PERIODIC TABLE WITH ATOMIC WEIGHTS
and Atomic Numbers as Reported by the Committee of the International Union of Chemistry, 1949

Group		O	I	II	III	IV	V	VI	VII	VIII		
Type Formula			R_2O	RO	R_2O_3	RO_2 H_4R	R_2O_5 H_3R	RO_3 H_2R	R_2O_7 HR	RO_4		
Period	Series 1		1 H-1.0080									
1	2	2 He-4.003	3 Li-6.940	4 Be-9.02	5 B-10.82	6 C-12.010	7 N-14.008	8 O-16.0000	9 F-19.00			
2	3	10 Ne-20.183	11 Na-22.997	12 Mg-24.32	13 Al-26.97	14 Si-28.06	15 P-30.98	16 S-32.066	17 Cl-35.457			
3	4	18 A-39.944	19 K-39.096	20 Ca-40.08	21 Sc-45.10	22 Ti-47.90	23 V-50.95	24 Cr-52.01	25 Mn-54.93	26 Fe-55.85	27 Co-58.94	28 Ni-58.69
	5		29 Cu-63.54	30 Zn-65.38	31 Ga-69.72	32 Ge-72.60	33 As-74.91	34 Se-78.96	35 Br-79.916			
4	6	36 Kr-83.7	37 Rb-85.48	38 Sr-87.63	39 Y-88.92	40 Zr-91.22	41 Cb-92.91	42 Mo-95.95	43 Tc-99 (?)	44 Ru-101.7	45 Rh-102.91	46 Pd-106.7
	7		47 Ag-107.880	48 Cd-112.41	49 In-114.76	50 Sn-118.70	51 Sb-121.76	52 Te-127.61	53 I-126.92			
5	8	54 Xe-131.3	55 Cs-132.91	56 Ba-137.36	57 La-138.92	58 Ce-140.13						
	9				Rare Earths Atomic Nos. 59-72 See Inside Back Cover							
6	10						73 Ta-180.88	74 W-183.92	75 Re-186.31	76 Os-190.2	77 Ir-193.1	78 Pt-195.23
	11		79 Au-197.2	80 Hg-200.61	81 Tl-204.39	82 Pb-207.21	83 Bi-209.00	84 Po-210 (?)	85 At-211 (?)			
7	12	86 Rn-222	87 Fr-223 (?)	88 Ra-226.05	89 Ac-227 (?)	90 Th-232.12	91 Pa-231	92 U-238.07	93 Np-237	94 Pu-239	95 Am-241	96 Cm-242

Chemical Engineers' Handbook

McGRAW-HILL SERIES IN CHEMICAL ENGINEERING

SIDNEY D. KIRKPATRICK, *Consulting Editor*

BUILDING FOR THE FUTURE OF A PROFESSION

Fifteen prominent chemical engineers first met in New York more than thirty years ago to plan a continuing literature for their rapidly growing profession. From industry came such pioneer practitioners as Leo H. Baekeland, Arthur D. Little, Charles L. Reese, John V. N. Dorr, M. C. Whitaker, and R. S. McBride. From the universities came such eminent educators as William H. Walker, Alfred H. White, D. D. Jackson, J. H. James, J. F. Norris, Warren K. Lewis, and Harry A. Curtis. H. C. Parmelee, then editor of *Chemical & Metallurgical Engineering*, served as chairman and was joined subsequently by S. D. Kirkpatrick as consulting editor.

After several meetings, this Editorial Advisory Committee submitted its report to the McGraw-Hill Book Company in September, 1925. In it were detailed specifications for a correlated series of more than a dozen text and reference books, including a chemical engineers' handbook and basic textbooks on the elements and principles of chemical engineering, on industrial applications of chemical synthesis, on materials of construction, on plant design, on chemical-engineering economics. Broadly outlined, too, were plans for monographs on unit operations and processes and on other industrial subjects to be developed as the need became apparent.

From this prophetic beginning has since come the McGraw-Hill Series in Chemical Engineering, which now numbers about thirty-five books. More are always in preparation to meet the ever-growing needs of chemical engineers in education and in industry. In the aggregate these books represent the work of literally hundreds of authors, editors, and collaborators. But no small measure of credit is due the pioneering members of the original committee and those engineering educators and industrialists who have succeeded them in the task of building a permanent literature for the classical engineering profession.

THE SERIES

ARIES AND NEWTON—*Chemical Engineering Cost Estimation*
BADGER AND BANCHERO—*Introduction to Chemical Engineering*
BADGER AND MCCABE—*Elements of Chemical Engineering*
CLARKE—*Manual for Process Engineering Calculations*
DODGE—*Chemical Engineering Thermodynamics*
GRISWOLD—*Fuels, Combustion, and Furnaces*
GROGGINS—*Unit Processes in Organic Synthesis*
HUNTINGTON—*Natural Gas and Natural Gasoline*
KIRKBRIDE—*Chemical Engineering Fundamentals*
LEE—*Materials of Construction*
LEWIS, RADASCH AND LEWIS—*Industrial Stoichiometry*
MANTELL—*Adsorption*
MANTELL—*Industrial Electrochemistry*
MCADAMS—*Heat Transmission*
NELSON—*Petroleum Refinery Engineering*
PERRY (EDITOR)—*Chemical Business Handbook*
PERRY (EDITOR)—*Chemical Engineers' Handbook*
PIERCE—*Chemical Engineering for Production Supervision*
RHODES, F. H.—*Technical Report Writing*
RHODES, T. J.—*Industrial Instruments for Measurement and Control*
ROBINSON AND GILLILAND—*Elements of Fractional Distillation*
SCHMIDT AND MARLIES—*Principles of High-polymer Theory and Practice*
SCHWEYER—*Process Engineering Economics*
SHERWOOD AND PIGFORD—*Absorption and Extraction*
SHERWOOD AND REED—*Applied Mathematics in Chemical Engineering*
SHREVE—*The Chemical Process Industries*
SMITH—*Introduction to Chemical Engineering Thermodynamics*
STEPHENSON—*Introduction to Nuclear Engineering*
TREYBAL—*Liquid Extraction*
TREYBAL—*Mass-transfer Operations*
TYLER—*Chemical Engineering Economics*
VILBRANDT—*Chemical Engineering Plant Design*
WALKER, LEWIS, MCADAMS, AND GILLILAND—*Principles of Chemical Engineering*
WILSON AND RIES—*Principles of Chemical Engineering Thermodynamics*
WILSON AND WELLS—*Coal, Coke, and Coal Chemicals*

CHEMICAL ENGINEERING SERIES

Chemical Engineers' Handbook

PREPARED BY A STAFF OF SPECIALISTS

JOHN H. PERRY, Ph.D., Editor

THIRD EDITION

New York, Toronto, and London
McGRAW-HILL BOOK COMPANY, INC.
1950

CHEMICAL ENGINEERS' HANDBOOK

V

THE MAPLE PRESS COMPANY, YORK, PA.

LIST OF CONTRIBUTORS

Robert Ammon, B.S., Chief Metallurgist, American Zinc, Lead and Smelting Co., St. Louis, Mo.; (Concentration of Ores by Sink-and-float Methods)

Anthony Anable, B.S., Engineer, The Dorr Co., Inc., New York. (Classification, Sedimentation)

***Evald Anderson, B.S.,** Late Technical Director, Western Precipitation Corp. (Separation of Dusts and Mists)

Robert S. Aries, M.Ch.E., M.A. (Econ.), M.Sc.D. Ch.E., Consulting Engineer and Economist; Adjunct Professor of Chemical Engineering, Polytechnic Institute of Brooklyn. (Plant Location)

H. R. Arnold, M.S., Chemist, E. I. du Pont de Nemours & Co. (Physical and Chemical Data)

W. L. Badger, M.S., Consulting Engineer, Ann Arbor, Mich., Formerly, Professor of Chemical Engineering, University of Michigan. (Evaporation)

Lawrence H. Bailey, B.S., M.S., Chief Engineer, F. J. Stokes Machine Co., Philadelphia. (Compacting)

***E. M. Baker, B.S.,** Late Professor of Chemical Engineering, University of Michigan. (Gas Absorption)

Theodore Baumeister, B.S., M.E., Professor of Mechanical Engineering, Columbia University. (Power and Power Machinery)

Hugh W. Bellas, B.S., Engineer, E. I. du Pont de Nemours & Co. (Filtration)

***A. F. Benning, B.S.,** Engineer, E. I. du Pont de Nemours & Co. (Materials of Construction)

Olaf P. Bergelin, Sc.D., Associate Professor of Chemical Engineering, University of Delaware. (Miscellaneous Over-all Coefficients of Heat Transfer)

C. E. Berry, B.S., Chemical Engineer, E. I. du Pont de Nemours & Co. (Miscellancous Methods of Mechanical Separation and Concentration, Size Reduction and Size Enlargement)

***A. D. Blake, M.E.,** Editor, *Combustion.* (Power Generation)

Harding Bliss, Ph.D., Professor of Chemical Engineering, Yale University. (Engineering Thermodynamic Properties)

P. W. Bridgman, Ph.D., Hollis Professor of Mathematics and Natural Philosophy, Harvard University. (Dimensional Analysis)

W. M. D. Bryant, B.S., Chemist, E. I. du Pont de Nemours & Co. (Physical and Chemical Principles)

C. P. Cabell, B.S., Chemical Engineer, General Electric Co. (Screening)

W. S. Calcott, Ch.E., Ll.D., Assistant Director (Development) Chemical Division, Organic Chemicals Dept., E. I. du Pont de Nemours & Co. (Sweating)

James S. Carey, Sc.D., Chemical Engineer, E. B. Badger & Sons Co. (Distillation)

W. H. Carrier, M.E., D.E., Chairman of the Board of Carrier Corp. (Humidification, Dehumidification, and Spray Ponds)

Allan P. Colburn, Ph.D., Assistant to the President and Professor of Chemical Engineering, University of Delaware. (Miscellaneous Over-all Coefficients of Heat Transfer, General Theory of Diffusional Operations, Plate Efficiencies for Distillation Columns, Packed Distillation Towers, Gas Absorption)

C. N. Collard, M.S., Chemical Engineer, E. B. Badger & Sons Co. (Plate Calculations for Multicomponent Mixtures)

Douglas M. Considine, B.S., Manager Technical Section, Sales Dept., Brown Instrument Co. (Process Control)

Carl S. Cragoe, A.B., Assistant Chief, Heat and Power Division, National Bureau of Standards. (Conversion Tables)

P. W. Crane, Ch.E., M.S., Assistant Director of Sales, Polychemicals Department, E. I. du Pont de Nemours & Co. (Extrusion)

***C. W. Cuno, Ph.D.,** Chemical Engineer and Consultant, The Lehon Co. (Economic Factors in Chemical Plant Location)

Harry A. Curtis, Ph.D., Director, Tennessee Valley Authority. (Solid Fuels)

D. S. Davis, M.S., Professor of Chemical Engineering, Virginia Polytechnic Institute. (Mathematics—Alignment Charts)

James G. DeFlon, B.S., Technical Advisor of Sales Department, Fluor Corp., Ltd. (Cooling Towers)

Fred D. DeVaney, E.M., M.S., Research Metallurgist, Picklands, Mather & Co. (Jigging, Tabling, Elutriation, Froth Flotation, and Magnetic Separation)

Barnett F. Dodge, D.Sc., Professor and Chairman of Department of Chemical Engineering, Yale University. (High-pressure Technique, Refrigeration—Low Temperature Processes)

***K. H. Donaldson, B.S.,** Professor and Head of Department of Mining Engineering, Case School of Applied Science. (Screening, Jigging, Tabling, Elutriation)

Thos. B. Dorris, Ph.D., Chief Chemical Engineer, Sprout, Waldron & Co. (Cutting)

T. B. Drew, M.S., Professor of Chemical Engineering, Columbia University. (Flow of Fluids)

Raymond W. Dull, Late Consulting Engineer, Chicago, Illinois. (Mathematics)

H. H. Dunkle, Engineer, Goetze Division, Johns-Manville Sales Corp. (Gaskets, Chemical Resistance of Gasket Materials)

Hans C. Duus, Ph.D., Chemical Engineer, E. I. du Pont de Nemours & Co. (Refrigeration, Thermodynamic Tables)

Joseph C. Elgin, Ph.D., Professor of Chemical Engineering, Princeton University. (Solvent Extraction)

C. H. Evans, B.S., Chemical Engineer, E. I. du Pont de Nemours & Co. (Furnaces and Kilns)

E. C. Fetter, B.S., Managing Editor, *Chemical Engineering,* McGraw-Hill Publishing Company. (Chemical Resistance of Gasket Materials)

Charles M. Fields, A.B., Technical Product Engineer, Shellmar Products Corp. (Extrusion)

A. E. Flowers, M.E. in E.E., M.M.E., Ph.D., Late Engineer-in-charge of Development, The DeLaval Separator Co. (Centrifuges)

* Contributions by these authors were made for previous editions and have been revised or rewritten by others for this edition. The stated professional position in these cases is that held by the author at the time of his contribution.

J. H. Foote, Assistant Research Director, Pulverizing Machinery Co. (Machining)

Samuel J. Friedman, M.S., Chemical Engineer, E. I. du Pont de Nemours & Co. (Drying)

I. H. Fullmer, B.S., Assistant Chief, Gage Section, National Bureau of Standards. (Wire and Sheet Metal Gages)

***J. F. C. Gartshore, B.Sc., Ph.D.,** Chemical Engineer, British Dyestuff Corp. (Sublimation)

R. P. Genereaux, Ch.E., Chemical Engineer, E. I. du Pont de Nemours & Co. (Flow of Fluids, Mathematics—Alignment Charts)

J. L. Gillson, Sc.D., Geologist, E. I. du Pont de Nemours & Co. (Electrostatic Methods of Concentration)

***S. L. Godshalk, Ch.E.,** Chemist, E. I. du Pont de Nemours & Co. (Spray Painting)

***G. D. Graves, Ph.D.,** Chemist, E. I. du Pont de Nemours & Co. (Organic Chemistry)

C. R. Groves, Ch.E., Engineer, E. I. du Pont de Nemours & Co. (Safety and Fire Protection)

C. Fred Gurnham, B.S., M.Ch.E., D.Eng.Sc., Consulting Engineer; Professor and Head of Department of Chemical Engineering, Tufts College, Medford, Mass. (Expression)

Selden H. Hall, M.E., Consulting Engineer and Registered Patent Agent, Poughkeepsie, N. Y. (Centrifuges)

***C. R. Harte, Jr., Ph.D.,** Chemical Engineer, Standard Oil Co. (Indiana). (Gas Absorption)

Carl V. Herrmann, B.S., M.A., E. I. du Pont de Nemours & Co. (Properties of Commercial Acids and Salts in Aqueous Solution)

K. C. D. Hickman, Ph.D., Consulting Engineer, Rochester, N. Y. (Molecular Distillation)

W. G. Hillen, M.E., Director, Sales Training, Carrier Corporation. (Humidification, Dehumidification, and Spray Ponds)

Hoyt C. Hottel, B.A., S.M., Professor of Fuel Engineering and Director, Fuels Research Laboratory, Massachusetts Institute of Technology. (Radiant-heat Transmission)

H. G. Houghton, B.S., S.M., D.Sc., Professor of Meteorology and Head of Department, Massachusetts Institute of Technology. (Spray Nozzles)

Wilbur G. Hudson, M.E., Consulting Engineer, Chicago, Illinois. (Material-handling Equipment)

Wilbert J. Huff, Ph.D., D.Sc., Chairman, Department of Chemical Engineering, University of Maryland. (Gaseous Fuels)

Reed W. Hyde, Metallurgical Engineer; President, Sintering Machinery Corp. (Feeders and Feeding Mechanisms)

Donald F. Irvin, B.S., Engineer, Oliver United Filters, Inc. (Filtration)

Lewis V. Judson, Ph.D., Chief of Length Section, National Bureau of Standards. (Conversion Tables, Specific Gravity, Weights and Measures, and Wire and Sheet Metal Gages)

***S. B. Kanowitz, B.S.,** Chemical Engineer, Raymond Pulverizer Division, Combustion Engineering Co., Inc. (Crushing, Grinding, and Pulverizing)

M. S. Kharasch, Ph.D., Professor of Chemistry, University of Chicago. (Physical and Chemical Data)

R. V. Kleinschmidt, A.B., A.M., S.B., S.D., Professor of the Practice of Mechanical Engineering, Harvard Graduate School of Engineering. (Theory of Dispersion of Liquid Droplets)

R. W. Lahey, B.S., American Cyanamid Co. (Packaging Equipment)

James A. Lane, B.S., Chem., M.S., Ph.D., Chemical Engineer, Head of Design Department, Technical Division, Oak Ridge National Laboratory. (Dialysis)

Norbert A. Lange, Ph.D., Western Reserve University; Handbook Publishers, Inc. (Physical Properties of Inorganic and Organic Compounds)

C. E. Lapple, Chem. E., Chemical Engineer, E. I. du Pont de Nemours & Co. (Dust and Mist Collection)

James A. Lee, B.S., B.A., M.A., Southwestern Representative, *Chemical Engineering,* McGraw-Hill Publishing Company. (Materials of Construction)

***E. T. Lessig, Ph.D.,** Physical Chemist, Goodrich Rubber Co. (Physical and Chemical Data)

F. L. Lucker, M.E., Professional Engineer of Pennsylvania; formerly, Special Representative to Chemical Industry, Ingersoll-Rand Co. (Movement of Liquids and Gases, Compression of Gas)

Gordon MacLean, Ch.E., Technical Director, Allied Foods Co. (Mixing of Material)

C. L. Mantell, Ph.D., Consulting Chemical Engineer, New York. (Adsorption, Electrochemistry)

W. R. Marshall, Jr., Ph.D., Associate Professor, Chemical Engineering Department, University of Wisconsin. (Drying)

H. L. Maxwell, Ph.D., Metallurgist, Engineering Department, E. I. du Pont de Nemours & Co. (Materials of Construction)

William H. McAdams, Sc.D., Professor of Chemical Engineering, Massachusetts Institute of Technology. (Heat Transmission by Conduction and Convection)

Warren L. McCabe, Ph.D., Director of Research, The Flintkote Co. (Crystallization)

F. T. McNamara, Ph.B., E.E., Associate Professor of Electrical Engineering, Yale University. (Electricity and Electrical Engineering)

E. J. Meyers, E.E., Assistant Manager, Safety and Fire Protection Division, E. I. du Pont de Nemours & Co. (Safety and Fire Protection)

Fred A. Miller, Materials Handling Consultant, E. I. du Pont de Nemours & Co. (Industrial Trucks, Tractors, and Trailers)

***G. H. Miller,** Industrial Relations Division, Service Department, E. I. du Pont de Nemours & Co. (Safety and Fire Protection)

S. A. Miller, B.S., Ph.D., Associate Professor of Chemical Engineering, University of Kansas. (Gas Dispersion)

H. L. Miner, Manager, Safety and Fire Protection Division, E. I. du Pont de Nemours & Co. (Safety and Fire Protection)

***G. L. Montgomery, M.E.,** Associate Editor, *Food Industries.* (Movement and Storage of Materials)

***L. H. Morrison, B.S. in M.E.,** *Business Journals.* (Refrigeration)

A. C. Mueller, Ph.D., Chemical Engineer, E. I. du Pont de Nemours & Co. (Furnaces and Kilns)

W. A. Myers, B.S.E., M.S.E., Assistant Manager, Research and Development Department, Atlantic Refining Co. (Liquid Fuels)

Harlan W. Nelson, Ph.D., Supervisor, Fuels Research, Battelle Memorial Institute. (Solid Fuels)

***T. R. Olive, A.B.**, Associate Editor, *Chemical Engineering.* (Measurement and Control of Process Variables)

E. L. Peffer, B.S., A.M., Late Chief of Capacity and Density Section, National Bureau of Standards. (Specific Gravity)

John H. Perry, Ph.D., Development Department, E. I. du Pont de Nemours & Co. (Mathematics, Physical and Chemical Data, Physical and Chemical Principles)

Robert H. Perry, A.B., B.S., Graduate Student, Massachusetts Institute of Technology; Formerly, Chemical Engineer, The Texas Co. (Mathematical Tables and Weights and Measures)

Robert L. Pigford, Ph.D., Professor and Head, Chemical Engineering Department, University of Delaware. (General Theory of Diffusional Operations, Distillation—Plate Efficiencies for Distillation Columns, Packed Distillation Towers, Gas Absorption)

Richard W. Porter, B.S., Editorial Director, *The Paper Industry and Paper World.* (Process Control)

***E. J. Prindle, M.E., LL.M., LL.D.**, Patent Lawyer, Senior Member, Prindle, Bean, and Mann. (Patents and Patent Law)

George A. Prochazka, Jr., E.M., Management Consultant; formerly, Chemical Economist, E. I. du Pont de Nemours & Co. (Accounting and Cost Finding)

***R. B. Purdy, M.E.**, Formerly, Associate Editor of *Power.* (Power Generation)

W. E. Rahm, Technical Field Service, E. I. du Pont de Nemours & Co. (Molding of Plastics)

A. F. Randolph, B.S., Ch.E., Assistant to Director of Sales, Polychemicals Department, E. I. du Pont de Nemours & Co. (Molding of Plastics)

Francis W. Reichelderfer, A.B., Sc.D., Chief, U.S. Weather Bureau. (Weather)

***E. E. Reid, Ph.D.**, Chemical Consultant, Baltimore, Md. (Physical and Chemical Data)

Frederick D. Rossini, Ph.D., Professor and Head of the Department of Chemistry, Carnegie Institute of Technology. Formerly, Chief of Section on Thermochemistry and Constitution of Petroleum, National Bureau of Standards. (Mathematical Tables and Weights and Measures, Physical and Chemical Data, Physical and Chemical Principles)

Percy H. Royster, A.B., A.M., Technical Advisor, Metallurgical Division, Bureau of Mines, U.S. Department of the Interior. (Furnaces and Kilns)

T. R. Running, Ph.D., Professor Emeritus of Mathematics, University of Michigan. (Mathematics—Graphical Calculus)

***W. P. Ryan, S.M.**, Late Professor of Chemical Engineering, Massachusetts Institute of Technology. (Industrial Stoichiometry)

Gilbert E. Seil, M.S., Ph.D., Late Technical Consultant, Day & Zimmermann, Inc. (Size Enlargement by Fusion)

W. M. Sheldon, B.S., Director of Research, Pulverizing Machinery Co. (Machining)

***T. K. Sherwood, Ph.D.**, Assistant Professor of Chemical Engineering, Massachusetts Institute of Technology. (Drying)

C. K. Sloan, A.B., M.S., Ph.D., Research Chemist, E. I. du Pont de Nemours & Co. (Emulsification, Flocculation)

Paul Sollenberger, Principal Astronomer, U.S. Naval Observatory. (Time)

William Staniar, M.E., Consulting Engineer. (Mechanical Power Transmission)

***W. O. Stauffer, M.S.**, Manager, Research Section of Technical Division, Remington Arms Co. (Physical and Chemical Data)

Jesse W. Stillman, Ph.D., E. I. du Pont de Nemours & Co. (Indicators, Sampling)

Daniel R. Stull, Ph.D., Dow Chemical Co (Physical and Chemical Data)

***H. S. Taylor, D.Sc.**, Research Professor of Chemistry, Princeton University. (Catalysis)

J. E. Teagarden, Ph.D., E. I. du Pont de Nemours & Co. (Molding of Plastics)

***P. V. Tilden, B.S. in M.E.**, Safety and Fire Protection Division, E. I. du Pont de Nemours & Co. (Safety and Fire Protection)

W. Trinks, Professor Emeritus, Carnegie Institute of Technology. (Furnaces and Kilns)

Louis B. Tuckerman, Ph.D., Assistant Chief, Mechanics Division, National Bureau of Standards. (Gravity)

Kenneth S. Valentine, A.B., Ch.E., New York Manager, The Patterson Foundry and Machine Co. (Mixing of Material)

D. J. Van Marle, Ch.E., Chemical Engineer, Buflovak Equipment Division of Blaw-Knox Co. (Flaking)

***J. V. Vaughen, Ph.D.**, Chemist, E. I. du Pont de Nemours & Co. (Physical and Chemical Data)

H. C. Vernon, M.S., Engineering Department, E. I. du Pont de Nemours & Co. (Mathematics, Physical and Chemical Data, Sublimation)

Donald D. Wagman, B.S., M.A., National Bureau of Standards. (Physical and Chemical Data)

R. G. Warner, Ph.B., E.E., Vice President, United Illuminating Co. (Electricity and Electrical Engineering)

H. M. Weir, B.Ch.E., Ph.D., Consulting Engineer, Philadelphia. (Liquid Fuels)

***F. L. Whitney, Jr., B.S.**, Engineer, E. I. du Pont de Nemours & Co. (Materials of Construction)

John C. Whitwell, Ch.E., Professor of Chemical Engineering, Princeton University. (Physical and Chemical Principles)

Richard Wiebe, Ph.D., U.S. Bureau of Agricultural and Industrial Chemistry. (Physical and Chemical Data)

***H. S. Winnicki, B.S.**, Head of Chemistry Section, South Charleston Research Division, Westvaco Chlorine Products Corp. (Measurement and Control of Process Variables)

F. W. Woodfield, Jr., M.S., Chemical Engineer, General Electric Co. (Distillation)

***Hood Worthington, M.S.**, Chemical Engineer, E. I. du Pont de Nemours & Co. (High Pressure Technique)

Raymond Wynkoop, Ph.D., Chemical Engineer, Standard Oil Co. (Indiana.) (Solvent Extraction)

J. I. Yellott, B.S., M.M.E., Director of Research, Locomotive Development Committee, Bituminous Coal Research, Inc. (Explosive Disintegration)

***H. L. Young, B.S.**, Sales Department, Aridye Corp. (Measurement and Control of Process Variables)

Stanley B. Zdonik, S.B., S.M., Chemical Engineer, E. B. Badger & Sons Co. (Azeotropic and Extractive Distillation)

***F. C. Zeisberg**, Late Technical Investigator, E. I. du Pont de Nemours & Co. (Reports and Report Writing)

ADVISORY BOARD

The following chemical engineers, not included in the List of Contributors, have consented to act as an advisory board to the editor of this handbook. All coauthors of this handbook actively assist the editor as advisors and are, therefore, ex-officio members of this Advisory Board.

L. W. Bass, Ph.D., Vice President, U.S. Industrial Chemicals, Inc.

G. G. Brown, Ph.D., Ch.E., Dean of Engineering and Professor of Chemical Engineering, University of Michigan.

T. H. Chilton, Ch.E., D.Sc., Technical Director, Development Engineering Division, Engineering Department, E. I. du Pont de Nemours & Company.

C. R. Downs, Ph.D., Consulting Chemical Engineer.

F. C. Frary, Ph.D., Retired. Formerly Director of Research, The Aluminum Company of America.

W. N. Jones, Ph.D., Director, College of Engineering, Carnegie Institute of Technology.

D. B. Keyes, Ph.D., Director, Heyden Chemical Corporation.

C. G. Kirkbride, President, The Houdry Process Corporation.

S. D. Kirkpatrick, B.S., Sc.D., D.Eng., Editorial Director, *Chemical Engineering* and *Chemical Week*, McGraw-Hill Publishing Company.

J. C. Lawrence, A.B., B.S., in Ch.E., Engineering Department, E. I. du Pont de Nemours & Company.

W. E. Lobo, B.S., Director of Chemical Engineering Department, The M. W. Kellogg Company.

D. F. Othmer, Professor and Head of the Department of Chemical Engineering, Polytechnic Institute of Brooklyn.

W. R. Veazey, Ph.D., D.E., Director, The Dow Chemical Company.

W. G. Whitman, S.M., Head, Department of Chemical Engineering, Massachusetts Institute of Technology.

Lincoln T. Work, Ph.D., Consulting Engineer.

PREFACE TO THE THIRD EDITION

The favorable reception of the first and second editions of this handbook by chemical and other engineers and by others in both industrial and academic fields has furnished the incentive for a thorough revision and an expansion in the present edition. It has been our continued purpose to keep pace with the theory and the practice of chemical engineering experiment and experience and thus to provide a handbook that is adequate in this and closely allied branches of engineering. The normal peacetime rate of progress in the theory and practice of chemical engineering was greatly accelerated during the late prewar years and especially during the war years. Since the war, a great deal of previously "classified" information has been released. The progress during prewar, war, and postwar years has been so great that complete, or almost complete, rewriting of most major sections of the handbook has been necessary.

The new page size of this edition will be welcome to the users of this handbook. The two previous editions had become unduly bulky for the original intention of a pocketbook or handbook. The present larger page size has enabled the thickness to be reduced and, even more importantly, permits larger sizes of graphs and illustrations which give needed clarity and accuracy of reading. All illustrations have been newly prepared and have greatly increased legibility. Every effort has been made by the editor, the authors, and the consultants to make the index of this new edition as thorough as possible by using not only the more common reference words but also their synonyms and equivalents. Suggestions for further entries for the index will be welcomed by the editor.

The following sections *have been thoroughly revised:* Flow of Fluids; Heat Transmission; Solvent Extraction; Mixing; Adsorption; Physical and Chemical Data; Physical and Chemical Principles; Mathematics; Mathematical Tables and Weights and Measures; Humidification, Dehumidification, Spray Ponds, and Cooling Towers; Fuels; Electrochemistry; Refrigeration; Electricity and Electrical Engineering; Materials of Construction; Mechanical Separations; Safety and Fire Protection; and Accounting and Cost Finding.

The following sections *have been rewritten and expanded:* Size Reduction; Power Generation; Gas Absorption; Distillation; Plant Location; Drying; Movement and Storage of Materials; Process Control; and High Pressure Technique. In addition, the following chapters have been rewritten and expanded: Low-temperature Refrigeration and Processes; Miscellaneous Methods of Mechanical Separations; and Sublimation.

The following new sections *have been added* in this edition: General Theory of Diffusional Operations; Furnaces and Kilns; Size Enlargement; Azeotropic Distillation; Multi-component Distillation; Extractive Distillation; Molecular Distillation; and Dialysis.

The following sections or chapters *have been deleted* in the present edition in order to use the space for material believed to be more valuable in this handbook: Reports and Report Writing; Indicators, Qualitative Analysis, Organic Chemistry. The material on indicators, however, has been revised and placed in another section. As a result of all these changes, some of the handbook sections have had to be renumbered.

The editor wishes to express here his appreciation of the spirit of cooperation shown by the contributors who have been extremely patient in furnishing several revisions of their material in order to assure the inclusion of the most up-to-date data and other information in their sections. The Advisory Board has also been very helpful to the editor in the preparation of this new edition.

It is a pleasure to express my indebtedness to the following persons, only a few of whom are authors or coauthors of specific sections of the handbook, for their assistance in their specialty fields: E. Bagnoli, E. P. Bartlett, C. H. Biesterfeld, F. C. Bond, D. F. Boucher, J. R. Boyer, C. O. Brown, G. G. Brown, M. H. Brown, W. D. Busch, J. R. Caddell, H. C. Carlson, T. H. Chilton, C. M. Cooper, Walter Coopey, D. S. Davis, R. L. Dodge, C. B. Dutton, Henry Eckhardt, J. F. Eversole, Leo Friend, J. F. C. Gartshore, W. M. Gaylord, J. L. Gillson, H. P. Grace, L. M. Greasel, W. P. G. Hall, V. F. Hanson, A. W. Hixson, O. A. Hougen, S. S. Hubard, R. M. Hunter, H. B. Irwin, R. H. Johnson, W. H. Kaiser, H. J. Kamack, H. S. Kemp, S. D. Kirkpatrick, F. E. Klutey, K. Kobe, J. H. Koffolt, N. A. Lange, A. W. Larchar, A. E. Lawrence, W. E. Lobo, E. M. Mahla, W. R. D. Manning, B. C. Mariacher, J. P. Martel, F. H. McBerty, S. A. Miller, M. C. Molstad, G. P. Monet, T. R. Olive, John Oliver, D. F. Othmer, R. H. Perry, Cyrus Pyle, P. G. Reynolds, G. W. Rigby, E. M. Rogers, F. D. Rossini, C. R. Schwartz, T. K. Sherwood, R. N. Shreve, S. Skowronski, Matt Souders, Jr., A. P. Spooner, J. W. Stillman, R. K. Taylor, J. B. Tepe, R. E. Treybal, Chaplin Tyler, S. L. Tyler, G. D. Van Arsdale, C. E. Waring, R. M. Watson, R. A. Weiman, N. P. Wescott, L. T. Work, F. W. Woodfield, Hood Worthington, and H. S. Young. The editor and the authors are also greatly indebted to a large number of firms and individuals for their assistance in the collection and correlation of data and other information, for the use of illustration copy, and for the reviewing of manuscripts and proofs. Finally, credit must be given to the host of chemical engineers, chemists, and others who have contributed materially to this new edition through their friendly criticisms and suggestions.

JOHN H. PERRY

WILMINGTON, DELAWARE
December, 1949

PREFACE TO THE FIRST EDITION

This handbook is intended to supply both the practicing engineer and the student with an authoritative reference work that covers comprehensively the field of chemical engineering as well as important related fields. To ensure the highest degree of reliability the cooperation of a large number of specialists has been necessary; this handbook presents the efforts of 60 contributing specialists. In addition, this book represents the experience and opinions of over 150 chemists and engineers who have aided with advice and suggestions. Each contributor is to be regarded as responsible for the statements in his section. Although this volume has been prepared primarily as a reference book, it is believed that many of the sections will be found suitable for courses in technical schools. Certainly, it should prove useful as a synopsis for students and practicing engineers.

The bibliography given with most sections is intended to direct the reader to more extended information in treatises and current periodicals. The space available for these bibliographies has made it necessary to omit many works and technical articles. The references are usually those which the writer of the section has found most useful; they are, therefore, in no sense complete, but will be found helpful in the search for additional information.

It is believed that the cost data given in the various sections will prove particularly helpful when properly used. These data usually represent the ordinary range of cost. It has been the experience of the contributors to this handbook that students and recent technical graduates lack even an approximate idea of the cost of apparatus and structures. It is primarily to supply approximate figures that the cost data are given in various sections. These data should also be found suitable for preliminary eatimates; but for close estimating, current prices, of course, should be obtained from manufacturers or purchasing agents.

Some duplications will be found among the various sections, but after due consideration it was decided to allow these duplications to remain in order to facilitate the use of any particular section and to avoid the necessity for continued reference to other parts of the handbook.

The editor-in-chief wishes to acknowledge his indebtedness to the editors of "International Critical Tables," who have permitted the use of a large amount of I.C.T. data. These data, for the purposes of this handbook, have, in a large number of cases, been rearranged and in many instances, have been recalculated to units commonly used by engineers. Acknowledgment is also here given to the various experts who initially correlated these data for the "International Critical Tables."

In addition to the authors of the individual sections, the editor-in-chief is indebted to a large number of individuals and firms for their assistance in the collection and correlation of data, for the use of illustration copy, and for the reviewing of manuscripts and proofs. Special thanks are due to T. H. Chilton, A. P. Colburn, R. P. Genereaux, and W. H. McAdams for the careful reading of a number of the individual sections, and for advice and suggestions throughout the preparation of this handbook. Every contribution, after examination by the editor-in-chief, was submitted to one or more specialists for criticisms and suggestions. The editor-in-chief is indebted to the following: T. Baker, W. D. Bancroft, E. P. Bartlett, W. M. D. Bryant, C. A. Butcher, C. H. Butcher, H. J. M. Creighton, L. T. Cummings, C. David, W. W. Heckert, C. V. Herrmann, J. A. Henricks, M. S. Kharasch, N. W. Krase, H. J. Macintire, T. A. Rickard, C. S. Robinson, G. G. Shannonhouse, G. C. Smith, E. F. Steinbring, R. S. Taylor, Chaplin Tyler, S. K. Varnes, W. M. Whitten, Jr., R. Wiebe, and L. T. Work.

The editor-in-chief wishes especially to thank H. C. Parmelee, Chairman of the Committee of the Chemical Engineering Series; S. D. Kirkpatrick, Editor of *Chemical and Metallurgical Engineering;* and the other members of the Committee for their aid and criticisms.

JOHN H. PERRY

CLEVELAND, OHIO
December, 1933

CONTENTS

For the detailed contents of any section consult the title page of that section; this can be located immediately by use of the thumb tabs.

For the alphabetical index see p. 1885.

	Section
Mathematical Tables and Weights and Measures	1
Mathematics	2
Physical and Chemical Data	3
Physical and Chemical Principles	4
Flow of Fluids	5
Heat Transmission	6
Evaporation	7
General Theory of Diffusional Operations	8
Distillation and Sublimation	9
Gas Absorption	10
Solvent Extraction and Dialysis	11
Humidification, Dehumidification, and Cooling Towers and Spray Ponds	12
Drying	13
Adsorption	14
Mechanical Separations	15
Size Reduction and Size Enlargement	16
Mixing of Material	17
High-pressure Technique	18
Process Control	19
Movement and Storage of Materials	20
Materials of Construction	21
Fuels	22
Furnaces and Kilns	23
Power Generation and Mechanical Power Transmission	24
Refrigeration	25
Plant Location	26
Electricity and Electrical Engineering	27
Electrochemistry	28
Accounting and Cost Finding	29
Safety and Fire Protection	30

CODE OF ETHICS
OF
THE AMERICAN INSTITUTE OF CHEMICAL ENGINEERS

The Institute expects that the following rules shall guide the acts of its members:

SECTION 1. They shall be guided in all their relations by the highest standards of integrity and fair dealing.

SECTION 2. They shall uphold before the public at all times the dignity of the profession generally and the reputation of the Institute.

SECTION 3. They shall avoid and discourage sensationalism, exaggeration, and unwarranted statements. In making the first publication concerning inventions or chemical discoveries or improvements, these should be made through scientific societies or technical publications.

SECTION 4. They shall refuse to undertake, for compensation, work which they believe will be unprofitable to clients without first advising clients as to the improbability of successful results.

SECTION 5. They shall uphold the principle that unreasonably low charges for professional work tend toward inferior and unreliable work.

SECTION 6. They shall refuse to lend their names to any questionable enterprise.

SECTION 7. They shall be conservative in all estimates, reports, testimony, etc., especially in connection with the promotion of business enterprises.

SECTION 8. They shall not engage in any occupation which is contrary to law or the public welfare.

SECTION 9. When a chemical engineer undertakes for others, work in connection with which he may make improvements, inventions, plans, designs, or other records, he shall preferably enter into a written agreement regarding their ownership. In a case where an agreement is not made or does not cover a point at issue, the following rules shall normally apply:

a. If a chemical engineer uses information which is not common knowledge or public property, but which he obtains from a client or employer, any results in the form of plans, designs, or other records shall not be regarded as his property, but the property of his client or employer.

b. If a chemical engineer uses only his own knowledge, or information or data which by prior publication or otherwise are public property, and obtains no chemical engineering data from a client or employer except performance specifications or routine information, then the results in the form of inventions, plans, designs, or other records should be regarded as the property of the engineer, and the client or employer should be entitled to their use only in the case for which the engineer was retained.

c. All work and results accomplished by the chemical engineer in the form of inventions, plans, designs, or other records outside of the field for which a client or employer has retained him, should be regarded as the chemical engineer's property.

d. When a chemical engineer participates in the building of apparatus from designs supplied him by a client, the designs remain the property of the client and should not be duplicated by the engineer or anyone representing him for others without express permission.

e. Chemical engineering data or information which a chemical engineer obtains from his client or employer or which he creates as a result of such information must be considered confidential by the engineer; and while he is justified in using such data or information in his own practice as forming part of his professional experience, its publication without express permission is improper.

f. Designs, data, records, and notes made by an employee other than a consultant, and referring to his employer's work, should be regarded as his employer's property.

g. A client does not acquire any exclusive right to plans or apparatus made or constructed by a consulting chemical engineer except for the specific case for which they were made.

SECTION 10. A chemical engineer cannot honorably accept compensation, financial or otherwise, from more than one interested party, without the consent of all parties; and whether consulting, designing, installing, or operating, must not accept compensation directly or indirectly from parties dealing with his client or employer except with the knowledge and consent of his client or employer.

When called upon to decide on the use of inventions, apparatus, processes, etc., in which he has a financial interest, he should make his status in the matter clearly understood before engagement.

SECTION 11. The chemical engineer should endeavor at all times to give credit for work to those who, so far as his knowledge goes, are the real authors of such work.

SECTION 12. Undignified, sensational, or misleading advertising is not permitted. The use of members' names or pictures as an aid to any such advertising is not permitted, and the use of the name of the Institute in connection with any such advertising will not be tolerated.

ABBREVIATIONS

Table 1. Symbols and Nomenclature of Chemical Engineering*

Arranged alphabetically by symbols

a	Acceleration	(ft./sec.)/sec., (ft./hr.)/hr.
	Activity	
	Aperture (also A)	in.
	Surface per unit volume	sq. ft./cu. ft.
A	Aperture (also a)	in.
	Area	sq. ft. (S is used for cross-sectional area available for flow)
	Free energy, Helmholtz ($U - TS$)	B.t.u., B.t.u./lb.-mole, p.c.u., p.c.u./lb.-mole
b	Breadth, width	ft.
B	Film thickness, effective	ft.
	Residue, waste, bottoms	lb.-moles/hr.
c	Concentration, volumetric	lb./cu. ft., lb.-moles/cu. ft.
	Specific heat	B.t.u./(lb.) (°F.), p.c.u./(lb.) (°C.)
c_p	Specific heat, at constant pressure	B.t.u./(lb.) (°F.), p.c.u./(lb.) (°C.)
c_s	Humid heat	B.t.u./(lb. dry air) (°F.), p.c.u. (lb. dry air) (°C.)
c_v	Specific heat, at constant volume	B.t.u./(lb.) (°F.), p.c.u./(lb.) (°C.)
C	Coefficient, of discharge, etc.	
	Coefficient of resistance	B.t.u./(hr.) (°F.), p.c.u./(hr.)(°C.)
	Conductance (see $1/R$)	B.t.u./(hr.) (°F.), p.c.u./(hr.)(°C.)
d	Differential operator	
D	Diameter	ft.
	Distillate rate	lb.-moles/hr.
D_v	Diffusivity of vapor	sq. ft./hr.
e	Base of natural logarithms	
E	Energy, in general	B.t.u., B.t.u./lb.-mole, p.c.u., p.c.u./lb.-mole
	Energy, internal	B.t.u., B.t.u./lb., p.c.u., p.c.u./lb.
	Entrainment ratio	lb./lb., lb.-moles/lb.-mole
	Evaporation	lb.
f	Coefficient of friction	
	Friction factor, Fanning	($F = 2fLV^2/g_cD$)
	Fugacity	lb. force/sq. ft., atm.
	Activity coefficient, molal basis, see γ (gamma)	
F	Feed rate	lb.-moles/hr.
	Force, total load	lb. force
	Friction in energy balance	(ft.) (lb. force)/lb.
	See G, Free energy, Gibbs	
g	Acceleration of gravity	(ft./sec.)/sec., (ft./hr.)/hr.
g_c	Newton law of motion, conversion factor in	(lb.)(ft.)/(sec.)^2(lb. force)
g_0	Acceleration of gravity, standard value	(ft./sec.)/sec., (ft./hr.)/hr.
G	Free energy, Gibbs ($H - TS$)	(F also is used)
	Mass velocity	lb./(hr.) (sq. ft.), lb./(sec.) (sq. ft.)
h	Coefficient of heat transfer, individual	B.t.u./(hr.) (sq. ft.) (°F.), p.c.u./(hr.) (sq. ft.) (°C.)
	Enthalpy, per unit weight	B.t.u./lb., p.c.u./lb. (i is used when necessary to distinguish)
h_{fg}	Latent heat. See λ (lambda)	B.t.u., B.t.u./lb., p.c.u., p.c.u./lb.
H	Enthalpy	B.t.u., B.t.u./lb.-mole p.c.u., p.c.u./lb.-mole
	Henry's law constant c/p	(lb.-moles/cu. ft.)/atm.
	Humidity	lb./lb. dry air
H_o	Solvent present	lb.
H_p	Height equivalent to a theoretical plate, "*H.E.T.P.*"	ft.
H_R	Humidity, relative	
H_t	Height of transfer unit, "*H.T.U.*"	ft.
i	See under Enthalpy, per unit weight	
I	Moment of inertia	(ft.)^4
j	Heat transfer factor	$(h/cG)\,\phi(c\mu/k)$
J	Mechanical equivalent of heat	(ft.) (lb. force)/B.t.u., (ft.) (lb. force)/p.c.u.
k	Mass transfer coefficient, individual	lb.-moles/(hr.) (sq. ft.) (atm.)
	Specific heats, ratio of	c_p/c_v, κ (kappa) and γ (gamma) also are used
	Thermal conductivity	B.t.u./(hr.) (sq. ft.) (°F./ft.), p.c.u./(hr.) (sq. ft.) (°C./ft.)
k_G	Mass transfer coefficient, gas film	lb.-moles/(hr.) (sq. ft.) (atm.)
k_L	Mass transfer coefficient, liquid film	lb.-moles/(hr.) (sq. ft.) (lb.-mole/cu. ft.)
K	Equilibrium constant, $y = Kx$	
	Mass transfer coefficient, over-all	lb.-moles/(hr.) (sq. ft.) (atm.)
K_G	Mass transfer coefficient, on gas film basis	
K_L	Mass transfer coefficient, on liquid film basis	
L	Length	ft.
	Liquid rate	lb.-moles/hr.
	Mass velocity of liquid	(lb./(hr.) (sq. ft.)
L_m	Liquid rate, below feed	lb.-moles/hr.
L_n	Liquid rate, above feed	lb.-moles/hr.
m	Mass	lb.
	Mesh (also M)	1/in.
	Slope of equilibrium curve	dy^*/dx
M	Mesh (also m)	1/in.
	Molecular weight	
n	Rate of rotation	r.p.m.
N	Number in general	
	Rate of transfer	lb.-moles/hr.
	Radiation, intensity of	B.t.u./(hr.) (sq. ft.) p.c.u./(hr.) (sq. ft.)
N_p	Plates, number of	
N_t	"Transfer units," number of	
p	Pressure	lb. force/sq. ft., atm.
P	Power	(ft.) (lb. force)/sec.
q	Rate of flow, volumetric	cu. ft./sec., cu. ft./hr.
	Rate of heat transfer	B.t.u./hr., p.c.u./hr.
	Thermal condition of feed $(L_m - L_n)/F$	
Q	Quantity of heat transferred	B.t.u., p.c.u.
r	Equivalent resistance of cloth	
	Radius	ft.
R	Gas constant	Where necessary to distinguish, use R_0
	Production rate	lb./hr.
	Reflux ratio	Use R_D for L/D, and R_V for L/V
	Resistance, thermal	°F./(B.t.u./hr.), °C./(p.c.u./hr.)
R_H	Hydraulic radius	ft., sq. ft./ft.
R_R	Reduction ratio	
s	Entropy, per unit weight	B.t.u./(lb.)(°R.), p.c.u./(lb.) (°K.)
	Exponent of compressibility of cake	
	Specific surface	sq. ft./lb., sq. cm./g.
S	Cross section	sq. ft.
	Entropy	B.t.u./(lb.) (°R.) or B.t.u./(lb.-mole) (°R.), p.c.u./(°K.) or p.c.u./(lb.-mole) (°K.)
	Solubility	lb./100 lb. solvent
t	Temperature	°F. or °C. (θ is used in some lists)
	Time (see τ (tau))	sec., hr. (θ also has been used)

* From *Trans. Am. Inst. Chem. Engrs.*, Apr. 25, 1944.

Table 1. Symbols and Nomenclature of Chemical Engineering*—(*Concluded*)

T	Temperature, absolute	°K. or °R. (Rankine)
u	Velocity, local	ft./sec., ft./hr.
	Internal energy per unit weight	B.t.u./lb., p.c.u./lb.
U	Energy, in general	B.t.u., B.t.u./lb.-mole, p.c.u., p.c.u./lb.-mole
	Heat transfer coefficient, over-all	B.t.u./(hr.) (sq. ft.) (°F.), p.c.u./(hr.) (sq. ft.) (°C.)
	Internal energy	B.t.u., B.t.u./lb.,p.c.u., p.c.u./lb.
v	Specific volume	cu. ft./lb.
v_H	Humid volume	cu. ft./lb. dry air
V	Vapor rate	lb.-moles/hr.
	Velocity, average	ft./sec., ft./hr.
	Volume, total or per mole	cu. ft., cu. ft./lb.-mole
V_a	Velocity, acoustic	ft./sec. (c is also used)
w	Mass flow rate	lb./sec., lb./hr.
W	Free moisture content	lb./lb.
	Residue, waste, bottoms	lb.-moles/hr.
	Work	B.t.u., p.c.u. Where necessary to distinguish, use W_k
	Weight, quantity of matter	lb.
W_e	Work, external	B.t.u., p.c.u.
W_k	See W	
x	Distance in direction of flow	ft.
	Mole fraction, in liquid	
x_v	Fraction by volume	
x_w	Fraction by weight	
X	Mole ratio, in liquid	
y	Depth	ft.
	Mole fraction, in vapor	
y^*	Mole fraction, in vapor, equilibrium value	
Y	Mole ratio, in vapor	
z	Compressibility	pV/RT
Z	Distance above datum plane	ft.
α (alpha)	Angle	Degrees, radians (θ and ϕ are also used)
	Coefficient of expansion, linear	(ft./ft.)/°F., (ft./ft.)/°C.
	Diffusivity, thermal	sq. ft./hr.
	Absorptivity (for radiation)	
	Relative volatility	
	Specific cake resistance	
β (beta)	Coefficient of expansion, volumetric	(cu. ft./cu. ft.)/°F., (cu. ft./cu. ft.)/°C.
γ (gamma)	Activity coefficient, molal basis. f also is used	
Γ (capital gamma)	Weight rate of flow per unit of breadth	lb./(sec.) (ft.), (lb./(hr.) (ft.)
Δ (capital delta)	Difference, finite; often that causing flow	
ϵ (epsilon)	Emissivity (for radiation)	
η (eta)	Efficiency	
θ (theta)	See τ (tau), α (alpha)	
κ (kappa)	Specific heats, ratio of: c_p/c_v	
λ (lambda)	Latent heat of evaporation	B.t.u./lb., p.c.u./lb.
μ (mu)	Viscosity, absolute	lb./(sec.) (ft.), lb./(hr.) (ft.)
$1/\mu$	Fluidity	(sec.) (ft.)/lb.
μ/μ_w	Viscosity, relative to water	
ν (nu)	Viscosity, kinematic η (eta) also is used	sq. ft./sec., sq. ft./hr.
ρ (rho)	Density	lb./cu. ft.
σ (sigma)	Surface tension	lb. force/ft., dynes/cm.
	Stefan-Boltzmann constant	
τ (tau)	See t (Time) (Θ (theta) also has been used)	sec., hr.
	Tractive force per unit area	lb. force/sq. ft.
ϕ, ψ, χ (phi, psi, chi)	Function	
	See α (alpha)	
ω (omega)	Solid angle	
	Velocity, angular	

Table 2. Common Abbreviations of Technical Journals

Mostly from the standard abbreviations of *Chemical Abstracts*, published by the American Chemical Society

Am. J. Sci.	American Journal of Science
Anales soc. españ. fís. y quim.	Anales fís. y quim (Madrid)
Angew. chem.	Angewandte chemie
Ann.	Liebig's Annalen der Chemie
Ann. chim.	Annales de chimie
Ann. phys.	Annales de physique
Ann. Physik	Annalen der Physik
Ber.*	Berichte der deutschen chemischen Gesellschaft
Bull. soc. chim. Belges	Bulletin de la sociétés chimique Belges
Chem. Absts. or, C. A.	Chemical Abstracts
Chem. Ber.	Chemische Berichte
Chem. Eng.	Chemical Engineering (formerly Chemical & Metallurgical Engineering)
Chem. Eng. Progress	Chemical Engineering Progress
Chemistry & Industry	Chemistry and Industry
Chem. Revs.	Chemical Reviews
Chem. Zentr.	Chemisches Zentralblatt
Chem.-Ztg.	Chemiker-Zeitung
Chimie & industrie	Chimie et industrie
Comm. Leiden	Communs. Kamerlingh Onnes Lab. Univ. Leiden
Compt. rend.	Comptes rendus de l'académie des Sciences
Elektrochem. Z.	Elektrochemische Zeitschrift (discontinued in 1922)
Elektrotech. Z.	Elektrotechnische Zeitschrift
Engineering	Engineering, London
Eng. Mining J.	Engineering and Mining Journal
Gazz. chim. ital.	Gazzetta chimica italiana
Helv. Chim. Acta	Helvetica Chimica Acta
Industrie chimique	L'Industrie chimique
Ind. Eng. Chem.	Industrial and Engineering Chemistry
J. Am. Chem. Soc.	Journal of the American Chemical Society
J. Applied Mechanics	Journal of Applied Mechanics
J. Applied Phys.	Journal of Applied Physics
J. Chem. Soc.	Journal of the Chemical Society (London)
J. Chem. Phys.	Journal of Chemical Physics
J. chim. phys.	Journal de chimie physique et de physico-chimie biologique
J. Eng. Education	Journal of Engineering Education
J. Phys. & Colloid Chem.	The Journal of Physical and Colloid Chemistry
J. prakt. Chem.	Journal für praktische Chemie
J. Research Nat. Bur. Standards	Journal of Research of the National Bureau of Standards
J. Russ. phys.-chem. Soc.	Journal of the Russian Physical-Chemical Society, Leningrad
Kolloid-Z.	Kolloid-Zeitschrift
Monatsh.	Monatshefte für Chemie und verwandte Teile anderer Wissenschaften
Phil. Mag.	The Philosophical Magazine
Trans. Roy. Soc. (London)	Philosophical Transactions of the Royal Society, London
Phys. Rev.	The Physical Review
Product Eng.	Product Engineering
Trans. A. I. Ch. E., or Trans. Am. Inst. Chem. Engrs.	Transactions of the American Institute of Chemical Engineers†
Trans. Electrochem. Soc.	Transactions of the Electrochemical Society‡
Trans. Am. Inst. Mining Met. Engrs.	Transactions of the American Institute of Mining & Metallurgical Engineers
Trans. Faraday Soc.	Transactions of the Faraday Society
Z. anal. Chem.	Zeitschrift für analytische Chemie
Z. anorg. Chem.	Zeitschrift für anorganische Chemie
Z. Elektrochem.	Zeitschrift für Elektrochemie und angewandte physikalische Chemie
Z. Physik	Zeitschrift für Physik
Z. physik. Chem.	Zeitschrift für physikalische Chemie
Z. Ver. deut. Ing.	Zeitschrift des Vereines deutscher Ingenieure

* Name discontinued. Now called Chem. Ber.

** Transactions of the American Institute of Chemical Engineers ceased publication with vol. 42 and Chemical Engineering Progress is now the official publication of the American Institute of Chemical Engineers.

*** Now appearing in J. of the Electrochem. Soc.

Table 3. Mathematical Signs, Symbols, and Abbreviations

$\pm$ ($\mp$)	plus or minus (minus or plus)
:	divided by, ratio sign
::	proportional sign
$<$	less than
$\nless$	not less than
$>$	greater than
$\ngtr$	not greater than
$\cong$	approximately equals, congruent
$\sim$	similar to
$\Leftrightarrow$	equivalent to
$\neq$	not equal to
$\doteq$	approaches, is approximately equal to
$\propto$	varies as
∞	infinity
$\therefore$	therefore
$\sqrt{\ }$	square root
$\sqrt[3]{\ }$	cube root
$\sqrt[n]{\ }$	nth root
$\angle$	angle
$\perp$	perpendicular to
$\parallel$	parallel to
$\|x\|$	numerical value of x
log or $\log_{10}$	common logarithm or Briggsian logarithm
$\log_e$ or ln	natural logarithm or hyperbolic logarithm or Naperian logarithm
e	base (2.718) of natural system of logarithms
$a°$	an angle a degrees
a'	a prime, an angle a minutes
a''	a double prime, an angle a seconds, a second
sin	sine
cos	cosine
tan	tangent
ctn or cot	cotangent
sec	secant
csc	cosecant
vers	versed sine
covers	coversed sine
exsec	exsecant
$\sin^{-1}$	anti sine or angle whose sine is
sinh	hyperbolic sine
cosh	hyperbolic cosine
tanh	hyperbolic tangent
$\sinh^{-1}$	anti hyperbolic sine or angle whose hyperbolic sine is
$f(x)$ or $\phi(x)$	function of x
Δx	increment of x
Σ	summation of
dx	differential of x
dy/dx or y'	derivative of y with respect to x
d^2y/dx^2 or y''	second derivative of y with respect to x
d^ny/dx^n	nth derivative of y with respect to x
$\partial y/\partial x$	partial derivative of y with respect to x
$\partial^n y/\partial x^n$	nth partial derivative of y with respect to x
$\dfrac{\partial^n y}{\partial x\,\partial y}$	nth partial derivative with respect to x and y
$\int$	integral of
$\int_a^b$	integral between the limits a and b
$\dot{y}$	first derivative of y with respect to time
$\ddot{y}$	second derivative of y with respect to time
Δ or ∇^2	the "Laplacian" $\left(\dfrac{\partial^2}{\partial x^2}+\dfrac{\partial^2}{\partial y^2}+\dfrac{\partial^2}{\partial z^2}\right)$
δ	sign of a variation
$\oint$	sign for integration around a closed path

LOGARITHMS

Table 4. Five-place Common Logarithms of Numbers†

100—170

No.	L	0	1	2	3	4	5	6	7	8	9
100	00	000	043	087	130	173	217	260	303	346	389
101		432	475	518	561	604	647	689	732	775	817
102		860	903	945	988	*030	*072	*115	*157	*199	*242
103	01	284	326	368	410	452	494	536	578	620	662
104		703	745	787	828	870	912	953	995	*036	*078
105	02	119	160	202	243	284	325	366	408	449	490
106		531	572	612	653	694	735	776	816	857	898
107		938	979	*019	*060	*100	*141	*181	*222	*262	*302
108	03	342	383	423	463	503	543	583	623	663	703
109		743	782	822	862	902	941	981	*021	*060	*100
110	04	139	179	218	258	297	336	376	415	454	493
111		532	571	610	650	689	727	766	805	844	883
112		922	961	999	*038	*077	*115	*154	*192	*231	*269
113	05	308	346	385	423	461	500	538	576	614	652
114		690	729	767	805	843	881	918	956	994	*032
115	06	070	108	145	183	221	258	296	333	371	408
116		446	483	521	558	595	633	670	707	744	781
117		819	856	893	930	967	*004	*041	*078	*115	*151
118	07	188	225	262	298	335	372	408	445	482	518
119		555	591	628	664	700	737	773	809	846	882
120		918	954	990	*027	*063	*099	*135	*171	*207	*243
121	08	279	314	350	386	422	458	493	529	565	600
122		636	672	707	743	778	814	849	884	920	955
123		991	*026	*061	*096	*132	*167	*202	*237	*272	*307
124	09	342	377	412	447	482	517	552	587	621	656
125		691	726	760	795	830	864	899	934	968	*003
126	10	037	072	106	140	175	209	243	278	312	346
127		380	415	449	483	517	551	585	619	653	687
128		721	755	789	823	857	890	924	958	992	*025
129	11	059	093	126	160	193	227	261	294	327	361
130		394	428	461	494	528	561	594	628	661	694
131		727	760	793	826	860	893	926	959	992	*024
132	12	057	090	123	156	189	222	254	287	320	353
133		385	418	450	483	516	548	581	613	646	678
134		710	743	775	808	840	872	905	937	969	*001
135	13	033	066	098	130	162	194	226	258	290	322
136		354	386	418	450	481	513	545	577	609	640
137		672	704	735	767	799	830	862	893	925	956
138		988	*019	*051	*082	*114	*145	*176	*208	*239	*270
139	14	301	333	364	395	426	457	489	520	551	582
140		613	644	675	706	737	768	799	829	860	891
141		922	953	983	*014	*045	*076	*106	*137	*168	*198
142	15	229	259	290	320	351	381	412	442	473	503
143		534	564	594	625	655	685	715	746	776	806
144		836	866	897	927	957	987	*017	*047	*077	*107
145	16	137	167	197	227	256	286	316	346	376	406
146		435	465	495	524	554	584	613	643	673	702
147		732	761	791	820	850	879	909	938	967	997
148	17	026	056	085	114	143	173	202	231	260	289
149		319	348	377	406	435	464	493	522	551	580
150	17	609	638	667	696	725	754	782	811	840	869
151		898	926	955	984	*013	*041	*070	*099	*127	*156
152	18	184	213	241	270	299	327	355	384	412	441
153		469	498	526	554	583	611	639	667	696	724
154		752	780	808	837	865	893	921	949	977	*005
155	19	033	061	089	117	145	173	201	229	257	285
156		312	340	368	396	424	451	479	507	535	562
157		590	618	645	673	700	728	756	783	811	838
158		866	893	921	948	976	*003	*030	*058	*085	*112
159	20	140	167	194	222	249	276	303	330	358	385
160		412	439	466	493	520	548	575	602	629	656
161		683	710	737	763	790	817	844	871	898	925
162		952	978	*005	*032	*059	*085	*112	*139	*165	*192
163	21	219	245	272	299	325	352	378	405	431	458
164		484	511	537	564	590	617	643	669	696	722
165		748	775	801	827	854	880	906	932	958	985
166	22	011	037	063	089	115	141	168	194	220	246
167		272	298	324	350	376	401	427	453	479	505
168		531	557	583	608	634	660	686	712	737	763
169		789	814	840	866	891	917	943	968	994	*019
170	23	045	070	096	121	147	172	198	223	249	274
No.	L	0	1	2	3	4	5	6	7	8	9

Proportional parts

	44	43	42
1	4.4	4.3	4.2
2	8.8	8.6	8.4
3	13.2	12.9	12.6
4	17.6	17.2	16.8
5	22.0	21.5	21.0
6	26.4	25.8	25.2
7	30.8	30.1	29.4
8	35.2	34.4	33.6
9	39.6	38.7	37.8

	41	40	39
1	4.1	4.0	3.9
2	8.2	8.0	7.8
3	12.3	12.0	11.7
4	16.4	16.0	15.6
5	20.5	20.0	19.5
6	24.6	24.0	23.4
7	28.7	28.0	27.3
8	32.8	32.0	31.2
9	36.9	36.0	35.1

	38	37	36
1	3.8	3.7	3.6
2	7.6	7.4	7.2
3	11.4	11.1	10.8
4	15.2	14.8	14.4
5	19.0	18.5	18.0
6	22.8	22.2	21.6
7	26.6	25.9	25.2
8	30.4	29.6	28.8
9	34.2	33.3	32.4

	35	34	33
1	3.5	3.4	3.3
2	7.0	6.8	6.6
3	10.5	10.2	9.9
4	14.0	13.6	13.2
5	17.5	17.0	16.5
6	21.0	20.4	19.8
7	24.5	23.8	23.1
8	28.0	27.2	26.4
9	31.5	30.6	29.7

	32	31	30
1	3.2	3.1	3.0
2	6.4	6.2	6.0
3	9.6	9.3	9.0
4	12.8	12.4	12.0
5	16.0	15.5	15.0
6	19.2	18.6	18.0
7	22.4	21.7	21.0
8	25.6	24.8	24.0
9	28.8	27.9	27.0

	29	28
1	2.9	2.8
2	5.8	5.6
3	8.7	8.4
4	11.6	11.2
5	14.5	14.0
6	17.4	16.8
7	20.3	19.6
8	23.2	22.4
9	26.1	25.2

	27	26
1	2.7	2.6
2	5.4	5.2
3	8.1	7.8
4	10.8	10.4
5	13.5	13.0
6	16.2	15.6
7	18.9	18.2
8	21.6	20.8
9	24.3	23.4

† For use of logarithm tables, see p. 50. * Indicates change in the first two decimal places.

Table 4. Five-place Common Logarithms of Numbers—(*Continued*)

170—240

No.	L	0	1	2	3	4	5	6	7	8	9
170	23	045	070	096	121	147	172	198	223	249	274
171		300	325	350	376	401	426	452	477	502	528
172		553	578	603	629	654	679	704	729	754	776
173		805	830	855	880	905	930	955	980	*005	*030
174	24	055	080	105	130	155	180	204	229	254	279
175		304	329	353	378	403	428	452	477	502	527
176		551	576	601	625	650	674	699	724	748	773
177		797	822	846	871	895	920	944	969	993	*018
178	25	042	066	091	115	139	164	188	212	237	261
179		285	310	334	358	382	406	431	455	479	503
180		527	551	575	600	624	648	672	696	720	744
181		768	792	816	840	864	888	912	935	959	983
182	26	007	031	055	079	102	126	150	174	198	221
183		245	269	293	316	340	364	387	411	435	458
184		482	505	529	553	576	600	623	647	670	694
185		717	741	764	788	811	834	858	881	905	928
186		951	975	998	*021	*045	*068	*091	*114	*138	*161
187	27	184	207	231	254	277	300	323	346	370	393
188		416	439	462	485	508	531	554	577	600	623
189		646	669	692	715	738	761	784	807	830	853
190		875	898	921	944	967	990	*012	*035	*058	*081
191	28	103	126	149	172	194	217	240	262	285	308
192		330	353	375	398	421	443	466	488	511	533
193		556	578	601	623	646	668	691	713	735	758
194		780	803	825	847	870	892	914	937	959	981
195	29	003	026	048	070	092	115	137	159	181	203
196		226	248	270	292	314	336	358	380	403	425
197		447	469	491	513	535	557	579	601	623	645
198		667	688	710	732	754	776	798	820	842	863
199		885	907	929	951	973	994	*016	*038	*060	*081
200	30	103	125	146	168	190	211	233	255	276	298
201		320	341	363	384	406	428	449	471	492	514
202		535	557	578	600	621	643	664	685	707	728
203		750	771	792	814	835	856	878	899	920	942
204		963	984	*006	*027	*048	*069	*091	*112	*133	*154
205	31	175	197	218	239	260	281	302	323	345	366
206		387	408	429	450	471	492	513	534	555	576
207		597	618	639	660	681	702	723	744	765	785
208		806	827	848	869	890	911	931	952	973	994
209	32	015	035	056	077	098	118	139	160	181	201
210		222	243	263	284	305	325	346	366	387	408
211		428	449	469	490	511	531	552	572	593	613
212		634	654	675	695	715	736	756	777	797	818
213		838	858	879	899	919	940	960	980	*001	*021
214	33	041	062	082	102	122	143	163	183	203	224
215		244	264	284	304	325	345	365	385	405	425
216		445	465	486	506	526	546	566	586	606	626
217		646	666	686	706	726	746	766	786	806	826
218		846	866	885	905	925	945	965	985	*005	*025
219	34	044	064	084	104	124	143	163	183	203	223
220		242	262	282	301	321	341	361	380	400	420
221		439	459	479	498	518	537	557	577	596	616
222		635	655	674	694	713	733	753	772	792	811
223		830	850	869	889	908	928	947	967	986	*005
224	35	025	044	064	083	102	122	141	160	180	199
225		218	238	257	276	295	315	334	353	372	392
226		411	430	449	468	488	507	526	545	564	583
227		603	622	641	660	679	698	717	736	755	774
228		793	813	832	851	870	889	908	927	946	965
229		984	*003	*021	*040	*059	*078	*097	*116	*135	*154
230	36	173	192	211	229	248	267	286	305	324	342
231		361	380	399	418	436	455	474	493	511	530
232		549	568	586	605	624	642	661	680	698	717
233		736	754	773	791	810	829	847	866	884	903
234		922	940	959	977	996	*014	*033	*051	*070	*088
235	37	107	125	144	162	181	199	218	236	254	273
236		291	310	328	346	365	383	401	420	438	457
237		475	493	511	530	548	566	585	603	621	639
238		658	676	694	712	731	749	767	785	803	822
239		840	858	876	894	912	931	949	967	985	*003
240	38	021	039	057	075	093	112	130	148	166	184
No.	L	0	1	2	3	4	5	6	7	8	9

Proportional parts

	25	24	23	22	21	20	19	18
1	2.5	2.4	2.3	2.2	2.1	2.0	1.9	1.8
2	5.0	4.8	4.6	4.4	4.2	4.0	3.8	3.6
3	7.5	7.2	6.9	6.6	6.3	6.0	5.7	5.4
4	10.0	9.6	9.2	8.8	8.4	8.0	7.6	7.2
5	12.5	12.0	11.5	11.0	10.5	10.0	9.5	9.0
6	15.0	14.4	13.8	13.2	12.6	12.0	11.4	10.8
7	17.5	16.8	16.1	15.4	14.7	14.0	13.3	12.6
8	20.0	19.2	18.4	17.6	16.8	16.0	15.2	14.4
9	22.5	21.6	20.7	19.8	18.9	18.0	17.1	16.2

* Indicates change in the first two decimal places.

Table 4. Five-place Common Logarithms of Numbers—*(Continued)*

240—310

No.	L	0	1	2	3	4	5	6	7	8	9	Proportional parts
240	38	021	039	057	075	093	112	130	148	166	184	
241		202	220	238	256	274	292	310	328	346	364	
242		382	399	417	435	453	471	489	507	525	543	
243		561	579	596	614	632	650	668	686	703	721	
244		739	757	775	792	810	828	846	863	881	899	18
245		917	934	952	970	987	*005	*023	*041	*058	*076	1 1.8
246	39	094	111	129	146	164	182	199	217	235	252	2 3.6
247		270	287	305	322	340	358	375	393	410	428	3 5.4
248		445	463	480	498	515	533	550	568	585	602	4 7.2
249		620	637	655	672	690	707	724	742	759	777	5 9.0
												6 10.8
250	39	794	811	829	846	863	881	898	915	933	950	7 12.6
251		967	985	*002	*019	*037	*054	*071	*088	*106	*123	8 14.4
252	40	140	157	175	192	209	226	243	261	278	295	9 16.2
253		312	329	346	364	381	398	415	432	449	466	
254		483	500	518	535	552	569	586	603	620	637	
255		654	671	688	705	722	739	756	773	790	807	
256		824	841	858	875	892	909	926	943	960	976	
257		993	*010	*027	*044	*061	*078	*095	*111	*128	*145	
258	41	162	179	196	212	229	246	263	280	296	313	17
259		330	347	364	380	397	414	430	447	464	481	1 1.7
												2 3.4
260		497	514	531	547	564	581	597	614	631	647	3 5.1
261		664	681	697	714	731	747	764	780	797	814	4 6.8
262		830	847	863	880	896	913	929	946	963	979	5 8.5
263		996	*012	*029	*045	*062	*078	*095	*111	*127	*144	6 10.2
264	42	160	177	193	210	226	243	259	275	292	308	7 11.9
												8 13.6
265		325	341	357	374	390	406	423	439	456	472	9 15.3
266		488	504	521	537	553	570	586	602	619	635	
267		651	667	684	700	716	732	749	765	781	797	
268		813	830	846	862	878	894	911	927	943	959	
269		975	991	*008	*024	*040	*056	*072	*088	*104	*120	
270	43	136	152	169	185	201	217	233	249	265	281	
271		297	313	329	345	361	377	393	409	425	441	
272		457	473	489	505	521	537	553	569	584	600	16
273		616	632	648	664	680	696	712	727	743	759	1 1.6
274		775	791	807	823	838	854	870	886	902	917	2 3.2
												3 4.8
275		933	949	965	981	996	*012	*028	*044	*059	*075	4 6.4
276	44	091	107	122	138	154	170	185	201	217	232	5 8.0
277		248	264	279	295	311	326	342	358	373	389	6 9.6
278		404	420	436	451	467	483	498	514	529	545	7 11.2
279		560	576	592	607	623	638	654	669	685	700	8 12.8
												9 14.4
280		716	731	747	762	778	793	809	824	840	855	
281		871	886	902	917	932	948	963	979	994	*010	
282	45	025	040	056	071	086	102	117	133	148	163	
283		179	194	209	225	240	255	271	286	301	317	
284		332	347	362	378	393	408	423	439	454	469	
												15
285		484	500	515	530	545	561	576	591	606	621	
286		637	652	667	682	697	712	728	743	758	773	1 1.5
287		788	803	818	834	849	864	879	894	909	924	2 3.0
288		939	954	969	984	*000	*015	*030	*045	*060	*075	3 4.5
289	46	090	105	120	135	150	165	180	195	210	225	4 6.0
												5 7.5
290		240	255	270	285	300	315	330	345	359	374	6 9.0
291		389	404	419	434	449	464	479	494	509	523	7 10.5
292		538	553	568	583	598	613	627	642	657	672	8 12.0
293		687	702	716	731	746	761	776	790	805	820	9 13.5
294		835	850	864	879	894	909	923	938	953	967	
295		982	997	*012	*026	*041	*056	*070	*085	*100	*115	
296	47	129	144	159	173	188	202	217	232	246	261	
297		276	290	305	319	334	349	363	378	392	407	
298		422	436	451	465	480	494	509	524	538	553	14
299		567	582	596	611	625	640	654	669	683	698	1 1.4
300	47	712	727	741	756	770	784	799	813	828	842	2 2.8
301		857	871	886	900	914	929	943	958	972	986	3 4.2
302	48	001	015	029	044	058	073	087	101	116	130	4 5.6
303		144	159	173	187	202	216	230	245	259	273	5 7.0
304		287	302	316	330	344	359	373	387	402	416	6 8.4
												7 9.8
305		430	444	458	473	487	501	515	530	544	558	8 11.2
306		572	586	601	615	629	643	657	671	686	700	9 12.6
307		714	728	742	756	770	785	799	813	827	841	
308		855	869	883	897	911	926	940	954	968	982	
309		996	*010	*024	*038	*052	*066	*080	*094	*108	*122	
310	49	136	150	164	178	192	206	220	234	248	262	
No.	L	0	1	2	3	4	5	6	7	8	9	Proportional parts

* Indicates change in the first two decimal places.

Table 4. Five-place Common Logarithms of Numbers—(*Continued*)

310—380

No.	L	0	1	2	3	4	5	6	7	8	9	Proportional parts
310	49	136	150	164	178	192	206	220	234	248	262	
311		276	290	304	318	332	346	360	374	388	402	
312		415	429	443	457	471	485	499	513	527	541	
313		554	568	582	596	610	624	638	651	665	679	
314		693	707	721	734	748	762	776	790	803	817	
315		831	845	859	872	886	900	914	927	941	955	
316		969	982	996	*010	*024	*037	*051	*065	*079	*092	
317	50	106	120	133	147	161	174	188	202	215	229	
318		243	256	270	284	297	311	325	338	352	365	
319		379	393	406	420	433	447	461	474	488	501	
320		515	529	542	556	569	583	596	610	623	637	
321		651	664	678	691	705	718	732	745	759	772	
322		786	799	813	826	840	853	866	880	893	907	
323		920	934	947	961	974	987	*001	*014	*028	*041	14
324	51	055	068	081	095	108	121	135	148	162	175	1 1.4
												2 2.8
325		188	202	215	228	242	255	268	282	295	308	3 4.2
326		322	335	348	362	375	388	402	415	428	441	4 5.6
327		455	468	481	495	508	521	534	548	561	574	5 7.0
328		587	601	614	627	640	654	667	680	693	706	6 8.4
329		720	733	746	759	772	786	799	812	825	838	7 9.8
												8 11.2
330		851	865	878	891	904	917	930	943	957	970	9 12.6
331		983	996	*009	*022	*035	*048	*061	*075	*088	*101	
332	52	114	127	140	153	166	179	192	205	218	231	
333		244	257	271	284	297	310	323	336	349	362	
334		375	388	401	414	427	440	453	466	479	492	
335		504	517	530	543	556	569	582	595	608	621	
336		634	647	660	673	686	699	711	724	737	750	
337		763	776	789	802	815	827	840	853	866	879	
338		892	905	917	930	943	956	969	982	994	*007	
339	53	020	033	046	058	071	084	097	110	122	135	
340		148	161	173	186	199	212	224	237	250	263	
341		275	288	301	314	326	339	352	365	377	390	
342		403	415	428	441	453	466	479	491	504	517	
343		529	542	555	567	580	593	605	618	631	643	13
344		656	668	681	694	706	719	732	744	757	769	1 1.3
												2 2.6
345		782	795	807	820	832	845	857	870	883	895	3 3.9
346		908	920	933	945	958	970	983	995	*008	*020	4 5.2
347	54	033	045	058	070	083	095	108	120	133	145	5 6.5
348		158	170	183	195	208	220	233	245	258	270	6 7.8
349		283	295	307	320	332	345	357	370	382	394	7 9.1
												8 10.4
350	54	407	419	432	444	456	469	481	494	506	518	9 11.7
351		531	543	555	568	580	593	605	617	630	642	
352		654	667	679	691	704	716	728	741	753	765	
353		777	790	802	814	827	839	851	864	876	888	
354		900	913	925	937	949	962	974	986	998	*011	
355	55	023	035	047	060	072	084	096	108	121	133	
356		145	157	169	182	194	206	218	230	242	255	
357		267	279	291	303	315	328	340	352	364	376	
358		388	400	413	425	437	449	461	473	485	497	
359		509	522	534	546	558	570	582	594	606	618	
360		630	642	654	666	678	691	703	715	727	739	
361		751	763	775	787	799	811	823	835	847	859	
362		871	883	895	907	919	931	943	955	967	979	
363		991	*003	*015	*027	*038	*050	*062	*074	*086	*098	
364	56	110	122	134	146	158	170	182	194	205	217	12
												1 1.2
365		229	241	253	265	277	289	301	313	324	336	2 2.4
366		348	360	372	384	396	407	419	431	443	455	3 3.6
367		467	478	490	502	514	526	538	549	561	573	4 4.8
368		585	597	608	620	632	644	656	667	679	691	5 6.0
369		703	714	726	738	750	761	773	785	797	808	6 7.2
												7 8.4
370		820	832	844	855	867	879	891	902	914	926	8 9.6
371		937	949	961	972	984	996	*008	*019	*031	*043	9 10.8
372	57	054	066	078	089	101	113	124	136	148	159	
373		171	183	194	206	217	229	241	252	264	276	
374		287	299	310	322	334	345	357	368	380	392	
375		403	415	426	438	449	461	473	484	496	507	
376		519	530	542	553	565	577	588	600	611	623	
377		634	646	657	669	680	692	703	715	726	738	
378		749	761	772	784	795	807	818	830	841	852	
379		864	875	887	898	910	921	933	944	956	967	
380		978	990	*001	*013	*024	*035	*047	*058	*070	*081	
No.	L	0	1	2	3	4	5	6	7	8	9	Proportional parts

* Indicates change in the first two decimal places.

Table 4. Five-place Common Logarithms of Numbers—(*Continued*)

380—450

No.	L	0	1	2	3	4	5	6	7	8	9	Proportional parts
380	57	978	990	*001	*013	*024	*035	*047	*058	*070	*081	
381	58	093	104	115	127	138	149	161	172	184	195	
382		206	218	229	240	252	263	275	286	297	309	
383		320	331	343	354	365	377	388	399	411	422	
384		433	444	456	467	478	490	501	512	524	535	
385		546	557	569	580	591	602	614	625	636	647	
386		659	670	681	692	704	715	726	737	749	760	
387		771	782	794	805	816	827	838	850	861	872	
388		883	894	906	917	928	939	950	961	973	984	
389		995	*006	*017	*028	*040	*051	*062	*073	*084	*095	
390	59	106	118	129	140	151	162	173	184	195	207	
391		218	229	240	251	262	273	284	295	306	318	
392		329	340	351	362	373	384	395	406	417	428	
393		439	450	461	472	483	494	506	517	528	539	
394		550	561	572	583	594	605	616	627	638	649	
395		660	671	682	693	704	715	726	737	748	759	
396		770	780	791	802	813	824	835	846	857	868	
397		879	890	901	912	923	934	945	956	966	977	
398		988	999	*010	*021	*032	*043	*054	*065	*076	*086	
399	60	097	108	119	130	141	152	163	173	184	195	
400	60	206	217	228	239	249	260	271	282	293	304	
401		314	325	336	347	358	369	379	390	401	412	
402		423	433	444	455	466	477	487	498	509	520	
403		531	541	552	563	574	584	595	606	617	627	
404		638	649	660	670	681	692	703	713	724	735	
405		746	756	767	778	788	799	810	821	831	842	
406		853	863	874	885	895	906	917	927	938	949	
407		959	970	981	991	*002	*013	*023	*034	*045	*055	
408	61	066	077	087	098	109	119	130	140	151	162	
409		172	183	194	204	215	225	236	247	257	268	
410		278	289	300	310	321	331	342	352	363	374	
411		384	395	405	416	426	437	448	458	469	479	
412		490	500	511	521	532	542	553	563	574	584	
413		595	606	616	627	637	648	658	669	679	690	
414		700	711	721	731	742	752	763	773	784	794	
415		805	815	826	836	847	857	868	878	888	899	
416		909	920	930	941	951	962	972	982	993	*003	
417	62	014	024	034	045	055	066	076	086	097	107	
418		118	128	138	149	159	170	180	190	201	211	
419		221	232	242	252	263	273	284	294	304	315	
420		325	335	346	356	366	377	387	397	408	418	
421		428	439	449	459	469	480	490	500	511	521	
422		531	542	552	562	572	583	593	603	614	624	
423		634	644	655	665	675	685	696	706	716	726	
424		737	747	757	767	778	788	798	808	818	829	
425		839	849	859	870	880	890	900	910	921	931	
426		941	951	961	972	982	992	*002	*012	*022	*033	
427	63	043	053	063	073	083	094	104	114	124	134	
428		144	155	165	175	185	195	205	215	225	236	
429		246	256	266	276	286	296	306	317	327	337	
430		347	357	367	377	387	397	407	417	428	438	
431		448	458	468	478	488	498	508	518	528	538	
432		548	558	568	579	589	599	609	619	629	639	
433		649	659	669	679	689	699	709	719	729	739	
434		749	759	769	779	789	799	809	819	829	839	
435		849	859	869	879	889	899	909	919	929	939	
436		949	959	969	979	988	998	*008	*018	*028	*038	
437	64	048	058	068	078	088	098	108	118	128	137	
438		147	157	167	177	187	197	207	217	227	237	
439		246	256	266	276	286	296	306	316	326	335	
440		345	355	365	375	385	395	404	414	424	434	
441		444	454	464	473	483	493	503	513	523	532	
442		542	552	562	572	582	591	601	611	621	631	
443		640	650	660	670	680	689	699	709	719	729	
444		738	748	758	768	777	787	797	807	816	826	
445		836	846	856	865	875	885	895	904	914	924	
446		933	943	953	963	972	982	992	*002	*011	*021	
447	65	031	040	050	060	070	079	089	099	108	118	
448		128	137	147	157	167	176	186	196	205	215	
449		225	234	244	254	263	273	283	292	302	312	
450		321	331	341	350	360	369	379	389	398	408	
No.	L	0	1	2	3	4	5	6	7	8	9	Proportional parts

Proportional parts:

	11
1	1.1
2	2.2
3	3.3
4	4.4
5	5.5
6	6.6
7	7.7
8	8.8
9	9.9

	10
1	1.0
2	2.0
3	3.0
4	4.0
5	5.0
6	6.0
7	7.0
8	8.0
9	9.0

	9
1	0.9
2	1.8
3	2.7
4	3.6
5	4.5
6	5.4
7	6.3
8	7.2
9	8.1

* Indicates change in the first two decimal places.

Table 4. Five-place Common Logarithms of Numbers—*(Continued)*

450—520

No.	L	0	1	2	3	4	5	6	7	8	9
450	65	321	331	341	350	360	369	379	389	398	408
451		418	427	437	447	456	466	475	485	495	504
452		514	523	533	543	552	562	571	581	591	600
453		610	619	629	639	648	658	667	677	686	696
454		706	715	725	734	744	753	763	773	782	792
455		801	811	820	830	839	849	858	868	877	887
456		896	906	916	925	935	944	954	963	973	982
457		992	*001	*011	*020	*030	*039	*049	*058	*068	*077
458	66	087	096	106	115	124	134	143	153	162	172
459		181	191	200	210	219	229	238	247	257	266
460		276	285	295	304	314	323	332	342	351	361
461		370	380	389	398	408	417	427	436	445	455
462		464	474	483	492	502	511	521	530	539	549
463		558	567	577	586	596	605	614	624	633	642
464		652	661	671	680	689	699	708	717	727	736
465		745	755	764	773	783	792	801	811	820	829
466		839	848	857	867	876	885	894	904	913	922
467		932	941	950	960	969	978	987	997	*006	*015
468	67	025	034	043	052	062	071	080	090	099	108
469		117	127	136	145	154	164	173	182	191	201
470		210	219	228	238	247	256	265	274	284	293
471		302	311	321	330	339	348	357	367	376	385
472		394	403	413	422	431	440	449	459	468	477
473		486	495	504	514	523	532	541	550	560	569
474		578	587	596	605	614	624	633	642	651	660
475		669	679	688	697	706	715	724	733	742	752
476		761	770	779	788	797	806	815	825	834	843
477		852	861	870	879	888	897	906	916	925	934
478		943	952	961	970	979	988	997	*006	*015	*024
479	68	034	043	052	061	070	079	088	097	106	115
480		124	133	142	151	160	169	178	187	196	205
481		215	224	233	242	251	260	269	278	287	296
482		305	314	323	332	341	350	359	368	377	386
483		395	404	413	422	431	440	449	458	467	476
484		485	494	502	511	520	529	538	547	556	565
485		574	583	592	601	610	619	628	637	646	655
486		664	673	682	690	699	708	717	726	735	744
487		753	762	771	780	789	797	806	815	824	833
488		842	851	860	869	878	886	895	904	913	922
489		931	940	949	958	966	975	984	993	*002	*011
490	69	020	028	037	046	055	064	073	082	090	099
491		108	117	126	135	144	152	161	170	179	188
492		197	205	214	223	232	241	249	258	267	276
493		285	294	302	311	320	329	338	346	355	364
494		373	381	390	399	408	417	425	434	443	452
495		461	469	478	487	496	504	513	522	531	539
496		548	557	566	574	583	592	601	609	618	627
497		636	644	653	662	671	679	688	697	705	714
498		723	732	740	749	758	767	775	784	793	801
499		810	819	827	836	845	854	862	871	880	888
500	69	897	906	914	923	932	940	949	958	966	975
501		984	992	*001	*010	*018	*027	*036	*044	*053	*062
502	70	070	079	088	096	105	114	122	131	140	148
503		157	165	174	183	191	200	209	217	226	234
504		243	252	260	269	278	286	295	303	312	321
505		329	338	346	355	364	372	381	389	398	406
506		415	424	432	441	449	458	467	475	484	492
507		501	509	518	526	535	544	552	561	569	578
508		586	595	603	612	621	629	638	646	655	663
509		672	680	689	697	706	714	723	731	740	749
510		757	766	774	783	791	800	808	817	825	834
511		842	851	859	868	876	885	893	902	910	919
512		927	935	944	952	961	969	978	986	995	*003
513	71	012	020	029	037	046	054	063	071	079	088
514		096	105	113	122	130	139	147	155	164	172
515		181	189	198	206	214	223	231	240	248	257
516		265	273	282	290	299	307	315	324	332	341
517		349	357	366	374	383	391	399	408	416	425
518		433	441	450	458	467	475	483	492	500	508
519		517	525	533	542	550	559	567	575	584	592
520		600	609	617	625	634	642	650	659	667	675
No.	L	0	1	2	3	4	5	6	7	8	9

Proportional parts

	10
1	1.0
2	2.0
3	3.0
4	4.0
5	5.0
6	6.0
7	7.0
8	8.0
9	9.0

	9
1	0.9
2	1.8
3	2.7
4	3.6
5	4.5
6	5.4
7	6.3
8	7.2
9	8.1

	8
1	0.8
2	1.6
3	2.4
4	3.2
5	4.0
6	4.8
7	5.6
8	6.4
9	7.2

* Indicates change in the first two decimal places.

Table 4. Five-place Common Logarithms of Numbers—(*Continued*)

520—590

No.	L	0	1	2	3	4	5	6	7	8	9	Proportional parts
520	71	600	609	617	625	634	642	650	659	667	675	
521		684	692	700	709	717	725	734	742	750	759	
522		767	775	784	792	800	809	817	825	834	842	
523		850	858	867	875	883	892	900	908	917	925	
524		933	941	950	958	966	975	983	991	999	*008	
525	72	016	024	032	041	049	057	066	074	082	090	
526		099	107	115	123	132	140	148	156	165	173	
527		181	189	198	206	214	222	230	239	247	255	
528		263	272	280	288	296	305	313	321	329	337	
529		346	354	362	370	378	387	395	403	411	419	
530		428	436	444	452	460	469	477	485	493	501	
531		509	518	526	534	542	550	559	567	575	583	
532		591	599	607	616	624	632	640	648	656	665	
533		673	681	689	697	705	713	722	730	738	746	
534		754	762	770	779	787	795	803	811	819	827	9
535		835	844	852	860	868	876	884	892	900	908	1 0.9
536		916	925	933	941	949	957	965	973	981	989	2 1.8
537		997	*006	*014	*022	*030	*038	*046	*054	*062	*070	3 2.7
538	73	078	086	094	102	111	119	127	135	143	151	4 3.6
539		159	167	175	183	191	199	207	215	223	231	5 4.5
540		239	247	255	264	272	280	288	296	304	312	6 5.4
541		320	328	336	344	352	360	368	376	384	392	7 6.3
542		400	408	416	424	432	440	448	456	464	472	8 7.2
543		480	488	496	504	512	520	528	536	544	552	9 8.1
544		560	568	576	584	592	600	608	616	624	632	
545		640	648	656	664	672	679	687	695	703	711	
546		719	727	735	743	751	759	767	775	783	791	
547		799	807	815	823	830	838	846	854	862	870	
548		878	886	894	902	910	918	926	934	941	949	
549		957	965	973	981	989	997	*005	*013	*020	*028	
550	74	036	044	052	060	068	076	084	092	099	107	
551		115	123	131	139	147	155	162	170	178	186	
552		194	202	210	218	225	233	241	249	257	265	
553		273	280	288	296	304	312	320	327	335	343	
554		351	359	367	374	382	390	398	406	414	421	8
555		429	437	445	453	461	468	476	484	492	500	1 0.8
556		507	515	523	531	539	547	554	562	570	578	2 1.6
557		586	593	601	609	617	624	632	640	648	656	3 2.4
558		663	671	679	687	695	702	710	718	726	733	4 3.2
559		741	749	757	764	772	780	788	796	803	811	5 4.0
560		819	827	834	842	850	858	865	873	881	889	6 4.8
561		896	904	912	920	927	935	943	950	958	966	7 5.6
562		974	981	989	997	*005	*012	*020	*028	*035	*043	8 6.4
563	75	051	059	066	074	082	089	097	105	113	120	9 7.2
564		128	136	143	151	159	166	174	182	189	197	
565		205	213	220	228	236	243	251	259	266	274	
566		282	289	297	305	312	320	328	335	343	351	
567		358	366	374	381	389	397	404	412	420	427	
568		435	442	450	458	465	473	481	488	496	504	
569		511	519	526	534	542	549	557	565	572	580	
570		587	595	603	610	618	626	633	641	648	656	
571		664	671	679	686	694	702	709	717	724	732	
572		740	747	755	762	770	778	785	793	800	808	
573		815	823	831	838	846	853	861	868	876	884	
574		891	899	906	914	921	929	937	944	952	959	7
575		967	974	982	989	997	*005	*012	*020	*027	*035	1 0.7
576	76	042	050	057	065	072	080	087	095	103	110	2 1.4
577		118	125	133	140	148	155	163	170	178	185	3 2.1
578		193	200	208	215	223	230	238	245	253	260	4 2.8
579		268	275	283	290	298	305	313	320	328	335	5 3.5
580		343	350	358	365	373	380	388	395	403	410	6 4.2
581		418	425	433	440	448	455	462	470	477	485	7 4.9
582		492	500	507	515	522	530	537	545	552	559	8 5.6
583		567	574	582	589	597	604	612	619	626	634	9 6.3
584		641	649	656	664	671	678	686	693	701	708	
585		716	723	730	738	745	753	760	768	775	782	
586		790	797	805	812	819	827	834	842	849	856	
587		864	871	879	886	893	901	908	916	923	930	
588		938	945	953	960	967	975	982	989	997	*004	
589	77	012	019	026	034	041	048	056	063	070	078	
590		085	093	100	107	115	122	129	137	144	151	
No.	L	0	1	2	3	4	5	6	7	8	9	Proportional parts

* Indicates change in the first two decimal places.

Table 4. Five-place Common Logarithms of Numbers—(*Continued*)

590—660

No.	L	0	1	2	3	4	5	6	7	8	9
590	77	085	093	100	107	115	122	129	137	144	151
591		159	166	173	181	188	195	203	210	218	225
592		232	240	247	254	262	269	276	283	291	298
593		305	313	320	327	335	342	349	357	364	371
594		379	386	393	401	408	415	422	430	437	444
595		452	459	466	474	481	488	495	503	510	517
596		525	532	539	546	554	561	568	576	583	590
597		597	605	612	619	627	634	641	648	656	663
598		670	677	685	692	699	706	714	721	728	735
599		743	750	757	764	772	779	786	793	801	808
600	77	815	822	830	837	844	851	859	866	873	880
601		887	895	902	909	916	924	931	938	945	952
602		960	967	974	981	989	996	*003	*010	*017	*025
603	78	032	039	046	053	061	068	075	082	089	097
604		104	111	118	125	132	140	147	154	161	168
605		176	183	190	197	204	211	219	226	233	240
606		247	254	262	269	276	283	290	297	305	312
607		319	326	333	340	347	355	362	369	376	383
608		390	398	405	412	419	426	433	440	447	455
609		462	469	476	483	490	497	505	512	519	526
610		533	540	547	554	561	569	576	583	590	597
611		604	611	618	625	633	640	647	654	661	668
612		675	682	689	696	704	711	718	725	732	739
613		746	753	760	767	774	781	789	796	803	810
614		817	824	831	838	845	852	859	866	873	880
615		888	895	902	909	916	923	930	937	944	951
616		958	965	972	979	986	993	*000	*007	*014	*021
617	79	029	036	043	050	057	064	071	078	085	092
618		099	106	113	120	127	134	141	148	155	162
619		169	176	183	190	197	204	211	218	225	232
620		239	246	253	260	267	274	281	288	295	302
621		309	316	323	330	337	344	351	358	365	372
622		379	386	393	400	407	414	421	428	435	442
623		449	456	463	470	477	484	491	498	505	512
624		518	525	532	539	546	553	560	567	574	581
625		588	595	602	609	616	623	630	637	644	651
626		657	664	671	678	685	692	699	706	713	720
627		727	734	741	748	754	761	768	775	782	789
628		796	803	810	817	824	831	837	844	851	858
629		865	872	879	886	893	900	906	913	920	927
630		934	941	948	955	962	969	975	982	989	996
631	80	003	010	017	024	030	037	044	051	058	065
632		072	079	085	092	099	106	113	120	127	134
633		140	147	154	161	168	175	182	188	195	202
634		209	216	223	229	236	243	250	257	264	271
635		277	284	291	298	305	312	318	325	332	339
636		346	353	359	366	373	380	387	393	400	407
637		414	421	428	434	441	448	455	462	468	475
638		482	489	496	502	509	516	523	530	536	543
639		550	557	564	570	577	584	591	598	604	611
640		618	625	632	638	645	652	659	665	672	679
641		686	693	699	706	713	720	726	733	740	747
642		754	760	767	774	781	787	794	801	808	814
643		821	828	835	841	848	855	862	868	875	882
644		889	895	902	909	916	922	929	936	943	949
645		956	963	969	976	983	990	996	*003	*010	*017
646	81	023	030	037	043	050	057	064	070	077	084
647		090	097	104	111	117	124	131	137	144	151
648		158	164	171	178	184	191	198	204	211	218
649		224	231	238	245	251	258	265	271	278	285
650	81	291	298	305	311	318	325	331	338	345	351
651		358	365	371	378	385	391	398	405	411	418
652		425	431	438	445	451	458	465	471	478	485
653		491	498	505	511	518	525	531	538	544	551
654		558	564	571	578	584	591	598	604	611	618
655		624	631	637	644	651	657	664	671	677	684
656		690	697	704	710	717	723	730	737	743	750
657		757	763	770	776	783	790	796	803	809	816
658		823	829	836	842	849	856	862	869	875	882
659		889	895	902	908	915	921	928	935	941	948
660		954	961	968	974	981	987	994	*000	*007	*014
No.	L	0	1	2	3	4	5	6	7	8	9

Proportional parts

	8
1	0.8
2	1.6
3	2.4
4	3.2
5	4.0
6	4.8
7	5.6
8	6.4
9	7.2

	7
1	0.7
2	1.4
3	2.1
4	2.8
5	3.5
6	4.2
7	4.9
8	5.6
9	6.3

	6
1	0.6
2	1.2
3	1.8
4	2.4
5	3.0
6	3.6
7	4.2
8	4.8
9	5.4

* Indicates change in the first two decimal places.

Table 4. Five-place Common Logarithms of Numbers—(*Continued*)

660—730

No.	L	0	1	2	3	4	5	6	7	8	9
660	81	954	961	968	974	981	987	994	*000	*007	*014
661	82	020	027	033	040	046	053	060	066	073	079
662		086	092	099	105	112	119	125	132	138	145
663		151	158	164	171	178	184	191	197	204	210
664		217	223	230	236	243	250	256	263	269	276
665		282	289	295	302	308	315	321	328	334	341
666		347	354	360	367	374	380	387	393	400	406
667		413	419	426	432	439	445	452	458	465	471
668		478	484	491	497	504	510	517	523	530	536
669		543	549	556	562	569	575	582	588	595	601
670		607	614	620	627	633	640	646	653	659	666
671		672	679	685	692	698	705	711	718	724	730
672		737	743	750	756	763	769	776	782	789	795
673		802	808	814	821	827	834	840	847	853	860
674		866	872	879	885	892	898	905	911	918	924
675		930	937	943	950	956	963	969	975	982	988
676		995	*001	*008	*014	*020	*027	*033	*040	*046	*052
677	83	059	065	072	078	085	091	097	104	110	117
678		123	129	136	142	149	155	161	168	174	181
679		187	193	200	206	213	219	225	232	238	245
680		251	257	264	270	276	283	289	296	302	308
681		315	321	327	334	340	347	353	359	366	372
682		378	385	391	398	404	410	417	423	429	436
683		442	448	455	461	468	474	480	487	493	499
684		506	512	518	525	531	537	544	550	556	563
685		569	575	582	588	594	601	607	613	620	626
686		632	639	645	651	658	664	670	677	683	689
687		696	702	708	715	721	727	734	740	746	753
688		759	765	771	778	784	790	797	803	809	816
689		822	828	835	841	847	853	860	866	872	879
690		885	891	898	904	910	916	923	929	935	942
691		948	954	960	967	973	979	986	992	998	*004
692	84	011	017	023	029	036	042	048	055	061	067
693		073	080	086	092	098	105	111	117	123	130
694		136	142	148	155	161	167	173	180	186	192
695		198	205	211	217	223	230	236	242	248	255
696		261	267	273	280	286	292	298	305	311	317
697		323	330	336	342	348	354	361	367	373	379
698		386	392	398	404	410	417	423	429	435	442
699		448	454	460	466	473	479	485	491	497	504
700	84	510	516	522	528	535	541	547	553	559	566
701		572	578	584	590	597	603	609	615	621	628
702		634	640	646	652	658	665	671	677	683	689
703		696	702	708	714	720	726	733	739	745	751
704		757	763	770	776	782	788	794	800	807	813
705		819	825	831	837	844	850	856	862	868	874
706		880	887	893	899	905	911	917	924	930	936
707		942	948	954	960	967	973	979	985	991	997
708	85	003	009	016	022	028	034	040	046	052	059
709		065	071	077	083	089	095	101	107	114	120
710		126	132	138	144	150	156	163	169	175	181
711		187	193	199	205	211	217	224	230	236	242
712		248	254	260	266	272	278	285	291	297	303
713		309	315	321	327	333	339	345	352	358	364
714		370	376	382	388	394	400	406	412	418	425
715		431	437	443	449	455	461	467	473	479	485
716		491	497	503	510	516	522	528	534	540	546
717		552	558	564	570	576	582	588	594	600	606
718		612	618	625	631	637	643	649	655	661	667
719		673	679	685	691	697	703	709	715	721	727
720		733	739	745	751	757	763	769	775	781	788
721		794	800	806	812	818	824	830	836	842	848
722		854	860	866	872	878	884	890	896	902	908
723		914	920	926	932	938	944	950	956	962	968
724		974	980	986	992	998	*004	*010	*016	*022	*028
725	86	034	040	046	052	058	064	070	076	082	088
726		094	100	106	112	118	124	130	136	141	147
727		153	159	165	171	177	183	189	195	201	207
728		213	219	225	231	237	243	249	255	261	267
729		273	279	285	291	297	303	308	314	320	326
730		332	338	344	350	356	362	368	374	380	386
No.	L	0	1	2	3	4	5	6	7	8	9

Proportional parts

	7
1	0.7
2	1.4
3	2.1
4	2.8
5	3.5
6	4.2
7	4.9
8	5.6
9	6.3

	6
1	0.6
2	1.2
3	1.8
4	2.4
5	3.0
6	3.6
7	4.2
8	4.8
9	5.4

Proportional parts

* Indicates change in the first two decimal places.

Table 4. Five-place Common Logarithms of Numbers—(*Continued*)

730—800

No.	L	0	1	2	3	4	5	6	7	8	9	Proportional parts
730	86	332	338	344	350	356	362	368	374	380	386	
731		392	398	404	410	416	421	427	433	439	445	
732		451	457	463	469	475	481	487	493	499	504	
733		510	516	522	528	534	540	546	552	558	564	
734		570	576	581	587	593	599	605	611	617	623	
735		629	635	641	646	652	658	664	670	676	682	
736		688	694	700	705	711	717	723	729	735	741	
737		747	753	759	764	770	776	782	788	794	800	
738		806	812	817	823	829	835	841	847	853	859	
739		864	870	876	882	888	894	900	906	911	917	
740		923	929	935	941	947	953	958	964	970	976	
741		982	988	994	999	*005	*011	*017	*023	*029	*035	
742	87	040	046	052	058	064	070	075	081	087	093	
743		099	105	111	116	122	128	134	140	146	151	
744		157	163	169	175	181	186	192	198	204	210	
745		216	221	227	233	239	245	251	256	262	268	
746		274	280	286	291	297	303	309	315	320	326	
747		332	338	344	350	355	361	367	373	379	384	
748		390	396	402	408	413	419	425	431	437	442	
749		448	454	460	466	471	477	483	489	495	500	
750	87	506	512	518	523	529	535	541	547	552	558	
751		564	570	576	581	587	593	599	604	610	616	
752		622	628	633	639	645	651	656	662	668	674	
753		680	685	691	697	703	708	714	720	726	731	
754		737	743	749	754	760	766	772	777	783	789	
755		795	800	806	812	818	823	829	835	841	846	
756		852	858	864	869	875	881	887	892	898	904	
757		910	915	921	927	933	938	944	950	955	961	
758		967	973	978	984	990	996	*001	*007	*013	*018	
759	88	024	030	036	041	047	053	059	064	070	076	
760		081	087	093	099	104	110	116	121	127	133	
761		138	144	150	156	161	167	173	178	184	190	
762		196	201	207	213	218	224	230	235	241	247	
763		252	258	264	270	275	281	287	292	298	304	
764		309	315	321	326	332	338	343	349	355	360	
765		366	372	378	383	389	395	400	406	412	417	
766		423	429	434	440	446	451	457	463	468	474	
767		480	485	491	497	502	508	514	519	525	530	
768		536	542	547	553	559	564	570	576	581	587	
769		593	598	604	610	615	621	627	632	638	643	
770		649	655	660	666	672	677	683	689	694	700	
771		705	711	717	722	728	734	739	745	750	756	
772		762	767	773	779	784	790	795	801	807	812	
773		818	824	829	835	840	846	852	857	863	868	
774		874	880	885	891	897	902	908	913	919	925	
775		930	936	941	947	953	958	964	969	975	981	
776		986	992	997	*003	*009	*014	*020	*025	*031	*037	
777	89	042	048	053	059	064	070	076	081	087	092	
778		098	104	109	115	120	126	131	137	143	148	
779		154	159	165	170	176	182	187	193	198	204	
780		209	215	221	226	232	237	243	248	254	260	
781		265	271	276	282	287	293	298	304	310	315	
782		321	326	332	337	343	348	354	360	365	371	
783		376	382	387	393	398	404	409	415	421	426	
784		432	437	443	448	454	459	465	470	476	481	
785		487	493	498	504	509	515	520	526	531	537	
786		542	548	553	559	564	570	575	581	586	592	
787		597	603	609	614	620	625	631	636	642	647	
788		653	658	664	669	675	680	686	691	697	702	
789		708	713	719	724	730	735	741	746	752	757	
790		763	768	774	779	785	790	796	801	807	812	
791		818	823	829	834	840	845	851	856	862	867	
792		873	878	883	889	894	900	905	911	916	922	
793		927	933	938	944	949	955	960	966	971	977	
794		982	988	993	998	*004	*009	*015	*020	*026	*031	
795	90	037	042	048	053	059	064	069	075	080	086	
796		091	097	102	108	113	119	124	129	135	140	
797		146	151	157	162	168	173	179	184	189	195	
798		200	206	211	217	222	227	233	238	244	249	
799		255	260	266	271	276	282	287	293	298	304	
800		309	314	320	325	331	336	342	347	352	358	
No.	L	0	1	2	3	4	5	6	7	8	9	Proportional parts

Proportional parts:

	6
1	0.6
2	1.2
3	1.8
4	2.4
5	3.0
6	3.6
7	4.2
8	4.8
9	5.4

	5
1	0.5
2	1.0
3	1.5
4	2.0
5	2.5
6	3.0
7	3.5
8	4.0
9	4.5

* Indicates change in the first two decimal places.

Table 4. Five-place Common Logarithms of Numbers—(*Continued*)

800—870

No.	L	0	1	2	3	4	5	6	7	8	9	Proportional parts
800	90	309	314	320	325	331	336	342	347	352	358	
801		363	369	374	380	385	390	396	401	407	412	
802		417	423	428	434	439	445	450	455	461	466	
803		472	477	482	488	493	499	504	509	515	520	
804		526	531	536	542	547	553	558	563	569	574	
805		580	585	590	596	601	607	612	617	623	628	
806		634	639	644	650	655	660	666	671	677	682	
807		687	693	698	704	709	714	720	725	730	736	
808		741	747	752	757	763	768	773	779	784	789	
809		795	800	806	811	816	822	827	832	838	843	
810		849	854	859	865	870	875	881	886	891	897	
811		902	907	913	918	924	929	934	940	945	950	
812		956	961	966	972	977	982	988	993	998	*004	
813	91	009	014	020	025	030	036	041	046	052	057	
814		062	068	073	078	084	089	094	100	105	110	
815		116	121	126	132	137	142	148	153	158	164	
816		169	174	180	185	190	196	201	206	212	217	
817		222	228	233	238	243	249	254	259	265	270	
818		275	281	286	291	297	302	307	312	318	323	
819		328	334	339	344	350	355	360	365	371	376	6
820		381	387	392	397	403	408	413	418	424	429	1 | 0.6
821		434	440	445	450	455	461	466	471	477	482	2 | 1.2
822		487	492	498	503	508	514	519	524	529	535	3 | 1.8
823		540	545	551	556	561	566	572	577	582	587	4 | 2.4
824		593	598	603	609	614	619	624	630	635	640	5 | 3.0
825		645	651	656	661	666	672	677	682	687	693	6 | 3.6 7 | 4.2 8 | 4.8
826		698	703	709	714	719	724	730	735	740	745	9 | 5.4
827		751	756	761	766	772	777	782	787	793	798	
828		803	808	814	819	824	829	834	840	845	850	
829		855	861	866	871	876	882	887	892	897	903	
830		908	913	918	924	929	934	939	944	950	955	
831		960	965	971	976	981	986	991	997	*002	*007	
832	92	012	018	023	028	033	038	044	049	054	059	
833		065	070	075	080	085	091	096	101	106	111	
834		117	122	127	132	137	143	148	153	158	163	
835		169	174	179	184	189	195	200	205	210	215	
836		221	226	231	236	241	247	252	257	262	267	
837		273	278	283	288	293	298	304	309	314	319	
838		324	330	335	340	345	350	355	361	366	371	
839		376	381	387	392	397	402	407	412	418	423	
840		428	433	438	443	449	454	459	464	469	474	
841		480	485	490	495	500	505	511	516	521	526	
842		531	536	542	547	552	557	562	567	572	578	
843		583	588	593	598	603	609	614	619	624	629	
844		634	639	645	650	655	660	665	670	675	681	5
845		686	691	696	701	706	711	717	722	727	732	1 | 0.5
846		737	742	747	752	758	763	768	773	778	783	2 | 1.0
847		788	793	799	804	809	814	819	824	829	834	3 | 1.5
848		840	845	850	855	860	865	870	875	881	886	4 | 2.0
849		891	896	901	906	911	916	921	927	932	937	5 | 2.5
850	92	942	947	952	957	962	967	973	978	983	988	6 | 3.0 7 | 3.5
851		993	998	*003	*008	*013	*018	*024	*029	*034	*039	8 | 4.0
852	93	044	049	054	059	064	069	075	080	085	090	9 | 4.5
853		095	100	105	110	115	120	125	131	136	141	
854		146	151	156	161	166	171	176	181	186	192	
855		197	202	207	212	217	222	227	232	237	242	
856		247	252	258	263	268	273	278	283	288	293	
857		298	303	308	313	318	323	328	334	339	344	
858		349	354	359	364	369	374	379	384	389	394	
859		399	404	409	414	420	425	430	435	440	445	
860		450	455	460	465	470	475	480	485	490	495	
861		500	505	510	515	520	526	531	536	541	546	
862		551	556	561	566	571	576	581	586	591	596	
863		601	606	611	616	621	626	631	636	641	646	
864		651	656	661	666	671	677	682	687	692	697	
865		702	707	712	717	722	727	732	737	742	747	
866		752	757	762	767	772	777	782	787	792	797	
867		802	807	812	817	822	827	832	837	842	847	
868		852	857	862	867	872	877	882	887	892	897	
869		902	907	912	917	922	927	932	937	942	947	
870		952	957	962	967	972	977	982	987	992	997	
No.	L	0	1	2	3	4	5	6	7	8	9	Proportional parts

* Indicates change in the first two decimal places.

Table 4. Five-place Common Logarithms of Numbers—(*Continued*)

870—940

No.	L	0	1	2	3	4	5	6	7	8	9	Proportional parts
870	93	952	957	962	967	972	977	982	987	992	997	
871	94	002	007	012	017	022	027	032	037	042	047	
872		052	057	062	067	072	077	082	087	091	096	
873		101	106	111	116	121	126	131	136	141	146	
874		151	156	161	166	171	176	181	186	191	196	
875		201	206	211	216	221	226	231	236	240	245	
876		250	255	260	265	270	275	280	285	290	295	
877		300	305	310	315	320	325	330	335	340	345	
878		349	354	359	364	369	374	379	384	389	394	
879		399	404	409	414	419	424	429	433	438	443	
880		448	453	458	463	468	473	478	483	488	493	
881		498	503	507	512	517	522	527	532	537	542	
882		547	552	557	562	567	571	576	581	586	591	
883		596	601	606	611	616	621	626	630	635	640	
884		645	650	655	660	665	670	675	680	685	689	
885		694	699	704	709	714	719	724	729	734	738	
886		743	748	753	758	763	768	773	778	783	787	
887		792	797	802	807	812	817	822	827	832	836	
888		841	846	851	856	861	866	871	876	880	885	
889		890	895	900	905	910	915	919	924	929	934	5
890		939	944	949	954	959	963	968	973	978	983	1 0.5
891		988	993	998	*002	*007	*012	*017	*022	*027	*032	2 1.0
892	95	036	041	046	051	056	061	066	071	075	080	3 1.5
893		085	090	095	100	105	109	114	119	124	129	4 2.0
894		134	139	143	148	153	158	163	168	173	177	5 2.5
895		182	187	192	197	202	207	211	216	221	226	6 3.0
896		231	236	240	245	250	255	260	265	270	274	7 3.5
897		279	284	289	294	299	303	308	313	318	323	8 4.0
898		328	332	337	342	347	352	357	361	366	371	9 4.5
899		376	381	386	390	395	400	405	410	415	419	
900	95	424	429	434	439	444	448	453	458	463	468	
901		472	477	482	487	492	497	501	506	511	516	
902		521	525	530	535	540	545	550	554	559	564	
903		569	574	578	583	588	593	598	602	607	612	
904		617	622	626	631	636	641	646	650	655	660	
905		665	670	674	679	684	689	694	698	703	708	
906		713	718	722	727	732	737	742	746	751	756	
907		761	766	770	775	780	785	789	794	799	804	
908		809	813	818	823	828	832	837	842	847	852	
909		856	861	866	871	875	880	885	890	895	899	
910		904	909	914	918	923	928	933	938	942	947	
911		952	957	961	966	971	976	980	985	990	995	
912		999	*004	*009	*014	*019	*023	*028	*033	*038	*042	
913	96	047	052	057	061	066	071	076	080	085	090	
914		095	099	104	109	114	118	123	128	133	137	4
915		142	147	152	156	161	166	171	175	180	185	1 0.4
916		190	194	199	204	209	213	218	223	227	232	2 0.8
917		237	242	246	251	256	261	265	270	275	280	3 1.2
918		284	289	294	298	303	308	313	317	322	327	4 1.6
919		332	336	341	346	350	355	360	365	369	374	5 2.0
920		379	384	388	393	398	402	407	412	417	421	6 2.4
921		426	431	435	440	445	450	454	459	464	468	7 2.8
922		473	478	483	487	492	497	501	506	511	515	8 3.2
923		520	525	530	534	539	544	548	553	558	563	9 3.6
924		567	572	577	581	586	591	595	600	605	609	
925		614	619	624	628	633	638	642	647	652	656	
926		661	666	670	675	680	685	689	694	699	703	
927		708	713	717	722	727	731	736	741	745	750	
928		755	759	764	769	774	778	783	788	792	797	
929		802	806	811	816	820	825	830	834	839	844	
930		848	853	858	862	867	872	876	881	886	890	
931		895	900	904	909	914	918	923	928	932	937	
932		942	946	951	956	960	965	970	974	979	984	
933		988	993	997	*002	*007	*011	*016	*021	*025	*030	
934	97	035	039	044	049	053	058	063	067	072	077	
935		081	086	090	095	100	104	109	114	118	123	
936		128	132	137	142	146	151	155	160	165	169	
937		174	179	183	188	192	197	202	206	211	216	
938		220	225	230	234	239	243	248	253	257	262	
939		267	271	276	280	285	290	294	299	304	308	
940		313	317	322	327	331	336	341	345	350	354	
No.	L	0	1	2	3	4	5	6	7	8	9	Proportional parts

* Indicates change in the first two decimal places.

Table 4. Five-place Common Logarithms of Numbers—(*Concluded*)
940—1000

No.	L	0	1	2	3	4	5	6	7	8	9	Proportional parts
940	97	313	317	322	327	331	336	341	345	350	354	
941		359	364	368	373	377	382	387	391	396	400	
942		405	410	414	419	424	428	433	437	442	447	
943		451	456	460	465	470	474	479	483	488	493	
944		497	502	506	511	516	520	525	529	534	539	
945		543	548	552	557	562	566	571	575	580	585	
946		589	594	598	603	607	612	617	621	626	630	
947		635	640	644	649	653	658	663	667	672	676	
948		681	685	690	695	699	704	708	713	717	722	
949		727	731	736	740	745	750	754	759	763	768	
950		772	777	782	786	791	795	800	804	809	813	
951		818	823	827	832	836	841	845	850	855	859	
952		864	868	873	877	882	887	891	896	900	905	
953		909	914	918	923	928	932	937	941	946	950	
954		955	959	964	968	973	978	982	987	991	996	5
955	98	000	005	009	014	019	023	028	032	037	041	1 | 0.5
956		046	050	055	059	064	069	073	078	082	087	2 | 1.0
957		091	096	100	105	109	114	118	123	127	132	3 | 1.5
958		137	141	146	150	155	159	164	168	173	177	4 | 2.0
959		182	186	191	195	200	205	209	214	218	223	5 | 2.5
960		227	232	236	241	245	250	254	259	263	268	6 | 3.0
961		272	277	281	286	290	295	299	304	308	313	7 | 3.5
962		318	322	327	331	336	340	345	349	354	358	8 | 4.0
963		363	367	372	376	381	385	390	394	399	403	9 | 4.5
964		408	412	417	421	426	430	435	439	444	448	
965		453	457	462	466	471	475	480	484	489	493	
966		498	502	507	511	516	520	525	529	534	538	
967		543	547	552	556	561	565	570	574	579	583	
968		588	592	597	601	605	610	614	619	623	628	
969		632	637	641	646	650	655	659	664	668	673	
970		677	682	686	691	695	700	704	709	713	717	
971		722	726	731	735	740	744	749	753	758	762	
972		767	771	776	780	785	789	793	798	802	807	
973		811	816	820	825	829	834	838	843	847	851	
974		856	860	865	869	874	878	883	887	892	896	
975		900	905	909	914	918	923	927	932	936	941	
976		945	949	954	958	963	967	972	976	981	985	
977		989	994	998	*003	*007	*012	*016	*021	*025	*029	
978	99	034	038	043	047	052	056	061	065	069	074	4
979		078	083	087	092	096	100	105	109	114	118	1 | 0.4
980		123	127	131	136	140	145	149	154	158	162	2 | 0.8
981		167	171	176	180	185	189	193	198	202	207	3 | 1.2
982		211	216	220	224	229	233	238	242	247	251	4 | 1.6
983		255	260	264	269	273	277	282	286	291	295	5 | 2.0
984		300	304	308	313	317	322	326	330	335	339	6 | 2.4
985		344	348	352	357	361	366	370	374	379	383	7 | 2.8
986		388	392	397	401	405	410	414	419	423	427	8 | 3.2
987		432	436	441	445	449	454	458	463	467	471	9 | 3.6
988		476	480	484	489	493	498	502	506	511	515	
989		520	524	528	533	537	542	546	550	555	559	
990		564	568	572	577	581	585	590	594	599	603	
991		607	612	616	621	625	629	634	638	642	647	
992		651	656	660	664	669	673	677	682	686	691	
993		695	699	704	708	712	717	721	726	730	734	
994		739	743	747	752	756	760	765	769	774	778	
995		782	787	791	795	800	804	808	813	817	822	
996		826	830	835	839	843	848	852	856	861	865	
997		870	874	878	883	887	891	896	900	904	909	
998		913	917	922	926	930	935	939	944	948	952	
999		957	961	965	970	974	978	983	987	991	996	
1000	00	000	004	009	013	017	022	026	030	035	039	
No.	L	0	1	2	3	4	5	6	7	8	9	Proportional parts

* Indicates change in the first two decimal places.

Table 5. Natural Trigonometric Functions and Their Logarithms

Degrees	Radians	Sines	Log sines	Cosines	Log cosines	Tangents	Log tangents	Cotangents	Log cotangents	Radians	Degrees
		Nat.		Nat.		Nat.		Nat.			
0° 00′	0.0000	0.0000	......	1.0000	0.0000	0.0000	......	....	......	1.5708	90° 00′
10	.0029	.0029	7.4637	1.0000	.0000	.0029	7.4637	343.77	2.5363	1.5679	50
20	.0058	.0058	7.7648	1.0000	10.0000	.0058	7.7648	171.89	2.2352	1.5650	40
30	.0087	.0087	7.9408	1.0000	10.0000	.0087	7.9409	114.59	2.0591	1.5621	30
40	.0116	.0116	8.0658	0.9999	10.0000	.0116	8.0658	85.940	1.9342	1.5592	20
50	.0145	.0145	8.1627	.9999	10.0000	.0146	8.1627	68.750	1.8373	1.5563	10
1° 00′	.0175	.0175	8.2419	.9999	9.9999	.0175	8.2419	57.290	1.7581	1.5533	89° 00′
10	.0204	.0204	8.3088	.9998	9.9999	.0204	8.3089	49.104	1.6911	1.5504	50
20	.0233	.0233	8.3668	.9997	9.9999	.0233	8.3669	42.964	1.6331	1.5475	40
30	.0262	.0262	8.4179	.9997	9.9999	.0262	8.4181	38.188	1.5819	1.5446	30
40	.0291	.0291	8.4637	.9996	9.9998	.0291	8.4639	34.368	1.5362	1.5417	20
50	.0320	.0320	8.5050	.9995	9.9998	.0320	8.5053	31.242	1.4947	1.5388	10
2° 00′	.0349	.0349	8.5428	.9994	9.9997	.0349	8.5431	28.636	1.4569	1.5359	88° 00′
10	.0378	.0378	8.5776	.9993	9.9997	.0378	8.5779	26.432	1.4221	1.5330	50
20	.0407	.0407	8.6097	.9992	9.9996	.0408	8.6101	24.542	1.3899	1.5301	40
30	.0436	.0436	8.6397	.9991	9.9996	.0437	8.6401	22.904	1.3599	1.5272	30
40	.0465	.0465	8.6677	.9989	9.9995	.0466	8.6682	21.470	1.3318	1.5243	20
50	.0495	.0494	8.6940	.9988	9.9995	.0495	8.6945	20.206	1.3055	1.5213	10
3° 00′	.0524	.0523	8.7188	.9986	9.9994	.0524	8.7194	19.081	1.2806	1.5184	87° 00′
10	.0553	.0552	8.7423	.9985	9.9993	.0553	8.7429	18.075	1.2571	1.5155	50
20	.0582	.0581	8.7645	.9983	9.9993	.0582	8.7653	17.169	1.2348	1.5126	40
30	.0611	.0611	8.7857	.9981	9.9992	.0612	8.7865	16.350	1.2135	1.5097	30
40	.0640	.0640	8.8059	.9980	9.9991	.0641	8.8067	15.605	1.1933	1.5068	20
50	.0669	.0669	8.8251	.9978	9.9990	.0670	8.8261	14.924	1.1739	1.5039	10
4° 00′	.0698	.0698	8.8436	.9976	9.9989	.0699	8.8446	14.301	1.1554	1.5010	86° 00′
10	.0727	.0727	8.8613	.9974	9.9989	.0729	8.8624	13.727	1.1376	1.4981	50
20	.0756	.0756	8.8783	.9971	9.9988	.0758	8.8795	13.197	1.1205	1.4952	40
30	.0785	.0785	8.8946	.9969	9.9987	.0787	8.8960	12.706	1.1040	1.4923	30
40	.0814	.0814	8.9104	.9967	9.9986	.0816	8.9119	12.251	1.0882	1.4893	20
50	.0844	.0843	8.9256	.9964	9.9985	.0846	8.9272	11.826	1.0728	1.4864	10
5° 00′	.0873	.0872	8.9403	.9962	9.9983	.0875	8.9420	11.430	1.0581	1.4835	85° 00′
10	.0902	.0901	8.9545	.9959	9.9982	.0904	8.9563	11.059	1.0437	1.4806	50
20	.0931	.0930	8.9683	.9957	9.9981	.0934	8.9701	10.712	1.0299	1.4777	40
30	.0960	.0959	8.9816	.9954	9.9980	.0963	8.9836	10.385	1.0164	1.4748	30
40	.0989	.0987	8.9945	.9951	9.9979	.0992	8.9966	10.078	1.0034	1.4719	20
50	.1018	.1016	9.0070	.9948	9.9978	.1022	9.0093	9.7882	0.9907	1.4690	10
6° 00′	.1047	.1045	9.0192	.9945	9.9976	.1051	9.0216	9.5144	.9784	1.4661	84° 00′
10	.1076	.1074	9.0311	.9942	9.9975	.1081	9.0336	9.2553	.9664	1.4632	50
20	.1105	.1103	9.0426	.9939	9.9973	.1110	9.0453	9.0098	.9547	1.4603	40
30	.1134	.1132	9.0539	.9936	9.9972	.1139	9.0567	8.7769	.9433	1.4573	30
40	.1164	.1161	9.0648	.9932	9.9971	.1169	9.0678	8.5556	.9323	1.4544	20
50	.1193	.1190	9.0755	.9929	9.9969	.1198	9.0786	8.3450	.9214	1.4515	10
7° 00′	.1222	.1219	9.0859	.9926	9.9968	.1228	9.0891	8.1443	.9109	1.4486	83° 00′
10	.1251	.1248	9.0961	.9922	9.9966	.1257	9.0995	7.9530	.9005	1.4457	50
20	.1280	.1276	9.1060	.9918	9.9964	.1287	9.1096	7.7704	.8904	1.4428	40
30	.1309	.1305	9.1157	.9914	9.9963	.1317	9.1194	7.5958	.8806	1.4399	30
40	.1338	.1334	9.1252	.9911	9.9961	.1346	9.1291	7.4287	.8709	1.4370	20
50	.1367	.1363	9.1345	.9907	9.9959	.1376	9.1385	7.2687	.8615	1.4341	10
8° 00′	.1396	.1392	9.1436	.9903	9.9958	.1405	9.1478	7.1154	.8522	1.4312	82° 00′
10	.1425	.1421	9.1525	.9899	9.9956	.1435	9.1569	6.9682	.8431	1.4283	50
20	.1454	.1449	9.1612	.9894	9.9954	.1465	9.1658	6.8269	.8342	1.4254	40
30	.1484	.1478	9.1697	.9890	9.9952	.1495	9.1745	6.6912	.8255	1.4224	30
40	.1513	.1507	9.1781	.9886	9.9950	.1524	9.1831	6.5606	.8169	1.4195	20
50	.1542	.1536	9.1863	.9881	9.9948	.1554	9.1915	6.4348	.8085	1.4166	10
9° 00′	.1571	.1564	9.1943	.9877	9.9946	.1584	9.1997	6.3138	.8003	1.4137	81° 00′
10	.1600	.1593	9.2022	.9872	9.9944	.1614	9.2078	6.1970	.7922	1.4108	50
20	.1629	.1622	9.2100	.9868	9.9942	.1644	9.2158	6.0844	.7842	1.4079	40
30	.1658	.1651	9.2176	.9863	9.9940	.1673	9.2236	5.9758	.7764	1.4050	30
40	.1687	.1679	9.2251	.9858	9.9938	.1703	9.2313	5.8708	.7687	1.4021	20
50	.1716	.1708	9.2324	.9853	9.9936	.1733	9.2389	5.7694	.7611	1.3992	10
10° 00′	.1745	.1737	9.2397	.9848	9.9934	.1763	9.2463	5.6713	.7537	1.3963	80° 00′
10	.1774	.1765	9.2468	.9843	9.9931	.1793	9.2536	5.5764	.7464	1.3934	50
20	.1804	.1794	9.2538	.9838	9.9929	.1823	9.2609	5.4845	.7391	1.3904	40
30	.1833	.1822	9.2606	.9833	9.9927	.1853	9.2680	5.3955	.7320	1.3875	30
40	.1862	.1851	9.2674	.9827	9.9924	.1884	9.2750	5.3093	.7250	1.3846	20
50	.1891	.1880	9.2741	.9822	9.9922	.1914	9.2819	5.2257	.7181	1.3817	10
11° 00′	.1920	.1908	9.2806	.9816	9.9920	.1944	9.2887	5.1446	.7114	1.3788	79° 00′
10	.1949	.1937	9.2871	.9811	9.9917	.1974	9.2954	5.0658	.7047	1.3759	50
20	.1978	.1965	9.2934	.9805	9.9915	.2004	9.3020	4.9894	.6981	1.3730	40
30	.2007	.1994	9.2997	.9799	9.9912	.2035	9.3085	4.9152	.6915	1.3701	30
40	.2036	.2022	9.3058	.9793	9.9909	.2065	9.3149	4.8430	.6851	1.3672	20
50	.2065	.2051	9.3119	.9788	9.9907	.2095	9.3212	4.7729	.6788	1.3643	10
12° 00′	.2094	.2079	9.3179	.9782	9.9904	.2126	9.3275	4.7046	.6725	1.3614	78° 00′
		Nat.		Nat.		Nat.		Nat.			
Degrees	Radians	Cosines	Log cosines	Sines	Log sines	Cotangents	Log cotangents	Tangents	Log tangents	Radians	Degrees

Table 5. Natural Trigonometric Functions and Their Logarithms—*(Continued)*

Degrees	Radians	Sines	Log sines	Cosines	Log cosines	Tangents	Log tangents	Cotangents	Log cotangents	Radians	Degrees
		Nat.		Nat.		Nat.		Nat.			
12° 00′	0.2094	0.2079	9.3179	0.9782	9.9904	0.2126	9.3275	4.7046	0.6725	1.3614	78° 00′
10	.2123	.2108	9.3238	.9775	9.9901	.2156	9.3337	4.6383	.6664	1.3584	50
20	.2153	.2136	9.3296	.9769	9.9899	.2186	9.3397	4.5736	.6603	1.3555	40
30	.2182	.2164	9.3353	.9763	9.9896	.2217	9.3458	4.5107	.6542	1.3526	30
40	.2211	.2193	9.3410	.9757	9.9893	.2248	9.3517	4.4494	.6483	1.3497	20
50	.2240	.2221	9.3466	.9750	9.9890	.2278	9.3576	4.3897	.6424	1.3468	10
13° 00′	.2269	.2250	9.3521	.9744	9.9887	.2309	9.3634	4.3315	.6366	1.3439	77° 00′
10	.2298	.2278	9.3575	.9737	9.9884	.2339	9.3691	4.2747	.6309	1.3410	50
20	.2327	.2306	9.3629	.9730	9.9881	.2370	9.3748	4.2193	.6252	1.3381	40
30	.2356	.2335	9.3682	.9724	9.9878	.2401	9.3804	4.1653	.6197	1.3352	30
40	.2385	.2363	9.3734	.9717	9.9875	.2432	9.3859	4.1126	.6141	1.3323	20
50	.2414	.2391	9.3786	.9710	9.9872	.2462	9.3914	4.0611	.6086	1.3294	10
14° 00′	.2443	.2419	9.3837	.9703	9.9869	.2493	9.3968	4.0108	.6032	1.3265	76° 00′
10	.2473	.2447	9.3887	.9696	9.9866	.2524	9.4021	3.9617	.5979	1.3235	50
20	.2502	.2476	9.3937	.9689	9.9863	.2555	9.4074	3.9136	.5926	1.3206	40
30	.2531	.2504	9.3986	.9682	9.9859	.2586	9.4127	3.8667	.5873	1.3177	30
40	.2560	.2532	9.4035	.9674	9.9856	.2617	9.4178	3.8208	.5822	1.3148	20
50	.2589	.2560	9.4083	.9667	9.9853	.2648	9.4230	3.7760	.5770	1.3119	10
15° 00′	.2618	.2588	9.4130	.9659	9.9849	.2680	9.4281	3.7321	.5720	1.3090	75° 00′
10	.2647	.2616	9.4177	.9652	9.9846	.2711	9.4331	3.6891	.5669	1.3061	50
20	.2676	.2644	9.4223	.9644	9.9843	.2742	9.4381	3.6471	.5619	1.3032	40
30	.2705	.2672	9.4269	.9636	9.9839	.2773	9.4430	3.6059	.5570	1.3003	30
40	.2734	.2700	9.4314	.9629	9.9836	.2805	9.4479	3.5656	.5521	1.2974	20
50	.2763	.2728	9.4359	.9621	9.9832	.2836	9.4527	3.5261	.5473	1.2945	10
16° 00′	.2793	.2756	9.4403	.9613	9.9828	.2868	9.4575	3.4874	.5425	1.2915	74° 00′
10	.2822	.2784	9.4447	.9605	9.9825	.2899	9.4622	3.4495	.5378	1.2886	50
20	.2851	.2812	9.4491	.9596	9.9821	.2931	9.4669	3.4124	.5331	1.2857	40
30	.2880	.2840	9.4533	.9588	9.9817	.2962	9.4716	3.3759	.5284	1.2828	30
40	.2909	.2868	9.4576	.9580	9.9814	.2994	9.4762	3.3402	.5238	1.2799	20
50	.2938	.2896	9.4618	.9572	9.9810	.3026	9.4808	3.3052	.5192	1.2770	10
17° 00′	.2967	.2924	9.4659	.9563	9.9806	.3057	9.4853	3.2709	.5147	1.2741	73° 00′
10	.2996	.2952	9.4701	.9555	9.9802	.3089	9.4898	3.2371	.5102	1.2712	50
20	.3025	.2979	9.4741	.9546	9.9798	.3121	9.4943	3.2041	.5057	1.2683	40
30	.3054	.3007	9.4781	.9537	9.9794	.3153	9.4987	3.1716	.5013	1.2654	30
40	.3083	.3035	9.4821	.9528	9.9790	.3185	9.5031	3.1397	.4969	1.2625	20
50	.3113	.3063	9.4861	.9520	9.9786	.3217	9.5075	3.1084	.4925	1.2595	10
18° 00′	.3142	.3090	9.4900	.9511	9.9782	.3249	9.5118	3.0777	.4882	1.2566	72° 00′
10	.3171	.3118	9.4939	.9502	9.9778	.3281	9.5161	3.0475	.4839	1.2537	50
20	.3200	.3145	9.4977	.9492	9.9774	.3314	9.5203	3.0178	.4797	1.2508	40
30	.3229	.3173	9.5015	.9483	9.9770	.3346	9.5245	2.9887	.4755	1.2479	30
40	.3258	.3201	9.5052	.9474	9.9765	.3378	9.5287	2.9600	.4713	1.2450	20
50	.3287	.3228	9.5090	.9465	9.9761	.3411	9.5329	2.9319	.4672	1.2421	10
19° 00′	.3316	.3256	9.5126	.9455	9.9757	.3443	9.5370	2.9042	.4630	1.2392	71° 00′
10	.3345	.3283	9.5163	.9446	9.9752	.3476	9.5411	2.8770	.4589	1.2363	50
20	.3374	.3311	9.5199	.9436	9.9748	.3509	9.5451	2.8502	.4549	1.2334	40
30	.3403	.3338	9.5235	.9426	9.9744	.3541	9.5492	2.8239	.4509	1.2305	30
40	.3432	.3366	9.5271	.9417	9.9739	.3574	9.5532	2.7980	.4469	1.2275	20
50	.3462	.3393	9.5306	.9407	9.9734	.3607	9.5571	2.7725	.4429	1.2246	10
20° 00′	.3491	.3420	9.5341	.9397	9.9730	.3640	9.5611	2.7475	.4389	1.2217	70° 00′
10	.3520	.3448	9.5375	.9387	9.9725	.3673	9.5650	2.7228	.4350	1.2188	50
20	.3549	.3475	9.5409	.9377	9.9721	.3706	9.5689	2.6985	.4311	1.2159	40
30	.3578	.3502	9.5443	.9367	9.9716	.3739	9.5727	2.6746	.4273	1.2130	30
40	.3607	.3529	9.5477	.9357	9.9711	.3772	9.5766	2.6511	.4234	1.2101	20
50	.3636	.3557	9.5510	.9346	9.9706	.3805	9.5804	2.6279	.4196	1.2072	10
21° 00′	.3665	.3584	9.5543	.9336	9.9702	.3839	9.5842	2.6051	.4158	1.2043	69° 00′
10	.3694	.3611	9.5576	.9325	9.9697	.3872	9.5879	2.5826	.4121	1.2014	50
20	.3723	.3638	9.5609	.9315	9.9692	.3906	9.5917	2.5605	.4083	1.1985	40
30	.3752	.3665	9.5641	.9304	9.9687	.3939	9.5954	2.5387	.4046	1.1956	30
40	.3782	.3692	9.5673	.9294	9.9682	.3973	9.5991	2.5172	.4009	1.1926	20
50	.3811	.3719	9.5704	.9283	9.9677	.4007	9.6028	2.4960	.3972	1.1897	10
22° 00′	.3840	.3746	9.5736	.9272	9.9672	.4040	9.6064	2.4751	.3936	1.1868	68° 00′
10	.3869	.3773	9.5767	.9261	9.9667	.4074	9.6100	2.4545	.3900	1.1839	50
20	.3898	.3800	9.5798	.9250	9.9661	.4108	9.6136	2.4342	.3864	1.1810	40
30	.3927	.3827	9.5828	.9239	9.9656	.4142	9.6172	2.4142	.3828	1.1781	30
40	.3956	.3854	9.5859	.9228	9.9651	.4176	9.6208	2.3945	.3792	1.1752	20
50	.3985	.3881	9.5889	.9216	9.9646	.4211	9.6243	2.3750	.3757	1.1723	10
23° 00′	.4014	.3907	9.5919	.9205	9.9640	.4245	9.6279	2.3559	.3722	1.1694	67° 00′
10	.4043	.3934	9.5948	.9194	9.9635	.4279	9.6314	2.3369	.3687	1.1665	50
20	.4072	.3961	9.5978	.9182	9.9629	.4314	9.6348	2.3183	.3652	1.1636	40
30	.4102	.3988	9.6007	.9171	9.9624	.4348	9.6383	2.2998	.3617	1.1606	30
40	.4131	.4014	9.6036	.9159	9.9619	.4383	9.6418	2.2817	.3583	1.1577	20
50	.4160	.4041	9.6065	.9147	9.9613	.4418	9.6452	2.2637	.3548	1.1548	10
24° 00′	.4189	.4067	9.6093	.9136	9.9607	.4452	9.6486	2.2460	.3514	1.1519	66° 00′
		Nat.		Nat.		Nat.		Nat.			
Degrees	Radians	Cosines	Log cosines	Sines	Log sines	Cotangents	Log cotangents	Tangents	Log tangents	Radians	Degrees

Table 5. Natural Trigonometric Functions and Their Logarithms—(*Continued*)

Degrees	Radians	Sines	Log sines	Cosines	Log cosines	Tangents	Log tangents	Cotangents	Log cotangents	Radians	Degrees
		Nat.		Nat.		Nat.		Nat.			
24° 00′	0.4189	0.4067	9.6093	0.9136	9.9607	0.4452	9.6486	2.2460	0.3514	1.1519	66° 00′
10	.4218	.4094	9.6121	.9124	9.9602	.4487	9.6520	2.2286	.3480	1.1490	50
20	.4247	.4120	9.6149	.9112	9.9596	.4522	9.6554	2.2113	.3447	1.1461	40
30	.4276	.4147	9.6177	.9100	9.9590	.4557	9.6587	2.1943	.3413	1.1432	30
40	.4305	.4173	9.6205	.9088	9.9584	.4592	9.6620	2.1775	.3380	1.1403	20
50	.4334	.4200	9.6232	.9075	9.9579	.4628	9.6654	2.1609	.3346	1.1374	10
25° 00′	.4363	.4226	9.6260	.9063	9.9573	.4663	9.6687	2.1445	.3313	1.1345	65° 00′
10	.4392	.4253	9.6287	.9051	9.9567	.4699	9.6720	2.1283	.3280	1.1316	50
20	.4422	.4279	9.6313	.9038	9.9561	.4734	9.6752	2.1123	.3248	1.1286	40
30	.4451	.4305	9.6340	.9026	9.9555	.4770	9.6785	2.0965	.3215	1.1257	30
40	.4480	.4331	9.6366	.9013	9.9549	.4806	9.6817	2.0809	.3183	1.1228	20
50	.4509	.4358	9.6392	.9001	9.9543	.4841	9.6850	2.0655	.3150	1.1199	10
26° 00′	.4538	.4384	9.6418	.8988	9.9537	.4877	9.6882	2.0503	.3118	1.1170	64° 00′
10	.4567	.4410	9.6444	.8975	9.9530	.4913	9.6914	2.0353	.3086	1.1141	50
20	.4596	.4436	9.6470	.8962	9.9524	.4950	9.6946	2.0204	.3054	1.1112	40
30	.4625	.4462	9.6495	.8949	9.9518	.4986	9.6977	2.0057	.3023	1.1083	30
40	.4654	.4488	9.6521	.8936	9.9512	.5022	9.7009	1.9912	.2991	1.1054	20
50	.4683	.4514	9.6546	.8923	9.9505	.5059	9.7040	1.9768	.2960	1.1025	10
27° 00′	.4712	.4540	9.6571	.8910	9.9499	.5095	9.7072	1.9626	.2928	1.0996	63° 00′
10	.4741	.4566	9.6595	.8897	9.9492	.5132	9.7103	1.9486	.2897	1.0966	50
20	.4771	.4592	9.6620	.8884	9.9486	.5169	9.7134	1.9347	.2866	1.0937	40
30	.4800	.4618	9.6644	.8870	9.9479	.5206	9.7165	1.9210	.2835	1.0908	30
40	.4829	.4643	9.6668	.8857	9.9473	.5243	9.7196	1.9074	.2805	1.0879	20
50	.4858	.4669	9.6692	.8843	9.9466	.5280	9.7226	1.8940	.2774	1.0850	10
28° 00′	.4887	.4695	9.6716	.8830	9.9459	.5317	9.7257	1.8807	.2743	1.0821	62° 00′
10	.4916	.4720	9.6740	.8816	9.9453	.5355	9.7287	1.8676	.2713	1.0792	50
20	.4945	.4746	9.6763	.8802	9.9446	.5392	9.7318	1.8546	.2683	1.0763	40
30	.4974	.4772	9.6787	.8788	9.9439	.5430	9.7348	1.8418	.2652	1.0734	30
40	.5003	.4797	9.6810	.8774	9.9432	.5467	9.7378	1.8291	.2622	1.0705	20
50	.5032	.4823	9.6833	.8760	9.9425	.5505	9.7408	1.8165	.2592	1.0676	10
29° 00′	.5061	.4848	9.6856	.8746	9.9418	.5543	9.7438	1.8041	.2563	1.0647	61° 00′
10	.5091	.4874	9.6878	.8732	9.9411	.5581	9.7467	1.7917	.2533	1.0617	50
20	.5120	.4899	9.6901	.8718	9.9404	.5619	9.7497	1.7796	.2503	1.0588	40
30	.5149	.4924	9.6923	.8704	9.9397	.5658	9.7526	1.7675	.2474	1.0559	30
40	.5178	.4950	9.6946	.8689	9.9390	.5696	9.7556	1.7556	.2444	1.0530	20
50	.5207	.4975	9.6968	.8675	9.9383	.5735	9.7585	1.7438	.2415	1.0501	10
30° 00′	.5236	.5000	9.6990	.8660	9.9375	.5774	9.7614	1.7321	.2386	1.0472	60° 00′
10	.5265	.5025	9.7012	.8646	9.9368	.5812	9.7644	1.7205	.2357	1.0443	50
20	.5294	.5050	9.7033	.8631	9.9361	.5851	9.7673	1.7090	.2328	1.0414	40
30	.5323	.5075	9.7055	.8616	9.9353	.5891	9.7702	1.6977	.2299	1.0385	30
40	.5352	.5100	9.7076	.8602	9.9346	.5930	9.7730	1.6864	.2270	1.0356	20
50	.5381	.5125	9.7097	.8587	9.9338	.5969	9.7759	1.6753	.2241	1.0327	10
31° 00′	.5411	.5150	9.7118	.8572	9.9331	.6009	9.7788	1.6643	.2212	1.0297	59° 00′
10	.5440	.5175	9.7139	.8557	9.9323	.6048	9.7816	1.6534	.2184	1.0268	50
20	.5469	.5200	9.7160	.8542	9.9315	.6088	9.7845	1.6426	.2155	1.0239	40
30	.5498	.5225	9.7181	.8526	9.9308	.6128	9.7873	1.6319	.2127	1.0210	30
40	.5527	.5250	9.7201	.8511	9.9300	.6168	9.7902	1.6213	.2099	1.0181	20
50	.5556	.5275	9.7222	.8496	9.9292	.6208	9.7930	1.6107	.2070	1.0152	10
32° 00′	.5585	.5299	9.7242	.8481	9.9284	.6249	9.7958	1.6003	.2042	1.0123	58° 00′
10	.5614	.5324	9.7262	.8465	9.9276	.6289	9.7986	1.5900	.2014	1.0094	50
20	.5643	.5348	9.7282	.8450	9.9268	.6330	9.8014	1.5798	.1986	1.0065	40
30	.5672	.5373	9.7302	.8434	9.9260	.6371	9.8042	1.5697	.1958	1.0036	30
40	.5701	.5398	9.7322	.8418	9.9252	.6412	9.8070	1.5597	.1930	1.0007	20
50	.5730	.5422	9.7342	.8403	9.9244	.6453	9.8098	1.5497	.1903	0.9977	10
33° 00′	.5760	.5446	9.7361	.8387	9.9236	.6494	9.8125	1.5399	.1875	.9948	57° 00′
10	.5789	.5471	9.7381	.8371	9.9228	.6536	9.8153	1.5301	.1847	.9919	50
20	.5818	.5495	9.7400	.8355	9.9219	.6577	9.8180	1.5204	.1820	.9890	40
30	.5847	.5519	9.7419	.8339	9.9211	.6619	9.8208	1.5108	.1792	.9861	30
40	.5876	.5544	9.7438	.8323	9.9203	.6661	9.8235	1.5013	.1765	.9832	20
50	.5905	.5568	9.7457	.8307	9.9194	.6703	9.8263	1.4919	.1737	.9803	10
34° 00′	.5934	.5592	9.7476	.8290	9.9186	.6745	9.8290	1.4826	.1710	.9774	56° 00′
10	.5963	.5616	9.7494	.8274	9.9177	.6788	9.8317	1.4733	.1683	.9745	50
20	.5992	.5640	9.7513	.8258	9.9169	.6830	9.8344	1.4641	.1656	.9716	40
30	.6021	.5664	9.7531	.8241	9.9160	.6873	9.8371	1.4550	.1629	.9687	30
40	.6050	.5688	9.7550	.8225	9.9151	.6916	9.8398	1.4460	.1602	.9657	20
50	.6080	.5712	9.7568	.8208	9.9143	.6959	9.8425	1.4370	.1575	.9628	10
35° 00′	.6109	.5736	9.7586	.8192	9.9134	.7002	9.8452	1.4282	.1548	.9599	55° 00′
10	.6138	.5760	9.7604	.8175	9.9125	.7046	9.8479	1.4193	.1521	.9570	50
20	.6167	.5783	9.7622	.8158	9.9116	.7089	9.8506	1.4106	.1494	.9541	40
30	.6196	.5807	9.7640	.8141	9.9107	.7133	9.8533	1.4020	.1467	.9512	30
40	.6225	.5831	9.7657	.8124	9.9098	.7177	9.8559	1.3934	.1441	.9483	20
50	.6254	.5854	9.7675	.8107	9.9089	.7221	9.8586	1.3848	.1414	.9454	10
36° 00′	.6283	.5878	9.7692	.8090	9.9080	.7265	9.8613	1.3764	.1387	.9425	54° 00′
		Nat.		Nat.		Nat.		Nat.			
Degrees	Radians	Cosines	Log cosines	Sines	Log sines	Cotangents	Log cotangents	Tangents	Log tangents	Radians	Degrees

Table 5. Natural Trigonometric Functions and Their Logarithms—(*Concluded*)

Degrees	Radians	Sines	Log sines	Cosines	Log cosines	Tangents	Log tangents	Cotangents	Log cotangents	Radians	Degrees
		Nat.		Nat.		Nat.		Nat.			
36° 00′	0.6283	0.5878	9.7692	0.8090	9.9080	0.7265	9.8613	1.3764	0.1387	0.9425	54° 00′
10	.6312	.5901	9.7710	.8073	9.9070	.7310	9.8639	1.3680	.1361	.9396	50
20	.6341	.5925	9.7727	.8056	9.9061	.7355	9.8666	1.3597	.1334	.9367	40
30	.6370	.5948	9.7744	.8039	9.9052	.7400	9.8692	1.3514	.1308	.9338	30
40	.6400	.5972	9.7761	.8021	9.9042	.7445	9.8719	1.3432	.1282	.9308	20
50	.6429	.5995	9.7778	.8004	9.9033	.7490	9.8745	1.3351	.1255	.9279	10
37° 00′	.6458	.6018	9.7795	.7986	9.9024	.7536	9.8771	1.3270	.1229	.9250	53° 00′
10	.6487	.6041	9.7811	.7969	9.9014	.7581	9.8797	1.3190	.1203	.9221	50
20	.6516	.6065	9.7828	.7951	9.9004	.7627	9.8824	1.3111	.1176	.9192	40
30	.6545	.6088	9.7845	.7934	9.8995	.7673	9.8850	1.3032	.1150	.9163	30
40	.6574	.6111	9.7861	.7916	9.8985	.7720	9.8876	1.2954	.1124	.9134	20
50	.6603	.6134	9.7877	.7898	9.8975	.7766	9.8902	1.2876	.1098	.9105	10
38° 00′	.6632	.6157	9.7893	.7880	9.8965	.7813	9.8928	1.2799	.1072	.9076	52° 00′
10	.6661	.6180	9.7910	.7862	9.8955	.7860	9.8954	1.2723	.1046	.9047	50
20	.6690	.6202	9.7926	.7844	9.8946	.7907	9.8980	1.2647	.1020	.9018	40
30	.6720	.6225	9.7942	.7826	9.8935	.7954	9.9006	1.2572	.0994	.8988	30
40	.6749	.6248	9.7957	.7808	9.8925	.8002	9.9032	1.2497	.0968	.8959	20
50	.6778	.6271	9.7973	.7790	9.8915	.8050	9.9058	1.2423	.0942	.8930	10
39° 00′	.6807	.6293	9.7989	.7772	9.8905	.8098	9.9084	1.2349	.0916	.8901	51° 00′
10	.6836	.6316	9.8004	.7753	9.8895	.8146	9.9110	1.2276	.0891	.8872	50
20	.6865	.6338	9.8020	.7735	9.8884	.8195	9.9135	1.2203	.0865	.8843	40
30	.6894	.6361	9.8035	.7716	9.8874	.8243	9.9161	1.2131	.0839	.8814	30
40	.6923	.6383	9.8050	.7698	9.8864	.8292	9.9187	1.2059	.0813	.8785	20
50	.6952	.6406	9.8066	.7679	9.8853	.8342	9.9213	1.1988	.0788	.8756	10
40° 00′	.6981	.6428	9.8081	.7660	9.8843	.8391	9.9238	1.1918	.0762	.8727	50° 00′
10	.7010	.6450	9.8096	.7642	9.8832	.8441	9.9264	1.1847	.0736	.8698	50
20	.7039	.6472	9.8111	.7623	9.8821	.8491	9.9289	1.1778	.0711	.8668	40
30	.7069	.6495	9.8125	.7604	9.8811	.8541	9.9315	1.1709	.0685	.8639	30
40	.7098	.6517	9.8140	.7585	9.8800	.8591	9.9341	1.1640	.0659	.8610	20
50	.7127	.6539	9.8155	.7566	9.8789	.8642	9.9366	1.1572	.0634	.8581	10
41° 00′	.7156	.6561	9.8169	.7547	9.8778	.8693	9.9392	1.1504	.0608	.8552	49° 00′
10	.7185	.6583	9.8184	.7528	9.8767	.8744	9.9417	1.1436	.0583	.8523	50
20	.7214	.6604	9.8198	.7509	9.8756	.8796	9.9443	1.1369	.0557	.8494	40
30	.7243	.6626	9.8213	.7490	9.8745	.8847	9.9468	1.1303	.0532	.8465	30
40	.7272	.6648	9.8227	.7470	9.8733	.8899	9.9494	1.1237	.0507	.8436	20
50	.7301	.6670	9.8241	.7451	9.8722	.8952	9.9519	1.1171	.0481	.8407	10
42° 00′	.7330	.6691	9.8255	.7431	9.8711	.9004	9.9544	1.1106	.0456	.8378	48° 00′
10	.7359	.6713	9.8269	.7412	9.8699	.9057	9.9570	1.1041	.0430	.8348	50
20	.7389	.6734	9.8283	.7392	9.8688	.9110	9.9595	1.0977	.0405	.8319	40
30	.7418	.6756	9.8297	.7373	9.8676	.9163	9.9621	1.0913	.0380	.8290	30
40	.7447	.6777	9.8311	.7353	9.8665	.9217	9.9646	1.0850	.0354	.8261	20
50	.7476	.6799	9.8324	.7333	9.8653	.9271	9.9671	1.0786	.0329	.8232	10
43° 00′	.7505	.6820	9.8338	.7314	9.8641	.9325	9.9697	1.0724	.0303	.8203	47° 00′
10	.7534	.6841	9.8351	.7294	9.8630	.9380	9.9722	1.0661	.0278	.8174	50
20	.7563	.6862	9.8365	.7274	9.8618	.9435	9.9747	1.0599	.0253	.8145	40
30	.7592	.6884	9.8378	.7254	9.8606	.9490	9.9773	1.0538	.0228	.8116	30
40	.7621	.6905	9.8391	.7234	9.8594	.9545	9.9798	1.0477	.0202	.8087	20
50	.7650	.6926	9.8405	.7214	9.8582	.9601	9.9823	1.0416	.0177	.8058	10
44° 00′	.7679	.6947	9.8418	.7193	9.8569	.9657	9.9848	1.0355	.0152	.8029	46° 00′
10	.7709	.6968	9.8431	.7173	9.8557	.9713	9.9874	1.0295	.0126	.7999	50
20	.7738	.6988	9.8444	.7153	9.8545	.9770	9.9899	1.0236	.0101	.7970	40
30	.7767	.7009	9.8457	.7133	9.8532	.9827	9.9924	1.0176	.0076	.7941	30
40	.7796	.7030	9.8469	.7112	9.8520	.9884	9.9950	1.0117	.0051	.7912	20
50	.7825	.7051	9.8482	.7092	9.8507	.9942	9.9975	1.0058	.0025	.7883	10
45° 00′	.7854	.7071	9.8495	.7071	9.8495	1.0000	0.0000	1.0000	0.0000	.7854	45° 00′
		Nat.		Nat.		Nat.		Nat.			
Degrees	Radians	Cosines	Log cosines	Sines	Log sines	Cotangents	Log cotangents	Tangents	Log tangents	Radians	Degrees

NUMBERS

Table 6. Squares, Square Roots, Cubes, Cube Roots

N	N^2	$\sqrt{N}$	N^3	$\sqrt[3]{N}$	N	N^2	$\sqrt{N}$	N^3	$\sqrt[3]{N}$	N	N^2	$\sqrt{N}$	N^3	$\sqrt[3]{N}$
					75	5 625	8.660	421 875	4.217	150	22 500	12.247	3 375 000	5.313
1	1	1.000	1	1.000	76	5 776	8.718	438 976	4.236	151	22 801	12.288	3 442 951	5.325
2	4	1.414	8	1.260	77	5 929	8.775	456 533	4.254	152	23 104	12.329	3 511 808	5.337
3	9	1.732	27	1.442	78	6 084	8.832	474 552	4.273	153	23 409	12.369	3 581 577	5.348
4	16	2.000	64	1.587	79	6 241	8.888	493 039	4.291	154	23 716	12.410	3 652 264	5.360
5	25	2.236	125	1.710	80	6 400	8.944	512 000	4.309	155	24 025	12.450	3 723 875	5.372
6	36	2.449	216	1.817	81	6 561	9.000	531 441	4.327	156	24 336	12.490	3 796 416	5.383
7	49	2.646	343	1.913	82	6 724	9.055	551 368	4.344	157	24 649	12.530	3 869 893	5.395
8	64	2.828	512	2.000	83	6 889	9.110	571 787	4.362	158	24 964	12.570	3 944 312	5.406
9	81	3.000	729	2.080	84	7 056	9.165	592 704	4.380	159	25 281	12.610	4 019 679	5.418
10	100	3.162	1 000	2.154	85	7 225	9.220	614 125	4.397	160	25 600	12.649	4 096 000	5.429
11	121	3.317	1 331	2.224	86	7 396	9.274	636 056	4.414	161	25 921	12.689	4 173 281	5.440
12	144	3.464	1 728	2.289	87	7 569	9.327	658 503	4.431	162	26 244	12.728	4 251 528	5.451
13	169	3.606	2 197	2.351	88	7 744	9.381	681 472	4.448	163	26 569	12.767	4 330 747	5.463
14	196	3.742	2 744	2.410	89	7 921	9.434	704 969	4.465	164	26 896	12.806	4 410 944	5.474
15	225	3.873	3 375	2.466	90	8 100	9.487	729 000	4.481	165	27 225	12.845	4 492 125	5.485
16	256	4.000	4 096	2.520	91	8 281	9.539	753 571	4.498	166	27 556	12.884	4 574 296	5.496
17	289	4.123	4 913	2.571	92	8 464	9.592	778 688	4.514	167	27 889	12.923	4 657 463	5.507
18	324	4.243	5 832	2.621	93	8 649	9.644	804 357	4.531	168	28 224	12.961	4 741 632	5.518
19	361	4.359	6 859	2.668	94	8 836	9.695	830 584	4.547	169	28 561	13.000	4 826 809	5.529
20	400	4.472	8 000	2.714	95	9 025	9.747	857 375	4.563	170	28 900	13.038	4 913 000	5.540
21	441	4.583	9 261	2.759	96	9 216	9.798	884 736	4.579	171	29 241	13.077	5 000 211	5.550
22	484	4.690	10 648	2.802	97	9 409	9.849	912 673	4.595	172	29 584	13.115	5 088 448	5.561
23	529	4.796	12 167	2.844	98	9 604	9.899	941 192	4.610	173	29 929	13.153	5 177 717	5.572
24	576	4.899	13 824	2.884	99	9 801	9.950	970 299	4.626	174	30 276	13.191	5 268 024	5.583
25	625	5.000	15 625	2.924	100	10 000	10.000	1 000 000	4.642	175	30 625	13.229	5 359 375	5.593
26	676	5.099	17 576	2.962	101	10 201	10.050	1 030 301	4.657	176	30 976	13.266	5 451 776	5.604
27	729	5.196	19 683	3.000	102	10 404	10.100	1 061 208	4.672	177	31 329	13.304	5 545 233	5.615
28	784	5.292	21 952	3.037	103	10 609	10.149	1 092 727	4.688	178	31 684	13.342	5 639 752	5.625
29	841	5.385	24 389	3.072	104	10 816	10.198	1 124 864	4.703	179	32 041	13.379	5 735 339	5.636
30	900	5.477	27 000	3.107	105	11 025	10.247	1 157 625	4.718	180	32 400	13.416	5 832 000	5.646
31	961	5.568	29 791	3.141	106	11 236	10.296	1 191 016	4.733	181	32 761	13.454	5 929 741	5.657
32	1 024	5.567	32 768	3.175	107	11 449	10.344	1 225 043	4.747	182	33 124	13.491	6 028 568	5.667
33	1 089	5.745	35 937	3.208	108	11 664	10.392	1 259 712	4.762	183	33 489	13.528	6 128 487	5.677
34	1 156	5.831	39 304	3.240	109	11 881	10.440	1 295 029	4.777	184	33 856	13.565	6 229 504	5.688
35	1 225	5.916	42 875	3.271	110	12 100	10.488	1 331 000	4.791	185	34 225	13.601	6 331 625	5.698
36	1 296	6.000	46 656	3.302	111	12 321	10.536	1 367 631	4.806	186	34 596	13.638	6 434 856	5.708
37	1 369	6.083	50 653	3.332	112	12 544	10.583	1 404 928	4.820	187	34 969	13.675	6 539 203	5.718
38	1 444	6.164	54 872	3.362	113	12 769	10.630	1 442 897	4.835	188	35 344	13.711	6 644 672	5.729
39	1 521	6.245	59 319	3.391	114	12 996	10.677	1 481 544	4.849	189	35 721	13.748	6 751 269	5.739
40	1 600	6.325	64 000	3.420	115	13 225	10.724	1 520 875	4.863	190	36 100	13.784	6 859 000	5.749
41	1 681	6.403	68 921	3.448	116	13 456	10.770	1 560 896	4.877	191	36 481	13.820	6 967 871	5.759
42	1 764	6.481	74 088	3.476	117	13 689	10.817	1 601 613	4.891	192	36 864	13.856	7 077 888	5.769
43	1 849	6.557	79 507	3.503	118	13 924	10.863	1 643 032	4.905	193	37 249	13.892	7 189 057	5.779
44	1 936	6.663	85 184	3.530	119	14 161	10.909	1 685 159	4.919	194	37 636	13.928	7 301 384	5.789
45	2 025	6.708	91 125	3.557	120	14 400	10.954	1 728 000	4.932	195	38 025	13.964	7 414 875	5.799
46	2 116	6.782	97 336	3.583	121	14 641	11.000	1 771 561	4.946	196	38 416	14.000	7 529 536	5.809
47	2 209	6.856	103 823	3.609	122	14 884	11.045	1 815 848	4.960	197	38 809	14.036	7 645 373	5.819
48	2 304	6.928	110 592	3.634	123	15 129	11.091	1 860 867	4.973	198	39 204	14.071	7 762 392	5.828
49	2 401	7.000	117 649	3.659	124	15 376	11.136	1 906 624	4.987	199	39 601	14.107	7 880 599	5.838
50	2 500	7.071	125 000	3.684	125	15 625	11.180	1 953 125	5.000	200	40 000	14.142	8 000 000	5.848
51	2 601	7.141	132 651	3.708	126	15 876	11.225	2 000 376	5.013	201	40 401	14.177	8 120 601	5.858
52	2 704	7.211	140 608	3.733	127	16 129	11.269	2 048 383	5.027	202	40 804	14.213	8 242 408	5.867
53	2 809	7.280	148 877	3.756	128	16 384	11.314	2 097 152	5.040	203	41 209	14.248	8 365 427	5.877
54	2 916	7.348	157 464	3.780	129	16 641	11.358	2 146 689	5.053	204	41 616	14.283	8 489 664	5.887
55	3 025	7.416	166 375	3.803	130	16 900	11.402	2 197 000	5.066	205	42 025	14.318	8 615 125	5.896
56	3 136	7.483	175 616	3.826	131	17 161	11.446	2 248 091	5.079	206	42 436	14.353	8 741 816	5.906
57	3 249	7.550	185 193	3.849	132	17 424	11.489	2 299 968	5.092	207	42 849	14.387	8 869 743	5.915
58	3 364	7.616	195 112	3.871	133	17 689	11.533	2 352 637	5.104	208	43 264	14.422	8 998 912	5.925
59	3 481	7.681	205 379	3.893	134	17 956	11.576	2 406 104	5.117	209	43 681	14.457	9 129 329	5.934
60	3 600	7.746	216 000	3.915	135	18 225	11.619	2 460 375	5.130	210	44 100	14.491	9 261 000	5.944
61	3 721	7.810	226 981	3.936	136	18 496	11.662	2 515 456	5.143	211	44 521	14.526	9 393 931	5.953
62	3 844	7.874	238 328	3.958	137	18 769	11.705	2 571 353	5.155	212	44 944	14.560	9 528 128	5.963
63	3 969	7.937	250 047	3.979	138	19 044	11.747	2 628 072	5.168	213	45 369	14.595	9 663 597	5.972
64	4 096	8.000	262 144	4.000	139	19 321	11.790	2 685 619	5.180	214	45 796	14.629	9 800 344	5.981
65	4 225	8.062	274 625	4.021	140	19 600	11.832	2 744 000	5.192	215	46 225	14.663	9 938 375	5.991
66	4 356	8.124	287 496	4.041	141	19 881	11.874	2 803 221	5.205	216	46 656	14.697	10 077 696	6.000
67	4 489	8.185	300 763	4.062	142	20 164	11.916	2 863 288	5.217	217	47 089	14.731	10 218 313	6.009
68	4 624	8.246	314 432	4.082	143	20 449	11.958	2 924 207	5.229	218	47 524	14.765	10 360 232	6.018
69	4 761	8.307	328 509	4.102	144	20 736	12.000	2 985 984	5.241	219	47 961	14.799	10 503 459	6.028
70	4 900	8.367	343 000	4.121	145	21 025	12.042	3 048 625	5.254	220	48 400	14.832	10 648 000	6.037
71	5 041	8.426	357 911	4.141	146	21 316	12.083	3 112 136	5.266	221	48 841	14.866	10 793 861	6.046
72	5 184	8.485	373 248	4.160	147	21 609	12.124	3 176 523	5.278	222	49 284	14.900	10 941 048	6.055
73	5 329	8.554	389 017	4.179	148	21 904	12.166	3 241 792	5.290	223	49 729	14.933	11 089 567	6.064
74	5 476	8.602	405 224	4.198	149	22 201	12.207	3 307 949	5.301	224	50 176	14.967	11 239 424	6.073

Table 6. Squares, Square Roots, Cubes, Cube Roots—(*Continued*)

N	N^2	$\sqrt{N}$	N^3	$\sqrt[3]{N}$	N	N^2	$\sqrt{N}$	N^3	$\sqrt[3]{N}$	N	N^2	$\sqrt{N}$	N^3	$\sqrt[3]{N}$
225	50 625	15.000	11 390 625	6.082	300	90 000	17.321	27 000 000	6.694	375	140 625	19.365	52 734 375	7.211
226	51 076	15.033	11 543 176	6.091	301	90 601	17.349	27 270 901	6.702	376	141 376	19.391	53 157 376	7.218
227	51 529	15.067	11 697 083	6.100	302	91 204	17.378	27 543 608	6.709	377	142 129	19.416	53 582 633	7.224
228	51 984	15.100	11 852 352	6.109	303	91 809	17.407	27 818 127	6.717	378	142 884	19.442	54 010 152	7.230
229	52 441	15.133	12 008 989	6.118	304	92 416	17.436	28 094 464	6.724	379	143 641	19.468	54 439 939	7.237
230	52 900	15.166	12 167 000	6.127	305	93 025	17.464	28 372 625	6.731	380	144 400	19.494	54 872 000	7.243
231	53 361	15.199	12 326 391	6.136	306	93 636	17.493	28 652 616	6.739	381	145 161	19.519	55 306 341	7.250
232	53 824	15.232	12 487 168	6.145	307	94 249	17.521	28 934 443	6.746	382	145 924	19.545	55 742 968	7.256
233	54 289	15.264	12 649 337	6.153	308	94 864	17.550	29 218 112	6.753	383	146 689	19.570	56 181 887	7.262
234	54 756	15.297	12 812 904	6.162	309	95 481	17.578	29 503 629	6.761	384	147 456	19.596	56 623 104	7.268
235	55 225	15.330	12 977 875	6.171	310	96 100	17.607	29 791 000	6.768	385	148 225	19.621	57 066 625	7.275
236	55 696	15.362	13 144 256	6.180	311	96 721	17.635	30 080 231	6.775	386	148 996	19.647	57 512 456	7.281
237	56 169	15.395	13 312 053	6.188	312	97 344	17.664	30 371 328	6.782	387	149 769	19.672	57 960 603	7.287
238	56 644	15.427	13 481 272	6.197	313	97 969	17.692	30 664 297	6.790	388	150 544	19.698	58 411 072	7.294
239	57 121	15.460	13 651 919	6.206	314	98 596	17.720	30 959 144	6.797	389	151 321	19.723	58 863 869	7.300
240	57 600	15.492	13 824 000	6.214	315	99 225	17.748	31 255 875	6.804	390	152 100	19.748	59 319 000	7.306
241	58 081	15.524	13 997 521	6.223	316	99 856	17.776	31 554 496	6.811	391	152 881	19.774	59 776 471	7.312
242	58 564	15.556	14 172 488	6.232	317	100 489	17.804	31 855 013	6.818	392	153 664	19.799	60 236 288	7.319
243	59 049	15.588	14 348 907	6.240	318	101 124	17.833	32 157 432	6.826	393	154 449	19.824	60 698 457	7.325
244	59 536	15.620	14 526 784	6.249	319	101 761	17.861	32 461 759	6.833	394	155 236	19.849	61 162 984	7.331
245	60 025	15.652	14 706 125	6.257	320	102 400	17.889	32 768 000	6.840	395	156 025	19.875	61 629 875	7.337
246	60 516	15.684	14 886 936	6.266	321	103 041	17.916	33 076 161	6.847	396	156 816	19.900	62 099 136	7.343
247	61 009	15.716	15 069 223	6.274	322	103 684	17.944	33 386 248	6.854	397	157 609	19.925	62 570 773	7.350
248	61 504	15.748	15 252 992	6.283	323	104 329	17.972	33 698 267	6.861	398	158 404	19.950	63 044 792	7.356
249	62 001	15.780	15 438 249	6.291	324	104 976	18.000	34 012 224	6.868	399	159 201	19.975	63 521 199	7.362
250	62 500	15.811	15 625 000	6.300	325	105 625	18.028	34 328 125	6.875	400	160 000	20.000	64 000 000	7.368
251	63 001	15.843	15 813 251	6.308	326	106 276	18.055	34 645 976	6.882	401	160 801	20.025	64 481 201	7.374
252	63 504	15.875	16 003 008	6.316	327	106 929	18.083	34 965 783	6.889	402	161 604	20.050	64 964 808	7.380
253	64 009	15.906	16 194 277	6.325	328	107 584	18.111	35 287 552	6.896	403	162 409	20.075	65 450 827	7.386
254	64 516	15.937	16 387 064	6.333	329	108 241	18.138	35 611 289	6.903	404	163 216	20.100	65 939 264	7.393
255	65 025	15.969	16 581 375	6.341	330	108 900	18.166	35 937 000	6.910	405	164 025	20.125	66 430 125	7.399
256	65 536	16.000	16 777 216	6.350	331	109 561	18.193	36 264 691	6.917	406	164 836	20.149	66 923 416	7.405
257	66 049	16.031	16 974 593	6.358	332	110 224	18.221	36 594 368	6.924	407	165 649	20.174	67 419 143	7.411
258	66 564	16.062	17 173 512	6.366	333	110 889	18.248	36 926 037	6.931	408	166 464	20.199	67 917 312	7.417
259	67 081	16.093	17 373 979	6.374	334	111 556	18.276	37 259 704	6.938	409	167 281	20.224	68 417 929	7.423
260	67 600	16.125	17 576 000	6.383	335	112 225	18.303	37 595 375	6.945	410	168 100	20.248	68 921 000	7.429
261	68 121	16.155	17 779 581	6.391	336	112 896	18.330	37 933 056	6.952	411	168 921	20.273	69 426 531	7.435
262	68 644	16.186	17 984 728	6.399	337	113 569	18.358	38 272 753	6.959	412	169 744	20.298	69 934 528	7.441
263	69 169	16.217	18 191 447	6.407	338	114 244	18.385	38 614 472	6.966	413	170 569	20.322	70 444 997	7.447
264	69 696	16.248	18 399 744	6.415	339	114 921	18.412	38 958 219	6.973	414	171 396	20.347	70 957 944	7.453
265	70 225	16.279	18 609 625	6.423	340	115 600	18.439	39 304 000	6.980	415	172 225	20.372	71 473 375	7.459
266	70 756	16.310	18 821 096	6.431	341	116 281	18.466	39 651 821	6.986	416	173 056	20.396	71 991 296	7.465
267	71 289	16.340	19 034 163	6.439	342	116 964	18.493	40 001 688	6.993	417	173 889	20.421	72 511 713	7.471
268	71 824	16.371	19 248 832	6.447	343	117 649	18.520	40 353 607	7.000	418	174 724	20.445	73 034 632	7.477
269	72 361	16.401	19 465 109	6.455	344	118 336	18.547	40 707 584	7.007	419	175 561	20.469	73 560 059	7.483
270	72 900	16.432	19 683 000	6.463	345	119 025	18.574	41 063 625	7.014	420	176 400	20.494	74 088 000	7.489
271	73 441	16.462	19 902 511	6.471	346	119 716	18.601	41 421 736	7.020	421	177 241	20.518	74 618 461	7.495
272	73 984	16.492	20 123 648	6.479	347	120 409	18.628	41 781 923	7.027	422	178 084	20.543	75 151 448	7.501
273	74 529	16.523	20 346 417	6.487	348	121 104	18.655	42 144 192	7.034	423	178 929	20.567	75 686 967	7.507
274	75 076	16.553	20 570 824	6.495	349	121 801	18.682	42 508 549	7.041	424	179 776	20.591	76 225 024	7.513
275	75 625	16.583	20 796 875	6.503	350	122 500	18.708	42 875 000	7.047	425	180 625	20.616	76 765 625	7.518
276	76 176	16.613	21 024 576	6.511	351	123 201	18.735	43 243 551	7.054	426	181 476	20.640	77 308 776	7.524
277	76 729	16.643	21 253 933	6.519	352	123 904	18.762	43 614 208	7.061	427	182 329	20.664	77 854 483	7.530
278	77 284	16.673	21 484 952	6.527	353	124 609	18.788	43 986 977	7.067	428	183 184	20.688	78 402 752	7.536
279	77 841	16.703	21 717 639	6.534	354	125 316	18.815	44 361 864	7.074	429	184 041	20.712	78 953 589	7.542
280	78 400	16.733	21 952 000	6.542	355	126 025	18.841	44 738 875	7.081	430	184 900	20.736	79 507 000	7.548
281	78 961	16.763	22 188 041	6.550	356	126 736	18.868	45 118 016	7.087	431	185 761	20.761	80 062 991	7.554
282	79 524	16.793	22 425 768	6.558	357	127 449	18.894	45 499 293	7.094	432	186 624	20.785	80 621 568	7.560
283	80 089	16.823	22 665 187	6.565	358	128 164	18.921	45 882 712	7.101	433	187 489	20.809	81 182 737	7.565
284	80 656	16.852	22 906 304	6.573	359	128 881	18.947	46 268 279	7.107	434	188 356	20.833	81 746 504	7.571
285	81 225	16.882	23 149 125	6.581	360	129 600	18.974	46 656 000	7.114	435	189 225	20.857	82 312 875	7.577
286	81 796	16.912	23 393 656	6.589	361	130 321	19.000	47 045 881	7.120	436	190 096	20.881	82 881 856	7.583
287	82 369	16.941	23 639 903	6.596	362	131 044	19.026	47 437 928	7.127	437	190 969	20.905	83 453 453	7.589
288	82 944	16.971	23 887 872	6.604	363	131 769	19.053	47 832 147	7.133	438	191 844	20.928	84 027 672	7.594
289	83 521	17.000	24 137 569	6.611	364	132 496	19.079	48 228 544	7.140	439	192 721	20.952	84 604 519	7.600
290	84 100	17.029	24 389 000	6.619	365	133 225	19.105	48 627 125	7.147	440	193 600	20.976	85 184 000	7.606
291	84 681	17.059	24 642 171	6.627	366	133 956	19.131	49 027 896	7.153	441	194 481	21.000	85 766 121	7.612
292	85 264	17.088	24 897 088	6.634	367	134 689	19.157	49 430 863	7.160	442	195 364	21.024	86 350 888	7.617
293	85 849	17.117	25 153 757	6.642	368	135 424	19.183	49 836 032	7.166	443	196 249	21.048	86 938 307	7.623
294	86 436	17.146	25 412 184	6.649	369	136 161	19.209	50 243 409	7.173	444	197 136	21.071	87 528 384	7.629
295	87 025	17.176	25 672 375	6.657	370	136 900	19.235	50 653 000	7.179	445	198 025	21.095	88 121 125	7.635
296	87 616	17.205	25 934 336	6.664	371	137 641	19.261	51 064 811	7.186	446	198 916	21.119	88 716 536	7.640
297	88 209	17.234	26 198 073	6.672	372	138 384	19.287	51 478 848	7.192	447	199 809	21.142	89 314 623	7.646
298	88 804	17.263	26 463 593	6.679	373	139 129	19.313	51 895 117	7.198	448	200 704	21.166	89 915 392	7.652
299	89 401	17.292	26 730 899	6.687	374	139 876	19.339	52 313 624	7.205	449	201 601	21.190	90 518 849	7.657

Table 6. Squares, Square Roots, Cubes, Cube Roots—(*Continued*)

N	N^2	$\sqrt{N}$	N^3	$\sqrt[3]{N}$	N	N^2	$\sqrt{N}$	N^3	$\sqrt[3]{N}$	N	N^2	$\sqrt{N}$	N^3	$\sqrt[3]{N}$
450	202 500	21.213	91 125 000	7.663	525	275 625	22.913	144 703 125	8.067	600	360 000	24.495	216 000 000	8.434
451	203 401	21.237	91 733 851	7.669	526	276 676	22.935	145 531 576	8.072	601	361 201	24.515	217 081 801	8.439
452	204 304	21.260	92 345 408	7.674	527	277 729	22.956	146 363 183	8.077	602	362 404	24.536	218 167 208	8.444
453	205 209	21.284	92 959 677	7.680	528	278 784	22.978	147 197 952	8.082	603	363 609	24.556	219 256 227	8.448
454	206 116	21.307	93 576 664	7.686	529	279 841	23.000	148 035 889	8.088	604	364 816	24.576	220 348 864	8.453
455	207 025	21.331	94 196 375	7.691	530	280 900	23.022	148 877 000	8.093	605	366 025	24.597	221 445 125	8.458
456	207 936	21.354	94 818 816	7.697	531	281 961	23.043	149 721 291	8.098	606	367 236	24.617	222 545 016	8.462
457	208 849	21.378	95 443 993	7.703	532	283 024	23.065	150 568 768	8.103	607	368 449	24.637	223 648 543	8.467
458	209 764	21.401	96 071 912	7.708	533	284 089	23.087	151 419 437	8.108	608	369 664	24.658	224 755 712	8.472
459	210 681	21.424	96 702 579	7.714	534	285 156	23.108	152 273 304	8.113	609	370 881	24.678	225 866 529	8.476
460	211 600	21.448	97 336 000	7.719	535	286 225	23.130	153 130 375	8.118	610	372 100	24.698	226 981 000	8.481
461	212 521	21.471	97 972 181	7.725	536	287 296	23.152	153 990 656	8.123	611	373 321	24.718	228 099 131	8.486
462	213 444	21.494	98 611 128	7.731	537	288 369	23.173	154 854 153	8.128	612	374 544	24.739	229 220 928	8.490
463	214 369	21.517	99 252 847	7.736	538	289 444	23.195	155 720 872	8.133	613	375 769	24.759	230 346 397	8.495
464	215 296	21.541	99 897 344	7.742	539	290 521	23.216	156 590 819	8.138	614	376 996	24.779	231 475 544	8.499
465	216 225	21.564	100 544 625	7.747	540	291 600	23.238	157 464 000	8.143	615	378 225	24.799	232 608 375	8.504
466	217 156	21.587	101 194 696	7.753	541	292 681	23.259	158 340 421	8.148	616	379 456	24.819	233 744 896	8.509
467	218 089	21.610	101 847 563	7.758	542	293 764	23.281	159 220 088	8.153	617	380 689	24.839	234 885 113	8.513
468	219 024	21.633	102 503 232	7.764	543	294 849	23.302	160 103 007	8.158	618	381 924	24.860	236 029 032	8.518
469	219 961	21.656	103 161 709	7.769	544	295 936	23.324	160 989 184	8.163	619	383 161	24.880	237 176 659	8.522
470	220 900	21.679	103 823 000	7.775	545	297 025	23.345	161 878 625	8.168	620	384 400	24.900	238 328 000	8.527
471	221 841	21.703	104 487 111	7.780	546	298 116	23.367	162 771 336	8.173	621	385 641	24.920	239 483 061	8.532
472	222 784	21.726	105 154 048	7.786	547	299 209	23.388	163 667 323	8.178	622	386 884	24.940	240 641 848	8.536
473	223 729	21.749	105 823 817	7.791	548	300 304	23.409	164 566 592	8.183	623	388 129	24.960	241 804 367	8.541
474	224 676	21.772	106 496 424	7.797	549	301 401	23.431	165 469 149	8.188	624	389 376	24.980	242 970 624	8.545
475	225 625	21.794	107 171 875	7.802	550	302 500	23.452	166 375 000	8.193	625	390 625	25.000	244 140 625	8.550
476	226 576	21.817	107 850 176	7.808	551	303 601	23.473	167 284 151	8.198	626	391 876	25.020	245 314 376	8.554
477	227 529	21.840	108 531 333	7.813	552	304 704	23.495	168 196 608	8.203	627	393 129	25.040	246 491 883	8.559
478	228 484	21.863	109 215 352	7.819	553	305 809	23.516	169 112 377	8.208	628	394 384	25.060	247 673 152	8.564
479	229 441	21.886	109 902 239	7.824	554	306 916	23.537	170 031 464	8.213	629	395 641	25.080	248 858 189	8.568
480	230 400	21.909	110 592 000	7.830	555	308 025	23.558	170 953 875	8.218	630	396 900	25.100	250 047 000	8.573
481	231 361	21.932	111 284 641	7.835	556	309 136	23.580	171 879 616	8.223	631	398 161	25.120	251 239 591	8.577
482	232 324	21.954	111 980 168	7.841	557	310 249	23.601	172 808 693	8.228	632	399 424	25.140	252 435 968	8.582
483	233 289	21.977	112 678 587	7.846	558	311 364	23.622	173 741 112	8.233	633	400 689	25.159	253 636 137	8.586
484	234 256	22.000	113 379 904	7.851	559	312 481	23.643	174 676 879	8.238	634	401 956	25.179	254 840 104	8.591
485	235 225	22.023	114 084 125	7.857	560	313 600	23.664	175 616 000	8.243	635	403 225	25.199	256 047 875	8.595
486	236 196	22.045	114 791 256	7.862	561	314 721	23.685	176 558 481	8.247	636	404 496	25.219	257 259 456	8.600
487	237 169	22.068	115 501 303	7.868	562	315 844	23.707	177 504 328	8.252	637	405 769	25.239	258 474 853	8.604
488	238 144	22.091	116 214 272	7.873	563	316 969	23.728	178 453 547	8.257	638	407 044	25.259	259 694 072	8.609
489	239 121	22.113	116 930 169	7.878	564	318 096	23.749	179 406 144	8.262	639	408 321	25.278	260 917 119	8.613
490	240 100	22.136	117 649 000	7.884	565	319 225	23.770	180 362 125	8.267	640	409 600	25.298	262 144 000	8.618
491	241 081	22.159	118 370 771	7.889	566	320 356	23.791	181 321 496	8.272	641	410 881	25.318	263 374 721	8.622
492	242 064	22.181	119 095 488	7.894	567	321 489	23.812	182 284 263	8.277	642	412 164	25.338	264 609 288	8.627
493	243 049	22.204	119 823 157	7.900	568	322 624	23.833	183 250 432	8.282	643	413 449	25.357	265 847 707	8.631
494	244 036	22.226	120 553 784	7.905	569	323 761	23.854	184 220 009	8.286	644	414 736	25.377	267 089 984	8.636
495	245 025	22.249	121 287 375	7.910	570	324 900	23.875	185 193 000	8.291	645	416 025	25.397	268 336 125	8.640
496	246 016	22.271	122 023 936	7.916	571	326 041	23.896	186 169 411	8.296	646	417 316	25.417	269 586 136	8.645
497	247 009	22.293	122 763 473	7.921	572	327 184	23.917	187 149 248	8.301	647	418 609	25.436	270 840 023	8.649
498	248 004	22.316	123 505 992	7.926	573	328 329	23.937	188 132 517	8.306	648	419 904	25.456	272 097 792	8.653
499	249 001	22.338	124 251 499	7.932	574	329 476	23.958	189 119 224	8.311	649	421 201	25.475	273 359 449	8.658
500	250 000	22.361	125 000 000	7.937	575	330 625	23.979	190 109 375	8.316	650	422 500	25.495	274 625 000	8.662
501	251 001	22.383	125 751 501	7.942	576	331 776	24.000	191 102 976	8.320	651	423 801	25.515	275 894 451	8.667
502	252 004	22.405	126 506 008	7.948	577	332 929	24.021	192 100 033	8.325	652	425 104	25.534	277 167 808	8.671
503	253 009	22.428	127 263 527	7.953	578	334 084	24.042	193 100 552	8.330	653	426 409	25.554	278 445 077	8.676
504	254 016	22.450	128 024 064	7.958	579	335 241	24.062	194 104 539	8.335	654	427 716	25.573	279 726 264	8.680
505	255 025	22.472	128 787 625	7.963	580	336 400	24.083	195 112 000	8.340	655	429 025	25.593	281 011 375	8.685
506	256 036	22.494	129 554 216	7.969	581	337 561	24.104	196 122 941	8.344	656	430 336	25.612	282 300 416	8.689
507	257 049	22.517	130 323 843	7.974	582	338 724	24.125	197 137 368	8.349	657	431 649	25.632	283 593 393	8.693
508	258 064	22.539	131 096 512	7.979	583	339 889	24.145	198 155 287	8.354	658	432 964	25.652	284 890 312	8.698
509	259 081	22.561	131 872 229	7.984	584	341 056	24.166	199 176 704	8.359	659	434 281	25.671	286 191 179	8.702
510	260 100	22.583	132 651 000	7.990	585	342 225	24.187	200 201 625	8.363	660	435 600	25.690	287 496 000	8.707
511	261 121	22.605	133 432 831	7.995	586	343 396	24.207	201 230 056	8.368	661	436 921	25.710	288 804 781	8.711
512	262 144	22.627	134 217 728	8.000	587	344 569	24.228	202 262 003	8.373	662	438 244	25.729	290 117 528	8.715
513	263 169	22.650	135 005 697	8.005	588	345 744	24.249	203 297 472	8.378	663	439 569	25.749	291 434 247	8.720
514	264 196	22.672	135 796 744	8.010	589	346 921	24.269	204 336 469	8.382	664	440 896	25.768	292 754 944	8.724
515	265 225	22.694	136 590 875	8.016	590	348 100	24.290	205 379 000	8.387	665	442 225	25.788	294 079 625	8.729
516	266 256	22.716	137 388 096	8.021	591	349 281	24.310	206 425 071	8.392	666	443 556	25.807	295 408 296	8.733
517	267 289	22.738	138 188 413	8.026	592	350 464	24.331	207 474 688	8.397	667	444 889	25.826	296 740 963	8.737
518	268 324	22.760	138 991 832	8.031	593	351 649	24.352	208 527 857	8.401	668	446 224	25.846	298 077 632	8.742
519	269 361	22.782	139 798 359	8.036	594	352 836	24.372	209 584 584	8.406	669	447 561	25.865	299 418 309	8.746
520	270 400	22.804	140 608 000	8.041	595	354 025	24.393	210 644 875	8.411	670	448 900	25.884	300 763 000	8.750
521	271 441	22.825	141 420 761	8.047	596	355 216	24.413	211 708 736	8.416	671	450 241	25.904	302 111 711	8.755
522	272 484	22.847	142 236 648	8.052	597	356 409	24.434	212 776 173	8.420	672	451 584	25.923	303 464 448	8.759
523	273 529	22.869	143 055 667	8.057	598	357 604	24.454	213 847 192	8.425	673	452 929	25.942	304 821 217	8.763
524	274 576	22.891	143 877 824	8.062	599	358 801	24.474	214 921 799	8.430	674	454 276	25.962	306 182 024	8.768

Table 6. Squares, Square Roots, Cubes, Cube Roots—(*Continued*)

N	N^2	$\sqrt{N}$	N^3	$\sqrt[3]{N}$	N	N^2	$\sqrt{N}$	N^3	$\sqrt[3]{N}$	N	N^2	$\sqrt{N}$	N^3	$\sqrt[3]{N}$
675	455 625	25.981	307 546 875	8.772	750	562 500	27.386	421 875 000	9.086	825	680 625	28.723	561 515 625	9.379
676	456 976	26.000	308 915 776	8.776	751	564 001	27.404	423 564 751	9.090	826	682 276	28.740	563 559 976	9.383
677	458 329	26.019	310 288 733	8.781	752	565 504	27.423	425 259 008	9.094	827	683 929	28.758	565 609 283	9.386
678	459 684	26.038	311 665 752	8.785	753	567 009	27.441	426 957 777	9.098	828	685 584	28.775	567 663 552	9.390
679	461 041	26.058	313 046 839	8.789	754	568 516	27.459	428 661 064	9.102	829	687 241	28.792	569 722 789	9.394
680	462 400	26.077	314 432 000	8.794	755	570 025	27.477	430 368 875	9.106	830	688 900	28.810	571 787 000	9.398
681	463 761	26.096	315 821 241	8.798	756	571 536	27.495	432 081 216	9.110	831	690 561	28.827	573 856 191	9.402
682	465 124	26.115	317 214 568	8.802	757	573 049	27.514	433 798 093	9.114	832	692 224	28.844	575 930 368	9.405
683	466 489	26.134	318 611 987	8.807	758	574 564	27.532	435 519 512	9.118	833	693 889	28.862	578 009 537	9.409
684	467 856	26.153	320 013 504	8.811	759	576 081	27.550	437 245 479	9.122	834	695 556	28.879	580 093 704	9.413
685	469 225	26.173	321 419 125	8.815	760	577 600	27.568	438 976 000	9.126	835	697 225	28.896	582 182 875	9.417
686	470 596	26.192	322 828 856	8.819	761	579 121	27.586	440 711 081	9.130	836	698 896	28.914	584 277 056	9.420
687	471 969	26.211	324 242 703	8.824	762	580 644	27.604	442 450 728	9.134	837	700 569	28.931	586 376 253	9.424
688	473 344	26.230	325 660 672	8.828	763	582 169	27.622	444 194 947	9.138	838	702 244	28.948	588 480 472	9.428
689	474 721	26.249	327 082 769	8.832	764	583 696	27.641	445 943 744	9.142	839	703 921	28.965	590 589 719	9.432
690	476 100	26.268	328 509 000	8.837	765	585 225	27.659	447 697 125	9.146	840	705 600	28.983	592 704 000	9.435
691	477 481	26.287	329 939 371	8.841	766	586 756	27.677	449 455 096	9.150	841	707 281	29.000	594 823 321	9.439
692	478 864	26.306	331 373 888	8.845	767	588 289	27.695	451 217 663	9.154	842	708 964	29.017	596 947 688	9.443
693	480 249	26.325	332 812 557	8.849	768	589 824	27.713	452 984 832	9.158	843	710 649	29.034	599 077 107	9.447
694	481 636	26.344	334 255 384	8.854	769	591 361	27.731	454 756 609	9.162	844	712 236	29.052	601 211 584	9.450
695	483 025	26.363	335 702 375	8.858	770	592 900	27.749	456 533 000	9.166	845	714 025	29.069	603 351 125	9.454
696	484 416	26.382	337 153 536	8.862	771	594 441	27.767	458 314 011	9.170	846	715 716	29.086	605 495 736	9.458
697	485 809	26.401	338 608 873	8.866	772	595 984	27.785	460 099 648	9.174	847	717 409	29.103	607 645 423	9.462
698	487 204	26.420	340 068 392	8.871	773	597 529	27.803	461 889 917	9.178	848	719 104	29.120	609 800 192	9.465
699	488 601	26.439	341 532 099	8.875	774	599 076	27.821	463 684 824	9.182	849	720 801	29.138	611 960 049	9.469
700	490 000	26.458	343 000 000	8.879	775	600 625	27.839	465 484 375	9.185	850	722 500	29.155	614 125 000	9.473
701	491 401	26.476	344 472 101	8.883	776	602 176	27.857	467 288 576	9.189	851	724 201	29.172	616 295 051	9.476
702	492 804	26.495	345 948 408	8.887	777	603 729	27.875	469 097 433	9.193	852	725 904	29.189	618 470 208	9.480
703	494 209	26.514	347 428 927	8.892	778	605 284	27.893	470 910 952	9.197	853	727 609	29.206	620 650 477	9.484
704	495 616	26.533	348 913 664	8.896	779	606 841	27.911	472 729 139	9.201	854	729 316	29.223	622 835 864	9.488
705	497 025	26.552	350 402 625	8.900	780	608 400	27.928	474 552 000	9.205	855	731 025	29.240	625 026 375	9.491
706	498 436	26.571	351 895 816	8.904	781	609 961	27.946	476 379 541	9.209	856	732 736	29.257	627 222 016	9.495
707	499 849	26.589	353 393 243	8.909	782	611 524	27.964	478 211 768	9.213	857	734 449	29.275	629 422 793	9.499
708	501 264	26.608	354 894 912	8.913	783	613 089	27.982	480 048 687	9.217	858	736 164	29.292	631 628 712	9.502
709	502 681	26.627	356 400 829	8.917	784	614 656	28.000	481 890 304	9.221	859	737 881	29.309	633 839 779	9.506
710	504 100	26.646	357 911 000	8.921	785	616 225	28.018	483 736 625	9.225	860	739 600	29.326	636 056 000	9.510
711	505 521	26.665	359 425 431	8.925	786	617 796	28.036	485 587 656	9.229	861	741 321	29.343	638 277 381	9.513
712	506 944	26.683	360 944 128	8.929	787	619 369	28.054	487 443 403	9.233	862	743 044	29.360	640 503 928	9.517
713	508 369	26.702	362 467 097	8.934	788	620 944	28.071	489 303 872	9.237	863	744 769	29.377	642 735 647	9.521
714	509 796	26.721	363 994 344	8.938	789	622 521	28.089	491 169 069	9.240	864	746 496	29.394	644 972 544	9.524
715	511 225	26.739	365 525 875	8.942	790	624 100	28.107	493 039 000	9.244	865	748 225	29.411	647 214 625	9.528
716	512 656	26.758	367 061 696	8.946	791	625 681	28.125	494 913 671	9.248	866	749 956	29.428	649 461 896	9.532
717	514 089	26.777	368 601 813	8.950	792	627 264	28.142	496 793 088	9.252	867	751 689	29.445	651 714 363	9.535
718	515 524	26.796	370 146 232	8.955	793	628 849	28.160	498 677 257	9.256	868	753 424	29.462	653 972 032	9.539
719	516 961	26.814	371 694 959	8.959	794	630 436	28.178	500 566 184	9.260	869	755 161	29.479	656 234 909	9.543
720	518 400	26.833	373 248 000	8.963	795	632 025	28.196	502 459 875	9.264	870	756 900	29.496	658 503 000	9.546
721	519 841	26.851	374 805 361	8.967	796	633 616	28.213	504 358 336	9.268	871	758 641	29.513	660 776 311	9.550
722	521 284	26.870	376 367 048	8.971	797	635 209	28.231	506 261 573	9.272	872	760 384	29.530	663 054 848	9.554
723	522 729	26.889	377 933 067	8.975	798	636 804	28.249	508 169 592	9.275	873	762 129	29.547	665 338 617	9.557
724	524 176	26.907	379 503 424	8.979	799	638 401	28.267	510 082 399	9.279	874	763 876	29.563	667 627 624	9.561
725	525 625	26.926	381 078 125	8.984	800	640 000	28.284	512 000 000	9.283	875	765 625	29.580	669 921 875	9.565
726	527 076	26.944	382 657 176	8.988	801	641 601	28.302	513 922 401	9.287	876	767 376	29.597	672 221 376	9.568
727	528 529	26.963	384 240 583	8.992	802	643 204	28.320	515 849 608	9.291	877	769 129	29.614	674 526 133	9.572
728	529 984	26.981	385 828 352	8.996	803	644 809	28.337	517 781 627	9.295	878	770 884	29.631	676 836 152	9.576
729	531 441	27.000	387 420 489	9.000	804	646 416	28.355	519 718 464	9.299	879	772 641	29.648	679 151 439	9.579
730	532 900	27.019	389 017 000	9.004	805	648 025	28.373	521 660 125	9.302	880	774 400	29.665	681 472 000	9.583
731	534 361	27.037	390 617 891	9.008	806	649 636	28.390	523 606 616	9.306	881	776 161	29.682	683 797 841	9.586
732	535 824	27.055	392 223 168	9.012	807	651 249	28.408	525 557 943	9.310	882	777 924	29.698	686 128 968	9.590
733	537 289	27.074	393 832 837	9.016	808	652 864	28.425	527 514 112	9.314	883	779 689	29.715	688 465 387	9.594
734	538 756	27.092	395 446 904	9.021	809	654 481	28.443	529 475 129	9.318	884	781 456	29.732	690 807 104	9.597
735	540 225	27.111	397 065 375	9.025	810	656 100	28.460	531 441 000	9.322	885	783 225	29.749	693 154 125	9.601
736	541 696	27.129	398 688 256	9.029	811	657 721	28.478	533 411 731	9.326	886	784 996	29.766	695 506 456	9.605
737	543 169	27.148	400 315 553	9.033	812	659 344	28.496	535 387 328	9.329	887	786 769	29.783	697 864 103	9.608
738	544 644	27.166	401 947 272	9.037	813	660 969	28.513	537 367 797	9.333	888	788 544	29.799	700 227 072	9.612
739	546 121	27.185	403 583 419	9.041	814	662 596	28.531	539 353 144	9.337	889	790 321	29.816	702 595 369	9.615
740	547 600	27.203	405 224 000	9.045	815	664 225	28.548	541 343 375	9.341	890	792 100	29.833	704 969 000	9.619
741	549 081	27.221	406 869 021	9.049	816	665 856	28.566	543 338 496	9.345	891	793 881	29.850	707 347 971	9.623
742	550 564	27.240	408 518 488	9.053	817	667 489	28.583	545 338 513	9.348	892	795 664	29.866	709 732 288	9.626
743	552 049	27.258	410 172 407	9.057	818	669 124	28.601	547 343 432	9.352	893	797 449	29.883	712 121 957	9.630
744	553 536	27.276	411 830 784	9.061	819	670 761	28.618	549 353 259	9.356	894	799 236	29.900	714 516 984	9.633
745	555 025	27.295	413 493 625	9.065	820	672 400	28.636	551 368 000	9.360	895	801 025	29.917	716 917 375	9.637
746	556 516	27.313	415 160 936	9.069	821	674 041	28.653	553 387 661	9.364	896	802 816	29.933	719 323 136	9.641
747	558 009	27.331	416 832 723	9.073	822	675 684	28.671	555 412 248	9.368	897	804 609	29.950	721 734 273	9.644
748	559 504	27.350	418 508 992	9.078	823	677 329	28.688	557 441 767	9.371	898	806 404	29.967	724 150 792	9.648
749	561 001	27.368	420 189 749	9.082	824	678 976	28.705	559 476 224	9.375	899	808 201	29.983	726 572 699	9.651

Table 6. Squares, Square Roots, Cubes, Cube Roots—(*Concluded*)

N	N^2	$\sqrt{N}$	N^3	$\sqrt[3]{N}$	N	N^2	$\sqrt{N}$	N^3	$\sqrt[3]{N}$	N	N^2	$\sqrt{N}$	N^3	$\sqrt[3]{N}$
900	810 000	30.000	729 000 000	9.655	935	874 225	30.578	817 400 375	9.778	970	940 900	31.145	912 673 000	9.899
901	811 801	30.017	731 432 701	9.658	936	876 096	30.594	820 025 856	9.782	971	942 841	31.161	915 498 611	9.902
902	813 604	30.033	733 870 808	9.662	937	877 969	30.610	822 656 953	9.785	972	944 784	31.177	918 330 048	9.906
903	815 409	30.050	736 314 327	9.666	938	879 844	30.627	825 293 672	9.789	973	946 729	31.193	921 167 317	9.909
904	817 216	30.067	738 763 264	9.669	939	881 721	30.643	827 936 019	9.792	974	948 676	31.209	924 010 424	9.913
905	819 025	30.083	741 217 625	9.673	940	883 600	30.659	830 584 000	9.796	975	950 625	31.225	926 859 375	9.916
906	820 836	30.100	743 677 416	9.676	941	885 481	30.676	833 237 621	9.799	976	952 576	31.241	929 714 176	9.919
907	822 649	30.116	746 142 643	9.680	942	887 364	30.692	835 896 888	9.803	977	954 529	31.257	932 574 833	9.923
908	824 464	30.133	748 613 312	9.683	943	889 249	30.708	838 561 807	9.806	978	956 484	31.273	935 441 352	9.926
909	826 281	30.150	751 089 429	9.687	944	891 136	30.725	841 232 384	9.810	979	958 441	31.289	938 313 739	9.930
910	828 100	30.166	753 571 000	9.691	945	893 025	30.741	843 908 625	9.813	980	960 400	31.305	941 192 000	9.933
911	829 921	30.183	756 058 031	9.694	946	894 916	30.757	846 590 536	9.817	981	962 361	31.321	944 076 141	9.936
912	831 744	30.199	758 550 528	9.698	947	896 809	30.773	849 278 123	9.820	982	964 324	31.337	946 966 168	9.940
913	833 569	30.216	761 048 497	9.701	948	898 704	30.790	851 971 392	9.824	983	966 289	31.353	949 862 087	9.943
914	835 396	30.232	763 551 944	9.705	949	900 601	30.806	854 670 349	9.827	984	968 256	31.369	952 763 904	9.946
915	837 225	30.249	766 060 875	9.708	950	902 500	30.822	857 375 000	9.830	985	970 225	31.385	955 671 625	9.950
916	839 056	30.265	768 575 296	9.712	951	904 401	30.838	860 085 351	9.834	986	972 196	31.401	958 585 256	9.953
917	840 889	30.282	771 095 213	9.715	952	906 304	30.854	862 801 408	9.837	987	974 169	31.417	961 504 803	9.956
918	842 724	30.299	773 620 632	9.719	953	908 209	30.871	865 523 177	9.841	988	976 144	31.432	964 430 272	9.960
919	844 561	30.315	776 151 559	9.722	954	910 116	30.887	868 250 664	9.844	989	978 121	31.448	967 361 669	9.963
920	846 400	30.332	778 688 000	9.726	955	912 025	30.903	870 983 875	9.848	990	980 100	31.464	970 299 000	9.967
921	848 241	30.348	781 229 961	9.729	956	913 936	30.919	873 722 816	9.851	991	982 081	31.480	973 242 271	9.970
922	850 084	30.364	783 777 448	9.733	957	915 849	30.935	876 467 493	9.855	992	984 064	31.496	976 191 488	9.973
923	851 929	30.381	786 330 467	9.736	958	917 764	30.952	879 217 912	9.858	993	986 049	31.512	979 146 657	9.977
924	853 776	30.397	788 889 024	9.740	959	919 681	30.968	881 974 079	9.861	994	988 036	31.528	982 107 784	9.980
925	855 625	30.414	791 453 125	9.743	960	921 600	30.984	884 736 000	9.865	995	990 025	31.544	985 074 875	9.983
926	857 476	30.430	794 022 776	9.747	961	923 521	31.000	887 503 681	9.868	996	992 016	31.559	988 047 936	9.987
927	859 329	30.447	796 597 983	9.750	962	925 444	31.016	890 277 128	9.872	997	994 009	31.575	991 026 973	9.990
928	861 184	30.463	799 178 752	9.754	963	927 369	31.032	893 056 347	9.875	998	996 004	31.591	994 011 992	9.993
929	863 041	30.480	801 765 089	9.758	964	929 296	31.048	895 841 344	9.879	999	998 001	31.607	997 002 999	9.997
930	864 900	30.496	804 357 000	9.761	965	931 225	31.064	898 632 125	9.882					
931	866 761	30.512	806 954 491	9.764	966	933 156	31.081	901 428 696	9.885					
932	868 624	30.529	809 557 568	9.768	967	935 089	31.097	904 231 063	9.889					
933	870 489	30.545	812 166 237	9.771	968	937 024	31.113	907 039 232	9.892					
934	872 356	30.561	814 780 504	9.775	969	938 961	31.129	909 853 209	9.896					

Table 7. Reciprocals of Numbers*

N	$\frac{1}{N}\times 10^3$	N	$\frac{1}{N}\times 10^3$	N	$\frac{1}{N}\times 10^3$	N	$\frac{1}{N}\times 10^3$	N	$\frac{1}{N}\times 10^3$	N	$\frac{1}{N}\times 10^3$	N	$\frac{1}{N}\times 10^3$	N	$\frac{1}{N}\times 10^3$	N	$\frac{1}{N}\times 10^3$	N	$\frac{1}{N}\times 10^3$
10	100.0000	45	22.2222	80	12.5000	115	8.69565	150	6.66667	185	5.40541	220	4.54546	255	3.92157	290	3.44828	325	3.07692
11	90.9091	46	21.7391	81	12.3457	116	8.62069	151	6.62252	186	5.37634	221	4.52489	256	3.90625	291	3.43643	326	3.06749
12	83.3333	47	21.2766	82	12.1951	117	8.54701	152	6.57895	187	5.34759	222	4.50451	257	3.89105	292	3.42466	327	3.05810
13	76.9231	48	20.8333	83	12.0482	118	8.47458	153	6.53595	188	5.31915	223	4.48431	258	3.87597	293	3.41297	328	3.04878
14	71.4286	49	20.4082	84	11.9048	119	8.40336	154	6.49351	189	5.29101	224	4.46429	259	3.86100	294	3.40136	329	3.03951
15	66.6667	50	20.0000	85	11.7647	120	8.33333	155	6.45161	190	5.26316	225	4.44444	260	3.84615	295	3.38983	330	3.03030
16	62.5000	51	19.6078	86	11.6279	121	8.26446	156	6.41026	191	5.23560	226	4.42478	261	3.83142	296	3.37838	331	3.02115
17	58.8235	52	19.2308	87	11.4943	122	8.19672	157	6.36943	192	5.20833	227	4.40529	262	3.81679	297	3.36700	332	3.01205
18	55.5556	53	18.8679	88	11.3636	123	8.13008	158	6.32911	193	5.18135	228	4.38597	263	3.80228	298	3.35571	333	3.00300
19	52.6316	54	18.5185	89	11.2360	124	8.06452	159	6.28931	194	5.15464	229	4.36681	264	3.78788	299	3.34448	334	2.99401
20	50.0000	55	18.1818	90	11.1111	125	8.00000	160	6.25000	195	5.12821	230	4.34783	265	3.77359	300	3.33333	335	2.98508
21	47.6190	56	17.8571	91	10.9890	126	7.93651	161	6.21118	196	5.10204	231	4.32900	266	3.75940	301	3.32226	336	2.97619
22	45.4545	57	17.5439	92	10.8696	127	7.87402	162	6.17284	197	5.07614	232	4.31035	267	3.74532	302	3.31126	337	2.96736
23	43.4783	58	17.2414	93	10.7527	128	7.81250	163	6.13497	198	5.05051	233	4.29185	268	3.73134	303	3.30033	338	2.95858
24	41.6667	59	16.9492	94	10.6383	129	7.75194	164	6.09756	199	5.02513	234	4.27350	269	3.71747	304	3.28947	339	2.94985
25	40.0000	60	16.6667	95	10.5263	130	7.69231	165	6.06061	200	5.00000	235	4.25532	270	3.70370	305	3.27869	340	2.94118
26	38.4615	61	16.3934	96	10.4167	131	7.63359	166	6.02410	201	4.97512	236	4.23729	271	3.69004	306	3.26797	341	2.93255
27	37.0370	62	16.1290	97	10.3093	132	7.57576	167	5.98802	202	4.95050	237	4.21941	272	3.67647	307	3.25733	342	2.92398
28	35.7143	63	15.8730	98	10.2041	133	7.51880	168	5.95238	203	4.92611	238	4.20168	273	3.66300	308	3.24675	343	2.91545
29	34.4828	64	15.6250	99	10.1010	134	7.46269	169	5.91716	204	4.90196	239	4.18410	274	3.64964	309	3.23625	344	2.90698
30	33.3333	65	15.3846	100	10.00000	135	7.40741	170	5.88235	205	4.87805	240	4.16667	275	3.63636	310	3.22581	345	2.89855
31	32.2581	66	15.1515	101	9.90099	136	7.35294	171	5.84795	206	4.85437	241	4.14938	276	3.62319	311	3.21543	346	2.89017
32	31.2500	67	14.9254	102	9.80392	137	7.29927	172	5.81395	207	4.83092	242	4.13223	277	3.61011	312	3.20513	347	2.88184
33	30.3030	68	14.7059	103	9.70874	138	7.24638	173	5.78035	208	4.80769	243	4.11523	278	3.59712	313	3.19489	348	2.87356
34	29.4118	69	14.4928	104	9.61539	139	7.19425	174	5.74713	209	4.78469	244	4.09836	279	3.58423	314	3.18471	349	2.86533
35	28.5714	70	14.2857	105	9.52381	140	7.14286	175	5.71429	210	4.76191	245	4.08163	280	3.57143	315	3.17460	350	2.85714
36	27.7778	71	14.0845	106	9.43396	141	7.09220	176	5.68182	211	4.73934	246	4.06504	281	3.55872	316	3.16456	351	2.84900
37	27.0270	72	13.8889	107	9.34579	142	7.04225	177	5.64972	212	4.71698	247	4.04858	282	3.54610	317	3.15457	352	2.84091
38	26.3158	73	13.6986	108	9.25926	143	6.99301	178	5.61798	213	4.69484	248	4.03226	283	3.53357	318	3.14465	353	2.83286
39	25.6410	74	13.5135	109	9.17431	144	6.94444	179	5.58659	214	4.67290	249	4.01606	264	3.52113	319	3.13480	354	2.82486
40	25.0000	75	13.3333	110	9.09091	145	6.89655	180	5.55556	215	4.65116	250	4.00000	285	3.50877	320	3.12500	355	2.81690
41	24.3902	76	13.1579	111	9.00901	146	6.84932	181	5.52486	216	4.62963	251	3.98406	286	3.49650	321	3.11527	356	2.80899
42	23.8095	77	12.9870	112	8.92857	147	6.80272	182	5.49451	217	4.60830	252	3.96825	287	3.48432	322	3.10559	357	2.80112
43	23.2558	78	12.8205	113	8.84956	148	6.75676	183	5.46448	218	4.58716	253	3.95257	288	3.47222	323	3.09598	358	2.79330
44	22.7273	79	12.6582	114	8.77193	149	6.71141	184	5.43478	219	4.56621	254	3.93701	289	3.46021	324	3.08642	359	2.78552

* For a more precise table of reciprocals, see "Six-place Tables," McGraw-Hill Book Company, Inc.

Table 7. Reciprocals of Numbers—(*Concluded*)

N	$\frac{1}{N} \times 10^3$	N	$\frac{1}{N} \times 10^3$	N	$\frac{1}{N} \times 10^3$	N	$\frac{1}{N} \times 10^3$	N	$\frac{1}{N} \times 10^3$	N	$\frac{1}{N} \times 10^3$	N	$\frac{1}{N} \times 10^3$	N	$\frac{1}{N} \times 10^3$	N	$\frac{1}{N} \times 10^3$	N	$\frac{1}{N} \times 10^3$
360	2.77778	425	2.35294	490	2.04082	555	1.80180	620	1.61290	685	1.45985	750	1.33333	815	1.22699	880	1.13636	945	1.05820
361	2.77008	426	2.34742	491	2.03666	556	1.79856	621	1.61031	686	1.45773	751	1.33156	816	1.22549	881	1.13507	946	1.05708
362	2.76243	427	2.34192	492	2.03252	557	1.79533	622	1.60772	687	1.45560	752	1.32979	817	1.22399	882	1.13379	947	1.05597
363	2.75482	428	2.33645	493	2.02840	558	1.79212	623	1.60514	688	1.45349	753	1.32802	818	1.22249	883	1.13250	948	1.05485
364	2.74725	429	2.33100	494	2.02429	559	1.78891	624	1.60256	689	1.45138	754	1.32626	819	1.22100	884	1.13122	949	1.05374
365	2.73973	430	2.32558	495	2.02020	560	1.78571	625	1.60000	690	1.44928	755	1.32450	820	1.21951	885	1.12994	950	1.05263
366	2.73224	431	2.32019	496	2.01613	561	1.78253	626	1.59744	691	1.44718	756	1.32275	821	1.21803	886	1.12867	951	1.05153
367	2.72480	432	2.31482	497	2.01207	562	1.77936	627	1.59490	692	1.44509	757	1.32100	822	1.21655	887	1.12740	952	1.05042
368	2.71739	433	2.30947	498	2.00803	563	1.77620	628	1.59236	693	1.44300	758	1.31926	823	1.21507	888	1.12613	953	1.04932
369	2.71003	434	2.30415	499	2.00401	564	1.77305	629	1.58983	694	1.44092	759	1.31752	824	1.21359	889	1.12486	954	1.04822
370	2.70270	435	2.29885	500	2.00000	565	1.76991	630	1.58730	695	1.43885	760	1.31579	825	1.21212	890	1.12360	955	1.04712
371	2.69542	436	2.29358	501	1.99601	566	1.76678	631	1.58479	696	1.43678	761	1.31406	826	1.21065	891	1.12233	956	1.04603
372	2.68817	437	2.28833	502	1.99203	567	1.76367	632	1.58228	697	1.43472	762	1.31234	827	1.20919	892	1.12108	957	1.04493
373	2.68097	438	2.28311	503	1.98807	568	1.76056	633	1.57978	698	1.43267	763	1.31062	828	1.20773	893	1.11982	958	1.04384
374	2.67380	439	2.27790	504	1.98413	569	1.75747	634	1.57729	699	1.43062	764	1.30890	829	1.20627	894	1.11857	959	1.04275
375	2.66667	440	2.27273	505	1.98020	570	1.75439	635	1.57480	700	1.42857	765	1.30719	830	1.20482	895	1.11732	960	1.04167
376	2.65957	441	2.26757	506	1.97629	571	1.75131	636	1.57233	701	1.42653	766	1.30548	831	1.20337	896	1.11607	961	1.04058
377	2.65252	442	2.26244	507	1.97239	572	1.74825	637	1.56986	702	1.42450	767	1.30378	832	1.20192	897	1.11483	962	1.03950
378	2.64550	443	2.25734	508	1.96850	573	1.74520	638	1.56740	703	1.42248	768	1.30208	833	1.20048	898	1.11359	963	1.03842
379	2.63852	444	2.25225	509	1.96464	574	1.74216	639	1.56495	704	1.42046	769	1.30039	834	1.19904	899	1.11235	964	1.03734
380	2.63158	445	2.24719	510	1.96078	575	1.73913	640	1.56250	705	1.41844	770	1.29870	835	1.19761	900	1.11111	965	1.03627
381	2.62467	446	2.24215	511	1.95695	576	1.73611	641	1.56006	706	1.41643	771	1.29702	836	1.19617	901	1.10988	966	1.03520
382	2.61780	447	2.23714	512	1.95313	577	1.73310	642	1.55763	707	1.41443	772	1.29534	837	1.19474	902	1.10865	967	1.03413
383	2.61097	448	2.23214	513	1.94932	578	1.73010	643	1.55521	708	1.41243	773	1.29366	838	1.19332	903	1.10742	968	1.03306
384	2.60417	449	2.22717	514	1.94553	579	1.72712	644	1.55280	709	1.41044	774	1.29199	839	1.19190	904	1.10620	969	1.03199
385	2.59740	450	2.22222	515	1.94175	580	1.72414	645	1.55039	710	1.40845	775	1.29032	840	1.19048	905	1.10497	970	1.03093
386	2.59067	451	2.21730	516	1.93798	581	1.72117	646	1.54799	711	1.40647	776	1.28866	841	1.18906	906	1.10375	971	1.02987
387	2.58398	452	2.21239	517	1.93424	582	1.71821	647	1.54560	712	1.40449	777	1.28700	842	1.18765	907	1.10254	972	1.02881
388	2.57732	453	2.20751	518	1.93050	583	1.71527	648	1.54321	713	1.40253	778	1.28535	843	1.18624	908	1.10132	973	1.02775
389	2.57069	454	2.20264	519	1.92678	584	1.71233	649	1.54083	714	1.40056	779	1.28370	844	1.18483	909	1.10011	974	1.02669
390	2.56410	455	2.19780	520	1.92308	585	1.70940	650	1.53846	715	1.39860	780	1.28205	845	1.18343	910	1.09890	975	1.02564
391	2.55755	456	2.19298	521	1.91939	586	1.70649	651	1.53610	716	1.39665	781	1.28041	846	1.18203	911	1.09770	976	1.02459
392	2.55102	457	2.18818	522	1.91571	587	1.70358	652	1.53374	717	1.39470	782	1.27877	847	1.18064	912	1.09649	977	1.02354
393	2.54453	458	2.18341	523	1.91205	588	1.70068	653	1.53139	718	1.39276	783	1.27714	848	1.17925	913	1.09529	978	1.02250
394	2.53807	459	2.17865	524	1.90840	589	1.69779	654	1.52905	719	1.39082	784	1.27551	849	1.17786	914	1.09409	979	1.02145
395	2.53165	460	2.17391	525	1.90476	590	1.69492	655	1.52672	720	1.38889	785	1.27389	850	1.17647	915	1.09290	980	1.02041
396	2.52525	461	2.16920	526	1.90114	591	1.69205	656	1.52439	721	1.38696	786	1.27227	851	1.17509	916	1.09170	981	1.01937
397	2.51889	462	2.16450	527	1.89753	592	1.68919	657	1.52207	722	1.38504	787	1.27065	852	1.17371	917	1.09051	982	1.01833
398	2.51256	463	2.15983	528	1.89394	593	1.68634	658	1.51976	723	1.38313	788	1.26904	853	1.17233	918	1.08933	983	1.01729
399	2.50627	464	2.15517	529	1.89036	594	1.68350	659	1.51745	724	1.38122	789	1.26743	854	1.17096	919	1.08814	984	1.01626
400	2.50000	465	2.15054	530	1.88679	595	1.68067	660	1.51515	725	1.37931	790	1.26582	855	1.16959	920	1.08696	985	1.01523
401	2.49377	466	2.14592	531	1.88324	596	1.67785	661	1.51286	726	1.37741	791	1.26422	856	1.16822	921	1.08578	986	1.01420
402	2.48756	467	2.14133	532	1.87970	597	1.67504	662	1.51057	727	1.37552	792	1.26263	857	1.16686	922	1.08460	987	1.01317
403	2.48139	468	2.13675	533	1.87617	598	1.67224	663	1.50830	728	1.37363	793	1.26103	858	1.16550	923	1.08342	988	1.01215
404	2.47525	469	2.13220	534	1.87266	599	1.66945	664	1.50602	729	1.37174	794	1.25945	859	1.16414	924	1.08225	989	1.01112
405	2.46914	470	2.12766	535	1.86916	600	1.66667	665	1.50376	730	1.36986	795	1.25786	860	1.16279	925	1.08108	990	1.01010
406	2.46305	471	2.12314	536	1.86567	601	1.66389	666	1.50150	731	1.36799	796	1.25628	861	1.16144	926	1.07991	991	1.00908
407	2.45700	472	2.11864	537	1.86220	602	1.66113	667	1.49925	732	1.36612	797	1.25471	862	1.16009	927	1.07875	992	1.00807
408	2.45098	473	2.11417	538	1.85874	603	1.65838	668	1.49701	733	1.36426	798	1.25313	863	1.15875	928	1.07759	993	1.00705
409	2.44499	474	2.10971	539	1.85529	604	1.65563	699	1.49477	734	1.36240	799	1.25156	864	1.15741	929	1.07643	994	1.00604
410	2.43902	475	2.10526	540	1.85185	605	1.65289	670	1.49254	735	1.36054	800	1.25000	865	1.15607	930	1.07527	995	1.00503
411	2.43309	476	2.10084	541	1.84843	606	1.65017	671	1.49031	736	1.35870	801	1.24844	866	1.15473	931	1.07411	996	1.00402
412	2.42718	477	2.09644	542	1.84502	607	1.64745	672	1.48810	737	1.35685	802	1.24688	867	1.15340	932	1.07296	997	1.00301
413	2.42131	478	2.09205	543	1.84162	608	1.64474	673	1.48588	738	1.35501	803	1.24533	868	1.15207	933	1.07181	998	1.00200
414	2.41546	479	2.08768	544	1.83824	609	1.64204	674	1.48368	739	1.35318	804	1.24378	869	1.15075	934	1.07066	999	1.00100
415	2.40964	480	2.08333	545	1.83486	610	1.63934	675	1.48148	740	1.35135	805	1.24224	870	1.14943	935	1.06952		
416	2.40385	481	2.07900	546	1.83150	611	1.63666	676	1.47929	741	1.34953	806	1.24070	871	1.14811	936	1.06838		
417	2.39808	482	2.07469	547	1.82815	612	1.63399	677	1.47711	742	1.34771	807	1.23916	872	1.14679	937	1.06724		
418	2.39234	483	2.07039	548	1.82482	613	1.63132	678	1.47493	743	1.34590	808	1.23762	873	1.14548	938	1.06610		
419	2.38664	484	2.06612	549	1.82149	614	1.62866	679	1.47275	744	1.34409	809	1.23609	874	1.14417	939	1.06496		
420	2.38095	485	2.06186	550	1.81818	615	1.62602	680	1.47059	745	1.34228	810	1.23457	875	1.14286	940	1.06383		
421	2.37530	486	2.05761	551	1.81488	616	1.62338	681	1.46843	746	1.34048	811	1.23305	876	1.14155	941	1.06270		
422	2.36967	487	2.05339	552	1.81159	617	1.62075	682	1.46628	747	1.33869	812	1.23153	877	1.14025	942	1.06157		
423	2.36407	488	2.04918	553	1.80832	618	1.61812	683	1.46413	748	1.33690	813	1.23001	878	1.13895	943	1.06045		
424	2.35849	489	2.04499	554	1.80505	619	1.61551	684	1.46199	749	1.33511	814	1.22850	879	1.13766	944	1.05932		

CIRCLES

Table 8. Circles: Diameters, Areas, Circumferences

Diameters in hundredths

Diam.	Area	Cir.	Diam.	Area	Cir.	Diam.	Area	Cir.	Diam.	Area	Cir.	Diam.	Area	Cir.	Diam.	Area	Cir.
1.00	0.78540	3.1416	1.75	2.40528	5.4978	2.50	4.9087	7.8540	3.25	8.2958	10.2102	4.00	12.5664	12.5664	4.75	17.7205	14.9226
1.01	.80119	3.1730	1.76	2.43285	5.5292	2.51	4.9481	7.8854	3.26	8.3469	10.2416	4.01	12.6293	12.5978	4.76	17.7952	14.9540
1.02	.81713	3.2044	1.77	2.46057	5.5606	2.52	4.9876	7.9168	3.27	8.3982	10.2730	4.02	12.6923	12.6292	4.77	17.8701	14.9854
1.03	.83323	3.2358	1.78	2.48846	5.5920	2.53	5.0273	7.9482	3.28	8.4496	10.3044	4.03	12.7556	12.6606	4.78	17.9451	15.0168
1.04	.84949	3.2673	1.79	2.51649	5.6235	2.54	5.0671	7.9796	3.29	8.5012	10.3358	4.04	12.8190	12.6920	4.79	18.0203	15.0482
1.05	.86590	3.2987	1.80	2.54469	5.6549	2.55	5.1071	8.0111	3.30	8.5530	10.3673	4.05	12.8825	12.7235	4.80	18.0956	15.0796
1.06	.88247	3.3301	1.81	2.57304	5.6863	2.56	5.1472	8.0425	3.31	8.6049	10.3987	4.06	12.9462	12.7549	4.81	18.1711	15.1111
1.07	.89920	3.3615	1.82	2.60155	5.7177	2.57	5.1875	8.0739	3.32	8.6570	10.4301	4.07	13.0100	12.7863	4.82	18.2467	15.1425
1.08	.91609	3.3929	1.83	2.63022	5.7491	2.58	5.2279	8.1053	3.33	8.7092	10.4615	4.08	13.0741	12.8177	4.83	18.3225	15.1739
1.09	.93313	3.4243	1.84	2.65904	5.7805	2.59	5.2685	8.1367	3.34	8.7616	10.4929	4.09	13.1382	12.8491	4.84	18.3984	15.2053
1.10	.95033	3.4558	1.85	2.68803	5.8119	2.60	5.3093	8.1681	3.35	8.8141	10.5243	4.10	13.2025	12.8805	4.85	18.4745	15.2367
1.11	.96769	3.4872	1.86	2.71716	5.8434	2.61	5.3502	8.1996	3.36	8.8668	10.5558	4.11	13.2670	12.9119	4.86	18.5508	15.2681
1.12	.98520	3.5186	1.87	2.74646	5.8748	2.62	5.3913	8.2310	3.37	8.9197	10.5872	4.12	13.3317	12.9434	4.87	18.6272	15.2996
1.13	1.00288	3.5500	1.88	2.77591	5.9062	2.63	5.4325	8.2624	3.38	8.9727	10.6186	4.13	13.3965	12.9748	4.88	18.7038	15.3310
1.14	1.02070	3.5814	1.89	2.80552	5.9376	2.64	5.4739	8.2938	3.39	9.0259	10.6500	4.14	13.4614	13.0062	4.89	18.7805	15.3624
1.15	1.03869	3.6128	1.90	2.83529	5.9690	2.65	5.5155	8.3252	3.40	9.0792	10.6814	4.15	13.5265	13.0376	4.90	18.8574	15.3938
1.16	1.05683	3.6442	1.91	2.86521	6.0004	2.66	5.5572	8.3566	3.41	9.1327	10.7128	4.16	13.5918	13.0690	4.91	18.9345	15.4252
1.17	1.07513	3.6757	1.92	2.89529	6.0319	2.67	5.5990	8.3881	3.42	9.1863	10.7442	4.17	13.6572	13.1004	4.92	19.0117	15.4566
1.18	1.09359	3.7071	1.93	2.92553	6.0633	2.68	5.6410	8.4195	3.43	9.2401	10.7757	4.18	13.7228	13.1319	4.93	19.0890	15.4881
1.19	1.11220	3.7385	1.94	2.95593	6.0947	2.69	5.6832	8.4509	3.44	9.2941	10.8071	4.19	13.7885	13.1633	4.94	19.1665	15.5195
1.20	1.13097	3.7699	1.95	2.98648	6.1261	2.70	5.7256	8.4823	3.45	9.3482	10.8385	4.20	13.8544	13.1947	4.95	19.2442	15.5509
1.21	1.14990	3.8013	1.96	3.01719	6.1575	2.71	5.7680	8.5137	3.46	9.4025	10.8699	4.21	13.9205	13.2261	4.96	19.3221	15.5823
1.22	1.16899	3.8327	1.97	3.04805	6.1889	2.72	5.8107	8.5451	3.47	9.4569	10.9013	4.22	13.9867	13.2575	4.97	19.4000	15.6137
1.23	1.18823	3.8642	1.98	3.07908	6.2204	2.73	5.8535	8.5765	3.48	9.5115	10.9327	4.23	14.0531	13.2889	4.98	19.4782	15.6451
1.24	1.20763	3.8956	1.99	3.11026	6.2518	2.74	5.8965	8.6080	3.49	9.5662	10.9642	4.24	14.1196	13.3204	4.99	19.5565	15.6765
1.25	1.22719	3.9270	2.00	3.14159	6.2832	2.75	5.9396	8.6394	3.50	9.6211	10.9956	4.25	14.1863	13.3518	5.00	19.6350	15.7080
1.26	1.24690	3.9584	2.01	3.17309	6.3146	2.76	5.9828	8.6708	3.51	9.6762	11.0270	4.26	14.2531	13.3832	5.01	19.7136	15.7394
1.27	1.26677	3.9898	2.02	3.20474	6.3460	2.77	6.0263	8.7022	3.52	9.7314	11.0584	4.27	14.3201	13.4146	5.02	19.7923	15.7708
1.28	1.28680	4.0212	2.03	3.23655	6.3774	2.78	6.0699	8.7336	3.53	9.7868	11.0898	4.28	14.3872	13.4460	5.03	19.8713	15.8022
1.29	1.30698	4.0527	2.04	3.26851	6.4088	2.79	6.1136	8.7650	3.54	9.8423	11.1212	4.29	14.4545	13.4774	5.04	19.9504	15.8336
1.30	1.32732	4.0841	2.05	3.30064	6.4403	2.80	6.1575	8.7965	3.55	9.8980	11.1527	4.30	14.5220	13.5088	5.05	20.0296	15.8650
1.31	1.34782	4.1155	2.06	3.33292	6.4717	2.81	6.2016	8.8279	3.56	9.9538	11.1841	4.31	14.5896	13.5403	5.06	20.1090	15.8965
1.32	1.36848	4.1469	2.07	3.36535	6.5031	2.82	6.2458	8.8593	3.57	10.0098	11.2155	4.32	14.6574	13.5717	5.07	20.1886	15.9279
1.33	1.38929	4.1783	2.08	3.39795	6.5345	2.83	6.2902	8.8907	3.58	10.0660	11.2469	4.33	14.7254	13.6031	5.08	20.2683	15.9593
1.34	1.41026	4.2097	2.09	3.43070	6.5659	2.84	6.3347	8.9221	3.59	10.1223	11.2783	4.34	14.7934	13.6345	5.09	20.3482	15.9907
1.35	1.43139	4.2412	2.10	3.46361	6.5973	2.85	6.3794	8.9535	3.60	10.1788	11.3097	4.35	14.8617	13.6659	5.10	20.4282	16.0221
1.36	1.45267	4.2726	2.11	3.49667	6.6288	2.86	6.4242	8.9850	3.61	10.2354	11.3411	4.36	14.9301	13.6973	5.11	20.5084	16.0535
1.37	1.47411	4.3040	2.12	3.52989	6.6602	2.87	6.4692	9.0164	3.62	10.2922	11.3726	4.37	14.9987	13.7288	5.12	20.5887	16.0850
1.38	1.49571	4.3354	2.13	3.56327	6.6916	2.88	6.5144	9.0478	3.63	10.3491	11.4040	4.38	15.0674	13.7602	5.13	20.6692	16.1164
1.39	1.51747	4.3668	2.14	3.59681	6.7230	2.89	6.5597	9.0792	3.64	10.4062	11.4354	4.39	15.1363	13.7916	5.14	20.7499	16.1478
1.40	1.53938	4.3982	2.15	3.63050	6.7544	2.90	6.6052	9.1106	3.65	10.4635	11.4668	4.40	15.2053	13.8230	5.15	20.8307	16.1792
1.41	1.56145	4.4296	2.16	3.66435	6.7858	2.91	6.6508	9.1420	3.66	10.5209	11.4982	4.41	15.2745	13.8544	5.16	20.9117	16.2106
1.42	1.58368	4.4611	2.17	3.69836	6.8173	2.92	6.6966	9.1735	3.67	10.5784	11.5296	4.42	15.3439	13.8858	5.17	20.9928	16.2420
1.43	1.60606	4.4925	2.18	3.73253	6.8487	2.93	6.7426	9.2049	3.68	10.6362	11.5611	4.43	15.4134	13.9173	5.18	21.0741	16.2734
1.44	1.62860	4.5239	2.19	3.76685	6.8801	2.94	6.7887	9.2363	3.69	10.6941	11.5925	4.44	15.4830	13.9487	5.19	21.1556	16.3049
1.45	1.65130	4.5553	2.20	3.8013	6.9115	2.95	6.8349	9.2677	3.70	10.7521	11.6239	4.45	15.5528	13.9801	5.20	21.2372	16.3363
1.46	1.67416	4.5867	2.21	3.8360	6.9429	2.96	6.8813	9.2991	3.71	10.8103	11.6553	4.46	15.6228	14.0115	5.21	21.3189	16.3677
1.47	1.69717	4.6181	2.22	3.8708	6.9743	2.97	6.9279	9.3305	3.72	10.8687	11.6867	4.47	15.6930	14.0429	5.22	21.4008	16.3991
1.48	1.72034	4.6496	2.23	3.9057	7.0058	2.98	6.9747	9.3619	3.73	10.9272	11.7181	4.48	15.7633	14.0743	5.23	21.4829	16.4305
1.49	1.74366	4.6810	2.24	3.9408	7.0372	2.99	7.0215	9.3934	3.74	10.9858	11.7496	4.49	15.8337	14.1058	5.24	21.5651	16.4619
1.50	1.76715	4.7124	2.25	3.9761	7.0686	3.00	7.0686	9.4248	3.75	11.0447	11.7810	4.50	15.9043	14.1372	5.25	21.6475	16.4934
1.51	1.79079	4.7438	2.26	4.0115	7.1000	3.01	7.1158	9.4562	3.76	11.1036	11.8124	4.51	15.9751	14.1686	5.26	21.7301	16.5248
1.52	1.81458	4.7752	2.27	4.0471	7.1314	3.02	7.1631	9.4876	3.77	11.1628	11.8438	4.52	16.0460	14.2000	5.27	21.8128	16.5562
1.53	1.83854	4.8066	2.28	4.0828	7.1628	3.03	7.2107	9.5190	3.78	11.2221	11.8752	4.53	16.1171	14.2314	5.28	21.8956	16.5876
1.54	1.86265	4.8381	2.29	4.1187	7.1942	3.04	7.2583	9.5504	3.79	11.2815	11.9066	4.54	16.1883	14.2628	5.29	21.9787	16.6190
1.55	1.88692	4.8695	2.30	4.1548	7.2257	3.05	7.3062	9.5819	3.80	11.3411	11.9381	4.55	16.2597	14.2942	5.30	22.0618	16.6504
1.56	1.91135	4.9009	2.31	4.1910	7.2571	3.06	7.3542	9.6133	3.81	11.4009	11.9695	4.56	16.3313	14.3257	5.31	22.1452	16.6819
1.57	1.93593	4.9323	2.32	4.2273	7.2885	3.07	7.4023	9.6447	3.82	11.4608	12.0009	4.57	16.4030	14.3571	5.32	22.2287	16.7133
1.58	1.96067	4.9637	2.33	4.2638	7.3199	3.08	7.4506	9.6761	3.83	11.5209	12.0323	4.58	16.4748	14.3885	5.33	22.3123	16.7447
1.59	1.98557	4.9951	2.34	4.3005	7.3513	3.09	7.4991	9.7075	3.84	11.5812	12.0637	4.59	16.5468	14.4199	5.34	22.3961	16.7761
1.60	2.01062	5.0265	2.35	4.3374	7.3827	3.10	7.5477	9.7389	3.85	11.6416	12.0951	4.60	16.6190	14.4513	5.35	22.4801	16.8075
1.61	2.03583	5.0580	2.36	4.3744	7.4142	3.11	7.5965	9.7704	3.86	11.7021	12.1265	4.61	16.6914	14.4827	5.36	22.5642	16.8389
1.62	2.06120	5.0894	2.37	4.4115	7.4456	3.12	7.6454	9.8018	3.87	11.7628	12.1580	4.62	16.7639	14.5142	5.37	22.6484	16.8704
1.63	2.08672	5.1208	2.38	4.4488	7.4770	3.13	7.6945	9.8332	3.88	11.8237	12.1894	4.63	16.8365	14.5456	5.38	22.7329	16.9018
1.64	2.11241	5.1522	2.39	4.4863	7.5084	3.14	7.7437	9.8646	3.89	11.8847	12.2208	4.64	16.9093	14.5770	5.39	22.8175	16.9332
1.65	2.13825	5.1836	2.40	4.5239	7.5398	3.15	7.7931	9.8960	3.90	11.9459	12.2522	4.65	16.9823	14.6084	5.40	22.9022	16.9646
1.66	2.16424	5.2150	2.41	4.5617	7.5712	3.16	7.8427	9.9274	3.91	12.0072	12.2836	4.66	17.0554	14.6398	5.41	22.9871	16.9960
1.67	2.19040	5.2465	2.42	4.5996	7.6027	3.17	7.8924	9.9588	3.92	12.0687	12.3150	4.67	17.1287	14.6712	5.42	23.0722	17.0274
1.68	2.21671	5.2779	2.43	4.6377	7.6341	3.18	7.9423	9.9903	3.93	12.1304	12.3465	4.68	17.2021	14.7027	5.43	23.1574	17.0588
1.69	2.24318	5.3093	2.44	4.6759	7.6655	3.19	7.9923	10.0217	3.94	12.1922	12.3779	4.69	17.2757	14.7341	5.44	23.2428	17.0903
1.70	2.26980	5.3407	2.45	4.7144	7.6969	3.20	8.0425	10.0531	3.95	12.2542	12.4093	4.70	17.3494	14.7655	5.45	23.3283	17.1217
1.71	2.29658	5.3721	2.46	4.7529	7.7283	3.21	8.0928	10.0845	3.96	12.3163	12.4407	4.71	17.4234	14.7969	5.46	23.4140	17.1531
1.72	2.32352	5.4035	2.47	4.7916	7.7597	3.22	8.1433	10.1159	3.97	12.3786	12.4721	4.72	17.4974	14.8283	5.47	23.4998	17.1845
1.73	2.35062	5.4350	2.48	4.8305	7.7911	3.23	8.1940	10.1473	3.98	12.4410	12.5035	4.73	17.5716	14.8597	5.48	23.5858	17.2159
1.74	2.37787	5.4664	2.49	4.8695	7.8226	3.24	8.2448	10.1788	3.99	12.5036	12.5350	4.74	17.6460	14.8911	5.49	23.6720	17.2473

Table 8. Circles: Diameters, Areas, Circumferences—(*Concluded*)

Diameters in hundredths

Diam.	Area	Cir.	Diam.	Area	Cir.	Diam.	Area	Cir.	Diam.	Area	Cir.	Diam.	Area	Cir.	Diam.	Area	Cir.
5.50	23.7583	17.2788	6.25	30.6796	19.6350	7.00	38.4845	21.9911	7.75	47.1730	24.3473	8.50	56.7450	26.7035	9.25	67.2006	29.0597
5.51	23.8448	17.3102	6.26	30.7779	19.6664	7.01	38.5945	22.0226	7.76	47.2948	24.3788	8.51	56.8786	26.7350	9.26	67.3460	29.0911
5.52	23.9314	17.3416	6.27	30.8763	19.6978	7.02	38.7047	22.0540	7.77	47.4168	24.4102	8.52	57.0124	26.7664	9.27	67.4915	29.1226
5.53	24.0182	17.3730	6.28	30.9748	19.7292	7.03	38.8151	22.0854	7.78	47.5389	24.4416	8.53	57.1463	26.7978	9.28	67.6372	29.1540
5.54	24.1051	17.4044	6.29	31.0736	19.7606	7.04	38.9256	22.1168	7.79	47.6612	24.4730	8.54	57.2803	26.8292	9.29	67.7831	29.1854
5.55	24.1922	17.4358	6.30	31.1725	19.7920	7.05	39.0363	22.1482	7.80	47.7836	24.5044	8.55	57.4146	26.8606	9.30	67.9291	29.2168
5.56	24.2795	17.4673	6.31	31.2715	19.8235	7.06	39.1471	22.1796	7.81	47.9062	24.5358	8.56	57.5490	26.8920	9.31	68.0753	29.2482
5.57	24.3669	17.4987	6.32	31.3707	19.8549	7.07	39.2580	22.2111	7.82	48.0290	24.5673	8.57	57.6835	26.9234	9.32	68.2216	29.2796
5.58	24.4545	17.5301	6.33	31.4700	19.8863	7.08	39.3692	22.2425	7.83	48.1519	24.5987	8.58	57.8182	26.9549	9.33	68.3680	29.3111
5.59	24.5422	17.5615	6.34	31.5696	19.9177	7.09	39.4805	22.2739	7.84	48.2750	24.6301	8.59	57.9530	26.9863	9.34	68.5147	29.3425
5.60	24.6301	17.5929	6.35	31.6692	19.9491	7.10	39.5919	22.3053	7.85	48.3982	24.6615	8.60	58.0880	27.0177	9.35	68.6615	29.3739
5.61	24.7181	17.6243	6.36	31.7690	19.9805	7.11	39.7035	22.3367	7.86	48.5216	24.6929	8.61	58.2232	27.0491	9.36	68.8084	29.4053
5.62	24.8063	17.6558	6.37	31.8690	20.0119	7.12	39.8153	22.3681	7.87	48.6451	24.7243	8.62	58.3585	27.0805	9.37	68.9555	29.4367
5.63	24.8947	17.6872	6.38	31.9692	20.0434	7.13	39.9272	22.3996	7.88	48.7688	24.7558	8.63	58.4940	27.1119	9.38	69.1028	29.4681
5.64	24.9832	17.7186	6.39	32.0695	20.0748	7.14	40.0393	22.4310	7.89	48.8927	24.7872	8.64	58.6297	27.1434	9.39	69.2502	29.4996
5.65	25.0719	17.7500	6.40	32.1699	20.1062	7.15	40.1515	22.4624	7.90	49.0167	24.8186	8.65	58.7655	27.1748	9.40	69.3978	29.5310
5.66	25.1607	17.7814	6.41	32.2705	20.1376	7.16	40.2639	22.4938	7.91	49.1409	24.8500	8.66	58.9014	27.2062	9.41	69.5455	29.5624
5.67	25.2497	17.8128	6.42	32.3713	20.1690	7.17	40.3765	22.5252	7.92	49.2652	24.8814	8.67	59.0375	27.2376	9.42	69.6934	29.5938
5.68	25.3388	17.8442	6.43	32.4722	20.2004	7.18	40.4892	22.5566	7.93	49.3897	24.9128	8.68	59.1738	27.2690	9.43	69.8415	29.6252
5.69	25.4281	17.8757	6.44	32.5733	20.2319	7.19	40.6020	22.5881	7.94	49.5143	24.9442	8.69	59.3102	27.3004	9.44	69.9897	29.6566
5.70	25.5176	17.9071	6.45	32.6745	20.2633	7.20	40.7150	22.6195	7.95	49.6391	24.9757	8.70	59.4468	27.3319	9.45	70.1380	29.6881
5.71	25.6072	17.9385	6.46	32.7759	20.2947	7.21	40.8282	22.6509	7.96	49.7641	25.0071	8.71	59.5835	27.3633	9.46	70.2865	29.7195
5.72	25.6970	17.9699	6.47	32.8775	20.3261	7.22	40.9416	22.6823	7.97	49.8892	25.0385	8.72	59.7204	27.3947	9.47	70.4352	29.7509
5.73	25.7869	18.0013	6.48	32.9792	20.3575	7.23	41.0550	22.7137	7.98	50.0145	25.0699	8.73	59.8575	27.4261	9.48	70.5840	29.7823
5.74	25.8770	18.0327	6.49	33.0810	20.3889	7.24	41.1687	22.7451	7.99	50.1399	25.1013	8.74	59.9947	27.4575	9.49	70.7330	29.8137
5.75	25.9672	18.0642	6.50	33.1831	20.4204	7.25	41.2825	22.7765	8.00	50.2655	25.1327	8.75	60.1320	27.4889	9.50	70.8822	29.8451
5.76	26.0576	18.0956	6.51	33.2853	20.4518	7.26	41.3965	22.8080	8.01	50.3912	25.1642	8.76	60.2696	27.5204	9.51	71.0315	29.8765
5.77	26.1482	18.1270	6.52	33.3876	20.4832	7.27	41.5106	22.8394	8.02	50.5171	25.1956	8.77	60.4073	27.5518	9.52	71.1810	29.9080
5.78	26.2389	18.1584	6.53	33.4901	20.5146	7.28	41.6248	22.8708	8.03	50.6432	25.2270	8.78	60.5451	27.5832	9.53	71.3306	29.9394
5.79	26.3298	18.1898	6.54	33.5927	20.5460	7.29	41.7393	22.9022	8.04	50.7694	25.2584	8.79	60.6831	27.6146	9.54	71.4803	29.9708
5.80	26.4208	18.2212	6.55	33.6955	20.5774	7.30	41.8539	22.9336	8.05	50.8958	25.2898	8.80	60.8212	27.6460	9.55	71.6303	30.0022
5.81	26.5120	18.2527	6.56	33.7985	20.6088	7.31	41.9686	22.9650	8.06	51.0223	25.3212	8.81	60.9595	27.6774	9.56	71.7804	30.0336
5.82	26.6033	18.2841	6.57	33.9016	20.6403	7.32	42.0835	22.9965	8.07	51.1490	25.3527	8.82	61.0980	27.7088	9.57	71.9306	30.0650
5.83	26.6948	18.3155	6.58	34.0049	20.6717	7.33	42.1986	23.0279	8.08	51.2758	25.3841	8.83	61.2366	27.7403	9.58	72.0810	30.0965
5.84	26.7865	18.3469	6.59	34.1084	20.7031	7.34	42.3138	23.0593	8.09	51.4028	25.4155	8.84	61.3754	27.7717	9.59	72.2316	30.1279
5.85	26.8783	18.3783	6.60	34.2119	20.7345	7.35	42.4292	23.0907	8.10	51.5300	25.4469	8.85	61.5143	27.8031	9.60	72.3823	30.1593
5.86	26.9703	18.4097	6.61	34.3157	20.7659	7.36	42.5447	23.1221	8.11	51.6573	25.4783	8.86	61.6534	27.8345	9.61	72.5332	30.1907
5.87	27.0624	18.4411	6.62	34.4196	20.7973	7.37	42.6604	23.1535	8.12	51.7848	25.5097	8.87	61.7927	27.8659	9.62	72.6842	30.2221
5.88	27.1547	18.4726	6.63	34.5237	20.8288	7.38	42.7762	23.1850	8.13	51.9124	25.5411	8.88	61.9321	27.8973	9.63	72.8354	30.2535
5.89	27.2471	18.5040	6.64	34.6279	20.8602	7.39	42.8922	23.2164	8.14	52.0402	25.5726	8.89	62.0717	27.9288	9.64	72.9867	30.2850
5.90	27.3397	18.5354	6.65	34.7323	20.8916	7.40	43.0084	23.2478	8.15	52.1681	25.6040	8.90	62.2114	27.9602	9.65	73.1382	30.3164
5.91	27.4325	18.5668	6.66	34.8368	20.9230	7.41	43.1247	23.2792	8.16	52.2962	25.6354	8.91	62.3513	27.9916	9.66	73.2899	30.3478
5.92	27.5254	18.5982	6.67	34.9415	20.9544	7.42	43.2412	23.3106	8.17	52.4245	25.6668	8.92	62.4913	28.0230	9.67	73.4417	30.3792
5.93	27.6184	18.6296	6.68	35.0464	20.9858	7.43	43.3578	23.3420	8.18	52.5529	25.6982	8.93	62.6315	28.0544	9.68	73.5937	30.4106
5.94	27.7117	18.6611	6.69	35.1514	21.0173	7.44	43.4746	23.3734	8.19	52.6814	25.7296	8.94	62.7718	28.0858	9.69	73.7458	30.4420
5.95	27.8051	18.6925	6.70	35.2565	21.0487	7.45	43.5916	23.4049	8.20	52.8102	25.7611	8.95	62.9124	28.1173	9.70	73.8981	30.4734
5.96	27.8986	18.7239	6.71	35.3618	21.0801	7.46	43.7087	23.4363	8.21	52.9391	25.7925	8.96	63.0530	28.1487	9.71	74.0506	30.5049
5.97	27.9923	18.7553	6.72	35.4673	21.1115	7.47	43.8259	23.4677	8.22	53.0681	25.8239	8.97	63.1938	28.1801	9.72	74.2032	30.5363
5.98	28.0862	18.7867	6.73	35.5730	21.1429	7.48	43.9433	23.4991	8.23	53.1973	25.8553	8.98	63.3348	28.2115	9.73	74.3559	30.5677
5.99	28.1802	18.8181	6.74	35.6788	21.1743	7.49	44.0609	23.5305	8.24	53.3267	25.8867	8.99	63.4760	28.2429	9.74	74.5088	30.5991
6.00	28.2743	18.8496	6.75	35.7847	21.2058	7.50	44.1786	23.5619	8.25	53.4562	25.9181	9.00	63.6173	28.2743	9.75	74.6619	30.6305
6.01	28.3687	18.8810	6.76	35.8908	21.2372	7.51	44.2965	23.5934	8.26	53.5858	25.9496	9.01	63.7587	28.3058	9.76	74.8151	30.6619
6.02	28.4631	18.9124	6.77	35.9971	21.2686	7.52	44.4146	23.6248	8.27	53.7157	25.9810	9.02	63.9003	28.3372	9.77	74.9685	30.6934
6.03	28.5578	18.9438	6.78	36.1035	21.3000	7.53	44.5328	23.6562	8.28	53.8456	26.0124	9.03	64.0421	28.3686	9.78	75.1221	30.7248
6.04	28.6526	18.9752	6.79	36.2101	21.3314	7.54	44.6511	23.6876	8.29	53.9758	26.0438	9.04	64.1840	28.4000	9.79	75.2758	30.7562
6.05	28.7475	19.0066	6.80	36.3168	21.3628	7.55	44.7697	23.7190	8.30	54.1061	26.0752	9.05	64.3261	28.4314	9.80	75.4296	30.7876
6.06	28.8426	19.0381	6.81	36.4237	21.3942	7.56	44.8883	23.7504	8.31	54.2365	26.1066	9.06	64.4683	28.4628	9.81	75.5837	30.8190
6.07	28.9379	19.0695	6.82	36.5308	21.4257	7.57	45.0072	23.7819	8.32	54.3671	26.1381	9.07	64.6107	28.4942	9.82	75.7378	30.8504
6.08	29.0333	19.1009	6.83	36.6380	21.4571	7.58	45.1262	23.8133	8.33	54.4979	26.1695	9.08	64.7533	28.5257	9.83	75.8922	30.8819
6.09	29.1289	19.1323	6.84	36.7453	21.4885	7.59	45.2453	23.8447	8.34	54.6288	26.2009	9.09	64.8960	28.5571	9.84	76.0466	30.9133
6.10	29.2247	19.1637	6.85	36.8528	21.5199	7.60	45.3646	23.8761	8.35	54.7599	26.2323	9.10	65.0388	28.5885	9.85	76.2013	30.9447
6.11	29.3206	19.1951	6.86	36.9605	21.5513	7.61	45.4841	23.9075	8.36	54.8912	26.2637	9.11	65.1818	28.6199	9.86	76.3561	30.9761
6.12	29.4166	19.2265	6.87	37.0684	21.5827	7.62	45.6037	23.9389	8.37	55.0226	26.2951	9.12	65.3250	28.6513	9.87	76.5111	31.0075
6.13	29.5128	19.2580	6.88	37.1764	21.6142	7.63	45.7234	23.9704	8.38	55.1541	26.3265	9.13	65.4684	28.6827	9.88	76.6662	31.0389
6.14	29.6092	19.2894	6.89	37.2845	21.6456	7.64	45.8434	24.0018	8.39	55.2858	26.3580	9.14	65.6118	28.7142	9.89	76.8214	31.0704
6.15	29.7057	19.3208	6.90	37.3928	21.6770	7.65	45.9635	24.0332	8.40	55.4177	26.3894	9.15	65.7555	28.7456	9.90	76.9769	31.1018
6.16	29.8024	19.3522	6.91	37.5013	21.7084	7.66	46.0837	24.0646	8.41	55.5497	26.4208	9.16	65.8993	28.7770	9.91	77.1325	31.1332
6.17	29.8992	19.3836	6.92	37.6099	21.7398	7.67	46.2041	24.0960	8.42	55.6819	26.4522	9.17	66.0433	28.8084	9.92	77.2882	31.1646
6.18	29.9962	19.4150	6.93	37.7187	21.7712	7.68	46.3247	24.1274	8.43	55.8142	26.4836	9.18	66.1874	28.8398	9.93	77.4441	31.1960
6.19	30.0934	19.4465	6.94	37.8276	21.8027	7.69	46.4454	24.1588	8.44	55.9467	26.5150	9.19	66.3317	28.8712	9.94	77.6002	31.2274
6.20	30.1907	19.4779	6.95	37.9367	21.8341	7.70	46.5663	24.1903	8.45	56.0794	26.5465	9.20	66.4761	28.9027	9.95	77.7564	31.2588
6.21	30.2882	19.5093	6.96	38.0459	21.8655	7.71	46.6873	24.2217	8.46	56.2122	26.5779	9.21	66.6207	28.9341	9.96	77.9128	31.2903
6.22	30.3858	19.5407	6.97	38.1554	21.8969	7.72	46.8085	24.2531	8.47	56.3452	26.6093	9.22	66.7654	28.9655	9.97	78.0693	31.3217
6.23	30.4836	19.5721	6.98	38.2649	21.9283	7.73	46.9298	24.2845	8.48	56.4783	26.6407	9.23	66.9103	28.9969	9.98	78.2260	31.3531
6.24	30.5815	19.6035	6.99	38.3746	21.9597	7.74	47.0513	24.3159	8.49	56.6116	26.6721	9.24	67.0554	29.0283	9.99	78.3828	31.3845

Table 9a. Circular Segments

Angle in degrees

Central angle, degrees	Height / R	Chord / R	Height / Chord	Area / R^2	Central angle, degrees	Height / R	Chord / R	Height / Chord	Area / R^2	Central angle, degrees	Height / R	Chord / R	Height / Chord	Area / R^2
1	0.000038	0.017453	0.002177	0.0000005	61	0.138371	1.01508	0.136315	0.0950155	121	0.507576	1.74071	0.291591	0.6273404
2	.000151	.034905	.004326	.0000035	62	.142833	1.03008	.138662	.0995782	122	.515190	1.74924	.294522	.6406267
3	.000343	.052354	.006552	.0000119	63	.147360	1.04500	.141014	.1042754	123	.522841	1.75763	.297469	.6540421
4	.000609	.069799	.008725	.0000283	64	.151952	1.05984	.143373	.1091083	124	.530528	1.76590	.300429	.6675852
5	.000952	.087239	.010913	.0000554	65	.156609	1.07460	.145737	.1140780	125	.538251	1.77402	.303411	.6812546
6	.001371	.104672	.013098	.0000956	66	.161329	1.08928	.148106	.1191858	126	.546010	1.78201	.306401	.6950488
7	.001865	.122097	.015275	.0001519	67	.166114	1.10387	.150483	.1244328	127	.553802	1.78987	.309409	.7089613
8	.002436	.139513	.017461	.0002266	68	.170962	1.11839	.152864	.1298199	128	.561629	1.79759	.312434	.7230052
9	.003083	.156918	.019647	.0003226	69	.175874	1.13281	.155255	.1353483	129	.569489	1.80517	.315477	.7371642
10	.003805	.174311	.021829	.0004423	70	.180848	1.14715	.157650	.1410188	130	.577382	1.81262	.318534	.7514417
11	.004604	.191692	.024018	.0005886	71	.185885	1.16140	.160053	.1468325	131	.585307	1.81992	.321611	.7658357
12	.005479	.209057	.026207	.0007639	72	.190983	1.17557	.162460	.1527902	132	.593263	1.82709	.324704	.7803448
13	.006428	.226406	.028391	.0009708	73	.196143	1.18965	.164875	.1588927	133	.601251	1.83412	.327814	.7949670
14	.007454	.243739	.030582	.0012121	74	.201365	1.20363	.167298	.1651409	134	.609269	1.84101	.330943	.8097006
15	.008555	.261052	.032771	.0014901	75	.206647	1.21752	.169728	.1715355	135	.617317	1.84776	.334089	.8245437
16	.009732	.278346	.034963	.0018076	76	.211989	1.23132	.172164	.1780773	136	.625393	1.85436	.337255	.8394945
17	.010984	.295619	.037156	.0021671	77	.217392	1.24503	.174608	.1847666	137	.633499	1.86084	.340437	.8545511
18	.012312	.312867	.039352	.0025711	78	.222854	1.25864	.177059	.1916045	138	.641632	1.86716	.343641	.8697117
19	.013714	.330095	.041547	.0030222	79	.228375	1.27216	.179518	.1985914	139	.649793	1.87334	.346863	.8849742
20	.015192	.347296	.043744	.0035229	80	.233956	1.28558	.181985	.2057277	140	.657980	1.87939	.350103	.9003667
21	.016745	.364471	.045943	.0040756	81	.239594	1.29890	.184459	.2130141	141	.666193	1.88528	.353366	.9157968
22	.018373	.381618	.048145	.0046829	82	.245290	1.31212	.186942	.2204508	142	.674432	1.89104	.356646	.9313529
23	.020075	.398736	.050347	.0053473	83	.251044	1.32524	.189433	.2280384	143	.682695	1.89665	.359948	.9470027
24	.021852	.415823	.052551	.0060712	84	.256855	1.33826	.191932	.2357772	144	.690983	1.90211	.363272	.9627442
25	.023704	.432879	.054759	.0068570	85	.262723	1.35118	.194440	.2436676	145	.699294	1.90743	.366616	.9785754
26	.025630	.449902	.056968	.0077072	86	.268646	1.36400	.196955	.2517094	146	.707628	1.91261	.369980	.9944937
27	.027630	.466891	.059178	.0086242	87	.274626	1.37671	.199481	.2599034	147	.715985	1.91764	.373368	1.0104973
28	.029704	.483844	.061392	.0096103	88	.280660	1.38932	.202012	.2682494	148	.724363	1.92252	.376778	1.0265840
29	.031852	.500760	.063607	.0106679	89	.286750	1.40182	.204556	.2767476	149	.732762	1.92726	.380209	1.0427512
30	.034074	.517638	.065826	.0117993	90	.292893	1.41421	.207107	.2853982	150	.741181	1.93185	.383664	1.0589969
31	.036370	.534477	.068048	.0130069	91	.299091	1.42650	.209668	.2942010	151	.749620	1.93630	.387140	1.0753188
32	.038738	.551275	.070270	.0142930	92	.305342	1.43868	.212238	.3031559	152	.758078	1.94059	.390643	1.0917144
33	.041180	.568031	.072496	.0156598	93	.311645	1.45075	.214816	.3122632	153	.766555	1.94474	.394168	1.1081816
34	.043695	.584743	.074725	.0171095	94	.318002	1.46271	.217406	.3215226	154	.775049	1.94874	.397718	1.1247180
35	.046283	.601412	.076957	.0186444	95	.324410	1.47456	.220005	.3309339	155	.783560	1.95259	.401293	1.1413210
36	.048944	.618034	.079193	.0202666	96	.330869	1.48629	.222614	.3404970	156	.792088	1.95630	.404891	1.1579885
37	.051676	.634609	.081430	.0219784	97	.337380	1.49791	.225234	.3502115	157	.800632	1.95985	.408517	1.1747179
38	.054481	.651136	.083671	.0237818	98	.343941	1.50942	.227863	.3600772	158	.809191	1.96325	.412169	1.1915068
39	.057359	.667614	.085917	.0256790	99	.350552	1.52081	.230503	.3700937	159	.817765	1.96651	.415845	1.2083528
40	.060307	.684040	.088163	.0276720	100	.357212	1.53209	.233153	.3802606	160	.826352	1.96962	.419549	1.2252533
41	.063328	.700415	.090415	.0297629	101	.363922	1.54325	.235815	.3905775	161	.834952	1.97257	.423281	1.2422059
42	.066420	.716736	.092670	.0319538	102	.370680	1.55429	.238488	.4010440	162	.843566	1.97537	.427042	1.2592082
43	.069582	.733002	.094927	.0342465	103	.377485	1.56522	.241171	.4116594	163	.852191	1.97803	.430828	1.2762575
44	.072816	.749213	.097190	.0366432	104	.384339	1.57602	.243867	.4224232	164	.860827	1.98054	.434643	1.2933512
45	.076121	.765367	.099457	.0391456	105	.391239	1.58671	.246572	.4333348	165	.869474	1.98289	.438488	1.3104871
46	.079495	.781462	.101730	.0417558	106	.398185	1.59727	.249299	.4443935	166	.878131	1.98509	.442363	1.3276623
47	.082940	.797498	.104000	.0444755	107	.405177	1.60771	.252021	.4555999	167	.886797	1.98714	.446268	1.3448744
48	.086455	.813473	.106278	.0473066	108	.412215	1.61803	.254764	.4669494	168	.895472	1.98904	.450203	1.3621207
49	.090039	.829386	.108561	.0502508	109	.419297	1.62823	.257517	.4784450	169	.904154	1.99079	.454169	1.3793987
50	.093692	.845237	.110847	.0533100	110	.426424	1.63830	.260284	.4900846	170	.912844	1.99238	.458165	1.3967057
51	.097415	.861022	.113139	.0564859	111	.433594	1.64825	.263063	.5018674	171	.921541	1.99383	.462196	1.4140393
52	.101206	.876742	.115434	.0597801	112	.440807	1.65808	.265854	.5137923	172	.930244	1.99513	.466257	1.4313966
53	.105067	.892396	.117736	.0631944	113	.448063	1.66777	.268660	.5258585	173	.938952	1.99627	.470358	1.4487751
54	.108994	.907981	.120040	.0667303	114	.455361	1.67734	.271478	.5380648	174	.947664	1.99726	.474482	1.4661721
55	.112989	.923497	.122349	.0703895	115	.462700	1.68678	.274310	.5504103	175	.956381	1.99810	.478645	1.4835852
56	.117052	.938943	.124664	.0741733	116	.470081	1.69610	.277154	.5628938	176	.965101	1.99878	.482845	1.5010115
57	.121182	.954318	.126983	.0780835	117	.477501	1.70528	.280013	.5755142	177	.973823	1.99931	.487080	1.5184484
58	.125380	.969619	.129308	.0821214	118	.484962	1.71433	.282887	.5882703	178	.982548	1.99970	.491348	1.5358933
59	.129644	.984847	.131639	.0862884	119	.492462	1.72326	.285773	.6011611	179	0.991274	1.99992	.495657	1.5533435
60	.133975	1.000000	.133975	.0905860	120	.500000	1.73205	.288684	.6141847	180	1.00000	2.00000	.500000	1.5707963

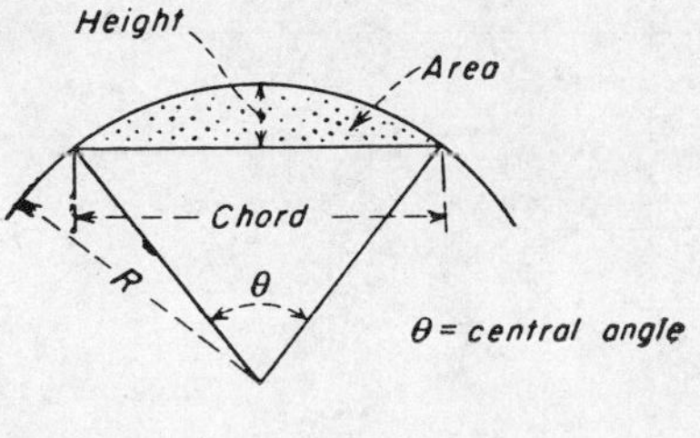

Table 9b. Circles: Areas of Segments

Height = h; Diameter = D; Area = A

h/D	A	h/D	A	h/D	A	h/D	A	h/D	A	h/D	A	h/D	A	h/D	A	h/D	A	h/D	A
		0.050	0.01468	0.100	0.04087	0.150	0.07387	0.200	0.11182	0.250	0.15355	0.300	0.19817	0.350	0.24498	0.400	0.29337	0.450	0.34278
0.001	0.00004	.051	.01512	.101	.04148	.151	.07459	.201	.11262	.251	.15441	.301	.19908	.351	.24593	.401	.29435	.451	.34378
.002	.00012	.052	.01556	.102	.04208	.152	.07531	.202	.11343	.252	.15528	.302	.20000	.352	.24689	.402	.29533	.452	.34477
.003	.00022	.053	.01601	.103	.04269	.153	.07603	.203	.11423	.253	.15615	.303	.20092	.353	.24784	.403	.29631	.453	.34577
.004	.00034	.054	.01646	.104	.04330	.154	.07675	.204	.11504	.254	.15702	.304	.20184	.354	.24880	.404	.29729	.454	.34676
.005	.00047	.055	.01691	.105	.04391	.155	.07747	.205	.11584	.255	.15789	.305	.20276	.355	.24976	.405	.29827	.455	.34776
.006	.00062	.056	.01737	.106	.04452	.156	.07819	.206	.11665	.256	.15876	.306	.20368	.356	.25071	.406	.29926	.456	.34876
.007	.00078	.057	.01783	.107	.04514	.157	.07892	.207	.11746	.257	.15964	.307	.20460	.357	.25167	.407	.30024	.457	.34975
.008	.00095	.058	.01830	.108	.04576	.158	.07965	.208	.11827	.258	.16051	.308	.20553	.358	.25263	.408	.30122	.458	.35075
.009	.00113	.059	.01877	.109	.04638	.159	.08038	.209	.11908	.259	.16139	.309	.20645	.359	.25359	.409	.30220	.459	.35175
.010	.00133	.060	.01924	.110	.04701	.160	.08111	.210	.11990	.260	.16226	.310	.20738	.360	.25455	.410	.30319	.460	.35274
.011	.00153	.061	.01972	.111	.04763	.161	.08185	.211	.12071	.261	.16314	.311	.20830	.361	.25551	.411	.30417	.461	.35374
.012	.00175	.062	.02020	.112	.04826	.162	.08258	.212	.12153	.262	.16402	.312	.20923	.362	.25647	.412	.30516	.462	.35474
.013	.00197	.063	.02068	.113	.04889	.163	.08332	.213	.12235	.263	.16490	.313	.21015	.363	.25743	.413	.30614	.463	.33573
.014	.00220	.064	.02117	.114	.04953	.164	.08406	.214	.12317	.264	.16578	.314	.21108	.364	.25839	.414	.30712	.464	.35673
.015	.00244	.065	.02166	.115	.05016	.165	.08480	.215	.12399	.265	.16666	.315	.21201	.365	.25936	.415	.30811	.465	35773
.016	.00268	.066	.02215	.116	.05080	.166	.08554	.216	.12481	.266	.16755	.316	.21294	.366	.26032	.416	.30910	.466	.35873
.017	.00294	.067	.02265	.117	.05145	.167	.08629	.217	.12563	.267	.16843	.317	.21387	.367	.26128	.417	.31008	.467	.35972
.018	.00320	.068	.02315	.118	.05209	.168	.08704	.218	.12646	.268	.16932	.318	.21480	.368	.26225	.418	.31107	.468	.36072
.019	.00347	.069	.02366	.119	.05274	.169	.08779	.219	.12729	.269	.17020	.319	.21573	.369	.26321	.419	.31205	.469	.36172
.020	.00375	.070	.02417	.120	.05338	.170	.08854	.220	.12811	.270	.17109	.320	.21667	.370	.26418	.420	.31304	.470	.36272
.021	.00403	.071	.02468	.121	.05404	.171	.08929	.221	.12894	.271	.17198	.321	.21760	.371	.26514	.421	.31403	.471	.36372
.022	.00432	.072	.02520	.122	.05469	.172	.09004	.222	.12977	.272	.17287	.322	.21853	.372	.26611	.422	.31502	.472	.36471
.023	.00462	.073	.02571	.123	.05535	.173	.09080	.223	.13060	.273	.17376	.323	.21947	.373	.26708	.423	.31600	.473	.36571
.024	.00492	.074	.02624	.124	.05600	.174	.09155	.224	.13144	.274	.17465	.324	.22040	.374	.26805	.424	.31699	.474	.36671
.025	.00523	.075	.02676	.125	.05666	.175	.09231	.225	.13227	.275	.17554	.325	.22134	.375	.26901	.425	.31798	.475	.36771
.026	.00555	.076	.02729	.126	.05733	.176	.09307	.226	.13311	.276	.17644	.326	.22228	.376	.26998	.426	.31897	.476	.36871
.027	.00587	.077	.02782	.127	.05799	.177	.09384	.227	.13395	.277	.17733	.327	.22322	.377	.27095	.427	.31996	.477	.36971
.028	.00619	.078	.02836	.128	.05866	.178	.09460	.228	.13478	.278	.17823	.328	.22415	.378	.27192	.428	.32095	.478	.37071
.029	.00653	.079	.02889	.129	.05933	.179	.09537	.229	.13562	.279	.17912	.329	.22509	.379	.27289	.429	.32194	.479	.37171
.030	.00687	.080	.02943	.130	.06000	.180	.09613	.230	.13646	.280	.18002	.330	.22603	.380	.27386	.430	.32293	.480	.37270
.031	.00721	.081	.02998	.131	.06067	.181	.09690	.231	.13731	.281	.18092	.331	.22697	.381	.27483	.431	.32392	.481	.37370
.032	.00756	.082	.03053	.132	.06135	.182	.09767	.232	.13815	.282	.18182	.332	.22792	.382	.27580	.432	.32491	.482	.37470
.033	.00791	.083	.03108	.133	.06203	.183	.09845	.233	.13900	.283	.18272	.333	.22886	.383	.27678	.433	.32590	.483	.37570
.034	.00827	.084	.03163	.134	.06271	.184	.09922	.234	.13984	.284	.18362	.334	.22980	.384	.27775	.434	.32689	.484	.37670
.035	.00864	.085	.03219	.135	.06339	.185	.10000	.235	.14069	.285	.18452	.335	.23074	.385	.27872	.435	.32788	.485	.37770
.036	.00901	.086	.03275	.136	.06407	.186	.10077	.236	.14154	.286	.18542	.336	.23169	.386	.27969	.436	.32887	.486	.37870
.037	.00938	.087	.03331	.137	.06476	.187	.10155	.237	.14239	.287	.18633	.337	.23263	.387	.28067	.437	.32987	.487	.37970
.038	.00976	.088	.03387	.138	.06545	.188	.10233	.238	.14324	.288	.18723	.338	.23358	.388	.28164	.438	.33086	.488	.38070
.039	.01015	.089	.03444	.139	.06614	.189	.10312	.239	.14409	.289	.18814	.339	.23453	.389	.28262	.439	.33185	.489	.38170
.040	.01054	.090	.03501	.140	.06683	.190	.10390	.240	.14494	.290	.18905	.340	.23547	.390	.28359	.440	.33284	.490	.38270
.041	.01093	.091	.03559	.141	.06753	.191	.10469	.241	.14580	.291	.18996	.341	.23642	.391	.28457	.441	.33384	.491	.38370
.042	.01133	.092	.03616	.142	.06822	.192	.10547	.242	.14666	.292	.19086	.342	.23737	.392	.28554	.442	.33483	.492	.38470
.043	.01173	.093	.03674	.143	.06892	.193	.10626	.243	.14751	.293	.19177	.343	.23832	.393	.28652	.443	.33582	.493	.38570
.044	.01214	.094	.03732	.144	.06963	.194	.10705	.244	.14837	.294	.19268	.344	.23927	.394	.28750	.444	.33682	.494	.38670
.045	.01255	.095	.03791	.145	.07033	.195	.10784	.245	.14923	.295	.19360	.345	.24022	.395	.28848	.445	.33781	.495	.38770
.046	.01297	.096	.03850	.146	.07103	.196	.10864	.246	.15009	.296	.19451	.346	.24117	.396	.28945	.446	.33880	.496	.38870
.047	.01339	.097	.03909	.147	.07174	.197	.10943	.247	.15095	.297	.19542	.347	.24212	.397	.29043	.447	.33980	.497	.38970
.048	.01382	.098	.03968	.148	.07245	.198	.11023	.248	.15182	.298	.19634	.348	.24307	.398	.29141	.448	.34079	.498	.39070
.049	.01425	.099	.04028	.149	.07316	.199	.11102	.249	.15268	.299	.19725	.349	.24403	.399	.29239	.449	.34179	.499	.39170
																		.500	.39270

Rules for Using Table: (1) Divide height of segment by the diameter; multiply the area in the table corresponding to the quotient, height/diameter, by the diameter squared. When segment exceeds a semicircle, its area is: Area of circle minus the area of a segment whose height is the circle diameter minus the height of the given segment. (2) To find the diameter when given the chord and the segment height: the diameter = $[(\frac{1}{2}\text{ chord})^2/\text{height}] + \text{height}$.

Table 10a. Circles: Lengths of Circular Arcs

[Given chord and arc height (*H*)]

H	*L*	*H*	*L*	*H*	*L*	*H*	*L*	*H*	*L*	*H*	*L*	*H*	*L*	*H*	*L*	*H*	*L*	*H*	*L*
0.001	1.00002	0.106	1.02970	0.15	1.05896	0.194	1.09752	0.238	1.14480	0.282	1.20014	0.326	1.26288	0.37	1.33234	0.414	1.40788	0.458	1.48889
.005	1.00007	.108	1.03082	.152	1.06051	.196	1.09949	.24	1.14714	.284	1.20284	.328	1.26588	.372	1.33564	.416	1.41145	.46	1.49269
.01	1.00027	.11	1.03196	.154	1.06209	.198	1.10147	.242	1.14951	.286	1.20555	.33	1.26892	.374	1.33896	.418	1.41503	.462	1.49651
.015	1.00061	.112	1.03312	.156	1.06368	.20	1.10347	.244	1.15189	.288	1.20827	.332	1.27196	.376	1.34229	.42	1.41861	.464	1.50033
.02	1.00107	.114	1.03430	.158	1.06530	.202	1.10548	.246	1.15428	.29	1.21102	.334	1.27502	.378	1.34563	.422	1.42221	.466	1.50416
.025	1.00167	.116	1.03551	.16	1.06693	.204	1.10752	.248	1.15670	.292	1.21377	.336	1.27810	.38	1.34899	.424	1.42583	.468	1.50800
.03	1.00240	.118	1.03672	.162	1.06858	.206	1.10958	.25	1.15912	.294	1.21654	.338	1.28118	.382	1.35237	.426	1.42945	.47	1.51185
.035	1.00327	.12	1.03797	.164	1.07025	.208	1.11165	.252	1.16156	.296	1.21933	.34	1.28428	.384	1.35575	.428	1.43309	.472	1.51571
.04	1.00426	.122	1.03923	.166	1.07194	.21	1.11374	.254	1.16402	.298	1.22213	.342	1.28739	.386	1.35914	.43	1.43673	.474	1.51958
.045	1.00539	.124	1.04051	.168	1.07365	.212	1.11584	.256	1.16650	.30	1.22495	.344	1.29052	.388	1.36254	.432	1.44039	.476	1.52346
.05	1.00665	.126	1.04181	.17	1.07537	.214	1.11796	.258	1.16899	.302	1.22778	.346	1.29366	.39	1.36596	.434	1.44405	.478	1.52736
.055	1.00805	.128	1.04313	.172	1.07711	.216	1.12011	.26	1.17150	.304	1.23063	.348	1.29681	.392	1.36939	.436	1.44773	.48	1.53126
.06	1.00957	.13	1.04447	.174	1.07888	.218	1.12225	.262	1.17403	.306	1.23349	.35	1.29997	.394	1.37283	.438	1.45142	.482	1.53518
.065	1.01123	.132	1.04584	.176	1.08066	.22	1.12444	.264	1.17657	.308	1.23636	.352	1.30315	.396	1.37628	.44	1.45512	.484	1.53910
.07	1.01302	.134	1.04722	.178	1.08246	.222	1.12664	.266	1.17912	.31	1.23926	.354	1.30634	.398	1.37974	.442	1.45883	.486	1.54302
.075	1.01493	.136	1.04862	.18	1.08428	.224	1.12885	.268	1.18169	.312	1.24216	.356	1.30954	.40	1.38322	.444	1.46255	.488	1.54696
.08	1.01698	.138	1.05003	.182	1.08611	.226	1.13108	.27	1.18429	.314	1.24507	.358	1.31276	.402	1.38671	.446	1.46628	.49	1.55091
.085	1.01916	.14	1.05147	.184	1.08797	.228	1.13331	.272	1.18689	.316	1.24801	.36	1.31599	.404	1.39021	.448	1.47002	.492	1.55487
.09	1.02146	.142	1.05293	.186	1.08984	.23	1.13557	.274	1.18951	.318	1.25095	.362	1.31923	.406	1.39372	.45	1.47377	.494	1.55854
.095	1.02389	.144	1.05441	.188	1.09174	.232	1.13785	.276	1.19214	.32	1.25391	.364	1.32249	.408	1.39724	.452	1.47753	.496	1.56282
.10	1.02646	.146	1.05591	.19	1.09365	.234	1.14015	.278	1.19479	.322	1.25689	.366	1.32577	.41	1.40077	.454	1.48131	.498	1.56681
.102	1.02752	.148	1.05743	.192	1.09557	.236	1.14247	.28	1.19746	.324	1.25988	.368	1.32905	.412	1.40432	.456	1.48509	.50	1.57080
.104	1.02860																		

Rule. Divide height of arc by the cord; find number corresponding to this quotient in the table of heights (*H*); take from the (*L*) table the value corresponding; multiply this *L* value by the length of the given chord; the answer is the required length of arc. If the arc is larger than a semicircle: Find the diameter from the relation: diameter = (square of half the chord ÷ height) + height; find the circumference. From the diameter subtract the arc height; the remainder will be the height of the smaller arc; find its length by the rule and subtract it from the circumference.

Table 10b. Circles: Lengths of Circular Arcs

Degrees

Deg.	Factor	Deg.	Factor	Deg.	Factor	Deg.	Factor	Deg.	Factor	Deg.	Factor	Deg.	Factor	Deg.	Factor	Deg.	Factor	Deg.	Factor	Deg.	Factor	Deg.	Factor
1	0.01745	16	0.27925	31	0.54105	46	0.80285	61	1.06465	76	1.32645	91	1.58825	106	1.85005	121	2.11185	136	2.37365	151	2.63545	166	2.89725
2	.03491	17	.29671	32	.55851	47	.82030	62	1.08210	77	1.34390	92	1.60570	107	1.86750	122	2.12930	137	2.39110	152	2.65290	167	2.91470
3	.05236	18	.31416	33	.57596	48	.83776	63	1.09956	78	1.36136	93	1.62316	108	1.88496	123	2.14676	138	2.40855	153	2.67035	168	2.93215
4	.06981	19	.33162	34	.59342	49	.85521	64	1.11701	79	1.37881	94	1.64061	109	1.90241	124	2.16421	139	2.42601	154	2.68781	169	2.94961
5	.08727	20	.34907	35	.61087	50	.87266	65	1.13446	80	1.39626	95	1.65806	110	1.91986	125	2.18166	140	2.44346	155	2.70526	170	2.96706
6	.10472	21	.36652	36	.62832	51	.89012	66	1.15192	81	1.41372	96	1.67552	111	1.93732	126	2.19911	141	2.46091	156	2.72271	171	2.98451
7	.12217	22	.38397	37	.64577	52	.90757	67	1.16937	82	1.43117	97	1.69297	112	1.95477	127	2.21657	142	2.47837	157	2.74017	172	3.00197
8	.13963	23	.40143	38	.66323	53	.92502	68	1.18682	83	1.44862	98	1.71042	113	1.97222	128	2.23402	143	2.49582	158	2.75762	173	3.01942
9	.15708	24	.41888	39	.68068	54	.94248	69	1.20428	84	1.46608	99	1.72788	114	1.98968	129	2.25147	144	2.51327	159	2.77507	174	3.03687
10	.17453	25	.43633	40	.69813	55	.95993	70	1.22173	85	1.48353	100	1.74533	115	2.00713	130	2.26893	145	2.53073	160	2.79253	175	3.05433
11	.19199	26	.45379	41	.71559	56	.97738	71	1.23918	86	1.50098	101	1.76278	116	2.02458	131	2.28638	146	2.54818	161	2.80998	176	3.07178
12	.20944	27	.47124	42	.73304	57	.99484	72	1.25664	87	1.51844	102	1.78024	117	2.04204	132	2.30383	147	2.56563	162	2.82743	177	3.08923
13	.22689	28	.48869	43	.75049	58	1.01229	73	1.27409	88	1.53589	103	1.79769	118	2.05949	133	2.32129	148	2.58309	163	2.84489	178	3.10669
14	.24435	29	.50615	44	.76794	59	1.02974	74	1.29154	89	1.55334	104	1.81514	119	2.07694	134	2.33874	149	2.60054	164	2.86234	179	3.12414
15	.26180	30	.52360	45	.78540	60	1.04720	75	1.30900	90	1.57080	105	1.83260	120	2.09440	135	2.35619	150	2.61799	165	2.87979	180	3.14159

Minutes

Min.	Factor	Min.	Factor	Min.	Factor	Min.	Factor	Min.	Factor	Min.	Factor	Min.	Factor	Min.	Factor	Min.	Factor	Min.	Factor	Min.	Factor	Min.	Factor
1	0.00029	6	0.00175	11	0.00320	16	0.00465	21	0.00611	26	0.00756	31	0.00902	36	0.01047	41	0.01193	46	0.01338	51	0.01484	56	0.01629
2	.00058	7	.00204	12	.00349	17	.00495	22	.00640	27	.00785	32	.00931	37	.01076	42	.01222	47	.01367	52	.01513	57	.01658
3	.00087	8	.00233	13	.00378	18	.00524	23	.00669	28	.00814	33	.00960	38	.01105	43	.01251	48	.01396	53	.01542	58	.01687
4	.00116	9	.00262	14	00407	19	.00553	24	.00698	29	.00844	34	.00989	39	.01134	44	.01280	49	.01425	54	.01571	59	.01716
5	.00145	10	.00291	15	.00436	20	.00582	25	.00727	30	.00873	35	.01018	40	.01164	45	.01309	50	.01454	55	.01600		

Radius = 1.
Length of arc = factor × radius.
Example. Find length of curve of radius 27 ft. and an angle of 43° 53′.

Factor for 43° (from top table) = 0.75049
Factor for 53′ (from bottom table) = 0.01542
Total factor = 0.76591
Total factor (0.76591) × 27′ = 20.6796 ft. length of arc required.

Table 11. Circles: Diameters of Circles with Sides of Squares of Equal Areas

Diameter of circle = 1.12838 × side of square of equal area
Side of square = 0.88623 × diameter of circle of same area

SPHERES

Table 12. Spheres: Diameters, Volumes*

Diameters by hundredths
Surface = 3.14159 × (diameter)²; volume = 0.523598 × (diameter)³

D	0.00	0.01	0.02	0.03	0.04	0.05	0.06	0.07	0.08	0.09
1.0	0.5236	0.5395	0.5556	0.5722	0.5890	0.6061	0.6236	0.6414	0.6596	0.6781
.1	.6969	.7161	.7356	.7555	.7757	.7963	.8173	.8386	.8603	.8823
.2	.9048	.9276	.9508	.9743	.9983	1.023	1.047	1.073	1.098	1.124
.3	1.150	1.177	1.204	1.232	1.260	1.288	1.317	1.346	1.376	1.406
.4	1.437	1.468	1.499	1.531	1.563	1.596	1.630	1.663	1.697	1.732
1.5	1.767	1.803	1.839	1.875	1.912	1.950	1.988	2.026	2.065	2.105
.6	2.145	2.185	2.226	2.268	2.310	2.352	2.395	2.439	2.483	2.527
.7	2.572	2.618	2.664	2.711	2.758	2.806	2.855	2.903	2.953	3.003
.8	3.054	3.105	3.157	3.209	3.262	3.315	3.369	3.424	3.479	3.535
.9	3.591	3.648	3.706	3.764	3.823	3.882	3.942	4.003	4.064	4.126
2.0	4.189	4.252	4.316	4.380	4.445	4.511	4.577	4.644	4.712	4.780
.1	4.849	4.919	4.989	5.060	5.131	5.204	5.277	5.350	5.425	5.500
.2	5.575	5.652	5.729	5.806	5.885	5.964	6.044	6.125	6.206	6.288
.3	6.371	6.454	6.538	6.623	6.709	6.795	6.882	6.970	7.059	7.148
.4	7.238	7.329	7.421	7.513	7.606	7.700	7.795	7.890	7.986	8.083
2.5	8.181	8.280	8.379	8.479	8.580	8.682	8.785	8.888	8.992	9.097
.6	9.203	9.309	9.417	9.525	9.634	9.744	9.855	9.966	10.079	10.19
.7	10.31	10.42	10.54	10.65	10.77	10.89	11.01	11.13	11.25	11.37
.8	11.49	11.62	11.74	11.87	11.99	12.12	12.25	12.38	12.51	12.64
.9	12.77	12.90	13.04	13.17	13.31	13.44	13.58	13.72	13.86	14.00
3.0	14.14	14.28	14.42	14.57	14.71	14.86	15.00	15.15	15.30	15.45
.1	15.60	15.75	15.90	16.06	16.21	16.37	16.52	16.68	16.84	17.00
.2	17.16	17.32	17.48	17.64	17.81	17.97	18.14	18.31	18.48	18.65
.3	18.82	18.99	19.16	19.33	19.51	19.68	19.86	20.04	20.22	20.40
.4	20.58	20.76	20.94	21.13	21.31	21.50	21.69	21.88	22.07	22.26
3.5	22.45	22.64	22.84	23.03	23.23	23.43	23.62	23.82	24.02	24.23
.6	24.43	24.63	24.84	25.04	25.25	25.46	25.67	25.88	26.09	26.31
.7	26.52	26.74	26.95	27.17	27.39	27.61	27.83	28.06	28.28	28.50
.8	28.73	28.96	29.19	29.42	29.65	29.88	30.11	30.35	30.58	30.82
.9	31.06	31.30	31.54	31.78	32.02	32.27	32.52	32.76	33.01	33.26
4.0	33.51	33.76	34.02	34.27	34.53	34.78	35.04	35.30	35.56	35.82
.1	36.09	36.35	36.62	36.88	37.15	37.42	37.69	37.97	38.24	38.52
.2	38.79	39.07	39.35	39.63	39.91	40.19	40.48	40.76	41.05	41.34
.3	41.63	41.92	42.21	42.51	42.80	43.10	43.40	43.70	44.00	44.30
.4	44.60	44.91	45.21	45.52	45.83	46.14	46.45	46.77	47.08	47.40
4.5	47.71	48.03	48.35	48.67	49.00	49.32	49.65	49.97	50.30	50.63
.6	50.97	51.30	51.63	51.97	52.31	52.65	52.99	53.33	53.67	54.02
.7	54.36	54.71	55.06	55.41	55.76	56.12	56.47	56.83	57.19	57.54
.8	57.91	58.27	58.63	59.00	59.37	59.73	60.10	60.48	60.85	61.22
.9	61.60	61.98	62.36	62.74	63.12	63.51	63.89	64.28	64.67	65.06
5.0	65.45	65.84	66.24	66.64	67.03	67.43	67.83	68.24	68.64	69.05
.1	69.46	69.87	70.28	70.69	71.10	71.52	71.94	72.36	72.78	73.20
.2	73.62	74.05	74.47	74.90	75.33	75.77	76.20	76.64	77.07	77.51
.3	77.95	78.39	78.84	79.28	79.73	80.18	80.63	81.08	81.54	81.99
.4	82.45	82.91	83.37	83.83	84.29	84.76	85.23	85.70	86.17	86.64
5.5	87.11	87.59	88.07	88.55	89.03	89.51	90.00	90.48	90.97	91.46
.6	91.95	92.45	92.94	93.44	93.94	94.44	94.94	95.44	95.95	96.46
.7	96.97	97.48	97.99	98.51	99.02	99.54	100.06	100.6	101.1	101.6
.8	102.2	102.7	103.2	103.8	104.3	104.8	105.4	105.9	106.4	107.0
.9	107.5	108.1	108.6	109.2	109.7	110.3	110.9	111.4	112.0	112.5
6.0	113.1	113.7	114.2	114.8	115.4	115.9	116.5	117.1	117.7	118.3
.1	118.8	119.4	120.0	120.6	121.2	121.8	122.4	123.0	123.6	124.2
.2	124.8	125.4	126.0	126.6	127.2	127.8	128.4	129.1	129.7	130.3
.3	130.9	131.5	132.2	132.8	133.4	134.1	134.7	135.3	136.0	136.6
.4	137.3	137.9	138.5	139.2	139.8	140.5	141.2	141.8	142.5	143.1
6.5	143.8	144.5	145.1	145.8	146.5	147.1	147.8	148.5	149.2	149.8
.6	150.5	151.2	151.9	152.6	153.3	154.0	154.7	155.4	156.1	156.8
.7	157.5	158.2	158.9	159.6	160.3	161.0	161.7	162.5	163.2	163.9
.8	164.6	165.4	166.1	166.8	167.6	168.3	169.0	169.8	170.5	171.3
.9	172.0	172.8	173.5	174.3	175.0	175.8	176.5	177.3	178.1	178.8
7.0	179.6	180.4	181.1	181.9	182.7	183.5	184.3	185.0	185.8	186.6
.1	187.4	188.2	189.0	189.8	190.6	191.4	192.2	193.0	193.8	194.6
.2	195.4	196.2	197.1	197.9	198.7	199.5	200.4	201.2	202.0	202.9
.3	203.7	204.5	205.4	206.2	207.1	207.9	208.8	209.6	210.5	211.3
.4	212.2	213.0	213.9	214.8	215.6	216.5	217.4	218.3	219.1	220.0
7.5	220.9	221.8	222.7	223.6	224.4	225.3	226.2	227.1	228.0	228.9
.6	229.8	230.8	231.7	232.6	233.5	234.4	235.3	236.3	237.2	238.1
.7	239.0	240.0	240.9	241.8	242.8	243.7	244.7	245.6	246.6	247.5
.8	248.5	249.4	250.4	251.4	252.3	253.3	254.3	255.2	256.2	257.2
.9	258.2	259.1	260.1	261.1	262.1	263.1	264.1	265.1	266.1	267.1

*See footnote at end of table, p. 35.

Table 12. Spheres: Diameters, Volumes*—(*Concluded*)

D	0.00	0.01	0.02	0.03	0.04	0.05	0.06	0.07	0.08	0.09
8.0	268.1	269.1	270.1	271.1	272.1	273.1	274.2	275.2	276.2	277.2
.1	278.3	279.3	280.3	281.4	282.4	283.4	284.5	285.5	286.6	287.6
.2	288.7	289.8	290.8	291.9	292.9	294.0	295.1	296.2	297.2	298.3
.3	299.4	300.5	301.6	302.6	303.7	304.8	305.9	307.0	308.1	309.2
.4	310.3	311.4	312.6	313.7	314.8	315.9	317.0	318.2	319.3	320.4
8.5	321.6	322.7	323.8	325.0	326.1	327.3	328.4	329.6	330.7	331.9
.6	333.0	334.2	335.4	336.5	337.7	338.9	340.1	341.2	342.4	343.6
.7	344.8	346.0	347.2	348.4	349.6	350.8	352.0	353.2	354.4	355.6
.8	356.8	358.0	359.3	360.5	361.7	362.9	364.2	365.4	366.6	367.9
.9	369.1	370.4	371.6	372.9	374.1	375.4	376.6	377.9	379.2	380.4
9.0	381.7	383.0	384.3	385.5	386.8	388.1	389.4	390.7	392.0	393.3
.1	394.6	395.9	397.2	398.5	399.8	401.1	402.4	403.7	405.1	406.4
.2	407.7	409.1	410.4	411.7	413.1	414.4	415.7	417.1	418.4	419.8
.3	421.2	422.5	423.9	425.2	426.6	428.0	429.4	430.7	432.1	433.5
.4	434.9	436.3	437.7	439.1	440.5	441.9	443.3	444.7	446.1	447.5
9.5	448.9	450.3	451.8	453.2	454.6	456.0	457.5	458.9	460.4	461.8
.6	463.2	464.7	466.1	467.6	469.1	470.5	472.0	473.5	474.9	476.4
.7	477.9	479.4	480.8	482.3	483.8	485.3	486.8	488.3	489.8	491.3
.8	492.8	494.3	494.8	497.3	498.9	500.4	501.9	503.4	505.0	506.5
.9	508.0	509.6	511.1	512.7	514.2	515.8	517.3	518.9	520.5	522.0
10.0	523.6									

* *Explanation.* Moving the decimal point one place in column D is equivalent to moving it three places in the body of the table (D = diameter).

$$\text{Volume of sphere} = \frac{\pi}{6} \times (\text{diam.})^3 = 0.523599 \times (\text{diam.})^3$$

Conversely,

$$\text{Diameter} = \sqrt[3]{\frac{6}{\pi}} \times \sqrt[3]{\text{vol.}} = 1.2407 \times \sqrt[3]{\text{vol.}}$$

Table 13. Spheres: Segments*

h = segment height; D = sphere diameter

h/D	Segment vol. / D^3	Segment vol. / Sphere vol.	h/D	Segment vol. / D^3	Segment vol. / Sphere vol.	h/D	Segment vol. / D^3	Segment vol. / Sphere vol.	h/D	Segment vol. / D^3	Segment vol. / Sphere vol.
0.01	0.000156	0.000298	0.16	0.035923	0.068608	0.31	0.119756	0.228718	0.41	0.191877	0.366458
.02	.000619	.001184	.17	.040251	.076874	.32	.126534	.241664	.42	.199503	.381024
.03	.001385	.002646	.18	.044787	.085536	.33	.133426	.254826	.43	.207180	.395686
.04	.002446	.004672	.19	.049522	.094582	.34	.140425	.268192	.44	.214901	.410432
.05	.003796	.007250	.20	.054454	.104000	.35	.147524	.281750	.45	.222660	.425250
.06	.005429	.010368	.21	.059573	.113778	.36	.154717	.295488	.46	.230450	.440128
.07	.007338	.014014	.22	.064875	.123904	.37	.161998	.309394	.47	.238265	.455054
.08	.009517	.018176	.23	.070353	.134366	.38	.169361	.323456	.48	.246099	.470016
.09	.011960	.022842	.24	.076001	.145152	.39	.176799	.337662	.49	.253946	.485002
.10	.014661	.028000	.25	.081812	.156250	.40	.184306	.352000	.50	.261799	.500000
.11	.017613	.033638	.26	.087780	.167648						
.12	.020809	.039744	.27	.093900	.179334						
.13	.024246	.046306	.28	.100160	.191296						
.14	.027914	.053312	.29	.106560	.203522						
.15	.031809	.060750	.30	.113097	.216000						

* Given the segment height h and the sphere diameter D, first form the ratio h/D, and find from the table the value of (segment volume/D^3); then multiply this latter value by D^3, that is: (segment volume/D^3) $\times D^3$ = segment volume.

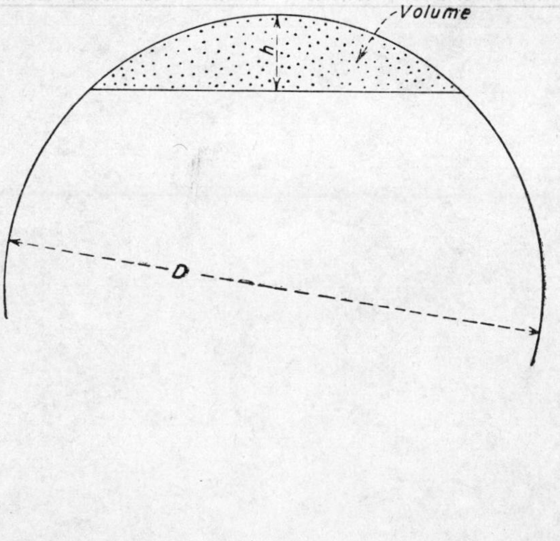

CONVERSION TABLES AND EQUIVALENTS

Table 14. Degrees to Radians

One radian = 57.29578 deg.

Deg.	Radians	Deg.	Radians	Deg.	Radians	Deg.	Radians	Deg.	Radians	Deg.	Radians	Deg.	Radians	Deg.	Radians	Deg.	Radians	Deg.	Radians	Deg.	Radians	Deg.	Radians
1	0.01745	16	.27925	31	0.54105	46	.80285	61	1.06465	76	1.32645	91	1.58825	106	1.85005	121	2.11185	136	2.37365	151	2.63545	166	2.89725
2	.03491	17	.29671	32	.55851	47	.82031	62	1.08210	77	1.34390	92	1.60570	107	1.86750	122	2.12930	137	2.39110	152	2.65290	167	2.91470
3	.05236	18	.31416	33	.57596	48	.83776	63	1.09956	78	1.36136	93	1.62316	108	1.88496	123	2.14676	138	2.40855	153	2.67035	168	2.93215
4	.06981	19	.33161	34	.59341	49	.85521	64	1.11701	79	1.37881	94	1.64061	109	1.90241	124	2.16421	139	2.42601	154	2.68781	169	2.94961
5	.08727	20	.34907	35	.61087	50	.87267	65	1.13446	80	1.39626	95	1.65806	110	1.91986	125	2.18166	140	2.44346	155	2.70526	170	2.96706
6	.10472	21	.36652	36	.62832	51	.89012	66	1.15192	81	1.41372	96	1.67552	111	1.93732	126	2.19912	141	2.46091	156	2.72271	171	2.98451
7	.12217	22	.38397	37	.64577	52	.90757	67	1.16937	82	1.43117	97	1.69297	112	1.95477	127	2.21657	142	2.47837	157	2.74017	172	3.00197
8	.13963	23	.40143	38	.66323	53	.92502	68	1.18682	83	1.44862	98	1.71042	113	1.97222	128	2.23402	143	2.49582	158	2.75762	173	3.01942
9	.15708	24	.41888	39	.68068	54	.94248	69	1.20428	84	1.46608	99	1.72788	114	1.98968	129	2.25147	144	2.51327	159	2.77507	174	3.03687
10	.17453	25	.43633	40	.69813	55	.95993	70	1.22173	85	1.48353	100	1.74533	115	2.00713	130	2.26893	145	2.53073	160	2.79253	175	3.05433
11	.19199	26	.45379	41	.71559	56	.97738	71	1.23918	86	1.50098	101	1.76278	116	2.02458	131	2.28638	146	2.54818	161	2.80998	176	3.07178
12	.20944	27	.47124	42	.73304	57	0.99484	72	1.25664	87	1.51844	102	1.78024	117	2.04204	132	2.30383	147	2.56563	162	2.82743	177	3.08923
13	.22689	28	.48869	43	.75049	58	1.01299	73	1.27409	88	1.53589	103	1.79769	118	2.05949	133	2.32129	148	2.58309	163	2.84489	178	3.10669
14	.24435	29	.50615	44	.76794	59	1.02974	74	1.29154	89	1.55334	104	1.81514	119	2.07694	134	2.33874	149	2.60054	164	2.86234	179	3.12414
15	.26180	30	.52360	45	.78540	60	1 04720	75	1.30900	90	1.57080	105	1.83260	120	2.09440	135	2.35619	150	2.61800	165	2.87979	180	3.14159

Minutes to Radians

Min.	Radians	Min.	Radians	Min.	Radians	Min.	Radians	Min.	Radians	Min.	Radians	Min.	Radians	Min.	Radians	Min.	Radians	Min.	Radians	Min.	Radians	Min.	Radians
1	0.00029	6	0.00175	11	0.00320	16	0.00465	21	0.00611	26	0.00756	31	0.00902	36	0.01047	41	0.01193	46	0.01338	51	0.01484	56	0.01629
2	.00058	7	.00204	12	.00349	17	.00495	22	.00640	27	.00785	32	.00931	37	.01076	42	.01222	47	.01367	52	.01513	57	.01658
3	.00087	8	.00233	13	.00378	18	.00524	23	.00669	28	.00814	33	.00960	38	.01105	43	.01251	48	.01396	53	.01542	58	.01687
4	.00116	9	.00262	14	.00407	19	.00553	24	.00698	29	.00844	34	.00989	39	.01134	44	.01280	49	.01425	54	.01571	59	.01716
5	.00145	10	.00291	15	.00436	20	.00582	25	.00727	30	.00873	35	.01018	40	.01164	45	.01309	50	.01454	55	.01600	60	.01745

Radians to Degrees

Radians	Hundredths of a radian: 0.00	.01	.02	.03	.04	.05	.06	.07	.08	.09
	Degrees									
0.0	0.0000	0.5730	1.1459	1.7189	2.2918	2.8648	3.4377	4.0107	4.5837	5.1566
.1	5.7296	6.3025	6.8755	7.4485	8.0214	8.5944	9.1673	9.7403	10.3132	10.8862
.2	11.4591	12.0321	12.6051	13.1780	13.7510	14.3239	14.8969	15.4699	16.0428	16.6158
.3	17.1887	17.7617	18.3346	18.9076	19.4806	20.0535	20.6265	21.1994	21.7724	22.3454
.4	22.9183	23.4913	24.0642	24.6372	25.2101	25.7831	26.3561	26.9290	27.5020	28.0749
.5	28.6479	29.2208	29.7938	30.3668	30.9397	31.1527	32.0856	32.6586	33.2316	33.8045
.6	34.3775	34.9504	35.5234	36.0963	36.6693	37.2423	37.8152	38.3882	38.9611	39.5341
.7	40.1070	40.6800	41.2530	41.8259	42.3989	42.9718	43.5448	44.1178	44.6907	45.2637
.8	45.8366	46.4096	46.9825	47.5555	48.1285	48.7014	49.2744	49.8473	50.4203	50.9932
.9	51.5662	52.1392	52.7121	53.2851	53.8580	54.4310	55.0039	55.5769	56.1499	56.7228

Radians	1	2	3	4	5	6	7
Degrees	57.2958	114.5916	171.8873	229.1831	286.4789	343.7747	401.0705

Table 15. Decimal Equivalents

From seconds and minutes to decimal parts of a degree				From decimal parts of a degree to minutes and seconds				Common fractions 8ths	16ths	32ds	64ths	Exact decimal values
0″	0°.0000	0′	0°.0000	0°.00	0′	0°.50	30′	..	..	..	1	0.01563
1	.0003	1	.0167	1	0′36″	1	30′36″	..	..	1	2	.03125
2	.0006	2	.0333	2	1′12″	2	31′12″	..	..	..	3	.04688
3	.0008	3	.05	3	1′48″	3	31′48″	..	1	2	4	.06250
4	.0011	4	.0667	4	2′24″	4	32′24″	..	..	..	5	.07813
5″	0°.0014	5′	0°.0833	0°.05	3′	0°.55	33′	..	..	3	6	.09375
6	.0017	6	.10	6	3′36″	6	33′36″	..	..	..	7	.10938
7	.0019	7	.1167	7	4′12″	7	34′12″	1	2	4	8	.12500
8	.0022	8	.1333	8	4′48″	8	34′48″	..	..	..	9	.14063
9	.0025	9	.15	9	5′24″	9	35′24″	..	..	5	10	.15625
10″	0°.0028	10′	0°.1667	0°.10	6′	0°.60	36′	..	..	..	11	.17188
1	.0031	1	.1833	1	6′36″	1	36′36″	..	3	6	12	.18750
2	.0033	2	.20	2	7′12″	2	37′12″	..	..	..	13	.20313
3	.0036	3	.2167	3	7′48″	3	37′48″	..	..	7	14	.21875
4	.0039	4	.2333	4	8′24″	4	38′24″	..	..	..	15	.23438
15″	0°.0042	15′	0°.25	0°.15	9′	0°.65	39′	2	4	8	16	.25000
6	.0044	6	.2667	6	9′36″	6	39′36″	..	..	..	17	.26563
7	.0047	7	.2833	7	10′12″	7	40′12″	..	..	9	18	.28125
8	.0050	8	.30	8	10′48″	8	40′48″	..	..	..	19	.29688
9	.0053	9	.3167	9	11′24″	9	41′24″	..	5	10	20	.31250
20″	0°.0056	20′	0°.3333	0°.20	12′	0°.70	42′	..	..	..	21	.32813
1	.0058	1	.35	1	12′36″	1	42′36″	..	..	11	22	.34375
2	.0061	2	.3667	2	13′12″	2	43′12″	..	..	..	23	.35938
3	.0064	3	.3833	3	13′48″	3	43′48″	3	6	12	24	.37500
4	.0067	4	.40	4	14′24″	4	44′24″	..	..	..	25	.39063
25″	0°.0069	25′	0°.4167	0°.25	15′	0°.75	45′	..	..	13	26	.40625
6	.0072	6	.4333	6	15′36″	6	45′36″	..	..	..	27	.42188
7	.0075	7	.45	7	16′12″	7	46′12″	..	7	14	28	.43750
8	.0078	8	.4667	8	16′48″	8	46′48″	..	..	..	29	.45313
9	.0081	9	.4833	9	17′24″	9	47′24″	..	..	15	30	.46875

From seconds and minutes to decimal parts of a degree				From decimal parts of a degree to minutes and seconds				Common fractions 8ths	16ths	32ds	64ths	Exact decimal values
30″	0°.0083	30′	0°.50	0°.30	18′	0°.80	48′	..	..	..	31	.48438
1	.0086	1	.5167	1	18′36″	1	48′36″	4	8	16	32	.50000
2	.0089	2	.5333	2	19′12″	2	49′12″	..	..	..	33	.51563
3	.0092	3	.55	3	19′48″	3	49′48″	..	..	17	34	.53125
4	.0094	4	.5667	4	20′24″	4	50′24″	..	..	..	35	.54688
35″	0°.0097	35′	0°.5833	0°.35	21′	0°.85	51′	..	9	18	36	.56250
6	.0100	6	.60	6	21′36″	6	51′36″	..	..	..	37	.57813
7	.0103	7	.6167	7	22′12″	7	52′12″	..	..	19	38	.59375
8	.0106	8	.6333	8	22′48″	8	52′48″	..	..	..	39	.60938
9	.0108	9	.65	9	23′24″	9	53′24″	5	10	20	40	.62500
40″	0°.0111	40′	0°.6667	0°.40	24′	0°.90	54′	..	..	..	41	.64063
1	.0114	1	.6833	1	24′36″	1	54′36″	..	..	21	42	.65625
2	.0117	2	.70	2	25′12″	2	55′12″	..	..	..	43	.67188
3	.0119	3	.7167	3	25′48″	3	55′48″	..	11	22	44	.68750
4	.0122	4	.7333	4	26′24″	4	56′24″	..	..	..	45	.70313
45″	0°.0125	45′	0°.75	0°.45	27′	0°.95	57′	..	..	23	46	.71875
6	.0128	6	.7667	6	27′36″	6	57′36″	..	..	..	47	.73438
7	.0131	7	.7833	7	28′12″	7	58′12″	6	12	24	48	.75000
8	.0133	8	.80	8	28′48″	8	58′48″	..	..	..	49	.76563
9	.0136	9	.8167	9	29′24″	9	59′24″	..	..	25	50	.78125
50″	0°.0139	50′	0°.8333	0°.50	30′	1°.00	60′	..	..	..	51	0.79688
1	.0142	1	.8500					..	13	26	52	.81250
2	.0144	2	.8667			0°.000	0″.0	..	..	..	53	.82813
3	.0147	3	.8833			1	3″.6	..	..	27	54	.84375
4	.015	4	.9000			2	7″.2	..	..	..	55	.85938
						3	10″.8	7	14	28	56	.87500
						4	14″.4	..	..	..	57	.89063
55″	0°.0153	55′	0°.9167									
6	.0156	6	.9333			0°.005	18″	..	..	29	58	.90625
7	.0158	7	.95			6	21″.6	..	..	..	59	.92188
8	.0161	8	.9667			7	25″.2	..	15	30	60	.93750
9	.0164	9	.9833			8	28″.8	..	..	..	61	.95313
60″	0°.0167	60′	1°.00			9	32″.4	..	..	31	62	.96875
								..	..	..	63	.98438

Decimal Equivalents

Decimals of a foot equivalent to inches and fractions of an inch

Inch	Inches 0	1	2	3	4	5	6	7	8	9	10	11
	Feet											
0	0.0000	0.08333	0.16667	0.25000	0.33333	0.41667	0.50000	0.58333	0.66667	0.75000	0.83333	0.91667
1/32	.00260	.08594	.16927	.25260	.33594	.41927	.50260	.58594	.66927	.75260	.83594	.91927
1/16	.00521	.08854	.17188	.25521	.33854	.42188	.50521	.58854	.67188	.75521	.83854	.92188
3/32	.00781	.09115	.17448	.25781	.34115	.42448	.50781	.59115	.67448	.75781	.84115	.92448
1/8	.01042	.09375	.17708	.26042	.34375	.42708	.51042	.59375	.67708	.76042	.84375	.92708
5/32	.01302	.09635	.17969	.26302	.34635	.42969	.51302	.59635	.67969	.76302	.84635	.92969
3/16	.01563	.09896	.18229	.26563	.34896	.43229	.51563	.59896	.68229	.76563	.84896	.93229
7/32	.01823	.10156	.18490	.26823	.35156	.43490	.51823	.60156	.68490	.76823	.85156	.93490
1/4	.02083	.10417	.18750	.27083	.35417	.43750	.52083	.60417	.68750	.77083	.85417	.93750
9/32	.02344	.10677	.19010	.27344	.35677	.44010	.52344	.60677	.69010	.77344	.85677	.94010
5/16	.02604	.10937	.19271	.27604	.35937	.44271	.52604	.60937	.69271	.77604	.85937	.94271
11/32	.02865	.11198	.19531	.27865	.36198	.44531	.52865	.61198	.69531	.77865	.86198	.94531
3/8	.03125	.11458	.19792	.28125	.36458	.44792	.53125	.61458	.69792	.78125	.86458	.94792
13/32	.03385	.11719	.20052	.28385	.36719	.45052	.53385	.61719	.70052	.78385	.86719	.95052
7/16	.03646	.11979	.20313	.28646	.36979	.45313	.53646	.61979	.70313	.78646	.86979	.95313
15/32	.03906	.12240	.20573	.28906	.37240	.45573	.53906	.62240	.70573	.78906	.87240	.95573
1/2	.04167	.12500	.20833	.29167	.37500	.45833	.54167	.62500	.70833	.79167	.87500	.95833
17/32	.04427	.12760	.21094	.29427	.37760	.46094	.54427	.62760	.71094	.79427	.87760	.96094
9/16	.04688	.13021	.21354	.29688	.38021	.46354	.54688	.63021	.71354	.79688	.88021	.96354
19/32	.04948	.13281	.21615	.29948	.38281	.46615	.54948	.63281	.71615	.79948	.88281	.96615
5/8	.05208	.13542	.21875	.30208	.38542	.46875	.55208	.63542	.71875	.80208	.88542	.96875
21/32	.05469	.13802	.22135	.30469	.38802	.47135	.55469	.63802	.72135	.80469	.88802	.97135
11/16	.05729	.14062	.22396	.30729	.39062	.47396	.55729	.64062	.72396	.80729	.89062	.97396
23/32	.05990	.14323	.22656	.30990	.39323	.47656	.55990	.64323	.72656	.80990	.89323	.97656
3/4	.06250	.14583	.22917	.31250	.39583	.47917	.56250	.64583	.72917	.81250	.89583	.97917
25/32	.06510	.14844	.23177	.31510	.39844	.48177	.56510	.64844	.73177	.81510	.89844	.98177
13/16	.06771	.15104	.23438	.31771	.40104	.48438	.56771	.65104	.73438	.81771	.90104	.98438
27/32	.07031	.15365	.23698	.32031	.40365	.48698	.57031	.65365	.73698	.82031	.90365	.98698
7/8	.07292	.15625	.23958	.32292	.40625	.48958	.57292	.65625	.73958	.82292	.90625	.98958
29/32	.07552	.15885	.24219	.32552	.40885	.49219	.57552	.65885	.74219	.82552	.90885	.99219
15/16	.07813	.16146	.24479	.32813	.41146	.49479	.57813	.66146	.74479	.82813	.91146	.99479
31/32	.08073	.16406	.24740	.33073	.41406	.49740	.58073	.66406	.74740	.83073	.91406	.99740

Examples. 1 in. (third column) = 0.08333 ft.; 1 1/32 in. = 0.08594 ft.; 4 13/32 in. = 0.36719 ft.

Table 16. Specific Gravity, Degrees Baumé, Degrees A.P.I., Degrees Twaddell, Pounds per Gallon, Pounds per Cubic Foot*

$°\text{Bé.} = 145 - \frac{145}{\text{sp. gr.}}$ (heavier than H_2O); $°\text{Bé.} = \frac{140}{\text{sp. gr.}} - 130$ (lighter than H_2O); $°\text{Tw.} = \frac{\text{sp. gr. } 60°/60°\text{F.} - 1}{0.005}$

Sp. gr. 60°/60°	°Bé.	°A.P.I.	Lb. per gal. at 60°F. wt. in air	Lb. per cu. ft. at 60°F. wt. in air	Sp. gr. 60°/60°	°Bé.	°A.P.I.	Lb. per gal. at 60°F. wt. in air	Lb. per cu. ft. at 60°F. wt. in air	Sp. gr. 60°/60°	°Bé.	°A.P.I.	Lb. per gal. at 60°F. wt. in air	Lb. per cu. ft. at 60°F. wt. in air	Sp. gr. 60°/60°	°Bé.	°A.P.I.	Lb. per gal. at 60°F. wt. in air	Lb. per cu. ft. at 60°F. wt. in air
0.600	103.33	104.33	4.9929	37.350	.700	70.00	70.64	5.8268	43.587	0.800	45.00	45.38	6.6606	49.825	.900	25.56	25.72	7.4944	56.062
.605	101.40	102.38	5.0346	37.662	.705	68.58	69.21	5.8685	43.899	.805	43.91	44.28	6.7023	50.137	.905	24.70	24.85	7.5361	56.374
.610	99.51	100.47	5.0763	37.973	.710	67.18	67.80	5.9101	44.211	.810	42.84	43.19	6.7440	50.448	.910	23.85	23.99	7.5777	56.685
.615	97.64	98.58	5.1180	38.285	.715	65.80	66.40	5.9518	44.523	.815	41.78	42.12	6.7857	50.760	.915	23.01	23.14	7.6194	56.997
.620	95.81	96.73	5.1597	38.597	.720	64.44	65.03	5.9935	44.834	.820	40.73	41.06	6.8274	51.072	.920	22.17	22.30	7.6612	57.310
.625	94.00	94.90	5.2014	39.910	.725	63.10	63.67	6.0352	45.146	.825	39.70	40.02	6.8691	51.384	.925	21.35	21.47	7.7029	57.622
.630	92.22	93.10	5.2431	39.222	.730	61.78	62.34	6.0769	45.458	.830	38.67	38.98	6.9108	51.696	.930	20.54	20.65	7.7446	57.934
.635	90.47	91.33	5.2848	39.534	.735	60.48	61.02	6.1186	45.770	.835	37.66	37.96	6.9525	52.008	.935	19.73	19.84	7.7863	58.246
.640	88.75	89.59	5.3265	39.845	.740	59.19	59.72	6.1603	46.082	.840	36.67	36.95	6.9941	52.320	.940	18.94	19.03	7.8280	58.557
.645	87.05	87.88	5.3682	40.157	.745	57.92	58.43	6.2020	46.394	.845	35.68	35.96	7.0358	52.632	.945	18.15	18.24	7.8697	58.869
.650	85.38	86.19	5.4098	40.468	.750	56.67	57.17	6.2437	46.706	.850	34.71	34.97	7.0775	52.943	.950	17.37	17.45	7.9114	59.181
.655	83.74	84.53	5.4515	40.780	.755	55.43	55.92	6.2854	47.018	.855	33.74	34.00	7.1192	53.255	.955	16.60	16.67	7.9531	59.493
.660	82.12	82.89	5.4932	41.092	.760	54.21	54.68	6.3271	47.330	.860	32.79	33.03	7.1609	53.567	.960	15.83	15.90	7.9947	59.805
.665	80.53	81.28	5.5349	41.404	.765	53.01	53.47	6.3688	47.642	.865	31.85	32.08	7.2026	53.879	.965	15.08	15.13	8.0364	60.117
.670	78.96	79.69	5.5766	41.716	.770	51.82	52.27	6.4104	47.953	.870	30.92	31.14	7.2443	54.191	.970	14.33	14.38	8.0780	60.428
.675	77.41	78.13	5.6183	42.028	.775	50.65	51.08	6.4521	47.265	.875	30.00	30.21	7.2860	54.503	.975	13.59	13.63	8.1197	60.740
.680	75.88	76.59	5.6600	42.340	.780	49.49	49.91	6.4938	48.577	.880	29.09	29.30	7.3277	54.815	.980	12.86	12.89	8.1615	61.052
.685	74.38	75.07	5.7017	42.652	.785	48.34	48.75	6.5355	48.889	.885	28.19	28.39	7.3694	55.127	.985	12.13	12.15	8.2032	61.364
.690	72.90	73.57	5.7434	42.963	.790	47.22	47.61	6.5772	49.201	.890	27.30	27.49	7.4111	55.438	.990	11.41	11.43	8.2449	61.676
.695	71.44	72.10	5.7851	43.275	.795	46.10	46.49	6.6189	49.513	.895	26.42	26.60	7.4528	55.750	.995	10.70	10.71	8.2866	61.988
															1.000	10.00	10.00	8.3283	62.300

Sp. gr. 60°/60°	°Bé.	°Tw.	Lb. per gal. at 60°F. wt. in air	Lb. per cu. ft. at 60°F. wt. in air	Sp. gr. 60°/60°	°Bé.	°Tw.	Lb. per gal. at 60°F. wt. in air	Lb. per cu. ft. at 60°F. wt. in air	Sp. gr. 60°/60°	°Bé.	°Tw.	Lb. per gal. at 60°F. wt. in air	Lb. per cu. ft. at 60°F. wt. in air	Sp. gr. 60°/60°	°Bé.	°Tw.	Lb. per gal. at 60°F. wt. in air	Lb. per cu. ft. at 60°F. wt. in air
1.005	0.72	1	8.3700	62.612	1.255	29.46	51	10.4546	78.206	1.505	48.65	101	12.5392	93.800	1.755	62.38	151	14.6238	109.394
1.010	1.44	2	8.4117	62.924	1.260	29.92	52	10.4963	78.518	1.510	48.97	102	12.5809	94.112	1.760	62.61	152	14.6655	109.705
1.015	2.14	3	8.4534	63.236	1.265	30.38	53	10.5380	78.830	1.515	49.29	103	12.6226	94.424	1.765	62.85	153	14.7072	110.017
1.020	2.84	4	8.4950	63.547	1.270	30.83	54	10.5797	79.141	1.520	49.61	104	12.6643	94.735	1.770	63.08	154	14.7489	110.329
1.025	3.54	5	8.5367	63.859	1.275	31.27	55	10.6214	79.453	1.525	49.92	105	12.7060	95.047	1.775	63.31	155	14.7906	110.641
1.030	4.22	6	8.5784	64.171	1.280	31.72	56	10.6630	79.765	1.530	50.23	106	12.7477	95.359	1.780	63.54	156	14.8323	110.953
1.035	4.90	7	8.6201	64.483	1.285	32.16	57	10.7047	80.077	1.535	50.54	107	12.7894	95.671	1.785	63.77	157	14.8740	111.265
1.040	5.58	8	8.6618	64.795	1.290	32.60	58	10.7464	80.389	1.540	50.84	108	12.8310	95.983	1.790	63.99	158	14.9157	111.577
1.045	6.24	9	8.7035	65.107	1.295	33.03	59	10.7881	80.701	1.545	51.15	109	12.8727	96.295	1.795	64.22	159	14.9574	111.889
1.050	6.91	10	8.7452	65.419	1.300	33.46	60	10.8298	81.013	1.550	51.45	110	12.9144	96.606	1.800	64.44	160	14.9990	112.200
1.055	7.56	11	8.7869	65.731	1.305	33.89	61	10.8715	81.325	1.555	51.75	111	12.9561	96.918	1.805	64.67	161	15.0407	112.512
1.060	8.21	12	8.8286	66.042	1.310	34.31	62	10.9132	81.636	1.560	52.05	112	12.9978	97.230	1.810	64.89	162	15.0824	112.824
1.065	8.85	13	8.8703	66.354	1.315	34.73	63	10.9549	81.948	1.565	52.35	113	13.0395	97.542	1.815	65.11	163	15.1241	113.136
1.070	9.49	14	8.9120	66.666	1.320	35.15	64	10.9966	82.260	1.570	52.64	114	13.0812	97.854	1.820	65.33	164	15.1658	113.448
1.075	10.12	15	8.9537	66.978	1.325	35.57	65	11.0383	82.572	1.575	52.94	115	13.1229	98.166	1.825	65.55	165	15.2075	113.760
1.080	10.74	16	8.9954	67.290	1.330	35.98	66	11.0800	82.884	1.580	53.23	116	13.1646	98.478	1.830	65.77	166	15.2492	114.072
1.085	11.36	17	9.0371	67.602	1.335	36.39	67	11.1217	83.196	1.585	53.52	117	13.2063	98.790	1.835	65.98	167	15.2909	114.384
1.090	11.97	18	9.0787	67.914	1.340	36.79	68	11.1634	83.508	1.590	53.81	118	13.2480	99.102	1.840	66.20	168	15.3326	114.696
1.095	12.58	19	9.1204	68.226	1.345	37.19	69	11.2051	83.820	1.595	54.09	119	13.2897	99.414	1.845	66.41	169	15.3743	115.007
1.100	13.18	20	9.1621	68.537	1.350	37.59	70	11.2467	84.131	1.600	54.38	120	13.3313	99.725	1.850	66.62	170	15.4160	115.318
1.105	13.78	21	9.2038	68.849	1.355	37.99	71	11.2884	84.443	1.605	54.66	121	13.3730	100.037	1.855	66.83	171	15.4577	115.630
1.110	14.37	22	9.2455	69.161	1.360	38.38	72	11.3301	84.755	1.610	54.94	122	13.4147	100.349	1.860	67.04	172	15.4993	115.943
1.115	14.96	23	9.2872	69.473	1.365	38.77	73	11.3718	85.067	1.615	55.22	123	13.4564	100.661	1.865	67.25	173	15.5410	116.255
1.120	15.54	24	9.3289	69.785	1.370	39.16	74	11.4135	85.379	1.620	55.49	124	13.4981	100.973	1.870	67.46	174	15.5827	116.567
1.125	16.11	25	9.3706	70.097	1.375	39.55	75	11.4552	85.691	1.625	55.77	125	13.5398	101.285	1.875	67.67	175	15.6244	116.879
1.130	16.68	26	9.4123	70.409	1.380	39.93	76	11.4969	86.003	1.630	56.04	126	13.5815	101.597	1.880	67.87	176	15.6661	117.191
1.135	17.25	27	9.4540	70.721	1.385	40.31	77	11.5386	86.315	1.635	56.32	127	13.6232	101.909	1.885	68.08	177	15.7078	117.503
1.140	17.81	28	9.4957	71.032	1.390	40.68	78	11.5803	86.626	1.640	56.59	128	13.6649	102.220	1.890	68.28	178	15.7495	117.814
1.145	18.36	29	9.5374	71.344	1.395	41.06	79	11.6220	86.938	1.645	56.85	129	13.7066	102.532	1.895	68.48	179	15.7912	118.126
1.150	18.91	30	9.5790	71.656	1.400	41.43	80	11.6637	87.250	1.650	57.12	130	13.7483	102.844	1.900	68.68	180	15.8329	118.438
1.155	19.46	31	9.6207	71.968	1.405	41.80	81	11.7054	87.562	1.655	57.39	131	13.7900	103.156	1.905	68.88	181	15.8746	118.740
1.160	20.00	32	9.6624	72.280	1.410	42.16	82	11.7471	87.874	1.660	57.65	132	13.8317	103.468	1.910	69.08	182	15.9163	119.062
1.165	20.54	33	9.7041	72.592	1.415	42.53	83	11.7888	88.186	1.665	57.91	133	13.8734	103.780	1.915	69.28	183	15.9580	119.374
1.170	21.07	34	9.7458	72.904	1.420	42.89	84	11.8304	88.498	1.670	58.17	134	13.9150	104.092	1.920	69.48	184	15.9996	119.686
1.175	21.60	35	9.7875	73.216	1.425	43.25	85	11.8721	88.810	1.675	58.43	135	13.9567	104.404	1.925	69.68	185	16.0413	119.998
1.180	22.12	36	9.8292	73.528	1.430	43.60	86	11.9138	89.121	1.680	58.69	136	13.9984	104.715	1.930	69.87	186	16.0830	120.309
1.185	22.64	37	9.8709	73.840	1.435	43.95	87	11.9555	89.433	1.685	58.95	137	14.0401	105.027	1.935	70.06	187	16.1247	120.621
1.190	23.15	38	9.9126	74.151	1.440	44.31	88	11.9972	89.745	1.690	59.20	138	14.0818	105.339	1.940	70.26	188	16.1664	120.933
1.195	23.66	39	9.9543	74.463	1.445	44.65	89	12.0389	90.057	1.695	59.45	139	14.1235	105.651	1.945	70.45	189	16.2081	121.245
1.200	24.17	40	9.9960	74.775	1.450	45.00	90	12.0806	90.369	1.700	59.71	140	14.1652	105.963	1.950	70.64	190	16.2498	121.557
1.205	24.67	41	10.0377	75.087	1.455	45.34	91	12.1223	90.681	1.705	59.96	141	14.2069	106.275	1.955	70.83	191	16.2915	121.869
1.210	25.17	42	10.0793	75.399	1.460	45.68	92	12.1640	90.993	1.710	60.20	142	14.2486	106.587	1.960	71.02	192	16.3332	122.181
1.215	25.66	43	10.1210	75.711	1.465	46.02	93	12.2057	91.305	1.715	60.45	143	14.2903	106.899	1.965	71.21	193	13.3749	122.493
1.220	26.15	44	10.1627	76.022	1.470	46.36	94	12.2473	91.616	1.720	60.70	144	14.3320	107.210	1.970	71.40	194	16.4166	122.804
1.225	26.63	45	10.2044	76.334	1.475	46.69	95	12.2890	91.928	1.725	60.94	145	14.3737	107.522	1.975	71.58	195	16.4583	123.116
1.230	27.11	46	10.2461	76.646	1.480	47.03	96	12.3307	92.240	1.730	61.18	146	14.4153	107.834	1.980	71.77	196	16.5000	123.428
1.235	27.59	47	10.2878	76.958	1.485	47.36	97	12.3724	92.552	1.735	61.34	147	14.4570	108.146	1.985	71.95	197	16.5417	123.740
1.240	28.06	48	10.3295	77.270	1.490	47.68	98	12.4141	92.864	1.740	61.67	148	14.4987	108.458	1.990	72.14	198	16.5833	124.052
1.245	28.53	49	10.3712	77.582	1.495	48.01	99	12.4558	93.176	1.745	61.91	149	14.5404	108.770	1.995	72.32	199	16.6250	124.364
1.250	29.00	50	10.4129	77.894	1.500	48.33	100	12.4975	93.488	1.750	62.14	150	14.5821	109.082	2.000	72.50	200	16.6667	124.676

* Prepared by Lewis V. Judson, Ph.D., Chief of Length Section of National Bureau of Standards with the advice and assistance of E. L. Peffer, B.S., A.M., late Chief of Capacity and Density Section, National Bureau of Standards.

Table 17. Wire and Sheet Metal Gages*

Values in approximate decimals of an inch

As a number of gages are in use for various shapes and metals, it is **advisable to state the thickness in thousandths when specifying gage number.**

Gage number	**American** (A.W.G.) or **Brown & Sharpe** (B. & S.) (for nonferrous wire and sheet)†	**U.S. Steel Wire** (Stl. W.G.) or **Washburn & Moen** or **Roebling** or **Am. Steel & Wire Co.** [A. (steel) W.G.] (for steel wire)	**Birmingham** (B.W.G.) (for steel wire) or **Stubs Iron Wire** (for iron or brass wire)‡	**U.S. Standard** (for sheet and plate metal, wrought iron)	**Standard Birmingham** (B.G.) (for sheet and hoop metal)	**Imperial Standard Wire Gage** (S.W.G.) (British legal standard)	Gage number
0000000	......	0.4900	.....	0.500	0.6666	0.500	**0000000**
000000	......	.4615	.....	.469	.6250	.464	**000000**
00000	......	.4305	.....	.438	.5883	.432	**00000**
0000	0.460	.3938	0.454	.406	.5416	.400	**0000**
000	.410	.3625	.425	.375	.5000	.372	**000**
00	.365	.3310	.380	.344	.4452	.348	**00**
0	.325	.3065	.340	.312	.3964	.324	**0**
1	.289	.2830	.300	.281	.3532	.300	**1**
2	.258	.2625	.284	.266	.3147	.276	**2**
3	.229	.2437	.259	.250	.2804	.252	**3**
4	.204	.2253	.238	.234	.2500	.232	**4**
5	.182	.2070	.220	.219	.2225	.212	**5**
6	.162	.1920	.203	.203	.1981	.192	**6**
7	.144	.1770	.180	.188	.1764	.176	**7**
8	.128	.1620	.165	.172	.1570	.160	**8**
9	.114	.1483	.148	.156	.1398	.144	**9**
10	.102	.1350	.134	.141	.1250	.128	**10**
11	.091	.1205	.120	.125	.1113	.116	**11**
12	.081	.1055	.109	.109	.0991	.104	**12**
13	.072	.0915	.095	.094	.0882	.092	**13**
14	.064	.0800	.083	.078	.0785	.080	**14**
15	.057	.0720	.072	.070	.0699	.072	**15**
16	.051	.0625	.065	.062	.0625	.064	**16**
17	.045	.0540	.058	.056	.0556	.056	**17**
18	.040	.0475	.049	.050	.0495	.048	**18**
19	.036	.0410	.042	.0438	.0440	.040	**19**
20	.032	.0348	.035	.0375	.0392	.036	**20**
21	.0285	.0317	.032	.0344	.0349	.032	**21**
22	.0253	.0286	.028	.0312	.0313	.028	**22**
23	.0226	.0258	.025	.0281	.0278	.024	**23**
24	.0201	.0230	.022	.0250	.0248	.022	**24**
25	.0179	.0204	.020	.0219	.0220	.020	**25**

Gage number	**American** (A.W.G.) or **Brown & Sharpe** (B. & S.) (for nonferrous wire and sheet)†	**U.S. Steel Wire** (Stl. W.G.) or **Washburn & Moen** or **Roebling** or **Am. Steel & Wire Co.** [A. (steel) W.G] (for steel wire)	**Birmingham** (B.W.G.) (for steel wire) or **Stubs Iron Wire** (for iron or brass wire)‡	**U.S. Standard** (for sheet and plate metal, wrought iron)	**Standard Birmingham** (B.G.) (for sheet and hoop metal)	**Imperial Standard Wire Gage** (S.W.G.) (British legal standard)	Gage number
26	0.0159	0.0181	0.018	0.0188	0.0196	0.018	**26**
27	.0142	.0173	.016	.0172	.0175	.0164	**27**
28	.0126	.0162	.014	.0156	.0156	.0148	**28**
29	.0113	.0150	.013	.0141	.0139	.0136	**29**
30	.0100	.0140	.012	.0125	.0123	.0124	**30**
31	.0089	.0132	.010	.0109	.0110	.0116	**31**
32	.0080	.0128	.009	.0102	.0098	.0108	**32**
33	.0071	.0118	.008	.0094	.0087	.0100	**33**
34	.0063	.0104	.007	.0086	.0077	.0092	**34**
35	.0056	.0095	.005	.0078	.0069	.0084	**35**
36	.0050	.0090	.004	.0070	.0061	.0076	**36**
37	.0045	.0085	.....	.0066	.0054	.0068	**37**
38	.0040	.0080	.....	.0062	.0048	.0060	**38**
39	.0035	.0075	.....	......	.0043	.0052	**39**
40	.0031	.0070	.....	......	.0039	.0048	**40**
41	......	.0066	.....	......	.0034	.0044	**41**
42	......	.0062	.....	......	.0031	.0040	**42**
43	......	.0060	.....	......	.0027	.0036	**43**
44	......	.0058	.....	......	.0024	.0032	**44**
45	......	.0055	.....	......	.0022	.0028	**45**
46	......	.0052	.....	......	.0019	.0024	**46**
47	......	.0050	.....	......	.0017	.0020	**47**
48	......	.0048	.....	......	.0015	.0016	**48**
49	......	.0046	.....	......	.0014	.0012	**49**
50	......	.0044	.....	......	.0012	.0010	**50**

Metric wire gage is ten times the diameter in millimeters.

* Courtesy of Dr. Lewis V. Judson with I. H. Fullmer, National Bureau of Standards.

† Sometimes used for iron wire.

‡ Sometimes used for copper plate and for steel plate 12 gage and heavier and for steel tubes.

Table 18. Conversion Tables or Equivalents*

Volume and Capacity Equivalents

Cu. in.	Cu. ft.	Cu. yd.	U. S. fluid oz.	U. S. qt.		U. S. gal.	U. S. bushels	Liters	Cu. cm.	Cu. m.
				Liquid	Dry					
1	$0.0_{3}5787$	$0.0_{4}2143$	0.5541	0.01732	0.01488	$0.0_{2}4329$	$0.0_{3}4650$	0.01639	16.39	1.639×10^{-5}
1,728	1	0.03704	957.5	29.92	25.71	7.481	0.8036	28.32	28,320	0.02832
46,656	27	1	25,853	807.9	694.3	202.0	21.70	764.6	764,559	0.7646
1.805	0.001044	$0.0_{4}3868$	1	0.03125	0.02686	0.007812	$0.0_{3}8392$	0.02957	29.57	2.957×10^{-5}
57.75	0.03342	0.001238	32	1	0.8594	0.25	0.02686	0.9463	946.3	9.463×10^{-4}
67.20	0.03889	0.00144	37.24	1.164	1	0.2909	0.03125	1.101	1,101	11.01×10^{-4}
231	0.1337	0.004951	128	4	3.437	1	0.1074	3.785	3,785	3.785×10^{-3}
2,150.42	1.244	0.04609	1,192	....	32	....	1	35.24	35,238	3.524×10^{-2}
61.03	0.03531	0.001308	33.81	1.057	0.9081	0.2642	0.02838	1	1,000	1×10^{-3}

Density Equivalents

G. per cc.	Kg. per cu. m.	Lb. per cu. in.	Lb. per cu. ft.	Lb. per cu. yd.	Lb. per U. S. gal.	Tons (2000 lb.) per cu. yard	Tons (2240 lb.) per cu. yard	Tons (metric) per cu. m.
1	1,000	0.03613	62.43	1,686	8.345	0.8428	0.7525	1.0000
0.001	1	0.00003613	0.06243	1.686	0.008345	0.8428×10^{-3}	0.7525×10^{-3}	0.001
27.68	27,680	1	1,728	46,656	231	23.33	20.83	27.68
0.01602	16.02	0.0005787	1	27	0.1337	0.01350	0.012054	0.01602
0.005933	0.59327	0.00002143	0.03704	1	0.004951	0.0005	0.0004464	0.005933
0.1198	119.8	0.004329	7.481	202.0	1	0.1010	0.09017	0.1198
1.187	1,187	0.04287	74.07	2,000	9.902	1	0.8929	1.187
1.329	1,329	0.0401	82.96	2,240	11.09	1.12	1	1.329

Mass Equivalents

Kg.	Grains	Ounces		Pounds		Tons			Grams
		Troy and apoth.	Avoir.	Troy and apoth.	Avoir.	Short	Long	Metric	
1	15,432	32.15	35.27	2.6792	2.205	0.001102	$0.0_{3}9842$	0.001	1,000.0
$0.0_{4}6480$	1	0.002083	0.002286	0.000174	0.0001429	$0.0_{7}7143$	$0.0_{7}6378$	$0.0_{7}6480$	0.0648
0.03110	480	1	1.09714	0.08333	0.06857	$0.0_{4}3429$	$0.0_{4}3061$	$0.0_{4}3110$	31.1
0.02835	437.5	0.9115	1	0.07595	0.0625	$0.0_{4}3125$	$0.0_{4}2790$	$0.0_{4}2835$	28.35
0.3732	5,760	12	13.17	1	0.8229	$0.0_{3}4114$	$0.0_{3}3673$	$0.0_{3}3732$	373.2
0.4536	7,000	14.58	16	1.215	1	0.0005	$0.0_{3}4464$	$0.0_{3}4536$	453.59
907.2	140_{6}	29,167	32,000	2,431	2,000	1	0.8929	0.9072	907,184
1,016	15,680,000	32,667	35,840	2,722	2,240	1.12	1	1.016	1,016,047
1,000	15,432,356	32,151	35,274	2,679	2,205	1.102	0.9842	1	1,000,000
0.001	15.432	0.03215	0.03527	0.00268	0.00220	$0.0_{5}1102$	$0.0_{6}984$	$0.0_{5}1$	1

Heat, Energy, or Work Equivalents†

Joules = 10^{7} ergs	Kg-m.	Ft.-lb.	Kw.-hr.	Hp.-hr.	Liter-atms.	Gram-calorie†	B.t.u.	Cal.††
1	0.10197	0.7376	$0.0_{6}2773$	$0.0_{6}3725$	0.009869	$0.0_{3}2390$	$0.0_{3}9478$	0.2390
9.80665	1	7.233	$0.0_{5}2724$	$0.0_{5}3653$	0.09678	0.002344	0.009296	2.3438
1.356	0.1383	1	$0.0_{6}3766$	$0.0_{6}50505$	0.01338	$0.0_{3}324$	0.001285	0.3241
3.6×10^{6}	3.671×10^{5}	2.655×10^{6}	1	1.341	35,534.3	860.57	3,412.76	860,565
2.6845×10^{6}	2.7375×10^{5}	1.98×10^{6}	0.7455	1	26,494	641.62	2,545	641,615
101.33	10.333	74.73	$0.0_{4}2815$	$0.0_{4}3774$	1	0.02422	0.09604	24.218
4,184	426.7	3,086	0.001162	0.001558	41.29	1	3.9657	1,000
1,055	107.58	778.16	$0.0_{3}2930$	$0.0_{3}3930$	10.41	0.252	1	252
4.184	0.4267	3.086	$0.0_{5}1162$	$0.0_{5}1558$	0.04129	0.001	0.00397	1

1 therm = 100,000 B.t.u.

Pressure Equivalents†

Megabars or megadynes per sq. cm.	Kg. per cm.^{2}	Lb. per sq. in.	Short tons per sq. ft.	Atm.	Columns of mercury at 0°C.		Columns of water at 15°C.			Lb. per sq. ft.	Kg. per m.^{2}	Mm. of mercury
					M.	In.	M.	In.	Ft.			
1	1.0197	14.50	1.044	0.9869	0.7500	29.53	10.21	401.8	33.48	2,088.55	10,197	750.062
0.9807	1	14.22	1.024	0.9678	0.7355	28.96	10.01	394.05	32.84	2,048.16	10,000	735.559
0.06895	0.07031	1	0.072	0.06804	0.05171	2.036	0.7037	27.70	2.309	144	703.1	51.7147
0.9576	0.9765	13.89	1	0.9450	0.7182	28.28	9.773	384.8	32.06	2,000	9,765	718.26
1.0133	1.0332	14.696	1.058	1	0.76	29.92	10.34	407.14	33.93	2,116.2	10,333	760
1.3333	1.3596	19.34	1.392	1.316	1	39.37	13.61	535.7	44.64	2,784.50	13,596	1,000
0.03386	0.03453	0.4912	0.03536	0.03342	0.02540	1	0.3456	13.61	1.134	70.7266	345.3	25.400
0.09798	0.09991	1.421	0.1023	0.0967	0.07349	2.893	1	39.37	3.281	204.633	999.1	73.4898
0.002489	0.002538	0.0361	0.002599	0.002456	0.001867	0.07349	0.02540	1	0.08333	5.19768	25.38	1.867
0.02986	0.03045	0.4332	0.03119	0.02947	0.0224	0.8819	0.3048	12	1	62.3722	304.5	22.3997

1 g. per sq. cm. = 980.655 dynes per sq. cm. = 0.45762 poundal per sq. in.
1 dyne per sq. cm. = 0.001019716 g. per sq. cm. = 0.000466642 poundal per sq. in.
1 poundal per sq. in. = 2142.97 dynes per sq. cm. = 2.18536 g. per sq. cm. = 0.031081 lb. per sq. in.

Equivalents of Weights or Masses per Unit Lengths

For wires, pipes, rails, etc.

Grams per cm.	Kg. per km.	Kg. per m.	Grains per in.	Lb. per ft.	Lb. per yd.	Lb. per mile
1	100	0.1	39.1983	0.067197	0.201591	354.80
0.01	1	0.001	0.391983	0.00067197	0.00201591	3.54800
10	1,000	1	391.983	0.67197	2.01591	3,548.00
0.025511	2.5511	0.0025511	1	0.00171429	0.00514286	9.0514
14.8816	1,488.16	1.48815	583.333	1	3	5,280
4.96054	496.054	0.49605	194.444	0.33333	1	1,760
0.0028185	0.28185	0.00028185	0.11048	0.00018939	0.00056818	1

* Dr. Lewis V. Judson, National Bureau of Standards.
‡ Dr. Lewis V. Judson and Carl S. Cragoe, Natural Bureau of Standards.
† Gram-calorie. Thermochemical calorie is defined as 4.1840 absolute joules.
†† Kilogram-calorie.

Table 18. Conversion Tables or Equivalents—(*Continued*)

Linear Measure Equivalents

Kilometer (km.)	Meter (m.)	Centimeter (cm.)	Millimeter (mm.)	Inch (in.)	Foot (ft.)	Yard (yd.)	Micron (μ)	Millimicron (mμ)	Rods	Chains	Miles	Nautical miles
1	1000	10^5	10^6	39,370	3280.83	1093.61	10^9	10^{12}	198.838	49.710	0.62137	0.5396
0.001	1	100	1000	39.37	3.28083	1.09361	10^6	10^9	0.19884	0.049710	0.0006214	0.0005396
10^{-5}	0.01	1	10	0.3937	0.032808	0.010936	10^4	10^7	0.0019884	0.0004971	0.0000062	0.0000054
10^{-6}	0.001	0.1	1	0.03937	0.0032808	0.0010936	10^3	10^6	0.000199	0.0000497	0.00000062	0.00000054
2.54×10^{-5}	0.0254	2.540	25.40005	1	0.08333	0.02778	25,400	2.54×10^7	0.00505	0.0012626	0.0000158	0.0000137
3.048×10^{-4}	0.30480	30.480	304.801	12	1	0.33333	304,801	3.048×10^8	0.0606	0.015152	0.00018939	0.00016447
9.144×10^{-4}	0.914402	91.440	914.4	36	3	1	914,402	9.144×10^8	0.1818	0.045455	0.0005682	0.0004934
10^{-9}	10^{-6}	10^{-4}	10^{-3}	3.937×10^{-5}	3.2808×10^{-6}	1.0936×10^{-6}	1	10^3				
10^{-12}	10^{-9}	10^{-7}	10^{-6}	3.937×10^{-8}	3.2808×10^{-9}	1.0936×10^{-9}	10^{-3}	1				

Surface and Area Equivalents

Sq. m.	Sq. in.	Sq. ft.	Sq. yd.	Sq. rods	Sq. chains	Acres	Sq. miles or sections	Sq. cm.	Sq. mm.
1	1,550	10.76	1.196	0.0395	0.002471	0.0002471	0.0000003861	10,000	10^6
0.0006452	1	0.006944	0.0007716	0.00002551	0.000001594	0.0000001594	$0.0_9 2491$	6.452	645.2
0.09290	144	1	0.1111	0.003673	$0.0_3 2296$	$0.0_4 2296$	$0.0_7 3587$	929.0	92,903
0.8361	1,296	9	1	0.03306	0.002066	0.0002066	$0.0_6 3228$	8,361	836,131
25.29	39,204	272.25	30.25	1	0.0625	0.00625	$0.0_5 9766$	252,930	25.29×10^6
404.7	627,264	4,356	484	16	1	0.1	0.0001562	4,046,873	404.7×10^6
4,047	6,272,640	43,560	4,840	160	10	1	0.001562	40,468,726	$4,047 \times 10^6$
2,589,998	....	27,878,400	3,097,600	102,400	6,400	640	1	25.9×10^9	25.9×10^{11}

Power Equivalents‡

Hp. (550 standard ft.-lb. per sec.)	Metric hp.	Kw. (1000 joules per sec.)	M.kg. per sec.	Ft.-lb. per sec.	Cal.† per sec.	B.t.u. per sec.	Cal.† per sec.
1	1.0138	0.7457	76.04	550	0.1782	0.7068	178.23
0.9863	1	0.7355	75	542.5	0.1758	0.6971	175.79
1.341	1.3596	1	101.97	737.56	0.2390	0.9478	239.01
0.01315	0.01333	0.009807	1	7.233	0.00234	0.009295	2.3438
0.001818	0.001843	0.001356	0.1383	1	0.000324	0.001285	0.3240
5.611	5.689	4.1840	426.7	3086	1	3.966	1000
1.415	1.434	1.055	107.58	778.16	0.2522	1	252.16
0.005611	0.005689	0.004184	0.4267	3.086	0.001	0.00397	1

1 boiler hp. = 33,475 B.t.u. per hr.; 1 ton refrigeration = 200 B.t.u. per min.

Velocity Equivalents

Cm. per sec.	M. per sec.	M. per min.	Km. per hr.	Ft. per sec.	Ft. per min.	Miles per hr.	Knots (nautical miles per hr.)
1	0.01	0.6	0.036	0.03281	1.9685	0.02237	0.01943
100	1	60	3.6	3.281	196.85	2.237	1.943
1.667	0.01667	1	0.06	0.0547	3.281	0.03728	0.03238
27.78	0.2778	16.67	1	0.9113	54.68	0.6214	0.53960
30.48	0.3048	18.29	1.097	1	60	0.6818	0.59209
0.5080	0.005080	0.3048	0.01829	0.01667	1	0.01136	0.00987
44.70	0.4470	26.82	1.609	1.467	88	1	0.86839
51.48	0.5148	30.887	1.8532	1.6889	101.337	1.15155	1

† Thermochemical calorie is defined as 4.1840 absolute joules.
‡ Dr. Lewis V. Judson and Carl S. Cragoe, National Bureau of Standards.

Angular Measure Equivalents

Circle	Degrees	Grades	Minutes	Seconds	Radians
1	360	400	21,600	1,296,000	6.283185
0.00278	1	1.11111	60	3,600	0.017453
0.0025	0.9000	1	54	3,240	0.015708
0.0000463	0.01667	0.01852	1	60	0.0002909
7.7×10^{-7}	0.00028	0.00031	0.01667	1	0.0000048481
0.159155	57.2958	63.662	3,437.75	206,265	1

Thermal Conductivity Equivalents‡

Cal.† per sec. per sq. cm. per cm. per °C.	International watts per sq. cm. per cm. per °C.	Cal. per hr. per sq. cm. per cm. per °C.†	B.t.u. per hr. per sq. ft. per in. per °F.	B.t.u. per day per sq. ft. per in. per °F.
1	4.1833	3,600	2,901.0	69,624
0.2390	1	860.6	693.5	16,643
0.000278	0.001162	1	0.8058	19.34
0.0003447	0.001441	1.241	1	24.0
0.0000144	0.0000601	0.05171	0.04167	1

Heat Flow Equivalents‡

Cal. per sec. per sq. cm.†	Cal. per hr. per sq. cm.†	B.t.u. per hr. per sq. ft.	B.t.u. per day per sq. ft.	Watts per sq. cm.
1	3,600	13,263	318,322	4.183
0.0002778	1	3.684	88.42	0.001162
0.0000754	0.2714	1	24	0.0003154
0.00000314	0.01131	0.04167	1	0.00001314
0.2390	860.6	3,171	76,094	1

Table 18. Conversion Tables or Equivalents—(*Concluded*)

Temperature Conversion Table

General formula: °F. = (°C. × 9/5) + 32; °C. = (°F. − 32) × 5/9

C.		F.
−273.1	**−459.4**	
−268	**−450**	
−262	**−440**	
−257	**−430**	
−251	**−420**	
−246	**−410**	
−240	**−400**	
−234	**−390**	
−229	**−380**	
−223	**−370**	
−218	**−360**	
−212	**−350**	
−207	**−340**	
−201	**−330**	
−196	**−320**	
−190	**−310**	
−184	**−300**	
−179	**−290**	
−173	**−280**	
−169	**−273**	−459.4
−168	**−270**	−454
−162	**−260**	−436
−157	**−250**	−418
−151	**−240**	−400
−146	**−230**	−382
−140	**−220**	−364
−134	**−210**	−346
−129	**−200**	−328
−123	**−190**	−310
−118	**−180**	−292
−112	**−170**	−274
−107	**−160**	−256
−101	**−150**	−238
− 95.6	**−140**	−220
− 90.0	**−130**	−202
− 84.4	**−120**	−184
− 78.9	**−110**	−166
− 73.3	**−100**	−148
− 67.8	**− 90**	−130
− 62.2	**− 80**	−112
− 56.7	**− 70**	− 94
− 51.1	**− 60**	− 76
− 45.6	**− 50**	− 58
− 40.0	**− 40**	− 40
− 34.4	**− 30**	− 22
− 28.9	**− 20**	− 4
− 23.3	**− 10**	14
− 17.8	**0**	32

C.		F.
−17.8	**0**	32
−17.2	**1**	33.8
−16.7	**2**	35.6
−16.1	**3**	37.4
−15.6	**4**	39.2
−15.0	**5**	41.0
−14.4	**6**	42.8
−13.9	**7**	44.6
−13.3	**8**	46.4
−12.8	**9**	48.2
−12.2	**10**	50.0
−11.7	**11**	51.8
−11.1	**12**	53.6
−10.6	**13**	55.4
−10.0	**14**	57.2
− 9.44	**15**	59.0
− 8.89	**16**	60.8
− 8.33	**17**	62.6
− 7.78	**18**	64.4
− 7.22	**19**	66.2
− 6.67	**20**	68.0
− 6.11	**21**	69.8
− 5.56	**22**	71.6
− 5.00	**23**	73.4
− 4.44	**24**	75.2
− 3.89	**25**	77.0
− 3.33	**26**	78.8
− 2.78	**27**	80.6
− 2.22	**28**	82.4
− 1.67	**29**	84.2
− 1.11	**30**	86.0
− 0.56	**31**	87.8
0	**32**	89.6
0.56	**33**	91.4
1.11	**34**	93.2
1.67	**35**	95.0
2.22	**36**	96.8
2.78	**37**	98.6
3.33	**38**	100.4
3.89	**39**	102.2
4.44	**40**	104.0
5.00	**41**	105.8
5.56	**42**	107.6
6.11	**43**	109.4
6.67	**44**	111.2
7.22	**45**	113.0
7.78	**46**	114.8
8.33	**47**	116.6
8.89	**48**	118.4
9.44	**49**	120.2

C.		F.
10.0	**50**	122.0
10.6	**51**	123.8
11.1	**52**	125.6
11.7	**53**	127.4
12.2	**54**	129.2
12.8	**55**	131.0
13.3	**56**	132.8
13.9	**57**	134.6
14.4	**58**	136.4
15.0	**59**	138.2
15.6	**60**	140.0
16.1	**61**	141.8
16.7	**62**	143.6
17.2	**63**	145.4
17.8	**64**	147.2
18.3	**65**	149.0
18.9	**66**	150.8
19.4	**67**	152.6
20.0	**68**	154.4
20.6	**69**	156.2
21.1	**70**	158.0
21.7	**71**	159.8
22.2	**72**	161.6
22.8	**73**	163.4
23.3	**74**	165.2
23.9	**75**	167.0
24.4	**76**	168.8
25.0	**77**	170.6
25.6	**78**	172.4
26.1	**79**	174.2
26.7	**80**	176.0
27.2	**81**	177.8
27.8	**82**	179.6
28.3	**83**	181.4
28.9	**84**	183.2
29.4	**85**	185.0
30.0	**86**	186.8
30.6	**87**	188.6
31.1	**88**	190.4
31.7	**89**	192.2
32.2	**90**	194.0
32.8	**91**	195.8
33.3	**92**	197.6
33.9	**93**	199.4
34.4	**94**	201.2
35.0	**95**	203.0
35.6	**96**	204.8
36.1	**97**	206.6
36.7	**98**	208.4
37.2	**99**	210.2
37.8	**100**	212.0

C.		F.
38	**100**	212
43	**109**	230
49	**120**	248
54	**130**	266
60	**140**	284
66	**150**	302
71	**160**	320
77	**170**	338
82	**180**	356
88	**190**	374
93	**200**	392
99	**210**	410
100	**212**	413
104	**220**	428
110	**230**	446
116	**240**	464
121	**250**	482
127	**260**	500
132	**270**	518
138	**280**	536
143	**290**	554
149	**300**	572
154	**310**	590
160	**320**	608
166	**330**	626
171	**340**	644
177	**350**	662
182	**360**	680
188	**370**	698
193	**380**	716
199	**390**	734
204	**400**	752
210	**410**	770
216	**420**	788
221	**430**	806
227	**440**	824
232	**450**	842
238	**460**	860
243	**470**	878
249	**480**	896
254	**490**	914

C.		F.
260	**500**	932
266	**510**	950
271	**520**	968
277	**530**	986
282	**540**	1004
288	**550**	1022
293	**560**	1040
299	**570**	1058
304	**580**	1076
310	**590**	1094
316	**600**	1112
321	**610**	1130
327	**620**	1148
332	**630**	1166
338	**640**	1184
343	**650**	1202
349	**660**	1220
354	**670**	1238
360	**680**	1256
366	**690**	1274
371	**700**	1292
377	**710**	1310
382	**720**	1328
388	**730**	1346
393	**740**	1364
399	**750**	1382
404	**760**	1400
410	**770**	1418
416	**780**	1436
421	**790**	1454
427	**800**	1472
432	**810**	1490
438	**820**	1508
443	**830**	1526
449	**840**	1544
454	**850**	1562
460	**860**	1580
466	**870**	1598
471	**880**	1616
477	**890**	1634
482	**900**	1652
488	**910**	1670
493	**920**	1688
499	**930**	1706
504	**940**	1724
510	**950**	1742
516	**960**	1760
521	**970**	1778
527	**980**	1796
532	**990**	1814

C.		F.
538	**1000**	1832
543	**1010**	1850
549	**1020**	1868
554	**1030**	1886
560	**1040**	1904
566	**1050**	1922
571	**1060**	1940
577	**1070**	1958
582	**1080**	1976
588	**1090**	1994
593	**1100**	2012
599	**1110**	2030
604	**1120**	2048
610	**1130**	2066
616	**1140**	2084
621	**1150**	2102
627	**1160**	2120
632	**1170**	2138
638	**1180**	2156
643	**1190**	2174
649	**1200**	2192
654	**1210**	2210
660	**1220**	2228
666	**1230**	2246
671	**1240**	2264
677	**1250**	2282
682	**1260**	2300
688	**1270**	2318
693	**1280**	2336
699	**1290**	2354
704	**1300**	2372
710	**1310**	2390
716	**1320**	2408
721	**1330**	2426
727	**1340**	2444
732	**1350**	2462
738	**1360**	2480
743	**1370**	2498
749	**1380**	2516
754	**1390**	2534
760	**1400**	2552
766	**1410**	2570
771	**1420**	2588
777	**1430**	2606
782	**1440**	2624
788	**1450**	2642
793	**1460**	2660
799	**1470**	2678
804	**1480**	2696
810	**1490**	2714

C.		F.
816	**1500**	2732
821	**1510**	2750
827	**1520**	2768
832	**1530**	2786
838	**1540**	2804
843	**1550**	2822
849	**1560**	2840
854	**1570**	2858
860	**1580**	2876
866	**1590**	2894
871	**1600**	2912
877	**1610**	2930
882	**1620**	2948
888	**1630**	2966
893	**1640**	2984
899	**1650**	3002
904	**1660**	3020
910	**1670**	3038
916	**1680**	3056
921	**1690**	3074
927	**1700**	3092
932	**1710**	3110
938	**1720**	3128
943	**1730**	3146
949	**1740**	3164
954	**1750**	3182
960	**1760**	3200
966	**1770**	3218
971	**1780**	3236
977	**1790**	3254
982	**1800**	3272
988	**1810**	3290
993	**1820**	3308
999	**1830**	3326
1004	**1840**	3344
1010	**1850**	3362
1016	**1860**	3380
1021	**1870**	3398
1027	**1880**	3416
1032	**1890**	3434
1038	**1900**	3452
1043	**1910**	3470
1049	**1920**	3488
1054	**1930**	3506
1060	**1940**	3524
1066	**1950**	3542
1071	**1960**	3560
1077	**1970**	3578
1082	**1980**	3596
1088	**1990**	3614
1093	**2000**	3632

C.		F.
1093	**2000**	3632
1099	**2010**	3650
1104	**2020**	3668
1110	**2030**	3686
1116	**2040**	3704
1121	**2050**	3722
1127	**2060**	3740
1132	**2070**	3758
1138	**2080**	3776
1143	**2090**	3794
1149	**2100**	3812
1154	**2110**	3830
1160	**2120**	3848
1166	**2130**	3866
1171	**2140**	3884
1177	**2150**	3902
1182	**2160**	3920
1188	**2170**	3938
1193	**2180**	3956
1199	**2190**	3974
1204	**2200**	3992
1210	**2210**	4010
1216	**2220**	4028
1221	**2230**	4046
1227	**2240**	4064
1232	**2250**	4082
1238	**2260**	4100
1243	**2270**	4118
1249	**2280**	4136
1254	**2290**	4154
1260	**2300**	4172
1266	**2310**	4190
1271	**2320**	4208
1277	**2330**	4226
1282	**2340**	4244
1288	**2350**	4262
1293	**2360**	4280
1299	**2370**	4298
1304	**2380**	4316
1310	**2390**	4334
1316	**2400**	4352
1321	**2410**	4370
1327	**2420**	4388
1332	**2430**	4406
1338	**2440**	4424
1343	**2450**	4442
1349	**2460**	4460
1354	**2470**	4478
1360	**2480**	4496
1366	**2490**	4514

C.		F.
1371	**2500**	4532
1377	**2510**	4550
1382	**2520**	4568
1388	**2530**	4586
1393	**2540**	4604
1399	**2550**	4622
1404	**2560**	4640
1410	**2570**	4658
1416	**2580**	4676
1421	**2590**	4694
1427	**2600**	4712
1432	**2610**	4730
1438	**2620**	4748
1443	**2630**	4766
1449	**2640**	4784
1454	**2650**	4802
1460	**2660**	4820
1466	**2670**	4838
1471	**2680**	4856
1477	**2690**	4874
1482	**2700**	4892
1488	**2710**	4910
1493	**2720**	4928
1499	**2730**	4946
1504	**2740**	4964
1510	**2750**	4982
1516	**2760**	5000
1521	**2770**	5018
1527	**2780**	5036
1532	**2790**	5054
1538	**2800**	5072
1543	**2810**	5090
1549	**2820**	5108
1554	**2830**	5126
1560	**2840**	5144
1566	**2850**	5162
1571	**2860**	5180
1577	**2870**	5198
1582	**2880**	5216
1588	**2890**	5234
1593	**2900**	5252
1599	**2910**	5270
1604	**2920**	5288
1610	**2930**	5306
1616	**2940**	5324
1621	**2950**	5342
1627	**2960**	5360
1632	**2970**	5378
1638	**2980**	5396
1643	**2990**	5414
1649	**3000**	5432

Note.—The numbers in bold-face type refer to the temperature (in either centigrade or Fahrenheit degrees) which it is desired to convert into the other scale. If converting from Fahrenheit degrees to centigrade degrees the equivalent temperature is in the left column, while if converting from degrees centigrade to degrees Fahrenheit, the equivalent temperature is in the column on the right. This table, made by Albert Sauveur, is published by permission of Mrs. Albert Sauveur.

Interpolation Factors

C.		F.	C.		F.
0.56	**1**	1.8	3.33	**6**	10.8
1.11	**2**	3.6	3.89	**7**	12.6
1.67	**3**	5.4	4.44	**8**	14.4
2.22	**4**	7.2	5.00	**9**	16.2
2.78	**5**	9.0	5.56	**10**	18.0

WEIGHTS AND MEASURES

Table 19. Weights and Measures of Various Systems*

United States Customary System

Linear Measure

12 inches (in.) or (″)	= 1 foot (ft.) or (′)
3 feet	= 1 yard (yd.)
16.5 feet } 5.5 yards }	= 1 rod (rd.)
5280 feet } 320 rods }	= 1 mile (mi.)
1 mil	= 0.001 inch

Nautical:

6080.2 feet	= 1 nautical mile
6 feet	= 1 fathom
120 fathoms	= 1 cable length
1 knot	= 1 nautical mile per hour
60 knots	= 1° (measured at equator)

Square Measure

144 sq. inches (sq. in.) or (in.^{2}) or (□″)	= 1 sq. foot (ft.^{2}) or (□′)
9 sq. feet (ft.^{2}) (□′)	= 1 sq. yard (yd.^{2})
30.25 sq. yards	= 1 sq. rod, pole, or perch
160 sq. rods = {10 sq. chains / 43,560 sq. ft.}	= 1 acre
640 acres = 1 sq. mile	= 1 section
1 circular inch (area of circle of 1 inch diameter)	= 0.7854 sq. inch
1 sq. inch	= 1.2732 circular inch
1 circular mil	= area of circle of 0.001 inch diameter
1,000,000 circular mils	= 1 circular inch

Circular Measure

60 seconds (″) (sec.)	= 1 minute (min.) or (′)
60 minutes (′)	= 1 degree (°)
90 degrees (°)	= 1 quadrant
360 degrees (°)	= 1 circumference
57.29578 degrees	{ = 1 radian (rad.) / = 57° 17′ 44.81″ }

Volume Measure

Solid:

1728 cubic in. (cu. in.) (in.^{3})	= 1 cubic foot (cu. ft.) (ft.^{3})
27 cu. ft.	= 1 cubic yard (cu. yd.)

Dry Measure:

2 pints	= 1 quart
8 quarts	= 1 peck
4 pecks	= 1 bushel
1 United States Winchester bushel	= 2150.42 cubic inches

Liquid:

4 gills	= 1 pint (pt.)
2 pints	= 1 quart (qt.)
4 quarts	= 1 gallon (gal.)
7.4805 gallons	= 1 cubic foot

Apothecaries' Liquid:

60 minims (min. or ♏)	= 1 fluid dram or drachm
8 drams (ʒ)	= 1 fluid ounce
16 ounces (oz. ℥)	= 1 pint

Water Measure:

1 miner's inch = amount of water flowing through an orifice of 1 sq. in. cross section under a head varying from 4 to 6.5 in. (as fixed by state law). Units now most generally used are: 1 cu. ft. per sec. (1 ft.^{3}/sec.) or 1 gal. per sec. (1 gal./sec.).

---

* By Dr. Lewis V. Judson, National Bureau of Standards.

Avoirdupois Weight

16 drams =	437.5 grains	= 1 ounce (oz.)
16 ounces =	7000 grains	= 1 pound (lb.)
100 pounds =	1 hundredweight (cwt.)	
2000 pounds =	1 short ton; 2240 pounds	= 1 long ton

Troy Weight

24 grains	= 1 pennyweight (dwt.)
20 pennyweights	= 1 ounce (oz.)
12 ounces	= 1 pound (lb.)

Apothecaries' Weight

20 grains (gr.)	= 1 scruple (℈)
3 scruples	= 1 dram (ʒ)
8 drams	= 1 ounce (℥)
12 ounces	= 1 pound (lb.)

Board Measure (B.M.)

1 board foot = product of 1 foot length, 1 foot breadth, and 1 inch thickness.

Metric System

Linear Measure

1 micromicron ($\mu\mu$)	= 0.000001 micron
1 Ångström unit (Å.)	= 0.0001 micron
1 millimicron (mμ)	= 0.001 micron (μ)
1 micron (μ)	= 0.001 millimeter (mm.)
10 millimeters (mm.)	= 1 centimeter (cm.)
10 centimeters (cm.)	= 1 decimeter (dm.)
10 decimeters (dm.) } 100 centimeters (cm.) }	= 1 meter (m.) = 39.37 inches
1 dekameter (dkm.)	= 10 meters (m.)
1 hectometer (hm.)	= 100 meters (m.)
1 kilometer (km.)	= 1,000 meters (m.)
1 myriameter	= 10,000 meters (m.)
1 megameter	= 1,000,000 meters (m.)

Square Measure

1 sq. millimeter (mm.^{2})	= 0.01 sq. centimeter (sq. cm.)
1 sq. centimeter (cm.^{2})	= 0.01 sq. decimeter (sq. dm.)
1 sq. decimeter (dm.^{2})	= 0.01 sq. meter (sq. m.)
1 sq. meter (centiare) or (m.^{2})	= 0.01 sq. dekameter (sq. dkm.)
1 sq. dekameter (are)	= 0.01 sq. hectometer (sq. hm.)
1 hectare (ha.)	= 10,000 sq. meters
1 sq. kilometer (km.^{2})	= 100 sq. hectometers

Volume Measure

1 cubic millimeter	= 10^{-6} liter = 0.001 cu. cm.
1 cubic centimeter	= 10^{-3} liter = 0.001 cu. dm.
1 cubic decimeter	= 1 liter† = 0.001 cu. m.
1 decistere	= 0.1 cubic meter
1 stere	= 1 cubic meter
1 dekastere	= 10 cubic meters
1 microliter(μl. or λ)	= 10^{-6} liter = 0.000001 liter
1 milliliter(ml.)	= 10^{-3} liter = 0.001 liter
1 centiliter(cl.)	= 10^{-2} liter = 0.01 liter
1 deciliter(dl.)	= 10^{-1} liter = 0.1 liter
1 dekaliter(dkl.)	= 10 liters
1 hectoliter(hl.)	= 10^2 liters = 100 liters

Mass Measure

1 microgram(μg. or γ)	= 10^{-6} gram
1 milligram(mg.)	= 10^{-3} gram
1 centigram(cg.)	= 10^{-2} gram
1 decigram(dg.)	= 10^{-1} gram = 0.1 gram
15.432 grains(gr.)	= 1 gram
1 dekagram(dkg.)	= 10 grams
1 hectogram(hg.)	= 10^2 grams = 100 grams
1 kilogram (kg.)	= 10^3 grams = 1,000 grams
1 metric ton	= 10^6 grams = 1,000,000 grams
1 metric carat	= 200 milligrams

---

† Accurately: 1 liter = 1.000028 cubic decimeters.

Table 20. Weights and Measures of Different Countries

1. Metric system (compulsory in all civilized countries except the British Commonwealth of Nations and the United States; optional in these two countries. Old non-metric units still in use in some countries. Compulsory use of the metric system in Japan effective Dec. 31, 1958):

1 meter (m.) = 443.284 Paris lignes = 3.280833 United States feet = 3.18620 Prussian feet
1 kilometer (km.) = 10 hectometers (hm.) = 0.6214 United States mile = 0.1328 Prussian mile = 0.9374 Russian verst = 0.5390 nautical mile
1 hectare (ha.) = 100 ares (a.) = 10,000 sq. m. = 0.01 sq. km. = 2.471 United States acres
1 liter (l.) = 0.001 cu. m. = 1000 ml. = 0.2642 United States gallons
1 hectoliter (hl.) = 0.1 cu. m. = 100 l. = 26.42 United States gallons
1 kilogram (kg.) = 1000 g. = weight of water at +4°C. = 2 German and Swiss pounds (zollpfund) = 2.2046 pounds avoirdupois = 1.7857 Austrian pounds = 2.3525 Swedish pounds = 2.4419 Russian pounds
1 gram (g.) = 15.432 grains (English and United States)
1 quintal = 100 kg. = 220.46 lb. avoirdupois = 1 cwt. 3 qrs. 0.84 lb.
1 metric ton = 1000 kg. = 0.9842 English ton and 0.9842 United States long ton = 1.1023 United States short tons (at 2000 lb.)

2. Great Britain and Ireland:

1 inch = 25.39998 mm.
1 foot = 0.3047997 m.
1 yard = 3 feet = 0.9143992 m.
1 fathom = 2 yards = 1.829 m.
1 rod (pole, perch) = 5½ yards = 5.0292 m.
1 chain = 22 yards 80 chains = 1 mile
1 statute mile = 8 furlongs = 320 poles = 1760 yards = 5280 feet = 1.6093 kilometers (km.)
1 nautical mile = 6080 feet = 1853.2 m.
1 acre = 160 square rods = 0.40468 ha. = 34,560 square feet = 4046.8 square meters
1 square mile = 640 acres = 258.998 ha.
1 gallon = 4 quarts = 8 pints = 277.42 cubic inches = 4.536 liters = 10 lb. water = 70,000 grains water = 4.535924 kg. water = 1.20094 United States gallons
1 cubic foot = 1728 cubic inches = 28.3162 l.
1 cubic inch = 16.3865 ml.
1 quarter = 8 bushels = 32 pecks = 64 gallons = 2.909 hl.
1 bushel = 8 gallons = 1.03205 United States bushels
1 fluid ounce = 1/20 pint = 28.412 ml.
1 pound avoirdupois (lb.) = 16 ounces (oz.) = 7000 grains = 0.4535924 kg.
1 ounce avoirdupois = 437½ grains = 28.35 g.
1 hundredweight (cwt.) = 4 quarters (qr.) = 8 stones = 112 lb. = 50.8024 kg.
1 ton = 20 cwt. = 2240 lb. = 1016.047 kg.

Apothecaries' Weight:

1 ounce apothecaries' = 8 drams = 24 scruples = 480 grains = 31.1035 g.
1 ounce troy (for gold and precious stones) = 20 pennyweight (dwt.) = 480 grains = 31.1035 g.
1 pennyweight (dwt.) = 1.5552 g.
1 grain (common to avoirdupois, apothecaries', and troy weight) = 0.06479892 g.

3. Russia: Metric measure and weight compulsory. The following units of the old system are sometimes still employed:

1 foute (foot) = 1 English or United States foot
1 sashen = 7 feet = 3 arshin = 12 tchetvert = 48 vershok = 2.1336 m.
1 verst = 500 sashen = 1066.80 m.
1 dessatine = 2400 square sashens = 10.925 sq. m.
1 vedro = 10 krushky (stoof) = 12.299 l. = 3.249 United States gallons
1 tchetvert = 2 osmini = 4 payok = 8 tchetverik = 209.9 l.
1 funt (pound) = 32 loth = 96 solotnik = 9216 doli = 0.9028 United States lb. = 409.512 g.
1 berkovetz = 10 poods = 400 lb. = 163.80 kg.
1 pood = 40 pounds = 36.113 United States lb. = 16.3805 kg.

4. Sweden: Metric measure and weight compulsory. The following units of the old system are sometimes still employed.

1 fot (foot) = 10 turn (inches) = 100 lines = 0.97408 United States foot = 0.2969 m.
1 famn (fathom) = 3 alnar (ells) = 6 feet = 5.8445 United States feet = 1.7814 m.
1 mil (mile) = 6000 fathoms = 6.6415 United States statute miles = 10.6884 km.
1 kanna = 0.69135 United States gallon = 2.617 l.
1 skålpund = 100 ort = 1000 korn = 0.9371 United States lb. = 425.076 g.
1 centner = 100 skålpund
1 skippund = 20 liespund = 400 skålpund

5. Switzerland: Metric measure and weight compulsory. The following units of the old system are sometimes still employed:

1 fuss = 0.3000 m. = 0.9842 United States foot
1 juchart = 36 are = 0.88957 United States acre
1 maass = 1.5 l.
1 saum = 100 maass = 150 l.

6. Central and South America (Bolivia, Chile, Costa Rica, Cuba, Dominican Republic, Ecuador, Guatemala, Honduras, Nicaragua, Peru, Venezuela): 1 quintal = 46.0093 kg.

Country	Quintal	Kilograms
Argentina	1	45.94
Brazil	1	58.752
Colombia	1	50.00
Mexico	1	46.0246
Paraguay	1	45.94
San Salvador	1	45.94
Uruguay	1	45.94

Square Feet, Square Meter:

1 square meter (sq. m.) = 10.764 square feet (United States, English, and Russian) = 10.008 square feet (Austrian) = 10.152 square feet (Prussian) = 11.344 square feet (Swedish)
1 square foot (United States, English, and Russian) = 0.09290 square meter

Cubic Feet, Cubic Meter:

1 cubic meter (cu. m.) = 35.314 cubic feet (English, United States, and Russian)
1 cubic meter (cu. m.) = 31.661 cubic feet (Austrian)
1 cubic meter (cu. m.) = 32.346 cubic feet (Prussian)
1 cubic meter (cu. m.) = 38.209 cubic feet (Swedish)
1 cubic foot (United States, English, and Russian) = 0.028315 cubic meter
1 kilogram per running meter = 0.6720 United States and English pound per linear foot = 0.6277 zollpfund per Prussian foot
1 United States and English pound per 1 United States and English foot = 1.4882 kg. per running meter
1 kilogram per square centimeter (for steam pressure) = 14.223 United States and English pounds per square inch = 13.681 zollpfund per Prussian square inch = 13.878 zollpfund per Austrian square inch

Table 20a. Horsepower (Per Second)

Kg.-cm.	Austria,* foot-pounds	Prussia, foot-pounds	England,† foot-pounds	Sweden, foot-pounds	Russia, foot-pounds
75	474.53	477.93	542.47	594.27	600.87
76.04	481.11	484.56	550	602.51	609.20

* Zollpfund. † Also United States.

75 kilogram-meters taken as unit,
550 English foot-pounds taken as unit
= 1 horsepower per second; or
33,000 foot-pounds per minute.

TIME

BY PAUL SOLLENBERGER

21. Time and Its Measurement. The standard by which time is usually measured is the period of rotation of the earth. That period is measured by observing the times of passage of celestial objects across a line passing north and south through the zenith. That line is called the **meridian.**

Sidereal time is based on the period of the earth's rotation with respect to the vernal equinox, which is the point among the stars where the sun appears to be at the beginning of spring. "Sidereal" means pertaining to the stars. Since the vernal equinox moves slowly among the stars, it would be more appropriate to designate this kind of time as **equinoctial** time.

Solar time is based on the period of the earth's rotation with respect to the sun. Since the sun appears to move westward among the stars from day to day, the solar day is longer than the sidereal day. The rate of the sun's apparent motion varies from day to day. This causes the **apparent** solar day to be of variable length. In order to overcome the handicaps which would result from a variable time standard, **mean solar** time has been devised. This time is based on a fictitious sun which moves uniformly along the equator. It is sometimes ahead of, and sometimes behind, the real sun, but its average position is just the same as that of the real sun. The ratio between the length of a sidereal and mean solar day is 1 to 1.002738.

The time at any place, as determined by transits of the fictitious sun across the meridian of that place, is called **local mean solar time,** and the difference between the local times of any two points is equal to their difference in longitude. Longitudes are frequently expressed in time instead of by the ordinary angular units. In order to avoid the use of different times at all different longitudes on the earth, 24 standard time zones have been established. The time in each of these zones differs from the time in the adjacent zones by exactly 1 hr. On land the zone boundaries are usually irregular in shape, since they have been made to conform with political or geographical boundaries, or they have been placed so as to cause minimum interference to commerce. There are still a number of countries where standard time is not yet used. Some of these use time differing by ½ hr. from the time of the nearest zone, while other places use the local times of important cities as their standard, or have no standard at all.

The tropical year is the period during which the sun appears to move among the stars from the vernal equinox, around the sky, and back again to the vernal equinox. The sidereal year is the period of the same motion with respect to the stars. The sidereal day corresponds not to the sidereal year but to the tropical year, both the sidereal day and the tropical year being equinoctial units. The anomalistic year is the period between successive passages of the earth through **perihelion,** which is the point of closest approach of the earth to the sun. The lengths of the three kinds of years, in mean solar days, are as follows:

Tropical	365.24220
Sidereal	365.25636
Anomalistic	365.25964

22. Standard Time Belts of the United States (see figure below). The United States is divided into four times zones: Eastern, Central, Mountain, and Western. The time of the Eastern zone is that of the 75° meridian or 5 hr. slower than Greenwich time; that of the Central zone is that of the 90° meridian; that of the Mountain zone is that of the 105° meridian; and that of the Western zone is that of the 120° meridian. Alaska has four time zones, using the time of the 120°, 135°, 150°, and 165° meridians. The time in one zone differs by 1 hr. from that of the adjacent zone. The time on the line dividing these zones is that of the zone most easterly. See Time Zone Chart, Fig. 1.

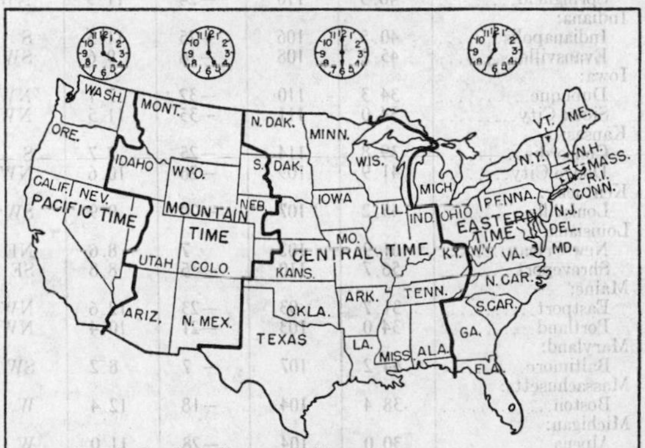

Drawn from the Time Zone Chart supplied by The National Bureau of Standards

FIG. 1.

GRAVITY

Table 23a. Theoretical Acceleration of Gravity. International Formula*,†

Latitude, $\varphi°$	$g_{(\varphi,0)}$, gal. = cm./sec.^{2}	$g_{(\varphi,0)}$, ft./sec.^{2}	$\frac{g_{(\varphi,0)}}{g_{(45°,0)}}$
0	978.0	32.088	0.9974
10	978.2	32.093	0.9975
20	978.7	32.108	0.9980
30	979.3	32.130	0.9987
40	980.2	32.158	0.9995
45	980.6	32.173	1.0000
50	981.1	32.188	1.0005
60	981.9	32.215	1.0013
70	982.6	32.238	1.0020
80	983.1	32.253	1.0025
90	983.2	32.258	1.0026

NOTE: Actual values for stations within the United States may differ from the values calculated from this table with free air reduction by ±0.3 gal. and for some foreign stations by as much as ±0.6 gal. If more accurate values are needed, inquire of the U.S. Coast and Geodetic Survey, Washington, 25, D.C.

* Adopted 1930 by the International Geodetic Association. Rounded to 4 and 5 significant figures. Free air reduction for altitude above sea level —0.309 gal. for each 1000 m.; and −0.00309 ft./sec.^{2} for each 1000 ft.

† Courtesy of L. B. Tuckerman, Assistant Chief, Mechanics Division, National Bureau of Standards.

Table 23b. Normal Gravity*

The conventional "normal value" of gravity to which all measurements involving local accelerations of gravity should be reduced is g_0 = 980.665 gal.† (cm./sec.^{2}) or 32.1740 ft./sec.^{2}. This was adopted by the International Bureau of Weights and Measures (1892) and approved as a *conventional constant* by the Fifth International Conference on Weights and Measures (1913). It is frequently called "standard gravity." It should not be called "the sea level value at 45° latitude," since it no longer has any reference to any particular location on the earth.

* Courtesy of L. B. Tuckerman, Assistant Chief, Mechanics Division, National Bureau of Standards.

† Contraction of Galileo.

MISCELLANEOUS

Table 24. Weather Data—U.S. Weather Bureau*

State and city	Average temperature, Oct. 1 to May 1	Highest temperature ever occurred, °F.	Lowest temperature ever occurred, °F.	Average wind velocity December, January, February, m.p.h.	Direction of prevailing wind December, January, February
Alabama:					
Mobile	59.2	103	−1	9.9	N
Birmingham	54.1	107	−10	8.0	N
Arizona:					
Phoenix	60.0	118	16	5.4	SW
Flagstaff	34.9	99	−30	6.7	SW
Arkansas:					
Fort Smith	50.9	113	−15	8.3	E
Little Rock	51.8	110	−12	8.3	NW
California:					
San Francisco	54.6	101	27	7.5	N
Los Angeles	59.3	109	28	6.4	NE
Colorado:					
Denver	39.3	105	−29	7.5	
Grand Junction	39.7	105	−21	4.4	S
Connecticut:					SE
New Haven	38.6	101	−15	9.4	
District of Columbia:					N
Washington	44.0	106	−15	7.7	NW
Florida:					
Jacksonville	62.2	104	10	9.0	NE
Georgia:					
Atlanta	53.3	103	−8	11.7	NW
Savannah	58.8	105	8	9.5	NW
Idaho:					
Lewiston	42.6	...	−23	4.7	E
Pocatello	36.5	105	−28	8.9	SW
Illinois:					
Chicago	36.9	105	−23	12.0	SW
Springfield	40.5	110	−24	11.9	NW
Indiana:					
Indianapolis	40.5	106	−25	11.3	S
Evansville	45.2	108	−16	9.6	SW
Iowa:					
Dubuque	34.3	110	−32	7.1	NW
Sioux City	34.0	111	−35	11.5	NW
Kansas:					
Concordia	39.8	114	−25	7.7	S
Dodge City	41.9	109	−26	10.6	NW
Kentucky:					
Louisville	45.2	107	−20	9.9	SW
Louisiana:					
New Orleans	61.9	102	7	8.6	NE
Shreveport	56.7	110	−5	8.8	SE
Maine:					
Eastport	31.7	93	−23	12.6	NW
Portland	34.0	103	−21	10.4	NW
Maryland:					
Baltimore	44.2	107	−7	8.2	SW
Massachusetts:					
Boston	38.4	104	−18	12.4	W
Michigan:					
Alpena	30.0	104	−28	11.0	W
Detroit	35.9	105	−24	12.0	SW
Marquette	31.2	108	−27	10.6	NW
Minnesota:					
Duluth	24.3	106	−41	13.4	NW
Minneapolis	29.9	108	−34	11.3	NW
Mississippi:					
Vicksburg	56.9	104	−1	8.3	SE
Missouri:					
St. Joseph	41.2	110	−24	9.3	NW
Springfield	44.5	106	−29	10.9	SE
Montana:					
Billings	35.2	112	−49	12.4	SW
Havre	28.5	112	−57	9.4	SW
Nebraska:					
Lincoln	37.9	115	−29	10.6	S
North Platte	36.3	109	−35	7.9	W
Nevada:					
Tonopah	39.4	98	−15	9.9	SE
Winnemucca	38.0	108	−36	8.1	NE
New Hampshire:					
Concord	31.2	102	−37	6.2	NW
New Jersey:					
Atlantic City	42.2	104	−9	15.8	NW
New Mexico:					
Santa Fe	38.5	97	−13	7.0	N
New York:					
Albany	35.3	104	−24	10.5	W
Buffalo	35.0	97	−20	17.1	W
New York	40.9	102	−14	16.9	NW
North Carolina:					
Raleigh	50.2	104	−2	7.9	SW
Wilmington	54.5	103	5	9.4	N
North Dakota:					
Bismarck	25.4	114	−45	9.1	NW
Devil's Lake	21.9	112	−46	10.1	NW
Ohio:					
Cleveland	36.1	103	−17	14.7	SW
Columbus	40.3	106	−20	11.6	SW
Oklahoma:					
Oklahoma City	48.6	113	−17	11.5	S
Oregon:					
Baker	35.6	104	−25	6.6	SE
Portland	46.1	107	−2	7.3	S
Pennsylvania:					
Philadelphia	42.7	106	−11	11.0	NW
Pittsburgh	40.7	103	−20	11.7	W
Rhode Island:					
Providence	38.3	100	−17	12.2	NW
South Carolina:					
Charleston	57.5	104	7	10.5	N
Columbia	54.3	106	−2	8.1	NE
South Dakota:					
Huron	29.3	111	−43	10.7	NW
Rapid City	34.0	106	−34	7.9	W
Tennessee:					
Knoxville	48.1	104	−16	7.2	SW
Memphis	51.3	106	−9	9.3	NW
Texas:					
El Paso	53.8	106	−6	9.0	NW
Fort Worth	55.5	112	−8	10.4	NW
San Antonio	60.9	107	4	8.3	NE
Utah:					
Modena	36.7	101	−32	8.4	W
Salt Lake City	32.7	105	−30	7.9	SE
Vermont:					
Burlington	31.2	101	−29	11.7	S
Virginia:					
Norfolk	49.6	105	2	12.1	N
Lynchburg	46.6	106	−7	8.1	NW
Richmond	47.5	107	−3	8.1	SW
Washington:					
Seattle	45.8	100	3	9.7	SE
Spokane	37.9	108	−30	6.2	SW
West Virginia:					
Elkins	39.7	99	−28	6.1	W
Parkersburg	42.7	106	−27	7.2	SW
Wisconsin:					
Green Bay	30.6	104	−36	10.5	SW
LaCrosse	32.0	108	−43	2.2	NW
Milwaukee	33.7	105	−25	12.1	W
Wyoming:					
Sheridan	31.0	106	−41	5.4	NW
Lander	30.3	102	−40	3.9	SW

* By F. W. Reichelderfer, Chief Weather Bureau, U.S. Department of Commerce.

Table 25. Greek Alphabet

Alpha	= Α, α	= A, a	Nu	= Ν, ν	= N, n
Beta	= Β, β	= B, b	Xi	= Ξ, ξ	= X, x
Gamma	= Γ, γ	= G, g	Omicron	= Ο, ο	= O, o
Delta	= Δ, δ	= D, d	Pi	= Π, π	= P, p
Epsilon	= Ε, ε	= E, e	Rho	= Ρ, ρ	= R, r
Zeta	= Ζ, ζ	= Z, z	Sigma	= Σ, σ	= S, s
Eta	= Η, η	= E, e	Tau	= Τ, τ	= T, t
Theta	= Θ, θ	= Th, th	Upsilon	= Υ, υ	= U, u
Iota	= Ι, ι	= I, i	Phi	= Φ, φ	= Ph, ph
Kappa	= Κ, κ	= K, k	Chi	= Χ, χ	= Ch, ch
Lambda	= Λ, λ	= L, l	Psi	= Ψ, ψ	= Ps, ps
Mu	= Μ, μ	= M, m	Omega	= Ω, ω	= O, o

Table 26. Numerical Constants

Constant	Value	Constant	Value	Constant	Value
π	3.141593	$1/\pi^3$	0.032252	$1/e^2$	1.135335
$1/\pi$	0.318310	$\sqrt[3]{\pi}$	1.464592	$\sqrt{e}$	1.648721
π^2	9.869604	$1/\sqrt[3]{\pi}$	0.682784	1 radian	57.295780 deg.
$1/\pi^2$	0.101321	$\log_e \pi$	1.144730	1 radian	3,437.7468 min.
$\sqrt{\pi}$	1.772454	$\log_{10} \pi$	0.497150	1 radian	206,264.81 sec.
$1/\sqrt{\pi}$	0.564190	e	2.718282	1 deg.	0.017453 radian
π^3	31.00628	$1/e$	0.367879	1 min.	0.0002909 radian
		e^2	7.389056	1 sec.	0.00000485 radian

Table 27. Fundamental Physical Constants*

1 sec. = 1.00273791 sidereal second	sec. = mean solar second
g_0 = 980.665 cm./sec.^{2}	Definition: g_0 = standard gravity
1 liter = 1000.028 ± 0.004 cu. cm.	
1 atm. = 1,013,250 dynes/sq. cm.	Definition: atm. = standard atmosphere
1 mm. Hg (pressure) = (1/760) atm.	mm. Hg (pressure) = standard millimeter mercury
= 1333.2237 dynes/sq. cm.	
1 int. ohm = 1.000495 ± 0.000015 abs. ohm†	int. = international; abs. = absolute
1 int. amp. = 0.999835 ± 0.000025 abs. amp.†	amp. = ampere
1 int. coul. = 0.999835 ± 0.000025 abs. coul.†	coul. = coulomb
1 int. volt = 1.000330 ± 0.000029 abs. volt†	
1 int. watt = 1.000165 ± 0.000052 abs. watt†	
1 int. joule = 1.000165 ± 0.000052 abs. joule†	
1 cal. = 4.1840 abs. joule	Definition: cal. = thermochemical calorie
= 4.1833 int. joule	
= 41.2929 ± 0.0020 cu. cm. atm.	
= 0.0412917 ± 0.0000020 liter atm.	
$T_{0°C.}$ = 273.160 ± 0.010°K.	Absolute temperature of the ice point, 0°C.
$(PV)_{0°C.}^{P=0} = (RT)_{0°C.}$ = 2271.16 ± 0.04 abs. joule/mole	
= 22,414.6 ± 0.4 cu. cm. atm./mole	*PV* product for ideal gas at 0°C.
= 22.4140 ± 0.0004 liter atm./mole	
R = 8.31439 ± 0.00034 abs. joule/deg. mole	R = gas constant per mole
= 1.98719 ± 0.00013 cal./deg. mole	
= 82.0567 ± 0.0034 cu. cm. atm./deg. mole	
= 0.0820544 ± 0.0000034 liter atm./deg. mole	
ln 10 = 2.302585	ln = natural logarithm (base *e*)
R ln 10 = 19.14460 ± 0.00078 abs. joule/deg. mole	
= 4.57567 ± 0.00030 cal./deg. mole	
N = (6.02283 ± 0.0022) × 10^{23}/mole	N = Avogadro number
$k = (R/N)$ = (1.38048 ± 0.00050) × 10^{-16} erg/deg.	k = Boltzmann constant
h = (6.6242 ± 0.0044) × 10^{-27} erg sec.	h = Planck constant
c = (2.99776 ± 0.00008) × 10^{10} cm./sec.	c = velocity of light
$(h^2/8\pi^2 k)$ = (4.0258 ± 0.0037) × 10^{-39} g. sq. cm. deg.	Constant in rotational partition function of gases
$(h/8\pi^2 c)$ = (2.7986 ± 0.0018) × 10^{-39} g. cm.	Constant relating wave number and moment of inertia
$Z = Nhc$ = 11.9600 + 0.0036 abs. joule cm./mole	Z = constant relating wave number and energy per mole
= 2.85851 ± 0.0009 cal. cm./mole	
$(Z/R) = (hc/k) = c_2$ = 1.43847 ± 0.00045 cm. deg.	c_2 = second radiation constant
$\mathfrak{F}$ = 96,501.2 ± 10.0 int. coul./g.-equiv. or int. joule/int. volt g.-equiv.	$\mathfrak{F}$ = Faraday constant
= 96,485.3 ± 10.0 abs. coul./g.-equiv. or abs. joule/abs. volt g.-equiv.	
= 23,068.1 ± 2.4 cal./int. volt g.-equiv.	
= 23,060.5 ± 2.4 cal./abs. volt g.-equiv.	
e = (1.60199 ± 0.00060) × 10^{-19} abs. coul.	e = electronic charge
= (1.60199 ± 0.00060) × 10^{-20} abs. e.m.u.	
= (4.80239 ± 0.00180) × 10^{-10} abs. e.s.u.	
1 int. electron-volt/molecule = 96,501.2 ± 10 int. joule/mole	
= 23,068.1 ± 2.4 cal./mole	
1 abs. electron-volt/molecule = 96,485.3 ± 10. abs. joule/mole	
= 23,060.5 ± 2.4 cal./mole	
1 int. electron-volt = (1.60252 ± 0.00060) × 10^{-12} erg	
1 abs. electron-volt = (1.60199 ± 0.00060) × 10^{-12} erg	
hc = (1.23916 ± 0.00032) × 10^{-4} int. electron-volt cm.	Constant relating wave number and energy per molecule
= (1.23957 ± 0.00032) × 10^{-4} abs. electron-volt cm.	
k = (8.61442 ± 0.00100) × 10^{-5} int. electron-volt/deg.	k = Boltzmann constant
= (8.61727 ± 0.00100) × 10^{-5} abs. electron-volt/deg.	
1 I.T. cal. = (1/860) = 0.00116279 int. watt-hr.	Definition of I.T. cal.: I.T. = International steam tables
= 4.18605 int. joule	
= 4.18674 abs. joule	
= 1.000654 cal.	cal. = thermochemical calorie
1 I.T. cal./g. = 1.8 B.t.u./lb.	Definition of B.t.u.: B.t.u. = I.T. British Thermal Unit
1 B.t.u. = 251.996 I.T. cal.	
= 0.293018 int. watt-hr.	
= 1054.866 int. joule	
= 1055.040 abs. joule	
= 252.161 cal.	cal. = thermochemical calorie
1 horsepower = 550 ft.-lb. (wt.)/sec.	Definition of horsepower (mechanical): lb. (wt.) = weight of 1 lb. at standard gravity
= 745.578 int. watt	
= 745.70 abs. watt	
1 in. = (1/0.3937) = 2.54000508 cm.	Definition of in.: in. = U.S. inch
1 ft. = 30.4800610 cm.	ft. = U.S. foot (1 ft. = 12 in.)
1 lb. = 453.5924277 g.	Definition; lb. = avoirdupois pound
1 gal. = 231 cu. in.	Definition; gal. = U.S. gallon
= 0.133680555 cu. ft.	
= 3785.43449 cu. cm.	
= 3.785329 liter	

* From the tables of Selected Values of Properties of Hydrocarbons of the American Petroleum Institute Research Project 44 and the tables of Selected Values of Chemical Thermodynamic Properties of the National Bureau of Standards, as of Dec. 31, 1947.

† The electrical units used in these tables are those in terms of which certification of standard cells, standard resistances, etc., is made by the National Bureau of Standards. Since January 1, 1948, all such certifications are in terms of absolute volts and absolute resistances.

SECTION 2

MATHEMATICS

BY

Raymond W. Dull, Late Consulting Engineer, Chicago, Ill.; Member, American Society of Mechanical Engineers, Western Society of Engineers.

T. R. Running, Ph. D., Professor Emeritus of Mathematics, University of Michigan; Fellow, American Association for the Advancement of Science; Member, American Mathematical Society, American Mathematical Association of America, Michigan Academy of Science, Arts, and Letters. (Graphical Calculus)

John H. Perry, Ph. D., Development Department, E. I. du Pont de Nemours & Co.; Member, American Institute of Chemical Engineers. American Chemical Society, American Association for the Advancement of Science, American Society for Engineering Education.

P. W. Bridgman, Ph. D., Hollis Professor of Mathematics and Natural Philosophy, Harvard University; Member, American Physical Society, American Academy of Arts and Sciences, Philosophical Society of Philadelphia, Washington Academy of Sciences, National Academy of Sciences. (Dimensional Analysis)

D. S. Davis, M. S., Professor of Chemical Engineering, Virginia Polytechnic Institute; Member, American Chemical Society. (Alignment Charts)

R. P. Genereaux, Ch. E., Chemical Engineer, Engineering Department, E. I. du Pont de Nemours & Co.; Member, American Institute of Chemical Engineers. (Alignment Charts)

H. C. Vernon, M. S. Chemical Engineer, Engineering Department, E. I. du Pont de Nemours & Co.; Member, American Institute of Chemical Engineers, American Chemical Society. (Graphical Integration)

CONTENTS

GENERAL

PAGE

Rules for Significant Figures 50
Logarithms 50
The Slide Rule 50

GEOMETRY AND MENSURATION

Theorems of Plane Geometry 53
Theorems of Solid Geometry 55
Theorems of Spherical Geometry 57

GEOMETRICAL CONSTRUCTIONS

Geometrical Constructions 59

ALGEBRA

Notation 62
Factoring 63
Binomial Theorem 64
Infinite Series 65
Progressions 65
Linear Equations 66
Quadratic Equations 66
Cubic Equations 67
Quartic and Higher Equations 68
Determinants 68

TRIGONOMETRY

Plane Trigonometry 70
Spherical Trigonometry 74

ANALYTICAL GEOMETRY

Conics 77
The Circle 78
The Ellipse 78
The Parabola 79
The Hyperbola 79
Miscellaneous Curves 80

DIFFERENTIAL AND INTEGRAL CALCULUS

PAGE

Differential Calculus 81
Differential Equations 84
Integral Calculus 85
Substitutes for Integration 89
Graphical Calculus 90

DIMENSIONAL ANALYSIS*

Dimensional Analysis 93

GRAPHS AND GRAPH PAPER

Methods of Representing Technical Data 97
Common Types of Graph Papers 98

ALIGNMENT CHARTS†

Alignment Charts 100

REFERENCES: Besides the elementary texts, the chemical engineer may find some of the following books helpful: Byerly, "Fourier's Series and Spherical Harmonics," Ginn, Boston, 1893. "Integral Calculus," Ginn, Boston, 1890. Carslaw, "Fourier's Series and Integrals," Macmillan, New York, 1921. "Six-place Tables," McGraw-Hill, New York, 1941. Chauvenet, "Plane and Spherical Trigonometry," Lippincott, Philadelphia, 1892. Dickson, "Elementary Theory of Equations." Dull, "Mathematics for Engineers," McGraw-Hill, New York, 1941; "Mathematical Aids for Engineers," McGraw-Hill, New York, 1946. Fry, "Elementary Differential Equations," Van Nostrand, New York, 1929. Granville and Smith, "Elements of Differential and Integral Calculus," Ginn, Boston, 1911. Lipka, "Graphical and Mechanical Computations," Wiley, New York, 1921. Mellor, "Higher Mathematics for Students of Chemistry and Physics," Longmans, London, 1916. Peirce, "Table of Integrals," Ginn, Boston, 1910. Running, "Empirical Formulas," Wiley, New York, 1917; "Graphical Mathematics," Wiley, New York, 1927. Sherwood and Reed, "Applied Mathematics in Chemical Engineering," McGraw-Hill, New York, 1939. Wilson, "Advanced Calculus," Ginn, Boston, 1912.

* By P. W. Bridgman. † By D. S. Davis and R. P. Genereaux.

GENERAL

RULES FOR SIGNIFICANT FIGURES

1. In rejecting superfluous figures, increase by one the last figure retained, if the following figure (that rejected) is 5 or greater.

2. In all measurements of deviation and precision, retain only two significant figures.

3. Retain as many places of figures in an average result and in data in general as correspond to the second place of significant figures in the measurements of deviation or precision.

4. The sum or difference of two or more quantities cannot be more precise numerically than the quantity having the largest deviation. Therefore, in adding or subtracting a large number of quantities, first find the average deviation of each and then retain in each quantity as many places of figures as correspond to the second place of significant figures in the largest deviation.

5. In multiplication or division, the **percentage** precision of the product or quotient cannot be greater than the **percentage** precision of the least precise factor entering into the computation. Therefore, in computations involving these operations, the number of significant figures to be retained in each factor is determined by the number properly retained under 3 above, in the factor which has the largest percentage deviation.

6. In performing operations of multiplication and division by logarithms, retain as many figures in the mantissa of the logarithm of each factor as are properly retained in the factors themselves under 5 above.

When any of the quantities to be multiplied or divided can be trusted no closer than 0.01 per cent, use a five-place logarithm table; when any of them can be trusted no closer than 0.1 per cent, use a four-place table; when any of them can be trusted no closer than 1 per cent, use a slide rule.

LOGARITHMS

The **logarithm of a number**, to a given base, is that power of the base which is equal to the number. For example, if $a^x = N$, then x is the logarithm to the base a of the number N, or $x = \log_a N$. The base a may be any positive number not zero. Logarithms to the base 10 are called **common** or **Briggsian logarithms.** The other base in most frequent use is e, the sum of the progression $1 + 1/1! + 1/2! + 1/3! + 1/4! + \cdots = 2.7182818 + \cdots$ equals the base of the **natural, hyperbolic,** or **Napierian logarithms.** If the subscript 10 or e is omitted, the base must be inferred from the context, the base 10 being used in numerical computations, and the base e in theoretical work such as mathematical derivations, etc. The two systems of logarithms are interrelated as follows:

$$\log_{10} e = M = 0.434294 \qquad \log_e 10 = 1/M = 2.302585$$
$$\log_{10} x = 0.4343 \log_e x \qquad \log_e x = 2.3026 \log_{10} x$$

$\log_e x$ is often abbreviated $\ln x$, especially in contexts where the common omission of 10 from $\log_{10} x$ might cause misunderstanding.

Every logarithm consists of two parts: an integral part called the **characteristic** and a decimal part called the **mantissa.** The mantissa only is given in ordinary logarithm tables, and usually without the decimal point. The most certain way of obtaining the characteristic is to set the number as a unit times a power of 10. The power of ten is then the characteristic; *e.g.*, $808 = 8.08 \times 10^2$, $\log 808 = 2.90741$, $0.000343 = 3.43 \times 10^{-4}$, $\log 0.000343 = \bar{4}.53529$. The latter expression is explained by the fact that when a logarithm consists of a negative characteristic and a positive mantissa it is usual to write the negative sign over the characteristic, or to add 10 to the characteristic and at the same time indicate the subtraction of 10 from the logarithm, *e.g.*, $\log 0.2 = \bar{1}.30103, = 9.30103 - 10$.

Regardless of the base used, the following relations are valid:

$$\log (ab) = \log a + \log b$$
$$\log \left(\frac{1}{n}\right) = -\log n = \text{colog } n$$
$$\log (\text{base}) = 1$$
$$\log 1 = 0$$
$$\log \left(\frac{a}{b}\right) = \log a - \log b = \log a + \text{colog } b$$
$$\log a^n = n \log a$$
$$\log \sqrt[n]{a} = \frac{1}{n} \log a$$
$$\log \sqrt[n]{a^m} = \frac{m}{n} \log a$$

$a^x = b$ may be transformed into $x \log a = \log b$, whence $x = (\log b)/\log a$.

Division, which introduces negative logarithms, may be simplified by the use of cologarithms, defined by the expression $\log (1/n) = -\log n = \text{colog } n$. Since the addition of a colog is the same as the subtraction of a log, the operation of division is reduced to addition. The following examples will make clear how the characteristics and mantissas of cologarithms are found: $\text{colog } 808 = -\log 808 = -2 - 0.90741 = -3 + 0.09259$, or $\bar{3}.09259$; $\text{colog } 0.000343 = -\log 0.000343, = -(-4 + 0.53529) = +4 - 0.53529 = 3.46471$.

Arrange the logarithm always to make the mantissa positive.

When a computer is making frequent changes from one system of logarithms to another or solving many exponential equations, the use of a table of logarithms becomes convenient. For example, $\log_e x = 2.3026 \log_{10} x$. Then $\log (\log_e x) = \log 2.3026 + \text{loglog } x$. Again, if $a^x = b$, then $\log x = \text{loglog } b - \text{loglog } a$. For a full discussion of this matter, together with tables of loglogs and antiloglogs, see "Six-Place Tables" (McGraw-Hill, New York, 1941).

THE SLIDE RULE

The slide rule is merely a mechanical equivalent of a logarithm table and consists of several scales which are arranged so that the distances, corresponding to logarithms, may be either added or subtracted. For most problems, the slide rule gives results more rapidly than a table of logarithms.

There are many different slide rules on the market. The basic form of slide rule is that known as the Mannheim which has been further developed in rules known as the Polyphase Duplex Trig* and Log Log Duplex

* Registered in U.S. Patent Office by Keuffel & Esser, New York.

Decitrig.* Others for more restricted uses include the Stadia, Electrical, Surveyor's, Trigonometrical, and Chemist's slide rules. In addition to the above straight rules there are a number of circular and cylindrical slide rules. Of the straight slide rules the Polyphase Duplex Trig* and Log Log Duplex Decitrig* are generally considered the best for engineers' use, since they combine the features of the original Polyphase* and Duplex* rules, with the addition of several special scales which are so arranged that necessary operations are most easily and quickly carried out. The Polyphase Duplex Trig* rule was chosen for the following discussion and the examples given refer directly to it. All the information given in regard to the Polyphase Duplex Trig* rule is also applicable to the Log Log Duplex Trig,* the Log Log Duplex Decitrig* and the Log Log Duplex Vector* slide rules.

The Polyphase Duplex Trig Slide Rule

This type of rule has 15 scales; *DF*, *CF*, *CIF*, *CI*, *C*, and *D* scales on one side; and the *K*, *A*, *B*, *T*, *ST*, *S*, *D*, and *DI* scales on the reverse side. Of these, the two *D* and the *C* scales are exactly alike; the *L* or log scale is divided into equal parts; the *CI* (*C* inverted) and *DI* (*D* inverted) scales are like the *C* and *D* scales except that they start from the right-hand side of the rule; the *S*, *ST*, and *T* scales are the sine and tangent scales, respectively; the *DF* (*D*-folded) and *CF* (*C*-folded) scales are exactly alike; and the *CIF* scales is an inverted and folded scale.

The *C* and *D* Scales. The graduations of these scales correspond to the logarithms of the numbers found on the *L* (log) scale. We can add the distances for 3 and 4, on the *C* and *D* scales, and the sum of these distances is 12, if we wish to multiply 3 and 4. The procedure for these operations is: Set 1 on the *C* scale opposite 3 on the *D* scale, move the glass indicator to 4 on the *C* scale; directly below this 4 (on the *D* scale) will be found 12, the answer. It will be noted in this example that, if the left-hand index (1) is set opposite 3 on the *D* scale, the slide will be extended to the right of the main part of the rule and no answer is possible for this particular example with this setting. In this case the right-hand index (1) is used. Both of these indexes have the same value in operations, even though the left-hand index really is 1, and the right-hand index might be considered to be 10. Multiplication can also be performed on the inverted scale, but, since the above method can be used for multiplying reciprocals, it is called the regular multiplication setting.

When more than two numbers are to be multiplied, it is advantageous to combine the two methods of multiplication as shown in the following example.

Example. Multiply 7.2 × 3.2 × 0.25 × 5.4. Beginning with the *CI* method and alternately using this and the *C* scale proceed as follows:

To 7.2 on *D*, set 3.2 on *CI* (by means of indicator). The result of this multiplication could be read opposite the slide index on *D* as 23, but we are not interested in this intermediate result and disregard it. With the slide in the above position, set the hairline of the indicator to the third factor, 0.25 on *C*. Again disregarding the intermediate result, move slide, and set the last factor, 5.4 on *CI*, under the indicator. Opposite the slide index read 311 on the *D* scale. The decimal point is placed by inspection of magnitude of the result, which appears to be about 30, and therefore the correct answer is 31.1.

The **general rule for multiplication**: To find the product of two factors: find one factor on the *D* scale, opposite this set the index 1; on the *C* scale find the second factor, and opposite this read the product on the *D* scale if possible. If the product falls beyond the scale, start over again, using the right-hand index of the *C* scale.

* Registered in U.S. Patent Office by Keuffel & Esser, New York.

To divide one number by another on the slide rule, we simply reverse the order of the work done in multiplication.

Example. Divide 8 by 56. Set the indicator to 8 on the *D* scale, move the slide (containing the *C* scale), so as to set 56 on the *C* scale exactly to the hairline of the indicator; in other words, directly over 8. Under the slide index we find 143 on the *D* scale and by inspection, place the decimal point before the 1 of the answer; that is, 0.143.

If both **multiplication and division** are to be performed in a series of operations, they should be carried out so that the minimum amount of settings is necessary, which will automatically increase the accuracy of, and decrease the labor in obtaining, the answer.

Example. (27 × 17)/15. First divide 27 by 15, by setting 15 on the *C* scale to 27 on the *D* scale, then multiply by 17, by moving the indicator to 17 on the *C* scale, then on the *D* scale opposite the indicator line and directly under 17 will be found the result, which by inspection must be about 30, and hence the reading which is 306 will be pointed off as 30.6.

Example. In a more complicated example, such as

$$\frac{18 \times 6 \times 14 \times 0.25}{3 \times 0.7 \times 78 \times 9.4},$$

the operations of multiplication and division are best performed alternately; that is, 18 ÷ 3 × 6 ÷ 0.7 × 14 ÷ 78 × 0.25 ÷ 9.4. Set indicator to 18, set 3 to indicator, set indicator to 6, set 7 to indicator, set indicator to right-hand index, set left-hand index to indicator, set indicator to 14, set 78 to indicator, set indicator to 25, set 94 to indicator, and under the right-hand index read the answer on the *D* scale, which is, by inspection, equal to 0.2155.

Methods of Placing the Decimal Point. Rough Calculation. In the above example,

$$\frac{18 \times 6 \times 14 \times 0.25}{3 \times 0.7 \times 78 \times 9.4},$$

the mental process of placing the decimal point by this method might be: 18 ÷ 3—6 × 6—36 ÷ 0.7—50 × 14—700 ÷ 78—9 × 0.25—2 ÷ 9—0.2, approx.

Method of Tens. Factor each number into two factors, one factor being 10 or a multiple of 10 as 10^2 or 10^{-2}, and the other factor being the first left-hand figure of the number; *e.g.*, 40,000 = 4×10^4, or 0.042 = 4.2×10^{-2}. In other words, the exponent of 10 is equal to the number of decimal places through which the decimal point has been moved.

Example.

$$\frac{4687 \times 0.031}{0.056 \times 4 \times 72} = \frac{4 \times 10^3 \times 3 \times 10^{-2}}{5 \times 10^{-2} \times 4 \times 7 \times 10} = \frac{4 \times 3}{5 \times 4 \times 7} \times \frac{10^3 \times 10^{-2}}{10^{-2} \times 10} = 0.1 \times 10^2, \text{ approx.}$$

Multiplying by Pi. Use the *C* and *D* scales as usual, but instead of reading the answer on the *D* scale, read it on the *DF* scale.

Example. Multiply: 5 × π. Set the right-hand index of the *C* scale on the 5 of the *D* scale; on the *DF* scale read the answer which by inspection is 15.7+.

Dividing by Pi. Use the *DF* and *CF* scales and read the answer on the *D* scale.

Example. (11 × 24)/π; set index of *CF* scale to 11 on *DF* scale; set runner to 24 on *CF* scale; read the answer on *D* scale which is 84.1.

Example. $\frac{(\pi \times 54 \times 3)}{14}$; mentally rearrange to the form 54 × 3 × 1/14 × π; set the indicator to 54 on *D* scale; set 3 on *CI* scale to indicator; set indicator to 14 on *CIF* scale; read answer under indicator on *D* scale, which is 36.8.

Reciprocals.

Example. $1/51$; set the indicator on 51 of *CI* or *DI* scale; read the answer on *C* or *D* scale, which is 0.0196.

Example. $\dfrac{1}{(3.1 \times 0.39 \times 81)}$. To 3.1 on *DI* scale set 0.39 on *C* scale. At 81 on *CI* scale read answer on *D* scale which is 0.0102. The problem can also be solved as $3.1 \times 0.39 \times 81$ by reading the result on the reciprocal scale *DI* instead of the *D* scale.

Example. $336/243$; $192/243$; $208/243$. Set slide index to 243 on *D* scale. At 336 on *D* scale read 1.382 on *C* scale; at 192 on *DF* scale read 0.79 on *CF* scale; at 208 on *DF* scale read 0.856 on *CF* scale.

Squares.

Example. $(14)^2$; set indicator to 14 on *D* scale; read answer on *A* scale, which is 196.0.

Example. $\dfrac{3.1 \times (14.1)^2 \times 28}{13.4 \times (56)^2}$; set indicator to 31 on *A* scale; move slide so as to set 134 on *B* scale over 31; set indicator to 141 on *C* scale; set 56 to *C* scale to indicator; set indicator to 28 on *B* scale; read answer on *A* scale, which is 0.41.

Square Roots.

Example. $\sqrt{54}$; set indicator to 54 on *A* scale; read the answer on *D* scale, which is 7.35.

Example. $1/(\sqrt{54})$; set indicator to 54 on *B* scale; read answer 0.136 on *CI* scale.

Example. $\dfrac{42.5 \times \sqrt{56} \times 0.38}{0.43 \times 126 \times \sqrt{419}}$; set indicator to 42.5 of *D* scale; set 43 of *C* scale to indicator; set indicator to 56 of *B* scale; set 126 of *C* scale to indicator; set indicator to 38 of *C* scale; set 419 of *B* scale to indicator; set indicator to index of *C* scale; read answer 0.109 on *D* scale at slide index.

Example. $(\sqrt{54})/8$. To 54 on *A* scale set 8 on *C* scale. At slide index read 0.92 on *D* scale.

Example. $8\sqrt{54}$. To 8 on *D* scale set slide index. At 54 on *B* scale read 58.8 on *D* scale.

Example. $1/(8\sqrt{54})$. To 54 on *A* scale set 8 on *CI* scale. At index of *D* scale read 0.017 on *C* scale.

Multiplication of xy = a Constant. Setting the slide index to the constant of the right-hand side of the equation above, on the *D* scale, the values of y on the *D* or *CI* scale for all corresponding values of x on *CI* or *D* scale, respectively, may be read without resetting the rule.

Squares and Square Roots.

Example. $\sqrt{53}/(8)^2$. To 53 on *A* scale set 8 on *C* scale. At 8 on *CI* scale read answer 0.1138 on *D* scale.

Example. $(8)^2/\sqrt{53}$. To 8 on *D* scale set 53 on *B* scale. At 8 on *C* scale read answer 8.79 on *D* scale.

Cubes and Cube Roots. Since the *D* scale is three times the length of each of the three scales on the *K* scale, the values of the readings on the *K* scale are cubes of the coinciding values on the *D* scale, and, conversely, the values on the *D* scale are the cube roots of the coinciding values of the *K* scale.

To find the cube of a number, set the runner at the number on the *D* scale and, under the hairline of the runner on *K*, read the answer. For cube root use the left-hand scale of the *K* scale for numbers between 1 and 10, the middle scale for numbers between 10 and 100, and the right-hand scale for numbers between 100 and 1000. The operation is to set the indicator on the number on the *K* scale (selecting the proper portion of the scale as noted above) under the hairline of the indicator on the *D* scale, and find the cube root.

Examples

$1/(8)^3$: To 8 on *DI* set indicator; read answer on *K*.

$1/\sqrt[3]{8}$: To 8 on *K* set indicator; read answer on *DI*.

$(8)^2/(41)^3$; set 41 on *B* to 8 on *D*, over 41 on *CI*; read answer on *A*.

$(8)^2(41)^3$; set 41 on *CI* to 8 on *D*, over 41 on *B*; read answer on *A*.

$8\sqrt[3]{41}$; set index of *C* to 41 on *K*, under 8 on *C*; read answer on *D*.

$1/(8\sqrt[3]{41})$; set 8 on *CI* to 41 on *K*, over index of *D*; read answer on *C*.

$\sqrt[3]{8/41}$; set 41 on *C* to 8 on *K*, under index of *C*; read answer on *D*.

$(8)^3(41)^3$; set index of *C* to 41 on *D*, under 8 on *C*; read answer on *K*.

$(8)^3/(41)^3$; set 41 on *C* to 8 on *D*, under index of *C*; read answer on *K*.

Logarithms. The logarithm of any number (in lieu of logarithm tables) is easily found on the slide rule.

Example. Log 7.8; with indicator set at 7.8 on the *D* scale read .892 on the *L* scale; the characteristic 0 being added according to the general rule. log 7.8 = 0.892.

Powers. Numbers raised to powers other than squares and cubes are solved as follows:

Example. $(2.3)^{4.38}$; 4.38 log 2.3 = log of the number desired.

Example. $(187)^{1/5}$; $1/5 \times \log 187$ = log of the number desired.

For this type of work, *e.g.*, higher and fractional powers and roots as well as for logarithms to the natural or any other base, the loglog scales are ideal, since they handle these types of problems as simply and as quickly as ordinary multiplication.

Sines. The *ST* and *S* scales are sine scales for angles from 35° to 5° 44′ and from 5° 44′ to 90°. Opposite any angle on *ST* or *S* scale, its sine is found on the *C* or *D* scale.

Example. Sine of 35°. Opposite 35° on *S* scale is found its sine on *C* (or, when slide is lined up, on *D* scale); sin 35° = 0.574. The decimal point in a sine of an angle appearing on *S* scale (except sin 90°) is directly preceding the first significant figure; while the sine for an angle appearing on *ST* scale will have one zero between the decimal point and its first significant figure.

Example. (sin 1° 25′.) Set indicator to 1° 25′ on *ST* scale and read answer 0.0247 on *C* or *D* scale.

Example. 56/sin 9°. To 56 on *D* scale set 9° on *S* scale. At slide index read answer 358 on *D* scale.

Example. (sin 6° 35′ × 43.) To 43 on *D* scale set slide index. At 6° 35′ on *S* scale read 4.93 on *D* scale.

Example. $\dfrac{\sqrt{2.66} \sin 10°}{14.2}$. To 2.66 on *A* scale set 14.2 on *C* scale. At 10° on *S* scale read answer 0.0199 on *D* scale.

Cosines. The cosine of an angle is equal to the sine of the complement of the angle $\cos\theta = \sin(90° - \theta)$. The *S* scale is numbered in red for complements. Therefore it is possible to read the cosines directly.

Example. Cosine 30°. Set indicator to 30° on *S* (red) and read cos 30° = 0.866 on *C* or *D* scale.

Example. Similarly, cos 15° 30′ = 0.964.

Tangents. The *T* scale is a scale of tangents for angles from 5° 43′ to 84° 17′. Opposite any angle between 5° 43′ and 45° on *T* scales (black numbering) its tangent is found on the *C* or *D* scale; and opposite any angle between 45° and 84° 17′ on *T* scale (red numbering) its tangent is found on *CI* or *DI* scale.

Example. Tangent 20°. Set indicator to 20° on *T* scale and read tan 20° = 0.364 on *C* or *D* scale. The decimal point in a tangent of an angle appearing on *T* scale (black) is directly preceding the first significant figure; while in the tangent of angle appearing on the *T* scale (red) the decimal point will be immediately behind the first significant figure.

Example. Tangent 75°. Set indicator to 75° on *T* scale and read tan 75° = 3.73 on *CI* or *DI* scale.

Example. 256/tan 10° 30′. To 256 on *D* scale set 10° 30′ on *T* scale, and at slide index read answer 1380 on *D* scale.

Since the tangents of small angles are practically equal to their sine, the *ST* scale can also be used to find tangents of angles between 35′ and 5° 44′ for most practical problems. Thus, tan 1° 30′ = sin 1° 30′ = 0.0262.

Example. $\frac{24 \tan 15°}{1.65 \sin 20°}$. To 24 on *D* scale set 1.65 on *C* scale. Indicator to 15° on *T* scale; 20° on *S* scale (black) to indicator; at slide index read answer 11.40 on *D* scale.

Cotangents. Since cot $\theta = 1/\tan \theta$, the cotangents of angles can be found on the slide rule by reading the reciprocal of its tangent.

Example. Cotangent 65°. To 65° on *T* scale set indicator. The tangent is read on *CI* or *DI* scale; therefore its reciprocal appears on *C* or *D* scale; cot 65° = 0.466.

Example. Cotangent 18° 10′. Opposite 18° 10′ on *T* scale read 3.045 on *CI* or *DI* scale.

Secants and Cosecants. These trigonometric functions are found as the reciprocals of cosine and sine, respectively; secant $\theta = 1/\cos\theta$ and cosecant $\theta = 1/\sin\theta$. Secants therefore are read from *S* scale (red) to *C* or *D* scale and cosecants are read from *S* scale (black) to *CI* or *DI* scale.

Rule. All direct functions (sin, tan, and secant) are read on scales of like colors (black to black or red to red), and all cofunctions (cos, cot, and csc) are read on scales of unlike colors (black to red or red to black).

Right Triangle and Vectors. The arrangement of the trigonometric scales allows a simple solution for the right triangle, vectors, and complex expressions.

Example. The two sides of a right-angle triangle are 3 and 4 in., respectively. Find the hypotenuse and the two acute angles. To the larger of the two sides, 4 on *D* scale, set the slide index. At 3 (the shorter side) on *D* scale read 36° 52′ on *T* scale (black) and 53° 08′ on *T* scale (red) for the two angles. Move slide to bring 36° 52′ on *S* scale (black) to the shorter side, 3 on *D* scale. At slide index read 5 on *D* scale for the hypotenuse.

Example. A force of 26.8 lb. acts at an angle of 38° with a given direction. Find the components of this force along the given direction and at right angle to it, respectively. To 26.8 on *D* scale set slide index. At 38° on *S* scale (black) read y = 16.50 lb. on *D* scale; at 38° on *S* scale (red) read x = 21.10 lb. on *D* scale.

Example. Find the polar coordinate for the rectangular notation 1.63 + j3.6. To 3.6 on *D* scale set slide index. At 1.63 on *D* scale read θ = 65° 38′ on *T* scale. Move 65° 38′ on *S* scale (red) to 1.63 on *D* scale. At the slide index read 3.95.

$$1.63 + j3.6 = 3.95/65° 38'$$

Additional examples of the use of the slide rule can be obtained from the manual of instructions by L. M. Kells, W. F. Kern, and J. R. Bland published by Keuffel & Esser Company, the manufacturer of the Polyphase Duplex Trig* slide rule; as well as from the manuals of other manufacturers.

Scales on Straight Slide Rules. Mannheim Slide Rules. One face has two single (*C* and *D*) and two double logarithmic (*A* and *B*) scales: the reverse side of the slide has scales of sines and of tangents and a scale for finding logarithms.

Polyphase Slide Rule.* In addition to the scales of the Mannheim rules this rule also has a scale for cubes and an inverted *C* scale for reciprocals.

Polyphase Duplex Trig and Polyphase Duplex Decitrig* Slide Rules.* These rules have all the scales of the Polyphase* rule with the addition of the *CF*, *DF*, and *CIF* scales, which are so-called "folded scales." The function of the folded scales is to enable factors to be read without resetting of the slide when such factors would be off the regular *C* and *D* scales. Also the trigonometric scales are rearranged and expanded for simplified solutions of the right triangle and other problems in trigonometry. The trigonometric scales are divided in degrees and minutes in the Polyphase Duplex Trig* slide rule and in degrees and decimal fractions thereof in the Polyphase Duplex Decitrig* slide rule.

Log Log Duplex Trig and Log Log Duplex Decitrig* Slide Rules.* These rules have all the scales of the Polyphase Duplex Trig* and the Polyphase Duplex Decitrig* rules, and in addition they carry five loglog scales arranged for multiplication and division of the logarithms of the numbers thereon and also giving logarithms to any base. The use of these rules is particularly recommended for determining powers and extracting roots. The solution of these problems is thereby simplified to the ease of multiplying and dividing with ordinary rules.

Log Log Duplex Vector Slide Rule.* This rule has all the scales of the Log Log Duplex Decitrig* rule, except the *DI* and *K* scales. In their place are three full-length hyperbolic scales giving hyperbolic sines and hyperbolic tangents as easily as the circular functions. The evaluation of hyperbolic functions of complex numbers and the inverse hyperbolic functions of complex numbers are tremendously simplified and the time for their solution reduced.

In addition to rules with scales as above described, there are many rules for special uses and necessarily with special scales some of which were mentioned at the beginning of this section.

* Trade marks of Keuffel & Esser Company, New York. Registered U.S. Patent Office.

GEOMETRY AND MENSURATION

Angles. Two angles are **complementary**, or complements, when their sum is equal to a right angle. Two angles are **supplementary**, or supplements, when their sum is equal to 180°. Two angles which have the sides of one perpendicular to the sides of the other, are either equal or supplementary. Two angles whose sides are parallel, each to each, are either equal or supplementary.

THEOREMS OF PLANE GEOMETRY

(1) Vertical angles are equal. (2) Two triangles are congruent: (*a*) when a side and the two adjacent angles of one are equal, respectively, to a side and the adjacent angles of the other; and (*b*) if two sides and the included angle of the one are equal, respectively, to two sides and the included angle of the other. (3) An exterior angle of a triangle is greater than either remote interior angle. (4) Two straight lines which are parallel to a third straight line are parallel to each other. (5) Two lines are parallel: (*a*) if a transversal to these lines makes the alternate interior angles equal; and (*b*) if a transversal to these lines makes the interior angles on the same side of the transversal supplementary. (6) If two parallels are cut by a transversal, the alternate interior angles are equal, also the corresponding angles are equal, and the interior angles on the same side of the transversal are supplementary. (7) Angles whose corresponding sides are parallel are either equal or supplementary. (8) The sum of the angles of a triangle is equal to 180°. (9) An exterior angle of a triangle is equal to the sum of the two remote interior angles. (10) The base angles of an isosceles triangle are equal. (11) If two angles of a triangle are equal, the sides opposite these angles are equal.

(12) Two triangles are equal (*a*) if three sides of the one are respectively equal to three sides of the other and (*b*) if they have equal bases and equal altitudes. (13) Two right triangles are equal if the hypotenuse and a side of the one are respectively equal to the hypotenuse and a side of the other. (14) The line that joins the vertices of two isosceles triangles on the same base bisects the common base at right angles. (15) Every point in the bisector of an angle is equidistant from the sides of the angle. (16) The sum of the two sides of a triangle is greater, and their difference is less, than the third side. (17) If two sides of a triangle are unequal, the angles opposite are unequal, and the greater angle is opposite the greater side. (18) If two angles of a triangle are unequal, the sides opposite are unequal, and the greater side is opposite the greater angle. (19) If two triangles have two sides of the one equal respectively to two sides of the other but the included angle of the first is greater than the included angle of the second, then the third side of the first is greater than the third side of the second. (20) If two triangles have two sides of the one equal respectively to two sides of the other but the third side of the first greater than the third side of the second, then the included angle of the first is greater than the included angle of the second. (21) A line which joins the mid-points of two sides of a triangle is parallel to the third side and equal to half of it. (22) If a line divides two sides of a triangle proportionally, it is parallel to the third side. (23) The bisector of an angle of a triangle divides the opposite side into segments having the same ratio as the two other sides. (24) The bisector of an exterior angle of a triangle divides the opposite side externally into segments which are proportional to the two other sides. (25) In a right triangle the altitude upon the hypotenuse is the mean proportional between the segments of the hypotenuse, and either arm is the mean proportional between the hypotenuse and the adjacent segment. (26) In any triangle the square of a side opposite an acute angle is equal to the sum of the squares of the two other sides diminished by twice the product of one of those sides and the projection of the other side upon it. (27) In any obtuse triangle, the square of the side opposite the obtuse angle is equal to the sum of the squares of the two other sides, increased by twice the product of one of these sides and the projection of the other side upon it. (28) In any triangle, (*a*) the square of one side plus four times the square of the corresponding median is equal to twice the sum of the squares of the other sides; (*b*) the product of two sides is equal to the square of the bisector of the included angle plus the product of the segments of the third side; (*c*) the product of two sides is equal to the altitude upon the third side multiplied by the diameter of the circumscribed circle.

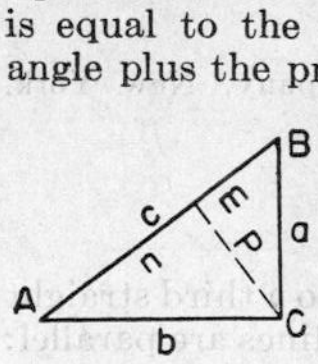

Fig. 1.

Right Triangles (Fig. 1). If a and b are the lengths of the legs and c is the length of the hypotenuse: $a^2 + b^2 = c^2$; $a = \sqrt{(c+b)(c-b)}$.

$$a = c \sin A = b \tan A = \sqrt{mc}$$
$$b = \sqrt{(c+a)(c-a)} = c \cos A = a/\tan A = \sqrt{nc}$$
$$c = \sqrt{a^2 + b^2} = a/\sin A = b/\cos A = m + n$$
$$\text{Area} = \tfrac{1}{2}ab = \tfrac{1}{2}a^2 \cot A = \tfrac{1}{2}b^2 \tan A$$
$$= \tfrac{1}{4}c^2 \sin 2A = \tfrac{1}{2}bc \sin A = \tfrac{1}{2}ac \sin B$$
$$= a^2/2 \tan A$$
$$p^2 = mn$$

Any triangle inscribed in a semicircle is a right triangle.

Equilateral Triangle (Fig. 2). $A = B = C = 60°$

$$\text{Area} = \tfrac{1}{2}ah = \tfrac{1}{4}a^2\sqrt{3} = 0.43301a^2$$
$$h = \tfrac{1}{2}a\sqrt{3} = 0.866a$$

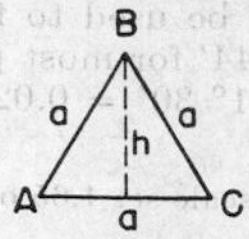

Fig. 2.

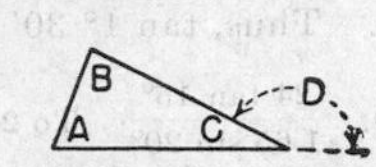

Fig. 3.

Any Plane Triangle (Fig. 3). $\angle A + \angle B + \angle C = 180° = \pi$ radians. An exterior angle, as $\angle D$ = the sum of the two opposite interior angles, as $\angle A + \angle B$. Where s is one-half the sum of the sides; $s = \tfrac{1}{2}(a + b + c)$; the radius r of the inscribed circle is $\sqrt{\frac{(s-a)(s-b)(s-c)}{s}}$ and the radius R of the circumscribed triangle is $\frac{a}{2 \sin A} = \frac{b}{2 \sin B} = \frac{c}{2 \sin C}$. The area of the triangle is

$$\tfrac{1}{2}bh = \sqrt{s(s-a)(s-b)(s-c)}.$$

The bisectors of the angles of a triangle meet in the center of the inscribed circle.

The medians are the lines joining the vertices with the centers of the opposite sides and intersect at the center of gravity G, which is one-third the altitude above the base.

Lines drawn perpendicular to the sides at their centers intersect at a point which is the center of the circumscribed circle.

A line parallel to one side of a triangle divides the two other sides into segments having equal ratios.

Congruent Triangles. (1) Two triangles are congruent if two sides and the included angle of one are equal respectively to the corresponding two sides and the included angle of the other; (2) two triangles are congruent if two angles and the side included between the vertices of one triangle are equal respectively to the corresponding two angles and the included side of the other; (3) two triangles are congruent if three sides of one triangle are equal respectively to the three sides of another triangle.

Similar Triangles. Two triangles are similar if (1) the angles of one are equal to the angles of the other, and their corresponding sides are proportional; (2) a line parallel to one side of a triangle forms with the two other sides a triangle; (3) two angles of one are equal respectively to the corresponding angles of the other.

Graphical Multiplication and Division. To multiply graphically, $a \times b$:

Draw preferably a right triangle whose base is equal to one of the numbers and whose altitude equals 1. From this triangle, draw a second similar triangle by extending the sides. Make the other number the altitude of this second triangle. The base of the second triangle is the required product.

To divide graphically, a/b.

Draw a right triangle with base equal to the divisor and altitude equal to 1. Draw a similar triangle by extending the sides, and make the base of the second triangle equal to the dividend. The altitude of this triangle is the quotient.

Lengths and Areas of Plane Figures. Rectangle (Fig. 4). Area $= ab = \tfrac{1}{2}D^2 \sin \theta$, where θ is the angle between diagonals, and the length of the diagonal $= \sqrt{a^2 + b^2}$ and $\theta = 2 \tan^{-1} \frac{b}{a}$.

Rhombus (Fig. 5). A quadrilateral having oblique angles but equal sides. Area = $a^2 \sin C = \frac{1}{2}D_1D_2$, where C = angle between two adjacent sides.

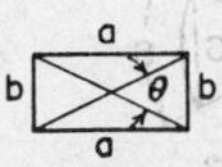
FIG. 4.

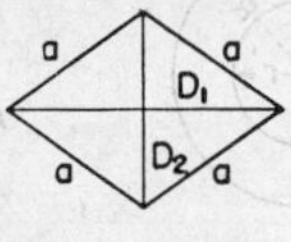
FIG. 5.

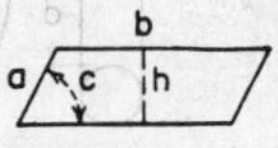
FIG. 6.

Parallelogram (Fig. 6). Area $= bh = ab \sin C = \frac{1}{2}D_1D_2 \sin \theta$, where D_1 and D_2 are the diagonals and θ the angle between them.

The opposite sides are equal and parallel. A diagonal divides a parallelogram into congruent triangles. The consecutive angles are supplementary. The diagonals bisect each other, and the center of gravity is at the intersection of the diagonals.

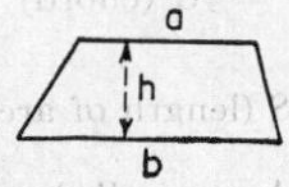
FIG. 7.

Trapezoid (Fig. 7). A quadrilateral with one pair of opposite sides parallel. Area $= \frac{(a+b)h}{2} = \frac{1}{2}D_1D_2 \sin \theta$, where D_1 and D_2 are the diagonals and θ the angle between them. A line drawn through the mid-points of the nonparallel sides is the median of the trapezoid. The length of the median is equal to one-half the sum of the bases. To find the center of gravity of a trapezoid, divide the trapezoid into triangles and find the centers of gravity of these triangles. The lines joining the centers of the parallel sides and the line joining the centers of gravity of the triangles intersect at the center of gravity of the trapezoid.

Quadrilateral. Any figure of four sides is a quadrilateral. The sum of the interior angles is 360°. Area $= \frac{1}{2}lh$ where l is the length of the diagonal and h is the altitude perpendicular to the diagonal. The most common method of finding the area is to divide the quadrilateral into two triangles and find the sum of the areas of these triangles. The area is also equal to $\frac{1}{2}l_1l_2 \sin \theta$, where θ is the angle between the diagonals.

The Circle. Circumference = diameter × 3.1416 = 2π × radius. Area $= \pi \times (\text{radius})^2 = \frac{1}{2}$ (circumference × radius) $= \frac{1}{4}\pi \times (\text{diameter})^2 = 0.7854 \times (\text{diameter})^2$.

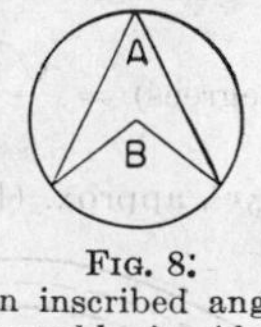
FIG. 8.

FIG. 9.

FIG. 8. An inscribed angle, as A, is measured by one-half the arc intercepted by its sides, or angle A is one-half of B.

FIG. 9. All inscribed angles subtended by the same arc are equal. $\angle A = \angle B = \angle C = \angle D$.

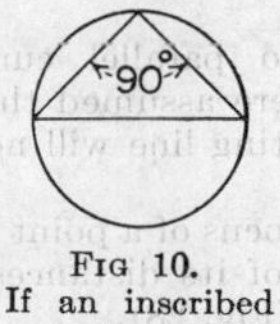
FIG 10.

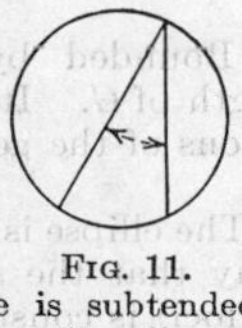
FIG. 11.

FIG. 10. If an inscribed angle is subtended by one-half the circumference, the angle is 90°, or $\pi/2$ radians.

FIG. 11. If the subtended arc is less than one-half the circumference, the angle is acute.

An angle inscribed in a semicircle is a **right angle.** An angle inscribed in a circle is measured by one-half the intercepted arc, which is also equal to the angle between a chord and a tangent. A **tangent** to a circle is a line perpendicular to the radius drawn to the point of contact. A **dihedral angle** between two planes is measured by the angle formed by two lines, one in each plane, perpendicular to the edge.

FIG. 12.

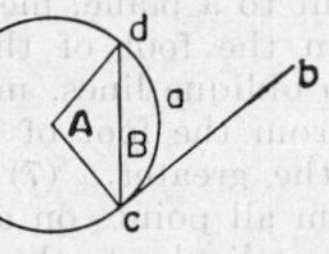
FIG. 13.

FIG. 12. If the subtended arc is greater than one-half the circumference the angle is obtuse.

FIG. 13. The angle B between the chord cd and the tangent cb is measured by one-half the arc dac, or is equal to one-half the angle A.

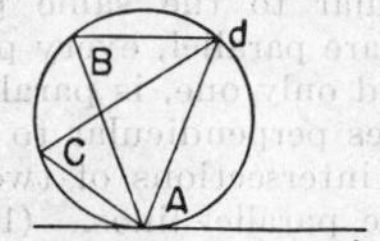
FIG. 14.

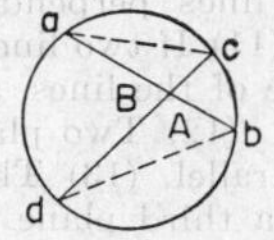
FIG. 15.

FIG. 14. The angle A between a tangent cb and a chord cd drawn from the point of tangency is equal to any inscribed angle, as B or C, subtending the same chord cd.

FIG. 15. If two chords intersect within a circle, either angle formed is measured by one-half the sum of the intercepted arcs.

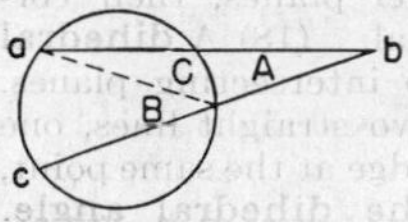
FIG. 16.

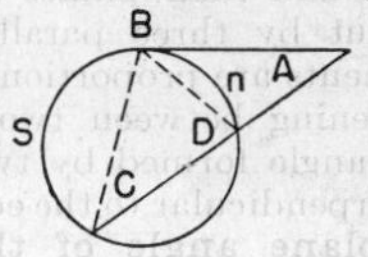
FIG. 17.

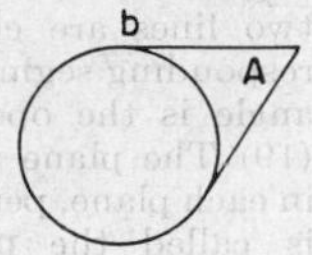
FIG. 18.

FIG. 16. If two secants, as ab and cb, meet outside a circle, the angle formed is measured by one-half the difference of the intercepted arcs $\angle A = \angle B - \angle C$.

FIG. 17. The angle formed by a tangent and a secant meeting outside of a circle is measured by one-half the difference of the intercepted arcs. Also, $\angle A$ is measured by one-half arc s minus, one-half arc n or $\angle A = \angle D - \angle C$.

FIG. 18. The angle formed by two tangents to a circle is equal to one-half the difference of the intercepted arcs.

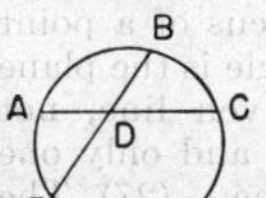
FIG. 19.

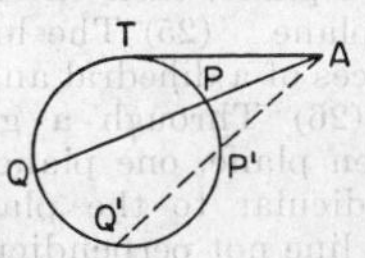
FIG. 20.

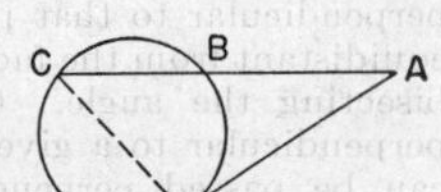
FIG. 21.

FIG. 19. If two chords of a circle intersect, the product of the segments of one is equal to the product of the segments of the other. $AD \times DC = ED \times DB$.

FIG. 20. If a variable line through A, located outside the circle, cuts the circle of P and Q, then $AP \times AQ = (AT)^2$, where AT is the tangent from A.

FIG. 21. If from a point A without a circle, a tangent and a secant be drawn, the tangent AD is a mean proportional between the entire secant AC and its external segment, $AC/AD = AD/AB$.

THEOREMS OF SOLID GEOMETRY

(1) If two planes cut each other, their intersection is a straight line. (2) If a line is perpendicular to each of two other lines at their point of intersection, it is perpendicular to the plane of the two lines. (3) All the perpendiculars that can be drawn to a given line at a given point lie in a plane which is perpendicular to the

given line at the given point. (4) Through a given point in a plane there can be drawn one line perpendicular to the plane, and only one. (5) Through a given external point, there can be drawn one, and only one, line perpendicular to a given plane. (6) Oblique lines drawn from a point to a plane, meeting the plane at equal distances from the foot of the perpendicular, are equal; and of two oblique lines, meeting the plane at unequal distances from the foot of the perpendicular, the more remote is the greater. (7) The locus of a point equidistant from all points on a circle is a line through the center, perpendicular to the plane of the circle. (8) The locus of a point equidistant from the vertices of a triangle is a line through the center of the circumscribed circle, perpendicular to the plane of the triangle. (9) The locus of a point equidistant from two given points is the plane perpendicular to the line joining them at its mid-point. (10) Two lines perpendicular to the same plane are parallel. (11) If two lines are parallel, every plane containing one of the lines, and only one, is parallel to the other line. (12) Two planes perpendicular to the same line are parallel. (13) The intersections of two parallel planes by a third plane are parallel lines. (14) A line perpendicular to one of two parallel planes is perpendicular to the other also. (15) If two intersecting lines are each parallel to a plane, the plane of these lines is parallel to that plane. (16) If two angles not in the same plane have their sides respectively parallel and lie on the same side of the straight line joining their vertices, the angles are equal and their planes are parallel. (17) If two lines are cut by three parallel planes, their corresponding segments are proportional. (18) A **dihedral** angle is the opening between two intersecting planes. (19) The plane angle formed by two straight lines, one in each plane, perpendicular to the edge at the same point, is called the **plane angle of the dihedral angle.** (20) Two dihedral angles are equal if their plane angles are equal. (21) Two dihedral angles have the same ratio as their plane angles, whether the plane angles are commensurable or incommensurable. (22) If two planes are perpendicular to each other, a line drawn in one of them perpendicular to their intersection is perpendicular to the other. (23) If a line is perpendicular to a plane, every plane passed through this line is perpendicular to the plane. (24) If two intersecting planes are each perpendicular to a third plane, their intersection is also perpendicular to that plane. (25) The locus of a point equidistant from the faces of a dihedral angle is the plane bisecting the angle. (26) Through a given line, not perpendicular to a given plane, one plane and only one can be passed perpendicular to the plane. (27) The projection of a straight line not perpendicular to a plane, upon that plane, is a straight line. (28) The acute angle which a line makes with its projection upon a plane is the least angle which it makes with any line of the plane. (29) Between two lines not in the same plane there can be one and only one common perpendicular. (30) The opening of three or more planes which meet at a common point is a **polyhedral angle,** the size of which depends only upon the relative position of the faces and not upon their extent. (31) The sum of any two face angles of a trihedral angle is greater than the third face angle. (32) The sum of the face angles of any convex polyhedral angle is less than four right angles. (33) Two trihedral angles are equal or symmetrical when the face angles of the one are equal respectively to the three face angles of the other.

Annulus. Area $= \pi(R^2 - r^2) = \dfrac{\pi(D^2 - d^2)}{4} = 2\pi R'b$, where R' = mean radius $= \frac{1}{2}(R + r)$ and $b = R - r$ (Fig. 22).

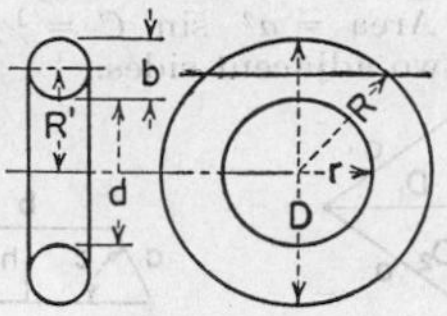

FIG. 22.

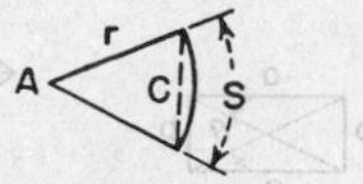

FIG. 23.

Sector (Fig. 23).

$$C \text{ (chord)} = 2r \sin \frac{A}{2} \quad (A \text{ in degrees or radians})$$

$$C \text{ (chord)} = 2r \sin \left(\frac{90S}{\pi r}\right)$$

S (length of arc) $= (\pi r A)/180$ (A in degrees) $= rA$ (A in radians)

Area $= rS/2 = (\pi r^2 A)/360$ (A in degrees) $= r^2A/2$ (A in radians)

Area of Sector of Annulus. Area $= \dfrac{\pi A(R^2 - r^2)}{360}$ (A in degrees) (Fig. 24)

FIG. 24.

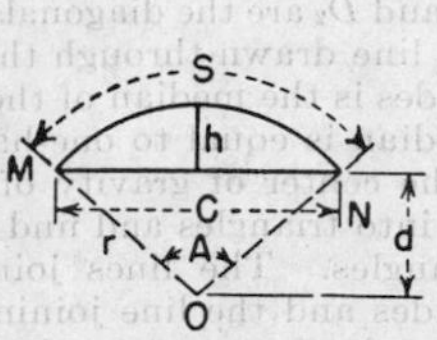

FIG. 25.

Segment (Fig. 25.)

$$C \text{ (chord)} = 2r \sin \frac{A}{2} \quad (A \text{ in degrees or radians})$$

$$= 2r \sin \left(\frac{90S}{\pi r}\right)^{\circ} = 2\sqrt{2hr - h^2}$$

$$S \text{ (length of arc)} = \frac{\pi r A}{180} \quad (A \text{ in degrees})$$

$$= 2r \frac{A}{2} \text{ or } rA \quad (A \text{ in radians})$$

$$h = r - d = r\left(1 - \cos \frac{A}{2}\right) \quad (A \text{ in degrees or radians})$$

$$d = r \cos \frac{A}{2} \quad (A \text{ in radians or degrees}) = \frac{\sqrt{4r^2 - C^2}}{2}$$

Fillet. Area $= 0.215r^2 = \frac{1}{5}r^2$, approx. (Fig. 26).

FIG. 26.

FIG. 27.

Ribbon. Bounded by two parallel curves. Area $= AB \times$ length of G. It is here assumed that the successive positions of the generating line will not intersect (Fig. 27).

Ellipse. The ellipse is the locus of a point that moves in such a way that the sum of its distances from two fixed points (foci) is constant (Fig. 28).

$$\text{Area of ellipse} = \pi ab$$

$$\text{Area of shaded segment} = xy + ab \sin^{-1} \frac{x}{a}$$

$$\text{Perimeter of ellipse} = \pi(a + b)k$$

where

$$k = 1 + \tfrac{1}{4}m^2 + \tfrac{1}{64}m^4 + \tfrac{1}{256}m^6 + \cdots$$

and

$$m = \frac{(a - b)}{(a + b)}.$$

Hyperbola. The hyperbola is the locus of a point which moves in such a way that the difference of its distances from two fixed points, called the foci, is con-

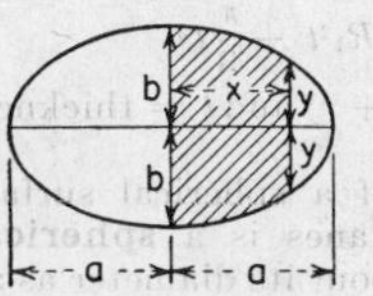

FIG. 28.

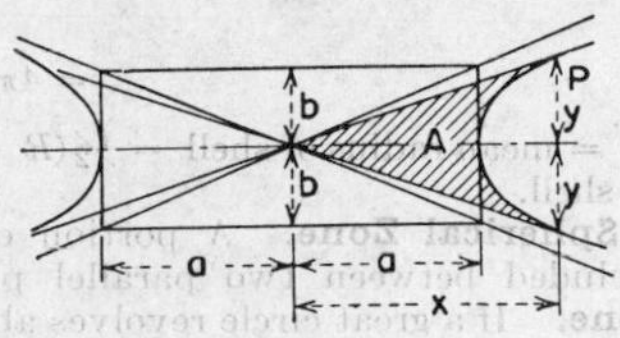

FIG. 29.

stant (Fig. 29). Shaded area $= ab \log_e \left(\frac{x}{a} + \frac{y}{b}\right)$. In an equilateral hyperbola, $a = b$. Shaded area $= a^2 \log$

$$\left(\frac{x + y}{a}\right) = a^2 \log \left(\frac{a}{x - y}\right)$$

$$= a^2 \sinh^{-1} \frac{y}{a} \text{ or } a^2 \cosh^{-1} \frac{x}{a}$$

Parabola. A parabola is the locus of a point which moves in such a way as to keep its distance from a fixed line called the directrix equal to its distance from a fixed point, called the focus. Shaded area $A = \frac{2}{3}ch$ or, in other words, area equals two-thirds the area of a rectangle having sides equal to c and h (Fig. 30).

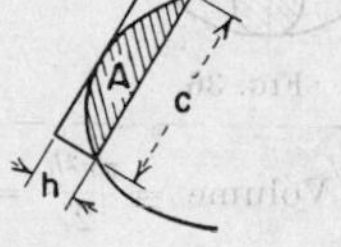

FIG. 30.

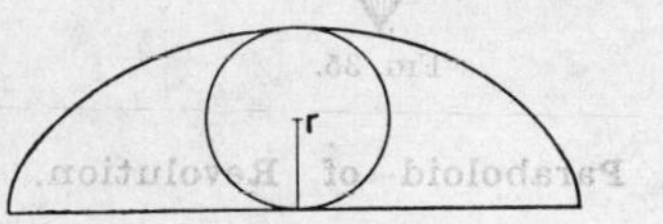

FIG. 31.

Cycloid. A curve described by a point on a circle which rolls along a fixed straight line. Length of arc $= S = 8r$; area $= 3\pi r^2$ (Fig. 31).

Irregular Areas. (Simpson's Rule). Divide the given area into n sections (where n is an even number) by means of $n + 1$ parallel lines, called ordinates, drawn at a constant distance apart. Then the area $= \frac{1}{3}h[(y_0 + y_n) + 4(y_1 + y_3 + y_5) + 2(y_2 + y_4 + y_6)]$, approx. (Fig. 32).

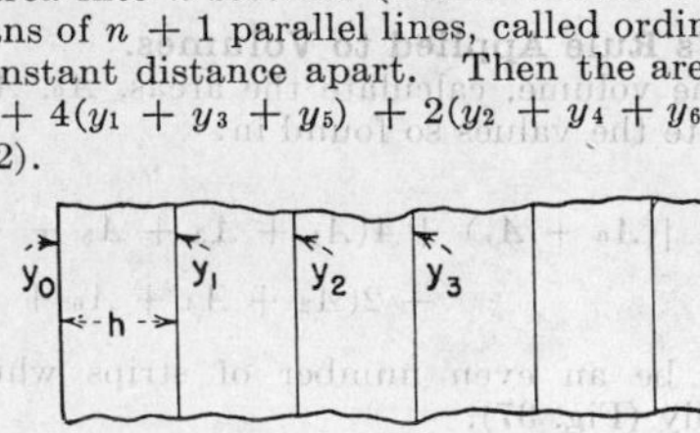

FIG. 32.

Surfaces and Volumes of Solids

Cubes. Volume $= a^3$; total surface area $= 6a^2$; diagonal $= a\sqrt{3}$, where a is the length of one side of the cube.

Rectangular Prism.

Volume = area of base × height $= blh$.

Total surface area $= 2(bl + lh + bh)$.

Diagonal $= d = \sqrt{b^2 + l^2 + h^2}$ = distance between opposite corners of parallel planes.

Regular Prism.

Volume $= \frac{1}{2}n \times r \times a \times h = B \times h$, where n = number of sides.

Lateral surface area $= n \times a \times h = P \times h$, where P = perimeter of base, h is the height, B is area of the base, and a is the length of a side.

Cylinder, Right Circular.

Volume $= \pi r^2 h = 0.7854 d^2 h$.

Lateral surface area $= 2\pi rh$. (Area of base $= \pi r^2$).

Total surface area $= 2\pi r(h + r)$, where r is the internal radius, h is the height, and d is the internal diameter.

Truncated Right Circular Cylinder (Fig. 33).

Volume $= \pi r^2 h = Bh$.

Lateral area $= 2\pi rh = Ph$, where h is the average height $= \frac{1}{2}(h_1 + h_2)$; B = area of base, P = perimeter of base.

FIG. 33.

Hollow Cylinder. Volume $= \pi h(R^2 - r^2)$, where r and R are the internal and external radii of the cylinder, and h is the height of the cylinder.

External surface area $= 2\pi Rh$

Internal surface area $= 2\pi rh$

Any Prism or Cylinder. Volume $= h \times$ area of base. Lateral surface area = length of cylinder × perimeter of normal section.

Any Truncated Prism or Cylinder. Volume = distance between centers of gravity of the two bases × area of normal section. Lateral surface area = perimeter of normal section × distance between centers of gravity of the two perimeters.

Regular Pyramid or Cone. Volume $= \frac{1}{3}$ (area of base × height). Lateral area of regular figure $= \frac{1}{2}$ (perimeter of base × slant height).

Any Pyramid or Cone. Volume $= \frac{1}{3}$ (area of base × distance from vertex to plane of base).

Frustum of Any Pyramid or Cone.

$$\text{Volume} = \frac{h}{3}(A_1 + A_2 + \sqrt{A_1 \times A_2}),$$

where A_1 and A_2 are the areas of bases made by parallel planes.

THEOREMS OF SPHERICAL GEOMETRY

Every intersection of a spherical surface by a plane is a circle. The intersection of a spherical surface by a plane (1) passing through the center is a **great circle** of the sphere, and (2) that not passing through the center is a **small circle.** If a diameter of a sphere is perpendicular to the plane of a circle of the sphere, the extremities are the **poles.** Parallel circles have the same poles. All great circles of a sphere are equal. Every great circle bisects the spherical surface. Two great circles bisect each other. If the planes of two great circles are perpendicular each circle passes through the poles of the other. Through two given points on the surface of a sphere an arc of a great circle may always be drawn. One and only one circle can be drawn through three given points on the surface of a sphere. The **spherical distance** between two points on the surface of a sphere is the length of the smaller arc of the great circle joining the points. The spherical distances of all points on a circle of a sphere from either pole of the circle are equal. The spherical distance from the nearer pole of a circle to any point on the circle is the **polar distance** of the circle. One-fourth of a great circle is a **quadrant,** and the polar distance of a great circle is a **quadrant.** A point on a sphere, which is at the distance of a quadrant from each of two other points, not the extremities of a diameter, is a pole of the great circle passing through these points. A plane perpendicular to a radius at its extremity is tangent to the sphere. A sphere may be inscribed in or circumscribed about any tetrahedron. Through four points not in the same plane one and only one spherical surface can be passed. The intersection of two spherical surfaces is a circle whose plane is perpendicular to the line which joins the centers of the spheres and whose center is in that line. The opening between two great circle arcs that intersect is a **spherical angle.** A spherical angle is measured by the arc of the great circle described from its vertex as a pole and included between its sides, which are produced if necessary. A **spherical polygon** is a portion of a spherical surface bounded by three or more arcs of great circles. An arc of a great circle joining two non-consecutive vertices of a spherical polygon is a **diagonal.** A **spherical triangle** is a spherical polygon of three sides. Each side of a spherical triangle is less than the sum of the two other sides. The sum of the sides of a spherical polygon is less than 360°. If, from the vertices of a spherical triangle as poles, arcs of great circles are described, another spherical triangle is formed which is the **polar triangle** of the first. If one spherical triangle is the polar triangle of another, then reciprocally the opposite is true. In two polar triangles each angle of one is the supplement of the opposite side of the other. The sum of the angles of a spherical triangle is greater than 180° and less than 540°. Two triangles on the same or equal spheres are congruent or symmetrical if (1) two sides and the included angle of one are respectively equal to the corresponding parts of the other, or (2) if two angles and the included side of one are respectively equal to the corresponding parts of the other. Two mutually equilateral triangles on the same or equal spheres are mutually equiangular and hence are congruent or symmetrical. In an isosceles spherical triangle the angles opposite the equal sides are equal. The shortest line that can be drawn on the surface of a sphere between two points is the arc of a great circle joining the two points, not greater than a semicircle. A portion of a spherical surface bounded by the halves of two great circles is a **lune.** The angle between the semicircles bounding a lune is the **angle of the lune.** The spherical excess is the excess of the sum of the angles of a spherical triangle over 180°. The **area of a lune** is to the area of the surface of a sphere as the angle of the lune is to four right angles. A spherical triangle is equivalent to a lune whose angle is half the spherical excess of the triangle. The **spherical excess of a polygon** is the excess of the sum of the angles of a spherical polygon of n sides over $(n - 2) \times 180°$. A spherical polygon is equivalent to a lune whose angle is half the spherical excess of the polygon. A portion of a sphere bounded by a spherical polygon and the planes of its sides is a **spherical pyramid.** A portion of a sphere generated by the revolution of a circular sector about any diameter of the circle of which the sector is a part is a **spherical sector.** A portion of a sphere contained between two parallel planes is a **spherical segment** (Fig. 34). A portion of a sphere bounded by a lune and the planes of two great circles is a **spherical wedge.**

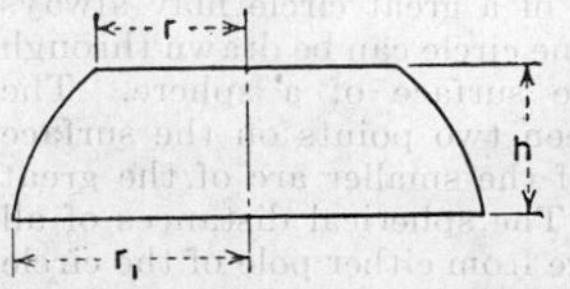

FIG. 34.

Sphere. Volume $= \frac{4}{3}\pi r^3 = 4.1888r^3 = \frac{\pi}{6} D^3 = 0.5236D^3$. The volumes of two spheres are to each other as the cubes of their radii. Area $= 4\pi r^2 = 12.5664r^2 = \pi D^2$ = area of four great circles. The area of a sphere is the same as the lateral area of a circumscribed cylinder. The areas of the surfaces of two spheres are to each other as the squares of their radii or as the squares of their diameters.

Hollow Sphere. Volume $= \frac{4}{3}\pi(R^3 - r^3)$

$$= 4.1888(R^3 - r^3)$$

$$= \frac{\pi}{6}(D^3 - d^3)$$

$$= 0.5236(D^3 - d^3)$$

$$= 4\pi R_1{}^2 t + \frac{\pi}{3} t^3$$

R_1 = mean radius of shell $= \frac{1}{2}(R + r)$ and t = thickness of shell.

Spherical Zone. A portion of a spherical surface included between two parallel planes is a **spherical zone.** If a great circle revolves about its diameter as an axis, any arc of the circle generates a **zone.**

$$\text{Area} = 2\pi R(R - \sqrt{R^2 - r^2}) = 2\pi Rh,$$

where R is the radius of the sphere.

Spherical Sector (Fig. 35). Volume $= \frac{2}{3}\pi r^2 h$.

$$\text{Total area} = 2\pi rh + \pi ra = \pi r(2h + a).$$

Ellipsoid. Volume $= \frac{4}{3}abc\pi = 4.1888abc$, where a, b, and c are the radii of the ellipsoid.

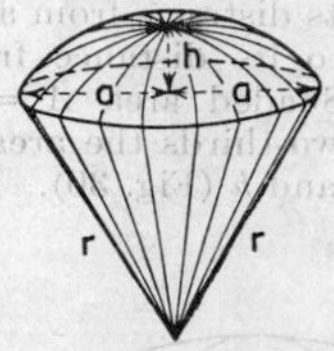

FIG. 35.

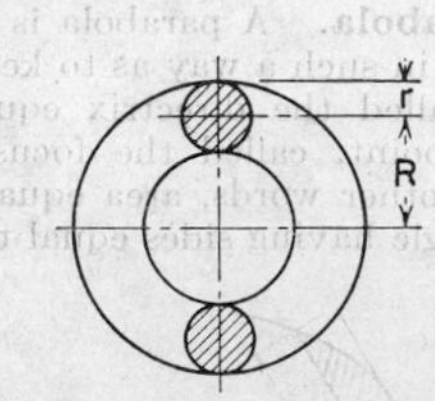

FIG. 36.

Paraboloid of Revolution. Volume $= \frac{\pi r^2 h}{2} = \frac{1}{2}$ volume of circumscribed cylinder.

Torus (Fig. 36). Volume $= 2\pi^2 Rr^2$. Surface $= 4\pi^2 Rr$.

Prismoidal Solid. Volume $= \frac{L}{6}(A + B + 4M)$, where L is the distance between parallel sides, A and B are the areas of end sections, and M is the average middle section.

Simpson's Rule Applied to Volumes.

To find the volume, calculate the areas, A_0, A_1, etc., and substitute the values so found in:

$$\text{Volume} = \frac{h}{3}\,[(A_0 + A_n) + 4(A_1 + A_3 + A_5 + \cdots) + 2(A_2 + A_4 + A_6 + \cdots)]$$

There must be an even number of strips which are spaced equally (Fig. 37).

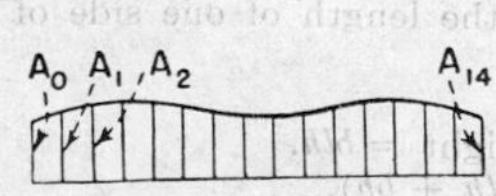

FIG. 37.

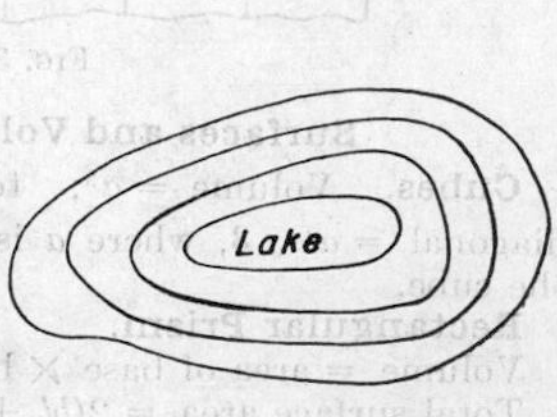

FIG. 38.

The volume of water can be computed by getting the areas of the contours and then using Simpson's rule to obtain the volume (Fig. 38).

GEOMETRICAL CONSTRUCTIONS*

To Bisect a Line. From either end as centers, and with equal radii, describe arcs which intersect at a point P

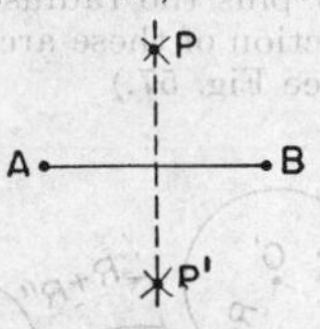

FIG. 39.

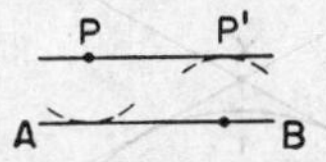

FIG. 40.

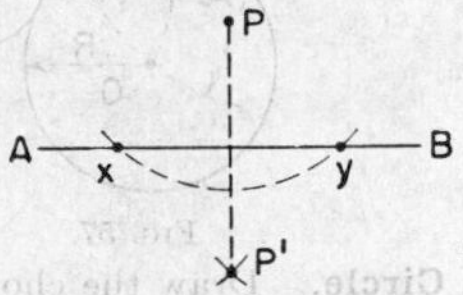

FIG. 41.

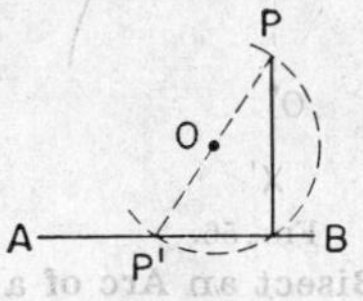

FIG. 42.

above the line AB, and also at a point P' below the line. Connect these points of intersection by a straight line PP' which will bisect the line AB. (See Fig. 39.)

To Draw a Parallel to a Given Line through Any Point. With the point P as a center, draw an arc which just touches the line AB; then with any point on the line AB as a center, draw an arc of the same radius. A line joining the original point P and touching the latter arc at P' will be the required parallel. (See Fig. 40.)

To Draw a Perpendicular to a Given Line from a Given Point, outside the Line. (1) With the point P as a center, describe an arc which cuts the line AB at two points X and Y; then bisect the line between these two points. The point of intersection is the foot of the perpendicular, PP'. (2) If the point P is nearly opposite one end of the line AB, take any point O above the line as a center and with a radius $= OP$ draw a circle cutting the line AB at P' and B, then connect P and B. (3) Use a triangle and a straightedge. (See Figs. 41 and 42.)

To Draw a Perpendicular to a Given Line at a Given Point on the Line. (1) With the point P on the line as a center, describe an arc cutting the line AB on either side of the point. Then with these points as

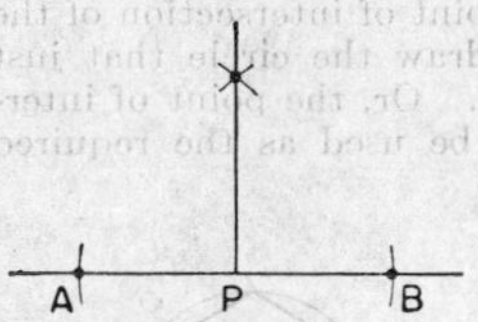

FIG. 43.

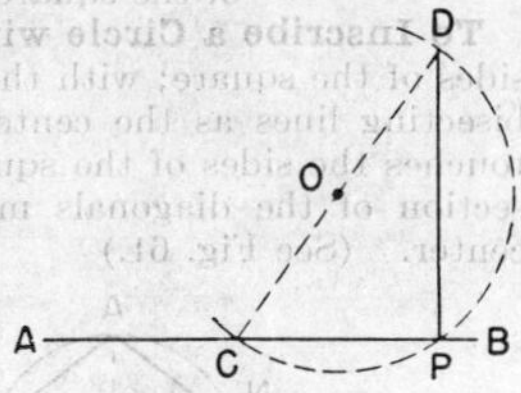

FIG. 44.

centers, and any radius greater than half the distance between them, describe arcs intersecting at a point above the line; connect this intersection of arcs with the original point P on the line by a straight line. (2) If the point P is near the end of the line, take any point O above the line as a center, and, with a radius OP, describe an arc cutting the line at a second point C; draw a straight line between the point O and the second point of intersection, and extend until it intersects the arc at a point D; connect this point D and the first point P to obtain the required perpendicular. (3) Use a straightedge and triangle. (See Figs. 43 and 44.)

To Divide a Line into n Equal Parts. Through one end of the line AB, draw a second line AX at any angle to the first line, preferably longer than the first line; lay off n equal divisions on this second line; connect the last of these divisions with the other end of the first line, and draw parallels through the other divisions of the second line. These parallels will divide the first line into the required n parts. (See Fig. 45.)

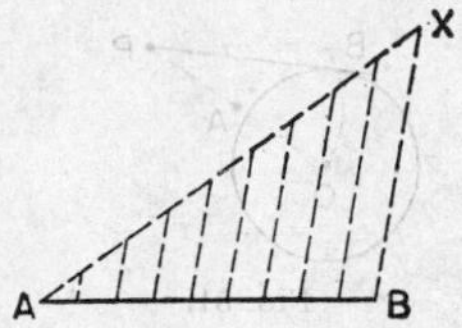

FIG. 45.

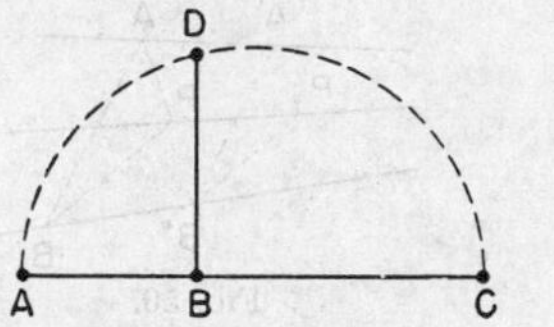

FIG. 46.

To Construct a Geometric Mean between Two Lengths. Lay off the lengths on a straight line, end to end; then construct a semicircle, with the combined lengths as the diameter; erect a perpendicular at the meeting point B of the lengths; this perpendicular will meet the circumference at a point D. The line BD will then be the geometric mean of the two lengths AB and BC. It can easily be proved that the **geometric mean** of two lengths is the square root of their product. (See Fig. 46.)

To Divide a Line in Extreme and Mean Ratio. At one end of the given line AB, erect a perpendicular QB equal to one-half the length of the given line; connect the other end of the given line to the upper end of the perpendicular; with the upper end of the perpendicular Q as a center and its length as the radius, describe an arc cutting the end of the line at B and the second line AQ at D. With A as center and radius AD, describe an arc cutting AB at C. The line AB is then divided into extreme and mean ratio by the point C. $AC = \sqrt{AB \cdot CB} = 0.618AB$. (See Fig. 47.)

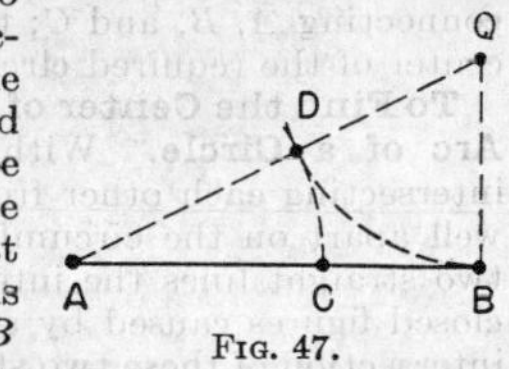

FIG. 47.

To Bisect an Angle. Lay off equal lengths on the sides of the angle; then with these points as centers and any radius, describe arcs which intersect; connect this intersection and the vertex of the angle with a straight line,

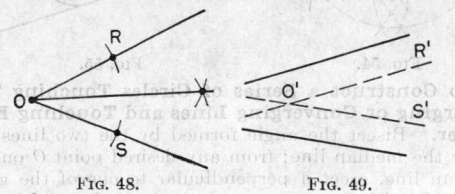

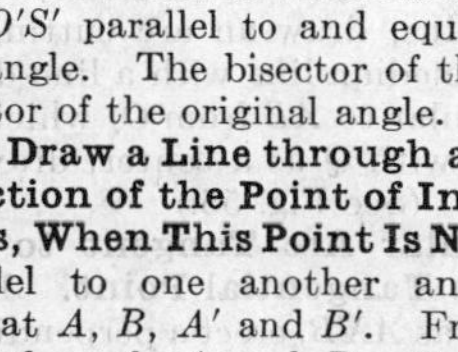

FIG. 48.

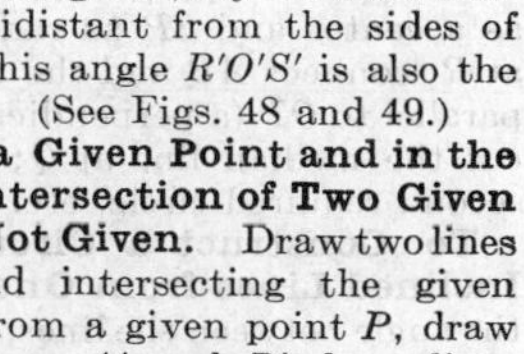

FIG. 49.

which will bisect the given angle. In order to bisect an angle, the vertex of which is not given, lay off lines $O'R'$ and $O'S'$ parallel to and equidistant from the sides of this angle. The bisector of this angle $R'O'S'$ is also the bisector of the original angle. (See Figs. 48 and 49.)

To Draw a Line through a Given Point and in the Direction of the Point of Intersection of Two Given Lines, When This Point Is Not Given. Draw two lines parallel to one another and intersecting the given lines at A, B, A' and B'. From a given point P, draw lines through A and B. From A' and B' draw lines parallel to PA and PB intersecting at P'. The line through P and P' is the desired line. (See Fig. 50.)

* The drawings for this portion of the Mathematics Section were made by C. B. Shepherd, E. I. duPont de Nemours & Co.

To Draw a Tangent to a Circle from an External Point. Bisect the line PO connecting the external point P and the center of the circle O; with the point of bisection A (at the center) as a center and a radius equal to one-half the length of the line PO, describe an arc intersecting the circumference of the circle; the point of intersection B is the required point of tangency. Connect the external point P with the point of tangency B by a straight line PB. (See Fig. 51.)

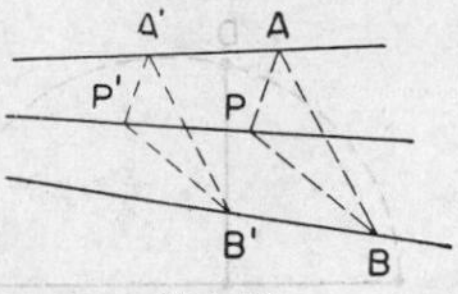

FIG. 50.

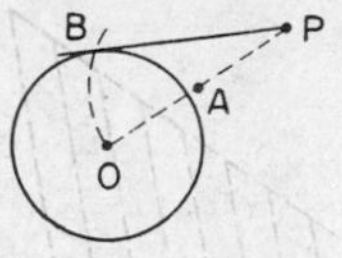

FIG. 51.

To Draw a Tangent to a Circle from a Given Point on the Circumference. Draw a radius PO of the circle through the given point P, and at the point P construct a perpendicular to this line which gives the required tangent. (See Fig. 52.)

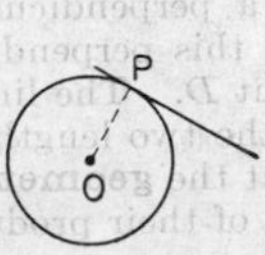

FIG. 52.

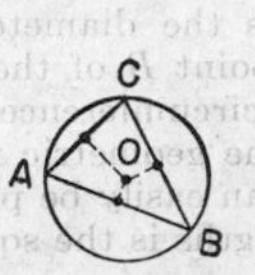

FIG. 53.

To Draw a Circle through Three Different Points A, B, and C or to Circumscribe a Triangle by a Circle. Draw the perpendicular bisectors of the lines connecting A, B, and C; these lines will intersect at the center of the required circle. (See Fig. 53.)

To Find the Center of a Circle, or the Center of an Arc of a Circle. With the same radius, draw arcs intersecting each other from any three points which are well apart on the circumference of a circle; connect by two straight lines the intersecting points of each of the closed figures caused by such intersections; the point of intersection of these two straight lines is the center of the circle or the center of the arc of the circle. (See Fig. 54.)

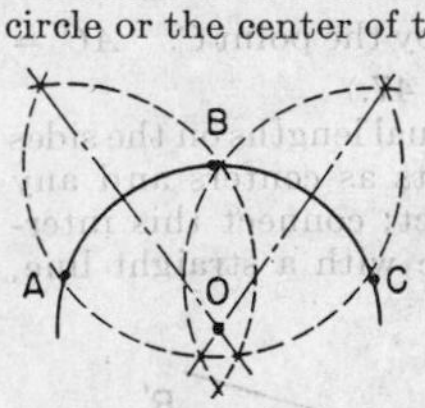

FIG. 54.

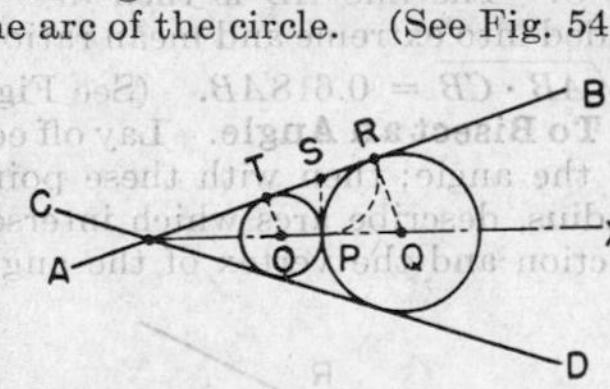

FIG. 55.

To Construct a Series of Circles Touching Two Diverging or Converging Lines and Touching Each Other. Bisect the angle formed by the two lines and draw the median line; from any desired point O on this median line, erect a perpendicular to one of the given lines AB; with O as a center, draw the circle touching AB and CD which will intersect the median line at a point P; from P, erect a perpendicular PS to the line AX; with S as a center and SP as radius, draw an arc cutting AB at R; connect AB and the median line with a line drawn parallel to OT (a perpendicular to AB from O) which will cut the median line at Q; with Q as a center, draw the second required circle, etc. (See Fig. 55.)

To Construct a Circular Arc Tangent to Two Inclined Lines from One Tangential Point. Bisect the angle between the lines, as AXB, erect a perpendicular PO to one line from the given tangential point P which will intersect the median line XX' at a point O which is the required center of the arc. (See Fig. 56.)

To Construct a Circle of Known Radius (R), Tangent to Two Given Circles. From the centers of the given circles draw two arcs, one of which has a radius equal to (R) plus the radius of the first circle, the other arc having a radius equal to (R) plus the radius of the second circle; the point of intersection of these arcs is the center of the required circle. (See Fig. 57.)

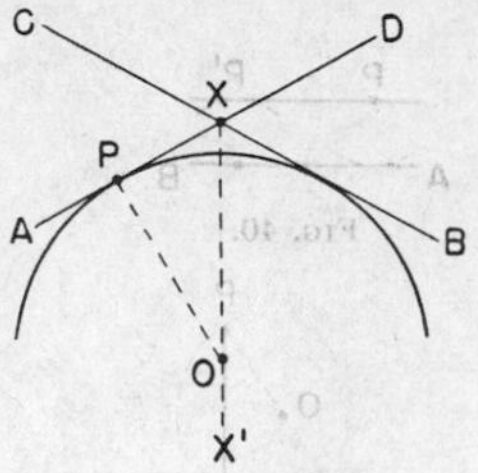

FIG. 56.

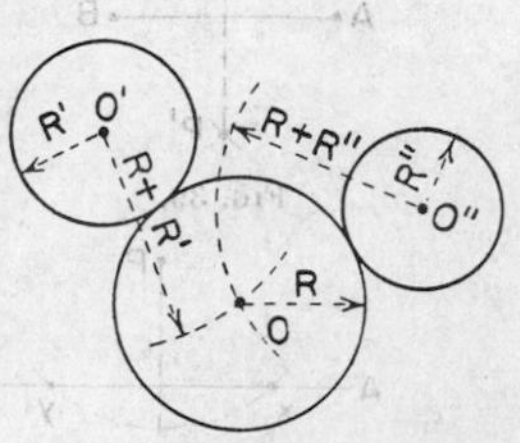

FIG. 57.

To Bisect an Arc of a Circle. Draw the chord and bisect it. From the center point D, erect the perpendicular DC which will bisect the arc. (See Fig. 58.)

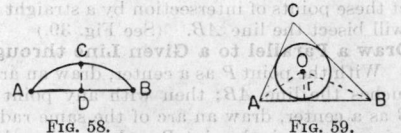

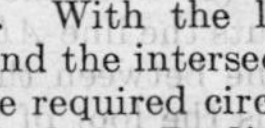

FIG. 58.

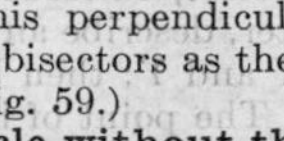

FIG. 59.

To Inscribe a Circle in a Given Triangle. Bisect two of the angles of the triangle; from the intersection of these bisectors, erect a perpendicular to one side of the triangle. With the length of this perpendicular as a radius and the intersection of the bisectors as the center, draw the required circle. (See Fig. 59.)

To Find the Radius of a Circle without the Center. The diameter is obtained by squaring half the chord, adding to this the square of the height and dividing the whole sum by the height.

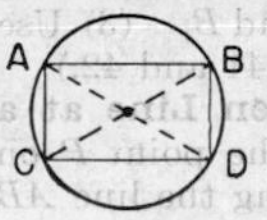

FIG. 60.

To Circumscribe a Square or a Rectangle by a Circle. Draw the diagonals of the square; with the point of intersection of the diagonals as the center, draw the circle that passes through the corners of the square. (See Fig. 60.)

To Inscribe a Circle within a Square. Bisect two sides of the square; with the point of intersection of the bisecting lines as the center, draw the circle that just touches the sides of the square. Or, the point of intersection of the diagonals may be used as the required center. (See Fig. 61.)

FIG. 61.

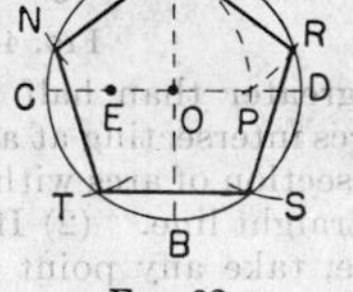

FIG. 62.

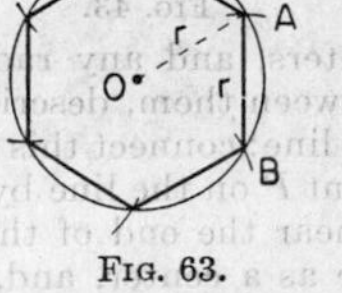

FIG. 63.

To Inscribe a Regular Pentagon in a Circle. Draw two diameters intersecting at right angles (at the circle center O); bisect line CO to give E; with radius EA and center at E, draw arc cutting CD at P; from A, with radius AP, cut the circle circumference at N and R; with the same radius and centers R and N, cut the circle at S and T; connect points A, N, T, S, and R to form the required pentagon. (See Fig. 62.)

To Inscribe a Regular Hexagon in a Circle. Using the radius of the circle as a chord, step off the circumference of the circle and connect the intersections of the circle with the resulting arcs by straight lines. Alternate method: Use a 60° triangle. (See Fig. 63.)

To Circumscribe a Regular Hexagon about a Circle. Draw a chord equal to the radius; bisect the arc projected by the chord; draw the tangent to the point of intersection of the bisector and the circumference; draw radii through the ends of the chord to intersect the tangent; draw a circle of a radius equal to the distance from the center of the circle to the point of intersection of the radius and the tangent, and inscribe a hexagon in this second circle. (See Fig. 64.)

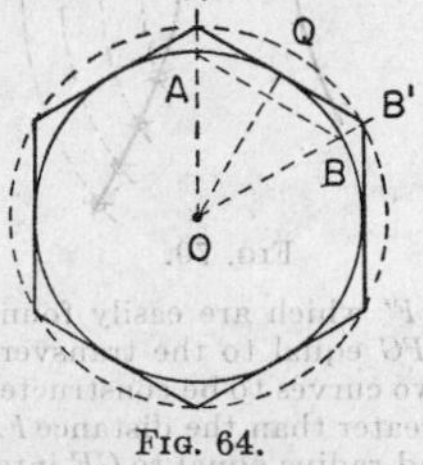

FIG. 64.

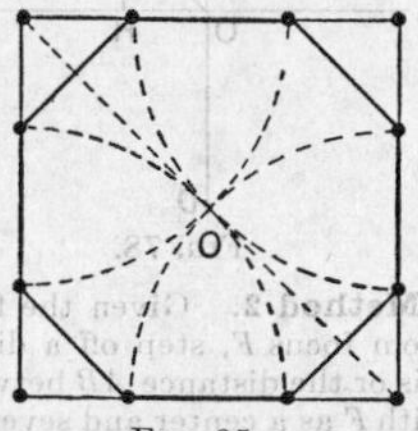

FIG. 65.

To Inscribe a Regular Octagon in a Square. With the corners of the square as centers, and a radius equal to one-half the length of a diagonal, draw four arcs, cutting the sides in eight points. These points are the vertices of the octagon. (See Fig. 65.)

To Inscribe a Regular Octagon in a Circle. Draw two perpendicular diameters, and bisect each of the quadrant arcs. (See Fig. 66.)

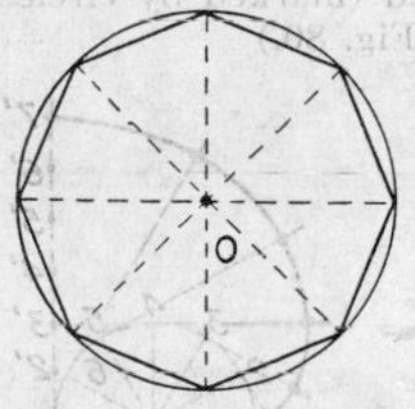

FIG. 66.

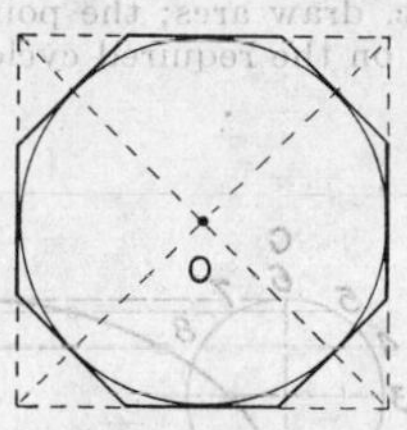

FIG. 67.

To Circumscribe a Regular Octagon about a Circle. Draw a square about the circle, then draw the tangents to the circle at the points where the circle is cut by the diagonals of the square. (See Fig. 67.)

To Construct a Regular Polygon of n Sides, When One Side Is Given. With one end of the given side as a center, and the length of the side as a radius, draw a semicircle and divide it into n equal parts, of which $n - 2$ parts are to be used. Draw lines from the end of the given side through these points of division of the semicircle. With radius equal to given side and end of given side as center, describe arc intersecting extension of nearest radius. With point of intersection as new center and same radius, describe arc cutting the extension of second radius. Repeat until center of circle is reached. (See Fig. 68.)

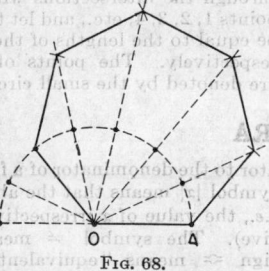

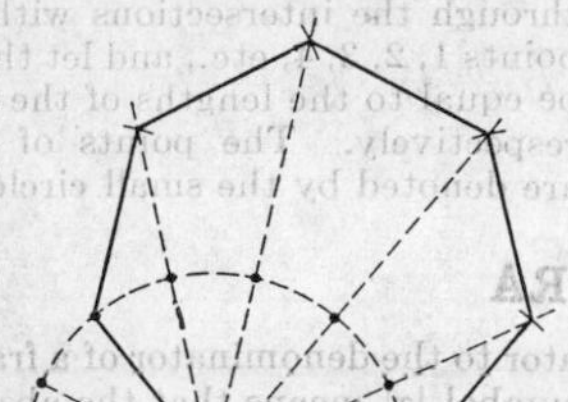

FIG. 68.

To Circumscribe Any Regular Polygon (with an Even Number of Sides) with a Circle. Find the center of the polygon by drawing one line between opposite points of the polygon, and drawing one line which bisects opposite sides of the polygon. The center of the circle is at the intersection of these two lines. (See Fig. 69.)

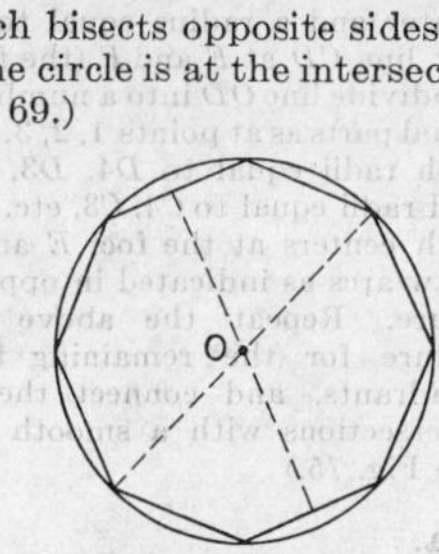

FIG. 69.

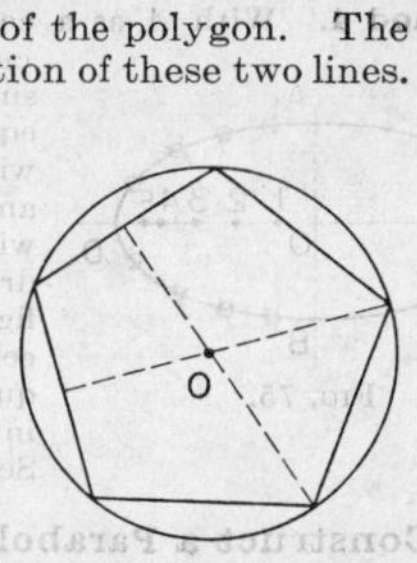

FIG. 70.

To Circumscribe Any Regular Polygon (with an Uneven Number of Sides) with a Circle. Find the center by drawing two lines each of which is drawn from one corner of the polygon to the middle point of the opposite side. The intersection of these lines is the center of the required circle. (See Fig. 70.)

To Draw an Annulus Which Shall Contain a Given Number of Equal Contiguous Circles. An annulus is defined as a ring-shaped area enclosed between two concentric circles. Let r be the radius of each of the n circles; and $R + r$ and $R - r$ will be the inner and outer radii of the annulus. The relation between these quantities is, $r = R \sin\ (180°/n)$, or $r = (R + r) \times \frac{\sin\ (180°/n)}{H \sin\ (180°/n)}$; therefore, $R = r \csc \frac{180°}{n}$. (See Fig. 71.)

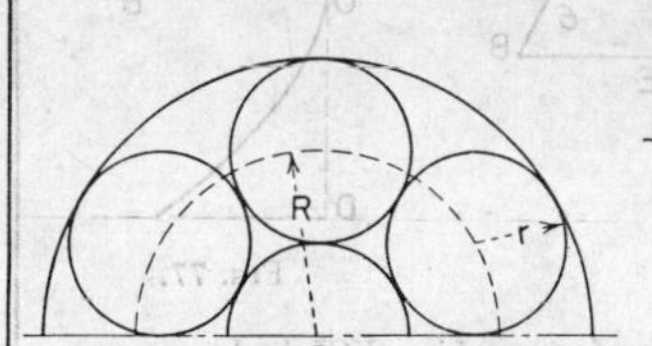

FIG. 71.

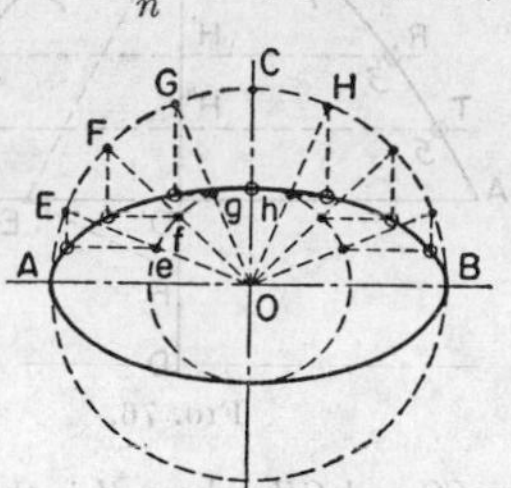

FIG. 72.

To Construct an Ellipse (Given the Two Diameters).

Method 1. With the intersection of the diameters as the center, draw two circles with the length of the ellipse diameters as the diameters of the circles. From a number of points on the circumference of the larger circle as E, F, G, H, etc., draw radii cutting the inner circle at e, f, g, etc. Drop perpendiculars to line AB from E, F, G, etc., and draw perpendiculars to line CD from e, f, g, etc. The intersections of these perpendiculars are points on the required ellipse. (See Fig. 72.)

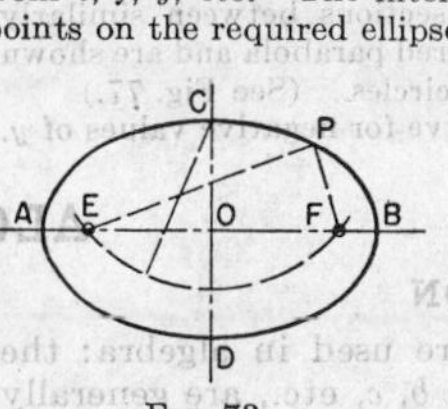

FIG. 73.

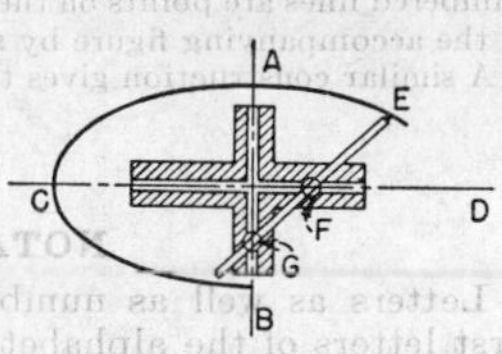

FIG. 74.

Method 2. From C as a center and with a radius equal to AO, describe an arc cutting the axis AB at E and F (the foci); insert pins at E and F and fasten each end of a string at these points, the length of the string being equal to the distance AB; place a pencil inside the string and guide it through A, B, C, and D thus completing the ellipse. (See Fig. 73.)

Method 3. A method similar to that of Method 2 above is to place a slotted right-angled cross over the axes of the ellipse to be made. The ellipse is then traced by means of a pointer or pencil inserted at the end of a bar at E by moving the bar keeping the pegs F and G of the bar in their respective slots. (See Fig. 74.)

Method 4. With A as a center and a radius equal to CO, cut line CD at E and F (the foci); subdivide line OD into a number of equal parts as at points 1, 2, 3, etc.; with radii equal to $D4$, $D3$, etc., and radii equal to $C4$, $C3$, etc., and with centers at the foci E and F, draw arcs as indicated in opposite figure. Repeat the above procedure for the remaining three quadrants, and connect the arc intersections with a smooth line. See Fig. 75.)

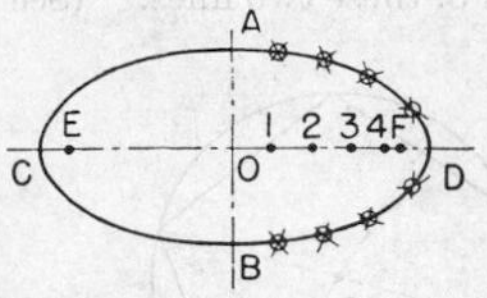

FIG. 75.

To Construct a Parabola.

Method 1. From the equation $X^2 = 2PY$, where X and Y are values of the abscissa and ordinate, respectively, and P = one-half the parameter = the distance from the focus to the directrix, substitute arbitrary values of X in the equation and plot the ordinate so obtained.

A similar method is applicable from the polar equation $r = P/(1 - \cos \theta)$.

Method 2. Given the vertex, an abscissa, and its ordinate for a parabola of the above type; bisect the abscissa OB at E, draw CE, construct a perpendicular to CE at E, and produce to strike line OD at F; mark off on line CD, distances equal to OF,

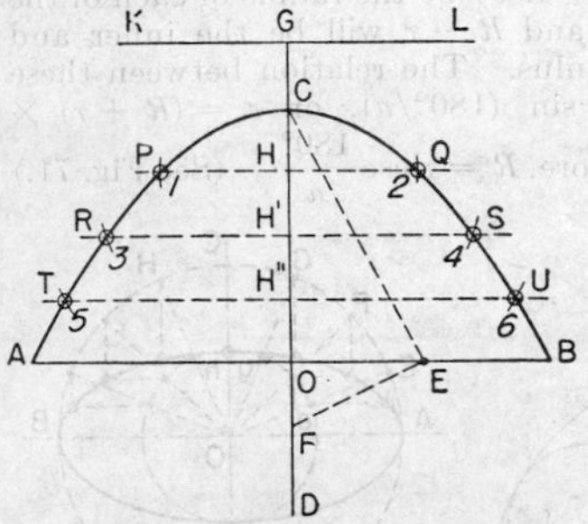

FIG. 76.

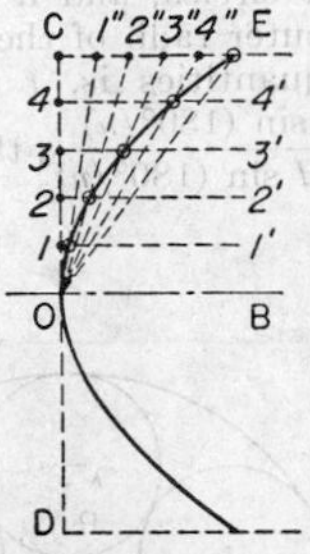

FIG. 77.

as CG and CH, where H is the focus. Line KGL is drawn perpendicular to line GO. Through H and any number of points, draw lines PQ, RS, and TU, parallel to the axis. With H as a center, draw arcs of radii HG, $H'G$, $H''G$, etc., to cut PQ, RS, TU at 1, 2, 3, 4, 5, 6, etc. Connect these intersections to obtain the required parabola. (See Fig. 76.)

Method 3. Erect CO perpendicular to X-axis at vertex O, with $CO = Y$ (ordinate of given point). Draw CE parallel to X-axis; given vertex, one point E, and equation $Y^2 = 2PX$. Divide line CO into several equal parts and divide CE into a like number of equal parts. Draw lines 1-1′, 2-2′, 3-3′, etc., parallel to CE and to OB. Draw lines connecting O with points 1″, 2″, 3″, 4″, etc., of line CE. The intersections between similarly numbered lines are points on the required parabola and are shown in the accompanying figure by small circles. (See Fig. 77.)

A similar construction gives the curve for negative values of y.

To Construct an Hyperbola

Method 1. From F (any point in, say, the upper right-hand quadrant) on the ordinate line GH, draw OF produced into the quadrant which meets the line GJ at K. Draw FL perpendicular to line GH, and KM perpendicular to line GJ and meeting FL at point M. Point M is then one point on the required hyperbola. Repeat the above operation for different ordinate lines for successive points on the required hyperbola. (See Fig. 78.)

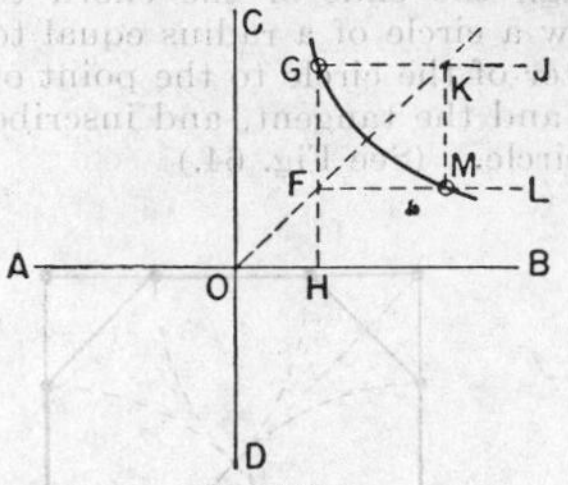

FIG. 78.

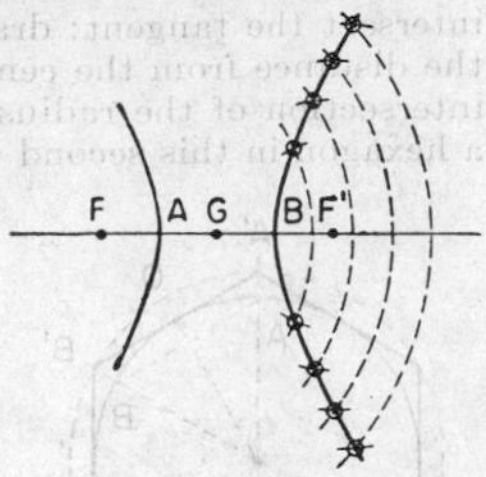

FIG. 79.

Method 2. Given the foci F and F' which are easily found. From focus F, step off a distance as FG equal to the transverse axis or the distance AB between the two curves to be constructed. With F as a center and several radii greater than the distance FB, draw arcs; then with F' as a center and radius equal to GE intersecting the arcs just described. The points of intersection are then points on the required hyperbola and are shown as small circles. (See Fig. 79.)

To Construct a Cycloid. Divide the semicircumference of the generating circle into a number of equal parts (say 6 parts); step off these arcs on the base line AB. With successive centers at 1, 2, and 3, etc., on AB and with radii equal to the distances 0-1′, 0-2′, and 0-3′, etc. draw arcs; the points obtained (marked by circles) lie on the required cycloid. (See Fig. 80.)

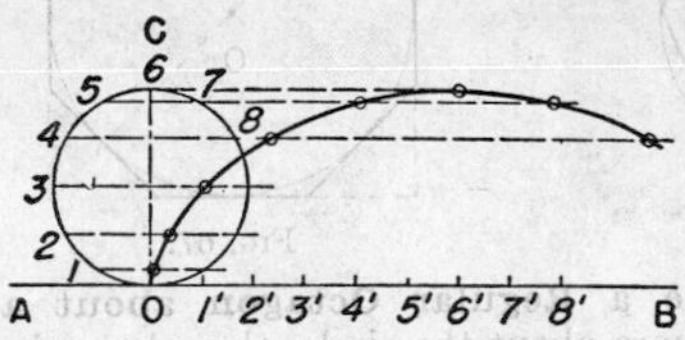

FIG. 80.

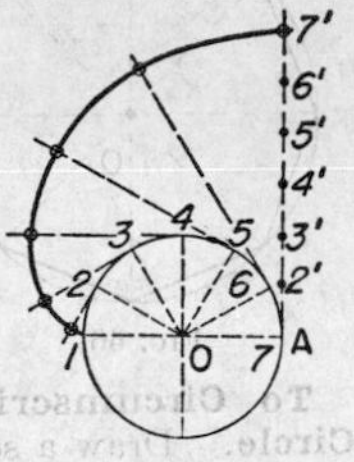

FIG. 81.

To Construct the Involute of a Circle. Divide the base circle into say 12 equal parts and draw tangents through the intersections with the circumference at the points 1, 2, 3, 4, etc., and let the lengths of these tangents be equal to the lengths of the arcs A-2′, A-3′, A-4′, etc., respectively. The points of the involute so obtained are denoted by the small circles. (See Fig. 81.)

ALGEBRA

NOTATION

Letters as well as numbers are used in algebra: the first letters of the alphabet as a, b, c, etc., are generally used to denote known quantities or numbers, while the last letters of the alphabet as x, y, z, are generally used to denote unknown quantities. As in arithmetic: $+$ indicates addition; $-$, subtraction; $\times$, multiplication; and $\div$, division. **Subscripts** as x_1, x_2, etc., denote different but unknown values of x. **Exponents, or superscripts,** denote the power to which the quantity specified is raised, *i.e.*, a^b denotes that a is to be raised to the bth power; here although the positive sign is omitted before the b, $a^b = a^{+b}$; a^{-b} is equivalent to $1/a^b$, since the sign of a superscript is changed when moved from the numerator to the denominator of a fraction, or vice versa. The symbol $|x|$ means that the absolute value is to be taken, *i.e.*, the value of x irrespective of sign (positive or negative). The symbol $\neq$ means "not equal to." The sign $\approx$ means "equivalent to." The sign $\doteq$ means "approaches" or "approximates to." $4 > 2$ or $a > b$ means "four is greater than two" or "a is greater than b"; similarly $2 < 4$ or $a < b$ means "two is less than four" or "a is less than b."

Collective Symbols. The parenthesis ( ), the bracket [ ], the brace { }, the bar | |, the vinculum ——, are all devices by which groups of quantities may be segregated. A minus sign before such a collective device requires the change of each sign therein when the collective device is removed. A plus sign does not affect the signs within such a device when the latter is removed, but a minus sign changes the signs of the terms within the device. For instance, $-(x^2 + y^2 - z^2) = -x^2 - y^2 + z^2$, and $+(x^2 + y^2 - z^2) = x^2 + y^2 - z^2$. If a collective device is located within another collective device, they are removed by removing innermost first and thus working to the outermost. For instance, $-[x^2 + y^2 - 2(x - y) + z] = -(x^2 + y^2 - 2x + 2y + z) = -x^2 - y^2 + 2x - 2y - z$.

The vinculum is used as follows: $\overline{x^2 + y^2} = (x^2 + y^2)$.

Addition and Subtraction. Only like terms can be added or subtracted in algebra, as $-5x + 3x + 4x^2 + 2x^2 + 5$ may be simplified by adding or subtracting like terms: $-2x + 6x^2 + 5$; *i.e.*, the x terms are like, and the x^2 terms are like. Or, $x^2 + 4x^2 + 5ax - 2ax - 3bx^4 + 2bx^4 - cy^5 = 5x^2 + 3ax - bx^4 - cy^5$.

Multiplication. Multiplication of factors with like signs gives plus; unlike signs give minus; *e.g.*, $(-x^2)(-2) = 2x^2$ (plus sign assumed when not given), and $(-x^2)(2) = -2x^2$. Other examples are: $ab + axb = bxa + ba$; $a(b + c) = ab + ac$; $ax(-c) = -acx$; $(a + b)(a - b) = a^2 - b^2$; $(a + b)(a + b) = a^2 + 2ab + b^2$; $(a - b)(a - b) = a^2 - 2ab + b^2$; $(a + b)^3 = a^3 + 3a^2b + 3ab^2 + b^3$; $(a - b)^3 = a^3 - 3a^2b + 3ab^2 - b^3$; $(x + a)(b + x) = x^2 + (a + b)x + ab$; $(a)(a) = a^2$; $(a^2)(a) = a^3$; $(a^m)a^n = a^{m+n}$. The operation of multiplication is usually carried out when the answer is not readily obtainable by inspection:

$(2a + 4x + 3y^2 + z)(a + x + y^2 + 2z)$:

$$\begin{array}{llllllllll}
2a & +4x & +3y^2 & +z \\
a & +x & +y^2 & +2z \\
\hline
2a^2 & +4ax & +3ay^2 & +az \\
 & +2ax & & & +4x^2 & +3xy^2 & +xz \\
 & & +2ay^2 & & & +4xy^2 & & +3y^4 & +y^2z \\
 & & & +4az & & & +8xz & & +6y^2z & +2z^2 \\
\hline
2a^2 & +6ax & +5ay^2 & +5az & +4x^2 & +7xy^2 & +9xz & +3y^4 & +7y^2z & +2z^2
\end{array}$$

Division. As in arithmetic the dividend is the quantity divided, the divisor is the quantity divided by, and the answer is the quotient. The quotient is positive if the signs of both the dividend and the divisor are alike; and the quotient is negative if the signs of the dividend and the divisor are unlike. $a^2b^3c^4/ab^4c^5 = a/bc$. Division is ordinarily performed by longhand as follows:

Divide $a^4 + 2a^3b + a^4b^4 + a^2b^2 + a^3b^5$ by $a + b$:

$$\begin{array}{l|l|l}
a + b & a^4 + 2a^3b + a^4b^4 + a^2b^2 + a^3b^5 & a^3 + a^2b + a^3b^4 \text{ (answer)} \\
 & a^4 + a^3b & \\
 & \quad a^3b + a^4b^4 + a^2b^2 + a^3b^5 & \\
 & \quad a^3b \qquad\quad + a^2b^2 & \\
 & \qquad a^4b^4 + a^3b^5 & \\
 & \qquad a^4b^4 + a^3b^5 &
\end{array}$$

Operations with Zero. All numerical operations (with the exception of division) can be made with zero: $a + 0 = a$; $a - 0 = a$; $(a)(0) = 0$; $a^0 = 1$; $0/a = 0$; $0/0$ and ∞/∞ are indeterminate.

Fractional Operations. $\frac{x}{y} = \frac{ax}{ay} = \frac{a^nx}{a^ny}$ or $\frac{x^2}{2} = \frac{ax^2}{2a} = \frac{a^nx^2}{2a^n}$; $-\left(\frac{x}{y}\right) = -\left(\frac{-x}{-y}\right) = -\left(\frac{+x}{y}\right)$; $\left(\frac{x}{y}\right) = \left(\frac{-x}{-y}\right)$; $\left(\frac{x}{y}\right)^n = \left(\frac{x^n}{y^n}\right)$; $\left(\frac{x}{y}\right) + \left(\frac{z}{y}\right) = \frac{x + z}{y}$; $\left(\frac{x}{y}\right) - \left(\frac{z}{y}\right) = \frac{x - z}{y}$; $\left(\frac{x}{y}\right)\left(\frac{z}{4}\right) = \frac{xz}{4y}$; $\frac{(a/b)}{(x/y)} = \left(\frac{a}{b}\right)\left(\frac{y}{x}\right) = \frac{ay}{bx}$.

Ratio and Proportion. The quotient of two quantities, as x/y, is called a **ratio** and may be written $x{:}y$ meaning "the ratio of x to y." Such a ratio is treated exactly like a fraction, and all the rules applying to fractions also apply to ratios. When two ratios are stated to be equal, the equality is called a **proportion.** Such a proportion is written $x/y = a/b$ or $x{:}y{::}a{:}b$, reading x is to y as a is to b. The first and last terms, x and b, are called the **extremes** of the ratio; and the second and third terms, y and a, are called the **means** of the ratio. The products of the corresponding terms of two or more proportions are themselves in proportion; *e.g.*, $a/b = c/d$ and $e/f = g/h$; then $ae/bf = cg/dh$. Product of extremes = product of means.

FACTORING

The process of dividing a given expression into two or more expressions which when multiplied together will produce the given expression is called **factoring.** The "two or more expressions" so obtained are called the **factors** of the given expression.

The first general rule is to factor out all monomials that are common to any terms of the expression as:

$$ax^2 + ay^2 + bx^2 + by^2 = a(x^2 + y^2) + b(x^2 + y^2) = (a + b)(x^2 + y^2).$$

The next step is to apply one of the following rules to any and all polynomials remaining in the expression:

(1) $(x^2 - y^2) = (x - y)(x + y)$

(2) $x^2 + 2xy + y^2 = (x + y)(x + y) = (x + y)^2$

(3) $x^2 + ax + c = (x + b)(x + d)$, where $b + d = a$ and $(b)(d) = c$

(4) $ey + by^2 + c = (dy + e)(fy + g)$, where $(d)(f) = b$, $(g)(e) = c$, and $(ef) + (gd) = a$

(5) $x^2 + y^2 + z^2 + 2xy + 2xz + 2yz = (x + y + z)(x + y + z) = (x + y + z)^2$

(6) $x^2 - y^2 - z^2 - 2yz = (x - y - z)(x + y + z)$

(7) $x^2 + y^2 + z^2 - 2xy - 2xz + 2yz = (x - y - z)^2$

(8) $(x^n - y^n) = (x + y)(x^{n-1} - x^{n-2}y + x^{n-3}y^2 - \cdots y^{n-1})$ when n is even $= (x - y)(x^{n-1} + x^{n-2}y + x^{n-3}y^2 + \cdots y^{n-1})$ always

(9) $x^3 - y^3 = (x - y)(x^2 + xy + y^2)$

(10) $x^3 + y^3 = (x + y)(x^2 - xy + y^2)$

(11) $x^4 - y^4 = (x^2 - y^2)(x^2 + y^2) = (x - y)(x + y)(x^2 + y^2)$

(12) $x^5 - y^5 = (x - y)(x^4 + x^3y + x^2y^2 + xy^3 + y^4)$

(13) $x^5 + y^5 = (x + y)(x^4 - x^3y + x^2y^2 - xy^3 + y^4)$

(14) $x^6 - y^6 = (x^3 - y^3)(x^3 + y^3) = (x - y)(x + y)(x^2 + xy + y^2)(x^2 - xy + y^2)$

Variation or Change. The variation sign is $\propto$, as $x \propto y$, meaning x varies with y. The variation sign may be replaced by an equality sign if we also place a sign as K denoting a constant on the right-hand side; *e.g.*, $x \propto y$ may be written $x = Ky$. The value of K is readily determined if we know any pair of values of the variables x and y. Suppose $x = 5$ and $y = 10$, then $K = x/y$ or $K = 5/10 = 0.5$.

Powers, Exponents, and Roots.

$\sqrt{x^2} = \pm x$; $\sqrt[n]{x^n} = x$ if n is odd, or $\pm x$ if n is even. $\sqrt[n]{x^m} = x^{\frac{m}{n}}$ [the nth root of (x^m)]: $(\sqrt[n]{x})^m = \sqrt[n]{x^m}$, $x^2/x^2 = x^{2-2} = x^0 = 1$, $x^{-2} = 1/x^2$, $1/x^{-2} = x^2$, $\sqrt{xy} = (\sqrt{x})(\sqrt{y})$, $(\sqrt{x^2y}) = x\sqrt{y}$. Any term like $\sqrt{-x}$ is *imaginary*, if x is *positive.*

$$(\sqrt{-x})^2 = -x; \quad \sqrt[n]{\frac{1}{x}} = \frac{1}{\sqrt[n]{x}} = \frac{1}{x^{\frac{1}{n}}} = x^{-\frac{1}{n}}$$

Approximations and Approximate Forms. The sign $\doteq$ means "approximates." $(1 + x)(1 + y) = 1 + x + y + xy$ approximates $1 + x + y$ when x and y are very small. $(1+x)(1-y) = 1 + x - y - xy$ approximates $1 + x - y$ when x and y are very small. $(1+x)(1+y)(1+z) = 1 + x + y + z + xy + xz + yz$ approximates $1 + x + y + z$ when x, y, and z are very small.

$$\frac{1}{(1-x)} \doteq 1 + x; \quad \frac{(1+x)}{(1-y)} \doteq 1 + x + y; \quad \frac{1}{(1+x)} \doteq 1 - x$$

$$(1 \pm x)^n \doteq 1 \pm nx$$
$$(x+a)^2 \doteq x^2 + 2xa$$
$$(1-x)^2 \doteq 1 - 2x$$
$$(1+x)^2 \doteq 1 + 2x; \quad \sqrt{1+x} \doteq 1 + \frac{x}{2}; \quad \sqrt{1-x} \doteq 1 - \frac{x}{2}$$
$$\frac{1}{\sqrt{1+x}} \doteq 1 - \frac{x}{2}; \quad \frac{1}{(1+x)^n} \doteq 1 - nx;$$
$$\frac{1}{\sqrt{1-x}} \doteq 1 + \frac{x}{2}; \quad \frac{1}{(1-x)^n} \doteq 1 + nx,$$

$(x \pm y)^n \doteq x^n \pm nx^{n-1}y$ when y is small

$\frac{1}{(1+x)} \doteq 1 - x$, where the error is less than the value of x^2 when $x > 0$ and < 1

$\frac{1}{(1+x)} \doteq 1 - x + x^2$, where the error is less than the value of x^3 when $x > 0$ and < 1

$\frac{1}{(1-x)} \doteq 1 + x$, where the error is less than $(x^2 + 2x^3)$ if $x < \frac{1}{2}$ and > 0

$\frac{1}{(1-x)} \doteq 1 + x + x^2$, where the error is less than $(x^3 + 2x^4)$ if $x < \frac{1}{2}$ and > 0

$$e^a \doteq 1 + a$$

THE BINOMIAL THEOREM

Positive Integral Exponents. This theorem is used to develop $(a+b)^n$ into the expanded form. The expansion of $(a+b)^n$ follows certain definite rules as is seen from the following:

$$(a+b)^n = a^n + na^{n-1}b + \frac{n(n-1)}{1 \times 2} a^{n-2}b^2 + \frac{n(n-1)(n-2)}{1 \times 2 \times 3} a^{n-3}b^3 + \cdots + \frac{n(n-1)(n-2) \cdots (n-r+2)}{1 \times 2 \times 3 \times \cdots x(r-1)} a^{n-r+1}b^{r-1} + \cdots + b^n$$

The **rules** are:

1. The exponent of b is less by one than the number of the term.
2. The exponent of a is n minus the exponent of b.
3. The last factor in the numerator of the coefficient is greater by one than the exponent of a.
4. The last factor of the denominator is the same as the exponent of b.
5. In the rth term the exponent of b will be $r - 1$.
6. The exponent of a will be $n - (r - 1)$ or $n - r + 1$.
7. The last factor of the numerator will be $n - r + 2$.
8. The last factor of the denominator will be $r - 1$.
9. The rth term

$$= \frac{n(n-1)(n-2) \cdots (n-r+2)}{1 \times 2 \times 3 \cdots (r-1)} a^{n-r+1}b^{r-1}.$$

This formula holds for positive, negative, or fractional values of n.

Any series developed from $(a+b)^n$ is infinite for fractional or negative values of n and convergent when b is numerically less than a. The series is divergent when b is numerically greater than a.

Some Binomial Series.

$$(1 \pm x)^n = 1 \pm nx + \frac{n(n-1)}{1 \times 2} x^2 \pm \frac{n(n-1)(n-2)}{1 \times 2 \times 3} x^3 + \cdots \text{ (convergent if } x^2 < 1)$$

$$(1 \pm x)^{-n} = 1 \mp nx + \frac{n(n+1)}{1 \times 2} x^2 \mp \frac{n(n+1)(n+2)}{1 \times 2 \times 3} x^3 + \cdots \text{ (convergent if } x^2 < 1)$$

$$(a - bx)^{-1} = \frac{1}{a}\left(1 + \frac{bx}{a} + \frac{b^2x^2}{a^2} + \frac{b^3x^3}{a^3} + \frac{b^4x^4}{a^4} + \cdots\right) \text{ (convergent if } b^2x^2 < a^2)$$

$$(1 \pm x)^{-1} = 1 \mp x + x^2 \mp x^3 + x^4 \mp x^5 + x^6 \mp x^7 + \cdots \text{ (convergent if } x^2 < 1)$$

$$(1 \pm x)^{-2} = 1 \mp 2x + 3x^2 \mp 4x^3 + 5x^4 \mp 6x^5 + \cdots \text{ (convergent if } x^2 < 1)$$

$$(1 \pm x)^{1/2} = 1 \pm 1/2\,x - 1/2 \cdot 1/4\,x^2 \pm 1/2 \cdot 1/4 \cdot 3/6\,x^3 - \cdots \text{ (convergent if } x^2 < 1)$$

$$(1 + x)^{-1/2} = 1 \mp 1/2\,x + 1/2 \cdot 3/4\,x^2 \mp 1/2 \cdot 3/4 \cdot 5/6\,x^3 + \cdots \text{ (convergent if } x^2 < 1)$$

$$(1 \pm x)^{1/3} = 1 \pm 1/3\,x - 1/3 \cdot 2/6\,x^2 \pm 1/3 \cdot 2/6 \cdot 5/9\,x^3 - \cdots \text{ (convergent if } x^2 < 1)$$

Binomial Approximations.

$$\log_e x = (x-1) - \frac{1}{2}(x-1)^2 + \frac{1}{3}(x-1)^3 + \cdots \frac{(-1)n - 1(x-1)^n}{n}; \; 2 \geqq x \geqq 0$$

$$\log_e x = \frac{x-1}{x} + \frac{1}{2}\frac{(x-1)^2}{(x)} + \frac{1}{3}\frac{(x-1)^3}{(x)} + \cdots + \frac{1}{n}\frac{(x-1)^n}{(x)}; \; x > \frac{1}{2}n$$

If a is small compared with 1, and n is reasonably small, say between the limits of $+2$ and -2, the following approximations are valid:

$$(1+a)^n = 1 + na \qquad (1-a)^n = 1 - na$$
$$(1+a)^{-n} = 1 - na \qquad (1-a)^{-n} = 1 + na$$

Logarithmic Series. This series is the expansion of $\log_e (1+x)$ in ascending powers of x.

$$\log_e (1 \pm x) = \pm x - \frac{x^2}{2} \pm \frac{x^3}{3} - \frac{x^4}{4} \pm \cdots$$

$$\log_e \left(\frac{1+x}{1-x}\right) = 2\left(x + \frac{x^3}{3} + \frac{x^5}{5} + \frac{x^7}{7} + \cdots\right) \text{ if } -1 < x < +1$$

$$\log_e \left(\frac{x+1}{x-1}\right) = 2\left(\frac{1}{x} + \frac{1}{3x^3} + \frac{1}{5x^5} + \frac{1}{7x^7} + \cdots\right) \text{ if } -1 > x \text{ or } x > +1$$

This series is convergent if $x < 1$.

Trigonometric Series. Where the angle (θ) is given in radians.

$\sin \theta = \theta - \frac{\theta^3}{3!} + \frac{\theta^5}{5!} + \frac{\theta^7}{7!} + \cdots$, where (!) means a factorial, *e.g.*, $3! = 3 \times 2 \times 1$ and $5! = 5 \times 4 \times 3 \times 2 \times 1$, etc.

$$\cos \theta = 1 - \frac{\theta^2}{2!} + \frac{\theta^4}{4!} - \frac{\theta^6}{6!} + \cdots$$

$$\tan \theta = \theta + \frac{\theta^3}{3} + \frac{2\theta^5}{15} + \frac{17\theta^7}{315} + \cdots$$

Exponential Series.

$$e = 1 + \frac{1}{1} + \frac{1}{2!} + \frac{1}{3!} + \frac{1}{4!} + \frac{1}{5!} + \cdots \frac{1}{(n-1)!}$$

$$e^x = 1 + \frac{x}{1} + \frac{x^2}{2!} + \frac{x^3}{3!} + \cdots + \frac{x^n}{n!};\ a^x = 1 + \frac{x \log a}{1} + \frac{(x \log a)^2}{2!} + \frac{(x \log a)^3}{3!} + \cdots + \frac{(x \log a)^n}{n!}$$

INFINITE SERIES

An infinite series is one in which there is no finite number of terms, *i.e.*, one in which the number of terms, n, can increase without limit.

If we have a series composed of terms that are positive, negative, or both, and we call the sum of the first n terms, S_n; that is, $S_n = U_1 + U_2 + U_3 + \cdots + U_n$, then if n increases without limit either (1) S_n approaches some finite number as a limit or (2) S_n does not approach a limit. In (1), the series is said to be convergent to S, the sum; whereas in (2) the series is non-convergent.

Non-convergent Series. This type of series embraces two classes: (1) a divergent series, in which S_n increases in absolute value without limit as n increases without limit; and (2) an **oscillating** series, in which S_n does not become infinite in absolute value as n increases without limit and which does not converge but **oscillates.** For example, $S_n = 1 - 1 + 1 - 1 + \cdots + (-1)^{n-1}$. A series of terms which are all positive cannot oscillate; S_n will always increase in such a series.

Example of Divergent Series. Consider the arithmetic series: $S_n = 1 + 2 + 3 + 4 + \cdots + n = \frac{n}{2}(1 + n)$. $S_1 = 1$; $S_2 = 1 + 2 = 3$; $S_3 = 6$; $S_4 = 10$, thus it is evident that S_n increases without limit as n increases without limit, and the series is divergent.

Comparison Test for Divergence. If we can find a series of terms already known to be divergent, whose terms are never greater than the corresponding terms of the series being tested, then the series being tested is a divergent series.

Example. Test the series: (1) $\frac{1}{\sqrt{2}} + \frac{1}{\sqrt{3}} + \frac{1}{\sqrt{4}} + \cdots + \frac{1}{\sqrt{n}}$ for divergence; compare with the divergent series: $\frac{1}{2} + \frac{1}{3} + \frac{1}{4} + \cdots + \frac{1}{n}$. (2) Since the denominator of each term in (1) is less than the corresponding denominator in (2), each term of (1) being greater than the corresponding term of (2), then (1) is divergent.

The following series are important in testing a series for divergence: the geometric series $a + ar + ar^2 + ar^3 + \cdots ar^{n-1}$, where $r \geqq 1$; and the harmonic series $1 + \frac{1}{2} + \frac{1}{3} + \frac{1}{4} + \cdots + \frac{1}{n}$.

Example of Convergent Series. Take the **geometric series,**

$$S_n = a + ar + ar^2 + ar^3 + \cdots ar^{n-1} = \frac{a(1 - r^n)}{1 - r}$$

where

$$-1 < r < 1.$$

Let $a = 1$, and $r = \frac{1}{2}$.
Then the series becomes

$$1 + \frac{1}{2} + \frac{1}{4} + \frac{1}{8} + \cdots + \frac{1}{2^{n-1}} \text{ and } S_n = 2 - \frac{1}{2^{n-1}}.$$

As n increases without limit, $\frac{1}{2^{n-1}}$ approaches zero as a limit, and S_n therefore approaches 2. As n increases, S_n can be made to differ from 2 by any assignable value, however small.

Comparison Test for Convergence. If we have a series of positive terms (1) which we wish to test for convergence, and we know another series of terms (2) which we know to be convergent, then if each of the terms of (1) is less than the corresponding term of series (2), the series (2) is convergent and its limiting value cannot be greater than the limiting value of (1).

Example. Prove that the series, $2 + 1 + \frac{1}{2^2} + \frac{1}{3^3} + \cdots + \frac{1}{(n-1)^{n-1}}$ (1) is convergent. Comparing with the geometrical series, $2 + 1 + \frac{1}{2^2} + \frac{1}{2^3} + \frac{1}{2^4} + \cdots + \frac{1}{2^{n-1}}$ (2), we see that, excluding the first three terms, each term of (1) is less than the corresponding term of (2), and further, by examining the nth terms of each series, that, if $n > 3$, then $1/(n-1)^{n-1} < 1/2^{n-1}$, so that series 1 is convergent. Other useful series for testing convergence are:

$$\frac{1}{1 \times 2} + \frac{1}{2 \times 3} + \frac{1}{3 \times 4} + \frac{1}{4 \times 5} + \cdots + \frac{1}{n(n+1)}$$

$$1 + \frac{1}{2^m} + \frac{1}{3^m} + \frac{1}{4^m} + \cdots + \frac{1}{n^m}$$

$$a + ar + ar^2 + ar^3 + \cdots + ar^{n-1}.$$

Ratio Test for Convergence. The ratio of the $(n+1)$st term of a series to the nth term is called the **ratio of convergence.** The absolute value of this ratio, as n increases without limit, will generally approach a definite limiting value, or it will increase without limit. If Q, the limit, < 1, the series is convergent; if $Q > 1$, the series is divergent; and if $Q = 1$, the test is inconclusive and the series may be either convergent or divergent.

Example. Test the series: $\frac{1}{2} + \frac{2}{2^2} + \frac{3}{2^3} + \frac{4}{2^4} + \cdots + \frac{n}{2^n}$.

The $(n+1)$st term is equal to $\frac{n+1}{2^{n+1}}$ and the nth term is $\frac{n}{2^n}$.

Therefore, $\frac{n+1}{2^{n+1}} \div \frac{2}{n} = \frac{1}{2} + \frac{1}{2n}$ and the limit approaches $\frac{1}{2}$ as n increases without limit. Since $\frac{1}{2} < 1$, the above series is convergent.

An infinite series which is composed of an infinite number of positive and an infinite number of negative terms is convergent if the series formed by taking the absolute values of all the terms is convergent.

PROGRESSIONS

Arithmetical Progression. A **series** is a succession of terms so related that each may be derived from one or more of the preceding terms in accordance with some fixed law. An **arithmetical progression** is a series, each term of which, except the first, is derived from the preceding by the addition of a constant number, a, $a + d$, $a + 2d$, etc., where d is the common difference. The **arithmetic mean** of two numbers a and b is equal to $(a + b)/2$.

Letting a = first term, l = last term, d = the common difference, n = the number of terms, and s = the sum of the terms, we have the following relations:

$$l = a + (n-1)d = -\left(\frac{1}{2}\right)d + \sqrt{2ds + \left(a - \frac{1}{2}d\right)^2} = \left(\frac{2s}{n}\right) - a = \left(\frac{s}{n}\right) + \frac{(n-1)d}{2}$$

$$s = \frac{1}{2}n[2a + (n-1)d] = \frac{(l+a)}{2} + \frac{(l^2 - a^2)}{2d} = \frac{(l+a)n}{2} = \frac{1}{2}n[2l - (n-1)d]$$

$$a = l - (n-1)d = \frac{s}{n} - \frac{(n-1)d}{2} = \frac{1}{2}d + \sqrt{\left(l + \frac{1}{2}d\right)^2 - 2ds} = \frac{2s}{n} - l$$

$$d = \frac{(l - a)}{(n - 1)} = \frac{2(s - an)}{n(n - 1)} = \frac{(l^2 - a^2)}{(2s - l - a)} = \frac{2(nl - s)}{n(n - 1)}$$

$$n = \frac{l - a}{d} + 1 = \frac{d - 2a + \sqrt{(2a - d)^2 + 8ds}}{2d}$$

$$= \frac{2s}{(l + a)} = \frac{2l + d + \sqrt{(2 + d)^2 - 8ds}}{2d}$$

Geometrical Progression. This type of progression is a progressive increase or decrease in each successive number by the same multiplier at each step, 1, 2, 4, 8, 16. The common multiplier is called the **ratio.**

Letting a = the first term, l = the last term, r = the ratio, n = the number of terms, m = any term, and s = the sum of the terms, we have the following relations:

$$l = ar^{n-1} = \frac{[a + (r - 1)s]}{r} = \frac{[(r - 1)sr^{n-1}]}{r^n - 1}$$

$$\log l = \log a + (n - 1) \log r$$

$$s = \frac{a(r^n - 1)}{(r - l)} = \frac{(rl - a)}{(r - 1)} = \frac{(lr^n - l)}{(r^n - r^{n-1})}$$

$$a = \frac{l}{r^{n-1}} = \frac{(r - 1)s}{(r^n - 1)} \qquad \log a = \log l - (n - 1) \log r$$

$$r = \sqrt[n-1]{\frac{l}{a}} = \frac{(s - a)}{(s - l)} \qquad \log r = \frac{(\log l - \log a)}{(n - 1)}$$

$$r^n - \frac{s}{a} r + \frac{(s - a)}{a} = 0 \qquad r^n - \frac{s}{s - l} r^{n-1} + \frac{l}{(s - l)} = 0$$

$$n = \frac{\log l - \log a}{\log r} + 1 = \frac{\log [a + (r - 1)s] - \log a}{\log r}$$

The **geometric mean** between two numbers equals the square root of their product, *i.e.*, $\sqrt{ab}$.

Example. The geometric mean of 9 and 4 = $\sqrt{9 \times 4} = \sqrt{36} = 6$.

Infinite Geometric Progression. If the ratio r of such a progression is less than unity, the value of r^n decreases as n increases. The general expression for this type of series is

$$a + ar + ar^2 + ar^3 + \cdot \cdot \cdot + ar^{n-1} = \frac{a}{(1 - r)} - \frac{ar^n}{(1 - r)}$$

As n increases without limit the right-hand member of the equation approaches $a/(1 - r)$ as a limit.

Combined Arithmetic and Geometric Progression. Such a progression is represented by the terms: a, $(a + d)r$, $(a + 2d)r^2$, $(a + 3d)r^3$, etc., and the sum of the first n terms is

$$\frac{a - [a + (n - 1)d]r^n}{1 - r} + \frac{rd(1 - r^{n-1})}{(1 - r)^2}$$

If r is less than unity and n is infinite, the sum is $a/(1 - r) + rd/(1 - r)^2$.

Harmonical Progression. The terms a, b, c, etc., form a harmonic series if their reciprocals $1/a$, $1/b$, $1/c$, etc., form an arithmetical series.

Harmonical Mean. The harmonical mean between two numbers is equal to twice their product divided by their sum; *e.g.*, $2ab/(a + b)$.

Geometrical Mean. The geometrical mean between two numbers is also the geometrical mean between their arithmetical and harmonical means.

Permutations and Combinations. Each separate order in which objects can be placed in a row is called a **permutation.** Each separate selection of objects that is possible irrespective of the order in which they are aligned is called a **combination.**

Example. Consider the letters a, b, and c, taken together; there are six permutations abc, bac, cab, acb, bca, and cba, but there is only one combination, *i.e.*, abc. The number of permutations of n different things, taken all at a time, is equal to the product $n(n - 1)(n - 2)(n - 3) \cdot \cdot \cdot$ or n-factorial ($n!$). The number of permutations of n things, taken r at a time, is equal to $n!/(n - r)!$

The number of combinations of n things, taken r at a time, is equal to

$$\frac{n(n - 1)(\cdot \cdot \cdot)(n - r + 1)}{r!} = \frac{n!}{r!(n - r)!}$$

In how many positions can five objects be placed in a row? 5! (five factorial) or $1 \times 2 \times 3 \times 4 \times 5 = 120$.

How many combinations of nine objects can be made, taking three at a time? $\dfrac{9 \times 8 \times 7}{1 \times 2 \times 3} = 84$.

LINEAR EQUATIONS

A linear equation is one of the first degree (*i.e.*, only the first powers of the variables are involved) and the process of obtaining definite values for the unknowns is called solving the equation. Terms of either side of the equation may be transposed to the other side of the equality if the sign of the term is changed, *e.g.*, $ax^2 + by = cz + d$ may be altered $ax^2 - cz - d = -by$.

General Types of Linear Equations.

Example 1. Solve the equations (1) $x + 2y = 4$ and (2) $3x - 5y = 10$.

A. Multiply Eq. (1) by 3 to give

$$\begin{aligned} x + 2y &= 4 \\ & \ 3 \\ \hline 3x + 6y &= 12 \end{aligned}$$

then subtract

$$\begin{aligned} 3x - 5y &= 10 \\ \hline 11y &= 2 \\ y &= 2/11 \end{aligned}$$

then substituting this value of $y = 2/11$ in Eq. (1) we have $x + 4/11 = 4$ or $x = 40/11$.

B. Equation (1) may be written $x = 4 - 2y$; substitute this value of x in Eq. (2) to give $3(4 - 2y) - 5y = 10$, or $12 - 6y - 5y = 10$, or $-11y = -2$, or $y = 2/11$.

Example 2. Solve the three equations (1) $x + y + z = 10$, (2) $x + 2y - 3z = 12$, and (3) $2x + 4y + z = 40$.

Solution. Subtract Eq. (1) from Eq. (2):

$$\begin{aligned} x + 2y - 3z &= 12 \\ x + y + z &= 10 \\ \hline y - 4z &= 2 \text{ (Eq. 4)} \end{aligned}$$

Multiply Eq. (2) by 2 and subtract the resulting equation from Eq. (3):

$$2(x + 2y - 3z) = 2x + 4y - 6z = 24$$

$$\begin{aligned} 2x + 4y + z &= 40 \\ 2x + 4y - 6z &= 24 \\ \hline 7z &= 16 \\ z &= 16/7 \end{aligned}$$

Substitute this value of z in Eq. (4) to get the value of y. Then substitute the values of y and z in one of the three original equations.

QUADRATIC EQUATIONS

General Equation. $Ax^2 + Bx + C = 0$ for which there are two solutions; *i.e.*, two values (roots) of x which satisfy the equation.

Solutions.

1. By Formula:

$$x = \frac{-B \pm \sqrt{B^2 - 4AC}}{2A}.$$

The sum of these values of x (the roots) $= -B/A$; and the product of the values of $x = C/A$.

Example. Solve the equation: $x^2 - 6x + 9 = 0$.

$$x = \frac{6 \pm \sqrt{36 - 36}}{2} = 3$$

2. By Completing the Square. Transfer each term containing the unknown or any power of the unknown to the left-hand side of the equation and all other members to the right-hand side of the equation; divide each term by the coefficient of the term containing the square of the unknown; add to each side of the equation the square of one-half the coefficient of the term containing the unknown to the first power; take the square root of each side of the equation, placing the sign $\pm$ before the right-hand side of the equation.

Example. $4x^2 + 3x - 16 = 0$; $4x^2 + 3x = 16$; $x^2 + \frac{3}{4}(x) = 4$; $x^2 + \frac{3}{4}x + (\frac{3}{8})^2 = 4 + (\frac{3}{8})^2$; $(x + \frac{3}{8})^2 = \frac{265}{64} = (2 + \frac{3}{8})^2 = (1\frac{9}{8})^2$; $x + \frac{3}{8} = \pm 2\frac{3}{8}$; $x = 2$ or $-2\frac{3}{4}$ *Ans.*

3. By Factoring. The two roots of a quadratic equation of the form $x^2 + 2ax + b = 0$ are $x_1 = -a + \sqrt{a^2 - b}$ and $x_2 = -a - \sqrt{a^2 - b}$.

Systems of Equations in Two Unknowns, One Equation of the First Degree and One of the Second Degree

Solutions. Solve the first-degree equation for one unknown: substitute the value of this unknown in the equation of the second degree; then solve the resulting quadratic equation.

Example. (1) $x^2 + 5xy - 15 = 0$; and (2) $x + 2y = 10$. $x + 2y = 10$ or $x = 10 - 2y$; substituting this value of x in (1) above, we have: $(10 - 2y)^2 + 5y(10 - 2y) - 15 = 0$; or $100 - 40y + 4y^2 + 50y - 10y^2 - 15 = 0$; collecting terms: $-6y^2 + 10y + 85 = 0$ or $6y^2 - 10y - 85 = 0$. This quadratic equation can now be solved by one of the methods given above.

Graphical Solution. Plot the equations of both curves on cross-sectional paper; the values of the variables at the points of intersection of the two curves will be the values of the variables which satisfy both equations. If the two curves do not intersect the values are imaginary.

CUBIC EQUATIONS

A cubic equation (an equation of the third degree) contains the third power of the unknown quantity and may also contain the second power or the second and first powers of the unknown. A cubic equation has three roots, two of which may be imaginary.

Solutions of Cubic Equations. Cardan's Method. (1) The **general cubic equation** $x^3 + ax^2 + bx + c = 0$, where a is not equal to zero, may be transformed to another equation, the x^2 term of which is missing, by substituting for x the term $\left(y - \frac{a}{3}\right)$, which reduces the general cubic equation to the type: $x^3 + ax + b = 0$.

To solve the equation $x^3 + 3x^2 - 6x + 20 = 0$. Substituting in this equation, the value $x = y - \frac{a}{3} = y - 1$: $y^3 - 3y^2 + 3y - 1 + 3y^2 - 6y + 3 - 6y + 6 + 20 = 0$, or $y^3 - 9y + 28 = 0$.

A Cubic Equation with the x^2 Term Missing. (2) $x^3 + ax + b = 0$.

Let $x = y + z$ and the above equation becomes $y^3 + 3yz(y + z) + z^3 + a(y + z) + b = 0$, or, by rearranging: $y^3 + z^3 + (3yz + a)(y + z) + b = 0$. Then, let $3yz + a = 0$ or $z = -a/3y$ and, substituting this value in the last equation, we obtain: $y^3 - \frac{a^3}{27y^3} + b = 0$ or $y^6 + by^3 = a^3/27$ which is a type of quadratic solved by the following formula:

$$y^3 = -\frac{b}{2} \pm \sqrt{\frac{b^2}{4} + \frac{a^3}{27}}$$

and hence,

$$z^3 = -y^3 - b = -\frac{b}{2} \mp \sqrt{\frac{b^2}{4} + \frac{a^3}{27}}$$

Therefore,

$$x = \sqrt[3]{\left(-\frac{b}{2} + \sqrt{\frac{b^2}{4} + \frac{a^3}{27}}\right)} + \sqrt[3]{\left(-\frac{b}{2} - \sqrt{\frac{b^2}{4} + \frac{a^3}{27}}\right)}$$

Cubic Equation with x-term Omitted. Trigonometric Solutions of Cubic Equation. $Ax^3 + Bx^2 + D = 0$.

Let $x = y - \frac{B}{3}$; substitute this value in the type equation, to give an equation of the form $x^3 + Ex + F = 0$. This type equation may then be solved in one of the following three ways:

(1) $x^3 + Ex + F = 0$.

Let

$$\tan\theta = \left(\frac{2E}{3F}\right)\sqrt{\frac{E}{3}} \text{ and } \tan\left(\frac{\mu}{2}\right) = \sqrt[3]{\tan\left(\frac{\theta}{2}\right)}$$

then

$$\text{The first root} = \left(-2\sqrt{\frac{E}{3}}\right)\cot\mu$$

$$\text{The second root} = \left(\sqrt{\frac{E}{3}}\right)\cot\mu + \csc\mu\sqrt{E}\sqrt{-1}$$

$$\text{The third root} = \left(\sqrt{\frac{E}{3}}\right)\cot\mu - \csc\mu\sqrt{E}\sqrt{-1}$$

(2) $x^3 - Ex + F = 0$, and where $\frac{E^3}{27} < \frac{F^2}{4}$.

Let

$$\sin\theta = \frac{2E}{3F}\sqrt{\frac{E}{3}} \text{ and } \tan\left(\frac{\mu}{2}\right) = \sqrt[3]{\tan\frac{\theta}{2}}$$

then

$$\text{The first root} = \left(-2\sqrt{\frac{E}{3}}\right)\csc\mu$$

$$\text{The second root} = \left(\sqrt{\frac{E}{3}}\right)\csc\mu + \cot\mu\sqrt{E}\sqrt{-1}$$

$$\text{The third root} = \left(\sqrt{\frac{E}{3}}\right)\csc\mu - \cot\mu\sqrt{E}\sqrt{-1}$$

(3) $x^3 - Ex + F = 0$, and where $\frac{E^3}{27} > \frac{F^2}{4}$.

Let

$$\sin\theta = \left(\frac{3F}{2E}\right)\sqrt{\frac{3}{E}}$$

then

$$\text{The first root} = \left(2\sqrt{\frac{E}{3}}\right)\sin\frac{\theta}{3}$$

$$\text{The second root} = \left(2\sqrt{\frac{E}{3}}\right)\sin\left(60° - \frac{\theta}{3}\right)$$

$$\text{The third root} = \left(-2\sqrt{\frac{E}{3}}\right)\sin\left(60° + \frac{\theta}{3}\right)$$

Newton's Method of Approximation. Find two numbers, one greater and the other less than a root of the equation by the rule. If two numbers, when substituted for the unknown quantity in an equation, give results having a different sign, at least one root lies between those numbers. Let a be one of these numbers, the nearest to the root if it can be ascertained. Then substitute $a + y$ for x in the given equation; then y is small and, by omitting the powers of y, greater than 1, in the new equation, a value of y is obtained which, added to a, gives b which is a closer approximation to the value of x. Next, substitute $b + z$ for x in the original equation and a second approximation is obtained as before.

Example. Find the real root of $x^3 - 2x - 5 = 0$. When 2 and 3 are substituted for x in the equation, the results are -1 and 16, respectively; therefore, the real root lies between 2 and 3 and nearer 2. Substitute $2 + y$ for x and $y^3 + 6y^2 + 10y - 1 = 0$ is obtained by neglecting the y^3 and y^2 terms, $y = -0.1$. Next, substitute $2.1 + z$ for x to obtain $0.061 \times 11.23z = 0$, or $z = -0.0054$ and $x = 2.1 - 0.0054 = 2.0946$, approx.

QUARTIC AND HIGHER EQUATIONS

(From A. de F. Palmer, "The Theory of Measurements," p. 220, McGraw-Hill.) Reduce the quartic equation to the form, $x^4 + 4ax^3 + 6bx^2 + 4cx + d = 0$. Then calculate the values of $g = a^2 - b$, $h = b^3 + c^2 - 2abc + dg$,

and $k = \frac{4}{3}ac - b^2 - \frac{1}{3}d$; $l = \frac{1}{2}(h + \sqrt{h^2 + k^3})^{1/3} + \frac{1}{2}(h - \sqrt{h^2 + k^3})^{1/3}$; $u = g + l$; $v = 2g - l$; $w = 4u^2 + 3k - 12gl$.

The four roots are:

$$x_1 = -a + \sqrt{u} + \sqrt{v + \sqrt{w}};\quad x_2 = -a + \sqrt{u} - \sqrt{v + \sqrt{w}};$$
$$x_3 = -a - \sqrt{u} + \sqrt{v - \sqrt{w}};\quad x_4 = -a - \sqrt{u} - \sqrt{v - \sqrt{w}}$$

in which the signs are to be used as written provided that $2a^3, - 3ab + c$ is a negative number; but if this is positive all radicals except $\sqrt{w}$ are to be changed in sign. The above expressions are irreducible when $h^2 + k^3$ is a negative number. In this case the given equation has either four imaginary roots or four real roots that can be determined numerically only by some method of approximation.

Simultaneous Equations of the First Degree. When two unknown quantities are involved in a problem, two statements or conditions must be found, and this results in two simultaneous linear equations. There are three accepted methods for analytical solutions. The one most frequently used is the method of addition and subtraction. One equation is multiplied by some number that will make the coefficients of one unknown the same in both equations; then, by adding or subtracting from the other equation, an equation with one unknown results, which is readily solved. As an example, if $6x - 9 = y$ (1) and $9x - 2y = 12$ (2) are the equations, multiply (1) by 2 and subtract from (2). Then,

$$\begin{aligned} 12x - 2y &= 18 \\ 9x - 2y &= 12 \\ \hline 3x &= 6 \\ x &= 2 \end{aligned}$$

Substituting $x = 2$ in (1) or (2) gives $y = 3$.

Another method is to eliminate by comparison as follows:

From (1), $x = (y + 9)/6$, and from (2), $x = (2y + 12)/9$. Then by comparison

$$\frac{y + 9}{6} = \frac{2y + 12}{9}$$

From which $y = 3$ and, when substituted in either (1) or (2), gives $x = 2$.

The third method is elimination by substitution, as follows:

If $x = (y + 9)/9$ found from (1) be substituted for x in (2), then x is eliminated in the resulting equation and y is readily found.

This same procedure may also be used for solving a system of three or more equations involving as many unknowns as there are independent equations.

The following method of procedure is recommended: Continue to eliminate the **same unknown** in the given equations until there is obtained a group of one less than the original number of equations with one less unknown. Next eliminate a second unknown from the new group in the same manner. Continue these operations until only two simultaneous equations remain, which can readily be solved. The other unknowns can be solved by substituting those found in some of the reduction equations.

DETERMINANTS

The use of determinants greatly facilitates the solution of certain linear algebraic operations, by making these operations more or less of a routine character so that the solution is obtained more or less mechanically. In the equation, $ay - bx = \begin{vmatrix} ab \\ xy \end{vmatrix}$, the right-hand term is called a **determinant,** while the left-hand term is called the **development** or **expansion** of the determinant. The individual members, a, b, x, and y are called the **elements.** When each term in the expansion is the product of **two** elements, the determinant is said to be of the **second order.**

The expansion of a determinant of the **second order** is formed by taking the product of the elements on the diagonal passing downward from left to right and subtracting from it the product of the elements on the other diagonal. For example,

$$\begin{vmatrix} a_1 b_1 \\ a_2 b_2 \end{vmatrix} = a_1b_2 - a_2b_1; \text{ or } \begin{vmatrix} 4 & 7 \\ 3 & -6 \end{vmatrix} = (4)(-6) - (3)(7) = -24 - 21 = -45$$

Each term of the expansion contains only one element from each column.

Solution of Simultaneous Equations. In Two Unknowns. Two linear simultaneous equations of the form

$$a_1x + b_1y = c_1$$
$$a_2x + b_2y = c_2$$

when solved in the usual analytical way may be put in the form

$$x = \frac{c_1b_2 - c_2b_1}{a_1b_2 - a_2b_1},\quad y = \frac{a_1c_2 - a_2c_1}{a_1b_2 - a_2b_1}$$

Writing both numerator and denominator in the determinate form

$$x = \frac{\begin{vmatrix} c_1b_1 \\ c_2b_2 \end{vmatrix}}{\begin{vmatrix} a_1b_1 \\ a_2b_2 \end{vmatrix}},\quad y = \frac{\begin{vmatrix} a_1c_1 \\ a_2c_2 \end{vmatrix}}{\begin{vmatrix} a_1b_1 \\ a_2b_2 \end{vmatrix}}$$

These expressions show that the unknowns may be found by setting up determinates found from the coefficients of the unknowns, together with the constant terms, arranged as a quotient. The denominator, which is the principal determinate, is formed from the coefficients of the unknown and is the same for both unknowns. The determinate for the numerators is formed by replacing the coefficients of the unknowns in the principal determinate by the constant terms.

Example 1. Solve the equations: $8x - y = -25$ and $5x + 2y = 29$. Rearranging them, $8x - y + 25 = 0$ and $5x + 2y - 29 = 0$. Each determinant in the numerator is formed from the denominator by replacing the coefficient of the unknown sought by the constant term. To find the numerator of x_1, replace a_1

and a_2 (the coefficients of x) by c_1 and c_2 (the constant terms); also, substitute c_1 and c_2 for b_1 and b_2 to find the value of y.

We have then from the above rearranged equations:

$$x = \frac{\begin{vmatrix}-25 & -1\\ 29 & 2\end{vmatrix}}{\begin{vmatrix}8 & -1\\ 5 & 2\end{vmatrix}} = \frac{-50+29}{16+5} = -1$$

$$y = \frac{\begin{vmatrix}8 & -25\\ 5 & 29\end{vmatrix}}{\begin{vmatrix}8 & -1\\ 5 & 2\end{vmatrix}} = \frac{232+125}{16+5} = 17$$

from which $x = -1$, and $y = 17$.

Example 2. Solve: $2x - y = 1$ and $3x + 2y = 3$.

The determinant for the denominator is $\begin{vmatrix}2 & -1\\ 3 & 2\end{vmatrix}$ for both x and y; while for the numerator of x, we have to replace the coefficients of x by the constants 1 and 3, to obtain $\begin{vmatrix}1 & -1\\ 3 & 2\end{vmatrix}$ for the numerator.

$$\therefore x = \frac{\begin{vmatrix}1 & -1\\ 2 & 2\end{vmatrix}}{\begin{vmatrix}2 & -1\\ 3 & 2\end{vmatrix}} = 5/7 \text{ and } y = \frac{\begin{vmatrix}2 & 1\\ 3 & 3\end{vmatrix}}{\begin{vmatrix}2 & -1\\ 3 & 2\end{vmatrix}} = 3/7$$

Third-order Determinants. This determinant is composed of nine elements arranged in three rows and three columns; *e.g.*,

$$\begin{vmatrix}a_1b_1c_1\\ a_2b_2c_2\\ a_3b_3c_3\end{vmatrix}$$

which is a convenient symbol for the expression, $a_1b_2c_3 + a_2b_3c_1 + a_3b_1c_2 - a_1b_3c_2 - a_2b_1c_3 - a_3b_2c_1$. Each term is the product of three elements: one only from each row, and one only from each column. The developed expression may be written $a_1(b_2c_3 - b_3c_2) - a_2(b_1c_3 - b_3c_1) + a_3(b_1c_2 - b_2c_1)$, where the terms in the parentheses are the expressions of second-order determinants. The latter can be written, then

$$\begin{vmatrix}a_1b_1c_1\\ a_2b_2c_2\\ a_3b_3c_3\end{vmatrix} = a_1\begin{vmatrix}b_2c_2\\ b_3c_3\end{vmatrix} - a_2\begin{vmatrix}b_1c_1\\ b_3c_3\end{vmatrix} + a_3\begin{vmatrix}b_1c_1\\ b_2c_2\end{vmatrix}$$

Solve:

$$\begin{vmatrix}48 & 3 & 3\\ 18 & 6 & -3\\ 21 & -3 & 2\end{vmatrix} = 48\begin{vmatrix}6 & -3\\ -3 & 2\end{vmatrix} - 18\begin{vmatrix}3 & 3\\ -3 & 2\end{vmatrix} + 21\begin{vmatrix}3 & 3\\ 6 & -3\end{vmatrix}$$
$$= 48(12 - 9) - 18(6 + 9) + 21(-9 - 18)$$
$$= -69.3$$

Solve the simultaneous equations:

$$5x + 3y + 3z - 48 = 0$$
$$2x + 6y - 3z - 18 = 0$$
$$8x - 3y + 2z - 21 = 0$$

whence the determinant of the system which is the determinant of the denominator is:

$$\begin{vmatrix}5 & 3 & 3\\ 2 & 6 & -3\\ 8 & -3 & 2\end{vmatrix} \text{ which is}$$

$$= 5\begin{vmatrix}6 & -3\\ -3 & 2\end{vmatrix} - 2\begin{vmatrix}3 & 3\\ -3 & 2\end{vmatrix} + 8\begin{vmatrix}3 & 3\\ 6 & -3\end{vmatrix}$$
$$= 5(12 - 9) - 2(6 + 9) + 8(-9 - 18)$$
$$= 15 - 30 - 216$$
$$= -231$$

$$x = \frac{\begin{vmatrix}48 & 3 & 3\\ 18 & 6 & -3\\ 21 & -3 & 2\end{vmatrix}}{-231}$$
$$= \frac{48\begin{vmatrix}6 & -3\\ -3 & 2\end{vmatrix} - 18\begin{vmatrix}3 & 3\\ -3 & 2\end{vmatrix} + 21\begin{vmatrix}3 & 3\\ 6 & -3\end{vmatrix}}{-231}$$
$$= \frac{48(12 - 9) - 18(6 + 9) + 21(-9 - 18)}{-231}$$
$$= \frac{-693}{-231} = 3$$

Similarly

$$y = \frac{\begin{vmatrix}5 & 48 & 3\\ 2 & 18 & -3\\ 8 & 21 & 2\end{vmatrix}}{-231} = 5 \text{ and } z = \frac{\begin{vmatrix}5 & 3 & 48\\ 2 & 6 & 18\\ 8 & -3 & 21\end{vmatrix}}{-231} = 6$$

Properties of Determinants. 1. The value of a determinant is not altered by changing the columns into rows, or the rows into columns.

$$\begin{vmatrix}a_1b_1\\ a_2b_2\end{vmatrix} = \begin{vmatrix}a_1a_2\\ b_1b_2\end{vmatrix} \text{ and } \begin{vmatrix}a_1b_1c_1\\ a_2b_2c_2\\ a_3b_3c_3\end{vmatrix} = \begin{vmatrix}a_1a_2a_3\\ b_1b_2b_3\\ c_1c_2c_3\end{vmatrix}$$

2. The sign and not the numerical value of a determinant is changed by the interchange of any two columns, or any two rows. Since,

$$\begin{vmatrix}a_1b_1\\ a_2b_2\end{vmatrix} = -\begin{vmatrix}b_1a_1\\ b_2a_2\end{vmatrix} \text{ and } \begin{vmatrix}a_1b_1c_1\\ a_2b_2c_2\\ a_3b_3c_3\end{vmatrix} = -\begin{vmatrix}b_1a_1c_1\\ b_2a_2c_2\\ b_3a_3c_3\end{vmatrix}$$

3. If two rows or two columns of a determinant are identical, the determinant is equal to zero. The expansion of

$$\begin{vmatrix}a_1a_1c_1\\ a_2a_2c_2\\ a_3a_3c_3\end{vmatrix} \text{ (will be proved) } = 0$$

4. When the constituents of two rows or two columns differ by a constant factor, the determinant equals zero. Thus, by expansion:

$$\begin{vmatrix}4 & 1 & 5\\ 8 & 2 & 6\\ 12 & 3 & 7\end{vmatrix} = 4\begin{vmatrix}1 & 1 & 5\\ 2 & 2 & 6\\ 3 & 3 & 7\end{vmatrix} = 4 \times 0 = 0$$

5. If a determinant has a row or column of zeros it is equal to zero. Expansion of

$$\begin{vmatrix}0 & b_1 & c_1\\ 0 & b_2 & c_2\\ 0 & b_3 & c_3\end{vmatrix} = 0$$

6. In order to multiply a determinant by any factor, multiply each constituent in one row or in one column by the factor, thus:

$$m\begin{vmatrix}a_1b_1c_1\\ a_2b_2c_2\\ a_3b_3c_3\end{vmatrix} = \begin{vmatrix}ma_1b_1c_1\\ ma_2b_2c_2\\ ma_3b_3c_3\end{vmatrix}$$

7. In order to divide a determinant by any factor divide each constituent in one row or in one column by

the factor. This operation will simplify the original determinant.

$$\frac{\begin{vmatrix} a_1b_1c_1 \\ a_2b_2c_2 \\ a_3b_3c_3 \end{vmatrix}}{m} = \begin{vmatrix} a_1/m & b_1 & c_1 \\ a_2/m & b_2 & c_2 \\ a_3/m & b_3 & c_3 \end{vmatrix}$$

8. If the sign of every constituent in a row or a column is changed, the sign of the determinant is changed. For example,

$$\begin{vmatrix} a_1b_1c_1 \\ a_2b_2c_2 \\ a_3b_3c_3 \end{vmatrix} = -\begin{vmatrix} -a_1b_1c_1 \\ -a_2b_2c_2 \\ -a_3b_3c_3 \end{vmatrix} = +\begin{vmatrix} -a_1 & -b_1c_1 \\ -a_2 & -b_2c_2 \\ -a_3 & -b_3c_3 \end{vmatrix}$$

9. If each constituent of a row or column can be expressed as the sum or difference of two or more terms, the determinant can be expressed as the sum or difference of two other determinants.

$$\begin{vmatrix} a_1\pm & o_1; & b_1 & c_1 \\ a_2\pm & p_1; & b_2 & c_2 \\ a_3\pm & q_1; & b_3 & c_3 \end{vmatrix} = \begin{vmatrix} a_1 & b_1 & c_1 \\ a_2 & b_2 & c_2 \\ a_3 & b_3 & c_3 \end{vmatrix} \pm \begin{vmatrix} o_1 & b_1 & c_1 \\ p_1 & b_2 & c_2 \\ q_1 & b_3 & c_3 \end{vmatrix}$$

10. The value of a determinant is not changed by adding or subtracting the constituents of any row or column from the corresponding constituents of one or more of the other rows or columns.

11. If all the constituents of a determinant on one side of the diagonal from the top left-hand corner are zeros, the determinant reduces to the leading term.

$$\begin{vmatrix} a_1 & b_1 & c_1 \\ 0 & b_2 & c_2 \\ 0 & 0 & c_3 \end{vmatrix} = a_1 \begin{vmatrix} b_2 & c_2 \\ 0 & c_3 \end{vmatrix} = a_1\ b_2\ c_3$$

12. If all but one of the constituents of a row or a column are zeros, the determinant can be reduced to the product of the one constituent, not zero, into a determinant whose order is one less than the original determinant.

$$\begin{vmatrix} 1 & a & b \\ 0 & a & b_1 \\ 0 & a_2 & b_2 \end{vmatrix} = 1 \begin{vmatrix} a_1 & b_1 \\ a_2 & b_2 \end{vmatrix}$$

or

$$\begin{vmatrix} 0 & 0 & -7 \\ 5 & 6 & -2 \\ -3 & 1 & 8 \end{vmatrix} = -7 \begin{vmatrix} 5 & 6 \\ -3 & 1 \end{vmatrix}$$

Multiplication of Determinants.

$$\begin{vmatrix} a_1 & b_1 \\ a_2 & b_2 \end{vmatrix} \times \begin{vmatrix} d_1 & e_1 \\ d_2 & e_2 \end{vmatrix} = \begin{vmatrix} a_1d_1 + b_1e_1; & a_1d_2 + b_1e_2 \\ a_2d_1 + b_2e_1; & a_2d_2 + b_2e_2 \end{vmatrix}$$

TRIGONOMETRY

PLANE TRIGONOMETRY

Trigonometric Functions. An angle is generated by the rotation of a line about a fixed center. If the rotation is clockwise, the angle is negative: if it be counterclockwise, the angle is positive. The line from which the rotation starts is the **initial line**, and the line at which the rotation stops is the **terminal line**. Angular magnitude is unlimited in size, since the rotation of the initial line may continue indefinitely, in either a positive or negative direction. Two angles are **complementary** if their sum is 90°, and are **supplementary** if their sum is 180°, and are **congruent** if they may be superposed so that their initial and terminal lines coincide.

Quadrants (Fig. 82). If the initial line runs horizontally to the right, the angle is in the first, second,

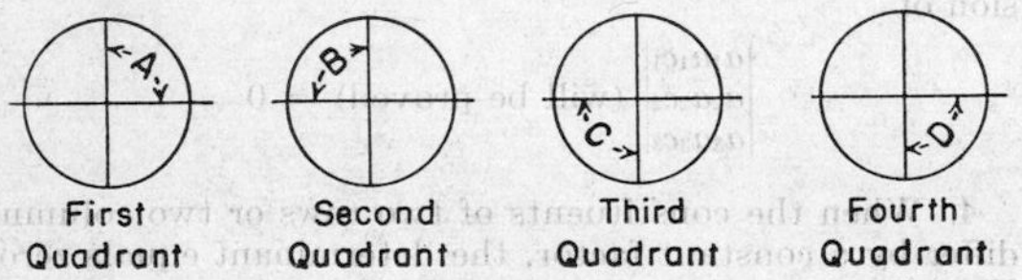

FIG. 82.

third, or fourth quadrant, according as the terminal line lies in the regions marked *A*, *B*, *C*, or *D*, respectively, in the accompanying diagram. The angles 0°, 90°, 180°, and 270° are called the quadrantal angles.

Angular Measurements. Sexagesimal. In this system, there are 360 degrees in one complete revolution, one degree being defined as 1/90th of a right angle. The degree is subdivided into 60 equal parts called minutes ('), and each minute is similarly subdivided into 60 equal parts called seconds ("). For most purposes it is usual to divide the degrees into decimal parts.

Radian or Circular. In this system, π radians (3.14159 radians) = 180°. One **radian** is the angle at the center of a circle subtended by an arc whose length is equal to the length of the radius. In practice, the radian is divided into decimals: 1 radian = 57.29578°, or 57.3°; 1° = 0.01745 radian; 1' = 0.00029089 radian; and 1" = 0.000004848 radian.

Protractor. A protractor is an instrument for measuring angles. It generally consists of a semicircular piece of cardboard, celluloid, or metal. The arc is divided on its margin into 180°. The center of the circular arc is marked by a hole. Protractors are sometimes made in the form of rectangles.

Functions of Trigonometry. The trigonometric functions of angles are the ratios between the various sides of the reference triangle of the angles. The reference triangle is the right-angled triangle formed by the initial line (or the initial line extended in the positive direction), the terminal line, and the perpendicular drawn from any point in the terminal line upon the initial line.

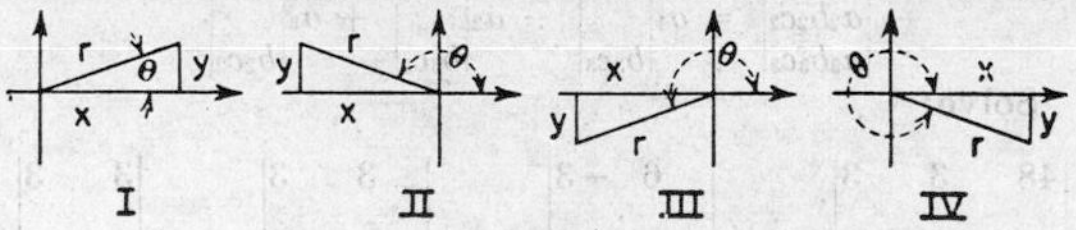

FIG. 83.

Figure 83 shows the reference triangles for the different quadrants.

Coordinates. A point is located in the rectangular system of coordinates by means of the values of x and y (the abscissa and ordinate), or by means of polar coordinates.

The sign of the hypotenuse is always taken to be positive. If the angle θ is placed with its vertex and its initial line along the positive end of the x-axis, allowing the terminal line to fall in each of the quadrants successively, and if the point $P(x,y)$, is any point on the terminal line, and r is its distance from the origin, the ratios of any two of the three values r, x, and y are called the trigonometric functions of the angle θ. The sign (positive or negative) of the function is determined by the signs of x and y; or, in other words, by the quadrant in which x and y are located. The following trigonometric definitions must be memorized—there is no substitute for the memorization.

$$\text{Sine of angle } \theta = \sin\theta = \frac{y}{r}$$

$$\text{Tangent of angle } \theta = \tan\theta = \frac{y}{x}$$

$$\text{Secant of angle } \theta = \sec\theta = \frac{r}{x}$$

$$\text{Cosine of angle } \theta = \cos\theta = \frac{x}{r}$$

$$\text{Cotangent of angle } \theta = \cot\theta = \frac{x}{y}$$

$$\text{Cosecant of angle } \theta = \csc\theta = \frac{r}{y}$$

It is convenient to remember that the last three are the reciprocals of the first three, *i.e.*:

$$\cot\theta = \frac{1}{\tan\theta}, \quad \sec\theta = \frac{1}{\cos\theta}, \quad \csc\theta = \frac{1}{\sin\theta}$$

When $r = 1$, the following ratios are valid:

$$\sin\theta = y, \quad \cos\theta = x, \quad \sec\theta = \frac{1}{x}, \quad \csc\theta = \frac{1}{y}$$

Since we know that $r^2 = x^2 + y^2$ by geometry, the following relations are valid.

Magnitude of Trigonometric Functions, When the Angle Varies from 0° to 360°

	0° to 90°	90° to 180°	180° to 270°	270° to 360°
$\sin\theta$	$+0$ to $+1$	$+1$ to $+0$	-0 to -1	-1 to -0
$\csc\theta$	$+\infty$ to $+1$	$+1$ to $+\infty$	$-\infty$ to -1	-1 to $-\infty$
$\cos\theta$	$+1$ to $+0$	-0 to -1	-1 to -0	$+0$ to $+1$
$\sec\theta$	$+1$ to $+\infty$	$-\infty$ to -1	-1 to $-\infty$	$+\infty$ to $+1$
$\tan\theta$	$+0$ to $+\infty$	$-\infty$ to -0	$+0$ to $+\infty$	$-\infty$ to -0
$\cot\theta$	$+\infty$ to $+0$	-0 to $-\infty$	$+\infty$ to $+0$	-0 to $-\infty$

Other functions are: versed sin θ = vers sin $\theta = 1 - \cos\theta$; coversed sin θ = covers sin $\theta = 1 - \sin\theta$; exterior sec θ = exsec $\theta = \sec\theta - 1$. These functions change with the quadrants as given below:

Quadrants

	First 0° to 90°	Second 90° to 180°	Third 180° to 270°	Fourth 270° to 360°
vers θ	$+0$ to $+1$	$+1$ to $+2$	$+2$ to $+1$	$+1$ to $+0$
covers θ	$+1$ to $+0$	$+0$ to $+1$	$+1$ to $+2$	$+2$ to $+1$

Trigonometrical Tables

In a right triangle, the side opposite the 30° angle is one-half the hypotenuse. Assuming $y = 1$, then $x = \sqrt{3}$, and the trigonometric functions of 30° are:

$\sin 30° = \frac{1}{2}$ $\cos 30° = \frac{1}{2}\sqrt{3}$ $\tan 30° = \frac{1}{3}\sqrt{3}$

$\cot 30° = \sqrt{3}$ $\sec 30° = \frac{2}{3}\sqrt{3}$ $\csc 30° = 2$

Similarly, the functions of the other angles may be calculated.

A convenient summary of the trigonometric functions for a number of angles is herewith given.

$\theta°$	θ radians	$\sin\theta$	$\cos\theta$	$\tan\theta$	cotan θ	$\sec\theta$	$\csc\theta$
0	0	0	1	0	∞	1	∞
30	$\pi/6$	$\frac{1}{2}$	$\sqrt{3}/2$	$\sqrt{3}/3$	$\sqrt{3}$	$2/\sqrt{3}$	2
45	$\pi/4$	$\sqrt{2}/2$	$\sqrt{2}/2$	1	1	$\sqrt{2}$	$\sqrt{2}$
60	$\pi/3$	$\sqrt{3}/2$	$\frac{1}{2}$	$\sqrt{3}$	$\sqrt{3}/3$	2	$2/\sqrt{3}$
90	$\pi/2$	1	0	∞	0	∞	1
120	$2\pi/3$	$\sqrt{3}/2$	$-\frac{1}{2}$	$-\sqrt{3}$	$-\sqrt{3}/3$	-2	$2\sqrt{3}/3$
135	$3\pi/4$	$\sqrt{2}/2$	$-\sqrt{2}/2$	-1	-1	$-\sqrt{2}$	$\sqrt{2}$
150	$5\pi/6$	$\frac{1}{2}$	$-\sqrt{3}/2$	$-\sqrt{3}/3$	$-\sqrt{3}$	$-\frac{2}{3}\sqrt{3}$	2
180	π	0	-1	0	∞	-1	∞
210	$7\pi/6$	$-\frac{1}{2}$	$-\sqrt{3}/2$	$\sqrt{3}/3$	$\sqrt{3}$	$-2\sqrt{3}/3$	-2
225	$5\pi/4$	$-\sqrt{2}/2$	$-\sqrt{2}/2$	1	1	$-\sqrt{2}$	$-\sqrt{2}$
240	$4\pi/3$	$-\sqrt{3}/2$	$-\frac{1}{2}$	$\sqrt{3}$	$\sqrt{3}/3$	-2	$-2\sqrt{3}/3$
270	$3\pi/2$	-1	0	∞	0	∞	-1
300	$5\pi/3$	$-\sqrt{3}/2$	$\frac{1}{2}$	$-\sqrt{3}$	$-\sqrt{3}/3$	2	$-2\sqrt{3}/3$
315	$7\pi/4$	$-\sqrt{2}/2$	$\sqrt{2}/2$	-1	-1	$\sqrt{2}$	$-\sqrt{2}$
330	$11\pi/6$	$-\frac{1}{2}$	$\sqrt{3}/2$	$-\sqrt{3}/3$	$-\sqrt{3}$	$2\sqrt{3}/3$	-2
360	2π	0	1	0	∞	1	∞

To Find a Function of a Given Angle. From a table which includes only the trigonometric functions of angles between 0° and 90°. Reduce the angle to the first quadrant by means of the following table from Marks, "Mechanical Engineers' Handbook":

When θ is between:	90° and 180°	180° to 270°	270° to 360°
Subtract:	90° from θ	180° from θ	270° from θ
Then			
$\sin\theta =$	$+\cos(\theta - 90)$	$-\sin(\theta - 180)$	$-\cos(\theta - 270)$
$\csc\theta =$	$+\sec(\theta - 90)$	$-\csc(\theta - 180)$	$-\sec(\theta - 270)$
$\cos\theta =$	$-\sin(\theta - 90)$	$-\cos(\theta - 180)$	$+\sin(\theta - 270)$
$\sec\theta =$	$-\csc(\theta - 90)$	$-\sec(\theta - 180)$	$+\csc(\theta - 270)$
$\tan\theta =$	$-\cot(\theta - 90)$	$+\tan(\theta - 180)$	$-\cot(\theta - 270)$
$\cot\theta =$	$-\tan(\theta - 90)$	$+\cot(\theta - 180)$	$-\tan(\theta - 270)$
vers $\theta =$	$1 + \sin(\theta - 90)$	$1 + \cos(\theta - 180)$	$1 - \sin(\theta - 270)$
covers $\theta =$	$1 - \cos(\theta - 90)$	$1 + \sin(\theta - 180)$	$1 + \cos(\theta - 270)$

The functions can then be found in a table which includes angles between 0° and 90°.

To Find the Angle When a Function of It Is Given. There will generally be two angles between 0° and 360° which correspond to the given function.

Given	Find from the tables an acute angle θ, such that	Required angles are
$\sin\theta = +a$	$\sin\theta_0 = a$	θ_0 and $(180° - \theta_0)$
$\cos\theta = +a$	$\cos\theta_0 = a$	θ_0 and $(360° - \theta_0)$
$\tan\theta = +a$	$\tan\theta_0 = a$	θ_0 and $(180° + \theta_0)$
$\cot\theta = +a$	$\cot\theta_0 = a$	θ_0 and $(180° + \theta_0)$
$\sin\theta = -a$	$\sin\theta_0 = a$	$180° + \theta_0$ and $(360° - \theta_0)$
$\cos\theta = -a$	$\cos\theta_0 = a$	$180° - \theta_0$ and $(180° + \theta_0)$
$\tan\theta = -a$	$\tan\theta_0 = a$	$180° - \theta_0$ and $(360° - \theta_0)$
$\cot\theta = -a$	$\cot\theta_0 = a$	$180° - \theta_0$ and $(360° - \theta_0)$

Relations between the Functions of a Single Angle.

$$\sin^2\theta + \cos^2\theta = 1;$$

$$\tan\theta = \frac{\sin\theta}{\cos\theta}; \quad \cot\theta = \frac{1}{\tan\theta} = \frac{\cos\theta}{\sin\theta};$$

$$1 + \tan^2\theta = \sec^2\theta = \frac{1}{\cos^2\theta};$$

$$1 + \cot^2\theta = \csc^2\theta = \frac{1}{\sin^2\theta}.$$

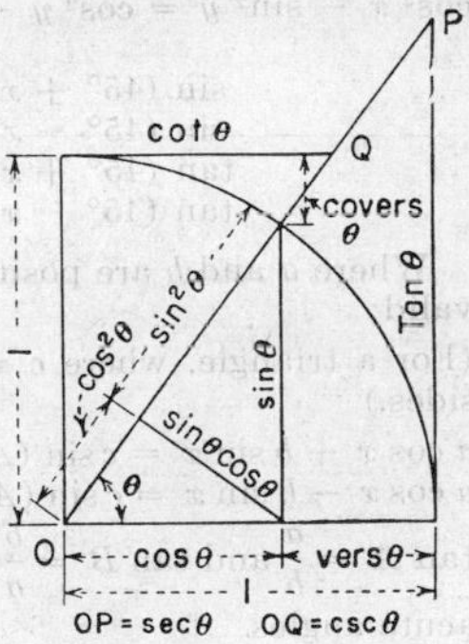

Fig. 84.

$\sin\theta < \theta < \tan\theta$ where $0 < \theta < 90°$

$$\sin\theta = \frac{\cos\theta}{\cot\theta} = \sqrt{1 - \cos^2\theta} = \frac{1}{\csc\theta} = \cos\theta\tan\theta$$

$$= \frac{\tan\theta}{\sqrt{(1 + \tan^2\theta)}} = \frac{1}{\sqrt{(1 + \cot^2\theta)}} = \frac{\sqrt{(\sec^2\theta - 1)}}{\sec\theta}$$

$$= 2\sin\left(\frac{\theta}{2}\right)\cos\left(\frac{\theta}{2}\right) = \pm\sqrt{\frac{(1 - \cos^2\theta)}{2}}$$

$$\cos\theta = \sqrt{1 - \sin^2\theta} = \frac{1}{\sqrt{(1 + \tan^2\theta)}} = \frac{\cot\theta}{\sqrt{(1 + \cot^2\theta)}}$$

$$= \frac{\sin\theta}{\tan\theta} = \frac{1}{\sec\theta} = \sqrt{1 - \sin\theta} = \frac{1}{\sqrt{1 \times \tan^2\theta}}$$

$$= \sin\theta\cot\theta = \frac{\sqrt{\csc^2\theta - 1}}{\csc\theta} = \cos^2\left(\frac{\theta}{2}\right) - \sin^2\frac{\theta}{2}$$

Figure 84 gives the values of the trigonometrical ratios as lengths of line segments.

Functions of Negative Angles.

$$\sin(-\theta) = -\sin\theta \qquad \cos(-\theta) = \cos\theta$$
$$\tan(-\theta) = -\tan\theta \qquad \cot(-\theta) = -\cot\theta$$
$$\sec(-\theta) = \sec\theta \qquad \csc(-\theta) = -\csc\theta$$

Trigonometric Functions of the Sum and Difference of Two Angles Represented Here by x and y.

$$\sin(x+y) = \sin x\cos y + \cos x\sin y,\ \cos(x+y) = \cos x\cos y - \sin x\sin y,$$
$$\sin(x-y) = \sin x\cos y - \cos x\sin y,\ \cos(x-y) = \cos x\cos y + \sin x\sin y,$$
$$\tan(x+y) = \left(\frac{\tan x + \tan y}{1 - \tan x\tan y}\right),\ \cot(x+y) = \left(\frac{\cot x\cot y - 1}{\cot x + \cot y}\right)$$
$$\tan(x-y) = \left(\frac{\tan x - \tan y}{1 + \tan x\tan y}\right),\ \cot(x-y) = \left(\frac{\cot x\cot y + 1}{\cot y - \cot x}\right)$$
$$\sin x + \sin y = 2\sin \tfrac{1}{2}(x+y)\cos\tfrac{1}{2}(x-y)$$
$$\sin x - \sin y = 2\cos\tfrac{1}{2}(x+y)\sin\tfrac{1}{2}(x-y)$$
$$\cos x + \cos y = 2\cos\tfrac{1}{2}(x+y)\cos\tfrac{1}{2}(x-y)$$
$$\cos x - \cos y = -2\sin\tfrac{1}{2}(x+y)\sin\tfrac{1}{2}(x-y)$$
$$\tan x + \tan y = \frac{\sin(x+y)}{\cos x\cos y},\ \cot x + \cot y = \frac{\sin(x+y)}{\sin x\sin y}$$
$$\tan x - \tan y = \frac{\sin(x-y)}{\cos x\cos y},\ \cot x - \cot y = \frac{\sin(y-x)}{\sin x\sin y}$$
$$\sin^2 x - \sin^2 y = \cos^2 y - \cos^2 x = \sin(x+y)\sin(x-y)$$
$$\cos^2 x - \sin^2 y = \cos^2 y - \sin^2 x = \cos(x+y)\cos(x-y)$$
$$\sin(45° + x) = \cos(45° - x)$$
$$\sin(45° - x) = \cos(45° + x)$$
$$\tan(45° + x) = \cot(45° - x)$$
$$\tan(45° - x) = \cot(45° + x)$$

Where a and b are positive, the following relations are valid:

(For a triangle, where $c = \sqrt{a^2 + b^2}$; a, b, c, being the sides.)

$$\left.\begin{aligned} a\cos x + b\sin x &= c\sin(A+x) = c\cos(B-x)\\ a\cos x - b\sin x &= c\sin(A-x) = c\cos(B+x)\end{aligned}\right\}\text{ where}$$

$\tan A = \dfrac{a}{b}$ and $\tan B = \dfrac{b}{a}$ and both A and B are positive acute angles.

Trigonometric Functions of Multiple Angles and Half Angles (From Marks, "Mechanical Engineers' Handbook").

$$\sin 2x = 2\sin x\cos x; \qquad \sin x = 2\sin\tfrac{1}{2}x\cos\tfrac{1}{2}x$$
$$\cos 2x = \cos^2 x - \sin^2 x = 1 - 2\sin^2 x = 2\cos^2 x - 1$$
$$\tan 2x = \left(\frac{2\tan x}{1 - \tan^2 x}\right)$$
$$\cot 2x = \left(\frac{\cot^2 x - 1}{2\cot x}\right)$$
$$\sin 3x = 3\sin x - 4\sin^3 x = 3\sin x\cos^2 x - \sin^3 x$$
$$\cos 3x = 4\cos^3 x - 3\cos x$$
$$\tan 3x = \left(\frac{3\tan x - \tan^3 x}{1 - 3\tan^2 x}\right)$$
$$\sin 4x = 4\sin x\cos x - 8\sin^3 x\cos x$$
$$\sin 5x = 5\sin x - 20\sin^3 x + 16\sin^5 x$$
$$\sin 6x = 6\sin x\cos x - 32\sin^3 x\cos x + 32\sin^5 x\cos x$$
$$\cos 3x = 4\cos^3 x - 3\cos x$$
$$\cos 4x = 8\cos^4 x - 8\cos^2 x$$
$$\cos 5x = 16\cos^5 x - 20\cos^3 x + 5\cos x$$
$$\cos 6x = 32\cos^6 x - 48\cos^4 x + 18\cos^2 x - 1$$

$$\sin\tfrac{1}{2}x = \sqrt{\tfrac{1}{2}(1 - \cos x)}; \qquad 1 - \cos x = 2\sin^2\tfrac{1}{2}x$$
$$\cos\tfrac{1}{2}x = \sqrt{\tfrac{1}{2}(1 + \cos x)}; \qquad 1 + \cos x = 2\cos^2\tfrac{1}{2}x$$
$$\tan\frac{1}{2}x = \sqrt{\frac{(1 - \cos x)}{(1 + \cos x)}} = \frac{\sin x}{1 + \cos x} = \frac{1 - \cos x}{\sin x}$$
$$\tan\left(\frac{x}{2} + 45°\right) = \pm\sqrt{\frac{(1 + \sin x)}{(1 - \sin x)}} = \pm\frac{1 + \sin x}{\cos x}$$

Trigonometric Relations between Three Angles Whose Sum Is 180°.

$$\sin A + \sin B + \sin C = 4\cos\frac{A}{2}\cos\frac{B}{2}\cos\frac{C}{2}$$
$$\cos A + \cos B + \cos C = 4\sin\frac{A}{2}\sin\frac{B}{2}\sin\frac{C}{2} + 1$$
$$\sin A + \sin B - \sin C = 4\sin\frac{A}{2}\sin\frac{B}{2}\cos\frac{C}{2}$$
$$\cos A + \cos B - \cos C = 4\cos\frac{A}{2}\cos\frac{B}{2}\cos\frac{C}{2} - 1$$
$$\sin^2 A + \sin^2 B + \sin^2 C = 2\cos A\cos B\cos C + 2$$
$$\sin^2 A + \sin^2 B - \sin^2 C = 2\sin A\sin B\cos C$$
$$\tan A + \tan B + \tan C = \tan A\tan B\tan C$$
$$\cot\frac{A}{2} + \cot\frac{B}{2} + \cot\frac{C}{2} = \cot\frac{A}{2}\cot\frac{B}{2}\cot\frac{C}{2}$$
$$\cot A\cot B + \cot A\cot C + \cot B\cot C = 1$$
$$\sin 2A + \sin 2B + \sin 2C = 4\sin A\sin B\sin C$$
$$\sin 2A + \sin 2B - \sin 2C = 4\cos A\cos B\sin C$$

Inverse Trigonometric Functions. $\sin^{-1} x$ is the antisine of x; *i.e.*, the inverse sine of x, and is sometimes written arc sin x and means the principal angle whose sine is x. The other trigonometric functions are treated in a similar way. The **principal angle** means an angle between +90° and −90° for $\sin^{-1} x$ and $\tan^{-1} x$, and 0° and 180° in the case of $\cos^{-1} x$.

Approximations. (1) When θ is small and measured in radians, sin θ, tan θ, and angle θ are approximately equal. This relation is very useful, for the angle (in radians) can be substituted if either the sine or the tangent is given which will avoid troublesome and time-consuming operations with decimals. (2) When θ (the angle) is small and given in radians, $\cos\theta = 1 - \tfrac{1}{2}\theta^2$.

THE SOLUTION OF PLANE TRIANGLES BY TRIGONOMETRY

Right Triangles (Fig. 85). Of the three sides and two acute angles of a right triangle, two of the parts (one of which is a side) must be known; the trigonometric function selected should involve two of the given parts together with one unknown part. The fundamental formulas are:

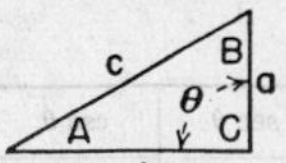

FIG. 85.

$$c^2 = a^2 + b^2$$
$$\text{Angle } A + \text{angle } B = 90°$$
$$\sin A = \cos B = \frac{a}{c} \qquad \cos A = \sin B = \frac{b}{c}$$
$$\tan A = \cot B = \frac{a}{b} \qquad \cot A = \tan B = \frac{b}{a}$$

The trigonometric ratios above are defined as follows:

1. The sine of either of the acute angles (those less than 90°) is equal to the quotient of the opposite side divided by the hypotenuse.

2. The cosine of either of the acute angles is equal to the quotient of the adjacent side divided by the hypotenuse.

3. The tangent of either of the acute angles is equal to the quotient of the opposite side divided by the adjacent side.

4. The cotangent of either of the acute angles is equal to the quotient of the adjacent side divided by the opposite side.

The solution of right triangles is made more accurate and rapid if logarithms are used:

Example 1. Given $a = 5$, and $b = 10$.

$\tan A = \frac{a}{b} = 5/10 = 0.5$; from table: $A = 26°34'$;

$B = 90° - 26°34' = 63°26'$.

$c = \sqrt{a^2 + b^2} = \sqrt{25 + 100} = 11.18$.

Example 2. Given $a = 5$, and $A = 30°$.

$B = 90° - A = 90° - 30° = 60°$;

$\sin A = \frac{a}{c}$; therefore $c = \frac{a}{\sin A}$; $\sin 30° = 0.5$;

therefore $c = \frac{5}{0.5} = 10$

$\cot B = \frac{b}{a}$; therefore $b = a \cot B = 5 \times 1.7321 = 8.6605$.

Example 3. Given $b = 10$, and $A = 40°$.

$\tan A = \frac{a}{b}$; therefore $a = 10 \times \tan 40° = 10 \times 0.8391 = 8.391$

$\cos A = \frac{b}{c}$; therefore $c = \frac{b}{\cos A} = 10 \div \cos 40° = 10 \div 0.766 = 13.055$.

$B = 90° - A = 90° - 40° = 50°$.

Example 4. Given $c = 10$, and $B = 75°$.

$\sin B = \frac{b}{c}$; therefore $b = c \sin B = 10 \sin 75° = (10)(0.9659) = 9.659$;

$\cos B = \frac{a}{c}$; therefore $a = c \cos B = 10 \cos 75° = 10(0.2588) = 2.588$;

$A = 90° - B = 90° - 75° = 15°$.

Example 5. The known parts are indicated in Fig. 86; find a, c, x, and y.

$$\frac{\tan 50°}{\tan 30°} = \frac{a}{x} \div \frac{a}{106 + x}.$$

$$2.0613 = \frac{106 + x}{x}, \quad (x = 99.6)$$

$$\tan 50° = \frac{a}{99.6}, \quad (a = 118.7)$$

$$y = \frac{118.7}{\sin 50°} = \frac{118.7}{0.766} = 154.96.$$

$$c = \frac{118.7}{\sin 30°} = \frac{118.7}{0.5} = 237.4$$

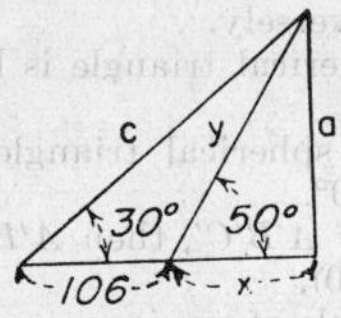

Fig. 86.

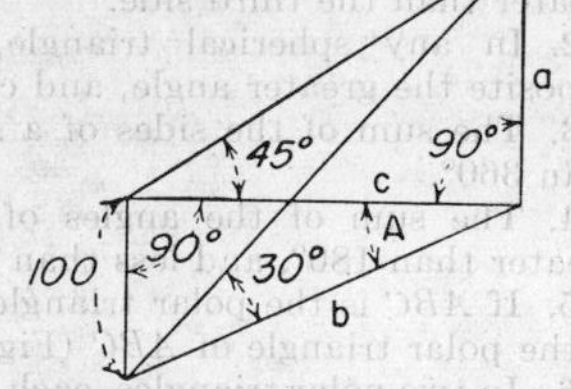

Fig. 87.

Example 6 (Fig. 87). From the given parts in the figure find a; a is perpendicular to the plane in which b and c lie.

$$\frac{a}{c} = \tan 45° = 1.$$

$$\frac{a}{b} = \tan 30° = \frac{\sqrt{3}}{3}$$

$$\frac{\tan 30°}{\tan 45°} = \frac{c}{b} = \frac{\sqrt{3}}{3} = \cos A,$$

$$\sin A = \frac{\sqrt{6}}{3}, \quad b = \frac{100}{\sin A} = 122.47.$$

$$\frac{a}{122.47} = \tan 30° = \frac{\sqrt{3}}{3}$$

$$a = \frac{122.47}{3}\sqrt{3} = 70.7$$

Oblique Triangles. Formulas for the solution of oblique triangles: $A + B + C = 180°$.

Law of Sines

$$(1)\ \frac{a}{b} = \frac{\sin A}{\sin B}$$

$$(2)\ \frac{a}{c} = \frac{\sin A}{\sin C}$$

$$(3)\ \frac{b}{c} = \frac{\sin B}{\sin C}$$

Law of Cosines

$$(4)\ a^2 = b^2 + c^2 - 2bc \cos A$$

$$(5)\ b^2 = a^2 + c^2 - 2ac \cos B$$

$$(6)\ c^2 = a^2 + b^2 - 2ab \cos C$$

$$(7)\ \sin \frac{A}{2} = \sqrt{\frac{(s - b)(s - c)}{bc}}$$

$$(8)\ \sin \frac{B}{2} = \sqrt{\frac{(s - c)(s - a)}{ca}}$$

$$(9)\ \sin \frac{C}{2} = \sqrt{\frac{(s - a)(s - b)}{ab}}$$

$$(10)\ \cos \frac{A}{2} = \sqrt{\frac{s(s - a)}{bc}}$$

$$(11)\ \cos \frac{B}{2} = \sqrt{\frac{s(s - b)}{ac}}$$

$$(12)\ \cos \frac{C}{2} = \sqrt{\frac{s(s - c)}{ab}}$$

$$(13)\ \tan \frac{A}{2} = \sqrt{\frac{(s - b)(s - c)}{s(s - a)}}$$

$$(14)\ \tan \frac{B}{2} = \sqrt{\frac{(s - a)(s - c)}{s(s - b)}}$$

$$(15)\ \tan \frac{C}{2} = \sqrt{\frac{(s - a)(s - b)}{s(s - c)}}$$

$$(16)\ \frac{a + b}{a - b} = \frac{\tan \frac{1}{2}(A + B)}{\tan \frac{1}{2}(A - B)}$$

$$(17)\ \frac{b + c}{b - c} = \frac{\tan \frac{1}{2}(B + C)}{\tan \frac{1}{2}(B - C)}$$

$$(18)\ \frac{a + c}{a - c} = \frac{\tan \frac{1}{2}(A + C)}{\tan \frac{1}{2}(A - C)}$$

In formulas (7) to (15) $a + b + c = 2s$. The area is represented by K. $2K = bc \sin A = ac \sin B = ab \sin C$, $K = \sqrt{s(s - a)(s - b)(s - c)}$. $2K = \frac{a^2 \sin B \sin C}{\sin A} = \frac{b^2 \sin A \sin C}{\sin B} = \frac{c^2 \sin A \sin B}{\sin C}$

To solve an oblique triangle select those formulas which contain the known parts and one of the unknown parts.

The laws of sines and cosines are valid for all triangles. The following rules are the most important laws in the solution of oblique-angle triangles:

Before starting the solution of such a triangle, it is advisable to solve graphically by drawing the triangle to scale. Use five- or six-place logarithm tables when great accuracy is required. Many oblique-angle triangles can be divided profitably into two right angles which are then readily solved by the trigonometric functions and the equation $c^2 = a^2 + b^2$.

A valuable rule in solving triangles is to find each part from the given parts rather than to find one part and use it to find another part. Also, use the law of cosines when three sides, or two sides and their included angle are given; in all other cases use the law of sines.

In solving a triangle, if the cosine is negative, the angle in question is obtuse. For example, if $\cos \theta = -0.7056$, then θ is obtuse and its supplement θ' has a cosine equal to $+0.7056$. From the table, $\theta' = 45°7'$; therefore $\theta = 180° - 45°7' = 134°53'$.

The solutions of oblique-angle triangles may be divided into four general cases:

Case 1. Three Sides Are Given. The longest side is less than the sum of the two other sides. Use the law of cosines and the law of sines.

1. Find the largest angle A (either acute or obtuse) from the relation $\cos A = (b^2 + c^2 - a^2)/2bc$; then find B and C, both of which must be acute, from the relation, $\sin B = b \sin \dfrac{A}{a}$ and $\sin C = c \sin \dfrac{A}{a}$. As a check, the three angles should add up to 180°.

2. Find A, B, and C, from the equations $\tan (A/2) = r/(s - a)$, $\tan (B/2) = r/(s - b)$, and $\tan (C/2) = r/(s - c)$, where $s = (a + b + c)/2$ and

$$r = \sqrt{\frac{(s - a)(s - b)(s - c)}{s}}.$$

3. If only one angle is required, say A, use one of the following equations: $\sin \left(\dfrac{A}{2}\right) = \sqrt{\dfrac{(s - b)(s - c)}{(bc)}}$, or $\cos \left(\dfrac{A}{2}\right) = \sqrt{\dfrac{s(s - a)}{(bc)}}$, depending upon whether $(A/2)$ is nearer 0° or nearer 90°, respectively.

Case 2. Two Sides and an Included Angle Are Given. Assume a and b are given, together with angle C, and assume that a is greater than b.

1. Divide the oblique triangle into two right triangles and solve.

2. Find $\dfrac{(A - B)}{2}$ from the equation, $\tan \frac{1}{2}(A - B) = \left[\dfrac{(a - b)}{(a + b)}\right] \dfrac{\cot C}{2}$; and $\frac{1}{2}(A + B)$ from the equation $\dfrac{(A + B)}{2} = 90° - \dfrac{c}{2}$, from which $A = \dfrac{(A + B)}{2} + \dfrac{(A - B)}{2}$ and $B = \dfrac{(A + B)}{2} - \dfrac{(A - B)}{2}$; then find c from the equation $c = a \sin C/\sin A$, or $c = b \sin C/\sin B$. The result may be checked by the equation $a \cos B + b \cos A = c$.

3. Find c from the relation $c^2 = a^2 + b^2 - 2ab \cos C$; then find the smaller angle B from $\sin B = (b/c) \sin C$; and, finally, find A from $A = 180° - (B + C)$. This method may be checked by the method given under Item 2 above.

Case 3. Two Angles and One Side Are Given. The sum of the two angles is less than 180°. The third angle is known, since $A + B + C = 180°$.

1. Divide the oblique triangle into right triangles and solve. The perpendicular should not be dropped to the given side as a base.

2. To find the remaining sides use the following equations:

$$b = \frac{a \sin B}{\sin A}$$

and

$$c = \frac{a \sin C}{\sin A}$$

(assuming that a is the side given).

Case 4. Two Sides and the Angle Opposite One of Them Are Given. Assume that b, c, and B are given. This is known as the "ambiguous" case, because it may either be impossible to construct the triangle or it may be possible to construct two triangles from the solutions obtained.

As the first test, construct the triangle. Then try to find C from the equation: $\sin C = \dfrac{c \sin B}{b}$. If $\sin C$ is greater than one, there is no solution. If, however, $\sin C = 1$, then $C = 90°$ and the triangle is a right triangle. If $\sin C$ is less than one, this shows that C may be either an acute or an obtuse angle; *i.e.*, C_1 or $C_2 = 180° - C_1$. If this is the case, C_1 is possible as a solution only when $(C_2 + B)$ is less than 180°.

Miscellaneous Properties of Triangles. Area $= \frac{1}{2}ab \sin C = \sqrt{s(s - a)(s - b)(s - c)} = rs$, where $s = \frac{1}{2}(a + b + c)$ and $r =$ the radius of the inscribed circle $= \sqrt{\dfrac{(s - a)(s - b)(s - c)}{s}}$. $2R = \dfrac{a}{\sin A} = \dfrac{b}{\sin B} = \dfrac{c}{\sin C}$, where R is the radius of the circumscribed circle and $r = 4R \sin (A/2) \sin (B/2) \sin (C/2) = (abc)/4Rs$; $r/R = \cos A + \cos B + \cos C - 1$.

SPHERICAL TRIGONOMETRY

Spherical Triangles. Definitions. A great circle is formed by the intersection of the surface of the sphere by a plane passing through the center of the sphere. In Fig. 88 the three great circles $ACDE$, CBE, and ABD form eight spherical triangles. Each of the spherical triangles is bounded by arcs of great circles. In the

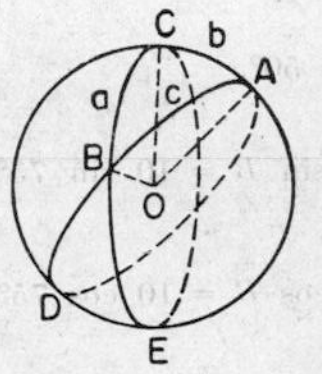

FIG. 88.

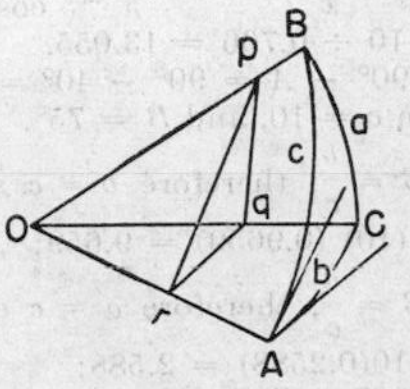

FIG. 89.

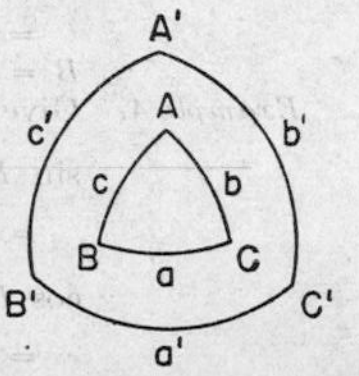

FIG. 90.

triangle ABC, the arcs a, b, and c are the sides. The solid O-ABC of Fig. 88 is drawn in Fig. 89. By the angle A is meant the angle between the planes of the sides b and c, or the angle at A between the tangents to the two great circles passing through A.

The triangle pqr lies in a plane perpendicular to OA. The trigonometric functions of the angle A are then the functions of the angle prq. It is to be observed that the angles are designated by capital letters and the sides opposite the angles by the corresponding small letters.

The following properties are proved in **solid geometry:**

1. The sum of any two sides of a spherical triangle is greater than the third side.

2. In any spherical triangle, the greater side lies opposite the greater angle, and conversely.

3. The sum of the sides of a spherical triangle is less than 360°.

4. The sum of the angles of a spherical triangle is greater than 180°, and less than 540°.

5. If ABC is the polar triangle of $A'B'C'$, then $A'B'C'$ is the polar triangle of ABC (Fig. 90).

6. In two polar triangles, each angle of one is measured by the supplement of the side lying opposite the homologous angle of the other (Fig. 90); *i.e.*,

$$a' = 180° - A \quad b' = 180° - B \quad c' = 180° - C$$
$$A' = 180° - a \quad B' = 180° - b \quad C' = 180° - c$$

A **zone** is that portion of the surface of a sphere which is included between two parallel planes.

A **spherical segment** is a portion of a sphere included between two parallel planes.

A **spherical sector** is the portion of a sphere generated by the revolution of a circular sector about a diameter.

Solution of Right Spherical Triangles. To solve a right spherical triangle, two elements must be given in addition to the right angle. There may be six cases. Given: (1) hypotenuse and an adjacent angle; (2) an angle and its opposite side; (3) an angle and its adjacent side; (4) the hypotenuse and another side; (5) two sides a and b; (6) two angles A and B.

The formulas for the solution of these six cases are:

$$\cos c = \cos a \cos b \qquad \tan A = \frac{\tan a}{\sin b}$$

$$\cos A = \frac{\tan b}{\tan c} \qquad \cos B = \frac{\tan a}{\tan c}$$

$$\sin B = \frac{\sin b}{\sin c} \qquad \sin A = \frac{\sin a}{\sin c}$$

$$\sin B = \frac{\cos A}{\cos a} \qquad \sin A = \frac{\cos B}{\cos b}$$

$$\cos c = \cot A \cot B \qquad \tan B = \frac{\tan b}{\sin a}$$

Example 1. Given $A = 50°43'$, $B = 122°18'$, C is the right angle. Find a, b, and c. The formulas to use are: $\cos c = \cot A \cot B$, $\sin A = \dfrac{\cos B}{\cos b}$, and $\sin B = \dfrac{\cos A}{\cos a}$.

$$\begin{aligned} \log \cot A &= 9.91276 - 10 & \log \cos B &= 9.72783 - 10n \\ \log \cot B &= 9.80084 - 10n & \log \sin A &= 9.88876 - 10 \\ \log \cos c &= 9.71360 - 10n & \log \cos b &= 9.83907 - 10n \\ c &= 121°8'.4 & b &= 133°39'.5 \end{aligned}$$

$$\begin{aligned} \log \cos A &= 9.80151 - 10 \\ \log \sin B &= 9.92699 - 10 \\ \log \cos a &= 9.87452 - 10 \\ a &= 41°29'.5 \end{aligned}$$

n, written at the right of a logarithm, indicates that the corresponding number is negative.

Example 2. Given $b = 157°41'$, $c = 136°21'$. Find a, A, and B.

$$\cos A = \frac{\tan b}{\tan c} \qquad \cos a = \frac{\cos c}{\cos b}$$

$$\begin{aligned} \log \tan b &= 9.61328 - 10n & \log \cos c &= 9.85948 - 10 \\ \log \tan c &= 9.97953 - 10n & \log \cos b &= 9.96619 - 10 \\ \log \cos A &= 9.63375 - 10 & \log \cos a &= 9.89329 - 10 \\ A &= 64°30'.9 & a &= 38°32'.5 \end{aligned}$$

$$\sin B = \frac{\sin b}{\sin c}$$

$$\begin{aligned} \log \sin b &= 9.57947 - 10 \\ \log \sin c &= 9.83901 - 10 \\ \log \sin B &= 9.74046 - 10 \\ \text{supplement } B &= 33°22'.5 \\ B &= 146°37'.5 \end{aligned}$$

B is the largest angle for it lies opposite the longest side; therefore the supplement of $33°22'.5$.

Example 3. Given $c = 127°9'$, $B = 80°51'$. Find a, b, and A.

$$\sin b = \sin B \sin c \qquad \tan c = \cos B \tan c$$

$$\begin{aligned} \log \sin B &= 9.99444 - 10 & \log \cos B &= 9.20145 - 10 \\ \log \sin c &= 9.90149 - 10 & \log \tan c &= 0.12052 - 10n \\ \log \sin b &= 9.89593 - 10 & \log \tan a &= 9.32197 - 10n \\ b &= 51°53'.9 & a &= 168°8'.8 \end{aligned}$$

$$\cot A = \frac{\cos c}{\cot B}$$

$$\begin{aligned} \log \cos c &= 9.78097 - 10n \\ \log \cot B &= 9.20701 - 10 \\ \log \cot A &= 0.57396 - 10n \\ A &= 165°4' \end{aligned}$$

tan a being negative shows that a is in the second quadrant. cot A is negative; therefore, A is in the second quadrant. This is also evident from the fact that a is the greatest side, and, therefore, A must be the greatest angle.

Solution of Oblique Spherical Triangles. There are six cases: (1) Given a side and the adjacent angles, (a, A, B); (2) given two sides and their included angle, (a, b, C); (3) given the three sides, (a, b, c); (4) given the three angles, (A, B, C); (5) given two sides and the angle opposite one of them, (a, b, A); (6) given two angles and the side opposite one of them, (A, B, a).

The formulas for solving oblique spherical triangles are given below.

$$(1)\quad \frac{\sin A}{\sin B} = \frac{\sin a}{\sin b}$$

$$(2)\quad \frac{\sin B}{\sin C} = \frac{\sin b}{\sin c}$$

$$(3)\quad \frac{\sin A}{\sin C} = \frac{\sin a}{\sin c}$$

$$(4)\quad \cos a = \cos b \cos c + \sin b \sin c \cos A$$

$$(5)\quad \cos b = \cos c \cos a + \sin c \sin a \cos B$$

$$(6)\quad \cos c = \cos a \cos b + \sin a \sin b \cos C$$

$$(7)\quad \cos A = -\cos B \cos C + \sin B \sin C \cos a$$

$$(8)\quad \cos B = -\cos C \cos A + \sin C \sin A \cos b$$

$$(9)\quad \cos C = -\cos A \cos B + \sin A \sin B \cos c$$

$$(10)\quad \sin \frac{1}{2} A = \sqrt{\frac{\sin (s-b) \sin (s-c)}{\sin b \sin c}}$$

$$(11)\quad \sin \frac{1}{2} B = \sqrt{\frac{\sin (s-c) \sin (s-a)}{\sin c \sin a}}$$

$$(12)\quad \sin \frac{1}{2} C = \sqrt{\frac{\sin (s-a) \sin (s-b)}{\sin a \sin b}}$$

$$(13)\quad \cos \frac{1}{2} A = \sqrt{\frac{\sin s \sin (s-a)}{\sin b \sin c}}$$

$$(14)\quad \cos \frac{1}{2} B = \sqrt{\frac{\sin s \sin (s-b)}{\sin a \sin c}}$$

$$(15)\quad \cos \frac{1}{2} C = \sqrt{\frac{\sin s \sin (s-c)}{\sin a \sin b}}$$

$$(16)\quad \tan \frac{1}{2} A = \sqrt{\frac{\sin (s-b) \sin (s-c)}{\sin s \sin (s-a)}}$$

$$(17)\quad \tan \frac{1}{2} B = \sqrt{\frac{\sin (s-c) \sin (s-a)}{\sin s \sin (s-b)}}$$

$$(18)\quad \tan \frac{1}{2} C = \sqrt{\frac{\sin (s-a) \sin (s-b)}{\sin s \sin (s-c)}}$$

$$(19)\quad \sin \frac{1}{2} a = \sqrt{\frac{-\cos S \cos (S-A)}{\sin B \sin C}}$$

$$(20)\quad \sin \frac{1}{2} b = \sqrt{\frac{-\cos S \cos (S-B)}{\sin C \sin A}}$$

$$(21)\quad \sin \frac{1}{2} c = \sqrt{\frac{-\cos S \cos (S-C)}{\sin A \sin B}}$$

$$(22)\quad \cos \frac{1}{2} a = \sqrt{\frac{\cos (S-B) \cos (S-C)}{\sin B \sin C}}$$

$$(23)\quad \cos \frac{1}{2} b = \sqrt{\frac{\cos (S-C) \cos (S-A)}{\sin C \sin A}}$$

$$(24)\quad \cos \frac{1}{2} c = \sqrt{\frac{\cos (S-A) \cos (S-B)}{\sin A \sin B}}$$

$$(25)\quad \tan \frac{1}{2} a = \sqrt{\frac{-\cos S \cos (S-A)}{\cos (S-B) \cos (S-C)}}$$

$$(26)\quad \tan \frac{1}{2} b = \sqrt{\frac{-\cos S \cos (S-B)}{\cos (S-C) \cos (S-A)}}$$

$$(27)\quad \tan \frac{1}{2} c = \sqrt{\frac{-\cos S \cos (S-C)}{\cos (S-A) \cos (S-B)}}$$

$$(28)\quad \frac{\sin \frac{1}{2}(A+B)}{\sin \frac{1}{2}(A-B)} = \frac{\tan \frac{1}{2}c}{\tan \frac{1}{2}(a-b)}$$

$$(29)\quad \frac{\cos \frac{1}{2}(A+B)}{\cos \frac{1}{2}(A-B)} = \frac{\tan \frac{1}{2}c}{\tan \frac{1}{2}(a+b)}$$

$$(30)\quad \frac{\sin \frac{1}{2}(a+b)}{\sin \frac{1}{2}(a-b)} = \frac{\cot \frac{1}{2}C}{\tan \frac{1}{2}(A-B)}$$

$$(31)\quad \frac{\cos \frac{1}{2}(a+b)}{\cos \frac{1}{2}(a-b)} = \frac{\cot \frac{1}{2}C}{\tan \frac{1}{2}(A+B)}$$

The sum of the angles $A + B + C$ is denoted by $2S$, $S = \frac{1}{2}(A + B + C)$.

Case 1. Given $A = 180°12'$, $B = 145°46'$, $C = 126°32'$. From formulas (28) and (29) are obtained the following:

$$\tan \tfrac{1}{2}(b - a) = \sin \tfrac{1}{2}(B - A) \csc \tfrac{1}{2}(B + A) \tan \tfrac{1}{2}c$$
$$\tan \tfrac{1}{2}(b + a) = \cos \tfrac{1}{2}(B - A) \sec \tfrac{1}{2}(B + A) \tan \tfrac{1}{2}c$$

$$\begin{aligned}
\log \sin \tfrac{1}{2}(B - A) &= 9.50784 - 10 \\
\log \csc \tfrac{1}{2}(B + A) &= 0.09756 \\
\log \tan \tfrac{1}{2}c &= \underline{0.29785} \\
\log \tan \tfrac{1}{2}(b - a) &= 9.90325 - 10 \\
\tfrac{1}{2}(b - a) &= 38°\ 40'.2 \\
\log \cos \tfrac{1}{2}(B - A) &= 9.97623 - 10 \\
\log \sec \tfrac{1}{2}(B + A) &= 0.22070n \\
\log \tan \tfrac{1}{2}c &= \underline{0.29785} \\
\log \tan \tfrac{1}{2}(b + a) &= 0.49478n \\
\tfrac{1}{2}(b + a) &= 107°\ 44'.8 \\
\tfrac{1}{2}(b + a) + \tfrac{1}{2}(b - a) &= b = 146°25' \\
\tfrac{1}{2}(b + a) - \tfrac{1}{2}(b - a) &= 2 = 69°4'.6
\end{aligned}$$

By (30),
$$\begin{aligned}
\log \sin \tfrac{1}{2}(b + a) &= 9.97883 - 10 \\
\operatorname{colog} \sin \tfrac{1}{2}(a - b) &= 0.20424n \\
\log \tan \tfrac{1}{2}(A - B) &= \underline{9.53161 - 10n} \\
\log \cot \tfrac{1}{2}C &= 9.71468 - 10 \\
\tfrac{1}{2}C &= 62°35.8' \\
C &= 125°11'.6
\end{aligned}$$

Case 2. Given $a = 62°20'$, $b = 54°10'$, $c = 97°50'$. Find A, B, and C.

By (16),
$$\begin{aligned}
\log \sin (s - b) &= 9.90235 - 10 \\
\log \sin (s - c) &= 9.20999 - 10 \\
\operatorname{colog} \sin (s - a) &= 0.15178 \\
\operatorname{colog} \sin a &= \underline{0.01979} \\
\log \tan^2 (\tfrac{1}{2}A) &= 9.28391 - 10 \\
\log \tan \tfrac{1}{2}A &= 9.64195 - 10 \\
\tfrac{1}{2}A &= 23°40'.6 \\
A &= 47°21'.2
\end{aligned}$$

B and C are found in the same way from formulas (17) and (18). The other cases are solved by the use of properly selected formulas.

Area of a Spherical Triangle. Given the three angles. K represents area, $2S = A + B + C$, R = radius of the sphere, $K = R^2(2S - \pi)$. It is to be remembered that $2S$ must be expressed in radians.

For an extensive treatment, see Chauvenet's "Trigonometry."

Hyperbolic Trigonometric Functions. Hyperbolic functions are certain combinations of the sum and difference of two exponential functions such as $e^u + e^{-u}$ and $e^u - e^{-u}$, and are defined by the statement that $(e^u - e^{-u})/2$ = the hyperbolic sine of u and is abbreviated as $\sinh u$. Similarly, $(e^u + e^{-u})/2$ is the hyperbolic cosine of u and is abbreviated as $\cosh u$. The other hyperbolic functions are:

$$\tanh u = \frac{\sinh u}{\cosh u} = \frac{e^u - e^{-u}}{e^u + e^{-u}}$$
$$\operatorname{sech} u = \frac{1}{\cosh u} = \frac{2}{(e^u + e^{-u})}$$
$$\coth u = \frac{1}{\tanh u} = \frac{e^u + e^{-u}}{e^u - e^{-u}}$$
$$\operatorname{csch} u = \frac{1}{\sinh u} = \frac{2}{(e^u - e^{-u})}$$

The following formulas can be derived from the definitions given above:

$$\cosh^2 u - \sinh^2 u = 1$$
$$\operatorname{sech}^2 u + \tanh^2 u = 1$$
$$\coth^2 u - \operatorname{csch}^2 u = 1$$
$$\sinh^2 u = \cosh^2 u - 1 = \frac{(\tanh^2 u)}{(1 - \tanh^2 u)} = \frac{1}{\coth^2 u - 1} = \frac{1 - \operatorname{sech}^2 u}{\operatorname{sech}^2 u} = \frac{1}{\operatorname{csch}^2 u}$$
$$\cosh^2 u = \sinh^2 u + 1 = \frac{1}{(1 - \tanh^2 u)} = \frac{\coth^2 u}{(\coth^2 u - 1)} = \operatorname{csch}^2 u + \frac{1}{\operatorname{csch} u} = \frac{1}{\operatorname{sech} u}$$
$$\sinh(-u) = -\sinh u, \quad \cosh(-u) = \cosh u, \quad \tanh(-u) = -\tanh u.$$
$$\tanh u = \frac{(\sinh u)}{(\sqrt{\sinh^2 u + 1})} = \frac{(\sqrt{\cosh^2 u - 1})}{\cosh u} = \sqrt{1 - \operatorname{sech}^2 u} = \frac{1}{(\sqrt{\operatorname{csch}^2 u + 1})} = \frac{1}{\coth u}$$
$$\sinh(u \pm v) = \frac{e^{u \pm v} - e^{-(u \pm v)}}{2} = \sinh u \cosh v \pm \cosh u \sinh v$$
$$\cosh(u \pm v) = \frac{e^{(u \pm v)} + e^{-(u \pm v)}}{2} = \cosh u \cosh v \pm \sinh u \sinh v$$
$$\tanh(u \pm v) = \frac{e^{(u \pm v)} - e^{-(u \pm v)}}{e^{(u \pm v)} + e^{-(u \pm v)}} = \frac{\tanh u \pm \tanh v}{1 \pm \tanh u \tanh v}$$
$$\sinh 2u = 2 \sinh u \cosh u = \frac{(2 \tanh u)}{(1 - \tanh^2 u)}$$
$$\cosh 2u = \cosh^2 u + \sinh^2 u = 1 + 2 \sinh^2 u = 2 \cosh^2 u - 1 = \frac{(1 + \tanh^2 u)}{(1 - \tanh^2 u)}$$
$$\tanh 2u = \frac{(2 \tanh u)}{(1 + \tanh^2 u)} \qquad \cosh u + 1 = 2 \cosh^2 \left(\frac{u}{2}\right)$$
$$\cosh u - 1 = 2 \sinh^2 \left(\frac{u}{2}\right)$$

When $x = a \cosh u$ and $y = a \sinh u$, then: $(x^2 - y^2) = a^2 (\cosh^2 u - \sinh^2 u)$; and since $\cosh^2 u = \sinh^2 u + 1$, then $x^2 - y^2 = a^2$. In other words, the hyperbolic functions in the parametric equations $x = a \cosh u$ and $y = a \sinh u$ have the same relation to the rectangular hyperbola: $x^2 - y^2 = a^2$ that the equations $x = a \cos \theta$ and $y = a \sin \theta$ have to the circle $x^2 + y^2 = a^2$.

$$\cosh u + \sinh u = \frac{e^u + e^{-u}}{2} + \frac{e^u - e^{-u}}{2} = e^u$$
$$\cosh u - \sinh u = e^{-u}$$

$\sinh u = (\tfrac{1}{2})(e^u - e^{-u})$ and, since $e^u = 1 + u + \frac{u^2}{\lfloor 2} + \frac{u^3}{\lfloor 3} + \frac{u^4}{\lfloor 4} + \cdots$ it follows that $\sinh u = u + \frac{u^3}{\lfloor 3} + \frac{u^5}{\lfloor 5} + \frac{u^7}{\lfloor 7} \cdots$ and, similarly, that $\cosh u = 1 + \frac{u^2}{\lfloor 2} + \frac{u^4}{\lfloor 4} + \frac{u^6}{\lfloor 6} + \cdots$

The above series representing $\sinh u$ and $\cosh u$ are convergent for all real values of u and may be used to calculate the hyperbolic functions.

Inverse Hyperbolic Functions. If $\sinh^{-1} u = v$, then $u = \sinh v$. For real values of u, the negative sign being excluded from $\sinh^{-1} u$:

$$\sinh^{-1} v = \log [v + \sqrt{(v^2 + 1)}]$$
$$\tanh^{-1} v = \frac{1}{2} \log \left[\frac{(1 + v)}{(1 - v)}\right]$$
$$\operatorname{sech}^{-1} v = \log \frac{1 + \sqrt{1 - v^2}}{v}$$
$$\cosh^{-1} v = \log [v + \sqrt{(v^2 - 1)}]$$
$$\coth^{-1} v = \frac{1}{2} \log \left[\frac{(v + 1)}{(v - 1)}\right]$$
$$\operatorname{csch}^{-1} v = \log \frac{1 + \sqrt{1 + v^2}}{v}$$

ANALYTICAL GEOMETRY

Coordinates. Two intersecting straight lines are generally used to determine the location of a point or curve. These lines, called coordinate axes, may intersect at right angles, as in Fig. 91; or at angles different from 90°, as in Fig. 92. A point is designated by (x, y), where x is the distance of the point from the y-axis, measured parallel to the y-axis, positive if to the right, negative if to the left. Y is the distance of the point from the x-axis, measured parallel to the y-axis, positive if upward, negative if downward. X is the abscissa and y the ordinate. O is the origin.

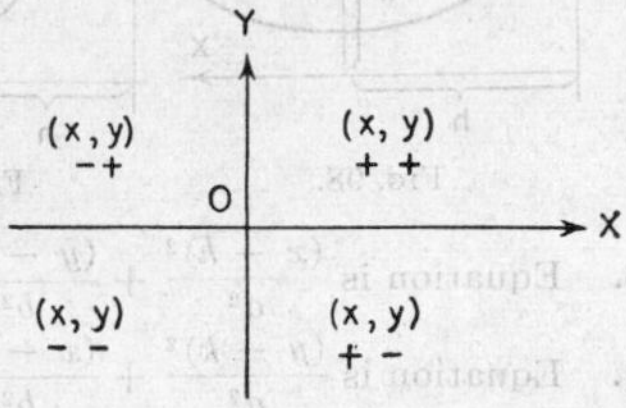

FIG. 91.

Polar Coordinates (Fig. 93). In this system of coordinates the position of a point is fixed by the coordinates (r, θ). r is the distance of the point from the origin, θ the angle that the line r makes with the positive direction of the x-axis or polar axis. From Fig. 93 it is seen that $x^2 + y^2 = r^2$, $x = r \cos \theta$, $y = r \sin \theta$, and $\theta = \tan^{-1} \frac{y}{x}$.

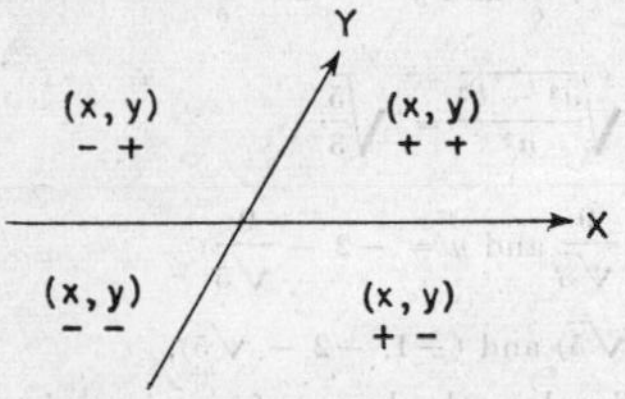

FIG. 92.

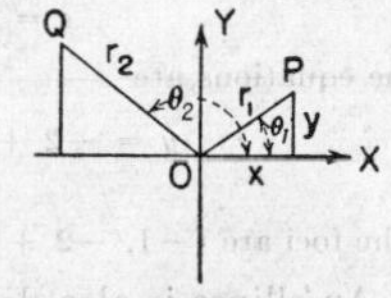

FIG. 93.

The distance between two points, (x_1, y_1) and (x_2, y_2) is represented by the formula $l = \sqrt{(x_2 - x_1)^2 + (y_2 - y_1)^2}$. In polar coordinates, $l = \sqrt{r_2^2 + r_1^2 - 2r_2r_1 \cos(\theta_1 - \theta_2)}$.

The Slope of a Line. The slope of a line, m, is the tangent of the angle (θ) (inclination angle) formed by the x-axis (abscissa), and the line, or the line produced to meet the axis where $m = \tan \theta$. m may be either positive or negative. The slope of a line parallel to the x-axis is zero, and that of a line parallel to the y-axis is infinity. In the case of two perpendicular lines, the slope of one line is the negative reciprocal of the slope of the second line. The slope of a straight line is positive when an increase in x causes an increase in y, and is negative when an increase in x causes a decrease in y. The slope of a line may be determined from the coordinates of two points on the line by the relation: $m = (y_1 - y_2)/(x_1 - x_2)$. The coordinates of a point in the middle of a line are: $x = \frac{1}{2}(x_1 + x_2)$ and $y = \frac{1}{2}(y_1 + y_2)$; and the coordinates of a point $1/n$th the way from (x_1, y_1) to (x_2, y_2) are $x = x_1 + \frac{1}{n}(x_2 - x_1)$ and $y = y_1 + \frac{1}{n}(y_2 - y_1)$.

The Angle One Line Makes with Another. Given two lines of slopes m_1 and m_2, respectively, to find the angle (θ) between the lines: $\tan \theta = \frac{m_2 - m_1}{m_2m_1 + 1}$.

Equations of the Straight Line. Every equation of the type $Ax + By = C$ represents a straight line of slope $-A/B$. A straight line is determined by the following cases:

1. Point Slope. A point on the line and the slope of the line are given. If the coordinates of a point be (x_1, y_1), the slope m and any other point on the line is assumed to be (x_2, y_2), then $m = (y_2 - y_1)/(x_2 - x_1)$ or $y_2 - y_1 = m(x_2 - x_1)$.

2. Intercept Slope. If the intercept of a line on the y-axis is A and the slope is m, its equation is $y = mx + A$.

3. Two Points. Let (x_1, y_1) and (x_2, y_2) be the two given points and (x, y) be any other point on the line. The equation of the line is $y - y_1 = \frac{y_2 - y_1}{x_2 - x_1}(x - x_1)$.

4. Two Intercepts. If the intercept of a line on the x-axis is a and the intercept on the y-axis is b, its equation is $\frac{x}{a} + \frac{y}{b} = 1$.

5. A line is determined if the length and direction of the perpendicular to it (**the normal**) from the origin are given (Fig. 94). Let p represent the perpendicular and θ the angle that this perpendicular makes with the positive direction of the x-axis. The equation of the line is $x \cos \theta + y \sin \theta = p$.

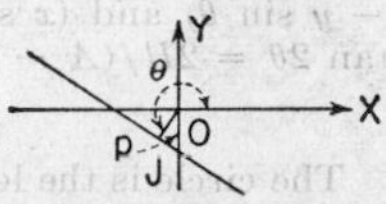

FIG. 94.

6. The general equation of a straight line is $Ax + By = -C$. If $A = 0$, the line is parallel to the x-axis. If $B = 0$, the line is parallel to the y-axis. Dividing by $-C$, the equation becomes:

$$\frac{x/-C}{A} + \frac{y/-C}{B} = 1$$

$-C/A$ is then the intercept on the x-axis and $-C/B$ the intercept on the y-axis. If $C = 0$, the line passes through the origin.

7. Equation of a Line through a Point (x_1, y_1) and Perpendicular to a Straight Line $Ax + By + C = 0$. Since the slope of this line is $= -A/B$, and the required line is perpendicular to this line, the slope of the latter $= B/A$; hence, the equation of the required line is $y - y_1 = (B/A)(x - x_1)$.

8. Distance from a Point (x_1, y_1) to a Straight Line Whose Equation Is $Ax + By + C = 0$. Let d represent the distance. Then $d = \frac{Ax_1 + By_1 + C}{\sqrt{A^2 + B^2}}$. d will be positive, if the point (x_1, y_1) and the origin are on the same side of the line; negative, if on opposite sides of the line.

CONICS

The term **conic** is derived from the fact that the curves included in this group are merely the various cross sections of a cone. The ellipse is produced by a plane intersecting the cone obliquely to the axis and to the surface. The parabola, by a plane intersecting parallel to an element of a surface, the hyperbola by a plane intersecting parallel to the axis, and with a circle the intersecting plane is perpendicular to the axis.

A **conic** is the locus of a point whose distance from a fixed point, called the **focus,** is in a constant ratio to its distance from a fixed line, called the **directrix.** This constant ratio is the eccentricity (e). If $e = 1$, the conic is a parabola; if $e < 1$, the curve is an ellipse; and if $e > 1$, the curve is an hyperbola.

The general equation of the second degree in two variables

$$Ax^2 + 2Hxy + By^2 + 2Gx + 2Fy + C = 0$$

represents a conic. The various criteria of the nature of the conic represented by such an equation are given in the following forms.

$\Delta = ABC + 2FGH - AF^2 - BG^2 - HG^2 =$ the discriminate of the quadratic, and $D = H^2 - AB =$ the characteristic.

1. When $H^2 - AB < 0$,
   $A\Delta < 0$ and $A \neq B$, an ellipse
   $A\Delta < 0$ and $A = B$, a circle
   $A\Delta > 0$, no locus
   $\Delta = 0$, a point
2. When $H^2 - AB > 0$,
   $\Delta \neq 0$, an hyperbola
   $\Delta = 0$, two intersecting lines
3. When $H^2 - AB = 0$,
   $\Delta \neq 0$, a parabola
   $\Delta = 0$, two parallel lines, one line, or no locus

To move the origin to the point (h, k), substitute for x and y in the original equation $(x + h)$ and $(y + k)$, respectively. To turn the axes through an angle θ, substitute for x and y in the original equation $(x \cos \theta - y \sin \theta)$ and $(x \sin \theta + y \cos \theta)$, respectively, where $\tan 2\theta = 2H/(A - B)$.

THE CIRCLE

The circle is the locus of a point at a constant distance from a fixed point.

General Equation. $x^2 + y^2 = r^2$, where r is the radius of the circle, and the center of the circle is at the origin. If the center is at (x_1, y_1), the equation is $(x - x_1)^2 + (y - y_1)^2 = r^2$. An equation of the type $x^2 + 2Ax + y^2 + 2By + C = 0$ represents a circle, with the center at $(-A, -B)$ and the radius equal to $\sqrt{A^2 + B^2 - C}$.

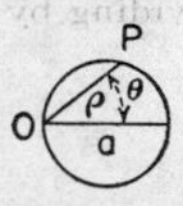

FIG. 95.

Polar Equation of Circle (Fig. 95). $\rho = a \cos \theta$, where $a =$ the diameter of the circle and the origin is on the circumference of the circle.

Equation of the Circle Passing through Three Points. Substitute the pairs of coordinates of the three points in the equation $x^2 + 2Ax + y^2 + 2By + C = 0$ and solve simultaneously for A, B, and C.

Equation of Tangent to a Circle. At a point (x_1, y_1 on $x^2 + y^2 = r^2$) the equation of the tangent is $xx_1 + yy_1 = r^2$. The tangent to the circle $x^2 + 2Ax + y^2 + 2By + C = 0$ at the point (x_1, y_1) is $(x_1 + A)x + (y_1 + B)y + Ax_1 + By_1 + C = 0$.

THE ELLIPSE

The ellipse is the locus of a point the sum of whose distances from two fixed points (foci) is constant. The foci are represented by F_1 and F_2. $F_1P + F_2P = 2a$. The major axis $= 2a$; the minor axis $= 2b$. The eccentricity is represented by e. If $a = b$, the ellipse becomes a circle. A circle is therefore an ellipse whose eccentricity is zero. In the ellipse $C = ae$, $b^2 = a^2(1 - e^2)$. The latus rectum is the double ordinate through F_1 or F_2 (Fig. 96), or the double abscissa through F_1 or F_2 (Fig. 97), and is equal to $2b^2/a$.

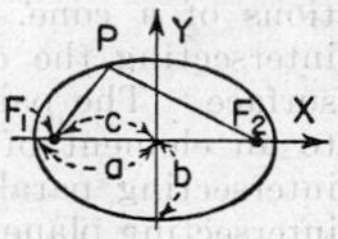

FIG. 96.

Figure 96. Equation is $\dfrac{x^2}{a^2} + \dfrac{y^2}{b^2} = 1$. $F_1 \equiv (-ae, 0)$, $F_2 \equiv (ae, 0)$. Equation of the normal at (x_1, y_1) is $y - y_1 = \dfrac{a^2 y_1}{b^2 x_1}(x - x_1)$. The equation of the tangent at (x_1, y_1) is $y - y_1 = \dfrac{b^2 x_1}{a^2 y_1}(x - x_1)$. The equation of the tangent in terms of its slope is $y = ms \pm \sqrt{m^2a^2 + b^2}$.

Figure 97. Equation is $\dfrac{y^2}{a^2} + \dfrac{x^2}{b^2} = 1$. $F_1 \equiv (0, ae)$, $F_2 \equiv (0, -ae)$. Equation of the tangent at (x_1, y_1) is $y - y_1 = -\dfrac{a^2 x_1}{b^2 y_1}(x - x_1)$. Equation of the normal is $y - y_1 = \dfrac{b^2 y_1}{a^2 x_1}(x - x_1)$.

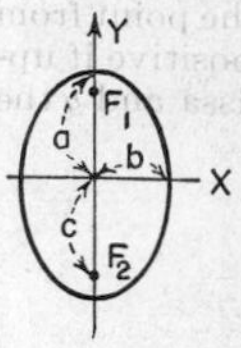

FIG. 97.

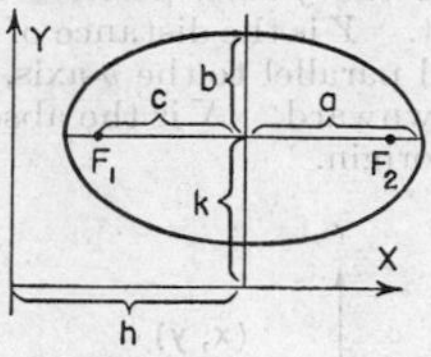

FIG. 98.

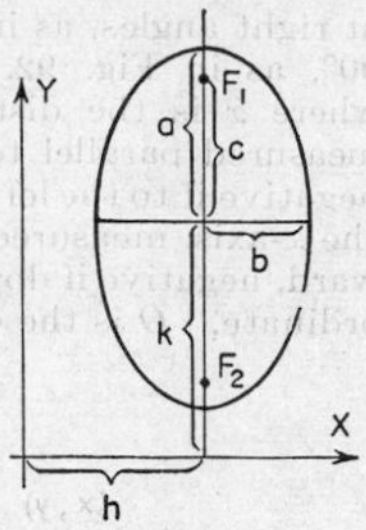

FIG. 99.

Figure 98. Equation is $\dfrac{(x - h)^2}{a^2} + \dfrac{(y - k)^2}{b^2} = 1$.

Figure 99. Equation is $\dfrac{(y - k)^2}{a^2} + \dfrac{(x - h)^2}{b^2} = 1$.

Example. Reduce to standard form:

$$9x^2 + 4y^2 + 18x + 16y - 11 = 0$$
$$9(x + 1)^2 + 4(y + 2)^2 = 36,$$

or

$$\frac{(x + 1)^2}{4} + \frac{(y + 2)^2}{9} = 1$$

The center is at $(-1, -2)$, $a = 3$, $b = 2$. The equations of the directrices are $y = -2 + \dfrac{a}{e}$, and $y = -2 - \dfrac{a}{e}$.

Since

$$e = \sqrt{\frac{a^2 - b^2}{a^2}} = \sqrt{\frac{5}{3}},$$

the equations are

$$y = -2 + \frac{9}{\sqrt{5}} \text{ and } y = -2 - \frac{9}{\sqrt{5}}$$

The foci are $(-1, -2 + \sqrt{5})$ and $(-1, -2 - \sqrt{5})$.

An ellipse is also defined as the locus of a point whose distance from a fixed point (focus) is in a constant ratio to its distance from a fixed straight line (directrix). The two lines parallel to the minor axis, QA and $Q'A'$ (Fig. 100), are the directrices. The distance of each directrix from the minor axis is a/e. The distance of each focus from the origin is ae $PF_1/PQ = PF_2/PQ' = e$.

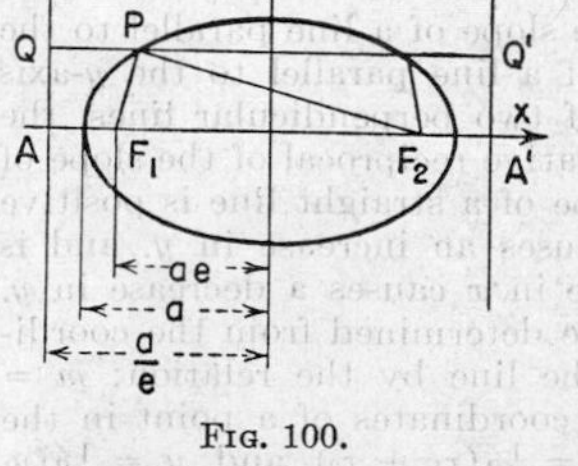

FIG. 100.

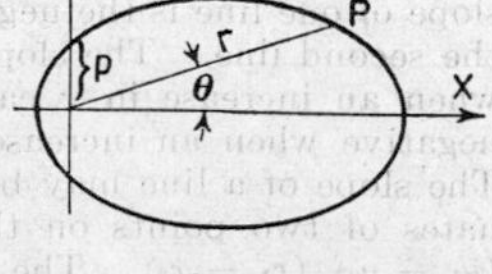

FIG. 101.

The equation of the diameter which bisects all chords of slope m is $y = -\dfrac{b^2}{a^2 m}x$. With the pole at the left-hand focus and $p = \dfrac{b^2}{a}$, the polar equation of the ellipse is $r = \dfrac{p}{l - e \cos \theta}$ (Fig. 101). The parametric equation of the ellipse with the center at the origin is $x = a \cos \theta$ and $y = b \sin \theta$ (Fig. 102).

Figure 102. A method for constructing the ellipse by points is as follows: Draw a circle with radius a, and with the same center and radius b another circle (Fig. 97). Draw the line OQ; through the point R draw the line SP parallel to the x-axis. Draw QT parallel to the y-axis. The intersection of QT with SP will be a point on the ellipse whose semiaxes are a and b.

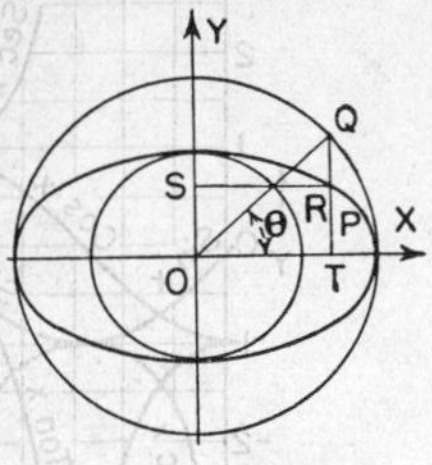

FIG. 102.

THE PARABOLA

The parabola is the locus of a point equidistant from a fixed point (focus) and a fixed straight line (directrix). The vertex is the point midway between the focus and directrix.

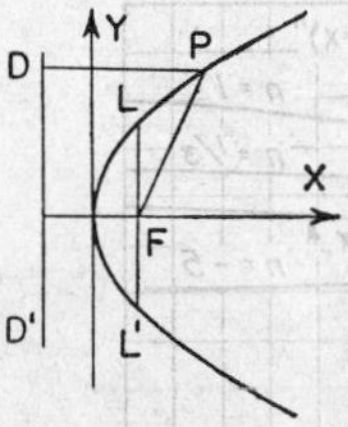

FIG. 103.

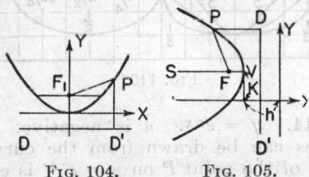

Figure 103. F is the focus, the line DD' is the directrix. $FP = PD$. The chord (double ordinate) through F, parallel to the y-axis, is the latus rectum. The x-axis is the axis of the parabola. The equation of the curve is $y^2 = 4px$. When p is positive, the parabola opens in the positive direction of the x-axis; when p is negative, the parabola opens in the negative direction of the x-axis. The length of the latus rectum is $4p$.

Figure 104. F is the focus, the line DD' is the directrix. $PD' = PF$. The chord through F, parallel to the x-axis, is the latus rectum $= 4p$. The equation of the curve is $x^2 = 4py$. When p is positive, the curve opens in the positive direction of the y-axis; when p is negative, it opens in the negative direction of the y-axis. The y-axis is the axis of the parabola.

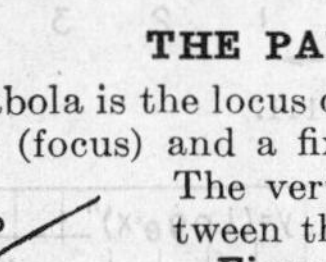

FIG. 104.

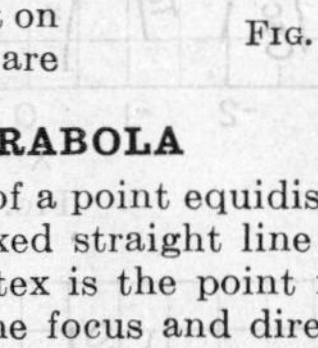

FIG. 105.

Figure 105. F is the focus, the line DD' the directrix. $DP = FP$. The vertex of the parabola is at the point (h, k). The equation of the curve is $(y - k)^2 = 4p(x - h)$. The equation of the directrix is $x = h - p$. If p is positive, the curve opens toward the right; if negative, toward the left. The axis of the parabola is parallel to the x-axis. An equation of the form $y^2 + Ax + By + C = 0$ represents a parabola whose axis is parallel to the x-axis, for it may be put in the form $(y + \frac{1}{2}B)^2 = -A\left(x + \frac{C}{A} - \frac{B^2}{4A}\right)$. The vertex is at the point $\left(\frac{B^2}{4A} - \frac{C}{A}, -\frac{B}{2}\right)$; $p = -\frac{A}{4}$. If A is positive, the curve opens up in the negative direction of the x-axis; if negative, in the positive direction of the x-axis. The equation of the axis is $y = -\frac{B}{2}$; the equation of the directrix is $x = \frac{B^2}{4A} - \frac{C}{A} + \frac{A}{4}$.

Figure 106. F is the focus, the line DD' the directrix. $DP = PF$. The vertex of the parabola is at the point (h, k). The equation of the curve is $(x - h)^2 = 4p(y - k)$. The equation of the directrix is $y = k - p$. If p is positive, the curve opens upward; if negative, downward as in the figure. The axis of the parabola is parallel to the y-axis. An equation of the form $x^2 + Ax + By + C = 0$ represents a parabola whose axis is parallel to the y-axis, for it may be put in the form $\left(x + \frac{1}{2A}\right)^2 = -B\left(y + \frac{C}{B} - \frac{A^2}{4A}\right)$. $p = -\frac{B}{4}$; equation of axis $x = -\frac{A}{2}$. The equation of the directrix is $y = \frac{A^2}{4A} - \frac{C}{B} + \frac{B}{4}$.

The diameter of a parabola is any line parallel to its axis. A diameter bisects all chords which are parallel to the tangent at the point where the diameter intersects the parabola.

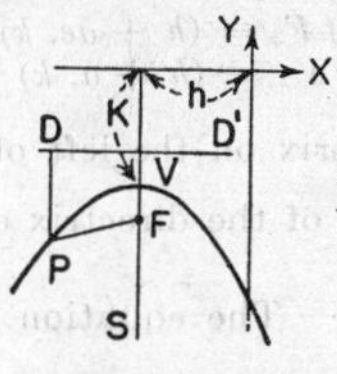

FIG. 106.

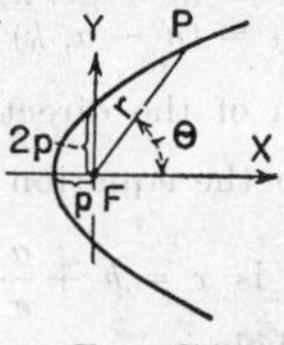

FIG. 107.

Figure 107. The **polar equation** of the parabola is $r = 2p/(1 - \cos \theta)$, where p has the same value as in the equations above. The focus is at the origin, the vertex at $(-p, \theta)$.

THE HYPERBOLA

An hyperbola is defined (1) as the locus of a point such that the difference of its distances from two fixed points (foci) is a constant, equal to $2a$, the length of the transverse axis; (2) as the locus of a point such that the ratio of its distance from a fixed point (focus) to a fixed straight line (directrix) is constant. This constant is equal to e (eccentricity). e is always greater than unity.

Figure 108. The equation of the hyperbola, whose vertices are V_1 and V_2, is $\frac{x^2}{a^2} - \frac{y^2}{b^2} = 1$, where a is the semitransverse axis, b the semiconjugate axis. The foci are F_1 and F_2. The latus rectum is the chord through a focus parallel to the y-axis and is equal to $2b^2/a$. The hyperbola is symmetrical with respect to each of the coordinate axes and also with respect to the origin. The lines dd' and DD' are the directrices. e (eccentricity) is equal to $\sqrt{(a^2 + b^2)}/a$; $b^2 = a^2(e^2 - 1)$. The lines AA' and A_1A_2 are the asymptotes to the hyperbola; their equations are $y = b/ax$ and $y = -b/ax$. The asymptotes approach the tangent to the hyperbola at infinity. The distance from the origin to either of the foci is ae, the distance from the y-axis to either of the directrices is a/e.

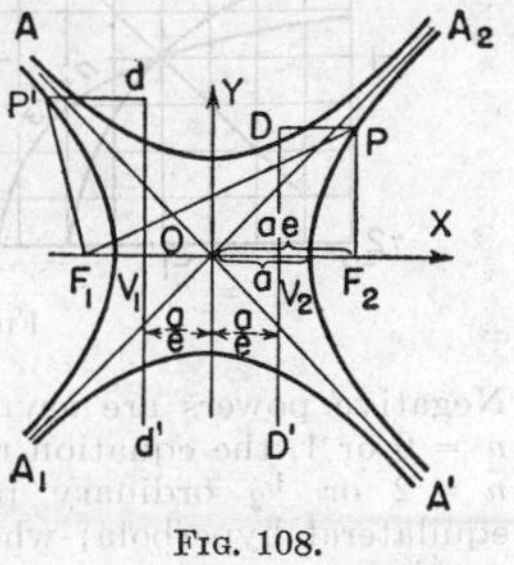

FIG. 108.

Conjugate Hyperbola. A curve bearing very close relations to the hyperbola, $\frac{x^2}{a^2} - \frac{y^2}{b^2} = 1$, is that represented by the equation $\frac{x^2}{a^2} - \frac{y^2}{b^2} = -1$, the hyperbola having its vertices on the y-axis (Fig. 91). The two hyperbolas have the same asymptotes.

The focal radii of an hyperbola whose equation is $\frac{x^2}{a^2} - \frac{y^2}{b^2} = 1$ to the point P (Fig. 108) are $F_1P = (ex + a)$ and $F_2P = (ex - a)$. The polar equation of the hyperbola with the center at the origin is

$$r = \pm \frac{ab}{\sqrt{b^2 \cos^2 \theta - a^2 \sin^2 \theta}}$$

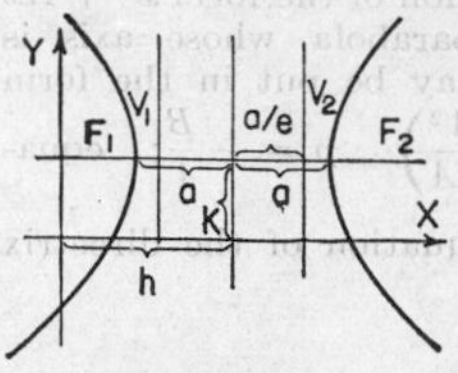

FIG. 109.

Figure 109. The equation of the hyperbola with center at the point (h, k) and axis parallel to the coordinate axes is

$$\frac{(x - h)^2}{a^2} - \frac{(y - k)^2}{b^2} = 1$$

$$F_1 \equiv (h - ae, k) \; F_2 \equiv (h + ae, k)$$
$$V_1 \equiv (h - a, k) \; V_2 \equiv (h + a, k)$$

The equation of the directrix on the left of the y-axis is $x = h - \frac{a}{e}$, the equation of the directrix on the right of the y-axis is $x = h + \frac{a}{e}$. The equation of the conjugate hyperbola is

$$\frac{(x - h)^2}{a^2} - \frac{(y - k)^2}{b^2} = -1$$

MISCELLANEOUS CURVES

Figure 110. Equation $y = x^n$. All the curves pass through (1, 1); positive powers also through (0, 0).

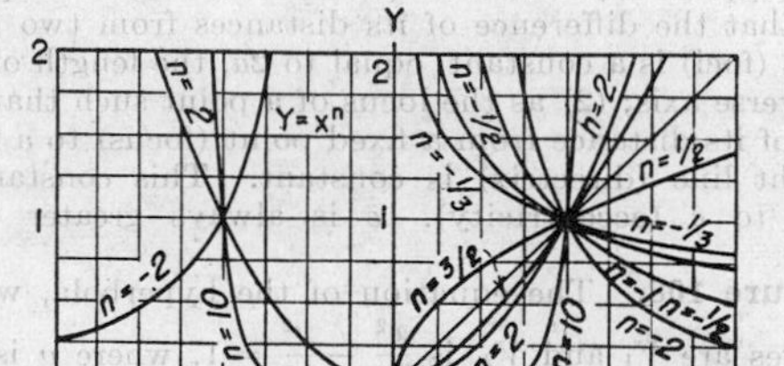

FIG. 110.

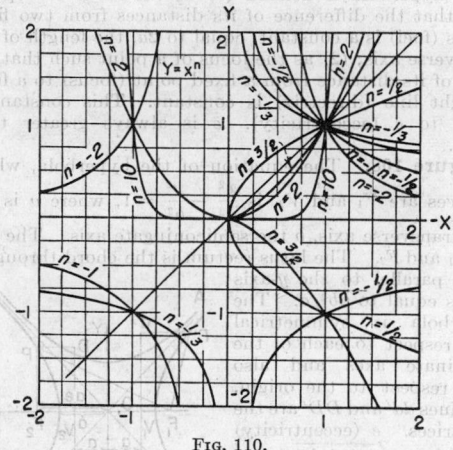

Negative powers are asymptotic to the y-axis. When $n = 0$ or 1, the equation represents straight lines; when $n = 2$ or $\frac{1}{2}$ ordinary parabolas; when $n = -1$ an equilateral hyperbola; when $n = \frac{3}{2}$ or $\frac{2}{3}$ semicubical parabolas.

Figure 111. Trigonometric Functions. The inverse functions are obtained as follows: From $y = \sin x$, read $x = \text{arc} \sin y$; from $y = \cos x$, read $x = \text{arc} \cos y$.

Figure 112. $y = (\log_e x)^n$; $y = \log_{10} x$. Note the form of the curve for values of x less than 1, when $n = \frac{1}{3}$.

Figure 113. $y = (\sin x)^n$. n is positive. When $n = 1$, the slope at the origin is 1; slope at the origin is zero, when n is greater than 1; infinite, when n is less than 1.

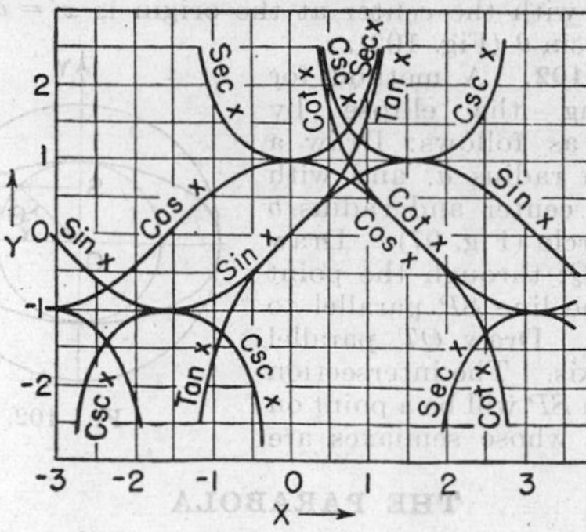

FIG. 111.

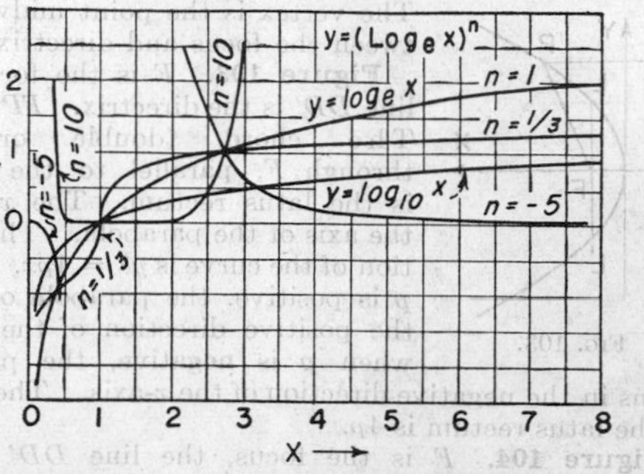

FIG. 112.

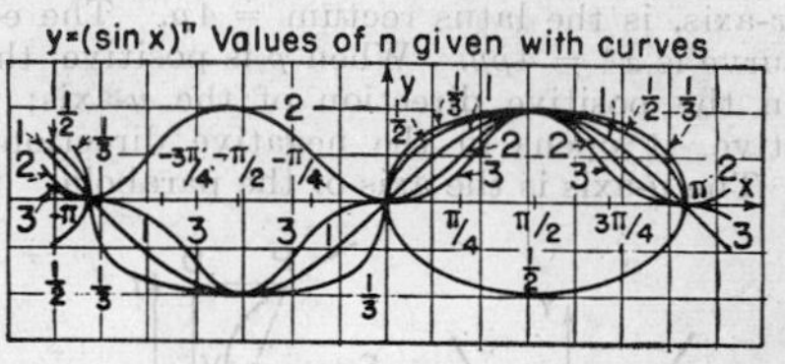

FIG. 113.

Figure 114. $y = e^{nx}$. n is negative. Note that all the curves can be drawn from the curve $y = e^{-x}$. The ordinate of the point P on $y = e^{-2x}$ is equal to the ordinate of the point Q on $y = e^{-x}$; the abscissa of Q is twice the abscissa of P. This relation holds true for

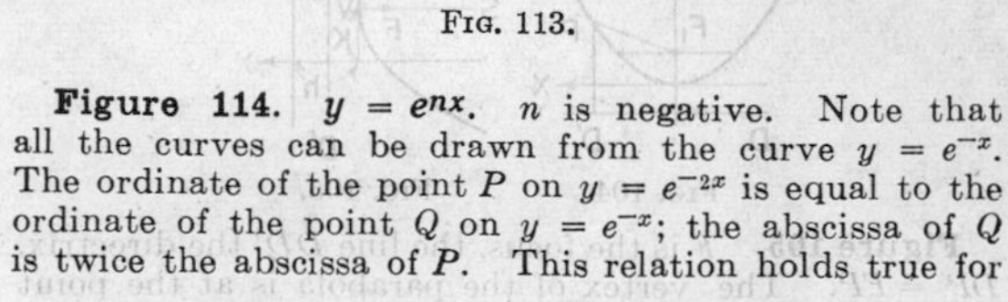

FIG. 114.

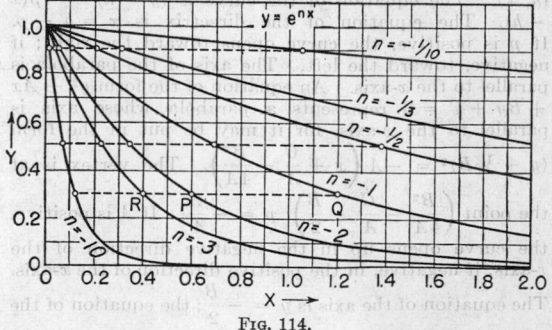

all the points on the two curves. The ordinate of R is equal to the ordinate of Q, and the abscissa of R is one-third of the abscissa of Q. Similar relations hold for the other curves.

Figure 115. Exponential and Hyperbolic Functions. The curves $y = e^x$ and $y = e^{-x}$ are first drawn, and the catenary (hyperbolic cosine) $y = \cosh x = (e^x + e^x)/2$ obtained by taking half the sum of the ordinates

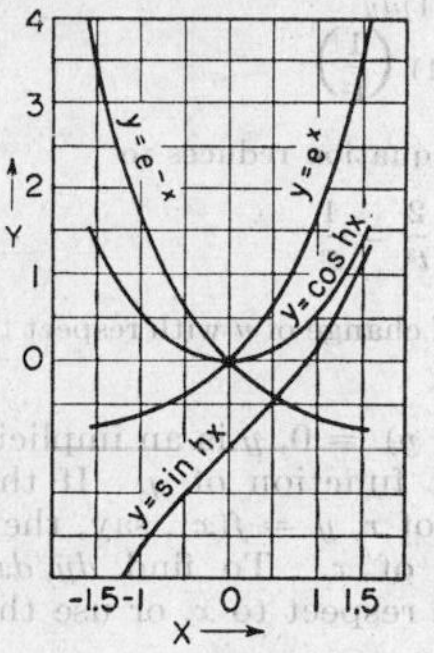

Fig. 115.

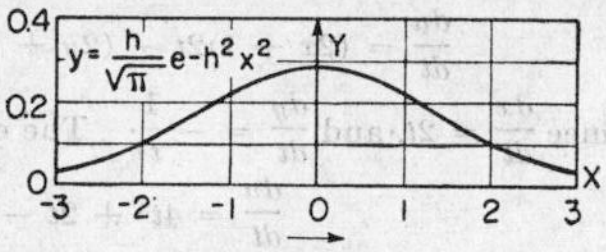

Fig. 116.

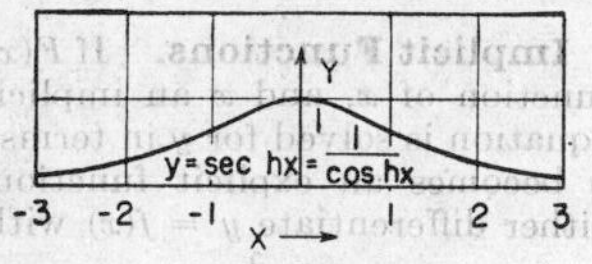

Fig. 117.

of the two curves. The hyperbolic sine, $y = (e^x - e^x)/2$, is obtained by taking half the difference of the ordinates. Both hyperbolic curves are asymptotic to $y = e^x/2$. The general catenary is $y = (a/2)(e^{\frac{x}{a}} + e^{\frac{x}{a}})$ $= a \cosh (x/a)$. It is obtained by changing the scale 1 to a on both axes. It is the curve in which a flexible inelastic cord will hang.

Figure 116. Probability Curves or Normal Curve of Error. $y = \dfrac{h}{\sqrt{\pi}} e^{-h^2x^2}$. In the figure, h (measure of precision) has been taken equal to one-half.

Figure 117. $y = \operatorname{sech} x = \dfrac{1}{\cosh x} = \dfrac{2}{(e^x + e^{-x})}$

Figure 118. The Witch of Agnesi. $y = \dfrac{8a^3}{(x^2 + 4a^2)}$.

Any one of the last three curves drawn to a proper scale

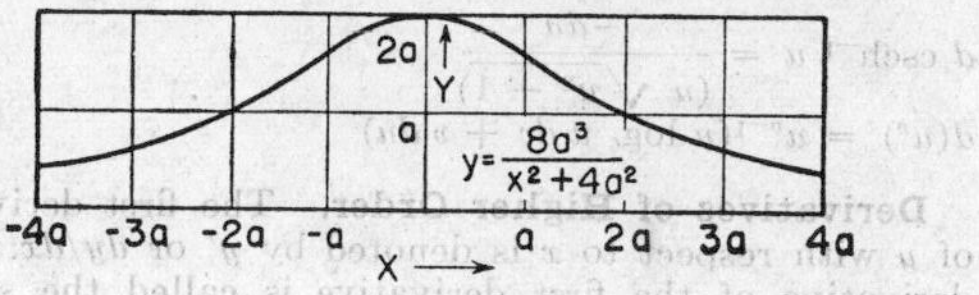

Fig. 118.

gives fairly good approximation to the probable distribution of accidental data which tend to distribute themselves about a mean.

DIFFERENTIAL AND INTEGRAL CALCULUS

DIFFERENTIAL CALCULUS

Derivative of a Function of x. A function of a single variable x is denoted by $f(x)$, $F(x)$, or by similar symbols. The value of $f(x)$ when x has the value a is represented by $f(a)$, and $F(x)$ when x has the value a is represented by $F(a)$. The derivative of $y = f(x)$ is $dy/dx = f'(x)$. When $x = a$, the derivative is $f'(a)$. This means the rate of change of the function at the point whose abscissa is a, or, geometrically, the slope of the curve at that point. By the slope of a curve at a given point is meant the slope of the tangent to the curve at that point.

The increment, Δy (read "delta y"), of y is the change produced in y by increasing x by Δx; that is, $\Delta y = f(x + \Delta x) - f(x)$, where x has any assigned value.

Differential dy. The differential, dy, of y is the value which Δy would have if the rate of change of the function became constant (coincides with the tangent). The

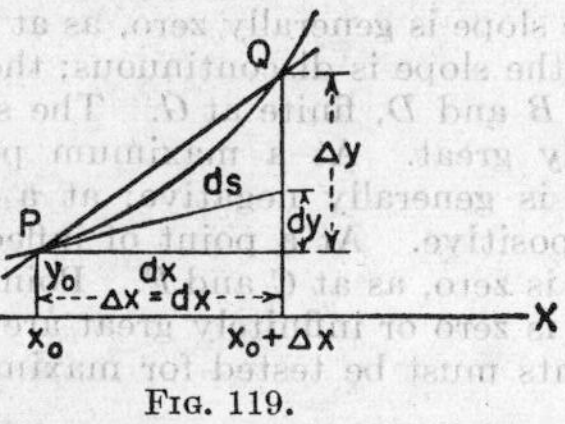

Fig. 119.

value of the derivative at any point depends only on the value of x at that point, while Δy and dy depend upon the values of both x and Δx. $\Delta y/\Delta x$ represents the slope of the secant, and dy/dx represents the slope of the tangent. Figure 119 shows geometrically the relations just stated. As Δx approaches zero, the secant approaches the tangent as a limiting position.

$$f'(x) = \frac{dy}{dx} = \lim_{\Delta x \to 0} \frac{\Delta y}{\Delta x} = \lim_{\Delta x \to 0} \frac{f(x + \Delta x) - f(x)}{\Delta x};$$

$$dy = f'(x)dx$$

Derivative at a Given Point. In the case of a tabulated function or a function given by a curve see Graphical Calculus.

When the function is given as a mathematical expression, use the following formulas.

Differentiation Formulas. $u, v, w, \ldots$ are functions of x. c is a constant; e is the base of the natural or Napierian logarithms and is equal to 2.71828.

$d(c + u) = du$

$d(cu) = c\,du$

$d(uv) = u\,dv + v\,du$

$d(uvw \cdots) = (uvw \cdots)\left(\dfrac{du}{u} + \dfrac{dv}{v} + \dfrac{dw}{w} + \cdots\right)$

$d(u + v + w + \cdots) = du + dv + dw + \cdots$

$d\dfrac{u}{v} = \dfrac{v\,du - u\,dv}{v^2}$

$d(u^n) = nu^{n-1}\,du$ $\quad d(u^{-1}) = -u^{-2}\,du$

$d(u^{1/2}) = \frac{1}{2}u^{-1/2}\,du$ $\quad d \sin u = \cos u\,du$

$d \log_e u = \dfrac{du}{u}$ $\quad d \cos u = -\sin u\,du$

$d(a^u) = (\log_e a)a^u\,du$ $\quad d \tan u = \sec^2 u\,du$

$d(e^u) = e^u\,du$ $\quad d \cot u = -\csc^2 u\,du$

$d \log_{10} u = \log_{10} e \dfrac{du}{u}$ $\quad d \sec u = \tan u \sec u\,du$

$= (0.4343 \cdots) \dfrac{du}{u}$ $\quad d \csc u = -\cot u \csc u\,du$

$d \sin^{-1} u = \dfrac{du}{\sqrt{1 - u^2}}$ $\quad d \log_e \sin u = \cot u\,du$

$d \cos^{-1} u = -\dfrac{du}{\sqrt{1 - u^2}}$ $\quad d \log_e \cos u = -\tan u\,du$

$d \tan^{-1} u = \dfrac{du}{1 + u^2}$ $\quad d \log_e \tan u = \dfrac{2\,du}{\sin 2u}$

$d \cot^{-1} u = -\dfrac{du}{1 + u^2}$ $\quad d \log_e \cot u = \dfrac{-2\,du}{\sin 2u}$

$d \sec^{-1} u = \dfrac{du}{u\sqrt{u^2 - 1}}$ $\quad d \log_e \sec u = \tan u\,du$

$d \csc^{-1} u = -\dfrac{du}{u\sqrt{u^2 - 1}}$ $\quad d \log_e \csc u = -\cot u\,du$

$d \sinh u = \cosh u\, du \qquad d \sinh^{-1} u = \dfrac{du}{\sqrt{u^2 + 1}}$

$d \cosh u = \sinh u\, du \qquad d \cosh^{-1} u = \dfrac{du}{\sqrt{u^2 - 1}}$

$d \tanh u = \text{sech}^2\, u\, du \qquad d \tanh^{-1} u = \dfrac{du}{(1 - u^2)}$

$d \coth u = -\text{csch}^2\, u\, du \qquad d \coth^{-1} u = \dfrac{du}{(1 + u^2)}$

$d\, \text{sech}\, u = -\text{sech}\, u \tanh u\, du$

$d\, \text{sech}^{-1} u = \dfrac{-du}{(u\sqrt{1 - u^2})}$

$d\, \text{csch}\, u = -\text{csch}\, u \coth u\, du$

$d\, \text{csch}^{-1} u = \dfrac{-du}{(u\sqrt{u^2 + 1})}$

$d(u^v) = u^{v-1}(u \log_e u\, dv + v\, du)$

Derivatives of Higher Order. The first derivative of y with respect to x is denoted by y' or dy/dx. The derivative of the first derivative is called the second derivative of y with respect to x and is denoted by y'' or d^2y/dx^2. y''' or d^3y/dx^3 is the derivative of the second derivative.

Example. $y = x \sin x + x^2$, $y' = dy/dx = \sin x + x \cos x + 2x$, $y'' = d^2y/dx^2 = 2 \cos x - x \sin x + 2$, $y''' = d^3y/dx^3 = -3 \sin x - x \cos x$.

Figure 120 shows the relation between the original curve and the first and second derivatives. In the

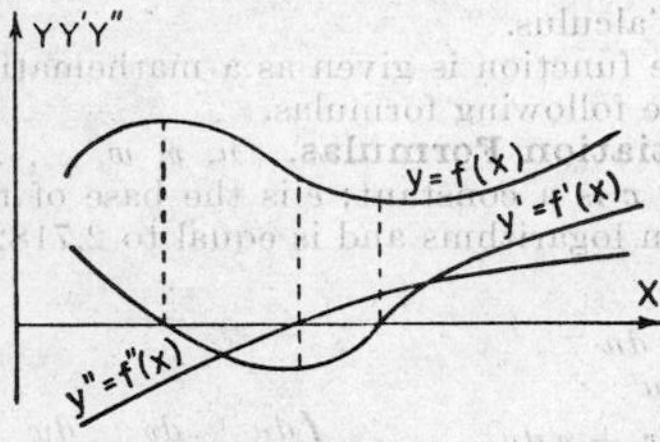

FIG. 120.

interval where the original curve is rising, y' is positive; where the original curve is falling, y' is negative. In the interval where the original curve is concave downward, y' is diminishing algebraically and y'' is negative; where it is concave upward, y' is increasing algebraically and y'' is positive. When y'' is equal to zero, the y-curve generally has a point of inflection. Exception: $y = x^4$, $y' = 4x^3$, $y'' = 12x^2$. Here $y'' = 0$ at $x = 0$, but the curve has no inflection at that point.

Partial Derivatives. If u is a function of several variables, $u = f(x, y, z, \cdots)$, the derivative of u on the assumption that x is the only variable, the other quantities, $y, z, \ldots$ being for the moment considered constants, is represented by $\partial y/\partial x$ or $f_x(x, y, z, \ldots)$. It is called the **partial derivative of *u* with respect to *x*.** Partial derivatives of the second order are denoted by $\partial^2 u/\partial x^2$, $\partial^2 u/\partial x \partial y$, $\partial^2 u/\partial y^2$.

If increments Δx, Δy, (or dx, dy) are assigned to x and y, in $u = f(x, y)$, the following equations are obtained:

$\Delta u = f(x + \Delta x, y + \Delta y) - f(x, y) =$ total increment of u

$du = \dfrac{\partial u}{\partial x}\, dx + \dfrac{\partial u}{\partial y}\, dy =$ total differential of u

If x and y are functions of a third variable t, then

$\dfrac{du}{dt} = \dfrac{\partial u}{\partial x}\dfrac{dx}{dt} + \dfrac{\partial u}{\partial y}\dfrac{dy}{dt} =$ total derivative of u with respect to t.

Example.

$$u = x^2 + y^2 + x + y,\ x = t^2,\ y = \frac{1}{t}$$

$$\Delta u = (2x + 1 + \Delta x)\Delta x + (2y + 1 + \Delta y)\Delta y$$

$$du = (2x + 1)dx + (2y + 1)dy$$

$$\frac{du}{dt} = (2x + 1)2t - (2y + 1)\left(\frac{1}{t^2}\right)$$

since $\dfrac{dx}{dt} = 2t$, and $\dfrac{dy}{dt} = -\dfrac{1}{t^2}$. The equation reduces to

$$\frac{du}{dt} = 4t^3 + 2t - \frac{2}{t^3} - \frac{1}{t^2}$$

The last equation expresses the rate of change of u with respect to t as a function of t.

Implicit Functions. If $F(x, y) = 0$, y is an implicit function of x, and x an implicit function of y. If the equation is solved for y in terms of x, $y = f(x)$, say, then y becomes an explicit function of x. To find dy/dx, either differentiate $y = f(x)$ with respect to x, or use the formula $\dfrac{dy}{dx} = -\dfrac{\dfrac{\partial}{\partial x}F(x, y)}{\dfrac{\partial}{\partial y}F(x, y)}$

Maxima and Minima. When a continuous function of x ceases to increase and begins to decrease, as at A, D, and G (Fig. 121), it is said to have a **maximum** value;

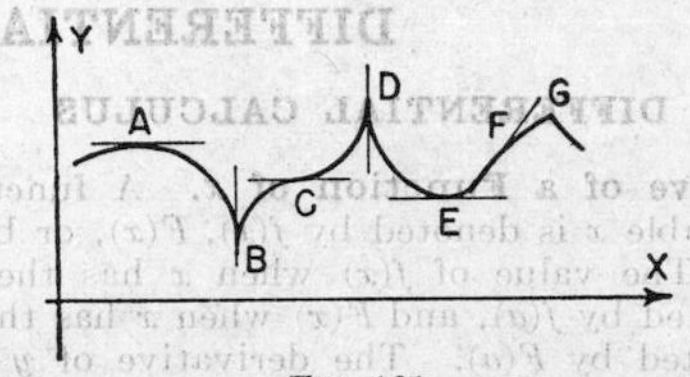

FIG. 121.

when the function ceases to decrease and begins to increase, as at B and E, it is said to have a **minimum** value.

A maximum point of a curve is one whose ordinate is algebraically greater than the ordinate of any other point in the immediate neighborhood. A minimum point of a curve is one whose ordinate as algebraically less than the ordinate of any other point in the immediate neighborhood.

At a maximum point the slope changes from positive to negative; at a minimum point the slope changes from negative to positive. The change may take place abruptly as at B, D, and G. At maximum and minimum points, the slope is generally zero, as at A and E. At B, D, and G, the slope is discontinuous; the discontinuity is infinite at B and D, finite at G. The slope at B and D is infinitely great. At a maximum point, the second derivative is generally negative; at a minimum point, generally positive. At a point of inflection, the second derivative is zero, as at C and F. Points where the first derivative is zero or infinitely great are **critical points.** These points must be tested for maximum or minimum values.

Example 1. $y = x^{2/3}$, $y' = \frac{2}{3}x^{-1/3}$. y' is infinitely great at $x = 0$, a critical point. The slope changes from negative to positive, hence a minimum point.

Example 2. $y = (x - 1)^3 + 2$. $y' = 3(x - 1)^2$, $y'' = 6(x - 1)$. Critical point at $x = 1$, but this is a point of inflection since $y'' = 0$. y' is positive on both sides of the critical point.

Example 3. $y = (x - 2)^{1/3} + 3$, $y' = \frac{1}{3}(x - 2)^{-2/3}$. Critical point at $x = 2$, y' is infinitely great, and does not change sign; hence, a point of inflection.

Example 4.

$$y = \tfrac{1}{4}x^4 - 2x^3 + (5.5)x^2 - 6x$$

$$y' = x^3 - 6x^2 + 11x - 6$$

$$y'' = 3x^2 - 12x + 11$$

Critical points at $x = 1, x = 2, x = 3$:

At $x = 1$, y'' is positive, minimum point
At $x = 2$, y'' is negative, maximum point
At $x = 3$, y'' is positive, minimum point

Formulas of Maclaurin and Taylor. If $f(x)$ and all of its derivatives are continuous in the neighborhood of the points $x = 0$ and $x = a$, then $f(x)$ can be developed into a power series arranged according to ascending powers of x or $x - a$ in the neighborhood of these points. The series are:

$$f(x) = f(0) + \frac{f'(0)}{1!}x + \frac{f''(0)}{2!}x^2 + \frac{f'''(0)}{3!}x^3 + \frac{f^4(0)}{4!}x^4 + \cdots R_1 \text{ (Maclaurin)}$$

$$f(x) = f(a) + \frac{f'(a)}{1!}(x-a) + \frac{f''(a)}{2!}(x-a)^2 + \frac{f'''(a)}{3!}(x-a)^3 + \cdots R_2 \text{ (Taylor)}$$

In using these series for computation, care must be exercised in taking enough terms so that the remainders R_1 and R_2 become negligible. When Maclaurin's series converges rapidly, the sum of the first few terms gives a good approximation to $f(x)$ for values of x near $x = 0$. When Taylor's series converges rapidly, the sum of the first few terms gives a good approximation to $f(x)$ for values of x near $x = a$.

Example. Expand $e^{nx}(= a^x)$ into a series of ascending powers of x.

$$f(x) = e^{nx}, \quad f(0) = 1 \quad f'''(x) = n^3e^{nx}, \quad f'''(0) = n^3$$
$$f'(x) = ne^{nx}, \quad f'(0) = n \quad f^4(x) = n^4e^{nx}, \quad f^4(0) = n^4$$
$$f''(x) = n^2e^{nx}, \quad f''(0) = n^2 \quad f^5(x) = n^5e^{nx}, \quad f^5(0) = n^5$$

$$e^{nx} = 1 + \frac{n}{1!}x + \frac{n^2}{2!}x^2 + \frac{n^3}{3!}x^3 + \frac{n^4}{4!}x^4 + \frac{n^5}{5!}x^5 + \cdots$$

When x and n are less than unity this series converges rapidly. Let $x = 1/2$ and $n = 1/3$, and the value of $e^{1/6}$ is found as follows ($n! = 1 \cdot 2 \cdot 3 \cdot 4 \cdot 5 \cdot 6 \cdots n$):

First term	1.0000000
Second term	0.1666667
Third term	0.0138889
Fourth term	0.0007716
Fifth term	0.0000321
Sixth term	0.0000011
$e^{1/6}$	1.1813604

Functions Expanded into Series. The range of values of x for which each series is convergent is given.

$$e^x = 1 + \frac{x}{1!} + \frac{x^2}{2!} + \frac{x^3}{3!} + \frac{x^4}{4!} + \frac{x^5}{5!} + \cdots \quad -\infty < x < +\infty$$

$$e^{-x} = 1 - \frac{x}{1!} + \frac{x^2}{2!} - \frac{x^3}{3!} + \frac{x^4}{4!} - \frac{x^5}{5!} + \cdots \quad -\infty < x < +\infty$$

$$\log_e(1+x) = x - \frac{x^2}{2} + \frac{x^3}{3} - \frac{x^4}{4} + \frac{x^5}{5} - \cdots \quad -1 < x < +1$$

$$\log_e(x) = (x-1) - \frac{(x-1)^2}{2} + \frac{(x-1)^3}{3} - \frac{(x-1)^4}{4} + \cdots \quad 2 > x > 0$$

$$\sin x = x - \frac{x^3}{3!} + \frac{x^5}{5!} - \frac{x^7}{7!} + \frac{x^9}{9!} - \cdots \quad -\infty < x < +\infty$$

$$\cos x = 1 - \frac{x^2}{2!} + \frac{x^4}{4!} - \frac{x^6}{6!} + \frac{x^8}{8!} - \cdots \quad -\infty < x < +\infty$$

$$\tan x = x + \frac{x^3}{3} + \frac{2x^5}{15} + \frac{17x^7}{315} + \frac{62x^9}{2835} + \cdots \quad -\frac{\pi}{2} < x < +\frac{\pi}{2}$$

$$\sin^{-1} x = x + \frac{x^3}{6} + \frac{3x^5}{40} + \frac{5x^7}{112} + \cdots \quad -1 \leq x \leq +1$$

$$\cos^{-1} x = \tfrac{1}{2}\pi - \sin^{-1} x$$

$$\tan^{-1} x = x - \frac{x^3}{3} + \frac{x^5}{5} - \frac{x^7}{7} + \cdots \quad -1 < x < +1$$

$$\cot^{-1} x = \tfrac{1}{2}\pi - \tan^{-1} x$$

$$\sinh x = x + \frac{x^3}{3!} + \frac{x^5}{5!} + \frac{x^7}{7!} + \cdots \quad -\infty < x < +\infty$$

$$\cosh x = 1 + \frac{x^2}{2!} + \frac{x^4}{4!} + \frac{x^6}{6!} + \frac{x^8}{8!} + \cdots \quad -\infty < x < +\infty$$

$$\sinh^{-1} x = x - \frac{x^3}{6} + \frac{3x^5}{40} - \frac{5x^7}{112} + \cdots \quad -1 < x < +1$$

$$\tanh^{-1} x = x + \frac{x^3}{3} + \frac{x^5}{5} + \frac{x^7}{7} + \cdots \quad -1 < x < +1$$

$$\coth^{-1} x = \frac{1}{x} + \frac{1}{3x^3} + \frac{1}{5x^5} + \cdots \quad -1 < x < +1$$

Indeterminate Forms. Indeterminate forms are functions of a variable which for certain values of the variable take on one of the following forms: $0/0$, ∞/∞, $0 \cdot \infty$, 0^0, 1^∞, ∞^0, and $\infty - \infty$. If two functions $f(x)$ and $F(x)$ both become zero for $x = a$, $f(a) = 0$, $F(a) = 0$, their quotient $\frac{f(x)}{F(x)}$ takes the form $0/0$ at $x = a$ and is undefined at that point. However, $\lim_{x \to a} \frac{f(x)}{F(x)}$ may exist. To find the limit at this point apply the theorem, $\lim_{x \to a} \frac{f(x)}{F(x)} = \lim_{x \to a} \frac{f'(x)}{F'(x)}$. If $f'(x) = 0$ and $F'(x) = 0$, for $x = a$ apply the theorem a second time.

Example 1. Value of $\frac{\sin x}{x}$ for $x = 0$. $\lim_{x \to 0} \frac{\sin x}{x} = \lim_{x \to 0} \frac{\cos x}{1} = 1.$

If $\frac{f(x)}{F(x)}$ takes on the form $\frac{\infty}{\infty}$ for $x = a$, then $\lim_{x \to a} \frac{f(x)}{F(x)} = \lim_{x \to a} \frac{f'(x)}{F'(x)}$ as before.

Example 2. $\lim_{x \to \infty} \frac{x - 1}{x^2 + 1} = \lim_{x \to \infty} \frac{1}{2x} = 0.$

If $f(x) \cdot F(x)$ takes on the form $0 \cdot \infty$, write the product in the form $\frac{F(x)}{\frac{1}{f(x)}}$ and proceed as above.

Example 3. $\lim_{x \to \frac{\pi}{2}} \tan x \left(x - \frac{\pi}{2}\right) = \lim_{x \to \frac{\pi}{2}} \frac{x - \frac{\pi}{2}}{\cot x} = \lim_{x \to \frac{\pi}{2}} \frac{1}{-\csc^2 x} = -1.$

If $[F(x)]^{f(x)}$ takes the form 0^0 for $x = a$, let $y = [F(x)]^{f(x)}$, $\log y = f(x) \log F(x)$. This is the form $0 \cdot \infty$.

Example 4. $\lim_{x \to 0} (\tan x)^{\sin x}$. Let $y = (\tan x)^{\sin x}$, $\log y = \sin x \log \tan x$. $\lim_{x\to 0} \sin x \log \tan x = \lim_{x \to 0} \frac{\log \tan x}{\csc x} = \lim_{x\to 0} \frac{\sec x \csc x}{-\csc x \cos x} = \lim_{x \to 0} \frac{\tan x}{-\cos x} = 0$. Therefore, $\lim_{x\to 0} (\tan x)^{\sin x} = 1$.

If $[F(x)]^{f(x)}$ takes the form 1^∞ for $x = a$, let $y = F(x)^{f(x)}$, $\log y = f(x) \log F(x) = \dfrac{\log F(x)}{\dfrac{1}{f(x)}}$. This is the form $\frac{0}{0}$.

Example 5. $\lim_{x \to \frac{\pi}{2}} (\sin x)^{\tan x}$. Let $y = (\sin x)^{\tan x}$, $\log y = \tan x \log \sin x = \frac{\log \sin x}{\cot x}$. $\lim_{x\to\frac{\pi}{2}} \frac{\log \sin x}{\cot x} = \lim_{x\to\frac{\pi}{2}} \frac{\cot x}{-\csc^2 x} = 0$, hence $(\sin x)^{\tan x} = 1$.

If $[F(x)]^{f(x)}$ takes the form ∞^0 for $x = a$, let $y = [F(x)]^{f(x)}$, $\log y = f(x) \log F(x) = \dfrac{\log F(x)}{1/fx}$. This is the form ∞/∞.

Example 6. $\lim_{x\to\frac{\pi}{2}} (\tan x)^{\cos x}$. Let $y = (\tan x)^{\cos x}$, $\log y = \cos x \log \tan x = \frac{\log \tan x}{\sec x}$. $\lim_{x\to\frac{\pi}{2}} \frac{\log \tan x}{\sec x} = \lim_{x\to\frac{\pi}{2}} \frac{\cos x}{\sin^2 x} = 0$, hence $(\tan x)^{\cos x} = 1$.

If $F(x) - f(x)$ takes the form $\infty - \infty$ for $x = a$, proceed as in the next example.

Example 7. $\lim_{x \to 1} \left(\frac{1}{x^2 - 1} - \frac{1}{x - 1}\right) = \lim_{x\to 1} \frac{-x^2 + x}{x^3 - x^2 - x + 1} = \lim_{x \to 1} \frac{-2x + 1}{3x^2 - 2x - 1} = -\infty$.

DIFFERENTIAL EQUATIONS

A differential equation is an equation which involves derivatives or differentials. An **ordinary differential equation** is one which contains a single independent variable and a single dependent variable with derivatives. A **partial differential equation** is one which contains several independent variables and partial derivatives.

The **order** of a differential equation is that of the highest derivative it contains. The **degree** of a differential equation is the power to which the derivative of the highest order in the equation is raised.

A **solution** of a differential equation is a relation between the variables, which, when substituted in the equation, will satisfy it. The general solution of an ordinary differential equation of the nth order will contain n arbitrary constants.

To find the differential equation whose general solution is $c(x + y) = xy$, differentiate the equation, giving $c\left(1 + \frac{dy}{dx}\right) = x\left(\frac{dy}{dx}\right) + y$. Eliminate c between these two equations, giving the equation $x^2\,dy + y^2\,dx = 0$. If an equation contains two independent constants, differentiate it twice and eliminate the two constants between the three equations.

Ordinary Differential Equations of the First Order. Separation of Variables. Every differential equation of the first order and of the first degree can be written $M\,dx + N\,dy = 0$, where M and N are functions of x and y. If the equation can be transformed so that M is a function of x alone and N a function of y alone, the variables are said to be separated. Such an equation can be solved by simple integration.

Example. $(1 - x)dy = (1 + y)dx$. Separating the variables this becomes $dy/(1 + y) = dx(1 - x)$. The solution is $\log (1 + y) + \log (1 - x) = A$ or $xy + x - y = B$.

Homogeneous Equations. An equation is said to be homogeneous if the coefficients of all terms are of the same order, considering all the variables; *e.g.*, $x^3\,dx + 3x^2y\,dy = 0$ is homogeneous, whereas $x^3\,dx + 3x^3y\,dy = 0$, is not. Let $y = vx$, $dy = v\,dx + x\,dv$. This substitution transforms the equation into one in v and x in which the variables can be separated.

Example 1. $(x^2 + 2y^2)dx = xy\,dy$. This becomes $(x^2 + 2v^2x^2)dx = vx^2(v\,dx + x\,dv)$ or $dx/x = v\,dv/(1 + v^2)$. This gives for the solution $x^4 = c(x^2 + y^2)$, since $v = y/x$.

Example 2. $dy/dx = (y + \sqrt{x^2 + y^2})/x$. This becomes $x\,dv/dx = \sqrt{1 + v^2}$ or $dv/(\sqrt{1 + v^2}) = dx/x$. The solution is $y + \sqrt{x^2 + y^2} = cx^2$.

Exact Differential Equations. If in $M\,dx + N\,dy = 0$, $\partial M/\partial y = \partial N/\partial x$, the equation is said to be exact and can be integrated directly.

Example. $(x^2 + 2y^2 + x)dx + (4xy + 1)dy = 0$ is exact; for, $M = x^2 + 2y^2 + x$, $N = 4xy + 1$. $\partial M/\partial y = 4y$, $\partial N/\partial x = 4y$. The solution is $\frac{1}{3}x^3 + 2xy^2 + \frac{1}{2}x^2 + y = C$.

Linear Equations. A differential equation of the first order is linear if it is of the first degree in y and dy/dx. Such an equation may be written $\frac{dy}{dx} + Py = Q$, where P and Q are functions of x alone. The equation becomes exact by multiplying each member by $e^{\int P dx}$.

Example 1. $\frac{dy}{dx} - \left(\frac{3}{x}\right)y = x$. Multiplying by $e^{-3\int\frac{dx}{x}} = e^{-\log x^3} = \frac{1}{x^3}$ it becomes $\frac{1}{x^3}\frac{dy}{dx} - \left(\frac{3}{x^4}\right)y = \frac{1}{x^2}$. The solution is $\frac{y}{x^3} = -\frac{1}{x} + c$ or $y = -x^2 + cx^3$.

Example 2. $\frac{dy}{dx} + y\cos x = \sin 2x$. Multiplying by $e^{\int \cos x\,dx} = e^{\sin x}$, $e^{\sin x}\frac{dy}{dx} + e^{\sin x}(\cos x)y = 2e^{\sin x}\sin x \cos x$. The solution is $e^{\sin x}y = 2e^{\sin x}\sin x - 2e^{\sin x} + c$ or $y = 2\sin x - 2 + ce^{-\sin x}$.

Equations Linear in $f(y)$. $f'(y)\frac{dy}{dx} + Pf(y) = Q$, where P and Q are functions of x alone, is linear in $f(y)$ and $f'(y)$.

Example. $\frac{3y^2\,dy}{dx} - 2y^3 = x + 1$. Let $y^3 = v$, $\frac{3y^2\,dy}{dx} = \frac{dv}{dx}$. The equation becomes $\frac{dv}{dx} - 2v = x + 1$, and this is transformed into an exact equation by multiplying by $e^{-2\int dx} = e^{-2x}$. $e^{-2x}\frac{dv}{dx} - 2e^{-2x}v = xe^{-2x} + e^{-2x}$. The solution is $e^{-2x}v = -\frac{1}{2}xe^{-2x} - \frac{3}{4}e^{-2x} + c$ or $y^3 = -\frac{1}{2}x - \frac{3}{4} + ce^{2x}$.

Clairaut's Equation. $y = xp + f(p)$, where $p = dy/dx$ is called Clairaut's equation. By differentiating with respect to x, it becomes $p = p + \frac{x\,dp}{dx} + \frac{f'(p)dp}{dx}$ or $\frac{[x + f'(p)]dp}{dx} = 0$. Elimination between this and the original equation gives as the solution the family of straight lines $y = cx + f(c)$, and the curve, obtained by eliminating p between $x + f'(p) = 0$, $y = xp + f(p)$.

Example. $y = xp + p^2$, $p = p + \frac{x\,dp}{dx} + \frac{2p\,dp}{dx}$ or $(x + 2p)\frac{dp}{dx} = 0$. $dp/dx = 0$ gives $p = c$. Replacing p by c in the original equation, $y = cx + c^2$. Replacing p by $-\frac{1}{2}x$ gives $y = -\frac{1}{2}x^2 + \frac{1}{4}x^2 = -\frac{1}{4}x^2$.

Differential Equations of the Second Order. The Equation $y'' = f(x)$. $y'' = d^2y/dx^2$, $y' = dy/dx$. The equation is solved by two integrations and will contain two constants of integration. It is evident that the equation $y^n = f(x)$ can be solved by n successive integrations.

Dependent Variable Absent. Let $y' = v$, $y'' = v'$.

Example. $xy'' = y' + x$. This becomes $xv' - v = x$. Dividing by x gives $v' - (1/x)v = 1$, which is the form for the linear equation. Multiplying by $e^{\int P\,dx(-1/x)}$ $(1/x)v' - (1/x^2)v = 1/x$. The first integration gives $(1/x)y' = \log x + c_1$ or $y' = x \log x + cx$. The solution of this equation is $y = \frac{1}{2}x^2 \log x + c_1x^2 + c_2$, where $c_1 = \frac{1}{2}c - \frac{1}{4}$.

Independent Variable Absent. Let $y' = v$, $y'' = v' = v(dv/dy)$.

Example. $yy'' + 2(y')^2 = 0$. $yv(dv/dy) + 2v^2 = 0$ or $y(dv/dy) + 2v = 0$. The first integration gives $vy^2 = c_1$; and the second integration, $y^3 = c_1x + c_2$.

Linear Differential Equations, Right-hand Member Zero. The solution of $y'' + ay' + by = 0$ depends upon the nature of the roots of the **characteristic equation** $m^2 + am + b = 0$.

Distinct Roots. If the roots of the characteristic equation are real and distinct, m_1 and m_2 say, the solution is $y = Ae^{m_1x} + Be^{m_2x}$, where A and B are arbitrary constants.

Example. $y'' + 11y' + 28y = 0$. The characteristic equation is $m^2 + 11m + 28 = 0$. The roots are -4 and -7, and the solution is $y = Ae^{-4x} + Be^{-7x}$.

Equal Roots. If the roots of the characteristic equation are equal, that is $m_1 = m_2$, the solution is $y = Ae^{m_1x} + Be^{m_1x}x$.

Example. $y'' - 6y' + 9y = 0$. The roots of $m^2 - 6m + 9 = 0$ are both equal to 3. The solution of the equation is $y = Ae^{3x} + Bxe^{3x}$.

Roots Complex. If the roots of the characteristic equation are complex, $p \pm qi$ say, the solution of the equation is $y = e^{px}(A \cos qx + B \sin qx)$.

Example. $y'' + 4y' + 8y = 0$. The roots of the characteristic equation $m^2 + 4m + 8 = 0$ are $-2 \pm 2i$, and the solution is $y = e^{-2x}(A \cos 2x + B \sin 2x)$. It is to be observed that in order to determine the values of the constants A and B, two conditions must be given.

Right-hand Member of Linear Differential Equations $f(x)$

Example. $y'' + y = e^{2x}$. Differentiating the equation, $y''' + y' = 2e^{2x}$ is obtained, and eliminating e^{2x} between the two equations, the homogeneous equation $y''' - 2y'' + y' - 2y = 0$ results. The corresponding characteristic equation is $m^3 - 2m^2 + m - 2 = 0$. The roots of this equation are 2 and $\pm i$, and the solution is $y = A \cos x + B \sin x + Ce^{2x}$. But this solution has three constants, one more than is required in the solution of a differential equation of the second order. By differentiating the solution twice and substituting in the original equation it is found that C is one-fifth. Therefore, the true solution is $y = A \cos x + B \sin x + \frac{1}{5}e^{2x}$.

INTEGRAL CALCULUS

If $f(x)dx$ is the differential of $F(x)$, the integral of $f(x)dx$ is $F(x)$. The derivative of $F(x)$ is $f(x)$. In the language of the calculus this is stated thus:

$$dF(x) = f(x)dx, \qquad \frac{dF(x)}{dx} = f(x)$$

$$\int f(x)dx = F(x) + C$$

where C is an arbitrary constant.

Fundamental Forms

u and v are functions of x. C is the constant of integration. Logarithms are to the base e. For a more complete table of integrals see Peirce's "Table of Integrals."

$$\int a\,du = au + C$$

$$\int \frac{du}{u} = \log_e u + C$$

$$\int u^n\,du = \frac{u^{n+1}}{n+1} + C$$

$$\int \epsilon^u\,du = \epsilon^u + C$$

$$\int \sin u\,du = -\cos u + C$$

$$\int \tan u\,du = -\log \cos u + C$$

$$\int \csc^2 u\,du = -\cot u + C$$

$$\int \cot u \csc u\,du = -\csc u + C$$

$$\int \csc u\,du = \log(\csc u - \cot u) + C$$

$$\int a^u\,du = \frac{a^u}{\log a} + C$$

$$\int \frac{du}{a^2 + u^2} = \frac{1}{a}\tan^{-1}\frac{u}{a} + C_1 \text{ or } -\frac{1}{a}\cot^{-1}\frac{u}{a} + C_2$$

$$\int \frac{du}{\sqrt{a^2 - u^2}} = \sin^{-1}\frac{u}{a} + C_1 \text{ or } -\cos^{-1}\frac{u}{a} + C_1$$

$$\int \cos u\,du = \sin u + C_1$$

$$\int \cot u\,du = \log \sin u + C_1$$

$$\int \sec^2 u\,du = \tan u + C_1$$

$$\int \tan u \sec u\,du = \sec u + C$$

$$\int \sec u\,du = \log(\sec u + \tan u) + C$$

$$\int (u + v)dx = \int u\,dx + \int v\,dx$$

$$\int u\,dv = uv - \int v\,du$$

In the following the constant of integration, C, is to be added.

Functions Involving $a + bx$.

$$\int (a + bx)^n\,dx = \frac{1}{b(n+1)}(a + bx)^{n+1} \quad (n \neq -1)$$

$$\int x(a + bx)^n\,dx = \frac{1}{b^2(n+2)}(a + bx)^{n+2} - \frac{a}{b^2(n+1)}(a + bx)^{n+1}$$

or

$$\frac{x}{b(n+1)}(a + bx)^{n+1} - \frac{1}{b^2(n+1)(n+2)}(a + bx)^{n+2} \quad (n \neq -1, -2)$$

$$\int x^2(a + bx)^n\,dx = \frac{(a + bx)^{n+1}}{b^3}\left[\frac{(a + bx)^2}{n+3} - \frac{2a(a + bx)}{n+2} + \frac{a^2}{n+1}\right] \quad (n \neq -1, -2, -3)$$

$$\int \frac{dx}{a + bx} = \frac{1}{b}\log(a + bx)$$

$$\int \frac{dx}{(a+bx)^2} = -\frac{1}{b}\frac{1}{a+bx}$$

$$\int \frac{x\,dx}{a+bx} = \frac{x}{b} - \frac{a}{b^2}\log(a+bx)$$

$$\int \frac{x^2\,dx}{a+bx} = \frac{1}{b^3}\left[\frac{1}{2}b^2x^2 - abx + a^2\log(a+bx)\right]$$

$$\int \frac{x\,dx}{(a+bx)^2} = \frac{a}{b^2(a+bx)} + \frac{1}{b^2}\log(a+bx)$$

$$\int \frac{x^2\,dx}{(a+bx)^2} = \frac{1}{b^3}\left[a+bx - \frac{a^2}{a+bx} - 2a\log(a+bx)\right]$$

$$\int \frac{dx}{x(a+bx)} = \frac{1}{a}\log\frac{x}{a+bx}$$

$$\int \frac{dx}{x^2(a+bx)} = -\frac{1}{ax} + \frac{b}{a^2}\log\frac{a+bx}{x}$$

$$\int \frac{dx}{x(a+bx)^2} = \frac{1}{a(a+bx)} - \frac{1}{a^2}\log\frac{a+bx}{x}$$

$$\int \frac{dx}{x^2(a+bx)^2} = -\frac{a+2bx}{a^2x(a+bx)} + \frac{2b}{a^3}\log\frac{a+bx}{x}$$

Functions Involving $a + bx^2$.

$$\int \frac{dx}{a+bx^2} = \frac{1}{\sqrt{ab}}\tan^{-1}\left(x\sqrt{\frac{b}{a}}\right) \quad (a \text{ and } b \text{ pos.})$$

$$= \frac{1}{2\sqrt{-ab}}\log\frac{\sqrt{a}+x\sqrt{-b}}{\sqrt{a}-x\sqrt{-b}} \quad (a \text{ pos., } b \text{ neg.})$$

$$\int \frac{dx}{(a+bx^2)^2} = \frac{x}{2a(a+bx^2)} + \frac{1}{2a}\int\frac{dx}{a+bx^2}$$

$$\int \frac{x\,dx}{a+bx^2} = \frac{1}{2b}\log\left(x^2 + \frac{a}{b}\right)$$

$$\int \frac{dx}{x(a+bx^2)} = \frac{1}{2a}\log\frac{x^2}{a+bx^2}$$

$$\int \frac{x^2\,dx}{a+bx^2} = \frac{x}{b} - \frac{a}{b}\int\frac{dx}{a+bx^2}$$

$$\int \frac{dx}{x^2(a+bx^2)} = -\frac{1}{ax} - \frac{b}{a}\int\frac{dx}{a+bx^2}$$

Functions Involving $a + bx + cx^2$.

$$\int \frac{dx}{a+bx+cx^2} = \frac{1}{\sqrt{b^2-4ac}}\log\frac{2cx+b-\sqrt{b^2-4ac}}{2cx+b+\sqrt{b^2-4ac}} \quad (b^2 > 4ac)$$

$$\int \frac{dx}{a+bx+cx^2} = \frac{2}{\sqrt{4ac-b^2}}\tan^{-1}\frac{2cx+b}{\sqrt{4ac-b^2}} \quad (b^2 < 4ac)$$

$$\int \frac{dx}{a+bx+cx^2} = -\frac{2}{2cx+b} \quad (b^2 = 4ac)$$

$$\int \frac{x\,dx}{a+bx+cx^2} = \frac{1}{2c}\log(a+bx+cx^2) - \frac{b}{2c}\int\frac{dx}{a+bx+cx^2}$$

$$\int \frac{x^2\,dx}{a+bx+cx^2} = \frac{x}{c} - \frac{b}{2c^2}\log(a+bx+cx^2) + \frac{b^2-2ac}{2c^2}\int\frac{dx}{a+bx+cx^2}$$

Functions Involving $\sqrt{a + bx}$.

$$\int \sqrt{a+bx}\,dx = \frac{2}{3b}(a+bx)^{3/2}$$

$$\int x\sqrt{a+bx}\,dx = \frac{2(3bx-2a)(a+bx)^{3/2}}{15b^2}$$

$$\int x^2\sqrt{a+bx}\,dx = \frac{2(8a^2-12abx+15b^2x^2)(a+bx)^{3/2}}{105b^3}$$

$$\int \frac{\sqrt{a+bx}}{x}dx = 2\sqrt{a+bx} + a\int\frac{dx}{x\sqrt{a+bx}}$$

$$\int \frac{dx}{\sqrt{a+bx}} = \frac{2\sqrt{a+bx}}{b}$$

$$\int \frac{x\,dx}{\sqrt{a+bx}} = \frac{2(bx-2a)}{3b^2}\sqrt{a+bx}$$

$$\int \frac{x^2\,dx}{\sqrt{a+bx}} = \frac{2(8a^2-4abx+3b^2x^2)}{15b^3}\sqrt{a+bx}$$

$$\int \frac{dx}{x^2\sqrt{a+bx}} = -\frac{\sqrt{a+bx}}{ax} - \frac{b}{2a}\int\frac{dx}{x\sqrt{a+bx}}$$

$$\int \frac{dx}{x\sqrt{a+bx}} = \frac{1}{\sqrt{a}}\log\frac{\sqrt{a+bx}-\sqrt{a}}{\sqrt{a+bx}+\sqrt{a}} \quad (a \text{ pos.})$$

$$\int \frac{dx}{x\sqrt{a+bx}} = \frac{2}{\sqrt{-a}}\tan^{-1}\sqrt{\frac{a+bx}{-a}} \quad (a \text{ neg.})$$

Functions Involving $\sqrt{a^2 + x^2}$.

$$\int \sqrt{a^2+x^2}\,dx = \tfrac{1}{2}\left[x\sqrt{a^2+x^2} + a^2\log\left(x+\sqrt{a^2+x^2}\right)\right]$$

$$\int \frac{dx}{\sqrt{a^2+x^2}} = \log\left(x+\sqrt{a^2+x^2}\right)$$

$$\int \frac{dx}{x\sqrt{a^2+x^2}} = -\frac{1}{a}\log\left(\frac{a+\sqrt{a^2+x^2}}{x}\right)$$

$$\int \frac{\sqrt{a^2+x^2}}{x} = \sqrt{a^2+x^2} - a\log\left(\frac{a+\sqrt{a^2+x^2}}{x}\right)$$

$$\int \frac{x\,dx}{\sqrt{a^2+x^2}} = \sqrt{a^2+x^2}$$

$$\int x\sqrt{a^2+x^2} = \tfrac{1}{3}(a^2+x^2)^{3/2}$$

Functions Involving $\sqrt{a^2 - x^2}$.

$$\int \sqrt{a^2-x^2}\,dx = \frac{1}{2}\left(x\sqrt{a^2-x^2} + a^2\sin^{-1}\frac{x}{a}\right)$$

$$\int \frac{dx}{x\sqrt{a^2-x^2}} = -\frac{1}{a}\log\left(\frac{a+\sqrt{a^2-x^2}}{x}\right)$$

$$\int \frac{\sqrt{a^2-x^2}}{x}dx = \sqrt{a^2-x^2} - a\log\left(\frac{a+\sqrt{a^2-x^2}}{x}\right)$$

$$\int \frac{x\,dx}{\sqrt{a^2-x^2}} = -\sqrt{a^2-x^2}$$

$$\int x\sqrt{a^2-x^2}\,dx = -\tfrac{1}{3}(a^2-x^2)^{3/2}$$

$$\int (a^2-x^2)^{3/2}dx = \frac{1}{4}\left[x(a^2-x^2)^{3/2} + \frac{3a^2x}{2}\sqrt{a^2-x^2} + \frac{3a^4}{2}\sin^{-1}\frac{x}{a}\right]$$

$$\int x^2\sqrt{a^2-x^2}\,dx = -\frac{x}{4}(a^2-x^2)^{3/2} + \frac{a^2}{8}\left(x\sqrt{a^2-x^2} + a^2\sin^{-1}\frac{x}{a}\right)$$

$$\int \frac{x^2\,dx}{\sqrt{a^2-x^2}} = -\frac{x}{2}\sqrt{a^2-x^2} + \frac{a^2}{2}\sin^{-1}\frac{x}{a}$$

$$\int \frac{dx}{x^2\sqrt{a^2-x^2}} = -\frac{\sqrt{a^2-x^2}}{a^2x}$$

$$\int \frac{\sqrt{a^2-x^2}}{x^2}\,dx = -\frac{\sqrt{a^2-x^2}}{x} - \sin^{-1}\frac{x}{a}$$

Functions Involving $\sqrt{x^2-a^2}$.

$$\int \sqrt{x^2-a^2} = \tfrac{1}{2}[x\sqrt{x^2-a^2} - a^2\log(x+\sqrt{x^2-a^2})]$$

$$\int \frac{dx}{\sqrt{x^2-a^2}} = \log(x+\sqrt{x^2-a^2})$$

$$\int \frac{dx}{x\sqrt{x^2-a^2}} = \frac{1}{a}\cos^{-1}\frac{a}{x},\ \text{or}\ \frac{1}{a}\sec^{-1}\frac{x}{a}$$

$$\int \frac{\sqrt{x^2-a^2}}{x}\,dx = \sqrt{x^2-a^2} - a\cos^{-1}\frac{a}{x}$$

$$\int \frac{x\,dx}{\sqrt{x^2-a^2}} = \sqrt{x^2-a^2}$$

$$\int x\sqrt{x^2-a^2}\,dx = \tfrac{1}{3}(x^2-a^2)^{3/2}$$

$$\int \frac{x^2\,dx}{\sqrt{x^2-a^2}} = \frac{x}{2}\sqrt{x^2-a^2} + \frac{a^2}{2}\log(x+\sqrt{x^2-a^2})$$

Transcendental Functions.

$$\int xe^{ax}\,dx = \frac{e^{ax}}{a^2}(ax-1)$$

$$\int x^ne^{ax}\,dx = \frac{x^ne^{ax}}{a} - \frac{n}{a}\int x^{n-1}e^{ax}\,dx$$

$$\int \frac{e^{ax}}{x^n}\,dx = \frac{1}{n-1}\left[-\frac{e^{ax}}{x^{n-1}} + a\int\frac{e^{ax}\,dx}{x^{n-1}}\right]$$

$$\int a^{bx}\,dx = \frac{a^{bx}}{b\log a}$$

$$\int x^na^x\,dx = \frac{x^na^x}{\log a} - \frac{n}{\log a}\int x^{n-1}a^x\,dx$$

$$\int \frac{dx}{1+e^x} = \log\frac{e^x}{1+e^x}$$

$$\int \frac{dx}{a+be^{nx}} = \frac{1}{an}[nx - \log(a+be^{nx})]$$

$$\int \frac{dx}{ae^{nx}+be^{-nx}} = \frac{1}{n\sqrt{ab}}\tan^{-1}\left(e^{nx}\sqrt{\frac{a}{b}}\right)$$

$$\int \log x\,dx = x\log x - x$$

$$\int \frac{(\log x)^n}{x}\,dx = \frac{1}{n+1}(\log x)^{n+1}$$

$$\int \sin^2 x\,dx = -\tfrac{1}{4}\sin 2x + \tfrac{1}{2}x = -\tfrac{1}{2}\sin x\cos x + \tfrac{1}{2}x$$

$$\int \cos^2 x\,dx = \tfrac{1}{4}\sin 2x + \tfrac{1}{2}x = \tfrac{1}{2}\sin x\cos x + \tfrac{1}{2}x$$

$$\int \sin nx\,dx = -\frac{\cos nx}{n}$$

$$\int \cos nx\,dx = \frac{\sin nx}{n}$$

$$\int \sin mx\cos nx\,dx = -\frac{\cos(m+n)x}{2(m+n)} - \frac{\cos(m-n)x}{2(m-n)}$$

$$\int \sin mx\sin nx\,dx = \frac{\sin(m-n)x}{2(m-n)} - \frac{\sin(m+n)x}{2(m+n)}$$

$$\int \cos mx\cos nx\,dx = \frac{\sin(m-n)x}{2(m-n)} + \frac{\sin(m+n)x}{2(m+n)}$$

$$\int \frac{dx}{\sin x} = \log\tan\frac{x}{2}$$

$$\int \frac{dx}{\cos x} = \log\tan\left(\frac{\pi}{4}+\frac{x}{2}\right)$$

$$\int \frac{dx}{1+\cos x} = \tan\frac{x}{2}$$

$$\int \frac{dx}{1-\cos x} = -\cot\frac{x}{2}$$

$$\int \sin x\cos x = \tfrac{1}{2}\sin^2 x$$

$$\int \frac{dx}{\sin x\cos x} = \log\tan x$$

$$\int \frac{dx}{a+b\cos x} = \frac{2}{\sqrt{a^2-b^2}}\tan^{-1}\left(\sqrt{\frac{a-b}{a+b}}\tan\frac{1}{2}x\right) \qquad (a^2>b^2)$$

$$= \frac{1}{\sqrt{b^2-a^2}}\log\frac{b+a\cos x+(\sin x)\sqrt{b^2-a^2}}{a+b\cos x} \qquad (a^2<b^2)$$

$$\int \frac{\cos x\,dx}{a+b\cos x} = \frac{x}{b} - \frac{a}{b}\int\frac{dx}{a+b\cos x}$$

$$\int \frac{\sin x\,dx}{a+b\cos x} = -\frac{1}{b}\log(a+b\cos x)$$

$$\int e^{ax}\sin bx\,dx = \frac{a\sin bx - b\cos bx}{a^2+b^2}e^{ax}$$

$$\int e^{ax}\cos bx\,dx = \frac{a\cos bx + b\sin bx}{a^2+b^2}e^{ax}$$

$$\int \sin^{-1}x\,dx = x\sin^{-1}x + \sqrt{1-x^2}$$

$$\int \cos^{-1}x\,dx = x\cos^{-1}x - \sqrt{1-x^2}$$

$$\int \tan^{-1}x\,dx = x\tan^{-1}x - \tfrac{1}{2}\log(1+x^2)$$

$$\int \cot^{-1}x\,dx = x\cot^{-1}x + \tfrac{1}{2}\log(1+x^2)$$

$$\int \frac{dx}{\sin^n x} = -\frac{\cos x}{(n-1)\sin^{n-1}x} + \frac{n-2}{n-1}\int\frac{dx}{\sin^{n-2}x}$$

$$\int \frac{dx}{\cos^n x} = \frac{\sin x}{(n-1)\cos^{n-1}x} + \frac{n-2}{n-1}\int\frac{dx}{\cos^{n-2}x}$$

Integration by Substitution. The following substitutions will often transform expressions into others which can be integrated.

In expressions involving fractional powers of $a + bx$, put $a + bx = z$.

Example 1. $\int \frac{x^2\,dx}{(1+2x)^{1/3}}$. Let $1+2x = z$, $x = \frac{1}{2}(z-1)$, $dx = \frac{1}{2}dz$.

$$\int\frac{x^2\,dx}{(1+2x)^{1/3}} = \frac{1}{2}\int\frac{z^2-2z+1}{4z^{1/3}}\,dz$$

$$= \frac{1}{8}\int(z^{5/3} - 2z^{2/3} + z^{-1/3})\,dz = \tfrac{1}{8}(\tfrac{3}{8}z^{8/3} - \tfrac{6}{5}z^{5/3} + \tfrac{3}{2}z^{2/3})$$

$$= \tfrac{3}{320}(1+2x)^{2/3}(9 - 12x + 20x^2)$$

In expressions involving fractional powers of $a^2 - x^2$, put $x = a\sin z$.

Example 2. $\int\sqrt{4-x^2}\,dx$. Let $x = 2\sin z$, $dx = 2\cos z\,dz$.

$$\int\sqrt{4-x^2}\,dx = 4\int\cos^2 z\,dz = 2\int(1+\cos 2z)\,dz = 2z + \sin 2z + C = 2\sin^{-1}\frac{x}{2} + \frac{x}{2}\sqrt{4-x^2} + C$$

$$= \frac{1}{2}\left(x\sqrt{4-x^2} + 4\sin^{-1}\frac{x}{2}\right) + C$$

In expressions involving fractional powers of $x^2 + a^2$, put $x = a\tan z$, $dx = a\sec^2 z\,dz$.

Example 3. $$\int\frac{dx}{(x^2+a^2)^{3/2}} = \int\frac{a\sec^2 z\,dz}{a^3\sec^3 z} = \frac{1}{a^2}\int\cos z\,dz$$

$$= \frac{1}{a^2}\sin z = \frac{x}{a^2\sqrt{x^2+a^2}},\ \text{since}\ \sin z = \frac{x}{\sqrt{x^2+a^2}}$$

In expressions involving fractional powers of $x^2 - a^2$, let $x = a \sec z$, $dx = a \sec z \tan z\, dz$.

Example 4. $\int x^3 \sqrt{x^2 - a^2}\, dx = \int a^3 \sec^3 za \tan za \sec z \tan z\, dz$
$= a^5 \int \sec^4 z \tan^2 z\, dz = a^5 \int (\sec^2 z \tan^2 z) \sec^2 z\, dz$
$= a^5 \int (\tan^4 z + \tan^2 z) \sec^2 z\, dz = a^5(\tfrac{1}{5} \tan^5 z + \tfrac{1}{3} \tan^3 z)$
$= a^5 \left[\frac{1}{5}\frac{(x^2 - a^2)^{5/2}}{a^5} + \frac{1}{3}\frac{(x^2 - a^2)^{3/2}}{a^3}\right] = \frac{1}{15}(x^2 - a^2)^{3/2}(2a^2 + 3x^2)$

Definite Integrals. If $\int f(x)dx = F(x)$, the definite integral $\int_a^b f(x)dx = F(b) - F(a)$. This is read the integral of $f(x)dx$ from $x = a$ to $x = b$ is equal to $F(b) - F(a)$. For example,

$$\int_3^5 x^2\, dx = \left[\frac{x^3}{3}\right]_3^5 = \frac{125}{3} - 9 = \frac{98}{3}.$$

The numbers a and b are the limits of integration. $\int_3^5 x^2\, dx$ represents the area bounded by $y = x^2$, the x-axis, and the two ordinates $x = 3$ and $x = 5$.

The value of the definite integral, $\int_a^b f(x)dx$, can be found approximately by drawing the curve $y = f(x)$ and measuring the area under it from $x = a$ to $x = b$, either by means of a planimeter or by "counting squares." Simpson's rule can be applied to advantage.

If an expression to be integrated is of such a nature that the ordinary processes do not apply, one of the methods just given will approximate the correct result.

Illustration: $\int_1^2 \sqrt{x + \frac{1}{x}}\, dx =$ the area bounded by $y = \sqrt{x + \frac{1}{x}}$, $y = 0$, $x = 1$, and $x = 2$. Divide the area into an even number of strips (in this case six) of equal width. The following table shows the process.

x	$x + \frac{1}{x}$	$\sqrt{x + \frac{1}{x}}$	
1	2.0000	1.414	1.414
7/6	2.0238	1.423	5.692
8/6	2.0833	1.443	2.886
9/6	2.1667	1.472	5.888
10/6	2.2667	1.506	3.012
11/6	2.3788	1.542	6.168
2	2.5000	1.581	1.581
		18	26.641
			1.4801

If the points represented by the corresponding numbers in columns 1 and 3 be plotted and a smooth curve drawn through them, the area under it can be estimated as shown in graphical integration.

By using Simpson's $\frac{1}{3}$ rule,
Area $= \frac{1}{3}h(y_0 + 4y_1 + 2y_2 + 4y_3 + 2y_4 + 4y_5 + y_6)$
a better approximation is obtained. The numbers in column 4 are obtained by multiplying the values of y by the corresponding coefficients in Simpson's rule. These products are added and the sum multiplied by $\frac{1}{3}h$, where h is the width of each strip. The value of the definite integral is found to be 1.4801. There is some doubt about the figure in the last place.

1. Single integration, double integration, and **triple integration** are illustrated by the following problem. Find the volume enclosed by the surface $z = 4 - x^2 - y^2$ above the xy-plane.

Single Integration. Since the cross section of the volume parallel to the xy-plane is a circle whose radius is $\sqrt{4 - z}$ (Fig. 122), the element of volume is $\pi(4 - z)dz$. The volume is equal to $\pi \int_0^4 (4 - z)dz = \pi\left[4z - \tfrac{1}{2}z^2\right]_0^4 = 8\pi$.

2. Double Integration. The volume is represented by

$$4 \int_0^2 \left[\int_0^{\sqrt{4-x^2}} (4 - x^2 - y^2)dy\right] dx.$$

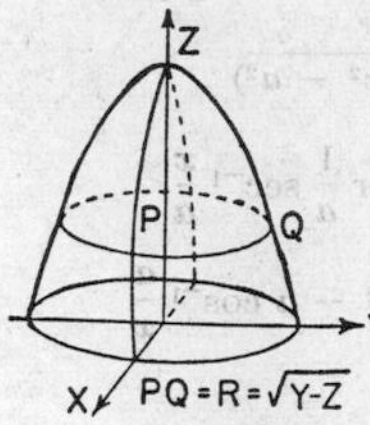

FIG. 122.

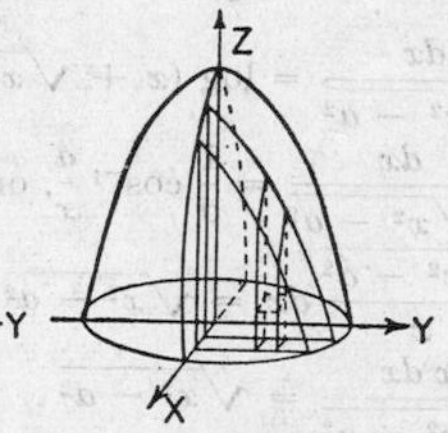

FIG. 123.

The expression in the brackets is the value of z, and the inner integral gives the area of the section parallel to the yz-plane in the first octant (Fig. 123). This area times dx gives the element of volume. The outer definite integral from $x = 0$ to $x = 2$ gives the volume in the first octant. The factor 4 is used to give the volume in the four octants. x and dx are constants while integrating with respect to y.

$$V = 4 \int_0^2 \left[\int_0^{\sqrt{4-x^2}} (4 - x^2 - y^2)dy\right] dx$$
$$= 4 \int_0^2 \left[4y - x^2y - \tfrac{1}{3}y^3\right]_0^{\sqrt{4-x^2}}$$
$$= \tfrac{8}{3} \int_0^2 \sqrt{4 - x^2}\,(4 - x^2)dx = 8\pi$$

3. Triple Integration.

$$V = 4 \int_0^2 \left[\int_0^{\sqrt{4-x^2}} \left(\int_0^{4-x^2-y^2} dz\right) dy\right] dx$$
$$= 4 \int_0^2 \left[\int_0^{\sqrt{4-x^2}} (4 - x^2 - y^2)dy\right] dx,$$

which is the integral above.

Simpson's Rule. The volume can also be obtained by the application of Simpson's rule. The areas of the sections parallel to the xy-plane at unit distance apart (Fig. 124) are, respectively, 0, π, 2π, 3π, and 4π. The

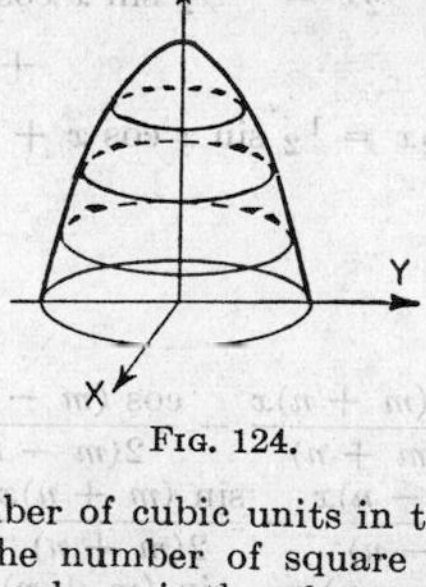

FIG. 124.

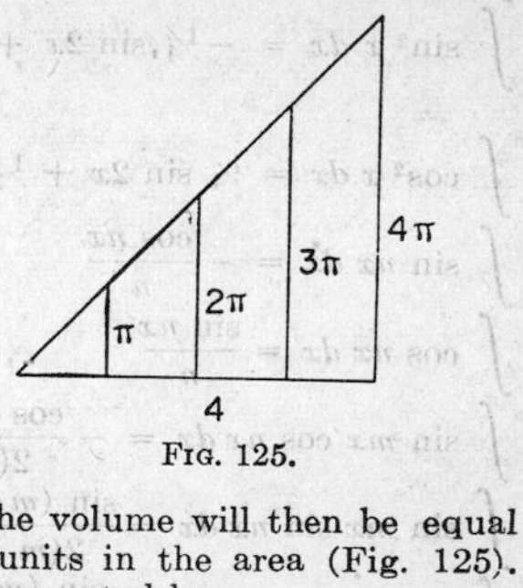

FIG. 125.

number of cubic units in the volume will then be equal to the number of square units in the area (Fig. 125). The volume is therefore represented by

$$V = \tfrac{1}{3}\pi(0 + 4 + 4 + 12 + 4) = 8\pi$$

Simpson's rule furnishes a convenient way of finding the approximate volume of any solid when the areas of the proper cross sections can be obtained.

For the evaluation of elliptic integrals, the probability integral, and gamma functions, see Peirce's "Table of Integrals."

SUBSTITUTES FOR INTEGRATION

I. When it is not convenient or it is impossible to integrate a differential equation, some less exact method may be used as a substitute, which, however, for many cases, will give a sufficiently exact answer.

Example 1. Find the value of $dx/dt = k(a - x)$, where x is the amount of substance transformed in time t, and a is the initial concentration, *i.e.*, a is a constant; dt is a unit time interval. Let Δx be the difference between the initial and final quantities of substance transformed in unit interval of time; then $\frac{1}{2}(\Delta x)$ is the average amount of substance transformed in the same time. For the first interval of time, $\Delta x = k_1(a - \frac{1}{2}\Delta x)$, which is by algebra $\Delta x = (k_1a)/(1 + \frac{1}{2}k_1)$; and for the next interval, $\Delta x = k_1(a - x - \frac{1}{2}\Delta x)$. Suppose $k_1 = 0.1$ and $a = 100$, then $\Delta x = 0.1 \times 100/1.05 = 9.52$ (first interval); and for the second interval, $k_1 = 0.1$ and $a = 100 - 9.52 = 90.48$; therefore, $\Delta x = 0.1 \times 90.48/1.05 = 8.62$.

Example 2. Assume the second-order rate of reaction: $dx/dt = k_2(a - x)^2$; in the same way as in Example 1, we get: $\Delta x = (k_2a^2)/(1 + k_2a)$; and $\Delta x = [k_2(a - x)^2]/[1 + k_2(a - x)]$, etc. Assume $k_2 = 0.1$ and $a = 100$, then $\Delta x = \dfrac{0.1 \times 10{,}000}{1 + 0.1(100)}$ $= 90.99$ (for first interval); and for the second interval, $k_2 = 0.1$ and $a = 100 - 90.99 = 9.01$, and $\Delta x = \dfrac{(9.01)^2 \times 0.1}{1 \times 0.1 \times 9.01} = 4.33$.

By direct integration these intervals are 90.99 and 4.33, respectively, which shows for this problem that the substitute used here is accurate.

II. The area enclosed by a curve can be estimated by determining the value of a definite integral; we may reverse the procedure and determine the numerical value of a definite integral by measuring the area enclosed by the curve.

Example 3. If the integral $[f(x)dx]$ is unknown, the value of $\int_b^a f(x)dx$ can be found by plotting the curve $y = f(x)$; then erecting ordinates to the curve on the points $x = a$ and $x = b$; and then measuring the surface by one of the following methods.

A. The area may be measured with a planimeter, which automatically measures and registers the area of a plane figure when a tracer is passed around the boundary lines. The **Amsler polar planimeter** is used for small areas, as for a few square inches, and consists of two principal parts: the *tracer arm* carrying the tracing point and the carriage with the measuring wheel, and the **pole arm,** which is fixed to the pole around which the instrument revolves. The area of any figure is readily and accurately obtained by tracing its boundary line with the tracing point, whereupon the result is computed from the reading of the graduated measuring wheel. Since all planimeters revolve around a fixed point, their scope is limited by the lengths of the instrument arms, which necessitates measuring large areas in sections. The use of the planimeter is restricted to definite integrals such as an area between two curves, or to the area between a curve and the X- or Y-axis, *i.e.*, between definite, known limits.

For measuring figures of any length but of a breadth limited only by the length of the tracer arm, roller planimeters must be used. The tracer arm moves in a straight line, on broad, heavy rollers, and is especially suited for measuring the areas of profiles, etc.

Integrators. These instruments determine the area and moments relative to any axis of any figure, by tracing its outline. They greatly facilitate the finding of the displacement, moments of stability and inertia, center of gravity of solids in fluids, the tensile strength, resistance, safe load, etc., of beams, cables, tracks, etc., and the volumes of embankments, etc.

B. The area may be measured by plotting the curve or data on cross-section paper, of uniform material and then cutting the figure out; let W_1 be the weight of a known area a_1; and W_2 be the weight of the piece cut out. The desired area x then can be obtained by direct proportion; *i.e.*, $W_1/a = W_2/x$ or $x = W_2a/W_1$.

C. The area, after plotting on cross-section paper, may be obtained approximately, by counting the number of squares between the known ordinates, and then adding to this sum the number of fractional parts of squares that are cut by the boundary of the curve. The total sum is the area in units of the dimensions of the squares. The finer the ruling of the cross-section paper, the more accurate is the result.

Integrators for Circular Charts. These instruments, which may be secured from such firms as the Bristol Co. and Builders Iron Foundry, are for measuring the average ordinate of circular charts having a constant radial scale.

Mechanical Integraph. This type of instrument draws, on paper, the curve $y = \int f(x)dx$, when the curve $y = f(x)$ is given. It also facilitates the solution of certain types of differential equations. The detailed directions for the use of all such instruments are given with each instrument.

Ordinate Methods. D. Trapezoidal Rule (Fig. 126). Let a corresponding series of values of x and y

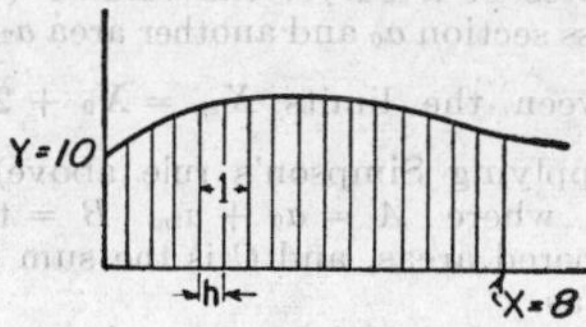

FIG. 126.

be plotted and the curve drawn through the points. To find the area bounded by the curve, the x-axis, and the two given ordinates, divide the area into n strips of equal width (k) by drawing ($n + 1$) ordinates (n being so large that each strip may be regarded as a trapezium). The required area under the curve is then the sum of the areas of the n trapeziums. This total area may be found by the following method. Assume that the data are as shown in the above figure; *i.e.*,

x	0	2	3	4	5	7	8
y	10	28	33	35	32	24	16

Since there are 8 pairs of values of x and y, we have arbitrarily divided the area into 16 strips, by drawing ordinates at intervals of 0.5 unit. The trapezoidal formula is written: $\frac{1}{2}h(y_0 + y_1) + \frac{1}{2}h(y_1 + y_2) + \cdots + \frac{1}{2}h(y_{n-1} + y_n)$, where h is the width of each trapezium. The above formula may be written in the following way, which facilitates its use: $= h[\frac{1}{2}(y_0 + y_n) + y_1 + y_2 + y_3 + \cdots + y_{n-1}]$; the sum of the two ordinates in the above problem is $(y_0 + y_n) = (10 + 16)$ and $\frac{1}{2}(10 + 16) = 13$; the other values of y, which are used from the graph, are: $y_1 = 15$, $y_2 = 19$; $y_3 = 23.5$, $y_4 = 28$, $y_5 = 31$, $y_6 = 33$, $y_7 = 34.5$, $y_8 = 35$, $y_9 = 34$, $y_{10} = 32$, $y_{11} = 29.5$, $y_{12} = 27.5$, $y_{13} = 24.5$, $y_{14} = 24$, $y_{n-1} = y_{15} = 19$, the sum of which is 409.5. Then, substituting in the above rearranged formula, area $= h(13 + 409.5)$, and, since $h = 0.5$ unit, the area $= \frac{1}{2}(13 + 409.5) = 211.25$.

E. Simpson's Rule. Here again, let $y = f(x)$ be a continuous function defined by a series of points through which a smooth curve is drawn. The area bounded by the curve, the x-axis, and the ordinates $y_0, y_1, y_2, \ldots$, is then divided into $2n$ strips of equal width, and, for convenience here, let y_0 coincide with the y-axis. Let $y_0 = a$, where $x_0 = 0$; and $y_1 = a + bh + ch^2$, where $h = x_1$; and $y_2 = a + 2bh + 4ch^2$, where $x_2 = 2h$. Solving the last two equations for b and c, we obtain: $b = (4y_1 - 3y_0 - y_2)/(2h)$ and $c = (y_2 - 2y_1 + y_0)/(2h^2)$. Then the area between y_0 and $y_2 = \int_0^{2h} (a + bx + cx^2)dx = \frac{2h(a + bh + 4ch^2)}{3} = \frac{1}{3}h(y_0 + 4y_1 + y_2)$.

The *rule* is then *divide the area to be calculated into $2n$ strips of equal width h by drawing $2n + 1$ ordinates, then the required area* $= \frac{1}{3}h(A + 4B + 2C)$, where A = sum of first and last ordinates, B = sum of the odd ordinates, and C = sum of the even ordinates; h is the ordinate interval.

Example. Use the data as given under the trapezoidal rule. h is therefore 0.5, since there must be $2n$ or $(2 \times 8) = 16$ ordinates. Then $A = 10 + 16 = 26$, $B = 211$, and $C = 214.5$; and hence $4B = 844$ and $2C = 397$; therefore, the area $= \frac{1}{3}(0.5)(26 + 858 + 422) = \frac{1}{3}(653) = 211.1$.

The Determination of Volume by Integration: Prismoidal Formula. This method is of use if the approximate volume of an irregular object, such as a tree trunk, must be determined. Here the cross-sectional areas perpendicular to the long axis are known at regular intervals along the axis. Then letting the areas be the ordinates of a curve, the volume (V), between an area of cross section a_0 and another area a_{2n}, is equal to $\int a\,dx$, between the limits $X_{2n} = X_0 + 2nh$ and x_0. This area (applying Simpson's rule above) $= \frac{1}{3}(h)(A + 4B + 2C)$, where $A = a_0 + a_{2n}$, B = the sum of the odd-numbered areas, and C is the sum of the even-numbered areas.

Similarly, the volume of water can be computed from a contour chart, by obtaining the areas of the contours and using Simpson's rule as given above.

Durand's Rule. This rule is more accurate but more cumbersome than the trapezoidal rule. The formula is

$$\text{Area} = h[0.4(y_0 + y_n) + 1.1(y_1 + y_{n-1}) + y_2 + y_3 + \cdots + y_{n-1}]$$

Weddle's Rule. The formula for obtaining the area under the curve represented by the integral $\int_0^5 f(x)dx$ is

$$\text{Area} = \frac{3}{10}h(y_0 + 5y_1 + y_2 + 6y_3 + y_4 + 5y_5 + y_6)$$

for the case when we take seven ordinates. For more detailed information, see *Philosophical Transactions*, vol. 167, i, p. 1 (1877).

Of the above formulas the trapezoidal and Simpson's rules are the most important for areas, and the latter is the more accurate; the prismoidal rule for volumes is the more important.

GRAPHICAL CALCULUS

When dealing with tabulated functions the following operations can be performed to a fair degree of approximation: integration, differentiation, interpolation, and smoothing or graduating the data. All these operations are made to depend upon one single principle of the formal calculus. Let

$$y = F(x)$$
$$y' = f(x)$$

Then

$$\int_a^b f(x)dx = F(b) - F(a) \qquad (1)$$

Equation (1) states that the number of square units under the derived curve between any two ordinates is equal to the number of linear units in the difference between the corresponding ordinates on the integral curve. In taking the difference, it is understood that the ordinate at the left is to be subtracted from the ordinate at the right.

Integration. Table 1, below, shows the process of integration; t is time in seconds and V is velocity in feet per second at any time t. It is required to find the distance passed over by a point moving according to the law expressed in the first two columns of the table.

The points represented by corresponding values of t and V in Table 1 are plotted in Fig. 127 and a smooth curve drawn through them. This curve must be drawn with great care. Upper boundaries of rectangles are drawn so that their areas are equal to the corresponding areas under the curve. For example, the line ae is the upper boundary of the rectangle whose area is equal to the area under the arc bd. This is done by making the triangular area abc equal to the triangular area cde. The eye is a very good judge of the equality of small areas. The height of each of these rectangles, represented by $\Delta v/\Delta t$ in Table 1, is the average velocity over the corresponding range. The areas of these rectangles are given in column 4. From Eq. (1) it is evident that the values of S, distance, will be obtained by adding successively these areas. In the figure it was assumed that s is zero when t is zero. The S curve is also drawn in Fig. 127.

Table 1

(1) t	(2) V	(3) $\Delta v/\Delta t$	(4) Areas	(5) S
0	0	....	.....	0
		2.0	2.0	
1.0	3.5	....	.....	2.0
		5.5	11.0	
3.0	8	....	.....	13.0
		9.5	9.5	
4.0	11	....	.....	22.5
		13.4	20.1	
5.5	16	....	.....	42.5
		18.0	14.4	
6.3	20	....	.....	57.0
		21.6	17.28	
7.1	23	....	.....	74.28
		24.0	21.6	
8.0	25	....	.....	95.88
		25.7	25.7	
9.0	26	....	.....	121.58
		25.4	25.4	
10.0	24	....	.....	146.98

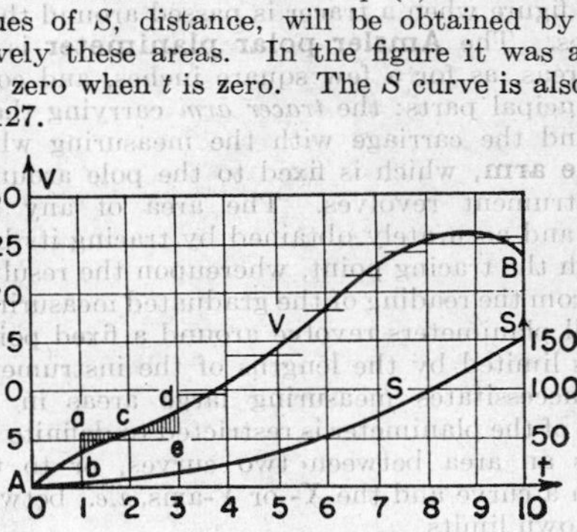

FIG. 127.

Differentiation. In the first two columns of Table 2 are given corresponding values of time t, in seconds, and θ, the excess in temperature of the body over the temperature of its surroundings. In column 5 are given the values of $\Delta\theta/\Delta t$, the average changes of temperature over the given ranges. The upper boundaries of the rectangles are drawn in Fig. 128. A smooth curve is now drawn so that the areas under the arcs shall be as nearly equal to the areas of the corresponding rectangles

Table 2

(1) t	(2) θ	(3) $\Delta\theta$	(4) Δt	(5) $\Delta\theta/\Delta t$	(6) $d\theta/dt$	(7) $\Delta\theta'$	(8) $\Delta\theta'/\Delta t$
0	99.9				−0.2985		
		−1.0	3.45	−0.2899		0.0151	0.004377
3.45	98.9				−0.2834		
		−2.0	7.40	−0.2703		0.0298	0.004027
10.85	96.9				−0.2536		
		−2.0	8.45	−0.2367		0.0301	0.003562
19.30	94.9				−0.2235		
		−2.0	9.50	−0.2105		0.0301	0.003126
28.80	92.9				−0.1938		
		−2.0	11.30	−0.1769		0.0302	0.002673
40.10	90.9				−0.1636		
		−2.0	13.65	−0.1465		0.0303	0.002212
53.75	88.9				−0.1333		
		−2.0	17.20	−0.1163		0.0303	0.001762
70.95	86.9				−0.1030		

as possible. This is the derived curve, and the values of the derivative read from this curve are given in column 6. A check on the accuracy of the differentiation is obtained by drawing the second derived curve. In column 7 are given the values of θ'. The values of

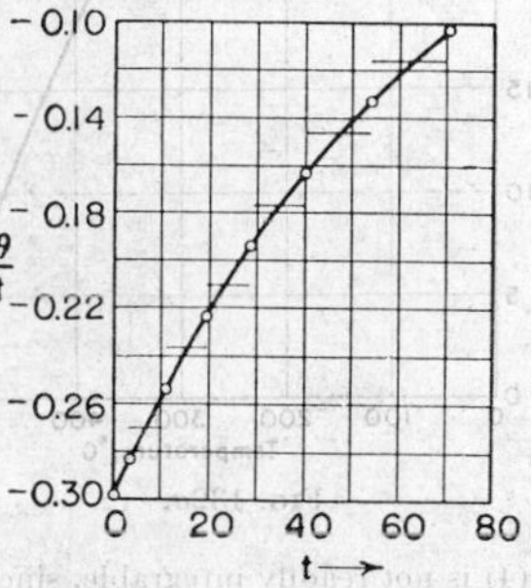

FIG. 128.

$\Delta\theta'/\Delta t$ given in column 8 are obtained by dividing the values of $\Delta\theta'$ by the corresponding values of Δt. Figure 129 gives the construction. It is seen that the second derived curve is quite smooth, which shows that the differentiation has been carried out with a high degree of accuracy. The values of the second derivative can be read from Fig. 129.

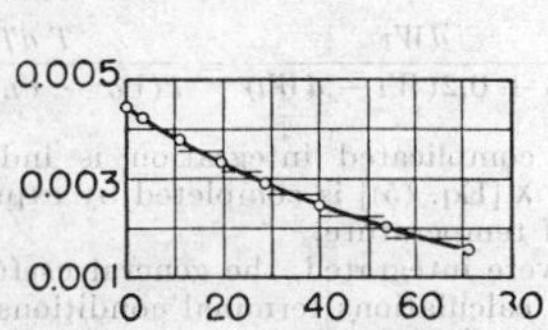

FIG. 129.

Smoothing of Data. Interlaced Parabolas. In smoothing data by the method of interlaced parabolas, the derived curve is drawn as a broken line, each segment being the derived curve of a parabolic arc.

In columns 1 and 2 of Table 3 are given the brush-contact resistance, R, in ohms, between the brushes and the metallic segments of the interrupter, and x, in revolutions per minute.

Data will be graduated when a smooth curve is constructed passing as nearly through the points represented by the data as possible. This is done by constructing the derived curve as a broken line and by the application of Eq. (1) computing the graduated values of the dependent variable. The values of $\Delta R/\Delta x$ in column 5, Table 3, are drawn as the upper boundaries of rectangles in Fig. 130. The broken line is now drawn to conform to the principle of areas; *i.e.*, so that the areas of the rectangles

Table 3

(1) R	(2) x	(3) ΔR	(4) Δx	(5) $\Delta R/\Delta x$	(6) R'	(7) Areas	(8) Sums	(9) Graduated R
0	0				0			0
		1.6	120	0.01333		1.152	1.152	
1.6	120				0.0192			1.152
		4.5	110	0.04091		3.080	4.232	
6.1	230				0.0368			4.232
		3.3	120	0.02750		5.568	9.800	
9.4	350				0.0560			9.800
		9.5	110	0.08636		8.515	18.315	
18.9	460				0.0988			18.315
		4.8	40	0.12000		4.265	22.580	
23.7	500				0.1144			22.580
		17.1	140	0.12214		19.836	42.416	
40.8	640				0.1689			42.416
		32.8	160	0.20500		32.016	74.432	
73.6	800				0.2313			74.432
		31.4	120	0.26167		30.558	104.990	
105.0	920				0.2780			104.990
279.1							277.917	277.917

will be as nearly equal to the corresponding areas under the broken line as possible. This is an approximation to the derived curve. The values of R' are read from the figure and given in column 6. The computed areas of the trapezoids given in column 7 are the differences between the corresponding graduated values of R. These areas added successively are given in column 8. The sum of column 8 subtracted from the sum of column 1 and this difference divided by 9 gives the first graduated value of R. This quotient is 0.131. But it is known that the first value is 0, so this value is taken. The other values in the last column are obtained by applying Eq. (1). The sum of the last column is slightly less than the sum of the first one, because the first value was taken as zero.

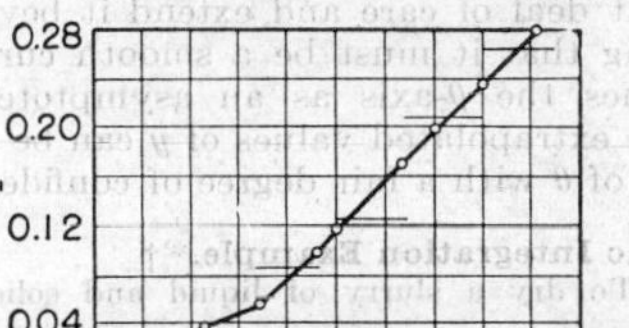

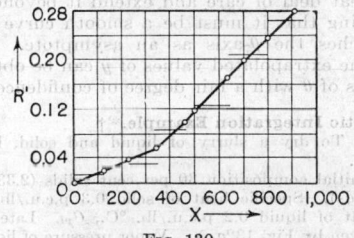

FIG. 130.

The curve of graduated data is made up of two parabolas, one covering the range from $x = 0$ to $x = 350$, the other from $x = 350$ to $x = 920$. At the point where they meet they have the same slope.

In the first two columns of Table 4 are given data taken from "Differential Equations" by Hitchcock and

Table 4

θ	y	Δy	$\Delta\theta$	$\Delta y/\Delta\theta$	y'	Areas	Sums	Graduated y
10	0.0348				0.002600			0.034813
		0.0091	4	0.002275		0.00908	0.00908	
14	0.0439				0.001940			0.043893
		0.0064	4	0.001600		0.00644	0.01552	
18	0.0503				0.001280			0.050333
		0.0044	4	0.001100		0.00444	0.01996	
22	0.0547				0.000980			0.054773
		0.0059	8	0.000738		0.00582	0.02578	
30	0.0606				0.000475			0.060593
		0.0036	11	0.000327		0.00360	0.02938	
41	0.0642				0.000180			0.064193
		0.0017	11	0.000154		0.00161	0.03099	
52	0.0659				0.000112			0.065803
		0.0009	16	0.000056		0.00100	0.03199	
68	0.0668				0.0000125			0.066803
	0.4412					0.03199	0.16270	0.441204
	0.1627							
8	0.2785							
	0.034813							

Robinson. The process of constructing the table is similar to the preceding. Figure 131 shows the construction. The sum of the numbers in column 8 subtracted from the sum of column 2 and this difference divided by 8 gives the first graduated value of y. The other values of y are obtained as before. In this example the smooth curve is made up of four interlaced parabolas.

For a better graduation the derived curve in Fig. 131 must be drawn as a smooth curve and checked by means of the second derivative.

Interpolation. In order to find the value of y corresponding to $\theta = 35$, it is only necessary to compute the area under the straight line from $\theta = 30$ to $\theta = 35$ and add this area to the value of y corresponding to $\theta = 30$. This gives for the value of y, 0.06223. The process depends upon Eq. (1).

Extrapolation. For extrapolating the data in Table 4 it is necessary to draw the derivative in Fig. 131

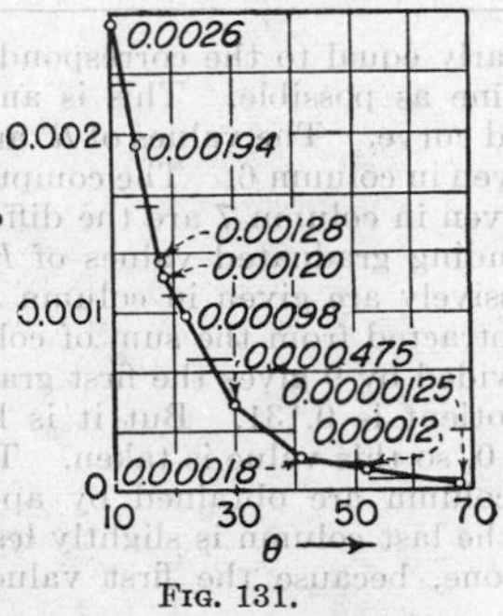

Fig. 131.

with a great deal of care and extend it beyond $\theta = 68$, remembering that it must be a smooth curve and that it approaches the θ-axis as an asymptote. Applying Eq. (1), the extrapolated values of y can be obtained for a few units of θ with a fair degree of confidence.

Arithmetic Integration Example.*†

Problem. To dry a slurry of liquid and solid, by a flash evaporation.

Data. Initial composition 30 per cent solids (2.33 lb. liquid per lb. solids). Specific heat of solid 0.3 p.c.u./lb. °C., C_{ps}. Specific heat of liquid 0.2 p.c.u./lb. °C., C_{pl}. Latent heat of liquid as given by Fig. 132*a*, λ. Vapor pressure of liquid, p.

Process. Heat slurry under pressure until there is sufficient sensible heat in the mixture to evaporate the liquid completely. Vent the pressure vessel to a condenser delivering to a closed receiver.

Required. Temperature and pressure before flash. Pressure at end of flash.

Calculation. This is a differential process whereby the quantity of heat removed in the vapor phase balances the sensible heat removed from the solid-liquid phase.

By heat and material balance,

$$\lambda\, dW_l = \text{heat removed by the vapor} \tag{1}$$

$$C_{ps}W_s\, dT = \text{heat removed from the solid} \tag{2}$$

$$C_{pl}(W_l - dW_l)dT = \text{heat removed from the liquid} \tag{3}$$

where λ = latent heat, p.c.u./lb.
W_s = weight of solid, lb.
W_l = weight of liquid, lb. (present at start of increment vaporized).
dW_l = incremental weight of liquid vaporized, lb.
C_{ps} = specific heat of solid, p.c.u./lb. °C.
C_{pl} = specific heat of liquid, p.c.u./lb. °C.
T = temperature, °K.

* The use of Gauss's method of numerical integration is a valuable tool for some chemical engineering calculations. See Kroll, *Chem. Eng.*, September, 1946, pp. 102–104; also, see Dull, "Mathematical Aids for Engineers," McGraw-Hill, New York, 1946.

† By H. C. Vernon.

Heat removed by flashing, dQ, must equal $\lambda\, dW_l$. This heat comes from the liquid and solid and is equal to

$$dQ = C_{ps}W_s\, dT + C_{pl}(W_l - dW_l)dT$$

Select a unit weight of solids as the basis of calculation, and since $C_{ps} = 0.3$, $C_{pl} = 0.2$*

$$\lambda\, dW_l = [0.3 + 0.2(W_l - dW_l)]dT \tag{4}$$

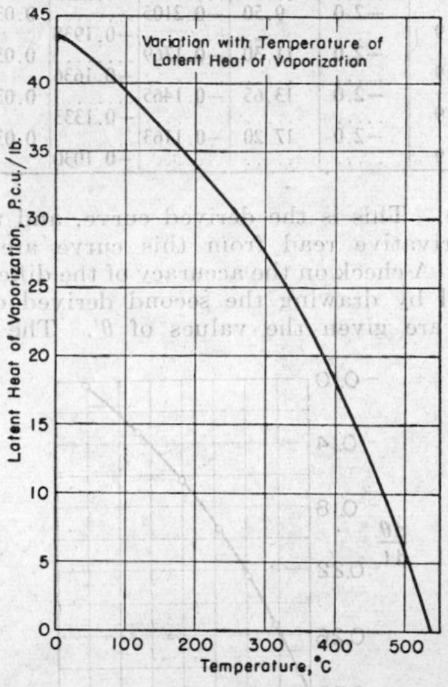

Fig. 132*a*.

However, Eq. (4) is not readily integrable, since λ is also a function of temperature and is given by an equation of the form

$$\lambda = k(V_G - V_L)\frac{e^{B-A/T}}{T} \tag{5}$$

where k, B, and A are constants and V_G and V_L are the molal volumes of the vapor and liquid, respectively, at any temperature, T.

Combining Eqs. (4) and (5),

$$\frac{dW_l}{0.3 + 0.2(W_l - dW_l)} = \frac{T\,dT}{k(V_G - V_L)e^{B-A/T}} \tag{6}$$

A still more complicated integration is indicated when this expression for λ [Eq. (5)] is completed by expressing V_G and V_L as functions of temperature.

If Eq. (6) were integrated, the general usefulness is doubtful, since in many calculations terminal conditions are being sought and no intermediate values exist for evaluation of the integration constant.

Proceeding by arithmetical integration, Eq. (4) is rewritten by substituting a finite change in variables for the differential change, *viz.*,

$$\lambda\,\Delta W_l = [0.3 + 0.2(W_l - \tfrac{1}{2}\Delta W_l)]\Delta T\dagger \tag{7}$$

* For simplification, C_{ps} and C_{pl} are assumed constant over the temperature range of the example. This is not exact, and engineering accuracy may require consideration of the variation with temperature.

† This includes an average value for the quantity of liquid existing during the finite interval taken as an approximation to dW_l. The limiting values are obtained by saying that:

(*a*) The increment vaporized takes its heat from the liquid remaining after the total increment is removed; Eq. (3) should then be written as $C_{pl}(W_l - \Delta W_l)\Delta T$. This leads to low values for ΔW_l.

(*b*) The increment vaporized takes its heat from the total quantity of liquid present at the start of the vaporization; Eq. (3) should then be written $C_{pl}W_l\Delta T$. This leads to high values for ΔW_l and assumes that the total quantity of liquid present at the start of the incremental vaporization is available for the whole increment.

Selecting 20°C. intervals, or $\Delta T = 20$ and rearranging,

$$\Delta W_l = \frac{6 + 4W_l}{\lambda + 2} \tag{8}$$

Assuming an initial temperature of 300°C. and solving for terminal conditions:

$T_1 - T_2$	$T_{avg.}$	λ	$\lambda + 2$	W_l	$4W_l$	$6 + 4W_l$	ΔW_l
30–0280	290	28	30	2.33	9.32	15.32	0.51
28–0260	270	29.5	31.5	1.82	7.28	13.28	0.42
26–0240	250	31	33	1.40	5.60	11.60	0.35
24–0220	230	32.5	34.5	1.05	4.20	10.20	0.30
22–0200	210	33.5	35.5	0.75	3.00	9.00	0.25
20–0180	190	34.5	36.5	0.50	2.00	8.00	0.22
18–0160	170	35.5	37.5	0.28	1.12	7.12	0.20
16–0140	150	36.5	38.5	0.08	0.32	6.32	(0.16)

By plotting the values (Fig. 132*b*), the final temperature can be obtained by extrapolation, and thus trial and error methods can be eliminated for the last arithmetic value.

By assuming various initial flash temperatures, curves similar to those in Fig. 132*b* can be obtained which cover the range of temperature from the critical temperature of the liquid down to the lowest final temperature desired. Then, by cross plotting, values of initial pressure (temperature) vs. final temperature for varying final compositions can be obtained. By again cross plotting, values of final composition vs. initial pressure (temperature) for varying final temperatures can be obtained as desired. The final pressure in the system (assuming no inert gases) will be the vapor pressure of the liquid corresponding to the cooling water temperature.

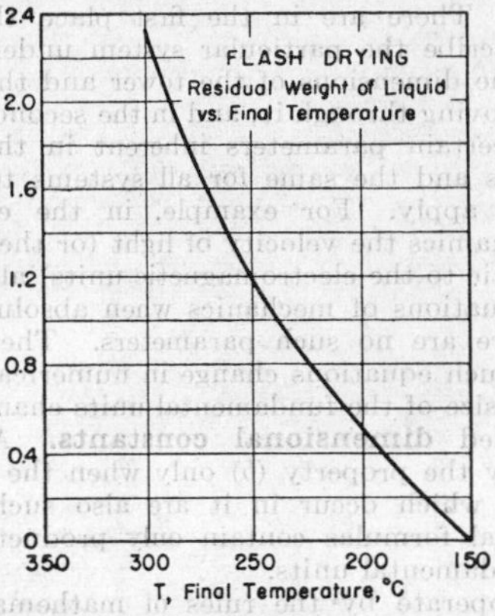

FIG. 132*b*.

DIMENSIONAL ANALYSIS

BY P. W. BRIDGMAN

REFERENCES: *Theory and General:* Bridgman, "Dimensional Analysis," Yale University Press, New Haven, 1922. 2d ed., 1931. Bridgman, Dimensional Analysis, "Encyclopaedia Britannica." Buckingham, *Phys. Rev.*, **4,** October, 1914; *J. Wash. Acad. Sci.*, vol. 14, July 19, 1914; *Phil. Mag.*, **42,** 696 (1921). Rayleigh, *Nature*, **95,** 202, 644 (1915). Robertson, *Gen. Elec. Rev.*, **33,** 207 (1930).

Various Practical Applications: Blake, *Trans. Am. Inst. Chem. Engrs.* **14,** 415 (1921–1922). Buckingham, *Engineering*, vol. 97, Mar. 13, 1914; *Bur. Standards Sci. Paper* 359, 1920; *J. Am. Soc. Nav. Eng.*, vol. 33, May, 1921; *Engineering*, **115,** 225 (Feb. 23, 1923). Greenewalt, *Ind. Eng. Chem.*, **18,** 1291 (1926). Hersey, *J. Wash. Acad. Sci.*, **6** (September and October) 1916; *Am. Soc. Mech. Eng.*, vol. 37, July, 1916; *Bur. Standards Sci. Paper*, 331, Sept. 25, 1919; *Bur. Standards, Aeronautic Instruments Circ.*, §30 and 32, 1918–1919. Rayleigh, *Phil. Mag.*, **34,** 59 (1892); **8,** 66 (1904).

Dimensional analysis is primarily a method by which partial information may be obtained about the relations which must hold between the variables that characterize definite physical systems. The advantage of the method is that it may be applied when only partial knowledge of the system is available, and the partial information thus obtained may often be of great practical value in limiting the number of special experiments or measurements necessary to obtain complete information. In addition to this most important use of dimensional analysis, there are other uses, as in changing units or in checking the correctness of equations, which will also be discussed.

In general, our knowledge of a physical system or situation is sufficient to permit an application of dimensional methods only when we understand it enough to see what is the nature of the steps that we would have to make in order to obtain a complete explanation of its behavior.

The reason that such partial information is possible is that the measured quantities and the parameters which enter the equations as well as the equations themselves have certain simple and very general properties.

Thus, in the problem of the absorption of water vapor by sulfuric acid in an absorption tower treated by Greenewalt (see reference above), we are convinced that if we could write out in detail the complicated hydrodynamic equations that govern the motion of the gas and the diffusion in the liquid phase we could obtain a complete description of the behavior of the system, from which we could isolate any particular piece of information that we need, such, for example, as the amount of water vapor absorbed in terms of the velocity of the gas passing through the tower. The amount absorbed may obviously be a most complicated function of the velocity and of *all the other variables which enter the equations by which the motion is determined.* The determination of the exact form of the relation between velocity and the other variables is hopelessly complicated, but dimensional analysis can tell us certain necessary connections that must be satisfied by any such relation. This is possible because: (*a*) Any system of measurement in scientific use satisfies the requirement that the number which expresses the ratio of the measures of any two concrete things (as, for example, the ratio of the masses of two objects, or the viscosities of two liquids) shall be independent of the size of the fundamental units at the basis of the system of measurement. (*b*) All the systems of equations in scientific use are written in such a form as to be independent of the size of the particular system of units employed, as, for example, the fundamental equation of mechanics: Force = mass × acceleration.

The property (*a*) places a most important restriction on the possible dimensional formulas of measured quantities. A dimensional formula expresses the way in which the fundamental units enter into the operations by which the property in question is measured. Thus, to obtain the number which represents the velocity of a given object, we divide the number which measures the distance it has passed over by the time required to pass over that distance, *i.e.*, velocity has the dimensions L/T. Or, force is defined as mass times acceleration and has the dimensions $M \times (LT^{-1}/T) = MLT^{-2}$. Or, pressure is defined as force per unit area and has dimensions $ML^{-1}T^{-2}$. Now the property (*a*) imposes the restriction that the *dimensional formulas of all measured quantities must have the form of products of powers of the kinds of unit chosen as fundamental.*

The equations which govern the behavior of the system, as, for example, the equations of hydrodynamics in the absorption-tower problem, contain two kinds of variable quantity. There are in the first place the variables which describe the particular system under discussion, such as the dimensions of the tower and the density of the gas moving through it; and in the second place there may be certain parameters inherent in the equations themselves and the same for all systems to which the equations apply. For example, in the equations of electrodynamics the velocity of light (or the ratio of the electrostatic to the electromagnetic units) always enters. In the equations of mechanics when absolute units are used, there are no such parameters. The parameters entering such equations change in numerical magnitude when the size of the fundamental units changes, and are often called **dimensional constants.** An equation can satisfy the property (*b*) only when the dimensional constants which occur in it are also such that their dimensional formulas contain only products of powers of the fundamental units.

If we operate by the rules of mathematics on the equations which completely determine the behavior of the system and which have the property (*b*) in such a way as to display any relations in which we are especially interested which are implied in the equations, then the results of such mathematical manipulation must also have the property (*b*), for the rules of mathematical manipulation imply no knowledge of the size of the units and are valid in all systems.

Hence, any relation of this sort between the physical variables and the dimensional constants of the fundamental equations has the property that it is independent of the size of the fundamental units. This statement, together with the fact that physical variables and dimensional constants are expressible as products of powers of the units, may be shown mathematically to lead to the so-called Π **theorem,** which states that any relation satisfying this requirement can be thrown into such a form that it involves only all the independent dimensionless products of **all** the physical variables and dimensional constants. The Π theorem contains the gist of all the information that can be obtained, and the problem of making a dimensional analysis of any situation is merely the problem of applying the Π theorem.

The applications of the Π theorem will presently be illustrated by a number of detailed examples. Two steps are obviously necessary in applying the theorem after the situation has been sufficiently analyzed from a physical point of view to show the nature of the governing equations: The dimensional formulas of the various quantities must be written down, and then all the dimensionless combinations of these must be found. In making the second step the following theorem is useful: In general, the number of dimensionless products is the difference between the number of arguments (physical variables plus dimensional constants) and the number of fundamental units. In special cases, there may be fewer or more products, but these can usually be recognized by inspection.

In taking the first step, *i.e.*, in writing out the dimensional formulas of the variables, it is sufficient merely to know the definitions of the various quantities in terms of the fundamental units. It is particularly to be emphasized that there is nothing absolute about the dimensions of any quantity, that they are entirely determined by the choice of the fundamental units, and that these may be chosen as convenient to suit the requirements of any particular problem, not even the number of kinds of units being determinate. In dealing with ordinary mechanical problems it is usually most convenient to use absolute units in which force is defined as mass times acceleration, and to express all quantities in terms of mass, length, and time as fundamental. The following list of the dimensions of ordinary mechanical quantities will be convenient.

Quantity	Dimensions	Quantity	Dimensions
Area	L^2	Moments of inertia	ML^2
Volume	L^3	Force	MLT^{-2}
Angle	O	Torque	ML^2T^{-2}
Frequency	T^{-1}	Work, energy	ML^2T^{-2}
Velocity	LT^{-1}	Power	ML^2T^{-3}
Angular velocity	T^{-1}	Pressure	$ML^{-1}T^{-2}$
Acceleration	LT^{-2}	Strain	O
Angular acceleration	T^{-2}	Compressibility	$M^{-1}LT^2$
Density	ML^{-3}	Elastic modulus	$ML^{-1}T^{-2}$
Momentum	MLT^{-1}	Viscosity	$ML^{-1}T^{-1}$
Moments of momentum	ML^2T^{-1}	Capillary constant	MT^{-2}

The dimensional formulas of simple quantities can be put to important use in finding how the numerical measure of any concrete physical thing changes when the size of the fundamental units is changed. Thus suppose that it is required to change a velocity of 88 ft. per second to miles per hour. In general, to get the velocity, we measure a distance, and the time required to pass over the distance, and divide one by the other. To get velocity in feet per second, we measure the distance in feet and the time in seconds, and divide the one number so obtained by the other. To get the velocity of the same object in miles per hour, we measure the **same** distance in miles and the **same** time in hours and divide the two numbers. Let us choose the distance arbitrarily as 88 ft.; the time will then be 1 sec. The operations just described can now be put in symbolic form:

$$\text{Velocity of given object} = \frac{88 \text{ ft.}}{1 \text{ sec.}} = \frac{88/5280 \text{ miles}}{1/3600 \text{ hr.}} = 60 \frac{\text{miles}}{\text{hr.}}$$

i.e., in the dimensional formula we replace the dimensional symbols by the names of the concrete units, and to change from one system of units to another express the old units numerically in terms of the new. This procedure is evidently general. For example:

$$g \text{ (acceleration of gravity)} = 32.17 \frac{\text{ft.}}{\text{sec.}^2} = 32.17 \frac{30.48 \text{ cm.}}{\text{sec.}^2} = 980.7 \frac{\text{cm.}}{\text{sec.}^2} \quad (1)$$

$$6.2 \frac{\text{g.}}{\text{cm.}^3} = 6.2 \frac{1/453.9 \text{ lb.}}{(1/30.48 \text{ ft.})^3} = 386.8 \frac{\text{lb.}}{\text{ft.}^3} \quad (2)$$

Another important use of simple dimensional formulas is in checking the correctness of various equations. Most equations that have a rational basis and can be deduced from general consideration are true in all systems of units. Such equations are called "complete" equations. It may be proved that any complete equation can be put into such form that it is dimensionally homogeneous, *i.e.*, all the terms have the same dimensions. If, on checking an equation which from its method of derivation should be complete, it is found that all the terms have not the same dimensions, then there has obviously been some error. For example, if the equation for the distance traveled by a body falling freely from rest should appear in the form $s = \frac{1}{2}gt$, it is obvious that there is an error, for the dimensions of the left-hand side are L, and those of the right, $LT^{-2}T = L/T$.

This principle is often useful in checking the equations of thermodynamics. For this purpose it is convenient to choose for the fundamental units for the thermodynamic equations, p, v, and τ (absolute temperature); time not entering the equations. In these units, energy, for example, is of the dimensions pv, and specific heat of the dimensions $pv\tau^{-1}$. As an example, suppose that the formula for the rise of temperature accompanying adiabatic compression appeared in the form

$$\left(\frac{\delta\tau}{\delta p}\right)_s = \frac{1}{C_p}\left(\frac{\delta v}{\delta\tau}\right)_p$$

The dimensions of the left-hand side are τp^{-1}; and those of the right,

$$(pv\tau^{-1})^{-1}v\tau^{-1} = p^{-1}$$

and there is therefore an error. The correct expression is

$$\left(\frac{\delta\tau}{\delta p}\right)_s = \frac{\tau}{C_p}\left(\frac{\delta v}{\delta\tau}\right)_p$$

Another consequence of the principle of dimensional homogeneity is that it is always possible to write a complete equation in such a form that the arguments of any transcendental functions are dimensionless. For example, logarithmic terms can be so combined with other logarithmic terms that either the logarithmic function disappears from the result, or the argument of the logarithm is dimensionless.

Often equations in engineering, which are obtained by purely empirical methods with measurements in a particular system of units, are written in a form not dimensionally homogeneous and not complete. Such equations can always be converted into complete equations and therefore be made dimensionally homogeneous by introducing enough dimensional constants of the proper numerical magnitude. Thus suppose the pressure head required to drive a fluid through a certain tower is $p = 0.001\,\frac{M^{1.8}\mu^{0.2}}{d}$, where p is the pressure head in inches of water per foot of height; M is the mass velocity in thousands of pounds per hour per square foot of sectional area; μ is the viscosity in centipoises; and d is the density in pounds per cubic foot. (Example of Blake; see reference, page 93.) This equation may now be put in complete form by writing

$$\frac{p}{p_0} = 0.001\,\frac{(M/M_0)^{1.8}(\mu/\mu_0)^{0.2}}{d/d_0}$$

where the dimensions of the dimensional constants p_0, M_0, μ_0, and d_0 are the same as those of p, M, μ, and d; and their numerical magnitudes are so chosen as to be unity in the systems of units specified for the four quantities, respectively.

We now consider a few illustrative examples of the more general use of the method to obtain permissible relations between physical variables.

1. The Problem of a Body Falling Freely from Rest in a Uniform Gravitational Field. This is obviously a problem in pure mechanics, and a complete solution can be obtained by integrating the equations of motion after substituting into the equations the characteristics of the particular situation. The equations of motion themselves contain no dimensional constants. The only property of the body which is to be considered in the equations of motion is its mass m; the gravitational field is completely characterized by the acceleration g, produced by it. We are to find a relation involving the two quantities m and g; the distance fallen, s; and the time of fall t. We must first find all the dimensionless products that can be made from these four quantities. The first step is to write their dimensional formulas. We may obviously use the ordinary M, L, T system of units.

Name of quantity	Symbol	Dimensions
Mass	m	M
Acceleration of gravity	g	LT^{-2}
Time of fall	t	T
Distance of fall	s	L

We have four quantities and three kinds of unit, so that, according to the general rule, there is one dimensionless product. This is obvious on inspection, or it may be found by a perfectly general procedure as follows: We must so choose the exponents α, β, γ, δ that $m^\alpha g^\beta t^\gamma s^\delta$ shall be dimensionless. Expressing these quantities in terms of their dimensional formulas, this means that $M^\alpha(LT^{-2})^\beta T^\gamma L^\delta$ shall be dimensionless, or the collected exponents of M, L, and T shall all vanish. The condition on the exponent of M is $\alpha = 0$; the condition on the exponent of L is $\beta + \delta = 0$; and the condition on the exponent of T is $-2\beta + \gamma = 0$. The solution of these two equations for β, γ, and δ is indeterminate and may be expressed as $\gamma = 2\beta$, $\delta = -\beta$. The dimensionless product may therefore be written as $m^0 g^\beta t^{2\beta} s^{-\beta}$ or $(gt^2/s)^\beta$. The mass drops from the result, a most important result not known to Aristotle. The Π theorem now states that the most general relation possible between these quantities is an arbitrary function of all the possible dimensionless products put equal to a constant. An arbitrary function of $(gt^2/s)^\beta$ is the same as an arbitrary function of gt^2/s itself, and, if this is to be constant, gt^2/s itself must be constant, or inverting and solving for s, $s = \text{const.}\ gt^2$ which checks with the known result, the value of the constant being given by the detailed work of integration.

2. To Find an Expression for the Pressure Exerted by a Perfect Gas. The molecules are supposed of negligible size, and perfectly elastic. This is evidently a problem in pure mechanics, although exceedingly complicated, and the pressure can be obtained in terms of the momentum of the molecules colliding with the walls. The molecules individually are characterized only by their mass, and the gas itself is characterized by the number of molecules per unit volume (N) and by the average energy of the molecules. Temperature is introduced into the result by means of the relation between average energy and temperature, and this involves the gas constant k. In the final result, the average energy may be replaced by the temperature and gas constant. k is a dimensional constant; there are obviously no other dimensional constants in the equations of motion themselves. We may now make the following schedule:

Name of quantity	Symbol	Dimensions
Pressure	p	$ML^{-1}T^{-2}$
Mass of molecule	m	M
Number of molecules (per unit volume)	N	L^{-3}
Absolute temperature	τ	τ
Gas constant (energy per degree)	k	$ML^2T^{-2}\tau^{-1}$

There are five quantities and four independent units, and hence one dimensionless product. Write this, as $p^\alpha m^\beta N^\gamma \tau^\delta k^\epsilon$, or substituting the dimensional formulas $(ML^{-1}T^{-2})^\alpha M^\beta (L^{-3})^\gamma \tau^\delta (ML^2T^{-2}\tau^{-1})^\epsilon$ is to be dimensionless. This gives

$$\begin{aligned} \alpha + \beta + \epsilon &= 0, \text{ condition on exponent of } M \\ -\alpha - 3\gamma + 2\epsilon &= 0, \text{ condition on exponent of } L \\ -2\alpha - 2\epsilon &= 0, \text{ condition on exponent of } T \\ \delta - \epsilon &= 0, \text{ condition on exponent of } \tau \end{aligned}$$

The solution is: $\beta = 0$, $\epsilon = -\alpha$, $\gamma = -\alpha$, $\delta = -\alpha$, and the dimensionless product is $(pN^{-1}\tau^{-1}k^{-1})^\alpha$. An arbitrary function of this must be a constant, which means that it itself must be a constant, or, solving for p, $p = \text{const.}\ Nk\tau$. Since N is inversely proportional to v, this may also be written, pv = const. τ, retaining only the physical variables p, v, and τ.

It is obvious that the dimensional constant k plays an essential part in this discussion and that, if it had been omitted, incorrect results would have been obtained.

3. Rayleigh's Problem of Heat Transfer. A solid body of given shape is fixed in a stream of liquid which flows past the solid. The stream of liquid is to be regarded as so extensive that it can be treated as infinite. The body is to be maintained at a definite temperature higher than the stream at points remote from the body, and it is required to find the rate of heat transfer from the body to the fluid. This problem discussed rigorously would be excessively complicated. We shall therefore make certain simplifying assumptions about the physical situation. The fluid is supposed to be in a state of steady motion, and the velocity is supposed to be so low that there is no turbulence. The motion is therefore streamline motion and is determined only by the geometry of the situation. The density of the fluid does not enter into the equations of motion as it would if there were turbulence, and the viscosity of the fluid has no effect on the velocity distribution and has therefore no effect on the heat transfer, although, of course, it affects the force required to hold the body in position in the stream. The boundary conditions between liquid and solid we suppose to be very simple; the liquid is at rest at the surface of the solid, and heat passes from the solid to the liquid by conduction. The solid is supposed to be a so much better heat conductor than the liquid that its temperature throughout is constant, so that the amount of heat transferred is not limited in any way by the thermal conductivity of the solid. The amount of heat which is transferred from the solid to the liquid depends immediately on the temperature gradient in the liquid directly in contact with the solid, and this depends on its thermal conductivity, thermal capacity, and the speed with which new fluid is brought up by the motion. We suppose furthermore all temperature differences, etc., to be so small that the variations of thermal conductivity, etc., with temperature are to be disregarded. It is obvious that in this system there is no conversion of thermal into mechanical energy, so that the mechanical equivalent of heat does not enter, and heat may be measured in thermal units. The body may be supposed defined geometrically in terms of a single absolute dimension and the ratios of other characteristic dimensions to this length. As, for example, if the body were an ellipsoid, it would be sufficient to give the absolute length of the longest axis, and the ratio of the two shorter axes to this axis. With all these simplifications we now have the following formulation:

Name of quantity	Symbol	Dimensions
Rate of heat transfer	h	HT^{-1}
Linear dimensions of body	a	L
Velocity of fluid	v	LT^{-1}
Temperature difference	τ	τ
Thermal conductivity of liquid	κ	$HL^{-1}T^{-1}\tau^{-1}$
Heat capacity of liquid per unit volume	c	$HL^{-3}\tau^{-1}$
Shape factors	r_1, r_2, etc.	0

Except for the shape factors, which are of zero dimensions and may be disregarded for the present, there are here six quantities expressed in terms of four kinds of units, and hence we expect two independent dimensionless products. These two independent products may be chosen in many ways, since, in general, a product may be formed from any five of the six quantities. Since we are primarily interested in h, we choose one of the products as containing h, and the second leaving h out, since in this way we will be able to solve our resultant functional equation for h. Take for the product containing h one involving the first five quantities. Then

$$h^\alpha a^\beta v^\gamma \tau^\delta \kappa^\epsilon$$

or

$$(HT^{-1})^\alpha L^\beta (LT^{-1})^\gamma \tau^\delta (HL^{-1}T^{-1}\tau^{-1})^\epsilon$$

must be dimensionless. As always, the system of equations for the exponents is homogeneous, so that one of the exponents may be assigned at pleasure (for any power of a dimensionless product is also dimensionless). We choose α as 1, and have the equations:

$$\begin{aligned} 1 + \epsilon &= 0 \\ \beta + \gamma - \epsilon &= 0 \\ -1 - \gamma - \epsilon &= 0 \\ \delta - \epsilon &= 0 \end{aligned}$$

the solution of which is

$$\epsilon = -1 \qquad \delta = -1 \qquad \gamma = 0 \qquad \beta = -1$$

and the dimensionless product is

$$ha^{-1}\tau^{-1}\kappa^{-1}$$

As the second dimensionless product choose the five quantities without h, or $a^\alpha v^\beta \tau^\gamma \kappa^\delta c^\epsilon$. Again put $\alpha = 1$ and work exactly like the preceding to obtain for the dimensionless product $av\kappa^{-1}c$. The II theorem now states that

$$f(ha^{-1}\tau^{-1}\kappa^{-1}, av\kappa^{-1}c)$$

is equal to a constant, where f is arbitrary. Solve this equation for the first argument and multiply up by $a\tau\kappa$, and an equivalent statement is

$$h = a\tau\kappa\varphi\left(\frac{avc}{\kappa}\right)$$

where φ is again arbitrary. The function φ involves the shape factors, which are also dimensionless products; we might have written φ in the expanded form, $\varphi'(avc/\kappa, r_1, r_2, \ldots)$, but, as long as the r's are maintained constant, φ' may be treated as a function of a single argument.

This is the first example in which an arbitrary function has remained in the final result. Physical systems that are so complicated that undetermined functions remain in the result of dimensional analysis can often be treated by making experiments on models, and it is here that the principal technical use of dimensional analysis lies. For example, if in the problem above a model one-tenth the actual dimensions is made, then the function φ will have the same value for the model as for the full-size example when the fluid in the model is moving with ten times the velocity of the fluid in the actual example, and hence h for the full-size example can be obtained from h for the model at another velocity. The result above shows in another way the advantages of dimensional analysis. Before applying the analysis we had to be prepared to investigate physically the effect of varying all the parameters, which involved studying different liquids with different thermal conductivities and different specific heats. After making the analysis, we find that we can get all the information by varying a single parameter, the velocity, of a single liquid.

It is particularly to be noticed that in applying the analysis we must include all the physically pertinent factors. Thus, if we had omitted to include the heat capacity of the liquid, which might easily be done, there would have been only one dimensionless product, and our result would have been h = const. $a\tau\kappa$, which is absurd because it is independent of the velocity. This is probably the easiest way in which to go wrong in applying dimensional analysis.

If the physical conditions change, the whole analysis

must be modified. Thus, if the motion is rapid enough to produce turbulence, the density of the liquid must be included, there are three dimensionless products, and the whole problem becomes much more complicated.

4. Consider an Application to Theoretical Physics in Which a Negative Result Is Obtained. Let us suppose that we suspect that it would be possible to develop a theoretical explanation of thermal conductivity in which there are no features of the mechanism which do not equally manifest themselves in determining the thermodynamic behavior. If there is a relation of this kind, then the Π theorem applies, since the relation must be dimensionally homogeneous. Since the thermodynamic behavior under ordinary conditions is fixed by the temperature, the compressibility, the thermal expansion, and the specific heat, we have the following formulation.

Name of quanti y	Symbol	Dimensions
Thermal conductivity	κ	$MLT^{-3}\tau^{-1}$
Compressibi ity per unit mass	a	$M^{-2}L^{1}T^{2}$
Thermal expansion per unit mass	b	$M^{-1}L^{3}\tau^{-1}$
Specific heat	c	$L^{2}T^{-2}\tau^{-1}$
Absolute temperature	τ	τ

Here we have five quantities and four units, and hence we expect one dimensionless product. Choose the exponent of κ as unity, and write the dimensionless product as

$$\kappa a^{\alpha} b^{\beta} c^{\gamma} \tau^{\delta}$$

or

$$MLT^{-3}\tau^{-1}(M^{-2}L^{4}T^{2})^{\alpha}(M^{-1}L^{3}\tau^{-1})^{\beta}(L^{2}T^{-2}\tau^{-1})^{\gamma}\tau^{\delta}$$

is to be dimensionless. This gives the conditions:

$$\begin{aligned} 1 - 2\alpha - \beta &= 0, \text{ on } M \\ 1 + 4\alpha + 3\beta + 2\gamma &= 0, \text{ on } L \\ -3 + 2\alpha - 2\gamma &= 0, \text{ on } T \\ -1 - \beta - \gamma + \delta &= 0, \text{ on } \tau \end{aligned}$$

Now these equations have no solution, as may be seen at once on eliminating γ between the second and third equations, when a condition will be found on the combination of terms $2\alpha + \beta$ inconsistent with that contained in the first equation. Another way of proving this is to notice that the determinant of the coefficients of the four quantities α, β, γ, δ vanishes, and hence there is no solution.

The conclusion to be drawn in this particular example is that there can be no relation of the kind suspected.

The above example shows that there are exceptions, and in fact the exceptions to the general rules laid down above are not infrequent. Sometimes there are fewer dimensionless products than the rule indicates, and sometimes there are more. Such exceptional conditions will always reveal themselves on working out the detailed solutions of the equations. Of course, a case where there are more dimensionless products than the rule predicts is more likely to escape notice than one where there are fewer; such cases can usually be discovered by inspection.

GRAPHS AND GRAPH PAPER

BY RAYMOND W. DULL, T. R. RUNNING, AND JOHN H. PERRY

METHODS OF REPRESENTING TECHNICAL DATA

In the following paragraphs there are given a number of typical curves, the points of which are represented by x and y coordinates on the bottom and left-hand scales, respectively; and other types of curves from the same data so plotted as to give straight or nearly straight lines, the points on which are represented by x and y coordinates on the top and right-hand scales, respectively. The chief advantages of using data so plotted as to give straight or nearly straight lines are (1) to increase the accuracy of drawing a representative curve through the points representing the data (a straight line may be drawn more accurately and more easily than a more complicated curve as a parabola, hyperbola, etc.) and (2) to increase the accuracy of extrapolation and interpolation of data.

Parabolic or Hyperbolic Curve. $Y = AX^B$. This type of curve approximates the form of a large number of sets of empirical data. If the constant B of the above equation type is positive, the curve is a parabola; if B is negative, the curve is an hyperbola. If the given data (*e.g.*, the pressure and volume of a saturated gas; or the solubility of a salt in a given solvent as a function of the temperature) are represented by an equation of this type, a plot of the data in a graph of log x vs. log y will approximate a straight line.

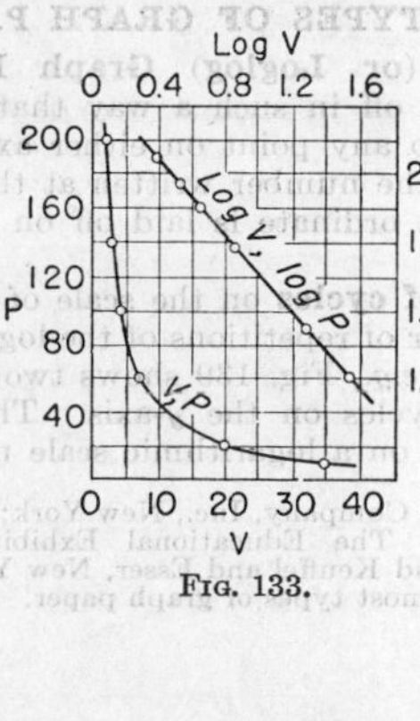

FIG. 133.

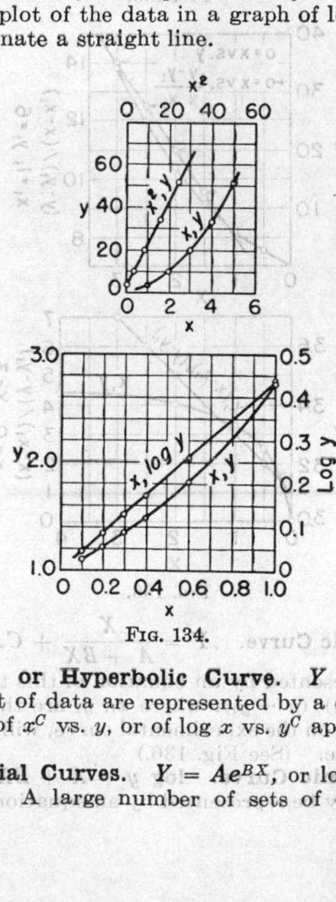

FIG. 134.

Parabolic or Hyperbolic Curve. $Y = A + BX^C$. If a given set of data are represented by a curve of this type, a plot of x^C vs. y, or of log x vs. y^C approximates a straight line.

Exponential Curves. $Y = Ae^{BX}$, or $\log y = \log A + Bx \log e$. A large number of sets of experimental data approximate a simple exponential curve of the above equation. If the given data are represented by a curve of this type, a plot of the data in a graph of x vs. $\log y$ will approximate a straight line. (See Fig. 134.)

Hyperbolic Curve. $Y = \dfrac{X}{A + BX}$. If a given set of data can be represented by an equation of this form (or its equal $x/y = A + BX$), a plot of x vs. x/y, or of $1/x$ vs. $1/y$, will approximate a straight line. (See Fig. 135.)

Exponential Curve. $Y = Ae^{BX} + C$. If the data can be represented by a curve of the form of $y = Ae^{BX} + C$, a plot x vs. $\log (y^{-c})$ will approximate a straight line. (See Fig. 135.)

Parabolic Curve. $Y = A + BX + CX^2$. If the data can be represented by a curve of the type $y = A + BX + CX^2$, a plot of x vs. $(y - y_1)/(x - x_1)$, where (x_1, y_1) is any point on the experimental curve, will give an approximate straight line. (See Fig. 136.)

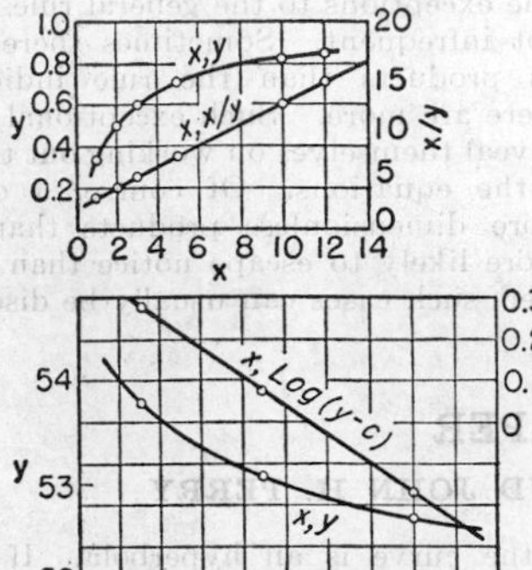

FIG. 135.

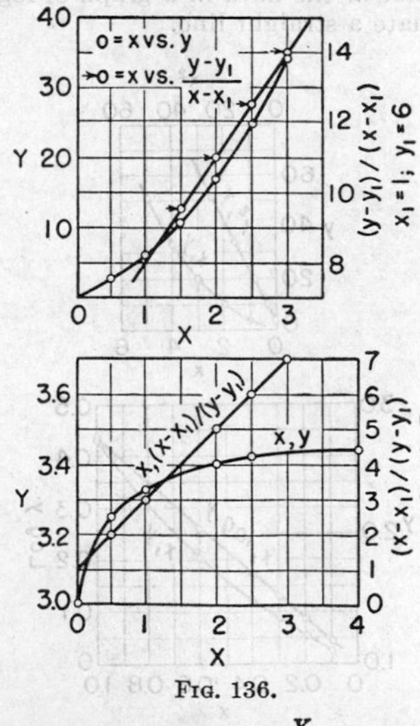

FIG. 136.

Hyperbolic Curve. $Y = \dfrac{X}{A + BX} + C$. If the data may be represented by an equation of this type, a plot of x vs. $(x - x_1)/(y - y_1)$, where x_1, y_1 are the coordinates of any point on the experimental curve, will approximate a straight line. (See Fig. 136.)

Logarithmic Curve. $\log y = A + BX + CX^2$. If the data may be represented by an equation of this type a plot of x vs. $(\log y - \log y_1)/(x - x_1)$ will approximate a straight line. (See Fig. 137.)

Vapor-pressure Temperature Curves. Vapor-pressure temperature data are conveniently represented by plotting the logarithm of the vapor pressure vs. the reciprocal of the absolute temperature to produce a straight or nearly straight line. Greater accuracy of extrapolation is obtained with straight lines than with other curves. The logarithmic reciprocal temperature graph paper referred to on page 99 is recommended for this type of curve. (See Fig. 137.)

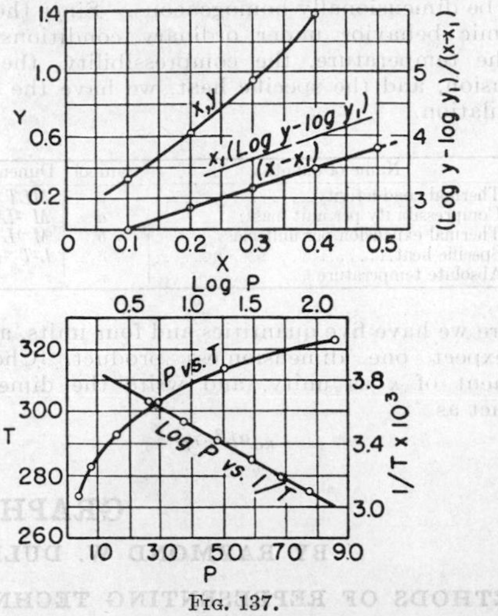

FIG. 137.

A method of compressing a large number of data over a large range while retaining great accuracy is that shown in Fig. 138. This figure shows millivolts

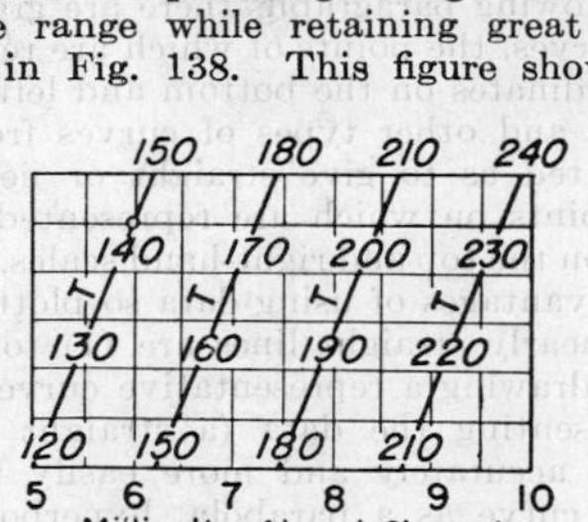

FIG. 138.

plotted against temperature (°C.). In the figure, the circled datum corresponds to the coordinates, 6 millivolts and 145°C. (for an alumel-chromel thermocouple).

COMMON TYPES OF GRAPH PAPERS*

Logarithmic (or Loglog) Graph Paper. Each coordinate is laid off in such a way that the distance from the origin to any point on either axis is equal to the logarithm of the number written at that point. In other words, each ordinate is laid off on a logarithmic scale.

The **number of cycles** on the scale of either coordinate is the number of repetitions of the logarithmic scale on that ordinate, *e.g.*, Fig. 139 shows two cycles on the x-axis and two cycles on the y-axis. The position of the decimal point on a logarithmic scale may be placed

* The Codex Book Company, Inc., New York; C. Schleicher & Schull, New York; The Educational Exhibition Company, Providence, R. I.; and Keuffel and Esser, New York, are among the firms supplying most types of graph paper.

at the discretion of the user; *i.e.*, in the x-axis of the figure, instead of values 1, 2, 3, 4, . . . , 9, 10, these may be regarded as 0.1, 0.2, 0.3, . . . , 0.9, 1.0.

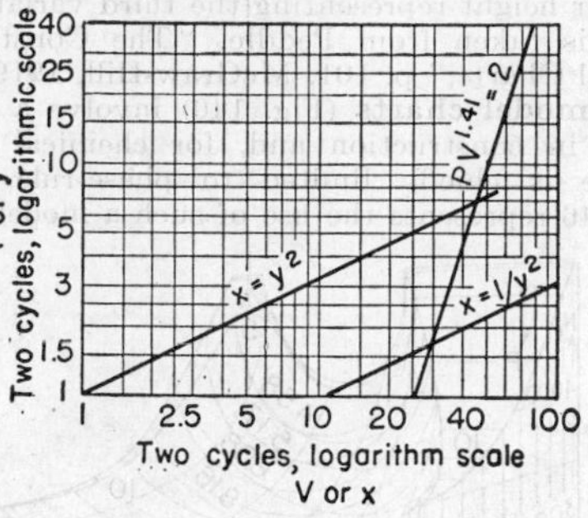

FIG. 139.

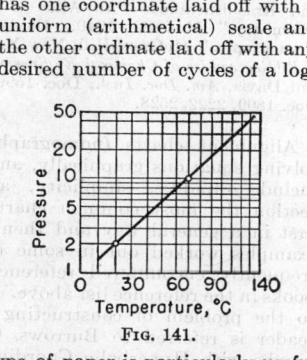

This type of paper is useful in plotting power functions of the general type $x = Ay^n$; the type of curve obtained from such a power function is shown for various values of n in the above formula. The method of plotting is easily seen by assuming $A = 1$ and by substituting various values of y for a constant value of, say, $n = 2$, whereby the curve $x = y^2$, in the above figure, is obtained. If A is not equal to 1, but is, say, 11, the equation is rewritten as $\log x = \log 11 + n \log y$; and, if $n = 2$, the curve as shown is produced, wherein A is the intercept on the x-axis and n is the slope of the curve as usual. Such a type of power curve is illustrated by Fig. 139. (1) The equation for the adiabatic expansion of perfect gases, $PV^{1.41}$ = const. (2) Fluid velocity = f (orifice manometer reading). Logarithmic paper can be secured from dealers in draftsmen's supplies or can be constructed easily by copying the logarithmic scale from the C or D scales of a slide rule. Inasmuch as this method of construction gives, for a definite length of slide rule, a scale of only the one length, a general method of constructing a scale of any desired length is: Make the desired length (x) of the scale equal to log 10 or 1.00; then the distance of point 9 from the origin (1) is equal to: the logarithm of 9 × logarithm of 10 or 0.9542 × 1.0 = 0.9542 × length x. In this way, logarithmic paper of any desired number of cycles of a definite length may be constructed.

Semilogarithmic (arith-log) graph paper shows percentage increases, inasmuch as y varies at a constant percentage rate. It has one coordinate laid off with a uniform (arithmetical) scale and the other ordinate laid off with any desired number of cycles of a log-

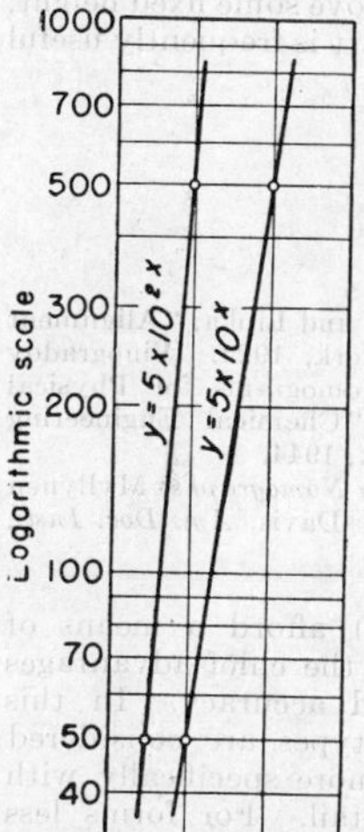

FIG. 140.

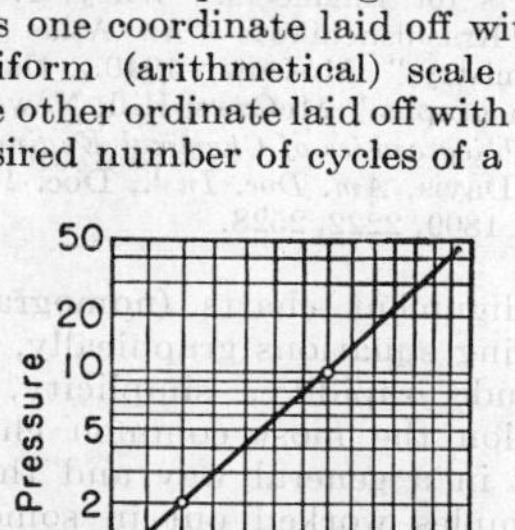

FIG. 141.

arithmic scale. This type of paper is particularly useful for plotting the graphs of an equation of the type, $y = n \times 10^{mx}$, where m and n are constants (see Fig. 140).

Logarithmic-reciprocal graph paper is particularly useful for plotting vapor pressures against temperatures and consists (Fig. 141) of one coordinate laid off in any desired number of cycles of a logarithmic scale, and the other coordinate laid off in a reciprocal scale. In Fig. 141, the y-axis is laid off in two cycles of a logarithmic scale and the x-axis is laid off in such a way that the divisions represent the reciprocals of the variable printed on the x-axis. Obviously the y-axis could be laid off to represent $\frac{1}{x_1^2}$, $\frac{1}{\log x}$, or for any other desired reciprocal. Graph paper of this general type may be secured through The Educational Exhibition Company, of Providence, R. I., and is laid off with the logarithmic scale on the y-axis and with the x-axis laid off as in Fig. 141, such that the points marked 20°C., 30°C., etc., represent the reciprocals of the absolute temperature corresponding to these temperatures, or $1/(273 + 20)$, $1/(273 + 30)$, etc.

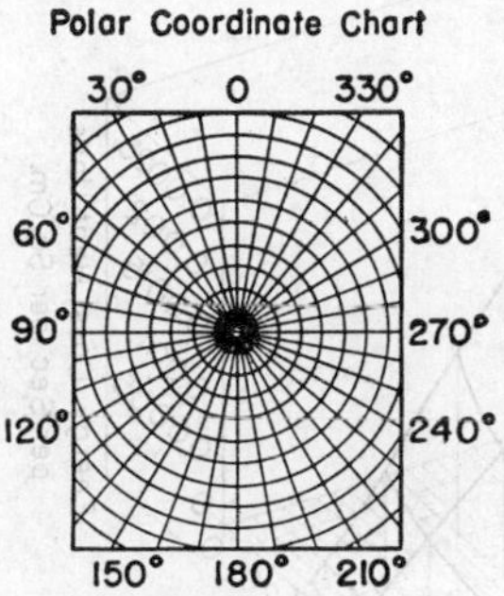

FIG. 142.

FIG. 143.

Polar coordinate graph paper consists of a series of concentric circles (1/10″ or other convenient unit apart) divided into 360° and numbered at 10° (or smaller) intervals in both a clockwise and counterclockwise manner. This type of chart is seldom required for chemical engineering work, although it is very useful as a paper protractor and for hourly (or fractions thereof) variations. Its great field of usefulness is in civil, mechanical, and electrical engineering (Fig. 142).

Triangular (trilinear) graph paper consists of an equilateral triangle, with scale lines at small intervals parallel to each side. From geometry it is known that in any equilateral triangle the sum of the perpendicular lines from any given point in that triangle to its sides is equal to the triangle altitude. In this type of graph the altitude usually represents 100 per cent (or unity); hence the sum of the perpendiculars from the three sides to a point in the triangle must always be equal to 100 per cent (or unity). Figure 143 indicates a typical triangular chart. This type of chart is extremely useful for plotting three-element relations, such as the properties of ternary alloys or of other ternary systems. Other variables, such as temperature and pressure, are shown on this type of diagram as contour lines.

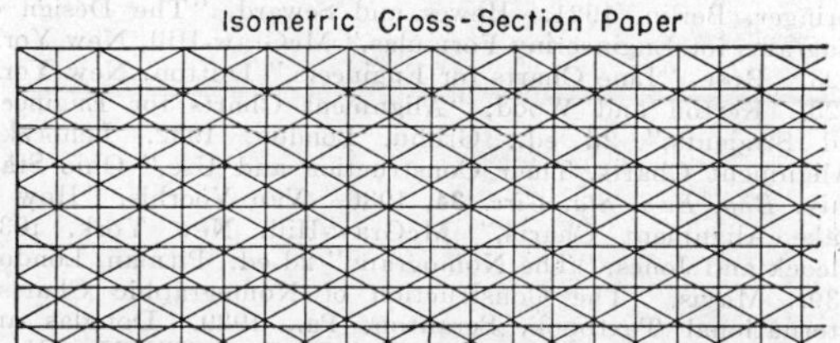

FIG. 144.

Miscellaneous Types of Graph Papers. Circular Percentage (Pie) Chart. Charts are those in which the circle circumference is divided into 100 equal parts, and is used for showing, as sectors, proportional parts.

Time charts are available in many different forms and show some unit of time as year, month, week, etc., divided into equal parts. **Isometric charts** (Fig. 144) combine the principles of perspective and mechanical drawing and eliminate the necessity for T squares and 60° to 30° triangles. An **erasable-line arithmetic graph** paper is also available which is useful for graphs or for design work. Upon completion of the design or graph, all uninked lines may be removed by a damp cloth leaving only those lines which have been inked on a white background. **Profile paper** is also of great utility and may be secured from all dealers supplying the ordinary arithmetic graph paper. It is drawn to a definite scale, such as 1 in. = 1 ft., and the like. For a treatise on **probability theory and charts**, the reader is referred to Fry, "Probability and Its Engineering Uses," Van Nostrand, New York, 1928.

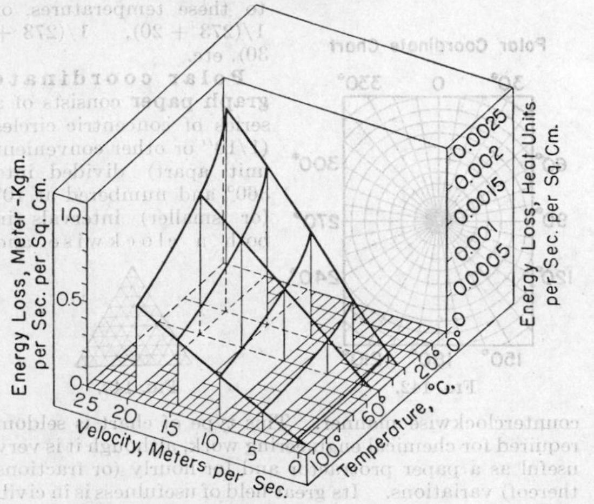

FIG. 145.

Axonometric charts (Fig. 145) are three-dimensional diagrams represented on two-dimensional space. The three axes and their planes are first drawn and suitable graduations of the planes are drawn in. From the graduations of the ground plane perpendiculars are drawn which give a cross-patched surface on which points may be obtained. At points on this ground plane, perpendiculars are drawn to this plane at suitable intervals, their height representing the third variable. (Figure 145 is taken from Peddle, "The Construction of Graphical Charts," p. 101, McGraw-Hill, 1919.)

Solid model charts (Fig. 146) involve a great deal of labor in construction and, for chemical engineers, their use is chiefly limited to phase-rule diagrams. Figure 146 represents the use of such a model involving

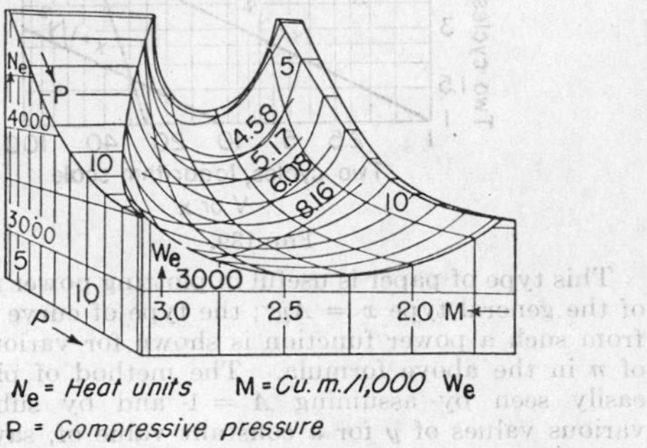

FIG. 146.

a relation between thermal units per hour per brake horsepower, compression pressure, and volume of gas mixture for a gas engine. (For details, see *Z. Ver. deut. Ing.*, Sept. 14, 1907; or Peddle, *op. cit.*, p. 105.) One suitable method of construction is by the use of strings stretched on suitable frames. To make a plaster-of-Paris model, a sheet of cross-section paper is placed on a board, and points corresponding to two of the three variables are laid off thereon. Vertical wires are then inserted at heights equivalent to the third variable. After building a box around the form, wet plaster of Paris is poured in until all the wires are just covered.

Contour-line charts are illustrated by a typical weather chart in which the contour lines show the barometric pressure, or by topographical charts wherein the contour lines show the elevation above some fixed height, usually sea level. This type of chart is frequently useful to the chemical engineer.

ALIGNMENT CHARTS

BY D. S. DAVIS AND R. P. GENEREAUX

REFERENCES: *Textbooks on Nomography:* Brodetsky, "A First Course in Nomography," G. Bell, London, 1920. Lipka, "Graphical and Mechanical Computation," Wiley, New York, 1921. d'Ocagne, "Traité de Nomographie," 2d ed., Gauthiers-Villars, Paris, 1921. Schwerdt, "Lehrbuch der Nomographie," Springer, Berlin, 1921. Hewes and Seward, "The Design of Diagrams for Engineering Formulas," McGraw-Hill, New York, 1923. Rose, "Line Charts for Engineers," Dutton, New York, 1923. Kerton and Wood, "Alignment Charts for Engineers and Students," 2d ed., Griffin, London, 1932. Lehoczky, "Alignment Charts, Their Construction and Use," Ohio State Univ. *Eng. Exp. Sta. Circ.* **34,** 1936. Van Voorhis, "How to Make Alignment Charts," McGraw-Hill, New York, 1937. Allcock and Jones, "The Nomogram," 2d ed., Pitman, London, 1939. Mavis, "The Construction of Nomographic Charts," International Textbook, Scranton, Pa., 1939. Douglas and Adams, "Elements of Nomography," McGraw-Hill, New York, 1947. Levens, "Nomography," Wiley, New York, 1948. Davis, "Empirical Equations and Nomography," McGraw-Hill, New York, 1943. Kraitchik, "Alignment Charts, Their Construction and Use," Van Nostrand, New York, 1945. [A novel method of design is given by Burrows, "Construction of Nomographs with Hyperbolic Coordinates," *Ind. Eng. Chem.*, **38,** 472 (1946).]

Collections of Nomographs: Cummings and Lipka, "Alignment Charts for Engineers," Wiley, New York, 1924. Vinogradov and Krasiilshchikov, "An Atlas of Nomograms for Physical Chemistry," Moscow, 1940. Davis, "Chemical Engineering Nomographs," McGraw-Hill, New York, 1944.

Bibliographies of Chemical Engineering Nomographs: Myllynen and Davis, *Am. Doc. Inst.*, Doc. 1599. Davis, *Am. Doc. Inst.*, Doc. 1809, 2222, 2528.

Alignment charts (nomography) afford a means of solving equations graphically, and the chief advantages include rapidity, simplicity, and accuracy. In this section the most common chart types are considered first in a general way and then more specifically with examples worked out in some detail. For forms less frequently encountered reference should be made to books in the reference list above. For a unique approach to the problem of constructing alignment charts, the reader is referred to Burrows, Construction of Nomographs with Hyperbolic Coordinates, *Ind. Eng. Chem.*, **38,** 472 (1946).

For use in laying off scales, one should first construct

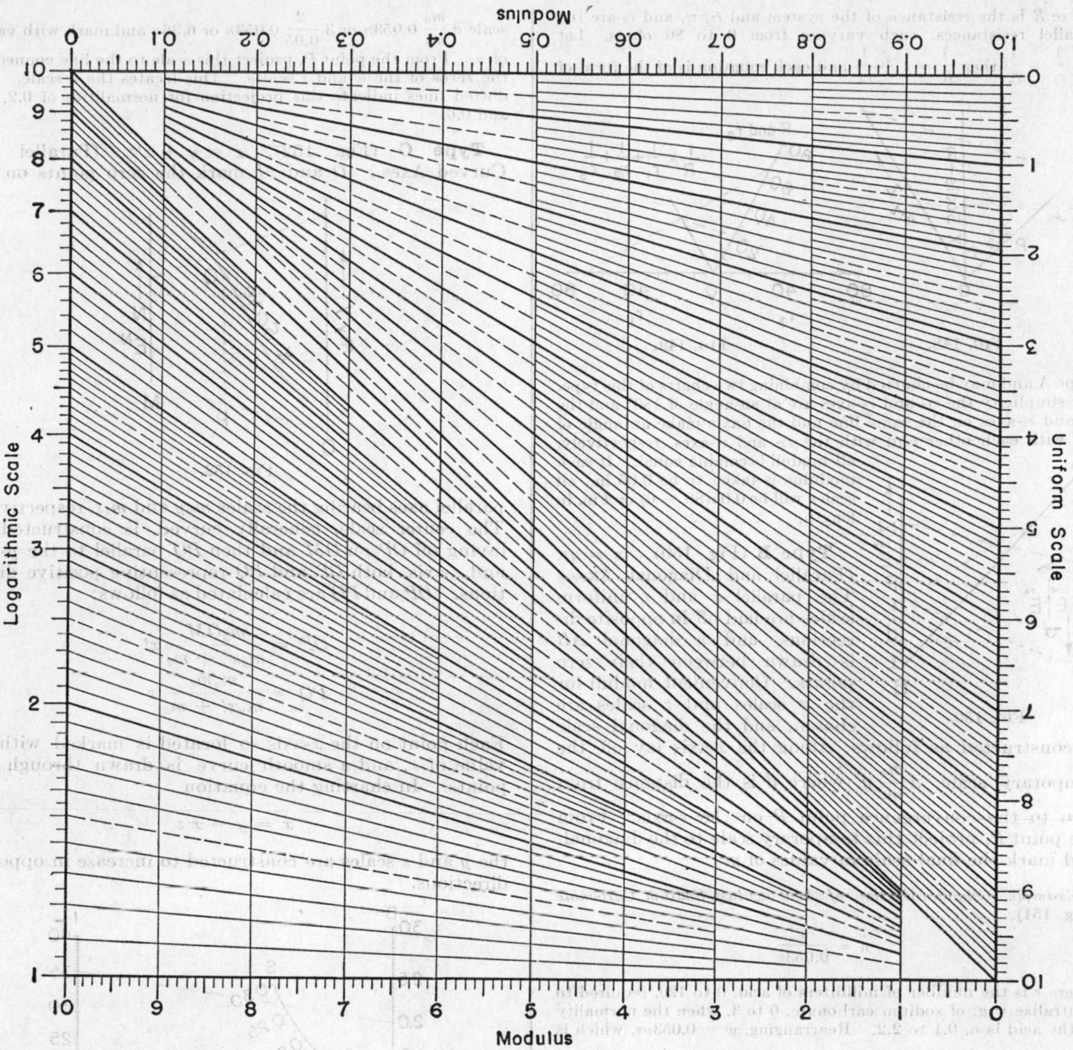

FIG. 147.

a modulus chart (Fig. 147), modulus being defined as the length of line corresponding to one unit of the scale. Thus for a uniform scale, if the distance between the 20 and 60 points is 4 in., the modulus is $\frac{4}{60 - 20}$ or 0.1 in.; and for a logarithmic scale with similar data, $\frac{4}{\log 60/20}$, or $\frac{4}{\log 3}$, or 8.38 in. To construct the logarithmic half of the modulus chart, use an engineer's decimal scale and a table of logarithms, making the cycle from 1 to 10 a distance of 10 in. along the left edge of the chart at modulus 10. Lay off the horizontal modulus scale so that 0 is at a distance of 10 in. from modulus 10, and draw the vertical modulus line. A pencil of lines from the modulus 10-line to the modulus 0 point will cut the vertical modulus lines in scales of moduli between 0 and 10. To lay off a logarithmic scale with a modulus of 8, fold the modulus chart along the 8-line, place along the line to be graduated, and mark off the desired points. The uniform half of the modulus chart is constructed and used in a similar manner.

In this section x, x', y, z, and w indicate variables or their functions and m_x, m_x', m_y, and m_z refer to the moduli of the various scales.

Chart Types—Type A (Fig. 148). $\frac{1}{x} = \frac{1}{y} + \frac{1}{z}$. (Three Convergent Axes.) The angles α and δ between the center and outer axes are such that $\sin \alpha/\sin \delta = m_y/m_z$. To construct, locate P and R so that $OP/PR = m_y/m_z$. On the y-axis lay off the temporary scale $m_y x$, and project to the center or x-axis by lines parallel to the z-axis. For the special case where $\alpha = \delta = 60°$, the moduli for all three scales are equal, and the above construction is unnecessary.

Example. Resistance of a System of Parallel Resistances (Fig. 149).

$$\frac{1}{R} = \frac{1}{r_1} + \frac{1}{r_2} + \frac{1}{r_3}$$

where R is the resistance of the system and r_1, r_2, and r_3 are the parallel resistances, each varying from 0 to 80 ohms. Let $t = \frac{1}{r_1} + \frac{1}{r_2}$, then $\frac{1}{R} = t + \frac{1}{r_3}$, and each equation is of the form of

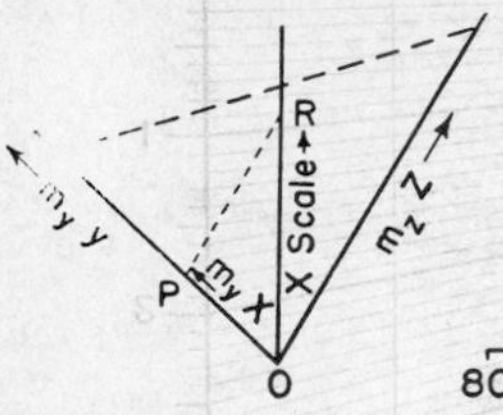

FIG. 148.

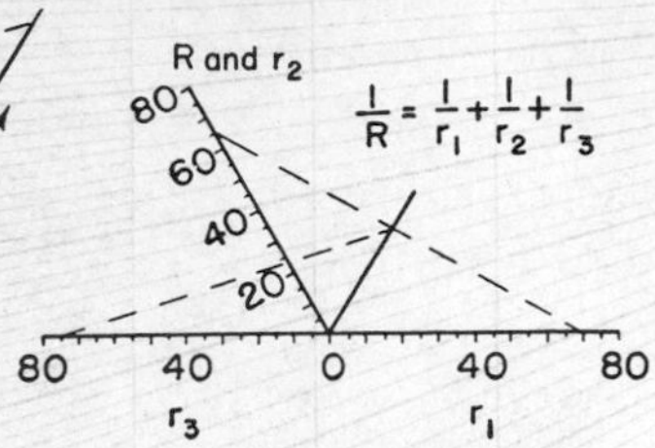

FIG. 149.

Type A and may be charted by combining two charts of this type. For simplicity the r_1- and r_3-axes are at an angle of 180° and the R- and r_2-axes on the same line and the t-axis make an angle of 60° with each other and with the r_3- and r_1-axes, respectively, all moduli becoming equal. If each modulus is taken to be 0.05 in., all scales will be 0.05(80 − 0) or 4 in. in length.

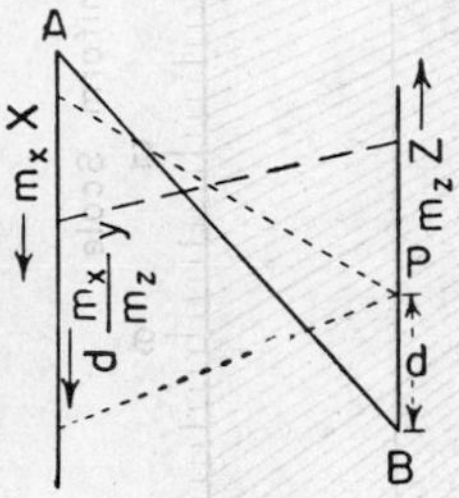

FIG. 150.

Type B (Fig. 150). $x = yz$. (Parallel and Diagonal Axes.) The parallel x and z uniform scales are laid off in opposite directions, and a diagonal AB is drawn between their zero points. Convenient moduli for the x scales and z scales are chosen, and the diagonal scale is constructed as follows: Along the x-axis lay off the temporary scale, $d \frac{m_x}{m_z} y$, where d is the distance from zero to the convergence point P on the z-axis. From the point P, project this temporary scale to the diagonal, and mark the final scale in values of y.

Example. Standardization of Acid against Sodium Carbonate (Fig. 151).

$$n = \frac{w}{0.053v}$$

where v is the number of milliliters of acid, 0 to 120, required to neutralize w g. of sodium carbonate, 0 to 3, when the normality of the acid is n, 0.1 to 2.2. Rearranging, $w = 0.053nv$, which is

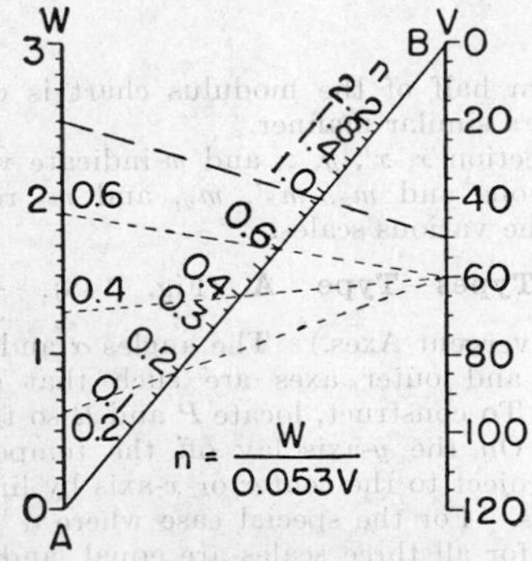

FIG. 151.

of the form of Type B. Choosing as the moduli, $m_w = 2$ and $m_v = 0.05$, lay off the scales $2w$ and $0.05v$ on parallel axes so that the scales increase in opposite directions. At a distance $d = 3$ in. from the zero of the v-scale, locate the point P. Along the w-axis starting at the zero point, construct the temporary scale $d \frac{m_w}{m_v}$ 0.053n or $3 \frac{2}{0.05}$ 0.053n or 6.36n and mark with values of n. From the point P, project this scale to the line connecting the zeros of the w and v scales. This locates the n scale. The dotted lines indicate this projection for normalities of 0.2, 0.4, and 0.6.

Type C (Fig. 152). $x = y + x'z$. (Parallel and Curved Axes.) O and M mark the zero points on the

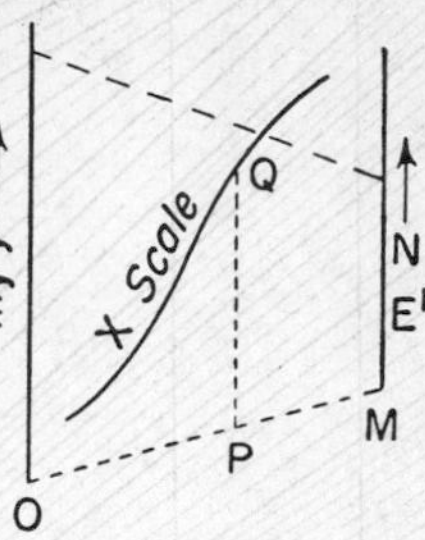

FIG. 152.

parallel axes bearing the scales $m_y y$ and $m_z z$, respectively. The center scale, generally curved, is constructed by laying off OP on OM and then PQ parallel to the y-axis and z-axis, both OP and PQ representing positive directions. OP and PQ are calculated as follows:

$$OP = \frac{m_y OM}{m_y x' + m_z} x'$$

$$PQ = \frac{m_y m_z}{m_y x' + m_z} x$$

Each point on the x-axis so located is marked with its value of x, and a smooth curve is drawn through the points. In charting the equation

$$x = y - x'z$$

the y and z scales are constructed to increase in opposite directions.

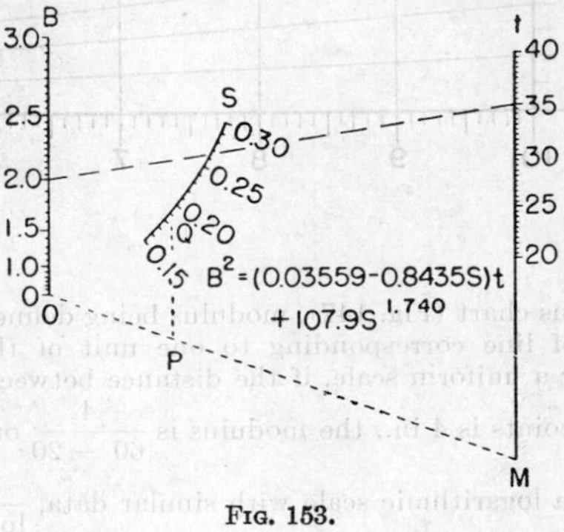

FIG. 153.

Example. Hydrometric Solids Determinations (Fig. 153).

$$B^2 = (0.03559 - 0.8435S)t + 107.9S^{1.740}$$

where, for a certain solute, B is the Baumé reading, 0.5° to 3.0°; t is the temperature, 20° to 40°C.; and S is the solids, 0.15 to 0.30 lb. per gallon. Rearranging,

$$107.9S^{1.740} = B^2 + (0.8435S - 0.03559)t$$

which is of the form

$$x = y + x'z$$

The B and t scales are laid off in the same direction on parallel axes using the moduli $m_B = 0.7$ and $m_t = 0.25$, the length of the

B scale being $0.7[(3.0)^2 - (0.5)^2]$ or 6.13 in. and the length of the *t* scale being $0.25(40 - 20)$ or 5 in. From *O* the line *OM* connecting the zero points of the two scales is laid off for values of *S* in steps of 0.01 so that

$$OP = \frac{m_B OM(0.8435S - 0.03559)}{m_B(0.8435S - 0.03559) + m_t}$$

and

$$PQ = \frac{m_B m_t(107.9S^{1.740})}{m_B(0.8435S - 0.03559) + m_t}$$

If *OM* is 12 in.,

$$OP = \frac{0.7(12)(0.8435S - 0.03559)}{0.7(0.8435S - 0.03559) + 0.25}$$

and

$$PQ = \frac{0.7(0.25)(107.9S^{1.740})}{0.7(0.8435S - 0.03559) + 0.25}$$

The locus of point *Q* is the curved *S* scale. The original equation may also be written

$$107.9S^{1.740} = B^2 - (0.03559 - 0.8435S)t$$

which is of the form

$$x = y - x'z$$

The *B* and *t* scales then increase in opposite directions, *OP* is always negative, *PQ* is always positive, and the *B* scale lies between the *t* and *S* scales. If *OM* be retained as 12 in. the *S* scale becomes nearly three times as long as before (Fig. 154).

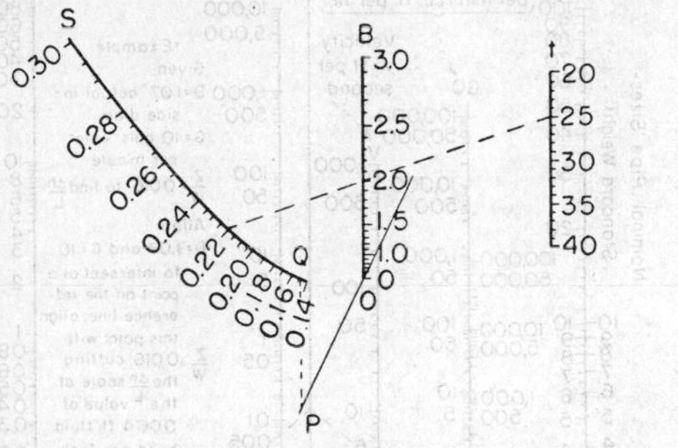

FIG. 154.

Type D (Fig. 155). $x^a = Ky^bz^c$. (Three Parallel Logarithmic Scales.) Taking logarithms,

$$a \log x = \log K + b \log y + c \log z$$

Two parallel axes *y* and *z* are drawn at a convenient distance, d_{yz}, apart and of convenient lengths, L_y and L_z. Logarithmic scales are constructed upon them to represent the ranges of values desired. The number of cycles, *N*, required for a range of 0.1 to 100 is log 100 − log 0.1 or 3. Values are determined for $m_y = L_y/N_yb$ and $m_z = L_z/N_zc$. The position of the *x*-axis is dependent upon m_y, m_z, and d_{yz}, and upon the position of the terms in the logarithmic form of the equation which is written so that all signs are positive. In cases where the equation is

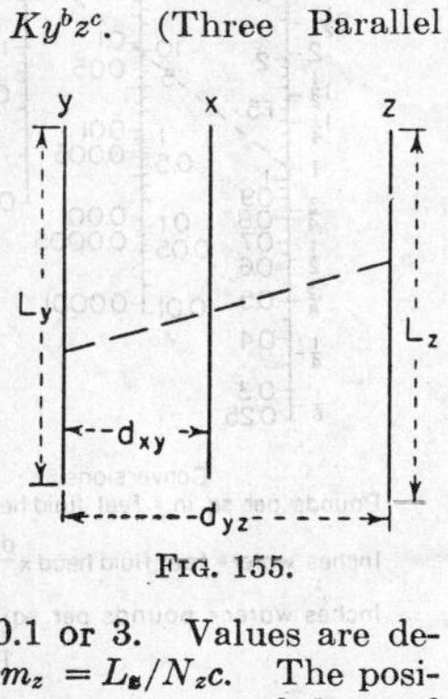

FIG. 155.

$$x^a = \frac{Ky^b}{z^c}$$

reciprocals can be used to give the terms positive signs in the logarithmic form:

$$a \log x = \log K + b \log y + c \log \frac{1}{z}$$

The axis for the single term *x* falls parallel to and between the axes for the two terms *y* and *z*. The distance between the x_x and *y*-axes is

$$d_{xy} = \frac{m_y}{m_y + m_z} d_{yz}$$

A starting point for the *x* scale is determined by a line connecting values of *y* and *z*. Substitution of these values in the original equation gives the value for the *x* point, the constant *K* having been included by this use of the equation. The length of a cycle of the *x* scale is am_x where

$$m_x = \frac{m_y m_z}{m_y + m_z}$$

In using the chart, a straight line through the values of any two variables will pass through the value of the third variable.

Example. Heat Transfer, Gases across Tubes (Fig. 156).*

$$\frac{h}{C_p} = \frac{16M^{0.7}}{D^{0.3}}$$

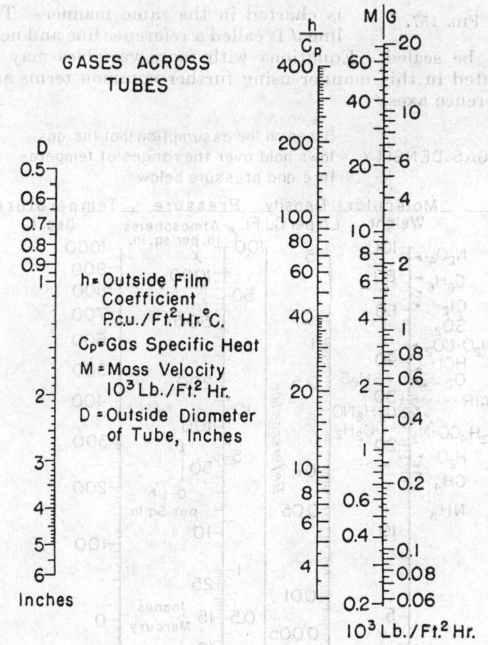

FIG. 156.*

where *h* is the outside film coefficient in p.c.u./(sq. ft.) (hr.) (°C.); C_p is the specific heat of the gas at constant pressure; *M* is the mass velocity in thousands of pounds per square foot per hour, 0.2 to 70 ($N_M = 2.544$); *D* is the outside diameter of tube in inches, 0.5 to 6 ($N_D = 1.08$). Denoting h/C_p by *H* and taking logarithms,

$$\log H = \log 16 + 0.7 \log M + 0.3 \log \frac{1}{D}$$

Draw the *M*- and *D*-axes parallel and choose $d_{MD} = 2.5$ in., $L_M = 4.46$ in. and $L_D = 3.24$ in. Scales are construc'ed, the *D* scale being inverted with respect to the others since *D* is in the

* This formula is used here only to illustrate construction of a specific type of alignment chart. For formula recommended for heat transfer problems (gases across tubes), see p. 473.

reciprocal form. $m_M = \dfrac{4.46}{2.544(0.7)} = 2.5$ in. and $m_D = \dfrac{3.24}{1.08(0.3)} = 10$ in. The H scale falls between the M and D scales with $d_{MH} = \dfrac{2.5(2.5)}{2.5 + 10} = 0.5$. Substituting $H = 10$ and $M = 1$ in the original equation gives $D = 4.8$. A line from $D = 4.8$ to $M = 1$ will cut the H scale at 10. The length of a cycle on H is $2(1) = 2$ in. where $m_H = \dfrac{2.5(10)}{2.5 + 10} = 2$ in. The H scale is constructed to cover a range of 3 to 400.

Type E (Fig. 157). $x^a = Ky^bz^cw^e$. (Four or More Parallel Logarithmic Scales.) Taking logarithms,

$$a \log x = \log K + b \log y + c \log z + e \log w$$

By placing two terms on each side of the equation, neglecting the constant, log K, and equating each side to a common term J, equations of three variables (Type D) are formed:

$$a \log x - b \log y = J \quad (1)$$
$$c \log z + e \log w = J \quad (2)$$

Equation (1) may be written as

$$a \log x + b \log \frac{1}{y} = J$$

and charted as described under Type D. Using the same line for J, Eq. (2) is charted in the same manner. The line J is called a reference line and need not be scaled. Equations with more variables may be treated in this manner using further common terms and reference axes.

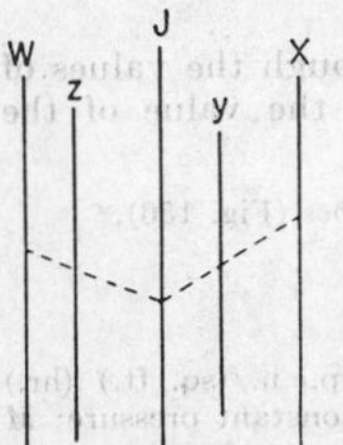

Fig. 157.

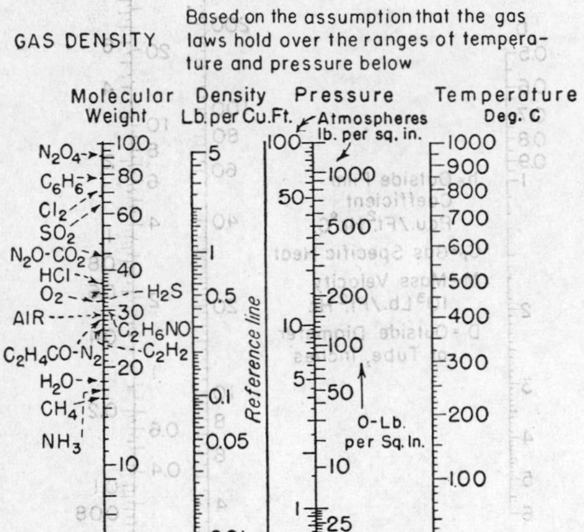

Fig. 158.

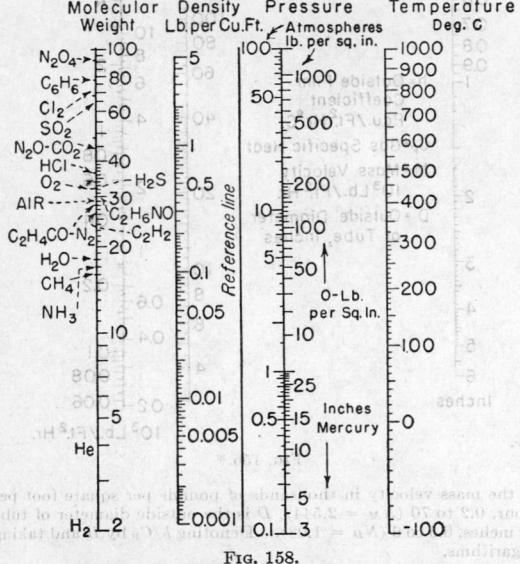

Example 1. Gas Density Chart (Fig. 158). The equation is

$$W = \frac{MP}{1.315T}$$

where W is the density in pounds per cubic foot; M is the molecular weight, 2 to 100 ($N_M = 1.699$); P is the pressure in atmospheres, 0.1 to 100 ($N_P = 3$); T is the centigrade absolute temperature, 173° to 1273° ($N_T = 0.8668$). Taking logarithms,

$$\log W + \log 1.315 + \log T = \log M + \log P$$

Rearranging and neglecting the term log 1.315 which is covered in locating the last scale,

$$\log P = \log T + J \quad (1)$$
$$\log W = \log M + J \quad (2)$$

Taking Eq. (1), draw two parallel lines, T and J, choosing $d_{JT} = 1.5$ in., $L_T = 5$ in., and $L_P = 5$ in. The temperature scale is constructed as in degrees centigrade absolute but is marked in degrees centigrade. $m_T = 5/0.8668 = 5.77$ in. and $m_P = 5/3 = 1.667$ in. while $m_J = \dfrac{5.77(1.667)}{5.77 - 1.667} = 2.34$ in. The negative sign is due to log P being on the opposite side of the equation from log T. $d_{TP} = \dfrac{5.77(1.5)}{5.77 + 2.34} = 1.07$ in. The line P falls between J and T. The pressure scale is marked in atmospheres on one side and on the other in pounds per square inch gage above 1 atm. and in inches of mercury below 1 atm. Taking Eq. (2) and choosing $d_{JM} = 1.5$ in., $L_M = 5$ in., m_M would be equal to $5/1.699 = 2.94$ in. and $m_W = \dfrac{2.94(2.34)}{2.94 + 2.34} = 1.31$ in., while $d_{MW} = \dfrac{2.94(1.5)}{2.94 + 2.34} = 0.833$ in. The W-axis falls between the

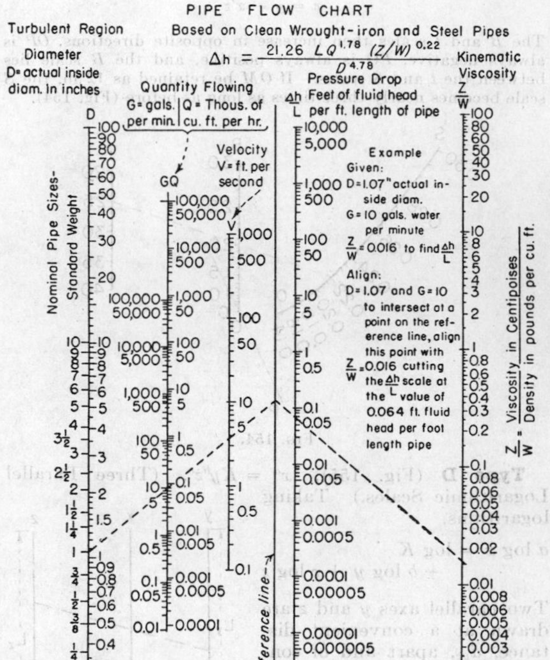

Fig. 159.

J- and M-axes. A line from a T value to a P value crosses J at a point. From this point a line to a value of M crosses W at a value determined by substituting the values of T, P, and M in the original equation. The length of a cycle on W is 1.31 in.

Example 2. Pipe Flow Chart (Fig. 159).* The equation is

$$\Delta h = \frac{21.26LQ^{1.78}(w/W)^{0.22}}{D^{4.78}}$$

* This formula used here only to illustrate construction of specific type of alignment chart. For formula and chart recommended for flow of fluids problems, see p. 379.

where Δh is the pressure drop in feet of fluid head; L is the length of pipe in feet; Q is the quantity of fluid flowing in thousands of cubic feet per hour; Z is the viscosity in centipoises; W is the density in pounds per cubic foot; and D is the inside diameter of the pipe in inches. This equation is made one of four variables by combining terms $\Delta h/L = P$, and $Z/W = K$. Then

$$P = 21.26Q^{1.78}K^{0.22}D^{-4.78}$$

Taking logarithms,

$$\log P = \log 21.26 + 1.78 \log Q + 0.22 \log K - 4.78 \log D$$

Separating terms and neglecting the constant

$$J + 0.22 \log K = \log P \qquad (1)$$
$$J + 4.78 \log D = 1.78 \log Q \qquad (2)$$

Ranges desired are D, 0.25 to 100, $N_D = 2.6$; K, 0.001 to 100, $N_K = 5$; P, 0.000001 to 10,000, $N_P = 10$. Drawing K with $L_K = 25$ in. and J with $d_{KJ} = 8$ in.; scale K is constructed with $m_K = \frac{25}{5(0.22)} = 22.7$ in. Choosing $L_P = 24$ in., $m_P = 24/10 = 2.4$ in. and $m_J = \frac{22.7(2.4)}{22.7 - 2.4} = 2.68$ in. Then $d_{KP} = \frac{22.7(8)}{22.7 + 2.68} = 7.16$ in. The P-axis is drawn in and the scale is constructed, starting at any desired point.

Taking Eq. (2) and using the same axis for J, the D-axis is drawn, $d_{JD} = 8$ in., $L_D = 24$ in.; then $m_D = \frac{24}{2.6(4.78)} = 1.93$ in. and $m_J = 2.68$; then $m_Q = \frac{1.93(2.68)}{1.93 + 2.68} = 1.12$ in. and $d_{DQ} = \frac{1.93(8)}{1.93 + 2.68} = 3.35$ in. The D scale is constructed and a starting point for the Q scale is obtained by substituting values in the original equation. The length of a cycle on the Q scale is $1.12(1.78) = 1.99$ in. Two additional scales have been added, one on the Q-axis to read gallons per minute and one between the Q-axis and the reference line, to read V (velocity in feet per second), the position and modulus of the latter being determined by the relationship between the variables D, Q, and V.

Line Coordinate Charts. Data may be represented on charts which consist of two lines and a point. This type of chart is constructed from a straight line representing data plotted on coordinate paper. If the scales used on the coordinate paper are constructed on two parallel axes, corresponding values of the variables when connected with straight lines will cross at a point. A line from the point thus established to a value on one of the parallel scales will cross the other scale at a value in accordance with the original plot.

The viscosity chart (see Sec. 5, p. 371) was constructed in this manner. The data for each fluid were plotted on logarithmic paper with viscosity in centipoises against degrees centigrade absolute and were found to vary little from a straight line. The parallel axes for viscosity and temperature were then constructed with logarithmic scales. The positions of the points were determined by corresponding values of viscosity and temperature. The viscosity of liquids decreases with temperature rise and in order that the points might fall between the axes the scales are constructed so as to increase in the same direction. For gases the viscosity increases with temperature rise. With scales constructed as in the liquid-viscosities chart, the points therefore fall outside the lines.

SECTION 3

PHYSICAL AND CHEMICAL DATA

BY

John H. Perry, Ph. D., E. I. du Pont de Nemours & Co.; Member, American Institute of Chemical Engineers, American Chemical Society, American Association for the Advancement of Science, American Society for Engineering Education.

AND

Frederick D. Rossini, Ph. D., Professor and Head of the Department of Chemistry, Carnegie Institute of Technology; formerly Chief of Section on Thermochemistry and Constitution of Petroleum, National Bureau of Standards; Member, American Institute of Chemical Engineers, American Chemical Society, American Physical Society, American Petroleum Institute, Washington Academy of Sciences, American Association for the Advancement of Science, Philosophical Society of Washington.

Harding Bliss, Ph. D., Professor of Chemical Engineering, Yale University; Member, American Institute of Chemical Engineers, American Chemical Society, American Association for the Advancement of Science, American Society for Engineering Education. (Engineering Thermodynamic Properties)

Hans C. Duus, Ph. D., E. I. du Pont de Nemours & Co.; Member, American Institute of Chemical Engineers, American Chemical Society. (Thermodynamic Tables)

WITH

Carl V. Herrmann, B. S., M. A., E. I. du Pont de Nemours & Co.; Member, American Institute of Chemical Engineers, American Chemical Society. (Properties of Commercial Acids and Salts in Aqueous Solution)

Norbert A. Lange, Ph. D., Western Reserve University, Handbook Publishers, Inc.; Member, American Chemical Society. (Physical Properties of Inorganic and Organic Compounds)

Jesse W. Stillman, Ph. D., E. I. du Pont de Nemours & Co.; Member, American Chemical Society, The Society of Public Analysts and Other Analytical Chemists. (Indicators and Hydrogen-ion Concentration)

Daniel R. Stull, Ph. D., Dow Chemical Co.; Member, American Chemical Society, American Association for the Advancement of Science. (Vapor Pressures of Inorganic and Organic Compounds)

H. C. Vernon, Ch. E., E. I. du Pont de Nemours & Co.; Member, American Institute of Chemical Engineers, American Chemical Society.

Donald D. Wagman, B. S., M. A., National Bureau of Standards; Member, American Chemical Society. (Heats and Free Energies of Formation, Heats of Combustion)

Richard Wiebe, Ph. D., U.S. Bureau of Agricultural and Industrial Chemistry; Member, American Chemical Society, Society of Automotive Engineers. (Compressibilities, *P-V-T* Relations)

CONTENTS

PHYSICAL PROPERTIES OF PURE SUBSTANCES

Table		Page
1.	Elements and Inorganic Compounds	110
2.	Organic Compounds	129

VAPOR PRESSURES OF PURE SUBSTANCES

Table		Page
3.	Water Ice, from −15° to 0°C	149
4.	Liquid Water, from −16° to 0°C	149
5.	Liquid Water, from 0° to 100°C	149
6.	Inorganic Compounds, above 1 atm	149
7.	Inorganic Compounds, up to 1 atm	150
8.	Organic Compounds, up to 1 atm	153
9.	Organic Compounds, above 1 atm	165

VAPOR PRESSURES OF SOLUTIONS

Table		Page
10.	H_2O over Aqueous HCl	166
11.	HCl over Aqueous HCl	167
12.	H_2O and SO_2 over Aqueous SO_2	167
	Charts for H_2O for Aqueous H_3PO_4 (Figs. 1 and 2)	167
13.	Total Pressure over Aqueous H_2SO_4	168
14.	H_2O and H_2SO_4 over Aqueous H_2SO_4	169
15.	SO_3 over Fuming H_2SO_4	169
16.	H_2O and HNO_3 over Aqueous HNO_3	169
17.	H_2O and HBr over Aqueous HBr	170
18.	HI over Aqueous HI	170
19.	System: Water-Sulfuric Acid–Nitric Acid	170
	Chart for Aqueous Diethylene Glycol (Fig. 3)	170
20.	Total Pressure over Aqueous CH_3COOH	170
21.	H_2O over Aqueous NH_3	171
22.	H_2O (in Mole Per Cent) over Aqueous NH_3	171
23.	NH_3 over Aqueous NH_3	172
24.	Total Pressure over Aqueous NH_3	172
25.	H_2O over Aqueous Na_2CO_3	172
26.	H_2O and CH_3OH over Aqueous CH_3OH	172
27.	H_2O over Aqueous NaOH	173

WATER-VAPOR CONTENT OF GASES

Charts for Gases at High Pressures (Figs. 4, 5, 5*A* and 6) 173

DISSOCIATION PRESSURES

Table		Page
28.	Barium Hydroxide	174
29.	Barium Peroxide	174
30.	Cadmium Carbonate	174
31.	Calcium Carbonate	174
32.	Calcium Cyanamide	174
33.	Calcium Oxalate	174
34.	Cobalt Sulfate	174
35.	Ferrous Sulfate	174
36.	Manganese Dioxide	174
37.	Mercuric Oxide	174
38.	Potassium Bicarbonate	174
39.	Potassium Carbonate	174
40.	Potassium Dihydrogen Phosphate	174

TABLE PAGE

41. Potassium Hydride 174
42. Silver Carbonate 174
43. Sodium Carbonate 174
44. Sodium Dihydrogen Phosphate 174

DENSITIES OF PURE SUBSTANCES

45. Water, 0° to 40°C 175
46. Water, 40° to 100°C 175
47. Water, below 0°C 176
48. Mercury 176
49. Gases, at 0°C. and 1 atm 176

SPECIFIC VOLUMES OF PURE SUBSTANCES

50. Water, 0° to 40°C 177
51. Water, 40° to 100°C 177

DENSITIES OF AQUEOUS INORGANIC SOLUTIONS

52. Aluminum Sulfate 177
53. Ammonia 178
54. Ammonium Acetate 178
55. Ammonium Bichromate 178
56. Ammonium Chloride 178
57. Ammonium Chromate 178
58. Ammonium Nitrate 178
59. Ammonium Sulfate 178
60. Arsenic Acid 178
61. Barium Chloride 178
62. Cadmium Nitrate 178
63. Calcium Chloride 178
64. Calcium Hydroxide 178
65. Calcium Hypochlorite 178
66. Calcium Nitrate 178
67. Chromic Acid 178
68. Chromium Chloride 179
69. Copper Nitrate 179
70. Copper Sulfate 179
71. Cuprous Chloride 179
72. Ferric Chloride 179
73. Ferric Sulfate 179
74. Ferric Nitrate 179
75. Ferrous Sulfate 179
76. Hydrogen Bromide 179
77. Hydrogen Cyanide 179
78. Hydrogen Chloride 179
79. Hydrogen Fluoride 179
80. Hydrogen Peroxide 179
81. Hydrofluosilicic Acid 179
82. Magnesium Chloride 179
83. Magnesium Sulfate 179
84. Nickel Chloride 179
85. Nickel Nitrate 179
86. Nickel Sulfate 179
87. Nitric Acid 180
88. Perchloric Acid 181
89. Phosphoric Acid 181
90. Potassium Bicarbonate 181
91. Potassium Bromide 181
92. Potassium Carbonate 181
93. Potassium Chromate 181
94. Potassium Chlorate 181
95. Potassium Chloride 181
96. Potassium Chrome Alum 182
97. Potassium Hydroxide 182
98. Potassium Nitrate 182
99. Potassium Dichromate 182
100. Potassium Sulfate 182
101. Potassium Sulfite 182
102. Sodium Acetate 182
103. Sodium Arsenate 182
104. Sodium Bichromate 182
105. Sodium Bromide 182
106. Sodium Formate 182
107. Sodium Carbonate 182
108. Sodium Chlorate 182
109. Sodium Chloride 182
110. Sodium Chromate 182
111. Sodium Hydroxide 182
112. Sodium Nitrate 182
113. Sodium Nitrite 183
114. Sodium Silicate 183
115. Sodium Sulfate 183
116. Sodium Sulfide 183
117. Sodium Sulfite 183
118. Sodium Thiosulfate 183
119. Sodium Thiosulfate Pentahydrate 183
120. Stannic Chloride 183
121. Stannous Chloride 183
122. Sulfuric Acid 184
123. Zind Bromide 185
124. Zinc Chloride 185
125. Zinc Nitrate 185
126. Zinc Sulfate 185

DENSITIES OF AQUEOUS ORGANIC SOLUTIONS

127. Formic Acid 186
128. Acetic Acid 186
129. Oxalic Acid 187
130. Methyl Alcohol 187
131. Ethyl Alcohol 188
132. Ethyl Alcohol 189
133. Ethyl Alcohol 190
134. *n*-Propyl Alcohol 190
135. Isopropyl Alcohol 191
136. Glycerol 191
137. Hydrazine 192
138. Miscellaneous Organic Compounds 192

DENSITIES OF MISCELLANEOUS MATERIALS

139. Approximate Values for Miscellaneous Solids and Liquids 194

SOLUBILITIES

140. Inorganic Compounds in Water 196

THERMAL EXPANSION

141. Expansion of Gases 200
142. Linear Expansion of the Solid Elements 200
143. Linear Expansion of Miscellaneous Substances 201
144. Cubical Expansion of Liquids 202
145. Cubical Expansion of Solids 202

JOULE-THOMSON EFFECT

146. Air 203
147. Carbon Dioxide 203
148. Methane 203
149. Ethyl Chloride 203

CRITICAL CONSTANTS

150. Elements and Inorganic and Organic Compounds 204

COMPRESSIBILITIES

151. Miscellaneous Gases at 0°C. and 1 atm 205
152. Oxygen 205
153. Hydrogen 205
154. Nitrogen 206
155. Carbon Dioxide 206
156. Air 207
157. Carbon Monoxide 207
158. Methane 207
159. Ethane 207
160. Ethylene 208
161. Acetylene 208
162. Methyl Chloride 208
163. Ammonia 208
164. Mixture of Hydrogen and Nitrogen, 3 to 1 209
165. Liquids 209

LATENT HEATS

166. Heats of Fusion and Vaporization of the Elements and Inorganic Compounds 210
167. Heats of Fusion of Organic Compounds 213
168. Heats of Vaporization of Organic Compounds 215
169. Heats of Fusion of Miscellaneous Materials 217
Alignment Chart of Heats of Vaporization of Water, Sulfur Dioxide, Ammonia, and Organic Compounds (Fig. 7) 217

TABLE PAGE

Alignment Chart of Heats of Vaporization of Hydrocarbons (Fig. 8) 218

SPECIFIC HEATS OF PURE COMPOUNDS

170. Elements and Inorganic Compounds 219
171. Water 225
172. Organic Liquid Compounds 225
Alignment Chart for Liquids (Fig. 9) 228
Alignment Chart for Gases (Fig. 10) 228
173. Organic Solid Compounds 230
174. Liquefied Gases 233
175. Air at High Pressures 233
Nitrogen, Charts Showing Variation with Pressure (Figs. 11 and 12) 233
176. Ratios of Specific Heats for Gases at 1 atm 233
177. Ratios of Specific Heats for Air at High Pressures 233

SPECIFIC HEATS OF AQUEOUS SOLUTIONS

178. Hydrochloric Acid 234
179. Sulfuric Acid 234
180. Nitric Acid 234
181. Phosphoric Acid 234
182. Acetic Acid 234
183. Sodium Hydroxide 234
184. Potassium Hydroxide 234
185. Ammonia 234
186. Sodium Carbonate 234
187. Sodium Chloride 234
188. Potassium Chloride 234
189. Zinc Sulfate 234
190. Copper Sulfate 235
191. Methyl Alcohol 235
192. Ethyl Alcohol 235
193. Normal Propyl Alcohol 235
194. Glycerol 235
195. Aniline 235

SPECIFIC HEATS OF MISCELLANEOUS MATERIALS

196. Miscellaneous Liquids and Solids 235

HEATS AND FREE ENERGIES OF FORMATION

197. Inorganic and Organic Compounds 237

HEATS OF COMBUSTION

198. Organic Compounds 244

HEATS OF SOLUTION

199. Inorganic Compounds in Water 246
200. Organic Compounds in Water 248

ENGINEERING THERMODYNAMIC PROPERTIES

Explanation of Tables 249
201. Air, Saturated 250
202. Air, Superheated 250
203. Ammonia, Saturated 250
204. Ammonia, Superheated 250
Ammonia, Aqueous; Enthalpy-concentration Chart (Fig. 13) 252
205. 1,3-Butadiene, Saturated 253
206. n-Butane, Saturated 254
207. n-Butane, Saturated and Superheated 254
208. Carbon Dioxide, Saturated 254
209. Carbon Dioxide, Superheated 255
Carbon Monoxide, Mollier Chart (Fig. 14) 256
210. Dowtherm, Saturated 257
211. Ethane, Saturated 257

TABLE PAGE

212. Ethane, Superheated 257
213. Ethylamine, Saturated 258
214. Ethyl Chloride, Saturated 258
215. Ethylene, Saturated 258
216. Ethylene, Superheated 259
217. F-11, Trichloromonofluoromethane, Saturated 259
218. F-11, Trichloromonofluoromethane, Superheated 260
219. F-12, Dichlorodifluoromethane, Saturated 261
Ethyl Alcohol, Enthalpy-concentration Chart (Fig. 15) 261
220. F-12, Dichlorodifluoromethane, Superheated 262
221. F-21, Dichloromonofluoromethane, Saturated 263
222. F-21, Dichloromonofluoromethane, Superheated 263
223. Isobutane, Saturated 264
224. Isobutane, Superheated 264
225. Mercury, Saturated 265
226. Methane, Saturated 265
Hydrogen, Temperature (14°–100°K.)-entropy Diagram (Fig. 16) 266
Hydrogen, Temperature (100°–300°K.)-entropy Diagram (Fig. 17) 267
Hydrogen Chloride, Enthalpy-concentration Chart (Fig. 18) 268
227. Methane, Superheated 270
228. Methylamine, Saturated 271
229. Methyl Chloride, Saturated 271
230. Methyl Chloride, Superheated 271
231. Methyl Formate, Saturated 272
232. Methylene Chloride, Saturated 272
233. Nitrogen, Saturated 272
Oxygen, Enthalpy-concentration Chart (Fig. 19) 272
234. Nitrogen, Superheated 273
235. Oxygen, Saturated 273
236. Oxygen, Superheated 273
237. Propane, Saturated 274
238. Propane, Superheated 274
Sodium Hydroxide, Enthalpy-concentration Chart (Fig. 20) 275
239. Sulfur Dioxide, Saturated 275
240. Sulfur Dioxide, Superheated 276
Sulfuric Acid, Aqueous, Enthalpy-concentration Chart (Fig. 21) 277
Sulfuric and Nitric Acids, Enthalpy-concentration Chart (Fig. 22) 277
241. Steam, Saturated: Temperature Table 277
242. Steam, Saturated: Pressure Table 278
243. Steam, Superheated 279
244. Trichloroethylene, Saturated 281

FREEZING POINTS OF AQUEOUS SOLUTIONS

245. Some Aqueous Solutions 281

FUSION TEMPERATURES

246. Alloys 281
247. Refractories 281

ELEVATION OF BOILING POINT

248. Aqueous Solutions of Inorganic Salts 282

CONSTANT TEMPERATURES

249. Production and Maintenance 282

INDICATORS AND HYDROGEN-ION CONCENTRATION

Indicators and Hydrogen-ion Concentration 283
Hydrogen-ion Indicator Chart 284
Table of Recommended Indicators 285
Electrometric Method 285

PHYSICAL PROPERTIES OF PURE SUBSTANCES

Table 1. Physical Properties of the Elements and Inorganic Compounds*

Abbreviations Used in the Table

a., acid
A., specific gravity with reference to air = 1
abs., absolute
ac., acetic acid
act., acetone
al., 95 per cent ethyl alcohol
alk, alkali (*i.e.*, aq. NaOH or KOH)
am., amyl (C_5H_{11})
amor., amorphous
anh., anhydrous
aq., aqueous or water
aq. reg., aqua regia
atm., atmosphere or 760 mm. of mercury pressure
bk., black
brn., brown
bz., benzene
c., cold
cb., cubic
cc, cubic centimeter
chl., chloroform
col., colorless or white
conc., concentrated
cr., crystals or crystalline
d., decomposes
D., specific gravity with reference to hydrogen = 1
d. 50, decomposes at 50°C.; 50 d., melts at 50°C. with decomposition
delq., deliquescent
dil., dilute
dk., dark
eff., effloresces or efflorescent
et., ethyl ether
expl., explodes
gel., gelatinous
gly., glycerol, (glycerin)
gn., green
h., hot
hex., hexagonal
hyg., hygroscopic
i., insoluble
ign., ignites
lq., liquid
lt., light
m. al., methyl alcohol
mn., monoclinic
nd., needles
NH_3, liquid ammonia
NH_4OH, ammonium hydroxide solution
oct., octahedral
or., orange
pd., powder
pl., plates
pr., prisms or prismatic
pyr., pyridine
rhb., rhombic (orthorhombic)
s., soluble
satd., saturated
sl., slightly
soln., solution
subl., sublimes
sulf., sulfides
tart. a., tartaric acid
tet., tetragonal
tr., transition
tri., triclinic
trig., trigonal
v., very
vac., in vacuo
vl., violet
volt., volatile or volatilizes
wh., white
yel., yellow
∞, soluble in all proportions
<, less than
>, greater than
42±, about or near 42
$-3H_2O$, 100, loses 3 moles of water per formula weight at 100°C.

Formula weights are based upon the International Atomic Weights of 1941 and are computed to the nearest hundredth.

Refractive index, where given for a uniaxial crystal, is for the ordinary (ω) ray; where given for a biaxial crystal, the index given is for the median (β) value. Unless otherwise specified, the index is given for the sodium D-line (λ = 589.3 mμ).

Specific gravity values are given at room temperatures (15° to 20°C.) unless otherwise indicated by the small figures which follow the value: thus, "$5.6\frac{18°}{4}$" indicates a specific gravity of 5.6 for the substance at 18°C. referred to water at 4°C. In this table the values for the specific gravity of gases are given with reference to air (A) = 1, or hydrogen (D) = 1.

Melting point is recorded in a certain case as "82 d." and in some other case as "d. 82," the distinction being made in this manner to indicate that the former is a melting point with decomposition at 82°C., while in the latter decomposition only occurs at 82°C. Where a value such as "$-2H_2O$, 82" is given it indicates loss of 2 moles of water per formula weight of the compound at a temperature of 82°C.

Boiling point is given at atmospheric pressure (760 mm, of mercury) unless otherwise indicated; thus, "$82^{15 mm.}$" indicates the boiling point is 82°C. when the pressure is 15 mm.

Solubility is given in parts by weight (of the formula shown at the extreme left) per 100 parts by weight of the solvent; the small superscript indicates the temperature. In the case of gases the solubility is often expressed in some manner as "$5^{10°}$ cc" which indicates that at 10°C., 5 cc. of the gas are soluble in 100 g. of the solvent. The symbols of the common mineral acids: H_2SO_4, HNO_3, HCl, etc., represent dilute aqueous solutions of these acids. See also special tables on Solubility.

REFERENCES: The information given in this table has been collected mainly from the following sources: Mellor, "A Comprehensive Treatise on Inorganic and Theoretical Chemistry," Longmans, New York, 1922. Abegg, "Handbuch der anorganischen Chemie," S. Hirzel, Leipzig, 1905. Gmelin-Kraut, "Handbuch der anorganischen Chemie," 7th ed., Carl Winter, Heidelberg; 8th ed., Verlag Chemie, Berlin, 1924. Friend, "Textbook of Inorganic Chemistry," Griffin, London, 1914. Winchell, "Microscopic Character of Artificial Inorganic Solid Substances or Artificial Minerals," Wiley, New York, 1931. "International Critical Tables," McGraw-Hill, New York, 1926. "Tables annuelles internationales de constants et donnes numeriques," McGraw-Hill, New York. "Annual Tables of Physical Constants and Numerical Data," National Research Council, Princeton, N.J., 1943. Comey and Hahn, "A Dictionary of Chemical Solubilities," Macmillan, New York, 1921. Seidell, "Solubilities of Inorganic and Meta. Organic Compounds," Van Nostrand, New York, 1940.

Name	Formula	Formula weight	Color, crystalline form and refractive index	Specific gravity	Melting point, °C.	Boiling point, °C.	Solubility in 100 parts: Cold water	Hot water	Other reagents
Aluminum	Al	26.97	silv., cb.	$2.70^{20°}$	660	2056	i.	i.	s. HCl, H_2SO_4, alk.
acetate, normal	$Al(C_2H_3O_2)_3$	204.10	wh. pd.	...	d. 200	...	s.	d.	
acetate, basic	$Al(OH)(C_2H_3O_2)_2$	162.07	wh., amor.	...	d.	...	i.	...	s.a.; i. NH_4 salts
bromide	$AlBr_3$	266.72	trig.	$3.01\frac{25°}{4}$	97.5	268	s.	...	s. al., act., CS_2
bromide	$AlBr_3.6H_2O$	374.82	col., delq. cr.	...	d. 100	...	s.	s.	s. al., CS_2
carbide	Al_4C_3	143.91	yel., hex., 2.70	2.95	d. >2200	...	d. to CH_4	...	s. a.; i. act.
chloride	$AlCl_3$	133.34	wh., delq., hex.	$2.44\frac{25°}{4}$	$194^{5.2atm.}$	182.7^{752mm}; subl. 178	$69.87^{15°}$	s. d.	s. et., chl., CCl_4; i. bz.
chloride	$AlCl_3.6H_2O$	241.44	col., delq., trig., 1.560	...	...	...	400	v. s.	50 al.; s. et.
fluoride (fluellite)	$AlF_3.H_2O$	101.99	col., rhb., 1.490	2.17	d.	...	sl. s.		
fluoride	$Al_2F_6.7H_2O$	294.05	wh., cr. pd.	...	$-4H_2O$, 120	$-6H_2O$, 250	i.	sl. s.	
hydroxide	$Al(OH)_3$	77.99	wh., mn.	2.42	$-2H_2O$, 300	...	$0.000104^{18°}$	i.	s. a., alk.; i. a.
nitrate	$Al(NO_3)_3.9H_2O$	375.14	rhb., delq.	...	73	d. 134	v. s.	v. s. d.	s. al., CS_2
nitride	Al_2N_2	81.96	yel., hex.	$3.05\frac{25°}{4}$	$2150^{4atm.}$	d. >1400	d. slowly	...	s. alk. d.
oxide	Al_2O_3	101.94	col., hex., 1.67–8	3.99	1999 to 2032	...	i.	i.	v. sl. s. a., alk.
oxide (corundum)	Al_2O_3	101.94	wh., trig., 1.768	4.00	1999 to 2032	2210	i.	i.	v. sl. s. a., alk.
phosphate	$AlPO_4$	121.95	col., hex.	2.59	...	...	i.	i.	s. a., alk.; i. ac.
potassium silicate (muscovite)	$3Al_2O_3.K_2O.6SiO_2.2H_2O$	796.40	mn., 1.590	2.9	d.	...	i.		
potassium silicate (orthoclase)	$Al_2O_3.K_2O.6SiO_2$	556.49	col., mn., 1.524	2.56	1450 (1150)	...	i.		

Aluminum potassium tartrate	$AlK(C_4H_4O_6)_2$	362.21	col.	...		...	s.	s.	i. HCl
sodium fluoride (cryolite)	$AlF_3.3NaF$	209.96	wh., mn., 1.3389	2.90	1000	...	sl. s.	...	d. a.
sodium silicate	$Al_2O_3.Na_2O.6SiO_2$	524.29	col., tri., 1.529	2.61	1100	...	i.	i.	
sulfate	$Al_2(SO_4)_3$	342.12	wh. cr.	2.71	d. 770		$31.3^{0°}$	$89^{100°}$	
Alum, ammonium (tschermigite)	$Al_2(SO_4)_3.(NH_4)_2SO_4.24H_2O$	906.64	col., oct., 1.4594	$1.64\frac{20°}{4}$	93.5	$-20H_2O$, 120; $-24H_2O$, 200	$3.9^{0°}$	∞ 100°	i. al.
ammonium chrome	$Cr_2(SO_4)_3.(NH_4)_2SO_4.24H_2O$	956.72	gn. or vl., oct., 1.4842	1.72	100 d.	...	$21.2^{25°}$	...	s. al.
ammonium iron	$Fe_2(SO_4)_3.(NH_4)_2SO_4.24H_2O$	964.40	vl., oct., 1.485	1.71	40	...	$124^{25°}$	...	i. al.
potassium (kalinite)	$Al_2(SO_4)_3.K_2SO_4.24H_2O$	948.76	col., mn., 1.4564	$1.76\frac{26°}{4}$	92	$-18H_2O$, 64.5	$5.7^{0°}$	∞ 93°	
potassium chrome	$Cr_2(SO_4)_3.K_2SO_4.24H_2O$	998.84	red or gn., cb., 1.4814	1.83	89	...	20	50	i. al.
sodium	$Al_2(SO_4)_3.Na_2SO_4.24H_2O$	916.56	col., oct., 1.4388	$1.675\frac{20°}{4}$	61	...	$106.4^{0°}$	$121.7^{45°}$	i. al.
Ammonia†	NH_3	17.03	col. gas, 1.325 (lq.)	$0.817^{-79°}$ 0.5971 (A)	−77.7	−33.4	$89.9^{0°}$	$7.4^{96°}$	$14.8^{20°}$ al.; s. et.
Ammonium acetate	$NH_4C_2H_3O_2$	77.08	wh., hyg. cr.	1.073	114	d.	$148^{4°}$	...	s. al.; sl. s. act.
auricyanide	$NH_4CN.Au(CN)_3.H_2O$	337.33	pl.	...	d. 200	...	s.	v. s.	i. al.
bicarbonate	NH_4HCO_3	79.06	mn. or rhb., 1.5358	1.573	d. 35–60	...	$11.9^{0°}$	$27^{30°}$	i. al.
bromide	NH_4Br	97.96	col., cb., 1.7108	$2.327\frac{15°}{4}$	subl. 542	...	$68^{10°}$	$145.6^{100°}$	s. al., et., act.
carbonate	$(NH_4)_2CO_3.H_2O$	114.11	col. pl.	...	d. 58	...	$100^{15°}$	...	i. al., CS_2, NH_3
carbonate, carbamate	$NH_4HCO_3.NH_2CO_2NH_4$‡	157.11	wh. cr.	...	subl.	...	$25^{15°}$	$67^{65°}$	
carbonate, sesqui-	$(NH_4)_2CO_3.2NH_4HCO_3.H_2O$	272.22	wh.	...	d.	...	$20^{15°}$	$50^{49°}$	
chloride (salammoniac)	NH_4Cl	53.50	wh., cb., 1.639, 1.6426	$1.53^{17°}$	d. 350	subl. 520	$29.4^{0°}$	$77.3^{100°}$	s. NH_3; sl. s. al., m. al.
chloroplatinate	$(NH_4)_2PtCl_6$	444.05	yel., cb.	3.065	d.	...	$0.7^{15°}$	$1.25^{100°}$	0.005 al.
chloroplatinite	$(NH_4)_2PtCl_4$	373.14	tet.	...	d.	...	s.	v. s.	
chlorostannate	$(NH_4)_2SnCl_6$	367.52	pink., cb.	2.4	...	...	$33.3^{15°}$		
chromate	$(NH_4)_2CrO_4$	152.09	yel., mn.	$1.917^{12°}$	d. 180	...	$40.5^{30°}$	d.	sl. s. act., NH_3; i. al.
cyanide	NH_4CN	44.06	col., cb.	$0.79^{100°}$ (A)	36	...	s.	v. s.	s. al.
dichromate	$(NH_4)_2Cr_2O_7$	252.10	or., mn.	2.15	d. 185	...	$47.2^{30°}$	v. s.	s. al.; i. act.
ferrocyanide	$(NH_4)_4Fe(CN)_6.6H_2O$	392.21	mn.	...	d.	...	s.	...	i. al.
fluoride	NH_4F	37.04	wh., hex.	...	...	...	v. s.	d.	s. al.; i. NH_3
fluoride, acid	$NH_4F.HF$	57.05	wh., rhb., 1.390	$2.21\frac{12°}{12}$	...	...	v. s.		
formate	HCO_2NH_4	63.06	col., mn., delq.	1.266	114–116	d. 180; subl. in vac.	$102^{0°}$	$531^{80°}$	s. al.
hydrosulfide	NH_4HS	51.11	col., rhb.	...	d.	subl. 120	v. s.	...	s. al.
hydroxide	NH_4OH	35.05	in soln. only	...	...	...	s.		
molybdate	$(NH_4)_2MoO_4$	196.03	mn.	2.27	d.	...	d.	d.	i. al., NH_3
molybdate, hepta-	$(NH_4)_6Mo_7O_{24}.4H_2O$‡	1235.95	col., mn.	...	...	...	$44^{25°}$	...	i. al.
nitrate (α), stable −16° to 32°	NH_4NO_3	80.05	col., tet., 1.611	$1.66\frac{25°}{4}$	169.6	d. 210	$118.3^{0°}$	$241.8^{30°}$	
nitrate (β), stable 32° to 84°	NH_4NO_3	80.05	col., rhb. or mn.	$1.725\frac{25°}{4}$	...	d. 210	$365.8^{35°}$	$580^{80°}$	$3.8^{20°}$ al., $17.1^{20°}$ m. al.; v. s. NH_3
nitrite	NH_4NO_2	64.05	wh. nd.	1.69	expl.	...	s.	d.	s. al.
osmochloride	$(NH_4)_2OsCl_6$	439.02	cb.	$2.93\frac{20°}{4}$					
oxalate	$(NH_4)_2C_2O_4.H_2O$	142.12	col., rhb.	1.501	...	...	$2.5^{0°}$	$11.8^{50°}$	sl. s. al.; i. NH_3
oxalate, acid	$NH_4HC_2O_4.H_2O$	125.08	col., trimetric	1.556	d.	...	s.		
perchlorate	NH_4ClO_4	117.50	col., rhb., 1.4833	1.95	d.	...	$10.9^{0°}$	$46.9^{100°}$	$2^{20°}$ al.; s. act.; i. et.
persulfate	$(NH_4)_2S_2O_8$	228.20	wh., mn., 1.5016	1.98	d. 120	...	$58.2^{0°}$	d.	
phosphate, monobasic	$NH_4H_2PO_4$	115.04	col., tet., 1.5246	$1.803\frac{19°}{4}$	...	...	$22.7^{0°}$	$173.2^{100°}$	i. ac.
phosphate, dibasic	$(NH_4)_2HPO_4$	132.07	col., mn., 1.53	1.619	...	...	$131^{15°}$	...	i. act.
phosphate, meta-	$(NH_4)_4P_4O_{12}$	388.08	col., mn.	2.21	...	...	s.		

*By N. A. Lange, Ph.D., Handbook Publishers, Inc., Sandusky, Ohio. Abridged from table of Physical Constants of Inorganic Compounds in Lange, "Handbook of Chemistry."

† See special tables.

‡ Usual commercial form.

Table 1. Physical Properties of the Elements and Inorganic Compounds—(*Continued*)

Name	Formula	Formula weight	Color, crystalline form and refractive index	Specific gravity	Melting point, °C.	Boiling point, °C.	Solubility in 100 parts: Cold water	Hot water	Other reagents
Ammonium phosphomolybdate	$(NH_4)_3PO_4.12MoO_3.3H_2O$ (?)	1930.55	yel.	...	d.	...	$0.03^{15°}$	i.	s. alk.; i. al., HNO_3
silicofluoride	$(NH_4)_2SiF_6$	178.14	cb., 1.3696	2.01	...	subl.	$18.5^{17.5°}$	55.5	s. al.; i. act.
sulfamate	$NH_4.SO_3NH_2$	114.12	col. pl.	...	132	d. 160	$134^{0°}$	$357^{50°}$	
sulfate (mascagnite)	$(NH_4)_2SO_4$	132.14	col., rhb., 1.5230	$1.769\frac{20°}{4}$	513 d.	...	$70.6^{0°}$	$103.3^{100°}$	i. al., act., CS_2
sulfate, acid	NH_4HSO_4	115.11	col., rhb., 1.480	1.78	146.9	490	100	...	v. sl. s. al.; i. act.
sulfide	$(NH_4)_2S$	68.14	yel.-wh.	...	d.	...	v. s.	...	$120^{25°}$ NH_3
sulfide, penta-	$(NH_4)_2S_5$	196.38	or.-red pr.	...	...	...	s.	...	
sulfite	$(NH_4)_2SO_3.H_2O$	134.16	col., mn.	1.41	d.	...	$100^{12°}$	...	i. al., act.
sulfite, acid	NH_4HSO_3	99.11	rhb.	$2.03\frac{12°}{4}$	d.	...	s.		
tartrate	$(NH_4)_2C_4H_4O_6$	184.15	col., mn.	1.60	d.	...	$45^{0°}$	$87^{60°}$	sl. s. al.
thiocyanate	NH_4CNS	76.12	col., mn., 1.685±	1.305	149.6	d. 170	$120^{0°}$	$170^{20°}$	s. al., act., NH_3, SO_2
vanadate, meta-	NH_4VO_3	116.99	col. cr.	2.326	d.	...	$0.44^{18°}$	$3.05^{70°}$	i. al., NH_4Cl
Antimony	Sb	121.76	tin wh., trig.	$6.684^{25°}$	630.5	1380	i.	i.	s. aq. reg., h. conc. H_2SO_4
chloride, tri- (butter of antimony)*	$SbCl_3$	228.13	col., rhb., delq.	$3.14\frac{20°}{4}$	73.4	220.2	$601.6^{0°}$	$\infty^{72°}$	s. al., HCl, HBr, $H_2C_4H_4O_6$
oxide, tri- (valentinite)	Sb_2O_3	291.52	rhb., 2.35	5.67	656	1570	v. sl. s.	sl. s.	s. HCl, KOH, $H_2C_4H_4O_6$
oxide, tri- (senarmontite)	Sb_2O_3	291.52	cb., 2.087	5.2	652				
sulfide, tri- (stibnite)	Sb_2S_3	339.70	bk., rhb., 4.046	4.64	550	...	$0.00017^{18°}$	d.	s. HCl; alk., NH_4HS, K_2S; i. ac.
sulfide, penta-	Sb_2S_5	403.82	golden	$4.120^{0°}$	−2S, 135	...	i.	i.	s. HCl, alk., NH_4HS
telluride, tri-	Sb_2Te_3	626.35	gray	...	629				
Antimonyl potassium tartrate (tartar emetic)	$(SbO)KC_4H_4O_6.\frac{1}{2}H_2O$	333.94	wh., rhb.	2.60	$-\frac{1}{2}H_2O$, 100	...	$5.26^{8.7°}$	$35.7^{100°}$	s. gly.; i. al.
sulfate, normal	$(SbO)_2SO_4$	371.58	wh. pd.	4.89	...	...	d.	d.	
sulfate, basic	$(SbO)_2SO_4.Sb_2(OH)_4$	683.13	wh. pd.	...	...	...	i.	d.	$5.15^{15°}$ gly.
Argon	A	39.94	col. gas	$1.65^{-233°}$; $1.402^{-185.7°}$; 1.38 (A)	−189.2	−185.7	$5.6^{0°}$ cc	$2.23^{50°}$ cc	$24^{25°}$ cc al.
Arsenic (crystalline)(α)	As_4	299.64	met., hex.	$5.727^{14°}$	$814^{36atm.}$	subl. 615	i.	i.	s. HNO_3
Arsenic (black)(β)	As_4	299.64	bk., amor.	$4.7^{20°}$	...	...	i.	i.	s. HNO_3, aq. reg., aq. Cl_2, h. alk.
Arsenic (yellow)(γ)	As_4	299.64	yel., cb.	$2.0^{20°}$	d. 358				
acid, ortho-	$H_3AsO_4.\frac{1}{2}H_2O$	150.94	col., hyg.	2.0–2.5	35.5	$-H_2O$, 160	16.7	50	s. alk.
acid, meta-	$HAsO_3$	123.92	wh., hyg.	...	d.	...	d. to form	H_3AsO_4	
acid, pyro-	$H_4As_2O_7$	265.85	col.	...	d. 206	...	d. to form	H_3AsO_4	
pentoxide	As_2O_5	229.82	wh., amor.	4.086	...	d.	$59.5^{0°}$	$76.7^{100°}$	s. alk., al.
sulfide, di- (realgar)	As_2S_2	213.94	red, mn., 2.68	(α)$3.506^{19°}$; (β)$3.254^{19°}$	(α)tr. 267; (β)307	565	i.	d.	s. K_2S, $NaHCO_3$
sulfide, penta-	As_2S_5	310.12	yel.	...	...	d. 500	$0.000136^{0°}$	i.	s. HNO_3, alk.
Arsenious chloride (butter of arsenic)	$AsCl_3$	181.28	oily lq.	lq. 2.163	−18	130	d.	d.	s. HCl, HBr, PCl_3
hydride (arsine)	AsH_3	77.93	col. gas	2.695 (A)	−113.5	−55; d. 230	20 cc	sl. s.	sl. s. alk.
oxide (arsenolite)	As_2O_3	197.82	col., cb., fibrous, 1.755	$3.865\frac{25°}{4}$	subl.	...	sl. s.	sl. s.	i. al., et.
oxide (claudetite)	As_2O_3	197.82	col., mn., 1.92	3.85	subl.	...	sl. s.	sl. s.	i. al., et.
oxide	As_2O_3	197.82	amor. or vitreous	3.738	315	...	$1.21^{0°}$	$2.93^{40°}$	s. HCl, alk., Na_2CO_3; i. al., et.
Auric chloride	$AuCl_3.2H_2O$	339.60	or. cr.	...	d.	...	v. s.	v. s.	s. HCl, al., et.; sl. s. NH_3
cyanide	$Au(CN)_3.6H_2O$	383.35	...	...	d. 50	...	v. s.	v. s.	s. al.
Aurous chloride	AuCl	232.66	yel. cr.	7.4	$AuCl_3$, 170	d. 290	d.	d.	s. HCl, HBr; d. al.
cyanide	AuCN	223.22	yel. cr.	...	d.	...	i.	i.	s. KCN; i. al., et.
Cf. also under *Gold*									
Barium	Ba	137.36	silv. met.	3.5	850	1140	d.	d.	s. a.; d. al.
acetate	$Ba(C_2H_3O_2)_2$	255.45	col.	2.468	...	...	$58.8^{0°}$	$75.0^{100°}$	
acetate	$Ba(C_2H_3O_2)_2.H_2O$	273.46	wh., tri. pr., 1.517	2.19	$-H_2O$, 41	...	$75^{30°}$(anh.)	$79^{40°}$(anh.)	i. al.
bromide	$BaBr_2$	297.19	col.	$4.781\frac{24°}{4}$	847	d.	$98^{0°}$	$149^{100°}$	v. s. m. al.; v. sl. s. act.
bromide	$BaBr_2.2H_2O$	333.22	col., mn., 1.7266	3.69	$-2H_2O$, 100	d.	v. s.	v. s.	s. al.
carbonate (witherite)	$BaCO_3$	197.37	wh., rhb., 1.676	4.29	tr. 811 to α	d. 1450	$0.0022^{18°}$	$0.0065^{100°}$	s. a.; i. al.
carbonate (α)	$BaCO_3$	197.37	wh., hex.	...	tr. 982 to β				
carbonate (β)	$BaCO_3$	197.37	wh.	...	1740^{90atm}	...	$0.0022^{18°}$	$0.0065^{100°}$	s. a.; i. al.

Barium chlorate	$Ba(ClO_3)_2$	304.27	col.	....	414	....	$20.35^{0°}$	$84.8^{80°}$	
chlorate	$Ba(ClO_3)_2.H_2O$†	322.29	col., mn., 1.577	3.179	d. 120	....	s.	s.	sl. s. al., act.
chloride	$BaCl_2$	208.27	col., mn., 1.7361	$3.856^{24°}_{4}$	tr. 925	1560	$31^{0°}$	$59^{100°}$	sl. s. HCl, HNO_3; i. al.
chloride	$BaCl_2$	208.27	col., cb.	....	962	1560			
chloride	$BaCl_2.2H_2O$†	244.31	col., mn., 1.646	$3.097^{24°}_{4}$	$-2H_2O$, 100	....	$39.3^{0°}$	$76.8^{100°}$	sl. s. HCl, HNO_3; i. al.
hydroxide	$Ba(OH)_2$	171.38	col., mn.	4.495	....	....	$1.67^{0°}$	$101.4^{80°}$	
hydroxide	$Ba(OH)_2.8H_2O$	315.50	col., mn., 1.5017	$2.188^{16°}$	77.9	$-8H_2O$, 550	$5.6^{15°}$	....	v. sl. s. al.; i. et.
nitrate (nitrobarite)	$Ba(NO_3)_2$	261.38	col., cb., 1.572	$3.244^{23°}$	592	d.	$5.0^{0°}$	$34.2^{100°}$	sl. s. a.; i. al.
oxalate	BaC_2O_4	225.38	wh. cr.	2.658	....	....	$0.0016^{8°}$	$0.0024^{24°}$	s. a., NH_4Cl; i. al.
oxide	BaO	153.36	col., cb., 1.98	5.72	1923	2000±	$1.5^{0°}$	$90.8^{80°}$	s. HCl, HNO_3, abs. al.; i. NH_3, act.
peroxide	BaO_2†	169.36	gray or wh. pd.	4.958	−O, 800	....	v. sl. s.	d.	s. dil. a.; i. act.
peroxide	$BaO_2.8H_2O$	313.49	pearly sc.	....	$-8H_2O$, 100	....	0.168	d.	s. dil. a.; i. al., et., act.
phosphate, monobasic	$BaH_4(PO_4)_2$	331.35	tri.	$2.9^{4°}$	....	....	d.	d.	s. a.
phosphate, dibasic	$BaHPO_4$	233.35	wh., rhb. nd., 1.635	$4.165^{15°}$	....	....	0.015	....	s. a., NH_4 salts
phosphate, tribasic	$Ba_3(PO_4)_2$	602.04	wh., cb.	$4.1^{16°}$	....	....	i.	....	s. a.
phosphate, pyro-	$Ba_2P_2O_7$	448.68	wh., rhb.	$3.9^{20°}$	....	....	0.01	....	s. a., NH_4 salts
silicofluoride	$BaSiF_6$	279.42	pr.	$4.279^{15°}$	....	....	$0.026^{17°}$	$0.09^{100°}$	sl. s. HCl, NH_4Cl; i. al.
sulfate (barite, barytes)	$BaSO_4$	233.42	col., rhb., 1.636	$4.499^{15°}$	1580 d.	tr. to mn. 1149	$0.000115^{0°}$	$0.000285^{30°}$	s. conc. H_2SO_4; 0.006, 3% HCl
sulfide, mono-	BaS	169.42	col., cb., 2.155	$4.25^{15°}$	....	....	d.	d.	d. HCl; i. al.
sulfide, tri-	BaS_3	233.54	yel.-gn.	....	d. 400	....	s.	s.	
sulfide, tetra-	$BaS_4.2H_2O$	301.63	red, rhb.	$2.988^{20°}$	d. 200	....	$41^{15°}$	v. s.	i. al., CS_2
Beryllium (glucinum)	Be(Gl)	9.02	gray, met., hex.	1.816	1284	2767	i.	sl. s. d.	s. dil. a., alk.
Bismuth	Bi	209.00	silv. wh. or reddish, hex.	$9.80^{20°}$	271	1450	i.	i.	s. aq. reg., conc. H_2SO_4, HNO_3
carbonate, sub-	$Bi_2O_3.CO_2.H_2O$	528.03	wh. pd.	6.86	d.	....	i.	i.	s. a.
chloride, di-	$BiCl_2$(?)	279.91	bk. nd.	4.86	163	d. 300	d.		
chloride, tri-	$BiCl_3$†	315.37	wh. cr.	4.75	230	447	d.	....	s. al.
nitrate	$Bi(NO_3)_3.5H_2O$	485.10	col., tri.	2.82	d. 30	$-5H_2O$, 80	d.	....	$42^{19°}$ act.; s. a.; i. al.
nitrate, sub-	$BiONO_3.H_2O$	305.02	hex. pl.	$4.928^{15°}$	d. 260	....	i.	i.	s. a.
oxide, tri-	Bi_2O_3	466.00	yel., rhb.	8.9	820	1900±	i.	i.	s. a.
oxide, tri-	Bi_2O_3	466.00	yel., tet.	8.55	860	....	i.	i.	s. a.
oxide, tri-	Bi_2O_3	466.00	yel., cb.	8.20	tr. 704	....	i.	i.	s. a.
oxychloride	BiOCl	260.46	wh., amor.	$7.72^{15°}$	....	....	sl. s.	sl. s.	s. a.; i. act., NH_3, $H_2C_4H_4O_6$
Boric acid	H_3BO_3	61.84	wh., tri.	$1.435^{15°}$	185 d.	....	$2.66^{0°}$	$40.2^{100°}$	$22.2^{20°}$ gly., $0.24^{25°}$ et.; s. al.
Boron	B	10.82	gray or bk., amor. or mn.	2.32	2300	2550	i.	i.	s. HNO_3; i. al.
carbide	B_4C	55.29	bk. cr.	2.54	2450	>3500	i.	i.	i. a.
oxide	B_2O_3	69.64	col. glass, 1.459	1.85	577	>1500	$1.1^{0°}$	$15.7^{100°}$	s. a., al., gly.
oxide (sassolite)	$B_2O_3.3H_2O$	123.69	tri., 1.456	1.49	d.	....	sl. s.	s.	
Bromic acid	$HBrO_3$	128.92	col.; in soln. only	....	d. 100	....	v. s.	d.	
Bromine	Br_2	159.83	rhb., or red lq.	$3.119^{20°}$; 5.87 (A)	−7.2	58.78	$4.22^{0°}$	$3.13^{30°}$	s. al., et., alk., CS_2
hydrate	$Br_2.10H_2O$	339.99	red, oct.	....	d. 6.8	....	s.		
Cadmium	Cd	112.41	silv. met., hex.	$8.65^{20°}$	320.9	767	i.	i.	s. a., NH_4NO_3
acetate	$Cd(C_2H_3O_2)_2$	230.50	col.	2.341	256	d.	v. s.	....	s. m. al.
acetate	$Cd(C_2H_3O_2)_2.2H_2O$†	266.53	col., mn.	2.01	$-H_2O$, 130	....	v. s.	....	s. al.
carbonate	$CdCO_3$	172.42	wh., trig.	$4.258^{4°}$	d. <500	....	i.	i.	s. a., KCN, NH_4 salts; i. NH_3
chloride	$CdCl_2$	183.32	wh., cb.	$4.047^{25°}_{4}$	568	960	$90^{0°}$	$147^{100°}$	$1.52^{15°}$ al.; i. et., act.
chloride	$CdCl_2.2\frac{1}{2}H_2O$	228.36	col., mn., 1.6513	3.327	tr. 34	....	$168^{20°}$	$180^{100°}$	$2.05^{15°}$ m. al.
cyanide	$Cd(CN)_2$	164.45	....	....	d. >200	....	$0.0247^{18°}$	....	s. a.; NH_4OH, KCN
hydroxide	$Cd(OH)_2$	146.43	wh., trig.	$4.79^{15°}_{4}$	d. 300	....	$0.00026^{25°}$	....	s. a., NH_4 salts; i. alk.
nitrate	$Cd(NO_3)_2$	236.43	col.	....	350	....	$109.7^{0°}$	$326^{59.5°}$	v. s. a.
nitrate	$Cd(NO_3)_2.4H_2O$†	308.49	col. nd.	$2.455^{17°}_{4}$	59.4	132	$215^{0°}$	....	s. al., NH_3; i. HNO_3
oxide	CdO	128.41	brn., cb.	8.15	....	....	i.	i.	s. a., NH_4 salts; i. alk.
oxide	CdO	128.41	brn., amor, 2.49	6.95	d. 900–1000	....	i.	i.	s. a., NH_4 salts; i. alk.
oxide, sub-	Cd_2O	240.82	gn., amor.	$8.192^{18°}_{4}$	d.	....	....	....	d. a., alk.

* Usually the solution.
† Usual commercial form.

Table 1. Physical Properties of the Elements and Inorganic Compounds—(*Continued*)

Name	Formula	Formula weight	Color, crystalline form and refractive index	Specific gravity	Melting point, °C.	Boiling point, °C.	Solubility in 100 parts: Cold water	Hot water	Other reagents
Cadmium sulfate	$CdSO_4$	208.47	rhb.	$4.691^{24°}_{4}$	1000	...	$76.5^{0°}$	$60.8^{100°}$	i. act., NH_3
sulfate	$CdSO_4.H_2O$	226.49	mn.	$3.786^{20°}$	tr. 108	...	s.	s.	
sulfate	$3CdSO_4.8H_2O$*	769.54	col., mn., 1.565	3.09	tr. 41.5	...	$114.2^{0°}$	$127.6^{60°}$	i. al.
sulfate	$CdSO_4.4H_2O$	280.53	col.	3.05	...	...	s.	s.	i. al.
sulfate	$CdSO_4.7H_2O$	334.58	mn.	$2.48^{20°}_{4}$	tr. 4	...	$350^{-5°}$	...	i. al.
sulfide (greenockite)	CdS	144.47	yel.-or., hex., 2.506	4.58	1750^{100atm}	subl. in N_2, 980	0.000001	Colloidal	s. a.; v. s. NH_4OH
Calcium	Ca	40.08	silv. met., cb.	$1.55^{20°}$	810	1200 ± 30	d.	d.	s. a.; sl. s. al.
acetate	$Ca(C_2H_3O_2)_2.H_2O$	176.18	wh. nd.	...	d.	...	$52^{0°}$	$45.5^{80°}$	sl. s. al.
aluminate	$Ca(AlO_2)_2$	158.02	col., rhb. or mn.	$3.67^{20°}$	1600	...	d.	...	s. HCl
aluminum silicate (anorthite)	$CaO.Al_2O_3.2SiO_2$	278.14	tri., 1.5832	2.765	1551				
arsenate	$Ca_3(AsO_4)_2$	398.06	wh. pd.	...	...	...	$0.013^{25°}$	i.	s. dil. a.
bromide	$CaBr_2$	199.91	delq. nd.	$3.353^{25°}_{4}$	760	810	$125^{0°}$	$312^{105°}$	s. al., act.; sl. s. NH_3
carbonate (aragonite)	$CaCO_3$	100.09	col., rhb., 1.6809	2.93	d. 825	...	$0.0012^{20°}$†	$0.002^{100°}$	s. a., NH_4Cl
carbonate (calcite)	$CaCO_3$	100.09	col., hex., 1.550	$2.711^{25°}_{4}$	$1339^{103atm.}$	...	$0.0014^{25°}$	$0.002^{100°}$	s. a., NH_4Cl
chloride (hydrophilite)	$CaCl_2$*	110.99	wh., delq., cb, 1.52	$2.152^{15°}_{4}$	772	>1600	$59.5^{0°}$	$347^{260°}$	s. al.
chloride	$CaCl_2.H_2O$	129.01	col., delq.	...	...	...	s.	s.	s. al.
chloride	$CaCl_2.6H_2O$	219.09	col., trig., 1.417	$1.68^{17°}$	29.92	$-6H_2O$, 200	v. s.	v. s.	s. al.
citrate	$Ca_3(C_6H_5O_7)_2.4H_2O$	570.50	col. nd.	...	$-2H_2O$, 130	$-4H_2O$, 185	$0.085^{18°}$	$0.096^{25°}$	$0.0065^{18°}$ al.
cyanamide	$CaCN_2$	80.11	col., rhombohedral	...	...	...	s. d.	d.	
ferrocyanide	$Ca_2Fe(CN)_6.12H_2O$	508.31	yel., tri., 1.5818	1.7	...	...	s.	$150^{90°}$	i. al.
fluoride (fluorite)	CaF_2	78.08	wh., cb., 1.4339	$3.180^{20°}$	1330	...	$0.0016^{18°}$	$0.0017^{26°}$	sl. s. a.
formate	$Ca(HCO_2)_2$	130.12	col., rhb.	2.015	d.	...	$16.1^{0°}$	$18.4^{100°}$	i. al., et.
hydride	CaH_2	42.10	wh. cr. or pd.	1.7	d. 675	...	d.	...	d. a.; i. bz.
hydroxide	$Ca(OH)_2$	74.10	col., hex., 1.574	2.2	$-H_2O$, 580	...	$0.185^{0°}$	$0.077^{100°}$	s. NH_4Cl
hypochlorite	$Ca(ClO)_2.4H_2O$	215.06	wh., feathery cr.	...	d.	...	delq.; d.	d.	d. a.
hypophosphate	$Ca_2P_2O_6.2H_2O$	274.15	granular	...	$-2H_2O$, 200	...	i.	...	s. HCl, $H_4P_2O_6$
lactate	$Ca(C_3H_5O_3)_2.5H_2O$	308.30	col., eff.	...	$-3H_2O$, 100	...	10.5	∞	∞ h. al.; i. et.
magnesium carbonate (dolomite)	$CaO.MgO.2CO_2$	184.42	trig., 1.68174	2.872	d. 730–760	...	$0.032^{18°}$		
magnesium silicate (diopside)	$CaO.MgO.2SiO_2$	216.52	wh., mn.	3.3	1391	...	i.	i.	
nitrate (nitrocalcite)	$Ca(NO_3)_2$	164.10	col., cb.	2.36	561	...	$102^{0°}$	$376^{151°}$	$14^{15°}$ al.; s. amyl al., NH_3
nitrate	$Ca(NO_3)_2.4H_2O$*	236.16	col., mn., 1.498	1.82	42.7	...	$266^{0°}$	v. s.	
nitride	Ca_3N_2	148.26	brn. cr.	$2.63^{17°}$	900	...	d.	d.	s. dil. a.; i. abs. al.
nitrite	$Ca(NO_2)_2.H_2O$	150.11	delq., hex.	$2.23^{34°}$	...	...	$77^{0°}$	$417^{90°}$	s. 90% al.
oxalate	CaC_2O_4	128.10	col., cb.	$2.2^{4°}$	d.	...	$0.00067^{13°}$	$0.0014^{95°}$	s. a.; i. ac.
oxalate	$CaC_2O_4.H_2O$	146.12	col.	2.2	$-H_2O$, 200	...	i.	i.	s. a.; i. ac.
oxide	CaO	56.08	col., cb., 1.837	3.32	2570	2850	Forms $Ca(OH)_2$	...	s. a.; i. al.
peroxide	$CaO_2.8H_2O$	216.21	pearly, tet.	...	$-8H_2O$, 100	expl. 275	sl. s.	d.	s. a. d.; i. al., et.
phosphate, monobasic	$CaH_4(PO_4)_2.H_2O$	252.09	wh., tri.	$2.220^{16°}_{4}$	$-H_2O$, 100	d. 200	...	d.	
phosphate, dibasic	$CaHPO_4.2H_2O$	172.10	wh., mn. pl.	$2.306^{16°}_{4}$	d.	...	$0.02^{24.5°}$	$0.075^{100°}$	
phosphate, tribasic	$Ca_3(PO_4)_2$	310.20	wh., amor.	3.14	1670	...	0.0025	d.	s. a.; i. al., ac.
phosphate, meta-	$Ca(PO_3)_2$	198.04	wh., tet., 1.588	2.82	975	...	i.	i.	i. a.
phosphate, pyro-	$Ca_2P_2O_7$	254.12	col., biaxial, 1.60	3.09	1230	...	i.	...	s. a.
phosphate, pyro- (brushite)	$Ca_2P_2O_7.5H_2O$	344.20	wh., mn.	2.25	...	...	sl. s.	...	s. a.; i. NH_4Cl
phosphide	Ca_3P_2	182.20	red cr.	$2.51^{15°}$	>1600	...	d.	...	s. dil. a.; i. al., et.
silicate (α) (pseudowollastonite)	$CaSiO_3$	116.14	col., pseudo hex., 1.6150 or mn.(?)	2.905	1540	...	$0.0095^{17°}$	...	s. HCl
silicate (β) (wollastonite)	$CaSiO_3$	116.14	col., mn., 1.610	2.915	tr. 1190 to α				
sulfate (anhydrite)	$CaSO_4$	136.14	col., rhb., 1.576, or mn, 1.50	2.96	1450(mn.)	tr. 1193 to rhb.	$0.298^{20°}$	$0.1619^{100°}$	s. a., $Na_2S_2O_3$, NH_4 salts
sulfate (gypsum)	$CaSO_4.2H_2O$	172.17	col., mn., 1.5226	2.32	$-1½H_2O$, 128	$-2H_2O$, 163	$0.223^{0°}$	$0.257^{50°}$	s. a., gly., $Na_2S_2O_3$, NH_4 salts
sulfhydrate	$Ca(SH)_2.6H_2O$	214.31	col. pr.	...	d. 15	...	v. s.	v. s.	s. al.
sulfide (oldhamite)	CaS	72.14	col., cb.	$2.8^{15°}$	...	...	d.	d.	s. a.
sulfite	$CaSO_3.2H_2O$	156.17	wh., cr., 1.595	...	$-2H_2O$, 100	d. 650	$0.0043^{18°}$	$0.0027^{90°}$	s. H_2SO_3
tartrate	$CaC_4H_4O_6.4H_2O$	260.22	col., rhb.	...	d.	...	$0.037^{0°}$	$0.22^{85°}$	sl. s. al.
thiocyanate	$Ca(CNS)_2.3H_2O$	210.28	wh., delq. cr.	...	...	...	s.	v. s.	v. s. al.
thiosulfate	$CaS_2O_3.6H_2O$	260.30	col., tri., 1.56	$1.873^{16°}$	d.	...	$71.2^{9°}$	d.	i. al.
tungstate (scheelite)	$CaWO_4$	288.00	wh., tet., 1.9200	6.06	...	...	0.2	...	s. NH_4Cl; i. a.

Carbon, *Cf.* table of organic compounds									
Carbon, amorphous	C	12.01	bk., amor.	1.8–2.1	>3500	4200	i.	i.	i. a., alk.
Carbon, diamond	C	12.01	col., cb., 2.4195	$3.51^{20°}$	>3500	4200	i.	i.	i. a., alk.
Carbon, graphite	C	12.01	bk., hex.	$2.26^{20°}$	>3500	4200	i.	i.	i. a., alk.
dioxide	CO_2	44.01	col. gas	lq. $1.101^{-37°}$; 1.53 (A); solid $1.56^{-79°}$	$-56.6^{5.2 atm.}$	subl. −78.5	179.7^0° cc	$90.1^{20°}$ cc	s. a., alk.
disulfide	CS_2	76.13	col. lq.	lq. $1.261\frac{22°}{20}$; 2.63 (A)	−108.6	46.3	$0.2^{0°}$	$0.014^{50°}$	s. al., et.
monoxide	CO	28.01	col., poisonous, odorless gas	lq. $0.814^{-195°}/4$; 0.968 (A)	−207	−192	$0.0044^{0°}$; $3.5^{0°}$ cc	$0.0018^{50°}$; $2.32^{20°}$ cc	s. al., Cu_2Cl_2
oxychloride (phosgene)	$COCl_2$	98.92	poisonous gas	$1.392\frac{19°}{4}$	−104	8.2^{756mm}	v. s. sl. d.	d.	s. ac., CCl_4, bz.; d.a.
oxysulfide	COS	60.07	gas	lq. $1.24^{-87°}$; 2.10 (A)	−138.2	-50.2^{760mm}	$133^{0°}$ cc	$40.3^{30°}$ cc	v. s. alk., al.
suboxide	C_3O_2	68.03	gas	lq. $1.114^{0°}$	−107	7^{761mm}	d.	.....	s. et.
thionyl chloride	$CSCl_2$	114.98	yel.-red lq.	$1.509^{15°}$	.....	73.5			
Ceric hydroxide	$2CeO_2.3H_2O$	398.31	yel., gelatinous	.....	.....	.....	.....	.....	s. a.; sl. s. alk. carb.; i. alk
hydroxynitrate	$Ce(OH)(NO_3)_3.3H_2O$	397.21	red, mn.	.....	.....	.....	d.		
oxide	CeO_2	172.13	wh. or pa. yel., cb.	7.3	1950	.....	i.	i.	s. H_2SO_4, HCl
sulfate	$Ce(SO_4)_2.4H_2O$	404.31	yel., rhb.	3.91	.....	.....	s. d.	.....	s. dil. H_2SO_4
Cerium	Ce	140.13	steel gray, cb. or hex.	$6.9^{20°}$ cb.; 6.7 hex.	645	1400	i.	Slowly oxidized	s. dil. a.; i. al.
Cerous sulfate	$Ce_2(SO_4)_3$	568.44	wh., mn. or rhb.	3.91	.....	.....	$18.98^{0°}$	$0.4^{100°}$	
sulfate	$Ce_2(SO_4)_3.8H_2O$	712.57	tri.	$2.886^{17°}$	$-8H_2O$, 630	.....	$25^{0°}$	$7.6^{40°}$	
Cesium	Ce	132.91	silv. met., hex.	$1.90^{20°}$	28.5	670	d.	.....	s. a., al., NH_3
Chloric acid	$HClO_3.7H_2O$	210.58	lq.	$1.282^{14.2°}$	<−20	d. 40	v. s.		
Chlorine	Cl_2	70.91	rhb., or gn.-yel. gas	lq. $1.56^{-33.6°}$; $2.49^{0°}$ (A)	−101.6	−34.6	$1.46^{0°}$; $310^{10°}$ cc	$0.57^{30°}$; $177^{30°}$ cc	s. alk.
hydrate	$Cl_2.8H_2O$	215.04	rhb.	1.23	d. 9.6	.....	s.	.....	s. alk.
Chloroplatinic acid	$H_2PtCl_6.6H_2O$	518.08	red-brn., delq.	2.431	60	.....	v. s.	v. s.	s. al., et.
Chlorostannic acid	$H_2SnCl_6.6H_2O$	441.55	delq.	$1.971^{28°}$	19.2	.....	s.		
Chlorosulfonic acid	$HO.SO_2.Cl$	116.52	col. lq.	$1.787^{25°}$	−80	151.5^{765mm}	d.	.....	d. al.; i. CS_2
Chromic acetate	$Cr_2(C_2H_3O_2)_6.2H_2O$	494.32	gn.	.....	.....	.....	s.	.....	$4.76^{15°}$ m. al.
chloride	$CrCl_3$	158.38	pink, trig.	$2.757^{15°}$	.....	1200–1500 d.	i.§	sl. s.	i. a., act., CS_2
chloride	$CrCl_3.6H_2O$*	266.48	vl. or gn., hex. pl.	$1.835\frac{25°}{4}$	subl. 83	.....	v. s. d.	.....	s. al.; i. et.
fluoride	CrF_3	109.01	gn., rhb.	3.8	>1000	d.	i.	.....	sl. s. a.; i. al., NH_3
hydroxide	$Cr(OH)_3$	103.03	gn. or blue, gelatinous	.....	.....	.....	i.	.....	s. a., alk.; sl. s. NH_3
hydroxide	$Cr(OH)_3.2H_2O$	139.07	gn.	.....	$-2H_2O$, 100	.....	i.	i.	s. a., alk.
nitrate	$Cr(NO_3)_3.9H_2O$*	400.18	purple pr.	.....	36.5	d. 100	s.	s.	s. a., alk., al., act.
nitrate	$Cr(NO_3)_3.7\frac{1}{2}H_2O$	373.15	purple, mn.	.....	100	d.	s.	s.	
oxide	Cr_2O_3	152.02	dark gn., hex.	5.21	1900	.....	i.	i.	sl. s. a.
sulfate	$Cr_2(SO_4)_3$	392.20	rose pd.	3.012	.....	.....	i.§	.....	i. a.
sulfate	$Cr_2(SO_4)_3.5H_2O$	482.28	gn.	.....	.....	.....	s.	.....	s. al., H_2SO_4
sulfate	$Cr_2(SO_4)_3.15H_2O$	662.44	vl.	$1.867^{17°}$	100	$-10H_2O$, 100	s.	d. 67°	sl. s. al.
sulfate	$Cr_2(SO_4)_3.18H_2O$	716.49	vl., cb., 1.564	$1.7^{22°}$	.....	$-12H_2O$, 100	$120^{20°}$	d.	s. al.
sulfide	Cr_2S_3	200.02	brn.-bk. pd.	$3.77^{19°}$	−S, 1350	.....	i.	d.	s. h. HNO_3
Chromium	Cr	52.01	gray, met., cb.	7.1	1615	2200	i.	i.	s. HCl, dil. H_2SO_4; i. HNO_3
trioxide (chromic acid)	CrO_3	100.01	red, rhb.	2.70	197 d.	.....	$164.9^{0°}$	$206.7^{100°}$	s. H_2SO_4, al., et.
Chromous chloride	$CrCl_2$	122.92	wh., delq.	2.75	.....	.....	v. s.	v. s.	sl. s. al.; i. et.
hydroxide	$Cr(OH)_2$	86.03	yel.-brn.	.....	d.	.....	d.	.....	s. conc. a.
oxide	CrO	68.01	bk. pd.	.....	.....	.....	i.	i.	i. dil. HNO_3
sulfate	$CrSO_4.7H_2O$	274.18	blue	.....	.....	.....	12.35^{0}	.....	sl. s. al.
sulfide (daubrelite)	CrS	84.07	bk. pd.	3.97	1550	.....	i.	.....	v. s. a.
Chromyl chloride	CrO_2Cl_2	154.92	dark red lq.	1.92	−96.5	117.6	d.	.....	s. et.
Cobalt	Co	58.94	silv. met., cb.	$8.9^{20°}$	1480	2900	i.	i.	s. a.
carbonyl	$Co(CO)_4$	170.98	or. cr.	$1.73^{18°}$	51	d. 52	i.	d.	s. al., et., CS_2
sulfide, di-	CoS_2	123.06	bk., cb.	4.269	.....	.....	i.	.....	s. HNO_3, aq. reg.
Cobaltic chloride	$CoCl_3$	165.31	red cr.	2.94	subl.	.....	s.	s.	
chloride, dichro	$Co(NH_3)_3Cl_3.H_2O$	234.42	.....	.....	.....	.....	s.	.....	s. a., al.
chloride, luteo	$Co(NH_3)_6Cl_3$	267.50	or., mn.	$1.7016^{20°}$	.....	.....	$4.26^{0°}$	$12.74^{46.5°}$	i. al., NH_4OH
chloride, praseo	$Co(NH_3)_4Cl_3.H_2O$	251.46	gn., rhb.	1.847	.....	.....	v. s.	.....	s. a.; i. al.

* Usual commercial form.
† The solubility of $CaCO_3$ in H_2O is greatly increased by increasing the amount of CO_2 in the H_2O.
§ Also a soluble modification.

Table 1. Physical Properties of the Elements and Inorganic Compounds—(*Continued*)

Name	Formula	Formula weight	Color, crystalline form and refractive index	Specific gravity	Melting point, °C.	Boiling point, °C.	Solubility in 100 parts: Cold water	Hot water	Other reagents
Cobaltic chloride, purpureo	$Co(NH_3)_5Cl_3$	250.47	rhb.	$1.819^{25°}_{25}$			$0.232^{0°}$	$1.031^{46.5°}$	i. al.
chloride, roseo	$Co(NH_3)_5Cl_3.H_2O$	268.49	brick red		d. 100		$16.12^{0°}$	$24.87^{16°}$	sl. s. HCl
hydroxide	$Co(OH)_3$	109.96	bk.		$-1\frac{1}{2}H_2O$, 100		i.	i.	s. a.; i. al.
oxide	Co_2O_3	165.88	bk.	5.18	d. 900		i.	i.	s. a.
sulfate	$Co_2(SO_4)_3$	406.06	blue cr.				d.		s. H_2SO_4
sulfide	Co_2S_3	214.06	bk. cr.	4.8			i.		d. a.
Cobalto-cobaltic oxide	Co_3O_4	240.82	bk., cb.	6.07			i.	i.	s. H_2SO_4; i. HCl, HNO_3
Cobaltous acetate	$Co(C_2H_3O_2)_2.4H_2O$	249.09	red-vl., mn., 1.542	$1.705^{18.7°}$	$-4H_2O$, 140		s.	s.	s. a., al.
chloride	$CoCl_2$	129.85	blue cr.	3.356	subl.	1049	$45^{7°}$	$105^{96°}$	31 al.; 8.6 act.
chloride	$CoCl_2.6H_2O$*	237.95	red, mn.	$1.924^{25°}_{25}$	86	$-6H_2O$, 110	$116.5^{0°}$	$177^{30°}$	v. s. et., act.
nitrate	$Co(NO_3)_2.6H_2O$	291.05	red, mn., 1.4	$1.883^{25°}_{25}$	<100	d.	$84.03^{0°}$(anh.)	$334.9^{90°}$(anh.)	$100^{12.5°}$ al.; s. act.; sl. s. NH_3
oxide	CoO	74.94	brn., cb.	5.68	d. 1800		i.	i.	s. a., NH_4OH; i. al.
sulfate	$CoSO_4$	155.00	red pd.	$3.710^{25°}$	d. 880		$25.6^{0°}$	$83^{100°}$	$1.04^{18°}$ m. al.; i. NH_3
sulfate	$CoSO_4.H_2O$	173.02	red pd., mn.(?), 1.639	3.13	d.		s.	s.	
sulfate (biebeorite)	$CoSO_4.7H_2O$*	281.11	red, mn., 1.483	$1.948^{25°}_{25}$	96.8	$-7H_2O$, 420	$33^{20°}$	s.	$2.5^{3°}$ al.
sulfide (syeporite)	CoS	91.00	brn. nd.	$5.45^{18°}$	>1100		$0.00038^{18°}$		s. a., aq. reg.
Copper	Cu	63.57	yel.-red met., cb.	$8.92^{20°}$	1083	2300	i.	i.	s. HNO_3, h. H_2SO_4
Cupric acetate	$Cu(C_2H_3O_2)_2$	181.66		$1.930^{20°}_{4}$			s.		
acetate	$Cu(C_2H_3O_2)_2.H_2O$	199.67	dark gn., mn.	1.882	115	240 d.	7.2	20	7 al.; s. et., gly.
aceto-arsenite (Paris green)	$(CuOAs_2O_3)_3.Cu(C_2H_3O_2)_2$*	1013.83	gn.				i.		s. a., NH_4OH
ammonium chloride	$CuCl_2.2NH_4Cl.2H_2O$	277.51	blue, tet., 1.670, 1.744	1.98	d. 110		$33.8^{0°}$	$99.3^{80°}$	s. a.
ammonium sulfate	$CuSO_4.4NH_3.H_2O$	245.77	blue, rhb.	1.81	d. 150		$18.05^{21.5°}$	d.	i. al.
carbonate, basic (azurite)	$2CuCO_3.Cu(OH)_2$	344.75	blue, mn., 1.758	3.88	d. 220		i.	d.	s. NH_4OH, h. aq. $NaHCO_3$
carbonate, basic (malachite)	$CuCO_3.Cu(OH)_2$	221.17	dark gn., mn., 1.875	3.9	d.		i.	d.	s. KCN; 0.03 aq. CO_2
chloride (eriochalcite)	$CuCl_2$	134.48	brn.-yel. pd.	3.054	498	Forms Cu_2Cl_2 993	$70.7^{0°}$	$107.9^{100°}$	$53^{15°}$ al.; $68^{15°}$ m. al.
chloride	$CuCl_2.2H_2O$	170.52	gn., rhb., 1.684	$2.39^{22.4°}$	$-2H_2O$, 110	d.	$110.4^{0°}$	$192.4^{100°}$	s. al., et., NH_4Cl
chromate, basic	$CuCrO_4.2CuO.2H_2O$	374.75	yel.-brn.		$-2H_2O$, 260		i.		s. HNO_3, NH_4OH
cyanide	$Cu(CN)_2$	115.61	yel.-gn.		d.		i.		s. KCN, C_5H_5N
dichromate	$CuCr_2O_7.2H_2O$	315.62	bk., tri.	$2.286^{18°}$	$-2H_2O$, 100		sl. s.	d.	s. a., NH_4OH
ferricyanide	$Cu_3[Fe(CN)_6]_2$	614.63	yel.-gn.				i.		s. NH_4OH; i. HCl
ferrocyanide	$Cu_2Fe(CN)_6.7H_2O$	465.21	red-brn.				i.	i.	s. NH_4OH; i. a., NH_3
formate	$Cu(HCO_2)_2$	153.61	blue, mn.	1.831			12.5	d.	0.25 al.
hydroxide	$Cu(OH)_2$	97.59	blue, gelatinous	3.368	$-H_2O$		i.	d.	s. a., NH_4OH, KCN, al.
lactate	$Cu(C_3H_5O_3)_2.2H_2O$	277.74	dark blue, mn.				16.7	$45^{100°}$	sl. s. al.
nitrate	$Cu(NO_3)_2.3H_2O$*	241.63	blue, delq.	$2.047^{3.9°}$	114.5	$-HNO_3$, 170	$381^{40°}$	$666^{80°}$	$100^{12.5°}$ al.
nitrate	$Cu(NO_3)_2.6H_2O$	295.68	blue, rhb.	2.074	$-3H_2O$, 26.4		$243.7^{0°}$	∞	s. al.
oxide (paramelaconite)	CuO	79.57	bk., cb.	6.40	d. 1026		i.	i.	s. a., KCN, NH_4Cl
oxide (tenorite)	CuO	79.57	bk., tri., 2.63	6.45	d. 1026		i.	i.	s. a., KCN, NH_4Cl
oxychloride	$CuCl_2.2CuO.4H_2O$	365.69	blue-gn.		$-3H_2O$, 140		i.		s. a.
phosphide	Cu_3P_2	252.67	bk.	6.35	d.		i.		s. HNO_3; i. HCl
sulfate (hydrocyanite)	$CuSO_4$	159.63	gn.-wh., rhb., 1.733	$3.606^{15°}$	d. >600	Forms CuO, 650	$14.3^{0°}$	$75.4^{100°}$	i. al.
sulfate (blue vitriol or chalcanthite)	$CuSO_4.5H_2O$*	249.71	blue, tri., 1.5368	$2.286^{15.6°}_{4}$	$-4H_2O$, 110	$-5H_2O$, 250	$24.3^{0°}$	$205^{100°}$	$1.1^{3°}$ al.
sulfide (covellite)	CuS	95.63	blue, hex. or mn., 1.45	4.6	tr. 103	d. 220	$0.000033^{18°}$		s. HNO_3, KCN
tartate	$CuC_4H_4O_6.3H_2O$	265.69	lt. gn. pd.		d.		$0.02^{15°}$	$0.14^{85°}$	s. a., KOH
Cuprous ammonium iodide	$CuI.NH_4I.H_2O$	353.47	rhb. pl.				d.		s. NH_4I
carbonate	Cu_2CO_3	187.15	yel.	4.4	d.		i.	i.	s. a., NH_4OH
chloride (nantokite)	Cu_2Cl_2	198.05	wh., cb., 1.973	3.53	422	1366	$1.52^{25°}$		s. HCl, NH_4OH, al.
cyanide	$Cu_2(CN)_2$	179.16	wh., mn.	2.9	474.5	d.	i.	i.	s. KCN, HCl, NH_4OH; sl. s. NH_3
ferricyanide	$Cu_3Fe(CN)_6$	402.67	brn.-red				i.		s. NH_4OH; i. HCl
ferrocyanide	$Cu_4Fe(CN)_6$	466.24	brn.-red				i.		s. NH_4OH; i. NH_4Cl
fluoride	Cu_2F_2	165.14	red cr.		908	subl. 1100	i.		s. HF, HCl, HNO_3; i. al.
hydroxide	CuOH	80.58	yel.	3.4	$-\frac{1}{2}H_2O$, 360		i.	i.	s. a., NH_4OH
oxide (cuprite)	Cu_2O	143.14	red, cb., 2.705	6.0	1235	$-O$, 1800	i.	i.	s. HCl, NH_4Cl, NH_4OH

Cuprous phosphide	Cu_6P_2	443.38	gray-bk.	6.4 to 6.8			i.		s. HNO_3; i. HCl
sulfide (chalcocite)	Cu_2S	159.20	bk., rhb.	5.6	1100		$0.0005^{18°}$		s. HNO_3, NH_4OH; i. act.
sulfide	Cu_2S	159.20	bk., cb.	5.80	1130		$0.0005^{18°}$		s. HNO_3, NH_4OH; i. act.
Cyanogen	C_2N_2	52.02	poisonous gas	lq. $0.866^{-17.2°}$; 1.806 (A)	−34.4	−20.5	$450^{20°}$ cc		$2300^{20°}$ cc al.; $500^{18°}$ cc et.
Cyanogen compounds, *Cf.* table of organic compounds									
Ferric acetate, basic	$Fe(OH)(C_2H_3O_2)_2$	190.95	brn., amor.				i.		s. a., al.
ammonium sulfate, *Cf.* Alum									
chloride (molysite)	$FeCl_3$	162.22	bk.-brn., hex. delq.	$2.804^{11°}$	282	315	$74.4^{0°}$	$535.8^{100°}$	v. s. al., et. +HCl
chloride	$FeCl_3.6H_2O$*	270.32	red-yel., delq.		37	280	$246^{0°}$	∞	s. al., act., gly.
ferrocyanide (Prussian blue)	$Fe_4[Fe(CN)_6]_3$	859.27	dark blue		d.		i.	d.	s. HCl, conc. H_2SO_4; i. al., et.
hydroxide	$Fe(OH)_3$	106.87	red-brn.	3.4 to 3.9	$-1\frac{1}{2}H_2O$, 500		i.	i.	s. a.; i. al., et.
lactate	$Fe(C_3H_5O_3)_3$	323.06	brn., amor., delq.				v. s.	v. s.	i. et.
nitrate	$Fe(NO_3)_3.6H_2O$	349.97	rhb., delq.	$1.684^{20°}$	35	d.	$150^{0°}$	∞	s. al., act.
oxide (hematite)	Fe_2O_3	159.70	red or bk., trig., 3.042	5.12	1560 d.		i.		s. HCl
sulfate	$Fe_2(SO_4)_3$	399.88	rhb., 1.814	$3.097^{18°}$	d. 480		sl. s.	d.	i. H_2SO_4, NH_3
sulfate (coquimbite)	$Fe_2(SO_4)_3.9H_2O$	562.02	yel., trig.	2.1			440	d.	s. abs. al.
Ferroso-ferric chloride	$FeCl_2.2FeCl_3.18H_2O$	775.49	yel., delq.		d. 50		s.	s.	
ferricyanide (Prussian green)	$Fe'''_4Fe''_3[Fe(CN)_6]_6$	1662.70	gn.		d. 180		i.		s. d. h. HCl
oxide (magnetite; magnetic iron oxide)	Fe_3O_4	231.55	bk., cb., 2.42	5.2	1538 d.		i.	i.	i. al.
oxide, hydrated	$Fe_3O_4.4H_2O$	303.61	bk.		d.		i.	i.	s. a.
Ferrous ammonium sulfate	$FeSO_4.(NH_4)_2SO_4.6H_2O$	392.15	blue-gn., mn., 1.4915	1.864	d.		$18^{0°}$	$100^{75°}$	i. al.
chloride (lawrencite)	$FeCl_2$	126.76	gn.-yel., hex., 1.567	2.7		delq.	$64.4^{10°}$	$105.7^{100°}$	100 al.; s. act.; i. et.
chloroplatinate	$FePtCl_6.6H_2O$	571.92	yel., hex.	2.714			v. s.	v. s.	
ferricyanide (Turnbull's blue)	$Fe_3[Fe(CN)_6]_2$	591.47	dark blue		d.		i.		i. dil. a., al.
ferrocyanide	$Fe_2Fe(CN)_6$	323.66	blue-wh., amor.				i.		
formate	$Fe(HCO_2)_2.2H_2O$	181.92			d.		sl. s.		
hydroxide	$Fe(OH)_2$	89.87	lt. gn.	3.4			0.00067		s. a., NH_4Cl
nitrate	$Fe(NO_3)_2.6H_2O$	287.96	cr.		60.5		$200^{0°}$	$300^{25°}$	
oxide	FeO	71.85	bk.	5.7	1420		i.	i.	s. a.; i. alk.
phosphate (vivianite)	$Fe_3(PO_4)_2.8H_2O$	501.64	blue, mn., 1.592, 1.603	2.58			i.	i.	s. a.; i. ac.
silicate	$FeSiO_3$	131.91	mn.	3.5	1550				
sulfate (siderotilate)	$FeSO_4.5H_2O$	241.99	gn., tri., 1.536	2.2		$-5H_2O$, 300	s.	s.	i. al.
sulfate (copperas)	$FeSO_4.7H_2O$*	278.02	blue-gn., mn.	$1.899^{14.8°}$	64	$-7H_2O$, 300	$32.8^{0°}$	$149^{50°}$	i. al.
sulfide	FeS	87.91	bk., hex.	4.84	1193	d.	$0.000616^{18°}$		s. a.; i. NH_3
Cf. also under iron									
Fluoboric acid	HBF_4	87.83	col. lq.			130 d.	∞	∞	s. al.
Fluorine	F_2	38.00	gn.-yel. gas	lq. $1.11^{-137°}$; $1.31^{15°}$ (A)	−223	−187	d.		
Fluosilicic acid	H_2SiF_6	144.08					s.	s.	
Gadolinium	Gd	156.9							
Gallium bromide	$GaBr_3$	309.47	delq. cr.				s.	s.	
Glucinum *Cf.* Beryllium									
Gold	Au	197.20	yel. met., cb.	$19.3^{20°}$	1063	2600	i.	i.	s. aq. reg., KCN; i. a.
Gold, colloidal	Au	197.20	blue to vi.				s.		s. aq. reg., KCN; i. a.
Gold salts *Cf.* under Auric and Aurous									
Hafnium	Hf	178.6	hex.	12.1	>1700	>3200(?)			
Helium	He	4.00	col. gas	0.1368 (A)	<−272.2	−268.9	$0.97^{0°}$ cc	$1.08^{50°}$ cc	Absorbed by Pt
Hydrazine	N_2H_4	32.05	col. lq.	$1.011\frac{15°}{4}$	1.4	113.5	∞	∞	s. al.
formate	$N_2H_4.2HCO_2H$	124.10	cb.		128		s.		
hydrate	$N_2H_4.H_2O$	50.06	col.	$1.03^{21°}$	−40	$118.5^{739.5mm}$	∞	∞	∞ al.; i. et.
hydrochloride	$N_2H_4.HCl$	68.51	yel. lq.				v. s.	v. s.	sl. s. al.
hydrochloride, di-	$N_2H_4.2HCl$	104.98	wh., cb.	1.42	198		s.	v. s.	s. al.
nitrate	$N_2H_4.HNO_3$	95.06	cr.		70.7	subl. 140	$174.9^{10°}$	v. s.	
nitrate, di-	$N_2H_4.2HNO_3$	158.08	nd.		104	d.	v. s.		
sulfate	$N_2H_4.\frac{1}{2}H_2SO_4$	81.09	delq. pl.		85		v. s.		i. al.
sulfate	$N_2H_4.H_2SO_4$	130.12	rhb.	1.378	254		$3.055^{22°}$	$27.65^{60°}$	v. sl. s. abs. al.
Hydrazoic acid (azoimide)	HN_3	43.03	col. lq.		−80	37	∞	∞	∞ al.
Hydriodic acid	HI	127.93	col. gas	$4.4^{0°}$ (A)	−50.8	−35.5	$42500^{10°}$ cc	v. s.	s. al.
Hydriodic acid	$HI.H_2O$	145.94	col. lq.	$1.7^{15°}$		127^{774mm}	∞		∞ al.
Hydriodic acid	$HI.2H_2O$	163.96	col. lq.		−43		∞		s. al.
Hydriodic acid	$HI.3H_2O$	181.98	col. lq.		−48		∞		s. al.
Hydriodic acid	$HI.4H_2O$	199.99	col. lq.		−36.5		∞		s. al.
Hydrobromic acid	HBr	80.92	col. gas; 1.325 (lq.)	$2.71^{0°}$ (A)	−86	−67	$221^{0°}$	$130^{100°}$	s. al.

* Usual commercial form.

Table 1. Physical Properties of the Elements and Inorganic Compounds—(*Continued*)

Name	Formula	Formula weight	Color, crystalline form and refractive index	Specific gravity	Melting point, °C.	Boiling point, °C.	Solubility in 100 parts: Cold water	Hot water	Other reagents
Hydrobromic acid	$HBr.H_2O$	98.94	col. lq.	1.78					Stable at −15.5° and 1 atm., and at −11.3° and 2.5 atm.
Hydrobromic acid	HBr (47.8% in H_2O)	80.92	col. lq.	1.486		126	∞		s. al.
Hydrobromic acid	$HBr.2H_2O$	116.96	wh. cr.	$2.11^{-15°}$	−11		s.	s.	
Hydrochloric acid	HCl*	36.47	col. gas; 1.256 (lq.)	$1.268^{0°}$ (A)	−111	−85	$82.3^{0°}$	$56.1^{60°}$	s. al., et.
Hydrochloric acid	HCl (45.2% in H_2O)	36.47	col. lq.	1.48	−15.35		∞		s. al.
Hydrochloric acid	$HCl.2H_2O$	72.50	col. lq.	$1.46^{-18.3°}_{4}$	0	d.	∞		s. al.
Hydrochloric acid	$HCl.3H_2O$	90.51	col. lq.		−24.4	d.	∞		s. al.
Hydrocyanic acid (prussic acid)	HCN	27.03	poisonous gas or col. lq., 1.254	$0.697^{18°}$	−14	26	∞		∞ al., et.
Hydrofluoric acid	HF	20.01	gas or col. lq.	$0.988^{13.6°}$	−83	19.4	∞ 0° to 19.4°	v. s.	
Hydrofluoric acid	HF (35.35% in H_2O)	20.01	col. lq.	1.15	−35	120	v. s.		
Hydrogen	H_2	2.016	col. gas or cb.	lq. $0.0709^{-252.7°}$ 0.06948 (A)	−259.1	−252.7	$2.1^{0°}$ cc	$0.85^{80°}$ cc	sl. s. Fe, Pd, Pt
peroxide	H_2O_2†	34.02	col. lq., 1.333	$1.438^{20°}_{4}$	−0.89	151.4^{760mm}	∞		s. a., et.; i. petr. et
selenide	H_2Se	81.22	col. gas	$2.12^{-42°}$	−64	−42	$377^{4°}$ cc	$270^{22.5°}$ cc	s. CS_2, $COCl_2$
sulfide	H_2S	34.08	col. gas	1.1895 (A)	−82.9	−59.6	$437^{0°}$ cc	$186^{40°}$ cc	$9.54^{15°}$ cc al.; s. CS_2
Hydroxylamine	NH_2OH	33.03	rhb., delq.	$1.35^{18°}$	34	56.5^{22mm}	s.	d.	s. a., al.
hydrochloride	$NH_2OH.HCl$	69.50	col., mn.	$1.67^{17°}$	151	d.	$83.3^{17°}$	v. s.	s. al.; i. et.
nitrate	$NH_2OH.HNO_3$	96.05	col. cr.		48	d. <100	v. s.	d.	v. s. abs. al.
sulfate	$NH_2OH.\frac{1}{2}H_2SO_4$	82.07	col., mn.		170 d.		$32.9^{0°}$	$68.5^{90°}$	v. sl. s. al.; i. et., abs. al.
Hypobromous acid	HBrO	96.92	yel.			40^{50mm}	s.	d.	
Illinium	Il	146(?)							
Indium	In	114.76	soft, tet. met.	$7.3^{20°}$	155	1450	i.	i.	s. a.
Iodic acid	HIO_3	175.93	col., rhb.	$4.629^{0°}$	110 d.		$286^{0°}$	$576^{101°}$	v. s. 87% al.; i. abs. al, et., chl.
Iodine	I_2	253.84	blue-bk., rhb.	$4.93^{20°}$	113.5	184.35	$0.0162^{0°}$	$0.09566^{60°}$	s. al., KI, et.
oxide, penta-	I_2O_5	333.84	wh., trimetric	$4.799^{25°}_{4}$	d. 300		$187.4^{13°}$		i. abs. al., et., chl.
Iodoplatinic acid	$H_2PtI_6.9H_2O$	1120.91	brn., delq. mn.				s. d.		
Iridium	Ir	193.10	wh. met., cb.	$22.4^{20°}$	2350	>4800	i.	i.	sl. s. aq. reg., aq. Cl_2
Iron, cast‡	Fe	55.85	gray	7.03	1275		i.	i.	s. a.; i. alk.
pure	Fe	55.85	silv. met., cb.	$7.86^{20°}$	1535	3000	i.	i.	s. a.; i. alk.
steel	Fe	55.85	silv. gray	7.6 to 7.8	1375		i.	i.	s. a.; i. alk.
white pig	Fe	55.85	gray	7.6 to 7.8	1075		i.	i.	s. a.; i. alk.
wrought	Fe	55.85	gray	7.86	1505		i.	i.	s. a.; i. alk.
carbide (cementite)	Fe_3C	179.56	pseudo hex.	7.4	1837		i.	i.	s. a.
carbonyl	$Fe(CO)_5$	195.90	pa. yel. lq.	$1.457^{21°}$	−21	102.5^{760mm}	i.		s. al., H_2SO_4, alk.
nitride	Fe_2N	125.71	gray	6.35	d. >560		d.		s. HCl, H_2SO_4
silicide	FeSi	83.91	yel.-gray, oct.	$6.1^{20°}_{4}$			i.	i.	i. aq. reg.
sulfide, di- (marcasite)	FeS_2	119.97	yel., rhb.	4.87	tr. 450	d.	0.00049		i. dil. a.
sulfide, di- (pyrite)	FeS_2	119.97	yel., cb.	5.0	1171	d.	0.0005		i. dil. a.
sulfide (pyrrhotite)	Fe_7S_8	647.43	hex.	$4.6^{20°}_{4}$	d. >700		i.		
Cf. also under ferric and ferrous									
Krypton	Kr	83.70	col. gas	2.818 (A)	−169	−151.8	$11.05^{0°}$ cc	$3.57^{60°}$ cc	sl. s. al., bz.
Lanthanum	La	138.92	lead gray	$6.15^{20°}$	826	1800	d.		s. a.
Lead	Pb	207.21	silv. met., cb.	$11.337^{20°}_{20}$	327.5	1620	i.	i.	s. HNO_3; i. c. HCl, H_2SO_4
acetate	$Pb(C_2H_3O_2)_2$	325.30	wh. cr.	$3.251^{20°}_{4}$	280		$19.7^{0°}$	$221^{50°}$	s. gly.; v. sl. s. al.
acetate (sugar of lead)	$Pb(C_2H_3O_2)_2.3H_2O$§	379.35	wh., mn.	2.55	$-3H_2O$, 75		$45.64^{15°}$	$200^{100°}$	s. gly.; sl. s. al.
acetate	$Pb(C_2H_3O_2)_2.10H_2O$	505.46	wh., rhb.	1.689	22		s.	s.	
acetate, basic	$Pb_2(C_2H_3O_2)_3OH$	608.56	wh.				v. s.		sl. s. al.
acetate, basic	$Pb(C_2H_3O_2)_2.Pb(OH)_2.H_2O$	584.54	wh. nd.				v. s.		s. al.
acetate, basic	$Pb(C_2H_3O_2)_2.2Pb(OH)_2$	807.75	wh. nd.				5.55	18.2	s. al.
arsenate, monobasic	$PbH_4(AsO_4)_2$	489.06	tri., 1.82	$4.46^{15°}$	d. 140		d.		s. HNO_3
arsenate, dibasic (schultenite)	$PbHAsO_4$	347.13	wh., mn., 1.9097	5.94	d. >200	$-H_2O$, 280	i.	sl. s.	s. HNO_3, NaOH
arsenate, meta-	$Pb(AsO_3)_2$	453.03	hex.	$6.42^{15°}$			d.		s. HNO_3
arsenate, pyro-	$Pb_2As_2O_7$	676.24	rhb., 2.03	$6.85^{15°}_{15}$	802		i.	d.	s. HCl, HNO_3; i. ac.

Lead azide	PbN_6	291.26	col. nd.	....	expl. 350	....	i.	$0.05^{100°}$	v. s. ac.; i. NH_4OH
bromide	$PbBr_2$	367.05	col., rhb.	6.66	373	918	$0.4554^{0°}$	$4.75^{100°}$	s. a., KBr.; sl. s. NH_3; i. al.
carbonate (cerussite)	$PbCO_3$	267.22	wh., rhb., 2.0763	6.6	d. 315	....	$0.00011^{20°}$	d.	s. a., alk.; i. NH_3, al.
carbonate, basic (hydrocerussite; white lead)	$2PbCO_3.Pb(OH)_2$ §	775.67	wh., hex.	6.14	d. 400	....	i.	i.	s. ac.; sl. s. aq. CO_2
chloride (cotunnite)	$PbCl_2$	278.12	wh., rhb., 2.2172	5.80	501	954^{760mm}	$0.673^{0°}$	$3.34^{100°}$	sl. s. dil. HCl, NH_3; i. al.
chromate (crocoite)	$PbCrO_4$	323.22	yel., mn., 2.42	6.12	844	d.	$0.000007^{20°}$	i.	s. a., alk.; i. NH_3, ac.
chromate, basic	$PbCrO_4.PbO$	546.43	or.-yel. nd.	....	....	....	i.	i.	s. a., alk.
formate	$Pb(HCO_2)_2$	297.25	wh., rhb.	4.56	d. 190	....	$1.6^{16°}$	$18^{100°}$ d.	i. al.
hydroxide	$3PbO.H_2O$	687.65	cb.	7.592	$-H_2O$, 130	....	0.014	....	s. a., alk.
nitrate	$Pb(NO_3)_2$	331.23	col., cb. or mn., 1.7815	4.53	d. 470	....	$38.8^{0°}$	$138.8^{100°}$	$8.8^{22°}$ al.
oxide, sub-	Pb_2O	430.42	bk., amor.	8.34	d. red heat	....	i.	i.	s. a., alk.
oxide, mono- (litharge)	PbO	223.21	yel., tet.	9.53	888	....	$0.0068^{18°}$	....	s. alk., PbAc, NH_4Cl, $CaCl_2$
oxide, mono (massicotite)	PbO	223.21	yel., rhb., 2.61	8.0					
oxide, mono-	PbO	223.21	amor.	9.2 to 9.5	....	....	i.	i.	s. alk., PbAc, NH_4Cl, $CaCl_2$
oxide, red (minium)	Pb_3O_4	685.63	red, amor.	9.1	d. 500	....	i.	i.	s. ac., h. HCl
oxide, sesqui-	Pb_2O_3	462.42	red-yel., amor.	....	d. 360	....	i.	i.	s. a., alk.
oxide, di- (plattnerite)	PbO_2	239.21	brn., tet., 2.229	9.375	d. 290	....	i.	i.	s. ac., h. alk.; i. al.
silicate	$PbSiO_3$	283.27	col., mn., 1.961	6.49	766	....	i.	....	s. a.
sulfate (anglesite)	$PbSO_4$	303.27	wh., mn. or rhb 1.8823	6.2	1170	....	$0.0028^{0°}$	$0.0056^{40°}$	s. conc. a., NH_4 salts; i. al.
sulfate, acid	$Pb(HSO_4)_2.H_2O$	419.36	cr.	....	d.	....	$0.0001^{18°}$	....	sl. s. H_2SO_4
sulfate, basic (lanarkite)	$PbSO_4.PbO$	526.48	col., mn.	6.92	977	....	$0.0044^{18°}$	....	sl. s. H_2SO_4
sulfide (galena)	PbS	239.27	lead gray, cb., 3.912	7.5	1120	....	$0.00009^{18°}$	i.	s. a.; i. alk.
thiocyanate	$Pb(CNS)_2$	323.37	col., mn.	3.82	d. 190	....	$0.05^{20°}$	s.	s. KCNS, HNO_3
Lithium	Li	6.94	silv. met. cb.	$0.53^{20°}$	186	1336 ± 5	d.	d.	s. a., NH_3
benzoate	$LiC_7H_5O_2$	128.05	wh. leaflets	....	....	....	$33^{25°}$	$40^{100°}$	$7.7^{25°}$, $10^{78°}$ al.
bromide	LiBr	86.86	wh., delq., cb., 1.784	$3.464^{25°}_{4}$	547	1265	$143^{0°}$ ($2H_2O$)	$266^{100°}$ ($1H_2O$)	s. al., act.
bromide	$LiBr.2H_2O$	122.89	wh. pr.	....	44	....	$246^{20°}$	....	s. al.
carbonate	Li_2CO_3	73.89	col., mn., 1.567	$2.11^{0°}$	618	d.	$1.54^{0°}$	$0.72^{100°}$	s. dil. a.; i. al., act., NH_3
chloride	LiCl	42.40	wh., delq., cb., 1.662	$2.068^{25°}_{4}$	614	1360	$67^{0°}$	$127.5^{100°}$	$2.48^{15°}$ al.; s. et.
citrate	$Li_3C_6H_5O_7.4H_2O$	281.98	wh. cr.	....	d.	....	$61.2^{15°}$	$66.7^{100°}$	sl. s. al., et.
fluoride	LiF	25.94	wh., cb., 1.3915	$2.295^{21.5°}$	870	1670	$0.27^{18°}$	$0.135^{35°}$	s. HF; i. act.
formate	$LiHCO_2.H_2O$	69.97	col., rhb.	1.46	$-H_2O$, 94	....	$49.2^{0°}$	$346.6^{104°}$	sl. s. al., et.
hydride	LiH	7.95	wh., cb.	0.820	680	....	d.	....	i. et.
hydroxide	LiOH	23.95	wh. cr.	2.54	445	925±	$12.7^{0°}$	$17.5^{100°}$	sl. s. al.
hydroxide	$LiOH.H_2O$	41.96	col., mn.	1.83	....	d.	$22.3^{10°}$	$26.8^{80°}$	sl. s. al.
nitrate	$LiNO_3$	68.95	col., trig., 1.735	2.38	261	....	$53.4^{0°}$	$194^{70°}$	s. al., NH_3
nitrate	$LiNO_3.3H_2O$	123.00	col.	....	29.88	....	v. s.	∞	
oxide	Li_2O	29.88	col., 1.644	$2.013^{25°}_{4}$	....	subl. <1000	forms LiOH		
phosphate, monobasic	LiH_2PO_4	103.94	col.	2.461	>100				
phosphate, tribasic	Li_3PO_4	115.80	wh., rhb.	$2.537^{17.5°}$	837	....	$0.034^{18°}$	v. sl. s.	s. a., NH_4Cl; i. act.
phosphate, tribasic	$Li_3PO_4.12H_2O$	331.99	wh., trig.	1.645	100	....	v. sl. s.	v. sl. s.	
salicylate	$LiC_7H_5O_3$	144.05	col.	....	d.	....	$128^{25°}$	....	v. s. al.
sulfate	Li_2SO_4	109.94	col., mn., 1.465	2.22	860	....	$35.34^{0°}$	$29.9^{100°}$	i. act., 80% al.
sulfate	$Li_2SO_4.H_2O$ §	127.96	col., mn., 1.477	2.06	$-H_2O$, 130	....	$43.6^{0°}$	$35^{100°}$	i. 80% al.
sulfate, acid	$LiHSO_4$	104.01	pr.	$2.123^{13°}$	170.5	....	d.		
Lutecium	Lu	174.99							
Magnesium	Mg	24.32	silv. met., hex.	$1.74^{20°}$	651	1110	i.	sl. s. d.	s. a., NH_4 salts
acetate	$Mg(C_2H_3O_2)_2$	142.41	wh.	1.42	323	....	v. s.	v. s.	$5.25^{15°}$ m. al.
acetate	$Mg(C_2H_3O_2)_2.4H_2O$ §	214.47	wh., mn. pr., 1.491	1.454	80	....	v. s.	v. s.	v. s. al.
aluminate (spinel)	$MgO.Al_2O_3$	142.26	col. cb., 1.718–23	3.6	2135	....	i.	....	v. sl. s. dil. HCl; i. dil. HNO_3
ammonium chloride	$MgCl_2.NH_4Cl.6H_2O$	256.83	wh., rhb., delq.	1.456	$-4H_2O$, 195	....	16.7	s.	
ammonium phosphate (struvite)	$MgNH_4PO_4.6H_2O$	245.44	col., rhb., 1.496	1.715	d. 100	....	$0.0231^{0°}$	$0.0195^{80°}$	s. a.; i. al.
ammonium sulfate (boussingaultite)	$MgSO_4.(NH_4)_2SO_4.6H_2O$	360.62	col., mn.	1.72	>120	....	$16.86^{0°}$	$130^{100°}$	
benzoate	$Mg(C_7H_5O_2)_2.3H_2O$	320.59	wh. pd.	....	$-3H_2O$, 110	....	$4.5^{25°}$ (anh.)	s.	s. act.
carbonate (magnesite)	$MgCO_3$	83.43	wh., trig. 1.700	3.037	d. 350	....	0.0106	....	s. a., aq. CO_2; i. act., NH_3
carbonate (nesquehonite)	$MgCO_3.3H_2O$	138.38	col., rhb., 1.501	1.852	$-H_2O$, 100	....	$0.1518^{19°}$	d.	s. a., aq. CO_2
carbonate, basic (hydromagnesite)	$3MgCO_3.Mg(OH)_2.3H_2O$	365.37	wh., rhb., 1.530	2.16	d.	....	0.04	0.011	s. a., NH_4 salts; i. al.

* Usual commercial form about 31 per cent.
† Usual commercial forms 3 or 30 per cent.
‡ See also a table of alloys.
§ Usual commercial form.

Table 1. Physical Properties of the Elements and Inorganic Compounds—(*Continued*)

Name	Formula	Formula weight	Color, crystalline form and refractive index	Specific gravity	Melting point, °C.	Boiling point, °C.	Solubility in 100 parts: Cold water	Hot water	Other reagents
Magnesium chloride (chloromagnesite)	$MgCl_2$	95.23	col., hex., 1.675	$2.325^{25°}$	712	1412	$52.8^{0°}$	$73^{100°}$	50 al.
chloride (bischofite)	$MgCl_2.6H_2O$*	203.33	wh., delq., mn., 1.507	1.56	118 d.	d.	$281^{0°}$	$918^{100°}$	50 al.
hydroxide (brucite)	$Mg(OH)_2$	58.34	wh., trig., 1.5617	2.4	d.		$0.0009^{18°}$		s. NH_4 salts, dil. a.
nitride	Mg_3N_2	100.98	gn.-yel., amor.		d.		i.	d.	s. a.; i. al.
oxide (magnesia; periclase)	MgO	40.32	col., cb., 1.7364	3.65	2800	3600	0.00062		s. a., NH_4 salts; i. al.
perchlorate	$Mg(ClO_4)_2$*	223.23	wh., delq.	$2.60^{25°}$	d.		$99.6^{25°}$	v. s.	24^{25} al., $51.8^{25°}$ m. al., 0.29 et.
peroxide	MgO_2	56.32	wh. pd.		expl. 275		i.	i.	s. a.
phosphate, pyro-	$Mg_2P_2O_7$	222.60	col., mn., 1.604	$2.598^{22°}$	1383		i.	i.	s. a.; i. alk.
phosphate, pyro-	$Mg_2P_2O_7.3H_2O$	276.65	wh., amor.	2.56	$-3H_2O$, 100		i.	sl. s.	s. a.; i. al.
potassium chloride (carnallite)	$MgCl_2.KCl.6H_2O$	277.88	delq., rhb., 1.475	$1.60\frac{19.4°}{4}$	265		$64.5^{19°}$ d.	d.	d. al.
potassium sulfate (picromerite)	$MgSO_4.K_2SO_4.6H_2O$	402.73	mn., 1.4629	2.15	d. 72		$19.26^{0°}$	$81.7^{75°}$	
silicofluoride	$MgSiF_6.6H_2O$	274.48	col., trig., 1.3439	$1.788\frac{17.5°}{4}$	d.		$64.8^{17.5°}$	s.	d. HF
sodium chloride	$MgCl_2.NaCl.H_2O$	171.70	col.				s.	s.	
sulfate	$MgSO_4$	120.38	col.	2.66	1185		$26.9^{0°}$	$68.3^{100°}$	s. al.
sulfate (epsom salt; epsomite)	$MgSO_4.7H_2O$*	246.49	col., rhb., 1.4554	1.68	70 d.		$72.4^{0°}$	$178^{40°}$	s. al.
Manganese	Mn	54.93	gray-pink met.	$7.2^{20°}$	1260	1900	d.		s. dil. a.
acetate	$Mn(C_2H_3O_2)_2$	173.02		$1.74\frac{20°}{4}$			s.	s.	
acetate	$Mn(C_2H_3O_2)_2.4H_2O$*	245.08	pa. pink, mn.	1.589			s.	$64.5^{50°}$	s. al., m. al.
carbonate (rhodocrosite)	$MnCO_3$	114.94	rose, trig., 1.817	3.125	d.		$0.0065^{25°}$		s. aq. CO_2, dil. a., i. NH_3, al.
chloride (scacchite)	$MnCl_2$	125.84	rose, delq., cb.	$2.977\frac{25°}{4}$	650	1190	$63.4^{0°}$	$123.8^{106°}$	s. al.; i. et., NH_3
chloride	$MnCl_2.4H_2O$*	197.91	rose red, delq., mn. 1.575	2.01	58.0	$-H_2O$, 106; $-4H_2O$, 200	$151^{8°}$	∞	s. al.; i. et.
chloride, per-	$MnCl_4$	196.76	gn.				s.	s.	s. al., et.
hydroxide (ous) (pyrochroite)	$Mn(OH)_2$	88.95	wh., trig.	$3.258^{13°}$	d.		$0.002^{20°}$	i.	s. a., NH_4 salts; i. alk.
hydroxide (ic) (manganite)	$Mn_2O_3.H_2O$	175.88	brn., rhb., 2.24	3.258	d.		i.	i.	s. h. H_2SO_4
nitrate	$Mn(NO_3)_2.6H_2O$	287.04	rose red, mn.	$1.82^{21°}$	25.8	129.5	$426^{0°}$	∞	v. s. al.
oxide (ous) (manganosite)	MnO	70.93	gray-gn., cb., 2.16	5.18	1650		i.	i.	s. a., NH_4Cl
oxide (ic)	Mn_2O_3	157.86	brn.-bk., cb.	4.81	−O, 1080		i.	i.	s. a.; i. act.
oxide, di- (pyrolusite; polianite)	MnO_2*	86.93	bk., rhb.	5.026	−O, >230		i.	i.	s. HCl; i. HNO_3, act.
sulfate (ous)	$MnSO_4$	150.99	red-wh.	3.235	700	d. 850	$53^{0°}$	$73^{50°}$	s. al.; i. et.
sulfate (ous) (szmikite)	$MnSO_4.H_2O$	169.01	pa. pink, mn., 1.595	2.87	Stable 57 to 117		$98.47^{48°}$	$79.77^{100°}$	
sulfate (ous)	$MnSO_4.2H_2O$	187.02		$2.526^{15°}$	Stable 40 to 57		$85.27^{35°}$	$106.8^{55°}$	
sulfate (ous)	$MnSO_4.3H_2O$	205.04		$2.356^{15°}$	Stable 30 to 40		$74.22^{5°}$	$99.31^{57°}$	
sulfate (ous)	$MnSO_4.4H_2O$*	223.05	pink, rhb. or mn., 1.518	2.107	Stable 18 to 30	$-4H_2O$, 450	$136^{16°}$	$169^{50°}$	i. al.
sulfate (ous)	$MnSO_4.5H_2O$	241.07	pink, tri., 1.508	$2.103^{15°}$	Stable 8 to 18		$142^{5°}$	$200^{35°}$	
sulfate (ous)	$MnSO_4.6H_2O$	259.09			Stable −5 to +8		$204^{0°}$	$247^{9°}$	
sulfate (ous)	$MnSO_4.7H_2O$	277.10	pink, mn. or rhb.	2.092	Stable −10 to −5; 19 d.	$-7H_2O$, 280	$176^{0°}$	$251^{14°}$	
sulfate (ic)	$Mn_2(SO_4)_3$	398.04	gn., delq. cr.	3.24	d. 160		v. s.	d.	s. HCl, dil. H_2SO_4; i. conc. H_2SO_4, HNO_3
Masurium	Ma	98–99.5		11.5	2300 (?)				
Mercuric acetate	$Hg(C_2H_3O_2)_2$	318.70	wh. pl.	3.270	d.		$25^{10°}$	$100^{100°}$	s. al. sl. d.
bromide	$HgBr_2$	360.44	wh., rhb.	6.053	237	322	$0.5^{20°}$	$25^{100°}$	$25.2^{0°}$ al.; v. sl. s. et.
carbonate, basic	$HgCO_3.2HgO$	693.84	brn.-red				i.		s. aq. CO_2, NH_4Cl
chloride (corrosive sublimate)	$HgCl_2$	271.52	wh., rhb., 1.859	5.44	277	304	$3.6^{0°}$	$61.3^{100°}$	$33^{25°}$ 99% al.; 33 et.
fulminate	$Hg(CNO)_2$	284.65	cb.	4.42	expl.		sl. s.		s. NH_4OH, al.
hydroxide	$Hg(OH)_2$	234.63			$-H_2O$, 175		i.	i.	s. a.
oxide (montroydite)	HgO	216.61	yel. or red, rhb., 2.5	11.14	d. 100		$0.0052^{25°}$	$0.041^{100°}$	s. a.; i. al.
oxychloride (kleinite)	$HgCl_2.3HgO$	921.35	yel., hex.	7.93	d. 260		i.	d.	s. HCl
silicofluoride, basic	$HgSiF_6.HgO.3H_2O$	613.33	yel. nd.				d.		s. a.
sulfate	$HgSO_4$	296.67	wh., rhb.	6.47	d.		d.		s. a.; i. al., act., NH_3
sulfate, basic (turpeth)	$HgSO_4.2HgO$	729.89	yel., tet.	6.44			0.005	$0.167^{100°}$	s. a.; i. al.
Mercurous acetate	$HgC_2H_3O_2$	259.65	wh. sc.		d.		$0.75^{13°}$	d.	s. H_2SO_4, HNO_3; i. al.
bromide	HgBr	280.53	wh., tet.	7.307	subl. 345		7×10^{-9}	i.	s. a.; i. al., act.
carbonate	Hg_2CO_3	461.23	yel. pd.		d. 130		i.	d.	s. NH_4Cl
chloride (calomel)	HgCl	236.07	wh., tet., 1.9733	7.150	302	383.7	$0.0014^{0°}$	$0.0007^{43°}$	s. aq. reg., $Hg(NO_3)_2$; sl. s. HNO_3, HCl; i. al., et.
iodide	HgI	327.53	yel., tet.	7.70	290 d.	subl. 140; 310d.	2×10^{-8}	v. sl. s.	s. KI; i. al.
nitrate	$HgNO_3.H_2O$	280.63	wh. mn.	$4.785^{3.9°}$	70	expl.	v. s.	d.	s. HNO_3; i. al., et.

Mercurous oxide	Hg_2O	417.22	bk.	9.8	d. 100	....	i.	0.0007	s. h. ac.; i. alk., dil. HCl, NH_3
sulfate	Hg_2SO_4	497.28	wh., mn.	7.56	d.	....	$0.055^{16.5°}$	$0.092^{100°}$	s. H_2SO_4, HNO_3
Mercury†	Hg	200.61	silv. lq. or hex.(?)	$13.546^{20°}$	−38.87	356.9	i.	i.	s. HNO_3; i. HCl
Molybdenum	Mo	95.95	gray, cb.	10.2	2620 ± 10	3700	i.	i.	s. h. conc. H_2SO_4; i. HCl, HF, NH_3, dil. H_2SO_4, Hg
chloride, di-	$MoCl_2$	166.85	yel., amor.	$3.714\frac{25°}{4}$	d.	....	i.	i.	s. HCl, H_2SO_4, NH_4OH, al., et.
chloride, tri-	$MoCl_3$	202.32	dark red pd.	$3.578\frac{25°}{4}$	d.	....	i.	d.	s. HNO_3, H_2SO_4; v. sl. s. al., et.
chloride, tetra-	$MoCl_4$	237.78	brn., delq.	....	volt.	d.	s.	d.	s. HNO_3, H_2SO_4; sl. s. al., et.
chloride, penta-	$MoCl_5$	273.24	bk. cr.	$2.928\frac{25°}{4}$	194	268	s.	d.	s. HNO_3, H_2SO_4; i. abs. al., et.
oxide, tri- (molybdite)	MoO_3	143.95	col., rhb.	$4.50^{19.5°}$	795	subl.	$0.107^{18°}$	$2.106^{79°}$	s. a., NH_4OH
sulfide, di- (molybdenite)	MoS_2	160.07	bk., hex., 4.7	$4.80^{14°}$	1185	....	i.	i.	s. H_2SO_4, aq. reg.
sulfide, tri-	MoS_3	192.13	red-brn.	....	d.	....	sl. s.	s.	s. alk. sulfides
sulfide, tetra-	MoS_4	224.19	brn. pd.	....	d.	....	i.	i.	s. alk. sulfides; i. NH_3
Molybdic acid	H_2MoO_4	161.97	yel-wh., hex.	....	d. 115	....	v. sl. s.	sl. s.	s. NH_4OH, H_2SO_4; i. NH_3
Molybdic acid	$H_2MoO_4.H_2O$	179.98	yel., mn.	$3.124^{15°}$	$-H_2O$, 70	$-2H_2O$, 200	$0.133^{18°}$	$2.13^{70°}$	s. a., NH_4OH, NH_4, salts
Neodymium	Nd	144.27	yellowish	$6.9^{20°}$	840		d.		
Neon	Ne	20.18	col. gas	lq. $1.204^{-245.9°}$ 0.674 (A)	−248.67	−245.9	$2.6^{0°}$ cc	$1.1^{45°}$ cc	s. lq. O_2, al., act., bz.
Neptunium	Np^{239}	239	....	....	Produced by Neutron bombardment of U^{238}				
Nickel	Ni	58.69	silv. met., cb.	8.90^{20}	1452	2900	i.	i.	s. dil. HNO_3; sl. s. H_2SO_4, HCl; i. NH_3
acetate	$Ni(C_2H_3O_2)_2$	176.78	gn. pr.	1.798	d.		16.6	....	i. al.
ammonium chloride	$NiCl_2.NH_4Cl.6H_2O$	291.20	gn., delq., mn.	1.645	....	....	$150^{25°}$	v. s.	
ammonium sulfate	$NiSO_4.(NH_4)_2SO_4.6H_2O$	394.99	blue-gn., mn., 1.5007	1.923	....	....	$2.5^{3.5°}$	$39.2^{85°}$	v. sl. s. $(NH_4)_2SO_4$
bromate	$Ni(BrO_3)_2.6H_2O$	422.62	gn., cb.	2.575	d.	....	28	....	s. NH_4OH
bromide	$NiBr_2$	218.52	yel., delq.	$4.64\frac{28°}{4}$	d.	....	$112.8^{0°}$	$156^{100°}$	s. al., et., NH_4OH
bromide	$NiBr_2.3H_2O$	272.57	gn., delq.	....	$-3H_2O$, 200	....	$199^{0°}$	$316^{100°}$	s. al., et., NH_4OH
bromide, ammonia	$NiBr_2.6NH_3$	320.71	vl. pd.	1.837	....		v. s.	d.	i. c. NH_4OH
bromoplatinate	$NiPtBr_6.6H_2O$	841.51	trig.	3.715					
carbonate	$NiCO_3$	118.70	lt. gn., rhb.		d.	....	$0.0093^{25°}$	i.	s. a.
carbonate, basic	$2NiCO_3.3Ni(OH)_2.4H_2O$	587.58	lt. gn.		d.	....	i.	d.	s. a., NH_4 salts
carbonyl	$Ni(CO)_4$	170.73	lq.	$1.31^{17°}$	−25	43^{751mm}	$0.018^{9.8°}$	i.	s. aq. reg., HNO_3, al., et
chloride	$NiCl_2$	129.60	yel., delq.	3.544	subl.	973	$53.8^{0°}$	$87.6^{100°}$	s. NH_4OH, al.; i. NH_3
chloride	$NiCl_2.6H_2O$*	237.70	gn., delq., mn., 1.57±	....	....	....	180	v. s.	v. s. al.
chloride, ammonia	$NiCl_2.6NH_3$	231.80	....				s.	d.	s. NH_4OH; i. al.
cyanide	$Ni(CN)_2.4H_2O$	182.79	gn. pl.		$-4H_2O$, 200	d.	i.	i.	s. KCN; i. dil. KCl
dimethylglyoxime	$NiC_8H_{14}O_4N_4$	288.91	scarlet red cr.	....	subl. 250	....	i.	i.	s. abs. al., a.; i. ac., NH_4OH
formate	$Ni(HCO_2)_2.2H_2O$	184.76	gn. cr.	2.154	d.	....	s.		
hydroxide (ic)	$Ni(OH)_3$	109.71	bk.		d.	....	i.	i.	s. a., NH_4OH, NH_4Cl
hydroxide (ous)	$Ni(OH)_2.\frac{1}{4}H_2O$	97.21	lt. gn.	4.36	d.	....	v. sl. s.	v. sl. s.	s. a., NH_4OH; i. alk.
nitrate	$Ni(NO_3)_2.6H_2O$	290.80	gn., mn.	2.05	56.7	136.7	$243.0^{0°}$	$\infty^{56.7°}$	s. NH_4OH; i. abs. al.
nitrate, ammonia	$Ni(NO_3)_2.4NH_3.2H_2O$	286.87	....		....	....	v. s.	....	i. al.
oxide, mono- (bunsenite)	NiO	74.69	gn.-bk., cb., 2.37	7.45	Forms Ni_2O_3 at 400		i.	i.	s. a. NH_4OH
potassium cyanide	$Ni(CN)_2.2KCN.H_2O$	258.97	red yel., mn.	$1.875^{11°}$	$-H_2O$, 100		s.		d. a.
sulfate	$NiSO_4$	154.75	yel., cb.	3.68	$-SO_3$, 840	....	$27.2^{0°}$	$76.7^{100°}$	i. al., et., act.
sulfate	$NiSO_4.6H_2O$*	262.85	gn. mn. or blue, tet., 1.5109	2.07	tr. 53.3	$-6H_2O$, 280	$131^{50°}$	$280^{100°}$	v. s. NH_4OH, al.
sulfate (morenosite)	$NiSO_4.7H_2O$	280.86	gn., rhb., 1.4893	1.948	98–100	$-6H_2O$, 103	$63.5^{0°}$	$117.8^{30°}$	s. al.
Nitric acid	HNO_3	63.02	col. lq.	1.502	−42	86	∞	∞	expl. with al.
Nitric acid	$HNO_3.H_2O$	81.03	col. lq.	....	−38		∞	∞	d. al.
Nitric acid	$HNO_3.3H_2O$	117.06	col. lq.	....	−18.5		$263^{-20°}$	∞	d. al.
Nitro acid sulfite	NO_2HSO_3	127.08	col., rhb.	....	73 d.		d.		s. H_2SO_4
Nitrogen	N_2	28.02	col. gas or cb. cr.	$1.026^{-252.5°}$ $0.808^{-195.8°}$ 12.5° (D)	−209.86	−195.8	$2.35^{0°}$ cc	$1.55^{20°}$ cc	sl. s. al.

* Usual commercial form.
† See also special table 48.

Table 1. Physical Properties of the Elements and Inorganic Compounds—(*Continued*)

Name	Formula	Formula weight	Color, crystalline form and refractive index	Specific gravity	Melting point, °C.	Boiling point, °C.	Solubility in 100 parts: Cold water	Solubility in 100 parts: Hot water	Solubility in 100 parts: Other reagents
Nitrogen oxide, mono- (ous)	N_2O	44.02	col. gas	lq. $1.226^{-89°}$ 1.530 (A)	−102.3	−90.7	$130.52^{0°}$ cc	$60.82^{24°}$ cc	s. H_2SO_4, al.
oxide, di- (ic)	NO or $(NO)_2$	30.01 (60.02)	col. gas	lq. $1.269^{-150.2°}$ 1.0367 (A)	−161	−151	$7.34^{0°}$ cc	$0.0^{100°}$	26.6 cc al.; 3.5 cc H_2SO_4; s. aq. $FeSO_4$
oxide, tri-	N_2O_3	76.02	red-brn. gas or blue lq. or solid	$1.447^{2°}$	−102	3.5	s.	...	s. a., et.
oxide, tetra- (per- or di-)	NO_2 or $(NO_2)_2$	46.01 (92.02)	yel. lq., col. solid, red-brn. gas	$1.448^{20°}$	−9.3	21.3	d.	...	s. HNO_3, H_2SO_4, chl., CS_2
oxide, penta-	N_2O_5	108.02	wh., rhb.	$1.63^{18°}$	30	47	s.	Forms HNO_3	
oxybromide	NOBr	109.92	brn. lq.	>1.0	−55.5	−2	d.		
oxychloride	NOCl	65.47	red-yel. lq. or gas	$1.417^{-12°}$ 2.31 (A)	−64.5	−5.5	d.	...	s. fuming H_2SO_4
Nitroxyl chloride	NO_2Cl	81.47	yel.-brn. gas	lq. $1.32^{14°}$	<−30	5	d		
Osmium	Os	190.2	blue, hex.	$22.48^{20°}$	2700	>5300	i.	i.	sl. s. aq. reg., HNO_3; i. NH_3
chloride, di-	$OsCl_2$	261.11	gn., delq.	...	...	...	s. d.	...	s. NaCl, al., et.
chloride, tri-	$OsCl_3$	296.57	brn., cb.	...	d. 560–600	...	sl. s.	...	s. a., alk., al.; sl. s. et.
chloride, tetra-	$OsCl_4$	332.03	red-yel. nd.	...	...	...	s. d.	...	s. HCl, al.
Oxygen	O_2	32.00	col. gas or hex. solid	$1.14^{-183°}$ $1.426^{-252.5°}$ 1.1053 (A)	−218.4	−183	$4.89^{0°}$ cc	$2.6^{30°}$ cc $1.7^{100°}$ cc	sl. s. al., s. fused Ag
Ozone	O_3	48.00	col. gas	$1.71^{-183°}$ $3.03^{-80°}$ 1.658 (A)	−251	−112	$0.494^{0°}$ cc	$0^{60°}$ cc	s. oil turp., oil cinn.
Palladium	Pd	106.70	silv. met., cb.	$12.0^{20°}$ $11^{1550°}$	1555	2200	i.	i.	s. aq. reg., h. H_2SO_4; i. NH_3
bromide (ous)	$PdBr_2$	266.53	brn.	...	...	...	i.	i.	s. HBr
chloride	$PdCl_2$	177.61	brn., cb.	...	500 d.	...	s.	s.	s. HCl, act., al.
chloride	$PdCl_2.2H_2O$	213.65	brn. pr.	...	...	...	s.	s.	s. HCl, act., al.
cyanide	$Pd(CN)_2$	158.74	yel.	...	d.	...	i.	i.	s. HCN, KCN, NH_4OH; i. dil. a.
hydride	Pd_2H	214.41	met.	11.06	d.				
Palladous dichlorodiammine	$Pd(NH_3)_2Cl_2$	211.68	red or yel., tet.	2.5	...	...	s.	...	s. a., NH_4OH
Perchloric acid	$HClO_4$	100.46	unstable, col. lq	$1.768_{4}^{22°}$	−112	16^{18mm}	s.		
Perchloric acid	$HClO_4.H_2O$	118.48	fairly stable nd.	1.88	50	d.	s.		
Perchloric acid	$HClO_4.2H_2O$* 73.6% anh.	136.50	stable lq., col.	$1.71_{4}^{25°}$	−17.8	200	v. s.	...	s. al.
Periodic acid	HIO_4	191.93	wh. cr.	...	d. 138	subl. 110	s.		
Periodic acid	$HIO_4.2H_2O$	227.96	delq., mn.	...	d. 110	...	v. s.	v. s.	sl. s. al., et.
Permanganic acid	$HMnO_4$	119.94	exists only in solution	...	...	...	v. s.	d.	d. al.
Permolybdic acid	$HMoO_4.2H_2O$	196.99	wh. cr.	...	...	...	v. s.	v. s.	
Persulfuric acid	$H_2S_2O_8$	194.14	hyg. cr.	...	<60	...	v. s.	d.	
Phosphamic acid	$PONH_2.(OH)_2$	97.02	cb.	...	d.	...	v. s.	v. s. d.	i. al.
Phosphatomolybdic acid	$H_7P(Mo_2O_7)_6.28H_2O$	2365.88	yel. cb.	...	78	$-25H_2O$, 140	s.	...	s. HNO_3
Phosphine	PH_3	34.00	col. gas	lq. $0.746^{-90°}$ 1.146 (A)	−132.5	−85	$26^{17°}$ cc	$i.^{100°}$	s. Cu_2Cl_2, al., et.
Phosphonium chloride	PH_4Cl	70.47	wh., cb.	...	28^{46atm}	subl.	d.		
Phosphoric acid, hypo-	$H_4P_2O_6$	161.99	cr.	...	55	d. 70	s.	$450^{62°}$	
Phosphoric acid, meta-	HPO_3	79.99	vitreous, delq.	2.2–2.5	subl.	...	s.	Forms H_3PO_4	i. lq. CO_2
Phosphoric acid, ortho-	H_3PO_4†	98.00	col., rhb.	$1.834^{18.2°}$	42.35	$-½H_2O$, 213	$2340^{26°}$	v. s.	s. al.
Phosphoric acid, pyro-	$H_4P_2O_7$	177.99	wh. nd.	...	61	...	$800^{23°}$	Forms H_3PO_4	v. s. al., et.
Phosphorous acid, hypo-	H_3PO_2	66.00	syrupy	$1.493^{18.8°}$	26.5	d.	∞	∞	
Phosphorous acid, ortho-	H_3PO_3	82.00	col.	$1.651^{21.2°}$	74	d. 200	$307.3^{0°}$	$730^{40°}$	
Phosphorous acid, pyro-	$H_4P_2O_5$	145.99	nd.	...	38	d. 130	d.		
Phosphorus, black	P_4	123.92	rhombohedral	2.69	...	ign. in air, 400	i.	i.	i. CS_2
Phosphorus, red	P_4	123.92	red, cb.	$2.20^{20°}$	590^{43atm}	ign. in air, 725	i.	i.	s. alk.; i. CS_2, NH_3, et.
Phosphorus, yellow	P_4	123.92	yel., hex., 2.1168	$1.82^{20°}$; lq. $1.745^{44.5°}$	44.1; ign. 34	280	0.0003	sl. s.	0.4 al.; $1000^{10°}$ CS_2; $1.5^{0°}$, $10^{81°}$ bz.; s. NH_3
chloride, tri-	PCl_3	137.35	col., fuming lq.	$1.574_{4}^{20.8°}$	−111.8	75.95^{760mm}	d.	...	s. et., chl., CS_2
chloride, penta-	PCl_5	208.27	delq., tet.	solid 1.6; $3.60^{295°}$ (A)	148 under pressure	subl. 160	d.	...	s. CS_2, C_6H_5COCl
oxide, penta-	P_2O_5	141.96	wh., delq., amor.	2.387	subl. 250	...	Forms H_3PO_4	v. s.	s. H_2SO_4; i. NH_3, act.
oxychloride	$POCl_3$	153.35	col., fuming lq.	1.675	2	107.2^{760mm}	d.		d. al.

Phosphotungstic acid	$P_2O_5.2WO_3.42H_2O$	3681.67	yel.-gn. cr.				s.		s. al., et.
Platinum	Pt	195.23	silv. met., cb.	$21.45^{20^{\circ}}$; lq. $19^{1755^{\circ}}$	1755	4300	i.	i.	s. aq. reg., fused alk.
chloride (ic)	$PtCl_4$	337.06	brn.		d. 370		$140^{25^{\circ}}$	v. s.	s. al., act.; sl. s. NH_3; i. et.
chloride (ous)	$PtCl_2$	266.14	brn.	$5.87^{11^{\circ}}$	d. 581		i.	i.	s. HCl, NH_4OH; sl. s. NH_3; i. al., et.
chloride (ic)	$PtCl_4.8H_2O$	481.19	red, mn.	2.43	$-4H_2O$, 100		v. s.	v. s.	s. al., et.
cyanide (ous)	$Pt(CN)_2$	247.27	yel.-brn.				i.	i.	i. alk.
Plutonium	Pu	238		Produced by deuteron bombardment on U^{238}					
Plutonium	Pu	239		Produced by neutron bombardment on U^{238}					
Potassium	K	39.10	silv. met., cb.	$0.86^{20^{\circ}}$; lq. $0.83^{62^{\circ}}$	62.3	760	d.	Forms KOH	s. a., al., Hg
acetate	$KC_2H_3O_2$	98.14	wh. pd.	1.8	292		$217^{0^{\circ}}$	$396^{90^{\circ}}$	33 al.; i. et.
acetate, acid	$KH(C_2H_3O_2)_2$	158.19	delq. nd. or pl.		148	d. 200	d.		s. ac.
aluminate	$K_2(AlO_2)_2.3H_2O$	250.18	cr.				s.	d.	s. alk.; i al.
amide	KNH_2	55.12	yel.-grn.		338	subl. 400	d.		d. al.; $3.6^{25^{\circ}}$ NH_3
arsenate (monobasic)	KH_2AsO_4	180.02	col., tet., 1.5674	2.867	288		$18.87^{6^{\circ}}$	v. s.	i. al.
auricyanide	$KAu(CN)_4.1.5H_2O$	367.39	pl.		d. 200		s.	v. s.	s. al.
aurocyanide	$KAu(CN)_2$	288.33	rhb.				14.3	$200^{100^{\circ}}$	sl. s. al.; i. et.
bicarbonate	$KHCO_3$	100.11	mn., 1.482	2.17	d. 100–200		$22.4^{0^{\circ}}$	$60^{60^{\circ}}$	i. satd. K_2CO_3, al.
bisulfate	$KHSO_4$	136.16	rhb., or mn., 1.480	2.35	210	d.	$36.3^{0^{\circ}}$	$121.6^{100^{\circ}}$	d. al.
bromate	$KBrO_3$	167.01	trig.	$3.27^{17.5^{\circ}}$	370 d.		$3.11^{0^{\circ}}$	$49.75^{100^{\circ}}$	sl. s. al.; i. act.
bromide	KBr	119.01	col., cb., 1.5594	$2.75^{25^{\circ}}$	730	1380	$53.5^{0^{\circ}}$	$104^{100^{\circ}}$	sl. s. al., et.
carbonate	K_2CO_3	138.20	wh., delq. pd., 1.531	2.29	891	d.	$105.5^{0^{\circ}}$	$156^{100^{\circ}}$	i. al.
carbonate	$K_2CO_3.2H_2O$	174.23	rhb.	2.043			$183^{0^{\circ}}$	$331^{100^{\circ}}$	
carbonate	$2K_2CO_3.3H_2O$	330.45	mn.	2.13			$129.4^{0^{\circ}}$	$268^{100^{\circ}}$	
chlorate	$KClO_3$	122.56	col., mn., 1.5167	2.32	368	d. 400	$3.3^{0^{\circ}}$	$57^{100^{\circ}}$	0.83 al.; s. alk.
chloride (sylvite)	KCl	74.56	col., cb., 1.4904	1.988	790	1500	$27.6^{0^{\circ}}$	$56.7^{100^{\circ}}$	s. a., alk.
chloroplatinate	K_2PtCl_6	486.16	yel., cb., 1.825±	3.499	d. 250		$0.74^{0^{\circ}}$	$5.2^{100^{\circ}}$	i. al., et.
chromate (tarapacaite)	K_2CrO_4	194.20	yel., rhb., 1.7261	$2.732^{18^{\circ}}$	975		$58.0^{0^{\circ}}$	$75.6^{100^{\circ}}$	i. al.
cyanate	KCNO	81.11	wh., tet.	2.048			s.	d.	v. sl. s. al.
cyanide	KCN	65.11	wh., cb., delq., 1.410	$1.52^{16^{\circ}}$	634.5		s.	$122.2^{103.3^{\circ}}$	s. gly.; $0.9^{19.5^{\circ}}$ al.; 1.3 h. al.
dichromate	$K_2Cr_2O_7$	294.21	red, tri.	2.69	398	d.	$4.9^{0^{\circ}}$	$80^{100^{\circ}}$	i. al.
ferricyanide	$K_3Fe(CN)_6$	329.25	red, mn. pr., 1.5689	1.84	d.		$33^{4.4^{\circ}}$	$77.5^{100^{\circ}}$	s. act.; sl. s. al.; i. NH_3
ferrocyanide	$K_4Fe(CN)_6.3H_2O$	422.39	yel., mn., 1.5772	$1.853^{17^{\circ}}$	$-3HO_2$, 70		$27.8^{12.2^{\circ}}$	$90.6^{96.3^{\circ}}$	s. act.; i. NH_3, al., et.
formate	$KHCO_2$	84.11	col., rhb.	1.91	167.5	d.	$331^{18^{\circ}}$	$657^{90^{\circ}}$	sl. s. al.; i. et.
hydride	KH	40.10	cb., 1.453	0.80	d.		d.		i. et., bz., CS_2
hydrosulfide	KHS	72.16	wh., delq., rhb.	2.0	455		s.	s. d.	s. al.
hydroxide	KOH	56.10	wh., delq., rhb.	2.044	380	1320	$97^{0^{\circ}}$	$178^{100^{\circ}}$	v. s. al., et.; i. NH_3
iodate	KIO_3	214.02	col., mn.	3.89	560		$4.73^{0^{\circ}}$	$32.2^{100^{\circ}}$	s. KI; i. al., NH_3
iodide	KI	166.02	wh., cb., 1.6670	3.13	723	1330	$127.5^{0^{\circ}}$	$208^{100^{\circ}}$	$4^{20^{\circ}}$ al.; s. NH_3; sl. s. et.
iodide, tri-	KI_3	419.86	dark blue, delq., mn.	3.498	45	d. 225	v. s.		s. KI, al.
iodoplatinate	K_2PtI_6	1034.94	cb.	5.18			s.		
manganate	K_2MnO_4	197.12	gn., rhb.		d. 190		d.		s. KOH
metabisulfite	$K_2S_2O_5$	222.31	mn., pl.		d. 150		$25^{0^{\circ}}$	$120^{94^{\circ}}$	sl. s. al.; i. et.
nitrate (saltpeter)	KNO_3	101.10	col., rhb., 1.5038	$2.11^{10.6^{\circ}}$	tr. 129; 333	d. 400	$13.3^{0^{\circ}}$	$246^{100^{\circ}}$	$0.1^{30^{\circ}}$ al.; i. et.
nitrite	KNO_2	85.10	pr.	1.915	297	d. 350	$281^{0^{\circ}}$	$413^{100^{\circ}}$	v. s. NH_3; sl. s. al.
oxalate	$K_2C_2O_4.H_2O$	184.23	wh., mn.	2.13	d.		$28.7^{0^{\circ}}$	$83.2^{100^{\circ}}$	
oxalate, acid	KHC_2O_4‡	128.12	mn., 1.545	2.0	d.		$14.3^{50^{\circ}}$	$48.1^{100^{\circ}}$	
oxalate, acid	$KHC_2O_4.\frac{1}{2}H_2O$	137.13	trimetric		d.		$2.2^{0^{\circ}}$	$51.5^{100^{\circ}}$	
oxide	K_2O	94.19	wh., cb.	$2.32\frac{20^{\circ}}{4}$			Forms KOH	v. s.	s. al., et.
perchlorate	$KClO_4$	138.55	col., rhb., 1.4737	$2.524\frac{11^{\circ}}{4}$	d. 400		$0.75^{0^{\circ}}$	$21.8^{100^{\circ}}$	$0.105^{20^{\circ}}$ m. al.; i. et.
permanganate	$KMnO_4$	158.03	purple, rhb.	2.703	d. <240		$2.83^{0^{\circ}}$	$32.35^{75^{\circ}}$	s. H_2SO_4; d. al.
persulfate	$K_2S_2O_8$	270.31	wh., tri., 1.4669		d. <100		$1.77^{0^{\circ}}$	$10^{40^{\circ}}$	i. al.
phosphate, monobasic	KH_2PO_4	136.09	col., delq., tet., 1.5095	2.338	256		$14.8^{0^{\circ}}$	$83.5^{90^{\circ}}$	i. al.
phosphate, dibasic	K_2HPO_4	174.18	wh., delq.		d.		$33^{25^{\circ}}$	v. s.	sl. s. al.
phosphate, tribasic	K_3PO_4	212.27	wh., rhb.	$2.564^{17^{\circ}}$	1340		$193.1^{25^{\circ}}$	v. s.	i. al.
phosphate, meta-	KPO_3	118.08	wh. pd.	$2.258^{14.5^{\circ}}$	tr. 450; 798	1320	s.	s.	
phosphate, meta-	$K_4P_4O_{12}.2H_2O$	508.34	amor.	$2.264^{14.5^{\circ}}$	$-2H_2O$, 100	d.	s.	83	s. a.
phosphate, pyro-	$K_4P_2O_7.3H_2O$	384.39	delq.	2.33	$-2H_2O$, 180	$-3H_2O$, 300	s.	v. s.	i. al.
phthalate, acid	$KHC_8H_4O_4$	204.22	wh. cr.	1.63	d.		$10.2^{25^{\circ}}$	36	
platinocyanide	$K_2Pt(CN)_4.3H_2O$	431.54	yel., rhb., 1.62±	$2.45^{16^{\circ}}$			sl. s.	v. s.	s. al., et.
silicate	K_2SiO_3	154.25	hyg. 1.521±		976		s.	s.	i. al.
silicate, tetra-	$K_2Si_4O_9.H_2O$	352.45	rhb., 1.530	2.417	d. 400		s.	s.	i. al.
sulfate (arcanite)	K_2SO_4	174.25	col., rhb., 1.4947	2.662	tr. 588		$7.35^{0^{\circ}}$	$24.1^{100^{\circ}}$	i. al., act., CS_2

* One commercial form 70 to 72 per cent.
† Common commercial form 85 per cent H_3PO_4 in aqueous solution.
‡ Usual commercial form.

Table 1. Physical Properties of the Elements and Inorganic Compounds—(*Continued*)

Name	Formula	Formula weight	Color, crystalline form and refractive index	Specific gravity	Melting point, °C.	Boiling point, °C.	Solubility in 100 parts: Cold water	Hot water	Other reagents
Potassium sulfate, pyro-	$K_2S_2O_7$	254.31	col.	2.277	300	...	s.	d.	
sulfide, mono-	$K_2S.5H_2O$	200.33	rhb., delq.	...	60	$-3H_2O$, 150	s.	...	s. al., gly.; i. et.
sulfite	$K_2SO_3.2H_2O$	194.28	wh., rhb.	...	d.	...	100	>100	sl. s. al.; i. NH_3
sulfite, acid	$KHSO_3$	120.16	wh., mn.	...	d. 190	...	$45.5^{15°}$	$91.5^{75°}$	i. abs. al.
tartrate	$K_2C_4H_4O_6.\frac{1}{2}H_2O$	235.27	col., mn., 1.526	1.98	...	d.	$12.5^{17.5°}$	$278^{100°}$	sl. s. al.
tartrate, acid	$KHC_4H_4O_6$*	188.18	col., rhb.	1.956	...	...	$0.37^{0°}$	$6.1^{100°}$	s. a., alk.; i. al., ac.
thiocyanate	KCNS	97.17	col.,delq., mn., 1.660±	1.886	172.3	d. 500	$177^{0°}$	$217^{20°}$	$20.8^{22°}$ act.; s. al.
thiosulfate	$K_2S_2O_3$	190.31	col., cb.	...	d. 400	...	$96.1^{0°}$	$311.2^{90°}$	
thiosulfate	$3K_2S_2O_3.H_2O$	588.95	delq., mn.	2.23	$-H_2O$, 180	d.	...	...	i. al.
Praseodymium	Pr	140.92	yel.	$6.5^{20°}$	940	...	d.		
Radium	Ra	226.05	wh., met.	5?	960	1140	d. $+H_2$	...	d. a.
bromide	$RaBr_2$	385.88	wh., mn.	5.79	728	subl. 900	$70^{20°}$	s.	s. al.
Radon (Niton)	Rn	222.0	gas	lq. 5.5; 111 (D)	−71	−62	$51^{0°}$ cc	$8.5^{60°}$ cc	
Rhenium	Re	186.31	hex.	...	3440	...	...	...	i. HF, HCl; s. H_2SO_4, HNO_3
Rhodium	Rh	102.91	gray-wh., cb.	12.5	1955	>2500	i.	i.	sl. s. aq. reg., a.
chloride	$RhCl_3$	209.28	red	...	d. 450	subl. 800±	i.	i.	v. sl. s. alk.; i. aq. reg., a.
chloride	$RhCl_3.4H_2O$	281.35	dark red	...	...	...	v. s.	...	s. HCl, al.; i. et.
Rubidium	Rb	85.48	silv. wh.	lq. $1.475^{38.5}$; $1.53^{20°}$	38.5	700	d.	...	s. a., al.
Ruthenium	Ru	101.70	bk., porous	8.6	>1950	...	i.	i.	sl. s. aq. reg., a.
Ruthenium	Ru	101.70	gray, hex.	$12.2^{20°}$	2450	>2700	i.	i.	
Samarium	Sm (also Sa)	150.43	...	7.7	>1300				
Scandium	Sc	45.10	...	2.5?	1200	2400			
Selenic acid	H_2SeO_4	144.98	hex. pr.	$2.950\frac{15°}{4}$	58	260	$1300^{30°}$	$\infty^{80°}$	s. H_2SO_4; d. al.; i. NH_3
Selenic acid	$H_2SeO_4.H_2O$	162.99	nd.	$2.627\frac{15°}{4}$	26	205	v. s.		
Selenium	Se_8	631.68	red pd., amor., 2.92	$4.26^{25°}$	50	688	i.	i.	s. CS_2, H_2SO_4, CH_2I_2
Selenium	Se_8	631.68	gray, trig., 3.00; red, hex.	4.80; 4.50	220	688	i.	i.	s. CS_2, H_2SO_4
Selenium	Se_8	631.68	steel gray	$4.8^{25°}$	217	688	i.	i.	i. CS_2; s. H_2SO_4
Selenous acid	H_2SeO_3	128.98	hex.	$3.004\frac{15°}{4}$	d.	...	$90^{0°}$	$400^{90°}$	v. s. al.; i. NH_3
Silicic acid, meta-	H_2SiO_3	78.08	amor., 1.41	2.1–2.3	...	...	i.	i.	s. alk.; i. NH_4Cl
Silicic acid, ortho-	H_4SiO_4	96.09	amor.	$1.576^{17°}$			sl. s.	sl. s.	s. alk.; i. NH_4Cl
Silicon, crystalline	Si	28.06	gray, cb., 3.736	$2.4^{20°}$	1420	2600	i.	i.	s. HNO_3 + HF, Ag; sl. s. Pb, Zn; i. HF
Silicon, graphitic	Si	28.06	cr.	2.0–2.5	...	2600	i.	i.	s. HNO_3 + HF, fused alk.; i. HF.
Silicon, amorphous	Si	28.06	brn., amor.	2	...	2600	i.	i.	s. HF, KOH
carbide	SiC	40.07	blue-bk., trig., 2.654	3.17	>2700	subl. 2200	i.	i.	s. fused alk.; i. a.
chloride, tri-	Si_2Cl_6	268.86	lf. or lq.	$1.58^{0°}$	−1	144^{760mm}	d.		d. alk.
chloride, tetra-	$SiCl_4$	169.89	col., fuming lq., 1.412	1.50	−70	57.6	d.	...	d. conc. H_2SO_4, al.
fluoride	SiF_4	104.06	gas	3.57 (A)	−95.7	-65^{1810mm}	v. s. d.	...	s. HNO_3, al., et.
hydride (silane)	SiH_4	32.09	col. gas	lq. $0.68^{-185°}$	−185	-112^{760mm}	i.	...	i. al., et.; d. KOH
oxide, di- (opal)	$SiO_2.xH_2O$	...	iridescent, amor.	2.2	1600–1750	subl. 1750	i.	i.	s. HF, h. alk., fused $CaCl_2$
oxide, di- (cristobalite)	SiO_2	60.06	col., cb. or tet., 1.487	2.32	1710	2230	i.	i.	s. HF; i. alk.
oxide, di- (lechatelierite)	SiO_2	60.06	...	2.20	...	2230	i.	i.	s. HF; i. alk.
oxide, di- (quartz)	SiO_2	60.06	hex., 1.5442	$2.650^{20°}$	tr. <1425	2230	i.	i.	s. HF; i. alk.
oxide, di- (tridymite)	SiO_2	60.06	trig., rhb., 1.469	2.26	tr. 1670	2230	i.	...	s. HF; i. alk.
Silver	Ag	107.88	silv. met., cb.	$10.5^{20°}$	960.5	1950	i.	i.	s. HNO_3, h. H_2SO_4; i. alk.
bromide (bromyrite)	AgBr	187.80	pa. yel., cb., 2.252	$6.473\frac{25°}{4}$	434	d. 700	$0.00002^{20°}$	$0.00037^{100°}$	$0.51^{18°}$ NH_4OH; s. KCN, $Na_2S_2O_3$
carbonate	Ag_2CO_3	275.77	yel. pd.	6.077	218 d.	...	$0.003^{20°}$	$0.05^{100°}$	s. NH_4OH, $Na_2S_2O_3$; i. al.
chloride (cerargyrite)	AgCl	143.34	wh., cb., 2.071	5.56	455	1550	$0.000089^{10°}$	$0.00217^{100°}$	s. NH_4OH, KCN; sl. s. HCl
cyanide	AgCN	133.90	wh., 1.685±	3.95	$-(CN)_2$, 320	...	$0.000022^{20°}$	...	s. NH_4OH, KCN, HNO_3
nitrate (lunar caustic)	$AgNO_3$	169.89	col., rhb., 1.744	$4.352\frac{19°}{4}$	212	444 d.	$122^{0°}$	$952^{100°}$	s. gly.; v. sl. s. al.
Sodium	Na	22.997	silv. met., cb.	$0.97^{20°}$	97.5	880	d., forms NaOH	...	i. bz.; d. al.
acetate	$NaC_2H_3O_2$	82.04	wh., mn., 1.464	1.528	324	...	$46.5^{20°}$	$170^{100°}$	$2.1^{18°}$ al.
acetate	$NaC_2H_3O_2.3H_2O$	136.09	wh., mn.	1.45	58	$-3H_2O$, 120	v. s.	v. s.	$7.8^{25°}$ abs. al.
aluminate	$NaAlO_2$	81.97	amor.	...	1650	...	s.	v. s.	i. al.
amide	$NaNH_2$	39.02	olive gn.	...	210	400	d	...	d. al.

Sodium ammonium phosphate	$NaNH_4HPO_4.4H_2O$	209.09	col., mn.	1.574	79 d.	......	16.7	100	i. al.
antimonate, meta-	$2NaSbO_3.7H_2O$	511.63	cb.	......		......	$0.031^{12.3°}$	......	sl. s. al., NH_4 salts; i. ac.
arsenate	$Na_3AsO_4.12H_2O$	424.09	hex., 1.4589	1.759	86.3	......	$26.7^{17°}$	......	1.67 al., $50^{15°}$ gly.
arsenate, acid (monobasic)	$NaH_2AsO_4.H_2O$	181.94	rhb., 1.5535	2.535	d. 100	......	s.		
arsenate, acid (dibasic)	$Na_2HAsO_4.7H_2O$*	312.02	col., mn., 1.4658	1.871	125	$-7H_2O$, 100	$61^{15°}$	v. s.	sl. s. al.
arsenate, acid (dibasic)	$Na_2HAsO_4.12H_2O$	402.10	mn., 1.4496	1.72	28	$-12H_2O$, 100	$5.59^{0.1°}$	$140.7^{30°}$	sl. s. al.
arsenite, acid	Na_2HAsO_3	169.91	col.	1.87	......	......	v. s.		
benzoate	$NaC_7H_5O_2$	144.11	col. cr.	......		......	$62.5^{25°}$	$76.9^{100°}$	$2.3^{25°}$, $8.3^{78°}$ al.
bicarbonate	$NaHCO_3$	84.01	wh., mn., 1.500	2.20	$-CO_2$, 270	......	$6.9^{0°}$	$16.4^{60°}$	i. al.
bifluoride	$NaHF_2$	62.00	col. cr.	......	d.	......	$3.7^{20°}$	s.	
bisulfate	$NaHSO_4$	120.06	col., tri.	2.742	>315	d., $-H_2O$	$50^{0°}$	$100^{100°}$	d. al.; i. NH_3
bisulfite	$NaHSO_3$	104.06	col., mn., 1.526	1.48	d.	......	sl. s.	s.	i. al., act.
borate, tetra-	$Na_2B_4O_7$	201.27	......	2.367	741	......	$1.3^{0°}$	$8.79^{40°}$	i. al.
borate, tetra	$Na_2B_4O_7.5H_2O$	291.35	col., rhb., 1.461	1.815	......	......	$22^{62°}$ (anh.)	$52.3^{100°}$ (anh.)	
borate, tetra- (borax)	$Na_2B_4O_7.10H_2O$*	381.43	wh., mn., 1.4694	1.73	75	$-10H_2O$, 200	$1.3^{0.5}$ (anh).	$20.3^{30°}$ (anh.)	s. gly.; i. abs. al.
bromate	$NaBrO_3$	150.91	col., cb.	$3.339^{17.5°}$	381	......	$27.5^{0°}$	$90.9^{100°}$	i. al.
bromide	$NaBr$	102.91	col., cb., 1.6412	$3.205^{17.5°}$	755	1390	$90^{20°}$	$121^{100°}$	sl. s. al.
bromide	$NaBr.2H_2O$	138.95	col., mn.	2.176	50.7	......	$79.5^{0°}$ (anh.)	$118.3^{80°}$ (anh.)	sl. s. al.
carbonate (soda ash)	Na_2CO_3	106.00	wh. pd., 1.535	2.533	851	d.	$7.1^{0°}$	$48.5^{104°}$	i. al., et.
carbonate	$Na_2CO_3.H_2O$	124.02	wh., rhb., 1.506–1.509	1.55	$-H_2O$, 100	......	s.	s.	s. gly.; i. al., et.
carbonate	$Na_2CO_3.7H_2O$	232.12	rhb. or trig.	1.51	d. 35.1	......	s.	s.	
carbonate (sal soda)	$Na_2CO_3.10H_2O$	286.16	wh., mn., 1.425	1.46	......	......	$21.5^{0°}$	$238^{30°}$	i. al.
carbonate, sesqui- (trona)	$Na_3H(CO_3)_2.2H_2O$	226.05	wh., mn., 1.5073	2.112	d.	......	$13^{0°}$	$42^{100°}$	
chlorate	$NaClO_3$	106.45	wh., cb., or trig., 1.5151	$2.490^{15°}$	248	d.	$79^{0°}$	$230^{100°}$	s. al.
chloride	$NaCl$	58.45	col., cb., 1.5443	2.163	800.4	1413	$35.7^{0°}$	$39.8^{100°}$	sl. s. al.; i. conc. HCl
chromate	Na_2CrO_4	162.00	yel., rhb.	2.723	392	......	$32^{0°}$	$126^{100°}$	
chromate	$Na_2CrO_4.10H_2O$	342.16	yel., delq., mn.	1.483	19.9	......	v. s.	∞	sl. s. al.
citrate	$2Na_3C_6H_5O_7.11H_2O$	714.36	wh., rhb.	$1.857^{\frac{23.5°}{4}}$	$-11H_2O$, 150	d.	$91^{25°}$	$250^{100°}$	i. al.
cyanide	$NaCN$	49.02	wh., cb., 1.452	......	563.7	1496	$48^{10°}$	$82^{35°}$	s. NH_3; sl. s. al.
dichromate	$Na_2Cr_2O_7.2H_2O$	298.05	red, mn., 1.6994	$2.52^{13°}$	$-2H_2O$, 84.6; 356 (anh.)	d. 400	$238^{0°}$	$508^{80°}$	
ferricyanide	$Na_3Fe(CN)_6.H_2O$	298.97	red, delq.	......	......	......	$18.9^{0°}$	$67^{100°}$	i. al.
ferrocyanide	$Na_4Fe(CN)_6.10H_2O$	484.11	yel., mn.	1.458	......	......	$17.9^{20°}$ (anh.)	$63^{98.5°}$ (anh.)	i. al.
fluoride (villiaumite)	NaF	42.00	tet., 1.3258	2.79	992	......	$4^{0°}$	$5^{100°}$	v. sl. s. al.
formate	$NaHCO_2$	68.01	wh., mn.	1.919	253	......	$44^{0°}$	$160^{100°}$	sl. s. al.; i. et.
hydride	NaH	24.005	silv. nd., 1.470	0.92	d. 800	......	d.	......	i. bz., CS_2, CCl_4, NH_3; s. molten metal
hydrosulfide	$NaSH.2H_2O$	92.10	col., delq., nd.	......	d.	......	s.	s.	s. al.; d. a.
hydrosulfide	$NaSH.3H_2O$	110.11	rhb.	......	22	d.	s.	s.	s. al.; d. a.
hydrosulfite	$Na_2S_2O_4.2H_2O$	210.15	col. cr.	......	d.	......	$22^{20°}$	d.	i. al.
hydroxide	$NaOH$	40.00	wh., delq.	2.130	318.4	1390	$42^{0°}$	$347^{100°}$	v. s. al., et., gly.; i. act.
hydroxide	$NaOH.3\frac{1}{2}H_2O$	103.06	mn.	......	15.5	......	s.	v. s.	
hypochlorite	$NaOCl$	74.45	pa. yel., in soln. only	......	d.	......	$26^{0°}$	$158^{56°}$	
iodide	NaI*	149.92	col., cb., 1.7745	$3.667^{0°}$	651	1300	$158.7^{0°}$	$302^{100°}$	v. s. al., act.
iodide	$NaI.2H_2O$	185.95	col., mn.	2.448	......	......	v. s.	v. s.	v. s. NH_3
lactate	$NaC_3H_5O_3$	112.07	col., amor.	......	d.	......	v. s.	v. s.	s. al.; i. et.
nitrate (soda niter)	$NaNO_3$	85.01	col., trig., 1.5874	2.257	308	d. 380	$73^{0°}$	$180^{100°}$	s. NH_3; sl. s. gly., al.
nitrite	$NaNO_2$	69.01	pa. yel., rhb.	$2.168^{0°}$	271	d. 320	$72.1^{0°}$	$163.2^{100°}$	$0.3^{20°}$ et.; 0.3 abs. al.; $4.4^{20°}$ m. al.; v. s. NH_3
oxide	Na_2O	61.99	wh., delq.	2.27	subl.	......	Forms NaOH	......	d. al.
perborate	$NaBO_3.H_2O$	99.83	wh. pd.	......	d. 40	......	sl. s.	d.	s. gly., alk.
perchlorate	$NaClO_4$	122.45	rhb., 1.4617	......	482 d.	......	$170^{0°}$	$320^{100°}$	s. al.; 51 m. al.; 52 act.; i. et.
perchlorate	$NaClO_4.H_2O$	140.47	hex.	2.02	d. 130	......	$209^{15°}$	$284^{50°}$	s. al.
peroxide	Na_2O_2*	77.99	yel.-wh. pd.	2.805	d.	......	s. d.	d.	s. dil. a.
peroxide	$Na_2O_2.8H_2O$	222.12	wh., hex.	......	d. 30	......	s. d.	d.	
phosphate, monobasic	$NaH_2PO_4.H_2O$*	138.01	col., rhb., 1.4852	2.040	$-H_2O$, 100	d. 200	$71^{0°}$	$390^{83°}$	i. al.
phosphate, monobasic	$NaH_2PO_4.2H_2O$	156.03	col., rhb., 1.4629	1.91	60	......	$91.1^{0°}$	$308^{40°}$	
phosphate, dibasic	$Na_2HPO_4.7H_2O$	268.09	col., mn., 1.4424	1.679	d.	......	$185^{40°}$	$2000^{100°}$	
phosphate, dibasic	$Na_2HPO_4.12H_2O$	358.17	col., mn., 1.4361	1.52	34.6	$-12H_2O$, 180	$4.3^{0°}$	$76.7^{30°}$	i. al.
phosphate, tribasic	Na_3PO_4	163.97	wh.	$2.537^{17.5°}$	1340	......	$4.5^{0°}$	$77^{100°}$	
phosphate, tribasic	$Na_3PO_4.12H_2O$*	380.16	wh., trig., 1.4458	1.62	73.4	$-11H_2O$, 100	$28.3^{15°}$	∞	i. CS_2
phosphate, meta-	$Na_4P_4O_{12}$	407.91	col.	2.476	616 d.	......	s.	s.	s. a., alk.
phosphate, pyro-	$Na_4P_2O_7$*	265.95	wh.	2.45	988	......	$2.26^{0°}$	$45^{96°}$	d. a.
phosphate, pyro-	$Na_4P_2O_7.10H_2O$	446.11	mn., 1.4525	1.82	d.	......	$5.4^{0°}$	$93^{100°}$	i. al., NH_3
phosphate (pyrodisodium)	$Na_2H_2P_2O_7$	221.97	col., mn., 1.510	1.862	d. 220	......	$4.5^{0°}$	$21^{40°}$	
phosphate (pyrodisodium)	$Na_2H_2P_2O_7.6H_2O$	330.07	col., mn., 1.4645	1.848	......	......	$6.9^{0°}$	$36^{40°}$	
potassium tartrate	$NaKC_4H_4O_6.4H_2O$	282.23	rhb., 1.493	1.790	70 to 80	$-4H_2O$, 215	$26^{0°}$	$66^{26°}$	sl. s. al.
silicate, meta-	Na_2SiO_3	122.05	col., rhb., 1.520	......	1088	......	s.	s. d.	i. Na or K salts, al.

* Usual commercial form.

Table 1. Physical Properties of the Elements and Inorganic Compounds—*(Continued)*

Name	Formula	Formula weight	Color, crystalline form and refractive index	Specific gravity	Melting point, °C.	Boiling point, °C.	Solubility in 100 parts: Cold water	Hot water	Other reagents
Sodium silicate, meta-	$Na_2SiO_3.9H_2O$	284.20	rhb.	...	47	$-6H_2O$, 100	v. s.	v. s.	$29^{18°}$, ½N NaOH
silicate, ortho-	Na_4SiO_4	184.05	col., hex., 1.530	...	1018	...	s.	s.	
silicofluoride	Na_2SiF_6	188.05	wh., hex., 1.312	2.679	d.	...	$0.44^{0°}$	$2.45^{100°}$	i. al.
stannate	$Na_2SnO_3.3H_2O$	266.74	hex. tablets	...	d. 140	...	$50^{0°}$	$67^{50°}$	i. al., act.
sulfate (thenardite)	Na_2SO_4	142.05	col., rhb., 1.477	2.698	tr. 100 to mn.	...	$5^{0°}$	$42^{100°}$	i. al.
sulfate	Na_2SO_4	142.05	col., mn.	...	tr. 500 to hex.	...	$48.8^{40°}$	$42.5^{100°}$	d. HI; s. H_2SO_4
sulfate	Na_2SO_4	142.05	col., hex.	...	884	...	$19.4^{20°}$	$45.3^{60°}$	
sulfate	$Na_2SO_4.7H_2O$	268.17	tet.	...	...	...	$44.9^{0°}$	$202.6^{26°}$	
sulfate (Glauber's salt)	$Na_2SO_4.10H_2O$	322.21	col., mn., 1.396	1.464	32.4	$-10H_2O$, 100	$36^{15°}$	$412^{34°}$	i. al.
sulfide, mono-	Na_2S	78.05	pink or wh., amor.	1.856	...	...	$15.4^{10°}$	$57.3^{90°}$	sl. s. al.; i. et.
sulfide, tetra-	Na_2S_4	174.23	yel., cb.	...	275	...	s.	s.	s. al.
sulfide, penta-	Na_2S_5	206.29	yel.	...	251.8	...	s.	s.	s. al.
sulfite	Na_2SO_3	126.05	hex. pr., 1.565	$2.633\frac{15}{4}°$	d.	...	$13.9^{0°}$	$28.3^{84°}$	i. al., NH
sulfite	$Na_2SO_3.7H_2O$	252.17	mn.	1.561	$-7H_2O$, 150	d.	$34.7^{2°}$	$67.8^{18°}$	i. al.
tartrate	$Na_2C_4H_4O_6.2H_2O$	230.10	rhb.	1.818	...	...	$29^{6°}$	$66^{43°}$	i. al.
thiocyanate	NaCNS	81.08	delq., rhb., 1.625±	...	287	...	$110^{10°}$	$225^{100°}$	v. s. al.
thiosulfate	$Na_2S_2O_3$	158.11	mn.	1.667	...	...	$50^{0°}$	$231^{80°}$	
thiosulfate (hypo)	$Na_2S_2O_3.5H_2O$*	248.19	mn. pr., 1.5079	1.685	d. 48.0	...	$74.7^{0°}$	$301.8^{60°}$	s. NH_3; v. sl. s. al.
tungstate	Na_2WO_4	293.91	wh., rhb.	4.179	692	...	$57.58^{0°}$	$97^{100°}$	
tungstate	$Na_2WO_4.2H_2O$*	329.95	wh., rhb.	3.245	$-2H_2O$, 100	...	$88^{0°}$	$123.5^{100°}$	sl. s. NH_3; i. a., al.
tungstate, para-	$Na_6W_7O_{24}.16H_2O$	2097.68	wh., tri.	$3.987^{14°}$	$-16H_2O$, 300	...	8	d.	
uranate	Na_2UO_4	348.06	yel.	...	...	...	i.	i.	s. alk. carb., dil. a.
vanadate	$Na_3VO_4.16H_2O$	472.20	col. nd.	...	866 (anh.)	...	v. s.	d.	i. al.
vanadate, pyro-	$Na_4V_2O_7$	305.89	hex.	...	654	...	s.	...	i. al.
Stannic chloride	$SnCl_4$	260.53	col., fuming lq.	2.226	−30.2	114.1	s.	d.	s. abs. al., act., NH_3; s. ∞ CS_2
oxide (cassiterite)	SnO_2	150.70	wh., tet., 1.9968	7.0	1127	...	i.	i.	s. conc. H_2SO_4; i. alk., NH_4OH, NH_3
sulfate	$Sn(SO_4)_2.2H_2O$	346.85	col., delq., hex.	...	...	...	v. s.	d.	s. dil. H_2SO_4, HCl; d. abs. al.
Stannous bromide	$SnBr_2$	278.53	yel., rhb.	$5.12^{17°}$	215.5	620	s.	d.	s. C_5H_5N
chloride	$SnCl_2$	189.61	wh., rhb.	...	246.8	623	$83.9^{0°}$	$269.8^{15°}$	s. alk., abs. al., et.
chloride (tin salt)	$SnCl_2.2H_2O$*	225.65	wh., tri.	$2.71^{15.5°}$	37.7	d.	$118.7^{0°}$	∞	s. tart. a., alk., al.
sulfate	$SnSO_4$	214.76	wh. cr.	...	$-SO_2$, 360	...	$19^{19°}$	$18^{100°}$	s. H_2SO_4
Strontium	Sr	87.63	silv. met.	2.6	800	1150	d.	Forms $Sr(OH)_2$	s. al., a.
acetate	$Sr(C_2H_3O_2)_2$	205.72	wh. cr.	2.099	...	d.	$36.9^{0°}$	$36.4^{97°}$	$0.26^{15°}$ m. al.
carbonate (strontianite)	$SrCO_3$	147.64	wh., rhb., 1.664	3.70	1497^{60atm}	$-CO_2$, 1350	$0.0011^{18°}$	$0.065^{100°}$	s. a., NH_4 salts, aq. CO_2
chloride	$SrCl_2$	158.54	wh., cb., 1.6499	3.052	873	...	$43.5^{0°}$	$100.8^{100°}$	v. sl. s. act., abs. al.; i. NH_3
chloride	$SrCl_2.6H_2O$*	266.64	wh., rhb., 1.5364	$1.933^{17°}$	$-4H_2O$, 61	$-6H_2O$, 100	$104^{0°}$	$198^{40°}$	
hydroxide	$Sr(OH)_2$	121.65	wh., delq.	3.625	375	...	$0.41^{0°}$	$21.83^{100°}$	s. NH_4Cl
hydroxide	$Sr(OH)_2.8H_2O$*	265.77	col., tet., 1.499	1.90	$-7H_2O$ in dry air	...	$0.90^{0°}$	$47.7^{100°}$	s. NH_4Cl; i. act.
nitrate	$Sr(NO_3)_2$*	211.65	col., cb., 1.5878	2.986	570	...	$40^{0°}$	$100^{89°}$	s. NH_3; 0.012 abs. al.
nitrate	$Sr(NO_3)_2.4H_2O$	283.71	wh., mn.	2.2	...	...	$62.2^{0°}$	$124^{20°}$	i. HNO_3
oxide (strontia)	SrO	103.63	col., cb., 1.870	4.7	2430	...	Forms $Sr(OH)_2$	...	sl. s. al.; i. et.
peroxide	SrO_2	119.63	wh. pd.	...	d.	...	$0.008^{20°}$	d.	s. al., NH_4Cl; i. act.
peroxide	$SrO_2.8H_2O$	263.76	wh. cr.	...	$-8H_2O$, 100	d.	$0.018^{20°}$	d.	s. al.; i. NH_4OH
sulfate (celestite)	$SrSO_4$	183.69	col., rhb., 1.6237	3.96	1580 d.	...	$0.0113^{0°}$	$0.0114^{32°}$	sl. s. a.; i. dil. H_2SO_4, al.
sulfate, acid	$Sr(HSO_4)_2$	281.77	col., granular	...	d.	...	d.	...	$14^{70°}$ H_2SO_4
Sulfamic acid	NH_2SO_3H	97.09	wh., rhb.	$2.03\frac{12}{4}°$	205 d.	...	$20^{0°}$	$40^{70°}$	sl. s. al., act.; i. et.
Sulfur, amorphous	S	32.06	pa. yel. pd., 2.0–2.9	2.046	120	444.6	i.	i.	sl. s. CS_2
Sulfur, monoclinic	S_8	256.48	pa. yel., mn.	1.96	119.0	444.6	i.	i.	s. CS_2, al.
Sulfur, rhombic	S_8	256.48	pa. yel., rhb.	2.07	112.8	444.6	i.	i.	$24^{0°}$, $181^{55°}$ CS_2
Sulfur bromide, mono-	S_2Br_2	223.95	red, fuming lq.	2.635	−46	$54^{0.18mm}$	d.		
chloride, mono-	S_2Cl_2	135.03	red-yel. lq.	1.687	−80	138	d.	...	s. CS_2, et., bz.
chloride, di-	SCl_2	102.97	dark red fuming lq.	$1.621\frac{15}{15}°$	−78	59	d.	...	d. al.
chloride, tetra-	SCl_4	173.89	yel.-brn. lq.	...	−30	d. > −20	d.		
oxide, di-	SO_2	64.06	col. gas	lq., $1.434^{0°}$; 2.264 (A)	−75.5	−10.0	$22.8^{0°}$	$4.5^{50°}$	s. H_2SO_4; al., ac.
oxide, tri-(α)	SO_3	80.06	col. pr.	lq., 1.923; 2.75 (A)	16.83	44.6	d.	...	s. H_2SO_4
oxide, tri-(β)	$(SO_3)_2$	160.12	col., silky, nd	$1.97^{20°}$	50	...	Forms H_2SO_4	...	s. H_2SO_4

Sulfuric acid	H_2SO_4*	98.08	col., viscous lq.	$1.834\frac{18°}{4}$	10.49	d. 340	∞	∞	d. al.
Sulfuric acid	$H_2SO_4.H_2O$	116.09	pr. or lq.	$1.842\frac{15°}{4}$	8.62	290	∞	∞	d. al.
Sulfuric acid	$H_2SO_4.2H_2O$	134.11	col. lq.	$1.650\frac{0°}{4}$	−38.9	167	∞	∞	d. al.
Sulfuric acid, pyro-	$H_2S_2O_7$	178.14	cr.	$1.9^{20°}$	35	d.	d.	....	d. al.
Sulfuric oxychloride	SO_2Cl_2	134.97	col. lq.	$1.667\frac{20°}{4}$	−54.1	69.1^{760mm}	d.	....	s. ac.; d. al.
Sulfurous oxybromide	$SOBr_2$	207.89	or.-yel. lq.	$2.68^{18°}$	−50	68^{40mm}	d.	....	s. bz., CS_2, CCl_4; d. act.
oxychloride	$SOCl_2$	118.97	col. lq.	1.638	−104.5	78.8	d.	....	s. bz., chl.
Tantalum	Ta	180.88	bk.-gray, cb.	16.6	2850	>4100	i.	i.	s. fused alk., HF; i. HCl, HNO_3, H_2SO_4
Tellurium	Te	127.61	met., hex.	(α) 6.24; (β) 6.00	452	1390	i.	i.	s. H_2SO_4, HNO_3, KCN, KOH, aq. reg.; i. CS_2
Terbium	Tb	159.20							
Thallium	Tl	204.39	blue-wh., tet.	11.85	303.5	1650	i.	i.	s. HNO_3, H_2SO_4; i. NH_3
acetate	$TlC_2H_3O_2$	263.43	silky nd.	3.68	110	....	v. s.	....	v. s. al.
chloride, mono-	TlCl	239.85	wh., cb.	7.00	430	806	$0.21^{0°}$	$1.8^{100°}$	sl. s. HCl; i. al., NH_4OH
chloride, sesqui-	Tl_2Cl_3	515.15	yel., hex.	5.9	400–500	d.	$0.26^{15°}$	$1.9^{100°}$	
chloride, tri-	$TlCl_3$	310.76	hex. pl.	....	25	d.	v. s.	....	s. al., et.
chloride, tri-	$TlCl_3.4H_2O$	382.83	nd.	....	37	$-4H_2O$, 100	$86.2^{17°}$	d.	s. al., et.
sulfate (ic)	$Tl_2(SO_4)_3.7H_2O$	823.07	lf.	....	$-6H_2O$, 200	d.	d.	d.	s. dil. H_2SO_4
sulfate (ous)	Tl_2SO_4	504.84	col., rhb., 1.8671	6.77	632	d.	$2.70^{0°}$	$18.45^{100°}$	
sulfate, acid	$TlHSO_4$	301.46	trimorphous	....	115 d.	....	....	....	v. sl. s. dil. H_2SO_4
Thio, *Cf.* sulfo or sulfur									
Thorium	Th	232.12	cb.	11.2	1845	>3000	i.	i.	s. HCl, H_2SO; sl. s. HNO_3; i. HF, alk.
oxide, di- (thorianite)	ThO_2	264.12	wh., cb.	9.69	>2800	4400	i.	....	s. h. H_2SO_4; i. alk.
sulfate	$Th(SO_4)_2$	424.24	....	$4.225^{17°}$	....	....	$0.74^{0°}$	$5.22^{50°}$	
sulfate	$Th(SO_4)_2.9H_2O$	586.38	mn. pr.	2.77	$-9H_2O$, 400	....	sl. s.	sl. s.	
Thulium	Tm	169.40	....	....	....	....	i.	i.	
Tin	Sn	118.70	silv. met., tet.	7.31	231.85	2260	i.	i.	s. HCl, H_2SO_4, dil. HNO_3 h. aq KOH
Tin	Sn	118.70	gray, cb.	5.750	Stable −163 to +18	2260	i.	i.	s. a., h. alk. solns.
Tin salts, *Cf.* stannic and stannous									
Titanic acid	H_2TiO_3	97.92	wh. pd.	....	....	....	i.	i.	s. alk.; v. sl. s. dil. a.; i. al.
Titanium	Ti	47.90	dark gray, cb.	$4.50^{17.5°}$	1800	>3000	i.	d.	s. a.
chloride, di-	$TiCl_2$	118.81	bk., delq.	....	Unstable in air	....	d.	....	i. CS_2, et., chl.
chloride, tri-	$TiCl_3$	154.27	vl. delq.	....	d. 440	....	s.	s.	
chloride, tetra-	$TiCl_4$*	189.73	col. lq.	lq., 1.726	−30	136.4	s.	d.	s. dil. HCl
oxide, di- (anatase)	TiO_2	79.90	brn. or bk., tet., 2.534–2.564	3.84	....	....	i.	i.	sl. s. alk.
oxide, di- (brookite)	TiO_2	79.90	brn. or bk., rhb., 2.586	4.17	....	....	i.	i.	
oxide, di- (rutile)	TiO_2	79.90	col. if pure, tet., 2.615	4.26	1640 d.	<3000	i.	i.	s. H_2SO_4, alk.
Tungsten	W	183.92	gray-bk., cb.	19.3	3370	5900	i.	i.	s. h. conc. KOH; sl. s. NH_3, HNO_3, aq. reg.
carbide	WC	195.93	gray pd., cb.	$15.7^{18°}$	2777	6000	i.	i.	s. F_2; i. a.
carbide	W_2C	379.85	iron gray	$16.06^{18°}$	2877	6000	i.	i.	s. h. HNO_3; sl. s. HCl, H_2SO_4
oxide, tri-	WO_3	231.92	yel., rhb.	7.16	>2130	....	i.	i.	s. alk.; i. a.
Tungstic acid (tungstite)	H_2WO_4	249.94	yel., rhb. 2.24	5.5	$-\frac{1}{2}H_2O$, 100; 1473	....	i.	sl. s.	s. HF, alk., NH_3
Uranic acid	H_2UO_4	304.09	yel. pd.	$5.926^{15°}$	$-H_2O$, 250 to 300	....	i.	i.	s. a., alk. carb.; i. alk.
Uranium	U	238.07	wh. cr.	$18.485\frac{13°}{4}$	1133	3500	i.	i.	s. a.; i. alk.
carbide	U_2C_3	512.14	cr.	11.28	2400	....	d.	....	d. a.
oxide, di- (uraninite)	UO_2	270.07	bk., rhb.	10.9	2176	....	i.	i.	s. HNO_3, conc. H_2SO_4
oxide (pitchblende)	U_3O_8	842.21	olive gn.	7.31	d.	....	i.	i.	s. HNO_3, H_2SO_4
sulfate (ous)	$U(SO_4)_2.4H_2O$	502.25	gn., rhb.	....	$-4H_2O$, 300	....	$23^{11°}$	$9^{63°}$	s. dil. a.
Uranyl acetate	$UO_2(C_2H_3O_2)_2.2H_2O$	424.19	yel., rhb.	$2.89^{15°}$	$-2H_2O$, 110	....	$9.2^{17°}$	d.	s. al., act.
carbonate (rutherfordine)	UO_2CO_3	330.08	tet.	5.6					
nitrate	$UO_2(NO_3)_2.6H_2O$	502.18	yel., rhb., 1.4967	2.807	60.2	118	$170.3^{0°}$	$\infty^{60°}$	v. s. ac., al., et.; i. dil. alk.
sulfate	$UO_2SO_4.3H_2O$	420.18	yel. cr.	$3.28^{16.5°}$	d. 100	....	$18.9^{13.2°}$	$230^{25°}$	4 al.; s. a.
Vanadic acid, meta-	HVO_3	99.96	yel. scales	....	....	....	i.	....	s. a., alk.; i. NH_3

* Usual commercial form.

Table 1. Physical Properties of the Elements and Inorganic Compounds—(*Concluded*)

Name	Formula	Formula weight	Color, crystalline form and refractive index	Specific gravity	Melting point, °C.	Boiling point, °C.	Solubility in 100 parts: Cold water	Solubility in 100 parts: Hot water	Solubility in 100 parts: Other reagents
Vanadic acid, pyro-	$H_4V_2O_7$	217.93	pa. yel., amor.				i.		s. a., alk., NH_4OH
Vanadium	V	50.95	lt. gray, cb.	5.96	1710	3000	i.	i.	s. HNO_3, H_2SO_4; i. aq. alk.
chloride, di-	VCl_2	121.86	gn., hex., delq.	$3.23^{18°}$			s.	d.	s. al., et.
chloride, tri-	VCl_3	157.23	pink, tabular, delq.	$3.00^{18°}$	d.		s.	d.	s. abs. al., et.
chloride, tetra-	VCl_4	192.78	red lq.	$1.816^{30°}$	−109	148.5^{755mm}	s. d.		s. abs. al., et., chl., ac.
oxide, di-	V_2O_2	133.90	lt. gray cr.	3.64	ign.		i.	i.	s. a.
oxide, tri-	V_2O_3	149.90	bk. cr.	$4.87\frac{18°}{4}$	1970		sl. s.	s.	s. HNO_3, HF, alk.
oxide, tetra-	V_2O_4	165.90	blue cr.	4.399	1967		i.	i.	s. a., alk.
oxide, penta-	V_2O_5	181.90	red-yel., rhb.	$3.357\frac{18°}{4}$	800	d. 1750	$0.8^{20°}$		s. a., alk.; i. abs. al.
oxychloride, mono-	VOCl	102.41	brn. pd.	2.824			i.		v. s. HNO_3
Vanadyl chloride	$(VO)_2Cl$	169.36	yel. cr.	3.64	d. in air		i.		s. HNO_3
chloride, di-	$VOCl_2$	137.86	gn., delq.	$2.88^{13°}$			d.		s. abs. al., dil. HNO_3
chloride, tri-	$VOCl_3$	173.32	yel. lq.	1.829	<−15	127.19	s. d.		s. al., et., ∞ Br_2
Water†	H_2O	18.016	col. lq., $1.33300^{20°}$; hex. solid, 1.309	$1.00^{4°}$ (lq.); $0.915^{0°}$ (ice)	0	100			∞ al.; sl. s. et.
Water, heavy	D_2O	20.029	col. lq., $1.32844^{20°}$	$1.107^{20°}$	3.82	101.42	∞	∞	∞ al.; sl. s. et.
Xenon	Xe	131.30	col. gas	lq., $3.06^{-109.1}$; $2.7^{-140°}$; 4.53 (A)	−140	−109.1	24.2° cc	$7.3^{50°}$ cc	
Ytterbium	Yb	173.04							
Yttrium	Y	88.92	dark gray, hex.	5.51	1490	2500	sl. d.	d.	v. s. dil. a., h. KOH
Zinc	Zn	65.38	silv. met., hex.	7.140	419.4	907	i.	i.	s. a., ac., alk.
acetate	$Zn(C_2H_3O_2)_2$	183.47	mn.	1.840	242	subl. in vac.	$30^{25°}$	$44.6^{100°}$	$2.8^{25°}$, $166^{79°}$ al.
acetate	$Zn(C_2H_3O_2)_2.2H_2O$*	219.50	wh., mn., 1.494	1.735	237	$-2H_2O$, 100	$40^{25°}$	$66.6^{100°}$	v. s. al.
bromide	$ZnBr_2$	225.21	rhb.	$4.219^{4°}$	394	650	$390^{0°}$	$670^{100°}$	v. s. NH_4OH, al., et.
carbonate	$ZnCO_3$	125.39	wh., trig., 1.818	4.42	$-CO_2$, 300		$0.001^{15°}$		s. a., alk., NH_4 salts; i. act., NH_3
chloride	$ZnCl_2$	136.29	wh., delq., 1.687, uniaxial	$2.91\frac{25°}{4}$	283	732	$432^{25°}$	$615^{100°}$	$100^{12.5°}$ al.; v. s. et.; i. NH_3
cyanide	$Zn(CN)_2$	117.42	col., rhb.		d. 80		$0.0005^{18°}$	sl. s.	s. KCN, NH_3, alk.; i. al
hydroxide	$Zn(OH)_2$	99.40	col., rhb.	3.053	d. 125		$0.00052^{18°}$		s. a., alk., NH_4OH
iodide	ZnI_2	319.22	cb.	$4.666\frac{14.2°}{4}$	446	624	$430^{0°}$	$510^{100°}$	s. a., al., NH_3, aq. $(NH_4)_2CO_3$
nitrate	$Zn(NO_3)_2.6H_2O$	297.49	col., tet.	$2.065\frac{14°}{4}$	36.4	$-6H_2O$, 105	324.5	$\infty^{36.4°}$	v. s. al.
oxide (zincite)	ZnO	81.38	wh., hex., 2.004	5.606	>1800		$0.00042^{18°}$		s. a., alk., NH_4Cl; i. NH_3
oxide	ZnO	81.38	wh., amor.	5.47	>1800		$0.00042^{18°}$		
peroxide	ZnO_2	97.38	yel.	1.571	expl. 212		0.0022		i. NH_4OH; d. a.
phosphide	Zn_3P_2	258.10	steel gray, cb.	$4.55\frac{13°}{4}$	>420	1100	i.		s. dil. a.
silicate	$ZnSiO_3$	141.44	hex. or rhb.; glass, 1.650	3.52	1437		i.		
sulfate (zincosite)	$ZnSO_4$	161.44	wh., rhb., 1.669	$3.74\frac{15°}{4}$	d.740		$42^{0°}$	$61^{100°}$	sl. s al., s. gly.
sulfate	$ZnSO_4.H_2O$	179.46	col.	$3.28\frac{15°}{4}$	d. 238		s.	$89.5^{100°}$	
sulfate	$ZnSO_4.6H_2O$	269.54	mn.	$2.072\frac{15°}{4}$	$-5H_2O$, 70		s.	s.	sl. s. al.; i. act., NH_3
sulfate (goslarite)	$ZnSO_4.7H_2O$*	287.55	rhb., 1.4801	$1.966^{16.5°}$	tr. 39	$-7H_2O$, 280	115.2°°	$653.6^{100°}$	sl. s. al.; i. act., NH_3
sulfide (α) (wurzite)	ZnS	97.44	wh., hex., 2.356	4.087	1850^{150atm}	subl. 1185	$0.00069^{18°}$	i.	v. s. a.; i. ac.
sulfide (β) (sphalerite)	ZnS	97.44	wh., cb.; glass (?) 2.18–2.25	$4.102\frac{25°}{4}$	tr. 1020		i.	i.	s. a.
sulfide (blende)	ZnS	97.44	wh., granular	4.04			i.	i.	v. s. a.; i. ac.
sulfite	$ZnSO_3.2\frac{1}{2}H_2O$	190.48	mn.		$-2\frac{1}{2}H_2O$, 100	d. 200	0.16	d.	s. H_2SO_3, NH_4OH; i. al.
Zirconium	Zr	91.22	cb., pd. ign. easily	6.4	1700	>2900	i.	i.	s. HF, aq. reg.; sl. s. a.
oxide, di- (baddeleyite)	ZrO_2	123.22	yel. or brn., mn., 2.19	5.49	2700		i.	i.	s. H_2SO_4, HF
oxide, di- (free from Hf)	ZrO_2	123.22	wh., mn.	5.73		4300	i.	i.	s. H_2SO_4, HF

* Usual commercial form.

† *Cf.* special tables on water and steam, Nos. 3, 4, 5, 45, 46, 47.

Table 2. Physical Properties of Organic Compounds*

Abbreviations Used in the Table

(A), density referred to air	cr., crystalline	i-, iso-, containing the group $(CH_3)_2CH$-	nd., needles	s-, sec-, secondary	v. s., very soluble
al., ethyl alcohol	d., decomposes	i., insoluble	o-, ortho	silv., silvery	v. sl. s., very slightly soluble
amor., amorphous	d-, dextrorotatory	ign., ignites	or., orange	sl., slightly	wh., white
aq., aqua, water	dl-, dextro-laevorotatory	l-, laevorotatory	p-, para	subl., sublimes	yel., yellow
brn., brown	et., ethyl ether	lf., leaflets	pd., powder	sym., symmetrical	(+), right rotation
bz., benzene	expl., explodes	lq., liquid	pet., petroleum ether	t-, tertiary	>, greater than
c., cubic	gn., green	m-, meta	pl., plates	tet., tetragonal	<, less than
cc., cubic centimeter	h., hot	mn., monoclinic	pr., prisms	tri., triclinic	∞, infinitely
chl., chloroform	hex., hexagonal	n-, normal	rhb., rhombic	uns., unsymmetrical	
col., colorless			s., soluble	v., very	

This table of the physical properties includes the organic compounds of most general interest. For the properties of other organic compounds, reference must be made to larger tables in Lange's "Handbook of Chemistry" (Handbook Publishers), "Handbook of Chemistry and Physics" (Chemical Rubber Publishing Co.), Van Nostrand's "Chemical Annual," "International Critical Tables" (McGraw-Hill), and similar works.

The **molecular weights** are based on the 1941 atomic weight values. The **densities** are given for the temperature indicated and are usually referred to water at 4°C., *e.g.*, $1.028^{95/4}$ a density of 1.028 at 95°C. referred to water at 4°C., the 4 being omitted when it is not clear whether the reference is to water at 4°C. or at the temperature indicated by the upper figure. The melting and boiling points given have been selected from available data as probably the most accurate. The **solubility** is given in grams of the substance in 100 g. of the solvent. In the case of gases, the solubility is often expressed in some manner as "5^{10} cc." which indicates that, at 10°C., 5 cc. of the gas are soluble in 100 g. of the solvent.

Name	Synonym	Formula	Formula weight	Form and color	Specific gravity	Melting point, °C.	Boiling point, °C.	Solubility in 100 parts: Water	Alcohol	Ether
Abietic acid	sylvic acid, abietinic acid	$C_{20}H_{30}O_2$	302.44	lf.	....	182	....	i.	v. s.	v. s.
Acenaphthene	naphthylene ethylene	$C_{10}H_6(CH_2)_2$	154.20	rhb./al.	$1.069^{95/95}$	95	278–9	i.	s. h.	s. chl.
Acetal	acetaldehyde diethylacetal	$CH_3CH(OC_2H_5)_2$	118.17	lq.	$0.821^{22/4}$	....	102.2	6^{25}	∞	∞
Acet-aldehyde	ethanal	CH_3CHO	44.05	col. lq.	$0.783^{18/4}$	−123.5	20.2	∞	∞	∞
-aldehyde, par-	paraldehyde	$(C_2H_4O)_3$	132.16	col. cr.	$0.994^{20/4}$	10.5–12	124.4^{752}	12^{13}	∞	∞
-aldehyde ammonia	....	$CH_3CHOHNH_2$	61.08	col. cr.	....	97	100–10 d.	v. s.	v. s.	sl. s.
-amide	ethanamide	CH_3CONH_2	59.07	col. cr.	1.159	81(69.4)	222	s.	s.	v. sl. s.
-anilide	antifebrin	$C_6H_5NHCOCH_3$	135.16	rhb./al.	1.21^4	113–4	305	0.5^6	21^{20}	7^{25}
-phenetidide (o-)	o-ethoxyacetanilide	$CH_3CONHC_6H_4OC_2H_5$	179.21	lf./al.	....	79	>250	i.	s.	
(m-)	acetyl-m-phenetidine	$CH_3CONHC_6H_4OC_2H_5$	179.21	lf./al.	....	96–7	....	sl. s.	s.	
-toluidide (o-)	N-tolylacetamide	$CH_3C_6H_4NHCOCH_3$	149.19	rhb.	1.168^{15}	110	296	0.86^{19}	s.	s.
(p-)	N-tolylacetamide	$CH_3C_6H_4NHCOCH_3$	149.19	rhb. or mn.	1.212^{15}	153	306–7	0.09^{22}	10^{25}	s.
Acetic acid	ethanoic acid, vinegar acid	CH_3CO_2H	60.05	col. lq.	$1.049^{20/4}$	16.7	118.1	∞	∞	∞
anhydride	acetyl oxide, acetic oxide	$(CH_3CO)_2O$	102.09	col. lq.	$1.082^{20/4}$	−73	139.6	12 c.	∞	∞
nitrile	methyl cyanide	CH_3CN	41.05	col. lq.	$0.783^{20/4}$	−41	81.6–2.0	∞	∞	∞
Acetone dimethylketal	propanone, dimethyl ketone,	CH_3COCH_3	58.08	col. lq.	$0.792^{20/4}$	−94.6	56.5	∞	∞	∞
Acetonyl urea	dimethyl hydantoin	$<NHCONHCOC>(CH_3)_2$	128.13	tri./al.	....	175	subl.	s.	s.	s.
Acetophenone benzoyl hydride	methyl-phenyl ketone	$CH_3COC_6H_5$	120.14	lf.	$1.033^{15/15}$	20.5	202.3^{749}	i.	s.	s.
Acetyl-chloride	ethanoyl chloride	CH_3COCl	78.50	col. lq.	$1.105^{20/4}$	−112.0	51–2	d.	d.	∞
-phenylenediamine (-p)	amino-acetanilide (p)	$C_2H_3ONHC_6H_4NH_2$	150.18	nd./aq.	....	162	....	s. h.	v. s.	v. s.
Acetylene	ethyne; ethine	$HC:CH$	26.04	col. gas	(A) 0 906	-81.5^{891}	-84^{760}	100 cc.[18]	600 cc.[18]	
dichloride (cis)	1,2-dichloroethene	$CHCl:CHCl$	96.95	col. lq.	$1.291^{15/4}$	−80.5	60.3	0.35^{20}	∞	∞
(trans)	dioform	$CHCl:CHCl$	96.95	col. lq.	$1.265^{15/4}$	−50	48.4	0.63^{20}	∞	∞
Aconitic acid	equisetic acid; citridic acid	$C_3H_3(CO_2H)_3$	174.11	cr./aq.	....	192 d.	....	33^{15}	sl. s.	v. sl. s.
Acridine	....	$C_6H_4<(CH)(N)>C_6H_4$	179.21	rhb./aq. al.	....	110–1	346	sl. s. h.	s.	s.
Acrolein ethylene aldehyde	acrylic aldehyde, propenal	$CH_2:CH.CHO$	56.06	col. lq.	$0.841^{20/4}$	−87.7	52.5	40	s.	s.
Acrylic acid	propenoic acid	$CH_2:CH.CO_2H$	72.06	col. lq.	$1.062^{16/4}$	12–13	141–2	∞	∞	
nitrile	vinyl cyanide	$CH_2:CH.CN$	53.06	col. lq.	0.811^{20}	−82	78–9	s.		
Adipic acid	hexandioc acid, adipinic acid	$(CH_2CH_2CO_2H)_2$	146.14	mn. pr.	$1.360^{25/4}$	151–3	265^{10}	1.4^{15}	v. s.	0.6^{15}
amide	....	$(CH_2CH_2CONH_2)_2$	144.17	cr. pd.	....	226–7	....	0.4^{12}		
nitrile	tetramethylene	$(CH_2CH_2CN)_2$	108.14	col. oil	$0.951^{19/19}$	1	295	v. sl. s.	s.	v. sl. s.
Adrenaline (1-) (3,4,1)	1-suprarenine	$C_6H_3(OH)_2(CHOHCH_2NHCH_3)$	183.20	col. pd.	....	d. 207–11	....	$0\ 03^{20}$	v. sl. s.	i.
Alanine (α) (dl-)	....	$CH_3CH(NH_2)CO_2H$	89.09	nd./aq.	....	295 d.	subl. >200	22^{17}	v. sl. s.	i.
Aldol acetaldol,	2-hydroxybutyraldehyde	$CH_3CH(OH)CH_2CO_2H$	88.10	col. lq.	$1.103^{20/4}$	....	83^{20}	∞	∞	s.
Alizarin	Anthraquinoic acid,	$C_6H_4(CO)_2C_6H_2(OH)_2$	240.20	red rhb.	....	289–90	430	$0\ 03^{100}$	v. s.	v. s.
Allyl alcohol	propen-1-ol-3,propenyl alcohol	$CH_2:CH.CH_2OH$	58.08	col. lq.	$0.854^{20/4}$	−129	96.6	∞	∞	∞
bromide	3-bromo-propene-1	$CH_2:CH.CH_2Br$	120.99	lq.	$1.398^{20/4}$	−119.4	$70–1^{753}$	i.	∞	∞
chloride	3-chloro-propene-1	$CH_2:CH.CH_2Cl$	76.53	col. lq.	$0.938^{20/4}$	−136.4	44.6	<0.1	∞	∞
thiocyanate (i)	mustard oil	$CH_2:CH.CH_2NCS$	99.15	col. oil	$1.013^{20/4}$	−80	152	0.2	∞	∞
thiourea	thiosinamide	$CH_2:CH.CH_2NHCSNH_2$	116.18	col. pr.	$1.219^{20/20}$	77–8	....	3^0	s.	v. sl. s.
Aluminum ethoxide	....	$Al(OCH_2CH_3)_3$	164.15	pd.	$1.142^{20/0}$	150–60	$200–5^{10}$	d.	i.	v. sl. s.
Amino-anthraquinone (α)	....	$C_6H_4(CO)_2C_6H_3NH_2$	223.22	red nd.	....	256	subl.	i.	s.	s.
(β)	....	$C_6H_4(CO)_2C_6H_3NH_2$	223.22	red nd.	....	302	subl.	i.	s.	i.
-azobenzene	....	$C_6H_5.N:N.C_6H_4NH_2$	197.23	yel. mn.	....	126–7	225^{120}	sl. s. h.	s. h.	s.
-benzoic acid (m-)	....	$H_2N.C_6H_4CO_2H$	137.13	nd./aq.	$1.511^{4°}$	173–4	....	v. sl. s.	2^{10}	1.8^6
(p-)	aminodracylic acid	$H_2N.C_6H_4CO_2H$	137.13	mn. pr.	....	187–8	....	0.3^{13}	11^{10}	$8\ 2^6$

* By N. A. Lange, Ph.D , Handbook Publishers, Inc., Sandusky, Ohio. Abridged from table of Physical Constants of Organic Compounds in Lange's "Handbook of Chemistry."

Table 2. Physical Properties of Organic Compounds—(*Continued*)

Name	Synonym	Formula	Formula weight	Form and color	Specific gravity	Melting point, °C.	Boiling point, °C.	Solubility in 100 parts: Water	Alcohol	Ether
Amino-diphenylamine (p-)		$H_2N.C_6H_4NH.C_6H_5$	184.23	nd./aq. al.		67	354	sl. s.	s.	s.
-G-acid (2-)(6-,8-), Na_2 salt		$C_{10}H_5(NH_2)(SO_3Na)_2$	347.28					v. sl. s.		
-mono-potassium salt		$C_{10}H_5(NH_2)S_2O_6HK$	341.39					12.8^{20}		
-sodium salt		$C_{10}H_5(NH_2)S_2O_6HNa$	325.29					2.7^{18}		
-J-acid (2-)(5-,7-)		$C_{10}H_5(NH_2)(SO_3H)_2$	303.30					10.0^{20}		
-mono-potassium salt		$C_{10}H_5(NH_2)S_2O_6HK$	341.39					3.4^{18}		
-naphthol sulfonic (1-,2-,4-)(α-)		$C_{10}H_5OHNH_2SO_3H\frac{1}{2}H_2O$	248.25					v. s.		
(1-,8-,4-)		$NH_2(OH)C_{10}H_5SO_3H$	239.24					v. sl. s.		
-phenol (o-)	2-aminophenol	$H_2N.C_6H_4.OH$	109.12	col. nd.		173	subl.	1.7^0	4.3^0	v. s.
(m-)	3-aminophenol	$H_2N.C_6H_4.OH$	109.12	pr.		122-3		2.6^0	s.	sl. s.
(p-)	p-hydroxyaniline	$H_2N.C_6H_4.OH$	109.12	lf.		184-6 d.	subl.	1.1^0	4^0	i. bz.
-toluene sulfonic acid (1-,2-,3-)		$C_6H_3(CH_3)(NH_2)SO_3H$	187.21	nd.				0.97^{11}		
(1-,4-,2-)		$C_6H_3(CH_3)(NH_2)SO_3H.H_2O$	205.23	mn.		d.		0.5^{20}	i.	
(1-,4-,3-)		$C_6H_3(CH_3)(NH_2)SO_3H.\frac{1}{2}H_2O$	196.22	nd.				0.47		
(1-,2-,5-)		$C_6H_3(CH_3)(NH_2)SO_3H.H_2O$	205.23	tri./aq.		$-H_2O$, 120		3^{11}	i.	
Amyl acetate (n-)		$CH_3CO_2CH_2(CH_2)_3CH_3$	130.18	col. lq.	$0.879^{20/20}$	−70.8	148.4^{737}	v. sl. s.	∞	∞
(i-)	common amyl acetate	$CH_3CO_2CH_2CH_2CH(CH_3)_2$	130.18	col. lq.	$0.876^{15/4}$		142^{757}	0.3^{15}	∞	∞
		$CH_3CO_2CH_2CH(CH_3)C_2H_5$	130.18	col. lq.	0.880^{13}		141–2	v. sl. s.	∞	∞
(s-)	α-Me-Bu-acetate	$CH_3CO_2CH(CH_3)CH_2C_2H_5$	130.18	col. lq.	0.922^0		133.5	sl. s.	∞	∞
(s-)	di Et-carbinol acetate	$CH_3CO_2CH(C_2H_5)_2$	130.18	col. lq.	$0.871^{20/4}$		133	sl. s.	∞	∞
(t-)		$CH_3CO_2C(CH_3)_2C_2H_5$	130.18	col. lq.	0.874^{19}		124.5^{749}	v. sl. s.	∞	∞
alcohol (n-) fusel oil,	pentanol-1	$CH_3(CH_2)_3CH_2OH$	88.15	col. lq.	$0.817^{20/20}$	−78.5	137.9	2.7^{22}	∞	∞
(s-,n-) methyl-propyl carbinol,	pentanol-2	$C_2H_5CH_2CH(OH)CH_3$	88.15	col. lq.	$0.810^{20/20}$		119.5	4^{20}	∞	∞
(prim.-,i-) isobutyl carbinol,	2-methyl-butanol-4	$(CH_3)_2CHCH_2CH_2OH$	88.15	col. lq.	$0.813^{15/4}$	−117.2	132.0	2^{14}	∞	∞
		$(C_2H_5)_2CHOH$	88.15	col. lq.	$0.815^{25/4}$		115.6	5.5^{30}	∞	∞
(s-,i-)	2-methyl-butanol-3	$(CH_3)_2CHCH(OH)CH_3$	88.15	col. lq.	0.819^{19}		113–4	2.8^{30}	∞	∞
(t-)	2-methyl-butanol-2	$(CH_3)_2C(OH)C_2H_5$	88.15	col. lq.	$0.809^{20/4}$	−11.9	102	sl. s.	s.	s.
		$(CH_3)_3CCH_2OH$	88.15	cr.		52–3	113–4	sl. s.	∞	∞
(d-)	active amyl alcohol	$C_2H_5CH(CH_3)CH_2OH$	88.15	col. lq.	$0.816^{20/4}$		128	3.6^{30}	∞	∞
-amine (n-)		$CH_3(CH_2)_4NH_2$	87.16	col. lq.	0.766^{19}	−55	103–4	s.	s.	s.
(s-,n-)		$(C_3H_7)(CH_3)CHNH_2$	87.16	col. lq.	$0.749^{20/4}$		91–2	∞	∞	∞
(i-)		$(CH_3)_2CH(CH_2)_2NH_2$	87.16	col. lq.	$0.751^{18/4}$		95	∞	∞	∞
(t-)		$(C_2H_5)(CH_3)_2CNH_2$	87.16	col. lq.	$0.731^{25/4}$	−105	77–8	∞	∞	∞
	1-NH_2-2-Me-butane	$C_2H_5CH(CH_3)CH_2NH_2$	87.16	col. lq.	0.755^{18}		95–6	∞	∞	∞
	3-amino pentane	$(C_2H_5)_2CHNH_2$	87.16	col. lq.	$0.749^{20/4}$		90–1	∞	∞	∞
	3-NH_2-2-Me-butane	$(CH_3)_2CHCH(CH_3)NH_2$	87.16	col. lq.	0.757^{18}		83–4	∞	∞	∞
aniline (i-)		$C_6H_5NHC_5H_{11}$	163.25	lq.	$0.928^{15/4}$		254.5	i.		
benzoate (i-)		$C_6H_5CO_2C_5H_{11}$	192.25	col. lq.	$0.992^{14/14}$		261^{746}	i.	∞	∞
bromide (n-)	1-bromopentane	$CH_3(CH_2)_3CH_2Br$	151.05	col. lq.	$1.218^{20/4}$	−95	129.7	i.	s.	
(i-)	4-Br-2-Me-butane	$(CH_3)_2CH(CH_2)_2Br$	151.05	col. lq.	$1.220^{17/15}$		120^{745}	0.02^{16}	s.	s.
(t-)	2-Br-2-Me-butane	$(CH_3)_2C(Br)C_2H_5$	151.05	lq.	$1.216^{19/0}$		108^{765}	i.	s.	s.
n-butyrate (n-)		$C_2H_5CH_2CO_2(CH_2)_4CH_3$	158.23	col. lq.	$0.871^{15/4}$	−73.2	186.4	0.05^{50}	∞	∞
(i-)		$C_2H_5CH_2CO_2.C_5H_{11}$	158.23	col. lq.	$0.866^{19/15}$		178.6	i.	∞	∞
(t-)		$C_3H_7CO_2C(CH_3)_2C_2H_5$	158.23	col. lq.	$0.865^{15/0}$		164	sl. s.	∞	∞
i-butyrate (i-)		$(CH_3)_2CHCO_2C_5H_{11}$	158.23	lq.	$0.876^{0/4}$		168.8	i.	s.	s.
chloride (n-)	1-chloropentane	$CH_3(CH_2)_3CH_2Cl$	106.60	col. lq.	$0.878^{20/4}$	−99	108.4	i.	s.	s.
(s-)	2-chloropentane	$C_2H_5CH_2CHClCH_3$	106.60	lq.	$0.870^{20/4}$		96.7	i.	s.	s.
(s-)	3-chloropentane	$(C_2H_5)_2CHCl$	106.60	col. lq.	0.895^{21}		97.3	i.	∞	∞
(i-)	4-Cl-2-Me-butane	$(CH_3)_2CH(CH_2)_2Cl$	106.60	col. lq.	$0.893^{20/4}$		99.7^{758}	i.	∞	∞
(s-,i-)	3-Cl-2-Me-butane	$(CH_3)_2CHCHClCH_3$	106.60	lq.	0.883^0		91^{753}	i.	s.	s.
(t-)	2-Cl-2-Me-butane	$(CH_3)_2CClC_2H_5$	106.60	lq.	$0.871^{20/4}$	−72.9	85.7	i.	s.	s.
	1-Cl-2-Me-butane	$(CH_3)(C_2H_5)CHCH_2Cl$	106.60	lq.	$0.881^{17.5}$		98–9	i.	s.	s.
i-cyanide (i-)	iso-caproic iso-nitrile	$(CH_3)_2CH(CH_2)_2NC$	97.16	lq.			137–9	i.	s.	s.
formate (n-)		$HCO_2CH_2(CH_2)_3CH_3$	116.16	lq.	0.902^0	−73.5	132	v. sl. s.	∞	∞
(i-)		$HCO_2CH_2CH_2CH(CH_3)_2$	116.16	lq.	$0.882^{20/4}$	−93.5	123.5	0.3^{22}	∞	∞
iodide (n-)	1-iodopentane	$CH_3(CH_2)_3CH_2I$	198.06	lq.	$1.510^{20/4}$	−86	157.0	i.	s.	∞
(i-)	4-I-2-Me-butane	$(CH_3)_2CHCH_2CH_2I$	198.06	lq	$1.515^{18/4}$		147^{765}	i.	∞	∞
(s-,n-)	2-iodopentane	$C_2H_5CH_2CHICH_3$	198.06	lq.	$1.507^{17/4}$		144–5	i.	∞	∞
(t-)	2-I-2-Me-butane	$(CH_3)_2CIC_2H_5$	198.06	lq.	$1.471^{19/15}$		127^{765}	i.	∞	∞
		$C_2H_5CH(CH_3)CH_2I$	198.06	lq.	$1.524^{20/4}$		148	i.	∞	∞
mercaptan (n-)	pentanthiol-1	$CH_3(CH_2)_3CH_2SH$	104.21	lq.	0.857^{20}		126^{767}	i.	∞	∞
(n-)	pentanthiol-3	$(C_2H_5)_2CHSH$	104.21	col. lq.			105	i.	∞	∞
(i-)	2-Me-butanthiol-4	$(CH_3)_2CH(CH_2)_2SH$	104.21	lq.	$0.835^{20/4}$		120	i.	∞	∞
phenol (t-)(p-)	pentaphen	$C_5H_{11}.C_6H_4OH$	164.24	cr.		93	265–7	sl. s.	s.	s.

propionate (n-)		$C_2H_5CO_2(CH_2)_4CH_3$	144.21	lq.	$0.876^{15/4}$	−73.1	168.7	i.	∞	∞
(i-)		$C_2H_5CO_2(CH_2)_2CH(CH_3)_2$	144.21	col. lq.	$0.870^{20/4}$	....	160.2	0.1^{25}	∞	∞
(act.)		$C_2H_5CO_2C_5H_{11}$	144.21	col. lq.	$0.866^{20/4}$	....	58^{16}	v. sl. s.	∞	∞
salicylate (n-)		$HOC_6H_4CO_2C_5H_{11}$	208.25	lq.	1.065^{15}	....	265	i.	∞	∞
Amyl i-valerate (i)		$C_4H_9CO_2C_5H_{11}$	172.26	col. lq.	$0.858^{20/15}$	....	194	v. sl. s.	∞	∞
(t-)		$C_4H_9CO_2C_5H_{11}$	172.26	col. lq.	$0.861^{14/0}$	....	173–4	sl. s.	s.	s.
Amylene (n-)(α-)	pentene-1	$C_2H_5CH_2CH{:}CH_2$	70.13	lq.	0.644^{20}	....	30–1	i.	∞	∞
(i-)	2-methyl-butene-3	$(CH_3)_2CHCH{:}CH_2$	70.13	col. lq.	0.632^{15}	−135	20.5^{771}	i.	∞	∞
(α-)	2-methyl-butene-1	$(C_2H_5)(CH_3)C{:}CH_2$	70.13	col. lq.	$0.667^{0/0}$	....	$31–2^{758}$	i.	∞	∞
-n)(β-)	pentene-2	$C_2H_5CH{:}CHCH_3$	70.13	col. lq.	$0.650^{20/4}$	−139	36.4	v. sl. s.	∞	∞
(i-)(β-)	2-methyl-butene-2	$(CH_3)_2C{:}CHCH_3$	70.13	col. lq.	$0.663^{19/4}$	−124	37–8	i.	s.	∞
Anethole (p-)	p-propenyl anisole	$CH_3CH{:}CH.C_6H_4OCH_3$	148.20	lf./al.	$0.991^{20/20}$	22.5	235.3	v. sl. s.	s.	∞
Anhydroformald-aniline	methylene aniline	$(CH_2NC_6H_5)_3$	315.40	pr./al.	....	143	185	i.	sl. s.	s.
Aniline	amino benzene, phenyl amine, cyanol	$C_6H_5NH_2$	93.12	col. oil	$1.022^{20/4}$	−6.2	184.4	3.6^{18}	∞	∞
hydrochloride	aniline salt, aniline chloride	$C_6H_5NH_2.HCl$	129.59	cr.	1.222^{4}	198	245	18^{15}	s.	i.
nitrate		$C_6H_5NH_2.HNO_3$	156.14	rhb.	1.356^{4}	d. 190	....	s.	s.	sl. s.
sulfate		$(C_6H_5NH_2)_2.H_2SO_4$	284.32	lf./al.	1.377^{4}	d.	....	5^{14}	sl. s.	i.
Anisal-acetone (p-)	MeO-benzalacetone	$CH_3OC_6H_4CH{:}CHCOCH_3$	176.22	lf./et.	....	73–4	....	i.	v. s.	v. s.
Anisic acid (p-)		$CH_3OC_6H_4CO_2H$	152.14	mn./aq.	1.385^{4}	184.2	275–80	0.03^{19}	v. s.	v. s.
aldehyde (p-)		$CH_3OC_6H_4CHO$	136.14	col. oil	$1.123^{20/4}$	2.5	247–8	v. sl. s.	∞	∞
Anisidine (o-)	2-amino-anisole	$CH_3OC_6H_4NH_2$	123.15	col. lq.	$1.098^{15/15}$	5.2	225	v. sl. s.	∞	∞
(m-)	MeO-aniline(m)	$CH_3OC_6H_4NH_2$	123.15	oil	$1.096^{20/4}$	<−12	251	v. sl. s.	s.	s.
(p-)	4-amino anisole	$CH_3OC_6H_4NH_2$	123.15	pl./aq.	$1.089^{55/55}$	57.2	243	s. h.	s.	s.
Anisole	methyl phenyl ether	$CH_3OC_6H_5$	108.13	col. lq.	$0.990^{22/4}$	−37.3	154–5	i.	s.	s.
Anthracene	paranaphthalene, anthracin green oil	$C_6H_4{:}(CH)_2{:}C_6H_4$	178.22	col. mn.	$1.25^{27/4}$	217–8	340–2	i.	1.5^{20}	
Anthramine (α)	α-amino-anthracene	$C_6H_4{:}(CH)_2{:}C_6H_3NH_2$	193.24	yel./al.	....	130±	....	i.	s.	
(β)	β-amino-anthracene	$C_6H_4{:}(CH)_2{:}C_6H_3NH_2$	193.24	yel./al.	....	238	subl.	i.	sl. s.	sl. s.
Anthranil		$C_6H_4{:}(NH)CO$	119.12	col. oil	$1.187^{15/4}$	<−18	d. >215	sl. s. h.	s.	s.
Anthranilic acid (o-)		$H_2NC_6H_4CO_2H$	137.13	col. rhb.	....	144–5	subl.	0.35^{14}	11^{10}	16^{7}
Anthrapurpurin (1-,2-,7-)		$C_{14}H_5O_2(OH)_3$	256.20	or. nd./al.	....	369	462	sl. s. h.	v. s. h.	sl. s.
Anthraquinone	diphenyleneketone, dihydrodiketoanthracene	$C_6H_4{:}(CO)_2{:}C_6H_4$	208.20	yel. rhb.	$1.438^{20/4}$	286	379–81	i.	0.05^{18}	v. sl. s.
disulfonate Na_2 (1-,5-)	ρ-anthraquinone disulfonate	$C_{14}H_6O_2(SO_3Na)_2.5H_2O$	502.38	yel. lf.	....	....	....	v. s.	i.	i.
(1-,8-)	x-anthraquinone disulfonate	$C_{14}H_6O_2(SO_3Na)_2.4H_2O$	484.37	yel. pr.	....	....	....	sl. s.		
(2-,6-)		$C_{14}H_6O_2(SO_3Na)_2.7H_2O$	538.41	col. cr.	....	....	....	3.9^{20}		
(2-,7-)		$C_{14}H_6O_2(SO_3Na)_2.4H_2O$	484.37	cr.	....	....	....	30.5^{20}	v. sl. s.	i.
sulfonate Na (1-)		$C_{14}H_7O_2SO_3Na$	310.25	yel. lf.	....	....	....	0.53^{20}	i.	i.
(2-)		$C_{14}H_7O_2SO_3Na$	310.25	silv. lf.	....	....	....	0.84^{25}	i.	i.
Anthrarufin (1-,5-)		$C_{14}H_6O_2(OH)_2$	240.20	yel. lf.	....	280	subl.	i.	sl. s.	s.
Antipyrene	1-ph-2,3-diMepyrazolone-5	$C_{11}H_{12}ON_2$	188.22	mn./aq.	$1.088^{113/4}$	113(109)	319^{174}	100^{25}	100	sl. s.
Apiole	1-allyl-2, 5-diMeO-3,4 methylenedioxybenzene	$C_{12}H_{14}O_4$	222.23	col. nd.	$1.02^{20/4}$	30	294	i	s.	s.
Arabinose (α)(d- or l-)		$CH_2OH(CHOH)_3CHO$	150.13	rhb. pr.	$1.585^{20/4}$	159.5	....	46^{0}	$0.5^{9°}$	i.
(dl-)		$CH_2OH(CHOH)_3CHO$	150.13	....	....	164.5	....	16.9^{10}		
Arachidic acid	eicosanoic acid	$CH_3(CH_2)_{18}CO_2H$	312.52	col. lf.	....	77	328	i.	s. h.	v. s.
Arsanilic acid (p-)		$H_2N.C_6H_4.AsO_3H_2$	217.04	nd./aq.	....	232	....	v. s. h.	v. s. h.	i.
Asparagine (l-)		$HO_2C.C_2H_3(NH_2).CONH_2$	132.12	rhb.	$1.543^{15/4}$	227–35	d. 235	3.1^{28}	i. c.	
Aspirin (o-)		$CH_3CO_2.C_6H_4.OH$	180.15	nd./aq.	....	135–6	....	1^{37}	s.	5^{20}
Atropic acid	α-phenyl acrylic acid	$C_6H_5C({:}CH_2).CO_2H$	148.15	nd./aq.	....	106–7	267 d.	0.1 c.	s.	s.
Auramine	4,4′-dimethylaminobenzophenomide	$[(CH_3)_2NC_6H_4]_2C{:}NH$	267.36	col./al.	....	136	....	i.	7^{20}	2.3^{20}
Aurine, coralline (4-,4′-)		$(HOC_6H_4)_2C{:}C_6H_4{:}O$	290.30	red	....	310 d.	....	i.	s.	s.
Azo-anisole (2-,2′-)	diMeO-azobenzene	$(CH_3O.C_6H_4N{:})2$	242.27	or. pr.	....	153	....	i.	s.	s.
benzene	diphenyldiimide	$C_6H_5N{:}N.C_6H_5$	182.22	or. mn.	$1.203^{20/4}$	68	297	i.	4.2^{20}	
Azoxybenzene		$(C_6H_5)_2N_2O$	198.22	yel. rhb.	$1.248^{20/20}$	36	d.	i.	11.4^{15}	
Barbituric acid	malonyl urea	$CO{:}(NHCO)_2{:}CH_2.2H_2O$	164.12	col./aq.	....	d. 245	....	s. h.	sl. s.	s.
Benzal acetone	Me-cinnamyl ketone	$C_6H_5CH{:}CHCOCH_3$	146.18	pl.	$1.035^{20/20}$	41–2	260–2	i.	s.	s.
Benzaldehyde	artificial almond oil	C_6H_5CHO	106.12	col. lq.	$1.046^{20/4}$	−26	179	0.3	∞	∞
Benzamide		$C_6H_5CONH_2$	121.13	col. pr.	1.341	130	290	1.35^{25}	17^{25}	sl. s.
Benzanilide		$C_6H_5CONHC_6H_5$	197.23	lf./al.	1.31^{4}	163	$117–9^{10}$	i.	4^{30}	sl. s.
Benzene	benzol, phenyl hydride, cyclohexatriene	C_6H_6	78.11	col. lq.	$0.879^{20/4}$	5.5	80.1	0.07^{22}	s.	∞
sulfinic acid		$C_6H_5SO_2H$	142.17	pr./aq.	....	83–4	d. >100	v. s. h.	v. s.	v. s.
sulfonic acid		$C_6H_5SO_3H$	158.17	col. nd.	....	65–6	d.	v. s.	v. s.	i.
sulfonic amide	benzene sulfonamide	$C_6H_5SO_2NH_2$	157.18	mn./aq.	....	156	....	0.43^{16}	v. s.	v. s.
sulfonic chloride	benzene sulfonyl chloride	$C_6H_5SO_2Cl$	176.62	cr.	$1.384^{15/15}$	14.5	251.5	i.	v. s.	s.
Benzidine (4-,4′-)		$NH_2.C_6H_4.C_6H_4.NH_2$	184.23	cr./aq.	....	128–9	400^{740}	1 h.	1 h.	2
disulfonic acid (2-,2′-)		$(.C_6H_3(NH_2)SO_3H)_2.3H_2O$	398.40	pr./aq.	....	d. >175	....	0.09^{25}	i.	i.
(3-,3′-)		$(.C_6H_3(NH_2)SO_3H)_2$	344.35	....	....	....	....	v. sl. s.		
Benzil	dibenzoyl	$C_6H_5CO.COC_6H_5$	210.22	pr.	1.23^{15}	95	348 d.	i.	v. s.	v. s.
Benzoic acid		$C_6H_5CO_2H$	122.12	mn. pr.	$1.266^{15/4}$	121.7	249.2	0.2^{17}	46^{15}	66^{15}
anhydride		$(C_6H_5CO)_2O$	226.22	rhb./et.	$1.199^{15/4}$	42	360	i.	s.	s.
nitrile	phenyl cyanide	C_6H_5CN	103.12	col. lq.	$1.001^{25/6}$	−12.9	190.7	1^{100}	∞	∞

Table 2. Physical Properties of Organic Compounds—(*Continued*)

Name	Synonym	Formula	Formula weight	Form and color	Specific gravity	Melting point, °C.	Boiling point, °C.	Solubility in 100 parts: Water	Alcohol	Ether
Benzoin (dl-)		$C_6H_5CO.CHOHC_6H_5$	212.24	mn.	....	133–7	344^{768}	v. sl. s.	s. h.	sl. s.
Benzophenone	diphenyl ketone	$C_6H_5COC_6H_5$	182.21	col. rhb.	1.083^{54}	48.5	305.4	i.	6.5^{15}	15^{13}
Benzotrichloride	phenyl chloroform	$C_6H_5CCl_3$	195.48	col. lq.	1.380^{14}	−4.75	220.7	i.	s.	s.
Benzoyl-benzoic acid (o-)		$C_6H_5COC_6H_4CO_2H.H_2O$	244.24	tri./aq.	....	93(128)	....	sl. s.		
-chloride		C_6H_5COCl	140.57	col. lq.	$1.212^{20/4}$	−0.5	197.2	d.	d. h.	∞
-peroxide		$(C_6H_5CO)_2O_2$	242.22	rhb./et.	....	108 d.	expl.	i.	s. h.	s.
Benzyl acetate		$CH_3CO_2CH_2C_6H_5$	150.17	col. lq.	1.057^{17}	−51.5	213.5	i.	∞	∞
alcohol	phenyl carbinol	$C_6H_5CH_2OH$	108.13	col. lq.	$1.043^{20/4}$	−15.3	204.7	4^{17}	∞	∞
amine	ω-amino toluene	$C_6H_5CH_2NH_2$	107.15	lq.	$0.982^{20/4}$	....	184.5	∞	∞	∞
aniline	phenyl-benzylamine	$C_6H_5CH_2NHC_5H_5$	183.24	mn. pr.	$1.065^{25/25}$	37–8	306^{759}	i.	....	s.
benzoate		$C_6H_5CO_2CH_2C_6H_5$	212.24	nd.	$1.12^{20/4}$	21	323–4	i.	∞	∞
butyrate		$C_2H_5CH_2CO_2CH_2C_6H_5$	178.22	col. lq.	$1.016^{16/18}$	....	238–40	i.	v. s.	v. s.
chloride	ω-chlorotoluene	$C_6H_5CH_2Cl$	126.58	col. lq.	$1.100^{20/20}$	−39	179.4	i.	∞	∞
ether	dibenzyl ether	$(C_6H_5CH_2)_2O$	198.25	lq.	1.036^{16}	....	295–8	i.	s. h.	s.
formate		$HCO_2CH_2C_6H_5$	136.14	col. lq.	1.081^{23}	3.6	$202–3^{747}$	i.	s.	∞
propionate		$C_2H_5CO_2CH_2C_6H_5$	164.20	lq.	$1.036^{16/17}$	....	220–2	i.		
Berberonic acid (2-,4-,5-)		$C_5H_2N(CO_2H)_3.2H_2O$	247.16	tri.	....	243	....	v. sl. s.	sl. s. h.	i.
Biuret	allophanamide	$NH(CONH_2)_2$	103.08	nd./al.	....	192–3 d.	....	1.3^0	s.	
Borneol (dl-)		$C_{10}H_{17}OH$	154.24	col. cr.	$1.011^{20/4}$	210.5	subl.	v. sl. s.		
(d- or l-)		$C_{10}H_{17}OH$	154.24	col. cr.	$1.011^{20/4}$	208–9	212–3	v. sl. s.	v. s.	v. s.
(iso-)		$C_{10}H_{17}OH$	154.24	col. cr.	....	212	....	i.		
Bornyl acetate (d-)		$CH_3CO_2C_{10}H_{17}$	196.28	rhb./pet.	0.991^{15}	29	226–7	i.	s.	s.
Bromo-aniline (p-)		$BrC_6H_4NH_2$	172.03	rhb.	1.8^{20}	63–4	....	i. c.	v. s.	v. s.
-benzene	phenyl bromide	C_6H_5Br	157.02	col. lq.	$1.495^{20/4}$	−30.6	156.2	i.	s.	∞
-camphor (3-)(d-)	α-bromocamphor	$BrC_{10}H_{15}O$	231.14	cr.	$1.449^{20/4}$	77–8	274	i.	20^{26}	v. s.
-diphenyl (p-)		$BrC_6H_4.C_6H_5$	233.11	cr./al.	....	90–1	310	i.	s.	34^{25}
-naphthalene (α-)	α-naphthyl bromide	$C_{10}H_7Br$	207.07	col. oil	$1.482^{20/4}$	5–6	281.1	i.	s.	∞
(β-)	β-naphthyl bromide	$C_{10}H_7Br$	207.07	lf./al.	1.605^0	59	281–2	i.	6^{20}	v. s.
-phenol (o-)		BrC_6H_4OH	173.02	col. lq.	1.553^{80}	5.6	194–5	s.	s.	∞
(m-)		BrC_6H_4OH	173.02	cr.	....	32–3	236–7	....	s.	s.
(p-)		BrC_6H_4OH	173.02	tet. cr.	1.588^{80}	63.5	238	1.4^{15}	v. s.	v. s.
-styrene (ω)(1)		$C_6H_5CH:CHBr$	183.05	lq.	$1.422^{20/4}$	7	221	i.	∞	∞
(2)		$C_6H_5CH:CHBr$	183.05	lq.	$1.427^{20/4}$	−7.5	108^{26}	i.	∞	∞
-toluene (o-)	o-tolyl bromide	$CH_3.C_6H_4Br$	171.04	col. lq.	$1.422^{20/4}$	−28	181.8	i.	s.	$∞^{25}$
(m-)		$CH_3.C_6H_4Br$	171.04	col. lq.	$1.410^{20/4}$	−39.8	183.7	i.	s.	s.
(p-)		$CH_3.C_6H_4Br$	171.04	cr./al.	$1.390^{20/4}$	28.5	184–5	i.	s.	$∞^{25}$
Bromoform	tribromo-methane	$CHBr_3$	252.77	col. lq.	$2.890^{20/4}$	8–9	150.5	0.1 c.	∞	∞
Butadiene (1-,2-)	methyl-allene	$CH_3CH:C:CH_2$	54.09	lq.	....	....	18–9	i.	∞	∞
(1-,3-)	erythrene	$CH_2:CHCH:CH_2$	54.09	col. gas	$0.621^{20/4}$	−108.9	−4.41	i.	∞	∞
Butadienyl acetylene		$CH_2:(CH)_2:CH.C:CH$	78.11	col. lq.	$0.773^{20/4}$	....	83–6	i.		
Butane	diethyl	$CH_3CH_2CH_2CH_3$	58.12	col. gas	0.60^0	−135	−0.6	i.	s.	s.
(i-)	trimethyl-methane	$(CH_3)_2CHCH_3$	58.12	col. gas	0.60^0	−145	−10	i.	s.	s.
Butyl acetate (n-)		$CH_3CO_2(CH_2)_2C_2H_5$	116.16	col. lq.	0.882^{20}	−76.3	125^{740}	0.7	∞	∞
(s-)		$CH_3CO_2CH(CH_3)C_2H_5$	116.16	col. lq.	$0.865^{25/4}$	....	112^{744}	i.	∞	∞
(i-)		$CH_3CO_2CH_2CH(CH_3)_2$	116.16	col. lq.	$0.871^{20/4}$	−98.9	118	0.6^{25}	∞	∞
(tert-)		$CH_3CO_2C(CH_3)_3$	116.16	col. lq.	$0.866^{20/4}$	....	$95–6^{750}$	i.	∞	∞
alcohol (n-)	butanol-1	$C_2H_5CH_2CH_2OH$	74.12	col. lq.	$0.810^{20/4}$	−79.9	117	9^{15}	∞	∞
(s-)	butanol-2	$C_2H_5CH(OH)CH_3$	74.12	col. lq.	$0.808^{20/4}$	−114.7	99.5	12.5^{20}	∞	∞
(i-)	2-methyl-propanol-1	$(CH_3)_2CHCH_2OH$	74.12	col. lq.	$0.805^{17.5}$	−108	107–8	10^{15}	∞	∞
(tert-)	2-methyl-propanol-2	$(CH_3)_3COH$	74.12	lq.	0.779^{26}	25.5	82.9	∞	∞	∞
amine (n-)		$C_2H_5CH_2CH_2NH_2$	73.14	col. lq.	$0.739^{25/4}$	−50	77.8	∞	∞	∞
(s-)		$C_2H_5CH(NH_2)CH_3$	73.14	col. lq.	$0.724^{20/4}$	−104	66^{772}	∞	∞	∞
(i-)		$(CH_3)_2CHCH_2NH_2$	73.14	col. lq.	$0.732^{20/20}$	−85	68–9	∞	∞	∞
(t-)		$(CH_3)_3CNH_2$	73.14	col. lq.	$0.698^{18/4}$	−67.5	45.2	....	∞	
p-aminophenol (N)(n)		$C_4H_9NH.C_6H_4.OH$	165.23	....	....	71	....	i.		
(N)(i-)		$C_4H_9NH.C_6H_4.OH$	165.23	....	....	79	....	i.		
aniline (n-)		$C_4H_9NHC_6H_5$	149.23	lq.	....	....	235^{720}	i.	v. s.	v. s.
(i-)		$C_4H_9NHC_6H_5$	149.23	oil	$0.940^{20/4}$	....	231–2	0.01^{15}	v. s.	v. s.
arsonic acid (n-)		$C_4H_9AsO(OH)_2$	182.04	col. lf.	....	158–9	....	s.	s.	i.
benzoate (n-)		$C_6H_5CO_2C_4H_9$	178.22	col. oil	$1.005^{25/25}$	−22	249–50	i.	s.	s.
(i-)		$C_6H_5CO_2C_4H_9$	178.22	col. oil	$0.997^{25/25}$	....	241.5	i.	∞	∞
bromide (n-)	1-bromo-butane	$C_2H_5CH_2CH_2Br$	137.03	lq.	$1.277^{20/4}$	−112.4	101.6	0.06^{16}	∞	∞
(s-)	2-bromo-butane	$C_2H_5CH(Br)CH_3$	137.03	lq.	$1.251^{25/4}$	−112	91.3	i.		
(i-)	1-Br-2-Me-propane	$(CH_3)_2CHCH_2Br$	137.03	lq.	$1.258^{25/4}$	−118.5	91.5	0.06^{18}	∞	∞
(t-)	2-Br-2-Me-propane	$(CH_3)_3CBr$	137.03	lq.	$1.211^{20/4}$	−16.2	73.3	i.	∞	∞

butyrate (n-)(n-)	...	$C_2H_5CH_2CO_2CH_2CH_2C_2H_5$	144.21	col. lq.	$0.872^{20/20}$	...	165.7^{736}	i.	∞	∞
(n-)(i-)	...	$C_2H_5CH_2CO_2CH_2CH(CH_3)_2$	144.21	col. lq.	$0.863^{18/4}$	...	156.9	i.	∞	∞
(i-)(i-)	...	$(CH_3)_2CHCO_2CH_2CH(CH_3)_2$	144.21	col. lq.	$0.875^{0/4}$	−80.7	148–9	i.	∞	∞
caproate	...	$CH_3(CH_2)_4CO_2C_4H_9$	172.26	col. lq.	$0.882^{0/0}$	...	204.3	i.	∞	∞
carbamate (i-)	...	$NH_2CO_2CH_2CH(CH_3)_2$	117.15	col. lf.	$0.956^{76/4}$	65	206–7	i.	s.	s.
cellosolve (n-)	2-BuO-ethanol-1	$C_4H_9OCH_2CH_2OH$	118.17	col. lq.	$0.903^{20/4}$	...	171.2	∞	∞	∞
chloride (n-)	1-chloro-butane	$C_2H_5CH_2CH_2Cl$	92.57	col. lq.	0.887^{20}	−123.1	77.9^{763}	0.07^{13}	∞	∞
(s-)	2-chloro-butane	$C_2H_5.CHCl.CH_3$	92.57	col. lq.	$0.871^{20/4}$	−131	67.8^{767}	i.	∞	∞
(i-)	1-Cl-2-Me-propane	$(CH_3)_2CHCH_2Cl$	92.57	col. lq.	0.884^{15}	−131.2	68.9	i.	∞	∞
(t-)	2-Cl-2-Me-propane	$(CH_3)_3CCl$	92.57	col. lq.	0.847^{15}	−26.5	51–2	i.	∞	∞
dimethylbenzene (t-)(1-,3-,5-)	...	$(CH_3)_3C.C_6H_3:(CH_3)_2$	162.26	col. lq.	...	...	$200–2^{147}$	i.		
formate (n-)	...	$HCO_2CH_2CH_2C_2H_5$	102.13	lq.	0.911^{0}	...	106.9	v. sl. s.	∞	∞
(s-)	...	$HCO_2CH(CH_3)C_2H_5$	102.13	lq.	$0.882^{20/4}$	...	97	sl. s.	∞	∞
(i-)	...	$HCO_2CH_2CH(CH_3)_2$	102.13	lq.	$0.885^{20/4}$	−95.3	98.2	1.1^{22}	∞	∞
furoate (n-)	...	$OC_4H_3CO_2C_4H_9$	168.19	col. lq.	$1.056^{20/4}$	...	$118–20^{25}$	i.	∞	∞
iodide (n-)	1-iodo-butane	$C_2H_5CH_2CH_2I$	184.03	lq.	$1.617^{20/4}$	−103.5	129.9	i.	∞	∞
(s-)	2-iodo-butane	$C_2H_5CHICH_3$	184.03	lq.	1.595^{20}	−104	118–9	i.	∞	∞
(i-)	1-iodo-2-Me-propane	$(CH_3)_2CHCH_2I$	184.03	lq.	$1.606^{20/4}$	−90.7	120	i.	∞	∞
(t-)	2-iodo-2-Me-propane	$(CH_3)_3CI$	184.03	lq.	$1.370^{19/15}$	−34	99	i.	∞	∞
lactate (n-)	...	$CH_3CH(OH)CO_2C_4H_9$	146.18	col. lq.	0.968	...	$75–6^{6}$	sl. s.	∞	∞
mercaptan (n-)	butanthiol-1	$C_2H_5CH_2CH_2SH$	90.18	col. lq.	$0.837^{25/4}$	−116	97–8	sl. s.	v. s.	v. s.
(i-)	2-Me-propanthiol-1	$(CH_3)_2CHCH_2SH$	90.18	lq.	$0.836^{20/4}$	<−79	88	v. sl. s.	s.	s.
(t-)	...	$(CH_3)_3CSH$	90.18	lq	...	...	65–7			
methacrylate (n-)	...	$CH_2:C(CH_3)CO_2C_4H_9$	142.19	lq.	$0.889^{15.6}$	...	155	i		
(i-)	...	$CH_2:C(CH_3)CO_2C_4H_9$	142.19	lq.	$0.889^{15.6}$	...	155	i.		
phenol (p-)(t-)	...	$(CH_3)_3C.C_6H_4.OH$	150.21	nd./aq.	$0.908^{112/4}$	99	236–8	sl. s.	s.	s.
propionate (n-)	...	$C_2H_5CO_2C_4H_9$	130.18	col. lq.	0.883^{15}	−89.55	146	i.	∞	∞
(s-)	...	$C_2H_5CO_2C_4H_9$	130.18	col. lq.	$0.866^{20/4}$	...	132.5	i.	∞	∞
(i-)	...	$C_2H_5CO_2C_4H_9$	130.18	col. lq.	$0.888^{0/4}$	−71.4	136.8	i.	∞	∞
stearate (n-)	...	$CH_3(CH_2)_{16}CO_2C_4H_9$	340.57	col. lq.	$0.855^{25/25}$	27.5	$220–5^{25}$	0.3^{25}	s.	s.
(i-)	...	$CH_3(CH_2)_{16}CO_2C_4H_9$	340.57	wax	...	25	...	i.		
iso-thiocyanate (n-)	butyl mustard oil	$C_2H_5CH_2CH_2.N:CS$	115.19	lq.	0.956^{11}	...	165^{724}	i.	s.	s.
(i-)	iso-Bu mustard oil	$(CH_3)_2CHCH_2.N:CS$	115.19	lq.	$0.964^{14/4}$	...	162	i.	s.	s.
(s-)(d-)	...	$C_4H_9.N:CS$	115.19	lq.	$0.943^{20/4}$	...	159–63	i.	s.	s.
(t-)	...	$(CH_3)_3C.N:CS$	115.19	lq.	0.919^{10}	10.5	140^{770}	i.	s.	s.
valerate (n-)(n-)	...	$CH_3(CH_2)_3CO_2(CH_2)_3CH_3$	158.23	lq.	$0.870^{15/4}$	−93	186	v. sl. s.	∞	∞
(i-)(n-)	...	$(CH_3)_2CHCH_2CO_2(CH_2)_3CH_3$	158.23	lq.	$0.862^{25/4}$	...	168.8	i.	∞	∞
(i-)(s-)	...	$(CH_3)_2CHCH_2CO_2C_4H_9$	158.23	col. lq.	$0.848^{20/4}$	...	$163–4^{752}$	i.	∞	∞
(i-)(i-)	...	$C_4H_9CO_2C_4H_9$	158.23	col. lq.	$0.874^{0/4}$	...	168.7	i.	∞	∞
Butylene (α-)	butene-1	$C_2H_5CH:CH_2$	56.10	col. gas	0.6^{9}	−130	-5^{758}	i.	v. s.	v. s.
(β-)	butene-2	$CH_3CH:CHCH_3$	56.10	col. gas	...	−127	3^{746}			
Butyraldehyde (n-)	...	$CH_3CH_2CH_2CHO$	72.10	col. lq.	$0.817^{20/4}$	−99	75.7	4	∞	∞
(i-)	2-Me-propanol	$(CH_3)_2CHCHO$	72.10	col. lq.	$0.794^{20/4}$	−65.9	64^{757}	11^{20}	∞	∞
Butyric acid (n-)	butanoic acid	$C_2H_5CH_2CO_2H$	88.10	col. lq.	$0.964^{20/4}$	−4.7	163.5^{757}	∞	∞	∞
(i-)	2-Me-propanoic acid	$(CH_3)_2CHCO_2H$	88.10	col. lq.	$0.949^{20/4}$	−47	154.5	20^{20}	∞	∞
amide (n-)	n-butyramide	$C_2H_5CH_2CONH_2$	87.12	rhb.	1.032	115–6	216	16.3^{15}	s.	sl. s.
(i-)	iso-butyramide	$(CH_3)_2CHCONH_2$	87.12	mn. pl.	1.013	129–30	216–20	v. s.	s.	sl. s.
anhydride (n-)	...	$(C_2H_5CH_2CO)_2O$	158.19	col. lq.	$0.968^{20/20}$	−75	199.5	d.	d	∞
(i-)	...	$[(CH_3)_2CHCO]_2O$	158.19	col. lq.	$0.950^{25/4}$	−53.5	181.5^{734}	d.	d	∞
anilide (n-)	n-butyranilide	$C_3H_7CONHC_6H_5$	163.21	mn. pr.	1.134	92	189^{15}	i.	s.	s.
Caffeic acid (3-,4-)	...	$(HO)_2C_6H_3C_2H_2CO_2H$	180.15	yel./aq.	...	195–213	d.	s. h.	s.	sl. s.
Caffeine	...	$C_8H_{10}O_2N_4.H_2O$	212.21	nd./al.	1.23^{19}	237	subl.	2	2	0.3
Camphene (dl-)	...	$C_{10}H_{16}$	136.23	cr.	0.822^{78}	50	160	i.	s.	s.
(d- or l-)	...	$C_{10}H_{16}$	136.23	cr.	$0.845^{50/4}$	42.7	159.6	i.	s.	s.
Camphor (d-)	...	$C_{10}H_{16}O$	152.23	trig.	$0.999^{9/9}$	178–9	209.1^{759}	0.1	120^{12}	v. s.
Camphoric acid (d-)	...	$C_8H_{14}(CO_2H)_2$	200.23	mn.	1.186	187	...	0.6^{12}	s.	
Cantharidine	...	$C_{10}H_{12}O_4$	196.20	cr.	...	212	...	0.003		
Capric acid	decanoic acid	$CH_3(CH_2)_8CO_2H$	172.26	col. nd.	0.889^{37}	31.5	268–70	0.003	s.	s.
Caproic acid (n-)	hexanoic acid	$CH_3(CH_2)_4CO_2H$	116.16	oily lq.	$0.922^{20/4}$	−1.5	202^{761}	1.1^{20}	s.	s.
(i-)	2-Me-pentanoic-5 acid	$(CH_3)_2CH(CH_2)_2.CO_2H$	116.16	col. oil	$0.925^{20/4}$	−35	207.7	v. sl. s.	s.	s.
Caprylic acid (n-)	octanoic acid	$CH_3(CH_2)_6CO_2H$	144.21	col. lf.	$0.910^{20/4}$	16	237.5	0.07^{15}	s.	s.
Carbazole	diphenylenelimine, dibenzopyrrole	$(C_6H_4)_2NH$	167.20	lf.	...	244.8	354.8	i.	0.92^{14}	sl. s.
Carbitol	diethylene glycol mono-Et ether	$C_2H_5O(CH_2)_2O(CH_2)_2OH$	134.17	col. lq.	$0.990^{20/20}$	...	201.9	∞	v. s.	s.
Carbon disulfide	...	CS_2	76.13	col. lq.	$1.263^{20/4}$	−108.6	46.3	0.2^{0}	∞	∞
monoxide	...	CO	28.01	col. gas	$0.81^{-195/4}$	−207	−192	3.5^{0} cc.	s.	
suboxide	...	$OC:C:CO$	68.03	gas	1.114^{0}	−107	7^{761}	d.	...	s.
tetrabromide	tetrabromomethane	CBr_4	331.67	col. mn.	3.42	90.1(48)	189.5	0.02^{30}	s.	s.
tetrachloride	tetrachloromethane	CCl_4	153.84	col. lq.	$1.595^{20/4}$	−22.6	76.8	0.08^{20}	∞	∞
tetrafluoride	tetrafluoromethane	CF_4	88.01	gas	...	...	−128	sl. s.		

Table 2. Physical Properties of Organic Compounds—*(Continued)*

Name	Formula	Formula weight	Form and color	Specific gravity	Melting point, °C.	Boiling point, °C.	Solubility in 100 parts: Water	Alcohol	Ether
Carbonyl sulfide	COS	60.07	col. gas	1.24^{-87}	−138.2	-50.2^{760}	80^{14} cc.	s.	s.
Carminic acid	$C_{22}H_{20}O_{13}$	492.40	red pd.	......	d. 136	......	s.	s.	v. sl. s.
Carvacrol (1-,2-,4-)	$CH_3C_6H_3(OH)CH(CH_3)_2$	150.21	col. lq.	$0.977^{20/4}$	0.5	238	v. sl. s.	∞	∞
Carvacrylamine (2-,1-,4-)	$H_2NC_6H_3(CH_3)C_3H_7$	149.23	oil	0.994^{20}	−16	241	v. sl. s.	s.	s.
Carvone (d-)	$C_{10}H_{14}O$	150.21	col. lq.	$0.961^{20/4}$	......	230^{755}	i.	∞	∞
Cellosolve	$C_2H_5O(CH_2)_2OH$	90.12	col. lq.	$0.931^{20/4}$	−70	135.1	∞	∞	∞
acetate	$CH_3CO_2CH_2CH_2OC_2H_5$	132.16	col. lq.	$0.975^{20/4}$	......	156.3	22	∞	∞
Cellulose	$(C_6H_{10}O_5)x$	162.14	amor.	1.3–1.4	......	......	i.	i.	i.
Cetyl acetate	$CH_3CO_2(CH_2)_{15}CH_3$	284.47	nd.	0.858^{20}	22–3	200^{15}	i.	v. sl. s. c.	
alcohol	$CH_3(CH_2)_{14}CH_2OH$	242.43	lf.	$0.818^{50/4}$	49–50	189.5^{15}	i.	s.	s.
Chloral	$CCl_3.CHO$	147.40	col. lq.	$1.505^{25/4}$	−57±	97.6^{768}	v. s.	∞	∞
hydrate	$CCl_3.CH(OH)_2$	165.42	mn. pr.	$1.619^{50/4}$	51.7±	d. 98	474^{17}	v. s.	s.
Chloranil	$OC:(CCl.CCl)_2:CO$	245.89	yel./bz.	......	290	subl.	i.	i. c.	i. c.
Chloretone	$Cl_3C.C(OH)(CH_3)_2$	177.47	col. cr.	......	97	167	0.8 c.	111	s.
Chloro-acetanilide (p-)	$CH_3CO_2NHC_6H_4Cl$	169.61	rhb.	1.385^{22}	175–6	......	sl. s.	s.	v. s.
-acetic acid	$ClCH_2CO_2H$	94.50	col. cr.	$1.58^{20/20}$	61.2	189.5	v. s.	s.	s.
-acetone	CH_3COCH_2Cl	92.53	col. lq.	1.162^{16}	−44.5	121	∞	∞	∞
-acetophenone (ω-)	$C_6H_5COCH_2Cl$	154.59	rhb.	1.324^{15}	58–9	245–7	0.11	v. s.	v. s.
-acetyl chloride	$ClCH_2COCl$	112.95	col. lq.	$1.498^{20/20}$	......	105	d.	d.	
-aniline (o-)	$ClC_6H_4NH_2$	127.57	lq.	$1.213^{20/4}$	0	210.5	i.	......	s.
(m-)	$ClC_6H_4NH_2$	127.57	lq.	$1.216^{20/4}$	−10.4	230^{767}	i.	......	s.
(p-)	$ClC_6H_4NH_2$	127.57	rhb.	1.427^{19}	70–1	230–1	s. h.	s.	s.
-anthraquinone (1-)	$C_6H_4(CO)_2C_6H_3Cl$	242.65	yel. nd.	......	162	subl.	i.	sl. s. h.	
(2-)	$C_6H_4(CO)_2C_6H_3Cl$	242.65	nd./al.	......	208–9	......	i.		
-benzaldehyde (o-)	ClC_6H_4CHO	140.57	nd.	1.29^{8}	11	208^{748}	v. sl. s.	v. s.	v. s.
(m-)	ClC_6H_4CHO	140.57	pr.	1.250^{15}	17–8	213–4	v. sl. s.	v. s.	v. s.
(p-)	ClC_6H_4CHO	140.57	pr.	1.196^{61}	47.8	213^{748}	s. h.	v. s.	v. s.
-benzene	C_6H_5Cl	112.56	col. lq.	$1.107^{20/4}$	−45.2	132.1	0.049^{20}	∞	∞
-benzoic acid (o-)	$ClC_6H_4CO_2H$	156.57	mn./aq.	$1.544^{25/4}$	141–2	......	0.208^{25}	s.	s.
(m-)	$ClC_6H_4CO_2H$	156.57	pr.	$1.496^{25/4}$	158	......	0.041^{25}	s.	s.
(p-)	$ClC_6H_4CO_2H$	156.57	tri.	1.541^{24}	242–3	subl.	0.008^{25}	s.	s.
-buta-1,3-diene (2-)	$CH_2:CCl.CH:CH_2$	88.54	col. lq.	$0.958^{20/20}$	......	59.4	v. sl. s.	∞	∞
(1-)	$CH_2:CH.CH:CHCl$	88.54	col. lq.	$0.965^{20/20}$	......	69	v. sl. s.	∞	∞
-buta-1,2-diene (4-)	$CH_2:C:CH.CH_2Cl$	88.54	col. lq.	$0.991^{20/20}$	......	88	d.		
-dimethylhydantoin	$—C(CH_3)_2N(Cl)CON(Cl)CO—$	197.03	......	$1.5^{20/20}$	130	......	0.21^{25}		
-dinitrobenzene (α)(1-,2-)(4-)	$ClC_6H_3(NO_2)_2$	202.56	cr./et.	......	39(36)	315 d.	i.	v. s. h.	v. s.
(α)(1-,3-)(4-)	$ClC_6H_3(NO_2)_2$	202.56	rhb./et.	1.697^{22}	53(43)	315 d.	i.	s. h.	s.
-diphenyl (o-)	$C_6H_5.C_6H_4Cl$	188.65	cr.	......	34	267–8	i.		
(m-)	$C_6H_5.C_6H_4Cl$	188.65	cr.	......	89	284–5	i.		
(p-)	$C_6H_5.C_6H_4Cl$	188.65	lf.	......	77.5	282	i.		
-hydroquinone	$ClC_6H_3(OH)_2$	144.56	mn.	......	106	263 sl. d.	v. s.	v. s.	v. s.
-naphthalene (α-)	$C_{10}H_7Cl$	162.61	col. lq.	$1.194^{20/4}$	−20	259.3	i.	s.	∞
(β-)	$C_{10}H_7Cl$	162.61	lf./al.	1.266^{16}	56–7	264^{751}	i.	v. s.	v. s.
-nitrobenzene (o-)	$ClC_6H_4NO_2$	157.56	mn. nd.	$1.305^{80/4}$	32.5	245.5^{753}	i.	s. h.	s.
(m-)	$ClC_6H_4NO_2$	157.56	yel./al.	$1.343^{50/4}$	44.4(24)	235.6	i.	v. s. h.	v. s.
(p-)	$ClC_6H_4NO_2$	157.56	mn. pr.	1.298^{91}	83–4	242^{761}	i.	v. s. h.	v. s.
-nitrotoluene (2-,4-)	$CH_3C_6H_3(NO_2)(Cl)$	171.56	cr.	1.256^{80}	38.2	240^{718}	i.		
(2-,6-)	$CH_3C_6H_3(NO_2)(Cl)$	171.56	cr.	......	37.5	238	i.		
-phenol (o-)	ClC_6H_4OH	128.56	col. lq.	$1.241^{18/15}$	7(0)	175–6	2.85^{20}	s.	s.
(m-)	ClC_6H_4OH	128.56	nd.	1.268^{25}	32–3	214	2.60^{20}	s.	s.
(p-)	ClC_6H_4OH	128.56	nd.	$1.306^{20/4}$	41–3	217	2.71^{20}	v. s.	v. s.
-propionic acid (α)(dl-)	$CH_3.CHCl.CO_2H$	108.53	col. lq.	1.306^{9}	<−20	186	∞	∞	∞
-toluene (o-)	$CH_3.C_6H_4Cl$	126.58	col. lq.	$1.082^{20/4}$	−34	159.5	i.	s.	∞
(m-)	$CH_3.C_6H_4Cl$	126.58	col. lq.	$1.072^{20/4}$	−47.8	161.6	i.	s.	∞
(p-)	$CH_3.C_6H_4Cl$	126.58	col. lq.	$1.070^{20/4}$	7.5	162.2	i.	s.	∞
Chloroform	$CHCl_3$	119.39	col. lq.	1.489^{20}	−63.5	61.2	0.82^{20}	∞	∞
Chlorophyll (α-)	$C_{55}H_{72}O_5N_4Mg$	893.48	......	......		d.	i.	s.	
Chloropicrin	Cl_3CNO_2	164.39	lq.	$1.651^{23/4}$	−64	112.3^{766}	0.17^{18}	s.	s.
Cholesterol	$C_{27}H_{45}OH.H_2O$	404.65	rhb./al.	1.067	149–51	subl.	0.26^{20}	1.1^{17}	18
Chrysene	$C_{18}H_{12}$	228.28	col. rhb.	......	253–4	448	i.	0.1^{16}	v. sl. s.
Chrysoidine (2-,4-)	$C_6H_5.N:N.C_6H_3(NH_2)_2$	212.25	yel. cr.	......	117.5	......	sl. s. h.	s.	s.
Chrysophanic acid	$C_{14}H_5(OH)_2(CH_3)O_2$	254.23	yel./al.	......	195	subl.	i. c.	s. h.	sl. s.
Cinchomeronic acid (3-,4-)	$C_5H_3N(CO_2H)_2$	167.12	cr./HCl	......	258–9 d.	subl. d.	v. sl. s.	sl. s.	i.
Cineole, eucalyptole	$C_{10}H_{18}O$	154.24	col. oil	0.927^{20}	1.5	176–7	1.9^{15}	∞	∞
Cinnamic acid (cis-)	$C_6H_5CH:CHCO_2H$	148.15	mn. pr.	1.284^{4}	68	125^{19}			
(trans-)	$C_6H_5CH:CHCO_2H$	148.15	mn. pr.	1.245	133	300	0.04^{18}	24^{20}	v. s.

aldehyde	$C_6H_5CH:CHCHO$	132.15	lq.	$1.110^{20/20}$	−7.5	252 sl. d.	v. sl. s.	s.	∞
Cinnamyl alcohol	$C_6H_5CH:CHCH_2OH$	134.17	nd.	$1.040^{35/35}$	33	257.5	sl. s.	v. s.	v. s.
cinnamate	$C_3H_7CO_2C_9H_9$	264.31	nd. or pr.	$1.085^{16.5}$	44	....	i.	4 c.	33
Citraconic acid (cis-)	$CH_3C(CO_2H):CHCO_2H$	130.10	nd.	1.617	92–3	....	360^{25}	s.	s.
Citral (α)	$C_9H_{15}CHO$	152.23	col. oil	$0.890^{17/4}$	....	229	i.	∞	∞
Citric acid	$C_3H_4(OH)(CO_2H)_3$	192.12	cr.	$1.542^{20/4}$	153	d.	207.7^{25}	76^{15}	2^{15}
Citronellal (d-)	$C_9H_{17}.CHO$	154.24	col. oil	$0.855^{17.5}$	....	204–8	v. sl. s.	∞	∞
Citronellol (d-)	$C_{10}H_{20}O$	156.26	col. oil	$0.848^{20/4}$	....	224–5	v. sl. s.	∞	∞
Coniine (d-)(2-)	$C_3H_7.C_5H_{10}N$	127.22	col. lq.	0.847^{17}	−2	166–7	1.1	v. s.	v. s.
Coumaric acid (o-)	$HOC_6H_4CH:CHCO_2H$	164.15	nd./aq.	....	207–8	subl.	sl. s. c.	s.	v. sl. s.
(p-)	$HOC_6H_4CH:CHCO_2H$	164.15	cr./aq.	....	206–7 d.	....	s. h.	v. s. h.	v. s.
Coumarin	$C_9H_6O_2$	146.14	rhb./et.	$0.935^{20/4}$	70	290–1	0.3 c.	v. s.	s.
Coumarone	C_8H_6O	118.13	oil	$1.078^{15/15}$	<−18	173–4	i.	....	s.
Creatine	$C_4H_9N_3O_2.H_2O$	149.15	mn./aq.	....	295	....	1.4^{18}	0.01^{17}	i.
Creatinine	$C_4H_7N_3O$	113.12	mn.	....	260 d.	....	8.7^{16}	1^{16}	
Creosol (3-,1-,4-)	$CH_3O.C_6H_3(CH_3)OH$	138.16	pr.	$1.092^{20/20}$	5.5	$221–2^{765}$	v. sl. s.	∞	∞
Cresidine (1-,2-,4-)	$CH_3(NH_2)C_6H_3.OCH_3$	137.18	nd./pet.	....	93–4	235	v. sl. s.	s.	s.
Cresol (o-)	$CH_3C_6H_4OH$	108.13	cr.	$1.048^{20/4}$	30.8	190.8	2.5	∞^{30}	∞^{30}
(m-)	$CH_3C_6H_4OH$	108.13	lq.	$1.034^{20/4}$	10.9	202.8	0.5	∞	∞
(p-)	$CH_3C_6H_4OH$	108.13	pr.	$1.035^{20/4}$	35–6	202	1.8	∞^{36}	∞^{36}
Cresyl benzoate (o-)	$C_6H_5CO_2C_6H_4CH_3$	212.24	lq.	....	....	308	i.		
(m-)	$C_6H_5CO_2C_6H_4CH_3$	212.24	cr.	....	55	314	i.		
(p-)	$C_6H_5CO_2C_6H_4CH_3$	212.24	cr.	....	71.5	316	i.		
Crotonic acid (α-)	$CH_3CH:CHCO_2H$	86.09	col. mn.	$0.964^{79.7}$	72	189	8.3^{15}		
acid (β-)(cis-)	$CH_3CH:CHCO_2H$	86.09	nd.	$1.031^{15/4}$	15.5	170–1 d.	∞^{25}	s.	
aldehyde (α)	$CH_3CH:CHCHO$	70.09	col. lq.	$0.853^{20/20}$	−69	102.2	18	∞	∞
Cumene	$C_6H_5CH(CH_3)_2$	120.19	col. lq.	$0.862^{20/4}$	−96.9	152.5	i.	∞	∞
Cumic acid (p-)	$(CH_3)_2CH.C_6H_4CO_2H$	164.20	tri.	1.162^{4}	116–7	subl.	0.02^{25}	s.	s.
Cumidine (p-)	$(CH_3)_2CH.C_6H_4NH_2$	135.20	lq.	0.953	<−20	225^{761}	i.		
Cyanamide	$H_2N.CN$	42.04	col. nd.	$1.073^{48/4}$	44–5	140^{19}	v. s.	v. s.	v. s.
Cyanic acid	HOCN or HNCO	43.03	gas	1.140^{0}	−80	-64^{0}	sl. s.	....	s.
Cyanoacetic acid	$CH_2(CN)CO_2H$	85.06	col. lq.	....	65–6	$108^{0.2}$	s.	s.	s.
Cyanogen	$(CN)_2$	52.04	col. gas	0.866^{17}	−34.4	−21	450^{20} cc.	2300^{20} cc.	500^{20} cc.
bromide	BrCN	105.93	nd.	$2.015^{20/4}$	52	61.3^{750}	s.	s.	s.
chloride	ClCN	61.48	gas	1.222^{0}	−6.5	12.5–13	2500^{20} cc.	v. s.	5000^{20} cc.
Cyanuric acid	$C_3H_3O_3N_3.2H_2O$	165.11	mn./aq.	$1.768^{0/4}$	>360	d.	0.27^{17}	0.1^{22}	
Cyclo-butane	$CH_2 < (CH_2)_2 > CH_2$	56.10	col. gas	$0.703^{0/4}$	−50	$11–12^{726}$	i.	v. s.	
-heptane	$CH_2 < (CH_2CH_2CH_2)_2 >$	98.18	oil	$0.810^{20/4}$	−12	118–20	i.		
-hexane	$CH_2 < (CH_2CH_2)_2 > CH_2$	84.16	col. lq.	$0.779^{20/4}$	6.5	80–1	i.	∞	∞
-hexanol	$CH_2 < (CH_2CH_2)_2 > CHOH$	100.16	col. nd.	$0.962^{20/4}$	23.9	160–1	3.6^{20}	s.	s.
-hexanone	$CH_2 < (CH_2CH_2)_2 > CO$	98.14	col. oil	$0.947^{19/4}$	−45	155–6	s.	s.	s.
-hexene	$(.CH_2.CH_2CH:)_2$	82.14	lq.	$0.810^{20/4}$	−103.7	83.3	v. sl. s.	v. s.	v. s.
-hexyl acetate	$CH_3CO_2C_6H_{11}$	142.19	oil	$0.985^{0/4}$	....	174^{750}	i.	∞	∞
amine	$CH_2 < (CH_2CH_2)_2 > CHNH_2$	99.17	col. lq.	$0.865^{20/0}$	....	134	i.	s.	
bromide	$CH_2 < (CH_2CH_2)_2 > CHBr$	163.06	col. lq.	$1.324^{20/20}$	....	165^{714}	i.	s.	s.
chloride	$CH_2 < (CH_2CH_2)_2 > CHCl$	118.61	col. lq.	$0.977^{18/4}$	−43.9	142	i.	....	∞
-pentadiene (1-,3-)	$CH_2 < (CH:CH)_2 >$	66.10	col. lq.	$0.805^{19/4}$	−85	41–2	i.	∞	∞
-pentane	$CH_2 < (CH_2CH_2)_2 >$	70.13	col. oil	$0.745^{20/4}$	−93.3	49–50	i.		
-pentanone	$< (CH_2CH_2)_2 > CO$	84.11	col. oil	0.948^{20}	−58.2	129–30	v. sl. s.		
-propane	$< CH_2CH_2CH_2 >$	42.08	col. gas	0.720^{-79}	−126.6	-34^{749}	i.	s.	s.
Cymene (o-)	$CH_3.C_6H_4CH(CH_3)_2$	134.21	col. lq.	$0.875^{20/4}$	....	177	i.	s.	s.
(m-)	$CH_3.C_6H_4CH(CH_3)_2$	134.21	col. lq.	0.862^{20}	<−25	175–6	i.	s.	s.
(p-)	$CH_3.C_6H_4CH(CH_3)_2$	134.21	col. lq.	$0.857^{20/4}$	−73.5	176–7	i.	s.	s.
Cystine (l-)	$[.SCH_2CH(NH_2)CO_2H]_2$	240.29	pl.	....	d. 258–61	....	0.01^{19}	i.	
Dambose	$C_6H_6(OH)_6$	180.16	mn./aq.	1.752	253	319^{15}	2^{12}	i.	i.
Decahydronaphthalene (cis-)	$C_{10}H_{18}$	138.24	lq.	$0.895^{18/4}$	−51	193.3	i.	s.	s.
(trans-)	$C_{10}H_{18}$	138.24	lq.	$0.872^{20/4}$	−32	185.3	i.	s.	s.
Decane (n-)	$CH_3(CH_2)_8CH_3$	142.28	col. lq.	0.730^{2}	−29.7	174.0	i.	∞	∞
Decyl alcohol	$CH_3(CH_2)_8CH_2OH$	158.28	col. oil	$0.830^{20/4}$	7	232.9	i.	s.	
Dextrin	$(C_6H_{10}O_5)x$	162.14	amor.	1.038	....	....	s.	i.	i.
Diacetone alcohol	$(CH_3)_2C(OH).CH_2COCH_3$	116.16	lq.	0.931^{25}	−47	167.9	∞	∞	∞
Diamino-benzophenone (4-,4′-)	$H_2NC_6H_4COC_6H_4NH_2$	212.24	yel. nd.	....	237–9	....	sl. s. h.	s.	s.
-diphenylamine (4-,4′-)	$H_2NC_6H_4NHC_6H_4NH_2$	199.25	lf./aq.	....	158	d.	sl. s.	s.	s.
-diphenylmethane (4-,4′-)	$H_2NC_6H_4CH_2C_6H_4NH_2$	198.26	nd./aq.	....	93–4	$249–53^{15}$	sl. s. c.	s.	s.
-diphenylurea (4-,4′-)	$(H_2NC_6H_4NH)_2CO$	242.28	cr.	....	subl. 310	....	v. sl. s.		
Diamyl-amine (i-)	$[(CH_3)_2CHCH_2CH_2]_2NH$	157.29	col. lq.	$0.767^{21/4}$	−44	188–90	sl. s.	s.	∞
ether (n-)	$(C_2H_5CH_2CH_2CH_2)_2O$	158.28	col. lq.	$0.774^{20/4}$	−69	190	i.	∞	∞
(i-)	$[(CH_3)_2CH(CH_2)_2]_2O$	158.28	col. lq.	$0.777^{20/4}$	....	173.4	i.	∞	∞

Table 2. Physical Properties of Organic Compounds—*(Continued)*

Name	Formula	Formula weight	Form and color	Specific gravity	Melting point, °C.	Boiling point, °C.	Solubility in 100 parts Water	Alcohol	Ether
Diamyl ketone (i-)	$[(CH_3)_2CHCH_2CH_2]_2CO$	170.29	yel. oil	$0.821^{25/4}$	14.6	228	i.	s.	s.
phthalate (n-)	$C_6H_4(CO_2C_5H_{11})_2$	306.39	col. lq.	....	....	$204\text{–}6^{11}$			
(i-)	$C_6H_4(CO_2C_5H_{11})_2$	306.39	col. lq.	1.03	....	225^{40}	i.	s.	s.
tartrate (i-)	$(HOCH.CO_2C_5H_{11})_2$	290.35	lq.	$1.063^{15/4}$	....	195^{16}	i.		
Dianisidine (o-)(4-,3-)$_2$	$[NH_2(OCH_3)C_6H_3.]_2$	244.28	col. lf.	....	131.5	....	i.	s.	s.
Diazo-aminobenzene	$C_6H_5N{:}N.NHC_6H_5$	197.23	yel. lf.	....	96–8	expl.	i.	s. h.	v. s.
-aminotoluene (2-,2′-)	$C_7H_7N{:}N.NHC_7H_7$	225.28	or. cr.	....	51	....	0.05		
-methane	$CH_2{:}N_2$	42.04	gas	....	−145	−23	d.	....	s.
Dibenzothiazyl-disulfide (2-,2′-)	$(C_6H_4NSC)_2S_2$	232.46	cr.	1.50	180	d.	i.		
Dibenzoyl methane	$(C_6H_5CO)_2CH_2$	224.25	rhb./al.	....	78	$219\text{–}21^{18}$	i.	4.4^{20}	s.
Dibenzyl-amine	$(C_6H_5CH_2)_2NH$	197.27	col. oil	$1.028^{25/25}$	−26	$268\text{–}71^{250}$	i.	s.	s.
-aniline	$C_6H_5N(CH_2C_6H_5)_2$	273.36	pr./al.	....	70–1	>300	i.	v. s. h.	s.
ketone	$(C_6H_5CH_2)_2CO$	210.26	cr.	....	34–5	330.6	i.		
phthalate (o-)	$C_6H_4(CO_2CH_2C_6H_5)_2$	346.36	pr./al.	....	42–3	274^{12}	v. sl. s.	s.	s.
succinate	$(.CH_2CO_2CH_2C_6H_5)_2$	298.32	lf./al.	....	45–6	238^{14}	i.	s.	s.
Dibromo-benzene (o-)	$C_6H_4Br_2$	235.92	col. lq.	$1.956^{20/4}$	1.8	221–2	i.	s.	s.
(m-)	$C_6H_4Br_2$	235.92	col. lq.	$1.952^{20/4}$	−6.9	219^{755}	i.	s.	s.
(p-)	$C_6H_4Br_2$	235.92	pl./al.	2.261^{18}	87–8	218.6^{758}	i.	1.6	71^{25}
-diphenyl (4-,4′-)	$BrC_6H_4.C_6H_4Br$	312.02	mn. pr.	1.897	164–5	355–60	i.	v. sl. s. h.	
Dibutyl-adipate (n-)	$(.CH_2CH_2CO_2C_4H_9)_2$	258.35	col. lq.	$0.965^{20/4}$	−38	183^{14}	i.	∞	∞
(i-)	$(.CH_2CH_2CO_2C_4H_9)_2$	258.35	col. lq.	0.950^{25}	−20	278–80	i.		
-amine (n-)	$(C_2H_5CH_2CH_2)_2NH$	129.24	col. lq.	$0.768^{20/20}$	....	159^{761}	∞	∞	∞
(i-)	$[(CH_3)_2CHCH_2]_2NH$	129.24	col. lq.	$0.741^{25/4}$	−70	139–40	v. sl. s.	s.	s.
-p-aminophenol (s-)	$(C_4H_9)_2N.C_6H_4OH$	221.33	lq.	....	....	170^{10}	i.		
-aniline (n-)	$C_6H_5N(C_4H_9)_2$	205.33	lq.	....	....	262.8	i.	∞	∞
carbonate (n-)	$CO(OC_4H_9)_2$	174.23	col. lq.	$0.924^{20/4}$	....	207^{740}	i.	s.	
(i-)	$CO(OC_4H_9)_2$	174.23	col. lq.	0.919^{15}	....	190	i.		
(s-)	$CO(OC_4H_9)_2$	174.23	col. lq.		....	178–80			
ether (n-)	$(C_2H_5CH_2CH_2)_2O$	130.22	lq.	$0.769^{20/20}$	−98	142.4	<0.05	∞	∞
(i-)	$[(CH_3)_2CHCH_2]_2O$	130.22	lq.	0.762^{15}	....	122.5	i.	∞	∞
(s-)	$[C_2H_5(CH_3)CH]_2O$	130.22	lq.	0.756^{21}	....	121	i.	∞	∞
ketone (n-)	$(C_2H_5CH_2CH_2)_2CO$	142.23	lq.	$0.827^{13/4}$	−5.9	187.7	i.	s.	v. s.
(i-)	$[(CH_3)_2CHCH_2]_2CO$	142.23	oil	$0.805^{21/4}$	....	168.1	<0.06	∞	∞
malate (l-)(n-)	$C_2H_4O(CO_2C_4H_9)_2$	246.30	lq.	$1.038^{20/4}$	....	$170\text{–}1^{13}$	v. sl. s.		
oxalate (n-)	$(.CO_2C_4H_9)_2$	202.24	col. lq.	$0.986^{20/4}$	−29.6	245.5	i.	s.	s.
phthalate (n-)	$C_6H_4(CO_2C_4H_9)_2$	278.34	col. lq.	1.045^{21}	....	340	0.04^{25}	∞	∞
tartrate (d-)(n-)	$(CHOHCO_2C_4H_9)_2$	262.30	pr.	1.098^{15}	22–2.5	$200\text{–}3^{18}$	i.		
(d-)(i-)	$(CHOHCO_2C_4H_9)_2$	262.30	cr.	$1.031^{75/4}$	73–4	323–5	v. sl. s.		
Dichloro-acetic acid	$Cl_2CH.CO_2H$	128.95	lq.	$1.560^{25/25}$	9.7(−4)	194.4	∞	∞	∞
-acetone (αα-)	$Cl_2CHCOCH_3$	126.98	lq.	1.234^{15}	....	120	v. sl. s.	s.	s.
-aniline (2-,5-)	$Cl_2C_6H_3NH_2$	162.02	nd.	....	50	251	v. sl. s.	s.	s.
-anthraquinone (1-,3-)	$C_6H_4{:}(CO)_2{:}C_6H_2Cl_2$	277.10	yel. nd.	....	208–9	....	i.	i.	
(1-,4-)	$C_6H_4{:}(CO)_2{:}C_6H_2Cl_2$	277.10	yel. nd.	....	187.5	....	i.	v. sl. s.	v. sl. s.
(1-,5-)	$C_6H_3Cl{:}(CO)_2{:}C_6H_3Cl$	277.10	yel. nd.	....	251	....	i.	sl. s.	
(1-,6-)	$C_6H_3Cl{:}(CO)_2{:}C_6H_3Cl$	277.10	yel. nd.	....	203–4	....	i.		
(1-,8-)	$C_6H_3Cl{:}(CO)_2{:}C_6H_3Cl$	277.10	yel. nd.	....	202–3	....	i.	sl. s.	
(2-,3-)	$C_6H_4{:}(CO)_2{:}C_6H_2Cl_2$	277.10	yel. nd.	....	268–70	....	i.	sl. s.	
(2-,6-)	$C_6H_3Cl{:}(CO)_2{:}C_6H_3Cl$	277.10	yel. nd.	....	282	....	i.		
(2-,7-)	$C_6H_3Cl{:}(CO)_2{:}C_6H_3Cl$	277.10	yel. nd.	....	210–11	....	i.		
-benzene (o-)	$C_6H_4Cl_2$	147.01	col. lq.	$1.305^{20/4}$	−17.6	179	i.	∞	∞
(m-)	$C_6H_4Cl_2$	147.01	col. lq.	$1.288^{20/4}$	−24.8	172^{766}	i.	s.	s.
(p-)	$C_6H_4Cl_2$	147.01	col. mn.	1.458^{21}	53	174^{764}	i.	v. s.	v. s.
-butane (n-)(1-,4-)	$ClCH_2(CH_2)_2CH_2Cl$	127.02	lq.	....	−38.7	161–3			
-diphenyl (4-,4′-)	$ClC_6H_4.C_6H_4Cl$	223.10	pr.	$1.442^{0/4}$	148	315–9	i.	v. sl. s.	4^{25}
-ethane (1-,2-)	$ClCH_2.CH_2Cl$	98.97	col. lq.	$1.256^{20/20}$	−35.3	83.7	0.9^{0}	∞	∞
-naphthalene (β-)(1-,4-)	$C_{10}H_6Cl_2$	197.06	nd./al.	$1.300^{76/4}$	67–8	$286\text{–}7^{740}$	i.	v. sl. s.	
(γ-)(1-,5-)	$C_{10}H_6Cl_2$	197.06	lf./al.	....	107	subl.	i.	s.	s.
-nitrobenzene (2-,5-)	$Cl_2C_6H_3NO_2$	192.01	tri./al.	1.669^{22}	54.6	266	i.	v. s. h.	
-pentane (1-,5-)	$ClCH_2(CH_2)_3CH_2Cl$	141.04	col. lq.	$1.094^{25/4}$	....	180–1	i.	s.	s.
-phenol (2-,4-)	$Cl_2C_6H_3OH$	163.01	nd.	$1.383^{60/25}$	45	209–10	0.45^{20}	v. s.	v. s.
Dichloramine T (p-)	$CH_3C_6H_4SO_2NCl_2$	240.11	cr.	....	83	....	sl. s.		
Dicyandiamide	$H_2N.C({:}NH).NH.CN$	84.08	mn. pl.	1.40^{14}	207–8	d.	2.3^{13}	1.3^{13}	0.01^{13}
Diethanolamine	$HN(CH_2CH_2OH)_2$	105.14	pr.	$1.097^{20/4}$	28	270^{748}	∞	∞	v. sl. s.
Diethyl adipate	$(.CH_2CH_2CO_2C_2H_5)_2$	202.24	col. lq.	$1.009^{20/4}$	−21	$239\text{–}41^{761}$	0.43^{30}	s.	s.
-amine	$(C_2H_5)_2NH$	73.14	col. lq.	$0.712^{15/15}$	−38.9	55.5^{759}	v. s.	∞	∞
-aminophenol (m-)	$(C_2H_5)_2N.C_6H_4.OH$	165.23	rhb.	....	78	276–80	s.		

-aniline	$(C_2H_5)_2NC_6H_5$	149.23	oil	$0.934^{20/4}$	−34.4	216	1.4^{12}	s.	s.
sulfonic acid (m-)	$(C_2H_5)_2NC_6H_4SO_3H$	229.29	cr.	....	270 d.	....	s.		
carbonate	$OC(OC_2H_5)_2$	118.13	col. lq.	$0.975^{20/4}$	−43	126^{759}	i.	∞	∞
diethyl malonate	$(C_2H_5)_2C(CO_2C_2H_5)_2$	216.27	col. lq.	$0.985^{20/4}$	....	230	i.	∞	∞
Diethyl dimethyl malonate	$(CH_3)_2C(CO_2C_2H_5)_2$	188.22	col. lq.	$0.994^{25/25}$	....	196.7	i.	∞	∞
glutarate	$CH_2(CH_2CO_2C_2H_5)_2$	188.22	syrup	1.025^{21}	−24	237	0.88^{20}	v. s.	s.
ketone	$(C_2H_5)_2CO$	86.13	col. lq.	$0.816^{19/4}$	−42	101.7	4.7^{20}	∞	∞
malonate	$CH_2(CO_2C_2H_5)_2$	160.17	col. lq.	$1.055^{20/4}$	−49.8	198.9	2.08^{20}	∞	∞
-malonic acid	$(C_2H_5)_2C(CO_2H)_2$	160.17	pr./aq.	....	125	d. 170–80	65^{16}	v. s.	v. s.
-naphthylamine (α-)	$C_{10}H_7N(C_2H_5)_2$	199.28	col. oil	1.005	....	285–90	i.	∞	∞
(β-)	$C_{10}H_7N(C_2H_5)_2$	199.28	col. oil	1.026	....	318	i.	∞	∞
oxalate	$(.CO_2C_2H_5)_2$	146.14	col. lq.	$1.079^{20/4}$	−40.6	186	v. sl. s.	∞	∞
phthalate (o-)	$C_6H_4(CO_2C_2H_5)_2$	222.23	col. lq.	$1.121^{25/25}$	....	298–9	i.	∞	∞
sulfate	$O_2S(OC_2H_5)_2$	154.18	col. lq.	$1.172^{25/4}$	−25	210	i.	s.	∞
sulfide	$(C_2H_5)_2S$	90.18	col. lq.	$0.837^{20/4}$	−99.5	$92–3^{754}$	0.31^{20}	∞	∞
tartrate (d-)	$(CHOH.CO_2C_2H_5)_2$	206.19	lq.	$1.204^{20/4}$	17	280	sl. s.	∞	∞
-toluidine (o-)	$CH_3.C_6H_4.N(C_2H_5)_2$	163.25	lq.	....	....	$208–9^{755}$	i.	s.	s.
(m-)	$CH_3.C_6H_4.N(C_2H_5)_2$	163.25	lq.	....	....	231–2	i.	s.	s.
(p-)	$CH_3.C_6H_4.N(C_2H_5)_2$	163.25	lq.	$0.924^{15.5}$	....	228–9	i.		
Diethyleneglycol dinitrate	$O(CH_2CH_2ONO_2)_2$	196.12	lq.	$1.377^{25/4}$	−11.3	....	i.		
Difluorodichloromethane	F_2CCl_2	120.92	gas	1.486^{-30}	−155	−29.2	5.7 cc.[26]	s.	s.
Diglycerol	$[(HO)_2C_3H_5]_2O$	166.17	lq.	....	....	$220–30^{10}$	s. h.	....	i.
Dihydroxy-dinaphthyl (α-)	$(HO.C_{10}H_6.)_2$	286.31	pl./al.	....	300	....	i.	s.	v. s.
(-2,-2′,-1,-1′)	$(HO.C_{10}H_6.)_2$	286.31	nd./al.	....	218	subl.	i.	s.	v. s.
-diphenyl (4-,4′-)	$(HO.C_6H_4.)_2$	186.20	rhb./al.	1.25	270–2	subl.	sl. s.	v. s.	v. s.
-ethyl formal (β-)	$CH_2(OCH_2CH_2OH)_2$	136.15	lq.	1.154^{25}	−5.3	264	∞		
-naphthalene (1-,5-)	$C_{10}H_6(OH)_2$	160.16	pr./aq.	....	258–60	d.	sl. s.	s.	v. s.
(1-,8-)	$C_{10}H_6(OH)_2$	160.16	nd.	....	140	....	sl. s. h.	....	v. s.
Dimethoxy-benzene (p-)	$(CH_3O)_2C_6H_4$	138.16	lf.	$1.053^{55/55}$	56	212.6	v. sl. s.	v. s.	v. s.
-diphenylamine (4-,4′-)	$HN(C_6H_4OCH_3)_2$	229.26	cr.	....	103	....	i.		
-ethyl adipate	$(CH_2)_4(CO_2C_2H_4OCH_3)_2$	262.30	lq.	$1.075^{15.6}$	....	$145–50^{2}$	5		
Dimethyl adipate	$[(CH_2)_2CO_2CH_3]_2$	174.19	col. lq.	$1.063^{20/4}$	10–1	115^{13}	i.		
-amine	$(CH_3)_2NH$	45.08	col. lq.	$0.680^{0/4}$	−96	7.4	v. s.	s.	s.
-aminoazobenzene (p-)	$C_6H_5N:N.C_6H_4N(CH_3)_2$	225.28	yel./al.	....	116–7	d.	i.	s.	s.
-aminoethanol	$(CH_3)_2NCH_2CH_2OH$	89.14	col. lq.	$0.887^{20/4}$	....	135^{758}	∞		
-aminophenol (m-)	$(CH_3)_2NC_6H_4OH$	137.18	nd.	....	85	265–8	sl. s. h.	s.	s.
-aniline	$(CH_3)_2NC_6H_5$	121.18	yel. lq.	$0.956^{20/4}$	2.5	193	i.	s.	s.
sulfonic acid (m-)	$(CH_3)_2NC_6H_4SO_3H$	201.24	cr.	....	d. 266	....	s.		
(p-)	$(CH_3)_2NC_6H_4SO_3H.H_2O$	219.25	pr.	....	257	....	s. h.	v. sl. s.	v. sl. s.
carbonate	$OC(OCH_3)_2$	90.08	col. lq.	$1.070^{20/4}$	0.5	89–90	i.	∞	∞
ether	CH_3OCH_3	46.07	gas	....	−138.5	−23.7	3700 cc.[18]	s.	s.
-formamide	$HCON(CH_3)_2$	73.09	lq.	0.945^{25}	−58.3	152.8	∞		
fumarate	$(:CHCO_2CH_3)_2$	144.12	col. tri.	....	102	192	i.	sl. s.	sl. s.
glutarate	$(CH_2)_3(CO_2CH_3)_2$	160.17	lq.	$1.089^{15.6}$	−37	130^{50}			
glyoxime	$(CH_3.C:NOH)_2$	116.12	col. cr.	....	240–6	....	0.06^{20}	v. s.	v. s.
-naphthalene (1-,4-)	$C_{10}H_6(CH_3)_2$	156.22	lq.	$1.016^{20/4}$	<−18	264–6	i.		
(2-,3-)	$C_{10}H_6(CH_3)_2$	156.22	lf./al.	....	104	265^{767}	i.	sl. s.	
-naphthylamine (α-)	$C_{10}H_7N(CH_3)_2$	171.23	col. oil	1.042^{20}	....	274.5^{711}	i.	s.	s.
(β-)	$C_{10}H_7N(CH_3)_2$	171.23	col. cr.	$1.039^{70/70}$	46	304–5	i.	s.	s.
oxalate	$(.CO_2CH_3)_2$	118.09	col. mn.	1.148^{54}	54	163.3	6	s.	s.
phthalate (o-)	$C_6H_4(CO_2CH_3)_2$	194.18	col. lq.	$1.189^{25/25}$	....	280^{734}	0.43		
sulfate	$(CH_3O)_2SO_2$	126.13	col. oil	$1.352^{0/4}$	−26.8	188.3	v. sl. s.	∞	∞
sulfide	$(CH_3)_2S$	62.13	oil	$0.846^{21/4}$	−83.2	37.3	i.	s.	s.
tartrate (d-)	$(CHOH.CO_2CH_3)_2$	178.14	cr.	$1.328^{20/4}$	61.5	280	s.	200^{15}	
-vinyl-ethenyl carbinol	$(CH_3)_2COH.C:C.CH:CH_2$	110.15	lq.	$0.887^{20/4}$	....	150	6^{29}		
Dinaphthyl (αα′-)	$C_{10}H_7.C_{10}H_7$	254.31	lf./al.	....	160	$240–4^{12}$	i.	s. h.	s.
-methane (αα′-)	$(C_{10}H_7)_2CH_2$	268.34	pr./al.	....	109	>360	i.	0.8 c.	v. s.
(β,β′-)	$(C_{10}H_7)_2CH_2$	268.34	nd./al.	....	92	....	i.	s.	
Dinitro-anisole (1-)(2-,4-)	$CH_3OC_6H_3(NO_2)_2$	198.13	col. mn.	1.341^{20}	94–5	....	sl. s. h.	1.5^{20}	
-benzene (o-)	$C_6H_4(NO_2)_2$	168.11	col. mn.	1.59^{18}	117–8	319^{774}	0.01 c.	1.9^{21}	
(m-)	$C_6H_4(NO_2)_2$	168.11	col. rhb.	$1.575^{20/4}$	89.8	300–2	0.3^{99}	3^{20}	
(p-)	$C_6H_4(NO_2)_2$	168.11	col. mn.	1.625^{18}	173–4	299^{777}	0.18^{100}	0.18^{21}	
sulfonic acid (2-,4-)(1-)	$(NO_2)_2C_6H_3SO_3H.3H_2O$	302.22	pr.	....	106–8	....	s.	s.	v. sl. s.
-benzoic acid (2-,4-)	$(NO_2)_2C_6H_3CO_2H$	212.12	cr./aq.	....	179–80	....	1.85^{25}	s.	
(3-,5-)	$(NO_2)_2C_6H_3CO_2H$	212.12	mn. pr.	....	204–5	subl.	s h.	v. s.	sl. s.
-benzophenone (4-,4′-)	$(NO_2C_6H_4)_2CO$	272.21	col. nd.	....	189	....	i.		
-diphenyl (4-,4′-)	$(NO_2C_6H_4)_2$	244.20	nd./al.	1.445	233	....	i.	1.5^{20}	
(2-,4′-)	$(NO_2C_6H_4)_2$	244.20	mn.	1.474	93.5	....	i.	v. s. h.	
-naphthalene (1-,5-)	$C_{10}H_6(NO_2)_2$	218.16	nd.	....	216	subl.	i.		
(1-,8-)	$C_{10}H_6(NO_2)_2$	218.16	rhb.	....	170–2	d.	i.	0.2 c.	

Table 2. Physical Properties of Organic Compounds—(*Continued*)

Name	Formula	Formula weight	Form and color	Specific gravity	Melting point, °C.	Boiling point, °C.	Solubility in 100 parts		
							Water	Alcohol	Ether
Dinitro-phenol (2-,3-)	$(NO_2)_2C_6H_3OH$	184.11	yel. mn.	1.681^{20}	144–5	....	sl. s.	v. s. h.	v. s.
(2-,4-)	$(NO_2)C_6H_3OH$	184.11	yel. rhb.	1.683^{24}	114–5	subl.	0.5 c.	4^{20}	v. s. h.
(2-,6-)	$(NO_2)C_6H_3OH$	184.11	yel. rhb.	....	63–4	....	s. h.	s. h.	s.
-salicylic acid (3-,5-)	$(NO_2)_2C_6H_2(OH)CO_2H.H_2O$	246.13	pl./aq.	....	173 d.	....	s. c.	v. s.	v. s.
-stilbene (4-,4′-)	$(NO_2C_6H_4CH:)_2$	270.24	yel. lf.	....	210–6	....	i.	v. sl. s.	v. sl. s.
-toluene (2-,4-)	$(NO_2)_2C_6H_3CH_3$	182.13	nd.	1.321^{71}	70	300	0.03^{22}	1.2^{15}	9^{15}
(3-,4-)	$(NO_2)_2C_6H_3CH_3$	182.13	nd.	1.259^{111}	60–1	....	i.		
(3-,5-)	$(NO_2)_2C_6H_3CH_3$	182.13	mn. pr.	1.277^{111}	92–3	subl.	sl. s.	s. h.	s.
Dioxane	$O < (CH_2.CH_2)_2 > O$	88.10	col. lq.	$1.033^{20/4}$	9.5–10.5	101.1	∞	s.	s.
Dipentene	$C_{13}H_{10}$	136.23	col. lq.	0.865^{18}	....	178	i.		
Diphenyl	$C_6H_5.C_6H_5$	154.20	col. mn.	$0.992^{73/4}$	69–70	254.9	i.	10^{20}	6.6^{20}
-amine	$C_6H_5NHC_6H_5$	169.22	col. mn.	$1.160^{20/20}$	52.9	302	0.03^{25}	$56^{19.5}$	s.
carbonate	$O(COC_6H_5)_2$	214.21	nd./al.	1.272^{14}	80	302–6	i.	v. s.	s.
-chloroarsine	$(C_6H_5)_2AsCl$	264.57	rhb.	1.583^{40}	43–4	d. 327	0.2 d.	20	s.
-ethane	$(C_6H_5CH_2.)_2$	182.25	col. pr.	$0.978^{50/50}$	52–3	284	i.	s.	v. s.
ether	$C_6H_5OC_6H_5$	170.20	col. rhb.	1.073^{20}	27	259	v. sl. s.	s.	∞
guanidine	$(C_6H_5NH)_2C:NH$	211.26	mn./al.	....	147–8	d. >170	v. sl. s.	9^{20}	sl. s.
-methane	$(C_6H_5)_2CH_2$	168.23	col. pr.	$1.001^{26/4}$	26–7	265	i.	v. s.	v. s.
phenylenediamine (p-)	$(C_6H_5NH)_2C_6H_4$	260.32	cr.	....	152	....	i.		
succinate	$(.CH_2CO_2C_6H_5)_2$	270.27	lf./al.	....	122–3	330	i.	....	s.
sulfide	$(C_6H_5)_2S$	186.26	col. lq.	$1.119^{15/15}$	<−40	296–7	i.	s. h.	∞
sulfone	$(C_6H_5)_2SO_2$	218.26	nd./aq.	$1.248^{25/4}$	128–9	379	sl. s. h.	s. h.	
urea (uns.)	$(C_6H_5)_2NCONH_2$	212.24	rhb.	1.276	189	....	v. sl. s.	s.	s.
Diphenylene oxide	$< (C_6H_4)_2O$	168.18	lf./al.	....	86–7	287–8	i.	s. h.	v. s.
Dipropyl adipate (n-)	$(.CH_2CH_2CO_2C_3H_7)_2$	230.30	col. lq.	$0.979^{20/4}$	−20.3	$143–5^{10}$	i.	s.	s.
-amine (n-)	$(C_2H_5CH_2)_2NH$	101.19	col. lq.	$0.739^{20/4}$	−39.6	110–1	s.	∞	∞
(i-)	$[(CH_3)_2CH]_2NH$	101.19	col. lq.	0.722^{22}	−61	83.5^{743}	s.	s.	
aniline (n-)	$C_6H_5N(C_3H_7)_2$	177.28	yel. oil	0.910^{20}	....	245.4	i.	s.	s.
carbonate (n-)	$O(COCH_2C_2H_5)_2$	146.18	col. lq.	0.968^{22}	....	168.2	v. sl. s.		
ether (n-)	$(C_2H_5CH_2)_2O$	102.17	col. lq.	$0.744^{21/0}$	−122	91	sl. s.	∞	∞
(i-)	$[(CH_3)_2CH]_2O$	102.17	col. lq.	$0.725^{21/0}$	−60	69	0.2	∞	∞
ketone (n-)	$(C_2H_5CH_2)_2CO$	114.18	col. lq.	$0.822^{20/4}$	−32.6	144.2	0.43	∞	∞
(i-)	$[(CH_3)_2CH]_2CO$	114.18	col. lq.	$0.806^{20/4}$	....	123.7	v. sl. s.	∞	∞
oxalate (n-)	$(CO_2CH_2C_2H_5)_2$	174.19	col. lq.	$1.038^{0/0}$	−51.7	213.5	d. h.		
(i-)	$[CO_2CH(CH_3)_2]_2$	174.19	col. lq.	....	....	190			
Disalicylal ethylenediamine	$[HOC_6H_4CH:NCH_2.]_2$	268.30	cr.	1.34	125–6	....	0.03^{28}		
Ditolyl guanidine (o-)	$(C_7H_7NH)_2C:NH$	239.31	cr.	$1.10^{20/4}$	178–9	....	v. sl. s.	s. h.	s.
Divinyl acetylene	$(H_2C:CH.C:)_2$	78.11	lq.	$0.776^{20/4}$	....	85	i.		
Docosane (n-)	$CH_3(CH_2)_{20}CH_3$	310.59	cr.	$0.778^{44/4}$	44.5	224.5^{15}	i.	4 h.	v. s.
Dodecane (n-)	$CH_3(CH_2)_{10}CH_3$	170.33	lq.	$0.751^{20/4}$	−9.6	214.5	i.	v. s.	v. s.
Dulcitol	$CH_2OH(CHOH)_4CH_2OH$	182.17	mn.	1.466^{15}	189	$290–5^{3}$	3.2^{15}	v. sl. s.	i.
Durene (1-,2-,4-,5-)	$(CH_3)_4C_6H_2$	134.21	mn.	$0.838^{81/4}$	79–80	193–5	i.	s.	s.
Elaidic acid	$C_8H_{17}CH:CH(CH_2)_7CO_2H$	282.45	lf./al.	$0.851^{79/4}$	51–2	288^{100}	i.	v. s.	v. s.
Eosine	$C_{20}H_8O_5Br_4$	647.93	col. cr.	....	....	....	i.	s.	
Ephedrine (l-)	$C_6H_5CHOHCH(CH_3)NHCH_3$	165.23	cr./et.	....	40	255	5	500	s.
Epichlorhydrin (α-)	$C_2H_3O.CH_2Cl$	92.53	lq.	$1.183^{25/25}$	−25.6	117^{756}	<5	∞	∞
Epidichlorohydrin (α-)	$CH_2:CCl.CH_2Cl$	110.98	col. lq.	1.204^{25}	....	94	i.	∞	∞
Erythritol (dl-)	$CH_2OH(CHOH)_2CH_2OH$	122.12	tet. pr.	$1.451^{20/4}$	126	329–31	60	sl. s. c.	i.
tetranitrate	$C_4H_6(ONO_2)_4$	302.12	lf./al.	....	61	expl.	i. c.	s.	s.
Ethane	CH_3CH_3	30.07	col. gas	0.546^{-88}	−172	−88.6	4.7 cc.^{20}	150 cc.	
Ethanol-amine	$HOCH_2CH_2NH_2$	61.08	col. oil	1.022^{20}	10.5	171^{757}	∞	∞	1
formamide	$HCONHCH_2CH_2OH$	89.09	lq.	1.169^{25}	<−40	d.	∞		
Ether	$(CH_3CH_2)_2O$	74.12	col. lq.	$0.708^{25/4}$	−116.3	34.6	7.5^{20}	∞	
Ethyl abietate	$C_{19}H_{29}CO_2C_2H_5$	330.49	lq.	$1.020^{20/20}$	....	200^{4}	i.		
acetate	$CH_3CO_2C_2H_5$	88.10	col. lq.	$0.901^{20/4}$	−82.4	77.1	8.5^{15}	∞	∞
acetoacetate	$CH_3COCH_2CO_2C_2H_5$	130.14	col. lq.	$1.025^{20/4}$	−45	180^{755}	13^{17}	∞	∞
alcohol	CH_3CH_2OH	46.07	col. lq.	$0.789^{20/4}$	−112	78.4	∞	....	∞
-amine	$C_2H_5NH_2$	45.08	col. lq.	$0.689^{15/15}$	−80.6	16.6	∞	∞	∞
hydrochloride	$C_2H_5NH_2.HCl$	81.55	mn.	1.216	108–9	....	240^{17}	v. s.	i.
aniline	$C_6H_5NHC_2H_5$	121.18	lq.	$0.963^{20/4}$	−63.5	204	i.	∞	∞
sulfonic acid (m-)	$C_2H_5NHC_6H_4SO_3H$	201.24	nd./aq.	....	d. 294	....	2.15^{15}		
anisate (p-)	$CH_3OC_6H_4CO_2C_2H_5$	180.20	lq.	$1.103^{25/25}$	7–8	269–70	i.	s.	s.
anthranilate (o-)	$NH_2C_6H_4CO_2C_2H_5$	165.19	cr.	$1.117^{20/4}$	13	266–8	v. sl. s.	s.	s.
benzene	$C_6H_5.C_2H_5$	106.16	col. lq.	$0.867^{20/4}$	−94.4	136.2	0.01^{15}	∞	∞
benzoate	$C_6H_5CO_2C_2H_5$	150.17	col. lq.	$1.052^{15/15}$	−34.6	211–2	0.08^{20}	∞	∞
-benzyl-aniline	$C_6H_5N(C_2H_5)CH_2C_6H_5$	211.29	yel. oil	$1.034^{18.5}$	....	285^{10}	i.	18	∞

bromide	C_2H_5Br	108.98	col. lq.	1.431$^{20/4}$	−117.8	38.4	1.06^{0}	∞	∞
butyrate (n-)	$C_2H_5CH_2CO_2C_2H_5$	116.16	col. lq.	0.879$^{20/4}$	−93.3	120–1	0.68^{25}	∞	∞
(i-)	$(CH_3)_2CHCO_2C_2H_5$	116.16	col. lq.	0.871$^{20/4}$	−88.2	110–1	sl. s.	∞	∞
caprate (n-)	$CH_3(CH_2)_8CO_2C_2H_5$	200.31	lq.	0.859^{28}	−20	244.6^{758}	0.002^{20}	∞	∞
Ethyl caproate (n-)	$CH_3(CH_2)_4CO_2C_2H_5$	144.21	col. lq.	0.873$^{20/20}$	−67.5	165–6^{736}	i.	∞	∞
caprylate (n-)	$CH_3(CH_2)_6CO_2C_2H_5$	172.26	col. lq.	0.878^{17}	−45	207–8^{753}	i.	∞	∞
chloride	CH_3CH_2Cl	64.52	col. lq.	0.917$^{6/6}$	−139	13	0.45^{0}	∞	∞
chloroacetate	$ClCH_2CO_2C_2H_5$	122.55	col. lq.	1.159$^{20/4}$	−26	144	i.	∞	∞
chlorocarbonate	$ClCO_2CH_2CH_3$	108.53	col. lq.	1.138$^{20/4}$	−80.6	94–5	d.	∞	∞
cinnamate (trans-)	$C_6H_5CH{:}CHCO_2H$	176.21	col. lq.	1.049$^{20/4}$	12	271	i.	∞	∞
cyanoacetate	$CH_2(CN)CO_2C_2H_5$	113.11	col. lq.	1.062$^{20/4}$	−22.5	208^{753}	2^{25}	∞	∞
formate	$HCO_2CH_2CH_3$	74.08	col. lq.	0.923$^{20/4}$	−79	54^{760}	11^{18}	∞	∞
furoate (α)	$OC_4H_3CO_2C_2H_5$	140.13	lf.	1.117$^{21/4}$	34	195^{765}	i.	∞	∞
heptoate	$CH_3(CH_2)_5CO_2C_2H_5$	158.23	col. lq.	0.872$^{20/20}$	−66.1	187–8	0.029^{20}	∞	∞
hypochlorite	$ClOCH_2CH_3$	80.52	yel. lq.	1.013$^{-6/4}$	expl.	36^{752}	....	....	∞
iodide	CH_3CH_2I	155.98	col. lq.	1.933$^{20/4}$	−105	72.4	0.4^{20}	∞	∞
lactate	$CH_3CH(OH)CO_2C_2H_5$	118.13	oil	1.030$^{25/4}$	....	155	∞	∞	∞
laurate	$CH_3(CH_2)_{10}CO_2C_2H_5$	228.36	oil	0.868$^{13/4}$	−10.7	269	i.	s.	∞
mercaptan	CH_3CH_2SH	62.13	lq.	0.839$^{20/4}$	−121	36–7	1.5	s.	s.
methacrylate	$CH_2{:}C(CH_3)CO_2C_2H_5$	114.14	col. lq.	0.913$^{15.6}$	....	118	i.	s.	s.
naphthylamine (α-)	$C_{10}H_7NHC_2H_5$	171.23	oil	1.060$^{20/4}$	....	303^{723}	i.	s.	s.
naphthyl ether (α-)	$C_{10}H_7OC_2H_5$	172.22	cr.	1.061$^{20/20}$	5.5	276.4	i.	s.	s.
nitrate	$C_2H_5ONO_2$	91.07	col. lq.	1.100$^{25/4}$	−102	87–8	1.3^{55}	∞	∞
nitrite	C_2H_5ONO	75.07	lq.	0.900$^{15.5}$	....	17	v. sl. s.	∞	∞
oleate	$C_{17}H_{33}CO_2C_2H_5$	310.50	oil	0.867^{25}	<−15	216–8^{15}	i.	∞	∞
palmitate	$CH_3(CH_2)_{14}CO_2C_2H_5$	284.47	col. nd.	0.858$^{25/4}$	24–5	191^{10}	i.	s.	s.
pelargonate	$CH_3(CH_2)_7CO_2C_2H_5$	186.29	col. lq.	0.866$^{17.5}$	−44.5	227–8^{757}	i.	∞	∞
propionate	$CH_3CH_2CO_2C_2H_5$	102.13	col. lq.	0.891$^{20/4}$	−72.6	99.1	2.4^{20}	∞	∞
salicylate (o-)	$HOC_6H_4CO_2C_2H_5$	166.17	col. lq.	1.136$^{15/4}$	1.3	233–4	i.	∞	∞
stearate	$CH_3(CH_2)_{16}CO_2C_2H_5$	312.52	col. cr.	0.848$^{36.3}$	33.4(31)	201^{10}	i.	s.	s.
toluate (o-)	$CH_3.C_6H_4CO_2C_2H_5$	164.20	lq.	1.032$^{25/25}$	<−10	227	i.	∞	∞
(m-)	$CH_3.C_6H_4CO_2C_2H_5$	164.20	lq.	1.030$^{20/20}$	....	231^{750}	i.	∞	∞
toluene sulfonate (p-)	$CH_3.C_6H_4SO_3C_2H_5$	200.25	pr./al.	1.166$^{48/4}$	33–4	221.3	i.	s.	s.
toluidine (o-)	$CH_3.C_6H_4NHC_2H_5$	135.20	lq.	0.948$^{25/4}$	<−15	215–6	i.		
(p-)	$CH_3.C_6H_4NHC_2H_5$	135.20	lq.	0.942$^{25/4}$	....	217	i.		
urea	$C_2H_5NH.CO.NH_2$	88.11	nd.	1.213^{18}	92	....	v. s.	80	i.
valerate (n-)	$CH_3(CH_2)_3CO_2C_2H_5$	130.18	col. lq.	0.877^{20}	−91.2	145.5	0.24^{25}	∞	∞
(i-)	$(CH_3)_2CH(CH_2)_2CO_2C_2H_5$	130.18	col. lq.	0.867$^{20/4}$	−99.3	135	0.17^{20}	∞	∞
Ethylal	$CH_2(OC_2H_5)_2$	104.15	lq.	0.824$^{25/4}$	−66.5	89	9^{18}	∞	∞
Ethylene	$H_2C{:}CH_2$	28.05	col. gas	0.57$^{-102/4}$	−169	−103.9	26 cc.^{0}	360 cc.	s.
bromide	$BrCH_2.CH_2Br$	187.88	col. lq.	2.180$^{20/4}$	10	131.5	0.43^{30}	∞	∞
bromohydrin	$BrCH_2.CH_2OH$	124.98	col. lq.	1.772$^{20/4}$	....	150.3	sl. s.	s.	
chlorobromide	$ClCH_2.CH_2Br$	143.43	lq.	1.689^{19}	−16.6	106.7	0.69^{30}		
chlorohydrin	$ClCH_2.CH_2OH$	80.52	col. lq.	1.213$^{20/4}$	−69	128.8	∞	∞	∞
diamine	$H_2NCH_2.CH_2NH_2$	60.10	col. lq.	0.900$^{20/20}$	8.5	117.2	∞	∞	0.3
oxide	$<(CH_2)_2>O$	44.05	lq.	0.887$^{7/4}$	−111.3	13.5^{747}	∞	∞	v. s.
Ethylidene diacetate	$CH_3CH(O_2CCH_3)_2$	146.14	col. lq.	1.061^{12}	18.85	168^{740}	sl. s.	∞	
Eugenol (1-,4-,3-)	$C_3H_5.C_6H_3(OH)OCH_3$	164.20	oil	1.070$^{15/15}$	10.3	253.5	v. sl. s.	∞	∞
(i-)(1-,3-,4-)	$C_3H_5.C_6H_3(OCH_3)OH$	164.20	oil	1.091$^{15/15}$	−10	267.5	v. sl. s.	∞	∞
Fenchyl alcohol (dl-)	$C_{10}H_{17}OH$	154.24	col. cr.	0.935^{40}	35	201	sl. s.		
(d-)(α-)	$C_{10}H_{17}OH$	154.24	col. pr.	0.964$^{20/4}$	45–7	201–2	sl. s.	s.	s.
(i-)(l-)	$C_{10}H_{17}OH$	154.24	col. cr.	0.961	61–2	201–2	i.		
Ferric dimethyl-dithiocarbamate	$Fe[SSCN(CH_3)_2]_3$	416.47	cr.	....	d. 100–30	ign. >150	v. sl. s.		
Fluorene	$(C_6H_4)_2>CH_2$	166.21	cr./al.	1.203$^{0/4}$	115–6	293–5	i.	s. h.	s.
Fluorescein	$C_{20}H_{12}O_5$	332.30	yel. red	....	d. >290	....	v. sl. s. h.	s. h.	
Fluoro-dichloromethane	$FCHCl_2$	102.93	gas	1.426^{0}	−127	14.5	i.	s.	s.
-trichloromethane	Cl_3CF	137.38	col. lq.	1.494$^{17.2}$	....	24.9	i.	∞	∞
Formaldehyde	$HCHO$	30.03	gas	0.815^{-20}	−92	−21	v. s.	v. s.	v. s.
(m-)	$(CH_2O)_3$	90.08	wh.	1.17^{65}	64	114.5^{759}	21^{25}	s.	s.
(p-)	$(CH_2O)_x.xH_2O$	(30.03)	amor.	....	150–60	subl.	20–30^{18}	i.	i.
Formamide	$HCONH_2$	45.04	lq.	1.139$^{20/4}$	2	193	∞	∞	v. sl. s.
Formanilide	$HCONHC_6H_5$	121.13	mn.	1.147$^{15/15}$	47	216^{120}	sl. s.	v. s.	s.
Formic acid	HCO_2H	46.03	col. lq.	1.220$^{20/4}$	8.6	100.8	∞	∞	∞
Fructose	$CH_2OH(CHOH)_3COCH_2OH$	180.16	nd./aq.	1.669$^{17.5}$	95–105	....	v. s.	8^{18}	
Fuchsin	$C_{20}H_{19}N_3HCl$	337.84	red	1.22	d. >200	....	0.3	s.	i.
Fulminic acid	$C{:}NOH$	43.03							
Fumaric acid (trans-)	$HO_2CCH{:}CHCO_2H$	116.07	col. pr.	1.635$^{20/4}$	286–7	290	0.7^{17}	5.8^{30}	0.7^{25}
Furfural	$C_4H_3O.CHO$	96.08	lq.	1.159$^{20/4}$	−38.7	161.7^{760}	9.1^{13}	∞	∞
Furfuran	C_4H_4O	68.07	col. lq.	0.937$^{20/4}$	....	31–2^{756}	i.	s.	s.

Table 2. Physical Properties of Organic Compounds—(*Continued*)

Name	Formula	Formula weight	Form and color
Furfuryl acetate	$CH_3CO_2CH_2C_4H_3O$	140.13	col. oil
alcohol	$C_4H_3O.CH_2OH$	98.10	oil
butyrate	$C_3H_7CO_2CH_2.C_4H_3O$	168.19	col. lq.
propionate	$C_2H_5CO_2CH_2.C_4H_3O$	154.16	col. lq.
Furoic acid	$C_4H_3O.CO_2H$	112.08	mn. pr.
G-acid, K salt (2-)(6-,8-)	$HOC_{10}H_5(SO_3K)_2$	380.46	cr.
Na salt (2-)(6-,8-)	$HOC_{10}H_5(SO_3Na)_2$	348.26	cr.
Galactose (d-)(α-)	$C_5H_{11}O_5.CHO$	180.16	pr.
Gallic acid (3-,4-,5-)	$(HO)_3C_6H_2CO_2H.H_2O$	188.13	mn./aq.
Gamma acid (2-,8-,6-)	$C_{10}H_5(NH_2)(OH)SO_3H$	239.24	cr.
Geraniol	$C_9H_{15}CH_2OH$	154.24	col. lq.
Glucose (d-)(α-)	$C_5H_{11}O_5.CHO$	180.16	rhb.
(d-)(β-)	$C_6H_{12}O_6.H_2O$	198.17	cr.
Glucuronic acid	$CHO(CHOH)_4CO_2H$	194.14	cr.
Glutam(in)ic acid (dl-)	$[.CHNH_2(CH_2)_2.](CO_2H)_2$	147.13	cr./aq.
Glutaric acid	$CH_2(CH_2CO_2H)_2$	132.11	col. cr.
Glycerol	$CH_2OH.CHOH.CH_2OH$	92.09	col. lq.
acetate (mono-)	$C_5H_{10}O_4$	134.13	col. oil
(di-)	$(CH_3CO_2)_2C_3H_5OH$	176.17	col. lq.
nitrate (mono-) (α-)	$CH_2OH.CHOH.CH_2NO_3$	137.09	col. pr.
(β-)	$CH_2OH.CHNO_3.CH_2OH$	137.09	lf.
dinitrate (1-,3-)	$CHOH(CH_2ONO_2)_2$	182.09	oil
Glyceryl triacetate	$(CH_3CO_2)_3C_3H_5$	218.20	col. lq.
tribenzoate	$(C_6H_5CO_2)_3C_3H_5$	404.40	nd.
tributyrate	$(C_2H_5CH_2CO_2)_3C_3H_5$	302.36	col. lq.
tricaprate	$[CH_3(CH_2)_8CO_2]_3C_3H_5$	554.83	col. cr.
tricaproate	$[CH_3(CH_2)_4CO_2]_3C_3H_5$	386.51	col. lq.
tricaprylate	$[CH_3(CH_2)_6CO_2]_3C_3H_5$	470.67	col. lq.
trilaurate	$[CH_3(CH_2)_{10}CO_2]_3C_3H_5$	638.98	col. nd.
trimyristate	$[CH_3(CH_2)_{12}CO_2]_3C_3H_5$	723.14	lf.
trinitrate	$CH_2NO_3.CHNO_3.CH_2NO_3$	227.09	yel. oil
trinitrite	$CH_2NO_2.CHNO_2.CH_2NO_2$	179.09	yel. lq.
trioleate	$(C_{17}H_{33}CO_2)_3C_3H_5$	885.40	col. oil
tripalmitate	$[CH_3(CH_2)_{14}CO_2]_3C_3H_5$	807.29	col. nd.
tristearate	$[CH_3(CH_2)_{16}CO_2]_3C_3H_5$	891.45	col pr.
Glycide	$C_2H_3O.CH_2OH$	74.08	col. lq.
Glycine, Glycocoll	$NH_2CH_2.CO_2H$	75.07	mn.
Glycol	$CH_2OH.CH_2OH$	62.07	col. lq.
diacetate	$(CH_3CO_2CH_2-)_2$	146.14	col. lq.
dibenzoate	$(C_6H_5CO_2CH_2-)_2$	270.27	rhb./et.
dibutyrate	$(C_3H_7CO_2CH_2-)_2$	202.24	col. lq.
dicaprylate	$(C_7H_{15}CO_2CH_2-)_2$	314.45	lq.
diformate	$(HCO_2CH_2-)_2$	118.09	lq.
dilaurate	$(C_{11}H_{23}CO_2CH_2-)_2$	426.66	amor.
dinitrate	$(O_2NO.CH_2-)_2$	152.07	yel. lq.
dinitrite	$(ONO.CH_2-)_2$	120.07	lq.
dipalmitate	$(C_{15}H_{31}CO_2CH_2-)_2$	538.87	nd.
dipropionate	$(C_2H_5CO_2CH_2-)_2$	174.19	lq.
ether	$(HO.CH_2CH_2)_2O$	106.12	lq.
formal	$< O.CH_2CH_2OCH_2 >$	74.08	lq.
formate (mono-)	$HCO_2CH_2CH_2OH$	90.08	lq.
Glycolic acid	$HOCH_2CO_2H$	76.05	nd./aq.
Guaiacol (o-)	$CH_3O.C_6H_4OH$	124.13	pr.
Guanidine	$NH:C(NH_2)_2$	59.07	col. cr.
H-acid, Na salt (1-,8-,3-,6-)	$C_{10}H_8O_7NS_2Na.1\frac{1}{2}H_2O$	368.31	cr.
Heptacosane (n-)	$CH_3(CH_2)_{25}CH_3$	380.72	col. cr.
Heptane (n-)	$CH_3(CH_2)_5CH_3$	100.20	col. lq.
(i-)	$(CH_3)_2CH(CH_2)_3CH_3$	100.20	col. lq.
	$C_3H_7.CH(CH_3).C_2H_5$	100.20	col. lq.
	$(CH_3)_3C.CH_2.C_2H_5$	100.20	col. lq.
	$[(CH_3)_2CH]_2CH_2$	100.20	col. lq.
	$(CH_3)_2C(C_2H_5)_2$	100.20	col. lq.
	$(C_2H_5)_3CH$	100.20	col. lq.
	$(CH_3)_3C.CH(CH_3)_2$	100.20	col. lq.
Heptoic acid	$CH_3(CH_2)_5CO_2H$	130.18	col. lq.
aldehyde	$CH_3(CH_2)_5CHO$	114.18	col. lq.

Name	Specific gravity	Melting point, °C.	Boiling point, °C.	Solubility in 100 parts: Water	Alcohol	Ether
Furfuryl acetate	$1.118^{20/4}$	....	175–7	i.	s.	s.
alcohol	$1.129^{25/4}$	....	169.5^{752}	∞	s.	s.
butyrate	$1.053^{20/4}$	....	212–3	v. sl. s.	s.	∞
propionate	$1.109^{20/4}$	....	195–6	v. sl. s.	s.	∞
Furoic acid	....	133–4	230–2	3.6^{15}	s.	s.
G-acid, K salt (2-)(6-,8-)	....	....	....	8^{25}		
Na salt (2-)(6-,8-)	....	....	....	34^{20}		
Galactose (d-)(α-)	....	165.5	....	10.3^{0}	0.6^{40}	
Gallic acid (3-,4-,5-)	$1.694^{4/4}$	d. 220	....	1^{13}	28^{15}	2.5^{15}
Gamma acid (2-,8-,6-)						
Geraniol	0.883^{15}	<−15	230	i.	∞	∞
Glucose (d-)(α-)	1.544^{25}	146	....	$82^{17.5}$	sl. s.	i.
(d-)(β-)	$1.562^{18/4}$	150	....	154^{15}		
Glucuronic acid	....	154	d.	v. s.		
Glutam(in)ic acid (dl-)	1.460	199 d.	....	1.5^{20}	v. sl. s.	v. sl. s.
Glutaric acid	1.429^{15}	97.5	200^{20}	63.9^{20}	v. s.	v. s.
Glycerol	$1.260^{50/4}$	17.9	290	∞	∞	i.
acetate (mono-)	$1.20^{20/4}$	....	158^{165}	v. s.	v. s.	sl. s.
(di-)	$1.178^{15/15}$	40	$175–6^{40}$	s.	s.	sl. s.
nitrate (mono-) (α-)	1.40^{15}	58–9	155–60	70^{15}	v. s.	v. sl. s.
(β-)	1.40^{15}	54	155–60	....	v. s.	sl. s.
dinitrate (1-,3-)	1.47^{15}	<−30	$146–8^{15}$	....	v. s.	v. s.
Glyceryl triacetate	$1.161^{17/4}$	−78	258–9	7.17^{15}	∞	∞
tribenzoate	1.228^{12}	75–6	d.	i.	s. h.	s.
tributyrate	$1.032^{20/4}$	<−75	305–9	i.	s.	s.
tricaprate	$0.921^{40/4}$	31(25)	....	i.	s. h.	v. s.
tricaproate	$0.987^{20/4}$	−25	....	i.	s.	s.
tricaprylate	$0.954^{20/4}$	8.3(−21)	....	i.	s.	s.
trilaurate	$0.894^{60/4}$	45–6	....	i.	sl. s. c.	v. s.
trimyristate	$0.885^{60/6}$	56.5	....	i.	s.	
trinitrate	1.601^{15}	13.3(2)	160^{15}	0.18^{20}	50^{20}	∞
trinitrite	$1.291^{10/16}$	....	150 sl. d.	d.	d.	s.
trioleate	0.915^{15}	−4	240^{18}	i.	sl. s.	v. s.
tripalmitate	$0.866^{80/4}$	65.1	$310–20^{0.1}$	i.	0.004^{21}	v. s.
tristearate	$0.862^{80/4}$	70.8(55)	....	i.	s. h.	s. h.
Glycide	$1.114^{16/16}$	....	166 sl. d.	∞	∞	∞
Glycine, Glycocoll	1.161	232–6 d.	....	23 c.	0.1 c.	i.
Glycol	$1.113^{19/4}$	−15.6	197.4	∞	∞	1.0
diacetate	$1.109^{14/4}$	−31	190.5	14.3^{22}	∞	∞
dibenzoate	....	73–4	>360	i.	....	s.
dibutyrate	1.024^{0}	....	240	i.	v. s.	v. s.
dicaprylate	....	22	....	i.		
diformate	....	....	174	v. sl. s.		
dilaurate	....	52–4	188^{20}	i.	v. s.	v. s.
dinitrate	$1.482^{21/2}$	−20	expl. 114	0.92^{25}	s.	∞
dinitrite	1.216^{0}	<−15	96–8	i.	s. d.	s.
dipalmitate	....	71–2	$260^{0.1}$	i.	s.	s.
dipropionate	1.045^{25}	....	211–2	sl. s.	∞	∞
ether	$1.118^{20/20}$	−10.5	244.8	∞	∞	i.
formal	$1.060^{20/4}$	....	75–6	∞		
formate (mono-)	$1.199^{15/4}$	....	180	∞		
Glycolic acid	....	79(63)	d.	∞	90^{25}	v. s.
Guaiacol (o-)	$1.140^{15/15}$	28.3	205	1.7^{15}	v. s.	v. s.
Guanidine	....	50	....	v. s.	s.	
H-acid, Na salt (1-,8-,3-,6-)	....	....	....	0.17^{20}		
Heptacosane (n-)	$0.780^{60/4}$	59.5	270^{15}	i.		
Heptane (n-)	$0.684^{20/4}$	−90.6	98.4^{760}	0.005^{15}	sl. s.	∞
(i-)	$0.679^{20/4}$	−118.2	90.0	i.	s.	∞
	$0.687^{20/4}$	−119.4	91.8	i.	s.	∞
	$0.674^{20/4}$	−125	79.1	i.	s.	∞
	$0.675^{20/4}$	−119.4	80.8	i.	s.	∞
	$0.693^{20/4}$	−135.0	86.0	i.	s.	∞
	$0.698^{20/4}$	−118.7	93.5	i.	s.	∞
	$0.690^{20/4}$	−25	80.8	i.	s.	∞
Heptoic acid	0.918^{20}	−10	221–2	0.25^{15}	s.	s.
aldehyde	$0.850^{20/4}$	−42	155	0.02^{20}	∞	∞

Heptyl acetate (n-)	$CH_3CO_2CH_2(CH_2)_5CH_3$	158.24	col. lq.	$0.874^{16/16}$	......	191.5^{759}	i.	s.	s.
alcohol (n-)	$CH_3(CH_2)_5CH_2OH$	116.20	col. lq.	$0.824^{20/4}$	34.6	175^{756}	0.18^{25}	∞	∞
	$[(CH_3)_2CH]_2CHOH$	116.20	col. lq.	$0.829^{20/4}$	......	140	v. sl. s.	∞	∞
	$(C_2H_5.CH_2.CH_2)_2CHOH$	116.20	lq.	$0.820^{20/4}$	−37	156	i.	s.	s.
mercaptan	$CH_3CH(SH).C_5H_{11}$	132.26	lq.	0.835^{20}	......	$174–5^{765}$	i.		
Hexachloro-benzene	C_6Cl_6	284.80	mn.	2.044^{24}	228–31	309^{742}	i.	v. sl. s. h.	s. h.
-ethane	$CCl_3.CCl_3$	236.76	rhb.	$2.091^{20/4}$	186–7	186^{777}	0.005^{22}	v. s.	v. s.
Hexacosane (n-)	$CH_3(CH_2)_{24}CH_3$	366.69	cr.	$0.779^{57/4}$	56.6	262^{15}	i.	v. sl. s.	
Hexadecane (n-)	$CH_3(CH_2)_{14}CH_3$	226.43	lf.	$0.774^{20/4}$	18.5	287.5	i.	∞	∞
Hexaethylbenzene	$C_6(C_2H_5)_6$	246.42	pr./al.	$0.831^{130/4}$	130	298.3	i.	0.75^{25}	8^{25}
Hexamethylbenzene	$C_6(CH_3)_6$	162.26	pl./al.	......	166	265	i.	0.2^{0}	v. s.
Hexamethylene-diamine	$NH_2(CH_2)_6NH_2$	116.20	lf.	......	42	204–5	v. s.	s.	
-diisocyanate	$OCN(CH_2)_6NCO$	168.19	lq.	1.04^{28}	......	$143–4^{20}$	d.	d.	
-glycol	$HO(CH_2)_6OH$	118.17	nd./aq.	......	42	250	s.	s.	sl. s. h.
tetramine	$(CH_2)_6N_4$	140.19	col. rhb.	......	subl.	......	81^{12}	3	v. sl. s.
Hexane (n-)	$CH_3(CH_2)_4CH_3$	86.17	col. lq.	$0.659^{20/4}$	−94	69	0.014^{15}	50^{33}	∞
(i-)	$(CH_3)_2CH(CH_2)_2CH_3$	86.17	lq.	$0.654^{20/4}$	−153.7	60.2	i.	......	s.
(neo-)	$(CH_3)_3C.C_2H_5$	86.17	lq.	$0.649^{20/20}$	−98.2	49.7	i.	......	s.
	$(CH_3)_2CH.CH(CH_3)_2$	86.17	lq.	$0.662^{20/4}$	−129.8	58.0^{760}	i.	......	s.
	$(C_2H_5)_2CHCH_3$	86.17	lq.	$0.664^{20/4}$	−118	63.2	i.	......	s.
Hexyl acetate (n-)	$CH_3CO_2(CH_2)_5CH_3$	144.21	col. lq.	$0.890^{0/0}$	......	169.2	i.	v. s.	v. s.
alcohol (n-)	$CH_3(CH_2)_4CH_2OH$	102.17	col. lq.	$0.820^{20/20}$	−51.6	157.2	0.6^{20}	∞	∞
	$(CH_3)_2CH.C(CH_3)_2OH$	102.17	lq.	$0.821^{20/0}$	−14	120–1	v. sl. s.	∞	∞
	$(CH_3)_2COH.CH_2C_2H_5$	102.17	lq.	$0.809^{20/4}$	−107	123^{762}	v. sl. s.	∞	∞
formate (n-)	$HCO_2CH_2(CH_2)_4CH_3$	130.18	lq.	0.898^{0}	......	153.6	......	∞	∞
resorcinol (2-,4-)	$CH_3(CH_2)_5C_6H_3(OH)_2$	194.26	col. nd.	......	68–70	179^{7}	0.05	v. s.	s.
Hippuric acid	$C_6H_5CONHCH_2CO_2H$	179.17	rhb.	$1.371^{20/4}$	187–8	d.	0.4^{20}	s. h.	0.25^{18}
Histidine (l-)	$C_6H_9O_2N_3$	155.16	lf./aq.	......	d. 287	......	s.	v. sl. s.	i.
Homophthalic acid (o-)	$HO_2C.C_6H_4.CH_2CO_2H$	180.15	cr./aq.	......	175–80	......	s. h.	v. s.	sl. s.
Hydracrylic acid	$HOCH_2CH_2CO_2H$	90.08	syrup	......	......	d.			
Hydro-cyanic acid	HCN	27.03	lq.	0.697^{18}	−12	25–6	∞	∞	∞
-quinone (p-)	$C_6H_4(OH)_2$	110.11	cr.	1.332^{15}	170.3	285^{730}	6^{15}	v. s.	v. s.
Hydroxy-benzaldehyde (p-)	$HO.C_6H_4.CHO$	122.12	nd./aq.	1.129^{130}	116–7	subl.	1.38^{31}		
-benzanilide (o-)	$HO.C_6H_4.CONHC_6H_5$	213.23	pr./al.	......	135	d.	v. sl. s. h.	s.	s.
-quinoline (2-)(α-)	$C_9H_6N.OH$	145.15	pr./al.	......	199–200	subl.	s. h.	v. s.	v. s.
(8-)(o-)	$C_9H_6N.OH$	145.15	pr.	......	75–6	266.6^{752}	v. sl. s. c.	s.	sl. s.
Indigo	$[C_6H_4(CO)(NH)C:]_2$	262.26	cr.	1.35	390–2	subl.	i.	i.	i.
White	$C_{16}H_{12}O_2N_2$	264.27	gray	......	......	......	i.	s.	s.
Indole	C_8H_7N	117.14	lf./aq.	......	52	253–4	s. h.	s. h.	s.
Indoxyl	C_8H_6NOH	133.14	yel. pr.	......	85	110	s.	s.	s.
Iodo-benzene	C_6H_5I	204.02	col. lq.	$1.824^{25/4}$	−28.5	188.6	0.034^{20}	s.	∞
-phenol (p-)	IC_6H_4OH	220.02	nd./aq.	1.857^{112}	93–4	d.	sl. s.	v. s.	v. s.
Iodoform	HCI_3	393.78	yel. hex.	4.008^{17}	119	subl.	0.01^{25}	1.5^{17}	13.6^{25}
Ionone (α-)	$C_{10}H_{16}:CHCOCH_3$	192.29	col. oil	0.930^{20}	......	136.1^{17}	sl. s.	∞	∞
(β-)	$C_{10}H_{16}:CHCOCH_3$	192.29	col. oil	0.944^{20}	......	140^{18}	sl. s.	∞	∞
Irone (β-)	$C_{14}H_{22}O$	206.32	col. oil	0.939^{20}	......	144^{16}	v. sl. s.	v. s.	v. s.
Isatin	$C_6H_4 < (CO)(N) > COH$	147.13	yel. red	......	200–1	subl.	s. h.	v. s. h.	sl. s.
Isoprene	$CH_2:CH.C(CH_3):CH_2$	68.11	col. lq.	$0.681^{20/4}$	−120	34	i.	∞	∞
Ketene	$H_2C:CO$	42.04	col. gas	......	−151	−56	d.	d.	s.
Koch acid (1-)(3-,6-,8-)	$C_{10}H_4(NH_2)S_3O_9HNa_2$	427.34	cr.	......	......	......	7.2^{20}		
Lactic acid (dl-)	$CH_3CH(OH)CO_2H$	90.08	hyg.	$1.249^{15/4}$	16.8	122^{14}	∞	∞	∞
anhydride	$C_6H_{10}O_5$	162.14	yel. oil	......	......	d. 250	v. sl. s.	s.	s.
Lactide (dl-)	$C_6H_8C_4$	144.12	tri./al.	$0.862^{10/4}$	124.5	255^{757}	v. sl. s.	v. sl. s. c.	
Lactose	$C_{12}H_{22}O_{11}.H_2O$	360.31	col. rhb.	1.525^{20}	202	d.	17^{10}	i.	i.
Lauric acid	$CH_3(CH_2)_{10}CO_2H$	200.31	col. nd.	$0.869^{50/4}$	48(44)	225^{100}	i.	s.	s.
Laurone	$[CH_3(CH_2)_{10}]_2CO$	338.60	pl.	$0.809^{69/4}$	69–70	......	i.	i. c.	
Lauryl alcohol	$CH_3(CH_2)_{10}CH_2OH$	186.33	lf.	$0.831^{24/4}$	24	255–9	i.	s.	s.
Lead tetraethyl	$Pb(CH_2CH_3)_4$	323.45	col. lq.	$1.659^{18/4}$	−136	152^{291}	i.	sl. s.	∞
tetramethyl	$Pb(CH_3)_4$	267.35	col. lq.	$1.995^{20/4}$	−27.5	110^{760}	i.		∞
Lecithin (protagon)	$C_{42}H_{48}O_9PN$	778.08	wax	......	150–200 d.	......	i.	s. h.	s. h.
Lepidine (py-4)	$C_9H_6N.CH_3$	143.18	lq.	1.086^{20}	9–10	261–3	sl. s.	∞	∞
Leucine (l-)	$(CH_3)_2CHCH_2CH(NH_2)CO_2H$	131.17	cr.	1.293^{18}	295	subl.	2.2^{18}		
Levulinic acid	$CH_3CO(CH_2)_2CO_2H$	116.11	lf.	$1.140^{20/20}$	33.5	245–6	v. s.	v. s.	v. s.
Limonene (d- or l-)	$C_{10}H_{16}$	136.23	lq.	$0.842^{20/4}$	−96.9	177	i.	∞	∞
Linalool (d- or l-)	$C_{10}H_{17}OH$	154.24	col. oil	0.868^{20}	......	198–200	v. sl. s.	s.	∞
Linalyl acetate	$CH_3CO_2C_{10}H_{17}$	196.28	col. lq.	0.895^{20}	......	220^{762} d.	v. sl. s.	∞	∞
Linoleic acid	$C_{17}H_{31}CO_2H$	280.44	yel. oil	$0.903^{18/4}$	−9.5	$229–30^{16}$	i.	∞	∞
Maleic acid	$HO_2C.CH:CH.CO_2H$	116.07	mn.	1.609	130.5	135 d.	79^{25}	70^{30}	8^{25}
anhydride	$< (.CHCO)_2 > O$	98.06	cr.	1.5	57–60	202	16.3^{30}		
Malic acid (dl-)	$HO_2CCH_2CH(OH)CO_2H$	134.09	col. cr.	$1.601^{20/4}$	128–9	150 d.	144^{26}	v. s.	v. s.
(d- or l-)	$HO_2CCH_2CH(OH)CO_2H$	134.09	col. cr.	$1.595^{20/4}$	99–100	140 d.	v. s.	v. s.	8.4^{15}
Malonic acid	$H_2C(CO_2H)_2$	104.06	col. tri.	1.631^{15}	130–5 d.	......	138^{16}	42^{25}	8^{15}

Table 2. Physical Properties of Organic Compounds—(*Continued*)

Name	Formula	Formula weight	Form and color	Specific gravity	Melting point, °C.	Boiling point, °C.	Solubility in 100 parts		
							Water	Alcohol	Ether
Maltose	$C_{12}H_{22}O_{11}.H_2O$	360.31	col. nd.	1.540^{17}	d.	....	108^{25}	v. sl. s. c.	i.
Mandelic acid (dl-)	$C_6H_5CH(OH)CO_2H$	152.14	rhb./aq.	$1.300^{20/4}$	118.1	d.	16^{20}	s.	s.
Mannitol (d-)	$CH_2OH(CHOH)_4CH_2OH$	182.17	col. rhb.	$1.489^{20/4}$	166	$290–5^{3}$	13^{14}	0.01^{14}	i.
Mannose (d-)	$CH_2OH(CHOH)_4CHO$	180.16	rhb.	$1.539^{20/4}$	132	....	248^{17}	v. sl. s.	i.
Margaric acid	$CH_3(CH_2)_{15}CO_2H$	270.44	col. pl.	0.853^{60}	60–1	227^{100}	i.	32^{28}	v. s.
Mellitic acid	$C_6(CO_2H)_6$	342.17	nd./al.	....	286–8	d.	v. s.	v. s.	
Menthol (l-)(α-)	$C_{10}H_{19}OH$	156.26	col. cr.	$0.890^{15/15}$	42–3	212	0.04 c.	v. s.	v. s.
Mercapto-benzothiazole (2-)	$< C_6H_4N:C(SH)S >$	167.24	nd.	$1.42^{20/4}$	179	d.	i.	s.	sl. s.
-thiazoline (2-)	$< CH_2N:C(SH)SCH_2 >$	119.20	cr.	1.50	106	....	1.6^{60}		
Mercuric cyanide	$Hg(CN)_2$	252.65	cr.	4.003^{22}	d. 320	....	12.5^{15}		
fulminate	$Hg(ONC)_2.\frac{1}{2}H_2O$	293.65	cr./aq.	4.4	expl.	....	0.07^{12}	s.	
Mesityl oxide	$(CH_3)_2C:CHCOCH_3$	98.14	lq.	$0.858^{20/4}$	−59	130^{750}	3^{20}	∞	∞
Mesitylene (1-,3-,5-)	$C_6H_3(CH_3)_3$	120.19	col. lq.	$0.865^{20/4}$	−45(−52)	164.8	i.	s.	∞
Metanilic acid (m-)	$H_2NC_6H_4SO_3H$	173.18	col. nd.	....	d.	....	2^{15}	v. sl. s.	v. sl. s.
Methane	CH_4	16.04	gas	0.415^{-164}	−182.6	−161.4	0.4^{20} cc.	47^{20} cc.	104^{10} cc.
Methoxy-methoxyethanol	$CH_3(OCH_2)_2CH_2OH$	106.12	lq.	1.038^{25}	<−70	167.5	∞		
Methyl acetate	$CH_3CO_2CH_3$	74.08	col. lq.	$0.924^{20/4}$	−98.7	57.1	33^{22}	∞	∞
acrylic acid (α-)	$CH_2:C(CH_3)CO_2H$	86.09	pr.	$1.015^{20/4}$	15–16	161–3	s. h.	∞	∞
alcohol	CH_3OH	32.04	col. lq.	$0.792^{20/4}$	−97–8	64.7	∞	∞	∞
-amine	CH_3NH_2	31.06	col. gas	0.699^{-11}	−92.5	-6.7^{758}	v. s.	v. s.	
-amine hydrochloride	$CH_3NH_2.HCl$	67.52	pl./al.	1.23	226–8	230^{15}	v. s.	23 h.	i.
aniline	$C_6H_5NHCH_3$	107.15	lq.	$0.989^{20/4}$	−57	195.5	0.01^{25}	s.	∞
anthracene (α-)	$C_6H_4:(CH)_2:C_6H_3CH_3$	192.25	lf./al.	$1.047^{99.4}$	86	....	i.		
(β-)	$C_6H_4:(CH)_2:C_6H_3CH_3$	192.25	col. lf.	$1.181^{0/4}$	207	....	i.	v. sl. s.	v. sl. s.
anthranilate (o-)	$NH_2C_6H_4CO_2CH_3$	151.16	col. lq.	$1.168^{19/4}$	24	135.5^{15}	sl. s.	s.	s.
anthraquinone (2-)	$C_6H_4:(CO)_2:C_6H_3CH_3$	222.23	col. nd.	....	176–7	subl.	i.	s.	s.
benzoate	$C_6H_5CO_2CH_3$	136.14	col. lq.	$1.087^{25/25}$	−12.5	198–9	0.02^{30}	∞	∞
benzylaniline	$C_6H_5N(CH_3)CH_2C_6H_5$	197.27	lq.	....	9.2	305–6	i.	s.	s.
bromide	CH_3Br	94.95	gas	$1.732^{0/0}$	−93	4.5^{758}	v. sl. s.	s.	s.
butyrate (n-)	$CH_3(CH_2)_2CO_2CH_3$	102.13	col. lq.	$0.898^{20/4}$	<−95	102.3	1.7	∞	∞
(i-)	$(CH_3)_2CHCO_2CH_3$	102.13	col. lq.	$0.891^{20/4}$	−84.7	92.6	v. sl. s.	∞	∞
caprate	$CH_3(CH_2)_8CO_2CH_3$	186.29	lq.	....	−18	223–4	i.	∞	∞
caproate (n-)	$CH_3(CH_2)_4CO_2CH_3$	130.18	col. lq.	$0.904^{0/0}$	....	149.5	i.	∞	∞
caprylate	$CH_3(CH_2)_6CO_2CH_3$	158.23	col. lq.	0.887^{18}	−40	192–4	i.	∞	∞
cellosolve	$CH_3OCH_2CH_2OH$	76.09	col. lq.	$0.965^{20/4}$	....	124–5	∞	∞	∞
chloride	CH_3Cl	50.49	gas	0.952^{0}	−97.7	−24	280^{16} cc.	v. s.	v. s.
chloroacetate	$ClCH_2CO_2CH_3$	108.53	col. lq.	$1.236^{20/4}$	−32.7	130^{740}	v. sl. s.	∞	∞
chloroformate	$ClCO_2CH_3$	94.50	col. lq.	1.236^{15}	....	71–2	d.	∞	∞
cinnamate	$C_6H_5CH:CHCO_2CH_3$	162.18	cr.	$1.042^{36/0}$	33.4	263	i.	v. s.	v. s.
cyclohexane	$CH_2 < (CH_2CH_2)_2 > CHCH_3$	98.18	col. lq.	$0.769^{20/4}$	−126.3	101	i.	s.	s.
ethyl carbonate	$CH_3O.CO.OC_2H_5$	104.10	lq.	1.002^{27}	−14.5	109.2	i.	∞	∞
ethyl ketone	$CH_3.CO.C_2H_5$	72.10	col. lq.	$0.805^{20/4}$	−85.9	79.6	35^{10}	∞	∞
ethyl oxalate	$CH_3OCO.CO_2C_2H_5$	132.11	lq.	$1.156^{0/0}$	....	173.7	i.	v. s.	v. s.
formate	HCO_2CH_3	60.05	lq.	$0.974^{20/4}$	−99.8	32	30^{20}	∞	
furoate	$C_4H_3O.CO_2CH_3$	126.11	col. lq.	$1.179^{21/4}$	....	181.3	i.	∞	∞
glucamine	$CH_2OH(CHOH)_4CH_2NHCH_3$	195.21							
glycolate	$HOCH_2CO_2CH_3$	90.08	lq.	1.168^{18}	....	151.2			
heptoate	$CH_3(CH_2)_5CO_2CH_3$	144.21	lq.	$0.881^{15/4}$	....	172–3	i.		
hypochlorite	$ClOCH_3$	66.49	gas	....	....	12^{726}			
iodide	CH_3I	141.95	col. lq.	$2.279^{20/4}$	−64.4	42.4	1.8^{15}	∞	∞
lactate	$CH_3CH(OH)CO_2CH_3$	104.10	lq.	1.090^{19}	....	144.8	∞	s.	s.
laurate	$CH_3(CH_2)_{10}CO_2CH_3$	214.34	lq.	....	5	148^{18}	i.		
mercaptan	CH_3SH	48.10	gas	0.896^{0}	−121	5.8^{752}	s.	v. s.	v. s.
methacrylate	$CH_2:C(CH_3)CO_2CH_3$	100.11	lq.	$0.950^{15.6}$	−48	100.3	i.		
myristate	$CH_3(CH_2)_{12}CO_2CH_3$	242.39	cr./al.	....	18–9	295^{715}	i.		
naphthalene (α-)	$C_{10}H_7CH_3$	142.19	oil	$1.025^{14/4}$	−19	244.6	i.	v. s.	v. s.
(β-)	$C_{10}H_7CH_3$	142.19	mn.	$0.994^{40/4}$	35–6	241–2	i.	v. s.	v. s.
nitrate	CH_3ONO_2	77.04	lq.	1.203^{25}	expl.	65	sl. s.	s.	s.
nitrite	CH_3ONO	61.04	gas	0.991^{15}	....	−12	....	s.	s.
nonyl ketone (n-)	$CH_3(CH_2)_8COCH_3$	170.29	col. oil	$0.828^{20/20}$	13.5	228	i.	s.	s.
oleate	$C_{17}H_{33}CO_2CH_3$	296.48	oil	0.879^{18}	....	$190–1^{10}$	i.	∞	∞
orange	$(CH_3)_2NC_6H_4N_2C_6H_4SO_3Na$	327.33	red pd.	....	....	....	0.2 c.		
palmitate	$CH_3(CH_2)_{14}CO_2CH_3$	270.44	col. cr.	....	30–1	196^{15}	i.	s.	s.
phosphine	CH_3PH_2	48.03	gas	....	....	-14^{759}	i.	sl. s.	
propionate	$CH_3CH_2CO_2CH_3$	88.10	col. lq.	$0.915^{20/4}$	−87.5	79.7	0.5^{20}	∞	∞
propyl ketone (n-)	$CH_3COCH_2CH_2CH_3$	86.13	col. lq.	$0.812^{15/15}$	−77.8	102	v. sl. s.	∞	∞
salicylate (o-)	$HO.C_6H_4CO_2CH_3$	152.14	col. lq.	$1.182^{25/25}$	−8.3	222.2	0.07^{30}	∞	∞

stearate	$CH_3(CH_2)_{16}CO_2CH_3$	298.49	col. cr.		38–9	215^{15}	i.	s.	s.
toluate (o-)	$CH_3.C_6H_4CO_2CH_3$	150.17	col. lq.	1.073^{15}	<-50	213	i.	∞	∞
(m-)	$CH_3.C_6H_4CO_2CH_3$	150.17	col. lq.	1.066^{15}		215	i.		
(p-)	$CH_3.C_6H_4CO_2CH_3$	150.17	cr.		33–4	217	i.	v. s.	v. s.
Methyl toluidine (o-)	$CH_3.C_6H_4NHCH_3$	121.18	lq.	0.973^{15}		206–7	i.	∞	∞
(m-)	$CH_3.C_6H_4NHCH_3$	121.18	lq.			206–7	i.	∞	∞
(p-)	$CH_3.C_6H_4NHCH_3$	121.18	lq.	$0.935^{55/4}$		211^{761}	i.	∞	∞
valerate (n-)	$CH_3(CH_2)_3CO_2CH_3$	116.16	lq.	$0.895^{15/4}$	−91	127.3	v. sl. s.	∞	∞
(i-)	$(CH_3)_2CHCH_2CO_2CH_3$	116.16	col. lq.	$0.881^{20/4}$		$116–7^{764}$	v. sl. s.	∞	∞
vinyl ketone	$CH_3COCH:CH_2$	70.09	lq.	$0.836^{20/4}$		81	>85		
Methylal	$HCH(OCH_3)_2$	76.09	col. lq.	$0.866^{15/4}$	−104.8	42–3	33	∞	∞
Methylene-bis-(phenyl-4-isocyanate)	$(OCN.C_6H_4)_2CH_2$	250.25	lq.	1.222^{30}		$210–2^{13}$	d.	d.	
bromide	CH_2Br_2	173.86	col. lq.	$2.495^{20/4}$	−52.8	98.5^{756}	1.17^{0}	∞	∞
chloride	CH_2Cl_2	84.94	col. lq.	$1.336^{20/4}$	−96.7	40–1	2^{20}	∞	∞
dianiline	$(C_6H_5NH)_2CH_2$	198.26	cr.		65	208–9 d.	i.	s.	s.
iodide	CH_2I_2	267.87	col. lq.	$3.325^{20/4}$	5.7	180 d.	1.4^{20}	∞	∞
Michler's hydrol (p-,p′-)	$[(CH_3)_2NC_6H_4]_2CHOH$	270.36	gn.		96–7		i.	s. h.	s.
ketone	$[(CH_3)_2NC_6H_4]_2CO$	268.35	lf./al.		174	>360 d.	i.	sl. s.	v. sl. s.
Morphine	$C_{17}H_{19}O_3N.H_2O$	303.35	pr./al.	1.317	254 d.		0.02^{20}	sl. s.	s.
Mucic acid	$(.CHOHCHOHCO_2H)_2$	210.14	pd.		206–14		0.33^{14}	i.	i.
Mustard gas	$(ClCH_2.CH_2)_2S$	159.08	oil	$1.275^{20/4}$	13–4	217	0.07^{25}	s.	s.
Myricyl alcohol	$C_{31}H_{63}OH(?)$	452.82	cr.	0.777^{95}	88		i.	v. sl. s.	v. s.
Myristic acid	$CH_3(CH_2)_{12}CO_2H$	228.36	col. lf.	$0.853^{70/4}$	57–8	250.5^{100}	i.	v. s.	v. s.
Myristyl alcohol	$CH_3(CH_2)_{12}CH_2OH$	214.38	cr.	$0.824^{38/4}$	38	167^{15}	<0.02	sl. s.	s.
Naphthalene	$C_{10}H_8$	128.16	pl./al.	$1.145^{20/4}$	80.2	217.9	0.003^{25}	9.5^{20}	v. s.
disulfonic acid (1-,5-)	$C_{10}H_6(SO_3H)_2$	288.28	lf.		d.		102^{20}	s.	i.
(1-,6-)	$C_{10}H_6(SO_3H)_2$	288.28	cr.		d. 125		164^{20}	s.	i.
sulfonic acid (α-)	$C_{10}H_7SO_3H.2H_2O$	244.26	cr.		90		v. s.	v. s.	sl. s.
(β-)	$C_{10}H_7SO_3H.H_2O$	226.24	cr.		125		77^{30}		
Naphthasultam (1-,8-)	$C_{10}H_7O_2NS$	205.22	nd.		177–8		s. h.	sl. s.	s.
disulfonate Na (1-,8-)	$C_{10}H_5O_8NS_3Na_2.2H_2O$	445.35	cr.				v. s.		
(2-,4-)	$C_{10}H_4O_8NS_3Na_3.8\frac{1}{2}H_2O$	584.45	lf.				v. s.	sl. s.	
Naphthoic acid (α-)	$C_{10}H_7CO_2H$	172.17	nd.		160–1	300	v. sl. s. h.	s. h.	s.
(β-)	$C_{10}H_7CO_2H$	172.17	mn.	$1.077^{100/4}$	184	>300	0.007^{25}	s.	s.
Naphthol (α-)	$C_{10}H_7OH$	144.16	mn.	1.224^{4}	96	278–80	sl. s. h.	v. s.	v. s.
(β-)	$C_{10}H_7OH$	144.16	mn.	1.217^{4}	122–3	285–6	0.074^{25}	v. s.	v. s.
sulfonic acid (α-)(1-,2-)	$HO.C_{10}H_6SO_3H$	224.22	pl./aq.		>250		v. s. h.		i.
(β-)(2-,6-)	$HO.C_{10}H_6SO_3H$	224.22	lf.		125		v. s.	v. s.	
Naphthyl acetate (α-)	$CH_3CO_2C_{10}H_7$	186.20	nd./al.		46–9		sl. s. h.	s.	s.
(β-)	$CH_3CO_2C_{10}H_7$	186.20	nd./al.		69–70		i.	s.	s.
amine (α-)	$C_{10}H_7NH_2$	143.18	rhb.	$1.123^{25/25}$	50	300.8	0.17 c.	v. s.	v. s.
(β-)	$C_{10}H_7NH_2$	143.18	lf./aq.	$1.061^{98/4}$	111–2	306.1	v. s. h.	s.	s.
amine hydrochloride (α-)	$C_{10}H_7NH_2.HCl$	179.65	nd.			subl.	3.8^{20}	s.	s.
(β-)	$C_{10}H_7NH_2.HCl$	179.65	lf.				v. s.	v. s.	
amine sulfonic acid (1-,4-)	$NH_2.C_{10}H_6.SO_3H$	223.24	nd.		d.		0.2^{100}	i.	i.
(1-,5-)	$NH_2.C_{10}H_6.SO_3H.H_2O$	241.26	cr.				sl. s.		
(1-,7-)	$NH_2.C_{10}H_6.SO_3H.H_2O$	241.26	cr.				0.46^{25}		
(1-,8-)	$NH_2.C_{10}H_6.SO_3H.H_2O$	241.26	cr.				0.42^{100}		
(2-,5-)	$NH_2.C_{10}H_6.SO_3H$	223.24	cr.				0.08		
(2-,6-)	$NH_2.C_{10}H_6.SO_3H.H_2O$	241.26	cr.				0.38^{100}		
(2-,7-)	$NH_2.C_{10}H_6.SO_3H.H_2O$	241.26	cr.				0.28^{100}		
isocyanate (α-)	$C_{10}H_7N:CO$	169.17	col. lq.	1.18		269–70	d.	s.	s.
Nicotine	$C_{10}H_{14}N_2$	162.23	oil	$1.009^{20/4}$	<-80	246^{730}	s.	∞	∞
Nicotinic acid (3-)	$C_5H_4NCO_2H$	123.11	nd./al.		235.2	subl.	s. h.	s. h.	v. sl. s.
(i-)(4-)	$C_5H_4NCO_2H$	123.11	nd./aq.		317	d.	s. h.	sl. s. h.	v. sl. s.
Nitro-acetanilide (p-)	$CH_3CONHC_6H_4NO_2$	180.16	rhb.		215–6		s. h.	s.	s.
-acetophenone (m-)	$CH_3COC_6H_4NO_2$	165.14	nd.		80–1	202	i.	s.	
-aminoanisole (4-,1-,2-)	$NO_2.C_6H_3(OCH_3)NH_2$	168.15	red nd.	1.207^{156}	118		i.	s.	
(5-,1-,2-)	$NO_2.C_6H_3(OCH_3)NH_2$	168.15	yel. nd.	1.211^{156}	139–40				
(3-,1-,4-)	$NO_2.C_6H_3(OCH_3)NH_2$	168.15	red		123		sl. s.	s.	s.
-aminophenol (4-,2-,1-)	$NO_2.C_6H_3(NH_2)OH$	154.12	or. pr.		142–3		sl. s. c.	v. s.	v. s.
-aniline (o-)	$NO_2.C_6H_4NH_2$	138.12	yel. rhb.	1.442^{15}	71.5	284.1	s. h.	v. s.	v. s.
(m-)	$NO_2.C_6H_4NH_2$	138.12	yel. rhb.	1.43	114	306.4	0.11^{20}	7.1^{20}	7.9^{20}
(p-)	$NO_2.C_6H_4NH_2$	138.12	yel. mn.	1.437^{14}	146–7	331.7	0.08^{19}	5.8^{20}	6.1^{20}
-anisole (o-)	$CH_3OC_6H_4NO_2$	153.13	col. cr.	$1.254^{20/4}$	9.4	272–3	0.17^{30}	∞	∞
(p-)	$CH_3OC_6H_4NO_2$	153.13	pr./al.	1.233^{20}	54	274	0.06^{30}	v. s.	v. s.
-anthraquinone (α-)	$C_6H_4:(CO)_2:C_6H_3NO_2$	253.20	nd.		230	270^{7}	i.	sl. s.	v. sl. s.
-anthraquinone sulfonic acid (1-,5-)	$NO_2.C_{14}H_6O_2.SO_3H$	333.26	yel. cr.				s	i.	i.
-benzal chloride (m-)	$NO_2.C_6H_4.CHCl_2$	206.03	mn.		65		i.	v. s. h.	v. s.
-benzaldehyde (m-)	$NO_2.C_6H_4CHO$	151.12	nd./aq.		58	164^{23}	1.95^{112}	v. s. h.	v. s.

Table 2. Physical Properties of Organic Compounds—(*Continued*)

Name	Formula	Formula weight	Form and color	Specific gravity	Melting point, °C.	Boiling point, °C.	Solubility in 100 parts: Water	Alcohol	Ether
Nitro-benzene	$C_6H_5NO_2$	123.11	yel. lq.	$1.205^{18/4}$	5.7	210.9	0.19^{20}	v. s.	∞
-benzidine (2-)	$NH_2C_6H_4C_6H_3(NH_2)NO_2$	229.23	red nd.		143		sl. s. h.		
-benzoic acid (o-)	$NO_2.C_6H_4.CO_2H$	167.12	tri./aq.	$1.575^{4/4}$	147.5		0.65^{20}	28^{11}	22^{11}
(m-)	$NO_2.C_6H_4.CO_2H$	167.12	mn.	$1.494^{4/4}$	140–1		0.24^{165}	$3.^{12}$	25^{10}
(p-)	$NO_2.C_6H_4.CO_2H$	167.12	yel. mn.	$1.550^{32/4}$	240–2	subl.	0.02^{15}	0.9^{10}	2.2^{13}
-benzyl alcohol (m-)	$NO_2.C_6H_4.CH_2OH$	153.13	cr.		27	$175–80^{3}$			
-benzyl bromide (p-)	$NO_2.C_6H_4CH_2Br$	216.04	nd./al.		99–100		i.	2^{19}	v. s.
-chlorotoluene (1-,2-,6-)	$CH_3.C_6H_3(NO_2)Cl$	171.58	cr.		37.5	238	i.		
-cresol (1-,3-,4-)	$CH_3.C_6H_3(NO_2)OH$	153.13	yel.	$1.240^{39/4}$	32	125^{22}	v. sl. s.	v. s.	v. s.
-cymene (1-,2-,4-)	$CH_3.C_6H_3(NO_2)CH(CH_3)_2$	179.21	oil	$1.067^{20/4}$		152^{15}	i.		
-dimethylaniline (o-)	$NO_2.C_6N_4NHCH_3$	166.18	yel. oil	$1.179^{20/4}$		$151–3^{30}$	v. sl. s.	v. s.	v. s.
(m-)	$NO_2.C_6H_4NHCH_3$	166.18	red mn.	1.313^{17}	60–1	280–5	i.	s.	s.
(p-)	$NO_2.C_6H_4NHCH_3$	166.18	yel. nd.		163–4		i.	s. h.	
-diphenyl (o-)	$C_6H_5.C_6H_4NO_2$	199.20	rhb.	1.44	37	320	i.	s.	v. s.
(p-)	$C_6H_5.C_6H_4NO_2$	199.20	nd./al.		113–4	340	i.	sl. s. c.	v. s.
-diphenylamine (o-)	$C_6H_5.NH.C_6H_4NO_2$	214.22	or. cr.		75–6				
-guanidine	$H_2NC(NH)NHNO_2$	104.07	nd./aq.		246–7		9^{100}	sl. s.	v. sl. s.
-naphthalene (α-)	$C_{10}H_7NO_2$	173.16	yel./al.	1.223^{62}	59–60	304	i.	s.	s.
(β-)	$C_{10}H_7NO_2$	173.16	col./al.		79	165^{15}	i.	v. s.	v. s.
-phenol (o-)	$NO_2.C_6H_4.OH$	139.11	yel. mn.	1.295^{45}	44–5	214.5	1.08^{100}	v. s.	v. s.
(m-)	$NO_2.C_6H_4.OH$	139.11	col. mn.	1.485^{20}	96–7	194^{70}	1.35^{20}	v. s.	s.
(p-)	$NO_2.C_6H_4.OH$	139.11	yel. pr.	1.479^{20}	113–4	subl.	1.6^{25}	v. s.	v. s.
-phenol sulfonic acid (1-,4-,2-)	$HO.C_6H_3(NO_2)SO_3H.3H_2O$	273.22	nd.		d. 110		v. s.	v. s.	sl. s
(1-,2-,4-)	$HO.C_6H_3(NO_2)SO_3H.3H_2O$	273.22	nd./aq.		51.5		v. s.	v. s.	
-phthalic acid (3-)	$NO_2.C_6H_3(CO_2H)_2$	211.13	yel./aq.		222		2.05^{25}	v. s. h.	sl. s.
(4-)	$NO_2.C_6H_3(CO_2H)_2$	211.13	yel. cr.		164–5		v. s.	v. s.	s.
-toluene (o-)	$CH_3.C_6H_4NO_2$	137.13	yel. lq.	$1.163^{20/4}$	−4.1	222.3	0.07^{30}	∞	∞
(m-)	$CH_3.C_6H_4NO_2$	137.13	lq.	$1.160^{18/4}$	15–16	230–1	0.05^{30}	∞	∞
(p-)	$CH_3.C_6H_4NO_2$	137.13	rhb.	$1.139^{55/55}$	51.9	237.7	0.04^{30}	8.6^{15}	80.8^{15}
-toluene sulfonic acid (1-,4-,2-)	$CH_3.C_6H_3(NO_2)SO_3H.2H_2O$	253.23	pl./aq.		130		47.7^{23}	v. s.	v. s.
-toluidine (4-,1-,2-)	$NO_2.C_6H_3(CH_3)NH_2$	152.15	yel. mn.	1.365^{15}	105–7		v. sl. s.	s.	s.
(3-,1-,4-)	$NO_2.C_6H_3(CH_3)NH_2$	152.15	red mn.	1.312^{17}	116–7		sl. s. h.	s.	
Nitron	$C_{20}H_{16}N_4$	312.36	yel. lf.		189–90 d.		i.	s. h.	v. sl. s.
Nitroso-dimethylaniline (p-)	$ON.C_6H_4N(CH_3)_2$	150.18	gn. tri.		86–7		i.	s.	s.
-naphthol (β-)(1-)	$ON.C_{10}H_6OH$	173.16	brn. pr.		109.5		0.1^{20}	2.4^{13}	
Nonadecane (n-)	$CH_3(CH_2)_{17}CH_3$	268.51	cr.	$0.777^{32/4}$	32	330	i.	sl. s.	s.
Nonane (n-)	$CH_3(CH_2)_7CH_3$	128.25	col. lq.	$0.718^{20/4}$	−53.7	150.5^{759}	i.	sl. s.	s.
Octadecane (n-)	$CH_3(CH_2)_{16}CH_3$	254.48	cr.	$0.775^{28/4}$	28	317	i.	sl. s.	s.
Octane (n-)	$CH_3(CH_2)_6CH_3$	114.22	col. lq.	$0.703^{20/4}$	−56.5	125.7	0.002^{16}	sl. s.	s.
(iso-)	$(CH_3)_3CCH_2CH(CH_3)_2$	114.22	col. lq.	$0.692^{20/4}$	−107.4	99.3^{760}	i.	sl. s.	s.
Octyl acetate (n-)	$CH_3CO_2CH_2(CH_2)_6CH_3$	172.26	col. lq.	$0.885^{0/4}$	−38.5	210	i.	s.	s.
(sec-)	$CH_3CO_2CH(CH_3)C_6H_{13}$	172.26	col. lq.	$0.863^{14/4}$		195	i.	s.	s.
alcohol (n-)	$CH_3(CH_2)_6CH_2OH$	130.22	col. lq.	$0.827^{20/4}$	−16	194–5	0.054^{25}	∞	∞
(sec-)	$CH_3(CH_2)_5CH(OH)CH_3$	130.22	col. lq.	$0.822^{20/4}$	−38.6	179–80	0.096^{25}	∞	∞
Octylene (n-)	$CH_3(CH_2)_5CH{:}CH_2$	112.21	lq.	$0.721^{18/4}$		126	i.	∞	∞
Oleic acid	$C_8H_{17}CH{:}CH(CH_2)_7CO_2H$	282.45	col. nd.	$0.854^{78/4}$	14	$285–6^{100}$	i.	∞	∞
Orcinol (1-,3-,5-)	$(HO)_2C_6H_3.CH_3$	124.13	pr./bz.	1.290^{4}	107–8	287–90	v. s.	v. s.	v. s.
Oxalic acid	$HO_2C.CO_2H.2H_2O$	126.07	col. mn.	$1.653^{19/4}$	101.5	subl.	s.	s.	1.3
Palmitic acid	$CH_3(CH_2)_{14}CO_2H$	256.42	col. pl.	$0.849^{70/4}$	63–4	271.5^{100}	i.	9^{20}	s.
Pelargonic acid	$CH_3(CH_2)_7CO_2H$	158.23	col. oil	$0.906^{20/4}$	12.5	253–4	v. sl. s.	s.	s.
Penta-chloroethane	$CHCl_2.CCl_3$	202.31	col. lq.	$1.671^{25/4}$	−22	162	0.05^{20}	∞	∞
-decane (n-)	$CH_3(CH_2)_{13}CH_3$	212.41	col. lq.	$0.770^{20/4}$	10	270.5	i.	v. s.	v. s.
-erythritol	$C(CH_2OH)_4$	136.15	cr.		262	276^{30}	5.6^{15}	v. sl. s.	i.
Pentandiol	$HOCH_2(CH_2)_3CH_2OH$	104.15	lq.	$0.994^{20/4}$		239.4	∞		
Pentane (n-)	$CH_3(CH_2)_3CH_3$	72.15	col. lq.	$0.630^{18/4}$	−129.7	36.3	0.036^{16}	∞	∞
(i-)	$(CH_3)_2CHCH_2CH_3$	72.15	col. lq.	0.621^{19}	−160.0	27.95	i.	∞	∞
(neo-)	$(CH_3)_2C(CH_3)_2$	72.15	col. lq.	$0.613^{20/4}$	−20	9.5	i.	s.	s.
Phenacetin	$C_2H_5OC_6H_4NHCOCH_3$	179.21	col. mn.		134–5	d.	0.7^{20}	40 h.	1.6^{25}
Phenanthrene	$<(C_6H_4CH)_2>$	178.22	pl./al.	1.179^{25}	99–100	340	i.	10 h.	v. s.
Phenetidine (o-)	$C_2H_5O.C_6H_4.NH_2$	137.18	oil		<−21	228–9	i.	s.	s.
(p-)	$C_2H_5O.C_6H_4.NH_2$	137.18	lq.	1.061^{15}	3–4	254–5	i.	s.	s.
Phenetole	$C_2H_5O.C_6H_5$	122.16	col. lq.	$0.967^{20/4}$	−30.2	172	i.	∞	∞
Phenol	C_6H_5OH	94.11	col. nd.	$1.071^{25/4}$	42–3	181.4	8.2^{15}	∞	∞
-phthalein	$C_{20}H_{14}O_4$	318.31	col. rhb.	$1.299^{25/4}$	261–2		0.2^{20}	10^{25}	5.9 c.
-sulfonic acid (o-)	$HO.C_6H_4SO_3H.\frac{3}{4}H_2O$	187.68	cr.		50 d.		v. s.	v. s.	
Phenyl acetaldehyde	$C_6H_5CH_2CHO$	120.14	lq.	1.025^{20}		193–4	v. sl. s.	∞	∞
acetic acid	$C_6H_5CH_2CO_2H$	136.14	lf.	$1.081^{30/4}$	76–7	265.5	1.66^{20}	v. s.	v. s.

-acetylene	$C_6H_5C{:}CH$	102.13	col. lq.	0.930$^{20/4}$	−43	142–3	i.	∞	∞
aniline (o-)	$C_6H_5.C_6H_4.NH_2$	169.22	cr.	....	45–6	299^{750}	v. sl. s.	s.	s.
(p-)	$C_6H_5.C_6H_4.NH_2$	169.22	lf.	....	50–2	302	s. h.	s.	s.
Phenyl-ethyl alcohol	$C_6H_5CH_2CH_2OH$	122.16	col. oil	1.023$^{13/4}$	....	219–21^{750}	1.6^{20}	s.	∞
-glycine	$C_6H_5NHCH_2CO_2H$	151.16	cr.	....	127	....	s.	s.	sl. s.
-hydrazine	$C_6H_5NH.NH_2$	108.14	yel. oil	1.097$^{23/4}$	19.6	243.5	sl. s. h.	∞	∞
-hydrazine sulfonic acid (p-)	$H_2NNHC_6H_4SO_3H$	188.20	cr./al.	....	286	....	0.6^{12}	sl. s.	
isocyanate	$C_6H_5N{:}CO$	119.12	lq.	1.096$^{20/4}$	....	166^{769}	d.	d.	v. s.
-methylpyrazolone (3-)(N-)	$C_4H_5ON_2.C_6H_5$	174.20	pr./aq.	....	128	191^{17}	I^{20}	v. s. h.	v. sl. s.
-mustard oil	$C_6H_5N{:}CS$	135.18	col. lq.	1.138$^{15/15}$	−21	219–20	i.	s.	s.
naphthalene (α-)	$C_{10}H_7.C_6H_5$	204.26	waxy	....	45	336–7	i.	v. s.	v. s.
(β-)	$C_{10}H_7.C_6H_5$	204.26	lf./al.	....	102.5	345–6	i.	sl. s.	sl. s.
naphthylamine (α-)	$C_{10}H_7NHC_6H_5$	219.27	pr./al.	1.17	62	335^{258}	0.08^{60}	s.	s.
(β-)	$C_{10}H_7NHC_6H_5$	219.27	rhb.	1.18	107–8	399.5	0.4^{60}	v. s. h.	v. s. h.
phenol (o-)	$C_6H_5.C_6H_4OH$	170.20	nd.	....	56–7	275	i.	s.	s.
(p-)	$C_6H_5.C_6H_4OH$	170.20	nd.	....	164–5	305–8	i.	s.	s.
propyl alcohol (γ-)	$C_6H_5(CH_2)_3OH$	136.19	oil	1.008$^{20/4}$	< −18	235–7	sl. s.	∞	∞
quinoline (2-)(α-)	$C_6H_5.C_9H_6N$	205.25	nd.	....	86	363	sl. s.	s. h.	s.
(8-)(0-)	$C_6H_5.C_9H_6N$	205.25	lq.	....	....	283^{187}	sl. s.	s.	s.
salicylate, salol	$HO.C_6H_4CO_2C_6H_5$	214.21	rhb./al.	1.250$^{20/4}$	42–3	172–3^{12}	0.015^{25}	v. s. h.	s.
stearate	$CH_3(CH_2)_{16}CO_2C_6H_5$	360.56	cr.	....	52	267^{15}	i.		
urethane	$C_6H_5NHCO_2C_2H_5$	165.19	pl./al.	1.106$^{30/4}$	52–3	237–8	i. c.	s.	s.
Phenylene-diamine (o-)	$C_6H_4(NH_2)_2$	108.14	lf./aq.	....	103–4	256–8	733^{81}	v. s.	v. s.
(m-)	$C_6H_4(NH_2)_2$	108.14	rhb.	1.139$^{15/15}$	62.8	284–7	35.1^{25}	v. s.	s.
(p-)	$C_6H_4(NH_2)_2$	108.14	mn.	....	140	267	669^{107}	s.	s.
Phloroglucinol (1-,3-,5-)	$C_6H_3(OH)_3.2H_2O$	162.14	rhb.	....	117	subl.	1.13^{25}	v. s.	v. s.
Phorone	$[(CH_3)_2C{:}CH]_2CO$	138.20	yel. pr.	0.885$^{20/4}$	28	197.2^{743}	0.1^{50}	s.	s.
Phosgene	$OCCl_2$	98.92	gas	1.392$^{19/4}$	−104	8.2^{756}	v. sl. s.		
Phthalic acid (o-)	$C_6H_4(CO_2H)_2$	166.13	mn./aq.	1.593$^{20/4}$	208	d.	0.70^{25}	12^{18}	0.68^{15}
(m-)(iso-)	$C_6H_4(CO_2H)_2$	166.13	nd./aq.	....	330	subl.	0.2^{100}	s.	
anhydride (o-)	$C_6H_4 < (CO)_2 > O$	148.11	rhb.	1.527^{4}	130.8	284.5	v. sl. s.	s.	sl. s.
nitrile (o-)	$C_6H_4(CN)_2$	128.13	cr.	....	141	....	sl. s. c.		
Phthalide	$C_6H_4(CH_2)(CO) > O$	134.13	nd./aq.	1.164$^{99/4}$	73(65)	290	v. sl. s.	s.	
Phthalimide (o-)	$C_6H_4 < (CO)_2 > NH$	147.13	cr./et.	....	238	subl.	0.04^{25}	5	s. h.
Picoline (α-)	$C_5H_4N.CH_3$	93.12	col. lq.	0.950$^{15/4}$	−70	128.8	v. s.	∞	∞
(β-)	$C_5H_4N.CH_3$	93.12	col. lq.	0.961$^{15/4}$	....	143.5	∞	∞	∞
(γ-)	$C_5H_4N.CH_3$	93.12	lq.	0.957$^{15/4}$	....	143.1	∞	∞	∞
Picramic acid (1-,2-,4-,6-)	$HO.C_6H_2(NH_2)(NO_2)_2$	199.12	red nd.	....	169	....	0.14^{22}	s.	sl. s.
Picric acid (2-,4-,6-)	$HO.C_6H_2(NO_2)_3$	229.11	yel. rhb.	1.763$^{20/4}$	121.8	expl.	1.23^{20}	6^{20}	1^{13}
Picryl chloride (2-,4-,6-)	$ClC_6H_2(NO_2)_3$	247.56	yel. mn.	1.797^{20}	83	d.	0.018^{15}	4.8^{17}	7^{17}
Pinacol	$[(CH_3)_2C.OH]_2$	118.17	col. nd.	0.967^{15}	43(38)	171–2^{739}	sl. s. c.	v. s.	v. s.
Pinacoline	$CH_3COC(CH_3)_3$	100.16	col. lq.	0.800^{16}	−52.5	106.2	2.5^{15}	s.	s.
Pinene (α-)(dl-)	$C_{10}H_{16}$	136.23	col. lq.	0.878$^{20/4}$	−55	154–6	v. sl. s.	s.	∞
hydrochloride	$C_{10}H_{17}Cl$	172.69	lf.	....	131–2	207–8	i.	33	s.
Pinol (dl-)	$C_{10}H_{16}O$	152.23	lq.	0.953$^{20/20}$	....	183–4	....	s.	s.
Piperidine	$CH_2 < (CH_2CH_2)_2 > NH$	85.15	lq.	0.860$^{20/4}$	−9	106	∞	∞	
carboxylic acid (α-)(dl-)	$HO_2C.CH < (CH_2CH_2)_2 > NH$	129.16	cr.	....	264	....	s.		
Piperidinium pentamethylene dithiocarbamate	$(CH_2)_5CS_2H.HN(CH_2)_5$	232.41	cr.	1.13	175	....	6^{28}		
Propane	$CH_3CH_2CH_3$	44.09	gas	0.585$^{-45/4}$	−187.1	−42.2	6.5^{18} cc.	s.	v. s.
Propionic acid	$CH_3CH_2CO_2H$	74.08	col. lq.	0.992$^{20/4}$	−22	141.1	∞	∞	∞
aldehyde	CH_3CH_2CHO	58.08	col. lq.	0.807$^{20/4}$	−81	49.5^{740}	20^{20}	∞	∞
anhydride	$(CH_3CH_2CO)_2O$	130.14	col. lq.	1.012$^{20/4}$	−45	168.8^{780}	d.	d.	
Propyl acetate (n-)	$CH_3COCH_2CH_2CH_3$	102.13	col. lq.	0.886$^{20/4}$	−92.5	101.6	1.6^{16}	∞	∞
(i-)	$CH_3CO_2CH(CH_3)_2$	103.13	col. lq.	0.874$^{20/20}$	−73.4	88.4	3^{20}	∞	∞
alcohol (n-)	$CH_3CH_2CH_2OH$	60.09	col. lq.	0.804$^{20/4}$	−127	97.8	∞	∞	∞
(i-)	$(CH_3)_2CHOH$	60.09	col. lq.	0.789$^{20/4}$	−85.8	82.5	∞	∞	∞
amine (n-)	$CH_3CH_2CH_2NH_2$	59.11	col. lq.	0.718$^{20/20}$	−83	49–50^{761}	∞	∞	∞
(i-)	$(CH_3)_2CHNH_2$	59.11	col. lq.	0.694$^{15/4}$	−101	33–4	∞	∞	∞
aniline (n-)	$C_6H_5NHCH_2CH_2CH_3$	135.20	lq.	0.949^{18}	....	222	i.	v. s.	v. s.
benzoate (n-)	$C_6H_5CO_2CH_2CH_2CH_3$	164.20	col. lq.	1.021$^{25/25}$	−51.6	231	i.	s.	s.
(i-)	$C_6H_5CO_2CH(CH_3)_2$	164.20	col. lq.	1.010$^{25/25}$	....	218.5	i.	s.	s.
bromide (n-)	$CH_3CH_2CH_2Br$	123.00	col. lq.	1.353$^{20/4}$	−109.9	70.8	0.25^{20}	∞	∞
(i-)	$(CH_3)_2CHBr$	123.00	col. lq.	1.310$^{20/4}$	−89	60	0.32^{20}	∞	∞
n-butyrate (n-)	$C_2H_5CH_2CO_2CH_2C_2H_5$	130.18	col. lq.	0.879^{15}	−95.2	142.7	0.17^{17}	∞	∞
i-butyrate (n-)	$(CH_3)_2CHCO_2CH_2C_2H_5$	130.18	col. lq.	0.884$^{0/4}$	....	134–5	v. sl. s.		
n-butyrate (i-)	$C_2H_5CH_2CO_2CH(CH_3)_2$	130.18	col. lq.	0.865^{13}	....	128	v. sl. s.		
i-butyrate (i-)	$(CH_3)_2CHCO_2CH(CH_3)_2$	130.18	col. lq.	0.869$^{0/4}$	....	120.8	v. sl. s.		
chloride (n-)	$CH_3CH_2CH_2Cl$	78.54	col. lq.	0.890$^{20/4}$	−122.8	46.4	0.27^{20}	∞	∞
(i-)	$(CH_3)_2CHCl$	78.54	col. lq.	0.859^{20}	−117	36.5	0.31^{20}	∞	∞

Table 2. Physical Properties of Organic Compounds—(*Continued*)

Name	Formula	Formula weight	Form and color	Specific gravity	Melting point, °C.	Boiling point, °C.	Solubility in 100 parts: Water	Alcohol	Ether
Propyl formate (n-)	$HCO_2CH_2CH_2CH_3$	88.10	col. lq.	$0.901^{20/4}$	−92.9	81.3	2.2^{22}	∞	∞
(i-)	$HCO_2CH(CH_3)_2$	88.10	col. lq.	$0.873^{20/4}$	....	$68\text{-}71^{751}$	2.1^{22}	∞	∞
furoate (n-)	$C_4H_3O.CO_2C_3H_7$	154.16	col. lq.	$1.075^{26/4}$	....	211	v. sl. s.	s.	∞
lactate (n-)	$CH_3CH(OH)CO_2CH_2C_2H_5$	132.16	col. lq.	....	....	$122\text{-}3^{150}$	s.	s.	s.
(i-)	$CH_3CH(OH)CO_2CH(CH_3)_2$	132.16	col. lq.	....	....	167.5	s.	s.	s.
mercaptan (n-)	$CH_3CH_2CH_2SH$	76.15	lq.	$0.836^{25/4}$	−112	67–8	v. sl. s.	s.	s.
(i-)	$(CH_3)_2CHSH$	76.15	lq.	$0.809^{25/4}$	−130.7	58–60	v. sl. s.	∞	∞
propionate (n-)	$C_2H_5CO_2CH_2C_2H_5$	116.16	col. lq.	$0.883^{20/4}$	−76	122–3	0.56^{25}	∞	∞
(i-)	$C_2H_5CO_2CH(CH_3)_2$	116.16	col. lq.	0.893^{0}	....	$109\text{-}11^{750}$	0.6^{25}	∞	∞
thiocyanate (i-)	$(CH_3)_2CH.CNS$	101.16	lq.	0.963^{20}	....	$152\text{-}3^{754}$	i.	∞	∞
n-valerate (n-)	$CH_3(CH_2)_3CO_2CH_2C_2H_5$	144.21	lq.	0.874^{15}	−70.7	67.5	i.	∞	∞
i-valerate (n-)	$(CH_3)_2CHCH_2CO_2C_3H_7$	144.21	col. lq.	$0.863^{20/4}$	....	155.9	i.	∞	∞
i-valerate (i-)	$(CH_3)_2CHCH_2CO_2C_3H_7$	144.21	col. lq.	0.854^{17}	....	142^{756}			
Propylene	$CH_3CH:CH_2$	42.08	gas	$0.609^{-47/4}$	−185	-48^{749}	44.6 cc.	1200 cc.	
bromide	$CH_3CHBrCH_2Br$	201.91	col. lq.	$1.933^{20/4}$	−55.5	141.6	0.25^{20}	s.	v. s.
chlorohydrin	$CH_3CHClCH_2OH$	94.54	col. lq.	1.103^{20}	....	133–4	s.	s.	s.
chloride	$CH_3CHClCH_2Cl$	112.99	col. lq.	$1.159^{20/20}$	< −70	96.8	0.27^{20}	v. s.	v. s.
glycol	$CH_3CH(OH)CH_2OH$	76.09	col. oil	$1.040^{19.4}$	....	188–9	∞	∞	8
oxide	$CH_3(CHCH_2)O$	58.08	col. lq.	$0.831^{20/20}$	....	35	33^{20}	∞	∞
Protocatechuic acid (3-,4-)	$(HO)_2C_6H_3CO_2H.H_2O$	172.13	nd./aq.	$1.542^{4/4}$	199 d.	....	1.82^{14}	v. s.	s.
Pulegol (iso-)(d-)	$C_{10}H_{17}OH$	154.24	col. lq.	$0.911^{20/4}$	....	$86\text{-}9^{10}$	v. sl. s.		
Pulegone	$C_{10}H_{16}O$	152.23	col. lq.	$0.932^{20/20}$	....	224^{754}	i.	∞	∞
Pyrazole	—NH.N:CH.CH:CH—	68.08	nd./et.	....	70	186–8	s.	s.	s.
Pyrazoline	$—NH.N:CH.CH_2CH_2—$	70.09	lq.	....	....	144	∞	∞	sl. s.
Pyrazolone	$—NH.CO.CH_2CH:N—$	84.08	nd.	....	165	subl. d.	s.	v. s.	v. sl. s.
Pyrene	$C_{16}H_{10}$	202.24	yel. pr.	$1.277^{0/4}$	149–50	>360	i.	3 h.	v. s.
Pyridazine	$N_2 < (CHCH)_2 >$	80.09	lq.	$1.107^{20/4}$	−8	208	∞	s.	s.
Pyridine	$CH < (CHCH)_2 > N$	79.10	col. lq.	$0.982^{20/4}$	−42	115–6	∞	∞	s.
Pyrocatechol (o-)	$C_6H_4(OH)_2$	110.11	nd./aq.	1.344^{4}	104–5	240–5	45.1^{20}	v. s.	v. s.
Pyrogallol (1-,2-,3-)	$C_6H_3(OH)_3$	126.11	nd.	1.453^{4}	133–4	309	40^{13}	s.	s.
Pyrone	$CO < (CHCH)_2 > O$	96.08	cr.	$1.190^{40.3}$	32.5	215–7	v. sl. s.	s.	v. s.
Pyrrole	$< (CH:CH)_2 > NH$	67.09	lq.	$0.948^{20/4}$	....	131	i.	s.	s.
Pyrrolidine	$< (CH_2.CH_2)_2 > NH$	71.12	lq.	$0.852^{22.5}$	....	87–8	∞	∞	∞
Pyrroline	$< (CH.CH_2)_2 > NH$	69.10	lq.	$0.910^{20/4}$	....	90–1	v. s.	∞	∞
Pyruvic acid	CH_3COCO_2H	88.06	col. lq.	$1.267^{20/4}$	13.6	165	∞	∞	∞
Quercitrin	$C_{21}H_{20}O_{11}.2H_2O$	484.40	yel. nd.	....	182–5	....	0.04^{20}	s.	sl. s.
Quinaldine (py-2)	$CH_3.C_9H_6N$	143.18	lq.	$1.059^{20/4}$	−1	$244\text{-}5^{750}$	v. sl. s.	....	s.
Quinoline	C_9H_7N	129.15	lq.	1.095^{20}	−15	237.1^{747}	6	∞	∞
(iso-)	C_9H_7N	129.15	pl.	$1.099^{21/4}$	24.6	240.5^{763}	sl. s.	....	s.
-diol (1-,3-)	$—C_6H_4CH:C(OH)N:C(OH)—$	161.15	cr.	....	237	....	v. sl. s.		
Quinone (p-)	$CO < (CHCH)_2 > CO$	108.09	yel. mn.	$1.318^{20/4}$	115.7	subl.	sl. s. h.	s.	s.
R-acid Ca salt (2-)(3-,6-)	$HOC_{10}H_5(SO_3)_2Ca$	342.35	cr.	....	....	....	30.6^{25}		
K salt	$HOC_{10}H_5(SO_3K)_2$	380.46	cr.	....	....	....	29.5^{25}		
Na salt	$HOC_{10}H_5(SO_3Na)_2$	348.26	cr.	....	....	....	25.2^{25}		
Raffinose	$C_{18}H_{32}O_{16}.5H_2O$	594.52	cr./aq.	1.465^{0}	119	d. 130	14.3^{20}	0.1^{20}	
Resorcinol (m-)	$C_6H_4(OH)_2$	110.11	col. rhb.	1.272^{15}	110.7	276.5	147^{12}	v. s.	v. s.
Retene	$C_{18}H_{18}$	234.32	lf./al.	1.13^{16}	98–9	390–4	i.	69 h.	v. s. h.
Rhamnose (β-)	$CH_3(CHOH)_4CHO.H_2O$	182.17	col. mn.	$1.471^{20/4}$	126	....	60.8^{21}	....	i.
Ricinoleic acid	$C_{17}H_{32}(OH)CO_2H$	298.45	lq.	0.954^{16}	4–5	$226\text{-}8^{10}$	i.	∞	∞
Rosaniline	$C_{20}H_{21}ON_3$	319.39	col. nd.	....	186 d.	....	v. sl. s.	sl. s.	i.
Rosolic acid	$C_{20}H_{16}O_3$	304.33	red lf.	....	308–10 d.	....	0.12^{25}	v. s. h.	sl. s.
Saccharin	$C_6H_4(CO)(SO_2) > NH$	183.18	mn.	....	225–8	subl.	0.4^{25}	3.1 c.	1.05 c.
Safrole (1-,3-,4-)	$CH_2:CHCH_2.C_6H_3:O_2CH_2$	162.18	col. mn.	$1.100^{20/4}$	11.2	233–4	i.	s.	∞
(iso-)(1-,3-,4-)	$CH_3.CH:CH.C_6H_3:O_2CH_2$	162.18	col. lq.	$1.122^{20/4}$	6–7	252–3	i.	∞	∞
Salicylic acid (o-)	$HO.C_6H_4.CO_2H$	138.12	mn.	$1.443^{20/4}$	159	211^{20}	0.2^{23}	49^{15}	51^{15}
aldehyde (o-)	$HO.C_6H_4.CHO$	122.12	col. oil	$1.153^{25/4}$	−7	196.5	1.7^{86}	∞	∞
Saligenin	$HO.C_6H_4.CH_2OH$	124.13	rhb./aq.	1.161^{25}	86–7	subl.	6.6^{15}	v. s.	v.s.
Schaeffer's salt, Ca	$(HOC_{10}H_6SO_3)_2Ca.5H_2O$	576.59	cr.	....	....	....	4.76^{20}		
K	$HOC_{10}H_6SO_3K$	262.31	cr.	....	....	....	3.46^{25}		
Na	$HOC_{10}H_6SO_3Na$	246.21	cr.	....	....	....	6.29^{25}		
Semicarbazide	$NH_2.CO.NH.NH_2$	75.07	pr./al.	....	96	....	v. s.	v. s.	i.
hydrochloride	$NH_2.CO.NH.NH_3Cl$	111.54	pr.	....	173 d.	....	v. s.	sl. s.	i.
Skatole (3-)	$CH_3.C_8H_6N$	131.17	lf.	....	95	$265\text{-}6^{755}$	0.05 c.	s.	s.
Sodium methylate	CH_3ONa	54.03	pd.	....	d. 300	....	d.		
Sorbitol	$[CH_2OH(CHOH)_2]_2$	182.17	cr.	....	110–2	....	v. s.	v. s. h.	
Sorbose (d- or l-)	$C_6H_{12}O_6$	180.16	rhb.	1.654^{15}	165	....	55^{17}	sl. s.	
Starch	$(C_6H_{10}O_5)x$	162.14	amor.	1.50^{21}	d.	....	i.	i.	i.

Stearic acid	$CH_3(CH_2)_{16}CO_2H$	284.47	mn.	$0.847^{69.3}$	70–1	291^{110}	0.03^{25}	2^{20}	6^{5}
amide	$CH_3(CH_2)_{16}CONH_2$	283.48	col. cr.	....	108–9	251^{12}	i.	s. h.	s. h.
Styrene	$C_6H_5CH{:}CH_2$	104.14	col. lq.	$0.903^{20/4}$	−31	145–6	v. sl. s.	∞	∞
Suberic acid	$HO_2C(CH_2)_6CO_2H$	174.19	nd./aq.	$1.266^{25/4}$	140–4	279^{100}	0.14^{16}	s.	0.8^{15}
Succinic acid	$HO_2C(CH_2)_2CO_2H$	118.09	col. mn.	$1.572^{25/4}$	189–90	235 d.	6.8^{20}	9.9^{15}	1.2^{15}
Sucrose	$C_{12}H_{22}O_{11}$	342.30	col. mn.	1.588^{15}	170–86 d.	....	179^{0}	0.9	i.
Sulfanilic acid (p-)	$H_2N.C_6H_4.SO_3H$	173.18	col. cr.	....	d. > 280	....	0.8^{10}	v. sl. s.	v. sl. s.
Sylvestrene (d-)	$C_{10}H_{16}$	136.23	lq.	$0.863^{20/4}$	....	176–7			
Tartaric acid (meso-)	$(CHOHCO_2H)_2$	150.09	cr.	1.737	159–60	....	120^{15}		
(racemic)	$(CHOHCO_2H)_2.H_2O$	168.10	tri.	$1.697^{20/4}$	205–6	....	20.6^{20}	2^{0}	0.09
(d- or l-)	$(CHOHCO_2H)_2$	150.09	mn.	$1.760^{20/4}$	168–70	d.	139^{20}	25^{15}	0.4^{15}
Tartronic acid	$CH(OH)(CO_2H)_2.\frac{1}{2}H_2O$	129.07	pr./aq.	....	d. 155–8	subl.	v. s.	v. s.	i.
Terephthalic acid (p-)	$C_6H_4(CO_2H)_2$	166.13	cr.	1.510	subl.	....	0.001 c.	sl. s. h.	i.
Terpin hydrate (cis-)	$C_{10}H_{20}O_2.H_2O$	190.28	rhb.	....	117	d.	0.4^{15}	10^{15}	1^{15}
Terpineol (α-)(d- or l-)	$C_{10}H_{18}O$	154.24	col. cr.	0.935^{15}	38–40	219–21	i.	v. s.	v. s.
(dl-)	$C_{10}H_{18}O$	154.24	col. cr.	$0.935^{20/20}$	35	$218–9^{752}$	i.	v. s.	v. s
Terpinyl acetate (α-)(dl-)	$CH_3CO_2.C_{10}H_{17}$	196.28	lq.	$0.966^{20/4}$	< −50	220 d.	i.	20	
Tetrabromo-ethane (sym)	$Br_2CH.CHBr_2$	345.70	col. lq.	$2.964^{20/4}$	−1.0	151^{54}	i.	∞	∞
(uns)	$Br_3C.CH_2Br$	345.70	col. lq.	$2.875^{20/4}$	0	104^{13}	....	s.	
Tetrachloro-ethane (sym)	$Cl_2CH.CHCl_2$	167.86	col. lq.	$1.600^{20/4}$	−36	146.3	0.29^{20}	∞	∞
(uns)	$Cl_3C.CH_2Cl$	167.86	lq.	$1.588^{20/4}$	....	129–30	i.	∞	∞
-ethylene	$Cl_2C{:}CCl_2$	165.85	col. lq.	$1.624^{15/4}$	−19	120.8	0.02^{20}	∞	∞
Tetracosane (n-)	$CH_3(CH_2)_{22}CH_3$	338.64	cr.	$0.779^{51/4}$	51.1	324	....	....	s.
Tetradecane (n-)	$CH_3(CH_2)_{12}CH_3$	198.38	col. lq.	$0.765^{20/4}$	5.5	252.5	i.	v. s.	v. s.
Tetraethyl-thiuram disulfide	$[(C_2H_5)_2NCS]_2S_2$	296.52	cr.	1.17	70	....	i.		
Tetrafluoro-ethylene	$F_2C{:}CF_2$	100.02	gas	1.58^{-78}	−142.5	−76.3	0.01^{30}		
Tetrahydro-furan	$—CH_2(CH_2)_2CH_2.O—$	72.10	col. lq.	$0.888^{21/4}$	−65	65–6	s.	s.	s.
-furfuryl alcohol	$C_4H_7O.CH_2OH$	102.13	col. lq.	$1.050^{20/4}$	....	$177–8^{743}$	∞	∞	∞
-pyran	$—CH_2(CH_2)_3CH_2.O—$	86.13	lq.	$0.881^{20/4}$	....	88	s.		
Tetralin	$—C_6H_4CH_2(CH_2)_2CH_2—$	132.20	col. lq.	$0.973^{18/4}$	−31	206^{754}	i.	s.	s.
Tetramethyl-thiuram disulfide	$[(CH_3)_2NCS]_2S_2$	240.41	cr.	1.29	155–6	....	i.		
Tetryl (2-,4-,6-)	$(NO_2)_3C_6H_2.N(CH_3)NO_2$	287.15	yel. mn.	1.57^{19}	130.5	expl.	i.	s. h.	s.
Theobromine	$C_7H_8O_2N_4$	180.17	rhb.	....	330	....	0.06^{15}	0.06 c.	0.03 h.
Thio-acetic acid	$CH_3.CO.SH$	76.11	yel. lq.	1.074^{10}	< −17	93	s.	∞	∞
-aniline (4-, 4′-)	$(NH_2.C_6H_4)_2S$	216.29	nd./aq.	....	108	....	sl. s. h.	s.	s.
-carbanilide	$(C_6H_5.NH)_2CS$	228.30	rhb./al.	1.3^{24}	154	d.	i.	v. s.	v. s.
-naphthol (β-)	$C_{10}H_7.SH$	160.22	cr./al.	....	81	286–8	v. sl. s.	v. s.	v. s.
-phenol	$C_6H_5.SH$	110.17	col. lq.	$1.074^{23/4}$	....	168–9	v. sl. s.	v. s.	∞
-salicylic acid (o-)	$HS.C_6H_4.CO_2H$	154.18	yel. nd.	....	164	subl.	sl. s. h.	s.	
-urea	$NH_2.CS.NH_2$	76.12	rhb./al.	$1.405^{20/4}$	180–2	d.	9.2^{13}	s.	sl. s.
Thiophene	$<(CH{:}CH)_2>S$	84.13	col. lq.	$1.070^{15/4}$	−30	84	i.	s.	
Thymol (5-,2-,1-)	$(CH_3)(C_3H_7)C_6H_3OH$	150.21	cr.	$0.972^{25/25}$	51.5	232^{752}	0.09^{19}	v. s.	v. s.
Tolidine (0-)(3-,3′-,4-,4′-)	$[CH_3(NH_2)C_6H_3]_2$	212.28	lf.	....	128–9	....	v. sl. s.	s.	s.
Toluene	$C_6H_5.CH_3$	92.13	col. lq.	$0.866^{20/4}$	−95	110.8	0.05^{16}	s.	∞
sulfonic acid (o-)	$CH_3.C_6H_4SO_3H.2H_2O$	208.23	cr.	....	d.	128.8^{0}	v. s.	s.	
(p-)	$CH_3.C_6H_4SO_3H.H_2O$	190.21	mn.	....	104–5	$146–7^{0}$	v. s.	s.	
sulfonic amide (p-)	$CH_3.C_6H_4SO_2NH_2$	171.21	mn.	....	137	....	0.2^{9}	7.4^{5}	
sulfonic chloride (p-)	$CH_3.C_6H_4.SO_2Cl$	190.64	tri.	....	69	134.5^{10}	i.	s.	s.
Toluic acid (o-)	$CH_3.C_6H_4.CO_2H$	136.14	cr./aq.	$1.062^{115/4}$	104–5	259^{751}	2.17^{100}	v. s.	
(m-)	$CH_3.C_6H_4.CO_2H$	136.14	pr./aq.	$1.054^{112/4}$	110–1	263	1.6^{100}	v. s.	v s.
(p-)	$CH_3.C_6H_4.CO_2H$	136.14	cr./aq.	....	179–80	274–5	1.3^{100}	v. s.	v. s.
Toluidine (o-)	$CH_3.C_6H_4.NH_2$	107.15	col. lq.	$0.999^{20/4}$	−16.3	199.7	1.5^{25}	∞	∞
(m-)	$CH_3.C_6H_4.NH_2$	107.15	col. lq.	$0.989^{20/4}$	−31.5	203.3	sl. s.	∞	∞
(p-)	$CH_3.C_6H_4.NH_2$	107.15	cr.	$1.046^{20/4}$	44–5	200.3	0.74^{21}	v s.	v. s.
hydrochloride (o-)	$CH_3.C_6H_4.NH_3Cl$	143.62	mn. pr.	....	218–20	242	s.	sl. s.	
sulfonic acid (1-,2-,3-)	$CH_3(NH_2)C_6H_3SO_3H$	187.21	cr.	....	....	....	0.97^{11}		
Toluylenediamine (1-,2-,4-)	$CH_3.C_6H_3(NH_2)_2$	122.17	rhb.	....	99	283–5	s. h.	s.	s.
Tolylene diisocyanate (1-,2-,4-)	$CH_3.C_6H_3(NCO)_2$	174.15	lq.	1.23^{28}	....	134.5^{20}	d.	d.	
Trehalose	$C_{12}H_{22}O_{11}.2H_2O$	378.33	rhb./al.	....	97	....	s. h.	sl. s. h.	i.
Triamylamine (n-)	$[CH_3(CH_2)_3CH_2]_3N$	227.42	lq.	....	....	240–5	i.		
(i-)	$[(CH_3)_2CH(CH_2)_2]_3N$	227.42	col. lq.	$0.786^{20/4}$	....	235	i.		
Tributyl-amine (n-)	$[CH_3(CH_2)_2CH_2]_3N$	185.34	col. lq.	$0.778^{20/20}$	....	216.5^{761}	i.	s.	∞
phosphite	$[CH_3(CH_2)_3O]_3P$	250.32	lq.	$0.925^{20/4}$	....	$122–3^{12}$	i.		
Trichloro-acetic acid	$Cl_3C.CO_2H$	163.40	cr.	$1.617^{46/15}$	58	195.5^{754}	120^{25}	s.	s.
-benzene (s-)(1-,3-,5-)	$C_6H_3Cl_3$	181.46	nd.	....	63.5	208.5^{764}	i.	sl. s.	
-ethane (1-,1-,1-)	$Cl_3C.CH_3$	133.42	lq.	$1.325^{26/4}$	....	74.1	i.	∞	∞
-ethylene	$Cl_2C{:}CHCl$	131.40	col. lq.	$1.466^{20/20}$	−73	87.2	0.1^{25}	∞	∞
-phenol	$Cl_3C_6H_2OH$	197.46	nd.	$1.490^{75/4}$	68–9	246	0.09^{25}	v. s.	v. s.
Tricosane (n-)	$CH_3(CH_2)_{21}CH_3$	324.61	lf.	$0.779^{48/4}$	47.7	234^{15}	i.		
Tricresyl phosphate (o-)	$OP(OC_6H_4CH_3)_3$	368.36	lq.	....	....	....	i.		
Tridecane (n-)	$CH_3(CH_2)_{11}CH_3$	184.35	col. lq.	$0.757^{20/4}$	−6.2	234	i.	v. s.	v. s.
Triethanol amine	$(HOCH_2CH_2)_3N$	149.19	col. lq.	$1.126^{20/20}$	20–1	$277–9^{150}$	∞	∞	sl. s.

Table 2. Physical Properties of Organic Compounds—(*Concluded*)

Name	Formula	Formula weight	Form and color	Specific gravity	Melting point, °C.	Boiling point, °C.	Solubility in 100 parts		
							Water	Alcohol	Ether
Triethyl-amine	$(CH_3CH_2)_3N$	101.19	col. oil	$0.729^{20/20}$	−114.8	89.4	∞ > 19^0	∞	∞
-benzene (1-,3-,5-)	$(C_2H_5)_3C_6H_3$	162.26	lq.	$0.861^{20/4}$	....	215	i.	s.	s.
(1-,2-,4-)	$(C_2H_5)_3C_6H_3$	162.26	lq.	$0.882^{17/4}$	....	$217–8^{755}$	i.	s.	s.
borate	$B(OCH_2CH_3)_3$	146.00	lq.	$0.864^{20/20}$	....	120	d.		
citrate	$HOC_3H_4(CO_2C_2H_5)_3$	276.28	oil	$1.137^{20/4}$	....	294	i.	∞	∞
Triethylene glycol	$(.CH_2OCH_2CH_2OH)_2$	150.17	col. lq.	$1.125^{20/20}$	−5	290	∞	∞	v. sl. s.
Trifluoro-chloromethane	CF_3Cl	104.47	gas	1.726^{-130}	−182	−80			
chloroethylene	$F_2C{:}CFCl$	116.48	gas	....	−157.5	−27.9	d.		
-trichloroethane	$Cl_2CF.CClF_2$	187.39	lq.	$1.576^{20/4}$	−35	47.6	i.	∞	∞
Trimethoxybutane (1-,3-,3-)	$CH_2(OCH_3)CH_2C(OCH_3)_2CH_3$	148.20	lq.	0.932	....	$63–5^{25}$	d.		
Trimethylamine	$(CH_3)_3N$	59.11	gas	0.662^{-5}	−124	3.5	41^{19}	s.	s.
Trimethylene bromide	$BrCH_2CH_2CH_2Br$	201.91	lq.	$1.987^{15/4}$	−34.4	167.5	0.17^{30}	s.	s.
chloride	$ClCH_2CH_2CH_2Cl$	112.99	lq.	1.201^{15}	....	123–5	0.27^{25}	s.	s.
glycol	$HOCH_2CH_2CH_2OH$	76.09	oil	$1.060^{20/4}$	....	214	∞	∞	
Trinitro-benzene (1-,3-,5-)	$C_6H_3(NO_2)_3$	213.11	col. rhb.	$1.688^{20/4}$	121	d.	0.03^{15}	1.9^{18}	1.5^{18}
-benzoic acid (2-,4-,6-)	$(NO_2)_3C_6H_2CO_2H$	257.12	rhb./aq.	....	210–20 d.	....	2.05^{24}		
-tert-butylxylene	$(NO_2)_3C_6(CH_3)_2C_4H_9$	297.26	nd./al.	....	110	....	i.	sl. s.	s.
-naphthalene (α-)(1-,3-,5-)	$C_{10}H_5(NO_2)_3$	263.16	rhb.	....	122–3	....	i.	s.	
(β-)(1-,3-,8-)	$C_{10}H_5(NO_2)_3$	263.16	cr./al.	....	218–9	....	0.02^{100}	0.05^{23}	0.13^{15}
(γ-)(1-,4-,5-)	$C_{10}H_5(NO_2)_3$	263.16	yel. cr.	....	148–9	....	i.	0.11^{19}	0.4^{19}
-phenol (2-,3-,6-)	$(NO_2)_3C_6H_2OH$	229.11	nd.	....	117–8	....	s. h.	v. s.	v. s.
-toluene (β-)(2-,3-,4-)	$CH_3C_6H_2(NO_2)_3$	227.13	cr.	$1.620^{20/4}$	112	expl.	i.	sl. s. c.	s.
(γ-)(2-,4-,5-)	$CH_3C_6H_2(NO_2)_3$	227.13	yel. pl.	$1.620^{20/4}$	104	expl.	i.	s. h.	v. s.
(α-)(2-,4-,6-)	$CH_3C_6H_3(NO_2)_3$	227.13	cr./al.	1.654	80.8	expl.	0.01^{20}	1.5^{22}	5^{33}
Trional	$C_2H_5(CH_3)C(SO_2C_2H_5)_3$	242.34	pl./al.	$1.199^{85/4}$	76	d.	0.3^{15}	5^0	6.6^{15}
Triphenyl-arsine	$(C_6H_5)_3As$	306.21	pl.	1.306	59–60	>360	i.	s.	v. s.
carbinol	$(C_6H_5)_3COH$	260.32	cr.	$1.188^{20/4}$	162.5	>360	i.	v. s.	v. s.
guanidine (α-)	$C_6H_5N{:}C(NHC_6H_5)_2$	287.35	rhb./al.	1.13	144–5	d.	i.	4^0	
methane	$(C_6H_5)_3CH$	244.32	cr.	$1.014^{99/4}$	93.4	359^{754}	i.	v. s. h.	v. s.
methyl	$(C_6H_5)_3C$ . . .	243.31	col. cr.	....	145–7	d.	i.	sl. s. h.	
phosphate	$OP(OC_6H_5)_3$	326.28	pr./al.	$1.206^{58/4}$	49–50	245^{11}	i.	155^{25}	v. s.
Tripropylamine (n-)	$(CH_3CH_2CH_2)_3N$	143.27	col. lq.	$0.757^{20/4}$	−93.5	156.5	v. sl. s.	∞	∞
Undecane (n-)	$CH_3(CH_2)_9CH_3$	156.30	col. lq.	$0.741^{20/4}$	−25.6	194.5	i.	∞	∞
Urea	$H_2N.CO.NH_2$	60.06	col. pr.	$1.335^{20/4}$	132.7	d.	100^{17}	20^{20}	sl. s.
nitrate	$CO(NH_2)_2.HNO_3$	123.07	col. mn.	....	152 d.	....	v. s. h.	s.	
Uric acid	$C_5H_4O_3N_4$	168.11	cr.	1.893^{20}	d.	....	0.06 h.	i.	i.
Valeric acid (n-)	$C_2H_5CH_2CH_2CO_2H$	102.13	col. lq.	$0.939^{20/4}$	−34.5	187	3.3^{16}	∞	∞
(i-)	$(CH_3)_2CHCH_2CO_2H$	102.13	col. lq.	$0.931^{20/20}$	−37.6	176	4.2^{20}	∞	∞
aldehyde (n-)	$C_2H_5CH_2CH_2CHO$	86.13	lq.	0.819^{11}	−92	103.4	v. sl. s.	s.	s.
(i-)	$(CH_3)_2CHCH_2CHO$	86.13	col. lq.	0.803^{17}	−51	92.5	sl. s.	s.	s.
amide (n-)	$C_2H_5CH_2CH_2CONH_2$	101.15	mn. pl.	1.023	106	....	v. s.	v. s.	v. s.
(i-)	$(CH_3)_2CHCH_2CONH_2$	101.15	mn.	$0.965^{20/4}$	135–7	232	s.	s.	s.
Vanillic acid (3-,4-,1-)	$CH_3O(OH)C_6H_3CO_2H$	168.14	nd./aq.	....	207	subl.	0.12^{14}	v. s.	v. s.
alcohol (3-,4-,1-)	$CH_3O(OH)C_6H_3CH_2OH$	154.16	mn./aq.	....	115	d.	v. s. h.	v. s.	v. s.
Vanillin (3-,4-,1-)	$CH_3O(OH)C_6H_3CHO$	152.14	mn.	1.056	81–2	285	1^{14}	v. s.	v. s.
Veratrole (o-)	$C_6H_4(OCH_3)_2$	138.16	cr.	$1.091^{15/15}$	22.5	207.1	v. sl. s.	s.	s.
Vinyl acetate	$CH_3CO_2CH{:}CH_2$	86.09	col. lq.	$0.932^{20/4}$	<−60	72–3	2^{20}	∞	∞
(poly-)	$(CH_3CO_2CH{:}CH_2)x$	(86.09)	....	1.19^{20}	100–25	....	i.		
acetic acid	$CH_2{:}CH.CH_2CO_2H$	86.09	col. lq.	$1.013^{15/15}$	−39	163	s.	∞	∞
acetylene	$CH_2{:}CH.C{:}CH$	52.07	gas	$0.705^{1.5}$	....	5.5	$0.67^{0.6}$		
alcohol	$CH_2{:}CHOH$	44.06							
(poly-)	$(CH_2{:}CHOH)x$	(44.06)	....	1.3^{20}	d. >200	....	s.		
chloride	$CH_2{:}CHCl$	62.50	gas	$0.908^{25/25}$	−160	−12	sl. s.	s.	v. s.
propionate	$C_2H_5CO_2CH{:}CH_2$	100.11	lq.	....	....	93–5	v. sl. s.		
Xylene (o-)	$C_6H_4(CH_3)_2$	106.16	col. lq.	$0.881^{20/4}$	−25	144	i.	s.	∞
(m-)	$C_6H_4(CH_3)_2$	106.16	col. lq.	$0.867^{17/4}$	−47.4	139.3	i.	s.	∞
(p-)	$C_6H_4(CH_3)_2$	106.16	col. lq.	$0.861^{20/4}$	13.2	138.5	i.	s.	v. s.
sulfonic acid (1-,4-,2-)	$(CH_3)_2C_6H_3SO_3H.2H_2O$	222.25	col. lf.	....	86	$149^{0.1}$	s.		
Xylidine (1:2)(3-)	$(CH_3)_2C_6H_3NH_2$	121.18	lq.	0.991^{15}	<−15	223	v. sl. s.	s.	s.
(1:2)(4-)	$(CH_3)_2C_6H_3NH_2$	121.18	pr.	$1.076^{17.5}$	49–50	224–6	v. sl. s.		
(1:3)(2-)	$(CH_3)_2C_6H_3NH_2$	121.18	lq.	0.980^{15}	....	216–7	v. sl. s.		
(1:3)(4-)	$(CH_3)_2C_6H_3NH_2$	121.18	lq.	$0.978^{20/4}$	....	213–4	v. sl. s.		
(1:3)(5-)	$(CH_3)_2C_6H_3NH_2$	121.18	oil	$0.972^{20/4}$	....	221–2	v. sl. s.		
(1:4)(2-)	$(CH_3)_2C_6H_3NH_2$	121.18	oil	$0.979^{21/4}$	15.5	215^{739}	v. sl. s.		
Xylose (l-)(+)	$CH_2OH(CHOH)_3CHO$	150.13	nd.	1.535^0	153–4	....	117^{20}	v. sl. s.	i.
Xylylene dichloride (p-)	$C_6H_4(CH_2Cl)_2$	175.06	mn.	1.417^0	100.5	240–5 d.	i.	s.	v. sl. s.
Zinc diethyl	$Zn(CH_2CH_3)_2$	123.50	col. lq.	1.182^{18}	−28	118	d.	d.	
dimethyl	$Zn(CH_3)_2$	95.45	col. lq.	1.386^{11}	−40	46	d.	d.	
dimethyl-dithiocarbamate	$Zn[S_2CN(CH_3)_2]_2$	305.79	....	$2.00^{40/4}$	248–50	....	i.		

VAPOR PRESSURES OF PURE SUBSTANCES

Table 3. Vapor Pressure of Water Ice from −15° to 0°C.

Mm. Hg

t, °C.	0.0	0.1	0.2	0.3	0.4	0.5	0.6	0.7	0.8	0.9
−14	1.361	1.348	1.336	1.324	1.312	1.300	1.288	1.276	1.264	1.253
−13	1.490	1.477	1.464	1.450	1.437	1.424	1.411	1.399	1.386	1.373
−12	1.632	1.617	1.602	1.588	1.574	1.559	1.546	1.532	1.518	1.504
−11	1.785	1.769	1.753	1.737	1.722	1.707	1.691	1.676	1.661	1.646
−10	1.950	1.934	1.916	1.899	1.883	1.866	1.849	1.833	1.817	1.800
− 9	2.131	2.112	2.093	2.075	2.057	2.039	2.021	2.003	1.985	1.968
− 8	2.326	2.306	2.285	2.266	2.246	2.226	2.207	2.187	2.168	2.149
− 7	2.537	2.515	2.493	2.472	2.450	2.429	2.408	2.387	2.367	2.346
− 6	2.765	2.742	2.718	2.695	2.672	2.649	2.626	2.603	2.581	2.559
− 5	3.013	2.987	2.962	2.937	2.912	2.887	2.862	2.838	2.813	2.790
− 4	3.280	3.252	3.225	3.198	3.171	3.144	3.117	3.091	3.065	3.039
− 3	3.568	3.539	3.509	3.480	3.451	3.422	3.393	3.364	3.336	3.308
− 2	3.880	3.848	3.816	3.785	3.753	3.722	3.691	3.660	3.630	3.599
− 1	4.217	4.182	4.147	4.113	4.079	4.045	4.012	3.979	3.946	3.913
− 0	4.579	4.542	4.504	4.467	4.431	4.395	4.359	4.323	4.287	4.252

Table 4. Vapor Pressure of Liquid Water from −16° to 0°C.*

Mm. Hg

t, °C.	0.0	0.1	0.2	0.3	0.4	0.5	0.6	0.7	0.8	0.9
−15	1.436	1.425	1.414	1.402	1.390	1.379	1.368	1.356	1.345	1.334
−14	1.560	1.547	1.534	1.522	1.511	1.497	1.485	1.472	1.460	1.449
−13	1.691	1.678	1.665	1.651	1.637	1.624	1.611	1.599	1.585	1.572
−12	1.834	1.819	1.804	1.790	1.776	1.761	1.748	1.734	1.720	1.705
−11	1.987	1.971	1.955	1.939	1.924	1.909	1.893	1.878	1.863	1.848
−10	2.149	2.134	2.116	2.099	2.084	2.067	2.050	2.034	2.018	2.001
− 9	2.326	2.307	2.289	2.271	2.254	2.236	2.219	2.201	2.184	2.167
− 8	2.514	2.495	2.475	2.456	2.437	2.418	2.399	2.380	2.362	2.343
− 7	2.715	2.695	2.674	2.654	2.633	2.613	2.593	2.572	2.553	2.533
− 6	2.931	2.909	2.887	2.866	2.843	2.822	2.800	2.778	2.757	2.736
− 5	3.163	3.139	3.115	3.092	3.069	3.046	3.022	3.000	2.976	2.955
− 4	3.410	3.384	3.359	3.334	3.309	3.284	3.259	3.235	3.211	3.187
− 3	3.673	3.647	3.620	3.593	3.567	3.540	3.514	3.487	3.461	3.436
− 2	3.956	3.927	3.898	3.871	3.841	3.813	3.785	3.757	3.730	3.702
− 1	4.258	4.227	4.196	4.165	4.135	4.105	4.075	4.045	4.016	3.986
− 0	4.579	4.546	4.513	4.480	4.448	4.416	4.385	4.353	4.320	4.289

* Computed from the above table with the aid of the thermodynamic equation

$$\log_{10} \frac{p_w}{p_i} = \frac{-1.1489t}{273.1 + t} - 1.330 \times 10^{-5}t^2 + 9.084 \times 10^{-8}t^3$$

Table 5. Vapor Pressure of Liquid Water from 0° to 100°C.*

Mm. Hg

t, °C.	0.0	0.1	0.2	0.3	0.4	0.5	0.6	0.7	0.8	0.9
0	4.579	4.613	4.647	4.681	4.715	4.750	4.785	4.820	4.855	4.890
1	4.926	4.962	4.998	5.034	5.070	5.107	5.144	5.181	5.219	5.256
2	5.294	5.332	5.370	5.408	5.447	5.486	5.525	5.565	5.605	5.645
3	5.685	5.725	5.766	5.807	5.848	5.889	5.931	5.973	6.015	6.058
4	6.101	6.144	6.187	6.230	6.274	6.318	6.363	6.408	6.453	6.498
5	6.543	6.589	6.635	6.681	6.728	6.775	6.822	6.869	6.917	6.965
6	7.013	7.062	7.111	7.160	7.209	7.259	7.309	7.360	7.411	7.462
7	7.513	7 565	7.617	7.669	7.722	7.775	7.828	7.882	7.936	7.990
8	8.045	8.100	8.155	8.211	8.267	8.323	8.380	8.437	8.494	8.551
9	8.609	8.668	8.727	8.786	8.845	8.905	8.965	9.025	9.086	9.147

* From th Physikalisch-technische Reichsanstalt, Holborn, Scheel and Henning, "Wärmetabellen," Friedrich Vieweg & Sohn, Brunswick, 1909. By permission.

Table 5. Vapor Pressure of Liquid Water from 0° to 100°C.*—(*Concluded*)

t, °C.	0.0	0.1	0.2	0.3	0.4	0.5	0.6	0.7	0.8	0.9
10	9.209	9.271	9.333	9.395	9.458	9.521	9.585	9.649	9.714	9.779
11	9.844	9.910	9.976	10.042	10.109	10.176	10.244	10.312	10.380	10.449
12	10.518	10.588	10.658	10.728	10.799	10.870	10.941	11.013	11.085	11.158
13	11.231	11.305	11.379	11.453	11.528	11.604	11.680	11 756	11.833	11.910
14	11.987	12.065	12.144	12.223	12.302	12.382	12.462	12.543	12.624	12.706
15	12.788	12.870	12.953	13.037	13.121	13.205	13.290	13.375	13.461	13.547
16	13.634	13.721	13.809	13.898	13.987	14.076	14.166	14.256	14.347	14.438
17	14.530	14.622	14.715	14.809	14.903	14.997	15.092	15.188	15.284	15.380
18	15.477	15.575	15.673	15.772	15.871	15.971	16.071	16.171	16.272	16.374
19	16.477	16.581	16.685	16.789	16.894	16.999	17.105	17.212	17.319	17.427
20	17.535	17.644	17.753	17.863	17.974	18.085	18.197	18.309	18.422	18.536
21	18.650	18.765	18.880	18.996	19.113	19.231	19.349	19.468	19.587	19.707
22	19.827	19.948	20.070	20.193	20.316	20.440	20.565	20.690	20.815	20.941
23	21.068	21.196	21.324	21.453	21.583	21.714	21.845	21.977	22.110	22.243
24	22.377	22.512	22.648	22.785	22.922	23.060	23.198	23.337	23.476	23.616
25	23.756	23.897	24.039	24.182	24.326	24.471	24.617	24.764	24.912	25.060
26	25.209	25.359	25.509	25.660	25.812	25.964	26.117	26.271	26.426	26.582
27	26.739	26.897	27.055	27.214	27.374	27.535	27.696	27.858	28.021	28.185
28	28.349	28.514	28.680	28.847	29.015	29.184	29.354	29.525	29.697	29.870
29	30.043	30.217	30.392	30.568	30.745	30.923	31.102	31.281	31.461	31.642
30	31.824	32.007	32.191	32.376	32.561	32.747	32.934	33.122	33 312	33.503
31	33.695	33.888	34.082	34.276	34.471	34.667	34.864	35.062	35.261	35.462
32	35.663	35.865	36.068	36.272	36.477	36.683	36.891	37.099	37.308	37.518
33	37 729	37.942	38.155	38.369	3 .584	38.801	39.018	39.237	39.457	39.677
34	39.898	40.121	40.344	40.569	40.796	41.023	41.251	41.480	41.710	41.942
35	42.175	42.409	42.644	42.880	43.117	43.355	43.595	43.836	44.078	44.320
36	44.563	44.808	45 054	45.301	45.549	45.799	46.050	46.302	46.556	46.811
37	47.067	47.324	47.582	47.841	48.102	48.364	48.627	48.891	49.157	49.424
38	49.692	49.961	50.231	50.502	50.774	51.048	51.323	51.600	51.879	52.160
39	52.442	52.725	53.009	53.294	53.580	53.867	54.156	54.446	54.737	55.030
40	55.324	55.61	55.91	56.21	56.51	56.81	57.11	57.41	57.72	58.03
41	58.34	58.65	58.96	59.27	59.58	59.90	60.22	60.54	60.86	61.18
42	61.50	61.82	62.14	62.47	62.80	63.13	63.46	63.79	64.12	64.46
43	64.80	65.14	65.48	65.82	66.16	66.51	66.86	67.21	67.56	67.91
44	68.26	68.61	68.97	69.33	69.69	70.05	70.41	70.77	71.14	71.51
45	71.88	72.25	72.62	72.99	73.36	73.74	74.12	74.50	74.88	75.26
46	75.65	76.04	76.43	76.82	77.21	77.60	78.00	78.40	78.80	79.20
47	79.60	80.00	80.41	80.82	81.23	81.64	82.05	82.46	82.87	83.29
48	83.71	84.13	84.56	84.99	85.42	85.85	86.28	86.71	87.14	87.58
49	88.02	88.46	88.90	89.34	89.79	90.24	90.69	91.14	91.59	92.05

t, °C.	0	1	2	3	4	5	6	7	8	9
50	92.51	97.20	102.09	107.20	112.51	118.04	123.80	129.82	136.08	142.60
60	149.38	156.43	163.77	171.38	179.31	187.54	196.09	204.96	214.17	223.73
70	233.7	243.9	254.6	265.7	277.2	289.1	301.4	314.1	327.3	341.0
80	355.1	369.7	384.9	400.6	416.8	433.6	450.9	468.7	487.1	506.1
90	525.76	527.76	529.77	531.78	533.80	535.82	537.86	539.90	541.95	544.00
91	546.05	548.11	550.18	552.26	554.35	556.44	558.53	560.64	562.75	564.87
92	566.99	569.12	571.26	573.40	575.55	577.71	579.87	582.04	584.22	586.41
93	588.60	590.80	593.00	595.21	597.43	599.66	601.89	604.13	606.38	608.64
94	610.90	613.17	615.44	617.72	620.01	622.31	624.61	626.92	629.24	631.57
95	633.90	636.24	638.59	640.94	643.30	645.67	648.05	650.43	652.82	655.22
96	657.62	660.03	662.45	664.88	667.31	669.75	672.20	674.66	677.12	679.69
97	682.07	684.55	687.04	689.54	692.05	694.57	697.10	699.63	702.17	704.71
98	707.27	709.83	712.40	714.98	717.56	720 15	722.75	725.36	727.98	730.61
99	733.24	735.88	738.53	741.18	743.85	746.52	749.20	751.89	754.58	757.29
100	760.00	762.72	765.45	768.19	770.93	773.68	776.44	779.22	782.00	784.78
101	787.57	790.37	793.18	796.00	798.82	801.66	804.50	807.35	810.21	813.08

Table 6. Vapor Pressures of Inorganic Compounds, above 1 Atm.*

Compound		Pressure, atm.									Critical point	
		1	2	5	10	20	30	40	50	60		
Name	Formula	Temperature, °C.									t_c, °C.	P_c, atm.
Ammonia	NH_3	−33.6	−18.7	+4.7	25.7	50.1	66.1	78.9	89.3	98.3	132.4	111.5
Carbon monoxide	CO	−191.3	−183.5	−170.7	−161.0	−149.7	−141.9	....	....	....	−138.7	34.6
dioxide	CO_2	−78.2	−69.1	−56.7	−39.5	−18.9	−5.3	+5.9	14.9	22.4	31.1	73.0
disulfide	CS_2	46.5	69.1	104.8	136.3	175.5	201.5	222.8	240.0	256.0	273.0	72.9
Chlorine	Cl_2	−33.8	−16.9	+10.3	35.6	65.0	84.8	101.6	115.2	127.1	144.0	76.1
para-Hydrogen	H_2	−252.5	−250.2	−246.0	−241.8	....	....	....	....	....	−240.0	12.80
Hydrogen bromide	HBr	−66.5	−51.5	−29.1	−8.4	+16.8	33.9	48.1	60.0	70.6	90.0	84.4
chloride	HCl	−84.8	−71.4	−50.5	−31.7	−8.8	+5.9	17.8	27.9	36.2	51.4	81.6
cyanide	HCN	25.9	45.8	75.8	102.7	135.0	153.8	169.9	183.5	....	183.5	50.0
Water	H_2O	100.0	120.1	152.4	180.5	213.1	234.6	251.1	264.7	276.5	374.2	218.0
Hydrogen sulfide	H_2S	−60.4	−45.9	−22.3	−0.4	+25.5	41.9	55.8	66.7	76.3	100.3	88.9
Krypton	Kr	−152.0	−143.5	−130.0	−118.0	−101.7	−88.8	−78.4	−66.5	....	−63	54
Nitrogen	N_2	−195.8	−189.2	−179.1	−169.8	−157.6	−148.3	....	....	....	−147.2	33.5
Oxygen	O_2	−183.1	−176.0	−164.5	−153.2	−140.0	−130.7	−124.1	....	....	−118.9	49.7
Sulfur dioxide	SO_2	−10.0	+6.3	32.1	55.5	83.8	102.6	118.0	130.2	141.7	157.2	77.7
trioxide	SO_3	44.8	60.0	82.5	104.0	138.0	157.8	175.0	187.8	198.0	218.3	83.6

* Compiled from the extended tables published by D. R. Stull in *Ind. Eng. Chem.*, **39**, 517 (1947).

Table 7. Vapor Pressures of Inorganic Compounds, up to 1 Atm.*

Compound: Name	Compound: Formula	Pressure, mm. Hg: 1	5	10	20	40	60	100	200	400	760	Melting point, °C.
		Temperature, °C.										
Aluminum	Al	1284	1421	1487	1555	1635	1684	1749	1844	1947	2056	660
borohydride	$Al(BH_4)_3$	......	−52.2	−42.9	−32.5	−20.9	−13.4	−3.9	+11.2	28.1	45.9	−64.
bromide	$AlBr_3$	81.3	103.8	118.0	134.0	150.6	161.7	176.1	199.8	227.0	256.3	97.
chloride	Al_2Cl_6	100.0	116.4	123.8	131.8	139.9	145.4	152.0	161.8	171.6	180.2	192.4
fluoride	AlF_3	1238	1298	1324	1350	1378	1398	1422	1457	1496	1537	1040
iodide	AlI_3	178.0	207.7	225.8	244.2	265.0	277.8	294.5	322.0	354.0	385.5	
oxide	Al_2O_3	2148	2306	2385	2465	2549	2599	2665	2766	2874	2977	2050
Ammonia	NH_3	−109.1	−97.5	−91.9	−85.8	−79.2	−74.3	−68.4	−57.0	−45.4	−33.6	−77.7
heavy	ND_3	......	......	......	......	......	−74.0	−67.4	−57.0	−45.4	−33.4	−74.0
Ammonium bromide	NH_4Br	198.3	234.5	252.0	270.6	290.0	303.8	320.0	345.3	370.8	396.0	
carbamate	$N_2H_6CO_2$	−26.1	−10.4	−2.9	+5.3	14.0	19.6	26.7	37.2	48.0	58.3	
chloride	NH_4Cl	160.4	193.8	209.8	226.1	245.0	256.2	271.5	293.2	316.5	337.8	520
cyanide	NH_4CN	−50.6	−35.7	−28.6	−20.9	−12.6	−7.4	−0.5	+9.6	20.5	31.7	36
hydrogen sulfide	NH_4HS	−51.1	−36.0	−28.7	−20.8	−12.3	−7.0	0.0	+10.5	21.8	33.3	
iodide	NH_4I	210.9	247.0	263.5	282.8	302.8	316.0	331.8	355.8	381.0	404.9	
Antimony	Sb	886	984	1033	1084	1141	1176	1223	1288	1364	1440	630.5
tribromide	$SbBr_3$	93.9	126.0	142.7	158.3	177.4	188.1	203.5	225.7	250.2	275.0	96.6
trichloride	$SbCl_3$	49.2	71.4	85.2	100.6	117.8	128.3	143.3	165.9	192.2	219.0	73.4
pentachlcride	$SbCl_5$	22.7	48.6	61.8	75.8	91.0	101.0	114.1	......	......	......	2.8
triiodide	SbI_3	163.6	203.8	223.5	244.8	267.8	282.5	303.5	333.8	368.5	401.0	167
trioxide	Sb_4O_6	574	626	666	729	812	873	957	1085	1242	1425	656
Argon	A	−218.2	−213.9	−210.9	−207.9	−204.9	−202.9	−200.5	−195.6	−190.6	−185.6	−189.2
Arsenic	As	372	416	437	459	483	498	518	548	579	610	814
Arsenic tribromide	$AsBr_3$	41.8	70.6	85.2	101.3	118.7	130.0	145.2	167.7	193.6	220.0	
trichloride	$AsCl_3$	−11.4	+11.7	+23.5	36.0	50.0	58.7	70.9	89.2	109.7	130.4	−18
trifluoride	AsF_3	......	......	......	......	−2.5	+4.2	13.2	26.7	41.4	56.3	−5.9
pentafluoride	AsF_5	−117.9	−108.0	−103.1	−98.0	−92.4	−88.5	−84.3	−75.5	−64.0	−52.8	−79.8
trioxide	As_2O_3	212.5	242.6	259.7	279.2	299.2	310.3	332.5	370.0	412.2	457.2	312.8
Arsine	AsH_3	−142.6	−130.8	−124.7	−117.7	−110.2	−104.8	−98.0	−87.2	−75.2	−62.1	−116.3
Barium	Ba	......	984	1049	1120	1195	1240	1301	1403	1518	1638	850
Beryllium borohydride	$Be(BH_4)_2$	+1.0	19.8	28.1	36.8	46.2	51.7	58.6	69.0	79.7	90.0	123
bromide	$BeBr_2$	289	325	342	361	379	390	405	427	451	474	490
chloride	$BeCl_2$	291	328	346	365	384	395	411	435	461	487	405
iodide	BeI_2	283	322	341	361	382	394	411	435	461	487	488
Bismuth	Bi	1021	1099	1136	1177	1217	1240	1271	1319	1370	1420	271
tribromide	$BiBr_3$	......	261	282	305	327	340	360	392	425	461	218
trichloride	$BiCl_3$	......	242	264	287	311	324	343	372	405	441	230
Diborane hydrobromide	B_2H_5Br	−93.3	−75.3	−66.3	−56.4	−45.4	−38.2	−29.0	−15.4	0.0	+16.3	−104.2
Borine carbonyl	BH_3CO	−139.2	−127.3	−121.1	−114.1	−106.6	−101.9	−95.3	−85.5	−74.8	−64.0	−137.0
triamine	$B_3N_3H_6$	−63.0	−45.0	−35.3	−25.0	−13.2	−5.8	+4.0	18.5	34.3	50.6	−58.2
Boron hydrides												
dihydrodecaborane	$B_{10}H_{14}$	60.0	80.8	90.2	100.0	117.4	127.8	142.3	163.8	......	......	99.6
dihydrodiborane	B_2H_6	−159.7	−149.5	−144.3	−138.5	−131.6	−127.2	−120.9	−111.2	−99.6	−86.5	−169
dihydropentaborane	B_5H_9	......	−40.4	−30.7	−20.0	−8.0	−0.4	+9.6	24.6	40.8	58.1	−47.0
tetrahydropentaborane	B_5H_{11}	−50.2	−29.9	−19.9	−9.2	+2.7	10.2	20.1	34.8	51.2	67.0	
tetrahydrotetraborane	B_4H_{10}	−90.9	−73.1	−64.3	−54.8	−44.3	−37.4	−28.1	−14.0	+0.8	16.1	−119.9
Boron tribromide	BBr_3	−41.4	−20.4	−10.1	+1.5	14.0	22.1	33.5	50.3	70.0	91.7	−45
trichloride	BCl_3	−91.5	−75.2	−66.9	−57.9	−47.8	−41.2	−32.4	−18.9	−3.6	+12.7	−107
trifluoride	BF_3	−154.6	−145.4	−141.3	−136.4	−131.0	−127.6	−123.0	−115.9	−108.3	−100.7	−126.8
Bromine	Br_2	−48.7	−32.8	−25.0	−16.8	−8.0	−0.6	+9.3	24.3	41.0	58.2	−7.3
pentafluoride	BrF_5	−69.3	−51.0	−41.9	−32.0	−21.0	−14.0	−4.5	+9.9	25.7	40.4	−61.4
Cadmium	Cd	394	455	484	516	553	578	611	658	711	765	320.9
chloride	$CdCl_2$	......	618	656	695	736	762	797	847	908	967	568
fluoride	CdF_2	1112	1231	1286	1344	1400	1436	1486	1561	1651	1751	520
iodide	CdI_2	416	481	512	546	584	608	640	688	742	796	385
oxide	CdO	1000	1100	1149	1200	1257	1295	1341	1409	1484	1559	
Calcium	Ca	......	926	983	1046	1111	1152	1207	1288	1388	1487	851
Carbon (graphite)	C	3586	3828	3946	4069	4196	4273	4373	4516	4660	4827	
dioxide	CO_2	−134.3	−124.4	−119.5	−114.4	−108.6	−104.8	−100.2	−93.0	−85.7	−78.2	−57.5
disulfide	CS_2	−73.8	−54.3	−44.7	−34.3	−22.5	−15.3	−5.1	+10.4	28.0	46.5	−110.8
monoxide	CO	−222.0	−217.2	−215.0	−212.8	−210.0	−208.1	−205.7	−201.3	−196.3	−191.3	−205.0
oxyselenide	$COSe$	−117.1	−102.3	−95.0	−86.3	−76.4	−70.2	−61.7	−49.8	−35.6	−21.9	
oxysulfide	COS	−132.4	−119.8	−113.3	−106.0	−98.3	−93.0	−85.9	−75.0	−62.7	−49.9	−138.8
selenosulfide	$CSeS$	−47.3	−26.5	−16.0	−4.4	+8.6	17.0	28.3	45.7	65.2	85.6	−75.2
subsulfide	C_3S_2	14.0	41.2	54.9	69.3	85.6	96.0	109.9	130.8	......	......	+0.4
tetrabromide	CBr_4	......	......	......	......	96.3	106.3	119.7	139.7	163.5	189.5	90.1
tetrachloride	CCl_4	−50.0	−30.0	−19.6	−8.2	+4.3	12.3	23.0	38.3	57.8	76.7	−22.6
tetrafluoride	CF_4	−184.6	−174.1	−169.3	−164.3	−158.8	−155.4	−150.7	−143.6	−135.5	−127.7	−183.7
Cesium	Cs	279	341	375	409	449	474	509	561	624	690	28.5
bromide	$CsBr$	748	838	887	938	993	1026	1072	1140	1221	1300	636
chloride	$CsCl$	744	837	884	934	989	1023	1069	1139	1217	1300	646
fluoride	CsF	712	798	844	893	947	980	1025	1092	1170	1251	683
iodide	CsI	738	828	873	923	976	1009	1055	1124	1200	1280	621
Chlorine	Cl_2	−118.0	−106.7	−101.6	−93.3	−84.5	−79.0	−71.7	−60.2	−47.3	−33.8	−100.7
fluoride	ClF	......	−143.4	−139.0	−134.3	−128.8	−125.3	−120.8	−114.4	−107.0	−100.5	−145
trifluoride	ClF_3	......	−80.4	−71.8	−62.3	−51.3	−44.1	−34.7	−20.7	−4.9	+11.5	−83
monoxide	Cl_2O	−98.5	−81.6	−73.1	−64.3	−54.3	−48.0	−39.4	−26.5	−12.5	+2.2	−116
dioxide	ClO_2	......	......	−59.0	−51.2	−42.8	−37.2	−29.4	−17.8	−4.0	+11.1	−59
heptoxide	Cl_2O_7	−45.3	−23.8	−13.2	−2.1	+10.3	+18.2	29.1	44.6	62.2	78.8	−91
Chlorosulfonic acid	HSO_3Cl	32.0	53.5	64.0	75.3	87.6	95.2	105.3	120.0	136.1	151.0	−80
Chromium	Cr	1616	1768	1845	1928	2013	2067	2139	2243	2361	2482	1615
carbonyl	$Cr(CO)_6$	36.0	58.0	68.3	79.5	91.2	98.3	108.0	121.8	137.2	151.0	
oxychloride	CrO_2Cl_2	−18.4	+3.2	13.8	25.7	38.5	46.7	58.0	75.2	95.2	117.1	
Cobalt chloride	$CoCl_2$	......	......	......	......	770	801	843	904	974	1050	735
nitrosyl tricarbonyl	$Co(CO)_3NO$	......	......	......	−1.3	+11.0	18.5	29.0	44.4	62.0	80.0	−11
Columbium fluoride	CbF_5	......	......	86.3	103.0	121.5	133.2	148.5	172.2	198.0	225.0	75.5
Copper	Cu	1628	1795	1879	1970	2067	2127	2207	2325	2465	2595	1083
Cuprous bromide	Cu_2Br_2	572	666	718	777	844	887	951	1052	1189	1355	504
chloride	Cu_2Cl_2	546	645	702	766	838	886	960	1077	1249	1490	422
iodide	Cu_2I_2	......	610	656	716	786	836	907	1018	1158	1336	605
Cyanogen	C_2N_2	−95.8	−83.2	−76.8	−70.1	−62.7	−57.9	−51.8	−42.6	−33.0	−21.0	−34.4
bromide	$CNBr$	−35.7	−18.3	−10.0	−1.0	+8.6	14.7	22.6	33.8	46.0	61.5	58
chloride	$CNCl$	−76.7	−61.4	−53.8	−46.1	−37.5	−32.1	−24.9	−14.1	−2.3	+13.1	−6.5
fluoride	CNF	−134.4	−123.8	−118.5	−112.8	−106.4	−102.3	−97.0	−89.2	−80.5	−72.6	

* Compiled from the extended tables published by D. R. Stull in *Ind. Eng. Chem.*, **39**. 517 (1947).

Table 7. Vapor Pressures of Inorganic Compounds, up to 1 Atm.—(*Continued*)

Compound		Pressure, mm. Hg										Melting point, °C.
Name	Formula	1	5	10	20	40	60	100	200	400	760	
		Temperature, °C.										
Deuterium cyanide	DCN	−68.9	−54.0	−46.7	−38.8	−30.1	−24.7	−17.5	−5.4	+10.0	26.2	−12
Fluorine	F_2	−223.0	−216.9	−214.1	−211.0	−207.7	−205.6	−202.7	−198.3	−193.2	−187.9	−223
oxide	F_2O	−196.1	−186.6	−182.3	−177.8	−173.0	−170.0	−165.8	−159.0	−151.9	−144.6	−223.9
Germanium bromide	$GeBr_4$	....	43.3	56.8	71.8	88.1	98.8	113.2	135.4	161.6	189.0	26.1
chloride	$GeCl_4$	−45.0	−24.9	−15.0	−4.1	+8.0	16.2	27.5	44.4	63.8	84.0	−49.5
hydride	GeH_4	−163.0	−151.0	−145.3	−139.2	−131.6	−126.7	−120.3	−111.2	−100.2	−88.9	−165
Trichlorogermane	$GeHCl_3$	−41.3	−22.3	−13.0	−3.0	+8.8	16.2	26.5	41.6	58.3	75.0	−71.1
Tetramethylgermane	$Ge(CH_3)_4$	−73.2	−54.6	−45.2	−35.0	−23.4	−16.2	−6.3	+8.8	26.0	44.0	−88
Digermane	Ge_2H_6	−88.7	−69.8	−60.1	−49.9	−38.2	−30.7	−20.3	−4.7	+13.3	31.5	−109
Trigermane	Ge_3H_8	−36.9	−12.8	−0.9	+11.8	26.3	35.5	47.9	67.0	88.6	110.8	−105.6
Gold	Au	1869	2059	2154	2256	2363	2431	2521	2657	2807	2966	1063
Helium	He	−271.7	−271.5	−271.3	−271.1	−270.7	−270.6	−270.3	−269.8	−269.3	−268.6	
para-Hydrogen	H_2	−263.3	−261.9	−261.3	−260.4	−259.6	−258.9	−257.9	−256.3	−254.5	−252.5	−259.1
Hydrogen bromide	HBr	−138.8	−127.4	−121.8	−115.4	−108.3	−103.8	−97.7	−88.1	−78.0	−66.5	−87.0
chloride	HCl	−150.8	−140.7	−135.6	−130.0	−123.8	−119.6	−114.0	−105.2	−95.3	−84.8	−114.3
cyanide	HCN	−71.0	−55.3	−47.7	−39.7	−30.9	−25.1	−17.8	−5.3	+10.2	25.9	−13.2
fluoride	H_2F_2	....	−74.7	−65.8	−56.0	−45.0	−37.9	−28.2	−13.2	+2.5	19.7	−83.7
iodide	HI	−123.3	−109.6	−102.3	−94.5	−85.6	−79.8	−72.1	−60.3	−48.3	−35.1	−50.9
oxide (water)	H_2O	−17.3	+1.2	11.2	22.1	34.0	41.5	51.6	66.5	83.0	100.0	0.0
sulfide	H_2S	−134.3	−122.4	−116.3	−109.7	−102.3	−97.9	−91.6	−82.3	−71.8	−60.4	−85.5
disulfide	HSSH	−43.2	−24.4	−15.2	−5.1	+6.0	12.8	22.0	35.3	49.6	64.0	−89.7
selenide	H_2Se	−115.3	−103.4	−97.9	−91.8	−84.7	−80.2	−74.2	−65.2	−53.6	−41.1	−64
telluride	H_2Te	−96.4	−82.4	−75.4	−67.8	−59.1	−53.7	−45.7	−32.4	−17.2	−2.0	−49.0
Iodine	I_2	38.7	62.2	73.2	84.7	97.5	105.4	116.5	137.3	159.8	183.0	112.9
heptafluoride	IF_7	−87.0	−70.7	−63.0	−54.5	−45.3	−39.4	−31.9	−20.7	−8.3	+4.0	5.5
Iron	Fe	1787	1957	2039	2128	2224	2283	2360	2475	2605	2735	1535
pentacarbonyl	$Fe(CO)_5$	....	−6.5	+4.6	16.7	30.3	39.1	50.3	68.0	86.1	105.0	−21
Ferric chloride	Fe_2Cl_6	194.0	221.8	235.5	246.0	256.8	263.7	272.5	285.0	298.0	319.0	304
Ferrous chloride	$FeCl_2$	....	....	700	737	779	805	842	897	961	1026	
Krypton	Kr	−199.3	−191.3	−187.2	−182.9	−178.4	−175.7	−171.8	−165.9	−159.0	−152.0	−156.7
Lead	Pb	973	1099	1162	1234	1309	1358	1421	1519	1630	1744	327.5
bromide	$PbBr_2$	513	578	610	646	686	711	745	796	856	914	373
chloride	$PbCl_2$	547	615	648	684	725	750	784	833	893	954	501
fluoride	PbF_2	....	861	904	950	1003	1036	1080	1144	1219	1293	855
iodide	PbI_2	479	540	571	605	644	668	701	750	807	872	402
oxide	PbO	943	1039	1085	1134	1189	1222	1265	1330	1402	1472	890
sulfide	PbS	852	928	975	1005	1048	1074	1108	1160	1221	1281	1114
Lithium	Li	723	828	881	940	1003	1042	1097	1178	1273	1372	186
bromide	LiBr	748	840	888	939	994	1028	1076	1147	1226	1310	547
chloride	LiCl	783	880	932	987	1045	1081	1129	1203	1290	1382	614
fluoride	LiF	1047	1156	1211	1270	1333	1372	1425	1503	1591	1681	870
iodide	LiI	723	802	841	883	927	955	993	1049	1110	1171	446
Magnesium	Mg	621	702	743	789	838	868	909	967	1034	1107	651
chloride	$MgCl_2$	778	877	930	988	1050	1088	1142	1223	1316	1418	712
Manganese	Mn	1292	1434	1505	1583	1666	1720	1792	1900	2029	2151	1260
chloride	$MnCl_2$	....	736	778	825	879	913	960	1028	1108	1190	650
Mercury	Hg	126.2	164.8	184.0	204.6	228.8	242.0	261.7	290.7	323.0	357.0	−38.9
Mercuric bromide	$HgBr_2$	136.5	165.3	179.8	194.3	211.5	221.0	237.8	262.7	290.0	319.0	237
chloride	$HgCl_2$	136.2	166.0	180.2	195.8	212.5	222.2	237.0	256.5	275.5	304.0	277
iodide	HgI_2	157.5	189.2	204.5	220.0	238.2	249.0	261.8	291.0	324.2	354.0	259
Molybdenum	Mo	3102	3393	3535	3690	3859	3964	4109	4322	4553	4804	2622
hexafluoride	MoF_6	−65.5	−49.0	−40.8	−32.0	−22.1	−16.2	−8.0	+4.1	17.2	36.0	17
oxide	MoO_3	734	785	814	851	892	917	955	1014	1082	1151	795
Neon	Ne	−257.3	−255.5	−254.6	−253.7	−252.6	−251.9	−251.0	−249.7	−248.1	−246.0	−248.7
Nickel	Ni	1810	1979	2057	2143	2234	2289	2364	2473	2603	2732	1452
carbonyl	$Ni(CO)_4$	....	....	....	....	−23.0	−15.9	−6.0	+8.8	25.8	42.5	−25
chloride	$NiCl_2$	671	731	759	789	821	840	866	904	945	987	1001
Nitrogen	N_2	−226.1	−221.3	−219.1	−216.8	−214.0	−212.3	−209.7	−205.6	−200.9	−195.8	−210.0
Nitric oxide	NO	−184.5	−180.6	−178.2	−175.3	−171.7	−168.9	−166.0	−162.3	−156.8	−151.7	−161
Nitrogen dioxide	NO_2	−55.6	−42.7	−36.7	−30.4	−23.9	−19.9	−14.7	−5.0	+8.0	21.0	−9.3
Nitrogen pentoxide	N_2O_5	−36.8	−23.0	−16.7	−10.0	−2.9	+1.8	7.4	15.6	24.4	32.4	30
Nitrous oxide	N_2O	−143.4	−133.4	−128.7	−124.0	−118.3	−114.9	−110.3	−103.6	−96.2	−85.5	−90.9
Nitrosyl chloride	NOCl	....	....	....	....	−60.2	−54.2	−46.3	−34.0	−20.3	−6.4	−64.5
fluoride	NOF	−132.0	−120.3	−114.3	−107.8	−100.3	−95.7	−88.8	−79.2	−68.2	−56.0	−134
Osmium tetroxide (yellow)	OsO_4	3.2	22.0	31.3	41.0	51.7	59.4	71.5	89.5	109.3	130.0	56
(white)	OsO_4	−5.6	+15.6	26.0	37.4	50.5	59.4	71.5	89.5	109.3	130.0	42
Oxygen	O_2	−219.1	−213.4	−210.6	−207.5	−204.1	−201.9	−198.8	−194.0	−188.8	−183.1	−218.7
Ozone	O_3	−180.4	−168.6	−163.2	−157.2	−150.7	−146.7	−141.0	−132.6	−122.5	−111.1	−251
Phosgene	$COCl_2$	−92.9	−77.0	−69.3	−60.3	−50.3	−44.0	−35.6	−22.3	−7.6	+8.3	−104
Phosphorus (yellow)	P	76.6	111.2	128.0	146.2	166.7	179.8	197.3	222.7	251.0	280.0	44.1
(violet)	P	237	271	287	306	323	334	349	370	391	417	590
tribromide	PBr_3	7.8	34.4	47.8	62.4	79.0	89.8	103.6	125.2	149.7	175.3	−40
trichloride	PCl_3	−51.6	−31.5	−21.3	−10.2	+2.3	10.2	21.0	37.6	56.9	74.2	−111.8
pentachloride	PCl_5	55.5	74.0	83.2	92.5	102.5	108.3	117.0	131.3	147.2	162.0	
Phosphine	PH_3	....	....	....	....	−129.4	−125.0	−118.8	−109.4	−98.3	−87.5	−132.5
Phosphonium bromide	PH_4Br	−43.7	−28.5	−21.2	−13.3	−5.0	+0.3	7.4	17.6	28.0	38.3	
chloride	PH_4Cl	−91.0	−79.6	−74.0	−68.0	−61.5	−57.3	−52.0	−44.0	−35.4	−27.0	−28.5
iodide	PH_4I	−25.2	−9.0	−1.1	+7.3	16.1	21.9	29.3	39.9	51.6	62.3	
Phosphorus trioxide	P_4O_6	....	39.7	53.0	67.8	84.0	94.2	108.3	129.0	150.3	173.1	22.5
pentoxide	P_4O_{10}	384	424	442	462	481	493	510	532	556	591	569
oxychloride	$POCl_3$	....	....	2.0	13.6	27.3	35.8	47.4	65.0	84.3	105.1	2
thiobromide	$PSBr_3$	50.0	72.4	83.6	95.5	108.0	116.0	126.3	141.8	157.8	175.0	38
thiochloride	$PSCl_3$	−18.3	+4.6	16.1	29.0	42.7	51.8	63.8	82.0	102.3	124.0	−36.2
Platinum	Pt	2730	3007	3146	3302	3469	3574	3714	3923	4169	4407	1755
Potassium	K	341	408	443	483	524	550	586	643	708	774	62.3
bromide	KBr	795	892	940	994	1050	1087	1137	1212	1297	1383	730
chloride	KCl	821	919	968	1020	1078	1115	1164	1239	1322	1407	790
fluoride	KF	885	988	1039	1096	1156	1193	1245	1323	1411	1502	880
hydroxide	KOH	719	814	863	918	976	1013	1064	1142	1233	1327	380
iodide	KI	745	840	887	938	995	1030	1080	1152	1238	1324	723
Radon	Rn	−144.2	−132.4	−126.3	−119.2	−111.3	−106.2	−99.0	−87.7	−75.0	−61.8	−71
Rhenium heptoxide	Re_2O_7	212.5	237.5	248.0	261.0	272.0	280.0	289.0	307.0	336.0	362.4	296

Table 7. Vapor Pressures of Inorganic Compounds, up to 1 Atm.—(*Concluded*)

Compound		Pressure, mm. Hg										Melting point, °C.
Name	Formula	1	5	10	20	40	60	100	200	400	760	
		Temperature, °C.										
Rubidium	Rb	297	358	389	422	459	482	514	563	620	679	38.5
bromide	RbBr	781	876	923	975	1031	1066	1114	1186	1267	1352	682
chloride	RbCl	792	887	937	990	1047	1084	1133	1207	1294	1381	715
fluoride	RbF	921	982	1016	1052	1096	1123	1168	1239	1322	1408	760
iodide	RbI	748	839	884	935	991	1026	1072	1141	1223	1304	642
Selenium	Se	356	413	442	473	506	527	554	594	637	680	217
dioxide	SeO_2	157.0	187.7	202.5	217.5	234.1	244.6	258.0	277.0	297.7	317.0	340
hexafluoride	SeF_6	−118.6	−105.2	−98.9	−92.3	−84.7	−80.0	−73.9	−64.8	−55.2	−45.8	−34.7
oxychloride	$SeOCl_2$	34.8	59.8	71.9	84.2	98.0	106.5	118.0	134.6	151.7	168.0	8.5
tetrachloride	$SeCl_4$	74.0	96.3	107.4	118.1	130.1	137.8	147.5	161.0	176.4	191.5	
Silicon	Si	1724	1835	1888	1942	2000	2036	2083	2151	2220	2287	1420
dioxide	SiO_2	......	......	1732	1798	1867	1911	1969	2053	2141	2227	1710
tetrachloride	$SiCl_4$	−63.4	−44.1	−34.4	−24.0	−12.1	−4.8	+5.4	21.0	38.4	56.8	−68.8
tetrafluoride	SiF_4	−144.0	−134.8	−130.4	−125.9	−120.8	−117.5	−113.3	−170.2	−100.7	−94.8	−90
Trichlorofluorosilane	$SiFCl_3$	−92.6	−76.4	−68.3	−59.0	−48.8	−42.2	−33.2	−19.3	−4.0	+12.2	−120.8
Iodosilane	SiH_3I	......	−53.0	−47.7	−33.4	−21.8	−14.3	−4.4	+10.7	27.9	45.4	−57.0
Diiodosilane	SiH_2I_2	......	3.8	18.0	34.1	52.6	64.0	79.4	101.8	125.5	149.5	−1.0
Disiloxan	$(SiH_3)_2O$	−112.5	−95.8	−88.2	−79.8	−70.4	−64.2	−55.9	−43.5	−29.3	−15.4	−144.2
Trisilane	Si_3H_8	−68.9	−49.7	−40.0	−29.0	−16.9	−9.0	+1.6	17.8	35.5	53.1	−117.2
Trisilazane	$(SiH_3)_3N$	−68.7	−49.9	−40.4	−30.0	−18.5	−11.0	−1.1	+14.0	31.0	48.7	−105.7
Tetrasilane	Si_4H_{10}	−27.7	−6.2	+4.3	15.8	28.4	36.6	47.4	63.6	81.7	100.0	−93.6
Octachlorotrisilane	Si_3Cl_8	46.3	74.7	89.3	104.2	121.5	132.0	146.0	166.2	189.5	211.4	
Hexachlorodisiloxane	$(SiCl_3)_2O$	−5.0	17.8	29.4	41.5	55.2	63.8	75.4	92.5	113.6	135.6	−33.2
Hexachlorodisilane	Si_2Cl_6	+4.0	27.4	38.8	51.5	65.3	73.9	85.4	102.2	120.6	139.0	−1.2
Tribromosilane	$SiHBr_3$	−30.5	−8.0	+3.4	16.0	30.0	39.2	51.6	70.2	90.2	111.8	−73.5
Trichlorosilane	$SiHCl_3$	−80.7	−62.6	−53.4	−43.8	−32.9	−25.8	−16.4	−1.8	+14.5	31.8	−126.6
Trifluorosilane	$SiHF_3$	−152.0	−142.7	−138.2	−132.9	−127.3	−123.7	−118.7	−111.3	−102.8	−95.0	−131.4
Dibromosilane	SiH_2Br_2	−60.9	−40.0	−29.4	−18.0	−5.2	+3.2	14.1	31.6	50.7	70.5	−70.2
Difluorosilane	SiH_2F_2	−146.7	−136.0	−130.4	−124.3	−117.6	−113.3	−107.3	−98.3	−87.6	−77.8	
Monobromosilane	SiH_3Br	......	−85.7	−77.3	−68.3	−57.8	−51.1	−42.3	−28.6	−13.3	+2.4	−93.9
Monochlorosilane	SiH_3Cl	−117.8	−104.3	−97.7	−90.1	−81.8	−76.0	−68.5	−57.0	−44.5	−30.4	
Monofluorosilane	SiH_3F	−153.0	−145.5	−141.2	−136.3	−130.8	−127.2	−122.4	−115.2	−106.8	−98.0	
Tribromofluorosilane	$SiFBr_3$	−46.1	−25.4	−15.1	−3.7	+9.2	17.4	28.6	45.7	64.6	83.8	−82.5
Dichlorodifluorosilane	SiF_2Cl_2	−124.7	−110.5	−102.9	−94.5	−85.0	−78.6	−70.3	−58.0	−45.0	−31.8	−139.7
Trifluorobromosilane	SiF_3Br	......	......	......	......	......	......	......	−69.8	−55.9	−41.7	−70.5
Trifluorochlorosilane	SiF_3Cl	−144.0	−133.0	−127.0	−120.5	−112.8	−108.2	−101.7	−91.7	−81.0	−70.0	−142
Hexafluorodisilane	Si_2F_6	−81.0	−68.8	−63.1	−57.0	−50.6	−46.7	−41.7	−34.2	−26.4	−18.9	−18.6
Dichlorofluorobromosilane	$SiFCl_2Br$	−86.5	−68.4	−59.0	−48.8	−37.0	−29.0	−19.5	−3.2	+15.4	35.4	−112.3
Dibromochlorofluorosilane	$SiFClBr_2$	−65.2	−45.5	−35.6	−24.5	−12.0	−4.7	+6.3	23.0	43.0	59.5	−99.3
Silane	SiH_4	−179.3	−168.6	−163.0	−156.9	−150.3	−146.3	−140.5	−131.6	−122.0	−111.5	−185
Disilane	Si_2H_6	−114.8	−99.3	−91.4	−82.7	−72.8	−66.4	−57.5	−44.6	−29.0	−14.3	−132.6
Silver	Ag	1357	1500	1575	1658	1743	1795	1865	1971	2090	2212	960.5
chloride	AgCl	912	1019	1074	1134	1200	1242	1297	1379	1467	1564	455
iodide	AgI	820	927	983	1045	1111	1152	1210	1297	1400	1506	552
Sodium	Na	439	511	549	589	633	662	701	758	823	892	97.5
bromide	NaBr	806	903	952	1005	1063	1099	1148	1220	1304	1392	755
chloride	NaCl	865	967	1017	1072	1131	1169	1220	1296	1379	1465	800
cyanide	NaCN	817	928	983	1046	1115	1156	1214	1302	1401	1497	564
fluoride	NaF	1077	1186	1240	1300	1363	1403	1455	1531	1617	1704	992
hydroxide	NaOH	739	843	897	953	1017	1057	1111	1192	1286	1378	318
iodide	NaI	767	857	903	952	1005	1039	1083	1150	1225	1304	651
Strontium	Sr	......	847	898	953	1018	1057	1111	1192	1285	1384	800
Strontium oxide	SrO	2068	2198	2262	2333	2410	......	......	......	......	......	2430
Sulfur	S	183.8	223.0	243.8	264.7	288.3	305.5	327.2	359.7	399.6	444.6	112.8
monochloride	S_2Cl_2	−7.4	+15.7	27.5	40.0	54.1	63.2	75.3	93.5	115.4	138.0	−80
hexafluoride	SF_6	−132.7	−120.6	−114.7	−108.4	−101.5	−96.8	−90.9	−82.3	−72.6	−63.5	−50.2
Sulfuryl chloride	SO_2Cl_2	......	−35.1	−24.8	−13.4	−1.0	+7.2	17.8	33.7	51.3	69.2	−54.1
Sulfur dioxide	SO_2	−95.5	−83.0	−76.8	−69.7	−60.5	−54.6	−46.9	−35.4	−23.0	−10.0	−73.2
trioxide (α)	SO_3	−39.0	−23.7	−16.5	−9.1	−1.0	+4.0	10.5	20.5	32.6	44.8	16.8
trioxide (β)	SO_3	−34.0	−19.2	−12.3	−4.9	+3.2	8.0	14.3	23.7	32.6	44.8	32.3
trioxide (γ)	SO_3	−15.3	−2.0	+4.3	11.1	17.9	21.4	28.0	35.8	44.0	51.6	62.1
Tellurium	Te	520	605	650	697	753	789	838	910	997	1087	452
chloride	$TeCl_4$	......	......	233	253	273	287	304	330	360	392	224
fluoride	TeF_6	−111.3	−98.8	−92.4	−86.0	−78.4	−73.8	−67.9	−57.3	−48.2	−38.6	−37.8
Thallium	Tl	825	931	983	1040	1103	1143	1196	1274	1364	1457	3035
Thallous bromide	TlBr	......	490	522	559	598	621	653	703	759	819	460
chloride	TlCl	......	487	517	550	589	612	645	694	748	807	430
iodide	TlI	440	502	531	567	607	631	663	712	763	823	440
Thionyl bromide	$SOBr_2$	−6.7	+18.4	31.0	44.1	58.8	68.3	80.6	99.0	119.2	139.5	−52.2
Thionyl chloride	$SOCl_2$	−52.9	−32.4	−21.9	−10.5	+2.2	10.4	21.4	37.9	56.5	75.4	−104.5
Tin	Sn	1492	1634	1703	1777	1855	1903	1968	2063	2169	2270	231.9
Stannic bromide	$SnBr_4$	......	58.3	72.7	88.1	105.5	116.2	131.0	152.8	177.7	204.7	31.0
Stannous chloride	$SnCl_2$	316	366	391	420	450	467	493	533	577	623	246.8
Stannic chloride	$SnCl_4$	−22.7	−1.0	+10.0	22.0	35.2	43.5	54.7	72.0	92.1	113.0	−30.2
iodide	SnI_4	......	156.0	175.8	196.2	218.8	234.2	254.2	283.5	315.5	348.0	144.5
hydride	SnH_4	−140.0	−125.8	−118.5	−111.2	−102.3	−96.6	−89.2	−78.0	−65.2	−52.3	−149.9
Tin tetramethyl	$Sn(CH_3)_4$	−51.3	−31.0	−20.6	−9.3	+3.5	11.7	22.8	39.8	58.5	78.0	
trimethyl-ethyl	$Sn(CH_3)_3.C_2H_5$	−30.0	−7.6	+3.8	16.1	30.0	38.4	50.0	67.3	87.6	108.8	
trimethyl-propyl	$Sn(CH_3)_3.C_3H_7$	−12.0	+10.7	21.8	34.0	48.5	57.5	69.8	88.0	109.6	131.7	
Titanium chloride	$TiCl_4$	−13.9	+9.4	21.3	34.2	48.4	58.0	71.0	90.5	112.7	136.0	−30
Tungsten	W	3990	4337	4507	4690	4886	5007	5168	5403	5666	5927	3370
Tungsten hexafluoride	WF_6	−71.4	−56.5	−49.2	−41.5	−33.0	−27.5	−20.3	−10.0	+1.2	17.3	−0.5
Uranium hexafluoride	UF_6	−38.8	−22.0	−13.8	−5.2	+4.4	10.4	18.2	30.0	42.7	55.7	69.2
Vanadyl trichloride	$VOCl_3$	−23.2	+0.2	12.2	26.6	40.0	49.8	62.5	82.0	103.5	127.2	
Xenon	Xe	−168.5	−158.2	−152.8	−147.1	−141.2	−137.7	−132.8	−125.4	−117.1	−108.0	−111.6
Zinc	Zn	487	558	593	632	673	700	736	788	844	907	419.4
chloride	$ZnCl_2$	428	481	508	536	566	584	610	648	689	732	365
fluoride	ZnF_2	970	1055	1086	1129	1175	1207	1254	1329	1417	1497	872
diethyl	$Zn(C_2H_5)_2$	−22.4	0.0	+11.7	24.2	38.0	47.2	59.1	77.0	97.3	118.0	−28
Zirconium bromide	$ZrBr_4$	207	237	250	266	281	289	301	318	337	357	450
chloride	$ZrCl_4$	190	217	230	243	259	268	279	295	312	331	437
iodide	ZrI_4	264	297	311	329	344	355	369	389	409	431	499

Table 8. Vapor Pressures of Organic Compounds, up to 1 Atm.*

Compound		Pressure, mm. Hg										Melting point, °C.
Name	Formula	1	5	10	20	40	60	100	200	400	760	
		Temperature, °C.										
Acenaphthalene	$C_{12}H_{10}$	....	114.8	131.2	148.7	168.2	181.2	197.5	222.1	250.0	277.5	95
Acetal	$C_6H_{14}O_2$	−23.0	−2.3	+8.0	19.6	31.9	39.8	50.1	66.3	84.0	102.2	
Acetaldehyde	C_2H_4O	−81.5	−65.1	−56.8	−47.8	−37.8	−31.4	−22.6	−10.0	+4.9	20.2	−123.5
Acetamide	C_2H_5NO	65.0	92.0	105.0	120.0	135.8	145.8	158.0	178.3	200.0	222.0	81
Acetanilide	C_8H_9NO	114.0	146.6	162.0	180.0	199.6	211.8	227.2	250.5	277.0	303.8	113.5
Acetic acid	$C_2H_4O_2$	−17.2	+6.3	17.5	29.9	43.0	51.7	63.0	80.0	99.0	118.1	16.7
anhydride	$C_4H_6O_3$	1.7	24.8	36.0	48.3	62.1	70.8	82.2	100.0	119.8	139.6	−73
Acetone	C_3H_6O	−59.4	−40.5	−31.1	−20.8	−9.4	−2.0	+7.7	22.7	39.5	56.5	−94.6
Acetonitrile	C_2H_3N	−47.0	−26.6	−16.3	−5.0	+7.7	15.9	27.0	43.7	62.5	81.8	−41
Acetophenone	C_8H_8O	37.1	64.0	78.0	92.4	109.4	119.8	133.6	154.2	178.0	202.4	20.5
Acetyl chloride	C_2H_3OCl	−50.0	−35.0	−27.6	−19.6	−10.4	−4.5	+3.2	16.1	32.0	50.8	−112.0
Acetylene	C_2H_2	−142.9	−133.0	−128.2	−122.8	−116.7	−112.8	−107.9	−100.3	−92.0	−84.0	−81.5
Acridine	$C_{13}H_9N$	129.4	165.8	184.0	203.5	224.2	238.7	256.0	284.0	314.3	346.0	110.5
Acrolein (2-propenal)	C_3H_4O	−64.5	−46.0	−36.7	−26.3	−15.0	−7.5	+2.5	17.5	34.5	52.5	−87.7
Acrylic acid	$C_3H_4O_2$	+3.5	27.3	39.0	52.0	66.2	75.0	86.1	103.3	122.0	141.0	14
Adipic acid	$C_6H_{10}O_4$	159.5	191.0	205.5	222.0	240.5	251.0	265.0	287.8	312.5	337.5	152
Allene (propadiene)	C_3H_4	−120.6	−108.0	−101.0	−93.4	−85.2	−78.8	−72.5	−61.3	−48.5	−35.0	−136
Allyl alcohol (propen-1-ol-3)	C_3H_6O	−20.0	+0.2	10.5	21.7	33.4	40.3	50.0	64.5	80.2	96.6	−129
chloride (3-chloropropene)	C_3H_5Cl	−70.0	−52.0	−42.9	−32.8	−21.2	−14.1	−4.5	10.4	27.5	44.6	−136.4
isopropyl ether	$C_6H_{12}O$	−43.7	−23.1	−12.9	−1.8	+10.9	18.7	29.0	44.3	61.7	79.5	
isothiocyanate	C_4H_5NS	−2.0	+25.3	38.3	52.1	67.4	76.2	89.5	108.0	129.8	150.7	−80
n-propyl ether	$C_6H_{12}O$	−39.0	−18.2	−7.9	+3.7	16.4	25.0	35.8	52.6	71.4	90.5	
4-Allylveratrole	$C_{11}H_{14}O_2$	85.0	113.9	127.0	142.8	158.3	169.6	183.7	204.0	226.2	248.0	
iso-Amyl acetate	$C_7H_{14}O_2$	0.0	+23.7	35.2	47.8	62.1	71.0	83.2	101.3	121.5	142.0	
n-Amyl alcohol	$C_5H_{12}O$	+13.6	34.7	44.9	55.8	68.0	75.5	85.8	102.0	119.8	137.8	
iso-Amyl alcohol	$C_5H_{12}O$	+10.0	30.9	40.8	51.7	63.4	71.0	80.7	95.8	113.7	130.6	−117.2
sec-Amyl alcohol (2-pentanol)	$C_5H_{12}O$	+1.5	22.1	32.2	42.6	54.1	61.5	70.7	85.7	102.3	119.7	
tert-Amyl alcohol	$C_5H_{12}O$	−12.9	+7.2	17.2	27.9	38.8	46.0	55.3	69.7	85.7	101.7	−11.9
sec-Amylbenzene	$C_{11}H_{16}$	29.0	55.8	69.2	83.8	100.0	110.4	124.1	145.2	168.0	193.0	
iso-Amyl benzoate	$C_{12}H_{16}O_2$	72.0	104.5	121.6	139.7	158.3	171.4	186.8	210.2	235.8	262.0	
bromide (1-bromo-3-methylbutane)	$C_5H_{11}Br$	−20.4	+2.1	13.6	26.1	39.8	48.7	60.4	78.7	99.4	120.4	
n-butyrate	$C_9H_{18}O_2$	21.2	47.1	59.9	74.0	90.0	99.8	113.1	133.2	155.3	178.6	
formate	$C_6H_{12}O_2$	−17.5	+5.4	17.1	30.0	44.0	53.3	65.4	83.2	102.7	123.3	
iodide (1-iodo-3-methylbutane)	$C_5H_{11}I$	−2.5	+21.9	34.1	47.6	62.3	71.9	84.4	103.8	125.8	148.2	
isobutyrate	$C_9H_{18}O_2$	14.8	40.1	52.8	66.6	81.8	91.7	104.4	124.2	146.0	168.8	
Amyl isopropionate	$C_8H_{16}O_2$	+8.5	33.7	46.3	60.0	75.5	85.2	97.6	117.3	138.4	160.2	
iso-Amyl isovalerate	$C_{10}H_{20}O_2$	27.0	54.4	68.6	83.8	100.6	110.3	125.1	146.1	169.5	194.0	
n-Amyl levulinate	$C_{10}H_{18}O_3$	81.3	110.0	124.0	139.7	155.8	165.2	180.5	203.1	227.4	253.2	
iso-Amyl levulinate	$C_{10}H_{18}O_3$	75.6	104.0	118.8	134.4	151.7	162.6	177.0	198.1	222.7	247.9	
nitrate	$C_5H_{11}NO_3$	+5.2	28.8	40.3	53.5	67.6	76.3	88.6	106.7	126.5	147.5	
4-tert-Amylphenol	$C_{11}H_{16}O$	....	109.8	125.5	142.3	160.3	172.6	189.0	213.0	239.5	266.0	93
Anethole	$C_{10}H_{12}O$	62.6	91.6	106.0	121.8	139.3	149.8	164.2	186.1	210.5	235.3	22.5
Angelonitrile	C_5H_7N	−8.0	+15.0	28.0	41.0	55.8	65.2	77.5	96.3	117.7	140.0	
Aniline	C_6H_7N	34.8	57.9	69.4	82.0	96.7	106.0	119.9	140.1	161.9	184.4	−6.2
2-Anilinoethanol	$C_8H_{11}NO$	104.0	134.3	149.6	165.7	183.7	194.0	209.5	230.6	254.5	279.6	
Anisaldehyde	$C_8H_8O_2$	73.2	102.6	117.8	133.5	150.5	161.7	176.7	199.0	223.0	248.0	2.5
o-Anisidine (2-methoxyaniline)	C_7H_9NO	61.0	88.0	101.7	116.1	132.0	142.1	155.2	175.3	197.3	218.5	5.2
Anthracene	$C_{14}H_{10}$	145.0	173.5	187.2	201.9	217.5	231.8	250.0	279.0	310.2	342.0	217.5
Anthraquinone	$C_{14}H_8O_2$	190.0	219.4	234.2	248.3	264.3	273.3	285.0	314.6	346.2	379.9	286
Azelaic acid	$C_9H_{16}O_4$	178.3	210.4	225.5	242.4	260.0	271.8	286.5	309.6	332.8	356.5	106.5
Azelaldehyde	$C_9H_{18}O$	33.3	58.4	71.6	85.0	100.2	110.0	123.0	142.1	163.4	185.0	
Azobenzene	$C_{12}H_{10}N_2$	103.5	135.7	151.5	168.3	187.9	199.8	216.0	240.0	266.1	293.0	68
Benzal chloride (α,α-Dichlorotoluene)	$C_7H_6Cl_2$	35.4	64.0	78.7	94.3	112.1	123.4	138.3	160.7	187.0	214.0	−16.1
Benzaldehyde	C_7H_6O	26.2	50.1	62.0	75.0	90.1	99.6	112.5	131.7	154.1	179.0	−26
Benzanthrone	$C_{17}H_{10}O$	225.0	274.5	297.2	322.5	350.0	368.8	390.0	426.5	......		174
Benzene	C_6H_6	−36.7	−19.6	−11.5	−2.6	+7.6	15.4	26.1	42.2	60.6	80.1	+5.5
Benzenesulfonylchloride	$C_6H_5ClO_2S$	65.9	96.5	112.0	129.0	147.7	158.2	174.5	198.0	224.0	251.5	14.5
Benzil	$C_{14}H_{10}O_2$	128.4	165.2	183.0	202.8	224.5	238.2	255.8	283.5	314.3	347.0	95
Benzoic acid	$C_7H_6O_2$	96.0	119.5	132.1	146.7	162.6	172.8	186.2	205.8	227.0	249.2	121.7
anhydride	$C_{14}H_{10}O_3$	143.8	180.0	198.0	218.0	239.8	252.7	270.4	299.1	328.8	360.0	42
Benzoin	$C_{14}H_{12}O_2$	135.6	170.2	188.1	207.0	227.6	241.7	258.0	284.4	313.5	343.0	132
Benzonitrile	C_7H_5N	28.2	55.3	69.2	83.4	99.6	109.8	123.5	144.1	166.7	190.6	−12.9
Benzophenone	$C_{13}H_{10}O$	108.2	141.7	157.6	175.8	195.7	208.2	224.4	249.8	276.8	305.4	48.5
Benzotrichloride (α,α,α-Trichlorotoluene)	$C_7H_5Cl_3$	45.8	73.7	87.6	102.7	119.8	130.0	144.3	165.6	189.2	213.5	−21.2
Benzotrifluoride (α,α,α-Trifluorotoluene)	$C_7H_5F_3$	−32.0	−10.3	+0.4	12.2	25.7	34.0	45.3	62.5	82.0	102.2	−29.3
Benzoyl bromide	C_7H_5BrO	47.0	75.4	89.8	105.4	122.6	133.4	147.7	169.2	193.7	218.5	0
chloride	C_7H_5ClO	32.1	59.1	73.0	87.6	103.8	114.7	128.0	149.5	172.8	197.2	−0.5
nitrile	C_8H_5NO	44.5	71.7	85.5	100.2	116.6	127.0	141.0	161.3	185.0	208.0	33.5
Benzyl acetate	$C_9H_{10}O_2$	45.0	73.4	87.6	102.3	119.6	129.8	144.0	165.5	189.0	213.5	+51.5
alcohol	C_7H_8O	58.0	80.8	92.6	105.8	119.8	129.3	141.7	160.0	183.0	204.7	−15.3
Benzylamine	C_7H_9N	29.0	54.8	67.7	81.8	97.3	107.3	120.0	140.0	161.3	184.5	
Benzyl bromide (α-bromotoluene)	C_7H_7Br	32.2	59.6	73.4	88.3	104.8	115.6	129.8	150.8	175.2	198.5	−4
chloride (α-chlorotoluene)	C_7H_7Cl	22.0	47.8	60.8	75.0	90.7	100.5	114.2	134.0	155.8	179.4	−39
cinnamate	$C_{16}H_{14}O_2$	173.8	206.3	221.5	239.3	255.8	267.0	281.5	303.8	326.7	350.0	39
Benzyldichlorosilane	$C_7H_8Cl_2Si$	45.3	70.2	83.2	96.7	111.8	121.3	133.5	152.0	173.0	194.3	
Benzyl ethyl ether	$C_9H_{12}O$	26.0	52.0	65.0	79.6	95.4	105.5	118.9	139.6	161.5	185.0	
phenyl ether	$C_{13}H_{12}O$	95.4	127.7	144.0	160.7	180.1	192.6	209.2	233.2	259.8	287.0	
isothiocyanate	C_8H_7NS	79.5	107.8	121.8	137.0	153.0	163.8	177.7	198.0	220.4	243.0	
Biphenyl	$C_{12}H_{10}$	70.6	101.8	117.0	134.2	152.5	165.2	180.7	204.2	229.4	254.9	69.5
1-Biphenyloxy-2,3-epoxypropane	$C_{15}H_{14}O_2$	135.3	169.9	187.2	205.8	226.3	239.7	255.0	280.4	309.8	340.0	
d-Bornyl acetate	$C_{12}H_{20}O_2$	46.9	75.7	90.2	106.0	123.7	135.7	149.8	172.0	197.5	223.0	29
Bornyl n-butyrate	$C_{14}H_{24}O_2$	74.0	103.4	118.0	133.8	150.7	161.8	176.4	198.0	222.2	247.0	
formate	$C_{11}H_{18}O_2$	47.0	74.8	89.3	104.0	121.2	131.7	145.8	166.4	190.2	214.0	
isobutyrate	$C_{14}H_{24}O_2$	70.0	99.8	114.0	130.0	147.2	157.6	172.2	194.2	218.2	243.0	
propionate	$C_{13}H_{22}O_2$	64.6	93.7	108.0	123.7	140.4	151.2	165.7	187.5	211.2	235.0	
Brassidic acid	$C_{22}H_{42}O_2$	209.6	241.7	256.0	272.9	290.0	301.5	316.2	336.8	359.6	382.5	61.5
Bromoacetic acid	$C_2H_3BrO_2$	54.7	81.6	94.1	108.2	124.0	133.8	146.3	165.8	186.7	208.0	49.5
4-Bromoanisole	C_7H_7BrO	48.8	77.8	91.9	107.8	125.0	136.0	150.1	172.7	197.5	223.0	12.5

* Compiled from the extended tables published by D. R. Stull in *Ind. Eng. Chem.*, **39**, 517 (1947).

Table 8. Vapor Pressures of Organic Compounds, up to 1 Atm.—(*Continued*)

Compound		Pressure, mm. Hg										Melting point, °C.
Name	Formula	1	5	10	20	40	60	100	200	400	760	
		Temperature, °C.										
Bromobenzene	C_6H_5Br	+2.9	27.8	40.0	53.8	68.6	78.1	90.8	110.1	132.3	156.2	−30.7
4-Bromobiphenyl	$C_{12}H_9Br$	98.0	133.7	150.6	169.8	190.8	204.5	221.8	248.2	277.7	310.0	90.5
1-Bromo-2-butanol	C_4H_9BrO	23.7	45.4	55.8	67.2	79.5	87.0	97.6	112.1	128.3	145.0	
1-Bromo-2-butanone	C_4H_7BrO	+6.2	30.0	41.8	54.2	68.2	77.3	89.2	107.0	126.3	147.0	
cis-1-Bromo-1-butene	C_4H_7Br	−44.0	−23.2	−12.8	−1.4	+11.5	19.8	30.8	47.8	66.8	86.2	
trans-1-Bromo-1-butene	C_4H_7Br	−38.4	−17.0	−6.4	+5.4	18.4	27.2	38.1	55.7	75.0	94.7	−100.3
2-Bromo-1-butene	C_4H_7Br	−47.3	−27.0	−16.8	−5.3	+7.2	15.4	26.3	42.8	61.9	81.0	−133.4
cis-2-Bromo-2-butene	C_4H_7Br	−39.0	−17.9	−7.2	+4.6	17.7	26.2	37.5	54.5	74.0	93.9	−111.2
trans-2-Bromo-2-butene	C_4H_7Br	−45.0	−24.1	−13.8	−2.4	+10.5	18.7	29.9	46.5	66.0	85.5	−114.6
1,4-Bromochlorobenzene	C_6H_4BrCl	32.0	59.5	72.7	87.8	103.8	114.8	128.0	149.5	172.6	196.9	
1-Bromo-1-chloroethane	C_2H_4BrCl	−36.0	−18.0	−9.4	0.0	+10.4	17.0	28.0	44.7	63.4	82.7	16.6
1-Bromo-2-chloroethane	C_2H_4BrCl	−28.8	−7.0	+4.1	16.0	29.7	38.0	49.5	66.8	86.0	106.7	−16.6
2-Bromo-4,6-dichlorophenol	$C_6H_3BrCl_2O$	84.0	115.6	130.8	147.7	165.8	177.6	193.2	216.5	242.0	268.0	68
1-Bromo-4-ethyl benzene	C_8H_9Br	30.4	42.5	74.0	90.2	108.5	121.0	135.5	156.5	182.0	206.0	−45.0
(2-Bromoethyl)-benzene	C_8H_9Br	48.0	76.2	90.5	105.8	123.2	133.8	148.2	169.8	194.0	219.0	
2-Bromoethyl 2-chloroethyl ether	C_4H_8BrClO	36.5	63.2	76.3	90.8	106.6	116.4	129.8	150.0	172.3	195.8	
(2-Bromoethyl)-cyclohexane	$C_8H_{15}Br$	38.7	66.6	80.5	95.8	113.0	123.7	138.0	160.0	186.2	213.0	
1-Bromoethylene	C_2H_3Br	−95.4	−77.8	−68.8	−58.8	−48.1	−41.2	−31.9	−17.2	−1.1	+15.8	−138
Bromoform (tribromomethane)	$CHBr_3$	......	22.0	34.0	48.0	63.6	73.4	85.9	106.1	127.9	150.5	8.5
1-Bromonaphthalene	$C_{10}H_7Br$	84.2	117.5	133.6	150.2	170.2	183.5	198.8	224.2	252.0	281.1	5.5
2-Bromo-4-phenylphenol	$C_{12}H_9BrO$	100.0	135.4	152.3	171.8	193.8	207.0	224.5	251.0	280.2	311.0	95
3-Bromopyridine	C_5H_4BrN	16.8	42.0	55.2	69.1	84.1	94.1	107.8	127.7	150.0	173.4	
2-Bromotoluene	C_7H_7Br	24.4	49.7	62.3	76.0	91.0	100.0	112.0	133.6	157.3	181.8	−28
3-Bromotoluene	C_7H_7Br	14.8	50.8	64.0	78.1	93.9	104.1	117.8	138.0	160.0	183.7	39.8
4-Bromotoluene	C_7H_7Br	10.3	47.5	61.1	75.2	91.8	102.3	116.4	137.4	160.2	184.5	28.5
3-Bromo-2,4,6-trichlorophenol	$C_6H_2BrCl_3O$	112.4	146.2	163.2	181.8	200.5	213.0	229.3	253.0	278.0	305.8	
2-Bromo-1,4-xylene	C_8H_9Br	37.5	65.0	78.8	94.0	110.6	121.6	135.7	156.4	181.0	206.7	+9.5
1,2-Butadiene (methyl allene)	C_4H_6	−89.0	−72.7	−64.2	−54.9	−44.3	−37.5	−28.3	−14.2	+1.8	18.5	
1,3-Butadiene	C_4H_6	−102.8	−87.6	−79.7	−71.0	−61.3	−55.1	−46.8	−33.9	−19.3	−4.5	−108.9
n-Butane	C_4H_{10}	−101.5	−85.7	−77.8	−68.9	−59.1	−52.8	−44.2	−31.2	−16.3	−0.5	−135
iso-Butane (2-methylpropane)	C_4H_{10}	−109.2	−94.1	−86.4	−77.9	−68.4	−62.4	−54.1	−41.5	−27.1	−11.7	−145
1,3-Butanediol	$C_4H_{10}O_2$	22.2	67.5	85.3	100.0	117.4	127.5	141.2	161.0	183.8	206.5	77
1,2,3-Butanetriol	$C_4H_{10}O_3$	102.0	132.0	146.0	161.0	178.0	188.0	202.5	222.0	243.5	264.0	
1-Butene	C_4H_8	−104.8	−89.4	−81.6	−73.0	−63.4	−57.2	−48.9	−36.2	−21.7	−6.3	−130
cis-2-Butene	C_4H_8	−96.4	−81.1	−73.4	−64.6	−54.7	−48.4	−39.8	−26.8	−12.0	+3.7	−138.9
trans-2-Butene	C_4H_8	−99.4	−84.0	−76.3	−67.5	−57.6	−51.3	−42.7	−29.7	−14.8	+0.9	−105.4
3-Butenenitrile	C_4H_5N	−19.6	+2.9	14.1	26.6	40.0	48.8	60.2	78.0	98.0	119.0	
iso-Butyl acetate	$C_6H_{12}O_2$	−21.2	+1.4	12.8	25.5	39.2	48.0	59.7	77.6	97.5	118.0	−98.9
n-Butyl acrylate	$C_7H_{12}O_2$	−0.5	+23.5	35.5	48.6	63.4	72.6	85.1	104.0	125.2	147.4	−64.6
alcohol	$C_4H_{10}O$	−1.2	+20.0	30.2	41.5	53.4	60.3	70.1	84.3	100.8	117.5	−79.9
iso-Butyl alcohol	$C_4H_{10}O$	−9.0	+11.6	21.7	32.4	44.1	51.7	61.5	75.9	91.4	108.0	−108
sec-Butyl alcohol	$C_4H_{10}O$	−12.2	+7.2	16.9	27.3	38.1	45.2	54.1	67.9	83.9	99.5	−114.7
tert-Butyl alcohol	$C_4H_{10}O$	−20.4	−3.0	+5.5	14.3	24.5	31.0	39.8	52.7	68.0	82.9	25.3
iso-Butyl amine	$C_4H_{11}N$	−50.0	−31.0	−21.0	−10.3	+1.3	8.8	18.8	32.0	50.7	68.6	−85.0
n-Butylbenzene	$C_{10}H_{14}$	22.7	48.8	62.0	76.3	92.4	102.6	116.2	136.9	159.2	183.1	−88.0
iso-Butylbenzene	$C_{10}H_{14}$	14.1	40.5	53.7	67.8	83.3	93.3	107.0	127.2	149.6	172.8	−51.5
sec-Butylbenzene	$C_{10}H_{14}$	18.6	44.2	57.0	70.6	86.2	96.0	109.5	128.8	150.3	173.5	−75.5
tert-Butylbenzene	$C_{10}H_{14}$	13.0	39.0	51.7	65.6	80.8	90.6	103.8	123.7	145.8	168.5	−58
iso-Butyl benzoate	$C_{11}H_{14}O_2$	64.0	93.6	108.6	124.2	141.8	152.0	166.4	188.2	212.8	237.0	
n-Butyl bromide (1-bromobutane)	C_4H_9Br	−33.0	−11.2	−0.3	+11.6	24.8	33.4	44.7	62.0	81.7	101.6	−112.4
iso-Butyl n-butyrate	$C_8H_{16}O_2$	+4.6	30.0	42.2	56.1	71.7	81.3	94.0	113.9	135.7	156.9	
carbamate	$C_5H_{11}NO_2$	......	83.7	96.4	110.1	125.3	134.6	147.2	165.7	186.0	206.5	65
Butyl carbitol (diethylene glycol butyl ether)	$C_8H_{18}O_3$	70.0	95.7	107.8	120.5	135.5	146.0	159.8	181.2	205.0	231.2	
n-Butyl chloride (1-chlorobutane)	C_4H_9Cl	−49.0	−28.9	−18.6	−7.4	+5.0	13.0	24.0	40.0	58.8	77.8	−123.1
iso-Butyl chloride	C_4H_9Cl	−53.8	−34.3	−24.5	−13.8	−1.9	+5.9	16.0	32.0	50.0	68.9	−131.2
sec-Butyl chloride (2-Chlorobutane)	C_4H_9Cl	−60.2	−39.8	−29.2	−17.7	−5.0	+3.4	14.2	31.5	50.0	68.0	−131.3
tert-Butyl chloride	C_4H_9Cl	......	......	......	......	−19.0	−11.4	−1.0	+14.6	32.6	51.0	−26.5
sec-Butyl chloroacetate	$C_6H_{11}ClO_2$	17.0	41.8	54.6	68.2	83.6	93.0	105.5	124.1	146.0	167.8	
2-tert-Butyl-4-cresol	$C_{11}H_{16}O$	70.0	98.0	112.0	127.2	143.9	153.7	167.0	187.8	210.0	232.6	
4-tert-Butyl-2-cresol	$C_{11}H_{16}O$	74.3	103.7	118.0	134.0	150.8	161.7	176.2	197.8	221.8	247.0	
iso-Butyl dichloroacetate	$C_6H_{10}Cl_2O_2$	28.6	54.3	67.5	81.4	96.7	106.6	119.8	139.2	160.0	183.0	
2,3-Butylene glycol (2,3-butanediol)	$C_4H_{10}O_2$	44.0	68.4	80.3	93.4	107.8	116.3	127.8	145.6	164.0	182.0	22.5
2-Butyl-2-ethylbutane-1,3-diol	$C_{10}H_{22}O_2$	94.1	122.6	136.8	151.2	167.8	178.0	191.9	212.0	233.5	255.0	
2-tert-Butyl-4-ethylphenol	$C_{12}H_{18}O$	76.3	106.2	121.0	137.0	154.0	165.4	179.0	200.3	223.8	247.8	
n-Butyl formate	$C_5H_{10}O_2$	−26.4	−4.7	+6.1	18.0	31.6	39.8	51.0	67.9	86.2	106.0	
iso-Butyl formate	$C_5H_{10}O_2$	−32.7	−11.4	−0.8	+11.0	24.1	32.4	43.4	60.0	79.0	98.2	−95.3
sec-Butyl formate	$C_5H_{10}O_2$	−34.4	−13.3	−3.1	+8.4	21.3	29.6	40.2	56.8	75.2	93.6	
sec-Butyl glycolate	$C_6H_{12}O_3$	28.3	53.6	66.0	79.8	94.2	104.0	116.4	135.5	155.6	177.5	
iso-Butyl iodide (1-iodo-2-methylpropane)	C_4H_9I	−17.0	+5.8	17.0	29.8	42.8	51.8	63.5	81.0	100.3	120.4	−90.7
isobutyrate	$C_8H_{16}O_2$	+4.1	28.0	39.9	52.4	67.2	75.9	88.0	106.3	126.3	147.5	−80.7
isovalerate	$C_9H_{18}O_2$	16.0	41.2	53.8	67.7	82.7	92.4	105.2	124.8	146.4	168.7	
levulinate	$C_9H_{16}O_3$	65.0	92.1	105.9	120.2	136.2	147.0	160.2	181.8	205.5	229.9	
naphthylketone (1-isovaleronaphthone)	$C_{15}H_{16}O$	136.0	167.9	184.0	201.6	219.7	231.5	246.7	269.7	294.0	320.0	
2-sec-Butylphenol	$C_{10}H_{14}O$	57.4	86.0	100.8	116.1	133.4	143.9	157.3	179.7	203.8	228.0	
2-tert-Butylphenol	$C_{10}H_{14}O$	56.6	84.2	98.1	113.0	129.2	140.0	153.5	173.8	196.3	219.5	
4-iso-Butylphenol	$C_{10}H_{14}O$	72.1	100.9	115.5	130.3	147.2	157.0	171.2	192.1	214.7	237.0	
4-sec-Butylphenol	$C_{10}H_{14}O$	71.4	100.5	114.8	130.3	147.8	157.9	172.4	194.3	217.6	242.1	
4-tert-Butylphenol	$C_{10}H_{14}O$	70.0	99.2	114.0	129.5	146.0	156.0	170.2	191.5	214.0	238.0	99
2-(4-tert-Butylphenoxy)ethyl acetate	$C_{14}H_{20}O_3$	118.0	150.0	165.8	183.3	201.5	212.8	228.0	250.3	277.6	304.4	
4-tert-Butylphenyl dichlorophosphate	$C_{10}H_{13}Cl_2O_2P$	96.0	129.6	146.0	164.0	184.3	197.2	214.3	240.0	268.2	299.0	
tert-Butyl phenyl ketone (pivalophenone)	$C_{11}H_{14}O$	57.8	85.7	99.0	114.3	130.4	140.8	154.0	175.0	197.7	220.0	
iso-Butyl propionate	$C_7H_{14}O_2$	−2.3	+20.9	32.3	44.8	58.5	67.6	79.5	97.0	116.4	136.8	−71
4-tert-Butyl-2,5-xylenol	$C_{12}H_{18}O$	88.2	119.8	135.0	151.0	169.8	180.3	195.0	217.5	241.3	265.3	
4-tert-Butyl-2,6-xylenol	$C_{12}H_{18}O$	74.0	103.9	119.0	135.0	152.2	163.6	176.0	196.0	217.8	239.8	
6-tert-Butyl-2,4-xylenol	$C_{12}H_{18}O$	70.3	100.2	115.0	131.0	148.5	158.2	172.0	192.3	214.2	236.5	
6-tert-Butyl-3,4-xylenol	$C_{12}H_{18}O$	83.9	113.6	127.0	143.0	159.7	170.0	184.0	204.5	226.7	249.5	
Butyric acid	$C_4H_8O_2$	25.5	49.8	61.5	74.0	88.0	96.5	108.0	125.5	144.5	163.5	−74

Table 8. Vapor Pressures of Organic Compounds, up to 1 Atm.—(*Continued*)

Compound: Name	Formula	1	5	10	20	40	60	100	200	400	760	Melting point, °C.
		Pressure, mm. Hg / Temperature, °C.										
iso-Butyric acid	$C_4H_8O_2$	14.7	39.3	51.2	64.0	77.8	86.3	98.0	115.8	134.5	154.5	−47
Butyronitrile	C_4H_7N	−20.0	+2.1	13.4	25.7	38.4	47.3	59.0	76.7	96.8	117.5	
iso-Valerophenone	$C_{11}H_{14}O$	58.3	87.0	101.4	116.8	133.8	144.6	158.0	180.1	204.2	228.0	
Camphene	$C_{10}H_{16}$	...	...	47.2	60.4	75.7	85.0	97.9	117.5	138.7	160.5	50
Campholenic acid	$C_{10}H_{16}O_2$	97.6	125.7	139.8	153.9	170.0	180.0	193.7	212.7	234.0	256.0	
d-Camphor	$C_{10}H_{16}O$	41.5	68.6	82.3	97.5	114.0	124.0	138.0	157.9	182.0	209.2	178.5
Camphylamine	$C_{10}H_{19}N$	45.3	74.0	83.7	97.6	112.5	122.0	134.6	153.0	173.8	195.0	
Capraldehyde	$C_{10}H_{20}O$	51.9	78.8	92.0	106.3	122.2	132.0	145.3	164.8	186.3	208.5	
Capric acid	$C_{10}H_{20}O_2$	125.0	142.0	152.2	165.0	179.9	189.8	200.0	217.1	240.3	268.4	31.5
n-Caproic acid	$C_6H_{12}O_2$	71.4	89.5	99.5	111.8	125.0	133.3	144.0	160.8	181.0	202.0	−1.5
iso-Caproic acid	$C_6H_{12}O_2$	66.2	83.0	94.0	107.0	120.4	129.6	141.4	158.3	181.0	207.7	−35
iso-Caprolactone	$C_6H_{10}O_2$	38.3	66.4	80.3	95.7	112.3	123.2	137.2	157.8	182.1	207.0	
Capronitrile	$C_6H_{11}N$	9.2	34.6	47.5	61.7	76.9	86.8	99.8	119.7	141.0	163.7	
Capryl alcohol (2-octanol)	$C_8H_{18}O$	32.8	57.6	70.0	83.3	98.0	107.4	119.8	138.0	157.5	178.5	−38.6
Caprylaldehyde	$C_8H_{16}O$	73.4	92.0	101.2	110.2	120.0	126.0	133.9	145.4	156.5	168.5	
Caprylic acid (octanoic acid)	$C_8H_{16}O_2$	92.3	114.1	124.0	136.4	150.6	160.0	172.2	190.3	213.9	237.5	16
Caprylonitrile	$C_8H_{15}N$	43.0	67.6	80.4	94.6	110.6	121.2	134.8	155.2	179.5	204.5	
Carbazole	$C_{12}H_9N$	...	...	...	...	...	248.2	265.0	292.5	323.0	354.8	244.8
Carbon dioxide	CO_2	−134.3	−124.4	−119.5	−114.4	−108.6	−104.8	−100.2	−93.0	−85.7	−78.2	−57.5
disulfide	CS_2	−73.8	−54.3	−44.7	−34.3	−22.5	−15.3	−5.1	+10.4	28.0	46.5	−110.8
monoxide	CO	−222.0	−217.2	−215.0	−212.8	−210.0	−208.1	−205.7	−201.3	−196.3	−191.3	−205.0
oxyselenide (carbonyl selenide)	$COSe$	−117.1	−102.3	−95.0	−86.3	−76.4	−70.2	−61.7	−49.8	−35.6	−21.9	
oxysulfide (carbonyl sulfide)	COS	−132.4	−119.8	−113.3	−106.0	−98.3	−93.0	−85.9	−75.0	−62.7	−49.9	−138.8
tetrabromide	CBr_4	...	...	...	...	96.3	106.3	119.7	139.7	163.5	189.5	90.1
tetrachloride	CCl_4	−50.0	−30.0	−19.6	−8.2	+4.3	12.3	23.0	38.3	57.8	76.7	−22.6
tetrafluoride	CF_4	−184.6	−174.1	−169.3	−164.3	−158.8	−155.4	−150.7	−143.6	−135.5	−127.7	−183.7
Carvacrol	$C_{10}H_{14}O$	70.0	98.4	113.2	127.9	145.2	155.3	169.7	191.2	213.8	237.0	+0.5
Carvone	$C_{10}H_{14}O$	57.4	86.1	100.4	116.1	133.0	143.8	157.3	179.6	203.5	227.5	
Chavibetol	$C_{10}H_{12}O_2$	83.6	113.3	127.0	143.2	159.8	170.7	185.5	206.8	229.8	254.0	
Chloral (trichloroacetaldehyde)	C_2HCl_3O	−37.8	−16.0	−5.0	+7.2	20.2	29.1	40.2	57.8	77.5	97.7	−57
hydrate (trichloroacetaldehyde hydrate)	$C_2H_3Cl_3O_2$	−9.8	+10.0	19.5	29.2	39.7	46.2	55.0	68.0	82.1	96.2	51.7
Chloranil	$C_6Cl_4O_2$	70.7	89.3	97.8	106.4	116.1	122.0	129.5	140.3	151.3	162.6	290
Chloroacetic acid	$C_2H_3ClO_2$	43.0	68.3	81.0	94.2	109.2	118.3	130.7	149.0	169.0	189.5	61.2
anhydride	$C_4H_4Cl_2O_3$	67.2	94.1	108.0	122.4	138.2	148.0	159.8	177.8	197.0	217.0	46
2-Chloroaniline	C_6H_6ClN	46.3	72.3	84.8	99.2	115.6	125.7	139.5	160.0	183.7	208.8	0
3-Chloroaniline	C_6H_6ClN	63.5	89.8	102.0	116.7	133.6	144.1	158.0	179.5	203.5	228.5	−10.4
4-Chloroaniline	C_6H_6ClN	59.3	87.9	102.1	117.8	135.0	145.8	159.9	182.3	206.6	230.5	70.5
Chlorobenzene	C_6H_5Cl	−13.0	+10.6	22.2	35.3	49.7	58.3	70.7	89.4	110.0	132.2	−45.2
2-Chlorobenzotrichloride (2-α,α,α-tetrachlorotoluene)	$C_7H_4Cl_4$	69.0	101.8	117.9	135.8	155.0	167.8	185.0	208.0	233.0	262.1	28.7
2-Chlorobenzotrifluoride (2-chloro-α,α,α-trifluorotoluene)	$C_7H_4ClF_3$	0.0	24.7	37.1	50.6	65.9	75.4	88.3	108.3	130.0	152.2	−6.0
2-Chlorobiphenyl	$C_{12}H_9Cl$	89.3	109.8	134.7	151.2	169.9	182.1	197.0	219.6	243.8	267.5	34
4-Chlorobiphenyl	$C_{12}H_9Cl$	96.4	129.8	146.0	164.0	183.8	196.0	212.5	237.8	264.5	292.9	75.5
α-Chlorocrotonic acid	$C_4H_5ClO_2$	70.0	95.6	108.0	121.2	135.6	144.4	155.9	173.8	193.2	212.0	
Chlorodifluoromethane	$CHClF_2$	−122.8	−110.2	−103.7	−96.5	−88.6	−83.4	−76.4	−65.8	−53.6	−40.8	−160
Chlorodimethylphenylsilane	$C_8H_{11}ClSi$	29.8	56.7	70.0	84.7	101.2	111.5	124.7	145.5	168.6	193.5	
1-Chloro-2-ethoxybenzene	C_8H_9ClO	45.8	72.8	86.5	101.5	117.8	127.8	141.8	162.0	185.5	208.0	
2-(2-Chloroethoxy) ethanol	$C_4H_9ClO_2$	53.0	78.3	90.7	104.1	118.4	127.5	139.5	157.2	176.5	196.0	
bis-2-Chloroethyl acetacetal	$C_6H_{12}Cl_2O_2$	56.2	83.7	97.6	112.2	127.8	138.0	150.7	169.8	190.5	212.6	
1-Chloro-2-ethylbenzene	C_8H_9Cl	17.2	43.0	56.1	70.3	86.2	96.4	110.0	130.2	152.2	177.6	−80.2
1-Chloro-3-ethylbenzene	C_8H_9Cl	18.6	45.2	58.1	73.0	89.2	99.6	113.6	133.8	156.7	181.1	−53.3
1-Chloro-4-ethylbenzene	C_8H_9Cl	19.2	46.4	60.0	75.5	91.8	102.0	116.0	137.0	159.8	184.3	−62.6
2-Chloroethyl chloroacetate	$C_4H_6Cl_2O_2$	46.0	72.1	86.0	100.0	116.0	126.2	140.0	159.8	182.2	205.0	
2-Chloroethyl 2-chloroisopropyl ether	$C_5H_{10}Cl_2O$	24.7	50.1	63.0	77.2	92.4	102.2	115.8	135.7	156.5	180.0	
2-Chloroethyl 2-chloropropyl ether	$C_5H_{10}Cl_2O$	29.8	56.5	70.0	84.8	101.5	111.8	125.6	146.3	169.8	194.1	
2-Chloroethyl α-methylbenzyl ether	$C_{10}H_{13}ClO$	62.3	91.4	106.0	121.8	139.6	150.0	164.8	186.3	210.8	235.0	
Chloroform (trichloromethane)	$CHCl_3$	−58.0	−39.1	−29.7	−19.0	−7.1	+0.5	10.4	25.9	42.7	61.3	−63.5
1-Chloronaphthalene	$C_{10}H_7Cl$	80.6	104.8	118.6	134.4	153.2	165.6	180.4	204.2	230.8	259.3	−20
4-Chlorophenethyl alcohol	C_8H_9ClO	84.0	114.3	129.0	145.0	162.0	173.5	188.1	210.0	234.5	259.3	
2-Chlorophenol	C_6H_5ClO	12.1	38.2	51.2	65.9	82.0	92.0	106.0	126.4	149.8	174.5	7
3-Chlorophenol	C_6H_5ClO	44.2	72.0	86.1	101.7	118.0	129.4	143.0	164.8	188.7	214.0	32.5
4-Chlorophenol	C_6H_5ClO	49.8	78.2	92.2	108.1	125.0	136.1	150.0	172.0	196.0	220.0	42
2-Chloro-3-phenylphenol	$C_{12}H_9ClO$	118.0	152.2	169.7	186.7	207.4	219.6	237.0	261.3	289.4	317.5	+6
2-Chloro-6-phenylphenol	$C_{12}H_9ClO$	119.8	153.7	170.7	189.8	208.2	220.0	237.1	261.6	289.5	317.0	
Chloropicrin (trichloronitromethane)	CCl_3NO_2	−25.5	−3.3	+7.8	20.0	33.8	42.3	53.8	71.8	91.8	111.9	−64
1-Chloropropene	C_3H_5Cl	−81.3	−63.4	−54.1	−44.0	−32.7	−25.1	−15.1	+1.3	18.0	37.0	−99.0
2-Chloropyridine	C_5H_4ClN	13.3	38.8	51.7	65.8	81.7	91.6	104.6	125.0	147.7	170.2	
3-Chlorostyrene	C_8H_7Cl	25.3	51.3	65.2	80.0	96.5	107.2	121.2	142.2	165.7	190.0	
4-Chlorostyrene	C_8H_7Cl	28.0	54.5	67.5	82.0	98.0	108.5	122.0	143.5	166.0	191.0	−15.0
1-Chlorotetradecane	$C_{14}H_{29}Cl$	98.5	131.8	148.2	166.2	187.0	199.8	215.5	240.3	267.5	296.0	+0.9
2-Chlorotoluene	C_7H_7Cl	+5.4	30.6	43.2	56.9	72.0	81.8	94.7	115.0	137.1	159.3	
3-Chlorotoluene	C_7H_7Cl	+4.8	30.3	43.2	57.4	73.0	83.2	96.3	116.6	139.7	162.3	
4-Chlorotoluene	C_7H_7Cl	+5.5	31.0	43.8	57.8	73.5	83.3	96.6	117.1	139.8	162.3	+7.3
Chlorotriethylsilane	$C_6H_{15}ClSi$	−4.9	+19.8	32.0	45.5	60.2	69.5	82.3	101.6	123.6	146.3	
1-Chloro-1,2,2-trifluoroethylene	C_2ClF_3	−116.0	−102.5	−95.9	−88.2	−79.7	−74.1	−66.7	−55.0	−41.7	−27.9	−157.5
Chlorotrifluoromethane	$CClF_3$	−149.5	−139.2	−134.1	−128.5	−121.9	−117.3	−111.7	−102.5	−92.7	−81.2	
Chlorotrimethylsilane	C_3H_9ClSi	−62.8	−43.6	−34.0	−23.2	−11.4	−4.0	+6.0	21.9	39.4	57.9	
trans-Cinnamic acid	$C_9H_8O_2$	127.5	157.8	173.0	189.5	207.1	217.8	232.4	253.3	276.7	300.0	133
Cinnamyl alcohol	$C_9H_{10}O$	72.6	102.5	117.8	133.7	151.0	162.0	177.8	199.8	224.6	250.0	33
Cinnamylaldehyde	C_9H_8O	76.1	105.8	120.0	135.7	152.2	163.7	177.7	199.3	222.4	246.0	−7.5
Citraconic anhydride	$C_5H_4O_3$	47.1	74.8	88.9	103.8	120.3	131.3	145.4	165.8	189.8	213.5	
cis-α-Citral	$C_{10}H_{16}O$	61.7	90.0	103.9	119.4	135.9	146.3	160.0	181.8	205.0	228.0	
d-Citronellal	$C_{10}H_{18}O$	44.0	71.4	84.8	99.8	116.1	126.2	140.1	160.0	183.8	206.5	
Citronellic acid	$C_{10}H_{18}O_2$	99.5	127.3	141.4	155.6	171.9	182.1	195.4	214.5	236.6	257.0	
Citronellol	$C_{10}H_{20}O$	66.4	93.6	107.0	121.5	137.2	147.2	159.8	179.8	201.0	221.5	
Citronellyl acetate	$C_{12}H_{22}O_2$	74.7	100.2	113.0	126.0	140.5	149.7	161.0	178.8	197.8	217.0	
Coumarin	$C_9H_6O_2$	106.0	137.8	153.4	170.0	189.0	200.5	216.5	240.0	264.7	291.0	70

Table 8. Vapor Pressures of Organic Compounds, up to 1 Atm.—(*Continued*)

Compound		Pressure, mm. Hg										Melting point, °C.
Name	Formula	1	5	10	20	40	60	100	200	400	760	
		Temperature, °C.										
o-Cresol (2-cresol; 2-methylphenol)	C_7H_8O	38.2	64.0	76.7	90.5	105.8	115.5	127.4	146.7	168.4	190.8	30.8
m-Cresol (3-cresol; 3-methylphenol)	C_7H_8O	52.0	76.0	87.8	101.4	116.0	125.8	138.0	157.3	179.0	202.8	10.9
p-Cresol (4-cresol; 4-methylphenol)	C_7H_8O	53.0	76.5	88.6	102.3	117.7	127.0	140.0	157.3	179.4	201.8	35.5
cis-Crotonic acid	$C_4H_6O_2$	33.5	57.4	69.0	82.0	96.0	104.5	116.3	133.9	152.2	171.9	15.5
trans-Crotonic acid	$C_4H_6O_2$			80.0	93.0	107.8	116.7	128.0	146.0	165.5	185.0	72
cis-Crotononitrile	C_4H_5N	−29.0	−7.1	+4.0	16.4	30.0	38.5	50.1	68.0	88.0	108.0	
trans-Crotononitrile	C_4H_5N	−19.5	+3.5	15.0	27.8	41.8	50.9	62.8	81.1	101.5	122.8	
Cumene	C_9H_{12}	+2.9	26.8	38.3	51.5	66.1	75.4	88.1	107.3	129.2	152.4	−96.0
4-Cumidene	$C_9H_{13}N$	60.0	88.2	102.2	117.8	134.2	145.0	158.0	180.0	203.2	227.0	
Cuminal	$C_{10}H_{12}O$	58.0	87.3	102.0	117.9	135.2	146.0	160.0	182.8	206.7	232.0	
Cuminyl alcohol	$C_{10}H_{14}O$	74.2	103.7	118.0	133.8	150.3	161.7	176.2	197.9	221.7	246.6	
2-Cyano-2-n-butyl acetate	$C_7H_{11}NO_2$	42.0	68.7	82.0	96.2	111.8	121.5	133.8	152.2	173.4	195.2	
Cyanogen	C_2N_2	−95.8	−83.2	−76.8	−70.1	−62.7	−57.9	−51.8	−42.6	−33.0	−21.0	−34.4
bromide	$CBrN$	−35.7	−18.3	−10.0	−1.0	+8.6	14.7	22.6	33.8	46.0	61.5	58
chloride	$CClN$	−76.7	−61.4	−53.8	−46.1	−37.5	−32.1	−24.9	−14.1	−2.3	+13.1	−6.5
iodide	CIN	25.2	47.2	57.7	68.6	80.3	88.0	97.6	111.5	126.1	141.1	
Cyclobutane	C_4H_8	−92.0	−76.0	−67.9	−58.7	−48.4	−41.8	−32.8	−18.9	−3.4	+12.9	−50
Cyclobutene	C_4H_6	−99.1	−83.4	−75.4	−66.6	−56.4	−50.0	−41.2	−27.8	−12.2	+2.4	
Cyclohexane	C_6H_{12}	−45.3	−25.4	−15.9	−5.0	+6.7	14.7	25.5	42.0	60.8	80.7	+6.6
Cyclohexaneethanol	$C_8H_{16}O$	50.4	77.2	90.0	104.0	119.8	129.8	142.7	161.7	183.5	205.4	
Cyclohexanol	$C_6H_{12}O$	21.0	44.0	56.0	68.8	83.0	91.8	103.7	121.7	141.4	161.0	23.9
Cyclohexanone	$C_6H_{10}O$	+1.4	26.4	38.7	52.5	67.8	77.5	90.4	110.3	132.5	155.6	−45.0
2-Cyclohexyl-4,6-dinitrophenol	$C_{12}H_{14}N_2O_5$	132.8	161.8	175.9	191.2	206.7	216.0	229.0	248.7	269.8	291.5	
Cyclopentane	C_5H_{10}	−68.0	−49.6	−40.4	−30.1	−18.6	−11.3	−1.3	+13.8	31.0	49.3	−93.7
Cyclopropane	C_3H_6	−116.8	−104.2	−97.5	−90.3	−82.3	−77.0	−70.0	−59.1	−46.9	−33.5	−126.6
Cymene	$C_{10}H_{14}$	17.3	43.9	57.0	71.1	87.0	97.2	110.8	131.4	153.5	177.2	−68.2
cis-Decalin	$C_{10}H_{18}$	22.5	50.1	64.2	79.8	97.2	108.0	123.2	145.4	169.9	194.6	−43.3
trans-Decalin	$C_{10}H_{18}$	−0.8	+30.6	47.2	65.3	85.7	98.4	114.6	136.2	160.1	186.7	−30.7
Decane	$C_{10}H_{22}$	16.5	42.3	55.7	69.8	85.5	95.5	108.6	128.4	150.6	174.1	−29.7
Decan-2-one	$C_{10}H_{20}O$	44.2	71.9	85.8	100.7	117.1	127.8	142.0	163.2	186.7	211.0	+3.5
1-Decene	$C_{10}H_{20}$	14.7	40.3	53.7	67.8	83.3	93.5	106.5	126.7	149.2	172.0	
Decyl alcohol	$C_{10}H_{22}O$	69.5	97.3	111.3	125.8	142.1	152.0	165.8	186.2	208.8	231.0	+7
Decyltrimethylsilane	$C_{13}H_{30}Si$	67.4	96.4	111.0	126.5	144.0	154.3	169.5	191.0	215.5	240.0	
Dehydroacetic acid	$C_8H_8O_4$	91.7	122.0	137.3	153.0	171.0	181.5	197.5	219.5	244.5	269.0	
Desoxybenzoin	$C_{14}H_{12}O$	123.3	156.2	173.5	192.0	212.0	224.5	241.3	265.2	293.0	321.0	60
Diacetamide	$C_4H_7NO_2$	70.0	95.0	108.0	122.6	138.2	148.0	160.6	180.8	202.0	223.0	78.5
Diacetylene (1,3-butadiyne)	C_4H_2	−82.5	−68.0	−61.2	−53.8	−45.9	−41.0	−34.0	−20.9	−6.1	+9.7	−34.9
Diallyldichlorosilane	$C_6H_{10}Cl_2Si$	+9.5	34.8	47.4	61.3	76.4	86.3	99.7	119.4	142.0	165.3	
Diallyl sulfide	$C_6H_{10}S$	−9.5	+14.4	26.6	39.7	54.2	63.7	75.8	94.8	116.1	138.6	−83
Diisoamyl ether	$C_{10}H_{22}O$	18.6	44.3	57.0	70.7	86.3	96.0	109.6	129.0	150.3	173.4	
oxalate	$C_{12}H_{22}O_4$	85.4	116.0	131.4	147.7	165.7	177.0	192.2	215.0	240.0	265.0	
sulfide	$C_{10}H_{22}S$	43.0	73.0	87.6	102.7	120.0	130.6	145.3	166.4	191.0	216.0	
Dibenzylamine	$C_{14}H_{15}N$	118.3	149.8	165.6	182.2	200.2	212.2	227.3	249.8	274.3	300.0	−26
Dibenzyl ketone (1,3-diphenyl-2-propanone)	$C_{15}H_{14}O$	125.5	159.8	177.6	195.7	216.6	229.4	246.6	272.3	301.7	330.5	34.5
1,4-Dibromobenzene	$C_6H_4Br_2$	61.0	79.3	87.7	103.6	120.8	131.6	146.5	168.5	192.5	218.6	87.5
1,2-Dibromobutane	$C_4H_8Br_2$	7.5	33.2	46.1	60.0	76.0	86.0	99.8	120.2	143.5	166.3	−64.5
dl-2,3-Dibromobutane	$C_4H_8Br_2$	+5.0	30.0	41.6	56.4	72.0	82.0	95.3	115.7	138.0	160.5	
meso-2,3-Dibromobutane	$C_4H_8Br_2$	+1.5	26.6	39.3	53.2	68.0	78.0	91.7	111.8	134.2	157.3	−34.5
1,2-Dibromodecane	$C_{10}H_{20}Br_2$	95.7	123.6	137.3	151.0	167.4	177.5	190.2	209.6	229.8	250.4	
Di(2-bromoethyl) ether	$C_4H_8Br_2O$	47.7	75.3	88.5	103.6	119.8	130.0	144.0	165.0	188.0	212.5	
α,β-Dibromomaleic anhydride	$C_4H_2Br_2O_3$	50.0	78.0	92.0	106.7	123.5	133.8	147.7	168.0	192.0	215.0	
1,2-Dibromo-2-methylpropane	$C_4H_8Br_2$	−28.8	−3.0	+10.5	25.7	42.3	53.7	68.8	92.1	119.8	149.0	−70.3
1,3-Dibromo-2-methylpropane	$C_4H_8Br_2$	14.0	40.0	53.0	67.5	83.5	93.7	107.4	117.8	150.6	174.6	
1,2-Dibromopentane	$C_5H_{10}Br_2$	19.8	45.4	58.0	72.0	87.4	97.4	110.1	130.2	151.8	175.0	
1,2-Dibromopropane	$C_3H_6Br_2$	−7.0	+17.3	29.4	42.3	57.2	66.4	78.7	97.8	118.5	141.6	−55.5
1,3-Dibromopropane	$C_3H_6Br_2$	+9.7	35.4	48.0	62.1	77.8	87.8	101.3	121.7	144.1	167.5	−34.4
2,3-Dibromopropene	$C_3H_4Br_2$	−6.0	+17.9	30.0	43.2	57.8	67.0	79.5	98.0	119.5	141.2	
2,3-Dibromo-1-propanol	$C_3H_6Br_2O$	57.0	84.5	98.2	113.5	129.8	140.0	153.0	173.8	196.0	219.0	
Diisobutylamine	$C_8H_{19}N$	−5.1	+18.4	30.6	43.7	57.8	67.0	79.2	97.6	118.0	139.5	−70
2,6-Ditert-butyl-4-cresol	$C_{15}H_{24}O$	85.8	116.2	131.0	147.0	164.1	175.2	190.0	212.8	237.6	262.5	
4,6-Ditert-butyl-2-cresol	$C_{15}H_{24}O$	86.2	117.3	132.4	149.0	167.4	179.0	194.0	217.5	243.4	269.3	
4,6-Ditert-butyl-3-cresol	$C_{15}H_{24}O$	103.7	135.2	150.0	167.0	185.3	196.1	211.0	233.0	257.1	282.0	
2,6-Ditert-butyl-4-ethylphenol	$C_{16}H_{26}O$	89.1	121.4	137.0	154.0	172.1	183.9	198.0	220.0	244.0	268.6	
4,6-Ditert-butyl-3-ethylphenol	$C_{16}H_{26}O$	111.5	142.6	157.4	174.0	192.3	204.4	218.0	241.7	264.6	290.0	
Diisobutyl oxalate	$C_{10}H_{18}O_4$	63.2	91.2	105.3	120.3	137.5	147.8	161.8	183.5	205.8	229.5	
2,4-Ditert-butylphenol	$C_{14}H_{22}O$	84.5	115.4	130.0	146.0	164.3	175.8	190.0	212.5	237.0	260.8	
Dibutyl phthalate	$C_{16}H_{22}O_4$	148.2	182.1	198.2	216.2	235.8	247.8	263.7	287.0	313.5	340.0	
sulfide	$C_8H_{18}S$	+21.7	51.8	66.4	80.5	96.0	105.8	118.6	138.0	159.0	182.0	−79.7
Diisobutyl d-tartrate	$C_{12}H_{22}O_6$	117.8	151.8	169.0	188.0	208.5	221.6	239.5	264.7	294.0	324.0	73.5
Dicarvacryl-mono-(6-chloro-2-xenyl) phosphate	$C_{32}H_{34}ClO_4P$	204.2	234.5	249.3	264.5	280.5	290.7	304.9	323.8	342.0	361.0	
Dicarvacryl-2-tolyl phosphate	$C_{27}H_{33}O_4P$	180.2	209.3	221.8	237.0	251.5	260.3	272.5	290.0	309.8	330.0	
Dichloroacetic acid	$C_2H_2Cl_2O_2$	44.0	69.8	82.6	96.3	111.8	121.5	134.0	152.3	173.7	194.4	9.7
1,2-Dichlorobenzene	$C_6H_4Cl_2$	20.0	46.0	59.1	73.4	89.4	99.5	112.9	133.4	155.8	179.0	−17.6
1,3-Dichlorobenzene	$C_6H_4Cl_2$	12.1	39.0	52.0	66.2	82.0	92.2	105.0	125.9	149.0	173.0	−24.2
1,4-Dichlorobenzene	$C_6H_4Cl_2$			54.8	69.2	84.8	95.2	108.4	128.3	150.2	173.9	53.0
1,2-Dichlorobutane	$C_4H_8Cl_2$	−23.6	−0.3	+11.5	24.5	37.7	47.8	60.2	79.7	100.8	123.5	
2,3-Dichlorobutane	$C_4H_8Cl_2$	−25.2	−3.0	+8.5	21.2	35.0	43.9	56.0	74.0	94.2	116.0	−80.4
1,2-Dichloro-1,2-difluoroethylene	$C_2Cl_2F_2$	−82.0	−65.6	−57.3	−48.3	−38.2	−31.8	−23.0	−10.0	+5.0	20.9	−112
Dichlorodifluoromethane	CCl_2F_2	−118.5	−104.6	−97.8	−90.1	−81.6	−76.1	−68.6	−57.0	−43.9	−29.8	
Dichlorodiphenyl silane	$C_{12}H_{10}Cl_2Si$	109.6	142.4	158.0	176.0	195.5	207.5	223.8	248.0	275.5	304.0	
Dichlorodiisopropyl ether	$C_6H_{12}Cl_2O$	29.6	55.2	68.2	82.2	97.3	106.9	119.7	139.0	159.8	182.7	
Di(2-chloroethoxy) methane	$C_5H_{10}Cl_2O_2$	53.0	80.4	94.0	109.5	125.5	135.8	149.6	170.0	192.0	215.0	
Dichloroethoxymethylsilane	$C_3H_8Cl_2OSi$	−33.8	−12.1	−1.3	+11.3	24.4	32.6	44.1	61.0	80.3	100.6	
1,2-Dichloro-3-ethylbenzene	$C_8H_8Cl_2$	46.0	75.0	90.0	105.9	123.8	135.0	149.8	172.0	197.0	222.1	−40.8
1,2-Dichloro-4-ethylbenzene	$C_8H_8Cl_2$	47.0	77.2	92.3	109.6	127.5	139.0	153.3	176.0	201.7	226.6	−76.4
1,4-Dichloro-2-ethylbenzene	$C_8H_8Cl_2$	38.5	68.0	83.2	99.8	118.0	129.0	144.0	166.2	191.5	216.3	−61.2
cis-1,2-Dichloroethylene	$C_2H_2Cl_2$	−58.4	−39.2	−29.9	−19.4	−7.9	−0.5	+9.5	24.6	41.0	59.0	−80.5
trans-1,2-Dichloro ethylene	$C_2H_2Cl_2$	−65.4	−47.2	−38.0	−28.0	−17.0	−10.0	−0.2	+14.3	30.8	47.8	−50.0

Table 8. Vapor Pressures of Organic Compounds, up to 1 Atm.—(*Continued*)

Compound: Name	Compound: Formula	Pressure, mm. Hg: 1	5	10	20	40	60	100	200	400	760	Melting point, °C.
		Temperature, °C.										
Di(2-chloroethyl) ether	$C_4H_8Cl_2O$	23.5	49.3	62.0	76.0	91.5	101.5	114.5	134.0	155.4	178.5	
Dichlorofluoromethane	$CHCl_2F$	−91.3	−75.5	−67.5	−58.6	−48.8	−42.6	−33.9	−20.9	−6.2	+8.9	−135
1,5-Dichlorohexamethyltrisiloxane	$C_6H_{18}Cl_2O_2Si_3$	26.0	52.0	65.1	79.0	94.8	105.0	118.2	138.3	160.2	184.0	−53.0
Dichloromethylphenylsilane	$C_7H_8Cl_2Si$	35.7	63.5	77.4	92.4	109.5	120.0	134.2	155.5	180.2	205.5	
1,1-Dichloro-2-methylpropane	$C_4H_8Cl_2$	−31.0	−8.4	+2.6	14.6	28.2	37.0	48.2	65.8	85.4	106.0	
1,2-Dichloro-2-methylpropane	$C_4H_8Cl_2$	−25.8	−4.2	+6.7	18.7	32.0	40.2	51.7	68.9	87.8	108.0	
1,3-Dichloro-2-methylpropane	$C_4H_8Cl_2$	−3.0	+20.6	32.0	44.8	58.6	67.5	78.8	96.1	115.4	135.0	
2,4-Dichlorophenol	$C_6H_4Cl_2O$	53.0	80.0	92.8	107.7	123.4	133.5	146.0	165.2	187.5	210.0	45.0
2,6-Dichlorophenol	$C_6H_4Cl_2O$	59.5	87.6	101.0	115.5	131.6	141.8	154.6	175.5	197.7	220.0	
α,α-Dichlorophenylacetonitrile	$C_8H_5Cl_2N$	56.0	84.0	98.1	113.8	130.0	141.0	154.5	176.2	199.5	223.5	
Dichlorophenylarsine	$C_6H_5AsCl_2$	61.8	100.0	116.0	133.1	151.0	163.2	178.9	202.8	228.8	256.5	
1,2-Dichloropropane	$C_3H_6Cl_2$	−38.5	−17.0	−6.1	+6.0	19.4	28.0	39.4	57.0	76.0	96.8	
2,3-Dichlorostyrene	$C_8H_6Cl_2$	61.0	90.1	104.6	120.5	137.8	149.0	163.5	185.7	210.0	235.0	
2,4-Dichlorostyrene	$C_8H_6Cl_2$	53.5	82.2	97.4	111.8	129.2	140.0	153.8	176.0	200.0	225.0	
2,5-Dichlorostyrene	$C_8H_6Cl_2$	55.5	83.9	98.2	114.0	131.0	142.0	155.8	178.0	202.5	227.0	
2,6-Dichlorostyrene	$C_8H_6Cl_2$	47.8	75.7	90.0	105.5	122.4	133.3	147.6	169.0	193.5	217.0	
3,4-Dichlorostyrene	$C_8H_6Cl_2$	57.2	86.0	100.4	116.2	133.7	144.6	158.2	181.5	205.7	230.0	
3,5-Dichlorostyrene	$C_8H_6Cl_2$	53.5	82.2	97.4	111.8	129.2	140.0	153.8	176.0	200.0	225.0	
1,2-Dichlorotetraethylbenzene	$C_{14}H_{20}Cl_2$	105.6	138.7	155.0	172.5	192.2	204.8	220.7	245.6	272.8	302.0	
1,4-Dichlorotetraethylbenzene	$C_{14}H_{20}Cl_2$	91.7	126.1	143.8	162.0	183.2	195.8	212.0	238.5	265.8	296.5	
1,2-Dichloro-1,1,2,2-tetrafluoroethane	$C_2Cl_2F_4$	−95.4	−80.0	−72.3	−63.5	−53.7	−47.5	−39.1	−26.3	−12.0	+3.5	−94
Dichloro-4-tolylsilane	$C_7H_8Cl_2Si$	46.2	71.7	84.2	97.8	113.2	122.6	135.5	153.5	175.2	196.3	
3,4-Dichloro-α,α,α-trifluorotoluene	$C_7H_3Cl_2F_3$	11.0	38.3	52.2	67.3	84.0	95.0	109.2	129.0	150.5	172.8	−12.1
Dicyclopentadiene	$C_{10}H_8$	...	34.1	47.6	62.0	77.9	88.0	101.7	121.8	144.2	166.6	32.9
Diethoxydimethylsilane	$C_6H_{16}O_2Si$	−19.1	+2.4	13.3	25.3	38.0	46.3	57.6	74.2	93.2	113.5	
Diethoxydiphenylsilane	$C_{16}H_{20}O_2Si$	111.5	142.8	157.6	174.3	193.2	205.0	220.0	243.8	259.7	296.0	
Diethyl adipate	$C_{10}H_{18}O_4$	74.0	106.6	123.0	138.3	154.6	165.8	179.0	198.2	219.1	240.0	−21
Diethylamine	$C_4H_{11}N$	...	...	−33.0	−22.6	−11.3	−4.0	+6.0	21.0	38.0	55.5	−38.9
N-Diethylaniline	$C_{10}H_{15}N$	49.7	78.0	91.9	107.2	123.6	133.8	147.3	168.2	192.4	215.5	−34.4
Diethyl arsanilate	$C_{10}H_{16}AsNO_3$	38.0	62.6	74.8	88.0	102.6	111.8	123.8	141.9	161.0	181.0	
1,2-Diethylbenzene	$C_{10}H_{14}$	22.3	48.7	62.0	76.4	92.5	102.6	116.2	136.7	159.0	183.5	−31.4
1,3-Diethylbenzene	$C_{10}H_{14}$	20.7	46.8	59.9	74.5	90.4	100.7	114.4	134.8	156.9	181.1	−83.9
1,4-Diethylbenzene	$C_{10}H_{14}$	20.7	47.1	60.3	74.7	91.1	101.3	115.3	136.1	159.0	183.8	−43.2
Diethyl carbonate	$C_5H_{10}O_3$	−10.1	+12.3	23.8	36.0	49.5	57.9	69.7	86.5	105.8	125.8	−43
cis-Diethyl citraconate	$C_9H_{14}O_4$	59.8	88.3	103.0	118.2	135.7	146.2	160.0	182.3	206.5	230.3	
Diethyl dioxosuccinate	$C_8H_{10}O_6$	70.0	98.0	112.0	126.8	143.8	153.7	167.7	188.0	210.8	233.5	
Diethylene glycol	$C_4H_{10}O_3$	91.8	120.0	133.8	148.0	164.3	174.0	187.5	207.0	226.5	244.8	
Diethyleneglycol-bis-chloroacetate	$C_8H_{12}Cl_2O_5$	148.3	180.0	195.8	212.0	229.0	239.5	252.0	271.5	291.8	313.0	
Diethylene glycol dimethyl ether Di(2-methoxyethyl) ether	$C_6H_{14}O_3$	13.0	37.6	50.0	63.0	77.5	86.8	99.5	118.0	138.5	159.8	
glycol ethyl ether	$C_6H_{14}O_3$	45.3	72.0	85.8	100.3	116.7	126.8	140.3	159.0	180.3	201.9	
Diethyl ether	$C_4H_{10}O$	−74.3	−56.9	−48.1	−38.5	27.7	−21.8	−11.5	+2.2	17.9	34.6	−116.3
ethylmalonate	$C_9H_{16}O_4$	50.8	77.8	91.6	106.0	122.4	132.4	146.0	166.0	188.7	211.5	
fumarate	$C_8H_{12}O_4$	53.2	81.2	95.3	110.2	126.7	137.7	151.1	172.2	195.8	218.5	+0.6
glutarate	$C_9H_{16}O_4$	65.6	94.7	109.7	125.4	142.8	153.2	167.8	189.5	212.8	237.0	
Diethylhexadecylamine	$C_{20}H_{43}N$	139.8	175.8	194.0	213.5	235.0	248.5	265.5	292.8	324.6	355.0	
Diethyl itaconate	$C_9H_{14}O_4$	51.3	80.2	95.2	111.0	128.2	139.9	154.3	177.5	203.1	227.9	
ketone (3-pentanone)	$C_5H_{10}O$	−12.7	+7.5	17.2	27.9	39.4	46.7	56.2	70.6	86.3	102.7	−42
malate	$C_8H_{14}O_5$	80.7	110.4	125.3	141.2	157.8	169.0	183.9	205.3	229.5	253.4	
maleate	$C_8H_{12}O_4$	57.3	85.6	100.0	115.3	131.8	142.4	156.0	177.8	201.7	225.0	
malonate	$C_7H_{12}O_4$	40.0	67.5	81.3	95.9	113.3	123.0	136.2	155.5	176.8	198.9	−49.8
mesaconate	$C_9H_{14}O_4$	62.8	91.0	105.3	120.3	137.3	147.9	161.6	183.2	205.8	229.0	
oxalate	$C_6H_{10}O_4$	47.4	71.8	83.8	96.8	110.6	119.7	130.8	147.9	166.2	185.7	−40.6
phthalate	$C_{12}H_{14}O_4$	108.8	140.7	156.0	173.6	192.1	204.1	219.5	243.0	267.5	294.0	
sebacate	$C_{14}H_{26}O_4$	125.3	156.2	172.1	189.8	207.5	218.4	234.4	255.8	280.3	305.5	1.3
2,5-Diethylstyrene	$C_{12}H_{16}$	49.7	78.4	92.6	108.5	125.8	136.8	151.0	173.2	198.0	223.0	
Diethyl succinate	$C_8H_{14}O_4$	54.6	83.0	96.6	111.7	127.8	138.2	151.1	171.7	193.8	216.5	−20.8
isosuccinate	$C_8H_{14}O_4$	39.8	66.7	80.0	94.7	111.0	121.4	134.8	155.1	177.7	201.3	
sulfate	$C_4H_{10}O_4S$	47.0	74.0	87.7	102.1	118.0	128.6	142.5	162.5	185.5	209.5	−25.0
sulfide	$C_4H_{10}S$	−39.6	−18.6	−8.0	+3.5	16.1	24.2	35.0	51.3	69.7	88.0	−99.5
sulfite	$C_4H_{10}O_3S$	10.0	34.2	46.4	59.7	74.2	83.8	96.3	115.8	137.0	159.0	
d-Diethyl tartrate	$C_8H_{14}O_6$	102.0	133.0	148.0	164.2	182.3	194.0	208.5	230.4	254.8	280.0	17
dl-Diethyl tartrate	$C_8H_{14}O_6$	100.0	131.7	147.2	163.8	181.7	193.2	208.0	230.0	254.3	280.0	
3,5-Diethyltoluene	$C_{11}H_{16}$	34.0	61.5	75.3	90.2	107.0	117.7	131.7	152.4	176.5	200.7	
Diethylzinc	$C_4H_{10}Zn$	−22.4	0.0	+11.7	24.2	38.0	47.2	59.1	77.0	97.3	118.0	−28
l-Dihydrocarvone	$C_{10}H_{16}O$	46.6	75.5	90.0	106.0	123.7	134.7	149.7	171.8	197.0	223.0	
Dihydrocitronellol	$C_{10}H_{22}O$	68.0	91.7	103.0	115.0	127.6	136.7	145.9	160.2	176.8	193.5	
1,4-Dihydroxyanthraquinone	$C_{14}H_8O_4$	196.7	239.8	259.8	282.0	307.4	323.3	344.5	377.8	413.0	450.0	194
Dimethylacetylene (2-butyne)	C_4H_6	−73.0	−57.9	−50.5	−42.5	−33.9	−27.8	−18.8	−5.0	+10.6	27.2	−32.5
Dimethylamine	C_2H_7N	−87.7	−72.2	−64.6	−56.0	−46.7	−40.7	−32.6	−20.4	−7.1	+7.4	−96
N,N-Dimethylaniline	$C_8H_{11}N$	29.5	56.3	70.0	84.8	101.6	111.9	125.8	146.5	169.2	193.1	+2.5
imethyl arsanilate	$C_8H_{12}AsNO_3$	15.0	39.6	51.8	65.0	79.7	88.6	101.0	119.8	140.3	160.5	
Di(α-methylbenzyl) ether	$C_{16}H_{18}O$	96.7	128.3	144.0	160.3	179.6	191.5	206.8	229.7	254.8	281.0	
2,2-Dimethylbutane	C_6H_{14}	−69.3	−50.7	−41.5	−31.1	−19.5	−12.1	−2.0	+13.4	31.0	49.7	−99.8
2,3-Dimethylbutane	C_6H_{14}	−63.6	−44.5	−34.9	−24.1	−12.4	−4.9	+5.4	21.1	39.0	58.0	−128.2
Dimethyl citraconate	$C_7H_{10}O_4$	50.8	78.2	91.8	106.5	122.6	132.7	145.8	165.8	188.0	210.5	
1,1-Dimethylcyclohexane	C_8H_{16}	−24.4	−1.4	+10.3	23.0	37.3	45.7	57.9	76.2	97.2	119.5	−34
cis-1,2-Dimethylcyclohexane	C_8H_{16}	−15.9	+7.3	18.4	31.1	45.3	54.4	66.8	85.6	107.0	129.7	−50.0
crans-1,2-Dimethylcyclohexane	C_8H_{16}	−21.1	+1.7	13.0	25.6	39.7	48.7	61.0	79.6	100.9	123.4	−88.0
crans-1,3-Dimethylcyclohexane	C_8H_{16}	−19.4	+3.4	14.9	27.4	41.4	50.4	62.5	81.0	102.1	124.4	−92.0
cis-1,3-Dimethylcyclohexane	C_8H_{16}	−22.7	0.0	+11.2	23.6	37.5	46.4	58.5	76.9	97.8	120.1	−76.2
cis-1,4-Dimethylcyclohexane	C_8H_{16}	−20.0	+3.2	14.5	27.1	41.1	50.1	62.3	80.8	101.9	124.3	−87.4
trans-1,4-Dimethylcyclohexane	C_8H_{16}	−24.3	−1.7	+10.1	22.6	36.5	45.4	57.6	76.0	97.0	119.3	−36.9
Dimethyl ether	C_2H_6O	−115.7	−101.1	−93.3	−85.2	−76.2	−70.4	−62.7	−50.9	−37.8	−23.7	−138.5
2,2-Dimethylhexane	C_8H_{18}	−29.7	−7.9	+3.1	15.0	28.2	36.7	48.2	65.7	85.6	106.8	
2,3-Dimethylhexane	C_8H_{18}	−23.0	−1.1	+9.9	22.1	35.6	44.2	56.0	73.8	94.1	115.6	
2,4-Dimethylhexane	C_8H_{18}	−26.9	−5.3	+5.2	17.2	30.5	39.0	50.6	68.1	88.2	109.4	
2,5-Dimethylhexane	C_8H_{18}	−26.7	−5.5	+5.3	17.2	30.4	38.9	50.5	68.0	87.9	109.1	−90.7

Table 8. Vapor Pressures of Organic Compounds, up to 1 Atm.—(*Continued*)

Compound: Name	Formula	1	5	10	20	40	60	100	200	400	760	Melting point, °C.
		Pressure, mm. Hg — Temperature °C.										
3,3-Dimethylhexane	C_8H_{18}	−25.8	−4.4	+6.1	18.2	31.7	40.4	52.5	70.0	90.4	112.0	
3,4-Dimethylhexane	C_8H_{18}	−22.1	+0.2	11.3	23.5	37.1	45.8	57.7	75.6	96.0	117.7	
Dimethyl itaconate	$C_7H_{10}O_4$	69.3	94.0	106.6	119.7	133.7	142.6	153.7	171.0	189.8	208.0	38
l-Dimethyl malate	$C_6H_{10}O_5$	75.4	104.0	118.3	133.8	150.1	160.4	175.1	196.3	219.5	242.6	
Dimethyl maleate	$C_6H_8O_4$	45.7	73.0	86.4	101.3	117.2	127.1	140.4	160.0	182.2	205.0	
malonate	$C_5H_8O_4$	35.0	59.8	72.0	85.0	100.0	109.7	121.9	140.0	159.8	180.7	−62
trans-Dimethyl mesaconate	$C_7H_{10}O_4$	46.8	74.0	87.8	102.1	118.0	127.8	141.5	161.0	183.5	206.0	
2,7-Dimethyloctane	$C_{10}H_{22}$	+6.3	30.5	42.3	55.8	71.2	80.8	93.9	114.0	136.0	159.7	−52.8
Dimethyl oxalate	$C_4H_6O_4$	20.0	44.0	56.0	69.4	83.6	92.8	104.8	123.3	143.3	163.3	
2,2-Dimethylpentane	C_7H_{16}	−49.0	−28.7	−18.7	−7.5	+5.0	13.0	23.9	40.3	59.2	79.2	−123.7
2,3-Dimethylpentane	C_7H_{16}	−42.0	−20.8	−10.3	+1.1	13.9	22.1	33.3	50.1	69.4	89.8	−135
2,4-Dimethylpentane	C_7H_{16}	−48.0	−27.4	−17.1	−5.9	+6.5	14.5	25.4	41.8	60.6	80.5	−119.5
3,3-Dimethylpentane	C_7H_{16}	−45.9	−25.0	−14.4	−2.9	+9.9	18.1	29.3	46.2	65.5	86.1	−135.0
2,3-Dimethylphenol (2,3-xylenol)	$C_8H_{10}O$	56.0	83.8	97.6	112.0	129.2	139.5	152.2	173.0	196.0	218.0	75
2,4-Dimethylphenol (2,4-xylenol)	$C_8H_{10}O$	51.8	78.0	91.3	105.0	121.5	131.0	143.0	161.5	184.2	211.5	25.5
2,5-Dimethylphenol (2,5-xylenol)	$C_8H_{10}O$	51.8	78.0	91.3	105.0	121.5	131.0	143.0	161.5	184.2	211.5	74.5
3,4-Dimethylphenol (3,4-xylenol)	$C_8H_{10}O$	66.2	93.8	107.7	122.0	138.0	148.0	161.0	181.5	203.6	225.2	62.5
3.5-Dimethylphenol (3,5-xylenol)	$C_8H_{10}O$	62.0	89.2	102.4	117.0	133.3	143.5	156.0	176.2	197.8	219.5	68
Dimethylphenylsilane	$C_8H_{12}Si$	+5.3	30.3	42.6	56.2	71.4	81.3	94.2	114.2	136.4	159.3	
Dimethyl phthalate	$C_{10}H_{10}O_4$	100.3	131.8	147.6	164.0	182.8	194.0	210.0	232.7	257.8	283.7	
3,5-Dimethyl-1,2-pyrone	$C_7H_8O_2$	78.6	107.6	122.0	136.4	152.7	163.8	177.5	198.0	221.0	245.0	51.5
4,6-Dimethylresorcinol	$C_8H_{10}O_2$	49.0	76.8	90.7	105.8	122.5	133.2	147.3	167.8	192.0	215.0	
Dimethyl sebacate	$C_{12}H_{22}O_4$	104.0	139.8	156.2	175.8	196.0	208.0	222.6	245.0	269.6	293.5	38
2,4-Dimethylstyrene	$C_{10}H_{12}$	34.2	61.9	75.8	90.8	107.7	118.0	132.3	153.2	177.5	202.0	
2,5-Dimethylstyrene	$C_{10}H_{12}$	29.0	55.9	69.0	84.0	100.2	110.7	124.7	145.6	168.7	193.0	
α,α-Dimethylsuccinic anhydride	$C_6H_8O_3$	61.4	88.1	102.0	116.3	132.3	142.4	155.3	175.8	197.5	219.5	
Dimethyl sulfide	C_2H_6S	−75.6	−58.0	−49.2	−39.4	−28.4	−21.4	−12.0	+2.6	18.7	36.0	−83.2
d-Dimethyl tartrate	$C_6H_{10}O_6$	102.1	133.2	148.2	164.3	182.4	193.8	208.8	230.5	255.0	280.0	61.5
dl-Dimethyl tartrate	$C_6H_{10}O_6$	100.4	131.8	147.5	164.0	182.4	193.8	209.5	232.3	257.4	282.0	89
N,N-Dimethyl-2-toluidine	$C_9H_{13}N$	28.8	54.1	66.2	80.2	95.0	105.2	118.1	138.3	161.5	184.8	−61
N,N-Dimethyl-4-toluidine	$C_9H_{13}N$	50.1	74.3	86.7	100.0	116.3	126.4	140.3	161.6	185.4	209.5	
Di(nitrosomethyl) amine	$C_2H_5N_3O_2$	+3.2	27.8	40.0	53.7	68.2	77.7	90.3	110.0	131.3	153.0	
Diosphenol	$C_{10}H_{16}O_2$	66.7	95.4	109.0	124.0	141.2	151.3	165.6	186.2	209.5	232.0	
1,4-Dioxane	$C_4H_8O_2$	−35.8	−12.8	−1.2	+12.0	25.2	33.8	45.1	62.3	81.8	101.1	10
Dipentene	$C_{10}H_{16}$	14.0	40.4	53.8	68.2	84.3	94.6	108.3	128.2	150.5	174.6	
Diphenylamine	$C_{12}H_{11}N$	108.3	141.7	157.0	175.2	194.3	206.9	222.8	247.5	274.1	302.0	52.9
Diphenyl carbinol (benzhydrol)	$C_{13}H_{12}O$	110.0	145.0	162.0	180.9	200.0	212.0	227.5	250.0	275.6	301.0	68.5
chlorophosphate	$C_{12}H_{10}ClPO_3$	121.5	160.5	182.0	203.8	227.9	244.2	265.0	299.5	337.2	378.0	
disulfide	$C_{12}H_{10}S_2$	131.6	164.0	180.0	197.0	214.8	226.2	241.3	262.6	285.8	310.0	61
1,2-Diphenylethane (dibenzyl)	$C_{14}H_{14}$	86.8	119.8	136.0	153.7	173.7	186.0	202.8	227.8	255.0	284.0	51.5
Diphenyl ether	$C_{12}H_{10}O$	66.1	97.8	114.0	130.8	150.0	162.0	178.8	203.3	230.7	258.5	27
1,1-Diphenylethylene	$C_{14}H_{12}$	87.4	119.6	135.0	151.8	170.8	183.4	198.6	222.8	249.8	277.0	
trans-Diphenylethylene	$C_{14}H_{12}$	113.2	145.8	161.0	179.8	199.0	211.5	227.4	251.7	278.3	306.5	124
1,1-Diphenylhydrazine	$C_{12}H_{12}N_2$	126.0	159.3	176.1	194.0	213.5	225.9	242.5	267.2	294.0	322.2	44
Diphenylmethane	$C_{13}H_{12}$	76.0	107.4	122.8	139.8	157.8	170.2	186.3	210.7	237.5	264.5	26.5
Diphenyl sulfide	$C_{12}H_{10}S$	96.1	129.0	145.0	162.0	182.8	194.8	211.8	236.8	263.9	292.5	
Diphenyl-2-tolyl thiophosphate	$C_{19}H_{17}O_3PS$	159.7	179.8	201.6	215.5	230.6	240.4	252.5	270.3	290.0	310.0	
1,2-Dipropoxyethane	$C_8H_{18}O_2$	−38.8	−10.3	+5.0	22.3	42.3	55.8	74.2	103.8	140.0	180.0	
1,2-Diisopropylbenzene	$C_{12}H_{18}$	40.0	67.8	81.8	96.8	114.0	124.3	138.7	159.8	184.3	209.0	
1,3-Diisopropylbenzene	$C_{12}H_{18}$	34.7	62.3	76.0	91.2	107.9	118.2	132.3	153.7	177.6	202.0	−105
Dipropylene glycol	$C_6H_{14}O_3$	73.8	102.1	116.2	131.3	147.4	156.5	169.9	189.9	210.5	231.8	
Dipropyleneglycol monobutyl ether	$C_{10}H_{22}O_3$	64.7	92.0	106.0	120.4	136.3	146.3	159.8	180.0	203.8	227.0	
isopropyl ether	$C_9H_{20}O_3$	46.0	72.8	86.2	100.8	117.0	126.8	140.3	160.0	183.1	205.6	
Di-n-propyl ether	$C_6H_{14}O$	−43.3	−22.3	−11.8	0.0	+13.2	21.6	33.0	50.3	69.5	89.5	−122
Diisopropyl ether	$C_6H_{14}O$	−57.0	−37.4	−27.4	−16.7	−4.5	+3.4	13.7	30.0	48.2	67.5	−60
Di-n-propyl ketone (4-heptanone)	$C_7H_{14}O$	23.0	44.4	55.0	66.2	78.1	85.8	96.0	111.2	127.3	143.7	−32.6
Di-n-propyl oxalate	$C_8H_{14}O_4$	53.4	80.2	93.9	108.6	124.6	134.8	148.1	168.0	190.3	213.5	
Diisopropyl oxalate	$C_8H_{14}O_4$	43.2	69.0	81.9	95.6	110.5	120.0	132.6	151.2	171.8	193.5	
Di-n-propyl succinate	$C_{10}H_{18}O_4$	77.5	107.6	122.2	138.0	154.8	166.0	180.3	202.5	226.5	250.8	
Di-n-propyl d-tartrate	$C_{10}H_{18}O_6$	115.6	147.7	163.5	180.4	199.7	211.7	227.0	250.1	275.6	303.0	
Diisopropyl d-tartrate	$C_{10}H_{18}O_6$	103.7	133.7	148.2	164.0	181.8	192.6	207.3	228.2	251.8	275.0	
Divinyl acetylene (1,5-hexadiene-3-yne)	C_6H_6	−45.1	−24.4	−14.0	−2.8	+10.0	18.1	29.5	46.0	64.4	84.0	
1,3-Divinylbenzene	$C_{10}H_{10}$	32.7	60.0	73.8	88.7	105.5	116.0	130.0	151.4	175.2	199.5	−66.9
Docosane	$C_{22}H_{46}$	157.8	195.4	213.0	233.5	254.5	268.3	286.0	314.2	343.5	376.0	44.5
n-Dodecane	$C_{12}H_{26}$	47.8	75.8	90.0	104.6	121.7	132.1	146.2	167.2	191.0	216.2	−9.6
1-Dodecene	$C_{12}H_{24}$	47.2	74.0	87.8	102.4	118.6	128.5	142.3	162.2	185.5	208.0	−31.5
n-Dodecyl alcohol	$C_{12}H_{26}O$	91.0	120.2	134.7	150.0	167.2	177.8	192.0	213.0	235.7	259.0	24
Dodecylamine	$C_{12}H_{27}N$	82.8	111.8	127.8	141.6	157.4	168.0	182.1	203.0	225.0	248.0	
Dodecyltrimethylsilane	$C_{15}H_{34}Si$	91.2	122.1	137.7	153.8	172.1	184.2	199.5	222.0	248.0	273.0	
Elaidic acid	$C_{18}H_{34}O_2$	171.3	206.7	223.5	242.3	260.8	273.0	288.0	312.4	337.0	362.0	51.5
Epichlorohydrin	C_3H_5ClO	−16.5	+5.6	16.6	29.0	42.0	50.6	62.0	79.3	98.0	117.9	−25.6
1,2-Epoxy-2-methylpropane	C_4H_8O	−69.0	−50.0	−40.3	−29.5	−17.3	−9.7	+1.2	17.5	36.0	55.5	
Erucic acid	$C_{22}H_{42}O_2$	206.7	239.7	254.5	270.6	289.1	300.2	314.4	336.5	358.8	381.5	33.5
Estragole (p-methoxy allyl benzene)	$C_{10}H_{12}O$	52.6	80.0	93.7	108.4	124.6	135.2	148.5	168.7	192.0	215.0	
Ethane	C_2H_6	−159.5	−148.5	−142.9	−136.7	−129.8	−125.4	−119.3	−110.2	−99.7	−88.6	−183.2
Ethoxydimethylphenylsilane	$C_{10}H_{16}OSi$	36.3	63.1	76.2	91.0	107.2	127.5	131.4	151.5	175.0	199.5	
Ethoxytrimethylsilane	$C_5H_{14}OSi$	−50.9	−31.0	−20.7	−9.8	+3.7	11.5	22.1	38.1	56.3	75.7	
Ethoxytriphenylsilane	$C_{20}H_{20}OSi$	167.0	198.2	213.5	230.0	247.0	258.3	273.5	295.0	319.5	344.0	
Ethyl acetate	$C_4H_8O_2$	−43.4	−23.5	−13.5	−3.0	+9.1	16.6	27.0	42.0	59.3	77.1	−82.4
acetoacetate	$C_6H_{10}O_3$	28.5	54.0	67.3	81.1	96.2	106.0	118.5	138.0	158.2	180.8	−45
Ethylacetylene (1-butyne)	C_4H_6	−92.5	−76.7	−68.7	−59.9	−50.0	−43.4	−34.9	−21.6	−6.9	+8.7	−130
Ethyl acrylate	$C_5H_8O_2$	−29.5	−8.7	+2.0	13.0	26.0	33.5	44.5	61.5	80.0	99.5	−71.2
α-Ethylacrylic acid	$C_5H_8O_2$	47.0	70.7	82.0	94.4	108.1	116.7	127.5	144.0	160.7	179.2	
α-Ethylacrylonitrile	C_5H_7N	−29.0	−6.4	+5.0	17.7	31.8	40.6	53.0	71.6	92.2	114.0	
Ethyl alcohol (ethanol)	C_2H_6O	−31.3	−12.0	−2.3	+8.0	19.0	26.0	34.9	48.4	63.5	78.4	−112
Ethylamine	C_2H_7N	−82.3	−66.4	−58.3	−48.6	−39.8	−33.4	−25.1	−12.3	+2.0	16.6	−80.6
4-Ethylaniline	$C_8H_{11}N$	52.0	80.0	93.8	109.0	125.7	136.0	149.8	170.6	194.2	217.4	−4
N-Ethylaniline	$C_8H_{11}N$	38.5	66.4	80.6	96.0	113.2	123.6	137.3	156.9	180.8	204.0	−63.5

Table 8. Vapor Pressures of Organic Compounds, up to 1 Atm.—(*Continued*)

Compound		Pressure, mm. Hg										Melting point, °C.
Name	Formula	1	5	10	20	40	60	100	200	400	760	
		Temperature, °C.										
2-Ethylanisole	$C_9H_{12}O$	29.7	55.9	69.0	83.1	98.8	109.0	122.3	142.1	164.2	187.1	
3-Ethylanisole	$C_9H_{12}O$	33.7	60.3	73.9	88.5	104.8	115.5	129.2	149.7	172.8	196.5	
4-Ethylanisole	$C_9H_{12}O$	33.5	60.2	73.9	88.5	104.7	115.4	128.4	149.2	172.3	196.5	
Ethylbenzene	C_8H_{10}	−9.8	+13.9	25.9	38.6	52.8	61.8	74.1	92.7	113.8	136.2	−94.9
Ethyl benzoate	$C_9H_{10}O_2$	44.0	72.0	86.0	101.4	118.2	129.0	143.2	164.8	188.4	213.4	−34.6
benzoylacetate	$C_{11}H_{12}O_3$	107.6	136.4	150.3	166.8	181.8	191.9	205.0	223.8	244.7	265.0	
bromide	C_2H_5Br	−74.3	−56.4	−47.5	−37.8	−26.7	−19.5	−10.0	+4.5	21.0	38.4	−117.8
α-bromoisobutyrate	$C_6H_{11}BrO_2$	10.6	35.8	48.0	61.8	77.0	86.7	99.8	119.7	141.2	163.6	
n-butyrate	$C_6H_{12}O_2$	−18.4	+4.0	15.3	27.8	41.5	50.1	62.0	79.8	100.0	121.0	−93.3
isobutyrate	$C_6H_{12}O_2$	−24.3	−2.4	+8.4	20.6	33.8	42.3	53.5	71.0	90.0	110.0	−88.2
Ethylcamphoronic anhydride	$C_{11}H_{16}O_5$	118.2	149.8	165.0	181.8	199.8	211.5	226.6	248.5	272.8	298.0	
Ethyl isocaproate	$C_8H_{16}O_2$	11.0	35.8	48.0	61.7	76.3	85.8	98.4	117.8	139.2	160.4	
carbamate	$C_3H_7NO_2$	......	65.8	77.8	91.0	105.6	114.8	126.2	144.2	164.0	184.0	49
carbanilate	$C_9H_{11}NO_2$	107.8	131.8	143.7	155.5	168.8	177.3	187.9	203.8	220.0	237.0	52.5
Ethylcetylamine	$C_{18}H_{39}N$	133.2	168.2	186.0	205.5	226.5	239.8	256.8	283.3	313.0	342.0	
Ethyl chloride	C_2H_5Cl	−89.8	−73.9	−65.8	−56.8	−47.0	−40.6	−32.0	−18.6	−3.9	+12.3	−139
chloroacetate	$C_4H_7ClO_2$	+1.0	25.4	37.5	50.4	65.2	74.0	86.0	103.8	123.8	144.2	−26
chloroglyoxylate	$C_4H_5ClO_3$	−5.1	+18.0	29.9	42.0	56.0	65.2	76.6	94.5	114.7	135.0	
α-chloropropionate	$C_5H_9ClO_2$	+6.6	30.2	41.9	54.3	68.2	77.3	89.3	107.2	126.2	146.5	
trans-cinnamate	$C_{11}H_{12}O_2$	87.6	108.5	134.0	150.3	169.2	181.2	196.0	219.3	245.0	271.0	12
3-Ethylcumene	$C_{11}H_{16}$	28.3	55.5	68.8	83.6	99.9	110.2	124.3	145.4	168.2	193.0	
4-Ethylcumene	$C_{11}H_{16}$	31.5	58.4	72.0	86.7	103.3	113.8	127.2	148.3	171.8	195.8	
Ethyl cyanoacetate	$C_5H_7NO_2$	67.8	93.5	106.0	119.8	133.8	142.1	152.8	169.8	187.8	206.0	
Ethylcyclohexane	C_8H_{16}	−14.5	+9.2	20.6	33.4	47.6	56.7	69.0	87.8	109.1	131.8	−111.3
Ethylcyclopentane	C_7H_{14}	−32.2	−10.8	−0.1	+11.7	25.0	33.4	45.0	62.4	82.3	103.4	−138.6
Ethyl dichloroacetate	$C_4H_6Cl_2O_2$	9.6	34.0	46.3	59.5	74.0	83.6	96.1	115.2	135.9	156.5	
N,N-diethyloxamate	$C_8H_{15}NO_3$	76.0	106.3	121.7	137.7	154.4	166.0	180.3	202.8	226.5	252.0	
N-Ethyldiphenylamine	$C_{14}H_{15}N$	98.3	130.2	146.0	162.8	182.0	193.7	209.8	233.0	258.8	286.0	
Ethylene	C_2H_4	−168.3	−158.3	−153.2	−147.6	−141.3	−137.3	−131.8	−123.4	−113.9	−103.7	−169
Ethylene-bis-(chloroacetate)	$C_6H_8Cl_2O_4$	112.0	142.4	158.0	173.5	191.0	201.8	215.0	237.3	259.5	283.5	
Ethylene chlorohydrin (2-chloroethanol)	C_2H_5ClO	−4.0	+19.0	30.3	42.5	56.0	64.1	75.0	91.8	110.0	128.8	−69
diamine (1,2-ethanediamine)	$C_2H_8N_2$	−11.0	+10.5	21.5	33.0	45.8	53.8	62.5	81.0	99.0	117.2	8.5
dibromide (1,2-dibromethane)	$C_2H_4Br_2$	−27.0	+4.7	18.6	32.7	48.0	57.9	70.4	89.8	110.1	131.5	10
dichloride (1,2-dichloroethane)	$C_2H_4Cl_2$	−44.5	−24.0	−13.6	−2.4	+10.0	18.1	29.4	45.7	64.0	82.4	−35.3
glycol (1,2-ethanediol)	$C_2H_6O_2$	53.0	79.7	92.1	105.8	120.0	129.5	141.8	158.5	178.5	197.3	−15.6
glycol diethyl ether (1,2-diethoxyethane)	$C_6H_{14}O_2$	−33.5	−10.2	+1.6	14.7	29.7	39.0	51.8	71.8	94.1	119.5	
glycol dimethyl ether (1,2-dimethoxyethane)	$C_4H_{10}O_2$	−48.0	−26.2	−15.3	−3.0	+10.7	19.7	31.8	50.0	70.8	93.0	
glycol monomethyl ether (2-methoxyethanol)	$C_3H_8O_2$	−13.5	+10.2	22.0	34.3	47.8	56.4	68.0	85.3	104.3	124.4	
oxide	C_2H_4O	−89.7	−73.8	−65.7	−56.6	−46.9	−40.7	−32.1	−19.5	−4.9	+10.7	−111.3
Ethyl α-ethylacetoacetate	$C_8H_{14}O_3$	40.5	67.3	80.2	94.6	110.3	120.6	133.8	153.2	175.6	198.0	
fluoride	C_2H_5F	−117.0	−103.8	−97.7	−90.0	−81.8	−76.4	−69.3	−58.0	−45.5	−32.0	
formate	$C_3H_6O_2$	−60.5	−42.2	−33.0	−22.7	−11.5	−4.3	−5.4	20.0	37.1	54.3	−79
2-furoate	$C_7H_8O_3$	37.6	63.8	77.1	91.5	107.5	117.5	130.4	150.1	172.5	195.0	34
glycolate	$C_4H_8O_3$	14.3	38.8	50.5	63.9	78.1	87.6	99.8	117.8	138.0	158.2	
3-Ethylhexane	C_8H_{18}	−20.0	+2.1	12.8	25.0	38.5	47.1	58.9	76.7	97.0	118.5	
2-Ethylhexyl acrylate	$C_{11}H_{20}O_2$	50.0	77.7	91.8	106.3	123.7	134.0	147.9	168.2	192.2	216.0	
Ethylidene chloride (1,1-dichloroethane)	$C_2H_4Cl_2$	−60.7	−41.9	−32.3	−21.9	−10.2	−2.9	+7.2	22.4	39.8	57.4	−96.7
fluoride (1,1-difluoroethane)	$C_2H_4F_2$	−112.5	−98.4	−91.7	−84.1	−75.8	−70.4	−63.2	−52.0	−39.5	−26.5	−117
Ethyl iodide	C_2H_5I	−54.4	−34.3	−24.3	−13.1	−0.9	+7.2	18.0	34.1	52.3	72.4	−105
Ethyl l-leucinate	$C_8H_{17}NO_2$	27.8	57.3	72.1	88.0	106.0	117.8	131.8	149.8	167.3	184.0	
Ethyl levulinate	$C_7H_{12}O_3$	47.3	74.0	87.3	101.8	117.7	127.6	141.3	160.2	183.0	206.2	
Ethyl mercaptan (ethanethiol)	C_2H_6S	−76.7	−59.1	−50.2	−40.7	−29.8	−22.4	−13.0	+1.5	17.7	35.0	−121
Ethyl methylcarbamate	$C_4H_9NO_2$	26.5	51.0	63.2	76.1	91.0	100.0	112.0	130.0	149.8	170.0	
Ethyl methyl ether	C_3H_8O	−91.0	−75.6	−67.8	−59.1	−49.4	−43.3	−34.8	−22.0	−7.8	+7.5	
1-Ethylnaphthalene	$C_{12}H_{12}$	70.0	101.4	116.8	133.8	152.0	164.1	180.0	204.6	230.8	258.1	−27
Ethyl α-naphthyl ketone (1-propionaphthone)	$C_{13}H_{12}O$	124.0	155.5	171.0	188.1	206.9	218.2	233.5	255.5	280.2	306.0	
Ethyl 3-nitrobenzoate	$C_9H_9NO_4$	108.1	140.2	155.0	173.6	192.6	205.0	220.3	244.6	270.6	298.0	47
3-Ethylpentane	C_7H_{16}	−37.8	−17.0	−6.8	+4.7	17.5	25.7	36.9	53.8	73.0	93.5	−118.6
4-Ethylphenetole	$C_{10}H_{14}O$	48.5	75.7	89.5	103.8	119.8	129.8	143.5	163.2	185.7	208.0	
2-Ethylphenol	$C_8H_{10}O$	46.2	73.4	87.0	101.5	117.9	127.9	141.8	161.6	184.5	207.5	−45
3-Ethylphenol	$C_8H_{10}O$	60.0	86.8	100.2	114.5	130.0	139.8	152.0	171.8	193.3	214.0	−4
4-Ethylphenol	$C_8H_{10}O$	59.3	86.5	100.2	115.0	131.3	141.7	154.2	175.0	197.4	219.0	46.5
Ethyl phenyl ether (phenetole)	$C_8H_{10}O$	18.1	43.7	56.4	70.3	86.6	95.4	108.4	127.9	149.8	172.0	−30.2
Ethyl propionate	$C_5H_{10}O_2$	−28.0	−7.2	+3.4	14.3	27.2	35.1	45.2	61.7	79.8	99.1	−72.6
Ethyl propyl ether	$C_5H_{12}O$	−64.3	−45.0	−35.0	−24.0	−12.0	−4.0	+6.8	23.3	41.6	61.7	
Ethyl salicylate	$C_9H_{10}O_3$	61.2	90.0	104.2	119.3	136.7	147.6	161.5	183.7	207.0	231.5	1.3
3-Ethylstyrene	$C_{10}H_{12}$	28.3	55.0	68.3	82.8	99.2	109.6	123.2	144.0	167.2	191.5	
4-Ethylstyrene	$C_{10}H_{12}$	26.0	52.7	66.3	80.8	97.3	107.6	121.5	142.0	165.0	189.0	
Ethylisothiocyanate	C_3H_5NS	−13.2	+10.6	22.8	36.1	50.8	59.8	71.9	90.0	110.1	131.0	−5.9
2-Ethyltoluene	C_9H_{12}	9.4	34.8	47.6	61.2	76.4	86.0	99.0	119.0	141.4	165.1	
3-Ethyltoluene	C_9H_{12}	7.2	32.3	44.7	58.2	73.3	82.9	95.9	115.5	137.8	161.3	−95.5
4-Ethyltoluene	C_9H_{12}	7.6	32.7	44.9	58.5	73.6	83.2	96.3	116.1	136.4	162.0	
Ethyl trichloroacetate	$C_4H_5Cl_3O_2$	20.7	45.5	57.7	70.6	85.5	94.4	107.4	125.8	146.0	167.0	
Ethyltrimethylsilane	$C_5H_{14}Si$	−60.6	−41.4	−31.8	−21.0	−9.0	−1.2	+9.2	25.0	42.8	62.0	
Ethyltrimethyltin	$C_5H_{14}Sn$	−30.0	−7.6	+3.8	16.1	30.0	38.4	50.0	67.3	87.6	108.8	
Ethyl isovalerate	$C_7H_{14}O_2$	−6.1	+17.0	28.7	41.3	55.2	64.0	75.9	93.8	114.0	134.3	−99.3
2-Ethyl-1,4-xylene	$C_{10}H_{14}$	25.7	52.0	65.6	79.8	96.0	106.2	120.0	140.2	163.1	186.9	
4-Ethyl-1,3-xylene	$C_{10}H_{14}$	26.3	53.0	66.4	80.6	97.2	107.4	121.2	141.8	164.4	188.4	
5-Ethyl-1,3-xylene	$C_{10}H_{14}$	22.1	48.8	62.1	76.5	92.6	103.0	116.5	137.4	159.6	183.7	
Eugenol	$C_{10}H_{12}O_2$	78.4	108.1	123.0	138.7	155.8	167.3	182.2	204.7	228.3	253.5	
iso-Eugenol	$C_{10}H_{12}O_2$	86.3	117.0	132.4	149.0	167.0	178.2	194.0	217.2	242.3	267.5	−10
Eugenyl acetate	$C_{12}H_{14}O_3$	101.6	132.3	148.0	164.2	183.0	194.0	209.7	232.5	257.4	282.0	295
Fencholic acid	$C_{10}H_{16}O_2$	101.7	128.7	142.3	155.8	171.8	181.5	194.0	215.0	237.8	264.1	19
d-Fenchone	$C_{10}H_{16}O$	28.0	54.7	68.3	83.0	99.5	109.8	123.6	144.0	166.8	191.0	5
dl-Fenchyl alcohol	$C_{10}H_{18}O$	45.8	70.3	82.1	95.6	110.8	120.2	132.3	150.0	173.2	201.0	35
Fluorene	$C_{13}H_{10}$	......	129.3	146.0	164.2	185.2	197.8	214.7	240.3	268.6	295.0	113
Fluorobenzene	C_6H_5F	−43.4	−22.8	−12.4	−1.2	+11.5	19.6	30.4	47.2	65.7	84.7	−42.1

Table 8. Vapor Pressures of Organic Compounds, up to 1 Atm.—(*Continued*)

Compound: Name	Compound: Formula	Pressure, mm. Hg: 1	5	10	20	40	60	100	200	400	760	Melting point, °C.
		Temperature, °C.										
2-Fluorotoluene	C_7H_7F	−24.2	−2.2	+8.9	21.4	34.7	43.7	55.3	73.0	92.8	114.0	−80
3-Fluorotoluene	C_7H_7F	−22.4	−0.3	+11.0	23.4	37.0	45.8	57.5	75.4	95.4	116.0	−110.8
4-Fluorotoluene	C_7H_7F	−21.8	+0.3	11.8	24.0	37.8	46.5	58.1	76.0	96.1	117.0	
Formaldehyde	CH_2O	...	...	−88.0	−79.6	−70.6	−65.0	−57.3	−46.0	−33.0	−19.5	−92
Formamide	CH_3NO	70.5	96.3	109.5	122.5	137.5	147.0	157.5	175.5	193.5	210.5	
Formic acid	CH_2O_2	−20.0	−5.0	+2.1	10.3	24.0	32.4	43.8	61.4	80.3	100.6	8.2
trans-Fumaryl chloride	$C_4H_2Cl_2O_2$	+15.0	38.5	51.8	65.0	79.5	89.0	101.0	120.0	140.0	160.0	
Furfural (2-furaldehyde)	$C_5H_4O_2$	18.5	42.6	54.8	67.8	82.1	91.5	103.4	121.8	141.8	161.8	
Furfuryl alcohol	$C_5H_6O_2$	31.8	56.0	68.0	81.0	95.7	104.0	115.9	133.1	151.8	170.0	
Geraniol	$C_{10}H_{18}O$	69.2	96.8	110.0	125.6	141.8	151.5	165.3	185.6	207.8	230.0	
Geranyl acetate	$C_{12}H_{20}O_2$	73.5	102.7	117.9	133.0	150.0	160.3	175.2	196.3	219.8	243.3	
Geranyl n-butyrate	$C_{14}H_{24}O_2$	96.8	125.2	139.0	153.8	170.1	180.2	193.8	214.0	235.0	257.4	
Geranyl isobutyrate	$C_{14}H_{24}O_2$	90.9	119.6	133.0	147.9	164.0	174.0	187.7	207.6	228.5	251.0	
Geranyl formate	$C_{11}H_{18}O_2$	61.8	90.3	104.3	119.8	136.2	147.2	160.7	182.6	205.8	230.0	
Glutaric acid	$C_5H_8O_4$	155.5	183.8	196.0	210.5	226.3	235.5	247.0	265.0	283.5	303.0	97.5
Glutaric anhydride	$C_5H_6O_3$	100.8	133.3	149.5	166.0	185.5	196.2	212.5	236.5	261.0	287.0	
Glutaronitrile	$C_5H_6N_2$	91.3	123.7	140.0	156.5	176.4	189.5	205.5	230.0	257.3	286.2	
Glutaryl chloride	$C_5H_6Cl_2O_2$	56.1	84.0	97.8	112.3	128.3	139.1	151.8	172.4	195.3	217.0	
Glycerol	$C_3H_8O_3$	125.5	153.8	167.2	182.2	198.0	208.0	220.1	240.0	263.0	290.0	17.9
Glycerol dichlorohydrin (1,3-dichloro-2-propanol)	$C_3H_6Cl_2O$	28.0	52.2	64.7	78.0	93.0	102.0	114.8	133.3	153.5	174.3	
Glycol diacetate	$C_6H_{10}O_4$	38.3	64.1	77.1	90.8	106.1	115.8	128.0	147.8	168.3	190.5	−31
Glycolide (1,4-dioxane-2,6-dione)	$C_4H_4O_4$	...	103.0	116.6	132.0	148.6	158.2	173.2	194.0	217.0	240.0	97
Guaicol (2-methoxyphenol)	$C_7H_8O_2$	52.4	79.1	92.0	106.0	121.6	131.0	144.0	162.7	184.1	205.0	28.3
Heneicosane	$C_{21}H_{44}$	152.6	188.0	205.4	223.2	243.4	255.3	272.0	296.5	323.8	350.5	40.4
Heptacosane	$C_{27}H_{56}$	211.7	248.6	266.8	284.6	305.7	318.3	333.5	359.4	385.0	410.6	59.5
Heptadecane	$C_{17}H_{36}$	115.0	145.2	160.0	177.7	195.8	207.3	223.0	247.8	274.5	303.0	22.5
Heptaldehyde (enanthaldehyde)	$C_7H_{14}O$	12.0	32.7	43.0	54.0	66.3	74.0	84.0	102.0	125.5	155.0	−42
n-Heptane	C_7H_{16}	−34.0	−12.7	−2.1	+9.5	22.3	30.6	41.8	58.7	78.0	98.4	−90.6
Heptanoic acid (enanthic acid)	$C_7H_{14}O_2$	78.0	101.3	113.2	125.6	139.5	148.5	160.0	179.5	199.6	221.5	−10
1-Heptanol	$C_7H_{16}O$	42.4	64.3	74.7	85.8	99.8	108.0	119.5	136.6	155.6	175.8	34.6
Heptanoyl chloride (enanthyl chloride)	$C_7H_{13}ClO$	34.2	54.6	64.6	75.0	86.4	93.5	102.7	116.3	130.7	145.0	
2-Heptene	C_7H_{14}	−35.8	−14.1	−3.5	+8.3	21.5	30.0	41.3	58.6	78.1	98.5	
Heptylbenzene	$C_{13}H_{20}$	64.0	94.6	110.0	126.0	144.0	154.8	170.2	193.3	217.8	244.0	
Heptyl cyanide (enanthonitrile)	$C_7H_{13}N$	21.0	47.8	61.6	76.3	92.6	103.0	116.8	137.7	160.0	184.6	
Hexachlorobenzene	C_6Cl_6	114.4	149.3	166.4	185.7	206.0	219.0	235.5	258.5	283.5	309.4	230
Hexachloroethane	C_2Cl_6	32.7	49.8	73.5	87.6	102.3	112.0	124.2	143.1	163.8	185.6	186.6
Hexacosane	$C_{26}H_{54}$	204.0	240.0	257.4	275.8	295.2	307.8	323.2	348.4	374.6	399.8	56.6
Hexadecane	$C_{16}H_{34}$	105.3	135.2	149.8	164.7	181.3	193.2	208.5	231.7	258.3	287.5	18.5
1-Hexadecene	$C_{16}H_{32}$	101.6	131.7	146.2	162.0	178.8	190.8	205.3	226.8	250.0	274.0	4
n-Hexadecyl alcohol (cetyl alcohol)	$C_{16}H_{34}O$	122.7	158.3	177.8	197.8	219.8	234.3	251.7	280.2	312.7	344.0	49.3
n-Hexadecylamine (cetylamine)	$C_{16}H_{35}N$	123.6	157.8	176.0	195.7	215.7	228.8	245.8	272.2	300.4	330.0	
Hexaethylbenzene	$C_{18}H_{30}$	...	134.3	150.3	168.0	187.7	199.7	216.0	241.7	268.5	298.3	130
n-Hexane	C_6H_{14}	−53.9	−34.5	−25.0	−14.1	−2.3	+5.4	15.8	31.6	49.6	68.7	−95.3
1-Hexanol	$C_6H_{14}O$	24.4	47.2	58.2	70.3	83.7	92.0	102.8	119.6	138.0	157.0	−51.6
2-Hexanol	$C_6H_{14}O$	14.6	34.8	45.0	55.9	67.9	76.0	87.3	103.7	121.8	139.9	
3-Hexanol	$C_6H_{14}O$	+2.5	25.7	36.7	49.0	62.2	70.7	81.8	98.3	117.0	135.5	
1-Hexene	C_6H_{12}	−57.5	−38.0	−28.1	−17.2	−5.0	+2.8	13.0	29.0	46.8	66.0	−98.5
n-Hexyl levulinate	$C_{11}H_{20}O_3$	90.0	120.0	134.7	150.2	167.8	179.0	193.6	215.7	241.0	266.8	
n-Hexyl phenyl ketone (enanthophenone)	$C_{13}H_{18}O$	100.0	130.3	145.5	161.0	178.9	189.8	204.2	225.0	248.3	271.3	
Hydrocinnamic acid	$C_9H_{10}O_2$	102.2	133.5	148.7	165.0	183.3	194.0	209.0	230.8	255.0	279.8	48.5
Hydrogen cyanide (hydrocyanic acid)	CHN	−71.0	−55.3	−47.7	−39.7	−30.9	−25.1	−17.8	−5.3	+10.2	25.9	−13.2
Hydroquinone	$C_6H_6O_2$	132.4	153.3	163.5	174.6	192.0	203.0	216.5	238.0	262.5	286.2	170.3
4-Hydroxybenzaldehyde	$C_7H_6O_2$	121.2	153.2	169.7	186.8	206.0	217.5	233.5	256.8	282.6	310.0	115.5
α-Hydroxyisobutyric acid	$C_4H_8O_3$	73.5	98.5	110.5	123.8	138.0	146.4	157.7	175.2	193.8	212.0	79
α-Hydroxybutyronitrile	C_5H_9NO	41.0	65.8	77.8	90.7	104.8	113.9	125.0	142.0	159.8	178.8	
4-Hydroxy-3-methyl-2-butanone	$C_5H_{10}O_2$	44.6	69.3	81.0	94.0	108.2	117.4	129.0	146.5	165.5	185.0	
4-Hydroxy-4-methyl-2-pentanone	$C_6H_{12}O_2$	22.0	46.7	58.8	72.0	86.7	96.0	108.2	126.8	147.5	167.9	−47
3-Hydroxypropionitrile	C_3H_5NO	58.7	87.8	102.0	117.9	134.1	144.7	157.7	178.0	200.0	221.0	
Indene	C_9H_8	16.4	44.3	58.5	73.9	90.7	100.8	114.7	135.6	157.8	181.6	−2
Iodobenzene	C_6H_5I	24.1	50.6	64.0	78.3	94.4	105.0	118.3	139.8	163.9	188.6	−28.5
Iodononane	$C_9H_{19}I$	70.0	96.2	109.0	123.0	138.1	147.7	159.8	179.0	199.3	219.5	
2-Iodotoluene	C_7H_7I	37.2	65.9	79.8	95.6	112.4	123.8	138.1	160.0	185.7	211.0	
α-Ionone	$C_{13}H_{20}O$	79.5	108.8	123.0	139.0	155.6	166.3	181.2	202.5	225.2	250.0	
Isoprene	C_5H_8	−79.8	−62.3	−53.3	−43.5	−32.6	−25.4	−16.0	−1.2	+15.4	32.6	−146.7
Lauraldehyde	$C_{12}H_{24}O$	77.7	108.4	123.7	140.2	157.8	168.7	184.5	207.8	231.8	257.0	44.5
Lauric acid	$C_{12}H_{24}O_2$	121.0	150.6	166.0	183.6	201.4	212.7	227.5	249.8	273.8	299.2	48
Levulinaldehyde	$C_5H_8O_2$	28.1	54.9	68.0	82.7	98.3	108.4	121.8	142.0	164.0	187.0	
Levulinic acid	$C_5H_8O_3$	102.0	128.1	141.8	154.1	169.5	178.0	190.2	208.3	227.4	245.8	33.5
d-Limonene	$C_{10}H_{16}$	14.0	40.4	53.8	68.2	84.3	94.6	108.3	128.5	151.4	175.0	−96.9
Linalyl acetate	$C_{12}H_{20}O_2$	55.4	82.5	96.0	111.4	127.7	138.1	151.8	173.3	196.2	220.0	
Maleic anhydride	$C_4H_2O_3$	44.0	63.4	78.7	95.0	111.8	122.0	135.8	155.9	179.5	202.0	58
Menthane	$C_{10}H_{20}$	+9.7	35.7	48.3	62.7	78.3	88.6	102.1	122.7	146.0	169.5	
1-Menthol	$C_{10}H_{20}O$	56.0	83.2	96.0	110.3	126.1	136.1	149.4	168.3	190.2	212.0	42.5
Menthyl acetate	$C_{12}H_{22}O_2$	57.4	85.8	100.0	115.4	132.1	143.2	156.7	178.8	202.8	227.0	
benzoate	$C_{17}H_{24}O_2$	123.2	154.2	170.0	186.3	204.3	215.8	230.4	253.2	277.1	301.0	54.5
formate	$C_{11}H_{20}O_2$	47.3	75.8	90.0	105.8	123.0	133.8	148.0	169.8	194.2	219.0	
Mesityl oxide	$C_6H_{10}O$	−8.7	+14.1	26.0	37.9	51.7	60.4	72.1	90.0	109.8	130.0	−59
Methacrylic acid	$C_4H_6O_2$	25.5	48.5	60.0	72.7	86.4	95.3	106.6	123.9	142.5	161.0	15
Methacrylonitrile	C_4H_5N	−44.5	−23.3	−12.5	−0.6	+12.8	21.5	32.8	50.0	70.3	90.3	
Methane	CH_4	−205.9	−199.0	−195.5	−191.8	−187.7	−185.1	−181.4	−175.5	−168.8	−161.5	−182.5
Methanethiol	CH_4S	−90.7	−75.3	−67.5	−58.8	−49.2	−43.1	−34.8	−22.1	−7.9	+6.8	−121
Methoxyacetic acid	$C_3H_6O_3$	52.5	79.3	92.0	106.5	122.0	131.8	144.5	163.5	184.2	204.0	
N-Methylacetanilide	$C_9H_{11}NO$	...	103.8	118.6	135.1	152.2	164.2	179.8	202.3	227.4	253.0	102
Methyl acetate	$C_3H_6O_2$	−57.2	−38.6	−29.3	−19.1	−7.9	−0.5	+9.4	24.0	40.0	57.8	−98.7
acetylene (propyne)	C_3H_4	−111.0	−97.5	−90.5	−82.9	−74.3	−68.8	−61.3	−49.8	−37.2	−23.3	−102.7
acrylate	$C_4H_6O_2$	−43.7	−23.6	−13.5	−2.7	+9.2	17.3	28.0	43.9	61.8	80.2	
alcohol (methanol)	CH_4O	−44.0	−25.3	−16.2	−6.0	+5.0	12.1	21.2	34.8	49.9	64.7	−97.8
Methylamine	CH_5N	−95.8	−81.3	−73.8	−65.9	−56.9	−51.3	−43.7	−32.4	−19.7	−6.3	−93.5

Table 8. Vapor Pressures of Organic Compounds, up to 1 Atm.—(*Continued*)

Compound		Pressure, mm. Hg										Melting point, °C.
Name	Formula	1	5	10	20	40	60	100	200	400	760	
		Temperature, °C.										
N-Methylaniline	C_7H_9N	36.0	62.8	76.2	90.5	106.0	115.8	129.8	149.3	172.0	195.5	−57
Methyl anthranilate	$C_8H_9NO_2$	77.6	109.0	124.2	141.5	159.7	172.0	187.8	212.4	238.5	266.5	24
benzoate	$C_8H_8O_2$	39.0	64.4	77.3	91.8	107.8	117.4	130.8	151.4	174.7	199.5	−12.5
2-Methylbenzothiazole	C_8H_7NS	70.0	97.5	111.2	125.5	141.2	150.4	163.9	183.2	204.5	225.5	15.4
α-Methylbenzyl alcohol	$C_8H_{10}O$	49.0	75.2	88.0	102.1	117.8	127.4	140.3	159.0	180.7	204.0	
Methyl bromide	CH_3Br	−96.3	−80.6	−72.8	−64.0	−54.2	−48.0	−39.4	−26.5	−11.9	+3.6	−93
2-Methyl-1-butene	C_5H_{10}	−89.1	−72.8	−64.3	−54.8	−44.1	−37.3	−28.0	−13.8	+2.5	20.2	−135
2-Methyl-2-butene	C_5H_{10}	−75.4	−57.0	−47.9	−37.9	−26.7	−19.4	−9.9	+4.9	21.6	38.5	−133
Methyl isobutyl carbinol (2-methyl-4-pentanol)	$C_6H_{14}O$	−0.3	+22.1	33.3	45.4	58.2	67.0	78.0	94.9	113.5	131.7	
n-butyl ketone (2-hexanone)	$C_6H_{12}O$	+7.7	28.8	38.8	50.0	62.0	69.8	79.8	94.3	111.0	127.5	−56.9
isobutyl ketone (4-methyl-2-pentanone)	$C_6H_{12}O$	−1.4	+19.7	30.0	40.8	52.8	60.4	70.4	85.6	102.0	119.0	−84.7
n-butyrate	$C_5H_{10}O_2$	−26.8	−5.5	+5.0	16.7	29.6	37.4	48.0	64.3	83.1	102.3	
isobutyrate	$C_5H_{10}O_2$	−34.1	−13.0	−2.9	+8.4	21.0	28.9	39.6	55.7	73.6	92.6	−84.7
caprate	$C_{11}H_{22}O_2$	63.7	93.5	108.0	123.0	139.0	148.6	161.5	181.6	202.9	224.0	−18
caproate	$C_7H_{14}O_2$	+5.0	30.0	42.0	55.4	70.0	79.7	91.4	109.8	129.8	150	
caprylate	$C_9H_{18}O_2$	34.2	61.7	74.9	89.0	105.3	115.3	128.0	148.1	170.0	193.0	−40
chloride	CH_3Cl	......	−99.5	−92.4	−84.8	−76.0	−70.4	−63.0	−51.2	−38.0	−24.0	−97.7
chloroacetate	$C_3H_5ClO_2$	−2.9	19.0	30.0	41.5	54.5	63.0	73.5	90.5	109.5	130.3	−31.9
cinnamate	$C_{10}H_{10}O_2$	77.4	108.1	123.0	140.0	157.9	170.0	185.8	209.6	235.0	263.0	33.4
α-Methylcinnamic acid	$C_{10}H_{10}O_2$	125.7	155.0	169.8	185.2	201.8	212.0	224.8	245.0	266.8	288.0	
Methylcyclohexane	C_7H_{14}	−35.9	−14.0	−3.2	+8.7	22.0	30.5	42.1	59.6	79.6	100.9	−126.4
Methylcyclopentane	C_6H_{12}	−53.7	−33.8	−23.7	−12.8	−0.6	+7.2	17.9	34.0	52.3	71.8	−142.4
Methylcyclopropane	C_4H_8	−96.0	−80.6	−72.8	−64.0	−54.2	−48.0	−39.3	−26.0	−11.3	+4.5	
Methyl n-decyl ketone (n-dodecan-2-one)	$C_{12}H_{24}O$	77.1	106.0	120.4	136.0	152.4	163.8	177.5	199.0	222.5	246.5	
dichloroacetate	$C_3H_4Cl_2O_2$	3.2	26.7	38.1	50.7	64.7	73.6	85.4	103.2	122.6	143.0	
N-Methyldiphenylamine	$C_{13}H_{13}N$	103.5	134.0	149.7	165.8	184.0	195.4	210.1	232.8	257.0	282.0	−7.6
Methyl n-dodecyl ketone (2-tetradecanone)	$C_{14}H_{28}O$	99.3	130.0	145.5	161.3	179.8	191.4	206.0	228.2	253.3	278.0	
Methylene bromide (dibromomethane)	CH_2Br_2	−35.1	−13.2	−2.4	+9.7	23.3	31.6	42.3	58.5	79.0	98.6	−52.8
chloride (dichloromethane)	CH_2Cl_2	−70.0	−52.1	−43.3	−33.4	−22.3	−15.7	− 6.3	+8.0	24.1	40.7	−96.7
Methyl ethyl ketone (2-butanone)	C_4H_8O	−48.3	−28.0	−17.7	−6.5	+6.0	14.0	25.0	41.6	60.0	79.6	−85.9
2-Methyl-3-ethylpentane	C_8H_{18}	−24.0	−1.8	+9.5	21.7	35.2	43.9	55.7	73.6	94.0	115.6	−114.5
3-Methyl-3-ethylpentane	C_8H_{18}	−23.9	−1.4	+9.9	22.3	36.2	45.0	57.1	75.3	96.2	118.3	−90
Methyl fluoride	CH_3F	−147.3	−137.0	−131.6	−125.9	−119.1	−115.0	−109.0	− 99.9	−89.5	−78.2	
formate	$C_2H_4O_2$	−74.2	−57.0	−48.6	−39.2	−28.7	−21.9	−12.9	+0.8	16.0	32.0	−99.8
α-Methylglutaric anhydride	$C_6H_8O_3$	93.8	125.4	141.8	157.7	177.5	189.9	205.0	229.1	255.5	282.5	
Methyl glycolate	$C_3H_6O_3$	+9.6	33.7	45.3	58.1	72.3	81.8	93.7	111.8	131.7	151.5	
2-Methylheptadecane	$C_{18}H_{38}$	119.8	152.0	168.7	186.0	204.8	216.3	231.5	254.5	279.8	306.5	
2-Methylheptane	C_8H_{18}	−21.0	+1.3	12.3	24.4	37.9	46.6	58.3	76.0	96.2	117.6	−109.5
3-Methylheptane	C_8H_{18}	−19.8	+2.6	13.3	25.4	38.9	47.6	59.4	77.1	97.4	118.9	−120.8
4-Methylheptane	C_8H_{18}	−20.4	+1.5	12.4	24.5	38.0	46.6	58.3	76.1	96.3	117.7	−121.1
2-Methyl-2-heptene	C_8H_{16}	−16.1	+6.7	17.8	30.4	44.0	52.8	64.6	82.3	102.2	122.5	
6-Methyl-3-hepten-2-ol	$C_8H_{16}O$	41.6	65.0	76.7	89.3	102.7	111.5	122.6	139.5	156.6	175.5	
6-Methyl-5-hepten-2-ol	$C_8H_{16}O$	41.9	66.0	77.8	90.4	104.0	112.8	123.8	140.0	156.6	174.3	
2-Methylhexane	C_7H_{16}	−40.4	−19.5	−9.1	+2.3	14.9	23.0	34.1	50.8	69.8	90.0	−118.2
3-Methylhexane	C_7H_{16}	−39.0	−18.1	−7.8	+3.6	16.4	24.5	35.6	52.4	71.6	91.9	
Methyl iodide	CH_3I	......	−55.0	−45.8	−35.6	−24.2	−16.9	−7.0	+8.0	25.3	42.4	−64.4
laurate	$C_{13}H_{26}O_2$	87.8	117.9	133.2	149.0	166.0	176.8	190.8	......	......	......	5
levulinate	$C_6H_{10}O_3$	39.8	66.4	79.7	93.7	109.5	119.3	133.0	153.4	175.8	197.7	
methacrylate	$C_5H_8O_2$	−30.5	−10.0	+1.0	11.0	25.5	34.5	47.0	63.0	82.0	101.0	
myristate	$C_{15}H_{30}O_2$	115.0	145.7	160.8	177.8	195.8	207.5	222.6	245.3	269.8	295.8	18.5
α-naphthyl ketone (1-acetonaphthone)	$C_{12}H_{10}O$	115.6	146.3	161.5	178.4	196.8	208.6	223.8	246.7	270.5	295.5	
β-naphthyl ketone (2-acetonaphthone)	$C_{12}H_{10}O$	120.2	152.3	168.5	185.7	203.8	214.7	229.8	251.6	275.8	301.0	55.5
n-nonyl ketone (undecan-2-one)	$C_{11}H_{22}O$	68.2	95.5	108.9	123.1	139.0	148.6	161.0	181.2	202.3	224.0	15
palmitate	$C_{17}H_{34}O_2$	134.3	166.8	184.3	202.0	......	......	......	......	......	......	30
n-pentadecyl ketone (2-heptdecanone)	$C_{17}H_{34}O$	129.6	161.6	178.0	196.4	214.3	226.7	242.0	265.8	291.7	319.5	
2-Methylpentane	C_6H_{14}	−60.9	−41.7	−32.1	−21.4	−9.7	−1.9	+8.1	24.1	41.6	60.3	−154
3-Methylpentane	C_6H_{14}	−59.0	−39.8	−30.1	−19.4	−7.3	+0.1	10.5	26.5	44.2	63.3	−118
2-Methyl-1-pentanol	$C_6H_{14}O$	15.4	38.0	49.6	61.6	74.7	83.4	94.2	111.3	129.8	147.9	
2-Methyl-2-pentanol	$C_6H_{14}O$	−4.5	+16.8	27.6	38.8	51.3	58.8	69.2	85.0	102.6	121.2	−103
Methyl n-pentyl ketone (2-heptanone)	$C_7H_{14}O$	19.3	43.6	55.5	67.7	81.2	89.8	100.0	116.1	133.2	150.2	
phenyl ether (anisole)	C_7H_8O	+5.4	30.0	42.2	55.8	70.7	80.1	93.0	112.3	133.8	155.5	−37.3
2-Methylpropene	C_4H_8	−105.1	−96.5	−81.9	−73.4	−63.8	−57.7	−49.3	−36.7	−22.2	−6.9	−140.3
Methyl propionate	$C_4H_8O_2$	−42.0	−21.5	−11.8	−1.0	+11.0	18.7	29.0	44.2	61.8	79.8	−87.5
4-Methylpropiophenone	$C_{10}H_{12}O$	59.6	89.3	103.8	120.2	138.0	149.3	164.2	187.4	212.7	238.5	
2-Methylpropionyl bromide	C_4H_7BrO	13.5	38.4	50.6	64.1	79.4	88.8	101.6	120.5	141.7	163.0	
Methyl propyl ether	$C_4H_{10}O$	−72.2	−54.3	−45.4	−35.4	−24.3	−17.4	−8.1	+6.0	22.5	39.1	
n-propyl ketone (2-pentanone)	$C_5H_{10}O$	−12.0	+8.0	17.9	28.5	39.8	47.3	56.8	71.0	86.8	103.3	−77.8
isopropyl ketone (3-Methyl-2-butanone)	$C_5H_{10}O$	−19.9	−1.0	+8.3	18.3	29.6	36.2	45.5	59.0	73.8	88.9	−92
2-Methylquinoline	$C_{10}H_9N$	75.3	104.0	119.0	134.0	150.8	161.7	176.2	197.8	211.7	246.5	−1
Methyl salicylate	$C_8H_8O_3$	54.0	81.6	95.3	110.0	126.2	136.7	150.0	172.6	197.5	223.2	−8.3
α-Methyl styrene	C_9H_{10}	7.4	34.0	47.1	61.8	77.8	88.3	102.2	121.8	143.0	165.4	−23.2
4-Methyl styrene	C_9H_{10}	16.0	42.0	55.1	69.2	85.0	95.0	108.6	128.7	151.2	175.0	
Methyl n-tetradecyl ketone (2-hexadecanone)	$C_{16}H_{32}O$	109.8	151.5	167.3	184.6	203.7	215.0	230.5	254.4	279.8	307.0	
thiocyanate	C_2H_3NS	−14.0	+9.8	21.6	34.5	49.0	58.1	70.4	89.8	110.8	132.9	−51
isothiocyanate	C_2H_3NS	−34.7	−8.3	+5.4	20.4	38.2	47.5	59.3	77.5	97.8	119.0	35.5
undecyl ketone (2-tridecanone)	$C_{13}H_{26}O$	86.8	117.0	131.8	147.8	165.7	176.6	191.5	214.0	238.3	262.5	28.5
isovalerate	$C_6H_{12}O_2$	−19.2	+2.9	14.0	26.4	39.8	48.2	59.8	77.3	96.7	116.7	
Monovinylacetylene (butenyne)	C_4H_4	−93.2	−77.7	−70.0	−61.3	−51.7	−45.3	−37.1	−24.1	−10.1	+5.3	
Myrcene	$C_{10}H_{16}$	14.5	40.0	53.2	67.0	82.6	92.6	106.0	126.0	148.3	171.5	
Myristadehyde	$C_{14}H_{28}O$	99.0	132.0	148.3	166.2	186.0	198.3	214.5	240.4	267.9	297.8	23.5
Myristic acid (tetradecanoic acid)	$C_{14}H_{28}O_2$	142.0	174.1	190.8	207.6	223.5	237.2	250.5	272.3	294.6	318.0	57.5
Naphthalene	$C_{10}H_8$	52.6	74.2	85.8	101.7	119.3	130.2	145.5	167.7	193.2	217.9	80.2
1-Naphthoic acid	$C_{11}H_8O_2$	156.0	184.0	196.8	211.2	225.0	234.5	245.8	263.5	281.4	300.0	160.5
2-Naphthoic acid	$C_{11}H_8O_2$	160.8	189.7	202.8	216.9	231.5	241.3	252.7	270.3	289.5	308.5	184
1-Naphthol	$C_{10}H_8O$	94.0	125.5	142.0	158.0	177.8	190.0	206.0	229.6	255.8	282.5	96
2-Naphthol	$C_{10}H_8O$	......	128.6	145.5	161.8	181.7	193.7	209.8	234.0	260.6	288.0	122.5
1-Naphthylamine	$C_{10}H_9N$	104.3	137.7	153.8	171.6	191.5	203.8	220.0	244.9	272.2	300.8	50

Table 8. Vapor Pressures of Organic Compounds, up to 1 Atm.—(*Continued*)

Compound		Pressure, mm. Hg										Melting point, °C.
Name	Formula	1	5	10	20	40	60	100	200	400	760	
		Temperature, °C.										
2-Naphthylamine	$C_{10}H_9N$	108.0	141.6	157.6	175.8	195.7	208.1	224.3	249.7	277.4	306.1	111.5
Nicotine	$C_{10}H_{14}N_2$	61.8	91.8	107.2	123.7	142.1	154.7	169.5	193.8	219.8	247.3	
2-Nitroaniline	$C_6H_6N_2O_2$	104.0	135.7	150.4	167.7	186.0	197.8	213.0	236.3	260.0	284.5	71.5
3-Nitroaniline	$C_6H_6N_2O_2$	119.3	151.5	167.8	185.5	204.2	216.5	232.1	255.3	280.2	305.7	114
4-Nitroaniline	$C_6H_6N_2O_2$	142.4	177.6	194.4	213.2	234.2	245.9	261.8	284.5	310.2	336.0	146.5
2-Nitrobenzaldehyde	$C_7H_5NO_3$	85.8	117.7	133.4	150.0	168.8	180.7	196.2	220.0	246.8	273.5	40.9
3-Nitrobenzaldehyde	$C_7H_5NO_3$	96.2	127.4	142.8	159.0	177.7	189.5	204.3	227.4	252.1	278.3	58
Nitrobenzene	$C_6H_5NO_2$	44.4	71.6	84.9	99.3	115.4	125.8	139.9	161.2	185.8	210.6	+5.7
Nitroethane	$C_2H_5NO_2$	−21.0	+1.5	12.5	24.8	38.0	46.5	57.8	74.8	94.0	114.0	−90
Nitroglycerin	$C_3H_5N_3O_9$	127	167	188	210	235	251	......	......			11
Nitromethane	CH_3NO_2	−29.0	−7.9	+2.8	14.1	27.5	35.5	46.6	63.5	82.0	101.2	−29
2-Nitrophenol	$C_6H_5NO_3$	49.3	76.8	90.4	105.8	122.1	132.6	146.4	167.6	191.0	214.5	45
2-Nitrophenyl acetate	$C_8H_7NO_4$	100.0	128.0	142.0	155.8	172.8	181.7	194.1	213.0	233.5	253.0	
1-Nitropropane	$C_3H_7NO_2$	−9.6	+13.5	25.3	37.9	51.8	60.5	72.3	90.2	110.6	131.6	−108
2-Nitropropane	$C_3H_7NO_2$	−18.8	+4.1	15.8	28.2	41.8	50.3	62.0	80.0	99.8	120.3	−93
2-Nitrotoluene	$C_7H_7NO_2$	50.0	79.1	93.8	109.6	126.3	137.6	151.5	173.7	197.7	222.3	−4.1
3-Nitrotoluene	$C_7H_7NO_2$	50.2	81.0	96.0	112.8	130.7	142.5	156.9	180.3	206.8	231.9	15.5
4-Nitrotoluene	$C_7H_7NO_2$	53.7	85.0	100.5	117.7	136.0	147.9	163.0	186.7	212.5	238.3	51.9
4-Nitro-1,3-xylene (4-nitro-m-xylene)	$C_8H_9NO_2$	65.6	95.0	109.8	125.8	143.3	153.8	168.5	191.7	217.5	244.0	+2
Nonacosane	$C_{29}H_{60}$	234.2	269.8	286.4	303.6	323.2	334.8	350.0	373.2	397.2	421.8	63.8
Nonadecane	$C_{19}H_{40}$	133.2	166.3	183.5	200.8	220.0	232.8	248.0	271.8	299.8	330.0	32
n-Nonane	C_9H_{20}	+1.4	25.8	38.0	51.2	66.0	75.5	88.1	107.5	128.2	150.8	−53.7
1-Nonanol	$C_9H_{20}O$	59.5	86.1	99.7	113.8	129.0	139.0	151.3	170.5	192.1	213.5	−5
2-Nonanone	$C_9H_{18}O$	32.1	59.0	72.3	87.2	103.4	113.8	127.4	148.2	171.2	195.0	−19
Octacosane	$C_{28}H_{58}$	226.5	260.3	277.4	295.4	314.2	326.8	341.8	364.8	388.9	412.5	61.6
Octadecane	$C_{18}H_{38}$	119.6	152.1	169.6	187.5	207.4	219.7	236.0	260.6	288.0	317.0	28
n-Octane	C_8H_{18}	−14.0	+8.3	19.2	31.5	45.1	53.8	65.7	83.6	104.0	125.6	−56.8
n-Octanol (1-octanol)	$C_8H_{18}O$	54.0	76.5	88.3	101.0	115.2	123.3	135.2	152.0	173.8	195.2	−15.4
2-Octanone	$C_8H_{16}O$	23.6	48.4	60.9	74.3	89.8	99.0	111.7	130.4	151.0	172.9	−16
n-Octyl acrylate	$C_{11}H_{20}O_2$	58.5	87.7	102.0	117.8	135.6	145.6	159.1	180.2	204.0	227.0	
iodide (1-Iodooctane)	$C_8H_{17}I$	45.8	74.8	90.0	105.9	123.8	135.4	150.0	173.3	199.3	225.5	−45.9
Oleic acid	$C_{18}H_{34}O_2$	176.5	208.5	223.0	240.0	257.2	269.8	286.0	309.8	334.7	360.0	14
Palmitaldehyde	$C_{16}H_{32}O$	121.6	154.6	171.8	190.0	210.0	222.6	239.5	264.1	292.3	321.0	34
Palmitic acid	$C_{16}H_{32}O_2$	153.6	188.1	205.8	223.8	244.4	256.0	271.5	298.7	326.0	353.8	64.0
Palmitonitrile	$C_{16}H_{31}N$	134.3	168.3	185.8	204.2	223.8	236.6	251.5	277.1	304.5	332.0	31
Pelargonic acid	$C_9H_{18}O_2$	108.2	126.0	137.4	149.8	163.7	172.3	184.4	203.1	227.5	253.5	12.5
Pentachlorobenzene	C_6HCl_5	98.6	129.7	144.3	160.0	178.5	190.1	205.5	227.0	251.6	276.0	85.5
Pentachloroethane	C_2HCl_5	+1.0	27.2	39.8	53.9	69.9	80.0	93.5	114.0	137.2	160.5	−22
Pentachloroethylbenzene	$C_8H_5Cl_5$	96.2	130.0	148.0	166.0	186.2	199.0	216.0	241.8	269.3	299.0	
Pentachlorophenol	C_6HCl_5O	......	......	......	192.2	211.2	223.4	239.6	261.8	285.0	309.3	188.5
Pentacosane	$C_{25}H_{52}$	194.2	230.0	248.2	266.1	285.6	298.4	314.0	339.0	365.4	390.3	53.3
Pentadecane	$C_{15}H_{32}$	91.6	121.0	135.4	150.2	167.7	178.4	194.0	216.1	242.8	270.5	10
1,3-Pentadiene	C_5H_8	−71.8	−53.8	−45.0	−34.8	−23.4	−16.5	−6.7	+8.0	24.7	42.1	
1,4-Pentadiene	C_5H_8	−83.5	−66.2	−57.1	−47.7	−37.0	−30.0	−20.6	−6.7	+8.3	26.1	
Pentaethylbenzene	$C_{16}H_{26}$	86.0	120.0	135.8	152.4	171.9	184.2	200.0	224.1	250.2	277.0	
Pentaethylchlorobenzene	$C_{16}H_{25}Cl$	90.0	123.8	140.7	158.1	178.2	191.0	208.0	230.3	257.2	285.0	
n-Pentane	C_5H_{12}	−76.6	−62.5	−50.1	−40.2	−29.2	−22.2	−12.6	+1.9	18.5	36.1	−129.7
iso-Pentane (2-methylbutane)	C_5H_{12}	−82.9	−65.8	−57.0	−47.3	−36.5	−29.6	−20.2	−5.9	+10.5	27.8	−159.7
neo-Pentane (2,2-dimethylpropane)	C_5H_{12}	−102.0	−85.4	−76.7	−67.2	−56.1	−49.0	−39.1	−23.7	−7.1	+9.5	−16.6
2,3,4-Pentanetriol	$C_5H_{12}O_3$	155.0	189.3	204.5	220.5	239.6	249.8	263.5	284.5	307.0	327.2	
1-Pentene	C_5H_{10}	−80.4	−63.3	−54.5	−46.0	−34.1	−27.1	−17.7	−3.4	+12.8	30.1	
α-Phellandrene	$C_{10}H_{16}$	20.0	45.7	58.0	72.1	87.8	97.6	110.6	130.6	152.0	175.0	
Phenanthrene	$C_{14}H_{10}$	118.2	154.3	173.0	193.7	215.8	229.9	249.0	277.1	308.0	340.2	99.5
Phenethyl alcohol (phenyl cellosolve)	$C_8H_{10}O_2$	58.2	85.9	100.0	114.8	130.5	141.2	154.0	175.0	197.5	219.5	
2-Phenetidine	$C_8H_{11}NO$	67.0	94.7	108.6	123.7	139.9	149.8	163.5	184.0	207.0	228.0	
Phenol	C_6H_6O	40.1	62.5	73.8	86.0	100.1	108.4	121.4	139.0	160.0	181.9	40.6
2-Phenoxyethanol	$C_8H_{10}O_2$	78.0	106.6	121.2	136.0	152.2	163.2	176.5	197.6	221.0	245.3	11.6
2-Phenoxyethyl acetate	$C_{10}H_{12}O_3$	82.6	113.5	128.0	144.5	162.3	174.0	189.2	211.3	235.0	259.7	−6.7
Phenyl acetate	$C_8H_8O_2$	38.2	64.8	78.0	92.3	108.1	118.1	131.6	151.2	173.5	195.9	
Phenylacetic acid	$C_8H_8O_2$	97.0	127.0	141.3	156.0	173.6	184.5	198.2	219.5	243.0	265.5	76.5
Phenylacetonitrile	C_8H_7N	60.0	89.0	103.5	119.4	136.3	147.7	161.8	184.2	208.5	233.5	−23.8
Phenylacetyl chloride	C_8H_7ClO	48.0	75.3	89.0	103.6	119.8	129.8	143.5	163.8	186.0	210.0	
Phenyl benzoate	$C_{13}H_{10}O_2$	106.8	141.5	157.8	177.0	197.6	210.8	227.8	254.0	283.5	314.0	70.5
4-Phenyl-3-buten-2-one	$C_{10}H_{10}O$	81.7	112.2	127.4	143.8	161.3	172.6	187.8	211.0	235.4	261.0	41.5
Phenyl isocyanate	C_7H_5NO	10.6	36.0	48.5	62.5	77.7	87.7	100.6	120.8	142.7	165.6	
isocyanide	C_7H_5N	12.0	37.0	49.7	63.4	78.3	88.0	01.0	120.8	142.3	165.0	
Phenylcyclohexane	$C_{12}H_{16}$	67.5	96.5	111.3	126.4	144.0	154.2	169.3	191.3	214.6	240.0	+7.5
Phenyl dichlorophosphate	$C_6H_5Cl_2O_2P$	66.7	95.9	110.0	125.9	143.4	153.6	168.0	189.8	213.0	239.5	
m-Phenylene diamine (1,3-phenylenediamine)	$C_6H_8N_2$	99.8	131.2	147.0	163.8	182.5	194.0	209.9	233.0	259.0	285.5	62.8
Phenylglyoxal	$C_8H_6O_2$	......	75.0	87.8	100.7	115.5	124.2	136.2	153.8	173.5	193.5	73
Phenylhydrazine	$C_6H_8N_2$	71.8	101.6	115.8	131.5	148.2	158.7	173.5	195.4	218.2	243.5	19.5
N-Phenyliminodiethanol	$C_{10}H_{15}NO_2$	145.0	179.2	195.8	213.4	233.0	245.3	260.6	284.5	311.3	337.8	
1-Phenyl-1,3-pentanedione	$C_{11}H_{12}O_2$	98.0	128.5	144.0	159.9	178.0	189.8	204.5	226.7	251.2	276.5	
2-Phenylphenol	$C_{12}H_{10}O$	100.0	131.6	146.2	163.3	180.3	192.2	205.9	227.9	251.8	275.0	56.5
4-Phenylphenol	$C_{12}H_{10}O$	......	......	176.2	193.8	213.0	225.3	240.9	263.2	285.5	308.0	164.5
3-Phenyl-1-propanol	$C_9H_{12}O$	74.7	102.4	116.0	131.2	147.4	156.8	170.3	191.2	212.8	235.0	
Phenyl isothiocyanate	C_7H_5NS	47.2	75.6	89.8	115.5	122.5	133.3	147.7	169.6	194.0	218.5	−21.0
Phorone	$C_9H_{14}O$	42.0	68.3	81.5	95.6	111.3	121.4	134.0	153.5	175.3	197.2	28
iso-Phorone	$C_9H_{14}O$	38.0	66.7	81.2	96.8	114.5	125.6	140.6	163.3	188.7	215.2	
Phosgene (carbonyl chloride)	CCl_2O	−92.9	−77.0	−69.3	−60.3	−50.3	−44.0	−35.6	−22.3	−7.6	+8.3	−104
Phthalic anhydride	$C_8H_4O_3$	96.5	121.3	134.0	151.7	172.0	185.3	202.3	228.0	256.8	284.5	130.8
Phthalide	$C_8H_6O_2$	95.5	127.7	144.0	161.3	181.0	193.5	210.0	234.5	261.8	290.0	73
Phthaloyl chloride	$C_8H_4Cl_2O_2$	86.3	118.3	134.2	151.0	170.0	182.2	197.8	222.0	248.3	275.8	88.5
2-Picoline	C_6H_7N	−11.1	+12.6	24.4	37.4	51.2	59.9	71.4	89.0	108.4	128.8	−70
Pimelic acid	$C_7H_{12}O_4$	163.4	196.2	212.0	229.3	247.0	258.2	272.0	294.5	318.5	342.1	103
α-Pinene	$C_{10}H_{16}$	−1.0	+24.6	37.3	51.4	66.8	76.8	90.1	110.2	132.3	155.0	−55
β-Pinene	$C_{10}H_{16}$	+4.2	30.0	42.3	58.1	71.5	81.2	94.0	114.1	136.1	158.3	
Piperidine	$C_5H_{11}N$	......	−7.0	+3.9	15.8	29.2	37.7	49.0	66.2	85.7	106.0	−9

Table 8. Vapor Pressures of Organic Compounds, up to 1 Atm.—(*Continued*)

Compound		Pressure, mm. Hg										Melting point, °C.
Name	Formula	1	5	10	20	40	60	100	200	400	760	
		Temperature, °C.										
Piperonal	$C_8H_6O_3$	87.0	117.4	132.0	148.0	165.7	177.0	191.7	214.3	238.5	263.0	37
Propane	C_3H_8	−128.9	−115.4	−108.5	−100.9	−92.4	−87.0	−79.6	−68.4	−55.6	−42.1	−187.1
Propenylbenzene	C_9H_{10}	17.5	43.8	57.0	71.5	87.7	97.8	111.7	132.0	154.7	179.0	−30.1
Propionamide	C_3H_7NO	65.0	91.0	105.0	119.0	134.8	144.3	156.0	174.2	194.0	213.0	79
Propionic acid	$C_3H_6O_2$	4.6	28.0	39.7	52.0	65.8	74.1	85.8	102.5	122.0	141.1	−22
anhydride	$C_6H_{10}O_3$	20.6	45.3	57.7	70.4	85.6	94.5	107.2	127.8	146.0	167.0	−45
Propionitrile	C_3H_5N	−35.0	−13.6	−3.0	+8.8	22.0	30.1	41.4	58.2	77.7	97.1	−91.9
Propiophenone	$C_9H_{10}O$	50.0	77.9	92.2	107.6	124.3	135.0	149.3	170.2	194.2	218.0	21
n-Propyl acetate	$C_5H_{10}O_2$	−26.7	−5.4	+5.0	16.0	28.8	37.0	47.8	64.0	82.0	101.8	−92.5
iso-Propyl acetate	$C_5H_{10}O_2$	−38.3	−17.4	−7.2	+4.2	17.0	25.1	35.7	51.7	69.8	89.0	
n-Propyl alcohol (1-propanol)	C_3H_8O	−15.0	+5.0	14.7	25.3	36.4	43.5	52.8	66.8	82.0	97.8	−127
iso-Propyl alcohol (2-propanol)	C_3H_8O	−26.1	−7.0	+2.4	12.7	23.8	30.5	39.5	53.0	67.8	82.5	−85.8
n-Propylamine	C_3H_9N	−64.4	−46.3	−37.2	−27.1	−16.0	−9.0	+0.5	15.0	31.5	48.5	−83
Propylbenzene	C_9H_{12}	6.3	31.3	43.4	56.8	71.6	81	94.0	113.5	135.7	159.2	−99.5
Propyl benzoate	$C_{10}H_{12}O_2$	54.6	83.8	98.0	114.3	131.8	143.3	157.4	180.1	205.2	231.0	−51.6
n-Propyl bromide (1-bromopropane)	C_3H_7Br	−53.0	−33.4	−23.3	−12.4	−0.3	+7.5	18.0	34.0	52.0	71.0	−109.9
iso-Propyl bromide (2-bromopropane)	C_3H_7Br	−61.8	−42.5	−32.8	−22.0	−10.1	−2.5	+8.0	23.8	41.5	60.0	−89.0
n-Propyl n-butyrate	$C_7H_{14}O_2$	−1.6	+22.1	34.0	47.0	61.5	70.3	82.6	101.0	121.7	142.7	−95.2
isobutyrate	$C_7H_{14}O_2$	−6.2	+16.8	28.3	40.6	54.3	63.0	73.9	91.8	112.0	133.9	
iso-Propyl isobutyrate	$C_7H_{14}O_2$	−16.3	+5.8	17.0	29.0	42.4	51.4	62.3	80.2	100.0	120.5	
Propyl carbamate	$C_4H_9NO_2$	52.4	77.6	90.0	103.2	117.7	126.5	138.3	155.8	175.8	195.0	
n-Propyl chloride (1-chloropropane)	C_3H_7Cl	−68.3	−50.0	−41.0	−31.0	−19.5	−12.1	−2.5	+12.2	29.4	46.4	−122.8
iso-Propyl chloride (2-chloropropane)	C_3H_7Cl	−78.8	−61.1	−52.0	−42.0	−31.0	−23.5	−13.7	+1.3	18.1	36.5	−117
iso-Propyl chloroacetate	$C_5H_9ClO_2$	+3.8	28.1	40.2	53.9	68.7	78.0	90.3	108.8	128.0	148.6	
Propyl chloroglyoxylate	$C_5H_7ClO_3$	9.7	32.3	43.5	55.6	68.8	77.2	88.0	104.7	123.0	150.0	
Propylene	C_3H_6	−131.9	−120.7	−112.1	−104.7	−96.5	−91.3	−84.1	−73.3	−60.9	−47.7	−185
Propylene glycol (1,2-Propanediol)	$C_3H_8O_2$	45.5	70.8	83.2	96.4	111.2	119.9	132.0	149.7	168.1	188.2	
Propylene oxide	C_3H_6O	−75.0	−57.8	−49.0	−39.3	−28.4	−21.3	−12.0	+2.1	17.8	34.5	−112.1
n-Propyl formate	$C_4H_8O_2$	−43.0	−22.7	−12.6	−1.7	+10.8	18.8	29.5	45.3	62.6	81.3	−92.9
iso-Propyl formate	$C_4H_8O_2$	−52.0	−32.7	−22.7	−12.1	−0.2	+7.5	17.8	33.6	50.5	68.3	
4,4′-iso-Propylidenebisphenol	$C_{15}H_{16}O_2$	193.0	224.2	240.8	255.5	273.0	282.9	297.0	317.5	339.0	360.5	
n-Propyl iodide (1-iodopropane)	C_3H_7I	−36.0	−13.5	−2.4	+10.0	23.6	32.1	43.8	61.8	81.8	102.5	−98.8
iso-Propyl iodide (2-iodopropane)	C_3H_7I	−43.3	−22.1	−11.7	0.0	+13.2	21.6	32.8	50.0	69.5	89.5	−90
n-Propyl levulinate	$C_8H_{14}O_3$	59.7	86.3	99.9	114.0	130.1	140.6	154.0	175.6	198.0	221.2	
iso-Propyl levulinate	$C_8H_{14}O_3$	48.0	74.5	88.0	102.4	118.1	127.8	141.8	161.6	185.2	208.2	
Propyl mercaptan (1-propanethiol)	C_3H_8S	−56.0	−36.3	−26.3	−15.4	−3.2	+4.6	15.3	31.5	49.2	67.4	−112
2-iso-Propylnaphthalene	$C_{13}H_{14}$	76.0	107.9	123.4	140.3	159.0	171.4	187.6	211.8	238.5	266.0	
iso-Propyl β-naphthyl ketone (2-isobutyronaphthone)	$C_{14}H_{14}O$	133.2	165.4	181.0	197.7	215.6	227.0	242.3	264.0	288.2	313.0	
2-iso-Propylphenol	$C_9H_{12}O$	56.6	83.8	97.0	111.7	127.5	137.7	150.3	170.1	192.6	214.5	15.5
3-iso-Propylphenol	$C_9H_{12}O$	62.0	90.3	104.1	119.8	136.2	146.6	160.2	182.0	205.0	228.0	26
4-iso-Propylphenol	$C_9H_{12}O$	67.0	94.7	108.0	123.4	139.8	149.7	163.3	184.0	206.1	228.2	61
Propyl propionate	$C_6H_{12}O_2$	−14.2	+8.0	19.4	31.6	45.0	53.8	65.2	82.7	102.0	122.4	−76
4-iso-Propylstyrene	$C_{11}H_{14}$	34.7	62.3	76.0	91.2	108.0	118.4	132.8	153.9	178.0	202.5	
Propyl isovalerate	$C_8H_{16}O_2$	+8.0	32.8	45.1	58.0	72.8	82.3	95.0	113.9	135.0	155.9	
Pulegone	$C_{10}H_{16}O$	58.3	82.5	94.0	106.8	121.7	130.2	143.1	162.5	189.8	221.0	
Pyridine	C_5H_5N	−18.9	+2.5	13.2	24.8	38.0	46.8	57.8	75.0	95.6	115.4	−42
Pyrocatechol	$C_6H_6O_2$	......	104.0	118.3	134.0	150.6	161.7	176.0	197.7	221.5	245.5	105
Pyrocaltechol diacetate (1,2-phenylene diacetate)	$C_{10}H_{10}O_4$	98.0	129.8	145.7	161.8	179.8	191.6	206.5	228.7	253.3	278.0	
Pyrogallol	$C_6H_6O_3$	......	151.7	167.7	185.3	204.2	216.3	232.0	255.3	281.5	309.0	133
Pyrotartaric anhydride	$C_5H_6O_3$	69.7	99.7	114.2	130.0	147.8	158.6	173.8	196.1	221.0	247.4	
Pyruvic acid	$C_3H_4O_3$	21.4	45.8	57.9	70.8	85.3	94.1	106.5	124.7	144.7	165.0	13.6
Quinoline	C_9H_7N	59.7	89.6	103.8	119.8	136.7	148.1	163.2	186.2	212.3	237.7	−15
iso-Quinoline	C_9H_7N	63.5	92.7	107.8	123.7	141.6	152.0	167.6	190.0	214.5	240.5	24.6
Resorcinol	$C_6H_6O_2$	108.4	138.0	152.1	168.0	185.3	195.8	209.8	230.8	253.4	276.5	110.7
Safrole	$C_{10}H_{10}O_2$	63.8	93.0	107.6	123.0	140.1	150.3	165.1	186.2	210.0	233.0	11.2
Salicylaldehyde	$C_7H_6O_2$	33.0	60.1	73.8	88.7	105.2	115.7	129.4	150.0	173.7	196.5	−7
Salicylic acid	$C_7H_6O_3$	113.7	136.0	146.2	156.8	172.2	182.0	193.4	210.0	230.5	256.0	159
Sebacic acid	$C_{10}H_{18}O_4$	183.0	215.7	232.0	250.0	268.2	279.8	294.5	313.2	332.8	352.3	134.5
Selenophene	C_4H_4Se	−39.0	−16.0	−4.0	+9.1	24.1	33.8	47.0	66.7	89.8	114.3	
Skatole	C_9H_9N	95.0	124.2	139.6	154.3	171.9	183.6	197.4	218.8	242.5	266.2	95
Stearaldehyde	$C_{18}H_{36}O$	140.0	174.6	192.1	210.6	230.8	244.2	260.0	285.0	313.8	342.5	63.5
Stearic acid	$C_{18}H_{36}O_2$	173.7	209.0	225.0	243.4	263.3	275.5	291.0	316.5	343.0	370.0	69.3
Stearyl alcohol (1-octadecanol)	$C_{18}H_{38}O$	150.3	185.6	202.0	220.0	240.4	252.7	269.4	293.5	320.3	349.5	58.5
Styrene	C_8H_8	−7.0	+18.0	30.8	44.6	59.8	69.5	82.0	101.3	122.5	145.2	−30.6
Styrene dibromide [(1,2-dibromoethyl) benzene]	$C_8H_8Br_2$	86.0	115.6	129.8	145.2	161.8	172.2	186.3	207.8	230.0	254.0	
Suberic acid	$C_8H_{14}O_4$	172.8	205.5	219.5	238.2	254.6	265.4	279.8	300.5	322.8	345.5	142
Succinic anhydride	$C_4H_4O_3$	92.0	115.0	128.2	145.3	163.0	174.0	189.0	212.0	237.0	261.0	119.6
Succinimide	$C_4H_5NO_2$	115.0	143.2	157.0	174.0	192.0	203.0	217.4	240.0	263.5	287.5	125.5
Succinyl chloride	$C_4H_4Cl_2O_2$	39.0	65.0	78.0	91.8	107.5	117.2	130.0	149.3	170.0	192.5	17
α-Terpineol	$C_{10}H_{18}O$	52.8	80.4	94.3	109.8	126.0	136.3	150.1	171.2	194.3	217.5	35
Terpenoline	$C_{10}H_{16}$	32.3	58.0	70.6	84.8	100.0	109.8	122.7	142.0	163.5	185.0	
1,1,1,2-Tetrabromoethane	$C_2H_2Br_4$	58.0	83.3	95.7	108.5	123.2	132.0	144.0	161.5	181.0	200.0	
1,1,2,2-Tetrabromoethane	$C_2H_2Br_4$	65.0	95.5	110.0	126.0	144.0	155.1	170.0	192.5	217.5	243.5	
Tetraisobutylene	$C_{16}H_{32}$	63.8	93.7	108.5	124.5	142.2	152.6	167.5	190.0	214.6	240.0	
Tetracosane	$C_{24}H_{50}$	183.8	219.6	237.6	255.3	276.3	288.4	305.2	330.5	358.0	386.4	51.1
1,2,3,4-Tetrachlorobenzene	$C_6H_2Cl_4$	68.5	99.6	114.7	131.2	149.2	160.0	175.7	198.0	225.5	254.0	46.5
1,2,3,5-Tetrachlorobenzene	$C_6H_2Cl_4$	58.2	89.0	104.1	121.6	140.0	152.0	168.0	193.7	220.0	246.0	54.5
1,2,4,5-Tetrachlorobenzene	$C_6H_2Cl_4$	......	......	......	......	146.0	157.7	173.5	196.0	220.5	245.0	139
1,1,2,2-Tetrachloro-1,2-difluoroethane	$C_2Cl_4F_2$	−37.5	−16.0	−5.0	+6.7	19.8	28.1	38.6	55.0	73.1	92.0	26.5
1,1,1,2-Tetrachloroethane	$C_2H_2Cl_4$	−16.3	17.4	19.3	32.1	46.7	56.0	68.0	87.2	108.2	130.5	−68.7
1,1,2,2-Tetrachloroethane	$C_2H_2Cl_4$	−3.8	+20.7	33.0	46.2	60.8	70.0	83.2	102.2	124.0	145.9	−36
1,2,3,5-Tetrachloro-4-ethylbenzene	$C_8H_6Cl_4$	77.0	110.0	126.0	143.7	162.1	175.0	191.6	215.3	243.0	270.0	
Tetrachloroethylene	C_2Cl_4	−20.6	+2.4	13.8	26.3	40.1	49.2	61.3	79.8	100.0	120.8	−19.0
2,3,4,6-Tetrachlorophenol	$C_6H_2Cl_4O$	100.0	130.3	145.3	161.0	179.1	190.0	205.2	227.2	250.4	275.0	69.5
3,4,5,6-Tetrachloro-1,2-xylene	$C_8H_6Cl_4$	94.4	125.0	140.3	156.0	174.2	185.8	200.5	223.0	248.3	273.5	
Tetradecane	$C_{14}H_{30}$	76.4	106.0	120.7	135.6	152.7	164.0	178.5	201.8	226.8	252.5	5.5

Table 8. Vapor Pressures of Organic Compounds, up to 1 Atm.—(*Continued*)

Compound: Name	Formula	1	5	10	20	40	60	100	200	400	760	Melting point, °C.
		Pressure, mm. Hg — Temperature, °C.										
Tetradecylamine	$C_{14}H_{31}N$	102.6	135.8	152.0	170.0	189.0	200.2	215.7	239.8	264.6	291.2	
Tetradecyltrimethylsilane	$C_{17}H_{38}Si$	120.0	150.7	166.2	183.5	201.5	213.3	227.8	250.0	275.0	300.0	
Tetraethoxysilane	$C_8H_{20}O_4Si$	16.0	40.3	52.6	65.8	81.1	90.7	103.6	123.5	146.2	168.5	
1,2,3,4-Tetraethylbenzene	$C_{14}H_{22}$	65.7	96.2	111.6	127.7	145.8	156.7	172.4	196.0	221.4	248.0	11.6
Tetraethylene glycol	$C_8H_{18}O_5$	153.9	183.7	197.1	212.3	228.0	237.8	250.0	268.4	288.0	307.8	
Tetraethylene glycol chlorohydrin	$C_8H_{17}ClO_4$	110.1	141.8	156.1	172.6	190.0	200.5	214.7	236.5	258.2	281.5	
Tetraethyllead	$C_8H_{20}Pb$	38.4	63.6	74.8	88.0	102.4	111.7	123.8	142.0	161.8	183.0	−136
Tetraethylsilane	$C_8H_{20}Si$	−1.0	+23.9	36.3	50.0	65.3	74.8	88.0	108.0	130.2	153.0	
Tetralin	$C_{10}H_{12}$	38.0	65.3	79.0	93.8	110.4	121.3	135.3	157.2	181.8	207.2	−31.0
1,2,3,4-Tetramethylbenzene	$C_{10}H_{14}$	42.6	68.7	81.8	95.8	111.5	121.8	135.7	155.7	180.0	204.4	−6.2
1,2,3,5-Tetramethylbenzene	$C_{10}H_{14}$	40.6	65.8	77.8	91.0	105.8	115.4	128.3	149.9	173.7	197.9	−24.0
1,2,4,5-Tetramethylbenzene	$C_{10}H_{14}$	45.0	65.0	74.6	88.0	104.2	114.8	128.1	149.5	172.1	195.9	79.5
2,2,3,3-Tetramethylbutane	C_8H_{18}	−17.4	+3.2	13.5	24.6	36.8	44.5	54.8	70.2	87.4	106.3	−102.2
Tetramethylene dibromide (1,4-dibromobutane)	$C_4H_8Br_2$	32.0	58.8	72.4	87.6	104.0	115.1	128.7	149.8	173.8	197.5	−20
Tetramethyllead	$C_4H_{12}Pb$	−29.0	−6.8	+4.4	16.6	30.3	39.2	50.8	68.8	89.0	110.0	−27.5
Tetramethyltin	$C_4H_{12}Sn$	−51.3	−31.0	−20.6	−9.3	+3.5	11.7	22.8	39.8	58.5	78.0	
Tetrapropylene glycol monoisopropyl ether	$C_{15}H_{32}O_5$	116.6	147.8	163.0	179.8	197.7	209.0	223.3	245.0	268.3	292.7	
Thioacetic acid (mercaptoacetic acid)	$C_2H_4O_2S$	60.0	87.7	101.5	115.8	131.8	142.0	154.0	...	...	...	−16.5
Thiodiglycol (2,2′-thiodiethanol)	$C_4H_{10}O_2S$	42.0	96.0	128.0	165.0	210.0	240.5	285				
Thiophene	C_4H_4S	−40.7	−20.8	−10.9	0.0	+12.5	20.1	30.5	46.5	64.7	84.4	−38.3
Thiophenol (benzenethiol)	C_6H_6S	18.6	43.7	56.0	69.7	84.2	93.9	106.6	125.8	146.7	168.0	
α-Thujone	$C_{10}H_{16}O$	38.3	65.7	79.3	93.7	110.0	120.2	134.0	154.2	177.8	201.0	
Thymol	$C_{10}H_{14}O$	64.3	92.8	107.4	122.6	139.8	149.8	164.1	185.5	209.2	231.8	51.5
Tiglaldehyde	C_5H_8O	−25.0	−1.6	+10.0	23.2	37.0	45.8	57.7	75.4	95.5	116.4	
Tiglic acid	$C_5H_8O_2$	52.0	77.8	90.2	103.8	119.0	127.8	140.5	158.0	179.2	198.5	64.5
Tiglonitrile	C_5H_7N	−25.5	−2.4	+9.2	22.1	36.7	46.0	58.2	77.8	99.7	122.0	
Toluene	C_7H_8	−26.7	−4.4	+6.4	18.4	31.8	40.3	51.9	69.5	89.5	110.6	−95.0
Toluene-2,4-diamine	$C_7H_{10}N_2$	106.5	137.2	151.7	167.9	185.7	196.2	211.5	232.8	256.0	280.0	99
2-Toluic nitrile (2-tolunitrile)	C_8H_7N	36.7	64.0	77.9	93.0	110.0	120.8	135.0	156.0	180.0	205.2	−13
4-Toluic nitrile (4-tolunitrile)	C_8H_7N	42.5	71.3	85.8	101.7	109.5	130.0	145.2	167.3	193.0	217.6	29.5
2-Toluidine	C_7H_9N	44.0	69.3	81.4	95.1	110.0	119.8	133.0	153.0	176.2	199.7	−16.3
3-Toluidine	C_7H_9N	41.0	68.0	82.0	96.7	113.5	123.8	136.7	157.6	180.6	203.3	−31.5
4-Toluidine	C_7H_9N	42.0	68.2	81.8	95.8	111.5	121.5	133.7	154.0	176.9	200.4	44.5
2-Tolyl isocyanide	C_8H_7N	25.2	51.0	64.0	78.2	94.0	104.0	117.7	137.8	159.9	183.5	
4-Tolylhydrazine	$C_7H_{10}N_2$	82.4	110.0	123.8	138.6	154.1	165.0	178.0	198.0	219.5	242.0	65.5
Tribromoacetaldehyde	C_2HBr_3O	18.5	45.0	58.0	72.1	87.8	97.5	110.2	130.0	151.6	174.0	
1,1,2-Tribromobutane	$C_4H_7Br_3$	45.0	73.5	87.8	103.2	120.2	131.6	146.0	167.8	192.0	216.2	
1,2,2-Tribromobutane	$C_4H_7Br_3$	41.0	69.0	83.2	98.6	116.0	127.0	141.8	163.5	188.0	213.8	
2,2,3-Tribromobutane	$C_4H_7Br_3$	38.2	66.0	79.8	94.6	111.8	122.2	136.3	157.8	182.2	206.5	
1,1,2-Tribromoethane	$C_2H_3Br_3$	32.6	58.0	70.6	84.2	100.0	110.0	123.5	143.5	165.4	188.4	−26
1,2,3-Tribromopropane	$C_3H_5Br_3$	47.5	75.8	90.0	105.8	122.8	134.0	148.0	170.0	195.0	220.0	16.5
Triisobutylamine	$C_{12}H_{27}N$	32.3	57.4	69.8	83.0	97.8	107.3	119.7	138.0	157.8	179.0	−22
Triisobutylene	$C_{12}H_{24}$	18.0	44.0	56.5	70.0	86.7	96.7	110.0	130.2	153.0	179.0	
2,4,6-Tritertbutylphenol	$C_{18}H_{30}O$	95.2	126.1	142.0	158.0	177.4	188.0	203.0	226.2	250.6	276.3	
Trichloroacetic acid	$C_2HCl_3O_2$	51.0	76.0	88.2	101.8	116.3	125.9	137.8	155.4	175.2	195.6	57
Trichloroacetic anhydride	$C_4Cl_6O_3$	56.2	85.3	99.6	114.3	131.2	141.8	155.2	176.2	199.8	223.0	
Trichloroacetyl bromide	C_2BrCl_3O	−7.4	+16.7	29.3	42.1	57.2	66.7	79.5	98.4	120.2	143.0	
2,4,6-Trichloroaniline	$C_6H_4Cl_3N$	134.0	157.8	170.0	182.6	195.8	204.5	214.6	229.8	246.4	262.0	78
1,2,3-Trichlorobenzene	$C_6H_3Cl_3$	40.0	70.0	85.6	101.8	119.8	131.5	146.0	168.2	193.5	218.5	52.5
1,2,4-Trichlorobenzene	$C_6H_3Cl_3$	38.4	67.3	81.7	97.2	114.8	125.7	140.0	162.0	187.7	213.0	17
1,3,5-Trichlorobenzene	$C_6H_3Cl_3$	...	63.8	78.0	93.7	110.8	121.8	136.0	157.7	183.0	208.4	63.5
1,2,3-Trichlorobutane	$C_4H_7Cl_3$	+0.5	27.2	40.0	55.0	71.5	82.0	96.2	118.0	143.0	169.0	
1,1,1-Trichloroethane	$C_2H_3Cl_3$	−52.0	−32.0	−21.9	−10.8	+1.6	9.5	20.0	36.2	54.6	74.1	−30.6
1,1,2 Trichloroethane	$C_2H_3Cl_3$	−24.0	−2.0	+8.3	21.6	35.2	44.0	55.7	73.3	93.0	113.9	−36.7
Trichloroethylene	C_2HCl_3	−43.8	−22.8	−12.4	−1.0	+11.9	20.0	31.4	48.0	67.0	86.7	−73
Trichlorofluoromethane	CCl_3F	−84.3	−67.6	−59.0	−49.7	−39.0	−32.3	−23.0	−9.1	+6.8	23.7	
2,4,5-Trichlorophenol	$C_6H_3Cl_3O$	72.0	102.1	117.3	134.0	151.5	162.5	178.0	201.5	226.5	251.8	62
2,4,6-Trichlorophenol	$C_6H_3Cl_3O$	76.5	105.9	120.2	135.8	152.2	163.5	177.8	199.0	222.5	246.0	68.5
Tri-2-chlorophenylthiophosphate	$C_{18}H_{12}Cl_3O_3PS$	188.2	217.2	231.2	246.7	261.7	271.5	283.8	302.8	322.0	341.3	
1,1,1-Trichloropropane	$C_3H_5Cl_3$	−28.8	−7.0	+4.2	16.2	29.9	38.3	50.0	67.7	87.5	108.2	−77.7
1,2,3-Trichloropropane	$C_3H_5Cl_3$	+9.0	33.7	46.0	59.3	74.0	83.6	96.1	115.6	137.0	158.0	−14.7
1,1,2-Trichloro-1,2,2-trifluoroethane	$C_2Cl_3F_3$	−68.0	−49.4	−40.3	−30.0	−18.5	−11.2	−1.7	+13.5	30.2	47.6	−35
Tricosane	$C_{23}H_{48}$	170.0	206.3	223.0	242.0	261.3	273.8	289.8	313.5	339.8	366.5	47.7
Tridecane	$C_{13}H_{28}$	59.4	98.3	104.0	120.2	137.7	148.2	162.5	185.0	209.4	234.0	−6.2
Tridecanoic acid	$C_{13}H_{26}O_2$	137.8	166.3	181.0	195.8	212.4	222.0	236.0	255.2	276.5	299.0	41
Triethoxymethylsilane	$C_7H_{18}O_3Si$	−1.5	+22.8	34.6	47.2	61.7	70.4	82.7	101.0	121.8	143.5	
Triethoxyphenylsilane	$C_{12}H_{20}O_3Si$	71.0	98.8	112.6	127.2	143.5	153.2	167.5	188.0	210.5	233.5	
1,2,4-Triethylbenzene	$C_{12}H_{18}$	46.0	74.2	88.5	104.0	121.7	132.2	146.8	168.3	193.7	218.0	
1,3,4-Triethylbenzene	$C_{12}H_{18}$	47.9	76.0	90.2	105.8	122.6	133.4	147.7	168.3	193.2	217.5	
Triethylborine	$C_6H_{15}B$	...	...	−148.0	−140.6	−131.4	−125.2	−116.0	−101.0	−81.0	−56.2	
Triethyl camphoronate	$C_{15}H_{26}O_6$	...	150.2	166.0	183.6	201.8	213.5	228.6	250.8	276.0	301.0	135
citrate	$C_{12}H_{20}O_7$	107.0	138.7	144.0	171.1	190.4	202.5	217.8	242.2	267.5	294.0	
Triethyleneglycol	$C_6H_{14}O_4$	114.0	144.0	158.1	174.0	191.3	201.5	214.6	235.2	256.6	278.3	
Triethylheptylsilane	$C_{13}H_{30}Si$	70.0	99.8	114.6	130.3	148.0	158.2	174.0	196.0	221.0	247.0	
Triethyloctylsilane	$C_{14}H_{32}Si$	73.7	104.8	120.6	137.7	155.7	168.0	184.3	208.0	235.0	262.0	
Triethyl orthoformate	$C_7H_{16}O_3$	+5.5	29.2	40.5	53.4	67.5	76.0	88.0	106.0	125.7	146.0	
phosphate	$C_6H_{15}O_4P$	39.6	67.8	82.1	97.8	115.7	126.3	141.6	163.7	187.0	211.0	
Triethylthallium	$C_6H_{15}Tl$	+9.3	37.6	51.7	67.7	85.4	95.7	112.1	136.0	163.5	192.1	−63.0
Trifluorophenylsilane	$C_6H_5F_3Si$	−31.0	−9.7	+0.8	12.3	25.4	33.2	44.2	60.1	78.7	98.3	
Trimethallyl phosphate	$C_{12}H_{21}PO_4$	93.7	131.0	149.8	169.8	192.0	207.0	225.7	255.0	288.5	324.0	
2,3,5-Trimethylacetophenone	$C_{11}H_{14}O$	79.0	108.0	122.3	137.5	154.2	165.7	179.7	201.3	224.3	247.5	
Trimethylamine	C_3H_9N	−97.1	−81.7	−73.8	−65.0	−55.2	−48.8	−40.3	−27.0	−12.5	+2.9	−117.1
2,4,5-Trimethylaniline	$C_9H_{13}N$	68.4	95.9	109.0	123.7	139.8	149.5	162.0	182.3	203.7	234.5	67
1,2,3-Trimethylbenzene	C_9H_{12}	16.8	42.9	55.9	69.9	85.4	95.3	108.8	129.0	152.0	176.1	−25.5
1,2,4-Trimethylbenzene	C_9H_{12}	13.6	38.3	50.7	64.5	79.8	89.5	102.8	122.7	145.4	169.2	−44.1
1,3,5-Trimethylbenzene	C_9H_{12}	9.6	34.7	47.4	61.0	76.1	85.8	98.9	118.6	141.0	164.7	−44.8
2,2,3-Trimethylbutane	C_7H_{16}	...	...	−18.8	−7.5	+5.2	13.3	24.4	41.2	60.4	80.9	−25.0
Trimethyl citrate	$C_9H_{14}O_7$	106.2	146.2	160.4	177.2	194.2	205.5	219.6	241.3	264.2	287.0	78.5

Table 8. Vapor Pressures of Organic Compounds, up to 1 Atm.—(*Concluded*)

Compound: Name	Formula	Pressure, mm. Hg — Temperature, °C.: 1	5	10	20	40	60	100	200	400	760	Melting point, °C.
Trimethyleneglycol (1,3-propanediol)	$C_3H_8O_2$	59.4	87.2	100.6	115.5	131.0	141.1	153.4	172.8	193.8	214.2	
1,2,4-Trimethyl-5-ethylbenzene	$C_{11}H_{16}$	43.7	71.2	84.6	99.7	106.0	126.3	140.3	160.3	184.5	208.1	
1,3,5-Trimethyl-2-ethylbenzene	$C_{11}H_{16}$	38.8	67.0	80.5	96.0	113.2	123.8	137.9	158.4	183.5	208.0	
2,2,3-Trimethylpentane	C_8H_{18}	−29.0	−7.1	+3.9	16.0	29.5	38.1	49.9	67.8	88.2	109.8	−112.3
2,2,4-Trimethylpentane	C_8H_{18}	−36.5	−15.0	−4.3	+7.5	20.7	29.1	40.7	58.1	78.0	99.2	−107.3
2,3,3-Trimethylpentane	C_8H_{18}	−25.8	−3.9	+6.9	19.2	33.0	41.8	53.8	72.0	92.7	114.8	−101.5
2,3,4-Trimethylpentane	C_8H_{18}	−26.3	−4.1	+7.1	19.3	32.9	41.6	53.4	71.3	91.8	113.5	−109.2
2,2,4-Trimethyl-3-pentanone	$C_8H_{16}O$	14.7	36.0	46.4	57.6	69.8	77.3	87.6	102.2	118.4	135.0	
Trimethyl phosphate	$C_3H_9O_4P$	26.0	53.7	67.8	83.0	100.0	110.0	124.0	145.0	167.8	192.7	
2,4,5-Trimethylstyrene	$C_{11}H_{14}$	48.1	77.0	91.6	107.1	124.2	135.5	149.8	171.8	196.1	221.2	
2,4,6-Trimethylstyrene	$C_{11}H_{14}$	37.5	65.7	79.7	94.8	111.8	122.3	136.8	157.8	182.3	207.0	
Trimethylsuccinic anhydride	$C_7H_{10}O_3$	53.5	82.6	97.4	113.8	131.0	142.2	156.5	179.8	205.5	231.0	
Triphenylmethane	$C_{19}H_{16}$	169.7	188.4	197.0	206.8	215.5	221.2	228.4	239.7	249.8	259.2	93.4
Triphenylphosphate	$C_{18}H_{15}O_4P$	193.5	230.4	249.8	269.7	290.3	305.2	322.5	349.8	379.2	413.5	49.4
Tripropyleneglycol	$C_9H_{20}O_4$	96.0	125.7	140.5	155.8	173.7	184.6	199.0	220.2	244.3	267.2	
Tripropyleneglycol monobutyl ether	$C_{13}H_{28}O_4$	101.5	131.6	147.0	161.8	179.8	190.2	204.4	224.4	247.0	269.5	
Tripropyleneglycol monoisopropyl ether	$C_{12}H_{26}O_4$	82.4	112.4	127.3	143.7	161.4	173.2	187.8	209.7	232.8	256.6	
Tritolyl phosphate	$C_{21}H_{21}O_4P$	154.6	184.2	198.0	213.2	229.7	239.8	252.2	271.8	292.7	313.0	
Undecane	$C_{11}H_{24}$	32.7	59.7	73.9	85.6	104.4	115.2	128.1	149.3	171.9	195.8	−25.6
Undecanoic acid	$C_{11}H_{22}O_2$	101.4	133.1	149.0	166.0	185.6	197.2	212.5	237.8	262.8	290.0	29.5
10-Undecenoic acid	$C_{11}H_{20}O_2$	114.0	142.8	156.3	172.0	188.7	199.5	213.5	232.8	254.0	275.0	24.5
Undecan-2-ol	$C_{11}H_{24}O$	71.1	99.0	112.8	127.5	143.7	153.7	167.2	187.7	209.8	232.0	
n-Valeric acid	$C_5H_{10}O_2$	42.2	67.7	79.8	93.1	107.8	116.6	128.3	146.0	165.0	184.4	−34.5
iso-Valeric acid	$C_5H_{10}O_2$	34.5	59.6	71.3	84.0	98.0	107.3	118.9	136.2	155.2	175.1	−37.6
γ-Valerolactone	$C_5H_8O_2$	37.5	65.8	79.8	95.2	101.9	122.4	136.5	157.7	182.3	207.5	
Valeronitrile	C_5H_9N	−6.0	+18.1	30.0	43.3	57.8	66.9	78.6	97.7	118.7	140.8	
Vanillin	$C_8H_8O_3$	107.0	138.4	154.0	170.5	188.7	199.8	214.5	237.3	260.0	285.0	81.5
Vinyl acetate	$C_4H_6O_2$	−48.0	−28.0	−18.0	−7.0	+5.3	13.0	23.3	38.4	55.5	72.5	
2-Vinylanisole	$C_9H_{10}O$	41.9	68.0	81.0	94.7	110.0	119.8	132.3	151.0	172.1	194.0	
3-Vinylanisole	$C_9H_{10}O$	43.4	69.9	83.0	97.2	112.5	122.3	135.3	154.0	175.8	197.5	
4-Vinylanisole	$C_9H_{10}O$	45.2	72.0	85.7	100.0	116.0	126.1	139.7	159.0	182.0	204.5	
Vinyl chloride (1-chloroethylene)	C_2H_3Cl	−105.6	−90.8	−83.7	−75.7	−66.8	−61.1	−53.2	−41.3	−28.0	−13.8	−153.7
cyanide (acrylonitrile)	C_3H_3N	−51.0	−30.7	−20.3	−9.0	+3.8	11.8	22.8	38.7	58.3	78.5	−82
fluoride (1-fluoroethylene)	C_2H_3F	−149.3	−138.0	−132.2	−125.4	−118.0	−113.0	−106.2	−95.4	−84.0	−72.2	−160.5
Vinylidene chloride (1,1-dichloroethene)	$C_2H_2Cl_2$	−77.2	−60.0	−51.2	−41.7	−31.1	−24.0	−15.0	−1.0	+14.8	31.7	−122.5
4-Vinylphenetole	$C_{10}H_{12}O$	64.0	91.7	105.6	120.3	136.3	146.4	159.8	180.0	202.8	225.0	
2-Xenyl dichlorophosphate	$C_{12}H_9Cl_2PO$	138.2	171.1	187.0	205.0	223.8	236.0	251.5	275.3	301.5	328.5	
2,4-Xyaldehyde	$C_9H_{10}O$	59.0	85.9	99.0	114.0	129.7	139.8	152.2	172.3	194.1	215.5	75
2-Xylene (2-xylene)	C_8H_{10}	−3.8	+20.2	32.1	45.1	59.5	68.8	81.3	100.2	121.7	144.4	−25.2
3-Xylene (3-xylene)	C_8H_{10}	−6.9	+16.8	28.3	41.1	55.3	64.4	76.8	95.5	116.7	139.1	−47.9
4-Xylene (4-xylene)	C_8H_{10}	−8.1	+15.5	27.3	40.1	54.4	63.5	75.9	94.6	115.9	138.3	+13.3
2,4-Xylidine	$C_8H_{11}N$	52.6	79.8	93.0	107.6	123.8	133.7	146.8	166.4	188.3	211.5	
2,6-Xylidine	$C_8H_{11}N$	44.0	72.6	87.0	102.7	120.2	131.5	146.0	168.0	193.7	217.9	

Table 9. Vapor Pressures of Organic Compounds, above 1 Atm.*

Compound: Name	Formula	Pressure, atm. — Temperature, °C.: 1	2	5	10	20	30	40	50	60	Critical point: t_c, °C.	P_c, atm.
Acetic acid	$C_2H_4O_2$	118.1	143.5	180.3	214.0	252.0	276.5	297.0	312.5	......	321.6	57.2
anhydride	$C_4H_6O_3$	139.6	162.0	194.0	221.5	253.0	272.8	288.5	......	......	296	46
Acetone	C_3H_6O	56.5	78.6	113.0	144.5	181.0	205.0	214.5	......	......	235.0	47.0
Acetylene	C_2H_2	−84.0	−71.6	−50.2	−32.7	−10.0	+4.8	16.8	26.8	34.8	36.0	62.0
Allene (propadiene)	C_3H_4	−35.0	−18.4	+8.0	33.2	64.5	85.5	103.5	118.0	......	120.7	51.8
Aniline	C_6H_7N	184.4	212.8	254.8	292.7	342.0	375.5	400.0	422.4	......	426	52.4
Benzene	C_6H_6	80.1	103.8	142.5	178.8	221.5	249.5	272.3	290.3	......	290.5	50.1
Bromobenzene	C_6H_5Br	156.2	186.2	232.5	274.5	327.0	359.8	387.5	......	......	397	44.6
1,3-Butadiene	C_4H_6	−4.5	+15.3	47.0	76.0	114.0	139.8	158.0	......	......	161.8	42.6
iso-Butane (2-methylpropane)	C_4H_{10}	−11.7	+7.5	39.0	66.8	99.5	120.5	......	......	......	134.0	37.0
n-Butane	C_4H_{10}	−0.5	+18.8	50.0	79.5	116.0	140.6	......	......	......	152.8	36.0
iso-Butyl alcohol (2-methylpropanol-1)	$C_4H_{10}O$	108.0	127.3	156.2	182.0	212.5	232.0	251.0	......	......	265	48
n-Butyl alcohol (1-butanol)	$C_4H_{10}O$	117.5	139.8	172.5	203.0	237.0	259.0	277.0	......	......	287	48.4
sec-Butyl alcohol (2-butanol)	$C_4H_{10}O$	99.5	118.2	147.5	172.0	204.0	230.0	251.0	......	......	265	48
tert-Butyl alcohol (trimethyl carbinol)	$C_4H_{10}O$	82.9	102.0	130.0	154.2	184.5	207.0	222.5	......	......	235	49
iso-Butyl formate	$C_5H_{10}C_2$	98.2	121.8	157.8	192.4	234.0	261.0	......	......	......	278.0	38.0
Butyric acid	$C_4H_8O_2$	163.5	188.3	225.0	257.0	295.0	319.0	338.0	352.0	......	355	52.0
iso-Butyric acid	$C_4H_8O_2$	154.5	179.8	217.0	250.0	289.0	315.0	336.0	......	......	336	40.0
Carbon dioxide	CO_2	−78.2	−69.1	−56.7	−39.5	−18.9	−5.3	+5.9	14.9	22.4	31.1	73.0
disulfide	CS_2	46.5	69.1	104.8	136.3	175.5	201.5	222.8	240.0	256.0	273.0	72.9
monoxide	CO	−191.3	−183.5	−170.7	−161.0	−149.7	−141.9	......	......	......	−138.7	34.6
tetrachloride	CCl_4	76.7	102.0	141.7	178.0	222.0	251.2	276.0	......	......	283.1	45.0
Chlorobenzene	C_6H_5Cl	132.2	160.2	205.0	245.3	292.8	324.4	349.8	......	......	359.2	44.6
Chlorodifluoromethane	$CHClF_2$	−40.8	−24.7	+0.3	24.0	52.0	70.3	85.3	......	......	96	48.7
Chloroform (trichloromethane)	$CHCl_3$	61.3	83.9	120.0	152.3	191.8	216.5	237.5	254.0	......	260	54.9
1-Chloro-1,2,2-trifluoroethylene	C_2ClF_3	−27.9	−11.1	+15.5	40.0	71.1	91.9	......	......	......	107.0	39.0
Chlorotrifluoromethane	$CClF_3$	−81.2	−66.7	−42.7	−18.5	+12.0	34.8	52.8	......	......	53	40.3
Cyanogen	C_2N_2	−21.0	−4.4	+21.4	44.6	72.6	91.6	106.5	118.2	......	126.6	58.2
Cyclohexane	C_6H_{12}	80.7	106.0	146.4	184.0	228.4	257.5	......	......	......	279.9	39.8
1,2-Dibromoethane	$C_2H_4Br_2$	131.5	157.7	200.0	237.0	269.0	286.0	295.0	300.0	304.5	309.8	70.6
Dichlorodifluoromethane	CCl_2F_2	−29.8	−12.2	+16.1	42.4	74.0	95.6	......	......	......	111.5	39.6
1,1-Dichloroethane	$C_2H_4Cl_2$	57.3	80.2	117.3	150.3	192.7	220.0	243.0	261.5	......	261.5	50.0
1,2-Dichloroethane	$C_2H_4Cl_2$	83.7	108.1	147.8	183.5	226.5	254.0	272.0	285.0	......	288.4	53.0
cis-1,2-Dichloroethylene	$C_2H_2Cl_2$	59.0	82.1	119.3	152.3	194.0	221.5	244.5	260.0	......	271.0	57.9
trans-1,2-Dichloroethylene	$C_2H_2Cl_2$	47.8	69.8	104.0	135.7	174.0	199.8	220.0	236.5	......	243.3	54.5
Dichlorofluoromethane	$CHCl_2F$	8.9	28.4	59.0	87.0	121.2	144.0	162.6	177.5	......	178.5	51.0
1,2-Dichloro-1,1,2,2-tetrafluoroethane	$C_2Cl_2F_4$	3.5	22.8	54.0	82.3	117.5	140.9	......	......	......	145.7	32.3
Diethylamine	$C_4H_{11}N$	55.5	77.8	113.0	145.3	184.5	210.0	......	......	......	223.3	36.6
Diethyl ether	$C_4H_{10}O$	34.6	56.0	90.0	122.0	159.0	183.3	......	......	......	193.8	35.5
sulfide	$C_4H_{10}S$	88.0	112.0	153.8	190.2	234.0	263.0	......	......	......	283.8	39.1

* Compiled from the extended tables published by D. R. Stull in *Ind. Eng. Chem.*, **39**, 517 (1947).

Table 9. Vapor Pressures of Organic Compounds, above 1 Atm.—(*Concluded*)

Compound		Pressure, atm.									Critical point	
Name	Formula	1	2	5	10	20	30	40	50	60	t_c, °C.	P_c, atm.
		Temperature, °C.										
Dimethylamine	C_2H_7N	7.4	25.0	53.9	80.0	111.7	132.2	149.8	162.6	......	164.5	52.4
2,3-Dimethylbutane	C_6H_{14}	58.0	82.0	120.3	155.7	198.7	225.5	......	......	......	227.4	30.7
Dimethyl ether	C_2H_6O	−23.7	−6.4	+20.8	45.5	75.7	96.0	112.1	125.2	......	126.9	52.0
oxalate	$C_4H_6O_4$	163.3	189.6	228.7	......	......	......	......	......	......	260	9.5
sulfide	C_2H_6S	36.0	57.8	92.3	124.5	163.8	188.5	209.0	224.5	......	229.9	54.6
n-Dodecane	$C_{12}H_{26}$	216.2	249.2	300.0	345.8	......	......	......	......	......	385	17.5
Ethane	C_2H_6	−88.6	−75.0	−52.8	−32.0	−6.4	+10.0	23.6	......	......	32.3	48.2
Ethyl acetate	$C_4H_8O_2$	77.1	100.6	136.6	169.7	209.5	235.0	......	......	......	250.1	37.9
alcohol (ethanol)	C_2H_6O	78.4	97.5	126.0	151.8	183.0	203.0	218.0	230.0	242.0	243.5	63.1
Ethylamine	C_2H_7N	16.6	35.7	65.3	91.8	124.0	146.0	163.0	176.0	......	183.2	55.5
Ethyl benzene	C_8H_{10}	136.2	163.5	207.5	246.3	294.5	326.5	......	......	......	346.4	38.1
bromide	C_2H_5Br	38.4	60.2	95.0	126.8	164.3	188.0	206.5	220.0	229.5	230.8	61.5
chloride	C_2H_5Cl	12.3	32.5	64.0	92.6	127.3	149.5	167.0	180.5	......	187.2	52.0
fluoride	C_2H_5F	−32.0	−16.7	+7.7	30.2	57.5	75.7	90.0	......	......	102.2	49.6
formate	$C_3H_6O_2$	54.3	76.0	110.5	142.2	180.0	205.0	225.0	......	......	235.3	46.8
isobutyrate	$C_6H_{12}O_2$	110.1	135.5	174.2	210.0	253.0	280.0	......	......	......	280.0	30.0
mercaptan (ethanethiol)	C_2H_6S	35.0	56.6	90.7	121.9	159.5	184.3	204.7	220.0	......	225.5	54.2
methyl ether	C_3H_8O	7.5	26.5	56.4	84.0	108.0	141.4	160.0	......	......	164.7	43.4
propionate	$C_5H_{10}O_2$	99.1	123.8	162.7	197.8	240.0	264.5	......	......	......	272.8	33.2
propyl ether	$C_5H_{12}O$	61.7	85.3	123.1	156.2	197.2	223.0	......	......	......	227.4	32.1
Ethylene	C_2H_4	−103.7	−90.8	−71.1	−52.8	−29.1	−14.2	−1.5	+8.9	......	9.6	50.7
Fluorobenzene	C_6H_5F	84.7	109.9	148.5	184.4	227.6	257.0	279.3	......	......	286.5	44.7
n-Heptane	C_7H_{16}	98.4	124.8	165.7	202.8	247.5	......	......	......	......	266.8	26.9
n-Hexane	C_6H_{14}	68.7	93.0	131.7	166.6	209.4	......	......	......	......	234.8	29.6
Hydrogen cyanide (hydrocyanic acid)	CHN	25.9	45.8	75.8	102.7	135.0	153.8	169.9	183.5	......	183.5	50.0
Iodobenzene	C_6H_5I	188.6	220.0	270.0	315.7	371.5	406.0	437.2	......	......	448	44.7
Methane	CH_4	−161.5	−152.3	−138.3	−124.8	−108.5	−96.3	−86.3	......	......	−82.1	45.8
Methyl acetate	$C_3H_6O_2$	57.8	79.5	113.1	144.2	181.0	205.0	225.0	......	......	233.7	46.3
acetylene (propyne)	C_3H_4	−23.3	−7.1	+19.5	43.8	74.0	94.0	111.5	125.0	......	128	52.8
alcohol	CH_4O	64.7	84.0	112.5	138.0	167.8	186.5	203.5	214.0	224.0	240.0	78.7
Methylamine	CH_5N	−6.3	+10.1	36.0	59.5	87.8	106.3	121.8	133.7	144.6	156.9	73.6
Methyl bromide	CH_3Br	3.6	23.3	54.8	84.0	121.7	147.5	170.2	190.0	......	194	51.6
butyrate	$C_5H_{10}O_2$	102.3	127.5	166.7	203.0	244.5	272.0	......	......	......	281.2	34.2
chloride	CH_3Cl	−24.0	−6.4	+22.0	47.3	77.3	97.5	113.8	126.0	137.5	143.8	65.8
fluoride	CH_3F	−78.2	−64.5	−42.0	−21.0	+2.6	15.5	26.5	36.0	43.5	44.9	62.0
formate	$C_2H_4O_2$	32.0	51.9	83.5	112.0	147.2	169.7	188.5	213.0	......	214.0	59.1
iodide	CH_3I	42.4	65.5	101.8	138.0	176.5	206.0	228.5	248.0	......	255	54.6
isobutyrate	$C_5H_{10}O_2$	92.6	116.7	155.2	190.2	232.0	259.5	......	......	......	267.5	33.9
mercaptan (methanethiol)	CH_4S	6.8	26.1	55.9	83.4	117.5	140.0	157.7	172.0	185.0	196.8	71.4
propionate	$C_4H_8O_2$	79.8	103.0	139.8	172.6	212.5	239.0	......	......	......	257.4	39.3
n-Octane	C_8H_{18}	125.6	152.7	196.2	235.8	281.4	......	......	......	......	296.2	24.7
iso Pentane (2-methylbutane)	C_5H_{12}	27.8	48.8	82.8	114.5	154.0	180.3	......	......	......	187.8	32.8
n-Pentane	C_5H_{12}	36.1	58.0	92.4	124.7	164.3	191.3	......	......	......	197.2	33.0
neo-Pentane (2,2-dimethylpropane)	C_5H_{12}	+9.5	29.5	61.1	90.7	127.6	152.5	......	......	......	159.0	33.0
Phenol	C_6H_6O	181.9	208.0	248.2	283.8	328.7	358.0	382.1	400.0	418.7	419	60.5
Phosgene (carbonyl chloride)	CCl_2O	8.3	27.3	57.2	85.0	119.0	141.8	159.8	174.0	......	181.7	56.0
Propane	C_3H_8	−42.1	−25.6	+1.4	26.9	58.1	78.7	94.8	......	......	96.8	42.0
Propionic acid	$C_3H_6O_2$	141.1	160.0	186.0	203.5	220.0	228.0	233.0	238.0	......	239.5	53.0
Propyl acetate	$C_5H_{10}O_2$	101.8	126.8	165.7	200.5	242.8	269.0	......	......	......	276.2	33.2
iso-Propyl alcohol (2-propanol)	C_3H_8O	82.5	101.3	130.2	155.7	186.0	205.0	220.2	232.0	......	235	53
n-Propyl alcohol (1-propanol)	C_3H_8O	97.8	117.0	149.0	177.0	210.8	232.3	250.0	......	......	263.7	49.9
Propylamine	C_3H_9N	48.5	69.8	102.8	133.4	170.0	194.3	214.5	......	......	223.8	46.8
Propyl formate	$C_4H_8O_2$	81.3	104.3	142.0	176.4	217.5	245.0	......	......	......	264.8	39.5
Propylene	C_3H_6	−47.7	−31.4	−4.8	+19.8	49.5	70.0	85.0	......	......	91.4	45.4
Tetramethylsilane	$C_4H_{12}Si$	27.0	48.0	82.0	113.0	152.0	178.0	......	......	......	185	33
Toluene	C_7H_8	110.6	136.5	178.0	215.8	262.5	292.8	319.0	......	......	320.6	41.6
Trichlorofluoromethane	CCl_3F	23.7	44.1	77.3	108.2	146.7	172.0	194.0	......	......	198.0	43.2
1,1,2-Trichloro-1,2,2-trifluoroethane	$C_2Cl_3F_3$	47.6	70.0	105.5	138.0	177.7	205.0	......	......	......	214.1	33.7

VAPOR PRESSURES OF SOLUTIONS*

Table 10. Partial Pressures of Water over Aqueous Solutions of HCl

$\log_{10} p\text{mm.} = A - \frac{B}{T}$, which, however, agrees only approximately with the table. The table is more nearly correct.

% HCl	A	B	Partial pressure of H_2O, mm. Hg, °C. 0°	5°	10°	15°	20°	25°	30°	35°	40°	45°	50°	60°	70°	80°	90°	100°	110°
6	8.99156	2282	4.18	6.04	8.45	11.7	15.9	21.8	29.1	39.4	50.6	66.2	86.0	139	220	333	492	715	
10	8.99864	2295	3.84	5.52	7.70	10.7	14.6	20.0	26.8	35.5	47.0	61.5	80.0	130	204	310	463	677	960
14	8.97075	2300	3.39	4.91	6.95	9.65	13.1	18.0	24.1	31.9	42.1	55.3	72.0	116	185	273	425	625	892
18	8.98014	2323	2.87	4.21	5.92	8.26	11.3	15.4	20.6	27.5	36.4	47.9	62.5	102	162	248	374	550	783
20	8.97877	2334	2.62	3.83	5.40	7.50	10.3	14.1	19.0	25.1	33.3	43.6	57.0	93.5	150	230	345	510	729
22	9.02708	2363	2.33	3.40	4.82	6.75	9.30	12.6	17.1	22.8	30.2	39.8	52.0	85.6	138	211	317	467	670
24	8.96022	2356	2.05	3.04	4.31	6.03	8.30	11.4	15.4	20.4	27.1	35.7	46.7	77.0	124	194	290	426	611
26	9.01511	2390	1.76	2.60	3.71	5.21	7.21	9.95	13.5	18.0	24.0	31.7	41.5	69.0	112	173	261	387	555
28	8.97611	2395	1.50	2.24	3.21	4.54	6.32	8.75	11.8	15.8	21.1	27.9	36.5	60.7	99.0	154	234	349	499
30	9.00117	2422	1.26	1.90	2.73	3.88	5.41	7.52	10.2	13.7	18.4	24.3	32.0	53.5	87.5	136	207	310	444
32	9.03317	2453	1.04	1.57	2.27	3.25	4.55	6.37	8.70	11.7	15.7	21.0	27.7	46.5	76.5	120	184	275	396
34	9.07143	2487	0.85	1.29	1.87	2.70	3.81	5.35	7.32	9.95	13.5	18.1	24.0	40.5	66.5	104	161	243	355
36	9.11815	2526	0.68	1.03	1.50	2.19	3.10	4.41	6.08	8.33	11.4	15.4	20.4	34.8	57.0	90.0	140	212	311
38	9.20783	2579	0.53	0.81	1.20	1.75	2.51	3.60	5.03	6.92	9.52	13.0	17.4	29.6	49.1	77.5	120	182	266
40	9.33923	2647	0.41	0.63	0.94	1.37	2.00	2.88	4.09	5.68	7.85	10.7	14.5	25.0	42.1	67.3	105	158	230
42	9.44953	2709	0.31	0.48	0.72	1.06	1.56	2.30	3.28	4.60	6.45	8.90	12.1	21.2	35.8	57.2	89.2	135	195

* Accuracy, *ca.* 2 per cent for solutions of 15 to 30 per cent HCl between 0 and 100°; for solutions of > 30 per cent HCl the accuracy is *ca.* 5 per cent at the lower temperatures and *ca.* 15 per cent at the higher temperatures. Below 15 per cent HCl, the accuracy is *ca.* 5 per cent at the lower temperatures and higher strengths to *ca.* 15 to 20 per cent at the lower strengths and perhaps 15 to 20 per cent at the higher temperatures and lower strengths.

Table 11. Partial Pressures of HCl over Aqueous Solutions of HCl

Mm. Hg, °C.

%HCl	A	B	0°	5°	10°	15°	20°	25°	30°	35°	40°	45°	50°	60°	70°	80°	90°	100°	110°
2	11.8037	4736	......	......	0.0000117	0.000023	0.000044	0.000084	0.000151	0.000275	0.00047	0.00083	0.00140	0.00380	0.0100	0.0245	0.058	0.132	0.280
4	11.6400	4471	0.000018	0.000036	.000069	.000131	.00024	.00044	.00077	.00134	.0023	.00385	.0064	.0165	.0405	.095	.21	.46	.93
6	11.2144	4202	.000066	.000125	.000234	.000425	.00076	.00131	.00225	.0038	.0062	.0102	.0163	.040	.094	.206	.44	.92	1.78
8	11.0406	4042	.000118	.000323	.000583	.00104	.00178	.0031	.00515	.0085	.0136	.022	.0344	.081	.183	.39	.82	1.64	3.10
10	10.9311	3908	.00042	.00075	.00134	.00232	.00395	.0067	.0111	.0178	.0282	.045	.069	.157	.35	.73	1.48	2.9	5.4
12	10.7900	3765	.00099	.00175	.00305	.0052	.0088	.0145	.0234	.037	.058	.091	.136	.305	.66	1.34	2.65	5.1	9.3
14	10.6954	3636	.0024	.00415	.0071	.0118	.0196	.0316	.050	.078	.121	.185	.275	.60	1.25	2.50	4.8	9.0	16.0
16	10.6261	3516	.0056	.0095	.016	.0265	.0428	.0685	.106	.163	.247	.375	.55	1.17	2.40	4.66	8.8	16.1	28
18	10.4957	3376	.0135	.0225	.037	.060	.095	.148	.228	.345	.515	.77	1.11	2.3	4.55	8.6	15.7	28	48
20	10.3833	3245	.0316	.052	.084	.132	.205	.32	.48	.72	1.06	1.55	2.21	4.4	8.5	15.6	28.1	49	83
22	10.3172	3125	.0734	.119	.187	.294	.45	.68	1.02	1.50	2.18	3.14	4.42	8.6	16.3	29.3	52	90	146
24	10.2185	2995	.175	.277	.43	.66	1.00	1.49	2.17	3.14	4.5	6.4	8.9	16.9	31.0	54.5	94	157	253
26	10.1303	2870	.41	.64	.98	1.47	2.17	3.20	4.56	6.50	9.2	12.7	17.5	32.5	58.5	100	169	276	436
28	10.0115	2732	1.0	1.52	2.27	3.36	4.90	7.05	9.90	13.8	19.1	26.4	35.7	64	112	188	309	493	760
30	9.8763	2593	2.4	3.57	5.23	7.60	10.6	15.1	21.0	28.6	39.4	53	71	124	208	340	542	845	
32	9.7523	2457	5.7	8.3	11.8	16.8	23.5	32.5	44.5	60.0	81	107	141	238	390	623	970		
34	9.6061	2316	13.1	18.8	26.4	36.8	50.5	68.5	92	122	161	211	273	450	720				
36	9.5262	2229	29.0	41.0	56.4	78	105.5	142	188	246	322	416	535	860					
38	9.4670	2094	63.0	87.0	117	158	210	277	360	465	598	758	955						
40	9.2156	1939	130	176	233	307	399	515	627	830									
42	8.9925	1800	253	332	430	560	709	900											
44	8.8621	1681	510	655	840														
46	.....	.....	940																

Table 12. Partial Pressures of H_2O and SO_2 over Aqueous Solutions of Sulfur Dioxide*

Partial Pressures of H_2O and SO_2, mm. Hg, °C.

Grams SO_2 per 100 g. water	10°C.		20°C.		30°C.		40°C.		50°C.		60°C.		70°C.		80°C.		90°C.		100°C.		110°C.		120°C.		130°C.	
	H_2O	SO_2	H_2O	SO_2	H_2O	SO_2	H_2O	SO_2	H_2O	SO_2	H_2O	SO_2	H_2O	SO_2	H_2O	SO_2	H_2O	SO_2	H_2O	SO_2	H_2O	SO_2	H_2O	SO_2	H_2O	SO_2
0.0	9.2	...	17.5	...	31.8	...	55.3	...	92.5	...	149.5	...	234	...	355	...	526	...	760	...	1074	....	1488	..	2026	
0.5	9.2	21	17.5	29	31.7	42	55.2	60	92.3	83	149.2	111	234	144	354	182	525	225	758	274	1072	326	1486	377	2024	420
1.0	9.2	42	17.4	59	31.7	85	55.1	120	92.2	164	149.0	217	233	281	354	356	524	445	757	548	1071	661	1484	775	2022	879
1.5	9.2	64	17.4	90	31.6	129	55.0	181	92.0	247	148.8	328	233	426	353	543	523	684	756	850	1070	1032				
2.0	9.1	86	17.4	123	31.6	176	55.0	245	91.9	333	148.6	444	233	581	353	746	523	940								
2.5	9.1	168	17.4	157	31.5	224	54.9	311	91.8	421	148.3	562	232	739	352	956										
3.0	9.1	130	17.3	191	31.5	273	54.7	378	91.6	511	148.1	682	232	897												
3.5	9.1	153	17.3	227	31.5	324	54.7	447	91.5	603	147.9	804														
4.0	9.1	176	17.3	264	31.4	376	54.6	518	91.4	698																
4.5	9.1	199	17.3	300	31.4	428	54.5	588	91.2	793																
5.0	9.1	223	17.2	338	31.3	482	54.4	661																		
5.5	9.0	247	17.2	375	31.3	536	54.4	733																		
6.0	9.0	271	17.2	411	31.2	588	54.3	804																		
6.5	9.0	295	17.2	448	31.2	642																				
7.0	9.0	320	17.1	486	31.1	698																				
7.5	9.0	345	17.1	524	31.1	752																				
8.0	9.0	370	17.1	562	31.0	806																				
8.5	9.0	395	17.0	600																						
9.0	9.0	421	17.0	638																						
9.5	8.9	447	17.0	676																						
10.0	8.9	473	17.0	714																						
10.5	8.9	499	17.0	751																						
11.0	8.9	526	16.9	789																						
11.5	8.9	553																								
12.0	8.9	580																								
12.5	8.9	608																								
13.0	8.8	635																								
13.5	8.8	662																								
14.0	8.8	689																								
14.5	8.8	716																								
15.0	8.8	743																								
15.5	8.8	771																								
16.0	8.8	799																								

* From "International Critical Tables," vol. 3, p. 302, McGraw-Hill.

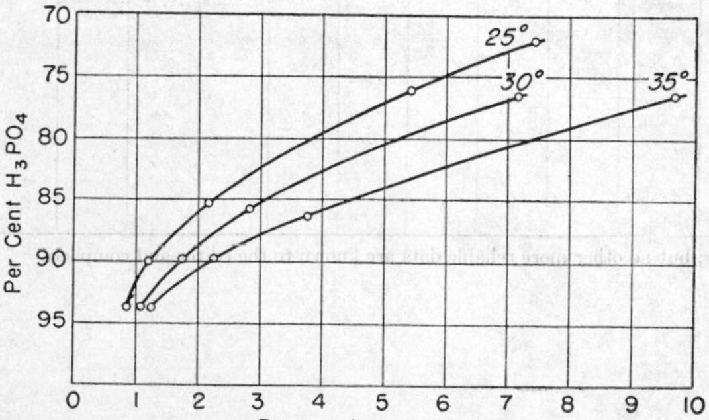

FIG. 1. Vapor pressures of H_3PO_4 aqueous. Partial pressure of H_2O vapor. *(Courtesy of Victor Chemical Works; measurements by W. H. Woodstock.)*

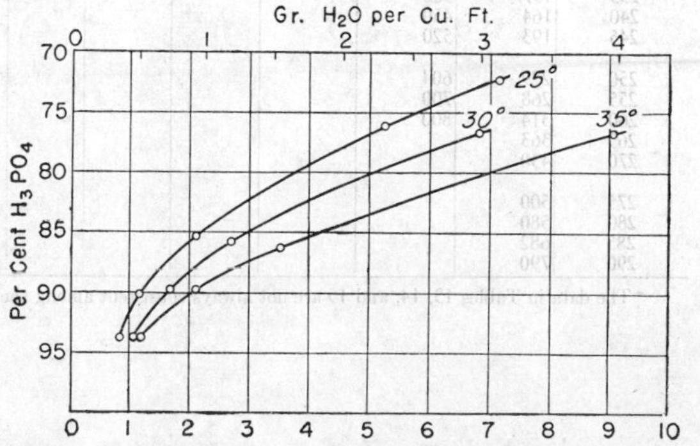

FIG. 2. Vapor pressures of H_3PO_4 aqueous. Weight of H_2O in saturated air. *(Courtesy of Victor Chemical Works; measurements by W. H. Woodstock.)*

Table 13. Vapor Pressures, Normal Boiling Points, and Latent Heats of Vaporization for Aqueous Solutions of H_2SO_4*

Percentages are wt. % H_2SO_4 in the solution

A and B are constants in the equation $\log_{10} p_{mm.} = A - \frac{B}{T}$

l = total heat of vaporization in g.-cal. per g. of water evaporated

B. P. = normal boiling point, °C.

For bibliography and discussion of data, see Greenewalt, *Ind. Eng. Chem.* **17**: 522, 1925.

Per cent	95	90	85	80	75	70	65	60	55	50	45	40	35	30	25	20	10
A	9.790	9.255	9.239	9.293	9.034	9.032	8.853	8.841	8.827	8.832	8.809	8.844	8.873	8.864		8.922	8.925
B	3888	3390	3175	3040	2810	2688	2533	2458	2400	2357	2322	2299	2286	2271		2268	2259
l	987	861	806	772	713	682	643	624	609	598	590	584	580	577		576	574
B. P.	290	255	225	202	182	165	151	140	130	123	118	114	110	108	106	104	102
°C.	Total vapor pressure, mm. Hg																
0	....	....	0.00418	0.0144	0.0550	0.154	0.377	0.686	1.08	1.55	2.07	2.55	3.06	3.43	3.72	4.02	4.38
5	....	0.00118	.00680	.0230	.0867	.235	.558	1.03	1.60	2.26	2.99	3.69	4.40	4.94	5.33	5.87	6.30
10	....	.00196	.0108	.0358	.128	.342	.800	1.46	2.26	3.19	4.19	5.22	6.23	6.91	7.46	8.05	8.80
15	....	.00318	.0169	.0555	.195	.506	1.15	2.05	3.19	4.50	5.85	7.27	8.65	9.65	10.5	11.3	12.3
20	....	.00497	.0257	.0835	.284	.723	1.61	2.87	4.43	6.20	8.10	9.95	11.8	13.2	14.3	15.4	16.6
25	....	.00765	.0390	.124	.408	1.03	2.24	3.97	6.15	8.45	10.9	13.5	15.8	17.8	19.4	20.8	22.4
30	....	.0117	.0585	.183	.580	1.44	3.09	5.41	8.29	11.3	14.7	18.0	21.2	23.8	26.0	27.8	30.0
35	0.00150	.0179	.0860	.265	.822	2.00	4.23	7.39	11.2	15.4	19.7	24.3	28.6	31.9	35.0	37.2	40.1
40	.00235	.0265	.125	.381	1.14	2.75	5.66	9.85	14.8	20.3	26.0	31.8	37.3	41.7	45.6	48.6	52.9
45	.00370	.0395	.181	.540	1.57	3.73	7.60	13.0	19.5	26.7	33.0	41.0	48.6	54.7	59.0	63.3	68.1
50	.00580	.0580	.260	.770	2.20	5.17	10.2	17.5	26.0	35.2	44.7	53.9	63.0	71.3	76.7	82.2	88.5
55	.00877	.0840	.367	1.06	2.95	6.89	13.4	22.7	33.7	45.5	57.5	69.0	80.2	91.0	98.2	106	113
60	.0133	.120	.411	1.47	3.98	9.12	18.6	29.3	43.0	58.0	73.0	87.3	102	116	124	133	143
65	.0196	.169	.707	2.00	5.30	10.2	22.7	37.7	55.1	73.7	92.3	110	127	145	156	167	178
70	.0288	.236	.960	2.68	7.02	15.6	29.0	48.0	69.6	92.5	116	138	159	180	195	207	223
75	.0415	.327	1.31	3.60	9.26	20.3	37.0	60.2	87.0	115	144	171	198	222	240	256	274
80	.0606	.450	1.77	4.77	12.0	26.0	47.0	75.3	108	143	179	211	244	273	295	314	337
85	.0879	.618	2.37	6.35	15.6	33.4	59.7	94.3	136	178	211	261	300	333	360	385	413
90	.123	.823	3.14	8.30	20.0	42.5	74.6	117	167	217	271	319	369	404	437	468	498
95	.172	1.12	4.18	10.8	25.7	53.9	92.7	144	205	268	335	390	450	493	531	580	608
100	.237	1.49	5.39	13.9	32.0	67.0	114	178	253	326	405	474	540	590	637	678	720
105	.321	1.93	6.95	17.6	40.0	82.3	140	213	302	393	484	568	642	702	758	812	
110	.437	2.52	9.00	22.5	50.0	103	172	260	367	471	580	679	768				
115	.590	3.23	11.4	28.3	62.0	126	207	313	435	562	684	800					
120	.788	4.19	14.5	35.6	76.5	153	251	377	522	670	812						
125	1.07	5.43	18.3	44.7	94.5	188	304	452	625	797							
130	1.42	6.97	23.2	56.0	117	230	370	544	744								
135	1.87	8.85	29.1	69.0	142	277	440	647									
140	2.40	11.2	36.3	85.5	173	332	525	760									
145	3.11	13.9	44.3	104	208	397	622										
150	4.02	17.5	54.6	127	248	471	730										
155	5.13	21.9	68.2	157	299	564											
160	6.47	27.7	82.0	188	354	665											
165	8.39	33.2	99.5	226	422	790											
170	10.3	39.8	119	267	496												
175	12.9	48.4	143	319	585												
180	15.9	59.0	169	378	685												
185	20.2	71.2	206	450	810												
190	24.8	85.0	245	535													
195	30.7	102	291	637													
200	36.7	120	340	735													
205	45.3	143	402														
210	55.0	170	472														
215	66.9	203	557														
220	79.8	240	647														
225	95.5	279	750														
230	115	326															
235	137	380															
240	164	450															
245	193	520															
250	229	604															
255	268	700															
260	314	800															
265	363																
270	430																
275	500																
280	580																
285	682																
290	790																

* The data in Tables 13, 14, and 15 are not always consistent among themselves, but no other more reliable data are known to the editor and compilers

Table 14. Partial Pressures of H_2SO_4 and H_2O over Sulfuric Acid Solutions

Mm. Hg

t, °C.	$p_{H_2SO_4}$	p_{H_2O}	t, °C.	$p_{H_2SO_4}$	p_{H_2O}	t, °C.	$p_{H_2SO_4}$	p_{H_2O}	t, °C.	$p_{H_2SO_4}$	p_{H_2O}	t, °C.	$p_{H_2SO_4}$	p_{H_2O}
89.25% H_2SO_4			91.26% H_2SO_4			95.06% H_2SO_4			98.06% H_2SO_4			99.23% H_2SO_4		
183.0	0.5	78.8	191.0	0.6	50.7	180.0	2.1	10.1	204.0	5.9	0.0	211.0	33.2	
197.5	1.3	116.9	205.0	1.9	84.7	200.0	4.8	21.2	218.5	9.8	1.5	225.0	49.9	
216.5	2.1	233.1	222.0	4.5	158.5	215.5	8.5	46.5	234.5	14.7	3.2	227.0	55.4	
230.0	3.6	306.3	242.5	6.4	271.6	232.0	13.4	91.9	249.0	28.5	2.6	244.0	84.1	<0.1
241.5	5.3	414.8	252.5	11.3	385.3	244.5	19.9	120.1	261.0	38.8	5.0	261.0	163.8	
			258.0	13.6	448.7	252.0	20.0	156.5	273.0	61.9	5.3	270.0	229.8	
			262.5	16.3	411.1	261.0	27.9	180.7	285.0	91.6	11.8	281.0	272.3	
						270.0	39.9	254.9	295.0	132.3	14.7	290.0	381.5	
						280.5	52.0	310.0						
						282.0	52.6	350.2						

Table 15. Partial Pressures of SO_3 over Fuming Sulfuric Acid

Mm. Hg, °C.

Total % H_2SO_4	% SO_3	20°	25°	30°	35°	40°	45°	50°	55°	60°	65°	70°	75°	80°	85°	90°
102.0	83.265	.....	.....	.....	0.4	0.6	1.0	1.6	2.5	3.8	5.7	8.5	12.5	18.2	26.3	37.5
103.0	84.081	.....	.....	.....	0.5	0.9	1.3	2.1	3.2	4.8	6.8	10.5	15.3	22.0	31.4	44.4
104.0	84.897	0.2	0.3	0.5	0.8	1.3	2.0	3.0	4.5	6.7	9.8	14.2	20.4	29.0	40.8	56.8
104.5	85.305	0.3	0.4	0.7	1.1	1.7	2.6	4.0	6.0	8.9	12.9	18.8	26.7	37.9	53.5	73.9
105.0	85.714	0.4	0.7	1.1	1.7	2.6	4.0	6.0	9.0	13.1	19.0	27.2	38.6	54.1	75.2	103.7
105.5	86.122	0.8	1.2	1.9	2.9	4.4	6.6	9.8	14.3	20.7	29.5	42.1	58.1	81.5	112.6	153.1
106.0	86.530	1.4	2.1	3.2	4.9	7.3	10.8	15.7	22.6	32.1	45.2	63.0	87.0	119.0	161.5	217.2
106.5	86.938	2.4	3.6	5.5	8.2	12.0	17.4	25.0	35.4	49.8	69.0	95.5	129.3	171.2	230.1	311.9
107.0	87.346	4.0	6.0	8.9	13.0	18.8	26.9	38.1	53.4	74.0	101.6	138.2	186.4	249.2	330.5	434.9
107.5	87.754	6.9	10.0	14.6	20.7	29.2	40.7	56.4	77.1	104.6	140.2	167.3	246.5	323.7	422.2	547.7
108.0	88.163	....	15.9	22.4	31.4	43.4	59.4	80.6	108.3	144.2	190.4	249.4	324.2			
108.5	88.571	.....	.....	35.1	47.8	64.8	86.9	115.7	152.3	199.4	257.9	333.6	425.1			
109.0	88.979	.....	.....	.....	64.5	86.6	115.3	152.1	199.0	258.3	332.5	425.0	539.5			
110.0	89.795	.....	.....	.....	100.5	133.7	176.1	230.1	298.2	383.5	489.5					
111.0	90.612	.....	.....	105.2	140.5	185.7	243.4	316.3	407.9							
112.0	91.428	76.1	103.2	138.5	184.0	242.4	316.5	409.9	526.6							
113.0	92.244	96.9	130.9	175.1	232.1	304.9	397.1	512.9	657.4							
114.0	93.060	119.1	160.7	214.9	284.5	373.3	485.8	626.9	802.8							
115.0	93.877	144.2	194.3	259.2	342.6	448.8	583.0	751.2	960.4							

Table 16. Partial Pressures of HNO_3 and H_2O over Aqueous Solutions of HNO_3

Mm. Hg

Percentages are wt. % HNO_3 in solution

°C.	20%		25%		30%		35%		40%		45%		50%	
	HNO_3	H_2O	HNO_3	H_2O	HNO_3	H_2O	HNO_3	H_2O	HNO_3	H_2O	HNO_3	H_2O	HNO_3	H_2O
0	....	4.1	....	3.8	....	3.6	.....	3.3	.....	3.0	.....	2.6	.....	2.1
5	....	5.7	....	5.4	....	5.0	.....	4.6	.....	4.2	.....	3.6	.....	3.0
10	....	8.0	....	7.6	....	7.1	.....	6.5	.....	5.8	.....	5.0	0.12	4.2
15	....	10.9	....	10.3	....	9.7	.....	8.9	.....	8.0	0.10	6.9	.18	5.8
20	....	15.2	....	14.2	....	13.2	.....	12.0	.....	10.8	.15	9.4	.27	7.9
25	....	20.6	....	19.2	....	17.8	.....	16.2	0.12	14.6	.23	12.7	.39	10.7
30	....	27.6	....	25.7	....	23.8	0.09	21.7	.17	19.5	.33	16.9	.56	14.4
35	....	36.5	....	33.8	....	31.1	.13	28.3	.25	25.5	.48	22.3	.80	19.0
40	....	47.5	....	44	0.11	41	.20	37.7	.36	33.5	.68	29.3	1.13	25.0
45	....	62	0.09	57.5	.17	53	.28	48	.52	43	.96	38.0	1.57	32.5
50	....	80	.13	75	.25	69	.42	63	.75	56	1.35	49.5	2.18	42.5
55	0.09	100	.18	94	.35	87	.59	79	1.04	71	1.83	62.5	2.95	54
60	.13	128	.28	121	.51	113	.85	102	1.48	90	2.54	80	4.05	70
65	.19	162	.40	151	.71	140	1.18	127	2.05	114	3.47	100	5.46	88
70	.27	200	.54	187	1.00	174	1.63	159	2.80	143	4.65	126	7.25	110
75	.38	250	.77	234	1.38	217	2.26	198	3.80	178	6.20	158	9.6	138
80	.53	307	1.05	287	1.87	267	3.07	243	5.10	218	8.15	195	12.5	170
85	.74	378	1.44	352	2.53	325	4.15	297	6.83	268	10.7	240	16.3	211
90	1.01	458	1.95	426	3.38	393	5.50	359	9.0	325	13.7	292	20.9	258
95	1.37	555	2.62	517	4.53	478	7.32	436	11.7	394	17.8	355	26.8	315
100	1.87	675	3.50	628	6.05	580	9.7	530	15.5	480	23.0	430	34.2	383
105	2.50	800	4.65	745	7.90	690	12.7	631	20.0	573	29.2	520	43.0	463
110	....	.....	....	.....	....	.....	16.5	755	25.7	688	37.0	625	54.5	560
115	....	.....	....	.....	....	.....	.....	.....	32.5	810	46	740	67	665
120	....	.....	....	.....	....	.....	.....	.....	.....	.....	.....	.....	84	785

Table 16. Partial Pressures of HNO_3 and H_2O over Aqueous Solutions of HNO_3—(*Concluded*)

°C.	55%		60%		65%		70%		80%		90%		100%
	HNO_3	H_2O	HNO_3	H_2O	HNO_3	H_2O	HNO_3	H_2O	HNO_3	H_2O	HNO_3	H_2O	HNO_3
0	......	1.8	0.19	1.5	0.41	1.3	0.79	1.1	2	.....	5.5	....	11
5	0.14	2.5	.28	2.1	.60	1.8	1.12	1.6	3	.....	8	....	15
10	.21	3.5	.41	3.0	.86	2.6	1.58	2.2	4	1.2	11	....	22
15	.31	4.9	.59	4.1	1.21	3.5	2.18	3.0	6	1.7	15	....	30
20	.45	6.7	.84	5.6	1.68	4.9	3.00	4.1	8	2.4	20	....	42
25	.66	9.1	1.21	7.7	2.32	6.6	4.10	5.5	10.5	3.2	27	1	57
30	.93	12.2	1.66	10.3	3.17	8.8	5.50	7.4	14	4	36	1.3	77
35	1.30	16.1	2.28	13.6	4.26	11.6	7.30	9.8	18.5	5.5	47	1.8	102
40	1.82	21.3	3.10	18.1	5.70	15.5	9.65	12.8	24.5	7	62	2.4	133
45	2.50	28.0	4.20	23.7	7.55	20.0	12.6	16.7	32	9.5	80	3	170
50	3.41	36.3	5.68	31	10.0	26.0	16.5	21.8	41	12	103	4	215
55	4.54	46	7.45	39	12.8	33.0	21.0	27.3	52	15	127	5	262
60	6.15	60	9.9	51	16.8	43.0	27.1	35.3	67	20	157	6.5	320
65	8.18	76	13.0	64	21.7	54.5	34.5	44.5	85	25	192	8	385
70	10.7	95	16.8	81	27.5	68	43.3	56	106	31	232	10	460
75	13.9	120	21.8	102	35.0	86	54.5	70	130	38	282	13	540
80	18.0	148	27.5	126	43.5	106	67.5	86	158	48	338	16	625
85	23.0	182	34.8	156	54.5	131	83	107	192	60	405	20	720
90	29.4	223	43.7	192	67.5	160	103	130	230	73	480	24	820
95	37.3	272	55.0	233	83.5	195	125	158	278	89	570	29	
100	47	331	69.5	285	103	238	152	192	330	108	675	35	
105	58.5	400	84.5	345	124	288	183	231	392	129	790	42	
110	73	485	103	417	152	345	221	270	465	155			
115	90	575	126	495	181	410	262	330	545	185			
120	110	685	156	590	218	490	312	393	640	219			
125	.....	.....	187	700	260	580	372	469					

Table 17. Partial Pressures of H_2O and HBr over Aqueous Solutions of HBr at 20° to 55°C.

Mm. Hg

% HBr	20°C.		25°C.		50°C.		55°C.	
	HBr	H_2O	HBr	H_2O	HBr	H_2O	HBr	H_2O
32	.....	...	0.0016					
34	.....	...	.0022					
36	.....	...	.0033					
38	.....	...	.0061					
40	.....	...	.011					
42	.....	...	.023					
44	.....	...	.048					
46	.....	...	.10					
48	0.09	6.2	.13	8.2	1.3	30.2	2.0	38
50	.23	4.5	.37	6.1	3.2	24.3	4.6	31
52	.71	3.3	1.1	4.5	7.2	19.3	10.2	25
54	2.2	2.4	3.2	3.3	17	16.0	23.0	21
56	6.8	1.7	9.3	2.4	40	13.3	51	18
58	21	1.3	27	1.9	91	10.4	115	14
60	.....	...	....	...	....	....	260	11.4

Table 18. Partial Pressures of HI over Aqueous Solutions of HI at 25°C.

Mm. Hg

%HI.....	4	46	48	50	52	54	56
p_{HI}......	0.00064	0.0010	0.0022	0.0050	0.013	0.035	0.10

Table 19. Vapor Pressures of the System: Water—Sulfuric Acid—Nitric Acid

For these data, reference must be made to the graphs of "International Critical Tables," vol. 3, pp. 306–308.

Table 20. Total Vapor Pressures of Aqueous Solutions of CH_3COOH

Percentages are wt. % acetic acid in the solution
Mm. Hg

°C.	25%	50%	75%
20	16.3	15.7	15.3
25	22.1	21.4	20.8
30	29.6	28.8	27.8
35	39.4	38.3	36.6
40	51.7	50.2	48.1
45	67.0	65.0	62.0
50	87.2	85.0	80.1
55	110	107	102
60	141	138	130
65	178	172	162
70	223	216	203
75	277	269	251
80	342	331	310
85	419	407	376
90	510	497	458
95	618	602	550
100	743	725	666

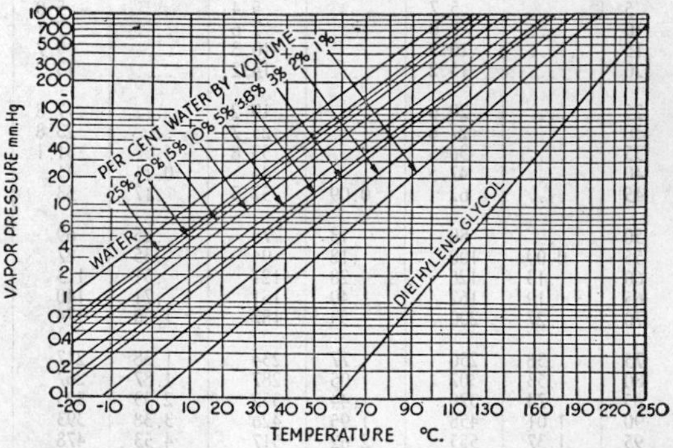

FIG. 3. Vapor pressure of aqueous diethylene glycol solutions. (*Courtesy of Carbide and Carbon Chemicals Corp.*)

Table 21. Partial Pressures of H_2O over Aqueous Solutions of NH_3*

Pressures are in pounds per square inch absolute

Molal concentration of ammonia in the solutions in percentages (Weight concentration of ammonia in the solution in percentages)

t, °F.	0 (0)	5 (4.74)	10 (9.50)	15 (14.29)	20 (19.10)	25 (23.94)	30 (28.81)	35 (33.71)	40 (38.64)	45 (43.59)	50 (48.57)	55 (53.58)	60 (58.62)	65 (63.69)	70 (68.79)	75 (73.91)	80 (79.07)	85 (84.26)	90 (89.47)	95 (94.72)
32	0.09	0.084	0.079	0.074	0.070	0.065	0.060	0.056	0.051	0.047	0.042	0.038	0.034	0.030	0.025	0.021	0.017	0.013	0.008	0.004
40	.12	.115	.108	.101	.095	.089	.083	.076	.070	.064	.058	.052	.046	.040	.035	.029	.023	.015	.012	.006
50	.18	.17	.16	.15	.14	.13	.12	.11	.10	.094	.085	.076	.068	.059	.051	.042	.034	.025	.017	.008
60	.26	.24	.23	.21	.20	.19	.17	.16	.15	.13	.12	.11	.097	.085	.073	.061	.049	.037	.024	.012
70	.36	.34	.32	.30	.28	.26	.25	.23	.21	.19	.17	.15	.14	.12	.10	.086	.069	.052	.034	.017
80	.51	.48	.45	.42	.40	.37	.34	.32	.29	.27	.24	.22	.19	.17	.14	.12	.096	.072	.048	.024
90	.70	.66	.63	.58	.55	.51	.47	.44	.40	.37	.33	.30	.26	.23	.20	.16	.13	.10	.066	.033
100	.95	.90	.85	.79	.74	.69	.64	.59	.55	.50	.45	.41	.36	.31	.27	.22	.18	.13	.090	.045
110	1.27	1.20	1.14	1.07	1.00	.93	.86	.80	.73	.67	.60	.54	.48	.42	.36	.30	.24	.18	.120	.061
120	1.69	1.60	1.51	1.42	1.33	1.24	1.15	1.06	.97	.89	.80	.72	.64	.56	.48	.40	.32	.24	.160	.081
130	2.22	2.10	1.98	1.86	1.74	1.62	1.51	1.39	1.28	1.17	1.05	.95	.84	.74	.63	.53	.42	.32	.210	.100
140	2.89	2.73	2.57	2.42	2.26	2.11	1.96	1.81	1.66	1.52	1.37	1.23	1.10	.96	.82	.69	.55	.41	.270	.140
150	3.72	3.51	3.31	3.11	2.91	2.72	2.52	2.33	2.14	1.95	1.76	1.59	1.41	1.24	1.06	.88	.71	.53	.350	.180
160	4.74	4.48	4.22	3.97	3.71	3.46	3.22	2.97	2.73	2.49	2.25	2.02	1.80	1.58	1.35	1.12	.90	.67	.450	.220
170	5.99	5.66	5.34	5.02	4.70	4.38	4.07	3.75	3.45	3.15	2.84	2.56	2.28	1.99	1.71	1.42	1.13	1.85	.570	.300
180	7.51	7.10	6.69	6.30	5.89	5.49	5.10	4.71	4.33	3.94	3.57	3.21	2.85	2.50	2.14	1.77	1.42	1.06		
190	9.34	8.83	8.32	7.82	7.32	6.83	6.34	5.86	5.38	4.91	4.44	3.99	3.55	3.10	2.65					
200	11.53	10.90	10.27	9.65	9.04	8.43	7.83	7.23	6.64	6.06	5.48	4.93	4.38	3.81						
210	14.12	13.35	12.58	11.82	11.07	10.32	9.59	8.86	8.13	7.42	6.71	6.04	5.34							
220	17.19	16.25	15.32	14.39	13.48	12.57	11.67	10.78	9.90	9.03	8.17	7.31								
230	20.78	19.64	18.51	17.40	16.29	15.19	14.11	13.03	11.97	10.91	9.87									
240	24.97	23.60	22.25	20.91	19.58	18.26	16.95	15.66	14.38	13.12	11.86									
250	29.83	28.20	26.58	25.00	23.39	21.82	20.25	18.71	17.18	15.67										

* Wilson, *Univ. Ill., Eng. Exp. Sta. Bull.* 146.

Table 22. Mole Percentages of H_2O over Aqueous Solutions of NH_3*

Molal concentration of ammonia in the solutions in percentages (Weight concentration of ammonia in the solutions in percentages)

t, °F.	0 (0)	5 (4.74)	10 (9.50)	15 (14.29)	20 (19.10)	25 (23.94)	30 (28.81)	35 (33.71)	40 (38.64)	45 (43.59)	50 (48.57)	55 (53.58)	60 (58.62)	65 (63.69)	70 (68.79)	75 (73.91)	80 (79.07)	85 (84.26)	90 (89.47)	95 (94.72)	100 (100.00)
32	100	24.3	13.2	7.63	4.43	2.50	1.43	0.856	0.514	0.335	0.216	0.151	0.109	0.0816	0.0585	0.0457	0.0345	0.0249	0.0146	0.00689	0.00
40	100	25.3	14.1	8.15	4.73	2.74	1.59	.943	.581	.372	.248	.172	.124	.0914	.0706	.0533	.0395	.0243	.0185	.00879	
50	100	26.6	15.2	9.09	5.24	3.03	1.78	1.060	.652	.434	.290	.202	.148	.1095	.0838	.0630	.0477	.0332	.0215	.00959	
60	100	27.9	16.2	9.50	5.69	3.42	1.97	1.210	.777	.481	.331	.238	.172	.1290	.0986	.0754	.0566	.0406	.0251	.01125	
70	100	29.1	17.4	10.30	6.14	3.65	2.27	1.390	.873	.569	.383	.266	.205	.1510	.112	.0882	.0656	.0474	.0296	.0135	
80	100	31.6	18.5	11.20	6.89	4.08	2.45	1.550	.978	.659	.444	.323	.230	.1750	.130	.103	.0772	.0528	.0351	.0167	
90	100	32.7	20.0	12.00	7.40	4.47	2.73	1.730	1.100	.742	.505	.366	.267	.2020	.157	.115	.0884	.0647	.0408	.0194	
100	100	34.4	21.0	12.90	7.92	4.85	3.00	1.890	1.250	.834	.574	.420	.307	.2290	.179	.135	.104	.0714	.0473	.0226	
110	100	35.9	22.2	13.80	8.59	5.29	3.30	2.110	1.370	.932	.644	.466	.347	.2640	.208	.157	.118	.0846	.0540	.0262	
120	100	37.5	23.4	14.70	9.22	5.75	3.63	2.320	1.520	1.044	.714	.529	.395	.3020	.233	.180	.135	.0970	.0619	.0300	
130	100	39.0	24.5	15.60	9.85	6.18	3.95	2.550	1.690	1.160	.811	.596	.444	.3430	.263	.205	.154	.1117	.0703	.0339	
140	100	40.7	25.8	16.50	10.50	6.69	4.28	2.790	1.860	1.286	.906	.663	.501	.3840	.297	.232	.175	.124	.0786	.0385	
150	100	42.3	27.1	17.50	11.20	7.19	4.63	3.080	2.040	1.410	1.004	.741	.558	.4320	.334	.257	.197	.140	.0892	.0439	
160	100	44.1	28.3	18.40	11.90	7.69	5.01	3.300	2.230	1.550	1.110	.818	.617	.4800	.372	.287	.218	.154	.1005	.0499	
170	100	45.6	29.6	19.40	12.70	8.22	5.38	3.580	2.430	1.700	1.220	.904	.689	.5300	.414	.320	.242	.174	.112	.0567	
180	100	47.3	30.9	20.40	13.40	8.76	5.78	3.870	2.640	1.850	1.340	.994	.756	.5860	.456	.352	.268	.192			
190	100	48.7	32.2	21.40	14.10	9.31	6.18	4.160	2.860	2.020	1.460	1.087	.830	.6420	.501						
200	100	50.4	33.4	22.30	14.90	9.88	6.59	4.470	3.080	2.190	1.580	1.187	.907	.7010							
210	100	52.1	34.7	23.40	15.70	10.45	7.03	4.780	3.310	2.360	1.720	1.272	.983								
220	100	53.7	36.1	24.40	16.40	11.05	7.48	5.100	3.560	2.540	1.860	1.390									
230	100	55.2	37.3	25.40	17.30	11.63	7.91	5.440	3.810	2.730	2.000										
240	100	56.8	38.6	26.50	18.00	12.24	8.36	5.780	4.060	2.920	2.150										
250	100	58.4	39.8	27.50	18.80	12.88	8.82	6.120	4.340	3.120											

* Wilson, *Univ. Ill., Eng. Exp. Sta. Bull.* 146.

Table 23. Partial Pressures of NH_3 over Aqueous Solutions of NH_3*

Pressures are in pounds per square inch absolute

t, °F.	Molal concentration of ammonia in the solutions in percentages (Weight concentration of ammonia in the solutions in percentages)																	
	5 (4.74)	10 (9.50)	15 (14.29)	20 (19.10)	25 (23.94)	30 (28.81)	35 (33.71)	40 (38.64)	45 (43.59)	50 (48.57)	55 (53.58)	60 (58.62)	65 (63.69)	70 (68.79)	75 (73.91)	80 (79.07)	85 (84.26)	90 (89.47)	95 (94.72)
32	0.26	0.52	0.90	1.51	2.67	4.27	6.54	8.93	14.13	19.36	25.12	31.13	36.74	42.69	45.92	49.26	52.13	54.89	58.01
40	.33	.66	1.14	1.92	3.16	5.13	7.98	11.98	17.14	23.33	30.15	37.15	43.69	49.56	54.40	58.31	61.62	64.77	68.31
50	.47	.89	1.50	2.53	4.16	6.63	10.24	15.24	21.56	29.17	37.46	45.86	53.79	60.82	66.63	71.26	75.22	79.05	83.40
60	.62	1.19	2.00	3.21	5.36	8.48	13.06	19.15	26.92	36.14	46.12	56.22	65.81	73.99	80.90	86.44	91.04	95.67	100.65
70	.83	1.52	2.60	4.28	6.87	10.76	16.33	23.84	33.20	44.25	56.29	68.32	79.42	89.26	97.42	104.01	109.55	114.83	120.61
80	1.04	1.98	3.34	5.45	8.69	13.52	20.29	29.40	40.69	53.84	67.97	82.36	95.52	107.06	116.42	124.20	130.57	136.35	143.70
90	1.36	2.52	4.25	6.88	10.89	16.76	25.04	35.94	49.45	64.99	81.61	98.35	113.79	127.22	138.18	147.02	154.46	161.74	169.73
100	1.72	3.20	5.34	8.60	13.53	20.68	30.57	43.57	59.49	77.85	97.27	116.81	134.70	150.23	162.94	173.22	181.97	190.13	199.17
110	2.14	4.00	6.65	10.64	16.65	25.21	37.01	52.43	71.20	92.59	115.16	137.62	158.42	176.18	190.85	203.02	212.71	222.22	232.79
120	2.67	4.95	8.21	13.09	20.30	30.54	44.56	62.62	84.44	109.40	135.48	161.44	185.14	205.81	222.28	236.05	247.14	258.24	270.02
130	3.28	6.09	10.05	15.93	24.58	36.74	53.16	74.27	99.69	128.45	158.45	188.16	215.14	238.70	257.87	272.88	286.08	298.46	311.80
140	3.97	7.41	12.21	19.23	29.43	43.77	62.97	87.53	116.72	149.93	184.17	218.18	248.70	275.33	297.12	314.45	328.99	342.93	358.46
150	4.78	8.92	14.70	23.09	35.09	51.91	74.28	102.51	136.15	173.64	212.91	251.24	286.00	316.24	340.82	360.39	376.57	392.45	409.62
160	5.68	10.70	17.57	27.45	41.56	61.03	86.91	119.37	157.71	200.45	244.98	288.38	327.82	361.75	389.08	411.30	429.73	447.35	466.38
170	6.75	12.67	20.85	32.41	48.89	71.48	101.09	138.30	181.95	230.36	280.54	329.42	373.61	411.59	442.28	466.67	487.85	507.63	528.50
180	7.90	14.96	24.56	38.13	57.19	83.07	116.97	159.37	208.66	263.43	319.89	374.25	424.10	466.26	500.63	528.08	551.24		
190	9.23	17.55	28.78	44.49	66.49	96.22	134.89	182.72	238.39	299.86	363.11	424.15	479.40	526.15					
200	10.70	20.45	33.49	51.58	76.90	110.85	154.58	208.56	270.94	340.02	410.17	478.62	539.79						
210	12.26	23.68	38.76	59.65	88.48	126.83	176.24	236.97	307.08	383.99	462.36	537.56							
220	14.02	27.15	44.61	68.43	101.24	144.74	200.46	268.30	346.07	431.43	518.19								
230	15.95	31.09	51.06	78.14	115.45	164.17	226.67	302.53	389.29	483.53									
240	17.92	35.40	58.00	89.02	130.94	185.79	255.26	339.72	435.78	540.44									
250	20.12	40.09	65.74	100.69	147.66	209.37	286.89	380.42	486.73										

* Wilson, *Univ. Ill., Eng. Exp. Sta. Bull.* 146.

Table 24. Total Vapor Pressures of Aqueous Solutions of NH_3*

Pressures are in pounds per square inch absolute

t, °F.	Molal concentration of ammonia in the solutions in percentages (Weight concentration of ammonia in the solutions in percentages)																			
	0 (0)	5 (4.74)	10 (9.50)	15 (14.29)	20 (19.10)	25 (23.94)	30 (28.81)	35 (33.71)	40 (38.64)	45 (43.59)	50 (48.57)	55 (53.58)	60 (58.62)	65 (63.69)	70 (68.79)	75 (73.91)	80 (79.07)	85 (84.26)	90 (89.47)	95 (94.72)	100 (100.00)
32	0.09	0.34	0.60	0.97	1.58	2.60	4.20	6.54	9.93	14.18	19.40	25.16	31.16	36.77	42.72	45.94	49.28	52.14	54.90	58.01	62.29
40	.12	.45	.77	1.24	2.01	3.25	5.21	8.06	12.05	17.20	23.39	30.20	37.20	43.73	49.60	54.43	58.33	61.64	64.78	68.32	73.32
50	.18	.64	1.05	1.65	2.67	4.29	6.75	10.35	15.34	21.65	29.26	37.54	45.93	53.85	60.87	66.67	71.29	75.25	79.07	83.41	89.19
60	.26	.86	1.42	2.21	3.51	5.55	8.65	13.22	19.30	27.05	36.26	46.23	56.32	65.90	74.06	80.96	86.49	91.08	95.69	100.66	107.6
70	.36	1.17	1.84	2.90	4.56	7.13	11.01	16.56	24.05	33.39	44.42	56.44	68.46	79.54	89.36	97.51	104.08	109.60	114.86	120.63	128.8
80	.51	1.52	2.43	3.76	5.85	9.06	13.86	20.61	29.69	40.96	54.08	68.19	82.55	95.69	107.20	116.54	124.30	130.64	136.40	143.72	153.0
90	.70	2.02	3.15	4.83	7.43	11.40	17.23	25.48	36.34	49.82	65.32	81.91	98.61	114.02	127.42	138.34	147.15	154.56	161.81	169.76	180.6
100	.95	2.62	4.05	6.13	9.34	14.22	21.32	31.16	44.12	59.99	78.30	97.68	117.17	135.01	150.50	163.16	173.40	182.10	190.22	199.22	211.9
110	1.27	3.34	5.14	7.72	11.64	17.58	26.07	37.81	53.16	71.87	93.19	115.7	138.10	158.84	176.54	191.15	203.26	212.89	222.34	232.85	247.0
120	1.69	4.27	6.46	9.63	14.42	21.54	31.69	45.62	63.59	85.33	110.2	136.2	162.08	185.70	206.29	222.68	236.37	247.38	258.40	270.1	286.4
130	2.22	5.38	8.07	11.91	17.67	26.20	38.25	54.55	75.55	100.86	129.5	159.	189.00	215.88	239.33	258.40	273.3	286.4	298.67	311.9	330.3
140	2.89	6.70	9.98	14.63	21.49	31.54	45.73	64.78	89.19	118.24	151.3	185.4	219.28	249.66	276.15	297.81	315.0	329.4	343.2	358.6	379.1
150	3.72	8.29	12.23	17.81	26.00	37.81	54.43	76.61	104.65	138.1	175.4	214.5	252.65	287.24	317.3	341.7	361.1	377.1	392.8	409.8	432.2
160	4.74	10.16	14.92	21.54	31.16	45.02	64.25	89.88	122.10	160.2	202.7	247.0	290.18	329.4	363.1	390.2	412.2	430.4	447.8	466.6	492.8
170	5.99	12.41	18.01	25.87	37.11	53.27	75.55	104.84	141.75	185.1	233.2	283.1	331.7	375.6	413.3	443.7	467.8	488.7	508.2	528.8	558.4
180	7.51	15.00	21.65	30.86	44.02	62.68	88.17	121.68	163.7	212.6	267.0	323.1	377.1	426.6	468.4	502.4	529.5	552.3			
190	9.34	18.06	25.87	36.60	51.81	73.32	102.56	140.75	188.1	243.3	304.3	367.1	427.7	482.5	528.8						
200	11.53	21.60	30.72	43.14	60.62	85.33	118.68	161.81	215.2	277.0	345.5	415.1	483.0	543.6							
210	14.12	25.61	36.26	50.58	70.72	98.80	136.42	185.10	245.1	314.5	390.7	468.4	542.9								
220	17.19	30.27	42.47	59.00	81.91	113.81	156.41	211.24	278.2	355.1	439.6	525.5									
230	20.78	35.59	49.60	68.46	94.43	130.64	178.28	239.70	314.5	400.2	493.4										
240	24.97	41.52	57.65	78.91	108.60	149.20	202.74	270.92	354.1	448.9	552.3										
250	29.83	48.32	66.67	90.74	124.08	169.48	229.62	305.60	397.6	502.4											

* Wilson, *Univ. Ill., Eng. Exp. Sta. Bull.* 146.

Table 25. Partial Pressures of H_2O over Aqueous Solutions of Sodium Carbonate

Mm. Hg

t, °C.	%Na_2CO_3						
	0	5	10	15	20	25	30
0	4.5	4.5					
10	9.2	9.0	8.8				
20	17.5	17.2	16.8	16.3			
30	31.8	31.2	30.4	29.6	28.8	27.8	26.4
40	55.3	54.2	53.0	57.6	50.2	48.4	46.1
50	92.5	90.7	88.7	86.5	84.1	81.2	77.5
60	149.5	146.5	143.5	139.9	136.1	131.6	125.7
70	239.8	235	230.5	225	219	211.5	202.5
80	355.5	348	342	334	325	315	301
90	526.0	516	506	494	482	467	447
100	760.0	746	731	715	697	676	648

Table 26. Partial Pressures of H_2O and CH_3OH over Aqueous Solutions of Methyl Alcohol*

39.9°C.			59.4°C.		
Mole fraction CH_3OH	P_{H_2O}, mm. Hg	P_{CH_3OH}, mm. Hg	Mole fraction CH_3OH	P_{H_2O}, mm. Hg	P_{CH_3OH}, mm. Hg
0	54.7	0	0	145.4	0
14.99	39.2	66.1	22.17	106.9	210.1
17.85	38.5	75.5	27.40	102.2	240.2
21.07	37.2	85.2	33.24	96.6	272.1
27.31	35.8	100.6	39.80	91.7	301.9
31.06	34.9	108.8	47.08	84.8	335.6
40.1	32.8	127.7	55.5	76.9	373.7
47.0	31.5	141.6	69.2	57.8	439.4
55.8	27.3	158.4	78.5	43.8	486.6
68.9	20.7	186.6	85.9	30.1	526.9
86.0	10.1	225.2	100.0	0	609.3
100.0	0	260.7			

* "International Critical Tables," vol. 3, p. 290, McGraw-Hill.

Table 27. Partial Pressures of H_2O over Aqueous Solutions of Sodium Hydroxide

Mm. Hg

Conc. g. NaOH/ 100 g H_2O	Temperature, °C.										
	0	20	40	60	80	100	120	160	200	250	300	350
0	4.6	17.5	55.3	149.5	355.5	760.0	1,489	4,633	11,647	29,771	64,200	123,600
5	4.4	16.9	53.2	143.5	341.5	730.0	1,430	4,450	11,200	28,600	61,800	118,900
10	4.2	16.0	50.6	137.0	325.5	697.0	1,365	4,260	10,750	27,500	59,300	114,100
20	3.6	13.9	44.2	120.5	288.5	621.0	1,225	3,860	9,800	25,300	54,700	105,400
30	2.9	11.3	36.6	101.0	246.0	537.0	1,070	3,460	8,950	23,300	50,800	98,000
40	2.2	8.7	28.7	81.0	202.0	450.0	920	3,090	8,150	21,500	47,200	91,600
50	...	6.3	20.7	62.5	160.5	368.0	770	2,690	7,400	19,900	44,100	85,800
60	...	4.4	15.5	47.0	124.0	294.0	635	2,340	6,750	18,400	41,200	80,700
70	...	3.0	10.9	34.5	94.0	231.0	515	2,030	6,100	17,100	38,700	76,000
80	...	2.0	7.6	24.5	70.5	179.0	415	1,740	5,500	15,800	36,300	71,900
90	...	1.3	5.2	17.5	53.0	138.0	330	1,490	5,000	14,700	34,200	68,100
100	...	0.9	3.6	12.5	38.5	105.0	262	1,300	4,500	13,650	32,200	64,600
120	...	....	1.7	6.3	20.5	61.0	164	915	3,650	11,800	28,800	58,600
140	...	....	....	3.0	11.0	35.5	102	765	2,980	10,300	25,900	53,400
160	...	....	....	1.5	6.0	20.5	63	470	2,430	8,960	23,300	49,000
180	...	....	....	.....	3.5	12.0	40	340	1,980	7,830	21,200	45,100
200	...	....	....	.....	2.0	7.0	25	245	1,620	6,870	19,200	41,800
250	...	....	....	.....	0.5	2.0	8	110	985	5,000	15,400	35,000
300	...	....	....	.....	0.1	0.5	2.7	50	610	3,690	12,500	29,800
350	...	....	....	.....	.....	....	0.9	23	380	2,750	10,300	25,700
400	...	....	....	.....	.....	....	....	11	240	2,080	8,600	22,400
500	...	....	....	.....	.....	....	.....	.....	100	1,210	6,100	17,500
700	...	....	....	.....	.....	....	.....	.....	.....	440	3,300	11,500
1000	...	....	....	.....	.....	....	.....	.....	.....	.....	1,470	6,800
2000	...	....	....	.....	.....	....	.....	.....	.....	.....	150	1,760
4000	...	....	....	.....	.....	....	.....	.....	.....	.....	.....	120
8000	...	....	....	.....	.....	....	.....	.....	.....	.....	.....	7

WATER-VAPOR CONTENT OF GASES

CHARTS FOR GASES AT HIGH PRESSURES

The accompanying figures are useful in determining the water-vapor content of gases at high pressure in contact with liquid water. Figure 4 shows the water-vapor content of hydrogen and nitrogen in contact with liquid water at high pressures. For additional experimental values of the water content of compressed nitrogen in contact with water at 100, 200, and 300 atm. and up to 230°C., see Saddington and Krase, *J. Am. Chem. Soc.* **56,** 360 (1934). Results to 100°C. are shown in Fig. 6, and comparisons with Bartlett's values at 50°C. are included. Figure 5 shows the water-vapor content of compressed gases in contact with liquid water. Figure 5*A* shows the volume percentage of water vapor in gases expanded from high-pressure contact with liquid water.

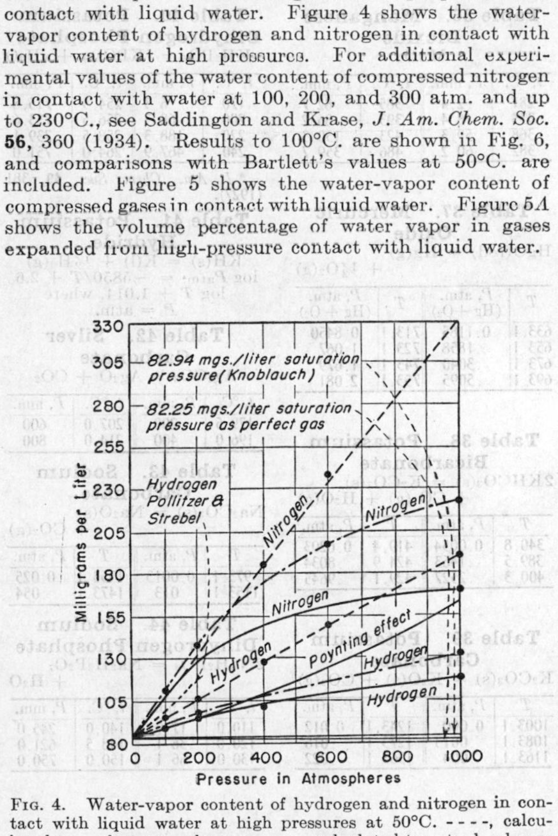

FIG. 4. Water-vapor content of hydrogen and nitrogen in contact with liquid water at high pressures at 50°C. - - - -, calculated to perfect gas volume; ———, calculated to actual volume; — · — ·, calculated to free space. [*Bartlett J. Am. Chem. Soc.,* **49,** 65 (1927).]

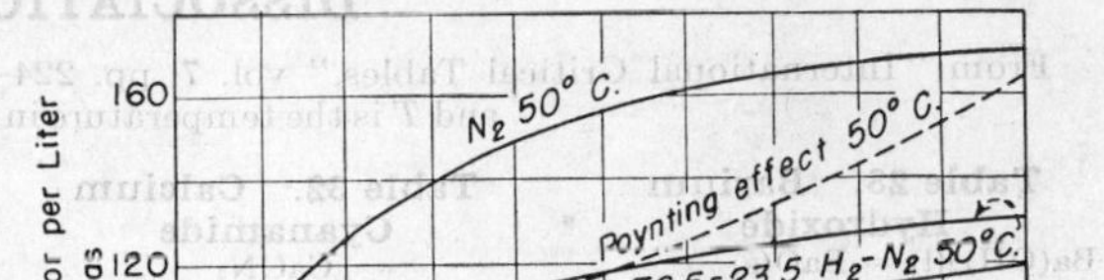

FIG. 5. Water-vapor content of compressed gases in contact with liquid water, 25.0°, 37.5°, 50.0°C. [*Bartlett, J. Am. Chem. Soc.,* **49,** 65 (1927).]

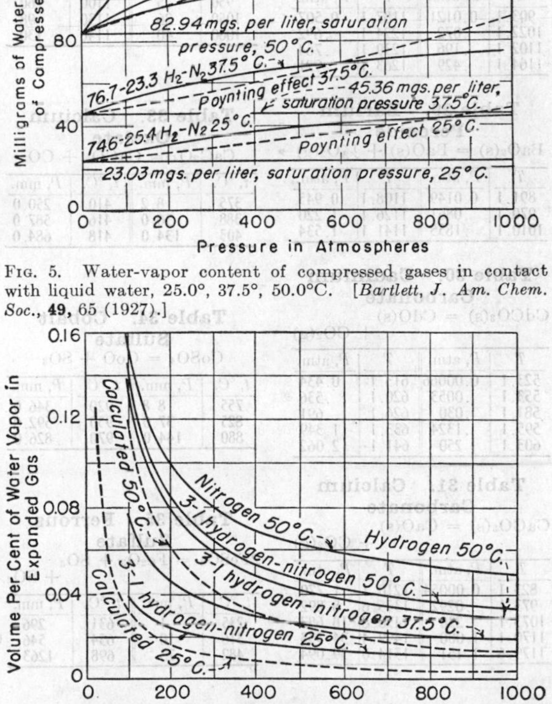

FIG. 5*A*. Volume percentage of water vapor in gases expanded from high-pressure contact with liquid water, 25.0°, 37.5°, 50.0°C. [*Bartlett, J. Am. Chem. Soc.,* **49,** 65 (1927).]

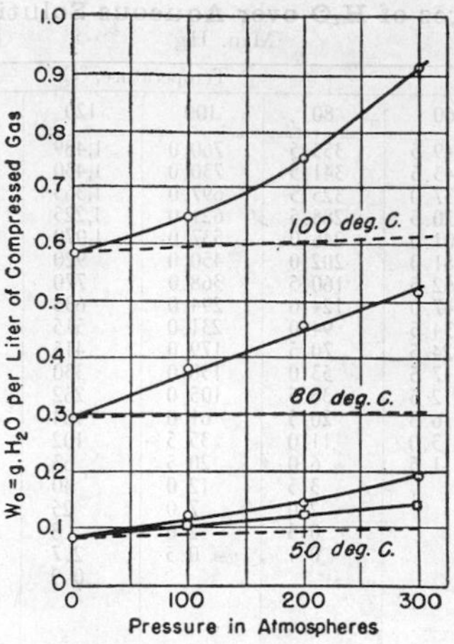

FIG. 6. Effect of pressure on the water-vapor content of compressed N_2 gas; - - -, Poynting relation; □, Bartlett; ◯, experimental. [*Saddington and Krase, J. Am. Chem. Soc.*, **56**, 360 (1934).]

DISSOCIATION PRESSURES

From "International Critical Tables," vol. 7, pp. 224–313, where P is the pressure in atmospheres or millimeters and T is the temperature in degrees Kelvin (°C. + 273.1).

Table 28. Barium Hydroxide

$Ba(OH)_2(l) = BaO(s) + H_2O(g)$

T	P, atm.	T	P, atm.
903.1	0.0121	1187.1	0.507
1022.1	.072	1224.1	.692
1102.1	.196	1239.1	.764
1164.1	.429	1263.1	.921

Table 29. Barium Peroxide

$BaO_2(s) = BaO(s) + \frac{1}{2}O_2(g)$

T	P, atm.	T	P, atm.
891.1	0.0149	1108.1	0.945
970.1	.0861	1126.1	1.220
1010.1	.1855	1141.1	1.534

Table 30. Cadmium Carbonate

$CdCO_3(s) = CdO(s) + CO_2(g)$

T	P, atm.	T	P, atm.
523.1	0.00066	613.1	0.434
553.1	.0053	620.1	.536
581.1	.030	626.1	.691
595.1	.1324	633.1	1.349
603.1	.250	641.1	2.062

Table 31. Calcium Carbonate

$CaCO_3(s) = CaO(s) + CO_2(g)$

T	P, atm.	T	P, atm.
823.1	0.00054	1210.1	1.770
973.1	.0292	1355.6	8.892
1073.1	.220	1430.8	18.687
1170.1	1.000	1499.4	34.333
1179.6	1.151	1514.0	39.094

Table 32. Calcium Cyanamide

$CaCN_2$

t, °C.	P, mm.	t, °C.	P, mm.
950	5	1100	60
1050	26	1130	98
1080	40	1146	130

Table 33. Calcium Oxalate

$CaC_2O_4 = CaCO_3 + CO$

t, °C.	P, mm.	t, °C.	P, mm.
375	8.2	410	250.0
388	30.0	416	587.0
403	134.0	418	684.0

Table 34. Cobalt Sulfate

$CoSO_4 = CoO + SO_3$

t, °C.	P, mm.	t, °C.	P, mm.
755	8.8	920	346.0
825	37.0	950	592.0
880	144.0	970	826.0

Table 35. Ferrous Sulfate

$2FeSO_4 = Fe_2O_3 + SO_3 + SO_2$

t, °C.	P, mm.	t, °C.	P, mm.
235	1	631	296
316	10	654	546
482	73	698	1263

Table 36. Manganese Dioxide

MnO_2

t, °C.	P, mm.	t, °C.	P, mm.
281	2.4	385	82.7
333	18.4	393	86.2
368	59.3	423	150.3
382	60.2	486	359.7

Table 37. Mercuric Oxide

$HgO(red) = Hg(g) + \frac{1}{2}O_2(g)$

T	P, atm. ($Hg + O_2$)	T	P, atm. ($Hg + O_2$)
633.1	0.1185	713.1	0.8450
653.1	.1858	723.1	1.067
673.1	.3040	743.1	1.679
693.1	.5095	753.1	2.081

Table 38. Potassium Bicarbonate

$2KHCO_3(s) = K_2CO_3(s) + CO_2(g) + H_2O(g)$

T	P, atm.	T	P, atm.
340.8	0.0054	419.4	0.6203
389.5	.1463	424.9	.8034
400.3	.2527	429.1	.9645

Table 39. Potassium Carbonate

$K_2CO_3(s) = K_2O(s) + CO_2(g)$

T	P, atm.	T	P, atm.
1003.1	0.000	1243.1	0.012
1083.1	.0013	1273.1	.016
1163.1	.004	1363.1	.022

Table 40. Potassium Dihydrogen Phosphate*

$xKH_2PO_4 = (KPO_3)x + xH_2O$

t, °C.	P, mm.	t, °C.	P, mm.
170	6.3	254.5	715.8
200	68.3	256.5	733.6
220	188.3	258.5	739.6
240	467.9	264.0	751.0

* *J. Am. Chem. Soc.*, **49**, 381 (1927).

Table 41. Potassium Hydride

$KH(s) = K(l) + \frac{1}{2}H_2(g)$

$\log P_{atm.} = -5850/T + 2.6 \log T + 1.014$, where P = atm.

Table 42. Silver Carbonate

$Ag_2CO_3 = Ag_2O + CO_2$

t, °C.	P, mm.	t, °C.	P, mm.
179.5	200	207.0	600
196.0	400	214.0	800

Table 43. Sodium Carbonate

$Na_2CO_3(s) = Na_2O(s) + CO_2(g)$

T	P, atm.	T	P, atm.
973.1	0.0013	1353.1	0.025
1153.1	.013	1473.1	.054

Table 44. Sodium Dihydrogen Phosphate

$2NaH_2PO_4 = Na_2H_2P_2O_7 + H_2O$

t, °C.	P, mm.	t, °C.	P, mm.
110.0	17.9	140.0	245.0
120.0	36.1	148.5	621.0
130.0	66.1	150.0	750.0

DENSITIES OF PURE SUBSTANCES

Table 45. Density of Water in Grams per Milliliter between 0° and 40°C.

°C.	0.0	0.1	0.2	0.3	0.4	0.5	0.6	0.7	0.8	0.9
0	0.9998679	0.9998746	0.9998811	0.9998874	0.9998935	0.9998995	0.9999053	0.9999109	0.9999163	0.9999216
1	9267	9315	9363	9408	9452	9494	9534	9573	9610	9645
2	9679	9711	9741	9769	9796	9821	9844	9866	9887	9905
3	9922	9937	9951	9962	9973	9981	9988	9994	9998	*0000
4	1.0000000	*9999	*9996	*9992	*9986	*9979	*9970	*9960	*9947	*9934
5	0.9999919	.9999902	.9999883	.9999864	.9999842	.9999819	.9999795	.9999769	.9999741	.9999712
6	9681	9649	9616	9581	9544	9506	9467	9426	9384	9340
7	9295	9248	9200	9150	9099	9046	8992	8936	8879	8821
8	8762	8701	8638	8574	8509	8442	8374	8305	8234	8162
9	8088	8013	7936	7859	7780	7699	7617	7534	7450	7364
10	7277	7189	7099	7008	6915	6820	6724	6627	6529	6428
11	6328	6225	6121	6017	5911	5803	5694	5585	5473	5361
12	5247	5132	5016	4898	4780	4660	4538	4415	4291	4166
13	4040	3913	3784	3655	3524	3391	3258	3123	2987	2850
14	2712	2572	2432	2290	2147	2003	1858	1711	1564	1415
15	1265	1113	0961	0808	0653	0497	0340	0182	0023	*9862
16	.9989701	.9989538	.9989374	.9989209	.9989043	.9988876	.9988707	.9988538	.9988367	.9988195
17	8022	7848	7673	7497	7319	7141	6961	6781	6599	6416
18	6232	6046	5861	5673	5485	5295	5105	4913	4720	4526
19	4331	4136	3938	3740	3541	3341	3140	2937	2733	2529
20	2323	2117	1909	1701	1490	1280	1068	0855	0641	0426
21	0210	*9993	*9775	*9556	*9335	*9114	*8892	*8669	*8444	*8219
22	.9977993	.9977765	.9977537	.9977308	.9977077	.9976846	.9976613	.9976380	.9976145	.9975910
23	5674	5437	5198	4959	4718	4477	4235	3991	3747	3502
24	3256	3009	2760	2511	2261	2010	1758	1505	1250	0995
25	0739	0482	0225	*9966	*9706	*9445	*9184	*8921	*8657	*8393
26	.9968128	.9967861	.9967594	.9967326	.9967057	.9966786	.9966515	.9966243	.9965970	.9965696
27	5421	5146	4869	4591	4313	4033	3753	3472	3190	2907
28	2623	2338	2052	1766	1478	1190	0901	0610	0319	0027
29	.9959735	.9959440	.9959146	.9958850	.9958554	.9958257	.9957958	.9957659	.9957359	.9957059
30	6756	6454	6151	5846	5541	5235	4928	4620	4312	4002
31	3692	3380	3068	2755	2442	2127	1812	1495	1178	0861
32	.9950542	.9950222	.9949901	.9949580	.9949258	.9948935	.9948612	.9948286	.9947961	.9947635
33	.9947308	.9946980	.9946651	.9946321	.9945991	.9945660	.9945328	.9944995	.9944661	.9944327
34	3991	3655	3319	2981	2643	2303	1963	1622	1280	0938
35	0594	0251	*9906	*9560	*9214	*8867	*8518	*8170	*7820	*7470
36	.9937119	.9936767	.9936414	.9936061	.9935707	.9935351	.9934996	.9934639	.9934282	.9933924
37	3565	3206	2846	2484	2123	1760	1397	1032	0668	0302
38	.9929936	.9929568	.9929201	.9928833	.9928463	.9928093	.9927722	.9927351	.9926978	.9926605
39	6232	5857	5482	5106	4730	4352	3974	3595	3216	2836
40	2455									

* Indicates change in the first three decimal places.

Table 46. Density of Water in Grams per Milliliter between 40° and 100°C.

°C.	0.0	1.0	2.0	3.0	4.0	5.0	6.0	7.0	8.0	9.0
40	0.99224	0.99186	0.99147	0.99107	0.99066	0.99024	0.98982	0.98940	0.98896	0.98852
50	.98807	.98762	.98715	.98669	.98621	.98573	.98525	.98475	.98425	.98375
60	.98324	.98272	.98220	.98167	.98113	.98059	.98005	.97950	.97894	.97838
70	.97781	.97723	.97666	.97607	.97548	.97489	.97429	.97368	.97307	.97245
80	.97183	.97121	.97057	.96994	.96930	.96865	.96800	.96734	.96668	.96601
90	.96534	.96467	.96399	.96330	.96261	.96192	.96122	.96051	.95981	.95909
100	.95838									

Table 47. Density and Specific Volume of Water below 0°C.

Grams per milliliter and milliliters per gram

t, °C.	Density	Volume	t, °C.	Density	Volume
0	0.9999	1.0001	− 8	0.9987	1.0013
−1	.9998	1.0002	− 9	.9984	1.0016
−2	.9997	1.0003	−10	.9982	1.0019
−3	.9996	1.0004	−11	.9976	1.0024
−4	.9995	1.0006	−12	.9973	1.0027
−5	.9993	1.0007	−13	.9969	1.0031
−6	.9991	1.0009			
−7	.9989	1.0011			

Table 48. Density of Mercury*

Density or mass in grams per cubic centimeter, and the volume in cubic centimeters of 1 g. mercury

t, °C.	Mass, g./cc.	Volume of 1 g. in cc.	t, °C.	Mass, g./cc.	Volume of 1 g. in cc.
−10	13.6198	0.0734225	30	13.5213	0.0739572
−9	6173	4358	31	5189	9705
−8	6148	4492	32	5164	9839
−7	6124	4626	33	5140	9973
−6	6099	4759	34	5116	40107
−5	6074	4893	35	5091	.0740241
−4	6050	5026	36	5066	0374
−3	6025	5160	37	5042	0508
−2	6000	5293	38	5018	0642
−1	5976	5427	39	4994	0776
0	5951	5560	40	4969	0910
1	5926	5694	50	4725	2250
2	5901	5828	60	4482	3592
3	5877	5961	70	4240	4936
4	5852	6095	80	3998	6282
5	5827	6228	90	3757	7631
6	5803	6362	100	3515	8981
7	5778	6496	110	3279	.0750305
8	5754	6629	120	3040	1653
9	5729	6763	130	2801	3002
10	5704	6893	140	2563	4354
11	5680	7030	150	2326	5708
12	5655	7164	160	2090	7064
13	5630	7298	170	1853	8422
14	5606	7431	180	1617	9784
15	5581	7565	190	1381	.0761149
16	5557	7699	200	1145	2516
17	5532	7832	210	0910	3886
18	5507	7966	220	0677	5260
19	5483	8100	230	0440	6637
20	5458	8233	240	13.0206	8017
21	5434	8367	250	12.9972	9402
22	5409	8501	260	9738	.077099
23	5385	8635	270	9504	2182
24	5360	8768	280	9270	3579
25	5336	8902	290	9036	4979
26	5311	9036	300	8803	6385
27	5287	9170	310	8569	7795
28	5262	9304	320	6338	9210
29	5238	9437	330	8102	.0780630
30	5213	9571	340	7869	2054
			350	7635	3485
			360	7402	4921

* "Smithsonian Tables."

Table 49. Densities of Gases at Standard Conditions (0°C., 1 atm.)

Gas	Formula	Mol. wt.	Density G./l.	Density Lb./cu. ft.
Acetylene	C_2H_2	26.02	1.1708	0.0732
Air			1.2928	0.0808
Ammonia	NH_3	17.03	0.7708	0.0482
Argon	A	39.91	1.7828	0.1114
Bromine	Br_2	159.83	7.1388	0.4460
Butane	C_4H_{10}	58.08	2.5985	0.1623
Carbon dioxide	CO_2	44.00	1.9768	0.1235
Carbon monoxide	CO	28.00	1.2501	0.0781
Carbon oxychloride	$COCl_2$	98.91	4.5313	0.2830
Carbon oxysulfide	COS	60.06	2.7201	0.1700
Chlorine	Cl_2	70.91	3.2204	0.2011
Chlorine monoxide	Cl_2O	86.91	3.8874	0.2428
Cyanogen	C_2N_2	52.02	2.3348	0.1459
Ethane	C_2H_6	30.05	1.3567	0.0848
Ethyl chloride	C_2H_5Cl	64.50	2.8700	0.1793
Ethylene	C_2H_4	28.03	1.2644	0.0783
Fluorine	F_2	38.00	1.6354	0.1022
Helium	He	4.00	0.1769	0.0111
Hydrogen	H_2	2.016	0.0898	0.0056
Hydrogen chloride	HCl	36.47	1.6394	0.1024
Hydrogen fluoride	HF	20.01	0.9218	0.0576
Hydrogen iodide	HI	127.94	5.7245	0.3576
Hydrogen selenide	H_2Se	81.22	3.6134	0.2258
Hydrogen sulfide	H_2S	34.08	1.5392	0.0961
Hydrogen telluride	H_2Te	129.52	5.8034	0.3625
Krypton	Kr	82.90	3.6431	0.2275
Methane	CH_4	16.03	0.7167	0.0448
Methyl chloride	CH_3Cl	50.48	2.3044	0.1440
Neon	Ne	20.40	0.8713	0.0544
Nitric oxide	NO	30.01	1.3401	0.0837
Nitrogen	N_2	28.02	1.2507	0.0782
Nitrous oxide	N_2O	44.02	1.9781	0.1235
Nitrosyl chloride	NOCl	65.47	2.9864	0.1865
Oxygen	O_2	32.00	1.4289	0.0892
Phosphine	PH_4	34.05	1.5293	0.0955
Silicon fluoride	SiF_2	104.06	4.6541	0.2907
Sulfur dioxide	SO_2	64.06	2.9268	0.1828
Xenon	X	130.20	5.7168	0.3570

SPECIFIC VOLUMES OF PURE SUBSTANCES

Table 50. Volume, in Milliliters, of 1 G. of Water between 0° and 40°C.*

°C.	0.0	0.1	0.2	0.3	0.4	0.5	0.6	0.7	0.8	0 9
0	1.0001322	1.0001255	1.0001190	1.0001127	1.0001065	1.0001005	1.0000948	1.0000892	1.0000837	1.0000785
1	0734	0685	0637	0592	0548	0506	0466	0427	0390	0355
2	0320	0289	0259	0230	0204	0179	0156	0134	0113	0095
3	0078	0063	0049	0038	0027	0019	0012	0006	0003	0000
4	0000	0001	0004	0008	0014	0022	0030	0040	0053	0067
5	0081	0098	0117	0137	0159	0182	0206	0232	0259	0288
6	0319	0351	0384	0419	0456	0494	0533	0574	0617	0661
7	0706	0753	0801	0851	0902	0954	1008	1064	1121	1179
8	1239	1300	1362	1426	1491	1558	1626	1696	1767	1839
9	1913	1988	2064	2142	2221	2302	2384	2467	2551	2637
10	2724	2813	2905	2993	3086	3181	3278	3374	3473	3573
11	3674	3776	3881	3985	4091	4199	4308	4417	4529	4641
12	4755	4871	4987	5105	5223	5343	5465	5588	5712	5837
13	5963	6091	6219	6349	6480	6613	6747	6882	7018	7155
14	7293	7433	7574	7716	7859	8003	8149	8296	8444	8592
15	8743	8895	9047	9201	9356	9512	9669	9828	9987	*0149
16	1.0010309	1.0010473	1.0010638	1.0010803	1.0010970	1.0011137	1.0011306	1.0011476	1.0011647	1.0011819
17	1993	2167	2343	2519	2697	2876	3056	3237	3419	3603
18	3787	3973	4160	4348	4537	4727	4918	5110	5303	5498
19	1.0015694	1.0015890	1.0016088	1.0016286	1.0016487	1.0016687	1.0016889	1.0017093	1.0017296	1.0017502
20	7708	7916	8124	8333	8544	8756	8968	9182	9397	9612
21	9830	*0048	*0266	*0486	*0708	*0930	*1153	*1377	*1603	*1828
22	1.0022056	1.0022285	1.0022514	1.0022744	1.0022976	1.0023208	1.0023442	1.0023676	1.0023912	1.0024148
23	4386	4624	4864	5104	5346	5589	5832	6077	6322	6569
24	6816	7065	7314	7565	7817	8069	8322	8577	8833	9089
25	9346	9605	9864	*0124	*0387	*0649	*0912	*1176	*1442	*1708
26	1.0031974	1.0032243	1.0032512	1.0032782	1.0033052	1.0033325	1.0033598	1.0033871	1.0034146	1.0034422
27	4699	4976	5255	5535	5815	6097	6379	6662	6947	7232
28	7517	7804	8093	8381	8671	8962	9253	9546	9839	*0134
29	1.0040428	1.0040725	1.0041022	1.0041320	1.0041619	1.0041918	1.0042219	1.0042521	1.0042824	1.0043127
30	3432	3736	4043	4350	4658	4966	5277	5587	5898	6211
31	6523	6838	7153	7469	7786	8104	8422	8741	9062	9382
32	9704	*0028	*0351	*0676	*1001	*1327	*1654	*1983	*2311	*2641
33	1.0052972	1.0053303	1.0053636	1.0053969	1.0054303	1.0054637	1.0054973	1.0055310	1.0055647	1.0055985
34	6324	6664	7005	7346	7688	8032	8376	8711	9067	9413
35	9761	*0109	*0457	*0807	*1158	*1510	*1862	*2215	*2570	*2924
36	1.0063279	1.0063636	1.0063993	1.0064351	1.0064710	1.0065070	1.0065430	1.0065791	1.0066153	1.0066516
37	6879	7244	7609	7975	8341	8709	9077	9447	9817	*0187
38	1.0070559	1.0070932	1.0071304	1.0071678	1.0072052	1.0072428	1.0072804	1.0073181	1.0073559	1.0073938
39	4317	4697	5077	5460	5841	6225	6609	6993	7379	7765
40	8152									

* See also steam tables, pp. 277–279.

Table 51. Volume, in Milliliters, of 1 G. of Water, between 40° and 100°C.*

°C.	0.0	1.0	2.0	3.0	4.0	5.0	6.0	7.0	8.0	9.0
40	1.00782	1.00821	1.00861	1.00901	1.00943	1.00985	1.01028	1.01072	1.01116	1.01161
50	1.01207	1.01254	1.01301	1.01349	1.01398	1.01448	1.01498	1.01548	1.01600	1.01652
60	1.01705	1.01758	1.01813	1.01867	1.01923	1.01979	1.02036	1.02093	1.02154	1.02210
70	1.02270	1.02330	1.02390	1.02452	1.02513	1.02576	1.02639	1.02703	1.02768	1.02833
80	1.02899	1.02965	1.03032	1.03099	1.03168	1.03237	1.03306	1.03376	1.03447	1.03518
90	1.03590	1.03663	1.03736	1.03810	1.03884	1.03959	1.04035	1.04111	1.04188	1.04265
100	1.04343									

* See also steam tables, pp. 277–279.

DENSITIES OF AQUEOUS INORGANIC SOLUTIONS

For more detailed data on densities see "International Critical Tables": tabular index, vol. 3, p. 1; abrasives, vol. 2, p. 87; air, moist, vol. 1, p. 71; building stones, vol. 2, p. 52; clays, vol. 2, p. 56; coals, vol. 2, p. 135; compounds, vol. 1, pp. 106, 176, 313, 341; elements, vol. 1, pp. 102, 340; fibers, vol. 2, p. 237; gases and vapors, vol. 3, pp. 3, 345; glass, vol. 2, p. 93; liquids and vitreous solids, vol. 3, p. 22; vol. 1, pp. 102, 340; vol. 2, pp. 456, 463; vol. 3, pp. 20, 35; temperatures of maximum solubility, vol. 3, p. 107; metals, vol. 2, p. 463; oils, fats, and waxes, vol. 2, p. 201; orthobaric, vol. 3, pp. 202, 228, 237, 244; petroleums, vol. 2, pp. 137, 144; plastics, vol. 2, p. 296; porcelains, vol. 2, pp. 68, 75; refrigerating brines, vol. 2, p. 327; rubber, vol. 2, pp. 255, 259; soaps, vol. 5, p. 447; metallic solid solutions, vo . 2, p. 358; solids, vol. 3, pp. 43, 45; vol. 2, p. 456; vol. 3, p. 21; solutions and mixtures, vol. 3, pp. 17, 51, 95, 104, 107, 111, 125, 130; woods, vol. 2, p. 1.

Table 52. Aluminum Sulfate [$Al_2(SO_4)_3$]

%	d_4^{15}	%	d_4^{15}
1	1.0093	16	1.1770
2	1.0195	20	1.2272
4	1.0404	24	1.2803
8	1.0837	26	1.3079
12	1.1293		

Table 53. Ammonia (NH_3)

%	−15°C.	−10°C.	−5°C.	0°C.	5°C.	10°C.	20°C.	25°C.	%	d_4^{15}
1	.....	0.9943	0.9954	0.9959	0.9958	0.9955	0.9939	0.993	32	0.889
2	.....	.9906	.9915	.9919	.9917	.9913	.9895	.988	36	.877
4	.....	.9834	.9840	.9842	.9837	.9832	.9811	.980	40	.865
8	0.970	.9701	.9701	.9695	.9686	.9677	.9651	.964	45	.849
12	.958	.9576	.9571	.9561	.9548	.9534	.9501	.948	50	.832
16	.947	.9461	.9450	.9435	.9420	.9402	.9362	.934	60	.796
20	.....	.9353	.9335	9316	.9296	.9275	.9229		70	.755
24	.....	.9249	.9226	.9202	.9179	.9155	.9101		80	.711
28	.....	.9150	.9122	.9094	.9067	.9040	.8980		90	.665
30	.....	.9101	.9070	.9040	.9012	.8983	.8920		100	.618

Table 54. Ammonium Acetate CH_3COONH_4

%	d_4^{25}
1	0.9992
2	1.0013
4	1.0055
8	1.0136
12	1.0216
16	1.0294
20	1.0368
24	1.0439
28	1.0507
30	1.0540
35	1.0618
40	1.0691
45	1.0760

Table 55. Ammonium Bichromate $(NH_4)_2Cr_2O_7$

%	d_4^{12}
1	1.0051
2	1.0108
4	1.0223
8	1.0463
12	1.0715
16	1.0981
20	1.1263

Table 56. Ammonium Chloride (NH_4Cl)

%	0°C.	10°C.	20°C.	30°C.	50°C.	80°C.	100°C.
1	1.0033	1.0029	1.0013	0.9987	0.9910	0.9749	0.9617
2	1.0067	1.0062	1.0045	1.0018	.9940	.9780	.9651
4	1.0135	1.0126	1.0107	1.0077	.9999	.9842	.9718
8	1.0266	1.0251	1.0227	1.0195	1.0116	.9963	.9849
12	1.0391	1.0370	1.0344	1.0310	1.0231	1.0081	.9975
16	1.0510	1.0485	1.0457	1.0422	1.0343	1.0198	1.0096
20	1.0625	1.0596	1.0567	1.0532	1.0454	1.0312	1.0213
24	1.0736	1.0705	1.0674	1.0641	1.0564	1.0426	1.0327

Table 57. Ammonium Chromate $(NH_4)_2CrO_4$

%	°C.	d_4^t
3.80	20	1.0219
10.52	13	1.0627
19.75	13.7	1.1189
28.04	19.6	1.1707

Table 58. Ammonium Nitrate (NH_4NO_3)

%	0°C.	10°C.	25°C.	40°C.	60°C.	80°C.
1.0	1.0043	1.0039	1.0011	0.9961	0.9870	0.9755
2.0	1.0088	1.0082	1.0051	1.0000	.9908	.9793
4.0	1.0178	1.0168	1.0132	1.0079	.9985	.9869
8.0	1.0358	1.0340	1.0297	1.0238	1.0142	1.0024
12.0	1.0539	1.0515	1.0464	1.0400	1.0301	1.0181
16.0	1.0721	1.0691	1.0633	1.0565	1.0462	1.0342
20.0	1.0905	1.0870	1.0806	1.0734	1.0627	1.0506
24.0	1.1090	1.1051	1.0982	1.0907	1.0796	1.0673
28.0	1.1277	1.1234	1.1161	1.1082	1.0968	1.0844
30.0	1.1371	1.1327	1.1252	1.1171	1.1055	1.0931
40.0	1.1862	1.1810	1.1727	1.1640	1.1515	1.1385
50.0	1.2380	1.2320	1.2229	1.2136	1.2006	1.1868

Table 59. Ammonium Sulfate [$(NH_4)_2SO_4$]

%	0°C.	20°C.	40°C.	80°C.	100°C.
1	1.0061	1.0041	0.9980	0.9777	0.9644
2	1.0124	1.0101	1.0039	.9836	.9705
4	1.0248	1.0220	1.0155	.9953	.9826
8	1.0495	1.0456	1.0387	1.0187	1.0066
12	1.0740	1.0691	1.0619	1.0421	1.0303
16	1.0980	1.0924	1.0849	1.0653	1.0539
20	1.1215	1.1154	1.1077	1.0883	1.0772
24	1.1448	1.1383	1.1304	1.1111	1.1003
28	1.1677	1.1609	1.1529	1.1338	1.1232
35	1.2072	1.2800	1.1919	1.1731	1.1629
40	1.2350	1.2277	1.2196	1.2011	1.1910
50	1.2899	1.2825	1.2745	1.2568	1.2466

Table 60. Arsenic Acid ($H_3A_3O_4$)

%	d_4^{15}	%	d_4^{15}
1	1.0057	20	1.1447
2	1.0124	30	1.2331
6	1.0398	40	1.3370
10	1.0681	50	1.4602
16	1.1128	60	1.6070
		70	1.7811

Table 61. Barium Chloride ($BaCl_2$)

%	0°C.	20°C.	40°C.	60°C.	80°C.	100°C.
2	1.0181	1.0159	1.0096	1.0004	0.9890	0.9755
4	1.0368	1.0341	1.0275	1.0181	1.0066	.9931
8	1.0760	1.0721	1.0648	1.0551	1.0434	1.0299
12	1.1178	1.1128	1.1047	1.0948	1.0827	1.0692
16	1.1627	1.1564	1.1478	1.1373	1.1249	1.1113
20	1.2105	1.2031	1.1938	1.1828	1.1702	1.1563
24	......	1.2531	1.2430	1.2316	1.2186	1.2045
26	......	1.2793	1.2688	1.2571	1.2440	1.2298

Table 62. Cadmium Nitrate [$Cd(NO_3)_2$]

%	d_4^{18}	%	d_4^{18}
2	1.0154	20	1.1904
4	1.0326	25	1.2488
8	1.0683	30	1.3124
12	1.1061	40	1.4590
16	1.1468	50	1.6356

Table 63. Calcium Chloride ($CaCl_2$)

%	−5°C.	0°C.	20°C.	30°C.	40°C.	60°C.	80°C.	100°C.	120°C.*	140°C.
2	......	1.0171	1.0148	1.0120	1.0084	0.9994	0.9881	0.9748	0.9596	0.9428
4	......	1.0346	1.0316	1.0286	1.0249	1.0158	1.0046	.9915	.9765	.9601
8	1.0708	1.0703	1.0659	1.0626	1.0586	1.0492	1.0382	1.0257	1.0111	.9954
12	1.1083	1.1072	1.1015	1.0978	1.0937	1.0840	1.0730	1.0610	1.0466	1.0317
16	1.1471	1.1454	1.1386	1.1345	1.1301	1.1202	1.1092	1.0973	1.0835	1.0691
20	1.1874	1.1853	1.1775	1.1730	1.1684	1.1581	1.1471	1.1352	1.1219	1.1080
25	......	1.2376	1.2284	1.2236	1.2186	1.2079	1.1965	1.1846		
30	......	1.2922	1.2816	1.2764	1.2709	1.2597	1.2478	1.2359		
35	......	......	1.3373	1.3316	1.3255	1.3137	1.3013	1.2893		
40	......	......	1.3957	1.3895	1.3826	1.3700	1.3571	1.3450		

* Corrected to atmospheric pressure.

Table 64. Calcium Hydroxide [$Ca(OH)_2$]

%	d_4^{15}	d_4^{25}
0.05	0.99979	0.99773
.10	1.00044	.99838
.15	1.00110	.99904

Table 65. Calcium Hypochlorite* ($CaOCl_2$)

% total salt	d_4^{15}
2	1.0169
4	1.0345
6	1.0520
8	1.0697
10	1.0876
12	1.1060

* $CaOCl_2$ = 89.15%;
$CaCl_2$ = 7.31%;
$Ca(ClO_3)_2$ = 0.26%;
$Ca(OH)_2$ = 2.92%.

Table 66. Calcium Nitrate [$Ca(NO_3)_2$]

%	6°C.	18°C.	25°C.	30°C.
2*	1.0157	1.0137	1.0120	1.0105
4	1.0316	1.0291	1.0272	1.0256
8	1.0641	1.0608	1.0585	1.0565
12	1.0979	1.0937	1.0911	1.0887
16	1.1330	1.1279	1.1250	1.1224
20	1.1694	1.1636	1.1602	1.1575
25	1.2168	1.2106	1.2065	1.2032
30	......	1.260		
35	......	1.311		
40	......	1.365		
45	......	1.422		
68*	......	1.747	1.741	1.736

* Supercooled tetrahydrate (m.p. 41.4°C.).

Table 67. Chromic Acid (CrO_3)

%	d_4^{15}	%	d_4^{15}
1	1.006	20	1.163
2	1.014	26	1.220
6	1.045	30	1.260
10	1.076	40	1.371
16	1.127	50	1.505
		60	1.663

Table 68. Chromium Chloride ($CrCl_3$)

%	d_4^{18} Violet	d_4^{18} Green	d_4^{18} Equilibrium mixture of violet and green
1	1.0076	1.0071	1.0075
2	1.0166	1.0157	1.0165
4	1.0349	1.0332	1.0347
8	1.0724	1.0691	1.0722
12	1.1114	1.1065	1.1111
14	1.1316		

Table 69. Copper Nitrate [$Cu(NO_3)_2$]

%	d_4^{20}	%	d_4^{20}
1	1.007	12	1.107
2	1.015	16	1.147
4	1.032	20	1.189
8	1.069	25	1.248

Table 70. Copper Sulfate ($CuSO_4$)

%	0°C.	20°C.	40°C
1	1.0104	1.0086	1.0024
4	1.0429	1.0401	1.0332
8	1.0887	1.084	1.0764
12	1.1379	1.1308	1.1222
16	......	1.180	
18	......	1.206	

Table 71. Cuprous Chloride (Cu_2Cl_2)

%	0°C.	20°C.	40°C.
1	1.0095	1.0072	1.002
4	1.0387	1.036	1.0305
8	1.0788	1.0754	1.0682
12	1.1208	1.1165	1.107
16	1.1653	1.1595	1.151
20	1.2121	1.2052	1.1953

Table 72. Ferric Chloride ($FeCl_3$)

%	0°C.	10°C.	20°C.	30°C.
1	1.0086	1.0084	1.0068	1.0040
2	1.0174	1.0168	1.0152	1.0122
4	1.0347	1.0341	1.0324	1.0292
8	1.0703	1.0692	1.0669	1.0636
12	1.1088	1.1071	1.1040	1.1006
16	1.1475	1.1449	1.1418	1.1386
20	1.1870	1.1847	1.1820	1.1786
25	1.2400	1.2380	1.2340	1.2290
30	1.2970	1.2950	1.2910	1.2850
35	1.3605	1.3580	1.3530	1.3475
40	1.4280	1.4235	1.4175	1.4115
45	......	1.4920	1.4850	
50	......	1.5610	1.5510	

Table 73. Ferric Sulfate [$Fe_2(SO_4)_3$]

%	$d_4^{17.5}$
1	1.0072
2	1.0157
4	1.0327
8	1.0670
12	1.1028
16	1.1409
20	1.1811
30	1.3073
40	1.4487
50	1.6127
60	1.7983

Table 74. Ferric Nitrate [$Fe(NO_3)_3$]

%	d_4^{18}
1	1.0065
2	1.0144
4	1.0304
8	1.0636
12	1.0989
16	1.1359
20	1.1748
25	1.2281

Table 75. Ferrous Sulfate ($FeSO_4$)

%	15°C.	18°C.	20°C.
0.2	......	1.00068	1.0002
0.4	......	1.00275	1.0022
0.8	......	1.00645	1.0062
1.0	1.0090	1.0085	1.0082
4.0	1.0380	1.0375	
8.0	1.0790	1.0785	
12.0	1.1235	1.1220	
16.0	1.1690	1.1675	
20.0	1.2150	1.2135	

Table 76. Hydrogen Bromide (HBr)

%	d_4^4	d_4^{10}	d_4^{25}
1.0	1.0073	1.0068	1.0041
2.0	1.0146	1.0139	1.0111
4.0	1.0295	1.0285	1.0255
6.0	1.0448	1.0435	1.0402
8.0	1.0604	1.0589	1.0552
10.0	1.0764	1.0747	1.0707
12.0	1.0928	1.0910	1.0867
14.0	1.1097	1.1078	1.1032
16.0	1.1272	1.1251	1.1202
18.0	1.1453	1.1430	1.1377
20.0	1.1640	1.1615	1.1557
22.0	1.1832	1.1806	1.1743
24.0	1.2030	1.2003	1.1935
26.0	1.2235	1.2206	1.2134
28.0	1.2446	1.2415	1.2340
30.0	1.2663	1.2630	1.2552
40.0	1.3877	1.3838	1.3736
50.0	1.5305	1.5257	1.5127
60.0	1.6950	1.6892	1.6731
65.0	1.7854	1.7792	1.7613

Table 77. Hydrogen Cyanide (HCN)

%	d_4^{15}
1	0.998
2	.996
4	.993
8	.984
12	.971
16	.956
82	.752
90	.724
100	.691

Table 78. Hydrogen Chloride (HCl)

%	−5°C.	0°C.	10°C.	20°C.	40°C.	60°C.	80°C.	100°C.
1	1.0048	1.0052	1.0048	1.0032	0.9970	0.9881	0.9768	0.9636
2	1.0104	1.0106	1.0100	1.0082	1.0019	.9930	0.9819	.9688
4	1.0213	1.0213	1.0202	1.0181	1.0116	1.0026	0.9919	.9791
6	1.0321	1.0319	1.0303	1.0279	1.0211	1.0121	1.0016	.9892
8	1.0428	1.0423	1.0403	1.0376	1.0305	1.0215	1.0111	.9992
10	1.0536	1.0528	1.0504	1.0474	1.0400	1.0310	1.0206	1.0090
12	1.0645	1.0634	1.0607	1.0574	1.0497	1.0406	1.0302	1.0188
14	1.0754	1.0741	1.0711	1.0675	1.0594	1.0502	1.0398	1.0286
16	1.0864	1.0849	1.0815	1.0776	1.0692	1.0598	1.0494	1.0383
18	1.0975	1.0958	1.0920	1.0878	1.0790	1.0694	1.0590	1.0479
20	1.1087	1.1067	1.1025	1.0980	1.0888	1.0790	1.0685	1.0574
22	1.1200	1.1177	1.1131	1.1083	1.0986	1.0886	1.0780	1.0668
24	1.1314	1.1287	1.1238	1.1187	1.1085	1.0982	1.0874	1.0761
26	1.1426	1.1396	1.1344	1.1290	1.1183	1.1076	1.0967	1.0853
28	1.1537	1.1505	1.1449	1.1392	1.1280	1.1169	1.1058	1.0942
30	1.1648	1.1613	1.1553	1.1493	1.1376	1.1260	1.1149	1.1030
32	......	......	......	1.1593				
34	......	......	......	1.1691				
36	......	......	......	1.1789				
38	......	......	......	1.1885				
40	......	......	......	1.1980				

Table 79. Hydrogen Fluoride (HF)

%	d_4^{20}	d_4^0
5	1.020	1.017
10	1.040	1.035
20	1.080	1.070
30	1.119	1.101
40	1.159	1.130
50	1.198	1.155
60	1.235	
70	1.258	
80	1.259	
90	1.178	
95	1.089	
100	1.0005	

Table 80. Hydrogen Peroxide (H_2O_2)

%	d_4^{18}	%	d_4^{18}
1	1.0022	26	1.0959
2	1.0058	28	1.1040
4	1.0131	30	1.1122
6	1.0204	35	1.1327
8	1.0277	40	1.1536
10	1.0351	45	1.1749
12	1.0425	50	1.1966
14	1.0499	55	1.2188
16	1.0574	60	1.2416
18	1.0649	70	1.2897
20	1.0725	80	1.3406
22	1.0802	90	1.3931
24	1.0880	100	1.4465

Table 81. Hydrofluosilicic Acid (H_2SiF_6)

%	$d_4^{17.5}$	%	$d_4^{17.5}$
1	1.0080	16	1.1373
2	1.0161	20	1.1748
4	1.0324	25	1.2235
8	1.0661	30	1.2742
12	1.1011	34	1.3162

Table 82. Magnesium Chloride ($MgCl_2$)

%	0°C.	20°C.	40°C.	60°C.	80°C.	100°C.
2	1.0168	1.0146	1.0084	0.9995	0.9883	0.9753
4	1.0338	1.0311	1.0248	1.0159	1.0050	.9923
8	1.0683	1.0646	1.0580	1.0493	1.0388	1.0269
12	1.1035	1.0989	1.0921	1.0836	1.0735	1.0622
16	1.1395	1.1342	1.1272	1.1188	1.1092	1.0984
20	1.1764	1.1706	1.1635	1.1552	1.1460	1.1359
25	1.2246	1.2184	1.2111	1.2031	1.1942	1.1847
30	1.2754	1.2688	1.2614	1.2535	1.2451	1.2360

Table 83. Magnesium Sulfate ($MgSO_4$)

%	0°C.	20°C.	30°C.	40°C.	50°C.	60°C.	80°C.
2	1.0210	1.0186	1.0158	1.0123	1.0081	1.0032	0.9916
4	1.0423	1.0392	1.0362	1.0326	1.0283	1.0234	1.0118
8	1.0858	1.0816	1.0782	1.0743	1.0700	1.0650	1.0534
12	1.1309	1.1256	1.1220	1.1179	1.1135	1.1083	1.0968
16	1.1777	1.1717	1.1679	1.1637	1.1592		
20	1.2264	1.2198	1.2159	1.2117	1.2072		
26	1.3032	1.2961	1.2922	1.2879	1.2836		

Table 84. Nickel Chloride ($NiCl_2$)

%	d_4^{13}
1	1.0082
2	1.0179
4	1.0375
8	1.0785
12	1.1217
16	1.1674
20	1.2163
30	1.353

Table 85. Nickel Nitrate [$Ni(NO_3)_2$]

%	d_4^{20}
1	1.0065
2	1.0150
4	1.0325
8	1.0688
12	1.1070
16	1.1480
20	1.191
30	1.311
35	1.377

Table 86. Nickel Sulfate ($NiSO_4$)

%	d_4^{18}
1	1.0091
2	1.0198
4	1.0415
8	1.0852
12	1.1325
16	1.1825
18	1.2090

Table 87. Nitric Acid (HNO_3)

%	0°C.	5°C.	10°C.	15°C.	20°C.	25°C.	30°C.	40°C.	50°C.	60°C.	80°C.	100°C.
1	1.0058	1.00572	1.00534	1.00464	1.00364	1.00241	1.0009	0.9973	0.9931	0.9882	0.9767	0.9632
2	1.0117	1.01149	1.01099	1.01018	1.00909	1.00778	1.0061	1.0025	.9982	.9932	.9816	.9681
3	1.0176	1.01730	1.01668	1.01576	1.01457	1.01318	1.0114	1.0077	1.0033	.9982	.9865	.9730
4	1.0236	1.02315	1.02240	1.02137	1.02008	1.01861	1.0168	1.0129	1.0084	1.0033	.9915	.9779
5	1.0296	1.02904	1.02816	1.02702	1.02563	1.02408	1.0222	1.0182	1.0136	1.0084	.9965	.9829
6	1.0357	1.03497	1.03397	1.03272	1.03122	1.02958	1.0277	1.0235	1.0188	1.0136	1.0015	.9879
7	1.0418	1.0410	1.0399	1.0385	1.0369	1.0352	1.0333	1.0289	1.0241	1.0188	1.0066	.9929
8	1.0480	1.0471	1.0458	1.0443	1.0427	1.0409	1.0389	1.0344	1.0295	1.0241	1.0117	.9980
9	1.0543	1.0532	1.0518	1.0502	1.0485	1.0466	1.0446	1.0399	1.0349	1.0294	1.0169	1.0032
10	1.0606	1.0594	1.0578	1.0561	1.0543	1.0523	1.0503	1.0455	1.0403	1.0347	1.0221	1.0083
11	1.0669	1.0656	1.0639	1.0621	1.0602	1.0581	1.0560	1.0511	1.0458	1.0401	1.0273	1.0134
12	1.0733	1.0718	1.0700	1.0681	1.0661	1.0640	1.0618	1.0567	1.0513	1.0455	1.0326	1.0186
13	1.0797	1.0781	1.0762	1.0742	1.0721	1.0699	1.0676	1.0624	1.0568	1.0509	1.0379	1.0238
14	1.0862	1.0845	1.0824	1.0803	1.0781	1.0758	1.0735	1.0681	1.0624	1.0564	1.0432	1.0289
15	1.0927	1.0909	1.0887	1.0865	1.0842	1.0818	1.0794	1.0739	1.0680	1.0619	1.0485	1.0341
16	1.0992	1.0973	1.0950	1.0927	1.0903	1.0879	1.0854	1.0797	1.0737	1.0675	1.0538	1.0393
17	1.1057	1.1038	1.1014	1.0989	1.0964	1.0940	1.0914	1.0855	1.0794	1.0731	1.0592	1.0444
18	1.1123	1.1103	1.1078	1.1052	1.1026	1.1001	1.0974	1.0913	1.0851	1.0787	1.0646	1.0496
19	1.1189	1.1168	1.1142	1.1115	1.1088	1.1062	1.1034	1.0972	1.0908	1.0843	1.0700	1.0547
20	1.1255	1.1234	1.1206	1.1178	1.1150	1.1123	1.1094	1.1031	1.0966	1.0899	1.0754	1.0598
21	1.1322	1.1300	1.1271	1.1242	1.1213	1.1185	1.1155	1.1090	1.1024	1.0956	1.0808	1.0650
22	1.1389	1.1366	1.1336	1.1306	1.1276	1.1247	1.1217	1.1150	1.1083	1.1013	1.0862	1.0701
23	1.1457	1.1433	1.1402	1.1371	1.1340	1.1310	1.1280	1.1210	1.1142	1.1070	1.0917	1.0753
24	1.1525	1.1501	1.1469	1.1437	1.1404	1.1374	1.1343	1.1271	1.1201	1.1127	1.0972	1.0805
25	1.1594	1.1569	1.1536	1.1503	1.1469	1.1438	1.1406	1.1332	1.1260	1.1185	1.1027	1.0857
26	1.1663	1.1638	1.1603	1.1569	1.1534	1.1502	1.1469	1.1394	1.1320	1.1244	1.1083	1.0910
27	1.1733	1.1707	1.1670	1.1635	1.1600	1.1566	1.1533	1.1456	1.1381	1.1303	1.1139	1.0963
28	1.1803	1.1777	1.1738	1.1702	1.1666	1.1631	1.1597	1.1519	1.1442	1.1362	1.1195	1.1016
29	1.1874	1.1847	1.1807	1.1770	1.1733	1.1697	1.1662	1.1582	1.1503	1.1422	1.1251	1.1069
30	1.1945	1.1917	1.1876	1.1838	1.1800	1.1763	1.1727	1.1645	1.1564	1.1482	1.1307	1.1122
31	1.2016	1.1988	1.1945	1.1906	1.1867	1.1829	1.1792	1.1708	1.1625	1.1542	1.1363	1.1175
32	1.2088	1.2059	1.2014	1.1974	1.1934	1.1896	1.1857	1.1772	1.1687	1.1602	1.1419	1.1228
33	1.2160	1.2131	1.2084	1.2043	1.2002	1.1963	1.1922	1.1836	1.1749	1.1662	1.1476	1.1281
34	1.2233	1.2203	1.2155	1.2113	1.2071	1.2030	1.1988	1.1901	1.1812	1.1723	1.1533	1.1335
35	1.2306	1.2275	1.2227	1.2183	1.2140	1.2098	1.2055	1.1966	1.1876	1.1784	1.1591	1.1390
36	1.2375	1.2344	1.2294	1.2249	1.2205	1.2163	1.2119	1.2028	1.1936	1.1842	1.1645	1.1440
37	1.2444	1.2412	1.2361	1.2315	1.2270	1.2227	1.2182	1.2089	1.1995	1.1899	1.1699	1.1490
38	1.2513	1.2479	1.2428	1.2381	1.2335	1.2291	1.2245	1.2150	1.2054	1.1956	1.1752	1.1540
39	1.2581	1.2546	1.2494	1.2446	1.2399	1.2354	1.2308	1.2210	1.2112	1.2013	1.1805	1.1589
40	1.2649	1.2613	1.2560	1.2511	1.2463	1.2417	1.2370	1.2270	1.2170	1.2069	1.1858	1.1638
41	1.2717	1.2680	1.2626	1.2576	1.2527	1.2480	1.2432	1.2330	1.2229	1.2126	1.1911	1.1687
42	1.2786	1.2747	1.2692	1.2641	1.2591	1.2543	1.2494	1.2390	1.2287	1.2182	1.1963	1.1735
43	1.2854	1.2814	1.2758	1.2706	1.2655	1.2606	1.2556	1.2450	1.2345	1.2238	1.2015	1.1783
44	1.2922	1.2880	1.2824	1.2771	1.2719	1.2669	1.2618	1.2510	1.2403	1.2294	1.2067	1.1831
45	1.2990	1.2947	1.2890	1.2836	1.2783	1.2732	1.2680	1.2570	1.2461	1.2350	1.2119	1.1879
46	1.3058	1.3014	1.2955	1.2901	1.2847	1.2795	1.2742	1.2630	1.2519	1.2406	1.2171	1.1927
47	1.3126	1.3080	1.3021	1.2966	1.2911	1.2858	1.2804	1.2690	1.2577	1.2462	1.2223	1.1976
48	1.3194	1.3147	1.3087	1.3031	1.2975	1.2921	1.2867	1.2750	1.2635	1.2518	1.2275	1.2024
49	1.3263	1.3214	1.3153	1.3096	1.3040	1.2984	1.2929	1.2811	1.2693	1.2575	1.2328	1.2073
50	1.3327	1.3277	1.3215	1.3157	1.3100	1.3043	1.2987	1.2867	1.2748	1.2628	1.2377	1.2118
51	1.3391	1.3339	1.3277	1.3218	1.3160	1.3102	1.3045	1.2923	1.2802	1.2680	1.2425	1.2163
52	1.3454	1.3401	1.3338	1.3278	1.3219	1.3160	1.3102	1.2978	1.2856	1.2731	1.2473	1.2208
53	1.3517	1.3462	1.3399	1.3338	1.3278	1.3218	1.3159	1.3033	1.2909	1.2782	1.2521	1.2252
54	1.3579	1.3523	1.3459	1.3397	1.3336	1.3275	1.3215	1.3087	1.2961	1.2833	1.2568	1.2296
55	1.3640	1.3583	1.3518	1.3455	1.3393	1.3331	1.3270	1.3141	1.3013	1.2883	1.2615	1.2339
56	1.3700	1.3642	1.3576	1.3512	1.3449	1.3386	1.3324	1.3194	1.3064	1.2932	1.2661	1.2382
57	1.3759	1.3700	1.3634	1.3569	1.3505	1.3441	1.3377	1.3246	1.3114	1.2981	1.2706	1.2424
58	1.3818	1.3757	1.3691	1.3625	1.3560	1.3495	1.3430	1.3298	1.3164	1.3029	1.2751	1.2466
59	1.3875	1.3813	1.3747	1.3680	1.3614	1.3548	1.3482	1.3348	1.3213	1.3077	1.2795	1.2507
60	1.3931	1.3868	1.3801	1.3734	1.3667	1.3600	1.3533	1.3398	1.3261	1.3124	1.2839	1.2547
61	1.3986	1.3922	1.3855	1.3787	1.3719	1.3651	1.3583	1.3447	1.3308	1.3169	1.2881	1.2587
62	1.4039	1.3975	1.3907	1.3838	1.3769	1.3700	1.3632	1.3494	1.3354	1.3213	1.2922	1.2625
63	1.4091	1.4027	1.3958	1.3888	1.3818	1.3748	1.3679	1.3540	1.3398	1.3255	1.2962	1.2661
64	......	1.4078	1.4007	1.3936	1.3866	1.3795	1.3725					
65	......	1.4128	1.4055	1.3984	1.3913	1.3841	1.3770					
66	......	1.4177	1.4103	1.4031	1.3959	1.3887	1.3814					
67	......	1.4224	1.4150	1.4077	1.4004	1.3932	1.3857					
68	......	1.4271	1.4196	1.4122	1.4048	1.3976	1.3900					
69	......	1.4317	1.4241	1.4166	1.4091	1.4019	1.3942					
70	......	1.4362	1.4285	1.4210	1.4134	1.4061	1.3983					
71	......	1.4406	1.4328	1.4252	1.4176	1.4102	1.4023					
72	......	1.4449	1.4371	1.4294	1.4218	1.4142	1.4063					
73	......	1.4491	1.4413	1.4335	1.4258	1.4182	1.4103					
74	......	1.4532	1.4454	1.4376	1.4298	1.4221	1.4142					

Table 87. Nitric Acid (HNO_3)—(*Concluded*)

%	0°C.	5°C.	10°C.	15°C.	20°C.	25°C.	30°C.	40°C.	50°C.	60°C.	80°C.	100°C.
75	......	1.4573	1.4494	1.4415	1.4337	1.4259	1.4180					
76	......	1.4613	1.4533	1.4454	1.4375	1.4296	1.4217					
77	......	1.4652	1.4572	1.4492	1.4413	1.4333	1.4253					
78	......	1.4690	1.4610	1.4529	1.4450	1.4369	1.4288					
79	......	1.4727	1.4647	1.4565	1.4486	1.4404	1.4323					
80	......	1.4764	1.4683	1.4601	1.4521	1.4439	1.4357					
81	......	1.4800	1.4718	1.4636	1.4555	1.4473	1.4391					
82	......	1.4835	1.4753	1.4670	1.4589	1.4507	1.4424					
83	......	1.4869	1.4787	1.4704	1.4622	1.4540	1.4456					
84	......	1.4903	1.4820	1.4737	1.4655	1.4572	1.4487					
85	......	1.4936	1.4852	1.4769	1.4686	1.4603	1.4518					
86	......	1.4968	1.4883	1.4799	1.4716	1.4633	1.4548					
87	......	1.4999	1.4913	1.4829	1.4745	1.4662	1.4577					
88	......	1.5029	1.4942	1.4858	1.4773	1.4690	1.4605					
89	......	1.5058	1.4970	1.4885	1.4800	1.4716	1.4631					
90	......	1.5085	1.4997	1.4911	1.4826	1.4741	1.4656					
91	......	1.5111	1.5023	1.4936	1.4850	1.4766	1.4681					
92	......	1.5136	1.5048	1.4960	1.4873	1.4789	1.4704					
93	......	1.5156	1.5068	1.4979	1.4892	1.4807	1.4722					
94	......	1.5177	1.5088	1.4999	1.4912	1.4826	1.4741					
95	......	1.5198	1.5109	1.5019	1.4932	1.4846	1.4761					
96	......	1.5220	1.5130	1.5040	1.4952	1.4867	1.4781					
97	......	1.5244	1.5152	1.5062	1.4974	1.4889	1.4802					
98	......	1.5278	1.5187	1.5096	1.5008	1.4922	1.4835					
99	......	1.5327	1.5235	1.5144	1.5056	1.4969	1.4881					
100	......	1.5402	1.5310	1.5217	1.5129	1.5040	1.4952					

Table 88. Perchloric Acid ($HClO_4$)

%	d_4^{15}	d_4^{20}	d_4^{25}	d_4^{50}	%	d_4^{15}	d_4^{20}	d_4^{50}
1	1.0050	......	1.0020	0.9933	28	1.1900	1.1851	1.1645
2	1.0109	......	1.0070	0.9986	30	1.2067	1.2013	1.1800
4	1.0228	......	1.0169	0.9906	32	1.2239	1.2183	1.1960
6	1.0348	......	1.0270	1.0205	34	1.2418	1.2359	1.2130
8	1.0471	......	1.0372	1.0320	36	1.2603	1.2542	1.2310
10	1.0597	......	1.0475	1.0440	38	1.2794	1.2732	1.2490
12	1.0726	......	......	1.0560	40	1.2991	1.2927	1.2680
14	1.0589	......	......	1.0680	45	1.3521	1.3450	1.3180
16	1.0995	......	......	1.0810	50	1.4103	1.4018	1.3730
18	1.1135	......	......	1.0940	55	1.4733	1.4636	1.4320
20	1.1279	......	......	1.1070	60	1.5389	1.5298	1.4950
22	1.1428	......	......	1.1205	65	1.6059	1.5986	1.5620
24	1.1581	......	......	1.1345	70	1.6736	1.6680	1.6290
26	1.1738	1.1697	......	1.1490				

Table 89. Phosphoric Acid (H_3PO_4)

°C.	2%	6%	14%	20%	26%	35%	50%	75%	100%
0	1.0113	1.0339	1.0811	1.1192					
10	1.0109	1.0330	1.0792	1.1167	1.1567	1.221	1.341		
20	1.0092	1.0309	1.0764	1.1134	1.1529	1.216	1.335	1.579	1.870
30	1.0065	1.0279	1.0728	1.1094	1.1484	1.211	1.329	1.572	1.862
40	1.0029	1.0241	1.0685	1.1048					

Table 90. Potassium Bicarbonate ($KHCO_3$)

°C.	1%	2%	4%	6%	8%	10%
0	1.0066	1.0134	1.0270			
10	1.0064	1.0132	1.0268			
15	1.0058	1.0125	1.0260	1.0396	1.0534	1.0674
20	1.0049	1.0117	1.0252			
30	1.0024	1.0092	1.0228			
40	0.9990	1.0058	1.0195			
50	.9949	1.0017	1.0154			
60	.9901	0.9969	1.0106			
80	.9786	.9855	0.9993			
100	.9653	.9722	9860			

Table 91. Potassium Bromide (KBr)

%	d_4^{20}
1	1.0054
2	1.0127
6	1.0426
12	1.0903
20	1.1601
30	1.2593
40	1.3746

Table 92. Potassium Carbonate (K_2CO_3)

%	0°C.	10°C.	20°C.	40°C.	60°C.	80°C.	100°C.
1	1.0094	1.0089	1.0072	1.0010	0.9919	0.9803	0.9670
2	1.0189	1.0182	1.0163	1.0098	1.0005	.9889	.9756
4	1.0381	1.0369	1.0345	1.0276	1.0180	1.0063	.9931
8	1.0768	1.0746	1.0715	1.0640	1.0538	1.0418	1.0291
12	1.1160	1.1131	1.1096	1.1013	1.0906	1.0786	1.0663
16	1.1562	1.1530	1.1490	1.1399	1.1290	1.1170	1.1049
20	1.1977	1.1941	1.1898	1.1801	1.1690	1.1570	1.1451
24	1.2405	1.2366	1.2320	1.2219	1.2106	1.1986	1.1869
28	1.2846	1.2804	1.2756	1.2652	1.2538	1.2418	1.2301
30	1.3071	1.3028	1.2979	1.2873	1.2759	1.2640	1.2522
35	1.3646	1.3600	1.3548	1.3440	1.3324	1.3206	1.3089
40	1.4244	1.4195	1.4141	1.4029	1.3913	1.3795	1.3678
45	1.4867	1.4815	1.4759	1.4644	1.4528	1.4408	1.4290
50	1.5517	1.5462	1.5404	1.5285	1.5169	1.5048	1.4928

Table 93. Potassium Chromate (K_2CrO_4)

%	d_4^{15}	d_4^{18}
1	1.0073	1.0066
2	1.0155	1.0147
4	1.0321	1.0311
8	1.0659	1.0647
12	1.1009	1.0999
16	......	1.1366
20	......	1.1748
24	......	1.2147
28	......	1.2566
30	......	1.2784

Table 94. Potassium Chlorate ($KClO_3$)

°C.	1%	2%	3%	4%
0	1.0061	1.0124	1.0189	1.0256
10	1.0059	1.0122	1.0187	1.0254
20	1.0045	1.0109	1.0174	1.0241
30	1.0020	1.0085	1.0151	1.0218
40	0.9986	1.0051	1.0116	1.0183
60	.9895	0.9959	1.0024	1.0091
80	.9781	.9845	0.9910	0.9977
100	.9646	.9709	.9774	.9840

Table 95. Potassium Chloride (KCl)

%	0°C.	20°C.	25°C.	40°C.	60°C.	80°C.	100°C.
1.0	1.00661	1.00462	1.00342	0.99847	0.9894	0.9780	0.9646
2.0	1.01335	1.01103	1.00977	1.00471	.9956	.9842	.9708
4.0	1.02690	1.02391	1.02255	1.01727	1.0080	.9966	.9834
8.0	1.05431	1.05003	1.04847	1.04278	1.0333	1.0219	1.0088
12.0	1.08222	1.07679	1.07506	1.06897	1.0592	1.0478	1.0350
16.0	1.11068	1.10434	1.10245	1.09600	1.0861	1.0746	1.0619
20.0	1.13973	1.13280	1.13072	1.12399	1.1138	1.1024	1.0897
24.0	....	1.16226	1.15995	1.15299	1.1425	1.1311	1.1185
28.0	....	....	....	1.18304	1.1723	1.1609	1.1483

%	110°C.	120°C.	130°C.	140°C.
3.79	0.9733	0.9663	0.9583	0.9502
7.45	.9978	.9899	.9827	.9745
13.62	1.0388	1.0313	1.0238	1.0159

Table 96. Potassium Chrome Alum [$K_2Cr_2(SO_4)_4$]

%	d_4^{15}
1	1.007
2	1.016
6	1.052
10	1.089
14	1.129
20	1.193
30	1.315
40	1.456
50	1.615

Table 97. Potassium Hydroxide (KOH)

%	d_4^{15}
1.0	1.0083
2.0	1.0175
4.0	1.0359
6.0	1.0544
8.0	1.0730
10.0	1.0918
15.0	1.1396
20.0	1.1884
25.0	1.2387
30.0	1.2905
35.0	1.3440
40.0	1.3991
45.0	1.4558
50.0	1.5143
51.7	1.5355 (sat'd. soln.)

Table 98. Potassium Nitrate (KNO_3)

%	0°C.	10°C.	20°C.	40°C.	60°C.	80°C.	100°C.
1	1.00654	1.00615	1.00447	0.99825	0.9890	0.9776	0.9641
2	1.01326	1.01262	1.01075	1.00430	.9949	.9834	.9699
4	1.02677	1.02566	1.02344	1.01652	1.0068	.9951	.9816
8	1.05419	1.05226	1.04940	1.04152	1.0313	1.0192	1.0056
12	1.08221	1.07963	1.07620	1.06740	1.0567	1.0442	1.0304
16	....	....	1.10392	1.09432	1.0831	1.0703	1.0562
20	....	....	1.13261	1.12240	1.1106	1.0974	1.0831
24	....	....	1.16233	1.15175	1.1391	1.1256	1.1110

Table 99. Potassium Dichromate ($K_2Cr_2O_7$)

%	d_4^{20}
1	1.0052
2	1.0122
4	1.0264
6	1.0408
8	1.0554
10	1.0703

Table 100. Potassium Sulfate (K_2SO_4)

%	d_4^{20}
1	1.0063
2	1.0145
4	1.0310
6	1.0477
8	1.0646
10	1.0817

Table 101. Potassium Sulfite (K_2SO_3)

%	d_4^{15}
1	1.0073
2	1.0155
4	1.0322
8	1.0667
12	1.1026
16	1.1402
20	1.1793
24	1.2197
26	1.2404

Table 102. Sodium Acetate ($NaC_2H_3O_2$)

%	d_4^{20}
1	1.0033
2	1.0084
4	1.0186
8	1.0392
12	1.0598
18	1.0807
20	1.1021
26	1.1351
23	1.1462

Table 103. Sodium Arsenate (Na_3AsO_4)

%	d_4^{17}
1	1.0097
2	1.0207
4	1.0431
8	1.0892
10	1.1130
12	1.1373

Table 104. Sodium Bichromate ($Na_2Cr_2O_7$)

%	d_4^{15}
1	1.006
2	1.013
4	1.027
8	1.056
12	1.084
16	1.112
20	1.140
24	1.166
28	1.193
30	1.207
35	1.244
40	1.279
45	1.312
50	1.342

Table 105. Sodium Bromide (NaBr)

%	d_4^{17}
1	1.0060
2	1.0139
4	1.0298
8	1.0631
10	1.0803
12	1.0981
20	1.1745
30	1.2841
40	1.4138

Table 106. Sodium Formate (HCOONa)

%	d_4^{25}
1	1.003
2	1.009
4	1.022
8	1.048
12	1.074
16	1.100
20	1.127
24	1.155
28	1.184
30	1.199
35	1.236
40	1.274

Table 107. Sodium Carbonate (Na_2CO_3)

%	0°C.	10°C.	20°C.	30°C.	40°C.	60°C.	80°C.	100°C.
1	1.0109	1.0103	1.0086	1.0058	1.0022	0.9929	0.9814	0.9683
2	1.0219	1.0210	1.0190	1.0159	1.0122	1.0027	.9910	.9782
4	1.0439	1.0423	1.0398	1.0363	1.0323	1.0223	1.0105	.9980
8	1.0878	1.0850	1.0816	1.0775	1.0732	1.0625	1.0503	1.0380
12	1.1319	1.1284	1.1244	1.1200	1.1150	1.1039	1.0914	1.0787
14	1.1543	1.1506	1.1463	1.1417	1.1365	1.1251	1.1125	1.0996
16	......	......	......	1.1636				
18	......	......	......	1.1859				
20	......	......	......	1.2086				
24	......	......	......	1.2552				
28	......	......	......	1.3031				
30	......	......	......	1.3274				

Table 108. Sodium Chlorate ($NaClO_3$)

%	d_4^{18}	%	d_4^{18}
1	1.0053	18	1.1288
2	1.0121	20	1.1449
4	1.0258	22	1.1614
6	1.0397	24	1.1782
8	1.0538	26	1.1953
10	1.0681	28	1.2128
12	1.0827	30	1.2307
14	1.0977	32	1.2491
16	1.1131	34	1.2680

Table 109. Sodium Chloride (NaCl)

%	0°C.	10°C.	25°C.	40°C.	60°C.	80°C.	100°C.
1	1.00747	1.00707	1.00409	0.99908	0.9900	0.9785	0.9651
2	1.01509	1.01442	1.01112	1.00593	.9967	.9852	.9719
4	1.03038	1.02920	1.02530	1.01977	1.0103	.9988	.9855
8	1.06121	1.05907	1.05412	1.04798	1.0381	1.0264	1.0134
12	1.09244	1.08946	1.08365	1.07699	1.0667	1.0549	1.0420
16	1.12419	1.12056	1.11401	1.10688	1.0962	1.0842	1.0713
20	1.15663	1.15254	1.14533	1.13774	1.1268	1.1146	1.1017
24	1.18999	1.18557	1.17776	1.16971	1.1584	1.1463	1.1331
26	1.20709	1.20254	1.19443	1.18614	1.1747	1.1626	1.1492

Table 110. Sodium Chromate (Na_2CrO_4)

%	d_4^{18}
1	1.0074
2	1.0164
4	1.0344
8	1.0718
12	1.1110
16	1.1518
20	1.1942
24	1.2383
26	1.2611

Table 111. Sodium Hydroxide (NaOH)

%	0°C.	15°C.	20°C.	40°C.	60°C.	80°C.	100°C.
1	1.0124	1.01065	1.0095	1.0033	0.9941	0.9824	0.9693
2	1.0244	1.02198	1.0207	1.0139	1.0045	.9929	.9797
4	1.0482	1.04441	1.0428	1.0352	1.0254	1.0139	1.0009
8	1.0943	1.08887	1.0869	1.0780	1.0676	1.0560	1.0432
12	1.1399	1.13327	1.1309	1.1210	1.1101	1.0983	1.0855
16	1.1849	1.17761	1.1751	1.1645	1.1531	1.1408	1.1277
20	1.2296	1.22183	1.2191	1.2079	1.1960	1.1833	1.1700
24	1.2741	1.26582	1.2629	1.2512	1.2388	1.2259	1.2124
28	1.3182	1.3094	1.3064	1.2942	1.2814	1.2682	1.2546
32	1.3614	1.3520	1.3490	1.3362	1.3232	1.3097	1.2960
36	1.4030	1.3933	1.3900	1.3768	1.3634	1.3498	1.3360
40	1.4435	1.4334	1.4300	1.4164	1.4027	1.3889	1.3750
44	1.4825	1.4720	1.4685	1.4545	1.4405	1.4266	1.4127
48	1.5210	1.5102	1.5065	1.4922	1.4781	1.4641	1.4503
50	1.5400	1.5290	1.5253	1.5109	1.4967	1.4827	1.4690

Table 112. Sodium Nitrate ($NaNO_3$)

%	0°C.	20°C.	40°C.	60°C.	80°C.	100°C.
1	1.0071	1.0049	0.9986	0.9894	0.9779	0.9644
2	1.0144	1.0117	1.0050	.9956	.9840	.9704
4	1.0290	1.0254	1.0180	1.0082	.9964	.9826
8	1.0587	1.0532	1.0447	1.0340	1.0218	1.0078
12	1.0891	1.0819	1.0724	1.0609	1.0481	1.0340
16	1.1203	1.1118	1.1013	1.0892	1.0757	1.0614
20	1.1526	1.1429	1.1314	1.1187	1.1048	1.0901
24	1.1860	1.1752	1.1629	1.1496	1.1351	1.1200
28	1.2204	1.2085	1.1955	1.1816	1.1667	1.1513
30	1.2380	1.2256	1.2122	1.1980	1.1830	1.1674
35	1.2834	1.2701	1.2560	1.2413	1.2258	1.2100
40	1.3316	1.3175	1.3027	1.2875	1.2715	1.2555
45	......	1.3683	1.3528	1.3371	1.3206	1.3044

Table 113. Sodium Nitrite ($NaNO_2$)

%	d_4^{15}
1	1.0058
2	1.0125
4	1.0260
8	1.0535
12	1.0816
16	1.1103
20	1.1394

Table 114. Sodium Silicate

	Concentration, %												
	1	2	4	8	10	14	20	24	30	36	40	45	50
	Density$_4^{20}$												
$Na_2O/3.9SiO_2$	1.006	1.014	1.030	1.063	1.080	1.116	1.172	1.211	1.275				
$Na_2O/3.36SiO_2$	1.006	1.014	1.030	1.065	1.083	1.120	1.179	1.222	1.290	1.365			
$Na_2O/2.40SiO_2$	1.007	1.016	1.034	1.071	1.090	1.130							
$Na_2O/2.44SiO_2$	.....	.....	.....	.....	.....	.....	.....	.....	1.309	1.387	1.445		
$Na_2O/2.06SiO_2$	1.007	1.016	1.035	1.073	1.093	1.134	1.200	1.247	1.321	1.397	1.450	1.520	1.594
$Na_2O/1.69SiO_2$	1.007	1.017	1.036	1.077	1.098	1.141	1.210	1.259	1.337	1.424			

Table 115. Sodium Sulfate (Na_2SO_4)

%	0°C.	20°C.	30°C.	40°C.	60°C.	80°C.	100°C.
1	1.0094	1.0073	1.0046	1.0010	0.9919	0.9805	0.9671
2	1.0189	1.0164	1.0135	1.0098	1.0007	.9892	.9758
4	1.0381	1.0348	1.0315	1.0276	1.0184	1.0068	.9934
8	1.0773	1.0724	1.0682	1.0639	1.0544	1.0426	1.0292
12	1.1174	1.1109	1.1062	1.1015	1.0915	1.0795	1.0661
16	1.1585	1.1506	1.1456	1.1406	1.1299	1.1176	1.1042
20	1.2008	1.1915	1.1865	1.1813	1.1696	1.1569	
24	1.2443	1.2336	1.2292	1.2237			

Table 116. Sodium Sulfide (Na_2S)

%	d_4^{18}
1	1.0098
2	1.0211
4	1.0440
8	1.0907
12	1.1388
16	1.1885
18	1.2140

Table 117. Sodium Sulfite (Na_2SO_3)

%	d_4^{19}
1	1.0078
2	1.0172
4	1.0363
8	1.0751
12	1.1146
16	1.1549
18	1.1755

Table 118. Sodium Thiosulfate ($Na_2S_2O_3$)

%	d_4^{20}
1	1.0065
2	1.0148
4	1.0315
8	1.0654
12	1.1003
16	1.1365
20	1.1740
24	1.2128
28	1.2532
30	1.2739
35	1.3273
40	1.3827

Table 119. Sodium Thiosulfate Pentahydrate ($Na_2S_2O_3.5H_2O$)

%	d_4^{19}
1	1.0052
2	1.0105
4	1.0211
8	1.0423
12	1.0639
16	1.0863
20	1.1087
24	1.1322
28	1.1558
30	1.1676
40	1.2297
50	1.2954

Table 120. Stannic Chloride ($SnCl_4$)

%	d_4^{15}
1	1.007
2	1.015
4	1.031
8	1.064
12	1.099
16	1.135
20	1.173
24	1.212
28	1.255
30	1.278
35	1.337
40	1.403
45	1.475
50	1.555
55	1.644
60	1.742
65	1.851
70	1.971

Table 121. Stannous Chloride ($SnCl_2$)

%	d_4^{15}
1	1.0068
2	1.0146
4	1.0306
8	1.0638
12	1.0986
16	1.1353
20	1.1743
24	1.2159
28	1.2603
30	1.2837
35	1.3461
40	1.4145
45	1.4897
50	1.5729
55	1.6656
60	1.7695
65	1.8865

Table 122. Sulfuric Acid (H_2SO_4)

%	0°C.	10°C.	15°C.	20°C.	25°C.	30°C.	40°C.	50°C.	60°C.	80°C.	100°C.
1	1.0074	1.0068	1.0060	1.0051	1.0038	1.0022	0.9986	0.9944	0.9895	0.9779	0.9645
2	1.0147	1.0138	1.0129	1.0118	1.0104	1.0087	1.0050	1.0006	.9956	.9839	.9705
3	1.0219	1.0206	1.0197	1.0184	1.0169	1.0152	1.0113	1.0067	1.0017	.9900	.9766
4	1.0291	1.0275	1.0264	1.0250	1.0234	1.0216	1.0176	1.0129	1.0078	.9961	.9827
5	1.0364	1.0344	1.0332	1.0317	1.0300	1.0281	1.0240	1.0192	1.0140	1.0022	.9888
6	1.0437	1.0414	1.0400	1.0385	1.0367	1.0347	1.0305	1.0256	1.0203	1.0084	.9950
7	1.0511	1.0485	1.0469	1.0453	1.0434	1.0414	1.0371	1.0321	1.0266	1.0146	1.0013
8	1.0585	1.0556	1.0539	1.0522	1.0502	1.0481	1.0437	1.0386	1.0330	1.0209	1.0076
9	1.0660	1.0628	1.0610	1.0591	1.0571	1.0549	1.0503	1.0451	1.0395	1.0273	1.0140
10	1.0735	1.0700	1.0681	1.0661	1.0640	1.0617	1.0570	1.0517	1.0460	1.0338	1.0204
11	1.0810	1.0773	1.0753	1.0731	1.0710	1.0686	1.0637	1.0584	1.0526	1.0403	1.0269
12	1.0886	1.0846	1.0825	1.0802	1.0780	1.0756	1.0705	1.0651	1.0593	1.0469	1.0335
13	1.0962	1.0920	1.0898	1.0874	1.0851	1.0826	1.0774	1.0719	1.0661	1.0536	1.0402
14	1.1039	1.0994	1.0971	1.0947	1.0922	1.0897	1.0844	1.0788	1.0729	1.0603	1.0469
15	1.1116	1.1069	1.1045	1.1020	1.0994	1.0968	1.0914	1.0857	1.0798	1.0671	1.0537
16	1.1194	1.1145	1.1120	1.1094	1.1067	1.1040	1.0985	1.0927	1.0868	1.0740	1.0605
17	1.1272	1.1221	1.1195	1.1168	1.1141	1.1113	1.1057	1.0998	1.0938	1.0809	1.0674
18	1.1351	1.1298	1.1271	1.1243	1.1215	1.1187	1.1129	1.1070	1.1009	1.0879	1.0744
19	1.1430	1.1375	1.1347	1.1318	1.1290	1.1261	1.1202	1.1142	1.1081	1.0950	1.0814
20	1.1510	1.1453	1.1424	1.1394	1.1365	1.1335	1.1275	1.1215	1.1153	1.1021	1.0885
21	1.1590	1.1531	1.1501	1.1471	1.1441	1.1410	1.1349	1.1288	1.1226	1.1093	1.0957
22	1.1670	1.1609	1.1579	1.1548	1.1517	1.1486	1.1424	1.1362	1.1299	1.1166	1.1029
23	1.1751	1.1688	1.1657	1.1626	1.1594	1.1563	1.1500	1.1437	1.1373	1.1239	1.1102
24	1.1832	1.1768	1.1736	1.1704	1.1672	1.1640	1.1576	1.1512	1.1448	1.1313	1.1176
25	1.1914	1.1848	1.1816	1.1783	1.1750	1.1718	1.1653	1.1588	1.1523	1.1388	1.1250
26	1.1996	1.1929	1.1896	1.1862	1.1829	1.1796	1.1730	1.1665	1.1599	1.1463	1.1325
27	1.2078	1.2010	1.1976	1.1942	1.1909	1.1875	1.1808	1.1742	1.1676	1.1539	1.1400
28	1.2160	1.2091	1.2057	1.2023	1.1989	1.1955	1.1887	1.1820	1.1753	1.1616	1.1476
29	1.2243	1.2173	1.2138	1.2104	1.2069	1.2035	1.1966	1.1898	1.1831	1.1693	1.1553
30	1.2326	1.2255	1.2220	1.2185	1.2150	1.2115	1.2046	1.1977	1.1909	1.1771	1.1630
31	1.2409	1.2338	1.2302	1.2267	1.2232	1.2196	1.2126	1.2057	1.1988	1.1849	1.1708
32	1.2493	1.2421	1.2385	1.2349	1.2314	1.2278	1.2207	1.2137	1.2068	1.1928	1.1787
33	1.2577	1.2504	1.2468	1.2432	1.2396	1.2360	1.2289	1.2218	1.2148	1.2008	1.1866
34	1.2661	1.2588	1.2552	1.2515	1.2479	1.2443	1.2371	1.2300	1.2229	1.2088	1.1946
35	1.2746	1.2672	1.2636	1.2599	1.2563	1.2526	1.2454	1.2383	1.2311	1.2169	1.2027
36	1.2831	1.2757	1.2720	1.2684	1.2647	1.2610	1.2538	1.2466	1.2394	1.2251	1.2109
37	1.2917	1.2843	1.2805	1.2769	1.2732	1.2695	1.2622	1.2550	1.2477	1.2334	1.2192
38	1.3004	1.2929	1.2891	1.2855	1.2818	1.2780	1.2707	1.2635	1.2561	1.2418	1.2276
39	1.3091	1.3016	1.2978	1.2941	1.2904	1.2866	1.2793	1.2720	1.2646	1.2503	1.2361
40	1.3179	1.3103	1.3065	1.3028	1.2991	1.2953	1.2880	1.2806	1.2732	1.2589	1.2446
41	1.3268	1.3191	1.3153	1.3116	1.3079	1.3041	1.2967	1.2893	1.2819	1.2675	1.2532
42	1.3357	1.3280	1.3242	1.3205	1.3167	1.3129	1.3055	1.2981	1.2907	1.2762	1.2619
43	1.3447	1.3370	1.3332	1.3294	1.3256	1.3218	1.3144	1.3070	1.2996	1.2850	1.2707
44	1.3538	1.3461	1.3423	1.3384	1.3346	1.3308	1.3234	1.3160	1.3086	1.2939	1.2796
45	1.3630	1.3553	1.3515	1.3476	1.3437	1.3399	1.3325	1.3251	1.3177	1.3029	1.2886
46	1.3724	1.3646	1.3608	1.3569	1.3530	1.3492	1.3417	1.3343	1.3269	1.3120	1.2976
47	1.3819	1.3740	1.3702	1.3663	1.3624	1.3586	1.3510	1.3435	1.3362	1.3212	1.3067
48	1.3915	1.3835	1.3797	1.3758	1.3719	1.3680	1.3604	1.3528	1.3455	1.3305	1.3159
49	1.4012	1.3931	1.3893	1.3854	1.3814	1.3775	1.3699	1.3623	1.3549	1.3399	1.3253
50	1.4110	1.4029	1.3990	1.3951	1.3911	1.3872	1.3795	1.3719	1.3644	1.3494	1.3348
51	1.4209	1.4128	1.4088	1.4049	1.4009	1.3970	1.3893	1.3816	1.3740	1.3590	1.3444
52	1.4310	1.4228	1.4188	1.4148	1.4109	1.4069	1.3991	1.3914	1.3837	1.3687	1.3540
53	1.4412	1.4329	1.4289	1.4248	1.4209	1.4169	1.4091	1.4013	1.3936	1.3785	1.3637
54	1.4515	1.4431	1.4391	1.4350	1.4310	1.4270	1.4191	1.4113	1.4036	1.3884	1.3735
55	1.4619	1.4535	1.4494	1.4453	1.4412	1.4372	1.4293	1.4214	1.4137	1.3984	1.3834
56	1.4724	1.4640	1.4598	1.4557	1.4516	1.4475	1.4396	1.4317	1.4239	1.4085	1.3934
57	1.4830	1.4746	1.4703	1.4662	1.4621	1.4580	1.4500	1.4420	1.4342	1.4187	1.4035
58	1.4937	1.4852	1.4809	1.4768	1.4726	1.4685	1.4604	1.4524	1.4446	1.4290	1.4137
59	1.5045	1.4959	1.4916	1.4875	1.4832	1.4791	1.4709	1.4629	1.4551	1.4393	1.4240
60	1.5154	1.5067	1.5024	1.4983	1.4940	1.4898	1.4816	1.4735	1.4656	1.4497	1.4344
61	1.5264	1.5177	1.5133	1.5091	1.5048	1.5006	1.4923	1.4842	1.4762	1.4602	1.4449
62	1.5375	1.5287	1.5243	1.5200	1.5157	1.5115	1.5031	1.4950	1.4869	1.4708	1.4554
63	1.5487	1.5398	1.5354	1.5310	1.5267	1.5225	1.5140	1.5058	1.4977	1.4815	1.4660
64	1.5600	1.5510	1.5465	1.5421	1.5378	1.5335	1.5250	1.5167	1.5086	1.4923	1.4766
65	1.5714	1.5623	1.5578	1.5533	1.5490	1.5446	1.5361	1.5277	1.5195	1.5031	1.4873
66	1.5828	1.5736	1.5691	1.5646	1.5602	1.5558	1.5472	1.5388	1.5305	1.5140	1.4981
67	1.5943	1.5850	1.5805	1.5760	1.5715	1.5671	1.5584	1.5499	1.5416	1.5249	1.5089
68	1.6059	1.5965	1.5920	1.5874	1.5829	1.5785	1.5697	1.5611	1.5528	1.5359	1.5198
69	1.6176	1.6081	1.6035	1.5989	1.5944	1.5899	1.5811	1.5724	1.5640	1.5470	1.5307
70	1.6293	1.6198	1.6151	1.6105	1.6059	1.6014	1.5925	1.5838	1.5753	1.5582	1.5417
71	1.6411	1.6315	1.6268	1.6221	1.6175	1.6130	1.6040	1.5952	1.5867	1.5694	1.5527
72	1.6529	1.6433	1.6385	1.6338	1.6292	1.6246	1.6155	1.6067	1.5981	1.5806	1.5637
73	1.6648	1.6551	1.6503	1.6456	1.6409	1.6363	1.6271	1.6182	1.6095	1.5919	1.5747
74	1.6768	1.6670	1.6622	1.6574	1.6526	1.6480	1.6387	1.6297	1.6209	1.6031	1.5857
75	1.6888	1.6789	1.6740	1.6692	1.6644	1.6597	1.6503	1.6412	1.6322	1.6142	1.5966
76	1.7008	1.6908	1.6858	1.6810	1.6761	1.6713	1.6619	1.6526	1.6435	1.6252	1.6074
77	1.7128	1.7026	1.6976	1.6927	1.6878	1.6829	1.6734	1.6640	1.6547	1.6361	1.6181
78	1.7247	1.7144	1.7093	1.7043	1.6994	1.6944	1.6847	1.6751	1.6657	1.6469	1.6286
79	1.7365	1.7261	1.7209	1.7158	1.7108	1.7058	1.6959	1.6862	1.6766	1.6575	1.6390

Table 122. Sulfuric Acid (H_2SO_4)—(*Concluded*)

%	0°C.	10°C.	15°C.	20°C.	25°C.	30°C.	40°C.	50°C.	60°C.	80°C.	100°C.
80	1.7482	1.7376	1.7323	1.7272	1.7221	1.7170	1.7069	1.6971	1.6873	1.6680	1.6493
81	1.7597	1.7489	1.7435	1.7383	1.7331	1.7279	1.7177	1.7077	1.6978	1.6782	1.6594
82	1.7709	1.7599	1.7544	1.7491	1.7437	1.7385	1.7281	1.7180	1.7080	1.6882	1.6692
83	1.7815	1.7704	1.7649	1.7594	1.7540	1.7487	1.7382	1.7279	1.7179	1.6979	1.6787
84	1.7916	1.7804	1.7748	1.7693	1.7639	1.7585	1.7479	1.7375	1.7274	1.7072	1.6878
85	1.8009	1.7897	1.7841	1.7786	1.7732	1.7678	1.7571	1.7466	1.7364	1.7161	1.6966
86	1.8095	1.7983	1.7927	1.7872	1.7818	1.7763	1.7657	1.7552	1.7449	1.7245	1.7050
87	1.8173	1.8061	1.8006	1.7951	1.7897	1.7842	1.7736	1.7632	1.7529	1.7324	1.7129
88	1.8243	1.8132	1.8077	1.8022	1.7968	1.7914	1.7809	1.7705	1.7602	1.7397	1.7202
89	1.8306	1.8195	1.8141	1.8087	1.8033	1.7979	1.7874	1.7770	1.7669	1.7464	1.7269
90	1.8361	1.8252	1.8198	1.8144	1.8091	1.8038	1.7933	1.7829	1.7729	1.7525	1.7331
91	1.8410	1.8302	1.8248	1.8195	1.8142	1.8090	1.7986	1.7883	1.7783	1.7581	1.7388
92	1.8453	1.8346	1.8293	1.8240	1.8188	1.8136	1.8033	1.7932	1.7832	1.7633	1.7439
93	1.8490	1.8384	1.8331	1.8279	1.8227	1.8176	1.8074	1.7974	1.7876	1.7681	1.7485
94	1.8520	1.8415	1.8363	1.8312	1.8260	1.8210	1.8109	1.8011	1.7914		
95	1.8544	1.8439	1.8388	1.8337	1.8286	1.8236	1.8137	1.8040	1.7944		
96	1.8560	1.8457	1.8406	1.8355	1.8305	1.8255	1.8157	1.8060	1.7965		
97	1.8569	1.8466	1.8414	1.8364	1.8314	1.8264	1.8166	1.8071	1.7977		
98	1.8567	1.8463	1.8411	1.8361	1.8310	1.8261	1.8163	1.8068	1.7976		
99	1.8551	1.8445	1.8393	1.8342	1.8292	1.8242	1.8145	1.8050	1.7958		
100	1.8517	1.8409	1.8357	1.8305	1.8255	1.8205	1.8107	1.8013	1.7922		

%	$d_4^{5.96}$	%	$d_4^{13.00}$	$d_4^{18.00}$
0.005	1.000 0140	0.05	0.999 810	0.999 028
.01	1.000 0576	.1	1.000 185	.999 400
.02	1.000 1434	.2	1.000 912	1.000 119
.03	1.000 2276	.3	1.001 623	1.000 820
.04	1.000 3104	4	1.002 326	1.001 512
.05	1.000 3920	.5	1.003 023	1.002 197
.06	1.000 4726	.6	1.003 716	1.002 877
.07	1.000 5523	.8	1.005 090	1.004 227
.08	1.000 6313	1.0	1.006 452	1.005 570
.09	1.000 7098	1.2	1.007 807	1.006 909
.10	1.000 7880	1.4	1.009 159	1.008 247
.15	1.001 1732	1.6	1.010 510	1.009 583
.20	1.001 5514	1.8	1.011 860	1.010 918
.25	1.001 9254	2.0	1.013 209	1.012 252
.30	1.002 2961	2.2	1.014 557	1.013 586
.35	1.002 6639	2.4	1.015 904	1.014 919
.40	1.003 0292			
.45	1.003 3923			
.50	1.003 7534			

Table 123. Zinc Bromide ($ZnBr_2$)

%	0°C.	20°C.	40°C.	60°C.	80°C.	100°C.
2	1.0188	1.0167	1.0102	1.0008	0.9890	0.9751
4	1.0381	1.0354	1.0285	1.0187	1.0065	0.9921
8	1.0777	1.0738	1.0660	1.0554	1.0422	1.0270
12	1.1186	1.1135	1.1046	1.0932	1.0789	1.0629
16	1.1609	1.1544	1.1445	1.1320	1.1169	1.1000
20	1.2043	1.1965	1.1855	1.1720	1.1560	1.1382
30	1.3288	1.3170	1.3030	1.2868	1.2688	1.2489
40	1.477	1.462	1.445	1.427	1.406	1.385
50	1.661	1.643	1.623	1.602	1.579	1.555
60	1.891	1.869	1.845	1.822	1.797	1.771
65	2 026	2 002	1.976	1.951	1.924	1.898

Table 124. Zinc Chloride ($ZnCl_2$)

%	0°C.	20°C.	40°C.	60°C.	80°C.	100°C.
2	1.0192	1.0167	1.0099	1.0003	0.9882	0.9739
4	1.0384	1.0350	1.0274	1.0172	1.0044	.9894
8	1.0769	1.0715	1.0624	1.0508	1.0369	1.0211
12	1.1159	1.1085	1.0980	1.0853	1.0704	1.0541
16	1.1558	1.1468	1.1350	1.1212	1.1055	1.0888
20	1.1970	1.1866	1.1736	1.1590	1.1428	1.1255
30	1.3062	1.2928	1.2778	1.2614	1.2438	1.2252
40	1.4329	1.4173	1.4003	1.3824	1.3637	1.3441
50	1.5860	1.5681	1.5495	1.5300	1.5097	1.4892
60	......	1.749				
70	......	1.962				

Table 125. Zinc Nitrate [$Zn(NO_3)_2$]

%	18°C.	%	18°C.
2	1.0154	18	1.1652
4	1.0322	20	1.1865
6	1.0496	25	1.2427
8	1.0675	30	1.3029
10	1.0859	35	1.3678
12	1.1048	40	1.4378
14	1.1244	45	1.5134
16	1.1445	50	1.5944

Table 126. Zinc Sulfate ($ZnSO_4$)

%	20°C.
2	1.019
4	1.0403
6	1.0620
8	1.0842
10	1.1071
12	1.1308
14	1.1553
16	1.1806

DENSITIES OF AQUEOUS ORGANIC SOLUTIONS

From "International Critical Tables," vol. 3, pp. 115–129. All compositions are in weight per cent *in vacuo*. All density values are d_4^t = g./ml. *in vacuo*.

Table 127. Formic Acid (HCOOH)

%	0°C.	15°C.	20°C.	30°C.	%	0°C.	15°C.	20°C.	30°C.	%	0°C.	15°C.	20°C.	30°C.	%	0°C.	15°C.	20°C.	30°C.
0	0.9999	0.9991	0.9982	0.9957	25	1.0706	1.0627	1.0609	1.0540	50	1.1349	1.1225	1.1207	1.1098	75	1.1953	1.1794	1.1769	1.1636
1	1.0028	1.0019	1.0019	0.9980	26	1.0733	1.0652	1.0633	1.0564	51	1.1374	1.1248	1.1223	1.1120	76	1.1976	1.1816	1.1785	1.1656
2	1.0059	1.0045	1.0044	1.0004	27	1.0760	1.0678	1.0656	1.0587	52	1.1399	1.1271	1.1244	1.1142	77	1.1999	1.1837	1.1801	1.1676
3	1.0090	1.0072	1.0070	1.0028	28	1.0787	1.0702	1.0681	1.0609	53	1.1424	1.1294	1.1269	1.1164	78	1.2021	1.1859	1.1818	1.1697
4	1.0120	1.0100	1.0093	1.0053	29	1.0813	1.0726	1.0705	1.0632	54	1.1448	1.1318	1.1295	1.1186	79	1.2043	1.1881	1.1837	1.1717
5	1.0150	1.0124	1.0115	1.0075	30	1.0839	1.0750	1.0729	1.0654	55	1.1472	1.1341	1.1320	1.1208	80	1.2065	1.1902	1.1806	1.1737
6	1.0179	1.0151	1.0141	1.0101	31	1.0866	1.0774	1.0753	1.0676	56	1.1497	1.1365	1.1342	1.1230	81	1.2088	1.1924	1.1876	1.1758
7	1.0207	1.0177	1.0170	1.0125	32	1.0891	1.0798	1.0777	1.0699	57	1.1523	1.1388	1.1361	1.1253	82	1.2110	1.1944	1.1896	1.1778
8	1.0237	1.0204	1.0196	1.0149	33	1.0916	1.0821	1.0800	1.0721	58	1.1548	1.1411	1.1381	1.1274	83	1.2132	1.1965	1.1914	1.1798
9	1.0266	1.0230	1.0221	1.0173	34	1.0941	1.0844	1.0823	1.0743	59	1.1573	1.1434	1.1401	1.1295	84	1.2154	1.1985	1.1929	1.1817
10	1.0295	1.0256	1.0246	1.0197	35	1.0966	1.0867	1.0847	1.0766	60	1.1597	1.1458	1.1424	1.1317	85	1.2176	1.2005	1.1953	1.1837
11	1.0324	1.0281	1.0271	1.0221	36	1.0993	1.0892	1.0871	1.0788	61	1.1621	1.1481	1.1448	1.1338	86	1.2196	1.2025	1.1976	1.1856
12	1.0351	1.0306	1.0296	1.0244	37	1.1018	1.0916	1.0895	1.0810	62	1.1645	1.1504	1.1473	1.1360	87	1.2217	1.2045	1.1994	1.1875
13	1.0379	1.0330	1.0321	1.0267	38	1.1043	1.0940	1.0919	1.0832	63	1.1669	1.1526	1.1493	1.1382	88	1.2237	1.2064	1.2012	1.1893
14	1.0407	1.0355	1.0345	1.0290	39	1.1069	1.0964	1.0940	1.0854	64	1.1694	1.1549	1.1517	1.1403	89	1.2258	1.2084	1.2028	1.1910
15	1.0435	1.0380	1.0370	1.0313	40	1.1095	1.0988	1.0963	1.0876	65	1.1718	1.1572	1.1543	1.1425	90	1.2278	1.2102	1.2044	1.1927
16	1.0463	1.0405	1.0393	1.0336	41	1.1122	1.1012	1.0990	1.0898	66	1.1742	1.1595	1.1565	1.1446	91	1.2297	1.2121	1.2059	1.1945
17	1.0491	1.0430	1.0417	1.0358	42	1.1148	1.1036	1.1015	1.0920	67	1.1766	1.1618	1.1584	1.1467	92	1.2316	1.2139	1.2078	1.1961
18	1.0518	1.0455	1.0441	1.0381	43	1.1174	1.1060	1.1038	1.0943	68	1.1790	1.1640	1.1604	1.1489	93	1.2335	1.2157	1.2099	1.1978
19	1.0545	1.0480	1.0464	1.0404	44	1.1199	1.1084	1.1062	1.0965	69	1.1813	1.1663	1.1628	1.1510	94	1.2354	1.2174	1.2117	1.1994
20	1.0571	1.0505	1.0488	1.0427	45	1.1224	1.1109	1.1085	1.0987	70	1.1835	1.1685	1.1655	1.1531	95	1.2372	1.2191	1.2140	1.2008
21	1.0598	1.0532	1.0512	1.0451	46	1.1249	1.1133	1.1108	1.1009	71	1.1858	1.1707	1.1677	1.1552	96	1.2390	1.2208	1.2158	1.2022
22	1.0625	1.0556	1.0537	1.0473	47	1.1274	1.1156	1.1130	1.1031	72	1.1882	1.1729	1.1702	1.1573	97	1.2408	1.2224	1.2170	1.2036
23	1.0652	1.0580	1.0561	1.0496	48	1.1299	1.1179	1.1157	1.1053	73	1.1906	1.1751	1.1728	1.1595	98	1.2425	1.2240	1.2183	1.2048
24	1.0679	1.0604	1.0585	1.0518	49	1.1324	1.1202	1.1185	1.1076	74	1.1929	1.1773	1.1752	1.1615	99	1.2441	1.2257	1.2202	1.2061
															100	1.2456	1.2273	1.2212	1.2073

Table 128. Acetic Acid (CH_3COOH)

%	0°C.	10°C.	15°C.	20°C.	25°C.	30°C.	40°C.	%	0°C.	10°C.	15°C.	20°C.	25°C.	30°C.	40°C.
0	0.9999	0.9997	0.9991	0.9982	0.9971	0.9957	0.9922	40	1.0621	1.0557	1.0522	1.0488	1.0450	1.0416	1.0338
1	1.0016	1.0013	1.0006	.9996	.9987	.9971	.9934	41	1.0633	1.0568	1.0532	1.0498	1.0460	1.0425	1.0346
2	1.0033	1.0029	1.0021	1.0012	1.0000	.9984	.9946	42	1.0644	1.0578	1.0542	1.0507	1.0469	1.0433	1.0353
3	1.0051	1.0044	1.0036	1.0025	1.0013	.9997	.9958	43	1.0656	1.0588	1.0551	1.0516	1.0477	1.0441	1.0361
4	1.0070	1.0060	1.0051	1.0040	1.0027	1.0011	.9970	44	1.0667	1.0598	1.0561	1.0525	1.0486	1.0449	1.0368
5	1.0088	1.0076	1.0066	1.0055	1.0041	1.0024	.9982	45	1.0679	1.0608	1.0570	1.0534	1.0495	1.0456	1.0375
6	1.0106	1.0092	1.0081	1.0069	1.0055	1.0037	.9994	46	1.0689	1.0618	1.0579	1.0542	1.0503	1.0464	1.0382
7	1.0124	1.0108	1.0096	1.0083	1.0068	1.0050	1.0006	47	1.0699	1.0627	1.0588	1.0551	1.0511	1.0471	1.0389
8	1.0142	1.0124	1.0111	1.0097	1.0081	1.0063	1.0018	48	1.0709	1.0636	1.0597	1.0559	1.0518	1.0479	1.0395
9	1.0159	1.0140	1.0126	1.0111	1.0094	1.0076	1.0030	49	1.0720	1.0645	1.0605	1.0567	1.0526	1.0486	1.0402
10	1.0177	1.0156	1.0141	1.0125	1.0107	1.0089	1.0042	50	1.0729	1.0654	1.0613	1.0575	1.0534	1.0492	1.0408
11	1.0194	1.0171	1.0155	1.0139	1.0120	1.0102	1.0054	51	1.0738	1.0663	1.0622	1.0582	1.0542	1.0499	1.0414
12	1.0211	1.0187	1.0170	1.0154	1.0133	1.0115	1.0065	52	1.0748	1.0671	1.0629	1.0590	1.0549	1.0506	1.0421
13	1.0228	1.0202	1.0184	1.0168	1.0146	1.0127	1.0077	53	1.0757	1.0679	1.0637	1.0597	1.0555	1.0512	1.0427
14	1.0245	1.0217	1.0199	1.0182	1.0159	1.0139	1.0088	54	1.0765	1.0687	1.0644	1.0604	1.0562	1.0518	1.0432
15	1.0262	1.0232	1.0213	1.0195	1.0172	1.0151	1.0099	55	1.0774	1.0694	1.0651	1.0611	1.0568	1.0525	1.0438
16	1.0278	1.0247	1.0227	1.0209	1.0185	1.0163	1.0110	56	1.0782	1.0701	1.0658	1.0618	1.0574	1.0531	1.0443
17	1.0295	1.0262	1.0241	1.0223	1.0198	1.0175	1.0121	57	1.0790	1.0708	1.0665	1.0624	1.0580	1.0536	1.0448
18	1.0311	1.0276	1.0255	1.0236	1.0210	1.0187	1.0132	58	1.0798	1.0715	1.0672	1.0631	1.0586	1.0542	1.0453
19	1.0327	1.0291	1.0269	1.0250	1.0223	1.0198	1.0142	59	1.0805	1.0722	1.0678	1.0637	1.0592	1.0547	1.0458
20	1.0343	1.0305	1.0283	1.0263	1.0235	1.0210	1.0153	60	1.0813	1.0728	1.0684	1.0642	1.0597	1.0552	1.0462
21	1.0358	1.0319	1.0297	1.0276	1.0248	1.0222	1.0164	61	1.0820	1.0734	1.0690	1.0648	1.0602	1.0557	1.0466
22	1.0374	1.0333	1.0310	1.0288	1.0260	1.0233	1.0174	62	1.0826	1.0740	1.0696	1.0653	1.0607	1.0562	1.0470
23	1.0389	1.0347	1.0323	1.0301	1.0272	1.0244	1.0185	63	1.0833	1.0746	1.0701	1.0658	1.0612	1.0566	1.0473
24	1.0404	1.0361	1.0336	1.0313	1.0283	1.0256	1.0195	64	1.0838	1.0752	1.0706	1.0662	1.0616	1.0571	1.0477
25	1.0419	1.0375	1.0349	1.0326	1.0295	1.0267	1.0205	65	1.0844	1.0757	1.0711	1.0666	1.0621	1.0575	1.0480
26	1.0434	1.0388	1.0362	1.0338	1.0307	1.0278	1.0215	66	1.0850	1.0762	1.0716	1.0671	1.0624	1.0578	1.0483
27	1.0449	1.0401	1.0374	1.0349	1.0318	1.0289	1.0225	67	1.0856	1.0767	1.0720	1.0675	1.0628	1.0582	1.0486
28	1.0463	1.0414	1.0386	1.0361	1.0329	1.0299	1.0234	68	1.0860	1.0771	1.0725	1.0678	1.0631	1.0585	1.0489
29	1.0477	1.0427	1.0399	1.0372	1.0340	1.0310	1.0244	69	1.0865	1.0775	1.0729	1.0682	1.0634	1.0588	1.0491
30	1.0491	1.0440	1.0411	1.0384	1.0350	1.0320	1.0253	70	1.0869	1.0779	1.0732	1.0685	1.0637	1.0590	1.0493
31	1.0505	1.0453	1.0423	1.0395	1.0361	1.0330	1.0262	71	1.0874	1.0783	1.0736	1.0687	1.0640	1.0592	1.0495
32	1.0519	1.0465	1.0435	1.0406	1.0372	1.0341	1.0272	72	1.0877	1.0786	1.0738	1.0690	1.0642	1.0594	1.0496
33	1.0532	1.0477	1.0446	1.0417	1.0382	1.0351	1.0281	73	1.0881	1.0789	1.0741	1.0693	1.0644	1.0595	1.0497
34	1.0545	1.0489	1.0458	1.0428	1.0392	1.0361	1.0289	74	1.0884	1.0792	1.0743	1.0694	1.0645	1.0596	1.0498
35	1.0558	1.0501	1.0469	1.0438	1.0402	1.0371	1.0298	75	1.0887	1.0794	1.0745	1.0696	1.0647	1.0597	1.0499
36	1.0571	1.0513	1.0480	1.0449	1.0412	1.0380	1.0306	76	1.0889	1.0796	1.0746	1.0698	1.0648	1.0598	1.0499
37	1.0584	1.0524	1.0491	1.0459	1.0422	1.0390	1.0314	77	1.0891	1.0797	1.0747	1.0699	1.0648	1.0598	1.0499
38	1.0596	1.0535	1.0501	1.0469	1.0432	1.0399	1.0322	78	1.0893	1.0798	1.0747	1.0700	1.0648	1.0598	1.0498
39	1.0608	1.0546	1.0512	1.0479	1.0441	1.0408	1.0330	79	1.0894	1.0798	1.0747	1.0700	1.0648	1.0597	1.0497

Table 128. Acetic Acid (CH_3COOH)—(*Concluded*)

%	0°C.	10°C.	15°C.	20°C.	25°C.	30°C.	40°C.	%	0°C.	10°C.	15°C.	20°C.	25°C.	30°C.	40°C.
80	1.0895	1.0798	1.0747	1.0700	1.0647	1.0596	1.0495	90	1.0865	1.0766	1.0708	1.0661	1.0605	1.0549	1.0445
81	1.0895	1.0797	1.0745	1.0699	1.0646	1.0594	1.0493	91	1.0857	1.0758	1.0700	1.0652	1.0597	1.0541	1 0436
82	1.0895	1.0796	1.0743	1.0698	1.0644	1.0592	1.0490	92	1.0848	1.0749	1.0690	1.0643	1.0587	1.0530	1.0426
83	1.0895	1.0795	1.0741	1.0696	1.0642	1.0589	1.0487	93	1.0838	1.0739	1.0680	1.0632	1.0577	1.0518	1.0414
84	1.0893	1.0793	1.0738	1.0693	1.0638	1.0585	1.0483	94	1.0826	1.0727	1.0667	1.0619	1.0564	1.0506	1.0401
85	1 0891	1.0790	1.0735	1.0689	1.0635	1.0582	1.0479	95	1.0813	1.0714	1.0652	1.0605	1.0551	1.0491	1.0386
86	1.0887	1.0787	1.0731	1.0685	1.0630	1.0576	1.0473	96	1.0798	......	1.0632	1.0588	1.0535	1.0473	1.0368
87	1.0883	1.0783	1.0726	1.0680	1.0626	1.0571	1.0467	97	1.0780	......	1.0611	1.0570	1.0516	1.0454	1.0348
88	1.0877	1.0778	1.0721	1.0675	1.0620	1.0564	1.0460	98	1.0759	......	1.0590	1.0549	1.0495	1.0431	1.0325
89	1.0872	1.0773	1.0715	1.0668	1.0613	1.0557	1.0453	99	1.0730	......	1.0567	1.0524	1.0468	1.0407	1.0299
								100	1.0697	......	1.0545	1.0498	1.0440	1.0380	1.0271

Table 129. Oxalic Acid ($H_2C_2O_4$)

%	$d_4^{17.5}$	%	$d_4^{17.5}$
1	1.0035	8	1.0280
2	1.0070	10	1.0350
4	1.0140	12	1.0420

Table 130. Methyl Alcohol (CH_3OH)

%	0°C.	10°C.	15.56°C.	20°C.	15°C.	%	0°C.	10°C.	15.56°C	20°C.	15°C.	%	0°C.	10°C.	15.56°C.	20°C.	15°C.
0	0.9999	0.9997	0.9990	0.9982	0.99913	35	0.9534	0.9484	0.9456	0.9433	0.94570	70	0.8869	0.8794	0.8748	0.8715	0.87507
1	.9981	.9980	.9973	.9965	.99727	36	.9520	.9469	.9440	.9416	.94404	71	.8847	.8770	.8726	.8690	.87271
2	.9963	.9962	.9955	.9948	.99543	37	.9505	.9453	.9422	.9398	.94237	72	.8824	.8747	.8702	.8665	.87033
3	.9946	.9945	.9938	.9931	.99370	38	.9490	.9437	.9405	.9381	.94067	73	.8801	.8724	.8678	.8641	.86792
4	.9930	.9929	.9921	.9914	.99198	39	.9475	.9420	.9387	.9363	.93894	74	.8778	.8699	.8653	.8616	.86546
5	.9914	.9912	.9904	.9896	.99029	40	.9459	.9403	.9369	.9345	.93720	75	.8754	.8676	.8629	.8592	.86300
6	.9899	.9896	.9889	.9880	.98864	41	.9443	.9387	.9351	.9327	.93543	76	.8729	.8651	.8604	.8567	.86051
7	.9884	.9881	.9872	.9863	.98701	42	.9427	.9370	.9333	.9309	.93365	77	.8705	.8626	.8579	.8542	.85801
8	.9870	.9865	.9857	.9847	.98547	43	.9411	.9352	.9315	.9290	.93185	78	.8680	.8602	.8554	.8518	.85551
9	.9856	.9849	.9841	.9831	.98394	44	.9395	.9334	.9297	.9272	.93001	79	.8657	.8577	.8529	.8494	.85300
10	.9842	.9834	.9826	.9815	.98241	45	.9377	.9316	.9279	.9252	.92815	80	.8634	.8551	.8503	.8469	.85048
11	.9829	.9820	.9811	.9799	.98093	46	.9360	.9298	.9261	.9234	.92627	81	.8610	.8527	.8478	.8446	.84794
12	.9816	.9805	.9796	.9784	.97945	47	.9342	.9279	.9242	.9214	.92436	82	.8585	.8501	.8452	.8420	.84536
13	.9804	.9791	.9781	.9768	.97802	48	.9324	.9260	.9223	.9196	.92242	83	.8560	.8475	.8426	.8394	.84274
14	.9792	.9778	.9766	.9754	.97660	49	.9306	.9240	.9204	.9176	.92048	84	.8535	.8449	.8400	.8366	.84009
15	.9780	.9764	.9752	.9740	.97518	50	.9287	.9221	.9185	.9156	.91852	85	.8510	.8422	.8374	.8340	.83742
16	.9769	.9751	.9738	.9725	.97377	51	.9269	.9202	.9166	.9135	.91653	86	.8483	.8394	.8347	.8314	.83475
17	.9758	.9739	.9723	.9710	.97237	52	.9250	.9182	.9146	.9114	.91451	87	.8456	.8367	.8320	.8286	.83207
18	.9747	.9726	.9709	.9696	.97096	53	.9230	.9162	.9126	.9094	.91248	88	.8428	.8340	.8294	.8258	.82937
19	.9736	.9713	.9695	.9681	.96955	54	.9211	.9142	.9106	.9073	.91044	89	.8400	.8314	.8267	.8230	.82667
20	.9725	.9700	.9680	.9666	.96814	55	.9191	.9122	.9086	.9052	.90839	90	.8374	.8287	.8239	.8202	.82396
21	.9714	.9687	.9666	.9651	.96673	56	.9172	.9101	.9065	.9032	.90631	91	.8347	.8261	.8212	.8174	.82124
22	.9702	.9673	.9652	.9636	.96533	57	.9151	.9080	.9045	.9010	.90421	92	.8320	.8234	.8185	.8146	.81849
23	.9690	.9660	.9638	.9622	.96392	58	.9131	.9060	.9024	.8988	.90210	93	.8293	.8208	.8157	.8118	.81568
24	.9678	.9646	.9624	.9607	.96251	59	.9111	.9039	.9002	.8968	.89996	94	.8266	.8180	.8129	.8090	.81285
25	.9666	.9632	.9609	.9592	.96108	60	.9090	.9018	.8980	.8946	.89781	95	.8240	.8152	.8101	.8062	.80999
26	.9654	.9618	.9595	.9576	.95963	61	.9068	.8998	.8958	.8924	.89563	96	.8212	.8124	.8073	.8034	.80713
27	.9642	.9604	.9580	.9562	.95817	62	.9046	.8977	.8936	.8902	.89341	97	.8186	.8096	.8045	.8005	.80428
28	.9629	.9590	.9565	.9546	.95668	63	.9024	.8955	.8913	.8879	.89117	98	.8158	.8068	.8016	.7976	.80143
29	.9616	.9575	.9550	.9531	.95518	64	.9002	.8933	.8890	.8856	.88890	99	.8130	.8040	.7987	.7948	.79859
30	.9604	.9560	.9535	.9515	.95366	65	.8980	.8911	.8867	.8834	.88662	100	.8102	.8009	.7959	.7917	.79577
31	.9590	.9546	.9521	.9499	.95213	66	.8958	.8888	.8844	.8811	.88433						
32	.9576	.9531	.9505	.9483	.95056	67	.8935	.8865	.8820	.8787	.88203						
33	.9563	.9516	.9489	.9466	.94896	68	.8913	.8842	.8797	.8763	.87971						
34	.9549	.9500	.9473	.9450	.94734	69	.8891	.8818	.8771	.8738	.87739						

Table 131. Ethyl Alcohol (C_2H_5OH)

%	10°C.	15°C.	20°C.	25°C.	30°C.	35°C.	40°C.	%	10°C.	15°C.	20°C.	25°C.	30°C.	35°C.	40°C.
0	0.99973	0.99913	0.99823	0.99708	0.99568	0.99406	0.99225	50	0.92126	0.91776	0.91384	0.90985	0.90580	0.90168	0.89750
1	785	725	636	520	379	217	034	51	.91943	555	160	760	353	.89940	519
2	602	542	453	336	194	031	.98846	52	723	333	.90936	534	125	710	288
3	426	365	275	157	014	.98849	663	53	502	110	711	307	.89896	479	056
4	258	195	103	.98984	.98839	672	485	54	279	.90885	485	079	667	248	.88823
5	098	032	.98938	817	670	501	311	55	055	659	258	.89850	437	016	589
6	.98946	.98877	780	656	507	335	142	56	.90831	433	031	621	206	.88784	356
7	801	729	627	500	347	172	.97975	57	607	207	.89803	392	.88975	552	122
8	660	584	478	346	189	009	808	58	381	.89980	574	162	744	319	.87888
9	524	442	331	193	031	.97846	641	59	154	752	344	.88931	512	085	653
10	393	304	187	043	.97875	685	475	60	.89927	523	113	699	278	.87851	417
11	267	171	047	.97897	723	527	312	61	698	293	.88882	446	044	615	180
12	145	041	.97910	753	573	371	150	62	468	062	650	233	.87809	379	.86943
13	026	.97914	775	611	424	216	.96989	63	237	.88830	417	.87998	574	142	705
14	.97911	790	643	472	278	063	829	64	006	597	183	763	337	.86905	466
15	800	669	514	334	133	.96911	670	65	.88774	364	.87948	527	100	667	227
16	692	552	387	199	.96990	760	512	66	541	130	713	291	.86863	429	.85987
17	583	433	259	062	844	607	352	67	308	.87895	477	054	625	190	747
18	473	313	129	.96923	697	452	189	68	074	660	241	.86817	387	.85950	407
19	363	191	.96997	782	547	294	023	69	.87839	424	004	579	148	710	266
20	252	068	864	639	395	134	.95856	70	602	187	.86766	340	.85908	470	025
21	139	.96944	729	495	242	.95973	687	71	365	.86949	527	100	667	228	.84783
22	024	818	592	348	087	809	516	72	127	710	287	.85859	426	.84986	540
23	.96907	689	453	199	.95929	643	343	73	.86888	470	047	618	184	743	297
24	787	558	312	048	769	476	168	74	648	229	.85806	376	.84941	500	053
25	665	424	168	.95895	607	306	.94991	75	408	.85988	564	134	698	257	.83809
26	539	287	020	738	442	133	810	76	168	747	322	.84891	455	013	564
27	406	144	.95867	576	272	.94955	625	77	.85927	505	079	647	211	.83768	319
28	268	.95996	710	410	098	774	438	78	685	262	.84835	403	.83966	523	074
29	125	844	548	241	.94922	590	248	79	442	018	590	158	720	277	.82827
30	.95977	686	382	067	741	403	055	80	197	.84772	344	.83911	473	029	578
31	823	524	212	.94890	557	214	.93860	81	.84950	525	096	664	224	.82780	329
32	665	357	038	709	370	021	662	82	702	277	.83848	415	.82974	530	079
33	502	186	.94860	525	180	.93825	461	83	453	028	599	164	724	279	.81828
34	334	011	679	337	.93986	626	257	84	203	.83777	348	.82913	473	027	576
35	162	.94832	494	146	790	425	051	85	.83951	525	095	660	220	.81774	322
36	.94986	650	306	.93952	591	221	.92843	86	697	271	.82840	405	.81965	519	067
37	805	464	114	756	390	016	634	87	441	014	583	148	708	262	.80811
38	620	273	.93919	556	186	.92808	422	88	181	.82754	323	.81888	448	003	552
39	431	079	720	353	.92979	597	208	89	.82919	492	062	626	186	.80742	291
40	238	.93882	518	148	770	385	.91992	90	654	227	.81797	362	.80922	478	028
41	042	682	314	.92940	558	170	774	91	386	.81959	529	094	655	211	.79761
42	.93842	478	107	729	344	.91952	554	92	114	688	257	.80823	384	.79941	491
43	639	271	.92897	516	128	733	332	93	.81839	413	.80983	549	111	669	220
44	433	062	685	301	.91910	513	108	94	561	134	705	272	.79835	393	.78947
45	226	.92852	472	085	692	291	.90884	95	278	.80852	424	.79991	555	114	670
46	017	640	257	.91868	472	069	660	96	.80991	566	138	706	271	.78831	388
47	.92806	426	041	649	250	.90845	434	97	698	274	.79846	415	.78981	542	100
48	593	211	.91823	429	028	621	207	98	399	.79975	547	117	684	247	.77806
49	379	.91995	604	208	.90805	396	.89979	99	094	670	243	.78814	382	.77946	507
								100	.79784	360	.78934	506	075	641	203

Table 132. Densities of Mixtures of C_2H_5OH and H_2O at 20°C.

G./ml.

% alcohol by weight	Tenths of %: 0	1	2	3	4	5	6	7	8	9
0	0.99823	804	785	766	748	729	710	692	673	655
1	636	618	599	581	562	544	525	507	489	471
2	453	435	417	399	381	363	345	327	310	292
3	275	257	240	222	205	188	171	154	137	120
4	103	087	070	053	037	020	003	*987	*971	*954
5	.98938	922	906	890	874	859	843	827	811	796
6	780	765	749	734	718	703	688	673	658	642
7	627	612	597	582	567	553	538	523	508	493
8	478	463	449	434	419	404	389	374	360	345
9	331	316	301	287	273	258	244	229	215	201
10	187	172	158	144	130	117	103	089	075	061
11	047	033	019	006	*992	*978	*964	*951	*937	*923
12	.97910	896	883	869	855	842	828	815	801	788
13	775	761	748	735	722	709	696	683	670	657
14	643	630	617	604	591	578	565	552	539	526
15	514	501	488	475	462	450	438	425	412	400
16	387	374	361	349	336	323	310	297	284	272
17	259	246	233	220	207	194	181	168	155	142
18	129	116	103	089	076	063	050	037	024	010
19	.96997	984	971	957	944	931	917	904	891	877
20	864	850	837	823	810	796	783	769	756	742
21	729	716	702	688	675	661	647	634	620	606
22	592	578	564	551	537	523	509	495	481	467
23	453	439	425	411	396	382	368	354	340	326
24	312	297	283	269	254	240	225	211	196	182
25	168	153	139	124	109	094	080	065	050	035
26	020	005	*990	*975	*959	*944	*929	*914	*898	*883
27	.95867	851	836	820	805	789	773	757	742	726
28	710	694	678	662	646	630	613	597	581	565
29	548	532	516	499	483	466	450	433	416	400
30	382	365	349	332	315	298	281	264	247	230
31	212	195	178	161	143	126	108	091	074	056
32	038	020	003	*985	*967	*950	*932	*914	*896	*878
33	.94860	842	824	806	788	770	752	734	715	697
34	679	660	642	624	605	587	568	550	531	512
35	494	475	456	438	419	400	382	363	344	325
36	306	287	268	249	230	211	192	172	153	134
37	114	095	075	056	036	017	*997	*978	*958	*939
38	.93919	899	879	859	840	820	800	780	760	740
39	720	700	680	660	640	620	599	579	559	539
40	518	498	478	458	437	417	396	376	356	335
41	314	294	273	253	232	212	191	170	149	129
42	107	086	065	044	023	002	*981	*960	*939	*918
43	.92897	876	855	834	812	791	770	749	728	707
44	685	664	642	621	600	579	557	536	515	493
45	472	450	429	408	386	365	343	322	300	279
46	257	236	214	193	171	150	128	106	085	063
47	041	019	*997	*976	*954	*932	*910	*889	*867	*845
48	.91823	801	780	758	736	714	692	670	648	626
49	604	582	560	538	516	494	472	450	428	406

% alcohol by weight	Tenths of %: 0	1	2	3	4	5	6	7	8	9
50	0.91384	361	339	317	295	272	250	228	206	183
51	160	138	116	093	071	049	026	004	*981	*959
52	.90936	914	891	869	846	824	801	779	756	734
53	711	689	666	644	621	598	576	553	531	508
54	485	463	440	417	395	372	349	327	304	281
55	258	236	213	190	167	145	122	099	076	054
56	031	008	*985	*962	*939	*917	*894	*871	*848	*825
57	.89803	780	757	734	711	688	665	643	620	597
58	574	551	528	505	482	459	436	413	390	367
59	344	321	298	275	252	229	206	183	160	137
60	113	090	067	044	021	*998	*975	*951	*928	*905
61	.88882	859	836	812	789	766	743	720	696	673
62	650	626	603	580	557	533	510	487	463	440
63	417	393	370	347	323	300	277	253	230	206
64	183	160	136	113	089	066	042	019	*995	*972
65	.87948	925	901	878	854	831	807	784	760	737
66	713	689	666	642	619	595	572	548	524	501
67	477	454	430	406	383	359	336	312	288	265
68	241	218	194	170	147	123	099	075	052	028
69	004	*981	*957	*933	*909	*885	*862	*838	*814	*790
70	.86766	742	718	694	671	647	623	599	575	551
71	527	503	479	455	431	407	383	339	335	311
72	287	263	239	215	191	167	143	119	095	071
73	047	022	*998	*974	*950	*926	*902	*878	*854	*830
74	.85806	781	757	733	709	685	661	636	612	588
75	564	540	515	491	467	443	419	394	370	346
76	322	297	273	249	225	200	176	152	128	103
77	079	055	031	006	*982	*958	*933	*909	*884	*860
78	.84835	811	787	762	738	713	689	664	640	615
79	590	566	541	517	492	467	443	418	393	369
80	344	319	294	270	245	220	196	171	146	121
81	096	072	047	022	*997	*972	*947	*923	*898	*873
82	.83848	823	798	773	748	723	698	674	649	624
83	599	574	549	523	498	473	448	423	398	373
84	348	323	297	272	247	222	196	171	146	120
85	095	070	044	019	*994	*968	*943	*917	*892	*866
86	.82840	815	789	763	738	712	686	660	635	609
87	583	557	531	505	479	453	427	401	375	349
88	323	297	271	245	219	193	167	140	114	088
89	062	035	009	*983	*956	*930	*903	*877	*850	*824
90	.81797	770	744	717	690	664	637	610	583	556
91	529	502	475	448	421	394	366	339	312	285
92	257	230	203	175	148	120	093	066	038	010
93	.80983	955	928	900	872	844	817	789	761	733
94	705	677	649	621	593	565	537	509	480	452
95	424	395	367	338	310	281	253	224	195	166
96	138	109	080	051	022	*993	*963	*934	*905	*875
97	.79846	816	787	757	727	698	668	638	608	578
98	547	517	487	456	426	396	365	335	305	274
99	243	213	182	151	120	089	059	028	*997	*966
100	.78934									

* Indicates change in the first two decimal places.

Table 133. Specific Gravity (60°/60°F.) (15.56°/15.56°C.) of Mixtures (by Volume) of C_2H_5OH and H_2O

% alcohol by volume at 60°F.	Tenths of %: 0	1	2	3	4	5	6	7	8	9
0	1.00000	*985	*970	*955	*940	*925	*910	*895	*880	865
1	0.99850	835	820	806	791	776	761	747	732	717
2	703	688	674	659	645	630	616	602	587	573
3	559	545	531	516	502	488	474	460	446	432
4	419	405	391	378	364	350	336	323	309	296
5	282	269	255	242	228	215	202	189	176	163
6	150	137	124	111	098	085	073	060	047	035
7	022	009	*997	*984	*972	*960	*947	*935	*923	*911
8	.98899	887	875	863	851	838	826	814	803	791
9	779	767	755	743	731	720	708	696	684	672
10	661	649	637	625	614	602	590	579	567	556
11	544	532	521	509	498	487	475	464	452	441
12	430	419	408	396	385	374	363	352	341	330
13	319	308	297	286	275	264	254	243	232	221
14	210	200	190	179	168	157	147	136	125	115
15	104	093	083	072	062	051	040	030	019	009
16	.97998	988	977	967	956	946	936	925	915	905
17	895	885	875	864	854	844	834	824	814	804
18	794	784	774	764	754	744	734	724	714	704
19	694	684	674	664	654	645	635	625	615	605
20	596	586	576	566	556	546	536	526	516	506
21	496	486	476	466	456	446	436	425	415	405
22	395	385	375	365	354	344	334	324	313	303
23	293	283	272	262	252	241	231	221	210	200
24	189	179	168	158	147	137	126	116	105	095
25	084	073	063	052	042	031	020	010	*999	*988
26	.96978	967	957	946	935	924	914	903	892	881
27	870	859	848	837	826	815	804	793	782	771
28	760	749	738	727	715	704	693	682	671	659
29	648	637	625	614	603	591	580	568	557	546
30	534	522	511	499	488	476	464	453	441	429
31	418	406	394	382	370	358	346	334	321	309
32	296	284	271	259	246	234	221	209	196	183
33	170	157	144	132	119	106	093	080	067	054
34	041	028	015	002	*988	*975	*962	*948	*935	*921
35	.95908	894	881	867	854	840	826	812	798	784
36	770	756	742	728	714	700	685	671	657	643
37	628	614	599	585	570	556	541	526	512	497
38	482	467	452	437	423	408	393	378	362	347
39	332	317	302	286	271	256	240	225	209	194
40	178	162	147	131	115	100	084	068	052	036
41	020	004	*988	*972	*956	*940	*923	*907	*891	*875
42	.94858	842	825	809	792	776	759	743	726	710
43	693	676	660	643	626	609	592	575	558	541
44	524	507	490	473	455	438	421	403	386	369
45	351	334	316	298	281	263	245	228	210	192
46	174	156	138	120	102	084	066	048	030	011
47	.93993	975	956	938	920	901	883	864	845	827
48	808	789	771	752	733	714	695	676	657	638
49	619	600	581	562	543	523	504	485	465	446

% alcohol by volume at 60°F.	Tenths of %: 0	1	2	3	4	5	6	7	8	9
50	0.93426	407	387	368	348	328	309	289	270	250
51	230	210	190	171	151	131	111	091	071	051
52	031	011	*991	*971	*951	*931	*911	*890	*870	*850
53	.92830	810	789	769	749	728	708	688	667	647
54	626	605	585	564	544	523	502	482	461	440
55	419	398	377	357	336	315	294	273	252	231
56	210	189	168	147	126	105	084	062	041	020
57	.91999	978	956	935	914	892	871	849	827	806
58	784	762	741	719	697	675	653	631	610	588
59	565	543	521	499	477	455	433	410	388	366
60	344	322	299	277	255	232	210	188	165	143
61	120	097	075	052	030	007	*984	*962	*939	*916
62	.90893	870	847	825	802	779	756	733	710	687
63	664	641	618	595	572	549	526	503	480	457
64	434	411	388	365	341	318	295	272	249	225
65	202	179	155	132	108	085	061	038	014	*991
66	.89967	943	920	896	872	848	825	801	777	753
67	729	705	681	657	633	609	585	561	537	513
68	489	465	441	416	392	368	343	319	295	270
69	245	220	196	171	147	122	098	073	048	024
70	.88999	974	950	925	900	875	850	825	801	776
71	751	725	700	675	650	625	600	574	549	524
72	499	474	448	423	397	372	346	321	296	270
73	244	218	193	167	141	116	090	064	039	013
74	.87987	961	935	910	884	858	832	806	780	754
75	728	702	676	650	623	597	571	545	518	492
76	465	439	412	386	359	332	306	279	252	226
77	199	172	145	118	092	065	038	011	*984	*957
78	.86929	902	875	847	820	793	766	738	711	684
79	656	629	601	574	546	518	491	463	435	408
80	380	352	324	296	269	241	213	185	157	129
81	100	072	044	015	*987	*959	*931	*902	*874	*846
82	.85817	789	760	732	703	674	646	617	588	560
83	531	502	473	444	415	386	357	328	299	270
84	240	211	181	152	122	093	063	033	004	*974
85	.84944	914	884	854	824	794	764	734	703	673
86	642	612	581	551	520	490	459	428	398	367
87	336	305	274	243	212	181	150	119	088	056
88	025	*994	*962	*930	*899	*867	*835	*803	*771	*739
89	.83707	675	643	610	578	545	513	480	447	415
90	382	349	315	282	249	216	183	150	116	083
91	049	015	*981	*947	*913	*879	*845	*810	*776	*741
92	.82705	670	635	600	565	529	494	458	423	387
93	351	315	279	243	206	170	133	096	059	022
94	.81984	947	909	871	834	796	757	719	681	642
95	603	564	525	486	446	407	367	327	287	247
96	206	165	125	084	042	001	*960	*918	*876	*834
97	.80792	750	707	664	620	577	533	489	445	401
98	356	311	265	219	173	127	080	033	*985	*937
99	.79889	841	792	743	693	643	593	543	492	441
100	389									

* Indicates change in first two decimal places.

Table 134. n-Propyl Alcohol (C_3H_7OH)

%	0°C.	15°C.	30°C.	%	0°C.	15°C.	30°C.	%	0°C.	15°C.	30°C.	%	0°C.	15°C.	30°C.	%	0°C.	15°C.	30°C.
0	0.9999	0.9991	0.9957	20	0.9789	0.9723	0.9643	40	0.9430	0.9331	0.9226	60	0.9033	0.8922	0.8807	80	0.8634	0.8516	0.8394
1	.9982	.9974	.9940	21	.9776	.9705	.9622	41	.9411	.9310	.9205	61	.9013	.8902	.8786	81	.8614	.8496	.8373
2	.9967	.9960	.9924	22	.9763	.9688	.9602	42	.9391	.9290	.9184	62	.8994	.8882	.8766	82	.8594	.8475	.8352
3	.9952	.9944	.9908	23	.9748	.9670	.9583	43	.9371	.9269	.9164	63	.8974	.8861	.8745	83	.8574	.8454	.8332
4	.9939	.9929	.9893	24	.9733	.9651	.9563	44	.9352	.9248	.9143	64	.8954	.8841	.8724	84	.8554	.8434	.8311
5	.9926	.9915	.9877	25	.9717	.9633	.9543	45	.9332	.9228	.9122	65	.8934	.8820	.8703	85	.8534	.8413	.8290
6	.9914	.9902	.9862	26	.9700	.9614	.9522	46	.9311	.9207	.9100	66	.8913	.8800	.8682	86	.8513	.8393	.8269
7	.9904	.9890	.9848	27	.9682	.9594	.9501	47	.9291	.9186	.9079	67	.8894	.8779	.8662	87	.8492	.8372	.8248
8	.9894	.9877	.9834	28	.9664	.9576	.9481	48	.9272	.9165	.9057	68	.8874	.8759	.8641	88	.8471	.8351	.8227
9	.9883	.9864	.9819	29	.9646	.9556	.9460	49	.9252	.9145	.9036	69	.8854	.8739	.8620	89	.8450	.8330	.8206
10	.9874	.9852	.9804	30	.9627	.9535	.9439	50	.9232	.9124	.9015	70	.8835	.8719	.8600	90	.8429	.8308	.8185
11	.9865	.9840	.9790	31	.9608	.9516	.9418	51	.9213	.9104	.8994	71	.8815	.8700	.8580	91	.8408	.8287	.8164
12	.9857	.9828	.9775	32	.9589	.9495	.9396	52	.9192	.9084	.8973	72	.8795	.8680	.8559	92	.8387	.8266	.8142
13	.9849	.9817	.9760	33	.9570	.9474	.9375	53	.9173	.9064	.8952	73	.8776	.8659	.8539	93	.8364	.8244	.8120
14	.9841	.9806	.9746	34	.9550	.9454	.9354	54	.9153	.9044	.8931	74	.8756	.8639	.8518	94	.8342	.8221	.8098
15	.9833	.9793	.9730	35	.9530	.9434	.9333	55	.9132	.9023	.8911	75	.8736	.8618	.8497	95	.8320	.8199	.8077
16	.9825	.9780	.9714	36	.9511	.9413	.9312	56	.9112	.9003	.8890	76	.8716	.8598	.8477	96	.8296	.8176	.8054
17	.9817	.9768	.9698	37	.9491	.9392	.9289	57	.9093	.8983	.8869	77	.8695	.8577	.8456	97	.8272	.8153	.8031
18	.9808	.9752	.9680	38	.9471	.9372	.9269	58	.9073	.8963	.8849	78	.8675	.8556	.8435	98	.8248	.8128	.8008
19	.9800	.9739	.9661	39	.9450	.9351	.9247	59	.9053	.8942	.8828	79	.8655	.8536	.8414	99	.8222	.8104	.7984
																100	.8194	.8077	.7958

Table 135. Isopropyl Alcohol (C_3H_7OH)

%	0°C.	15°C.*	15°C.*	20°C.	30°C	%	0°C.	15°C.*	15°C.*	20°C.	30°C.	%	0°C.	15°C.*	15°C.*	20°C.	30°C.
0	0.9999	0.9991	0.99913	0.9982	0.9957	35	0.9557	......	0.9446	0.9419	0.9338	70	0.8761	0.8639	0.86346	0.8584	0.8511
1	.9980	.9973	.9972	.9962	.9939	36	.9536	......	.9424	.9399	.9315	71	.8738	.8615	.8611	.8560	.8487
2	.9962	.9956	.9954	.9944	.9921	37	.9514	......	.9401	.9377	.9292	72	.8714	.8592	.8588	.8537	.8464
3	.9946	.9938	.9936	.9926	.9904	38	.9493	......	.9379	.9355	.9269	73	.8691	.8568	.8564	.8513	.8440
4	.9930	.9922	.9920	.9909	.9887	39	.9472	......	.9356	.9333	.9246	74	.8668	.8545	.8541	.8489	.8416
5	.9916	.9906	.9904	.9893	.9871	40	.9450	......	.93333	.9310	.9224	75	.8644	.8521	.8517	.8464	.8392
6	.9902	.9892	.9890	.9877	.9855	41	.9428	......	.9311	.9287	.9201	76	.8621	.8497	.8493	.8439	.8368
7	.9890	.9878	.9875	.9862	.9839	42	.9406	......	.9288	.9264	.9177	77	.8598	.8474	.8470	.8415	.8344
8	.9878	.9864	.9862	.9847	.9824	43	.9384	......	.9266	.9239	.9154	78	.8575	.8450	.8446	.8391	.8321
9	.9866	.9851	.9849	.9833	.9809	44	.9361	......	.9243	.9215	.9130	79	.8551	.8426	.8422	.8366	.8297
10	.9856	.9838	.98362	.9820	.9794	45	.9338	......	.9220	.9191	.9106	80	.8528	.8403	.83979	.8342	.8273
11	.9846	.9826	.9824	.9808	.9778	46	.9315	......	.9197	.9165	.9082	81	.8503	.8379	.8374	.8317	.8248
12	.9838	.9813	.9812	.9797	.9764	47	.9292	......	.9174	.9141	.9059	82	.8479	.8355	.8350	.8292	.8224
13	.9829	.9802	.9800	.9876	.9750	48	.9270	......	.9150	.9117	.9036	83	.8456	.8331	.8326	.8268	.8200
14	.9821	.9790	.9788	.9776	.9735	49	.9247	......	.9127	.9093	.9013	84	.8432	.8307	.8302	.8243	.8175
15	.9814	.9779	.9777	.9765	.9720	50	.9224	......	.91043	.9069	.8990	85	.8408	.8282	.8278	.8219	.8151
16	.9806	.9768	.9765	.9754	.9705	51	.9201	......	.9081	.9044	.8966	86	.8384	.8259	.8254	.8194	.8127
17	.9799	.9756	.9753	.9743	.9690	52	.9178	......	.9058	.9020	.8943	87	.8360	.8234	.8229	.8169	.8201
18	.9792	.9745	.9741	.9731	.9675	53	.9155	......	.9035	.8996	.8919	88	.8336	.8209	.8205	.8145	.8078
19	.9784	.9730	.9728	.9717	.9658	54	.9132	......	.9011	.8971	.8895	89	.8311	.8184	.8180	.8120	.8053
20	.9777	.9719	.97158	.9703	.9642	55	.9109	......	.8988	.8946	.8871	90	.8287	.8161	.81553	.8096	.8029
21	.9768	.9704	.9703	.9688	.9624	56	.9086	......	.8964	.8921	.8847	91	.8262	.8136	.8130	.8072	.8004
22	.9759	.9690	.9689	.9669	.9606	57	.9063	......	.8940	.8896	.8823	92	.8237	.8110	.8104	.8047	.7979
23	.9749	.9675	.9674	.9651	.9587	58	.9040	......	.8917	.8874	.8800	93	.8212	.8085	.8079	.8023	.7954
24	.9739	.9660	.9659	.9634	.9569	59	.9017	......	.8893	.8850	.8777	94	.8186	.8060	.8052	.7998	.7929
25	.9727	.9643	.9642	.9615	.9549	60	.8994	......	.88690	.8825	.8752	95	.8160	.8034	.8026	.7973	.7904
26	.9714	.9626	.9624	.9597	.9529	61	.8970	......	.8845	.8800	.8728	96	.8133	.8008	.7999	.7949	.7878
27	.9699	.9608	.9605	.9577	.9509	62	.8947	0.8829	.8821	.8776	.8704	97	.8106	.7981	.7972	.7925	.7852
28	.9684	.9590	.9586	.9558	.9488	63	.8924	.8805	.8798	.8751	.8680	98	.8078	.7954	.7945	.7901	.7826
29	.9669	.9570	.9568	.9540	.9467	64	.8901	.8781	.8775	.8727	.8656	99	.8048	.7926	.7918	.7877	.7799
30	.9652	.9551	.95493	.9520	.9446	65	.8878	.8757	.8752	.8702	.8631	100	.8016	.7896	.78913	.7854	.7770
31	.9634	......	.9530	.9500	.9426	66	.8854	.8733	.8728	.8679	.8607						
32	.9615	......	.9510	.9481	.9405	67	.8831	.8710	.8705	.8656	.8583						
33	.9596	......	.9489	.9460	.9383	68	.8807	.8686	.8682	.8632	.8559						
34	.9577	......	.9468	.9440	.9361	69	.8784	.8662	.8658	.8609	.8535						

* Two different observers; see "International Critical Tables," vol. 3, p. 120.

Table 136. Density and Percentage of Glycerol*

Glycerol, %	Density					Glycerol, %	Density					Glycerol, %	Density				
	15°C.	15.5°C.	20°C.	25°C.	30°C.		15°C.	15.5°C.	20°C.	25°C.	30°C.		15°C.	15.5°C.	20°C.	25°C.	30°C.
100	1.26415	1.26381	1.26108	1.15802	1.25495	65	1.17030	1.17000	1.16750	1.16475	1.16195	30	1.07455	1.07435	1.07270	1.07070	1.06855
99	1.26160	1.26125	1.25850	1.25545	1.25235	64	1.16755	1.16725	1.16475	1.16200	1.15925	29	1.07195	1.07175	1.07010	1.06815	1.06605
98	1.25900	1.25865	1.25590	1.25290	1.24975	63	1.16480	1.16445	1.16205	1.15925	1.15650	28	1.06935	1.06915	1.06755	1.06560	1.06355
97	1.25645	1.25610	1.25335	1.25030	1.24710	62	1.16200	1.16170	1.15930	1.15655	1.15375	27	1.06670	1.06655	1.06495	1.06305	1.06105
96	1.25385	1.25350	1.25080	1.24770	1.24450	61	1.15925	1.15895	1.15655	1.15380	1.15100	26	1.06410	1.06390	1.06240	1.06055	1.05855
95	1.25130	1.25095	1.24825	1.24515	1.24190	60	1.15650	1.15615	1.15380	1.15105	1.14830	25	1.06150	1.06130	1.05980	1.05800	1.05605
94	1.24865	1.24830	1.24560	1.24250	1.23930	59	1.15370	1.15340	1.15105	1.14835	1.14555	24	1.05885	1.05870	1.05720	1.05545	1.05350
93	1.24600	1.24565	1.24300	1.23985	1.23670	58	1.15095	1.15065	1.14830	1.14560	1.14285	23	1.05625	1.05610	1.05465	1.05290	1.05100
92	1.24340	1.24305	1.24035	1.23725	1.23410	57	1.14815	1.14785	1.14555	1.14285	1.14010	22	1.05365	1.05350	1.05205	1.05035	1.04850
91	1.24075	1.24040	1.23770	1.23460	1.23150	56	1.14535	1.14510	1.14280	1.14015	1.13740	21	1.05100	1.05090	1.04950	1.04780	1.04600
90	1.23810	1.23775	1.23510	1.23200	1.22890	55	1.14260	1.14230	1.14005	1.13740	1.13470	20	1.04840	1.04825	1.04690	1.04525	1.04350
89	1.23545	1.23510	1.23245	1.22935	1.22625	54	1.13980	1.13955	1.13730	1.13465	1.13195	19	1.04590	1.04575	1.04440	1.04280	1.04105
88	1.23280	1.23245	1.22975	1.22665	1.22360	53	1.13705	1.13680	1.13455	1.13195	1.12925	18	1.04335	1.04325	1.04195	1.04035	1.03860
87	1.23015	1.22980	1.22710	1.22400	1.22095	52	1.13425	1.13400	1.13180	1.12920	1.12650	17	1.04085	1.04075	1.03945	1.03790	1.03615
86	1.22750	1.22710	1.22445	1.22135	1.21830	51	1.13150	1.13125	1.12905	1.12650	1.12380	16	1.03835	1.03825	1.03695	1.03545	1.03370
85	1.22485	1.22445	1.22180	1.21870	1.21565	50	1.12870	1.12845	1.12630	1.12375	1.12110	15	1.03580	1.03570	1.03450	1.03300	1.03130
84	1.22220	1.22180	1.21915	1.21605	1.21300	49	1.12600	1.12575	1.12360	1.12110	1.11845	14	1.03330	1.03320	1.03200	1.03055	1.02885
83	1.21955	1.21915	1.21650	1.21340	1.21035	48	1.12325	1.12305	1.12090	1.11840	1.11580	13	1.03080	1.03070	1.02955	1.02805	1.02640
82	1.21690	1.21650	1.21380	1.21075	1.20770	47	1.12055	1.12030	1.11820	1.11575	1.11320	12	1.02830	1.02820	1.02705	1.02560	1.02395
81	1.21425	1.21385	1.21115	1.20810	1.20505	46	1.11780	1.11760	1.11550	1.11310	1.11055	11	1.02575	1.02565	1.02455	1.02315	1.02150
80	1.21160	1.21120	1.20850	1.20545	1.20240	45	1.11510	1.11490	1.11280	1.11040	1.10795	10	1.02325	1.02315	1.02210	1.02070	1.01905
79	1.20885	1.20845	1.20575	1.20275	1.19970	44	1.11235	1.11215	1.11010	1.10775	1.10530	9	1.02085	1.02075	1.01970	1.01835	1.01670
78	1.20610	1.20570	1.20305	1.20005	1.19705	43	1.10960	1.10945	1.10740	1.10510	1.10265	8	1.01840	1.01835	1.01730	1.01600	1.01440
77	1.20335	1.20300	1.20030	1.19735	1.19435	42	1.10690	1.10670	1.10470	1.10240	1.10005	7	1.01600	1.01590	1.01495	1.01360	1.01205
76	1.20060	1.20025	1.19760	1.19465	1.19170	41	1.10415	1.10400	1.10200	1.09975	1.09740	6	1.01360	1.01350	1.01255	1.01125	1.00970
75	1.19785	1.19750	1.19485	1.19195	1.18900	40	1.10145	1.10130	1.09930	1.09710	1.09475	5	1.01120	1.01110	1.01015	1.00890	1.00735
74	1.19510	1.19480	1.19215	1.18925	1.18635	39	1.09875	1.09860	1.09665	1.09445	1.09215	4	1.00875	1.00870	1.00780	1.00655	1.00505
73	1.19235	1.19205	1.18940	1.18650	1.18365	38	1.09605	1.09590	1.09400	1.09180	1.08955	3	1.00635	1.00630	1.00540	1.00415	1.00270
72	1.18965	1.18930	1.18670	1.18380	1.18100	37	1.09340	1.09320	1.09135	1.08915	1.08690	2	1.00395	1.00385	1.00300	1.00180	1.00035
71	1.18690	1.18655	1.18395	1.18110	1.17830	36	1.09070	1.09050	1.08865	1.08655	1.08430	1	1.00155	1.00145	1.00060	0.99945	0.99800
70	1.18415	1.18385	1.18125	1.17840	1.17565	35	1.08800	1.08780	1.08600	1.08390	1.08165	0	0.99913	0.99905	0.99823	0.99708	0.99568
69	1.18135	1.18105	1.17850	1.17565	1.17290	34	1.08530	1.08515	1.08335	1.08125	1.07905						
68	1.17860	1.17830	1.17575	1.17295	1.17020	33	1.08265	1.08245	1.08070	1.07860	1.07645						
67	1.17585	1.17555	1.17300	1.17020	1.16745	32	1.07995	1.07975	1.07800	1.07600	1.07380						
66	1.17305	1.17275	1.17025	1.16745	1.16470	31	1.07725	1.07705	1.07535	1.07335	1.07120						

* Bosart and Snoddy, *Ind. Eng. Chem.*, **20**, 1378 (1928).

Table 137. Hydrazine (N_2H_4)

%	d_4^{15}	%	d_4^{15}
1	1.0002	30	1.0305
2	1.0013	40	1.038
4	1.0034	50	1.044
8	1.0077	60	1.047
12	1.0121	70	1.046
16	1.0164	80	1.040
20	1.0207	90	1.030
24	1.0248	100	1.011
28	1.0286		

Table 138. Densities of Aqueous Solutions of Miscellaneous Organic Compounds*

d (resp., d_w, d_s) = density of the solution [resp., water; resp., the pure liquid solute] in g. per ml. p_s (resp., p_w) = weight % of solute (resp., water) in the solution. "Range" = range of applicability of the equation.

Section A. $d = d_w + Ap_s + Bp_s^2 + Cp_s^3$

Name	Formula	t, °C.	Range, p_s	A	B	C
Acetaldehyde	C_2H_4O	18	0– 30	$+0.0_{3}255$	$-0.0_{5}16$	
Acetamide	C_2H_5NO	15	0– 6	$+0.0_{3}639$	$+0.0_{4}171$	
Acetone	C_3H_6O	0	0–100	$-0.0_{3}856$	$-0.0_{5}449$	$-0.0_{7}588$
		4	0–100	$-0.0_{3}7648$	$-0.0_{4}1193$	$+0.0_{8}272$
		15	0–100	$-0.0_{2}1009$	$-0.0_{5}9682$	$-0.0_{8}624$
		20	0–100	$-0.0_{2}1233$	$-0.0_{5}3529$	$-0.0_{7}5327$
		25	0–100	$-0.0_{2}1171$	$-0.0_{5}904$	$-0.0_{8}56$
Acetonitrile	C_2H_3N	15	0– 16	$-0.0_{2}1175$	$-0.0_{4}2024$	
Allyl alcohol	C_3H_6O	0	0– 89	$-0.0_{3}3729$	$-0.0_{4}1232$	$+0.0_{7}2984$
Benzenepentacarboxylic acid	$C_{11}H_6O_{10}$	25	0– 0.6	$+0.0_{2}5615$	$-0.0_{2}117$	
Butyl alcohol (n-)	$C_4H_{10}O$	20	0– 7.9	$-0.0_{2}1651$	$+0.0_{4}285$	
Butyric acid (n-)	$C_4H_8O_2$	18	0– 10	$+0.0_{3}414$	$+0.0_{4}131$	
		25	0– 62	$+0.0_{3}5135$	$-0.0_{4}166$	$+0.0_{6}11$
Chloral hydrate	$C_2H_3Cl_3O_2$	0	0– 70	$+0.0_{2}4489$	$+0.0_{4}2802$	$-0.0_{7}1291$
		15	0– 78	$+0.0_{2}4455$	$+0.0_{4}2198$	$+0.0_{7}4366$
		30	0– 90	$+0.0_{2}4401$	$+0.0_{4}1887$	$+0.0_{7}6549$
Chloroacetic acid	$C_2H_3ClO_2$	20	0– 32	$+0.0_{2}3648$	$+0.0_{5}302$	
		25	0– 86	$+0.0_{2}3602$	$+0.0_{5}552$	$+0.0_{7}22$
Citric acid (hydrate)	$C_6H_8O_7 + H_2O$	18	0– 50	$+0.0_{2}3824$	$+0.0_{4}1141$	$+0.0_{7}17$
Dichloroacetic acid	$C_2H_2Cl_2O_2$	20	0– 30	$+0.0_{2}4427$	$+0.0_{5}537$	$+0.0_{7}7534$
		25	0– 97	$+0.0_{2}4427$	$+0.0_{5}537$	$+0.0_{7}7534$
Diethylamine hydrochloride	$C_4H_{12}ClN$	21	0– 36	$+0.0_{3}34$	$+0.0_{6}76$	
Ethylamine hydrochloride	C_2H_8ClN	21	0– 65	$+0.0_{2}1193$	$-0.0_{5}307$	$-0.0_{7}47$
Ethylene glycol	$C_2H_6O_2$	0	0–100	$+0.0_{2}1483$	$+0.0_{5}2992$	$-0.0_{7}5248$
		15	0– 6	$+0.0_{2}133$	$-0.0_{5}108$	
Ethyl ether	$C_4H_{10}O$	20	0– 5	$-0.0_{2}221$	$+0.0_{4}48$	
		25	0– 4.5	$-0.0_{2}221$	$+0.0_{4}35$	
tartrate	$C_8H_{14}O_6$	15	0– 95	$+0.0_{2}2367$	$+0.0_{5}358$	$-0.0_{7}6005$
Formaldehyde	CH_2O	15	0– 40	$+0.0_{2}2518$	$-0.0_{5}658$	$+0.0_{6}542$
Formamide	CH_3NO	25	22– 96	$+0.0_{2}1217$	$+0.0_{5}3199$	$-0.0_{7}2529$
Furfural	$C_5H_4O_2$	20	0– 8	$+0.0_{2}1827$	$+0.0_{5}366$	
		25	0– 8	$+0.0_{2}1664$	$+0.0_{4}21$	
Isoamyl alcohol	$C_5H_{12}O$	20	0– 2.5	$+0.0_{2}155$	$+0.0_{4}3$	
Isobutyl alcohol	$C_4H_{10}O$	15	0– 8	$-0.0_{2}146$	$+0.0_{5}6$	
		20	0– 8	$-0.0_{2}169$	$+0.0_{4}38$	
Isobutyric acid	$C_4H_8O_2$	15	0– 9	$+0.0_{3}52$		
		18	0– 9	$+0.0_{3}45$		
		25	0– 12	$+0.0_{3}37$		
Isovaleric acid	$C_5H_{10}O_2$	25	0– 5	$+0.0_{3}253$	$-0.0_{4}282$	
Lactic acid	C_3H_6O	25	0– 9	$+0.0_{2}231$	$+0.0_{5}186$	
Maleic acid	$C_4H_4O_4$	25	0– 40	$+0.0_{2}34$	$+0.0_{5}75$	
Malic acid	$C_4H_6O_5$	20	0– 40	$+0.0_{2}3933$	$+0.0_{5}957$	
		25	0– 40	$+0.0_{2}3736$	$+0.0_{4}175$	
Malonic acid	$C_3H_4O_4$	20	0– 40	$+0.0_{2}389$	$+0.0_{4}1066$	
Methyl acetate	$C_3H_6O_2$	20	0– 20	$+0.0_{3}40$	$-0.0_{5}74$	
glucoside (α-)	$C_7H_{14}O_6$	0	26– 51	$+0.0_{2}3336$	$+0.0_{5}996$	$+0.01544$
		30	26– 51	$+0.0_{2}3151$	$+0.0_{5}975$	$+0.0_{8}978$
Nicotine	$C_{10}H_{14}N_2$	20	0– 60	$+0.0_{3}642$	$+0.0_{5}454$	$-0.0_{7}687$
Nitrophenol (p-)	$C_6H_5NO_3$	15	0– 1.5	$+0.0_{2}3216$	$-0.0_{4}55$	
Oxalic acid	$C_2H_2O_4$	0	0– 4	$+0.0_{2}5898$	$-0.0_{3}3185$	$+0.0_{4}41$
		15	0– 4	$+0.0_{2}494$	$-0.0_{5}8$	
		17.5	0– 9	$+0.0_{2}494$	$-0.0_{5}8$	
		20	0– 4	$+0.0_{2}5264$	$-0.0_{3}1996$	$+0.0_{4}254$
		25	0– 4	$+0.0_{2}5108$	$-0.0_{3}1607$	$+0.0_{4}208$
Phenol	C_6H_6O	15	0– 5	$+0.0_{2}111$	$-0.0_{4}283$	
		80	0– 65	$+0.0_{3}462$	$-0.0_{6}86$	
Phenylglycolic acid	$C_8H_8O_3$	25	0– 11	$+0.0_{2}207$	$+0.0_{4}23$	
Picoline (α-)	C_6H_7N	25	0– 70	$-0.0_{4}386$	$-0.0_{5}1405$	$-0.0_{7}4167$
(β-)	C_6H_7N	25	0– 60	$-0.0_{4}683$	$-0.0_{5}13$	
Propionic acid	$C_3H_6O_2$	18	0– 10	$+0.0_{3}95$	$-0.0_{4}172$	
		25	0– 40	$+0.0_{3}9245$	$-0.0_{5}99$	$+0.0_{7}361$
Pyridine	C_5H_5N	25	0– 60	$+0.0_{3}229$	$-0.0_{5}204$	$-0.0_{8}28$
Resorcinol	$C_6H_6O_2$	18	0– 52	$+0.0_{2}201$	$+0.0_{5}519$	$-0.0_{8}19$
Succinic acid	$C_4H_6O_4$	25	0– 5.5	$+0.0_{2}304$		
Tartaric acid (d., l., or dl.)	$C_4H_6O_6$	15	0– 15	$+0.0_{2}4482$	$+0.0_{4}185$	
		17.5	0– 50	$+0.0_{2}4455$	$+0.0_{4}185$	
		20	0– 50	$+0.0_{2}4432$	$+0.0_{4}1837$	
		30	0– 50	$+0.0_{2}4335$	$+0.0_{4}185$	
		40	0– 50	$+0.0_{2}4265$	$+0.0_{4}185$	
		50	0– 50	$+0.0_{2}4205$	$+0.0_{4}185$	
		60	0– 50	$+0.0_{2}4155$	$+0.0_{4}185$	

* From "International Critical Tables," vol. 3, pp. 111–114.

Table 138. Densities of Aqueous Solutions of Miscellaneous Organic Compounds—*(Concluded)*

Name	Formula	*t*, °C.	Range, p_s	*A*	*B*	*C*
Tetraethyl ammonium chloride	$C_8H_{20}ClN$	21	0– 63	$+0.0_{3}1884$	$+0.0_{5}6$	$+0.0_{7}122$
Thiourea	CH_4N_2S	15	0– 7	$+0.0_{2}2995$	$+0.0_{5}374$	
Trichloroacetic acid	$C_2HCl_3O_2$	12.5	0– 61	$+0.0_{2}499$	$+0.0_{4}153$	
		20	10– 30	$+0.0_{2}5053$	$+0.0_{4}1387$	
		25	0– 94	$+0.0_{2}5051$	$+0.0_{6}6119$	$+0.0_{6}1038$
Triethylamine hydrochloride	$C_6H_{16}ClN$	21	0– 54	$+0.0_{4}6$	$+0.0_{5}558$	$-0.0_{8}69$
Trimethyl carbinol	$C_4H_{10}O$	20	0–100	$-0.0_{2}117$	$-0.0_{4}1908$	$+0.0_{7}957$
		25	0–100	$-0.0_{2}1286$	$-0.0_{4}176$	$+0.0_{7}887$
Urea	CH_4N_2O	14.8	0– 12	$+0.0_{2}3213$	$-0.0_{4}4802$	$+0.0_{5}1216$
		18	0– 51	$+0.0_{2}2718$	$+0.0_{5}1552$	$+0.0_{7}2573$
		20	0– 35	$+0.0_{2}2702$	$+0.0_{5}3712$	$-0.0_{7}2285$
		25	0– 10	$+0.0_{2}2728$	$-0.0_{4}1817$	$+0.0_{5}1379$
Urethane	$C_3H_7NO_2$	20	0– 56	$+0.0_{2}1278$	$-0.0_{5}245$	$-0.0_{7}3437$
Valeric acid (n-)	$C_5H_{10}O_2$	25	0– 3	$+0.0_{3}34$	$-0.0_{4}27$	

Section B. $d = d_s + Ap_w + Bp_w^2 + Cp_w^3$

Name	Formula	d_s	*t*, °C.	Range, p_w	*A*	*B*	*C*
Butyl alcohol (n-)	$C_4H_{10}O$	0.8097	20	0–20	$+0.0_{2}2103$	$-0.0_{4}113$	
Butyric acid (n-)	$C_4H_8O_2$	0.9534	25	0–38	$+0.0_{2}1854$	$-0.0_{4}2314$	
Ethyl ether	$C_4H_{10}O$	0.7077	25	0– 1.1	$+0.0_{2}34$	$+0.0_{3}36$	
Isobutyl alcohol	$C_4H_{10}O$	0.8170	0	0–14	$+0.0_{2}2437$	$-0.0_{4}285$	
		0.8055	15	0–16	$+0.0_{2}224$	$-0.0_{4}129$	
Isobutyric acid	$C_4H_8O_2$	0.9425	26	0–80	$+0.0_{2}1808$	$-0.0_{4}2358$	$+0.0_{6}1253$
Nicotine	$C_{10}H_{14}N_2$	1.0093	20	0–40	$+0.0_{2}199$	$-0.0_{4}331$	$+0.0_{7}315$
Picoline (α-)	C_6H_7N	0.9404	25	0–30	$+0.0_{2}2715$	$-0.0_{4}393$	
(β-)	C_6H_7N	0.9515	25	0–40	$+0.0_{2}1925$	$-0.0_{4}352$	$+0.0_{6}25$
Pyridine	C_5H_5N	0.9776	25	0–40	$+0.0_{2}1157$	$-0.0_{5}536$	$-0.0_{6}2$
Trimethyl carbinol	$C_4H_{10}O$	0.7856	20	0–20	$+0.0_{2}2287$	$+0.0_{5}275$	

Section C. $d_t = d_o + At + Bt^2$

Name	Formula	p_s	d_0	Range, °C.	*A*	*B*
Allyl alcohol	C_3H_6O	76.60	0.9122	0–45	$-0.0_{3}8$	$-0.0_{5}27$
Butyl alcohol (n-)	$C_4H_{10}O$	80.95	0.8614	0–43	$-0.0_{3}7292$	$-0.0_{6}75$
Chloral hydrate	$C_2H_3Cl_3O_2$	2.00	1.0094	7–80	$-0.0_{4}2597$	$-0.0_{5}4313$
		10.00	1.0476	7–80	$-0.0_{4}7955$	$-0.0_{5}4253$
Ethyl tartrate	$C_7H_{14}O_6$	5.00	1.0150	15–80	$-0.0_{3}2103$	$-0.0_{5}2544$
		10.00	1.0270	15–80	$-0.0_{3}2116$	$-0.0_{6}2929$
		25.00	1.0665	15–80	$-0.0_{3}401$	$-0.0_{5}23$
Furfural	$C_5H_4O_2$	4.62	1.0125	22–74	$-0.0_{3}232$	$-0.0_{5}254$
		5.69	1.0140	22–74	$-0.0_{3}221$	$-0.0_{5}268$
		6.56	1.0155	22–74	$-0.0_{3}211$	$-0.0_{5}290$
Pyridine	C_5H_5N	9.34	1.0055	11–73	$-0.0_{3}171$	$-0.0_{6}3615$
		21.20	1.0115	14–73	$-0.0_{3}378$	$-0.0_{5}248$
		29.50	1.0145	12–72	$-0.0_{3}463$	$-0.0_{5}235$
		40.40	1.0182	9–74	$-0.0_{3}605$	$-0.0_{5}167$

DENSITIES OF MISCELLANEOUS MATERIALS

Table 139. Approximate Specific Gravities and Densities of Miscellaneous Solids and Liquids*

Water at 4°C. and normal atmospheric pressure taken as unity

For more detailed data on any material, see the section dealing with the properties of that material.

Substance	Sp. gr.	Aver. weight lb./cu. ft.
Metals, Alloys, Ores		
Aluminum, cast-hammered	2.55–2.80	165
bronze	7.7	481
Brass, cast-rolled	8.4–8.7	534
Bronze, 7.9 to 14% Sn	7.4–8.9	509
phosphor	8.88	554
Copper, cast-rolled	8.8–8.95	556
ore, pyrites	4.1–4.3	262
German silver	8.58	536
Gold, cast-hammered	19.25–19.35	1205
coin (U.S.)	17.18–17.2	1073
Iridium	21.78–22.42	1383
Iron, gray cast	7.03–7.13	442
cast, pig	7.2	450
wrought	7.6–7.9	485
spiegeleisen	7.5	468
ferro-silicon	6.7–7.3	437
ore, hematite	5.2	325
ore, limonite	3.6–4.0	237
ore, magnetite	4.9–5.2	315
slag	2.5–3.0	172
Lead	11.34	710
ore, galena	7.3–7.6	465
Manganese	7.42	475
ore, pyrolusite	3.7–4.6	259
Mercury	13.6	849
Monel metal, rolled	8.97	555
Nickel	8.9	537
Platinum, cast-hammered	21.5	1330
Silver, cast-hammered	10.4–10.6	656
Steel, cold-drawn	7.83	489
machine	7.80	487
tool	7.70–7.73	481
Tin, cast-hammered	7.2–7.5	459
cassiterite	6.4–7.0	418
Tungsten	19.22	1200
Zinc, cast-rolled	6.9–7.2	440
blende	3.9–4.2	253
Various Solids		
Cereals, oats, bulk	0.51	26
barley, bulk	0.62	39
corn, rye, bulk	0.73	45
wheat, bulk	0.77	48
Cork	0.22–0.26	15
Cotton, flax, hemp	1.47–1.50	93
Fats	0.90–0.97	58
Flour, loose	0.40–0.50	28
pressed	0.70–0.80	47
Glass, common	2.40–2.80	162
plate or crown	2.45–2.72	161
crystal	2.90–3.00	184
flint	3.2–4.7	247
Hay and straw, bales	0.32	20
Leather	0.86–1.02	59
Paper	0.70–1.15	58
Potatoes, piled	0.67	44
Rubber, caoutchouc	0.92–0.96	59
goods	1.0–2.0	94
Salt, granulated, piled	0.77	48
Saltpeter	1.07	67
Starch	1.53	96
Sulfur	1.93–2.07	125
Wool	1.32	82

Substance	Sp. gr.	Aver. weight lb./cu. ft.
Timber, Air-dry		
Apple	0.66–0.74	44
Ash, black	0.55	34
white	0.64–0.71	42
Birch, sweet, yellow	0.71–0.72	44
Cedar, white, red	0.35	22
Cherry, wild red	0.43	27
Chestnut	0.48	30
Cypress	0.45–0.48	29
Elm, white	0.56	35
Fir, Douglas	0.48–0.55	32
balsam	0.40	25
Hemlock	0.45–0.50	29
Hickory	0.74–0.80	48
Locust	0.67–0.77	45
Mahogany	0.56–0.85	44
Maple, sugar	0.68	43
white	0.53	33
Oak, chestnut	0.74	46
live	0.87	54
red, black	0.64–0.71	42
white	0.77	48
Pine, Norway	0.55	34
Oregon	0.51	32
red	0.48	30
Southern	0.61–0.67	38–42
white	0.43	27
Poplar	0.43	27
Redwood, California	0.42	26
Spruce, white, red	0.45	28
Teak, African	0.99	62
Indian	0.66–0.88	48
Walnut, black	0.59	37
Willow	0.42–0.50	28
Various Liquids		
Alcohol, ethyl (100%)	0.789	49
methyl (100%)	0.796	50
Acid, muriatic, 40%	1.20	75
nitric, 91%	1.50	94
sulfuric, 87%	1.80	112
Chloroform	1.500	95
Ether	0.736	46
Lye, soda, 66%	1.70	106
Oils, vegetable	0.91–0.94	58
mineral, lubricants	0.88–0.94	57
Turpentine	0.861–0.867	54
Water, 4°C. max. density	1.0	62.428
100°C.	0.9584	59.830
ice	0.88–0.92	56
snow, fresh fallen	0.125	8
sea water	1.02–1.03	64
Ashlar Masonry		
Bluestone	2.3–2.6	153
Granite, syenite, gneiss	2.4–2.7	159
Limestone	2.1–2.8	153
Marble	2.4–2.8	162
Sandstone	2.0–2.6	143
Rubble Masonry		
Bluestone	2.2–2.5	147
Granite, syenite, gneiss	2.3–2.6	153
Limestone	2.0–2.7	147
Marble	2.3–2.7	156
Sandstone	1.9–2.5	137

Substance	Sp. gr.	Aver. weight lb./cu. ft.
Dry Rubble Masonry		
Granite, syenite, gneiss	1.9–2.3	130
Limestone, marble	1.9–2.1	125
Sandstone, bluestone	1.8–1.9	110
Brick Masonry		
Hard brick	1.8–2.3	128
Medium brick	1.6–2.0	112
Soft brick	1.4–1.9	103
Sand-lime brick	1.4–2.2	112
Concrete Masonry		
Cement, stone, sand	2.2–2.4	144
slag, etc.	1.9–2.3	130
cinder, etc.	1.5–1.7	100
Various Building Materials		
Ashes, cinders	0.64–0.72	40–45
Cement, Portland, loose	1.5	94
Lime, gypsum, loose	0.85–1.00	53–64
Mortar, lime, set	1.4–1.9	103
Portland cement	2.08–2.25	94–135
Portland cement	3.1–3.2	196
Slags, bank slag	1.1–1.2	67–72
bank screenings	1.5–1.9	98–117
machine slag	1.5	96
slag sand	0.8–0.9	49–55
Earth, etc., Excavated		
Clay, dry	1.0	63
damp plastic	1.76	110
and gravel, dry	1.6	100
Earth, dry, loose	1.2	76
dry, packed	1.5	95
moist, loose	1.3	78
moist, packed	1.6	96
mud, flowing	1.7	108
mud, packed	1.8	115
Riprap, limestone	1.3–1.4	80–85
Riprap, sandstone	1.4	90
Riprap, shale	1.7	105
Sand, gravel, dry, loose	1.4–1.7	90–105
gravel, dry, packed	1.6–1.9	100–120
gravel, wet	1.89–2.16	126
Excavations in Water		
Clay	1.28	80
River mud	1.44	90
Sand or gravel	0.96	60
and clay	1.00	65
Soil	1.12	70
Stone riprap	1.00	65
Minerals		
Asbestos	2.1–2.8	153
Barytes	4.50	281
Basalt	2.7–3.2	184
Bauxite	2.55	159
Bluestone	2.5–2.6	159
Borax	1.7–1.8	109
Chalk	1.8–2.8	143
Clay, marl	1.8–2.6	137
Dolomite	2.9	181
Feldspar, orthoclase	2.5–2.7	162
Gneiss	2.7–2.9	175
Granite	2.6–2.7	165
Greenstone, trap	2.8–3.2	187
Gypsum, alabaster	2.3–2.8	159
Hornblende	3.0	187
Limestone	2.1–2.86	155
Marble	2.6–2.86	170
Magnesite	3.0	187
Phosphate rock, apatite	3.2	200
Porphyry	2.6–2.9	172

* From Marks, "Mechanical Engineers' Handbook," McGraw-Hill.

Table 139. Approximate Specific Gravities and Densities of Miscellaneous Solids and Liquids—(*Concluded*)

Substance	Sp. gr.	Aver. weight lb./cu. ft.
Minerals (*Cont'd*)		
Pumice, natural	0.37–0.90	40
Quartz, flint	2.5–2.8	165
Sandstone	2.0–2.6	143
Serpentine	2.7–2.8	171
Shale, slate	2.6–2.9	172
Soapstone, talc	2.6–2.8	169
Syenite	2.6–2.7	165
Stone, Quarried, Piled		
Basalt, granite, gneiss	1.5	96
Greenstone, hornblende	1.7	107
Limestone, marble, quartz	1.5	95
Sandstone	1.3	82
Shale	1.5	92

Substance	Sp. gr.	Aver. weight lb./cu. ft.
Bituminous Substances		
Asphaltum	1.1–1.5	81
Coal, anthracite	1.4–1.8	97
bituminous	1.2–1.5	84
lignite	1.1–1.4	78
peat, turf, dry	0.65–0.85	47
charcoal, pine	0.28–0.44	23
charcoal, oak	0.47–0.57	33
coke	1.0–1.4	75
Graphite	1.64–2.7	135
Paraffin	0.87–0.91	56

Substance	Sp. gr.	Aver. weight lb./cu. ft.
Bituminous Substances (*Cont'd*)		
Petroleum	0.87	54
refined (kerosene)	0.78–0.82	50
benzine	0.73–0.75	46
gasoline	0.70–0.75	45
Pitch	1.07–1.15	69
Tar, bituminous	1.20	75
Coal and Coke, Piled		
Coal, anthracite	0.75–0.93	47–58
bituminous, lignite	0.64–0.87	40–54
peat, turf	0.32–0.42	20–26
charcoal	0.16–0.23	10–14
coke	0.37–0.51	23–32

SOLUBILITIES

Table 140. Solubilities of Inorganic Compounds in Water at Various Temperatures*

This table shows the amount of anhydrous substance which is soluble in 100 g. of water at the temperature in degrees centigrade as indicated; where the formula is followed by † the value is expressed in grams of substance in 100 cc. of saturated solution. Solid phase gives the hydrated form in equilibrium with the saturated solution.

	Substance	Formula	Solid phase	0°C.	10°C.	20°C.	30°C.	40°C.	50°C.	60°C.	70°C.	80°C.	90°C.	100°C.	
1	Aluminum chloride	$AlCl_3$	$6H_2O$			69.86^15°									1
2	sulfate	$Al_2(SO_4)_3$	$18H_2O$	31.2	33.5	36.4	40.4	46.1	52.2	59.2	66.1	73.0	80.8	89.0	2
3	Ammonium aluminum sulfate	$(NH_4)_2Al_2(SO_4)_4$	$24H_2O$	2.1	4.99	7.74	10.94	14.88	20.10	26.70				109.7^95°	3
4	bicarbonate	NH_4HCO_3		11.9	15.8	21	27								4
5	bromide	NH_4Br		60.6	68	75.5	83.2	91.1	99.2	107.8	116.8	126	135.6	145.6	5
6	chloride	NH_4Cl		29.4	33.3	37.2	41.4	45.8	50.4	55.2	60.2	65.6	71.3	77.3	6
7	chloroplatinate	$(NH_4)_2PtCl_6$			0.7									1.25	7
8	chromate	$(NH_4)_2CrO_4$					40.4								8
9	chromium sulfate	$(NH_4)_2Cr_2(SO_4)_4$	$24H_2O$			10.78^25°									9
10	dichromate	$(NH_4)_2Cr_2O_7$					47.17								10
11	dihydrogen phosphite	$NH_4H_2PO_3$		171		190^14.5°	260^31°								11
12	hydrogen phosphate	$(NH_4)_2HPO_4$				131^15									12
13	iodide	NH_4I		154.2	163.2	172.3	181.4	190.5	199.6	208.9	218.7	228.8		250.3	13
14	magnesium phosphate	NH_4MgPO_4	$6H_2O$	0.023		0.052		0.036	0.030	0.040	0.016	0.019			14
15	manganese phosphate	NH_4MnPO_4	$7H_2O$			0		0		0	0.005	0.007			15
16	nitrate	NH_4NO_3		118.3		192	241.8	297.0	344.0	421.0	499.0	580.0	740.0	871.0	16
17	oxalate	$(NH_4)_2C_2O_4$	$1H_2O$	2.2	3.1	4.4	5.9	8.0	10.3						17
18	perchlorate†	NH_4ClO_4†		11.56		20.85		30.58		39.05		48.19		57.01	18
19	persulfate	$(NH_4)_2S_2O_8$		58.2											19
20	sulfate	$(NH_4)_2SO_4$		70.6	73.0	75.4	78.0	81.0		88.0		95.3		103.3	20
21	thiocyanate	NH_4CNS		119.8	144	170	207.7								21
22	vanadate (meta)	NH_4VO_3				0.48	0.84	1.32	1.78		3.05				22
23	Antimonious fluoride	SbF_3		384.7		444.7	563.6								23
24	sulfide	Sb_2S_3				0.000175^18°									24
25	Arsenic oxide	As_2O_5		59.5	62.1	65.8	69.5	71.2		73.0		75.1		76.7	25
26	Arsenious sulfide	As_2S_3		5.17 × 10^-5 at 18°											26
27	Barium acetate	$Ba(C_2H_3O_2)_2$	$3H_2O$	59	63	71									27
28	acetate	$Ba(C_2H_3O_2)_2$	$1H_2O$				75	79	77	74	74			75	28
29	carbonate	$BaCO_3$			0.0016^8°	0.0022^18°	0.0024 at 24.2°								29
30	chlorate	$Ba(ClO_3)_2$	$1H_2O$	20.34	26.95	33.80	41.70	49.61		66.81		84.84		104.9	30
31	chloride	$BaCl_2$	$2H_2O$	31.6	33.3	35.7	38.2	40.7	43.6	46.4	49.4	52.4		58.8	31
32	chromate	$BaCrO_4$		0.0002	0.00028	0.00037	0.00046								32
33	hydroxide	$Ba(OH)_2$	$8H_2O$	1.67	2.48	3.89	5.59	8.22	13.12	20.94		101.4			33
34	iodide	BaI_2	$6H_2O$	170.2	185.7	203.1	219.6								34
35	iodide	BaI_2	$2H_2O$					231.9		247.3		261.0		271.7	35
36	nitrate	$Ba(NO_3)_2$		5.0	7.0	9.2	11.6	14.2	17.1	20.3		27.0		34.2	36
37	nitrite	$Ba(NO_2)_2$	$1H_2O$			67.5						205.8		300	37
38	oxalate	BaC_2O_4			0.0016^8°	0.0022^18°	0.0024 at 24.2°								38
39	perchlorate	$Ba(ClO_4)_2$	$3H_2O$	205.8		289.1		358.7	426.3		495.2		562.3		39
40	sulfate	$BaSO_4$		1.15×10^{-4}	2.0×10^{-4}	2.4×10^{-4}	2.85×10^{-4}								40
41	Beryllium sulfate	$BeSO_4$	$6H_2O$				52		60.67						41
42	sulfate	$BeSO_4$	$4H_2O$				43.78	46.74			62		83	100	42
43	sulfate	$BeSO_4$	$2H_2O$									84.76	98	110	43
44	Boric acid	H_3BO_3		2.66	3.57	5.04	6.60	8.72	11.54	14.81	16.73	23.75	30.38	40.25	44
45	Boron oxide	B_2O_3		1.1	1.5	2.2		4.0		6.2		9.5		15.7	45
46	Bromine	Br_2		4.22	3.4	3.20	3.13								46
47	Cadmium chloride	$CdCl_2$	$4H_2O$	97.59	125.1										47
48	chloride	$CdCl_2$	$2\frac{1}{2}H_2O$	90.01			132.1								48
49	chloride	$CdCl_2$	$1H_2O$		135.1	134.5		135.3		136.5		140.4		147.0	49
50	cyanide	$Cd(CN)_2$				1.7^15°									50
51	hydroxide	$Cd(OH)_2$					2.6×10^{-4} at 25°								51
52	sulfate	$CdSO_4$		76.48	76.00	76.60		78.54		83.68			63.13	60.77	52
53	Calcium acetate	$Ca(C_2H_3O_2)_2$	$2H_2O$	37.4	36.0	34.7	33.8	33.2		32.7		33.5			53
54	acetate	$Ca(C_2H_3O_2)_2$	$1H_2O$										31.1	29.7	54

* By N. A. Lange, Handbook Publishers, Inc., Sandusky, Ohio. Abridged from table of Solubilities of Inorganic Compounds in Water at Various Temperatures in Lange's "Handbook of Chemistry."

Table 140. Solubilities of Inorganic Compounds in Water at Various Temperatures—(*Continued*)

	Substance	Formula	Solid phase	0°C.	10°C.	20°C.	30°C.	40°C.	50°C.	60°C.	70°C.	80°C.	90°C.	100°C.	
1	Calcium bicarbonate	$Ca(HCO_3)_2$	...	16.15	...	16.60	...	17.05	...	17.50	...	17.95	...	18.40	1
2	chloride	$CaCl_2$	$6H_2O$	59.5	65.0	74.5	102	...	...	...	...	...	...	...	2
3	chloride	$CaCl_2$	$2H_2O$	...	...	...	...	...	...	136.8	141.7	147.0	152.7	159	3
4	fluoride	CaF_2	...	...	...	$0.0016^{18°}$	$0.0017^{26°}$	...	...	...	...	...	...	...	4
5	hydroxide	$Ca(OH)_2$	...	0.185	0.176	0.165	0.153	0.141	0.128	0.116	0.106	0.094	0.085	0.077	5
6	nitrate	$Ca(NO_3)_2$	$4H_2O$	102.0	115.3	129.3	152.6	195.9	...	...	...	...	...	...	6
7	nitrate	$Ca(NO_3)_2$	$3H_2O$	...	...	...	...	237.5	281.5	...	...	...	...	...	7
8	nitrate	$Ca(NO_3)_2$	...	...	...	...	...	...	...	...	...	358.7	...	363.6	8
9	nitrite	$Ca(NO_2)_2$	$4H_2O$	62.07	...	76.68	...	...	...	...	...	...	...	...	9
10	nitrite	$Ca(NO_2)_2$	$2H_2O$	...	...	...	...	...	...	132.6	151.9	...	244.8	...	10
11	oxalate	CaC_2O_4	...	...	6.7×10^{-4} at 13°	6.8×10^{-4} at 25°	9.5×10^{-4} at 50°	14×10^{-4} at 95°	...	...	...	...	...	...	11
12	sulfate	$CaSO_4$	$2H_2O$	0.1759	0.1928	...	0.2090	0.2097	...	0.2047	0.1966	...	...	0.1619	12
13	Carbon dioxide, 760 mm	CO_2	...	0.3346	0.2318	0.1688	0.1257	0.0973	0.0761	0.0576	...	...	...	0	13
14	monoxide, 760 mm	CO	...	0.0044	0.0035	0.0028	0.0024	0.0021	0.0018	0.0015	0.0013	0.0010	0.0006	0	14
15	Cesium chloride	$CsCl$	...	161.4	174.7	186.5	197.3	208.0	218.5	229.7	239.5	250.0	260.1	270.5	15
16	nitrate	$CsNO_3$	...	9.33	14.9	23.0	33.9	47.2	64.4	83.8	107.0	134.0	163.0	197.0	16
17	sulfate	Cs_2SO_4	...	167.1	173.1	178.7	184.1	189.9	194.9	199.9	205.0	210.3	214.9	220.3	17
18	Chlorine, 760 mm	Cl_2	...	1.46	0.980	0.716	0.562	0.451	0.386	0.324	0.274	0.219	0.125	0	18
19	Chromic anhydride	CrO_3	...	164.9	...	...	...	174.0	182.1	...	...	...	217.5	206.8	19
20	Cupric chloride	$CuCl_2$	$2H_2O$	70.7	73.76	77.0	80.34	83.8	87.44	91.2	...	99.2	...	107.9	20
21	nitrate	$Cu(NO_3)_2$	$6H_2O$	81.8	95.28	125.1	...	...	...	...	...	...	...	...	21
22	nitrate	$Cu(NO_3)_2$	$3H_2O$	...	...	...	...	159.8	...	178.8	...	207.8	...	...	22
23	sulfate	$CuSO_4$	$5H_2O$	14.3	17.4	20.7	25	28.5	33.3	40	...	55	...	75.4	23
24	sulfide	CuS	...	...	...	3.3×10^{-5} at 18°	...	...	...	...	...	...	...	...	24
25	Cuprous chloride	$CuCl$	...	...	...	$1.52^{25°}$	...	...	...	...	...	...	...	...	25
26	Ferric chloride	$FeCl_3$	...	74.4	81.9	91.8	...	...	315.1	...	...	525.8	...	535.7	26
27	Ferrous chloride	$FeCl_2$	$4H_2O$	...	64.5	...	73.0	77.3	82.5	88.7	...	...	...	...	27
28	chloride	$FeCl_2$	...	...	...	...	...	...	...	...	...	100	105.3	105.8	28
29	nitrate	$Fe(NO_3)_2$	$6H_2O$	71.02	...	83.8	...	...	...	165.6	...	...	...	...	29
30	sulfate	$FeSO_4$	$7H_2O$	15.65	20.51	26.5	32.9	40.2	48.6	...	...	...	...	...	30
31	sulfate	$FeSO_4$	$1H_2O$	...	...	...	...	...	...	...	50.9	43.6	37.3	...	31
32	Hydrobromic acid, 760 mm	HBr	...	221.2	210.3	198	...	...	171.5	...	...	...	...	130	32
33	Hydrochloric acid, 760 mm	HCl	...	82.3	...	...	67.3	63.3	59.6	56.1	...	...	...	...	33
34	Iodine	I_2	...	...	...	0.029	0.04	0.056	0.078	...	...	...	...	...	34
35	Lead acetate	$Pb(C_2H_3O_2)_2$	$3H_2O$	...	...	...	$55.04^{25°}$	...	...	...	...	...	...	...	35
36	bromide	$PbBr_2$	...	0.4554	...	0.85	1.15	1.53	1.94	2.36	...	3.34	...	4.75	36
37	carbonate	$PbCO_3$	...	...	...	0.00011	...	...	...	...	...	...	...	...	37
38	chloride	$PbCl_2$	...	0.6728	...	0.99	1.20	1.45	1.70	1.98	...	2.62	...	3.34	38
39	chromate	$PbCrO_4$	...	...	...	7×10^{-6}	...	...	...	...	...	...	...	...	39
40	fluoride	PbF_2	...	...	0.060	0.064	0.068	...	...	...	...	...	...	...	40
41	nitrate	$Pb(NO_3)_2$	...	38.8	48.3	56.5	66	75	85	95	...	115	...	38.8	41
42	sulfate	$PbSO_4$	...	0.0028	0.0035	0.0041	0.0049	0.0056	...	...	...	...	...	...	42
43	Magnesium bromide	$MgBr_2$	$6H_2O$	91.0	94.5	96.5	99.2	101.6	104.1	107.5	...	113.7	...	120.2	43
44	chloride	$MgCl_2$	$6H_2O$	52.8	53.5	54.5	...	57.5	...	61.0	...	66.0	...	73.0	44
45	hydroxide	$Mg(OH)_2$	...	...	...	$0.0009^{18°}$	...	...	...	...	...	...	...	...	45
46	nitrate	$Mg(NO_3)_2$	$6H_2O$	66.55	...	...	...	84.74	...	...	...	...	137.0	...	46
47	sulfate	$MgSO_4$	$7H_2O$	...	30.9	35.5	40.8	45.6	...	...	...	...	...	...	47
48	sulfate	$MgSO_4$	$6H_2O$	40.8	42.2	44.5	45.3	...	50.4	53.5	59.5	64.2	69.0	74.0	48
49	sulfate	$MgSO_4$	$1H_2O$	...	...	...	...	...	...	...	...	62.9	...	68.3	49
50	Manganous sulfate	$MnSO_4$	$7H_2O$	53.23	60.01	...	...	...	...	...	...	...	...	...	50
51	sulfate	$MnSO_4$	$5H_2O$	...	59.5	62.9	67.76	...	...	...	...	...	...	...	51
52	sulfate	$MnSO_4$	$4H_2O$	...	...	64.5	66.44	68.8	72.6	...	...	...	...	...	52
53	sulfate	$MnSO_4$	$1H_2O$	...	...	...	...	...	58.17	55.0	52.0	48.0	42.5	34.0	53
54	Mercurous chloride	$HgCl$	...	0.00014	...	0.0002	...	0.0007	...	...	...	...	...	...	54
55	Molybdic oxide	MoO_3	$2H_2O$	...	...	0.138	0.264	0.476	0.687	1.206	2.055	2.106	...	...	55
56	Nickel chloride	$NiCl_2$	$6H_2O$	53.9	59.5	64.2	68.9	73.3	78.3	82.2	85.2	...	...	87.6	56
57	nitrate	$Ni(NO_3)_2$	$6H_2O$	79.58	...	96.31	...	122.2	...	...	...	...	...	...	57
58	nitrate	$Ni(NO_3)_2$	$3H_2O$	...	...	...	...	...	...	163.1	169.1	...	235.1	...	58
59	sulfate	$NiSO_4$	$7H_2O$	27.22	32	...	42.46	...	...	...	...	...	...	...	59
60	sulfate	$NiSO_4$	$6H_2O$	...	...	...	...	...	50.15	54.80	59.44	63.17	...	76.7	60
61	Nitric oxide, 760 mm	NO	...	0.00984	0.00757	0.00618	0.00517	0.00440	0.00376	0.00324	0.00267	0.00199	0.00114	0	61
62	Nitrous oxide	N_2O	...	...	0.1705	0.1211	...	...	...	...	...	...	...	...	62

Table 140. Solubilities of Inorganic Compounds in Water at Various Temperatures—*(Continued)*

	Substance	Formula	Solid phase	0°C.	10°C.	20°C.	30°C.	40°C.	50°C.	60°C.	70°C.	80°C.	90°C.	100°C.	
1	Potassium acetate	$KC_2H_3O_2$	$1\frac{1}{2}H_2O$	216.7	233.9	255.6	283.8	323.3							1
2	acetate	$KC_2H_3O_2$	$\frac{1}{2}H_2O$						337.3	350	364.8	380.1	396.3		2
3	alum	$K_2SO_4.Al_2(SO_4)_3$	$24H_2O$	3.0	4.0	5.9	8.39	11.70	17.00	24.75	40.0	71.0	109.0		3
4	bicarbonate	$KHCO_3$		22.4	27.7	33.2	39.1	45.4		60.0					4
5	bisulfate	$KHSO_4$		36.3		51.4		67.3						121.6	5
6	bitartrate	$KHC_4H_4O_6$		0.32	0.40	0.53	0.90	1.32	1.83	2.46		4.6		6.95	6
7	carbonate	K_2CO_3	$2H_2O$	105.5	108	110.5	113.7	116.9	121.2	126.8	133.1	139.8	147.5	155.7	7
8	chlorate	$KClO_3$		3.3	5	7.4	10.5	14	19.3	24.5		38.5		57	8
9	chloride	KCl		27.6	31.0	34.0	37.0	40.0	42.6	45.5	48.3	51.1	54.0	56.7	9
10	chromate	K_2CrO_4		58.2	60.0	61.7	63.4	65.2	66.8	68.6	70.4	72.1	73.9	75.6	10
11	dichromate	$K_2Cr_2O_7$		5	7	12	20	26	34	43	52	61	70	80	11
12	ferricyanide	$K_3Fe(CN)_6$		31	36	43	50	60		66				82.6^{104}	12
13	hydroxide	KOH	$2H_2O$	97	103	112	126								13
14	hydroxide	KOH	$1H_2O$						140					178	14
15	nitrate	KNO_3		13.3	20.9	31.6	45.8	63.9	85.5	110.0	138	169	202	246	15
16	nitrite	KNO_2		278.8		298.4		334.9						412.8	16
17	perchlorate	$KClO_4$		0.75	1.05	1.80	2.6	4.4	6.5	9	11.8	14.8	18	21.8	17
18	permanganate	$KMnO_4$		2.83	4.4	6.4	9.0	12.56	16.89	22.2					18
19	persulfate†	$K_2S_2O_8$†	†	1.62	2.60	4.49	7.19	9.89							19
20	sulfate	K_2SO_4		7.35	9.22	11.11	12.97	14.76	16.50	18.17	19.75	21.4	22.8	24.1	20
21	thiocyanate	$KCNS$		177.0		217.5									21
22	Silver cyanide	$AgCN$				2.2×10^{-5}									22
23	nitrate	$AgNO_3$		122	170	222	300	376	455	525		669		952	23
24	sulfate	Ag_2SO_4		0.573	0.695	0.796	0.888	0.979	1.08	1.15	1.22	1.30	1.36	1.41	24
25	Sodium acetate	$NaC_2H_3O_2$	$3H_2O$	36.3	40.8	46.5	54.5	65.5	83	139					25
26	acetate	$NaC_2H_3O_2$		119	121	123.5	126	129.5	134	139.5	146	153	161	170	26
27	bicarbonate	$NaHCO_3$		6.9	8.15	9.6	11.1	12.7	14.45	16.4					27
28	carbonate	Na_2CO_3	$10H_2O$	7	12.5	21.5	38.8								28
29	carbonate	Na_2CO_3	$1H_2O$				50.5	48.5		46.4		45.8		45.5	29
30	chlorate	$NaClO_3$		79	89	101	113	126	140	155	172	189		230	30
31	chloride	$NaCl$		35.7	35.8	36.0	36.3	36.6	37.0	37.3	37.8	38.4	39.0	39.8	31
32	chromate	Na_2CrO_4	$10H_2O$	31.70	50.17	88.7									32
33	chromate	Na_2CrO_4	$4H_2O$				88.7	95.96	104	114.6					33
34	chromate	Na_2CrO_4									123.0	124.8		125.9	34
35	dichromate	$Na_2Cr_2O_7$	$2H_2O$	163.0		177.8			244.8		316.7	376.2			35
36	dichromate	$Na_2Cr_2O_7$												426.3	36
37	dihydrogen phosphate	NaH_2PO_4	$2H_2O$	57.9	69.9	85.2	106.5	138.2							37
38	dihydrogen phosphate	NaH_2PO_4	$1H_2O$						158.6						38
39	dihydrogen phosphate	NaH_2PO_4								179.3	190.3	207.3	225.3	246.6	39
40	hydrogen arsenate	Na_2HAsO_4	$12H_2O$	7.3	15.5	26.5	37	47		65		85			40
41	hydrogen phosphate	Na_2HPO_4	$12H_2O$	1.67	3.6	7.7	20.8								41
42	hydrogen phosphate	Na_2HPO_4	$7H_2O$					51.8							42
43	hydrogen phosphate	Na_2HPO_4	$2H_2O$						80.2	82.9	88.1	92.4	102.9		43
44	hydrogen phosphate	Na_2HPO_4												102.2	44
45	hydroxide	$NaOH$	$4H_2O$	42											45
46	hydroxide	$NaOH$	$3\frac{1}{2}H_2O$		51.5										46
47	hydroxide	$NaOH$	$1H_2O$			109	119	129	145	174					47
48	hydroxide	$NaOH$											313	347	48
49	nitrate	$NaNO_3$		73	80	88	96	104	114	124		148		180	49
50	nitrite	$NaNO_2$		72.1	78.0	84.5	91.6	98.4	104.1			132.6		163.2	50
51	oxalate	$Na_2C_2O_4$				3.7								6.33	51
52	phosphate, tri-	Na_3PO_4	$12H_2O$	1.5	4.1	11	20	31	43	55		81		108	52
53	pyrophosphate	$Na_4P_2O_7$	$10H_2O$	3.16	3.95	6.23	9.95	13.50	17.45	21.83		30.04		40.26	53
54	sulfate	Na_2SO_4	$10H_2O$	5.0	9.0	19.4	40.8								54
55	sulfate	Na_2SO_4	$7H_2O$	19.5	30	44									55
56	sulfate	Na_2SO_4						48.8	46.7	45.3		43.7		42.5	56
57	sulfide	Na_2S	$9H_2O$		15.42	18.8	22.5	28.5							57
58	sulfide	Na_2S	$5\frac{1}{2}H_2O$						39.82	42.69	45.73	51.40	59.23		58
59	sulfide	Na_2S	$6H_2O$						36.4	39.1	43.31	49.14	57.28		59
60	sulfite	Na_2SO_3	$7H_2O$	13.9	20	26.9	36								60
61	sulfite	Na_2SO_3						28	28.2	28.8		28.3			61
62	tetraborate	$Na_2B_4O_7$	$10H_2O$	1.3	1.6	2.7	3.9		10.5	20.3					62
63	tetraborate	$Na_2B_4O_7$	$5H_2O$								24.4	31.5	41	52.5	63
64	vanadate (meta)	$NaVO_3$	$2H_2O$			$15.3^{25°}$		30.2		68.4					64

Table 140. Solubilities of Inorganic Compounds in Water at Various Temperatures—(*Concluded*)

	Substance	Formula	Solid phase	0°C.	10°C.	20°C.	30°C.	40°C.	50°C.	60°C.	70°C.	80°C.	90°C.	100°C.	
1	Sodium vanadate (meta)	$NaVO_3$				21.10[25°]		26.23		32.97	36.9	38.8[75°]			1
2	Stannous chloride	$SnCl_2$		83.9		269.8[15°]									2
3	sulfate	$SnSO_4$				19								18	3
4	Strontium acetate	$Sr(C_2H_3O_2)_2$	$4H_2O$	36.9	43.61										4
5	acetate	$Sr(C_2H_3O_2)_2$	$\frac{1}{2}H_2O$		42.95	41.6	39.5		37.35		36.24	36.10		36.4	5
6	chloride	$SrCl_2$	$6H_2O$	43.5	47.7	52.9	58.7	65.3	72.4	81.8					6
7	chloride	$SrCl_2$	$2H_2O$								85.9	90.5		100.8	7
8	nitrate	$Sr(NO_3)_2$	$1H_2O$	52.7		64.0			83.8	97.2			130.4	139	8
9	nitrate	$Sr(NO_3)_2$	$4H_2O$	40.1		70.5									9
10	nitrate	$Sr(NO_3)_2$					88.6	90.1		93.8	96	98	100		10
11	sulfate	$SrSO_4$		0.0113		0.0114	0.0114								11
12	Sulfur dioxide, 760 mm	SO_2		22.83	16.21	11.29	7.81	5.41	4.5						12
13	Thallium sulfate	Tl_2SO_4		2.70	3.70	4.87	6.16		9.21	10.92	12.74	14.61	16.53	18.45	13
14	Thorium sulfate	$Th(SO_4)_2$	$9H_2O$	0.74	0.98	1.38	1.995	2.998	5.22						14
15	sulfate	$Th(SO_4)_2$	$8H_2O$	1.0	1.25	1.62									15
16	sulfate	$Th(SO_4)_2$	$6H_2O$	1.50		1.90	2.45			6.64					16
17	sulfate	$Th(SO_4)_2$	$4H_2O$					4.04	2.54	1.63	1.09				17
18	Zinc chlorate	$ZnClO_3$	$6H_2O$	145.0	152.5										18
19	chlorate	$ZnClO_3$	$4H_2O$			200.3	209.2	223.2	273.1						19
20	nitrate	$Zn(NO_3)_2$	$6H_2O$	94.78		118.3									20
21	nitrate	$Zn(NO_3)_2$	$3H_2O$					206.9							21
22	sulfate	$ZnSO_4$	$7H_2O$	41.9	47	54.4									22
23	sulfate	$ZnSO_4$	$6H_2O$					70.1	76.8						23
24	sulfate	$ZnSO_4$	$1H_2O$									86.6	83.7	80.8	24

THERMAL EXPANSION

The tables given under this subject are reprinted by permission from the "Smithsonian Tables." For more detailed data on thermal expansion see "International Critical Tables": tabular index, vol. 3, p. 1; abrasives, vol. 2, p. 87; alloys, vol. 2, p. 463; building stones, vol. 2, p. 54; carbons, vol. 2, p. 303; elements, vol. 1, p. 102; enamels, vol. 2, p. 115; glass, vol. 2, p. 93; metals, vol. 2, p. 459; petroleums, vol. 2, p. 145; porcelains, vol. 2, pp. 70, 78; refractory materials, vol. 2, p. 83; solid insulators, vol. 2, p. 310.

Table 141. Coefficients of Thermal Expansion of Gases*

Coefficient at constant volume for temperatures from 0 to 100°C. $\alpha_p = \frac{1}{p_0}\left[\frac{dp}{dt}\right]_v$

Substance	Initial pressure, mm. Hg	$10^6\alpha_p$	Substance	Initial pressure, mm. Hg	$10^6\alpha_p$
Air	760	3671.6	Hydrogen	760	3662.7
Air	1000	3675	Hydrogen	1000	3662.6
Ammonia	760	3767.8	chloride	760	3721
Argon	517	3668	Krypton	1000	3689.9
Argon	760	3672	Methane	760	3679
Argon	1000	3675	Neon	760	3662.8
Carbon dioxide	760	3711	Neon	1362.8	3662.3
dioxide	1000	3726	Nitrogen	760	3672
monoxide	760	3673	Nitrogen	994	3674
Chlorine	760	3803	Nitrous oxide	760	3719
Cyanogen	760	3830	Oxygen	760	3673.5
Ethylene	760	3722	Oxygen	1000	3675.7
Helium	760	3661.3	Sulfur dioxide	760	3840
Helium	1000	3660.7	Xenon	1000	3720

Coefficient at constant pressure for temperatures from 0 to 100°C. $\alpha_v = \frac{1}{v_0}\left[\frac{dv}{dt}\right]_p$

Substance	Initial pressure, mm. Hg	$10^6\alpha_p$	Substance	Initial pressure, mm. Hg	$10^6\alpha_p$
Air	760	3671.1	Hydrogen chloride	760	3734
Air	1000	3674	Krypton	862	3691.6
Ammonia	760	3721	Krypton	1000	3696.7
Argon	760	3672.4	Methane	760	3682
Argon	1000	3676	Neon	760	3660.6
Carbon dioxide	760	3725	Neon	1007	3660.2
monoxide	760	3672	Nitrogen	760	3671
Chlorine	760	3830	Nitrogen	994	3673.4
Cyanogen	760	3870	Nitrous oxide	760	3732
Ethylene	760	3735	Nitrous oxide	1000	3706.7
Helium	760	3659.1	Oxygen	760	3674
Helium	994	3657.9	Oxygen	1000	3676.3
Hydrogen	760	3660.3	Sulfur dioxide	760	3880
Hydrogen	1095	3659.0	Sulfur hexafluoride	760	3808
			Xenon	1000	3739.5

* The data were taken from collected and calculated values of Coppock, *Phil. Mag.*, (7) **19**, 446 (1935).

Table 142. Linear Expansion of the Solid Elements*

C is the true expansion coefficient at the given temperature; M is the mean coefficient between given temperatures; where one temperature is given, the true coefficient at that temperature is indicated; α and β are coefficients in formula $l_t = l_0(1 + \alpha t + \beta t^2)$; l_0 is length at 0°C. (unless otherwise indicated, when, if x is the reference temperature, $l_t = l_x[1 + \alpha(t - t_x) + \beta(t - t_x)^2]$; l_t is length at t°C.).

Element	Temp., °C.	$C \times 10^4$	Temp. range, °C.	$M \times 10^4$	Temp. range, °C.	$\alpha \times 10^4$	$\beta \times 10^6$
Aluminum	20	0.224	100	0.235	0, 500	0.22	0.009
Aluminum	300	0.284	500	0.311			
Antimony	20	0.136‖	20	0.080⊥			
Arsenic	20	0.05					
Bismuth	20	0.014‖	20	0.103⊥			
Cadmium	0	0.54‖	−180, −140	0.59‖	20, 100	0.526‖	
Cadmium	0	0.20⊥	−180, −140	0.117⊥	20, 100	0.214⊥	
Carbon, diamond	50	0.012					
graphite	50	0.06					
Chromium	...	...	20, 100	0.068	20, 500	0.086	
Cobalt	20	0.123	...	...	6, 121	0.121	0.0064
Copper	20	0.162	100	0.166	0, 625	0.161	0.0040
Copper	200	0.170	300	0.175			
Gold	20	0.140	17, 100	0.143	0, 520	0.142	0.0022
Gold	...	...	−191, 17	0.132			
Indium	40	0.417					
Iodine	...	...	−190, 17	0.837			
Iridium	20	0.065	...	...	0, 80	0.0636	0.0032
Iridium	...	...	...	...	1070, 1720	0.0679	0.0011
Iron, soft	40	0.1210	0, 100	0.11			
cast	20	0.118	...	...	0, 750	0.1158	0.0053
wrought	20	0.119	...	...	0, 750	0.1170	0.0053
steel	20	0.114	...	...	0, 750	0.1118	0.0053
Lead (99.9)	...	...	20, 100	0.291	100, 240	0.269	0.011
	100	0.291	20, 200	0.300			
	280	0.343					
Magnesium	20	0.254	−100, + 20	0.240	+ 20, 500	0.2480	0.0096
			20, 100	0.260			
Manganese	20	0.233	0, 100	0.228			
			−190, 0	0.159	20, 300	0.216	0.0121
Molybdenum†	20	0.053	0, 100	0.052	−142, 19	0.0515	0.0057
			25, 100	0.049	19, +305	0.0501	0.0014
			25, 500	0.055			
Nickel	20	0.126	0, 100	0.130	−190, + 20	0.1308	0.0166
					+ 20, +300	0.1236	0.0066
					500, 1000	0.1346	0.0033
Osmium	40	0.066					
Palladium	20	0.1173	...	...	−190, +100	0.1152	0.00517
					0, 1000	0.1167	0.0022
Platinum	20	0.0887	...	...	−190, −100	0.0875	0.00314
	20	0.0893	...	...	0, + 80	0.0890	0.00121
					0, 1000	0.0887	0.00132
Potassium	...	...	0, 50	0.83			
Rhodium	40	0.0850	6, 21	0.0876	− 75, −112	0.0746	
Ruthenium	40	0.0963					
Selenium	0	0.439	0, 100	0.660			

Table 142. Linear Expansion of the Solid Elements*—(*Concluded*)

Element	Temp., °C.	$C \times 10^4$	Temp. range, °C.	$M \times 10^4$	Temp. range, °C.	$\alpha \times 10^4$	$\beta \times 10^6$
Silicon	40	0.0763	− 3, + 18	0.0249	− 75, − 67	0.0182	
Silver	20	0.1846	0, 100	0.197	0, 875	0.1827	0.00479
	20	0.195			20, 500	0.1939	0.00295
Sodium	....	......	−190, −17	0.622	0, 50	0.72	
Steel, 36.4Ni	....	......	20, 260	0.031	260, 500	0.144	
			20, 340	0.055	340, 500	0.136	
Tantalum†	20	0.065	− 78, 0	0.059	20, 400	0.0646	0.0009
			0, 100	0.0655			
Tellurium	20	0.016‖	20	0.272⊥			
Thallium	40	0.302					
Tin	20	0.214	....	......	8, 95	0.2033	0.0263
	20	0.305‖	20	0.154⊥			
Tungsten†	27	0.0444	0, 100	0.045	−105, +502	0.0428	0.00058
Zinc	20‡	0.643‖	−140, −100	0.656‖	+ 0, 400	0.354	0.010
	20‡	0.125⊥	+ 20, 100	0.639‖			
	20	0.358	+ 20, 100	0.141⊥			

* "Smithsonian Tables."
† Molybdenum, 300° to 2500°C.; $l_t = l_{300}[1 + 5.00 \times 10^{-6}(t - 300) + 10.5 \times 10^{-10}(t - 300)^2]$
Tantalum, 300° to 2800°C.; $l_t = l_{300}[1 + 6.60 \times 10^{-6}(t - 300) + 5.2 \times 10^{-10}(t - 300)^2]$
Tungsten, 300° to 2700°C.; $l_t = l_{300}[1 + 4.44 \times 10^{-6}(t - 300) + 4.5 \times 10^{-10}(t - 300)^2]$
Beryllium, 20° to 100°C.; 12.3×10^{-6} per °C.
Columbium, 0° to 100°C.; 7.2×10^{-6} per °C.
Tantalum, 20° to 100°C.; 6.6×10^{-6} per °C.
‡ Two errors in the data of zinc have been corrected. The new values were taken from Grüneisen and Goens, *Z. Physik.*, **29**, 141 (1924).

Table 143. Linear Expansion of Miscellaneous Substances*

The coefficient of cubical expansion may be taken as three times the linear coefficient. t is the temperature or range of temperature, C the coefficient of expansion.

Substance	t°C	$C \times 10^4$
Amber	0–30	0.50
	0–09	0.61
Bakelite, bleached	20–60	0.22
Brass:		
Cast	0–100	0.1875
Wire	0–100	0.1930
Wire	0–100	0.1783 to 0.193
71.5 Cu + 27.7 Zn + 0.3 Sn + 0.5 Pb	40	0.1859
71 Cu + 29 Zn	0–100	0.1906
Bronze:		
3 Cu + 1 Sn	16.6–100	0.1844
3 Cu + 1 Sn	16.6–350	0.2116
3 Cu + 1 Sn	16.6–957	0.1737
86.3 Cu + 9.7 Sn + 4 Zn	40	0.1782
97.6 Cu + 2.2 Sn + 0.2 P, hard	0–80	0.1713
97.6 Cu + 2.2 Sn + 0.2 P, soft	0–80	0.1708
Caoutchouc	....	0.657 to 0.686
Caoutchouc	16.7–25.3	0.770
Celluloid	20–70	1.00
Constantan	4–29	0.1523
Duralumin, 94Al	20–100	0.23
	20–300	0.25
Ebonite	25.3–35.4	0.842
Fluorspar. CaF_2	0–100	0.1950
German silver	0–100	0.1836
Gold-platinum, 2 Au + 1 Pt	0–100	0.1523
Gold-copper, 2 Au + 1 Cu	0–100	0.1552
Glass:		
Tube	0–100	0.0833
Tube	0–100	0.0828
Plate	0–100	0.0891
Crown (mean)	0–100	0.0897
Crown (mean)	50–60	0.0954
Flint	50–60	0.0788
Jena thermometer 16III normal	0–100	0.081

Substance	t°C	$C \times 10^4$
Jena thermometer 59III	0–100	0.058
Jena thermometer 59III	−191 to +16	0.424
Gutta percha	20	1.983
Ice	−20 to −1	0.51
Iceland spar:		
Parallel to axis	0–80	0.2631
Perpendicular to axis	0–80	0.0544
Lead tin (solder) 2 Pb + 1 Sn	0–100	0.2508
Limestone	25–100	0.09
Magnalium	12–39	0.238
Manganin	....	0.181
Marble	15–100	0.117
Monel metal	25–100	0.14
	25–600	0.16
Paraffin	0–16	1.0662
Paraffin	16–38	1.3030
Paraffin	38–49	4.7707
Platinum-iridium, 10 Pt + 1 Ir	40	0.0884
Platinum-silver, 1 Pt + 2 Ag	0–100	0.1523
Porcelain	20–790	0.0413
Porcelain Bayeux	1000–1400	0.0553
Quartz:		
Parallel to axis	0–80	0.0797
Parallel to axis	−190 to +16	0.0521
Perpend. to axis	0–80	0.1337
Quartz glass	−190 to +16	−0.0026
Quartz glass	16 to 500	0.0057
Quartz glass	16 to 1000	0.0058
Rock salt	40	0.4040
Rubber, hard	0	0.691
Rubber, hard	−160	0.300
Speculum metal	0–100	0.1933
Steel, 0.14 C, 34.5 Ni	25–100	0.037
	25–600	0.136

Substance	t°C	$C \times 10^4$
Topaz:		
Parallel to lesser horizontal axis	0–100	0.0832
Parallel to greater horizontal axis	0–100	0.0836
Parallel to vertical axis	0–100	0.0472
Tourmaline:		
Parallel to longitudinal axis	0–100	0.0937
Parallel to horizontal axis	0–100	0.0773
Type metal	16.6–254	0.1952
Vulcanite	0–18	0.6360
Wedgwood ware	0–100	0.0890
Wood:		
Parallel to fiber:		
Ash	0–100	0.0951
Beech	2.34	0.0257
Chestnut	2.34	0.0649
Elm	2.34	0.0565
Mahogany	2.34	0.0361
Maple	2.34	0.0638
Oak	2.34	0.0492
Pine	2.34	0.0541
Walnut	2.34	0.0658
Across the fiber:		
Beech	2.34	0.614
Chestnut	2.34	0.325
Elm	2.34	0.443
Mahogany	2.34	0.404
Maple	2.34	0.484
Oak	2.34	0.544
Pine	2.34	0.341
Walnut	2.34	0.484
Wax white	10–26	2.300
Wax white	26–31	3.120
Wax white	31–43	4.860
Wax white	43–57	15.227

* "Smithsonian Tables."

Table 144. Cubical Expansion of Liquids*

If V_0 is the volume at 0° then at t° the expansion formula is $V_t = V_0(1 + \alpha t + \beta t^2 + \gamma t^3)$. The table gives values of α, β, and γ and of C, the true coefficient of cubical expansion at 20° for some liquids and solutions. Δt is the temperature range of the observation.

Liquid	Range	$\alpha \times 10^3$	$\beta \times 10^6$	$\gamma \times 10^8$	$C \times 10^3$ at 20°
Acetic acid	16–107	1.0630	0.12636	1.0876	1.071
Acetone	0–54	1.3240	3.8090	− 0.87983	1.487
Alcohol:					
Amyl	−15–80	0.9001	0.6573	1.18458	0.902
Ethyl, 30% by volume	18–39	0.2928	10.790	−11.87	
Ethyl, 50% by volume	0–39	0.7450	1.85	0.730	
Ethyl, 99.3% by volume	27–46	1.012	2.20	....	1.12
Ethyl, 500 atm. pressure	0–40	0.866			
Ethyl, 3000 atm. pressure	0–40	0.524			
Methyl	0–61	1.1342	1.3635	0.8741	1.199
Benzene	11–81	1.17626	1.27776	0.80648	1.237
Bromine	0–59	1.06218	1.87714	− 0.30854	1.132
Calcium chloride:					
5.8% solution	18–25	0.07878	4.2742	....	0.250
40.9% solution	17–24	0.42383	0.8571	....	0.458
Carbon disulfide	−34–60	1.13980	1.37065	1.91225	1.218
500 atm. pressure	0–50	0.940			
3000 atm. pressure	0–50	0.581			
Carbon tetrachloride	0–76	1.18384	0.89881	1.35135	1.236
Chloroform	0–63	1.10715	4.66473	− 1.74328	1.273
Ether	−15–38	1.51324	2.35918	4.00512	1.656
Glycerin	....	0.4853	0.4895	....	0.505
Hydrochloric acid, 33.2% solution	0–33	0.4460	0.215	....	0.455
Mercury	0–100	0.18182	0.0078	....	0.18186
Olive oil	....	0.6821	1.1405	− 0.539	0.721
Pentane	0–33	1.4646	3.09319	1.6084	1.608
Potassium chloride, 24.3% solution	16–25	0.2695	2.080	....	0.353
Phenol	36–157	0.8340	0.10732	0.4446	1.090
Petroleum, 0.8467 density	24–120	0.8994	1.396	....	0.955
Sodium chloride, 20.6% solution	0–29	0.3640	1.237	....	0.414
Sodium sulfate, 24% solution	11–40	0.3599	1.258	....	0.410
Sulfuric acid:					
10.9% solution	0–30	0.2835	2.580	....	0.387
100.0%	0–30	0.5758	−0.432	....	0.558
Turpentine	− 9–106	0.9003	1.9595	− 0.44998	0.973
Water	0–33	−0.06427	8.5053	− 6.7900	0.207

* "Smithsonian Tables," Table 269.

Bromoform[1] 7.7 − 50°C.

$V_t = 0.34204[1 + 0.00090411(t - 7.7) + 0.0000006766(t - 7.7)^2]$

0.34204 is the specific volume of bromoform at 7.7°C.

Glycerin[2] −62 to 0°C.

$V_t = V_0(1 + 4.83 \times 10^{-4}t - 0.49 \times 10^{-6}t^2)$

0 − 80°C.

$V_t = V_0(1 + 4.83 \times 10^{-4}t + 0.49 \times 10^{-6}t^2)$

Mercury[3] 0 − 300°C.

$V_t = V_0[1 + 10^{-8}(18153.8t + 0.7548t^2 + 0.001533t^3 + 0.00000536t^4)]$

[1] Sherman and Sherman, *J. Am. Chem. Soc.*, **50**, 1119 (1928). (An obvious error in their equation has been corrected.)
[2] Samsoen, *Ann. phys.*, (10) **9**, 91 (1928).
[3] Harlow, *Phil. Mag.*, (7) **7**, 674 (1929).

Table 145. Cubical Expansion of Solids*

If v_2 and v_1 are the volumes at t_2 and t_1, respectively, then $v_2 = v_1(1 + C\Delta t)$, C being the coefficient of cubical expansion and Δt the temperature interval. Where only a single temperature is stated, C represents the true coefficient of cubical expansion at that temperature.

Substance	t or Δt	$C \times 10^4$
Antimony	0–100	0.3167
Beryl	0–100	0.0105
Bismuth	0–100	0.3948
Copper†	0–100	0.4998
Diamond	40	0.0354
Emerald	40	0.0168
Galena	0–100	0.558
Glass, common tube	0–100	0.276
hard	0–100	0.214
Jena, borosilicate 59 III	20–100	0.156
pure silica	0–80	0.0129
Gold	0–100	0.4411
Ice	−20 to −1	1.1250
Iron	0–100	0.3550
Lead†	0–100	0.8399
Paraffin	20	5.88
Platinum	0–100	0.265
Porcelain, Berlin	20	0.0814
chloride	0–100	1.094
nitrate	0–100	1.967
sulfate	20	0.0754
Quartz	0–100	0.3840
Rock salt	50–60	1.2120
Rubber	20	4.87
Silver	0–100	0.5831
Sodium	20	2.13
Stearic acid	33.8–45.4	8.1
Sulfur, native	13.2–50.3	2.23
Tin	0–100	0.6889
Zinc†	0–100	0.8928

* "Smithsonian Tables," Table 268.
† See additional data below.

Aluminum[1] 100 − 530°C.

$V = V_0(1 + 2.16 \times 10^{-5}t + 0.95 \times 10^{-8}t^2)$

Cadmium[1] 130 − 270°C.

$V = V_0(1 + 8.04 \times 10^{-5}t + 5.9 \times 10^{-8}t^2)$

Copper[1] 110 − 300°C.

$V = V_0(1 + 1.62 \times 10^{-5}t + 0.20 \times 10^{-8}t^2)$

Colophony[2] 0 − 34°C.

$V = V_0(1 + 2.21 \times 10^{-4}t + 0.31 \times 10^{-6}t^2)$

34 − 150°C.

$V = V_{34}[1 + 7.40 \times 10^{-4}(t - 34) + 5.91 \times 10^{-6}(t - 34)^2]$

Lead[1] 100 − 280°C.

$V = V_0(1 + 1.60 \times 10^{-5}t + 3.2 \times 10^{-8}t^2)$

Shellac[2] 0 − 46°C.

$V = V_0(1 + 2.73 \times 10^{-4}t + 0.39 \times 10^{-6}t^2)$

46 − 100°C.

$V = V_{46}[1 + 13.10 \times 10^{-4}(t - 46) + 0.62 \times 10^{-6}(t - 46)^2]$

Silica (vitreous)[3] 0 − 300°C.

$V_t = V_0[1 + 10^{-8}(93.6t + 0.7776t^2 - 0.003315t^2 + 0.000005244t^4)$

Sugar (cane, amorphous)[2] 0 − 67°C.

$V_t = V_0(1 + 2.34 \times 10^{-4}t + 0.14 \times 10^{-6}t^2)$

67 − 160°C.

$V_t = V_{67}[1 + 5.02 \times 10^{-4}(t - 67) + 0.43 \times 10^{-6}(t - 67)^2]$

Zinc[1] 120 − 360°C.

$V_t = V_0(1 + 8.50 \times 10^{-5}t + 3.9 \times 10^{-8}t^2)$

[1] Uffelmann, *Phil. Mag.*, (7) **10**, 633 (1930).
[2] Samsoen, *Ann. phys.*, (10) **9**, 83 (1928).
[3] Harlow, *Phil. Mag.*, (7) **7**, 674 (1929).

JOULE-THOMSON EFFECT*

Table 146. Air

μ = °C./atm.

Pressure, atm.	−150°C.	−140°C.	−130°C.	−120°C.	−100°C.	−25°C.	0°C.	50°C.	100°C.	200°C.	280°C.
1	1.100	0.936	0.807	0.710	0.576	0.317	0.266	0.189	0.133	0.0625	0.0297
20	1.200	0.967	0.819	0.710	0.562	0.297	0.249	0.178	0.124	0.0564	0.0246
40	0.052	0.245	0.776	0.577	0.534	0.276					
80	0.034	0.067	0.141	0.299	0.386	0.232					
100	0.021	0.043	0.087	0.158	0.284	0.211	0.178	0.128	0.089	0.0347	0.0078
140	0.00	0.017	0.038	0.069	0.142	0.164	0.145	0.105	0.072	0.0258	0.0011
180	−0.022	−0.008	0.008	0.028	0.075	0.125	0.113	0.083	0.058	0.0185	−0.0054
220	−0.042	−0.028	−0.015	−0.002	0.031	0.093	0.081	0.063	0.045	0.0127	−0.0110

* Rearranged from "International Critical Tables," vol. 5, p. 144. For more complete data see "International Critical Tables," vol. 5, pp. 144–146. Also Sec. 25 of this handbook, p. 1709. All values are positive values unless otherwise noted. Positive values denote cooling upon expansion.

Table 147. Carbon Dioxide

μ = °C./atm.

Vapor

Temp., °K.	0 atm.	1 atm.	10 atm.	40 atm.	60 atm.	80 atm.	100 atm.
220	2.2855	2.3035					
250	1.6885	1.6954	1.7570				
275	1.3455	1.3455	1.3470				
300	1.1070	1.1045	1.0840	1.0175	0.9675		
325	0.9425	0.9375	0.9075	0.8025	0.7230	0.6165	0.4220
350	0.8195	0.8150	0.7850	0.6780	0.6020	0.5210	0.4340
380	0.7080	0.7045	0.6780	0.5835	0.5165	0.5405	0.3855
400	0.6475	0.6440	0.6210	0.5375	0.4790	0.4225	0.3635

Liquid

Temp., °K.	10 atm.	40 atm.	60 atm.	80 atm.	100 atm.
220	−0.0294	−0.0323	−0.0341	−0.0359	−0.0375
250	....	+0.00073	+0.00073		
275	....	+0.0414	+0.0355		

Table 148. Methane

$\mu = (t_1 - t_2)/(P - 1)$ = °C./atm.

P, atm..............	17	25	27	55
t_1, °C..............	−78	−77	+10	+10
μ..............	0.75	0.75	0.35	0.40

Table 149. Ethyl Chloride*

μ = °C./atm.

°C..........	9	20	40	60	80	100
μ..........	5.22	4.51	3.86	3.31	2.84	2.43

* For initial pressures of 3 atm. or less see Jenkin and Shorthose, "International Critical Tables," vol. 5, p. 146.

CRITICAL CONSTANTS

Table 150. Critical Constants of Elements and Inorganic and Organic Compounds

For additional values of the critical temperature and pressure, see Table 7, p. 150, and Table 9, p. 165, on vapor pressures above 1 atm.

Name	Formula	t_c, °C.	P_c, atm.	d_c, g./cc.
Acetaldehyde	C_2H_4O	188.0		
Acetic acid	$C_2H_4O_2$	321.6	57.2	0.351
anhydride	$C_4H_6O_3$	296.0	46.0	
Acetone	C_3H_6O	235.0	47.0	0.268
Acetonitrile	C_2H_3N	274.7	47.7	0.240
Acetylene	C_2H_2	36.0	62.0	0.231
Air	.....	−140.7	37.2	0.35 (0.31)
Allyl alcohol	C_3H_6O	272.0		
Allylene	C_3H_4	128.0		
Allyl ethyl ether	$C_5H_{10}O$	245		
sulfide	$C_6H_{10}S$	380		
Ammonia	NH_3	132.4	111.5	0.235
Amyl alcohol (t-)	$C_5H_{12}O$	272		
Aniline	C_6H_7N	426	52.4	
Anisole	C_7H_8O	369	41.3	
Argon	A	−122	48.0	0.531
Arsenic	As	803	342.0	
Benzene	C_6H_6	288.5	47.7	0.304
Benzonitrile	C_7H_5N	426.0	41.6	
Boron tribromide	BBr_3	300	....	0.90
Bromine	Br_2	302		
Bromobenzene	C_6H_5Br	397	44.6	0.486
Butadiene-1,3	C_4H_6	152	42.7	0.245
Butane (n-)	C_4H_{10}	153	36.0	
Butyl acetate (n-)	$C_6H_{12}O_2$	306.0		
alcohol (n-)	$C_4H_{10}O$	287	48.4	
alcohol (s-)	$C_4H_{10}O$	265		
alcohol (t-)	$C_4H_{10}O$	235		
Butyric acid (n-)	$C_4H_8O_2$	355	....	0.302
Butyronitrile	C_4H_7N	309	37.4	
Capronitrile	$C_6H_{11}N$	349.0	32.2	
Carbon dioxide	CO_2	31.1	73.0	0.460
disulfide	CS_2	273.0	76.0	0.441
monoxide	CO	−139.0	35.0	0.311
oxysulfide	COS	105.0	61.0	
tetrachloride	CCl_4	283.1	45.0	0.558
Chlorine	Cl_2	144.0	76.1	0.573
Chlorobenzene	C_6H_5Cl	359.0	44.6	0.365
Chloroform	$CHCl_3$	263.0	....	0.516
Cresol (o-)	$CH_3.C_6H_4OH$	422.0	49.4	
(m-)	$CH_3.C_6H_4OH$	432.0	45.0	
(p-)	$CH_3.C_6H_4OH$	426.0	50.8	
Cyanogen	$(CN)_2$	128.0	59.0	
Cyclohexane	C_6H_{12}	281.0	40.4	0.270
Dichlordifluormethane	CCl_2F_2	111.5	39.56	0.555
Diethylamine	$(C_2H_5)_2NH$	223.5	36.2	0.246
Di-isobutyl	$(CH_3)_2CH(CH_2)_2CH(CH_3)_2$	277.0	24.5	0.237
isopropyl	C_6H_{14}	227.4	30.6	0.241
Dimethylamine	$(CH_3)_2NH$	164.6	51.7	
Dimethyl aniline	$C_6H_5N(CH_3)_2$	415.0	35.8	
toluidine (o-)	$C_8H_{11}N$	395.0	30.8	
Dipropylamine	$(C_3H_7)_2NH$	277.0	31.0	
Ethane	C_2H_6	32.1	48.8	0.21
Ethyl acetate	$CH_3COOC_2H_5$	250.1	37.8	0.308
alcohol	C_2H_5OH	243.1	63.1	0.2755
allyl ether	$C_2H_5OCH_2CHCH_2$	245.0		
amine	$C_2H_5NH_2$	183.2	55.5	
bromide	C_2H_5Br	231.0	....	0.513
butyrate	$C_3H_7COOC_2H_5$	293.0	30.0	0.276
caprylate	$C_7H_{15}COOC_2H_5$	386.0		
chloride	C_2H_5Cl	187.2	52.0	0.33
chloroformate	$ClCOOC_2H_5$	<235.0		
crotonate	$C_6H_{10}O_2$	326.0		
disulfide	$(C_2H_5)_2S_2$	369.0		
Ethylene	C_2H_4	9.7	50.9	0.22
chloride	$C_2H_4Cl_2$	290	....	0.45
oxide	C_2H_4O	192.0		
Ethyl ether	$(C_2H_5)_2O$	193.8	35.5	0.2625
formate	$HCOOC_2H_5$	235.3	46.65	0.323
isobutyrate	$(CH_3)_2CHCOOC_2H_5$	280.0	30.0	0.276
isovalerate	$C_7H_{14}O_2$	315.0		
mercaptan	C_2H_5SH	225.5	54.2	0.301
methyl ether	$CH_3C_2H_5O$	164.7	43.4	0.270
sulfide	$C_2H_5CH_3S$	260.0	42.0	
nonylate	$C_{11}H_{22}O_2$	400.0		
propionate	$C_2H_5COOC_2H_5$	272.9	33.0	0.2965
propyl ether	$C_2H_5C_3H_7O$	227.4	32.1	0.258
sulfide	$(C_2H_5)_2S$	283.8	39.1	0.279
valerate	$C_7H_{14}O_2$	297.0		
Fluorobenzene	C_6H_5F	286.0	44.6	0.354
Fluorine	F	−155.0	25.0	
Germanium tetrachloride	$GeCl_4$	277.0	38.0	
Helium	He	−267.9	2.26	0.0693
Heptane (n-)	C_7H_{16}	266.8	26.8	0.234
Heptyl alcohol (n-)	$C_7H_{15}OH$	365.0		
Hexane (n-)	C_6H_{14}	234.8	29.5	0.234
Hydrazine	N_2H_4	380.0	145.0	
Hydrogen	H_2	−239.9	12.8	0.0310
bromide	HBr	90.0	84.0	
chloride	HCl	51.4	81.6	0.42
cyanide	HCN	183.5	50.0	0.20
fluoride	HF	230.2		
iodide	HI	151.0	82.0	
selenide	H_2Se	138.0	88.0	
sulfide	H_2S	100.4	88.9	
Iodine	I_2	553.0		
Iodobenzene	C_6H_5I	448.0	44.6	0.581
Isoamyl acetate	$CH_3COOC_5H_{11}$	326.0		
alcohol	$C_5H_{11}OH$	307.0		
butyrate	$C_3H_7COOC_5H_{11}$	346.0		
formate	$HCOOC_5H_{11}$	303.0	34.0	0.282
mercaptan	$C_5H_{11}SH$	321.0		
propionate	$C_2H_5COOC_5H_{11}$	338.0		
sulfide	$(C_5H_{11})_2S$	391.0		
Isobutane	C_4H_{10}	134.0	37.0	
Isobutyl acetate	$CH_3COOC_4H_9$	288.0	31.0	0.281
alcohol	C_4H_9OH	265.0	48.0	
butyrate	$C_3H_7COOC_4H_9$	338.0		
formate	$HCOOC_4H_9$	278.0	38.0	0.288
isobutyrate	$C_3H_7COOC_4H_9$	329.0		
isovalerate	$C_9H_{18}O_2$	348.0		
propionate	$C_7H_{14}O_2$	319.0		
Isobutyric acid	C_3H_7COOH	336.0	....	0.304
Isopentane	C_5H_{12}	187.8	32.8	0.234
Isopropyl alcohol	C_3H_7OH	235.0	53.0	
Isovaleric acid	C_4H_9OH	361.0		
Krypton	Kr	−63	54	0.78
Mercury	Hg	>1550	>200	4–5
Methane	CH_4	−82.5	45.8	0.162
Methyl acetate	CH_3COOCH_3	233.7	46.3	0.325
Methylal	$CH_2(OCH_3)_2$	224.0		
Methyl alcohol	CH_3OH	240.0	78.7	0.272
amine	CH_3NH_2	156.9	73.6	
aniline	$C_6H_5NHCH_3$	429.0	51.3	
butyrate	$C_3H_7COOCH_3$	281.3	34.2	0.300
chloride	CH_3Cl	143.1	65.8	0.37
ether	$(CH_3)_2O$	126.9	52.0	0.271
ethyl ether	$CH_3OC_2H_5$	164.7	43.4	0.270
sulfide	$CH_3SC_2H_5$	260.0	42.0	
fluoride	CH_3F	44.9	62.0	
formate	$HCOOCH_3$	214.0	59.15	0.349
isobutyrate	$C_3H_7COOCH_3$	267.55	33.7	0.301
mercaptan	CH_3SH	196.8	71.4	0.323
oxalate	$(CH_3O_2C)_2$	260.0	9.48	
propionate	$C_2H_5COOCH_3$	257.4	39.3	0.312
sulfide	$(CH_3)_2S$	229.9	54.6	0.306
valerate	$C_6H_{12}O_2$	294.0(d)	32.0	0.279
Methyl diethyl ether	$C_5H_{12}O_2$	254.0		
Neon	Ne	−228.7	25.9	0.484
Niton	Nt	+104.5	62.5	
Nitric oxide	NO	−94.0	65.0	0.52
Nitrogen	N_2	−147.1	33.5	0.3110
tetroxide	N_2O_4	158.0	99.0	
Nitrous oxide	N_2O	36.5	71.7	0.45
Octane (n-)	C_8H_{18}	296.0	24.6	0.234
Octyl alcohol (n-)	$C_8H_{17}OH$	385.0		
(s-)	$C_8H_{17}OH$	364.0		
Oxygen	O_2	−118.8	49.7	0.430
Paraldehyde	$(C_2H_4O)_3$	290.0		
Pentane (n-)	C_5H_{12}	197.2	33.0	0.232
Phenetole	$C_6H_5OC_2H_5$	374.0	33.8	
Phenol	C_6H_5OH	419.0	60.5	
Phosgene	$COCl_2$	182.0	56.0	0.52
Phosphine	PH_3	51.0	64.0	0.30
Phosphonium chloride	PH_4Cl	49.0	73.0	0.226
Propane (n-)	C_3H_8	96.8	42.0	
Propionic acid	C_2H_5COOH	339.5	53.0	0.315
Propyl alcohol (n-)	C_3H_7OH	263.7	49.95	0.273
acetate	$CH_3COOC_3H_7$	276.2	32.9	0.296
Propylamine	$C_3H_7NH_2$	223.8	46.3	

* Rearranged from "International Critical Tables," vol. 3, pp. 248–249. Revised from literature, 1947.

Table 150. Critical Constants of Elements and Inorganic and Organic Compounds—(*Concluded*)

Name	Formula	t_c, °C.	P_c, atm.	d_c, g./cc.
Propyl butyrate	$C_3H_7COOC_3H_7$	327.0		
chloride (n-)	C_3H_7Cl	230.0	45.2	
formate	$HCOOC_3H_7$	264.85	40.1	0.309
Propylene	C_3H_6	92.3	45.0	
Propyl ethyl ether	$C_3H_7OC_2H_5$	227.4	32.1	0.258
isobutyrate	$C_3H_7COOC_3H_7$	316.0		
isovalerate	$C_8H_{16}O_2$	336.0		
Propionitrile	C_2H_5N	291.2	41.3	0.241
Propyl propionate	$C_2H_5COOC_3H_7$	305.0		
Pyridine	C_5H_5N	344.0	60.0	
Quinoline	C_9H_7N	520.0		
Radon	Rn	104.0	62.0	
Silicon tetrafluoride	SiF_4	−1.5	50.0	
tetrahydride	SiH_4	−3.5	48.0	
Stannic tetrachloride	$SnCl_4$	318.7	37.0	0.742
Steam	H_2O	374.0	217.7	0.4
Sulfur	S	1040		
dioxide	SO_2	157.2	77.7	0.52
trioxide	SO_3	218.3	83.6	0.630
Thiophene	C_4H_4S	317.0	48.0	
Thymol	$CH_3C_6H_3(OH)C_3H_7$	425.0		
Toluene	$C_6H_5CH_3$	320.6	41.6	0.292
Tolunitrile	$CH_3C_6H_4CN$	450.0		
Triethylamine	$(C_2H_5)_3N$	262.0	30.0	0.251
Trimethylamine	$(CH_3)_3N$	161.0	41.0	
Valeric acid (n-)	$C_5H_{10}O_2$	379.0		
Water	H_2O	374.15	218.4	0.323
Xenon	Xe	16.6	58.2	1.155

COMPRESSIBILITIES

The tables given under this subject, with the exception of those for methyl chloride, ethylene, acetylene, and those of liquids, were calculated and arranged by Dr. Richard Wiebe, of The Fixed Nitrogen Research Laboratory, Washington, D.C., for "Fixed Nitrogen" by Dr. Harry A. Curtis. The Chemical Catalog Company has kindly granted permission to reprint these data.

Table 151. Molal Volumes of Gases at 0°C. and 1 Atm.

Gas	$1 + \lambda = \frac{pv(0°, 0 \text{ atm.})}{pv(0°, 1 \text{ atm.})}$	$\frac{22,414.1}{1 + \lambda}$ (cc./mole)
Acetylene	1.010	22,192
Air	1.0006	22,401
Ammonia	1.015	22,081
Carbon dioxide	1.0067	22,264
Carbon monoxide	1.0005	22,404
Ethylene	1.0078	22,240
Hydrogen	0.99939	22,428
Methane	1.0024	22,360
Methyl chloride	1.0244	21,880
Nitrogen	1.00047	22,404
Oxygen	1.0009	22,393
3:1 hydrogen-nitrogen	0.99966	22,422

Example. The volume that 1 g. of hydrogen occupies at 100° and at a pressure of 1000 atm. is given

$$V = \frac{f}{p} \times \frac{V_0°_{,1 \text{ atm.}}}{\text{M.W.}} = \frac{2.0784}{1000} \times \frac{22,428}{2.0154} = 23.10 \text{ cc./g.}$$

where f is the compressibility factor, M.W. is the molecular weight, and $V_0°_{,1 \text{ atm.}}$ is the volume given in the table.

For more detailed data on Compressibilities see "International Critical Tables": tabular index, vol. 3, p. 1; building stones, vol. 2, p. 54; compounds, vol. 3, p. 49; elements, vol. 3, pp. 35, 46; gases, vol. 3, pp. 3, 17, 435; glass, vol. 2, p. 93; liquids and vitreous solids, vol. 3, pp. 35, 40, 41; metals, vol. 3, p. 46; minerals and rocks, vol. 3, p. 49; animal and vegetable oils, vol. 2, p. 208; petroleum, vol. 2, p. 146; porcelains, vol. 2, p. 68; rubber, vol. 2, p. 269; solutions, vol. 3, p. 439; woods, vol. 2, p. 1. For Heats of Compression, see vol. 5, p. 144.

For values of the densities of gases at standard conditions, 0°C. and 1 atm., see "Chemical Engineers' Handbook," 2d ed., Table 43, p. 411.

The following table is given to enable one to make calculations of densities, specific volumes, or the corresponding molal quantities from compressibility data.

Table 152. Compressibility Factors of Oxygen*

PV = 1.0000 at 1 atm. and 0°C.

P, atm.	0°C.	50°C.	100°C.
1	1.0000	1.1838	1.3674
10	0.9913	1.1796	1.3661
25	.9776	1.1732	1.3644
50	.9569	1.1642	1.3630
75	.9388	1.1571	1.3632
100	.9234	1.1520	1.3651

* Calculated from equations given by Holborn and Otto, *Z. Physik*, **33**, 1 (1925). For additional data see "International Critical Tables," vol. 3, p. 8.

Table 153. Compressibility Factors of Hydrogen*

PV = 1.0000 at 1 atm. and 0°C.

Pressure, atm.	−239.9 °C.	−207.9 °C.	−183 °C.	−150 °C.	−100 °C.	−70 °C.	−50 °C.	−25 °C.	0.0 °C.	20 °C.	50 °C.	100 °C.	200 °C.	300 °C.
1	......	0.2380	0.3297	0.4508	0.6340	0.7438	0.8170	0.9085	1.0000	1.0732	1.1830	1.3660	1.7317	2.0974
10	......	.2308	.3279	.4520	.6377	.7432	.8219	.9138	1.0057	1.0791	1.1891	1.3723	1.7380	2.1037
20	0.0286	.2239	.3265	.4541	.6421	.7535	.8275	.9187	1.0120	1.0855	1.1959	1.3792	1.7450	2.1108
30	.0389	.2183	.3260	.4564	.6466	.7583	.8331	.9257	1.0183	1.0920	1.2027	1.3862	1.7520	2.1178
40	.0595	.2144	.3262	.4591	.6513	.7642	.8389	.9318	1.0247	1.0985	1.2094	1.3931	1.7590	2.1249
50	.0717	.2126	.3271	.4623	.6562	.7695	.8447	.9378	1.0309	1.1051	1.2162	1.4001	1.7660	2.1319
60	......	.2127	.3289	.4658	.6613	.7752	.8506	.9441	1.0376	1.1116	1.2230	1.4070	1.7731	2.1392
80	......	.2187	.3346	.4740	.6720	.7870	.8628	.9567	1.0507	1.1249	1.2365	1.4209	1.7871	2.1530
100	......	.2301	.3434	.4839	.6834	.8003	.8754	.9700	1.0639	1.1388	1.2510	1.4356	1.8042	2.1733
200	......	......	......	......	......	.8640	.9411	1.0383	1.1336	1.2066	1.3203	1.5071	1.8756	2.2393
300	......	......	......	......	......	.9340	1.0112	1.1093	1.2045	1.2799	1.3915	1.5790		
400	......	......	......	......	......	1.0075	1.0832	1.1803	1.2775	1.3511	1.4635	1.6513	2.0206	2.3826
500	......	......	......	......	......	1.0804	1.1568	1.2542	1.3500	1.4240	1.5357	1.7235		
600	......	......	......	......	......	1.1555	1.2301	1.3272	1.4226	1.4958	1.6081	1.7955	2.1628	2.5246
800	......	......	......	......	......	1.3018	1.3755	1.4717	1.5665	1.6391	1.7512	1.9380	2.3043	2.6653
1000	......	......	......	......	......	1.4443	1.5185	1.6139	1.7101	1.7795	1.8917	2.0784	2.4568	2.8026

* Crommelin and Swallow, *Proc. 4th Intern. Congr. Refrig.*, **1**, 53*a* (1924); Holborn and Otto, *Z. Physik*, **23**, 77 (1924); *ibid.*, **33**, 1 (1925); Verschoyle, *Proc. Roy. Soc., London*, **111A**, 552 (1926); Holborn and Otto, *Z. Physik*, **38**, 359 (1926); Bartlett, Cupples, and Tremearne, *J. Am. Chem. Soc.*, **50**, 1275 (1928); Bartlett. Hetherington, Kvalnes, and Tremearne, *ibid.*, **52**, 1363 (1930). Michels, Niihoff, and Gerver, *Ann. Physik.*, **12**, 562 (1932); Wiebe and Gaddy, *J. Am. Chem. Soc.*, **60**, 2300 (1938). See also calculations of physical properties by Deming and Shupe, *Phys. Rev.*, **40**, 848 (1932); Deming and Deming, *ibid.*, **45**, 109 (1934).

Table 154. Compressibility Factors of Nitrogen*,†

PV = 1.0000 at 1 atm. and 0°C.

Pressure, atm.	−146.3°C.	−130°C.	−100°C.	−70°C.	−50°C.	−25°C.	0.0°C.	20°C.	50°C.	100°C.	200°C.	300°C.	400°C.
1	.....	0.5209	0.6319	0.7426	0.8162	0.9077	1.0000	1.0730	1.1835	1.3669	1.7335	2.1000	2.4663
10	.....	.4873	.6109	.7292	.8060	.9010	.9962	1.0705	1.1836	1.3695	1.7398	2.1083	2.4758
20	0.3539	.4465	.5874	.7130	.7951	.8940	.9925	1.0690	1.1842	1.3728	1.7469	2.1175	2.4864
30	.2670	.4005	.5637	.7010	.7851	.8886	.9894	1.0677	1.1851	1.3765	1.7542	2.1271	2.4971
40	.....	.3487	.5404	.6850	.7757	.8830	.9870	1.0668	1.1866	1.3805	1.7617	2.1366	2.5079
50	.....	.2943	.5180	.6716	.7672	.8790	.9848	1.0669	1.1884	1.3849	1.7694	2.1462	2.5189
60	.....	.2483	.4970	.6620	.7596	.8764	.9840	1.0670	1.1907	1.3896	1.7772	2.1559	2.5299
80	.....	.2986	.4632	.6432	.7476	.8700	.9835	1.0687	1.1906	1.4002	1.7935	2.1755	2.5522
100	.....	.....	.4471	.6362	.7424	.8676	.9848	1.0749	1.2046	1.4121	1.8111	2.1973	2.5751
200	.....	.....	.....	.6823	.7854	.9151	1.0355	1.1309	1.2742	1.4965	1.9119	2.3127	2.6971
300	.....	.....	.....	.8053	.8986	1.0179	1.1335	1.2293	1.3711	1.5978	2.0216	2.4287	2.8193
400	.....	.....	.....	.9477	1.0334	1.1445	1.2557	1.3467	1.4870	1.7119	2.1455	2.5506	2.9450
500	.....	.....	.....	1.0914	1.1748	1.2798	1.3885	1.4782	1.6171	1.8388	2.2708	2.6774	3.0714
600	.....	.....	.....	1.2331	1.3159	1.4186	1.5214	1.6098	1.7473	1.9657	2.3961	2.8042	3.1977
800	.....	.....	.....	1.5111	1.5928	1.6958	1.7959	1.8817	2.0155	2.2279	2.6557	3.0623	3.4587
1000	.....	.....	.....	1.7783	1.8573	1.9600	2.0641	2.1481	2.2825	2.4948	2.9212	3.3203	3.7224

* Holborn and Otto, *Z. Physik*, **23, 77** (1924); *ibid.*, **33**, 1 (1925); Verschoyle, *Proc. Roy. Soc., London*, **111A**, 552 (1926); Holborn and Otto, *Z. Physik*, **38**, 359 (1926); Bartlett, Cupples, and Tremearne, *J. Am. Chem. Soc.*, **50**, 1275 (1928); Bartlett, Hetherington, Kvalnes, and Tremearne, *ibid.*, **52**, 1363 (1930); Amagat, *Ann. chim. phys.*, **29**, 68 (1893); Onnes and Urk, *Proc. 4th Intern. Congr. Refr.*, p. 69, *Comm. phys. Lab. Leyden*, No. 169*d*.

† Recent data show some disagreement, particularly in the high-pressure range at 50° and 100°C. (0.5 per cent). If more accurate data are needed between 0° and 150°C. to 400 atm., it is recommended to use the following equation taken from Otto, Michels, and Wouters, *Physik. Z.*, **35**, 97 (1934).

$$pv = A + Bp + Cp^2 + Dp^4 + Ep^6 + Fp^8$$

t, °C.	A	$B10^3$	$C10^6$	$D10^{12}$	$E10^{18}$	$F10^{24}$	t, °C.	A	$B10^3$	$C10^6$	$D10^{12}$	$E10^{18}$	$F10^{24}$
0	1.00045	−0.45890	2.90964	23.1895	−587.22	4503.8	100	1.36671	+0.28848	1.69106	−1.68806	− 8.2022	47.065
25	1.09201	−0.22110	2.65436	5.06187	−159.266	1008.8	125	1.45828	+0.41812	1.37824	−0.56983	−11.4400	43.566
50	1.18358	−0.01616	2.23372	2.64600	− 95.758	535.08	150	1.54985	+0.52421	1.19192	−0.91694	− 4.5638	18.788
75	1.27515	+0.15299	1.92483	−0.46531	− 31.6196	163.97							

The following equation taken from Michels, Wouters, and DeBoer, *Physica*, **3**, 585 (1936) may be used for values to 3000 atm. and at temperatures between 0° and 150°C. if a possible deviation of 3×10^{-4} is permissible. For higher accuracy, the authors recommend use of table giving $(pv_{calc.} - Pv_{exp.}) \times 10^5$.

$$p(v - \alpha) = A + \beta d + \gamma d^2 + \delta d^3 + \epsilon d^4$$

In this equation, d is the Amagat density defined as P/PV.

t, °C.	$\alpha 10^3$	A	$\beta 10^3$	$\gamma 10^6$	$\delta 10^9$	$\epsilon 10^{12}$	t, °C.	$\alpha 10^3$	A	$\beta 10^3$	$\gamma 10^6$	$\delta 10^9$	$\epsilon 10^{12}$
0	0.48398	1.00045	−0.95241	3.38591	−2.22621	9.2040	100	0.62860	1.36671	−0.47050	3.12708	+0.69591	4.8982
25	.59626	1.09202	−0.88854	3.18775	−0.84709	6.6287	125	.65156	1.45828	−0.36509	3.17849	+0.84240	4.4999
50	.54734	1.18358	−0.67607	3.24421	−0.58091	7.0958	150	.77424	1.54985	−0.39933	2.82427	+2.03012	1.4968
75	.51846	1.27515	−0.48391	3.29668	−0.22363	7.2360							

See also Deming and Shupe, *Phys. Rev.*, **37,** 638 (1931); Deming and Deming, *ibid.*, **45,** 109 (1934).

Table 155. Compressibility Factors of Carbon Dioxide*

PV = 1.0000 at 1 atm. and 0°C.

Pressure, atm.	0°C.	10°C.	20°C.	30°C.	40°C.	50°C.	60°C.	70°C.	80°C.	90°C.	100°C.	137°C.	198°C.	258°C.
1	1.0000													
50	0.1050	0.1145	0.6800	0.7750	0.8500	0.9200	0.9840	1.0430	1.0960	1.1530	1.2065	1.3800		
75	.1530	.1630	.1800	.2190	.6200	.7470	.8410	0.9180	0.9880	1.0515	1.1180	1.3185	1.6150	1.8670
100	.2020	.2130	.2285	.2550	.3090	.4910	.6610	.7770	.8725	0.9535	1.0300	1.2590	1.5820	1.8470
125	.2490	.2620	.2785	.3000	.3350	.3950	.5100	.6430	.7590	.8580	0.9470	1.2050	1.5530	1.8310
150	.2950	.3090	.3260	.3460	.3770	.4190	.4850	.5750	.6805	.7815	.8780	1.1585	1.5295	1.8180
175	.3405	.3550	.3725	.3930	.4215	.4570	.5055	.5730	.6515	.7410	.8320	1.1230	1.5100	1.8095
200	.3850	.4010	.4190	.4400	.4675	.5000	.5425	.5955	.6600	.7315	.8145	1.0960	1.4960	1.8040
225	.4305	.4455	.4655	.4875	.5130	.5425	.5825	.6285	.6815	.7460	.8175	1.0835	1.4890	1.8035
250	.4740	.4900	.5100	.5335	.5580	.5865	.6250	.6670	.7135	.7690	.8355	1.0810	1.4870	1.8060
275	.5170	.5340	.5545	.5775	.6040	.6330	.6675	.7070	.7515	.8015	.8600	1.0885	1.4875	1.8115
300	.5595	.5775	.5985	.6225	.6485	.6765	.7100	.7485	.7900	.8375	.8900	1.1080	1.4935	1.8200
350	.6445	.6640	.6850	.7090	.7365	.7650	.7980	.8325	.8725	.9135	.9615	1.1565	1.5210	1.8465
400	.7280	.7475	.7710	.7950	.8230	.8515	.8840	.9180	.9560	.9660	1.0385	1.2175	1.5630	1.8830
450	.8090	.8310	.8550	.8800	.9075	.9365	.9690	1.0035	1.0400	1.0775	1.1190	1.2880	1.6160	1.9280
500	.8905	.9130	.9380	.9630	.9900	1.0210	1.0540	1.0880	1.1240	1.1610	1.2005	1.3620	1.6775	
550	.9700	.9935	1.0200	1.0465	1.0740	1.1035	1.1370	1.1720	1.2085	1.2430	1.2830	1.4400	1.7450	
600	1.0495	1.0730	1.0995	1.1275	1.1570	1.1865	1.2190	1.2540	1.2900	1.3265	1.3655	1.5180	1.8120	
650	1.1275	1.1530	1.1800	1.2075	1.2375	1.2680	1.3010	1.3360	1.3725	1.4085	1.4475	1.5960	1.8835	
700	1.2055	1.2320	1.2590	1.2890	1.3190	1.3500	1.3825	1.4170	1.4535	1.4900	1.5285	1.6760	1.9560	
800	1.3580	1.3870	1.4170	1.4475	1.4790	1.5105	1.5435	1.5780	1.6140	1.6505	1.6890	1.8355	2.1080	
850	1.4340	1.4625	1.4935	1.5245	1.5570	1.5885	1.6225	1.6575	1.6925	1.7285	1.7680	1.9150	2.1860	
900	1.5090	1.5385	1.5685	1.6000	1.6325	1.6650	1.6995	1 7345	1.7710	1.8075	1.8460	1.9944	2.2600	
950	1.5830	1.6115	1.6430	1.6740	1.7065	1.7395	1.7745	1.8100	1.8470	1.8845	1.9230	2.0720	2.3350	
1000	1.6560	1.6850	1.7160	1.7480	1.7800	1.8140	1.8475	1.8840	1.9210	1.9590	1.9990			

* Amagat, *Ann. chim. phys.*, **29**, 68 (1893). More recent work of Michels, Michels, and Wouters, *Proc. Roy. Soc.*, *London*, *A*, **153**, 201 (1935); Michels thesis (Amsterdam, 1937); Michels, Bijl, and Michels, *Proc. Roy. Soc.*, *London*, *A*, **160**, 376 (1937), has shown rather remarkable agreement with the work of Amagat. For fugacity calculations see Deming and Deming, *Phys. Rev.*, **56**, 108 (1939). Thermodynamic properties of carbon dioxide were calculated by Sweigert, Weber, and Allen, *Ind. Eng. Chem.*, **38**, 185 (1946).

Table 156. Compressibility Factors of Air*

PV = 1.0000 at 1 atm. and 0°C.

P, atm.	0°C.	50°C.	100°C.	150°C.	200°C.
1	1.0000	1.1837	1.3670	1.5504	1.7335
10	0.9949	1.1826	1.3686	1.5539	1.7385
25	.9874	1.1817	1.3720	1.5603	1.7473
50	.9779	1.1822	1.3792	1.5724	1.7630
75	.9724	1.1855	1.3888	1.5861	1.7800
100	.9695	1.1914	1.4006	1.6015	1.7984

* Calculated from equations of Holborn and Otto, *Z. Physik*, **33**, 1 (1925). For additional data see "International Critical Tables," vol. 3, p. 9.

Table 157. Compressibility Factors of Carbon Monoxide*

PV = 1.0000 at 1 atm. and 0°C.

Pressure, atm.	−70°C.	−50°C.	−25°C.	0°C.	25°C.	50°C.	100°C.	150°C.	200°C.
1	0.7427	0.8162	0.9082	1.0000	1.0918	1.1836	1.3671	1.5504	1.7336
10	.7275	.8061	.9025	0.9960	1.0885	1.1830	1.3700	1.5565	1.7423
20	.7115	.7947	.8960	.9912	1.0858	1.1827	1.3735	1.5630	1.7510
30	.6950	.7840	.8895	.9868	1.0836	1.1820	1.3769	1.5695	1.7595
40	.6792	.7730	.8833	.9825	1.0820	1.1821	1.3800	1.5755	1.7683
50	.6636	.7622	.8768	.9780	1.0814	1.1826	1.3837	1.5823	1.7758
60	.6495	.7530	.8712	.9755	1.0810	1.1842	1.3878	1.5880	1.7833
80	.6274	.7355	.8620	.9718	1.0820	1.1893	1.3967	1.6008	1.7985
100	.6147	.7264	.8592	.9725	1.0851	1.1955	1.4062	1.6151	1.8146
120	.6110	.7240	.8590	.9763	1.0917	1.2042	1.4176	1.6288	1.8310
140	.6142	.7270	.8627	.9832	1.1009	1.2139	1.4312	1.6450	1.8492
160	.6255	.7355	.8715	.9935	1.1125	1.2255	1.4456	1.6620	1.8683
180	.6431	.7485	.8856	1.0125	1.1255	1.2400	1.4615	1.6792	1.8883
200	.6631	.7656	.9022	1.0200	1.1415	1.2561	1.4794	1.6987	1.9090
250	.7247	.8205	.9520	1.0665	1.1885	1.3022	1.5280	1.7500	1.9632
300	.7955	.8872	1.0087	1.1211	1.2408	1.3521	1.5798	1.8054	2.0183
400	.9434	1.0285	1.1403	1.2487	1.3625	1.4716	1.6963	1.9178	2.1380
500	1.0920	1.1755	1.2831	1.3843	1.4940	1.6023	1.8235	2.0450	2.2627
600	1.2386	1.3225	1.4282	1.5256	1.6317	1.7378	1.9557	2.1757	2.3923
800	1.5236	1.6100	1.7153	1.8064	1.9115	2.0144	2.2244	2.4442	2.6602
1000	1.7992	1.8871	1.9935	2.0827	2.1857	2.2879	2.4935	2.7142	2.9264

* Bartlett, Hetherington, Kvalnes, and Tremearne, *J. Am. Chem. Soc.*, **52**, 1374 (1930); Goig, *Compt. rend.*, **189**, 246 (1929); Scott, *Proc. Roy. Soc., London*, **122A**, 283 (1929). For calculations of physical properties see Deming and Shupe, *Phys., Rev.*, **38**, 2245 (1931); Deming and Deming, *ibid.*, **45**, 109 (1934).

Table 158. Compressibility Factors of Methane*

PV = 1.0000 at 1 atm. and 0°C.

Pressure, atm.	−70°C.	−50°C.	−25°C.	0.0°C.	25°C.	50°C.	100°C.	150°C.	200°C.
1	0.7410	0.8150	0.9075	1.0000	1.0922	1.1845	1.3686	1.5525	1.7363
10	.6985	.7795	.8803	0.9785	1.0733	1.1780	1.3595	1.5470	1.7348
20	.6473	.7402	.8493	.9543	1.0549	1.1590	1.3500	1.5422	1.7330
30	.5910	.6991	.8183	.9297	1.0373	1.1412	1.3411	1.5370	1.7311
40	.5244	.6547	.7873	.9061	1.0198	1.1275	1.3335	1.5345	1.7309
50	.4425	.6069	.7558	.8830	1.0034	1.1152	1.3268	1.5319	1.7307
60	.3366	.5551	.7243	.8607	0.9871	1.1017	1.3200	1.5292	1.7308
80	.2556	.4604	.6651	.8192	.9569	1.0799	1.3098	1.5248	1.7322
100	.2808	.4088	.6167	.7845	.9319	1.0624	1.3018	1.5237	1.7357
120	.3175	.4095	.5877	.7604	.9126	1.0487	1.2965	1.5241	1.7414
140	.3543	.4304	.5801	.7457	.9003	1.0399	1.2939	1.5272	1.7485
160	.3915	.4601	.5891	.7425	.8949	1.0367	1.2952	1.5325	1.7570
180	.4288	.4924	.6079	.7482	.8970	1.0373	1.2997	1.5398	1.7668
200	.4656	.5269	.6319	.7631	.9048	1.0437	1.3076	1.5500	1.7774
250	.5567	.6142	.7066	.8184	.9469	1.0776	1.3364	1.5867	1.8126
300	.6458	.7025	.7879	.8886	1.0062	1.1286	1.3785	1.6234	1.8534
400	.8185	.8750	.9561	1.0468	1.1499	1.2608	1.4929	1.7268	1.9586
500	.9867	1.0433	1.1221	1.2086	1.3064	1.4106	1.6277	1.8542	2.0803
600	1.1487	1.2071	1.2862	1.3709	1.4659	1.5653	1.7729	1.9935	2.2131
800	1.4631	1.5246	1.6046	1.6894	1.7801	1.8781	2.0744	2.2828	2.4949
1000	1.7656	1.8287	1.9110	2.0000	2.0892	2.1845	2.3757	2.5797	2.7861

* Keyes, Smith, and Joubert, *J. Math. Physics*, **1**, 191 (1922); Kvalnes and Gaddy, *J. Am. Chem. Soc.*, **53**, 394 (1931). For extension to 1675 atm. and 300°C. and for data on methane-ammonia mixtures, see Kazarnovskii and Levchenko, *J. Phys. Chem.* (U.S.S.R.), **18**, 380 (1944). Isotherms of methane-ethane mixtures at 0, 25, and 50 to 60 atm. were measured by Michels and Nederbragt, *Physica*, **6**, 656 (1939).

Table 159. Compressibility of Ethane*

Density, moles/liter	0.5	1.0	1.5	2.0	2.5	3.0	3.5	4.0	4.5	5.0
Temp., °C.	Pressures, atm.									
25	11.11	20.14	27.34	32.84	36.88	39.66	41.35			
50	12.24	22.58	31.24	38.41	44.27	49.02	52.87	55.97	58.49	60.56
75	13.34	24.95	35.08	43.80	51.37	57.98	63.78	68.89	73.52	77.79
100	14.43	27.28	38.79	49.03	58.29	66.68	74.36	81.50	88.25	94.73
125	15.52	29.64	42.49	54.25	65.13	75.26	84.81	93.91	102.73	111.46
150	16.60	31.89	46.08	59.31	71.78	83.61	94.99	106.06	117.00	127.98
175	17.67	34.17	49.66	64.36	78.40	91.93	105.13	118.16	131.20	144.41
200	18.71	36.44	53.25	69.38	84.97	100.19	115.20	130.18	145.30	160.77
225	19.80	38.64	56.73	74.27	91.38	108.24	125.02	141.90	159.08	176.73
250	20.89	40.87	60.23	79.16	97.81	116.30	134.84	153.59	172.87	192.77

Density, moles/liter	5.5	6.0	6.5	7.0	7.5	8.0	9.0	10.0
Temp., °C.	Pressures, atm.							
50	62.40	63.99	65.52	67.08	68.82	70.88	76.90	87.76
75	81.97	85.95	90.00	94.29	99.01	104.38	118.32	139.23
100	101.36	107.86	114.60	121.83	129.74	138.64	160.77	191.98
125	120.56	129.64	139.19	149.41	160.60	173.10	203.63	244.97
150	139.68	151.38	163.70	176.99	191.52	207.66	246.57	298.02
175	158.68	172.98	188.20	204.54	222.44	242.24	289.49	350.95
200	177.62	194.53	212.59	232.03	253.29	276.76	332.56	
225	196.44	215.98	236.88	259.42	284.02	311.09		
250	215.21	237.38	261.13	286.74	314.67	345.38		
275	233.90	258.69	285.28	314.00	345.30			

* Data by Beattie, Hadlock, and Poffenberger, *J. Chem. Phys.*, **3**, 93 (1935); Beattie, Su, and Simard, *J. Am. Chem. Soc.*, **61**, 926 (1939). For thermal properties see Planck and Kambertz, *Ice and Cold Storage*, **39**, 159, 176 (1936).

For densities between 0.5 and 5.0 moles/liter and at temperatures from 25° to 250°C., the Beattie-Bridgeman equation of state may be used:

$$P = \left[\frac{RT(1-\epsilon)}{V^2}\right][V + B] - \frac{A}{V^2}$$

	$A = A_0(1 - a/V)$		$B = B_0(1 - b/V)$		$\epsilon = c/VT^3$	
R	A_0	a	B_0	b	c	Mol. wt.
Units: normal atmospheres, liters/mole, °K. (T°K. = t°C. + 273.13)						
0.08206	5.8800	0.05861	0.09400	0.01915	90.00×10^4	30.0462
Amagat units: normal atmospheres, V = 1 at 0°C. and 1 atm., °K.						
3.69658×10^{-3}	11.9320×10^{-3}	2.6402×10^{-3}	4.2344×10^{-3}	0.8627×10^{-3}	40.543×10^3	30.0462

Table 160. Compressibility Factors of Ethylene*

P, atm.	0°C.	20°C.	40°C.	60°C.	80°C.	100°C.
1	1.0000					
50	0.1755	0.6290	0.8140	0.9535	1.0770	1.1920
100	.3100	.3600	.4705	.6680	0.8465	1.0050
150	.4405	.4850	.5505	.6490	.7760	0.9240
200	.5650	.6095	.6690	.7440	.8380	.9460
300	.8055	.8520	.9075	.9720	1.0475	1.1330
500	1.2555	1.3075	1.3670	1.4310	1.5000	1.5775
1000	2.2890	2.3535	2.4215	2.4925	2.5660	2.6425

* For more detailed data see "International Critical Tables," vol. 3, p. 15.

See also recent data of Michels, De Gruyter, and Nilsen, *Physica*, **3**, 346 (1936).

For practical purposes, the results of Michels and Geldermans, *Physica*, **9**, 967 (1942) can be expressed in terms of Amagat densities P/PV at pressures to 3000 atm. and at temperatures from 0° to 150°C.

$$PV = A + Bd + Cd^2 + Zd^3 + Dd^4 + Yd^5 + Ed^6$$

t, °C.	A	$B \cdot 10^3$	$C \cdot 10^6$	$Z \cdot 10^9$	$D \cdot 10^{12}$	$Y \cdot 10^{15}$	$E \cdot 10^{18}$
0	1.007526	−7.547208	9.986890	169.8126	−1117.346	2574.578	−1678.789
25	1.096504	−6.595575	4.786190	168.9001	− 996.0618	2274.842	−1462.893
50	1.188234	−6.000185	6.552788	124.4417	− 749.2238	1796.149	−1144.402
75	1.279360	−5.391539	6.020634	110.2406	− 641.1927	1582.912	−1009.174
100	1.372791	−4.970870	7.634562	86.27056	− 509.2971	1332.537	− 847.2137
125	1.461910	−4.351752	5.178556	97.58054	− 519.1989	1344.884	− 868.2760
150	1.555980	−4.023550	7.215812	78.09239	− 422.7762	1175.214	− 767.8247

For greater accuracy, the table of deviations given by the authors in above reference may be consulted.

Table 161. Compressibility Factors of Acetylene*

$PV = 1.0000$ at 0°C. and 1 atm.

Pressure, atm.	*pv* at 0°C.	*pv* at 25°C.
0.5	1.0057	1.0989
1.0	1.0000	1.0937
2.0	0.9891	1.0841
4.0	.9708	1.0684
6.0	.9530	1.0531
8.0	.9360	1.0385
10.0	.9194	1.0255
12.0	.9026	1.0139

* Sameshima, *Bull. Chem. Soc. Japan*, **1**, 41 (1926).

Table 162. Compressibility Factors of Methyl Chloride

P, atm.	PV	P, atm.	PV
69.9°C.		84.95°C.	
15.065	1.0251	16.19	1.0909
15.425	1.0190	16.586	1.0834
15.76	1.0115	17.09	1.0734
16.10	1.0062	17.61	1.0665
16.44	0.9966	18.13	1.0561
16°C.		18.74	1.0477
		19.36	1.0368
760 mm.	1.0000	19.40	1.0367
1200	0.9796	20.00	1.0239
1650	.9648	20.68	1.0106
2100	.9533	21.49	0.9969
2800	.9335	22.24	.9813

Table 163. Compressibility Factors of Ammonia*,†

$PV = 1.0000$ at 1 atm. and 0°C.

Pressure, atm.	0°C.	25°C.	50°C.	100°C.	132.9°C.	150°C.	200°C.	250°C.	300°C.
1	1.000	1.095	1.191	1.379	1.503	1.567	1.754	1.940	2.126
2	0.986	1.085	1.182	1.373	1.498	1.563	1.750	1.937	2.124
5	.....	1.049	1.153	1.354	1.483	1.549	1.740	1.930	2.119
10	.....	.....	1.103	1.321	1.457	1.526	1.724	1.917	2.109
20	.....	.....	0.988	1.252	1.404	1.479	1.690	1.892	2.089
30	.....	.....	.....	1.176	1.349	1.430	1.656	1.867	2.071
40	.....	.....	.....	1.090	1.290	1.378	1.621	1.840	2.051
50	.....	.....	.....	0.995	1.227	1.326	1.586	1.816	2.032
60	.....	.....	.....	.....	1.160	1.272	1.551	1.791	2.013
80	.....	.....	.....	.....	.....	1.157	1.479	1.740	1.976
100	.....	.....	.....	.....	.....	.....	1.408	1.690	1.938
120	.....	.....	.....	.....	.....	.....	.....	.....	1.903

Pressure, atm.	0°C.‡	10°C.	20°C.	30°C.	40°C.	50°C.	60°C.	70°C.	80°C.	90°C.	100°C.
100	0.1196	0.1223	0.1250	0.1279	0.1305	0.1334	0.1373	0.1417	0.1465	0.1535	0.1606
200	.2437	.2457	.2484	.2526	.2570	.2628	.2693	.2769	.2859	.2963	.3081
300	.3533	.3594	.3657	.3722	.3796	.3886	.3980	.4082	.4190	.4309	.4441
400	.4720	.4777	.4843	.4913	.4998	.5094	.5206	.5322	.5449	.5593	.5705
500	.5839	.5914	.5995	.6084	.6186	.6300	.6419	.6552	.6696	.6853	.7019
600	.6942	.7027	.7113	.7213	.7321	.7443	.7576	.7728	.7898	.8076	.8261
700	.8060	.8137	.8222	.8330	.8453	.8608	.8754	.8916	.9094	.9279	.9472
800	.9171	.9256	.9340	.9441	.9564	.9711	.9873	1.0050	1.0243	1.0443	1.0659
900	1.0281	1.0351	1.0436	1.0544	1.0659	1.0814	1.0976	1.1168	1.1361	1.1577	1.1801
1000	1.1431	1.1489	1.1562	1.1654	1.1770	1.1917	1.2086	1.2279	1.2495	1.2719	1.2942
1100	1.2557	1.2603	1.2672	1.2757	1.2873	1.3012	1.3182	1.3382	1.3606	1.3829	1.4069

Pressure, atm.	110°C.	120°C.	130°C.	140°C.	150°C.	160°C.	170°C.	180°C.	190°C.	200°C.	210°C.
100	0.1724										
200	.3209	0.3363	0.3563	0.3845	0.4273	0.4921	0.5769	0.6841	0.8137	0.9680	
300	.4588	.4760	.4963	.5199	.5479	.5798	.6209	.6703	.7327	.8114	
400	.5917	.6097	.6308	.6541	.6801	.7090	.7421	.7790	.8237	.8700	
500	.7204	.7404	.7622	.7852	.8099	.8369	.8669	.8797	.9387	.9788	
600	.8469	.8677	.8901	.9132	.9379	.9634	.9927	1.0243	1.0567	1.0922	
700	.9680	.9896	1.0127	1.0366	1.0613	1.0875	1.1153	1.1423	1.1708	1.2001	
800	1.0875	1.1107	1.1346	1.1585	1.1839	1.2102	1.2372	1.2665	1.2958	1.3266	
900	1.2032	1.2271	1.2518	1.2773	1.3035	1.3297	1.3583	1.3868	1.4161	1.4477	1.4817
1000	1.3189	1.3436	1.3698	1.3961	1.4223	1.4493	1.4778	1.5056	1.5364	1.5673	1.5989
1100	1.4323	1.4585	1.4855	1.5118	1.5395	1.5673	1.5958	1.6259	1.6560	1.6861	1.7161

* Calculated from the Beattie-Bridgeman equation of state. See Beattie and Lawrence, *J. Am. Chem. Soc.*, **52**, 6 (1930).

† Calculated from experimental data of F. G. Keyes, *J. Am. Chem. Soc.*, **53**, 965 (1931).

‡ Data at 0° were extrapolated graphically.

Table 164. Compressibility Factors of 3:1 Hydrogen-nitrogen Mixture*

$PV = 1.0000$ at 1 atm. and 0°C

P, atm.	−70°C.	−50°C.	−25°C.	0°C.	25°C.	50°C.	100°C.	200°C.	300°C.
25	0.7506	0.8251	0.9187						
50	.7593	.8364	.9320	1.0263	1.1201	1.2133	1.4009	1.7704	2.1323
75	.7700	.8481	.9449						
100	.7816	.8615	.9601	1.0569	1.1527	1.2466	1.4377	1.8098	2.1759
125	.7947	.8750	.9760						
150	.8092	.8901	.9909						
200	.8430	.9256	1.0264	1.1273	1.2247	1.3207	1.5149	1.8912	2.2600
300	.9180	1.0003	1.1024						
400	1.0019	1.0833	1.1833	1.2864	1.3850	1.4825	1.6781	2.0601	2.4303
500	1.0897	1.1693	1.2679						
600	1.1771	1.2568	1.3561	1.4564	1.5549	1.6513	1.8472	2.2278	2.5993
800	1.3531	1.4306	2.5280	1.6295	1.7259	1.8217	2.0176	2.3962	2.7684
1000	1.5264	1.6024	1.6987	1.8009	1.8966	1.9916	2.1867	2.5654	2.9352

* From −70° to −25°C. smoothed values by Deming and Shupe from the work of Bartlett and Vershoyle (see Table 152). The other values were interpolated from Wiebe and Gaddy, *J. Am. Chem. Soc.*, **60,** 2300 (1938).

For data on some other hydrocarbons, see the following:
Propane: Sage, Schaafsma, and Lacey, *Ind. Eng. Chem.*, **26,** 1218 (1934). Beattie, Kay, and Kaminsky, *J. Am. Chem. Soc.*, **59,** 1589 (1937). For thermodynamic properties see Dana, Jenkins, Burdick, and Timm, *Refrig. Eng.*, **12,** 387 (1926).
Butane: (n- and isobutane) For thermodynamic properties see Dana, Jenkins, Burdick, and Timm, *Refrig. Eng.*, **12,** 387 (1926); Sage Webster, and Lacey, *J. Ind. Eng. Chem.*, **29,** 1188 (1937).
n-Pentane: Young, *Sci. Proc. Roy. Dublin, Soc.*, **13,** 310 (1912).
n-Heptane: Smith, Beattie, and Kay, *J. Am. Chem. Soc.*, **59,** 1587 (1937).

Table 165. Compressibilities of Liquids*

At the constant temperature t, the compressibility $\beta = (1/V_0)(dV/dP)$. In general as P increases, β decreases rapidly at first and then slowly; the change of β with t is large at low pressures but very small at pressures above 1000 to 2000 megabars. 1 megabar = 0.987 atm. = 10^6 dyne/cm.²

Substance	Temp., °C.	Pressure, megabars	Compressibility per megabar $\beta \times 10^6$	Substance	Temp., °C.	Pressure, megabars	Compressibility per megabar $\beta \times 10^6$	Substance	Temp., °C.	Pressure, megabars	Compressibility per megabar $\beta \times 10^6$
Acetone	14	23	111	Ethyl acetate	20	400	75	Methyl alcohol	15	23	103
Acetone	20	500	61	alcohol	14	23	100	alcohol	20	200	95
Acetone	20	1,000	52	alcohol	20	500	63	alcohol	20	400	80
Acetone	40	12,000	9	alcohol	20	1,000	54	alcohol	20	500	65
Amyl alcohol	14	23	88	alcohol	20	12,000	8	alcohol	20	1,000	54
alcohol, iso	20	200	84	bromide	20	200	100	alcohol	20	12,000	8
alcohol, iso	20	400	70	bromide	20	400	82	Nitric acid	0	17	32
alcohol, n	20	500	61	bromide	20	500	70	Oils:			
alcohol, n	20	1,000	46	bromide	20	1,000	54	Almond	15	5	53
alcohol, n	20	12,000	8	bromide	20	12,000	8	Castor	15	5	46
alcohol, n	40	12,000	8	chloride	15	23	151	Linseed	15	5	51
Benzene	17	5	89	chloride	20	500	102	Olive	15	5	55
Benzene	20	200	77	chloride	20	1,000	66	Rapeseed	20	......	59
Benzene	20	400	67	chloride	20	12,000	8	Phosphorus trichloride	10	250	71
Bromine	20	200	56	ether	25	23	188	trichloride	20	500	63
Bromine	20	400	51	ether	20	500	84	trichloride	20	1,000	47
Butyl alcohol, iso	18	8	97	ether	20	1,000	61	trichloride	20	12,000	8
alcohol, iso	20	200	81	ether	20	12,000	10	Propyl alcohol (n)	20	200	77
alcohol, iso	20	400	64	iodide	20	200	81	alcohol (n)	20	400	67
alcohol, iso	20	500	56	iodide	20	400	69	alcohol (n?)	20	500	65
alcohol, iso	20	1,000	46	iodide	20	500	64	alcohol (n?)	20	1,000	47
alcohol, iso	20	12,000	8	iodide	20	1,000	50	alcohol (n?)	20	12,000	7
Carbon bisulfide	16	21	86	iodide	20	12,000	8	Toluene	20	200	74
bisulfide	20	500	57	Gallium	30	300	3.97	Toluene	20	400	64
bisulfide	20	1,000	48	Glycerol	15	5	22	Turpentine	20	......	74
bisulfide	20	12,000	6	Hexane	20	200	117	Water	20	13	49
tetrachloride	20	200	86	Hexane	20	400	91	Water	20	200	43
tetrachloride	20	400	73	Kerosene	20	500	55	Water	20	400	41
Chloroform	20	200	83	Kerosene	20	1,000	45	Water	20	500	39
Chloroform	20	400	70	Kerosene	20	12,000	8	Water	40	500	38
Dichloroethylsulfide	32	1,000	34	Mercury	20	300	3.95	Water	40	1,000	33
Dichloroethylsulfide	32	2,000	24	Mercury	22	500	3.97	Water	40	12,000	9
Ethyl acetate	13	23	103	Mercury	22	1,000	3.91	Xylene, meta	20	200	69
acetate	20	200	90	Mercury	22	12,000	2.37	meta	20	400	60

* "Smithsonian Tables," Table 106.

LATENT HEATS

Table 166. Heats of Fusion and Vaporization of the Elements and Inorganic Compounds*

Unless stated otherwise, the values have been taken from the compilations by K. K. Kelley on Heats of Fusion of Inorganic Compounds, *U. S. Bur. Mines Bull.* 393 (1936), and The Free Energies of Vaporization and Vapor Pressures of Inorganic Substances, *U. S. Bur. Mines Bull.* 383 (1935).

Substance	M.p., °C.	Heat of fusion,[a,b] cal./mole	B.p. at 1 atm., °C.	Heat of vaporization,[a,b] cal./mole
Aluminum:				
Al	660.0	2,550	2057	61,020
Al_2Br_6	97.5	5,420	256.4	10,920
Al_2Cl_6	192.5	16,960	180.2[c]	26,750[c]
$AlF_3.3NaF$	1000	16,380		
Al_2I_6	191.0	7,960	385.5	15,360
Al_2O_3	2045	(26,000)	3000	
Antimony:				
Sb	630.5	4,770	1440	46,670
$SbBr_3$	97	3,510		
$SbCl_3$	73.4	3,030	219	10,360
$SbCl_5$	4	2,400	172[d]	11,570
Sb_4O_6	655	(27,000)	1425	17,820
Sb_4S_6	546	11,200		
Argon:				
A	−189.3	290	−185.8	1,590
Arsenic:				
As	814	(6,620)	610[c]	31,000[c]
$AsBr_3$	31	2,810		
$AsCl_3$	− 16	2,420	122	7,570
AsF_5	− 80.7	2,800	− 52.8	4,980
As_4O_6	313	8,000	457.2	14,300
Barium:				
Ba	704	(1,400)[e]	1638	35,670
$BaBr_2$	847	6,000		
$BaCl_2$	960	5,370		
BaF_2	1287	3,000		
$Ba(NO_3)_2$	595	(5,900)		
$Ba_3(PO_4)_2$	1730	18,600		
$BaSO_4$	1350	9,700		
Beryllium:				
Be	1280	2,500[e]		
Bismuth:				
Bi	271.3	2,505	1420	
$BiBr_3$			461	18,020
$BiCl_3$	224	2,600	441	17,350
Bi_2O_3	817	6,800		
Bi_2S_5	747	8,900		
Boron:				
BBr_3	......	......	91.3	7,300
BCl_3	......	......	12.5	5,680
BF_3	−128	480	−100.9	4,620
B_2H_6	−165.5	......	− 92.4	3,685
B_3H_{10}	−119.8	......	16	6,470
B_5H_9	− 46.9	......	58	7,700
B_5H_{11}	......	......	67	8,500
$B_{10}H_{14}$	99.7	7,800	[f]	11,600
B_2H_5Br	−104	......	16	6,230
$B_3N_3H_6$	− 58	......	50.4	7,670
Bromine:				
Br_2	− 7.2	2,580	58.0	7,420
BrF_5	− 61.3	1,355	40.4	7,470
Cadmium:				
Cd	320.9	1,460	765	23,870
$CdBr_2$	568	(5,000)		
$CdCl_2$	568	5,300	967	29,860
CdF_2	1110	(5,400)		
CdI_2	387	3,660	796	25,400
CdO	......	......	1559[c]	53,820[c]
$CdSO_4$	1000	4,790		
Calcium:				
Ca	851	2,230	1487	36,580
$CaBr_2$	730	4,180		
$CaCO_3$	1282	(12,700)		
$CaCl_2$	782	6,100		
CaF_2	1392	4,100		
$Ca(NO_3)_2$	561	5,120		
CaO	2707	(12,240)		
$CaO.Al_2O_3.2SiO_2$	1550	29,400		
$CaO.MgO.2SiO_2$	1392	(18,200)		
$CaO.SiO_2$	1512	13,400		
$CaSO_4$	1297	6,700		
Carbon:				
C (graphite)	3600	11,000[c]		
CBr_4	90	1,050		
CCl_4	− 24.0	644	77	7,280
CF_4	......	......	−127.9	3,110
CH_4	−182.5	224	−161.4	2,040
C_2N_2	− 27.8	1,938[u]	− 21.1	5,576[u]
CNBr	52	......	......	11,010[c]
CNCl	− 5	2,240	13	6,300
Carbon (*Cont.*):				
CNF	......	......	− 72.8	5,780[c]
CNI	......	......	141	13,980[c]
CO	−205.0	200	−191.5	1,444
CO_2	− 57.5	1,900	− 78.4[c]	6,030[c,j]
COS	−138.8	1,129[k]	− 50.2	4,423[k]
$COCl_2$	......	......	8.0	5,990
CS_2	−112.0	1,049[l]		
Cerium:				
Ce	775	2,120		
Cesium:				
Cs	28.4	500	690	16,320
CsBr	......	......	1300	35,990
CsCl	642	3,600	1300	35,690
CsF	715	(2,450)	1251	34,330
CsI	......	......	1280	35,930
$CsNO_3$	407	3,250		
Chlorine:				
Cl_2	−101.0	1,531[m]	− 34.1	4,878[m]
ClF	......	......	−101	
ClF_3	......	......	11.3	5,890
Cl_2O	......	......	2.0	6,280
ClO_2	......	......	10.9	7,100
Cl_2O_7	......	......	79	8,480
Chromium:				
Cr	1550	3,930	2475	
CrO_2Cl_2	......	......	117	8,250
Cobalt:				
Co	1490	3,660		
$CoCl_2$	727	7,390	1050	27,170
Copper:				
Cu	1083.0	3,110	2595	72,810
Cu_2Br_2	......	......	1355	16,310
Cu_2Cl_2	430	4,890	1490	11,920
CuI	......	......	1336	15,940
$Cu_2(CN)_2$	473	(5,400)		
Cu_2O	1230	(13,400)		
CuO	1447	2,820		
Cu_2S	1127	5,500		
Fluorine:				
F_2	−223	......	−188.2	1,640
F_2O	......	......	−144.8	2,650
Gallium:				
Ga	29.8	1,336	2071	
Germanium:				
Ge	959	(8,300)		
GeH_4	−165	......	− 89.1	3,580
Ge_2H_6	−109	......	31.4	5,900
Ge_3H_8	−105.6	......	110.6	7,550
$GeHCl_3$	− 71	......	75[g]	8,000
$GeBr_4$	26.1	......	189	8,560
$GeCl_4$	− 49.5	......	84	7,030
$Ge(CH_3)_4$	− 88	......	44	6,460
Gold:				
Au	1063.0	3,030	2966	81,800
Helium:				
He	−271.4		−268.4	22
Hydrogen:				
H_2	−259.2	28	−252.7	216
HBr	− 86.9	575	− 66.7	4,210
HCl	−114.2	476	− 85.0	3,860
HCN	− 13.2	2,009[i]	25.7	6,027[i]
HF	− 83.0	1,094	33.3	7,460
$(HF)_6$	......	......	51.2	5,020
HI	− 50.8	686		
H_2O	0.0	1,436	100.0	9,729[h,q]
$H^2{}_2O$ (= D_2O)	3.8	1,501[s]	101.4	9,945[r,q]
H_2O_2	− 2	2,520[c]	158	10,270
HNO_3	− 47	600		
H_3PO_2	17.4	2,310		
H_3PO_3	74	3,070		
H_3PO_4	42.4	2,520		
$H_4P_2O_6$	55	8,300		
H_2S	− 85.5	568[t]	− 60.3	4,463[t]
H_2S_2	− 87.6	1,805		
H_2SO_4	10.5	2,360		
H_2Se	......	......	− 41.3	4,880
H_2SeO_4	58	3,450		
H_2Te	− 48.9	1,670	− 2.2	5,650
Indium:				
In	156.4	781		

* For data on NH_3, CO_2, SO_2, see the section on engineering thermodynamic properties, Sec. 3, pp. 250, 252, 255, 256, 275, and 276.

Table 166. Heats of Fusion and Vaporization of the Elements and Inorganic Compounds—(*Continued*)

Substance	M.p., °C.	Heat of fusion,[a,b] cal./mole	B.p. at 1 atm., °C.	Heat of vaporization,[a,b] cal./mole
Iodine:				
I_2	113.0	3,650	183	10,390
$ICl(\alpha)$	17.2	2,660		
$ICl(\beta)$	13.9	2,270		
IF_7	......	......	4[c]	7,460[c]
Iron				
Fe	1530	3,560	2735	84,600
$FeCl_2$	677	7,800	1026	30,210
Fe_2Cl_6	304	20,590	319	12,040
$Fe(CO)_5$	− 21	3,250	105	9,000
FeO	1380	(7,700)		
FeS	1195	5,000		
Krypton:				
Kr	−157	360[e]	152.9	2,310[e]
Lead:				
Pb	327.4	1,224	1744	42,060
$PbBr_2$	488	4,290	914	27,700
$PbCl_2$	498	5,650	954	29,600
PbF_2	824	1,860	1293	38,300
PbI_2	412	5,970	872	24,850
$PbMoO_4$	1065	(25,800)		
PbO	890	2,820	1472	51,310
PbS	1114	4,150	1281	(50,000)
$PbSO_4$	1087	9,600		
$PbWO_4$	1123	(15,200)		
Lithium:				
Li	179	1,100	1372	32,250
$LiBO_2$	845	(5,570)		
LiBr	552	2,900	1310	35,420
LiCl	614	3,200	1382	35,960
LiF	847	(2,360)	1681	50,970
LiI	440	(1,420)	1171	40,770
LiOH	462	2,480		
Li_2MoO_4	705	4,200		
$LiNO_3$				
Li_2SiO_3	1177	7,210		
Li_4SiO_4	1249	7,430		
Li_2SO_4	857	3,040		
Li_2WO_4	742	(6,700)		
Magnesium:				
Mg	650	2,160	1107	32,520
$MgBr_2$	711	8,300		
$MgCl_2$	712	8,100	1418	32,690
MgF_2	1221	5,900		
MgO	2642	18,500		
$Mg_3(PO_4)_2$	1184	(11,300)		
$MgSiO_3$	1524	14,700		
$MgSO_4$	1127	3,500		
$MgZn_2$	589	(8,270)		
Manganese:				
Mn	1220	3,450	2152	55,150
$MnCl_2$	650	7,340	1190	29,630
$MnSiO_3$	1274	(8,200)		
$MnTiO_3$	1404	(7,960)		
Mercury:				
Hg	− 38.9	557	361	13,980
$HgBr_2$	241	3,960	319	14,080
$HgCl_2$	277	4,150	304	14,080
HgI_2	250	4,500	354	14,260
$HgSO_4$	850	(1,440)		
Molybdenum:				
Mo	2622	(6,660)	(4800)	(128,000)
MoF_6	17	2,500	36	6,000
MoO_3	745	(2,500)	1151	
Neon:				
Ne	−248.5	77	−246.0	440[e]
Nickel:				
Ni	1455	4,200	2730	87,300
$NiCl_2$	......	......	987[c]	48,360[c]
$Ni(CO)_4$	......	......	42.5	7,000
Ni_2S	645	(2,980)		
Ni_3S_2	790	5,800		
Nitrogen:				
N_2	−210.0	172	−195.8	1,336
NF_3	......	......	−129.0	3,000
NH_3	− 77.7	1,352[n]	− 33.4	5,581[n]
NH_4CNS	146	(4,700)		
NH_4NO_3	169.6	1,460		
N_2O	− 90.8	1,563	− 88.5	3,950
NO	−163.6	550	−151.7	3,307
N_2O_4	− 13	5,540	30	7,040
N_2O_5	......	......	32.4	13,800[c]
NOCl	......	......	− 6.4	6,140
Osmium:				
OsF_8	......	......	47.4	6,840
OsO_4(yellow)	56	4,060	130	9,450
OsO_4 (white)	42	2,340		
Oxygen				
O_2	−218.9	106	−183.0	1,629
O_3	......	......	−111	2,880
Palladium:				
Pd	1554	4,120		
Phosphorus:				
P_4 (yellow)	44.2	615	280	12,520
P_4 (violet)	......	......	417[c]	25,600[c]
P_4 (black)	......	......	453[c]	33,100
PCl_3	......	......	74.2	7,280
PH_3	−133.8	270[o]	− 87.7	3,489[o]
P_4O_6	23.8	3,360	174	10,380
$P_4O_{10}(\alpha)$	569	17,080	591	20,670
$P_4O_{10}(\beta)$	......	......	358[c]	
$POCl_3$	1.1	3,110	105.1	8,380
P_2S_3	......	......	508	
Platinum:				
Pt	1773.5	4,700	(4400)	(107,000)
Potassium:				
K	63.5	574	776	18,920
KBO_2	947	(5,700)		
KBr	742	5,000	1383	37,060
KCl	770	6,410	1407	38,840
KCN	623	(3,500)		
KCNS	179	2,250		
K_2CO_3	897	7,800		
K_2CrO_4	984	6,920		
$K_2Cr_2O_7$	398	8,770		
KF	857	6,500		
KI	682	4,100	1324	34,690
K_2MoO_4	922	(4,000)		
KNO_3	338	2,840		
KOH	360	(2,000)	1327	30,850
KPO_3	817	2,110		
K_3PO_4	1340	8,900		
$K_4P_2O_7$	1092	14,000		
K_2SO_4	1074	8,100		
K_2TiO_3	810	(10,600)		
K_2WO_4	927	(4,400)		
Praseodymium:				
Pr	932	2,700		
Radon:				
Rn	−71	......	− 61.8	4,010
Rhenium:				
Re	(3000)			
Re_2O_7	296	15,340	362.4	18,060
Re_2O_8	147	3,800		
Rubidium:				
Rb	39.1	525	679	18,110
RbBr	677	3,700	1352	37,120
RbCl	717	4,400	1381	36,920
RbF	833	4,130	1408	39,510
RbI	638	2,990	1304	35,960
$RbNO_3$	305	1,340		
Selenium:				
Se_2	217	1,220	753	25,490
Se_6			736	20,600
SeF_6			− 45.8[c]	6,350[c]
SeO_2			317[c]	20,900
$SeOCl_2$	10	1,010	168	
Silicon:				
Si	1427	9,470	2290	
$SiCl_4$	− 67.6	1,845	56.8	6,860
Si_2Cl_6	− 1		139	
Si_3Cl_8			211.4	12,340
$(SiCl_3)_2O$	− 33		135.6	8,820
SiF_4	......	......	− 94.8[c]	6,130[c]
Si_2F_6	− 18.5	3,900	− 18.9[c]	10,400[c]
SiF_3Cl	−138	......	− 70.1	4,460
SiF_2Cl_2	−144	......	− 31.5	5,080
SiH_4	−185	......	−111.6	2,960
Si_2H_6	−132.5	......	− 14.3	5,110
Si_3H_8	−117	......	53.1	6,780
Si_4H_{10}	− 93.5	......	100	8,890
SiH_3Br	− 93.8	......	2.4	5,650
SiH_2Br_2	− 70.0	......	70.5	6,840
$SiHCl_3$	−126.5	......	31.8	6,360
$(SiH_3)_3N$	−105.6	......	48.7	6,850
$(SiH_3)_2O$	−144	......	− 15.4	5,350
SiO_2 (quartz)	1470	3,400	2230	
SiO_2 (cristobalite)	1700	2,100		
Silver:				
Ag	960.5	2,700	2212	60,720
AgBr	430	2,180		
AgCl	455	3,155	1564	42,520
AgCN	350	2,750		
AgI	557	2,250	1506	34,450
$AgNO_3$	209	2,755		
Ag_2S	842	3,360		
Ag_2SO_4	657	(4,300)		
Sodium:				
Na	97.7	630	914	23,120
$NaBO_2$	966	8,660		

Table 166. Heats of Fusion and Vaporization of the Elements and Inorganic Compounds—(*Concluded*)

Substance	M.p., °C.	Heat of fusion,[a,b] cal./mole	B.p. at 1 atm., °C.	Heat of vaporization,[a,b] cal./mole
Sodium (*Cont.*):				
NaBr	747	6,140	1392	37,950
NaCl	800	7,220	1465	40,810
$NaClO_3$	255	5,290		
NaCN	562	(4,400)	1500	37,280
NaCNS	323	4,450		
Na_2CO_3	854	7,000		
NaF	992	7,000	1704	53,260
NaI	662	5,240		
Na_2MoO_4	687	3,600		
$NaNO_3$	310	3,760		
NaOH	322	2,000	1378	
$\frac{1}{2}Na_2O.\frac{1}{2}Al_2O_3.3SiO_2$	1107	13,150		
$NaPO_3$	988	(5,000)		
$Na_4P_2O_7$	970	(13,700)		
Na_2S	920	(1,200)		
Na_2SiO_3	1087	10,300		
$Na_2Si_2O_5$	884	8,460		
Na_2SO_4	884	5,830		
Na_2WO_4	702	5,800		
Strontium:				
Sr	757	2,190	1384	33,610
$SrBr_2$	643	4,780		
$SrCl_2$	872	4,100		
SrF_2	1400	4,260		
$Sr_3(PO_4)_2$	1770	18,500		
Sulfur:				
S (rhombic)	112.8	....	444.6	20,200
S (monoclinic)	119.2			
S_2Cl_2	....	....	138	8,720
SF_6	....	....	− 63.5[c]	5,600[c]
SO_2	− 75.5	1,769[p]	− 5.0	5,960[p]
$SO_3(\alpha)$	17	2,060	44.8	10,190
$SO_3(\beta)$	32.4	2,890		
$SO_3(\gamma)$	62.2	6,310		
$SOBr_2$	....	....	139.5	9,920
$SOCl_2$	....	....	75.4	7,600
SO_2Cl_2	....	....	69.2	7,760
Tellurium:				
Te	453	3,230	1090	
$TeCl_4$	....	....	392	16,830
TeF_6	....	....	− 38.6[c]	6,700[c]
Thallium:				
Tl	302.5	1,030	1457	38,810
TlBr	460	5,990	819	23,800
TlCl	427	4,260	807	24,420
Tl_2CO_3	273	4 400		
TlI	440	3,125	823	25,030
$TlNO_3$	207	2,290		
Tl_2S	449	3,000		
Tl_2SO_4	632	5,500		
Tin:				
Sn_4	231.8	1,720	2270	68,000
$SnBr_2$	232	(1,700)		
$SnBr_4$	30	3,000		
$SnCl_2$	247	3,050	623	20,740
$SnCl_4$	− 33.2	2,190	113	8,330
$Sn(CH_3)_4$	....	....	78.3	7,320
SnH_4	−149.8	....	− 52.3	4,420
SnI_4	143.5	(4,300)		
Titanium:				
$TiBr_4$	38.2	(2,060)		
$TiCl_4$	− 23	2,240	136	8,350
TiO_2	1825	(11,400)		
Tungsten:				
W	3390	(8,400)	(5900)	(176,000)
WF_6	− 0.4	1,800	17.3	6,350
Uranium:				
UF_6	....	....	55.1[c]	9,990[c]
Xenon:				
Xe	−111.5	740	−108.0	3,110
Zinc:				
Zn	419.5	1,595	907	27,430
$ZnCl_2$	283	(5,500)	732	28,710
$Zn(C_2H_5)_2$	....	....	118	8,960
ZnO	1975	4,470		
ZnS	1645	(9,000)		
Zirconium:				
$ZrBr_4$	....	....	357[c]	25,800[c]
$ZrCl_4$	....	....	311[c]	25,290[c]
ZrI_4	....	....	431[c]	29,030[c]
ZrO_2	2715	20,800		

[a] Values in parentheses are uncertain.
[b] For the freezing point or the normal boiling point unless otherwise stated.
[c] Sublimation.
[d] Decomposes at about 75°C.; value obtained by extrapolation.
[e] Bichowsky and Rossini, "Thermochemistry of the Chemical Substances," Reinhold, New York (1936).
[f] Decomposes before the normal boiling point is reached.
[g] Decomposes at about 40°C.; value obtained by extrapolation.
[h] See also pp. 277–279 on steam table.
[i] Giauque and Ruehrwein, *J. Am. Chem. Soc.*, **61**, 2626 (1939).
[j] Giauque and Egan, *J. Chem. Phys.*, **5**, 45 (1937).
[k] Kemp and Giauque, *J. Am. Chem. Soc.*, **59**, 79 (1937).
[l] Brown and Manov, *J. Am. Chem. Soc.*, **59**, 500 (1937).
[m] Giauque and Powell, *J. Am. Chem. Soc.*, **61**, 1970 (1939).
[n] Overstreet and Giauque, *J. Am. Chem. Soc.*, **59**, 254 (1937).
[o] Stephenson and Giauque, *J. Chem. Phys.*, **5**, 149 (1937).
[p] Giauque and Stephenson, *J. Am. Chem. Soc.*, **60**, 1389 (1938).
[q] Osborne, Stimson, and Ginnings, *Bur. Standards J. Research*, **23**, 197, 261 (1939).
[r] Miles and Menzies, *J. Am. Chem. Soc.*, **58**, 1067 (1936).
[s] Long and Kemp, *J. Am. Chem. Soc.*, **58**, 1829 (1936).
[t] Giauque and Blue, *J. Am. Chem. Soc.*, **58**, 831 (1936).
[u] Ruehrwein and Giauque, *J. Am. Chem. Soc.*, **61**, 2940 (1939).

Table 167. Heats of Fusion of Organic Compounds

The values for the hydrocarbons are from the tables of the American Petroleum Institute Research Project 44 at the National Bureau of Standards, with some from Parks and Huffman, *Ind. Eng. Chem.*, **23**, 1138 (1931).

The values for the non-hydrocarbon compounds were recalculated from data in "International Critical Tables," vol. 5.

Hydrocarbon compounds	Formula	M.p., °C.	Heat of fusion, cal./g.
Paraffins:			
Methane	CH_4	−182.48	14.03
Ethane	C_2H_6	−183.23	22.712
Propane	C_3H_8	−187.65	19.100
n-Butane	C_4H_{10}	−138.33	19.167
2-Methylpropane	C_4H_{10}	−159.60	18.668
n-Pentane	C_5H_{12}	−129.723	27.874
2-Methylbutane	C_5H_{12}	−159.890	17.076
2,2-Dimethylpropane	C_5H_{12}	− 16.6	10.786
n-Hexane	C_6H_{14}	− 95.320	36.138
2-Methylpentane	C_6H_{14}	−153.680	17.407
2,2-Dimethylbutane	C_6H_{14}	− 99.73	1.607
2,3-Dimethylbutane	C_6H_{14}	−128.41	2.251
n-Heptane	C_7H_{16}	− 90.595	33.513
2-Methylhexane	C_7H_{16}	−118.270	21.158
3-Ethylpentane	C_7H_{16}	−118.593	22.555
2,2-Dimethylpentane	C_7H_{16}	−123.790	13.982
2,4-Dimethylpentane	C_7H_{16}	−119.230	15.968
3,3-Dimethylpentane	C_7H_{16}	−134.46	16.856
2,2,3-Trimethylbutane	C_7H_{16}	− 24.96	5.250
n-Octane	C_8H_{18}	− 56.798	43.169
2-Methylheptane	C_8H_{18}	−109.04	21.458
3-Methylpentane	C_8H_{18}	−120.50	23.795
4-Methylheptane	C_8H_{18}	−120.955	22.692
2,2-Dimethylhexane	C_8H_{18}	−121.18	24.226
2,5-Dimethylhexane	C_8H_{18}	− 91.200	26.903
3,3-Dimethylhexane	C_8H_{18}	−126.10	14.9
2-Methyl-3-ethylpentane	C_8H_{18}	−114.960	23.690
3-Methyl-3-ethylpentane	C_8H_{18}	− 90.870	22.657
2,2,3-Trimethylpentane	C_8H_{18}	−112.27	18.061
2,2,4-Trimethylpentane	C_8H_{18}	−107.365	19.278
2,3,3-Trimethylpentane	C_8H_{18}	−100.70	3.204
2,3,4-Trimethylpentane	C_8H_{18}	−109.210	19.392
2,2,3,3-Tetramethylbutane	C_8H_{18}	+100.69	14.900
n-Nonane	C_9H_{20}	− 53.9	41.2
n-Decane	$C_{10}H_{22}$	− 30.0	48.3
n-Undecane	$C_{11}H_{24}$	− 25.9	34.1
n-Dodecane	$C_{12}H_{26}$	− 9.6	51.3
Eicosane	$C_{20}H_{42}$	+ 36.4	52.0
Pentacosane	$C_{25}H_{52}$	+ 53.3	53.6
Tritriacontane	$C_{33}H_{68}$	+ 71.1	54.0
Aromatics:			
Benzene	C_6H_6	+ 5.533	30.100
Methylbenzene (Toluene)	C_7H_8	− 94.991	17.171
Ethylbenzene	C_8H_{10}	− 94.950	20.629
o-Xylene	C_8H_{10}	− 25.187	30.614
m-Xylene	C_8H_{10}	− 47.872	26.045
p-Xylene	C_8H_{10}	+ 13.263	38.526
n-Propylbenzene	C_9H_{12}	− 99.500	16.97
Isopropylbenzene	C_9H_{12}	− 96.028	19.22
1-Methyl-2-ethylbenzene	C_9H_{12}	− 80.833	21.13
Aromatics—(*Cont.*):			
1-Methyl-3-ethylbenzene	C_9H_{12}	− 95.55	15.14
1-Methyl-4-ethylbenzene	C_9H_{12}	− 62.350	25.29
1,2,3-Trimethylbenzene	C_9H_{12}	− 25.375	16.64
1,2,4-Trimethylbenzene	C_9H_{12}	− 43.80	24.54
1,3,5-Trimethylbenzene	C_9H_{12}	− 44.720	18.97
Naphthalene	$C_{10}H_8$	+ 80.0	36.0
Camphene	$C_{10}H_{12}$	+ 51	57
Durene	$C_{10}H_{14}$	+ 79.3	37.4
Isodurene	$C_{10}H_{14}$	− 24.0	23.0
Prehnitene	$C_{10}H_{14}$	− 7.7	20.0
p-Cymene	$C_{10}H_{14}$	− 68.9	17.1
n-Butyl benzene	$C_{10}H_{14}$	− 88.5	19.5
tert-Butyl benzene	$C_{10}H_{14}$	− 58.1	14.9
β-Methyl naphthalene	$C_{11}H_{10}$	+ 34.1	20.1
Diphenyl	$C_{12}H_{10}$	+ 68.6	28.8
Hexamethyl benzene	$C_{12}H_{18}$	+165.5	30.4
Diphenyl methane	$C_{13}H_{12}$	+ 25.2	26.4
Anthracene	$C_{14}H_{10}$	+216.5	38.7
Phenanthrene	$C_{14}H_{10}$	+ 96.3	25.0
Tolane	$C_{14}H_{10}$	+ 60	28.7
Stilbene	$C_{14}H_{12}$	+124	40.0
Dibenzil	$C_{14}H_{14}$	+ 51.4	30.7
Triphenyl methane	$C_{19}H_{16}$	+ 92.1	21.1
Alkyl cyclohexanes:			
Cyclohexane	C_6H_{12}	+ 6.67	7 569
Methylcyclohexane	C_7H_{14}	−126.58	16.429
Alkyl cyclopentanes:			
Cyclopentane	C_5H_{10}	− 93.80	2.068
Methylcyclopentane	C_6H_{12}	−142.445	19.68
Ethylcyclopentane	C_7H_{14}	−138.435	11.10
1,1-Dimethylcyclopentane	C_7H_{14}	− 69.73	3.36
cis-1,2-Dimethylcyclopentane	C_7H_{14}	− 53.85	3.87
trans-1,2-Dimethylcyclopentane	C_7H_{14}	−117.57	15.68
trans-1,3-Dimethylcyclopentane	C_7H_{14}	−133.680	17.93
Monoolefins:			
Ethene (Ethylene)	C_2H_4	−169.15	28.547
Propene (Propylene)	C_3H_6	−185.25	17.054
1-Butene	C_4H_8	−185.35	16.393
cis-2-Butene	C_4H_8	−138.91	31.135
trans-2-Butene	C_4H_8	−105.55	41.564
2-Methylpropene (isobutene)	C_4H_8	−140.35	25.265
1-Pentene	C_5H_{10}	−165.27	16.82
cis-2-pentene	C_5H_{10}	−151.363	24.239
trans-2-pentene	C_5H_{10}	−140.235	26.536
2-Methyl-1-butene	C_5H_{10}	−137.560	26.879
3-Methyl-1-butene	C_5H_{10}	−168.500	18.009
2-Methyl-2-butene	C_5H_{10}	−133.780	25.738
Acetylenes:			
Acetylene	C_2H_2	− 81.5	23.04
2-Butyne (dimethylacetylene)	C_4H_6	−132.23	40.808

Non-hydrocarbon compounds	Formula	M.p., °C.	Heat of fusion, cal./g.
Acetic acid	$C_2H_4O_2$	16.7	46.68
Acetone	C_3H_6O	−95.5	23.42
Acrylic acid	$C_3H_4O_2$	12.3	37.03
Allo-cinnamic acid	$C_9H_8O_2$	68	27.35
Aminobenzoic acid (o-)	$C_7H_7NO_2$	145	35.48
(m-)	$C_7H_7NO_2$	179.5	38.03
(p-)	$C_7H_7NO_2$	188.5	36.46
Amyl alcohol	$C_5H_{12}O$	−78.9	26.65
Anethole	$C_{10}H_{12}O$	22.5	25.80
Aniline	$C_6H_5NH_2$	−6.3	27.09
Anthraquinone	$C_{14}H_8O_2$	284.8	37.48
Apiol	$C_{12}H_{14}O_4$	29.5	25.80
Azobenzene	$C_{12}H_{10}N_2$	67.1	28.91
Azoxybenzene	$C_{12}H_{10}N_2O$	36	21.62
Benzil	$C_{14}H_{10}O_2$	95.2	22.15
Benzoic acid	$C_7H_8O_2$	122.45	33.90
Benzophenone	$C_{13}H_{10}O$	47.85	23.53
Benzylaniline	$C_{13}H_{13}N$	32.37	21.86
Bromocamphor	$C_{10}H_{15}BrO$	78	41.57
Bromochlorbenzene (o-)	C_6H_4BrCl	−12.6	15.41
(m-)	C_6H_4BrCl	−21.2	15.29
(p-)	C_6H_4BrCl	64.6	23.41
Bromoiodobenzene (o-)	C_6H_4BrI	21	12.18
(m-)	C_6H_4BrI	9.3	10.27
(p-)	C_6H_4BrI	90.1	16.60
Bromol hydrate	$C_2H_3Br_3O_2$	46	16.90
Bromophenol (p-)	C_6H_5BrO	63.5	20.50
Bromotoluene (p-)	C_7H_7Br	28	20.86
Butyl alcohol (n-)	$C_4H_{10}O$	−89.2	29.93
(t-)	$C_4H_{10}O$	25.4	21.88
Butyric acid (n-)	$C_4H_8O_2$	−5.7	30.04
Capric acid (n-)	$C_{10}H_{20}O_2$	31.99	38.87
Caprylic acid (n-)	$C_8H_{16}O_2$	16.3	35.40
Carbazole	$C_{12}H_9N$	243	42.05
Carbon tetrachloride	CCl_4	−22.8	41.57
Carvoxime (d-)	$C_{10}H_{15}NO$	71.5	23.29
(l-)	$C_{10}H_{15}NO$	71	23.41
(dl-)	$C_{10}H_{15}NO$	91	24.61
Cetyl alcohol	$C_{16}H_{34}O$	49.27	33.80
Chloracetic acid (α-)	$C_2H_4ClO_2$	61.2	31.06
(β-)	$C_2H_4ClO_2$	56	35.12
Chloral alcoholate	$C_4H_7Cl_3O_2$	9	24.03
hydrate	$C_2H_3Cl_3O_2$	47.4	33.18
Chloroaniline (p-)	C_6H_6ClN	71	37.15
Chlorobenzoic acid (o-)	$C_7H_5ClO_2$	140.2	39.30
(m-)	$C_7H_5ClO_2$	154.25	36.41
(p-)	$C_7H_5ClO_2$	239.7	49.21
Chloronitrobenzene (m-)	$C_6H_4ClNO_2$	44.4	29.38
(p-)	$C_6H_4ClNO_2$	83.5	31.51
Cinnamic acid	$C_9H_8O_2$	133	36.50
anhydride	$C_{18}H_{14}O_3$	48	28.14
Cresol (p-)	C_7H_8O	34.6	26.28
Crotonic acid (α-)	$C_4H_6O_2$	72	25.32
(cis-)	$C_4H_6O_2$	71.2	34.90
Cyanamide	CH_2N_2	44	49.81
Cyclohexanol	$C_6H_{12}O$	25.46	4.19

Table 167. Heats of Fusion of Organic Compounds—(*Concluded*)

Non-hydrocarbon compounds	Formula	M.p., °C.	Heat of fusion, cal./g.
Dibromobenzene (o-)	$C_6H_4Br_2$	1.8	12.78
(m-)	$C_6H_4Br_2$	−6.9	13.38
(p-)	$C_6H_4Br_2$	86	20.55
Dibromophenol (2, 4-)	$C_6H_4Br_2O$	12	13.97
Dichloroacetic acid	$C_2H_2Cl_2O_2$	−4(?)	14.21
Dichlorobenzene (o-)	$C_6H_4Cl_2$	−16.7	21.02
(m-)	$C_6H_4Cl_2$	−24.8	20.55
(p-)	$C_6H_4Cl_2$	53.13	29.67
Dihydroxybenzene (o-)	$C_6H_6O_2$	104.3	49.40
(m-)	$C_6H_6O_2$	109.65	46.20
(p-)	$C_6H_6O_2$	172.3	58.77
Di-iodobenzene (o-)	$C_6H_4I_2$	23.4	10.15
(m-)	$C_6H_4I_2$	34.2	11.54
(p-)	$C_6H_4I_2$	129	16.20
Dimethyl tartrate (dl-)	$C_6H_{10}O_6$	87	35.12
(d-)	$C_6H_{10}O_6$	49	21.50
pyrone	$C_7H_8O_2$	132	56.14
Dinitrobenzene (o-)	$C_6H_4N_2O_4$	116.93	32.25
(m-)	$C_6H_4N_2O_4$	89.7	24.70
(p-)	$C_6H_4N_2O_4$	173.5	39.99
Dinitrotoluene (2, 4-)	$C_7H_6N_2O_4$	70.14	26.40
Dioxane	$C_4H_8O_2$	11.0	34.85
Diphenyl amine	$C_{12}H_{11}N$	52.98	25.23
Elaidic acid	$C_{18}H_{34}O_2$	44.4	52.08
Ethyl acetate	$C_4H_8O_2$	83.8	28.43
alcohol	C_2H_6O	−114.4	25.76
Ethylene dibromide	$C_2H_4Br_2$	10.012	13.52
Ethyl ether	$C_4H_{10}O$	−116.3	23.54
Formic acid	CH_2O_2	8.40	58.89
Glutaric acid	$C_5H_8O_4$	97.5	37.39
Glycerol	$C_3H_8O_3$	18.07	47.49
Glycol, ethylene	$C_2H_6O_2$	−11.5	43.26
Hydrazo benzene	$C_{12}H_{12}N_2$	134	22.89
Hydrocinnamic acid	$C_9H_{10}O_2$	48	28.14
Hydroxyacetanilide	$C_8H_9NO_2$	91.3	33.59
Iodotoluene (p-)	C_7H_7I	34	18.75
Isopropyl alcohol	C_3H_8O	−88.5	21.08
ether	$C_6H_{14}O$	−86.8	25.79
Lauric acid (n-)	$C_{12}H_{24}O_2$	43.22	43.72
Levulinic acid	$C_5H_8O_3$	33	18.97
Menthol (l-) (α)	$C_{10}H_{20}O$	43.5	18.63
Methyl alcohol	CH_4O	−97.8	23.7
Myristic acid	$C_{14}H_{28}O_2$	53.86	47.49
Methyl cinnamate	$C_{10}H_{10}O_2$	36	26.53
fumarate	$C_6H_8O_4$	102	57.93
oxalate	$C_4H_6O_4$	54.35	42.64
phenylpropiolate	$C_{10}H_8O_2$	18	22.86
succinate	$C_6H_{10}O_4$	19.5	35.72
Naphthol (α-)	$C_{10}H_8O$	95.0	38.94
(β-)	$C_{10}H_8O$	120.6	31.30
Naphthylamine (α-)	$C_{10}H_9N$	50	22.34
Nitroaniline (o-)	$C_6H_6N_2O_2$	71.2	27.88
(m-)	$C_6H_6N_2O_2$	114.0	40.97
(p-)	$C_6H_6N_2O_2$	147.3	36.46
Nitrobenzene	$C_6H_5NO_2$	5.85	22.52
Nitrobenzoic acid (o-)	$C_7H_5NO_4$	145.8	40.06
(m-)	$C_7H_5NO_4$	141.1	27.59
(p-)	$C_7H_5NO_4$	239.2	52.80
Nitronaphthalene	$C_{10}H_7NO_2$	56.7	25.44
Nitrophenol (o-)	$C_6H_5NO_3$	45.13	26.76
Palmitic acid	$C_{16}H_{32}O_2$	61.82	39.18
Paraldehyde	$C_6H_{12}O_3$	10.5	25.02
Pelargic acid (n-) (β-)	$C_9H_{18}O_2$	....	39.04
Pelargonic acid (n-) (α-)	$C_9H_{18}O_2$	12.35	30.63
Phenol	C_6H_6O	40.92	29.03
Phenylacetic acid	$C_8H_8O_2$	76.7	25.44
Phenylhydrazine	$C_6H_8N_2$	19.6	36.31
Propyl ether (n)	$C_6H_{14}O$	−126.1	20.66
Quinone	$C_6H_4O_2$	115.7	40.85
Stearic acid	$C_{18}H_{30}O_2$	68.82	47.54
Succinic anhydride	$C_4H_4O_3$	119	48.74
Succinonitrile	$C_4H_4N_2$	54.5	11.71
Tetrachloroxylene (o-)	$C_8H_6Cl_4$	86	21.02
(p-)	$C_8H_6Cl_4$	95	22.10
Thiophene	C_4H_4S	−39.4	14.11
Thiosinamine	$C_4H_8N_2S$	77	33.45
Thymol	$C_{10}H_{14}O$	51.5	27.47
Toluic acid (o-)	$C_8H_8O_2$	103.7	35.40
(m-)	$C_8H_8O_2$	108.75	27.59
(p-)	$C_8H_8O_2$	179.6	39.90
Toluidine (p-)	C_7H_9N	43.3	39.90
Tribromophenol (2, 4, 6-)	$C_6H_3Br_3O$	93	13.38
Trichloroacetic acid	$C_2HCl_3O_2$	57.5	8.60
Trinitroglycerol	$C_3H_5N_3O_9$	12.3	23.02
Trinitrotoluene (2, 4, 6-)	$C_7H_5N_3O_6$	80.83	22.34
Tristearin	$C_{57}H_{110}O_6$	70.8, 54.5	45.63
Undecylic acid (α-) (n-)	$C_{11}H_{22}O_2$	28.25	32.20
(β-) (n-)	$C_{11}H_{22}O_2$	....	42.91
Urethane	$C_3H_7NO_2$	48.7	40.85
Veratrol	$C_8H_{10}O_2$	22.5	27.45
Xylene dibromide (o-)	$C_8H_8Br_2$	95	24.25
(m-)	$C_8H_8Br_2$	77	21.45
dichloride (o-)	$C_8H_8Cl_2$	55	29.03
(m-)	$C_8H_8Cl_2$	34	26.64
(p-)	$C_8H_8Cl_2$	100	32.73

Table 168. Heats of Vaporization of Organic Compounds

The values for the hydrocarbons are from the tables of the American Petroleum Institute Research Project 44 at the National Bureau of Standards.

The values for the non-hydrocarbon compounds were recalculated from data in "International Critical Tables," vol. 5.

Hydrocarbon compounds	Formula	Temperature, °C.	ΔH_v, cal./g.
Paraffins:			
Methane	CH_4	−161.6	121.87
Ethane	C_2H_6	−88.9	116.87
Propane	C_3H_8	25	81.76
		−42.1	101.76
n-Butane	C_4H_{10}	25	86.63
		−0.50	92.09
2-Methylpropane (isobutane)	C_4H_{10}	25	78.63
		−11.72	87.56
n-Pentane	C_5H_{12}	25	87.54
		36.08	85.38
2-Methylbutane (isopentane)	C_5H_{12}	25	81.47
		27.86	80.97
2,2-Dimethylpropane (neopentane)	C_5H_{12}	25	72.15
		9.45	75.37
n-Hexane	C_6H_{14}	25	87.50
		68.74	80.48
2-Methylpentane	C_6H_{14}	25	82.83
		60.27	76.89
3-Methylpentane	C_6H_{14}	25	83.96
		63.28	78.42
2,2-Dimethylbutane	C_6H_{14}	25	76.79
		49.74	73.75
2,3-Dimethylbutane	C_6H_{14}	25	80.77
		57.99	76.53
n-Heptane	C_7H_{16}	25	87.18
		98.43	76.45
2-Methylhexane	C_7H_{16}	25	83.02
		90.05	73.4
3-Methylhexane	C_7H_{16}	25	83.68
		91.95	74.1
3-Ethylpentane	C_7H_{16}	25	84.02
		93.47	74.3
2,2-Dimethylpentane	C_7H_{16}	25	77.36
		79.20	69.7
2,3-Dimethylpentane	C_7H_{16}	25	81.68
		89.79	72.9
2,4-Dimethylpentane	C_7H_{16}	25	78.44
		80.51	70.9
3,3-Dimethylpentane	C_7H_{16}	25	78.76
		86.06	70.6
2,2,3-Trimethylbutane	C_7H_{16}	25	76.42
		80.88	69.3
n-Octane	C_8H_{18}	25	86.80
		125.66	73.19
2-Methylheptane	C_8H_{18}	25	83.02
		117.64	70.3
3-Methylheptane	C_8H_{18}	25	83.35
		118.92	71.3
4-Methylheptane	C_8H_{18}	25	83.01
		117.71	70.91
3-Ethylhexane	C_8H_{18}	25	82.95
		118.53	71.7
2,2-Dimethylhexane	C_8H_{18}	25	78.02
		106.84	67.7
2,3-Dimethylhexane	C_8H_{18}	25	81.17
		115.60	70.2
2,4-Dimethylhexane	C_8H_{18}	25	79.02
		109.43	68.5
2,5-Dimethylhexane	C_8H_{18}	25	79.21
		109.10	68.6
3,3-Dimethylhexane	C_8H_{18}	25	78.54
		111.97	68.5
3,4-Dimethylhexane	C_8H_{18}	25	81.55
		117.72	70.2
2-Methyl-3-ethylpentane	C_8H_{18}	25	80.60
		115.65	69.7
3-Methyl-3-ethylpentane	C_8H_{18}	25	79.49
		118.26	69.3
2,2,3-Trimethylpentane	C_8H_{18}	25	77.24
		109.84	67.3
2,2,4-Trimethylpentane	C_8H_{18}	25	73.50
		99.24	64.87
2,3,3-Trimethylpentane	C_8H_{18}	25	77.87
		114.76	68.1
2,3,4-Trimethylpentane	C_8H_{18}	25	78.90
		113.47	68.37
2,2,3,3-Tetramethylbutane	C_8H_{18}	106.30	66.2
Alkyl benzenes:			
Benzene	C_6H_6	25	103.57
		80.10	94.14
Methylbenzene (toluene)	C_7H_8	25	98.55
		110.62	86.8
Ethylbenzene	C_8H_{10}	25	95.11
		136.19	81.0
1,2-Dimethylbenzene (o-xylene)	C_8H_{10}	25	97.79
		144.42	82.9
1,3-Dimethylbenzene (m-xylene)	C_8H_{10}	25	96.03
		139.10	82.0
1,4-Dimethylbenzene (p-xylene)	C_8H_{10}	25	95.40
		138.35	81.2
n-Propylbenzene	C_9H_{12}	25	91.93
		159.22	76.0
Isopropylbenzene	C_9H_{12}	25	89.77
		152.40	74.6
1-Methyl-2-ethylbenzene	C_9H_{12}	25	94.9
		165.15	77.3
1-Methyl-3-ethylbenzene	C_9H_{12}	25	93.3
		161.30	76.6
1-Methyl-4-ethylbenzene	C_9H_{12}	25	92.7
		162.05	76.4
1,2,3-Trimethylbenzene	C_9H_{12}	25	97.56
		176.15	79.6
1,2,4-Trimethylbenzene (pseudocumene)	C_9H_{12}	25	95.33
		169.25	78.0
1,3,5 Trimethylbenzene (mesitylene)	C_9H_{12}	25	94.40
		164.70	77.6
Alkyl cyclopentanes:			
Cyclopentane	C_5H_{10}	25	97.1
		49.26	93.1
Methylcyclopentane	C_6H_{12}	25	89.83
		71.81	83.2
Ethylcyclopentane	C_7H_{14}	25	88.6
		103.45	78.3
1,1-Dimethylcyclopentane	C_7H_{14}	25	82.5
		87.5	74.6
cis-1,2-Dimethylcyclopentane	C_7H_{14}	25	86.4
		99.3	77.0
trans-1,2-Dimethylcyclopentane	C_7H_{14}	25	83.9
		91.9	75.5
trans-1,3-Dimethylcyclopentane	C_7H_{14}	25	83.6
		90.8	75.3
Alkyl cyclohexanes:			
Cyclohexane	C_6H_{12}	25	93.81
		80.74	85.6
Methylcyclohexane	C_7H_{14}	25	86.07
		100.94	76.9
Ethylcyclohexane	C_8H_{16}	25	86.21
		131.79	73.7
1,1-Dimethylcyclohexane	C_8H_{16}	25	80.9
		119.50	70.7
cis-1,2-Dimethylcyclohexane	C_8H_{16}	25	84.59
		129.73	72.9
trans-1,2-Dimethylcyclohexane	C_8H_{16}	25	81.70
		123.42	71.1
cis-1,3-Dimethylcyclohexane	C_8H_{16}	25	83.49
		124.45	72.1
trans-1,3-Dimethylcyclohexane	C_8H_{16}	25	81.42
		120.09	70.9
cis-1,4-Dimethylcyclohexane	C_8H_{16}	25	83.13
		124.32	71.9
trans-1,4-Dimethylcyclohexane	C_8H_{16}	25	80.67
		119.35	70.4
Monoolefins.			
Ethene (ethylene)	C_2H_4	−103.71	115.39
Propene (propylene)	C_3H_6	−47.70	104.62
1-Butene	C_4H_8	25	86.8
		−6.25	93.36
cis-2-Butene	C_4H_8	25	94.5
		3.72	99.46
trans-2-Butene	C_4H_8	25	91.8
		0.88	96.94
2-Methylpropene (isobutene)	C_4H_8	25	87.7
		−6.90	94.22

Non-hydrocarbon compounds	Formula	Temperature, °C.	ΔH_v, cal./g.
Acetal	$C_6H_{14}O_2$	102.9	66.18
Acetaldehyde	C_2H_4O	21	136.17
Acetic acid	$C_2H_4O_2$	118.3	96.75
		140	94.37
		220	81.23
		321	0
anhydride	$C_4H_6O_3$	137	92.2
Acetone	C_3H_6O	0	134.74
		20	131.87
		40	128.05
		60	123.51
		80	118.26
		100	112.76
		235	0

Table 168. Heats of Vaporization of Organic Compounds—(*Concluded*)

Non-hydrocarbon compounds	Formula	Temperature, °C.	ΔHv, cal./g.
Acetonitrile	C_2H_3N	80	173.68
Acetophenone	C_8H_8O	203.7	77.16
Acetyl chloride	C_3H_3ClO	51	78.84
Air	....	......	51.0
Allyl alcohol	C_3H_6O	96	163.41
Amyl alcohol (n-)	$C_5H_{11}OH$	131	120.17
alcohol (t-)	$C_5H_{11}OH$	102	105.83
amine (n-)	$C_5H_{13}N$	95	98.67
bromide (n-)	$C_5H_{11}Br$	129	48.26
ether (n-)	$C_{10}H_{22}O$	170	69.52
iodide (n-)	$C_5H_{11}I$	155	47.54
methyl ketone (n-)	$C_7H_{14}O$	149.2	82.66
Amylene	C_5H_{10}	12.5	75.01
Anethole (p-)	$C_{10}H_{12}O$	232	71.43
Aniline	C_6H_7N	183	103.68
Benzaldehyde	C_7H_6O	179	86.48
Benzonitrile	C_7H_5N	189	87.68
Benzyl alcohol	C_7H_8O	204.3	112.28
Butyl acetate (n-)	$C_6H_{12}O_2$	124	73.82
alcohol (n-)	$C_4H_{10}O$	116.8	141.26
alcohol (s-)	$C_4H_{10}O$	98.1	134.38
alcohol (t-)	$C_4H_{10}O$	83	130.44
formate	$C_4H_{10}O_2$	105.1	86.74
methyl ketone (n-)	$C_6H_{12}O$	127	82.42
propionate (n-)	$C_7H_{14}O_2$	144.9	71.74
Butyric acid (n-)	$C_4H_8O_2$	163.5	113.96
Butyronitrile (n-)	C_4H_7N	117.4	114.91
Bromobenzene	C_6H_5Br	155.9	57.60
Capronitrile	$C_6H_{11}N$	156	88.15
Carbon disulfide	CS_2	0	89.35
		46.25	84.09
		100	75.49
		140	67.37
tetrachloride	CCl_4	0	52.06
		76.75	46.42
		200	32.73
Carvacrol	$C_{10}H_{14}O$	237	68.09
Chloral	C_2HCl_3O	....	53.99
hydrate	$C_2H_3Cl_3O_2$	96	131.87
Chlorobenzene	C_6H_5Cl	130.6	77.59
Chloroethyl alcohol (2-)	C_2H_5ClO	126.5	122.94
acetate (β-)	$C_4H_7ClO_2$	141.5	80.75
Chloroform	$CHCl_3$	0	64.74
		40	60.92
		61.5	59.01
		100	55.19
		260	0
Chlorotoluene (o-)	C_7H_7Cl	158.1	72.63
(p-)	C_7H_7Cl	160.4	73.13
Cresol (m-)	C_7H_8O	202	100.58
Cyanogen	$(CN)_2$	0	102.97
chloride	$CNCl$	13	134.98
Cyclohexanol	$C_6H_{12}O$	161.1	108.22
Cycohexyl chloride	$C_6H_{11}Cl$	142.0	74.78
Dichloroacetic acid	$C_2H_2Cl_2O_2$	194.4	77.16
Dichlorodifluormethane	CCl_2F_2	−29.8	40.40
Diethylamine	$C_4H_{11}N$	58	91.02
carbonate	$C_5H_{10}O_3$	126	73.10
ketone	$C_5H_{10}O$	101	90.78
oxalate	$C_6H_{10}O_4$	185	67.61
Di-isobutylamine	$C_8H_{19}N$	134	65.70
Dimethyl aniline	$C_8H_{11}N$	193	80.75
carbonate	$C_3H_6O_3$	90	88.15
Dipropyl ketone	$C_7H_{14}O$	143.5	75.73
Dipropylamine (n-)	$C_6H_{15}N$	108	75.73
Ethyl acetate	$C_4H_8O_2$	0.0	102.01
alcohol	C_2H_6O	78.3	204.26
Ethylamine	C_2H_7N	15	145.97
Ethyl benzoate	$C_9H_{10}O_2$	213	64.50
bromide	C_2H_5Br	38.4	59.92
butyrate (n-)	$C_6H_{12}O_2$	118.9	74.68
caprylate	$C_{10}H_{20}O_2$	207	60.44
chloride	C_2H_5Cl	4.7	92.93
		15.0	92.45
		20.0	92.22
		25.0	91.98
Ethylene bromide	$C_2H_4Br_2$	130.8	46.23
chloride	$C_2H_4Cl_2$	0	85.29
		82.3	77.33
glycol	$C_2H_6O_2$	197	191.12
oxide	C_2H_4O	13	138.56
Ethyl ether	$C_4H_{10}O$	34.6	83.85
formate	$C_3H_6O_2$	53.3	97.18
iodide	C_2H_5I	71.2	45.61
Ethylidine chloride	$C_2H_4Cl_2$	0.0	76.69
		60	67.13
Ethyl isobutyl ether	$C_6H_{14}O$	79.0	74.78
isobutyrate	$C_6H_{12}O_2$	109.2	72.05
isovalerate	$C_7H_{14}O_2$	144	67.85
methyl ketone	C_4H_8O	78.2	105.93
methyl ketoxime	C_4H_9NO	182	115.87

Non-hydrocarbon compounds	Formula	Temperature, °C.	ΔHv, cal./g.
Ethyl nonylate	$C_{11}H_{22}O_2$	227	58.05
propionate	$C_5H_{10}O_2$	97.6	80.08
propyl ether	$C_5H_{12}O$	60	82.66
valerate (n-)	$C_7H_{14}O_2$	98	77.16
Formic acid	CH_2O_2	101	119.93
Furane	C_4H_4O	31	95.32
Furfural	$C_5H_4O_2$	160.5	107.51
Heptyl alcohol (n-)	$C_7H_{16}O$	176	104.88
Hexylmethyl ketone	$C_8H_{16}O$	173	74.06
Hydrogen cyanide	HCN	20	210.23
Isoamyl acetate	$C_7H_{14}O_2$	143.6	69.04
alcohol	$C_5H_{12}O$	130.2	119.78
butyrate (n-)	$C_9H_{18}O_2$	169	61.88
formate	$C_6H_{12}O_2$	123	73.58
isobutyrate	$C_9H_{18}O_2$	168	57.57
propionate	$C_8H_{16}O_2$	161	65.22
valerate (n-)	$C_{10}H_{20}O_2$	187	56.14
Isobutyl acetate	$C_6H_{12}O_2$	115.3	73.75
alcohol	$C_4H_{10}O$	106.9	138.08
butyrate (n-)	$C_8H_{16}O_2$	157	64.50
formate	$C_5H_{10}O_2$	97	78.50
isovalerate	$C_9H_{18}O_2$	169	60.44
isobutyrate	$C_8H_{16}O_2$	148	63.31
propionate	$C_7H_{14}O_2$	137	65.94
valerate (n-)	$C_9H_{18}O_2$	169	57.81
Isobutyric acid	$C_4H_8O_2$	154	111.57
Isopropyl alcohol	C_3H_8O	82.3	159.35
methyl ketone	$C_5H_{10}O$	92	89.83
Isovaleric acid	$C_5H_{10}O_2$	176.3	101.05
Limonene	$C_{10}H_{16}$	165	69.52
Mesityl oxide	$C_6H_{10}O$	128	85.77
Methyl acetate	$C_3H_6O_2$	0.0	113.96
		56.3	98.09
Methylal	$C_3H_8O_2$	42	89.83
Methyl alcohol	CH_4O	0	284.29
		64.7	262.79
		100	241.29
		160	193.51
		200	148.12
		220	109.89
		240	0
amyl ketone (n-)	$C_7H_{14}O$	149.2	82.66
aniline	C_7H_9N	194	95.56
butyl ketone (n-)	$C_6H_{12}O$	127	82.42
butyrate (n-)	$C_5H_{10}O_2$	102.6	79.79
chloride	CH_3Cl	−23.8	102.25
		+15.0	96.04
		20.0	95.32
		25.0	94.60
ethyl ketone	C_4H_8O	78.2	105.93
ethyl ketoxime	C_4H_9NO	182	115.87
formate	$C_2H_4O_2$	31.3	112.35
hexyl ketone	$C_8H_{16}O$	173	74.06
iodide	CH_3I	42	45.87
isobutyrate	$C_5H_{10}O_2$	91.1	78.12
isopropyl ketone	$C_5H_{10}O$	92	89.83
isovalerate	$C_6H_{12}O_2$	116	72.39
phenyl ether	C_7H_8O	153	81.46
propionate	$C_4H_8O_2$	79.0	87.56
valerate (n-)	$C_6H_{12}O_2$	116	70.00
Naphthalene	$C_{10}H_8$	218	75.49
Nitrobenzene	$C_6H_5NO_2$	210	79.08
Nitromethane	CH_3NO_2	99.9	134.98
Octyl alcohol (n-)	$C_8H_{18}O$	196	97.47
alcohol (dl-) (sec-)	$C_8H_{18}O$	180	94.37
Phenyl methyl ether	C_7H_8O	153	81.46
Picoline (α-)	C_6H_7N	129	90.78
Piperidine	$C_5H_{11}N$	106	89.35
Propionic acid	$C_3H_6O_2$	139.3	98.81
Propionitrile	C_3H_5N	97	134.26
Propyl acetate (n-)	$C_5H_{10}O_2$	100.4	80.27
alcohol (n-)	C_3H_8O	97.2	164.36
butyrate (n-)	$C_7H_{14}O_2$	143.6	68.33
formate (n-)	$C_4H_8O_2$	80.0	88.13
isobutyrate (n-)	$C_7H_{14}O_2$	134	63.79
isovalerate (n-)	$C_8H_{16}O_2$	156	64.50
propionate (n-)	$C_6H_{12}O_2$	120.6	73.15
Pyridine	C_5H_5N	114.1	107.36
Salicylaldehyde	$C_7H_6O_2$	196	74.78
Tetrachloroethane (1,1,2,2-)	$C_2H_2Cl_4$	145	55.07
Tetrachloroethylene	C_2Cl_4	120.7	50.05
Toluidine (o-)	C_7H_9N	198	95.08
Trichloroethylene	C_2HCl_3	85.7	57.24
Valeronitrile (n-)	C_5H_9N	129	96.28

Table 169. Heats of Fusion of Miscellaneous Materials

Material	M.p., °C.	Heat of fusion, cal./g.
Alloys		
30.5 Pb + 69.5 Sn	183	17
36.9 Pb + 63.1 Sn	179	15.5
63.7 Pb + 36.3 Sn	177.5	11.6
77.8 Pb + 22.2 Sn	176.5	9.54
1 Pb + 9 Sn	236	28
24 Pb + 27.3 Sn + 48.7 Bi	98.8	6.85
25.8 Pb + 14.7 Sn + 52.4 Bi + 7 Cd	75.5	8.4
Silicates		
Anorthite ($CaAl_2Si_2O_8$)	.....	100
Orthoclase ($KAlSi_2O_8$)	.....	100
Microcline ($KAlSi_3O_8$)	.....	83
Wollastonite ($CaSiO_3$)	.....	100
Malacolite ($Ca_3MgSi_4O_{12}$)	.....	94
Diopside ($CaMgSi_2O_6$)	.....	100
Olivine (Mg_2SiO_4)	.....	130
Fayalite (Fe_2SiO_4)	.....	85
Spermaceti	43.9	37.0
Wax (bees')	61.8	42.3

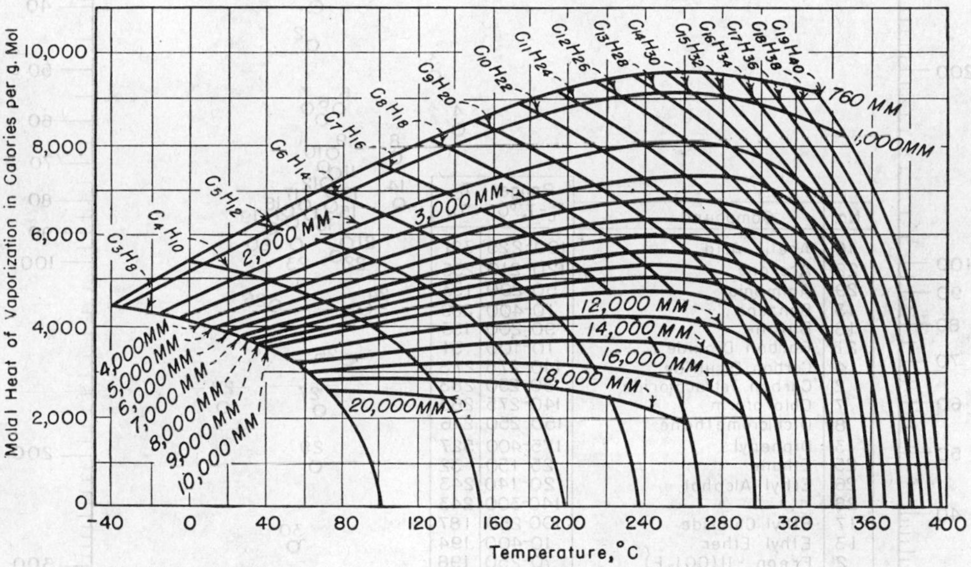

FIG. 7. Molal heats of vaporization of hydrocarbons. [*Schultz, Ind. Eng. Chem.*, **22**, 785 (1930).]

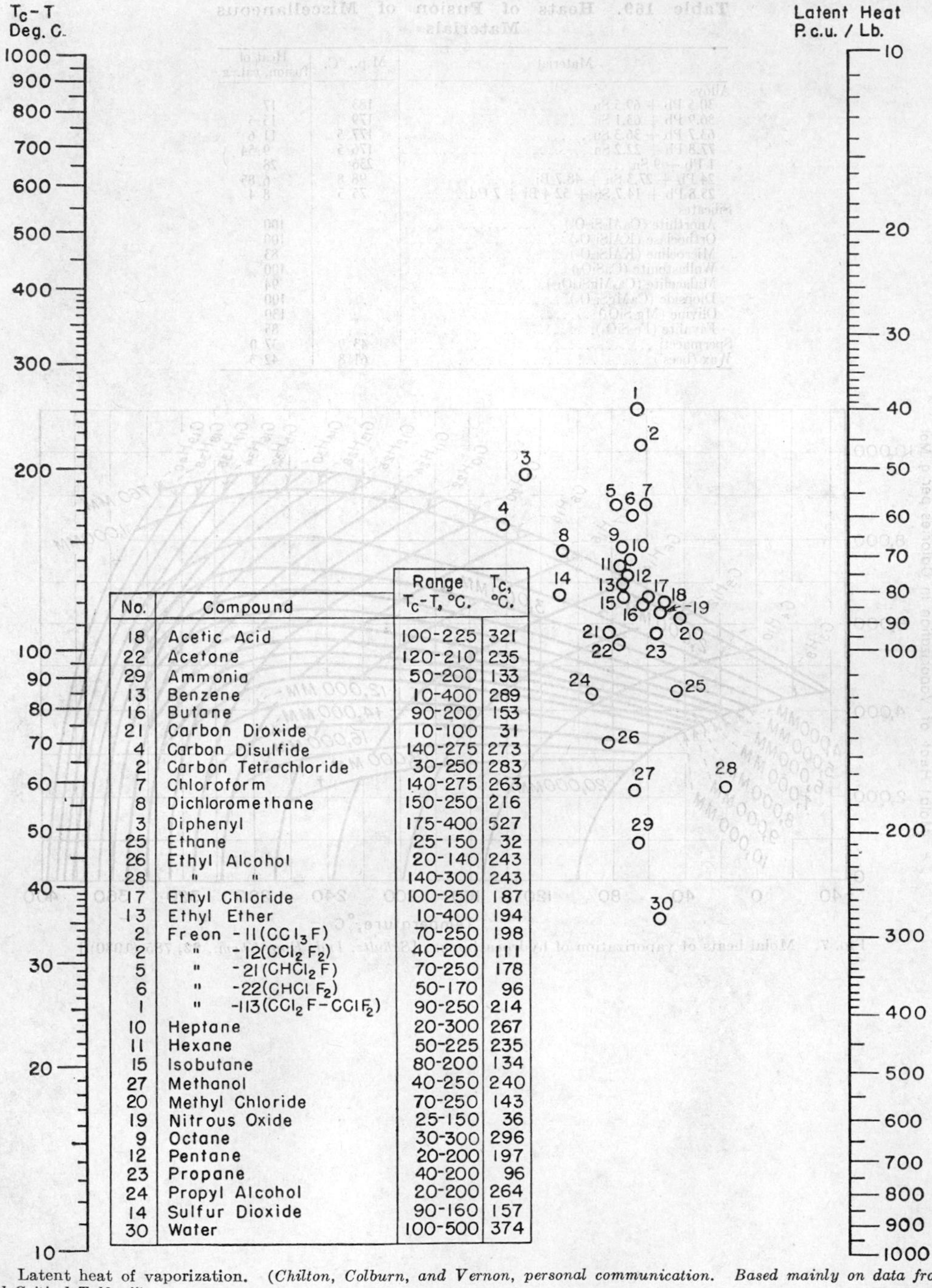

No.	Compound	Range T_c-T, °C.	T_c, °C.
18	Acetic Acid	100-225	321
22	Acetone	120-210	235
29	Ammonia	50-200	133
13	Benzene	10-400	289
16	Butane	90-200	153
21	Carbon Dioxide	10-100	31
4	Carbon Disulfide	140-275	273
2	Carbon Tetrachloride	30-250	283
7	Chloroform	140-275	263
8	Dichloromethane	150-250	216
3	Diphenyl	175-400	527
25	Ethane	25-150	32
26	Ethyl Alcohol	20-140	243
28	" "	140-300	243
17	Ethyl Chloride	100-250	187
13	Ethyl Ether	10-400	194
2	Freon -11 (CCl_3F)	70-250	198
2	" -12 (CCl_2F_2)	40-200	111
5	" -21 ($CHCl_2F$)	70-250	178
6	" -22 ($CHClF_2$)	50-170	96
1	" -113 ($CCl_2F-CClF_2$)	90-250	214
10	Heptane	20-300	267
11	Hexane	50-225	235
15	Isobutane	80-200	134
27	Methanol	40-250	240
20	Methyl Chloride	70-250	143
19	Nitrous Oxide	25-150	36
9	Octane	30-300	296
12	Pentane	20-200	197
23	Propane	40-200	96
24	Propyl Alcohol	20-200	264
14	Sulfur Dioxide	90-160	157
30	Water	100-500	374

FIG. 8. Latent heat of vaporization. (*Chilton, Colburn, and Vernon, personal communication. Based mainly on data from "International Critical Tables."*)

SPECIFIC HEATS OF PURE COMPOUNDS

Table 170. Heat Capacities of the Elements and Inorganic Compounds

Taken from the compilation by K. K. Kelley on high-temperature specific-heat equations for inorganic substances, *U.S. Bur. Mines Bull.* 371 (1934).

Substance	State*	Heat capacity at constant pressure (T = °K.; 0°C. = 273.1°K.), cal./deg. mol	Range of temperature, °K.	Uncertainty, %
Aluminum:				
Al	*c*	$4.80 + 0.00322T$	273–931	1
	l	7.00	931–1273	5
$AlBr_3$	*c*	$18.74 + 0.01866T$	273–370	3
	l	29.5	370–407	5
$AlCl_3$	*c*	$13.25 + 0.02800T$	273–465	3
	l	31.2	465–504	3
$AlCl_3.6H_2O$	*c*	76	288–327	?
AlF_3	*c*	19.3	288–326	?
$AlF_3.3\frac{1}{2}H_2O$	*c*	50.5	288–326	?
$AlF_3.3NaF$	*c*	$38.63 + 0.04760T - 449200/T^2$	273–1273	2
	l	142	1273–1373	?
AlI_3	*c*	$16.88 + 0.02266T$	273–464	3
	l	28.8	464–480	5
Al_2O_3	*c*	$22.08 + 0.008971T - 522500/T^2$	273–1973	3
$Al_2O_3.SiO_2$	*c*, sillimanite	$40.79 + 0.004763T - 992800/T^2$	273–1573	3
	c, disthene	$41.81 + 0.005283T - 1211000/T^2$	273–1673	2
	c, andalusite	$43.96 + 0.001923T - 1086000/T^2$	273–1573	3
$3Al_2O_3.2SiO_2$	*c*, mullite	$59.65 + 0.0670T$	273–576	5
$4Al_2O_3.3SiO_2$	*c*	$113.2 + 0.0652T$	273–575	3
$Al_2(SO_4)_3$	*c*	63.5	273–373	?
$Al_2(SO_4)_3.18H_2O$	*c*	235	288–325	?
Antimony:				
Sb	*c*	$5.51 + 0.00178T$	273–903	2
	l	7.15	903–1273	5
$SbBr_3$	*c*	$17.2 + 0.0293T$	273–370	?
$SbCl_3$	*c*	$10.3 + 0.0511T$	273–346	?
Sb_2O_3	*c*	$19.1 + 0.0171T$	273–929	?
Sb_2O_4	*c*	$22.6 + 0.0162T$	273–1198	?
Sb_2S_3	*c*	$24.2 + 0.0132T$	273–821	?
Argon:				
A	*g*	4.97	All	0
Arsenic:				
As	*c*	$5.17 + 0.00234T$	273–1168	5
$AsCl_3$	*l*	31.9	286–371	?
As_2O_3	*c*	$8.37 + 0.0486T$	273–548	?
As_2S_3	*c*	25.8	293–373	?
Barium:				
$BaCl_2$	*c*	$17.0 + 0.00334T$	273–1198	?
$BaCl_2.H_2O$	*c*	28.2	273–307	?
$BaCl_2.2H_2O$	*c*	37.3	273–307	?
$Ba(ClO_3)_2.H_2O$	*c*	51	289–320	?
$BaCO_3$	*c*, α	$17.26 + 0.0131T$	273–1083	5
	c, β	30.0	1083–1255	15
$BaMoO_4$	*c*	34	273–297	?
$Ba(NO_3)_2$	*c*	39.8	285–371	?
$BaSO_4$	*c*	$21.35 + 0.0141T$	273–1323	5
Beryllium:				
Be	*c*	$4.698 + 0.001555T - 121000/T^2$	273–1173	1
BeO	*c*	$8.69 + 0.00365T - 313000/T^2$	273–1175	5
$BeO.Al_2O_3$	*c*	25.4	273–373	?
$BeSO_4$	*c*	20.8	273–373	?
Bismuth:				
Bi	*c*	$5.38 + 0.00260T$	273–544	3
	l	7.60	544–1273	3
Bi_2O_3	*c*	$23.27 + 0.01105T$	273–777	2
Bi_2S_3	*c*	30.4	284–372	?
Boron:				
B	*c*	$1.54 + 0.00440T$	273–1174	5
B_2O_3	*gls*	$5.14 + 0.0320T$	273–513	3
	gls	30.4	513–623	3
BN	*c*	$1.61 + 0.00400T$	273–1173	5
Bromine:				
Br_2	*g*	9.00	300–2000	5
Cadmium:				
Cd	*c*	$5.46 + 0.002466T$	273–594	1
	l	7.13	594–973	5
CdO	*c*	$9.65 + 0.00208T$	273–2086	?
CdS	*c*	$12.9 + 0.00090T$	273–1273	?
$CdSO_4.\frac{8}{3}H_2O$	*c*	51.3	293	?
Calcium:				
Ca	*c*	$5.31 + 0.00333T$	273–673	2
	c	$6.29 + 0.00140T$	673–873	2
$CaCl_2$	*c*	$16.9 + 0.00386T$	273–1055	?
$CaCO_3$	*c*	$19.68 + 0.01189T - 307600/T^2$	273–1033	3
CaF_2	*c*	$14.7 + 0.00380T$	273–1651	?
$CaMg(CO_3)_2$	*c*	40.1	299–372	?
$CaMoO_4$	*c*	33	273–297	?
CaO	*c*	$10.00 + 0.00484T - 108000/T^2$	273–1173	2
$Ca(OH)_2$	*c*	21.4	276–373	?
$CaO.Al_2O_3.2SiO_2$	*c*, anorthite	$63.13 + 0.01500T - 1537000/T^2$	273–1673	1
	gls	$67.41 + 0.01048T - 1874000/T^2$	273–973	1
$CaO.MgO.2SiO_2$	*c*, diopside	$54.46 + 0.005746T - 1500000/T^2$	273–1573	1
	gls	$51.68 + 0.009724T - 1308000/T^2$	273–973	1
$CaO.SiO_2$	*c*, wollastonite	$27.95 + 0.002056T - 745600/T^2$	273–1573	1
	c, pseudowollastonite	$25.48 + 0.004132T - 488100/T^2$	273–1673	1
	gls	$23.16 + 0.009672T - 487100/T^2$	273–973	1

* The symbols in this column have the following meaning; *c*, crystal; *l*, liquid; *g*, gas; *gls*, glass.

Table 170. Heat Capacities of the Elements and Inorganic Compounds—(*Continued*)

Substance	State	Heat capacity at constant pressure (T = °K.; 0°C. = 273.1°K.), cal./deg. mol	Range of tempera- ture, °K.	Uncer- tainty, %
Calcium—(*Cont.*):				
CaP_2O_6	*c*	39.5	287–371	?
$CaSO_4$	*c*	$18.52 + 0.02197T - 156800/T^2$	273–1373	5
$CaSO_4.2H_2O$	*c*	46.8	282–373	?
$CaWO_4$	*c*	27.9	292–322	?
Carbon:				
C	*c*, graphite	$2.673 + 0.002617T - 116900/T^2$	273–1373	2
	c, diamond	$2.162 + 0.003059T - 130300/T^2$	273–1313	3
CH_4	*g*	$5.34 + 0.0115T$	273–1200	2
CO	*g*	$6.60 + 0.00120T$	273–2500	1½
CO_2	*g*	$10.34 + 0.00274T - 195500/T^2$	273–1200	1½
CS_2	*l*	18.4	293	?
Cerium:				
Ce	*c*	$5.88 + 0.00123T$	273–908	?
CeO_2	*c*	15.1	273–373	?
$Ce_2(MoO_4)_3$	*c*	96	273–297	?
$Ce_2(SO_4)_3$	*c*	66.4	273–373	?
$Ce_2(SO_4)_3.5H_2O$	*c*	131.6	273–319	?
Cesium:				
Cs	*c*	$1.96 + 0.0182T$	273–301	3
	l	8.00	302	3
	g	4.97	All	0
CsBr	*c*	$12.6 + 0.00259T$	273–909	?
CsCl	*c*	$11.7 + 0.00309T$	273–752	?
CsF	*c*	$11.3 + 0.00285T$	273–957	?
CsI	*c*	$11.6 + 0.00268T$	273–894	?
Chlorine:				
Cl_2	*g*	$8.28 + 0.00056T$	273–2000	1½
Chromium:				
Cr	*c*	$4.84 + 0.00295T$	273–1823	5
	l	9.70	1823–1923	10
$CrCl_3$	*c*	23	286–319	?
Cr_2O_3	*c*	$26.0 + 0.00400T$	273–2263	?
CrSb	*c*	$12.3 + 0.00120T$	273–1383	?
$CrSb_2$	*c*	$19.2 + 0.00184T$	273–949	?
$Cr_2(SO_4)_3$	*c*	67.4	273–373	?
Cobalt:				
Co	*c*	$5.12 + 0.00333T$	273–1763	5
	l	8.40	1763–1873	5
$CoAs_2.CoS_2$	*c*	32.9	283–373	?
CoSb	*c*	$11.7 + 0.00156T$	273–1464	?
Co_2Sn	*c*	$15.83 + 0.00950T$	273–903	2
CoS	*c*	$10.6 + 0.00251T$	273–1373	?
$CoSO_4.7H_2O$	*c*	96	286–303	?
Copper:				
Cu	*c*	$5.44 + 0.001462T$	273–1357	1
	l	7.50	1357–1573	3
CuAl	*c*	$9.88 + 0.00500T$	273–733	2
$CuAl_2$	*c*	$16.78 + 0.00366T$	273–773	2
Cu_3Al	*c*	$19.61 + 0.01054T$	273–775	2
CuI	*c*	$12.1 + 0.00286T$	273–675	?
CuI_2	*c*	20.1	274–328	?
CuO	*c*	$10.87 + 0.003576T - 150600/T^2$	273–810	2
$CuO.SiO_2.H_2O$	*c*	29	293–323	?
CuS	*c*	$10.6 + 0.00264T$	273–1273	?
Cu_2S	*c*, α	$9.38 + 0.0312T$	273–376	3
	c, β	20.9	376–1173	2
CuS.FeS	*c*	24	292–321	?
Cu_2Sb	*c*	$13.73 + 0.01350T$	273–573	2
Cu_3Sb	*c*	$21.79 + 0.00900T$	273–693	2
Cu_2Se	*c*, α	20.85	273–383	5
	c, β	20.35	383–488	5
Cu_3Si	*c*	$21.3 + 0.00587T$	273–1135	?
$CuSO_4$	*c*	24.1	282	?
$CuSO_4.H_2O$	*c*	31.3	282	?
$CuSO_4.3H_2O$	*c*	49.0	282	?
$CuSO_4.5H_2O$	*c*	67.2	282	?
Fluorine:				
F_2	*g*	$6.50 + 0.00100T$	300–3000	5
Gallium:				
Ga_2O_3	*c*	$18.2 + 0.0252T$	273–923	?
$Ga_2(SO_4)_3$	*c*	62.4	273–373	?
Gold:				
Au	*c*	$5.61 + 0.00144T$	273–1336	2
	l	7.00	1336–1573	5
$AuSb_2$	*c*, α	$17.12 + 0.00465T$	273–628	1
	c, βγ	$11.47 + 0.01756T$	628–713	?
Helium:				
He	*g*	4.97	All	0
Hydrogen:				
H	*g*	4.97	All	0
H_2	*g*	$6.62 + 0.00081T$	273–2500	2
HBr	*g*	$6.80 + 0.00084T$	273–2000	2
HCl	*g*	$6.70 + 0.00084T$	273–2000	1½
HI	*g*	$6.93 + 0.00083T$	273–2000	2
H_2O	*l*	See Table 171		
	g	$8.22 + 0.00015T + 0.00000134T^2$	300–2500	?
H_2S	*g*	$7.20 + 0.00360T$	300–600	8
$H_2S_2O_7$	*c*	27	281	?
	l	58	308	?
Iodine:				
I_2	*g*	9.00	300–2000	5

Table 170. Heat Capacities of the Elements and Inorganic Compounds—(*Continued*)

Substance	State	Heat capacity at constant pressure (T = °K.; 0°C. = 273.1°K.), cal./deg. mol	Range of temperature, °K.	Uncertainty, %
Iridium:				
Ir	*c*	$5.50 + 0.00148T$	273–1873	1
Iron:				
Fe	*c*, α	$4.13 + 0.00638T$	273–1041	3
	c, β	$6.12 + 0.00336T$	1041–1179	3
	c, γ	8.40	1179–1674	5
	c, δ	10.0	1674–1803	5
	l	8.15	1803–1873	5
$FeAs_2$	*c*	17.8	283– 373	?
Fe_3C	*c*	$25.17 + 0.00223T$	273–1173	10
$FeCO_3$	*c*	22.7	293– 368	?
FeO	*c*	$12.62 + 0.001492T - 76200/T^2$	273–1173	2
Fe_2O_3	*c*	$24.72 + 0.01604T - 423400/T^2$	273–1097	2
Fe_3O_4	*c*	$41.17 + 0.01882T - 979500/T^2$	273–1065	2
$Fe_2O_3.3H_2O$	*c*	47.8	286– 373	?
FeS	*c*, α	$2.03 + 0.0390T$	273– 411	5
	c, β	$12.05 + 0.00273T$	411–1468	3
FeS_2	*c*	$10.7 + 0.01336T$	273– 773	?
FeSi	*c*	$10.54 + 0.00458T$	273– 903	2
Fe_2SiO_4	*c*	$33.57 + 0.01907T - 879700/T^2$	273–1161	2
$FeSO_4$	*c*	22	293– 373	?
$Fe_2(SO_4)_3$	*c*	66.2	273– 373	?
$FeSO_4.4H_2O$	*c*	63.6	282	?
$FeSO_4.7H_2O$	*c*	96	291– 319	?
Krypton:				
Kr	*g*	4.97	All	0
Lanthanum:				
La	*c*	$5.91 + 0.00100T$	273–1099	?
La_2O_3	*c*	$22.6 + 0.00544T$	273–2273	?
$La_2(MoO_4)_3$	*c*	86	273– 307	?
$La_2(SO_4)_3$	*c*	66.9	273– 373	?
$La_2(SO_4)_3.9H_2O$	*c*	152	273– 319	?
Lead:				
Pb	*c*	$5.77 + 0.00202T$	273– 600	2
	l	6.8	600–1273	5
$Pb_3(AsO_4)_2$	*c*	65.5	286– 370	?
PbB_2O_4	*c*	26.5	288– 371	?
PbB_4O_7	*c*	41.4	289– 371	?
$PbBr_2$	*c*	$18.13 + 0.00310T$	273– 761	2
	l	27.4	761– 860	10
$PbCl_2$	*c*	$15.88 + 0.00835T$	273– 771	2
	l	27.2	771– 851	10
$2PbCl_2.NH_4Cl$	*c*	53.1	293	?
$PbCO_3$	*c*	21.1	286– 320	?
$PbCrO_4$	*c*	29.1	292– 323	?
PbF_2	*c*	$16.5 + 0.00412T$	273–1091	?
PbI_2	*c*	$18.66 + 0.00293T$	273– 648	2
	l	32.3	648– 776	20
$PbMoO_4$	*c*	30.4	292– 322	?
$Pb(NO_3)_2$	*c*	36.4	286– 320	?
PbO	*c*	$10.33 + 0.00318T$	273 544	2
PbO_2	*c*	$12.7 + 0.00780T$	273– ?	?
$Pb_2P_2O_7$	*c*	48.3	284– 371	?
PbS	*c*	$10.63 + 0.00401T$	273– 873	3
$PbSO_4$	*c*	26.4	293– 372	?
PbS_2O_3	*c*	29	293– 373	?
$PbWO_4$	*c*	35	273– 297	?
Lithium:				
Li	*c*	$0.68 + 0.0180T$	273– 459	10
	g	4.97	All	0
LiBr	*c*	$11.5 + 0.00302T$	273– 825	?
$LiBr.H_2O$	*c*	22.6	278– 318	?
LiCl	*c*	$11.0 + 0.00339T$	273– 887	?
$LiCl.H_2O$	*c*	23.6	279– 360	?
LiF	*c*	$8.20 + 0.00520T$	273–1117	?
LiI	*c*	$12.5 + 0.00208T$	273– 723	?
$LiI.H_2O$	*c*	23.6	277– 359	?
$LiI.2H_2O$	*c*	32.9	277– 345	?
$LiI.3H_2O$	*c*	43.2	277– 347	?
$LiNO_3$	*c*	$9.17 + 0.0360T$	273– 523	5
	l	26.8	523– 575	5
Magnesium:				
Mg	*c*	$6.20 + 0.00133T - 67800/T^2$	273– 923	1
	l	7.4	923–1048	10
MgAg	*c*	$10.58 + 0.00412T$	273– 905	2
Mg_4Al_3	*c*	$34.4 + 0.0198T$	273– 736	?
MgAu	*c*	$11.3 + 0.00189T$	273–1433	?
Mg_2Au	*c*	$16.2 + 0.00451T$	273–1073	?
Mg_3Au	*c*	$21.2 + 0.00614T$	273–1103	?
$MgCl_2$	*c*	$17.3 + 0.00377T$	273– 991	?
$MgCl_2.6H_2O$	*c*	77.1	292– 342	?
$MgCO_3$	*c*	16.9	290	?
$MgCu_2$	*c*	$14.96 + 0.00776T$	273– 903	3
Mg_2Cu	*c*	$15.5 + 0.00652T$	273– 843	?
$MgNi_2$	*c*	$15.87 + 0.00692T$	273– 903	2
MgO	*c*	$10.86 + 0.001197T - 208700/T^2$	273–2073	2
$MgO.Al_2O_3$	*c*	28	288– 319	?
$MgO.SiO_2$	*c*, amphibole	$25.60 + 0.004380T - 674200/T^2$	273–1373	1
	c, pyroxene	$23.35 + 0.008062T - 558800/T^2$	273– 773	1
	gls	$23.30 + 0.007734T - 542000/T^2$	273– 973	1
$6MgO.MgCl_2.8B_2O_3$	*c*, α	$58.7 + 0.408T$	273– 538	5
	c, β	$107.2 + 0.2876T$	538– 623	5
$Mg(OH)_2$	*c*	18.2	292– 323	?

Table 170. Heat Capacities of the Elements and Inorganic Compounds—(*Continued*)

Substance	State	Heat capacity at constant pressure (T = °K.; 0°C. = 273.1°K.), cal./deg. mol	Range of temperature, °K.	Uncertainty, %
Magnesium—(*Cont.*):				
Mg_3Sb_2	*c*	$28.2 + 0.00560T$	273–1234	?
Mg_2Si	*c*	$15.4 + 0.00415T$	273–1343	?
$MgSO_4$	*c*	26.7	296–372	?
$MgSO_4.H_2O$	*c*	33	282	?
$MgSO_4.6H_2O$	*c*	80	282	?
$MgSO_4.7H_2O$	*c*	89	291–319	?
Manganese:				
Mn	*c*, α	$3.76 + 0.00747T$	273–1108	5
	c, β	$5.06 + 0.00395T$	1108–1317	5
	c, γ	$4.80 + 0.00422T$	1317–1493	5
	l	11.0	1493–1673	10
$MnCl_2$	*c*	$16.2 + 0.00520T$	273–923	?
$MnCO_3$	*c*	$7.79 + 0.0421T + 0.0000090T^2$	273–773	?
MnO	*c*	$7.43 + 0.01038T - 0.00000362T^2$	273–1923	?
Mn_2O_3	*c*	$10.33 + 0.0530T - 0.0000257T^2$	273–1173	?
Mn_3O_4	*c*	$19.25 + 0.0538T - 0.0000209T^2$	273–1773	?
MnO_2	*c*	$1.92 + 0.0471T - 0.0000297T^2$	273–773	?
$Mn_2O_3.H_2O$	*c*	31	291–322	?
MnS	*c*	$10.21 + 0.00656T - 0.00000242T^2$	273–1883	?
$MnSO_4$	*c*	27.5	293–373	?
$MnSO_4.5H_2O$	*c*	78	290–319	?
Mercury:				
Hg	*l*	6.61	273–630	1
	g	4.97	All	0
Hg_2	*g*	9.00	300–2000	5
HgCl	*c*	$11.05 + 0.00370T$	273–798	?
$HgCl_2$	*c*	$15.3 + 0.0103T$	273–553	?
$Hg(CN)_2$	*c*	25	285–319	?
HgI	*c*	$11.4 + 0.00461T$	273–563	?
HgI_2	*c*, α	$17.4 + 0.004001T$	273–403	3
	c, β	20.2	403–523	3
HgO	*c*	11.5	278–371	?
HgS	*c*	$10.9 + 0.00365T$	273–853	?
Hg_2SO_4	*c*	31.0	273–307	?
Molybdenum:				
Mo	*c*	$5.69 + 0.00188T - 50300/T^2$	273–1773	5
MoO_3	*c*	$15.1 + 0.0121T$	273–1068	?
MoS_2	*c*	$19.7 + 0.00315T$	273–729	?
Neon:				
Ne	*g*	4.97	All	0
Nickel:				
Ni	*c*, α	$4.26 + 0.00640T$	273–626	2
	c, β	$6.99 + 0.000905T$	626–1725	5
	l	8.55	1725–1903	10
NiO	*c*	$11.3 + 0.00215T$	273–1273	?
NiS	*c*	$9.25 + 0.00640T$	273–597	3
Ni_2Si	*c*	$15.8 + 0.00329T$	273–1582	?
NiSi	*c*	$10.0 + 0.00312T$	273–1273	?
Ni_3Sn	*c*	$20.78 + 0.0102T$	273–904	2
$NiSO_4$	*c*	33.4	293–373	?
$NiSO_4.6H_2O$	*c*	82	291–325	?
NiTe	*c*	$11.00 + 0.00433T$	273–700	2
Nitrogen:				
N_2	*g*	$6.50 + 0.00100T$	300–3000	3
NH_3	*g*	$6.70 + 0.00630T$	300–800	1½
NH_4Br	*c*	22.8	274–328	?
NH_4Cl	*c*, α	$9.80 + 0.0368T$	273–457	5
	c, β	$5.0 + 0.0340T$	457–523	5
NH_4I	*c*	17.8	273–328	?
NH_4NO_3	*c*	31.8	273–293	?
$(NH_4)_2SO_4$	*c*	51.6	275–328	?
NO	*g*	$8.05 + 0.000233T - 156300/T^2$	300–5000	2
Osmium:				
Os	*c*	$5.686 + 0.000875T$	273–1877	1
Oxygen:				
O_2	*g*	$8.27 + 0.000258T - 187700/T^2$	300–5000	1
Palladium:				
Pd	*c*	$5.41 + 0.00184T$	273–1822	2
Phosphorus:				
P	*c*, yellow	5.50	273–317	5
	c, red	$0.21 + 0.0180T$	273–472	10
	l	6.6	317–373	10
PCl_3	*l*	28.7	284–371	?
P_4O_{10}	*c*	$15.72 + 0.1092T$	273–631	2
	g	73.6	631–1371	3
Platinum:				
Pt	*c*	$5.92 + 0.00116T$	273–1873	1
Potassium:				
K	*c*	$5.24 + 0.00555T$	273–336	5
	l	7.7	336–373	5
	g	4.97	All	0
K_2	*g*	9.00	300–2000	5
$KAsO_3$	*c*	25.3	290–372	?
KBO_2	*c*	$12.6 + 0.0126T$	273–1220	?
$K_2B_4O_7$	*c*	51.3	290–372	?
KBr	*c*	$11.49 + 0.00360T$	273–543	2
KCl	*c*	$10.93 + 0.00376T$	273–1043	2
$KClO_3$	*c*	25.7	289–371	?
$KClO_4$	*c*	26.3	287–318	?
$2KCl.CuCl_2.2H_2O$	*c*	63	292–323	?
$2KCl.PtCl_4$	*c*	55	286–319	?

Table 170. Heat Capacities of the Elements and Inorganic Compounds—(*Continued*)

Substance	State	Heat capacity at constant pressure (T = °K.; 0°C. = 273.1°K.), cal./deg. mol	Range of tempera- ture, °K.	Uncer- tainty, %
Potassium—(*Cont.*):				
$2KCl.SnCl_4$	c	54.5	292– 323	?
$2KCl.ZnCl_2$	c	43.4	279– 319	?
$2KCN.Zn(CN)_2$	c	57.4	277– 319	?
K_2CO_3	c	29.9	296– 372	?
K_2CrO_4	c	35.9	289– 371	?
$K_2Cr_2O_7$	c	$42.80 + 0.0410T$	273– 671	5
	l	96.9	671– 757	5
KF	c	$10.8 + 0.00284T$	273–1129	?
$K_4Fe(CN)_6$	c	80.1	273– 319	?
$K_4Fe(CN)_6.3H_2O$	c	114.5	273– 310	?
KH_2AsO_4	c	32	289– 319	?
KH_2PO_4	c	28.3	290– 320	?
$KHSO_4$	c	30	292– 324	?
$KMnO_4$	c	28	287– 318	?
KNO_3	c	$6.42 + 0.0530T$	273– 401	10
	c	28.8	401– 611	5
	l	29.5	611– 683	10
$K_2O.Al_2O_3.3SiO_2$	c, orthoclase	$69.26 + 0.00821T - 2331000/T^2$	273–1373	1½
	gls, orthoclase	$69.81 + 0.01053 - 2403000/T^2$	273–1373	1½
	c, microcline	$65.65 + 0.01102T - 1748000/T^2$	273–1373	1½
	gls, microcline	$64.83 + 0.01438T - 1641000/T^2$	273–1373	1½
$K_4P_2O_7$	c	63.1	290– 371	?
K_2SO_4	c	33.1	287– 371	?
$K_2S_2O_3$	c	37	293– 373	?
$K_2SO_4.Al_2(SO_4)_3.24H_2O$	c	352	292– 322	?
$K_2SO_4.Cr_2(SO_4)_3.24H_2O$	c	324	292– 324	?
$K_2SO_4.MgSO_4.6H_2O$	c	106	292– 323	?
$K_2SO_4.NiSO_4.6H_2O$	c	107	289– 319	?
$K_2SO_4.ZnSO_4.6H_2O$	c	120	293– 317	?
Radon:				
Rn	g	4.97	All	0
Rhenium:				
Re	c	$6.30 + 0.00053T$	273–2273	?
Rhodium:				
Rh	c	$5.40 + 0.00219T$	273–1877	2
Rubidium:				
Rb	c	$3.27 + 0.0131T$	273– 312	2
	l	7.85	312– 373	5
$RbBr$	c	$11.6 + 0.00255T$	273– 954	?
$RbCl$	c	$11.5 + 0.00249T$	273– 987	?
Rb_2CO_3	c	28.4	291– 320	?
RbF	c	$11.3 + 0.00256T$	273–1048	?
RbI	c	$11.6 + 0.00263T$	273– 913	?
Scandium:				
Sc_2O_3	c	21.1	273– 373	?
$Sc_2(SO_4)_3$	c	62.0	273– 373	?
Selenium:				
Se	c	$4.53 + 0.00550T$	273– 490	2
	l	8.35	490– 570	3
Silicon:				
Si	c	$5.74 + 0.000617T - 101000/T^2$	273–1174	2
SiC	c	$8.89 + 0.00291T - 284000/T^2$	273–1629	2
$SiCl_4$	l	32.4	293– 373	?
SiO_2	c, quartz, α	$10.87 + 0.008712T - 241200/T^2$	273– 848	1
	c, quartz, β	$10.95 + 0.00550T$	848–1873	3½
	c, cristobalite, α	$3.65 + 0.0240T$	273– 523	2½
	c, cristobalite, β	$17.09 + 0.000454T - 897200/T^2$	523–1973	2
	gls	$12.80 + 0.00447T - 302000/T^2$	273–1973	3½
Silver:				
Ag	c	$5.60 + 0.00150T$	273–1234	1
	l	8.2	1234–1573	3
Ag_3Al	c	$22.56 + 0.00570T$	273– 902	2
Ag_2Al	c	$16.85 + 0.00450T$	273– 903	2
$AgAl_{12}$	c	$58.62 + 0.0575T$	273– 768	5
$AgBr$	c	$8.58 + 0.0141T$	273– 703	6
	l	14.9	703– 836	5
$AgCl$	c	$9.60 + 0.00929T$	273– 728	2
	l	14.05	728– 806	5
$AgCNO$	c	18.7	273– 353	?
AgI	c, α	$8.58 + 0.0141T$	273– 423	6
$AgNO_3$	c, α	$18.83 + 0.0160T$	273– 433	2
	c, β	25.7	433– 482	5
	l	30.2	482– 541	5
Ag_3PO_4	c	37.5	293– 325	?
Ag_2S	c, α	18.8	273– 448	5
	c, β	21.8	448– 597	5
Ag_3Sb	c	$19.53 + 0.0160T$	273– 694	5
Ag_2Se	c, α	20.2	273– 406	5
	c, β	20.4	406– 460	5
Sodium:				
Na	c	$5.01 + 0.00536T$	273– 371	1½
	l	7.50	371– 451	2
	g	4.97	All	0
$NaBO_2$	c	$10.4 + 0.0199T$	273–1239	?
$Na_2B_4O_7$	c	47.9	289– 371	?
$Na_2B_4O_7.10H_2O$	c	147	292– 323	?
$NaBr$	c	$11.74 + 0.00233T$	273– 543	2
$NaCl$	c	$10.79 + 0.00420T$	273–1073	2
	l	15.9	1073–1205	3

Table 170. Heat Capacities of the Elements and Inorganic Compounds—(*Concluded*)

Substance	State	Heat capacity at constant pressure (T = °K.; 0°C. = 273.1°K.), cal./deg. mol	Range of tempera- ture, °K.	Uncer- tainty, %
Sodium—(*Cont.*):				
$NaClO_3$	*c*	$9.48 + 0.0468T$	273–528	3
	l	31.8	528–572	5
NaCNO	*c*	13.1	273–353	?
Na_2CO_3	*c*	28.9	288–371	?
NaF	*c*	$10.4 + 0.00289T$	273–1261	?
$Na_2HPO_4.7H_2O$	*c*	86.6	275–307	?
$Na_2HPO_4.12H_2O$	*c*	133.4	275–307	?
NaI	*c*	$12.5 + 0.00162T$	273–936	?
$NaNO_3$	*c*	$4.56 + 0.0580T$	273–583	5
	l	37.2	583–703	10
$Na_2O.Al_2O_3.3SiO_2$	*c*, albite	$63.78 + 0.01171T - 1678000/T^2$	273–1373	?
	gls	$61.25 + 0.01768T - 1545000/T^2$	273–1173	?
$NaPO_3$	*c*	22.1	290–319	?
$Na_4P_2O_7$	*c*	60.7	290–371	?
Na_2SO_4	*c*	32.8	289–371	?
$Na_2S_2O_3$	*c*	34.9	273–307	?
$Na_2S_2O_3.5H_2O$	*c*	86.2	273–307	?
Strontium:				
$SrBr_2$	*c*	$18.1 + 0.00311T$	273–923	?
$SrBr_2.H_2O$	*c*	28.9	277–370	?
$SrBr_2.6H_2O$	*c*	82.1	276–327	?
$SrCl_2$	*c*	$18.2 + 0.00244T$	273–1143	?
$SrCl_2.H_2O$	*c*	28.7	276–365	?
$SrCl_2.2H_2O$	*c*	38.3	277–366	?
$SrCO_3$	*c*	21.8	281–371	?
SrI_2	*c*	$18.6 + 0.00304T$	273–783	?
$SrI_2.H_2O$	*c*	28.5	276–363	?
$SrI_2.2H_2O$	*c*	39.1	275–336	?
$SrI_2.6H_2O$	*c*	84.9	275–333	?
$SrMoO_4$	*c*	37	273–297	?
$Sr(NO_3)_2$	*c*	38.3	290–320	?
$SrSO_4$	*c*	26.2	293–369	?
Sulfur:				
S	*c*, rhombic	$3.63 + 0.00640T$	273–368	3
	c, monoclinic	$4.38 + 0.00440T$	368–392	3
S_2	*g*	$8.58 + 0.00030T$	300–2500	?
S_2Cl_2	*l*	27.5	273–332	?
SO_2	*g*	$7.70 + 0.00530T - 0.00000083T^2$	300–2500	2½
Tantalum:				
Ta	*c*	$5.91 + 0.00099T$	273–1173	2
Tellurium:				
Te	*c*	$5.19 + 0.00250T$	273–600	3
Thallium:				
Tl	*c*, α	$5.32 + 0.00385T$	273–500	?
	c, β	8.12	500–576	1
	l	7.12	576–773	3
TlBr	*c*	$12.53 + 0.00100T$	273–733	10
	l	16.0	733–800	10
TlCl	*c*	$12.56 + 0.00088T$	273–700	5
	l	14.2	700–803	10
Thorium:				
Th	*c*	6.40	273–373	?
ThO_2	*c*	$14.6 + 0.00507T$	273–1273	?
$Th(SO_4)_2$	*c*	41.2	273–373	?
Tin:				
Sn	*c*	$5.05 + 0.00480T$	273–504	2
	l	6.6	504–1273	10
SnAu	*c*	$11.79 + 0.00233T$	273–581	1
$SnCl_2$	*c*	$16.2 + 0.00926T$	273–520	?
$SnCl_4$	*l*	38.4	286–371	?
SnO	*c*	$9.40 + 0.00362T$	273–1273	?
SnO_2	*c*	$13.94 + 0.00565T - 252000/T^2$	273–1373	?
SnPt	*c*	$11.49 + 0.00190T$	273–1318	1
SnS	*c*	$12.1 + 0.00165T$	273–1153	?
SnS_2	*c*	$20.5 + 0.00400T$	273–873	?
Titanium:				
Ti	*c*	$8.91 + 0.00114T - 433000/T^2$	273–713	3
$TiCl_4$	*l*	35.7	285–372	?
TiO_2	*c*	$11.81 + 0.00754T - 41900/T^2$	273–713	3
Tungsten:				
W	*c*	$5.65 + 0.00866$	273–2073	1
WO_3	*c*	$16.0 + 0.00774T$	273–1550	?
Uranium:				
U	*c*	6.64	273–372	?
U_3O_8	*c*	59.8	276–314	?
Vanadium:				
V	*c*	$5.57 + 0.00097T$	273–1993	?
Xenon:				
Xe	*g*	4.97	All	0
Zinc:				
Zn	*c*	$5.25 + 0.00270T$	273–692	1
	l	$7.59 + 0.00055T$	692–1122	3
$ZnCl_2$	*c*	$15.9 + 0.00800T$	273–638	?
ZnO	*c*	$11.40 + 0.00145T - 182400/T^2$	273–1573	1
ZnS	*c*	$12.81 + 0.00095T - 194600/T^2$	273–1173	5
ZnSb	*c*	$11.5 + 0.00313T$	273–810	?
$ZnSO_4$	*c*	28	293–373	?
$ZnSO_4.H_2O$	*c*	34.7	282	?
$ZnSO_4.6H_2O$	*c*	80.8	282	?
$ZnSO_4.7H_2O$	*c*	100.2	273–307	?
Zirconium:				
ZrO_2	*c*	$11.62 + 0.01046T - 177700/T^2$	273–1673	5
$ZrO_2.SiO_2$	*c*	26.7	297–372	?

Table 171. Heat Capacity of Water*

Air-free, at a constant pressure of 1 atm.

Temperature, °C.	Heat capacity, constant pressure, 1 atm. cal./g. °C.†	Temperature, °C.	Heat capacity, constant pressure, 1 atm., cal./g. °C.†	Temperature, °C.	Heat capacity, constant pressure, 1 atm., cal./g. °C.†	Temperature, °C.	Heat capacity, constant pressure, 1 atm., cal./g. °C.†
0	1.00803	25	.99892	50	0.99919	75	1.00208
1	1.00717	26	.99885	51	.99926	76	1.00225
2	1.00636	27	.99878	52	.99935	77	1.00241
3	1.00564	28	.99873	53	.99943	78	1.00258
4	1.00495	29	.99869	54	.99950	79	1.00277
5	1.00433	30	.99866	55	.99959	80	1.00294
6	1.00378	31	.99864	56	.99969	81	1.00313
7	1.00325	32	.99861	57	.99978	82	1.00332
8	1.00277	33	.99861	58	.99988	83	1.00351
9	1.00234	34	.99859	59	.99998	84	1.00373
10	1.00194	35	.99859	60	1.00007	85	1.00392
11	1.00158	36	.99861	61	1.00019	86	1.00414
12	1.00124	37	.99861	62	1.00029	87	1.00435
13	1.00093	38	.99864	63	1.00041	88	1.00457
14	1.00067	39	.99866	64	1.00053	89	1.00480
15	1.00041	40	.99869	65	1.00065	90	1.00502
16	1.00019	41	.99871	66	1.00079	91	1.00526
17	0.99998	42	.99876	67	1.00091	92	1.00550
18	.99978	43	.99880	68	1.00105	93	1.00574
19	.99962	44	.99883	69	1.00117	94	1.00600
20	.99947	45	.99890	70	1.00131	95	1.00626
21	.99933	46	.99895	71	1.00146	96	1.00653
22	.99921	47	.99900	72	1.00160	97	1.00684
23	.99912	48	.99907	73	1.00177	98	1.00705
24	.99902	49	.99912	74	1.00191	99	1.00734
						100	1.00763

* From the data of Osborne, Stimson, and Ginnings, *Bur. Standards J. Research*, **23**, 197 (1939).
† 1 calorie = 4.1833 NBS int. j (National Bureau of Standards international joule).

Table 172. Specific Heats of Organic Liquids*

From "International Critical Tables," vol. 5, pp. 107–113 and a few data from other sources

Compound	Formula	Temperature, °C.	Sp. ht., cal./g. °C.
Acetal	$C_6H_{14}O_2$	0	0.467
		19–99	.520
Acetic acid*	$C_2H_4O_2$	26–95	.522
Acetone*	C_3H_6O	3–22.6	.514
		0	.506
		24.2–49.4	.538
Acetonitrile	C_2H_3N	21–76	.541
Acetophenone	C_8H_8O	20–196	.450
Acetyl chloride	C_2H_3ClO	0	.339
Allyl acetate	$C_5H_8O_2$	0	.430
alcohol	C_3H_6O	0	.386
		21–96	.665
benzoate	$C_{10}H_{10}O$	20	.388
butyrate	$C_7H_{12}O_2$	20	.451
chloride	C_3H_5Cl	0	.313
chloroacetate	$C_5H_7ClO_2$	20	.396
dichloroacetate	$C_5H_6Cl_2O_2$	20	.332
isobutyrate	$C_7H_{12}O_2$	20	.448
propionate	$C_6H_{10}O_2$	20	.451
trichloroacetate	$C_5H_5Cl_3O_2$	20	.288
valerate	$C_8H_{14}O$	20	.451
Aminobenzoic acid (o-)	$C_7H_7NO_2$	M. P.	.435
(m-)	$C_7H_7NO_2$	M. P.	.435
(p-)	$C_7H_7NO_2$	M. P.	.444
Amyl alcohol (d-primary)	$C_5H_{12}O$	22–125	.711
(t-)	$C_5H_{12}O$	20–99	.753
Amylene	C_5H_{10}	0	.282
Anethole	$C_{10}H_{12}O$	23–233	.511
		22.48	.551
		24.59	.564
		25.23	.612
Aniline	C_6H_7N	8–82	.512
Benzaldehyde	C_7H_6O	22–172	.428
Benzene*	C_6H_6	6–60	.419
		10	.340
		65	.482
Benzonitrile	C_7H_5N	22–186	.441
Benzophenone (β-)	$C_{13}H_{10}O$	3–40	.382
		0	.346
Benzyl alcohol	C_7H_8O	20 100	.511
		22–200	.540
chloride	C_7H_7Cl	0	.323
ethylene	C_9H_{10}	0	.393
Bromobenzene	C_6H_5Br	0	.215
		20–100	.231
		16.9–65	.239

Compound	Formula	Temperature, °C.	Sp. ht., cal./g. °C.
Bromochlorobenzene (o-)	C_6H_4BrCl	0	.215
(m-)	C_6H_4BrCl	0	.212
Bromoiodobenzene (o-)	C_6H_4BrI	0	.153
		5–100	.160
		3.2–64.6	.157
		1.8–34	.157
(m-)	C_6H_4BrI	0	.152
		5–100	.158
		3.2–64.5	.156
		1.7–34.1	.154
		1.7–36.2	.149
Bromophenol	C_6H_5BrO	18–77	.316
Butane (n-)	C_4H_{10}	0	.549
Butyl alcohol (n-)	$C_4H_{10}O$	21–115	.687
		30	.582
		−76.2	.443
		−33.3	.453
		2.3	.526
		19.2	.563
chloride (n-)	C_4H_9Cl	20	.451
formate (n-)	$C_5H_{10}O_2$	20	.459
propionate	$C_7H_{14}O_2$	20	.459
valerate	$C_9H_{18}O_2$	20	.459
Butyric acid (n-)	$C_4H_8O_2$	0	.444
		40	.501
		20–100	.515
Butyronitrile (n-)	C_4H_7N	21–113	.547
Caproic acid	$C_6H_{12}O_2$	29–105	.531
Capronitrile	$C_6H_{11}N$	18–156	.541
Carbon tetrachloride*	CCl_4	0	.198
		20	.201
		30	.200
Carvacrol	$C_{10}H_{14}O$	24–233	.575
Chloral	C_2HCl_3O	17–53	.250
hydrate	$C_2H_3Cl_3O_2$	55–88	.470
Chlorobenzene*	C_6H_5Cl	0	.273
		10	.298
		20	.308
Chlorobenzoic acid (o-)	$C_7H_5ClO_2$	0	0.390
(m-)	$C_7H_5ClO_2$	0	.265
(p-)	$C_7H_5ClC_2$	M. P.	.545
Chloroform*	$CHCl_3$	0	.232
		15	.226
		30	.234
Chlorophenol	C_6H_5ClO	0–20	.399
Chlorotoluene	C_7H_7Cl	0	.315

* See line coordinate chart, Fig. 9, p. 228, for the specific heats of a number of substances as a function of the temperature.

Table 172. Specific Heats of Organic Liquids—(*Continued*)

Compound	Formula	Temperature, °C.	Sp. ht., cal./g. °C.
Cresol (o-)	C_7H_8O	0–20	.497
(m-)	C_7H_8O	21–197	.551
		0–20	.477
Cresyl methyl ether (p-)	$C_8H_{10}O$	0	.404
Crotonic acid	$C_4H_6O_2$	71.4	.500
Cyclohexanol	$C_6H_{12}O$	15–18	.416
Cyclohexanone	$C_6H_{10}O$	15–18	.431
o-Cymene	$C_{10}H_{14}$	0	.398
Decahydronaphthalene (cis-)	$C_{10}H_{18}$	15–18	.393
Decane	$C_{10}H_{22}$ b.p. = 159	21–154	.588
	$C_{10}H_{22}$ b.p. = 162	0–50	.493
	$C_{10}H_{22}$ b.p. = 172	0–50	.500
Decylene (γ-)	$C_{10}H_{20}$	0–50	.467
Diallyl oxalate	$C_8H_{10}O_4$	20	.424
succinate	$C_{10}H_{14}O_4$	20	.450
Diamylene	$C_{10}H_{20}$	20–130	.543
Dibromobenzene (o-)	$C_6H_4Br_2$	0	.179
(m-)	$C_6H_4Br_2$	0	.175
Dibutyl oxalate	$C_{10}H_{18}O_4$	20	.439
Dichlorodifluormethane	CCl_2F_2	−43	.21
Dichloroacetic acid	$C_2H_2Cl_2O_2$	21–106	.349
		21–196	.348
Dichlorobenzene (o-)	$C_6H_4Cl_2$	0	.269
(m-)	$C_6H_4Cl_2$	0	.269
(p-)	$C_6H_4Cl_2$	53–99	.297
Diethylamine	$C_4H_{11}N$	22.5	.516
Diethyl carbonate	$C_5H_{10}O_3$	0	.245
		20–100	.462
ether (see Ether)	$C_4H_{10}O$	20.2–123	.473
ketone	$C_5H_{10}O$	20–98.5	.555
malate	$C_8H_{14}O_5$	24–186	.473
malonate	$C_7H_{12}O_4$	20	.431
oxalate	$C_6H_{10}O_4$	20	.431
succinate	$C_8H_{14}O_4$	20	.450
Dihydronaphthalene	$C_{10}H_{10}$	18–28	.345
Di-iodobenzene (m-)	$C_6H_4I_2$	34.2–99.6	.139
Di-isoamyl	$C_{10}H_{22}$	21.5–155	.588
oxalate	$C_{12}H_{22}O_4$	20	.447
Di-isobutylamine	C_8H_9N	22–130	.569
Dimethyl aniline	$C_8H_{11}N$	0–20	.416
		0	.403
naphthalene (β-)	$C_{12}H_{12}$	0	.392
pyrone	$C_7H_8O_2$	166	.547
Dinitrobenzene (m-)	$C_8H_4N_2O_4$	M. P.	.404
Diphenylamine	$C_{12}H_{11}N$	54	.437
		56	.441
		66	.480
Dipropyl ketone	$C_7H_{14}O$	20–140	.550
malonate	$C_9H_{16}O_4$	20	.431
oxalate (n-)	$C_8H_{14}O_4$	20	.431
succinate	$C_{10}H_{18}O_4$	20	.450
Dodecane	$C_{12}H_{26}$	14–20	.505
		0–50	.498
Dodecylene	$C_{12}H_{24}$	0–50	.455
Ether*	$C_4H_{10}O$	−100	.511
		−50	.515
		−5	.525
		0	.521
		+30	.545
		80	.687
		120	.800
		140	.819
		180	1.037
Ethyl acetate*	$C_4H_8O_2$	20	0.457
		20	.476
acetoacetate	$C_6H_{10}O_3$	0	.428
		20–100	.475
alcohol*	C_2H_6O (100%)	−20	.505
		0–98	.680
benzene	C_8H_{10}	0	.392
		30	.407
benzoate	$C_9H_{10}O_2$	20	.387
bromide*	C_2H_5Br	−100	0.194
		−20	.206
		5–10	.216
		10–15	.213
		15–20	.214
butyrate	$C_6H_{12}O_2$	20	.457
chloride	C_2H_5Cl	−28 to +4	.426
		0	.367
chloroacetate	$C_4H_7ClO_2$	9–138	.416
		20	.397
cresyl ether (p-)	$C_9H_{12}O$	0	.427
dichloroacetate	$C_4H_6Cl_2O_2$	20	.328
ether	$C_4H_{10}O$	0	.521
formate	$C_3H_6O_2$	14–49	.508
		−20 to +14	.454
Ethyl (*Cont.*):			
iodide*	C_2H_5I	−30	.156
		0	.161
		60	.171
isobutyrate	$C_6H_{12}O_2$	20	.457
propionate	$C_5H_{10}O_2$	20	.457
silicate	$C_8H_{20}SiO_4$	15–98	.424
sulfide	$C_4H_{10}S$	0	.468
		5–10	.470
		10–15	.473
		20–70	.477
trichloroacetate	$C_4H_5Cl_3O_2$	10–81	.294
		9–139	.305
		20	.284
valerate	$C_7H_{14}O_2$	20	.457
Ethylene bromide	$C_2H_4Br_2$	8–95	.182
		13–106	.175
		20	.173
chloride	$C_2H_4Cl_2$	−30	.278
		+20	.299
		30	.304
		50	.313
		60	.318
dichloroacetate	$C_6H_6Cl_4O_4$	0	.321
glycol*	$C_2H_6O_2$	−11.1	.535
		0	.542
		+ 2.5	.550
		5.1	.554
		14.9	.569
		19.9	.573
Formamide	CH_3NO	19	.549
Formic acid	CH_2O_2	0	.436
		15.5	.509
		20–100	.524
Furfural	$C_5H_4O_2$	0	.367
		20–100	.416
Glycerol*	$C_3H_8O_3$	15–50	.576
Glycol (ethylene)*	$C_2H_6O_2$	(see ethylene	glycol)
Heptaldehyde	$C_7H_{14}O$	0	.364
Heptane (n-)*	C_7H_{16}	0–50	.507
		20	.490
		30	.518
Heptylene	C_7H_{14}	0–50	.486
Heptylic acid	$C_7H_{14}O_2$	9	.556
Hexadecane (n-)	$C_{16}H_{34}$	0–50	.496
Hexadiene (1,5-)	C_6H_{10}	0	.405
Hexahydrocresol (o-)	$C_7H_{14}O$	15–18	.416
(m-)	$C_7H_{14}O$	15–18	.420
(p-)	$C_7H_{14}O$	15–18	.421
Hexane (n-)	C_6H_{14}	0–50	.527
		20–100	.600
Hexylene	C_6H_{12}	0–50	.504
Isoamyl acetate	$C_7H_{14}O_2$	20	.459
alcohol	$C_5H_{12}O$	0	.502
		20	.535
		30	.570
		47.9	.662
		10–117	.693
		21–130	.695
		75.5	.688
amine	$C_5H_{13}N$	22–91	.614
butyrate	$C_9H_{18}O_2$	20	.459
formate	$C_6H_{12}O_2$	16–65	.509
isobutyrate	$C_9H_{18}O_2$	20	.459
propionate	$C_8H_{16}O_2$	20	.459
succinate	$C_{14}H_{26}O_4$	0	.449
valerate	$C_{10}H_{20}O_2$	20	.459
Isobutane	C_4H_{10}	0	.549
Isobutyl acetate	$C_6H_{12}O_2$	20	.459
alcohol	$C_4H_{10}O$	21–109	.716
		30	.603
butyrate	$C_8H_{16}O_2$	20	0.459
succinate	$C_{12}H_{22}O_4$	0	.442
Isobutyric acid	$C_4H_8O_2$	20	.450
Isoheptane	C_7H_{16}	0–50	.501
Isopentane	C_5H_{12}	0	.512
		8	.527
Isovaleric acid	$C_5H_{10}O_2$	20	.463
		23–93	.590
Lauric acid	$C_{12}H_{24}O_2$	40–100	0.572
		57	.515
Mesitylene	C_9H_{12}	0	.393
Mesityl oxide	$C_6H_{10}O$	21–12[illegible]	.521
Methyl acetate	$C_3H_6O_2$	15	.468
Methylal	$C_3H_8O_2$	15–41	.521

Table 172. Specific Heats of Organic Liquids—(*Concluded*)

Compound	Formula	Temperature, °C.	Sp. ht., cal./g. °C.
Methyl alcohol*	CH_4O	5–10	.590
		15–20	.601
aniline	C_7H_9N	20–197	.512
benzoate	$C_8H_8O_2$	0	.363
butyl ketone	$C_6H_{12}O$	21–127	.553
butyrate (n-)	$C_5H_{10}O_2$	20	.459
chloroacetate	$C_3H_5ClO_2$	20	.382
cyclohexanone (o-)	$C_7H_{12}O$	15–18	.436
(m-)	$C_7H_{12}O$	15–18	.441
(p-)	$C_7H_{12}O$	15–18	.441
dichloroacetate	$C_3H_4Cl_2O_2$	20	.311
ethylketone	C_4H_8O	20–78	.549
ethylketoxime	C_4H_9NO	21.8–151.5	.650
formate	$C_2H_4O_2$	13–29	.516
hexyl ketone	$C_8H_{16}O$	22–168	.552
isobutyl ketone	$C_6H_{12}O$	20	.459
isopropyl ketone	$C_5H_{10}O$	20–91	.525
propionate	$C_4H_8O_2$	20	.459
trichloroacetate	$C_3H_3Cl_3O_2$	20	.267
valerate	$C_6H_{12}O_2$	20	.459
Methylene chloride	CH_2Cl_2	15–40	.288
Myristic acid	$C_{14}H_{28}O_2$	56–100	.539
Naphthalene	$C_{10}H_8$	87.5	.402
Naphthylamine (α-)	$C_{10}H_9N$	53.2	.475
		94.2	.476
Nitrobenzene	$C_6H_5NO_2$	10	.358
		30	.339
		50	.330
		70	.330
		90	.343
		120	.394
Nitrobenzoic acid (p-)	$C_7H_5NO_4$	M. P.	.449
Nitromethane	CH_3NO_2	17	.412
Nitronaphthalene (α-)	$C_{10}H_7NO_2$	58.6	.365
		61.4	.378
		94.3	.390
Nonane	C_9H_{20}	0–50	.503
Nonylene	C_9H_{18}	0–50	.485
Octane (n-)*	C_8H_{18}	0–50	.505
		20–123	.578
Octylene	C_8H_{16}	0–50	.486
Palmitic acid	$C_{16}H_{32}O_2$	65–104	.653
Paraldehyde	$C_6H_{12}O_3$	0	.436
Pentadecane	$C_{15}H_{32}$	0–50	.497
Pentadecylene	$C_{15}H_{30}$	0–50	.471
Phenetole	$C_8H_{10}O$	20	.446
Phenyl methyl ether	C_7H_8O	0	.405
		20–152	.483
Picoline (α-)	C_6H_7N	22–124	.434
Piperidine	$C_5H_{11}N$	20–98	.523
Propane	C_3H_8	0	.576
Propionaldehyde	C_3H_6O	0	.522
Propionic acid	$C_3H_6O_2$	0	.444
		20–137	.560
Propionitrile	C_3H_5N	0	.508
		19–95	.538
Propyl acetate (n-)	$C_5H_{10}O_2$	20	.459
benzene	C_9H_{12}	0	.400
benzoate	$C_{10}H_{12}O_2$	20	.398
butyrate	$C_7H_{14}O_2$	20	.459

Compound	Formula	Temperature, °C.	Sp. ht., cal./g. °C.
Propyl (*Cont.*):			
chloroacetate	$C_5H_9ClO_2$	20	.414
dichloroacetate	$C_5H_8Cl_2O_2$	20	.341
formate (n-)	$C_4H_8O_2$	20	.459
isobutyrate	$C_7H_{14}O_2$	20	.459
phenyl ether	$C_9H_{12}O$	0	.429
propionate	$C_6H_{12}O_2$	20	.459
trichloroacetate	$C_5H_7Cl_3O_2$	20	.297
valerate	$C_8H_{16}O_2$	20	.459
Pseudocumene	C_9H_{12}	20	.414
Pyridine	C_5H_5N	20	.405
		21–108	.431
		0–20	.395
Quinoline	C_9H_7N	0–20	.352
Salicylaldehyde	$C_7H_6O_2$	18	.382
Salol	$C_{13}H_{10}O_3$	44.1	.391
Stearic acid	$C_{18}H_{36}O_2$	75–137	.550
Tetrachloroethane	$C_2H_2Cl_4$	20	.268
Tetrachloroethylene	C_2Cl_4	20	.216
		24	.211
Tetradecane	$C_{14}H_{30}$	0–50	0.497
Thymol (m-)	$C_{10}H_{14}O$	50	.566
Toluene*	C_7H_8	{ 10	.364
		85	.534
		12–99	.440
Toluidine (o-)	C_7H_9N	0	.454
		22–195	.598
		40.5	.498
(p-)	C_7H_9N	43	.524
		58	.634
		94	.533
Trichloroethane	$C_2H_3Cl_3$	20	0.266
Trichloroethylene	C_2HCl_3	20	.223
Tridecane	$C_{13}H_{28}$	0–50	.499
Tridecylene	$C_{13}H_{26}$	0–50	.457
Trinitrotoluene (2,4,6-)	$C_7H_5N_3O_6$	....	.335
Undecane	$C_{11}H_{24}$	0–50	.501
Undecylene	$C_{11}H_{22}$	0–50	.482
Valeronitrile	C_5H_9N	23–121	.520
Xylene (o-)*	C_8H_{10}	30	.411
		39	.450
(m-)*	C_8H_{10}	0	.383
		9–40	.400
		16–35	.387
		30	.401
(p-)*	C_8H_{10}	0	.383
		30	.397
		40.8	.428
dibromide (o-)	$C_8H_8Br_2$	15–40	.183
(m-)	$C_8H_8Br_2$	15–40	.184
(p-)	$C_8H_8Br_2$	15–40	.180
dichloride (o-)	$C_8H_8Cl_2$	15–40	.283
(m-)	$C_8H_8Cl_2$	15–40	.295
(p-)	$C_8H_8Cl_2$	15–40	.282
tetrachloride (o-)	$C_8H_6Cl_4$	15–40	.240
(p-)	$C_8H_6Cl_4$	15–40	.242
Xylyl ethyl ether (2,4-)	$C_{10}H_{14}O$	0	.417

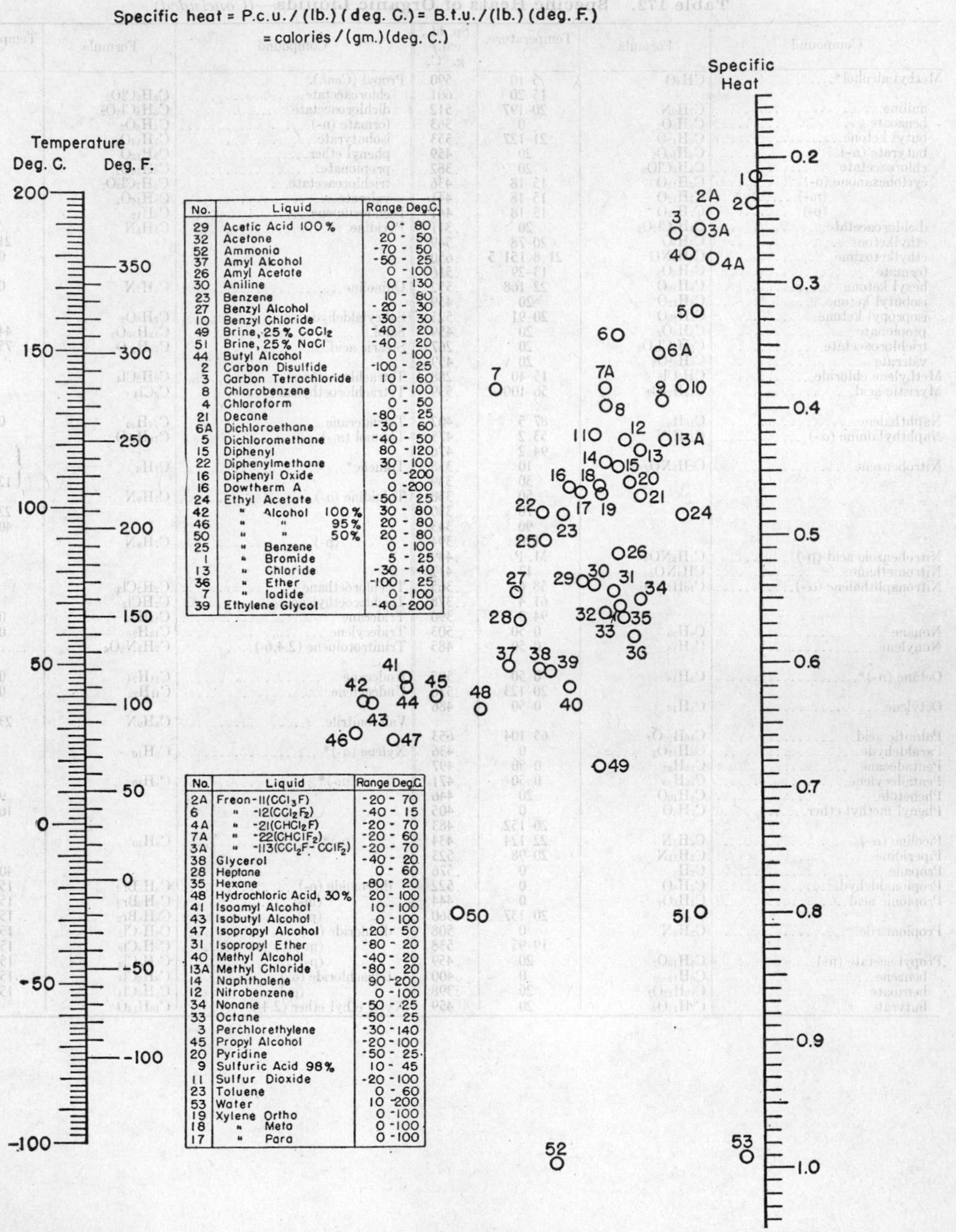

No.	Liquid	Range Deg.C.
29	Acetic Acid 100%	0 - 80
32	Acetone	20 - 50
52	Ammonia	-70 - 50
37	Amyl Alcohol	-50 - 25
26	Amyl Acetate	0 - 100
30	Aniline	0 - 130
23	Benzene	10 - 80
27	Benzyl Alcohol	-20 - 30
10	Benzyl Chloride	-30 - 30
49	Brine, 25% $CaCl_2$	-40 - 20
51	Brine, 25% NaCl	-40 - 20
44	Butyl Alcohol	0 - 100
2	Carbon Disulfide	-100 - 25
3	Carbon Tetrachloride	10 - 60
8	Chlorobenzene	0 - 100
4	Chloroform	0 - 50
21	Decane	-80 - 25
6A	Dichloroethane	-30 - 60
5	Dichloromethane	-40 - 50
15	Diphenyl	80 - 120
22	Diphenylmethane	30 - 100
16	Diphenyl Oxide	0 - 200
16	Dowtherm A	0 - 200
24	Ethyl Acetate	-50 - 25
42	" Alcohol 100%	30 - 80
46	" " 95%	20 - 80
50	" " 50%	20 - 80
25	" Benzene	0 - 100
1	" Bromide	5 - 25
13	" Chloride	-30 - 40
36	" Ether	-100 - 25
7	" Iodide	0 - 100
39	Ethylene Glycol	-40 - 200

No.	Liquid	Range Deg.C.
2A	Freon-11(CCl_3F)	-20 - 70
6	" -12(CCl_2F_2)	-40 - 15
4A	" -21($CHCl_2F$)	-20 - 70
7A	" -22($CHClF_2$)	-20 - 60
3A	" -113($CCl_2F-CClF_2$)	-20 - 70
38	Glycerol	-40 - 20
28	Heptane	0 - 60
35	Hexane	-80 - 20
48	Hydrochloric Acid, 30%	20 - 100
41	Isoamyl Alcohol	10 - 100
43	Isobutyl Alcohol	0 - 100
47	Isopropyl Alcohol	-20 - 50
31	Isopropyl Ether	-80 - 20
40	Methyl Alcohol	-40 - 20
13A	Methyl Chloride	-80 - 20
14	Naphthalene	90 - 200
12	Nitrobenzene	0 - 100
34	Nonane	-50 - 25
33	Octane	-50 - 25
3	Perchlorethylene	-30 - 140
45	Propyl Alcohol	-20 - 100
20	Pyridine	-50 - 25
9	Sulfuric Acid 98%	10 - 45
11	Sulfur Dioxide	-20 - 100
23	Toluene	0 - 60
53	Water	10 - 200
19	Xylene Ortho	0 - 100
18	" Meta	0 - 100
17	" Para	0 - 100

FIG. 9. Specific heats of liquids. (*Chilton, Colburn, and Vernon, personal communication. Based mainly on data from "International Critical Tables."*)

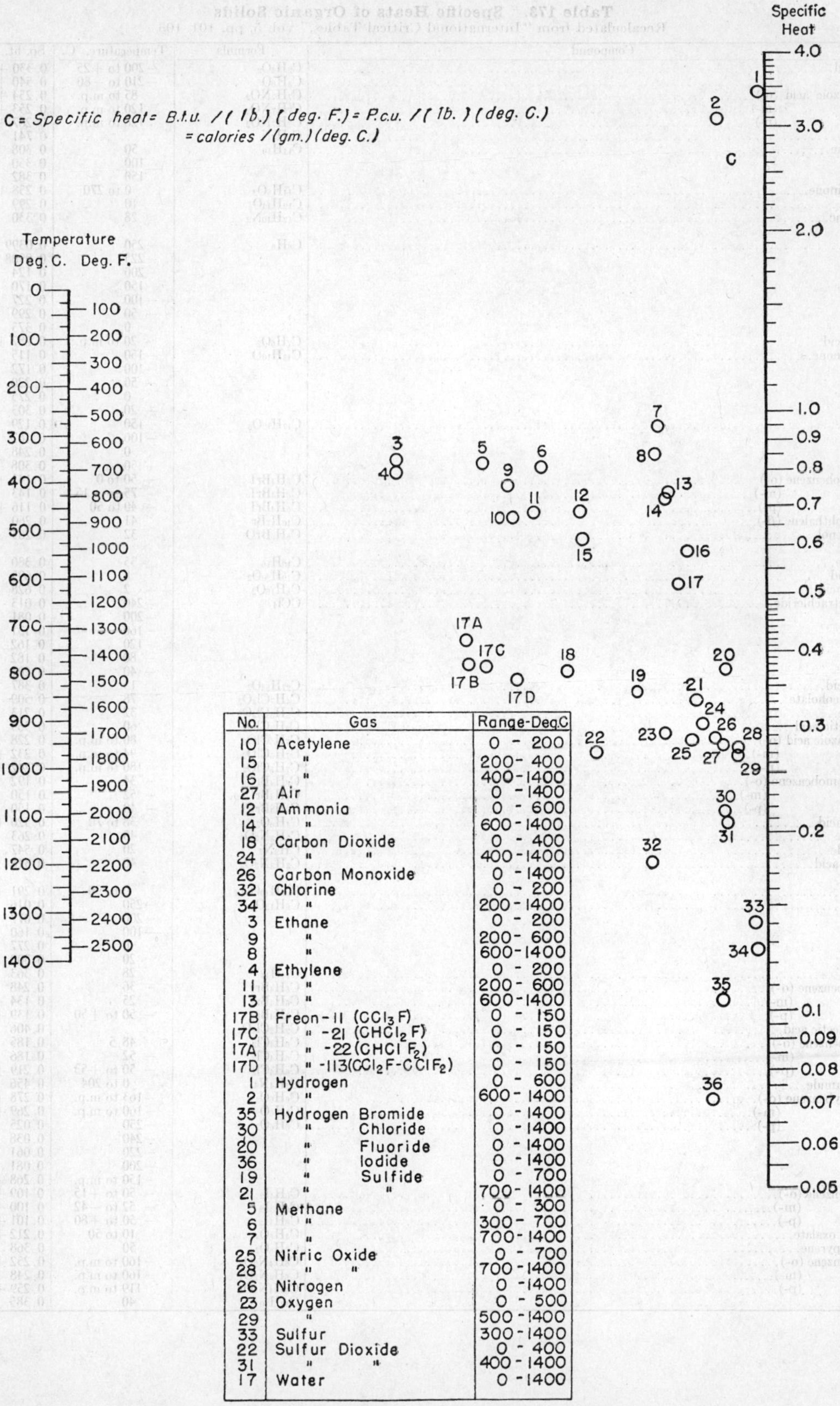

No.	Gas	Range-Deg.C
10	Acetylene	0 - 200
15	"	200 - 400
16	"	400 - 1400
27	Air	0 - 1400
12	Ammonia	0 - 600
14	"	600 - 1400
18	Carbon Dioxide	0 - 400
24	" "	400 - 1400
26	Carbon Monoxide	0 - 1400
32	Chlorine	0 - 200
34	"	200 - 1400
3	Ethane	0 - 200
9	"	200 - 600
8	"	600 - 1400
4	Ethylene	0 - 200
11	"	200 - 600
13	"	600 - 1400
17B	Freon-11 (CCl_3F)	0 - 150
17C	" -21 ($CHCl_2F$)	0 - 150
17A	" -22 ($CHClF_2$)	0 - 150
17D	" -113 (CCl_2F-$CClF_2$)	0 - 150
1	Hydrogen	0 - 600
2	"	600 - 1400
35	Hydrogen Bromide	0 - 1400
30	" Chloride	0 - 1400
20	" Fluoride	0 - 1400
36	" Iodide	0 - 1400
19	" Sulfide	0 - 700
21	" "	700 - 1400
5	Methane	0 - 300
6	"	300 - 700
7	"	700 - 1400
25	Nitric Oxide	0 - 700
28	" "	700 - 1400
26	Nitrogen	0 - 1400
23	Oxygen	0 - 500
29	"	500 - 1400
33	Sulfur	300 - 1400
22	Sulfur Dioxide	0 - 400
31	" "	400 - 1400
17	Water	0 - 1400

FIG. 10. Specific heats (C_p) of gases at 1 atm. pressure.

Table 173. Specific Heats of Organic Solids
Recalculated from "International Critical Tables," vol. 5, pp. 101–105

Compound	Formula	Temperature, °C.	Sp. ht., cal./g. °C.
Acetic acid	$C_2H_4O_2$	−200 to +25	0.330 + 0.00080t
Acetone	C_3H_6O	−210 to −80	0.540 + 0.0156t
Aminobenzoic acid (o-)	$C_7H_7NO_2$	85 to m.p.	0.254 + 0.00136t
(m-)	$C_7H_7NO_2$	120 to m.p.	0.253 + 0.00122t
(p-)	$C_7H_7NO_2$	128 to m.p.	0.287 + 0.00088t
Aniline	C_6H_7N	....	0.741
Anthracene	$C_{14}H_{10}$	50	0.308
		100	0.350
		150	0.382
Anthraquinone	$C_{14}H_8O_2$	0 to 270	0.258 + 0.00069t
Apiol	$C_{12}H_{14}O_4$	10	0.299
Azobenzene	$C_{12}H_{10}N_2$	28	0.330
Benzene	C_6H_6	−250	0.0399
		−225	0.0908
		−200	0.124
		−150	0.170
		−100	0.227
		− 50	0.299
		0	0.375
Benzoic acid	$C_7H_6O_2$	20 to m.p.	0.287 + 0.00050t
Benzophenone	$C_{13}H_{10}O$	−150	0.115
		−100	0.172
		− 50	0.220
		0	0.275
		+ 20	0.303
Betol	$C_{17}H_{12}O_3$	−150	0.129
		−100	0.167
		0	0.248
		+ 50	0.308
Bromoiodobenzene (o-)	C_6H_4BrI	− 50 to 0	0.143 + 0.00025t
(m-)	C_6H_4BrI	− 75 to −15	0.143
(p-)	C_6H_4BrI	− 40 to 50	0.116 + 0.00032t
Bromonaphthalene (β-)	$C_{10}H_7Br$	41	0.260
Bromophenol	C_6H_5BrO	32	0.263
Camphene	$C_{10}H_{16}$	35	0.380
Capric acid	$C_{10}H_{20}O_2$	8	0.695
Caprylic acid	$C_8H_{16}O_2$	− 2	0.628
Carbon tetrachloride	CCl_4	−240	0.013
		−200	0.081
		−160	0.131
		−120	0.162
		− 80	0.182
		− 40	0.201
Cerotic acid	$C_{27}H_{54}O_2$	15	0.387
Chloral alcoholate	$C_4H_7Cl_3O_2$	78	0.509
hydrate	$C_2H_3Cl_3O_2$	32	0.213
Chloroacetic acid	$C_2H_3ClO_2$	60	0.363
Chlorobenzoic acid (o-)	$C_7H_5ClO_2$	80 to m.p.	0.228 + 0.00084t
(m-)	$C_7H_5ClO_2$	94 to m.p.	0.232 + 0.00073t
(p-)	$C_7H_5ClO_2$	180 to m.p.	0.242 + 0.00055t
Chlorobromobenzene (o-)	C_6H_4BrCl	− 34	0.192
(m-)	C_6H_4BrCl	− 52	0.150
(p-)	C_6H_4BrCl	− 40	0.150
Crotonic acid	$C_4H_6O_2$	38 to 70	0.520 + 0.00020t
Cyamelide	$C_3H_3N_3O_3$	40	0.263
Cyanamide	CH_2N_2	20	0.547
Cyanuric acid	$C_3H_3N_3O_3$	40	0.318
Dextrin	$(C_6H_{10}O_5)_x$	0 to 90	0.291 + 0.00096t
Dextrose	$C_6H_{12}O_6$	−250	0.016
		−200	0.077
		−100	0.160
		0	0.277
		20	0.300
Dibenzyl	$C_{14}H_{14}$	28	0.363
Dibromobenzene (o-)	$C_6H_4Br_2$	− 36	0.248
(m-)	$C_6H_4Br_2$	− 25	0.134
(p-)	$C_6H_4Br_2$	− 50 to +50	0.139 + 0.00038t
Dichloroacetic acid	$C_2H_2Cl_2O_2$	....	0.406
Dichlorobenzene (o-)	$C_6H_4Cl_2$	− 48.5	0.185
(m-)	$C_6H_4Cl_2$	− 52	0.186
(p-)	$C_6H_4Cl_2$	− 50 to +53	0.219 + 0.0021t
Dicyandiamide	$C_2H_4N_4$	0 to 204	0.456
Dihydroxybenzene (o-)	$C_6H_6O_2$	−163 to m.p.	0.278 + 0.00098t
(m-)	$C_6H_6O_2$	−160 to m.p.	0.269 + 0.00118t
(p-)	$C_6H_6O_2$	−250	0.025
		−240	0.038
		−220	0.061
		−200	0.081
		−150 to m.p.	0.268 + 0.00093t
Di-iodobenzene (o-)	$C_6H_4I_2$	− 50 to +15	0.109 + 0.00026t
(m-)	$C_6H_4I_2$	− 52 to −42	0.100 + 0.00026t
(p-)	$C_6H_4I_2$	− 50 to +80	0.101 + 0.00026t
Dimethyl oxalate	$C_4H_6O_4$	10 to 50	0.212 + 0.0044t
Dimethylpyrene	$C_7H_8O_2$	50	0.368
Dinitrobenzene (o-)	$C_6H_4N_2O_4$	−160 to m.p.	0.252 + 0.00083t
(m-)	$C_6H_4N_2O_4$	−160 to m.p.	0.248 + 0.00077t
(p-)	$C_6H_4N_2O_4$	119 to m.p.	0.259 + 0.00057t
Diphenyl	$C_{12}H_{10}$	40	0.385

Table 173. Specific Heats of Organic Solids—(*Continued*)

Compound	Formula	Temperature, °C.	Sp. ht., cal./g. °C.
Diphenylamine	$C_{12}H_{11}N$	26	0.337
Dulcitol	$C_6H_{14}O_6$	20	0.282
Erythritol	$C_4H_{10}O_4$	60	0.351
Ethyl alcohol	C_2H_6O (crystalline)	−190	0.232
		−180	0.248
		−160	0.282
		−140	0.318
		−130	0.376
	(vitreous)	−190	0.260
		−180	0.296
		−175	0.380
		−170	0.399
Ethylene glycol	$C_2H_6O_2$	−190 to −40	$0.366 + 0.00110t$
Formic acid	CH_2O_2	− 22	0.387
		0	0.430
Glutaric acid	$C_5H_8O_4$	20	0.299
Glycerol	$C_3H_8O_3$	−265	0.009
		−260	0.022
		−250	0.047
		−220	0.085
		−200	0.115
		−100	0.217
		0	0.330
Hexachloroethane	C_2Cl_6	25	0.174
Hexadecane	$C_{16}H_{34}$	....	0.495
Hydroxyacetanilide	$C_8H_9NO_2$	41 to m.p.	$0.249 + 0.00154t$
Iodobenzene	C_6H_5I	40	0.191
Isopropyl alcohol	C_3H_8O	−200 to −160	$0.051 + 0.00165t$
Lactose	$C_{12}H_{22}O_{11}$	20	0.287
	$C_{12}H_{22}O_{11}.H_2O$	20	0.299
Lauric acid	$C_{12}H_{24}O_2$	− 30 to +40	$0.430 + 0.000027t$
Levoglucosane	$C_6H_{10}O_5$	40	0.607
Levulose	$C_6H_{12}O_6$	20	0.275
Malonic acid	$C_3H_4O_4$	20	0.275
Maltose	$C_{12}H_{22}O_{11}$	20	0.320
Mannitol	$C_6H_{14}O_6$	0 to 100	$0.313 + 0.00025t$
Melamine	$C_3H_6N_6$	40	0.351
Myristic acid	$C_{14}H_{28}O_2$	0 to 35	$0.381 + 0.00545t$
Naphthalene	$C_{10}H_8$	−130 to m.p.	$0.281 + 0.00111t$
Naphthol (α-)	$C_{10}H_8O$	50 to m.p.	$0.240 + 0.00147t$
(β-)	$C_{10}H_8O$	61 to m.p.	$0.252 + 0.00128t$
Naphthylamine (α-)	$C_{10}H_9N$	0 to 50	$0.270 + 0.0031t$
Nitroaniline (o-)	$C_6H_6N_2O_2$	−160 to m.p.	$0.269 + 0.000920t$
(m-)	$C_6H_6N_2O_2$	−160 to m.p.	$0.275 + 0.000946t$
(p-)	$C_6H_6N_2O_2$	−160 to m.p.	$0.276 + 0.001000t$
Nitrobenzoic acid (o-)	$C_7H_5NO_4$	−163 to m.p.	$0.256 + 0.00085t$
(m-)	$C_7H_5NO_4$	66 to m.p.	$0.258 + 0.00091t$
(p-)	$C_7H_5NO_4$	−160 to m.p.	$0.247 + 0.00077t$
Nitronaphthalene	$C_{10}H_7NO_2$	0 to 55	$0.236 + 0.00215t$
Oxalic acid	$C_2H_2O_4$	−200 to +50	$0.259 + 0.00076t$
	$C_2H_2O_4.2H_2O$	−200	0.117
		−100	0.239
		0	0.338
		+ 50	0.385
		100	0.416
Palmitic acid	$C_{16}H_{32}O_2$	−180	0.167
		−140	0.208
		−100	0.251
		− 50	0.306
		0	0.382
		+ 20	0.430
Phenol	C_6H_6O	14 to 26	0.561
Phthalic acid	$C_8H_6O_4$	20	0.232
Picric acid	$C_6H_3N_3O_7$	−100	0.165
		0	0.240
		+ 50	0.263
		100	0.297
		120	0.332
Propionic acid	$C_3H_6O_2$	− 33	0.726
Propyl alcohol (n-)	C_3H_8O	−200	0.170
		−175	0.363
		−150	0.471
		−130	0.497
Pyrotartaric acid	$C_6H_8O_4$	20	0.301
Quinhydrone	$C_{12}H_{10}O_4$	−250	0.017
		−225	0.061
		−200	0.098
		−100	0.191
		0	0.256

Table 173. Specific Heats of Organic Solids—(*Concluded*)

Compound	Formula	Temperature, °C.	Sp. ht., cal./g. °C.
Quinone	$C_6H_4O_2$	−250	0.031
		−225	0.082
		−200	0.113
		−150 to m.p.	0.282 + 0.00083t
Salol	$C_{13}H_{10}O_3$	32	0.289
Stearic acid	$C_{18}H_{36}O_2$	15	0.399
Succinic acid	$C_4H_6O_4$	0 to 160	0.248 + 0.00153t
Sucrose	$C_{12}H_{22}O_{11}$	20	0.299
Sugar (cane)	$C_{12}H_{22}O_{11}$	22 to 51	0.301
Tartaric acid	$C_4H_6O_6$	36	0.287
Tartaric acid	$C_4H_6O_6.H_2O$	−150	0.112
		−100	0.170
		− 50	0.231
		0	0.308
		+ 50	0.366
Tetrachloroethylene	C_2Cl_4	− 40 to 0	0.198 + 0.00018t
Tetryl	$C_7H_5N_5O_8$	−100	0.182
		− 50	0.199
		0	0.212
		+100	0.236
1 Tetryl + 1 picric acid	$C_{13}H_8N_8O_{15}$	−100 to +100	0.253 + 0.00072t
1 Tetryl + 2 TNT	$C_{21}H_{15}N_{11}O_{20}$	−100	0.172
		0	0.280
		+ 50	0.325
Thymol	$C_{10}H_{14}O$	0 to 49	0.315 + 0.0031t
Toluic acid (o-)	$C_8H_8O_2$	54 to m.p.	0.277 + 0.00120t
(m-)	$C_8H_8O_2$	54 to m.p.	0.239 + 0.00195t
(p-)	$C_8H_8O_2$	130 to m.p.	0.271 + 0.00106t
Toluidine (p-)	C_7H_9N	0	0.337
		20	0.387
		40	0.440
Trichloroacetic acid	$C_2H_2Cl_3O$	solid	0.459
Trimethyl carbinol	$C_4H_{10}O$	− 4	0.559
Trinitrotoluene	$C_7H_5N_3O_6$	−100	0.170
		− 50	0.253
		0	0.311
		+100	0.385
Trinitroxylene	$C_8H_7N_3O_6$	−185 to +23	0.241
		20 to 50	0.423
Triphenylmethane	$C_{19}H_{16}$	0 to 91	0.189 + 0.0027t
Urea	CH_4N_2O	20	0.320

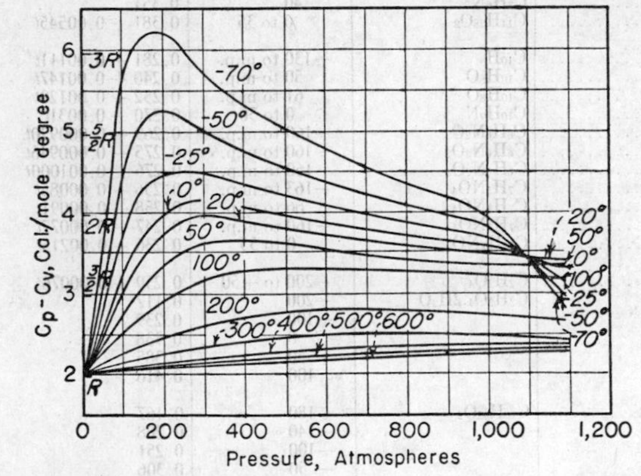

FIG. 11. The variation of $C_p - C_v$ for nitrogen with pressure at various temperatures. [*Deming and Shupe, Phys. Rev.*, **37**, 638 (1931).]

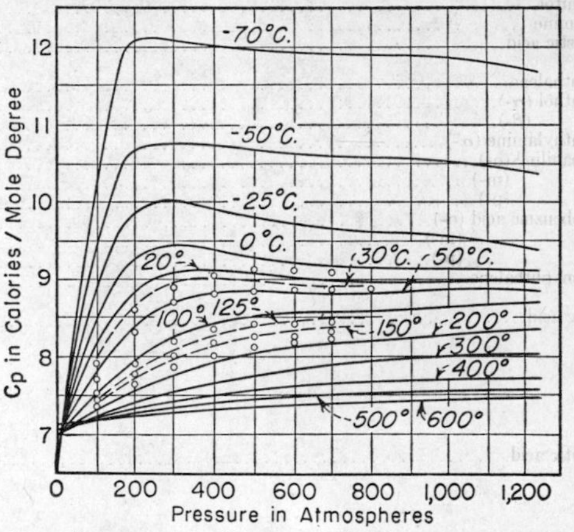

FIG. 12. The variation of heat capacity for nitrogen with pressure at various temperatures. [*Deming and Shupe, Phys. Rev.*, **37**, 638 (1931).]

Table 174. Liquefied Gases

Substance (liquid)	Temperature or range of temperature, °C.	Specific heat, cal./g. °C.
Ammonia	− 60	1.05
	0	1.10
	40	1.16
	80	1.29
	100	1.48
	110	1.61
Carbon dioxide (63 atm.)	−50 to −10	0.465 to 0.539
disulfide	−100 to 150	$0.235 + 0.00046t$
monoxide	−206 to −190	0.0615
Chlorine	−205	0.229
Hydrogen	−258 to −252	1.75 to 2.33
Nitric oxide	−158 to −156	0.580
Nitrogen	−209 to −197	.475
Oxygen	−216 to −200	.398
Sulfur dioxide	− 20	.313
	0	.318
	20	.328
	60	.361
	100	.419
	150	.846

Table 175. Specific Heat of Air at High Pressures

Cal./g. °C.

Temp., °C.	Atmospheres					
	1	10	20	40	70	100
100	0.237	0.239	0.240	0.245	0.250	0.258
0	.238	.242	.247	.251	.277	.298
−50	.238	.246	.257	.279	.332	.412
−100	.239	.259	.285	.370	.846	
−150	.240	.311	.505			

For the specific heats of other materials as a function of the pressure, see "International Critical Tables," vol. 5, pp. 82–83. See Figs. 9 and 12.

Table 176. C_p/C_v: Ratios of Specific Heats of Gases at 1 Atm. Pressure*

Compound	Formula	Temperature, °C.	Ratio of specific heats, $(\gamma) = C_p/C_v$
Acetaldehyde	C_2H_4O	30	1.14
Acetic acid	$C_2H_4O_2$	136	1.15
Acetylene	C_2H_2	15	1.26
		−71	1.31
Air		925	1.36
		17	1.403
		−78	1.408
		−118	1.415
Ammonia	NH_3	15	1.310
Argon	A	15	1.668
		−180	1.76 (?)
		0–100	1.67
Benzene	C_6H_6	90	1.10
Bromine	Br_2	20–350	1.32
Carbon dioxide	CO_2	15	1.304
		−75	1.37
disulfide	CS_2	100	1.21
monoxide	CO	15	1.404
		−180	1.41
Chlorine	Cl_2	15	1.355
Chloroform	$CHCl_3$	100	1.15
Cyanogen	$(CN)_2$	15	1.256
Cyclohexane	C_6H_{12}	80	1.08
Dichlorodifluormethane	CCl_2F_2	25	1.139
Ethane	C_2H_6	100	1.19
		15	1.22
		−82	1.28
Ethyl alcohol	C_2H_6O	90	1.13
ether	$C_4H_{10}O$	35	1.08
		80	1.086
Ethylene	C_2H_4	100	1.18
		15	1.255
		−91	1.35
Helium	He	−180	1.660
Hexane (n-)	C_6H_{14}	80	1.08
Hydrogen	H_2	15	1.410
		−76	1.453
		−181	1.597
bromide	HBr	20	1.42
chloride	HCl	15	1.41
		100	1.40
cyanide	HCN	65	1.31
		140	1.28
		210	1.24
Hydrogen (*Cont.*):			
iodide	HI	20–100	1.40
sulfide	H_2S	15	1.32
		−45	1.30
		−57	1.29
Iodine	I_2	185	1.30
Isobutane	C_4H_{10}	15	1.11
Krypton	Kr	19	1.68
Mercury	Hg	360	1.67
Methane	CH_4	600	1.113
		300	1.16
		15	1.31
		−80	1.34
		−115	1.41
Methyl acetate	$C_3H_6O_2$	15	1.14
alcohol	CH_4O	77	1.203
ether	C_2H_6O	6–30	1.11
Methylal	$C_3H_8O_2$	13	1.06
		40	1.09
Neon	Ne	19	1.64
Nitric oxide	NO	15	1.400
		−45	1.39
		−80	1.38
Nitrogen	N_2	15	1.404
		−181	1.47
Nitrous oxide	N_2O	100	1.28
		15	1.303
		−30	1.31
		−70	1.34
Oxygen	O_2	15	1.401
		−76	1.415
		−181	1.45
Pentane (n-)	C_5H_{12}	86	1.086
Phosphorus	P	300	1.17
Potassium	K	850	1.77
Sodium	Na	750–920	1.68
Sulfur dioxide	SO_2	15	1.29
Xenon	Xe	19	1.66

* From "International Critical Tables," vol. 5, pp. 80–82.

Table 177. Ratios of Specific Heats of Air at High Pressures

Pressure, atm.	Ratio of specific heats, $(\gamma) = C_p/C_v$ 0°C.	−79.4°C.	Pressure, atm.	Ratio of specific heats, $(\gamma) = C_p/C_v$ 0°C.	−79.4°C.
25	1.47	1.57	125	1.69	2.40
50	1.53	1.77	150	1.74	2.47
75	1.59	2.00	175	1.78	2.41
100	1.65	2.20	200	1.83	2.33

For additional data, see "International Critical Tables," vol. 5, pp. 115–116, and pp. 122–125.

SPECIFIC HEATS OF AQUEOUS SOLUTIONS

For additional data see "International Critical Tables," vol. 5, pp. 115–116, and pp. 122–125

Table 178. Hydrochloric Acid

Mole % HCl	Specific heat, cal./g. °C. 0°C.	10°C.	20°C.	40°C.	60°C.
0.0	1.00				
9.09	0.72	0.72	0.74	0.75	0.78
16.7	.61	.605	.631	.645	.67
20.0	.58	.575	.591	.615	.638
25.9	.55	.....	.....	.....	.61

Table 179. Sulfuric Acid*

$\%H_2SO_4$	C_p at 20°C., cal./g. °C.	$\%H_2SO_4$	C_p at 20°C., cal./g. °C.
0.34	0.9968	35.25	0.7238
0.68	.9937	37.69	.7023
1.34	.9877	40.49	.6770
2.65	.9762	43.75	.6476
3.50	.9688	47.57	.6153
5.16	.9549	52.13	.5801
9.82	.9177	57.65	.5420
15.36	.8767	64.47	.5012
21.40	.8339	73.13	.4628
22.27	.8275	77.91	.4518
23.22	.8205	81.33	.4481
24.25	.8127	82.49	.4467
25.39	.8041	84.48	.4408
26.63	.7945	85.48	.4346
28.00	.7837	89.36	.4016
29.52	.7717	91.81	.3787
30.34	.7647	94.82	.3554
31.20	.7579	97.44	.3404
33.11	.7422	100.00	.3352

* Vinal and Craig, *Bur. Standards J. Research*, **24**, 475 (1940).

Table 180. Nitric Acid Solutions

$\%HNO_3$ by Weight	Specific Heat at 20°C., Cal./g. °C.
0	1.000
10	0.900
20	.810
30	.730
40	.675
50	.650
60	.640
70	.615
80	.575
90	.515

Table 181. Phosphoric Acid*

$\%H_3PO_4$	C_p at 21.3°C. cal./g. °C.	$\%H_3PO_4$	C_p at 21.3°C. cal./g. °C.
2.50	0.9903	50.00	0.6350
3.80	.9970	52.19	.6220
5.33	.9669	53.72	.6113
8.81	.9389	56.04	.5972
10.27	.9293	58.06	.5831
14.39	.8958	60.23	.5704
16.23	.8796	62.10	.5603
19.99	.8489	64.14	.5460
22.10	.8300	66.13	.5349
24.56	.8125	68.14	.5242
25.98	.8004	69.97	.5157
28.15	.7856	69.50	.5160
29.96	.7735	71.88	.5046
32.09	.7590	73.71	.4940
33.95	.7432	75.79	.4847
36.26	.7270	77.69	.4786
38.10	.7160	79.54	.4680
40.10	.7024	80.00	.4686
42.08	.6877	82.00	.4593
44.11	.6748	84.00	.4500
46.22	.6607	85.98	.4419
48.16	.6475	88.01	.4359
49.79	.6370	89.72	.4206

* *Z. physik. Chem.*, **A167**, 42 (1933).

Table 182. Acetic Acid (at 38°C.)

Mole % acetic acid	0	6.98	30.9	54.5	100
Cal./g. °C.	1.0	0.911	0.73	0.631	0.535

Table 183. Sodium Hydroxide (at 20°C.)

Mole % NaOH	0	0.5	1.0	9.09	16.7	28.6	37.5
Cal./g. °C.	1.0	0.985	0.97	0.835	0.80	0.784	0.782

Table 184. Potassium Hydroxide (at 19°C.)

Mole % KOH	0	0.497	1.64	4.76	9.09
Cal./g. °C.	1.0	0.975	0.93	0.814	0.75

Table 185. Ammonia

Mole % NH_3	Specific heat, cal./g. °C. 2.4°C.	20.6°C.	41°C.	61°C.
0	1.01	1.0	0.995	1.0
10.5	0.98	0.995	1.06	1.02
20.9	.96	.99	1.03	
31.2	.956	1.0		
41.4	.985			

Table 186. Sodium Carbonate*

% Na_2CO_3 by weight	Temperature, °C. 17.6	30.0	76.6	98.0
0.000	0.9992	0.9986	1.0098	1.0084
1.498	.9807			
2.000		.9786		
2.901	.9597			
4.000	.....	.9594		
5.000	.9428		.9761	
6.000		.9392		
8.000	.9183			
10.000	.9086	.....	.9452	
13.790	.8924			
13.840		.8881		
20.000	.....	.8631	.8936	
25.000	.....	.....	.8615	0.8911

* *J. Chem. Soc.*, pp. 3062–3079 (1931).

Table 187. Sodium Chloride

Mole % NaCl	Specific heat, cal./g. °C. 6°C.	20°C.	33°C.	57°C.
0.249	.....	0.99		
.99	0.96	.97	0.97	
2.44	.91	.915	.915	0.923
9.09	.805	.81	.81	.82

Table 188. Potassium Chloride

Mole % KCl	Specific heat, cal./g. °C. 6°C.	20°C.	33°C.	40°C.
0.99	0.945	0.947	0.947	0.947
3.85	.828	.831	.835	.837
5.66	.77	.775	.778	.775
7.41	.....	.727		

Table 189. Zinc Sulfate

Composition	Temperature	Sp. ht., cal./g. °C.
$ZnSO_4 + 50H_2O$	20° to 52°C.	0.842
$ZnSO_4 + 200H_2O$	20° to 52°C.	.952

Table 190. Copper Sulfate

Composition	Temperature	Sp. ht., cal./g. °C.
$CuSO_4 + 50H_2O$	12° to 15°C.	0.848
$CuSO_4 + 200H_2O$	12° to 14°C.	.951
$CuSO_4 + 400H_2O$	13° to 17°C.	.975

Table 191. Methyl Alcohol

Mole % CH_3OH	Specific heat, cal./g. °C. 5°C.	20°C.	40°C.
5.88	1.02	1.0	0.995
12.3	0.975	0.982	.98
27.3	.877	.917	.92
45.8	.776	.811	.83
69.6	.681	.708	.726
100	.576	.60	.617

Table 192. Ethyl Alcohol

Mole % C_2H_5OH	Specific heat, cal./g. °C. 3°C.	23°C.	41°C.
4.16	1.05	1.02	1.02
11.5	1.02	1.03	1.03
37.0	0.805	0.86	0.875
61.0	.67	.727	.748
100.0	.54	.577	.621

Table 193. Normal Propyl Alcohol

Mole % C_3H_7OH	Specific heat, cal./g. °C. 5°C.	20°C.	40°C.
1.55	1.03	1.02	1.01
5.03	1.07	1.06	1.03
11.4	1.035	1.032	0.99
23.1	0.877	0.90	.91
41.2	.75	.78	.815
73.0	.612	.645	.708
100.0	.534	.57	.621

Table 194. Glycerol

Mole % $C_3H_5(OH)_3$	Specific heat, cal./g. °C. 15°C.	32°C.
2.12	0.961	0.960
4.66	.929	.924
11.5	.851	.841
22.7	.765	.758
43.9	.67	.672
100.0	.555	.576

NOTE.—For the specific heats of non-aqueous solutions, see "International Critical Tables," vol. 5, pp. 116, 125.

Table 195. Aniline (at 20°C.)

Mol % aniline	100	95	90.5	82.3	75.2
Cal./g. °C.	0.497	0.52	0.53	0.56	0.581

SPECIFIC HEATS OF MISCELLANEOUS MATERIALS

Table 196. Specific Heats of Miscellaneous Liquids and Solids

Material	Specific Heat, cal./g. °C.
Alumina	0.2 (100°C.); 0.274 (1500°C.)
Alundum	0.186 (100°C.)
Asbestos	0.25
Asphalt	0.22
Bakelite	0.3 to 0.4
Brickwork	About 0.2
Carbon	0.168 (26° to 76°C.)
	0.314 (40° to 892°C.)
	0.387 (56° to 1450°C.)
(gas retort)	0.204
(See under Graphite)	
Cellulose	0.32
Cement, Portland Clinker	0.186
Charcoal (wood)	0.242
Chrome brick	0.17
Clay	0.224
Coal	0.26 to 0.37
tar oils	0.34 (15° to 90°C.)
Coal tars	0.35 (40°C.); 0.45 (200°C.)
Coke	0.265 (21° to 400°C.)
	0.359 (21° to 800°C.)
	0.403 (21° to 1300°C.)
Concrete	0.156 (70° to 312°F.); 0.219 (72° to 1472°F.)
Cryolite	0.253 (16° to 55°C.)
Diamond	0.147
Fireclay brick	0.198 (100°C.); 0.298 (1500°C.)
Fluorspar	0.21 (30°C.)
Gasoline	0.53
Glass (crown)	0.16 to 0.20
(flint)	0.117
(pyrex)	0.20
(silicate)	0.188 to 0.204 (0 to 100°C.)
	0.24 to 0.26 (0 to 700°C.)
wool	0.157
Granite	0.20 (20° to 100°C.)
Graphite	0.165 (26° to 76°C.); 0.390 (56° to 1450°C.)
Gypsum	0.259 (16° to 46°C.)
Kerosene	0.47
Limestone	0.217
Litharge	0.055
Magnesia	0.234 (100°C.); 0.188 (1500°C.)
Magnesite brick	0.222 (100°C.); 0.195 (1500°C.)
Marble	0.21 (18°C.)
Pyrites (copper)	0.131 (19° to 50°C.)
(iron)	0.136 (15° to 98°C.)
Quartz	0.17 (0°C.); 0.28 (350°C.)
Sand	0.191
Silica	0.316
Steel	0.12
Stone	About 0.2
Turpentine	0.42 (18°C.)
Wood (oak)	0.570
Most woods vary between	0.45 and 0.65

Oils (animal, vegetable, mineral oils)

$$C_p(\text{cal./g. °C.}) = \frac{A}{\sqrt{d_4^{15}}} + B(t - 15)$$

Oils	A	B
Castor	0.500	0.0007
Citron	(0.438 at 54°C.)	
Fatty drying	0.440	0.0007
non-drying	0.450	0.0007
semidrying	0.445	0.0007
oils (except castor)	0.450	0.0007
Naphthene base	0.405	0.0009
Olive	(0.47 at 7°C.)	
Paraffin base	0.425	0.0009
Petroleum oils	0.415	0.0009

Porcelain	Average specific heat between 20°C. and 100°C.	300°C.	500°C.	1100°C.
Fired Berlin	0.189	0.203	0.222	0.337
Green Berlin	.185	.197	.228	
Fired Berlin (glaze)	.179	.189	.199	245
Green Berlin (glaze)	.170	.183	.208	
Fired earthenware	.186	.203	.223	324
Green earthenware	.181	.192	.215	

Material	Specific Heat, cal./g. °C.
Pyrex glass	0.20
Pyroxylin plastics	0.34 to 0.38
Rubber (vulcanized)	0.415
Silica brick	0.202 (100°C.); 0.195 (1500°C.)
Silicon carbide brick	0.202 (100°C.)
Silk	0.33
Stoneware (common)	0.185 to 0.191 (20° to 100°C.)
Wool	0.325
Zirconium oxide	0.11 (100°C.); 0.179 (1500°C.)

HEATS AND FREE ENERGIES OF FORMATION

Table 197. Heats and Free Energies of Formation of Inorganic and Organic Compounds

The values given in the following table for the heats and free energies of formation of inorganic compounds are derived from (*a*) Bichowsky and Rossini, "Thermochemistry of the Chemical Substances," Reinhold, New York, 1936; (*b*) Latimer, "Oxidation States of the Elements and Their Potentials in Aqueous Solution," Prentice-Hall, New York, 1938; (*c*) the tables of the American Petroleum Institute Research Project 44 at the National Bureau of Standards; and (*d*) the tables of Selected Values of Chemical Thermodynamic Properties of the National Bureau of Standards. The reader is referred to the preceding books and tables for additional details as to methods of calculation, standard states, etc. See Sec. 1, page 47, of this book, for the definition of the unit of energy.

The organic compounds in the following table are all given under the element carbon. The values for the non-hydrocarbons are largely from E. I. du Pont de Nemours & Co., Ammonia Department, Chemical Division, Experimental Station; and the values for the hydrocarbons are from the tables of the American Petroleum Institute Research Project 44 at the National Bureau of Standards.*

Compound	State†	Heat of formation‡§ ΔH (formation) at 25°C., kcal./mole	Free energy of formation‖¶ ΔF (formation) at 25°C., kcal./mole
Aluminum:			
Al	c	0.00	0.00
$AlBr_3$	c	−123.4	
	aq	−209.5	−189.2
Al_4C_3	c	−30.8	−29.0
$AlCl_3$	c	−163.8	
	aq, 600	−243.9	−209.5
AlF_3	c	−329	
	aq	−360.8	−312.6
AlI_3	c	−72.8	
	aq	−163.4	−152.5
AlN	c	−57.7	−50.4
$Al(NH_4)(SO_4)_2$	c	−561.19	−486.17
$Al(NH_4)(SO_4)_2.12H_2O$	c	−1419.36	−1179.26
$Al(NO_3)_3.6H_2O$	c	−680.89	−526.32
$Al(NO_3)_3.9H_2O$	c	−897.59	
Al_2O_3	c, corundum	−399.09	−376.87
$Al(OH)_3$	c	−304.8	−272.9
$Al_2O_3.SiO_2$	c, sillimanite	−648.7	
$Al_2O_3.SiO_2$	c, disthene	−642.4	
$Al_2O_3.SiO_2$	c, andalusite	−642.0	
$3Al_2O_3.2SiO_2$	c, mullite	−1874	
Al_2S_3	c	−121.6	
$Al_2(SO_4)_3$	c	−820.99	−739.53
	aq	−893.9	−759.3
$Al_2(SO_4)_3.6H_2O$	c	−1268.15	−1103.39
$Al_2(SO_4)_3.18H_2O$	c	−2120	
Antimony:			
Sb	c	0.00	0.00
$SbBr_3$	c	−59.9	
$SbCl_3$	c	−91.3	−77.8
$SbCl_5$	l	−104.8	
SbF_3	c	−216.6	
SbI_3	c	−22.8	
Sb_2O_3	c, I, orthorhombic	−165.4	−146.0
	c, II, octahedral	−166.6	
Sb_2O_4	c	−213.0	−186.6
Sb_2O_5	c	−230.0	−196.1
Sb_2S_3	c, black	−38.2	−36.9
Arsenic:			
As	c	0.00	0.00
$AsBr_3$	c	−45.9	
$AsCl_3$	l	−80.2	−70.5
AsF_3	l	−223.76	−212.27
AsH_3	g	43.6	37.7
AsI_3	c	−13.6	
As_2O_3	c	−154.1	−134.8
As_2O_5	c	−217.9	−183.9
As_2S_3	c	−20	−20
	amorphous	−34.76	
Barium:			
Ba	c	0.00	0.00
$BaBr_2$	c	−180.38	
	aq, 400	−185.67	−183.0
$BaCl_2$	c	−205.25	
	aq, 300	−207.92	−196.5
$Ba(ClO_3)_2$	c	−176.6	
	aq, 1600	−170.0	−134.4
$Ba(ClO_4)_2$	c	−210.2	
	aq, 800	....	−155.3
$Ba(CN)_2$	c	−48	
$Ba(CNO)_2$	c	−212.1	
	aq	....	−180.7
$BaCN_2$	c	−63.6	
$BaCO_3$	c, witherite	−284.2	−271.4
$BaCrO_4$	c	−342.2	
BaF_2	c	−287.9	
	aq, 1600	−284.6	−265.3
BaH_2	c	−40.8	−31.5
$Ba(HCO_3)_2$	aq	−459	−414.4
BaI_2	c	−144.6	
	aq, 400	−155.17	−158.52
$Ba(IO_3)_2$	c	−264.5	
	aq	−237.50	−198.35
$BaMoO_4$	c	−370	
Ba_3N_2	c	−90.7	
$Ba(NO_2)_2$	c	−184.5	
	aq	−179.05	−150.75
Barium (*Cont.*):			
$Ba(NO_3)_2$	c	−236.99	−189.94
	aq, 600	−227.74	
BaO	c	−133.0	
$Ba(OH)_2$	c	−225.9	
	aq, 400	−237.76	−209.02
$BaO.SiO_2$	c	−363	
$Ba_3(PO_4)_2$	c	−992	
$BaPtCl_6$	c	−284.9	
BaS	c	−111.2	
$BaSO_3$	c	−282.5	
$BaSO_4$	c	−340.2	−313.4
$BaWO_4$	c	−402	
Beryllium:			
Be	c	0.00	0.00
$BeBr_2$	c	−79.4	
	aq	−142	−127.9
$BeCl_2$	c	−112.6	
	aq	−163.9	−141.4
BeI_2	c	−39.4	
	aq	−112	−103.4
Be_3N_2	c	−134.5	−122.4
BeO	c	−145.3	−138.3
$Be(OH)_2$	c	−215.6	
BeS	c	−56.1	
$BeSO_4$	c	−281	
	aq	....	−254.8
Bismuth:			
Bi	c	0.00	0.00
$BiCl_3$	c	−90.5	−76.4
	aq	−101.6	
BiI_3	c	−24	
	aq	−27	
BiO	c	−49.5	−43.2
Bi_2O_3	c	−137.1	−117.9
$Bi(OH)_3$	c	−171.1	
Bi_2S_3	c	−43.9	−39.1
$Bi_2(SO_4)_3$	c	−607.1	
Boron:			
B	c	0.00	0.00
BBr_3	l	−52.7	
	g	−44.6	−50.9
BCl_3	g	−94.5	−90.8
BF_3	g	−265.2	−261.0
B_2H_6	g	7.5	19.9
BN	c	−32.1	−27.2
B_2O_3	c	−302.0	−282.9
	gls	−297.6	−280.3
$B(OH)_3$	c	−260.0	−229.4
B_2S_3	c	−56.6	
Bromine:			
Br_2	l	0.00	0.00
	g	7.47	0.931
BrCl	g	3.06	−0.63
Cadmium:			
Cd	c	0.00	0.00
$CdBr_2$	c	−75.8	−70.7
	aq, 400	−76.6	−67.6
$CdCl_2$	c	−92.149	−81.889
	aq, 400	−96.44	−81.2
$Cd(CN)_2$	c	36.2	
$CdCO_3$	c	−178.2	−163.2
CdI_2	c	−48.40	
	aq, 400	−47.46	−43.22
Cd_3N_2	c	39.8	
$Cd(NO_3)_2$	aq, 400	−115.67	−71.05
CdO	c	−62.35	−55.28
$Cd(OH)_2$	c	−135.0	−113.7
CdS	c	−34.5	−33.6
$CdSO_4$	c	−222.23	
	aq, 400	−232.635	−194.65
Calcium:			
Ca	c	0.00	0.00
$CaBr_2$	c	−162.20	
	aq, 400	−187.19	−181.86
CaC_2	c	−14.8	−16.0
$CaCl_2$	c	−190.6	−179.8
	aq	−209.15	−195.36

* For footnotes see p. 243.

Table 197. Heats and Free Energies of Formation of Inorganic and Organic Compounds—(*Continued*)

Compound	State†	Heat of formation‡§ ΔH (formation) at 25°C., kcal./mole	Free energy of formation‖¶ ΔF (formation) at 25°C., kcal./mole
Calcium (*Cont.*):			
$CaCN_2$	c	−85	
$Ca(CN)_2$	c	−43.3	
	aq	....	−54.0
$CaCO_3$	c, calcite	−289.5	−270.8
	c, aragonite	−289.54	−270.57
$CaCO_3.MgCO_3$	c	−558.8	
CaC_2O_4	c	−332.2	
$Ca(C_2H_3O_2)_2$	c	−356.3	
	aq	−364.1	−311.3
CaF_2	c	−290.2	
	aq	−286.5	−264.1
CaH_2	c	−46	−35.7
CaI_2	c	−128.49	
	aq, 400	−156.63	−157.37
Ca_3N_2	c	−103.2	−88.2
$Ca(NO_3)_2$	c	−224.05	−177.38
	aq, 400	−228.29	
$Ca(NO_3)_2.2H_2O$	c	−367.95	−293.57
$Ca(NO_3)_2.3H_2O$	c	−439.05	−351.58
$Ca(NO_3)_2.4H_2O$	c	−509.43	−409.32
CaO	c	−151.7	−144.3
$Ca(OH)_2$	c	−235.58	−213.9
	aq, 800	−239.2	−207.9
$CaO.SiO_2$	c, II, wollastonite	−377.9	−357.5
	c, I, pseudowollastonite	−376.6	−356.6
CaS	c	−114.3	−113.1
$CaSO_4$	c, insoluble form	−338.73	−311.9
	c, soluble form α	−336.58	−309.8
	c, soluble form β	−335.52	−308.8
$CaSO_4.\frac{1}{2}H_2O$	c	−376.13	
$CaSO_4.2H_2O$	c	−479.33	−425.47
$CaWO_4$	c	−387	
Carbon:			
C	c, graphite	0.00	0.00
	c, diamond	0.453	0.685
CO	g	−26.416	−32.808
CO_2	g	−94.052	−94.260
CH_4 methane	g	−17.889	−12.140
C_2H_6 ethane	g	−20.236	−7.860
C_3H_8 propane	g	−24.820	−5.614
C_4H_{10} n-butane	g	−29.812	−3.754
C_4H_{10} isobutane	g	−31.452	−4.296
C_5H_{12} n-pentane	g	−35.00	−1.96
	l	−41.36	−2.21
C_5H_{12} 2-methylbutane	g	−36.92	−3.50
	l	−42.85	−3.59
C_5H_{12} 2,2-dimethylpropane	g	−39.67	−3.64
C_6H_{14} n-hexane	g	−39.96	0.05
	l	−47.52	−0.91
C_6H_{14} 2-methylpentane	g	−41.66	−0.96
	l	−48.82	−1.73
C_6H_{14} 3-methylpentane	g	−41.02	−0.29
	l	−48.28	−1.12
C_6H_{14} 2,2-dimethylbutane	g	−44.35	−2.35
	l	−51.00	−2.88
C_6H_{14} 2,3-dimethylbutane	g	−42.49	−0.73
	l	−49.48	−1.44
C_7H_{16} n-heptane	g	−44.89	2.09
	l	−53.63	0.42
C_7H_{16} 2-methylhexane	g	−46.60	0.98
	l	−54.93	−0.47
C_7H_{16} 3-methylhexane	g	−45.96	1.10
	l	−54.35	−0.39
C_7H_{16} 3-ethylpentane	g	−45.34	2.59
	l	−53.77	1.06
C_7H_{16} 2,2-dimethylpentane	g	−49.29	0.09
	l	−57.05	−1.08
C_7H_{16} 2,3-dimethylpentane	g	−47.62	0.16
	l	−55.81	−1.27
C_7H_{16} 2,4-dimethylpentane	g	−48.30	0.72
	l	−56.17	−0.49
C_7H_{16} 3,3-dimethylpentane	g	−48.17	0.63
	l	−56.07	−0.69
C_7H_{16} 2,2,3-trimethylbutane	g	−48.96	0.76
	l	−56.63	−0.43
C_8H_{18} n-octane	g	−49.82	4.14
	l	−59.74	1.77
C_8H_{18} 2-methylheptane	g	−51.50	3.06
	l	−60.98	0.92
C_8H_{18} 3-methylheptane	g	−50.82	3.29
	l	−60.34	1.12
C_8H_{18} 4-methylheptane	g	−50.69	4.00
	l	−60.17	1.86
C_8H_{18} 3-ethylhexane	g	−50.40	3.95
	l	−59.88	1.80

Compound	State†	Heat of formation‡§ ΔH (formation) at 25°C., kcal./mole	Free energy of formation‖¶ ΔF (formation) at 25°C., kcal./mole
Carbon (*Cont.*):			
C_8H_{18} 2,2-dimethylhexane	g	−53.71	2.56
	l	−62.63	−0.72
C_8H_{18} 2,3-dimethylhexane	g	−51.13	4.23
	l	−60.40	2.17
C_8H_{18} 2,4-dimethylhexane	g	−52.44	2.80
	l	−61.47	0.89
C_8H_{18} 2,5-dimethylhexane	g	−53.21	2.50
	l	−62.26	0.59
C_8H_{18} 3,3-dimethylhexane	g	−52.61	3.17
	l	−61.58	1.23
C_8H_{18} 3,4-dimethylhexane	g	−50.91	4.97
	l	−60.23	2.86
C_8H_{18} 2-methyl-3-ethylpentane	g	−50.48	5.08
	l	−59.69	3.03
C_8H_{18} 3-methyl-3-ethylpentane	g	−51.38	4.76
	l	−60.46	2.69
C_8H_{18} 2,2,3-trimethylpentane	g	−52.61	4.09
	l	−61.44	2.22
C_8H_{18} 2,2,4-trimethylpentane	g	−53.57	3.13
	l	−61.97	1.51
C_8H_{18} 2,3,3,-trimethylpentane	g	−51.73	4.52
	l	−60.63	2.54
C_8H_{18} 2,3,4-trimethylpentane	g	−51.97	4.32
	l	−60.98	2.34
C_8H_{18} 2,2,3,3,-tetramethylbutane	g	−53.99	4.88
	c	−64.23	2.74
C_2H_4 ethylene	g	12.496	16.282
C_3H_6 propylene	g	4.879	14.964
C_4H_8 1-butene	g	0.280	17.217
C_4H_8 cis-2-butene	g	−1.362	16.007
C_4H_8 trans-2-butene	g	−2.405	15.323
C_4H_8 2-methyl-2-propene	g	−3.343	14.574
C_5H_{10} 1-pentene	g	−5.000	18.787
C_5H_{10} cis-2-pentene	g	−6.710	17.173
C_5H_{10} trans-2-pentene	g	−7.590	16.575
C_5H_{10} 2-methyl-1-butene	g	−8.680	15.509
C_5H_{10} 3-methyl-1-butene	g	−6.920	17.874
C_5H_{10} 2-methyl-2-butene	g	−10.170	14.267
C_2H_2 acetylene	g	54.194	50.000
C_3H_4 methylacetylene	g	44.319	46.313
C_4H_6 1-butyne	g	39.70	48.52
C_4H_6 2-butyne	g	35.374	44.725
C_5H_8 1-pentyne	g	34.50	50.17
C_5H_8 2-pentyne	g	30.80	46.41
C_5H_8 3-methyl-1-butyne	g	32.60	49.12
C_6H_6 benzene	g	19.820	30.989
C_7H_8 toluene	l	11.718	29.756
	g	11.950	29.228
	l	2.867	27.282
C_8H_{10} ethylbenzene	g	7.120	31.208
	l	−2.977	28.614
C_8H_{10} o-xylene	g	4.540	29.177
	l	−5.841	26.370
C_8H_{10} m-xylene	g	4.120	28.405
	l	−6.075	25.730
C_8H_{10} p-xylene	g	4.290	28.952
	l	−5.838	26.310
C_9H_{12} n-propylbenzene	g	1.870	32.810
	l	−9.178	29.600
C_9H_{12} isopropylbenzene	g	0.940	32.738
	l	−9.848	29.708
C_9H_{12} 1-methyl-2-ethylbenzene	g	0.290	31.323
	l	−11.110	27.973
C_9H_{12} 1-methyl-3-ethylbenzene	g	−0.460	30.217
	l	−11.670	26.977
C_9H_{12} 1-methyl-4-ethylbenzene	g	−0.780	30.281
	l	−11.920	27.041
C_9H_{12} 1,2,3-trimethylbenzene	g	−2.290	29.319
	l	−14.013	25.679
C_9H_{12} 1,2,4-trimethylbenzene	g	−3.330	27.912
	l	−14.785	24.462
C_9H_{12} 1,3,5-trimethylbenzene	g	−3.840	28.172
	l	−15.184	24.832
C_5H_{10} cyclopentane	g	−18.46	9.23
	l	−25.31	8.70
C_6H_{12} methylcyclopentane	g	−25.50	8.55
	l	−33.08	7.53
C_7H_{14} ethylcyclopentane	g	−30.38	10.59
	l	−39.09	8.84

Table 197. Heats and Free Energies of Formation of Inorganic and Organic Compounds—(*Continued*)

Compound	State†	Heat of formation‡§ ΔH (formation) at 25°C., kcal./mole	Free energy of formation‖¶ ΔF (formation) at 25°C., kcal./mole
Carbon (*Cont.*):			
C_6H_{12} cyclohexane	g	−29.43	7.59
	l	−37.34	6.39
C_7H_{14} methylcyclohexane	g	−37.00	6.52
	l	−45.46	4.86
C_8H_{16} ethylcyclohexane	g	−41.06	9.38
	l	−50.73	6.96
CH_4O methanol	g	−48.08	−38.62
	l	−57.04	−39.80
C_2H_6O ethanol	g	−52.23	−40.23
	l	−66.35	−41.76
C_3H_8O *n*-propanol	g	−61.17	−38.83
	l	−71.87	−39.84
C_3H_8O isopropanol	g	−62.41	−38.20
	l	−74.32	−38.83
$C_4H_{10}O$ *n*-butanol	g	−67.81	−38.88
	l	−79.61	−40.37
$C_4H_{10}O$ isobutanol	g	−69.05	−38.25
	l	−81.06	−39.36
$C_2H_6O_2$ ethylene glycol	g	−92.53	−71.26
	l	−107.91	−76.44
$C_3H_8O_3$ glycerol	g		
	l	−159.16	−113.65
C_6H_6O phenol	g	−21.71	−6.26
	l	−37.80	−11.02
C_7H_8O cresol	g	....	−13.17
C_2H_4O ethylene oxide	g	−16.1	−6.94
C_2H_6O dimethyl ether	g	−43.06	−26.06
	l	−51.3	
$C_4H_{10}O$ diethyl ether	l	−65.2	−27.75
CH_2O formaldehyde	g	−28.29	−26.88
C_2H_4O acetaldehyde	g	−39.72	−31.46
C_3H_4O acrolein	g	−20.50	−15.57
	l	−27.97	−16.17
C_3H_6O propionaldehyde	g	−49.15	−33.96
C_4H_8O *n*-butyraldehyde	g	−52.40	−73.24
C_7H_6O benzaldehyde	g	−9.57	5.85
	l	−21.23	2.24
C_8H_8O *p*-toluic aldehyde	g	−17.78	4.09
	l	−29.79	0.97
C_2H_2O ketene	g	−14.78	−14.30
	l	−18.78	−13.32
C_3H_6O acetone	g	−51.79	−36.45
	l	−59.32	−37.16
$C_5H_{10}O$ diethylketone	l	−73.8	
CH_2O_2 formic acid	g	−86.67	−80.24
	l	−97.8	−82.7
½$(CH_2O_2)_2$ bimolecular formic acid	g	−93.85	−81.90
$C_2H_4O_2$ acetic acid	g	−104.72	−91.24
	l	−116.2	−93.56
$C_3H_6O_2$ propionic acid	g	−108.75	−88.27
	l	−121.7	−91.65
$C_2H_4O_3$ hydroxyacetic acid	l	−155.33	−125.57
$C_6H_{10}O_4$ adipic acid	g	−216.19	−163.96
	l	−235.51	−177.17
$C_2H_4O_2$ methyl formate	g	−84.69	−71.37
	l	−95.26	−71.53
$C_4H_6O_2$ methyl acrylate	g	−70.10	−56.78
	l	−82.76	−58.13
$C_4H_8O_2$ ethyl acetate	g	−102.02	−74.93
	l	−110.72	−76.11
$C_5H_{10}O_2$ ethyl propionate	g	−112.36	−77.37
	l	−122.16	−79.16
$C_4H_6O_3$ acetic anhydride	g	−148.82	−119.29
	l	−155.16	−121.75
$C_6H_{10}O_3$ propionic anhydride	g	−147.32	−109.78
	l	−161.53	−113.66
CS_2 carbon disulfide	g	28.11	16.13
COS carbonyl sulfide	g	−33.83	−40.85
C_2N_2 cyanogen	g	73.82	71.02
HCN hydrogen cyanide	g	31.1	27.94
	l	25.2	29.0
	aq, 100	25.2	26.8
C_2H_3N acetonitrile	g	19.81	
CH_5N methylamine	g	−6.7	6.6
C_2H_7N ethylamine	g	−12.24	10.01
C_3H_9N propylamine	g	−16.45	14.38
$C_4H_{11}N$ butylamine	g	−15.60	19.55
$C_6H_{13}N$ hexamethyleneimine	g	−14.37	31.52
	l	−24.90	28.84
CH_2N_2 cyanamide	l	11.18	24.30
	c	9.15	24.18
$C_6H_8N_2$ adiponitrile	g	33.34	61.43
	l	19.19	54.63
$C_6H_{16}N_2$ hexamethyienediamine	g	−30.57	28.91
CH_5N_3 guanidine	l	−27.48	7.34
	c	−30.68	6.33

Compound	State†	Heat of formation‡§ ΔH (formation) at 25°C., kcal./mole	Free energy of formation‖¶ ΔF (formation) at 25°C., kcal./mole
Carbon (*Cont.*):			
$C_3H_6N_6$ melamine	l	−19.33	40.80
CH_3NO formamide	g	−44.64	−36.60
C_2H_7NO ethanolamine	l	−62.52	27.50
CH_4N_2O urea	l	−77.55	−46.45
	c	−79.634	−47.118
Cerium:			
Ce	c	0.00	0.00
CeN	c	−78.2	−70.8
Cesium:			
Cs	c	0.00	0.00
CsBr	c	−97.64	
	aq, 500	−91.39	−94.86
CsCl	c	−106.31	
	aq, 400	−102.01	−101.61
Cs_2CO_3	c	−271.88	
CsF	c	−131.67	
	aq, 400	−140.48	−135.98
CsH	c	−12	−7.30
$CsHCO_3$	c	−230.6	
	aq, 2000	−226.6	−210.56
CsI	c	−83.91	
	aq, 400	−75.74	−82.61
$CsNH_2$	c	−28.2	
$CsNO_3$	c	−121.14	
	aq, 400	−111.54	−96.53
Cs_2O	c	−82.1	
CsOH	c	−100.2	
	aq, 200	−117.0	−107.87
Cs_2S	c	−87	
Cs_2SO_4	c	−344.86	
	aq	−340.12	−316.66
Chlorine:			
Cl_2	g	0.00	0.00
ClF	g	−25.7	
ClO	g	33	
ClO_2	g	24.7	29.5
ClO_3	g	37	
Cl_2O	g	18.20	22.40
Cl_2O_7	g	63	
Chromium:			
Cr	c	0.00	0.00
$CrBr_3$	aq	....	−122.7
Cr_3C_2	c	−21.008	−21.20
Cr_4C	c	−16.378	−16.74
$CrCl_2$	c	−103.1	−93.8
	aq	....	−102.1
CrF_2	c	−152	
CrF_3	c	−231	
CrI_2	c	−63.7	
	aq	....	−64.1
CrO_3	c	−139.3	
Cr_2O_3	c	−268.8	−249.3
$Cr_2(SO_4)_3$	aq	....	−626.3
Cobalt:			
Co	c	0.00	0.00
$CoBr_2$	c	−55.0	
	aq	−73.61	−61.96
Co_3C	c	9.49	7.08
$CoCl_2$	c	−76.9	−66.6
	aq, 400	−95.58	−75.46
$CoCO_3$	c	−172.39	−155.36
CoF_2	aq	−172.98	−144.2
CoI_2	c	−24.2	
	aq	−43.15	−37.4
$Co(NO_3)_2$	c	−102.8	
	aq	−114.9	−65.3
CoO	c	−57.5	
Co_3O_4	c	−196.5	
$Co(OH)_2$	c	−131.5	−108.9
$Co(OH)_3$	c	−177.0	−142.0
CoS	c	−22.3	−19.8
Co_2S_3	c	−40.0	
$CoSO_4$	c	−216.6	
	aq, 400	....	−188.9
Columbium:			
Cb	c	0.00	0.00
Cb_2O_5	c	−462.96	
Copper:			
Cu	c	0.00	0.00
CuBr	c	−26.7	−23.8
$CuBr_2$	c	−34.0	
	aq	−42.4	−33.25
CuCl	c	−31.4	−24.13
$CuCl_2$	c	−48.83	
	aq, 400	−64.7	
$CuClO_4$	aq	−28.3	1.34
$Cu(ClO_3)_2$	aq, 400	....	15.4
$Cu(ClO_4)_2$	aq	....	−5.5

Table 197. Heats and Free Energies of Formation of Inorganic and Organic Compounds—*(Continued)*

Compound	State†	Heat of formation‡§ ΔH (formation) at 25°C., kcal./mole	Free energy of formation‖¶ ΔF (formation) at 25°C., kcal./mole
Copper (*Cont.*):			
CuI	c	−17.8	−16.66
CuI_2	c	−4.8	
	aq	−11.9	−8.76
Cu_3N	c	17.78	
$Cu(NO_3)_2$	c	−73.1	
	aq, 200	−83.6	−36.6
CuO	c	−38.5	−31.9
Cu_2O	c	−43.00	−38.13
$Cu(OH)_2$	c	−108.9	−85.5
CuS	c	−11.6	−11.69
Cu_2S	c	−18.97	−20.56
$CuSO_4$	c	−184.7	−158.3
	aq, 800	−200.78	−160.19
Cu_2SO_4	c	−179.6	
	aq	....	−152.0
Erbium:			
Er	c	0.00	0.00
$Er(OH)_3$	c	−326.8	
Fluorine:			
F_2	g	0.00	0.00
F_2O	g	5.5	9.7
Gallium:			
Ga	c	0.00	0.00
$GaBr_3$	c	−92.4	
$GaCl_3$	c	−125.4	
GaN	c	−26.2	
Ga_2O	c	−84.3	
Ga_2O_3	c	−259.9	
Germanium:			
Ge	c	0.00	0.00
Ge_3N_4	c	−15.7	
GeO_2	c	−128.6	
Gold:			
Au	c	0.00	0.00
$AuBr$	c	−3.4	
$AuBr_3$	c	−14.5	
	aq	−11.0	24.47
$AuCl$	c	−8.3	
$AuCl_3$	c	−28.3	
	aq	−32.96	4.21
AuI	c	0.2	−0.76
Au_2O_3	c	11.0	18.71
$Au(OH)_3$	c	−100.6	
Hafnium:			
Hf	c	0.00	0.00
HfO_2	c	−271.1	−258.2
Hydrogen:			
H_3AsO_3	aq	−175.6	−153.04
H_3AsO_4	c	−214.9	
	aq	−214.8	−183.93
HBr	g	−8.66	−12.72
	aq, 400	−28.80	−24.58
$HBrO$	aq	−25.4	−19.90
$HBrO_3$	aq	−11.51	5.00
HCl	g	−22.063	−22.778
	aq, 400	−39.85	−31.330
HCN	g	31.1	27.94
	aq, 100	24.2	26.55
$HClO$	aq, 400	−28.18	−19.11
$HClO_3$	aq	−23.4	−0.25
$HClO_4$	aq, 660	−31.4	−10.70
$HC_2H_3O_2$	l	−116.2	−93.56
	aq, 400	−116.74	−96.8
$H_2C_2O_4$	c	−196.7	
	aq, 300	−194.6	−165.64
$HCOOH$	l	−97.8	−82.7
	aq, 200	−98.0	−85.1
H_2CO_3	aq	−167.19	−149.0
HF	g	−64.2	−64.7
	aq, 200	−75.75	
HI	g	6.27	0.365
	aq, 400	−13.47	−12.35
HIO	aq	−38	−23.33
HIO_3	c	−56.77	
	aq	−54.8	−32.25
HN_3	g	70.3	78.50
HNO_3	g	−31.99	−17.57
	l	−41.35	−19.05
	aq, 400	−49.210	
$HNO_3.H_2O$	l	−112.91	−78.36
$HNO_3.3H_2O$	l	−252.15	−193.70
H_2O	g	−57.7979	−54.6351
	l	−68.3174	−56.6899
H_2O_2	l	−45.16	−28.23
	aq, 200	−45.80	−31.47
H_3PO_2	c	−145.5	
	aq	−145.6	−120.0
H_3PO_3	c	−232.2	
	aq	−232.2	−204.0

Compound	State†	Heat of formation‡§ ΔH (formation) at 25°C., kcal./mole	Free energy of formation‖¶ ΔF (formation) at 25°C., kcal./mole
Hydrogen (*Cont.*):			
H_3PO_4	c	−306.2	
	aq, 400	−309.32	−270.0
H_2S	g	−4.77	−7.85
	aq, 2000	−9.38	
H_2S_2	l	−3.6	
H_2SO_3	aq, 200	−146.88	−128.54
H_2SO_4	l	−193.69	
	aq, 400	−212.03	
H_2Se	g	20.5	17.0
	aq	18.1	18.4
H_2SeO_3	c	−126.5	
	aq	−122.4	−101.36
H_2SeO_4	c	−130.23	
	aq, 400	−143.4	
H_2SiO_3	c	−267.8	−247.9
H_4SiO_4	c	−340.6	
H_2Te	g	36.9	33.1
H_2TeO_3	c	−145.0	−115.7
	aq	−145.0	
H_2TeO_4	aq	−165.6	
Indium:			
In	c	0.00	0.00
$InBr_3$	c	−97.2	
	aq	−112.9	−97.2
$InCl_3$	c	−128.5	
	aq	−145.6	−117.5
InI_3	c	−56.5	
	aq	−67.2	−60.5
InN	c	−4.8	
In_2O_3	c	−222.47	
Iodine:			
I_2	c	0.00	0.00
	g	14.88	4.63
IBr	g	10.05	1.24
ICl	g	4.20	−1.32
ICl_3	c	−21.8	−6.05
I_2O_5	c	−42.5	
Iridium:			
Ir	c	0.00	0.00
$IrCl$	c	−20.5	−16.9
$IrCl_2$	c	−40.6	−32.0
$IrCl_3$	c	−60.5	−46.5
IrF_6	l	−130	
IrO_2	c	−40.14	
Iron:			
Fe	c, α	0.00	0.00
$FeBr_2$	c	−57.15	
	aq, 540	−78.7	−69.47
$FeBr_3$	aq	−95.5	−76.26
Fe_3C	c	5.69	4.24
$Fe(CO)_5$	l	−187.6	
$FeCO_3$	c, siderite	−172.4	−154.8
$FeCl_2$	c	−81.9	−72.6
	aq	−100.0	−83.0
$FeCl_3$	c	−96.4	
	aq, 2000	−128.5	−96.5
FeF_2	aq, 1200	−177.2	−151.7
FeI_2	c	−24.2	
	aq	−47.7	−45
FeI_3	aq	−49.7	−39.5
Fe_4N	c	−2.55	0.862
$Fe(NO_3)_2$	aq	−118.9	−72.8
$Fe(NO_3)_3$	aq, 800	−156.5	−81.3
FeO	c	−64.62	−59.38
Fe_2O_3	c	−198.5	−179.1
Fe_3O_4	c	−266.9	−242.3
$Fe(OH)_2$	c	−135.9	−115.7
$Fe(OH)_3$	c	−197.3	−166.3
$FeO.SiO_2$	c	−273.5	
Fe_2P	c	−13	
$FeSi$	c	−19.0	
FeS	c	−22.64	−23.23
FeS_2	c, pyrites	−38.62	−35.93
	c, marcasite	−33.0	
$FeSO_4$	c	−221.3	−195.5
	aq, 400	−236.2	−196.4
$Fe_2(SO_4)_3$	aq, 400	−653.3	−533.4
$FeTiO_3$	c, ilmenite	−295.51	−277.06
Lanthanum:			
La	c	0.00	0.00
$LaCl_3$	c	−253.1	
	aq	−284.7	
La_3H_8	c	−160	
LaN	c	−72.0	−64.6
La_2O_3	c	−539	
LaS_2	c	−148.3	
La_2S_3	c	−351.4	
$La_2(SO_4)_3$	aq	−972	

Table 197. Heats and Free Energies of Formation of Inorganic and Organic Compounds—(*Continued*)

Compound	State†	Heat of formation‡§ ΔH (formation) at 25°C., kcal./mole	Free energy of formation‖¶ ΔF (formation) at 25°C., kcal./mole
Lead:			
Pb	c	0.00	0.00
$PbBr_2$	c	−66.24	−62.06
	aq	−56.4	−54.97
$PbCO_3$	c, cerussite	−167.6	−150.0
$Pb(C_2H_3O_2)_2$	c	−232.6	
	aq, 400	−234.2	−184.40
PbC_2O_4	c	−205.3	
$PbCl_2$	c	−85.68	−75.04
	aq	−82.5	−68.47
PbF_2	c	−159.5	−148.1
PbI_2	c	−41.77	−41.47
$Pb(NO_3)_2$	c	−106.88	
	aq, 400	−99.46	−58.3
PbO	c, red	−51.72	−45.53
	c, yellow	−50.86	−43.88
PbO_2	c	−65.0	−52.0
Pb_3O_4	c	−172.4	−142.2
$Pb(OH)_2$	c	−123.0	−102.2
PbS	c	−22.38	−21.98
$PbSO_4$	c	−218.5	−192.9
Lithium:			
Li	c	0.00	0.00
$LiBr$	c	−83.75	
	aq, 400	−95.40	−95.28
$LiBrO_3$	aq	−77.9	−65.70
Li_2C_2	c	−13.0	
$LiCN$	aq	−31.4	−31.35
$LiCNO$	aq	−101.2	−94.12
$LiC_2H_3O_2$	aq	−183.9	−160.00
Li_2CO_3	c	−289.7	−269.8
	aq, 1900	−293.1	−267.58
$LiCl$	c	−97.63	
	aq, 278	−106.45	−102.03
$LiClO_3$	aq	−87.5	−70.95
$LiClO_4$	aq	−106.3	−81.4
LiF	c	−145.57	
	aq, 400	−144.85	−136.40
LiH	c	−22.9	
$LiHCO_3$	aq, 2000	−231.1	−210.98
LiI	c	−65.07	
	aq, 400	−80.09	−83.03
$LiIO_3$	aq	−121.3	−102.95
Li_3N	c	−47.45	−37.33
$LiNO_3$	c	−115.350	
	aq, 400	−115.88	−96.95
Li_2O	c	−142.3	
Li_2O_2	c	−151.9	−138.0
	aq	−159	
$LiOH$	c	−116.58	−106.44
	aq, 400	−121.47	−108.29
$LiOH.H_2O$	c	−188.92	
$Li_2O.SiO_2$	gls	−374	
Li_2Se	c	−84.9	
	aq	−95.5	−105.64
Li_2SO_4	c	−340.23	−314.66
	aq, 400	−347.02	
$Li_2SO_4.H_2O$	c	−411.57	−375.07
Magnesium:			
Mg	c	0.00	0.00
$Mg(AsO_4)_2$	c	−731.3	
	aq	−749	−630.14
$MgBr_2$	c	−123.9	
	aq, 400	−167.33	−156.94
$Mg(CN)_2$	aq	−39.7	−29.08
$MgCN_2$	c	−61	
$Mg(C_2H_3O_2)_2$	aq	−344.6	−286.38
$MgCO_3$	c	−261.7	−241.7
$MgCl_2$	c	−153.220	−143.77
	aq, 400	−189.76	
$MgCl_2.H_2O$	c	−230.970	−205.93
$MgCl_2.2H_2O$	c	−305.810	−267.20
$MgCl_2.4H_2O$	c	−453.820	−387.98
$MgCl_2.6H_2O$	c	−597.240	−505.45
MgF_2	c	−263.8	
MgI_2	c	−86.8	
	aq, 400	−136.79	−132.45
$MgMoO_4$	c	−329.9	
Mg_3N_2	c	−115.2	−100.8
$Mg(NO_3)_2$	c	−188.770	−140.66
	aq, 400	−209.927	−160.28
$Mg(NO_3)_2.2H_2O$	c	−336.625	
$Mg(NO_3)_2.6H_2O$	c	−624.48	−496.03
MgO	c	−143.84	−136.17
$MgO.SiO_2$	c	−347.5	−326.7
$Mg(OH)_2$	c, ppt.	−221.90	−200.17
	c, brucite	−223.9	−193.3
MgS	c	−84.2	
	aq	−108	

Compound	State†	Heat of formation‡§ ΔH (formation) at 25°C., kcal./mole	Free energy of formation‖¶ ΔF (formation) at 25°C., kcal./mole
Magnesium (*Cont.*):			
$MgSO_4$	c	−304.94	−277.7
	aq, 400	−325.4	−283.88
$MgTe$	c	−25	
$MgWO_4$	c	−345.2	
Manganese:			
Mn	c, α	0.00	0.00
$MnBr_2$	c	−91	
	aq	−106	−97.8
Mn_3C	c	1.1	1.26
$Mn(C_2H_3O_2)_2$	c	−270.3	
	aq	−282.7	−227.2
$MnCO_3$	c	−211	−192.5
MnC_2O_4	c	−240.9	
$MnCl_2$	c	−112.0	−102.2
	aq, 400	−128.9	
MnF_2	aq, 1200	−206.1	−180.0
MnI_2	c	−49.8	
	aq	−76.2	−73.3
Mn_5N_2	c	−57.77	−46.49
$Mn(NO_3)_2$	c	−134.9	
	aq, 400	−148.0	−101.1
$Mn(NO_3)_2.6H_2O$	c	−557.07	−441.2
MnO	c	−92.04	−86.77
MnO_2	c	−124.58	−111.49
Mn_2O_3	c	−229.5	−209.9
Mn_3O_4	c	−331.65	−306.22
$MnO.SiO_2$	c	−301.3	−282.1
$Mn(OH)_2$	c	−163.4	−143.1
$Mn(OH)_3$	c	−221	−190
$Mn_3(PO_4)_2$	c	−736	
$MnSe$	c	−26.3	−27.5
MnS	c, green	−47.0	−48.0
$MnSO_4$	c	−254.18	−228.41
	aq, 400	−265.2	
$Mn_2(SO_4)_3$	c	−635	
	aq	−657	
Mercury:			
Hg	l	0.00	0.00
$HgBr$	g	23	18
$HgBr_2$	c	−40.68	−38.8
	aq	−38.4	−9.74
$Hg(C_2H_3O_2)_2$	c	−196.3	
	aq	−192.5	−139.2
$HgCl_2$	c	−53.4	−42.2
	aq	−50.3	−23.25
$HgCl$	g	19	14
Hg_2Cl_2	c	−63.13	
$Hg(CN)_2$	c	62.8	
	aq, 1110	66.25	
HgC_2O_4	c	−159.3	
HgH	g	57.1	52.25
HgI_2	c, red	−25.3	−24.0
HgI	g	33	23
Hg_2I_2	c	−28.88	−26.53
$Hg(NO_3)_2$	aq	−56.8	−13.09
$Hg_2(NO_3)_2$	aq	−58.5	−15.65
HgO	c, red	−21.6	−13.94
	c, yellow ppt.	−20.8	
Hg_2O	c	−21.6	−12.80
HgS	c, black	−10.7	−8.80
$HgSO_4$	c	−166.6	
Hg_2SO_4	c	−177.34	−149.12
Molybdenum:			
Mo	c	0.00	0.00
Mo_2C	c	4.36	2.91
Mo_2N	c	−8.3	
MoO_2	c	−130	−118.0
MoO_3	c	−180.39	−162.01
MoS_2	c	−56.27	−54.19
MoS_3	c	−61.48	−57.38
Nickel:			
Ni	c	0.00	0.00
$NiBr_2$	c	−53.4	
	aq	−72.6	−60.7
Ni_3C	c	9.2	8.88
$Ni(C_2H_3O_2)_2$	aq	−249.6	−190.1
$Ni(CN)_2$	aq	230.9	66.3
$NiCl_2$	c	−75.0	
	aq, 400	−94.34	−74.19
NiF_2	c	−157.5	
	aq	−171.6	−142.9
NiI_2	c	−22.4	
	aq	−42.0	−36.2
$Ni(NO_3)_2$	c	−101.5	
	aq, 200	−113.5	−64.0
NiO	c	−58.4	−51.7
$Ni(OH)_2$	c	−129.8	−105.6
$Ni(OH)_3$	c	−163.2	

Table 197. Heats and Free Energies of Formation of Inorganic and Organic Compounds—(*Continued*)

Compound	State†	Heat of formation‡§ ΔH (formation) at 25°C., kcal./mole	Free energy of formation‖¶ ΔF (formation) at 25°C., kcal./mole
Nickel (*Cont.*):			
NiS	c	−20.4	
$NiSO_4$	c	−216	
	aq, 200	−231.3	−187.6
Nitrogen:			
N_2	g	0.00	0.00
NF_3	g	−27	
NH_3	g	−10.96	−3.903
	aq, 200	−19.27	
NH_4Br	c	−64.57	
	aq	−60.27	−43.54
$NH_4C_2H_3O_2$	c	−148.1	
	aq, 400	−148.58	−108.26
NH_4CN	c	−0.7	
	aq	3.6	20.4
NH_4CNS	c	−17.8	
	aq	−12.3	4.4
$(NH_4)_2CO_3$	aq	−223.4	−164.1
$(NH_4)_2C_2O_4$	c	−266.3	
	aq	−260.6	−196.2
NH_4Cl	c	−75.23	−48.59
	aq, 400	−71.20	
NH_4ClO_4	c	−69.4	
	aq	−63.2	−21.1
$(NH_4)_2CrO_4$	c	−276.9	
	aq	−271.3	−209.3
NH_4F	c	−111.6	
	aq	−110.2	−84.7
NH_4I	c	−48.43	
	aq	−44.97	−31.3
NH_4NO_3	c	−87.40	
	aq, 500	−80.89	
NH_4OH	aq	−87.59	
$(NH_4)_2S$	aq, 400	−55.21	−14.50
$(NH_4)_2SO_4$	c	−281.74	−215.06
	aq, 400	−279.33	−214.02
N_2H_4	l	12.06	
$N_2H_4.H_2O$	l	−57.96	
$N_2H_4.H_2SO_4$	c	−232.2	
N_2O	g	19.55	24.82
NO	g	21.600	20.719
NO_2	g	7.96	12.26
N_2O_4	g	2.23	23.41
N_2O_5	c	−10.0	
$NOBr$	l	11.6	19.26
$NOCl$	g	12.8	16.1
Osmium:			
Os	c	0.00	0.00
OsO_4	c	−93.6	−70.9
	g	−80.1	−68.1
Oxygen:			
O_2	g	0.00	0.00
O_3	g	33.88	38.86
Palladium:			
Pd	c	0.00	0.00
PdO	c	−20.40	
Phosphorus:			
P	c, white ("yellow")	0.00	0.00
	c, red ("violet")	−4.22	−1.80
P	g	150.35	141.88
P_2	g	33.82	24.60
P_4	g	13.2	5.89
PBr_3	l	−45	
PBr_5	c	−60.6	
PCl_3	g	−70.0	−65.2
	l	−76.8	−63.3
PCl_5	g	−91.0	−73.2
PH_3	g	2.21	−1.45
PI_3	c	−10.9	
P_2O_5	c	−360.0	
$POCl_3$	g	−138.4	−127.2
Platinum:			
Pt	c	0.00	0.00
$PtBr_4$	c	−40.6	
	aq	−50.7	
$PtCl_2$	c	−34	
$PtCl_4$	c	62.6	
	aq	−82.3	
PtI_4	c	−18	
$Pt(OH)_2$	c	−87.5	−67.9
PtS	c	−20.18	−18.55
PtS_2	c	−26.64	−24.28
Potassium:			
K	c	0.00	0.00
K_3AsO_3	aq	−323.0	
K_3AsO_4	aq	−390.3	−355.7
KH_2AsO_4	c	−271.2	−236.7
KBr	c	−94.06	−90.8
	aq, 400	−89.19	−92.0
Potassium (*Cont.*):			
$KBrO_3$	c	−81.58	−60.30
	aq, 1667	−71.68	
$KC_2H_3O_2$	c	−173.80	
	aq, 400	−177.38	−156.73
KCl	c	−104.348	−97.76
	aq, 400	−100.164	−98.76
$KClO_3$	c	−93.5	−69.30
	aq, 400	−81.34	
$KClO_4$	c	−103.8	−72.86
	aq, 400	−101.14	
KCN	c	−28.1	
	aq, 400	−25.3	−28.08
$KCNO$	c	−99.6	
	aq	−94.5	−90.85
$KCNS$	c	−47.0	
	aq, 400	−41.07	−44.08
K_2CO_3	c	−274.01	
	aq, 400	−280.90	−264.04
$K_2C_2O_4$	c	−319.9	
	aq, 400	−315.5	−293.1
K_2CrO_4	c	−333.4	
	aq, 400	−328.2	−306.3
$K_2Cr_2O_7$	c	−488.5	
	aq, 400	−472.1	−440.9
KF	c	−134.50	
	aq, 180	−138.36	−133.13
$K_3Fe(CN)_6$	c	−48.4	
	aq	−34.5	
$K_4Fe(CN)_6$	c	−131.8	
	aq	−119.9	
KH	c	−10	−5.3
$KHCO_3$	c	−229.8	
	aq, 2000	−224.85	−207.71
KI	c	−78.88	−77.37
	aq, 500	−73.95	−79.76
KIO_3	c	−121.69	−101.87
	aq, 400	−115.18	−99.68
KIO_4	aq	−98.1	
$KMnO_4$	c	−192.9	−169.1
	aq, 400	−182.5	−168.0
K_2MoO_4	aq, 880	−364.2	−342.9
KNH_2	c	−28.25	
KNO_2	aq	−86.0	−75.9
KNO_3	c	−118.08	−94.29
	aq, 400	−109.79	−93.68
K_2O	c	−86.2	
$K_2O.Al_2O_3.4H_2O$	c, leucite	−1379.6	
	gls	−1368.2	
$K_2O.Al_2O_3.6H_2O$	c, adularia	−1810.7	
	c, microcline	−1784.5	
	gls	−1747	
KOH	c	−102.02	
	aq, 400	−114.96	−105.0
K_3PO_3	aq	−397.5	
K_3PO_4	aq	−478.7	−443.3
KH_2PO_4	c	−362.7	−326.1
K_2PtCl_4	c	−254.7	
	aq	−242.6	−226.5
K_2PtCl_6	c	−299.5	−263.6
	aq, 9400	−286.1	
K_2Se	c	−74.4	
	aq	−83.4	−99.10
K_2SeO_4	aq	−267.1	−240.0
K_2S	c	−121.5	
	aq, 400	−110.75	−111.44
K_2SO_3	c	−267.7	
	aq	−269.7	−251.3
K_2SO_4	c	−342.65	−314.62
	aq, 400	−336.48	−310.96
$K_2SO_4.Al_2(SO_4)_3$	c	−1178.38	−1068.48
$K_2SO_4.Al_2(SO_4)_3.24H_2O$	c	−2895.44	−2455.68
$K_2S_2O_6$	c	−418.62	
Rhenium:			
Re	c	0.00	0.00
ReF_6	g	−274	
Rhodium:			
Rh	c	0.00	0.00
RhO	c	−21.7	
Rh_2O	c	−22.7	
Rh_2O_3	c	−68.3	
Rubidium:			
Rb	c	0.00	0.00
$RbBr$	c	−95.82	
	g	−45.0	−52.50
	aq, 500	−90.54	−93.38
$RbCN$	aq	−25.9	
Rb_2CO_3	c	−273.22	
	aq, 220	−282.61	−263.78

Table 197. Heats and Free Energies of Formation of Inorganic and Organic Compounds—(*Continued*)

Compound	State†	Heat of formation‡ § ΔH (formation) at 25°C., kcal./mole	Free energy of formation‖ ¶ ΔF (formation) at 25°C., kcal./mole
Rubidium (*Cont.*):			
RbCl	c	-105.06	-98.48
	g	-53.6	-57.9
	aq, ∞	-101.06	-100.13
RbF	c	-133.23	
	aq, 400	-139.31	-134.5
$RbHCO_3$	c	-230.01	
	aq, 2000	-225.59	-209.07
RbI	c	-81.04	
	g	-31.2	-40.5
	aq, 400	-74.57	-81.13
$RbNH_2$	c	-27.74	
$RbNO_3$	c	-119.22	
	aq, 400	-110.52	-95.05
Rb_2O	c	-82.9	
Rb_2O_2	c	-107	
RbOH	c	-101.3	
	aq, 200	-115.8	-106.39
Ruthenium:			
Ru	c	0.00	0.00
RuS_2	c	-46.99	-44.11
Selenium:			
Se	c, I, hexagonal	0.00	0.00
	c, II, red, monoclinic	0.2	
Se_2Cl_2	l	-22.06	-13.73
SeF_6	g	-246	-222
SeO_2	c	-56.33	
Silicon:			
Si	c	0.00	0.00
$SiBr_4$	l	-93.0	
SiC	c	-28	-27.4
$SiCl_4$	l	-150.0	-133.9
	g	-142.5	-133.0
SiF_4	g	-370	-360
SiH_4	g	-14.8	-9.4
SiI_4	c	-29.8	
Si_3N_4	c	-179.25	-154.74
SiO_2	c, cristobalite, 1600° form	-202.62	
	c, cristobalite, 1100° form	-202.46	
	c, quartz	-203.35	-190.4
	c, tridymite	-203.23	
Silver:			
Ag	c	0.00	0.00
AgBr	c	-23.90	-23.02
Ag_2C_2	c	84.5	
$AgC_2H_3O_2$	c	-95.9	
	aq	-91.7	-70.86
AgCN	c	33.8	38.70
Ag_2CO_3	c	-119.5	-103.0
$Ag_2C_2O_4$	c	-158.7	
AgCl	c	-30.11	-25.98
AgF	c	-48.7	
	aq, 400	-53.1	-47.26
AgI	c	-15.14	-16.17
$AgIO_3$	c	-42.02	-24.08
$AgNO_2$	c	-11.6	3.76
	aq	-2.9	9.99
$AgNO_3$	c	-29.4	-7.66
	aq, 6500	-24.02	-7.81
Ag_2O	c	-6.95	-2.23
Ag_2S	c	-5.5	-7.6
Ag_2SO_4	c	-170.1	-146.8
	aq	-165.8	-139.22
Sodium:			
Na	c	0.00	0.00
Na_3AsO_3	aq, 500	-314.61	
Na_3AsO_4	c	-366	
	aq, 500	-381.97	-341.17
NaBr	c	-86.72	
	aq, 400	-86.33	-87.17
NaBrO	aq	-78.9	
$NaBrO_3$	aq, 400	-68.89	-57.59
$NaC_2H_3O_2$	c	-170.45	
	aq, 400	-175.450	-152.31
NaCN	c	-22.47	
	aq, 200	-22.29	-23.24
NaCNO	c	-96.3	
	aq	-91.7	-86.00
NaCNS	c	-39.94	
	aq, 400	-38.23	-39.24
Na_2CO_3	c	-269.46	-249.55
	aq, 1000	-275.13	-251.36
$NaCO_2NH_2$	c	-142.17	
$Na_2C_2O_4$	c	-313.8	
	aq, 600	-309.92	-283.42
NaCl	c	-98.321	-91.894
	aq, 400	-97.324	-93.92
Sodium (*Cont.*):			
$NaClO_3$	c	-83.59	
	aq, 400	-78.42	-62.84
$NaClO_4$	c	-101.12	
	aq, 476	-97.66	-73.29
Na_2CrO_4	c	-319.8	
	aq, 800	-323.0	-296.58
$Na_2Cr_2O_7$	aq, 1200	-465.9	-431.18
NaF	c	-135.94	-129.0
	aq, 400	-135.711	-128.29
NaH	c	-14	-9.30
$NaHCO_3$	c	-226.0	-202.66
	aq	-222.1	-202.87
NaI	c	-69.28	
	aq, ∞	-71.10	-74.92
$NaIO_3$	aq, 400	-112.300	-94.84
Na_2MoO_4	c	-364	
	aq	-358.7	-333.18
$NaNO_2$	c	-86.6	
	aq	-83.1	-71.04
$NaNO_3$	c	-111.71	-87.62
	aq, 400	-106.880	-88.84
Na_2O	c	-99.45	-90.06
Na_2O_2	c	-119.2	-105.0
$Na_2O.SiO_2$	c	-383.91	-361.49
$Na_2O.Al_2O_3.3SiO_2$	c, natrolite	-1180	
$Na_2O.Al_2O_3.4SiO_2$	c	-1366	
NaOH	c	-101.96	-90.60
	aq, 400	-112.193	-100.18
Na_3PO_3	aq, 1000	-389.1	
Na_3PO_4	c	-457	
	aq, 400	-471.9	-428.74
Na_2PtCl_4	aq	-237.2	-216.78
Na_2PtCl_6	c	-272.1	
	aq	-280.9	
Na_2Se	c	-59.1	
	aq, 440	-78.1	-89.42
Na_2SeO_4	c	-254	
	aq, 800	-261.5	-230.30
Na_2S	c	-89.8	
	aq, 400	-105.17	-101.76
Na_2SO_3	c	-261.2	-240.14
	aq, 800	-264.1	-241.58
Na_2SO_4	c	-330.50	-302.38
	aq, 1100	-330.82	-301.28
$Na_2SO_4.10H_2O$	c	-1033.85	-870.52
Na_2WO_4	c	-391	
	aq	-381.5	-345.18
Strontium:			
Sr	c	0.00	0.00
$SrBr_2$	c	-171.0	
	aq, 400	-187.24	-182.36
$Sr(C_2H_3O_2)_2$	c	-358.0	
	aq	-364.4	-311.80
$Sr(CN)_2$	aq	-59.5	-54.50
$SrCO_3$	c	-290.9	-271.9
$SrCl_2$	c	-197.84	
	aq, 400	-209.20	-195.86
SrF_2	c	-289.0	
$Sr(HCO_3)_2$	aq	-459.1	-413.76
SrI_2	c	-136.1	
	aq, 400	-156.70	-157.87
Sr_3N_2	c	-91.4	-76.5
$Sr(NO_3)_2$	c	-233.2	
	aq, 400	-228.73	-185.70
SrO	c	-140.8	-133.7
$SrO.SiO_2$	gls	-364	
SrO_2	c	-153.3	-139.0
Sr_2O	c	-153.6	
$Sr(OH)_2$	c	-228.7	
	aq, 800	-239.4	-208.27
$Sr_3(PO_4)_2$	c	-980	
	aq	-985	-881.54
SrS	c	-113.1	
	aq	-120.4	-109.78
$SrSO_4$	c	-345.3	
	aq, 400	-345.0	-309.30
$SrWO_4$	c	-393	
Sulfur:			
S	c, rhombic	0.00	0.00
	c, monoclinic	-0.071	-0.023
	l, λ	0.257	0.072
	l, λμ equilibrium	....	0.071
	g	53.25	43.57
S_2	g	31.02	19.36
S_6	g	27.78	13.97
S_8	g	27.090	12.770
S_2Br_2	l	-4	
SCl_4	l	-13.7	

Table 197. Heats and Free Energies of Formation of Inorganic and Organic Compounds—(*Concluded*)

Compound	State†	Heat of formation‡ § ΔH (formation) at 25°C., kcal./mole	Free energy of formation‖ ¶ ΔF (formation) at 25°C., kcal./mole
Sulfur (*Cont.*):			
S_2Cl_2	l	−14.2	−5.90
S_2Cl_4	l	−24.1	
SF_6	g	−262	−237
SO	g	19.02	12.75
SO_2	g	−70.94	−71.68
SO_3	g	−94.39	−88.59
	l	−103.03	−88.28
	c, α	−105.09	−88.22
	c, β	−105.92	−88.34
	c, γ	−109.34	−88.98
SO_2Cl_2	g	−82.04	−74.06
	l	−89.80	−75.06
Tantalum:			
Ta	c	0.00	0.00
TaN	c	−51.2	−45.11
Ta_2O_5	c	−486.0	−453.7
Tellurium:			
Te	c	0.00	0.00
$TeBr_4$	c	−49.3	
$TeCl_4$	c	−77.4	−57.4
TeF_6	g	−315	−292
TeO_2	c	−77.56	−64.66
Thallium:			
Tl	c	0.00	0.00
TlBr	c	−41.5	−39.43
	aq	28.0	−32.34
TlCl	c	−49.37	−44.46
	aq	−38.4	−39.09
$TlCl_3$	c	−82.4	
	aq	−91.0	−44.25
TlF	aq	−77.6	−73.46
TlI	c	−31.1	−31.3
	aq	−12.7	−20.09
$TlNO_3$	c	−58.2	−36.32
	aq	−48.4	−34.01
Tl_2O	c	43.18	
Tl_2O_3	c	−120	
TlOH	c	−57.44	−45.54
	aq	−53.9	−45.35
Tl_2S	c	−22	
Tl_2SO_4	c	−222.8	−197.79
	aq, 800	−214.1	−191.62
Thorium:			
Th	c	0.00	0.00
$ThBr_4$	c	−281.5	
	aq	−352.0	−295.31
ThC_2	c	−45.1	
$ThCl_4$	c	−335	
	aq	−392	−322.32
ThI_4	aq	−292.0	−246.33
Th_3N_4	c	−309.0	−282.3
ThO_2	c	−291.6	−280.1
$Th(OH)_4$	c, "soluble"	−336.1	
$Th(SO_4)_2$	c	−632	
	aq	−668.1	−549.2
Tin:			
Sn	c, II, tetragonal	0.00	0.00
	c, III, "gray," cubic	0.6	1.1
$SnBr_2$	c	−61.4	
	aq	−60.0	−55.43
$SnBr_4$	c	−94.8	
	aq	−110.6	−97.66
$SnCl_2$	c	−83.6	
	aq	−81.7	−68.94
$SnCl_4$	l	−127.3	−110.4
	aq	−157.6	−124.67
SnI_2	c	−38.9	
	aq	−33.3	−30.95
Tin (*Cont.*):			
SnO	c	−67.7	−60.75
SnO_2	c	−138.1	−123.6
$Sn(OH)_2$	c	−136.2	−115.95
$Sn(OH)_4$	c	−268.9	−226.00
SnS	c	−18.61	
Titanium:			
Ti	c	0.00	0.00
TiC	c	−110	−109.2
$TiCl_4$	l	−181.4	−165.5
TiN	c	−80.0	−73.17
TiO_2	c, III, rutile	−225.0	−211.9
	amorphous	−214.1	−201.4
Tungsten:			
W	c	0.00	0.00
WO_2	c	−130.5	−118.3
WO_3	c	−195.7	−177.3
WS_2	c	−84	
Uranium:			
U	c	0.00	0.00
UC_2	c	−29	
UCl_3	c	−213	
UCl_4	c	−251	
U_3N_4	c	−274	−249.6
UO_2	c	−256.6	−242.2
$UO_2(NO_3)_2.6H_2O$	c	−756.8	−617.8
UO_3	c	−291.6	
U_3O_8	c	−845.1	
Vanadium:			
V	c	0.00	0.00
VCl_2	c	−147	
VCl_3	l	−187	
VCl_4	l	−165	
VN	c	−41.43	−35.08
V_2O_2	c	−195	
V_2O_3	c	−296	−277
V_2O_4	c	−342	−316
V_2O_5	c	−373	−342
Zinc:			
Zn	c	0.00	0.00
ZnSb	c	−3.6	−3.88
$ZnBr_2$	c	−77.0	−72.9
	aq, 400	−93.6	
$Zn(C_2H_3O_2)_2$	c	−259.4	
	aq, 400	−269.4	−214.4
$Zn(CN)_2$	c	17.06	
$ZnCO_3$	c	−192.9	−173.5
$ZnCl_2$	c	−99.9	−88.8
	aq, 400	−115.44	
ZnF_2	aq	−192.9	−166.6
ZnI_2	c	−50.50	−49.93
	aq	−61.6	
$Zn(NO_3)_2$	aq, 400	−134.9	−87.7
ZnO	c, hexagonal	−83.36	−76.19
$ZnO.SiO_2$	c	−282.6	
$Zn(OH)_2$	c, rhombic	−153.66	
ZnS	c, wurtzite	−45.3	−44.2
$ZnSO_4$	c	−233.4	
	aq, 400	−252.12	−211.28
Zirconium:			
Zr	c	0.00	0.00
ZrC	c	−29.8	−34.6
$ZrCl_4$	c	−268.9	
ZrN	c	−82.5	−75.9
ZrO_2	c, monoclinic	−258.5	−244.6
$Zr(OH)_4$	c	−411.0	
$ZrO(OH)_2$	c	−337	−307.6

† The physical state is indicated as follows: *c*, crystal (solid); *l*, liquid; *g*, gas; *gls*, glass or solid supercooled liquid; *aq*, in aqueous solution. A number following the symbol *aq* applies only to the values of the heats of formation (not to those of free energies of formation); and indicates the number of moles of water per mole of solute; when no number is given, the solution is understood to be dilute. For the free energy of formation of a substance in aqueous solution, the concentration is always that of the hypothetical solution of unit molality.

‡ The increment in heat content, ΔH, in the reaction of forming the given substance from its elements in their standard states. When ΔH is negative, heat is evolved in the process, and, when positive, heat is absorbed.

§ The heat of solution in water of a given solid, liquid, or gaseous compound is given by the difference in the value for the heat of formation of the given compound in the solid, liquid, or gaseous state and its heat of formation in aqueous solution. The following two examples serve as an illustration of the procedure: (1) For NaCl(*c*) and NaCl(*aq*, 400H_2O), the values of ΔH(formation) are, respectively, −98.321 and −97.324 kg.-cal. per mole. Subtraction of the first value from the second gives $\Delta H = 0.998$ kg.-cal. per mole for the reaction of dissolving crystalline sodium chloride in 400 moles of water. When this process occurs at a constant pressure of 1 atm., 0.998 kg.-cal. of energy are absorbed. (2) For HCl(*g*) and HCl(*aq*, 400H_2O), the values for ΔH(formation) are, respectively, −22.06 and −39.85 kg.-cal. per mole. Subtraction of the first from the second gives $\Delta H = -17.79$ kg.-cal per mole for the reaction of dissolving gaseous hydrogen chloride in 400 moles of water. At a constant pressure of 1 atm. 17.79 kg.-cal. of energy are evolved in this process.

‖ The increment in the free energy, ΔF, in the reaction of forming the given substance in its standard state from its elements in their standard states. The standard states are: for a gas, fugacity (approximately equal to the pressure) of 1 atm.; for a pure liquid or solid, the substance at a pressure of 1 atm.; for a substance in aqueous solution, the hypothetical solution of unit molality, which has all the properties of the infinitely dilute solution except the property of concentration.

¶ The free energy of solution of a given substance from its normal standard state as a solid, liquid, or gas to the hypothetical one molal state in aqueous solution may be calculated in a manner similar to that described in footnote § for calculating the heat of solution.

HEATS OF COMBUSTION

Table 198. Hydrogen, Carbon, Carbon Monoxide, and Hydrocarbons

Heats of combustion of additional compounds may be calculated from the heats of formation given in Table 197, p. 237.
The following values are taken from the tables of the American Petroleum Institute Research Project 44 of the National Bureau of Standards on the Collection, Analysis, Calculation, and Compilation of Data on the Properties of Hydrocarbons.

Compound	Formula	State	Heat of combustion, $-\Delta Hc°$, at 25°C. and constant pressure, to form					
			H_2O (liq.) and CO_2 (gas)			H_2O (gas) and CO_2 (gas)		
			Kcal./mole	Cal./g.	B.t.u./lb.	Kcal./mole	Cal./g.	B.t.u./lb.
Hydrogen	H_2	gas	68.3174	33,887.6	60,957.7	57.7979	28,669.6	51,571.4
Carbon	C	solid, graph.	94.0518	7,831.1	14,086.8			
Carbon monoxide	CO	gas	67.6361	2,414.7	4,343.6			
Paraffins								
Methane	CH_4	gas	212.798	13,265.1	23,861	191.759	11,953.6	21,502
Ethane	C_2H_6	gas	372.820	12,399.2	22,304	341.261	11,349.6	20,416
Propane	C_3H_8	gas	530.605	12,033.5	21,646	488.527	11,079.2	19,929
Propane	C_3H_8	liq.*	526.782	11,946.8	21,490	484.704	10,992.5	19,774
n-Butane	C_4H_{10}	gas	687.982	11,837.3	21,293	635.384	10,932.3	19,665
n-Butane	C_4H_{10}	liq.*	682.844	11,748.9	21,134	630.246	10,843.9	19,506
2-Methylpropane (Isobutane)	C_4H_{10}	gas	686.342	11,809.1	21,242	633.744	10,904.1	19,614
2-Methylpropane (Isobutane)	C_4H_{10}	liq.*	681.625	11,727.9	21,096	629.027	10,822.9	19,468
n-Pentane	C_5H_{12}	gas	845.16	11,714.6	21,072	782.04	10,839.7	19,499
n-Pentane	C_5H_{12}	liq.	838.80	11,626.4	20,914	775.68	10,751.5	19,340
2-Methylbutane (Isopentane)	C_5H_{12}	gas	843.24	11,688.0	21,025	780.12	10,813.1	19,451
2-Methylbutane (Isopentane)	C_5H_{12}	liq.	837.31	11,605.8	20,877	774.19	10,730.9	19,303
2,2-Dimethylpropane (Neopentane)	C_5H_{12}	gas	840.49	11,649.8	20,956	777.37	10,775.0	19,382
2,2-Dimethylpropane (Neopentane)	C_5H_{12}	liq.	835.18	11,576.2	20,824	772.06	10,701.4	19,250
n Hexane	C_6H_{14}	gas	1,002.57	11,634.5	20,928	928.93	10,780.0	19,391
n-Hexane	C_6H_{14}	liq.	995.01	11,546.8	20,771	921.37	10,692.2	19,233
2-Methylpentane	C_6H_{14}	gas	1,000.87	11,614.8	20,893	927.23	10,760.2	19,356
2-Methylpentane	C_6H_{14}	liq.	993.71	11,531.7	20,743	920.07	10,677.1	19,206
3-Methylpentane	C_6H_{14}	gas	1,001.51	11,622.2	20,906	927.87	10,767.6	19,369
3-Methylpentane	C_6H_{14}	liq.	994.25	11,538.0	20,755	920.61	10,683.4	19,218
2,2-Dimethylbutane	C_6H_{14}	gas	998.17	11,583.5	20,837	924.53	10,728.9	19,299
2,2-Dimethylbutane	C_6H_{14}	liq.	991.52	11,506.3	20,698	917.88	10,651.7	19,161
2,3-Dimethylbutane	C_6H_{14}	gas	1,000.04	11,605.2	20,876	926.40	10,750.6	19,338
2,3-Dimethylbutane	C_6H_{14}	liq.	993.05	11,524.0	20,730	919.41	10,669.5	19,192
n-Heptane	C_7H_{16}	gas	1,160.01	11,577.2	20,825	1,075.85	10,737.2	19,314
n-Heptane	C_7H_{16}	liq.	1,151.27	11,489.9	20,668	1,067.11	10,650.0	19,157
2-Methylhexane	C_7H_{16}	gas	1,158.30	11,560.1	20,795	1,074.14	10,720.2	19,284
2-Methylhexane	C_7H_{16}	liq.	1,149.97	11,477.0	20,645	1,065.81	10,637.0	19,134
3-Methylhexane	C_7H_{16}	gas	1,158.94	11,566.5	20,806	1,074.78	10,726.6	19,295
3-Methylhexane	C_7H_{16}	liq.	1,150.55	11,482.8	20,655	1,066.39	10,642.8	19,145
3-Ethylpentane	C_7H_{16}	gas	1,159.56	11,572.7	20,817	1,075.40	10,732.7	19,306
3-Ethylpentane	C_7H_{16}	liq.	1,151.13	11,488.6	20,666	1,066.97	10,648.6	19,155
2,2-Dimethylpentane	C_7H_{16}	gas	1,155.61	11,533.3	20,746	1,071.45	10,693.3	19,235
2,2-Dimethylpentane	C_7H_{16}	liq.	1,147.85	11,455.8	20,607	1,063.69	10,615.9	19,096
2,3-Dimethylpentane	C_7H_{16}	gas	1,157.28	11,549.9	20,776	1,073.12	10,710.0	19,265
2,3-Dimethylpentane	C_7H_{16}	liq.	1,149.09	11,468.2	20,629	1,064.93	10,628.3	19,118
2,4-Dimethylpentane	C_7H_{16}	gas	1,156.60	11,543.1	20,764	1,072.44	10,703.2	19,253
2,4-Dimethylpentane	C_7H_{16}	liq.	1,148.73	11,464.6	20,623	1,064.57	10,624.7	19,112
3,3-Dimethylpentane	C_7H_{16}	gas	1,156.73	11,544.4	20,766	1,072.57	10,704.5	19,255
3,3-Dimethylpentane	C_7H_{16}	liq.	1,148.83	11,465.6	20,625	1,064.67	10,625.7	19,114
2,2,3-Trimethylbutane	C_7H_{16}	gas	1,155.94	11,536.6	20,752	1,071.78	10,696.6	19,241
2,2,3-Trimethylbutane	C_7H_{16}	liq.	1,148.27	11,460.0	20,614	1,064.11	10,620.1	19,104
n-Octane	C_8H_{18}	gas	1,317.45	11,533.9	20,747	1,222.77	10,705.0	19,256
n-Octane	C_8H_{18}	liq.	1,307.53	11,447.1	20,591	1,212.85	10,618.2	19,100
2-Methylheptane	C_8H_{18}	gas	1,315.76	11,519.1	20,721	1,221.08	10,690.2	19,230
2-Methylheptane	C_8H_{18}	liq.	1,306.28	11,436.1	20,572	1,211.60	10,607.2	19,080
3-Methylheptane	C_8H_{18}	gas	1,316.44	11,525.1	20,732	1,221.76	10,696.2	19,240
3-Methylheptane	C_8H_{18}	liq.	1,306.92	11,441.7	20,582	1,212.24	10,612.8	19,091
4-Methylheptane	C_8H_{18}	gas	1,316.57	11,526.2	20,734	1,221.89	10,697.3	19,243
4-Methylheptane	C_8H_{16}	liq.	1,307.09	11,443.2	20,584	1,212.41	10,614.3	19,093
3-Ethylhexane	C_8H_{18}	gas	1,316.87	11,528.8	20,738	1,222.19	10,699.9	19,247
3-Ethylhexane	C_8H_{18}	liq.	1,307.39	11,445.8	20,589	1,212.71	10,616.9	19,098
2,2-Dimethylhexane	C_8H_{18}	gas	1,313.56	11,499.9	20,686	1,218.88	10,671.0	19,195
2,2-Dimethylhexane	C_8H_{18}	liq.	1,304.64	11,421.8	20,546	1,209.96	10,592.9	19,055
2,3-Dimethylhexane	C_8H_{18}	gas	1,316.13	11,522.4	20,727	1,221.45	10,693.5	19,236
2,3-Dimethylhexane	C_8H_{18}	liq.	1,306.86	11,441.2	20,581	1,212.18	10,612.3	19,090
2,4-Dimethylhexane	C_8H_{18}	gas	1,314.83	11,511.0	20,706	1,220.15	10,682.1	19,215
2,4-Dimethylhexane	C_8H_{18}	liq.	1,305.80	11,431.9	20,564	1,211.12	10,603.0	19,073
2,5-Dimethylhexane	C_8H_{18}	gas	1,314.05	11,504.2	20,694	1,219.37	10,675.3	19,203
2,5-Dimethylhexane	C_8H_{18}	liq.	1,305.00	11,424.9	20,551	1,210.32	10,596.0	19,060
3,3-Dimethylhexane	C_8H_{18}	gas	1,314.65	11,509.4	20,703	1,219.97	10,680.5	19,212
3,3-Dimethylhexane	C_8H_{18}	liq.	1,305.68	11,430.9	20,562	1,211.00	10,602.0	19,071
3,4-Dimethylhexane	C_8H_{18}	gas	1,316.36	11,524.4	20,730	1,221.68	10,695.5	19,239
3,4-Dimethylhexane	C_8H_{18}	liq.	1,307.04	11,442.8	20,583	1,212.36	10,613.9	19,092
2-Methyl-3-ethylpentane	C_8H_{18}	gas	1,316.79	11,528.1	20,737	1,222.11	10,699.2	19,246
2-Methyl-3-ethylpentane	C_8H_{18}	liq.	1,307.58	11,447.5	20,592	1,212.90	10,618.6	19,101
3-Methyl-3-ethylpentane	C_8H_{18}	gas	1,315.88	11,520.2	20,723	1,221.20	10,691.3	19,232
3-Methyl-3-ethylpentane	C_8H_{18}	liq.	1,306.80	11,440.7	20,580	1,212.12	10,611.8	19,089
2,2,3-Trimethylpentane	C_8H_{18}	gas	1,314.66	11,509.5	20,703	1,219.98	10,680.6	19,212
2,2,3-Trimethylpentane	C_8H_{18}	liq.	1,305.83	11,432.2	20,564	1,211.15	10,603.3	19,073
2,2,4-Trimethylpentane	C_8H_{18}	gas	1,313.69	11,501.0	20,688	1,219.01	10,672.1	19,197
2,2,4-Trimethylpentane	C_8H_{18}	liq.	1,305.29	11,427.5	20,556	1,210.61	10,598.6	19,065
2,3,3-Trimethylpentane	C_8H_{18}	gas	1,315.54	11,517.2	20,717	1,220.86	10,688.3	19,226
2,3,3-Trimethylpentane	C_8H_{18}	liq.	1,306.64	11,439.3	20,577	1,211.96	10,610.4	19,086
2,3,4-Trimethylpentane	C_8H_{18}	gas	1,315.29	11,515.0	20,713	1,220.61	10,686.1	19,222
2,3,4-Trimethylpentane	C_8H_{18}	liq.	1,306.28	11,436.1	20,572	1,211.60	10,607.2	19,080
2,2,3,3-Tetramethylbutane	C_8H_{18}	gas	1,313.27	11,497.3	20,682	1,218.59	10,668.4	19,191

Table 198. Hydrogen, Carbon, Carbon Monoxide, and Hydrocarbons—(*Continued*)

Compound	Formula	State	Heat of combustion, $-\Delta Hc°$, at 25°C. and constant pressure, to form: H_2O (liq.) and CO_2 (gas)			H_2O (gas) and CO_2 (gas)		
			Kcal./mole	Cal./g.	B.t.u./lb.	Kcal./mole	Cal./g.	B.t.u./lb.
2,2,3,3-Tetramethylbutane	C_8H_{18}	solid	1,303.03	11,407.7	20,520	1,208.35	10,578.8	19,029
n-Nonane	C_9H_{20}	gas	1,474.90	11,500.2	20,687	1,369.70	10,680.0	19,211
n-Nonane	C_9H_{20}	liq.	1,463.80	11,413.6	20,531	1,358.60	10,593.4	19,056
n-Decane	$C_{10}H_{22}$	gas	1,632.34	11,473.0	20,638	1,516.63	10,659.7	19,175
n-Decane	$C_{10}H_{22}$	liq.	1,620.06	11,386.7	20,483	1,504.35	10,573.4	19,020
n-Undecane	$C_{11}H_{24}$	gas	1,789.78	11,450.8	20,598	1,663.55	10,643.2	19,145
n-Undecane	$C_{11}H_{24}$	liq.	1,776.32	11,364.7	20,443	1,650.09	10,557.0	18,990
n-Dodecane	$C_{12}H_{26}$	gas	1,947.23	11,432.2	20,564	1,810.48	10,629.4	19,120
n-Dodecane	$C_{12}H_{26}$	liq.	1,932.59	11,346.3	20,410	1,795.84	10,543.4	18,966
n-Tridecane	$C_{13}H_{28}$	gas	2,104.67	11,416.5	20,536	1,957.40	10,617.6	19,099
n-Tridecane	$C_{13}H_{28}$	liq.	2,088.85	11,330.6	20,382	1,941.58	10,531.8	18,945
n-Tetradecane	$C_{14}H_{30}$	gas	2,262.11	11,402.9	20,512	2,104.32	10,607.5	19,081
n-Tetradecane	$C_{14}H_{30}$	liq.	2,245.11	11,317.2	20,358	2,087.32	10,521.8	18,927
n-Pentadecane	$C_{15}H_{32}$	gas	2,419.55	11,391.2	20,491	2,251.24	10,598.7	19,065
n-Pentadecane	$C_{15}H_{32}$	liq.	2,401.37	11,305.6	20,337	2,233.06	10,513.2	18,911
n-Hexadecane	$C_{16}H_{34}$	gas	2,577.00	11,380.9	20,472	2,398.17	10,591.1	19,052
n-Hexadecane	$C_{16}H_{34}$	liq.	2,557.64	11,295.4	20,318	2,378.81	10,505.6	18,898
n-Heptadecane	$C_{17}H_{36}$	gas	2,734.44	11,371.8	20,456	2,545.09	10,584.3	19,039
n-Heptadecane	$C_{17}H_{36}$	liq.	2,713.90	11,286.4	20,302	2,524.55	10,498.9	18,886
n-Octadecane	$C_{18}H_{38}$	gas	2,891.88	11,363.7	20,441	2,692.01	10,578.3	19,028
n-Octadecane	$C_{18}H_{38}$	liq.	2,870.16	11,278.4	20,288	2,670.29	10,493.0	18,875
n-Nonadecane	$C_{19}H_{40}$	gas	3,049.33	11,356.5	20,428	2,838.94	10,572.9	19,019
n-Nonadecane	$C_{19}H_{40}$	liq.	3,026.43	11,271.2	20,275	2,816.04	10,487.7	18,865
n-Eicosane	$C_{20}H_{42}$	gas	3,206.77	11,350.0	20,416	2,985.86	10,568.1	19,010
n-Eicosane	$C_{20}H_{42}$	liq.	3,182.69	11,264.7	20,263	2,961.78	10,482.8	18,857
Alkyl benzenes								
Benzene	C_6H_6	gas	789.08	10,102.4	18,172	757.52	9,698.4	17,446
Benzene	C_6H_6	liq.	780.98	9,998.7	17,986	749.42	9,594.7	17,259
Methylbenzene (toluene)	C_7H_8	gas	943.58	10,241.4	18,422	901.50	9,784.7	17,601
Methylbenzene (toluene)	C_7H_8	liq.	934.50	10,142.8	18,245	892.42	9,686.1	17,424
Ethylbenzene	C_8H_{10}	gas	1,101.13	10,372.4	18,658	1,048.53	9,876.9	17,767
Ethylbenzene	C_8H_{10}	liq.	1,091.03	10,277.2	18,487	1,038.43	9,781.7	17,596
1,2-Dimethylbenzene (o-xylene)	C_8H_{10}	gas	1,098.54	10,348.0	18,614	1,045.94	9,852.5	17,723
1,2-Dimethylbenzene (o-xylene)	C_8H_{10}	liq.	1,088.16	10,250.2	18,438	1,035.56	9,754.7	17,547
1,3-Dimethylbenzene (m-xylene)	C_8H_{10}	gas	1,098.12	10,344.0	18,607	1,045.52	9,848.5	17,716
1,3-Dimethylbenzene (m-xylene)	C_8H_{10}	liq.	1,087.92	10,247.9	18,434	1,035.32	9,752.4	17,543
1,4-Dimethylbenzene (p-xylene)	C_8H_{10}	gas	1,098.29	10,345.6	18,610	1,045.69	9,850.1	17,719
1,4-Dimethylbenzene (p-xylene)	C_8H_{10}	liq.	1,088.16	10,250.2	18,438	1,035.56	9,754.7	17,547
n-Propylbenzene	C_9H_{12}	gas	1,258.24	10,469.1	18,832	1,195.12	9,943.9	17,887
n-Propylbenzene	C_9H_{12}	liq.	1,247.19	10,377.2	18,667	1,184.07	9,852.0	17,722
Isopropylbenzene (cumene)	C_9H_{12}	gas	1,257.31	10,461.4	18,818	1,194.19	9,936.2	17,873
Isopropylbenzene (cumene)	C_9H_{12}	liq.	1,246.52	10,371.6	18,657	1,183.40	9,846.4	17,712
1-Methyl-2-ethylbenzene	C_9H_{12}	gas	1,256.66	10,456.0	18,808	1,193.54	9,930.8	17,864
1-Methyl-2-ethylbenzene	C_9H_{12}	liq.	1,245.26	10,361.1	18,638	1,182.14	9,835.9	17,693
1-Methyl-3-ethylbenzene	C_9H_{12}	gas	1,255.92	10,449.8	18,797	1,192.80	9,924.6	17,853
1-Methyl-3-ethylbenzene	C_9H_{12}	liq.	1,244.71	10,356.5	18,630	1,181.59	9,831.3	17,685
1-Methyl-4-ethylbenzene	C_9H_{12}	gas	1,255.59	10,447.1	18,792	1,192.47	9,921.9	17,848
1-Methyl-4-ethylbenzene	C_9H_{12}	liq.	1,244.45	10,354.4	18,626	1,181.33	9,829.2	17,681
1,2,3-Trimethylbenzene (hemimellitene)	C_9H_{12}	gas	1,254.08	10,434.5	18,770	1,190.96	9,909.3	17,825
1,2,3-Trimethylbenzene (hemimellitene)	C_9H_{12}	liq.	1,242.36	10,337.0	18,594	1,179.24	9,811.8	17,650
1,2,4-Trimethylbenzene (pseudocumene)	C_9H_{12}	gas	1,253.04	10,425.8	18,754	1,189.92	9,900.7	17,809
1,2,4-Trimethylbenzene (pseudocumene)	C_9H_{12}	liq.	1,241.58	10,330.5	18,583	1,178.46	9,805.3	17,638
1,3,5-Trimethylbenzene (mesitylene)	C_9H_{12}	gas	1,252.53	10,421.6	18,747	1,189.41	9,896.4	17,802
1,3,5-Trimethylbenzene (mesitylene)	C_9H_{12}	liq.	1,241.19	10,327.2	18,577	1,178.07	9,802.1	17,632
n-Butylbenzene	$C_{10}H_{14}$	gas	1,415.44	10,546.3	18,971	1,341.80	9,997.6	17,984
n-Butylbenzene	$C_{10}H_{14}$	liq.	1,403.46	10,457.0	18,810	1,329.82	9,908.4	17,823
Alkyl cyclopentanes								
Cyclopentane	C_5H_{10}	gas	793.39	11,313.1	20,350	740.79	10,563.1	19,001
Cyclopentane	C_5H_{10}	liq.	786.54	11,215.5	20,175	733.94	10,465.4	18,825
Methylcyclopentane	C_6H_{12}	gas	948.72	11,273.4	20,279	885.60	10,523.3	18,930
Methylcyclopentane	C_6H_{12}	liq.	941.14	11,183.3	20,117	878.02	10,433.2	18,768
Ethylcyclopentane	C_7H_{14}	gas	1,106.21	11,266.9	20,267	1,032.57	10,516.9	18,918
Ethylcyclopentane	C_7H_{14}	liq.	1,097.50	11,178.2	20,108	1,023.86	10,428.2	18,758
n-Propylcyclopentane	C_8H_{16}	gas	1,263.56	11,260.9	20,256	1,179.40	10,510.8	18,907
n-Propylcyclopentane	C_8H_{16}	liq.	1,253.74	11,173.4	20,099	1,169.58	10,423.3	18,750
n-Butylcyclopentane	C_9H_{18}	gas	1,421.10	11,257.7	20,250	1,326.42	10,507.6	18,901
n-Butylcyclopentane	C_9H_{18}	liq.	1,410.10	11,170.5	20,094	1,315.42	10,420.5	18,745
Alkyl cyclohexanes								
Cyclohexane	C_6H_{12}	gas	944.79	11,226.7	20,195	881.67	10,476.7	18,846
Cyclohexane	C_6H_{12}	liq.	936.88	11,132.7	20,026	873.76	10,382.7	18,676
Methylcyclohexane	C_7H_{14}	gas	1,099.59	11,199.5	20,146	1,025.95	10,449.5	18,797
Methylcyclohexane	C_7H_{14}	liq.	1,091.13	11,113.3	19,991	1,017.49	10,363.3	18,642
Ethylcyclohexane	C_8H_{16}	gas	1,257.90	11,210.4	20,166	1,173.74	10,460.4	18,816
Ethylcyclohexane	C_8H_{16}	liq.	1,248.23	11,124.3	20,011	1,164.07	10,374.3	18,661
n-Propylcyclohexane	C_9H_{18}	gas	1,415.12	11,210.3	20,165	1,320.44	10,460.3	18,816
n-Propylcyclohexane	C_9H_{18}	liq.	1,404.34	11,124.9	20,012	1,309.66	10,374.9	18,663
n-Butylcyclohexane	$C_{10}H_{20}$	gas	1,572.74	11,213.0	20,170	1,467.54	10,463.0	18,821
n-Butylcyclohexane	$C_{10}H_{20}$	liq.	1,560.78	11,127.8	20,017	1,455.58	10,377.8	18,668
Monoolefins								
Ethene (ethylene)	C_2H_4	gas	337.234	12,021.7	21,625	316.195	11,271.7	20,276
Propene (propylene)	C_3H_6	gas	491.987	11,692.3	21,032	460.428	10,942.3	19,683
1-Butene	C_4H_8	gas	649.757	11,581.3	20,833	607.679	10,831.3	19,484
cis-2-Butene	C_4H_8	gas	648.115	11,552.0	20,780	606.037	10,802.0	19,431
trans-2-Butene	C_4H_8	gas	647.072	11,533.4	20,747	604.994	10,783.4	19,397
2-Methylpropene (isobutene)	C_4H_8	gas	646.134	11,516.7	20,716	604.056	10,766.7	19,367
1-Pentene	C_5H_{10}	gas	806.85	11,505.1	20,696	754.25	10,755.1	19,346

Table 198. Hydrogen, Carbon, Carbon Monoxide, and Hydrocarbons—(*Concluded*)

Compound	Formula	State	Heat of combustion, $-\Delta Hc°$, at 25°C. and constant pressure, to form: H_2O (liq.) and CO_2 (gas)			H_2O (gas) and CO_2 (gas)		
			Kcal./mole	Cal./g.	B.t.u./lb.	Kcal./mole	Cal./g.	B.t.u./lb.
cis-2-Pentene	C_5H_{10}	gas	805.34	11,483.5	20,657	752.74	10,733.5	19,308
trans-2-Pentene	C_5H_{10}	gas	804.26	11,468.1	20,629	751.66	10,718.1	19,280
2-Methyl-1-butene	C_5H_{10}	gas	803.17	11,452.6	20,601	750.57	10,702.6	19,252
3-Methyl-1-butene	C_5H_{10}	gas	804.93	11,477.7	20,646	752.33	10,727.7	19,297
2-Methyl-2-butene	C_5H_{10}	gas	801.68	11,431.3	20,563	749.08	10,681.3	19,214
Acetylenes								
Ethyne (acetylene)	C_2H_2	gas	310.615	11,930.2	21,460	300.096	11,526.2	20,734
Propyne (methylacetylene)	C_3H_4	gas	463.109	11,559.8	20,794	442.070	11,034.6	19,849
1-Butyne (ethylacetylene)	C_4H_6	gas	620.86	11,478.7	20,648	589.302	10,895.2	19,599
2-Butyne (dimethylacetylene)	C_4H_6	gas	616.533	11,398.7	20,504	584.974	10,815.2	19,455
1-Pentyne	C_5H_8	gas	778.03	11,422.5	20,547	735.95	10,804.7	19,436
2-Pentyne	C_5H_8	gas	774.33	11,368.2	20,449	732.25	10,750.4	19,338
3-Methyl-1-butyne	C_5H_8	gas	776.13	11,394.6	20,497	734.05	10,776.8	19,386

* Saturation pressure.

HEATS OF SOLUTION

Table 199. Heats of Solution of Inorganic Compounds in Water

Heat evolved, in kilogram calories per gram formula weight, on solution in water at 18°C. Computed from data in Bichowsky and Rossini "Thermochemistry of Chemical Substances," Reinhold, New York, 1936

Substance	Dilution*	Formula	Heat, kg.-cal./g.-mole
Aluminum bromide	aq	$AlBr_3$	+85.3
chloride	600	$AlCl_3$	+77.9
	600	$AlCl_3.6H_2O$	+13.2
fluoride	aq	AlF_3	+31
	aq	$AlF_3.\frac{1}{2}H_2O$	+19.0
	aq	$AlF_3.3\frac{1}{2}H_2O$	−1.7
iodide	aq	AlI_3	+89.0
sulfate	aq	$Al_2(SO_4)_3$	+126
	aq	$Al_2(SO_4)_3.6H_2O$	+56.2
	aq	$Al_2(SO_4)_3.18H_2O$	+6.7
Ammonium bromide	aq	NH_4Br	−4.45
chloride	∞	NH_4Cl	−3.82
chromate	aq	$(NH_4)_2CrO_4$	−5.82
dichromate	600	$(NH_4)_2Cr_2O_7$	−12.9
iodide	aq	NH_4I	−3.56
nitrate	∞	NH_4NO_3	−6.47
perborate	aq	$NH_4BO_3.H_2O$	−9.0
sulfate	∞	$(NH_4)_2SO_4$	−2.75
sulfate, acid	800	NH_4HSO_4	+0.56
sulfite	aq	$(NH_4)_2SO_3$	−1.2
	aq	$(NH_4)_2SO_3.H_2O$	−4.13
Antimony fluoride	aq	SbF_3	−1.7
iodide	aq	SbI_3	−0.8
Arsenic acid	aq	H_3AsO_4	−0.4
Barium bromate	∞	$Ba(BrO_3)_2.H_2O$	−15.9
bromide	∞	$BaBr_2$	+5.3
	∞	$BaBr_2.H_2O$	−0.8
	∞	$BaBr_2.2H_2O$	−3.87
chlorate	∞	$Ba(ClO_3)_2$	−6.7
	∞	$Ba(ClO_3)_2.H_2O$	−10.6
chloride	∞	$BaCl_2$	+2.4
	∞	$BaCl_2.H_2O$	−2.17
	∞	$BaCl_2.2H_2O$	−4.5
cyanide	aq	$Ba(CN)_2$	+1.5
	aq	$Ba(CN)_2.H_2O$	−2.4
	aq	$Ba(CN)_2.2H_2O$	−4.9
iodate	∞	$Ba(IO_3)_2$	−9.1
	∞	$Ba(IO_3)_2.H_2O$	−11.3
iodide	∞	BaI_2	+10.5
	∞	$BaI_2.H_2O$	+2.7
	∞	$BaI_2.2H_2O$	+0.14
	∞	$BaI_2.2\frac{1}{2}H_2O$	−0.58
	∞	$BaI_2.7H_2O$	−6.61
nitrate	∞	$Ba(NO_3)_2$	−10.2
perchlorate	∞	$Ba(ClO_4)_2$	−2.8
	∞	$Ba(ClO_4)_2.3H_2O$	−10.5
sulfide	∞	BaS	+7.2
Beryllium bromide	aq	$BeBr_2$	+62.6
chloride	aq	$BeCl_2$	+51.1
iodide	aq	BeI_2	+72.6
sulfate	aq	$BeSO_4$	+18.1
	aq	$BeSO_4.H_2O$	+13.5
	aq	$BeSO_4.2H_2O$	+7.9
	aq	$BeSO_4.4H_2O$	+1.1
Bismuth iodide	aq	BiI_3	+3

Substance	Dilution*	Formula	Heat, kg.-cal./g.-mole
Boric acid	aq	H_3BO_3	−5.4
Cadmium bromide	400	$CdBr_2$	+0.4
	400	$CdBr_2.4H_2O$	−7.3
chloride	400	$CdCl_2$	+3.1
	400	$CdCl_2.H_2O$	+0.6
	400	$CdCl_2.2\frac{1}{2}H_2O$	−3.00
nitrate	400	$Cd(NO_3)_2.H_2O$	+4.17
	400	$Cd(NO_3)_2.4H_2O$	−5.08
sulfate	400	$CdSO_4$	+10.69
	400	$CdSO_4.H_2O$	+6.05
	400	$CdSO_4.2\frac{2}{3}H_2O$	+2.51
Calcium acetate	∞	$Ca(C_2H_3O_2)_2$	+7.6
	∞	$Ca(C_2H_3O_2)_2.H_2O$	+6.5
bromide	∞	$CaBr_2$	+24.86
	∞	$CaBr_2.6H_2O$	−0.9
chloride	∞	$CaCl_2$	+4.9
	∞	$CaCl_2.H_2O$	+12.3
	∞	$CaCl_2.2H_2O$	+12.5
	∞	$CaCl_2.4H_2O$	+2.4
	∞	$CaCl_2.6H_2O$	−4.11
formate	400	$Ca(CHO_2)_2$	+0.7
iodide	∞	CaI_2	+28.0
	∞	$CaI_2.8H_2O$	+1.8
nitrate	∞	$Ca(NO_3)_2$	+4.1
	∞	$Ca(NO_3)_2.H_2O$	+0.7
	∞	$Ca(NO_3)_2.2H_2O$	−3.2
	∞	$Ca(NO_3)_2.3H_2O$	−4.2
	∞	$Ca(NO_3)_2.4H_2O$	−7.99
phosphate, mono-	aq	$Ca(H_2PO_4)_2.H_2O$	−0.6
dibasic	aq	$CaHPO_4.2H_2O$	−1
sulfate	∞	$CaSO_4$	+5.1
	∞	$CaSO_4.\frac{1}{2}H_2O$	+3.6
	∞	$CaSO_4.2H_2O$	−0.18
Chromous chloride	aq	$CrCl_2$	+18.6
		$CrCl_2.3H_2O$	+5.3
		$CrCl_2.4H_2O$	+2.0
iodide	aq	CrI_2	+5.7
Cobaltous bromide	aq	$CoBr_2$	+18.4
	aq	$CoBr_2.6H_2O$	−1.25
chloride	400	$CoCl_2$	+18.5
	400	$CoCl_2.2H_2O$	+9.8
	400	$CoCl_2.6H_2O$	−2.9
iodide	aq	CoI_2	+18.8
sulfate	400	$CoSO_4$	+15.0
	400	$CoSO_4.6H_2O$	−1.4
	400	$CoSO_4.7H_2O$	−3.6
Cupric acetate	aq	$Cu(C_2H_3O_2)_2$	+2.4
formate	aq	$Cu(CHO_2)_2$	+0.5
nitrate	200	$Cu(NO_3)_2$	+10.3
	200	$Cu(NO_3)_2.3H_2O$	−2.6
	200	$Cu(NO_3)_2.6H_2O$	−10.7
sulfate	800	$CuSO_4$	+15.9
		$CuSO_4.H_2O$	+9.3
		$CuSO_4.3H_2O$	+3.65
		$CuSO_4.5H_2O$	−2.85
Cuprous sulfate	aq	Cu_2SO_4	+11.6

* The numbers represent moles of water used to dissolve 1 g. formula weight of substance; ∞ means "infinite dilution"; and aq means "aqueous solution of unspecified dilution."

Table 199. Heats of Solution of Inorganic Compounds in Water—(*Continued*)

Substance	Dilution*	Formula	Heat, kg.-cal./g.-mole
Ferric chloride	1000	$FeCl_3$	+31.7
	1000	$FeCl_3.2\frac{1}{2}H_2O$	+21.0
	1000	$FeCl_3.6H_2O$	+5.6
nitrate	800	$Fe(NO_3)_3.9H_2O$	−9.1
Ferrous bromide	aq	$FeBr_2$	+18.0
chloride	400	$FeCl_2$	+17.9
	400	$FeCl_2.2H_2O$	+8.7
	400	$FeCl_2.4H_2O$	+2.7
iodide	aq	FeI_2	+23.3
sulfate	400	$FeSO_4$	+14.7
	400	$FeSO_4.H_2O$	+7.35
	400	$FeSO_4.4H_2O$	+1.4
	400	$FeSO_4.7H_2O$	−4.4
Lead acetate	400	$Pb(C_2H_3O_2)_2$	+1.4
	400	$Pb(C_2H_3O_2)_2.3H_2O$	−5.9
bromide	aq	$PbBr_2$	−10.1
chloride	aq	$PbCl_2$	−3.4
formate	aq	$Pb(CHO_2)_2$	−6.9
nitrate	400	$Pb(NO_3)_2$	−7.61
Lithium bromide	∞	$LiBr$	+11.54
	∞	$LiBr.H_2O$	+5.30
	∞	$LiBr.2H_2O$	+2.05
	∞	$LiBr.3H_2O$	−1.59
chloride	∞	$LiCl$	+8.66
	∞	$LiCl.H_2O$	+4.45
	∞	$LiCl.2H_2O$	+1.07
	∞	$LiCl.3H_2O$	−1.98
fluoride	∞	LiF	−0.74
hydroxide	∞	$LiOH$	+4.74
	∞	$LiOH.\frac{1}{8}H_2O$	+4.39
	∞	$LiOH.H_2O$	+9.6
iodide	∞	LiI	+14.92
	∞	$LiI.\frac{1}{2}H_2O$	+10.08
	∞	$LiI.H_2O$	+6.93
	∞	$LiI.2H_2O$	+3.43
	∞	$LiI.3H_2O$	−0.17
nitrate	∞	$LiNO_3$	+0.466
	∞	$LiNO_3.3H_2O$	−7.87
sulfate	∞	Li_2SO_4	+6.71
	∞	$Li_2SO_4.H_2O$	+3.77
Magnesium bromide	∞	$MgBr_2$	+43.7
	∞	$MgBr_2.H_2O$	+35.9
	∞	$MgBr_2.6H_2O$	+19.8
chloride	∞	$MgCl_2$	+36.3
	∞	$MgCl_2.2H_2O$	+20.8
	∞	$MgCl_2.4H_2O$	+10.5
	∞	$MgCl_2.6H_2O$	+3.4
iodide	∞	MgI_2	+50.2
nitrate	∞	$Mg(NO_3)_2.6H_2O$	−3.7
phosphato	aq	$Mg_3(PO_4)_2$	+10.2
sulfate	∞	$MgSO_4$	+21.1
	∞	$MgSO_4.H_2O$	+14.0
	∞	$MgSO_4.2H_2O$	+11.7
	∞	$MgSO_4.4H_2O$	+4.9
	∞	$MgSO_4.6H_2O$	+0.55
	∞	$MgSO_4.7H_2O$	−3.18
sulfide	aq	MgS	+25.8
Manganic nitrate	400	$Mn(NO_3)_2$	+12.9
	400	$Mn(NO_3)_2.3H_2O$	−3.9
	400	$Mn(NO_3)_2.6H_2O$	−6.2
sulfate	aq	$Mn_2(SO_4)_3$	+22
Manganous acetate	aq	$Mn(C_2H_3O_2)_2$	+12.2
	aq	$Mn(C_2H_3O_2)_2.4H_2O$	+1.6
bromide	aq	$MnBr_2$	+15
	aq	$MnBr_2.H_2O$	+14.4
	aq	$MnBr_2.4H_2O$	+16.1
chloride	400	$MnCl_2$	+16.0
	400	$MnCl_2.2H_2O$	+8.2
	400	$MnCl_2.4H_2O$	+1.5
formate	aq	$Mn(CHO_2)_2$	+4.3
	aq	$Mn(CHO_2)_2.2H_2O$	−2.9
iodide	aq	MnI_2	+26.2
	aq	$MnI_2.H_2O$	+24.1
	aq	$MnI_2.2H_2O$	+22.7
	aq	$MnI_2.4H_2O$	+19.9
	aq	$MnI_2.6H_2O$	+21.2
sulfate	400	$MnSO_4$	+13.8
	400	$MnSO_4.H_2O$	+11.9
	400	$MnSO_4.7H_2O$	−1.7
Mercuric acetate	aq	$Hg(C_2H_3O_2)_2$	−4.0
bromide	aq	$HgBr_2$	−2.4
chloride	aq	$HgCl_2$	−3.3
nitrate	aq	$Hg(NO_3)_2.\frac{1}{2}H_2O$	−0.7
Mercurous nitrate	aq	$Hg_2(NO_3)_2.2H_2O$	−11.5
Nickel bromide	aq	$NiBr_2$	+19.0
	aq	$NiBr_2.3H_2O$	+0.2

Substance	Dilution*	Formula	Heat, kg.-cal./g.-mole
Nickel chloride	800	$NiCl_2$	+19.23
	800	$NiCl_2.2H_2O$	+10.4
	800	$NiCl_2.4H_2O$	+4.2
	800	$NiCl_2.6H_2O$	−1.15
iodide	aq	NiI_2	+19.4
nitrate	200	$Ni(NO_3)_2$	+11.8
	200	$Ni(NO_3)_2.6H_2O$	−7.5
sulfate	200	$NiSO_4$	+15.1
	200	$NiSO_4.7H_2O$	−4.2
Phosphoric acid, ortho-	400	H_3PO_4	+2.79
	400	$H_3PO_4.\frac{1}{2}H_2O$	−0.1
pyro-	aq	$H_4P_2O_7$	+25.9
	aq	$H_4P_2O_7.1\frac{1}{2}H_2O$	+4.65
Potassium acetate	∞	$KC_2H_3O_2$	+3.55
aluminum sulfate	600	$KAl(SO_4)_2$	+48.5
	600	$KAl(SO_4)_2.3H_2O$	+26.6
		$KAl(SO_4)_2.12H_2O$	−10.1
bicarbonate	2000	$KHCO_3$	−5.1
bromate	∞	$KBrO_3$	−10.13
bromide	∞	KBr	−5.13
carbonate	∞	K_2CO_3	+6.58
	∞	$K_2CO_3.\frac{1}{2}H_2O$	+4.25
		$K_2CO_3.1\frac{1}{2}H_2O$	−0.43
chlorate	∞	$KClO_3$	−10.31
chloride	∞	KCl	−4.404
chromate	2185	K_2CrO_4	−4.9
chrome sulfate	600	$KCr(SO_4)_2$	+55
		$KCr(SO_4)_2.H_2O$	+42
		$KCr(SO_4)_2.2H_2O$	+33
		$KCr(SO_4)_2.6H_2O$	+7
		$KCr(SO_4)_2.12H_2O$	−9.5
cyanide	200	KCN	−3.0
dichromate	1600	$K_2Cr_2O_7$	−17.8
fluoride	∞	KF	+3.96
	∞	$KF.2H_2O$	−1.85
	∞	$KF.4H_2O$	−6.05
hydrosulfide	∞	KHS	+0.86
	∞	$KHS.\frac{1}{4}H_2O$	+1.21
hydroxide	∞	KOH	+12.91
	∞	$KOH.\frac{3}{4}H_2O$	+4.27
	∞	$KOH.H_2O$	+3.48
	∞	$KOH.7H_2O$	+0.86
iodate	∞	KIO_3	−6.93
iodide	∞	KI	−5.23
nitrate	∞	KNO_3	−8.633
oxalate	400	$K_2C_2O_4$	−4.6
		$K_2C_2O_4.H_2O$	−7.5
perchlorate	∞	$KClO_4$	−12.94
permanganate	400	$KMnO_4$	−10.4
phosphate, dihydrogen	aq	KH_2PO_4	+4.7
pyrosulfite	aq	$K_2S_2O_5$	−11.0
	aq	$K_2S_2O_5.\frac{1}{2}H_2O$	−10.22
sulfate	∞	K_2SO_4	−6.32
sulfate, acid	800	$KHSO_4$	−3.10
sulfide	∞	K_2S	−11.0
sulfite	aq	K_2SO_3	+1.8
	aq	$K_2SO_3.H_2O$	+1.37
thiocyanate	∞	$KCNS$	−6.08
thionate, di-	aq	$K_2S_2O_6$	−13.0
thiosulfate	∞	$K_2S_2O_3$	−4.5
Silver acetate	aq	$AgC_2H_3O_2$	−5.4
nitrate	200	$AgNO_3$	−4.4
Sodium acetate	∞	$NaC_2H_3O_2$	+4.085
	∞	$NaC_2H_3O_2.3H_2O$	−4.665
arsenate	500	Na_3AsO_4	+15.6
	500	$Na_3AsO_4.12H_2O$	−12.61
bicarbonate	1800	$NaHCO_3$	−4.1
borate, tetra-	900	$Na_2B_4O_7$	+10.0
	900	$Na_2B_4O_7.10H_2O$	−16.8
bromide	∞	$NaBr$	−0.58
	∞	$NaBr.2H_2O$	−4.57
carbonate	∞	Na_2CO_3	+5.57
	∞	$Na_2CO_3.H_2O$	+2.19
	∞	$Na_2CO_3.7H_2O$	−10.81
	∞	$Na_2CO_3.10H_2O$	−16.22
chlorate	∞	$NaClO_3$	−5.37
chloride	∞	$NaCl$	−1.164
chromate	800	Na_2CrO_4	+2.50
	800	$Na_2CrO_4.4H_2O$	−7.52
	800	$Na_2CrO_4.10H_2O$	−16.0
cyanide	200	$NaCN$	−0.37
	200	$NaCN.\frac{1}{2}H_2O$	−0.92
	200	$NaCN.2H_2O$	−4.41
fluoride	∞	NaF	−0.27
hydrosulfide	∞	$NaHS$	+4.62
	∞	$NaHS.2H_2O$	−1.49

Table 199. Heats of Solution of Inorganic Compounds in Water—(*Concluded*)

Substance	Dilution*	Formula	Heat, kg.-cal./g.-mole
Sodium hydroxide	∞	$NaOH$	+10.18
	∞	$NaOH.\frac{1}{2}H_2O$	+8.17
	∞	$NaOH.\frac{2}{3}H_2O$	+7.08
	∞	$NaOH.\frac{3}{4}H_2O$	+6.48
	∞	$NaOH.H_2O$	+5.17
iodide	∞	NaI	+1.57
	∞	$NaI.2H_2O$	−3.89
metaphosphate	600	$NaPO_3$	+3.97
nitrate	∞	$NaNO_3$	−5.05
nitrite	aq	$NaNO_2$	−3.6
perchlorate	∞	$NaClO_4$	−4.15
phosphate, di-	1600	Na_2HPO_4	+5.21
tri-	1600	Na_3PO_4	+13
phosphate	1600	$Na_3PO_4.12H_2O$	−15.3
di-	1600	$Na_2HPO_4.2H_2O$	−0.82
	1600	$Na_2HPO_4.7H_2O$	−12.04
	1600	$Na_2HPO_4.12H_2O$	−23.18
phosphite, mono-	600	NaH_2PO_3	+0.90
	600	$NaH_2PO_3.2\frac{1}{2}H_2O$	−5.29
di-	800	Na_2HPO_3	+9.30
	800	$Na_2HPO_3.5H_2O$	−4.54
pyrophosphate	1600	$Na_4P_2O_7$	+11.9
	1600	$Na_4P_2O_7.10H_2O$	−11.7
di-	1200	$Na_2H_2P_2O_7$	−2.2
	1200	$Na_2H_2P_2O_7.6H_2O$	−14.0
sulfate	∞	Na_2SO_4	+0.28
	∞	$Na_2SO_4.10H_2O$	−18.74
sulfate, acid	800	$NaHSO_4$	+1.74
	800	$NaHSO_4.H_2O$	+0.15
sulfide	∞	Na_2S	+15.2
	∞	$Na_2S.4\frac{1}{2}H_2O$	+0.09
	∞	$Na_2S.5H_2O$	−6.54
	∞	$Na_2S.9H_2O$	−16.65
sulfite	∞	Na_2SO_3	+2.8
	∞	$Na_2SO_3.7H_2O$	−11.1
thiocyanate	∞	$NaCNS$	−1.83
thionate, di-	aq	$Na_2S_2O_6$	−5.80
	aq	$Na_2S_2O_6.2H_2O$	−11.86

Substance	Dilution*	Formula	Heat, kg.-cal./g.-mole
Sodium thiosulfate	aq	$Na_2S_2O_3$	+2.0
	aq	$Na_2S_2C_3.5H_2O$	−11.30
Stannic bromide	aq	$SnBr_4$	+15.5
Stannous bromide	aq	$SnBr_2$	−1.6
iodide	aq	SnI_2	−5.8
Strontium acetate	∞	$Sr(C_2H_3O_2)_2$	+6.2
	∞	$Sr(C_2H_3O_2)_2.\frac{1}{2}H_2O$	+5.9
bromide	∞	$SrBr_2$	+16.4
	∞	$SrBr_2.H_2O$	+9.25
	∞	$SrBr_2.2H_2O$	+6.5
	∞	$SrBr_2.4H_2O$	+0.4
	∞	$SrBr_2.6H_2O$	−6.1
chloride	∞	$SrCl_2$	+11.54
	∞	$SrCl_2.H_2O$	+6.4
	∞	$SrCl_2.2H_2O$	+2.95
	∞	$SrCl_2.6H_2O$	−7.1
iodide	∞	SrI_2	+20.7
	∞	$SrI_2.H_2O$	+12.65
	∞	$SrI_2.2H_2O$	+10.4
	∞	$SrI_2.6H_2O$	−4.5
nitrate	∞	$Sr(NO_3)_2$	−4.8
	∞	$Sr(NO_3)_2.4H_2O$	−12.4
sulfate	∞	$SrSO_4$	+0.5
Sulfuric acid, pyro-	∞	$H_2S_2O_7$	−18.08
Zinc acetate	400	$Zn(C_2H_3O_2)_2$	+9.8
	400	$Zn(C_2H_3O_2)_2.H_2O$	+7.0
	400	$Zn(C_2H_3O_2)_2.2H_2O$	+3.9
bromide	400	$ZnBr_2$	+15.0
chloride	400	$ZnCl_2$	+15.72
iodide	aq	ZnI_2	+11.6
nitrate	400	$Zn(NO_3)_2.3H_2O$	−5
	400	$Zn(NO_3)_2.6H_2O$	−6.0
sulfate	400	$ZnSO_4$	+18.5
	400	$ZnSO_4.H_2O$	+10.0
	400	$ZnSO_4.6H_2O$	−0.8
	400	$ZnSO_4.7H_2O$	−4.3

Table 200. Heats of Solution of Organic Compounds in Water (at Infinite Dilution and Approximately Room Temperature)

(Recalculated and rearranged from "International Critical Tables," vol. 5, pp. 148–150)

Solute	Heat of Solution, G.-cal./g.-mole Solute*
Acetic acid (solid), $C_2H_4O_2$	−2,251
Acetylacetone, $C_5H_8O_2$	−641
Acetylurea, $C_3H_6N_2O_2$	−6,812
Aconitic acid, $C_6H_6O_6$	−4,206
Ammonium benzoate, $C_7H_9NO_2$	−2,700
picrate	−8,700
succinate (n-)	−3,489
Aniline, hydrochloride, C_6H_8ClN	−2,732
Barium picrate	−4,708
Benzoic acid, $C_7H_6O_2$	−6,501
Camphoric acid, $C_{10}H_{16}O_4$	−502
Citric acid, $C_6H_8O_7$	−5,401
Dextrin, $C_{12}H_{20}O_{10}$	268
Fumaric acid, $C_4H_4O_4$	−5,903
Hexamethylenetetramine, $C_6H_{12}N_4$	4,780
Hydroxybenzamide (m-), $C_7H_7NO_2$	−4,161
(m-), (HCl)	−7,003
(o-), $C_7H_7NO_2$	−4,340
(p-)	−5,392
Hydroxybenzoic acid (o-), $C_7H_6O_3$	−6,350
(p-), $C_7H_6O_3$	−5,781
Hydroxybenzyl alcohol (o-), $C_7H_8O_2$	−3,203
Inulin, $C_{36}H_{62}O_{31}$	−96
Isosuccinic acid, $C_4H_6O_4$	−3,420
Itaconic acid, $C_5H_6O_4$	−5,922
Lactose, $C_{12}H_{22}O_{11}.H_2O$	−3,705
Lead picrate	−7,098
($2H_2O$)	−13,193
Magnesium picrate	14,699
($8H_2O$)	−15,894
Maleic acid, $C_4H_4O_4$	−4,441
Malic acid, $C_4H_6O_5$	−3,150
Malonic acid, $C_3H_4O_4$	−4,493
Mandelic acid, $C_8H_2O_3$	−3,090
Mannitol, $C_6H_{14}O_6$	−5,260
Menthol, $C_{10}H_{20}O$	0
Nicotine dihydrochloride, $C_{10}H_{16}Cl_2N_2$	6,561
Nitrobenzoic acid (m-), $C_7H_5NO_4$	−5,593
(o-), $C_7H_5NO_4$	−5,306
(p-), $C_7H_5NO_4$	−8,891
Nitrophenol (m-), $C_6H_5NO_3$	−5,210
(o-), $C_6H_5NO_3$	−6,310
(p-), $C_6H_5NO_3$	−4,493
Oxalic acid, $C_2H_2O_4$	−2,290
($2H_2O$)	−8,485
Phenol (solid), C_6H_6O	−2,605
Phthalic acid, $C_8H_6O_4$	−4,871
Picric acid, $C_6H_3N_3O_7$	−7,098
Piperic acid, $C_{12}H_{10}O_4$	−10,492
Piperonylic acid, $C_8H_6O_4$	−9,106
Potassium benzoate	−1,506
citrate	2,820
tartrate (n-) (0.5 H_2O)	−5,562
Pyrogallol, $C_6H_6O_3$	−3,705
Pyrotartaric acid	−5,019
Quinone	−3,991
Raffinose, $C_{18}H_{32}O_{16}$ ($5H_2O$)	−9,703
Resorcinol, $C_6H_6O_2$	−3,960
Silver malonate (n-)	−9,799
Sodium citrate (tri-)	5,270
picrate	−6,441
potassium tartrate	−1,817
($4H_2O$)	−12,342
succinate (n-)	2,390
($6H_2O$)	−10,994
tartrate (n-)	−1,121
($2H_2O$)	−5,882
Strontium picrate	7,887
($6H_2O$)	−14,412
Succinic acid, $C_4H_6O_4$	−6,405
Succinimide, $C_4H_5NO_2$	−4,302
Sucrose, $C_{12}H_{22}O_{11}$	−1,319
Tartaric acid (d-)	−3,451
Thiourea, CH_4N_2S	−5,330
Urea, CH_4N_2O	−3,609
acetate	−8,795
formate	−7,194
nitrate	−10,803
oxalate	−17,806
Vanillic acid	−5,160
Vanillin	−5,210
Zinc picrate	−11,496
($8H_2O$)	−15,894

* + denotes heat evolved, and − denotes heat absorbed. All values are positive unless otherwise noted. The data in the "International Critical Tables" were calculated by E. Anderson.

ENGINEERING THERMODYNAMIC PROPERTIES

EXPLANATION OF TABLES

Pure Substances and Mixtures of Invariant Composition

Tabular Presentation of Saturation Properties. The thermodynamic quantities with which the chemical engineer is most frequently concerned are

Pressure, p.
Temperature, t or T (if absolute).
Specific volume, v.
Specific internal energy, u or e.
Specific enthalpy, h.
Specific entropy, s.
Quality, fraction of vapor in a mixture of liquid and vapor, x.

The subscript f will be used to denote the saturated liquid and g the saturated vapor. If a quantity appears without a subscript, it refers to a mixture of liquid and vapor or to a superheated vapor.

It is well known that two thermodynamic properties define the state of a pure substance, which in turn defines all the remaining properties. Any two of the above list except the first two, which are mutually dependent, may be used for this definition of state. The ordinary way of presenting the properties of the saturated state is to list the specific volume, enthalpy, and entropy of both saturated liquid and vapor as functions of either pressure or temperature. The second coordinate for definition of the state is not shown in the table but is supplied by the user. Thus, if at a specified pressure (or temperature) the quality (the second coordinate) is fixed, one can determine the other properties by these relations:

$$v = (1 - x)v_f + xv_g$$
$$h = (1 - x)h_f + xh_g$$
$$s = (1 - x)s_f + xs_g$$

On the other hand, if at a specified pressure (or temperature) the specific volume (the second coordinate) is fixed, one can determine the quality by the first of the above equations and the other properties by the remaining ones.

It may be noted that the original sources of the tables to follow often give the latent heat of evaporation and latent entropy of evaporation. These have been deleted from these tables to save space, but one has only to observe that these figures are equal to the enthalpy or entropy of the saturated liquid subtracted from that of the saturated vapor.

Tabular Presentation of Superheated Properties. In this case, one is interested in the same properties as before except the quality, which has no meaning here. The definition of the state is in terms of two variables, nearly always pressure and temperature, although almost any other two could be used. The usual information presented is the specific volume, enthalpy, and entropy. These may then be utilized to compute the specific internal energy by the relation

$$u = h - \frac{pv}{J}$$

where J is the factor for converting the pv product to the same unit of energy as u and h.

A note of caution should be made here to emphasize that, in the solution of problems involving both the saturated and superheated regions, it is imperative that the basis of the compilation (*i.e.*, the state or states at which enthalpy and entropy are assigned arbitrary zero values) be the same for both regions. In the tabulations to follow, those substances for which this is not true will be so noted.

One particular case of a related nature is that of air, oxygen, and nitrogen. The properties of these three materials on the same basis are often required for the solution of problems pertaining to the separation of air. The bases of the two pure gases in the following tables are the same, but they differ from that for air. The difference is relatively easily resolved, however, as follows: One may assume that, at 500°R. and 1 atm., air is an ideal mixture of nitrogen and oxygen, *i.e.*, its enthalpy may be computed from the sum of the enthalpies of its components. According to Table 202, the enthalpy of air is 220.10 B.t.u./lb. or 220.10 × 28.85 = 6350 B.t.u./lb.-mole. Interpolating in Tables 234 and 236, the enthalpies of nitrogen and oxygen are at 500°F. and 1 atm. 2729 and 3095 cal./g.-mole. These, when converted to B.t.u./lb.-mole, become 4912 and 5571. Assuming ideality, the enthalpy of the mixture is

$$4912 \times 0.790 + 5571 \times 0.210 = 5050 \quad \text{B.t.u./lb.-mole.}$$

Therefore (6350 − 5050) or 1300 B.t.u./lb.-mole must be subtracted from the air values, converted to the pound-mole basis, in order to make them comparable with the oxygen and nitrogen values when these are converted to the pound-mole basis. The entropy is not computed in the same way as the enthalpy, since an entropy of mixing contributes to the entropy of even an ideal mixture. Ordinarily, however, there is little need of having the entropies on a common basis.

Graphical Presentation. Both the saturated and superheated properties are usually presented in one thermodynamic diagram, but the two coordinates used for definition of the state are much different from those found in tabular presentations. The most common pairs are temperature and entropy and enthalpy and entropy (Mollier), although several other coordinate sets are found. Among the figures presented here, two are enthalpy-entropy (steam and carbon monoxide), and the rest are temperature-entropy charts.

References. The tables and diagrams to follow are abridged from the published sources cited. Most of those sources give considerable detail as to the original data used and the means of preparation of the table or diagram. In addition, certain other literature references listed below should be consulted for the general method of preparation and utility of such compilations.

General References: Dodge, "Chemical Engineering Thermodynamics," McGraw-Hill, New York, 1944. Keenan, "Thermodynamics," Wiley, New York, 1941. Keenan and Keyes, "Thermodynamic Properties of Steam," Wiley, New York, 1936. Bošnjaković, "Technische Thermodynamik," T. Steinkopf, Leipzig, 1935. Goff, "Notes on Thermodynamics," 3d ed., Swift, St. Louis, 1939. Sage and Lacey, "Volumetric and Phase Behavior of Hydrocarbons," Stanford University Press, Stanford University, Calif. 1939. Weber, "Thermodynamics for Chemical Engineers," Wiley, New York, 1939.

Mixtures of Variable Composition

When attention is turned to mixtures of variable composition, a new quantity for each new component, namely, the concentration of that component, is added to the previous list. This so complicates the problem that most of the other thermodynamic properties are abandoned, and the diagram is usually limited to the representation of enthalpy, concentration, and temperature. Such charts are frequently called "Merkel charts" after the German engineer who used them extensively.

The relatively few Merkel charts that have been prepared fall into two groups. One is the case of the relatively nonvolatile solute, with which the chart affords its greatest utility in the solution of problems pertaining to the heat effects accompanying mixing and evaporation. The second group is the case of volatile solvent and solute with which the chart is usually applied to distillation problems. In the latter case, phase-equilibrium data are usually included.

The source of each chart to follow is listed, and most of such sources go into considerable detail about the original data and the method of preparation of the charts. The reference states, mentioned in the figures to follow, are elaborated there. In addition, however, a few general references pertaining to the preparation or use of such charts are listed below.

General References: *For enthalpy-concentration diagrams:* Dodge, "Chemical Engineering Thermodynamics," McGraw-Hill, New York, 1944. Merkel, *Z. Ver. deut. Ing.*, **72**, 109 (1928); *Z. ges. Kälte-Ind.*, **35**, 130 (1928). Ponchon, *Tech. Moderne*, **13**, 20–24, 55–58 (1921). Savarit, *Arts et métiers*, pp. 65, 142, 178, 241, 266, 307 (1922). Bosnjakovič, "Technische Thermodynamik," vols. 1 and 2, T. Steinkopf, Leipzig, 1935. Hougen and Watson, "Chemical Process Principles," Part I, Wiley, New York, 1943.

Table 201. Properties of Saturated Air*

Absolute pressure, atm.	Temperature		Enthalpy, B.t.u./lb.		Entropy, B.t.u./(lb.)(°R.)		Absolute pressure, atm.	Temperature		Enthalpy, B.t.u./lb.		Entropy, B.t.u./(lb.)(°R.)	
	Dew point, °R.	Bubble point, °R.	Liquid	Vapor	Liquid	Vapor		Dew point, °R.	Bubble point, °R.	Liquid	Vapor	Liquid	Vapor
p			h_f	h_g	s_f	s_g	p			h_f	h_g	s_f	s_g
1	147.17	141.70	46.33	134.55	0	0.6110	8	188.55	184.68	67.75	140.20	.1260	.5134
2	158.78	153.61	52.00	136.80	0.0363	0.5818	10	194.42	190.89	71.13	140.32	.1427	.5018
4	172.39	167.72	59.00	138.83	.0782	.5500	15	206.08	203.33	78.46	140.08	.1763	.4773
5	177.28	172.84	61.48	139.38	.0925	.5376	20	215.23	213.17	84.70	139.30	.2056	.4602
6	181.49	177.25	63.75	139.80	.1057	.5294	30	229.61	228.80	96.95	135.35	.2563	.4238
							37.25	238.34		118.75		0.3453	

* Williams, *Trans. Am. Inst. Chem. Engrs.*, **39**, 93 (1943).

Table 202. Properties of Superheated Air

h, enthalpy, B.t.u./lb.; s, entropy, B.t.u./(lb.)(°R.); pressure, atm.

Temp. °R.	1 atm.		5 atm.		10 atm.		20 atm.		60 atm.		100 atm.		140 atm.		180 atm.		220 atm.	
T	h	s	h	s	h	s	h	s	h	s	h	s	h	s	h	s	h	s
160	137.72	0.6316	......	......	......	......	......	......	......	......	......	......	......	......	......	......	......	......
180	142.60	.6603	140.10	0.5416	......	......	......	......	......	......	......	......	......	......	......	......	......	......
200	147.53	.6863	145.24	.5687	141.90	0.5098	......	......	......	......	......	......	......	......	......	......	......	......
220	152.40	.7095	150.42	.5934	147.66	.5373	140.90	0.4676	......	......	......	......	......	......	......	......	......	......
240	157.27	.7307	155.62	.6160	153.28	.5617	147.68	.4971	101.94	0.2670	98.23	0.2368	96.95	0.2194	97.21	0.2082	97.86	0.1990
260	162.13	.7502	160.70	.6364	158.76	.5836	154.15	.5230	122.74	.3501	110.35	.2853	107.60	.2620	107.02	.2475	107.15	.2361
280	167.00	.7682	165.70	.6549	164.11	.6035	160.38	.5461	140.83	.4173	123.79	.3351	118.67	.3031	116.90	.2841	116.38	.2703
300	171.90	.7851	170.77	.6724	169.36	.6216	166.30	.5665	151.75	.4551	136.78	.3799	129.50	.3404	126.74	.3180	125.48	.3017
320	176.75	.8007	175.85	.6888	174.52	.6382	171.99	.5849	160.24	.4825	147.97	.4160	140.03	.3744	136.42	.3493	134.42	.3306
340	181.60	.8154	180.73	.7035	179.60	.6536	177.43	.6014	167.45	.5043	157.39	.4446	149.75	.4039	145.73	.3775	143.16	.3571
360	186.50	.8294	185.70	.7177	184.63	.6680	182.69	.6164	173.93	.5229	165.43	.4676	158.68	.4294	154.64	.4029	151.69	.3814
380	191.30	.8424	190.60	.7310	189.61	.6815	187.83	.6303	180.04	.5394	172.76	.4874	166.83	.4514	162.88	.4252	159.93	.4037
400	196.13	.8548	195.46	.7435	194.56	.6942	192.90	.6433	185.91	.5544	179.60	.5049	174.30	.4706	170.55	.4449	167.80	.4239
420	200.90	.8664	200.30	.7553	199.43	.7060	197.92	.6555	191.62	.5684	186.07	.5207	181.25	.4876	177.76	.4625	175.17	.4419
440	205.73	.8777	205.15	.7665	204.37	.7175	202.90	.6671	197.22	.5814	192.29	.5352	187.90	.5030	184.61	.4784	182.10	.4580
460	210.53	.8884	209.90	.7771	209.20	.7283	207.90	.6782	202.78	.5937	198.36	.5487	194.30	.5172	191.21	.4931	188.78	.4729
480	215.35	.8986	214.66	.7872	214.07	.7386	212.83	.6887	208.29	.6055	204.35	.5614	200.53	.5305	197.70	.5069	195.39	.4869
500	220.10	.9083	219.45	.7970	218.96	.7486	217.92	.6991	213.76	.6166	210.29	.5736	206.84	.5434	204.15	.5201	201.96	.5003

* Williams, *Trans. Am. Inst. Chem. Engrs.*, **39**, 93 (1943).

Table 203. Properties of Saturated Ammonia*

Temp., °F.	Abs. pressure, lb./sq. in.	Volume, cu. ft./lb.		Enthalpy, B.t.u./lb.		Entropy, B.t.u./(lb.)(°R.)		Temp., °F.	Abs. pressure, lb./sq. in.	Volume, cu. ft./lb.		Enthalpy, B.t.u./lb.		Entropy, B.t.u./(lb.)(°R.)	
		Liquid	Vapor	Liquid	Vapor	Liquid	Vapor			Liquid	Vapor	Liquid	Vapor	Liquid	Vapor
t	p	v_f	v_g	h_f	h_g	s_f	s_g	t	p	v_f	v_g	h_f	h_g	s_f	s_g
−60	5.55	0.02278	44.73	−21.2	589.6	−0.0517	1.4769	24	52.59	....	5.443	69.1	618.9	.1528	1.2897
−50	7.67	.02299	33.08	−10.6	593.7	− .0256	1.4497	28	57.28	....	5.021	73.5	619.9	.1618	1.2825
−40	10.41	.02322	24.86	0.0	597.6	.0000	1.4242	32	62.29	....	4.637	77.9	621.0	.1708	1.2755
−30	13.90	....	18.97	10.7	601.4	.0250	1.4001	36	67.63	....	4.289	82.3	622.0	.1797	1.2686
−20	18.30	.02369	14.68	21.4	605.0	.0497	1.3774	40	73.32	.02533	3.971	86.8	623.0	.1885	1.2618
−16	20.34	....	13.29	25.6	606.4	.0594	1.3686	50	89.19	.02564	3.294	97.9	625.2	.2105	1.2453
−12	22.56	....	12.06	30.0	607.8	.0690	1.3600	60	107.6	.02597	2.751	109.2	627.3	.2322	1.2294
− 8	24.97	....	10.97	34.3	609.2	.0786	1.3516	70	128.8	.02632	2.312	120.5	629.1	.2537	1.2140
− 4	27.59	....	9.991	38.6	610.5	.0880	1.3433	80	153.0	.02668	1.955	132.0	630.7	.2749	1.1991
0	30.42	.02419	9.116	42.9	611.8	.0975	1.3352	90	180.6	.02707	1.661	143.5	632.0	.2958	1.1846
4	33.47	....	8.333	47.2	613.0	.1069	1.3273	100	211.9	.02747	1.419	155.2	633.0	.3166	1.1705
8	36.77	....	7.629	51.6	614.3	.1162	1.3195	110	247.0	.02790	1.217	167.0	633.7	.3372	1.1566
12	40.31	....	6.996	56.0	615.5	.1254	1.3118	120	286.4	.02836	1.047	179.0	634.0	.3576	1.1427
16	44.12	....	6.425	60.3	616.6	.1346	1.3043	125	307.8	.02860	0.973	183.9	634.0	.3659	1.1372
20	48.21	.02474	5.910	64.7	617.8	.1437	1.2969								

* *U. S. Bur. Standards Circ.* 142, 1923.

Table 204. Properties of Superheated Ammonia*

v, volume, cu. ft./lb.; h, enthalpy, B.t.u /lb.; s, entropy, B.t.u./(lb.)(°R.)

Absolute pressure, lb. per sq. in. (saturation temperature, °F., in parentheses)

Temp., °F.	5 (−63.11)			7 (−52.88°)			10 (−41.34)			14 (−29.76)			18 (−20.61)		
	v	h	s	v	h	s	v	h	s	v	h	s	v	h	s
Sat.	49.31	588.5	1.4857	36.01	592.5	1.4574	25.81	597.1	1.4276	18.85	601.4	1.3996	14.90	604.8	1.3787
−50	51.05	595.2	1.5025	36.29	594.0	1.4611									
−40	52.36	600.3	1.5149	37.25	599.3	1.4739									
−30	53.67	605.4	1.5269	38.19	604.5	1.4861	26.58	603.2	1.4420						
−20	54.97	610.4	1.5385	39.13	609.6	1.4979	27.26	608.5	1.4542	19.33	606.8	1.4119	14.93	605.1	1.3795
−10	56.26	615.4	1.5498	40.07	614.7	1.5094	27.92	613.7	1.4659	19.82	612.2	1.4241	15.32	610.7	1.3921
0	57.55	620.4	1.5608	41.00	619.8	1.5206	28.58	618.9	1.4773	20.30	617.6	1.4358	15.70	616.2	1.4042
10	58.84	625.4	1.5716	41.93	624.9	1.5314	29.24	624.0	1.4884	20.78	622.8	1.4472	16.08	621.6	1.4158
20	60.12	630.4	1.5821	42.85	629.9	1.5421	29.90	629.1	1.4992	21.26	628.0	1.4582	16.46	626.9	1.4270
30	61.41	635.4	1.5925	43.77	635.0	1.5525	30.55	634.2	1.5097	21.73	633.2	1.4688	16.83	632.2	1.4380
40	62.69	640.4	1.6026	44.69	640.0	1.5627	31.20	639.3	1.5200	22.20	638.4	1.4793	17.20	637.5	1.4486
50	63.96	645.5	1.6125	45.61	645.0	1.5727	31.85	644.4	1.5301	22.67	643.6	1.4896	17.57	642.7	1.4590
60	65.24	650.5	1.6223	46.53	650.1	1.5825	32.49	649.5	1.5400	23.14	648.7	1.4996	17.94	647.9	1.4691
70	66.51	655.5	1.6319	47.44	655.2	1.5921	33.14	654.6	1.5497	23.60	653.9	1.5094	18.30	653.1	1.4790
80	67.79	660.6	1.6413	48.36	660.2	1.6016	33.78	659.7	1.5593	24.06	659.0	1.5191	18.67	658.4	1.4887

* *U. S. Bur. Standards Circ.* 142, 1923.

Table 204. Properties of Superheated Ammonia*—(*Continued*)

Temp. °F.	5 (−63.11)			7 (−52.88°)			10 (−41.34)			14 (−29.76)			18 (−20.61)		
	v	*h*	*s*	*v*	*h*	*s*	*v*	*h*	*s*	*v*	*h*	*s*	*v*	*h*	*s*
90	69.06	665.6	1.6506	49.27	665.3	1.6110	34.42	664.8	1.5687	24.53	664.2	1.5285	19.03	663.6	1.4983
100	70.33	670.7	1.6598	50.18	670.4	1.6202	35.07	670.0	1.5779	24.99	669.4	1.5378	19.39	668.8	1.5077
110	71.60	675.8	1.6689	51.09	675.5	1.6292	35.71	675.1	1.5870	25.45	674.5	1.5470	19.75	674.0	1.5169
120	72.87	680.9	1.6778	52.00	680.7	1.6382	36.35	680.3	1.5960	25.91	679.7	1.5560	20.11	679.2	1.5260
130	74.14	686.1	1.6865	52.91	685.8	1.6470	36.99	685.4	1.6049	26.37	684.9	1.5649	20.47	684.4	1.5349
140	75.41	691.2	1.6952	53.82	691.0	1.6557	37.62	690.6	1.6136	26.83	690.1	1.5737	20.83	689.7	1.5438
150	76.68	696.4	1.7038	54.73	696.2	1.6643	38.26	695.8	1.6222	27.29	695.4	1.5824	21.19	694.9	1.5525
160	77.95	701.6	1.7122	55.63	701.4	1.6727	38.90	701.1	1.6307	27.74	700.6	1.5909	21.54	700.2	1.5610
170	79.21	706.8	1.7206	56.54	706.6	1.6811	39.54	706.3	1.6391	28.20	705.9	1.5993	21.90	705.5	1.5695
180	80.48	712.1	1.7289	57.45	711.9	1.6894	40.17	711.6	1.6474	28.66	711.2	1.6076	22.26	710.8	1.5778
190	.....	.....	......	.....	.....	......	40.81	716.9	1.6556	29.11	716.5	1.6159	22.61	716.1	1.5861
200	.....	.....	......	.....	.....	......	41.45	722.2	1.6637	29.57	721.8	1.6240	22.97	721.4	1.5943
220	.....	.....	......	.....	.....	......	.....	.....	......	.....	.....	......	23.68	732.2	1.6103

Temp., °F.	24 (−9.58)			30 (−0.57)			38 (9.42)			48 (19.80)		
	v	*h*	*s*	*v*	*h*	*s*	*v*	*h*	*s*	*v*	*h*	*s*
Sat.	11.39	608.6	1.3549	9.236	611.6	1.3364	7.396	614.7	1.3168	5.934	617.7	1.2973
0	11.67	614.1	1.3670	9.250	611.9	1.3371						
10	11.96	619.7	1.3791	9.492	617.8	1.3497	7.407	615.0	1.3175			
20	12.25	625.2	1.3907	9.731	623.5	1.3618	7.603	621.0	1.3301	5.937	617.8	1.2976
30	12.54	630.7	1.4019	9.966	629.1	1.3733	7.795	626.9	1.3422	6.096	624.0	1.3103
40	12.82	636.1	1.4128	10.20	634.6	1.3845	7.983	632.6	1.3538	6.251	630.0	1.3225
50	13.11	641.4	1.4234	10.43	640.1	1.3953	8.170	638.3	1.3650	6.404	635.9	1.3341
60	13.39	646.7	1.4337	10.65	645.5	1.4059	8.353	643.8	1.3758	6.554	641.6	1.3453
70	13.66	652.0	1.4438	10.88	650.9	1.4161	8.535	649.3	1.3863	6.702	647.3	1.3561
80	13.94	657.3	1.4537	11.10	656.2	1.4261	8.716	654.8	1.3965	6.848	652.9	1.3666
90	14.22	662.6	1.4634	11.33	661.6	1.4359	8.895	660.2	1.4065	6.993	658.5	1.3768
100	14.49	667.8	1.4729	11.55	666.9	1.4456	9.073	665.6	1.4163	7.137	664.0	1.3868
110	14.76	673.1	1.4822	11.77	672.2	1.4550	9.250	671.0	1.4258	7.280	669.5	1.3965
120	15.04	678.4	1.4914	11.99	677.5	1.4642	9.426	676.4	1.4352	7.421	675.0	1.4061
130	15.31	683.6	1.5004	12.21	682.9	1.4733	9.602	681.8	1.4444	7.562	680.5	1.4154
140	15.58	688.9	1.5093	12.43	688.2	1.4823	9.776	687.2	1.4534	7.702	685.9	1.4246
150	15.85	694.2	1.5180	12.65	693.5	1.4911	9.950	692.6	1.4623	7.842	691.4	1.4336
160	16.12	699.5	1.5266	12.87	698.8	1.4998	10.12	698.0	1.4711	7.981	696.8	1.4425
170	16.39	704.8	1.5352	13.08	704.2	1.5083	10.30	703.3	1.4797	8.119	702.3	1.4512
180	16.66	710.2	1.5436	13.30	709.6	1.5168	10.47	708.7	1.4883	8.257	707.7	1.4598
190	16.93	715.5	1.5518	13.52	714.9	1.5251	10.64	714.2	1.4966	8.395	713.2	1.4683
200	17.20	720.9	1.5600	13.73	720.3	1.5334	10.81	719.6	1.5049	8.532	718.7	1.4766
220	17.73	731.7	1.5761	14.16	731.1	1.5495	11.16	730.5	1.5212	8.805	729.6	1.4930
240	18.27	742.6	1.5919	14.59	742.0	1.5653	11.50	741.4	1.5371	9.077	740.6	1.5090
260	.....	.....	......	15.02	753.0	1.5808	11.84	752.4	1.5526	9.348	751.7	1.5246
280	.....	.....	......	......	.....	......	12.18	763.5	1.5678	9.619	762.9	1.5399
300	.....	.....	......	......	.....	......	......	.....	......	9.888	774.1	1.5548

Temp., °F.	60 (30.21)			80 (44.40)			100 (56.05)			120 (66.02)		
	v	*h*	*s*	*v*	*h*	*s*	*v*	*h*	*s*	*v*	*h*	*s*
Sat.	4.805	620.5	1.2787	3.655	624.0	1.2545	2.952	626.5	1.2356	2.476	628.4	1.2201
30												
40	4.933	626.8	1.2913	.....	.....	......	.....	.....	......	.....	.....	......
50	5.060	632.9	1.3035	3.712	627.7	1.2619	.....	.....	......	.....	.....	......
60	5.184	639.0	1.3152	3.812	634.3	1.2745	2.985	629.3	1.2409	.....	.....	......
70	5.307	644.9	1.3265	3.909	640.6	1.2866	3.068	636.0	1.2539	2.505	631.3	1.2255
80	5.428	650.7	1.3373	4.005	646.7	1.2981	3.149	642.6	1.2661	2.576	638.3	1.2386
90	5.547	656.4	1.3479	4.098	652.8	1.3092	3.227	649.0	1.2778	2.645	645.0	1.2510
100	5.665	662.1	1.3581	4.190	658.7	1.3199	3.304	655.2	1.2891	2.712	651.6	1.2628
110	5.781	667.7	1.3681	4.281	664.6	1.3303	3.380	661.3	1.2999	2.778	658.0	1.2741
120	5.897	673.3	1.3778	4.371	670.4	1.3404	3.454	667.3	1.3104	2.842	664.2	1.2850
130	6.012	678.9	1.3873	4.460	676.1	1.3502	3.527	673.3	1.3206	2.905	670.4	1.2956
140	6.126	684.4	1.3966	4.548	681.8	1.3598	3.600	679.2	1.3305	2.967	676.5	1.3058
150	6.239	689.9	1.4058	4.635	687.5	1.3692	3.672	685.0	1.3401	3.029	682.5	1.3157
160	6.352	695.5	1.4148	4.722	693.2	1.3784	3.743	690.8	1.3495	3.089	688.4	1.3254
170	6.464	701.0	1.4236	4.808	698.8	1.3874	3.813	696.6	1.3588	3.149	694.3	1.3348
180	6.576	706.5	1.4323	4.893	704.4	1.3963	3.883	702.3	1.3678	3.209	700.2	1.3441
190	6.687	712.0	1.4409	4.978	710.0	1.4050	3.952	708.0	1.3767	3.268	706.0	1.3531
200	6.798	717.5	1.4493	5.063	715.6	1.4136	4.021	713.7	1.3854	3.326	711.8	1.3620
210	6.909	723.1	1.4576	5.147	721.3	1.4220	4.090	719.4	1.3940	3.385	717.6	1.3707
220	7.019	728.6	1.4658	5.231	726.9	1.4304	4.158	725.1	1.4024	3.442	723.1	1.3793
230	.....	.....	......	5.315	732.5	1.4386	4.226	730.8	1.4108	3.500	729.2	1.3877
240	7.238	739.7	1.4819	5.398	738.1	1.4467	4.294	736.5	1.4190	3.557	734.9	1.3960
250	.....	.....	......	5.482	743.8	1.4547	4.361	742.2	1.4271	3.614	740.7	1.4042
260	7.457	750.9	1.4976	5.565	749.4	1.4626	4.428	747.9	1.4350	3.671	746.5	1.4123
270	.....	.....	......	.....	.....	......	.....	.....	......	3.727	752.2	1.4202
280	7.675	762.1	1.5130	5.730	760.7	1.4781	4.562	759.4	1.4507	3.783	758.0	1.4281
290	.....	.....	......	.....	.....	......	.....	.....	......	3.839	763.8	1.4359
300	7.892	773.3	1.5281	5.894	772.1	1.4933	4.695	770.8	1.4660	3.895	769.6	1.4435

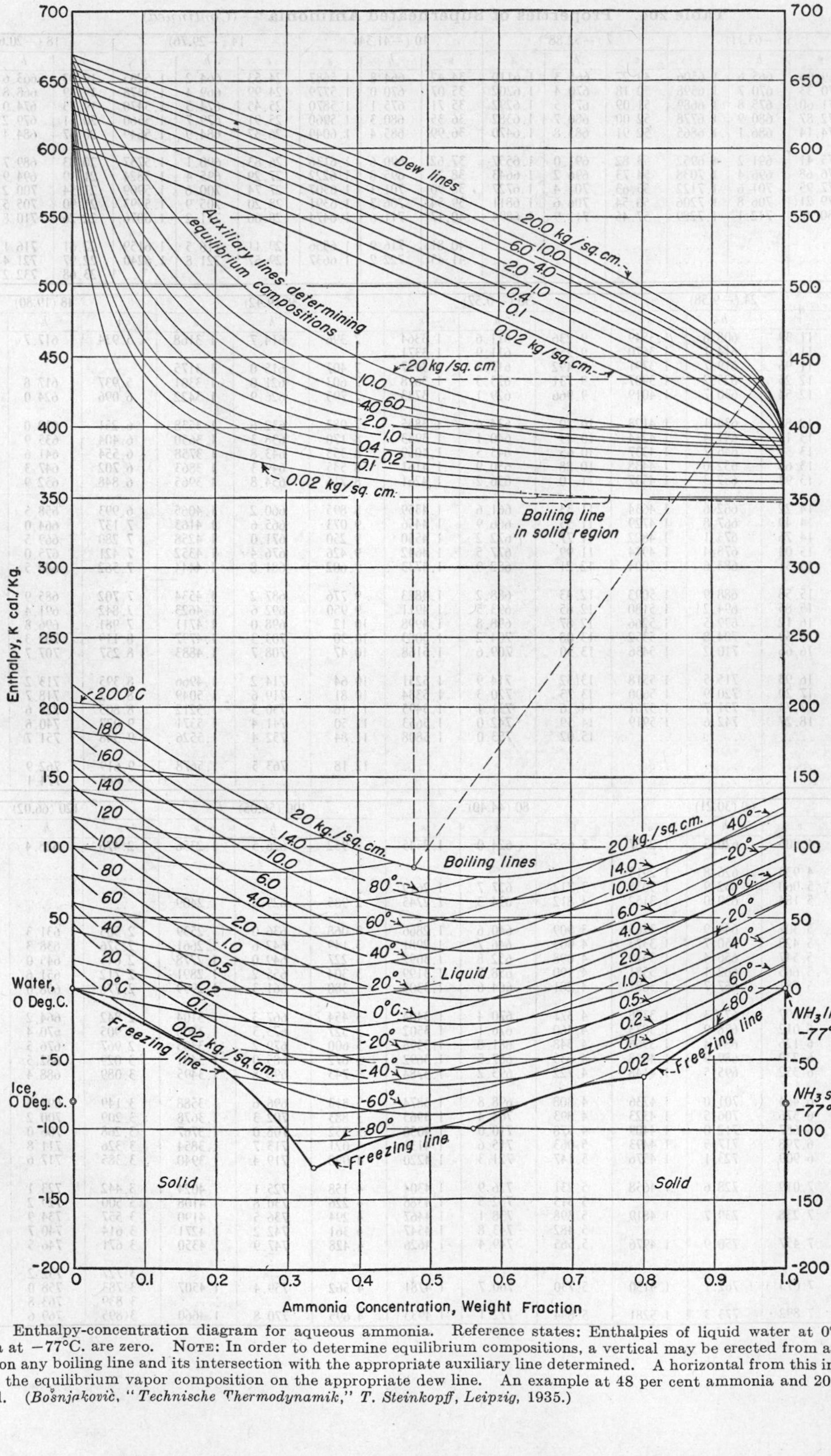

FIG. 13. Enthalpy-concentration diagram for aqueous ammonia. Reference states: Enthalpies of liquid water at 0°C. and liquid ammonia at −77°C. are zero. NOTE: In order to determine equilibrium compositions, a vertical may be erected from any liquid composition on any boiling line and its intersection with the appropriate auxiliary line determined. A horizontal from this intersection will establish the equilibrium vapor composition on the appropriate dew line. An example at 48 per cent ammonia and 20 kg./sq. cm. is indicated. (*Bošnjaković, "Technische Thermodynamik," T. Steinkopff, Leipzig,* 1935.)

Table 204. Properties of Superheated Ammonia*—(Concluded)

Temp., °F.	140 (74.79)			160 (82.64)			180 (89.78)		
	v	h	s	v	h	s	v	h	s
Sat.	2.132	629.9	1.2068	1.872	631.1	1.1952	1.667	632.0	1.1850
80	2.166	633.8	1.2140						
90	2.228	640.9	1.2272	1.914	636.6	1.2055	1.668	632.2	1.1853
100	2.288	647.8	1.2396	1.969	643.9	1.2186	1.720	639.9	1.1992
110	2.347	654.5	1.2515	2.023	651.0	1.2311	1.770	647.3	1.2123
120	2.404	661.1	1.2628	2.075	657.8	1.2429	1.818	654.4	1.2247
130	2.460	667.4	1.2738	2.125	664.4	1.2542	1.865	661.3	1.2364
140	2.515	673.7	1.2843	2.175	670.9	1.2652	1.910	668.0	1.2477
150	2.569	679.9	1.2945	2.224	677.2	1.2575	1.955	674.6	1.2586
160	2.622	686.0	1.3045	2.272	683.5	1.2859	1.999	681.0	1.2691
170	2.675	692.0	1.3141	2.319	689.7	1.2958	2.042	687.3	1.2792
180	2.727	698.0	1.3236	2.365	695.8	1.3054	2.084	693.6	1.2891
190	2.779	704.0	1.3328	2.411	701.9	1.3148	2.126	699.8	1.2987
200	2.830	709.9	1.3418	2.457	707.9	1.3240	2.167	705.9	1.3081
210	2.880	715.8	1.3507	2.502	713.9	1.3331	2.208	712.0	1.3172
220	2.931	721.6	1.3594	2.547	719.9	1.3419	2.248	718.1	1.3262
230	2.981	727.5	1.3679	2.591	725.8	1.3506	2.288	724.1	1.3350
240	3.030	733.3	1.3763	2.635	731.7	1.3591	2.328	730.1	1.3436
250	3.080	739.2	1.3846	2.679	737.6	1.3675	2.367	736.1	1.3521
260	3.129	745.0	1.3928	2.723	743.5	1.3757	2.407	732.0	1.3605
270	3.179	750.8	1.4008	2.766	749.4	1.3838	2.446	748.0	1.3687
280	3.227	756.7	1.4088	2.809	755.3	1.3919	2.484	753.9	1.3768
290	3.275	762.5	1.4166	2.852	761.2	1.3998	2.523	759.9	1.3847
300	3.323	768.3	1.4243	2.895	767.1	1.4076	2.561	765.8	1.3926
320	3.420	780.0	1.4395	2.980	778.9	1.4229	2.637	777.7	1.4081
340	.....	.....	.....	3.064	790.7	1.4379	2.713	789.6	1.4231

Temp., °F.	200 (96.34)			220 (102.42)			240 (108.09)			260 (113.42)			300 (123.21)		
	v	h	s	v	h	s	v	h	s	v	h	s	v	h	s
Sat.	1.502	632.7	1.1756	1.367	633.2	1.1671	1.253	633.6	1.1592	1.155	633.9	1.1518	0.999	634.0	1.1383
110	1.567	643.4	1.1947	1.400	639.4	1.1781	1.261	635.3	1.1621						
120	1.612	650.9	1.2077	1.443	647.3	1.1917	1.302	643.5	1.1764	1.182	639.5	1.1617			
130	1.656	658.1	1.2200	1.485	654.8	1.2045	1.342	651.3	1.1898	1.220	647.8	1.1757	1.023	640.1	1.1487
140	1.698	665.0	1.2317	1.525	662.0	1.2167	1.380	658.8	1.2025	1.257	655.6	1.1889	1.058	648.7	1.1632
150	1.740	671.8	1.2429	1.564	669.0	1.2281	1.416	666.1	1.2145	1.292	663.1	1.2014	1.091	656.9	1.1767
160	1.780	678.4	1.2537	1.601	675.8	1.2394	1.452	673.1	1.2259	1.326	670.4	1.2132	1.123	664.7	1.1894
170	1.820	684.9	1.2641	1.638	682.5	1.2501	1.487	680.0	1.2369	1.359	677.5	1.2245	1.153	672.2	1.2014
180	1.859	691.3	1.2742	1.675	689.1	1.2604	1.521	686.7	1.2475	1.391	684.4	1.2354	1.183	679.5	1.2129
190	1.897	697.7	1.2840	1.710	695.5	1.2704	1.554	693.3	1.2577	1.422	691.1	1.2458	1.211	686.5	1.2239
200	1.935	703.9	1.2935	1.745	701.9	1.2801	1.587	699.8	1.2677	1.453	697.7	1.2560	1.239	693.5	1.2344
210	1.972	710.1	1.3029	1.780	708.2	1.2896	1.619	706.2	1.2773	1.484	704.3	1.2658	1.267	700.3	1.2447
220	2.009	716.3	1.3120	1.814	714.4	1.2989	1.651	712.6	1.2867	1.514	710.7	1.2754	1.294	706.9	1.2546
230	2.046	722.4	1.3209	1.848	720.6	1.3079	1.683	718.9	1.2959	1.543	717.1	1.2847	1.320	713.5	1.2642
240	2.082	728.4	1.3296	1.881	726.8	1.3168	1.714	725.1	1.3049	1.572	723.4	1.2938	1.346	720.0	1.2736
250	2.118	734.5	1.3382	1.914	732.9	1.3255	1.745	731.3	1.3137	1.601	729.7	1.3027	1.372	726.5	1.2827
260	2.154	740.5	1.3467	1.947	739.0	1.3340	1.775	737.5	1.3224	1.630	736.0	1.3115	1.397	732.9	1.2917
270	2.189	746.5	1.3550	1.980	745.1	1.3424	1.805	743.6	1.3308	1.658	742.2	1.3200	1.422	739.2	1.3004
280	2.225	752.5	1.3631	2.012	751.1	1.3507	1.835	749.8	1.3392	1.686	748.4	1.3285	1.447	745.5	1.3090
290	2.260	758.5	1.3712	2.044	757.2	1.3588	1.865	755.9	1.3474	1.714	754.5	1.3367	1.472	751.8	1.3175
300	2.295	764.5	1.3791	2.076	763.2	1.3668	1.895	762.0	1.3554	1.741	760.7	1.3449	1.496	758.1	1.3257
320	2.364	776.5	1.3947	2.140	775.3	1.3825	1.954	774.1	1.3712	1.796	772.9	1.3608	1.544	770.5	1.3419
340	2.432	788.5	1.4099	2.203	787.4	1.3978	2.012	786.3	1.3866	1.850	785.2	1.3763	1.592	782.9	1.3576
360	2.500	800.5	1.4247	2.265	799.5	1.4127	2.069	798.4	1.4016	1.904	797.4	1.3914	1.639	795.3	1.3729
380	2.568	812.5	1.4392	2.327	811.6	1.4273	2.126	810.6	1.4163	1.957	809.6	1.4062	1.686	807.7	1.3878
400	.....	.....	.....	.....	.....	.....	.....	.....	.....	2.009	821.9	1.4206	1.732	820.1	1.4024

* *U. S. Bur. Standards Circ.* 142, 1923.

Table 205. Properties of Saturated 1,3-Butadiene*

Temp., °F.	Abs. pressure, lb/sq. in.	Volume, cu. ft./lb.		Enthalpy, B.t.u./lb.		Entropy, B.t.u./(lb.)(°R)		Temp., °F.	Abs. pressure, lb/sq. in.	Volume, cu ft./lb.		Enthalpy, B.t.u./lb.		Entropy, B.t.u./(lb.)(°R)	
		Liquid	Vapor	Liquid	Vapor	Liquid	Vapor			Liquid	Vapor	Liquid	Vapor	Liquid	Vapor
t	p	v_f	v_g	h_f	h_g	s_f	s_g	t	p	v_f	v_g	h_f	h_g	s_f	s_g
−164.05	0.010	0.02097	5706.	122.61	341.8	0.5904	1.3317	10	10.728	0.02429	8.441	204.97	386.7	.8092	1.1962
−160	.013	.02104	4504.	124.44	342.7	.5973	1.3256	20	13.45	.02453	6.840	210.05	389.6	.8199	1.1942
−140	.045	.02136	1406.	133.55	347.3	.6267	1.2953	30	16.68	.02478	5.595	215.19	392.4	.8305	1.1925
−120	.130	.02170	516.5	142.72	352.0	.6546	1.2707	40	20.49	.02503	4.617	220.40	395.3	.8410	1.1910
−100	.329	.02205	216.7	151.96	356.9	.6810	1.2509	50	24.94	.02529	3.840	225.66	398.2	.8514	1.1899
− 90	.500	.02224	146.4	156.61	359.5	.6938	1.2425	60	30.11	.02557	3.218	231.00	401.1	.8617	1.1890
− 80	.740	.02242	101.44	161.29	362.0	.7062	1.2350	70	36.05	.02585	2.715	236.40	404.0	.8719	1.1883
− 70	1.071	.02261	71.88	165.99	364.7	.7184	1.2283	80	42.84	.02614	2.305	241.88	406.8	.8821	1.1878
− 60	1.516	.02280	52.00	170.72	367.3	.7304	1.2223	90	50.57	.02645	1.968	247.43	409.7	.8922	1.1874
− 50	2.103	.02300	38.33	175.49	370.0	.7422	1.2170	100	59.30	.02678	1.689	253.0	412.5	.9023	1.1872
− 40	2.867	.02320	28.75	180.29	372.7	.7538	1.2123	120	80.11	.02747	1.262	264.6	418.2	.9223	1.1873
− 30	3.841	.02341	21.91	185.14	375.5	.7652	1.2081	140	105.93	.02823	0.9576	276.4	423.6	.9422	1.1877
− 20	5.068	.02362	16.94	190.02	378.2	.7764	1.2045	160	137.4	.02909	.7362	288.6	428.9	.9620	1.1883
− 10	6.592	.02384	13.27	194.96	381.0	.7875	1.2013	180	175.4	.03007	.5715	301.3	433.9	.9817	1.1891
0	8.461	.02406	10.525	199.94	383.9	.7984	1.1985	200	220.5	.03121	.4465	315.	439.	1.001	1.190

* C. H. Meyers, C. S. Cragoe, and E. F. Mueller, *J. Research N. B. S.* **39**, 507 (1947).

Table 206. Properties of Saturated n-Butane*

Abs. pressure, lb./sq. in. p	Temp., °F. t	Volume, cu. ft./lb. Liquid v_f	Volume, cu. ft./lb. Vapor v_g	Enthalpy, B.t.u./lb. Liquid h_f	Enthalpy, B.t.u./lb. Vapor h_g	Entropy, B.t.u./(lb.)(°R.) Liquid s_f	Entropy, B.t.u./(lb.)(°R.) Vapor s_g
30	67.6	0.02747	3.027	4.20	163.88	0.0106	0.3108
40	84.3	0.02802	2.301	13.80	169.11	0.0284	0.3116
50	98.0	0.02850	1.8568	22.09	173.51	0.0407	0.3124
60	109.7	0.02891	1.5556	29.29	177.22	0.0527	0.3132
70	120.1	0.02926	1.3377	35.65	180.49	0.0639	0.3142
80	129.3	0.02960	1.1728	41.50	183.38	0.0741	0.3152
90	137.7	0.02993	1.0433	46.80	186.00	0.0834	0.3161
100	145.5	0.03025	0.9393	51.89	188.42	0.0919	0.3172
125	162.6	0.03104	0.7492	63.70	193.77	0.1105	0.3196

Abs. pressure, lb./sq. in. p	Temp., °F. t	Volume, cu. ft./lb. Liquid v_f	Volume, cu. ft./lb. Vapor v_g	Enthalpy, B.t.u./lb. Liquid h_f	Enthalpy, B.t.u./lb. Vapor h_g	Entropy, B.t.u./(lb.)(°R.) Liquid s_f	Entropy, B.t.u./(lb.)(°R.) Vapor s_g
150	177.3	0.03183	0.6203	74.30	198.33	0.1267	0.3218
175	190.3	0.03264	0.5259	83.17	202.14	0.1408	0.3237
200	202.0	0.03342	0.4536	91.55	205.29	0.1534	0.3252
225	212.7	0.03422	0.3959	99.40	207.88	0.1646	0.3261
250	222.5	0.03497	0.3489	106.68	209.97	0.1755	0.3267
275	231.7	0.03580	0.3095	113.63	211.68	0.1856	0.3270
300	240.2	0.03671	0.2761	120.37	212.97	0.1950	0.3270

* Sage, Webster, and Lacey, *Ind. Eng. Chem.*, **29**, 1188 (1937); with permission.

Table 207. Properties of Saturated and Superheated n-Butane*

p, absolute pressure, lb./sq. in.; v, volume, cu. ft./lb.; h, enthalpy, B.t.u./lb.; s, entropy, B.t.u./(lb.)(°R.)
Parenthetic figures after temperatures are saturation pressures

p	70°F. (31.30) v	h	s	100°F. (51.62) v	h	s	130°F. (80.78) v	h	s	160°F. (120.99) v	h	s
Satd. vapor	2.907	164.65	0.3109	1.7999	174.18	0.3125	1.1617	183.52	0.3152	0.7745	192.94	0.3192
Satd. liquid	0.02754	5.67	.01075	0.02856	23.32	.04295	0.02963	42.00	.07518	.03090	61.84	.10764
14.696	6.451	166.97	.3396	6.847	178.48	.3607	7.237	190.38	.3814	7.627	202.67	.4017
20	4.680	166.36	.3279	4.981	177.98	.3492	5.273	189.94	.3701	5.564	202.29	.3905
30	3.043	164.92	.3125	3.256	176.96	.3345	3.458	189.07	.3556	3.659	201.51	.3762
40	....	....	....	2.390	175.78	.3232	2.549	188.11	.3447	2.704	200.71	.3655
50	....	....	....	1.8669	174.45	.3139	2.003	187.07	.3359	2.131	199.87	.3570
60	....	....	....	....	....	....	1.6358	186.00	.3283	1.7477	198.97	.3497
70	....	....	....	....	....	....	1.3735	184.83	.3215	1.4742	189.05	.3433
80	....	....	....	....	....	....	1.1751	183.62	.3157	1.2681	197.10	.3376
90	....	....	....	....	....	....	....	....	....	1.1072	196.11	.3323
100	....	....	....	....	....	....	....	....	....	0.9783	195.11	.3274

p	190°F. (174.4) v	h	s	220°F. (243.5) v	h	s	250°F. (330.4) v	h	s
Satd. vapor	0.5275	202.06	0.3237	0.3504	209.44	0.3266	0.2417	214.23	0.3267
Satd. liquid	.03262	82.94	.14033	.03476	105.11	.17311	.03797	128.21	.2055
14.696	8.013	215.33	.4217	8.398	228.38	.4414	8.780	241.82	.4607
20	5.851	215.00	.4105	6.136	228.10	.4302	6.419	241.59	.4496
30	3.855	214.35	.3964	4.048	227.54	.4163	4.240	241.10	.4358
40	2.855	213.65	.3859	3.003	226.96	.4060	3.149	240.59	.4256
50	2.255	212.94	.3776	2.377	226.35	.3978	2.4953	240.09	.4175
60	1.8548	212.18	.3706	1.9583	225.71	.3910	2.059	239.55	.4109
70	1.5688	211.38	.3647	1.6601	225.06	.3851	1.7485	238.98	.4051
80	1.3546	210.59	.3590	1.4365	224.37	.3799	1.5155	238.39	.4001
90	1.1867	209.76	.3541	1.2625	223.64	.3752	1.3340	237.78	.3955
100	1.0530	208.90	.3494	1.1232	222.89	.3708	1.1896	237.14	.3913
125	0.8095	206.68	.3390	0.8715	220.88	.3611	0.9286	235.45	.3819
150	.6461	204.36	.3305	.7022	218.72	.3525	.7535	233.62	.3740
175	....	....	....	.5791	216.42	.3445	.6270	231.61	.3669
200	....	....	....	.4844	213.99	.3370	.5303	229.41	.3600
225	....	....	....	.4082	211.41	.3306	.4530	227.03	.3533
250	....	....	....	....	....	....	.3901	224.46	.3468
275	....	....	....	....	....	....	.3366	221.66	.3405
300	....	....	....	....	....	....	.2904	218.56	.3343

* Sage, Webster, and Lacey, *Ind. Eng. Chem.*, **29**, 1188 (1937); with permission.

Table 208. Properties of Saturated Carbon Dioxide*†‡

Temp., °F. t	Abs. pressure, lb./sq. in. p	Volume, cu. ft./lb. Condensed phase* v_f	Volume, cu. ft./lb. Vapor v_g	Enthalpy, B.t.u./lb. Condensed phase* h_f	Enthalpy, B.t.u./lb. Vapor h_g	Entropy B.t.u./(lb.)(°R.) Condensed phase* s_f	Entropy B.t.u./(lb.)(°R.) Vapor s_g
−140	3.18	0.01008	24.320	−121.5	129.2	0.6065	1.3908
−120	8.90	.01018	9.179	−116.0	132.0	.6232	1.3636
−100	22.22	.01032	3.804	−110.1	134.3	.6403	1.3199
− 90	33.98	.01040	2.525	−106.7	135.1	.6499	1.3033
− 80	50.85	.01048	1.700	−102.5	135.7	.6607	1.2881
− 70	74.82	.01059	1.162	− 98.0	135.9	.6724	1.2726
− 69.9	75.10	.01059	1.157	− 97.9	135.9	.6725	1.2724
− 69.9	75.10	.01360	1.1570	− 13.7	135.9	.8885	1.2724
− 60	94.7	.01384	0.9270	− 9.2	136.6	.8997	1.2647
− 50	118.2	.01409	.7492	− 4.7	137.2	.9110	1.2572
− 40	145.8	.01437	.6113	.00	137.8	.9218	1.2503
− 30	177.8	.01466	.5029	4.5	138.2	.9325	1.2436

Temp., °F. t	Abs. pressure, lb./sq. in. p	Volume, cu. ft./lb. Condensed phase* v_f	Volume, cu. ft./lb. Vapor v_g	Enthalpy, B.t.u./lb. Condensed phase* h_f	Enthalpy, B.t.u./lb. Vapor h_g	Entropy B.t.u./(lb.)(°R.) Condensed phase* s_f	Entropy B.t.u./(lb.)(°R.) Vapor s_g
− 20	214.9	.01498	.4168	9.1	138.5	.9430	1.2372
− 10	257.3	.01532	.3472	13.9	138.7	.9532	1.2303
0	305.5	.01570	.2904	18.8	138.9	.9636	1.2247
10	360.2	.01614	.2437	24.0	138.7	.9744	1.2188
20	421.8	.01663	.2049	29.4	138.3	.9856	1.2127
30	490.8	.01719	.1722	35.4	137.8	.9976	1.2067
40	567.8	.01787	.1444	41.7	136.7	1.0092	1.1994
50	653.6	.01868	.1205	48.4	135.0	1.0218	1.1917
60	748.6	.01970	.0994	55.5	132.1	1.0353	1.1826
70	853.4	.02112	.08040	63.7	127.5	1.0500	1.1724
80	968.7	.02370	.06064	73.9	118.7	1.0694	1.1555
87.8	1069.4	.03454	.03454	97.0	97.0	1.1098	1.1098

* Above the solid line the condensed phase is solid; below the line it is liquid.
† "Refrigerating Data Book," 5th ed., American Society of Refrigerating Engineers, New York, 1942.
‡ s_f = 1.0 at 32°F.
h_f = 36.7 at 32°F.

Table 209. Properties of Superheated Carbon Dioxide*,†

v, volume, cu. ft./lb.; *h*, enthalpy, B.t.u./lb.; *s*, entropy, B.t.u/(lb.)(°R.); *p*, absolute pressure, lb./sq. in.

p	−75°F.	−50°F.	0°F.	50°F.	100°F.	150°F.	200°F.	300°F.	400°F.	600°F.	800°F.	1000°F.	1200°F.	1400°F.	1600°F.	1800°F.
v	93.90	100.0	112.2	124.4	136.6	148.8	161.0	185.4	209.7	258.5	307.2	356.0	404.8	453.6	502.3	551.0
1.00 *h*	283.2	288.0	297.8	307.7	318.0	328.4	339.1	361.4	384.7	434.4	487.1	542.4	599.6	658.6	718.8	780.0
s	1.4772	1.4892	1.5112	1.5316	1.5506	1.5684	1.5852	1.6165	1.6451	1.6969	1.7423	1.7829	1.8197	1.8533	1.8838	1.9123
v	9.280	9.902	11.15	12.38	13.61	14.84	16.06	18.51	20.96	25.85	30.73	35.61	40.49	45.36	50.24	55.11
10.0 *h*	282.6	287.5	297.3	307.3	317.7	328.2	339.0	361.3	384.6	434.4	487.1	542.4	599.6	658.6	718.8	780.0
s	1.3733	1.3853	1.4073	1.4277	1.4467	1.4645	1.4813	1.5126	1.5412	1.5930	1.6384	1.6790	1.7158	1.7494	1.7799	1.8084
v	4.586	4.904	5.542	6.119	6.778	7.407	8.016	9.247	10.47	12.92	15.36	17.80	20.24	22.68	25.11	27.55
20.0 *h*	281.9	287.0	296.8	306.8	317.3	327.9	338.8	361.1	384.5	434.3	487.1	542.4	599.6	658.6	718.8	780.0
s	1.3417	1.3538	1.3759	1.3964	1.4154	1.4332	1.4500	1.4813	1.5099	1.5617	1.6071	1.6477	1.6845	1.7181	1.7486	1.7771
v	2.239	2.404	2.738	3.053	3.363	3.688	3.993	4.615	5.230	6.458	7.688	8.901	10.12	11.37	12.56	13.78
40.0 *h*	280.6	285.9	295.8	305.9	316.5	327.4	338.4	360.9	384.3	434.2	487.0	542.4	599.6	658.6	718.8	780.0
s	1.3088	1.3211	1.3435	1.3642	1.3834	1.4014	1.4184	1.4499	1.4787	1.5305	1.5759	1.6165	1.6533	1.6869	1.7174	1.7459
v	......	1.154	1.335	1.498	1.657	1.828	1.982	2.298	2.608	3.226	3.839	4.448	5.060	5.670	6.281	6.887
80.0 *h*	......	283.8	293.8	304.1	315.1	326.4	337.7	360.2	383.9	434.0	486.9	542.3	599.5	658.6	718.8	780.0
s	......	1.2778	1.3044	1.3284	1.3490	1.3679	1.3855	1.4177	1.4468	1.4991	1.5446	1.5852	1.6220	1.6556	1.6861	1.7146
v	......	......	0.8665	0.9799	1.088	1.208	1.311	1.525	1.734	2.148	2.559	2.966	3.373	3.781	4.188	4.592
120 *h*	......	......	291.7	302.2	313.6	325.4	337.0	359.7	383.5	433.8	486.8	542.3	599.5	658.6	718.8	780.0
s	......	......	1.2833	1.3086	1.3297	1.3488	1.3666	1.3993	1.4285	1.4808	1.5263	1.5669	1.6037	1.6373	1.6678	1.6963
v	......	......	0.6305	0.7207	0.8033	0.8986	0.9760	1.139	1.297	1.610	1.918	2.224	2.530	2.836	3.141	3.445
160 *h*	......	......	289.7	300.4	312.1	324.4	336.3	359.1	383.1	433.6	486.6	542.2	599.5	658.6	718.8	780.0
s	......	......	1.2666	1.2928	1.3154	1.3350	1.3529	1.3857	1.4151	1.4675	1.5133	1.5539	1.5907	1.6243	1.6548	1.6833
v	......	......	0.4891	0.5652	0.6376	0.7125	0.7748	0.9075	1.035	1.287	1.534	1.779	2.024	2.269	2.513	2.757
200 *h*	......	......	287.7	298.6	310.6	323.4	335.6	358.5	382.7	433.4	486.5	542.2	599.5	658.5	718.8	780.0
s	......	......	1.2519	1.2805	1.3038	1.3239	1.3421	1.3753	1.4049	1.4574	1.5033	1.5439	1.5807	1.6143	1.6448	1.6733
v	......	......	0.3948	0.4614	0.5237	0.5886	0.6407	0.7532	0.8604	1.071	1.273	1.482	1.687	1.891	2.095	2.297
240 *h*	......	......	285.6	296.7	309.1	322.4	334.9	358.0	382.3	433.1	486.4	542.1	599.5	658.5	718.8	780.0
s	......	......	1.2395	1.2694	1.2940	1.3145	1.3330	1.3671	1.3963	1.4490	1.4948	1.5356	1.5724	1.6060	1.6365	1.6650
v	......	......	......	0.3563	0.4100	0.4636	0.5065	0.5985	0.6868	0.8556	1.021	1.186	1.349	1.513	1.676	1.838
300 *h*	......	......	......	294.0	306.9	320.9	333.9	357.1	381.6	432.8	486.2	542.0	599.4	658.5	718.7	780.0
s	......	......	......	1.2562	1.2813	1.3029	1.3219	1.3560	1.3862	1.4389	1.4848	1.5256	1.5624	1.5960	1.6265	1.6550
v	......	......	......	0.2858	0.3341	0.3780	0.4171	0.4958	0.5693	0.7212	0.8502	0.9874	1.125	1.261	1.397	1.533
360 *h*	......	......	......	291.2	304.6	319.4	332.8	356.3	381.0	432.5	486.0	541.9	599.4	658.5	718.7	779.9
s	......	......	......	1.2436	1.2699	1.2925	1.3124	1.3475	1.3779	1.4307	1.4766	1.5174	1.5542	1.5878	1.6183	1.6468
v	......	......	......	0.2216	0.2652	0.3040	0.3358	0.4022	0.4633	0.5817	0.6950	0.8079	0.9201	1.032	1.142	1.255
440 *h*	......	......	......	287.6	301.6	317.4	331.4	355.1	380.2	432.1	485.8	541.7	599.3	658.4	718.6	779.9
s	......	......	......	1.2282	1.2559	1.2797	1.3006	1.3370	1.3681	1.4215	1.4675	1.5083	1.5451	1.5787	1.6092	1.6377
v	......	......	......	0.1772	0.2174	0.2513	0.2795	0.3374	0.3901	0.4912	0.5881	0.6832	0.7785	0.8733	0.9672	1.062
520 *h*	......	......	......	283.9	298.7	315.4	330.0	354.0	379.4	431.7	485.5	541.5	599.2	658.3	718.6	779.9
s	......	......	......	1.2148	1.2438	1.2687	1.2905	1.3281	1.3599	1.4138	1.4599	1.5007	1.5375	1.5711	1.6010	1.6301
v	......	......	......	0.1452	0.1823	0.2123	0.2383	0.2898	0.3363	0.4250	0.5093	0.5921	0.6747	0.7571	0.8385	0.9202
600 *h*	......	......	......	280.3	295.7	313.4	328.6	352.8	378.6	431.1	485.3	541.4	599.0	658.2	718.6	779.8
s	......	......	......	1.2020	1.2323	1.2583	1.2809	1.3198	1.3525	1.4071	1.4534	1.4942	1.5310	1.5646	1.5951	1.6236
v	......	......	......	......	0.1196	0.1483	0.1712	0.2126	0.2489	0.3173	0.3812	0.4436	0.5060	0.5680	0.6292	0.6906
800 *h*	......	......	......	......	288.2	308.4	325.1	350.0	376.5	430.1	484.7	541.0	598.8	658.0	718.4	779.7
s	......	......	......	......	1.2111	1.2391	1.2631	1.3041	1.3380	1.3935	1.4404	1.4812	1.5180	1.5516	1.5821	1.6106
v	......	......	......	......	......	0.1101	0.1310	0.1663	0.1966	0.2526	0.3048	0.3547	0.4049	0.4545	0.5037	0.5526
1000 *h*	......	......	......	......	......	303.4	321.6	347.1	374.5	429.1	484.0	540.6	598.5	657.8	718.3	779.6
s	......	......	......	......	......	1.2218	1.2472	1.2903	1.3258	1.3828	1.4302	1.4712	1.5080	1.5416	1.5721	1.6006
v	......	......	......	......	......	......	0.1042	0.1356	0.1621	0.2096	0.2531	0.2953	0.3374	0.3789	0.4199	0.4609
1200 *h*	......	......	......	......	......	......	318.4	344.2	372.5	428.1	483.5	540.2	598.2	657.7	718.2	779.5
s	......	......	......	......	......	......	1.2343	1.2791	1.3158	1.3740	1.4216	1.4628	1.4996	1.5332	1.5637	1.5922
v	......	......	......	......	......	......	......	0.1136	0.1375	0.1788	0.2160	0.2529	0.2892	0.3249	0.3601	0.3551
1400 *h*	......	......	......	......	......	......	......	341.4	370.4	427.0	482.9	539.8	598.0	657.5	718.0	779.5
s	......	......	......	......	......	......	......	1.2703	1.3078	1.3668	1.4145	1.4558	1.4927	1.5263	1.5568	1.5853
v	......	......	......	......	......	......	......	......	0.1191	0.1557	0.1898	0.2211	0.2530	0.2843	0.3153	0.3461
1600 *h*	......	......	......	......	......	......	......	......	367.6	426.0	482.3	539.5	597.7	657.2	717.9	779.4
s	......	......	......	......	......	......	......	......	1.3002	1.3602	1.4083	1.4497	1.4867	1.5193	1.5508	1.5793
v	......	......	......	......	......	......	......	......	0.1047	0.1377	0.1675	0.1964	0.2249	0.2528	0.2804	0.3079
1800 *h*	......	......	......	......	......	......	......	......	364.0	424.9	481.7	539.1	597.4	657.0	717.8	779.3
s	......	......	......	......	......	......	......	......	1.2930	1.3539	1.4023	1.4440	1.4812	1.5148	1.5453	1.5738
v	......	......	......	......	......	......	......	......	......	0.1120	0.1368	0.1605	0.1840	0.2070	0.2296	0.2522
2200 *h*	......	......	......	......	......	......	......	......	......	421.7	480.5	538.4	596.9	656.6	717.5	779.2
s	......	......	......	......	......	......	......	......	......	1.3426	1.3925	1.4344	1.4720	1.5057	1.5362	1.5647
v	......	......	......	......	......	......	......	......	......	......	0.1156	0.1357	0.1557	0.1752	0.1945	0.2137
2600 *h*	......	......	......	......	......	......	......	......	......	......	479.3	537.6	596.4	656.3	717.2	779.1
s	......	......	......	......	......	......	......	......	......	......	1.3834	1.4260	1.4640	1.4981	1.5286	1.5571
v	......	......	......	......	......	......	......	......	......	......	0.1001	0.1176	0.1349	0.1519	0.1687	0.1856
3000 *h*	......	......	......	......	......	......	......	......	......	......	478.1	536.8	595.8	656.0	717.0	778.8
s	......	......	......	......	......	......	......	......	......	......	1.3752	1.4183	1.4569	1.4915	1.5220	1.5505

* Sweigert, Weber, and Allen, *Ind. Eng. Chem.*, **38**, 185 (1946); with permission.

† s_f = 1.0 at 32°F. h_f = 180 at 32°F. Therefore, according to the bases of Table 208, the entropies of these two CO_2 tables are consistent, but (180 − 36.7) or 143.3 B.t.u./lb. must be added to the enthalpies of saturated CO_2 to make them consistent with those of superheated CO_2.

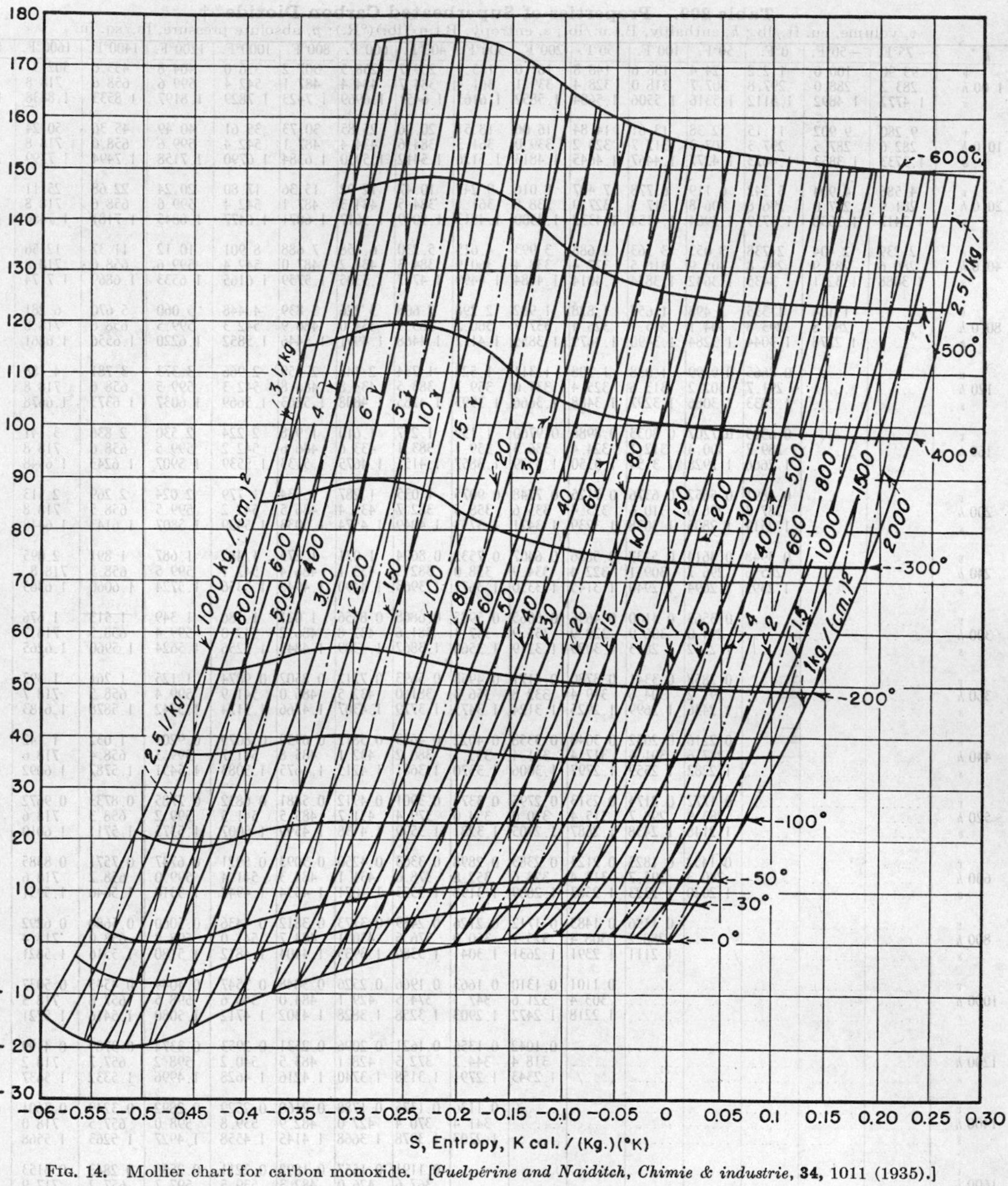

FIG. 14. Mollier chart for carbon monoxide. [*Guelpérine and Naiditch, Chimie & industrie*, **34**, 1011 (1935).]

Table 210. Properties of Saturated Dowtherm A*
(73.5 % diphenyloxide, 26.5 % diphenyl)

Temp., °F. t	Abs. pressure, lb./sq. in. p	Volume, cu. ft./lb. Liquid v_f	Volume, cu. ft./lb. Vapor v_g	Enthalpy, B.t.u./lb. Liquid h_f	Enthalpy, B.t.u./lb. Vapor h_g
53.6	......	......	......	0	164
100	......	......	......	18.0	176
150	......	......	......	38.4	192
200	......	0.0160	833	60.0	210
220	......	.0162	500	69.0	217
240	0.20	.0163	294	78.2	224
260	.29	.0165	179	87.7	232
280	.49	.0166	125	97.5	240
300	.74	.0168	83.3	108.0	250
320	1.1	.0170	52.6	118.0	258
340	1.6	.0171	38.5	128.0	266
360	2.2	.0173	27.8	138.0	275
380	3.0	.0175	20.0	150.0	286
400	4.1	.0176	14.7	162.0	296
420	5.4	.0178	11.4	174.0	306
440	6.9	.0180	8.40	186.0	316
460	8.8	.0181	6.54	198.0	326
480	12	.0183	5.13	210.0	334
500	15	.0185	4.08	222.0	345
520	19	.0188	3.23	234.0	354
540	24	.0190	2.44	247.0	365
560	30	.0193	2.00	260.0	375
580	36	.0195	1.67	274.0	386
600	43	.0198	1.47	288.0	398
620	51	.0201	1.20	302.0	409
640	62	.0204	0.980	316.0	421
660	74	.0207	.826	330.0	432
680	88	.0211	.694	344.0	443
700	103	.0213	.617	358.0	455
720	120	.0218	.526	372.0	465
750	150	.0225	.417	393.0	482

* Dow Chemical Co.

Table 211. Properties of Saturated Ethane*

Temp., °F. t	Abs. pressure, lb./sq. in. p	Volume, cu. ft./lb. Liquid v_f	Volume, cu. ft./lb. Vapor v_g	Enthalpy, B.t.u /lb. Liquid h_f	Enthalpy, B.t.u /lb. Vapor h_g	Entropy, B.t.u./(lb.)(°R.) Liquid s_f	Entropy, B.t.u./(lb.)(°R.) Vapor s_g
−220	0.27	0.02669	310.5	117.6	353.9	0.8249	1.8107
−210	.50	.02694	179.2	123.1	356.8	.8474	1.7833
−200	.85	.02720	107.8	128.7	359.7	.8691	1.7587
−190	1.40	.02746	68.43	134.3	362.6	.8901	1.7366
−180	2.20	.02774	44.90	139.9	365.5	.9133	1.7201
−170	3.36	.02802	30.32	145.5	368.3	.9332	1.7025
−160	4.97	.02831	21.11	151.1	371.2	.9523	1.6865
−150	7.14	.02861	15.10	156.8	373.9	.9710	1.6720
−140	9.97	.02893	11.05	162.3	376.6	.9891	1.6593
−130	13.68	.02923	8.282	168.3	379.3	1.0065	1.6464
−127.55	14.696	.02931	7.741	169.9	379.9	1.0111	1.6435
−120	18.33	.02957	6.316	174.3	381.8	1.0240	1.6346
−110	24.17	.02991	4.876	180.3	384.3	1.0407	1.6241
−100	31.32	.03029	3.830	186.2	386.8	1.0570	1.6143
− 90	39.98	.03067	3.043	192.9	389.0	1.0731	1.6055
− 80	50.34	.03108	2.451	198.2	391.2	1.0890	1.5974
− 70	62.63	.03152	1.994	204.4	393.2	1.1047	1.5896
− 60	77.02	.03199	1.638	210.5	395.2	1.1205	1.5825
− 50	93.76	.03249	1.355	216.8	397.0	1.1362	1.5758
− 40	113.1	.03303	1.127	223.2	398.6	1.1519	1.5697
− 30	135.0	.03359	0.9452	229.7	400.2	1.1672	1.5639
− 20	159.9	.03422	.7983	236.3	401.7	1.1824	1.5583
− 10	188.1	.03494	.6775	243.2	402.9	1.1977	1.5530
0	219.7	.03570	.5754	250.3	403.9	1.2132	1.5476
+ 10	254.9	.03655	.4909	257.5	404.5	1.2289	1.5420
20	294.0	.03754	.4198	265.1	404.9	1.2445	1.5361
30	337.1	.03866	.3595	272.8	404.9	1.2604	1.5301
40	385.0	.03990	.3062	281.0	404.5	1.2762	1.5234
50	437.5	.04144	.2596	298.7	403.4	1.2926	1.5158
60	494.2	.04358	.2164	299.3	401.3	1.3100	1.5064
70	558.3	.04625	.1795	309.8	397.6	1.3284	1.4940
80	630.7	.05063	.1411	323.7	391.4	1.3505	1.4751
90.11	716.0	.0755	.0755	362.0	362.0	1.4234	1.4234

* Barkelew, Valentine, and Hurd, *Trans. Amer. Inst. Chem. Engrs.*, **43**, 25 (1947).

Table 212. Properties of Superheated Ethane*
v, volume, cu. ft./lb.; *h*, enthalpy, B.t.u./lb.; *s*, entropy, B.t.u./(lb.)(°R.)
Parenthetic figures after pressures are saturation temperatures

Temp., °F.	10 lb./sq. in. abs. (−140°F.)			20 lb./sq. in. abs. (−117.2°F.)			30 lb./sq. in. abs. (−102.0°F.)			40 lb./sq. in. abs. (−89.9°F.)			60 lb./sq. in. abs. (−72.0°F.)			80 lb./sq. in. abs. (−57.8°F.)			100 lb./sq. in. abs. (−46.0°F.)		
t	v	h	s	v	h	s	v	h	s	v	h	s	v	h	s	v	h	s	v	h	s
−140	11.05	376.6	1.6593	......	......	......	......	......	......	......	......	......	......	......	......	......	......	......	......	......	......
−120	11.82	383.5	1.6795	......	......	......	......	......	......	......	......	......	......	......	......	......	......	......	......	......	......
−100	12.58	390.4	1.6995	6.149	388.7	1.6491	4.010	386.9	1.6177	......	......	......	......	......	......	......	......	......	......	......	......
− 80	13.32	397.5	1.7186	6.542	396.0	1.6691	4.279	394.5	1.6382	3.151	392.8	1.6157	......	......	......	......	......	......	......	......	......
− 60	14.06	404.7	1.7373	6.926	403.4	1.6885	4.546	402.1	1.6579	3.356	400.7	1.6362	2.164	397.8	1.6039	......	......	......	......	......	......
− 40	14.80	412.0	1.7554	7.306	410.9	1.7072	4.808	409.7	1.6772	3.559	408.5	1.6558	2.305	406.0	1.6248	1.678	403.4	1.6015	1.298	400.6	1.5809
− 20	15.53	419.5	1.7732	7.682	418.5	1.7254	5.066	417.4	1.6959	3.757	416.4	1.6749	2.447	414.2	1.6449	1.790	412.0	1.6224	1.394	409.6	1.6031
0	16.26	427.1	1.7903	8.054	426.2	1.7428	5.320	425.3	1.7139	3.952	424.4	1.6932	2.583	422.5	1.6638	1.897	420.5	1.6419	1.484	418.4	1.6234
20	16.98	434.9	1.8069	8.425	434.1	1.7597	5.572	433.3	1.7310	4.145	432.4	1.7108	2.717	430.7	1.6817	2.002	429.0	1.6600	1.573	427.1	1.6425
40	17.71	442.9	1.8230	8.795	442.1	1.7763	5.822	441.4	1.7477	4.336	440.6	1.7279	2.849	439.1	1.6990	2.105	437.5	1.6777	1.658	435.8	1.6605
60	18.43	451.0	1.8389	9.163	450.3	1.7921	6.070	449.6	1.7641	4.525	448.9	1.7444	2.979	447.6	1.7157	2.206	446.1	1.6949	1.741	444.6	1.6778
80	19.16	459.3	1.8545	9.529	458.7	1.8079	6.319	458.1	1.7802	4.713	457.4	1.7606	3.108	456.2	1.7320	2.305	454.9	1.7116	1.823	453.5	1.6946
90.1	19.52	463.6	1.8624	9.713	463.0	1.8159	6.443	462.4	1.7883	4.808	461.8	1.7687	3.173	460.5	1.7402	2.354	459.3	1.7199	1.863	458.0	1.7030
100	19.89	467.9	1.8701	9.894	467.3	1.8236	6.565	466.7	1.7961	4.900	466.1	1.7765	3.236	464.9	1.7481	2.402	463.8	1.7278	1.902	462.5	1.7111
120	20.61	476.6	1.8854	10.26	476.0	1.8389	6.810	475.5	1.8118	5.086	475.0	1.7920	3.362	473.9	1.7637	2.499	472.8	1.7436	1.982	471.7	1.7274
140	21.33	485.5	1.9006	10.62	485.1	1.8542	7.055	484.6	1.8271	5.271	484.1	1.8074	3.487	483.0	1.7793	2.595	482.0	1.7593	2.060	481.0	1.7430
160	22.05	494.7	1.9160	10.98	494.2	1.8697	7.299	493.7	1.8425	5.456	493.3	1.8228	3.612	492.4	1.7951	2.690	491.5	1.7749	2.137	490.5	1.7587
180	22.77	504.0	1.9312	11.35	503.6	1.8849	7.542	503.1	1.8577	5.639	502.7	1.8382	3.736	501.9	1.8104	2.785	501.0	1.7905	2.214	500.3	1.7744
200	23.49	513.6	1.9460	11.71	513.3	1.8997	7.784	512.9	1.8727	5.822	512.5	1.8530	3.860	511.7	1.8254	2.879	510.9	1.8056	2.290	510.0	1.7897
220	24.20	523.5	1.9606	12.07	523.2	1.9144	8.026	522.8	1.8874	6.005	522.4	1.8678	3.984	521.6	1.8403	2.973	520.9	1.8205	2.366	520.2	1.8046
240	24.92	533.6	1.9751	12.43	533.2	1.9289	8.268	532.8	1.9019	6.187	532.5	1.8823	4.107	531.9	1.8549	3.066	531.2	1.8352	2.442	530.5	1.8194
260	25.64	543.9	1.9895	12.79	543.6	1.9433	8.511	543.2	1.9163	6.370	542.9	1.8967	4.229	542.2	1.8692	3.159	541.6	1.8497	2.517	540.9	1.8341
280	26.35	554.5	2.0037	13.15	554.2	1.9575	8.752	553.9	1.9305	6.552	553.6	1.9112	4.352	553.0	1.8836	3.252	552.3	1.8645	2.592	551.7	1.8485
300	27.07	565.3	2.0181	13.51	565.1	1.9721	8.992	564.7	1.9449	6.734	564.4	1.9256	4.474	563.9	1.8982	3.344	563.3	1.8786	2.666	562.7	1.8632
320	27.79	576.3	2.0323	13.87	576.0	1.9861	9.233	575.8	1.9591	6.917	575.5	1.9400	4.596	575.0	1.9127	3.436	574.4	1.8935	2.740	573.9	1.8776
340	28.50	587.6	2.0469	14.23	587.4	2.0007	9.474	587.1	1.9737	7.097	586.9	1.9544	4.717	586.3	1.9274	3.528	585.8	1.9078	2.815	585.3	1.8924
360	29.22	599.1	2.0609	14.59	598.9	2.0148	9.714	598.7	1.9878	7.278	598.4	1.9686	4.839	597.9	1.9412	3.620	597.4	1.9217	2.888	596.9	1.9064
380	29.93	610.9	2.0752	14.95	610.6	2.0292	9.954	610.4	2.0022	7.459	610.2	1.9828	4.960	609.7	1.9557	3.711	609.3	1.9362	2.962	608.8	1.9209
400	30.65	622.9	2.0891	15.31	622.6	2.0431	10.19	622.4	2.0161	7.639	622.2	1.9968	5.082	621.7	1.9697	3.803	621.3	1.9502	3.036	620.9	1.9350
420	31.37	635.0	2.1029	15.67	634.8	2.0569	10.43	634.5	2.0299	7.819	634.3	2.0108	5.202	633.9	1.9831	3.894	633.5	1.9641	3.109	633.1	1.9488
440	32.08	647.3	2.1166	16.03	647.1	2.0706	10.67	646.9	2.0437	7.999	646.7	2.0246	5.323	646.3	1.9973	3.985	645.8	1.9778	3.183	645.4	1.9626
460	32.80	659.8	2.1301	16.38	659.6	2.0841	10.91	659.4	2.0572	8.179	659.2	2.0382	5.444	658.8	2.0109	4.076	658.4	1.9914	3.256	658.0	1.9763
480	33.51	672.5	2.1440	16.74	672.3	2.0980	11.15	672.1	2.0711	8.359	671.9	2.0518	5.565	671.5	2.0249	4.168	671.2	2.0054	3.329	670.8	1.9900
500	34.23	685.4	2.1575	17.10	685.2	2.1115	11.39	685.0	2.0846	8.539	684.9	2.0655	5.686	684.5	2.0384	4.258	684.2	2.0189	3.402	683.8	2.0040

* Barkelew, Valentine, and Hurd, *Trans. Am. Inst. Chem. Engrs.* **43**, 25 (1947).

Table 212. Properties of Superheated Ethane—(*Concluded*)

Temp., °F.	150 lb./sq. in. abs. (−24.2°F.)			200 lb./sq. in. abs. (−6.2°F.)			300 lb./sq. in. abs. (21.6°F.)			500 lb./sq. in. abs. (61.0°F.)			800 lb./sq. in. abs.			1000 lb./sq. in. abs.			1500 lb./sq. in. abs.		
t	v	h	s	v	h	s	v	h	s	v	h	s	v	h	s	v	h	s	v	h	s
−20	0.8639	403.0	1.5649	......	......	......	......	......	......	......	......	......	......	......	......	......	......	......	......	......	......
0	.9306	412.9	1.5870	0.6508	406.5	1.5582	......	......	......	......	......	......	......	......	......	......	......	......	......	......	......
20	.9964	422.3	1.6073	.7036	417.2	1.5805	......	......	......	......	......	......	......	......	......	......	......	......	......	......	......
40	1.059	431.6	1.6267	.7562	427.1	1.6013	0.4510	416.4	1.5586	......	......	......	......	......	......	......	......	......	......	......	......
60	1.120	440.9	1.6454	.8086	436.8	1.6208	.4921	427.9	1.5812	......	......	......	......	......	......	......	......	......	......	......	......
80	1.179	450.1	1.6633	.8578	446.5	1.6395	.5305	438.5	1.6019	0.2587	418.3	1.5386	......	......	......	......	......	......	......	......	......
90.1	1.207	454.8	1.6721	.8814	451.3	1.6487	.5489	443.9	1.6119	.2761	425.4	1.5526	0.05125	336.1	1.3687	0.04689	329.8	1.3535	0.04338	323.8	1.3282
100	1.235	459.5	1.6803	.9044	456.2	1.6574	.5662	449.3	1.6214	.2909	432.3	1.5648	.07366	362.4	1.4206	.05114	336.4	1.3694	.04488	326.8	1.3407
120	1.291	468.9	1.6966	.9481	466.0	1.6742	.5993	459.7	1.6396	.3169	445.3	1.5870	.1404	413.0	1.5103	.06852	375.9	1.4352	.04822	340.5	1.3651
140	1.346	478.2	1.7129	.9901	475.8	1.6909	.6310	470.2	1.6573	.3401	457.6	1.6077	.1691	432.9	1.5440	.1038	409.3	1.4964	.05401	360.2	1.4045
160	1.400	488.1	1.7289	1.032	485.7	1.7073	.6609	480.7	1.6746	.3625	469.3	1.6273	.1905	448.7	1.5705	.1302	431.5	1.5334	.0649	388.2	1.4497
180	1.454	497.9	1.7450	1.072	495.6	1.7235	.6903	491.0	1.6916	.3835	480.8	1.6459	.2089	463.0	1.5939	.1493	448.8	1.5615	.0779	413.2	1.4903
200	1.507	508.0	1.7606	1.112	505.9	1.7393	.7193	501.6	1.7080	.4039	492.4	1.6638	.2254	476.6	1.6149	.1652	464.8	1.5855	.0898	434.3	1.5223
220	1.559	518.2	1.7758	1.152	516.3	1.7547	.7480	512.3	1.7238	.4236	503.9	1.6808	.2405	489.7	1.6343	.1795	479.6	1.6076	.1007	453.1	1.5490
240	1.611	529.6	1.7909	1.192	526.9	1.7699	.7766	523.1	1.7393	.4433	515.2	1.6970	.2550	502.7	1.6528	.1925	493.5	1.6281	.1112	470.3	1.5735
260	1.662	539.2	1.8057	1.231	537.6	1.7849	.8048	534.1	1.7545	.4623	526.8	1.7135	.2687	515.3	1.6703	.2046	507.2	1.6472	.1210	486.4	1.5964
280	1.713	550.1	1.8204	1.270	548.6	1.7997	.8325	545.4	1.7698	.4804	538.6	1.7293	.2819	528.0	1.6877	.2161	520.7	1.6654	.1306	502.0	1.6182
300	1.763	561.1	1.8352	1.310	559.8	1.8145	.8593	556.8	1.7849	.4980	550.4	1.7450	.2949	540.6	1.7045	.2272	534.0	1.6830	.1398	517.0	1.6386
320	1.813	572.3	1.8496	1.350	571.0	1.8291	.8859	568.2	1.7998	.5156	562.4	1.7604	.3074	553.2	1.7206	.2382	547.0	1.7006	.1484	531.6	1.6573
340	1.863	583.8	1.8644	1.388	582.6	1.8441	.9124	580.0	1.8149	.5329	574.5	1.7760	.3196	565.9	1.7366	.2487	560.1	1.7170	.1564	545.9	1.6757
360	1.914	595.5	1.8785	1.427	594.4	1.8582	.9386	591.9	1.8292	.5499	586.8	1.7908	.3314	578.7	1.7524	.2589	573.3	1.7329	.1639	560.2	1.6928
380	1.964	607.4	1.8933	1.465	606.4	1.8731	.9650	603.9	1.8440	.5665	599.2	1.8061	.3428	591.6	1.7684	.2687	586.5	1.7480	.1712	574.3	1.7103
400	2.013	619.6	1.9073	1.503	618.7	1.8873	.9912	616.3	1.8584	.5830	611.6	1.8207	.3540	604.6	1.7833	.2783	599.9	1.7645	.1782	588.4	1.7269
420	2.063	631.8	1.9214	1.540	631.0	1.9013	1.017	628.7	1.8726	.5992	624.3	1.8352	.3653	617.6	1.7983	.2878	613.4	1.7797	.1853	602.5	1.7431
440	2.113	644.3	1.9352	1.577	643.5	1.9153	1.042	641.4	1.8867	.6152	637.2	1.8494	.3765	630.7	1.8130	.2973	626.8	1.7946	.1922	616.7	1.7588
460	2.162	657.0	1.9489	1.614	656.2	1.9290	1.067	654.2	1.9005	.6315	650.2	1.8634	.3877	644.0	1.8274	.3066	640.3	1.8092	.1992	630.8	1.7741
480	2.212	669.8	1.9629	1.651	669.1	1.9431	1.093	667.1	1.9147	.6475	663.5	1.8778	.3986	657.5	1.8421	.3160	653.9	1.8242	.2062	645.1	1.7895
500	2.261	682.9	1.9766	1.689	682.2	1.9568	1.119	680.2	1.9284	.6637	676.8	1.8918	.4095	671.6	1.8564	.3255	667.6	1.8387	.2132	659.6	1.8044

Table 213. Properties of Saturated Ethylamine*

($C_2H_5NH_2$)

Temp., °F. t	Abs. Pressure, lb./sq. in. p	Volume, cu. ft./lb. vapor v_g	Enthalpy, B.t.u./lb. Liquid h_f	Enthalpy, B.t.u./lb. Vapor h_g	Entropy, B.t.u./(lb.)(°R.) Liquid s_f	Entropy, B.t.u./(lb.)(°R.) Vapor s_g	Temp., °F. t	Abs. Pressure, lb./sq. in. p	Volume, cu. ft./lb. vapor v_g	Enthalpy, B.t.u./lb. Liquid h_f	Enthalpy, B.t.u./lb. Vapor h_g	Entropy, B.t.u./(lb.)(°R.) Liquid s_f	Entropy, B.t.u./(lb.)(°R.) Vapor s_g
− 58	0.355	270.8	− 7.82	284.78	−0.0263	0.7094	23	5.590	20.00	38.91	311.92	.0857	.6512
− 40	.740	134.5	0.00	290.93	.0000	.6931	41	8.960	12.88	50.36	317.70	.1091	.6430
− 22	1.408	72.87	10.93	296.84	.0253	.6788	68	16.896	7.156	68.18	326.50	.1437	.6332
− 4	2.546	41.71	23.01	303.05	.0498	.6664	86	24.72	5.039	80.50	332.46	.1664	.6281
+ 5	3.342	32.32	27.61	306.04	.0619	.6609	113	41.49	3.136	99.52	341.44	.1999	.6223

* "Refrigerating Data Book," 5th ed., American Society of Refrigerating Engineers, New York, 1942.

Table 214. Properties of Saturated Ethyl Chloride*

Temp., °F. t	Abs. pressure, lb./sq. in. p	Volume, cu. ft./lb. Liquid v_f	Volume, cu. ft./lb. Vapor v_g	Enthalpy, B.t.u./lb. Liquid h_f	Enthalpy, B.t.u./lb. Vapor h_g	Entropy, B.t.u./(lb.)(°R.) Liquid s_f	Entropy, B.t.u./(lb.)(°R.) Vapor s_g
−22	2.20	0.01657	34.4	−23.1	158.2	−0.0497	0.3642
−13	2.85	.01669	26.95	−19.2	160.7	− .0410	.3615
− 4	3.66	.01682	21.33	−15.4	163.1	− .0324	.3591
+ 5	4.65	.01695	17.06	−11.6	165.4	− .0241	.3566
14	5.85	.01708	13.77	− 7.7	167.8	− .0159	.3543
23	7.28	.01721	11.21	− 3.8	170.2	− .0079	.3523
32	8.99	.01735	9.21	0.0	172.5	.0000	.3506
41	11.01	.01749	7.62	+ 3.8	174.7	+ .0077	.3488
50	13.37	.01763	6.36	7.7	177.0	.0154	.3475
59	16.11	.01777	5.34	11.6	179.3	.0228	.3459
68	19.29	.01792	4.51	15.4	181.4	.0302	.3446
77	22.94	.01807	3.84	19.2	183.5	.0374	.3433
86	27.10	.01822	3.29	23.1	185.7	.0445	.3423
95	31.82	.01838	2.83	26.9	187.7	.0515	.3412
104	37.17	.01854	2.44	30.8	189.9	.0583	.3402
113	43.16	.01870	2.13	34.6	191.8	.0651	.3394
122	49.88	.01887	1.86	38.5	193.8	.0718	.3386
131	57.36	.01904	1.63	42.3	195.6	.0783	.3377

* *Am. Soc. Refrigerating Eng. Circ.* 9, 1926.

Table 215. Properties of Saturated Ethylene*

Temp., °F. t	Abs. pressure, atm. p	Volume, cu. ft./lb. Liquid v_f	Volume, cu. ft./lb. Vapor v_g	Enthalpy, B.t.u./lb. Liquid h_f	Enthalpy, B.t.u./lb. Vapor h_g	Entropy, B.t.u./(lb.)(°R.) Liquid s_f	Entropy, B.t.u./(lb.)(°R.) Vapor s_g
−272.47	0.0012	....	4064.0	− 68.19	176.5	−0.2826	1.024
−260.0	.0037	....	1405.0	− 60.84	180.1	− .2351	0.972
−240.0	.0177	....	328.6	− 49.16	185.7	− .1887	.880
−220.0	.0606	....	103.1	− 37.56	191.4	− .1382	.817
−200.0	.169	....	39.74	− 26.03	197.1	− .0922	.767
−180.0	.402	....	17.92	− 14.57	202.6	− .0498	.727
−160.0	.837	....	9.047	− 3.12	207.2	− .0103	.692
−154.66	1.0000	0.02818	7.6712	0.0	207.9	.0000	.6814
−140.00	1.5775	.02877	5.005	+ 8.6	210.5	+ .0249	.6563
−120.00	2.7376	.02964	2.987	20.8	214.2	.0648	.6340
−100.00	4.4616	.03122	1.879	32.8	217.2	.0995	.6121
− 80.00	6.8697	.03179	1.732	45.2	219.9	.1374	.5974
− 60.00	10.099	.03308	0.857	57.9	221.9	.1666	.5769
− 40.00	14.0338	.03468	.593	70.8	222.8	.1935	.5556
− 20.00	19.722	.03662	.419	84.7	222.7	.2245	.5383
0.00	26.397	.03912	.301	100.3	221.0	.2577	.5202
+ 20.00	34.55	.04292	.212	119.6	216.6	.2968	.4991
+ 40.00	44.54	.05035	.139	148.7	206.2	.3533	.4683
+ 49.82	50.50	.070	.070	171.8	171.8	.3964	.3964

* York and White, *Trans. Am. Inst. Chem. Engrs.*, **40**, 227 (1944).

Table 216. Properties of Superheated Ethylene*

v, volume, cu. ft./lb.; *h*, enthalpy, B.t.u./lb.; *s*, entropy, B.t.u./(lb.)(°R.)

Abs. pressure, atm.	Temperature, °F.																			
	−140°	−120°	−100°	−80°	−60°	−40°	−20°	0°	20°	40°	60°	100°	140°	180°	220°	260°	320°	380°	440°	500°
v	8.061	8.617	9.168	9.714	10.256	10.794	11.329	11.862	12.497	12.924	13.453	14.51	15.56	16.61	17.66	18.71	20.28	21.85	23.42	24.99
1 *h*	212.2	218.3	224.3	230.6	236.9	243.3	249.9	256.6	263.4	270.4	277.7	292.6	308.3	324.8	342.0	360.0	388.4	418.4	450.4	482 9
s	0.695	0.714	0.732	0.748	0.764	0.780	0.795	0.810	0.825	0.839	0.853	0.881	0.908	0.934	0.961	0.986	1.025	1.062	1.098	1.133
v	.....	4.190	4.479	4.765	5.047	5.324	5.599	5.873	6.144	6.414	6.683	7.217	7.749	8.280	8.808	9.34	10.12	10.91	11.70	12.48
2 *h*	.....	215.9	222.3	228.9	235.4	242.0	248.7	255.6	262.5	269.5	276.9	291.9	307.6	324.2	341.5	359.7	388.1	418.2	450.2	482.7
s	.....	0.659	0.677	0.693	0.710	0.726	0.741	0.756	0.771	0.785	0.799	0.828	0.855	0.881	0.909	0.934	0.974	1.011	1.048	1.083
v	.....	.....	2.130	2.287	2.439	2.588	2.733	2.878	3.019	3.159	3.297	3.571	3.843	4.113	4.380	4.647	5.046	5.443	5.838	6.234
4 *h*	.....	.....	218.2	225.5	232.4	239.3	246.3	253.4	260.6	267.8	275.3	290.6	306.6	323.3	340.6	358.9	387.4	417.6	449.7	482.3
s	.....	.....	0.623	0.641	0.658	0.675	0.691	0.706	0.722	0.737	0.751	0.780	0.807	0.833	0.860	0.886	0.925	0.963	1.000	1.035
v	.....	.....	.....	1.456	1.566	1.673	1.776	1.833	1.877	2.099	2.168	2.356	2.541	2.724	2.905	3.084	3.352	3.620	3.885	4.150
6 *h*	.....	.....	.....	221.6	229.1	236.6	243.9	251.2	258.6	266.0	273.7	289.3	305.4	322.3	339.8	358.1	386.8	417.0	449.2	481.8
s	.....	.....	.....	0.607	0.625	0.642	0.659	0.675	0.691	0.706	0.721	0.750	0.777	0.804	0.831	0.856	0.896	0.934	0.971	1.006
v	.....	.....	.....	.....	0.860	0.936	1.006	1.074	1.140	1.202	1.264	1.382	1.499	1.613	1.724	1.834	1.999	2.161	2.323	2.483
10 *h*	.....	.....	.....	.....	222.4	230.8	238.8	246.6	254.4	262.4	270.4	286.5	303.1	320.3	338.0	356.5	385.4	415.8	448.2	480.9
s	.....	.....	.....	.....	0.578	0.599	0.616	0.632	0.649	0.665	0.680	0.711	0.739	0.766	0.794	0.820	0.860	0.898	0.934	0.969
t	.....	.....	.....	.....	.....	.....	0.614	0.669	0.718	0.765	0.811	0.897	0.978	1.057	1.134	1.209	1.322	1.433	1.542	1.650
15 *h*	.....	.....	.....	.....	.....	.....	230.9	239.9	248.7	257.5	266.2	283.0	300.1	317.7	335.8	354.6	383.8	414.4	447.0	479.9
s	.....	.....	.....	.....	.....	.....	0.573	0.593	0.612	0.630	0.646	0.677	0.707	0.734	0.763	0.789	0.829	0.867	0.903	0.938
v	.....	.....	.....	.....	.....	.....	.....	0.459	0.503	0.545	0.583	0.652	0.717	0.778	0.838	0.897	0.983	1.068	1.151	1.233
20 *h*	.....	.....	.....	.....	.....	.....	.....	232.5	242.5	252.1	261.3	279.2	296.9	315.1	333.5	352.6	382.1	413.0	445.7	478.8
s	.....	.....	.....	.....	.....	.....	.....	0.560	0.582	0.591	0.618	0.651	0.682	0.710	0.740	0.766	0.807	0.845	0.882	0.917
v	.....	.....	.....	.....	.....	.....	.....	.....	0.282	0.318	0.350	0.405	0.454	0.498	0.543	0.585	0.645	0.704	0.760	0.817
30 *h*	.....	.....	.....	.....	.....	.....	.....	.....	226.9	239.5	250.8	271.2	290.2	309.5	328.7	348.3	378.6	410.1	443.3	476.7
s	.....	.....	.....	.....	.....	.....	.....	.....	0.528	0.552	0.574	0.613	0.646	0.676	0.706	0.734	0.775	0.814	0.851	0.887
v	.....	.....	.....	.....	.....	.....	.....	.....	.....	0.192	0.228	0.279	0.323	0.359	0.395	0.428	0.476	0.521	0.565	0.609
40 *h*	.....	.....	.....	.....	.....	.....	.....	.....	.....	221.1	237.3	262.3	283.4	303.7	323.8	344.1	375.1	407.2	440.8	474.5
s	.....	.....	.....	.....	.....	.....	.....	.....	.....	0.503	0.534	0.580	0.617	0.650	0.681	0.709	0.751	0.791	0.828	0.865
v	.....	.....	.....	.....	.....	.....	.....	.....	.....	.....	0.138	0.201	0.243	0.275	0.309	0.334	0.375	0.412	0.448	0.484
50 *h*	.....	.....	.....	.....	.....	.....	.....	.....	.....	.....	218.1	250.3	275.6	297.5	318.7	339.7	371.5	404.2	438.2	472.4
s	.....	.....	.....	.....	.....	.....	.....	.....	.....	.....	0.484	0.550	0.590	0.626	0.659	0.688	0.731	0.772	0.810	0.847
v	.....	.....	.....	.....	.....	.....	.....	.....	.....	.....	0.069	0.144	0.186	0.219	0.248	0.273	0.307	0.340	0.371	0.402
60 *h*	.....	.....	.....	.....	.....	.....	.....	.....	.....	.....	171.0	238.9	267.4	291.2	313.6	335.4	367.8	401.4	435.8	470.3
s	.....	.....	.....	.....	.....	.....	.....	.....	.....	.....	0.390	0.517	0.568	0.607	0.640	0.671	0.716	0.756	0.795	0.833
v	.....	.....	.....	.....	.....	.....	.....	.....	.....	.....	0.051	0.082	0.121	0.150	0.174	0.195	0.224	0.250	0.275	0.301
80 *h*	.....	.....	.....	.....	.....	.....	.....	.....	.....	.....	146.4	205.0	248.3	277.6	302.8	326.4	360.5	395.3	430.8	466.1
s	.....	.....	.....	.....	.....	.....	.....	.....	.....	.....	0.340	0.477	0.522	0.569	0.608	0.641	0.688	0.731	0.770	0.808
v	.....	.....	.....	.....	.....	.....	.....	.....	.....	.....	0.044	0.057	0.082	0.109	0.131	0.149	0.174	0.197	0.219	0.239
100 *h*	.....	.....	.....	.....	.....	.....	.....	.....	.....	.....	143.5	182.5	228.3	263.0	291.3	317.1	353.3	389.6	426.0	462.0
s	.....	.....	.....	.....	.....	.....	.....	.....	.....	.....	0.327	0.401	0.480	0.535	0.579	0.614	0.665	0.709	0.751	0.789
v	.....	.....	.....	.....	.....	.....	.....	.....	.....	.....	0.039	0.046	0.056	0.067	0.081	0.094	0.111	0.128	0.143	0.159
150 *h*	.....	.....	.....	.....	.....	.....	.....	.....	.....	.....	139.1	167.7	199.8	235.0	267.2	296.5	336.5	376.0	414.8	452.6
s	.....	.....	.....	.....	.....	.....	.....	.....	.....	.....	0.308	0.361	0.417	0.474	0.524	0.564	0.619	0.668	0.712	0.752
v	.....	.....	.....	.....	.....	.....	.....	.....	.....	.....	0.037	0.042	0.047	0.054	0.062	0.070	0.085	0.097	0.108	0.119
200 *h*	.....	.....	.....	.....	.....	.....	.....	.....	.....	.....	138.3	163.8	191.1	221.5	252.3	281.7	323.4	364.7	404.9	444.2
s	.....	.....	.....	.....	.....	.....	.....	.....	.....	.....	0.295	0.342	0.392	0.440	0.487	0.529	0.585	0.636	0.682	0.724
v	.....	.....	.....	.....	.....	.....	.....	.....	.....	.....	0.035	0.039	0.043	0.049	0.054	0.059	0.069	0.078	0.088	0.099
250 *h*	.....	.....	.....	.....	.....	.....	.....	.....	.....	.....	138.2	161.8	187.5	215.1	243.9	272.3	314.1	355.8	397.1	437.5
s	.....	.....	.....	.....	.....	.....	.....	.....	.....	.....	0.286	0.331	0.376	0.419	0.462	0.503	0.561	0.612	0.659	0.702
v	.....	.....	.....	.....	.....	.....	.....	.....	.....	.....	0.034	0.038	0.041	0.045	0.049	0.054	0.061	0.069	0.077	0.083
300 *h*	.....	.....	.....	.....	.....	.....	.....	.....	.....	.....	137.9	161.4	186.3	212.1	238.7	265.9	307.7	349.4	391.2	432.0
s	.....	.....	.....	.....	.....	.....	.....	.....	.....	.....	0.277	0.322	0.364	0.404	0.446	0.484	0.541	0.593	0.640	0.684

* York and White, *Trans. Am. Inst. Chem. Engrs.*, **40**, 227 (1944).

Table 217. Properties of Saturated Trichloromonofluoromethane (F-11)*

Temp., °F. t	Abs. pressure, lb./sq. in. p	Volume, cu. ft./lb. Liquid v_f	Volume, cu. ft./lb. Vapor v_g	Enthalpy, B.t.u./lb. Liquid h_f	Enthalpy, B.t.u./lb. Vapor h_g	Entropy, B.t.u./(lb.)(°R.) Liquid s_f	Entropy, B.t.u./(lb.)(°R.) Vapor s_g
−40	0.7391	0.00988	44.21	0.00	87.48	0.0000	0.2085
−30	1.034	.00995	32.33	1.97	88.67	.0046	.2064
−20	1.420	.01002	24.06	3.94	89.87	.0091	.2046
−10	1.920	.01010	18.17	5.91	91.07	.0136	.2030
5	2.931	.01022	12.27	8.88	92.88	.0201	.2009
10	3.352	.01026	10.83	9.88	93.48	.0222	.2003
20	4.342	.01034	8.519	11.87	94.69	.0264	.1991
30	5.557	.01042	6.776	13.88	95.91	.0306	.1981
40	7.032	.01051	5.447	15.89	97.11	.0346	.1972
50	8.804	.01060	4.421	17.92	98.32	.0386	.1964
60	10.90	.01069	3.626	19.96	99.53	.0426	.1958
70	13.40	.01079	2.993	22.02	100.73	.0465	.1951
80	16.31	.01088	2.492	24.09	101.93	.0504	.1947
86	18.28	.01094	2.242	25.34	102.65	.0527	.1944
90	19.69	.01098	2.091	26.18	103.12	.0542	.1942
100	23.60	.01109	1.765	28.27	104.30	.0580	.1938
110	28.09	.01119	1.499	30.40	105.47	.0617	.1935
120	33.20	.01130	1.281	32.53	106.63	.0654	.1933
130	38.96	.01142	1.101	34.67	107.78	.0691	.1931
140	45.50	.01154	0.9505	36.84	108.91	.0727	.1929
150	52.85	.01166	.8240	39.02	110.02	.0763	.1927
160	61.04	.01179	.7176	41.23	111.12	.0798	.1926

* Courtesy Kinetic Chemicals, Inc.

Table 218. Properties of Superheated Trichloromonofluoromethane (F-11)*

v, volume, cu. ft./lb.; *h*, enthalpy, B.t.u./lb.; *s*, entropy, B.t.u./(lb.)(°R.)
Parenthetic figures after pressures are saturation temperatures

Temp., °F. *t*	Abs. pressure 0.7 lb./sq. in. (−41.6°F.) *v*	*h*	*s*
Sat.	46.54	87.29	0.2090
−40	46.69	87.49	.2093
−30	47.81	88.69	.2121
−20	48.93	89.91	.2149
−10	50.06	91.14	.2177
0	51.18	92.38	.2204
10	52.30	93.63	.2231
20	53.42	94.89	.2258
30	54.54	96.16	.2284
40	55.66	97.44	.2310
50	56.77	98.73	.2335
60	57.89	100.03	.2361
70	59.01	101.34	.2386
80	60.13	102.67	.2411
90	61.25	104.00	.2435
100	62.37	105.34	.2459
110	63.49	106.69	.2483
120	64.60	108.06	.2507
130	65.72	109.43	.2531
140	66.84	110.82	.2554
150	67.96	112.22	.2577
160	69.08	113.62	.2600
170	70.20	115.04	.2622
180	71.32	116.47	.2645
190	72.43	117.90	.2667
200	73.55	119.35	.2689
210	74.67	120.81	.2711
220	75.79	122.27	.2733
230	76.90	123.75	.2754
240	78.02	125.24	.2776
250	79.14	126.74	.2797
260	80.26	128.25	.2818
270			

Temp., °F. *t*	Abs. pressure 2.0 lb./sq. in. (−8.6°F.) *v*	*h*	*s*
Sat.	17.50	91.24	0.2028
0	17.83	92.31	.2052
10	18.23	93.56	.2079
20	18.62	94.82	.2105
30	19.02	96.09	.2131
40	19.41	97.37	.2157
50	19.81	98.66	.2183
60	20.20	99.97	.2208
70	20.59	101.28	.2233
80	20.99	102.61	.2258
90	21.38	103.94	.2282
100	21.77	105.28	.2307
110	22.17	106.64	.2331
120	22.56	108.01	.2354
130	22.95	109.38	.2378
140	23.35	110.76	.2401
150	23.74	112.17	.2424
160	24.13	113.57	.2447
170	24.52	114.98	.2470
180	24.92	116.41	.2492
190	25.31	117.85	.2515
200	25.70	119.30	.2537
210	26.10	120.76	.2559
220	26.49	122.22	.2581
230	26.88	123.70	.2602
240	27.27	125.19	.2624
250	27.66	126.69	.2645
260	28.06	128.21	.2666
270	28.45	129.73	.2687
280	28.84	131.26	.2708
290	29.23	132.80	.2728
300	29.62	134.35	.2749

Temp., °F. *t*	Abs. pressure 5.0 lb./sq. in. (25.6°F.) *v*	*h*	*s*
Sat.	7.475	95.38	0.1985
30	7.543	95.94	.1997
40	7.703	97.22	.2023
50	7.863	98.51	.2048
60	8.023	99.82	.2074
70	8.182	101.14	.2099
80	8.341	102.47	.2124
90	8.500	103.80	.2148
100	8.659	105.16	.2173
110	8.818	106.51	.2197
120	8.977	107.88	.2220
130	9.135	109.26	.2244
140	9.294	110.65	.2267
150	9.452	112.05	.2290
160	9.610	113.45	.2313
170	9.769	114.87	.2336
180	9.927	116.30	.2359
190	10.09	117.74	.2381
200	10.25	119.19	.2403
210	10.40	120.65	.2425
220	10.56	122.12	.2447
230	10.72	123.60	.2469
240	10.88	125.09	.2490
250	11.04	126.59	.2511
260	11.19	128.11	.2533
270	11.35	129.63	.2554
280	11.51	131.16	.2574
290	11.67	132.71	.2595
300	11.82	134.27	.2616
310	11.98	135.83	.2636
320	12.14	137.40	.2656
330	12.30	138.98	.2677

Temp., °F. *t*	Abs. pressure 10.0 lb./sq. in. (55.9°F.) *v*	*h*	*s*
Sat.	3.928	99.04	0.1960
60	3.961	99.57	.1970
70	4.042	100.89	.1996
80	4.123	102.23	.2021
90	4.204	103.57	.2045
100	4.285	104.93	.2070
110	4.366	106.29	.2094
120	4.447	107.67	.2118
130	4.528	109.05	.2141
140	4.609	110.44	.2165
150	4.690	111.85	.2188
160	4.771	113.26	.2211
170	5.851	114.69	.2234
180	4.932	116.12	.2257
190	5.012	117.56	.2279
200	5.092	119.01	.2301
210	5.171	120.48	.2323
220	5.251	121.95	.2345
230	5.331	123.43	.2367
240	5.411	124.92	.2388
250	5.491	126.43	.2409
260	5.571	127.95	.2431
270	5.650	129.48	.2452
280	5.730	131.01	.2473
290	5.809	132.56	.2494
300	5.889	134.12	.2514
310	5.969	135.68	.2535
320	6.048	137.26	.2555
330	6.127	138.84	.2575
340	6.206	140.43	.2595
350	6.285	142.04	.2615
360	6.365	143.66	.2635

Temp., °F. *t*	Abs. pressure 20 lb./sq. in. (90.8°F.) *v*	*h*	*s*
Sat.	2.061	103.22	0.1942
100	2.099	104.47	.1964
110	2.140	105.84	.1989
120	2.182	107.23	.2013
130	2.224	108.63	.2036
140	2.266	110.03	.2060
150	2.307	111.44	.2083
160	2.349	112.87	.2107
170	2.390	114.31	.2130
180	2.431	115.75	.2152
190	2.473	117.19	.2175
200	2.514	118.65	.2197
210	2.555	120.12	.2219
220	2.596	121.60	.2241
230	2.637	123.09	.2263
240	2.678	124.59	.2285
250	2.719	126.11	.2306
260	2.760	127.63	.2328
270	2.800	129.16	.2349
280	2.841	130.70	.2370
290	2.881	132.25	.2391
300	2.922	133.82	.2411
310	2.962	135.39	.2432
320	3.003	136.97	.2452
330	3.043	138.56	.2472
340	3.084	140.16	.2493
350	3.124	141.77	.2513
360	3.164	143.40	.2532
370	3.205	145.03	.2552
380	3.245	146.68	.2572
390	3.286	148.33	.2592
400	3.326	149.99	.2611

Temp., °F. *t*	Abs. pressure 40 lb./sq. in. (131.6°F.) *v*	*h*	*s*
Sat.	1.074	107.96	0.1930
140	1.092	109.17	.1950
150	1.115	110.60	.1974
160	1.137	112.04	.1997
170	1.159	113.49	.2021
180	1.181	114.95	.2044
190	1.202	116.42	.2067
200	1.224	117.89	.2089
210	1.246	119.38	.2112
220	1.268	120.88	.2134
230	1.289	122.39	.2156
240	1.311	124.91	.2178
250	1.332	125.44	.2199
260	1.353	126.97	.2221
270	1.374	128.52	.2242
280	1.396	130.08	.2263
290	1.417	131.64	.2284
300	1.438	133.22	.2305
310	1.459	134.80	.2326
320	1.480	136.40	.2347
330	1.501	138.00	.2367
340	1.522	139.61	.2387
350	1.543	141.24	.2407
360	1.564	142.88	.2427
370	1.584	144.52	.2447
380	1.605	146.18	.2467
390	1.626	147.83	.2487
400	1.647	149.51	.2506
410	1.668	151.19	.2526
420	1.688	152.89	.2545
430	1.709	154.59	.2564
440	1.729	156.30	.2583

Temp., °F. *t*	Abs. pressure 60 lb./sq. in. (158.8°F.) *v*	*h*	*s*
Sat.	0.7297	110.99	0.1926
160	.7314	111.17	.1929
170	.7471	112.65	.1953
180	.7627	114.14	.1976
190	.7782	115.63	.1999
200	.7937	117.13	.2022
210	.8090	118.64	.2045
220	.8241	120.16	.2068
230	.8391	121.68	.2090
240	.8541	123.21	.2112
250	.8691	124.76	.2134
260	.8839	126.31	.2156
270	.8987	127.87	.2177
280	.9134	129.44	.2198
290	.9281	131.01	.2220
300	.9427	132.60	.2241
310	.9574	134.20	.2262
320	.9719	135.81	.2282
330	.9863	137.42	.2303
340	1.001	139.05	.2323
350	1.015	140.69	.2344
360	1.030	142.33	.2364
370	1.044	143.98	.2384
380	1.059	145.65	.2404
390	1.073	147.32	.2424
400	1.087	149.01	.2443
410	1.101	150.70	.2463
420	1.115	152.41	.2483
430	1.130	154.12	.2502
440	1.144	155.84	.2521
450	1.158	157.57	.2540
460	1.172	159.31	.2559

* Courtesy Kinetic Chemicals, Inc.

Table 219. Properties of Saturated Dichlorodifluoromethane (F-12)*

Temp., °F.	Abs. pressure, lb./sq. in.	Volume, cu. ft./lb.		Enthalpy, B.t.u./lb.		Entropy, B.t.u./(lb.)(°R.)		Temp., °F.	Abs. pressure, lb./sq. in.	Volume, cu. ft./lb.		Enthalpy, B.t.u./lb.		Entropy, B.t.u./(lb.)(°R.)	
		Liquid	Vapor	Liquid	Vapor	Liquid	Vapor			Liquid	Vapor	Liquid	Vapor	Liquid	Vapor
t	p	v_f	v_g	h_f	h_g	s_f	s_g	t	p	v_f	v_g	h_f	h_g	s_f	s_g
−40	9.32	0.0106	3.911	0	73.50	0	0.17517	32	44.77	.0115	.908	15.21	81.63	.03323	.16876
−30	12.02	.0107	3.088	2.03	74.70	0.00471	.17387	36	48.13	.0116	.848	16.10	82.27	.03502	.16854
−20	15.28	.0108	2.474	4.07	75.87	.00940	.17275	40	51.68	.0116	.792	17.00	82.71	.03680	.16833
−16	16.77	.0108	2.271	4.89	76.34	.01126	.17232	50	61.39	.0118	.673	19.27	83.78	.04126	.16785
−12	18.37	.0109	2.088	5.72	76.81	.01310	.17124	60	72.41	.0119	.575	21.57	84.82	.04568	.16741
− 8	20.08	.0109	1.922	6.57	77.29	.01496	.17158	70	84.82	.0121	.493	23.90	85.82	.05009	.16701
− 4	21.91	.0110	1.772	7.41	77.75	.01682	.17123	80	98.76	.0123	.425	26.28	86.80	.05446	.16662
0	23.87	.0110	1.637	8.25	78.21	.01869	.17091	90	114.3	.0125	.368	28.70	87.74	.05882	.16624
+ 4	25.96	.0111	1.514	9.10	78.67	.02052	.17060	100	131.6	.0127	.319	31.16	88.62	.06316	.16584
8	28.18	.0111	1.403	9.96	79.13	.02235	.17030	110	150.7	.0129	.277	33.65	89.43	.06749	.16542
12	30.56	.0112	1.301	10.82	79.59	.02419	.17001	120	171.8	.0132	.240	36.16	90.15	.07180	.16495
16	33.08	.0112	1.207	11.70	80.05	.02601	.16974	130	194.9	.0134	.208	38.69	90.76	.07607	.16438
20	35.75	.0113	1.121	12.55	80.49	.02783	.16949	140	220.2	.0138	.180	41.24	91.24	.08024	.16363
24	38.58	.0113	1.043	13.44	80.95	.02963	.16926								
28	41.59	.0114	0.973	14.32	81.39	.03143	.16900								

* From *Am. Soc. Refrigerating Eng. Circ.* 12, by permission.

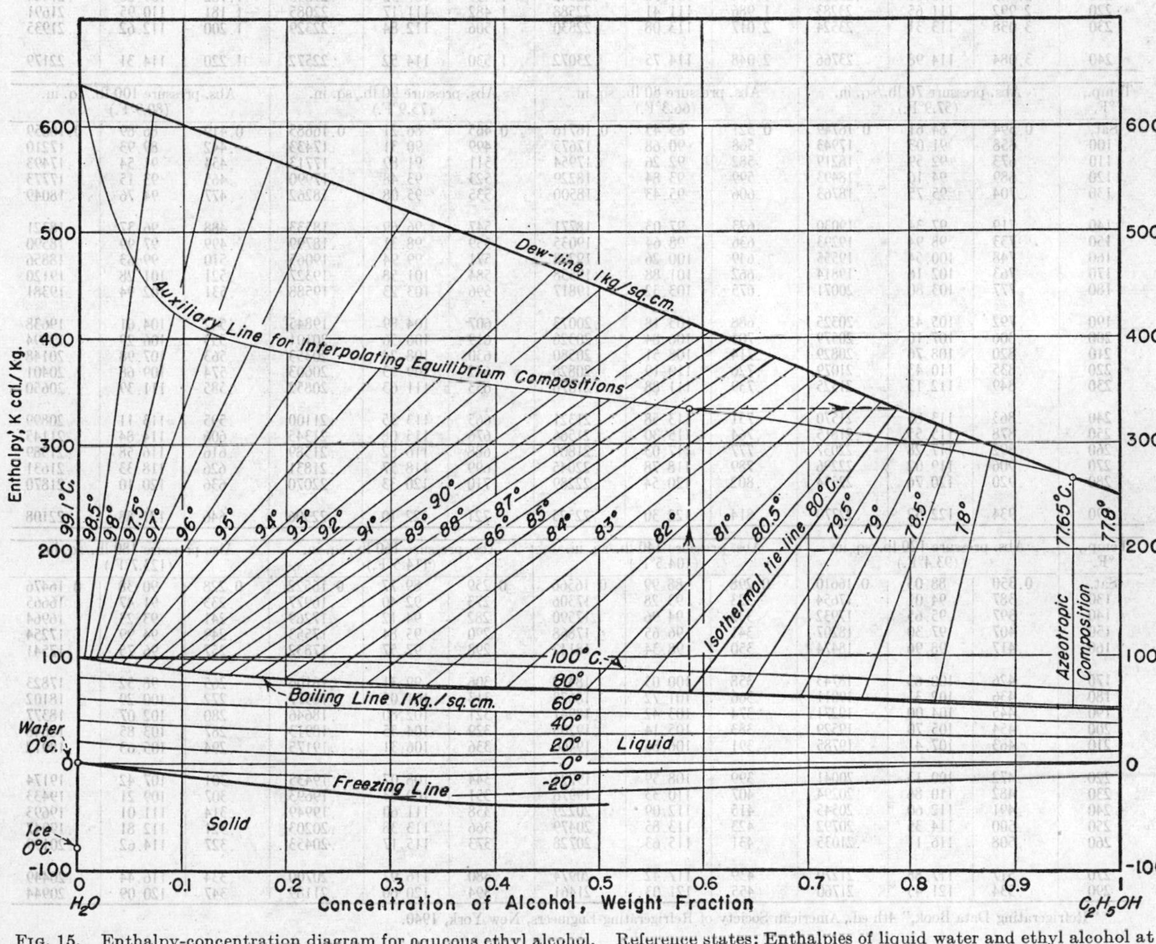

FIG. 15. Enthalpy-concentration diagram for aqueous ethyl alcohol. Reference states: Enthalpies of liquid water and ethyl alcohol at 0°C. are zero. NOTE: In order to interpolate equilibrium compositions, a vertical may be erected from any liquid composition on the boiling line and its intersection with the auxiliary line determined. A horizontal from this intersection will establish the equilibrium vapor composition on the dew line. (*Bošnjaković, "Technische Thermodynamik," T. Steinkopff, Leipzig,* 1935.)

Table 220. Properties of Superheated Dichlorodifluoromethane (F-12)*

v, volume, cu. ft./lb.; *h*, enthalpy, B.t.u./lb.; *s*, entropy, B.t.u./(lb.)(°R.)
Parenthetic figures after pressures are saturation temperatures

Temp., °F.	Abs. pressure 20 lb./sq. in. (−8.2°F.)			Abs. pressure 30 lb./sq. in. (11.1°F.)			Abs. pressure 40 lb./sq. in. (25.9°F.)			Abs. pressure 50 lb./sq. in. (38.3°F.)		
t	*v*	*h*	*s*	*v*	*h*	*s*	*v*	*h*	*s*	*v*	*h*	*s*
Sat.	1.925	77.27	0.17160	1.323	79.47	0.17008	1.009	81.16	0.16914	0.817	82.52	0.16841
50	2.203	85.40	.18853	1.448	85.03	.18138	1.070	84.65	.17612	.842	84.24	.17187
60	2.250	86.85	.19138	1.480	86.48	.18420	1.095	86.11	.17896	.863	85.72	.17475
70	2.297	88.31	.19415	1.512	87.95	.18690	1.120	87.60	.18178	.884	87.22	.17760
80	2.343	89.78	.19688	1.544	89.43	.18974	1.144	89.09	.18455	.904	88.72	.18040
90	2.390	91.26	.19959	1.576	90.91	.19249	1.169	90.58	.18731	.924	90.23	.18317
100	2.437	92.75	.20229	1.608	92.41	.19519	1.194	92.09	.19004	.944	91.75	.18591
110	2.483	94.26	.20494	1.640	93.93	.19787	1.218	93.62	.19272	.964	93.29	.18862
120	2.530	95.78	.20759	1.672	95.46	.20053	1.242	95.15	.19538	.984	94.83	.19132
130	2.577	97.31	.21020	1.703	97.00	.20315	1.267	96.70	.19803	1.004	96.39	.19397
140	2.623	98.85	.21280	1.735	98.54	.20577	1.291	98.26	.20066	1.024	97.96	.19662
150	2.669	100.40	.21537	1.767	100.11	.20836	1.315	99.83	.20325	1.044	99.54	.19923
160	2.716	101.97	.21792	1.799	101.69	.21092	1.340	101.42	.20583	1.064	101.14	.20182
170	2.762	103.56	.22045	1.829	103.28	.21344	1.364	103.02	.20838	1.084	102.75	.20439
180	2.808	105.15	.22297	1.860	104.88	.21597	1.388	104.63	.21092	1.103	104.36	.20694
190	2.854	106.76	.22545	1.891	106.49	.21846	1.412	106.25	.21343	1.123	105.98	.20946
200	2.901	108.38	.22794	1.923	108.12	.22096	1.435	107.88	.21592	1.142	107.62	.21196
210	2.947	110.01	.23039	1.954	109.76	.22342	1.459	109.52	.21840	1.162	109.28	.21444
220	2.992	111.65	.23283	1.986	111.41	.22588	1.482	111.17	.22085	1.181	110.95	.21691
230	3.038	113.31	.23524	2.017	113.08	.22830	1.506	112.84	.22329	1.200	112.62	.21935
240	3.084	114.98	.23766	2.048	114.75	.23072	1.530	114.52	.22572	1.220	114.31	.22179

Temp., °F.	Abs. pressure 70 lb./sq. in. (57.9°F.)			Abs. pressure 80 lb./sq. in. (66.3°F.)			Abs. pressure 90 lb./sq. in. (73.9°F.)			Abs. pressure 100 lb./sq. in. (80.9°F.)		
Sat.	0.594	84.61	0.16749	0.521	85.45	0.16716	0.465	86.21	0.16685	0.419	86.89	0.16659
100	.658	91.03	.17943	.568	90.68	.17675	.499	90.31	.17433	.442	89.93	.17210
110	.673	92.59	.18219	.582	92.26	.17954	.511	91.89	.17713	.454	91.54	.17493
120	.689	94.16	.18493	.599	93.84	.18229	.523	93.48	.17990	.465	93.15	.17773
130	.704	95.75	.18763	.606	95.43	.18500	.535	95.08	.18262	.477	94.76	.18049
140	.719	97.34	.19030	.623	97.03	.18771	.547	96.69	.18533	.488	96.37	.18321
150	.733	98.94	.19293	.636	98.64	.19035	.559	98.31	.18799	.499	97.99	.18590
160	.748	100.54	.19555	.649	100.26	.19298	.571	99.94	.19065	.510	99.63	.18856
170	.763	102.16	.19814	.662	101.88	.19558	.584	101.58	.19327	.521	101.28	.19120
180	.777	103.80	.20071	.675	103.52	.19817	.596	103.23	.19588	.531	102.94	.19381
190	.792	105.45	.20325	.688	105.18	.20073	.607	104.89	.19845	.542	104.61	.19638
200	.806	107.10	.20579	.701	106.84	.20328	.619	106.56	.20101	.553	106.29	.19894
210	.820	108.76	.20829	.714	108.51	.20580	.630	108.24	.20353	.563	107.98	.20148
220	.835	110.43	.21079	.726	110.19	.20828	.642	109.93	.20603	.574	109.68	.20401
230	.849	112.13	.21325	.739	111.88	.21076	.653	111.63	.20852	.585	111.39	.20650
240	.863	113.83	.21570	.751	113.58	.21321	.665	113.35	.21100	.595	113.11	.20899
250	.878	115.55	.21815	.764	115.30	.21566	.676	115.08	.21345	.606	114.84	.21145
260	.892	117.28	.22057	.777	117.03	.21809	.688	116.82	.21589	.616	116.58	.21389
270	.906	119.02	.22296	.789	118.78	.22045	.699	118.57	.21831	.626	118.33	.21631
280	.920	120.76	.22534	.802	120.54	.22289	.710	120.33	.22070	.636	120.10	.21870
290	.934	122.52	.22770	.814	122.30	.22525	.721	122.10	.22306	.646	121.88	.22108

Temp., °F.	Abs. pressure 120 lb./sq. in. (93.4°F.)			Abs. pressure 140 lb./sq. in. (104.5°F.)			Abs. pressure 160 lb./sq. in. (114.5°F.)			Abs. pressure 180 lb./sq. in. (123.7°F.)		
Sat.	0.350	88.05	0.16610	0.298	88.99	0.16566	0.259	89.77	0.16522	0.228	90.38	0.16476
130	.387	94.01	.17654	.323	93.28	.17306	.273	92.40	.16977	.233	91.47	.16665
140	.397	95.65	.17932	.332	94.96	.17590	.282	94.12	.17269	.241	93.23	.16964
150	.407	97.30	.18207	.341	96.65	.17868	.290	95.84	.17553	.249	94.99	.17254
160	.417	98.96	.18474	.350	98.34	.18142	.298	97.57	.17832	.257	96.75	.17541
170	.426	100.63	.18743	.358	100.03	.18412	.306	99.31	.18106	.265	98.52	.17823
180	.436	102.31	.19011	.366	101.72	.18678	.313	101.05	.18377	.272	100.29	.18102
190	.445	104.00	.19271	.374	103.42	.18941	.321	102.80	.18646	.280	102.07	.18377
200	.454	105.70	.19529	.383	105.14	.19205	.329	104.55	.18913	.287	103.85	.18648
210	.463	107.41	.19785	.391	106.86	.19466	.336	106.31	.19175	.294	105.63	.18912
220	.472	109.13	.20041	.399	108.59	.19724	.344	108.07	.19435	.301	107.42	.19174
230	.482	110.86	.20294	.407	110.33	.19976	.351	109.83	.19693	.307	109.21	.19433
240	.491	112.60	.20545	.415	112.09	.20229	.358	111.60	.19949	.314	111.01	.19693
250	.500	114.35	.20792	.423	113.85	.20479	.366	113.38	.20203	.321	112.81	.19947
260	.508	116.11	.21035	.431	115.63	.20728	.373	115.17	.20453	.327	114.62	.20199
270	.517	117.88	.21279	.439	117.42	.20974	.380	116.97	.20700	.334	116.44	.20449
290	.534	121.45	.21760	.455	121.03	.21461	.394	120.60	.21189	.347	120.09	.20944

* "Refrigerating Data Book," 4th ed., American Society of Refrigerating Engineers, New York, 1940.

Table 221. Properties of Saturated Dichloromonofluoromethane (F-21)*

Temp., °F.	Abs. pressure, lb./sq. in.	Volume, cu. ft./lb.		Enthalpy, B.t.u./lb.		Entropy, B.t.u./(lb.)(°R.)	
		Liquid	Vapor	Liquid	Vapor	Liquid	Vapor
t	p	v_f	v_g	h_f	h_g	s_f	s_g
−40	1.358	0.01058	32.09	0.00	114.56	0.0000	0.2730
−30	1.888	.01066	23.61	2.36	115.76	.0055	.2695
−20	2.578	.01075	17.66	4.71	116.96	.0109	.2663
−10	3.463	.01084	13.43	7.07	118.17	.0162	.2633
0	4.582	.01093	10.35	9.44	119.37	.0214	.2606
5	5.243	.01097	9.132	10.63	119.97	.0240	.2593
10	5.978	.01102	8.085	11.81	120.57	.0265	.2581
20	7.699	.01112	6.392	14.21	121.78	.0316	.2559
30	9.793	.01122	5.112	16.61	122.98	.0365	.2538
40	12.32	.01132	4.130	19.04	124.19	.0414	.2519
50	15.33	.01142	3.370	21.49	125.39	.0463	.2502
60	18.90	.01153	2.773	23.98	126.60	.0511	.2486
70	23.08	.01164	2.300	26.49	127.79	.0559	.2471
80	27.96	.01176	1.923	29.03	128.98	.0606	.2458
86	31.23	.01183	1.733	30.56	129.68	.0634	.2450
90	33.58	.01188	1.619	31.59	130.14	.0652	.2446
100	40.04	.01200	1.371	34.18	131.29	.0699	.2434
110	47.40	.01213	1.169	36.79	132.42	.0745	.2424
120	55.75	.01226	1.001	39.46	133.53	.0791	.2414
130	65.15	.01240	0.8623	42.13	134.61	.0837	.2405
140	75.72	.01254	.7457	44.86	135.66	.0882	.2396
150	87.51	.01269	.6476	47.62	136.68	.0927	.2388
160	100.6	.01284	.5646	50.43	137.69	.0972	.2381

* Courtesy Kinetic Chemicals, Inc.

Table 222. Properties of Superheated Dichloromonofluoromethane (F-21)*

v, volume, cu. ft./lb.; h, enthalpy, B.t.u./lb.; s, entropy, B.t.u./(lb.)(°R.)

Parenthetic figures after pressures are saturation temperatures

Temp., °F.	Abs. pressure 1.2 lb./sq. in. (−43.6°F.)		
t	v	h	s
Sat.	36.02	114.13	0.2744
−40	36.34	114.57	.2754
−30	37.21	115.80	.2783
−20	38.08	117.03	.2812
−10	38.95	118.28	.2840
0	39.83	119.54	.2867
10	40.70	120.80	.2894
20	41.57	122.08	.2921
30	42.44	128.38	.2948
40	43.32	124.69	.2975
50	44.19	126.01	.3001
60	45.06	127.35	.3027
70	45.93	128.71	.3053
80	46.80	130.09	.3079
90	47.67	131.48	.3104
100	48.54	132.88	.3129
110	49.42	134.29	.3154
120	50.29	135.71	.3179
130	51.16	137.14	.3203
140	52.03	138.59	.3228
150	52.90	140.06	.3252
160	53.77	141.55	.3276
170	54.64	143.05	.3300
180	55.51	144.56	.3324
190	56.38	146.08	.3348
200	57.25	147.61	.3371
210	58.12	149.16	.3394
220	58.99	150.73	.3418
230	59.86	152.31	.3441
240	60.73	153.90	.3464
250	61.60	155.51	.3486

Temp., °F.	Abs. pressure 2 lb./sq. in. (−28.2°F.)		
t	v	h	s
Sat.	22.37	115.98	0.2689
−20	22.80	116.99	.2712
−10	23.33	118.24	.2740
0	23.85	119.49	.2768
10	24.38	120.76	.2795
20	24.90	122.04	.2822
30	25.43	123.34	.2849
40	25.95	124.65	.2876
50	26.47	125.98	.2902
60	27.00	127.33	.2928
70	27.52	128.69	.2954
80	28.05	130.07	.2979
90	28.57	131.45	.3005
100	29.09	132.85	.3030
110	29.62	134.26	.3055
120	30.14	135.68	.3080
130	30.66	137.12	.3104
140	31.19	138.57	.3129
150	31.71	140.03	.3153
160	32.23	141.52	.3177
170	32.76	143.02	.3201
180	33.28	144.53	.3225
190	33.80	146.05	.3249
200	34.33	147.59	.3272
210	34.85	149.14	.3296
220	35.37	150.71	.3319
230	35.89	152.29	.3342
240	36.42	153.88	.3365
250	36.94	155.59	.3388

Temp., °F.	Abs. pressure 4 lb./sq. in. (−4.9°F.)		
t	v	h	s
Sat.	11.74	118.78	0.2619
0	11.87	119.40	.2633
10	12.14	120.66	.2660
20	12.40	121.95	.2687
30	12.66	123.25	.2714
40	12.93	124.57	.2741
50	13.19	125.90	.2767
60	13.45	127.25	.2793
70	13.72	128.61	.2819
80	13.98	129.99	.2845
90	14.24	131.37	.2870
100	14.51	132.77	.2895
110	14.77	134.18	.2920
120	15.03	135.61	.2945
130	15.29	137.04	.2970
140	15.56	138.49	.2994
150	15.82	139.96	.3019
160	16.08	141.45	.3043
170	16.35	142.95	.3067
180	16.61	144.46	.3091
190	16.87	145.98	.3114
200	17.13	147.52	.3138
210	17.39	149.08	.3161
220	17.66	150.65	.3184
230	17.92	152.23	.3207
240	18.18	153.82	.3230
250	18.44	155.43	.3253
260	18.70	157.05	.3276
270	18.97	158.68	.3298

Temp., °F.	Abs. pressure 10 lb./sq. in. (30.9°F.)		
t	v	h	s
Sat.	5.014	123.10	0.2536
40	5.112	124.30	.2561
50	5.219	125.63	.2587
60	5.326	126.99	.2613
70	5.443	128.36	.2639
80	5.540	129.74	.2665
90	5.646	131.13	.2691
100	5.753	132.54	.2716
110	5.860	133.95	.2741
120	5.966	135.38	.2766
130	6.072	136.82	.2791
140	6.178	138.28	.2815
150	6.285	139.75	.2840
160	6.391	141.24	.2864
170	6.497	142.75	.2888
180	6.603	144.27	.2912
190	6.709	145.79	.2936
200	6.815	147.33	.2959
210	6.921	148.89	.2983
220	7.027	150.46	.3006
230	7.133	152.05	.3029
240	7.239	153.65	.3052
250	7.344	155.26	.3075
260	7.450	156.88	.3098
270	7.555	158.52	.3120
290	7.761	160.17	.3143
290	7.767	161.84	.3165
300	7.872	163.52	.3187
310	7.978	165.21	.3210

* Courtesy of Kinetic Chemicals, Inc.

Table 222. Properties of Superheated Dichloromonofluoromethane (F-21)*—(*Concluded*)

Temp., °F.	Abs. pressure 20 lb./sq. in. (62.8°F.)			Temp., °F.	Abs. pressure 50 lb./sq. in. (113.3°F.)			Temp., °F.	Abs. pressure 100 lb./sq. in. (159.6°F.)		
t	*v*	*h*	*s*	*t*	*v*	*h*	*s*	*t*	*v*	*h*	*s*
Sat.	2.630	126.94	0.2482	Sat.	1.111	132.79	0.2421	Sat.	0.5680	137.66	0.2381
60				120	1.127	133.78	.2438	160	.5685	137.72	.2382
70	2.670	127.94	.2501	130	1.150	135.26	.2463	170	.5815	139.32	.2408
80	2.725	129.33	.2527					180	.5945	140.92	.2433
90	2.780	130.73	.2552	140	1.173	136.75	.2488	190	.6073	142.53	.2458
				150	1.197	138.26	.2513				
100	2.834	132.14	.2578	160	1.220	139.79	.2538	200	.6200	144.16	.2483
110	2.889	133.56	.2603	170	1.243	141.33	.2563	210	.6326	145.80	.2508
120	2.943	135.00	.2628	180	1.265	142.88	.2587	220	.6540	147.45	.2532
130	2.998	136.45	.2653					230	.6574	149.10	.2556
140	3.052	137.91	.2678	190	1.288	144.43	.2612	240	.6698	150.76	.2580
				200	1.311	146.00	.2636				
150	3.106	139.39	.2702	210	1.334	147.59	.2659	250	.6820	152.44	.2604
160	3.160	140.90	.2727	220	1.356	149.19	.2683	260	.6942	154.13	.2628
170	3.214	142.41	.2751	230	1.379	150.80	.2707	270	.7063	155.83	.2651
180	3.268	143.93	.2775					280	.7183	157.54	.2674
190	3.322	145.46	.2799	240	1.401	152.42	.2730	290	.7303	159.26	.2698
				250	1.424	154.06	.2753				
200	3.376	147.01	.2822	260	1.446	155.71	.2776	300	.7423	161.00	.2721
210	3.430	148.58	.2846	270	1.469	157.37	.2799	310	.7542	162.75	.2743
220	3.484	150.16	.2869	280	1.491	159.05	.2822	320	.7660	164.51	.2766
230	3.537	151.75	.2892					330	.7777	166.28	.2789
240	3.591	153.36	.2915	290	1.513	160.74	.2845	340	.7894	168.06	.2811
				300	1.535	162.44	.2867				
250	3.645	154.97	.2938	310	1.558	164.16	.2890	350	.8011	169.85	.2833
260	3.698	156.60	.2961	320	1.580	165.89	.2912	360	.8128	171.66	.2856
270	3.752	158.24	.2984	330	1.602	167.63	.2934	370	.8244	173.48	.2878
280	3.805	159.90	.3007					380	.8360	175.31	.2900
290	3.859	161.57	.3029	340	1.624	169.38	.2956	390	.8475	177.16	.2922
				350	1.646	171.14	.2978				
300	3.912	163.25	.3051	360	1.668	172.92	.3000	400	.8590	179.02	.2943
310	3.965	164.95	.3074	370	1.690	174.72	.3022				
320	4.019	166.67	.3096	380	1.712	176.52	.3044				
330	4.072	168.40	.3118								
340	4.125	170.14	.3140	390	1.734	178.34	.3065				
...	.....	.....	.....								

Table 223. Properties of Saturated Isobutane*

Abs. pressure, lb./sq. in.	Temp., °F.	Volume, cu. ft./lb.		Enthalpy, B.t.u./lb.		Entropy, B.t.u./(lb.)(°R.)		Abs. pressure, lb./sq. in.	Temp., °F.	Volume, cu. ft./lb.		Enthalpy, B.t.u./lb.		Entropy, B.t.u./(lb.)(°R.)	
		Liquid	Vapor	Liquid	Vapor	Liquid	Vapor			Liquid	Vapor	Liquid	Vapor	Liquid	Vapor
p	t	v_f	v_g	h_f	h_g	s_f	s_g	p	t	v	v	h	h	s	s
40	63.0	0.02838	2.210	1.64	146.4	0.0032	0.2803	200	178.3	.03412	.4305	73.94	181.0	.1259	.2938
50	76.5	.02888	1.7813	9.30	151.11	.0173	.2818	225	188.7	.03496	.3769	81.42	183.8	.1373	.2951
60	88.1	.02932	1.4904	16.01	154.82	.02957	.2831	250	198.3	.03578	.3327	88.51	185.8	.1478	.2957
70	98.2	.02973	1.2796	21.96	157.97	.0403	.2841	275	207.3	.03663	.2954	95.26	187.3	.1578	.2959
80	107.3	.03013	1.1198	27.34	160.81	.0499	.2852	300	215.6	.03748	.2633	101.7	188.7	.1671	.2959
90	115.5	.03049	0.9947	32.37	163.33	.0586	.2862	325	223.5	.03838	.2325	108.0	189.6	.1760	.2954
100	123.8	.03088	.8949	37.57	165.73	.0674	.2871	350	231.0	.03935	.2110	114.1	189.6	.1846	.2941
125	139.8	.03167	.7103	47.89	170.44	.0844	.2889	375	238.1	.04036	.1888	120.1	189.5	.1928	.2920
150	154.2	.03245	.5864	57.36	174.49	.0998	.2906	400	244.9	.04143	.1686	126.1	188.7	.2009	.2897
175	167.0	.03331	.4979	66.06	178.03	.1136	.2923								

* Sage and Lacey, *Ind. Eng. Chem.*, **30**, 673 (1938); with permission.

Table 224. Properties of Superheated Isobutane*

p, absolute pressure, lb./sq. in.; v, volume, cu. ft./lb.; h, enthalpy, B.t.u./lb.; s, entropy, B.t.u./(lb.)(°R.)
Parenthetic figures after temperatures are saturation pressures

p	70°F. (44.97)			100°F. (71.91)			130°F. (109.80)			160°F. (161.10)		
	v	h	s	v	h	s	v	h	s	v	h	s
Satd. vapor	1.975	148.94	0.2181	1.2463	158.51	0.2843	0.8123	167.57	0.2877	0.5437	176.10	0.2914
Satd. liquid	0.02863	5.58	.01040	0.02981	23.03	.04216	.03118	41.56	.07399	.03285	61.27	.10607
10	9.607	152.68	.3374	10.175	164.49	.3590	10.733	176.50	.3800	11.291	188.92	.4005
14.696	6.482	152.32	.3238	6.872	164.17	.3456	7.257	176.25	.3666	7.641	188.71	.3872
20	4.715	151.88	.3128	5.008	163.76	.3346	5.294	175.94	.3558	5.578	188.45	.3765
30	3.075	......	....	3.279	......	....	3.476	......	....	3.669	......	......
40	2.249	149.68	.2862	2.411	162.09	.3090	2.565	174.57	.3308	2.715	187.30	.3518
50	....	......	....	1.888	......	....	2.017	......	....	2.140	......	......
60	....	......	....	1.5379	159.93	.2945	1.6508	172.91	.3151	1.7575	185.91	.3366
80	....	......	....	....	......	....	1.1905	170.94	.3030	1.2781	184.30	.3251
100	....	......	....	....	......	....	0.9121	168.74	.2926	0.9888	182.64	.3154
125	....	......	....	....	......	....	....	......	....	.7558	180.28	.3048
150	....	......	....	....	......	....	....	......	....	.5983	177.48	.2953

Table 224. Propertise of Superheated Isobutane*—(*Concluded*)

p	190°F. (229.3)			220°F. (313.7)			250°F. (419.7)		
	v	h	s	v	h	s	v	h	s
Satd. vapor	0.3687	184.1	0.2953	0.2476	189.2	0.2958	0.154	187.9	0.2881
Satd. liquid	0.03506	82.37	.13867	.03799	105.2	.17207	.04228	130.4	.2072
10	11.854	201.70	.4207	12.410	214.85	.4404	12.965	228.39	.3600
14.696	8.023	201.52	.4074	8.403	214.71	.4273	8.783	228.28	.4468
20	5.860	201.30	.3968	6.141	214.53	.4167	6.422	228.14	.4363
30	3.861	......	....	4.051	......	....	4.240	......	......
40	2.861	200.35	.3724	3.006	213.66	.3925	3.150	227.49	.4123
50	2.260	......	....	2.379	......	....	2.497	......	......
60	1.8600	199.15	.3575	1.9619	212.66	.3779	2.061	226.66	.3979
80	1.3604	197.77	.3463	1.4404	211.49	.3670	1.5178	225.67	.3874
100	1.0600	196.23	.3371	1.1278	210.16	.3582	1.1925	224.54	.3788
125	0.8183	194.08	.3272	0.8771	208.31	.3487	0.9322	222.96	.3697
150	.6557	191.71	.3183	.7091	206.26	.3403	.7585	221.20	.3617
175	.5391	189.40	.3103	.5888	204.03	.3326	.6355	219.27	.3545
200	.4505	187.0	.3030	.4977	201.64	.3254	.5392	217.19	.3477
225	.3794	184.5	.2963	.4252	199.08	.3185	.4652	214.97	.3413
250	....	......	....	.3649	196.35	.3118	.4048	212.61	.3352
275	....	......	....	.3139	193.45	.3053	.3541	210.1	.3292
300	....	......	....	.2696	190.7	.2999	.3111	207.5	.3232
350	....	......	....	....	......	......	.2416	201.3	.3113
400	....	......	....	....	......	......	.187	192.6	.2963

* Sage and Lacey, *Ind. Eng. Chem.*, **30,** 673 (1938) with permission.

Table 225. Properties of Saturated Mercury*

Abs. pressure, lb./sq. in.	Temp., °F.	Volume, cu. ft./lb.	Enthalpy, B.t.u./lb.		Entropy, B.t.u./(lb.)(°R.)		Abs. pressure, lb./sq. in.	Temp., °F.	Volume, cu. ft./lb.	Enthalpy, B.t.u./lb.		Entropy, B.t.u./(lb.)(°R.)	
		Vapor	Liquid	Vapor	Liquid	Vapor			Vapor	Liquid	Vapor	Liquid	Vapor
p	t	v_f	h_f	h_g	s_f	s_g	p	t	v_f	h_f	h_g	s_f	s_g
0.4	402.3	114.5	13.81	141.9	0.02094	0.1696	25	730.4	2.429	26.05	146.9	.03297	.1345
.6	426.1	78.23	14.70	142.3	.02195	.1660	30	750.9	2.053	26.81	147.2	.03360	.1331
.8	443.8	59.71	15.36	142.6	.02269	.1635	35	769.0	1.781	27.49	147.5	.03416	.1319
1.0	458.1	48.45	15.89	142.8	.02328	.1615	40	784.8	1.576	28.08	147.8	.03464	.1308
1.5	485.1	33.14	16.90	143.2	.02436	.1580	45	799.3	1.414	28.62	148.0	.03507	.1299
2	505.2	25.31	17.65	143.5	.02514	.1556	50	812.5	1.284	29.11	148.2	.03546	.1291
3	535.4	17.34	18.78	144.0	.02629	.1521	60	836.1	1.086	29.99	148.6	.03614	.1276
4	558.0	13.26	19.62	144.3	.02714	.1497	70	856.6	0.9436	30.75	148.9	.03672	.1264
5	576.2	10.77	20.30	144.6	.02780	.1478	80	874.8	0.8349	31.43	149.1	.03725	.1254
6	591.4	9.096	20.87	144.8	.02834	.1462	90	891.6	0.7497	32.06	149.4	.03771	.1245
7	605.0	7.882	21.37	145.0	.02882	.1450	100	906.9	0.6811	32.63	149.6	.03813	.1237
8	616.8	6.963	21.81	145.2	.02923	.1439	120	934.4	0.5767	33.60	150.1	.03887	.1224
9	627.5	6.244	22.21	145.4	.02960	.1429	140	958.3	0.5012	34.55	150.4	.03951	.1212
10	637.3	5.661	22.58	145.5	.02993	.1420	160	979.9	0.4438	35.35	150.8	.04007	.1202
15	676.5	3.892	24.04	146.1	.03124	.1387	180	999.6	0.3990	36.09	151.1	.04058	.1193
20	706.2	2.983	25.15	146.6	.03220	.1363							

* Marks, "Mechanical Engineers' Handbook," 4th ed., McGraw-Hill, New York, 1941.

Table 226. Properties of Saturated Methane*

Temp., °F.	Abs. pressure, lb./sq. in.	Volume, cu. ft./lb.		Enthalpy, B.t.u./lb.		Entropy B.t.u./(lb.)(°R.)		Temp., °F.	Abs. pressure, lb./sq. in.	Volume, cu. ft./lb.		Enthalpy, B.t.u./lb.		Entropy B.t.u./(lb.)(°R.)	
		Liquid	Vapor	Liquid	Vapor	Liquid	Vapor			Liquid	Vapor	Liquid	Vapor	Liquid	Vapor
t	p	v_f	v_g	h_f	h_g	s_f	s_g	t	p	v_f	v_g	h_f	h_g	s_f	s_g
−280	4.90	0.03635	24.04	0	228.2	0	1.2699	−180	191.5	.04575	.773	87.8	257.0	.3767	.9816
−270	8.44	.03698	14.61	8.2	232.3	0.0423	1.2236	−170	240.0	.04745	.610	98.0	257.2	.4127	.9622
−260	13.80	.03766	9.31	16.6	236.4	.0823	1.1830	−160	297.0	.04944	.483	108.7	256.5	.4476	.9411
−250	21.71	.03839	6.13	25.0	240.3	.1201	1.1468	−150	364	.05197	.381	120.3	254.5	.4839	.9169
−240	32.4	.03915	4.24	33.3	243.9	.1578	1.1164	−140	440	.05224	.3008	133.2	251.2	.5214	.8905
−230	46.4	.03999	3.04	42.0	247.3	.1962	1.0900	−130	527	.05999	.2318	148.1	245.9	.5656	.8622
−220	64.5	.04092	2.23	50.6	250.2	.2333	1.0660	−120	627	.06961	.1613	171.8	231.4	.6329	.8083
−210	87.6	.04193	1.67	59.5	252.8	.2693	1.0434	−115.8	673	.0983	.0983	203.4	203.4	.7232	.7232
−200	115.7	.04306	1.281	68.8	254.8	.3062	1.0224								
−190	150.0	.04431	0.990	78.2	256.2	.3419	1.0019								

* Matthews and Hurd, *Trans. Am. Inst. Chem. Engrs.* **42,** 55 (1946).

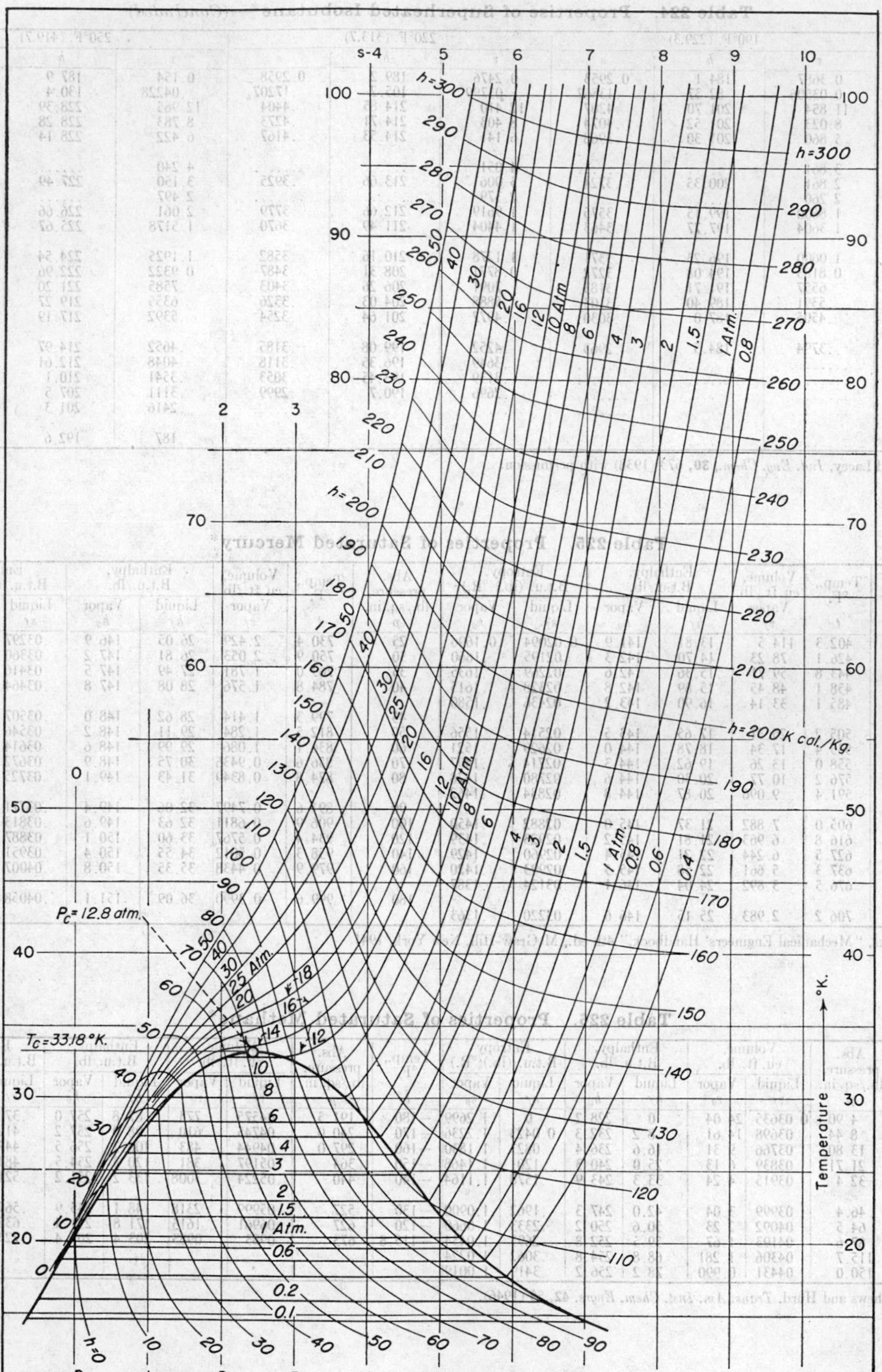

Fig. 16. Temperature-entropy diagram for hydrogen (14 to 100°K.). (*Keesom and Houthoff, Communs. Physical Lab. Univ. Leiden, Suppl. 65d,* 1926).

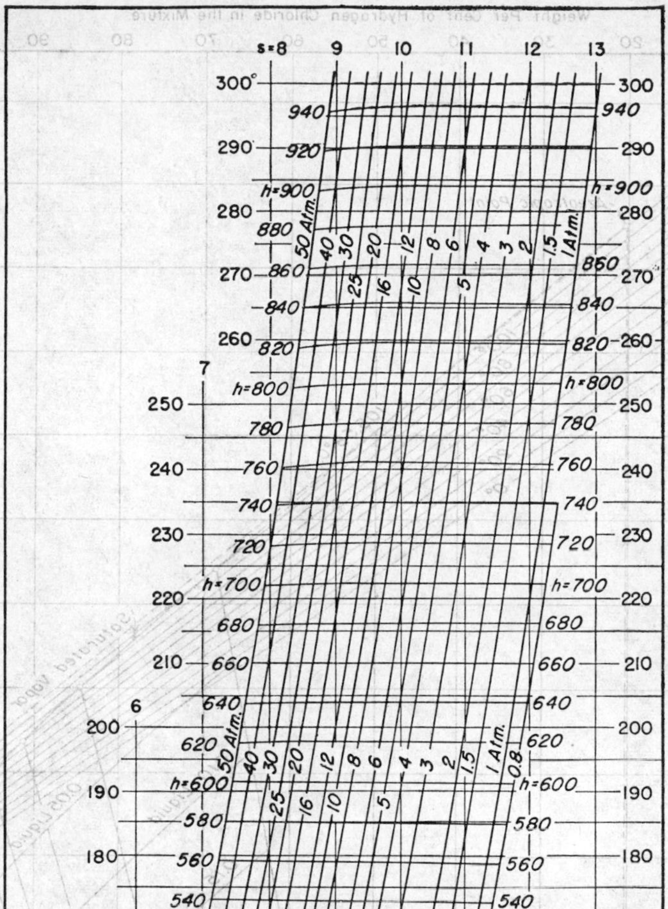

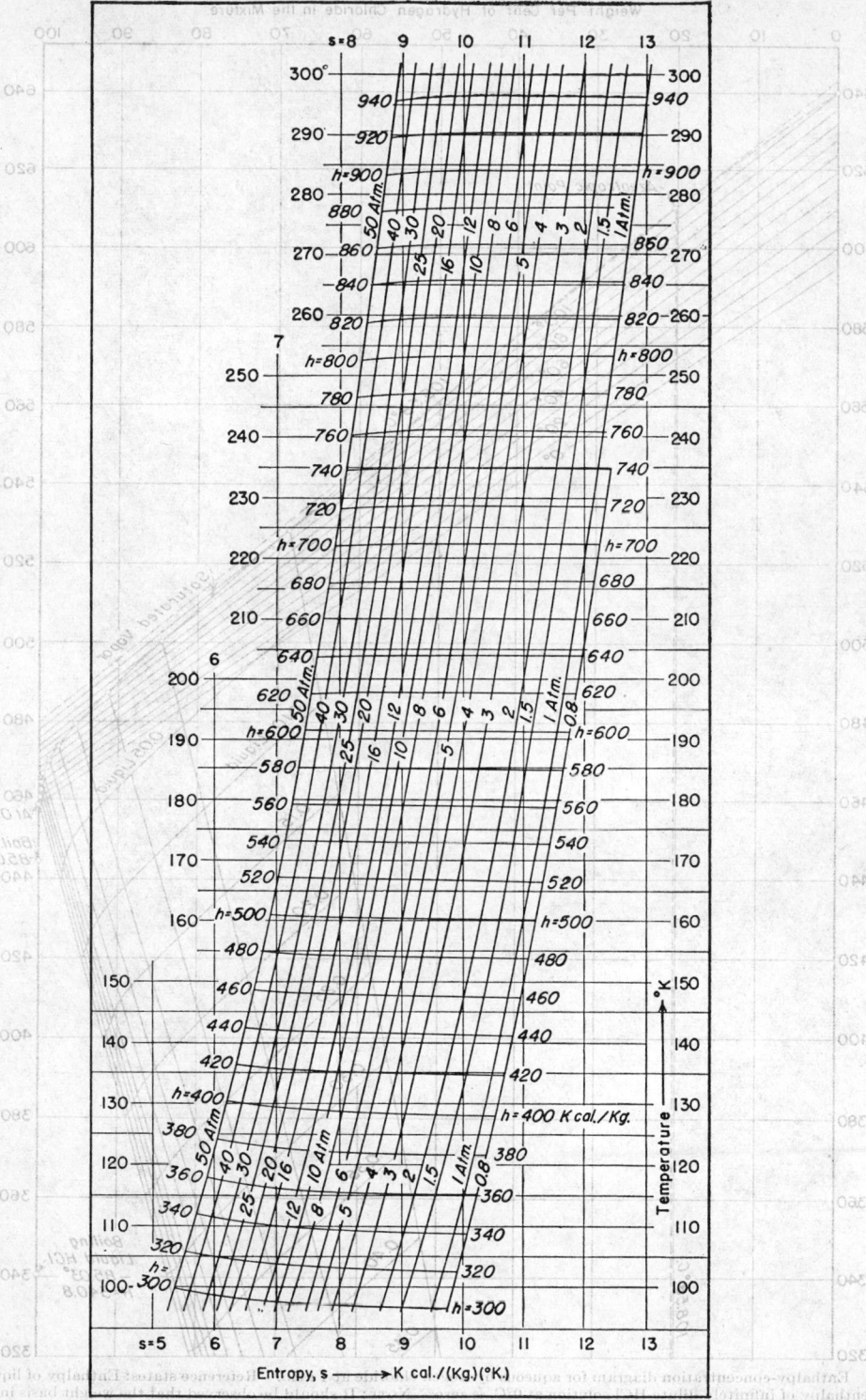

FIG. 17. Temperature-entropy diagram for hydrogen (100 to 300°K.). (*Keesom and Houthoff, Communs. Physical Lab. Univ. Leiden, Suppl.* 65*d*, 1926.)

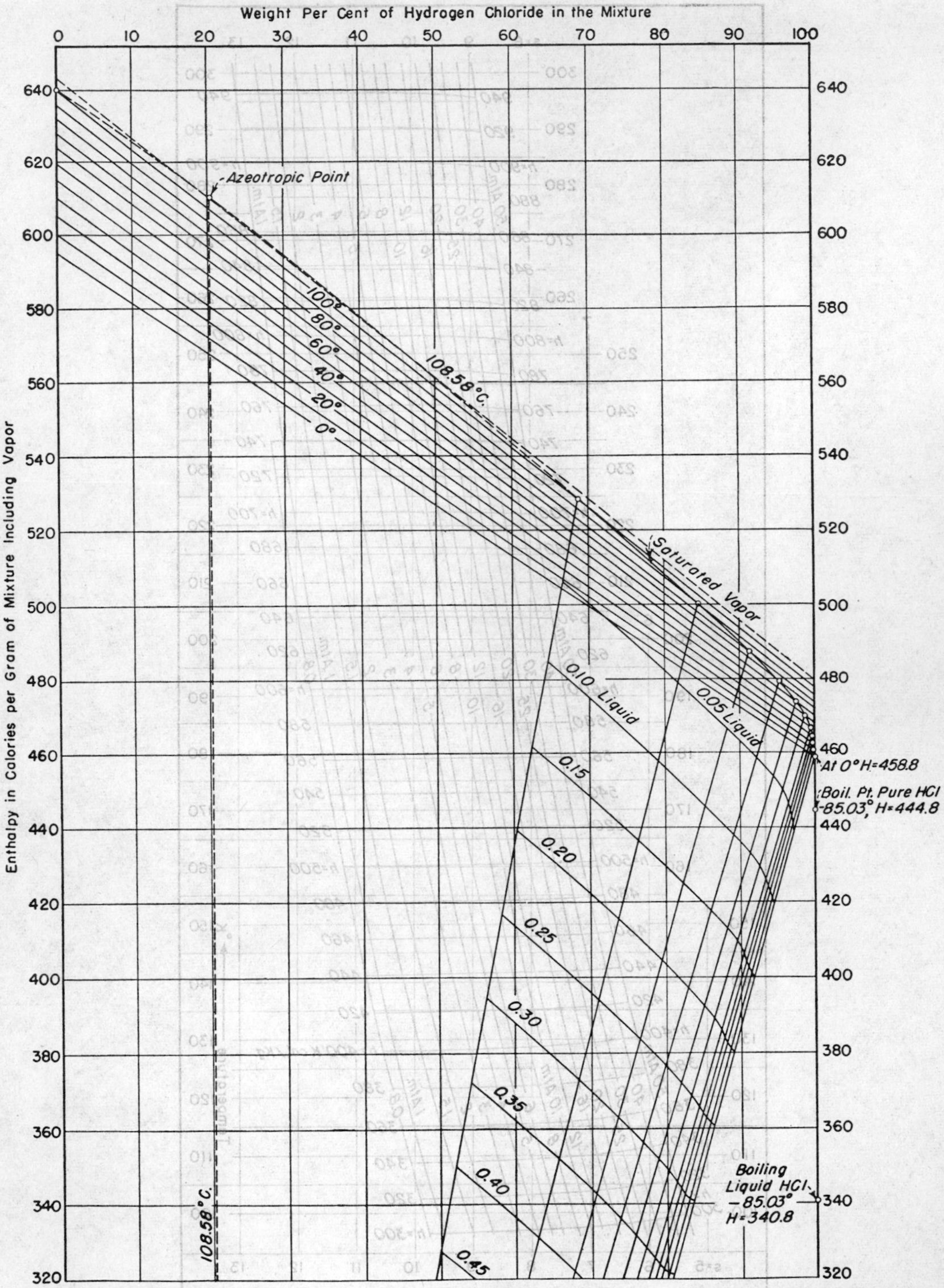

FIG. 18. Enthalpy-concentration diagram for aqueous hydrogen chloride at 1 atm. Reference states: Enthalpy of liquid water at 0°C. is zero; enthalpy of infinitely dilute HCl solution at 0°C. is zero. NOTE: It should be observed that the weight basis includes the vapor.

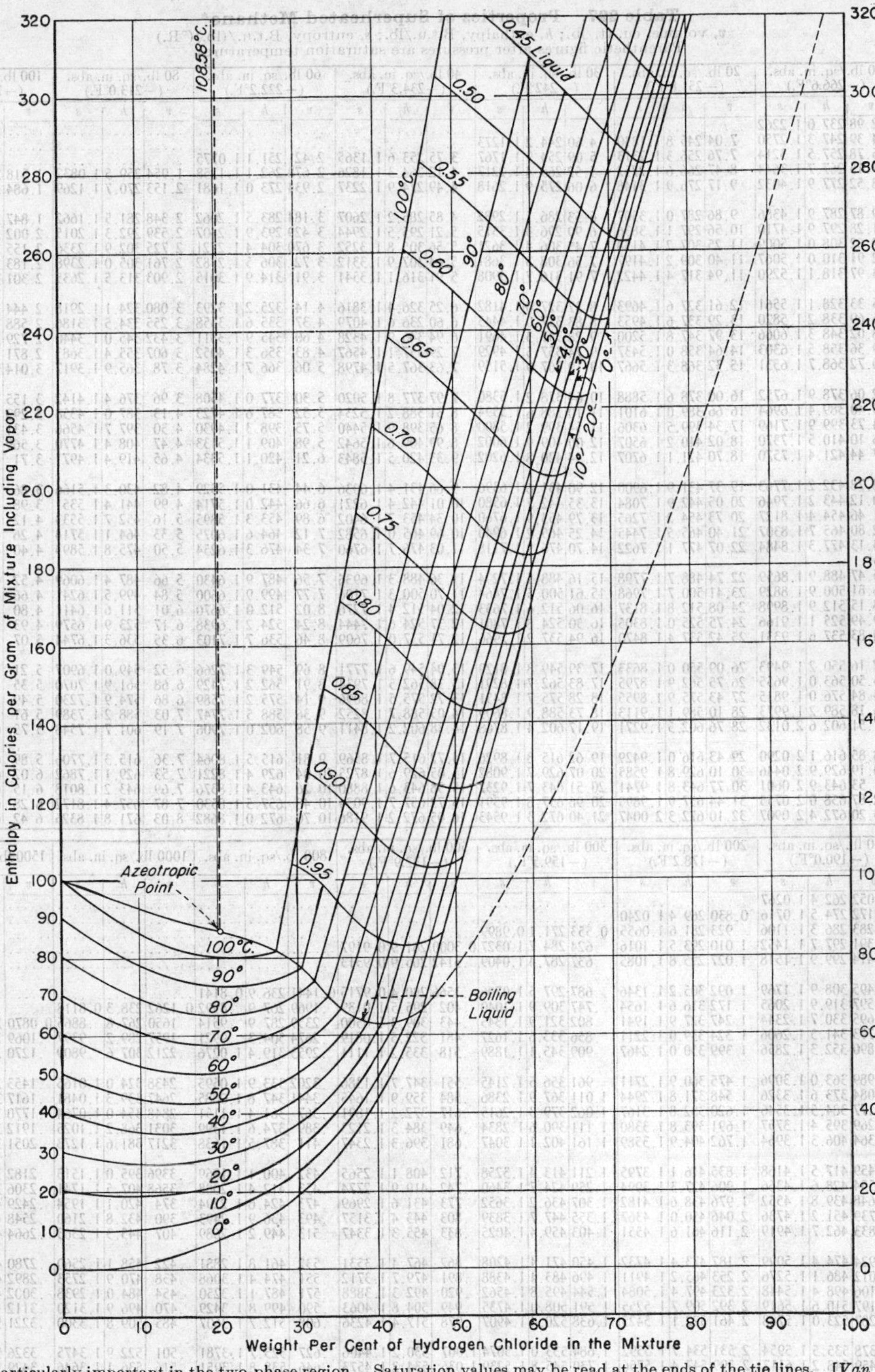

which is particularly important in the two-phase region. Saturation values may be read at the ends of the tie lines. [*Van Nuys, Trans. Am. Inst. Chem. Engrs.* **39**, 663 (1943).]

Table 227. Properties of Superheated Methane*

v, volume, cu. ft./lb.; *h*, enthalpy, B.t.u./lb.; *s*, entropy, B.t.u./(lb.)(°R.)
Parenthetic figures after pressures are saturation temperatures

Temp., °F.	10 lb./sq. in. abs. (−266.6°F.)			20 lb./sq. in. abs. (−251.8°F.)			30 lb./sq. in. abs. (−242°F.)			40 lb./sq. in. abs. (−234.3°F.)			60 lb./sq. in. abs. (−222.2°F.)			80 lb./sq. in. abs. (−213.0°F.)			100 lb./sq. in. abs. (−205.5°F.)		
t	v	h	s	v	h	s	v	h	s	i	h	s	v	h	s	v	h	s	v	h	s
−260	12.98	237.0	1.2262	.....	.....	.....	.....	.....	.....	.....	.....	.....	.....	.....	.....	.....	.....	.....	.....	.....	.....
−240	14.39	247.3	1.2750	7.04	245.8	1.1830	4.60	244.2	1.1273	.....	.....	.....	.....	.....	.....	.....	.....	.....	.....	.....	.....
−220	15.78	257.5	1.3214	7.76	256.3	1.2313	5.09	254.9	1.1767	3.75	253.6	1.1365	2.421	251.1	1.0775	.....	.....	.....	.....	.....	.....
−200	17.15	267.7	1.3644	8.47	266.6	1.2752	5.57	265.4	1.2217	4.12	264.3	1.1826	2.678	262.1	1.1258	1.954	259.5	1.0832	1.518	256.9	1.0476
−180	18.52	277.9	1.4032	9.17	276.9	1.3148	6.06	275.9	1.2618	4.49	274.9	1.2237	2.934	273.0	1.1681	2.153	270.7	1.1269	1.684	268.4	1.0935
−160	19.87	287.9	1.4386	9.86	287.0	1.3507	6.53	286.1	1.2982	4.85	285.2	1.2607	3.184	283.5	1.2062	2.348	281.5	1.1662	1.847	279.6	1.1339
−140	21.28	297.9	1.4710	10.56	297.1	1.3836	6.99	296.3	1.3315	5.21	295.5	1.2944	3.429	293.9	1.2407	2.539	292.3	1.2015	2.002	290.6	1.1701
−120	22.62	308.0	1.5009	11.25	307.2	1.4138	7.45	306.4	1.3621	5.56	305.8	1.3252	3.670	304.4	1.2721	2.725	302.9	1.2336	2.155	301.4	1.2029
−115.8	22.91	310.0	1.5067	11.40	309.2	1.4197	7.56	308.3	1.3681	5.64	307.9	1.3312	3.72	306.5	1.2782	2.761	305.0	1.2398	2.183	303.6	1.2092
−100	23.97	318.1	1.5290	11.94	317.4	1.4422	7.91	316.7	1.3908	5.91	316.1	1.3541	3.91	314.9	1.3015	2.903	313.5	1.2635	2.301	312.2	1.2332
− 80	25.33	328.1	1.5561	12.61	327.6	1.4693	8.37	327.0	1.4182	6.25	326.4	1.3816	4.14	325.2	1.3293	3.080	324.1	1.2918	2.444	322.9	1.2619
− 60	26.69	338.2	1.5820	13.29	337.6	1.4953	8.83	337.0	1.4443	6.60	236.6	1.4079	4.37	335.6	1.3558	3.255	334.5	1.3186	2.588	333.4	1.2891
− 40	28.02	348.3	1.6066	13.97	347.8	1.5200	9.28	347.3	1.4691	6.94	346.9	1.4328	4.60	345.9	1.3811	3.432	345.0	1.3440	2.729	344.0	1.3148
− 20	29.36	358.5	1.6303	14.64	358.0	1.5437	9.74	357.5	1.4929	7.29	357.1	1.4567	4.83	356.3	1.4052	3.607	355.4	1.3683	2.871	354.5	1.3393
0	30.72	368.7	1.6531	15.32	368.3	1.5667	10.19	367.9	1.5159	7.63	367.5	1.4798	5.06	366.7	1.4284	3.78	365.9	1.3917	3.014	365.2	1.3628
20	32.06	378.9	1.6752	16.00	378.6	1.5888	10.64	378.2	1.5380	7.97	377.8	1.5020	5.30	377.0	1.4508	3.96	376.4	1.4142	3.155	375.7	1.3854
40	33.40	389.4	1.6964	16.66	389.0	1.6101	11.10	388.6	1.5594	8.31	388.2	1.5234	5.52	287.6	1.4723	4.13	387.0	1.4358	3.293	386.3	1.4072
60	34.73	399.9	1.7169	17.34	399.5	1.6306	11.54	399.2	1.5800	8.65	398.9	1.5440	5.75	398.3	1.4930	4.30	397.7	1.4566	3.431	397.0	1.4280
80	36.10	410.5	1.7370	18.02	410.2	1.6507	12.00	409.9	1.6002	8.99	409.6	1.5642	5.98	409.1	1.5133	4.47	408.4	1.4770	3.569	407.9	1.4485
100	37.44	421.4	1.7570	18.70	421.1	1.6707	12.44	420.8	1.6202	9.33	420.5	1.5843	6.21	420.1	1.5334	4.65	419.4	1.4971	3.71	418.9	1.4687
120	38.78	432.2	1.7763	19.37	431.9	1.6900	12.90	431.6	1.6396	9.66	431.4	1.6036	6.44	431.0	1.5529	4.82	430.3	1.5166	3.86	429.8	1.4882
140	40.12	443.2	1.7946	20.05	442.9	1.7084	13.35	442.7	1.6580	10.01	442.4	1.6221	6.66	442.0	1.5714	4.99	441.4	1.5351	3.98	440.9	1.5068
160	41.46	454.4	1.8127	20.73	454.2	1.7265	13.79	453.9	1.6760	10.34	453.7	1.6402	6.89	453.3	1.5895	5.16	452.7	1.5533	4.12	452.2	1.5250
180	42.80	465.7	1.8307	21.40	465.5	1.7445	14.25	465.2	1.6940	10.69	465.0	1.6582	7.12	464.6	1.6075	5.33	464.1	1.5714	4.26	463.7	1.5432
200	44.13	477.3	1.8484	22.07	477.1	1.7622	14.70	476.9	1.7118	11.03	476.7	1.6760	7.34	476.3	1.6254	5.50	475.8	1.5893	4.40	475.4	1.5611
220	45.47	488.9	1.8659	22.74	488.7	1.7798	15.16	488.5	1.7294	11.36	488.3	1.6936	7.56	487.9	1.6430	5.66	487.4	1.6069	4.52	487.1	1.5788
240	46.81	500.9	1.8829	23.41	500.7	1.7968	15.61	500.5	1.7464	11.70	500.3	1.7106	7.77	499.9	1.6600	5.84	499.5	1.6241	4.66	499.2	1.5960
260	48.15	512.9	1.8998	24.08	512.8	1.8137	16.06	512.6	1.7633	12.04	512.4	1.7276	8.02	512.0	1.6670	6.01	511.6	1.6411	4.80	511.3	1.6130
280	49.49	525.1	1.9166	24.75	525.0	1.8305	16.50	524.8	1.7801	12.37	524.6	1.7444	8.24	524.2	1.6938	6.17	523.9	1.6579	4.93	523.5	1.6299
300	50.83	537.6	1.9331	25.42	537.4	1.8470	16.94	537.2	1.7966	12.71	537.0	1.7609	8.46	536.7	1.7103	6.35	536.3	1.6744	5.07	536.0	1.6464
320	52.16	550.2	1.9493	26.09	550.0	1.8633	17.39	549.8	1.8129	13.04	549.6	1.7771	8.69	549.3	1.7266	6.52	549.0	1.6907	5.21	548.6	1.6627
340	53.50	563.0	1.9655	26.75	562.9	1.8795	17.83	562.7	1.8291	13.38	562.5	1.7934	8.91	562.2	1.7429	6.68	561.9	1.7070	5.35	561.5	1.6790
360	54.84	576.0	1.9815	27.43	575.9	1.8955	18.28	575.7	1.8451	13.72	575.5	1.8096	9.14	575.2	1.7589	6.86	574.9	1.7230	5.49	574.6	1.6951
380	56.18	589.2	1.9973	28.10	589.1	1.9113	18.73	588.9	1.8609	14.05	588.7	1.8252	9.36	588.5	1.7747	7.03	588.2	1.7389	5.61	587.8	1.7110
400	57.51	602.6	2.0132	28.76	602.5	1.9271	19.17	602.4	1.8768	14.38	602.2	1.8411	9.58	602.0	1.7906	7.19	601.7	1.7548	5.75	601.3	1.7269
420	58.85	616.1	2.0290	29.43	616.0	1.9429	19.62	615.8	1.8926	14.72	615.7	1.8569	9.81	615.5	1.8064	7.36	615.3	1.7706	5.89	614.9	1.7427
440	60.19	629.9	2.0446	30.10	629.8	1.9585	20.07	629.7	1.9082	15.05	629.6	1.8725	10.04	629.4	1.8221	7.53	629.1	1.7862	6.02	628.8	1.7583
460	61.53	643.9	2.0601	30.77	643.8	1.9741	20.51	643.7	1.9237	15.38	643.6	1.8880	10.26	643.4	1.8376	7.69	643.2	1.8018	6.15	642.9	1.7739
480	62.87	658.0	2.0755	31.44	657.9	1.9895	20.96	657.8	1.9391	15.72	657.7	1.9034	10.48	657.5	1.8530	7.87	657.4	1.8172	6.28	657.0	1.7893
500	64.20	672.4	2.0907	32.10	672.3	2.0047	21.40	672.3	1.9543	16.05	672.2	1.9186	10.70	672.0	1.8682	8.03	671.8	1.8325	6.42	671.5	1.8046

Temp., °F.	150 lb./sq. in. abs. (−190.0°F.)			200 lb./sq. in. abs. (−178.2°F.)			300 lb./sq. in. abs. (−159.5°F.)			500 lb./sq. in. abs. (−132.9°F.)			800 lb./sq. in. abs.			1000 lb./sq. in. abs.			1500 lb./sq. in. abs.		
t	v	h	s	v	h	s	v	h	s	v	h	s	v	h	s	v	h	s	v	h	s
−180	1.052	262.4	1.0267	.....	.....	.....	.....	.....	.....	.....	.....	.....	.....	.....	.....	.....	.....	.....	.....	.....	.....
−160	1.172	274.5	1.0716	0.830	269.4	1.0240	.....	.....	.....	.....	.....	.....	.....	.....	.....	.....	.....	.....	.....	.....	.....
−140	1.283	286.3	1.1106	.923	281.6	1.0655	0.553	271.1	0.9899	.....	.....	.....	.....	.....	.....	.....	.....	.....	.....	.....	.....
−120	1.391	297.7	1.1452	1.010	293.5	1.1016	.624	284.7	1.0327	0.3000	261.9	0.9197	.....	.....	.....	.....	.....	.....	.....	.....	.....
−115.8	1.414	299.9	1.1518	1.027	295.8	1.1085	.637	287.4	1.0409	.3142	265.9	0.9313	.....	.....	.....	.....	.....	.....	.....	.....	.....
−100	1.495	308.9	1.1769	1.092	305.2	1.1346	.687	297.5	1.0706	.3566	280.4	0.9715	0.1441	236.9	0.8141	.....	.....	.....	.....	.....	.....
− 80	1.597	319.9	1.2065	1.172	316.6	1.1654	.747	309.9	1.1040	.402	295.5	1.0139	.1969	267.9	.8992	0.1262	238.3	0.8118	.....	.....	.....
− 60	1.695	330.7	1.2344	1.247	327.9	1.1941	.802	321.9	1.1345	.443	309.3	1.0500	.2359	287.9	.9514	.1650	267.8	.8865	0.0870	223.5	0.7723
− 40	1.793	341.5	1.2606	1.324	339.0	1.2211	.856	333.6	1.1627	.481	322.5	1.0819	.2674	304.3	.9921	.1957	289.2	.9397	.1069	253.3	.8373
− 20	1.890	352.3	1.2856	1.399	350.0	1.2467	.909	345.1	1.1889	.518	335.2	1.1111	.2953	319.4	1.0276	.2212	307.6	.9809	.1270	276.6	.8860
0	1.989	363.0	1.3096	1.475	360.9	1.2711	.961	356.5	1.2145	.551	347.7	1.1386	.3202	333.9	1.0595	.2438	324.0	1.0165	.1453	297.8	.9292
20	2.084	373.6	1.3326	1.548	371.8	1.2944	1.011	367.9	1.2386	.584	359.9	1.1645	.3441	347.8	1.0885	.2647	339.3	1.0481	.1617	317.3	.9657
40	2.177	384.5	1.3546	1.620	382.8	1.3167	1.062	379.3	1.2615	.617	372.2	1.1891	.367	361.4	1.1151	.2848	354.0	1.0764	.1770	335.5	.9983
60	2.269	395.4	1.3757	1.691	393.8	1.3380	1.111	390.6	1.2834	.649	384.3	1.2123	.389	374.6	1.1399	.3031	368.2	1.1025	.1912	352.4	1.0276
80	2.364	406.3	1.3964	1.762	404.9	1.3589	1.161	402.1	1.3047	.681	396.3	1.2347	.411	387.5	1.1638	.3217	381.6	1.1273	.2051	367.9	1.0554
100	2.459	417.5	1.4168	1.835	416.1	1.3795	1.211	413.4	1.3258	.712	408.1	1.2565	.432	400.1	1.1869	.3396	395.0	1.1513	.2182	382.5	1.0818
120	2.554	428.6	1.4366	1.906	427.3	1.3994	1.259	424.7	1.3460	.742	419.9	1.2774	.453	412.4	1.2088	.3568	407.5	1.1741	.2306	396.5	1.1069
140	2.648	439.8	1.4552	1.976	438.6	1.4182	1.307	436.2	1.3652	.773	431.6	1.2969	.473	424.6	1.2294	.374	420.1	1.1954	.2429	410.0	1.1302
160	2.739	451.2	1.4736	2.046	450.0	1.4367	1.355	447.7	1.3839	.803	443.4	1.3157	.493	436.9	1.2493	.390	432.8	1.2160	.2548	423.3	1.1524
180	2.833	462.7	1.4919	2.116	461.6	1.4551	1.403	459.4	1.4025	.833	455.3	1.3347	.513	449.2	1.2689	.407	445.3	1.2363	.2664	436.4	1.1741
200	2.924	474.4	1.5099	2.187	473.4	1.4732	1.450	471.4	1.4208	.862	467.4	1.3531	.532	461.8	1.2881	.422	458.1	1.2560	.2780	449.6	1.1948
220	3.012	486.1	1.5276	2.253	485.2	1.4911	1.496	483.4	1.4388	.891	479.7	1.3712	.551	474.4	1.3068	.438	470.9	1.2753	.2892	462.7	1.2150
240	3.106	498.4	1.5448	2.323	497.4	1.5084	1.544	495.8	1.4562	.920	492.3	1.3888	.571	487.1	1.3250	.454	484.0	1.2938	.3002	476.0	1.2343
260	3.197	510.6	1.5619	2.392	509.7	1.5255	1.591	508.0	1.4735	.949	504.8	1.4063	.590	499.8	1.3429	.470	496.9	1.3120	.3112	489.2	1.2531
280	3.287	523.0	1.5788	2.461	522.1	1.5425	1.638	520.4	1.4907	.978	517.4	1.4236	.608	512.7	1.3607	.485	509.8	1.3300	.3221	502.6	1.2716
300	3.378	535.5	1.5954	2.531	534.7	1.5592	1.684	533.0	1.5074	1.007	530.2	1.4406	.627	525.7	1.3781	.501	522.9	1.3475	.3326	516.2	1.2896
320	3.459	548.2	1.6117	2.599	547.4	1.5756	1.730	545.8	1.5239	1.036	534.2	1.4573	.646	538.8	1.3951	.516	536.1	1.3646	.3429	529.8	1.3072
340	3.559	561.1	1.6281	2.667	560.3	1.5919	1.776	558.8	1.5403	1.064	556.3	1.4740	.664	552.1	1.4120	.530	549.4	1.3817	.3534	543.5	1.3247
360	3.66	574.2	1.6442	2.736	573.4	1.6080	1.823	571.9	1.5566	1.093	569.6	1.4904	.682	565.6	1.4287	.545	563.0	1.3985	.364	557.3	1.3419
380	3.75	587.5	1.6601	2.805	586.7	1.6239	1.869	585.3	1.5726	1.121	583.1	1.5066	.700	579.2	1.4450	.560	576.7	1.4149	.374	571.3	1.3589

* Matthews and Hurd, *Trans. Am. Inst. Chem. Engrs.* **42**, 55 (1946).

Table 227. Properties of Superheated Methane*—(*Concluded*)

Temp., °F.	150 lb./sq. in. abs. (−190.0°F.)			200 lb./sq. in. abs. (−178.2°F.)			300 lb./sq. in. abs. (−159.5°F.)			500 lb./sq. in. abs. (−132.9°F.)			800 lb./sq. in. abs.			1000 lb./sq. in. abs.			1500 lb./sq. in. abs.		
t	v	h	s	v	h	s	v	h	s	v	h	s	v	h	s	v	h	s	v	h	s
400	3.83	601.0	1.6760	2.873	600.2	1.6389	1.915	598.9	1.5886	1.148	596.7	1.5228	.717	593.1	1.4614	.575	590.7	1.4314	.384	585.4	1.3757
420	3.93	614.5	1.6919	2.941	613.8	1.6557	1.961	612.6	1.6045	1.176	610.5	1.5389	.735	607.0	1.4777	.589	605.8	1.4479	.394	599.7	1.3926
440	4.02	628.4	1.7076	3.009	627.7	1.6714	2.007	626.5	1.6202	1.204	624.5	1.5547	.754	621.1	1.4937	.603	619.0	1.4641	.404	614.2	1.4092
460	4.10	642.5	1.7231	3.077	641.8	1.6874	2.052	640.6	1.6358	1.232	638.8	1.5704	.772	635.5	1.5096	.617	633.4	1.4802	.414	628.8	1.4256
480	4.19	656.6	1.7386	3.144	656.0	1.7025	2.098	654.9	1.6513	1.260	653.1	1.5860	.790	649.8	1.5253	.632	647.9	1.4962	.424	643.5	1.4420
500	4.28	671.1	1.7539	3.210	670.5	1.7178	2.142	669.4	1.6666	1.287	667.8	1.6013	.809	664.5	1.5409	.648	662.6	1.5120	.435	658.5	1.4581

Table 228. Properties of Saturated Methylamine
(CH_3NH_2)*

Temp., °F.	Abs. pressure, lb./sq. in.	Volume, cu. ft./lb., vapor	Enthalpy, B.t.u./lb.		Entropy, B.t.u./lb. (°R.)		Temp., °F.	Abs. pressure, lb./sq. in.	Volume, cu. ft./lb., vapor	Enthalpy, B.t.u./lb.		Entropy, B.t.u./lb. (°R.)	
			Liquid	Vapor	Liquid	Vapor				Liquid	Vapor	Liquid	Vapor
t	p	v_g	h_f	h_g	s_f	s_g	t	p	v_g	h_f	h_g	s_f	s_g
− 58	1.322	98.93	−12.4	374.8	−0.0503	0.9135	23	15.99	10.01	45.6	404.3	.1015	.8443
− 40	2.532	56.72	0.0	381.7	.0000	.9092	41	24.49	6.796	59.2	410.1	.1292	.8300
− 22	4.551	32.56	12.7	388.4	.0294	.8877	68	43.52	3.985	80.2	418.8	.1698	.8115
− 4	7.78	19.84	25.7	395.0	.0586	.8689	86	61.53	2.962	94.5	424.4	.1963	.8008
+ 5	10.03	15.54	32.6	398.5	.0731	.8603	113	98.76	1.867	116.9	432.6	.2360	.7873

*"Refrigerating Data Book," 5th ed., American Society of Refrigerating Engineers, New York, 1942.

Table 229. Properties of Saturated Methyl Chloride*

Temp., °F.	Abs. pressure, lb./sq. in.	Volume, cu. ft./lb.		Enthalpy, B.t.u./lb.		Entropy, B.t.u./(lb.)(°R.)	
		Liquid	Vapor	Liquid	Vapor	Liquid	Vapor
t	p	v_f	v_g	h_f	h_g	s_f	s_g
−40	6.878	0.01553	12.72	0.000	190.66	0.0000	0.4544
−30	9.036	.01568	9.873	3.562	192.08	.0084	.4472
−20	11.71	.01583	7.761	7.146	193.49	.0166	.4405
−10	14.96	.01598	6.176	10.75	194.87	.0247	.4343
0	18.90	.01613	4.969	14.39	196.23	.0327	.4284
5	21.15	.01622	4.471	16.21	196.92	.0367	.4257
10	23.60	.01631	4.038	18.04	197.58	.0406	.4229
20	29.16	.01647	3.312	21.73	198.84	.0484	.4177
30	35.68	.01665	2.739	25.44	200.03	.0560	.4126
40	43.25	.01684	2.286	29.17	201.17	.0636	.4079
50	51.99	.01704	1.920	32.93	202.28	.0710	.4034
60	62.00	.01724	1.624	36.71	203.33	.0784	.3991
70	73.41	.01744	1.382	40.52	204.34	.0856	.3950
80	86.26	.01764	1.183	44.36	205.27	.0928	.3910
86	94.70	.01778	1.081	46.47	205.80	.0970	.3887
90	100.6	.01786	1.018	48.21	206.13	.0998	.3872
100	116.7	.01808	0.8814	52.09	206.94	.1069	.3836
110	134.5	.01833	.7672	56.00	207.70	.1138	.3801
120	154.2	.01859	.6710	59.93	208.39	.1206	.3768
130	175.9	.01887	.5889	63.89	209.02	.1274	.3736
140	199.6	.01915	.5189	67.87	209.58	.1341	.3705
150	225.4	.01945	.4586	71.87	210.10	1407	.3674
160	253.5	.01978	.4070	75.90	210.56	.1473	.3646
170	283.9	.02015	.3613	79.97	210.93	.1538	.3618

* Tanner, Banning, and Matthewson, *Ind. Eng. Chem.*, **31**, 878 (1939). Copyright, 1939, E. I. du Pont de Nemours & Co., Inc.

Table 230. Properties of Superheated Methyl Chloride*

v, volume, cu. ft./lb.; h, enthalpy, B.t.u./lb.; s, entropy, B.t.u./(lb.)(°R.)

Parenthetic figures after pressures are saturation temperatures

Temp., °F.	Abs. pressure, 6 lb./sq. in. (−44.8°F.)			Temp., °F.	Abs. pressure, 10 lb./sq. in. (−26.1°F.)			Temp., °F.	Abs. pressure, 20 lb./sq. in. (2.5°F.)		
t	v	h	s	t	v	h	s	t	v	h	s
Sat.	14.45	189.96	0.4580	Sat.	8.993	192.64	0.4446	Sat.	4.710	196.58	0.4270
−40	14.62	190.77	.4599	−20	9.124	193.67	.4471	20	4.917	199.90	.4341
−20	15.36	194.27	.4681	0	9.567	197.32	.4552	40	5.146	203.75	.4420
0	16.09	197.84	.4760	20	10.01	201.04	.4630	60	5.373	207.66	.4496
20	16.82	201.48	.4838	40	10.45	204.78	.4707	80	5.599	211.65	.4572
40	17.55	205.19	.4914	60	10.89	208.62	.4782	100	5.823	215.69	.4645
60	18.27	209.01	.4989	80	11.33	212.53	.4856	120	6.046	219.80	.4717
80	18.99	212.88	.5061	100	11.77	216.50	.4928	140	6.268	223.99	.4788
100	19.71	216.82	.5133	120	12.21	220.54	.5000	160	6.489	228.24	.4858
120	20.42	220.84	.5204	140	12.65	224.67	.5069	180	6.709	232.56	.4927
140	21.14	224.94	.5274	160	13.08	228.86	.5138	200	6.929	236.94	.4994
160	21.86	229.11	.5342	180	13.52	233.13	.5206	220	7.147	241.40	.5061
180	22.57	233.36	.5410	200	13.95	237.47	.5273	240	7.365	245.96	.5127
200	23.29	237.69	.5476	220	14.38	241.89	.5339	260	7.583	250.55	.5192
220	24.00	242.09	.5542	240	14.81	246.42	.5045	280	7.801	255.23	.5256
240	24.71	246.60	.5607	260	15.24	250.98	.5469	300	8.019	259.96	.5319
260	25.42	251.15	.5672	280	15.67	255.66	.5532				

Temp., °F.	Abs. pressure, 50 lb./sq. in. (47.8°F.)			Temp., °F.	Abs. pressure, 100 lb./sq. in. (89.6°F.)			Temp., °F.	Abs. pressure, 200 lb./sq. in. (140.3°F.)		
t	v	h	s	t	v	h	s	t	v	h	s
Sat.	1.992	202.09	0.4043	Sat.	1.025	206.11	0.3872	Sat.	0.517	209.60	0.3702
60	2.054	204.65	.4094	100	1.055	208.58	.3920	160	.551	215.30	.3796
80	2.154	208.89	.4174	120	1.111	213.33	.4003	180	.582	220.87	.3886
100	2.252	213.18	.4252	140	1.165	218.15	.4084	200	.612	226.35	.3971
120	2.348	217.52	.4328	160	1.217	222.94	.4163	220	.641	231.75	.4052
140	2.443	221.88	.4402	180	1.268	227.71	.4239	240	.668	237.12	.4129
160	2.537	226.32	.4475	200	1.318	232.50	.4312	260	.695	242.39	.4204
180	2.630	230.79	.4546	220	1.367	237.32	.4384	280	.721	247.65	.4276
200	2.722	235.32	.4616	240	1.415	242.20	.4455	300	.747	252.93	.4346
220	2.813	239.90	.4684	260	1.463	247.06	.4523	320	.772	258.21	.4414

* Tanner, Banning, and Matthewson, *Ind. Eng. Chem.*, **31**, 878 (1939). Copyright, 1939, E. I. du Pont de Nemours & Co., Inc.

Table 230. Properties of Superheated Methyl Chloride*—(*Concluded*)

Temp., °F.	Abs. pressure, 50 lb./sq. in. (47.8°F.)			Temp., °F.	Abs. pressure, 100 lb./sq. in. (89.6°F.)			Temp., °F.	Abs. pressure, 200 b./sq. in. (140.3°F.)		
t	*v*	*h*	*s*	*t*	*v*	*h*	*s*	*t*	*v*	*h*	*s*
240	2.903	244.58	.4752	280	1.511	251.96	.4591	340	.797	263.51	.4481
260	2.993	249.27	.4818	300	1.557	256.92	.4657	360	.822	268.84	.4547
280	3.083	254.02	.4884	320	1.603	261.93	.4722	380	.847	274.19	.4612
300	3.173	258.83	.4948	340	1.649	266.97	.4786	400	.870	279.59	.4676
320	3.261	263.71	.5011	360	1.695	272.07	.4849	420	.895	285.05	.4738
340	3.349	268.65	.5074	380	1.739	277.25	.4911	440	.918	290.52	.4800

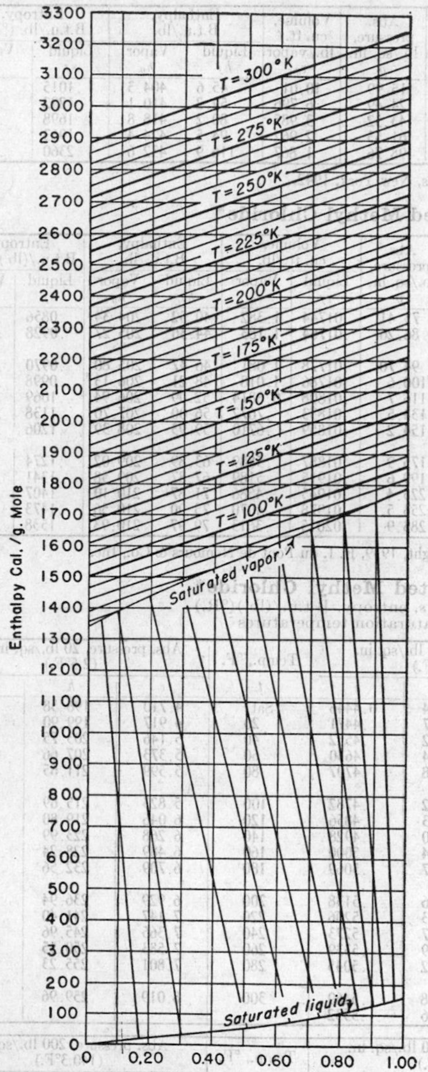

FIG. 19. Enthalpy-concentration diagram for oxygen-nitrogen mixture at 1 atm. Reference states: Enthalpies of liquid oxygen and liquid nitrogen at the normal boiling point of nitrogen are zero. (Dodge, "Chemical Engineering Thermodynamics," McGraw-Hill, New York, 1944.)

Table 231. Properties of Saturated Methyl Formate ($HCOOCH_3$)*

Temp., °F.	Abs. pressure, lb./sq. in.	Volume, cu. ft./lb., vapor	Enthalpy, B.t.u./lb.		Entropy, B.t.u./(lb.)(°R.)	
			Liquid	Vapor	Liquid	Vapor
t	*p*	v_g	h_f	h_g	s_f	s_g
0	1.50	54.0	0	232.5	0.0000	0.5055
20	2.70	31.0	10.3	236.6	.0219	.4934
40	4.66	18.9	20.6	240.7	.0432	.4837
60	7.61	12.0	30.9	244.8	.0633	.4748
80	12.07	7.98	41.2	248.8	.0825	.4670
100	18.26	5.38	51.5	252.9	.1015	.4615
120	27.24	3.74	61.8	257.0	.1192	.4559
140	38.41	2.65	72.1	261.0	.1375	.4525

* "Refrigerating Data Book," 5th ed., American Society of Refrigerating Engineers, New York, 1942.

Table 232. Properties of Saturated Methylene Chloride (CH_2Cl_2).*

Temp. °F.	Abs. pressure, lb./sq. in.	Volume, cu. ft./lb., vapor	Enthalpy, B.t.u./lb.		Entropy, B.t.u./(lb.)(°R.)	
			Liquid	Vapor	Liquid	Vapor
t	*p*	v_g	h_f	h_g	s_f	s_g
10	1.38	42.55	3.4	164.4	0.0072	0.3502
20	1.92	31.40	6.8	165.6	.0151	.3461
30	2.56	23.90	10.2	166.9	.0222	.3425
40	3.38	18.60	13.6	168.0	.0285	.3377
60	5.52	11.68	20.4	170.1	.0410	.3292
80	8.81	7.50	27.2	172.0	.0520	.3202
100	13.25	5.14	34.0	173.7	.0620	.3113
120	19.20	3.65	40.8	175.0	.0714	.3031
140	26.79	2.69	47.6	176.0	.0795	.2935

* "Refrigerating Data Book," 5th ed., American Society of Refrigerating Engineers, New York, 1942.

Table 233. Properties of Saturated Nitrogen*

Temp., °K.	Abs. pressure, atm.	Volume, cc./g.-mole		Enthalpy, cal./g.-mole		Entropy, cal./(g.-mole)(°K.)	
		Liquid	Vapor	Liquid	Vapor	Liquid	Vapor
T	*p*	v_f	v_g	h_f	h_g	s_f	s_g
77.4	1.00	34.7	6190	0	1335	0.0	17.25
80	1.36	35.2	4640	39	1352	.42	16.83
85	2.25	36.3	2880	115	1381	1.18	16.08
90	3.54	37.5	1872	189	1402	1.90	15.38
95	5.31	39.0	1257	257	1412	2.56	14.72
100	7.67	40.6	875	326	1412	2.21	14.07
105	10.71	42.6	624	393	1402	3.81	13.43
110	14.54	45.0	452	460	1379	4.39	12.74
115	19.28	48.3	324	534	1338	5.01	11.99
120	25.04	53.2	227	625	1269	5.78	11.14
125	31.94	64.9	137	781	1112	7.02	9.64
126	33.47	90.1	90.1	952	952	8.67	8.67

* Millar and Sullivan, *U.S. Bur. Mines Tech. Paper* 424, 1928.

Table 234. Properties of Superheated Nitrogen*

v, volume, cc./g.-mole; h, enthalpy, cal./g.-mole; s, entropy, cal./(g.-mole)(°K.); pressure, atm.; saturation temperature in °K. in parentheses after pressure

Temp., °K.	1 atm. (77.4)			5 atm. (94.1)			10 atm. (103.8)			20 atm. (115.7)			40 atm.			60 atm.		
	v	h	s	v	h	s	v	h	s	v	h	s	v	h	s	v	h	s
80	6,410	1,352	17.45	....	.....	.....	.....	.....	.....	.....	.....	.....	...	.....	.....	...	.....	.....
90	7,260	1,425	18.31	....	.....	.....	.....	.....	.....	.....	.....	.....	...	.....	.....	...	.....	.....
100	8,110	1,497	19.07	1,445	1,458	15.31	.....	.....	.....	.....	.....	.....	...	.....	.....	...	.....	.....
110	8,965	1,568	19.73	1,640	1,534	16.03	770	1,473	14.20	.....	.....	.....	...	.....	.....	...	.....	.....
120	9,800	1,638	20.34	1,825	1,610	16.69	870	1,564	14.99	345	1,422	12.70	...	.....	.....	...	.....	.....
130	10,635	1,708	20.90	2,005	1,684	17.28	965	1,646	15.65	417	1,555	13.76	95	1,120	9.43	...	784	6.75
140	11,475	1,778	21.42	2,190	1 757	17.82	1,060	1,724	16.23	476	1,660	14.55	181	1,461	11.96	...	1,150	9.45
150	12,290	1,848	21.90	2,370	1,829	18.32	1,155	1,800	16.76	527	1,745	15.13	229	1,602	13.03	112	1,400	11.19
160	13,125	1,918	22.36	2,540	1,901	18.78	1,245	1,875	17.24	577	1,825	15.64	265	1,708	13.73	147	1,562	12.23
170	13,960	1,988	22.78	2,710	1,971	19.21	1,335	1,949	17.69	626	1,903	16.12	296	1,802	14.28	176	1,694	13.04
180	14,775	2,058	23.18	2,880	2,042	19.61	1,420	2,022	18.11	675	1,979	16.55	324	1,893	14.79	203	1,803	13.69
190	15,600	2,126	23.55	3,050	2,112	19.99	1,510	2,094	18.50	722	2,055	16.96	350	1,980	15.26	224	1,910	14.25
200	16,410	2,194	23.89	3,220	2,182	20.35	1,590	2,166	18.87	769	2,131	17.36	376	2,064	15.70	245	2,002	14.73
210	17,230	2,262	24.23	3,390	2,251	20.69	1,675	2,238	19.22	816	2,206	17.73	401	2,147	16.11	263	2,090	15.15
220	18,060	2,330	24.56	3,560	2,321	21.02	1,760	2,310	19.55	864	2,281	18.08	426	2,228	16.48	281	2,178	15.56
230	18,880	2,400	24.87	3,720	2,391	21.33	1,845	2,380	19.86	910	2,353	18.40	450	2,306	16.83	296	2,263	15.93
240	19,700	2,470	25.16	3,890	2,461	21.63	1,930	2,450	20.16	958	2,426	18.71	474	2,383	17.16	314	2,345	16.29
250	20,510	2,540	25.44	4,050	2,531	21.91	2,020	2,520	20.45	1,010	2,498	19.01	498	2,461	17.48	328	2,425	16.62
260	21,330	2,608	25.71	4,220	2,600	22.18	2,110	2,591	20.72	1,063	2,573	19.29	522	2,536	17.77	345	2,503	16.92
270	22,145	2,676	25.97	4,390	2,669	22.44	2,200	2,660	20.98	1,118	2,643	19.56	546	2,610	18.05	362	2,579	17.21
280	22,960	2,744	26.22	4,560	2,737	22.70	2,290	2,730	21.24	1,174	2,714	19.82	569	2,684	18.32	378	2,655	17.49
290	23,785	2,814	26.46	4,730	2,8C8	22.94	2,375	2,800	21.48	1,231	2,785	20.08	591	2,758	18.58	393	2,731	17.76
300	24,620	2,884	26.70	4,900	2,878	23.18	2,465	2,870	21.72	1,290	2,857	20.32	612	2,832	18.83	418	2,807	18.01

* Millar and Sullivan, *U.S. Bur. Mines Tech. Paper* 424, 1928.

Table 235. Properties of Saturated Oxygen*

Temp., °K. T	Abs. pressure, atm. p	Volume, cc./g.-mole Liquid v_f	Volume, cc./g.-mole Vapor v_g	Enthalpy, cal./g.-mole Liquid h_f	Enthalpy, cal./g.-mole Vapor h_g	Entropy, cal./(g.-mole)(°K.) Liquid s_f	Entropy, cal./(g.-mole)(°K.) Vapor s_g	Temp., °K. T	Abs. pressure, atm. p	Volume, cc./g.-mole Liquid v_f	Volume, cc./g.-mole Vapor v_g	Enthalpy, cal./g.-mole Liquid h_f	Enthalpy, cal./g.-mole Vapor h_g	Entropy, cal./(g.-mole)(°K.) Liquid s_f	Entropy, cal./(g.-mole)(°K.) Vapor s_g
90.15	1.00	27.9	7223	158	1788	1.90	19.99	125	13.51	34.3	592	626	1864	6.13	16.02
95	1.60	28.6	4508	219	1800	2.55	19.21	130	17.52	35.7	460	700	1863	6.63	15.58
100	2.50	29.3	3028	284	1816	3.23	18.55	135	22.23	37.4	363	776	1854	7.19	15.18
105	3.73	30.1	2000	351	1831	3.86	17.96								
110	5.38	31.0	1456	420	1844	4.48	17.43	140	27.9	39.4	286	860	1833	7.78	14.73
								145	34.4	42.3	220	954	1783	8.41	14.13
115	7.51	32.0	1045	489	1854	5.08	16.95	150	42.2	47.9	157	1081	1682	9.24	13.25
120	10.20	33.1	784	557	1861	5.61	16.47	154.27	49.7	74.5	74.5	1393	1393	11.25	11.25

* Millar and Sullivan, *U.S. Bur. Mines Tech. Paper* 424, 1928.

Table 236. Properties of Superheated Oxygen*

v, volume, cc./g.-mole; h, enthalpy, cal./g.-mole; s, entropy, cal./(g.-mole)(°K.)
Pressure, atm.; saturation temperature, °K., in parentheses after pressure

Temp., °K. T	1 atm. (90.15)			5 atm. (108.9)			10 atm. (119.7)			20 atm. (132.7)			40 atm. (148.7)			60 atm.		
	v	h	s	v	h	s	v	h	s	v	h	s	v	h	s	v	h	s
95	7,630	1,824	20.38	....	....	.....	....	....	.....	...	.....	.....	...	.....	.....	...	.....	.....
100	8,050	1,861	20.76	....	....	.....	....	....	.....	...	.....	.....	...	.....	.....	...	.....	.....
110	8,890	1,933	21.44	1,560	1,853	17.66	....	....	.....	...	.....	.....	...	.....	.....	...	.....	.....
120	9,730	2,005	22.07	1,790	1,947	18.48	805	1,865	16.53	...	.....	.....	...	.....	.....	...	.....	.....
130	10,560	2,076	22.64	2,000	2,033	19.17	930	1,977	17.44	...	.....	.....	...	.....	.....	...	.....	.....
140	11,400	2,146	23.16	2 210	2,113	19.76	1,040	2,069	18.13	455	1,961	16.10	...	.....		...	.....	.....
150	12,240	2,216	23.65	2,400	2,188	20.28	1,150	2,151	18.69	516	2,072	16.86	184	1,780	13.90	...	.....	.....
160	13,070	2,286	24.10	2,590	2,262	20.76	1,250	2,230	19.20	572	2,163	17.45	228	1,993	15.28	...	1,570	12.25
170	13,900	2,356	24.52	2,770	2,334	21.20	1,350	2,307	19.67	626	2,249	17.97	266	2,116	16.03	139	1,927	14.42
180	14,740	2,426	24.92	2,940	2,406	21.61	1,450	2,381	20.09	677	2,328	18.42	298	2,219	16.62	168	2,093	15.37
190	15,570	2,496	25.30	3,110	2,477	22.00	1,540	2,452	20.47	725	2,404	18.84	328	2,311	17.11	194	2,219	16.06
200	16,390	2,565	25.66	3,280	2,547	22.36	1,625	2,523	20.84	772	2,479	19.22	356	2,397	17.56	220	2,322	16.59
210	17,210	2,633	25.99	3,450	2,615	22.69	1,715	2,593	21.18	818	2,552	19.58	382	2,478	17.95	243	2,409	17.01
220	18,030	2,700	26.30	3,610	2,684	23.01	1,800	2,664	21.51	864	2,627	19.93	408	2,558	18.32	264	2,493	17.40
230	18,860	2,767	26.60	3,770	2,752	23.32	1,880	2,735	21.83	910	2,701	20.26	434	2,638	18.68	284	2,577	17.77
240	19,680	2,834	26.88	3,940	2,822	23.61	1,960	2,806	22.13	956	2 775	20.57	460	2,717	19.01	303	2,661	18.13
250	20,500	2,901	27.16	4,100	2,892	23.90	2,050	2,877	22.42	1,001	2,849	20.88	484	2,796	19.33	321	2,745	18.47
260	21,330	2,971	27.43	4,260	2,962	24.17	2,130	2,948	22.70	1,046	2,922	21.16	508	2,874	19.64	338	2,828	18.80
270	22,150	3,041	27.68	4,430	3,032	24.43	2,210	3,019	22.96	1,090	2,996	21.44	532	2,952	19.93	353	2,910	19.11
280	22,970	3,111	27.93	4,590	3,102	24.68	2,290	3,090	23.22	1,136	3,068	21.70	556	3,028	20.21	367	2,991	19.40
290	23,790	3,180	28.18	4,750	3,172	24.93	2,370	3,161	23.47	1,180	3,140	21.96	580	3,104	20.47	381	3,072	19.69
300	24,610	3,251	28.41	4,920	3,242	25.17	2,460	3,232	23.71	1,226	3,212	22.20	604	3,179	20.73	395	3,152	19.96

* Millar and Sullivan, *U.S. Bur. Mines Tech. Paper* 424, 1928.

Table 237. Properties of Saturated Propane*

Temp., °F. t	Abs. pressure, lb./sq.in. p	Volume, cu. ft./lb. Liquid v_f	Vapor v_g	Enthalpy, B.t.u./lb. Liquid h_f	Vapor h_g	Entropy, B.t.u./(lb.)(°R.) Liquid s_f	Vapor s_g	Temp., °F. t	Abs. pressure, lb./sq. in. p	Volume, cu. ft./lb. Liquid v_f	Vapor v_g	Enthalpy, B.t.u./lb. Liquid h_f	Vapor h_g	Entropy, B.t.u./(lb.)(°R.) Liquid s_f	Vapor s_g
−80	5.65	0.0265	16.2	162.6	354.0	0.8794	1.3832	70	124.3	.03209	.854	245.7	394.4	1.0624	1.3427
−70	7.48	.0268	12.5	167.6	357.0	.8927	1.3781	80	143.6	.03269	.745	251.9	396.4	1.0737	1.3413
−60	9.78	.02703	9.77	172.7	360.0	.9060	1.3740	90	165.0	.03329	.643	258.2	398.3	1.0850	1.3400
−50	12.60	.02733	7.73	177.8	362.8	.9188	1.3702	100	188.7	.03390	.558	264.6	400.2	1.0963	1.3388
−40	16.00	.02763	6.16	183.0	365.7	.9315	1.3670	110	214.8	.03452	.487	271.1	401.9	1.1080	1.3378
−30	20.18	.02794	5.02	188.4	368.6	.9441	1.3640	120	243.4	.03532	.426	278.0	403.8	1.1195	1.3368
−20	25.05	.02826	4.06	193.8	371.5	.9568	1.3610	130	274.5	.03612	.370	285.2	405.4	1.1310	1.3356
−10	30.95	.02859	3.33	199.4	374.4	.9690	1.3582	140	308.4	.03702	.320	292.7	407.0	1.1430	1.3347
0	37.81	.02893	2.74	205.0	377.2	.9812	1.3555	150	345.4	.03817	.278	300.2	408.2	1.1552	1.3326
10	45.85	.02930	2.30	210.7	380.0	.9932	1.3531	160	385.0	.03962	.240	308.4	408.8	1.1680	1.3303
20	55.00	.02970	1.93	216.6	382.6	1.0050	1.3510	170	426.0	.04132	.208	317.5	408.6	1.1816	1.3272
30	65.70	.03011	1.60	222.3	385.1	1.0167	1.3491	180	473.2	.04367	.180	327.5	407.6	1.1970	1.3223
40	77.80	.03055	1.33	227.9	387.5	1.0283	1.3473	190	523.4	.04712	.149	339.2	404.6	1.2140	1.3156
50	91.50	.03101	1.14	233.8	389.9	1.0398	1.3456	200	575.0	.0521	.113	353.5	398.3	1.2360	1.3040
60	106.9	.03150	0.984	239.6	392.2	1.0511	1.3441								

* Stearns and George, *Ind. Eng. Chem.*, **35**, 602 (1943); with perm ssion.

Table 238. Properties of Superheated Propane*

v, volume, cu. ft./lb.; h, enthalpy, B.t.u./lb.; s, entropy, B.t.u./(lb.)(°R.); p, absolute pressure, lb./sq. in.
Parenthetic figures after pressures are saturation temperatures

Temp., °F.	7.35 (−70.7°F.) v	h	s	12.24 (−51°F.) v	h	s	14.696 (−43.708°F.) v	h	s	20 (−30.30°F.) v	h	s
Sat.	12.72	356.5	1.3785	7.90	362.75	1.3718	6.66	364.6	1.3681	5.050	368.4	1.3640
−60	13.11	360.3	1.3869									
−40	13.79	366.8	1.4023	8.182	366.2	1.3796	6.775	365.8	1.3710	.....	.....	.....
−20	14.47	373.4	1.4177	8.591	373.0	1.3948	7.123	372.8	1.3866	5.186	372.2	1.3719
0	15.14	380.2	1.4329	9.000	379.9	1.4100	7.471	379.8	1.4020	5.439	379.3	1.3873
20	15.82	387.4	1.4481	9.409	387.0	1.4252	7.816	386.8	1.4772	5.695	386.5	1.4025
40	16.50	394.8	1.4633	9.818	394.4	1.4404	8.160	394.3	1.4324	5.951	393.9	1.4177
60	17.17	402.4	1.4785	10.23	402.0	1.4556	8.503	401.9	1.4474	6.207	401.6	1.4327
80	17.85	410.2	1.4937	10.64	410.0	1.4706	8.844	409.8	1.4623	6.461	409.4	1.4477
100	18.52	418.4	1.5088	11.050	418.1	1.4854	9.187	418.0	1.4772	6.716	417.6	1.4625
120	19.20	426.6	1.5235	11.45	426.2	1.5002	9.528	426.2	1.4918	6.969	425.9	1.4771
140	19.88	435.2	1.5380	11.86	435.9	1.5148	9.869	434.8	1.5064	7.221	434.6	1.4917
160	20.55	444.1	1.5524	12.27	443.8	1.5293	10.21	443.7	1.5210	7.473	443.5	1.5063
180	21.23	453.1	1.5668	12.68	452.9	1.5437	10.51	452.9	1.5256	7.725	452.6	1.5209
200	21.90	462.5	1.5812	13.09	462.4	1.5581	10.84	462.4	1.5500	7.977	461.9	1.5355

Temp., °F.	30 (−11.52°F.) v	h	s	40 (+2.90°F.) v	h	s	60 (24.80°F.) v	h	s	80 (41.69°F.) v	h	s
Sat.	3.30	374.00	1.3588	2.61	378.2	1.3548	1.76	383.8	1.3501	1.32	387.8	1.3470
0	3.559	378.3	1.3678	.....	.....	.....	.....	.....	.....	.....	.....	.....
20	3.735	385.7	1.3830	2.753	384.7	1.3684	.....	.....	.....	.....	.....	.....
40	3.911	393.1	1.3980	2.889	392.2	1.3838	1.863	390.1	1.3630	.....	.....	.....
60	4.087	400.8	1.4130	3.025	400.0	1.3990	1.959	398.2	1.3789	1.424	396.1	1.3624
80	4.261	408.7	1.4280	3.159	408.0	1.4140	2.053	406.3	1.3948	1.500	404.6	1.3785
100	4.432	417.0	1.4428	3.289	416.2	1.4290	2.145	414.8	1.4102	1.573	413.2	1.3940
120	4.602	425.4	1.4576	3.419	424.6	1.4440	2.235	423.4	1.4254	1.644	421.8	1.4094
140	4.772	434.0	1.4724	3.549	433.4	1.4588	2.325	432.2	1.4404	1.714	430.9	1.4248
160	4.942	443.0	1.4872	3.679	442.5	1.4736	2.415	441.4	1.4554	1.784	440.3	1.4402
180	5.112	452.2	1.5020	3.809	451.8	1.4884	2.505	450.8	1.4703	1.852	449.8	1.4554
200	5.282	461.8	1.5168	3.939	461.3	1.5030	2.593	460.5	1.4851	1.920	459.7	1.4704

Temp., °F.	100 (55.62°F.) v	h	s	130 (73.20°F.) v	h	s	160 (87.71°F.) v	h	s
Sat.	1.06	391.2	1.3448	0.8165	395.0	1.3424	0.6685	397.9	1.3404
60	1.094	393.5	1.3488	.....	.....	.....	.....	.....	.....
80	1.164	402.6	1.3656	.8456	398.8	1.3486	.....	.....	.....
100	1.227	411.4	1.3816	.9045	408.3	1.3659	.6969	404.7	1.3521
120	1.289	420.3	1.3962	.9588	417.8	1.3822	.7464	414.6	1.3698
140	1.347	429.6	1.4130	1.006	427.3	1.3987	.7908	424.8	1.3867
160	1.400	439.1	1.4286	1.052	437.2	1.4150	.8319	434.9	1.4031
180	1.460	448.8	1.4440	1.098	447.0	1.4310	.8712	445.1	1.4195
200	1.516	458.8	1.4598	1.143	457.1	1.4468	.9093	455.4	1.4357

Temp., °F.	190 (100.50°F.) v	h	s	220 (111.85°F.) v	h	s	250 (122.12°F.) v	h	s	300 (137.55°F.) v	h	s
Sat.	0.5540	400.4	1.3388	0.4738	402.5	1.3375	0.4130	404.2	1.3355	0.3332	406.6	1.3345
120	.5995	411.6	1.3580	.4911	407.7	1.3460						
140	.6415	422.1	1.3759	.5314	419.2	1.3650	.4473	415.9	1.3550	.3392	408.7	1.3372
160	.6792	432.4	1.3930	.5673	429.8	1.3827	.4816	427.1	1.3732	.3745	422.0	1.3580
180	.7144	443.0	1.4096	.5998	440.6	1.3999	.5121	438.0	1.3908	.4037	433.2	1.3765
200	.7472	453.4	1.4255	.6302	451.2	1.4161	.5408	448.9	1.4074	.4303	444.9	1.3944

* Stearns and George, *Ind. Eng. Chem.*, **25**, 602 (1943); with permission.

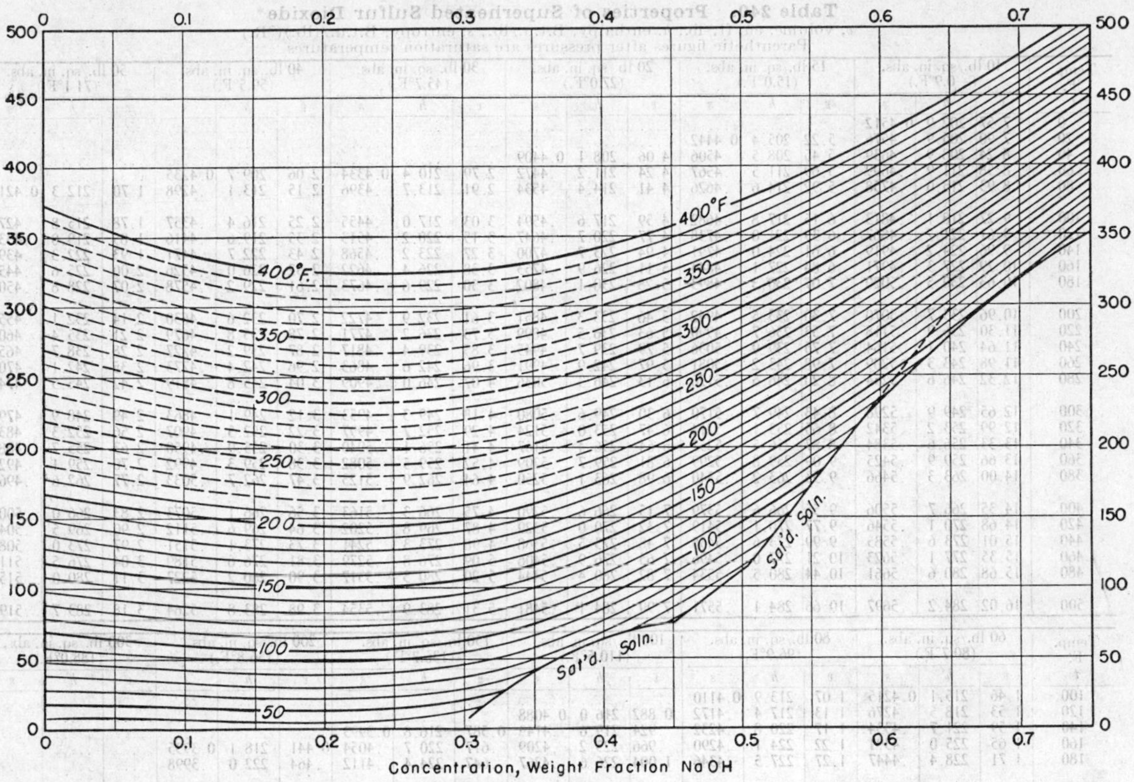

FIG. 20. Enthalpy-concentration diagram for aqueous sodium hydroxide at 1 Atm. Reference states: Enthalpy of liquid water at 32°F. and vapor pressure is zero; partial molal enthalpy of infinitely dilute NaOH solution at 64°F. and 1 atm. is zero. [McCabe, *Trans. Am. Inst. Chem. Engrs.* **31**, 129 (1935).]

Table 239. Properties of Saturated Sulfur Dioxide*

Temp., °F. t	Abs. pressure, lb./sq. in. p	Volume, cu. ft./lb. Liquid v_f	Volume, cu. ft./lb. Vapor v_g	Enthalpy, B.t.u./lb. Liquid h_f	Enthalpy, B.t.u./lb. Vapor h_g	Entropy, B.t.u./(lb.)(°R.) Liquid s_f	Entropy, B.t.u./(lb.)(°R.) Vapor s_g
−100	0.294	0.009856	204.7	0	190.1	0	0.5285
− 90	.465	.009954	132.9	3.3	191.4	0.0090	.5179
− 80	.710	.010054	89.3	6.5	192.7	.0177	.5081
− 70	1.056	.01015	61.0	9.8	194.0	.0262	.4990
− 60	1.550	.01025	42.6	13.0	195.3	.0344	.4905
− 50	2.225	.01034	30.4	16.3	196.7	.0424	.4827
− 40	3.12	.01044	22.2	19.6	198.0	.0502	.4753
− 30	4.33	.01053	16.4	22.8	199.2	.0579	.4684
− 20	5.88	.01062	12.5	26.0	200.4	.0654	.4620
− 10	7.83	.01072	9.48	29.3	201.7	.0726	.4560
0	10.26	.01082	7.35	32.5	202.8	.0797	.4503
10	13.3	.01092	5.77	35.7	204.0	.0867	.4450
20	16.9	.01103	4.59	39.0	205.3	.0935	.4402
30	21.3	.01114	3.70	42.2	206.4	.1002	.4356
40	26.6	.01125	3.02	45.5	207.7	.1067	.4313
50	32.9	.01137	2.48	48.7	208.7	.1132	.4271
60	40.3	.01149	2.05	52.0	209.8	.1195	.4232
70	49.1	.01162	1.70	55.3	210.8	.1258	.4194
80	59.3	.01175	1.42	58.6	211.7	.1320	.4158
90	71.0	.01189	1.20	61.9	212.6	.1382	.4123
100	84.1	.01204	1.02	65.3	213.5	.1443	.4090
110	99.1	.01219	0.868	68.8	214.5	.1503	.4060
120	116.3	.01235	.746	72.2	215.2	.1564	.4030
130	135.8	.01251	.646	75.8	215.8	.1624	.4001
140	157.7	.01269	.554	79.3	216.4	.1684	.3972
150	182	.01288	.481	83.0	217.2	.1744	.3944
160	209	.01309	.418	86.7	217.6	.1805	.3917
170	238	.01330	.364	90.5	218.1	.1865	.3892
180	272	.01350	.319	94.4	218.4	.1926	.3865
190	307	.01371	.281	98.3	218.6	.1988	.3840
200	347	.01396	.246	102.4	218.9	.2050	.3815
210	390	.01422	.217	106.5	218.9	.2112	.3792
220	437	.01453	.191	110.8	218.8	.2175	.3764
230	487	.01487	.168	115.2	218.7	.2239	.3737
240	543	.01527	.147	119.7	217.9	.2304	.3707
250	604	.01574	.129	124.4	216.9	.2370	.3674
260	669	.01628	.113	129.2	215.4	.2438	.3636
270	739	.01690	.0987	134.3	213.3	.2509	.3591
280	816	.01767	.0848	139.7	210.6	.2582	.3543
290	898	.01861	.0723	148.6	203.8	.2703	.3434
300	987	.02002	.0599	152.1	200.7	.2746	.3386
315.4	1143	.03070	.0307	176.2	176.2	.3062	.3062

* Rynning and Hurd, *Trans. Am. Inst. Chem. Engrs.* **41**, 265 (1945).

Table 240. Properties of Superheated Sulfur Dioxide*

v, volume, cu. ft./lb.; *h*, enthalpy, B.t.u./lb.; *s*, entropy, B.t.u./(lb.)(°R.)
Parenthetic figures after pressures are saturation temperatures

Temp. °F.	10 lb./sq. in. abs. (−0.9°F.)			15 lb./sq. in. abs. (15.0°F.)			20 lb./sq. in. abs. (27.0°F.)			30 lb./sq. in. abs. (45.7°F.)			40 lb./sq. in. abs. (59.5°F.)			50 lb./sq. in. abs. (71.1°F.)		
t	*v*	*h*	*s*	*v*	*h*	*s*	*v*	*h*	*s*	*v*	*h*	*s*	*v*	*h*	*s*	*v*	*h*	*s*
0	7.57	202.9	0.4512	.....	.....	.....	.....	.....	.....	.....	.....	.....	.....	.....	.....	.....	.....	.....
20	7.90	205.9	.4576	5.22	205.4	0.4442	.....	.....	.....	.....	.....	.....	.....	.....	.....	.....	.....	.....
40	8.24	208.9	.4639	5.45	208.5	.4506	4.06	208.1	0.4409	.....	.....	.....	.....	.....	.....	.....	.....	.....
60	8.58	211.9	.4699	5.69	211.5	.4567	4.24	211.2	.4472	2.79	210.4	0.4334	2.06	209.7	0.4235	.....	.....	.....
80	8.93	215.0	.4758	5.92	214.6	.4626	4.41	214.4	.4534	2.91	213.7	.4396	2.15	213.1	.4298	1.70	212.3	0.4217
100	9.27	218.1	.4813	6.15	217.8	.4682	4.59	217.6	.4594	3.03	217.0	.4455	2.25	216.4	.4357	1.78	215.8	.4278
120	9.61	221.2	.4867	6.39	221.0	.4738	4.77	220.7	.4647	3.15	220.2	.4513	2.35	219.6	.4416	1.85	219.0	.4339
140	9.96	224.2	.4919	6.61	224.0	.4791	4.94	223.7	.4700	3.27	223.2	.4568	2.43	222.7	.4471	1.93	222.3	.4396
160	10.29	227.3	.4971	6.84	227.1	.4844	5.11	226.9	.4753	3.38	226.4	.4622	2.52	226.0	.4526	2.00	225.6	.4451
180	10.63	230.5	.5020	7.07	230.3	.4893	5.28	230.1	.4802	3.50	229.6	.4673	2.61	229.2	.4578	2.07	228.8	.4505
200	10.96	233.7	.5070	7.30	233.5	.4942	5.46	233.3	.4851	3.61	232.9	.4722	2.70	232.6	.4630	2.14	232.1	.4558
220	11.30	236.9	.5118	7.53	236.7	.4990	5.63	236.5	.4899	3.73	236.2	.4771	2.78	235.8	.4679	2.21	235.4	.4608
240	11.64	240.1	.5164	7.75	239.9	.5036	5.79	239.7	.4945	3.85	239.4	.4817	2.87	239.2	.4727	2.28	238.7	.4656
260	11.98	243.3	.5209	7.98	243.2	.5081	5.97	242.9	.4991	3.96	242.6	.4863	2.96	242.4	.4773	2.35	242.1	.4703
280	12.32	246.6	.5254	8.20	246.5	.5126	6.13	246.3	.5036	4.07	246.0	.4909	3.04	245.8	.4819	2.42	245.5	.4749
300	12.65	249.9	.5298	8.43	249.7	.5170	6.30	249.6	.5080	4.18	249.3	.4953	3.12	249.1	.4863	2.49	248.9	.4794
320	12.99	253.2	.5342	8.65	253.1	.5214	6.47	253.0	.5124	4.29	252.7	.4997	3.22	252.5	.4907	2.56	252.3	.4838
340	13.33	256.6	.5384	8.87	256.5	.5257	6.64	256.4	.5167	4.41	256.1	.5040	3.30	255.9	.4950	2.63	255.7	.4881
360	13.66	259.9	.5425	9.09	259.8	.5299	6.81	259.7	.5209	4.52	259.5	.5082	3.38	259.3	.4992	2.70	259.1	.4923
380	14.00	263.3	.5466	9.33	263.2	.5340	6.98	263.1	.5250	4.64	262.9	.5123	3.47	262.7	.5033	2.77	262.6	.4964
400	14.35	266.7	.5506	9.55	266.6	.5379	7.15	266.6	.5290	4.75	266.3	.5163	3.56	266.1	.5073	2.83	266.0	.5003
420	14.68	270.1	.5546	9.77	270.1	.5419	7.32	270.0	.5329	4.87	269.8	.5202	3.64	269.6	.5112	2.90	269.5	.5041
440	15.01	273.6	.5585	9.99	273.6	.5458	7.48	273.5	.5368	4.98	273.3	.5241	3.73	273.1	.5151	2.97	273.0	.5081
460	15.35	277.1	.5623	10.21	277.0	.5496	7.65	276.9	.5406	5.09	276.8	.5279	3.81	276.6	.5189	3.04	276.5	.5119
480	15.68	280.6	.5661	10.44	280.5	.5534	7.82	280.4	.5444	5.20	280.3	.5317	3.90	280.2	.5227	3.11	280.0	.5157
500	16.02	284.2	.5697	10.66	284.1	.5571	7.99	284.1	.5481	5.31	283.9	.5354	3.98	283.8	.5264	3.18	283.7	.5194

Temp. F.	60 lb./sq. in. abs. (80.7°F.)			80 lb./sq. in. abs. (96.9°F.)			100 lb./sq. in abs. (110.5°F.)			150 lb./sq. in. abs. (136.7°F.)			200 lb./sq. in. abs. (156.8°F.)			300 lb./sq. in. abs. (188.0°F.)		
t	*v*	*h*	*s*	*v*	*h*	*s*	*v*	*h*	*s*	*v*	*h*	*s*	*v*	*h*	*s*	*v*	*h*	*s*
100	1.46	215.1	0.4215	1.07	213.9	0.4110	.....	.....	.....	.....	.....	.....	.....	.....	.....	.....	.....	.....
120	1.53	218.5	.4276	1.13	217.4	.4172	0.882	216.0	0.4088	.....	.....	.....	.....	.....	.....	.....	.....	.....
140	1.59	221.7	.4334	1.17	220.8	.4232	.924	219.6	.4149	0.587	216.8	0.3993	.....	.....	.....	.....	.....	.....
160	1.65	225.0	.4391	1.22	224.1	.4290	.966	223.2	.4209	.617	220.7	.4054	0.441	218.1	0.3935	.....	.....	.....
180	1.71	228.4	.4447	1.27	227.5	.4346	1.004	226.6	.4267	.647	224.4	.4112	.464	222.0	.3998	.....	.....	.....
200	1.78	231.8	.4501	1.32	230.9	.4400	1.042	230.2	.4323	.677	228.1	.4172	.487	226.0	.4058	0.299	221.3	0.3883
220	1.83	235.1	.4553	1.36	234.3	.4452	1.080	233.6	.4376	.707	231.7	.4227	.509	229.7	.4116	.316	225.5	.3946
240	1.89	238.5	.4603	1.41	237.8	.4502	1.118	237.1	.4427	.735	235.3	.4280	.531	233.5	.4172	.333	229.7	.4005
260	1.95	241.8	.4650	1.45	241.2	.4550	1.154	240.5	.4476	.760	238.9	.4330	.551	237.3	.4226	.348	233.6	.4060
280	2.01	245.2	.4694	1.49	244.7	.4598	1.190	244.1	.4524	.786	242.5	.4379	.571	241.1	.4277	.364	237.8	.4123
300	2.07	248.6	.4736	1.54	248.1	.4643	1.226	247.5	.4570	.810	246.1	.4427	.590	244.8	.4325	.378	241.8	.4174
320	2.12	252.0	.4778	1.59	251.5	.4687	1.263	251.0	.4614	.834	249.8	.4473	.610	248.5	.4372	.392	246.0	.4223
340	2.18	255.5	.4820	1.63	255.1	.4731	1.298	254.6	.4656	.858	253.4	.4517	.629	252.2	.4418	.405	249.8	.4270
360	2.24	258.9	.4862	1.67	258.5	.4773	1.333	258.1	.4697	.882	257.0	.4561	.652	255.8	.4462	.419	253.5	.4316
380	2.30	262.4	.4903	1.71	262.0	.4814	1.367	261.6	.4737	.905	260.5	.4603	.667	259.4	.4505	.433	257.2	.4360
400	2.36	265.9	.4943	1.76	265.5	.4854	1.403	265.1	.4777	.929	264.1	.4645	.685	263.1	.4548	.446	261.0	.4404
420	2.41	269.4	.4983	1.80	269.0	.4893	1.439	268.6	.4817	.952	267.6	.4686	.704	266.7	.4589	.458	264.7	.4447
440	2.47	272.9	.5023	1.85	272.5	.4932	1.474	272.1	.4857	.974	271.2	.4726	.727	270.3	.4630	.471	268.4	.4489
460	2.53	276.4	.5061	1.89	276.0	.4970	1.509	275.7	.4896	.996	274.8	.4766	.740	273.9	.4670	.484	272.1	.4531
480	2.59	279.9	.5099	1.93	279.5	.5008	1.543	279.2	.4935	1.018	278.4	.4805	.759	277.6	.4710	.497	275.9	.4572
500	2.64	283.5	.5136	1.97	283.1	.5045	1.578	282.8	.4973	1.039	282.1	.4843	.776	281.3	.4748	.509	279.7	.4611

Temp. °F.	400 lb./sq. in. abs. (212.0°F.)			500 lb./sq. in. abs. (232.5°F.)			600 lb./sq. in. abs. (249.5°F.)			800 lb./sq. in. abs. (278.0°F.)			1000 lb./sq. in. abs. (301.5°F.)		
t	*v*	*h*	*s*	*v*	*h*	*s*	*v*	*h*	*s*	*v*	*h*	*s*	*v*	*h*	*s*
220	0.216	221.1	0.3810	.....	.....	.....	.....	.....	.....	.....	.....	.....	.....	.....	.....
240	.232	225.6	.3875	0.168	220.2	0.3753	.....	.....	.....	.....	.....	.....	.....	.....	.....
260	.246	230.0	.3936	.182	225.2	.3820	0.138	220.4	0.3714	.....	.....	.....	.....	.....	.....
280	.255	234.4	.3994	.195	230.2	.3883	.150	226.3	.3783	0.0914	212.5	0.3560	.....	.....	.....
300	.269	238.8	.4049	.206	235.0	.3944	.162	231.4	.3849	.1035	221.8	.3649	.....	.....	.....
320	.282	243.1	.4102	.216	239.7	.4002	.172	236.5	.3913	.1140	229.1	.3734	0.0761	217.3	0.3504
340	.293	247.2	.4154	.226	244.2	.4059	.181	241.2	.3976	.1230	234.9	.3814	.0865	226.1	.3655
360	.305	251.0	.4203	.236	248.5	.4114	.189	245.8	.4035	.1308	240.0	.3887	.0944	232.7	.3760
380	.315	254.8	.4251	.244	252.6	.4165	.197	250.1	.4090	.1379	244.8	.3951	.1014	238.6	.3821
400	.325	258.8	.4298	.253	256.7	.4212	.205	254.4	.4138	.1446	249.5	.4004	.1076	244.2	.3888
420	.335	262.6	.4343	.261	260.6	.4257	.212	258.5	.4184	.1510	254.0	.4056	.1135	249.1	.3941
440	.346	266.4	.4386	.270	264.6	.4301	.220	262.5	.4229	.1572	258.3	.4106	.1190	253.9	.3998
460	.356	270.2	.4428	.279	268.5	.4345	.227	266.4	.4273	.1632	262.5	.4153	.1243	258.4	.4051
480	.366	274.1	.4469	.287	272.4	.4387	.234	270.5	.4317	.1691	266.8	.4200	.1296	262.8	.4101
500	.376	278.0	.4509	.295	276.3	.4428	.241	274.6	.4359	.1746	271.0	.4245	.1345	267.1	.4147

* Rynning and Hurd, *Trans. Am. Inst. Chem. Engrs.* **41**, 265 (1945).

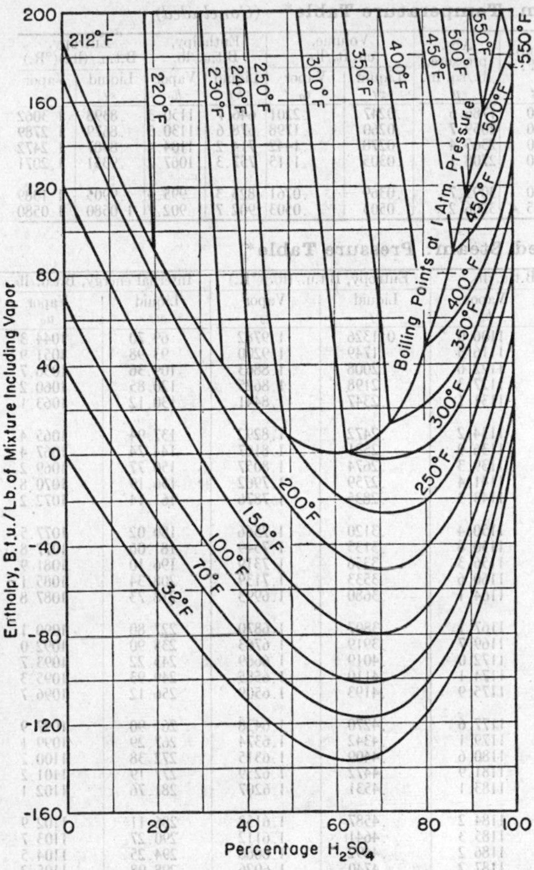

FIG. 21. Enthalpy-concentration diagram for aqueous sulfuric acid at 1 Atm. Reference states: Enthalpies of pure liquid components at 32°F. and vapor pressures are zero. NOTE: It should be observed that the weight basis includes the vapor, which is particularly important in the two-phase region. The upper ends of the tie lines in this region are assumed to be pure water. (*Hougen and Watson, "Chemical Process Principles," Part I, Wiley, New York*, 1943.)

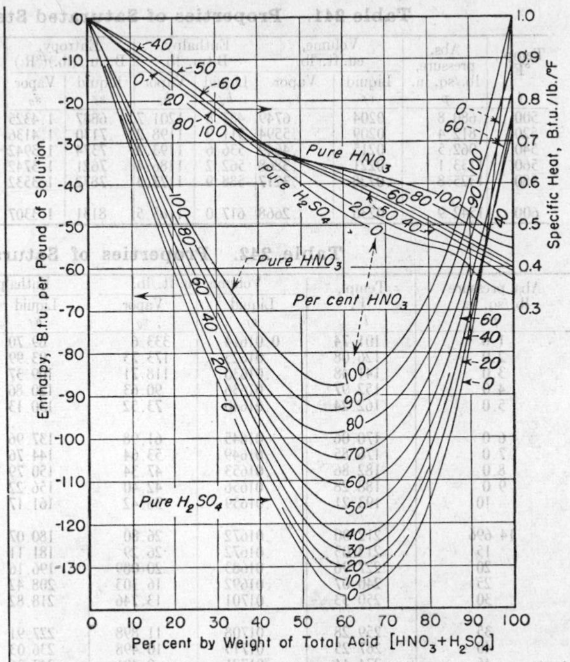

FIG. 22. Enthalpy-concentration diagram for aqueous sulfuric and nitric acids at 32°F. Reference states: Enthalpy of pure components at 32°F. is zero. NOTE: The per cent HNO_3 is computed on a water-free basis. Enthalpies at temperatures other than 32°F. may be computed by utilizing the specific heat data given, which may be assumed independent of temperature as a first approximation. [*McKinley and Brown, Chem. & Met. Eng.* **49,** 142 (1942).]

Table 241. Properties of Saturated Steam: Temperature Table*

Temp., °F.	Abs. pressure, lb./sq. in.	Volume, cu. ft./lb.		Enthalpy, B.t.u./lb.		Entropy, B.t.u./(lb.)(°R.)	
		Liquid	Vapor	Liquid	Vapor	Liquid	Vapor
t	p	v_f	v_g	h_f	h_g	s_f	s_g
32	0.08854	0.01602	3306	0.00	1075.8	0.0000	2.1877
35	.09995	.01602	2947	3.02	1077.1	.0061	2.1770
40	.12170	.01602	2444	8.05	1079.3	.0162	2.1597
45	.14752	.01602	2036.4	13.06	1081.5	.0262	2.1429
50	.17811	.01603	1703.2	18.07	1083.7	.0361	2.1264
60	.2563	.01604	1206.7	28.06	1088.0	.0555	2.0948
70	.3631	.01606	867.9	38.04	1092.3	.0745	2.0647
80	.5069	.01608	633.1	48.02	1096.6	.0932	2.0360
90	.6982	.01610	468.0	57.99	1100.9	.1115	2.0087
100	.9492	.01613	350.4	67.97	1105.2	.1295	1.9826
110	1.2748	.01617	265.4	77.94	1109.5	.1471	1.9577
120	1.6924	.01620	203.27	87.92	1113.7	.1645	1.9339
130	2.2225	.01625	157.34	97.90	1117.9	.1816	1.9112
140	2.8886	.01629	123.01	107.89	1122.0	.1984	1.8894
150	3.718	.01634	97.07	117.89	1126.1	.2149	1.8685
160	4.741	.01639	77.29	127.89	1130.2	.2311	1.8485
170	5.992	.01645	62.06	137.90	1134.2	.2472	1.8293
180	7.510	.01651	50.23	147.92	1138.1	.2630	1.8109
190	9.339	.01657	40.96	157.95	1142.0	.2785	1.7932
200	11.526	.01663	33.64	167.99	1145.9	.2938	1.7762
210	14.123	.01670	27.82	178.05	1149.7	.3090	1.7598
212	14.696	.01672	26.80	180.07	1150.4	.3120	1.7566
220	17.186	.01677	23.15	188.13	1153.4	.3239	1.7440
230	20.780	.01684	19.382	198.23	1157.0	.3387	1.7288
240	24.969	.01692	16.323	208.34	1160.5	.3531	1.7140
250	29.825	.01700	13.821	216.48	1164.0	.3675	1.6998
260	35.429	.01709	11.763	228.64	1167.3	.3817	1.6860
270	41.858	.01717	10.061	238.84	1170.6	.3958	1.6727
280	49.203	.01726	8.645	249.06	1173.8	.4096	1.6597
290	57.556	.01735	7.461	259.31	1176.8	.4234	1.6472
300	67.013	.01745	6.466	269.59	1179.7	.4369	1.6350
310	77.68	.01755	5.626	279.92	1182.5	.4504	1.6231
320	89.66	.01765	4.914	290.28	1185.2	.4637	1.6115
330	103.06	.01776	4.307	300.68	1187.7	.4769	1.6002
340	118.01	.01787	3.788	311.13	1190.1	.4900	1.5891
350	134.63	.01799	3.342	321.63	1192.3	.5029	1.5783
360	153.04	.01811	2.957	332.18	1194.4	.5158	1.5677
370	173.37	.01823	2.625	342.79	1196.3	.5286	1.5573
380	195.77	.01836	2.335	353.45	1198.1	.5413	1.5471
390	220.37	.01850	2.0836	364.17	1199.6	.5539	1.5371
400	247.31	.01864	1.8633	374.97	1201.0	.5664	1.5272
410	276.75	.01878	1.6700	385.83	1202.1	.5788	1.5174
420	308.83	.01894	1.5000	396.77	1203.1	.5912	1.5078
430	343.72	.01910	1.3499	407.79	1203.8	.6035	1.4982
440	381.59	.01926	1.2171	418.90	1204.3	.6158	1.4887
450	422.6	.0194	1.0993	430.1	1204.6	.6280	1.4793
460	466.9	.0196	0.9944	441.4	1204.6	.6402	1.4700
470	514.7	.0198	.9009	452.8	1204.3	.6523	1.4606
480	566.1	.0200	.8172	464.4	1203.7	.6645	1.4513
490	621.4	.0202	.7423	476.0	1202.8	.6766	1.4419

* Abridged from Keenan and Keyes, "Thermodynamic Properties of Steam," Wiley, New York, 1936. Copyright, 1937, by Joseph H. Keenan and Frederick G. Keyes.

Table 241. Properties of Saturated Steam: Temperature Table*—(*Concluded*)

Temp., °F.	Abs. pressure, lb./sq. in.	Volume, cu. ft./lb.		Enthalpy, B.t.u./lb.		Entropy, B.t.u./(lb.)(°R.)		Temp., °F.	Abs. pressure, lb./sq. in.	Volume, cu. ft./lb.		Enthalpy, B.t.u./lb.		Entropy, B.t.u./(lb.)(°R.)	
		Liquid	Vapor	Liquid	Vapor	Liquid	Vapor			Liquid	Vapor	Liquid	Vapor	Liquid	Vapor
t	p	v_f	v_g	h_f	h_g	s_f	s_g	t	p	v_f	v_g	h_f	h_g	s_f	s_g
500	680.8	.0204	.6749	487.8	1201.7	.6887	1.4325	620	1786.6	.0247	.2201	646.7	1150.3	.8398	1.3062
520	812.4	.0209	.5594	511.9	1198.2	.7130	1.4136	640	2059.7	.0260	.1798	678.6	1130.5	.8679	1.2789
540	962.5	.0215	.4649	536.6	1193.2	.7374	1.3942	660	2365.4	.0278	.1442	714.2	1104.4	.8987	1.2472
560	1133.1	.0221	.3868	562.2	1186.4	.7621	1.3742	680	2708.1	.0305	.1115	757.3	1067.2	.9351	1.2071
580	1325.8	.0228	.3217	588.9	1177.3	.7872	1.3532								
								700	3093.7	.0369	.0761	823.3	995.4	.9905	1.1389
600	1542.9	.0236	.2668	617.0	1165.5	.8131	1.3307	705.4	3206.2	.0503	.0503	902.7	902.7	1.0580	1.0580

Table 242. Properties of Saturated Steam: Pressure Table*

Abs. pressure lb./sq. in.	Temp., °F.	Volume, cu. ft./lb.		Enthalpy, B.t.u./lb.		Entropy, B.t.u./(lb.)(°R.)		Internal energy, B.t.u./lb.	
		Liquid	Vapor	Liquid	Vapor	Liquid	Vapor	Liquid	Vapor
p	t	v_f	v_g	h_f	h_g	s_f	s_g	u_f	u_g
1.0	101.74	0.01614	333.6	69.70	1106.0	0.1326	1.9782	69.70	1044.3
2.0	126.08	.01623	173.73	93.99	1116.3	.1749	1.9200	93.98	1051.9
3.0	141.48	.01630	118.71	109.37	1122.6	.2008	1.8863	109.36	1056.7
4.0	152.97	.01636	90.63	120.86	1127.3	.2198	1.8625	120.85	1060.2
5.0	162.24	.01640	73.52	130.13	1131.1	.2347	1.8441	130.12	1063.1
6.0	170.06	.01645	61.98	137.96	1134.2	.2472	1.8292	137.94	1065.4
7.0	176.85	.01649	53.64	144.76	1136.9	.2581	1.8167	144.74	1067.4
8.0	182.86	.01653	47.34	150.79	1139.3	.2674	1.8057	150.77	1069.2
9.0	188.28	.01656	42.40	156.22	1141.4	.2759	1.7962	156.19	1070.8
10	193.21	.01659	38.42	161.17	1143.3	.2835	1.7876	161.14	1072.2
14.696	212.00	.01672	26.80	180.07	1150.4	.3120	1.7566	180.02	1077.5
15	213.03	.01672	26.29	181.11	1150.8	.3135	1.7549	181.06	1077.8
20	227.96	.01683	20.089	196.16	1156.3	.3356	1.7319	196.10	1081.9
25	240.07	.01692	16.303	208.42	1160.6	.3533	1.7139	208.34	1085.1
30	250.33	.01701	13.746	218.82	1164.1	.3680	1.6993	218.73	1087.8
35	259.28	.01708	11.898	227.91	1167.1	.3807	1.6870	227.80	1090.1
40	267.25	.01715	10.498	236.03	1169.7	.3919	1.6763	235.90	1092.0
45	274.44	.01721	9.401	243.36	1172.0	.4019	1.6669	243.22	1093.7
50	281.01	.01727	8.515	250.09	1174.1	.4110	1.6585	249.93	1095.3
55	287.07	.01732	7.787	256.30	1175.9	.4193	1.6509	256.12	1096.7
60	292.71	.01738	7.175	262.09	1177.6	.4270	1.6438	261.90	1097.9
65	297.97	.01743	6.655	267.50	1179.1	.4342	1.6374	267.29	1099.1
70	302.92	.01748	6.206	272.61	1180.6	.4409	1.6315	272.38	1100.2
75	307.60	.01753	5.816	277.43	1181.9	.4472	1.6259	277.19	1101.2
80	312.03	.01757	5.472	282.02	1183.1	.4531	1.6207	281.76	1102.1
85	316.25	.01761	5.168	286.39	1184.2	.4587	1.6158	286.11	1102.9
90	320.27	.01766	4.896	290.56	1185.3	.4641	1.6112	290.27	1103.7
95	324.12	.01770	4.652	294.56	1186.2	.4692	1.6068	294.25	1104.5
100	327.81	.01774	4.432	298.40	1187.2	.4740	1.6026	298.08	1105.2
110	334.77	.01782	4.049	305.66	1188.9	.4832	1.5948	305.30	1106.5
120	341.25	.01789	3.728	312.44	1190.4	.4916	1.5878	312.05	1107.6
130	347.32	.01796	3.455	318.81	1191.7	.4995	1.5812	318.38	1108.6
140	353.02	.01802	3.220	324.82	1193.0	.5069	1.5751	324.35	1109.6
150	358.42	.01809	3.015	330.51	1194.1	.5138	1.5694	330.01	1110.5
160	363.53	.01815	2.834	335.93	1195.1	.5204	1.5640	335.39	1111.2
170	368.41	.01822	2.675	341.09	1196.0	.5266	1.5590	340.52	1111.9
180	373.06	.01827	2.532	346.03	1196.9	.5325	1.5542	345.42	1112.5
190	377.51	.01833	2.404	350.79	1197.6	.5381	1.5497	350.15	1113.1
200	381.79	.01839	2.288	355.36	1198.4	.5435	1.5453	354.68	1113.7
250	400.95	.01865	1.8438	376.00	1201.1	.5675	1.5263	375.14	1115.8
300	417.33	.01890	1.5433	393.84	1202.8	.5879	1.5104	392.79	1117.1
350	431.72	.01913	1.3260	409.69	1203.9	.6056	1.4966	408.45	1118.0
400	444.59	.0193	1.1613	424.0	1204.5	.6214	1.4844	422.6	1118.5
450	456.28	.0195	1.0320	437.2	1204.6	.6356	1.4734	435.5	1118.7
500	467.01	.0197	0.9278	449.4	1204.4	.6487	1.4634	447.6	1118.6
550	476.94	.0199	.8424	460.8	1203.9	.6608	1.4542	458.8	1118.2
600	486.21	.0201	.7698	471.6	1203.2	.6720	1.4454	469.4	1117.7
650	494.90	.0203	.7083	481.8	1202.3	.6826	1.4374	479.4	1117.1
700	503.10	.0205	.6554	491.5	1201.2	.6925	1.4296	488.8	1116.3
750	510.86	.0207	.6092	500.8	1200.0	.7019	1.4223	598.0	1115.4
800	518.23	.0209	.5687	509.7	1198.6	.7108	1.4153	506.6	1114.4
850	525.26	.0210	.5327	518.3	1197.1	.7194	1.4085	515.0	1113.3
900	531.98	.0212	.5006	526.6	1195.4	.7275	1.4020	523.1	1112.1
950	538.43	.0214	.4717	534.6	1193.7	.7355	1.3957	530.9	1110.8
1000	544.61	.0216	.4456	542.4	1191.8	.7430	1.3897	538.4	1109.4
1100	556.31	.0220	.4001	557.4	1187.8	.7575	1.3780	552.9	1106.4
1200	567.22	.0223	.3619	571.7	1183.4	.7711	1.3667	566.7	1103.0
1300	577.46	.0227	.3293	585.4	1178.6	.7840	1.3559	580.0	1099.4
1400	587.10	.0231	.3012	598.7	1173.4	.7963	1.3454	592.7	1095.4
1500	596.23	.0235	.2765	611.6	1167.9	.8082	1.3351	605.1	1091.2
2000	635.82	.0257	.1878	671.7	1135.1	.8619	1.2849	662.2	1065.6
2500	668.13	.0287	.1307	730.6	1091.1	.9126	1.2322	717.3	1030.6
3000	695.36	.0346	.0858	802.5	1020.3	.9731	1.1615	783.4	972.7
3206.2	705.40	.0503	.0503	902.7	902.7	1.0580	1.0580	872.9	872.9

* Abridged from Keenan and Keyes, "Thermodynamic Properties of Steam," Wiley, New York, 1936. Copyright, 1937, by Joseph H. Keenan and Frederick G. Keyes.

Table 243. Properties of Superheated Steam*

v, volume, cu. ft./lb.; *h*, enthalpy, B.t.u./lb.; *s*, entropy, B.t.u./(lb.)(°R.)

Abs. pressure, lb./sq. in. (sat. temp.)		Temp., °F. 200	300	400	500	600	700	800	900	1000	1100	1200	1400	1600
	v	392.6	452.3	512.0	571.6	631.2	690.8	750.4	809.9	869.5	929.1	988.7	1107.8	1227.0
1	*h*	1150.4	1195.8	1241.7	1288.3	1335.7	1383.8	1432.8	1482.7	1533.5	1585.2	1637.7	1745.7	1857.5
(101.74)	*s*	2.0512	2.1153	2.1720	2.2233	2.2702	2.3137	2.3542	2.3923	2.4283	2.4625	2.4952	2.5566	2.6137
	v	78.16	90.25	102.26	114.22	126.16	138.10	150.03	161.95	173.87	185.79	197.71	221.6	245.4
5	*h*	1148.8	1195.0	1241.2	1288.0	1335.4	1383.6	1432.7	1482.6	1533.4	1585.1	1637.7	1745.7	1857.4
(162.24)	*s*	1.8718	1.9370	1.9942	2.0456	2.0927	2.1361	2.1767	2.2148	2.2509	2.2851	2.3178	2.3792	2.4363
	v	38.85	45.00	51.04	57.05	63.03	69.01	74.98	80.95	86.92	92.88	98.84	110.77	122.69
10	*h*	1146.6	1193.9	1240.6	1287.5	1335.1	1383.4	1432.5	1482.4	1533.2	1585.0	1637.6	1745.6	1857.3
(193.21)	*s*	1.7927	1.8595	1.9172	1.9689	2.0160	2.0596	2.1002	2.1383	2.1744	2.2086	2.2413	2.3028	2.3598
	v	......	30.53	34.68	38.78	42.86	46.94	51.00	55.07	59.13	63.19	67.25	75.37	83.48
14.696	*h*	......	1192.8	1239.9	1287.1	1334.8	1383.2	1432.3	1482.3	1533.1	1584.8	1637.5	1745.5	1857.3
(212.00)	*s*	......	1.8160	1.8743	1.9261	1.9734	2.0170	2.0576	2.0958	2.1319	2.1662	2.1989	2.2603	2.3174
	v	......	22.36	25.43	28.46	31.47	34.47	37.46	40.45	43.44	46.42	49.41	55.37	61.34
20	*h*	......	1191.6	1239.2	1286.6	1334.4	1382.9	1432.1	1482.1	1533.0	1584.7	1637.4	1745.4	1857.2
(227.96)	*s*	......	1.7808	1.8396	1.8918	1.9392	1.9829	2.0235	2.0618	2.0978	2.1321	2.1648	2.2263	2.2834
	v	......	11.040	12.628	14.168	15.688	17.198	18.702	20.20	21.70	23.20	24.69	27.68	30.66
40	*h*	......	1186.8	1236.5	1284.8	1333.1	1381.9	1431.3	1481.4	1532.4	1584.3	1637.0	1745.1	1857.0
(267.25)	*s*	......	1.6994	1.7608	1.8140	1.8619	1.9058	1.9467	1.9850	2.0212	2.0555	2.0883	2.1498	2.2069
	v	......	7.259	8.357	9.403	10.427	11.441	12.449	13.452	14.454	15.453	16.451	18.446	20.44
60	*h*	......	1181.6	1233.6	1283.0	1331.8	1380.9	1430.5	1480.8	1531.9	1583.8	1636.6	1744.8	1856.7
(292.71)	*s*	......	1.6492	1.7135	1.7678	1.8162	1.8605	1.9015	1.9400	1.9762	2.0106	2.0434	2.1049	2.1621
	v	......	......	6.220	7.020	7.797	8.562	9.322	10.077	10.830	11.582	12.332	13.830	15.325
80	*h*	......	......	1230.7	1281.1	1330.5	1379.9	1429.7	1480.1	1531.3	1583.4	1636.2	1744.5	1856.5
(312.03)	*s*	......	......	1.6791	1.7346	1.7836	1.8281	1.8694	1.9079	1.9442	1.9787	2.0115	2.0721	2.1303
	v	......	......	4.937	5.589	6.218	6.835	7.446	8.052	8.656	9.259	9.860	11.060	12.258
100	*h*	......	......	1227.6	1279.1	1329.1	1378.9	1428.9	1479.5	1530.8	1582.9	1635.7	1744.2	1856.2
(327.81)	*s*	......	......	1.6518	1.7085	1.7581	1.8029	1.8443	1.8829	1.9193	1.9538	1.9867	2.0484	2.1056
	v	......	......	4.081	4.636	5.165	5.683	6.195	6.702	7.207	7.710	8.212	9.214	10.213
120	*h*	......	......	1224.4	1277.2	1327.7	1377.8	1428.1	1478.8	1530.2	1582.4	1635.3	1743.9	1856.0
(341.25)	*s*	......	......	1.6287	1.6869	1.7370	1.7822	1.8237	1.8625	1.8990	1.9335	1.9664	2.0281	2.0854
	v	......	......	3.468	3.954	4.413	4.861	5.301	5.738	6.172	6.604	7.035	7.895	8.752
140	*h*	......	......	1221.1	1275.2	1326.4	1376.8	1427.3	1478.2	1529.7	1581.9	1634.9	1743.5	1855.7
(353.02)	*s*	......	......	1.6087	1.6683	1.7190	1.7645	1.8063	1.8451	1.8817	1.9163	1.9493	2.0110	2.0683
	v	......	......	3.008	3.443	3.849	4.244	4.631	5.015	5.396	5.775	6.152	6.906	7.656
160	*h*	......	......	1217.6	1273.1	1325.0	1375.7	1426.4	1477.5	1529.1	1581.4	1634.5	1743.2	1855.5
(363.53)	*s*	......	......	1.5908	1.6519	1.7033	1.7491	1.7911	1.8301	1.8667	1.9014	1.9344	1.9962	2.0535
	v	......	......	2.649	3.044	3.411	3.764	4.110	4.452	4.792	5.129	5.466	6.136	6.804
180	*h*	......	......	1214.0	1271.0	1323.5	1374.7	1425.6	1476.8	1528.6	1581.0	1634.1	1742.9	1855.2
(373.06)	*s*	......	......	1.5745	1.6373	1.6894	1.7355	1.7776	1.8167	1.8534	1.8882	1.9212	1.9831	2.0404
	v	......	......	2.361	2.726	3.060	3.380	3.693	4.002	4.309	4.613	4.917	5.521	6.123
200	*h*	......	......	1210.3	1268.9	1322.1	1373.6	1424.8	1476.2	1528.0	1580.5	1633.7	1742.6	1855.0
(381.79)	*s*	......	......	1.5594	1.6240	1.6767	1.7232	1.7655	1.8048	1.8415	1.8763	1.9094	1.9713	2.0287
	v	......	......	2.125	2.465	2.772	3.066	3.352	3.634	3.913	4.191	4.467	5.017	5.565
220	*h*	......	......	1206.5	1266.7	1320.7	1372.6	1424.0	1475.5	1527.5	1580.0	1633.3	1742.3	1854.7
(389.86)	*s*	......	......	1.5453	1.6117	1.6652	1.7120	1.7545	1.7939	1.8308	1.8656	1.8987	1.9607	2.0181
	v	......	......	1.9276	2.247	2.533	2.804	3.068	3.327	3.584	3.839	4.093	4.597	5.100
240	*h*	......	......	1202.5	1264.5	1319.2	1371.5	1423.2	1474.8	1526.9	1579.6	1632.9	1742.0	1854.5
(397.37)	*s*	......	......	1.5319	1.6003	1.6546	1.7017	1.7444	1.7839	1.8209	1.8558	1.8889	1.9510	2.0084
	v	......	......	......	2.063	2.330	2.582	2.827	3.067	3.305	3.541	3.776	4.242	4.707
260	*h*	......	......	......	1262.3	1317.7	1370.4	1422.3	1474.2	1526.3	1579.1	1632.5	1741.7	1854.2
(404.42)	*s*	......	......	......	1.5897	1.6447	1.6922	1.7352	1.7748	1.8118	1.8467	1.8799	1.9420	1.9995
	v	......	......	......	1.9047	2.156	2.392	2.621	2.845	3.066	3.286	3.504	3.938	4.370
280	*h*	......	......	......	1260.0	1316.2	1369.4	1421.5	1473.5	1525.8	1578.6	1632.1	1741.4	1854.0
(411.05)	*s*	......	......	......	1.5796	1.6354	1.6834	1.7265	1.7662	1.8033	1.8383	1.8716	1.9337	1.9912
	v	......	......	......	1.7675	2.005	2.227	2.442	2.652	2.859	3.065	3.269	3.674	4.078
300	*h*	......	......	......	1257.6	1314.7	1368.3	1420.6	1472.8	1525.2	1578.1	1631.7	1741.0	1853.7
(417.33)	*s*	......	......	......	1.5701	1.6268	1.6751	1.7184	1.7582	1.7954	1.8305	1.8638	1.9260	1.9835
	v	......	......	......	1.4923	1.7036	1.8980	2.084	2.266	2.445	2.622	2.798	3.147	3.493
350	*h*	......	......	......	1251.5	1310.9	1365.5	1418.5	1471.1	1523.8	1577.0	1630.7	1740.3	1853.1
(431.72)	*s*	......	......	......	1.5481	1.6070	1.6563	1.7002	1.7403	1.7777	1.8130	1.8463	1.9086	1.9663
	v	......	......	......	1.2851	1.4770	1.6508	1.8161	1.9767	2.134	2.290	2.445	2.751	3.055
400	*h*	......	......	......	1245.1	1306.9	1362.7	1416.4	1469.4	1522.4	1575.8	1629.6	1739.5	1852.5
(444.59)	*s*	......	......	......	1.5281	1.5894	1.6398	1.6842	1.7247	1.7623	1.7977	1.8311	1.8936	1.9513

* Abridged from Keenan and Keyes, "Thermodynamic Properties of Steam," Wiley, New York, 1936.

Table 243. Properties of Superheated Steam*—(*Concluded*)

Abs. pressure, lb./sq. in. (sat. temp.)		Temp., °F. 500	550	600	620	640	660	680	700	800	900	1000	1200	1400	1600
	v	1.1231	1.2155	1.3005	1.3332	1.3652	1.3967	1.4278	1.4584	1.6074	1.7516	1.8928	2.170	2.443	2.714
450	h	1238.4	1272.0	1302.8	1314.6	1326.2	1337.5	1348.8	1359.9	1414.3	1467.7	1521.0	1628.6	1738.7	1851.9
(456.28)	s	1.5095	1.5437	1.5735	1.5845	1.5951	1.6054	1.6153	1.6250	1.6699	1.7108	1.7486	1.8177	1.8803	1.9381
	v	0.9927	1.0800	1.1591	1.1893	1.2188	1.2478	1.2763	1.3044	1.4405	1.5715	1.6996	1.9504	2.197	2.442
500	h	1231.3	1266.8	1298.6	1310.7	1322.6	1334.2	1345.7	1357.0	1412.1	1466.0	1519.6	1627.6	1737.9	1851.3
(467.01)	s	1.4919	1.5280	1.5588	1.5701	1.5810	1.5915	1.6016	1.6115	1.6571	1.6982	1.7363	1.8056	1.8683	1.9262
	v	0.8852	0.9686	1.0431	1.0714	1.0989	1.1259	1.1523	1.1783	1.3038	1.4241	1.5414	1.7706	1.9957	2.219
550	h	1223.7	1261.2	1294.3	1306.8	1318.9	1330.8	1342.5	1354.0	1409.9	1464.3	1518.2	1626.6	1737.1	1850.6
(476.94)	s	1.4751	1.5131	1.5451	1.5568	1.5680	1.5787	1.5890	1.5991	1.6452	1.6868	1.7250	1.7946	1.8575	1.9155
	v	0.7947	0.8753	0.9463	0.9729	0.9988	1.0241	1.0489	1.0732	1.1899	1.3013	1.4096	1.6208	1.8279	2.033
600	h	1215.7	1255.5	1289.9	1302.7	1315.2	1327.4	1339.3	1351.1	1407.7	1462.5	1516.7	1625.5	1736.3	1850.0
(486.21)	s	1.4586	1.4990	1.5323	1.5443	1.5558	1.5667	1.5773	1.5875	1.6343	1.6762	1.7147	1.7846	1.8476	1.9056
	v	......	0.7277	0.7934	0.8177	0.8411	0.8639	0.8860	0.9077	1.0108	1.1082	1.2024	1.3853	1.5641	1.7405
700	h	......	1243.2	1280.6	1294.3	1307.5	1320.3	1332.8	1345.0	1403.2	1459.0	1513.9	1623.5	1734.8	1848.8
(503.10)	s	......	1.4722	1.5084	1.5212	1.5333	1.5449	1.5559	1.5665	1.6147	1.6573	1.6963	1.7666	1.8299	1.8881
	v	......	0.6154	0.6779	0.7006	0.7223	0.7433	0.7635	0.7833	0.8763	0.9633	1.0470	1.2088	1.3662	1.5214
800	h	......	1229.8	1270.7	1285.4	1299.4	1312.9	1325.9	1338.6	1398.6	1455.4	1511.0	1621.4	1733.2	1847.5
(518.23)	s	......	1.4467	1.4863	1.5000	1.5129	1.5250	1.5366	1.5476	1.5972	1.6407	1.6801	1.7510	1.8146	1.8729
	v	......	0.5264	0.5873	0.6089	0.6294	0.6491	0.6680	0.6863	0.7716	0.8506	0.9262	1.0714	1.2124	1.3509
900	h	......	1215.0	1260.1	1275.9	1290.9	1305.1	1318.8	1332.1	1393.9	1451.8	1508.1	1619.3	1731.6	1846.3
(531.98)	s	......	1.4216	1.4653	1.4800	1.4938	1.5066	1.5187	1.5303	1.5814	1.6257	1.6656	1.7371	1.8009	1.8595
	v	......	0.4533	0.5140	0.5350	0.5546	0.5733	0.5912	0.6084	0.6878	0.7604	0.8294	0.9615	1.0893	1.2146
1000	h	......	1198.3	1248.8	1265.9	1281.9	1297.0	1311.4	1325.3	1389.2	1448.2	1505.1	1617.3	1730.0	1845.0
(544.61)	s	......	1.3961	1.4450	1.4610	1.4757	1.4893	1.5021	1.5141	1.5670	1.6121	1.6525	1.7245	1.7886	1.8474
	v	......	......	0.4532	0.4738	0.4929	0.5110	0.5281	0.5445	0.6191	0.6866	0.7503	0.8716	0.9885	1.1031
1100	h	......	......	1236.7	1255.3	1272.4	1288.5	1303.7	1318.3	1384.3	1444.5	1502.2	1615.2	1728.4	1843.8
(556.31)	s	......	......	1.4251	1.4425	1.4583	1.4728	1.4862	1.4989	1.5535	1.5995	1.6405	1.7130	1.7775	1.8363
	v	......	......	0.4016	0.4222	0.4410	0.4586	0.4752	0.4909	0.5617	0.6250	0.6843	0.7967	0.9046	1.0101
1200	h	......	......	1223.5	1243.9	1262.4	1279.6	1295.7	1311.0	1379.3	1440.7	1449.2	1613.1	1726.9	1842.5
(567.22)	s	......	......	1.4062	1.4243	1.4413	1.4568	1.4710	1.4843	1.5409	1.5879	1.6293	1.7025	1.7672	1.8263
	v	......	......	0.3174	0.3390	0.3580	0.3753	0.3912	0.4062	0.4714	0.5281	0.5805	0.6789	0.7727	0.8640
1400	h	......	......	1193.0	1218.4	1240.4	1260.3	1278.5	1295.5	1369.1	1433.1	1493.2	1608.9	1723.7	1840.0
(587.10)	s	......	......	1.3639	1.3877	1.4079	1.4258	1.4419	1.4567	1.5177	1.5666	1.6093	1.6836	1.7489	1.8083
	v	......	......	......	0.2733	0.2936	0.3112	0.3271	0.3417	0.4034	0.4553	0.5027	0.5906	0.6738	0.7545
1600	h	......	......	......	1187.8	1215.2	1238.7	1259.6	1278.7	1358.4	1425.3	1487.0	1604.6	1720.5	1837.5
(604.90)	s	......	......	......	1.3489	1.3741	1.3952	1.4137	1.4303	1.4964	1.5476	1.5914	1.6669	1.7328	1.7926
	v	......	......	......	......	0.2407	0.2597	0.2760	0.2907	0.3502	0.3986	0.4421	0.5218	0.5968	0.6693
1800	h	......	......	......	......	1185.1	1214.0	1238.5	1260.3	1347.2	1417.4	1480.8	1600.4	1717.3	1835.0
(621.03)	s	......	......	......	......	1.3377	1.3638	1.3855	1.4044	1.4765	1.5301	1.5752	1.6520	1.7185	1.7786
	v	......	......	......	......	0.1936	0.2161	0.2337	0.2489	0.3074	0.3532	0.3935	0.4668	0.5352	0.6011
2000	h	......	......	......	......	1145.6	1184.9	1214.8	1240.0	1335.5	1409.2	1474.5	1596.1	1714.1	1832.5
(635.82)	s	......	......	......	......	1.2945	1.3300	1.3564	1.3783	1.4576	1.5139	1.5603	1.6384	1.7055	1.7660
	v	......	......	......	......	......	......	0.1484	0.1686	0.2294	0.2710	0.3061	0.3678	0.4244	0.4784
2500	h	......	......	......	......	......	......	1132.3	1176.8	1303.6	1387.8	1458.4	1585.3	1706.1	1826.2
(668.13)	s	......	......	......	......	......	......	1.2687	1.3073	1.4127	1.4772	1.5273	1.6088	1.6775	1.7389
	v	......	......	......	......	......	......	......	0.0984	0.1760	0.2159	0.2476	0.3018	0.3505	0.3966
3000	h	......	......	......	......	......	......	......	1060.7	1267.2	1365.0	1441.8	1574.3	1698.0	1819.9
(695.36)	s	......	......	......	......	......	......	......	1.1966	1.3690	1.4439	1.4984	1.5837	1.6540	1.7163
	v	......	......	......	......	......	......	......	......	0.1583	0.1981	0.2288	0.2806	0.3267	0.3703
3206.2	h	......	......	......	......	......	......	......	......	1250.5	1355.2	1434.7	1569.8	1694.6	1817.2
(705.40)	s	......	......	......	......	......	......	......	......	1.3508	1.4309	1.4874	1.5742	1.6452	1.7080
	v	......	......	......	......	......	......	......	0.0306	0.1364	0.1762	0.2058	0.2546	0.2977	0.3381
3500	h	......	......	......	......	......	......	......	780.5	1224.9	1340.7	1424.5	1563.3	1689.8	1813.6
	s	......	......	......	......	......	......	......	0.9515	1.3241	1.4127	1.4723	1.5615	1.6336	1.6968
	v	......	......	......	......	......	......	......	0.0287	0.1052	0.1462	0.1743	0.2192	0.2581	0.2943
4000	h	......	......	......	......	......	......	......	763.8	1174.8	1314.4	1406.8	1552.2	1681.7	1807.2
	s	......	......	......	......	......	......	......	0.9347	1.2757	1.3827	1.4482	1.5417	1.6154	1.6795
	v	......	......	......	......	......	......	......	0.0276	0.0798	0.1226	0.1500	0.1917	0.2273	0.2602
4500	h	......	......	......	......	......	......	......	753.5	1113.9	1286.5	1388.4	1540.8	1673.5	1800.9
	s	......	......	......	......	......	......	......	0.9235	1.2204	1.3529	1.4253	1.5235	1.5990	1.6640
	v	......	......	......	......	......	......	......	0.0268	0.0593	0.1036	0.1303	0.1696	0.2027	0.2329
5000	h	......	......	......	......	......	......	......	746.4	1047.1	1256.5	1369.5	1529.5	1665.3	1794.5
	s	......	......	......	......	......	......	......	0.9152	1.1622	1.3231	1.4034	1.5066	1.5839	1.6499
	v	......	......	......	......	......	......	......	0.0262	0.0462	0.0880	0.1143	0.1516	0.1825	0.2106
5500	h	......	......	......	......	......	......	......	741.3	985.0	1224.1	1349.3	1518.2	1637.0	1788.1
	s	......	......	......	......	......	......	......	0.9090	1.1093	1.2930	1.3821	1.4908	1.5699	1.6369

Table 244. Properties of Saturated Trichloroethylene (C_2HCl_3)*

Temp., °F. t	Abs. pressure, lb./sq. in. p	Volume, cu. ft./lb. vapor v_g	Enthalpy, B.t.u./lb. Liquid h_f	Enthalpy, B.t.u./lb. Vapor h_g	Entropy, B.t.u./(lb.)(°R.) Liquid s_f	Entropy, B.t.u./(lb.)(°R.) Vapor s_g	Temp., °F. t	Abs. pressure, lb./sq. in. p	Volume, cu. ft./lb. vapor v	Enthalpy, B.t.u./lb. Liquid h_f	Enthalpy, B.t.u./lb. Vapor h_g	Entropy, B.t.u./(lb.)(°R.) Liquid s_f	Entropy, B.t.u./(lb.)(°R.) Vapor s_g
0	0.150	261.3	0.0	112.60	0.0000	0.2451	80	1.56	28.00	19.84	129.62	.0374	.2414
10	.194	197.6	1.86	114.18	.0039	.2433	90	1.98	22.46	23.22	132.54	.0428	.2422
20	.252	153.3	3.90	116.00	.0081	.2429							
30	.352	114.5	6.12	117.92	.0127	.2411	100	2.45	18.54	26.60	135.41	.0480	.2428
40	.492	85.8	8.42	119.83	.0178	.2412	110	3.15	14.62	29.92	138.22	.0530	.2432
							120	3.98	11.73	33.60	141.38	.0581	.2440
50	.672	63.2	11.10	122.15	.0227	.2405	130	5.05	9.20	37.70	144.90	.0632	.2450
60	.900	47.53	13.80	124.43	.0278	.2404	140	6.31	7.70	42.00	148.57	.0680	.2459
70	1.20	36.00	16.80	127.02	.0324	.2407							

* "Refrigerating Data Book," 5th ed., American Society of Refrigerating Engineers, New York, 1942.

FREEZING POINTS OF AQUEOUS SOLUTIONS

Table 245. Freezing Points of Some Aqueous Solutions

%	Freezing points, °C.					%	Freezing points, °C.				
	H_2SO_4	Na_2CO_3	NaOH	HNO_3	HCl		H_2SO_4	Na_2CO_3	NaOH	HNO_3	HCl
5	−2	−2	−12	−3		55	−29	....	+32	−19	−20
10	−5	+10	−15	−7	−22	60	−29	....	+52	−23	−20
15	−8	+18	−21	−11	−42	65	−40	....	+62	−29	−16
20	−12	+24	−27	−17	−65	70	<−40	....	+66	−40	−15
25	−22	+27	−17	−24	−78	75	−39	....	>100	−39	
30	−36	+30	0	−35	−47	80	−3	....	.....	−39	
35	−60	....	+11	−38	−30	85	+8	....	.....	−45	
40	−68	....	+14	−30	−24	90	−5	....	.....	−62	
45	−48	....	+10	−25	−26	95	−24	....	.....	−52	
50	−37	....	+10	−21	−18	100	+10	....	.....	−42	

FUSION TEMPERATURES

Table 246. Melting Points of Alloys

Metals		Melting points, °C. Percentage of metal in second column										
		0%	10%	20%	30%	40%	50%	60%	70%	80%	90%	100%
Pb	Sn	327.3	295	276	262	240	220	190	185	200	216	231.8
	Bi	327.3	290	....	......	179	145	126	168	205	....	271.3
	Te	327.3	710	790	880	917	760	600	480	410	425	453
	Ag	327.3	460	545	590	620	650	705	775	840	905	960.8
	Na	327.3	360	420	400	370	330	290	250	200	130	97.5
	Cu	327.3	870	920	925	945	950	955	985	1005	1020	1083
	Sb	327.3	250	275	330	395	440	490	525	560	600	630.5
Al	Sb	660.1	750	840	925	945	950	970	1000	1040	1010	630.5
	Cu	660.1	630	600	560	540	580	610	755	930	1055	1083
	Au	660.1	675	740	800	855	915	970	1025	1055	675	1063.0
	Ag	660.1	625	615	600	590	580	575	570	650	750	960.8
	Zn	660.1	640	620	600	580	560	530	510	475	425	419.5
	Fe	660.1	860	1015	1110	1145	1145	1220	1315	1425	1500	1530
	Sn	660.1	645	635	625	620	605	590	570	560	540	231.8
Sb	Bi	630.5	610	590	575	555	540	520	470	405	330	271.3
	Ag	630.5	595	570	545	520	500	505	545	680	850	960.8
	Sn	630.5	600	570	525	480	430	395	350	310	255	231.8
	Zn	630.5	555	510	540	570	565	540	525	510	470	419.5
Ni	Sn	1453	1380	1290	1200	1235	1290	1305	1230	1060	800	231.8
Na	Bi	97.5	425	520	590	645	690	720	730	715	570	271.3
	Cd	97.5	125	185	245	285	325	330	340	360	390	320.9
Cd	Ag	320.9	420	520	610	700	760	805	850	895	940	960.5
	Tl	320.9	300	285	270	262	258	245	230	210	235	302.5
	Zn	320.9	280	270	295	313	327	340	355	370	390	419.3
Au	Cl	1063.0	910	890	895	905	925	975	1000	1025	1060	1083
	Ag	1063.0	1062	1061	1058	1054	1049	1039	1025	1006	982	960.8
	Pt	1063.0	1125	1190	1250	1320	1380	1455	1530	1610	1685	1773.5
K	Na	62.3	17.5	−10	−3.5	5	11	26	41	58	77	97.5
	Hg	62.3	......	....		.....	90	110	135	162	265	
	Tl	62.3	133	165	188	205	215	220	240	280	305	302.5
Cu	Ni	1083	1180	1240	1290	1320	1335	1380	1410	1430	1440	1455
	Ag	1083	1035	990	945	910	870	830	788	814	875	960.8
	Sn	1083	1005	890	755	725	680	630	580	530	440	231.8
	Zn	1083	1040	995	930	900	880	820	780	700	580	419.5
Ag	Zn	960.8	850	755	705	690	660	630	610	570	505	419.5
	Sn	960.8	870	750	630	550	495	450	420	375	300	231.8
Na	Hg	97.7	90	80	70	60	45	22	55	95	215	

* "Smithsonian Tables." (Melting points of pure metals from Table 166, p. 210.)

Table 247. Fusion Temperatures of Refractories*

Material	Temp., °C.	Material	Temp., °C
Alumina	2000	Magnesia, sintered	2200–2600
Alundum	1750–2000	Magnesite brick	2150–2165
Bauxite brick	1565–1785	Silica (SiO_2)	1700
Bauxite clay	1750–2000	Silica brick	1685–1800
Carborundum	2200–2240	Sillimanite, $(Al_2O_3)_3.(SiO_2)_2$	1816
Chrome brick	1850–2050	Spinel, ($MgO.Al_2O_3$)	2135
Chromium oxide (Cr_2O_3)	1990	Zirconia (ZrO_2)	2500–2950
Fireclay brick	1500–1750	Zirconia brick	2000–2600
Magnesium oxide (MgO)	2800		

* From "International Critical Tables," vol. 2, p. 83.

ELEVATION OF BOILING POINT

Table 248. Aqueous Solutions of Inorganic Salts*

This table gives the number of grams of the salt which, when dissolved in 100 g. water, will raise the boiling point by the amount stated in the headings of the different columns. The pressure is supposed to be 76 cm. Hg.

Salt	Temperature, °C.									
	1	2	3	4	5	7	10	15	20	25
$BaCl_2 + 2H_2O$	15.0	31.1	47.3	63.5	(71.6 gives 4.5° rise of temp.)					
$CaCl_2$	6.0	11.5	16.5	21.0	25.0	32.0	41.5	55.5	69.0	84.5
$Ca(NO_3)_2 + 2H_2O$	12.0	25.5	39.5	53.5	68.5	101.0	152.5	240.0	331.5	443.5
KOH	4.7	9.3	13.6	17.4	20.5	26.4	34.5	47.0	57.5	67.3
$KC_2H_3O_2$	6.0	12.0	18.0	24.5	31.0	44.0	63.5	98.0	134.0	171.5
KCl	9.2	16.7	23.4	29.9	36.2	48.4	(57.4 gives a rise of 8.5°)			
K_2CO_3	11.5	22.5	32.0	40.0	47.5	60.5	78.5	103.5	127.5	152.5
$KClO_3$	13.2	27.8	44.6	62.2						
KI	15.0	30.0	45.0	60.0	74.0	99.5	134.	185.0	(220 gives 18.5°)	
KNO_3	15.2	31.0	47.5	64.5	82.0	120.5	188.5	338.5		
$K_2C_4H_4O_6 + \frac{1}{2}H_2O$	18.0	36.0	54.0	72.0	90.0	126.5	182.0	284.0		
$KNaC_4H_4O_6$	17.3	34.5	51.3	68.1	84.8	119.0	171.0	272.5	390.0	510.0
$KNaC_4H_4O_6 + 4H_2O$	25.0	53.5	84.0	118.0	157.0	266.0	554.0	5510.0		
$LiCl$	3.5	7.0	10.0	12.5	15.0	20.0	26.0	35.0	42.5	50.0
$LiCl + 2H_2O$	6.5	13.0	19.5	26.0	32.0	44.0	62.0	92.0	123.0	160.5
$MgCl_2 + 6H_2O$	11.0	22.0	33.0	44.0	55.0	77.0	110.0	170.0	241.0	334.5
$MgSO_4 + 7H_2O$	41.5	87.5	138.0	196.0	262.0					
$NaOH$	4.3	8.0	11.3	14.3	17.0	22.4	30.0	41.0	51.0	60.1
$NaCl$	6.6	12.4	17.2	21.5	25.5	33.5	(40.7 gives 8.8° rise)			
$NaNO_3$	9.0	18.5	28.0	38.0	48.0	68.0	99.5	156.0	222.0	
$NaC_2H_3O_2 + 3H_2O$	14.9	30.0	46.1	62.5	79.7	118.1	194.0	480.0	6250.0	
$Na_2S_2O_3$	14.0	27.0	39.0	49.5	59.0	77.0	104.0	152.0	214.5	311.0
Na_2HPO_4	17.2	34.4	51.4	68.4	85.3					
$Na_2C_4H_4O_6 + 2H_2O$	21.4	44.4	68.2	93.9	121.3	183.0	(237.3 gives 8.4° rise)			
$Na_2S_2O_3 + 5H_2O$	23.8	50.0	78.6	108.1	139.3	216.0	400.0	1765.0		
$Na_2CO_3 + 10H_2O$	34.1	86.7	177.6	369.4	1052.9					
$Na_2B_4O_7 + 10H_2O$	39.0	93.2	254.2	898.5	(5555.5 gives 4.5° rise)					
NH_4Cl	6.5	12.8	19.0	24.7	29.7	39.6	56.2	88.5		
NH_4NO_3	10.0	20.0	30.0	41.0	52.0	74.0	108.0	172.0	248.0	337.0
$(NH_4)_2SO_4$	15.4	30.1	44.2	58.0	71.8	99.1				
$SrCl_2 + 6H_2O$	20.0	40.0	60.0	81.0	103.0	150.0	234.0	524.0		
$Sr(NO_3)_2$	24.0	45.0	63.6	81.4	97.6					
$C_4H_6O_6$	17.0	34.4	52.0	70.0	87.0	123.0	177.0	272.0	374.0	484.0
$C_2H_2O_4 + 2H_2O$	19.0	40.0	62.0	86.0	112.0	169.0	262.0	540.0	1316.0	50,000.0
$C_6H_8O_7 + H_2O$	29.0	58.0	87.0	116.0	145.0	208.0	320.0	553.0	952.0	

Salt	Temperature, °C.									
	40	60	80	100	120	140	160	180	200	240
$CaCl_2$	137.5	222.0	314.0							
KOH	92.5	121.7	152.6	185.0	219.8	263.1	312.5	375.0	444.4	623.0
$NaOH$	93.5	150.8	230.0	345.0	526.3	800.0	1333.0	2353.0	6452.0	
NH_4NO_3	682.0	1370.0	2400.0	4099.0	8547.0	∞				
$C_4H_6O_6$	980.0	3774.0	(infinity gives 170)							

* "Smithsonian Tables."

CONSTANT TEMPERATURES

Table 249. Production and Maintenance

From "International Critical Tables"

Below Zero Degrees Centigrade

1. Bath Fluids Boiling at Constant Pressure (See "International Critical Tables," vol. 1, pp. 61–62.) The vapor pressure data of a number of suitable liquids are available in this data section (see pp. 149–174) and need not be duplicated here. For this type of temperature control, the following substances are among those most often used: ammonia, diphenyl, ethane, ethyl chloride, ethylene, hydrogen, hydrogen sulfide, methane, methyl chloride, nitrogen, nitrous oxide, oxygen, propane, propylene, sulfur dioxide.

2. Bath Fluids for Thermostatic Control. (See "International Critical Tables," vol. 1, pp. 61–67, for a more complete list of substances.)

Temperature Range	Recommended Substances
Below −150°C	Amylene, butane, isopentane, methyl chloride (25%) + methyl ether (75%), petroleum, propane, propylene
−150° to −125°C	Ethyl chloride, methylcyclohexane, pentane, petroleum distillate
−125° to −100°C	Carbon disulfide, chloroform (23%) + ethyl ether (77%), ethyl bromide, ethyl ether
−100° to −90°C	Acetone, chloroform (79%) + ethyl ether (21%), methyl chloride, methylene chloride, toluene
−90° to −80°C	Carbon tetrachloride (49%) + chloroform (51%), ethyl acetate, ethyl alcohol, trichloroethylene
−80° to −50°C	Chloroform, ether (80%) + ethyl alcohol (20%)
−50° to −25°C	Chlorobenzene, ethyl alcohol (25%) + glycerine (25%) + water (50%), gasoline + carbon tetrachloride
−25° to 0°C	Carbon tetrachloride, sodium chloride + water

Above Zero Degrees Centigrade

1. Bath Fluids Boiling at Constant Pressure. For this type of constant-temperature bath, the following are among the substances often used: Acetone, aniline, benzene, benzophenone, carbon disulfide, chloroform, diphenyl, ethanol, ethyl chloride, ethyl ether, methanol, m-xylene, naphthalene, quinoline, sulfur, toluene, water, zinc, etc. For accurate vapor pressure data see this section, or for more complete data see such standard reference books as, "International Critical Tables," vol. 5.

2. Bath Fluids for Thermostatic Control.

Temperature Range	Recommended Substances
0° to 90°C.	Water
Up to 20° or 30° of its flash point.	Mineral oil
From melting point to 300°C.	Paraffin
10°C. to 285°C.	Hydrogenated cottonseed oil

3. Transition Points of Hydrates (Solid-liquid Non-variant Points).

Substance	Hydration Temp., °C.
Sodium chromate	19.7
Sodium sulfate	32.38
Sodium carbonate	35.3
Sodium thiosulfate	48.0
Manganese chloride	57.8
Trisodium phosphate	73.4

4. Metal Blocks. According to Stähler (see "International Critical Tables," vol. 1, p. 66), metallic blocks of aluminum or copper have been used up to 600°C. with a constancy of temperature of 1°C.

5. High Temperatures. The following references (compiled by the Geophysical Laboratory) are given in "International Critical Tables," vol. 1, p. 66, bearing on the construction and the temperature regulation of high-temperature furnaces:

Kolovrat, *J. phys. et radium*, **8**, 495 (1909). Haughton and Hanson, *J. Inst. Metals*, **14**, 145 (1915); **18**, 173 (1917). White and Adams, *Phys. Rev.*, **14**, 44 (1919). Haagn, *Elektrotech. Z.*, **40**, 670 (1919). Roberts, *J. Wash. Acad. Sci.*, **11**, 409 (1921). *J. Optical Soc. Am.*, **6**, 965 (1922). Bunting, *J. Am. Ceram. Soc.*, **6**, 1209 (1923). Adams, *J. Optical Soc. Am.*, **9**, 599 (1924). Roberts, *J. Optical Soc. Am.*, **10**, 723 (1925).

INDICATORS AND HYDROGEN-ION CONCENTRATION

BY J. W. STILLMAN

The fundamental importance of indicators in analytical work is so obvious that a clear understanding of this subject is demanded of every analytical chemist. Indicators are used in analytical chemistry to show, usually by means of color change, when a reaction has reached that point desired in the analysis. For the purpose of this discussion, indicators in general may be divided into two classes:

1. Indicators for acidimetry and alkalimetry (H-ion concentration).
2. Miscellaneous special indicators (for precipitation, iodine titrations, other oxidation-reduction reactions, etc.).

In this brief survey, only the first class will be considered.

In the first place it cannot be too strongly emphasized that no one indicator so far known is applicable to all reactions of acids and bases. Each indicator has a particular H-ion range within which it gives satisfactory results. Also the hydrogen-ion concentration at the end point of a reaction when equivalent quantities of acid and base have been added varies with the particular acid and base used.

The **characteristics of a good indicator** are:

1. The color change should be sharp; *i.e.*, it should change color within a short range of H-ion concentration.
2. The change should be between contrasting colors which are easily distinguished.
3. It should change color at the H-ion concentration corresponding to the end point of the reaction being used.

It has been found that a change in the shade of a color is more easily detected by the eye than is the depth of a single shade. For this reason a two-color indicator is to be preferred to a one-color indicator.

Figure 23 gives a list of the commonly used indicators showing their color change and effective H-ion range.

A number of so-called "mixed indicators" have been described in the literature usually being recommended for some specific reaction. Mixed indicators may be of two kinds: (1) a mixture of two colored substances both of which are indicators and therefore change color, and (2) a mixture of two colored substances only one of which changes color with the reaction. Several so-called "universal indicators" have been devised which give different shades of color over the H-ion range. They are usually mixtures of several indicators and are of value in determining the approximate H-ion concentration corresponding to the end point of a reaction. The advantage of a mixed indicator is that by this means the color change is sharpened at a particular desired point or an indication is given when the end point has been overrun.

Choice of an Indicator. In titrating we are not primarily interested in obtaining a solution which is exactly neutral, but we are interested in knowing at what point equivalent quantities of acid and base have been added. For reactions involving only a strong acid and a strong base, any indicator changing color between pH = 3 and pH = 10 will be satisfactory (pH = logarithm of the reciprocal of the H-ion concentration expressed in moles per liter). If a weak acid is being titrated with a strong base, the pH at the end point will be above 7 (on the alkaline side of the neutral point) and an indicator must be selected which changes in this range. If a weak base is being titrated with a strong acid, the pH at the end point will be below 7 (on the acid side of the neutral point) and a corresponding indicator should be chosen. It should be kept in mind that unless precautions are taken to remove it, CO_2 (carbonic acid, H_2CO_3) will always be present in solutions and if an indicator changing above a pH of 4 is used the CO_2 present will be included in the titration. Since in titrations using weak acids and weak bases no sharp change in pH is obtained at any point, no indicator will give a satisfactory color change.

In the case of dibasic or tribasic acids there are end points corresponding to the neutralization of the successive H-ions. The indicator chosen must correspond in color change with the reaction desired. The same is true for bases.

The following table lists a number of common substances which may be determined by titration and gives the pT (pH at the end point) of the reaction and the proper indicator to use. It is assumed that a strong acid or base is used as the titrating agent.

Determination of Hydrogen-ion Concentration

Colorimetric Method. It has been shown in the chart below (Fig. 23) that the indicators change color at different concentration of hydrogen ion, and this makes possible another use for indicators in addition to the titration of acids or bases. The H-ion concentration of a sample can be determined by adding to it a measured amount of the indicator and comparing the color developed with the color obtained when the same amount of indicator is added to a standard buffer solution of known H-ion concentration. In practice a series of

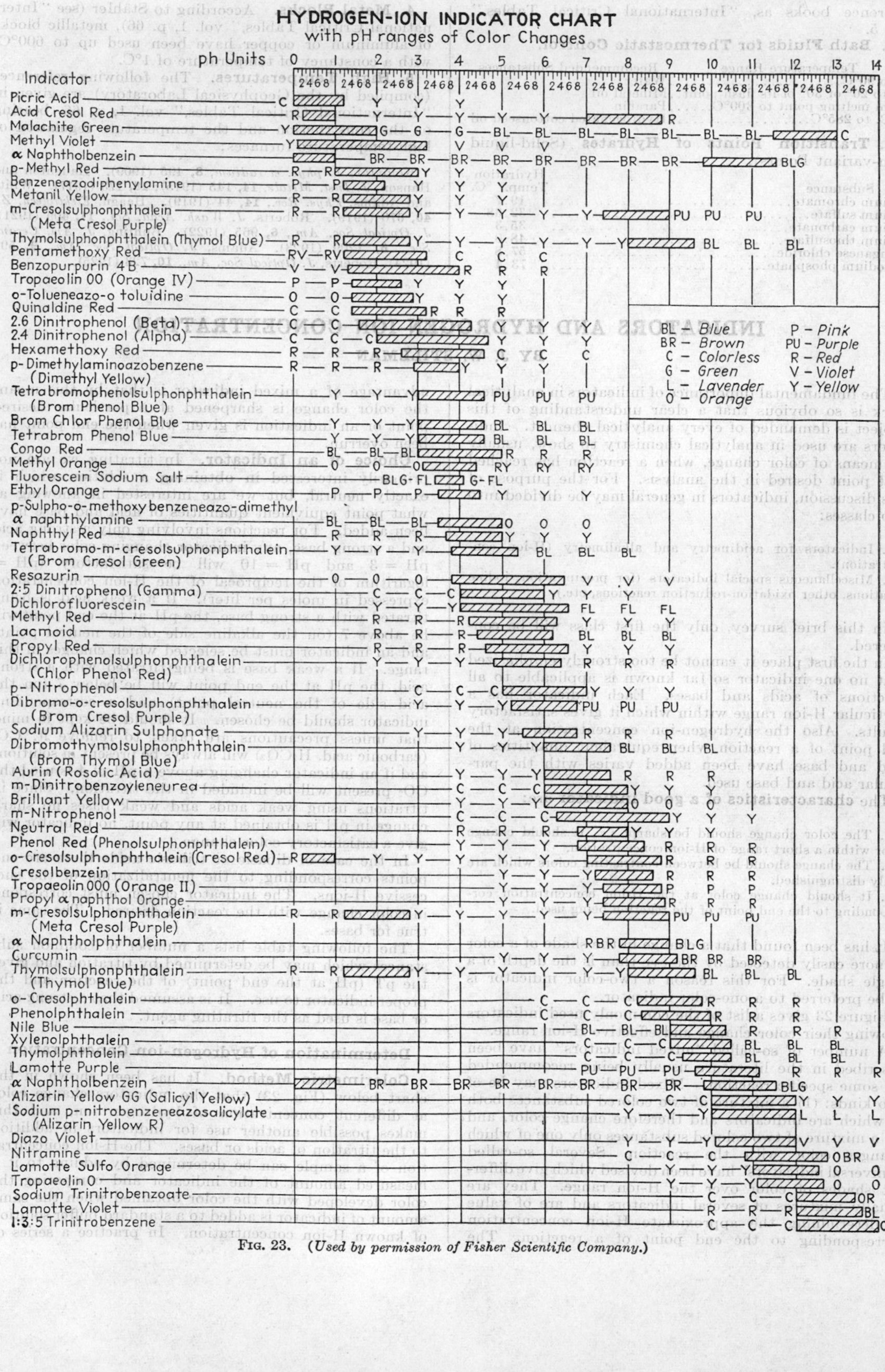
HYDROGEN-ION INDICATOR CHART
with pH ranges of Color Changes
ph Units
Indicator
Picric Acid
Acid Cresol Red
Malachite Green
Methyl Violet
α Naphtholbenzein
p-Methyl Red
Benzeneazodiphenylamine
Metanil Yellow
m-Cresolsulphonphthalein
(Meta Cresol Purple)
Thymolsulphonphthalein (Thymol Blue)
Pentamethoxy Red
Benzopurpurin 4B
Tropaeolin 00 (Orange IV)
o-Tolueneazo-o toluidine
Quinaldine Red
2.6 Dinitrophenol (Beta)
2.4 Dinitrophenol (Alpha)
Hexamethoxy Red
p-Dimethylaminoazobenzene
(Dimethyl Yellow)
Tetrabromophenolsulphonphthalein
(Brom Phenol Blue)
Brom Chlor Phenol Blue
Tetrabrom Phenol Blue
Congo Red
Methyl Orange
Fluorescein Sodium Salt
Ethyl Orange
p-Sulpho-o-methoxybenzeneazo-dimethyl
α naphthylamine
Naphthyl Red
Tetrabromo-m-cresolsulphonphthalein
(Brom Cresol Green)
Resazurin
2:5 Dinitrophenol (Gamma)
Dichlorofluorescein
Methyl Red
Lacmoid
Propyl Red
Dichlorophenolsulphonphthalein
(Chlor Phenol Red)
p-Nitrophenol
Dibromo-o-cresolsulphonphthalein
(Brom Cresol Purple)
Sodium Alizarin Sulphonate
Dibromothymolsulphonphthalein
(Brom Thymol Blue)
Aurin (Rosolic Acid)
m-Dinitrobenzoyleneurea
Brilliant Yellow
m-Nitrophenol
Neutral Red
Phenol Red (Phenolsulphonphthalein)
o-Cresolsulphonphthalein (Cresol Red)
Cresolbenzein
Tropaeolin 000 (Orange II)
Propyl α naphthol Orange
m-Cresolsulphonphthalein
(Meta Cresol Purple)
α Naphtholphthalein
Curcumin
Thymolsulphonphthalein
(Thymol Blue)
o-Cresolphthalein
Phenolphthalein
Nile Blue
Xylenolphthalein
Thymolphthalein
Lamotte Purple
α Naphtholbenzein
Alizarin Yellow GG (Salicyl Yellow)
Sodium p-nitrobenzeneazosalicylate
(Alizarin Yellow R)
Diazo Violet
Nitramine
Lamotte Sulfo Orange
Tropaeolin 0
Sodium Trinitrobenzoate
Lamotte Violet
1:3:5 Trinitrobenzene
BL - Blue
BR - Brown
C - Colorless
G - Green
L - Lavender
O - Orange
P - Pink
PU - Purple
R - Red
V - Violet
Y - Yellow

FIG. 23. *(Used by permission of Fisher Scientific Company.)*

Recommended Indicators

Acids	pT	Indicators	End color
ACIDS			
Hydrochloric Hydrobromic Hydriodic Sulfuric	7	Methyl-thymol blue Methyl red Bromphenol blue	Yellow Orange Green
Picric	7	Methyl red	Yellow
Saccharin	7.1	Phenol red	Orange
Salicylic	7.2	Phenol red	Orange
Nitric	7.5	Phenol red	Orange
Hippuric	7.5	Phenol red	Orange
Fumaric	7.5	Phenol red	Orange to neutral salt
Benzoic	7.6	Phenol red	Orange
Formic	7.8	Phenol red	Orange
Lactic	7.8	Phenol red	Orange
Cinnamic	8.0	Phenol red	Red
Oxalic	8.0	Phenol red	Red
Acetylsalicylic	8.0	Phenol red	Red
Tartaric	8.1	Phenol red	Red
Valeric	8.3	Thymol blue or phenol violet	Green
Carbonic	8.4	Thymol blue or phenol violet	Green to bicarbonate
Maleic	8.5	Thymol blue or phenol violet	Green to neutral salt
Malonic	8.5	Thymol blue or phenol violet	Green
Boric, with glycerin	8.6	Thymol blue or phenol violet	Green Blue
Acetic Phthalic Succinic Malic	8.8	Thymol blue or phenol violet or phenol-thymolphthalein	Pink
Citric	9.5	Phenol violet or phenolthymolphthalein	Violet
Oleic	9.5		Violet
Diethylbarbituric	10.2	Thymol violet	Green to standard color
Phosphoric, or glycerophosphoric	4.0 8.2	Bromphenol blue Phenol red	Maximum blue to acid salt (NaH_2PO_4), red to acid salt (Na_2HPO_4)
Hypophosphorous	5.5	Methyl red	Orange
Boric, without glycerin	11.1	Too weak for titration	
Phenol	12		
BASES			
Strong bases	7	Methyl-thymol blue Methyl red Bromphenol blue	Yellow Orange Green
Nicotine	5.5	Methyl red	Orange
Homatropine	5.5	Methyl red	Orange
Ammonia Morphine Codeine	5.2	Methyl red	Orange
Atropine Strychnine Brucine Ethylmorphine	5.0	Methyl red	Orange
Pyridine	3.6	Bromphenol blue	Green
Aniline	2.8	Bromphenol blue	Green, not sharp
SALTS			
Borax	5.2	Methyl red	Orange
Sodium carbonate to bicarbonate	8.1	Phenol violet or thymol blue	Yellow
to carbonic acid	4.1	Bromcresol green	Yellow
Sodium phenate	6.5	Phenol red	
Sodium dihydrogen phosphate to acid salt (Na_2HPO_4)	8.2	Phenol red	Red
Sodium arsenate	5.5	Methyl red	Orange to standard color

The above table was taken in the main from Lizius and Evers, *Analyst*, 47, 331 (1922), in which will be found a description of mixed indicators.

buffer solutions differing from each other in steps of 0.2 pH for the effective range of the indicator are prepared, a definite amount of indicator is added to each, and the color standard thus formed is sealed in an ampoule. These standards, if protected from direct light, are stable for a considerable time.

The advantages in using color indicators are that no special apparatus is required; the indicators covering the entire pH range are readily available; and the standard buffer solutions can be purchased or easily prepared.

Disadvantages encountered in the use of color indicators include the following:

The Acid Error. In slightly buffered solutions, the fact that indicators dissociate as acids and thus supply additional hydrogen ions will cause an error in the recorded pH.

The Salt Error. The presence of salts in the solution affects the solubility of indicators, which may already be low, and also influences the ionic equilibrium of the forms of the indicator upon which the color reactions are based.

The Protein Error. Proteins and other colloidal materials adsorb selectively acidic and basic indicators and thus affect the resultant color.

The Temperature Effect. As would be expected, a change in temperature has a direct influence on the ionization of the indicator.

Dichromatism. Certain indicators appear to have a distinctly different color when viewed through different thicknesses, and the use of such indicators in turbid solutions should be avoided.

Electrometric Method. The H-ion concentration of a solution can also be determined electrometrically if a suitable electrode system is set up. Such a system usually consists of two electrodes dipping in the solution, and, by means of a potentiometer, the potential developed by the cell can be measured. Each electrode exhibits a potential as the result of the reaction between the electrode and the constituents of the solution. If one of the electrodes is a standard, the potential of which is known, and if the potential of the cell can be measured, then the potential of the other electrode can be calculated. Such a standard or reference electrode is the calomel electrode, which consists of mercury in contact with calomel in a solution of potassium chloride saturated with calomel. Its potential at a given temperature depends upon the concentration of the potassium chloride, and the electrode is designated tenth normal, normal, or saturated, depending upon the concentration of potassium chloride used. The saturated calomel electrode is most readily prepared and is the one in common use.

Again, if the potential of the second or measuring electrode is dependent upon the H-ion concentration, a means is provided for the measurement electrometrically of the hydrogen-ion concentration. There are four types of such electrodes in common use, and a brief description of each will be given together with a tabulation of their advantages and disadvantages.

The hydrogen electrode is the recognized standard for H-ion measurements and consists of a platinum surface coated with platinum black which dips in the solution and is bathed with a stream of pure hydrogen gas. The potential depends upon the equilibrium between the hydrogen gas and the hydrogen ions in solution. The hydrogen electrode is used to standardize buffer solutions and the other methods for hydrogen-ion determination.

Quinhydrone is an equimolecular compound of quinone and hydroquinone which is almost completely dissociated when dissolved in water. The possibility of measuring H-ion concentration through the medium of quinhydrone depends upon reactions which, while not clearly established, can be set down as follows:

$$\text{Quinone} + 2\ \text{electrons} \rightleftharpoons \text{anion of hydroquinone}$$
$$\text{Anion of hydroquinone} + 2H^+ \rightleftharpoons \text{hydroquinone}$$

Electrodes for pH Measurements

Electrode	Advantages	Disadvantages
Hydrogen Useful range 0–14	No salt error The established method for standardization of buffers A measurement with a limit of error of 0.01 pH is possible over the total range of 0.0 to 14.0 pH Adaptable to turbid or colored solutions Is a low resistance system	Requires a special catalytic surface, which may be poisoned, and a supply of pure hydrogen gas Not easily adaptable to continuous measurements Inaccurate when dissolved gases are involved in the pH equilibrium Oxidation-reduction systems cause errors Inaccurate in the presence of metals below hydrogen in the "electromotive series" Sulfides, sulfites, calomel, arsenic compounds, etc., are possible sources of error Not adaptable to unbuffered solutions over the range of 5.0 to 8.5 pH
Quinhydrone Useful range 0–9	Does not require a special electrode surface or gas Not influenced by dissolved gases such as CO_2, air, etc. A limit of error as small as 0.02 pH is possible below 8.8 pH May be used in solutions of mild oxidizing or reducing intensity Adaptable to solutions containing nickel, copper, aluminum and to milk, beer, blood, soil solutions, etc. Adaptable to turbid or colored solutions Is a low resistance system	Addition of quinhydrone contaminates the solution Limited to the range of 0.0 to 9.0 pH Decomposition and oxidation at 9.0 pH and above causes erroneous readings Has certain salt errors if salt concentration is above 1M High concentrations of ferric salts, hydrogen sulfide, chromates, peroxides, permanganates, hypochlorites, and sulfites give errors ranging between 0.3 and 2.0 pH Sparingly buffered solutions may involve errors in solutions above 5.0 pH
Antimony Useful range 4–11.5	Very rugged construction Does not require catalytic surface or addition of auxiliary material May be standardized for specific solutions with a limit of error of ± 0.1 pH May be used in turbid or colored solutions Adaptable to measurements in heavy sludges, viscous fluids, and semisolids May be used for continuous recording May be used in the presence of acetates, alum, ammonia, borates, carbonates, clays, lime, phosphates, soap, soil suspensions, sugar, sulfites, wool scouring solutions, etc. Is a low resistance system	Salt errors of considerable magnitude An over-all limit of error of ± 0.2 pH for average conditions Calibration for still solution different from that of moving solutions Very susceptible to contamination by copper even in concentrations of 0.5 p.p.m. Susceptible to errors in presence of oxidation-reduction systems Calibration dependent upon whether solution is saturated with air or other gas
Glass Useful range 1–14	Does not require gas, catalytic surface, or addition of auxiliary material Adaptable to oxidizing and reducing solutions and unbuffered solutions May be used in colored solutions After standardization a limit of error of ± 0.03 pH may be obtained May be used for continuous recording Adaptable to semisolids and thick fluids In absence of sodium salts, may be used in presence of Ba, Ca, and NH_4OH of high concentrations	Inaccurate in the presence of sodium salts at ranges above 9.0 pH. The higher the pH and the greater the temperature above 20°C. the greater is the error An error may be introduced in pure water measurements due to the solubility of the glass in water The glass surface is highly absorptive When changing from a buffer solution to an unbuffered solution considerable washing is required. The thicker and more robust the glass electrode, the greater must be the degree of washing This is a high resistance system Susceptible to errors due to current pick-up and leakage

The setting up of the quinhydrone electrode is very simple. A certain amount of quinhydrone is dissolved in the solution under test and a platinum (or other "unattackable" metal) wire is dipped into the solution.

The antimony electrode is a metal-metal oxide electrode which is set up by immersing an antimony rod in the solution which contains dissolved oxygen. The antimony oxide in equilibrium with water forms antimony hydroxide which supplies the metal ions. Of the possible metals for a metal-metal oxide electrode all fail to meet all the necessary qualifications except antimony and silicon. To get the most out of the antimony electrode, special attention must be given to the conditions for its use.

The glass electrode consists of a thin membrane of glass which is blown or sealed to the end of a tube and separates the solution within the electrode from the solution under test in which the glass electrode is immersed. Authorities differ in their theories of the mechanism of the glass electrode, but its operation can be described in simple outline. The glass membrane containing water can be considered as a medium into which and out of which hydrogen ions can pass. If the composition of the glass remains essentially unchanged, the hydrogen activity should remain constant. Practically, the glass electrode behaves like a hydrogen electrode over a certain pH range. A number of different solutions have been used inside the glass membrane. Among these are calomel in 1.0*N* hydrochloric acid, silver-silver chloride in 0.1*N* hydrochloric acid, quinhydrone in 1 pH hydrochloric acid, etc. The glass electrode approaches more closely to the ideal universal electrode than do the others, and apparatus for the determination of H-ion concentration using the glass electrode is being produced by several manufacturers. However, the limitations of the glass electrode, given in the table on this page, must not be overlooked.

Choosing the Proper Electrode. The problem of determining H-ion concentration should be considered from a definitely scientific point of view, and, before selecting an electrode system, the conditions under which the determination is to be made should be examined, and the advantages and disadvantages of the different types of electrodes reviewed in the light of these conditions. If the problem is of a research nature where great accuracy is desired, the hydrogen electrode should probably be selected. If a continuous determination is desired, the antimony or glass electrode should be used. Both the antimony and the glass electrodes can be used in connection with continuous recording and controlling equipment, and commercial apparatus involving these principles is available on the market.

To assist in the selection of the proper electrode the advantages and disadvantages of the four types of electrode mentioned above have been listed in the table on this page. Acknowledgment is made to G. A. Perley of the Leeds & Northrup Company for permission to use the data and arrangement which he followed in his publications: The Measurement and Control of Hydrogen Ion Concentration in Industrial Plants, *Trans. Am. Inst. Chem. Engrs.* **29**, 257 (1933); and Modern Views of pH Measurement, *Am. Dyestuff Reptr.*, **26**, 832 (1937).

SECTION 4

PHYSICAL AND CHEMICAL PRINCIPLES

BY

W. M. D. Bryant, B. S., E. I. duPont de Nemours & Co.; Member, American Chemical Society, American Institute of Physics.

Joseph C. Elgin, Ph. D., Professor of Chemical Engineering, Princeton University; Member, American Institute of Chemical Engineers, American Chemical Society, American Institute of Mining and Metallurgical Engineers, American Society for Engineering Education.

John H. Perry, Ph. D., E. I. duPont de Nemours & Co.; Member, American Institute of Chemical Engineers, American Chemical Society, American Association for the Advancement of Science, American Society for Engineering Education.

Frederick D. Rossini, Ph. D., Professor and Head of the Department of Chemistry, Carnegie Institute of Technology; formerly Chief of Section on Thermochemistry and Constitution of Petroleum, National Bureau of Standards; Member, American Institute of Chemical Engineers, American Chemical Society, American Physical Society, Philosophical Society of Washington, Washington Academy of Sciences.

John C. Whitwell, Ch. E., Professor of Chemical Engineering, Princeton University; Member, American Institute of Chemical Engineers, American Chemical Society, American Society for Engineering Education, American Statistical Association, The Fiber Society.

CONTENTS

INTRODUCTION

PAGE

Introduction 288

GENERAL PRINCIPLES

Law of Conservation of Mass 289
Laws of Definite and Multiple Proportions 289
Molal Quantities 289
Mixture Law 289
Specific Volume and Density of Mixtures 289

GASES

Ideal Gas Laws 289
Actual Gases 290
Dalton's Law—Additive Pressures 290
Amagat's Law—Additive Volumes 291
Liquefaction—Critical Constants 291
Law of Cailletet and Mathias—Rectilinear Diameter 292
Theory of Corresponding States 292
Gaseous Diffusion 292
Lifting Power of a Gas 292

LIQUIDS

Vapor Pressure 293
Effect of Temperature on the Vapor Pressure 293
Effect of Indifferent Gas Pressure on the Vapor Pressure 293
Vapor Pressure of Small Drops (as Mists) 293
Estimation of Vapor Pressure 293

THERMODYNAMICS, THERMOCHEMISTRY, AND CHEMICAL EQUILIBRIUM

Introduction 294
Nomenclature 294
Units of Energy 295
Fundamental Constants and Conversion Factors 295
Energy and the First Law of Thermodynamics 295
Heat Content or Enthalpy 296
Calorimetric Determination of Heats of Reaction 296
Heat Capacity 296
Ideal Gas 296
Heat Capacity of Gases 297
Heat Capacity of Saturated Vapors 297
Heat Capacity of Liquids 297
Heat Capacity of Solids 298
Heats of Formation 298
Variation of Heat of Reaction with Temperature 299
Heats of Fusion, Vaporization, Sublimation, and Transition 299
Heat of Solution 299

PAGE

Estimation of Heat of Vaporization 299
Heats of Adsorption and Wetting 301
Variation of the Energy and Heat Content of Real Gases with Pressure or Volume 301
Reversible Process 302
Isothermal Expansion and Compression 302
Adiabatic Expansion and Compression 303
Entropy and the Second Law of Thermodynamics 303
Heat Engine 304
Refrigerating Machine 304
Heating Machine 304
System in a Reversible Process with Work Only of Expansion or Compression 304
Variation of Entropy with Temperature, Volume, and Pressure 304
Maximum Work 305
Maximum Useful Work or Free Energy 305
Variation of Free Energy with Pressure and Temperature .. 305
Galvanic Cells 306
Criteria of Equilibrium 306
Fugacity of Gases 306
Fugacity of Liquids and Solids 307
Activity 307
Activity Coefficient 307
Standard States 307
Ideal Solution 308
Real Solutions 308
Partial Molal Properties 308
Equilibrium Constant and the Standard Free Energy Change 309
Development of the Free Energy Equation for a Reaction Over a Given Range of Temperature 310
Effect of Temperature on the Equilibrium Constant and the Position of Equilibrium 311
Effect of Pressure on the Equilibrium Constant and the Position of Equilibrium 311
Effect of Indifferent Gases and Proportion of Reactants on the Equilibrium 312
Third Law of Thermodynamics 313
Thermodynamic Properties of Gases from Statistical Calculations 313
Free Energy Function 314
Tabulations of Thermodynamic Data 314
Summary of the More Important Thermodynamic Equations 314

PHASE EQUILIBRIUM AND SOLUTIONS

Phase Rule 315
Phase Relations and Diagrams 315

PAGE
Raoult's and Henry's Laws 317
Solubility of Gases in Liquids 317
Mutual Solubility of Liquids 318
Vapor Pressure and Boiling Points of Binary Liquid Mixtures 319
Solubility of Solids in Liquids 319
Elevation of the Boiling Point by Non-volatile Solutes 319
Lowering of the Freezing Point 319
"Salting-out" Effect 320
Distribution between Two Liquid Phases 320

CHEMICAL REACTION KINETICS

Introduction 321
Homogeneous Reactions 321
General Rate Equation 321
Order of Reaction 321
First-order Reaction 321
Constant Volume 321
Constant Pressure 322
Second-order Reaction 322
Constant Volume 322
Constant Pressure 323
Higher Order Reactions 323
Effect of Temperature on Reaction Velocity 323
Apparent Reaction Order 324
Simultaneous Reactions 324
Illustrative Cases of Simultaneous Reactions 325
Reversible Reactions 325
Consecutive Reactions 325
Side Reactions 327
Non-isothermal Reactions 327
Determination of Reaction Order 327
Heterogeneous Reactions 328
General Character 328
Heterogeneous Reaction Cases 329
Contact Catalytic Gas Reactions 329
Flow Systems. Space-time-yield 329

PAGE

GRAPHICAL REPRESENTATION OF THERMODYNAMIC FUNCTIONS

Pressure-volume Diagram 330
Pressure-volume-temperature Diagram 330
Pressure-volume vs. Pressure Diagram 330
Pressure-temperature Diagram 331
Temperature-energy Diagram 331
Temperature-entropy Diagram 331
Mollier Diagram 332
Joule-Thomson Effect or Throttling 333

TECHNICAL CALCULATIONS

Introduction 333
General Considerations 333
Units and Their Conversion 334
Ideal Gas Law 335
Material Balances 337
Sensible Heats 339
Latent Heats 341
Vaporization 341
Fusion 343
Total Enthalpy Change 344
Standard Heats of Reaction 344
From Heats of Formation 344
From Standard Heats of Combustion 345
Heats of Reaction at a Specified Temperature 346
Heat Balances 347
Equilibrium 348
Calculation of Equilibrium Constant 348
Use of Equilibrium Constants 350
High Pressure 352
Compressibility Charts for Pure Gases 352
Compressibility Chart for Mixtures 353
Fugacity Charts 355
Enthalpy Charts 356
Illustrative Problem 357

INTRODUCTION

Since they cannot be applied quantitatively under all conditions of physical and chemical change, the fundamental principles and generalizations of physical chemistry must frequently be regarded as limiting laws. In the field of chemical industry this is especially true. Nevertheless, these laws provide powerful theoretical and practical tools for dealing with such operations as are encountered in chemical-engineering practice.

In the first part of this section methods and principles of physical chemistry which are believed to be of most technical importance are presented. The latter part illustrates the use and applications of such principles in technical calculations. In some cases principles basic to particular operations will be found in other sections of the handbook.

As exact laws are frequently not available, empirical methods, particularly useful in the absence of necessary data and seldom regarded as fundamentally important in more theoretical treatments of the subject, are outlined in many cases. The scope of the book has necessitated the omission of a majority of the details, of much of the theoretical background and development, and of many subjects entirely. For the application and utilization of the material presented, a knowledge of elementary physical chemical principles is presumed. Where theoretical treatment and complete details are desired reference should be made to sources of information such as those listed below.

GENERAL REFERENCES: Getman and Daniels, "Outlines of Theoretical Chemistry," 6th ed., Wiley, New York, 1937. MacDougall, "Physical Chemistry," Macmillan, New York, 1936. Millard, "Physical Chemistry for Colleges," 6th ed., McGraw-Hill, New York, 1946. Noyes and Sherrill, "Chemical Principles," 2d ed., Macmillan, New York, 1938. Taylor and Taylor, "Elementary Physical Chemistry," 2d ed., Van Nostrand, New York, 1937. Taylor, "Treatise on Physical Chemistry," 2d ed., 2 vols., Van Nostrand, New York, 1931. Jellinek, "Lehrbuch der Physikalischen Chemie," 5 vols., Enke, Stuttgart, 1928–1937. Adam, "Physics and Chemistry of Surfaces," 2d ed., Oxford, New York, 1938. Hougen and Watson, "Industrial Chemical Calculations," 2d ed., Wiley, New York, 1936. Eucken-Jakob, "Der Chemie-Ingenieur," 3 vols. 16 parts, Akademische Verlagsgesellschaft, Leipzig, 1933–1938. Lewis and Randall, "Thermodynamics and the Free Energy of Chemical Substances," McGraw-Hill, New York, 1923. MacDougall, "Thermodynamics and Chemistry," 3d ed., Wiley, New York, 1939. Kiefer and Stewart, "Principles of Engineering Thermodynamics," Wiley, New York, 1930. Weber, "Thermodynamics for Chemical Engineers," Wiley, New York, 1939. Sage and Lacey, "Volumetric and Phase Behavior of Hydrocarbons," Stanford University Press, Stanford University, Calif., 1939. Sherwood and Reed, "Applied Mathematics in Chemical Engineering," McGraw-Hill, New York, 1939.

Physical, Chemical, and Thermodynamic Data. "International Critical Tables," McGraw-Hill, New York. Landolt-Börnstein, "Physikalisch-Chemischen Tabellen," 5th ed., Springer, Berlin, 1936. "International Annual Tables of Constants and Numerical Data," (International Council of Scientific Unions and International Union of Chemistry), McGraw-Hill, New York. Justi, "Spezifische Wärme, Enthalpie, Entropie, und Dissoziation technischer Gase," Springer, Berlin, 1938. Bichowsky and Rossini, "Thermochemistry of Chemical Substances," Reinhold, New York, 1936. Parks and Huffman, "Free Energies of Some Organic Compounds," Reinhold, New York, 1932. "Selected Values of Properties of Hydrocarbons," American Petroleum Institute Research Project 44, National Bureau of Standards. "Selected Values of Chemical Thermodynamic Properties," National Bureau of Standards. See also Sec. 3, p. 107, of this handbook.

GENERAL PRINCIPLES

Law of Conservation of Mass. In a chemical reaction the sum of the masses of the products is exactly equal to the sum of the masses of the reactants.

Laws of Definite and Multiple Proportions. A pure chemical compound always contains its elements in the same proportions by weight, and, when two elements unite to form more than one compound, the weights of the second element combining with a fixed weight of the first bear to each other a simple ratio.

Molal Quantities. A *mole* of any substance is that quantity of the substance whose weight in pounds, grams, or any other convenient unit is numerically equal to its molecular weight. If expressed in pounds, it is called the *pound-mole;* if in grams, the *gram-mole*. If the substance is an element the term *atom* is used instead of *mole*.

Substance	Mol. wts.	Lb.-mole, pounds	G.-mole, grams
NH_3	17	17	17
C_6H_6	78	78	78
$CaCO_3$	100	100	100

Molal Volume. According to the hypothesis of Avogadro, equal volumes of ideal gases at the same temperature and pressure contain an equal number of molecules. The molal volume of any gas at standard temperature and pressure (S.T.P.), 0°C. and 760 mm. Hg, or 32°F. and 29.92 in. Hg, is 22,412 cc./g.-mole or 359 cu. ft./lb.-mole.

Example. Calculate the volume occupied at 0°C. and 760 mm. Hg by (*a*) 10 lb. gaseous ammonia and (*b*) 10 lb. acetone vapor.

(*a*) Lb.-moles $NH_3 = 10/17 = 0.588$
Volume of $NH_3 = 0.588 \times 359 = 211.1$ cu. ft.
(*b*) Lb.-moles acetone (C_3H_6O) $= 10/58 = 0.172$
Volume of acetone vapor $= 0.172 \times 359 = 61.75$ cu. ft.

While the molal volume of any ideal gas under standard temperature and pressure is the same, the molal volume of liquids and solids varies. It depends upon their specific chemical nature and their density. The bulk density of solids depends upon their physical form and state of subdivision.

The Mole Fraction. In a homogeneous mixture or solution the **mole fraction** of any component is defined as the number of moles of that component divided by the sum of the number of moles of all components. If n_1, n_2, etc., are the number of moles, W_1, W_2, etc., the weights, and M_1, M_2, etc., the molecular weights of the components of a mixture, then

$$\text{Mole fraction of component } 1 = \frac{n_1}{n_1 + n_2 + \cdots} = \frac{W_1/M_1}{\dfrac{W_1}{M_1} + \dfrac{W_2}{M_2} + \cdots} \quad (1)$$

The sum of the mole fractions of all components equals unity. **Mole percentage** is the mole fraction multiplied by 100. The composition of solutions and mixtures is often more conveniently expressed in terms of mole percentage than in terms of weight or volume percentage. In the case of gaseous mixtures the mole percentage and the volume percentage are identical.

The average molecular weight of a mixture of gases or of a liquid mixture in which none of the components is dissociated or associated is the sum of the individual weights of each substance present divided by the sum of the number of moles of each.

Example 1. In a mixture of 24 lb. oxygen and 14 lb. nitrogen,

$$\text{Mole fraction, } O_2 = \frac{24/32}{24/32 + 14/28} = \frac{0.75}{0.75 + 0.5} = 0.6$$

Mole percentage, $O_2 = 0.6 \times 100 = 60$ per cent

$$\text{Volume percentage, } O_2 = \frac{0.75 \times 359}{0.75 \times 359 + 0.5 \times 359} \times 100 = 60 \text{ per cent}$$

$$\text{Average molecular weight} = \frac{24 + 14}{24/32 + 14/28} = 30.4$$

1 lb.-mole of the mixture = 30.4 lb.

Example 2. In a mixture containing 36 lb. pentane (C_5H_{12}), 43 lb. hexane (C_6H_{14}), and 25 lb. heptane (C_7H_{16}),

$$\text{Mole fraction, pentane} = \frac{36/72}{36/72 + 43/86 + 25/100} = \frac{0.5}{0.5 + 0.5 + 0.25} = 0.4$$

$$\text{Mole fraction, hexane} = \frac{43/86}{36/72 + 43/86 + 25/100} = \frac{0.5}{0.5 + 0.5 + 0.25} = 0.4$$

$$\text{Mole fraction, heptane} = \frac{25/100}{36/72 + 43/86 + 25/100} = \frac{0.25}{0.5 + 0.5 + 0.25} = 0.2$$

$$\text{Average molecular weight} = \frac{36 + 43 + 25}{0.5 + 0.5 + 0.25} = 83.2$$

1 lb.-mole of the mixture = 83.2 lb.

Example 3. In a 10 per cent aqueous solution of caustic soda,

$$\text{Mole fraction, NaOH} = \frac{10/40}{10/40 + 90/18} = \frac{0.25}{0.25 + 5.0} = 0.0476$$

Mole fraction, $H_2O = 1 - 0.0476 = 0.9524$

Mixture Law. Some of the physical properties of both liquid and solid solutions may frequently be calculated with useful accuracy from those of the components by the mixture law. This law assumes that the property of the mixture is an additive function of the same property of the components, an assumption which in many cases is far from accurate. If A_1, A_2, etc., be the physical property in question per unit weight, and if the weights are W_1, W_2, etc., respectively, then for a weight W of the mixture

$$WA = W_1A_1 + W_2A_2 + \cdots \quad (2)$$

Specific Volume and Density of Mixtures. In general, for engineering purposes, the density and the specific volume (properties referred to unit weight are customarily termed specific) of mixtures may be calculated from the general mixture law. If V_1, V_a, and V_b are the volumes of the mixture, component a, and component b, respectively; v, v_a, and v_b are the corresponding specific volumes; and W, W_a, and W_b, the corresponding weights, then

$$V_a = v_aW_a,\ V_b = v_bW_b,\ V = v(W_a + W_b) = vW$$

and

$$v = \frac{v_aW_a + v_bW_b}{W_a + W_b}$$

In general, in the case of mixtures of liquids the error involved in the use of this principle is less the more closely the liquids are related chemically. The mixture law does not hold as a rule for concentrated solutions of electrolytes.

GASES

Ideal Gas Laws. For any ideal gas or mixture of ideal gases at any definite temperature

$$pV = k$$

where V is the total volume, p is the total pressure, and k is a constant. This is **Boyle's law.**

According to **Charles's law,** at constant pressure the volume of a given mass of an ideal gas increases 1/273.16 of its volume at 0°C. for each degree centigrade increase in temperature. The **absolute temperature** is defined by the statement that it is directly proportional to the pressure-volume product of an ideal gas. On the centigrade scale the absolute temperature, hereafter denoted as T, is $273.16 + t°C$. A temperature of 25°C. is then equal to $273.16 + 25 = 298.16$ degrees centigrade absolute (°C. abs.) or °K. (Kelvin). On the Fahrenheit scale, $T = 460° + t°F.$, and is termed degrees Rankine (°R). Hence, at constant pressure,

$$V = k_1T$$

Boyle's and Charles's laws are combined to give a general equation, $pV = nRT$, in which p is the pressure, V the volume, n the number of moles, T the absolute temperature, and R is a universal constant called the gas constant. This relation is known as the **ideal gas law.** An ideal gas is that hypothetical gas which exactly obeys these laws.

The gas law is frequently used in the form

$$\frac{p_1V_1}{T_1} = \frac{p_2V_2}{T_2}$$

The ideal gas law is not strictly valid for any actual gas, but at moderate pressures and ordinary temperatures it is sufficiently accurate for most engineering purposes. Actual gases conform more closely to the law at low pressures and high temperatures.

In the following table the value of the constant R is given when p and V are expressed in various units.

Table 1. Values of the Gas Constant, R, in $pV = RT$ for 1 Mole of Ideal Gas*

Temp. units	Pressure units	Volume units	Energy units	R per gram-mole
°K			calories	1.9872
°K			abs. joules	8.3144
°K			international joules	8.3130
°K	atm.	cc.		82.057
°K	atm.	liters		0.082054
°K	mm. Hg	liters		62.361
°K	Bar.	liters		0.08314
°K	kg./sq. cm.	liters		0.08478

Temp. units	Pressure units	Volume units	Energy units	R per pound-mole
°R			B.t.u. (*I.T.*)	1.986
°R			hp.-hr.	0.0007805
°R			kw.-hr.	0.0005819
°R	atm.	cu. ft.		0.7302
°R	in. Hg	cu. ft.		21.85
°R	mm. Hg	cu. ft.		555.0
°R	lb./sq. in., abs.	cu. ft.		10.73
°R	lb./sq. ft. abs.	cu. ft.	ft.-lb.	1545.0
°K	atm.	cu. ft.		1.314
°K	mm. Hg	cu. ft.		998.9
°K			c.h.u.	1.986

* See p. 334 for methods of calculation involving changes of units.

Example 1. How many pounds of CO are contained in a vessel of 500 cu. ft. capacity if the pressure is 1 atm. and the temperature 70°F.?

$$pV = nRT \text{ and } n = \frac{W}{M}$$

Then

$$W = \frac{MpV}{RT} = \frac{(28)(14.7 \times 144)(500)}{(1545)(460 + 70)} = 36.2 \text{ lb.}$$

Example 2. Calculate the volume occupied at 110°C. and 2 atm. pressure by 11 cu. ft. of a gas measured at 0°C. and 760 mm. Hg.

$$\frac{p_2V_2}{p_1V_1} = \frac{T_2}{T_1}$$

Therefore,

$$V_{2(110°\text{C.},\ 3\ \text{atm.})} = \left(\frac{p_1}{p_2}\right)\left(\frac{t_2 + 273.2}{t_1 + 273.2}\right) V_{1(0°\text{C.},\ 760\ \text{mm.})}$$

$$V_2 = \frac{760}{2 \times 700} \times \frac{383.2}{273.2} \times 11 = 7.7 \text{ cu. ft.}$$

Actual Gases. The magnitudes of the deviations of actual gases from the requirements of the ideal gas law increase with increasing pressure and decreasing temperature and vary widely with the nature of the gas. Vapors conform more closely to the law the greater the superheat. For each gas there is a definite temperature, commonly referred to as the **Boyle Point,** at which the gas conforms exactly to the laws of Boyle and of Avogadro. The deviations differ in direction above and below this temperature.

Numerous equations of state* have been developed in the attempt to correlate the pressure-volume-temperature variables for actual gases with experimental data. These equations contain a number of arbitrary constants and in general are cumbersome to use. Although some are quite accurate for particular gases in definite temperature and pressure ranges, none is valid over all ranges. The simplest and best known of these equations is that of van der Waals,

$$\left(p + \frac{a}{v^2}\right)(v - b) = RT \qquad (3)$$

where v is specific volume and a and b are constants characteristic of the gas. The constants may be evaluated from critical data (see p. 291) or, with greater accuracy, from actual p-V-T data for the gas in the region in which van der Waals' equation is applicable. They have been determined for a number of gases. Van der Waals' equation is of limited accuracy over wide temperature and pressure ranges. It applies with an accuracy of 5 per cent, although sometimes less, under the following conditions: (1) if the temperature is well above its critical value down to a molal volume of approximately 5 cu. ft. minimum (*e.g.,* 150 atm. at 400°C.); (2) when the temperature is near or below its critical value down to a molal volume of approximately 50 cu. ft. minimum (*e.g.,* 12 atm. at 25°C.) for the more "imperfect" gases.

For detailed information concerning types of equations of state, their respective validities and methods of employing them, reference should be made to text and reference books and to the literature.[1] An extensive list of equations of state may be found in "Handbuch der Experimentalphysik," vol. 8, 2d half, p. 224, Akademische Verlagsgesellschaft, Leipzig, 1929.

For gas calculations at elevated pressures see pp. 352–357.

Dalton's Law—Additive Pressures. In a mixture of gases each component exerts that pressure which it would exert if present alone at the same temperature in the volume occupied by the mixture. The total pressure of a gas mixture is the sum of the partial pressures of the individual components

$$p = p_1 + p_2 + p_3 + \cdots$$

The partial pressure of each component is equal to the total pressure multiplied by its mole fraction in the mix-

* An equation connecting the variables, pressure, temperature, and volume.

[1] See for example Taylor, "Treatise on Physical Chemistry," 2d ed., vol. 1, pp. 236*ff.*, Van Nostrand, New York, 1931. Lewis, "A System of Physical Chemistry," 3d ed., vol. 1, pp., 68*ff.*, and vol. 2, pp. 59*ff.*, Longmans, Green, New York, 1920. Hougen and Watson, "Industrial Chemical Calculations," 2d ed., Wiley, New York, 1936. Weber, "Thermodynamics for Chemical Engineers," Wiley, New York, 1939. Beattie and Bridgman, *Proc. Am. Acad. Arts Sci.*, **63,** No. 5, p. 229 (1928).

ture. For example, in a mixture of 75 per cent oxygen and 25 per cent nitrogen at 1 atm. total pressure, the partial pressure of oxygen equals 75/(75 + 25) × 1 = 0.75 atm., or 0.75 × 760 = 570 mm.

Table 2. Deviations from Dalton's Law of Additive Pressures

a. Argon-ethylene mixtures

Calcd. pressure, atm.	Deviation from calcd. pressure, atm.		
	C_2H_4, 24.74 per cent	C_2H_4, 49.94 per cent	C_2H_4, 90.06 per cent
30	−0.75	−0.85	−0.45
70	−3.35	−5.45	−1.95
90	−5.2	−7.65	1.0
110	−6.60	−8.15	5.25

b. CO, H_2, N_2 mixtures

Calcd. pressure, atm.	Deviation from calcd. pressure, atm.	
	CO 28.23 per cent H_2 52.22 per cent N_2 19.55 per cent	CO 35.66 per cent H_2 39.06 per cent N_2 25.28 per cent
25	−0.40	0.00
100	2.67	2.35
200	7.70	8.46
500	29.75	33.98
1000	68.66	81.32

Dalton's law is rigidly accurate only for ideal gases. For actual gases at high pressures and low temperatures the deviations from this law become appreciable; at pressures of 1000 atm. the deviations are large. At pressures of a few atmospheres and below, gas mixtures may be regarded as ideal gases and the law employed for engineering calculations.

The order of magnitude of the deviations to be expected is indicated by the data of Masson and Dolley [*Proc. Roy. Soc. London*, **103A**, 524 (1923)] for argon-ethylene mixtures and by those of Bartlett (private communication) for carbon monoxide-hydrogen-nitrogen mixtures (Table 2).

The calculated pressure is that obtained by adding the pressures which each constituent would exert if present alone in the same volume. According to Masson and Dolley, the deviations for oxygen-ethylene mixtures are approximately the same as those for argon-ethylene mixtures, whereas oxygen-argon mixtures obey Dalton's law more closely.

Amagat's Law—Additive Volumes. The volume occupied by a gas mixture is equal to the sum of the volumes occupied separately by each constituent at the same temperature and pressure as the mixture, *i.e.*,

$$V = v_1 + v_2 + v_3 + \cdots$$

The partial volume of each constituent is equal to its mole fraction x multiplied by the total volume of the mixture; $v_1 = x_1V$. This rule is obeyed strictly only by ideal gases, but deviations from it are in some cases considerably smaller than from Dalton's law under like conditions.

Liquefaction—Critical Constants. Gases and vapors may be liquefied by the application of pressure, provided that the temperature is at or below a definite value characteristic for each. That definite temperature above which a gas cannot be liquefied by an increase in pressure is the critical temperature T_c of the gas. The pressure required to liquefy the gas at this critical temperature is the critical pressure p_c. The specific volume at the critical temperature and pressure is the critical volume v_c. The pressure necessary to produce liquefaction decreases as the temperature decreases below the critical temperature.

The critical constants (see pp. 150, 165, and 204) are known for many gases and vapors, but they are difficult to determine with accuracy experimentally. There are several methods which may be employed for their estimation in cases where they are unknown, but such approximations should be resorted to only when exact data are not at hand. The critical constants are related to the van der Waals constants a and b by the following equations:

$$v_c = 3b \qquad p_c = \frac{a}{27b^2} \qquad T_c = \frac{8a}{27bR}$$

and, according to the van der Waals equation, (3), they are connected by the relation $RT_c = 2.67p_cv_c$, from which it is possible to calculate any one of the critical constants if the other two are known.* It is found experimentally that the latter relation is more accurately $RT_c = 3.6p_cv_c$ and that v_c is more nearly equal to $4b$ than $3b$. The results obtained in the estimation of the critical constants from the above relations are of a low degree of accuracy because of the inexactness of the van der Waals equation.

The approximate relation of Guldberg, $T_c = 1.5T_b$, where T_b is the normal boiling point of the liquid, holds for a limited number of liquids. For **non-polar** liquids, Watson [*Ind. Eng. Chem.*, **23**, 360 (1931)] has proposed the approximate relation

$$\frac{T_e}{T_c} = 0.283\left(\frac{M}{d_s}\right)^{0.18} \tag{4}$$

where T_c is the critical temperature, M the molecular weight, d_s the density at the normal boiling point in grams per cubic centimeter, and T_e the temperature at which the liquid is in equilibrium with a molal vapor volume of 22.4 l. T_e is connected with the normal boiling point T_s by the relation

$$\ln T_e = 9.8\frac{T_e}{T_s} - 4.2 \tag{4a}$$

It is claimed that values of T_c may be calculated from these relations with a general accuracy of 2 per cent.

Example 1. The values of the constants a and b for nitrogen are 1.31×10^6 cc.^{2} × atm./mole and 37.3 cc./mole; and for ethane, 6.0×10^6 cc.^{2} × atm./mole and 69.9 cc./mole, respectively. Estimate the critical constants for these gases from the van der Waals relations given above.

Solution.

For nitrogen:

$$p_c = \frac{a}{27b^2} = \frac{1.31 \times 10^6}{27(37.3)^2} = 34.8 \text{ atm.}$$

$$T_c = \frac{8a}{27bR} = \frac{8 \times 1.31 \times 10^6}{27 \times 37.3 \times 82.07} = 127°\text{K.}$$

For ethane:

$$p_c = \frac{6.0 \times 10^6}{27(69.9)^2} = 45.6 \text{ atm.}$$

$$T_c = \frac{8 \times 6.0 \times 10^6}{27 \times 69.9 \times 82.07} = 310°\text{K.}$$

The relatively close agreement with accepted values found in these examples is not always to be expected.

Example 2. Calculate the critical temperature of benzene by the method of Watson from the following data: Mol. wt. = 78, d_s = 0.80, normal boiling point = 353°K.

Solution. Solving Eq. (4a), T_e is found to be 363°K. (In the original paper of Watson, *op. cit.*, a plot is given from which the values of T_e corresponding to T_s may be conveniently read.) Substitution of this value of T_e in Eq. (4) and solving gives T_c = 563°K. The critical temperature of benzene is found experimentally to be 561°K.

Since the critical density of benzene is known to be 0.305 g./cc., the critical pressure is found from the relation $RT_c = 3.6p_cv_c$ to be approximately 52 atm.

In general the critical temperature of a gaseous mixture lies between those of the individual components and

* As mentioned on p. 290, the constants a and b may likewise be calculated from experimentally determined values of the critical constants; thus $b = \frac{1}{3}v_c = \frac{1}{8}\frac{RT_c}{p_c}$ and $a = \frac{9RT_cv_c}{8}$.

varies according to the composition of the mixture. The behavior of mixtures of gases on liquefaction comprises a relatively complex and specialized subject whose treatment exceeds the scope of the present handbook. Explanations and discussions of phase behavior and relations in the critical region and methods of dealing with the condensation and vaporization of gaseous mixtures are to be found in Weber, "Thermodynamics for Chemical Engineers," pp. 36–42; Hougen and Watson, "Industrial Chemical Calculations," pp. 406–409; Sage and Lacey, "Volumetric and Phase Behavior of Hydrocarbons"; and Katz and Kurata, Retrograde Condensation, *Ind. Eng. Chem.*, **32**, 817 (1940). The mixtures for which critical constants and phase data in the critical region are available and their source are summarized in the latter reference; data pertaining to the liquefaction of mixtures are also to be found in the other references cited and in recent papers in *Ind. Eng. Chem.*, by G. G. Brown; Cummings; Gilliland; Katz; Kay; and Sage, Lacey, and coworkers. Watson and Smith [*Ind. Eng. Chem.*, **29**, 1408 (1937)] present a correlation of boiling point and critical properties of hydrocarbon mixtures.

See Maass, *et al.*, *Can. J. Research*, **18**, 103, 118 (1940), for experimental data and discussion, and Mayer, *et al.*, *J. Chem. Phys.*, **6**, 87, 101 (1938), for theoretical interpretation of the "critical dispersion temperature" and the distinction between this and the true critical temperature.

Law of Cailletet and Mathias—Rectilinear Diameter. According to this law the arithmetical average of the densities of a pure unassociated liquid and

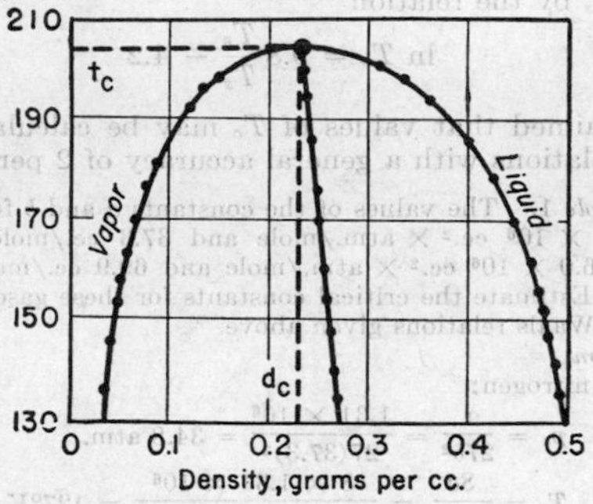

FIG. 1. Law of rectilinear diameter.

its saturated vapor (orthobaric densities) is a linear function of the temperature. The relation is illustrated in Fig. 1. Mathematically expressed,

$$\frac{d_L + d_v}{2} = A + Bt \tag{5}$$

where d_L and d_v are the densities of liquid and vapor, respectively; t is the temperature in degrees centigrade; and A and B are constants characteristic of the liquid, B being negative. At the critical temperature $d_L = d_v = d_c$. Therefore

$$d_c = A + Bt_c$$

and

$$d_c = d_t + B(t_c - t) \tag{5a}$$

where d_t is the average density at temperature t. Although not exact, this relation is quite accurate for many substances and provides a very useful method for calculating the critical density (and thus the critical volume), a quantity not readily measured directly. The constants may be evaluated from measurements of liquid and vapor densities at several temperatures below the critical, or d_c may be estimated graphically from such measurements by means of a plot similar to that of Fig. 1.

Theory of Corresponding States. If p, V, and T in the van der Waals equation be expressed as fractions of the corresponding critical pressure, volume, and temperature ($p_r = p/p_c$; $V_r = V/V_c$; $T_r = T/T_c$), and if a, b, and R also be expressed in terms of the critical constants, an equation

$$\left(p_r + \frac{3}{V_r}\right)(3V_r - 1) = 8T_r \tag{6}$$

results in which all constants characteristic of the substance have apparently disappeared and which, therefore, should be independent of the nature of the gas. This is termed the reduced equation of state; and the principle involved, the law of corresponding states. These reduced values may be used in conjunction with any equation of state, their accuracy being independent of the particular equation involved. If the characteristic constants can be related to the critical constants, the former may then disappear from the resulting equation. Theoretically, according to this principle, in terms of these reduced quantities there should be a universally valid equation of state which holds for both liquids and gases. The corresponding states principle is not entirely rigorous.

Two substances are said to be in corresponding or reduced states when their pressures are proportional to their critical pressures, their volumes to their critical volumes (or their densities to their critical densities), and their temperatures to their critical temperatures

$$\frac{p_1}{p_{c_1}} = \frac{p_2}{p_{c_2}}; \qquad \frac{V_1}{V_{c_1}} = \frac{V_2}{V_{c_2}}; \qquad \frac{T_1}{T_{c_1}} = \frac{T_2}{T_{c_2}}$$

When two substances have any two of these equal, the third must then be equal. It follows that, if the density-temperature relationship to the critical point is available for a substance, the density of a second related substance over the temperature range may be calculated from its critical temperature and a single known measurement of its density at some temperature, provided that the values have been measured at equivalent reduced pressures. Except at elevated temperatures this latter condition may be neglected for liquids.

This principle has proved useful in the systemization of data and has served as the basis for the development of generalized functions relating the physical and thermal properties of various substances. For example, see pp. 352–357.

Gaseous Diffusion. The rates at which gases diffuse vary inversely as the square roots of their densities, irrespective of the fluid in which diffusion occurs. The diffusion coefficient for one gas in another varies inversely as the pressure and directly as the 1.7 to 2.0 power of the absolute temperature. It is slightly dependent upon the composition of the gaseous mixture. Diffusion coefficients may be calculated according to the kinetic gas theory; for methods consult Gilliland, *Ind. Eng. Chem.*, **26**, 681 (1934). See also Sec. 8.

Lifting Power of a Gas. This depends upon the difference between the density of air and that of the lifting gas such as hydrogen or helium. If V is the volume of the balloon in cubic feet; ρ_A and ρ_H the densities (pounds per cubic feet) of air and lifting gas, respectively; L the lift in pounds per cubic foot of gas; and L_T the total lift in pounds; then,

$$L = \rho_A - \rho_H \qquad \text{and} \qquad L_T = VL$$

The effect of a change in temperature and pressure on the densities ρ_A and ρ_H, and thus upon the lifting power of the gas, may be calculated from the gas laws.

LIQUIDS

Vapor Pressure. At any temperature a pure liquid exerts a pressure of vapor, called the vapor pressure, which is characteristic of this liquid. This pressure of vapor in equilibrium with the liquid has, at constant temperature, a definite value in contrast to the pressure of the vapor in the absence of the liquid, which may have any value up to that of the saturated vapor pressure of the liquid. The normal boiling point of a liquid is that temperature at which its vapor pressure equals the standard external pressure of 760 mm. Hg. Normal boiling points are those generally recorded. In general, that temperature at which the vapor pressure of the liquid equals the external pressure is the boiling point under this pressure, and, when the vapor pressure exceeds this external pressure by an infinitesimal amount, the liquid boils. For further treatment of vapor pressure see Sec. 8, pp. 525*ff*.

Effect of Temperature on the Vapor Pressure. The vapor pressure of liquids (and of solids) increases rapidly with the temperature. Variation of vapor pressure with temperature is expressed quantitatively by the Clapeyron equation

$$\frac{dp}{dT} = \frac{L_v}{T\Delta v} \tag{7}$$

where T is the absolute temperature, L_v the molal latent heat of vaporization, and Δv the volume of 1 mole of vapor minus that of 1 mole of the liquid at this temperature. If the necessary data are available, the change in vapor pressure per degree may be calculated by this equation, and, since the external pressure and the vapor pressure are equal at the boiling point, the change of boiling point with unit change of pressure may also be calculated from the equation* in inverted form.

Based on the assumptions that the volume of the liquid is negligible compared with that of the vapor and that the vapor conforms to the ideal gas law, the more useful, approximate Clausius-Clapeyron equation

$$\frac{d \ln p}{dT} = \frac{L}{RT^2} \tag{8}$$

is derived. Equation (8) may be integrated over small temperature intervals without appreciable error on the assumption that L is constant, independent of temperature,

$$\ln \frac{p_1}{p_2} = \frac{L}{R} \frac{T_1 - T_2}{T_1 T_2} \tag{8a}$$

This approximate form of the equation is useful for calculating average heats of vaporization from vapor pressure-temperature data and for estimating vapor pressures over small temperature intervals.

The Clausius-Clapeyron equations also apply to solids. In this case L represents the molal heat of sublimation and Δv the difference in volume between vapor and solid. The exact form may also be used to calculate change of melting point with pressure. Here

$$dT = \frac{T(v_L - v_s)}{L_f}\, dp$$

where L_f is the molal heat of fusion, T the melting point, and v_L and v_s the molal volumes of liquid and solid, respectively. L_f, Δv, and T may be considered constant for moderate changes in pressure, say 15 atm.

Since the boiling point of liquids is the temperature at which the vapor pressure becomes equal to the external pressure, it is obvious that the boiling point varies with the external pressure. The approximate form of the Clapeyron equation [Eq. (8*a*)] may be conveniently employed to estimate boiling points under various pressures. In conjunction with Trouton's rule (see p. 300) it leads to the relation,

$$\Delta T_b = 0.00012 T_b\, \Delta p^*$$

where T_b is the normal boiling point and ΔT_b the change in the normal boiling point corresponding to the change in pressure, Δp, expressed in mm. Hg. This relation gives a close approximation of the effect of pressure on the boiling point for small pressure changes.

Effect of Indifferent Gas Pressure on the Vapor Pressure. The following relation has been derived expressing the change of the vapor pressure in the presence of an indifferent gas pressure p_i:

$$\left(\frac{dp}{dP}\right)_T = \frac{v_L}{v_g} \tag{9}$$

where p and P are the vapor pressure of the liquid and the total pressure on the liquid, $p + p_i$, and v_L and v_g the molal volumes of liquid and its own vapor under pressure p, respectively. Change of vapor pressure with external pressure is ordinarily negligibly small since v_L/v_g is a small number. Under high pressures it may become appreciable. For example, at 20°C. and under an inert gas pressure of 300 atm. the normal vapor pressure of water is increased by approximately 20 per cent. For the practical application of Eq. (9) the relation between p and v_g must be known. If the vapor conforms to the ideal gas law and on the assumption that v_L is independent of pressure, we have on integration

$$\ln \frac{p_2}{p_1} = \frac{v_L}{RT}(P_2 - P_1) \tag{9a}$$

where p_2 and p_1 are the vapor pressures under total pressures P_2 and P_1, respectively.†

Vapor Pressure of Small Drops (as Mists). If p_0 is the bulk vapor pressure and r the radius of a small droplet, the vapor pressure p of the droplet is given by

$$p = p_0\left(1 + \frac{2\sigma M}{rdRT}\right)$$

where σ is the surface tension, M the molecular weight, and d the density. This equation is approximate since σ decreases with the radius for small drops. For water at 20°C.,

$$\frac{p - p_0}{p_0} = \frac{1.08 \times 10^{-7}}{r}$$

and therefore a droplet of water of a radius of 10^{-5} cm. has a vapor pressure approximately 1 per cent greater than water in bulk.

Estimation of Vapor Pressure. Vapor pressure–temperature data are to be found in Sec. 3, pp. 149–173, Sec. 9, p. 563*ff*., and for refrigerants in Sec. 3, pp. 249–280. Data for the vapor pressure of liquids, particu-

* When employing this or any other similar equation, the quantities L, dp, and Δv must be expressed in consistent units; see p. 342, Example, Part *c*.

* MacDougall, "Thermodynamics and Chemistry," 3d ed., p. 365, Wiley, New York, 1939.

† Equation (9*a*) may be reduced to $\ln \frac{p_2}{p_s} = \frac{v_L}{RT}(P_2 - p_s)$, where p_s is the saturated vapor pressure of the liquid in the absence of an indifferent gas.

larly over a range of temperatures, are often lacking; consequently their estimation is of considerable importance. In contrast to an ordinary vapor pressure vs. temperature plot, a plot of the log of the vapor pressure vs. the reciprocal of the absolute temperature (as indicated by the Clausius-Clapeyron equation; see p. 293) gives a substantially straight line over narrow temperature intervals and lines of relatively small curvature over even wide ranges. Interpolation and extrapolation of data are thus facilitated. Cox [*Ind. Eng. Chem.*, **15**, 592 (1923)] and Calingaert and Davis [*Ind. Eng. Chem.*, **17**, 1287 (1925)] have developed a method of plotting vapor-pressure data of especial convenience for hydrocarbons (see Sec. 9, p. 563*ff.*).

Dühring's Rule. This principle is among the most convenient and useful for plotting and estimating vapor-pressure data. The rule states that the temperature at which one liquid exerts a given vapor pressure is a linear function of the temperature at which a second liquid exerts the identical vapor pressure. Accordingly, if the centigrade, Fahrenheit, or absolute temperatures at which two liquids exert the same vapor pressures be plotted against each other, a straight line is obtained. The rule holds more closely and the resulting plot more nearly approximates a straight line the greater the chemical and physical similarity between the two liquids; *e.g.*, where the two liquids concerned are chemically related as are benzene and toluene, the aliphatic ethers, organic hydroxyl compounds and water, and aliphatic hydrocarbons. The rule is a variation of the more accurate but less convenient Ramsay and Young relationship

$$\frac{T_{A}'}{T_{B}'} = \frac{T_A}{T_B} + C(T_{A}' - T_A) \tag{10}$$

which is more nearly applicable to all liquids. T_A' and T_B' and T_A and T_B are, respectively, the absolute temperatures corresponding to two points at which the substances A and B have identical vapor pressures and C is a constant which approaches zero as the two liquids are more closely related.

Using this method of plotting, if a suitable reference liquid whose vapor pressure–temperature curve is known over the desired temperature range is available, the entire vapor pressure–temperature curve of a second unknown liquid may be estimated from its vapor pressure at only two temperatures. It is evident that the results will be more accurate if the reference liquid is related closely to the unknown liquid, *e.g.*, for hydrocarbons another hydrocarbon should be employed as a reference. Water is ordinarily the most convenient reference liquid because its vapor-pressure curve is accurately known over the entire temperature range.

It follows that the Dühring plot can be employed for estimating boiling points under different pressures. It is applicable for the estimation of vapor pressures and boiling points of aqueous solutions, water being the reference liquid [see p. 319]. Perry and Smith [*Ind. Eng. Chem.*, **25**, 195 (1933)] extended the Dühring plot to many other physical properties.

Othmer's Method. For checking, correlating, and estimating vapor-pressure data, Othmer [*Ind. Eng. Chem.*, **32**, 841 (1940); see also p. 301] finds that *logarithmic plots* of the vapor pressures of liquids against those of another or reference liquid at corresponding values of the temperature give straight lines over wide temperature ranges. The lines are more nearly straight than those obtained by previously suggested methods of plotting, and in practice, for ordinary purposes, in the majority of cases may be considered as straight over wide ranges of temperature. As is usual, agreement is best in the case of two related substances, but water is a convenient reference liquid. Use of loglog paper is a convenience in plotting, and these plots are more readily constructed from experimental data than with other methods.

The Clausius-Clapeyron equation provides the theoretical background for the method. It is rearranged as

$$\frac{dp/p}{dp_R/P_R} = \frac{L}{L_R} \quad \text{or} \quad \frac{d \log p}{d \log p_R} = \frac{L}{L_R} \tag{11}$$

where p and p_R are the vapor pressures and L and L_R are the molal latent heats of two substances at the same temperature. The integrated form is

$$\log p = \frac{L}{L_R} \log p_R + C \tag{12}$$

The straightness of the line is due to the fact that the ratio L/L_R is essentially constant over wide temperature ranges. A differential plot according to Eq. (11), *i.e.*, $d \log p$ vs. $d \log p_R$ may be made. In this case the lines go through the origin, and with the data for a known reference liquid the entire vapor-pressure curve of a second liquid may be constructed from a known value of dp/p, $\Delta p/p$, $d \log p$, or L, the latter at one temperature. However, the integral plot, $\log p$ vs. $\log p_R$, as indicated by Eq. (12), is more convenient and accurate to plot and use. For this plot, in addition to data for the reference liquid, it is in general necessary to have either vapor pressures for the second liquid at two different temperatures or the vapor-pressure value at one temperature (*e.g.*, the normal boiling point) and either the latent heat or dp/p at the same or any other known temperature.

The reference cited gives Othmer plots for a large number of substances. The method is also extended to the estimation of latent heats (see p. 343) and to a variety of other physical properties.

THERMODYNAMICS, THERMOCHEMISTRY, AND CHEMICAL EQUILIBRIUM

By Frederick D. Rossini

REVISED BY FREDERICK D. ROSSINI AND W. M. D. BRYANT

Introduction. Thermodynamics is concerned with the transformation of energy from one form to another, with the availability of energy for doing useful work, and with the stability of, and equilibrium among, chemical substances. The whole science is founded upon the first law of thermodynamics (which was contained implicitly in the work of Carnot in 1832 and the work of Mayer in 1842 but first stated unambiguously by Helmholtz in 1847), upon the second law (which was implied in the work of Carnot as early as 1824, but first clearly enunciated by Clausius in 1850 and independently by Kelvin in 1851), and upon another principle (first stated by Nernst in 1906, but later clarified and made rigorous by G. N. Lewis and his associates) which has come to be known as the third law of thermodynamics. The deductions from these laws are numerous and may be applied to any chemical or physical change. Because of the limited nature of the present exposition, the reader is referred to the many texts and treatises on the subject for more detailed discussion and applications.

Nomenclature. A *system* is taken to mean that real or ideal space confined by known boundaries through which pass, in or out, the various forms of energy that are involved in the process in which the given system is participating.

Unless otherwise indicated, Q denotes the heat absorbed by the system from the surroundings and W denotes the work done by the system on the surroundings.

The *fundamental properties* of the system are as follows:

p = pressure. V = volume.
T = absolute temperature. S = entropy.
U = internal or intrinsic energy.

The *derived properties* of the system are as follows:

$H = U + pV$ = heat content or enthalpy.
$A = U - TS$ = maximum work.
$F = U + pV - TS = H - TS$ = maximum useful work or free energy.

The symbol Δ denotes the increment in any given property of the system as it passes from one state to another. For example, when the system goes from a state A to a state B, the increment in the internal energy is $\Delta U = U_B - U_A$, where the subscripts A and B refer, respectively, to the initial and final states of the system. A differential increment of any property Y is indicated by the usual notation dY.

A *numerical subscript* indicates the absolute temperature to which the given property applies, except when the subscripts 1 and 2 are attached to a partial molal property, in which case the 1 and 2 refer to the solvent and solute, respectively.

The *superscript zero*, as H^0, indicates the selected standard or reference state applicable to the given system.

The symbol x_i denotes the mole fraction of component i in a given solution, or the ratio of the number of moles of component i to the total number of moles in the solution.

The symbol m denotes, for aqueous solutions, the molality in terms of the number of moles of solute per 1000 g. of the solvent water.

Units of Energy. The actual unit of heat energy in modern thermodynamics is the electrical joule, which is based upon standards of electromotive force and resistance maintained at the various national standardizing laboratories. Prior to 1948, all accurate measurements of energy were made in terms of the international electrical joule, which differs from the absolute joule by a small amount. The present "best" relation between the international joule and the absolute joule is

$$1 \text{ int. joule} = 1.000165 \pm 0.000052 \text{ abs. joules} \quad (1)$$

By international agreement, beginning Jan. 1, 1948, the various national standardizing laboratories are certifying standard cells and resistances in terms of absolute volts and absolute ohms, so that measurements of energy are now made in terms of absolute joules.

Because chemists and physicists and engineers have for many decades been accustomed to thinking of heat energy in terms of calories, it is apparently desirable psychologically to retain the calorie as the name of the unit of energy. For this purpose, chemical thermodynamicists have in recent years been using an artificial thermochemical calorie defined in terms of international joules prior to Jan. 1, 1948 and in terms of absolute joules since then, with the conversion factors being taken so that the thermochemical calorie represents exactly the same quantity of energy now as prior to 1948. This artificial calorie bears no relation whatever to the heat capacity of water, except that its name is historically derived from, and its value is approximately equal to, the amount of heat required to raise the temperature of one gram of water through one degree centigrade. In this section, then, the unit of energy is taken as the thermochemical calorie, defined as

$$1 \text{ cal.} = 4.1840 \text{ abs. joules} \quad (2)$$

Prior to 1948, the thermochemical calorie was defined as

$$1 \text{ cal.} = 4.1833 \text{ int. joules} \quad (3)$$

The relation of this unit of energy to other units of energy is given by the following:

$$1 \text{ cal.} = 41.2929 \text{ cc. atm.} \quad (4)$$
$$1 \text{ cal.} = 0.0412917 \text{ l.-atm.} \quad (5)$$
$$1 \text{ cal.} = 0.99935 \text{ I.T. cal.*} \quad (6)$$
$$1 \text{ cal.} = 0.0039657 \text{ B.t.u.} \quad (7)$$
$$1 \text{ cal.} = 1.16203 \times 10^{-6} \text{ int. kw.-hr.} \quad (8)$$
$$1 \text{ cal.} = 1.55856 \times 10^{-6} \text{ hp.-hr.} \quad (9)$$
$$1 \text{ cal.} = 3.08595 \text{ ft.-lb.(wt.)} \quad (10)$$
$$1 \text{ cal.} = 0.021430 \text{ cu. ft.-lb.(wt.)/sq. in.} \quad (11)$$

Fundamental Constants and Conversion Factors. For a complete and self-consistent set of fundamental constants and conversion factors the reader is referred to Table 27 in Sec. 1.

Energy and the First Law of Thermodynamics. The first law is represented mathematically by the equation†

$$dU = dQ - dW \quad (12)$$

where, in infinitesimal amounts, dU is the increase in internal energy of the system, dQ is the heat absorbed by the system from the surroundings, and dW is the work done by the system on the surroundings. The first law means simply that, when any system takes part in any process whatsoever, the amount of the energy resident in that system changes only by the net amount of energy (of any form) which may be absorbed or given off by the system during the process.

When a system changes from a state A to a state B, one may write

$$U_B - U_A = \Delta U = Q - W \quad (13)$$

where ΔU represents the increase in internal energy of the system, Q is the heat absorbed by the system from the surroundings, and W is the work done by the system on the surroundings. For any such change the value of ΔU is determined completely by the initial and final states of the system and is independent of the path by which the system passes from the initial to the final state. (The state of a system is defined by its pressure, volume, temperature, mass, and chemical composition.) Whereas ΔU, and thus its equivalent $Q - W$, is dependent only upon the initial and final states, the quantities Q and W themselves depend upon the path by which the system travels from one state to another. The only restriction upon Q and W is that their difference, $Q - W$, must equal ΔU.

The work done by the system on the surroundings, dW (or the work done by the surroundings on the system, $-dW$), may consist of energy from sources mechanical, electrical, magnetic, etc. If p and V are the pressure and volume of the system, and the only work done is that of expansion or compression against a pressure differing from p by only a differential amount dp, then

$$dW = p\,dV \quad \text{or} \quad W = \int_{V_A}^{V_B} p\,dV \quad (14)$$

* The I.T. calorie is the International Steam Table calorie, defined as 1/860 international watt-hr., and is used in the steam tables.

† All the various forms of energy must, of course, be expressed in terms of the same units.

When the pressure is constant,

$$W = p(V_B - V_A) = p\,\Delta V \qquad (15)$$

Heat Content or Enthalpy. When a system takes part in a process occurring *at constant pressure*, with the only work done being that of expansion or compression, one has, by the first law,

$$\Delta U = Q - W = Q - p\,\Delta V \qquad (16)$$

Substituting $U_B - U_A$ and $V_B - V_A$, one has

$$U_B - U_A = Q - p(V_B - V_A) = Q - p_BV_B + p_AV_A \qquad (17)$$

or

$$Q = (U_B + p_BV_B) - (U_A + p_AV_A) = \Delta(U + pV) \qquad (18)$$

Because many such constant-pressure processes are encountered in thermodynamics, it is convenient to abbreviate the combination of properties, $U + pV$, by using the symbol H. This property H is called the *heat content* or *enthalpy*, and, for the process considered above,

$$Q = \Delta(U + pV) = \Delta H \qquad (19)$$

The increment in heat content of the system is equal to the heat absorbed by the system during a process occurring at constant pressure when the only work is that of compression or expansion.

Calorimetric Determination of Heats of Reaction. According to Eq. (19), when the only work involved is that of expansion against or compression by the prevailing pressure, the heat absorbed by a system taking part in a process occurring at constant pressure is

$$Q = \Delta H \qquad (20)$$

where ΔH is the heat content of the final state of the system (products of the reaction) less the heat content of the initial state of the system (reactants of the reaction). If a given reaction is permitted to take place at atmospheric pressure in an appropriate reaction vessel in a calorimeter, the heat evolved per mole of the reaction is the value of $-\Delta H$ for the given reaction with the reactants and products in the states in which they actually exist at the beginning and end, respectively, of the process. Examples of reactions or processes that may be carried out at constant pressure in a calorimeter include the following: (*a*) heat of combustion of a given substance in oxygen with the reaction taking place in a flame; (*b*) heat of solution of a gas, liquid, or solid in a liquid; (*c*) heat of reaction in the liquid phase, as neutralization of an acid with a base; (*d*) heat of hydrogenation of unsaturated organic compounds; (*e*) heat of fusion; (*f*) heat of vaporization.

When the process takes place with no change in volume of the given system and no other work is involved, then the heat absorbed by the system with the process taking place at constant volume is

$$Q = \Delta U \qquad (21)$$

where ΔU is the internal energy of the final state of the system (products of the reaction) less the internal energy of the initial state of the system (reactants of the reaction). For a given reaction, the value of ΔU so determined is to be assigned to the reaction with the reactants and products in the states in which they actually exist at the beginning and end, respectively, of the process. A familiar example of a reaction occurring at constant volume in a calorimeter is the combustion of a given substance in oxygen (at a pressure near 30 atm.) by explosion in a closed bomb.

In the best modern practice, the heat of any given process that evolves heat is determined calorimetrically in general by the substitution method, wherein the heat evolved by a measured amount of the given process or reaction is compared with the heat evolved by a measured amount of electrical energy, using the calorimeter as the comparator of the two energies. The procedure is to measure, in one kind of experiment, the change in a standard calorimeter system produced by a measured amount of the given process and, in another kind of experiment, the amount of electrical energy required to produce the same change in the same calorimeter. When the process is one that absorbs heat, the value of the heat of the process is usually determined by adding electrical energy to the calorimeter in an amount sufficient to balance the heat absorbed by the process. When, for any of the calorimetric experiments, apparatus is not available for measuring the electrical energy with sufficient accuracy, the investigator may compare the heat of the given process or reaction with the heat of a similar process or reaction that has been measured accurately in terms of electrical energy in a standardizing laboratory. Standard calibrating reactions include (*a*) the combustion of benzoic acid, for calorimeters used for measuring heats of combustion in a bomb at constant volume, and (*b*) the combustion of hydrogen in oxygen, for calorimeters used for measuring heats of combustion at constant atmospheric pressure with the reaction taking place in a flame.

Heat Capacity. Heat capacity is defined by the relation

$$C = \frac{dQ}{dT} \qquad (22)$$

where Q is the heat absorbed by the system. Since, by the first law, when the only work done is that of expansion or compression,

$$dQ = dU + p\,dV \qquad (23)$$

then

$$C = \frac{dU}{dT} + p\frac{dV}{dT} \qquad (24)$$

At constant pressure

$$C_p = \left[\frac{\delta(U + pV)}{\delta T}\right]_p = \left(\frac{\delta H}{\delta T}\right)_p \qquad (25)$$

At constant volume

$$C_V = \left(\frac{\delta U}{\delta T}\right)_V \qquad (26)$$

C_p is related to C_V by the formula

$$C_p - C_V = \left[p + \left(\frac{\delta U}{\delta V}\right)_T\right]\left(\frac{\delta V}{\delta T}\right)_p \qquad (27)$$

Ideal Gas. The ideal gas is defined by the two relations, $pV = nRT$ and $(\delta U/\delta V)_T = 0$, where n is the number of moles of gas. As any real gas is taken to lower and lower pressures (or to greater and greater volumes), its p-V-T relations approach those of the ideal gas. Since the internal energy of the ideal gas at constant temperature is independent of its volume, it follows that $(dU/dp)_T = 0$. For 1 mole of ideal gas,

$$pV = RT \qquad (28)$$

$$H = U + pV = U + RT \qquad (29)$$

and, therefore,

$$\left(\frac{\delta H}{\delta p}\right)_T = 0 \qquad (30)$$

For the ideal gas, Eq. (27) becomes

$$C_p - C_V = p\left(\frac{\delta V}{\delta T}\right)_p = R \qquad (31)$$

Heat Capacity of Gases. Values for the heat capacities of most inorganic gases are given in Table 170 of Sec. 3. Values for some organic gaseous compounds may be found in the "International Critical Tables," pp. 79–83, vol. 5. New data on organic gases have appeared in the literature within the past few years, especially in *Ind. Eng. Chem.* and *J. Am. Chem. Soc.* For a semiempirical calculation of the heat capacities of organic gases, see Dobratz, *Ind. Eng. Chem.*, **33**, 759 (1941); Stull and Mayfield, *Ind. Eng. Chem.*, **35**, 639 (1943); and Bennewitz and Rossner, *Z. physik. Chem.*, **39B**, 126 (1938).

For gases each degree of translational freedom and each degree of fully excited rotational freedom contribute an amount $\frac{1}{2}R$, and each degree of fully excited vibrational freedom contributes an amount R, to the heat capacity per mole. For all gases the translational degrees of freedom may be considered fully excited at all temperatures; for all except several gases, the rotational degrees of freedom are fully excited at all except very low temperatures; for most gases, the vibrational degrees of freedom are not fully excited except at very high temperatures, the vibrational contribution to the heat capacity being very small below room temperature. The following table illustrates the magnitude of the contributions to the heat capacity made by the various degrees of freedom, when fully excited, for molecules consisting of n atoms:

then, for the ideal gas,

$$\kappa = 1 + \frac{R}{C_V} \tag{34}$$

For the ideal monatomic gas, $\kappa = 1\frac{2}{3}$; for the ideal polyatomic linear molecule having its two degrees of rotational freedom fully excited but its vibrational degrees of freedom not excited at all, $\kappa = 1\frac{2}{5}$; for the ideal polyatomic non-linear molecule, having its three degrees of rotational freedom fully excited but its vibrational degrees of freedom not excited at all, $\kappa = 1\frac{1}{3}$. Some values of C_p/C_V for various real gases are given in Tables 176 to 177 of Sec. 3. For real gases, C_p/C_V varies with temperature and pressure as indicated by the values given in Table 2,* but for ordinary engineering calculations C_p/C_V may be assumed constant without appreciable error. It should be noted that the error may be considerable over wide ranges of pressure and temperature.

The heat capacities of mixtures of real gases may be taken, without appreciable error, to be that of the sum of the components, as

$$C_p = x_A C_{p_A} + x_B C_{p_B} \tag{35}$$

Heat Capacity of Saturated Vapors. The specific heat of a saturated vapor may be positive, negative, or zero, as may be seen on examination of Eq. (24). If the temperature T of 1 mole of vapor be raised to $T + dT$

Table 1. Degrees of Freedom of Gaseous Molecules and Their Contribution to the Heat Capacity
(n is the number of atoms per molecule)

Gaseous molecule	Degrees of freedom				Heat capacity, when fully excited			
	Translational	Rotational	Vibrational	Total	Translational	Rotational	Vibrational	Total
Monatomic	3	0	0	$3n$	$\frac{3}{2}R$	0	0	$\frac{3}{2}R$
Polyatomic, linear	3	2	$3n - 5$	$3n$	$\frac{3}{2}R$	R	$(3n - 5)R$	$(3n - \frac{5}{2})R$
Polyatomic, non-linear	3	3	$3n - 6$	$3n$	$\frac{3}{2}R$	$\frac{3}{2}R$	$(3n - 6)R$	$(3n - 3)R$

In many polyatomic molecules, degrees of freedom normally assigned to vibrational frequencies are associated with internal rotation of component radicals about single covalent bonds. Such rotation is usually restricted by potential-energy barriers and is not to be confused with rotation of the molecule as a whole. Where rotation about a bond is essentially free, the contribution may be considered constant and equal to $\frac{1}{2}R$ for each covalent bond acting as an axis of rotation. If the binding forces are very strong (as in bicovalent bond) and the torsional displacement is small, the motion may be treated as a vibration. Usually, however, the motion is of intermediate character and cannot at present be treated without recourse to experimental heat capacity or entropy data. The molecular heat due to internal rotation, as well as the corresponding entropy, heat content, and free-energy function, can be calculated by methods outlined by Pitzer, *J. Chem. Phys.*, **5**, 469 (1937); *ibid.*, **14**, 239 (1946); and by Pitzer and Gwinn, *ibid.*, **10**, 428 (1942).

The heat capacity of the ideal gas is independent of the pressure; whereas for real gases the heat capacity changes significantly with the pressure at constant temperature. The magnitude of this variation with pressure may be seen from Fig. 12, p. 232, Sec. 3, which gives values of C_p for N_2 from $-70°$ to 600°C. and from 0 to 1200 atm. See also Table 175, Sec. 3.

The ratio of the heat capacity at constant pressure to that at constant volume is

$$\frac{C_p}{C_V} = \kappa \tag{32}$$

Since by Eq. (31)

$$C_p - C_V = R \tag{33}$$

and its pressure p_s be raised to $p_s + dp_s$, in order to maintain the vapor at the saturation pressure, dV_s will be negative and the ratio of the heat added to the temperature rise, which is the heat capacity, will be positive if dU is greater than $p_s\,dV_s$, negative if the reverse is true, and zero if they are equal. For H_2O, CS_2, $CHCl_3$, CCl_4, and acetone the heat capacity of the saturated vapor is negative, while for ether it is positive.

Table 2. Effect of Temperature and Pressure on C_p/C_V for Air

Pressure, atm.	0°C.	−79.3°C.
1	1.405	1.405
25	1.473	1.569
50	1.530	1.767
100	1.646	2.200
150	1.739	2.469
200	1.828	2.333

Heat Capacity of Liquids. The heat capacity of liquids ordinarily, but not invariably, increases with the temperature. In the case of a few liquids, of which water is an important example, there may be an interval of temperature in which the heat capacity actually decreases. This phenomenon is apparently dependent upon association of the substance in the liquid state. For mixtures that constitute substantially ideal solutions the heat capacity may be calculated as the sum of the heat capacities of the components in the pure state. For most real solutions, however, and especially aqueous solutions of electrolytes, the additivity of the heat capacities does not hold. For dilute aqueous solutions of practically all strong electrolytes, the heat capacity

* Also see Edmister, *Ind. Eng. Chem.*, **32**, 373 (1940), for methods of estimating C_p/C_V ratios for hydrocarbons at any temperature and pressure.

of a given amount of the solution is slightly less than that of the pure water from which it was made. In the absence of experimental data the heat capacity of aqueous solutions that are not too concentrated may be estimated without serious error as being equal to the heat capacity of the pure water from which the given solution was made. Values for the heat capacities of inorganic liquids are given in Table 170 on p. 219, Sec. 3, values for organic liquids in Table 172 on p. 225, Sec. 3, and values for some solutions on pp. 233–234, Sec. 3.

Heat Capacity of Solids. Values for the heat capacities of most inorganic solids for which data exist are given on pp. 230–235, Sec. 3, while values for some organic solids may be found in the "International Critical Tables," pp. 101–105, vol. 5.

The heat capacities of practically all elements in the solid crystalline state, and of most compounds in the isotropic crystalline state, are fairly well described as a function of the temperature by the Debye heat-capacity equation:

$$C_V = 3R\left[3\left(\frac{1}{x_m}\right)^3 \int_0^{x_m} \frac{x^4 e^x}{(e^x - 1)^2}\,dx\right] \text{ per g.-atom} \tag{36}$$

where $x_m = \Theta/T = h\nu/kT$. Here Θ is a constant characteristic of the given substance, h is the Planck constant, k is the Boltzmann constant, and ν is the characteristic frequency of vibration for the given substance. In the Debye equation, as T approaches zero, C_V approaches zero. For small values of T the Debye equation reduces to

$$C_V = 3R\left(\frac{4\pi^4}{5\Theta^3}T^3\right) = aT^3 \text{ per g.-atom} \tag{37}$$

That is, from the absolute zero up to some small value of the absolute temperature C_V increases with the cube of the absolute temperature. For large values of T, the Debye equation becomes

$$C_V = 3R \text{ per g.-atom} \tag{38}$$

since the term in brackets in Eq. (36) approaches unity for large values of T and does so become for $T = \infty$.

Figure 2 gives a plot of C_V against T/Θ, or $1/x_m$, as given by the Debye equation [Eq. (36)]. The values

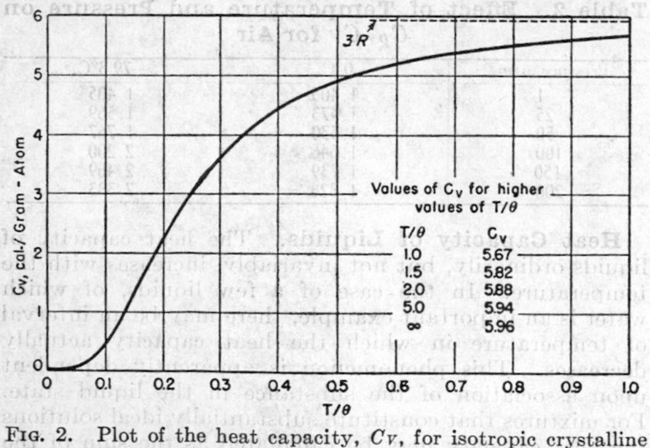

FIG. 2. Plot of the heat capacity, C_V, for isotropic crystalline solids, as a function of T/θ, as calculated by the Debye heat capacity equation (see text).

of C_V per g.-atom for most isotropic crystalline solids will fall on or near this curve. That is, if the value of Θ for a given solid is known, its heat capacity per gram-atom at any given temperature can be immediately estimated from the curve of Fig. 2. Similarly, if the value of the temperature is known at which the heat capacity of a given solid has a known value between about 1 and 5 cal. per g.-atom, the value of Θ can thereby be deduced, and the heat capacity for all other temperatures becomes known. The following are values of Θ for a number of elements and compounds in the solid state (taken from "A Treatise on Heat," by Saha and Srivastava, Indian Press, Calcutta, India, 1935): Cs, 68; Rb, 85; Pb, 88; Tl, 96; Hg, 97; I_2, 106; Bi, 111; Sn, 119; K, 126; Cd, 168; Au, 175; Na, 202; Ag, 218; Ca, 226; Zn, 235; Ge, 290; Mg, 290; W, 310; Cu, 315; Fe, 370; Ni, 375; Al, 396; Li, 510; Be, 900; C (diamond), 1860; KBr, 177; KCl, 230; NaCl, 281; CaF_2, 474; FeS_2, 645.

The utility of the values in Fig. 2 is illustrated by the following examples:

1. What is the value of C_V at 100°C. for a solid for which the Debye Θ has the value 235? *Answer:* $T/\Theta = (100 + 273)/235 = 1.59$. From the values of C_V vs. T/Θ tabulated in the lower right part of Fig. 2, C_V is estimated to be 5.8 cal./deg. g.-atom.
2. What is the value of C_V at 0°C. for a solid for which the Debye Θ has the value 1200? *Answer:* $T/\Theta = (0 + 273)/1200 = 0.228$. From the curve in Fig. 2, for this value of T/Θ, $C_V = 2.7$ cal./deg. g.-atom.

Examination of Eq. (36) and Fig. 2 shows that when $T/\Theta = 1$, *i.e.*, when $T = \Theta$, the heat capacity of crystalline solids will be approximately 5.67 cal./deg. g.-atom, which is 95 per cent of the Dulong and Petit value of $3R$ per g.-atom. It may be noted that, according to this relation, the heat capacities of the first 17 elements listed above, and of the first 3 compounds, will have reached 95 per cent of the Dulong and Petit value at or below room temperature, while for the remaining solids listed this value is not reached until higher temperatures—at 1860°K. for C (diamond). It is important to note also that solids of complicated crystalline structure may deviate appreciably from the Debye equation.

Equation (36) gives the value for C_V. Since it is C_p that is most frequently needed, it is necessary to know the value of $C_p - C_V$. Equation (27) may be transformed to

$$C_p - C_V = \frac{\alpha^2 VT}{\beta} \tag{39}$$

where C_p, C_V, and V refer to 1 mole of the substance, $\alpha = 1/V(dV/dT)$, the coefficient of expansion, and $\beta = -1/V(dV/dp)$, the coefficient of compression. For most solids, $C_p - C_V$ is small, though it becomes appreciable for substances having large values of the coefficient of expansion. The following values illustrate the magnitude of $C_p - C_V$ for a number of elements in the solid state at room temperature, in calories per degree gram-atom, as given by Lewis and Gibson [*J. Am. Chem. Soc.*, **39**, 2554 (1917)]: Li, 0.3; C, 0.0; Na, 0.5; Al, 0.2; Si, 0.1; S, 0.4; K, 0.6; Cr, 0.1; Mn, 0.1; Fe, 0.1; Co, 0.1; Ni, 0.2; Cu, 0.2; Zn, 0.3; As, 0.0; Ag, 0.3; Cd, 0.3; I, 0.9; W, 0.1; Pt, 0.2; Au, 0.3; Pb, 0.4; Th, 0.1.

Heats of Formation. The heat of formation of a chemical compound is the heat associated with the formation of the compound in its standard state from its elements in their standard states, all at a given temperature and pressure.* For example, for the formation of water at 25°C. and 1 atm.,

$$H_2\,(g) + \tfrac{1}{2}O_2\,(g) = H_2O\,(l)$$
$$\Delta H^0_{298.16} = -68.317 \text{ kg.-cal./mole} \tag{40}$$

That is, the formation of 1 g.-mole of liquid water from gaseous hydrogen and oxygen, at 25°C. and a constant pressure of 1 atm., is accompanied by a decrease in heat

* The symbols used to indicate the physical state of a substance have the following meaning: c, crystal (solid); l, liquid; g, gas; aq, aqueous.

content, or an evolution of heat energy, equal to 68,317 cal.

Values for the heats of formation of many chemical compounds are given in Table 197, Sec. 3 (see also Table 198). These values have been derived by first determining the heats of formation of those substances whose formation from the elements can take place in a reaction susceptible of calorimetric measurement. Then, in order to determine the heats of formation of those substances that cannot be formed directly from their elements in reactions susceptible of calorimetric measurement, there are measured the heats of reactions in which all the reactants and products, save the given "unknown," are either elements or substances whose heats of formation have already been determined.

Variation of Heat of Reaction with Temperature. Consider the reaction

$$bB + cC = mM + nN \qquad (41)$$

Application of Eq. (25) yields, for the reaction,

$$\frac{d(\Delta H)}{dT} = \Delta C_p \qquad (42)$$

For most substances in a given phase, over a more or less limited range of temperature, the heat capacity can be expressed as some function of the temperature, as

$$C_p = a + bT + cT^2 + \cdots \qquad (43)$$

or (see Kelley, *U.S. Bur. Mines Bull.* 371, 1934)

$$C_p = a + bT - cT^{-2} \qquad (44)$$

If for each of the reactants and products of the given reaction the heat capacity for the given range of temperature can be expressed by an equation of the form of Eq. (43), then

$$\Delta C_p = (\Delta a) + (\Delta b)T + (\Delta c)T^2 + \cdots \qquad (45)$$

Combination of Eqs. (42) and (45), followed by integration, yields

$$\Delta H = \Delta H_* + (\Delta a)T + \tfrac{1}{2}(\Delta b)T^2 + \tfrac{1}{3}(\Delta c)T^3 + \cdots \qquad (46)$$

where ΔH_* is the constant of integration and is the value of ΔH given by this equation when T has the value zero. This equation is not valid outside the stated interval of temperature for which the heat-capacity equations are valid, and it must be emphasized that ΔH_* is only the constant of integration and is not $\Delta H_0{}^0$, the heat of the reaction at the absolute zero of temperature. Three of the four constants in Eq. (46), Δa, Δb, and Δc, can be evaluated from the heat-capacity equations for the reactants and products and the fourth, ΔH_*, from a value for ΔH for some temperature within the range for which the heat-capacity equations are valid.

Heats of Fusion, Vaporization, Sublimation, and Transition. The heat of fusion of a given substance is the heat required to convert unit mass of it from the solid state to the liquid state at a given pressure (and temperature). Values for the heats of fusion of many inorganic substances are given on pp. 210–212, Sec. 3, and of many organic substances on p. 213, Sec. 3.

The heat of vaporization of a given substance is the heat required to convert unit mass of it from the liquid state to the gaseous state at a given pressure (and temperature). Values for the heats of vaporization of many inorganic substances are given on pp. 210–212, Sec. 3, and of many organic substances on p. 215, Sec. 3.

The heat of sublimation of a given substance is the heat required to convert unit mass of it from the solid state to the gaseous state. At a given temperature the heat of sublimation is equal to the sum of the heats of fusion and vaporization for that same temperature.

The heat of transition is the heat required to convert unit mass of a given substance from one crystalline form to another crystalline form. Values for the heats of transition of various solids that exist in more than one crystalline form may be calculated from the values for the heats of formation given on pp. 244–246, Sec. 3. For example, for the transition

$$\text{C (c, graphite)} = \text{C (c, diamond)} \qquad (47)$$

$$\Delta H = \underset{(0.45)}{\Delta Hf[\text{C (c, diamond)}]} - \underset{(0.00)}{\Delta Hf[\text{C (c, graphite)}]} = 0.45 \text{ kg.-cal./mole} \qquad (48)$$

That is, the conversion of 1 g.-mole of graphite into diamond at 18°C. and 1 atm. pressure is accompanied by an absorption of 0.45 kg.-cal., or 450 cal., of heat energy.

Heat of Solution. Values for a number of heats of solution may be calculated from the values for heats of formation given on pp. 236–243, Sec. 3. For example, the heat of solution of 1 g.-mole of sodium chloride in 400 moles of water at 18°C. is calculated as follows:

$$\text{NaCl (c)} = \text{NaCl (aq, } 400H_2O) \qquad (49)$$

$$\Delta H = \underset{(-97.105)}{\Delta Hf[\text{NaCl (aq, } 400H_2O)]} - \underset{(-98.330)}{\Delta Hf[\text{NaCl (c)}]} = 1.225 \text{ kg.-cal./mole} \qquad (50)$$

The foregoing is the total or integral heat of solution of 1 mole of solute dissolved in a definite amount of solvent. The partial or differential heat of solution is the heat, per mole of solute added, associated with the solution of a very small amount of solute in a large amount of solution of a given concentration, the relative amounts being such that the concentration is substantially unchanged. It is also the heat associated with the solution of 1 mole of solute in a very large amount of solution of a given concentration, the amount of solution taken being so large that the addition of 1 mole of solute leaves the concentration substantially unchanged. The partial molal heat of dilution is similarly the heat associated with the addition of a small amount of solvent to a large amount of solution of a given concentration, the concentration remaining substantially unchanged and the heat being calculated per mole of solvent added. In making thermodynamic calculations involving the components of solutions, it is convenient to use the partial molal equations and functions, which are described briefly on pp. 308–309 and in great detail by Lewis and Randall ("Thermodynamics and the Free Energy of Chemical Substances," McGraw-Hill, New York, 1923).

Estimation of Heat of Vaporization. Data on heats of vaporization (see pp. 210–212 and 215, Sec. 3), particularly over a range of temperatures, are often lacking for many substances. The majority of the methods proposed for their estimation are empirical and few are applicable over extended temperature and pressure ranges.

Trouton's rule states that the molal heat of vaporization of normal liquids at the boiling point under atmospheric pressure divided by the absolute boiling temperature is a constant which is approximately 22; $\Lambda_b/T_b = 22$. This constant varies considerably, however, with variation in the boiling temperature. Nernst has proposed the modified form

$$\frac{\Lambda}{T_b} = 9.5 \log T_b - 0.007T_b \qquad (51)$$

Neither of these relations is accurate over wide boiling ranges. The ratio, in general, increases with boiling temperature and the values of Λ calculated for high boiling points are too large.

Hildebrand has shown that the ratio Λ/T (molal entropy of vaporization) is approximately constant for substances having widely different boiling points if the comparison temperatures be those corresponding to an equal vapor concentration and if the pressure range be such that the vapor conforms to the ideal gas law. The value of Hildebrand's constant, calculated for temperatures at which the concentration of vapor is 0.005 mole/l., is 26.2 to 27.8 for many substances such as O_2, Cl_2, N_2, hexane, and mercury; and for NH_3, H_2O, and ethyl alcohol is approximately 32 to 33.

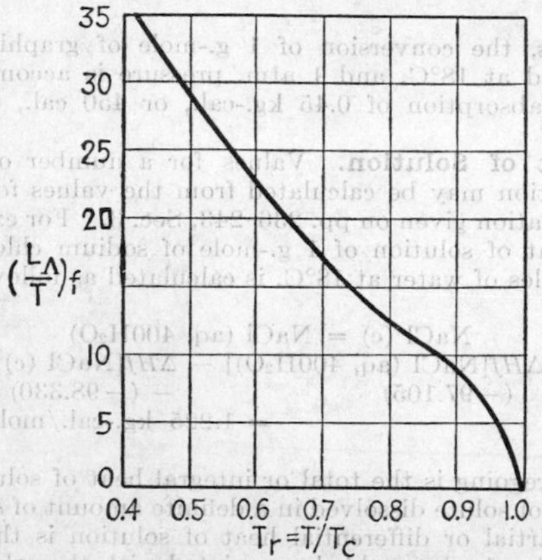

FIG. 3. Watson's plot showing form of variation of molal entropy of vaporization from temperature. Curve determined from experimental data for liquids: C_6H_6, C_8H_{18}, SO_2, CS_2, ether, C_6H_{14}, C_2H_5OH, H_2O. [*Ind. Eng. Chem.*, **23**, 362 (1931).]

In order to estimate latent heats over a range of temperature and pressure from a relatively few experimental measurements, Lewis and Weber [*Ind. Eng. Chem.*, **14**, 486 (1922)] and McAdams and Morrell [*Ind. Eng. Chem.*, **16**, 375 (1924)] utilize Hildebrand's rule in the following manner. Values of Λ/T (the Hildebrand function) over the range of temperature and pressure are plotted against $1000p/T$, p and T being the pressure and temperature at which vaporization occurs. Heats of vaporization at various desired temperatures and pressures may be obtained from this curve. The method does not hold at high temperatures and at pressures near the critical point.

Latent heats of vaporization may be calculated from vapor-pressure data by means of the Clausius-Clapeyron equation, but use of this equation is confined to narrow temperature intervals and to a pressure region where the vapor obeys the ideal gas law. The Clausius-Clapeyron equation may be conveniently employed in conjunction with a Dühring plot (see p. 293) to calculate heats of vaporization at any desired temperature (Lewis and Weber, *op. cit.*). The slope of the Dühring line is

$$\frac{\Delta T_r}{\Delta T} = \left(\frac{\Lambda}{\Lambda_r}\right)\left(\frac{T_r}{T}\right)^2 \tag{52}$$

where Λ_r is the latent heat of the reference liquid at temperature T_r (°K.), and Λ is the latent heat of the liquid in question at the temperature T, corresponding to an equal vapor pressure. If the slope of the line and the latent heat of the reference liquid be known, that of the second liquid may be calculated. The latent heat of the reference liquid employed should be accurately known over a wide temperature range. Although the assumption that the vapors obey the ideal gas law is involved in such a calculation, the method will be valid if deviations are in the same direction and of the same order of magnitude.

In reviewing the applicability of various methods proposed for estimating latent heats of vaporization, Watson [*Ind. Eng. Chem.*, **23**, 360 (1931)] found that the relation developed theoretically by Kistiakowsky [*Z. phys. Chem.*, **107**, 65–73 (1923)],

$$\frac{\Lambda}{T_b} = 8.75 + 4.575 \log T_b \tag{53}$$

where Λ is the latent heat of vaporization, cal. per mole, holds with surprising accuracy for **non-polar** liquids at atmospheric pressure over the entire range of experimentally measured values.

Latent heats of vaporization at any desired temperature and pressure may be calculated, if a value at one temperature T and the critical temperature T_c are known, by the following graphical method due to Watson (*op. cit.*, p. 363), which is applicable to all liquids. Watson has plotted values of $(\Lambda/T)f$ (where f is a characteristic constant for each liquid determined graphically from the available data) for a series of liquids against corresponding values of the reduced temperature $T_r = T/T_c$, such a plot resulting in a single curve for all liquids. This curve is reproduced in Fig. 3. From the curve of Fig. 3 the latent heat of vaporization may be calculated from the relationship

$$\left(\frac{\Lambda}{T}\right)_2 = \frac{Y_2}{Y_1}\left(\frac{\Lambda}{T}\right)_1 \tag{54}$$

where $(\Lambda/T)_1$ is the known ratio at temperature T; corresponding to a reduced temperature $T_1/T_c = T_{r_1}$; $(\Lambda/T)_2$ is the desired ratio at temperature T_2, corresponding to a reduced temperature $T_2/T_c = T_{r_2}$; and Y_1 and Y_2 are, respectively, the corresponding ordinates from the curve.

In the entire absence of experimental data the latent heats of vaporization of non-polar liquids at the normal boiling point may be calculated from the Kistiakowsky equation above, and the value at any desired temperature then estimated by the above method with a general accuracy better than 5 per cent.

A knowledge of latent heats of vaporization of water from aqueous solutions is of practical importance for engineering calculations involved in evaporator design. As a rule there is very little information concerning the heats of vaporization of solvents from even moderately concentrated solutions. The heat of vaporization of water from highly concentrated caustic soda solutions may, for example, differ greatly from that for pure water. The method using a Dühring plot in conjunction with the Clausius-Clapeyron equation, outlined on p. 293 or the method of Othmer below, may be advantageously employed for calculating heats of vaporization of water from moderately concentrated aqueous solutions where the solute is non-volatile, water being the reference liquid. The Dühring method is of doubtful value for very concentrated solutions since at higher concentrations the curves depart markedly from straight lines.

Where data are available for the heat of vaporization of the pure solvent and for the heat of dilution of the given solution with the solvent, the heat of vaporization of the solvent from the solution can be easily calculated, as illustrated in the following example:

Example. Calculate, for 18°C., the heat required to remove water from an aqueous solution of sodium hydroxide sufficient to reduce the concentration from 1 mole in 5 moles of water to

1 mole in 3 moles of water. The data given are the heat of vaporization of pure water at 18°C., 10,580 cal./mole, and the heat of diluting a solution of $NaOH.3H_2O$ with $2H_2O$ to form $NaOH.5H_2O$, where there is absorbed 1160 cal./g.-mole of H_2O added.

Solution.

$$(NaOH.5H_2O)\ (l) = (NaOH.3H_2O)\ (l) + 2H_2O\ (l); \quad \Delta H = -2320 \text{ cal.} \tag{55}$$

$$2H_2O\ (l) = 2H_2O\ (g); \quad \Delta H = 21{,}160 \text{ cal.} \tag{56}$$

$$(NaOH.5H_2O)\ (l) = (NaOH.3H_2O)\ (l) + 2H_2O\ (g); \quad \Delta H = 18{,}840 \text{ cal.} \tag{57}$$

A new and useful method for the estimation of latent heats of various changes of state from vapor-pressure data has recently been presented by Othmer [*Ind. Eng. Chem.*, **32**, 841 (1940)]. Writing the Clausius-Clapeyron relation (see p. 293) for each of two substances, dividing one by the other to eliminate temperature terms, rearranging and integrating, the following equation is obtained

$$\log p = \left(\frac{\Lambda}{\Lambda_R}\right) \log p_R + C \tag{58}$$

where p and p_R are vapor pressures, Λ and Λ_R molal latent heats, respectively, of the two substances, taken at the same temperature, and C is a constant.

If a loglog plot be made of the vapor pressure of any substance vs. the vapor pressure exerted by a reference substance at the same temperature, Othmer finds that the line is more nearly straight than that obtained by other common vapor-pressure extrapolation methods; hence, the constant slope, which from Eq. (58) is Λ/Λ_R, readily permits the calculation of values of Λ from values of Λ_R at the same temperature. Latent heats of vaporization from both pure liquids and from solutions, of sublimation, and of fusion are thus readily obtainable at a desired temperature.* It is, obviously, important to select a reference material for which sufficient data concerning vapor pressure and latent heats are available. Illustrations of the use of this method will be found on pp. 341–344.

Heats of Adsorption and Wetting. The absorption of gases and vapors on solid surfaces is accompanied by the evolution of heat. The heat of adsorption varies with the fraction of the surface covered by adsorbed gas. The **total or integral heat of adsorption** is the heat evolved in the adsorption of a definite quantity of gas on the bare surface, while the **differential heat** is that evolved when 1 mole of gas is adsorbed by a large quantity of adsorbent already having a specified quantity of adsorbed gas. Heats of adsorption normally decrease as successive amounts of gas are adsorbed, but in certain cases [Fryling, *J. Phys. Chem.*, **30**, 818 (1927); Taylor and Kistiakowsky, *Z. physik. Chem.*, **125**, 341 (1927)] low initial values increasing to a maximum and then decreasing have been found. Lamb and Coolidge [*J. Am. Chem. Soc.*, **42**, 1146 (1920)] obtained an almost linear relation between the heat of absorption and the amount adsorbed in the case of some organic vapors on charcoal. In general, the heat of adsorption is greater than the heat of liquefaction, the difference being particularly large in the case of active catalyst surfaces.

Heat changes attending the adsorption of gases and vapors by solids can be related to the temperature coefficient of adsorption by the Clausius-Clapeyron equation.

* Othmer's method may also be applied to the checking of experimental data and to the estimation of heat effects accompanying hydration, dissociation, dilution, or solution, and chemical reaction. Freezing points of salt solutions, enthalpy charts, steam-distillation relations, composition of vapors from boiling binary solutions, constant boiling mixtures, and other phenomena relating to the disengaging of vapors from a condensed phase may also be studied and evaluated by the method.

This equation is applied in the form

$$\ln \frac{p_2}{p_1} = \frac{\Delta H_{\text{ad}}}{R}\left(\frac{T_2 - T_1}{T_1 T_2}\right) \tag{59}$$

where p_2 and p_1 are the equilibrium pressures in the gas phase corresponding to the adsorption of the same definite quantity of gas at temperatures T_2 and T_1, respectively, and ΔH_{ad} is the isosteric heat of adsorption. The term "isosteric" signifies the fact that this heat is that calculated from the variation of pressure with temperature for a constant amount adsorbed and not from the variation of amount adsorbed with temperature at constant pressure. Details and a discussion of the significance of such calculations are to be found in Freundlich, "Colloid and Capillary Chemistry," pp. 134–137, Methuen, London, 1926. This method should not be indiscriminately employed for calculating heats of adsorption, particularly in cases where active catalyst surfaces are involved. Before it is used, reference should be made to a specialized treatment of the subject.

The heat evolved or adsorbed when a liquid and a solid surface are placed in contact is known as the **heat of wetting.** Data are available for some solids and liquids. The heat of wetting is proportional to the extent of surface involved and varies with the nature of this surface and of the liquid.

Variation of the Energy and Heat Content of Real Gases with Pressure or Volume. The variation, with temperature and volume, of the internal energy of a real gas is given, as for any thermodynamic system, by the relation

$$dU = \left(\frac{\delta U}{\delta T}\right)_V dT + \left(\frac{\delta U}{\delta V}\right)_T dV \tag{60}$$

or

$$dU = C_V\, dT + \left(\frac{\delta U}{\delta V}\right)_T dV \tag{61}$$

Similarly, the variation of heat content with temperature and pressure is given by

$$dH = \left(\frac{\delta H}{\delta T}\right)_p dT + \left(\frac{\delta H}{\delta p}\right)_T dp \tag{62}$$

or

$$dH = \left(\frac{\delta H}{\delta T}\right)_p dT - \left(\frac{\delta T}{\delta p}\right)_H \left(\frac{\delta H}{\delta T}\right)_p dp \tag{63}$$

or

$$dH = C_p\, dT - \mu C_p\, dp \tag{64}$$

where

$$\mu = \left(\frac{\delta T}{\delta p}\right)_H \tag{65}$$

is the Joule-Thomson coefficient. For an ideal gas, $\mu = 0$, because $(\delta H/\delta p)_T = -\mu C_p = 0$ and C_p is finite.

At constant temperature all real gases suffer some change in internal energy with change in pressure (or volume), and, while the effect per atmosphere is small, it becomes important for large changes of pressure (or volume). Calorimetric measurements have been made of both $(\delta U/\delta p)_T$, the change of internal energy with pressure at constant temperature, and of $(\delta H/\delta p)_T$, the change of heat content with pressure at constant temperature. The first experiment is one in which a given mass of gas in a pressure vessel in a calorimeter is permitted to expand against the atmosphere and the temperature of the calorimeter is maintained constant by adding the appropriate amount of electrical energy. When the easily calculated work of "pushing" back the atmosphere is subtracted from the energy added, the remainder represents the change in internal energy of the gas as it expands from the initial to the final pressure. The second experiment is one in which the gas is passed through a porous plug and the temperature of the gas (assuming that it gains internal energy on expansion) is maintained constant by adding the appropriate amount of electrical power. The energy added gives directly the change in heat content with pressure at constant temperature as the gas passes from the initial to the final pressure.

If, during the change in pressure, no heat is allowed to enter or leave the gas, its change in internal energy must be compensated for by a corresponding change in its temperature. These are the conditions that hold in the Joule-Thomson experiment, where a gas is passed through a porous plug that is thermally isolated from its surroundings and the change in energy of the gas is reflected by a change in its temperature as it passes through the plug. This experiment in adiabatic throttled expansion measures directly the Joule-Thomson coefficient $\mu = (\delta T/\delta p)_H$, because the heat content of the system remains constant and the change in temperature and the change in pressure are measured. As indicated above, the relation between the Joule-Thomson coefficient and the change of heat content with pressure at constant temperature involves the heat capacity at constant pressure:

$$\left(\frac{\delta H}{\delta p}\right)_T = -\mu C_p \tag{66}$$

The Joule-Thomson coefficient varies with the gas, the temperature, and the pressure. For all gases at the lowest temperatures, adiabatic throttled expansion results in cooling of the gas. As the temperature is raised at a given pressure, there is reached a temperature (different for different gases) where throttled expansion results in zero change in temperature. This represents, for the given gas and pressure, the inversion temperature, where $\mu = 0$. Above the inversion temperature all gases undergo heating on throttled expansion, below it, cooling. At room temperature most gases (including air, oxygen, nitrogen, and carbon dioxide) undergo cooling on throttled expansion, but a few (including hydrogen) undergo heating. For the latter group the inversion temperature lies below room temperature. At any given temperature the Joule-Thomson coefficient changes a small amount with change in pressure.

The Joule-Thomson effect is employed in the Linde process for liquefying air and may be similarly used in the liquefaction of other gases. Values of the coefficient have been experimentally determined for many gases, including air and carbon dioxide, though the existing data are far from complete. Relatively little is known concerning the effect in the case of gas mixtures, as related to the values for the pure components.*

Reversible Process. A process proceeding in a manner such that it may be reversed at will by an infinitesimal change of the external conditions is designated a reversible process. In a reversible process the opposing forces differ only by a differential amount. For example, in the reversible expansion of a gas whose pressure at any moment is p, the pressure resisting the expansion is $p - \Delta p$; for reversible compression of the same gas, the impressed pressure is $p + \Delta p$. The reversible process, while never actually realizable, is important because it is the limit to which all actual processes can be made to approach as frictional and analogous losses are made smaller and smaller.

Isothermal Expansion and Compression. When a gas undergoes any change, the first law

$$\Delta U = Q - W \tag{67}$$

gives the relation between the increase in internal energy of the gas, the heat absorbed by the gas from the surroundings, and the work done by the gas on the surroundings. In the case of the isothermal reversible expansion of a gas from an initial volume V_A to a final volume V_B, the work done by the gas on the surroundings is

$$W = \int_{V_A}^{V_B} p\,dV \tag{68}$$

Then the heat absorbed by the gas from the surroundings during the isothermal reversible expansion is

$$Q = \Delta U\Big]_{V_A}^{V_B} - \int_{V_A}^{V_B} p\,dV \tag{69}$$

where $\Delta U\Big]_{V_A}^{V_B}$ must be evaluated from experiments similar to those discussed on p. 301, and the integration of the last term is performed by expressing p as a function of V or V as a function of p.

If the gas is ideal and there is 1 mole of it, then for the isothermal reversible expansion

$$\Delta U = 0 \tag{70}$$

$$pV = RT \tag{71}$$

$$W = \int_{V_A}^{V_B} p\,dV = RT \ln \frac{V_B}{V_A} \tag{72}$$

and

$$Q = W = RT \ln \frac{V_B}{V_A} = RT \ln \frac{p_A}{p_B} \tag{73}$$

For the ideal gas in the isothermal reversible expansion, the heat absorbed from the surroundings is therefore exactly equal to the work done on the surroundings.

For the isothermal reversible compression of a gas from an initial volume V_A to a final volume V_B, the work done by the gas on the surroundings and the heat absorbed by the gas from the surroundings are given by the same equations as for expansion, the only difference being that the final volume is now smaller than the initial volume and the signs of each of the quantities are reversed. Since V_B is smaller than V_A, both Q and W will be negative, *i.e.*, a positive quantity of heat is absorbed by the surroundings from the gas and a positive quantity of work is done on the gas by the surroundings. For the same values of V_A and V_B these quantities of energy are equal and opposite in sign to the corresponding quantities in expansion.

In the case of real gases under conditions such that their p-V-T relations cannot be represented with sufficient accuracy by the ideal gas equation $pV = nRT$, the calculation of the work of expansion or compression offers some difficulty because, for the integration, V must be expressed explicitly in terms of p, or vice versa, and, in those cases where adequate p-V-T data are available, the relations obtained are very complex. In these circumstances the calculation of the work of expansion or compression is more conveniently accomplished by graphical methods, which will be described later.

Because of frictional losses, etc., the actual work done by a gas on the surroundings in expansion will be somewhat less, and the actual work done on a gas by the surroundings in compression will be somewhat more, than that calculated by the above equations, the degree of departure depending upon how closely the limiting conditions of true reversibility are approached. Also, for the expansion or compression of the gas to be isothermal, it is necessary to add heat to the gas in expansion and to remove heat from the gas in compression in order to maintain the temperature constant. In actual practice, unless provision otherwise is deliberately made, the

* Joule-Thomson coefficients for gaseous mixtures of methane with ethane and with butane have been reported by Budenholzer, Sage, and Lacey [*Ind. Eng. Chem.*, **31**, 1288 (1939); **32**, 384 (1940)].

sudden expansion or compression of a gas is more nearly adiabatic, because the rate of heat transfer from or to the surroundings is usually not sufficiently rapid to maintain the temperature constant.

Adiabatic Expansion and Compression. In an adiabatic process, there is no heat absorbed or given off by the system, and hence $Q = 0$. If the only work is that of reversible expansion or compression, $dW = p\,dV$. Then, by the first law,

$$dU = -p\,dV \tag{74}$$

For any system undergoing a thermodynamic change, Eq. (61) holds. When applied to the ideal gas, this becomes

$$dU = C_V\,dT \tag{75}$$

Therefore, for the ideal gas undergoing adiabatic reversible expansion or compression,

$$C_V\,dT = -p\,dV \tag{76}$$

This equation cannot be integrated without knowing how C_V varies with T. Remembering that for the ideal gas $C_p - C_V = R$, placing $C_p/C_V = \kappa$ and taking the special case where C_V is constant, one obtains

$$TV^{R/C_V} = TV^{(\kappa-1)} = \text{const.} \tag{77}$$

Substituting pV/R for T, one obtains

$$pV^{C_P/C_V} = pV^{\kappa} = \text{const.} \tag{78}$$

If under the same conditions Eq. (76) is integrated between two given states, one obtains the relation

$$\frac{T_B}{T_A} = \left(\frac{V_B}{V_A}\right)^{-(\kappa-1)} = \left(\frac{p_B}{p_A}\right)^{\frac{(\kappa-1)}{\kappa}} \tag{79}$$

where the subscripts A and B refer to the initial and final states, respectively.

The work done by the ideal gas on adiabatic reversible expansion or compression is

$$W = \int p\,dV = -\int C_V\,dT \tag{80}$$

Taking C_V constant, one obtains

$$W = -C_V(T_B - T_A) = C_V(T_A - T_B) \tag{81}$$

where T_A and T_B are the initial and final temperatures.

The value of C_p/C_V for gases is discussed on p. 297.

Qualitatively one can see that the energy imparted to the system during adiabatic reversible compression must be taken up by an increase in the temperature of the gas. Conversely, the energy given to the surroundings during adiabatic reversible expansion must be supplied by a decrease in the temperature of the gas. The above equations relating to adiabatic reversible expansion or compression are derived on the assumption that the gas is ideal, so that $pV = nRT$ and, further, that C_V is constant. (This latter assumption means also constancy of C_p and κ because for the ideal gas $C_p - C_V = R$.) For real gases over limited ranges of temperature and pressures (especially low pressure where for the real gas pV approaches nRT) these assumptions are approximately true, and the corresponding equations may be used in the limited sense. Actually, however, over wide ranges of temperature and pressure, real gases depart considerably from these assumptions and the above equations are hardly applicable. For such cases graphical methods and temperature-entropy diagrams (see p. 331) are to be used.

Entropy and the Second Law of Thermodynamics. Whereas the first law of thermodynamics is concerned with the accounting of the various kinds of energy involved in a given process, the second law is concerned with the availability of the energy of a given system for doing useful work.

Every system, if left to itself, changes, rapidly or slowly, in such a way as to approach a definite final state of rest or equilibrium. Examples of processes tending toward equilibrium are as follows: (*a*) the diffusion of material from a concentrated to a dilute solution, leading to uniform concentration; (*b*) the passage of heat from a hot body to a cold body, leading to a uniform temperature; (*c*) the oxidation of organic substances by the oxygen of the air; (*d*) the running down of a mechanical clock. As these processes pass toward equilibrium, they lose some measure of their capacity for spontaneous change. Useful work can be extracted from any process that occurs spontaneously. The further removed a given system is from its state of equilibrium, the greater the amount of useful work that can be obtained from it when appropriately harnessed. The passage of a system toward its state of equilibrium does not necessarily mean that its intrinsic energy is decreasing. This energy may actually remain constant. What is being lost by the system is the availability of its energy for doing useful work. In connection with the second law, therefore, one might better speak, not of the dissipation or degradation of the energy, but rather of the degradation of the system as a whole as measured by its ability to perform useful work. The second law gives a quantitative measure of the useful work that can be obtained from any process.

Various statements of the second law have been made, of which two are as follows:

1. When any actual process occurs, it is impossible to invent a means of restoring to its original condition every system participating in the process.
2. Every system that is left to itself will, on the average, change toward a condition of maximum probability.

Changes in entropy are quantitatively measured as follows: In any reversible process, the increase in entropy of any participating system is equal to the heat absorbed by that system divided by its absolute temperature. That is, for a system participating in a process carried on reversibly,*

$$dS_i = \frac{dQ_i}{T_i} \tag{82}$$

where, for the system i, dS_i is the increase in entropy, dQ_i is the heat absorbed, and T_i is its absolute temperature.

From the nature of the definition of the measure of entropy, it follows that for the ideal reversible process the sum of all the changes in entropy for all the participating systems is equal to zero. That is, for the reversible process

$$\sum dS_i = \sum \frac{dQ_i}{T_i} = 0 \tag{83}$$

Since all actual processes are irreversible to a greater or lesser extent because of friction, resistance, etc., the truly reversible process is the ideal one toward which all actual processes may be made to approach as a limit.

For any process that occurs irreversibly, the sum of all the changes in entropy for all the participating systems is positive, which is to say that every process occurring spontaneously is accompanied by an increase in entropy. For the irreversible process, then

$$\Sigma\, dS_i > 0 \tag{84}$$

* As already mentioned, a truly reversible process is that ideal process in which all friction, electrical resistance, etc., are eliminated.

The entropy of a system depends only upon the variables that define its state, and the entropy change accompanying the passage of a system from one state to another is determined entirely by the two states and not by the path. That is, the difference in the entropy of a system in the states A and B, which is $S_B - S_A$, is independent of the manner in which the system passes from state A to state B. In order to determine the magnitude of the change in the entropy of the system $S_B - S_A$, one may utilize an ideal reversible process to return the system from state B to state A by means of a coupling of the system with the standard "spring-reservoir" system. This ideal reversible process is carried on in such a way that the change in entropy of the standard spring-reservoir system during the process is calculable. Since the process is a reversible one with only the two systems involved, then the total change in entropy is zero, and the change in entropy of the standard spring reservoir is equal to the negative value of the change in entropy of the given system as it is returned from state B to state A. $S_A - S_B$, of course, is the negative value of $S_B - S_A$.

Heat Engine. The purpose of a heat engine is to take a quantity of heat from a "hot" reservoir, convert as much as possible of it into work, and discharge the remainder into a "cold" reservoir. If T_1 and T_2 are the absolute temperatures of the hot and cold reservoirs, respectively, Q_1 is the heat absorbed by the engine from the hot reservoir, W is the work done by the engine, and Q_2 is the heat given by the engine to the cold reservoir, then, taking the engine as the system and considering a complete cycle of reversible changes, one may write by the first law

$$\Sigma\, dU = 0 \qquad \text{or} \qquad Q_1 - Q_2 - W = 0 \tag{85}$$

and by the second law

$$\Sigma\, dS = 0 \qquad \text{or} \qquad \frac{Q_1}{T_1} - \frac{Q_2}{T_2} = 0 \tag{86}$$

Solving the two equations for W/Q_1, which gives the fractional part of the heat energy taken by the engine from the hot reservoir that is convertible into useful work, one finds

$$\frac{W}{Q_1} = \frac{T_1 - T_2}{T_1} \tag{87}$$

This important equation, which was deduced by Carnot over one hundred years ago, gives the maximum amount of work obtainable from a given quantity of heat by an engine operating between the temperatures T_1 and T_2. For example, the so-called *maximum thermal efficiency* of any heat engine receiving heat at 250°C. and discharging it at 80°C. is equal to $(T_1 - T_2)/T_1 = 0.325$ or 32.5 per cent.

Refrigerating Machine. The purpose of a refrigerating machine is to absorb a given amount of work energy, absorb heat from a cold reservoir, and discharge heat to a hot reservoir. That is, the refrigerating machine utilizes work energy to transfer a given quantity of heat energy from a refrigerated space to the warmer surroundings. If T_1 and T_2 are the absolute temperatures of the hot and cold reservoirs, respectively, W is the work energy absorbed by the machine, Q_2 is the heat absorbed by the machine from the cold reservoir, and Q_1 is the heat given by the machine to the hot reservoir, then, taking the machine as the system and considering a complete cycle of reversible changes, one may write by the first law

$$\Sigma\, dU = 0 \qquad \text{or} \qquad W + Q_2 - Q_1 = 0 \tag{88}$$

and by the second law

$$\Sigma\, dS = 0 \qquad \text{or} \qquad \frac{Q_2}{T_2} - \frac{Q_1}{T_1} = 0 \tag{89}$$

Solving for W/Q_2, one obtains

$$\frac{W}{Q_2} = \frac{T_1 - T_2}{T_2} \tag{90}$$

This equation tells how much work is required to transfer Q_2 units of heat from a refrigerated space (cold reservoir) at T_2 to the surroundings (hot reservoir) at T_1. For example, a minimum of 9.2 units of electrical energy is required to transfer 100 units of heat energy from a refrigerated space at 0°C. to surroundings at 25°C.

Heating Machine. The purpose of a heating machine is to absorb a given amount of work energy, absorb heat from a cold reservoir, and discharge heat into a hot reservoir. If T_1 and T_2 are the absolute temperatures of the hot and cold reservoirs, respectively, W is the work energy absorbed by the machine, Q_2 is the heat absorbed by the machine from the cold reservoir, and Q_1 is the heat given by the machine to the hot reservoir, then, taking the machine as the system and considering a complete cycle of reversible changes, one may write by the first law

$$\Sigma\, dU = 0 \qquad \text{or} \qquad W + Q_2 - Q_1 = 0 \tag{91}$$

and by the second law

$$\Sigma\, dS = 0 \qquad \text{or} \qquad \frac{Q_2}{T_2} - \frac{Q_1}{T_1} = 0 \tag{92}$$

Solving for W/Q_1, one obtains

$$\frac{W}{Q_1} = \frac{T_1 - T_2}{T_1} \tag{93}$$

This equation tells how much work energy is required to transfer Q_1 units of heat from the surroundings (cold reservoir) at T_2 into a heated space (hot reservoir) at T_1. For example, a minimum of 7.5 units of electrical energy are required to "pump" 100 units of heat into a heated space at 22°C. when the surroundings are at 0°C.

System in a Reversible Process with Work Only of Expansion or Compression. When a system participates in a reversible process in which the energy changes involved are those of heat and of work of expansion or compression, the behavior of the system may be described by a simple but powerful equation deduced by application of the first and second laws to the process. If dU and dS are, respectively, the increments in internal energy and entropy of the system, dQ is the heat absorbed by the system from the surroundings, dW is the work done by the system on the surroundings, then by the first law

$$dU = dQ - dW = dQ - p\, dV \tag{94}$$

and by the second law

$$dS = \frac{dQ}{T} \tag{95}$$

Eliminating dQ from the two equations, one obtains

$$dU = T\, dS - p\, dV \tag{96}$$

This equation, the most important in all thermodynamics, describes the behavior of any system that is subjected to a reversible change in which the work is only that of expansion or compression.

Variation of Entropy with Temperature, Volume, and Pressure. For a system undergoing a reversible change, one may write, by the second law.

$$dS = \frac{dQ}{T} \tag{97}$$

But, by the definition of heat capacity,

$$C = \frac{dQ}{dT} \tag{98}$$

Therefore

$$dS = \frac{C}{T}\,dT \tag{99}$$

For a process at constant pressure

$$dS = \frac{C_p}{T}\,dT = C_p\,d\ln T \tag{100}$$

For a process at constant volume

$$dS = \frac{C_V}{T}\,dT = C_V\,d\ln T \tag{101}$$

When the heat capacity is known, either analytically or graphically as a function of the temperature, then Eq. (99) may be integrated to give the change in entropy as a given system is taken from one temperature to another.

From Eq. (96) one obtains the following relation, giving the variation of entropy with volume at constant temperature:

$$\left(\frac{\delta U}{\delta V}\right)_T = T\left(\frac{\delta S}{\delta V}\right)_T - p \tag{102}$$

or

$$\left(\frac{\delta S}{\delta V}\right)_T = \frac{1}{T}\left[p + \left(\frac{\delta U}{\delta V}\right)_T\right] \tag{103}$$

From Eq. (96) one obtains the following relation giving the variation of entropy with pressure at constant temperature:

$$\left(\frac{\delta U}{\delta p}\right)_T = T\left(\frac{\delta S}{\delta p}\right)_T - p\left(\frac{\delta V}{\delta p}\right)_T \tag{104}$$

or

$$\left(\frac{\delta S}{\delta p}\right)_T = \frac{1}{T}\left[p\left(\frac{\delta V}{\delta p}\right)_T + \left(\frac{\delta U}{\delta p}\right)_T\right] \tag{105}$$

or

$$\left(\frac{\delta S}{\delta p}\right)_T = \frac{1}{T}\left[-V + \left(\frac{\delta H}{\delta p}\right)_T\right] \tag{106}$$

Maximum Work. When a system takes part in any reversible process at constant temperature, one may write, from Eqs. (12) and (82),

$$dU = T\,dS - dW = d(TS) - dW \tag{107}$$

or

$$dW = -d(U - TS) \tag{108}$$

or

$$W = -\Delta(U - TS) \tag{109}$$

That is, the work done by the system on the surroundings during the process is determined entirely by the initial and final states of the system, being given by the decrement in $(U - TS)$. Because of its recurrence in thermodynamic problems, this combination of properties has been abbreviated for convenience to the symbol A. This property A is called the *maximum work* because its increment is the maximum work obtainable from a system participating in a reversible process at constant temperature. Therefore, for the reversible process at constant temperature,

$$W = -\Delta(U - TS) = -\Delta A \tag{110}$$

Maximum Useful Work or Free Energy. When a system takes part in a reversible process at constant temperature and constant pressure, one may write, from Eqs. (12) and (82),

$$dU = T\,dS - dW = T\,dS - p\,dV - dW' \tag{111}$$

Here dW is replaced by the two terms $p\,dV$ and dW', the first representing the work of expansion (or compression) against the prevailing constant pressure (which is usually the atmosphere) and the second term including all other forms of work energy performed by the system on the surroundings, which can be applied to useful purposes. Rewriting Eq. (111), one obtains

$$-dW' = dU + p\,dV - T\,dS \tag{112}$$

or

$$-dW' = dU + d(pV) - d(TS) \tag{113}$$

or

$$dW' = -d(U + pV - TS) \tag{114}$$

Integrating, one obtains

$$W' = -\Delta(U + pV - TS) = -\Delta(H - TS) \tag{115}$$

That is, the maximum useful work obtainable from a system participating in a reversible process at constant temperature and constant pressure is determined entirely by the initial and final states of the system and is given by the decrement of $(U + pV - TS)$. Because of its frequent recurrence in thermodynamics, this combination of properties, $U + pV - TS$, is abbreviated by using the symbol F. This property F is called the *free energy* because, for this kind of process, it measures the energy that is free to be put to useful purposes. Therefore, for the reversible process at constant temperature and constant pressure,

$$W' = -\Delta(U + pV - TS) = -\Delta(H - TS) = -\Delta F \tag{116}$$

Variation of the Free Energy with Pressure and Temperature. The change of free energy with pressure and temperature is given by the general differential equation

$$dF = \left(\frac{\delta F}{\delta p}\right)_T dp + \left(\frac{\delta F}{\delta T}\right)_p dT \tag{117}$$

From the definition of the free energy,

$$F = U + pV - TS \tag{118}$$

one may obtain by differentiation

$$dF = dU + p\,dV + V\,dp - T\,dS - S\,dT \tag{119}$$

Remembering that $dU = T\,dS - p\,dV$ (from Eq. (96)), one obtains

$$dF = V\,dp - S\,dT \tag{120}$$

which describes the complete change in free energy of a system that is participating in a reversible process in which the only work is that of expansion or compression.

Comparing Eqs. (117) and (120), one sees that

$$\left(\frac{\delta F}{\delta p}\right)_T = V \tag{121}$$

and

$$\left(\frac{\delta F}{\delta T}\right)_p = -S \tag{122}$$

Equation (121) shows that the change of the free energy of the system with pressure at constant temperature is equal to its volume, and Eq. (122) shows that the change of the free energy of the system with temperature at constant pressure is equal to the negative of its entropy. When a system changes from a pressure A to a pressure

B, at constant temperature, one may write, from Eq. (121),

$$F_B - F_A = \int_{p_A}^{p_B} V\,dp \qquad (123)$$

Likewise, for the change from a temperature A to a temperature B at constant pressure, from Eq. (122), one has

$$F_B - F_A = -\int_{T_A}^{T_B} S\,dT \qquad (124)$$

It is useful to write Eq. (122) in another form by substituting $(H - F)/T$ for S, according to the definition of F, obtaining

$$\left(\frac{\delta F}{\delta T}\right)_p = -S = \frac{F - H}{T} \qquad (125)$$

Equation (125) may be rearranged to give

$$\left[\frac{\delta\left(\frac{F}{T}\right)}{\delta T}\right]_p = -\frac{H}{T^2} \qquad (126)$$

or

$$\left[\frac{\delta\left(\frac{F}{T}\right)}{\delta\left(\frac{1}{T}\right)}\right]_p = H \qquad (127)$$

Equation (126) shows that the change of F/T with temperature at constant pressure is equal to the negative of H/T^2. Equation (127) shows that the change of F/T with $1/T$ is equal to the heat content.

Galvanic Cells.* For any process taking place in a galvanic cell operating reversibly at constant temperature and pressure, the maximum useful work obtainable from the system is, as for any reversible process at constant temperature and pressure,

$$W' = -\Delta(U + pV - TS) = -\Delta(H - TS) = -\Delta F \qquad (128)$$

But the maximum useful work obtainable from such a galvanic cell is also equal to the quantity of electricity transferred through the cell multiplied by the potential difference through which the electricity is carried. If n is the number of chemical equivalents per mole of reaction in the cell, $\mathfrak{F}$ is the Faraday constant giving the quantity of electricity associated with one chemical equivalent, and $\mathcal{E}$ is the electromotive force of the cell, then per mole of reaction the quantity of electricity is $n\mathfrak{F}$ and the useful work obtainable from the cell is $n\mathfrak{F}\mathcal{E}$. We may write, therefore, that

$$W' = n\mathfrak{F}\mathcal{E} \qquad (129)$$

From Eqs. (90) and (91), it follows that

$$\Delta F = -n\mathfrak{F}\mathcal{E} \qquad (130)$$

For any galvanic cell to be significant thermodynamically, its e.m.f. must be constant and reproducible and the reaction actually taking place in the cell must be identified.

From Eqs. (125), (126), and (130) one may obtain the following relations giving, for the actual reaction that occurs in the cell, at constant pressure, the values of the increments of the entropy and heat content in terms of the e.m.f. and its temperature coefficient:

$$\Delta S = n\mathfrak{F}\left(\frac{\delta\mathcal{E}}{\delta T}\right)_p \qquad (131)$$

$$\Delta H = n\mathfrak{F}\left[-\mathcal{E} + T\left(\frac{\delta\mathcal{E}}{\delta T}\right)_p\right] \qquad (132)$$

* See also Sec. 28, p. 1788.

From Eqs. (121) and (130) one may obtain the relation giving the variation of the e.m.f. of the cell with pressure at constant temperature,

$$\left(\frac{\delta\mathcal{E}}{\delta p}\right)_T = \frac{\Delta V}{n\mathfrak{F}} \qquad (133)$$

where ΔV is the increment in volume for the reaction actually occurring in the cell.

Criteria of Equilibrium. When a given reaction proceeds in a reversible manner,

$$dU = T\,dS - p\,dV - dW' \qquad (134)$$

where, as before, W' includes all the work other than that of expansion or compression and represents the maximum useful work obtainable from the system. Rewriting Eq. (134), we have

$$dW' = -(dU + p\,dV - T\,dS) \qquad (135)$$

Now useful work can be obtained from the system only if it is removed from its state of equilibrium. In this case the process is a natural or spontaneous one and

$$dW' \text{ is positive} \quad \text{or} \quad dW' > 0 \qquad (136)$$

If the system is already in its state of equilibrium, then no work can be obtained from it, and

$$dW' = 0 \qquad (137)$$

The third situation is one in which energy is supplied to the system in order to move it away from its state of equilibrium. For this unnatural or not spontaneous process

$$dW' \text{ is negative} \quad \text{or} \quad dW' < 0 \qquad (138)$$

The general criterion of equilibrium is, then, that

$$dW' = -(dU + p\,dV - T\,dS) = 0 \qquad (139)$$

At constant pressure and constant temperature

$$dU + p\,dV - T\,dS = d(U + pV) - d(TS) = d(H - TS) = dF = 0 \qquad (140)$$

At constant volume and constant temperature

$$dU + p\,dV - T\,dS = dU - d(TS) = d(U - TS) = dA = 0 \qquad (141)$$

At constant pressure and constant entropy

$$dU + p\,dV - T\,dS = dU + d(pV) = d(U + pV) = dH = 0 \qquad (142)$$

At constant volume and constant entropy

$$dU + p\,dV - T\,dS = dU = 0 \qquad (143)$$

These criteria of equilibrium may be summarized thus:

At constant p and T, $dF = 0$ (144)
At constant V and T, $dA = 0$ (145)
At constant p and S, $dH = 0$ (146)
At constant V and S, $dU = 0$ (147)

Because most processes that we encounter are those at constant pressure and temperature, the most widely used criterion of equilibrium is that ΔF is zero. When ΔF is negative, one can say that, for constant p and T, the reaction will tend to proceed spontaneously or naturally in the direction indicated; whereas, when ΔF is positive, it will tend to proceed in the reverse direction.

Fugacity of Gases. Combination of Eq. (121) with that of the ideal gas yields the following relation, giving for 1 mole of the ideal gas (or of any gas in the ideal state) the variation of the free energy with pressure at constant temperature:

$$dF_{\text{ideal}} = V_{\text{ideal}} dp = RT d \ln p \qquad (148)$$

Integrating between the standard or reference state of one atmosphere (see section on standard states) and p atmospheres, and letting the superscript zero indicate the standard or reference state, one has for the ideal gas

$$F_{\text{ideal}} - F^0_{\text{ideal}} = RT \ln p \qquad (149)$$

For real gases, however, V is not exactly equal to RT/p, and the integration cannot be performed as simply as for the ideal gas. From appropriate p-V data at a given temperature one can tabulate values of the quantity α, defined as

$$\alpha = V_{\text{ideal}} - V_{\text{real}} = \frac{RT}{p} - V \qquad (150)$$

One may write then

$$d(F_{\text{real}} - F_{\text{ideal}}) = (V_{\text{real}} - V_{\text{ideal}})\,dp = -\alpha\,dp \qquad (151)$$

Integrating between the limits of $p = p$ and $p = 0$, and remembering that at zero pressure $(F_{\text{real}} - F_{\text{ideal}})$ is zero, one obtains

$$F_{\text{real}} - F_{\text{ideal}} = -\int_0^p \alpha\,dp \qquad (152)$$

But from Eq. (149)

$$F_{\text{ideal}} = F^0_{\text{ideal}} + RT \ln p \qquad (153)$$

Hence

$$F_{\text{real}} - F^0_{\text{ideal}} = RT \ln p - \int_0^p \alpha\,dp \qquad (154)$$

or, removing the subscripts,

$$F - F^0 = RT \ln p - \int_0^p \alpha\,dp \qquad (155)$$

As will be seen later, this is an inconvenient expression to use in thermodynamic equations involving the equilibrium constant, and, in order to retain the simplicity of the form of Eq. (149), a new function, f, the fugacity, is defined as follows:

$$RT \ln f = RT \ln p - \int_0^p \alpha\,dp \qquad (156)$$

Therefore Eq. (155) becomes

$$F - F^0 = RT \ln f \qquad (157)$$

Fugacity is thus seen to be a thermodynamic function (resembling a "corrected" pressure) that permits one to use thermodynamic equations in the simple form they possess when the system consists of ideal gases. Fugacity is the real measure of the escaping tendency of any component from a given system.

Fugacity of Liquids and Solids. The fugacity or escaping tendency of any liquid or solid substance (or of any component of any solution) is equal to the fugacity of the given substance in the vapor phase that is in equilibrium with the liquid or solid (or with the solution). Consider the equilibrium

$$X\ (l) = X\ (g) \qquad (158)$$

Now

$$F\ (g) = F^0\ (g) + RT \ln f\ (g) \qquad (159)$$

At equilibrium

$$F\ (g) = F\ (l) \quad \text{and} \quad f\ (g) = f\ (l) \qquad (160)$$

Hence

$$F\ (l) = F^0\ (g) + RT \ln f\ (l) \qquad (161)$$

That is, the free energy of the substance in the condensed phase is equal to the free energy of that substance in the standard or reference state in the gas phase plus RT times the natural logarithm of the fugacity of that substance in the condensed phase (which at equilibrium is equal to the fugacity of that substance in the gas phase).

Activity. In order to avoid a situation such as occurs in Eq. (161), where the free energy of the substance in the liquid phase is referred to a standard state in the gas phase, it is convenient to invent a new thermodynamic function called the *activity*, which permits one to use a standard state in the same phase. The activity is defined as the ratio of the fugacity of the substance in the given state to the fugacity in a selected standard or reference state in the same phase. For the case considered above, one has, as before,

$$F\ (l) = F^0\ (g) + RT \ln f\ (l) \qquad (162)$$

But, by the definition of activity,

$$a\ (l) = \frac{f\ (l)}{f^0\ (l)} \qquad (163)$$

Now

$$F^0\ (l) = F^0\ (g) + RT \ln f^0\ (l) \qquad (164)$$

Subtracting Eq. (164) from Eq. (162), one obtains

$$F\ (l) - F^0\ (l) = RT \ln \frac{f\ (l)}{f^0\ (l)} \qquad (165)$$

Substituting from Eq. (163), one has

$$F\ (l) - F^0\ (l) = RT \ln a\ (l) \qquad (166)$$

That is, the activity of a substance in a certain condition in a given phase is the ratio of the fugacity of that substance in the given condition to the fugacity of that substance in its standard or reference state in the same phase.

Activity Coefficient. In aqueous solutions the fugacity and molality in the very dilute range of concentration (see the article on Real Solutions, p. 308) bear the relation

$$f_2 = km \qquad (167)$$

This is the ideal line along which one extrapolates from a dilute real solution to the hypothetical standard or reference state of unit molality. When $m = 1$,

$$f_2 = k = f_2^0 \qquad (168)$$

and therefore in the very dilute range of concentration where Eq. (167) is valid, the activity of the solute is

$$a_2 = \frac{f_2}{f_2^0} = m \qquad (169)$$

The departure from equality of a_2 and m, or the departure from unity of the ratio a_2/m, is indicative of the departure of the solution from ideality. In problems involving aqueous solutions, it has been found convenient to use a new function, γ, called the *activity coefficient*, which is defined as the ratio of a_2 to m:

$$\gamma = \frac{a_2}{m} \qquad (170)$$

In the formulas involving activities in aqueous solutions, one may substitute for a_2 the product γm. The value of γ is, of course, unity over all the range of concentration in which Eq. (169) is valid.

Standard States. In the application of the principles of chemical thermodynamics to practical problems, it is convenient to select certain standard or reference states to which may be referred the fugacity or free energy of a substance in a given phase.

For the gaseous phase the standard state is taken as that of the ideal gas at a fugacity of 1 atm. At any given temperature the real gas can be brought to this standard state by expanding the gas to zero pressure along its own p-V isotherm and then compressing it to 1 atm. along the ideal gas isotherm of $pV = RT$. In the gas phase then $f^0 = 1$, and

$$a = \frac{f}{f^0} = f \tag{171}$$

That is, the activity of a gas is numerically equal to its fugacity. As has already been pointed out, the fugacity of a gas at low pressure is nearly equal to the pressure, and for any pressure the fugacity may be calculated from the appropriate p-V data at the given temperature by means of Eq. (156) (see also pp. 353–356).

For pure liquids and solids the standard state is taken as that of the pure liquid or solid at a pressure of 1 atm.

For aqueous solutions in the liquid phase the standard state for the solvent, water, is taken as that of pure liquid water at the given temperature; and the standard state for the solute is taken as that of the solute in a hypothetical one-molal solution. The hypothetical one-molal solution is that solution that has all the properties of the infinitely dilute solution except the property of concentration, which is one molal. A real solution may, theoretically, be taken to the hypothetical one-molal state by first diluting the real solution to infinite dilution and then bringing it back to the hypothetical one-molal solution along the ideal line, which is fixed by the relation of the fugacity to molality of the given component in the very dilute range of concentration.

For non-aqueous solutions in the liquid phase, the standard state for the solvent may be taken as that of the pure liquid solvent at the same temperature; and the standard state for the solute may be taken as that of the solute in the hypothetical standard state of unit mole fraction of solute, which is fixed by the relation of the fugacity to mole fraction in the very dilute range of concentration. An alternative plan is to take the standard state for each component to be that of the pure liquid state of that component at the given temperature.

Ideal Solution. The ideal solution is defined as one in which, for any given component i, the fugacity f_i is equal to the mole fraction x_i multiplied by the fugacity of the component i in the pure state in the same phase at the same temperature. That is, for any component i,

$$f_i = x_i f_i^0 \tag{172}$$

Since the activity is defined as

$$a_i = \frac{f_i}{f_i^0} \tag{173}$$

one sees that in the ideal solution the activity of any component is always equal to its mole fraction:

$$a_i = x_i \tag{174}$$

When the vapor phase behaves sufficiently like the ideal gas for the fugacity of the gas to be replaced by its partial pressure, the above equations assume the forms

$$p_i = x_i p_i^0 \tag{175}$$

$$a_i = \frac{p_i}{p_i^0} \tag{176}$$

From the definition of the ideal solution, it can be shown that the partial molal volume and partial molal heat content of any component are constant for all concentrations and are equal, respectively, to the volume and heat content of the given component in the pure state in the same phase at the same temperature. That is, there is no change in volume or heat content when the components are mixed to form an ideal solution in the same phase at the same temperature.

Figure 28, p. 577, shows graphically the relation between the mole fraction, the partial pressure of each component, and the total pressure, for any ideal binary solution at given temperatures.

From Eqs. (125), (130), (161), and (176) one may derive the following equations describing the behavior of an ideal liquid solution of two components in equilibrium with the solid phase of one of them (i):

$$\left(\frac{\delta \ln x_i}{\delta p}\right)_T = -\frac{\Delta V_i}{RT} \tag{177}$$

$$\left(\frac{\delta \ln x_i}{\delta T}\right)_p = \frac{\Delta H_i}{RT^2} \tag{178}$$

In Eq. (177), ΔV_i is for that component that exists in the two phases the volume of the given component in the liquid phase less its volume in the solid phase. In Eq. (178), ΔH_i is for that component that exists in the two phases the heat content of the given component in the liquid phase less that in the solid phase, *i.e.*, the heat of solution of the given component. Equation (177) describes the change of solubility with pressure at constant temperature; Eq. (178) describes the change of solubility with temperature at constant pressure.

Real Solutions. It has been found experimentally that real solutions in the dilute range of concentration follow the relation*

$$f_2 = k_2 x_2 \tag{179}$$

That is, in the dilute range of concentration, the fugacity of the solute is proportional to its mole fraction (though the proportionality factor is not necessarily the fugacity of the solute in the pure state in the same phase at the same temperature). From Eqs. (174) and (179) one can then show that, for this dilute solution, the fugacity of the solvent is equal to its mole fraction multiplied by the fugacity of the solvent in the pure state in the same phase at the same temperature, *i.e.*, for the solvent in the dilute solution,

$$f_1 = x_1 f_1^0 \tag{180}$$

Equation (179) is substantially Henry's law of the dilute solution; and Eq. (180) is substantially Raoult's law of the lowering of the vapor pressure of the solvent in a dilute solution. From Eqs. (126), (157), and (180) one may derive the equation that is substantially the van't Hoff law of the lowering of the freezing point of the solvent in a dilute solution,

$$\left(\frac{\delta T}{\delta x_2}\right)_p = -\frac{RT^2}{\Delta H} \tag{181}$$

In Eq. (181), ΔH is the heat of fusion of the solvent in the dilute solution.

Partial Molal Properties. One of the important fields of application of the laws of thermodynamics is that of solutions. In such problems it is desirable to know thermodynamically how a given property of a solution changes with change in the amount of one of the components, all the other variables and the amount of the other components being kept constant. The partial molal property $\bar{G}_1$ for component 1 in a given solution is

* Following the convention of Lewis and Randall, we use the subscripts 1 and 2 to refer to given properties of the solvent and solute, respectively, in a binary solution.

defined as

$$\bar{G}_1 = \left(\frac{\delta G}{\delta n_1}\right)_{p,T,n_2,n_3,\ldots} \quad (182)$$

where G is the value of the property for the entire solution, and the pressure, temperature, and the amounts of all the other components are kept constant as the number of moles of component 1 is changed. To facilitate the thermodynamic study of such systems, the following partial molal equations have been developed:

$$G = n_1\bar{G}_1 + n_2\bar{G}_2 + \cdots \quad (183)$$

$$n_1\,d\bar{G}_1 + n_2\,d\bar{G}_2 + \cdots = 0 \quad (184)$$

These equations, together with the first and second laws of thermodynamics, have been used to derive a number of important equations describing the behavior of solutions. All the thermodynamic equations previously derived for pure substances or for complete systems are applicable to the components of solutions. Examples are the following:

$$\bar{C}_{p1} = \left(\frac{\delta \bar{H}_1}{\delta T}\right)_p; \quad \left(\frac{\delta \bar{S}_1}{\delta T}\right)_p = \frac{\bar{C}_{p1}}{T} \quad (185)$$

$$\left(\frac{\delta \bar{F}_1}{\delta p}\right)_T = \bar{V}_1; \quad \left(\frac{\delta \bar{F}_1}{\delta T}\right)_p = -\bar{S}_1 \quad (186)$$

$$\left[\frac{\delta\left(\frac{\bar{F}_1}{T}\right)}{\delta T}\right]_p = -\frac{\bar{H}_1}{T^2}; \quad \left[\frac{\delta\left(\frac{\bar{F}_1}{T}\right)}{\delta\left(\frac{1}{T}\right)}\right]_p = \bar{H}_1 \quad (187)$$

Equilibrium Constant and the Standard Free Energy Change. Consider the reaction

$$bB + cC = mM + nN \quad (188)$$

taking place under certain conditions R. Then the increment in free energy is

$$\Delta F^R = mF_M{}^R + nF_N{}^R - bF_B{}^R - cF_C{}^R \quad (189)$$

Likewise, for the reaction taking place under certain conditions S, the increment in free energy is

$$\Delta F^S = mF_M{}^S + nF_N{}^S - bF_B{}^S - cF_C{}^S \quad (190)$$

Subtracting Eq. (190) from Eq. (189) gives the difference in the free energy increments for the two sets of conditions,

$$\Delta F^R - \Delta F^S = m(F_M{}^R - F_M{}^S) + n(F_N{}^R - F_N{}^S) - b(F_B{}^R - F_B{}^S) - c(F_C{}^R - F_C{}^S) \quad (191)$$

By applying the relation given by Eq. (166) for the change in free energy of a given substance between two states, as for substance M between the states R and S, one obtains

$$F_M{}^R - F_M{}^S = RT \ln (a_M{}^R)/(a_M{}^S) \quad (192)$$

Substituting appropriate equations of the form of Eq. (192) into Eq. (191), one obtains

$$\Delta F^R - \Delta F^S = mRT \ln (a_M{}^R)/(a_M{}^S) + nRT \ln (a_N{}^R)/(a_N{}^S) - bRT \ln (a_B{}^R)/(a_B{}^S) - cRT \ln (a_C{}^R)/(a_C{}^S) \quad (193)$$

Rearranging the terms on the right, one obtains

$$\Delta F^R - \Delta F^S = RT \ln \frac{(a_M{}^R)^m (a_N{}^R)^n}{(a_B{}^R)^b (a_C{}^R)^c} - RT \ln \frac{(a_M{}^S)^m (a_N{}^S)^n}{(a_B{}^S)^b (a_C{}^S)^c} \quad (194)$$

The two logarithmic terms on the right are the proper activity quotients for reaction (188). If the states R are taken to be the states at which the given reactants and products of reaction are in equilibrium at constant temperature and pressure, then

$$\Delta F^R = 0 \quad \text{and} \quad \frac{(a_M{}^R)^m (a_N{}^R)^n}{(a_B{}^R)^b (a_C{}^R)^c} = K \quad (195)$$

where K is the equilibrium constant for the given reaction, expressed in terms of activities (see p. 307). If the states S are taken to be the standard states at which the given reactants and products each have unit activity, then $a_M{}^S = 1$, $a_N{}^S = 1$, etc., and

$$RT \ln \frac{(a_M{}^S)^m (a_N{}^S)^n}{(a_B{}^S)^b (a_C{}^S)^c} = 0 \quad \text{also} \quad \Delta F^S = \Delta F^0 \quad (196)$$

Equation (194) then becomes

$$\Delta F^0 = -RT \ln K \quad (197)$$

which is the familiar equation relating the standard free energy change with the equilibrium constant.

It should be emphasized that the foregoing procedure follows Lewis and Randall, in whose system there is only one kind of equilibrium constant. The following examples will illustrate the method of setting up the equilibrium constant K for different kinds of equilibrium.

Example 1. Reaction between gases and pure liquids and solids:

$$bB\ (g) + dD\ (l) = eE\ (g) + gG\ (s) \quad (198)$$

$$K = \frac{(a_E)^e (a_G)^g}{(a_B)^b (a_D)^d} \quad (199)$$

For all substances (see pp. 307–308)

$$a_i = \frac{f_i}{f_i{}^0} \quad (200)$$

where $f_i{}^0$ is the fugacity of the given substance in its standard state. For gases $f_i{}^0 = 1$ atm., and $a_i = f_i/f_i{}^0 = f_i$; and for liquids and solids the standard state is the pure liquid or solid, so that $f_i/f_i{}^0 = 1 = a_i$. Therefore for the above reaction

$$a_D = 1 \quad \text{and} \quad a_G = 1 \quad (201)$$

If the gases are such that the fugacity may be replaced by the partial pressure, then

$$a_B = f_B = p_B \quad (202)$$

$$a_E = f_E = p_E \quad (203)$$

Then, for the reaction given by Eq. (199),

$$K = \frac{(p_E)^e}{(p_B)^b} \quad (204)$$

Example 2. Reaction between gases and aqueous liquid solutions where the concentration is expressed in molality:

$$bB\ (g) + dD\ (\text{aq soln.}, m_D) = eE\ (g) + gG\ (\text{aq soln.}, m_G) \quad (205)$$

For the solute in aqueous solutions of non-electrolytes, the standard state is taken as the hypothetical solution of unit molality, so that, if the solution is sufficiently ideal for the fugacity to be proportional to the molality, then

$$a_i = \frac{f_i}{f_i{}^0} = m_i \quad (206)$$

Therefore

$$a_D = m_D \quad \text{and} \quad a_G = m_G \quad (207)$$

And

$$K = \frac{(p_E)^e (m_G)^g}{(p_B)^b (m_D)^d} \quad (208)$$

Example 3. Reaction in a solution where the concentrations are measured in terms of mole fractions:

$$bB\ (\text{soln.}\ x_B) = dD\ (\text{soln.}, x_D) \quad (209)$$

For a solution when the concentrations are measured in mole fractions, the standard state for each component may be taken as the pure component in the liquid state, and, if the solution is sufficiently ideal for the fugacity of any component to be proportional to its mole fraction, one may write

$$a_i = \frac{f_i}{f_i^0} = x_i \tag{210}$$

so that

$$K = \frac{(x_D)^d}{(x_B)^b} \tag{211}$$

Example 4. Reaction involving gas, liquid, solid, and liquid solution phases:

$$bB\ (\mathrm{g}) + dD\ (\mathrm{l}) = eE\ (\mathrm{soln.},\ x_E) + gG\ (\mathrm{s}) \tag{212}$$

$$K = \frac{(x_E)^e}{(p_B)^b} \tag{213}$$

Example 5. Reaction in the gaseous phase at constant volume, where the concentration is given in moles, n, in the total volume, V:

$$bB\ (\mathrm{g}) + dD\ (\mathrm{g}) = eE\ (\mathrm{g}) + gG\ (\mathrm{g}) \tag{214}$$

Let $C_i = n_i/V$, the concentration in moles per liter. Then, if the ideal gas laws may be assumed to hold,

$$a_B = \frac{f_B}{f_B^0} = f_B = p_B = \frac{n_B}{V}RT = C_BRT \tag{215}$$

$$a_D = \frac{f_D}{f_D^0} = f_D = p_D = \frac{n_D}{V}RT = C_DRT \tag{216}$$

$$a_E = \frac{f_E}{f_E^0} = f_E = p_E = \frac{n_E}{V}RT = C_ERT \tag{217}$$

$$a_G = \frac{f_G}{f_G^0} = f_G = p_G = \frac{n_G}{V}RT = C_GRT \tag{218}$$

Then

$$K = \frac{(a_E)^e(a_G)^g}{(a_B)^b(a_D)^d} = \frac{(p_E)^e(p_G)^g}{(p_B)^b(p_D)^d} = \frac{(C_ERT)^e(C_GRT)^g}{(C_BRT)^b(C_DRT)^d} = \left[\frac{(C_E)^b(C_G)^g}{(C_B)^b(C_D)^d}\right](RT)^{e+g-b-d} \tag{219}$$

When a given reaction can be resolved into several other reactions whose equilibrium constants or standard free energy changes are known, these values for the separate reactions may be combined to give that for the given reaction. For example, consider the water-gas reaction

$$CO\ (\mathrm{g}) + H_2O\ (\mathrm{g}) = CO_2\ (\mathrm{g}) + H_2\ (\mathrm{g}) \tag{220}$$

for which

$$K = \frac{p_{CO_2}p_{H_2}}{p_{CO}p_{H_2O}} \tag{221}$$

and

$$\Delta F^0 = \Delta F^0_{\mathrm{Eq.}\ (220)} \tag{222}$$

The water-gas reaction may be obtained from the following two reactions:

$$CO\ (\mathrm{g}) + \tfrac{1}{2}O_2\ (\mathrm{g}) = CO_2\ (\mathrm{g}) \tag{223}$$

for which

$$K = \frac{p_{CO_2}}{p_{CO}p_{O_2}^{1/2}} \tag{224}$$

$$\Delta F^0 = \Delta F^0_{\mathrm{Eq.}\ (223)} \tag{225}$$

and

$$H_2O\ (\mathrm{g}) = H_2\ (\mathrm{g}) + \tfrac{1}{2}O_2\ (\mathrm{g}) \tag{226}$$

for which

$$K = \frac{p_{H_2}p_{O_2}^{1/2}}{p_{H_2O}} \tag{227}$$

$$\Delta F^0 = \Delta F^0_{\mathrm{Eq.}\ (226)} \tag{228}$$

Combination of Eqs. (223) and (226) gives Eq. (220), and likewise

$$\Delta F^0_{\mathrm{Eq.}\ (220)} = \Delta F^0_{\mathrm{Eq.}\ (223)} + \Delta F^0_{\mathrm{Eq.}\ (226)} \tag{229}$$

and

$$RT \ln K_{\mathrm{Eq.}\ (220)} = RT \ln K_{\mathrm{Eq.}\ (223)} + RT \ln K_{\mathrm{Eq.}\ (226)} \tag{230}$$

or

$$K_{\mathrm{Eq.}\ (220)} = (K_{\mathrm{Eq.}\ (223)})(K_{\mathrm{Eq.}\ (226)}) = \frac{p_{CO_2}p_{H_2}}{p_{CO}p_{H_2O}} \tag{231}$$

In a similar way values of K and ΔF^0 for many other reactions may be combined. This procedure makes it possible to determine values for K and ΔF^0 for many reactions for which these constants cannot be determined directly.

Development of the Free Energy Equation for a Reaction over a Given Range of Temperature. From Eq. (126), one can write for any given reaction at constant pressure:

$$\frac{d(\Delta F^0/T)}{dT} = -\frac{\Delta H^0}{T^2} \tag{232}$$

When the reactants and products of the given reaction are substances for which the heat capacity can be simply expressed as a function of the temperature over a given range of temperature, ΔH^0 can be expressed as a function of the temperature, as derived on p. 299, Eqs. (41) to (46) of this section:

$$\Delta H^0 = \Delta H_*{}^0 + (\Delta a)T + \tfrac{1}{2}(\Delta b)T^2 + \tfrac{1}{3}(\Delta c)T^3 + \cdots \tag{233}$$

As pointed out in the derivation, this equation is valid only for the range of temperature for which the equations of heat capacity are valid. Further, it is important to remember that $\Delta H_*{}^0$ is a constant of integration and is the value of ΔH^0 given by this equation when T has the value zero. $\Delta H_*{}^0$ is not $\Delta H_0{}^0$, the heat of the given reaction at 0°K. Substituting into Eq. (232) the value of ΔH^0 given by Eq. (233) and integrating, one obtains

$$\Delta F^0 = \Delta H_*{}^0 - (\Delta a)T \ln T - \tfrac{1}{2}(\Delta b)T^2 - \tfrac{1}{6}(\Delta c)T^3 + \cdots + IT \tag{234}$$

In this equation there are five constants, of which three, Δa, Δb, and Δc are evaluated from data on heat capacities, one $\Delta H_*{}^0$, is evaluated from the heat of reaction at some one temperature in the given range, and the other, I, is evaluated from the value of the standard free energy increment, ΔF^0, or the equilibrium constant, K, at some one temperature. Since, by Eq. (197),

$$\Delta F^0 = -RT \ln K \tag{235}$$

Eq. (234) may be expressed explicitly in terms of $R \ln K$, as follows:

$$R \ln K = -\frac{\Delta H_*{}^0}{T} + (\Delta a) \ln T + \frac{1}{2}(\Delta b)T + \frac{1}{6}(\Delta c)T^2 + \cdots - I \tag{236}$$

Equations (234) and (236) serve to evaluate ΔF^0 or K for the given reaction at any temperature within the range for which the equations are valid.

The following example illustrates the development of such a free energy equation:

Example. For the water-gas reaction

$$CO_2\ (\mathrm{g}) + H_2\ (\mathrm{g}) = CO\ (\mathrm{g}) + H_2O\ (\mathrm{g}) \tag{237}$$

$$\Delta H^0_{298.16} = 9838\ \mathrm{cal./mole} \tag{238}$$

$$\Delta F^0_{298.16} = 6817\ \mathrm{cal./mole} \tag{239}$$

The following heat capacity equations are given for the range 273° to 1500°K:

CO_2 (g); $C_p = 6.3957 + 10.1933 \times 10^{-3}T - 35.333 \times 10^{-7}T^2$ cal./(deg.)(mole) (240)

H_2 (g); $C_p = 6.9469 - 0.1999 \times 10^{-3}T + 4.808 \times 10^{-7}T^2$ cal./(deg.)(mole) (241)

CO (g); $C_p = 6.3424 + 1.8363 \times 10^{-3}T - 2.801 \times 10^{-7}T^2$ cal./(deg.)(mole) (242)

H_2O (g); $C_p = 7.1873 + 2.3733 \times 10^{-3}T + 2.084 \times 10^{-7}T^2$ cal./(deg.)(mole) (243)

Derive the equation giving ΔF^0 and $R \ln K$ for the water-gas reaction as a function of the temperature in the range 273° to 1500°K.

Solution. Appropriate combination of Eqs. (240) to (243) yields for the water-gas reaction [Eq. (237)]

$$\Delta C_p = 0.1871 - 5.7838 \times 10^{-3}T + 29.808 \times 10^{-7}T^2 \text{ cal./(deg.)(mole)} \quad (244)$$

Substituting this value of ΔC_p into the equation,

$$d(\Delta H) = \Delta C_p \, dT \quad (245)$$

and integrating, one obtains

$$\Delta H^0 = \Delta H_*{}^0 + 0.1871T - 2.8919 \times 10^{-3}T^2 + 9.936 \times 10^{-7}T^3 \text{ cal./mole} \quad (246)$$

When $T = 298.16$, $\Delta H^0 = 9838$ cal./mole. Substituting these values into Eq. (246) and solving for $\Delta H_*{}^0$, one obtains

$$\Delta H_*{}^0 = 10{,}013 \text{ cal./mole} \quad (247)$$

Therefore

$$\Delta H^0 = 10{,}013 + 0.1871T - 2.8919 \times 10^{-3}T^2 + 9.936 \times 10^{-7}T^3 \text{ cal./mole} \quad (248)$$

Substituting this value of ΔH^0 into Eq. (232) and integrating, one obtains

$$\frac{\Delta F^0}{T} - \frac{10{,}013}{T} = 0.1871 \ln T + 2.8919 \times 10^{-3}T - 4.968 \times 10^{-7}T^2 + I \quad (249)$$

When $T = 298.16$, $\Delta F^0 = 6817$ cal./mole. Substituting these values and solving for I, one obtains

$$I = -10.47 \quad (250)$$

Therefore,

$$\Delta F^0 = 10{,}013 - 0.1871T \ln T + 2.8919 \times 10^{-3}T^2 - 4.968 \times 10^{-7}T^3 - 10.47 \text{ cal./mole} \quad (251)$$

or, since

$$\Delta F^0 = -RT \ln K \quad (252)$$

$$R \ln K = -\frac{10{,}013}{T} + 0.1871 \ln T - 2.8919 \times 10^{-3}T + 4.968 \times 10^{-7}T^2 + 10.47 \quad (253)$$

In terms of $\log_{10} K$, which is $2.3026 \ln K$, with $R = 1.98718$ cal./(deg.)(mole), this equation becomes

$$\log_{10} K = -\frac{2188}{T} + 0.0941 \log_{10} T - 0.6320 \times 10^{-3}T + 1.086 \times 10^{-7}T^2 + 2.288 \quad (254)$$

These equations are valid for the range for which the heat capacity equations are valid, *viz.*, 273° to 1500°K.

Effect of Temperature on the Equilibrium Constant and the Position of Equilibrium. As shown in the foregoing section, ΔF^0 and $\log K$ can easily be expressed as a function of the temperature when the needed data are available. At any given temperature the change of the equilibrium constant with temperature at constant pressure is given by the relation [see Eqs. (126) and (197)]

$$\frac{d\left(\frac{\Delta F^0}{T}\right)}{dT} = -\frac{R \, d \ln K}{dT} = -\frac{\Delta H^0}{T^2} \quad (255)$$

If ΔH^0 is positive, *i.e.*, if heat is absorbed in the reaction at the given temperature, then with increasing temperature ΔF^0 will decrease and K will increase, *i.e.*, the equilibrium will be shifted to the right for the reaction as written.

Effect of Pressure on the Equilibrium Constant and the Position of Equilibrium. Since ΔF^0 is the increment in free energy for a given reaction with all the reactants and products in their standard states and since these standard states are fixed independently of the pressure, ΔF^0 is constant at a given temperature. Then since

$$\Delta F^0 = -RT \ln K \quad (256)$$

K is independent of the pressure at a given temperature. It must be emphasized, however, that this equilibrium constant is expressed in terms of the activities (see p. 307) of the reactants and products.

For liquids and solids the activity changes little with the pressure. For gases the activity is equal to the fugacity, which for low pressures, is substantially equal to the partial pressure. At higher pressures, however, the fugacity and the partial pressure may differ considerably, and in these cases, for a given reaction involving gases, the "apparent" equilibrium constant expressed in terms of partial pressure may change markedly with the pressure. The use of the proper values of the fugacities for the given pressures will result in the equilibrium constant being really constant. The method of evaluating fugacities is mentioned on pp. 307–308 of this section, and some correlation of values of fugacity with pressure for a number of gaseous substances is given on pp. 355–356 of this section. Larson and Dodge [*J. Am. Chem. Soc.*, **45**, 2918 (1923); **46**, 367 (1924)] found, for example, that for the ammonia synthesis at 450°C. the value of the "apparent" equilibrium constant expressed in terms of partial pressures varied from 0.0066 at 10 atm. to 0.0233 at 1000 atm. This difference is entirely due to the non-ideal change in the fugacities of the three gases with pressure as shown by the calculations of Gillespie and Beattie [*Phys. Rev.*, **36**, 743, 1008 (1930); **37**, 655 (1931); *J. Am. Chem. Soc.*, **52**, 4239 (1930)]. Through the use of the Beattie-Bridgeman equation of state they were able to achieve constancy of K over the entire experimental range of pressures.

It is important to distinguish the foregoing zero effect of pressure upon the true equilibrium constant from the effect of pressure in shifting the position of equilibrium with respect to the amounts of reactants and products. At any given temperature the effect of pressure is such that, with the true equilibrium constant remaining constant, the amounts of the reactants and products will change in such a way that the total volume of the system will decrease with increase in pressure. By Eq. (121) at constant temperature

$$\frac{d(\Delta F)}{dp} = \Delta V \quad (257)$$

which defines the variation of the increment of free energy with pressure for reactants and products in any initial and states, not the standard. If ΔV is negative for the reaction as written, *i.e.*, if the stoichiometric volume of the products is less than the stoichiometric volume of the reactants, $d(\Delta F)/dp$ will be negative and ΔF will decrease with increase in pressure. A decrease in ΔF means that the equilibrium will be shifted so as to produce more of the products, *i.e.*, the reaction will proceed to the right as written until new values of the equilibrium concentrations are established for the new pressure. The proper quotient of the activities, as expressed by the equilibrium constant, will have the same value for the new pressure as for the old.

In the case of reactions where the *sum of the molal volumes of the products is the same as that of the reactants, i.e.*, where $\Delta V = 0$, a change in pressure will affect the equilibrium concentrations only insofar as the change in pressure produces different changes in the fugacities of the reactants and products. If the gases are sufficiently ideal that the fugacities may be replaced by the partial pressures, then, for $\Delta V = 0$, change in pressure will produce zero change in the equilibrium amounts of the reactants and products.

In the case of reactions for which ΔV is not equal to zero, an increase of pressure will shift the equilibrium in the direction corresponding to a decrease in volume, as mentioned above. Examples of gaseous reactions for which ΔV is not equal to zero are as follows:

$$\tfrac{1}{2}N_2\ (g) + \tfrac{3}{2}H_2\ (g) = NH_3\ (g) \qquad (258)$$

$$CO\ (g) + 2H_2\ (g) = CH_3OH\ (g) \qquad (259)$$

$$CH_3OH\ (g) + CO\ (g) = CH_3COOH\ (g) \qquad (260)$$

In terms of molal volumes the value of ΔV for these three reactions is -1, -2, and -1, respectively, and with increase in pressure at a given temperature the amounts of NH_3 (g), CH_3OH (g), and CH_3COOH (g) at equilibrium will increase. If all the gases may be assumed sufficiently ideal for the pressures considered that the partial pressures may be used in place of fugacities, equations may be derived for the mole fraction of the respective products obtained at equilibrium, from stoichiometric mixtures of the reactants, as a function of the pressure at constant temperature. If y is the mole fraction of NH_3 (g) in the equilibrium mixture produced from nitrogen and hydrogen in stoichiometric amounts, then, assuming ideal gases at a total pressure p, we may proceed as follows:

Gas	Initial no. of moles	Change in no. of moles	No. of moles at equilibrium	Mole fraction	Partial pressure
N_2	$\frac{1}{2}$	$-\frac{1}{2}x$	$\frac{1}{2}-\frac{1}{2}x$	$\frac{1}{2}\left(\frac{1-x}{2-x}\right)$	$\frac{1}{2}\left(\frac{1-x}{2-x}\right)p$
H_2	$\frac{3}{2}$	$-\frac{3}{2}x$	$\frac{3}{2}-\frac{3}{2}x$	$\frac{3}{2}\left(\frac{1-x}{2-x}\right)$	$\frac{3}{2}\left(\frac{1-x}{2-x}\right)p$
NH_3	0	x	x	$\frac{x}{2-x}$	$\frac{x}{2-x}p$
Total......	2	$-x$	$2-x$	1	p

Then

$$K = \frac{p_{NH_3}}{(p_{N_2})^{1/2}(p_{H_2})^{3/2}}$$

$$= \frac{\left(\frac{x}{2-x}\right)p}{\left[\frac{1}{2}\left(\frac{1-x}{2-x}\right)p\right]^{1/2}\left[\frac{3}{2}\left(\frac{1-x}{2-x}\right)p\right]^{3/2}}$$

$$= \left(\frac{4}{3^{3/2}}\right)\frac{x(2-x)}{(1-x)^2 p} \qquad (261)$$

Let $y = \dfrac{x}{2-x}$, the mole fraction of NH_3 (g) in the equilibrium mixture. Then $x = \dfrac{2y}{1+y}$; $2 - x = \dfrac{2}{1+y}$; and $1 - x = \dfrac{1-y}{1+y}$. Substituting for x, one obtains K in terms of y and p:

$$K = \left(\frac{16}{3^{3/2}}\right)\frac{y}{(1-y)^2}\frac{1}{p} \qquad (262)$$

or

$$\frac{y}{(1-y)^2} = \left(\frac{3^{3/2}K}{16}\right)p \qquad (263)$$

That is, $y/(1-y)^2$ increases directly with the pressure; and, for values of y small in comparison with 1, the mole fraction of ammonia, y, is directly proportional to p.

Likewise, for the reaction expressed by Eq. (259), the synthesis of methanol from carbon monoxide and hydrogen, if y is the mole fraction of CH_3OH in the equilibrium mixture produced from stoichiometric amounts of CO and H_2 at a total pressure p, and assuming the gases sufficiently ideal as before, one finds for a given temperature

$$K = \left(\frac{27}{4}\right)\frac{y}{(1-y)^3}\frac{1}{p^2} \qquad (264)$$

or

$$\frac{y}{(1-y)^3} = \left(\frac{4}{27}K\right)p^2 \qquad (265)$$

That is, for all values of y, the quantity $y/(1-y)^3$ varies with p^2 and, for values of y small in comparison with 1, y varies with p^2.

In a similar way and with the same assumptions, one finds for the conversion of methanol to acetic acid with carbon monoxide, with y equal to the mole fraction of acetic acid in the equilibrium mixture,

$$K = 4\frac{y}{(1-y)^2}\frac{1}{p} \qquad (266)$$

or

$$\frac{y}{(1-y)^2} = \left(\frac{K}{4}\right)p \qquad (267)$$

So that here $y/(1-y)^2$ varies with p, and, for values of y small in comparison with 1, y varies with p.

It should be emphasized that with increasing pressure the fugacities will begin to deviate markedly from the partial pressures and the above formulas become less and less accurate. In such cases recourse must be had to the use of fugacities (see pp. 307–308 of this section).

Effect of Indifferent Gases and the Proportions of Reactants on the Equilibrium. Frequently in technical practice the reaction mixture contains as diluents materials which do not participate in the reaction. For example, nitrogen is present in the burner gases passing to the converter in the contact sulfuric acid process, and liquid-phase reactions are frequently carried out in the presence of an indifferent solvent. Even if indifferent gases are initially present in very small proportion, their concentration gradually increases if a circulatory system is employed, and "bleeding" must be resorted to in order to maintain the concentration of the indifferent gas at a permissible value. This latter is the case in the synthesis of ammonia where argon concentrates if a circulatory system is used.

The presence of indifferent gases has no effect upon the true equilibrium constant. Their presence does change the mole fraction and consequently, for a given total pressure, will reduce the fugacity and partial pressure of each of the active constituents. The addition of an indifferent gas, keeping the total pressure unchanged, corresponds substantially to a decrease in total pressure without the addition of the indifferent gas. Where the value of ΔV for a reaction (see pp. 311–312) is zero, the presence of the indifferent gas will be to alter the fugacity–partial-pressure relationships for the active constituents. At low pressures this effect will be small, and there will be substantially no change in the relative amounts of products and reactants, while at high pressures this alteration of the relation between fugacity and partial pressure may be very marked. For reactions where ΔV is not zero (see pp. 311–312), the effect of the indifferent gas becomes considerable, as in the case of change in total pressure (see p. 311), and the addition of the inert gas without change in total pressure is analogous to reduction of the total pressure with no addition of inert gas. For ΔV positive, addition of inert gas will shift the equilibrium to the right to give a greater amount of products; while for ΔV negative, the addition of inert gas will shift the equilibrium to the left, to give less of the products. In those cases where the gases are sufficiently ideal for the partial pressures to be used in place of fugacities, the exact effect of the inert gas can be easily calculated. The maximum equilibrium concentration of products, and therefore the maximum yield per unit volume, will in general be obtained when the reactants are used in the stoichiometric proportions demanded by the theoretical reaction equation. In practice, however, the proportions in which the reactants are employed are frequently governed by considerations other than a maximum yield of products. Thus, it may be desirable to convert one of the reactants as completely as possible, and this may be effected by use of another in large excess. In other instances the economical technical methods of producing the reactant mixture govern its composition. In general, the presence of a diluting material, either an indifferent substance or an excess of one of the reactants, results in a decreased concentration of products in the equilibrium mixture.

Third Law of Thermodynamics. When for a given reaction the temperature of the products is the same as that of the reactants, one may write, from the definition of free energy (which see),

$$\Delta F = \Delta H - T\,\Delta S \tag{268}$$

When values of ΔH and ΔS are known at some temperature for a given reaction, with the products and reactants in their standard states, this equation permits evaluation of ΔF^0 and the corresponding equilibrium constant [see Eq. (197)].

For any substance that normally exists in only one crystalline phase, the heat content and entropy of the gas at a temperature T are given by the following equations:

$$H = H_0 + \int_0^{T_{\text{m.p.}}} C_p\,(\text{c})dT + \Delta H\,(\text{fusion}) + \int_{T_{\text{m.p.}}}^{T_{\text{b.p.}}} C_p\,(\text{l})dT + \Delta H\,(\text{vaporization}) + \int_{T_{\text{b.p.}}}^{T} C_p\,(\text{g})dT \tag{269}$$

$$S = S_0 + \int_0^{T_{\text{m.p.}}} C_p\,(\text{c})d\ln T + \frac{\Delta H\,(\text{fusion})}{T_{\text{m.p.}}} + \int_{T_{\text{m.p.}}}^{T_{\text{b.p.}}} C_p\,(\text{l})d\ln T + \frac{\Delta H\,(\text{vaporization})}{T_{\text{b.p.}}} + \int_{T_{\text{b.p.}}}^{T} C_p\,(\text{g})d\ln T \tag{270}$$

Here H_0 and S_0 are the heat content and the entropy of the substance at the absolute zero of temperature. When the given substance normally exists in more than one crystalline phase, an extra term or terms involving the heat or entropy of transition appear in the above equations. If the given substance is normally a liquid or a solid at the temperature under consideration, then the appropriate number of terms are dropped from the equations. In more abbreviated form one may write the above equations as

$$H = H_0 + \int_0^T dQ \tag{271}$$

$$S = S_0 + \int_0^T \frac{dQ}{T} \tag{272}$$

where dQ is the heat absorbed by the substance at constant pressure. Combination of Eqs. (268), (271), and (272) yields

$$\Delta F = \Delta H_0 + \Delta \int_0^T dQ - T\left(\Delta S_0 + \Delta \int_0^T \frac{dQ}{T}\right) \tag{273}$$

or

$$\frac{\Delta F}{T} = \frac{\Delta H_0}{T} + \frac{1}{T}\Delta \int_0^T dQ - \Delta \int_0^T \frac{dQ}{T} - \Delta S_0 \tag{274}$$

In this last equation all the terms on the right except the last are capable of being evaluated from calorimetric measurements. If the last term ΔS_0 were known, one would have available here a powerful equation which would permit evaluation of ΔF^0 and the equilibrium constant K_p entirely from calorimetric measurements. It was a study of this problem of evaluating ΔS_0 for chemical reactions, by Le Chatelier, Lewis, Richards, Nernst, Planck, and others, that finally resulted in the principle that is today commonly called the third law of thermodynamics. A rigorous statement of the third law is the following:

For any substance in a single pure quantum state, the entropy at the absolute zero of temperature may be taken as zero.

No exceptions to this law are known. It has been found by experience that with some few exceptions most substances that can be obtained in a crystalline state conform to the requirements of the third law. For all substances falling into the category defined by the third law, $S_0 = 0$; and values of ΔF^0 and K for reactions involving these substances are deducible entirely from appropriate calorimetric measurements since $\Delta S_0 = 0$. The importance of the third law lies in the fact that for many reactions (especially organic ones) the values of ΔF^0 and K can be determined in no other way.

From tables of values of the heats of formation and the entropies of the chemical compounds one can easily compute the appropriate values of the free energy increment according to Eq. (268).

Thermodynamic Properties of Gases from Statistical Calculations. Within recent years statistical methods have been developed for calculating certain thermodynamic properties of gaseous molecules as a function of the temperature from spectroscopic and other molecular data. The properties so calculable include the following:

$U^0 - U_0^0$, the internal energy referred to the absolute zero.
$H^0 - H_0^0$, the heat content referred to the absolute zero.
C_V, the heat capacity at constant volume.
C_p, the heat capacity at constant pressure.
S^0, the entropy in the ideal standard state.
$(F^0 - H_0^0)/T$, the free energy function.

These statistical calculations were first made for monatomic gases only; then methods for the diatomic molecules were developed; and in the past few years the calculations have been extended to more complicated molecules.

For any monatomic gas in the ideal state,

$$U^0 - U_0^0 = \tfrac{3}{2}RT \tag{275}$$

$$C_V = \tfrac{3}{2}R \tag{276}$$

$$U_0^0 = H_0^0 \tag{277}$$

$$H^0 - H_0^0 = \tfrac{5}{2}RT \tag{278}$$

$$dS = \left(\frac{\delta S}{\delta V}\right)_T dV + \left(\frac{\delta S}{\delta T}\right)_V dT = \frac{p}{T}dV + \frac{C_V}{T}dT = Rd\ln V + \frac{3}{2}R\ln T \tag{279}$$

Integrating the last equation, one obtains

$$S = R\ln V + \tfrac{3}{2}R\ln T + c \tag{280}$$

The constant c has been evaluated from theoretical considerations, by the work of Sackur, Tetrode, Stern, and Ehrenfest to be

$$c = R \ln [e^{5/2}(2\pi k)^{3/2}N^{-5/2}M^{3/2}g_0] \qquad (281)$$

where e is the base of the natural logarithms, k is the Boltzmann gas constant per molecule N is Avogadro's number giving the number of atoms or molecules per mole, M is the molecular weight, grams per mole, and g_0 is the "statistical weight," or "multiplicity" of the ground level of the molecule. Substituting RT/p for V and inserting the appropriate values of the constants, one obtains for the entropy of 1 mole of a monatomic gas in the ideal state

$$S = R \ln (p^{-1}T^{5/2}M^{3/2}g_0) - 2.30 \text{ cal./(deg.)(mole)} \qquad (282)$$

The above equations give also the translational part of the energy and entropy of any polyatomic gas in the ideal state. For 1 mole of any given polyatomic gas in the ideal state, one may separate the total energy referred to the absolute zero, the heat capacity, and the entropy into the parts arising from the (1) translational, (2) rotational, (3) vibrational, (4) internal rotational, and (5) electronic degrees of freedom, thus:

$$U^0 - U_0{}^0 = U^0 \text{ (translation)} + U^0 \text{ (rotation)} + U^0 \text{ (vibration)} + U^0 \text{ (internal rotation)} + U^0 \text{ (electronic)} \qquad (283)$$

$$C = C \text{ (translation)} + C \text{ (rotation)} + C \text{ (vibration)} + C \text{ (internal rotation)} + C \text{ (electronic)} \qquad (284)$$

$$S = S \text{ (translation)} + S \text{ (rotation)} + S \text{ (vibration)} + S \text{ (internal rotation)} + S \text{ (electronic)} \qquad (285)$$

To ascertain the total energy (referred to the absolute zero), the heat capacity, and the entropy of the polyatomic gas, one needs to add to the contribution from translation the contributions from rotation, vibration, internal rotation, and electronic excitation. These contributions may be calculated statistically in a number of special ways [see, for example, Giauque, *J. Am. Chem. Soc.*, **52**, 4816 (1930); Johnston and Chapman, *J. Am. Chem. Soc.*, **55**, 153 (1933); Kassel, *Chem. Rev.*, **18**, 277 (1936); Wilson, *Chem. Rev.*, **27**, 17 (1940); Fowler and Guggenheim, "Statistical Thermodynamics," Cambridge, London, (1940)]. For most gases at room temperature the contributions from electronic excitation are negligible and those from vibrational degrees of freedom are small. On the other hand the contributions from internal rotation may be appreciable at or somewhat above room temperature. This is particularly true of many organic compounds.

Free Energy Function. With the development of methods for calculating certain thermodynamic properties of gaseous molecules statistically from spectroscopic and other molecular data, the thermodynamic values so calculated are becoming increasingly useful and important. In particular is this true of the free energy function, $(F^0 - H_0{}^0)/T$ or $(F^0 - U_0{}^0)/T$, the usefulness of which can be illustrated as follows:

$$\Delta \left(\frac{F^0 - H_0{}^0}{T}\right) = \frac{\Delta F^0}{T} - \frac{\Delta H_0{}^0}{T} \qquad (286)$$

or

$$\frac{\Delta F^0}{T} = -R \ln K = \frac{\Delta H_0{}^0}{T} + \Delta \left(\frac{F^0 - H_0{}^0}{T}\right) \qquad (287)$$

To calculate the value of $\Delta F^0/T$, or its equivalent, $-R \ln K$, for any given reaction at a temperature T, it is necessary to have only values of the free energy function for each of the reactants and products at the temperature T, together with the value of the heat of the reaction at the absolute zero. The free energy function is calculated statistically from spectroscopic and other molecular data. The heat of the reaction at the absolute zero, $\Delta H_0{}^0$, is easily obtained by correcting the calorimetrically determined heat of reaction at 25°C. to the absolute zero by appropriately combining with $\Delta H_{298.16}$, values of $H_{298.16} - H_0{}^0$ for each of the reactants and products. That is,

$$\Delta H_0{}^0 = \Delta H^0{}_{298.16} - \Delta(H^0{}_{298.16} - H_0{}^0) \qquad (288)$$

Tabulations of Thermodynamic Data. For the convenience of those interested in making chemical thermodynamic calculations of equilibrium, etc., there have been compiled by various workers tables giving: (1), values of the entropies of the chemical compounds, usually at the standard temperature of 25°C. [see, *e.g.*, Kelley, *U.S. Bur. Mines Bull.* 434 (1941); Parks and Huffman, "Free Energies of Some Organic Compounds," Reinhold, New York (1932); Wenner, "Thermochemical Calculations," McGraw-Hill, New York (1941)]; (2), values of the heats of formation of the chemical compounds from their elements, all in the appropriate standard states at the selected standard temperature [see, for example, Bichowsky and Rossini, "Thermochemistry of the Chemical Substances," Reinhold, New York (1936)]; (3), values of the free energy of formation of the chemical compounds from their elements, all in the appropriate standard states at the selected standard temperature [see, for example, Lewis and Randall, "Thermodynamics and the Free Energy of Chemical Substances," McGraw-Hill, New York (1923); Randall, "International Critical Tables," vol. 5, pp. 224–353, McGraw-Hill, New York (1930)]; (4), values of the free energy function,† $(F^0 - U_0{}^0)/T$ or $(F^0 - H_0{}^0)/T$, and the heat content referred to the absolute zero, $H - H_0{}^0$ or $H - U_0{}^0$,† for various gaseous molecules (see, for example, papers published from 1930 to 1946 in the *J. Am. Chem. Soc.* and the *J. Chem. Phys.* by Giauque, Johnston, Kassel, Gordon, Pitzer, and others, in the numerous papers from the National Bureau of Standards on hydrocarbons and other substances, the tabulation in *Landolt-Bornstein-Roth-Scheel Tabellen*, 5th ed., 3d suppl., Part III, pp. 2334–2344, Springer, Berlin, 1936), the tables of thermodynamic properties issued by American Petroleum Institute Research Project 44 at the National Bureau of Standards, and the tables of Selected Values of Chemical Thermodynamic Properties of the National Bureau of Standards.

Many values for the thermodynamic constants, such as heat of formation, free energy of formation, heat capacity, heat of fusion, and heat of vaporization, are given in Sec. 3.

Summary of the More Important Thermodynamics Equations*

$$Q = (V_1 - V_2)T\frac{dp}{dT} \qquad dQ = C_V dT + T\left(\frac{\partial p}{\partial T}\right)_V dV$$

$$dQ = C_p dT - T\left(\frac{\partial V}{\partial T}\right)_p dp$$

$$dQ = C_V\left(\frac{\partial T}{\partial P}\right)_V dp + C_p\left(\frac{\partial T}{\partial V}\right)_p dV$$

$$C_p = \left(\frac{\partial H}{\partial T}\right)_p = \left(\frac{\partial Q}{\partial T}\right)_p$$

$$C_V = \left(\frac{\partial U}{\partial T}\right)_V = \left(\frac{\partial U}{\partial p}\right)_V \left(\frac{\partial p}{\partial T}\right)_V$$

* See, also, Bridgman, *Phys. Rev.*, **3**, 273 (1914), for a condensed summary of thermodynamic formulas.

† In much of the earlier literature $F^0 - U_0{}^0/T$ appears as $F^0 - E_0{}^0/T$ and $H^0 - U_0{}^0$ as $H^0 - E_0{}^0$.

$$C_p = \left(\frac{\partial U}{\partial T}\right)_p + p\left(\frac{\partial V}{\partial T}\right)_p$$

$$C_p = \left[\left(\frac{\partial U}{\partial V}\right)_T + p\right]\left(\frac{\partial V}{\partial T}\right)_p + \left(\frac{\partial U}{\partial T}\right)_V$$

$$\left(\frac{\partial U}{\partial p}\right)_T = -p\left(\frac{\partial V}{\partial p}\right)_T - T\left(\frac{\partial V}{\partial T}\right)_p$$

$$\left(\frac{\partial U}{\partial V}\right)_T = T\left(\frac{\partial p}{\partial T}\right)_V - p$$

$$\left(\frac{\partial C_p}{\partial p}\right)_T = -T\left(\frac{\partial^2 V}{\partial T^2}\right)_p \qquad \left(\frac{\partial C_V}{\partial V}\right)_T = T\left(\frac{\partial^2 p}{\partial T^2}\right)_V$$

$$C_p dT + \left(\frac{\partial U}{\partial p}\right)_T + p\left(\frac{\partial V}{\partial p}\right)_T dp = 0$$

$$C_V dT + \left[\left(\frac{\partial U}{\partial V}\right)_T + p\right] dV = 0$$

$$C_p - C_v = T\left(\frac{\partial p}{\partial T}\right)_V\left(\frac{\partial V}{\partial T}\right)_p = -T\left(\frac{\partial p}{\partial V}\right)_T\left(\frac{\partial V}{\partial T}\right)_p^2$$

$$\left(\frac{\partial S}{\partial T}\right)_V = \frac{C_V}{T} \qquad \left(\frac{\partial S}{\partial V}\right)_T = \left(\frac{\partial p}{\partial T}\right)_V$$

$$\left(\frac{\partial S}{\partial T}\right)_p = \frac{C_p}{T} \qquad \left(\frac{\partial S}{\partial P}\right)_T = -\left(\frac{\partial V}{\partial T}\right)_p$$

$$\left(\frac{\partial T}{\partial p}\right)_u = \frac{p(\partial V/\partial p)_T + T(\partial V/\partial T)_p}{C_p - p(\partial V/\partial T)_p}$$

$$\left(\frac{\partial T}{\partial V}\right)_u = \frac{p - T(\partial p/\partial T)_V}{C_V}$$

$$H = U + pV \qquad dH = C_p dT - \left[T\left(\frac{\partial V}{\partial T}\right)_p - V\right] dp$$

$$\left(\frac{\partial T}{\partial p}\right)_s = \frac{T(\partial V/\partial T)_p}{C_p} \qquad \left(\frac{\partial T}{\partial V}\right)_s = -\frac{T(\partial p/\partial T)_V}{C_V}$$

$$\left(\frac{\partial p}{\partial V}\right)_s = -\frac{C_p}{C_V}\frac{(\partial T/\partial V)_p}{(\partial T/\partial p)_V}$$

$$C_s = \frac{dQ}{dT} = C_p - T\left(\frac{\partial V}{\partial T}\right)_p \frac{dp}{dt} = C_p - \frac{Q}{V_1 - V_2}\left(\frac{\partial V}{\partial T}\right)_p$$

$$C_s = \frac{dQ}{dT} = C_v + T\left(\frac{\partial p}{\partial T}\right)_V \frac{dV}{dT} = C_V\left(\frac{\partial T}{\partial p}\right)_V \frac{dp}{dT} + C_p\left(\frac{\partial T}{\partial V}\right)_p \frac{dV}{dT}$$

$$C_{s_1} - C_{s_2} = \frac{dQ}{dT} - \frac{Q}{T} = \frac{d(Q/T)}{dT}$$

$$\text{Joule-Thomson coefficient} = \mu = \left(\frac{dT}{dp}\right)_H = \frac{T\left(\frac{\partial V}{\partial T}\right)_p - V}{C_p}$$

PHASE EQUILIBRIUM AND SOLUTIONS

References: Findlay, "The Phase Rule," 6th ed., Longmans, New York, 1923. Bowden, "Phase Rule and Its Applications," Macmillan, London, 1938. Roozeboom, "Die heterogenen Gleichgewichte," Vieweg, Brunswick, 1901–1918. Taylor, "Treatise on Physical Chemistry," 2d ed., vol. 1, pp. 467–582, Van Nostrand, New York, 1926. Sage and Lacey, "Volumetric and Phase Behavior of Hydrocarbons," Stanford University Press, Stanford University, Calif., 1939. Hildebrand, "Solubility," 2d ed., Reinhold, New York, 1936. For further treatment and application of specific principles see Secs. 8, 9, and 10 of this handbook.

Phase Rule. The principle known as the **phase rule** is generally applicable to equilibriums between phases and provides a useful method for interpreting and classifying such equilibriums. The phase rule is expressed by the equation

$$P + F = C + 2 \tag{1}$$

in which P is the number of phases present at equilibrium, C is the number of components of the system, and F is the number of degrees of freedom. These terms must be exactly defined.

Any portion of a system which is homogeneous throughout, which is bounded by a surface, and which may be mechanically separated from the other portions (phases) is a **phase.** A gas or gaseous mixture is one phase. A pure liquid or a solution forms one phase, a mixture of two immiscible liquids is two phases. Every pure solid is a separate phase; if a solid solution is formed between two solids, however, these comprise a single phase. The number of solid phases which may coexist is unlimited.

The least number of independently variable constituents required to express the composition of each phase of a system of equilibrium is defined to be the number of **components** of the system. In practice the number of components is usually determined from information concerning the changes or reactions which occur in the system. Thus, in the system $CaCO_3$-CaO-CO_2, it is seen that the composition of any phase will be determined by specifying the relative total quantities of any two of the constituents* but not by a smaller number. This system has then two components. If the temperature range in question were one in which no decomposition of $CaCO_3$ occurred, three components would be necessary to define this system. The number of components is therefore determined in part by the equilibriums established as well as by the substances present.

Degrees of freedom are defined as the number of independently variable quantities such as temperature, pressure, and concentration which may be altered at will without producing an alteration in the number of phases, or, as the number of such variables which must be arbitrarily fixed in order to completely define the system.

Where additional restrictions must be made in applying the phase rule, the effect of each such restriction is to reduce by one the number of degrees of freedom obtained from Eq. (1). For example, the system NH_4Cl, solid and vapor, has only one component since the ratio of NH_3 to HCl is restricted to unity; the system NH_4Cl, solid, vapor, and excess NH_3, however, has two components. In the first case composition is arbitrarily fixed and the degrees of freedom are one, whereas in the latter case they are two.

The phase rule specifies only the number of phases at equilibrium and does not determine their nature, exact composition, or total quantity. Furthermore, it applies only to a system in stable equilibrium and states nothing regarding the rate at which this equilibrium is attained.

If a system has one component, the phase rule shows that for one phase it is divariant, for two phases univariant, and for three phases non-variant. The independent variables are temperature and pressure.

Phase Relations and Diagrams. In correlating relations between, and in predicting the number of, phases which can exist in equilibrium in a given system, the phase rule is a valuable guide. For the physical and chemical nature of these phases and their quantita-

* If $CaCO_3$ is selected as one of the components, negative or zero values for the composition of the other two phases in terms of $CaCO_3$ will result. This is, however, not excluded.

tive interrelationship with given variables over definite ranges of conditions one must, however, resort to experiment. For example, for a single pure substance, such as a one-component system, for any specified range of pressure, since $P = 3 - 1 = 2$, two phases may coexist in equilibrium if one does not specify arbitrarily the conditions of temperature. But the rule does not predict at what particular temperatures the two phases will be vapor-liquid, liquid-solid, or vapor-solid. Again in the case of two partially miscible liquids, one can predict that, if two liquid layers are present in equilibrium at a specified temperature and pressure, each phase must have one specific and definite composition. The value of this composition for this temperature and pressure cannot, however, be predicted.

Experimental data on phase relationships are usefully and commonly represented graphically in terms of phase diagrams. An almost unlimited variety of such diagrams is possible. In ordinary engineering and chemical work the variables to be considered are pressure, volume, temperature, and (for systems of more than one component) composition. By means of special expedients and solid figures more or less complete representation of phase relations in terms of all these variables is possible. This is, however, inconvenient for most practical uses, and it usually suffices to show the relationships of phases to two variables at a time by use of plane diagrams, while the others are neglected or maintained constant. It is ordinarily unnecessary to cover the entire range of even two selected variables on a single such diagram. Consequently, phase diagrams as usually found represent only one section of the complete phase equilibrium diagram for any particular system. It is common to find separate diagrams showing the one-phase region, two-phase, three-phase, etc., regions of a system. One particular region can usually be singled out as being that of significance to a given engineering operation.

Numerous pressure-temperature, pressure-volume, pressure-composition, volume-temperature, and temperature-composition diagrams for typical systems are to be found in the references, the handbook, and the literature. Sage and Lacey ("Volumetric and Phase Behavior of Hydrocarbons," Stanford University Press, Stanford University, Calif., 1939) give a number of such diagrams for hydrocarbons; Katz and Kurata [*Ind. Eng. Chem.*, **32**, 817 (1940)] have recently summarized and discussed vapor-liquid phase behavior for simple and complex systems.

The vapor-pressure–temperature curves of pure liquids, vapor-pressure–composition and boiling-point–composition curves of binary liquid mixtures, pressure-temperature curves for gases in the critical region, the familiar equilibrium-pressure–composition curves for the solubility of gases in liquids, freezing-point–composition curves for mixtures of two substances, and the solubility curves for solids in liquids are all common examples of simple phase diagrams. A frequently useful type of plane phase diagram shows the composition of one phase in equilibrium with the other, *e.g.*, the vapor-liquid (y-x) equilibrium curve used in distillation calculations. It should be noted that strictly speaking such curves are constant-pressure or constant-temperature diagrams but that in certain cases variations in these over reasonable ranges do not affect significantly the value at equilibrium. In dealing with equilibrium between liquid phases and liquid-solid phases (*i.e.*, condensed phases), the effect of pressure and the vapor phase need not be considered or can be regarded as constant. Its effect, unless the pressure is extremely high, is negligible for most purposes in considering equilibrium between solid and liquid phases. Only temperature and composition need be taken into account.

Where solid phases are concerned, the phase diagrams are usually of the temperature-composition type but may be exceedingly complex, especially where more than two components are involved, because of the large number of solid phases which may exist. For the most part such diagrams are encountered in the metallurgical, ceramic, and fractional crystallization fields.

Extended treatment of phase equilibriums and the many types of phase relationships and phase diagrams falls beyond the scope of this handbook. Some examples of phase diagrams are mentioned in subsequent pages.

System Water. Phase equilibriums are conveniently represented graphically by phase diagrams. In Fig. 4 the pressure-temperature relations of the system water are illustrated. ER is the equilibrium curve for ice and vapor; t_cEX is the equilibrium curve for liquid and vapor, *i.e.*, the vapor-pressure curve of water; E is the point at which water freezes under its own vapor pres-

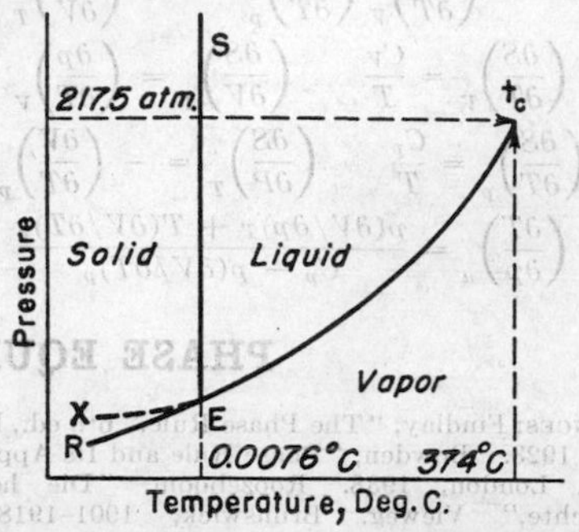

FIG. 4. Phase diagram for one-component system, water (not drawn to scale).

sure,* but it may be supercooled along the line EX. At E, solid, liquid, and vapor coexist in equilibrium. This is the so-called "triple point." ES represents the equilibrium, solid-liquid, in the absence of water vapor. Along any of these curves the system has one degree of freedom, *i.e.*, they represent univariant equilibriums. At E, the triple point, there are no degrees of freedom and the system is non-variant. The intersection of three boundary curves, such as ES, ER, and Et_c, represents in general a triple point at which three phases are in equilibrium. Bridgman has shown that at very high pressures other triple points are present in the water system, where different modifications of solid (ice) exist. A similar p-t phase diagram for the system sulfur shows four triple points: monoclinic sulfur-liquid-vapor, rhombic-monoclinic-liquid, rhombic-monoclinic-vapor, and the metastable rhombic-liquid-vapor.

Two-component Systems. A two-component system must be defined by specifying the compositions of all phases in terms of two constituents. Such a system is represented by Na_2SO_4 and H_2O and by benzene and toluene. Independent variables are temperature, pressure, and composition. Since, in this case, a complete phase diagram can be given only in three dimensions, it is more convenient for practical purposes arbitrarily to fix one variable at some constant value and use one of the three possible two-variable combinations, pressure-temperature, pressure-composition, or temperature-composition, representable on plane diagrams. The phase rule shows that for two-component systems four coexisting phases such as, for example, in the Na_2SO_4-H_2O system, vapor, solution, solid Na_2SO_4, ice, or a solid hydrate as $Na_2SO_4.H_2O$, are required for non-variant equilibrium: $F = 2 + 2 - 4 = 0$. Univariant equilib-

* The freezing point at atmospheric pressure is 0°C.

rium requires three phases; divariant, two phases; and trivariant, one phase.

In three-component systems of liquid and solid phases the existing relations are frequently represented by showing the relations between compositions at constant temperature on an equilateral triangular diagram (see Sec. 11).

Raoult's and Henry's Laws. Raoult's law states that the partial vapor pressure of the solvent in equilibrium with a solution is directly proportional to its mole fraction N_0, or $p = N_0 p_0$, where p_0 is the vapor pressure of the pure solvent at the same temperature. If the solute is non-volatile, the partial vapor pressure of the solvent evidently corresponds to the total vapor pressure over the solution. According to Raoult's law the fractional lowering of the vapor pressure of the pure solvent by a dissolved substance is

$$\frac{p_0 - p}{p_0} = \frac{N_1}{N_0 + N_1} = N_1 \tag{2}$$

where N_1 is the mole fraction of the solute and N_1 and N_0 are the moles of solute and solvent, respectively.

Henry's law states that the quantity of gas dissolved in a given quantity of solvent or solution is directly proportional to its partial pressure over the solution, *i.e.*, in terms of the mole fraction N_g of the gas in the solution,

$$p_g = kN_g \tag{3}$$

The law is expressed in other terms, *e.g.*, in terms of the concentration of the gas in solution, $p_g = k'C_g$. The Henry's law coefficient k varies with the temperature and the nature of the gas and solvent. Its numerical value depends upon the units in which pressure and concentration are expressed, and, in using data, particular note should be taken of the units in which the value of k is recorded. The law may be applied to the solution of gases in liquids or to the partial vapor pressure of a volatile liquid dissolved in another liquid in small concentration.

Henry's law applies for the individual components of a gaseous mixture as well as for a single pure gas, the concentration of each component dissolved being, according to the law, proportional to its own partial pressure and not the total gas pressure. Consequently, if the law holds, the equilibrium partial pressure over the liquid is unaffected by the presence of an inert gas, except possibly at very high pressures.

It will be noted that Raoult's law differs from Henry's in the value of the proportionality constant. In dilute solution the two laws bear a relation to each other; when Raoult's law holds for the solvent, Henry's law holds for the solute and vice versa.

While Raoult's and Henry's laws are closely followed by many substances and approximated by others with accuracy satisfactory for engineering calculations, in numerous cases the deviations are large. They are therefore to be regarded as ideal solution laws to which real solutions approximate more closely the more dilute they become. In both cases deviations increase with increasing concentration in either phase, *i.e.*, as the concentration of the solution or the pressure over it is raised. The usual effect of increasing the temperature is to cause a system to approximate more closely to Raoult's or Henry's law, as the case may be. As the pressure is raised, the solubility of a gas tends to deviate more widely (the law is more closely followed the more closely the ideal gas law is approximated), and at high pressures the deviations may be very great. In general, if the solute undergoes a molecular change in solution, such as dissociation, association, or combination with the solvent, divergences from both laws are likely to be large. Deviations may occur, however, even though this is not the case.

Raoult's law is likely to be approached more closely by mixtures of chemically similar liquids and when there is no significant heat effect on mixing. Certain mixtures of volatile liquids, for example, as benzene-toluene, carbon tetrachloride-toluene, deviate very little from Raoult's law over the entire composition range. As a rule deviations in the case of mixtures of hydrocarbons, members of homologous series, stereoisomers, and the like, are small and may be neglected in the majority of engineering calculations. Mixtures of aliphatic and aromatic organic compounds (*e.g.*, benzene-ethyl alcohol, ethyl alcohol-water, acetone-chloroform) deviate widely.

In concentrated solutions of non-volatile solutes and particularly with dissolved electrolytes, divergences from the law are large. The reduction of vapor pressure effected by dissolved electrolytes is always greater than that calculated from their molecular weights. Approximate corrections for the dissociation of the solute may be made if the degree of dissociation is known.

By making use of the methods of thermodynamics, Raoult's and Henry's laws are expressed in terms of the thermodynamic functions, *fugacity* and *activity* (see p. 307), so that they assume exact and generally valid form. In terms of fugacities, Raoult's law is

$$f_1 = f_1^0 N_1 \tag{4}$$

where f_1 is the fugacity of the solvent in the solution and f_1^0 its fugacity in pure form. Or, since the activity is defined to be the ratio of the fugacity of a substance in a system to its fugacity in an arbitrarily chosen standard state ($a = f/f^0$), if the pure liquid is chosen as the standard state, Raoult's law becomes

$$\frac{f_1}{f_1^0} = a_1 = N_1 \tag{5}$$

and Henry's law

$$a_2 = \frac{f_2}{f_2^0} = k_2 N_2$$

Hildebrand gives a number of rules and generalizations concerning solubility and deviations from Henry's and Raoult's laws which may be qualitatively useful as a guide in the absence of data.

Further discussion of these laws, and their utility and applications will be found in Secs. 8 and 9.

Solubility of Gases in Liquids. Gas solubility is expressed in a diversity of forms, one of which is in terms of the Henry's law coefficient; another, the Bunsen absorption coefficient, is the volume of gas at 0°C. and 1 atm., which at the experimental temperature is dissolved in unit volume of solvent when the partial pressure of the gas is 1 atm. Other absorption coefficients express the amount of solvent in grams. Data and detailed treatment are given in Sec. 8. For chemical engineering use it is most convenient to represent data on gas solubility graphically on a plot of partial pressure vs. concentration, or the mole ratio of the solute gas to inert gas in the gas phase vs. the mole ratio of dissolved gas to solvent in the liquid phase. If Henry's law holds, such a plot is a straight line through the origin.

In general, the solubility of a gas diminishes with increasing temperature. The Clausius-Clapeyron equation (p. 341), where it can be expected to hold, affords a possible means of estimating solubilities at various temperatures. It is also possible that the method of Othmer (p. 341) might be applied to this problem.

In general, gases which dissociate or combine with the solvent, *e.g.*, ammonia, sulfur dioxide, and hydrogen chloride in water, are more strongly soluble and show large deviations from Henry's law. The so-called permanent gases which exist in the same molecular form in solution, *e.g.*, hydrogen, oxygen, nitrogen, methane, in water, are usually more slightly soluble and approximate more closely to Henry's law in ordinary ranges of temperature and pressure. These factors, of course, vary with the solvent, and a gas may deviate from the law in one solvent and behave normally in

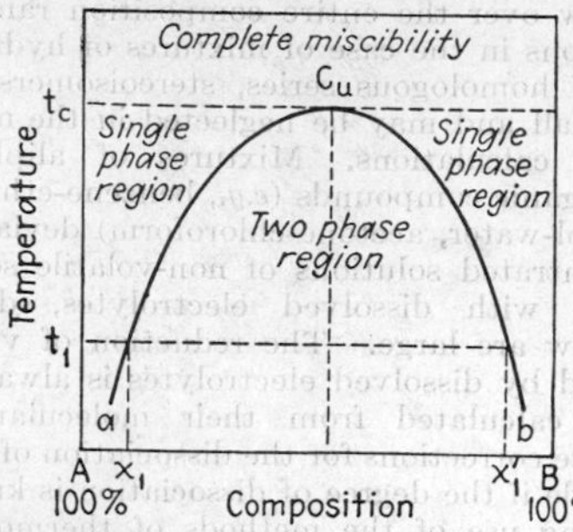

FIG. 5. Schematic *t-C* phase diagram for two liquids having an upper consolute temperature.

another, and its conformity with the law will depend upon the temperature.

Frolich and coworkers [*Ind. Eng. Chem.*, **23**, 548 (1931); see also Goodman and Krase, *Ind. Eng. Chem.*, **23**, 401 (1931)] have concluded from an experimental investigation that, if a gas does not combine chemically with the solvent, Henry's law may be assumed for calculating solubilities over a wide pressure range for engineering purposes. In the case of the permanent gases (H_2, N_2, CH_4, etc.) in a variety of solvents, such as water, ethyl alcohol, propane, benzene, and gas oil, solubilities at high pressures may be estimated approximately from Henry's law and the experimentally determined solubility at atmospheric pressure. [For recent data on the solubility of N_2 and CO_2 in C_6H_6, see Dodge and coworkers, *Ind. Eng. Chem.*, **32**, 95, 434 (1940).] Application of the law to gases of the vapor type, such as propane, ethylene, and hydrogen sulfide, gives satisfactory results only over pressure ranges up to one-half to two-thirds of the saturation point at any temperature. Data for the solubility of gases in liquids are given in Sec. 10.

Mutual Solubility of Liquids. In a given temperature range a pair of liquids may be miscible, *i.e.*, soluble in all proportions, partially miscible, or, for practical purposes, immiscible. Temperature has a pronounced effect upon mutual solubility, while the effect of pressure is, under ordinary conditions, small. On the basis of the effect of temperature on their mutual solubility two liquids exhibit varying types of behavior These may be represented by simple phase diagrams in which composition is plotted vs. temperature at constant pressure.

Type 1. Mutual solubility increases with temperature (Fig. 5). At lower temperatures, *e.g.*, t_1, on mixing, two layers are formed—a saturated solution of B in A whose composition is given by a point on the curve aCb, *i.e.*, x_1, and a saturated solution of A in B whose composition is given by x_1'. These are called *conjugate layers.* At any point on the curve, one *saturated layer* or phase exists. At any over-all composition represented by the region within the curve two mutually saturated liquid phases coexist whose respective compositions are given by points on the opposite sections of the curve corresponding to the given temperature. Outside of the curve below t_c, to the left, there exists a partly saturated layer of A and, to the right, a partly saturated solution of B. As the temperature is raised, mutual solubility increases, the compositions of the two saturated layers approach each other, and a point C_u is reached at which miscibility in all proportions occurs. This temperature is called the *critical solution* or *consolute temperature.* Aniline-H_2O, phenol-H_2O, methyl alcohol-cyclohexane, and aniline-hexane are examples of liquids exhibiting this behavior.

Type 2. Mutual solubility decreases with temperature (Fig. 6). In some cases the mutual solubility of two liquids diminishes with temperature, and a lower critical solution temperature C_L is exhibited, below which the two are miscible in all proportions. Figure 6 illustrates the *t-C* phase diagram for this type. Its characteristics are analogous to those of Fig. 5. Examples of liquid pairs of this type are triethylamine-H_2O dimethylaniline-H_2O.

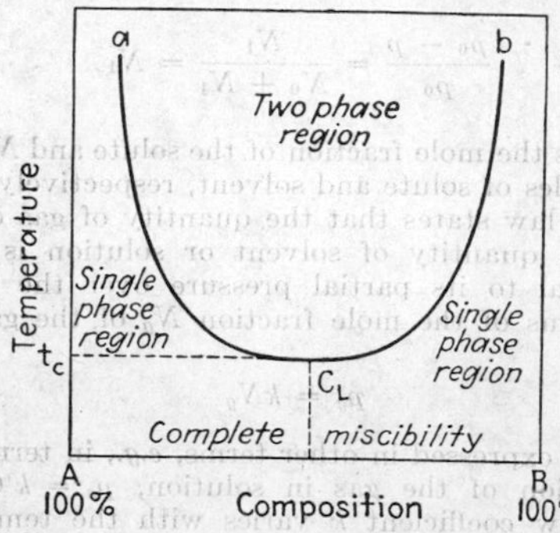

FIG. 6. Schematic *t-C* phase diagram for two liquids having a lower consolute temperature.

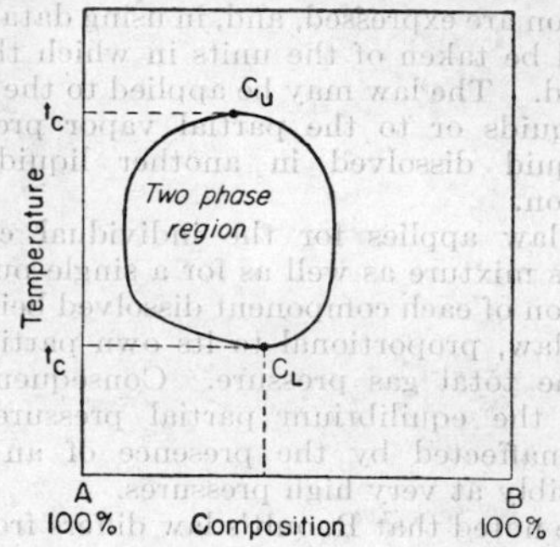

FIG. 7. Schematic *t-C* phase diagram for two liquids having both upper and lower consolute temperature.

Type 3. Closed solubility curve. A few cases are known, the classic example being nicotine-water, in which two liquids show both a higher and lower critical solution temperature (Fig. 7). Within the curve they are partially miscible. Above and below the consolute temperatures they are miscible in all proportions.

Other Cases. Many cases are known in which neither an upper nor lower consolute temperature is found to exist. In some cases, such as ethyl ether and water, the solubility of one liquid in the second diminishes with temperature while that of the second in the first increases with temperature. Theoretically, all liquid pairs may be considered to have both upper and lower consolute temperature, but either a solid phase appears or, on the other end, one liquid is converted to vapor before complete miscibility occurs.

The presence of impurities, *i.e.*, a third component, markedly alters the mutual solubility of two liquids, the shape of the saturation curve, and the consolute temperature. In some cases a relatively small amount of a third substance may at a given temperature suffice to render two partially miscible liquids completely miscible, while in other cases the opposite effect is obtained. Three component systems consisting of two liquids and a third substance are considered in Sec. 11.

Theoretical considerations regarding the solubilities of both liquids and solids in liquids which may offer some qualitative guide in complete absence of data are discussed by Hildebrand ("Solubility," 2d ed., Reinhold, New York, 1936).

Vapor Pressure and Boiling Points of Binary Liquid Mixtures. The phase-equilibrium diagrams and relationships for liquid mixtures are of fundamental importance in distillation and are comprehensively treated in Sec. 9.

Solubility of Solids in Liquids. No satisfactory method for estimating the extent to which a solid will dissolve in a given liquid has been developed. Attempts to predict the solubilities of non-electrolytes from vapor-pressure and thermal data have been made, but these are not of practical use and cannot be considered here.

Certain useful generalizations concerning the solubilities of solids may be pointed out. The solubility of a given solid is dependent upon the temperature, nature of the liquid solvent, and the nature of the solid. As a general rule the solubility of a solid in a liquid increases with rising temperature, but the rate of increase differs widely for different solids, and in some cases the solubility is diminished by increasing temperature in some ranges. For example, the solubility of calcium sulfate above 40°C., of sodium sulfate between 33° and 100°C., and of calcium hydroxide decreases with rising temperature. Solubility-temperature curves are generally continuous, but in some cases discontinuities occur. These are usually associated with the existence of different hydrated forms of the solid in solution, the different hydrates being stable in different temperature ranges and exhibiting different variations of solubility with temperature (*i.e.*, Na_2SO_4 and $Na_2SO_4 \cdot 10H_2O$).

Little is known concerning the specific influence of the nature of the solid and the liquid upon solubility. In general, substances closely related chemically show marked mutual solvent action. The solubility of electrolytes, such as inorganic salts, in organic liquids is normally small and considerably less than that in water. Unless the latent heats of fusion are considerably different, a solid having a higher melting point is usually less soluble at a given temperature than one of lower melting point.

The solubility of solids has been shown to increase somewhat with decreasing particle size (also the rate of solution).

The effect of pressure upon the solubility of a solid in a liquid is very small and may be either to increase or decrease the solubility. According to van't Hoff the solubility of ammonium chloride, which expands when dissolved, increases by 1 per cent for 160 atm., while copper sulfate, which contracts, increases by about 3 per cent for 60 atm. Pressures up to 2000 to 3000 atm., as a rule, have very little effect upon the solubility of inorganic substances. For organic substances the change of solubility with pressure is much greater. In some cases the solubility is decreased by 20 to 25 per cent at pressures around 1000 atm.

The solubility curve of a solid in a liquid or the freezing-point curve of any two-component system are both examples of phase diagrams of the *t-C* type. A wide and complex variety of phase relations in systems involving solid phases may exist since compound formation between the two, solid solutions, etc., may be formed and the number of solid phases which may coexist can be unlimited. Their discussion transcends the scope of the handbook. Consult the references and Sec. 11.

Elevation of the Boiling Point by Non-volatile Solutes. Since the addition of a non-volatile solute lowers the vapor pressure of a pure liquid, the boiling point of the resulting solution is higher than that of the pure solvent. Assuming Raoult's law, the following relation may be derived

$$\Delta T = T - T_0 = \frac{p_0}{dp_0/dT_0}(N_1) \tag{6}$$

or

$$\Delta T = KN_1 \tag{7}$$

where ΔT is the boiling-point rise, N_1 the mole fraction of solute, p_0 the vapor pressure of the pure solvent at its boiling point T_0, and K is a constant. For very dilute solutions, since $N_1 = n_1/n_0$ approximately.

$$\Delta T = \frac{K_0 n_1}{n_0} \tag{8}$$

The proportionality constant K_0 is characteristic of the solvent and independent of the nature of the solute. It represents the boiling-point rise per mole of solute in 1 mole of pure solvent. K_0 may be determined experimentally or calculated from the Clausius-Clapeyron equation. In the latter case

$$K_0 = \frac{RT_0^2}{L_v} \quad \text{and} \quad \Delta T = \frac{RT_0^2}{L_v}(N_1) \tag{9}$$

where L_v is the molal latent heat of vaporization of the pure solvent at its boiling temperature T_0. For convenience, a constant giving the molal boiling-point rise per 1000 g. solvent instead of per mole is customarily determined and recorded in the literature.

These rules are accurate only for very dilute solutions, and then, in the case of electrolytes, only when a proper correction for dissociation is made. For concentrated solutions in the ranges commonly met in practice, the boiling-point rise thus calculated may differ from that experimentally determined by several hundred per cent.

The boiling points of concentrated aqueous solutions may be estimated with considerable accuracy over ordinary ranges of pressure and temperature from a small number of experimental measurements by the use of the Dühring principle (see p. 294). A Dühring line, the boiling point (°C. or °F.) of water at two different pressures against the boiling point of the solution at the same pressures, is plotted for each of a number of individual concentrations covering the range of concentration desired. Knowing the boiling point of water at the desired pressure, that of the solution of the required concentration may be read from this plot. Concentrations intermediate between those for which lines are available may be determined by interpolation. Plots are available for the more commonly encountered solutions such as salt (Badger and McCabe, 'Elements of Chemical Engineering," 2d ed., p. 187, McGraw-Hill, New York, 1936) and caustic soda [Baker and Waite, *Chem. & Met.*, **25**, No. 25 (Dec. 21, 1921); Adams and Richards, *Ind. Eng. Chem.*, **20**, 470 (1928)]. The method proposed by Othmer (pp. 342 and 343) provides an alternate procedure for plotting and estimating boiling points which has some advantages.

Lowering of the Freezing Point. The freezing point of a solution is that temperature at which the solid solvent and the solution exist in equilibrium, and this freezing point is always lower than that of the pure solvent. Normally the solid separating from the solution on freezing is the pure solvent.

Since the vapor pressures of solid solvent and solvent in solution must be equal at the freezing point, the relation

$$\frac{p_0 - p}{T_0 - T} = \frac{L_f p}{RT^2} \tag{10}$$

is obtained from a consideration of the vapor-pressure temperature curves of liquid and solid solvent, solution, and from the Clausius-Clapeyron equation. Here p_0 is the vapor pressure of the pure solvent at its freezing point, T_0, p is the vapor pressure of the solution at T_0, T is the freezing point of the solution, and L_f is the molal latent heat of fusion of the solid solvent. From Raoult's law $(p_0 - p)/p = N_1$, hence

$$-\Delta T_f = T_0 - T = \frac{RT^2}{L_f} N_1 \tag{11}$$

or

$$-\Delta T = K_f N_1 \tag{12}$$

N_1 being the mole fraction of solute and K_f the freezing-point constant which is characteristic of the solvent. Freezing-point constants giving the depression in the freezing point produced by 1 mole of solute in 1000 g. solvent, $K_f = RT_f^2/1000L_f$, are ordinarily reported in the literature.

The above relations are strictly valid only in dilute solutions where Raoult's law and the other assumptions involved hold true. The freezing-point depressions of concentrated solutions cannot be calculated from them.

The boiling-point and freezing-point relationships are useful for determining the molecular weights of non-volatile dissolved substances from the observed changes in these temperatures in dilute solution. Molecular weight = $mK/\Delta T$, where m is the grams of solute in 1000 g. solvent, ΔT is the corresponding change in boiling or freezing point, and K the corresponding molal constant.

"Salting-out" Effect. When an electrolyte (salt) is added to an aqueous solution of a non-electrolyte, as hydrogen, ether, or succinic acid, the solubility of the latter is generally diminished. The mutual solubility of two liquids such as ether and water is reduced by the addition of such a salt as sodium chloride or sodium sulfate. This phenomenon is called the "salting-out effect." The effect is complicated and is a characteristic property of the electrolyte and of the substance salted out. The diminution in solubility produced by an electrolyte often follows the empirical relation

$$\ln\left(\frac{S}{S_0}\right) = k_s C \tag{13}$$

over a considerable concentration range (up to approximately 4.0 molal). S is the solubility in the solution of the electrolyte of concentration C, S_0 is the solubility in pure water, and k_s is a constant for each specific electrolyte when salting out a particular non-electrolyte.

Some high molecular weight organic salts are known which "salt in" rather than out, *e.g.*, the sodium and potassium salts of toluene, naphthalene, and xylene sulfonic acids. These substances are very highly soluble in water and have a marked effect in increasing the solubility of many organic substances in water. They are called *hydrotropic salts*.

Distribution between Two Liquid Phases. A third substance introduced as solute into a mixture of two partially miscible liquids, if soluble in both liquids, distributes itself between the two liquid layers in definite equilibrium concentrations, the distribution normally favoring one of the liquids in preference to the other. The simple or ideal distribution law states that at equilibrium at constant temperature the ratio of the concentrations of the solute in the two phases is a constant, irrespective of the total quantity in which it is present; *i.e.*,

$$\frac{C_b}{C_a} = D \tag{14}$$

C_b and C_a are the solute concentrations in the two liquid phases, respectively, and D is the "distribution" or "partition" coefficient. The latter is a constant independent of concentration if the law is obeyed. The numerical value of D depends upon the units of concentration. Any units may be used, but those most frequently found in the scientific literature are moles per liter of solution or solvent. Mole ratios are more convenient in engineering calculations. The solute may be either solid or another liquid.

This is an ideal or limiting law, rigorously exact only for ideal solutions, and is a special case of the ideal general distribution law which also embraces Henry's law (see p. 317). It is more closely approximated where the two solvents are essentially immiscible and their mutual solubility remains unaffected by the presence of the added substance. This is more nearly the case in dilute concentrations so that the quantity of the solute must usually be very small for the law to hold.

The law in its simple form, Eq. (14), can be expected to hold only when the solute has the same molecular formula in both phases, *e.g.*, if it neither dissociates, associates, nor combines with the solvent; *i.e.*, only the ratio between like molecules in the two phases can be said to be constant. If the type of molecular alteration which occurs is known, suitable corrections can be applied to the law to obtain a coefficient which may be expected to be constant. Thus, if a solute exists entirely as undissociated single molecules in one phase and is associated into double molecules in the other, the law can be expressed as

$$D' = \frac{C_b}{\sqrt{C_a}} \tag{15}$$

Or, if the solute partially dissociates in one phase and not in the other, then

$$D'' = \frac{C_a(1 - \alpha)}{C_b} \tag{16}$$

where α is the degree of dissociation. Analogous forms may be derived for other possible cases.

For the majority of practical chemical engineering calculations which would involve the use of the law, it is not a satisfactory approximation. In the relatively few cases where it is sufficiently accurate, it provides a useful generalization. In most engineering applications the added substance is present in such proportions as to affect materially the mutual solubility of the two liquid layers, and exact phase equilibrium and composition data covering the desired concentration range must be available.

Where the ideal law holds, the distribution of a solid between two liquids bears a definite relation to its solubility in each of the two. As the solubility of the solute varies with temperature and to a different extent for each solvent, temperature changes affect the distribution coefficient. Ordinarily, where the law is approximated, the effect of temperature is small, but it may be large. Pressure has little effect on the distribution ratio.

The distribution principle finds important application in the process of solvent extraction. Thus organic substances mixed with inorganic material in aqueous solution are readily separated therefrom by extracting with a suitable organic solvent in which the organic substance is preferentially soluble. Solvent extraction in recent years has assumed marked importance as a chemical engineering unit operation and is treated in Sec. 11.

CHEMICAL REACTION KINETICS

Introduction. The rates at which chemical processes proceed are of great theoretical and practical importance in chemical engineering. Marked progress has been made in recent years toward developing and establishing the basic scientific principles which govern chemical-reaction velocities. At present, chemical kinetics is not, however, an exact science as is thermodynamics. From a practical standpoint it is not yet possible to formulate simple generalized mathematical relations or rigorous laws quantitatively expressing all kinetic behavior. Quantitative treatment of reaction velocities is essentially complex and rests largely on an empirical basis, especially for the majority of industrially important reactions. The interpretation of experimental data and kinetic analysis is in most cases an individual problem. However, certain general principles are available on the basis of which mathematical relations have been developed for some simpler cases. These may serve as a guide in the treatment of kinetic problems.

REFERENCES: Hinshelwood, "Kinetics of Chemical Change in Gaseous Systems," 3d ed., Oxford, New York, 1933. Kassel, "Kinetics of Homogeneous Gas Reactions," Reinhold, New York, 1932. Daniels, "Chemical Kinetics," Cornell University Press, Ithaca, N. Y., 1938. Schumacher, "Chemische Gas Reaktionen," Steinkopff, Dresden, 1938. Moelwyn-Hughes, "Kinetics of Reactions in Solution," Oxford, New York, 1933. Sherwood and Reed, "Applied Mathematics in Chemical Engineering," pp. 53–67, McGraw-Hill, New York, 1939. Eucken-Jakob, "Der Chemie-Ingenieur," vol. 3, Akademische Verlagsgesellschaft, Leipzig, 1937. Schwab, Taylor, and Spence, "Catalysis from the Standpoint of Chemical Kinetics," Van Nostrand, New York, 1937. Rideal and Taylor, "Catalysis in Theory and Practice," 2d ed., Macmillan, London, 1926. Green, "Industrial Catalysis," Benn, London, 1928. Ipatieff, "Catalytic Reactions at High Pressures and Temperatures," Macmillan, New York, 1936. Marek and Hahn, "Catalytic Oxidation of Organic Compounds in the Vapor Phase," Reinhold, New York, 1932.

HOMOGENEOUS REACTIONS

General Rate Equation. According to the law of mass action the instantaneous velocity of chemical reaction is proportional to the concentrations of the reactants at any moment. The rate may be expressed in terms of the rate of disappearance of any reactant or of the rate of appearance of any of the products. Thus for the general irreversible reaction

$$aA + bB + dD \rightarrow rR + sS + qQ \tag{1}$$

in terms of the disappearance of A, the rate is mathematically formulated by the general differential equation (Sherwood and Reed, "Applied Mathematics in Chemical Engineering," p. 54)

$$\frac{-dn_A}{d\theta} = kn_A C_A{}^{a-1} C_B{}^b C_D{}^d \tag{2}$$

where n_A represents the number of moles of A present in unit mass; C_A, C_B, and C_D, the concentration of specie A, B, and D, respectively, as moles per unit volume; θ the time; and k the specific reaction velocity constant. k is characteristic of the reaction and a function of temperature. Except for a first-order reaction its numerical value depends upon the units of quantity, volume, and time employed. Reaction velocities are conveniently described by stating the numerical value of k.

For reaction *at constant volume* or *in the special case that there is no change in the number of moles at constant pressure*, since $C_A = n_A/V$, and $dn_A = VdC_A$ (V = volume of the reacting system), Eq. (2) reduces to the more familiar expression

$$\frac{-dC_A}{d\theta} = kC_A{}^a C_B{}^b C_D{}^d \tag{3}$$

In terms of the rate of formation of the product R, it becomes

$$\frac{dn_R}{d\theta} = k\left(\frac{r}{a}\right) n_A C_A{}^{a-1} C_B{}^b C_D{}^d \tag{4}$$

Order of Reaction. Reactions are classified into first, second, third, and higher orders according to the number of molecules which appear to enter into the reaction as determined by the relationship existing between the rate and the reactant concentrations. Mathematical rate equations for reactions of various orders are developed below.

The stoichiometric chemical equation does not necessarily determine the order of reaction. This can be decided only by experiment. Measurement of the rate of a reaction supplies the only practical information concerning the mechanism of the change taking place which can be used for determining the influence of operating variables and the optimum conditions for carrying out a reaction in practice.

While the kinetic behavior of many simple systems is satisfactorily represented by the basic reaction-rate equations subsequently presented, by far the greater number of reactions which have been studied have been found to follow more complex rate equations than those for the simple reaction orders. Frequently, empirical equations and approximate kinetic expressions are more useful than others. This is so in many cases of industrial importance. Intermediate products may first be formed which subsequently decompose or which react among themselves or with other molecules of the original reactants to form the final products. Or either or both the reactants or the intermediate products may undergo to an appreciable extent side reactions in addition to the principal reaction. A few common reactions which do not follow the simple reaction orders are as follows:

$$H_2 + Cl_2 \rightarrow 2HCl$$
$$2SO_2 + O_2 \rightarrow 2SO_3$$
$$2CO + O_2 \rightarrow 2CO_2$$
$$2H_2 + O_2 \rightarrow 2H_2O$$

First-order Reaction. For a truly unimolecular reaction, as

$$A \rightarrow \nu B \qquad \text{or} \qquad A \rightarrow B + D \tag{5}$$

under isothermal conditions, the rate equation is *for constant volume*

$$\frac{-dC_A}{d\theta} = k_1 C_A \tag{6}$$

or, in a more generally useful form applicable to either constant volume or constant pressure (since $C_A = n_A/V$),

$$\frac{-dn_A}{d\theta} = k_1 n_A \tag{7}$$

Constant Volume. Integration between limits for Eq. (6) gives

$$k_1 = \frac{2.303}{\theta} \log \frac{C_{A_0}}{C_A} \tag{8}$$

and for Eq. (7)

$$k_1 = \frac{2.303}{\theta} \log \frac{n_{A_0}}{n_A} \tag{9}$$

where C_{A_0} and n_{A_0} are the concentration and number of moles, respectively, initially present at zero time, and C_A and n_A, the concentration and the moles remaining after time θ. Or, if f is the fraction of n_{A_0} reacting in time θ

$$n_A = (1 - f)n_{A_0},$$

$$k_1 = \left(\frac{2.303}{\theta}\right) \log \left(\frac{1}{1-f}\right) \tag{10}$$

Any of these forms may be employed. The velocity constant k_1 for a first-order reaction is independent of the concentration units used. The time required for any given fraction of the reactant originally present to undergo change is independent of the initial concentration.

Constant Pressure. The preceding relations apply strictly where there is no volume change during reaction. This is the case for isothermal reaction in constant-volume "batch" or "closed" systems, or for gas reactions in continuous-flow systems in the special case that no change in volume takes place, and approximately for liquid phase reactions. Constant-pressure flow systems are more important industrially. The formulation of mathematical relations for such cases has been treated by Benton [*J. Am. Chem. Soc.*, **53**, 2984 (1931)] and by Sherwood and Reed ("Applied Mathematics in Chemical Engineering," McGraw-Hill, New York, 1939).

In flow systems the time θ refers to that during which the reactants are in the reaction zone and during which reaction is progressing, the so-called "time of contact," which is not usually directly measurable. It varies with the rate of flow, the volume of the reaction zone, and the extent to which the volume changes during the reaction. The extent of reaction during this time is determined by analysis of the inlet and exit streams. If there is no volume change, Eq. (9) applies directly, the time θ being replaced by V_R/V_0, where V_R is the volume of the reaction zone and V_0 the volume of the incoming stream per unit time at the temperature and pressure of the reaction zone.

Consider the general reaction

$$A \rightarrow \nu B \tag{11}$$

where ν is the number of moles of products formed from the reaction of 1 mole of A, and the ideal gas laws are assumed. Let n_{A_0} moles of A continuously enter the reaction zone per unit time, n_A be the number of moles unreacted after time θ and passing any given point in the zone per unit time, and V be the total volume of material (unchanged reactant plus products) corresponding to n_A after any reaction time, *i.e.*, the volume per unit time passing any point in the zone. Then

$$V_0 = \left(\frac{RT}{p}\right) n_{A_0} \quad \text{and} \quad V = \left(\frac{RT}{p}\right)[n_A + \nu(n_{A_0} - n_A)]$$

$$= \left(\frac{RT}{p}\right)[\nu n_{A_0} - (\nu - 1)n_A]$$

$$= V_0\left[\nu - \frac{(\nu-1)n_A}{n_{A_0}}\right] \tag{12}$$

Treating V_R as the variable,

$$d\theta = \frac{dV_R}{V} = \frac{dV_R}{V_0\left[\nu - \dfrac{(\nu-1)n_A}{n_{A_0}}\right]} \tag{13}$$

substituting for $d\theta$ in Eq. (7) and integrating,

$$\frac{k_1 V_R}{V_0} = 2.303\nu \log \frac{n_{A_0}}{n_A} - (\nu - 1)\left(1 - \frac{n_A}{n_{A_0}}\right) \tag{14}$$

Or, if f represents the fraction of A which has reacted

$$\frac{k_1 V_R}{V_0} = 2.303\nu \log \frac{1}{1-f} - (\nu - 1)f \tag{15}$$

In the above development, pressure drop through the reaction zone due to friction has been assumed negligible.

If the entering reactant is mixed with an inert diluent, the preceding relations must be modified slightly to take the volume of diluent into account.

Where the fractional conversion to products is not too large, a sufficiently accurate approximation may result if V_0 is calculated as the average of the inlet and exit rates of flow and Eq. (9) is used as in the case of no volume change.

Very few homogeneous truly unimolecular reactions are known.

Second-order Reaction. In a second-order or bimolecular change two molecules, either of the same or different molecular species, react:

Case 1. $$A + B \rightarrow D + E + \cdots \tag{16}$$

or

Case 2. $$2A \rightarrow D + E + \cdots \tag{17}$$

From Eq. (2) the general rate equations are

For case 1. $$\frac{-dn_A}{d\theta} = k_2 n_A C_B \tag{18}$$

and

For case 2. $$\frac{-dn_A}{d\theta} = k_2 n_A C_A \tag{19}$$

At *constant volume*

Case 1. $$\frac{-dC_A}{d\theta} = k_2 C_A C_B \tag{20}$$

and

Case 2. $$\frac{-dC_A}{d\theta} = k_2 C_A{}^2 \tag{21}$$

or, alternately, for the general case, since $C_A = n_A/V$,

Case 1. $$\frac{-dn_A}{d\theta} = \frac{k_2 n_A n_B}{V} \tag{22}$$

Case 2. $$\frac{-dn_A}{d\theta} = \frac{k_2 n_A{}^2}{V} \tag{23}$$

from which it is seen that the rate at which moles of A disappear is inversely dependent upon the volume of the reaction space.

If C_{A_0} and C_{B_0} are the initial concentrations at zero time and C_A and C_B the concentrations after time θ, of A and B, respectively, integration of Eq. (20) gives

Case 1. $$k_2 = \frac{2.303}{\theta(C_{A_0} - C_{B_0})} \log \left(\frac{C_{B_0}}{C_{A_0}}\right)\left(\frac{C_A}{C_B}\right) \tag{24}$$

and of Eq. (21)

Case 2. $$k_2 = \frac{1}{\theta C_{A_0}}\left(\frac{C_{A_0} - C_A}{C_A}\right) \tag{25}$$

To integrate in the form of the general Eq. (22), n_A must be related to V. This is readily done at constant volume since V is constant independent of n_A.

Case 1. $$k_2 = \frac{2.303V}{(n_{A_0} - n_{B_0})\theta} \log \left(\frac{n_{B_0}}{n_{A_0}}\right)\left(\frac{n_A}{n_B}\right) \tag{26}$$

Case 2. $$k_2 = \frac{V}{n_{A_0}\theta}\left(\frac{n_{A_0} - n_A}{n_A}\right) \tag{27}$$

Alternate integrated forms of the above equations are, in terms of f, the fractional conversion of reactant A

to product at time θ, $(n_{A0} - n_A)/n_{A0} = f$,

Case 1. $$k_2 = \frac{2.303V}{(n_{A0} - n_{B0})\theta} \log \frac{1}{1 - f\left(\frac{n_{A0}}{n_{B0}}\right)} - f \tag{28}$$

Case 2. $$k_2 = \frac{V}{n_{A0}\theta}\left(\frac{f}{1-f}\right) \tag{29}$$

and, in terms of n, the number of moles of A and of B which have reacted up to time θ

Case 1. $$k_2 = \frac{2.303V}{(n_{A0} - n_{B0})\theta} \log \left(\frac{n_{B0}}{n_{A0}}\right)\left(\frac{n_{A0} - n}{n_{B0} - n}\right) \tag{30}$$

Case 2. $$k_2 = \frac{V}{\theta}\left(\frac{n}{n_{A0}(n_{A0} - n)}\right) \tag{31}$$

In the event that in reaction (16) the initial concentrations of the two reactants are equal, the rate equations for this case 1 reduce to those of the simpler case 2.

The numerical value of the second-order velocity constant k_2 is dependent upon both the units of concentration and the units of time employed. In the second-order reaction the time required for any specified fraction of one of the reactants to be transformed into reaction product depends upon the initial concentrations, whereas in the first-order reaction such time is independent of initial concentration.

The greater number of homogeneous reactions are second order; examples are

$$H_2 + I_2 \rightarrow 2HI$$
$$2HI \rightarrow H_2 + I_2$$
$$2NO_2 \rightarrow 2NO + O_2$$
$$2CH_3CHO \rightarrow 2CH_4 + 2CO$$
$$CH_3COOC_2H_5 + NaOH \rightarrow CH_3COONa + C_2H_5OH$$

Constant Pressure. The rate of a bimolecular reaction proceeding in a flow system under constant pressure is from Eqs. (18) and (19) given by

Case 1. $$\frac{-dn_A}{d\theta} = k_2VC_AC_B = \frac{k_2n_An_B}{V} \tag{32}$$

Case 2. $$\frac{-dn_A}{d\theta} = k_2VC_A{}^2 = \frac{k_2n_A{}^2}{V} \tag{33}$$

Generalized treatment is more complicated. The analysis of the special case of a kinetically second-order irreversible reaction of the type

$$2A \rightarrow 2B + 2D$$

carried out by Sherwood and Reed (*ibid.*) illustrates the method of attack.

Let n_{A0} represent the moles of A entering the reaction zone per unit time, n_A the moles unreacted after any reaction time θ, V_R the volume of the reaction zone, V the volume of unchanged A plus products corresponding to the value of n_A; assume the ideal gas laws; and treat V_R as the variable. It may then be shown that

$$V = \frac{RT}{p}[n_{A0} + (n_{A0} - n_A)] \tag{34}$$

$$d\theta = \frac{dV_R}{V} \tag{35}$$

and

$$C_A = \frac{n_A}{V} = \frac{n_Ap}{RT(2n_{A0} - n_A)} \tag{36}$$

Substitution in Eq. (33) gives

$$\frac{dn_A}{dV_R} = -k_2\left(\frac{p}{RT}\right)^2\left(\frac{n_A}{2n_{A0} - n_A}\right)^2 \tag{37}$$

Separating variables,

$$\left(\frac{4n_{A0}}{n_A{}^2} - \frac{4n_{A0}}{n_A} + 1\right) dn_A = -k_2\left(\frac{p}{RT}\right)^2 dV_R \tag{38}$$

Letting $\alpha = n_A/n_{A0}$, the fraction of A unchanged, substituting and integrating (neglect the small change in p through the zone due to friction) between the limits of $\alpha = 1$, $V_R = 0$, and of $\alpha = \alpha$, $V_R = V_R$, there results the relation

$$4 - \frac{4}{\alpha} - 4 \ln \alpha + (\alpha - 1) = \frac{k_2}{n_{A0}}\left(\frac{p}{RT}\right)^2 V_R \tag{39}$$

From this relation k_2 may be calculated by experimentally measuring α, or conversely, k_2 having been established, α or V_R may be calculated for any stated conditions.

As an approximation where the fractional conversion of reactant to products is not too large, Eq. (33) may be written

$$\frac{\Delta n_A}{\Delta\theta} = k_2\bar{V}(\bar{C}_A)^2 \tag{40}$$

where $\bar{V}$ and $\bar{C}_A$ are the average volume and the average concentration of A based on inlet and exit streams, and $\Delta\theta$ is taken as V_R divided by the average rate of flow. Such an approximation may serve for some purposes.

Higher Order Reactions. Third-order or trimolecular reactions involving the simultaneous reaction of three molecules of the same or different species are rare and of relatively little industrial importance. The only simple ones known are the reactions of NO with O_2, H_2, Cl_2, and Br_2. Similar kinetic expressions may be developed for this case if necessary. Their rate is proportional to the third power of the concentration of reactants and inversely to the square of the volume, or directly to the square of the pressure.

No reactions of orders higher than the third are known, and the probability of molecular collisions of order above third is so small that reactions of higher orders should be extremely rare. Where a stoichiometric equation involves a larger number (greater than 3) of reactant molecules, the reaction proceeds in a succession of reactions of simpler order.

Effect of Temperature on Reaction Velocity. The rates of chemical reactions increase rapidly with rising temperature. Generally, for homogeneous reactions the rate increases two- to threefold for a 10°C. rise in temperature but sometimes even more rapidly. The relative effect of an increase of temperature diminishes as the temperature is raised.

The empirical Arrhenius equation

$$\frac{d \ln k}{dT} = \frac{A}{RT^2} \tag{41}$$

in most cases expresses satisfactorily the relation between the specific reaction velocity k and temperature. T is the absolute temperature, R is the gas-law constant, and A is an energy-quantity characteristic of the reaction which is commonly termed the *energy of activation.* Theoretically, A is interpreted to represent the quantity of energy required to activate 1 mole of reactants to a state sufficiently above the average energy level of all molecules such that reaction may occur. In general, the only means of determining A is empirically from the Arrhenius equation by measurement of the reaction-velocity constants at two temperatures. In recent years some success, however, has been attained in predicting from purely theoretical considerations energies of activation agreeing in order of magnitude with those calculated from experimental data by the Arrhenius equation.

Over wide temperature ranges the value of A usually diminishes somewhat. However, for the majority of homogeneous reactions it is usually relatively satisfactory to assume A constant, in which case Eq. (41) integrates to

$$\ln \frac{k_{t_2}}{k_{t_1}} = \frac{A}{R}\left(\frac{1}{T_1} - \frac{1}{T_2}\right) \tag{42}$$

If A is relatively constant, a plot of $\ln k$ vs. $1/T$ gives a substantially straight line of slope $-A/R$ and is a convenient method of representing reaction-velocity–temperature data. From such a plot or from Eq. (42), A, or the velocity constant, at any temperature may be calculated from experimental measurements at two temperatures.

Apparent Reaction Order. Many reactions appear to follow a reaction order to which they do not actually belong. This may arise, for example, because one or more of the reactants is initially present in very large concentration relative to the others, so that its concentration change is negligible compared with those present in small concentration. Thus, if in the bimolecular reaction of Eq. (16) the concentration of A is so large as to remain substantially constant while B is consumed, the second-order Eq. (24) approximates to

$$k_2(C_{A0} - C_{B0}) = \frac{2.303}{\theta} \log \frac{C_{B0}}{C_B} \tag{43}$$

which is that characteristic of a first-order reaction. $k_2(C_{A0} - C_{B0})$ remains substantially constant, and the reaction will appear to be first order. The true velocity constants can be determined by taking the concentrations of all reactants into account. An example is the hydrolysis of sucrose

$$\underset{}{C_{12}H_{22}O_{11}} + H_2O \rightarrow \underset{\text{fructose}}{C_6H_{12}O_6} + \underset{\text{glucose}}{C_6H_{12}O_6}$$

in which the water concentration being large with respect to sucrose is practically constant during reaction, and the rate appears to depend only on the sucrose concentration.

In other cases the true reaction order may be masked by the simultaneous occurrence of reactions following other mechanisms. In the event that one reaction of a series is slow compared with the others, the over-all rate will be given by the simple equation for its order. Again the rates of some reactions frequently appear to depend upon non-integral powers of the concentration of reactants. This almost invariably indicates the simultaneous occurrence of complicating simultaneous reactions other than that under consideration. Complex industrial reactions such as the cracking of petroleum are extremely complicated, and rigorous kinetic treatment is almost hopeless. Satisfactory treatments of the kinetics of such cases have been accomplished through the correlation of empirical test data in terms of simpler reaction orders under the guidance of the kinetic principles developed for simple systems.

Simultaneous Reactions. The simple kinetic equations previously presented apply strictly only to irreversible reactions proceeding by a single mechanism with the formation of a single group of products. Many industrially important systems involve a much more complex reaction process, a considerable number of reactions taking place simultaneously.

Where a series of reactions proceed simultaneously in the same system, each may be assumed to take place at its own specific rate independently of the others and to follow the simple reaction-rate equation for its own particular order. The net total rate of change is then the summation of the rates of all the independent reactions taking place, each governed by its own specific rate. Certain definitive classes of simultaneous reactions are commonly recognized: *reversible reactions*, *consecutive reactions*, and *concurrent* or *side reactions*.

Many industrially important cases are encountered in which the reaction proceeds only partially to completion in the desired direction, and the effect of the reverse reaction on the net rate of formation of product cannot be neglected. In this case the net velocity of reaction is given by the difference of the velocities of forward and reverse reactions, and eventually a state of equilibrium is attained.

In numerous other industrially important cases the initial reaction between the original reactants is followed by one or more subsequent or *consecutive* reactions, *e.g.*,

$$A \rightarrow B \rightarrow D + E$$

in which the products first formed decompose or react with each other or the original substances to form a new set of final products. The measured rate of formation of the final products depends upon the relative rates of the consecutive reactions and is frequently expressed by a complicated kinetic equation. Intermediate products are often difficult to isolate. If the velocity of one of the series of consecutive reactions is relatively much slower than those of the others, which is frequently the case, the rate of this slow reaction determines the rate of formation of the final products and the order of the entire reaction.

In still other important cases a given set of substances react in more than one way by different mechanisms giving rise to a series of *concurrent or side reactions* which form different sets of final products, *e.g.*,

$$2A \rightarrow B + D$$
$$A \rightarrow R + S$$

Each of these side reactions has its own characteristic rate, one usually being much faster than the others. The total rate of disappearance of the reactants is then the sum of the rates of the individual side reactions. The relative rates of such side reactions are determined by reaction conditions, and it is frequently possible to alter the relative amounts of the various sets of products by altering the conditions. This can often be accomplished with the aid of a selective catalyst.

Reactions progressing simultaneously may be of different order; for example, in a reversible reaction the forward reaction may be first order, the reverse reaction second order. Further, a given system may involve at the same time side and consecutive reactions, some of which are reversible, others irreversible, as well as combinations of different reaction orders. For example, an original reversible reaction of two molecules of reactant may be followed by combination of a molecule of the original reactant with one of the products in a consecutive reaction which is also reversible; thus

$$\begin{aligned} 2B &\rightleftarrows D + H \\ B + D &\rightleftarrows T + H \end{aligned} \tag{44}$$

Unless the situation is unusually complicated, if the reactions taking place and their respective orders are known or can be determined, it is usually possible to derive reaction-rate equations to express the kinetic behavior of the system. The general procedure is to set up the simple differential-rate equations of the proper order for each separate reaction in terms of the rate of disappearance of reactants and the rate of formation of products and then to combine these to express the net reaction rate.

Treatment of the numerous possible cases is outside the scope of this handbook. Each is an individual kinetic problem. Kinetic equations have been developed for a number of prestated simpler cases of reversible, consecutive, and side reactions. Consult the references at the front of this section (MacDougall, "Physical Chemistry," Macmillan, New York, 1936, treats a number of cases) and the literature. Integration of the differential kinetic equations and calculations of the specific velocity constants for complex cases which are required for practical use of the rigorous relations are frequently difficult and not feasible. Integration graphically will often be found most convenient. Empirically formulated equations may be more convenient for practical purposes.

As a usual thing in a complex case the exact reaction mechanisms involved and their orders are not directly measurable, and kinetic analysis of rate data offers the only method of approach. A common procedure is to postulate a number of hypothetical reaction mechanisms, assigning appropriate simple integral reaction orders to the several reaction equations, and then to analyze the experimental rate data in terms of the demands of the kinetic rate relations developed for each. It is not always possible to differentiate between two or more possible mechanisms. Practical kinetic analyses are illustrated for many cases in the literature [see, for example Russell and Hottel, Rate of Ethylene Polymerization, *Ind. Eng. Chem.*, **30**, 183 (1938); Egbert, Solution to the Student Problem Contest, 1937, *Trans. Am. Inst. Chem. Engrs.*, **34**, 435 (1938); Haas, McBee, and Weber, Chlorination of Paraffins, *Ind. Eng. Chem.*, **28**, 333 (1936); Huntington and Brown, Laboratory Cracking Data as a Basis for Plant Design, *Ind. Eng. Chem.*, **27**, 699 (1937)].

Illustrative Cases of Simultaneous Reactions. A few of many possible examples may be useful to the handbook reader in illustrating methods of attack.

Reversible Reactions. The simplest case occurs when both forward and reverse reactions are of the same order, either first or second, and the products initially absent from the reacting mixture. If both reactions are second order, according to the scheme,

$$A + B \underset{k_R}{\overset{k_F}{\rightleftarrows}} D + E \tag{45}$$

the net rate of disappearance of A (which, since 1 mole of D and of E result from reaction of 1 mole of A and of B, is equal to the net rate of formation of product, *i.e.*, $-dn_A/d\theta = +dn_D/d\theta$) is

$$\frac{-dn_A}{d\theta} = k_A\frac{n_A n_B}{V} - k_R\frac{n_D n_E}{V} \tag{46}$$

Or, if n_{A_0} and n_{B_0} are the number of moles of reactants originally present in unit mass of the mixture and n is the number of moles of each which have reacted in time θ

$$\frac{-dn_A}{d\theta} = \frac{+dn_D}{d\theta} = \frac{dn}{d\theta} = k_F\frac{(n_{A_0} - n)(n_{B_0} - n)}{V} - \frac{k_R n^2}{V} \tag{47}$$

At constant volume Eq. (47) is readily integrated (or, since no volume change is involved, also at constant pressure). In the general case where volume changes are involved, integration is most readily accomplished graphically.

In the case of reversible reactions of this type, at the equilibrium state, where $dn_A/d\theta = 0$, then

$$k_F\frac{n_A n_B}{V} = k_R\frac{n_D n_E}{V} \tag{48}$$

and

$$\frac{n_D n_E}{n_A n_B} = \frac{k_F}{k_R} \tag{49}$$

If the reacting substances are approximately ideal, *i.e.*, if conditions are such that activities are proportional to concentration,

$$\frac{n_D n_E}{n_A n_B} = \frac{k_F}{k_R} = K \tag{50}$$

where K is the equilibrium constant. From Eq. (50) and the expression resulting from integration of Eq. (47) both of the velocity constants can be calculated.

As a somewhat more complicated case, consider the reversible reaction

$$A \underset{k_R}{\overset{k_F}{\rightleftarrows}} B + C \tag{51}$$

in which the forward reaction is first order, the reverse second order, and where the reaction mixture contains initially at time zero n_{B_0} and n_{C_0} moles of the products in addition to n_{A_0} moles of the reactant A. (See Watson, paper presented before the Chemical Engineering Division, Soc. Promotion Eng. Ed., Pennsylvania State College, June, 1939.) Then after time θ

$$n_B = n_{B_0} + (n_{A_0} - n_A) \quad \text{and} \quad n_C = n_{C_0} + (n_{A_0} + n_A) \tag{52}$$

and the differential rate equation is

$$\frac{-dn_A}{d\theta} = k_F n_A - k_R\frac{(n_{B_0} + n_{A_0} - n_A)(n_{C_0} + n_{A_0} - n_A)}{V} \tag{53}$$

or integrating

$$\theta = \int_{n_{A_0}}^{n_A} \frac{dn_A}{-k_F n_A + (k_R/V)(n_{B_0} + n_{A_0} - n_A)(n_{C_0} + n_{A_0} - n_A)} \tag{54}$$

At constant volume, or at constant pressure by evaluating V as a function of n_A, the integration may be performed graphically by plotting the reciprocal of the denominator on the right-hand side as a function of n_A. The time required for a selected degree of completion is then the area under the curve between n_{A_0} and the selected value of n_A.

From the conditions at equilibrium, as above,

$$k_F n_A = \frac{k_R n_B n_C}{V} \tag{55}$$

and

$$\frac{n_B n_C}{V n_A} = \frac{k_F}{k_R} = \frac{(n_B/V)(n_C/V)}{(n_A/V)} = K \tag{56}$$

so that both velocity constants may be calculated from a knowledge of the equilibrium constant at the desired temperature and the integration of Eq. (54).

Consecutive Reactions. Consider the simple consecutive reaction scheme

$$A \xrightarrow{k_1} B \xrightarrow{k_2} D$$

in which an intermediate group of products represented by B is formed from the reactant A in an irreversible first step and subsequently in a consecutive irreversible reaction changes into the final products D. For simplicity, considering the case where volume is constant, if the two reactions are of the general mth and nth orders, respectively, we may write the kinetic equations for the rate of disappearance of A

$$\frac{-dC_A}{d\theta} = k_1 C_A{}^m \tag{57}$$

for the rate of formation of D

$$\frac{+dC_D}{d\theta} = k_2 C_B{}^n \tag{58}$$

or the net rate of appearance of B

$$\frac{+dC_B}{d\theta} = k_1 C_A{}^m = k_2 C_B{}^n \tag{59}$$

Equation (57) is readily integrated and arranged in form suitable for the calculation of the constant k_1 from experimental data. The others are not so readily handled, but in such cases it is often possible to devise mathematical manipulations and approximations to permit the integration and calculation of the several velocity constants.

It may be noted that if, for example, the velocity constant k_2 is very much larger than k_1, the system in the early stages of the reaction will attain a stationary state in which the net rate of formation of B is zero and which corresponds to very small, scarcely detectable, concentrations of B. The rate of formation of D then becomes practically identical with the rate of disappearance of A, and the reaction appears to be of the mth order throughout its course, *e.g.*, at the stationary state $+dC_B/d\theta = 0$

$$k_1 C_A{}^m = k_2 C_B{}^n \tag{60}$$

whence

$$\frac{-dC_A}{d\theta} = \frac{dC_D}{d\theta} = k_1 C_A{}^m \tag{61}$$

If k_1 and k_2 are of the same relative order of magnitude, the concentration of B in the reaction mixture continually increases in the early stages, passes through a maximum at some intermediate time, diminishes, and eventually vanishes leaving the substance D as the final product of reaction. The course of such consecutive reactions with respect to the concentration-time relation of the several substances is illustrated in Fig. 8. The exact shape of each curve depends upon the orders of the several reactions. The concentration of B at its maximum and the position of the maximum depend also upon the relative values of the velocity constants for the several reactions. If B is a desired product, it is evidently important to stop the reaction at a time near the maximum.

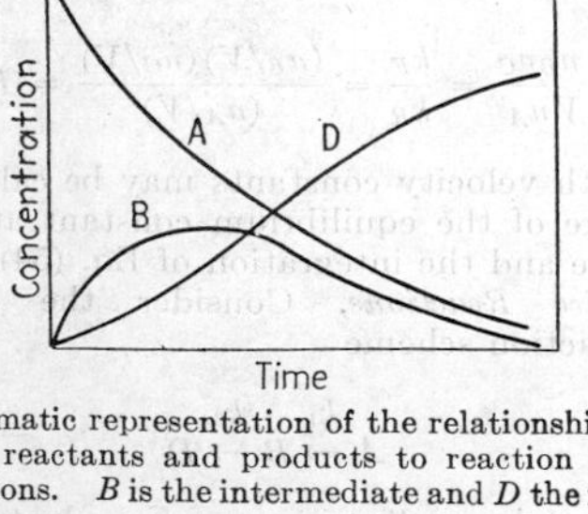

FIG. 8. Schematic representation of the relationship of the concentration of reactants and products to reaction time in consecutive reactions. B is the intermediate and D the final product.

Experimental data are often conveniently represented by plots such as Fig. 8. It is also frequently convenient in such cases to eliminate the time variable and to plot the percentage of a particular reactant consumed vs. the percentage of it which is converted into each of the products of reaction. Such curves can be calculated theoretically by developing the appropriate kinetic relations for postulated reaction schemes and orders and eliminating the time variable between the equations expressing rate of formation of products and disappearance of reactants. Comparison with experimental data may then be made by plotting analyses of the mixture at various times. Such a curve from the data of Mason *et al.* [*J. Chem. Soc.*, 1931, 3150] for the vapor-phase chlorination of benzene is illustrated in Fig. 9.

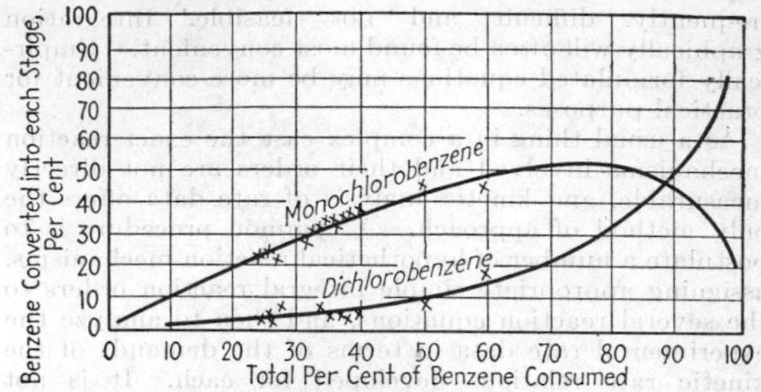

FIG. 9. Vapor-phase chlorination of benzene. Continuous curves = theoretical curves for two-stage reaction. Points marked X = experimental values. (*From Groggins, "Unit Processes in Organic Snythesis."*)

Watson (*op. cit.*) has indicated a method of treating a relatively complex case of consecutive reversible reactions of different order represented by the scheme of Eq. (44). This is exemplified by the pyrolysis of benzol to produce diphenyl. Triphenyl and hydrogen are also products.

$$\underset{(C_6H_6)}{2B} \underset{k_{R1}}{\overset{k_{F1}}{\rightleftarrows}} \underset{(C_{12}H_{10})}{D} + \underset{(H_2)}{H} \tag{62}$$

$$D + B \underset{k_{R2}}{\overset{k_{F2}}{\rightleftarrows}} \underset{(C_{18}H_{14})}{T} + \underset{(H_2)}{H} \tag{63}$$

If the benzol is initially free of the products, its rate of decomposition is given by

$$\frac{dn_B}{d\theta} = -k_{F1}\frac{n_B{}^2}{V} + k_{R1}\frac{n_D n_H}{V} - k_{F2}\frac{n_B n_D}{V} + k_{R2}\frac{n_T n_H}{V} \tag{64}$$

Similarly, for the rate of formation of diphenyl,

$$\frac{dn_D}{d\theta} = k_{F1}\frac{n_B{}^2}{V} - k_{R1}\frac{n_D n_H}{V} - k_{F2}\frac{n_B n_D}{V} + k_{R2}\frac{n_T n_H}{V} \tag{65}$$

Analogous equations may be written for the rate of formation of the other products.

To determine the several reaction-velocity constants, a series of experimental data covering a range of conversions at constant temperature is necessary. The moles of reactants and products per unit mass of system are plotted against reaction time. These curves may be graphically differentiated by taking the slope (this is the instantaneous reaction rate, *e.g.*, $-dn_B/d\theta$) at various time points and the corresponding concentration of benzene reactant and of products, respectively, read off for the same time. If the simultaneous values of rate $dn/d\theta$ and concentration are then substituted in the corresponding velocity equation, *e.g.*, Eqs. (64) and (65), four equations which may be solved simultaneously for the constants are obtained. In this specific case a plot of $dn_B/d\theta$ vs. n_B when extrapolated to n_{B0} (*i.e.*, zero time, thus obtaining the initial reaction rate) permits k_{F1} to be estimated. If equilibrium or the necessary thermodynamic data (see pp. 348–352) are available, k_{R1} may then be calculated from K and k_{F1}, since $K = k_{F1}/k_{R1}$. A similar procedure using the rate equation for the formation of triphenyl $dn_T/d\theta$ and the equilibrium constant of the second reaction permits the other constants to be estimated.

Side Reactions. The scheme

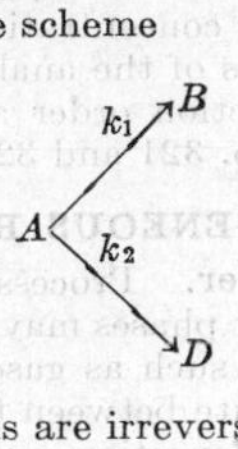

in which both reactions are irreversible is a simpler case. For general reactions of the mth and nth order at constant volume the kinetic relations are

$$\frac{+dC_B}{d\theta} = k_1 C_A{}^m \tag{66}$$

$$\frac{+dC_D}{d\theta} = k_2 C_A{}^n \tag{67}$$

$$\frac{-dC_A}{d\theta} = k_1 C_A{}^m + k_2 C_A{}^n \tag{68}$$

Also from Eqs. (66) and (67)

$$\frac{dC_B}{dC_D} = \frac{k_1}{k_2} C_A{}^{m-n} \tag{69}$$

Except in the simple case that both reactions are of the same order, integration is complicated. If this is the case,

$$\frac{-dC_A}{d\theta} = (k_1 + k_2)C_A{}^n \tag{70}$$

and

$$\frac{dC_B}{dC_D} = \frac{k_1}{k_2} \tag{71}$$

Equation (70) is readily integrated and the integral of (71) is

$$\frac{C_B}{C_D} = \frac{k_1}{k_2} = \text{const.}$$

From these relations and the appropriate experimental data k_1 and k_2 can then be calculated.

Non-isothermal Reactions. In technical practice heat effects attending chemical reaction or arising from the design of equipment frequently lead to reaction under non-isothermal conditions. To handle such cases modification of the simple kinetic equations previously given is necessary to take into account the varying temperature. Sherman [*Ind. Eng. Chem.*, **28**, 1026 (1936)] has developed integrated reaction rate equations involving temperature variations for a number of cases where the temperature variation is expressible as a relatively simple methematical function of time or composition.

Watson (*op. cit.*) suggests that the general case in which the relationship of temperature to time is complex and the underlying kinetic equation complicated may be handled through step-by-step integrations in which the reaction zone is considered to be divided into relatively small elements. His suggested procedure is to start from the terminal conditions at either end and to calculate successively the extent of conversion and the changes in conditions in each element. The arithmetic average values of such variables as temperature, pressure, and density may be used with satisfactory accuracy if sufficiently small elements are chosen. Although obviously laborious, complicated situations involving combined variation of temperature, pressure, rate of heat dissipation, and composition throughout the reaction zone can be solved without unreasonable expenditure of effort. It is sometimes necessary to proceed by trial, assuming as a first approximation the distribution of one variable, *e.g.*, pressure. The results of an analysis of the reaction zone on this basis are then used to obtain a corrected curve for the distribution of the variable as a second approximation, and this procedure is repeated until agreement is obtained. Watson states that such calculations rapidly converge to the correct solution and lead to satisfactory accuracy in two or three trials in the majority of problems.

In dealing with systems involving only a single reaction or reactions of approximately equal activation energies, Watson finds a concept of "equivalent time" useful in simplifying calculations. By this method a convenient intermediate temperature is arbitrarily selected as a base. Complete reaction-rate data are then compiled showing the time-conversion relationships at this base temperature, which are conveniently expressed graphically. In dealing with a given case all reaction times at different temperatures are calculated by the Arrhenius equation (41) to the equivalent time at the base temperature required to produce the same degree of reaction. Such equivalent times may then be added directly and all rate calculations made for the base temperature. To facilitate calculations of equivalent time, Watson develops a chart based on the Arrhenius equation which relates temperature and activation energy to a correction factor f by which reaction time at any temperature T must be multiplied to give the equivalent time at the base temperature T_B.

Determination of Reaction Order. *Method* 1. The change in concentration of either the reactants or products, whichever is most conveniently measured, is determined as a function of time during the progress of the reaction. From the sets of values thus obtained, series of k values are calculated from the integrated rate equations corresponding to various postulated reaction orders and mechanisms. The equation giving the most satisfactory constant series of k values represents the reaction order. In using this method it is important to make several sets of experimental measurements employing different initial concentrations in order to avoid erroneous conclusions. Rather than calculating velocity constants analytically, this method may be applied graphically by plotting the data in such manner that comparison with the postulated kinetic equation similarly plotted may be made. A method of plotting giving a straight line is usually preferable.

This method is more readily applied to relatively simple reaction cases; side and consecutive reactions, if they occur to a significant extent, may lead to erroneous conclusions unless they can be accounted for in the developed kinetic equation.

Method 2. The reaction order may be determined from the times required for equal fractional decreases of concentration in experiments where different initial concentrations of reactants are used, and, if more than one reactant is involved, all are present in equivalent amounts. Let two experiments be made with different initial concentrations C_0 and C_0' and let θ_1 and θ_2 be the respective time intervals required for the same definite fraction (usually taken as one-half or one-fourth of the initial concentration) of the reactants originally present to undergo reaction. In general, the time θ required for any selected fractional completion of reaction is proportional to $1/C^{n-1}$ where n represents the order of reaction. Therefore, if θ_1 and θ_2 are taken so that the same fraction f of the initial concentrations in the two experiments has reacted, then

$$\frac{\theta_1}{\theta_2} = \frac{C'_0{}^{n-1}}{C_0{}^{n-1}} \tag{72}$$

Hence

$$n = 1 + \frac{\log \theta_1 - \log \theta_2}{\log C_0' - \log C_0} \tag{73}$$

from which expression the value of n may be determined.

Instead of using two separate experiments, if the reaction follows the same course throughout, the time θ required for the fraction f of the initial concentration to react may be compared with the succeeding time interval required for the same fraction f of the concentration remaining at the end of time θ to undergo reaction. Thus, for example, if the reaction is bimolecular the time required for the initial concentration C_0 to decrease to $\frac{1}{2}C_0$ will be only half that required for the succeeding decrease from $\frac{1}{2}C_0$ to $\frac{1}{4}C_0$.

This method is the most accurate of the various methods which may be employed and is particularly useful where side and consecutive reactions occur to a small extent. Its accuracy and utility at constant pressure in flow systems are, however, limited.

Method 3. The number of molecules of each substance involved in a polymolecular reaction may be investigated by a series of experiments in which each reactant is in turn present in very small concentration compared with that of the other reactants. It is evident from the general velocity equation, $-dC/d\theta = kC_A^{n_1}C_B^{n_2}C_D^{n_3}$, that, if C_B and C_D are very large compared with C_A, they may be regarded as constants and the whole reaction will appear to be of the n_1th order.

$$\frac{-dC}{d\theta} = (kC_B^{n_2}C_D^{n_3})C_A^{n_1} = k'C_A^{n_1} \tag{74}$$

The value of n_1 can then be determined by one of the preceding methods. Similarly the values of n_2 and n_3 may be determined.

Method 4. In complex reactions the influence of the concentration of each reactant on the rate may be approximately investigated by varying the concentration of each in turn, maintaining the others constant, and measuring the *initial* reaction rate over a very short time interval such that the change in concentrations is not over 10 per cent. If C_0 and C_1 represent two different initial concentrations of the reactant being investigated, the concentrations of the others being practically constant, we may write over short time intervals

$$\frac{\Delta C_0/\Delta\theta}{\Delta C_1/\Delta\theta} = \left(\frac{C_0}{C_1}\right)^n \text{ approx.} \tag{75}$$

where n is the exponent of the concentration of this reactant in the velocity equation.

These methods are most readily applied to relatively simple kinetic cases. In complex cases it may be necessary to devise special procedures, and it is not always possible to determine the true reaction order to differentiate between several possible mechanisms. Unless the case is obviously a simple one, data should be tested by checking against several of the above methods.

Experiments to study reaction mechanism and kinetics are usually conducted isothermally at constant volume. Although reaction order may be investigated by these methods for gas reactions proceeding at constant pressure in flow systems, less precision is obtained since the time of contact (which measures the time interval during which the reaction is progressing) is not a directly measurable quantity and depends upon the fractional conversion when the reaction involves a change in volume. Since for small conversions the time of contact is approximately constant if the rate of flow is constant, the reaction order may be most satisfactorily studied by determining the effect on the fractional conversion of varying the entering concentration of reactants at constant rate of flow. Thus, for example, if only one reactant is involved and the fractional conversion is small, the latter should be approximately independent of the entering concentration of reactant if the reaction is unimolecular and should be approximately doubled by doubling the entering concentration if it is bimolecular.

Numerous examples of the analysis of kinetic data for determination of reaction order are to be found in the references cited on pp. 321 and 325.

HETEROGENEOUS REACTIONS

General Character. Processes occurring between substances in different phases may involve solely physical processes of solution, such as gases and solids in liquids or the transfer of solute between two immiscible liquids; vaporization or condensation, such as vapors into or from gases; or physical adsorption, such as vapors by charcoal or silica gel. The rates and mechanisms of such processes are treated in other sections of the handbook (see, *e.g.*, Secs. 8, 9, 10, 11, 12, and 13). The velocities of such processes are in general governed by the rates at which the particular substance in question is transferred to and away from the interface between the two phases by convection and diffusion through the laminar fluid films existing at such interfaces.

In numerous important cases chemical reaction takes place between substances in different phases and is frequently the process of principal significance. Since reactant from one phase must be brought into the presence of the other in the second phase as a preliminary to reaction, in the general case both the rate of this material transfer, or diffusion process, as well as that of the chemical reaction is involved. If the chemical reaction is slow relative to the diffusional process, the net rate of the entire process will be controlled by the rate of the chemical reaction and exhibit the corresponding kinetic behavior. On the other hand, if, relatively, the rate of the chemical reaction is rapid, the net rate will be controlled by the diffusional process and exhibit corresponding characteristics. Both situations are known to exist. It is not always easy to predict or to determine which is the fact. However, it is well established in many cases, *e.g.*, in reactions at the surface of heterogeneous contact catalysts, that the mechanism and rate of the process are governed by the reaction occurring on the catalyst surface.

The rates of both the diffusional process and chemical reaction vary with conditions, but the extent to which their rates are affected by conditions varies. Since such reactions take place at, or adjacent to, the interface between two phases, their rates are, in general, whether diffusion or reaction controls, directly proportional to the surface of contact between the two. Compared with the marked acceleration of the velocity of chemical reaction by increasing temperature, the velocity of diffusion is much less affected. The temperature coefficients and activation energies of surface catalytic reactions are, however, usually considerably smaller than those for homogeneous reactions. The nature and activity of the catalyst surface have, in contrast to the diffusional processes, a profound influence on both the velocity and course of chemical reaction. On the other hand, while the diffusional transfer process is speeded up by agitation of one or both phases, the velocity of the chemical reaction is unaffected.

It is thus evident that the nature and relative control in a heterogeneous process may be altered by varying the conditions and that a heterogeneous reaction occurring in the same system may under one set of conditions be governed by diffusion while under other conditions the velocity of the chemical reaction is of principal importance. Thus, for example, the speed of reaction of a dissolved substance with gaseous oxygen or hydrogen in a quiescent system may be controlled by diffusion, whereas under vigorous agitation typical chemical kinetic behavior is exhibited. A gas reaction proceeding in contact with a very active catalyst at high temperature may be controlled by diffusion, whereas with a less active catalyst at low temperature the kinetics of the chemical reaction predominate, exhibiting, for example, a temperature coefficient typical of such cases. Cases where both situations exist under different conditions in the same system have been observed.

The speed of the chemical reaction may be such that its rate controls the process over the range of conditions which it is normally possible to realize in practice. This appears undoubtedly to be the case in the majority of catalytic gas reactions. The reverse is also true. Where ionic reactions which are usually very rapid are involved, the rate is likely to be governed by the diffusion process. These differences in the dependence of velocities of diffusion and of chemical reaction on operating conditions are commonly utilized to discover which process controls in heterogeneous reactions. Knowledge thus gained determines the conditions most suitable for the reaction.

The influence of diffusion, fluid, and heat flow on the yield in chemical reactions has been treated by Damkoehler [Eucken-Jakob, "Der Chemie-Ingenieur" vol. 3, part 1, p. 359 (1937)]; Jakob [*Trans. Am. Inst. Chem. Engrs.*, **34**, 173 (1938); **35**, 563 (1939)] has discussed the relation of heat transfer to the true temperature of the catalyst surface in catalytic gas reactions.

Heterogeneous Reaction Cases. Many gas reactions, superficially homogeneous, in reality take place wholly or partly on the walls or surface of the reaction vessel. Liquid-phase reactions are also influenced by the catalytic effect of the walls of the container. Whether the reaction is heterogeneous and catalyzed by the walls is not always readily distinguished. A common test method is to increase the surface by changing the size or shape of the vessel or adding broken material of the same nature to the vessel, or to block the surface by covering it with a supposedly inert material. Many reactions of the so-called "chain" type, such as rapid combustions or polymerizations, may be either initiated or stopped by the catalytic action of solid surfaces.

In the reaction of solids with dissolved substances the quantity of solid reacting in unit time is proportional to the surface of the solid and to the concentration of the dissolved substance. The rate is not universally controlled by diffusion velocities. For example, in the solution of certain metals in acids (Sn, Cd, and Al in HCl) the velocity of the process is controlled by that of the chemical reaction at the interface and is independent of diffusion velocity. In these cases the velocity is affected very little by agitation and has a high temperature coefficient.

In the reaction of gases with solids such as the reduction of metallic oxides by hydrogen the chemical reaction appears in many cases to be accelerated by the catalytic effect of the reduced metal. Such reactions commence slowly and speed up rapidly with time as reduced metal appears. Such cases are termed *autocatalytic*.

Contact Catalytic Gas Reactions. The velocities of gas reactions at solid catalyst surfaces and the factors influencing them are of great importance technically, but no quantitative generalizations can be made concerning them. Many of the scientific principles involved have been established, but rigorous general theoretical treatment of their kinetics is not yet feasible. For engineering purposes they must still be approached on an empirical basis. Kinetic principles developed in connection with homogeneous reactions serve as a guide but require modification in application.

The rates of such processes and their reaction orders are determined by the nature and extent of the catalyst surface* and by the relative degree to which the reactants and products are adsorbed by the catalyst. Whereas in homogeneous reactions the determined order represents the number of molecules entering into the reaction, this is not necessarily the case in heterogeneous reactions. Thus a catalyzed reaction between two substances, which is in reality bimolecular, may appear to be first order due to the fact that one of the reactants is strongly adsorbed by the surface independently of its pressure. In many cases the rates of such reactions are inversely proportional to some power of the concentration of the products. The presence of foreign substances in the reactants even in minor amounts often exerts a powerful retarding or poisoning influence on the reaction speed owing to their preferential adsorption on the surface. The reaction rate may be an inverse function of the concentrations of such substances or stopped entirely by their presence in traces. The effect may be reversible, the activity of the catalyst being restored by their removal from the reactants, or it may be entirely irreversible.

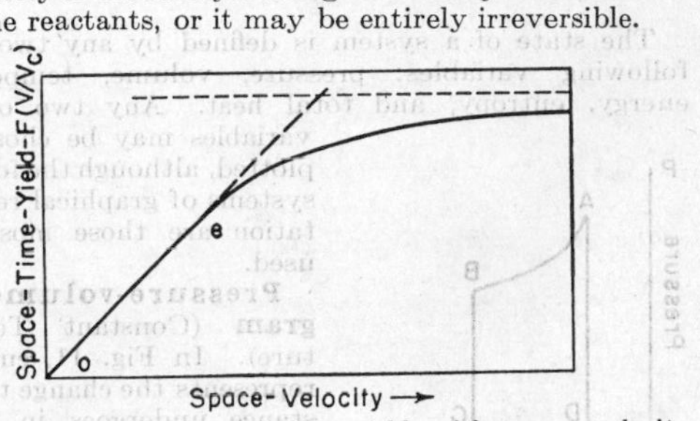

FIG. 10. Variation of space-time-yield with space velocity (schematic).

Comprehensive treatments of this subject are to be found in the references heading this topic and the general section.

Flow Systems. Space-time-yield. The analysis and correlation of reaction-velocity results in flow systems are very complicated and have not reached the state of development attained in the treatment of results obtained for scientific purposes in static or batch systems. The role played by diffusion, flow, and heat-transfer factors has not been definitely established, and nonisothermal conditions are the normal rule in large-scale industrial reactors. For methods of treating the kinetics of catalyzed gas reactions under flow conditions and development of relations for specific cases, see Benton *Ind. Eng. Chem.*, **19**, 494 (1927); *J. Am. Chem. Soc.*, **53**, 2984 (1931); Laupichler, Mass Transfer and Catalyst Activity in the Catalytic Water Gas Reaction, *Ind. Eng. Chem.*, **30**, 578 (1938); Kassel, Application of Reaction Kinetics to Process Design—Catalytic Hydrogenation of Isooctene, *Ind. Eng. Chem.*, **31**, 275 (1939).

In flow systems, other conditions being constant, the fractional conversion is determined by the time during which the reactants are in contact with the catalyst. The time of contact is not directly measurable but depends upon the rate of flow and the quantity of the catalyst. It is usually treated in practice in terms of space velocity. Space velocity is defined to be the volume of gas (S.T.P.)† or liquid passing through a given volume of catalyst space v_c in unit time divided by the latter, *i.e.*, space velocity = V/v_c. It may be referred to any principal constituent of the reacting mixture. The yield of desired product in unit time per

* This factor may be largely influenced by the method and conditions under which the catalyst is prepared.

† It would seem more correct to measure this volume at the temperature and pressure of the reaction space.

unit volume of catalyst per passage is the space-time-yield. This is the product of the fractional conversion F by the space velocity, *space-time-yield* $= F(V/v_c)$. In Fig. 10 the variation of space-time-yield with space velocity is illustrated schematically. The initial portion of the curve represents the low space-velocity region in which equilibrium conditions are always attained, the space-time-yield here being directly proportional to the space velocity. At higher space velocities, *i.e.*, above e, equilibrium conditions are not attained and the fractional conversion becomes a function of space velocity. It decreases with increasing space velocity, however, at a rate which is less than the increase of the latter. Consequently, the space-time-yield, F (*space velocity*), continues to increase, but at a slower rate. At very high space velocities the space-time-yield tends to become independent of space velocity and approaches a constant steady value determined by the nature of the catalyst surface and its temperature. Where consecutive or side reactions are involved, the desired product being formed as an intermediate step in the series of reactions, the space-time-yield–space-velocity curve referred to this product may pass through a maximum, the yield of this product diminishing at too high space velocities. This is due to the diverging effect of contact time on the relative over-all rates of the several competing reactions. This cannot be the case where only a single reaction is taking place. In cases where the reaction is retarded by the products, a relatively greater increase of space-time-yield with space velocity is to be expected than in cases where the reaction is not so retarded.

Space velocity plays a predominant role in determining the size of equipment required for a given production. In practice the optimum space velocity involves a balance between increased yield and the additional operating costs attending increases in space velocity.

GRAPHICAL REPRESENTATION OF THERMODYNAMIC FUNCTIONS

The state of a system is defined by any two of the following variables: pressure, volume, temperature, energy, entropy, and total heat. Any two of these variables may be chosen and plotted, although the following systems of graphical representation are those most often used.

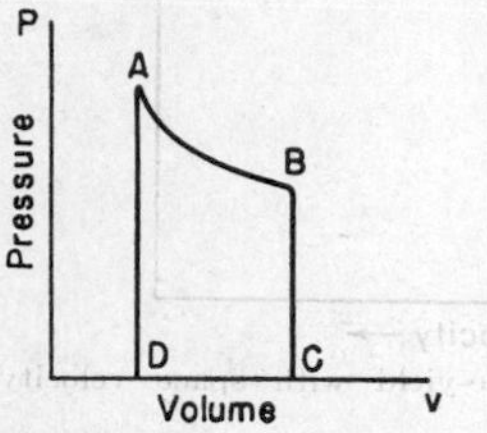

FIG. 11. Pressure-volume diagram.

Pressure-volume Diagram (Constant Temperature). In Fig. 11, curve AB represents the change the substance undergoes in passing from state A to state B at constant temperature. State A corresponds to p_1V_1; state B to p_2V_2; and the area between the curve and the x-axis (area $ABCDA$) is given by the integral $\int_{V_1}^{V_2} p\,dV$ and represents the external work done by the gas in going from state A (high pressure and small volume) to state B (low pressure and large volume).

Pressure-volume-temperature Diagram. Figure 12 represents the pressure-volume-temperature relations of a pure substance. Assume a gas to be in a state represented by a point (Ap_A, V_A) on the 150°C. isotherm. Let this gas be compressed isothermally (*i.e.*, along the 150°C. line AE). When state E is reached, the gas is saturated and this state (E) is represented by the coordinates p_sV_s, the pressure and volume, respectively, at saturation. If the compression is continued at constant temperature, the pressure remains constant and follows the line ECB until state B is reached. At B the material being compressed is entirely liquid. Curve V_1BN is the saturated liquid curve, and curve V_2EN is the saturated vapor curve. At any point on V_1BN the material is entirely liquid, and at any point on V_2EN the material is entirely vapor. Within the dome V_1BNEV_2 the material is a mixture of vapor and liquid. At point C the relative length of the lines CB/BE is the fraction of the mixture in the form of vapor, and this fraction is called the *quality* of the mixture and is usually denoted by x. CE/BE or $(1 - x)$ is the fraction of the material that is liquid. The region to the right of the curve V_2EN is the superheated region, and the region to the left of the curve V_2BN is the liquid region. Along the saturation line in the entire dome V_2EN the pressure is a function only of the temperature; and the volume depends upon the temperature and the quality x. In the superheated region the volume is a function of the pressure and the temperature, both of which may be varied independently. The point N represents the **critical point**, the coordinates of which are p_c and V_c, the critical pressure and critical volume, respectively; the temperature corresponding to this point is the critical temperature T_c and is defined as that temperature above which no pressure, however high, causes the separation

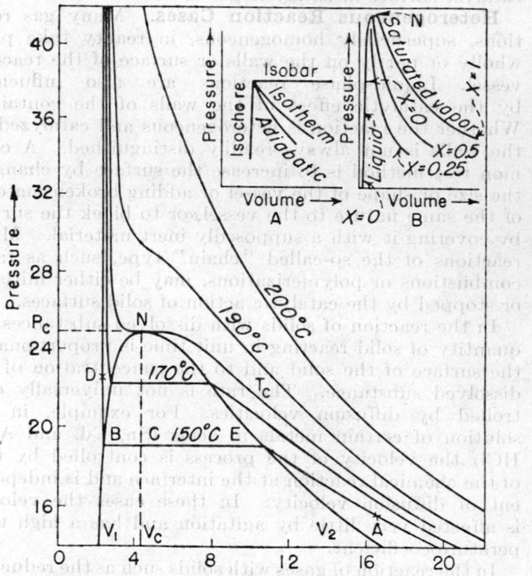

FIG. 12. Pressure-volume-temperature diagram.

of additional phases. The inserted small graph A gives the relative positions of the principal variables of such a system: **isobar** (constant-pressure), **isotherm** (constant-temperature), **adiabatic** and **isochore** (constant-volume) lines. The inserted graph B indicates the relative positions of the constant-quality (x) lines.

Pressure-volume vs. Pressure Diagram (Fig. 13). The pressure-volume relations of gases are often plotted as pV vs. p. A perfect gas would give a horizontal line at any definite pressure. The construction of this diagram is comparatively simple. Assume a given mass of material, as 1 lb. or 1 g. Given the p-V-T data for a definite gas, the product pV is (usually) plotted as the ordinate with p as the abscissa. An isothermal is easily obtained by assuming a definite value for the temperature, and assuming successive values of p to obtain corresponding successive values of pV. Similarly, the constant-volume lines are easily obtained from the pV-p relation ($V = pV/p$). The total heat at any condition of p-V-T may be obtained as follows: Fix some arbitrary temperature as a base temperature which is below the region in which it is desired to operate. The **total heat** i of the liquid is then at this base temperature, by convention, equal to zero. The amount of heat a to heat the liquid from this base temperature to a chosen boiling temperature is calculated by multiplying the mass of material assumed by its specific heat and by the difference between the temperature assumed and the base temperature. Next calculate the heat required to vaporize the mass of material to give a heat b. Next calculate the heat necessary to raise the temperature of the vaporized material to the chosen temperature to give heat c. The sum of a, b, and c will give the total heat of the superheated vapor above the base temperature chosen. Obviously, by a similar procedure, lines of constant total-heat content may be calculated. As a matter of convenience in the preparation of a diagram of this type, it is convenient to pursue the following order for construction: the vapor-liquid dome, the constant-volume lines, the constant-temperature lines, and finally the constant total-heat lines.

Temperature-entropy Diagram (Fig. 16). Assume some temperature below the range in which it is desired to work, and arbitrarily assume the entropy to be zero

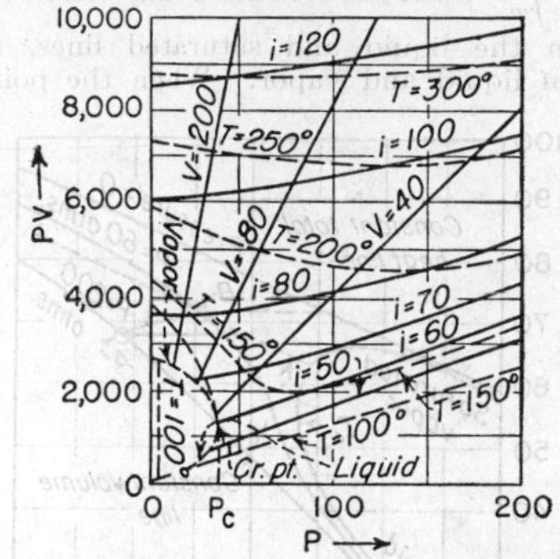

FIG. 13. Pressure-volume vs. pressure (for air). (*Wien-Harms, Handbuch der Experimentalphysik, vol.* 9, *Part* 1, "*Gasverflüssigung und ihre thermodynamischen Grundlagen,*" *by H. Lenz, pp.* 90, 106, *Akademische Verlagsgesellschaft, Leipzig,* 1929.)

at this temperature. The heat required to raise a unit weight, say, 1 lb.-mole of a liquid, divided by the temperature at which the heat is added, will give the entropy of the liquid (S_l) at that temperature; dS_l is then =

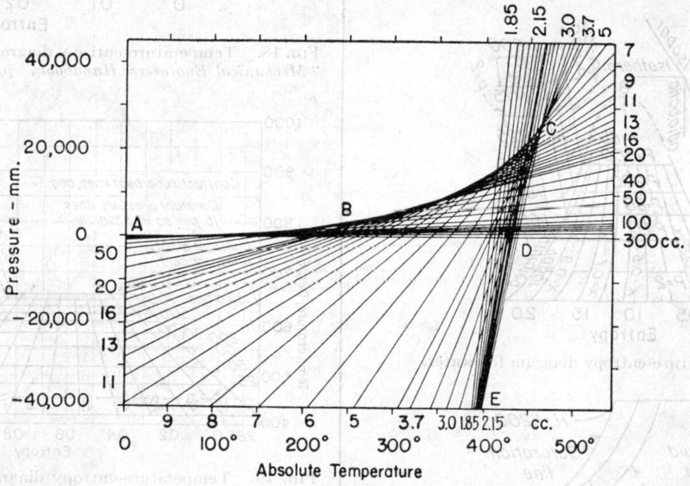

FIG. 14. Pressure-temperature diagram, showing isochores. (*Reproduced by permission of Longmans, Green, from Young, "Stoichiometry."*)

Pressure-temperature Diagram (Fig. 14). The lines of constant volume (isochores) are shown, which form a complicated three-dimensional surface. It will be noted that these isochores are nearly straight lines.

Temperature-energy Diagram (Fig. 15). The internal energy as a function of the temperature (°C. abs.) is shown in Fig. 15. This T-E type of diagram is useful in throttling (wire-drawing) processes or changes of state. Consider the change of state from A to B, and assume A to be defined by $p_1 = 200$ atm., $T_1 = 210°$, and by V_1, and E_1; and B to be defined by $p = 60$ atm., $T_2 = 175°$, and by V_2 and E_2. (1) Assuming E, the internal energy, to be constant and starting at A, we follow the constant-energy line E to the constant-volume line V_2 corresponding to the final pressure and temperature p_2 and T_2; then (2) follow the constant-volume line (V_2) to the (final pressure p_2) isobar to T_2. Hence $E_2 - E_1$ is obtained. Other applications will be obvious.

$C_p\,dT/T$ and $S_l = \int_{T_{\text{base}}}^{T} C_p\,dT/T$. The series of points obtained up to the critical temperature will give the so-called "liquid curve," as OB in Fig. 16.

During vaporization an amount of heat, L, is absorbed at the vaporization temperature; therefore, the increase in entropy is L/T. The entropy of the saturated vapor S_v is then at a constant temperature, the entropy of the liquid S_l plus the entropy of vaporization per unit mass (per pound-mole), *i.e.*, entropy of the saturated vapor = $S_l + \frac{T}{L}$. A sufficient number of values of S_v are calculated between the reference (base) temperature and the critical temperature to give the saturation curve.

If the saturated vapor is further heated, the material heated undergoes a further increase of entropy and goes into the superheated region. If T_1 and T_2 are the absolute temperatures of the initial and final states and C_p is the specific heat of the superheated vapor, the entropy increase is $\int_{T_1}^{T_2} C_p\,dT/T$. Points lying within the dome, *i.e.*, within the liquid and saturated lines, represent mixtures of liquid and vapor. When the point under

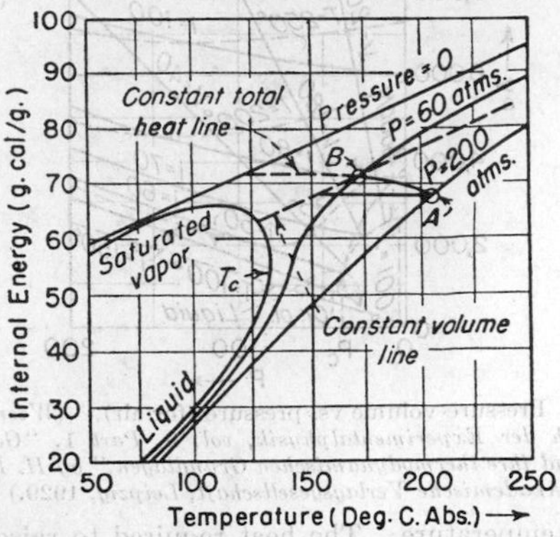

FIG. 15. Temperature-energy diagram. *Wien-Harms, Handbuch der Experimentalphysik, vol. 9, Part 1, "Gasverflüssigung und ihre thermodynamischen Grundlagen," by H. Lenz, p. 88, Akademische Verlagsgesellschaft, Leipzig, 1929.*)

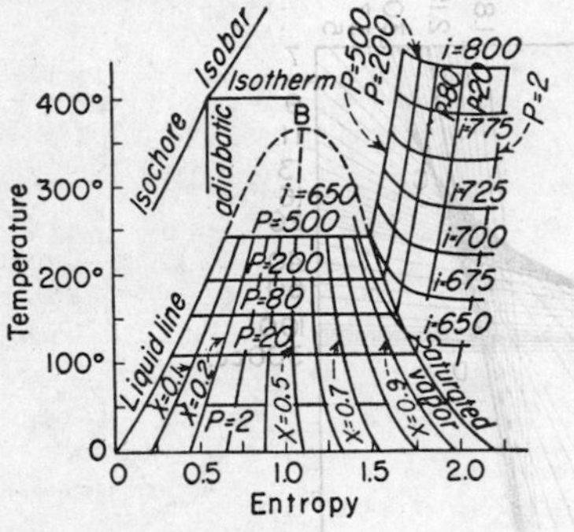

FIG. 16. Temperature-entropy diagram for water.

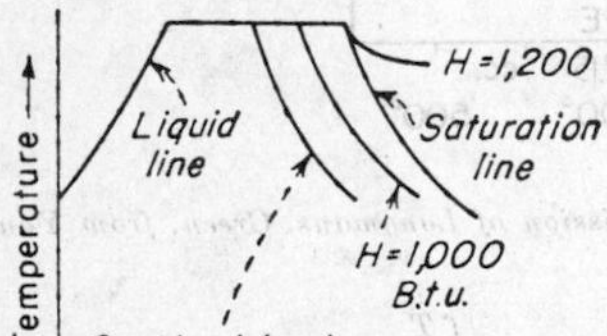

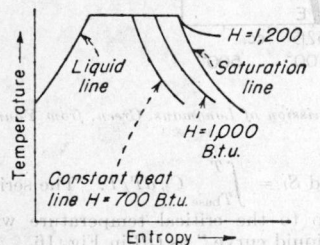

FIG. 17. Temperature-entropy diagram for steam showing constant heat lines. H = constant.

consideration is near the saturation curve, the values of the thermal factors may be found by subtracting from their values at the saturation limit. For points on the same adiabatic, the sum of the entropy of the liquid from the base temperature to any temperature and the product of the quality and the entropy of complete vaporization at that temperature are constant.

Chemical engineers frequently have need for the thermodynamic properties of other substances. For this reason we have included thermodynamic diagrams for the available substances and have based typical calculations therewith on the temperature-entropy diagram for hydrogen. See Figs. 16 and 17, Sec. 3.

Example 1. To superheat 1 kg. hydrogen from 60° to 90°K. at 1 atm. requires what amount of heat? At 60°K. and 1 atm. total heat = 209; while at 90°K. and 1 atm., total heat = 284; therefore heat to be added = 284 − 209 cal./kg.

Example 2. To compress 1 kg. hydrogen from 0.6 atm. to 50 atm. at a constant temperature of 45°K., what amount of heat must be abstracted? At 0.06 atm. and 45°K., entropy = 0.8; and at 50 atm. and 45°K., entropy = 0.24. Heat to be abstracted = $T\,ds$ = 45(0.8 − 0.24) = 25.2 cal.

Example 3. For 1 kg. hydrogen at 60°K. and 30 atm., what is the volume occupied; the total heat; and the entropy? At the intersection of the 30-atm. isobar and the 60°K. isothermal, the volume is 72; the total heat is 181 cal./kg.; and the entropy is 0.44.

Example 4. One kilogram liquid hydrogen at 4.5 atm. and 27°K. is expanded adiabatically to 1 atm. What are the final temperature and the final quality? Following the constant-entropy line vertically downward (S = about 0.08) to the 1-atm. isobar, we arrive at a temperature of slightly over 20°K.; and interpolating between the 0.1 and 0.2 quality lines at a constant quality of about 0.18.

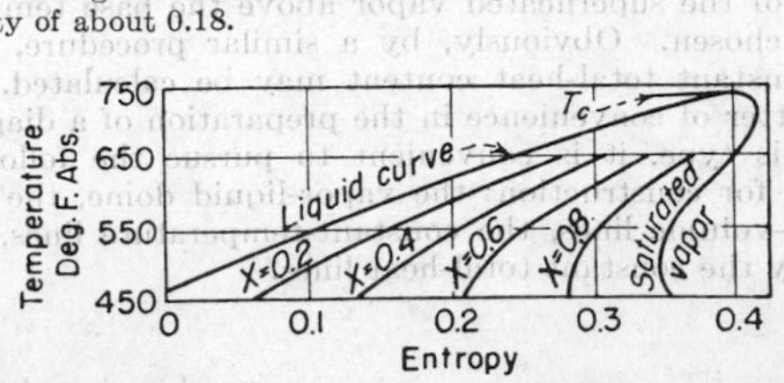

FIG. 18. Temperature-entropy diagram for isobutane. (*Marks, "Mechanical Engineers' Handbook," p. 329, McGraw-Hill, 1930.*)

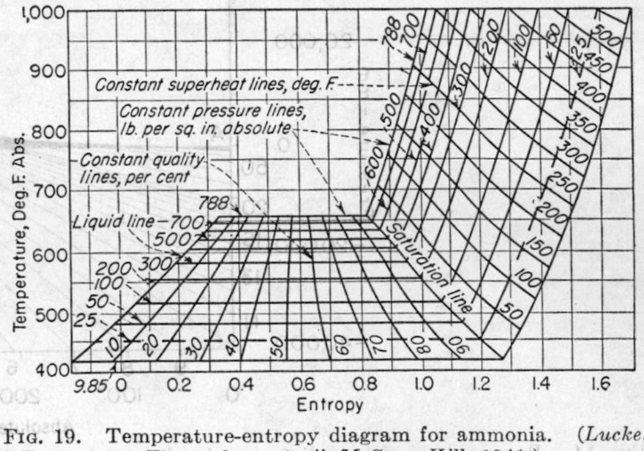

FIG. 19. Temperature-entropy diagram for ammonia. (*Lucke, "Engineering Thermodynamics," McGraw-Hill, 1941.*)

The temperature-entropy diagrams indicated are included here for the convenience of engineers, since most of these are available only in journals or books that are not readily obtainable. Figure 17, for steam, shows the relative positions of the liquid and saturation (vapor) lines, as well as the relative positions of total-constant-heat lines in a temperature-entropy diagram. Figure 18 shows a portion of the temperature-entropy diagram for isobutane, with four lines of constant quality (x = 0.2, 0.4, 0.6, and 0.8). Figure 19 is an abbreviated portion of the temperature-entropy diagram for ammonia. Figure 20 is a similar portion of the temperature-entropy diagram for carbon dioxide. Similar figures for other compounds are given in the following publication: *Commun. Kamerlingh Onnes Lab. Univ. Leiden, Suppl.* 65—ethylene, helium, hydrogen, methane, nitrogen.

Mollier Diagram (Heat-content vs. Entropy Chart). This type of chart is especially useful in problems involving the flow of fluids, such as wire-drawing or throttling, the main distinction over other types of chart being that heat and work quantities are represented by line segments instead of by areas. The slope of any isobar is a measure of the temperature, since at

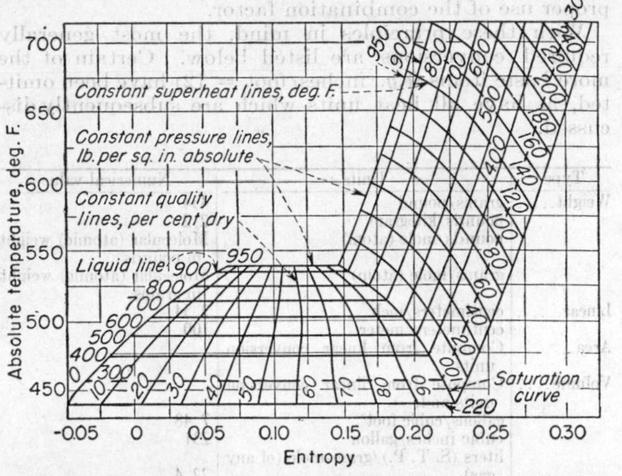

FIG. 20. Temperature-entropy diagram for carbon dioxide. (*Lucke, "Engineering Thermodynamics," McGraw-Hill*, 1941.)

constant pressure the change in total heat equals the product of the temperature and the change in entropy. Therefore, in the wet region the isobars are straight, but in the superheated vapor region the slope of the isobars changes as the amount of superheat increases. The most useful portions of this chart are the regions immediately below and immediately above the vapor-liquid boundary curve, and for this reason most Mollier charts are large-scale drawings of these regions and show little if any of the other regions of the complete total heat-entropy diagram. See pp. 249–280 for Mollier and other diagrams.

Joule-Thomson Effect or Throttling. If actual gases or vapors are expanded isenthalpically, a cooling effect is produced which is measured by the decrease of temperature per unit decrease of pressure with the total heat of the system remaining constant, *i.e.*, $(dT/dP)_i =$ the Joule-Thomson effect. Although the general effect of throttling is to cause decrease of temperature of the

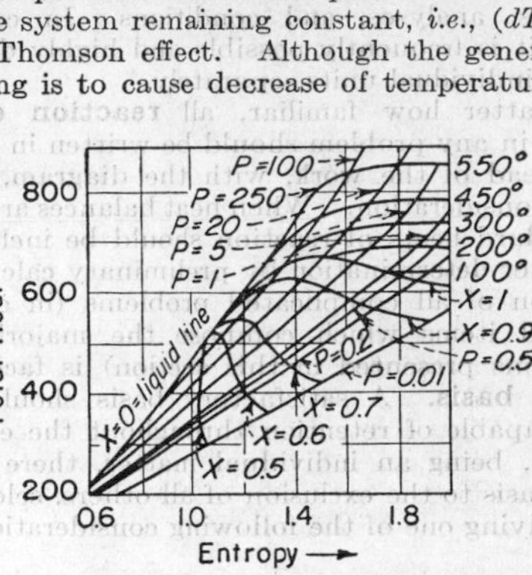

FIG. 21. Mollier diagram. Total heat entropy for water. (*Wien-Harms, "Handbuch der Experimentalphysik," vol.* 9, *Part* 2; *"Wärmekraft und Wärmearbeitsmaschinen," by A. Loschge, p.* 11, *Akademische Verlagsgesellschaft, Leipzig,* 1929.)

gas throttled, a gas may be heated during a throttling process.

If an initially saturated gas is throttled, thereby lowering its pressure, it becomes superheated. Similarly, a wet gas becomes drier. Horizontal lines (of constant total heat) on the Mollier diagram indicate the throttling process, and hence one can follow the amount of superheating a gas undergoes when it is throttled, and also one can determine the extent to which a liquid (or a wet mixture) will be vaporized. Since the Joule-Thomson effect $(dT/dp)_i = [T(dV/dT)c - V]/C_p$, the effect becomes zero when $V/T = (dV/dT)_p$. The temperature at which (at any definite pressure) $(dV/dT)_p = 0$ is called the **inversion temperature.** Above this temperature the throttling effect is positive, and the temperature is lowered. Obviously, the inversion temperature is a function of the pressure.

TECHNICAL CALCULATIONS*

GENERAL REFERENCES: Bichowsky and Rossini, "Thermochemistry of Chemical Substances," Reinhold, New York, 1936. Dodge, "Chemical Engineering Thermodynamics," McGraw-Hill, New York, 1944. Hodgman, "Handbook of Chemistry and Physics," Chemical Rubber Publishing Co., Cleveland, Ohio, 1946. Lewis and Randall, "Thermodynamics and the Free Energy of Chemical Substances," McGraw-Hill, New York, 1923. Lange, "Handbook of Chemistry," Handbook Publishers, Sandusky, Ohio, 1946. Rossini and coworkers, American Petroleum Institute Research Project 44, National Bureau of Standards; and *Bur. Standards J. Research*, **33**, 255 (1944); **34**, 59, 65, 143, 263, 403 (1945). Sage and Lacey, "Phase Equilibria in Hydrocarbon Systems," Stanford University Press, Stanford University, Calif., 1939. Sherwood and Reed, "Applied Mathematics in Chemical Engineering," McGraw-Hill, New York, 1939. Weber, "Thermodynamics for Chemical Engineers," Wiley, New York, 1939. Wenner, "Thermochemical Calculations," McGraw-Hill, New York, 1941.

Introduction. To present comprehensive examples covering every physical chemical problem which may arise is obviously impossible. It is the intention of this section to illustrate the use of the principles through the inclusion of calculations typifying an approach along general lines. The majority of illustrations are purposely kept simple to emphasize specific points; combination of many items into involved problems would tend to submerge the important detail, making them thoroughly inaccessible for reference.

Calculations and methods should not be considered as formulas through which analogous problems may be worked. It is, however, hoped that the material presented may be found valuable in the suggestion of methods of attack.

The human mind is extremely variable; thus, what might seem to the author to be the only logical method of solving one type of problem might well be completely at variance with another's ideas. With this fact under consideration, no attempt has been made to standardize problem solutions. Further, because of this situation, it is thought well to consider only certain basic principles of problem attack.

General Considerations in Problem Solution. A thorough knowledge of the problem as presented is obviously necessary. The accumulation of this knowledge may involve many readings and rereadings of the material at hand, or it may be completely accomplished by one careful consideration, if the work is simple.

Since visualization is almost inevitably an aid to clear analysis, **diagrams** of complicated problems are practical and, often timesaving. To fulfill its requirements, a diagram should be simple while still representing

* Acknowledgment is due to W. P. Ryan, whose work, entitled Industrial Stoichiometry, appearing in the first edition of this section has been freely drawn upon in the first revision.

the general features of the equipment. The sketch should be large enough to include basic data regarding quantities, analyses, and conditions. In complicated systems it is frequently possible and highly desirable to consider individual units separately.*

No matter how familiar, all **reaction equations** involved in any problem should be written in full, either at the head of the work, with the diagram, or at the point of consideration. When heat balances are involved, the standard heats of reaction should be included, generally after determination by preliminary calculation.

Solution of all complicated problems (in contrast to individual items which compose the majority of the illustrations presented in this section) is facilitated by use of a **basis.** A satisfactory basis should be of a nature capable of retention throughout the entire solution, but, being an individual matter, there is no one correct basis to the exclusion of all others, selection usually involving one of the following considerations:

1. Custom of an individual or concern, to provide easy analysis of methods or to check calculations.
2. Completeness of analytical data on a particular stream (*e.g.*, Orsat analysis and dew point of a flue gas).
3. Inertness of a component to subsequent chemical reaction or physical change (*e.g.*, nitrogen in combustion, non-volatile matter in evaporation, solvent in extraction or crystallization).

One common basis which is worthy of note is that of time, usually chosen when capacity is involved, and frequently implying a quantity basis as well.

A point which cannot be overlooked but which should be automatic with those dealing with chemical reactions and/or physical changes is the value of dealing with **quantities expressed as moles.** Not only is analysis easier, but the calculations are generally simplified, with resultant reduction of the chances of errors. These statements do not imply that the basis should invariably be expressed in moles, for analyses of solids and liquids are customarily reported in weight units; even in these cases, however, simplicity is generally found to result from the immediate conversion of weights to moles when reaction is involved.

One final point which cannot be overemphasized is the proper compilation and **labeling of results,** both intermediate and final, and orderly recording of all calculations performed. For clarity, ease of checking, and proper interpretation, such system cannot be neglected. Wherever possible, tabular or graphical representations aid general clarity.

Units and Their Conversion. Although industry operates in large part on the English system of units, scientific chemical and physical data are invariably reported in metric units. Thus the chemical engineer must be thoroughly conversant with both systems and with their interrelation (*i.e.*, conversion factors).

Conversions from one system to another may be performed directly by the use of handbooks; if one becomes thoroughly familiar with one such reference, it may be possible to open quickly to the proper section without recourse to the slow procedure of searching through an often inadequate index. However, when the factor is obtained, frequently the problem still remains as to the manner in which it should be applied, *i.e.*, multiplication or division. This problem arises in any case where absolute visualization of the relative sizes of the two quantities involved is difficult. It is the opinion of the author that only the most basic factors should be used; their number and the particular items chosen should be suitably varied in accordance with the type of problem most encountered by the individual. These factors will rapidly become a part of standard mental equipment, leaving troublesome handbook factors for infrequent reference. When a particular conversion requires the combination of several of these factors, reasoning replaces memory, with the result automatically indicating the proper use of the combination factor.

With these principles in mind, the most generally required conversions are listed below. Certain of the more basic items (*e.g.*, inches/foot = 12) have been omitted, as have all heat units which are subsequently discussed.

Type	Units	Numerical value
Weight.....	grams/pound	454
	pounds/kilogram	2.2
	pounds/mole (atom)	Molecular (atomic) weight in pounds
	grams/mole (atom)	Molecular (atomic) weight in grams
Linear......	centimeters/inch	2.54
	centimeters/meter	100
Area.......	Compute from linear conversion units	
Volume....	Compute from linear conversion units and,	
	gallons/cubic foot	7.48
	cubic inches/gallon	231
	liters (S. T. P.)/gram-mole (of any gas)	22.4
	cubic feet (S. T. P.)/pound-mole (of any gas)	359
Pressure....	pounds per square inch/atmosphere	14.7
	inches of mercury/atmosphere	29.92
	millimeters of mercury/atmosphere	760

Example 1. To how many cubic meters do X cu. ft. correspond?

Solution.

$$\text{Cubic feet} \times \left(\frac{\text{inches}}{\text{foot}}\right)^3 \times \left(\frac{\text{centimeters}}{\text{inch}}\right)^3 \times \left(\frac{\text{meters}}{\text{centimeter}}\right)^3 = \text{cubic meters}$$

$$\therefore \text{Cubic meters} = X \times (12)^3 \times (2.54)^3 \times \left(\frac{1}{100}\right)^3 = X\left[\frac{(12)(2.54)}{100}\right]^3 = \frac{X}{35.35}$$

Example 2. To how many lb./(hr.) (ft.) viscosity units do X centipoises correspond?

Solution. An additional factor, centipoises/poise = 100, is needed as well as the knowledge that the units of the poise are grams/(second)(centimeter). Then the procedure is as before.

$$\text{Centipoises} \times \frac{\text{poises}}{\text{centipoise}} \times \frac{\text{pounds}}{\text{gram}} \times \frac{\text{seconds}}{\text{hour}} \times \frac{\text{centimeters}}{\text{inch}} \times \frac{\text{inches}}{\text{foot}} = \text{English hourly units}$$

$$X \text{ centipoises} \times \tfrac{1}{100} \times \tfrac{1}{454} \times 3600 \times 2.54 \times 12 = 2.42X \text{ English units}$$

Temperature units are not directly convertible by ratios. They should be considered on the basis of a comparable range (*e.g.*, freezing water to boiling water). Thus there are 180 Fahrenheit degrees for 100 centigrade degrees, but the zero for the Fahrenheit scale is 32 degrees below the freezing point. Thus, for conversion, the ranges should first be made comparable and the conversion ratio then applied, where °F./°C. = 1.8.

Example 1. To how many °F. does 150°C. correspond?

Solution.

$$150 \times 1.8 = 270 \text{ Fahrenheit degrees.}$$

To position properly on Fahrenheit scale, add 32.

$$270 + 32 = 302\text{°F.}$$

Example 2. To how many °C. does −13°F. correspond?

* The 1936 Student Problem of the American Institute of Chemical Engineers illustrates such segregation, Staaterman, *Trans. Am. Inst. Chem. Engrs.*, **32**, 431 (1936).

Solution. First subtract 32° to get scales on equivalent range.

$$-13 - 32 = -45°$$

Then convert.

$$\frac{(-45)}{(1.8)} = -25°\text{C.}$$

The use of this method rather than conversion tables is most essential for all ordinary work. When large numbers of conversions are required, the use of tables may be substituted, *provided* that the engineer first understands the method.

Absolute scales can best be represented as follows:

°C. + 273 = °K. (Kelvin scale)
°F. + 460 = °R. (Rankine scale)

Heat units frequently give difficulty, not because of their complexity but because of an inherent difficulty in their visualization. Heat units may be considered to consist of three basic types:

1. A quantity of heat, as the calorie (cal.), or British thermal unit (B.t.u.).
2. A latent heat—a quantity of heat per unit mass.
3. A heat capacity—a quantity of heat/(unit mass × unit temperature difference).

Of the three, only the first introduces any new units over those just discussed. Even this one may be reduced to simpler terms for the normal usage, for the practical basic heat units are all expressed as the amount of heat to raise unit weight through one degree of temperature difference, generally expressed in consistent units. Thus all may be reduced to the simple terms: (mass) × (degrees). The common basic units are numerous.*

Heat unit	Definition	Units
calorie (cal.)	Heat to raise 1 g. water from 14° to 15°C.	g. × °C.
Calorie (Cal.) or kilogram-calorie (kg.-cal.)	Heat to raise 1 kg. water from 14° to 15°C.	kg. × °C.
British thermal unit (B.t.u.)	Heat to raise 1 lb. water from 58° to 59°F.	lb. × °F.
centigrade heat unit (c.h.u.) or pound centigrade unit (p.c.u.)	Heat to raise 1 lb. water from 14° to 15°C. This unit is a hybrid arising from the common use of the English weight and centigrade temperature	lb. × °C.

Conversions from one of these units to another may be made by simple consideration of the component weight and temperature scales. It is important to note that temperatures are used as differences (*i.e.*, from one temperature to another) in the definition and that the size of the relative units is therefore all that need be converted, since the range (or zero) differences have been eliminated by the taking of a difference.

Example. Convert X cal. (*a*) to B.t.u.; (*b*) to c.h.u.
Solution. (*a*) cal. = grams × Δ°C.

$$\text{B.t.u.} = \text{pounds} \times \Delta°\text{F.}$$

$$\text{cal.} \times \frac{\text{pounds}}{\text{gram}} \times \frac{\Delta°\text{F.}}{\Delta°\text{C.}} = \text{B.t.u.}$$

and

$$X \times \frac{1}{454} \times 1.8 = \frac{X}{252}\ \text{B.t.u.}$$

(*b*) c.h.u. = pounds × Δ°C.

$$\text{cal.} \times \frac{\text{pounds}}{\text{gram}} \times \frac{\Delta°\text{C.}}{\Delta°\text{C.}} = \text{c.h.u.}$$

$$X \times \frac{1}{454} \times 1 = \frac{X}{454}\ \text{c.h.u.}$$

In a similar manner the following conversions may be shown. They are not considered as fundamental and should not be committed to memory.

Original unit	Factor	Result
calorie	÷252	B.t.u.
calorie	÷454	c.h.u.
c.h.u.	×1.8	B.t.u.
kilogram-calorie	÷0.252	B.t.u.

The second unit, that of *latent heat*, is expressed as follows, or in equivalent consistent units:

Per Unit Molal Weight	Per Unit Weight
calories per gram-mole	calories per gram
c.h.u. per pound-mole	c.h.u. per pound
B.t.u. per pound-mole	B.t.u. per pound
kilogram-calories per kilogram-mole	kilogram-calories per kilogram

Division of any quantity, expressed in the units of the first column, by the molecular weight of the compound considered gives the equivalent quantity expressed in the units of the second column. Quantities in the second column are obviously of smaller magnitude.

Conversion between **latent heat units** is accomplished in the same manner as proposed above for the pure basic heat unit. However, in this case, since the weights are consistent in the basic heat unit and in the denominator, they cancel, leaving only temperature scale conversions summarized below:

Original unit	Factor	Result
calories per gram-mole	×1.8	B.t.u. per pound-mole
calories per gram	×1.8	B.t.u. per pound
calories per gram-mole	None	c.h.u. per pound-mole
calories per gram	None	c.h.u. per pound
c.h.u. per pound-mole	×1.8	B.t.u. per pound-mole
c.h.u. per pound	×1.8	B.t.u. per pound

Standard heats of formation and combustion are given in basic heat units per unit weight—customarily calories per gram-mole. This unit is of the same type as that just considered for latent heat; conversions are therefore identical to those in the immediately preceding table.

When *heat capacity* units* are studied, it is noted that they not only include the units just discussed in connection with latent heat but also another item—temperature difference in the denominator. By study of the conversions from one unit to any other it will be found, therefore, that heat capacities are numerically equivalent in all systems of units, as long as the unit weight is expressed in similar terms, *i.e.*, in molecular weights in both cases, *or* in unit weights in both cases. Thus

calories per (gram-mole × °C.) = c.h.u. per (pound-mole × °C.)
= B.t.u. per (pound-mole × °F.)
calories per (gram × °C.) = c.h.u. per (pound × °C.)
= B.t.u. per (pound × °F.)

Although a considerable number of heat unit conversion factors have been given above, it is not recommended that they be used as a conversion table. Those who have much concern with these units should make the conversions for themselves until the principles are thoroughly familiar, at which time the conversions become automatic.

No completely consistent system of units has been attempted in the illustrative problems subsequently presented. All types have been used indiscriminately, although in many cases the answers may be given in two systems.

Ideal Gas Law. The law relating pressure (p), specific volume per mole (v), and absolute temperature (T), where ideality of the gas has been assumed, was discussed on p. 296 and indicated to be of the form

$$pv = RT \qquad (1)$$

* The joule, and its relation to these more commonly employed units, has already been defined (see p. 295).

* For remarks concerning specific heats, see p. 341.

If other than 1 mole of gas is considered, that number is represented by n, and the total volume (V) of this number of moles is used in place of the specific volume (v), the relation becoming

$$pV = nRT \tag{2}$$

Also noted on p. 290 was the useful form

$$\frac{p_1V_1}{T_1} = \frac{p_2V_2}{T_2} \tag{3}$$

Illustrations are provided in the following material to indicate proper use of either Eq. (2) or (3), (1) being considered as a special case of (2) where $n = 1$. There is nothing difficult about the use of either, but the establishment of the value of R must be thoroughly understood. It does not have to be obtained from a table but may be calculated from material already supplied.

Units of R. By writing

$$R = \frac{pV}{nT}$$

it is apparent that the units are those of pressure times volume per mole divided by temperature, and that any of these quantities may be expressed in any desired units. R may be calculated if at standard conditions the molal volume is known in the system being considered and if p and T are known in the desired units at standard conditions. Standard conditions being 1 atm. and 0°C., or equivalent, these values may be obtained in any system by simple conversions previously discussed.

Calculation of Values of R.

Example 1. What is the value of the gas constant R when it is desired to calculate volume in cubic feet, using pressure in pounds per square inch and temperature in degrees Rankine?

Solution. At S.T.P.,

$$p = 14.7 \text{ lb./sq. in.}$$
$$T = 32°\text{F.} + 460 = 492°\text{R.}$$
$$\frac{V}{n} = v = 359 \text{ cu. ft./lb.-mole}$$
$$\therefore R = \frac{14.7 \times 359}{492} = 10.73$$

When the gas is non-ideal, the problem becomes somewhat more complicated. Methods and calculations are illustrated subsequently (pp. 352*ff.*).

Other Useful Relations. Using Dalton's law of additive volumes for mixtures ($p = p_1 + p_2 + \cdots + p_n$, where p_1, p_2, etc., are partial pressures of the individual components), the following relations will be found particularly useful:

$$\frac{p_1}{p_2} = \frac{n_1}{n_2} = \frac{V_1}{V_2}$$
$$\frac{p_1}{p} = \frac{n_1}{n} = \frac{V_1}{V}$$

Example 2. A mixture of hydrocarbon vapors contains 40 lb. of benzene, 40 lb. of toluene, and 30 lb. of xylenes. What volume will be occupied by this mixture at 300°F. and 443 mm. Hg? What is the density of the mixture under these conditions?

Solution.

Material	Molecular Weight
Benzene....................	78
Toluene....................	92
Xylene.....................	106

Gas law constant, $R = 0.729$ (atm. × cu. ft.)/(lb.-mole × °R.)

Basis: 110 lb. of mixture

$$B: \frac{40 \text{ lb.}}{78} = 0.513 \text{ mole}$$
$$T: \frac{40 \text{ lb.}}{92} = 0.435 \text{ mole}$$
$$X: \frac{30 \text{ lb.}}{106} = 0.283 \text{ mole}$$
$$\text{Total: } 110 \quad = 1.231 \text{ moles}$$

Method a. Using ideal gas law equation, and units in which R is noted,

$$pV = nRT$$
$$\tfrac{443}{760} V = (1.231)(0.729)(760)$$
$$V = 1170 \text{ cu. ft. at } 300°\text{F., 443 mm. Hg}$$
$$\text{Density} = \frac{W}{V} = \frac{110}{1170} = 0.094 \text{ lb./cu. ft. at } 300°\text{F., 443 mm. Hg}$$

Method b. By pV/T = const.

$$\text{Volume at S.T.P.} = 1.231 \times 359$$
$$\therefore V = 1.231 \times 359 \times \tfrac{760}{4\ 2} \times \tfrac{760}{443} = 1170 \text{ cu. ft. at } 300°\text{F., 443 mm. Hg}$$

Density is found as in the preceding method.

Example 3. Calculate the amount of water vapor in 1000 cu. ft. of "wet" air, measured and saturated at 70°F., 760 mm. Hg; under the same conditions of temperature and pressure but with a relative saturation of 50 per cent.

Solution. Part 1. Saturated

Data:

Vapor pressure H_2O at 70°F. = 18.6 mm. Hg*

$$\frac{\text{Pounds water}}{\text{pound air}} = 0.016 \text{ (chart, p. 811)}$$

$$\frac{\text{Moles water}}{\text{mole air}} = 0.0251 \text{ (chart, Hougen and Watson†)} = \frac{18.6}{760 - 18.6}$$

$$1000 \times \tfrac{492}{530} \times \tfrac{1}{359} = 2.59 \text{ moles wet gas}$$

Method a. Dalton's law corollary: $p_1 = N_1p$

N = mole fraction; subscript 1 refers to water vapor, 2 to air, and T to total

Let n_1 = moles water vapor, n_2 = moles air, and n_T = moles total

$$18.6 = \frac{n_1}{2.59}\,760$$
$$n_1 = 0.0634 \text{ moles } H_2O \text{ vapor present (1.14 lb.)}$$

Method b. Dalton's law (using direct pressure ratio)

$$n_1 = 2.59 \times \frac{18.6}{760} = 0.0634 \text{ moles } H_2O \text{ vapor present}$$

Method c. From molal humidity chart:

$$\frac{n_1}{n_2} = \frac{n_1}{2.59 - n_1} = 0.0251$$
$$\therefore n_1 = 0.0634 \text{ moles } H_2O \text{ vapor present}$$

Method d.

$$\frac{18n_1}{29(2.59 - n_1)} = 0.016$$
$$n_1 = 0.065 \text{ moles } H_2O \text{ vapor present}$$

(Error of 2.5 per cent due to difficulty in reading chart.)

Part 2. Relative saturation = 50 per cent

Methods a and b.

$$0.5 = \frac{p_1}{p_s} = \frac{p_1}{18.6}$$
$$p_1 = 9.3 \text{ mm.}$$
$$2.59 \times \frac{9.3}{760} = 0.0317 \text{ moles of water vapor present}$$

Methods c and d. Per cent relative saturation curves do not appear on either of the charts used. Ratios corresponding to

* For graphical vapor pressure representations and applications see pp. 341*ff.* on latent heats.

† Hougen and Watson, "Chemical Process Principles," p. 100, Wiley, New York, 1943.

those obtained from these charts may be readily calculated, however, if desired.

$$\frac{\text{Moles water}}{\text{mole air}} = \frac{9.3}{760 - 9.3} = 0.0124$$

$$\frac{\text{Pounds water}}{\text{pound air}} = 0.0124 \frac{18}{29} = 0.00769$$

Example 4. A gas is to be dried to 1 per cent moisture by compression. What final pressure is required if the gas is ultimately cooled to 80°C.? 20°C? Assume the validity of the ideal gas laws. (In this problem it is unnecessary to specify the initial condition unless the amount of water removed in the drying is desired.)

Solution.

t°C.	Vapor Pressure H_2O, mm. Hg
20	17.4
80	360

Applying the corollary of Dalton's law, stating

$$p_1 = N_1 p$$

∴ at 20°C.

$$\frac{17.4}{760} = 0.01p$$

$$p = 2.3 \text{ atm.}$$

and at 80°C.

$$\frac{360}{760} = 0.01p$$

$$p = 47.4 \text{ atm.}$$

Example 5. In the preceding problem (Example 4), assume that compression is truly adiabatic with cooling at constant volume. What are the pressure and the temperature *after* compression and *before* cooling, if the gas was originally at 20°C. and 1 atm.? Again assume the validity of the ideal gas law; further assume that the amount of water condensed is insufficient to invalidate the basic equation of adiabatic compression and constant volume cooling.

Solution. Case I, final temperature of 20°C.

Compression (see p. 303):

$$\left(\frac{T_2}{T_1}\right) = \left(\frac{p_2}{p_1}\right)^{\frac{\kappa-1}{\kappa}}$$

where κ is assumed equal to that for air $= \frac{C_p}{C_v} = 1.4$

$$\therefore \frac{T_2}{293} = \left(\frac{p_2}{1}\right)^{0.4/1} \quad (4)$$

Cooling (see p. 336):

$$\left(\frac{T_2}{T_3}\right) = \left(\frac{p_2}{p_3}\right) \quad (5)$$

$$T_2 = ? \qquad T_3 = 293°\text{K.}$$

$$p_2 = ? \qquad p_3 = 2.3 \text{ atm.}$$

$$\frac{T_2}{293} = \frac{p_2}{2.3} \quad (6)$$

Solving (6) for T_2 and substituting in (4)

$$\frac{293p_2}{(2.3)(293)} = \left(\frac{p_2}{1}\right)^{0.4/1.4}$$

$$\frac{1}{2.3} = \frac{(p_2)^{0.4/1.4}}{p_2} = \frac{1}{(p_2)^{1/1.4}}$$

$$\therefore p_2 = (2.3)^{1.4} = 3.21 \text{ atm.}$$

From Eq. (5)

$$T_2 = \frac{3.21}{2.3} 293 = 409°\text{K.} = 136°\text{C.}$$

Case II, final temperature, T_3, of 80°C.

By similar methods:

$$p_2 = 171 \text{ atm.}$$

$$T_2 = 1271°\text{K.} = 998°\text{C.}$$

Example 6. Repeat the preceding problem (Example 5) if the cooling is isobaric. Employ the same assumptions used in Example 5. Solve for Case I only.

Solution.

Compression:

$$\frac{T_2}{T_1} = \left(\frac{V_1}{V_2}\right)^{\kappa-1}$$

Cooling:

$$\frac{V_2}{V_3} = \frac{T_2}{T_3}$$

per mole,

$$V_1 = \frac{RT_1}{p_1} = \frac{(1.316)(293)}{1} = 385 \text{ cu. ft.}$$

and

$$V_3 = \frac{RT_3}{p_3} = \frac{(1.316)(293)}{2.3} = 167.5 \text{ cu. ft.}$$

$$\frac{V_2}{167.5} = \frac{T_2}{293}; \qquad T_2 = \frac{293}{167.5} V_2$$

and

$$\frac{T_2}{293} = \left(\frac{385}{V_2}\right)^{0.4} = \frac{(293)(V_2)}{(167.5)(293)}$$

$$\therefore \frac{385^{0.4}}{V_2^{1.4}} = \frac{1}{167.5}$$

$$V_2 = (1812)^{1/1.4} = 212 \text{ cu. ft.}$$

$$\therefore T_2 = (212)\left(\frac{293}{167.5}\right) = 371°\text{K.} = 98°\text{C.}$$

as a check, p_2, calculated from the equation

$$\frac{T_2}{T_1} = \left(\frac{p_2}{p_1}\right)^{\frac{\kappa-1}{\kappa}}$$

is found to be 2.3, which is correct.

Material Balances.* All material balances are based upon the law of conservation of mass (see p. 289). By this law every unit of mass entering a system (or process) must subsequently leave as mass; however, chemical and/or physical changes may have occurred in the system, so that form of feed and products may be expected to change. The introduction of the concepts of atomic energy modify this law that was previously considered exact but are not involved in the calculations in which this section is concerned.

When in physical processes, such as the unit operations, only the physical state changes, material balances may be made around each unit, or a group of units, or both; the balances possible include that on the total mass of every stream and those on the total mass of each component in every stream. Since there is no chemical change involved, the balance may also be performed in terms of moles, total or of each constituent. However, when chemical action occurs, it is not proper to balance in moles because of the possibility of changes resulting in increased or decreased numbers on this basis. In this case three courses are open, the pure mass balance, the atomic balance, or the balance of some compound artificially retained for the purposes of calculation (*e.g.*, SO_3 in gas and liquid phases, despite chemical combination in the latter to give H_2SO_4).

The material balance is invariably employed to complete, or to check, data. Since the checking of data is an experimental procedure, the following examples will show completion of essential material by calculation, in order to avoid direct measurements.

Example 1. Chlorine gas, analyzing 1.6 per cent oxygen, is generated in caustic cells. In the line carrying this gas, for purposes of measurement of rate of generation, oxygen is added, allowed to mix thoroughly, and the gas again analyzed, showing 3.6 per cent oxygen. In 5 min. and 33 sec., 10 lb. of oxygen were added. What is the rate of generation of the original gas?

Solution.

Basis: 10 lb. of oxygen added = 0.312 moles†

* See also Sec. 22, and various unit operations that use the material balance extensively in development of theoretical equations and in problem solutions.

† There being no chemical change in the mixing of these gases, and moles a convenient way of expressing quantities, this choice is legitimate.

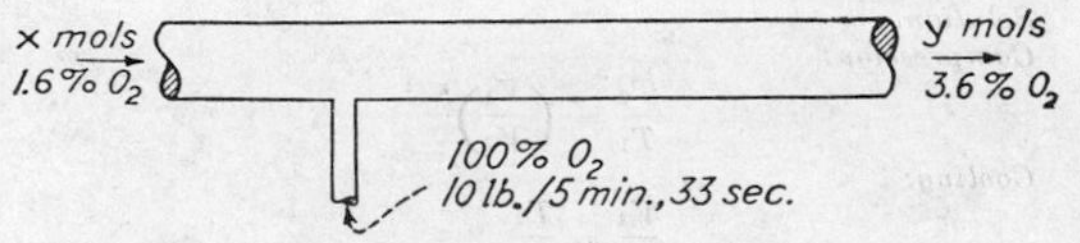

FIG. 22.

Time = 5 min. 33 sec. = 5.533 min.
Rate of oxygen flow = 0.0564 moles per min

Let x = moles of original gas per minute
y = moles of final gas per minute

Over-all balance: $x + 0.0564 = y$

Oxygen balance:
$$0.016x + 0.0564 = 0.036y$$
$$= 0.036(x + 0.0564)$$
$$= 0.036x + 0.00203$$
$$0.020x = 0.0544$$
$$x = 2.72 \text{ moles/min.}$$

Measured at S.T.P., this flow rate is 977 cu. ft./min.

Example 2. The feed to a continuous still analyzes by weight: 28 per cent benzene, 72 per cent toluene. Benzene is 52 per cent in the distillate, 5 per cent in the bottoms. Calculate the per cent recovery in the overhead stream.

Solution.

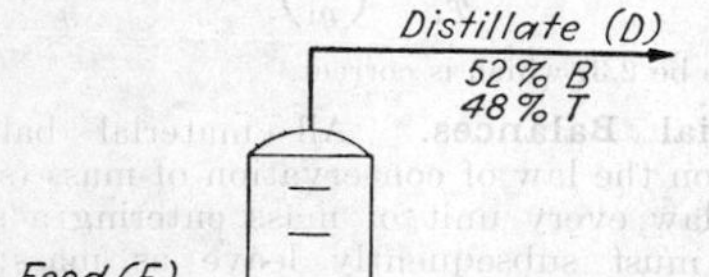

FIG. 23.

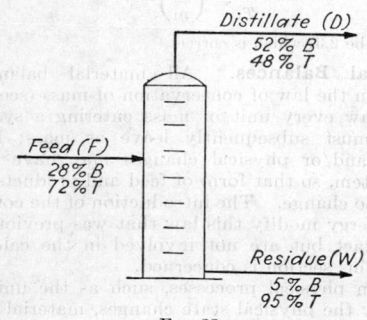

Basis: 100 lb.* of feed
28 lb. of benzene
72 lb. of toluene

Over-all balance: $100 = D + W$ (7)
Benzene balance: $28 = 0.52D + 0.05W$ (8)
Toluene balance: $72 = 0.48D + 0.95W$ (9)

There being but two unknowns, the simplest pair of the three equations is chosen for the solution. Take Eqs. 7 and 8 although Eqs. 7 and 9 are just as easy.

(7) × 0.05 $\quad 5 = 0.05D + 0.05W$ (10)
$\quad 28 = 0.52D + 0.05W$ (11)
(11) − (10) $\quad 23 = 0.47D$
D = 49 lb. of distillate
(49)(0.52) = 25.5 lb. of benzene in distillate
$\frac{25.5}{28}$ = 91.1 per cent recovery of benzene in distillate

Equation (9) serves as a useful check:

$$72 = (0.48)(49) + (0.95)(51)$$
$$= 23.5 + 48.5 = 72.0$$

Example 3. A furnace fired with a hydrocarbon fuel oil has a dry stack gas analysis as follows: 10.2 per cent CO_2; 8.3 per cent O_2; 81.5 per cent N_2. Calculate:

a. Composition of the original fuel oil.
b. Per cent excess air used.
c. Cubic feet (S.T.P.) of air per pound of fuel.

* Again, there being no chemical reaction, the unit of weight is not necessary for a correct solution. However, in this case, the data make it the only practical basis.

Solution. Since chemical change is involved in the combustion process, moles do not provide a proper basis for calculation. Further, too few data are available for the over-all weight to be of importance. The following solution will be seen to be based upon elementary materials. It will be seen that the calculations are in terms of moles and atoms, rather than weight, although the molal basis has been noted above to be incorrect. No inconsistency exists, however, since each stream is broken down into its elements. Such a method is equivalent to analysis of streams on a weight basis.

Basis: 100 moles of flue gas—dry
10.2 moles of CO_2
8.3 moles of O_2
81.3 moles of N_2

Essential reactions:
$$C + O_2 \rightarrow CO_2$$
$$H_2 + \tfrac{1}{2}O_2 \rightarrow H_2O$$

Part a.

O_2 in flue gas = 10.2 + 8.3 moles = 18.5 moles
O_2 supplied in air = O_2 calculated from N_2 in flue gas
$$= \frac{81.5}{79}\,21 = 21.69 \text{ moles}$$
O_2 balance: 21.69 − 18.5 = 3.19 moles O_2 not accounted for in flue gas

Unaccounted oxygen must have disappeared as water when gas is dried. Therefore,

3.19 moles O_2 burned H_2
6.38 moles H_2 were burned
12.76 atoms of H_2 were burned = 12.86 lb.
10.2 atoms of C were burned = 122.4 lb.
135.3 lb.
135.3 lb. of fuel per 100 moles dry flue gas

Final analysis:
$$C = \frac{122.4}{135.3} = 90.5 \text{ per cent}$$
$$H = \frac{12.9}{135.3} = 9.5 \text{ per cent}$$

Empirical formula:
$$(C_{10.2}H_{12.76})_{n'} = (CH_{1.25})_{n''} = (C_4H_5)_n$$

Part b.

O_2 supplied	=	21.69 moles	21.69
O_2 needed:			
For C	=	10.2 moles	
For H	=	3.19 moles	13.39
O_2 excess	=		8.30

$$\text{Per cent excess air} = \text{per cent excess } O_2 = 100\,\frac{8.30}{13.39} = 62.0 \text{ per cent}$$

Part c.

Fuel = 135.3 lb.
Air = 21.69 + 81.5 = 103.4 moles
$$\frac{103.4 \times 359}{135.3} = 274 \text{ cu. ft. air per pound oil}$$

Example 4. To form a solution saturated at 90°F., 500 lb. of crude salt cake are dissolved in water. In the neutralization treatment to remove iron impurities, 0.4 per cent of the original charge is removed chemically and 100 lb. of water are added to the process. Calculate:

a. Minimum temperature, to prevent crystallization, which must be maintained at the end of purification.

b. Maximum quantity of crystals if the purified solution be cooled to 68°F.

c. Increase in crystals produced if the solution be evaporated at a temperature not in excess of 140°F. and without entering the two-phase region, and finally cooled to 68°F.

Discussion. This problem is not purely physical in character, for the Glauber's salt crystallized represents a combination of Na_2SO_4 with $10H_2O$.

The possibility of predicting maximum yields by application of the material balance principle is here illustrated. Since it is felt that the reader is now acquainted with the mechanics of material balances, no attempt will be made to define sharply the balances employed.

Solution.

	Molecular Weights
Na_2SO_4	142
$Na_2SO_4.10H_2O$	322

Basis: 500 lb. of salt cake

0.004 — fraction lost
2.0 — weight lost
498 lb. of pure Na_2SO_4

From the accompanying phase diagram (Fig. 24) at 90°F., 47 parts of Na_2SO_4 will dissolve in 100 parts of water.

$^{498}/_{47} \times 100 = 1060$ lb. of water originally
100 lb. of water added
1160 lb. of water after purification
(498/1160)100 = 42.9 parts of Na_2SO_4 per 100 parts of water after purification

Part a. From the phase diagram, the minimum temperature after purification, if crystallization is to be prevented, is 87.5°F.

Part b. At 68°F., 20 parts Na_2SO_4 per 100 parts water.

Let x = pounds of Na_2SO_4 crystallizing

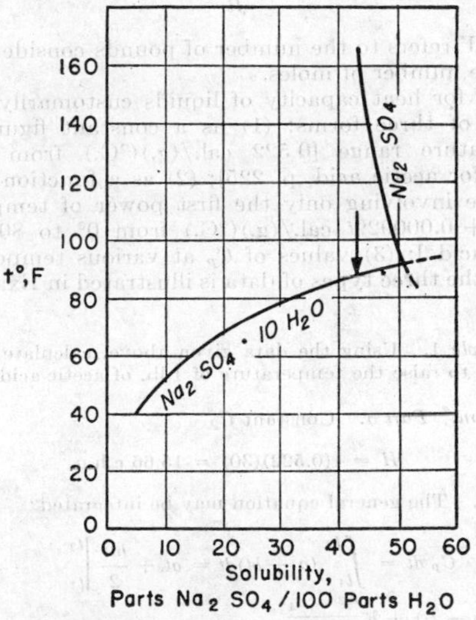

FIG. 24. Solubilty of sodium sulfate.

$322x/142$ = pounds of Glauber's salt ($Na_2SO_4.10H_2O$) crystallizing
$180x/142$ = pounds of water crystallizing
$498 - x$ = pounds of Na_2SO_4 left in solution
$1160 - 180x/142$ = pounds of H_2O left in solution

$$\therefore \left(\frac{498 - x}{1160 - 180x/142}\right) 100 = 20$$

$49{,}800 - 100x = 23{,}200 - 25.4x$
$74.6x = 26{,}600$
$x = 356$ lb. of Na_2SO_4 crystallized as 808 lb. of Glauber's salt
$1160 + 498 - 808 = 850$ lb. of saturated mother liquid remaining

Part c. At 140°F., the limiting concentration is 44 parts Na_2SO_4 per 100 parts of water—taken slightly on the single-phase side of the line.

Let y = water evaporated

$$100\left(\frac{498}{1160 - y}\right) = 44$$

$49{,}800 = 51{,}050 - 44y$
$y = 28.4$ lb. of water evaporated
1132 lb. of water remaining

Proceeding as in part *b*:

$$100\left(\frac{498 - z}{1132 - 180z/142}\right) = 20$$

$49{,}800 - 100z = 22{,}640 - 25.4z$
$z = 364$ lb. Na_2SO_4 crystallized as 825 lb. of Glauber's salt

$\frac{825 - 808}{808} = 2.1$ per cent increase in yield by inclusion of evaporation step

Sensible Heats.* *Gases.* Although many data (see p. 229)† are available expressing the heat capacity of gases, usually in the form

$$C_p = a + bt + ct^2 \quad (12)$$

or

$$C_p = a' + b'T + c'T^2 \quad (13)$$

the expression is not too convenient for calculating sensible heats since the following integration is required, H being enthalpy at t_2 relative to that at t_1, and n being number of moles of gas:

$$H = n \int_{t_1}^{t_2} C_p\,dt = n \int_{t_1}^{t_2} (a + bt + ct^2)dt$$
$$= n[a(t_2 - t_1) + \frac{b}{2}(t_2^2 - t_1^2) + \frac{c}{3}(t_2^3 - t_1^3)] \quad (14)$$

Calculations are greatly simplified if values of mean heat capacities *over a temperature range* are available. The latter are obtained as follows:

$$\underset{(t_2 - t_1)}{C_{pm}} = \frac{\int_{t_1}^{t_2} C_p\,dt}{(t_2 - t_1)} \quad (15)$$

To prevent the recording of an infinite number of values it is essential that t_1 become a base temperature. If this be zero, Eq. (15) becomes

$$\underset{(t_2 - 0)}{C_{pm}} = \frac{\int_0^{t_2} C_p\,dt}{t_2} \quad (16)$$

Values of C_{pm} from zero to t may then be recorded graphically as illustrated in Fig. 25, and

$$H \text{ (from 0 to } t) = n \underset{(0 - t)}{C_{pm}} t \quad (17)$$

If H is desired from t_1 to t_2, it is readily calculated as follows:

$$H \text{ (from } t_1 \text{ to } t_2) = n(C_{pm_2}t_2 - C_{pm_1}t_1) \quad (18)$$

In these calculations it is important that the C_{pm} be chosen in the proper temperature units, since the integration, Eq. (16), involves a C_p expressed in terms of temperature. Therefore, the data of Fig. 25 have been given for the centigrade scale *and* for heat units in the metric system. As noted, the same figure may be used with the Fahrenheit scale and heat units in the English system with the following changes:

1. Heat capacities are determined for $(t - 32)°$ on the Fahrenheit abscissa instead of $t°$.

* See also Sec. 22, p. 1588 for enthalpy curves. For tabular data on refrigerants see Sec. 3, pp. 249–280. Mean specific heat curves are also shown in Sec. 22, p. 1588.

† For hydrocarbons, see A.P.I. Research Project 44 National Bureau of Standards, *op. cit.;* Stull and Mayfield, *Ind. Eng. Chem.*, **35,** 639 (1943); and, for general estimates, Fugassi and Rudy, *Ind. Eng. Chem.*, **30,** 1029 (1938).

2. Equation (18) is converted to

$$H \text{ (from } t_1 \text{ to } t_2) = n[C_{pm_2}(t_2 - 32) - C_{pm_2}(t_1 - 32)]$$

Example 1. Calculate the heat input to raise 132 lb. of CO_2 from 100° to 1000°C. Perform the calculation in the following ways:

a. By integrating the expression for C_p in terms of t.
b. By calculating c.h.u. from the mean heat capacity chart.
c. By calculating B.t.u. from the mean heat capacity chart.

Solution. Part a.

$$132 \text{ lb.} = 3 \text{ moles } CO_2$$

$$C_p = 6.85 + 8.533 \times 10^{-3}T - 2.475 \times 10^{-6}T^2$$

$$H = 3\int_{373}^{1273} C_p\,dT = 3\left(6.85T + 4.266 \times 10^{-3}T^2 - 0.825 \times 10^{-6}T^3\right)_{373}^{1273}$$

$$= 3(8720 - 2555 + 6920 - 594 - 1705 + 43)$$
$$= 3(10{,}830) = 32{,}490 \text{ c.h.u.}$$

Part b.

$$C_{pm}\underset{1000^\circ}{} = 11.82 \qquad C_{pm}\underset{100^\circ}{} = 9.44$$

$$H = 3[(11.82)(1000) - (9.44)(100)]$$
$$= 3(10{,}876) = 32{,}430 \text{ c.h.u.}$$

Part c.

1000°C. = 1832°F.	100°C. = 212°F.
$(t - 32)$°F. = 1800	$(t - 32)$°F. = 180
C_{pm} = 11.82	C_{pm} = 9.44

$$H = 3[(11.82)(1800) - (9.44)(180)]$$
$$= (3)(1.8)(10{,}876) = 58{,}731 \text{ B.t.u.}$$
$$= 32{,}430 \text{ c.h.u.}$$

Maximum deviation of methods

$$\frac{32{,}490 - 32{,}430}{32{,}490} = \frac{60}{32{,}490} = 0.18\%$$

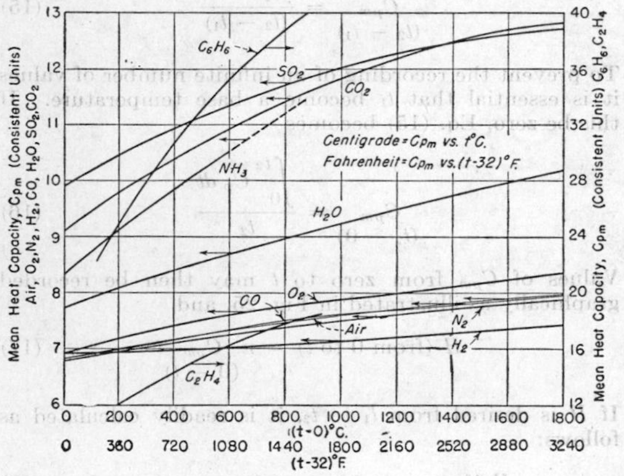

FIG. 25.

Obviously the method leads to calculations far more accurate than warranted by the data (2.5 per cent from 0° to 2200°C. in this case). However, it is usually as easy to use the charts as to attempt simple estimations.

Gaseous Mixtures. Although the mixture law (see p. 289) may be applied as follows:

$$(C_p)_{\text{mix}} = N_A C_{p_A} + N_B C_{p_B} + \cdots \tag{19}$$

it is frequently easier to divide the gas into its component parts and calculate, using the individual heat capacities, as shown in Example 2.

Example 2. The analysis of 13,600 cu. ft. of gas (S.T.P.) is as follows: 10 per cent SO_2, 12 per cent O_2, and 78 per cent N_2. How much heat must be added to this gas to raise its temperature from 30° to 425°C.?

Basis: 13,600 cu. ft. (S.T.P.) = 37.9 moles, 3.79 moles SO_2, 4.55 moles O_2, 29.56 moles N_2

From Fig. 25:

C_{pm}	30°C.	425°C.
SO_2	10.00	11.00
O_2	6.96	7.32
N	6.80	7.12

$$H_{SO_2} = 3.79[11.00(425) - 10.00(30)] = 16{,}600 \text{ c.h.u.}$$
$$H_{O_2} = 4.55[7.32(425) - 6.96(30)] = 13{,}200 \text{ c.h.u.}$$
$$H_{N_2} = 29.56[7.12(425) - 6.80(30)] = 83{,}400 \text{ c.h.u.}$$
$$\text{Total heat required} = 113{,}200 \text{ c.h.u.}$$

Liquids. With *pure liquids* it is customary to report heat capacity data per unit of weight rather than per mole. However, this does not change the method of treatment except that the equation for sensible heat change reads

$$H = W\int_{t_1}^{t_2} C_p\,dt \tag{20}$$

rather than

$$H = n\int_{t_1}^{t_2} C_p\,dt \tag{21}$$

where W refers to the number of pounds considered, and n to the number of moles.

Data for heat capacity of liquids customarily appear in one of three forms: (1) as a constant figure for a temperature range [0.522 cal./(g.)(°C.) from 26° to 95°C., for acetic acid, p. 225]; (2) as a function of temperature involving only the first power of temperature [$0.468 + 0.000929t$ cal./(g.)(°C.) from 0° to 80°C., for acetic acid*]; (3) values of C_p at various temperatures. Use of the three types of data is illustrated in Examples 1 and 2.

Example 1. Using the data given above, calculate the heat required to raise the temperature of 1 lb. of acetic acid from 30° to 60°C.

Solution. Part a. Constant C_p

$$H = 1(0.522)(30) = 15.66 \text{ c.h.u.}$$

Part b. The general equation may be integrated

$$H = \int_{t_2}^{t_1} C_p\,dt = \int_{t_1}^{t_2}(a + bt)dt = at + \frac{bt^2}{2}\Big]_{t_1}^{t_2}$$
$$= a(t_2 - t_1) + b\,\frac{t_2^2 - t_1^2}{2}$$
$$= (t_2 - t_1)\left(a + b\,\frac{t_2 + t_1}{2}\right) = \frac{(a + bt_2) + (a + bt_1)}{2}(t_2 - t_1)$$

The last expression shows that for *any* first-power heat capacity equation the average of the heat capacities at the terminal temperatures is mathematically correct. Therefore

$$C_{p_{\text{avg}}} = 0.468 + (0.000929)(45) = 0.510$$
$$H = 1(0.51)(30) = 15.3 \text{ c.h.u.}$$

Because of the simple first-power form of such data, alignment charts may frequently be found (see p. 217) for heat capacity of liquids.

Example 2. With the following data, estimate the sensible heat in 1 lb. of pure ethyl alcohol at 60°C., referred to that at 10°C. as a datum.

t°C.	c.h.u./(lb.)(°C.)
0	0.535
25	0.580
50	0.652

Solution. Part a. Using an estimated average C_p, assuming it to be approximately that at 35°C., estimate, by inspection, that the value is 0.605.

Then $H = (1)(0.605)(50) = 30.3$ c.h.u.

* Hougen and Watson, *op. cit.*

Part b. Graphical integration of the curve of C_p vs. t (see Fig. 26) from 10° to 60°C. will give an exact value of H.

Area under curve = 11.09 squares
Area not shown = 50.00 squares
(to $C_p = 0$)

61.09 squares
0.5 c.h.u. per square

30.5 c.h.u.

Error by method of *Part a:*

$$\frac{0.10}{30.5} = 0.33 \text{ per cent}$$

The method of part *b* is extremely laborious. Moreover, the added accuracy is rarely of use in engineering calculations. As seen, the method of part *a* is quite accurate if reasonable care is taken in choosing the value of $C_{p_{avg}}$.

Graphical integration is always laborious. The work may be considerably simplified by utilizing a method such as Moore's* for obtaining the average ordinate, rather than counting squares. Figure 26 illustrates the method for this problem, the whole curve here being assumed to be parabolic. If this assumption cannot be made in other cases, the integration can be performed in smaller subdivisions of the abscissa.

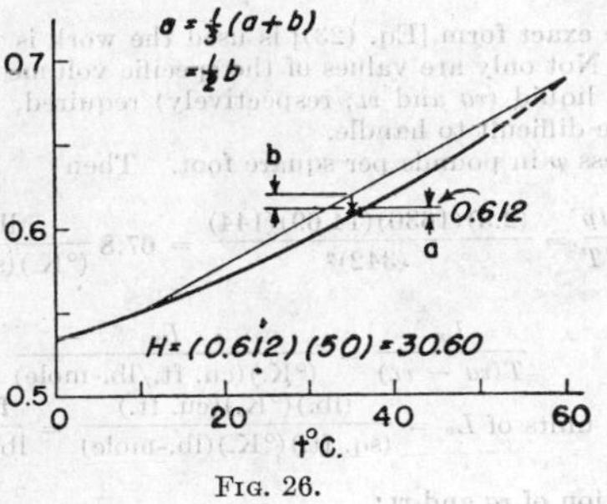

FIG. 26.

Solutions. The majority of data regarding solutions is reported in a manner similar to that used for pure liquids (*i.e.*, a value of C_p for some temperature, or temperature range, and for a given solution). Use of such data must involve careful consideration of the fact that C_p is not constant, generally increasing with temperature increase; moreover, the data reported are applicable only to the solution specified. Remarks have previously been made (see p. 297) on estimation of heat capacities of aqueous solutions of electrolytes. Calculation from these heat capacities is similar to some one of the methods outlined above.

In some few cases partial heat capacity data may be available. Their use is quite complicated, requiring several plots and dealing exclusively with constant-temperature data. The latter point eliminates any engineering value which might otherwise be attributed to their accuracy.

Solids. Heat capacities of solids are customarily reported in a manner similar to that employed with liquids; there will, therefore, be no need of repetition of examples on this class of compounds.

It should be noted that for solids specific heats rather than heat capacities are generally recorded. Specific heat *has no units* (although often erroneously labeled calories per degree centigrade) but is equivalent numerically to the heat capacity per unit of weight [*i.e.*, cal./(g.)(°C.), B.t.u./(lb.)(°F.), c.h.u./(lb.)(°C.), etc.].

Latent Heats. *Vaporization.* There are many available methods for the estimation of latent heats of vaporization. Several methods will be summarized in the examples.* To conserve space, material will be presented for the calculation of the latent heat of one compound only, hexane, and that at its normal boiling point.

Example. Calculate the latent heat of vaporization of n-hexane at its normal boiling point† by each of the following methods:

a. Kistiakowsky equation for non-polar liquids.
b. Dühring line method of estimation.
c. Clapeyron relation.
d. Othmer method [*Ind. Eng. Chem.*, **32,** 841 (1940)].

t°C.	Benzene p, mm. Hg	Hexane p, mm. Hg	Water p, mm. Hg
32.0	...	200	35.6
42.1	...	300	61.8
43.0	200	309	
50.1	...	400	93.0
53.0	300	448	
56.6	...	500	127.4
61.0	400	590	
61.9	...	600	160.8
66.3	...	700	199
67.6	500	725	
73.2	600	1000	
78.0	700	1012	

Latent heat of vaporization per unit weight $= \dfrac{L_v}{M}$

t°C.	Benzene, cal./g.	Water, cal./g.
60	97.47	
69	.....	558
80	94.17	
100	90.57	539

Solution. Part a. Kistiakowsky equation:

$$\frac{L_v}{T_s} = 8.75 + 4.571 \log T_s$$

where L_v = molal latent heat, metric units
T_s = normal boiling point, °K. = 342°

$$\therefore L_v = 8.75T_s + 4.571 T_s \log T_s$$
$$= 2995 + 3960 = 6955 \text{ cal./g.-mole}$$
$$\frac{L_v}{M} = \frac{6955}{86.1} = 80.8 \text{ cal./g. (c.h.u./lb.)}$$
$$\text{True value} = 79.2 \text{ cal./g.}$$
$$\text{Error} = 1.6$$
$$\frac{1.6}{79.2} = 2^+ \text{ per cent}$$

Part b. Dühring line method (see p. 294).

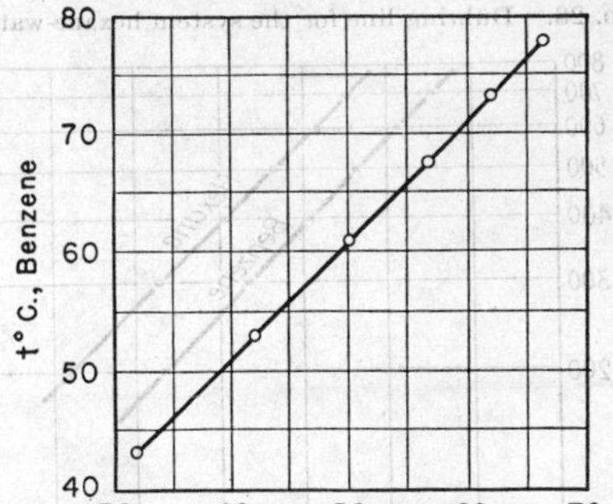

FIG. 27. Dühring line for the system hexane-benzene.

Figure 27 is a Dühring plot of boiling points of benzene *vs.* those of hexane at an equivalent pressure. The equation for

* Moore, *J. Eng. Education*, **31,** 452 (1941).

* For lack of space, other methods are necessarily omitted. The Antoine equation [Thomson, *Chem. Rev.*, **38,** 1 (1946)] and such modifications as that of Calingaert [*Ind. Eng. Chem.*, **17,** 1287 (1925)] should be noted to be useful.

† Several values are given in the literature, but 69°C. seems to be the consensus. 79.2 cal./g. is most frequently quoted as the value of the latent heat of vaporization at this temperature.

determining molal latent heat from this curve is

$$L_v = L_{v_R}\left(\frac{T}{T_R}\right)^2\left(\frac{dT_R}{dT}\right) \tag{22}$$

L_v, L_{v_R} = molal latent heats of the compound and the reference materials, respectively, at the same pressure.

T, T_R = boiling points of the compound and the reference material, at the same pressure, absolute scale

$\frac{dT_R}{dT}$ = slope of the Dühring line. Note that the line is not entirely straight.

$$\frac{dT_R}{dT} = \frac{39}{38} \text{ (over the whole line)} = 1.027$$

$$= 1.095 \text{ (for range of hexane's normal boiling point)}$$

$$L_v = (94.3)(78)(1.095)\left(\frac{342}{353.2}\right)^2 = 7540 \text{ cal./g.-mole}$$

$$\frac{L_v}{M} = 87.6 \text{ cal./g.}$$

True value = 79.2 cal./g.
Error = 8.4
Per cent error = 10.6

This error can be considerably reduced by using the lower value of dT_R/dT, L_v/M becoming 82.1, an error of less than 4 per cent. This calculation can be made equally well using water as a reference state. L_v/M is then calculated as 86.2 cal./g., an error of 8.8 per cent. Since the method is most likely of success when the reference material is of structure similar to that of the material in question, neither water nor benzene is ideal in this case. They were used as representative of the more difficult problem, frequently encountered when data for a similar material are not known in sufficient detail to make the calculation possible.

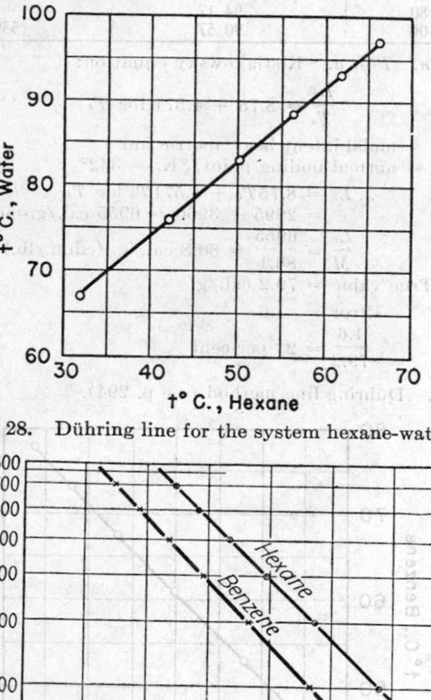

FIG. 28. Dühring line for the system hexane-water.

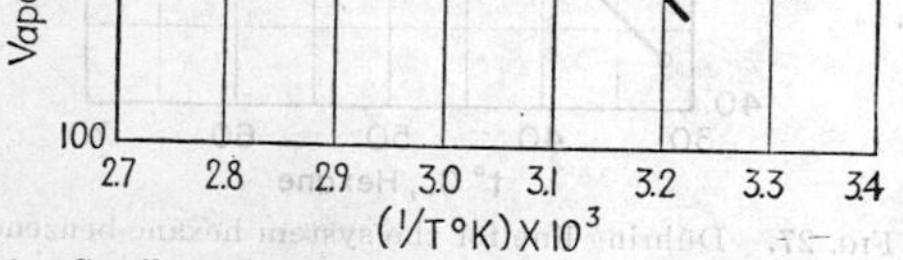

FIG. 29. Semilog plot of vapor pressure vs. reciprocal of temperature for hexane and benzene.

Part c. Clapeyron relation:

$$\frac{dp}{dT} = \frac{L_v}{T(v_G - v_L)} \quad \text{exact form} \tag{23}$$

$$\frac{dp}{dT} = \frac{pL_v}{RT^2} \text{ approximately, neglecting } v_L \text{ in comparison with } v_G \tag{24}$$

In these expressions L_v is assumed to be a constant, which immediately leads to errors. However, as an approximation, either expression may be used to estimate L_v. Necessary for the calculation is a value of dP/dT, best obtained by differentiating an equation for the vapor pressure vs. temperature relation. The latter may be obtained from a semilog plot, Fig. 29, as follows:

$$\log p = \frac{-1630}{T} + 7.63$$

[It is interesting to note that the equation reported in the literature is $\log p = (-1656/T) + 7.724$.]

Differentiating the experimental equation:

$$d(2.3 \log p) = d \ln p = \frac{dp}{p} = 2.3\frac{1630}{T^2}\,dT$$

$$\therefore \frac{dp}{dT} = \frac{(2.3)(1630)p}{T^2}$$

In the approximate form, Eq. (24), $\frac{dp}{dT} = \frac{pL_v}{RT^2}$

$$\frac{(2.3)(1630)p}{T^2} = \frac{pL_v}{RT^2}$$

$$L_v = (2.3)(1630)R = 7450 \text{ cal./g.-mole}$$

$$L_v/M = 86.5 \text{ cal./g.}$$

Per cent error = 9.2

If the exact form [Eq. (23)] is used the work is much complicated. Not only are values of the specific volumes of the vapor and the liquid (v_G and v_L, respectively) required, but the units are more difficult to handle.

Express p in pounds per square foot. Then

$$\frac{dp}{dT} = \frac{(2.3)(1630)(14.69)(144)}{(342)^2} = 67.8\,\frac{\text{lb.}}{(\text{°K.})(\text{sq. ft.})}$$

and

$$\frac{L_v}{T(v_G - v_L)} = \frac{L_v}{(\text{°K.})(\text{cu. ft./lb.-mole})}$$

$$\therefore \text{units of } L_v = \frac{(\text{lb.})(\text{°K.})(\text{cu. ft.})}{(\text{sq. ft.})(\text{°K.})(\text{lb.-mole})} = \frac{\text{ft.-lb.}}{\text{lb.-mole}}$$

Calculation of v_G and v_L:

Assume that the density of liquid hexane at 20°C. may be used since small changes in v_L will not greatly affect the term $(v_G - v_L)$.

$$\rho_{20°\text{C}} = 0.659(62.4) \text{ lb./cu. ft.}$$

$$v_L = \frac{86}{\rho} = 2.18 \text{ cu. ft./lb.-mole}$$

$$v_G = 359 \times {}^{342}\!/_{273} = 450 \text{ cu. ft./lb.-mole}$$

$$L_v = \frac{dp}{dT}(T)(v_G - v_L) \text{ in consistent units}$$

$$\therefore L_v = (67.8)(342)(450 - 2.18) = 103.8 \times 10^5 \text{ ft.-lb./lb.-mole}$$

$$L_v = \frac{103.8 \times 10^5}{(778)1.8} = 7410 \text{ cal./g.-mole}$$

$$\frac{L_v}{M} = 86 \text{ cal./g.}$$

Per cent error = 8.6

Part d. Othmer (*op. cit.*)* has developed the equation

$$\log p = \frac{L_v}{L_{vR}} \log p_R + C \tag{25}$$

* For the purposes of this section it has been convenient to represent the Othmer applications to vapor pressure and latent heats only. However, the same author with various coworkers has shown application of the principle to many other calculations. Other subjects covered, all published in *Ind. Eng. Chem.*, include heats of hydration, dissociation, dilution, solution [**32**, 84, (1940)]; gas solubilities, partial pressure, and partial molal heats of solution [**34**, 952 (1944)]; use of reduced conditions to reduce errors near critical regions [**34**, 1072 (1942)]; vapor pressure of liquids on solids [**35**, 1269 (1943)]; equilibrium constants for distillation and absorption systems [**36**, 669, 858 (1944)]; gas and liquid viscosities [**38**, 111 (1946); **37**, 1112 (1945), respectively]; and finally a paper on surface tension [**40**, 886 (1948)].

Whitwell and Toner have developed further uses for the method in relation to equilibrium moisture and solutions [*Textile Research J.*, **16**, 255 (1946); **16**, 307 (1946), the last paper with McCarthy] for use in drying and concentration processes.

Therefore the slope of a loglog line such as shown in Figs. 30 and 31 represents L_v/L_{vR}, both latent heats taken at the same temperature. From the benzene-hexane plot (Fig. 30) the slope is 0.941 and

$$\frac{L_v}{M} = \frac{(L_{vR})(0.941)}{86.1} = \frac{(0.941)(96)(78)}{96.1}$$
$$= 81.8$$
$$\text{Per cent error} = 3.3$$

If water is used as the reference, the slope is 0.692, and

$$\frac{L_v}{M} = \frac{(0.692)(1004.5)(18)}{(86.1)(1.8)} = 80.7 \quad \text{(see Fig. 31)}$$
$$\text{Per cent error} = 1.9$$

Note that the methods and errors are much similar to those obtained in the Dühring line method, but somewhat lower.

Note. Of the above, only the Dühring and Othmer methods make simple calculations of latent heat, in ordinary heat units, at various temperatures.

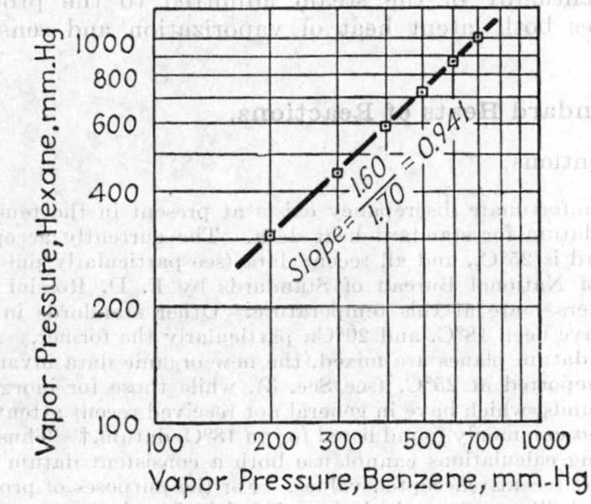

FIG. 30. Othmer plot for the system hexane-benzene.

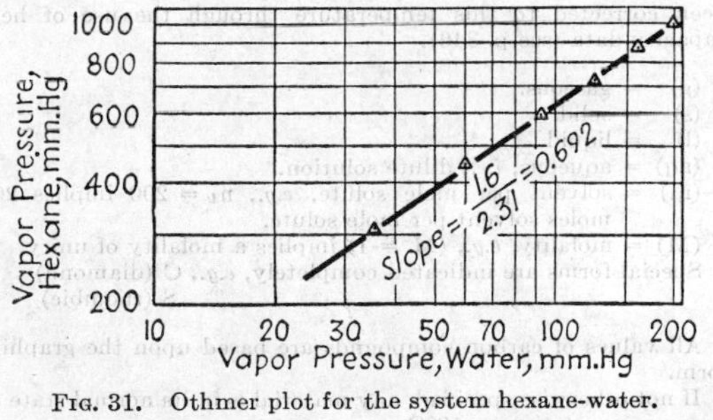

FIG. 31. Othmer plot for the system hexane-water.

Watson Method. Watson [*Ind. Eng. Chem.*, **23**, 360 (1931)] has formulated a chart for the estimation of latent heats of vaporization at high temperatures, duplicated here as Fig. 3, p. 300. His equation is [see p. 300, Eq. (54)]

$$\left(\frac{L_v}{T}\right)_2 = \frac{Y_2}{Y_1}\left(\frac{L_v}{T}\right)_1$$

Values of Y_2 and Y_1 are obtained from the chart, and L_v/T must be known at some temperature. Usually the most convenient reference L_v is that at the normal boiling point. It is also necessary to know the critical temperature T_c.

Example. Estimate the latent heat of vaporization of benzene at a temperature of 227°C. (500°K.).

Solution.

$$L_v = 94.2 \text{ cal./g. at } 80.1°\text{C.}$$
$$T_c = 561.7°\text{K.}$$

$$\left(\frac{T}{T_c}\right)_2 = \frac{500}{561.7} = 0.89$$
$$\left(\frac{T}{T_c}\right)_1 = \frac{80.1 + 273.1}{561.7} = \frac{353.2}{561.7} = 0.63$$

From chart

$$Y_1 = 21.3$$
$$Y_2 = 9.1$$
$$\left(\frac{L_v}{T}\right)_2 = \frac{9.1}{21.3}\left(\frac{94.2}{353}\right)$$
$$L_v = \frac{(9.1)(94.2)(500)}{(21.3)(353)} = 57 \text{ cal./g.}$$

This problem illustrates the fact that the latent heats may be expressed either per mole or per unit weight in this method.

Fusion. The common rule for latent heats of fusion is that data are to be supplanted by Trouton's rule only in cases of lack of the former. However, a recent paper by Othmer (*op. cit.*) has advanced a very convenient method of estimation for any compound, similar to that proposed for latent heats of vaporization. The principle is simple: the slope of the vapor pressure line for the solid state is indicative of the latent heat of fusion *plus* the latent heat of vaporization.

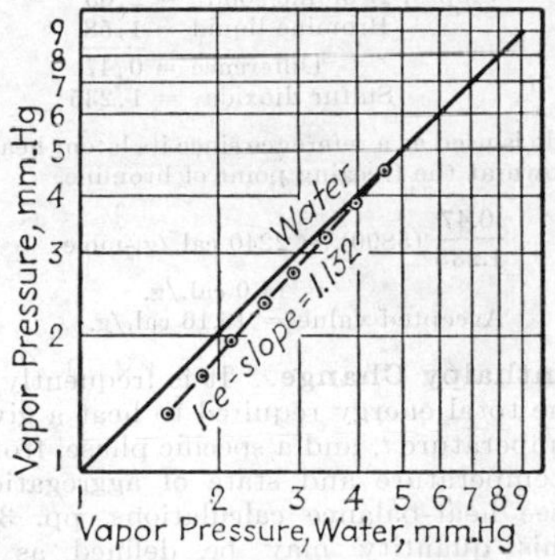

FIG. 32.

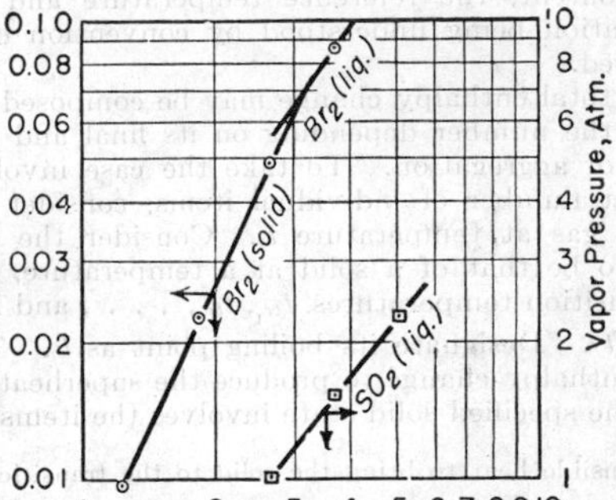

FIG. 33.

Example 1. Find the latent heat of fusion of water from the vapor-pressure data:

t°C.	Vapor pressures, mm. Hg: Ice	Vapor pressures, mm. Hg: Water
−14	1.361	1.560
−12	1.632	1.834
−10	1.950	2.149
− 8	2.326	2.514
− 6	2.765	2.931
− 4	3.280	3.410
− 2	3.880	3.956
0	4.579	4.579

Solution. A loglog plot of these data is given in Fig. 32. The slope of the two lines:

Ice.................................... 1.132
Water.................................. 1.000

$$L_v(0°C.) = 595.4 \text{ cal./g.}$$
$$L_v + L_f(0°C.) = (595.4)(1.132) = 674.0$$
$$L_f = 674.0 - 595.4 = 78.6 \text{ cal./g.}$$

or

$$1.132 - 1.000 = 0.132$$
$$L_f = (595.4)(0.132) = 78.6 \text{ cal./g.}$$
$$\text{True value} = 79.7 \text{ cal./g.}$$
$$\text{Per cent error} = 1.4$$

Example 2. Find the latent heat of fusion of bromine, given the following data:

t°C.	Vapor pressures		
	Cl_2, atm.	Br_2, atm.	SO_2, atm.
10	4.95	0.143	2.256
0	3.66	.0868	1.529
−10	2.63	.0481	0.9995
−20	1.84	.0225	.627
−30	1.23	.0098	.376

$$L_v(SO_2) \text{ at } -7.32°C. = 5890 \text{ cal./g.-mole}$$
$$\text{Freezing point, } Br_2 = -7.32°C.$$

Solution. A loglog plot of these data is given in Fig. 33.

Slopes: Bromine solid = 2.05
Bromine liquid = 1.58
Difference = 0.47
Sulfur dioxide = 1.235

Sulfur dioxide is used as a reference since its latent heat of vaporization is known at the freezing point of bromine.

$$\frac{0.47}{1.235}(5890) = 2240 \text{ cal./g.-mole}$$
$$= 14.0 \text{ cal./g.}$$
$$\text{Accepted value} = 16.16 \text{ cal./g.}$$

Total Enthalpy Change. It is frequently desirable to know the total energy required to heat a given material to a temperature t, and a specific phase, from a given reference temperature and state of aggregation. (For example, see heat-balance calculations, pp. 347*ff.* and 357*ff.*) This quantity may be defined as the *total enthalpy change*, often referred to as the total relative heat content, the reference temperature and state of aggregation being understood by convention or clearly indicated.

The total enthalpy change may be composed of many items, the number depending on its final and reference states of aggregation. To take the case involving the greatest number of individual items, consider a superheated gas at temperature t_s. Consider the reference state to be that of a solid at a temperature, t_c, below its transition temperatures, t_{tr_1}, t_{tr_2}, . . . , and its fusion point, t_f. Designate its boiling point as t_b. Then the total enthalpy change to produce the superheated vapor from the specified solid state involves the items:

1. Sensible heat to bring the solid to the transition temperature t_{tr_1}.
2. Latent heat of transition at the constant temperature t_{tr_1}.
3. Repetition of items 1 and 2 until all solid transition states are passed.
4. Sensible heat to bring the solid to its fusion temperature t_f.
5. Latent heat of fusion at the constant temperature t_f.
6. Sensible heat to bring the liquid to its boiling temperature t_b.
7. Latent heat of vaporization at the constant temperature t_b.
8. Sensible heat to bring the vapor to the temperature t_s.

Rarely will all these items be found necessary. If the reference state is liquid at its freezing point, items 1, 2, 3, 4, and 5 are omitted. This is the case with steam tables whose reference state is liquid water at 32°F. (0°C.). If material remains a gas throughout the problem, only item 8 remains. If the final state is that of liquid at the boiling point and the initial state solid above its last transition point, only items 4, 5, and 6 remain. Thus, as the occasion demands, items may be stripped from the top or bottom of the list to leave only those changes involved in the individual problem.

In heat balances involving chemical reaction, it should be carefully noted that the reference state is specified by the equation for which heat of reaction is calculated. Thus, if the catalytic water gas reaction is written

$$CO\ (g) + H_2O\ (g) \rightarrow CO_2\ (g) + H_2\ (g)$$

the reference states of all materials are indicated to be gaseous. But if the equation had been written

$$CO\ (g) + H_2O\ (l) \rightarrow CO_2\ (g) + H_2\ (g)$$

the reference state for water is the liquid state and the total enthalpy of the steam admitted to the process includes both latent heat of vaporization and sensible heat.

Standard Heats of Reactions.

Conventions

An unfortunate discrepancy exists at present in the temperature datum for standard heat data. The currently accepted standard is 25°C., and all recent data (see particularly publications of National Bureau of Standards by F. D. Rossini and coworkers*) are at this temperature. Other standards in the past have been 18°C. and 20°C., particularly the former. As a result, datum planes are mixed, the new organic data invaribly being reported at 25°C. (see Sec. 3), while those for inorganic compounds, which have in general not received recent attention, are most commonly found listed for an 18°C. datum.† Thus the following calculations cannot use both a consistent datum and the most recently accepted values. For the purposes of providing simple illustrative calculations, it is felt that the latter are the more justifiably sacrificed. An 18°C. datum has therefore been chosen and used consistently. Where possible new values have been corrected to this temperature through the use of heat capacity data (see p. 346).

(g) = gaseous.
(s) = solid.
(l) = liquid.
(aq) = aqueous, *i.e.*, dilute solution.
(n_1) = solvent per mole solute, *e.g.*, $n_1 = 200$ implies 200 moles solvent per mole solute.
(M) = molality, *e.g.*, (M = 1) implies a molality of unity.
Special forms are indicated completely, *e.g.*, C (diamond) S (rhombic)

All values of carbon compounds are based upon the graphite form.

If not otherwise specified, any material is in its normal state of aggregation at 1 atm., 18°C.

Positive values of ΔH indicate absorption of heat by the system, from the surroundings.

Negative values of ΔH indicate loss (generation) of heat by the system to the surroundings.

Standard heats of formation of all elements (in their basic form) are equal to zero.

All examples calculated in c.h.u., and c.h.u. per pound-mole; to convert to B.t.u. and B.t.u. per pound-mole, respectively, multiply by 1.8.

From Heats of Formation. Hess's law of constant heat summation may be used to formulate a general rule for the calculation of heats of reaction at a standard temperature. For example, take the reaction

$$KClO_3\ (s) \rightarrow KCl\ (s) + \tfrac{3}{2}O_2\ (g)\ (\Delta H_{291} = ?) \qquad (26)$$

From heat of formation tables

* *Op. cit.*
† Bichowsky and Rossini, *op. cit.*

$$K + \tfrac{1}{2}Cl_2 + \tfrac{3}{2}O_2 \rightarrow KClO_3\ (s) \qquad (\Delta H_{291} = -91{,}330 \text{ c.h.u.}) \quad (27)$$

$$K + \tfrac{1}{2}Cl_2 \rightarrow KCl\ (s) \qquad (\Delta H_{291} = -104{,}361 \text{ c.h.u.}) \quad (28)$$

(28) − (27)

$$KClO_3\ (s) + \cancel{K} + \cancel{\tfrac{1}{2}Cl_2} \rightarrow KCl\ (s) + \cancel{K} + \cancel{\tfrac{1}{2}Cl_2} + \tfrac{3}{2}O_2 (\Delta H_{291} = -13{,}031 \text{ c.h.u.}) \quad (29)$$

Two items are immediately apparent: (*a*) If the original equation is balanced and the heats of formation of the compounds are properly listed, the final equation [*e.g.*, Eq. (29)] must be the same as the first [*e.g.*, Eq. (26)]; and (*b*) the above work is equivalent to subtracting the standard heat of formation of the reactants from the standard heat of formation of the products. (In the above case ΔH_F for O_2, since it is an element, is zero.) It may be shown, therefore, that for the general equation

$$bB + cC + \cdot \cdot \cdot \rightarrow rR + sS + \cdot \cdot$$

the standard heat of reaction may be calculated as follows:

$$\Delta H_{291} = (r\,\Delta H_{FR} + s\,\Delta H_{FS} + \cdot \cdot \cdot - b\,\Delta H_{FB} - c\,\Delta H_{FC} - \cdot \cdot \cdot) \quad (30)$$

Examples 1 and 2 below illustrate this simple principle.

When infinitely dilute solutions are involved, the problem is complicated but little, for the total standard heat of formation consists of the two parts: standard heat of formation plus the standard heat of infinite solution. For HCl (aq)

$$\Delta H_{F\text{aq}} = \Delta H_F + \Delta H_{\text{soln}} = -22{,}060 - 17{,}470 = -39{,}687 \text{ c.h.u.}$$

This addition is incorporated into Examples 3 and 4.

Example 1.

$$3MnO_2\ (s) + 4Al\ (s) \rightarrow 3Mn\ (s) + 2Al_2O_3\ (s) \qquad (\Delta H_{291} = ?)$$

ΔH_F's: 3(−123,000) 0 0 2(−380,000)

$\therefore \Delta H_{291} = -760{,}000 + 369{,}000 = -391{,}000$ c.h.u.

ΔH_{291} per mole Al = −97,750 c.h.u. = −175,950 B.t.u.

Example 2.

$$Fe_3O_4\ (s) + 4H_2\ (g) \rightarrow 3Fe\ (s) + 4H_2O\ (g) \qquad (\Delta H_{291} = ?)$$

ΔH_F's: −266,900 0 0 4(−57,801)

$\therefore \Delta H_{291} = -231{,}204 + 266{,}900 = 35{,}696$ c.h.u.

ΔH_{291} per mole H_2 (or H_2O) = 8924 c.h.u.

Example 3.

$$ZnS\ (s) + 2HCl\ (aq) \rightarrow ZnCl_2\ (aq) + H_2S\ (g) \qquad (\Delta H_{291} = ?)$$

	ZnS (s)	2HCl (aq)	ZnCl₂ (aq)	H₂S (g)
ΔH_F's:	−44,000	2(−22,060)	−99,550	−5300
ΔH_{soln}		2(17,627)	15,720	
		−79,374	−115,270	

$\therefore \Delta H_{291} = -115{,}270 - 5300 + 79{,}374 + 44{,}000 = 2804$ c.h.u.

Example 4.

$$2Pb(C_2H_3O_2)_2\ (n_1 = 400) + K_2Cr_2O_7\ (n_1 = 800) + H_2O\ (l) \rightarrow 2PbCrO_4\ (s) + 2KC_2H_3O_2 (n_1 = 800) + 2CH_3COOH\ (n_1 = 800)$$

ΔH_F's:	2(−233,420)	−488,500	−68,370	2(−221,400)	2(−174,230)
ΔH_{soln}:	2(− 1,400)	+ 17,250			2(− 3,444)
	2(234,820)	−471,250			2(−177,674)
ΔH_F's:					2(−117,260)
ΔH_{soln}:					2(− 359)
					2(−117,619)

$\Delta H_{291} = 442{,}800 + 355{,}348 + 235{,}238 - 469{,}640 - 471{,}250 - 68{,}370$

$\Delta H_{291} = -1{,}033{,}386 + 1{,}009{,}260 = 24{,}126$ c.h.u.

This last example indicates clearly the fact that small errors in any of the standard heats of formation may introduce large errors in the standard heat of reaction.

Example. Suppose the error in ΔH_F for lead acetate were 1 per cent = 2330 c.h.u. This would produce an error of 4660 c.h.u. in the heat of reaction, or nearly 20 per cent.

It is therefore apparent that it is unwise to use a slide rule in calculating standard heats of reaction.

From Standard Heats of Combustion. Since all heats of combustion represent the evolution of heat by the system, all values of ΔH_c will be negative. Usually reported, however, as heat evolved, Q_c values may be expected to be listed with a positive sign ($Q = -\Delta H$).

In calculating standard heats of reaction, standard heats of combustion have two important applications: (*a*) when values for all reactants and all products are available, they may be used in a manner analogous to that employed with heats of formation, and developed below; (*b*) they may be used to calculate heats of formation, a less readily measured item for organic compounds.

a. Again applying Hess's law of constant heat summation, a general rule may be formulated for calculation of standard heats of reaction from standard heats of combustion. Take the reaction

$$CO_2\ (g) + C\ (\text{graphite}) \rightarrow 2CO\ (g) \qquad (\Delta H_{291} = ?) \quad (31)$$

ΔH_c's: 0 −94,230 2(−67,610)

Express the combustion reactions as follows:

$$2CO\ (g) + O_2 \rightarrow 2CO_2 \qquad (\Delta H_c = 2 \times -67{,}610) \quad (32)$$

$$C\ (\text{graphite}) + O_2 \rightarrow CO_2 \qquad (\Delta H_c = -94{,}230) \quad (33)$$

(33) − (32)

$$C + O_2 - 2CO - O_2 \rightarrow CO_2 - 2CO_2 \qquad (\Delta H_c = +40{,}990)$$

$$\therefore C + CO_2 \rightarrow 2CO \qquad (\Delta H_c = +40{,}990 \text{ c.h.u.})$$

This process is equivalent to subtracting the sum of the heats of combustion of the products from the sum of the heats of combustion of the reactants. Since this is the reverse of the case for heats of formation [in which the rule was products minus reactants, see Eq. (30)], it is well to reverse the statement with standard heats of combustion by a change of sign. The new rule then becomes as follows: *The standard heat of reaction* may be calculated by subtraction of the sum of the negative values of the heats of combustion of the reactants from the sum of the negative values of the standard heats of combustion of the products. Algebraically, for the reaction

$$bB + cC + \cdot \cdot \cdot \rightarrow rR + sS + \cdot \cdot \cdot$$

$$\Delta H_{291} = -(r\,\Delta H_{c_R} + s\,\Delta H_{c_S} + \cdot \cdot \cdot - b\,\Delta H_{c_B} - c\,\Delta H_{c_C} - \cdot \cdot \cdot) \quad (34)$$

Compare Eq. (30)

$$\Delta H_{291} = (r\,\Delta H_{FR} + s\,\Delta H_{FS} + \cdot \cdot \cdot - b\,\Delta H_{FB} - c\,\Delta H_{FC} - \cdot \cdot \cdot)$$

Example 1.

$$CO\ (g) + 2H_2\ (g) \rightarrow CH_3OH\ (l) \qquad (\Delta H_{291} = ?)$$

ΔH_c's: −67,610 2(−68,370) −173,740

$\therefore \Delta H_{291} = -(-173{,}740 + 136{,}740 + 67{,}610) = -30{,}670$ c.h.u.

Example 2.

$$(C_6H_{10}O_5)_n\ (s) + nH_2O \rightarrow 2nC_2H_5OH\ (l) + 2nCO_2\ (g) \qquad (\Delta H_{291} = ?)$$

starch
M = 162*n*

ΔH_c's: −4179 c.h.u./lb. $2n(-327{,}550)$ c.h.u./mole

$\therefore \Delta H_{291} = -n(655{,}100 + 676{,}998) = -21{,}898n$ c.h.u./mole starch

$\Delta H_{291} = -10{,}949$ c.h.u./lb.-mole of C_2H_5OH

In general, heats of combustion may be used with great saving in time and labor when organic reactions are involved; however, it should be emphasized that careful analysis of the product form is necessary. For example, in the reaction

$$CO\ (g) + H_2O\ (l) \rightarrow CO_2\ (g) + H_2\ (g)$$

the use of ΔH_c's will yield a correct result. But if it is of value to determine the standard heat of reaction of the same reaction when written

$$CO\ (g) + H_2O\ (g) \rightarrow CO_2\ (g) + H_2\ (g) \qquad (35)$$

a correction must be made for the fact that H_2O is in the gaseous state, for ΔH_c for H_2 is reported with a product of H_2O (l); the latter will not cancel H_2O (g). The heat of combustion of any compound containing H_b may be corrected to give a product of H_2O (g) by adding 5285b to the reported value. The heat of reaction of Eq. (35) would then be calculated as follows:

$$\Delta H_{291} = -(0 - 68{,}370 + 5285 \times 2 + 67{,}610 + 0) = -9810 \text{ c.h.u.}$$

The same caution should be employed in utilization of values of ΔH_c for compounds containing chlorine. Although it is convenient to measure values of ΔH_c with a product of HCl (aq), the latter may well complicate the solution if a free Cl_2 molecule appears elsewhere in the equation, as in direct chlorinations. Since the heat of oxidation of 2HCl (aq) to Cl_2 (g) is a ΔH = 10,690, compounds containing Cl_d and reported with products of HCl (aq) may be corrected to products in the form of Cl_2 (g) by addition of 5502d to the reported value.

Example 3.

$$C_2H_4\ (g) + Cl_2\ (g) \rightarrow C_2H_4Cl_2\ (g) \qquad (\Delta H_{291} = ?)$$

ΔH_c's: −336,640   0   −307,030

[products include HCl (aq)]

$$\therefore \Delta H_{291} = -[(-307{,}030 + 5502 \times 2) - (-336{,}640)] = -40{,}614 \text{ c.h.u.}$$

b. In some cases reactions are found combining both organic and inorganic compounds, *e.g.*,

$$CaC_2\ (s) + H_2O\ (l) \rightarrow C_2H_2\ (g) + CaO\ (s) \qquad (\Delta H_{291} = ?)$$

Neither ΔH_F's nor ΔH_c's are available for *all* these compounds, so that it must be made possible to calculate the former from the latter, or vice versa. In the above equation, ΔH_F for acetylene would complete the knowledge of the ΔH_F's for the equation. Again let Hess's law form the basis for treatment of the problem.

$$C_2H_2\ (g) + \tfrac{5}{2}O_2 \rightarrow 2CO_2\ (g) + H_2O\ (l) \qquad (\Delta H_{291} = -311{,}170) \quad (36)$$

$$2C\ (\text{graphite}) + 2O_2 \rightarrow 2CO_2\ (g) \qquad (\Delta H_{291} = -188{,}460) \quad (37)$$

$$H_2\ (g) + \tfrac{1}{2}O_2\ (g) \rightarrow H_2O\ (l) \qquad (\Delta H_{291} = -\ 68{,}370) \quad (38)$$

Reversing Eq. (36):

$$2CO_2\ (g) + H_2O\ (l) \rightarrow C_2H_2\ (g) + \tfrac{5}{2}O_2 \qquad (\Delta H_{291} = \ 311{,}170) \quad (39)$$

Adding Eqs. (37), (38), and (39), and simplifying

$$2C\ (\text{graphite}) + H_2\ (g) \rightarrow C_2H_2\ (g) \qquad (\Delta H_{291} = 54{,}340)$$

Using similar examples of standard heats of formation from standard heats of combustion, a general rule is readily formulated for hydrocarbons in the following form* for the compound, C_aH_b,

$$\Delta H_F = -\Delta H_c - 94{,}230a - 34{,}185b$$

* Adapted from Hougen and Watson, *op. cit.*, p. 171.

when the products of combustion are CO_2 (g) and H_2O (l).

Expanding this rule to include organics containing more elements than carbon and hydrogen, a more general expression may be obtained, as follows: for any compound, $C_aH_bBr_cCl_dF_eI_fN_gO_hS_i$, whose combustion products are reported as CO_2 (g), H_2O (l), Br (l), Cl_2 (g), HF (aq), I (s), N_2 (g), SO_2 (g), the general formula is

$$\Delta H_F = -\Delta H_c - 94{,}230a - 34{,}185b - 41{,}515e - 70{,}920i \quad (40)$$

For the same compound, whose products of combustion are CO_2 (g), H_2O (l), Br_2 (g), HCl (aq), HF (aq), I (s), HNO_3 (aq), H_2SO_4 (aq),

$$\Delta H_F = -\Delta H_c - 94{,}230a - 34{,}185b + 7{,}650c - 5502d - 41{,}515e - 15{,}005g - 147{,}430i \quad (41)$$

Obviously for different combinations of products of combustion, individual terms of Eqs. (40) and (41) may be used to form a new general equation. Such an equation is more rapid in use than the method employed in the determination of ΔH_F from the ΔH_c value for acetylene.

Example 1.

$$CaC_2\ (s) + H_2O\ (l) \rightarrow C_2H_2\ (g) + CaO\ (s) \qquad (\Delta H_{291} = ?)$$

ΔH_F's: −14,060   −68,370   +54,340   −151,700

$$\Delta H_{291} = (54{,}340 - 151{,}700) - (-14{,}060 - 68{,}370) = -14{,}930 \text{ c.h.u.}$$

Example 2.

$$C_6H_6\ (l) + 3HNO_3\ (l) \rightarrow C_6H_3(NO_2)_3\ (s) + 3H_2O\ (l) \qquad (\Delta H_{291} = ?)$$

Owing to the inclusion of the inorganic compound HNO_3, it is best to convert ΔH_c values to ΔH_F. For trinitrobenzene, the value of $\Delta H_c = -664{,}000$ is reported for a product of N_2 (g), so that formula (40) applies.

$$\Delta H_F = +664{,}000 - 565{,}380 - 102{,}555 = -3{,}935$$

$$C_6H_6: \Delta H_F = +783{,}400 - 565{,}380 - 205{,}110 = +12{,}910$$

$$\therefore \Delta H_{291} = (-3{,}935 - 205{,}110) - (12{,}910 - 124{,}980) = -96{,}975 \text{ c.h.u.}$$

Heats of Reaction at a Specified Temperature. At other than standard temperatures (18°C. in previous examples), the heat of reaction must be calculated by a heat balance. The latter may be stated in words as follows: Heat in reactants plus heat *evolved* in reaction plus heat from external source equals heat in products plus heat *abstracted* for any purpose.

In the case where reactants and products are at the same temperature, no heat is added from external sources, and heat abstracted is ΔH_t, that not required to maintain products at the temperature t, Kirchhoff's law (see p. 299) applies. This may be written in either of the following forms

$$\Delta H_t = \Delta H_{273} + \Delta at + \frac{\Delta b}{2}t^2 + \frac{\Delta c}{3}t^3 \quad (42)$$

$$\Delta H_t = \Delta H_0 + \Delta a'T + \frac{\Delta b'}{2}T^2 + \frac{\Delta c'}{3}T^3 \quad (43)$$

ΔH_{273} and ΔH_0 do *not* equal the heats of reaction at the absolute temperatures subscripted *unless* the integrated heat capacity equations employed are a complete analysis of heat input from the subscripted temperature to the temperature t (or T). Thus ΔH_{273} is often an accurate measure of heat of reaction at 0°C., but ΔH_0 could never be a measure of heat of reaction at 0°K. since latent heats of phase change are not taken into account in the heat capacity equation. Example 1 below illustrates the use of Eq. (42). Development of Eq. (43) has been illustrated on p. 299.

Example 1. If products of ammonia synthesis leave at 659°C., the reactants also having been charged at that temperature, what is the heat of reaction per mole of ammonia formed? The standard heat of formation of ammonia at 18°C. is 11,000 cal./g.-mole of NH_3.

Solution.

Reaction: $\frac{1}{2}N_2 + \frac{3}{2}H_2 = NH_3 \quad (\Delta H_{291} = -11,000 \text{ cal.})$

Heat capacities:

Per Mole

$$NH_3: C_p = 8.497 + 8.001 \times 10^{-3}t - 1.764 \times 10^{-6}t^2$$
$$N_2: C_p = 6.822 + 1.631 \times 10^{-3}t - 0.345 \times 10^{-6}t^2$$
$$H_2: C_p = 6.919 + 0.218 \times 10^{-3}t + 0.279 \times 10^{-6}t^2$$

In Reacting Quantities

$$NH_3: C_p = 8.479 + 8.001 \times 10^{-3}t - 1.764 \times 10^{-6}t^2$$
$$N_2: 0.5C_p = 3.411 + 0.816 \times 10^{-3}t - 0.172 \times 10^{-6}t^2$$
$$H_2: 1.5C_p = 10.379 + 0.327 \times 10^{-3}t + 0.418 \times 10^{-6}t^2$$
$$\Delta a = 8.497 - 3.411 - 10.374 = -5.293$$
$$\Delta b = (8.001 - 0.816 - 0.327) \times 10^{-3} = 6.858 \times 10^{-3}$$
$$\Delta c = (-1.764 + 0.172 - 0.418) \times 10^{-6} = -2.010 \times 10^{-6}$$
$$\Delta H_t = \Delta H_{273} + \Delta a t + \frac{\Delta b}{2} t^2 + \frac{\Delta c}{3} t^3$$

Calculating ΔH_{273} from known value at $t = 18$°C.:

$$-11,000 = \Delta H_{273} - 5.288(18) + (3.429)(18)^2 \times 10^{-3} - (2.010)(18)^3 \times 10^{-6}$$
$$\Delta H_{273} = -10,906$$

Calculating ΔH_{659}, using previously calculated value of ΔH_{273}:

$$\Delta H_{659} = -10,906 - 5.293(659) + 3.429(659)^2 \times 10^{-3} - (2.010)(659)^3 \times 10^{-6}$$
$$= -10,906 - 3488 + 1490 - 576$$
$$= -13,480 \text{ cal./g.-mole}$$

Heat Balances.* The principle of the heat balance is that of the conservation of energy (the first law of thermodynamics). The simplest statement of the principle for use in calculations centering around chemical reactions is "heat input = heat output." However, since heat is relative, a datum temperature is essential; in most calculations it is most conveniently taken as that at which standard heats of reaction may be determined, 18°C. for work in this section. On this basis, *heat input* consists of the following several parts:

1. The sum of the sensible and latent heats of the reactants (above that, at 18°C., of the phases specified in writing the chemical equation).
2. Energy from extraneous sources, *e.g.*, by direct fire, steam jacket, electrical input.
3. Heats, at 18°C., of all reactions involved $= -\Delta H_{291}$.

The several parts of the total *heat output* are as follows:

1. The sum of the sensible and latent heats of the products (above that, at 18°C., of the phases specified in writing the chemical equation).
2. Energy abstracted from the process for any purpose, *e.g.*, radiation, electrical work, mechanical work—the net heat of reaction under the conditions existing.

By the heat balance, the heat input may be equated to the heat output to obtain any unknown item in either list.

Example 1. For the reaction

$$\frac{1}{2}N_2 + \frac{3}{2}H_2 = NH_3 \text{ (g)}$$

calculate the heat produced per mole of ammonia formed when the reactants enter and the products leave, both at 659°C.

Solution. It will be noted that this problem is identical to that solved in the preceding section entitled Heats of Reaction at a Specified Temperature. Considerable simplification will be seen to be effected by the use of the following method:

* See also Sec. 7, pp. 515*ff.*, Sec. 9, pp. 589*ff.*; 604*ff.* Also, see Staaterman, *Trans. Am. Inst. Chem. Engrs.*, **32**, 431 (1936).

Sensible heats of reactants:

$$H_{N_2} = \frac{1}{2}[(659)(7.27) - (18)(6.78)] = 2334 \text{ c.h.u.}$$
$$H_{H_2} = \frac{3}{2}[(659)(7.01) - (18)(6.92)] = 6742 \text{ c.h.u.}$$
$$H_{\text{reactants}} = 9076 \text{ c.h.u.}$$

Sensible heats of products:

$$H_{NH_3} = (659)(10.72) - (18)(8.40) = 6919 \text{ c.h.u.}$$

Heat of reaction at 18°C. = 11,000 c.h.u. per mole of ammonia $= -\Delta H_{291}$

By a heat balance:

$$9076 + 11,000 = -\Delta H_{659} + 6919$$
$$\Delta H_{659} = -13,160 \text{ c.h.u.}$$

The error from the true value of −13,470 is 2.2% less than that from the heat capacity equations.

A further simplification may be effected by assuming that $\Delta H_{273} = \Delta H_{291}$. The second term of the enthalpy equations could be omitted in this case.

$$H_{N_2} = \frac{1}{2}(659)(7.27) = 2395$$
$$H_{H_2} = \frac{3}{2}(659)(7.01) = 6929 \qquad 9324$$
$$H_{NH_3} = (659)(10.72) = 7070$$
$$\Delta H_t = -13,260 \text{ (error} = 1.0\%)$$

Obviously the error introduced by this assumption is relatively large in cases of a large excess of reactants over products, or vice versa.

Example 2. A theoretical producer gas (34.7 per cent CO, 65.3 per cent N_2), at 18°C., is burned with 25 per cent excess air,

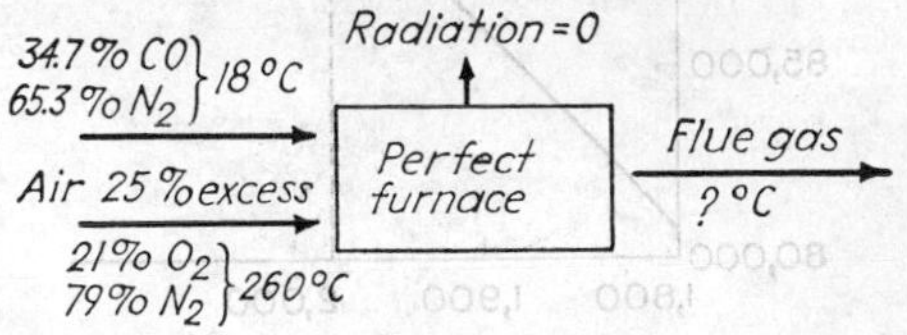

FIG. 34.

the latter preheated to 260°C. Calculate the theoretical flame temperature.

Solution. Basis: 1 mole CO; 1.88 moles N_2

Datum: 18°C.

Reaction: $CO + \frac{1}{2}O_2 \rightarrow CO_2$

Oxygen needed: 0.5 mole	
Oxygen supplied: (0.5)(1.25)	= 0.625 mole (0.125 excess)
Nitrogen in air	= 2.35 moles
Nitrogen in fuel	= 1.88 moles
Nitrogen total	4.23 moles

Flue gas:

CO_2 − 1.0 mole
O_2 − 0.13 mole
N_2 − 4.23 moles

Heat input:

1. In air
   (2.98)(7.1)(260) − (2.98)(6.7)(18) = 5,140 c.h.u.
2. In fuel—none
3. Reaction $-\Delta H_c$ = 67,610 c.h.u.

   72,750 c.h.u.

Heat Output:

1. Heat of radiation—none since *theoretical* flame temperature desired
2. Heat in O_2, N_2, CO_2

Per mole:

$$C_{p_{N_2}} = 6.30 + 1.819 \times 10^{-3}T - 0.345 \times 10^{-6}T^2$$
$$C_{p_{O_2}} = 6.13 + 2.99 \times 10^{-3}T - 0.806 \times 10^{-6}T^2$$

In product quantities:

$$C_{p_{CO_2}} = 6.85 + 8.533 \times 10^{-3}T - 2.475 \times 10^{-6}T^2$$
$$C_{p_{N_2}} = 26.65 + 7.690 \times 10^{-3}T - 1.461 \times 10^{-6}T^2$$
$$C_{p_{O_2}} = 0.80 + 0.389 \times 10^{-3}T - 0.105 \times 10^{-6}T^2$$
$$C_p\ (g) = 34.30 + 16.612 \times 10^{-3}T - 4.041 \times 10^{-6}T^2$$
$$H_{gas} = 34.30T + 8.306 \times 10^{-3}T^2 - 1.347 \times 10^{-6}T^3 - 9{,}980 - 7030 + 33$$

Balance:

$$72{,}750 = 34.30T + 8.306 \times 10^{-3}T^2 - 1.347 \times 10^{-6}T^3 - 17{,}000$$
$$89{,}750 = 34.30T + 8.306 \times 10^{-3}T^2 - 1.347 \times 10^{-6}T^3 = \varphi(T)$$

Trials

T	$\varphi(T)$	
2000	91,050	Plot for solution at $\varphi(T) = 89{,}750$
1900	85,920	$T = 1980°K.\ (1610°C.)$
1800	80,800	(see Fig. 35)

This problem may also be solved without use of the formal heat capacity equations by calculating the enthalpies of exit gases at assumed temperatures, plotting the values of enthalpies against temperature, as in the case illustrated.

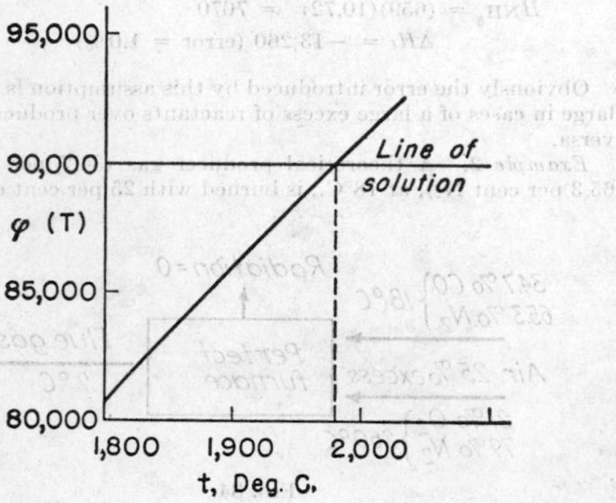

FIG. 35.

Example 3. A vertical kiln is charged with lime and coke at the top; air, dry, at 18°C., blown in the bottom, provides heat by burning the coke to CO_2. It is desired to estimate:

a. Required $CaCO_3$: coke ratio in charge.
b. Analysis of gases to be expected from the top of the kiln.

Assumptions:

Gases: exit temperature—600°F. No free oxygen.
Lime: exit temperature—950°F. Contains no carbon or unburned carbonate.
Coke: pure carbon. Charged at 18°C.
Limestone: pure $CaCO_3$. Charged at 18°C.
Radiation losses: none, giving maximum $CaCO_3$: C ratio.

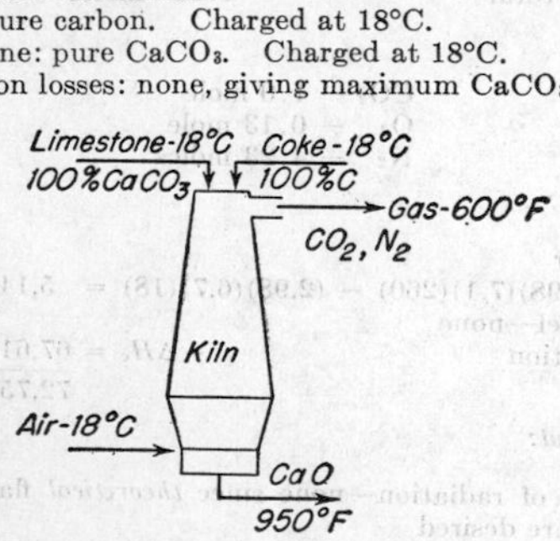

FIG. 36.

Solution.

Data and preliminary calculations:

18°C. = 64.4°F. 600°F. = 315.5°C. 950°F. = 510°C.

$$CaCO_3\ (s) \rightarrow CaO\ (s) + CO_2\ (g)$$

$$\Delta H_{291} = (-151{,}800 - 94{,}230) - (-289{,}500)$$
$$= +43{,}470 \text{ c.h.u./lb.-mole } CaCO_3 \text{ decomposed}$$
$$C\ (coke) + O_2 \rightarrow CO_2\ (g)$$
$$\Delta H_{291} = 96{,}530 \text{ c.h.u./lb.-mole C}$$

C_{pm} (for CaO from 18°C. to 510°C.) = 0.200
C_{pm} (for N_2, from 18°C. to 315.5°C.) = 7.05
C_{pm} (for CO_2, from 18°C. to 315.5°C.) = 10.1

Basis: 1 mole $CaCO_3$ decomposed
x moles C burned
x moles O_2 used

$$\frac{79}{21}x = 3.76x \text{ moles } N_2 \text{ used}$$

Heat Input

1. Reactants 0
2. Reaction $CaCO_3$: $-43{,}470$ c.h.u.
3. Reaction coke: $+96{,}530x$ c.h.u.

Heat Output

1. CaO: $(56)(0.200)(510 - 18)$ c.h.u.
2. CO_2: $(1 + x)(10.1)(316 - 18)$ c.h.u.
3. N_2: $(3.76x)(7.05)(316 - 18)$ c.h.u.

Balance

$$-43{,}470 + 96{,}530x = (56)(0.2)(492) + (1 + x)(10.1)(298) + (3.76x)(7.05)(298)$$
$$96{,}530x - 3010x - 7900x = 43{,}470 + 5510 + 3010$$
$$85{,}620x = 51{,}990$$
$$x = 0.603 \text{ mole C per mole } CaCO_3$$
$$\frac{0.603 \times 12}{100} = 0.0724 \text{ lb. C per pound } CaCO_3$$

GAS LEAVING—ANALYSIS

	Moles	Moles	%
CO_2	$1 + x$	1.603	41.4
N_2	$3.76x$	2.27	58.6
Total		3.87	100.0

This problem is considerably simplified, as compared to calculations of an actual case, by the specified data, *i.e.*, all reactants at 18°C., and pure $CaCO_3$ and carbon. Although material and heat balances would be more complicated with actual data, the principles would remain the same as illustrated in the simple case.

Equilibrium. *Calculation of Equilibrium Constant.* By the well-known equation (see p. 311)

$$-\Delta F_T° = RT \ln K \tag{44}$$

values of K, the equilibrium constant, may be calculated at the temperature for which the value of $\Delta F_T°$ is known. It therefore becomes essential to know the methods of calculating values of $\Delta F_T°$ for any reaction. Several such methods exist:

1. Calculation of $\Delta F°$ at a definite temperature from values of $\Delta F°$ for other reactions, or for formation of the compounds involved, all at one standard temperature.

2. Calculation of $\Delta F°$ at a definite temperature from values of ΔH and ΔS at the same temperature by equation

$$\Delta F = \Delta H - T\Delta S$$

(see p. 305).*

3. Calculation of $\Delta F°$ as a function of temperature by the relation (see p. 313).

$$\Delta F_T° = \Delta H_0 - \Delta aT \ln T - \tfrac{1}{2}\Delta bT^2 - \tfrac{1}{6}\Delta cT^3 - \cdots + IT \tag{45}$$

and

$$\Delta H_T = \Delta H_0 + \Delta aT + \tfrac{1}{2}\Delta bT^2 + \tfrac{1}{3}\Delta cT^3 + \cdots \tag{46}$$

Example 1. From the following data on free energies of formation at 25°C., calculate the equilibrium constant for the reaction

* Recent data on standard entropies and standard heats of formation of hydrocarbons are given in *Chem. Rev.*, **27**, 1–83 (1940).

$$CO\ (g) + H_2O\ (g) \rightleftarrows CO_2\ (g) + H_2\ (g)$$

	$\Delta F_F°$
CO	−32,808 cal./g.-mole
H_2O	−54,635 cal./g.-mole
CO_2	−94,260 cal./g.-mole

where $\Delta F_F°$ is the standard free energy of formation.

Solution.

$$\begin{aligned}\Delta F_{298}° &= \Sigma\Delta F_F°_{\text{products}} - \Sigma\Delta F_F°_{\text{reactants}}\\ &= (\Delta F_F°_{CO_2} + \Delta F_F°_{H_2}) - (\Delta F_F°_{CO} + \Delta F_F°_{H_2O})\\ &= (-94{,}260 + 0) - (-54{,}635 - 32{,}808)\\ &= -6817 \text{ cal.}\\ +6817 &= RT \ln K\\ \log K &= \frac{6817}{4.58T} = 4.99\\ K &= 9.77 \times 10^4 \text{ at } 25°\text{C.}\end{aligned}$$

Example 2. Calculate the value of the equilibrium constant at 25°C. from entropy and free energy data for the reaction

$$C_2H_4\ (g) + H_2O\ (g) \rightleftarrows C_2H_5OH\ (g)$$

Solution. The following data are available:

	S_{298}*	$\Delta S_{F_{298}}$†	$\Delta H_{F_{298}}$*	L_v
C_2H_4 (g)	52.75	−12.68	12,496	
H_2O (g)	45.17	−10.61	−57,798	
C_2H_5OH (g)	66.4		−56,350	
C_2H_5OH (l)		−82.4		10,000

Method a.

ΔS_{298} may be calculated from S_{298} data as follows:

$$\begin{aligned}\Delta S_{298} &= \Sigma S_{298}\ (\text{products}) - \Sigma S_{298}\ (\text{reactants})\\ \Delta S_{298} &= 66.4 - (52.75 + 45.17)\\ &= -31.52 \text{ E.U. (entropy units)}\\ T\Delta S_{298} &= (-31.52)(298) = -9390 \text{ cal.}\end{aligned}$$

ΔH_{298} is calculated from ΔH_F's in the usual manner (see p. 313)

$$\begin{aligned}\Delta H_{298} &= -56{,}350 - (-57{,}798 + 12{,}496)\\ &= -11{,}048 \text{ cal.}\\ \therefore \Delta F_{298}° &= -11{,}048 + 9390 = -1658 \text{ cal.}\\ RT \ln K &= +1658\\ \log K &= 1.22\\ K &= 16.6\end{aligned}$$

Method b. If the data on S_{298} were not available, it might be necessary to use those for $\Delta S_{F_{298}}$. From these data ΔS_{298} may be calculated by

$$\Delta S_{298} = \Sigma\ \Delta S_{F_{298}}\ (\text{products}) - \Sigma\ \Delta S_{F_{298}}\ (\text{reactants})$$

However, in this case $\Delta S_{F_{298}}$ is known for the liquid and desired for the vapor. At 25°C., ethyl alcohol boils under a pressure of 10.8 mm. (0.0142 atm.), with a latent heat of 10,000 cal./mole.

∴ for C_2H_5OH (l, 0.0142 atm.) → C_2H_5OH (g, 0.0142 atm.)‡

$$\Delta S_{298} = \frac{10{,}000}{298} = 33.55 \text{ E.U.}$$

and since the relation for change of pressure is

$$\Delta S_{298} = R \ln \frac{P_1}{P_2}$$

assuming the simple gas law,

for C_2H_5OH (g, 0.0142 atm.) → C_2H_5OH (g, 1 atm)

$$\begin{aligned}\Delta S_{298} &= R \ln \frac{0.0142}{1}\\ &= 4.58 \log 0.0142 = (4.58)(-1.848)\\ &= -8.45 \text{ E.U.}\end{aligned}$$

Adding the two results and neglecting the entropy change of the liquid with change of pressure, we get for

C_2H_5OH (l, 1 atm.) → C_2H_5OH (g, 1 atm.)

$$\Delta S_{298} = 25.1 \text{ E.U.}$$

∴ for C_2H_5OH (g, 1 atm.)

$$\begin{aligned}\Delta S_{F\,(g)} &= \Delta S_{F\,(l)} + \Delta S_{F\,(\text{vap})}\\ &= -82.4 + 25.1 = -57.3\end{aligned}$$

* Ewell, *Ind. Eng. Chem.*, **31**, 267 (1939).

† Calculated, from data on ΔF and ΔH, Sec. 3.

‡ See Lewis and Randall, "Thermodynamics and the Free Energy of Chemical Substances," p. 146, McGraw-Hill, 1923.

Then ΔS_{298}, for the reaction

$$\begin{aligned}&= -57.3 - (-12.68 \quad 10.61)\\ &= -34.01 \text{ E.U.}\end{aligned}$$

which is in quite good agreement with Ewell's data used in method *a*.

The calculation of $\Delta F_{298}°$ from this entropy value is exactly as in method *a*, and will not be repeated.

Example 3. For the reaction

$$C_2H_4\ (g) + H_2O\ (g) \rightarrow C_2H_5OH\ (g)$$

calculate expressions for $\Delta F°$ and $\log K$ as functions of the absolute temperature.

Solution. From the calculations of Example 2,

$$\Delta H_{298} = -11{,}048 \text{ cal.}$$

and

$$\Delta F_{298}° = -1658 \text{ cal.}$$

From A.P.I. data,* the following C_p is calculated:

$$C_2H_4\text{: } C_p = 9.085 - 0.164 \times 10^{-3}T + 0.153 \times 10^{-6}T^2$$

Combining this equation with that previously employed for water:

$$\begin{aligned}H_2O\ (g)\text{: } C_p &= 6.89 + 3.283 \times 10^{-3}T - 0.343 \times 10^{-6}T^2\\ C_2H_4\ (g)\text{: } C_p &= 9.09 - 0.164 \times 10^{-3}T + 0.153 \times 10^{-6}T^2\\ \hline C_p\ (\text{reactants}) &= 15.98 + 3.119 \times 10^{-3}T - 0.190 \times 10^{-6}T^2\end{aligned}$$

Calculating by the method of Fugassi and Rudy:

$$\begin{aligned}C_2H_5OH\ (g)\text{: } C_p &= 1.430 + 53.802 \times 10^{-3}T - 19.496 \times 10^{-6}T\\ &\qquad + 10.98 \times 10^{+6}T^{-3}\\ \hline \Delta C_p &= -14.550 + 50.683 \times 10^{-3}T - 19.306 \times 10^{-6}T\\ &\qquad + 10.98 \times 10^{+6}T^{-3}\end{aligned}$$

The equation for $\Delta F°$ as a function of T has been derived (see p. 310) as

$$\Delta F_T° = \Delta H_0 - \Delta a\ T \ln T - \tfrac{1}{2}\Delta b\ T^2 - \tfrac{1}{6}\Delta c\ T^3 + \cdots + IT \qquad (47)$$

in which the constant ΔH_0 must be obtained from a known value of ΔH_T ($\Delta H_{298} = -11{,}048$ in this problem) and the equation

$$\Delta H_T = \Delta H_0 + \Delta a\ T + \tfrac{1}{2}\Delta b\ T^2 + \tfrac{1}{3}\Delta c\ T^3 + \cdots \qquad (48)$$

The constant of integration I in Eq. (47) may then be obtained from one known value of $\Delta F_T°$ (= −1658 at 298°K.)

In Eq. (48)

$$\begin{aligned}-11{,}048 &= \Delta H_0 - 14.550(298) + 25.342 \times 10^{-3}(298)^2\\ &\qquad -6.435 \times 10^{-6}(298)^3 - 5.49 \times 10^{+6}T^{-2}\\ &= \Delta H_0 - 4335 + 2250 - 171 - 61.8\\ \Delta H_0 &= -8730\end{aligned}$$

Substituting this value in Eq. (47) as well as $\Delta F_{298}° = -1658$ and $T = 298$

$$\begin{aligned}-1658 &= -8730 + 14.550(298)(2.3)(\log_{10} 298) - 25.342\\ &\qquad \times 10^{-3}T^2 + 3.218 \times 10^{-6}T^3 - 2.77 \times 10^{+6}T^{-5} + 298I\end{aligned}$$

(The last T term is negligible)

$$\begin{aligned}-1658 &= -8730 + 24{,}674 - 2250 + 86 + 298I\\ I &= -51.8\\ \therefore \Delta F_T° &= -8730 + 14.550T \ln T - 25.342 \times 10^{-3}T^2\\ &\qquad + 3.218 \times 10^{-6}T^3 - 51.8T\end{aligned}$$

and since $\log K = -\dfrac{\Delta F_T°}{2.303RT}$

$$\log K = \frac{1907}{T} - 7.32 \log T + 5.53 \times 10^{-3}T - 0.702 \times 10^{-6}T^2 + 11.31$$

Values determined from the last two equations are plotted in Fig. 37, along with values of $\Delta F_T°$ calculated from the data of Example 2, on the assumptions that neither ΔH_{298} nor ΔS_{298} varies with change in temperature, a useful method of first

* API Research Project 44, National Bureau of Standards. The data are plotted and fitted to the second degree form of equation. The fit is not good.

approximation and probably not much more inaccurate than the use of uncertain heat capacity equations.

Figure 38 also includes data for $\log_{10} K$ as a function of T°K. for other common reactions.*

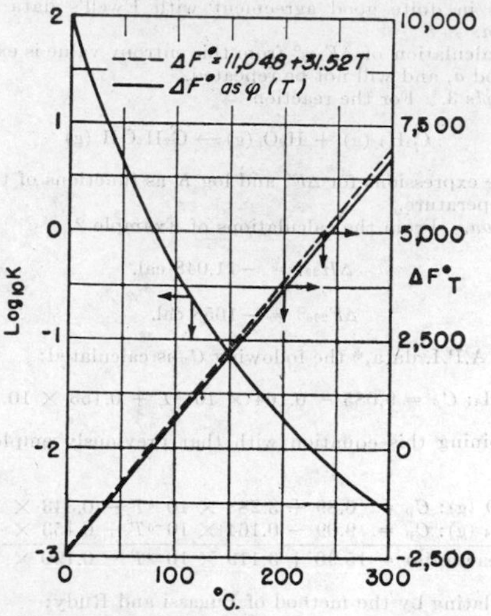

FIG. 37.

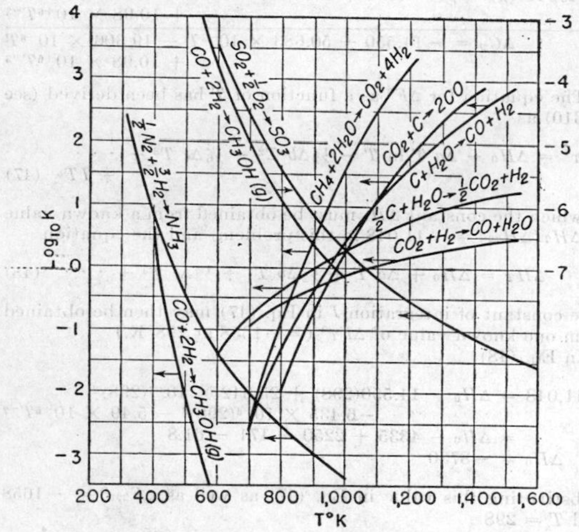

FIG. 38.

Use of Equilibrium Constants. Starting with the general equilibrium constant K for the expression $bB + cC + \cdots \rightleftarrows rR + sS + \cdots$, written in terms of activities, a, as follows,

$$K = \frac{(a_R)^r(a_S)^s \cdots}{(a_B)^b(a_C)^c \cdots} \tag{49}$$

considerable simplification may be performed for low-pressure work.†

First

$$a_A = \gamma_A N_A p$$

where N_A = mole fraction of material A.

γ_A = activity coefficient of material A.

* Lines revised since the second edition are based on data by Rossini and coworkers, *Bur. Standards J. Research*, **34**, 143 (1945).

† High-pressure equilibrium calculations are illustrated on pp. 355*ff*.

Further, at low pressures, $\gamma = 1$ and $a = p$.

$$\therefore\ a_A = p_A = N_A P$$

Therefore, the K may be written as K_p' (ideal gas law applying) according to the following

$$K_p' = \frac{(p_R)^r(p_S)^s \ldots}{(p_B)^b(p_C)^c \ldots} = \frac{(n_R)^r(n_S)^s \ldots}{(n_B)^b(n_C)^c \ldots}\left(\frac{p}{n_T}\right)^{r+s+\cdots-b-c\cdots} \tag{50}$$

where $n_R, n_S \ldots$ = number of moles of material R, S, ...

n_T = total gaseous moles.

Example 1. For the expression

$$N_2 + 3H_2 \rightleftarrows 2NH_3$$

$$K_p' = \frac{(n_{NH_3})^2}{n_{N_2}(n_{H_2})^3}\frac{(n_T)^2}{p^2}$$

It is frequently found that the values of K_p are given for the equation with different coefficients, that of at least one product (or reactant) having been made equal to unity.

$$\frac{b}{r}B + \frac{c}{r}C \rightleftarrows R + \frac{s}{r}S$$

then

$$K_p = \frac{(p_R)(p_S)^{s/r}}{(p_B)^{b/r}(p_C)^{c/r}} = \sqrt[r]{K_p'} \tag{51}$$

Example 2. For the expression

$$\tfrac{1}{2}N_2 + \tfrac{3}{2}H_2 \rightleftarrows NH_3$$

$$K_p = \frac{n_{NH_3}n_T}{(n_{N_2})^{1/2}(n_{H_2})^{3/2}\,p} = \sqrt{K_p'} \tag{52}$$

It is sometimes found that values of K_p will be reported for the reverse of the action desired, as

$$rR + sS \rightleftarrows bB + cC$$

Then

$$K_p'' = \frac{(p_B)^b(p_C)^c}{(p_R)^r(p_S)^s}$$

and

$$K_p = 1/K_p'' \tag{53}$$

Example 3. For the expression

$$NH_3 \rightleftarrows \tfrac{1}{2}N_2 + \tfrac{3}{2}H_2$$

$$K_p'' = \frac{(n_{N_2})^{1/2}(n_{H_2})^{3/2}\,p}{n_{NH_3}n_T} = \frac{1}{K_p'} \tag{54}$$

If some components are gaseous while others are liquid, or solid, it is permissible to eliminate the latter from the equilibrium expression. Thus, if component B is a solid and component S a liquid, since

$$a_S = 1 \quad \text{and} \quad a_B = 1$$

$$K_p = \frac{(p_R)^r}{(p_C)^c} = \frac{(n_R)^r}{(n_C)^c}\left(\frac{p}{n_T}\right)^{r+\cdots-c-\cdots} \tag{55}$$

(see Example 7).

In general, equilibrium expressions have two uses in calculations: (*a*) calculation of values of K_p from analysis of constituents after equilibrium is attained; and (*b*) prediction of equilibrium concentrations from known values of K_p. Below are given four examples, one (Example 4) of type *a* and three of type *b*. The latter include three types: Example 5—solution of a simple quadratic; Example 6—solution of equation of higher than second order; and Example 7—example of at least one constituent in liquid or solid phase; Example 8 shows the use of K for liquid phase reaction.

Example 4. If a stoichiometric mixture, 3 moles H_2 to 1 mole N_2, reacts under a constant pressure of 100 atm. to form at equilibrium 0.5 mole ammonia, calculate K for the reaction, $3/2H_2 + 1/2N_2 \rightleftarrows NH_3$, at this temperature, assuming the gases sufficiently ideal that the fugacity may be replaced by the partial pressure.

Solution. 0.5 mole N_2 and 1.5 moles H_2 disappear per mole of NH_3 formed. Therefore at equilibrium:

	Moles	Mole Fractions
N_2:	$1 - 0.25 = 0.75$	$0.75/3.5 = 0.214$
H_2:	$3 - 0.75 = 2.25$	$2.25/3.5 = 0.644$
NH_3:	$= 0.50$	$0.50/3.5 = 0.142$
Total moles	$= 3.50$	

Partial Pressures

$0.214 \times 100 = 21.4$ atm.
$0.644 \times 100 = 64.4$ atm.
$0.142 \times 100 = 14.2$ atm.

$$K = \frac{p_{NH_3}}{p_{H_2}^{3/2} \times p_{N_2}^{1/2}} = \frac{14.2}{(64.4)^{3/2}(21.4)^{1/2}} = 5.99 \times 10^{-3}$$

Example 5. A water gas is mixed with steam and fed to a catalytic water-gas converter, in which the following reaction occurs:

$$CO + H_2O \rightleftarrows CO_2 + H_2$$

Calculate: the equilibrium per cent conversion of CO to CO_2 if 600 per cent excess steam is used, and the operation is conducted at 932°F. and 1 atm.

Water-gas analysis:

4 per cent CO_2; 43 per cent CO; 51 per cent H_2; 2 per cent N_2

Solution. Basis: 100 moles dry water gas

43 moles CO
∴ 43 moles H_2O needed

at 600 per cent excess, 301 moles H_2O supplied

Let x = moles CO_2 formed at equilibrium.

Then at equilibrium

$CO_2 = 4 + x$ moles
$CO = 43 - x$ moles
$H_2 = 51 + x$ moles
$N_2 = 2$ moles
$H_2O = 301 - x$ moles
Total = 401 moles

From chart (see p. 350) at 932°F. for the reaction

$$CO_2 + H_2 \rightleftarrows CO + H_2O$$

$$\log K_p'' = -0.70 = \bar{1}.30$$
$$K_p'' = 0.1995$$
$$K_p = \frac{1}{K_p''} = 5.01 = \frac{n_{CO_2} n_{H_2}}{n_{CO} n_{H_2O}}$$

$$\therefore 5.01 = \frac{(4 + x)(51 + x)}{(43 - x)(301 - x)} = \frac{204 + 55x + x^2}{12{,}930 - 344x + x^2}$$

$$64{,}800 - 1725x + 5.01x^2 = 204 + 55x + x^2$$
$$4.01x^2 - 1780x + 64{,}600 = 0$$
$$x = \frac{1780 \pm \sqrt{3{,}165{,}000 - 1{,}037{,}000}}{8.02}$$
$$x = \frac{1780 \pm 1458}{8.02} = 40.1$$

$$\therefore \text{Per cent CO converted} = \frac{40.1}{43} = 93.4$$

The gases leaving may be readily analyzed, as follows:

	Moles	Per Cent (Wet)	Per Cent (Dry)
CO_2	44.1	11.0	31.5
CO	2.9	0.7	2.1
H_2	91.1	22.7	65.0
N_2	2.0	0.5	1.4
H_2O	260.9	65.1	
Total	401.0 / 140.1 (dry)	100.0	100.0

Example 6. A burner gas, analyzing 8 per cent SO_2, 12.9 per cent O_2, and 79.1 per cent N_2 is passed to a contact converter operating at 475°C. and 1 atm. pressure. Calculate: the per cent conversion of SO_2 to SO_3 if equilibrium were attained in the converter.

Solution. Basis: 100 moles gas fed to the converter

8 moles SO_2
12.9 moles O_2
79.1 moles N_2

Reaction: $SO_2 + \frac{1}{2}O_2 \rightleftarrows SO_3$

Let α = fractional conversion of SO_2 to SO_3*

Equilibrium Composition

$SO_2 = 8(1 - \alpha)$
$SO_3 = 8\alpha$
$O_2 = 12.9 - 4\alpha$
$N_2 = 79.1$
Total $= 100 - 4\alpha$

Equilibrium constant: $\log K_p = 1.92$ (see Fig. 38)
$K_p = 83.1$

$$83.1 = \frac{n_{SO_3}}{(n_{O_2})^{1/2}\, n_{SO_2}} \left(\frac{nT}{P}\right)^{1/2} = \frac{(8\alpha)(100 - 4\alpha)^{1/2}}{(12.9 - 4\alpha)^{1/2}(8 - 8\alpha)} = f(\alpha)$$

To solve, assume values of α, calculating the values of $f(\alpha)$. Plot the latter against the assumed values of α as abscissa. Where the curve crosses an ordinate value of 83.1, the equation is satisfied.

Results of trials.

α	$f(\alpha)$
0.99	324
.98	160
.97	106
.96	78.2
.95	61.8

From Fig. 39

$$\alpha = 0.962$$

∴ Per cent conversion = 96.2 per cent

Another method of effecting a graphical solution of the third power equation is also illustrated in Fig. 39 by the right-hand

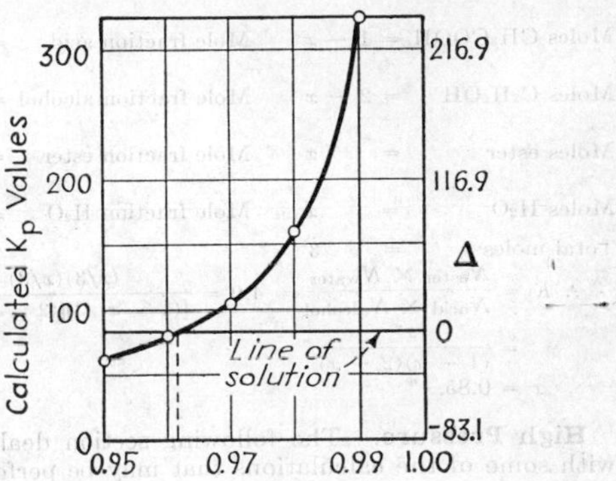

FIG. 39.

ordinates. For this case the equilibrium expression is rearranged to read

$$\frac{(8\alpha)(100 - 4\alpha)^{1/2}}{(12.9 - 4\alpha)^{1/2}(8 - 8\alpha)} - 83.1 = \Delta$$

Δ's are calculated for assumed values of α. The solution of the equation occurs when $\Delta = 0$, as shown by the coordinates at the right, Fig. 39.

Example 7. A stream of pure CO_2 contacts a bed of hot graphite, and the resulting reaction is allowed to progress to equilibrium at a temperature of 800°C. and a pressure of $\frac{1}{2}$ atm. Calculate: the analysis of the product gases.

Solution. Basis: 1 mole of CO_2 charged

Reaction: $CO_2 + C \rightleftarrows 2CO$

Equilibrium expression: $K = \dfrac{(a_{CO})^2}{(a_{CO_2})(a_C)}$ $\quad (a_C = 1)$

* This method is presented as an alternate to that in which the unknown is established as a number of moles converted (see Example 5). Choice of one of these methods is entirely a personal matter. The author believes that analysis of the constituents is most clean-cut when the unknown is expressed as a number of moles.

$$\therefore K_p = \frac{(p_{CO})^2}{(p_{CO_2})} = \frac{(n_{CO})^2 P}{n_{CO_2} n_T}$$

Let $2x$ = moles CO formed at equilibrium
$1 - x$ = moles CO_2 consumed
$1 + x$ = total moles

Equilibrium constant: $\log K_p = 0.89$ (see Fig. 38)

$$K_p = 7.86$$

$$\therefore 7.86 = \frac{(4x^2)(0.5)}{(1-x)(1+x)} = \frac{2x^2}{1-x^2}$$

$$7.86 - 7.86x^2 = 2x^2$$

$$x^2 = \frac{7.86}{9.86} = 0.797$$

$x = 0.89$ = fraction CO_2 converted, at equilibrium

CO_2 –	0.11 moles –	5.8 per cent
CO-	1.78 moles –	94.2 per cent
	1.89 moles	100.0

Example 8. Assume the equilibrium constant K for the esterification of ethyl alcohol in the liquid phase

$$C_2H_5OH + CH_3COOH \rightleftarrows CH_3OOC_2H_5 + H_2O$$

to be 4.0. Calculate the equilibrium conversion to ester if 2 moles of alcohol, 1 mole of acid, and no water are initially present in the reaction mixture and if the activity of each component may be taken as equal to its mole fraction.

Solution. Let x = moles ester = moles H_2O formed at equilibrium. Then, at equilibrium

Moles CH_3COOH	$= 1 - x$	Mole fraction acid	$= \frac{(1-x)}{3}$
Moles C_2H_5OH	$= 2 - x$	Mole fraction alcohol	$= \frac{(2-x)}{3}$
Moles ester	$= x$	Mole fraction ester	$= \frac{x}{3}$
Moles H_2O	$= x$	Mole fraction H_2O	$= \frac{x}{3}$
Total moles	$= 3$		

$$\therefore K = \frac{N_{\text{ester}} \times N_{\text{water}}}{N_{\text{acid}} \times N_{\text{alcohol}}} = 4.0 = \frac{(x/3)(x/3)}{[(1-x)/3][(2-x)/3]}$$

$$= \frac{x^2}{(1-x)(2-x)}$$

$$x = 0.85.$$

High Pressure. The following section deals briefly with some of the calculations that may be performed in regions where the application of the ideal gas law would lead to errors larger than can be tolerated. Two systems of applying these corrections exist. In the first, an empirical constant is introduced into the ideal gas law equation, so that it reads

$$pV = CnRT$$

In the second, a volume corrective term α is used

$$V_{\text{true}} = \frac{nRT}{p} - \alpha$$

In this section, only the use of the compressibility factor C will be illustrated. For information on α, which has some advantages for certain types of work, see Sage and Lacey, "Phase Equilibria in Hydrocarbon Systems," and Edminster, *Ind. Eng. Chem.*, **28**, 1112 (1936), and **30**, 352 (1938).

The importance of either C or α lies in the ability to correlate relatively accurately the values for different gases by use of the reduced conditions and the consequent elimination of use of complicated equations involving a multitude of constants or higher degree equations, with virial coefficients, such as found in the theoretical van der Waals and Beattie-Bridgeman or the empirical Maron and Turnbull equations.* For C, this general correlation is presented in Fig. 40, values of $C = pv/RT$ being plotted vs. p_r at constant values of T_r. As primarily discussed (see p. 292)

$$p_r = \frac{p}{p_c}\dagger \quad (56)$$

$$T_r = \frac{T}{T_c}\dagger \quad (57)$$

Also

$$V_r = \frac{V}{V_c}$$

Figure 40 represents *average* data for a number of gases and is exact for none of them.

Compressibility Chart for Pure Gases. There are three main problems involving pure gases, cases of volume, temperature, and pressure being the unknowns. Either of the following equations may be applied:

$$pV = CnRT$$

$$\frac{p_1V_1}{p_2V_2} = \frac{C_1T_1}{C_2T_2} \quad (58)$$

If volume is the unknown, both p_r and T_r may be determined, and C obtained from the compressibility chart, Fig. 40. But, if either p or T is unknown, graphical means must be employed for a solution.

The following example will illustrate each of the three cases. For the sake of simplicity, one set of conditions will be used for all three calculations, 100 lb. of ethane, at a pressure of 50 atm., a temperature of 100°C., and a calculated volume of 25.9 cu. ft.

Case I. Let volume be the unknown.
Solution. $n = {}^{100}/_{30} = 3.3$ moles

Data: $p_c = 48.2$ atm.
$T_c = 305.4°K.$

$$\left.\begin{aligned} p_r &= \frac{50.0}{48.2} = 1.04 \\ T_r &= \frac{373}{305.4} = 1.22 \end{aligned}\right\} \therefore \text{from Fig. 40, } C = 0.8$$

$$R = 1.317 \frac{(\text{atm.})(\text{cu. ft.})}{°\text{K.}}$$

$$\therefore V = \frac{CnRT}{p} = \frac{(0.8)(3.3)(1.317)(373)}{50} = 25.9 \text{ cu. ft.}$$

Case II. Let pressure be the unknown.
Solution. In this case C cannot be determined directly, since p is unknown. However, with the known quantities, n, T, and V, the general equation $pV = CnRT$ may be reduced to

$$C = a'p = a'p_rp_c$$

$$\therefore C = ap_r \quad \text{where} \quad a = \frac{p_cV}{nRT}$$

The equation is that of a straight line on either logarithmic or coordinate paper. The solution of the problem lies on this line

* Maron and Turnbull, *Ind. Eng. Chem.*, **34**, 544 (1942).
† For critical conditions of various materials, see Sec. 3.

For helium, hydrogen, and neon, Newton [*Ind. Eng. Chem.*, **27**, 302 (1935)] has shown that better correlation is obtained by the expression

$$p_r = \frac{p}{p_c + 8} \quad \text{pressures in atm.}$$

$$T_r = \frac{T}{T_c + 8} \quad \text{temperatures in °K.}$$

Many of the errors that result from use of reduced conditions are reported by Morgen and Childs [*Ind. Eng. Chem.*, **37**, 667 (1945)] to be eliminated by their equations for $p_r = \frac{p}{p_c + K}$ and $T_r = \frac{T}{T_c + K'}$, methods of calculating K and K' being given

Many methods of estimating critical conditions have been proposed. Meissner and Redding [*Ind. Eng. Chem.*, **34**, 521 (1942)] have provided a summary.

at its intersection with the known T_r. On the logarithmic coordinates of Fig. 40 the line $C = ap_r$ is one of slope equal to 1 (45°) and intercept $a(C = a \text{ at } p_r = 1.0)$. On a coordinate C vs. p_r chart, the line $C = ap_r$ has no intercept, a slope a, and represents the solution to the problem at the known value of T_r.

The results of Fig. 41 show

$$C = 0.8$$
$$T_r = 1.22$$
$$\therefore T = (1.22)(305.4) = 373°\text{K.} = 100°\text{C.}$$

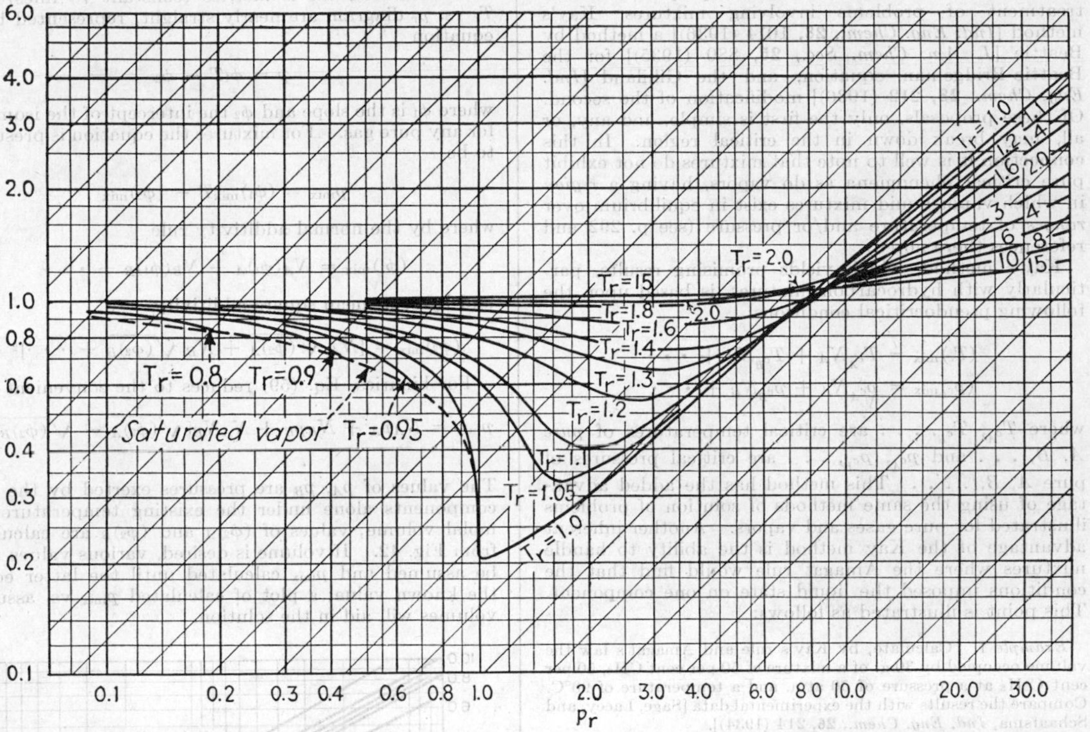

FIG. 40. Compressibility factors of gases and vapors. (*From Hougen and Watson, Industrial Chemical Calculations, 2d ed., Wiley, 1936.*)

In the specific problem here presented:

$$T_r = \frac{373}{305.4} = 1.22$$

$$a = \frac{(48.2)(25.9)}{(3.3)(1.317)(373)} = 0.77$$

Graphically $C = 0.8$; $p_r = 1.05$

$$p = 1.05p_r = 50.6 \text{ atm.}$$

Case III. Temperature unknown.

Solution. As in Case II, C cannot be determined directly but the general equation $pV = CnRT$ may be reduced to

$$C = \frac{b'}{T} = \frac{b'}{T_rT_c} = \frac{b}{T_r}$$

where $b = \dfrac{pV}{nRT_c}$

The solution of the problem lies at the intersection of two lines on an auxiliary plot:

$C = b/T_r$ for assumed values of T_r

C at constant known p_r for assumed values of T_r

In the specific problem:

$$p_r = 1.04$$

$$b = \frac{(50)(25.9)}{(3.3)(1.317)(305.4)} = 0.976$$

Solved graphically on Fig. 41, by the calculated data:

Assumed T_r	C	
	From b/T_r	At constant $p_r = 1.04$
1.0	0.976	0.20
1.2	.814	.78
1.4	.697	.89
1.6	.610	.93

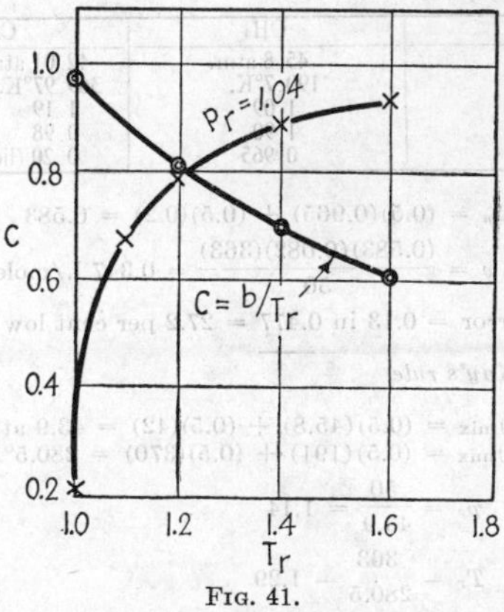

FIG. 41.

Compressibility Chart for Mixtures. Determination of the correct values of compressibility factors for mixtures is somewhat difficult. Dalton's and Amagat's laws (see pp. 290 and 291) may be applied, but either may lead to considerable errors when high pressures are involved. Amagat's law, however, leads to a simple relation

$$C_m = C_AN_A + C_BN_B + \cdots$$

where C_m is the mean compressibility factor for the

mixture, C_A, C_B, . . . are compressibility factors for the components A, B, . . . calculated at the total pressure and temperature conditions for the *mixture*, and N_A, N_B, . . . are mole fractions of components A, B, ...

Several other methods have been proposed for the treatment of problems involving mixtures, Kay's method [*Ind. Eng. Chem.*, **28,** 1014 (1936)] a method by Beattie [*J. Am. Chem. Soc.*, **25,** 880 (1935)] for the Beattie-Bridgeman equation, and the Gilliland [*Ind. Eng. Chem.*, **28,** 212 (1936)] modification of the second. Of these proposals, only the first is simple, and any, or all, may break down in the critical region. In this connection it is well to note that mixtures do not exhibit pure critical phenomena as do vapors, having a *region* in which vapor-liquid mixtures exist in equilibrium over *ranges* of temperature and/or pressure (see p. 292 and references there noted).

Kay's method, which yields promising results, particularly with hydrocarbon mixtures, is based upon the following pseudocritical conditions:

$$(T_c)_{\text{mix}} = T_{c_A}N_A + T_{c_B}N_B + \cdots$$
$$(p_c)_{\text{mix}} = p_{c_A}N_A + p_{c_B}N_B + \cdots$$

where T_{c_A}, T_{c_B}, . . . are critical temperatures of *pure* A, B, . . . and p_{c_A}, p_{c_B}, . . . are critical pressures of pure A, B, ... This method has the added advantage of using the same methods of solution of problems illustrated for pure gases and vapors. Another inherent advantage of the Kay method is the ability to handle mixtures where the Amagat rule would find that the conditions imposed the liquid state on one component. This point is illustrated as follows:

Example 1. Calculate, by Kay's rule and Amagat's law the volume occupied by 30 g. of a mixture of 50 per cent CH_4, 50 per cent C_3H_8 at a pressure of 50 atm. and a temperature of 90°C. Compare the results with the experimental data [Sage, Lacey, and Schaafsma, *Ind. Eng. Chem.*, **26,** 214 (1934)].

Solution. The average molecular weight of the mixture is

$$(0.5)(16) + (0.5)(44) = 30 \text{ g.}$$

The required volume is, therefore, the *molal* volume. Sage, Lacey, and Schaafsma found for these conditions, a density of 62.9 g./l., or a calculated molal volume of 0.477 l./mole.

Part a. Amagat's law.

	CH_4	C_3H_8
p_c	45.8 atm.	42.01 atm.
T_c	190.7°K.	369.97°K.
p_r	1.09	1.19
T_r	1.90	0.98
C	0.965	0.20 (liquid region)

$$C_m = (0.5)(0.965) + (0.5)(0.2) = 0.583$$
$$v = \frac{(0.583)(0.082)(363)}{50} = 0.347 \text{ l./mole}$$

Error = 0.13 in 0.477 = 27.2 per cent low

Part b. Kay's rule.

$$(p_c)_{\text{mix}} = (0.5)(45.8) + (0.5)(42) = 43.9 \text{ atm.}$$
$$(T_c)_{\text{mix}} = (0.5)(191) + (0.5)(370) = 280.5°\text{K.}$$
$$p_r = \frac{50}{43.9} = 1.14$$
$$T_r = \frac{363}{280.5} = 1.29$$
$$C = 0.82$$
$$pv = CRT$$
$$v = \frac{(0.82)(0.082)(363)}{50} = 0.489$$

Error = 2.5 per cent high

A third method, Gilliland's modification of the Beattie method, has been termed by the author "the additivity of intrinsic pressures" and is much more complicated than either of the methods proposed above. This proposal is based upon the good assumption that, for most mixtures, the isometrics (constant V_r lines) on a T_r vs. p_r diagram are nearly straight, represented by an equation

$$p = \phi_1 T - \phi_2$$

where ϕ_1 is the slope and ϕ_2 the intercept of the isometric for any pure gas. For mixtures the equation is presumed to be

$$p_{\text{mix}} = (\phi_1)_{\text{mix}}T - (\phi_2)_{\text{mix}} \tag{59}$$

where by the normal additivity rule

$$(\phi_1)_{\text{mix}} = N_A(\phi_1)_A + N_B(\phi_1)_B + \cdots$$

and by root mean square additivity

$$(\phi_2)_{\text{mix}} = [N_A\sqrt{(\phi_2)_A} + N_B\sqrt{(\phi_2)_B} + \cdots]^2 \tag{60}$$

For binaries, Eq. (59) reduces to the convenient form

$$p_{\text{mix}} = N_Ap_A + N_Bp_B + N_AN_B[\sqrt{(\phi_2)_A} - \sqrt{(\phi_2)_B}]^2 \tag{61}$$

The values of p_A, p_B are pressures exerted by the pure components alone under the existing temperature and molal volume; values of $(\phi_2)_A$ and $(\phi_2)_B$ are calculated from Fig. 42. If volume is desired, various values must be assumed and p_{mix} calculated until the latter equals the known value; a plot of calculated p_{mix} vs. assumed volumes will aid in the solution.

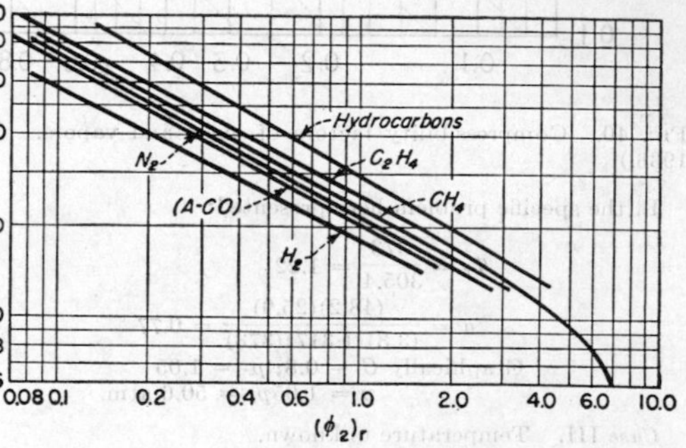

FIG. 42. [*From E. R. Gilliland, Ind. Eng. Chem.*, **28,** 212 (1936)].

Example 2. By the method of Gilliland repeat the calculation required in Example 1.

Solution. Assume various values of the volume (in this case the molal volume) from 0.4 to 0.5 l. In each case calculate the pressure exerted by each component if existing separately in this volume, at the temperature of 90°C.

$$p = C\frac{RT}{v}$$
$$p_r = C\frac{RT}{p_cv}$$
$$CH_4: p_r = C\frac{(0.082)(363)}{45.8v} = \frac{0.65C}{v}$$
$$C_3H_8: p_r = C\frac{(0.082)(363)}{42v} = \frac{0.709C}{v}$$

Substituting assumed values of v, determining p_r from the compressibility chart (Fig. 40), and calculating p for the pure component from the relation $p = p_cp_r$, the following table results:

	v, assumed		
	0.4	0.45	0.5
pCH_4	73.2	64.1	57.7
pC_3H_8	36.1	34.9	32.8

Similarly, calculating ϕ_2, values from Fig. 42 and the following v_c data:

$$CH_4: v_c = 0.099 \text{ l./mole}$$
$$C_3H_8: v_c = 0.195 \text{ l./mole}$$

	v, assumed		
	0.4	0.45	0.5
$CH_4: v_r$	4.04	4.54	5.05
$C_3H_8: v_r$	2.05	2.31	2.56
$CH_4: \phi_2/p_c$	0.34	0.26	0.23
$C_3H_8: \phi_2/p_c$	1.90	1.60	1.3
$CH_4: \phi_2$	15.6	11.9	10.5
$C_3H_8: \phi_2$	79.8	67.2	54.6

Applying Eq. (61)
at $v = 0.4$

$$p_{\text{mix}} = (0.5)(73.2) + (0.5)(36.1) + (0.5)(0.5)(\sqrt{15.6} - \sqrt{79.8})^2 = 61.2$$

at $v = 0.45$

$$p_{\text{mix}} = (0.5)(64.1) + (0.5)(34.9) + (0.25)(\sqrt{11.9} - \sqrt{67.2})^2 = 55.14$$

at $v = 0.5$

$$p_{\text{mix}} = (0.5)(57.7) + (0.5)(32.8) + (0.25)(\sqrt{10.5} - \sqrt{54.6})^2 = 49.30$$

From the graph of these values of p_{mix} vs. v (see Fig. 43), the calculated volume is 0.497 l./mole. Although the error here is 4.2 per cent, the use of more accurate compressibility charts would improve the calculations tremendously. The method is, however, extremely cumbersome, and therefore it is desirable to avoid it when possible.

It is advisable to note that, for the general case, values of ϕ_1 may be calculated from values of ϕ_2 and p for each component from the equation

$$p = \phi_1 T - \phi_2$$

Therefore, values of $(\phi_1)_{\text{mix}}$ and $(\phi_2)_{\text{mix}}$ may be calculated from Eqs. (59) and (60) for cases other than those of binaries. The steps involved are slightly altered, but the basic calculations are unchanged.

Fugacity Charts. From actual pressure-volume-temperature data it is possible to formulate a new chart of fugacity in terms of reduced conditions. This work has been done by many investigators* following two different methods which may be shown to be identical.

One method of derivation is from the compressibility chart by graphical integration. From

$$dF = dH - d(TS) = dH - T\,dS - S\,dT$$

and

$$dH = dE + d(pv) = dE + p\,dv + v\,dp$$
$$dE = dq - dw = T\,dS - p\,dv$$
$$\left(\frac{\partial F}{\partial p}\right)_T = v$$

By definition,

$$dF = RT\,d\ln f$$
$$\therefore \left(\frac{\partial \ln f}{\partial p}\right)_T = \frac{1}{RT}\left(\frac{\partial F}{\partial p}\right)_T = \frac{v}{RT} = \frac{C}{p}$$

* Selheimer, Souders, Smith, Brown, *Ind. Eng. Chem.*, **26**, 514 (1932); Lewis and Luke, *Trans. Am. Soc. Mech. Engrs.*, **54**, 55 (1932), and *Ind. Eng. Chem.*, **25**, 725 (1933); Newton, *Ind. Eng. Chem.*, **27**, 302 (1935); Watson and Smith, *National Petroleum News*, July, 1936; Morgen and Childs, *Ind. Eng. Chem.*, **37**, 667 (1945).

Isothermally

$$d\ln f = C\frac{dp}{p} = C\,d\ln p$$
$$d\ln\frac{f}{f_i} = d\ln\frac{f}{p} = (C-1)\,d\ln p = (C-1)\,d\ln p_r$$
$$\therefore \ln\frac{f}{p} = \ln\gamma = \int_0^p \frac{C-1}{p_r}\,dp_r \qquad (62)$$

From this development it is apparent that a fugacity chart may be obtained by graphical integration of data represented by the compressibility chart, along constant T_r lines. The curves of Newton are reproduced here. Some uses of this general fugacity correlation are given below.

A. Determination of $x - y^*$ relations:

This calculation is illustrated briefly in Sec. 8, pp. 525*ff*. For more complete material, see Sherwood ("Absorption and Extraction," McGraw-Hill, New York, 1937) and Dodge ("Chemical Engineering Thermodynamics," McGraw-Hill, New York, 1944).

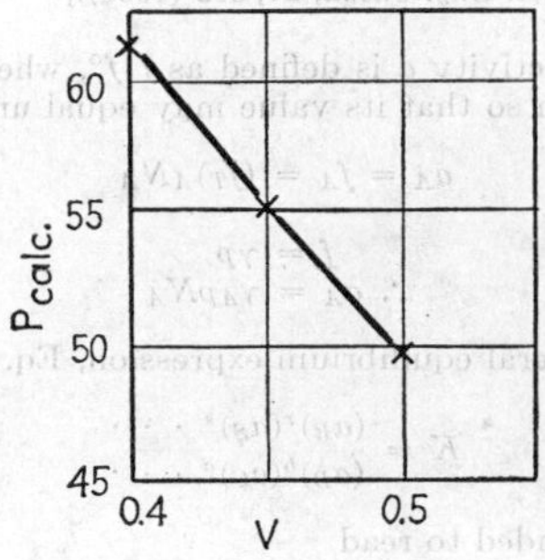

FIG. 43.

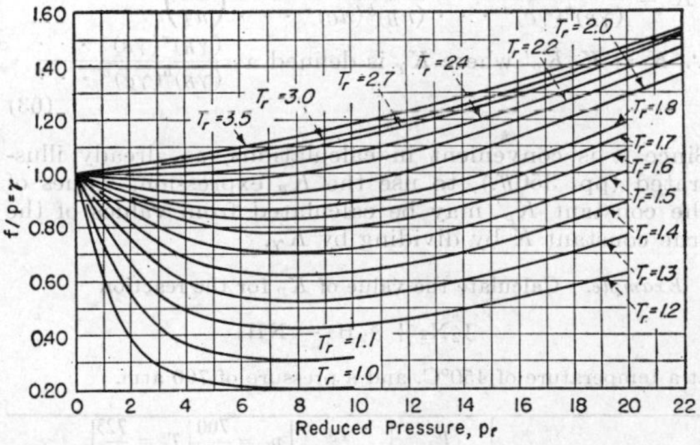

FIG. 44. Plot of f/p, the ratio of fugacity to pressure, as a function of the reduced pressure, p_r, for gases at a number of values of the reduced temperature, T_r, for the intermediate pressure range. [*Taken from Newton, Ind. Eng. Chem.*, **27**, 302 (1935).]

B. Calculation of equilibrium conditions:

In determinations of fugacity of components of a gaseous mixture, the Lewis and Randall fugacity rule is very useful, stating

$$f_A = (f_T)_A N_A$$

* Although not consistent with previous, or subsequent, nomenclature x and y are retained here to indicate mole fraction of more volatile component in liquid and vapor, respectively, owing to custom in this type of unit operation.

where f_A = fugacity of component A in the mixture.
$(f_T)_A$ = fugacity of *pure* component A under the total temperature and pressure of the mixture,
and N_A = the mole fraction of component A in the mixture.

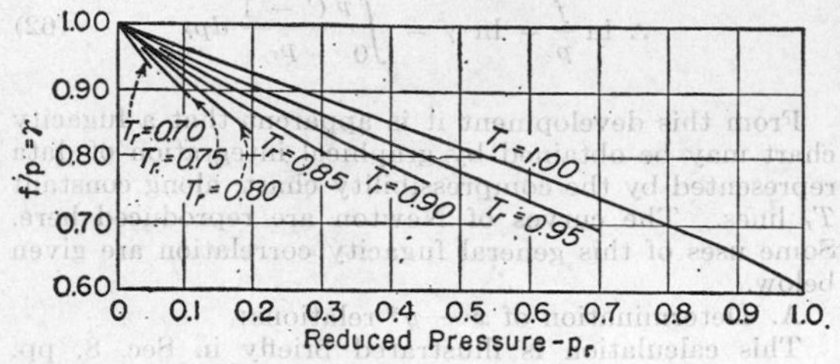

FIG. 45. Plot of f/p, the ratio of fugacity to pressure, as a function of the reduced pressure, p_r, for gases at a number of values of the reduced temperature, T_r, for the low pressure range. [*Taken from Newton, Ind. Eng. Chem.*, **27**, 302 (1935).]

Since the activity a is defined as $f/f°$, where the state of $f°$ is chosen so that its value may equal unity,

$$a_A = f_A = (f_T)_A N_A$$

But

$$f = \gamma p$$

$$\therefore a_A = \gamma_A p N_A$$

Thus the general equilibrium expression, Eq. (49),

$$K = \frac{(a_R)^r (a_S)^s \cdots}{(a_B)^b (a_C)^c \cdots}$$

may be expanded to read

$$K = \frac{(\gamma_R)^r (\gamma_S)^s \cdots}{(\gamma_B)^b (\gamma_C)^c \cdots} \cdot \frac{(n_R)^r (n_S)^s \cdots}{(n_B)^b (n_C)^c \cdots} \left(\frac{p}{n_T}\right)^{r+s+\cdots b-c-\cdots}$$

$$\therefore K = K_\gamma K_p' \text{ where } K_\gamma \text{ is defined as } \frac{(\gamma_R)^r (\gamma_S)^s \cdots}{(\gamma_B)^b (\gamma_C)^c \cdots} \quad (63)$$

Since it is convenient in calculations, as already illustrated (pp. 350*ff.*), to use the K_p expression, values of the constant K_p' may be calculated from values of the true constant K by dividing by K_γ.

Example. Calculate the value of K_p for the reaction

$$\tfrac{1}{2}N_2 + \tfrac{3}{2}H_2 \rightarrow NH_3$$

at a temperature of 450°C. and a pressure of 700 atm.

	p_c atm.	T_c °K.	$p_r = \frac{700}{p_c}$	$T_r = \frac{723}{T_c}$	γ
N_2	33.5	126	20.9	5.74	1.38
H_2	12.8	33.3	33.6*	17.5*	1.22
NH_3	111.5	405.6	6.27	1.78	0.86

* $p_r = \frac{700}{p_c + 8}$ and $T_r = \frac{723}{T_c + 8}$.

$$K\gamma = \frac{\gamma_{NH_3}}{\gamma_{N_2}^{1/2}\gamma_{H_2}^{3/2}} = \frac{0.86}{(1.38)^{1/2}(1.22)^{3/2}} = 0.55$$

at 723°K., $K = 0.00708$

$$\therefore K_p = \frac{K}{K\gamma} = \frac{0.00708}{0.55} = 0.0129$$

Enthalpy Charts. Consider two sets of conditions, the temperature being identical, with pressures varying as follows: (1) at *any* pressure condition, employing the thermodynamic symbols H, p, f, F, etc.; (2) a standard state, at sufficiently low pressure for application of the simple gas law, employing the symbols $H°$, $f° = p$, $F°$, etc.

$$F° - F = RT \ln \frac{f°}{f} \quad (64)$$

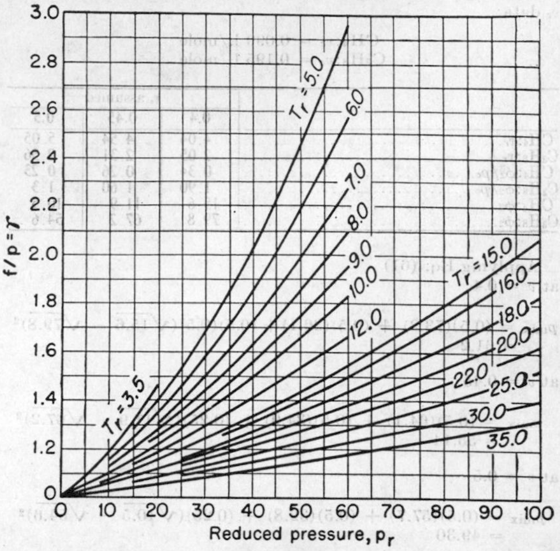

FIG. 46. Plot of f/p, the ratio of fugacity to pressure, as a function of the reduced pressure, p_r, for gases at a number of values of the reduced temperature, T_r, for the high pressure range. [*Taken from Newton, Ind. Eng. Chem.*, **27**, 302 (1935).]

Differentiating with respect to temperature at constant pressure,

$$\left(\frac{\partial F°}{\partial T}\right)_p - \left(\frac{\partial F}{\partial T}\right)_p = RT\left(\frac{\partial \ln f°}{\partial T}\right)_p - RT\left(\frac{\partial \ln f}{\partial T}\right)_p + R \ln \frac{f°}{f}$$

But $f° = p$, which latter does not change at constant pressure, so that

$$(\partial \ln f°)_p = (\partial \ln p)_p = 0$$

and from Eq. (64)

$$R \ln \frac{f°}{f} = \frac{F°}{T} - \frac{F}{T}$$

$$\therefore \left(\frac{\partial F°}{\partial T}\right)_p - \left(\frac{\partial F}{\partial T}\right)_p = \frac{F° - F}{T} - RT\left(\frac{\partial \ln f}{\partial T}\right)_p$$

Now limit the work performed to expansion. Then, at constant pressure,

$$\left(\frac{\partial F}{\partial T}\right)_p = -S = \frac{F - H}{T}$$

Thus from the last two equations

$$\frac{H° - H}{T} = \frac{\Delta H^*}{T} = +RT\left(\frac{\partial \ln f}{\partial T}\right)_p$$

Since, at any particular constant pressure,

$$\partial \ln f = \partial \ln \frac{f}{p} = \partial \ln \gamma$$

$$\frac{\Delta H^*}{RT^2} = \left(\frac{\partial \ln \gamma}{\partial T}\right)_p$$

and, for reduced conditions where $T = T_r T_c$ and $\partial T = T_c\, \partial T_r$,

$$\frac{\Delta H^*}{T} = R\left(\frac{\partial \ln \gamma}{\partial \ln T_r}\right)_p \qquad (65)$$

From the last equation a diagram of values of $\Delta H^*/T$ may be constructed by graphical methods from the γ vs. p_r data. This construction involves replotting of the latter data as γ vs. T_r, loglog, for lines of constant p_r. Then, along any one of these lines, slopes $= \left(\frac{\partial \ln \gamma}{\partial \ln T_r}\right)_{p_r}$ may be taken which, when multiplied by R, give values of $\Delta H^*/T$. This chart, as developed by Watson and Smith (*op. cit.*) is given in Fig. 47.* Its use is illustrated in the final problem.

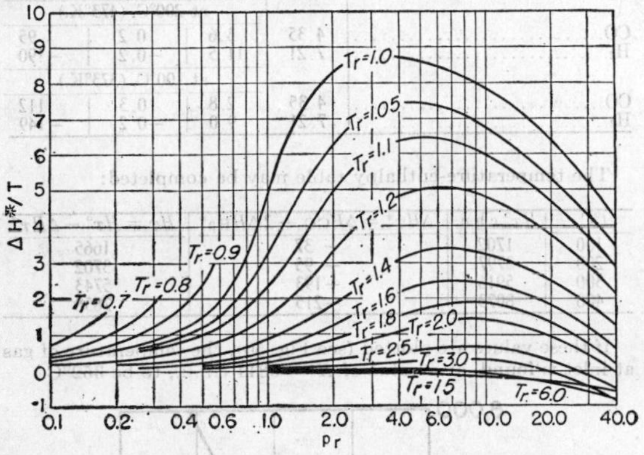

FIG. 47. Enthalpy-pressure relationship for gases and vapors. (*From Hougen and Watson, "Industrial Chemical Calculations," 2d ed., Wiley, 1936.*)

Illustrative Problem. A methanol converter is being designed to operate on a stoichiometric gas mixture at a constant pressure of 150 atm. The space velocity will be arranged to give 50 per cent of equilibrium conversion calculated for a temperature of 400°C. This temperature will be that of the fresh gas to the catalyst bed; but, uring passage over the catalyst, absorption of heat will be allowed to raise the gas temperature until, at exit, gases leave at not over 450°C. Cooling of the bed will be performed by an internal heat exchanger, fresh feed gas being the cooling medium.

Calculate:

1. The heat abstracted in the internal exchanger, assuming radiation losses to be zero. (Note: Exchanger designed to provide for transfer of this amount of heat will have maximum area.)
2. Temperature of the feed gases at inlet to the internal exchanger.

Solution:

a. Calculation of the Conversion:

$$\text{From Fig. 38, } \log_{10} K = -4.75 = \bar{5}.25,\quad K = 1.78 \times 10^{-5} \quad \Big\}\ T = 673°\text{K}.$$

With critical constants from p. 204, and values of $\gamma = f/p$ from Fig. 44.

	CO	H_2†	CH_3OH
p_c, atm	34.5	12.8	78.7
T_c, °K	133	33.3	513.2
p_r	4.35	7.21	1.91
T_r	5.06	16.3	1.31
γ	1.07	1.06	0.78

† $p_c = 12.8 + 8.0$ and $T_c = 33.3 + 8.0$ for calculation of reduced conditions.

* Weber and York, *Ind. Eng. Chem.*, **32**, 388 (1940), have developed, for a limited range of p_r and T_r values, a more accurate chart, involving a new but limited factor, ϕ, in the ordinate, $\Delta H^*/T\phi$. Values of ϕ are calculable from the material presented in the paper.

$$K = K_\gamma K_p = K_\gamma \frac{(n_{CH_3OH})(n\pi)^2}{(n_{CO})(n_{H_2})^2 p^2}$$

$$K_\gamma = \frac{\gamma_{CH_3OH}}{(\gamma_{CO})(\gamma_{H_2})^2} = \frac{0.78}{(1.07)(1.06)^2} = 0.65$$

$$K_p = \frac{K}{K_\gamma} = \frac{1.78 \times 10^{-5}}{0.65} = 2.74 \times 10^{-5}$$

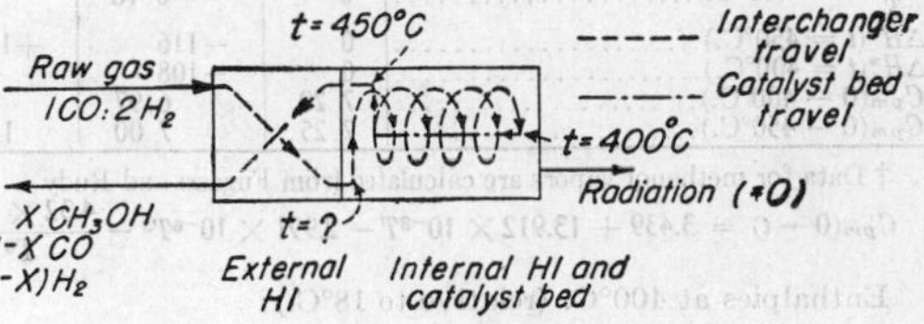

FIG. 48.

Basis: 3 moles gas feed
1 mole CO
2 moles H_2

Let x = moles CH_3OH at equilibrium
Then, the equilibrium quantities, in moles. are

CH_3OH	x
CO	$1 - x$
H_2	$2(1 - x)$
Total	$3 - 2x$

$$2.74 \times 10^{-5} = \frac{(x)(3 - 2x)}{(1 - x)(2^2)(1 - x)^2 150^2}$$

$$0.615 = \frac{(x)(3 - 2x)}{4(1 - x)^3} = \phi(x)$$

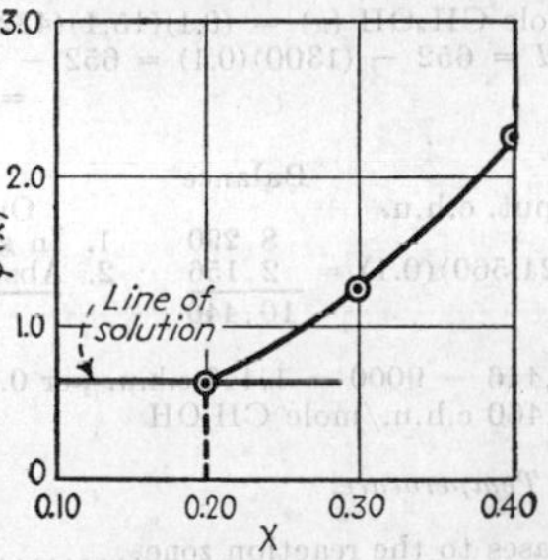

FIG. 49.

Solving by trial and error, with the accompanying plot of $\phi(x)$ vs. x (Fig. 49)

$$x = 0.20$$

At 50 per cent of the equilibrium conversion, 10 per cent of the CO is converted to methanol, and $x = 0.10$.

b. Heat Balance over Converter:

Heat input
1. Heat of reaction.
2. Heat content of gases.

1. Heat of Reaction.

From data of Bichowsky and Rossini ("Thermochemistry of Chemical Substances," Reinhold, New York, 1936), the heat of formation ΔH_{F291} for methanol in the vapor state is −48,400 cal./mole, using diamond as a reference state. Correcting this value for the β-graphite reference state and using other standard data:

$$CO\ (g) + 2H_2\ (g) \rightarrow CH_3OH\ (g)$$

$$\Delta H_{291} = (-48{,}180) - (-26{,}620 + 0) = -21{,}560 \text{ c.h.u./mole } CH_3OH$$

2. Enthalpy of Gases.

	Quantities, moles			
	Input	Formed	Consumed	Output
CO	1	0	0.10	0.90
H_2	2	0	0.20	1.80
CH_3OH	0	0.10	....	0.10

Enthalpies are figured with the aid of reduced pressure relations (Fig. 47).

	CO	H_2	CH_3OH
p_r	4.35	7.21	1.91
$T_r(t = 450°C.)$	5.44	17.5	1.41
$T_r(t = 400°C.)$	5.06	16.3	
$\frac{\Delta H^*}{T}(t = 450°C.)$	0	−0.16	1.8
$\frac{\Delta H^*}{T}(t = 400°C.)$	0	−0.16	
$\Delta H^*(t = 450°C.)$	0	−116	+1300
$\Delta H^*(t = 400°C.)$	0	−108	
$C_{pm}(0 - 400°C.)$	7.20	6.97	
$C_{pm}(0 - 450°C.)$	7.25	7.00	15.1†

† Data for methanol vapors are calculated from Fugassi and Rudy

$$C_{pm}(0 - t) = 3.439 + 13.912 \times 10^{-3}T - 2.991 \times 10^{-6}T^2 - \frac{4.32 \times 10^{+6}}{T^3}$$

Enthalpies at 400°C. (relative to 18°C.):

1. Heat in 1 mole CO (1 atm.) = 1(7.20)(382)
   = 2750 c.h.u.

   (No correction for pressure)
2. Heat in 2 moles H_2 (1 atm.) = 2(6.97)(382)
   = 5320 c.h.u.

at 150 atm. $\Delta H^* = H° - H$

$H = 5320 - (2)(-108) = 5320 + 220$
$= 5540$ c.h.u.

8290 c.h.u. total

Enthalpies at 450°C. (relative to 18°C.):

1. Heat in 0.9 mole CO (1 atm.) = (0.9)(7.25)(432)
   = 2820 c.h.u.

   (No correction for pressure)
2. Heat in 1.8 moles H_2 (1 atm.) = (1.8)(7.00)(432)
   = 5450 c.h.u.

at 150 atm., $H = 5450 - (1.8)(-116) = 5450 + 210$
$= 5660$ c.h.u.

Heat in 0.1 mole CH_3OH (g) = (0.1)(15.1)(432) = 652 moles
at 150 atm., $H = 652 - (1300)(0.1) = 652 - 130$
$= 520$ c.h.u.

9000 c.h.u. total

Balance

Input, c.h.u.		Output, c.h.u.	
1. Gases	8,290	1. In gases	9000
2. Reaction (21,560)(0.1) =	2,156	2. Abstracted	x
	10,446		$9000 + x$

$x = 10{,}446 - 9000 = 1{,}446$ c.h.u. per 0.1 mole CH_3OH
$= 14{,}460$ c.h.u./mole CH_3OH

c. Inlet Gas Temperature:

Enthalpy of gases to the reaction zone	8290 c.h.u.
Heat abstracted from reaction zone by heat exchange	1446 c.h.u.
Enthalpy of gases at entrance to interchanger	6844 c.h.u.

This solution is one of trial and error, since T_r cannot be calculated until the temperature of the gases is known. A logical method of attack would be as follows:

Enthalpies at temperature t, 1 atm., may be calculated from heat capacity equations, expressed

$$C_p = a + bt + ct^2$$

Using the proper equations in integrated form

$$H_2\text{:}\quad H° = 2(6.88T + 0.033 \times 10^{-3}T^2 + 0.093 \times 10^{-6}T^3)]_{291}^{T}$$
$$= 13.76T + 0.066 \times 10^{-3}T^2 + 0.186 \times 10^{-6}T^3]_{291}^{T}$$
$$CO\text{:}\quad H° = 6.25T + 1.046 \times 10^{-3}T^2 - 0.153 \times 10^{-6}T^3]_{291}^{T}$$
$$\text{Total:}\quad H_T° = 20.01T + 1.112 \times 10^{-3}T^2 + 0.033 \times 10^{-6}T^3]_{291}^{T}$$

Values of $H_T°$ may be calculated over a reasonable range, as 100° to 400°C., all relative to 18°C.

t, °C.	$H_T°$, c.h.u.
100	1702
200	3797
300	5916
400	8073

Values of ΔH^*, needed for correction of the above atmospheric enthalpies, are obtained as follows:

	p_r	T_r	$\Delta H^*/T$	$n\,\Delta H^*$
		at 400°C. (673°K.)		
CO	4.35	5.1	0	0
H_2	7.21	16.3	−0.16	−215
		at 300°C. (573°K.)		
CO	4.35	4.3	0.1	57
H_2	7.21	13.9	−0.2	−230
		at 200°C. (473°K.)		
CO	4.35	3.6	0.2	95
H_2	7.21	11.5	−0.2	−190
		at 100°C. (373°K.)		
CO	4.35	2.8	0.3	112
H_2	7.21	9.0	−0.2	−149

The temperature-enthalpy table may be completed:

t°C.	$H_T°$ c.h.u.	$\Delta H_T^* = \Delta H_{CO} + 2\,\Delta H_{H_2}^*$	$H_T = H_T° - \Delta H_T^*$
100	1702	− 37	1665
200	3797	− 95	3702
300	5916	−193	5743
400	8073	−215	7858

If these values are plotted (see Fig. 50), the temperature of gas at inlet is found, at a value of H = 6844 c.h.u., to be 352°C.

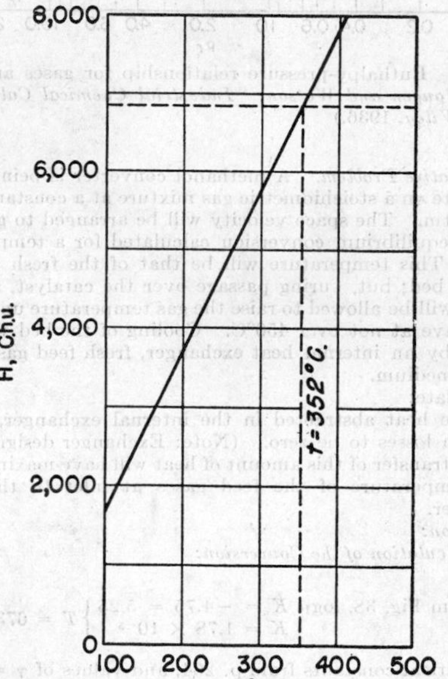

FIG. 50.

SECTION 5

FLOW OF FLUIDS*

BY

T. B. Drew, S. M., Professor of Chemical Engineering, Columbia University; Member, American Institute of Chemical Engineers, American Chemical Society, American Society of Mechanical Engineers; Fellow, New York Academy of Sciences.

H. H. Dunkle, Goetze Division, Johns-Manville Sales Corp. (Gaskets)

R. P. Genereaux, Ch. E., Chemical Engineer, E. I. duPont de Nemours & Co.; Member, American Institute of Chemical Engineers.

CONTENTS

FLUID STATICS

	Page
Hydrostatic Pressure	360
Buoyancy and Flotation	362
Capillarity and Surface Tension	363

PRESSURE GAGES AND MANOMETERS

Liquid-column Manometers	364
Multiplying Gages	365
Micromanometers	366
Mechanical Pressure Gages	367
Calibration of Gages	367
Gages for Location of Interfaces	367

FLUIDS IN MOTION

Substances Which Flow	369
Viscosity Data	369
Terminology in Fluid Dynamics	374
Bernoulli's Theorem	375
Flow in Pipes and Channels	377
General Formulas	377
Fanning Friction Factors	381
Economic Pipe Diameter	384
Isothermal Streamline Flow in Straight Passages	385
Flow of Suspensions	385
Miscellaneous Pressure Drops	387
Enlargement, Expansion, or Exit Loss	388
Contraction or Entrance Loss	388
Diaphragms, Orifices, etc	389
Pipe Fittings, Bends, etc	389
Flow through Equipment	390
Heat Exchangers	390
Packed Towers	393

FLOW MEASUREMENTS

Introductory Note	396
Static Pressure	396
Local Static Pressure	396
Average Static Pressure	397
Specifications for Piezometer Taps	397
Local Velocity	397

	Page
Impact and Pitot Tubes	397
Anemometers	398
Traversing for Mean Velocity	399
Direction Indicators	400
Orifices, Nozzles, and Venturi Meters	400
General Principles and Definitions	400
Theoretical Formulas	401
Nomenclature	401
Liquids	401
Gases	402
Maximum Discharge for Gases	402
Working Formulas	403
Orifices	404
Flow Nozzles	406
Venturi Meters	406
Miscellaneous Openings	407
Accuracy	407
Area Meters	408
General Principles	408
Rotameters	408
Weirs	408
General Principles and Definitions	408
Applicability of Weir Formulas	409
Application of Weir Formulas to Gases	410
Quantity Meters	410
Dilution Metering	412
Thermal Meters	412
Mixture Metering	412
Miscellaneous Mechanical Meters	412

PIPE AND FITTINGS

Ferrous Metal Pipe and Tubing	413
Non-ferrous Metal Pipe and Tubing	423
Non-metallic Pipe and Tubing	431
Pipe Joints, Flanges, and Fittings	441
Pipe Joints	441
Flanges and Fittings	442
Valves	447
Gaskets	451
Pipe Supports	453
Cements and Solders	453

*** Special thanks are due to D. F. Boucher for a critical review of this section, including the preparation of new material on compressible flow, pressure drop across tube banks, and pressure drop across packed solids.**

FLUID STATICS

REFERENCES: Hughes and Safford, "Hydraulics," Macmillan, New York, 1925. Gibson, "Hydraulics and Its Applications," Van Nostrand, New York, 1930. Minchin, "Treatise on Hydrostatics," Oxford, New York, 1912. Lea, "Hydraulics," Longmans, New York, 1930. Greenhill, "Treatise on Hydrostatics," Macmillan, London, 1894. Dodge and Thompson, "Fluid Mechanics," McGraw-Hill, New York, 1937. O'Brien and Hickox, "Applied Fluid Mechanics," McGraw-Hill, New York, 1937.

Hydrostatics treats of the properties and behavior of liquids when there is no relative motion within the fluid. In the case of gases the same problems lie in the province of **pneumatics.** This section considers both liquids and gases.

HYDROSTATIC PRESSURE

The **pressure (intensity of pressure)** at a point on a surface that sustains a distributed force is the intensity with which the force acts on the surface at that point; *e.g.*, if a force F (lb.) is distributed uniformly over a plane of area A (sq. ft.), the pressure is $p = F/A$ (lb. force/sq. ft.). If the surface is curved or if the force is not uniformly distributed, the pressure at a point on the surface is the limit approached by the ratio $\delta F/\delta A$, where δF is the force acting on an element of area δA which includes the point, as δA is taken smaller and smaller. The dimensions of pressure are those of (force ÷ area); *e.g.*, pounds *force* per square inch, dynes per square centimeter, etc. A **normal pressure** acts only in a direction perpendicular to the surface that sustains it. In common engineering usage, the unmodified word "pressure" sometimes means "resultant pressure" (see p. 361), but usually when used of fluids the meaning is "hydrostatic pressure."

A **gage pressure** is the difference between a given pressure and that of the atmosphere. The terms "inches of mercury (or other fluid) vacuum," "pounds (per square inch) suction," etc., indicate negative gage pressures, *i.e.*, less than atmospheric. The readings of pressure gages are commonly gage pressures.

Absolute pressure is used to denote the true pressure whenever the unqualified word might be confused with *gage pressure*. The absolute pressure in a perfect vacuum is zero. The sum of the gage pressure (taken with the proper sign) and the atmospheric pressure equals the absolute pressure.

The **hydrostatic, or static, pressure** at a point in a fluid is the compressive stress at that point. It is normal to any surface on which it acts, and at any given point has the same magnitude irrespective of the orientation of the surface. The pressure exerted on the containing walls by a fluid at rest is wholly normal and equals, at each point, the static pressure in the adjacent fluid. In a flowing fluid the static pressure is, by definition, the normal pressure exerted by the fluid on a stationary surface paralleling the direction of motion (see pp. 396*ff.*); the pressure on the surface in this case is not wholly normal.

The variation of the static pressure p (lb. *force*/sq. ft.) in any direction within a fluid is found by evaluating Newton's law, force = mass × acceleration/g_c, for a differential cube of fluid, chosen with its faces perpendicular to the coordinate axes. When there is no relative internal motion, the only forces on such a cube are attractions typified by gravity and the pressures acting normal to the faces. For example, the result for the direction of y (ft.) is

$$\rho Y - \frac{\partial p}{\partial y} = \frac{\rho a_y}{g_c} \tag{1}$$

where ρ is the density of the fluid, lb. *mass*/cu. ft.; a_y is the y-component of its acceleration, ft./sec.^{2}; Y is the y-component of "extraneous" forces, lb. *force*/lb. *mass*; and the dimensional constant g_c is 32.1740 (lb. *mass*)(ft.)/(lb. *force*)(sec.^{2}). If y is measured vertically downward, $Y = g/g_c$,* where g is the local acceleration due to gravity, ft./sec.^{2}; for a horizontal direction, say that of x, the corresponding force component X is ordinarily zero. Ordinarily, therefore, within an unaccelerated body of fluid the pressure varies with the depth y (ft.) according to the relation $dp/dy = \rho g/g_c$* and, provided that the density is uniform or varies only vertically, has the same value everywhere at the same level. For incompressible fluids, *e.g.*, liquids, these considerations lead to **Pascal's principle:** An increase in the pressure at any point in the fluid is attended by an equal increase at every point in the fluid.

The static pressure at the foot of a vertical column of a fluid of uniform density, ρ (lb. *mass*/cu. ft.), exceeds that at the top by $H\rho g/g_c$ (lb. *force*/sq. ft.), if H (ft.) is the height, or **head,** of the column. The magnitude of any pressure or pressure difference may be specified by stating the head of the column of fluid (of given density) that causes an increase in static pressure equal in magnitude to the pressure described. Generally, fluid heads measure gage pressures. When not otherwise stated, it is assumed that the fluid composing the column has throughout its height the density of the fluid in the apparatus at the point of measurement.

Example 1. The pressure corresponding to a head of 18 in. of Hg, of which the specific gravity is 13.6, is, if g/g_c is assumed unity,

$$\frac{18 \times 13.6 \times 62.3}{12 \times 144} = 8.83 \text{ lb./sq. in.}$$

Example 2. What depth of water is required to produce a pressure of 140 lb./sq. in. in excess of that on the free surface?

Solution. The head required is $140 \times 144/62.3 = 324$ ft. Each foot of increase in depth of liquid causes an increase in pressure of (density/144) lb./sq. in.

The **absolute static pressure** p_2 (lb. *force*/sq. ft.) **at the foot of a column of a perfect gas** ($p/\rho = RT/M$) at the uniform absolute temperature T [°F. + 459.5, or 1.8(273.1 + °C.)] is computed from the formula

$$\log_{10} p_2 = \log_{10} p_1 + \frac{HMg}{2.3RTg_c} \tag{2}$$ *

where H is the height of the column, ft.; p_1 is the static pressure, lb. *force*/sq. ft., at the top; M is the molecular weight of the gas; g is the local acceleration due to gravity, ft./sec.^{2}; g_c is 32.1740 (lb. *mass*)(ft.)/(lb. *force*)(sec.^{2}); and R is 1546 (ft.-lb. *force*)/(°F)(lb.-mole). If the temperature is not uniform and its manner of variation with the height is known, the relation between

* In most localities the variation of g from the standard value g_0 = 32.1740 ft./sec.^{2} [which is numerically equal to the dimensional constant g_c in (lb. *mass*) (ft.)/(lb. *force*) (sec.^{2})] is so slight that g/g_c is numerically very close to unity; see p. 375.

p_1 and p_2 may be found by integrating the equation $dp/p = Mg\,dH/RTg_c$. The absolute pressure and the density of the atmosphere at a height H (ft.) above sea level may be estimated by the formulas below, if H is less than 36,000 ft.:

$$p = p_1(1 - 0.00000687H)^{5.256} \qquad (3)$$

$$\rho = \rho_1(1 - 0.00000687H)^{4.256} \qquad (4)$$

Any units may be used to express p and ρ. The subscript ($_1$) indicates the conditions at sea level. These equations are based on the convention that the temperature is 15°C. at sea level and decreases at the rate of 1.98°C. per 1,000 ft. of elevation (cf. "International Critical Tables," vol. 1, p. 72).

The **resultant pressure** on a submerged surface is the resultant force arising from the pressure on the surface. If the surface is plane, the resultant pressure is sometimes called the total normal pressure and, for a fluid of uniform density, is the product of the density, the area of the surface, the static head at the center of gravity of the surface, and the ratio g/g_c.* For a surface of any shape in contact with a fluid at rest, the component of the resultant pressure, F_a, in any given direction is $F_a = \int p \cos\theta\, dA$, where θ is the angle between that direction and the normal to the surface at the element thereof, dA, that sustains the pressure normal, p. The coordinates $(\bar{x}, \bar{y})$ of the line of action of the component F_a, with respect to any convenient set of axes OXY in a plane perpendicular to the given direction, are found by the formulas

$$\bar{x} = \frac{\int px \cos\theta\, dA}{F_a} \qquad \bar{y} = \frac{\int py \cos\theta\, dA}{F_a} \qquad (5a, b)$$

where (x, y) are the coordinates with respect to OXY of the surface element dA. The vector sum

$$\sqrt{F_1^2 + F_2^2 + F_3^2}$$

of the pressure components F_1, F_2, and F_3 in three mutually perpendicular directions equals the resultant pressure F_R; and the angles between the lines of action of F_R and of F_1, F_2, and F_3 have the cosines F_1/F_R, F_2/F_R, and F_3/F_R, respectively. Figure 1 illustrates the results when only the hydrostatic pressure in a liquid at

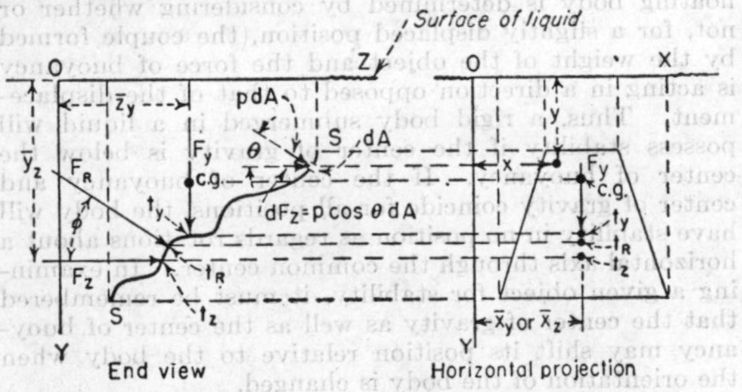

FIG. 1. Resultant pressure on surface SS. c.g. = center of gravity of liquid above SS.

rest is to be considered. In that diagram the "given direction" is taken to be the z-direction and, as is often convenient when there is a plane free surface, one of the reference planes OXZ has been taken coincident therewith. The points where F_y, F_z, and F_R act on the surface are t_y, t_z, and t_R, respectively.

The **center of pressure** of a submerged plane area is the common point where the lines of action of the resultant pressure and of its components meet the surface; hence a single supporting force acting at this point will balance all the pressure on the surface. The coordinates of the center of pressure may be found by locating the line of action of any convenient pressure component by the formulas of the preceding paragraph. To find the center of hydrostatic pressure (due to the weight of the liquid only) for a vertical plane surface immersed in a liquid at rest, it is simplest to take one axis, say OY, vertically downward, and the other, OX, coincident with the intersection of the plane of the

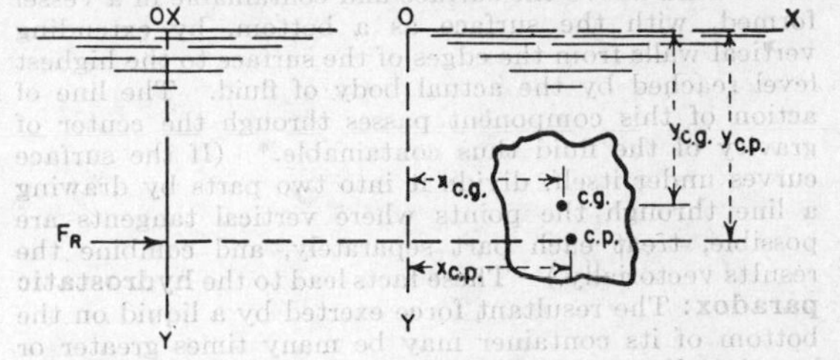

FIG. 2. Center of pressure—vertical plane surface.

vertical area and that of the free liquid surface (see Fig. 2). In such a case the vertical distance from OX to the center of pressure is

$$y_{c.p.} = \frac{\int y^2\, dA}{(\text{area})(y_{c.g.})} = \frac{\text{moment of inertia of the area about } OX}{(\text{area})(y_{c.g.})} \qquad (6)$$

where $y_{c.g.}$ is the vertical distance from OX to the center of gravity of the area. The other coördinate is

$$x_{c.p.} = \frac{\int xy\, dA}{(\text{area})(y_{c.g.})} \qquad (7)$$

NOTE. (1) The same method may be used when the liquid completely fills a closed vessel, if the axis OX be a horizontal line in the plane of the surface and pass through the highest point that is occupied by liquid in free communication with the liquid exerting the pressure. (2) It must be remembered that in addition to the hydrostatic pressure arising from the weight of the liquid, which acts at the center of pressure found by the above equations, there is usually acting on the surface in question a uniform pressure p_a, the center of pressure of which is at the center of gravity of the surface. If the liquid has a free surface, p_a is the gas pressure on the free surface; if the liquid fills a closed container, p_a is the static pressure in that part of the liquid that fills the highest point in the apparatus (cf. Pascal's principle, p. 360).

The moment of inertia of any area about any axis is the sum of the moment of inertia about the parallel axis through the center of gravity plus the product of the area times the square of the distance between the two axes. For certain common areas the moments of inertia I about the given axes through their centers of gravity are as follows: Rectangle, sides a and b, $I = ab^3/12$ about axis parallel to side a; triangle, base $= a$, altitude $= b$, $I = ab^3/36$ about axis parallel to base, c.g. is a distance $b/3$ above base; circle, diameter $= D$, $I = \pi D^4/64 = 0.0491D^4$; ellipse, axes $= a$ and b, $I = \pi a^3b/64 = 0.0491a^3b$ about the axis of length b. For less common areas see Marks, "Mechanical Engineers' Handbook," 4th ed., McGraw-Hill, New York, 1941.

From the above paragraphs, together with the fact that a submerged solid is subject to no unbalanced force if it has the same density as the fluid, it follows for fluids at rest that:

1. For any immersed surface, the pressure component in any horizontal direction has the same magnitude and line of action as the resultant pressure on the projection of that surface on a plane perpendicular to the given direction. (When the static pressure does not vary appreciably in the vertical direction, *e.g.*, as within a small region in a gas, the last sentence remains approximately true if the word "horizontal" be omitted.) Hence the center of pressure of an inclined plane surface has the same coordinates on any given vertical reference

* See footnote on p. 360.

plane as does the center of pressure of its projection on that reference plane.

2. For any immersed surface, the vertical component of the resultant pressure equals the weight of the fluid above the surface and containable in a vessel formed, with the surface as a bottom, by extending vertical walls from the edges of the surface to the highest level reached by the actual body of fluid. The line of action of this component passes through the center of gravity of the fluid thus containable.* (If the surface curves under itself, divide it into two parts by drawing a line through the points where vertical tangents are possible, treat each part separately, and combine the results vectorially.) These facts lead to the **hydrostatic paradox:** The resultant force exerted by a liquid on the bottom of its container may be many times greater or less than the weight of the liquid. If the tanks of water shown in Fig. 3 each have the same bottom area, the same resultant force will be sustained by each bottom because the hypothetical tanks (dotted side walls) formed in accordance with the rule here stated would contain the same weight of water if filled to the same level. The fact that one actual tank contains more water than the other has no bearing on the matter.

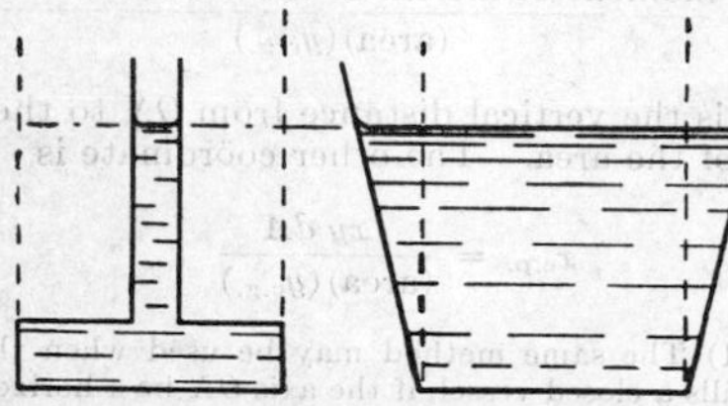

FIG. 3. Hydrostatic paradox.

3. The resultant pressure on a closed surface immersed in a fluid is a vertical force equal to the weight of the fluid that could be contained within the surface under such conditions that the contents would be in equilibrium with their surroundings if the surface were removed. This force acts through the center of gravity of the fluid thus supposedly enclosed, and it acts upward or downward according as the outer or inner side of the surface sustains the fluid pressure.

4. For an immersed plane surface, the pressure component in any given direction is as follows: The area of the projection of the given surface on a plane perpendicular to that direction times the pressure on the center of gravity of the given surface.

Example. Find the **hoop tension** at the bottom of a cylindrical metal tank 10 ft. in diameter if it is placed with its axis vertical and is filled with water to a depth of 100 ft.

Solution. Imagine a circular ring of metal to be cut out of the tank by two horizontal planes 1 in. apart, one of which is the bottom of the tank. Suppose this ring to be divided into two parts by a vertical plane through the axis of the tank. Either half of the metal ring will be in equilibrium only if the resultant pressure on that half is equal and opposite to the sum of the tensile forces at the point where the vertical plane cuts the ring. The resultant pressure F_R, according to paragraph 1, p. 361, is the same as the resultant pressure on that part of the assumed vertical plane that is cut out by the two horizontal planes and the inner surface of the ring. Therefore, if $g = g_c$ numerically,

$$F_R = \left(\frac{100 \times 62.3}{144}\right) \times (10 \times 12 \times 1) = 5200 \text{ lb.}$$

Therefore, the two tensions are equal to each other and to $F_R/2$ or 2600 lb.

BUOYANCY AND FLOTATION

Archimedes' Principle. The resultant pressure of a fluid on a body immersed in it acts vertically upward through the center of gravity of the displaced fluid and is equal to the weight of the fluid displaced. This resultant upward force exerted by the fluid on the body is called the **buoyancy,** or **force of buoyancy;** hence the center of gravity of the displaced fluid is called the **center of buoyancy.**

Example. What weight of lead must be added to the bottom of a buoy to assist in keeping it upright in sea water if a net downward pull of 500 lb. is required?

Lead weighs 708 lb./cu. ft.; 1 cu. ft. sea water weighs 64 lb.; the net downward pull is 708 − 64 = 644 lb./cu. ft. lead if $g = g_c$ numerically. There is therefore required: 500/644 = 0.7764 cu. ft. or 549.7 lb. lead.

The center of buoyancy of a rigid body wholly submerged in a fluid of uniform density has the same location relative to the body whatever the position of the body in the fluid. In such a case, if the density of an object is uniform, its center of buoyancy and center of gravity will coincide. On the other hand, when a body floats at the surface of a liquid (and, in general, whenever the surrounding fluid is of a non-uniform density), the center of buoyancy is usually moved relative to the body whenever the position of the body is altered. This effect results from the change in shape of the space from which the liquid is displaced. The locus of the center of buoyancy is called the **curve of buoyancy.**

Equilibrium and Stability of Floating Bodies. The only forces acting on a freely floating body are its weight, which acts downward, and the buoyancy, which acts upward. Hence, for equilibrium, these forces must be equal, and the center of gravity and the center of buoyancy of a floating body must lie in the same vertical line. A body floating in equilibrium is said to possess stability if a small displacement brings into action forces tending to restore the body to its original position of equilibrium. The displacements usually discussed are small rotations about a horizontal axis since the behavior for vertical and horizontal translations is generally obvious (see Greenhill, "Treatise on Hydrostatics," Macmillan, London, 1894).

Evidently the stability as regards rotation of a given floating body is determined by considering whether or not, for a slightly displaced position, the couple formed by the weight of the object and the force of buoyancy is acting in a direction opposed to that of the displacement. Thus, a rigid body submerged in a liquid will possess stability if the center of gravity is below the center of buoyancy. If the center of buoyancy and center of gravity coincide for all positions, the body will have stability in no position as regards rotations about a horizontal axis through the common center. In examining a given object for stability, it must be remembered that the center of gravity as well as the center of buoyancy may shift its position relative to the body when the orientation of the body is changed.

In the case of a rigid body floating at the interface between two fluids and subject to rotations in a plane perpendicular to a vertical plane of symmetry of the body, the point at which a vertical line through the center of buoyancy in a very slightly displaced position cuts the originally vertical line through the original center of buoyancy and center of gravity is called the metacenter. (For large displacements the point of intersection generally will not be at the metacenter.) If the metacenter lies above the center of gravity, the floating body possesses stability; if below, stability is lacking. For complicated cases, see Greenhill, *ibid.*

The **depth of flotation,** or **draft,** of an object floating at the interface between two fluids is the vertical distance of the lowest point of the body below the interface. This distance can be easily found if the weight, shape, and density of the body are known.

* See Fig. 1, but observe also part 2 of Note on p. 361.

Example. What should be the relation between the stem cross section and weight of a hydrometer if an increase of the specific gravity of a fluid of 1 part in 1000 is to cause the hydrometer to ride 0.1 in. higher? The volume of 0.1 in. length of stem must be 0.001 × total volume of hydrometer = 0.001 × total weight/density of fluid. Assuming the fluid is water, 1 cu. in. weighs 62.3/1728 = 0.0361 lb., 0.1 (area of stem) = 0.001 × total weight/0.0361, or area = pounds total weight/3.61.

Specific Gravity and Volume of Solids by Immersion. By weighing any substance in a liquid in which it is insoluble, and then weighing it in air, the difference in weight is the weight of an equal volume of the liquid. Letting W_a = weight in air and W_f = weight in fluid, we have the relation: $W_a - W_f$ = weight of equal volume of liquid (W_w). Hence, $W_a/(W_a - W_f)$ = relative density.

CAPILLARITY AND SURFACE TENSION

Interfacial Tension. As a consequence of intermolecular attraction the interface between two immiscible fluids tends to assume the minimum area consistent with the volume of fluid enclosed and with the external forces acting on the fluids. To extend the area A of the interface isothermally requires the expenditure of work $dW = \sigma\, dA$, where σ, the work done per unit increase of area, is called the interfacial tension. σ is the "free energy" of the interface per unit area and is a function of the state of the system. Its value generally varies markedly with the compositions of the fluids and with the temperature, while the variation with pressure is frequently small. Its effective value may be considerably altered by the accumulation of traces of foreign matter at the interface. The dimensions of σ are those of energy ÷ area which are the same as those of force ÷ length; tabulated values are customarily expressed in dynes per centimeter and must be multiplied by 6.8523×10^{-5} to convert to pounds *force* per foot.

When one of the fluids is a gas, a change in the nature of this gas, other things being equal, often produces no appreciable change in σ. Hence the interfacial tension for liquid-gas systems is usually treated as a property of the liquid alone and is generally called the **surface tension.** Except within 20 to 30°C. of the critical temperature t_c (°C.), the variation of the surface tension with the temperature t (°C.) is generally well approximated by the Eötvös equation:

$$\sigma = \left(\frac{\rho}{M}\right)^{2/3} (B - k_E t) \quad \text{dynes/cm.} \tag{8}$$

where ρ = density of the liquid, g./cc.; M = molecular weight; and B and k_E are experimentally determined constants. Approximately,

$$B = k_E(t_c - 6)$$

For normal (*i.e.*, unassociated) liquids, $k_E = 2.1$.

Contact Angle. Where the interface between two fluids meets a solid, the angle of contact θ is dependent on the nature of the solid surface and on that of each fluid. In practice, for a given system, θ varies between more or less definite limits, but the exact value appears to depend somewhat upon the history of the system. For a discussion of the phenomena and a bibliography see Haller, *Kolloid-Z.*, **53**, 247 (1930).

At room temperature, if θ is the angle subtended by the liquid at the junction of an air-liquid interface and a glass surface, approximate values are as follows: $\theta = 0°$ for water, most aqueous solutions, hydrogen peroxide, and most organic liquids, provided that the glass is covered with a film of the liquid; for mercury, $\theta = 140°$. For the system water-air-paraffin wax, $\theta = 105°$.

Table 1. Surface Tension of Common Liquids in Contact with Air*

Liquid	Surface tension, σ, dynes/cm.						Eötvös constant, k_E
	0°C.	20°C.	40°C.	60°C.	80°C.	100°C.	
Benzene, C_6H_6	31.6	28.9	26.3	23.7	21.3	18.8†	$2.22^{0-100°C.}$
Ethanol, C_2H_5OH	24.1	22.3	20.6	19.0	....	15.5†	$1.0^{20°C.}$, $1.3^{150°C.}$
Methanol, CH_3OH	24.5	22.6	20.9	....	....	15.7†	$1.0^{100°C.}$
Toluene, C_7H_8	30.7	28.4	26.1	23.8	21.5	19.4	
Water, H_2O	75.6	72.8	69.6	66.2	62.6	58.9	$1.03^{25°C.}$, $1.18^{100°C.}$

* For extensive data see "International Critical Tables," vol. 4, pp. 432–475, McGraw-Hill, New York.
† In contact with saturated vapor.

Table 2. Interfacial Tension of Mercury and Water in Contact with Certain Fluids*

Mercury

Second fluid	σ, dynes/cm.	Temp., °C.
Air, or O_2	437†	15
Hydrogen, H_2	470	19
Nitrogen, N_2	496	15
Vacuum	470–480	0
Benzene, C_6H_6	357	20
Ethanol, C_2H_5OH	364	20
Water, H_2O	375	20

Water

Second fluid	σ, dynes/cm.	Temp., °C.
Benzene, C_6H_6	35.0	20
Carbon tetrachloride, CCl_4	45.0	20
Cyclohexanol, $C_6H_{11}OH$	3.92	16.2
Ethyl ether, $C_4H_{10}O$	10.7	20
n-Hexane, C_6H_{14}	51.1	20
Oleic acid, $C_{18}H_{34}O_2$	15.6	20
Toluene, C_7H_8	36.1	25

* For extensive data see "International Critical Tables," vol. 4, pp. 432–475, McGraw-Hill, New York.
† Variable, and decreases with time.

Capillary Rise. In a small, vertical, open tube of circular section (Fig. 4) a column of liquid will stand at a height greater or less than that corresponding to the static head at the foot of the column by the amount

$$H = \frac{4\sigma \cos\theta}{gD(\rho_1 - \rho_2)} \quad \text{cm.} \tag{9}$$

Here σ is the surface tension, dynes/cm.; D is the diameter, cm.; ρ_1 and ρ_2 are the densities, g./cc., of the liquid and gas (or light liquid), respectively; g = 981

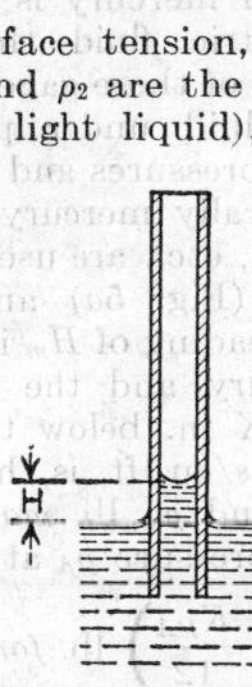

FIG. 4. Capillary rise.

cm./sec.², approximately; and θ is the contact angle subtended by the heavier fluid. If it is desired to express H and D in inches, the units of the other quantities being unchanged, the factor 4 of the above formula should be replaced by 0.62. This formula is fairly accurate only if the tube is small enough for the meniscus to be substantially spherical. For corrections for use in the case of larger tubes see "International Critical Tables," vol. 4, p. 435; also vol. 1, p. 70, for the case of mercury.

PRESSURE GAGES AND MANOMETERS

REFERENCES: Walker, Lewis, McAdams, and Gilliland, "Principles of Chemical Engineering," McGraw-Hill, New York, 1937. Pannell, "Measurement of Fluid Velocity and Pressure," Arnold, London, 1924.

Most common pressure gages actually show the difference between two pressures and therefore are **differential gages**, but usually this term is applied only to U-tube gages when indicating the difference between two pressures, both distinct from that of the atmosphere. In an **open gage** one of the pressures is atmospheric so that the reading is the **gage pressure** (see p. 360). When the pressure is that of a moving fluid, a physical interpretation of the gage readings cannot be given without specific knowledge as to the position and orientation of the pressure tap relative to the flow. In the formulas below, the pressure p is the actual pressure exerted on the opening at which the gage is attached.

LIQUID-COLUMN MANOMETERS

The **height**, or **head** (*cf.* p. 360), to which a fluid rises in an open vertical tube attached to an apparatus containing a liquid is a direct measure of the pressure at the point of attachment and is frequently used to show the level of liquids in tanks, etc. With U-tube gages (Fig. 5*a*) and equivalent devices such as that shown in Fig. 5*b*, the same principle can be used when it is desirable or necessary (as in the case of a gas pressure) that the pressure be measured as a head of some fluid other than that of which the pressure is wanted. Most of these gages may be used either as open or as differential manometers. Although mercury is shown in the illustrations, the manometric fluid that constitutes the measured liquid column of these gages may be any liquid immiscible with the fluid under pressure. For high vacuums and for large pressures and pressure differences the gage liquid is generally mercury; for low pressures, kerosene, water, alcohol, etc., are used.

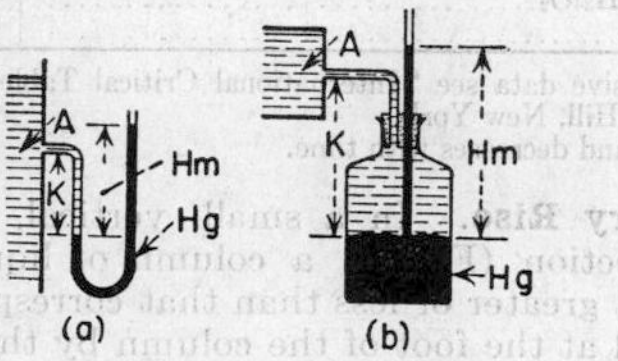

FIG. 5. Open manometers.

The **open U-tube** (Fig. 5*a*) and the **open gage** (Fig. 5*b*) each show a reading of H_m in. mercury, and the interface of the mercury and the fluid of which the pressure is wanted is K in. below the point of attachment A. If ρ_A lb. *mass*/cu. ft. is the weight density of the latter fluid at A, and ρ_m lb. *mass*/cu. ft. is that of the mercury, the gage pressure p_A at A is

$$p_A = \frac{1}{144}\left(\frac{H_m \rho_m}{12} - \frac{K\rho_A}{12}\right) \text{ lb. } force\text{/sq. in. gage*}$$

and the head H_A at A as feet of the fluid at that point is

$$H_A = \frac{H_m}{12}\left(\frac{\rho_m}{\rho_A}\right) - \frac{K}{12} \quad \text{ft.*}$$

* It is assumed that the lines leading from the respective pressure taps to the gage are filled with fluid of the same density as that in the apparatus at the location of the pressure taps; if this is not the case, ρ_A and ρ_B are the densities of the fluids actually filling the gage lines and the values given for H_A and ΔH must be multiplied by ρ_A/ρ where ρ is density of the fluid of which the head is being measured. Theoretically the factor $g/g_c \sim 1$ should appear in each formula for p_A or $p_A - p_B$ (see pp. 360, 375).

When a gas pressure is measured, unless it is very great, ρ_A is so much smaller than ρ_m that the terms involving K in the above formulas are negligible.

The **differential U-tube** (Fig. 6) shows the pressure difference between the taps A and B to be

$$p_A - p_B = \frac{1}{144}\left[\frac{H_m}{12}(\rho_m - \rho_A) + \frac{K_A\rho_A}{12} - \frac{K_B\rho_B}{12}\right] \text{ lb. } force\text{/sq. in.*}$$

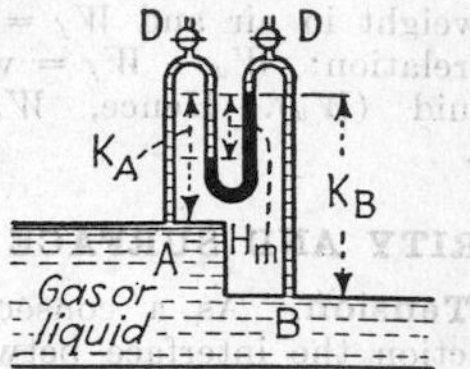

FIG. 6. Differential U-tube.

where H_m is in inches; K_A and K_B, both in inches, are the vertical distances of the upper mercury surface above A and B, respectively; and ρ_A and ρ_B are the weight densities, lb. *mass*/cu. ft., of the fluids at A and at B, respectively. If either pressure tap is above the higher mercury level, the corresponding K is taken to be negative. If the differential head is that caused by an orifice or similar device measuring the flow of a liquid, the indicated orifice differential ΔH in feet (see p. 401) is

$$\Delta H = (p_1v_1 - p_2v_2) + (Z_1 - Z_2)$$
$$= \frac{H_m}{12}\left(\frac{\rho_m}{\rho_A} - 1\right) \quad \text{ft.*}$$

For gases, except at very high pressures, ρ_A and ρ_B are so small compared with ρ_m that the formulas reduce to

$$(p_A - p_B) = \frac{H_m\rho_m}{144 \times 12} \text{ lb. } force\text{/sq. in.}$$

The **inverted differential U-tube** (Fig. 7), in which the fluid filling the U-tube may be a gas, is often used to measure liquid pressure differentials when open liquid columns would be inordinately high, or when the liquid under pressure cannot be exposed to the atmosphere. If ρ_m (lb. *mass*/cu. ft.) is the weight density of the fluid in the U-tube, K_A and K_B (in.) are as shown, and the remaining symbols mean the same as in the previous paragraph:

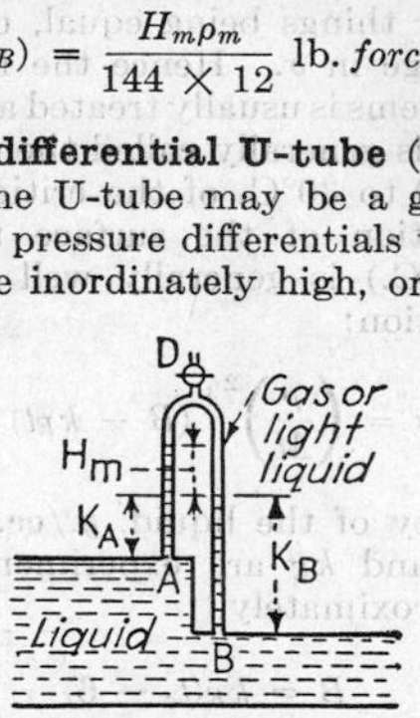

FIG. 7. Inverted differential U-tube.

$$p_A - p_B = \frac{1}{144}\left[\frac{H_m}{12}(\rho_A - \rho_m) + \frac{K_A\rho_A}{12} - \frac{K_B\rho_B}{12}\right] \text{ lb. } force\text{/sq. in.*}$$

If the gage is indicating the orifice differential of a head meter operating with a liquid, ΔH (ft.) (see p. 401) is

$$\Delta H = \frac{H_m}{12}\left(1 - \frac{\rho_m}{\rho_A}\right) \quad \text{ft.*}$$

Table 1. Reduction of Barometric Height to Standard Temperature

Corrections for brass scale and English measure				Corrections for brass scale and metric measure				Correction for glass scale and metric measure			
Height of barometer, in.	α, in. for temp., °F.	Height of barometer, in.	α, in. for temp., °F.	Height of barometer, mm.	α, mm. for temp., °C.	Height of barometer, mm.	α, mm. for temp., °C.	Height of barometer, mm.	α, mm. for temp., °C.	Height of barometer, mm.	α, mm. for temp., °C.
15.0	0.0014	26.0	0.0024	400	0.065	600	0.098	50	0.009	660	0.114
16.0	.0015	26.5	.0024	410	.067	610	.099	100	.017	670	.115
17.0	.0015	27.0	.0025	420	.068	620	.101	150	.026	680	.117
17.5	.0016	27.5	.0025	430	.070	630	.102	200	.035	690	.119
18.0	.0016	28.0	.0025	440	.072	640	.104	250	.043	700	.121
18.5	.0017	28.5	.0026	450	.073	650	.106	300	.052	710	.122
19.0	.0017	29.0	.0026	460	.075	660	.107	350	.060	720	.124
19.5	.0018	29.2	.0027	470	.077	670	.109	400	.069	730	.126
20.0	.0018	29.4	.0027	480	.078	680	.111	450	.078	740	.128
20.5	.0019	29.6	.0027	490	.080	690	.112	500	.086	750	.129
21.0	.0019	29.8	.0027	500	.081	700	.114	520	.090	760	.131
21.5	.0019	30.0	.0027	510	.083	710	.115	540	.093	770	.133
22.0	.0020	30.2	.0027	520	.085	720	.117	560	.097	780	.134
22.5	.0020	30.4	.0028	530	.086	730	.119	580	.100	790	.136
23.0	.0021	30.6	.0028	540	.088	740	.120	600	.103	800	.138
23.5	.0021	30.8	.0028	550	.089	750	.122	610	.105	850	.146
24.0	.0022	31.0	.0028	560	.091	760	.124	620	.107	900	.155
24.5	.0022	31.2	.0028	570	.093	770	.125	630	.109	950	.164
25.0	.0023	31.4	.0029	580	.094	780	.127	640	.110	1000	.172
25.5	.0023	31.6	.0029	590	.096	790	.128	650	.112		
						800	.130				

"Smithsonian Tables."

The height of the barometer is affected by the relative thermal expansion of the mercury and the glass, in the case of instruments graduated on the glass tube, and by the relative expansion of the mercury and the metallic enclosing case, usually of brass, in the case of instruments graduated on the brass case. This relative, expansion is practically proportional to the first power of the temperature. The above tables of values of the coefficient of relative expansion will be found to give corrections almost identical with those given in the "International Meteorological Tables." The numbers tabulated under α are the values of α in the equation $H_t = H_t' - \alpha(t' - t)$, where H_t is the height at the standard temperature, H_t' the observed height at the temperature t', and $\alpha(t' - t)$ the correction for temperature. The standard temperature is 0°C. for the metric system and 28.5°F. for the English system. The English barometer is correct for the temperature of melting ice at a temperature of approximately 28.5°F., because the brass scale is graduated so as to be standard at 62°F., while mercury has the standard density at 32°F.

Example.—A barometer having a brass scale gave $H = 765$ mm. at 25°C.; required, the corresponding reading at 0°C. Here the value of α is the mean of 0.124 and 0.125 or 0.1245; therefore $\alpha(t' - t) = 0.1245 \times 25 = 3.11$. Hence $H_0 = 765 - 3.11 = 761.89$.

Except at very high pressures, ρ_m is negligible if the fluid in the U-tube is a gas.

Valve D, which is kept closed when the gage is in use, makes it possible to vary the quantity of the light fluid in the gage. This is an especial convenience when the U of the gage is filled with gas.

Closed U-tubes (Fig. 8) serve to measure directly the absolute pressure p of a fluid. If ρ_m (lb. *mass*/cu. ft.) is the weight density of mercury, and H_m is in inches, $p = \dfrac{H_m \rho_m}{12 \times 144}$ lb. *force*/sq. in. abs., provided that the space above the mercury is substantially a perfect vacuum. For liquids, and for gases under very high pressures, the quantity $K\rho$ should be subtracted from the

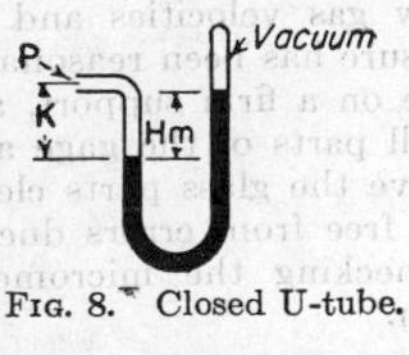

FIG. 8.* Closed U-tube.

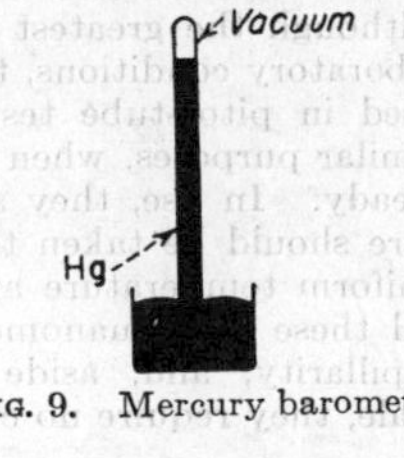

FIG. 9. Mercury barometer.

numerator of this equation.* K is the vertical distance in inches of the point of attachment above the lower mercury level. ρ (lb. *mass*/cu. ft.) is the density of the fluid under pressure.

The **mercury barometer** (Fig. 9) records directly the absolute pressure of the atmosphere in terms of the height of the mercury column. Normal (standard) barometric pressure is 760 mm. mercury (at 0°C.) by definition. Equivalents of this pressure in other units are 29.921 in. mercury (at 0°C.); 14.696 lb./sq. in.; 1 atm. Barometer readings, when expressed by the height of the mercury column, must be corrected to standard temperature (usually 0°C.) (see Table 1).

* See footnote on p. 364.

The **aneroid barometer** is a delicate diaphragm gage (see p. 367).

Technique in the Use of Manometers. The most serious common errors in manometric measurements arise as follows:

1. Leaky connections between the pressure taps and the instrument permit flow to take place through the tap opening and leads. Since these passages are usually small, even a minute leak may occasion a substantial pressure drop in the leads, by which amount the pressure effectively impressed on the instrument will be in error.

2. In small manometer tubing capillarity may cause marked displacement of the observed interface from its proper level. The extent of capillary rise or depression (p. 363) varies inversely as the diameter of the tube; *e.g.*, for a clean water-air interface at 20°C., the rise H is $0.0461/D$ in. if D is the inside diameter, in.; hence $H = 0.368$ in. for $D = 1/8$ in. and $H = 0.092$ in. for $D = 1/2$ in. A theoretical estimate of the correction is apt to be very unreliable because of local variations in bore and doubt as to the proper values of the interfacial tension and contact angles. Both these properties are markedly altered by slight contamination of the interface. Especially when the interface observed is that between two liquids, the shape of the meniscus and the extent of the capillary rise will be often found to depend upon the direction in which the interface has most recently moved and to settle back to normal only very slowly. Consequently it is preferable to avoid capillary error by the use of tubes sufficiently large and manometric fluids of such densities that the effect of capillarity is negligible in comparison with the gage reading. The minimum inside diameters consonant with good practice are, for U-tubes, 1/4 in. and, for devices such as that shown in Fig. 5*b*, 1/2 in. The smaller diameter is generally permissible for U-tubes because the capillary displacement in one leg tends to cancel that in the other. A good rule with ordinary manometers is to choose, if possible, a type in which the reading will be at least 2 in.

3. A false interpretation of the readings will result if the gage lines are not in fact wholly filled with the fluid they are supposed to contain. Gas pockets in lines supposedly filled with liquid are particularly troublesome if there are high points in the line, because a substantial flow of liquid may be needed to displace the bubbles.

MULTIPLYING GAGES

To attain the requisite precision in the measurement of small pressure differences by liquid-column manometers, it is often necessary to devise means of magnifying the reading. Of the schemes that follow, the second and third may give tenfold multiplication; the fourth, as much as thirtyfold. In general, the greater the multiplication, the more elaborate must be the precautions in the use of the gage, if the gain in precision is not to be illusory.

1. *A Change of the Manometric Fluid.* In open manometers choose a fluid of less density. In differential manometers, such as in Fig. 6, choose a fluid such that the difference between its density and that of the fluid under pressure is as small as possible.

2. *An Inclined U-tube* (Fig. 10). If the reading R in. is taken as shown, by making the substitution $H_m = (R - R_0) \sin \theta$ the formulas of preceding paragraphs give $(p_A - p_B)$ when the corresponding upright U-tube is

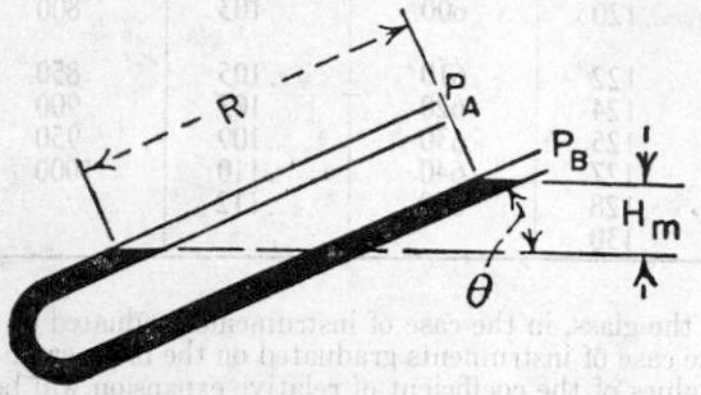

FIG. 10.

replaced by one inclined. For precise work it is advisable to calibrate the gage because of possible variations in tube diameter and in inclination. R_0 = zero reading.

3. *The draft gage* (Fig. 11), commonly used for low gas heads, has for one leg of the U a reservoir of much larger bore than the tubing that forms the inclined leg. Hence variations of the level in the inclined tube produce very little change in the level in the reservoir. Although H_m may be readily computed in terms of the reading R and the dimensions of the tube, it is preferable to calibrate the gage (p. 367), because often the changes of level in the reservoir are not negligible and, moreover, variations in tube diameter may introduce serious error into the computation. Commercial gages are often provided with a scale giving H_m directly in inches of water, provided that a particular liquid (often not water) fills the tube; failure to appreciate that the scale is incorrect unless the gage is filled with the specified liquid is a frequent source of error. If the scale reads correctly when the density of the gage liquid is ρ_0, then the reading must be multiplied by ρ/ρ_0 if the density of the fluid actually in use is ρ.

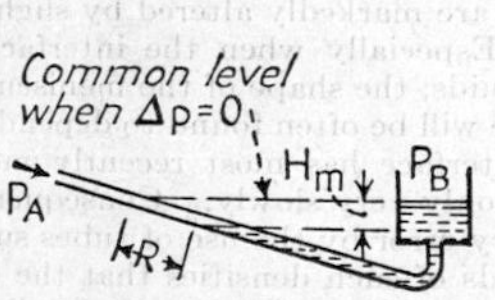

FIG. 11. Draft gage.

4. *Two-fluid U-tubes* (Fig. 12) are very sensitive instruments for measuring small gas heads. Let A be the cross-sectional area of each of the reservoirs, a that of the tube forming the U; let ρ_1 be the density of the lighter fluid, ρ_2 that of the heavier, and ρ_w the density of water, all in pounds per cubic foot. Then for Type I (Fig. 12)

$$p_A - p_B = \frac{R - R_0}{\rho_w}\left(\rho_2 - \rho_1 + \frac{a}{A}\rho_1\right) \quad \text{in. water}$$

where the reading R is in inches and R_0 is its value with zero pressure difference. For Type II, if the reading R (in.) is taken as shown in Fig. 12,

$$p_A - p_B = \frac{R}{\rho_w}\left[\rho_2 - \rho_1 + \frac{a}{A}(\rho_2 + \rho_1)\right] \quad \text{in. water}$$

When A/a is sufficiently large, the terms $(a/A)\rho_1$ and $(a/A)(\rho_2 + \rho_1)$ of the above formulas become negligible in comparison with the difference $(\rho_2 - \rho_1)$. However, *these terms should not be omitted without due consideration.* Since the magnification of the reading is inversely as the difference in densities, the tendency is to choose a liquid pair for which that difference is very small; and, consequently, to admit of the above omission, the reservoirs required may be of excessive diameter. In applying the formulas *the densities of the gage liquids may not be taken from tables without introducing serious*

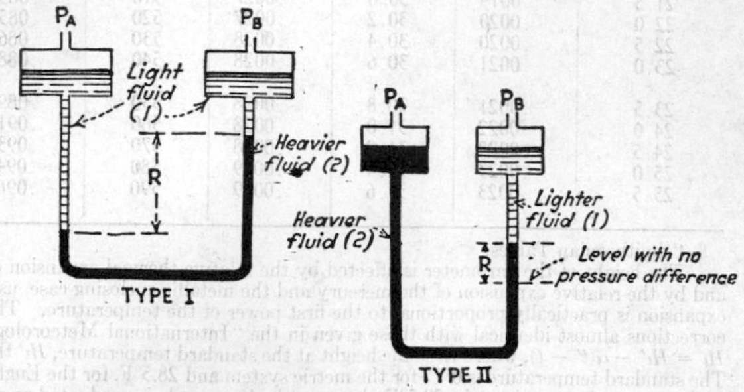

FIG. 12.

error, for each liquid will dissolve appreciable quantities of the other. Before the gage is filled, the liquids should be shaken together, and the densities of the two layers should be determined for the temperature at which the gage is to be used. When high magnification is being sought, it may be necessary to enclose the U-tube in a constant-temperature bath in order that $(\rho_2 - \rho_1)$ may be accurately known. In general, if accuracy is desired, the gage should be calibrated (see p. 367).

MICROMANOMETERS

Several micromanometers, based on the liquid-column principle, and of extreme precision and sensitivity, have been developed for measuring minute pressure differences and for calibrating low-range gages. Although the greatest accuracy can be had only under laboratory conditions, these gages have been successfully used in pitot-tube tests at low gas velocities and for similar purposes, when the pressure has been reasonably steady. In use, they should be on a firm support, and care should be taken to have all parts of the gage at a uniform temperature and to have the glass parts clean. All these micromanometers are free from errors due to capillarity, and, aside from checking the micrometer scale, they require no calibration.

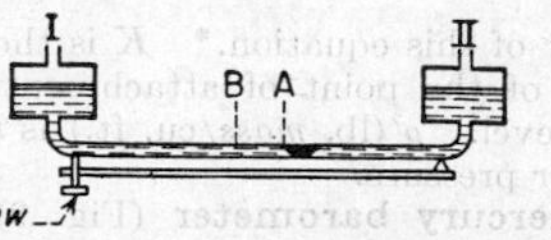

FIG. 13. Tilting manometer.

1. The simplest are merely upright U-tubes or vertical tubes provided with a delicate means, such as a hook gage (see p. 367) or microscope for determining the levels of the manometric liquid. The legs of such a gage are at least 2 or 3 in. in diameter so that errors due to capillarity are avoided.

2. In the **Chattock and Chattock-Fry tilting manometers,** a U-tube, with widely separated legs of

large diameter connected by a smaller tube, is mounted in a vertical plane on a rigid beam that can be raised at one end by a micrometer screw. The apparatus shown in Fig. 13 serves to illustrate the principle of operation. By applying a slight excess of pressure at opening II, the level of the fluid in the right arm will be slightly depressed. This will cause the bubble of immiscible liquid at *A* to move to some position *B*. The greater the diameter of the reservoirs compared with that of the connected tube, the greater is the distance *AB* for a given change in level. Tilting the gage by turning the micrometer screw returns the bubble to its zero position at *A*. In practice, a microscope mounted on the beam of the instrument is used to observe the position of the bubble. From the dimensions of the instrument and the micrometer readings the pressure difference can be calculated. Capillary errors are avoided because at the time of reading all interfaces are always at the same position relative to the U-tube. (For a good description of tilting gages see Pannell, "Measurement of Fluid Velocity and Pressure," Arnold, London, 1924.) Ower (Aeronautical Research Committee, Great Britain, *Reports and Memoranda*, No. 1308, 1930) has devised an instrument of this type with a sensitivity of 0.00001 in. of manometric liquid.

3. Similar in principle to the Chattock gage is the apparatus illustrated in Fig. 14, which is a U-tube with one flexible leg. A rigidly held reservoir is connected to an inclined glass tube *T* by a rubber tube. The tube *T* is carried by a sliding block that may be raised or lowered by micrometer screw *M*. A scratch is placed on the glass tube, and the readings are made by adjusting the micrometer until the meniscus is at the scratch. The difference between the zero reading of the micrometer

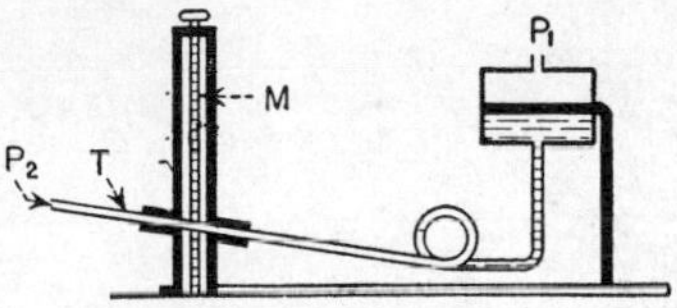

FIG. 14. Micromanometer.

and reading under pressure gives the pressure difference directly. Differential heads of less than 1/1000 in. of manometric fluid can be determined. The sensitivity may be varied by changing the inclination of tube *T*.

MECHANICAL PRESSURE GAGES

The **Bourdon ring gage**, commonly found on steam mains, etc., is operated by the flection under internal pressure of a tube of oval cross section, bent in an arc of a circle and closed at one end. An increase of internal pressure tends to straighten the bent tube, the movements of which are communicated to a pointer by a suitable mechanism. Bourdon gages are available for all pressures, below and above atmospheric. The pressure tube is usually made of brass, but commercial gages can be obtained with tubes of other materials (see p. 1279).

Diaphragm gages depend for their indications on the deflection of a diaphragm, usually metallic, when subjected to a difference of pressure between its two faces. They are available for the same general purposes as Bourdon gages but are not usually employed for high pressures. Very delicate gages of this type can be made by using optical or electrical methods for magnifying the movements of the diaphragm. The aneroid barometer is a type of diaphragm gage.

Bourdon and diaphragm gages that show both pressure and vacuum indications on the same dial are available. Such instruments are called **compound gages.**

Bourdon gages should not be exposed to temperatures over about 150°F. (65.6°C.), since above this temperature the metallic tube loses part of its elasticity. When the pressure of a hotter fluid is to be measured, some sort of liquid seal should be used to keep the hot fluid from the gage. In using either the Bourdon or diaphragm gage to measure gas pressures, if the gage be below the pressure tap of the apparatus so that water or other liquid can collect in the leads, the gage reading will be too high by an amount equal to the hydrostatic head of the accumulated fluid. In the case of condensed water the correction in pounds per square inch is 0.433 × the hydrostatic head in feet. Both types of gage require occasional calibration if accuracy is desired.

CALIBRATION OF GAGES

Simple liquid-column manometers do not require calibration if they are so constructed as to minimize errors due to capillarity (see p. 363). If the scales used to measure the reading have been checked against a standard, the accuracy of the gages depends solely upon the precision of determining the position of the liquid surfaces. It follows that liquid-column manometers are used to standardize other gages. For high pressures and, with commercial mechanical gages, even for quite moderate pressures, a dead-weight gage (see pp. 1254 and 1280) is commonly used as the primary standard because it is safer and more convenient than the use of manometers. When manometers are used as high-pressure standards, an inordinately high mercury column may be avoided by connecting a number of U-tubes in series. Multiplying gages are standardized by comparing them with any of the micromanometers described on p. 366. The procedure in the calibration of a gage consists merely of connecting it (in parallel with the standard instrument) to a reservoir wherein any desired constant pressure may be maintained. Readings of the unknown gage are then made for various reservoir pressures as determined by the standard.

GAGES FOR LOCATION OF INTERFACES

For precisely locating the position of a free liquid surface the following instruments are much used:

1. The **hook gage** (Fig. 15) utilizes the very distinct optical effect produced when a sharp point pierces a

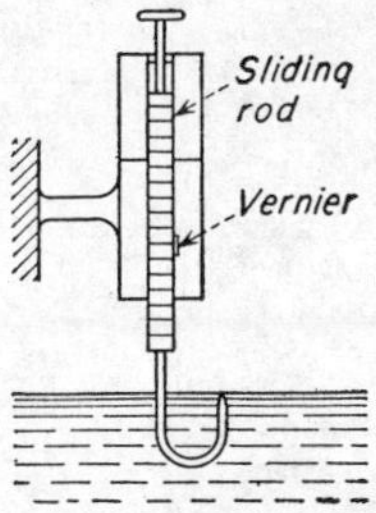

FIG. 15. Hook gage.

liquid surface from the underside. The upturned point should be a sharp cone of large vertex angle (45, 90, or even 120 deg).

2. The **point gage** resembles the hook gage, save that the point is directed downward. In use, the point is lowered until a "bubble" is formed in the surface. A plumb bob suspended by a metal tape measure is a serviceable point gage.

3. A **float gage** (Fig. 16), for exact measurements, is usually a hollow metal float surmounted by a slender vertical spindle that moves in guides along a scale. The float may be made to operate a pointer on a dial or the pen of a recording device, but, when precision is wanted, these additions are unwise on account of the possibility of lost motion in, or inertia of, the mechanism. When using a float gage, the bounding walls of the liquid surface should be everywhere at least 1 in. from the float in order to avoid errors due to capillarity.

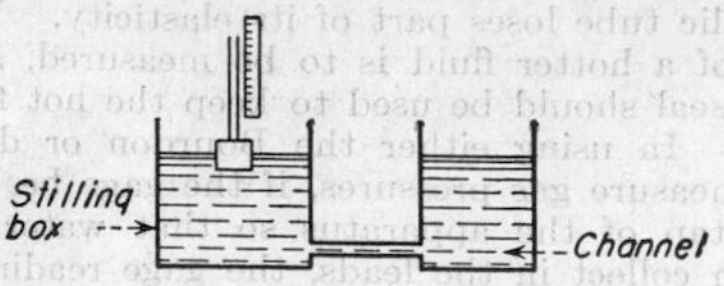

FIG. 16. Float gage.

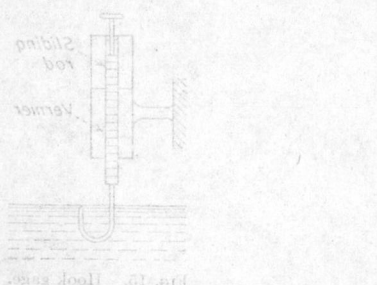

If the liquid surface is substantially stationary, there seems to be little choice between the hook gage and the float gage as far as accuracy is concerned. Both can be made to give correct readings within less than 0.01 in. If the level fluctuates even slightly, it is difficult to obtain satisfactory hook-gage readings, and the float gage then becomes the superior. The latter instrument has an advantage in that it is an indicating gage. When the head of a fluid flowing in an open channel is to be measured, if accuracy is desired, all these gages must be used in a stilling box (Fig. 16) which has free communication with the moving fluid by means of an opening flush with the channel wall (see pp. 396*ff.*).

Remote indication of the liquid level in a tank is commonly effected by utilizing the fact that the pressure at any point in liquid is proportional to its depth below the surface (see pp. 360, 1289). Thus, from the reading of any convenient type of pressure gage or manometer, so installed that it indicates the pressure at a point that is always submerged, the level in the tank may be inferred.

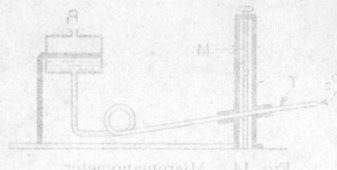

FLUIDS IN MOTION

REFERENCES: Lamb, "Hydrodynamics," Cambridge University Press, London, 1924. Gibson, "Hydraulics and Its Applications," Van Nostrand, New York, 1930. Stodola, "Steam Turbines," Loewenstein's transl., McGraw-Hill, New York, 1927. Walker, Lewis, McAdams, and Gilliland, "Principles of Chemical Engineering," McGraw-Hill, New York, 1937. Prandtl-Tietjens, "Applied Hydro- and Aeromechanics," McGraw-Hill, New York, 1934. National Research Council, Report of the Committee on Hydrodynamics, *Bull.* 84, February, 1932. Dodge and Thompson, "Fluid Mechanics," McGraw-Hill, New York, 1937. O'Brien and Hickox, "Applied Fluid Mechanics," McGraw-Hill, New York, 1937. Aeronautical Research Committee, Great Britain, Goldstein (ed.), "Modern Developments in Fluid Dynamics," Oxford, New York, 1938.

SUBSTANCES WHICH FLOW

A **fluid** is a substance which undergoes continuous deformation when subjected to a shear stress.* The resistance offered by a real fluid to such deformation is called its **consistency**. If τ (lb. *force*/sq. ft.) is the shear stress which maintains at a point in a fluid the time rate of shearing strain du/dy [*i.e.*, velocity gradient, (ft./sec.)/ft. perpendicular to the direction of τ], the numerical measure of the consistency is the quantity

$$\frac{\tau g_c}{du/dy} \quad \frac{\text{lb. } mass}{(\text{sec.})(\text{ft.})} \qquad (1)$$

in which g_c is the dimensional constant 32.1740 (lb. *mass*)(ft.)/(lb. *force*)(sec.^{2}). For **gases** and for **simple** (or **Newtonian**) **liquids** the consistency is constant if the static pressure and temperature are fixed; for such substances the consistency is usually called the **viscosity** (= absolute viscosity). If two parallel plates are separated by a thin layer, L ft. thick, of a fluid of viscosity μ lb. *mass*/(sec.)(ft.), the force τ per unit area (lb. *force*/sq. ft.) which must be exerted in order to keep one plate moving parallel to the other at a relative velocity of 1 ft./sec. is $\tau = \mu/g_cL$. If the consistency of a fluid is not constant but is, at constant temperature, a function only of the shear stress τ, the fluid is called a **complex** (or **non-Newtonian**) **fluid**.

An **ideal** or **perfect fluid** is a hypothetical liquid or gas which offers no resistance to shear and therefore has zero consistency. The imaginary perfect fluid is not to be confused with a "perfect gas," which is a fluid for which $pv = RT$ (*cf.* p. 289), where p is the absolute static pressure, v is the specific volume, R is the gas constant, and T is the absolute temperature. In many cases highly incorrect results are obtained if the viscosity is neglected, although in the same problem the relation $pv = RT$ may be safely used.

Plastic solids are substances which behave like elastic solids (*i.e.*, the strain in shear is a single valued function of the shear stress) unless the shear stress exceeds the yield stress τ_0 and, for shear stresses greater than τ_0, progressively suffer permanent deformation.

* This and most of the definitions immediately following are those recommended by the Committee on Definitions and Nomenclature of the Society of Rheology, *Rheology Leaflet*, No. 7, pp. 11–17, November, 1938.

A **plastico-viscous solid** or **ideal plastic** behaves like an elastic solid unless the shear stress exceeds the yield stress and, for a shear stress τ (lb. *force*/sq. ft.) greater than τ_0, deforms continuously at a rate of shearing strain du/dy ft./(sec.)(ft.) which is a function of τ. The **mobility** of a plastico-viscous solid is $(du/dy)/g_c(\tau - \tau_0)$ (sec.)(ft.)/lb. *mass*. The **rigidity** is the reciprocal of the mobility. Many **slurries**, even when quite thin, approximate the behavior of plastico-viscous solids.

A **thixotropic** substance is such that deformation reduces its consistency.

VISCOSITY DATA

Data on the viscosities of common fluids are discussed below. For more extensive tables of data, references should be made to the "International Critical Tables" as indicated in Table 1.

Table 1. Viscosity Data in "International Critical Tables"

	Volume	Page
Tabular index	5	1
Alloys	5	6
Elements	1	102
Elements	5	2, 6
Gases and vapors	5	1
Gelatins	2	233
Glass	2	94
Liquids	5	10
Liquids	7	211
Metals	5	6
Oils, fats, waxes	2	209
Petroleum	2	146
Refrigerating brines	2	328
Rubber	2	255, 259
Solutions	5	7, 12, 20
Solutions	5	21, 25, 447

The unit of viscosity (*i.e.*, absolute viscosity) in the c.g.s. system is the **poise** (p.) = 1 (dyne)(sec.)/sq. cm. = 1 g./(sec.)(cm.). Viscosities are usually tabulated in **centipoises** (cp.) = 0.01 poise. The dimensions of viscosity may with equal propriety be considered those of

$$\frac{(\text{force})(\text{time})}{(\text{length}^2)} \text{ or of } \frac{(\text{mass})}{(\text{time})(\text{length})}$$

In "absolute" systems of units, such as c.g.s., where the dimensional constant g_c of Newton's law is unity, the numerical value of the viscosity is the same on either basis, if the fundamental units of the system are used to evaluate the dimensions. In most engineering systems of units, in both Europe and America, this is not true, because g_c in such systems is usually other than unity. It is important, therefore, invariably to state the units in which a viscosity is expressed. Conversion factors are given in Table 2.

The **relative viscosity** of a fluid is the ratio of its viscosity to that (at 68°F.) of water. The viscosity of water at 68°F. is very nearly 1 centipoise; therefore, for practical purposes, the relative viscosity of a fluid is identical with the viscosity in centipoises.

A **specific viscosity** is the ratio of the viscosity of a fluid to that of a standard fluid (usually water, or the

Table 2. Conversion of Units for Viscosity

To convert centipoises to:	$\frac{\text{lb. } mass}{(\text{ft.})(\text{sec.})}$ or $\frac{(\text{poundal})(\text{sec.})}{\text{sq. ft.}}$	$\frac{\text{lb. } mass}{(\text{ft.})(\text{hr.})}$	$\frac{(\text{lb. } force)(\text{sec.})}{\text{sq. ft.}}$	$\frac{\text{kg. } mass}{(\text{m.})(\text{sec.})}$	$\frac{(\text{kg. } force)(\text{sec.})}{\text{sq. m.}}$
Multiply by	$0.672 \times 10^{-3} = 1/1488$	2.42	0.209×10^{-4}	1/1000	$1.02 \times 10^{-4} = 1/9810$

solvent in the case of solutions), both viscosities being taken at the *same* temperature.

The **kinematic viscosity** of a fluid of density ρ lb. *mass*/cu. ft. and viscosity μ lb. *mass*/(ft.)(sec.) is $\nu = \mu/\rho$, sq. ft./sec. The c.g.s. unit of kinematic viscosity is sometimes called the **stoke** and equals 1 sq. cm./sec. In several common commercial viscometers the kinematic viscosity is measured in terms of the time of efflux θ (sec.) of a fixed volume of liquid through a standard capillary tube. The term **Saybolt seconds,** for example, refers to the time of efflux in a Saybolt viscometer. The tube in these instruments is so short that the entrance and exit effects often constitute an important part of the resistance to flow. Consequently, the relation between the time of efflux θ and the kinematic viscosity is empirically determined. With liquids of low viscosity these instruments are unsatisfactory because turbulent flow occurs in the tube. The principal commercial viscometers of this type are designed to obey the empirical equation

$$\nu = A\theta - \frac{B}{\theta} \quad \text{sq. cm./sec.}$$

where θ is in seconds and the values of A and B are as follows:

Viscometer	A	B
Saybolt Universal	0.0022	1.8
Redwood	.0026	1.72
Redwood Admiralty	.027	20.
Engler	.00147	3.74

For a conversion chart see p. 1198.

The **fluidity** is the reciprocal of the viscosity. In the c.g.s. system the unit is sometimes called the **rhe** = 1 poise^{-1}.

Table 3. Viscosity Constants of Gases
At 1 atm.

Gas	Sutherland constant, C, in Eq. (3)	Temp. range, °C.	n in Eq. (2)	Temp. range, °C.	μ_0*, centipoise
Air	114	0–300+	0.768	0–100	0.01709
Ammonia	377	15–184	0.981	15–100	0.0096
Argon	169.9	15–184	0.816	15–100	0.0210
	147	20–100			
Carbon dioxide	239.7	−21–302	0.935	−20–140	0.0137
	274	15–100			
Carbon monoxide	118	15–100	0.758	15–100	0.0166
Chlorine	325	13–99	1.00	20–100	0.0129
Deuterium	.....	....	0.699	−183–22	0.01295 30°C.
Ethylene	225.9	−21–302	.....	....	0.0096
Helium	78.2	−61–184	0.647	−258–19	0.0187
	80.3	15–185	0.685	15–100	
Hydrogen	83	−60–185	0.695	−183–21	0.0084
	71.7	−21–302			
Hydrogen chloride	357	13–100	1.03	20–96	0.0137
Hydrogen sulfide	331	17–100	.....	....	0.0117
Krypton	188	16–100	.....	....	0.0232
Methane	198	17–100	0.873	17–100	0.0120
Methyl chloride	454	−15–302	.....	....	0.00988
Neon	61	20–100	0.657	20–100	0.0297
Nitric oxide	128	20–200	0.78	20–100	0.0179
Nitrogen	118	15–100	0.756	15–10C	0.0166
Nitrous oxide	102.7	−76–250			
	274	15–100	0.89	28–278	0.0135
Oxygen	260	28–278			
	138	17–186	0.814	17–100	0.0187
Sulfur dioxide	416	18–100	.....	....	0.0117
Xenon	252	15–100	.....	....	0.0210

* Viscosity at 0°C. and 1 atm. except as noted; "International Critical Tables" or "Smithsonian Tables."

Gases. The viscosities of gases commonly met may be found for atmospheric pressure from the alignment chart, Fig. 17, with an accuracy adequate for most engineering calculations. The chart is based on the formula

$$\frac{\mu}{\mu_0} = \left(\frac{T}{273.1}\right)^n \tag{2}$$

where μ_0 is the viscosity at 0°C., T is the absolute temperature (°K. = °C. + 273.1), and n has been empirically determined from the available data. This type of formula is in good agreement with experiment for most gases, especially in the lower temperature ranges. For hydrogen and helium over most of the experimental range it is preferable to **Sutherland's formula**

$$\frac{\mu}{\mu_0} = \frac{273.1 + C}{T + C}\left(\frac{T}{273.1}\right)^{3/2} \tag{3}$$

which is otherwise generally considered a more accurate expression, particularly above the critical temperature. Table 3 gives values of C and n taken from Chapman and Cowling, "The Mathematical Theory of Non-uniform Gases," pp. 223 and 225, Cambridge University Press, London, 1939. Their book contains the best account of the pertinent parts of kinetic theory.

Table 4. Viscosities of Gases
Coordinates for use with Fig. 17

No.	Gas	X	Y	No.	Gas	X	Y
1	Acetic acid	7.7	14.3	29	Freon-113	11.3	14.0
2	Acetone	8.9	13.0	30	Helium	10.9	20.5
3	Acetylene	9.8	14.9	31	Hexane	8.6	11.8
4	Air	11.0	20.0	32	Hydrogen	11.2	12.4
5	Ammonia	8.4	16.0	33	$3H_2 + 1N_2$	11.2	17.2
6	Argon	10.5	22.4	34	Hydrogen bromide	8.8	20.9
7	Benzene	8.5	13.2	35	Hydrogen chloride	8.8	18.7
8	Bromine	8.9	19.2	36	Hydrogen cyanide	9.8	14.9
9	Butene	9.2	13.7	37	Hydrogen iodide	9.0	21.3
10	Butylene	8.9	13.0	38	Hydrogen sulfide	8.6	18.0
11	Carbon dioxide	9.5	18.7	39	Iodine	9.0	18.4
12	Carbon disulfide	8.0	16.0	40	Mercury	5.3	22.9
13	Carbon monoxide	11.0	20.0	41	Methane	9.9	15.5
14	Chlorine	9.0	18.4	42	Methyl alcohol	8.5	15.6
15	Chloroform	8.9	15.7	43	Nitric oxide	10.9	20.5
16	Cyanogen	9.2	15.2	44	Nitrogen	10.6	20.0
17	Cyclohexane	9.2	12.0	45	Nitrosyl chloride	8.0	17.6
18	Ethane	9.1	14.5	46	Nitrous oxide	8.8	19.0
19	Ethyl acetate	8.5	13.2	47	Oxygen	11.0	21.3
20	Ethyl alcohol	9.2	14.2	48	Pentane	7.0	12.8
21	Ethyl chloride	8.5	15.6	49	Propane	9.7	12.9
22	Ethyl ether	8.9	13.0	50	Propyl alcohol	8.4	13.4
23	Ethylene	9.5	15.1	51	Propylene	9.0	13.8
24	Fluorine	7.3	23.8	52	Sulfur dioxide	9.6	17.0
25	Freon-11	10.6	15.1	53	Toluene	8.6	12.4
26	Freon-12	11.1	16.0	54	2, 3, 3-Trimethylbutane	9.5	10.5
27	Freon-21	10.8	15.3	55	Water	8.0	16.0
28	Freon-22	10.1	17.0	56	Xenon	9.3	23.0

The relatively few data available for the **effect of pressure** on the viscosities of gases have been correlated by Comings and Egly [*Ind. Eng. Chem.*, **32**, 714–718 (1940)] in terms of the reduced pressure and reduced temperature, as shown in Fig. 18. In most cases Fig. 18 appears to agree with the data as well as different observers agree with each other. The seemingly best data for steam (Table 5), however, diverge widely from the chart and from other available data. *Cf.* Hawkins, Sibbitt, and Solberg, "Review of Available Data on Viscosity of Water and Superheated Steam," Chicago Mtg., *Am. Soc. Mech. Engrs.*, June, 1947.

Table 5. Viscosity of Steam*
Values in centipoises

Temp.		Pressure, lb. force/sq. in. abs.					
°C.	°F.	100	200	400	500	600	800
204	400	0.0198	0.0230				
260	500	.0213	.0236	0.0272	0.0289	0.0311	
316	600	.0228	.0246	.0279	.0294	.0314	0.0350
371	700	.0243	.0259	.0290	.0304	.0321	.0357
427	800	.0260	.0275	.0304	.0318	.0334	.0370
482	900	.0278	.0292	.0320	.0335	.0352	.0390
538	1000	.0296	.0310	.0338	.0354	.0372	.0414

* Data of Hawkins, Solberg, and Potter, *Trans. Am. Soc. Mech. Engrs.*, **62**, p. 677 (1940).

According to kinetic theory, the viscosity of a gas should be independent of the pressure, provided that the gas is neither (1) so rarefied that the mean free path

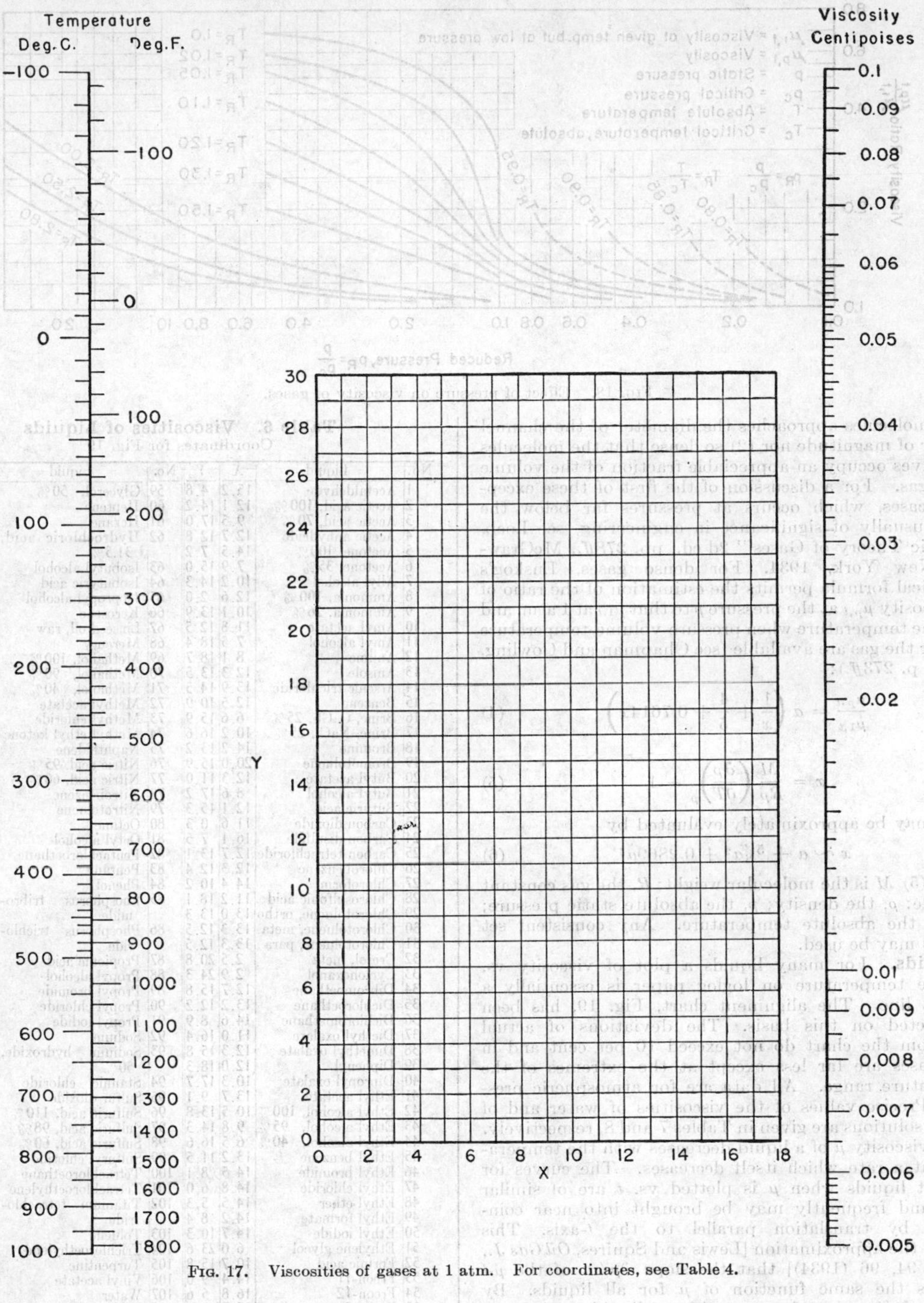

FIG. 17. Viscosities of gases at 1 atm. For coordinates, see Table 4.

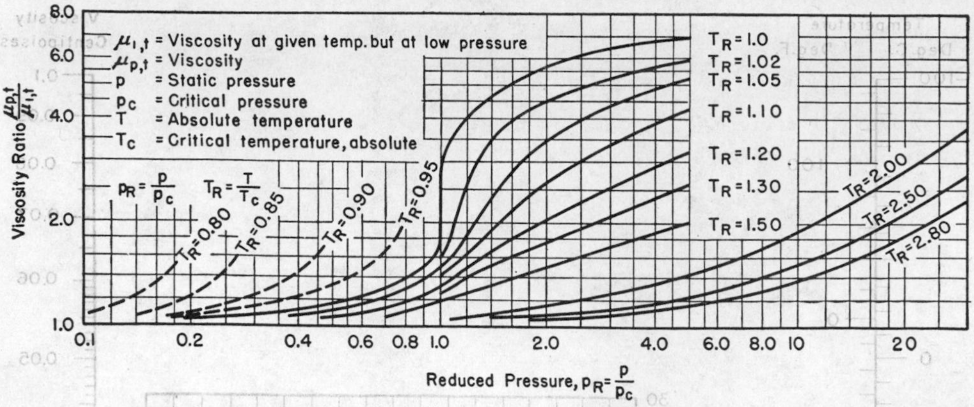

FIG. 18. Effect of pressure on viscosity of gases.

of the molecules approaches the diameter of the channel in order of magnitude nor (2) so dense that the molecules themselves occupy an appreciable fraction of the volume of the gas. For a discussion of the first of these exceptional cases, which occurs at pressures far below the ranges usually of significance in engineering, see Loeb, "Kinetic Theory of Gases," 2d ed., pp. 278*ff*., McGraw-Hill, New York, 1934. For dense gases, Enskog's theoretical formula permits the estimation of the ratio of the viscosity $\mu_{p,t}$ at the pressure p to that $\mu_{1,t}$ at 1 atm. and the same temperature when pressure-volume-temperature data for the gas are available (see Chapman and Cowling, *op. cit.*, p. 273*ff*.):

$$\frac{\mu_{p,t}}{\mu_{1,t}} = a\left(\frac{1}{x} + \frac{4}{5} + 0.7614x\right) \tag{4}$$

where

$$x = \frac{M}{R\rho}\left(\frac{\partial p}{\partial T}\right)_\rho - 1 \tag{5}$$

and a may be approximately evaluated by

$$x \sim a + \tfrac{5}{8}a^2 + 0.2869a^3 \tag{6}$$

In Eq. (5) M is the molecular weight; R, the gas constant per mole; ρ, the density; p, the absolute static pressure; and T, the absolute temperature. Any consistent set of units may be used.

Liquids. For many liquids a plot of viscosity vs. absolute temperature on loglog paper is essentially a straight line. The alignment chart, Fig. 19, has been constructed on this basis. The deviations of actual data from the chart do not exceed 10 per cent and in most cases are far less except at the extremes of the temperature range. All data are for atmospheric pressure. Precise values of the viscosities of water and of sucrose solutions are given in Tables 7 and 8, respectively.

The viscosity μ of a liquid decreases with the temperature t at a rate which itself decreases. The curves for different liquids when μ is plotted vs. t are of similar shape and frequently may be brought into near coincidence by translation parallel to the t-axis. This leads to the approximation [Lewis and Squires, *Oil Gas J.*, **33**, 92, 94, 96 (1934)] that the slope $\partial\mu/\partial t$ of the μ-t curve is the same function of μ for all liquids. By plotting $\partial t/\partial\mu$ vs. μ for a number of liquids and determining the integral of a representative curve through the points, a generalized viscosity curve is obtained which can serve as a first approximation for any liquid if the temperature scale is moved parallel to itself to a position such that some one value of the viscosity is correctly given. Such a curve is given in Fig. 20. Although substantial deviations are not uncommon,

Table 6. Viscosities of Liquids
Coordinates for Fig. 19

No.	Liquid	X	Y
1	Acetaldehyde	15.2	4.8
2	Acetic acid, 100%	12.1	14.2
3	Acetic acid, 70%	9.5	17.0
4	Acetic anhydride	12.7	12.8
5	Acetone, 100%	14.5	7.2
6	Acetone, 35%	7.9	15.0
7	Allyl alcohol	10.2	14.3
8	Ammonia, 100%	12.6	2.0
9	Ammonia, 26%	10.1	13.9
10	Amyl acetate	11.8	12.5
11	Amyl alcohol	7.5	18.4
12	Aniline	8.1	18.7
13	Anisole	12.3	13.5
14	Arsenic trichloride	13.9	14.5
15	Benzene	12.5	10.9
16	Brine, $CaCl_2$, 25%	6.6	15.9
17	Brine, NaCl, 25%	10.2	16.6
18	Bromine	14.2	13.2
19	Bromotoluene	20.0	15.9
20	Butyl acetate	12.3	11.0
21	Butyl alcohol	8.6	17.2
22	Butyric acid	12.1	15.3
23	Carbon dioxide	11.6	0.3
24	Carbon disulfide	16.1	7.5
25	Carbon tetrachloride	12.7	13.1
26	Chlorobenzene	12.3	12.4
27	Chloroform	14.4	10.2
28	Chlorosulfonic acid	11.2	18.1
29	Chlorotoluene, ortho	13.0	13.3
30	Chlorotoluene, meta	13.3	12.5
31	Chlorotoluene, para	13.3	12.5
32	Cresol, meta	2.5	20.8
33	Cyclohexanol	2.9	24.3
34	Dibromoethane	12.7	15.8
35	Dichloroethane	13.2	12.2
36	Dichloromethane	14.6	8.9
37	Diethyl oxalate	11.0	16.4
38	Dimethyl oxalate	12.3	15.8
39	Diphenyl	12.0	18.3
40	Dipropyl oxalate	10.3	17.7
41	Ethyl acetate	13.7	9.1
42	Ethyl alcohol, 100%	10.5	13.8
43	Ethyl alcohol, 95%	9.8	14.3
44	Ethyl alcohol, 40%	6.5	16.6
45	Ethyl benzene	13.2	11.5
46	Ethyl bromide	14.5	8.1
47	Ethyl chloride	14.8	6.0
48	Ethyl ether	14.5	5.3
49	Ethyl formate	14.2	8.4
50	Ethyl iodide	14.7	10.3
51	Ethylene glycol	6.0	23.6
52	Formic acid	10.7	15.8
53	Freon-11	14.4	9.0
54	Freon-12	16.8	5.6
55	Freon-21	15.7	7.5
56	Freon-22	17.2	4.7
57	Freon-113	12.5	11.4
58	Glycerol, 100%	2.0	30.0
59	Glycerol, 50%	6.9	19.6
60	Heptene	14.1	8.4
61	Hexane	14.7	7.0
62	Hydrochloric acid, 31.5%	13.0	16.6
63	Isobutyl alcohol	7.1	18.0
64	Isobutyric acid	12.2	14.4
65	Isopropyl alcohol	8.2	16.0
66	Kerosene	10.2	16.9
67	Linseed oil, raw	7.5	27.2
68	Mercury	18.4	16.4
69	Methanol, 100%	12.4	10.5
70	Methanol, 90%	12.3	11.8
71	Methanol, 40%	7.8	15.5
72	Methyl acetate	14.2	8.2
73	Methyl chloride	15.0	3.8
74	Methyl ethyl ketone	13.9	8.6
75	Naphthalene	7.9	18.1
76	Nitric acid, 95%	12.8	13.8
77	Nitric acid, 60%	10.8	17.0
78	Nitrobenzene	10.6	16.2
79	Nitrotoluene	11.0	17.0
80	Octane	13.7	10.0
81	Octyl alcohol	6.6	21.1
82	Pentachloroethane	10.9	17.3
83	Pentane	14.9	5.2
84	Phenol	6.9	20.8
85	Phosphorus tribromide	13.8	16.7
86	Phosphorus trichloride	16.2	10.9
87	Propionic acid	12.8	13.8
88	Propyl alcohol	9.1	16.5
89	Propyl bromide	14.5	9.6
90	Propyl chloride	14.4	7.5
91	Propyl iodide	14.1	11.6
92	Sodium	16.4	13.9
93	Sodium hydroxide, 50%	3.2	25.8
94	Stannic chloride	13.5	12.8
95	Sulfur dioxide	15.2	7.1
96	Sulfuric acid, 110%	7.2	27.4
97	Sulfuric acid, 98%	7.0	24.8
98	Sulfuric acid, 60%	10.2	21.3
99	Sulfuryl chloride	15.2	12.4
100	Tetrachloroethane	11.9	15.7
101	Tetrachloroethylene	14.2	12.7
102	Titanium tetrachloride	14.4	12.3
103	Toluene	13.7	10.4
104	Trichloroethylene	14.8	10.5
105	Turpentine	11.5	14.9
106	Vinyl acetate	14.0	8.8
107	Water	10.2	13.0
108	Xylene, ortho	13.5	12.1
109	Xylene, meta	13.9	10.6
110	Xylene, para	13.9	10.9

the curve is very valuable in the frequent circumstance that μ is known for only one temperature. By devising a special scale for the μ-axis, the plot in Fig. 20 can readily be made a straight line. That Dühring's rule should be

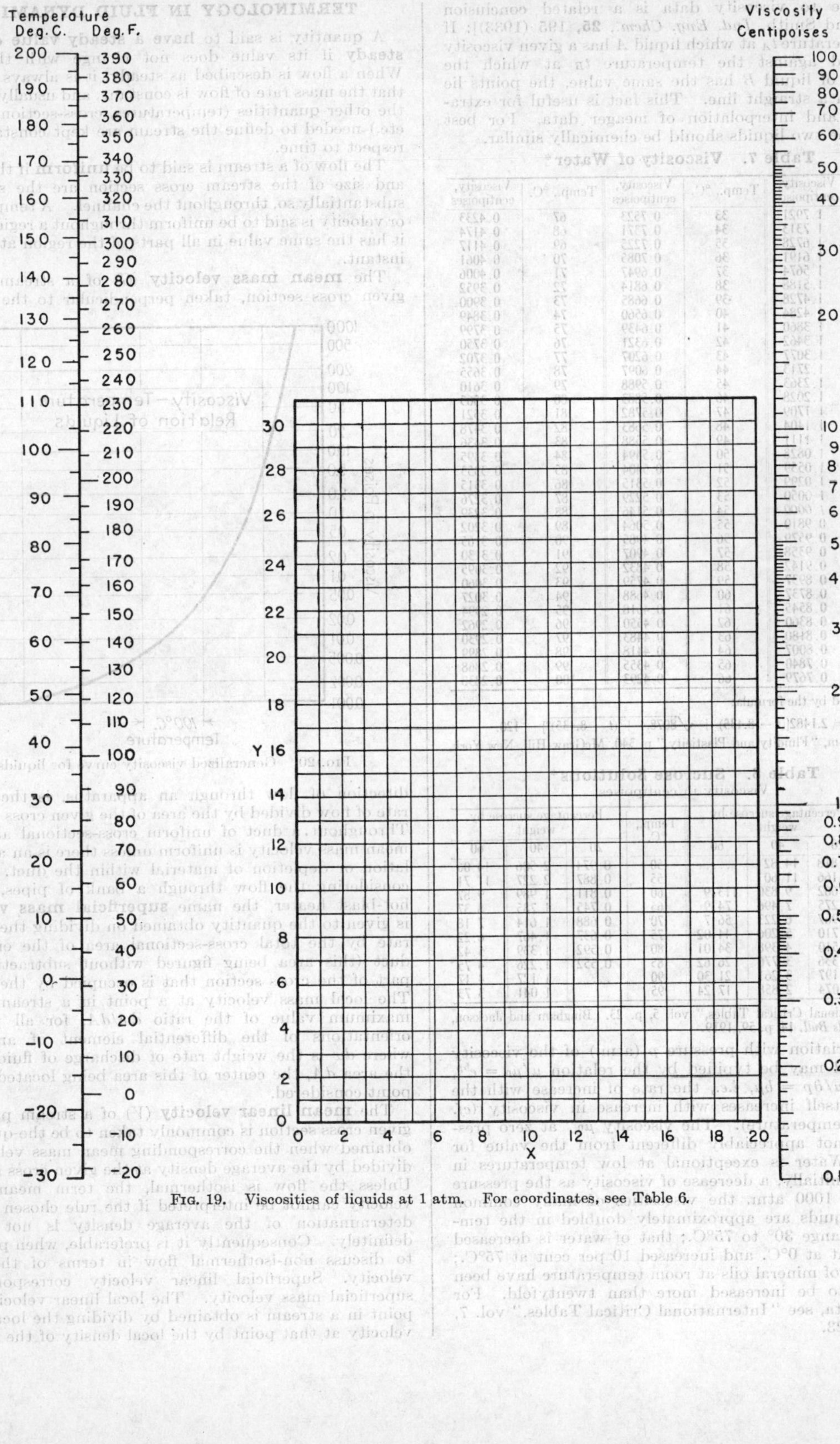

FIG. 19. Viscosities of liquids at 1 atm. For coordinates, see Table 6.

applicable to viscosity data is a related conclusion [Perry and Smith, *Ind. Eng. Chem.*, **25**, 195 (1933)]: If the temperature t_A at which liquid A has a given viscosity is plotted against the temperature t_B at which the viscosity of liquid B has the same value, the points lie nearly on a straight line. This fact is useful for extrapolation and interpolation of meager data. For best results the two liquids should be chemically similar.

Table 7. Viscosity of Water*

Temp., °C.	Viscosity, centipoises	Temp., °C.	Viscosity, centipoises	Temp., °C.	Viscosity, centipoises
0	1.7921	33	0.7523	67	0.4233
1	1.7313	34	0.7371	68	0.4174
2	1.6728	35	0.7225	69	0.4117
3	1.6191	36	0.7085	70	0.4061
4	1.5674	37	0.6947	71	0.4006
5	1.5188	38	0.6814	72	0.3952
6	1.4728	39	0.6685	73	0.3900
7	1.4284	40	0.6560	74	0.3849
8	1.3860	41	0.6439	75	0.3799
9	1.3462	42	0.6321	76	0.3750
10	1.3077	43	0.6207	77	0.3702
11	1.2713	44	0.6097	78	0.3655
12	1.2363	45	0.5988	79	0.3610
13	1.2028	46	0.5883	80	0.3565
14	1.1709	47	0.5782	81	0.3521
15	1.1404	48	0.5683	82	0.3478
16	1.1111	49	0.5588	83	0.3436
17	1.0828	50	0.5494	84	0.3395
18	1.0559	51	0.5404	85	0.3355
19	1.0299	52	0.5315	86	0.3315
20	1.0050	53	0.5229	87	0.3276
20.20	*1.0000*	54	0.5146	88	0.3239
21	0.9810	55	0.5064	89	0.3202
22	0.9579	56	0.4985	90	0.3165
23	0.9358	57	0.4907	91	0.3130
24	0.9142	58	0.4832	92	0.3095
25	0.8937	59	0.4759	93	0.3060
26	0.8737	60	0.4688	94	0.3027
27	0.8545	61	0.4618	95	0.2994
28	0.8360	62	0.4550	96	0.2962
29	0.8180	63	0.4483	97	0.2930
30	0.8007	64	0.4418	98	0.2899
31	0.7840	65	0.4355	99	0.2868
32	0.7679	66	0.4293	100	0.2838

* Calculated by the formula:

$$1/\mu = 2.1482[(t - 8.435) + \sqrt{8078.4 + (t - 8.435)^2}] - 120.$$

From Bingham, "Fluidity and Plasticity," p. 340, McGraw-Hill, New York, 1922.

Table 8. Sucrose Solutions*

Viscosity in centipoises

Temp., °C.	Percentage sucrose by weight			Temp., °C.	Percentage sucrose by weight		
	20	40	60		20	40	60
0	3.818	14.82	......	50	0.974	2.506	14.06
5	3.166	11.60	......	55	0.887	2.227	11.71
10	2.662	9.830	113.9	60	0.811	1.989	9.87
15	2.275	7.496	74.9	65	0.745	1.785	8.37
20	1.967	6.223	56.7	70	0.688	1.614	7.18
25	1.710	5.206	44.02	75	0.637	1.467	6.22
30	1.510	4.398	34.01	80	0.592	1.339	5.42
35	1.336	3.776	26.62	85	0.552	1.226	4.75
40	1.197	3.261	21.30	90	.....	1.127	4.17
45	1.074	2.858	17.24	95	.....	1.041	3.73

* "International Critical Tables," vol. 5, p. 23. Bingham and Jackson, *Bur. Standards Bull.* 14, p. 59, 1919.

The variation with pressure p (atm.) of the viscosity of a liquid may be typified by the relation $\mu/\mu_0 = e^{bp}$, whence $\partial\mu/\partial p = b\mu$, *i.e.*, the rate of increase with the pressure itself increases with increase in viscosity (*cf.* effect of temperature). The viscosity μ_0 "at zero pressure" is not appreciably different from the value for 1 atm. Water is exceptional at low temperatures in showing, initially, a decrease of viscosity as the pressure rises. At 1000 atm. the viscosities of many common organic liquids are approximately doubled in the temperature range 30° to 75°C.; that of water is decreased 10 per cent at 0°C. and increased 10 per cent at 75°C.; and those of mineral oils at room temperature have been reported to be increased more than twentyfold. For further data, see "International Critical Tables," vol. 7, pp. 222–223.

TERMINOLOGY IN FLUID DYNAMICS

A quantity is said to **have a steady value** or to **be steady** if its value does not change with the time. When a flow is described as steady, it is always inferred that the mass rate of flow is constant; and usually that all the other quantities (temperatures, cross-sectional areas, etc.) needed to define the stream are kept constant with respect to time.

The flow of a stream is said to be **uniform** if the shape and size of the stream cross section are the same, or substantially so, throughout the channel. A temperature or velocity is said to be uniform throughout a region when it has the same value in all parts of the region at a given instant.

The **mean mass velocity** (G) of a stream past a given cross section, taken perpendicular to the general

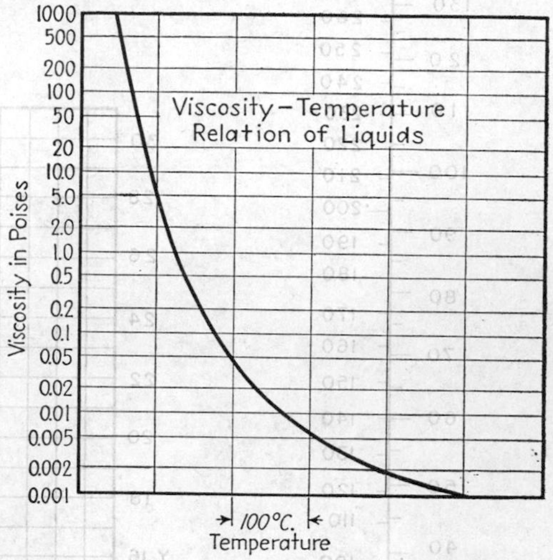

FIG. 20. Generalized viscosity curve for liquids.

direction of flow through an apparatus, is the weight rate of flow divided by the area of the given cross section. Throughout a duct of uniform cross-sectional area, the mean mass velocity is uniform unless there is an accumulation or depletion of material within the duct. When considering the flow through a bank of pipes, *e.g.*, a hot-blast heater, the name **superficial mass velocity** is given to the quantity obtained on dividing the weight rate by the total cross-sectional area of the enclosing duct (this area being figured without subtracting the part of the cross section that is occupied by the pipes). The local mass velocity at a point in a stream is the maximum value of the ratio dw/dA, for all possible orientations of the differential element of area dA, where dw is the weight rate of discharge of fluid across the area dA, the center of this area being located at the point considered.

The **mean linear velocity** (V) of a stream past any given cross section is commonly taken to be the quantity obtained when the corresponding mean mass velocity is divided by the average density at the given cross section. Unless the flow is isothermal, the term mean linear velocity cannot be interpreted if the rule chosen for the determination of the average density is not stated definitely. Consequently it is preferable, when possible, to discuss non-isothermal flow in terms of the mass velocity. Superficial linear velocity corresponds to superficial mass velocity. The local linear velocity at a point in a stream is obtained by dividing the local mass velocity at that point by the local density of the fluid.

The **acoustic velocity**, the velocity of sound, in a fluid of large extent or contained in a rigid walled vessel is given by the formulas

$$V_a = \sqrt{g_c \left(\frac{\partial p}{\partial \rho}\right)_S} = \sqrt{g_c \kappa \left(\frac{\partial p}{\partial \rho}\right)_T} = \sqrt{\frac{K g_c}{\rho}} \text{ ft./sec.}$$

where p is the pressure, lb. *force*/sq. ft.; ρ is the density, lb. *mass*/cu. ft.; K is the bulk modulus of elasticity, lb. *force*/sq. ft.; g_c = 32.1740 (lb. *mass*) (ft.)/(lb. *force*) (sec.^{2}); and $\kappa = c_p/c_v$, the ratio of the specific heats at constant pressure and volume, respectively. The subscript S denotes constancy of entropy; the subscript T, constancy of temperature. For perfect gases $(\partial p/\partial \rho)_T = p/\rho = RT/M$, where T is the absolute temperature (°R. = °F. + 459.5), R = 1546 (ft.-lb. *force*)/(°R.) (mole), and M is the molecular weight. Consequently, for a perfect gas $V_a = \sqrt{g_c \kappa RT/M}$.

For definition of the thermodynamic terms—internal energy, enthalpy, total heat, entropy, etc.—consult any textbook on thermodynamics (*e.g.*, Kiefer and Stuart, "Principles of Engineering Thermodynamics," Wiley, New York, 1930; Keenan, "Thermodynamics," Wiley, New York, 1941; Weber, "Thermodynamics for Chemical Engineers," Wiley, New York, 1939; Dodge "Chemical Engineering Thermodynamics," McGraw-Hill, New York, 1944.)

A **Reynolds number,** N_{Re}, is any of several dimensionless quantities, of the form $LV\rho/\mu$, that occur in the theory of fluid motion. Here L is a characteristic linear dimension of the apparatus through which the flow is taking place; V is the linear velocity; ρ is the density; and μ is the absolute viscosity—all either at some particular place in the apparatus or at some sort of average location therein. In discussions of flow through ducts or open channels, the term "Reynolds number," when used without special qualification, generally designates the quantity $N_{Re} = 4w/L_p\mu$, where w is the weight rate of flow, L_p is the wetted perimeter of a cross section of the passage, and μ is the viscosity of the fluid as it passes that cross section. For **circular pipes** running full, $L_p = \pi D$, where D is the diameter; consequently, $N_{Re} = 4w/\pi D\mu = DG/\mu = DV\rho/\mu$. For streams of **non-circular cross section,** the corresponding result is $N_{Re} = 4w/L_p\mu = 4R_H G/\mu$ where R_H is the hydraulic radius. In computing a Reynolds number, consistent units must be used; *e.g.*, L_p, R_H, and D in ft., G in lb. *mass*/(sq. ft.)(sec.), μ in lb. *mass*/(ft.)(sec.), w in lb. *mass*/sec.; or L_p, R_H, and D in cm., G in g./(sq. cm.)(sec.), μ in g./(cm.)(sec.), w in g./sec.

The **mean hydraulic radius** R_H of a channel is the cross-sectional area of that part of the channel that is filled with fluid divided by the length of the wetted perimeter. The hydraulic radius of a circular pipe is one-fourth of the diameter; hence for a non-circular duct the **equivalent diameter** is said to be four times the hydraulic radius. For formulas, see Table 9, p. 378.

Streamlines. A **line of motion or streamline** is an imaginary line drawn from point to point within a stream, so that its direction is everywhere that of the motion of the fluid at a given instant. In general, the course of the line of motion which passes through any given fixed point varies with time. However, if the motion is steady *at every point* in the stream (or in a continuous portion of the stream), the lines of motion are obviously fixed in space and represent the actual paths of the fluid "particles"; *i.e.*, the particles located on a given line of motion slide along that line like rings along a rigid rod. If this state of motion exists permanently throughout a stream, the fluid is said to move in steady **streamline flow.** The streamlines are not necessarily straight lines. In general, at sufficiently low velocities real fluids do move in streamline flow. This type of motion is also called **laminar flow** and **viscous flow.**

A **filament line** or **streak line** is a line drawn from a point in a stream through the instantaneous positions of the successive particles which have previously occupied that point. Such a line is recorded by a snapshot when coloring matter, for example, is continuously introduced at a fixed point.

Particle paths are paths of particular portions of the fluid; they are recorded as short streaks on the film when a time exposure is made of a stream containing a number of solid particles. A filament line is the locus of the present termini of the paths of particles which once passed through a given point.

In steady flow, streamlines, filament lines, and particle paths are identical.

Turbulent Flow. Critical Velocity. In a given case, if the mean linear velocity of a fluid exceeds a value known as the critical velocity, the motion is generally found not to be streamline throughout the stream. On the contrary, either locally or everywhere in the moving fluid, the particles are found to pursue erratic and continually varying courses. The flow is then said to be turbulent. It is possible to have turbulent flow in some portions of a stream which streamline flow exists in other parts. For further discussion of the critical velocity see p. 383. For a discussion of the modern theory of turbulence, of means of measuring the intensity of turbulence, etc., see Goldstein "Modern Developments in Fluid Dynamics," Oxford, New York, 1938.

BERNOULLI'S THEOREM*

The name Bernoulli's theorem is commonly applied to any of several forms in which the law of conservation of energy may appear when fluid motion is considered. If the terms W_0 and F are zero, Eq. (11), p. 376, is correctly so designated.

Total Energy Balance. Consider a unit weight (1 lb. *mass*) of fluid, and let

G = mass velocity, lb. *mass*/(sec.) (sq. ft. of cross section).
g = local acceleration due to gravity, ft./sec.^{2}
g_c = 32.1740 (lb. *mass*) (ft.)/(lb. *force*) (sec.^{2}), dimensional constant in Newton's law, force = mass × acceleration/g_c.
J = the mechanical equivalent of heat, 778 ft.-lb. *force*/B.t.u.
p = absolute static pressure, lb. *force*/sq. ft.
u = internal (or intrinsic) energy, B.t.u./lb. *mass*.
V = linear velocity, ft./sec.
v = specific volume, cu. ft./lb. *mass*.
Z = height above any arbitrary horizontal datum plane, ft.

Then the potential energy relative to the chosen reference level is Zg/g_c ft.-lb. *force*/lb. *mass*, and the kinetic energy is $V^2/2g_c$ ft.-lb. *force*/lb. *mass*, so that the total energy of the pound of fluid is $(Ju + Zg/g_c + V^2/2g_c)$ ft.-lb. *force*/lb. *mass*. The numerical value of g rarely differs from 32.1740 by as much as 3 parts per 1000, a deviation considerably less than the probable error of most engineering data for fluid flow. Consequently, the ratio g/g_c is regarded as unity in this section unless otherwise noted.

Suppose a fluid to be flowing steadily through an apparatus which itself is in no way altered by the flow, although the fluid in its passage may be changed chemically or otherwise. Let there be no accumulation or depletion of either energy or matter within the apparatus, and let all the conditions at the entrance and outlet be steady. Now, the total energy of a sample of matter can be altered only by having external work done on,

* REFERENCES: Stodola, *op. cit.;* Berry, *Mech. Eng.*, **51,** 816 (1929); Weber, "Thermodynamics for Chemical Engineers," Chaps. II and XI, Wiley, New York, 1939; Keenan, "Thermodynamics," Wiley, New York, 1941.

or by, the sample, or by permitting heat to flow into or out of the sample. Therefore, if the subscripts (1) and (2) indicate the conditions at the inlet and outlet respectively,

$$\left(Ju_2 + Z_2 + \frac{V_2^2}{2g_c}\right) - \left(Ju_1 + Z_1 + \frac{V_1^2}{2g_c}\right) = JQ + W$$

where Q is the heat received, B.t.u./lb. *mass*, from sources **external to the apparatus**, and W is the net external work (ft.-lb. *force*/lb. *mass*) **done on** the pound of fluid while in the apparatus.

A part of the work W is done **on** the pound of fluid as it is being pushed past the entrance by the fluid behind it; this amounts to p_1v_1. Similarly, on passing the outlet, an amount of work p_2v_2 is done **by** the pound of fluid on the fluid just ahead of it. Hence W may be replaced by $W = (p_1v_1 - p_2v_2) + W_0$, where W_0 is the work, ft.-lb. *force*/lb. *mass*, **delivered by outside machinery** to the average pound of fluid during its passage. The resulting expression of the first law of thermodynamics is often known as the **over-all energy balance** form of Bernoulli's theorem:

$$Ju_1 + Z_1 + p_1v_1 + \frac{V_1^2}{2g_c} + JQ + W_0 = Ju_2 + Z_2 + p_2v_2 + \frac{V_2^2}{2g_c} \quad (7)$$

No friction term appears in this form of the theorem. For adiabatic processes, $Q = 0$. If there is no pump, turbine, or similar device between cross sections 1 and 2, $W_0 = 0$. **For gases,** except when under high pressure and consequently of high density, the difference between the terms Z_1 and Z_2 is usually negligible compared with the other terms.

Equation (7) in differential form is

$$dZ + J\,du + p\,dv + v\,dp + \frac{V\,dV}{g_c} = J\,\delta Q + \delta W_0 \quad (8)$$

NOTE. In the practical application of the above equations and most of those derived from them, it is customary to give Z the value obtaining at the center of gravity of the cross section of the stream considered; to determine u and v from the reading of a thermometer (corrected for radiation errors, etc.) inserted in the fluid, together with a measurement of the static pressure; and to take $V = vw/A$, where w is the weight rate of discharge, lb. *mass*/sec., and A is the cross-sectional area, sq. ft., at the point in question. In the more usual cases the values thus found for u, v, and Z are in theory substantially the correct averages. The individual kinetic energy terms are theoretically incorrect if V is computed in the customary way unless the velocity is uniform over the cross section. In the case of viscous flow (p. 375) in a circular pipe, if $V = vw/A$, the true average kinetic energy per pound is easily shown to be V^2/g_c not $V^2/2g_c$. In turbulent flow, neglecting the kinetic energy of turbulence, it is estimated from typical velocity distributions that $V^2/2g_c$ is 3 to 8 per cent too low for flow in circular pipes. For a discussion which includes the effects of turbulence, see van Driest, *J. Applied Mechanics*, **13**, A-231, 1946. It frequently occurs that abnormal orifice coefficients and friction terms can be traced to a failure of the conventional method in evaluating the various terms of the above formulas. (*Cf.* Johansen, Aeronautical Research Committee, Great Britain, *Reports and Memoranda*, No. 1252, p. 15, June, 1929; *A.S.M.E. Fluid Meter Report*, Part I, 4th ed., 1937.)

If $i = u + (pv/J)$ (B.t.u./lb. *mass*) is the enthalpy (= total heat or heat content), Eq. (7) takes a form convenient for use with steam and other fluids for which the thermal properties are tabulated or calculable.

$$(Z_1 - Z_2) + J(i_1 - i_2) + JQ + W_0 = \frac{V_2^2 - V_1^2}{2g_c} \quad (9)$$

To calculate the heat effects in steady flow processes, Eq. (9) is solved for Q; in most such cases the terms $(Z_1 - Z_2)$ and $\dfrac{V_2^2 - V_1^2}{2g_c}$ are negligible. For **adiabatic flow of gases** in pipe lines of uniform cross section, $Q = W_0 = 0$ and $(Z_1 - Z_2)$ is usually negligible, so that Eq. (9) reduces to

$$\frac{G^2}{2g_c}(v_2^2 - v_1^2) = J(i_1 - i_2) \quad (9a)$$

which, since i is a function of p and v, gives the relationship between p and v throughout the pipe. The result differs substantially from that for constant entropy (*e.g.*, pv^k = const. in the case of a perfect gas); it is given for perfect gases by Eq. (19), p. 379. A **Fanno line** is a graph of Eq. (9a) drawn on a temperature-entropy diagram. For perfect gases (*i.e.*, if $pv = T \times$ const.),

$$J(i_1 - i_2) = J\int_{T_2}^{T_1} c_p\,dT = Jc_{p_{\text{avg}}}(T_1 - T_2) = \frac{\kappa}{\kappa - 1}(p_1v_1 - p_2v_2)$$

where $\kappa = c_p/c_v$, the ratio of the specific heats at constant pressure and volume, respectively, and T is the absolute temperature (°R. = °F. + 459.5).

Example 1. Dry saturated steam at 212°F. enters a laboratory superheater at a velocity of 100 ft./sec. Subsequently the steam is bled into a large duct where the velocity is negligible and pressure is 1 lb. *force*/sq. in. abs. If the temperature of the low-pressure steam is required to be 400°F., how much heat must be supplied by the superheater to each pound *mass* of steam?

Solution: Use Eq. (9). Assume $(Z_1 - Z_2)$ negligible. $W_0 = 0$, since there is no pump or similar device.

$$Q = \frac{V_2^2 - V_1^2}{2g_cJ} + (i_2 - i_1)$$

From pp. 277*ff*. i_1 = 1150.4 B.t.u. for dry saturated steam at 212°F. i_2 = 1240.6 B.t.u./lb. *mass* for steam at 1 lb. *force*/sq. in. abs. and 400°F.

$$Q = \frac{0^2 - (100)^2}{2 \times 32.17 \times 778} + (1240.6 - 1150.4) = -0.2 + 90.5 = 90.3 \text{ B.t.u./lb. } mass$$

Here, as is usual in heating and cooling moving gases, the kinetic energy effects are nearly negligible.

Mechanical Energy Balance. The change in internal energy $J\,du$ of the pound of fluid may, on thermodynamic grounds, be expressed either in terms of p, v, the specific entropy s, and the absolute temperature T as $JT\,ds - p\,dv$ or as $J\,\delta Q - \delta W$, where δW is the work done by the fluid on itself and its surroundings. Because there is friction in real flow, any path of states through which the fluid is presumed actually to pass must be irreversible. Consequently $JT\,ds > J\,\delta Q$ and $p\,dv > \delta W$. The discrepancy δF is the same in each case and is commonly described as the mechanical energy lost by frictional conversion into heat. By introducing δF, the second expression for $J\,du$ may be written $J\,\delta Q - p\,dv + \delta F$, which, if used in Eq. (8), leads to

$$dZ + v\,dp + \frac{V\,dV}{g_c} = \delta W_0 - \delta F \quad (10)$$

As before, all the terms are expressed in ft.-lb. *force*/lb. *mass*. The integral of this equation is the so-called mechanical energy balance form of Bernoulli's theorem:

$$Z_1 + \frac{V_1^2}{2g_c} + W_0 - \int_{p_1,v_1}^{p_2,v_2} v\,dp - F = Z_2 + \frac{V_2^2}{2g_c} \quad (11)^*$$

Generally v continually increases as p decreases, along a curve of moderate curvature. Hence it is evident from geometry that $(p_1 - p_2)v_{\text{avg}}$ is often a good approximation to $-\int_1^2 v\,dp$ if v_{avg} is the arithmetic mean of

* For flow in pipes, etc., special integrals of Eq. (10), which are more easy to evaluate accurately than is Eq. (11), are given on pp. 378*ff*.

the terminal specific volumes. For isothermal flow of liquids, which are practically incompressible, this simplification is very precise. In the case of isothermal flow of gases in pipes the error is not greater than the probable error of the friction data if the absolute pressure ratio, p_1/p_2, is less than 2. Thus in these and in many other practical cases Eq. (11) may be sufficiently precise in the form

$$Z_1 + \frac{V_1^2}{2g_c} + p_1 v_{avg} + W_0 = Z_2 + \frac{V_2^2}{2g_c} + p_2 v_{avg} + F \quad (12)$$

in which p_1 and p_2 may be taken as gage pressures, lb./sq. ft., if desired, rather than absolute pressures.

The expression of Eq. (12) in heads is common in hydraulics and in treatments of low-pressure gas flow (*e.g.*, as in ventilating ducts). The **velocity head,** $H_V = V^2/2g$ (ft.), is the height at which 1 lb. *mass* has a gravitational potential energy equal to $V^2/2g_c$; the **static** or **pressure head** H_p (p. 360), if expressed in terms of a column of fluid of density $1/v_{avg}$, is $H_p = pv_{avg}g_c/g$ (ft.); and, as noted in deriving Eq. (7), the Z's in Eqs. (7) to (12) are properly Zg/g_c. Therefore, if Eq. (12) is multiplied throughout by g_c/g, one obtains

$$Z_1 + H_{V_1} + H_{p_1} + \frac{W_0 g_c}{g} = Z_2 + H_{V_2} + H_{p_2} + \frac{F g_c}{g} \quad (13)$$

The factor g_c/g is, as before, customarily omitted because it is so nearly unity. Fg_c/g is called the **friction head** and is expressed in feet of a column of the fluid, as are all the terms in Eq. (13).

Example 2. A closed tank kept partly filled with oil (sp. gr. = 0.9) has a pressure in the gas space above the liquid of 10 lb. *force*/sq. in. gage. If the oil is discharging through a hose at a rate of 40 gal./min., estimate the static pressure before the nozzle which is located 12 ft. lower than the oil surface in the tank and has a 1-in. (i.d.) cylindrical entrance. Assume the total friction in the line amounts to 1 "ft. of oil" (*i.e.*, 1 ft.-lb. *force*/lb. *mass* of oil passing).

Solution. Use Eq. (11). Take section 1 and also the datum plane $Z = 0$ to be the oil surface in the tank. Let section 2 be at the nozzle entrance. Substituting the data

$$Z_1 + \frac{V_1^2}{2g_c} + W_0 - \int v\,dp - F = Z_2 + \frac{V_2^2}{2g_c}$$

$$0 + 0 + 0 + \frac{10 \times 144 - p_2}{0.9 \times 62.3} - 1 = -12 + \frac{\left[\frac{40}{7.48} \times \frac{1}{60} \times \frac{4 \times 144}{(1)^2\pi}\right]^2}{2 \times 32.17}$$

$$p_2 = 0.9 \times 62.3 \times 32.5 = 1822.3 \text{ lb. } force\text{/sq. ft. (gage)}$$

$$p_2 = \frac{0.9 \times 62.3 \times 32.5}{144} = 12.7 \text{ lb. } force\text{/sq. in. (gage)}$$

FLOW IN PIPES AND CHANNELS

General Formulas and Methods. The problem of finding one of the three quantities, rate of discharge, size of channel, pressure or head loss, when the other two are given is solved by substituting the data of the problem in an appropriate form of the mechanical energy balance (p. 376) after the term F or δF, the frictional loss of mechanical energy, has been evaluated. The part of F which arises from friction within the channel proper is considered below. The part due to fittings, bends, etc., which often constitutes a major part of the friction, is discussed on pp. 387 to 390. Serious errors are often made by overlooking the fact that most "flow formulas" merely give the value of F and make no allowance for the other terms in Bernoulli's theorem.

The **Fanning, or Darcy, equation** for steady flow in uniform circular pipes running full of liquid under isothermal conditions

$$F = \frac{4fLV^2}{2g_cD} = \frac{4fLG^2}{2g_cD\rho^2} = \frac{4fLH_V}{D} = \frac{32fLw^2}{\pi^2\rho^2g_cD^5} = \frac{32fLq^2}{\pi^2g_cD^5} \quad (14)$$

gives the friction loss F in ft.-lb. *force*/lb. *mass* of fluid passing (or ft. of fluid flowing), if L = duct length, ft.; D = diameter, ft.; ρ = density, lb. *mass*/cu. ft.; V = linear velocity, ft./sec.; g_c = conversion factor = 32.1740 (lb. *mass*)(ft.)/(lb. *force*)(sec.^{2}); G = mass velocity, lb. *mass*/(sec.) (sq. ft.); w = weight rate of discharge, lb. *mass*/sec.; q = volumetric discharge, cu. ft./sec.; H_V = velocity head, ft. of fluid flowing; and f = the Fanning *friction factor* which varies with $DV\rho/\mu$ (see pp. 381–383 and Fig. 23). This formula and the variant of it known as the **Chézy formula,**

$$V = C\sqrt{R_HF/L} \quad (14a)$$

(R_H = hydraulic radius, ft.; $C = \sqrt{2g_c/f}$) are widely used. The Chézy formula (14a) and the first three forms of the Fanning equation (14) are **applicable to ducts and open channels of any cross-sectional shape** whatever if D, wherever it occurs, is replaced by ($4R_H$) (see Table 9). The pressure drop due to friction is $\Delta p = F\rho$ (lb. *force*/sq. ft.) if ρ is the constant weight density of the liquid flowing. In general, Eq. (14) is applicable for the evaluation of F in Eqs. (12) and (13) whenever the variation in the density of the fluid is small, although it should be noted that if, in Eqs. (12) and (13) v_{avg} **is taken as** $(v_1 + v_2)/2$ the allowance for "stack effect" will be inexact [*cf.* Eq. (22a), p. 381]. More exact formulas for gases are given on pp. 378*ff*.

A rapid method of solving Eq. (14) is to use the **alignment chart** (Fig. 21) which was constructed on the assumption that $f = 0.04/N_{Re}^{0.16}$ where $N_{Re} = DG/\mu = 4R_HG/\mu$ is the Reynolds number. If, for the value of N_{Re} obtaining, some other value of f, say f', is preferred, the quantity sought, as given by the chart, should be multiplied by the following factors:

Quantity sought	$\frac{\Delta p}{L}$ or $\frac{\Delta H}{L}$	D_i	w or G
Factor	f'/f	$(f'/f)^{1/5}$	$\sqrt{f/f'}$

The **Williams and Hazen formula,** $V = C'R_H^{0.63}(F/L)^{0.54}0.001^{-0.04}$, is widely used to compute the flow of **water.** It is not so convenient for fluids in general as are formulas of the Fanning type.

$$C' = (\sqrt{2g_c/f})/[1.319R_H^{0.13}(F/L)^{0.04}]$$

Correlation of Friction Data. When passages of geometrically similar cross section are considered, it is readily shown by dimensional analysis [Rayleigh, *Phil. Mag.* (5), **34,** 59 (1892)] that for a fluid of absolute viscosity μ and density ρ, moving with the mean linear velocity V, in a duct of cross section proportional to D^2, the loss dF/dx of mechanical energy [*cf.* Bernoulli's theorem, Eq. (11), p. 376] per pound *mass* of fluid flowing per foot of pipe is expressible by an equation of the type

$$\frac{Dg_c}{V^2}\left(\frac{dF}{dx}\right) = 2f = \text{a function of } \left(\frac{DV\rho}{\mu}\right)$$

where g_c is a conversion factor = 32.1740 (lb. *mass*)(ft.)/(lb. *force*) (sec.)^2. The relation has been extensively tested. It is found to hold quite accurately for sections of a given duct insofar as variations in V, ρ, and μ, are concerned, provided that the region considered is sufficiently far from the inlet. If, however, there are considered ducts supposedly geometrically similar but of different cross-sectional areas, or even of the same cross-sectional area but of different materials of construction, the relation breaks down. The discrepancies have been universally attributed to the impossibility in practice of attaining geometrical similarity as regards the roughness of the wetted surfaces. There is confirmation for this point of view in the fact that large pipe frequently shows a lesser friction loss than that predicted from tests on smaller pipe made in the same manner from identical materials. In such cases the absolute roughness is the same, but its relative effect becomes less as the diameter becomes greater (see p. 381). Since it has, so far, proved impossible to devise any satisfactory measure of roughness and since the nature of the wetted surface changes during service on account of corrosion, etc., no friction formula can be entirely reliable. It is probably inadvisable to expect an accuracy of better than ±5 per cent for any formula. Precision as great as this can be had only when the wetted surfaces are such that they can be reproduced satisfactorily (*e.g.*, drawn tubing) and when appreciable corrosion or fouling is known to be absent.

In the case of gases, the drop in pressure as one proceeds along a passage occasions a decrease in density, and hence an increase in linear velocity. The foregoing formulas, therefore, will apply only to a differential length of duct, dx, throughout which the density may be considered constant. **In non-uniform ducts** a similar situation arises as a consequence of variation in D or R_H. Thus, in general, Fanning's equation must be used in differential form:

$$\frac{dF}{dx} = \frac{4fV^2}{2g_cD} = \frac{fV^2}{2g_cR_H} = \frac{fG^2}{2g_c\rho^2R_H} \qquad (14b)$$

The substitution of Eq. (14*b*) in the differential mechanical energy balance Eq. (10), p. 376 gives

$$v\,dp + \frac{V\,dV}{g_c} = -\left(\frac{fV^2}{2g_cR_H} + K\right)dx \qquad (15)$$

where $v = 1/\rho$ is the specific volume, cu. ft./lb. *mass*, of the fluid; $K\,dx = dZ$ is the vertical distance, ft., through which the fluid is raised when it moves the distance dx along the pipe, and δW_0 in Eq. (10) has been taken to be zero on the supposition that no pump is in the line. The other symbols are as in Eq. (14). In a uniform duct the mass velocity $G = V/v$ is constant so that, if p is known as a function of v only, $v\,dp$ can be written $\phi(V)\,dV$. Then, when the K is constant, the variables in Eq. (15) are separable, and exact integration is possible although graphical procedures may be needed.

Table 9. Values of Hydraulic Radius R_H for Various Cross Sections

$$R_H = \frac{\text{area of stream cross section}}{\text{wetted perimeter}}; \text{“equivalent diameter”} = 4R_H$$

Shape of Cross Section	R_H
Pipes and ducts, running full:	
Circle, diam. = D	$\frac{D}{4}$
Annulus, inner diam. = d, outer diam. = D	$\frac{(D-d)}{4}$
Square, side = D	$\frac{D}{4}$
Rectangle, sides a, b	$\frac{ab}{2(a+b)}$
Ellipse, major axis = $2a$, minor axis = $2b$	$\frac{ab}{K(a+b)}$*
Open channels or partly filled ducts:	
Rectangle, depth = y, width = b	$\frac{by}{b+2y}$
Semicircle, free surface on a diam. D	$\frac{D}{4}$
Wide shallow stream on flat plate, depth = y	y
Triangular trough, ∠ = 90°, bisector vertical, depth = y, slant depth = d	$\frac{d}{4} = \frac{y}{2\sqrt{2}}$
Trapezoid (depth = y, bottom width = b): Side slope 60° from horizontal	$y\left(\frac{b + y/\sqrt{3}}{b + 4y/\sqrt{3}}\right)$
Side slope 45°	$\frac{yb + y^2}{b + 2\sqrt{2}y}$
Film (thickness = t) on wall of vertical wetted wall tower of diameter = D	$t - t^2/D = t$ (approx.)

* Values of K. If $S = (a - b)/(a + b)$,

S =	0.2	0.3	0.4	0.5	0.6	0.7	0.8	0.9	1.0
K =	1.010	1.023	1.040	1.064	1.092	1.127	1.168	1.216	1.273

Directions for Use of Pipe Flow Chart, Fig. 21

Gas Example. Air at a pressure of 120 lb. *force*/sq. in. gage and a temperature of 30°C. is flowing at the rate of 500 lb. *mass*/hr. through a 2-in. standard steel pipe. What is the pressure drop per foot of pipe? The actual inside diameter is 2.067 in. The pressure of the air is $(120 + 14.7)/14.7 = 9.16$ atm. abs. Connect $D_1 = 2.067$ with $w = 0.5$, and extend the line to intersect the reference line at $A = 6.15$. Connect 30°C. on the gas temperature scale with molecular weight = 29 and intersect the $\mu_c^{0.16}/\rho$ line at 7.1. Join this last intersection with point A, intersecting the $\Delta p(P)/L$ line at 0.008. The pressure drop is then $0.008/9.16 = 0.00087$ lb. *force*/sq. in./ft. of pipe.

Molecular Weights of Gases

Acetylene	26.0	Helium	4.0
Air	29.0	Hexane	86.1
Ammonia	17.0	Hydrogen	2.0
Argon	39.9	Hydrogen bromide	80.9
Bromine vapor	159.8	Hydrogen chloride	36.5
Butane	58.1	Hydrogen cyanide	27.0
Butylene	56.1	Hydrogen fluoride	20.0
Carbon dioxide	44.0	Hydrogen sulfide	34.1
Carbon monoxide	28.0	Methane	16.0
Chlorine	70.9	Methyl chloride	50.5
Cyanogen	52.0	Nitric oxide	30.0
Ethane	30.1	Nitrogen	28.0
Ethylene	28.0	Oxygen	32.0
Fluorine	38.0	Pentane	72.1
“Freon-11”	137.4	Propane	44.1
“Freon-12”	120.9	Propylene	42.1
“Freon-21”	102.9	Sulfur dioxide	64.1
“Freon-22”	86.5	Water	18.0
“Freon-113”	187.4		

Liquid Example. A 25 per cent calcium chloride brine is to be pumped through a line at 250 gal./min. at a temperature of 0°C. If the allowable pressure drop is 0.006 lb. *force*/sq. in./ft. of pipe, what size of pipe is required? Connect 0°C. on the liquid temperature scale with the intersection of grid values $X = 2.6$ and $Y = 4.2$ shown in the table of Coordinates for Liquids. Extend the line to $\mu_c^{0.16}/\rho = 0.0179$ and connect that point to $\Delta p/L = 0.006$ and extend to the reference line at point $B = 11.65$. Connect point B through $w = 147$ (since at density of 73.3 lb. *mass*/cu. ft., 250 gal./min. = 147,000 lb. *mass*/hr.) to intersect at $D_i = 5.5$ in., indicating a 6-in. pipe.

Coordinates for Liquids and Aqueous Solutions

	X	Y		X	Y
Acetaldehyde	−0.3	3.7	“Freon 113”	0.9	6.2
Acetic acid, 100%	1.0	4.0	Glycerol, 100%	6.9	1.8
Acetic acid, 77%	2.6	3.8	Glycerol, 50%	3.0	3.7
Acetic anhydride	0.7	4.3	Hydrochloric acid, 31.5%	1.1	4.2
Acetone, 100%	0.9	3.4	Linseed oil, raw	3.4	1.8
Acetone, 35%	2.7	3.7	Mercury	See	chart
Ammonia, anhydrous	0.9	3.6	Methanol, 100%	0.8	3.3
Ammonia, 26%	1.9	3.6	Methanol, 40%	2.8	3.6
Aniline	2.5	3.4	Methyl acetate	0.0	4.2
Benzene	0.6	3.6	Methyl chloride	−0.8	4.3
Butanol	2.6	2.6	Nitric acid, 95%	0.8	5.8
Calcium chloride brine, 25%	2.6	4.2	Nitric acid, 60%	1.5	4.8
Carbon disulfide	0.0	5.6	Nitrobenzene	1.7	4.4
Carbon tetrachloride	0.7	6.0	Octane	0.4	2.7
Chloroform	0.0	6.0	Phenol	2.4	3.4
Chlorosulfonic acid	1.5	5.8	Propionic acid	0.6	3.8
Cyclohexanol	5.3	2.2	Sodium chloride brine, 25%	2.1	4.4
Diphenyl	0.0	3.5	Sodium hydroxide, 50%	5.3	3.7
Ether	−0.3	3.2	Sulfur dioxide	−0.2	6.1
Ethyl acetate	0.2	3.9	Sulfuric acid, 110%	3.7	4.7
Ethyl alcohol, 95%	1.9	3.0	Sulfuric acid, 98%	3.5	4.8
Ethyl alcohol, 45%	3.6	3.4	Sulfuric acid, 78%	3.2	4.8
Ethyl chloride	0.2	4.3	Tetrachlorethylene	0.3	6.2
Ethylene glycol	3.5	2.9	Toluene	0.4	3.6
Formic acid	1.5	4.5	Trichlorethylene	0.1	5.9
“Freon-11”	0.0	6.2	Turpentine	1.1	3.1
“Freon-12”	−1.2	5.9	Vinyl acetate	0.4	4.2
“Freon-21”	−0.4	5.9	Water	2.0	4.2
“Freon-22”	−1.7	5.5			

Allowance for Viscosity and Density

Gases. Temperature and molecular weight scales are given in Fig. 21 in the form of a line-coordinate chart by which values of $\mu_c^{0.16}/\rho$ at atmospheric pressure are determined directly. While it is true that the viscosities of gases and gas mixtures are not exactly proportional to the molecular weight, the error is small when the 0.16 power of the viscosity is used.

Liquids. A separate temperature scale and a grid are given on the chart. Coordinates given in the table above locate the point for a given liquid on the grid; a line through the point and the given temperature determines $\mu_c^{0.16}/\rho$ directly. Coordinates for liquids not given in the table may be determined by calculating values of $\mu_c^{0.16}/\rho$ for two temperatures and noting the intersection of lines connecting corresponding values of $\mu_c^{0.16}/\rho$ and temperature.

Integrated Flow Equations for Perfect Gases. The results below are obtained from Eq. (15) in the case of perfect gases ($pv = RT/M$).

Nomenclature for Perfect-gas Equations

D = diameter, ft.
f = friction factor in Eqs. (14) and (14*b*).
g_c = 32.1740 (lb. *mass*)(ft.)/(lb. *force*)(sec.^{2}).
$G = V\rho = V/v$ = mass velocity, lb. mass/(sec.)(sq. ft.).
G_c = critical or maximum mass velocity.
$G_{ci} = \sqrt{g_c p_0/(2.718 v_0)} = p_0\sqrt{g_c M/(2.718 R T_0)}$, see Fig. 22*A*.
$K = \sin\theta$, where θ = angle of inclination to horizontal [see Eq. (15)].
$\kappa = c_p/c_v$, ratio of specific heat at constant pressure to that at constant volume.
L = length, ft.
M = molecular weight, lb. *mass*/lb. mole.
μ = viscosity, lb. *mass*/(ft.)(sec.) = centipoises/1488.
$N = fL/R_H$ = frictional resistance in velocity heads.
p_1, p_2 = pressure at inlet and outlet, respectively, lb. *force*/sq. ft.
r_c = critical pressure ratio, p_2/p_0, see Fig. 22*A*.
R = gas constant = 1546 ft.-lb. *force*/(lb.-mole)(°R.)
= 2780 ft.-lb. *force*/(lb.-mole)(°K.).
T = absolute temperature, °R. = °F. + 459.6; or °K. = °C. + 273.1.
v = specific volume of gas, cu. ft./lb. *mass*.
V_1, V_2 = linear velocity at inlet and outlet, respectively, ft./sec.
w = weight rate of flow, lb. *mass*/sec.
x = distance measured in direction of flow, ft.

Isothermal Flow. If $B = f(RTG)^2/2g_cR_HM^2K$,

$$\frac{MKL}{RT} = 1.151\log_{10}\left(\frac{p_1^2+B}{p_2^2+B}\right) + \frac{2.30MKR_H}{RTf}\log_{10}\left[\frac{p_2^2}{p_1^2}\left(\frac{p_1^2+B}{p_2^2+B}\right)\right] \quad (16)$$

The last term is often negligible.

Isothermal Flow in Horizontal Ducts ($K = 0$).

$$p_1^2 - p_2^2 = \frac{fLG^2RT}{g_cR_HM}\left(1 + \frac{4.61R_H}{fL}\log_{10}\frac{p_1}{p_2}\right) \quad (17)$$

In ducts of appreciable length the last term in parentheses is generally negligible unless the pressure drop is very large. For example, if $L/D = 100$, $(p_1 - p_2)/p_1$ may be as great as 20 per cent before the error in omitting the last term becomes as great as the probable error in the friction factor. When the last term is omitted, Eq. (17) may be written

$$p_1 - p_2 = \frac{fLG^2}{2g_c\rho_{avg}R_H} \quad (17a)$$

where ρ_{avg} is the density at the average pressure $(p_1 + p_2)/2$. This formula is identical with that for a liquid of density ρ_{avg} flowing in a uniform horizontal duct. When solved for the weight rate of flow, Eq. (17*a*) becomes

$$w = \frac{\pi}{8}\sqrt{\frac{(p_1^2 - p_2^2)g_c\,D^5M}{fLRT}} \quad (17b)$$

In ordinary pipe lines the flow is commonly more nearly adiabatic than truly isothermal. However, as shown below, the formulas for adiabatic flow often deviate little from those for isothermal flow.

Adiabatic Flow in Horizontal Ducts. Let the initial acoustic velocity $\sqrt{\kappa g_c p_1 v_1} = \sqrt{\kappa g_c R T_1/M} = c_1$, then

$$\frac{fL}{R_H} = \frac{-2.30(\kappa+1)}{\kappa}\log_{10}\frac{V_2}{V_1} + \frac{1}{\kappa}\left[\frac{c_1^2}{V_1^2} + \frac{(\kappa-1)}{2}\right]\left(1 - \frac{V_1^2}{V_2^2}\right) \quad (18)$$

$$\frac{p_2}{p_1} = \frac{V_1}{V_2}\left[1 + \frac{(\kappa-1)V_1^2}{2c_1^2}\left(1 - \frac{V_2^2}{V_1^2}\right)\right] \quad (19)$$

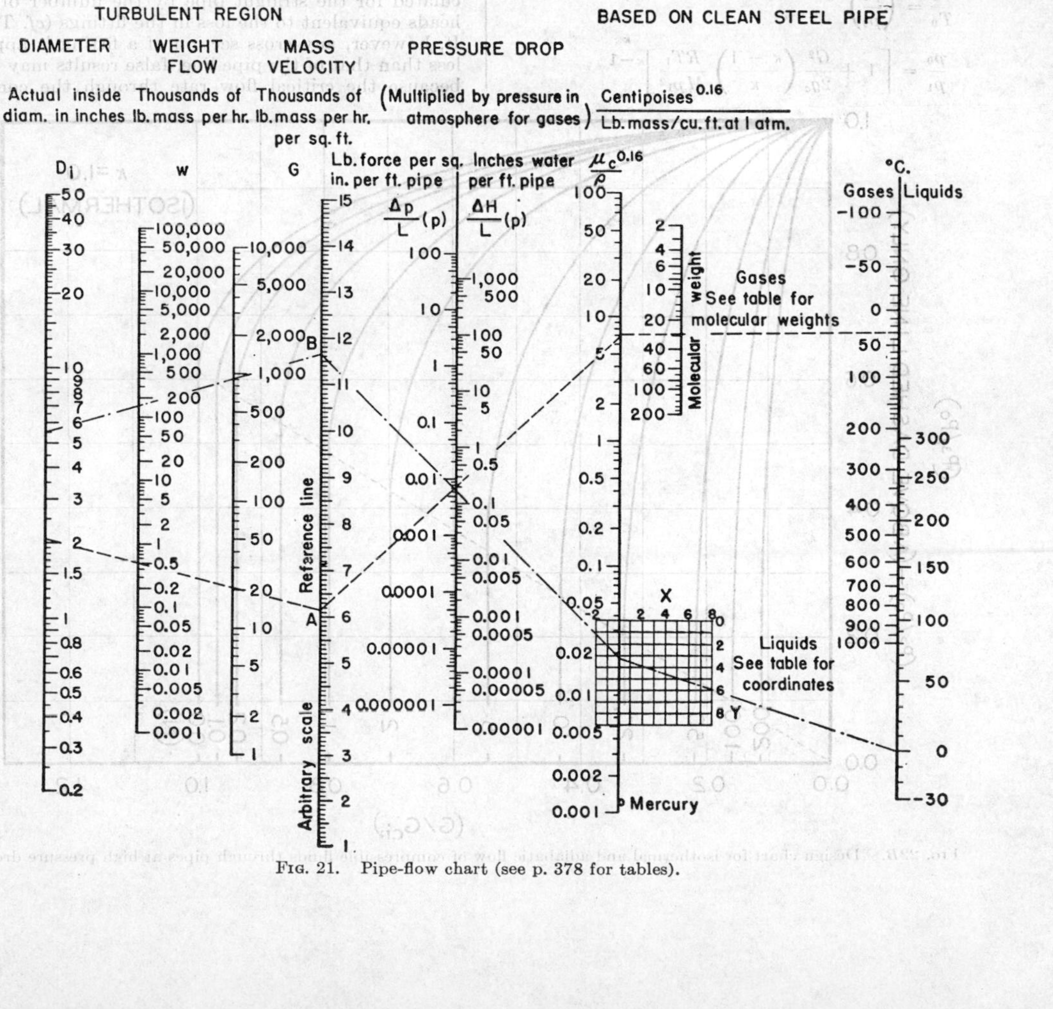

FIG. 21. Pipe-flow chart (see p. 378 for tables).

Equation (19) follows from Eq. (9), p. 376, when the specific heats c_p and c_v are assumed constant. Frequently the last term in the bracket of Eq. (19) is negligible so that the equation becomes $p_2v_2 = p_1v_1$; in such cases Eq. (18) may be replaced by Eq. (17) because the departure from isothermal flow will be negligible.

A convenient graphical method of solving Eqs. (18) and (19) simultaneously [Lapple, *Trans. Am. Inst. Chem. Engrs.*, **39**, 385 (1943)] results from the device of assuming that the conditions of flow at the inlet arise from the adiabatic expansion of the gas through a frictionless nozzle leading from a chamber where the velocity is negligible. Such a chamber frequently is present in fact; defects of the real entrance from a perfect nozzle may be approximately allowed for by supposing the length of pipe to be increased.

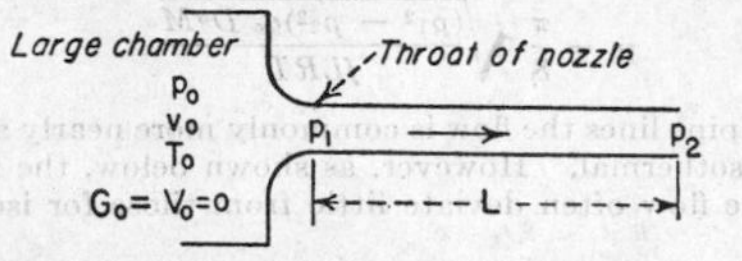

FIG. 22*A*. Pipe discharging from a large chamber.

Figure 22*A* shows the flow system and the usage of subscripts. For the system shown,

$$\frac{v_1}{v_0} = \left(\frac{p_0}{p_1}\right)^{\frac{1}{\kappa}}$$

$$\frac{T_1}{T_0} = \left(\frac{p_1}{p_0}\right)^{\frac{\kappa-1}{\kappa}}$$

$$\frac{p_0}{p_1} = \left[1 + \frac{G^2}{2g_c}\left(\frac{\kappa-1}{\kappa}\right)\frac{RT_1}{Mp_1^2}\right]^{\frac{\kappa}{\kappa-1}}$$

$$\frac{T_2}{T_1} = \left(\frac{p_2}{p_1}\right)\left(\frac{v_2}{v_1}\right)$$

The charts of Figs. 22*B*, *C*, and *D* show, for three values of the ratio of the specific heats κ and for various values of the frictional resistance ($N = fL/R_H$) in the duct, the relation between p_2/p_0 or p_3/p_0 and the ratio of the mass velocity G in the duct to a parameter

$$G_{ci} = \sqrt{\frac{g_c p_0}{(2.718 v_0)}} = p_0 \sqrt{\frac{g_c M}{(2.718 RT_0)}}.$$

The quantity G_{ci} is the maximum mass velocity hypothetically attainable on isothermal expansion through the system of Fig. 22*A* when $N = 0$. Such an isothermal expansion is not physically realizable. The ratios p_2/p_0 and p_3/p_0 are identical if, for a given N, the mass velocity is less than a certain maximum or critical value, G_c; this is true above the dashed line on the charts. If p_3/p_0 is less than the value of $p_2/p_0 = r_c$ corresponding to G_c, $p_2/p_0 = r_c$ and $G = G_c$; in other words, the flow is then independent of p_3 (*cf.* p. 402). The dashed line is a plot of r_c vs. G_c/G_{ci}.

The use of the charts is most easily understood by studying the illustrative example given below. Interpolation between charts is permissible for values of κ other than 1.0, 1.4, and 1.8. When G is being sought, assume $f = 0.0035$ to determine a rough value of G; then, using the approximate G to form the Reynolds number, find f from Fig. 23, and repeat the calculation using this revised estimate of f. When fittings are present in the pipe line, increase the value of N calculated for the straight pipe by the number of velocity heads equivalent to the loss in the fittings (*cf.* Table 14). If, however, any cross section of a fitting is appreciably less than that of the pipe line, false results may be found because the critical flow rate through the constriction

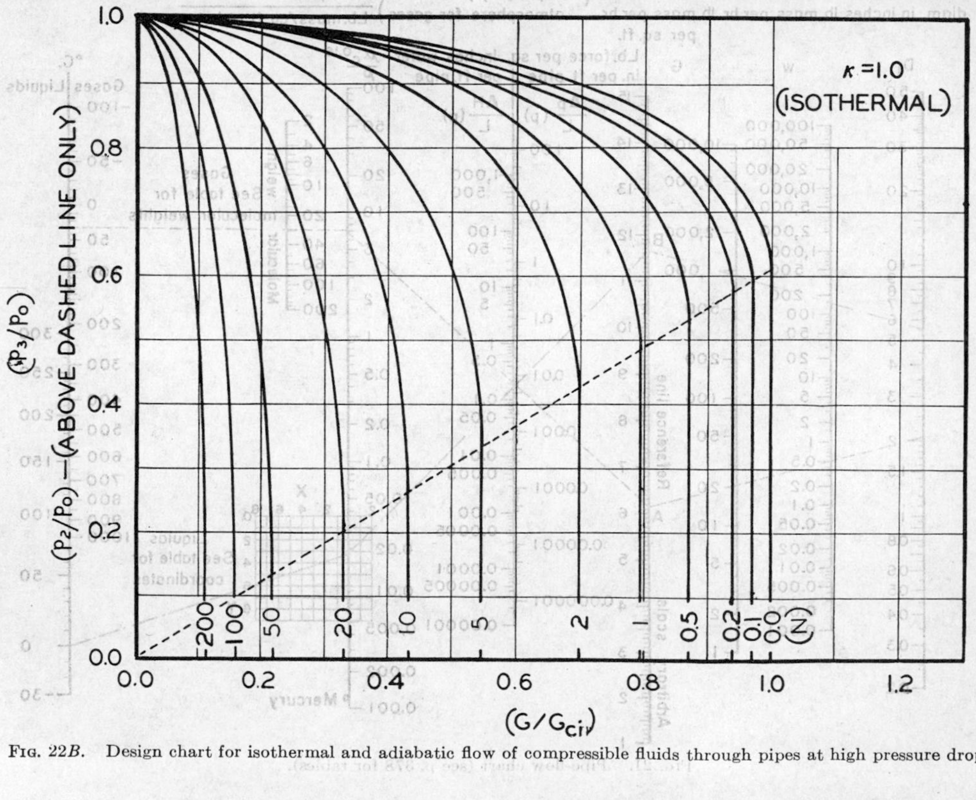

FIG. 22*B*. Design chart for isothermal and adiabatic flow of compressible fluids through pipes at high pressure drop.

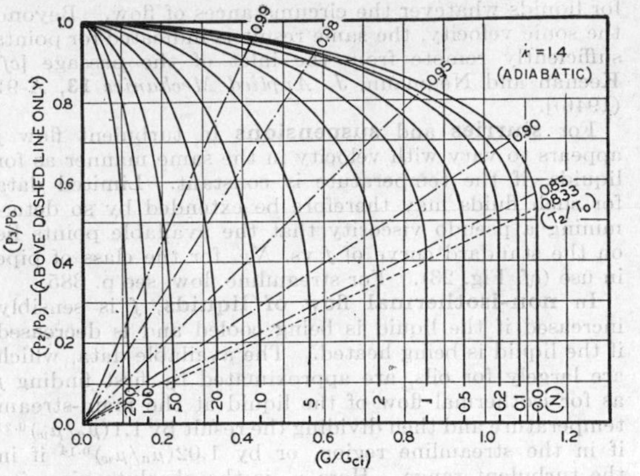

FIG. 22C. Design chart for isothermal and adiabatic flow of compressible fluids through pipes at high pressure drop.

may limit the capacity of the line; this occurs if the acoustic velocity (*cf.* p. 375) is approached at the constriction. For a sharp or abrupt inlet the charts lead to approximately correct results if 0.5 is added to the value of N for the duct. In this case, however, the formulas given above for v_1/v_0, p_0/p_1, and T_1/T_0 are inapplicable; they apply only to a rounded entrance.

Illustrative Example. It is desired to calculate the discharge rate of air to the atmosphere from a reservoir at 150 lb. *force*/sq. in. gage and 70°F. through 39 ft. of straight 2-in. Schedule 40 steel pipe (i.d. = 2.067 in.) and three standard elbows. The pipe inlet is abrupt.

To solve the problem, it is necessary to assume a value of the friction factor, f, and express all resistance in terms of N as follows:

Resistance	(L/D)	(N)
Inlet	...	0.50*
Straight pipe	226	3.16†
Elbows (three)	96	1.34†
		5.00

* Assumed equivalent values.
† Calculated, assuming $f = 0.0035$.

From the conditions of the problem,

$T_0 = 530°\text{R}$.
$p_0 = (150 + 14.7)(144) = 23{,}700$ lb. *force*/sq. ft.
$p_3 = (14.7)(144) = 2120$ lb. *force*/sq. ft.
$p_3/p_0 = 0.0893$.
$M = 29$.

$$G_{ci} = p_0\sqrt{\frac{g_cM}{2.718RT_0}} = 23{,}700\sqrt{\frac{32.17 \times 29}{2.718 \times 1546 \times 530}}$$

$G_{ci} = 486$ lb. *mass*/(sec.)(sq. ft.)

$$\text{Pipe cross section} = (0.785)\left(\frac{2.067}{12}\right)^2 = 0.0233 \text{ sq. ft.}$$

It is now possible to calculate the discharge by the use of Fig. 22C as shown by the following tabulation:

κ	1.4
(G/G_{ci}), from curve for $N = 5.0$ and $(p_3/p_0) = 0.0893$	0.565
G, lb. *mass*/(sec.)(sq. ft.)	275
Discharge rate, lb. *mass*/sec	6.41
(T_2/T_0), from dashed line, since (p_3/p_0) is below the dashed line	0.833
T_2, °R	442
Average gas temp. in pipe, °F	26
Viscosity, μ, at average gas temp., lb. *mass*/(ft.)(sec.)	1.14×10^{-5}
N_{Re} or (DG/μ)	4,160,000
f	0.0033

Since the value of f so obtained checks the assumed value reasonably well, it is not necessary in this case to repeat the determination of G.

Approximate integration of Eq. (15) may be accomplished for uniform ducts, for example, by dividing throughout by v^2 which gives

$$\frac{-dp}{v} = -\rho\, dp = \frac{G^2\, dv}{g_c v} + \left(\frac{fG^2}{2g_cR_H} + \frac{K}{v^2}\right) dx \qquad (20)$$

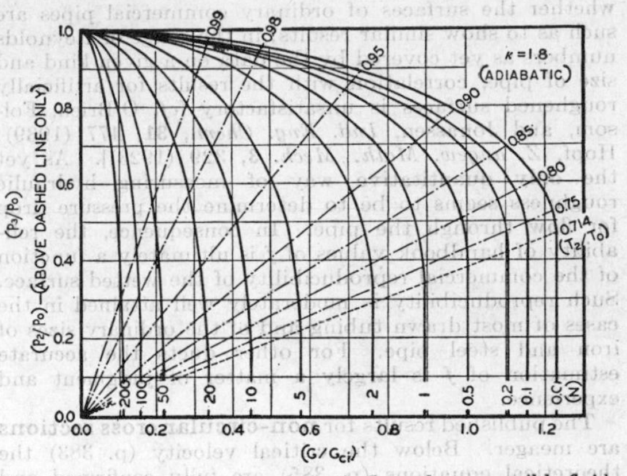

FIG. 22D. Design chart for adiabatic flow of compressible fluids through pipes at high pressure drop.

and noting that, **for gases**, $\rho_{\text{avg}}(p_1 - p_2) = (\rho_1 + \rho_2)(p_1 - p_2)/2$ is usually a good approximation for $-\int \rho\, dp$ because a plot of ρ vs. p is generally of small curvature. On this basis, the result is

$$p_1 - p_2 = 2.30\frac{G^2}{g_c\rho_{\text{avg}}}\log_{10}\frac{v_2}{v_1} + \frac{fG^2L}{2g_cR_H\rho_{\text{avg}}} + KL\rho_{\text{avg}} \qquad (21)$$

This equation is inexact for large values of K. For the isothermal flow of a perfect gas in a horizontal duct, this result agrees with Eq. (17). For **nonisothermal flow of liquids** (*e.g.*, in a heat exchanger), $\rho = 1/v$ varies with temperature, and so with x, but not with p. Hence a better result is obtained by dividing Eq. (15) throughout by v (rather than v^2) expressing ρ in terms of x and integrating:

$$p_1 - p_2 = \frac{G^2(v_2 - v_1)}{g_c} + \int_0^L \left(\frac{fG^2v}{2g_cR_H} + K\rho\right) dx \qquad (22)$$

$$= \frac{G^2(v_2 - v_1)}{g_c} + \frac{fG^2v_{\text{avg}}L}{2g_cR_H} + K\rho_{\text{avg}}L \qquad (22a)$$

Note that $v_{\text{avg}} \neq 1/\rho_{\text{avg}}$ and that both are here averaged with respect to x, not with respect to p as in Eq. (21). Generally the first term on the right is negligible.

Fanning Friction Factors. The dimensionless Fanning friction factor f discussed below is that to be used in the flow formulas, Eqs. (14) to (22a) pp. 377*ff*. It is related to the **Chézy coefficient** C which has the dimensions [(lb. *mass*)(ft.)/(lb. *force*)(sec.^{2})]$^{0.5}$ by the formula $C = \sqrt{2g_c/f}$. *Care must be exercised when values of f are taken from the literature because the same name and symbol are used by some writers to denote various multiples of the present f.*

For pipe and ducts of all cross-sectional shapes so far examined, the friction factor f has been found to be a function of DG/μ [D = diameter of circular pipe (or 4 × hydraulic radius R_H for non-circular ducts); G = mass velocity; μ = absolute viscosity] and of the relative roughness of the wetted surfaces. It has been shown by experiments with artificially roughened pipes [*e.g.*, Nikuradse, *Forschungsheft*, **361** (1933); Colebrook and White, *Proc. Roy. Soc.* (*London*), **A161**, 367 (1937)] that, for surfaces coated with sand or studded with rivets, f becomes a function of roughness alone at sufficiently large Reynolds numbers. It is not yet certain whether the surfaces of ordinary commercial pipes are such as to show similar results; in the range of Reynolds numbers as yet covered by the data on a given kind and size of pipe, correlation with the results for artificially roughened surfaces is unsatisfactory [*cf.* O'Brien, Folsom, and Jonassen, *Ind. Eng. Chem.*, **31,** 477 (1939); Hopf, *Z. angew. Math. Mech.* **3,** 329 (1923)]. As yet the only quantitative way of measuring hydraulic roughness seems to be to determine the pressure drop for flow through the pipe. In consequence, the reliability of handbook values of f is ultimately a function of the commercial reproducibility of the wetted surface. Such reproducibility is moderately well attained in the cases of most drawn tubing and of the ordinary sizes of iron and steel pipe. For other ducts the accurate estimation of f is largely a matter of judgment and experience.

The published results for **non-circular cross sections** are meager. Below the critical velocity (p. 383) the theoretical equations (p. 385) are fully confirmed and should be used; above the critical velocity f does not differ greatly from the results for circular pipes. Data for **open channels** are almost exclusively based on experiments with water, and the results are usually given in terms of Chézy coefficients. For other fluids it is recommended that open channels be treated as special cases of non-circular ducts; if desired C can be computed from the corresponding values of f.

In the case of **gases** most of the experimental determinations of f have been made with small pressure gradients and at moderate pressures. Usually the flow was substantially isothermal, and deviations from the perfect gas law were not large. In the moderate pressure range it has been demonstrated that up to the velocity of sound f for gases is the same function of DG/μ as for liquids whatever the circumstances of flow. Beyond the sonic velocity, the same result is indicated for points sufficiently remote from the inlet of the passage [*cf.* Keenan and Neumann, *J. Applied Mechanics* **13,** A-91 (1946)].

For **slurries** and **suspensions** in turbulent flow f appears to vary with velocity in the same manner as for liquids, if the temperature is constant. Limited data for such fluids may therefore be extended by so determining a pseudo viscosity that the available points lie on the standard curve of f vs. N_{Re} for the class of pipe in use (*cf.* Fig. 23). For streamline flow, see p. 385.

In **non-isothermal flow of liquids,** f is sensibly increased if the liquid is being cooled and is decreased if the liquid is being heated. The available data, which are largely for oils, are approximated by first finding f as for isothermal flow of the liquid at the main-stream temperature and then dividing the result by $1.1(\mu_a/\mu_w)^{0.25}$ if in the streamline region, or by $1.02(\mu_a/\mu_w)^{0.14}$ if in the turbulent range. Here μ_a is the absolute viscosity at the main-stream temperature and μ_w is that at the wall temperature [*cf.* Sieder and Tate, *Ind. Eng. Chem.*, **28,** 1429 (1936)].

For **circular pipes** Fig. 23 shows the friction factor f, as ordinates, plotted vs. the Reynolds number DG/μ, as abscissas. For small values of DG/μ, with all fluids (except possibly gases under high pressures or high-pressure gradients) and in any ordinary type of pipe

$$f = \frac{16\mu}{DG} \tag{23}$$

where μ = absolute viscosity lb. *mass*/(sec.)(ft.) = centipoises/1488; D = diameter, ft.; G = mass velocity, lb. *mass*/(sec.)(sq. ft.); V = linear velocity, ft./sec. This equation (curve A) is that theoretically deducible

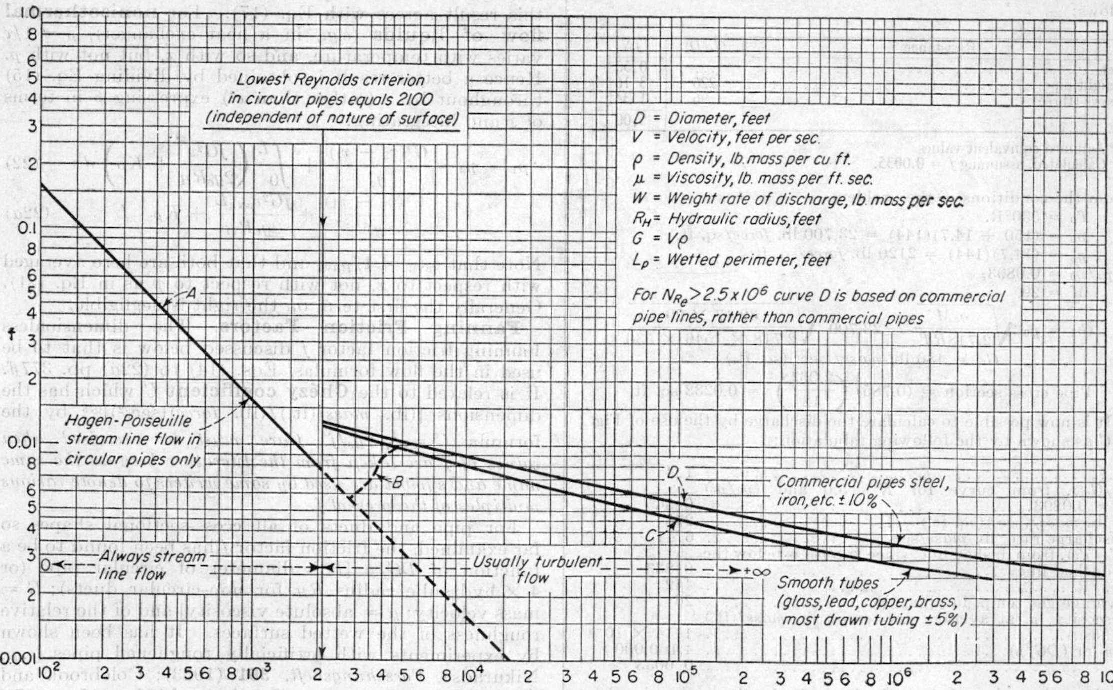

FIG. 23. Fanning friction factors f in Eqs. (14) to (22a). Straight ducts only.

on the assumption of rectilinear streamline (or viscous) flow (*cf.* pp. 375 and 385*ff.*). If $DG/\mu >$ about 2100, which value is known as the Reynolds criterion, f first rises rapidly as DG/μ increases (*e.g.*, curve B in Fig. 23) and then falls off along a curve (*e.g.*, curve C or D) of gradually decreasing slope. The location of the latter curve depends considerably upon the nature of the wetted surface.

If $DG/\mu > 2100$, all self-consistent data, for clean smooth tubes (glass, lead, copper, and most drawn tubing of any metal), both for gases at moderate pressures and for liquids, lie within a band of ± 5 per cent from curve C of Fig. 23. The equation of curve C is

$$f = 0.00140 + 0.125 \left(\frac{\mu}{DG}\right)^{0.32} \tag{24}$$

This formula, devised by Koo, is sometimes easier to use analytically than the von Kármán equation:

$$\frac{1}{\sqrt{f}} = 4.0 \log_{10} (N_{Re} \sqrt{f}) - 0.40 \tag{25}$$

which has a rational theoretical basis and is preferable for extrapolation [*cf.* Drew, Koo, and McAdams, *Trans. Am. Inst. Chem. Engrs.*, **28**, 56 (1933); von Kármán, National Advisory Committee for Aeronautics, *Tech. Mem.* No. 611 (1931); Nikuradse, *Forschungsheft*, **356** (1932)]. For no type of pipe or duct is f less than the value given by Eq. (25).

For clean commercial iron and steel pipe, Curve D on Fig. 23 represents the data (air, steam, water, oils) within ± 10 per cent. For $N_{Re} > 2.5 \times 10^6$ the curve is based on data for pipe lines rather than for "standard pipe." The equation of curve D is

$$f = 0.0014 + 0.090 \left(\frac{\mu}{DG}\right)^{0.27} \tag{26}$$

which is close to the recommended approximation of Drew and Genereaux [*Trans. Am. Inst. Chem. Engrs.*, **32**, 16–17 (1936)].

In the case of very small pipes, there are indications that the friction factor has values greater than those given by curves C and D [Eqs. (24) and (26)]. However, in the size range covered by the data on which curve D and Eq. (26) are based, there is no consistent trend relative to curve D with variation in diameter even when consideration is restricted to a single series of experiments. In general, a line representing the data for a particular pipe is found to be of less slope than curve D even though all the points lie in the ± 10 per cent band.

Except in the case of water, for which texts on hydraulics should be consulted, there are few consistent available data for other than smooth tubes and commercial iron pipe. As a possible guide Table 10 shows, for various pipes, roughness factors X by which f as read from curve D of Fig. 23 should be multiplied. For rough and irregular walls at high Reynolds numbers there is some reason to expect that the f's corresponding to Manning's n (Table 11) represent lower limits (*cf.* Weymouth formula). As an estimate for old iron pipe, f may be taken as 0.01 when $DG/\mu >$ about 10^4. For $DG/\mu < 10^4$, but > 2100, take f about 20 per cent greater than the values given by curve D of Fig. 23. For the purpose of design when moist gases, steam, etc., are to be handled in corrodible pipes, it is recommended that f be taken at not less than 0.006 for large values of Reynolds number (DG/μ).

The equivalent of the Weymouth formula, which is widely used to estimate the flow of natural gas, results when, in Eq. (17*b*) (p. 379), f is taken as $0.008/D_i^{1/3}$, where D_i is the pipe diameter, in.

The friction factor for long **curved pipes** of uniform radius of curvature (*e.g.*, helical coils) may be estimated as follows: In isothermal streamline flow $f = Cf_s$ where f_s is the friction factor in straight pipes, as given by curve A of Fig. 23 [or Eq. (23)] for the same DG/μ, and C is the function of $(DG/\mu)\sqrt{D/D_c}$ plotted in Fig. 24. D = pipe diameter, ft., D_c = diameter of curvature, ft. [White, *Proc. Roy. Soc.* (*London*), **A123**, 645 (1929)]. This result has been shown to be theoretically sound by Dean [*Phil. Mag.* (7), **4**, 208 (1927); **5**, 673 (1928)] and has been tested experimentally for values of D/D_c between $1/15$ and $1/2050$ using air, water, and oil. If f, thus calculated, is less than 0.009, the flow will be actually turbulent and the above rule will fail. White's results for turbulent flow in smooth tubes appear to indicate that the friction factor is substantially the same as in straight passages if D/D_c is not greater than $1/500$. For sharper curves the data available do not suffice to formulate a rule, but at least in the lower range of DG/μ the friction is markedly increased above that in straight pipes by increasing the curvature. For losses in bends as distinguished from long coils see p. 390.

Table 10

Condition of pipe	Roughness factor X	C' of Williams and Hazens formula (water)
Smooth brass, copper or lead pipe	0.9	140
New steel or cast-iron pipe	1.0	130
Smooth wooden, good covered pipe	1.2	120
Old cast-iron, new riveted steel pipe	1.4	110
Vitrified pipe, or old steel pipe	1.6	100
Old riveted steel pipe	2.0	90
Badly tuberculated cast-iron pipe	2.5	80

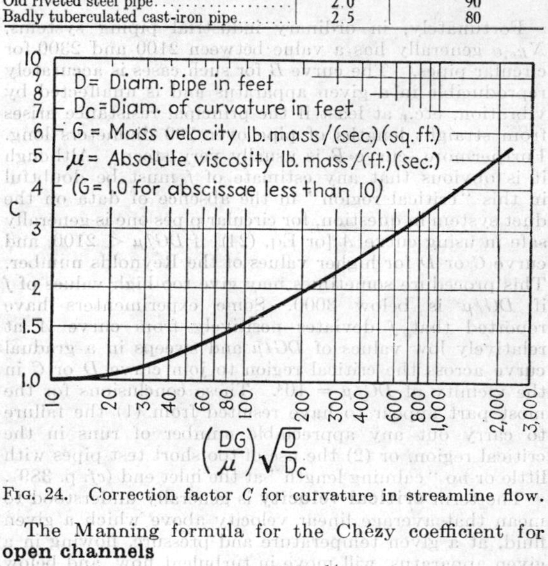

FIG. 24. Correction factor C for curvature in streamline flow.

The Manning formula for the Chézy coefficient for **open channels**

$$C = \frac{1.486 R_H^{1/6}}{n} \tag{27}$$

$$f = 29.1 \frac{n^2}{R_H^{1/3}} \tag{27a}$$

where R_H, ft., is the hydraulic radius and n is a roughness factor, appears to represent the data as well as, or better than, the widely used but complicated Ganguillet-Kutter formula. The value of n is so nearly identical with that of Kutter's n that commonly no distinction between them is made. Table 11 gives a number of typical values.

Critical Region. Critical Velocity. According to the most reliable experiments, the Reynolds criterion, $N_{Re,C}$ (= "critical value" of DG/μ at which the Fanning friction factor suddenly departs from curve A of Fig. 23), in round pipes apparently never has a value less than about 2100, whatever be the nature of the wetted surfaces. However, $N_{Re,C}$ may have values greater than 2100, in the absence of turbulence at the entrance, and insofar as experimental evidence is concerned, there seems to be no definite upper limit. Consequently, no accurate location can be assigned to curve B in Fig. 23 nor can the shape of that curve be given definitely. For a review of experimental work in this connection, see Schiller [*Physik. Z.*, **26**, 566 (1925)] who gives an extensive bibliography.

Table 11. Average Values of n for the Manning Formula, Eq. (27)*

Surface	n
Cast-iron pipe, fair condition	0.014
Galvanized wrought-iron pipe, fair condition	0.015
Smooth brass pipe	0.011
Riveted steel pipe	0.017
Vitrified sewer pipe	0.015
Concrete pipe and flumes, average construction	0.015
Planed plank flume	0.013
Semicircular metal flumes, smooth	0.013
Semicircular metal flumes, corrugated	0.0275
Canals and ditches:	
Earth, straight and uniform	0.0225
Winding sluggish channels	0.0275
Dredged-earth channels	0.030
Natural-stream channels:	
Clean, straight bank, full stage	0.030
Winding, some pools, and shoals	0.040
Same, but with stony sections	0.055
Sluggish reaches, very deep pools, very weedy	0.070–0.125

* King, "Handbook of Hydraulics," 5th ed., McGraw-Hill, New York, 1939.

Fortunately, in ordinary industrial piping systems, $N_{Re,C}$ generally has a value between 2100 and 2300 for circular pipes. The curve B for such cases is accurately reproducible in a given apparatus and is unaffected by vibration, etc., at least if the principal resistance arises from straight lengths of pipe over 200 diameters long. Furthermore, curve B is usually very steep. Although it is obvious that any estimate of f must be doubtful in this "critical region" in the absence of data on the duct system in question, for circular pipes one is generally safe in using curve A [or Eq. (24)] if $DG/\mu < 2100$, and curve C or D for higher values of the Reynolds number. This procedure sometimes may give too high values of f if DG/μ is below 3000. Some experimenters have reported that f deviates positively from curve A at relatively low values of DG/μ and sweeps in a gradual curve across the critical region to join curve D or C in the vicinity of $DG/\mu = 10^4$. These conclusions for the most part appear to have resulted from (1) the failure to carry out any appreciable number of runs in the critical region, or (2) the use of too short test pipes with little or no "calming length" at the inlet end (*cf.* p. 389).

The term **critical velocity** is generally understood to mean that average linear velocity above which a given fluid, at a given temperature and pressure, flowing in a given apparatus, will move in turbulent flow, and below which velocity the flow is streamline or viscous (*cf.* p. 375). The critical velocity is obviously associated with the value of DG/μ (Reynolds criterion) at which f begins to deviate from curve A of Fig. 23. If $N_{Re,C}$ is the value of Reynolds criterion for the duct and circumstances in question, ρ is the density (lb. *mass*/cu. ft.) and μ is the absolute viscosity [lb. *mass*/(ft.)(sec.)]; then

$$V_c = (N_{Re,C}) \frac{\mu}{D\rho} \text{ (ft. per sec.)} \quad (28)$$

is the rational expression for the critical velocity in a circular pipe of diameter D (ft.). Since the value of $N_{Re,C}$ cannot be definitely stated in the absence of tests on the system in question, there exists no way at present of surely evaluating V_c, in advance of such tests, in the sense inferred by the usual definition. The "lower critical velocity,"

$$V_{lc} = \frac{2100\mu}{D\rho} \quad (29)$$

in the sense of Schiller (*ibid.*) has a definite physical meaning, *i.e.*, below V_{lc} the flow *is* streamline; above V_{lc}, it *may be* turbulent.

Several other meanings have been assigned to the term "critical velocity" in the literature: (1) The velocities corresponding to the values of DG/μ at which curves C and D intersect curve A. (2) The velocity corresponding to the value of DG/μ at which curve B meets curve C or D. (3) The velocity corresponding to some supposed upper limit of $N_{Re,C}$. None of these meanings has any real physical significance, and only the first of these three kinds of critical velocities can be given any definite numerical value.

For non-circular passages neither the lower limiting values nor the usual range of values of the Reynolds criterion have been satisfactorily determined in any case, but at least for passages running full the indications are that the transition from viscous to turbulent flow occurs when $4R_HG/\mu$ is of the same order of magnitude as the value of $N_{Re,C}$ in circular pipes (*i.e.*, 2100 to 2300).

Taylor [*Proc. Roy. Soc.* (*London*), **A124**, 243 (1929)] has shown conclusively, by introducing bands of colored fluid into a liquid stream carried by a helical coil of glass tubing, that turbulence develops less readily in curved pipe than in straight pipe. The value $DG/\mu (= N_{Re,C})$ above which the friction is incalculable by the rule for viscous flow, given on p. 383, agrees well with the lowest value of DG/μ for which Taylor found fully developed turbulence (*cf.* Table 12). Even if the flow is turbulent in a straight section preceding the pipe coil the flow becomes streamline shortly after entering the coil if DG/μ is below the approximate limits shown in Table 12.

Table 12. Effect of Curvature on Critical Reynolds Number

D/D_c	$N_{Re,C}$	
	White (friction)	Taylor (color band)
1/15.15	7590	
1/18.7	....	7100
1/31.9	....	6350
1/50	6020	
1/2050	2270	

Economic Pipe Diameter. Turbulent Flow. Pipe diameters should be selected to afford the minimum total cost of owning and operating the fluid-handling system. The initial cost of the pipe and fittings is directly proportional to the diameter, as are the other factors of pipe-operating cost, depreciation, and maintenance, which are a constant percentage of initial pipe-line cost. The cost of pressure drop (cost of pumping or blowing), however, is inversely proportional to the diameter, so that a balance can be struck at a diameter which will give the minimum sum for these two costs.

The cost of pipe can be expressed approximately in terms of that of 1-in. pipe, multiplied by the actual diameter raised to a power, or $X D_i^n$. A logarithmic plot of cost per foot vs. diameter is essentially a straight line, and, if X is the cost in \$/ft. of 1-in. pipe, the cost of all sizes is expressed as $X D_i^n$. For steel pipe, $n = 1.5$. The total annual cost of the pipe line is

$$C_p = (a + b)(F + 1)X D^n \quad (30)$$

where a = amortization, fractional; b = maintenance, fractional; F = factor for fittings and erection (if the cost of fittings and erection were one-half the cost of the pipe, then $F = \frac{1}{2}$).

The second item in total annual cost, the cost of pressure drop, is taken as zero when it is not charged to operation, as in drawing water from a main. When pumping of liquids or compressing of gases is required, that cost must be included. In the case of liquids the volume of flow multiplied by the pressure drop expresses exactly the work done, but for gases this is not exact. Assuming it to be true introduces a percentage error equal to about half the percentage pressure drop, an error which is unimportant for small pressure drops. Hence this approximation is made. (This method cannot, of course, apply to steam, since the value of the pressure lost in a steam line depends on the temperature and pressure level.) Assuming that the cost of pressure drop is proportional to the product of flow volume and pressure drop, the hourly energy consumption is $1000(w/\rho)(144\Delta p)$ ft.-lb. *force* or $1000w\ (144\Delta p)/(2.654{,}200E\rho)$ kw.-hr. In dollars per year this cost becomes

$$\frac{0.0542w\ \Delta p\ YK}{E\rho}$$

or, substituting for Δp its value from the equation used to construct Fig. 21, the annual dollar cost of pressure drop per foot of pipe becomes, for turbulent flow,

$$C_{pd} = \frac{0.0072w^{2.84}\mu_c^{0.16}YK}{D_i^{4.84}\rho^2 E} \tag{31}$$

where w = thousands of lb. *mass*/hr.; μ_c = viscosity in centipoises; Y = hr. of operation per year; K = cost of electrical energy, \$/kw.-hr.; D_i = pipe diameter, in.; ρ = density, lb. *mass* per cu. ft.; E = efficiency of motor and pump; and Δp = pressure drop, lb. *force*/sq. in.

The total annual cost, C, is the sum of Eqs. (30) and (31); the minimum cost is determined by differentiating C with respect to D_i and setting the derivative equal to zero.

The resulting equation is the **general expression for any pipe:**

$$D_i^{4.84+n} = \frac{0.0348w^{2.84}\mu_c^{0.16}YK}{n(a+b)(F+1)XE\rho^2} \tag{32}$$

For steel pipe, with $n = 1.5$, Eq. (32) becomes

$$D_i = \frac{w^{0.448}\mu_c^{0.025}}{\rho^{0.312}}\left[\frac{0.0232YK}{(a+b)(F+1)XE}\right]^{0.158} \tag{33}$$

To simplify further, the "cost expression" in the brackets may be evaluated for extreme and normal values of the variables; a resultant value (after applying the 0.158 power) of 2.2 is used in Eq. (34). Since $\mu_c^{0.025}$ is nearly unity for most viscosity values, it is neglected in arriving at **the simplified equation:**

$$D_i = \frac{2.2w^{0.45}}{\rho^{0.31}} \tag{34}$$

Figure 25 has been constructed from Eq. (34); in using the chart it is more economical to select the next standard pipe size above the actual diameter determined.

After the economic pipe diameter has been obtained from the chart, the Reynolds number should be calculated to make sure that the flow is actually turbulent. See p. 383 for information on critical Reynolds numbers.

Economic Pipe Diameter. Streamline Flow. For the streamline flow region, the economic pipe diameter can be obtained from an alignment chart given by Sarchet and Colburn [*Ind. Eng. Chem.*, **32,** 1249–1252 (1940)]. The use of this chart is recommended for those cases where the chart for turbulent flow (Fig. 25) gives a pipe diameter that corresponds to streamline flow with the given flow rate.

Isothermal Streamline Flow in Straight Passages. In Table 13 are given the formulas derived by the methods of hydrodynamics for the streamline flow of liquids and gases in passages and channels of various cross sections. These formulas do not include end corrections. The pressure drop $(p_1 - p_2)$ is that measured between static pressure taps L ft. apart in the wall of a continuous duct when sufficient distance is allowed between the inlet and the upstream pressure tap to ensure the existence of the normal velocity distribution at the latter point (see p. 389). When short tubes are involved, these formulas will give highly incorrect results if the pressure drop is measured between terminal reservoirs without applying end corrections. For these corrections, see pp. 388*ff.*

The notation used in Table 13 is:

g_c = 32.1740 (lb. *mass*)(ft.)/(lb. *force*) (sec.)².
L = length of passage, ft.
M = molecular weight.
p_1, p_2 = upstream and downstream static pressures, lb. *force*/sq. ft.
R = 1543, the gas constant per mole, ft. lb. *force*/°R.
T = absolute temperature °R., *i.e.*, °F. + 459.5.
w = weight rate of discharge, lb. *mass*/sec.
z = compressibility factor, pvM/RT, where v = specific volume, cu. ft./lb. *mass*
α = angle between duct axis and horizontal, deg.
ρ = density of fluid, lb. *mass*/cu. ft.
μ = absolute viscosity, lb. *mass*/(ft.)(sec.).

The weight rate of discharge through a passage of any constant cross section is always expressible by $w = aN$, where a is a factor depending only on the cross-sectional shape, and N is the first or the second of the groups of variables shown at the head of Table 13 depending on whether a liquid or a gas is considered. The volumetric rate of flow is in all cases $q = w/\rho$, and the mean linear velocity is found by dividing q by the cross-sectional area. In the case of gases the volumetric rate and the linear velocity obviously vary from point to point along the duct.

For results applicable to less usual cross sections, see Boussinesq [*J. math. pures et appl.* (*Liouville's J.*) [2], **13,** 377 (1868)] for equilateral triangle, tubes of variable cross section, curved tubes, general theory; Graetz [*Z. Math. Physik,* **25,** 316, 375 (1880)] for curvilinear quadrilaterals, four-pointed star; Greenhill [*Proc. London Math. Soc.*, **13,** 43 (1881)] for sector of circle, cross section bounded by hyperbolas, best method of attack in general case; Piercy, Hooper, and Winny [*Phil. Mag.* [7], **15,** 647 (1933)] for pipes with eccentric cores.

The **velocity distribution** for the circular cross section is

$$u_r = \frac{2w}{\pi R^2\rho}\left[1 - \left(\frac{r}{R}\right)^2\right] \tag{35}$$

in which u_r is the linear velocity at the distance r from the pipe axis and R is the radius of the pipe; for the case of parallel plates

$$u_y = \frac{3}{4}\,\frac{w(b^2 - y^2)}{b^3\rho} \tag{36}$$

where u_y is the linear velocity at a distance y from a plane midway between the two plates and b is half the distance between the plates. Lamb ("Hydrodynamics," 5th ed., pp. 555, 556, Cambridge University Press, London, 1924) gives the velocity distribution for the annulus and ellipse; for other cases, see Graetz, *op. cit.*, and Greenhill, *op. cit.*

Flow of Suspensions. The discussion of flow in pipes given above concerned fluids whose viscosity at any given temperature remained constant with rate of flow (Newtonian fluids, see p. 369). While many suspensions of solids in liquids and gases and of liquids in gases are approximately Newtonian, many are not, especially concentrated suspensions. Several types of behavior have been observed, see p. 1197. For non-

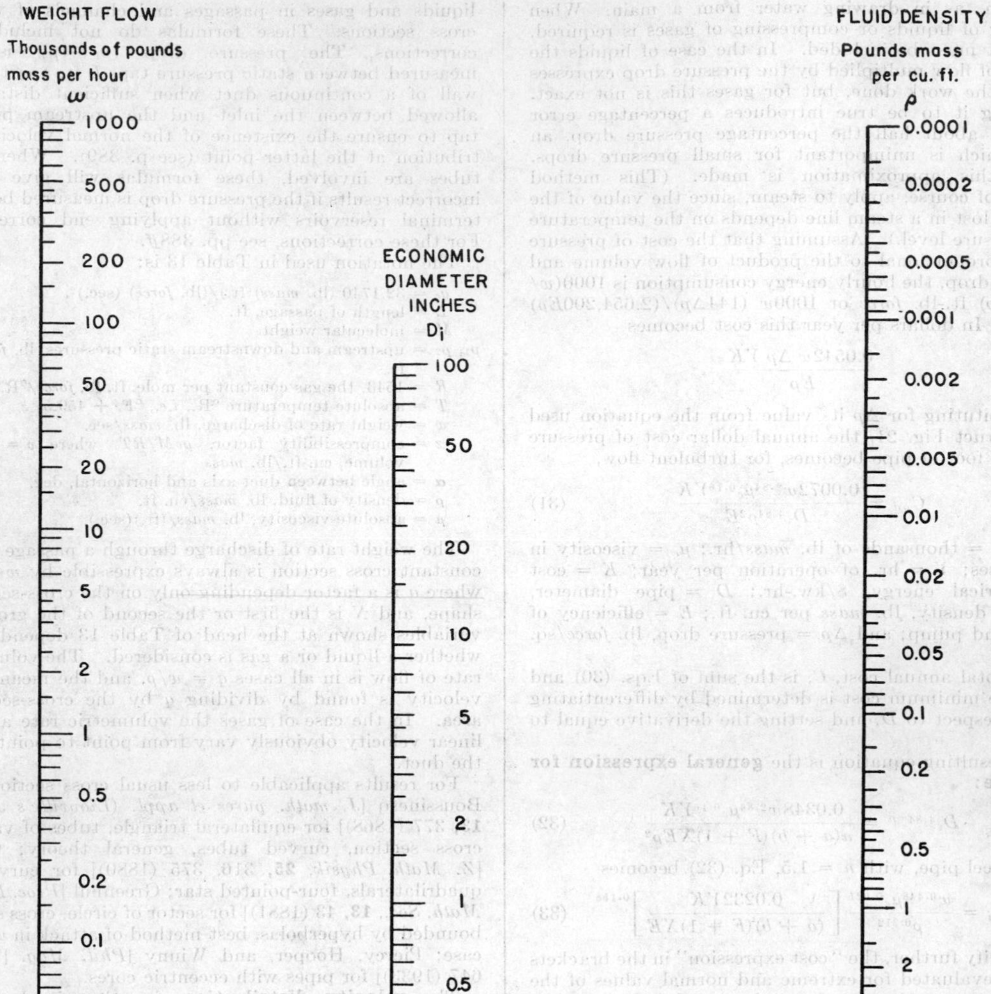

Fig. 25. Economic pipe diameter; chart constructed from Eq. (34). Connect values of w and ρ to obtain the economic diameter.

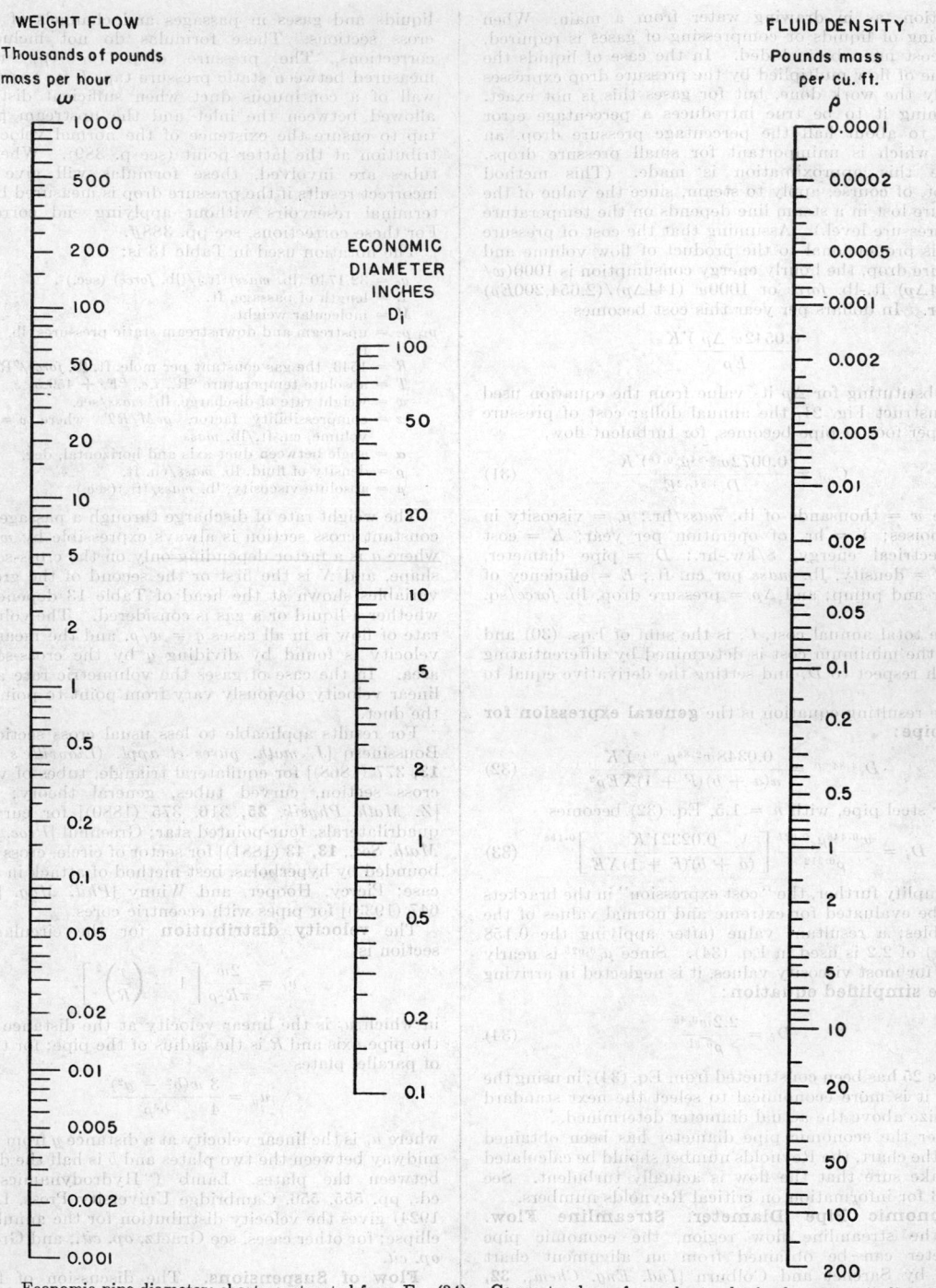

Newtonian fluids which exhibit the characteristics of a true plastic substance, Babbitt and Caldwell (*Univ. Ill. Eng. Exp. Sta. Bull.* 319, 1939) found the following equation applicable to laminar flow

$$\Delta p = \frac{16\tau_0 L}{3D} + \frac{32\eta VL}{g_c D^2} \tag{37}$$

where Δp = lb. *force*/sq. ft.; L = length of pipe, ft.; g_c = conversion factor = 32.1740 (lb. *mass*)(ft.)/(lb. *force*)(sec.)^2; τ_0 = shearing stress at the yield point (called yield value) lb. *force*/sq. ft.; η = rigidity, lb. *mass*/(ft.)(sec.); V = mean velocity, ft./sec.; and D = diameter of pipe, ft. For the sludges measured, values of τ_0 and η were found to be independent of diameter and roughness of the pipe in which they were measured. The quantities τ_0 and η may be found readily by the use of a viscometer of the Stormer type if the slurry does not settle too rapidly, or they may be deduced from test measurements of flow through a pipe [*cf.* Binder and Busher, *J. Appl. Mechanics*, **13**, A101(1946)]. A literature review on this subject is given by Babbitt and Caldwell (*ibid.*) and by Busher and Binder (*ibid.*).

For turbulent flow of suspensions, see p. 382.

Table 13. Streamline-flow Formulas

Tubes, etc., Running Full

For liquids, let $N = \frac{\rho g_c}{\mu}\left(\rho \sin\alpha + \frac{p_1 - p_2}{L}\right)$; for gases, let $N = \frac{g_c M}{2zRT\mu}\left(\frac{p_1^2 - p_2^2}{L}\right)$.*

Tube Cross Section	Theoretical Equation for Weight Rate of Flow
Circle, diam. = D	$w = \frac{\pi D^4 N}{128}$. For liquids this reduces to $p_1 - p_2 = \frac{32\mu LV}{g_c D^2}$, (*i.e.*, Poiseuille's law) if the tube is horizontal
Ellipse, semiaxes = a, b	$w = \frac{\pi a^3 b^3}{a^2 + b^2}\left(\frac{N}{4}\right)$
Rectangle, sides = a, b	$w = \frac{a^2 b^2 n N}{4}$, where $n = f\left(\frac{b}{a}\right)$ given by Fig. 26. The square has the greatest capacity of all rectangles of a given cross-sectional area
Broad parallel plates, spacing = $2b$ (*i.e.*, rectangle with one pair of sides infinite)	$w = \frac{2}{3} b^3 N$ per unit breadth
Annulus, outer diam. = D_2, inner daim. = D_1	$w = \frac{\pi(D_2^2 - D_1^2)N}{128}\left[D_2^2 + D_1^2 - \frac{D_2^2 - D_1^2)}{2.3 \log_{10}(D_2/D_1)}\right]$

Open Channels†

Let $N = \frac{\rho^2 g_c \sin\alpha}{\mu}$, since necessarily $p_1 = p_2$. The following equations are valid only when the variation in depth is negligible.

Channel Cross Section	Theoretical Equation for Weight Rate of Flow
Broad stream on flat plate, depth = B	$w = \frac{B^3 N}{3}$ per unit breadth. Applies accurately to flow down vertical wetted wall column at low liquid rates
Rectangle, depth = $\frac{b}{2}$, width = a	$w = \frac{a^2 b^2 n N}{8}$, where n is found from Fig. 26‡
V-trough, vertical $\angle = 90°$, bisector vertical, slant depth = a	$w = \frac{0.14 a^4 N‡}{8}$

* If the pressure drop is less than 10 per cent of the downstream absolute pressure, the approximate expression $N = \frac{\rho g_c (p_1 - p_2)}{\mu L}$ may be used in case of gases.

† Reynolds criterion is not well known for open channels. It appears to depend upon the surface tension and the width of channel. However, streamline flow does occur at very low rates of flow in open channels.

‡ Experimental confirmation lacking insofar as known.

Miscellaneous Pressure Drops. Despite the fact that the frictional losses of mechanical energy caused by the presence of valves, bends, and other obstructions in a piping system may often account for a considerable part of the total fluid friction, the actual data available are unfortunately neither complete nor entirely consistent. Many of the formulas commonly given are based upon scattered experiments by the older hydraulicians who naturally worked almost exclusively with water. Since the temperature of the fluid is frequently unrecorded, any effect of viscosity is often indeterminable. Moreover, possible scale effects usually cannot be certainly detected because of the experimenter's failure to follow the dictates of dynamical similarity in the construction of his models. There is, therefore, more or less disagreement among the various authorities.

Most of the losses discussed below (with a partial exception in the case of bends, *q.v.*) are theoretically reducible to expansion losses. Usually when a fluid passes an obstruction, some sort of *vena contracta*, or contraction of the stream, is formed (*cf.* p. 401), and the major frictional loss appears to arise from the excessive turbulence induced when that jet flows into the more slowly moving fluid downstream. From this point of view, on account of the results for orifices (pp. 400*ff.*) the resistances of geometrically similar systems, in general, would be expected to be functionally related to the Reynolds number $DV\rho/\mu$, in which D is any characteristic linear dimension of the type of apparatus considered, V is the linear velocity at some point therein, and ρ and μ are the density and absolute viscosity, respectively. That such is indeed the case has been found roughly true for pipe elbows and bends, which as a matter of fact are among the few devices that have been studied at all completely.

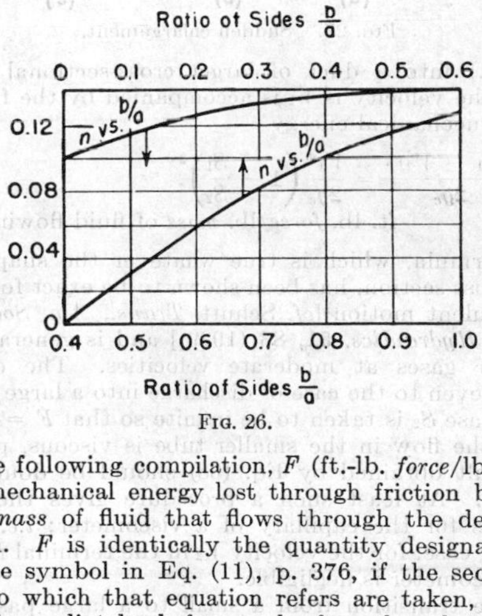

FIG. 26.

In the following compilation, F (ft.-lb. *force*/lb. *mass*) is the mechanical energy lost through friction by each pound *mass* of fluid that flows through the device in question. F is identically the quantity designated by the same symbol in Eq. (11), p. 376, if the sections 1 and 2 to which that equation refers are taken, respectively, immediately before and after the obstruction. Hence to calculate the pressure drop occasioned by the obstruction, Eq. (11), p. 376, must be solved using the appropriate F. Whereas the dimensions of F reduce to "feet," they should not, especially in dealing with gases, be regarded as feet of any particular fluid since such terminology sometimes leads to the often erroneous conclusion that multiplying F by the density of that fluid will give the total pressure drop. The symbol L_e designates the length (expressed in pipe diameters) of straight pipe (having the same diameter as the pipe to which the fitting or other device is designed to be attached), which would cause the same frictional loss of mechanical energy as does the fitting, *e.g.*, for a standard 90-deg. elbow of nominal size, $1\frac{1}{2}$ in.; $L_e = 32$ (see Table 14). The actual inner diameter of standard $1\frac{1}{2}$-in. pipe is 1.61 in. Therefore, the elbow has a frictional resistance equal to that of (32)(1.61)($\frac{1}{12}$) = 4.29 ft. of standard $1\frac{1}{2}$-in. pipe. The other symbols are listed below. Subscripts are interpreted in the sections wherein they are used.

D = inside diameter, ft.
g_c = conversion factor = 32.1740(lb. *mass*)(ft.)/(lb. *force*)(sec.^{2}).
G = mean mass velocity, lb. *mass*/(sq. ft.)(sec.).
L = length of pipe or duct, ft.
p = static pressure, lb. *force*/sq. ft.
S = cross-sectional area, sq. ft.
V = mean linear velocity, ft./sec.
w = weight rate of discharge, lb. *mass*/sec.
μ = absolute viscosity, lb. *mass*/(ft.)(sec.).
ρ = density, lb. *mass*/cu. ft.

Unless the formulas and data below are definitely stated to apply to viscous flow, they should be considered applicable only in the turbulent range. In the absence of specific statement to the contrary the original experiments were performed with liquids. Gases and vapors moving at moderate velocities show substantially the same results in most cases as do liquids. For high-velocity flow of gases (over a few hundred feet per second) treatises such as Stodola ("Steam Turbines," McGraw-Hill, New York, 1927) should be consulted.

Enlargement, Expansion, or Exit Loss. The **sudden expansion** (Fig. 27) of a duct of cross-sectional area S_1 (sq. ft.) where the mean linear velocity is V_1

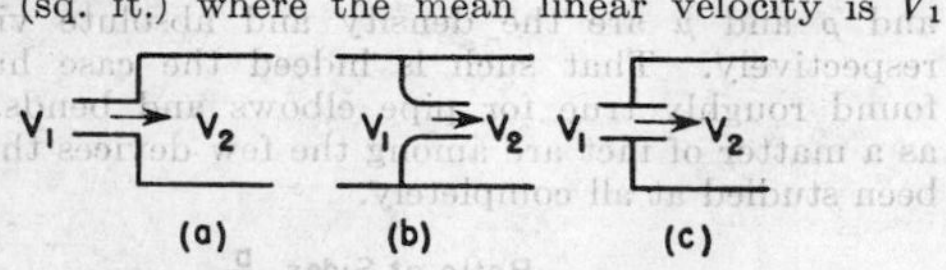

FIG. 27. Sudden enlargement.

(ft./sec.) into a duct of larger cross-sectional area S_2 where the velocity is V_2 is accompanied by the frictional loss of mechanical energy

$$F = \frac{(V_1 - V_2)^2}{2g_c} = \frac{V_1^2}{2g_c}\left(1 - \frac{S_1}{S_2}\right)^2$$

ft.-lb. *force*/lb. *mass* of fluid flowing (38)

This formula, which is true whatever the shape of the duct cross section, has been shown to be exact for liquids in turbulent motion [*cf.* Schutt, *Trans., Am. Soc. Mech. Engrs., Hydraulics*, **51**, 83 (1929)] and is generally used also for gases at moderate velocities. The equation applies even to the case of discharge into a large tank, in which case S_2 is taken to be infinite so that $F = V_1^2/2g_c$. When the flow in the smaller tube is viscous, probably the result obtained by Eq. (38) should be doubled (*cf.* p. 376). At least such a procedure gives the proper exit loss for the capillary of a viscometer; *i.e.*, in the limiting case, for the velocity V_2 in the terminal reservoir of a viscometer is negligible.

If the transition from a small to a large passage, of any cross-sectional shape, is accomplished by a **uniformly diverging duct** (Fig. 28) with a straight axis, the total pressure change Δp between the ends of the

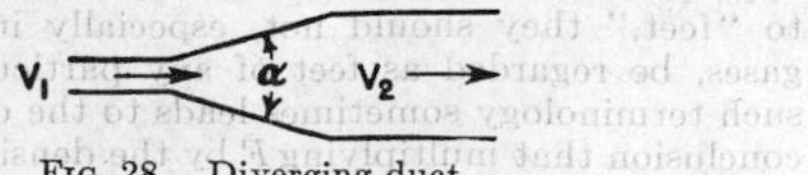

FIG. 28. Diverging duct.

diverging section may be estimated by integrating Eq. (15), p. 378, provided that the total angle α between the diverging walls is not greater than about 7 deg. For values of $\alpha >$ about 10 deg., the friction loss increases very rapidly, and above an α from 30 to 40 deg. the loss often exceeds considerably that accompanying a sudden expansion from the smaller to the larger pipe. Since (in addition to the angle of divergence and ratio of expansion) the value of D_1G_1/μ, the nature of the initial velocity distribution, and the length and cross-sectional shape of the **downstream passage** all appear to influence the measured losses, it is difficult, if not impossible, to give a general and simple formulation for $\alpha > 7$ deg. [For experimental data and theoretical discussion, see Nikuradse, *Forschungsarbeiten*, **289** (1929); Donch, *ibid.*, **282** (1926); Kröner, *ibid.*, **222** (1915).] Gibson ("Hydraulics and Its Applications") gives for α = 7.5 to 35 deg.

$$F = 3.50\left(\tan\frac{\alpha}{2}\right)^{1.22}\frac{(V_1 - V_2)^2}{2g_c}$$

ft.-lb. *force*/lb. *mass* of fluid (39)

which approximates his results for water flowing in turbulent flow through expanding cones that joined the following pairs of circular pipes: 1.5 to 3.0 in.; 2.0 to 3.0 in.; 0.5 to 1.5 in.; 1.0 to 3.0 in. The dimensions given are internal diameters. According to Gibson, **trumpet-shaped enlargements** so designed that there is a constant decrease per unit length of pipe axis in the velocity head $V^2/2g_c$ were found to give from 20 to 60 per cent less frictional loss than straight taper pipes of the same length.

Contraction or Entrance Loss. At a **sharp-edged entrance** to a pipe line or at a **sudden reduction** in the cross-sectional area of a duct (Fig. 28*A*), the frictional loss of mechanical energy is

$$F = K\frac{V_2^2}{2g_c} \quad \text{ft.-lb. } \textit{force}\text{/lb. } \textit{mass} \text{ of fluid} \qquad (40)$$

where V_2 is the linear velocity, ft./sec., in the smaller pipe; and K is a factor depending on the ratio of the smaller cross section S_2 to the larger cross section S_1. For $S_2/S_1 < 0.715$,

$$K = 0.4\left(1.25 - \frac{S_2}{S_1}\right)$$

for $S_2/S_1 > 0.715$,

$$K = 0.75\left(1 - \frac{S_2}{S_1}\right)$$

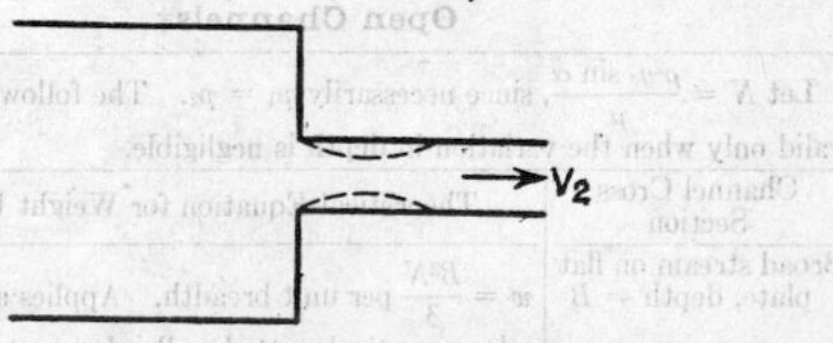

FIG. 28*A*. Sudden contraction.

If the entrance is **trumpet-shaped or rounded,** like a well-shaped measuring nozzle (*cf.* p. 406), or if it is **conical** like the upstream section of a venturi tube (*cf.* p. 406), K of Eq. (40) has a value of only about 0.05 whatever may be the ratio S_2/S_1, provided only that the flow in the smaller duct be turbulent. If, however, viscous motion obtains in the smaller passage, it is found that, although the entrance loss proper is negligible, there is an abnormally high rate of pressure drop for a distance along the tube, computed by Boussinesq [*Compt. rend.*, **113**, 9, 49 (1891)] to be 0.065 (DG/μ) pipe diameters for the case of a circular pipe. This is in good agreement with the experiments of Nikuradse (*op. cit.*). The cause of the abnormality lies in the work required to set up the parabolic velocity distribution obtaining in viscous flow through a pipe. An increased frictional resistance near the pipe inlet arises from the same cause in turbulent flow [*e.g.*, Schiller, "Vorträge aus dem

Gebiete der Aerodynamik . . . (Aachen, 1929)" Springer (1930)], but the data are as yet insufficient to give rules for its estimation. Fortunately the effect appears generally to be of less importance in the turbulent region than in streamline flow. It is, however, an important cause for discrepancies among various sets of pipe-friction measurements.

For a Reynolds number $DG/\mu = 2000$, the entrance effect would be felt for 130 diameters. Since, in a straight pipe with a rounded entrance leading from a quiet tank, streamline flow may be maintained without great difficulty up to very high Reynolds numbers [*e.g.*, Schiller, *Z. Physik*, **3**, 412 (1920) reports viscous flow in a very rough pipe up to a Reynolds number of 20,000] the deviation from Poiseuille's law near the inlet may be important in short lengths of pipe having rounded inlets. The deviation is of very considerable importance in viscometry.

For circular tubes the estimation of the pressure drop for viscous flow in the entrance length following a rounded inlet is readily accomplished by the use of the plot [Schiller, *Z. angew. Math. Mech.* **2**, 96 (1922)] given in Fig. 29. Although the plot is accurate only for the first half of the entrance length, it gives results in excellent agreement with experiment because the deviation from Poiseuille's law is most serious near the inlet. In Fig. 29, p_0 and p_1 are, respectively, the static pressures, lb. *mass*/sq. ft., within the tank and at a distance L (ft.) from the beginning of the straight tube walls; D is the inside diameter of the tube, ft.; and V is the mean linear velocity, ft./sec., inside the tube $= G/\rho$. The abscissas and ordinates of Fig. 29 are dimensionless;

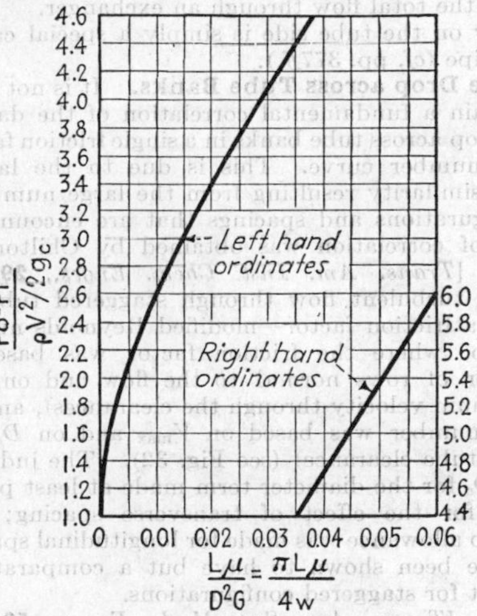

FIG. 29. Entrance effect in streamline flow.

therefore actually any consistent set of units may be used. When the tube is not horizontal the ordinates of Fig. 29 should be taken as

$$\left(\frac{p_0 - p_1}{\rho V^2/2g_c} + \frac{Z_0 - Z_1}{V^2/2g_c}\right)$$

where $(Z_0 - Z_1)$ is the difference in level, ft., between the points indicated.

Diaphragms, Orifices, Etc. The over-all frictional loss due to the presence of a **square-edged orifice** or of an equivalent diaphragm in a pipe line can be estimated from the data given on p. 404 if the flow is turbulent. If the flow is streamline in the upstream channel ($D_1G_1/\mu_1 < 2100$), the data, although very incomplete, indicate that the over-all friction loss F (ft.-lb. *force*/lb. *mass* of fluid) is equal to the orifice differential expressed in feet of fluid flowing. For **sharp-edged, *i.e.*, knife-edged, circular orifices** when

$$\left[\left(\frac{D_1G_1}{\mu}\right)\left(\frac{D_1}{D_2}\right)\right] < 10$$

$$L_e = \frac{\alpha^2}{64}\left(\frac{D_2}{D_1}\right)\left[\left(\frac{D_1}{D_2}\right)^4 - 1\right] \quad (41)$$

where $\alpha = 6.38 + 2.33(D_2/D_1)^2$ and the subscripts (1) and (2) denote, respectively, conditions in the upstream pipe and in the orifice (Johansen, Aeronautical Research Committee, Great Britain, *Reports and Memoranda*, No. 1252, 1929). For methods of computing the orifice differential, see pp. 401*ff*.

The over-all friction loss caused by a **well-shaped nozzle** or **rounded orifice** placed in a pipe line is found by applying the expansion-loss formula [Eq. (38), p. 388] to the expansion from the cross section of the throat to that of the downstream passage.

Pipe Fittings, Bends, Etc. Determinations of the resistance of fittings are ordinarily carried out by measur-

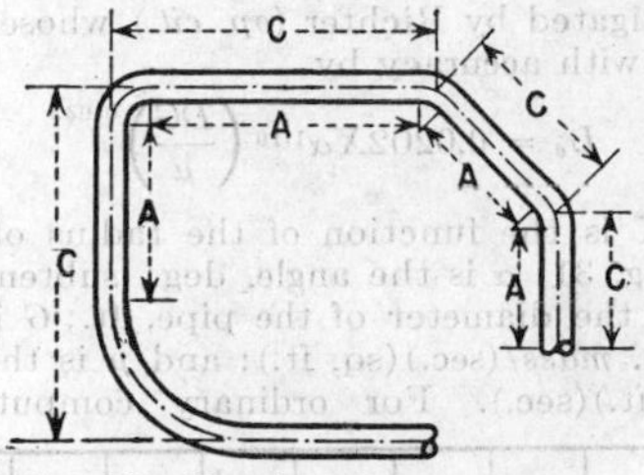

FIG. 30. Pipe-length measurement.

ing the over-all friction loss F_T in a system made up of two or more lengths of straight pipe connected in series by a suitable number of identical fittings. To obtain the loss due to the fittings themselves, the friction loss in the straight pipe is subtracted from F_T. There are three distinct conventions for computing the length L_s of the straight pipe in the test system: (*a*) $L_s = \Sigma A$ (see Fig. 30), the total length of the pieces of pipe that are actually straight; (*b*) $L_s = \Sigma C$, the sum of the distances between the intersections of the extended center lines of the successive straight pipes; (*c*) $L_s =$ the actual length of the center line of the entire system. Convention *a* is used in this handbook; convention *c* is quite commonly used in reference books on heating and ventilating, and on large-scale hydraulics. Obviously the differences among the three conventions are nearly negligible except in cases such as that of the long radius bend, but some of the apparent discrepancies among the data from different sources arise from the failure of the authors to state the convention used.

The data below are reported insofar as possible in terms of L_e, the number of diameters of straight pipe equivalent to the fitting (see p. 387). In the region of turbulent flow, L_e varies but slightly with the Reynolds number, $DG/\mu = 4w/\pi D\mu$, in the cases of bends and of elbows, which are the fittings most completely studied.

Table 14 has been prepared by comparing the data tabulated by various authorities, including the Crane Co., and is believed to give safe values of L_e. For references to the original experiments and for discussion thereof, see H. Richter, *Forschungsarbeiten*, **338** (1930), Gibson, "Hydraulics and Its Applications"; Wilson, McAdams, and Seltzer, *Ind. Eng. Chem.*, **14**, 105 (1922); Giesecke, Reming, and Knudsen, *Univ. Texas Bull.* 2712 (Eng. series No. 22) (1927). Most of the data were obtained for flow of liquids through so-called **standard screwed fittings**. The figures given for valves and water meters should be regarded as very rough approxi-

Table 14. Friction Loss of Screwed Fittings, Valves, Etc.

	Equivalent length in pipe diameters, L_e	Number of "velocity heads"
45-deg. elbows	15	0.3
90-deg. elbows, standard radius	32	0.74
90-deg. elbows, medium radius	26	0.60
90-deg. elbows, long sweep	20	0.46
90-deg. square elbow	60	1.3
180-deg. close return bends	75	1.7
180-deg. medium radius return bends	50	1.2
Tee (used as elbow, entering run)	60	1.3
Tee (used as elbow, entering branch)	90	1.9
Couplings	Negligible	
Unions	Negligible	
Gate valves, open	7	0.13
Globe valves, open	300	6.0
Angle valves, open	170	3.0
Water meters, disk	400	8.0
Water meters, piston	600	12.0
Water meters, impulse wheel	300	6.0

mations since there is much variation in equipment of the same type from different manufacturers.

Smooth bends of round copper pipe have been carefully investigated by Richter (*op. cit.*) whose results are expressible with accuracy by

$$L_e = 0.0202X\alpha^{1.10}\left(\frac{DG}{\mu}\right)^{0.032} \tag{42}$$

in which X is the function of the radius of curvature given in Fig. 31; α is the angle, deg., subtended by the bend; D is the diameter of the pipe, ft.; G is the mass velocity, lb. *mass*/(sec.)(sq. ft.); and μ is the viscosity, lb. *mass*/(ft.)(sec.). For ordinary computations the

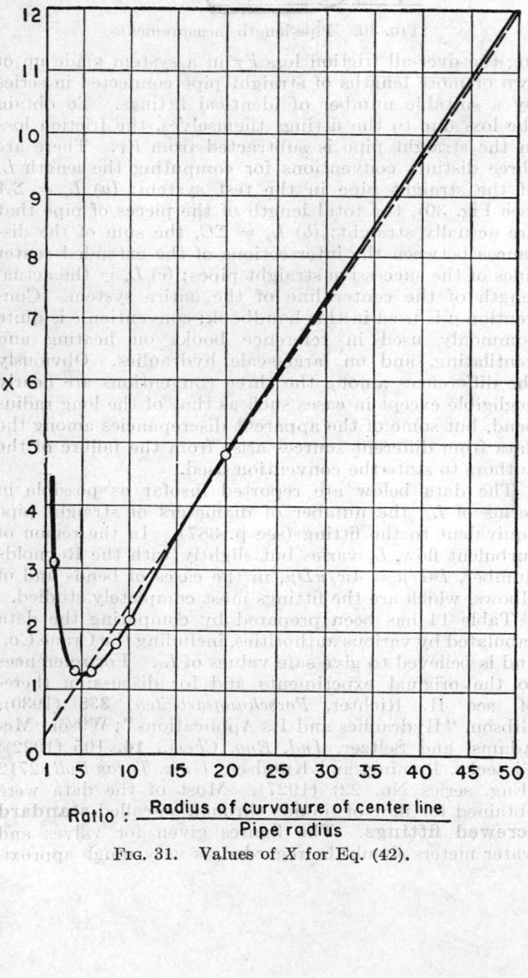

FIG. 31. Values of X for Eq. (42).

factor containing the Reynolds number DG/μ may be taken as unity. When the radius of curvature is greater than 8 pipe diameters, $X = 0.482 \times$ (radius of curvature of center line in pipe diameters), approximately, as shown by the dotted line in Fig. 31. In Richter's experiments 21 different bends of 40 mm. (1.57 in.) in diameter pipe were used having α's varying from 10.3 to 180 deg., and radii of curvature from 0.85 to 47.75 pipe diameters; and one 88.4-deg. bend of 20 mm. (0.788 in.) pipe was also tested. The Reynolds number was varied from 1660 to 72,000, but Eq. (42) represents the results for the turbulent region only. Equation (42), which is in fair agreement with the scattered results of other experimenters, should be applicable to **smooth welded bends** assembled from commercial fittings such as tube turns.

FLOW THROUGH EQUIPMENT

Heat Exchangers. The pressure drop on the shell side of a heat exchanger cannot be estimated adequately by a single equation because of the variety of baffle types and arrangements, the extent of the clearances between tubes, baffles, and shell, and the sizes of entrance and exit connections. The pressure drop is best approximated by calculating the individual losses as follows. The entrance and exit losses may be determined as indicated on pp. 388*ff*. The drop due to flow normal to the tubes can be calculated by Eqs. (43) and (44). Losses due to flow parallel to the tubes may be calculated by means of a hydraulic radius. The loss due to the baffles may be determined by Eq. (45). It should be emphasized that leakage through the clearances between tubes, baffles, and shell may amount to 40 per cent or more of the total flow through an exchanger.

The flow on the tube side is simply a special case of flow in a pipe (*cf.* pp. 377*ff*.).

Pressure Drop across Tube Banks. It is not possible to obtain a fundamental correlation of the data on pressure drop across tube banks in a single friction factor–Reynolds number curve. This is due to the lack of geometric similarity resulting from the large number of tube configurations and spacings that are encountered. A degree of correlation was obtained by Chilton and Genereaux [*Trans. Am. Inst. Chem. Engrs.*, **29**, 161 (1933)] for turbulent flow through staggered tubes by the use of a friction factor—modified Reynolds number relationship, where the friction factor was based on the number of rows normal to the flow and on V_{max} (the maximum velocity through the clearances), and the Reynolds number was based on V_{max} and on D_c (the transverse tube clearance) (see Fig. 32). The judicious choice of D_c for the diameter term made at least partial allowance for the effect of transverse spacing; and, although no allowance was made for longitudinal spacing, it has since been shown to have but a comparatively minor effect for staggered configurations.

Grimison [*Trans. Am. Soc. Mech. Engrs.*, **59**, 583 (1937)] has presented a friction factor–modified Reynolds number correlation for turbulent flow through banks of staggered and in-line tubes. Grimison employed the same friction-factor relation as Chilton and Genereaux (*op. cit.*) but based the Reynolds number on D (tube diameter), with transverse and longitudinal spacing as parameters on the friction factor–Reynolds number plot (see Figs. 33*A* and 33*B*). For the viscous-flow region, a correlation based on an equivalent diameter was found to be better.

For tube bundles in a circular shell, average values of N_c (the number of transverse tube clearances per row) should be used, as well as an average value of D_c, to allow for the variation in the number of tubes per row and in the space between the tube and shell. The value of N_r to use is the number of rows of tubes in the whole exchanger, which is the number of rows in a pass times the number of passes; except for the case in which some of the tubes pass through the baffle openings. In that case the number of rows per pass has to be reduced arbitrarily to allow for the fact that the fluid does not then effectively cross all the rows.

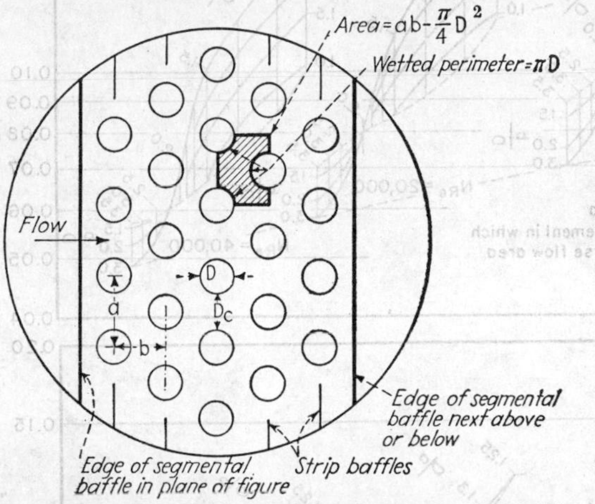

FIG. 32. Heat-exchange pressure-drop diagram for determining dimensions; one pass shown. $N_r = 5$; $N_c = 5, 6, 7, 6, 5$, respectively.

For the **turbulent-flow region,** the pressure drop may be calculated from the following equation using consistent units:

$$\Delta p = \frac{4fN_r\rho V^2_{\max}}{2g_c} \tag{43}$$

For banks of **staggered tubes, the friction factor** is given by Eq. (43*a*). This equation is recommended particularly for the common commercial spacings, *i.e.*, transverse pitches in the range of 1.25 to 1.50 tube diameters. The reliability in this range is probably within ±25 per cent. The flow may be considered to be turbulent for values of $(D_cV_{\max}\rho/\mu)$ greater than 40.

$$f = 0.75\left(\frac{D_cV_{\max}\rho}{\mu}\right)^{-0.2} \tag{43a}$$

where D_c = transverse clearance, ft.; f = friction factor, dimensionless; g_c = conversion factor, 32.1740 (lb. *mass*) (ft.)/(lb. *force*)(sec.²); N_r = number of rows of tubes normal to the flow, dimensionless; $V_{\max}$ = fluid velocity through the minimum area available for flow, ft./sec.; Δp = pressure drop, lb. *force*/sq. ft. ρ = density of fluid, lb. *mass*/cu. ft.; μ = viscosity, lb. *mass*/(ft.)(sec.).

It has been assumed in the foregoing discussion that the minimum area for flow occurs in the transverse openings. However, if this minimum area occurs in the diagonal openings, Eqs. (43) and (43*a*) can still be used by defining D_c as the diagonal clearance and N_r as the number of rows of tubes normal to the flow less one, and basing $V_{\max}$ on the area of the diagonal openings.

For banks of staggered tubes with **transverse pitches of 1.50 or more tube diameters,** or in general where greater accuracy is needed than is given by Eq. (43*a*), it is recommended that the friction-factor term for use in Eq. (43) be selected from the Grimison (*op. cit.*) chart given in Fig. 33*A*. Each "fence" represents a specific Reynolds number $(DV_{\max}\rho/\mu)$, where D = tube diameter, ft. For intermediate Reynolds numbers, it may be necessary to interpolate by drawing a curve through values obtained from two or more of the "fences," although in most cases mental interpolation should be adequate. Here again, if the minimum area occurs in the diagonal openings, N_r should be taken as the number of rows normal to the flow less one, $V_{\max}$ should be based on the diagonal clearance area, and f should be taken from the lower chart in Fig. 33*A*. $V_{\max}$ is always based on the minimum area available for flow. A recent comparison of this correlation with all the data now available has indicated an average deviation of the correlation from the data of only ±15 per cent.

For banks of **in-line tubes** (rectangular arrangement), the pressure drop can be calculated using Eq. (43) and taking the friction factor from the Grimison (*op. cit.*) chart in Fig. 33*B*. Each "fence" represents a specific Reynolds number, which is defined in the previous paragraph. The deviation of the correlation from all the now available data is on the order of ±15 per cent.

In the case of turbulent flow through **shallow banks** of tubes, the average friction factor per row will be somewhat higher than indicated by the equations and charts given up to this point which are based on banks 10 or more rows deep. The magnitude of this increase is dependent upon the tube spacing and arrangement. A 30 per cent increase per row for two rows, 15 per cent per row for three rows, and 7 per cent per row for four rows may be taken as about the maximum likely to be encountered.

For the case where a large temperature change occurs in a *gas* as it flows across a tube bundle, it is necessary to select a suitable mean temperature upon which to base the gas properties. The mean temperature for gases may be taken as

$$t_m = t_t + K_1\,\Delta t_m \tag{43b}$$

where K_1 = a constant, dimensionless; t_m = mean temperature for evaluation of gas properties, °F. or °C.; t_t = average tube-wall temperature, °F. or °C.; Δt_m = logarithmic mean temperature difference between the gas and the tubes, °F. or °C. Values of K_1, averaged from the recommendations of Chilton and Genereaux (*op. cit.*) and Grimison (*op. cit.*), are as follows:

	Gas cooling off		Gas heating up	
	In-line	Staggered	In-line	Staggered
K_1	0.9	0.75	−0.9	−0.8

In the case of non-isothermal flow of *liquids* across tube bundles, the friction factor is sensibly increased if the liquid is being cooled and decreased if the liquid is being heated. It is recommended that the factors given for non-isothermal flow of liquids in pipes (p. 382) be employed for this case.

In the **viscous-flow** region, the equation presented by Chilton and Genereaux (*op. cit.*) is recommended for the calculation of pressure drop:

$$\Delta p = \frac{53\mu LV_{\max}}{g_cD_e^2} \tag{44}$$

where D_e = equivalent diameter, equal to four times the hydraulic radius (see Eq. 44*a*), ft.; g_c = conversion factor, 32.17 (lb. *mass*)(ft.)/(lb. *force*)(sec.²); L = length of tube bank in the direction of flow, ft.; $V_{\max}$ = fluid velocity through the minimum area available for flow, ft./sec.; Δp = pressure drop, lb. *force*/sq. ft.; μ = fluid viscosity, lb. *mass*/(ft.)(sec.). The flow may be considered to be viscous for all values of $(D_eV_{\max}\rho/\mu)$ less than 150. The equivalent diameter for a tube bank is given by the following equation:

$$D_e = \frac{4[ab - (\pi/4)D^2]}{\pi D} \tag{44a}$$

where a = transverse pitch, ft.; b = longitudinal pitch, ft.; and D = tube diameter, ft.; and ρ = fluid density, lb. *mass*/cu. ft.

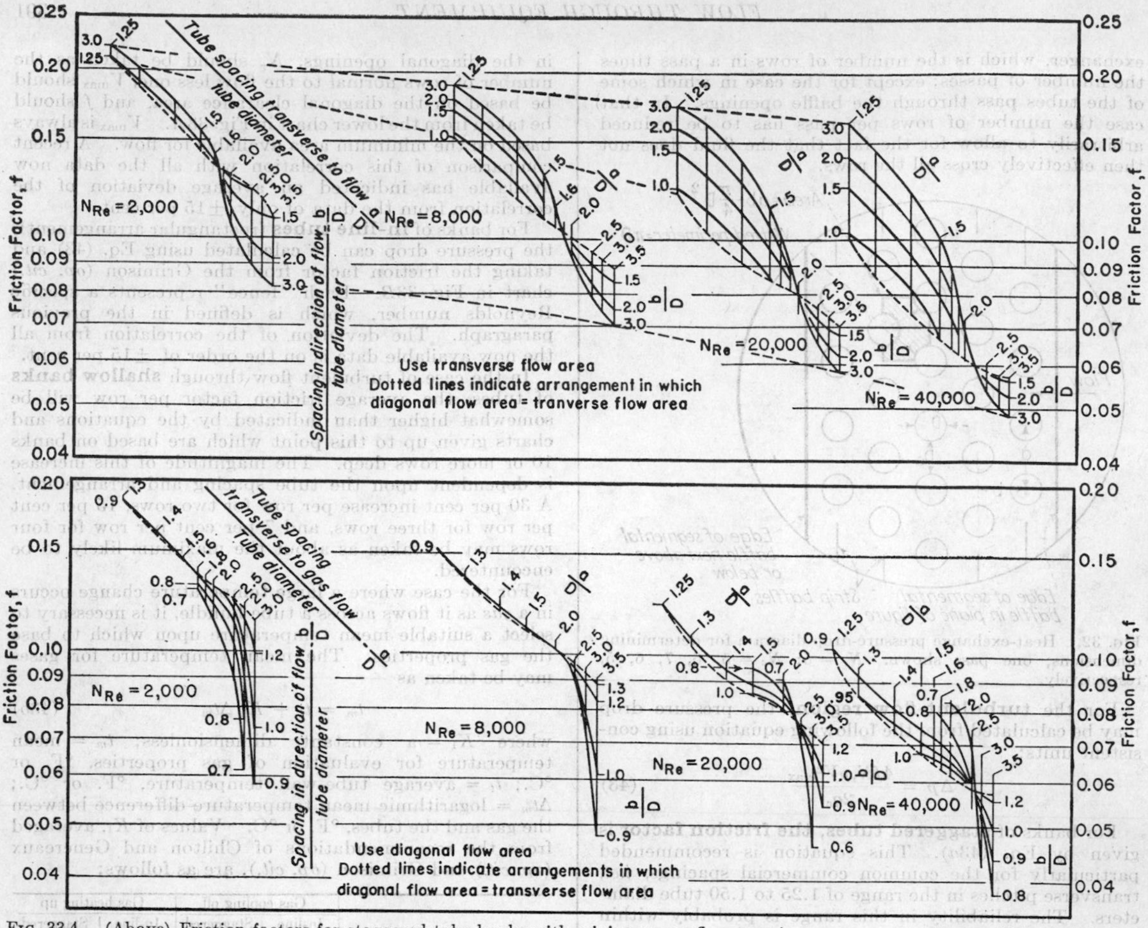

FIG. 33A. (Above) Friction factors for staggered-tube banks with minimum gas-flow area in transverse openings. (Below) Friction factors for staggered-tube banks with minimum gas-flow area in diagonal openings.

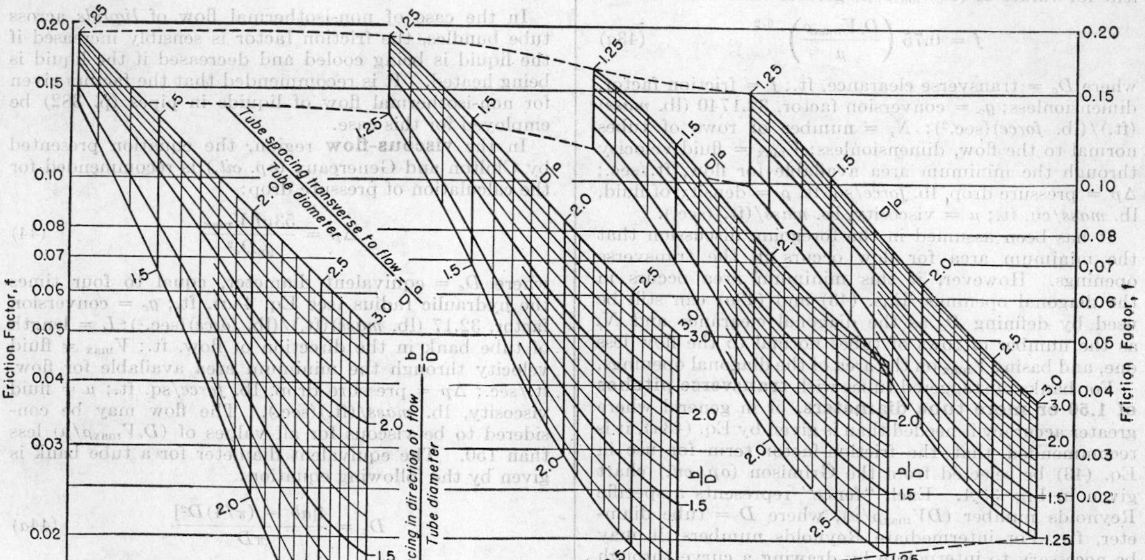

FIG. 33B. Friction factors for in-line tube banks.

This correlation was based only on data for two equilateral-staggered arrangements with transverse pitches of 1.25 and 1.59 tube diameters, and it should therefore be used with caution for spacings outside this range. Although no information is available on viscous flow through banks of in-line tubes, it is suggested that Eq. (44) be used as an approximation for this case.

Pressure Drop across Baffles. A tube-bundle baffle in the form of a segment of a circle may be treated as an orifice with an approximate coefficient of 0.7 to give in consistent units the equation

$$\Delta p = \frac{w^2 N_b}{\rho S_b^2 g_c} \tag{45}$$

where Δp = pressure drop, lb. *force*/sq. ft.; w = weight rate of flow, lb. *mass*/sec.; N_b = number of baffles in series; ρ = density of fluid, lb. *mass*/cu. ft.; S_b = area of free opening of baffle segment, sq. ft.; and g_c = conversion factor = 32.1740 (lb. *mass*) (ft.)/(lb. *force*) (sec.^{2}). In certain common engineering units, Eq. (45) can be expressed as

$$\Delta p_i = \frac{4.5 w^2 N_b}{S_{bi}^2 \rho} \tag{45a}$$

where Δp_i = lb. *force*/sq. in., S_{bi} = area of free opening of baffle segment, sq. in., and other units are as defined above.

Packed Towers. Data on the **flow of fluids through beds of granular solids** or other porous structures such as are used in absorption towers, distillation columns, sand beds, filters, driers, and catalyst beds are not readily correlated. This is due to the variety of packings, their arrangement, and the effect of countercurrent flow when more than one phase is present.

For the flow of a single fluid through uniform granular solid particles, Chilton and Colburn [*Trans. Am. Inst. Chem. Engrs.*, **26**, 178 (1931)] correlated the available data, see Fig. 34, on a friction factor f vs. a modified Reynolds number basis, using the following equations, all in *consistent units:*

$$f = \frac{2 g_c D_p \Delta p}{4 L \rho V_0^2 A_f} \tag{46}$$

which in the form of the Fanning equation is as follows:

$$\Delta p = \frac{4 f L \rho V_0^2 A_f}{2 g_c D_p} \tag{46a}$$

and

$$N_{Re} = \frac{D_p V_0 \rho}{\mu} \tag{46b}$$

Although the data form a curve without a sharp break, straight lines for the viscous and turbulent regions were drawn for convenience and expressed for the viscous region ($N_{Re} < 40$) as

$$f = \frac{850}{N_{Re}}$$

and for the turbulent region ($N_{Re} > 40$) as

$$f = \frac{38}{N_{Re}^{0.15}}$$

Expressions for calculating the pressure drop are, for the **viscous region,**

$$\Delta p = \frac{53 \mu L V_0 A_f}{D_p^2} \tag{47}$$

and for the **turbulent region**

$$\Delta p = \frac{2.36 \mu^{0.15} L \rho^{0.85} V_0^{1.85} A_f}{D_p^{1.15}} \tag{48}$$

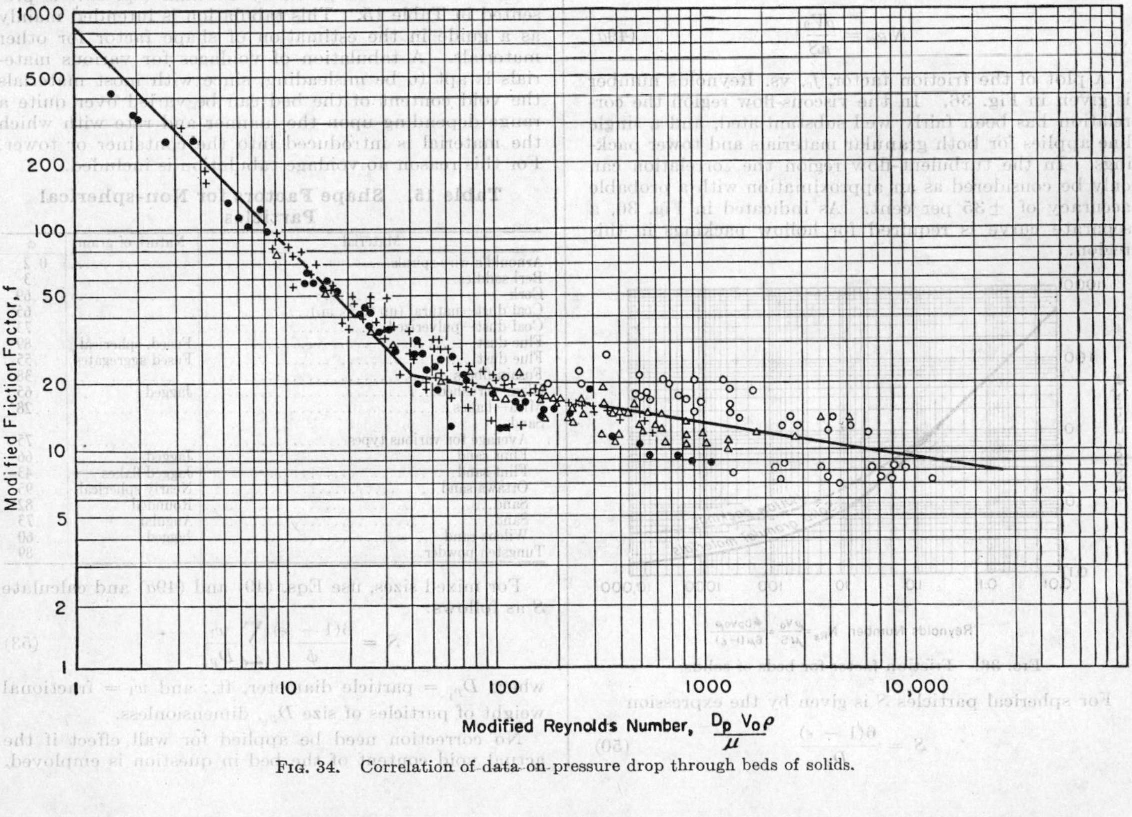

FIG. 34. Correlation of data on pressure drop through beds of solids.

where Δp = pressure drop, lb. *force*/sq. ft.; μ = viscosity of fluid, lb. *mass*/(ft.)(sec.); L = depth of packing, ft.; ρ = density of fluid, lb. *mass*/cu. ft.; V_0 = velocity based on the cross section of empty tower, ft. per sec.; A_f = wall-effect factor (see Fig. 35) dimensionless; D_p = nominal diameter of particles, ft.; and g_c = conversion factor = 32.1740 (lb. *mass*)(ft.)/(lb. *force*)(sec.^{2}).

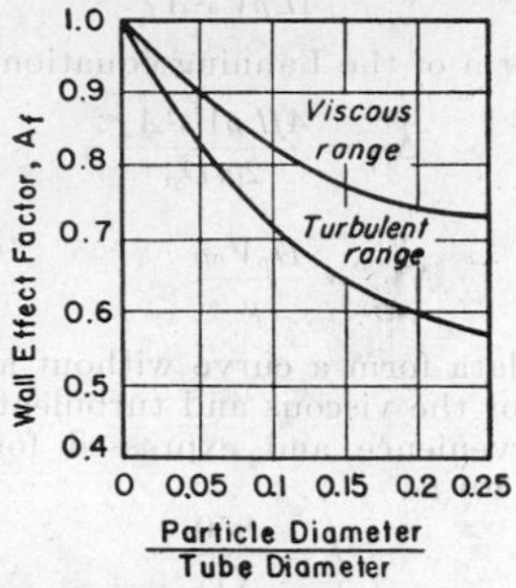

FIG. 35. Wall-effect factor. Computed by Furnas, *U.S. Bur. Mines Bull.* **307,** 1929.

For beds of **abnormal void content** (outside the range from 35 to 45 per cent), it is recommended that the correlation of Carman [*Trans. Inst. Chem. Engrs.* (*London*), **15,** 150 (1937)] be employed, particularly where the percentage voids and specific surface or shape factor of the material in question is available, or where a reasonable estimate can be made of these quantities.

$$\Delta p = \frac{f_c L S \rho V_0^2}{g_c \epsilon^3} \tag{49}$$

where f_c is a function of the Reynolds number, N_{Re}

$$N_{Re} = \frac{\rho V_0}{\mu S} \tag{49a}$$

A plot of the friction factor, f_c, vs. Reynolds number is given in Fig. 36. In the viscous-flow region the correlation has been fairly well substantiated, and a single line applies for both granular materials and tower packings. In the turbulent-flow region the correlation can only be considered as an approximation with a probable accuracy of ±35 per cent. As indicated in Fig. 36, a separate curve is required for hollow packings in this region.

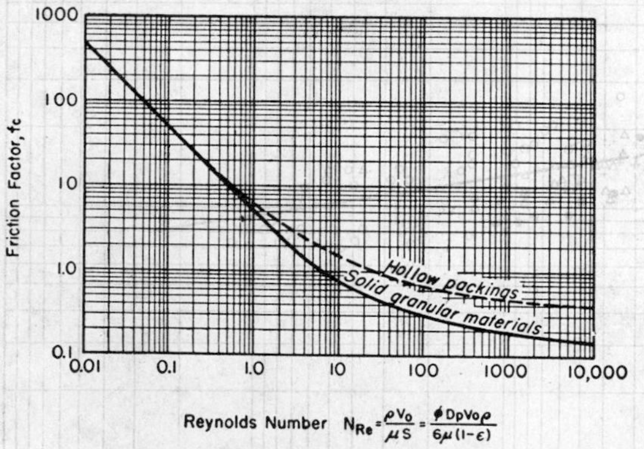

FIG. 36. Friction factor for beds of solids.

For spherical particles S is given by the expression

$$S = \frac{6(1-\epsilon)}{D_p} \tag{50}$$

For non-spherical particles, this may be written

$$S = \frac{6(1-\epsilon)}{\phi D_p} \tag{50a}$$

Thus

$$\Delta p = \frac{6 f_c L \rho V_0^2 (1-\epsilon)}{g_c \phi D_p \epsilon^3} \tag{51}$$

and

$$N_{Re} = \frac{\phi D_p V_0 \rho}{6\mu(1-\epsilon)} \tag{51a}$$

where D_p = average particle diameter, ft.; f_c = friction factor, dimensionless; g_c = conversion factor, 32.1740 (lb. *mass*)(ft.)/(lb. *force*)(sec.^{2}); L = depth of bed, ft.; N_{Re} = Reynolds number, dimensionless; S = area of particle surface per unit volume of bed, $S_0(1-\epsilon)$, sq. ft./cu. ft.; S_0 = area of particle surface per unit volume of solid, sq. ft./cu. ft.; V_0 = superficial velocity based on empty chamber cross section, ft./sec.; Δp = pressure drop, lb. *force*/sq. ft.; ϵ = voidage (fractional free volume), dimensionless; ϕ = shape factor (unity for spheres), dimensionless; ρ = fluid density, lb. *mass*/cu. ft.; μ = viscosity, lb. *mass*/(ft.)(sec.).

In the viscous-flow region ($N_{Re} < 2$ for solid granular materials, and $N_{Re} < 0.25$ for hollow packings)

$$f_c = \frac{5}{N_{Re}} \tag{52}$$

and the corresponding expression for pressure drop is

$$\Delta p = \frac{5\mu L S^2 V_0}{g_c \epsilon^3} = \frac{180 \mu L (1-\epsilon)^2 V_0}{g_c \epsilon^3 \phi^2 D_p^2} \tag{52a}$$

The value of the shape factor (unity for spheres) depends upon the evaluation of the diameter term for irregular particles. A tabulation of shape factors for various materials as given by Carman (*op. cit.*) is presented in Table 15. This tabulation is intended mainly as a guide in the estimation of shape factor for other materials. A tabulation of voidages for various materials is apt to be misleading, since with most materials the void content of the bed can be varied over quite a range depending upon the manner and rate with which the material is introduced into the container or tower. For this reason no voidage tabulation is included.

Table 15. Shape Factors for Non-spherical Particles

Material	Nature of grain	ϕ
Arnould's wire spirals		0.2
Berl saddles		.3
Cork		.69
Coal dust—natural (up to 3/8 in.)		.65
Coal dust—pulverized		.73
Flue dust	Fused, spherical	.89
Flue dust	Fused aggregates	.55
Fusain fibers		.38
Glass—crushed	Jagged	.65
Mica—flakes		.28
Sand		
Average for various types		.75
Flint sand	Jagged	.66
Flint sand	Jagged flakes	.43
Ottawa sand	Nearly spherical	.95
Sand	Rounded	.82
Sand	Angular	.73
Wilcox sand	Jagged	.60
Tungsten powder		.89

For mixed sizes, use Eqs. (49) and (49a) and calculate S as follows:

$$S = \frac{6(1-\epsilon)}{\phi} \sum \frac{w_1}{D_{p_1}} \tag{53}$$

where D_{p_1} = particle diameter, ft.; and w_1 = fractional weight of particles of size D_{p_1}, dimensionless.

No correction need be applied for wall effect if the actual void content of the bed in question is employed.

However, if the void content is known only for a bed of large cross section, multiply the pressure drop as calculated for this void content by the appropriate wall-effect factor as given by Fig. 35.

For **hollow shapes** such as Raschig and Lessing rings and Berl saddles (see p. 680 *et seq.* for descriptions), factors developed by Chilton by which to multiply the calculated pressure drop data of White [*Trans. Am. Inst. Chem. Engrs.*, **31**, 390 (1935)] are as follows:
For Raschig or Lessing rings,

$$F_h = \frac{0.24}{D_{pi}^{0.5}} \tag{54}$$

and for Berl saddles,

$$F_h = \frac{0.13}{D_{pi}^{0.5}} \tag{55}$$

where F_h is the factor by which to multiply Δp in Eqs. (47) and (48), and D_{pi} is the nominal packing diameter, in. Equations (54) and (55) are to be used without the wall-effect factor A_f, mentioned above, but represent the data only when the ratio of particle size to tower diameter is less than $\frac{1}{6}$.

In the case of gas flow, the pressure drop is increased due to reduction in free volume if a **liquid is wetting or circulating** through the packing. Factors to allow for wetting of solid packings (wet and drained) and for the effect of circulating liquid [water, at a rate of 0.18 lb. *mass*/(sec.) (sq. ft. tower cross section) or 660 lb. *mass*/(hr.)(sq. ft.)] were tentatively proposed by Chilton and Colburn (*op. cit.*) based on data of Zeisberg (*Trans. Am. Inst. Chem. Engrs.*, **12**, 231, 1919) as follows:
packing wet and drained,

$$F_w = 1 + \frac{0.22}{D_{pi}} \tag{56}$$

water circulating at rate of 0.18 lb. *mass*/(sec.)(sq. ft.),

$$F_c = 1 + \frac{0.47}{D_{pi}} \tag{57}$$

The pressure drop calculated for dry packing is to be multiplied by the factors given.

The effect of water circulation on pressure loss in gas flow through hollow packings is also considered by White (*op. cit.*), who gives curves showing the relative increase in pressure drop due to water circulation over that for the wet and drained Raschig rings. These curves have been replotted and added to by Sherwood ("Absorption and Extraction," McGraw-Hill, New York, 1937) as shown in Figs. 37 and 38. The ordinate is a factor, F_L, by which to multiply the pressure drop for dry packing.

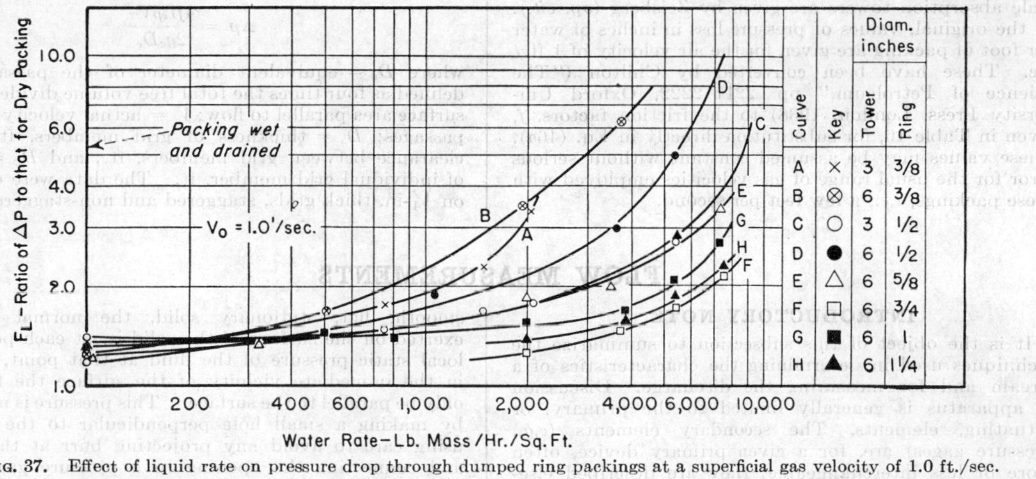

FIG. 37. Effect of liquid rate on pressure drop through dumped ring packings at a superficial gas velocity of 1.0 ft./sec.

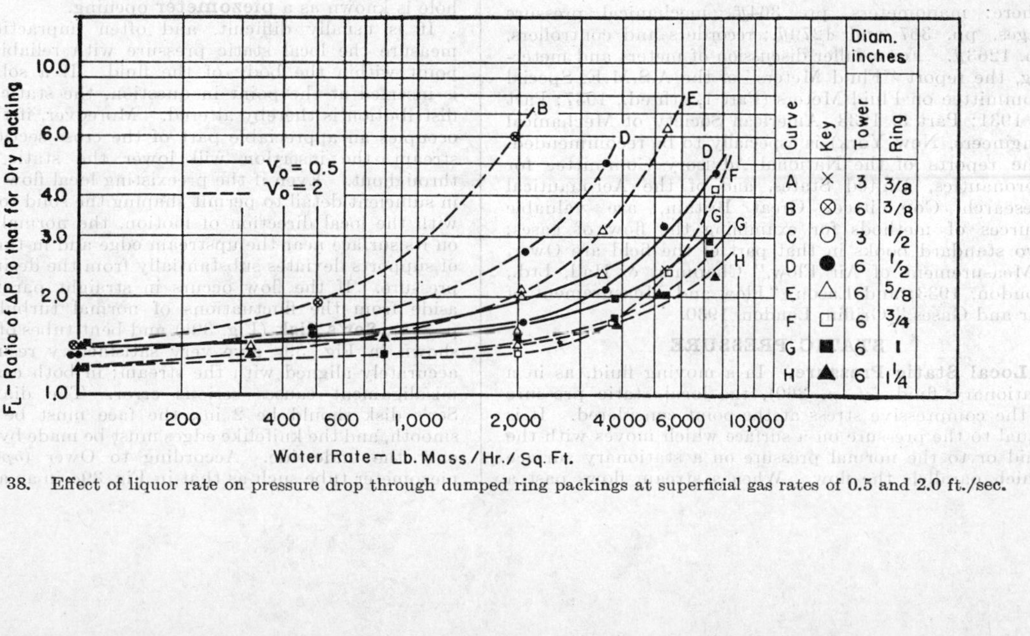

FIG. 38. Effect of liquor rate on pressure drop through dumped ring packings at superficial gas rates of 0.5 and 2.0 ft./sec.

Not enough data are available to generalize as to the effect of liquids other than water. It seems likely, however, that more viscous liquids would cause a greater increase in pressure drop, at the same rate of flow.

Considerable experimental work on pressure drop, flooding velocity, distribution, etc., is reported by Uchida and Fujita in the following references: *J. Soc. Chem. Ind. Japan*, **37**, 1578 (1934), also *Suppl. Binding*, p. 724B, pressure drop through dry lead shot and aluminum Lessing rings; *ibid.*, **37**, 1707 (1934), also 791B, pressure drop through 5, 7, and 10 mm. dry broken limestone; *ibid.*, **39**, 876 (1936), also 431B, pressure drop, flooding velocity, liquid hold-up, and liquid distribution in 15, 26, and 35 mm. Raschig rings, dumped and stacked, irrigated with water; *ibid.*, **40**, 538 (1937), also p. 238B, same as previous paper but at higher water rates, also some measurements on 16, 25, and 35 mm. broken solids; *ibid.*, **41**, 563 (1938), also p. 275B, same as two previous papers but irrigated with two viscous oils. Pressure drop measurements on irrigated Raschig rings, Berl saddles, Prym rings and prisms, in 8- to 30-mm. sizes are reported by Mach, *Forsch. Gebiete Ingenieurw.* 6, *Forschungsheft* No. 375, 9–20 (1935). The effect of gas density and liquid viscosity is also considered. More recently, Schoenborn and Dougherty have made measurements on the effect of liquor rate on pressure drops through tower packings using water and two oils; see *Trans. Am. Inst. Chem. Engrs.*, **40**, 51–77 (1944).

Data on the pressure loss in the **larger sized fabricated stoneware packing** shapes often used in large-scale absorption towers are given by Zeisberg (*op. cit.*). In the original, values of pressure loss in inches of water per foot of packing are given for the air velocity of 1 ft./sec. These have been converted by Chilton ("The Science of Petroleum," pp. 2211–2222, Oxford University Press, London, 1938) to the friction factors, f, given in Table 16, for substitution directly in Eq. (46*a*). These values may be assumed constant without serious error for the usual range of gas velocities employed with these packings, *i.e.*, a few feet per second.

Table 16. Friction Factors for Fabricated Packings,* for Use with Eq. (46*a*)

Packing	D_P, ft.	Dumped			Stacked			Packed		
		Dry	Wet	Circ.	Dry	Wet	Circ.	Dry	Wet	Circ.
4 × 3 in. smooth diaphragm rings	0.33	6.1	6.7	7.6	3.0	4.1	4.6	...	3.9	4.1
3 × 3 in. corrugated diaphragm rings	0.25	4.7	5.1	5.8	2.8	3.8	5.4	2.4	2.6	2.6
3 × 3 in. corrugated spiral rings	0.25	4.1	4.9	5.8	4.9	4.9	5.6			
6 × 6 in. corrugated spiral rings	0.50	...	...	...	6.9	8.6	9.1	4.1	5.7	6.4
6 in. corrugated Hechenbleikner blocks	0.50	2.9	6.6	6.6	6.4	6.9	6.9	4.7	5.1	5.1

* Data of Zeisberg.

"Dumped" signifies packing dumped at random.

"Stacked" signifies packing arranged regularly in layers, but with no attempt to have pieces in one layer in any way related to those in adjacent layers.

"Packed" signifies packing arranged regularly in layers, with axes of pieces in adjacent layers coinciding.

"Circ." signifies water circulated at rate of 0.18 lb. *mass*/(sec.)(sq. ft.) = 600 lb. *mass*/(hr.)(sq. ft.).

For **grid packings**, Johnstone and Singh [*Ind. Eng. Chem.*, **29**, 286 (1937)] propose the friction factor equation

$$f = 0.08\left(\frac{D_e V \rho}{\mu}\right)^{-0.2} + 0.52\left(\frac{D_b}{D_s}\right)^{1.5}\left(\frac{D_e}{D_g}\right)^{0.75}$$

for use in the Fanning equation (consistent units),

$$\Delta p = \frac{4fL\rho V^2}{2g_c D_e} \qquad (58)$$

where D_e = equivalent diameter of the passages, ft. defined as four times the total free volume divided by the surface area parallel to flow; V = actual velocity through passages; D_b = thickness of grid members, ft.; D_s = clearance between grid members, ft.; and D_g = height of individual grid member, ft. The data were obtained on ¼-in. thick grids, staggered and non-staggered.

FLOW MEASUREMENTS

INTRODUCTORY NOTE

It is the object of this subsection to summarize the techniques used in determining the characteristics of a stream and for measuring the discharge. Discussion of apparatus is generally limited to the primary, or actuating, elements. The secondary elements (*e.g.*, pressure gages) are, for a given primary device, often more or less interchangeable; they are described elsewhere: manometers, pp. 364*ff.*, mechanical pressure gages, pp. 367 and 1279*ff.*; recorders and controllers, pp. 1263*ff.* For a fuller discussion of meters and metering, the report "Fluid Meters" of the A.S.M.E. Special Committee on Fluid Meters (Part 1, 4th ed., 1937; Part 2, 1931; Part 3, 1933, American Society of Mechanical Engineers, New York) is especially to be recommended. The reports of the National Advisory Committee for Aeronautics, United States, and of the Aeronautical Research Committee, Great Britain, are valuable sources of methods for examining the flow of gases; two standard books in that part of the field are Ower, "Measurement of Air Flow," Chapman & Hall, Ltd., London, 1933; and Eason, "Flow and Measurement of Air and Gases," Griffin, London, 1930.

STATIC PRESSURE

Local Static Pressure. In a moving fluid, as in a stationary fluid (*cf.* p. 360), the local static pressure is the compressive stress at the point considered. It is equal to the pressure on a surface which moves with the fluid or to the normal pressure on a stationary surface which parallels the flow. When a stream flows past a smooth, flat, stationary solid, the normal pressure exerted on the surface of the solid is, at each point, the local static pressure of the fluid at that point, because in the immediate vicinity of the surface the flow can only be parallel to the surface. This pressure is measured by making a small hole perpendicular to the surface, using care to avoid any projecting burr at the edges, and connecting the opening to a pressure gage. The hole is known as a **piezometer** opening.

It is usually difficult, and often impracticable, to measure the local static pressure with reliability at a point within the body of the fluid. If a solid object is inserted at the point in question, the static pressure distribution is thereby altered. Moreover, if the object occupies an appreciable part of the cross section of the stream, the insertion will lower the static pressure throughout. Even if the preexisting local flow is known in sufficient detail to permit shaping the solid to conform with the local direction of motion, the normal pressure on its surface near the upstream edge and in the vicinity of supports deviates substantially from the desired static pressure. If the flow occurs in straight parallel lines, aside from the fluctuations of normal turbulence (*cf.* p. 375), **Ser's disk** (Fig. 39*b*) and bent tubes of the type shown in Fig. 39*c* give very satisfactory results when accurately aligned with the stream; in both cases slight misalignment causes serious error. The diameter of Ser's disk should be 2 in., the face must be flat and smooth, and the knifelike edges must be made by beveling only the underside. According to Ower (*op. cit.*), a piezometer tube such as that in Fig. 39*c*, in an air stream at 70 ft./sec., reads correctly because of a balancing of errors if seven static holes (0.04 in. diameter) are located in a ring 6 tube diameters downstream from the beginning of the straight walls at the closed end and 15 tube diameters upstream from the supporting tube. For water at 8 ft./sec. Hubbard [*Trans. Am. Soc. Mech. Engrs.*, **61,** 477–506 (1939)] found distances of 6 diameters from the tip and 8 diameters from the support to be adequate. The readings given by open straight tubes (Fig. 39*d*) are too low. The readings of closed tubes extending perpendicularly to the axis of the stream and provided with side openings (Fig. 39*e*) may be too low by as much as two velocity heads, but, if the openings face 30 deg. on either side of the upstream direction, the correct static pressure will be approximated. Tubes, constructed as just described, have been suggested for the exploration of non-parallel flows, but caution is advisable because possible baffle effects of the tube itself may alter the local motion.

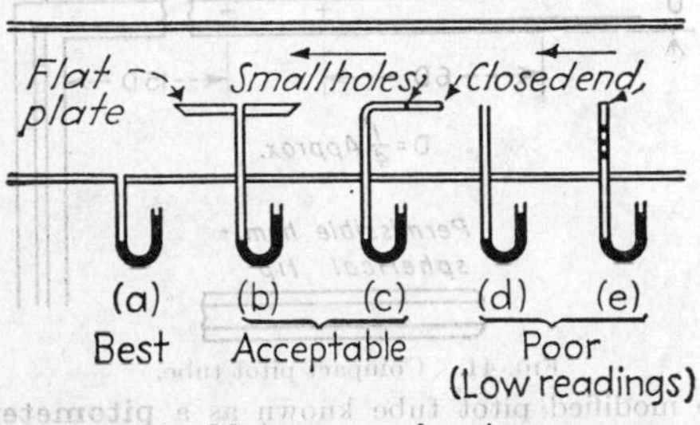

FIG. 39. Measurement of static pressure.

Average Static Pressure. In most cases the object of a static-pressure measurement is to obtain a suitable average value for substitution in Bernoulli's theorem or in an equivalent flow formula. This can be simply done only in case the flow is in straight lines parallel to the confining walls. Such flow occurs, for example, in straight ducts at sufficient distances beyond bends and other obstructions and in the throats of well-shaped nozzles. For streams of this class the sum of the static head and the gravitational potential head is the same at all points of a cross section taken perpendicular to the current. An adequate average static pressure can therefore be obtained by making a piezometer opening in the duct wall at any convenient point of the periphery of the cross section. The exact location of the opening is immaterial, provided that its elevation is known, because only the sum of the static and potential heads enters in flow formulas; it is customary in stating the static pressure to give the value at the center line of the stream.

In other than straight passages or for swirling flow in straight ducts, *e.g.*, in the tail pipe of a cyclone separator (pp. 1017 and 1023), the determination of a true average static pressure is, in general, impracticable. It would involve the estimation of a mean of a set of local static-pressure measurements weighted in proportion to the discharge through the corresponding parts of the cross section. In curved pipes the average static pressure at a given cross section has been arbitrarily taken by some authorities to be that impressed on a piezometer opening at the end of a pipe diameter drawn perpendicular to the radius of curvature. In metering, straightening vanes (see p. 407) preceding the upstream pressure tap are used to eliminate the effect of swirl.

Specifications for Piezometer Taps. Theoretically, the size of a static opening should be small compared with the cross section of the channel and large compared with the scale of the surface irregularities. For reliable results, it is essential that (*a*) the surface in which the hole is made be substantially smooth and parallel to the flow for some little distance on either side of the hole and (*b*) the opening be flush with the surface and possess no "burr" or other irregularity about its inner edges. To ensure the absence of a burr, it is advisable to round the edges slightly. The foregoing precautions are by no means merely academic; gross errors may be introduced by neglecting them.

For side-wall openings in pipes of 2½ in. diameter and less, the Report of the A.S.M.E. Research Committee on Fluid Meters ("Fluid Meters," Part 1, p. 15, 4th ed., 1937) recommends ⅛- to $\frac{3}{16}$-in. holes drilled perpendicular to the wall and rounded to a radius of $\frac{1}{32}$ in. For larger pipes up to 16 in., ¼- to ½-in. holes are permissible. The American Gas Association gives pressure tap diameters as one-eighth of the pipe diameter for 2- to 4-in. pipe, as ½-in. pipe size for 4- to 8-in. pipe, and as one-sixteenth of the pipe diameter for larger than 8-in. pipe. The data of Fuhrmann [*Jahrb. Motorluftschiffstudienges.*, **5,** 63 (1911–1912)] indicate that holes of 0.3 mm. (0.0118 in.) diameter and larger give too low indications by between 0.8 and 0.9 per cent of a velocity head. For smaller openings the error decreases almost linearly to zero as the diameter decreases.

A **piezometer ring** is constructed by connecting several side-wall static openings in the same cross section to a common manifold which usually encircles the pipe. The pressure gage is attached to the manifold. The cross section of the manifold is fairly large, and the size of the several static openings is small. The piezometer ring has been considerably used in the belief that it automatically averages the possibly slightly differing pressures on the various openings. However, if there is any real difference between the time averages of the pressures on the separate openings, it is prima facie evidence that the flow is not wholly parallel to the walls. Under such conditions, even supposing that the reading of the gage on the manifold is actually the mean of the readings that would be obtained by attaching separate gages to the several holes, the reading has no definite relation to the true average static pressure. The principal advantage of the ring is that the use of several holes in place of one hole reduces the possibility of completely plugging the static openings.

LOCAL VELOCITY

Impact and Pitot Tubes. An impact tube, usually a piece of tubing bent at right angles as shown in Fig. 40, has its pressure opening facing squarely upstream. Sometimes the instrument is called a **total head tube.** For gases (at ordinary velocities) and for liquids, the pressure (lb. *force*/sq. ft.) on such an opening has been shown both theoretically and experimentally to be $p_i = p_0 + \frac{\rho_0 V_0^2}{2g_c}$ where p_0 (lb. *force*/sq. ft.) is the static pressure; ρ_0 and V_0 are the density (lb. *mass*/cu. ft.) and velocity (ft./sec.), respectively, of the fluid at the point where the tip is located; and g_c = 32.1740 (lb. *mass*)(ft.)/(lb. *force*)(sec.²); *i.e.*, the impact head is the sum of the static head and velocity head. With gases at velocities above 200 ft./sec. the exact theoretical formula for a perfect gas is

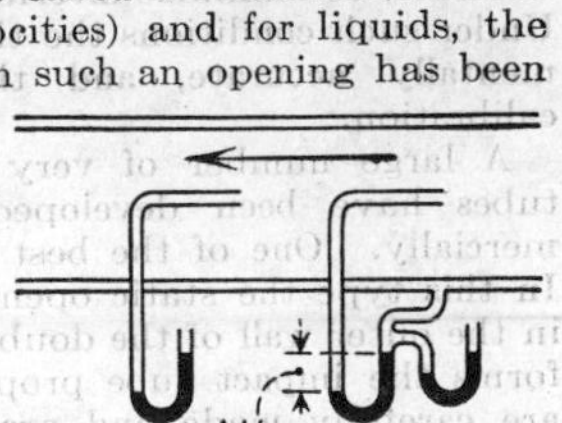

FIG. 40. Impact tubes.

$$V_0 = \sqrt{\left(\frac{2g_c\kappa}{\kappa - 1}\right)\frac{p_i}{\rho_i}\left[1 - \left(\frac{p_0}{p_i}\right)^{\frac{\kappa-1}{\kappa}}\right]} \tag{1}$$

where κ is the ratio of the specific heat at constant pressure to that at constant volume, and ρ_i is the density measured at the pressure p_i and at the temperature recorded by a thermometer inserted in the stream (*cf.* p. 402). For gas velocities near that of sound, see Rayleigh [*Proc. Roy. Soc.* (*London*), **A84**, 247, (1910)]. Only for exceedingly low velocities and for very small diameter tubes is there indication that the reading of an impact tube is affected by the viscosity of the fluid. Barker [*Proc. Roy. Soc.* (*London*), **A101**, 435 (1922)], using water as the fluid, found a tube 0.1 cm. in diameter to give readings agreeing with the usual formula until velocities less than 0.2 ft./sec. were reached. Such minute impact tubes, made of hypodermic syringe tubing, have been employed by Stanton and others to study the flow of a fluid near the pipe wall [see *Proc. Roy. Soc.* (*London*), **A85**, 366 (1911); **A97**, 413 (1920)].

The length and shape of the tip, so long as the opening faces upstream, have no effect upon the head indicated by an impact tube. The shape influences the behavior of the tube somewhat when the plane of the opening is inclined to its proper position, but usually even 10 to 15 deg. misalignment causes no significant error. The readings, in general, do not agree with those calculated by determining the component of the velocity, or of the impact pressure that acts perpendicular to the actual plane of the opening. Evidently an impact tube is an unsatisfactory instrument for finding the direction of flow at particular points in the fluid.

If the legs of a differential manometer are connected, one to a static opening and the other to an impact opening as in Fig. 40*b*, the reading evidently measures the velocity head directly. A **pitot tube** is an impact tube that is used in such a manner. When the reading ΔH is expressed in feet of the fluid flowing (see pp. 360 and 364) and the static opening is so located that it truly records the static pressure of the fluid at the tip of the impact tube, the velocity at that point is $V_0 = \sqrt{2g\,\Delta H} = \sqrt{2g_c(p_i - p_0)/\rho_0}$, which is easily derived from the formula for p_i on p. 397 and is subject to the same limitations. It should be clear from the discussion of impact and static openings that practically the only cause of a deviation in practice from the theoretical formula given above is an improper static opening. In a pipe or duct, if the point of measurement is preceded by at least 50 diameters of straight pipe, the greatest precision is obtained by the use of properly made side-wall static holes at the cross section where the impact tip is located. The impact tube, preferably streamlined, should be very small compared with the cross section of the duct, to minimize interference with the normal flow. Under such conditions the theoretical formula is experimentally accurate, and the apparatus requires no calibration.

A large number of very convenient compact pitot tubes have been developed and are available commercially. One of the best types is shown in Fig. 41. In this type the static openings are small holes drilled in the outer wall of the double pipe. The inner passage forms the impact tube proper. If the static openings are carefully made and are located 6 tube diameters from the point where the tube wall becomes parallel to the stream when the apparatus is correctly placed and at least 10 diameters from the support rod, the readings will agree closely with the theoretical equation. In general, however, it is advisable to take the formula for a commercial compact-type tube to be $V = C\sqrt{2g\,\Delta H}$ and to determine the coefficient C by calibration. The value of C for commercial pitot tubes is generally between 0.8 and 1.0, but, for a given tube, C may vary with the velocity, viscosity, and density of the fluid under investigation. The compact-type instrument is used almost wholly in practical engineering work on account of its convenience. When accuracy is desired, all pitot tubes should be preceded by as great a length of straight channel as possible. With compact pitot tubes, great care should be exercised to have the impact opening facing exactly upstream, for in the case of many such instruments a slight inclination of the tip will bring the static openings into an improper position and thus cause a serious error in the results. *It is most important to realize that the orientation for the maximum reading does not necessarily correspond with that for the correct reading;* for one of the best compact pitot tubes the maximum reading is 15 per cent greater than the true reading and occurs when the tube is 30 deg. out of alignment.

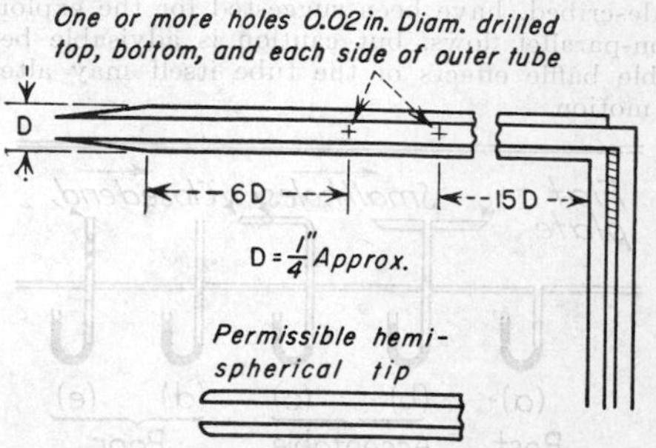

FIG. 41. Compact pitot tube.

The modified pitot tube known as a **pitometer** has one pressure opening facing upstream and the other facing downstream. The differential between the two openings is usually from 25 to 50 per cent more than that for a standard pitot tube, so that the instrument is of advantage in dealing with low velocities. There are commercially available very compact types of pitometers which require relatively small openings for their insertion into a duct. It is important that these instruments be calibrated under substantially the conditions of use.

Anemometers. An anemometer may be any instrument for the measurement of gas velocity, *e.g.*, a pitot tube, but usually the term refers to one of the types below.

The **vane anemometer** is a delicate revolution counter with jeweled bearings, which is actuated by a small windmill (usually 3 to 4 in. diameter) constructed of flat, or slightly curved, radially disposed vanes. Suitable means, similar to those of stop watches, are provided for stopping and starting the counting apparatus. In use, the device is placed with the axis of the vaned wheel parallel to the direction of flow; the counting mechanism and a stop watch are simultaneously started, and after a convenient interval both are stopped. The reading (ft.) of the anemometer divided by the reading of the watch (sec.) gives the apparent linear velocity V_a (ft./sec.) of the gas at the location of the instrument. If the calibration was carried out in a gas of density ρ_0 lb. *mass*/cu. ft. and the density of the gas stream being measured is ρ_1, the true wind speed V_t is found as follows: From the calibration data find the speed V_{0t} which, for the calibration conditions, corresponded to an apparent speed of $V_a\sqrt{\rho_1/\rho_0}$. Then the present true speed is $V_t = V_{0t}\sqrt{\rho_0/\rho_1}$ (*cf.* Ower, *op. cit.*). Correction for the effect of density is usually unimportant except at very low speeds, say below 5 ft./sec. for the usual atmospheric conditions. In all cases care must be taken to hold the anemometer well away from one's body or from any object not normally present in the stream.

The use of the vane anemometer is restricted, principally, to the estimation of gas velocities too low (less than 20 ft./sec. for air at 1 atm.) to give satisfactory readings on ordinary gages if a pitot tube be used. Although the instrument can be employed (accuracy of ±3 per cent, at best) for velocities up to 40 ft./sec., such use should not be made of it, except for rough work, if the case at hand admits of another meter. Most vane anemometers must be handled with considerable care, since a relatively slight shock is sufficient to alter their adjustment. No reliance can be placed on the indications unless a recent calibration has been made and the history of the instrument since the time of the calibration is known. Obviously anemometers of this type cannot be used with corrosive gases.

The **Robinson cup anemometer,** long used in meteorological work, consists of four hemispherical cups mounted on light arms radiating from a hub which turns freely about a vertical axis. The cups are so arranged that, when one is facing the wind, the one diametrically opposite is presenting its convex side. For the U.S. Weather Bureau type (cups 4 in. in diameter with their centers 6.72 in. from the axis), the following formula is given by the "International Critical Tables":

$$\log V = 0.079 + 0.9012 \log V_r$$

wherein V and V_r are the true air speed and the "recorded air speed," respectively, m.p.h. V_r = 3 times the linear velocity of cup centers = 0.120 × (r.p.m.).

The term **current meter** is applied to any of a large variety of mechanical devices for measuring the velocity of liquids in open channels. Most of these meters are substantially the same in principle as the vane or cup anemometer, but the construction is more rugged. For details a treatise on hydraulics should be consulted. The same name is sometimes given to certain types of commercial water meters (*q.v.*, p. 411).

The indications of most anemometers and current meters are affected considerably by the degree of turbulence in the stream. It is therefore most important that they be calibrated under substantially the conditions of normal use.

A **hot-wire anemometer** consists essentially of a fine, electrically heated wire (generally platinum) that is exposed to the stream of which the velocity is wanted. An increase of fluid speed, other things being equal, increases the rate of heat flow from wire to fluid (*cf.* Sec. 6) and consequently tends to cool the wire. Cooling the wire, however, alters its resistance, so that a measurable effect is produced in the electrical heating circuit. The differences between various types of hot-wire anemometers lie principally in the circuits and electrical instruments used. In the simplest type [King, *Trans. Roy. Soc. (London)*, **A214**, 373 (1914)], a constant voltage is maintained across the hot wire, so that the current flowing in the circuit is a measure of the fluid speed. For a description of different forms of the hot-wire anemometer see Pannell, "The Measurement of Fluid Velocity and Pressure," Arnold, London, 1924. In all cases the relation between the electrical indications and the fluid speed must be determined by calibration. Hot-wire anemometers are usually intended for use with gases, but they can be modified for liquid measurement.

Traversing for Mean Velocity. Any of the instruments for measuring local velocity may evidently be used to find the mean velocity in or the total discharge through a duct by dividing the cross section into a number of equal areas, finding the local velocity at a representative point of each, and averaging the results. Obviously the greater the subdivision of the cross section, the greater is the precision. When, as in the case of the pitot tube, the actual reading is not directly the velocity, it should be noted that the readings must be converted into velocities before averaging. In the case of **rectangular passages** the cross section is usually divided into small squares or rectangles, and the velocity is found at the center of each. In **circular pipes** the cross section is divided into several equal annular areas and a central circle. Readings of velocity are made at the intersections of a diameter and the set of circles which bisect the annuli and the central circle.

For an N-point traverse, make readings on each side of the cross section at

$$100 \times \sqrt{\frac{2n-1}{N}}\ \% \left(n = 1, 2, 3 \text{ to } \frac{N}{2}\right) \qquad (2)$$

of the pipe radius from the center. Figure 42 shows the layout for a 10-point traverse. When a circular pipe must be more carefully explored than by a single traverse, it is usual to traverse several diameters spaced at equal

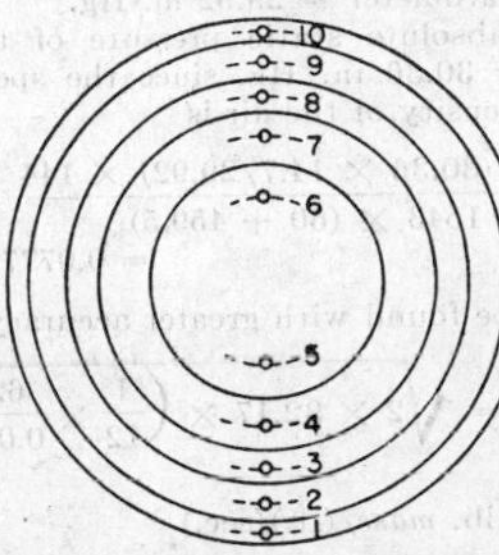

FIG. 42. Locations for pitot tube in 10-point traverse.

angles about the pipe. With the normal-velocity distribution in a circular pipe a single 10-point traverse gives a mean velocity theoretically 0.30 per cent high and a 20-point traverse gives a result 0.1 per cent high. A **3-point traverse** based on Gauss's method of numerical integration requires a reading at the center (V_c) and at points on each side of the center at a distance of 0.880 × radius of the pipe (V_a, V_b). The mean velocity is then $V = \frac{4}{9}V_c + \frac{5}{18}(V_a + V_b)$. The result is adequately accurate for most purposes if the velocity distribution is symmetrical about the diameter traversed.

For the normal velocity distribution in straight circular pipe at points preceded by a run of at least 50 diameters without pipe fittings or other obstructions, the graph in Fig. 43 shows the ratio of V (the mean

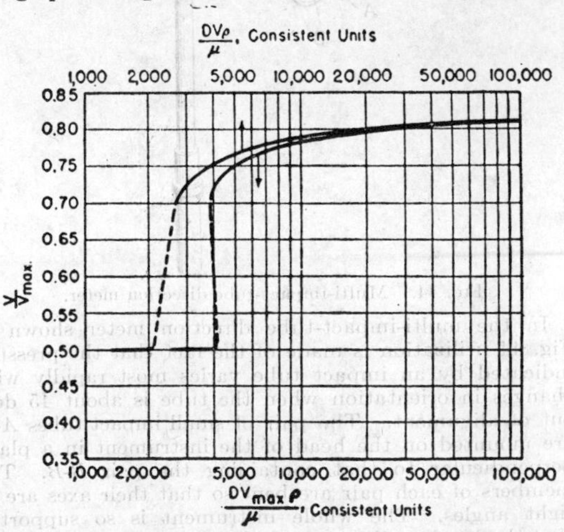

FIG. 43. Pipe factor in circular pipes. (*McAdams, "Heat Transmission," McGraw-Hill, New York,* 1933.)

velocity) to V_{max} (the actual velocity at the center line) plotted against $DV_{max}\rho/\mu$ and $DV\rho/\mu$. From this chart and a pitot reading at the center of a pipe the mean velocity is readily determined. For accuracy in all cases and whenever there is reason to suspect an abnormal velocity distribution, the values of the pipe factor V/V_{max} at the particular measuring station in use should be determined by making several traverses at various velocities. Thereafter, by using the experimentally determined ratio, one reading at the center of the pipe will be sufficient. The pipe factor should always be found by this method in case $DV_{max}\rho/\mu$ is found to be in the range 2000 to 5000; for values of Reynolds number less than 2000, V/V_{max} is 0.50.

Example A pitot tube inserted at the center of an air duct 1 ft. in diameter gives a differential reading of 1 in. water. The static pressure recorded by an open manometer is 6 in. water, and the air temperature is 60°F. If the air is dry, what is its mean velocity? Barometer = 29.92 in. Hg.

Solution. The absolute static pressure of the air is $p_0 = (6/13.6) + 29.92 = 30.36$ in. Hg, since the specific gravity of Hg is 13.6. The density of the air is

$$\rho_0 = \frac{Mp_0}{RT} = \frac{29 \times (30.36 \times 14.7/29.92) \times 144}{1543 \times (60 + 459.5)} = 0.0777 \text{ lb. } mass/\text{cu. ft.}$$

if desired, ρ_0 may be found with greater accuracy from tables.

$$V_{max} = \sqrt{2g_c\,\Delta H} = \sqrt{2 \times 32.17 \times \left(\frac{1}{12} \times \frac{62.3}{0.0777}\right)} = 65.6 \text{ ft./sec.}$$

$$\mu_{air} = 1.21 \times 10^{-5} \text{ lb. } mass/\text{(ft.)(sec.)}$$

$$\frac{DV_{max}\rho}{\mu} = \frac{1 \times 65.6 \times 0.0777}{1.21 \times 10^{-5}} = 421{,}250$$

In extrapolating the curve in Fig. 43, it should be continued nearly horizontally so V/V_{max} is about 0.81.

$$\therefore \text{ mean velocity} = 0.81 \times 65.6 = 53.1 \text{ ft./sec.}$$

Direction Indicators. Two useful flow-direction indicators which have been evolved in aeronautical work are shown in Figs. 44 and 45. For accuracy both types require calibration in a stream of known direction because of the impracticability of attaining exact symmetry in their construction.

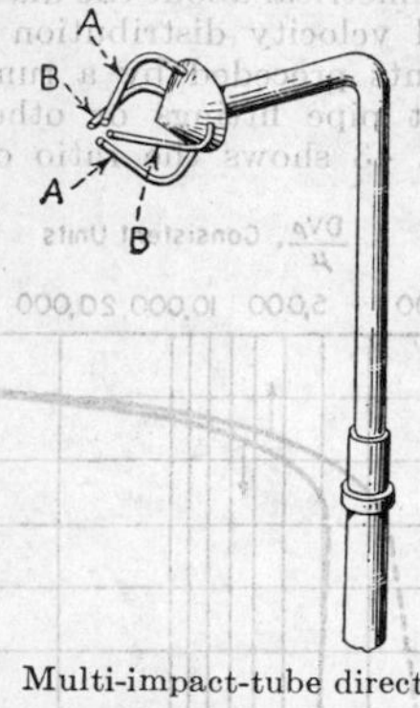

FIG. 44. Multi-impact-tube direction meter.

In the multi-impact-tube direction meter shown in Fig. 44 utilization is made of the fact that the pressure indicated by an impact tube varies most rapidly with changes in orientation when the tube is about 45 deg. out of alignment. The pair of small impact tubes *A-A* are mounted on the head of the instrument in a plane perpendicular to that containing the pair *B-B*. The members of each pair are bent so that their axes are at right angles. The whole instrument is so supported that the head can be rotated about each of two mutually perpendicular lines drawn through the common point of intersection of the axes of the impact tubes. In use, the position is varied until there is no difference in pressure between the mouths of the two members of either pair. In the instrument as developed by Lavender (Aeronautical Research Committee, Great Britain, *Reports and Memoranda*, No. 844, 1923) there is an extra tube opening into a hollow conical space behind the head. The differential pressure between this fifth opening and any of the impact tubes serves, after calibration, to measure the air velocity. The accuracy of this direction meter is ±0.1 deg.

The more accurate of two types of direction indicator based on the principle of the hot-wire anemometer is illustrated in Fig. 45. Three short platinum wires of equal length radiate at equal angles from one of the four conducting supports. The three wires will be equally cooled only if they lie at equal angles to the direction of flow. Thus, using a suitable Wheatstone-bridge circuit to show the position of balance, the instrument can be readily brought into a position such that the axis of the pyramid of wires is in the line of flow. For details see Simmons and Bailey, Aeronautical Research Committee, Great Britain, *Reports and Memoranda*, No. 1019, 1926. When the flow is known to be approximately parallel to some plane, a simplified instrument made with two wires in the form of a V is serviceable. The second type of hot-wire direction indicator consists of two parallel hot-wire anemometer elements mounted close together. The divergence of the indications of the two wires is a maximum when one wire is wholly in the wake of the other. Either type may be used to measure velocity if suitably calibrated.

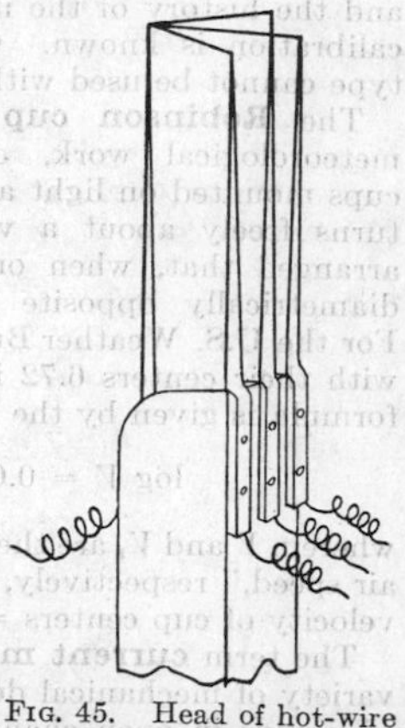

FIG. 45. Head of hot-wire direction indicator.

A good survey of research methods for the visualization of flow is given in "Modern Developments in Fluid Dynamics," Aeronautical Research Committees, Great Britain, vol. 1, pp. 280*ff*., Oxford, London, 1938.

ORIFICES, NOZZLES, AND VENTURI METERS

General Principles and Definitions. Orifices, nozzles, and venturi meters all measure rates of total discharge. They are identical in principle, and the same group of formulas apply to all of them. In the channel of a steady stream there is a constricted portion of fixed cross section, or throat, wherein, evidently, the mean linear velocity [and, hence, the kinetic energy $(V^2/2g_c)$] of the fluid must be greater than in the remainder of the channel. According to Bernoulli's theorem (p. 375), assuming no heat exchange with the surroundings, the total energy of the fluid must be the same at every cross section. Consequently any increase of the kinetic energy at any section is necessarily accompanied by a corresponding decrease in some or all of the other terms in the energy balance. This decrease in energy other than kinetic is theoretically calculable from static pressure and temperature measurements made in the wide and in the narrow parts of the channel. Hence the linear velocity in either of the cross sections where these measurements are made can be estimated if the ratio of the cross-sectional areas is known. Since, in general, actual flow fails to fulfill the assumptions made in the theory, the real discharge differs from the theoretical value. The ratio of the former quantity to the latter is called the **coefficient of discharge** *C*.

The term **orifice** may be applied to any aperture of less diameter than the supply main through which a fluid discharges, but in modern usage the square-edged orifice in a thin plate (see p. 404) is usually meant. In the terminology of flow measurement, a **nozzle** (see p. 406) is a discharge opening shaped on the upstream side to conform more or less to the streamlines of the contracting flow. A **venturi tube** (see Fig. 46) is a special type of nozzle immediately followed by a gradually expanding cone, an attachment which largely averts the substantial frictional loss of kinetic energy occasioned when the jet from a simple nozzle or orifice discharges into the slowly moving fluid downstream.

When used with a liquid, an orifice or a nozzle is said to be **submerged** if the liquid in the downstream channel stands above the opening. The term **free discharge** is used when the stream issues into a gas-filled space. For reliable performance as measuring devices, nozzles, and orifices should operate either fully submerged or with wholly free discharge.

Typically in the case of the thin plate orifice, but also for some other openings, the issuing jet contracts beyond the orifice to a minimum cross section called the *vena contracta*. The ratio of the cross section of the jet at the *vena contracta* to the smallest cross section of the orifice is the **coefficient of contraction** C_c. For the experimental determination of C_c for liquids see Gibson, "Hydraulics and Its Applications," Van Nostrand, New York, 1930.

Theoretical Formulas. The following summary outlines the derivations of the principal theoretical equations for computing the flow through orifices and nozzles. In present-day practical meter calculations a quasi-empirical formula (see p. 403) is chiefly used, and most recently determined coefficients of discharge are for use in that formula. The coefficients of discharge for the various formulas in the literature are not in general interchangeable.

Nomenclature.

C = coefficient of discharge.
C_c = coefficient of contraction.
C_v, C_v' = coefficients of velocity.
c_v, c_p = specific heats at constant volume and at constant pressure, respectively.
D = diameter (ft.); D_i = diameter, in.
g = acceleration due to gravity, ft./sec.^{2}
g_c = dimensional constant, 32.1740 (lb. *mass*)(ft.)/(lb. *force*)(sec.^{2}); see p. 375.
G = mass velocity, lb. *mass*/(sec.) (sq. ft.) = $V\rho$.
H_m = reading of differential U-tube, ft.
H_p = static head in terms of the fluid flowing, ft.; ΔH, ft. (see pp. 364 and 377).
i = $u + pv/J$ = enthalpy (= heat content, total heat, B.t.u./lb. *mass*).
M = molecular weight, lb. *mass*/mole.
p = absolute static pressure, lb. *force*/(sq. ft.).
q = volumetric discharge rate, cu. ft./sec.
R = gas constant per mole 1546 ft.-lb. *force*/(lb.-mole)(°F.), 1.99 B.t.u./(lb.-mole) (°F.), or 1.99 g.-cal./(g.-mole)(°C.).
S = cross-sectional area, sq. ft.
T = absolute temperature, in °Rankine (R.) = 459.5 + t°F.; in °Kelvin (K.) = 273.1 + t°C.
u = internal energy per unit weight, B.t.u./lb. *mass*.
V = average linear velocity, ft./sec. = q/A = G/ρ.
v = specific volume, cu. ft./lb. *mass* = $1/\rho$.
w = weight rate of discharge, lb. *mass*/sec.
W_0 = external work done per unit weight, ft.-lb. *force*/lb. *mass*.
Z = potential head, ft. (see p. 375*ff*).
β = D_2/D_1.
κ = c_p/c_v.
μ = absolute viscosity, lb. *mass*/(sec.)(ft.) = 0.0672 × absolute viscosity in poises.
ρ = density, lb. *mass*/cu. ft.

Liquids. For brevity, let

$$\left.\begin{aligned}\Delta H &= H_{p_1} - H_{p_2} + Z_1 - Z_2 = H_m\left(\frac{\rho_m}{\rho} - 1\right)\\ &= \frac{(p_1 - p_2)g_c}{\rho g} + Z_1 - Z_2\end{aligned}\right\} \quad (3)$$

Then Bernoulli's theorem in terms of heads [Eq. (13), p. 377] may be rewritten as

$$\sqrt{V_2^2 - V_1^2} = C_v\sqrt{2g\,\Delta H} \quad (4)$$

or

$$V_2 = C_v'\sqrt{2g\,\Delta H + C_1^2} \quad (4a)$$

since W_0 and $\int p\,dv$ are zero, and friction may be allowed for by a coefficient of velocity, C_v or C_v', rather than by an additive term. The subscripts ($_1$) and ($_2$), respectively differentiate between the conditions at an upstream and at a downstream cross section; *e.g.*, sections *I* and *II* of Fig. 46 which can serve to illustrate the entire discussion of formulas. The cross sections are assumed to be so chosen that the fluid there moves only in an axial direction. C_v and C_v' are not identical except when they equal unity or when the square of the **velocity of approach** V_1^2 is negligible compared with $2g\,\Delta H$. (For the experimental determination of the C_v's consult any treatise on hydraulics. The values range from 0.95 to 0.99 for most common meters but may be much less for devices such as the standard short tube or a nozzle with a long

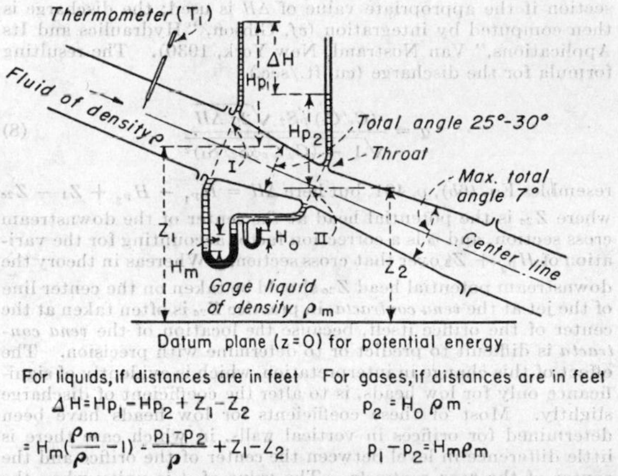

FIG. 46. Venturi tube.

throat. If the pressure taps are not really located at the points to which V_1 and V_2 are supposed to refer, the correction for this is included in these coefficients.) If S_1 is taken to be the upstream cross-sectional area of the stream and S_2, as is customary, to be that of the smallest cross section of the meter, $V_1 = C_cS_2V_2/S_1 = C_c\beta^2V_2$ where C_c is the ratio of the true cross-sectional area of the stream, at the point to which V_2 refers, to S_2. By substituting this relation in the above equations, pairs of formulas are obtained for the downstream linear velocity V_2 and for the volumetric discharge rate q;

$$V_2 = \frac{C_v\sqrt{2g\,\Delta H}}{\sqrt{1 - (C_cS_2/S_1)^2}} \qquad q = \frac{(C_vC_c)S_2\sqrt{2g\,\Delta H}}{\sqrt{1 - (C_cS_2/S_1)^2}} \quad (5, a \text{ and } b)$$

Using C_v',

$$V_2 = \frac{C_v'\sqrt{2g\,\Delta H}}{\sqrt{1 - (C_v'C_cS_2/S_1)^2}} \qquad q = \frac{(C_v'C_c)S_2\sqrt{2g\,\Delta H}}{\sqrt{1 - (C_v'C_cS_2/S_1)^2}} \quad (6, a \text{ and } b)$$

If C_c differs from unity (*e.g.*, square-edged orifice, free-running Borda mouthpiece, poorly made nozzles, etc.) only formula (6*b*) of those above is useful, since the others require separate values of C_v and of C_c, quantities which individually cannot be accurately measured.

When S_1/S_2 = 25 or more, the denominator of any of the above formulas may be taken equal to unity.

An impact tube, so located that it records the mean upstream impact head, sometimes replaces the upstream static hole. In such a case

$$q = (C_v'C_c)S_2\sqrt{2g\,\Delta H'} \quad (7)$$

where $\Delta H' = H_{p_1} - H_{p_2} + Z_1 - Z_2 + \frac{V_1^2}{2g} = H_m\left(\frac{\rho_m}{\rho} - 1\right)$ is the differential head that would be shown by the gages of Fig. 46 were an impact tube replacing the upstream static connection.

The coefficients in Eqs. (5), (6), and (7), are related to the coefficients of discharge C and K of the equation in practical use [Eq. (15), p. 403] by the following equations:

$$C = C_vC_c\sqrt{\frac{1-\beta^4}{1-C_c^2\beta^4}} = C_v'C_c\sqrt{\frac{1-\beta^4}{1-(C_v'C_c)^2\beta^4}} \tag{7a}$$

$$C_v'C_c = \frac{C}{\sqrt{1-(1-C^2)\beta^4}} = \frac{K}{\sqrt{1+K^2\beta^4}} \tag{7b}$$

$$C_v = C \quad \text{if} \quad C_c = 1 \tag{7c}$$

At the *vena contracta* of a free jet discharging into a gas space, the static pressure is substantially the same throughout the cross section and equals the pressure of the gas; *i.e.*, the static head H_p is the same, not $H_p + Z$ (Z being the potential head) as in the case of a submerged orifice. Consequently, in passing from the upstream to the downstream stations, the change ΔH in $H_p + Z$ for a given particle of fluid depends on the point at which the downstream cross section is traversed (unless the jet be vertical), instead of being the same for every particle as in submerged operation. It is customary to suppose that Eq. (4*a*) (p. 401) gives the velocity at any particular point of the downstream cross section if the appropriate value of ΔH is used; the discharge is then computed by integration (*cf.* Gibson, "Hydraulics and Its Applications," Van Nostrand, New York, 1930). The resulting formula for the discharge (cu. ft./sec.)

$$q = \frac{(C_v'C_c)\phi S_2\sqrt{2g\,\Delta H}}{\sqrt{1-(C_v'C_c\phi S_2/S_1)^2}} \tag{8}$$

resembles Eq. (6*b*), p. 401, but here $\Delta H = H_{p_1} - H_{p_2} + Z_1 - Z_{2c}$ where Z_{2c} is the potential head at the center of the downstream cross section, and ϕ is a correction factor accounting for the variation of $H_{p_2} + Z_2$ over that cross section. Whereas in theory the downstream potential head Z_{2c} should be taken on the center line of the jet at the *vena contracta*, in practice Z_{2c} is often taken at the center of the orifice itself, because the location of the *vena contracta* is difficult to predict or to determine with precision. The effect of this change in interpretation, which is evidently of significance only for low heads, is to alter the coefficient of discharge slightly. Most of these coefficients for low heads have been determined for orifices in vertical walls, in which case there is little difference in level between the center of the orifice and the center of the *vena contracta*. The value of ϕ is unity when the jet is projected vertically upward or downward. For circular openings in a vertical plane ϕ varies from 0.960 when the ratio $(\Delta H + V_1^2/2g)/D_2$ is 0.5 to 0.994 when that ratio is 1.2.

Gases. The flow through nozzles, venturi tubes, well-rounded orifices, standard short tubes, etc., may be computed by the following formula for the velocity of discharge (ft./sec.):

$$V_2 = C_v\sqrt{\frac{2g_c(i_1 - i_2)778}{1-(S_2v_1/S_1v_2)^2}} \tag{9}$$

if the upstream static pressure p_1 and temperature T_1 and the downstream static pressure p_2 are known. The formula is derived from Eq. (9) on p. 376. i_1 and i_2 are the total heats, B.t.u./lb. *mass* (= enthalpy, see p. 401), and v_1, v_2 are the specific volumes, cu. ft./lb. *mass*, of the fluid in the main channel and in the restricted section, respectively. S_1 and S_2 are the cross-sectional areas (sq. ft.). The weight discharge is $w = V_2S_2/v_2$ (lb. *mass*/sec.). Equation (9) applies to any fluid, but its practical use is restricted to those for which the thermal properties are tabulated (*e.g.*, steam, NH_3, etc.) or are calculable (*e.g.*, perfect gas). To determine i from tables, both the pressure and the temperature (or quality, in case of a wet vapor) must be known. When these quantities are correctly measurable at both sections *I* and *II* (see Fig. 46, p. 401), the coefficient C_v is nearly unity; a slight departure from unity occurs because the variation of velocity across the channel is neglected in the derivation. In practice, it is usually inconvenient, if not impossible, to measure the temperature (or quality) at the smaller cross section. Hence it is customary to obtain i_2 from a measurement of p_2 and the assumption of frictionless adiabatic flow (constant entropy). For this purpose a Mollier diagram (see p. 332) (entropy vs. total heat chart) for the fluid in question is convenient. When i_2 is found in this manner, the value of C_v will differ from unity on account of friction.

If a vapor is initially superheated, or dry and saturated, and the observed p_2, interpreted by ordinary tables or diagrams, indicates partial condensation at the throat, false results will be obtained by use of Eq. (9) when i_2 is found as above. In these cases it has been shown conclusively (*cf.* Stodola, "Steam Turbines," tr. by Loewenstein, McGraw-Hill, New York, 1927) that condensation does not occur until the throat is passed, so that the vapor there is dry and supersaturated. Special tables and Mollier diagrams that give the values of i for supersaturated vapor can be prepared (*cf.* Stodola, *ibid.*), but, in their absence, the use of Eq. (10) gives substantially correct results for nozzles if κ is evaluated as for superheated vapor.

The usual "theoretical" equation, in terms of the absolute static pressures, p_1 and p_2 and absolute temperature T_1 is derivable from Eq. (9) on the assumptions that the fluid is a perfect gas [$pv = RT/M$; c_p = const.; $(i_1 - i_2) = c_p(T_1 - T_2)$] and that the flow is frictionless as well as adiabatic. The latter assumption permits the use of the relation

$$\left(\frac{p_1}{p_2}\right)^{(\kappa-1)/\kappa} = \frac{T_1}{T_2}$$

where $\kappa = c_p/c_v$.

There results for the linear velocity (ft./sec.) through the constriction

$$V_2 = C_v\sqrt{\frac{2g_c\frac{\kappa}{\kappa-1}\frac{RT_1}{M}\left[1-\left(\frac{p_2}{p_1}\right)^{(\kappa-1)/\kappa}\right]}{1-(S_2/S_1)^2(p_2/p_1)^{2/\kappa}}} \tag{10}$$

where $R = 1546$; T_1 is in °F. abs.; M is the molecular weight.

For square-edged orifices and other devices with which p_2 is supposedly measured at the *vena contracta*, Eqs. (9) and (10) could be modified by replacing S_2 by C_cS_2 [*cf.* Eqs. (5*a*) and (6*a*), p. 401]. Usually Eq. (15), below, is used for square-edged orifices, since most of the published coefficients have been computed for it.

As in the case of liquids, some simplification results if one static pressure is replaced by an impact pressure measurement. Let p_i be the pressure, lb. *force*/sq. ft., on an impact tube facing upstream in the jet from a nozzle; let p_2 be the downstream static pressure, lb. *force*/sq. ft.; and let T_i be the temperature, °F. abs., attainable when the gas is compressed adiabatically and without friction from the upstream conditions to a static pressure equal to p_i. Then a set of formulas is obtainable differing from Eqs. (9) and (10) in that the denominators are lacking and the subscript ($_1$) is replaced everywhere by (i). The same results are found if the impact tube is so located in the upstream channel that a mean reading is recorded, for the friction loss through a good nozzle is so small that the difference between the upstream and downstream impact pressures is experimentally undetectable (*cf.* Bean, Buckingham, and Murphy, *Bur. Standards Research Paper* 49, 1929). The temperature T_i is taken to be that recorded by a thermometer inserted in the stream above the nozzle. This is the same measurement as that taken to be the T_1 of the formulas in the preceding paragraphs. Actually, if the upstream velocity is appreciable, the reading of such a thermometer (properly corrected for radiation errors) is more nearly equal to the theoretical value of T_i than it is to that of T_1 (*cf.* Kiefer and Stuart, "Principles of Engineering Thermodynamics," Wiley, New York, 1930). If the upstream velocity is small, T_i and T_1 are in theory nearly identical.

Maximum Discharge for Gases. Critical Pressure Ratio. For a given set of upstream conditions the discharge of a gas from a nozzle, venturi tube, or orifice will theoretically increase for a decrease in the value of absolute pressure ratio p_2/p_1 until the linear velocity in the throat reaches that of sound in the gas at that point. The value of p_2/p_1 for which the acoustic velocity (*cf.* p. 375) is just attained is called the critical pressure ratio r_c. The actual pressure in the true throat does not fall below p_1r_c, even if a much lower pressure exists downstream.

In the case of nozzles, venturi tubes, rounded orifices, etc., r_c satisfies quite well the following theoretical relation for perfect gases, derived by equating formula (10) to the acoustic velocity, taking $C_v = 1$,

$$\frac{\kappa+1}{r_c^{2/\kappa}} - \frac{2}{r_c^{(\kappa+1)/\kappa}} - \frac{\kappa-1}{(S_1/S_2)^2} = 0 \tag{11}$$

which reduces to

$$r_c = \left(\frac{2}{\kappa+1}\right)^{\kappa/(\kappa-1)} \tag{12}$$

if S_1/S_2 is 25 or more. The values of r_c at 15°C. computed from Eq. (12) are as follows: for monatomic gases, 0.49; for H_2, O_2, N_2, air, HCl, NO, 0.53; for Cl_2, H_2S, NH_3, CH_4, 0.54; for SO_2, N_2O, CO_2, C_2H_2, C_2H_4, 0.55. The theoretical maximum is $r_c = 0.608$. For imperfect gases and vapors, if the thermodynamic properties are known (*e.g.*, steam, NH_3, etc.), the value of r_c can be found by making a plot of the weight discharge w [calculated from Eq. (9)] vs. the pressure p_2 that corresponds to a set of assumed values of i_2, the downstream total heat. The maximum value of w will occur at a p_2 such that $r_c = p_2/p_1$. The value of r_c is about 0.55 for moderately superheated steam and about 0.58 for saturated steam.

Especially in the cases of rounded orifices, nozzles, etc., it may occur that the observed downstream pressure p_2 is less than the critical value. In such cases the downstream pressure tap is not showing the true throat pressure, and p_2 should be replaced by $p_1 r_c$ in all the above flow formulas. Thus, for $S_1/S_2 \geqq 25$, Eq. (10) leads to

$$w_{\max} = C_v S_2 p_1 \sqrt{\frac{g_c \kappa M}{R T_1} \left(\frac{2}{\kappa + 1}\right)^{(\kappa+1)/(\kappa-1)}} \quad \text{lb. } mass\text{/sec.} \tag{13}$$

which should be used when p_2, as measured, is less than $p_1 r_c$. Correspondingly, when Eq. (9) is used under these conditions the value of i_2 in the equation should be that corresponding to $p_1 r_c$, not to the observed p_2. **For air,** Eq. (13) reduces to Fliegner's equation for $p_2 < 0.53 p_1$ and $S_1/S_2 > 25$:

$$w = 0.533 \frac{C_v S_2 p_1}{\sqrt{T_1}} \quad \text{lb. } mass\text{/sec.} \tag{14}$$

In these two formulas T_1 is in °F. abs., $R = 1546$; and p is in lb. *force*/sq. ft., if S is in sq. ft.; or in lb. *force*/sq. in. if S is in sq. in. M is the molecular weight.

For square-edged orifices, the theoretical conclusions in regard to maximum discharge are found experimentally to be invalid. The discharge continues to increase as p_2/p_1 decreases below r_c as given by Eqs. (11) and (12) [see Schiller, *Forsch. Gebiete Ingenieurw.*, **4**, 128 (1933)]. The theoretical conclusions are confirmed for well-rounded orifices and nozzles. Such devices are extensively used under conditions of maximum discharge for the standardization of other meters. Before such use they should always be calibrated to determine the coefficient of discharge.

Working Formulas. The basic form of the practical equation for the weight rate of discharge adopted by the A.S.M.E. Special Research Committee on Fluid Meters (see p. 396) for use either with gases or with liquids is

$$w = q_1 \rho_1 = C Y S_2 \sqrt{\frac{2 g_c (p_1 - p_2) \rho_1}{1 - \beta^4}} = K Y S_2 \sqrt{2 g_c (p_1 - p_2) \rho_1} \tag{15*}$$

$$= C Y S_2 \sqrt{\frac{2 g \rho_1^2 \, \Delta H}{1 - \beta^4}} \tag{15a}$$

where C = coefficient of discharge (no dimensions).

g_c = dimensional constant = 32.1740 (lb. *mass*) (ft.)/(lb. *force*) (sec.^{2}).

g = local acceleration due to gravity, ft./sec.^{2}, numerically very nearly equal to g_c (see p. 375).

ΔH = orifice differential, ft. of fluid of upstream density [see Eq. (3), p. 401].

$K = C/\sqrt{1 - \beta^4}$ (no dimensions).

p_1, p_2 = pressures at upstream and downstream static pressure taps, respectively, lb. *force*/sq. ft.

q_1 = volumetric rate of discharge measured at upstream pressure and temperature, cu. ft./sec.

S_2 = cross-sectional area of the discharge opening, sq. ft.

w = weight rate of discharge, lb. *mass*/sec.

Y = expansion factor, see below (no dimensions); for liquids $Y = 1$.

β^2 = ratio of cross section of constriction to that of upstream channel; for circular openings in circular pipes, $\beta = D_2/D_1$.

ρ_1 = density at upstream temperature and pressure, lb. *mass*/cu. ft.

* When there is a difference in level between the pressure taps, $(p_1 - p_2)$ should be replaced by $(p_1 - p_2) + (Z_1 - Z_2)\rho_1 g/g_c$ (*cf.* Eq. (3), p. 401).

Comparison of Eq. (15*a*) when $Y = 1$ and the theoretical equation for liquids [Eq. (5*b*), p. 401] shows that the two equations are equivalent if there is no contraction of the jet. Comparison of Eq. (15) with the theoretical equation for gas flow through nozzles [Eq. (10), p. 402], shows them to be equivalent if

$$Y = \sqrt{r^{2/\kappa} \left(\frac{\kappa}{\kappa - 1}\right) \left(\frac{1 - r^{(\kappa-1)/\kappa}}{1 - r}\right) \left(\frac{1 - \beta^4}{1 - \beta^4 r^{2/\kappa}}\right)}$$

(for nozzles and venturi meters) (16)

where $r = p_2/p_1$. In both cases C becomes identical with C_v. It is found for **nozzles and venturi meters** that the rules for evaluating C are the same for gases as for liquids if for gases Y is evaluated by Eq. (16) and for liquids $Y = 1$. Table 1 and Fig. 47 give values of Y computed from Eq. (16) for several values of κ, r, and β. For centered circular **square-edged orifices** in pipe lines it is found that, when C is evaluated in the same way for gases as for liquids and Y is assumed unity for liquids, Y for gases is expressible as an empirical function of β and $(p_1 - p_2)/p_1\kappa$, where $\kappa = c_p/c_v$. Equation (17) and Fig. 47 give values of Y for orifices

Table 1. Values of Y for Nozzles and Venturi Meters

Calculated from Eq. (16) for use in Eq. (15)

κ	r \ β	0.25	0.4	0.5	0.6	0.65
1.40	0.98	0.9893	0.9890	0.9884	0.9872	0.9863
	.96	.9782	.9776	.9765	.9743	.9724
	.94	.9672	.9662	.9646	.9613	.9585
	.92	.9560	.9548	.9526	.9483	.9447
	.90	.9447	.9432	.9405	.9352	.9309
1.30	.98	.9883	.9880	.9874	.9861	.9851
	.96	.9767	.9759	.9747	.9723	.9703
	.94	.9647	.9636	.9620	.9584	.9555
	.92	.9526	.9514	.9491	.9445	.9407
	.90	.9405	.9389	.9361	.9305	.9259
1.20	.98	.9873	.9870	.9863	.9849	.9839
	.96	.9747	.9740	.9727	.9701	.9679
	.94	.9618	.9608	.9589	.9550	.9519
	.92	.9489	.9475	.9450	.9401	.9360
	.90	.9358	.9341	.9311	.9250	.9201

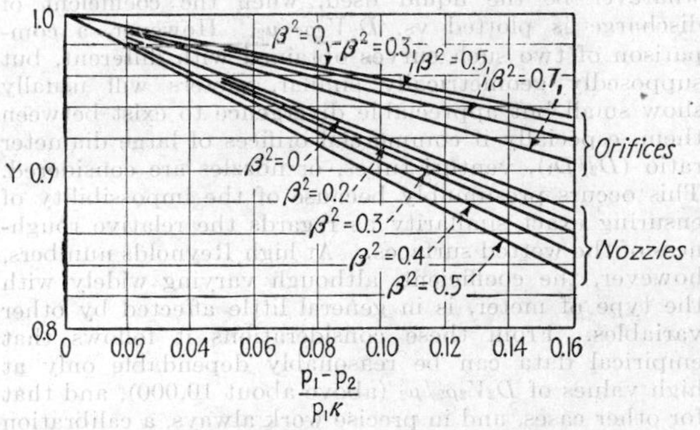

FIG. 47. Values of expansion factor Y for orifices and nozzles.

$$Y = 1 - \left(\frac{p_1 - p_2}{p_1 \kappa}\right)(0.41 + 0.35\beta^4) \quad \text{(for square-edged orifices)} \tag{17}$$

Probably results similar to those for centered circular orifices obtain for other types of opening which cause jet contraction; in the absence of specific data Eq. (17) is often used to approximate Y in such cases in order to convert a calibration for liquids into one for gases. The validity of such usage is open to question.

Convenient reduced forms of Eq. (15), expressed in commonly used units, are given below for circular openings of diameter D_{2i} (in.):

$$w = 0.525KYD_{2i}^2 \sqrt{\rho_1(p_1 - p_2)} \text{ lb. } mass/\text{sec.} \qquad (15b)$$

where ρ_1 is in lb. *mass*/cu. ft., and p_1 and p_2 are in lb. *force*/sq. in.

$$q_1 = XKYD_{2i}^2 \sqrt{\frac{H_m(\rho_m - \rho_A)}{\rho_1}} \qquad (15c)$$

Table 2. Values of X in Eq. (15c)

Volume units for q_1	Time units for q_1			
	Sec.	Min.	Hr.	24-hr. day
Cubic feet	0.0126	0.758	45.5	1090
United States gallons	.0945	5.67	340.0	8160

where H_m (in.) is the reading of a differential U-tube with a liquid of specific gravity ρ_m filling the U; ρ_A is the specific gravity of the fluid in the leads (seal liquid); and ρ_1 that upstream of the fluid flowing. Table 2 gives the values of the factor X which should be used to obtain the upstream volumetric rate of flow in any one of several sets of units.

The **discharge coefficients** C for use with Eqs. (15) and (15a) are given on the following pages. It can be shown by dimensional analysis that, for geometrically similar meters running submerged, any rationally defined discharge coefficient may be expected to be a function of Reynolds number, $D_2V_2\rho_2/\mu_2$, and of the roughness of the pipe and meter surfaces.

NOTE: μ = absolute viscosity; D_2 = a characteristic linear dimension of the orifice or throat; *e.g.*, the diameter of a circular orifice. For pipe orifices, etc., evidently $D_2V_2\rho_2/\mu_2$ can be replaced by $D_1V_1\rho_1/\mu_1$ if desired, since one Reynolds number can be expressed algebraically in terms of the other.

This expectation is partly confirmed by the fact that for a given meter the same experimental curve is obtained, whatever be the liquid used, when the coefficient of discharge is plotted vs. $D_2V_2\rho_2/\mu_2$. However, a comparison of two such curves obtained with different, but supposedly geometrically similar, meters will usually show small but appreciable divergence to exist between them, especially if commercial orifices of large diameter ratio (D_2/D_1), venturi tubes, or nozzles are considered. This occurs presumably because of the impossibility of ensuring exact similarity as regards the relative roughness of the wetted surfaces. At high Reynolds numbers, however, the coefficient, although varying widely with the type of meter, is in general little affected by other variables. From these considerations it follows that empirical data can be reasonably dependable only at high values of $D_2V_2\rho_2/\mu_2$ (above about 10,000), and that for other cases, and in precise work always, a calibration with some fluid of the particular instrument in use is desirable. Until recently most of the published values of meter coefficients have been for high ranges of $D_2V_2\rho_2/\mu_2$. When metering fluids of high kinematic viscosity (μ/ρ), *e.g.*, oils, hydrogen gas, etc., it is important to estimate $D_2V_2\rho_2/\mu_2$ before using such coefficients in order to make certain that they are applicable.

Practically the only sources of data on **unsubmerged operation** are experiments with water as the fluid. Consequently the effect of viscosity on the discharge coefficients is not well known. In general, as in submerged operation, the coefficient at high rates of flow and for large openings appears to depend principally upon the type of meter. At low rates of flow the data of the various investigators do not agree very well, especially for small openings. In general there appears to be little difference between the results for free discharge and those for submerged operation.

Orifices. A **square-edged, or sharp-edged, orifice** (Fig. 48) is a cleancut square-edged hole with straight

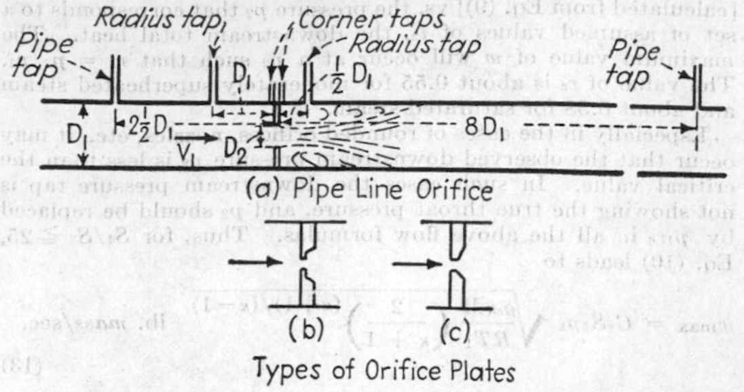

FIG. 48. Square-edged orifice.

walls perpendicular to the flat upstream surface of a thin plate which blocks a channel. The stream issuing from such an orifice attains its minimum cross section (*vena contracta*) beyond the orifice at a distance downstream that depends on the circumstances of operation. For centered circular orifices in pipe lines, the position of the *vena contracta*, as judged from the point of minimum static pressure, varies with the ratio β of orifice to pipe diameter D approximately as follows:

β	0.80	0.50	0.30
Distance from orifice	.33D	.66D	.80D

Beyond the *vena contracta* (in submerged orifices) the kinetic energy of the jet is almost completely destroyed by turbulence in mixing with the slowly moving fluid in the discharge line; hence, the over-all loss in static head is considerable as shown by Fig. 49 (*cf.* venturi tube).

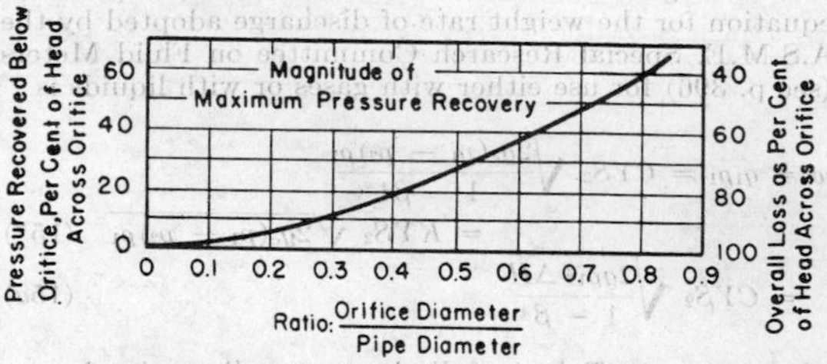

FIG. 49. Pressure recovery following an orifice.

If the calculated discharge is to be correct, it is most important that (1) the walls of the hole and the upstream surface of the plate meet sharply at right angles, (2) the area of the opening be precisely known, (3) the plate be not thicker than 1/30 of the pipe diameter, 1/8 of the orifice diameter, or 1/4 of the distance from the pipe wall to the edge of the opening, and (4) the upstream face of the plate be a smooth unpitted plane. Very slight departures from these specifications may cause very substantial errors; this is especially true of a slight rounding of the upstream edge of the opening, the result of which is to cause readings lower than the actual flow. When for reasons of strength a thick plate is required, the construction shown in Fig. 48b and 48c are used; the bevel in type c should be at 45 deg. or more to the pipe axis.

For a centered circular **orifice in a pipe**, the pressure drop, or *orifice differential* is customarily measured between one of the following pressure-tap pairs. Except in the case of flange taps all measurements of distance from the orifice are made from the *upstream* surface of the plate.

a. Corner Taps. Static holes drilled one in the upstream and one in the downstream flange with the openings as close as possible to the orifice plate.

b. Radius Taps. Static holes located 1 pipe diameter upstream and ½ pipe diameter downstream from the plate.

c. Pipe Taps. Static holes, one 2½ pipe diameters upstream and the other 8 diameters downstream.

d. Flange Taps. Static holes located 1 in. upstream and 1 in. downstream from the orifice plate.

e. Vena Contracta Taps. The upstream static hole is from ½ to 2 pipe diameters from the orifice plate. The downstream tap is located at the position of minimum pressure (See Fig. 51).

Radius taps are theoretically the best: the downstream pressure tap is located at about the mean position of the *vena contracta;* the upstream tap is sufficiently far upstream to be unaffected by the distortion of the flow in the immediate vicinity of the orifice (in practice, the upstream tap can be as much as 2 pipe diameters from the plate without affecting the results). Practically, *vena contracta* taps give the largest differential head for a given rate of flow; they make for precision of reading but are inconvenient if the size of the opening is changed from time to time. Corner taps offer the sometimes great advantage that the pressure taps can be built into the plate carrying the orifice. Thus the entire apparatus can be quickly inserted in a pipe line at any convenient flanged joint without the necessity for drilling holes in the pipe. Flange taps are similarly convenient. By merely replacing standard flanges with special orifice flanges* suitable pressure taps are made available. Pipe taps are, in general, to be avoided because of low differential pressure, although satisfactory results can be obtained with them.

The **coefficients of discharge** C for corner taps are plotted vs. the Reynolds number in Fig. 50 [Tuve and Sprenkle, *Instruments*, **6**, 201 (1933)]. In the range covered, the values for modified radius taps having the downstream tap ⅓ pipe diameter beyond the plate were found to differ from those for corner taps by less than 1½ per cent. For ordinary calculations the coefficients of Fig. 50 may be used, with a probable error usually less than 2 per cent, for all types of pressure tap pairs except pipe taps. There is little variation with Reynolds number above $N_{Re,2} = 30{,}000$. Above this value, K is given by Fig. 51 for various positions of the downstream tap; the variation with position of the upstream tap is very small. At these high Reynolds numbers, if $\beta = D_2/D_1 < 0.2$, C lies between 0.595 and 0.61 for all cases except that of pipe taps; for the latter, C is about 1 per cent higher. As Fig. 51 shows, the deviations among the types of taps is marked at higher values of β. Very precise values of C and K are given by the A.S.M.E. Special Research Committee, "Fluid Meters" (Part 1, p. 111, 4th ed., 1937) for flange taps, *vena contracta* taps, and pipe taps.

In pipe lines **non-circular square-edged orifices (and eccentric circular orifices)**, running submerged in the turbulent range, have discharge coefficients C ranging from 0.595 to 0.62 (*i.e.*, essentially identical with those for centered circular openings), provided that the channel walls are not nearer any edge of the orifice than 2½ times the smallest dimension of the opening and the pressure connections are substantially equivalent to *vena contracta* or flange taps. More precise data cannot be given unless it is known that the exact conditions of some previously recorded experiment are fulfilled. The variation of C with the Reynolds number is presumably similar to that found for centered circular openings, but the data are insufficient to permit definite conclusions. One of the principal reasons for using eccentric circular orifices is that foreign matter carried

* Some "special orifice flanges" available commercially have pressure taps so widely at variance with the rules for measurement of static pressure (p. 396) that even approximately reliable results may be had only after special calibration.

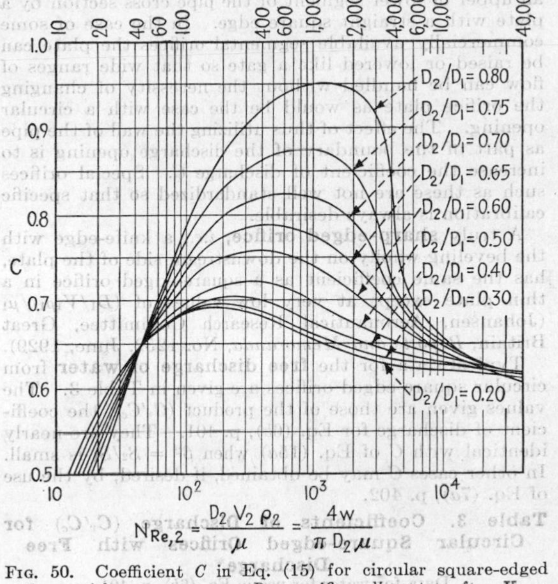

FIG. 50. Coefficient C in Eq. (15) for circular square-edged orifices with corner taps. D_2 = orifice diameter, ft., V_2 = velocity through orifice, ft./sec.; ρ_2 = density at downstream conditions, lb. *mass*/cu. ft.; w = weight rate of flow, lb. *mass*/sec.; μ_2 = viscosity, lb. *mass*/(ft.)(sec.) = 0.000672 × centipoises.

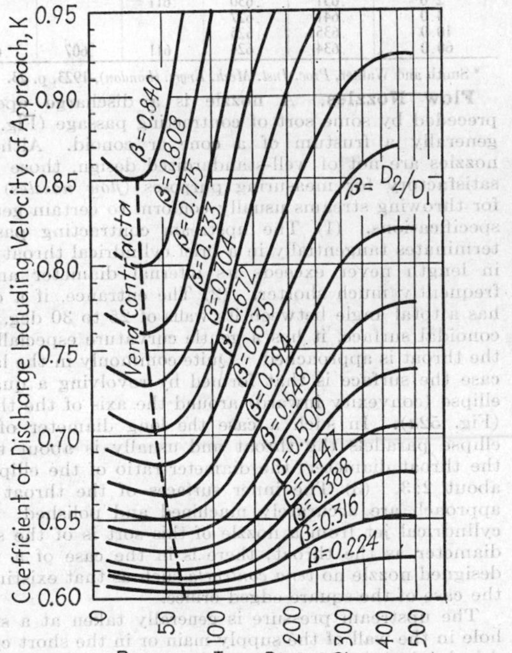

FIG. 51. Coefficient of discharge for square-edged circular orifices, $N_{Re,2} > 30{,}000$ upstream tap located between one and two pipe diameters from orifice plate.

by the fluid has less chance of fouling the orifice if the edge of the opening is placed tangent to the top or to the bottom of the pipe, as the nature of the suspended matter may require. The same end is accomplished by the use of a **segmental orifice** which is formed by blocking an upper or lower segment of the pipe cross section by a plate with a straight square edge. In the case of some commercially available segmental orifices the plate can be raised or lowered like a gate so that wide ranges of flow can be handled without the necessity of changing the orifice plate as would be the case with a circular opening. The effect of thus utilizing the wall of the pipe as part of the boundary of the discharge opening is to increase the coefficient of discharge C. Special orifices such as these are not well standardized so that specific calibration is always desirable.

A truly **sharp-edged orifice,** *i.e.*, a knife-edge with the beveling wholly on the downstream side of the plate, has the same coefficient as a square-edged orifice in a thin plate except at very low values of $(D_1/V_1\rho_1)/\mu_1$ (Johansen, Aeronautical Research Committee, Great Britain, *Reports and Memoranda*, No. 1252, June, 1929).

Typical data for the **free discharge of water** from circular square-edged orifices are given in Table 3. The values given are those of the product $(C_v'C_c)$, the coefficient of discharge for Eq. (6*b*), p. 401. They are nearly identical with C of Eq. (15*a*) when $\beta^2 = S_2/S_1$ is small. In other cases C may be obtained, if desired, by the use of Eq. (7*a*), p. 402.

Table 3. Coefficients of Discharge $(C_v'C_c)$ for Circular Square-edged Orifices with Free Discharge*

Data for water for use in Eq. (6*b*), p. 401

Head, ft. of water	Orifice diameter, in.				
	0.75	1.0	1.5	2.0	2.5
1.0	0.657	0.633	0.612	0.607	0.606
2.0	.651	.630	.611		
4.0	.641	.627			
10.0	.635	.625			
60.0	.634	.624	.611	.607	.606

* Smith and Walker, *Proc. Inst. Mech. Engrs. (London)*, 1923, p. 23.

Flow Nozzles. A nozzle is a discharge opening preceded by some sort of contracting passage (Fig. 52), generally a frustum of a cone or conoid. Although nozzles are not of well-standardized design, those most satisfactory for measuring purposes (*flow nozzles*) and for throwing streams usually conform to certain general specifications. (1) The upstream contracting passage terminates tangentially in a brief cylindrical throat that in length never exceeds its internal diameter and is frequently much shorter. (2) The entrance, if a cone, has a total angle between its walls of 25 to 30 deg.; if a conoidal surface, it has a gentle curvature especially as the throat is approached. Quite commonly in the latter case the surface is that formed by revolving a quarter ellipse (convexity inward) around the axis of the throat (Fig. 52*a*). In such a case the long diameter of the ellipse parallels the throat and usually is about twice the throat diameter; the diameter ratio of the ellipse is about 2:3. (3) The inner surfaces of the throat and approach are accurately machined and polished. The cylindrical jet from a nozzle of this sort is of the same diameter as the throat; there is in the case of a well-designed nozzle no *vena contracta* such as that existing in the case of the square-edged orifice.

The upstream pressure is generally taken at a static hole in the wall of the supply main or in the short cylindrical inlet provided before the contracting section in some nozzles. The downstream static pressure tap may be in the wall, either of the throat, or of the chamber or pipe into which the discharge takes place. The **coefficients of discharge** for the "long radius" or elliptical nozzle (Fig. 52*a*) have not yet been adequately correlated. In general they approximate those for venturi tubes (see below) and vary with the Reynolds number in a similar manner. Data on these nozzles are given by Folsom [*Trans. Am. Soc. Mech. Engrs.*, **61**, 233 (1939)]. For ordinary calculations the use of the venturi-tube data is recommended. Discharge coefficients and details of construction of the International Standard or German Standard flow nozzle which is extensively used abroad are given in "German Industrial Standards," No. 1952, 4th ed., 1937, and in the April, 1935, Reports of International Standards Association, Committee No. 30.

The regain in static pressure downstream from a nozzle in a pipe line has not been so carefully investigated as in the case of the square-edged orifice. The permanent loss in head is of the same order of magnitude as in the latter case, for the kinetic energy of the jet is, here also, largely lost in turbulent friction. An estimate of the over-all loss can be made by using the "sudden expansion loss" formula (p. 388) since the losses in the contracting section and throat of a well-made nozzle are negligible.

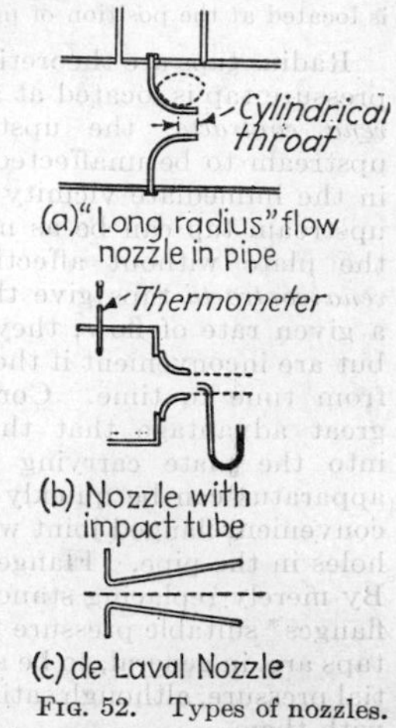

FIG. 52. Types of nozzles.

Sometimes, when the circumstances of operation permit, an impact tube is placed facing upstream in the center of the jet a short distance beyond the nozzle (Fig. 52*b*) [Moss, *Trans. Am. Soc. Mech. Engrs.*, **38**, 761 (1916)]. In such cases the differential pressure recorded is measured between the impact tube and the downstream static pressure tap; or if the discharge side is at atmospheric pressure, an open manometer attached to the impact tube gives the required differential. For gases the coefficient of velocity C_v for use in Eqs. (9) and (10), modified as noted on p. 402, varies from 0.99 to 0.995 in the turbulent range.

A **deLaval nozzle,** as shown in Fig. 52*c* (used mostly for attaining high gas velocities, as in turbines), is generally a rounded orifice followed by a diffusing cone. The term is sometimes used to denote any opening followed by such a cone. For the design and theory of flow in deLaval cones see Stodola, "Steam Turbines," Loewenstein tra., McGraw-Hill, New York, 1927.

Venturi Meters. The over-all pressure loss caused by a nozzle may be considerably reduced by adding, downstream, a diverging passage. The term venturi meter may denote any such modified nozzle, but it is generally synonymous with *venturi tube*. A venturi tube (Fig. 46, p. 401) has an upstream section consisting of (1) a cylindrical inlet I equaling the supply pipe in diameter; (2) a contracting portion that is a frustum of a cone with a vertex angle of 25 to 30 deg., and (3) a cylindrical throat II of diameter $D_2 = \frac{1}{4}$ to $\frac{1}{2}$ of the inlet diameter D_1. The length of the throat is usually about $\frac{1}{2}D_2$; in no case should it be over $1D_2$. Static pressure taps (usually piezometer rings cast as integral parts of the meter) are provided at the inlet and throat (see Fig. 46) between which a suitable differential gage is ordinarily connected. The downstream section is a cone diverging from the throat diameter to that of the discharge pipe at a constant total angle of 7 deg. or less. The junctions between the cones and cylinders are made by gradually curving surfaces. The over-all pressure loss is from 10 to 20 per cent of the reading.

For venturi tubes, **the coefficient of discharge** C may be estimated from Fig. 53 within the precision probable without special calibration [Smith, *Mech. Eng.*, **52**, 968 (1930)]. Venturi tubes are usually designed

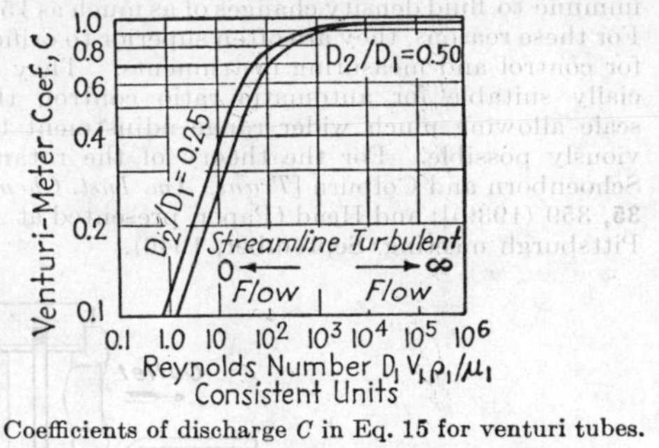

FIG. 53. Coefficients of discharge C in Eq. 15 for venturi tubes.

to operate at Reynolds numbers above 10,000 (based on conditions upstream); for such operation the commonly assumed value $C = 0.98$ is seen to be satisfactory.

Miscellaneous Openings. Except for well-made rounded orifices, the use of the following as measuring devices is inadvisable except in special circumstances after proper calibration.

A **Borda mouthpiece** is a short cylindrical reentrant tube (Figs. 54*a*, *d*) with a sharp, square upstream end.

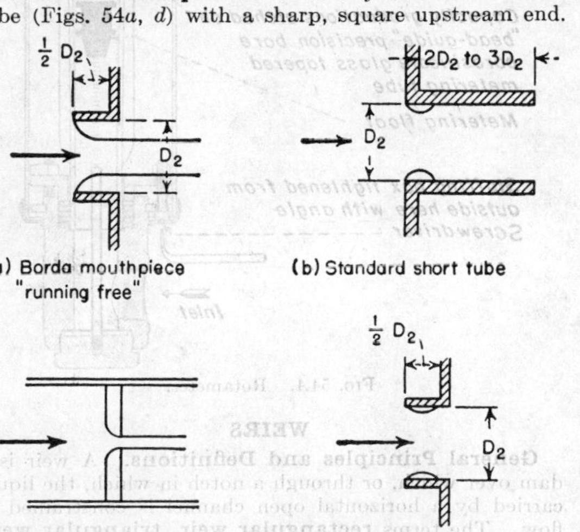

FIG. 54.—Miscellaneous discharge openings.

Under favorable conditions the jet does not touch the inner walls of the tube (Fig. 54*a*); the device is then said to *run free.* A Borda more than ½ a diameter long, or one discharging from other than a quiet pool, is apt to *run full, i.e.*, with the jet touching the tube walls (Fig. (54*d*). For free discharge of liquids the coefficient $(C_v'C_c)$ in Eq. (8) for a Borda mouthpiece running free is usually given as 0.53; when running full the values of $(C_v'C_c)$ customarily used are those given by Bilton (*Eng. News*, July 9, 1908) for reentrant tubes 2½ diameters long:

Diam., in.	⅛	¼	⅜	½	1	1½	2	2½
$(C_v'C_c)$.	0.91	0.87	0.85	0.83	0.79	0.77	0.76	0.75

For submerged operation C of Eq. (15) has been estimated to be 0.72.

A **standard short tube** (Fig. 54*b*) is a straight cylindrical tube two or three times as long as its internal diameter. Its entrance is like that of the square-edged orifice. When running free the short tube behaves like a square-edged orifice. When running-full C is commonly given as 0.82.

A **rounded orifice** (Fig. 54*c*) is a hole with rounded upstream edges in a flat plate. If the rounding is not too sharp, the device is readily a measuring nozzle (see above), and the remarks thereto pertaining are applicable. In other cases the coefficients of discharge will vary between those for square-edged orifices and those for well-made nozzles. Usually a square-edged orifice is safer.

Accuracy. Square-edged orifices and venturi tubes have been so well studied and standardized that one standard instrument may be expected to reproduce the results of another within from 1 to 2 per cent. This is, therefore, the order of reliability to be had, granting accurate measurement of the meter differential, when coefficients of discharge from the literature are used in estimating the flow from the indications of the meter. Usually, with meters of the class under consideration, relatively small constructional deviations cause substantial differences in the coefficient of discharge. Consequently strict adherence to the standard specifications is essential unless a specific calibration is to be made. A further consequence is that relatively slight corrosion or fouling may seriously influence the readings.

Except after calibration in place, none of these meters gives readily interpretable indications if there is swirling or helical flow or other abnormality of velocity distribution in the upstream channel. For example, the A.S.M.E. Special Research Committee (*Report on Fluid Meters,* Part 3) found that to obtain indications correct within 2 per cent pipe-line orifices and nozzles should not be located nearer to upstream fittings than the distances given in Table 4. The table shows also the reduction in spacing made possible by the use of straightening vanes between the fittings and the meter. Entirely adequate straightening vanes may be provided by fitting a bundle of short small tubes within the pipe.

Table 4. Locations of Orifices and Nozzles Relative to Pipe Fittings

Distances in pipe diameters, D

Kind of fitting, upstream	D_2/D	Distance, upstream fitting to orifice: Without straightening vanes	Distance, upstream fitting to orifice: With straightening vanes	Distance, vanes to orifice	Distance, nearest downstream fitting from orifice
Single 90-deg. ell, tee or cross used as ell	0.2	6			1
	.4	6			
	.6	8			
	.8	20	10	8	2
2 short radius 90-deg. ells in form of S	0.2	9*			1
	.4	9*	8		
	.6	14*	10*	6	
	.8	25*	16*	10	2
2 long or short radius 90-deg. ells in perpendicular planes	0.2	15*	5.5	5	1
	.4	18*	6		
	.6	25*	8	6	
	.8	40*	12	6.5	2
Contraction or enlargement	0.2	4	Vanes have no advantage		1
	.4	6			
	.6	9			
	.8	15			2
Globe valve	0.2	18	8	5	1
	.4	22	8	5	
	.6	30	9	6	
	.8	50	15	9	2
Gate valve, ⅕ open	0.2	10	Same as globe valve		1
	.4	12			
	.6	48			
	.8	>60			2

* A.G.A. Gas Measurement Committee *Report* No. 2.

The indications of orifices, nozzles, and venturi meters are undependable if the flow pulsates. Pulsations in the stream must be damped out by preceding the meter by air chambers in the case of liquids, or reservoirs in the case of gases, and by throttling the flow. A common error is to damp the fluctuations of the gages by placing restrictions in the gage lines. This may result in a steady reading which, in a given instance, may be interpretable in the control of an operation, but the relation of the reading to the actual flow will be indeterminate without calibration. Complete reproduction of operating conditions, including wave form, is necessary for an accurate calibration. Pulsation errors of about 40 per cent have been noted. The reading assumed by the gage will be a complicated function of the resistances in the gage connections and the inertia of the indicating mechanism. For discussions of the effects of pulsation see Bailey [*Trans. Am. Soc. Mech. Engrs.*, **61**, 301 (1939)], Beitler [*Trans. Am. Soc. Mech. Engrs.*, **61**, 309 (1939)] and Lindahl, [*Trans. Am. Soc. Mech. Engrs.*, **68**, 883–894 (1946)].

AREA METERS

General Principles. The underlying principle of an ideal area meter is the same as that of a head meter of the orifice type (*cf.* p. 400). The stream to be measured is throttled by a constriction, but, instead of observing the variation with the flow of the differential head across an orifice of fixed size, the constriction of an area meter is so arranged that its size is varied to accommodate the flow while the differential head is held constant. A simple example of an area meter is a gate valve of the rising stem type provided with static pressure taps before and after the gate. In most common types of area meters the variation of the opening is automatically brought about by the motion of a weighted piston or float supported by the fluid.

The calibration of area meters can be predicted precisely, without individual test, from Reynolds number versus coefficient curves similar to those used for orifice meters.

Rotameters. The rotameter (Fig. 54*A*) has justifiably become one of the most popular meters in chemical industry. It is especially useful for either manual or automatic control because it permits visualization of the flow process by the operator and comparison between primary and secondary instrument without shutting down. It consists essentially of a plummet which is free to move in a vertical, slightly tapered tube with the small end down. The fluid enters the lower end of the tube and causes the plummet to rise until the annular area between the plummet and the wall of the tube is such that the pressure drop across this constriction is just sufficient to support the plummet. Typically, the tapered tube is of glass and carries etched upon it a nearly linear scale on which the position of the floating plummet indicates the flow. Rotameters are available with pneumatic, electric, and electronic transmitters for actuating remote recorders, integrators, and automatic flow controllers.

Interchangeable precision-bore glass tubes and metal metering tubes are available. Rotameters have proved satisfactory both for gases and for liquids at high and at low pressures. A single instrument can readily cover a tenfold range of flow and by providing more than one float a two-hundredfold range is practicable. Rotameters require no straight runs of pipe before or after the point of installation. Pressure loss remains constant over the whole flow range.

In experimental work, for the greatest precision, a rotameter can be calibrated with the fluid which is to be metered. However, modern rotameters not only have predictable calibrations but are immune to effect of viscosity variations over a wide range of Reynolds numbers, so that viscosity variations usually cause no alteration of the calibration. Rotameters can also be rendered immune to fluid density changes of as much as 15 per cent. For these reasons, they are often superior to orifice meters for control and measuring instruments. They are especially suitable for automatic ratio control, the linear scale allowing much wider range adjustment than previously possible. For the theory of the rotameter see Schoenborn and Colburn [*Trans. Am. Inst. Chem. Engrs.*, **35**, 359 (1939)]; and Head (Paper, presented at A.S.M.E. Pittsburgh meeting, September, 1946).

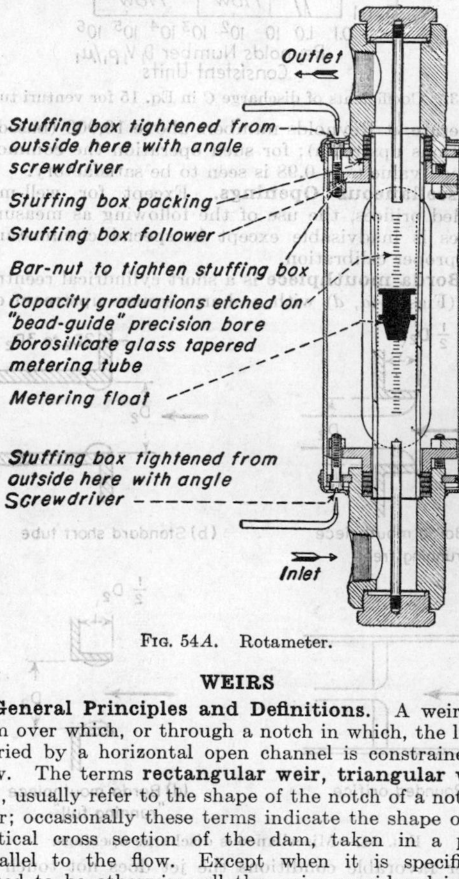

FIG. 54*A*. Rotameter.

WEIRS

General Principles and Definitions. A weir is a dam over which, or through a notch in which, the liquid carried by a horizontal open channel is constrained to flow. The terms **rectangular weir, triangular weir,** etc., usually refer to the shape of the notch of a notched weir; occasionally these terms indicate the shape of the vertical cross section of the dam, taken in a plane parallel to the flow. Except when it is specifically stated to be otherwise, all the weirs considered in this section have flat upstream faces that are perpendicular to the bed and walls of the channel.

Sharp-edged weirs have edges like those of square-edged orifices (see p. 404). Notched weirs are ordinarily sharp-edged. Weirs not in the sharp-edged class are, for the most part, those described as **broad-crested;** for a detailed discussion and description of them a treatise on hydraulics should be consulted.

The theory of the weir is very similar to that of an orifice that discharges into a gas-filled space; in fact, a weir may be regarded as such an orifice operating under a head so low that the opening is not filled. (For theory of weirs, see Gibson, O'Brien, and Hickox, or Dodge and Thompson, *op. cit.*) In operation, likewise, weirs have many of the characteristics of orifices; *e.g.*, the *nappe* (as the sheet of discharging liquid is called) generally contracts after leaving the edges of the opening as does the jet from an orifice; the amount of contraction may be decreased, with a consequent increase in the discharge under a given head, by rounding the upstream edges of the weir, etc. The distinction between a weir and a head meter (pp. 400*ff.*) is found in the following facts. In the case of the latter the area of the discharge opening is fixed, and independent of the head; in the case of the weir this area varies with the head. Roughly, Eq. (8) (p. 402) may be regarded as the type formula for a weir if A_2 is regarded as a function of ΔH.

A weir is said to be **submerged** if the liquid level in the downstream channel stands above the lowest edge of the weir. Submerged weirs are unsatisfactory as measuring devices and will not be considered here.

To measure the head on a weir the height of the surface of the liquid in the channel is determined at a point sufficiently far upstream to avoid the drop in level occasioned, immediately above the weir, by the overfall of liquid. For this purpose a stilling box, fitted with a hook gage or float gage (p. 367), is usually attached to the channel. The conditions requisite for the determination of the static pressure must be strictly observed (p. 396) in joining the stilling box to the channel.

Applicability of Weir Formulas. Nearly all weir formulas in current use have been devised empirically to fit certain sets of data for the discharge of water. None of them can be used with complete confidence unless the experimental conditions to which they apply are substantially duplicated. Sufficient data are not available to determine precisely whether or not other liquids obey the same formulas as does water. Dimensional analysis (except for V-notches) is not so helpful as it is in the case of submerged orifices [*cf.* Gibson, *op. cit.*, pp. 118*ff.*; also, Lindquist, *Trans. Am. Soc. Civil Engrs.*, **93**, 999 (1929) discussion of an article by Schoder and Turner]. The indications are that variations in viscosity produce only small effects on the rate of flow. When precision is desired, especially with very viscous fluids, it is highly advisable to calibrate all weirs. This is particularly true of small-scale apparatus. In all cases satisfactory measurements can be made only when the weir is preceded by a channel of sufficient length and breadth to ensure a regular and smooth flow above the weir plate. Weirs generally behave erratically if the nappe does not spring clear from the downstream side and from the top of the weir plate. Hence, for the case of a "clinging nappe," no great dependence can be placed upon formulas, and even after calibration the results cannot be regarded with confidence.

NOTE. For a review of weir data and formulas, see Schoder and Turner, *Trans., Am. Soc. Civil Engrs.*, **93**, 999 (1929), and the discussion of their article. For tests of various notches, see Cone, U.S. Dept. Agr., *J. Agr. Research*, **5**, 1051 (1916).

Rectangular weirs have one edge (the *crest*) of the opening horizontal. The head H is the vertical distance between the crest and the surface of the liquid in the upstream channel. For sharp-edged notches of this type the **Francis formula** is commonly used:

$$q = 3.33(L - 0.2H)H^{3/2} \quad \text{(cu. ft./sec.)} \qquad (18)$$
$$= 0.415(L - 0.2H)H^{3/2}\sqrt{2g} \qquad (18a)$$

where L = length of crest, ft.; and H = head, ft. [Equation (18*a*) is a dimensionally consistent form of Eq. (18).] If one of the vertical edges of the notch is so designed that there is substantially no contraction of the nappe on that side, replace $(L - 0.2H)$ by $(L - 0.1H)$. If this condition occurs at both sides of the nappe, replace the whole parenthesis by L. In the latter circumstance the device is called a **suppressed weir.** When the velocity of approach is appreciable, the term $H^{3/2}$ of the above formula is replaced by $\left[\left(H + \frac{V_1^2}{2g}\right)^{3/2} - \left(\frac{V_1^2}{2g}\right)^{3/2}\right]$, where V_1 is the mean linear velocity, ft./sec., in the upstream channel, and g = 32.17 ft./sec.² is the local acceleration due to gravity. The value of V_1 may usually be obtained sufficiently closely by making a preliminary computation without the correction for velocity of approach. The Francis formula agrees with experiment within 3 per cent or less, if (*a*) L is greater than $2H$; (*b*) velocity of approach is 1 ft./sec., or less; (*c*) height of crest above the bottom of the upstream channel is at least $3H$; and (*d*) H is not less than 0.3 ft. In the case of nappe contraction at a vertical edge, the contraction will be less than that allowed for in the formula, unless the nearest upstream channel wall is at a distance of at least $2H$ from the edge in question.

Narrow rectangular notches ($H \geq L$) have been found to give, quite closely, 90 per cent of the discharge computed by the Francis formula for a suppressed weir.

If the crest of a suppressed weir is a sufficiently broad ($3H$, at least) horizontal flat surface, the nappe, during the latter part of its passage across the crest, will form a horizontal-surfaced stream of substantially uniform depth t (ft.) (see Fig. 55). For a broad-crested weir of this type, with square upstream edges, the Francis suppressed-weir formula may be used to estimate the flow if the coefficient 3.33 is replaced by from 2.7 to 2.6. The higher value corresponds to a breadth of about $3H$, and smaller coefficients are found as the breadth increases. Theoretically, the depth of liquid above the crest is $t = \frac{2}{3}H$, at the point where the free surface of the nappe is horizontal.

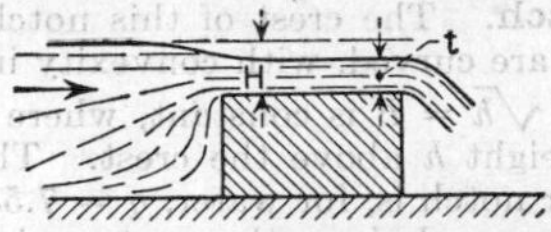

FIG. 55. Broad-crested weir.

Triangular or V-notch weirs are V-shaped notches cut in the weir plate with the vertex pointing downward and with each side inclined equally to the vertical. For such weirs of the sharp-edged type the measurements reported by Greve [*Eng. News-Record*, **105**, 166 (1930)] give for water

$$w \text{ (lb. } mass\text{/sec.)} = 156\left(\tan\frac{\alpha}{2}\right)^{0.996} H^{2.47} \qquad (19)$$
$$q \text{ (cu. ft./sec.)} = 2.505\left(\tan\frac{\alpha}{2}\right)^{0.996} H^{2.47} \qquad (20)$$

where α is the vertical angle of the notch and H is the head measured, ft., above the vertex. The range of α was from 25 to 118 deg.; H varied from 0.15 to 1.2 ft. The vertex of the V was always over 4 ft. from the bottom of the upstream channel. The velocity of approach was negligible. This result agrees with those of previous workers and, in the light of the dimensional analysis of Gibson (*op. cit.*), indicates the following dimensionally consistent formula, applicable to any fluid:

$$\frac{q\rho}{\mu H} = B\left(\frac{H^3 q_L \rho^2}{\mu^2}\right)^{0.490} \qquad (21)$$

where $B = 0.575\,(\tan\alpha/2)^{0.996}$; H and α are as stated above; and the other symbols agree with the table of nomenclature, p. 401. Hence, for liquids other than water, Eq. (19) may be used, if the results are multiplied by $(\mu/\mu_w s)^{0.02}$, where μ/μ_w is the relative viscosity and s is the specific gravity of the liquid in question. The correction is generally so small that it is negligible.

The **Cipolletti trapezoidal weir** is a sharp-edged, horizontal-crested notch of which the sides slope upward and outward (1 horizontal to 4 vertical). The slope has been so chosen that the relatively simple Francis suppressed-weir formula ($q = 3.33LH^{3/2}$, where L = length of crest) is substantially correct for this notch.

Circular or semicircular notches are not normally used; hence the data concerning them are meager. Need for such data occasionally arises when a square-edged orifice is operating under so low a head that the opening is not filled. Consequently the data available are presented in Fig. 56 [Cone, U.S. Dept. Agr., *J. Agr. Research*, **5**, 1051 (1916)]. The head H is measured above the lowest point of the circular opening. The velocity of approach was negligible in the experiments on which the plot is based. The lines are drawn only for the range covered by the data.

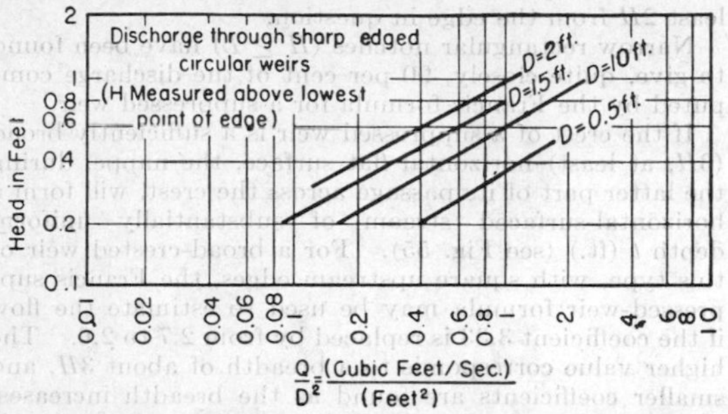

FIG. 56. Discharge through circular-notched weirs.

Weirs with the discharge directly proportional to the head can be constructed. The simplest is the **inverted notch.** The crest of this notch is horizontal and the sides are curved, with convexity inward, in such a way that $L\sqrt{h} = B$ is constant, where L is the notch width at a height h above the crest. The formula for this particular notch is, for water, $q = 7.55BH$, where H is the head measured above the crest. q is in cu. ft./sec., and h, L, and H are in ft. To construct this weir, the necessary value of B is found by substituting the desired maximum value of q and H in the above formula.

Curved sharp-edged weirs with the crest in a horizontal plane (*e.g.*, overflow pipes when not running full), as might be expected, appear to obey roughly the Francis formula for suppressed weirs if the radius of curvature is very large. For radii less than 13 in., the meager experimental data indicate that the Francis formula gives values of the discharge that are distinctly too high, whether the flow is radially inward or outward, unless the head is low. The most commonly quoted formula **for inward flow** (Gourley, *Proc. Inst. Civil Eng.*, Part 2, p. 297, 1910) is approximately $q = 3.0LH^{1.4}$ (cu. ft./sec.) which may be used for heads up to one-fifth the diameter of the crest curve. L is the perimeter of the crest, ft., and H is the head, ft., above the crest. In considering the discharge of an overflow pipe from any reservoir but a quiet tank, it should be remembered that the actual discharge is considerably dependent upon the kind and extent of agitation in the reservoir. Consequently formulas, other than those derived from previous data on equipment similar in all respects to that at hand, must be regarded as rough approximations.

For radially outward flow (*e.g.*, over the edge of a cylindrical tank or standpipe), $q = 2.0(LH)^{1.3}$ (cu. ft./sec.) is approximately correct for heads less than 0.03 of the tank diameter [*Proc. Am. Soc. Civil Engrs.*, **32**, 497 (1906)].

Application of Weir Formulas to Gases. If a gas A flows into a chamber filled with another gas B, the stream of A will generally remain distinct for an appreciable length of travel. If A has a density greater than B, gas A will settle to the floor of the chamber and flow along the floor much as a liquid flows in an open channel. In fact the behavior of such a gas stream on meeting weirs, etc., may be estimated by applying the foregoing formulas for liquids, interpreting all heads as feet of the gas flowing. Analogously, if gas A has a density less than B, gas A will flow along the roof of the chamber, and the conditions found on meeting an obstruction suspended from the ceiling can be visualized by inverting a diagram drawn for a liquid flowing past a similar obstruction that rises from the floor of its channel. For example, Fig. 55, if inverted, can represent the flow of hot gases along the roof of a furnace, *under* a "dam," and up a chimney. The familiar experiment of pouring hydrogen *up* into a bottle illustrates these phenomena.

Groume-Grjimailo ("Flow of Gases in Furnaces," tr. by Williams, Wiley, New York, 1923) has made brilliant use of these principles in furnace design. For a brief discussion see Haslam and Russell, "Fuels and Their Combustion," pp. 740*ff.*, McGraw-Hill, New York, 1925.

QUANTITY METERS

The meters described below are designed to measure the total quantity of fluid discharged during a period of time. In this respect they differ from many of the instruments previously discussed, most of which actually indicate rates of flow, although integrating devices can be sometimes provided to record the total discharge. Most of them have moving metal parts in contact with fluid so that their use for the measurement of corrosive fluids is generally impossible without special construction. Commercial mechanical flow meters are so numerous that no attempt can be made to discuss more than the principal types (see pp. 1279–1282).

Wet gas meters (Fig. 57) accurately measure the volumetric discharge of gases even at the smallest rates of flow. Within a cylindrical casing, partly filled with liquid, is a drum that rotates on a horizontal axis. The drum has a vertical cross section similar to that shown in the diagram. By referring to Fig. 57, it is seen that gas entering the internal drum through the hub can pass only into compartment A since the openings b, c, and d into the other compartments are sealed with liquid. Consequently, admission of gas causes the drum to rotate in the direction of the arrow. The rotation soon seals opening a, through which the gas entered compartment A; permits flow into compartment B; and eventually unseals the passage a' through which the gas in A escapes into the outer casing. The rotation of the drum evidently continues as long as gas is flowing and may be made to operate a suitable revolution counter. Since each revolution corresponds to the passage of a definite volume of gas (measured at the temperature and pressure inside the meter, and saturated with the vapor of the liquid therein), the dial may be readily made to read directly in units of volume.

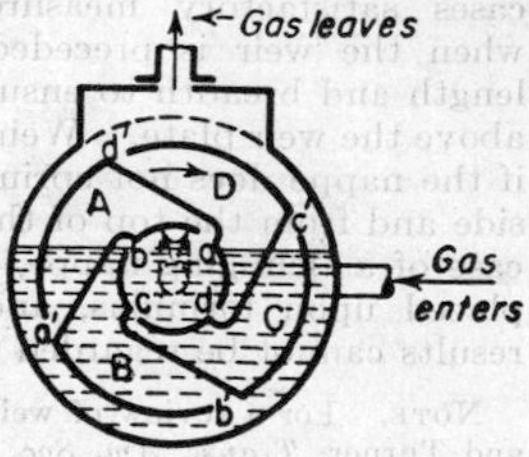

FIG. 57. Wet gas meter.

To obtain satisfactory results with a wet gas meter, the surface of the liquid inside must be, during use, in exactly the same position relative to the casing that it occupied during calibration. Consequently the meter is always provided with crossed spirit levels, or their equivalent, and with a sight glass, so that the location of the liquid surface can be accurately adjusted. A rapidly fluctuating rate of flow may cause error by producing surges of the liquid level and thus interfering with the proper sequence of sealing and unsealing the various openings. It is therefore necessary when such flows are met, to insert in the line some means (*e.g.*, a gas bag) of damping the variations. The sealing fluid in wet gas meters is usually water, although any noncorrosive liquid may be employed, provided that it is inert with respect to the gas being measured. When precision is wanted, the vapor pressure of the liquid must be accurately known so that volume of *dry* gas can be computed from the volumes of *wet* gas read by the meter.

Meters for liquids, the same in principle as the wet gas meters, are commercially available. The construction and mode of operation become obvious if Fig. 57 is turned upside down. In the inverted diagram the shaded portion is gas filled, and the unshaded part is filled with the liquid being measured.

Dry gas meters, bellows, or **diaphragm meters,** are those generally used to meter illuminating gas supplied for household or commercial purposes. Two flexible containers, connected by automatic valves to the gas main, are alternately filled and emptied as the gas flows through the meter. The reciprocating motion arising from the expansion, first of one gas compartment and then of the other, is communicated to a counting mechanism. The two containers are enclosed in a box which carries the dial of the meter. A typical meter is shown in Fig. 58. The temperature and static pressure of the gas at the meter must be recorded in order to compute weight rate of flow or the volumetric rate of flow for any set of standard conditions. The use of dry

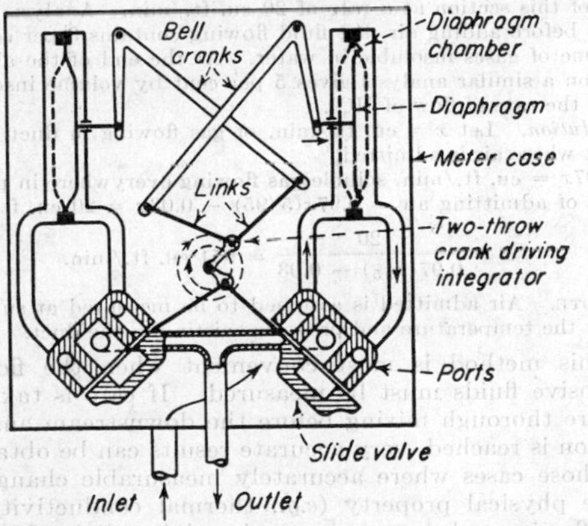

FIG. 58. Typical diaphragm gas-meter mechanism.

gas meters under conditions of temperature and pressure for which they have not been recently calibrated is highly inadvisable. When calibrated, these meters can be expected to have an accuracy of ±1 per cent within their rated range.

Cycloidal meters (Fig. 59) are much more compact. They are very similar to a two-lobed cycloidal blower of the Roots type. Passage of the gas causes the meter to revolve. This type permits a certain amount of slippage around the rotors but has the advantage that the volume is invariable, whereas there is a possibility of variation in the water-sealed meter. Depending on size, such meters operate on a differential of from ½ to 1 in. water. They are built for pressures to 500 lb./sq. in., in sizes from 100 to 1,000,000 cu. ft./hr. of gas corrected to atmospheric pressure. Meters are sold with charts for correcting for the slight leakage around the rotors, which varies in a known manner with the differential under which the meter is operating. Rotation of these meters is recorded as integrated flow on a counter. Such meters may also be provided with a recorder on which flow rate, temperature, and pressure are simultaneously recorded.

Rotary-disk water meters are those commonly used for measuring the household water supply in American cities. Inside an outer casing, which carries the dial and protects the gear train, is a chamber A (Fig. 60) that has an inwardly projecting, conical top and bottom. The chamber contains a movable, flat, hard-rubber disk B, affixed to the central ball C, any motion of which is communicated to the gears by the pin E. In the disk is a radial slot which straddles the fixed partition D-

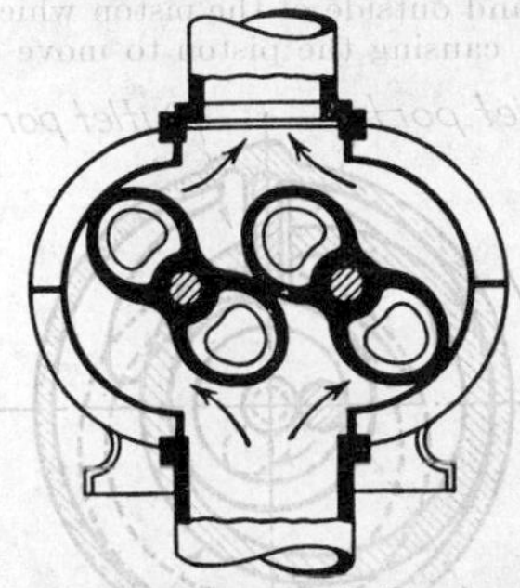

FIG. 59. Typical cycloidal gas meter.

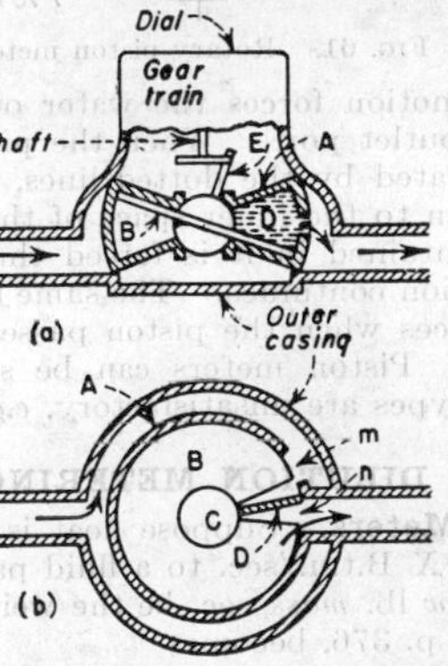

FIG. 60. Rotary-disk water meter.

Since the end of pin E is constrained to move in a circle of a fixed diameter about the shaft that leads to the gear train, and the partition D prevents any rotation of the disk about its axis, the only motion possible for the disk is like that of a disk top near the end of its spin. By properly choosing the angle of the cones forming the top and bottom of the inner chamber, some radius of the upper face of the disk can always be kept tangent to the upper cone, while the diametrically opposed radius of the lower face is tangent to the lower cone. In these circumstances, water entering the chamber A at opening m can reach the outlet n only by moving the rubber disk.

As their common name implies, disk water meters are usually restricted to the measurement of water, but they may be used with any liquid which is without action on the materials of construction. It is inadvisable to attempt the measurement of hot fluids with meters of this type since the rubber disk is liable to be damaged by heat. Reasonable accuracy may be expected of these meters if they are frequently calibrated under the exact conditions of use as to pressures, rate of flow, and temperature. However, if under operating conditions the rates of flow, or the pressures, vary greatly, little reliance can be placed on disk meters because of the consequent variation in leakage.

Piston meters resemble a pair of double-acting piston pumps working backward. Usually slide valves, that of one cylinder being operated by the motion of the piston of the other cylinder, control the alternate admission and expulsion of the liquid from either end of each cylinder. Only one piston moves at a time. A counting device is actuated by one of the pistons. In one type a circle of pistons is connected to a nutating disk or wobble plate which transmits motion to a dial. In still another type the piston is of the rotary type as shown in Fig. 61. Liquid flowing through the inlet port enlarges the spaces on the inside and outside of the piston which are open to the inlet port, causing the piston to move counterclock-

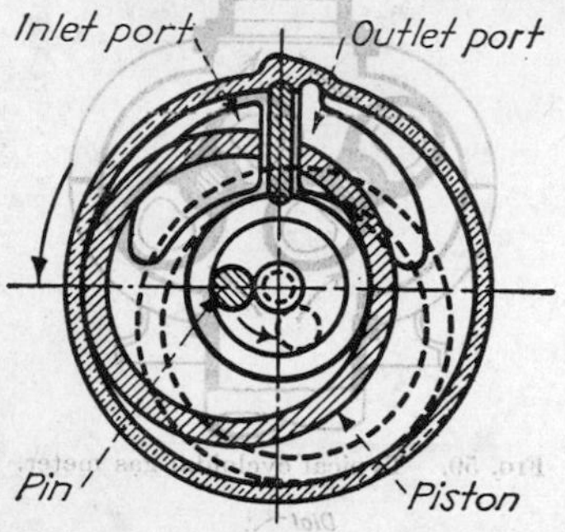

FIG. 61. Rotary piston meter.

wise. This motion forces the water out of the spaces open to the outlet port. When the piston reaches the position indicated by the dotted lines, the inlet port is no longer open to the inner space of the piston, and all the water contained in it is forced through the outlet piston as motion continues. The same thing happens to the outer spaces when the piston passes the upper half of its travel. Piston meters can be successfully used where other types are unsatisfactory, *e.g.*, in the case of hot water.

DILUTION METERING

Thermal Meters. Suppose heat is supplied at the constant rate X B.t.u./sec. to a fluid passing through a duct, and let w lb. *mass*/sec. be the weight rate of flow. Then Eq. (9), p. 376, becomes

$$J\frac{X}{w} = \frac{V_2^2 - V_1^2}{2g_c} + (Z_2 - Z_1) + J(i_2 - i_1)$$

for, since there is no pump, $W_0 = 0$. If the temperature, static pressure, and duct cross section are measurable on each side of the heater, the entire right-hand side of the equation can be expressed in terms of w and known quantities. ($V_1 = w/\rho_1 S_1$, etc., where ρ_1 and S_1 are the fluid density and duct cross-sectional area, respectively.) Therefore, if the rate of heat input X is measured, w can always be found by solving the above equation. If the temperature rise of the fluid is small (in the case of gases the static pressure drop must also be small), the term $(V_2^2 - V_1^2)/2g_c$ will be negligible, and usually $(Z_2 - Z_1)$ is negligible in all practical cases. Hence, the equation reduces to $w = X/(i_2 - i_1)$ or, for liquids and for perfect gases,

$$w \text{ (lb. } mass\text{/sec.)} = \frac{X \text{ (B.t.u./sec.)}}{c_p(t_2 - t_1)} \qquad (22)$$

where c_p is the specific heat at constant pressure, B.t.u./(lb. *mass*) (°F.) and $(t_2 - t_1)$ is the temperature rise of the fluid, °F.

It is obvious that any convenient means of supplying heat to, or removing heat from, the fluid at an accurately measurable rate may be made the basis of a thermal flowmeter. In any case, if the temperature and heat-flow measurements are correct, and specific heat of the fluid is known, the weight rate of flow will be strictly accurate.

The Thomas electric flowmeter for gases (Cutler-Hammer Manufacturing Co., Milwaukee) is a commercially available thermal flowmeter. Heat is supplied to the fluid by an electric heating coil wound in a grid that covers the cross section of the duct. The temperature rise is determined by placing resistance thermometers (also wound in grids) one before and one after the heater. It may operate in either of two ways: (*a*) The temperature rise may be kept constant by varying the direct current in the heater; then a wattmeter placed in the heating circuit gives a reading directly proportional to weight rate of flow. (*b*) The wattage supplied may be kept constant; then the temperature rise is inversely proportional to the rate of flow. In either case the electrical instruments used can be calibrated to read the flow directly.

Mixture Metering. Given any foreign substance that is soluble in the fluid flowing through a duct and is susceptible of ready quantitative determination when mixed therewith. If this foreign substance is added to the stream at a known constant rate and analyses of the flowing fluid are made on both the upstream and downstream sides of the point of admixture, then the rate of flow of the stream can be readily calculated.

Example. HCl gas is flowing through a stoneware duct under a slight vacuum. To measure the flow, a section of the duct is first made airtight, and then dry air is admitted at the upstream end of this section at a rate of 20 cu. ft./min. Analysis shows that, before adding air, the fluid flowing contains 3 per cent by volume of gases insoluble in water. At the end of the airtight section a similar analysis gives 5 per cent by volume insoluble. Find the rate of flow of HCl.

Solution. Let x = cu. ft./min. of gas flowing in duct before point where air is admitted.

$0.97x$ = cu. ft./min. soluble gas flowing everywhere in pipe.

Rate of admitting air $= 0.97x(5/95) - 0.03x = 20$ cu. ft./min.

$$x = \frac{20}{0.97(5/95) - 0.03} = 951 \text{ cu. ft./min.}$$

NOTE. Air admitted is assumed to be measured at substantially the temperature and pressure existing in the duct.

This method is most convenient when the flow of corrosive fluids must be measured. If care is taken to ensure thorough mixing before the downstream analysis station is reached, very accurate results can be obtained. In those cases where accurately measurable changes of some physical property (*e.g.*, thermal conductivity) of the flowing substance are produced by slight dilution, measurements of that property may replace the chemical analyses.

MISCELLANEOUS MECHANICAL METERS

Impulse Wheel or Turbine Meters. Several types of meters for liquids are in principle the result of mounting a vane anemometer in a pipe line. Variously constructed turbine wheels are mounted in casings which carry a suitable revolution-counting mechanism. These meters are generally used only for high rates of flow. Moderate accuracy is obtained if the meters are used in the range for which they are calibrated, but at low flows seriously low readings are usual.

Vane Meters. A small rod or vane suspended in a stream will be deflected by the force of the moving fluid against the resistance of a spring or against the force of gravity. By providing a suitable indicating mechanism the extent of the deflection may be used as a measure of the flow.

PIPE AND FITTINGS

Pipe, tubing, and fittings are available in many materials, metallic and non-metallic. Selection should be made according to the requirements of economic size, strength, and corrosion resistance. It is advisable to check the cost and availability of stock vs. mill material. In the case of non-ferrous and alloy-steel tubing, the extra cost for close tolerances may not be warranted in some installations. Data on the dimensions for those materials listed below are given in the pages and tables indicated.

	Page	Tables
General:		
Linear expansion of pipes	413	1
Metallic pipe: ferrous:		
Steel and wrought-iron pipe standards	414	2
Line-pipe and alloy-pipe standards	414	3
Wrought-iron and wrought-steel pipe	413	
Welded- and seamless-steel pipe	415	4
Steel-pipe dimensions, capacities, and weights	415	5
Steel pipe "double extra strong"	417	6
Steel tubing, tolerances	417	7
Stainless-steel pipe and tubing	417	8
Cast-iron pipe	418	
Pit-cast cast-iron pipe	419	9, 10
Standard bell-and-spigot pipe	421	11
Cast-iron flanged pipe	422	12
Centrifugally cast, bell and spigot	423	13
Duriron pipe	423	14
Metallic pipe: non-ferrous:		
Aluminum	424	15
Copper, tolerances	424	15*a*
Condenser and heat-exchanger tube	425	16
Copper water tubes	427	17
Copper and copper-alloy	428	18
Finned copper tubes	429	18*a*
Lead and lead-lined	429	19, 20

	Page	Tables
Magnesium	429	
Nickel and nickel-alloy	430	15
Tantalum and columbium	430	
Tin	430	21, 22
Bimetallic drawn tubing	430	
Flexible metal hose	431	
Non-metallic pipe and tubing:		
Asbestos-cement	431	23, 24
Carbon and graphite	431	25, 26
Cement-lined	434	27
Chemical ware	434	28, 29, 30
Clay sewer	435	31
Concrete	435	
Glass	436	32
Glass-lined	436	33
Plastic and fiber plastic	436	
Haveg	437	34
Lucite, methyl methacrylate	437	35
Laminated phenolic	438	36
Saran	438	37*a*, *b*
Porcelain	438	38
Rubber and rubber-lined	438	39*a*, 39*b*
Rubber-fabric hose	439	
Silica	440	
Wood	440	40
Wood-lined	440	

Reference can be made to the "Code for Pressure Piping," A.S.A. B31.1-1942 with supplements B31.1*a*-1944 and B31.1*b*-1947, which represents a standard of minimum safety requirements for materials, dimensions, design, erection, and test (American Standards Association, New York, N.Y.). Considerable information on piping is given in Crocker, "Piping Handbook," 4th ed. (McGraw-Hill, New York, 1945).

The A.S.A. recommended practice A13-1928 is a scheme for identification of piping systems above ground including conduits for transport of gases, liquids, semiliquids, and plastics. Special reference to personal hazards is given.

Table 1. Linear Expansion of Pipes

The expansion in inches for a pipe of any length between any two given temperatures is found by taking the difference in length at these temperatures, dividing by 100, and multiplying by the length of the pipe in feet.

Temp., °F.	Cast iron	Wrought iron	Steel	Brass and copper	Temp., °F.	Cast iron	Wrought iron	Steel	Brass and copper
0	0.00	0.00	0.00	0.00					
50	.36	.40	.38	.57	450	3.89	4.28	4.08	6.18
100	.72	.79	.76	1.14	475	4.20	4.62	4.41	6.68
125	.88	.97	.92	1.40	500	4.45	4.90	4.67	7.06
150	1.10	1.21	1.15	1.75	525	4.75	5.22	4.99	7.55
175	1.28	1.41	1.34	2.04	550	5.05	5.55	5.30	8.03
200	1.50	1.65	1.57	2.38	575	5.36	5.90	5.63	8.52
225	1.70	1.87	1.78	2.70	600	5.70	6.26	5.98	9.06
250	1.90	2.09	1.99	3.02	625	6.05	6.65	6.35	9.62
275	2.15	2.36	2.26	3.42	650	6.40	7.05	6.71	10.18
300	2.35	2.58	2.47	3.74	675	6.78	7.46	7.12	10.78
325	2.60	2.86	2.73	4.13	700	7.15	7.86	7.50	11.37
350	2.80	3.08	2.94	4.45	725	7.58	8.33	7.96	12.06
375	3.15	3.46	3.31	5.01	750	7.96	8.75	8.36	12.66
400	3.30	3.63	3.46	5.24	775	8.42	9.26	8.84	13.38
425	3.68	4.05	3.86	5.85	800	8.87	9.76	9.31	14.10

FERROUS METAL PIPE AND TUBING

Considerable standardization of specifications and dimensions of ferrous metal pipe and tubing has been accomplished by the American Standards Association. Tables 2 and 3 list those of interest to chemical engineers.

Wrought-iron and Wrought-steel Pipe

Dimensions and weights of wrought-iron and steel pipes from 1/8 to 30 in. have been standardized by the American Standards Association as A.S.A. Standard B36.10; the system of classes known as "standard weight," "extra strong," and "double extra strong," has been supplanted by 10 weight and thickness schedules of which the lightest wall is Schedule 10 and the heaviest is Schedule 160. Weights for steel and wrought iron are identical in sizes up to and including 12 in., although the nominal wall thicknesses for the two materials have been adjusted slightly to compensate for the small difference in density. The schedules and wall thicknesses for steel pipe are given in Table 4, and the thicknesses for wrought iron are about 2 per cent greater. For 14-in. pipe and larger, the wall thicknesses are the same; the necessary adjustment has been made in the weights. Schedule 80 is the largest so far contemplated for wrought iron. For dimensions, weights, capacities, etc., of steel pipe, see Table 5. For dimensions of the now obsolete "double extra strong" pipe, see Table 6.

Table 2. Steel and Wrought-iron Pipe Standards*

Specifications for Pipe, with the Tensile Strengths Called for, and the Corresponding Uses for Which Each Material Is Intended

A.S.A. designation	A.S.T.M. designation	Title standard specifications	Tensile strength, lb./sq. in.	Scope
B36.1-1945	A53-44	Welded and seamless steel pipe	Welded, Bessemer, 50,000 min. Welded, open-hearth, 45,000 min. Seamless, low carbon, 48,000 min. Seamless, medium, 62,000 min.	Commercial steel pipe for general uses, also for coiling, bending, flanging, and similar forming operations when so specified
B36.2-1939	A72-39	Welded wrought-iron pipe	40,000 min.	Commercial wrought-iron pipe for general uses, also for coiling, bending, flanging, and other special purposes
B36.3-1942	A106-42T	Lap-welded and seamless steel pipe for high-temperature service	Welded, open-hearth, 45,000 min. Seamless, grade A, 48,000 min. Seamless, grade B, 62,000 min. Seamless, grade C, 75,000 min.	Lap-welded and seamless steel pipe for high-temperature service. Suitable for bending, flanging, and similar forming operations
B36.4-1942	A134-42	Electric-fusion-welded steel pipe, sizes 30 in. and over	Material as specified from a list of A.S.T.M. Standards included in A 134	Covers pipe 30 in. diameter and over, in wall thicknesses up to ¾ in., inclusive, fabricated from steel plates by straight seam or spiral seam electric fusion welding
B36.5-1945	A135-44	Electric-resistance-welded steel pipe	Grade A, 48,000 min. Grade B, 60,000 min.	Pipe up to 30 in., inclusive, intended for conveying liquids, gas, or vapor at temperatures below 450°F. Adapted for flanging, bending, and similar forming operations in grade A class
B36.9-1942	A139-42	Electric-fusion-welded steel pipe, 8 in. to, but not including 30 in.	Grade A, 48,000 min. Grade B, 60,000 min. or other suitable material as specified from a list of A.S.T.M. Standards	Covers sizes 8 in. up to but not including 30 in. in wall thicknesses not over ⅝ in., fabricated from steel plates by straight seam or spiral seam electric fusion welding. Intended for conveying liquids, gas, or vapor at temperatures below 450°F. Adapted for flanging and bending
B36.10-1939	....	Wrought-iron and wrought-steel pipe	See Table 4	Sizes from ⅛ in. to 30 in., inclusive
B36.11-1942	A155-42	Electric-fusion-welded steel pipe for high-temperature and high-pressure service	Grade A, 45,000 min. Grade B, 50,000 min. Grade C, 55,000 min.	Electric-fusion-welded steel pipe having an outside diameter of 18 in. and over for high-temperature and high-pressure service. Suitable for bending, flanging, corrugating, and similar forming operations. Welding in accordance with Par. U-68 of the A.S.M.E. Code for Unfired Pressure Vessels
B36.16-1945	A211-44	Spiral-welded steel or iron pipe	Steel and iron of A.S.T.M. A78, A89, A70, A10, and A129	Sizes from 4 to 48 in., inclusive and 1/16 to 11/64 in. wall thickness
G8.7-1945	A120-44	Black and hot-dipped zinc-coated (galvanized) welded and seamless steel pipe for ordinary uses	†	Commercial steel pipe for ordinary uses such as low-pressure steam, liquid, or gas lines. Not intended for coiling or close bending, or for high-temperature service

* American Standards Association, New York, N.Y.

† No physical test requirements are called for in A.S.T.M. A 120 except the mill hydrostatic tests on each length, which are for the same pressures as called for in B36.1 for the respective types and sizes of pipe. As indicated in the title, the purpose of the society in framing A.S.T.M. A 120 is to provide a medium for the purchase of pipe from warehouse stocks for use under conditions where the physical and chemical properties of the material are of lesser moment.

Table 3. Line-pipe and Alloy-pipe Standards*

A.P.I. and A.S.T.M. Specifications for Pipe, with Tensile Strengths Called for, and the Corresponding Uses for Which Each Material Is Intended

Designation	Title	Tensile strength, lb./sq. in.	Scope
A.P.I. 5-L	Line-pipe specifications	Furnace welded: Bessemer, 50,000 min. Open-hearth, class I, 45,000 min. Open-hearth, class II, 48,000 min. Wrought iron, 42,000 min. Seamless or electric welded: Open-hearth, grade A, 48,000 min. Open-hearth, grade B, 60,000 min. Open-hearth, grade C, 75,000 min. Seamless or furnace welded: Open-hearth iron, 42,000 min.	Pipe for line-pipe purposes, to convey gas, water, or oil. Sizes covered, ⅛ in. nominal size to 24-in. outside diameter. Couplings for threaded line pipe are of special design. Threads on pipe and in couplings are subject to official gage limits
A.S.T.M. A206	Seamless carbon-molybdenum alloy steel pipe for service at temperatures from 750° to 1000°F.	55,000 min.	Seamless carbon—½ per cent molybdenum-alloy-steel pipe intended for service at metal temperatures from 750° to 1000°F. Suitable for bending, flanging, and similar forming operations, and for fusion welding
A.S.T.M. A158	Seamless alloy-steel pipe for service at temperatures from 750° to 1100°F.	Ferritic, 60,000 min. Austenitic, 75,000 min.	Seamless alloy-steel pipe intended for service at metal temperatures from 750° to 1100°F. Several classes of materials that have been rather extensively used are included. Choice from the respective steels should be made on the basis of requirements of design, service conditions, and the physical properties.

* These specifications do not constitute part of the American Standard but are included for information.

Table 4. Welded- and Seamless-steel Pipe*

Nominal pipe size, in.	Outside diam., in.	Nominal wall thickness, in. Schedule No.† 10	20	30	40	60	80	100	120	140	160
1/8	0.405				0.068		0.095				
1/4	.540				.088		.119				
3/8	.675				.091		.126				
1/2	.840				.109		.147				0.187
3/4	1.050				.113		.154				.218
1	1.315				.133		.179				.250
1 1/4	1.660				.140		.191				.250
1 1/2	1.900				.145		.200				.281
2	2.375				.154		.218				.343
2 1/2	2.875				.203		.276				.375
3	3.500				.216		.300				.437
3 1/2	4.000				.226		.318				
4	4.500				.237		.337		0.437		.531
5	5.563				.258		.375		.500		.625
6	6.625				.280		.432		.562		.718
8	8.625		0.250	0.277	.322	0.406	.500	0.593	.718	0.812	.906
10	10.75		.250	.307	.365	.500	.593	.718	.843	1.000	1.125
12	12.75		.250	.330	.406	.562	.687	.843	1.000	1.125	1.312
14 o.d.	14.0	0.250	.312	.375	.437	.593	.750	.937	1.062	1.250	1.406
16 o.d.	16.0	.250	.312	.375	.500	.656	.843	1.031	1.218	1.437	1.562
18 o.d.	18.0	.250	.312	.437	.562	.718	.937	1.156	1.343	1.562	1.750
20 o.d.	20.0	.250	.375	.500	.593	.812	1.031	1.250	1.500	1.750	1.937
24 o.d.	24.0	.250	.375	.562	.687	.937	1.218	1.500	1.750	2.062	2.312
30 o.d.	30.0	.312	.500	.625							

* A.S.A. Standards B36.10-1939.

† The schedule numbers indicate approximate values of the expression: $1000P/S$, where P = the internal pressure, lb./sq. in. and S = the allowable fiber stress, lb./sq. in.

Thicknesses shown in boldface type in schedules 30 and 40 are identical with thicknesses for "standard weight" pipe in former lists; those in schedules 60 and 80 are identical with thicknesses for "extra strong" pipe in former lists.

The decimal thicknesses listed for the respective pipe sizes represent their nominal or average wall dimensions. A mill tolerance of 12.5 per cent under the nominal thicknesses is permitted.

Table 5. Steel-pipe Dimensions, Capacities, and Weights*

Nominal pipe size, in.	Outside diam., in.	Schedule No.	Wall thickness, in.	Inside diam., in.	Cross sectional area metal, sq. in.	Inside sectional area, sq. ft.	Circumference, ft., or surface, sq. ft./ft., of length: Outside	Inside	Capacity at 1 ft./sec. velocity: U.S. gal./min.	Lb./hr. water	Weight of pipe per ft., lb.
1/8	0.405	40†	0.068	0.269	0.072	0.00040	0.106	0.0705	0.179	89.5	0.25
		80‡	.095	.215	.093	.00025	.106	.0563	.112	56.0	.32
1/4	0.540	40†	.088	.364	.125	.00072	.141	.0954	.323	161.5	.43
		80‡	.119	.302	.157	.00050	.141	.0792	.224	112.0	.54
3/8	0.675	40†	.091	.493	.167	.00133	.177	.1293	.596	298.0	.57
		80‡	.126	.423	.217	.00098	.177	.1110	.440	220.0	.74
1/2	0.840	40†	.109	.622	.250	.00211	.220	.1630	.945	472.5	.85
		80‡	.147	.546	.320	.00163	.220	.1430	.730	365.0	1.09
		160	.187	.466	.384	.00118	.220	.1220	.529	264.5	1.31
3/4	1.050	40†	.113	.824	.333	.00371	.275	.2158	1.665	832.5	1.13
		80‡	.154	.742	.433	.00300	.275	.1942	1.345	672.5	1.48
		160	.218	.614	.570	.00206	.275	.1610	0.924	462.0	1.94
1	1.315	40†	.133	1.049	.494	.00600	.344	.2745	2.690	1,345.	1.68
		80‡	.179	0.957	.639	.00499	.344	.2505	2.240	1,120.	2.17
		160	.250	0.815	.837	.00362	.344	.2135	1.625	812.5	2.85
1 1/4	1.660	40†	.140	1.380	.669	.01040	.435	.362	4.57	2,285.	2.28
		80‡	.191	1.278	.881	.00891	.435	.335	3.99	1,995.	3.00
		160	.250	1.160	1.107	.00734	.435	.304	3.29	1,645.	3.77
1 1/2	1.900	40†	.145	1.610	0.799	.01414	.498	.422	6.34	3,170.	2.72
		80‡	.200	1.500	1.068	.01225	.498	.393	5.49	2,745.	3.64
		160	.281	1.338	1.429	.00976	.498	.350	4.38	2,190.	4.86
2	2.375	40†	.154	2.067	1.075	.02330	.622	.542	10.45	5,225.	3.66
		80‡	.218	1.939	1.477	.02050	.622	.508	9.20	4,600.	5.03
		160	.343	1.689	2.190	.01556	.622	.442	6.97	3,485.	7.45
2 1/2	2.875	40†	.203	2.469	1.704	.03322	.753	.647	14.92	7,460.	5.80
		80‡	.276	2.323	2.254	.02942	.753	.609	13.20	6,600.	7.67
		160	.375	2.125	2.945	.02463	.753	.557	11.07	5,535.	10.0
3	3.500	40†	.216	3.068	2.228	.05130	.917	.804	23.00	11,500.	7.58
		80‡	.300	2.900	3.016	.04587	.917	.760	20.55	10,275.	10.3
		160	.437	2.626	4.205	.03761	.917	.688	16.90	8,450.	14.3
3 1/2	4.000	40†	.226	3.548	2.680	.06870	1.047	.930	30.80	15,400.	9.11
		80‡	.318	3.364	3.678	.06170	1.047	.882	27.70	13,850.	12.5
4	4.500	40†	.237	4.026	3.173	.08840	1.178	1.055	39.6	19,800.	10.8
		80‡	.337	3.826	4.407	.07986	1.178	1.002	35.8	17,900.	15.0
		120	.437	3.626	5.578	.07170	1.178	0.950	32.2	16,100.	19.0
		160	.531	3.438	6.621	.06447	1.178	0.901	28.9	14,450.	22.6
5	5.563	40†	.258	5.047	4.304	.1390	1.456	1.322	62.3	31,150.	14.7
		80‡	.375	4.813	6.112	.1263	1.456	1.263	57.7	28,850.	20.8
		120	.500	4.563	7.953	.1136	1.456	1.197	51.0	25,500.	27.1
		160	.625	4.313	9.696	.1015	1.456	1.132	45.5	22,750.	33.0
6	6.625	40†	0.280	6.065	5.584	0.2006	1.734	1.590	90.0	45,000.	19.0
		80‡	.432	5.761	8.405	.1810	1.734	1.510	81.1	40,550.	28.6
		120	.562	5.501	10.71	.1650	1.734	1.445	73.9	36,950.	36.4
		160	.718	5.189	13.32	.1469	1.734	1.360	65.8	32,900.	45.3

* Based on A.S.A. Standards B36.10-1939.

† Designates former "standard" sizes.

‡ Former "extra strong." For "double extra strong" (discontinued as a standard), see Table 6.

Table 5. Steel-pipe Dimensions, Capacities, and Weights*—(*Concluded*)

Nominal pipe size, in.	Outside diam., in.	Schedule No.	Wall thickness, in.	Inside diam., in.	Cross sectional area metal, sq. in.	Inside sectional area, sq. ft.	Circumference, ft., or surface, sq. ft./ft., of length: Outside	Circumference, ft., or surface, sq. ft./ft., of length: Inside	Capacity at 1 ft./sec. velocity: U.S. gal./min.	Capacity at 1 ft./sec. velocity: Lb./hr. water	Weight of pipe per ft., lb.
8	8.625	20	.250	8.125	6.570	.3601	2.258	2.130	161.5	80,750.	22.4
		30†	.277	8.071	7.260	.3553	2.258	2.115	159.4	79,700.	24.7
		40†	.322	7.981	8.396	.3474	2.258	2.090	155.7	77,850.	28.6
		60	.406	7.813	10.48	.3329	2.258	2.050	149.4	74,700.	35.7
		80‡	.500	7.625	12.76	.3171	2.258	2.000	142.3	71,150.	43.4
		100	.593	7.439	14.96	.3018	2.258	1.947	135.3	67,650.	50.9
		120	.718	7.189	17.84	.2819	2.258	1.883	126.5	63,250.	60.7
		140	.812	7.001	19.93	.2673	2.258	1.835	120.0	60,000.	67.8
		160	.906	6.813	21.97	.2532	2.258	1.787	113.5	56,750.	74.7
10	10.75	20	.250	10.250	8.24	.5731	2.814	2.685	257.0	128,500.	28.1
		30†	.307	10.136	10.07	.5603	2.814	2.655	252.0	126,000.	34.3
		40†	.365	10.020	11.90	.5475	2.814	2.620	246.0	123,000.	40.5
		60‡	.500	9.750	16.10	.5185	2.814	2.550	233.0	116,500.	54.8
		80	.593	9.564	18.92	.4989	2.814	2.503	224.0	112,000.	64.4
		100	.718	9.314	22.63	.4732	2.814	2.440	212.0	106,000.	77.0
		120	.843	9.064	26.24	.4481	2.814	2.373	201.0	100,500.	89.2
		140	1.000	8.750	30.63	.4176	2.814	2.290	188.0	93,750.	105.0
		160	1.125	8.500	34.02	.3941	2.814	2.230	177.0	88,500.	116.0
12	12.75	20	0.250	12.250	9.82	.8185	3.338	3.21	367.0	183,500.	33.4
		30†	.330	12.090	12.87	.7972	3.338	3.17	358.0	179,000.	43.8
		40	.406	11.938	15.77	.7773	3.338	3.13	349.0	174,500.	53.6
		60	.562	11.626	21.52	.7372	3.338	3.05	331.0	165,500.	73.2
		80	.687	11.376	26.03	.7058	3.338	2.98	317.0	158,500.	88.6
		100	.843	11.064	31.53	.6677	3.338	2.90	299.0	149,500.	108.0
		120	1.000	10.750	36.91	.6303	3.338	2.82	283.0	141,500.	126.0
		140	1.125	10.500	41.08	.6013	3.338	2.75	270.0	135,000.	140.0
		160	1.312	10.126	47.14	.5592	3.338	2.66	251.0	125,500.	161.0
14	14.0	10	0.250	13.500	10.80	.9940	3.665	3.54	446.0	223,000.	36.8
		20	.312	13.376	13.42	.9750	3.665	3.51	438.0	219,000.	45.7
		30	.375	13.250	16.05	.9575	3.665	3.47	430.0	215,000.	54.6
		40	.437	13.126	18.61	.9397	3.665	3.44	422.0	211,000.	63.3
		60	.593	12.814	24.98	.8956	3.665	3.36	402.0	201,000.	85.0
		80	.750	12.500	31.22	.8522	3.665	3.28	382.0	191,000.	107.0
		100	.937	12.126	38.45	.8020	3.665	3.18	360.0	180,000.	131.0
		120	1.062	11.876	43.17	.7693	3.665	3.11	345.0	172,500.	147.0
		140	1.250	11.500	50.07	.7213	3.665	3.01	324.0	162,000.	171.0
		160	1.406	11.188	55.63	.6827	3.665	2.93	306.0	153,000.	190.0
16	16.0	10	0.250	15.500	12.37	1.3104	4.189	4.06	587.0	293,500.	42.1
		20	.312	15.376	15.38	1.2895	4.189	4.03	578.0	289,000.	52.3
		30	.375	15.250	18.41	1.2680	4.189	4.00	568.0	284,000.	62.6
		40	.500	15.000	24.35	1.2272	4.189	3.93	550.0	275,000.	82.8
		60	.656	14.688	31.62	1.1766	4.189	3.85	528.0	264,000.	108.0
		80	.843	14.314	40.14	1.1175	4.189	3.76	500.0	250,000.	137.0
		100	1.031	13.938	48.48	1.0596	4.189	3.65	474.0	237,000.	165.0
		120	1.218	13.564	56.56	1.0035	4.189	3.56	450.0	225,000.	193.0
		140	1.437	13.126	65.74	0.9397	4.189	3.44	422.0	211,000.	224.0
		160	1.562	12.876	70.85	.9043	4.189	3.37	405.0	202,500.	241.0
18	18.0	10	0.250	17.50	13.94	1.6703	4.712	4.59	748.0	374,000.	47.4
		20	.312	17.376	17.34	1.6468	4.712	4.55	738.0	369,000.	59.0
		30	.437	17.126	24.11	1.5993	4.712	4.49	717.0	358,500.	82.0
		40	.562	16.876	30.79	1.5533	4.712	4.42	697.0	348,500.	105.0
		60	.718	16.564	38.98	1.4964	4.712	4.34	670.0	335,000.	133.0
		80	.937	16.126	50.23	1.4183	4.712	4.23	635.0	317,500.	171.0
		100	1.156	15.688	61.17	1.3423	4.712	4.11	602.0	301,000.	208.0
		120	1.343	15.314	70.28	1.2791	4.712	4.02	573.0	286,500.	239.0
		140	1.562	14.876	80.66	1.2070	4.712	3.90	540.0	270,000.	275.0
		160	1.750	14.500	89.34	1.1467	4.712	3.80	514.0	257,000.	304.0
20	20.0	10	0.250	19.500	15.51	2.0740	5.236	5.11	930.0	465,000.	52.8
		20	.375	19.250	23.12	2.0211	5.236	5.05	902.0	451,000.	78.6
		30	.500	19.000	30.63	1.9689	5.236	4.98	883.0	441,500.	105.0
		40	.593	18.814	36.15	1.9305	5.236	4.94	866.0	433,000.	123.0
		60	.812	18.376	48.95	1.8417	5.236	4.81	826.0	413,000.	167.0
		80	1.031	17.938	61.44	1.7550	5.236	4.70	787.0	393,500.	209.0
		100	1.250	17.500	73.63	1.6703	5.236	4.59	750.0	375,000.	251.0
		120	1.500	17.000	87.18	1.5762	5.236	4.46	707.0	353,500.	297.0
		140	1.750	16.500	100.3	1.4849	5.236	4.32	665.0	332,500.	342.0
		160	1.937	16.126	109.9	1.4183	5.236	4.22	635.0	317,500.	374.0
24	24.0	10	0.250	23.500	18.65	3.012	6.283	6.16	1350.	675,000.	63.5
		20	.375	23.250	27.83	2.948	6.283	6.09	1325.	662,500.	94.7
		30	.562	22.876	41.39	2.854	6.283	6.00	2380.	640,000.	141.0
		40	.687	22.626	50.31	2.792	6.283	5.94	1254.	627,000.	171.0
		60	.937	22.126	67.89	2.670	6.283	5.80	1200.	600,000.	231.0
		80	1.218	21.564	87.17	2.536	6.283	5.65	1136.	568,000.	297.0
		100	1.500	21.000	106.0	2.405	6.283	5.50	1080.	540,000.	361.0
		120	1.750	20.500	122.3	2.292	6.283	5.37	1030.	515,000.	416.0
		140	2.062	19.876	142.1	2.155	6.283	5.21	965.	482,500.	484.0
		160	2.312	19.376	157.5	2.048	6.283	5.08	918.	409,000.	536.0
30	30.0	10	0.312	29.376	29.10	4.707	0.7854	7.69	2110.	1,055,000.	99.0
		20	.500	29.000	46.34	4.587	0.7854	7.60	2055.	1,027,500.	158.0
		30	.625	28.750	57.68	4.508	0.7854	7.53	2020.	1,010,000.	197.0

* Based on A.S.A. Standards B36.10-1939.
† Designates former "standard" sizes.
‡ Former "extra strong." For "double extra strong" (discontinued as a standard), see Table 6.

Table 6. Steel Pipe "Double Extra Strong" Weight*

Nominal diam., in.	Outside, diam., in.	Inside diam., in.	Wall thickness, in.	Cross-sectional area metal, sq. in.	Inside sectional area, sq. ft.	Circumference, ft., or surface, sq. ft./ft. of length		Capacity at 1 ft./sec. velocity		Weight per ft., lb.
						Outside	Inside	U.S. gal./min.	Lb./hr. water	
½	0.840	0.252	0.294	0.504	0.000347	0.220	0.0660	0.156	78	1.714
¾	1.050	.434	.308	.718	.001027	.275	.1135	.461	231	2.440
1	1.315	.599	.358	1.076	.001958	.344	.1568	.879	440	3.659
1¼	1.660	.896	.382	1.534	.00438	.435	.2345	1.966	983	5.214
1½	1.900	1.100	.400	1.885	.00660	.497	.2880	2.962	1,481	6.408
2	2.375	1.503	.436	2.656	.01231	.622	.394	5.52	2,760	9.029
2½	2.875	1.771	.552	4.028	.01711	.753	.464	7.68	3,840	13.695
3	3.500	2.300	.600	5.466	.02885	.916	.602	12.95	6,480	18.583
3½	4.000	2.728	.636	6.721	.0406	1.047	.714	18.22	9,110	22.850
4	4.500	3.152	.674	8.101	.0542	1.178	.825	24.32	12,200	27.541
4½	5.000	3.580	.710	9.569	.0699	1.308	.937	31.37	15,700	32.530
5	5.563	4.063	.750	11.340	.0900	1.456	1.064	40.4	20,200	38.552
6	6.625	4.897	.864	15.637	.1308	1.734	1.282	48.7	29,400	53.160
7	7.625	5.875	.875	18.555	.1883	1.996	1.538	84.5	42,300	63.079
8	8.625	6.875	.875	21.304	.2578	2.258	1.800	115.7	57,900	72.424

* This weight has now been dropped as a standard; for present standard weights, see Tables 4 and 5.

Steel Tubing

Seamless steel tubing is made in an extremely wide range of shapes, sizes, and thicknesses, and in many metal compositions. Two common classifications of tubing are "pressure" and "mechanical," although there is some overlapping and some tubing does not come under either class. In the specification of wall thickness or gage, two conventions exist: the specified gage may denote either (1) the "average wall" thickness or (2) the "minimum wall" thickness. Since the standard sizes are available on both bases, that intended should always be stated. The permissible variations in the diameters and wall thicknesses of round cold-drawn seamless mechanical tubing are given in Table 7.

Hypodermic needle tubing in 0.70 to 0.80 carbon steel is available from 0.012 in. outside diameter and 0.00275 in. wall thickness to 0.109 in. outside diameter and 0.012 in. wall thickness; in 18-8 stainless, the wall thicknesses are 0.001 in. thicker.

Pressure tubing, for boilers, condensers, heat exchangers, etc., is usually made in outside diameter sizes ordinarily ¼ to 2 in. and thicknesses varying from 20 to 9 B.W.G. Dimensions, capacities, and weights of commonly used tubing are given in Table 16, condenser and heat-exchanger tube dimensions.

Mechanical tubing is available in many sizes and thicknesses from 1/16 in. outside diameter and 36 B.W.G. wall to 10¾ in. outside diameter and 1⅝ in. wall. The usual steels are S.A.E. 1015, 1020, 1025, 1035, and 1045.

Table 7. Permissible Tolerances of Cold-drawn Seamless Tubing*

Size range of outside diameter		Permissible variations from					
		Outside diameter, in.		Inside diameter, in.		Wall thickness, %	
From	Up to but not including	Over	Under	Over	Under	Over	Under
3/16	½	0.004	0	(a),(b)	(a),(b)	(a),(b)	(a),(b)
½	1½	.005(c)	0(c)	0(a),(b)	0.005(a),(b)	10(a),(b)	10(a),(b)
1½	3½	.010(c)	0(c)	0(a)	.010(a)	10(a)	10(a)
3½	5½	.015	0(c)	0.005(a)	.015(a)	10(a)	10(a)

(a) For tubes with inside diameter less than 50 per cent of outside diameter or with wall thickness more than 25 per cent of outside diameter, or with wall thickness over 1¼ in., or weighing more than 90 lb./ft. which cannot be successfully drawn over a mandrel, the inside diameter may vary over or under by an amount equal to 10 per cent of the wall thickness.

(b) For tubes with inside diameter less than ½ in. (or less than ⅝ in. when the wall thickness is more than 20 per cent of the outside diameter), which cannot be successfully drawn over a mandrel, the wall thickness may vary 10 per cent over or under that specified, and the inside diameter will be governed by the outside diameter and wall thickness variations.

(c) Tubing having a wall thickness less than 3 per cent of the outside diameter cannot be straightened properly without a certain amount of distortion. Consequently such tubes, while having an average outside diameter and inside diameter within the tolerances shown in this table, will require an ovality tolerance of ½ per cent over or under the nominal outside diameter and inside diameter, this being in addition to the tolerances indicated in the table.

* Summerill Tubing Co.

Stainless Steel Pipe and Tubing

Under this classification are analyses such as 12, 18, and 24 per cent chromium, 18 per cent chromium–8 per cent nickel, etc., which are made in pipe and tubing, usually seamless, but also welded in most sizes. For

Table 8. Stainless Steel Tubing
All dimensions in inches

Outside diam.	Wall thickness range, inclusive		Outside diam.	Wall thickness range, inclusive		Standard wall thickness	
	From	To		From	To	Decimal	B.W.G. or fraction
1/32	0.004	0.013	3⅛	0.035	1.000	0.004	36
1/16	.004	.028	3¼	.035	1.000	.005	35
3/32	.004	.035	3⅜	.035	1.000	.007	34
⅛	.004	.049	3½	.035	1.000	.008	33
3/16	.004	.065	3⅝	.035	1.000	.009	32
¼	.004	.083	3¾	.035	1.000	.010	31
5/16	.004	.109	3⅞	.035	1.000	.012	30
⅜	.005	.134	4	.035	1.000	.013	29
7/16	.008	.148	4⅛	.049	1.000	.014	28
½	.009	.165	4¼	.049	1.000	.016	27
9/16	.009	.165	4⅜	.049	1.000	.018	26
⅝	.012	.203	4½	.049	1.000	.020	25
11/16	.013	.203	4⅝	.058	1.000	.022	24
¾	.014	.238	4¾	.072	1.000	.025	23
13/16	.014	.238	4⅞	.072	1.000	.028	22
⅞	.016	.238	5	.109	1.000	.032	21
15/16	.016	.238	5⅛	.134	1.000	.035	20
1	.018	.313	5¼	.134	1.000	.042	19
1 1/16	.020	.313	5⅜	.134	1.000	.049	18
1⅛	.020	.313	5½	.134	1.000	.058	17
1 3/16	.022	.313	5¾	.134	1.000	.065	16
1¼	.022	.313	6	.134	1.000	.072	15
1 5/16	.022	.313	6¼	.134	1.000	.083	14
1⅜	.025	.375	6½	.134	1.000	.095	13
1 7/16	.025	.375	6⅝	.180	1.000	.109	12
1½	.025	.375	6¾	.188	1.000	.125	⅛
1⅝	.028	.375	7	.203	1.000	.134	10
1¾	.028	.500	7¼	.220	1.000	.148	9
1⅞	.028	.500	7½	.238	1.000	.156	5/32
2	.032	.625	7⅝	.250	1.000	.165	8
2⅛	.032	.625	7¾	.281	1.000	.180	7
2¼	.035	.625	8	.281	1.000	.188	3/16
2⅜	.035	.750	8¼	.313	1.000	.203	6
2½	.035	.750	8½	.313	1.000	.220	5
2⅝	.035	.750	8⅝	.313	1.000	.238	4
2¾	.035	.875	8¾	.375	1.000	.250	¼
2⅞	.035	.875	9	.375	1.000	.281	9/32
3	.035	1.000				.313	5/16
						.375	⅜
						.438	7/16
						.500	½
						.563	9/16
						.625	⅝
						.750	¾
						.875	⅞
						1.000	1

Example: 1-in. o.d. tubing can be obtained in the wall thicknesses shown in the two right-hand columns from 0.018 to 0.313 in., inclusive.

other compositions such as 25-12 and 25-20 Cr-Ni and modifications of 18-8 with molybdenum, columbium, and titanium, the producer should be consulted. The standards and costs are based on 18-8. Pipe is available in the "standard" and "extra heavy" weights from ⅛ to 8 in. and the "double extra strong" weight from ½ to 2 in. Dimensions are the same as for ordinary steel pipe (see Tables 5 and 6). Pipe is plain end unless ordered threaded.

Stainless tubing can be obtained in the standard sizes shown in Table 8. At a higher cost per pound, intermediate sizes of outside diameter, inside diameter, or wall thickness can be obtained.

Cast-iron Pipe

Cast-iron pipe is used for many fluids and is particularly adapted for use under ground and water. It is usually coated inside or out with such materials as pitch, asphalt, and cement when corrosion or nodulation conditions exist. Electrolysis can be reduced by using such non-conducting joints as cement or leadite. Cast-iron pipe, valves, and fittings are not recommended for use much above 500°F. or 260°C.

The pipe is available in various thicknesses with bell-and-spigot, flanged, bolt-lug, and plain ends. In general, the bell-and-spigot joint is used for subsurface installations; the flanged joint for exposed piping and when rigidity, strength, and tightness are required; and the plain ends joined by couplings or bronze welding for high-pressure gas lines. Two general types of pipe are made, pit-cast and centrifugally cast. The latter is stronger and freer from blowholes, thus permitting thinner wall sections.

The standards program of the American Standards Association and allied groups on cast-iron pipe is shown below. Those in preparation are marked with an asterisk.

A21.1-1939	Manual for the Computation of Strength and Thickness of Cast-iron pipe
A21.2-1939	Specifications for Cast-iron Pit-cast Pipe for Water or Other Liquids
A21.3-*	Cast-iron Pit-cast Pipe for Gas
A21.4-1939	Specifications for Cement Mortar Lining for Cast-iron Pipe and Fittings
A21.5-*	Coal-tar Dip Coating for Cast-iron Pipe and Fittings
A21.6-*	Specifications for Special Types of Cast-iron Pipe
A21.7-*	Specifications for Cast-iron Fittings

Wall thickness can be readily determined by use of the complete tables and curves in A21.1-1939. Consideration is given to methods of laying, internal pressure, external load, water hammer, and impact loads. Dimensions of **pit-cast bell-and-spigot joint pipe** from A21.2-1939 are given in Table 9. The inside diameter is very nearly the nominal diameter. Standard thicknesses for various pressure and laying conditions are given in Table 9*a*.

The Federal government has issued specification WW-P-421, "Pipe; Water, Cast-iron (Bell and Spigot, and Bolted Joint)"; 5 cts., which covers five types as follows:

Type	Casting method	Mold	Lengths, ft.	Joint
I	Centrifugal	Metal contact	12, 18	Bell and spigot
II	Centrifugal	Sand lined	16, 16½, 20	Bell and spigot
III	Horizontal	Green sand	12, 16	Bell and spigot
IV	Horizontal	Sand	6, 12	Bolt lugs and integral machine-tapered lugs
V	Vertical pit	Dry sand	12	Bell and spigot

The dimensions and weights of type V are given in Table 10. Some of the items in Types I to IV are given in Tables 11 and 13.

The American Water Works Association has established standards for pit-cast pipe with **bell-and-spigot** joints (Table 11) for 12-ft. lengths in Classes A to H, in steps of 100 ft. of water head or 43 lb./sq. in. pressure. Their standards for **flanged pipe**, Classes A to H, are given in Table 12. Standards for **centrifugally cast bell-and-spigot** pipe are given in Table 13 for 100 to 250 lb./sq. in. working pressures.

The American Gas Association has established standards for centrifugally cast bell-and-spigot pipe for carrying gases. Dimensions and weights are given in Table 13*a*. They correspond closely with A.W.W.A. Class A pipe but have two types of joint, type 1 for sealing with lead, and type 2 for sealing with cement. This pipe is also made with flanged ends and plain ends.

The **maximum bend** in laying bell-and-spigot A.G.A.

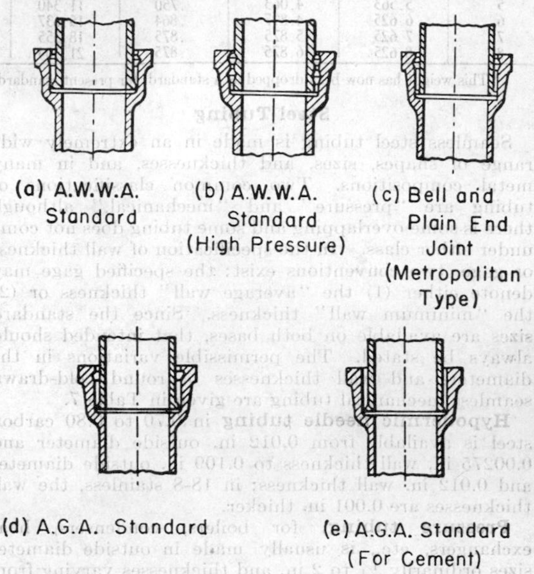

FIG. 62. Sections through typical cast-iron pipe joints.

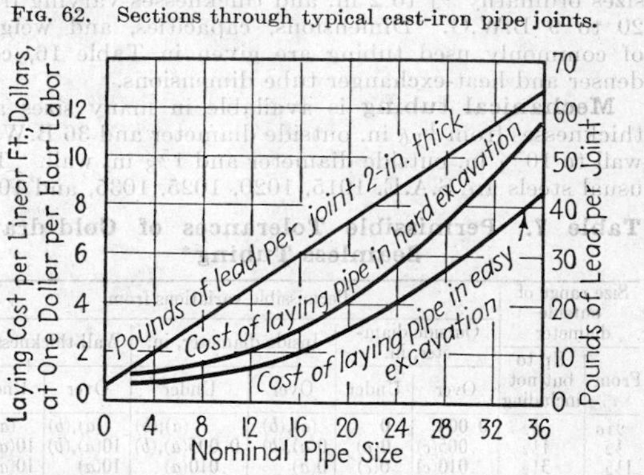

FIG. 63. Cost of laying cast-iron pipe, pounds of lead required per joint; standard bell-and-spigot joints.

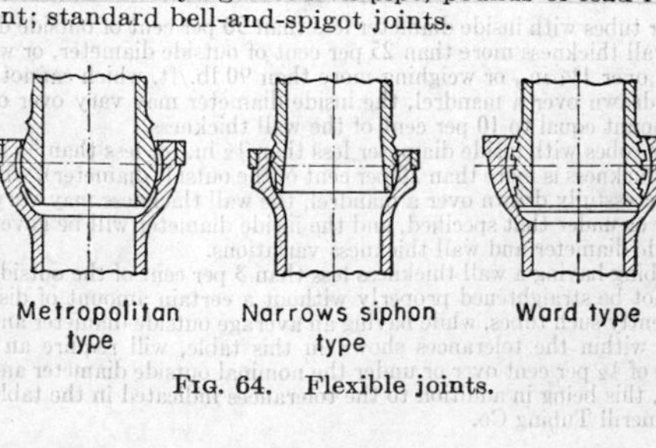

FIG. 64. Flexible joints.

pipe with the Anthony high-pressure bell is equivalent to 3°53′ per joint, or a deflection of 9 in. for 12-ft. pipe and 13.5 in. for 18-ft. pipe. These are equivalent to radii of curvature of 193 and 289 ft., respectively. For A.W.W.A. standard bell-and-spigot pipe, the following table indicates the maximum bends:

	Nominal diameter, in.								
	4	6	8	10	12	14	16	18	20
Deflection, in., 12-ft. pipe..	9.02	7.84	5.99	5.40	5.13	4.64	4.05	3.63	3.29
Deflection, in., 18-ft. pipe..	13.44	11.68	8.92	8.04	6.16				
Radius, ft., 12-ft. pipe.....	193	222	290	322	421	375	430	480	529
Radius, ft., 18-ft. pipe.....	289	333	436	483	631				

Joints available with cast-iron pipe are the bell-and-spigot, flanged, bolt lugs with integral machine-tapered joint, flexible, and welded. The **bell-and-spigot joint** is available in several designs, to allow for pressure and installation conditions. Figure 62 illustrates typical examples. Figure 62*a* is for A.W.W.A. Classes A to D, Fig. 62*b* for Classes E to H, termed high pressure. Integrally cast lugs are sometimes used to secure the joint. The joint is usually calked with hemp and then filled with one of the following: lead, lead wool, lead wool and cement, a lead alloy containing 2 per cent each of tin and antimony (for high-pressure gas), leadite, and cement. Rubber rings are also used. A properly made joint is generally tight and yet permits reasonable expansion and contraction. The cost of laying cast-iron pipe and the amount of lead required for joints are given in Fig. 63.

The **flexible joint** is used mainly for underwater work, where leaded or cemented joints would be too rigid. Three types of flexible joint are shown in Fig. 64. The **Metropolitan** type is the simplest and most common, as the lead is stationary and therefore readily calked. The **Narrows siphon** type is more effective for tightness, but calking of the lead is not possible. The **Ward** type is not generally recommended as the lead may be disturbed when the pipe is bent.

Table 9. Pit-cast Cast-iron Pipe—Dimensions*

Dimensions, in.					Weights, lb			
Nominal diam.	Thickness class	Thickness	Outside diam.	Inside diam.	Weight of barrel per ft.	Weight of bell	Weight based on 12-ft. length† Per length	Weight based on 12-ft. length† Average per foot
3	1	0.37	3.80	3.06	12.4	19	170	14.2
	2	.40	3.80	3.00	13.3	19	180	15.0
	3	.43	3.80	2.94	14.2	19	190	15.8
	4	.46	3.96	3.04	15.8	20	210	17.5
	5	.50	3.96	2.96	17.0	20	225	18.8
4	1	.40	4.80	4.00	17.3	23	230	19.2
	2	.43	4.80	3.94	18.4	23	245	20.4
	3	.46	5.00	4.08	20.5	24	270	22.5
	4	.50	5.00	4.00	22.1	24	290	24.2
	5	.54	5.00	3.92	23.6	24	310	25.8
6	1	.43	6.90	6.04	27.3	33	360	30.0
	2	.46	6.90	5.98	29.0	33	380	31.7
	3	.50	6.90	5.90	31.4	33	410	34.2
	4	.54	7.10	6.02	34.7	34	450	37.5
	5	.58	7.10	5.94	37.1	34	480	40.0
	6	.63	7.22	5.96	40.7	56	545	45.4
	7	.68	7.38	6.02	44.7	62	600	50.0
	8	.73	7.38	5.92	47.6	62	635	52.9
8	1	.46	9.05	8.13	38.7	48	515	42.9
	2	.50	9.05	8.05	41.9	48	550	45.8
	3	.54	9.05	7.97	45.0	48	590	49.2
	4	.58	9.30	8.14	49.6	49	645	53.8
	5	.63	9.30	8.04	53.5	49	690	57.5
	6	.68	9.30	7.94	57.5	49	740	61.7
	7	.73	9.42	7.96	62.2	75	825	68.8
	8	.79	9.60	8.02	68.2	84	905	75.4
	9	.85	9.60	7.90	72.9	84	960	80.0
10	1	.50	11.10	10.10	52.0	58	685	57.1
	2	.54	11.10	10.02	55.9	58	730	60.8
	3	.58	11.10	9.94	59.8	58	775	64.6
	4	.63	11.40	10.14	66.5	64	865	72.1
	5	.68	11.40	10.04	71.5	64	925	77.1
	6	.73	11.40	9.94	76.4	64	980	81.7
	7	.79	11.60	10.02	83.7	99	1,105	92.1
	8	.85	11.60	9.90	89.6	99	1,175	97.9
	9	.92	11.84	10.00	98.5	110	1,295	107.9
12	1	.54	13.20	12.12	67.0	72	880	73.3
	2	.58	13.20	12.04	71.8	72	935	77.9
	3	.63	13.20	11.94	77.6	72	1,005	83.8
	4	.68	13.50	12.14	85.5	78	1,105	92.1
	5	.73	13.50	12.04	91.4	78	1,175	97.9
	6	.79	13.50	11.92	98.4	78	1,260	105.0
	7	.85	13.78	12.08	107.7	125	1,420	118.3
	8	.92	13.78	11.94	116.0	125	1,520	126.7
	9	.99	14.08	12.10	127.0	144	1,670	139.2
	10	1.07	14.08	11.94	136.4	144	1,785	148.8
14	1	0.54	15.30	14.22	78.1	88	1,025	85.4
	2	.58	15.30	14.14	83.7	88	1,095	91.3
	3	.63	15.65	14.39	92.8	96	1,210	100.8
	4	.68	15.65	14.29	99.8	96	1,295	107.9
	5	0.73	15.65	14.19	106.8	96	1,380	115.0
	6	.79	15.65	14.07	115.1	96	1,480	123.3
	7	.85	15.65	13.95	123.3	96	1,580	131.7
	8	.92	15.98	14.14	135.8	148	1,780	148.3
	9	.99	15.98	14.00	145.5	148	1,895	157.9
	10	1.07	16.32	14.18	160.0	183	2,105	175.4
	11	1.16	16.32	14.00	172.4	183	2,255	187.9
16	1	0.58	17.40	16.24	95.6	114	1,265	105.4
	2	.63	17.40	16.14	103.6	114	1,360	113.3
	3	.68	17.80	16.44	114.1	128	1,500	125.0
	4	.73	17.80	16.34	122.1	128	1,595	132.9
	5	.79	17.80	16.22	131.7	128	1,710	142.5
	6	.85	17.80	16.10	141.2	128	1,825	152.1
	7	.92	17.80	15.96	152.2	128	1,955	162.9
	8	.99	18.16	16.18	166.6	180	2,180	181.7
	9	1.07	18.16	16.02	179.2	180	2,335	194.6
	10	1.16	18.54	16.22	197.6	224	2,600	216.7
	11	1.25	18.54	16.04	211.8	224	2,770	230.8
18	1	0.63	19.50	18.24	116.5	133	1,535	127.9
	2	.68	19.50	18.14	125.4	133	1,640	136.7
	3	.73	19.92	18.46	137.3	154	1,805	150.4
	4	.79	19.92	18.34	148.1	154	1,935	161.3
	5	.85	19.92	18.22	158.9	154	2,065	172.1
	6	.92	19.92	18.08	171.3	154	2,210	184.2
	7	.99	19.92	17.94	183.7	154	2,360	196.7
	8	1.07	20.34	18.20	202.1	222	2,650	220.8
	9	1.16	20.34	18.02	218.1	222	2,840	236.7
	10	1.25	20.78	18.28	239.3	283	3,155	262.9
	11	1.35	20.78	18.08	257.1	283	3,370	280.8
20	1	0.66	21.60	20.28	135.5	156	1,785	148.8
	2	.71	21.60	20.18	145.4	156	1,905	158.8
	3	.77	21.60	20.06	157.2	156	2,045	170.4
	4	.83	22.06	20.40	172.7	189	2,265	188.8
	5	.90	22.06	20.26	186.7	189	2,430	202.5
	6	.97	22.06	20.12	200.5	189	2,600	216.7
	7	1.05	22.06	19.96	216.2	189	2,785	232.1
	8	1.13	22.54	20.28	237.1	260	3,110	259.2
	9	1.22	22.54	20.10	255.0	260	3,325	277.1
	10	1.32	22.54	19.90	274.6	260	3,560	296.7
	11	1.43	23.02	20.16	302.6	326	3,960	330.0
24	1	0.74	25.80	24.32	181.8	199	2,385	198.8
	2	.80	25.80	24.20	196.0	199	2,555	212.9
	3	.86	25.80	24.08	210.2	199	2,725	227.1
	4	.93	26.32	24.46	231.5	250	3,030	252.5
	5	1.00	26.32	24.32	248.2	250	3,230	269.2
	6	1.08	26.32	24.16	267.2	250	3,460	288.3
	7	1.17	26.32	23.98	288.4	250	3,715	309.6
	8	1.26	26.90	24.38	316.7	349	4,155	346.2
	9	1.36	26.90	24.18	340.5	349	4,440	370.0
	10	1.47	26.90	23.96	366.4	349	4,750	395.8
	11	1.59	27.76	24.58	407.9	489	5,385	448.8
	12	1.72	27.76	24.32	439.0	489	5,760	480.0

* A.S.A. Standards A21.-2-1939.
† Including bell-and-spigot bead. Calculated weight of pipe rounded off to nearest 5 lb.

Table 9. Pit-cast Cast-iron Pipe—Dimensions*—(*Concluded*)

Dimensions, in.					Weights, lb.			
Nominal diam.	Thickness class	Thickness	Outside diam.	Inside diam.	Weight of barrel per ft.	Weight of bell	Weight based on 12-ft. length† Per length	Weight based on 12-ft. length† Average per foot
24	13	1.86	27.76	24.04	472.2	489	6,160	513.3
30	1	0.87	31.74	30.00	263.3	296	3,460	288.3
	2	.94	32.00	30.12	286.2	298	3,735	311.3
	3	1.02	32.00	29.96	309.7	298	4,020	335.0
	4	1.10	32.40	30.20	337.5	351	4,405	367.1
	5	1.19	32.74	30.36	368.0	416	4,835	402.9
	6	1.29	32.74	30.16	397.7	416	5,190	432.5
	7	1.39	32.74	29.96	427.1	416	5,545	462.1
	8	1.50	33.10	30.10	464.6	557	6,135	511.3
	9	1.62	33.46	30.22	505.6	626	6,695	557.9
	10	1.75	33.46	29.96	543.9	626	7,155	596.3
36	1	0.97	37.96	36.02	351.7	383	4,610	384.2
	2	1.05	38.30	36.20	383.4	446	5,050	420.8
	3	1.13	38.30	36.04	411.7	446	5,390	449.2
	4	1.22	38.70	36.26	448.2	512	5,895	491.3
	5	1.32	38.70	36.06	483.6	512	6,320	526.7
	6	1.43	39.16	36.30	528.9	586	6,940	578.3
	7	1.54	39.16	36.08	567.9	586	7,405	617.1
	8	1.66	39.60	36.28	617.3	770	8,180	681.7
	9	1.79	39.60	36.02	663.4	770	8,735	727.9
	10	1.93	40.04	36.18	720.9	876	9,530	794.2
42	1	1.07	44.20	42.06	452.3	539	5,970	497.5
	2	1.16	44.50	42.18	492.8	586	6,505	542.1
	3	1.25	44.50	42.00	529.9	586	6,950	579.2
	4	1.35	45.10	42.40	578.9	701	7,655	637.9
	5	1.46	45.10	42.18	624.5	701	8,200	683.3
	6	1.58	45.58	42.42	681.4	805	8,990	749.2
	7	1.71	45.58	42.16	735.3	805	9,635	802.9
48	1	1.18	50.50	48.14	570.4	660	7,510	625.8
	2	1.27	50.50	47.96	612.8	660	8,020	668.3
	3	1.37	50.80	48.06	663.8	745	8,715	726.3
	4	1.48	51.40	48.44	724.2	900	9,595	799.6
	5	1.60	51.40	48.20	781.0	900	10,280	856.7
	6	1.73	51.98	48.52	852.1	1046	11,280	940.0
	7	1.87	51.98	48.24	918.5	1046	12,075	1006.3
	8	2.02	51.98	47.94	989.2	1046	12,925	1077.1
54	1	1.30	56.66	54.06	705.4	855	9,325	777.1
	2	1.40	57.10	54.30	764.3	993	10,170	847.5
	3	1.51	57.10	54.08	822.8	993	10,875	906.3
	4	1.63	57.80	54.54	897.4	1189	11,965	997.1
	5	1.76	57.80	54.28	966.8	1189	12,800	1066.7
	6	1.90	58.40	54.60	1052.2	1391	14,025	1168.8
	7	2.05	58.40	54.30	1132.3	1391	14,985	1248.8
	8	2.21	58.40	53.98	1217.2	1391	16,005	1333.8
60	1	1.39	62.80	60.02	836.7	1021	11,070	922.5
	2	1.50	63.40	60.40	910.1	1145	12,075	1006.3
	3	1.62	63.40	60.16	981.0	1145	12,925	1077.1
	4	1.75	63.40	59.90	1057.5	1145	13,845	1153.8
	5	1.89	64.20	60.42	1154.3	1393	15,250	1270.9
	6	2.04	64.20	60.12	1242.9	1393	16,315	1359.6
	7	2.20	64.82	60.42	1350.3	1647	17,860	1488.3
	8	2.38	64.82	60.06	1456.6	1647	19,135	1594.6

* A.S.A. Standards A21.2-1939.

† Including bell-and-spigot bead. Calculated weight of pipe rounded off to nearest 5 lb.

Table 9*a*. Pit-cast Cast-iron Pipe—Standard Thicknesses for Various Pressure and Laying Conditions*

Thicknesses in inches (include allowances for water hammer, foundry practice, and corrosion).
Working pressure in pounds per square inch.
See Table 9 for other dimensions.
Laying condition *A*—flat-bottom trench, without blocks, untamped backfili.
Laying condition *B*—flat-bottom trench, without blocks, tamped backfill.
Laying condition *C*—pipe laid on blocks, untamped backfill.
Laying condition *D*—pipe laid on blocks, tamped backfill.

Size, in.	Working pressure	3½ ft. of cover, Laying condition A	B	C	D	5 ft. of cover, Laying condition A	B	C	D	8 ft. of cover, Laying condition A	B	C	D
3	50	0.37	0.37	0.37	0.37	0.37	0.37	0.37	0.37	0.37	0.37	0.43	0.37
	100	.37	.37	.37	.37	.37	.37	.37	.37	.37	.37	.43	.37
	150	.37	.37	.37	.37	.37	.37	.40	.37	.37	.37	.43	.37
	200	.37	.37	.37	.37	.37	.37	.40	.37	.37	.37	.43	.37
	250	.37	.37	.37	.37	.37	.37	.40	.37	.37	.37	.46	.37
	300	.37	.37	.37	.37	.37	.37	.40	.37	.37	.37	.46	.37
	350	.37	.37	.40	.37	.37	.37	.40	.37	.37	.37	.46	.37
4	50	.40	.40	.40	.40	.40	.40	.43	.40	.40	.40	.50	.40
	100	.40	.40	.40	.40	.40	.40	.43	.40	.40	.40	.50	.40
	150	.40	.40	.40	.40	.40	.40	.43	.40	.40	.40	.50	.40
	200	.40	.40	.43	.40	.40	.40	.43	.40	.40	.40	.50	.40
	250	.40	.40	.43	.40	.40	.40	.46	.40	.40	.40	.50	.40
	300	.40	.40	.43	.40	.40	.40	.46	.40	.40	.40	.54	.40
	350	.40	.40	.46	.40	.40	.40	.46	.40	.40	.40	.54	.40
6	50	.43	.43	.46	.43	.43	.43	.50	.43	.43	.43	.58	.43
	100	.43	.43	.46	.43	.43	.43	.50	.43	.43	.43	.58	.43
	150	.43	.43	.50	.43	.43	.43	.50	.43	.43	.43	.58	.43
	200	.43	.43	.50	.43	.43	.43	.54	.43	.43	.43	.58	.43
	250	.43	.43	.50	.43	.43	.43	.54	.43	.46	.46	.63	.46
	300	.46	.43	.54	.43	.46	.46	.54	.46	.50	.46	.63	.46
	350	.50	.50	.54	.46	.50	.50	.58	.50	.50	.50	.63	.50
8	50	.46	.46	.54	.46	.46	.46	.54	.46	.46	.46	.63	.46
	100	.46	.46	.54	.46	.46	.46	.54	.46	.46	.46	.63	.50
	150	.46	.46	.58	.46	.46	.46	.58	.46	.50	.50	.68	.50
	200	.46	.46	.58	.50	.50	.46	.58	.50	.54	.50	.68	.54
	250	.50	.50	.63	.50	.54	.50	.63	.54	.54	.54	.68	.58
	300	.54	.54	.63	.54	.58	.54	.63	.58	.58	.58	.73	.58
	350	.58	.58	.63	.58	.58	.58	.68	.58	.63	.63	.73	.63
10	50	.50	.50	.58	.50	.50	.50	.58	.50	.54	.50	.68	.54
	100	.50	.50	.58	.50	.50	.50	.63	.50	.58	.54	.73	.58
	150	.50	.50	.58	.50	.54	.54	.63	.54	.58	.58	.73	.58
	200	.54	.54	.63	.54	.58	.58	.68	.58	.63	.58	.79	.63
	250	.58	.58	.63	.58	.63	.63	.68	.63	.68	.63	.79	.68
	300	.63	.63	.68	.63	.68	.68	.73	.68	.68	.68	.79	.73
	350	.68	.68	.73	.68	.73	.73	.79	.73	.73	.73	.85	.73
12	50	.54	.54	.58	.54	.54	.54	.63	.54	.58	.58	.73	.63
	100	.54	.54	.63	.54	.54	.54	.63	.58	.63	.58	.79	.63
	150	.58	.54	.63	.58	.58	.58	.68	.63	.68	.63	.79	.68
	200	.63	.58	.68	.63	.63	.63	.73	.68	.68	.68	.85	.73
	250	.68	.68	.73	.68	.68	.68	.79	.73	.73	.73	.85	.73
	300	.73	.73	.79	.73	.73	.73	.79	.79	.79	.79	.92	.79
	350	.79	.79	.85	.79	.79	.79	.85	.85	.85	.85	.92	.85
14	50	0.54	0.54	0.63	0.58	0.58	0.54	0.68	0.58	0.68	0.63	0.79	0.68
	100	.58	.54	.68	.58	.63	.58	.73	.63	.68	.68	.85	.73
	150	.63	.63	.68	.63	.68	.63	.73	.68	.73	.73	.85	.79
	200	.68	.68	.73	.68	.73	.68	.79	.73	.79	.79	.92	.79
	250	.79	.73	.79	.79	.79	.79	.85	.79	.85	.79	.92	.85
	300	.85	.85	.85	.85	.85	.85	.92	.85	.92	.85	.99	.92
	350	.92	.92	.92	.92	.92	.92	.99	.92	.99	.92	1.07	.99
16	50	.58	.58	.68	.63	.63	.58	.73	.63	.73	.68	0.85	.73
	100	.63	.58	.73	.63	.68	.63	.73	.68	.73	.73	.85	.79
	150	.68	.68	.73	.68	.73	.68	.79	.73	.79	.79	.92	.85
	200	.73	.73	.79	.73	.79	.79	.85	.79	.85	.79	.92	.85
	250	.79	.79	.85	.85	.85	.85	.92	.85	.92	.85	.99	.92
	300	.92	.92	.92	.92	.92	.92	.99	.92	.99	.92	1.07	.99
	350	.99	.99	.99	.99	.99	.99	1.07	.99	1.07	.99	1.16	1.07
18	50	.63	.63	.73	.68	.68	.63	0.79	.68	0.79	.73	0.85	0.79
	100	.68	.83	.73	.68	.73	.68	.79	.73	.85	.79	.92	.85
	150	.73	.73	.79	.73	.79	.73	.85	.79	.85	.85	.99	.92
	200	.79	.79	.85	.85	.85	.85	.92	.85	.92	.85	.99	.99
	250	.92	.92	.92	.92	.92	.92	.99	.92	.99	.99	1.07	.99
	300	.99	.99	.99	.99	.99	.99	1.07	1.07	1.07	1.07	1.16	1.07
	350	1.07	1.07	1.07	1.07	1.07	1.07	1.16	1.07	1.16	1.16	1.25	1.16
20	50	0.66	0.66	0.77	0.71	0.71	0.66	0.83	0.77	0.83	0.77	0.90	0.83
	100	.71	.66	.83	.77	.77	.71	.83	.83	.83	.83	.97	.90
	150	.77	.77	.83	.77	.83	.83	.90	.83	.90	.90	1.05	.97
	200	.83	.83	.90	.90	.90	.90	.97	.90	.97	.97	1.05	1.05
	250	.97	.97	.97	.97	.97	.97	1.05	1.05	1.05	1.05	1.13	1.13
	300	1.05	1.05	1.05	1.05	1.13	1.05	1.13	1.13	1.13	1.13	1.22	1.13
	350	1.22	1.13	1.22	1.22	1.22	1.22	1.22	1.22	1.22	1.22	1.32	1.22

* A.S A. Standards A21.2-1939.

Table 9a. Pit-cast Cast-iron Pipe—Standard Thicknesses for Various Pressure and Laying Conditions*

(Concluded)

Size, in.	Working pressure	3½ ft. of cover, Laying condition A	B	C	D	5 ft. of cover, Laying condition A	B	C	D	8 ft. of cover, Laying condition A	B	C	D
24	50	0.74	0.74	0.86	0.80	0.80	0.74	0.86	0.86	0.93	0.86	1.00	0.93
	100	.80	.74	.86	.86	.86	.80	.93	.93	1.00	.93	1.08	1.00
	150	.86	.86	.93	.86	.93	.93	1.00	1.08	1.08	1.00	1.08	1.08
	200	1.00	1.00	1.00	1.00	1.08	1.00	1.08	1.08	1.17	1.08	1.17	1.17
	250	1.08	1.08	1.17	1.08	1.17	1.08	1.17	1.17	1.26	1.17	1.26	1.26
	300	1.26	1.26	1.26	1.26	1.26	1.26	1.26	1.26	1.26	1.26	1.36	1.36
	350	1.36	1.36	1.36	1.36	1.36	1.36	1.36	1.36	1.47	1.36	1.47	1.47
30	50	0.94	0.87	1.02	0.94	0.94	0.87	1.02	0.94	1.10	1.02	1.19	1.10
	100	.94	.87	1.02	.94	1.02	.94	1.10	1.02	1.19	1.10	1.29	1.19
	150	1.10	1.02	1.10	1.10	1.19	1.10	1.19	1.19	1.29	1.19	1.39	1.29
	200	1.19	1.19	1.19	1.19	1.29	1.19	1.29	1.29	1.39	1.29	1.39	1.39
	250	1.29	1.29	1.39	1.39	1.39	1.39	1.50	1.39	1.50	1.39	1.50	1.50
	300	1.50	1.50	1.50	1.50	1.50	1.50	1.62	1.50	1.62	1.50	1.62	1.62
	350	1.62	1.62	1.62	1.62	1.62	1.62	1.75	1.62	1.75	1.75	1.75	1.75
36	50	1.05	0.97	1.13	1.05	1.05	0.97	1.22	1.05	1.22	1.13	1.32	1.22
	100	1.13	.97	1.22	1.13	1.22	1.05	1.22	1.13	1.32	1.22	1.43	1.32
	150	1.22	1.13	1.22	1.22	1.32	1.22	1.32	1.32	1.43	1.32	1.54	1.43
	200	1.32	1.32	1.43	1.32	1.43	1.43	1.43	1.43	1.54	1.43	1.66	1.54
	250	1.54	1.54	1.54	1.54	1.66	1.54	1.66	1.66	1.66	1.66	1.79	1.66
	300	1.79	1.79	1.79	1.79	1.79	1.79	1.79	1.79	1.79	1.79	1.93	1.79
	350	1.93	1.93	1.93	1.93	1.93	1.93	2.08	1.93	2.08	1.93	2.08	2.08
42	50	1.16	1.07	1.25	1.16	1.25	1.07	1.35	1.25	1.35	1.25	1.46	1.35
	100	1.25	1.07	1.35	1.25	1.35	1.25	1.35	1.35	1.46	1.35	1.58	1.46
	150	1.35	1.25	1.46	1.35	1.46	1.35	1.46	1.46	1.58	1.46	1.71	1.58
	200	1.58	1.46	1.58	1.58	1.58	1.58	1.71	1.58	1.71	1.58	1.85	1.71
	250	1.71	1.71	1.85	1.71	1.85	1.71	1.85	1.85	1.85	1.85	2.00	1.85
48	50	1.27	1.18	1.37	1.27	1.37	1.18	1.48	1.37	1.60	1.37	1.60	1.48
	100	1.37	1.27	1.48	1.37	1.48	1.37	1.60	1.48	1.73	1.48	1.73	1.60
	150	1.48	1.48	1.60	1.48	1.60	1.48	1.73	1.60	1.87	1.60	1.87	1.73
	200	1.73	1.73	1.73	1.73	1.87	1.73	1.87	1.87	2.02	1.87	2.02	1.87
	250	2.02	2.02	2.02	2.02	2.02	2.02	2.02	2.02	2.18	2.02	2.18	2.18
54	50	1.40	1.30	1.51	1.40	1.51	1.30	1.63	1.40	1.63	1.51	1.76	1.63
	100	1.51	1.40	1.63	1.51	1.63	1.51	1.76	1.63	1.90	1.63	1.90	1.76
	150	1.63	1.63	1.76	1.63	1.76	1.63	1.90	1.76	2.05	1.76	2.05	1.90
	200	1.90	1.90	2.05	1.90	2.05	1.90	2.05	2.05	2.21	2.05	2.21	2.21
	250	2.21	2.21	2.21	2.21	2.21	2.21	2.39	2.21	2.39	2.21	2.39	2.39
60	50	1.50	1.39	1.62	1.50	1.62	1.39	1.75	1.50	1.89	1.62	1.89	1.75
	100	1.62	1.50	1.75	1.62	1.75	1.62	1.89	1.75	2.04	1.75	2.04	1.89
	150	1.89	1.75	1.89	1.75	2.04	1.89	2.04	1.89	2.20	2.04	2.20	2.04
	200	2.04	2.04	2.20	2.04	2.20	2.20	2.20	2.20	2.38	2.20	2.38	2.38
	250	2.38	2.38	2.38	2.38	2.57	2.38	2.57	2.57	2.57	2.57	2.78	2.57

* A.S.A. Standards A21.2-1939.

Table 10. Pit-cast Cast-iron Water Pipe*

(Dimensions in inches)

Nominal diam.	Outside diam.	Class 150† Wall thickness	Class 150† Inside diam.	Class 150† Weight, lb. per 12-ft. length	Class 250‡ Wall thickness	Class 250‡ Inside diam.	Class 250‡ Weight, lb. per 12-ft. length
4	4.80	0.40	4.00	230	0.40	4.00	230
6	6.90	0.43	6.04	360	0.43	6.04	360
8	9.05	0.46	8.13	515	0.50	8.05	550
10	11.10	0.54	10.04	730	0.63	9.84	835
12	13.20	0.58	12.04	935	0.68	11.84	1075
14	15.65	0.63	14.39	1210	0.79	14.07	1480
16	17.80	0.68	16.44	1500	0.85	16.10	1825
18	19.92	0.73	18.46	1805	0.92	18.08	2210
20	22.06	0.83	20.40	2265	0.97	20.12	2600
24	26.32	0.93	24.46	3030	1.08	24.16	3460

* Federal Specification WW-P-421: Type V.
† Class 150 is for 150 lb./sq. in. or 346 ft. head.
‡ Class 250 is for 250 lb./sq. in. or 576 ft. head.

Table 11. A.W.W.A. Standard Cast-iron Bell-and-spigot Pipe (Pit-cast)

(See Fig. 62*a* and *b*, p. 418)

Nominal diam., in.	Class	Outside diam., in.	Wall thickness, in.	Inside diam., in.	Wt. per ft.-excluding bell, lb.	Wt. of 12-ft. length including bell, lb.
3	A	3.80	0.39	3.02	14.5	175
	B	3.96	.42	3.12	16.2	194
	C	3.96	.45	3.06	17.1	205
	D	3.96	.48	3.00	18.0	216
4	A	4.80	.42	3.96	20.0	240
	B	5.00	.45	4.10	21.7	260
	C	5.00	.48	4.04	23.3	280
	D	5.00	.52	3.96	25.0	300
6	A	6.90	.44	6.02	30.8	370
	B	7.10	.48	6.14	33.3	400
	C	7.10	.51	6.08	35.8	430
	D	7.10	.55	6.00	38.3	460
	E	7.22	.58	6.06	42.5	510
	F	7.22	.61	6.00	44.3	531
	G	7.38	.65	6.08	48.1	577
	H	7.38	.69	6.00	50.5	606
8	A	9.05	.46	8.13	42.9	515
	B	9.05	.51	8.03	47.5	570
	C	9.30	.56	8.18	52.1	625
	D	9.30	.60	8.10	55.8	670
	E	9.42	.66	8.10	60.9	731
	F	9.42	.71	8.00	66.8	802
	G	9.60	.75	8.10	72.3	868
	H	9.60	.80	8.00	76.1	913
10	A	11.10	.50	10.10	57.1	685
	B	11.10	.57	9.96	63.8	765
	C	11.40	.62	10.16	70.8	850
	D	11.40	.68	10.04	76.7	920
	E	11.60	.74	10.12	86.9	1,043
	F	11.60	.80	10.00	92.8	1,114
	G	11.84	.86	10.12	101.4	2,217
	H	11.84	.92	10.00	107.3	1,288
12	A	13.20	0.54	12.12	72.5	870
	B	13.20	.62	11.96	82.1	985
	C	13.50	.68	12.14	91.7	1,100
	D	13.50	.75	12.00	100.0	1,200
	E	13.78	.82	12.14	114.6	1,375
	F	13.78	.89	12.00	122.8	1,474
	G	14.08	.97	12.14	136.2	1,634
	H	14.08	1.04	12.00	144.4	1,733
14	A	15.30	0.57	14.16	89.6	1,075
	B	15.30	.66	13.98	102.5	1,230
	C	15.65	.74	14.17	116.7	1,400
	D	15.65	.82	14.01	129.2	1,550
	E	15.98	.90	14.18	145.6	1,747
	F	15.98	.99	14.00	158.8	1,905
	G	16.32	1.07	14.18	175.1	2,101
	H	16.32	1.16	14.00	187.5	2,250
16	A	17.40	0.60	16.20	108.3	1,300
	B	17.40	.70	16.00	125.0	1,500
	C	17.80	.80	16.20	143.8	1,725
	D	17.80	.89	16.02	158.3	1,900
	E	18.16	.98	16.20	180.7	2,168
	F	18.16	1.08	16.00	196.5	2,358
	G	18.54	1.18	16.18	218.0	2,616
	H	18.54	1.27	16.00	233.8	2,805
18	A	19.50	0.64	18.22	129.2	1,550
	B	19.50	.75	18.00	150.0	1,800
	C	19.92	.87	18.18	175.0	2,100
	D	19.92	.96	18.00	191.7	2,300
	E	20.34	1.07	18.20	221.8	2,662
	F	20.34	1.17	18.00	239.3	2,872
	G	20.78	1.28	18.22	268.2	3,218
	H	20.78	1.39	18.00	287.8	3,453
20	A	21.60	0.67	20.26	150.0	1,800
	B	21.60	.80	20.00	175.0	2,100
	C	22.06	.92	20.22	208.3	2,500
	D	22.06	1.03	20.00	229.2	2,750
	E	22.54	1.15	20.24	265.8	3,190
	F	22.54	1.27	20.00	287.3	3,448
	G	23.02	1.39	20.24	321.8	3,862
	H	23.02	1.51	20.00	345.8	4,149

Table 11. A.W.W.A. Standard Cast-iron Bell-and-spigot Pipe (Pit-cast)—(*Concluded*)

Nominal diam., in.	Class	Outside diam., in.	Wall thickness, in.	Inside diam., in.	Wt. per ft.-excluding bell, lb.	Wt. of 12-ft. length including bell, lb.
24	A	25.80	0.76	24.28	204.2	2,450
	B	25.80	.89	24.02	233.3	2,800
	C	26.32	1.04	24.24	279.2	3,350
	D	26.32	1.16	24.00	306.7	3,680
	E	26.90	1.31	24.28	359.1	4,309
	F	26.90	1.45	24.00	392.3	4,707
	G	27.76	1.74	24.26	479.8	5,758
	H	27.76	1.88	24.00	510.6	6,127
30	A	31.74	0.88	29.98	291.7	3,500
	B	32.00	1.03	29.94	333.3	4,000
	C	32.40	1.20	30.00	400.0	4,800
	D	32.74	1.37	30.00	450.0	5,400
	E	33.10	1.55	30.00	530.9	6,371
	F	33.46	1.73	30.00	588.8	7,065
36	A	37.96	0.99	35.98	391.7	4,700
	B	38.30	1.15	36.00	454.2	5,450
	C	38.70	1.36	35.98	545.8	6,550
	D	39.16	1.58	36.00	625.0	7,500
	E	39.60	1.80	36.00	738.1	8,857
	F	40.04	2.02	36.00	821.0	9,852
42	A	44.20	1.10	42.00	512.5	6,150
	B	44.50	1.28	41.94	591.7	7,100
	C	45.10	1.54	42.02	716.7	8,600
	D	45.58	1.78	42.02	825.0	9,900
48	A	50.50	1.26	47.98	666.7	8,000
	B	50.80	1.42	47.96	750.0	9,000
	C	51.40	1.71	47.98	908.3	10,900
	D	51.98	1.96	48.06	1050.0	12,600
54	A	56.66	1.35	53.96	800.0	9,600
	B	57.10	1.55	54.00	933.0	11,200
	C	57.80	1.90	54.00	1141.7	13,700
	D	58.40	2.23	53.94	1341.7	16,100
60	A	62.80	1.39	60.02	916.7	11,000
	B	63.40	1.67	60.06	1104.2	13,250
	C	64.20	2.00	60.20	1341.7	16,100
	D	64.82	2.38	60.06	1583.3	19,000
72	A	75.34	1.62	72.10	1281.9	15,380
	B	76.00	1.95	72.10	1547.3	18,570
	C	76.88	2.39	72.10	1904.3	22,850
	D	77.74	2.82	72.10	2260.9	27,137
84	A	87.54	1.72	84.10	1635.8	19,630
	B	88.54	2.22	84.10	2104.1	25,250
	C	89.58	2.74	84.10	2596.4	31,165
	D	90.58	3.24	84.10	3084.6	37,023

Table 12. Cast-iron Flanged Pipe—Classes A to D

Nominal diameter, in.	Diameter of flange, in.	Diameter of bolt circle, in.	Number of bolts	Diameter of bolts, in.	Thickness of flange, in.	Class A, 100 ft. head, 43 lb. pressure: Thickness, in.	Class A: Weight, lb. per Foot	Class A: Weight, lb. per 12-ft. length	Class A: Weight, lb. per Single flange	Class B, 200 ft. head, 86 lb. pressure: Thickness, in.	Class B: Weight, lb. per Foot	Class B: Weight, lb. per 12-ft. length	Class B: Weight, lb. per Single flange	Class C, 300 ft. head, 130 lb. pressure: Thickness, in.	Class C: Weight, lb. per Foot	Class C: Weight, lb. per 12-ft. length	Class C: Weight, lb. per Single flange	Class D, 400 ft. head, 173 lb. pressure: Thickness, in.	Class D: Weight, lb. per Foot	Class D: Weight, lb. per 12-ft. length	Class D: Weight, lb. per Single flange	Nominal diameter, in.
3	7.50	6.00	4	5/8	0.75	0.39	13.0	169	6.4	0.42	14.6	188	6.2	0.45	15.5	198	6.2	0.48	16.4	209	6.2	3
4	9.00	7.50	8	5/8	.94	.42	18.0	238	11.1	.45	20.1	263	10.7	.48	21.3	277	10.7	.52	22.8	295	10.7	4
6	11.00	9.50	8	3/4	1.00	.44	27.9	365	15.0	.48	31.1	402	14.4	.51	32.9	424	14.4	.55	35.3	452	14.4	6
8	13.50	11.75	8	3/4	1.13	.46	38.7	511	23.1	.51	42.7	559	23.1	.56	48.0	620	22.0	.60	51.2	658	22.0	8
10	16.00	14.25	12	7/8	1.19	.50	51.9	687	32.2	.57	58.8	770	32.2	.62	65.5	847	30.6	.68	71.4	918	30.6	10
12	19.00	17.00	12	7/8	1.25	.54	67.0	899	47.7	.62	76.4	1,012	47.7	.68	85.4	1,116	45.6	.75	93.7	1,216	45.6	12
14	21.00	18.75	12	1	1.38	.57	82.3	1,104	58.1	.66	94.7	1,253	58.1	.74	108.1	1,407	55.1	.82	119.2	1,541	55.1	14
16	23.50	21.25	16	1	1.44	.60	98.8	1,332	73.2	.70	114.6	1,522	73.2	.80	133.3	1,738	69.1	.89	147.5	1,908	69.1	16
18	25.00	22.75	16	1 1/8	1.56	.64	118.3	1,576	78.1	.75	137.8	1,810	78.1	.87	162.4	2,094	72.8	.96	178.4	2,286	72.8	18
20	27.50	25.00	20	1 1/8	1.69	.67	137.4	1,848	99.8	.80	163.1	2,157	99.8	.92	190.6	2,473	92.9	1.03	212.3	2,733	92.9	20
24	32.00	29.50	20	1 1/4	1.88	.76	186.5	2,512	137.2	.89	217.3	2,882	137.2	1.04	257.6	3,345	126.8	1.16	286.0	3,686	126.8	24
30	38.75	36.00	28	1 3/8	2.13	.88	266.1	3,622	214.4	1.03	312.6	4,166	207.2	1.20	366.9	4,795	196.0	1.37	421.2	5,427	186.4	30
36	46.00	42.75	32	1 1/2	2.38	.99	358.7	4,959	327.4	1.15	418.7	5,654	314.8	1.36	497.7	6,572	299.9	1.58	581.9	7,548	282.5	36
40	50.75	47.25	36	1 5/8	2.50	1.06	427.2	5,940	406.6	1.23	497.0	6,753	394.5	1.48	601.6	7,965	372.7	1.72	703.4	9,143	351.1	40
42	53.00	49.50	36	1 5/8	2.63	1.10	464.6	6,492	458.5	1.28	542.2	7,395	444.2	1.54	657.4	8,720	415.4	1.78	764.1	9,953	392.1	42
48	59.50	56.00	44	1 5/8	2.75	1.26	608.0	8,408	555.9	1.42	687.2	9,324	538.9	1.71	832.7	11,001	504.4	1.96	960.8	12,471	470.8	48

Classes E to H

Nominal diameter, in.	Diameter of flange, in.	Thickness of flange, in.	Diameter of bolt circle, in.	Number of bolts	Diameter of bolts, in.	Class E, 500 ft. head, 217 lb. pressure: Thickness, in.	Class E: Weight, lb. per Foot	Class E: Weight, lb. per 12-ft. length	Class E: Weight, lb. per Single flange	Class F, 600 ft. head, 260 lb. pressure: Thickness, in.	Class F: Weight, lb. per Foot	Class F: Weight, lb. per 12-ft. length	Class F: Weight, lb. per Single flange	Class G, 700 ft. head, 304 lb. pressure: Thickness, in.	Class G: Weight, lb. per Foot	Class G: Weight, lb. per 12-ft. length	Class G: Weight, lb. per Single flange	Class H, 800 ft. head, 347 lb. pressure: Thickness, in.	Class H: Wieght, lb. per Foot	Class H: Wieght, lb. per 12-ft. length	Class H: Wieght, lb. per Single flange	Nominal diameter, in.
6	12.50	1 7/16	10.63	12	3/4	0.58	37.7	515	31.1	0.61	39.5	536	31.1	0.65	42.9	576	30.4	0.69	45.2	603	30.4	6
8	15.00	1 5/8	13.00	12	7/8	.66	54.7	748	45.9	.71	60.6	819	45.9	.75	65.1	871	44.7	.80	68.8	915	44.7	8
10	17.50	1 7/8	15.25	16	1	.74	78.8	1,079	66.5	.80	84.7	1.149	66.5	.86	92.5	1,239	64.3	.92	98.5	1,311	64.3	10
12	20.50	2	17.75	16	1 1/8	.82	104.2	1,441	95.1	.89	112.4	1,539	95.1	.97	124.6	1,678	91.6	1.04	132.9	1,778	91.6	12
14	23.00	2 1/8	20.25	20	1 1/8	.90	133.1	1,837	120.1	.99	146.2	1,995	120.1	1.07	160.2	2,153	115.2	1.16	172.6	2,302	115.2	14
16	25.50	2 1/4	22.50	20	1 1/4	.98	165.0	2,277	148.7	1.08	180.8	2,467	148.7	1.18	199.2	2,675	142.1	1.27	215.0	2,864	142.1	16
18	28.00	2 3/8	24.75	24	1 1/4	1.07	202.3	2,790	181.2	1.17	219.8	3,000	181.2	1.28	244.6	3,280	172.2	1.39	264.1	3,514	172.2	18
20	30.50	2 1/2	27.00	24	1 3/8	1.15	241.1	3,328	217.3	1.27	262.5	3,585	217.3	1.39	294.4	3,945	205.9	1.51	318.3	4,231	205.9	20
24	36.00	2 3/4	32.00	24	1 5/8	1.31	328.5	4,590	323.8	1.45	361.6	4,987	323.8	1.75	446.2	5,948	296.8	1.88	476.9	6,316	296.8	24
30	43.00	3	39.25	28	1 3/4	1.55	484.7	6,747	465.3	1.73	538.0	7,357	450.3	....	.....	.....	.....	....	.....	.....	.....	30
36	50.00	3 3/8	46.00	32	1 7/8	1.80	674.2	9,384	646.8	2.02	748.7	10,229	622.2	....	.....	.....	.....	....	.....	.....	.....	36

Table 13. Centrifugally Cast Cast-iron Pipe—Bell-and-spigot
(See Fig. 62*c*, p. 418)

Nominal diam., in.	Class or maximum working pressure	Outside diam., in.	Average thickness, in.	Inside diam., in.	Weight per ft. excluding bell, lb.	Weight of pipe, lb., including bell: 12 ft.	18 ft.
3	150	3.80	0.33	3.14	11	140	
	250	3.80	.36	3.08	12	155	
4	150	4.80	.34	4.12	15	195	285
	250	4.80	.38	4.04	17	220	325
6	150	6.90	.37	6.16	24	315	460
	250	6.90	.43	6.04	27	350	515
8	150	9.05	.42	8.21	36	475	690
	200	9.05	.46	8.13	39	510	745
	250	9.05	.50	8.05	42	545	800
10	150	11.10	.47	10.16	49	640	935
	200	11.10	.52	10.06	54	700	1025
	250	11.10	.57	9.96	59	760	1115
12	150	13.20	.50	12.20	62	810	1180
	200	13.20	.57	12.06	70	905	1325
	250	13.20	.62	11.96	77	990	1450
14	100	15.30	.48	14.34	70	920	1340
	150	15.65	.55	14.55	80	1060	1555
	200	15.65	.62	14.41	89	1190	1735
	250	15.65	.69	14.27	99	1320	1930
16	100	17.40	.52	16.36	86	1130	1645
	150	17.80	.60	16.60	99	1320	1935
	200	17.80	.68	16.44	111	1490	2175
	250	17.80	.75	16.30	121	1635	2390
18	100	19.50	.56	18.48	104	1365	1990
	150	19.92	.65	18.62	120	1595	2330
	200	19.92	.74	18.44	136	1810	2645
	250	19.92	.83	18.26	147	2015	2950
20	100	21.60	.58	20.44	120	1585	2305
	150	22.06	.68	20.70	139	1860	2720
	200	22.06	.78	20.50	159	2125	3105
	250	22.06	.88	20.30	175	2365	3465
24	100	25.80	.64	24.52	158	2085	3035
	150	26.32	.76	24.80	187	2480	3630
	200	26.32	.88	24.56	215	2855	4170
	250	26.32	1.00	24.32	242	3200	4690

Table 13*a*. Centrifugally Cast Cast-iron Pipe—Standard A.G.A. Bell-and-spigot—Types 1 and 2, and Anthony High Pressure
(See Fig. 62*d* and *e*, p. 418)

Nominal diam., in.	Class or maximum working pressure	Outside diam., in.	Average thickness, in.	Inside diam., in.	Weight per ft. excluding bell, lb.	Weight of pipe, lb., including bell*: 12 ft.	18 ft.
3	150	3.80	0.33	3.14	11	140	
4	150	4.80	0.34	4.12	15	195	285
6	150	6.90	0.37	6.16	24	315	460
8	150	9.05	0.42	8.21	36	475	690
10	150	11.10	0.47	10.16	49	640	935
12	150	13.20	0.50	12.20	62	810	1180
16	100	17.40	0.52	16.36	86	1130	1645
	150	17.40	0.60	16.20	99	1285	1880
20	100	21.60	0.58	20.44	120	1580	2300
	150	21.60	0.68	20.24	139	1805	2640
24	100	25.80	0.64	24.52	158	2100	3050
	150	25.80	0.76	24.28	187	2450	3570

* Weights are approximate and for type 1 bell; multiply by 1.02 for type 2 bell; multiply by 0.98 for Anthony high-pressure bell.

Duriron and Durichlor Pipe

Duriron is a high-silicon iron and contains approximately 14.5 per cent silicon and 0.85 per cent carbon. It is resistant to most chemicals, such as sulfuric, nitric, and acetic acids at any strength and temperature. Durichlor is a special high-silicon iron, containing an appreciable amount of molybdenum. It is resistant to the same corrosives for which Duriron is recommended and in addition is almost entirely resistant to hydrochloric acid at all concentrations and at all temperatures up to the boiling point. These alloys are available in the cast form only. Pipe and fittings are cast with upset ends being joined by split flanges. Integrally cast flanged pipe is also available. Allowable working pressures cannot be stated in the manner customary for other types of pipe because of such variables as thermal shock, pulsating pressures, and the corrosive being handled. Although rupture does not occur below 400 lb./sq. in. pressure in sizes up to and including 6 in., 50 lb./sq. in. is a normal recommendation even though the pipe has been used for pressure considerably in excess of that figure. Table 14 lists sizes 1 to 12 in., and larger sizes can be obtained. Bell-and-spigot pipe is produced in the weights and dimensions shown in Table 14; fittings are available. The coefficient of linear expansion of these alloys in the temperature range 20° to 100°C. is 12.2×10^{-6} per degree centigrade, which is slightly above that of cast iron (National Bureau of Standards). Since these alloys have practically no elasticity, it is necessary to use expansion joints in relatively short pipe lines. Connections for Duriron-flanged pipe, fittings, valves, and pumps are made to 125-lb. American Standard drilling.

Table 14. Duriron and Durichlor Pipe*

Size, inside diam., in.	Flanged or upset ends: Outside diam., in.	Wall thickness, in.	Standard† length, ft.	Weight per piece, lb.	Bell-and-spigot ends: Outside diam., in.	Wall thickness, in.	Standard† length, ft.	Weight per piece, lb.
1	1¾	⅜	3	16				
1½	2¼	⅜	3	24	2⅛	5⁄16	3	19
2	2¾	⅜	4	37	2⅝	5⁄16	4	30
2½	3¼	⅜	5	60				
3	3⅞	7⁄16	5	83	3⅝	5⁄16	5	56
4	4⅞	7⁄16	5	103	4⅝	5⁄16	5	91
6	7	½	5	210	6 11⁄16	11⁄32	5	135
8	9¼	⅝	6	350	8¾	⅜	5	230
10	11½	¾	6	510	11¼	⅝	5	380
12	14	1	6	850	13¼	⅝	5	460
15	....	.....	....	...	16¾	⅞	5	680

* The Duriron Co.
† Laying lengths; lengths less than standard are available.

NON-FERROUS METAL PIPE AND TUBING

Pipo and tubing dimensions for aluminum, columbium, copper, copper alloy, lead, nickel, nickel alloy, tantalum, and tin are given below, as well as for pipe lined or coated with some of these metals. Pipe is usually made in "iron pipe size" dimensions for outside diameter, but the wall thicknesses and consequently the inside diameters differ.

Aluminum Pipe and Tubing

Seamless pipe and tubing are available in about 17 aluminum alloys and tempers, the strength of each varying over the working temperature range (up to 500°F., 260°C. maximum). An ultimate strength of 12,000 lb./sq. in. at 75°F. is given for "2S-0" material, and a safety factor of 4 is recommended by the manufacturer. The **tubing** is made in sizes (outside diameter) from 1/8 to 11 1/4 in. and in wall thicknesses from 31 Stubs (= 31 B.W.G. = 0.010 in.) for the 1/8-in. size to 0.500 in. thickness for the 11 1/4-in. size. All alloys and tempers are not made in all sizes and thicknesses. For dimensions of common **tube** sizes, see Table 16. Aluminum **pipe**, Table 15, is made in iron pipe sizes, the outside and inside diameters being identical with iron and steel pipe; see Table 4. The sizes above 5 in. standard and 4 in. extra strong are not made in all tempers.

Table 15. Aluminum and Nickel Pipe

Size, in.	Outside diam., in.	Standard pipe: Inside diam., in.	Standard pipe: Wall thickness, in.	Standard pipe: Weight per ft., lb., Aluminum	Standard pipe: Weight per ft., lb., Nickel	Extra strong pipe: Inside diam., in.	Extra strong pipe: Wall thickness, in.	Extra strong pipe: Weight per ft., lb., Aluminum	Extra strong pipe: Weight per ft., lb., Nickel	Double extra strong pipe: Inside diam., in.	Double extra strong pipe: Wall thickness, in.	Double extra strong pipe: Weight per ft., lb., nickel
1/8	0.405	0.269	0.068	0.084	0.274	0.215	0.095	0.109	0.383			
1/4	.540	.364	.088	.147	.477	.302	.119	.185	.602			
3/8	.675	.493	.091	.196	.638	.423	.126	.255	.830			
1/2	.840	.622	.109	.294	.957	.546	.147	.376	1.223			
3/4	1.050	.824	.113	.390	1.272	.742	.154	.509	1.657	0.434	0.308	2.757
1	1.315	1.049	.133	.580	1.889	.957	.179	.750	2.442	.599	.358	4.133
1 1/4	1.660	1.380	.140	.785	2.558	1.278	.191	1.035	3.371	.896	.382	5.889
1 1/2	1.900	1.610	.145	.939	3.059	1.500	.200	1.254	4.085			
2	2.375	2.067	.154	1.262	4.112	1.939	.218	1.735	5.650			
2 1/2	2.875	2.469	.203	2.002	6.522	2.323	.276	2.647	8.619			
3	3.500	3.068	.216	2.617	8.529	2.900	.300	3.543	11.581			
3 1/2	4.000	3.548	.226	3.147	10.248	3.364	.318	4.321	14.125			
4	4.500	4.026	.237	3.729	12.139	3.826	.337	5.178	16.925			
4 1/2	5.000	4.506	.247	4.333	14.105	4.290	.355	6.086				
5	5.563	5.045	.259	5.051	16.572	4.813	.375	7.180				
6	6.625	6.065	.280	6.556	21.431	5.761	.432	9.874				
7	7.625	7.023	.301	8.136	......	6.625	.500	13.147				
8	8.625	7.981	.322	9.867	......	7.625	.500	14.993				
9	9.625	8.941	.342	11.800	......	8.625	.500	17.010				
10	10.750	10.020	.365	14.000	......	9.750	.500	19.103				

Copper and Copper-alloy Pipe and Tubing

Seamless copper, brass, commercial bronze, and other copper-alloy pipe and tubing can be obtained in a wide range of sizes and wall thicknesses. The tubing is available in many shapes other than circular: square, rectangular, oval, etc. **Circular seamless tubing** is fabricated in "o.d." (outside diameter) and in "i.d." (inside diameter) sizes ranging from 1/32 to 16 in., and in a range of wall thicknesses* (Stubs gage = B.W.G.; B.&S. gage; decimal) varying from 36 B.&S, gage (or 0.005 in.) for the smallest tubing to 0000 B.&S. gage (or 0.75 in.) for the 16 in. size; larger diameters and wall thicknesses can be obtained on order. The o.d. sizes are more commonly used. The cost per pound varies considerably with size and thickness. Tubing should always be specified by two of the following dimensions outside diameter, inside diameter, and wall thickness. The 1/8- and 1/4-in. tubes are commercially available in coils up to 100 ft. in length, although some mills can supply up to 300 ft. on special order. Straight lengths of 20 ft. are commercially available. Diameter tolerances are generally ±5 per cent.

* See p. 39 for table of sheet-metal and wire gages.

Condenser and heat exchanger tubes are held to closer tolerances, as shown in Table 15*a* as standard TUBE-5 of the Copper & Brass Research Association, New York, N.Y. Dimensions of the tubes in the nominal or outside-diameter sizes from 1/4 to 2 in. are given in Table 16. The Association also publishes Data 68 and 69 "Estimating Data" for tubing from 5/8 to 2 in.

Table 15*a*. Tolerances of Condenser and Heat Exchange Tubes*

Diameter

Applicable to straight lengths only

Diameter, in.	Outside diameter tolerances, in. (plus and minus): Alloys other than cupronickel	Outside diameter tolerances, in. (plus and minus): Cupronickel
Up to 0.500, inclusive	0.002	0.0025
Over 0.500–0.740, inclusive	.0025	.003
Over 0.740–1.000, inclusive	.003	.0035
Over 1.000–1.250, inclusive	.0035	.004
Over 1.250–1.500, inclusive	.004	.0045

Length

Applicable to straight lengths only

The length shall be not less than that specified, when measured at a temperature of 20°C. (68°F.), but may be more than that specified by the amounts in the following table:

Length, Ft.	Tolerances, In.
Up to 15, inclusive	1/16
Over 15–20, inclusive	3/32
Over 20	1/8

Wall Thickness

All tolerances plus, the upper limit being controlled by the weight tolerances.

Weight

Any lot of 600 tubes or any shipment of more than 600 tubes may exceed the theoretical weight by not more than 5 per cent. One cubic inch of the various alloys covered by these specifications shall be assumed to weigh:

Alloy	Weight, Lb./Cu. In.
Admiralty	0.308
Copper (all types)	.323
Cupronickel, 30%	.323
Cupronickel, 20%	.323
Aluminum brass	.301
Aluminum bronze, 5%	.295
Red brass, 85%	.316
Muntz metal	.303

* Copper & Brass Research Association, standard TUBE-5-1946.

Table 16. Condenser and Heat-exchanger Tube Dimensions

Outside diam., in.	Wall thickness: B.W.G. and Stubs' gage*	Wall thickness: In.	Inside diam., in.	Cross-sectional area metal, sq. in.	Inside sectional area, sq. ft.	Circumference, ft., or surface, sq. ft./ft. of length: Outside	Circumference, ft., or surface, sq. ft./ft. of length: Inside	Velocity, ft./sec. for 1 U.S. gal./min.	Capacity at 1 ft./sec. velocity: U.S. gal./min.	Capacity at 1 ft./sec. velocity: Lb./hr. water	Weight per ft., lb.†
¼	14	0.083	0.084	0.0435	0.000039	0.0654	0.0219	57.14	0.0175	8.75	0.161
	16	.065	.120	.0377	.000079	.0654	.0314	28.20	.0355	17.73	.140
	18	.049	.152	.0309	.000126	.0654	.0397	17.68	.0566	28.30	.115
	20	.035	.180	.0236	.000177	.0654	.0471	12.59	.0794	39.70	.0876
	22	.028	.194	.0195	.000205	.0654	.0507	10.869	.0920	46 00	.0724
	24	.022	.206	.0157	.000231	.0654	.0539	9.645	.1037	51.85	.0584
⅜	14	.083	.209	.0761	.000238	.0981	.0547	9.362	.1068	53.40	.282
	16	.065	.245	.0633	.000327	.0981	.0641	6.814	.1468	73.40	.235
	18	.049	.277	.0501	.000418	.0981	.0725	5 330	.1876	93.80	.186
	20	.035	.305	.0373	.000507	.0981	.0798	4.395	.2275	113.8	.139
	22	.028	.319	.0305	.000555	.0981	.0835	4.015	.2494	124.7	.113
	24	.022	.331	.0243	.000597	.0981	.0866	3.732	.2679	134.0	.0904
½	12	.109	.282	.1338	.000433	.1309	.0748	5.142	.1945	97.25	.493
	14	.083	.334	.1087	.000608	.1309	.0874	3.662	.2730	136.5	.403
	16	.065	.370	.0888	.000747	.1309	.0969	2.981	.3352	167.5	.329
	18	.049	.402	.0694	.000882	.1309	.1052	2.530	.3952	197.6	.258
	20	.035	.430	.0511	.001009	.1309	.1125	2.209	.4528	226.4	.190
⅝	10	.134	.357	.2067	.000695	.1636	.0935	3.206	.3119	156.0	.769
	11	.120	.385	.1904	.000808	.1636	.1008	2.758	.3626	181.3	.708
	12	.109	.407	.1767	.000903	.1636	.1066	2.468	.4053	202.7	.657
	13	.095	.435	.1582	.00103	.1636	.1139	2.163	.4623	231.2	.588
	14	.083	.459	.1460	.00115	.1636	.1202	1.938	.5161	258.1	.526
	15	.072	.481	.1250	.00126	.1636	.1259	1.768	.5655	258.9	.465
	16	.065	.495	.1143	.00134	.1636	.1296	1.663	.6014	300.7	.425
	17	.058	.509	.1033	.00141	.1636	.1333	1.580	.6328	316.4	.384
	18	.049	.527	.0887	.00151	.1636	.1380	1.476	.6777	338.9	.330
	19	.042	.541	.0596	.00160	.1636	.1469	1.393	.7181	359.1	.286
¾	10	.134	.482	.2593	.00127	.1963	.1262	1.754	.5700	235.0	.965
	11	.120	.510	.2375	.00142	.1963	.1335	1.569	.6373	318.7	.884
	12	.109	.532	2195	.00154	.1963	.1393	1.447	.6912	345.6	.817
	13	.095	.560	.1955	.00171	.1963	.1466	1.303	.7674	383.7	.727
	14	.083	.584	.1739	.00186	.1963	.1529	1.198	.8348	417.4	.647
	15	.072	.606	.1534	.00200	.1963	.1587	1.114	.8976	448.8	.571
	16	.065	.620	.1398	.00210	.1963	.1623	1.061	.9425	471.3	.520
	17	.058	.634	.1261	.00219	.1963	.1660	1.017	.9829	491.5	.469
	18	.049	.652	.1079	.00232	.1963	.1707	0.962	1.041	520.5	.401
	19	.042	.666	.0934	.00242	.1963	.1744	.920	1.086	543.0	.348
⅞	9	.148	.579	.3380	.00183	.2291	.1516	1.218	0.8213	410.7	1.26
	10	.134	.607	.3119	.00201	.2291	.1589	1.109	.9021	451.1	1.16
	11	.120	.635	.2846	.00220	.2291	.1662	1.012	.9874	493.7	1.06
	12	.109	.657	.2623	.00235	.2291	.1720	0.948	1.055	527.5	0.976
	13	.095	.685	.2328	.00256	.2291	.1793	.870	1.149	574.5	.866
	14	.083	.709	.2065	.00274	.2291	.1856	.813	1.230	615.0	.768
	15	.072	.731	.1816	.00291	.2291	.1914	.766	1.306	653.0	.676
	16	.065	.745	.1654	.00303	.2291	.1950	.735	1.360	680.0	.615
	17	.058	.759	.1488	.00314	.2291	.1987	.709	1.409	704.5	.554
	18	.049	.777	.1271	.00329	.2291	.2034	.678	1.477	738.5	.473
	19	.042	.791	.1099	.00341	.2291	.2071	.654	1.530	751.5	.409
1	7	.180	.640	.4637	.00223	.2618	.1676	.999	1.001	500.5	1.73
	8	.165	.670	.4328	.00245	.2618	.1754	.909	1.100	505.0	1.61
	9	.148	.704	.3962	.00270	.2618	.1843	.826	1.212	606.0	1.47
	10	.134	.732	.3654	.00292	.2618	.1916	.763	1.310	655.0	1.36
	11	.120	.760	.3318	.00315	.2618	.1990	.707	1.414	707.0	1.23
	12	.109	.782	.3051	.00334	.2618	.2048	.667	1.499	750.0	1.14
	13	.095	.810	.2701	.00358	.2618	.2121	.622	1.607	803.5	1.000
	14	.083	.834	.2391	.00379	.2618	.2183	.588	1.701	850.5	0.890
	15	.072	.856	.2099	.00400	.2618	.2241	.557	1.795	897.5	.781
	16	.065	.870	.1909	.00413	.2618	.2277	.538	1.854	927.0	.710
	17	.058	.884	.1716	.00426	.2618	.2314	.523	1.912	956.0	.639
	18	.049	.902	.1463	.00444	.2618	.2361	.501	1.993	996.5	.545
	19	.042	.916	.1264	.00458	.2618	.2398	.486	2.056	1028.	.470
1⅛	7	.180	.765	.5355	.00319	.2945	.2003	.698	1.432	716.0	1.99
	8	.165	.795	.4979	.00345	.2945	.2081	.646	1.548	774.0	1.85
	9	.148	.829	.4546	.00375	.2945	.2170	.594	1.683	841.5	1.69
	10	.134	.857	.4175	.00401	.2945	.2244	.556	1.800	900.0	1.55
	11	.120	.885	.3792	.00427	.2945	.2317	.521	1.916	958.	1.41
	12	.109	.907	.3479	.00449	.2945	.2375	.496	2.015	1008.	1.29
	13	.095	.935	.3074	.00477	.2945	.2448	.467	2.141	1071.	1.14
	14	.083	.959	.2717	.00502	.2945	.2511	.443	2.253	1127.	1.01
	15	.072	.981	.2381	.00525	.2945	.2568	.424	2.356	1178.	0.886
	16	.065	.995	.2165	.00540	.2945	.2605	.412	2.424	1212.	.805
	17	.058	1.009	.1944	.00555	.2945	.2642	.401	2.491	1246.	.723
	18	.049	1.029	.1624	.00575	.2945	.2694	.387	2.581	1291.	.616
	19	.042	1.041	.1429	.00591	.2945	.2725	.377	2.652	1326.	.532
1¼	7	.180	0.890	.6051	.00432	.3271	.2330	.516	1.939	969.5	2.25
	8	.165	.920	.5624	.00462	.3271	.2409	.482	2.073	1037.	2.09
	9	.148	.954	.5124	.00496	.3271	.2498	.449	2.226	1113.	1.91
	10	.134	.982	.4698	.00526	.3271	.2572	.424	2.361	1181.	1.75
	11	.120	1.010	.4260	.00556	.3271	.2644	.401	2.495	1248.	1.58
	12	.109	1.032	.3907	.00581	.3271	.2701	.384	2.608	1304.	1.45
	13	.095	1.060	.3447	.00613	.3271	.2775	.363	2.751	1376.	1.28
	14	.083	1.084	.3042	.00641	.3271	.2839	.348	2.877	1439.	1.13

* B.W.G. = "Birmingham wire gage" commonly used for ferrous tubing; it is identical with Stubs' = "Stubs' iron-wire gage," see p. 39.
† In brass, sp. gr. = 8.56; sp. gr. of steel = 7.8.

Table 16. Condenser and Heat-exchanger Tube Dimensions—(*Concluded*)

Outside diam., in.	Wall thickness: B.W.G. and Stubs' gage*	Wall thickness: In.	Inside diam., in.	Cross-sectional area metal, sq. in.	Inside sectional area, sq. ft.	Circumference, ft., or surface, sq. ft./ft. of length: Outside	Circumference, ft., or surface, sq. ft./ft. of length: Inside	Velocity, ft./sec. for 1 U.S. gal./min.	Capacity at 1 ft./sec. velocity: U.S. gal./min.	Capacity at 1 ft./sec. velocity: Lb./hr. water	Weight per ft., lb.†
1¼	15	0.072	1.106	0.2665	0.00667	0.3271	0.2896	0.334	2.993	1497.	0.991
	16	.065	1.120	.2419	.00684	.3271	.2932	.326	3.070	1535.	.900
	17	.058	1.134	.2172	.00701	.3271	.2969	.318	3.146	1573.	.808
	18	.049	1.152	.1848	.00724	.3271	.3015	.308	3.249	1625.	.688
	19	.042	1.166	.1590	.00742	.3271	.3053	.300	3.330	1665.	.593
1⅜	7	.180	1.015	.6758	.00562	.3620	.2657	.397	2.522	1261.	2.51
	8	.165	1.045	.6272	.00596	.3620	.2736	.374	2.675	1338.	2.33
	9	.148	1.079	.5705	.00635	.3620	.2825	.351	2.850	1425.	2.12
	10	.134	1.107	.5224	.00668	.3620	.2898	.334	2.998	1499.	1.94
	11	.120	1.135	.4731	.00703	.3620	.2971	.317	3.155	1578.	1.76
	12	.109	1.157	.4335	.00730	.3620	.3029	.305	3.276	1638.	1.61
	13	.095	1.185	.3820	.00766	.3620	.3102	.291	3.438	1719.	1.42
	14	.083	1.209	.3369	.00797	.3620	.3165	.280	3.577	1789.	1.25
	15	.072	1.231	.2947	.00827	.3620	.3223	.269	3.712	1856.	1.10
	16	.065	1.245	.2675	.00845	.3620	.3259	.264	3.792	1896.	0.995
	17	.058	1.259	.2399	.00865	.3620	.3296	.258	3.882	1941.	.893
	18	.049	1.277	.2041	.00889	.3620	.3343	.251	3.990	1995.	.759
	19	.042	1.291	.1759	.00909	.3620	.3380	.245	4.080	2040.	.654
1½	7	.180	1.140	.7464	.00709	.3925	.2985	.314	3.182	1591.	2.78
	8	.165	1.170	.6920	.00747	.3925	.3063	.298	3.353	1677.	2.57
	9	.148	1.204	.6286	.00791	.3925	.3152	.282	3.550	1775.	2.34
	10	.134	1.232	.5750	.00828	.3925	.3225	.269	3.716	1858.	2.14
	11	.120	1.260	.5202	.00866	.3925	.3299	.257	3.887	1944.	1.94
	12	.109	1.282	.4763	.00896	.3925	.3356	.249	4.021	2011.	1.77
	13	.095	1.310	.4193	.00936	.3925	.3430	.238	4.201	2101.	1.56
	14	.083	1.334	.3694	.00971	.3925	.3492	.229	4.358	2176.	1.37
	15	.072	1.358	.3187	.0100	.3925	.3555	.223	4.488	2244.	1.20
	16	.065	1.370	.2930	.0102	.3925	.3587	.218	4.578	2289.	1.09
	17	.058	1.384	.2627	.0104	.3925	.3623	.214	4.668	2334.	0.978
	18	.049	1.402	.2234	.0107	.3925	.3670	.208	4.802	2401.	.831
	19	.042	1.416	.1923	.0109	.3925	.3707	.204	4.892	2446.	.716
1⅝	7	.180	1.265	.8171	.00873	.4254	.3312	.255	3.918	1959.	3.04
	8	.165	1.295	.8354	.00915	.4254	.3390	.243	4.107	2054.	2.82
	9	.148	1.329	.6868	.00963	.4254	.3479	.231	4.322	2161.	2.55
	10	.134	1.357	.6198	.0100	.4254	.3553	.223	4.488	2244.	2.34
	11	.120	1.385	.5674	.0105	.4254	.3626	.212	4.712	2356.	2.11
	12	.109	1.407	.5191	.0108	.4254	.3684	.206	4.847	2424.	1.93
	13	.095	1.435	.4566	.0112	.4254	.3757	.199	5.027	2514.	1.70
	14	.083	1.459	.4020	.0116	.4254	.3820	.192	5.206	2603.	1.50
	15	.072	1.481	.3512	.0120	.4254	.3877	.186	5.386	2693.	1.31
	16	.065	1.495	.3186	.0122	.4254	.3914	.183	5.475	2738.	1.19
	17	.058	1.509	.2855	.0124	.4254	.3951	.180	5.565	2783.	1.06
	18	.049	1.527	.2426	.0127	.4254	.3998	.175	5.700	2850.	0.903
	19	.042	1.541	.2300	.0130	.4254	.4034	.171	5.834	2967.	.777
1¾	7	.180	1.390	.8878	.0105	.4582	.3639	.212	4.712	2356.	3.30
	8	.165	1.420	.8216	.0110	.4582	.3718	.203	4.937	2469.	3.06
	9	.148	1.454	.7449	.0115	.4582	.3807	.194	5.161	2581.	2.77
	10	.134	1.482	.6803	.0120	.4582	.3880	.186	5.386	2693.	2.53
	11	.120	1.510	.6145	.0124	.4582	.3953	.180	5.565	2783.	2.29
	12	.109	1.532	.5620	.0128	.4582	.4011	.174	5.745	2873.	2.09
	13	.095	1.560	.4939	.0133	.4582	.4084	.168	5.969	2985.	1.84
	14	.083	1.584	.4346	.0137	.4582	.4147	.163	6.149	3075.	1.62
	15	.072	1.606	.3796	.0141	.4582	.4205	.158	6.328	3169.	1.41
	16	.065	1.620	.3441	.0143	.4582	.4241	.156	6.418	3209.	1.28
	17	.058	1.634	.3083	.0146	.4582	.4278	.153	6.552	3276.	1.15
	18	.049	1.652	.2619	.0149	.4582	.4325	.150	6.687	3344.	0.974
	19	.042	1.666	.2253	.0151	.4582	.4362	.148	6.777	3389.	.838
1⅞	7	.180	1.515	.9585	.0125	.4909	.3966	.178	5.610	2805.	3.57
	8	.165	1.545	.8864	.0130	.4909	.4045	.171	5.834	2917.	3.30
	9	.148	1.579	.8030	.0136	.4909	.4134	.164	6.104	3052.	2.99
	10	.134	1.607	.7329	.0141	.4909	.4207	.158	6.328	3164.	2.73
	11	.120	1.635	.6616	.0146	.4909	.4280	.153	6.552	3276.	2.46
	12	.109	1.657	.6048	.0150	.4909	.4338	.149	6.732	3366.	2.25
	13	.095	1.685	.5312	.0155	.4909	.4411	.144	6.956	3478.	1.98
	14	.083	1.709	.4594	.0159	.4909	.4474	.140	7.136	3568.	1.74
	15	.072	1.731	.4078	.0163	.4909	.4532	.137	7.315	3658.	1.52
	16	.065	1.745	.3695	.0166	.4909	.4568	.134	7.450	3725.	1.38
	17	.058	1.759	.3310	.0169	.4909	.4605	.132	7.585	3793.	1.23
	18	.049	1.777	.2811	.0172	.4909	.4652	.130	7.719	3860.	1.05
	19	.042	1.791	.2418	.0175	.4909	.4689	.127	7.854	3927.	0.900
2	7	.180	1.640	1.0289	.0147	.5233	.4294	.152	6.597	3299.	3.83
	8	.165	1.670	0.9511	.0152	.5233	.4372	.147	6.822	3411.	3.54
	9	.148	1.704	.8608	.0158	.5233	.4461	.141	7.091	3546.	3.20
	10	.134	1.732	.7855	.0164	.5233	.4534	.136	7.360	3680.	2.92
	11	.120	1.760	.7084	.0169	.5233	.4608	.132	7.585	3793.	2.64
	12	.109	1.782	.6475	.0173	.5233	.4665	.129	7.764	3882.	2.41
	13	.095	1.810	.5686	.0179	.5233	.4739	.125	8.034	4017.	2.12
	14	.083	1.834	.4998	.0183	.5233	.4801	.122	8.213	4107.	1.86
	15	.072	1.856	.4359	.0188	.5233	.4859	.118	8.437	4219.	1.62
	16	.065	1.870	.3951	.0191	.5233	.4896	.117	8.572	4286.	1.47
	17	.058	1.884	.3542	.0194	.5233	.4932	.115	8.707	4354.	1.32
	18	.049	1.902	.3000	.0197	.5233	.4979	.113	8.841	4421.	1.12
	19	.042	1.916	.2584	.0200	.5233	.5016	.111	8.976	4488.	0.961

* B.W.G. = "Birmingham wire gage" commonly used for ferrous tubing; it is identical with Stubs' = "Stubs' iron-wire gage," see p. 39.

† In brass, sp. gr. = 8.56; sp. gr. of steel = 7.8.

Copper water tubes are more costly because they are made to one-half the tolerances of common copper tubes to permit the use of swage and solder fittings. The nominal size is 1/8 in. less than the outside diameter; dimensions are given in Table 17. The working pressures vary with wall thickness, diameter, alloy, and heat-treatment.

Type	K	K	L	L	M
Temper	Hard	Soft	Hard	Soft	Hard
Working pressure	400	250	250	150	250

Copper and copper-alloy pipe is made in so-called "iron pipe sizes" or "standard pipe sizes" and, although the outside diameters are identical with those of standard iron pipe, the inside diameters are not the same. Table 18 indicates the regular and extra strong sizes. The copper and copper-alloy pipe is in many cases more corrosion-resistant than iron pipe. When corrosion does occur, there is generally not the decrease in diameter that frequently occurs with iron pipe. It should be remembered that brass pipe is sometimes subject to dezincification and consequent disintegration when in contact with natural waters and with solutions. At 360°F., or 182°C., the strength of copper pipe is reduced 15 per cent. The maximum working pressures can be calculated by the formula

$$\frac{\text{Tensile strength} \times \text{wall thickness, in.}}{\text{Tube radius, in.} \times \text{safety factor}} = \text{safe limit of working pressure}$$

The tensile strength of copper is about 30,000 lb./sq. in. and of brass, about 40,000. The tube radius is half the inside diameter in inches. A safety factor of 6, based on the ultimate tensile strength, is common.

Table. 17 Copper Water Tube—Types K, L, M*

For compression or soldered fittings
All tolerances plus and minus except as otherwise indicated

Nominal size	Actual outside diam., in.	Mean outside diam. tolerances, in.		Wall thickness, in.						Theoretical weight, lb./ft.		
				Type K		Type L		Type M				
		Soft annealed	Hard drawn	Nominal	Tolerance	Nominal	Tolerance	Nominal	Tolerance	Type K	Type L	Type M
3/8	0.500	0.0025	0.001	0.049	0.004	0.035	0.0035	.....	.....	0.269	0.198	
1/2	.625	.0025	.001	.049	.004	.040	0035	.....	.....	.344	.285	
5/8	.750	.0025	.001	.049	.004	.042	.0035	.....	.....	.418	.362	
3/4	.875	.003	.001	.065	.0045	.045	.004	.....	.....	.641	.455	
1	1.125	.0035	.0015	.065	.0045	.050	.004	.....	.....	.839	.655	
1 1/4	1.375	.004	.0015	.065	.0045	.055	.0045	.....	.....	1.04	.884	
1 1/2	1.625	.0045	.002	.072	.005	.060	.0045	.....	.....	1.36	1.14	
2	2.125	.005	.002	.083	.007	.070	.006	.....	.....	2.06	1.75	
2 1/2	2.625	.005	.002	.095	.007	.080	.006	0.065	0.006	2.93	2.48	2.03
3	3.125	.005	.002	.109	.007	.090	.007	.072	.006	4.00	3.33	2.68
3 1/2	3.625	.005	.002	.120	.008	.100	.007	.083	.007	5.12	4.29	3.58
4	4.125	.005	.002	.134	.010	.110	.009	.095	.009	6.51	5.38	4.66
5	5.125	.005	.002	.160	.010	.125	.010	.109	.009	9.67	7.61	6.66
6	6.125	.005	.002	.192	.012	.140	.011	.122	.010	13.9	10.2	8.92
8	8.125	.006	+ .002 − .004	.271	.016	.200	.014	.170	.014	25.9	19.3	16.5
10	10.125	.008	.002 .006	338	.018	250	.016	212	.015	40.3	30.1	25.6
12	12.125	.008	.002 .006	.405	.020	.280	.018	.254	.016	57.8	40.4	36.7

* Copper & Brass Research Association, Standard TUBE-3-1946, corresponds to the National Bureau of Standards simplified practice recommendations R217-46.

Lengths

a. The standard length for tubes furnished straight is 20 ft.
b. The standard length for tubes furnished in coils is 60 ft.

Length Tolerance

Same as for regular seamless tube, as given in Table 15*a*.

Weight Tolerance

Tube shall not vary in weight by more than 5 per cent from the theoretical weight given above.

Tempers

Types K and L: hard and soft tempers.
Type M: hard temper only.

Table 18. Copper, Brass, and Bronze Pipe*

A. Dimensions and Weights of Regular Pipe

Nominal pipe size, in.	Nominal dimensions, in. Outside diam.	Inside diam.	Wall Thickness	Cross-sectional area of bore, sq. in.	Lb./ft. Red brass	Copper	Nominal pipe size, in.	Nominal dimensions, in. Outside diam.	Inside diam.	Wall thickness	Cross-sectional area of bore, sq. in.	Lb./ft. Red brass	Copper
1/8	0.405	0.281	0.062	0.062	0.253	0.259	2½	2.875	2.501	0.187	4.91	5.99	6.12
1/4	.540	.376	.082	.110	.447	.457	3	3.500	3.062	.219	7.37	8.56	8.75
3/8	.675	.495	.090	.192	.627	.641	3½	4.000	3.500	.250	9.62	11.2	11.4
½	.840	.626	.107	.307	.934	.955	4	4.500	4.000	.250	12.6	12.7	12.9
¾	1.050	.822	.114	.531	1.27	1.30	5	5.562	5.062	.250	20.1	15.8	16.2
1	1.315	1.063	.126	.887	1.78	1.82	6	6.625	6.125	.250	29.5	19.0	19.4
1¼	1.660	1.368	.146	1.47	2.63	2.69	8	8.625	8.001	.312	50.3	30.9	31.6
1½	1.900	1.609	.150	2.01	3.13	3.20	10	10.750	10.020	.365	78.8	45.2	46.2
2	2.375	2.063	.156	3.34	4.12	4.22	12	12.750	12.000	.375	113.	55.3	56.5

B. Dimensions and Weights of Extra Strong Pipe

Nominal pipe size, in.	Nominal dimensions, in. Outside diam.	Inside diam.	Wall Thickness	Cross-sectional area of bore, sq. in.	Lb./ft. Red brass	Copper	Nominal pipe size, in.	Nominal dimensions, in. Outside diam.	Inside diam.	Wall thickness	Cross-sectional area of bore, sq. in.	Lb./ft. Red brass	Copper
1/8	0.405	0.205	0.100	0.033	0.363	0.371	2½	2.875	2.315	0.280	4.21	8.66	8.85
1/4	.540	.294	.123	.068	.611	.625	3	3.500	2.892	.304	6.57	11.6	11.8
3/8	.675	.421	.127	.139	.829	.847	3½	4.000	3.358	.321	8.86	14.1	14.4
½	.840	.542	.149	.231	1.23	1.25	4	4.500	3.818	.341	11.5	16.9	17.3
¾	1.050	.736	.157	.425	1.67	1.71	5	5.562	4.812	.375	18.2	23.2	23.7
1	1.315	.951	.182	.710	2.46	2.51	6	6.625	5.751	.437	26.0	32.2	32.9
1¼	1.660	1.272	.194	1.27	3.39	3.46	8	8.625	7.625	.500	45.7	48.4	49.5
1½	1.900	1.494	.203	1.75	4.10	4.19	10	10.750	9.750	.500	74.7	61.1	62.4
2	2.375	1.933	.221	2.94	5.67	5.80							

C. Weight and Wall Thickness Tolerances

Standard pipe size	Weight per ft. tolerances Plus or minus, %	Wall thickness tolerances to nearest 0.001 in. Minus, %	Plus
Up to 6 in., inclusive	5	5	Limited only by weight tolerances
Over 6–8 in., inclusive	7	7	
Over 8 in.	8	8	

Length tolerances: Standard length 12 ft. plus or minus ½ in.
Bronze pipe weighs 0.96 times copper.
* Copper & Brass Research Association standard TUBE-4, corresponds to the National Bureau of Standards simplified practice recommendations R217-46.

Table 18*a*. Finned Copper Tubes*

Number of fins per in.	Root diameter D Nominal, in.	Actual ± 0.006 in., in.	Wall thickness W ± 0.003 in., in.	Fin height H, in.	Diameter over fins Do ± 3/64 in., in.	Mean fin thickness Fm, in.	Outside area, sq. ft./ft. Ao	Inside area, sq. ft./ft. Ai	Ratio Ao/Ai R	Approx. weight per ft. copper, lb.
4	½	0.516	0.065	21/64	1 11/64	0.034	0.715	0.101	7.08	0.787
6	7/8	.891	.058	21/64	1 35/64	.024	1.49	.203	7.33	1.26
6			.083	21/64	1 35/64	.024	1.49	.190	7.83	1.49
6	1	1.016	.072	21/64	1 43/64	.024	1.65	.228	7.25	1.57
6			.083	21/64	1 43/64	.024	1.65	.222	7.44	1.69
7			.042	21/64	1 11/64	.024	1.15	.113	10.2	0.762
7	½	0.516	.049	21/64	1 11/64	.024	1.15	.109	10.6	.798
7			.065	21/64	1 11/64	.024	1.15	.101	11.4	.876
7			.042	21/64	1 19/64	.024	1.33	.146	9.11	.907
7	5/8	.641	.049	21/64	1 19/64	.024	1.33	.142	9.37	.954
7			.065	21/64	1 19/64	.024	1.33	.134	9.93	1.06
7			.042	21/64	1 27/64	.024	1.52	.179	8.49	1.05
7	¾	.766	.049	21/64	1 27/64	.024	1.52	.175	8.69	1.11
7			.083	21/64	1 27/64	.024	1.52	.157	9.68	1.37

For 9 fins per inch and other fin spacings, consult mill.
In column 3, the actual root diameter is 1/64 in. larger than the nominal in each case. This allowance is to permit stripping of fins from the ends of the tube, resulting in a smooth finish at the nominal size.
Tube ends may be stripped of fins for fitting to female connectors or expanded without stripping for male connectors.
Standard length is 10 ft., but lengths to 30 ft. can be obtained.
* Calumet and Hecla Consolidated Copper Co., Wolverine Tube Division.

Finned copper tubes are available in various designs, with the fins integral with the tube or attached. Table 18*a* shows the dimensions of one integral type.

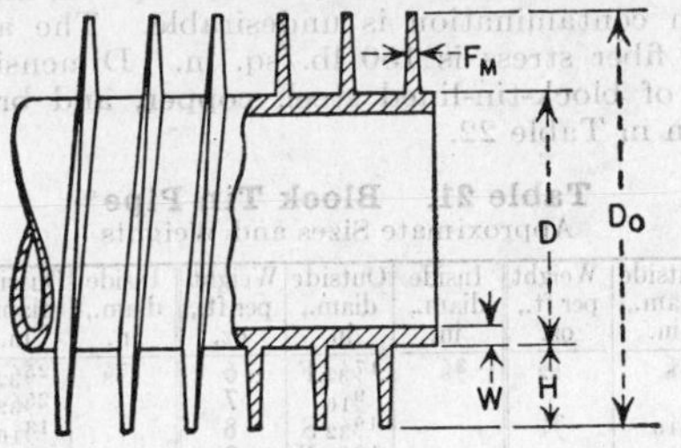

FIG. 65.

Lead and Lead-lined Pipe

Standard lead pipe sizes have been set by the Lead Industries Association as given in Table 19. The National Bureau of Standards has issued Commercial Standard CS95-41 "Lead Pipe," and the Federal government has adopted Federal Specification WW-P-325 for "Pipe, Bends and Traps; Lead (for) Plumbing and Water-distribution." These are identical with the standards of The Lead Industries Association. The nominal size is the actual inside diameter. For ⅜- to 2-in. size, the various wall thicknesses for a given nominal size are classified by different letters for East and West, the dividing line being roughly the Mississippi River. There were originally variations between East and West wall thicknesses, but they have now been eliminated. The West letters represent XL extra light, L light, M medium, S strong, XS extra strong, XXS double extra strong; the East letters are alphabetical. The weights are given in terms of a sp. gr. of 11.34 for acid, chemical, and soft lead. The specific gravity of antimonial lead is as follows:

% antimony	Specific gravity	Multiply weight given in Table 19 by
3	11.1	0.83
4	11.0	.83
6	10.9	.81
8	10.8	.80
10	10.6	.79
12	10.5	.79

The safe working pressures at various operating temperatures may be calculated from the formula: $P = 2ST/D$, where P = safe working pressure, lb./sq. in.; S = maximum allowable fiber stress, lb./sq. in. (Fig. 66), T = thickness of pipe wall, in.; and D = inside diameter of pipe, in.

Lead-lined steel pipe may be used where strength and support are required. Dimensions are given in Table 20 for lined standard steel pipe. Lead-lined brass pipe and cast-iron pipe are also available.

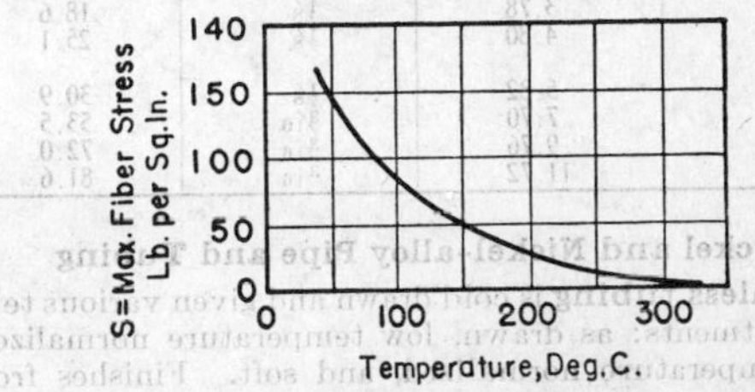

FIG. 66. Maximum allowable working stress for lead pipe.

Magnesium

Magnesium (99.9 per cent) and alloys with aluminum, manganese, and zinc are fabricated into tubing. The

Table 19. Lead Pipe

Size, inside diam., in.	Classification East	Classification West	Out-side diam., in.	Wall thick-ness, in.	Weight per ft., lb.
⅜	D	XL	0.549	0.087	0.63
	C	L	.577	.101	.75
	B	M	.631	.128	1.0
	A	S	.725	.175	1.5
	AA	XS	.811	.218	2.0
	AAA	XXS	.888	.257	2.5
½	D	XL	.666	.083	.75
	C	L	.712	.106	1.0
	B	M	.756	.128	1.3
	A	S	.798	.149	1.5
	AA	XS	.876	.188	2.0
	AAA	XXS	1.012	.256	3.0
⅝	D	XL	.803	.089	1.0
	C	L	.881	.128	1.5
	B	M	.953	.164	2.0
	A	S	1.019	.197	2.5
	AA	XS	1.082	.228	3.0
	AAA	XXS	1.137	.256	3.5
¾	D	XL	.940	.095	1.3
	C	L	1.006	.128	1.8
	B	M	1.068	.159	2.3
	A	S	1.156	.203	3.0
	AA	XS	1.212	.231	3.5
	AAA	XXS	1.336	.293	4.8
1	D	XL	1.232	.116	2.0
	C	L	1.284	.142	2.5
	B	M	1.356	.178	3.3
	A	S	1.428	.214	4.0
	AA	XS	1.492	.246	4.8
	AAA	XXS	1.596	.298	6.0
1¼	D	XL	1.486	.128	2.5
	C	L	1.528	.139	3.0
	B	M	1.592	.171	3.8
	A	S	1.670	.210	4.8
	AA	XS	1.765	.257	6.0
	AAA	XXS	1.889	.319	7.8
1½	D	XL	1.776	.138	3.5
	C	L	1.830	.165	4.3
	B	M	1.882	.191	5.0
	A	S	1.984	.242	6.5
	AA	XS	2.076	.288	8.0
	AAA	XXS	2.272	.386	11.3
1¾	D	XL	2.024	0.137	4.0
	C	L	2.086	.168	5.0
	B	M	2.146	.198	6.0
	A	S	2.193	.221	6.8
	AA	XS	2.404	.327	10.5
	AAA	XXS	2.624	.437	14.8
2	D	XL	2.284	.142	4.8
	C	L	2.354	.177	6.0
	B	M	2.410	.205	7.0
	A	S	2.503	.251	8.8
	AA	XS	2.751	.375	13.8
	AAA	XXS	3.008	.504	19.5
2½			2.75	.125	5.0
			3.00	.250	10.6
3			3.25	.125	6.0
			3.50	.250	12.5
			3.75	.375	20.0
3½			3.75	.125	7.0
			4.00	.250	14.5
4			4.25	.125	7.9
			4.50	.250	16.4
			4.75	.375	25.0
4½			4.75	.125	8.9
			5.00	.250	17.9
5			5.25	.125	9.9
			5.50	.250	20.3
5½			5.75	.125	10.9
			6.00	.250	22.3
6			6.25	.125	11.9
			6.50	.250	24.1
			6.75	.375	37.0
			7.00	.500	50.2
7			7.25	.125	13.8
			7.50	.250	28.0
			7.75	.375	42.8
			8.00	.500	58.0
8			8.25	.125	15.7
			8.50	.250	31.9
			8.75	.375	48.5
			9.00	.500	65.7
10			10.75	.375	60.1
			11.00	.500	81.1
			11.25	.625	102.6
			11.50	.750	124.6
12			12.75	.375	71.7
			13.00	.500	96.6
			13.25	.625	122.0
			13.50	.750	147.8
14			14.75	.375	83.3
			15.00	.500	112.0
			15.25	.625	141.3
			15.50	.750	171.0

outside-diameter range is 3/16 to 8 in. The wall thickness ranges from a minimum of 0.022 in. to a maximum of 0.028 in. for the 3/16 in. diameter; and from a minimum of 0.284 in. to a maximum of 1.0 in. for the 8 in. diameter. Intermediate diameters are available in a wide range of wall thicknesses. The specific gravity of magnesium is 1.74, the melting point 651°C., or 1204°F., the coefficient of thermal expansion between 68° and 932°C. is 0.0000166 in./in./°F., and the thermal conductivity is 1090 B.t.u./(hr.)(sq. ft.)(°F./ft.)

Table 20. Lead-lined Steel Pipe

Nominal pipe size, in.	Inside diam. lining, in.	Thickness of lining, in.	Weight pipe per ft., lb.
½	0.497	1/16	1.5
¾	.699	1/16	2.0
1	.924	1/16	2.7
1¼	1.20	3/32	5.0
1½	1.43	3/32	5.3
2	1.88	3/32	6.8
2½	2.22	⅛	10.0
3	2.82	⅛	13.4
4	3.78	⅛	18.6
5	4.80	⅛	25.1
6	5.82	⅛	30.9
8	7.70	3/16	53.5
10	9.76	3/16	72.0
12	11.72	3/16	81.6

Nickel and Nickel-alloy Pipe and Tubing

Seamless tubing is cold drawn and given various temper treatments: as drawn, low temperature normalized, high temperature normalized, and soft. Finishes from "satin" to "extra smooth" may be produced. Sizes are as follows: outside diameter ranging from ¼ to 5 in., and Stubs' gage (= B.W.G.) from 1 to 23. The sizes greater than 4½ in. are semicold drawn. The tensile strength varies from 60,000 to 105,000 lb./sq. in.

Seamless pipe is made in "iron pipe sizes"; the outside and inside diameters are identical with those of iron pipe; see Table 15. Pipe and tubing are usually tested hydraulically as follows:

Outside diam., in.	Gage	Pressure, lb./sq. in.
¼–1	18–20	400–600
¼–1	14–17	500–700
¼–1	13 and heavier	700–800
Over 1	All	700–1000
⅛–1	Iron pipe sizes	500–700
Over 1	Iron pipe sizes	700–1000

Square seamless tubing is available from ⅝ to 2¾ in. outside dimension in some of the thicknesses from 18 to 12 Stubs' gage.

Tantalum and Columbium Tubing

Welded and seamless tubing is manufactured by the Fansteel Metallurgical Corp. The diameter range of the **welded** type is ½ to 8 in. The wall thickness is 0.010 in. maximum for ½ to 1 in. diameters, and 0.015 in. for 4 to 8 in. diameters. The maximum lengths are 4 ft. for ½ to ⅝ in. diameters; 6 ft. for ¾ to 1 in.; 8 ft. for 1¼ to 2 in.; and 10 ft. above 2 in. The diameter range for the **seamless tubing** is 1/16 to 2 in. The wall thickness is 0.008 to 0.015 in. for 1/16 to ¼ in. diameters, and 0.015 to 0.025 in. for ¼ to 2 in. diameters. Lengths depend on the weight of a piece except in the smaller diameters, the maximum per piece being approximately 1250 g. for Ta and 625 g. for Cb. Extra heavy wall thicknesses can be made specially in certain diameter and length ranges.

Tin Pipe and Tubing

Pure high-grade block tin is made into pipe and tubing in straight lengths, coils, or on reels; the dimensions of the usual sizes are given in Table 21. The longest length of pipe is one weighing 50 lb. Because of the relative inactivity of tin, it is used principally for handling fluids intended for human consumption, and others of which contamination is undesirable. The allowable working fiber stress is 150 lb./sq. in. Dimensions and weights of block-tin-lined steel, copper, and brass pipe are given in Table 22.

Table 21. Block Tin Pipe*
Approximate Sizes and Weights

Inside diam., in.	Outside diam., in.	Weight per ft., oz.	Inside diam., in.	Outside diam., in.	Weight per ft., oz.	Inside diam., in.	Outside diam., in.	Weight per ft., oz.
1/16	⅛	½	⅜	17/32 F	6	⅝	25/32 S	8
				9/16	7		25/32	9
⅛	3/16	¾		19/32 S	8		13/16 S	10
				19/32 F	9		⅞ S	14
3/16	¼ F	1½		⅝	10			
	5/16	2½		21/32	12	¾	⅞	8
							29/32	10
¼	⅜	3	7/16	17/32 F	4		29/32 F	11
	13/32	4		19/32 S	6		15/16 S	12
	7/16	5		⅝	8		31/32 F	16
	15/32 S	6					1	17
	½ S	7	½	19/32 F	4½		1 1/32	20
	½ F	8		⅝ S	5			
				⅝	5½	1	1 3/32 F	9
5/16	7/16 F	4		⅝ F	6		1 5/32 F	14
	½ S	5½		21/32	7		1 3/16	16
	17/32	7½		21/32 F	8		1 3/16 F	18
	17/32 F	8		23/32 S	10		1 7/32 F	20
				¾ S	12		1¼	22
⅜	½ S	4						
	½	4½						
	17/32 S	5						

F = full; S = scant.
* National Lead Co.

Table 22. Block-tin-lined Pipe

Nominal pipe size, in.	Thickness of lining, in.	Tin-lined steel: Inside diam. lined pipe, in.	Tin-lined steel: Total weight per ft., lb.	Tin-lined copper and brass: Inside diam. lined pipe, in.	Total weight per ft.: Copper, lb.	Total weight per ft.: Brass, lb.
½	0.0556	0.511	1.14	0.514	1.25	1.19
¾	.0625	.699	1.56	0.697	1.73	1.68
1	.0625	.924	2.23	0.937	2.39	2.26
1¼	.0625	1.255	3.05	1.243	3.47	3.28
1½	.0781	1.454	3.84	1.444	4.33	4.13
2	.0938	1.880	5.37	1.875	5.95	5.72
2½	.125	2.219	8.59	2.250	8.93	8.63
3	.125	2.818	11.0	2.812	12.2	11.7
4	.125	3.776	15.5	3.750	17.6	17.0
5	.125	4.797	20.5	4.812	22.1	21.3
6	.125	5.815	26.1	5.875	26.6	25.6

Bimetallic Drawn Tubing

This type of tubing is made by drawing the outside tube down to the desired size while maintaining the inner tube as a liner by means of a mandrel. The following copper-base alloys can be combined with other metals or alloys, either inside or outside: admiralty, aluminum brass, aluminum bronze, copper, cupronickel (80-20 and 70-30), Muntz metal, naval brass, red brass (85-15), and yellow brass with other metals; and alloys such as aluminum, monel, nickel, steel (low-carbon), and stainless steel.

Under the trade name "Duplex Tubing," the Bridgeport Brass Co. supplies straight lengths up to 24 ft.; the outside diameters ranging from ⅝ to 2 in. The total wall thickness ranges from 0.049 to 0.080 in. On thinner gages, the wall thickness is generally divided equally between the two components. On heavier gages, however, the wall thickness of the copper-base alloy component may be from one-third to one-fifth of the total wall thickness. "Composite Tubing" combining steel, copper, nickel, brass, aluminum, etc., is available from the Summerill Tubing Co.

Table 23. Asbestos-cement Pressure Pipe*

Size, actual inside diam., in.	Length, ft.	Class 50†			Class 100†			Class 150†			Class 200†		
		Wall thickness, in.	Outside diam., in.	Weight per ft., lb.	Wall thickness, in.	Outside diam., in.	Weight per ft., lb.	Wall thickness, in.	Outside diam., in.	Weight per ft., lb.	Wall thickness, in.	Outside diam., in.	Weight per ft., lb.
3	13	0.33	3.66	3.6	0.35	3.70	3.8	0.44	3.88	4.6	0.60	4.20	6.6
4‡	13	.33	4.66	4.7	.35	4.70	5.0	.45	4.85	6.0	.60	5.20	8.4
6‡	13	.36	6.72	7.6	.38	6 76	7.8	.55	6.95	10.7	.75	7.50	15.4
8‡	13	.42	8.84	11.7	.44	8.88	11.9	.65	9.15	16.8	.88	9.76	23 7
10	13	.44	10.88	15.2	.59	11.18	19.8	.85	11.70	28.0	1.10	12.20	37.0
12	13	.48	12.96	19.8	.68	13.36	27.6	.98	13.96	38.6	1.24	14.48	49.6
14	13	.52	15.04	24.8	.78	15.56	36.6	1.13	16.26	51.6	1.44	16.88	67.0
16	13	.56	17.12	30.6	.88	17.76	47.0	1.25	18.50	65.0	1.65	19.30	87.8
18	13	.59	19.18	35.9	.97	19.94	58.2	1.39	20.78	81.2	1.87	21.74	112.0
20	13	.63	21.26	42.5	1.07	22.14	71.2	1.53	23.06	99.5	2.09	24.18	139.5
24	13	.69	25.38	55.5	1.25	26.50	99.3	1.82	27.64	141.5	2.48	28.96	199.0
30	13	.90	31.80	89.2	1.54	33.08	150.6	2.29	34.58	221.0	3.12	36.24	310.0
36	13	1.09	38.18	126.3	1.83	39.66	211.0	2.80	41.60	318.0	3.74	43.48	435.0

Weights given are exclusive of couplings and fittings.
* Johns-Manville Co.
† Class is equivalent to working pressure in pounds per square inch.
‡ Inside diameters for 4, 6, 8 in. Class 150 are 3.95, 5.85, 7.85 in., respectively.

Table 24. Asbestos-cement Sewer Pipe*

Pipe size, in.	Class 1			Class 2			Class 3			Class 4		
	Wall thickness, in.	Outside diam., in.	Weight per ft., lb.	Wall thickness, in.	Outside diam., in.	Weight per ft., lb.	Wall thickness, in.	Outside diam., in.	Weight per ft., lb.	Wall thickness, in.	Outside diam. in.,	Weight per ft., lb.
4	0.39	4.78	4.9									
5	.41	5.82	6.6									
6	.42	6.84	7.9									
8	.48	8.96	11.9									
10	.50	11.00	15.3	0.56	11.12	17.7	0.65	11.30	21.0			
12	.54	13.08	19.9	.64	13.28	23.6	.74	13.48	28.6			
14	.58	15.16	24.6	.73	15.46	31.0	.84	15.68	37.0			
16	.62	17.24	30.2	.82	17.64	40.6	.94	17.88	47.8			
18	.65	19.30	35.5	.90	19.80	51.0	1.03	20.06	58.0	1.12	20.24	66.0
20	.69	21.38	41.7	.94	21.88	57.5	1.13	22.26	70.0	1.25	22.50	84.0
24	.75	25.50	54.3	1.06	26.12	77.6	1.31	26.62	100.0	1.45	26.90	110.0
30	.96	31.92	86.8	1.24	32.48	113.2	1.64	33.28	155.0	1.85	33.70	175.0
36	1.15	38.30	124.8	1.41	38.82	154.3	1.93	39.86	215.0	2.18	40.36	248.0

Laying length of all sizes of pipe is 13 ft.
* Johns-Manville Co.

Flexible Metal Hose

Various metals such as brass, bronze, monel, aluminum, and steel are used in manufacturing flexible hose which may be had with metal or fabric coverings and with male-female and flanged joints. Threads are usually cut to standard "iron pipe sizes." Inside diameters range from ⅛ to 12 in. The maximum recommended temperature for bronze hose is approximately 450°F., or 230°C.

NON-METALLIC PIPE AND TUBING

A large number of non-metallic materials are manufactured into pipe and tubing. No general size standardization exists. Some materials and dimensions are given below; for further details the manufacturers' catalogues should be consulted. In many cases, fittings and valves are available.

Asbestos-cement Pipe

Asbestos and Portland cement are used to make a seamless pipe, generally with plain ends. Pipe of this material is corrosion-resistant and finds especial application in the conveyance of relatively corrosive fluids. It has a smooth interior surface and, being non-metallic, is free from tuberculation. The joints are usually made with sleeves, sealed with rubber or other elastic rings, or with cement or asphalt. Ends are sometimes machined or tapered for use with metal couplings. A variety of fittings is made. This material can withstand temperatures to 370°C., or 700°F., and has a low thermal conductivity.

Pressure pipe is made in four classes, corresponding to working pressures of 50, 100, 150, and 200 lb./sq. in. See Table 23.

Sewer pipe, as shown in Table 24, is made in four classes, which, in wall thickness, lie between classes 50 and 150 for pressure pipe.

Ducts, as shown in Table 25, are made with wall thicknesses slightly less than class 50 pressure pipe.

Table 25. Asbestos-cement Ducts*

Inside diam., in.	Length, ft.	Wall thickness, in.	Outside diam., in.	Weight† per ft., lb.
3	10	0.32	3.64	3.5
4	10	.32	4.64	4.5
5	10	.35	5.70	5.9
6	10	.35	6.70	7.3
7	10	.40	7.80	8.6
8	13	.40	8.80	11.2
10	13	.40	10.80	13.9
12	13	.45	12.90	18.4
14	13	.45	14.90	18.8
16	13	.50	17.00	23.6
18	13	.50	19.00	26.6
20	13	.55	21.10	32.6
24	13	.60	25.20	42.7
30	13	.67	31.34	59.4
36	13	.75	37.50	79.5

* Johns-Manville Co.
† Weights given are exclusive of couplings and fittings.

Carbon, Graphite, and "Karbate" Pipe and Fittings

These materials are resistant to practically all acids (including hydrofluoric), alkalies, salt solutions, and organic compounds except those of a highly oxidizing character. They combine reasonable strength with light weight and are highly resistant to thermal shock. Their coefficients of expansion are one-fourth to one-fifth that of steel. Carbon is a fairly good insulator, whereas graphite has a thermal conductivity higher than many metals. Carbon and graphite are inherently porous and are slightly permeable to fluids under pressure. "Karbate" materials, produced by National Carbon Co., Inc., are carbon and graphite made impervious to fluids by impregnation with chemically resistant, thermal-setting, synthetic resins. Carbon oxidizes in air at approximately 660°F. and graphite at about 840°F. "Karbate" materials may be used up to 338°F. Recommended maximum operating pressures for "Karbate" pipe and fittings are 75 lb./sq. in. hydrostatic pressure and 50 lb./sq. in. (gage) steam pressure. Graphite and "Karbate" graphite base materials are more widely used than carbon base materials.

Table 26 lists the standard sizes of **pipe**; ½-, ¾-, and ⅞-in. sizes are heat exchanger tubing, and standard fittings are not available for these sizes. Pipe lengths are joined by couplings, flanges, and other fittings made from the same material. Permanent connections are made by threaded joints cemented in place with chemically suitable cements. Table 26*a* shows standard thread details.

Table 26*b* illustrates the standard flanged **connection** and flexible pipe coupling employed with "Karbate" pipe. The flexible coupling simplifies final pipe-length adjustment and permits sufficient pipe movement in any direction to protect against the effects of vibration, misalignment, and differential expansion. Flexlock joints, as shown by Table 26*c*, are used with pipe assemblies requiring considerable flexibility, but submerged in corrosive fluids that prevent the use of the flexible pipe coupling. Dimensions of standard pipe fittings are detailed in Table 26*d*.

Table 26. Standard Sizes of Carbon, Graphite, and "Karbate" Pipe*

Nominal pipe size in.	Inside diam., in.	Outside diam., in.	Wall thickness, in.	Max. length, ft.	Average weight, lb./ft.: Carbon and graphite	Average weight, lb./ft.: "Karbate"	Inside cross-sectional area, sq. ft.	Circumference, ft., or surface sq. ft./ft. of length: Inside	Circumference, ft., or surface sq. ft./ft. of length: Outside
½	½	¾	⅛	6	0.17	0.19	0.00136	0.131	0.196
¾	¾	1	⅛	6	.24	.27	.00307	.196	.262
⅞	⅞	1¼	3/16	9	.44	.48	.00417	.229	.327
1	1	1½	¼	9	.66	.74	.00545	.262	.393
1½	1½	2	¼	9	.97	1.1	.01227	.393	.524
2	2	2¾	⅜	9	1.5	1.7	.0218	.524	.687
2½	2⅜	3	5/16	9	1.8	2.0	.0308	.622	.785
3	3	4	½	9	4.8	5.4	.0491	.785	1.047
4	4	5¼	⅝	9	7.2	8.1	.0873	1.047	1.374
6	6	7½	¾	9	13.8	15.6	.1965	1.571	1.964
8	7¾	9¾	1	6	21.6	24.3	.328	2.029	2.553
10	10	13	1½	6	40.3	45.1	.545	2.618	3.403

* National Carbon Co., Inc.

"Karbate" globe **valves** are produced in 1- and 2-in. sizes, as shown in Table 26*e*. All parts in contact with the fluid are "Karbate" material except the packing.

Table 26*a*. Standard Thread Details—Carbon, Graphite, and "Karbate" Pipe

Pipe size, in.	Dimensions, in.: A*	B	C
1	1	1.490	12
1½	1	1.960	12
2	1¼	2.670	8
2½	1¼	3.000	8
3	1½	3.980	8
4	1½	5.230	8
6	1¾	7.400	8
8	2⅛	9.507	8
10	2⅜	12.500	8

* *A* dimension for threaded couplings, elbows, and tees only.

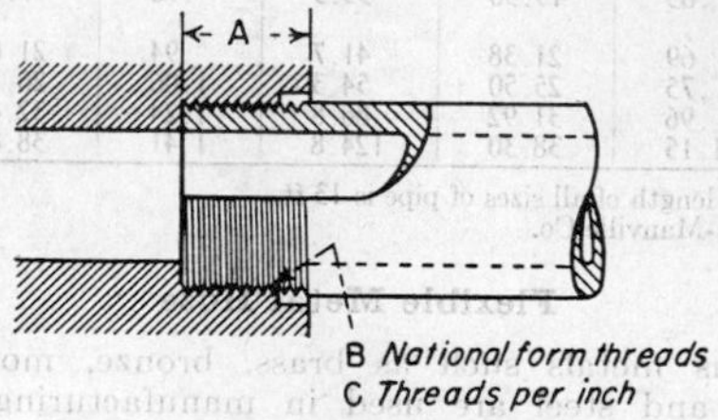

FIG. 67*A*.

Table 26*b*. Standard Flanged Connection and Flexible Pipe Coupling

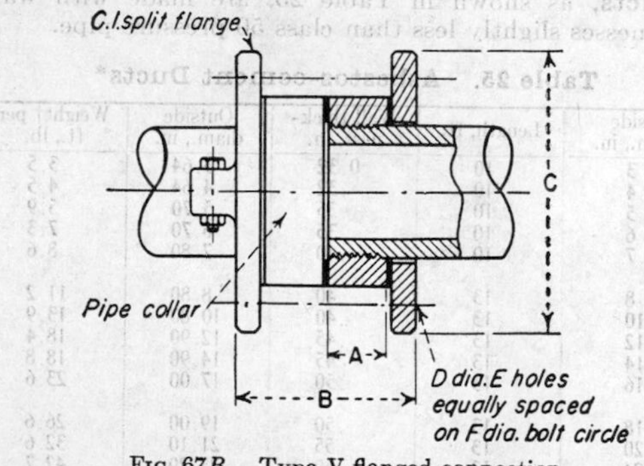

FIG. 67*B*. Type V flanged connection.

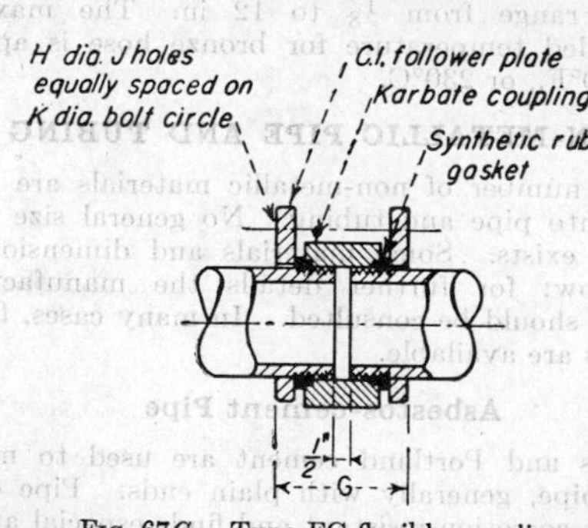

FIG. 67*C*. Type FC flexible coupling.

Pipe size, in.	Dimensions, in.: A	B	C	D	E	F	G	H	J	K
1	1	2 15/16	4⅝	⅝	4	3½	3⅜	⅝	2	3 11/16
1½	1⅛	3 5/16	5	⅝	4	3⅞	4⅜	⅝	2	4¾
2	1¼	3 11/16	6	¾	4	4¾	4⅜	⅝	3	5⅛
2½	1 7/16	4 1/16	7	¾	4	5½	4⅜	⅝	3	5⅞
3	1 9/16	4 7/16	7½	¾	4	6	5⅛	⅝	3	6⅞
4	1¾	4 15/16	9	¾	4	7½	5¼	¾	4	8⅞
6	2	5 11/16	12½	⅞	6	10⅝	5¾	⅞	6	12
8	2 3/16	6 1/16	15	⅞	6	13				
10	2⅜	6 11/16	17½	⅞	8	15¼				

Table 26c. Flexlock Joint for Use with Carbon, Graphite, and "Karbate" Pipe

Pipe size, in.	Dimensions, in.	
	A	*B*
1	1¼	1¾
1½	1¼	2⁵⁄₁₆
2	1½	2³¹⁄₃₂

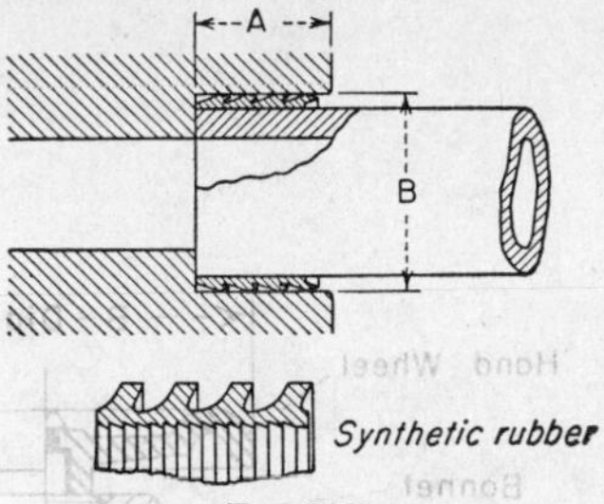

FIG. 67*D*.

Table 26d. Dimensions of Carbon, Graphite, and "Karbate" Pipe Fittings

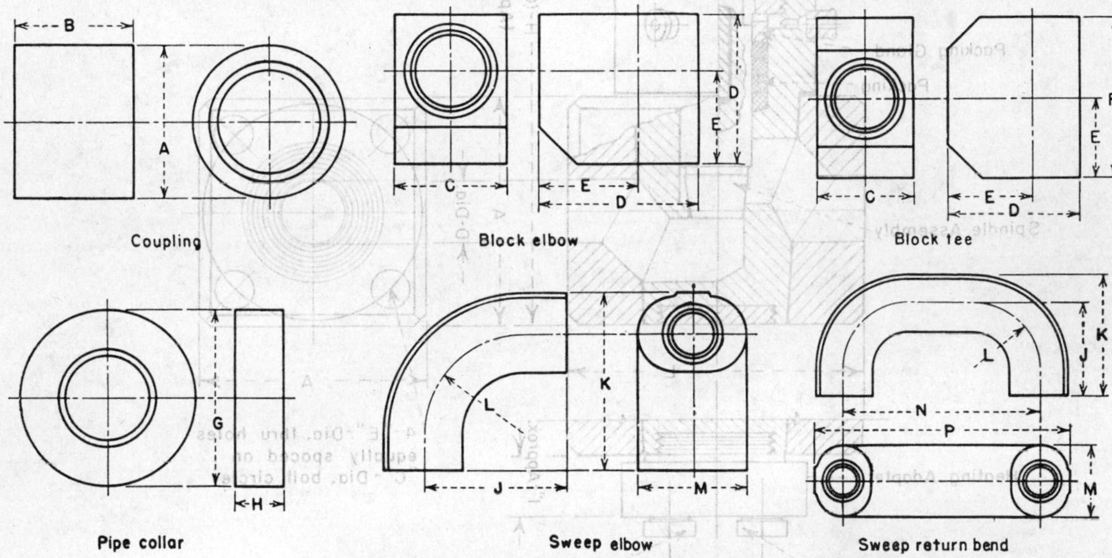

FIG. 67*E*.

Pipe size, in.	Dimensions, in.													
	A	*B*	*C*	*D*	*E*	*F*	*G*	*H*	*J*	*K*	*L*	*M*	*N*	*P*
Threaded joint														
1	2⅛	2	2⅛	2¹⁵⁄₁₆	1⅞	3¾	2⅞	1	3¹¹⁄₁₆	4⅞	2¾	2½	7¾	10⅛
1½	2⅝	2	2⅝	3⁷⁄₁₆	2⅛	4¼	3¼	1⅛	3¹¹⁄₁₆	5	3	3⅛	7¾	10⅜
2	3⅜	2½	3⅜	4⁷⁄₁₆	2¾	5½	4	1¼	4¹³⁄₁₆	6¹¹⁄₁₆	3½	4	10	13¾
2½	3¾	2½	3¾	4¾	2⅞	5¾	4¾	1⁷⁄₁₆						
3	5⅛	3	5⅛	6¹⁄₁₆	3⅝	7¼	5¼	1⁹⁄₁₆						
4	6⅝	3	6⅝	7⁹⁄₁₆	4¼	8½	6¾	1¾						
6	9	3½	9	10⅛	5⅝	11¼	9¾	2						
8	11⅜	4¼	12	18¼	7¼	14½	12⅛	2³⁄₁₆						
10	14⅝	4¾	14¾	16⅜	9	18	14⅜	2⅜						
"Flexlock" joint														
1	2⅜	2½	2⅜	3½	2⁵⁄₁₆	4⅝								
1½	3⅛	2½	3	4⅛	2⅜	5¼								
2	4	3	3¾	5	3⅛	6¼								

Table 26e. Dimensions of "Karbate" Globe Valves

Dimensions in inches

Dim.	1 in. size	2 in. size
A	$3\frac{5}{8}$	$5\frac{1}{2}$
B	$3\frac{5}{8}$	$6\frac{1}{2}$
C	$3\frac{1}{2}$	$5\frac{1}{2}$
D	1	2
E	$\frac{5}{8}$	$\frac{3}{4}$
H	$8\frac{1}{4}$	$12\frac{1}{2}$
L	$4\frac{1}{2}$	7

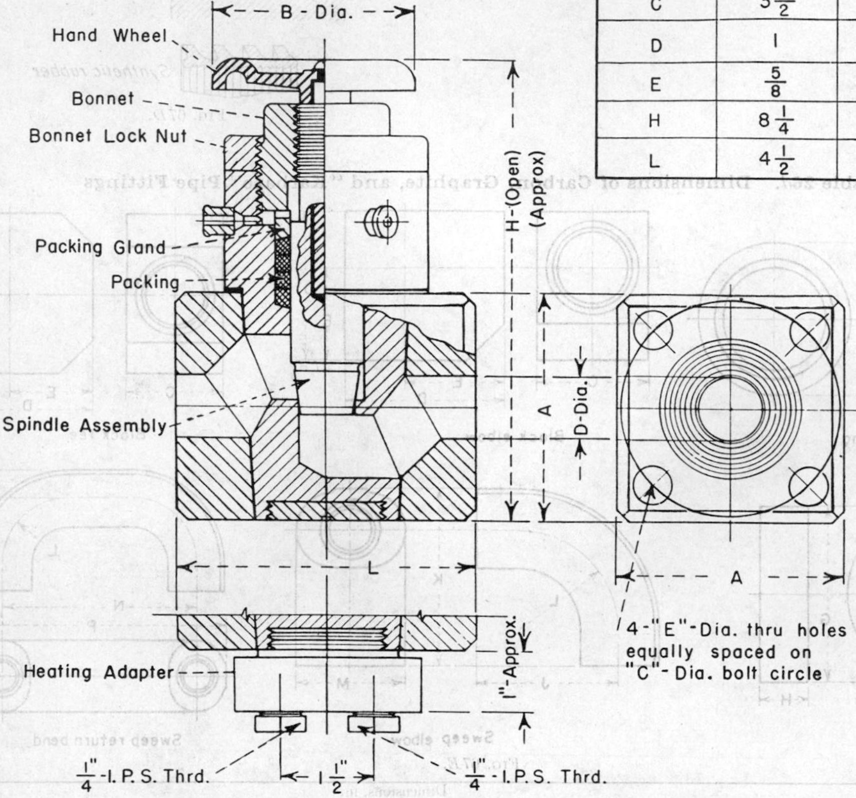

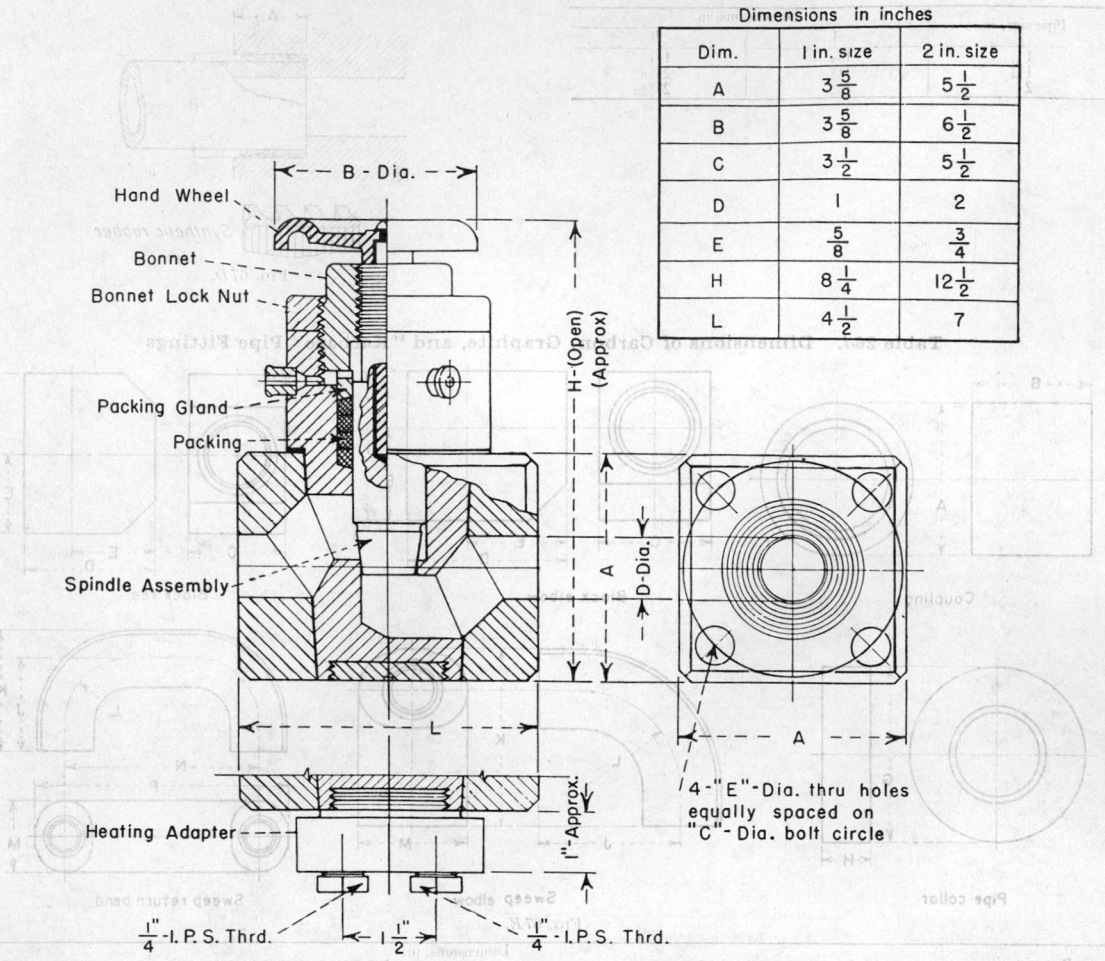

Section Showing Assembly of Heating Adapter

Fig. 6/F.

Cement-lined Pipe

Cement-lined pipe is made by lining iron or steel pipe with special cement. Threaded pipe in sizes from ¾ to 4 in. are stocked. Pipe of larger sizes can be obtained on order. The fittings are generally lead- or tin-lined to avoid chipping the lining in making up a joint and to provide smooth flow through the fitting. The coefficients of expansion of iron and cement are nearly alike. Table 27 gives dimensions of cement-lined pipe. Sizes ¾ to 6 in. are threaded. Larger sizes can be obtained for use with any type of joint.

Table 27. Cement-lined Pipe*

Standard pipe size, in.	Inside diam. after lining, in.	Thickness of lining, in.	Weight, per ft., lb.	Standard pipe size, in.	Inside diam. after lining, in.	Thickness of lining, in.	Weight per ft., lb.
¾	0.70	0.06	1.3	3	2.70	0.13	8.3
1	.90	.07	1.9	4	3.60	.16	12.0
1¼	1.20	.08	2.5	6	5.40	.25	24.0
1½	1.40	.09	3.0	8	7.40	.25	32.0
2	1.80	.10	4.1	10	9.40	.30	43.0
2½	2.20	.10	6.6	12	11.40	.30	55.0

* Cement Lined Pipe Co.

Chemical Ware Pipe

Acidproof chemical stoneware pipe and fittings withstand most acid, alkali, or other corrosives, the main exception being hydrofluoric acid. The widest range of sizes is made with the **bell-and-spigot joint** and with plain butt ends (Table 28), while the **conical flanged joint** can be obtained in the "low-pressure" class in sizes from 1 to 8 in. (Table 29). Pipe with a ring for joining with a steel band (for use in ventilating work where space is too limited for the bell-and-spigot joint) is given in Table 30, with the dimensions of **rectangular** bell-and-spigot pipe. These are in the "medium-pressure" class.

Ordinary chemical stoneware is sensitive to changes of temperature, and pipe of more than 3 or 4 in. inside diameter made of ordinary chemical ware should not be used when temperatures of more than 140°F. are involved. Special "heat-shock-resistant" stoneware (such as

Table 28. Chemical Ware—Bell-and-spigot and Plain Butt-end Pipe

Low pressure*				Medium pressure†					High pressure*				
Inside diam., in.	Outside diam., in.	Wall thick-ness, in.	Working pressure, lb./sq. in.	Inside diam., in.	Outside diam., in.	Wall thick-ness, in.	Max. length, ft.	Weight per ft., lb.	Nominal size, in.	Inside diam., in.	Outside diam., in.	Wall thick-ness, in.	Working pressure, lb./sq. in.
1	1¾	⅜		1	2	½	3	3	1	½	1¾	⅝	
1½	2½	½	30	1½	2½	½	3	4¾	1½	1	2½	¾	60
2	3	½	30	2	3	½	5	5	2	1½	3	¾	60
3	4	½	25	3	4	½	5	7	3	2	4	⅞	50
4	5	½	25	4	5	½	5	10	4	3	5	1	50
5	6¼	⅝	20	5	6¼	⅝	5	12½					
6	7¼	⅝	20	6	7½	¾	6	18	6	5	7¼	1⅛	45
8	9½	¾	20	8	9½	¾	6	23	8	7	9½	1¼	45
9	10¾	⅞		9	11	1	6	27					
10	11¾	⅞		10	12	1	6	32	10	9	11¾	1⅜	
12	13¾	⅞		12	14	1	6	40	12	10¾	13¾	1½	
14	16	1							14	12¾	16	1⅝	
15	17¼	1⅛		15	17	1	5	45					
16	18½	1¼							16	15	18½	1¾	
18	20½	1¼		18	20	1	5	70	18	16¾	20½	1⅞	
20	22¾	1⅜		20	22	1	5	80	20	18¾	22¾	2	
22	25	1½											
24	27	1½		24	26½	1¼	4	95	24	22¾	27	2⅛	
26	29	1½											
				27	29½	1¼	4	110					
28	31¼	1⅝											
30	33¼	1⅝		30	32½	1¼	4	120	30	28¼	33¼	2½	
36	39½	1¾											
42	45¾	1⅞											
48	52	2											

* Standard lengths up to 5 ft. for 1- to 12-in. pipe; above 12 in., 4 ft.; U.S. Stoneware Co.
† Maurice A. Knight Co.

Table 29. Chemical Ware—Flanged Pressure Pipe

Inside diam., in.	Outside diam., in.	Wall thickness, in.	Max. length, ft.	Inside diam., in.	Outside diam., in.	Wall thickness, in.	Max. length, ft.
1	1¾	⅜	3.3	3	4	½	5
1¼	2¼	½	3.3	4	5	½	5
1½	2½	½	4	5	6¼	⅝	5
2	3	½	5	6	7¼	⅝	5
2½	3½	½	5	8	9½	¾	5

Table 30. Chemical Ware—Ventilating Pipe

Round pipe with ring					Rectangular pipe—bell-and-spigot*					Round pipe with ring					Rectangular pipe—bell-and-spigot*				
Actual inside diam., in.	Out-side diam., in.	Wall thick-ness, in.	Max. length, ft.	Weight per ft., lb.	Inside width, in.	Inside length, in.	Wall thick-ness, in.	Max. length, ft.	Weight per ft., lb.	Actual inside diam., in.	Out-side diam., in.	Wall thick-ness, in.	Max. length, ft.	Weight per ft., lb.	Inside width, in.	Inside length, in.	Wall thick-ness, in.	Max. length, ft.	Weight per ft., lb.
6	8	1	5	14	8	12	⅞	5	33	12	14	1	5	34	18	24	1½	5	140
8	10	1	5	20	10	15	1⅛	5	45	15	17½	1¼	5	40	20	30	1⅝	4	165
9	11	1	5	23	12	18	1¼	5	60	18	20¼	1⅛	5	65					
10	12	1	5	26	15	20	1⅜	5	95	20	22½	1¼	5	80					

* Round pipe with bell-and-spigot joint is also made for ventilating work; see Table 28.

U. S. Stoneware Co. "Ceratherm-500," or General Ceramics Co. "SP-22") should be used above 130°F. and especially where pipe lines carry alternatively hot and cold gases or liquids.

Clay Sewer Pipe

The specifications are given in A.S.T.M. Designation C13-35, and the pipe is intended for the conveyance of sewage, industrial wastes, and storm water. The bell-and-spigot joint is used with a hemp or oakum gasket and with cement or other joining materials. Dimensions of pipe are given in Table 31.

Table 31. Clay Sewer Pipe

Inside diam., in.	Outside diam., in.	Wall thickness, in.	Laying lengths, ft.
4	5	½	2
6	7¼	⅝	2, 2½
8	9½	¾	2, 2½, 3
10	11¾	⅞	2, 2½, 3
12	14	1	2, 2½, 3
15	17½	1¼	2, 2½, 3
18	21	1½	2, 2½, 3
21	24½	1¾	2, 2½, 3
24	28	2	2, 2½, 3
27	31½	2¼	2½, 3
30	35	2½	2½, 3
33	38¼	2⅝	2½, 3
36	41½	2¾	2½, 3

Concrete Pipe

Two general classifications are reinforced and plain, and the nominal size is the inside diameter. For use at pressures from 50 to 600 ft. of water, the **steel-cylinder reinforced-concrete pressure pipe** is made in sizes from 16 in. to 12½ ft. and lengths up to 16 ft. A steel cylinder is used with the usual reinforcing bars. The joint is of the expansion type, employing a rubber or lead gasket, sealed in with cement mortar. **Reinforced-concrete pressure pipe** is intended for pressures up to 50 ft. water head and is made by pouring in molds or casting centrifugally. Expansion joints are also used. It is made in sizes from 16 to 108 in. in 12-ft. lengths. For **subaqueous** work the reinforced pipe is made with and without the steel cylinder, in lengths up to 24 ft., and a rubber gasket is used in the joint. Reinforced **culvert** and **sewer** pipes are made in sizes from 12 to 108 in. and vary in wall thickness depending on the strength of concrete used, see A.S.T.M. Designations C76-37 and C75-35, respectively. Bell, slip, and tongue-and-groove joints are used. **Plain sewer pipe** (not reinforced) is made with the bell end and in sizes from 4 to 24 in., A.S.T.M. Designation C14-35. Some manufacturers fabricate the concrete pipes to more stringent specifications than the A.S.T.M. calls for. Concrete pipe can be **lined** with special salt-glazed vitrified-clay liner plates, joined with a die-cast asphalt joint.

Glass Pipe and Tubing

Glass pipe and tubing are made from heat- and chemical-resistant borosilicate glass (*e.g.*, Corning Glass Works No. 774). This glass is highly stable in acids and resists attack by alkalies in solutions where the pH is 8 or less. It is attacked by hydrofluoric and glacial phosphoric acids. Some important physical properties are:

Linear coefficient of thermal expansion, in./in./°F. between 32° and 570°F.	0.00000184
Modulus of elasticity, lb./sq. in.	9,750,000
Specific gravity	2.23
Specific heat	0.20
Thermal conductivity at 75°F., B.t.u./(hr.) (sq. ft.) (°F./in. of thickness)	8.1

Conical flanged glass pipe is made in the sizes shown in Table 32 and in lengths from 6 in. to 10 ft. The normal working pressure is 50 lb./sq. in. A complete line of fittings is available, and special parts are made to order.

Conical flanged heat exchanger tubes are intended for use in drip-type or metal-jacketed return bend heat exchangers. Tube lengths from 4 to 10 ft., stuffing boxes, and glass return bends are available. Sizes, dimensions, and working pressures are shown in Table 32.

Bell-and-spigot pipe is used where large-diameter corrosion-resistant pipe is required for low-pressure gas lines, fume ducts, or packed columns. Dimensions are shown in Table 32.

Industrial glass tubing is available in medium and heavy wall in stock lengths of 60 in. ± 1/4 in. Dimensions are shown in Table 32.

Tubular gage glasses are of three types: (1) Low pressure, suitable for 200 lb./sq. in. in 10-in. length to 100 lb./sq. in. in 70-in. length, are made in nominal outside diameter sizes 1/2, 5/8, and 3/4 in. (2) High pressure and temperature, suitable for 500 lb./sq. in. in 10-in. length to 140 lb./sq. in. in 70-in. length, are made in nominal outside diameters from 1/2 to 1 1/2 in. (3) Heavy wall, suitable for 600 lb./sq. in. in 5/8-, 3/4-, and 7/8-in. sizes up to 30-in. length, and for 650 lb./sq. in. in 1-in. size.

FIG. 68. Ball-and-socket joint.

A **ball-and-socket joint** shown in Fig. 68 is manufactured for use with glass tubing to secure flexibility and ease of assembly. The joint is leakproof and interchangeable because of the precision grinding.

Glass-lined Pipe

Glass-lined steel pipe can be used at temperatures up to 600°F., provided that there are no excessive sudden temperature changes, and at pressures up to 75 lb./sq. in. for standard pipe and 300 lb./sq. in. for special pipe. The glass lining is usually 1/32 in. thick. Dimensions of 1 1/2- to 10-in. pipe are shown in Table 33. Larger sizes can be made to order. Complete fittings are available as glass-lined castings for pressures up to 75 lb./sq. in. and as glass-lined steel for higher pressures.

Table 32. Glass Pipe and Tubing*

Pipe, Conical Flanged Joint
(See Fig. 69*A*)

Inside diam., in.	Outside diam., in.	Wall thickness, in.	Weight per ft., lb.
1	1.31	0.156	0.55
1.5	1.84	.171	.87
2	2.34	.171	1.13
3	3.41	.202	1.97
4	4.50	.264	3.41

Heat Exchanger Tubing, Conical Flanged Joint

Nominal size	Inside diam., in.	Outside diam., in.	Wall thickness, in.	Internal cross-sectional area, sq. in.	Mean surface area, sq. ft./ft.	Max. internal pressure, lb./sq. in.
1	1.19	1.31	0.062	1.11	0.33	25
1.5	1.72	1.84	.062	2.32	.47	25
2	2.22	2.34	.062	3.87	.60	25
3	3.23	3.41	.090	8.19	.87	20
4	4.25	4.50	.125	14.19	1.14	15

Pipe, Bell-and-spigot Joint
Dimensions in inches
(See Fig. 69*B*)

Nominal size	A	B	C	D	E	F nominal
6 1/2	6 1/2	8 5/8	36	33	3	3/16–7/16
9	9 1/8	11 1/2	42	39	3	3/16–7/16
12	12	14 1/8	30	27	3	3/16–7/16
16	16	18	27	24	3	3/16–7/16
18	18	20	24	21	3	3/16–7/16

Industrial Tubing, Plain Ends

Outside diameter, in.	Medium wall		Heavy wall	
	Wall thickness, in.	Weight per ft., lb.	Wall thickness, in.	Weight per ft., lb.
1/4	3/64	0.029		
1/2	1/16	.083	3/32	0.11
3/4	1/16	.131	1/8	.24
1	3/32	.26	5/32	.40
1 1/4	3/32	.33	5/32	.53
1 1/2	3/32	.41	5/32	.64
1 3/4	3/32	.48	5/32	.76
2	1/8	.72	3/16	1.05
2 1/4	1/8	.82	3/16	1.19
2 1/2	1/8	.91	3/16	1.33
2 3/4	1/8	1.01	3/16	1.48
3	1/8	1.10	3/16	1.60
3 1/4	1/8	1.20	3/16	1.77
3 1/2	1/8	1.30	3/16	1.91
4	3/16	2.17	1/4	2.85
4 1/2	3/16	2.45	1/4	3.20

* Corning Glass Works.

Table 33. Glass-lined Steel Pipe*

Size, in.	Inside diam., in.	Outside diam., in.	Max. length, ft.	Weight per ft., lb.
1 1/2	1.500	1.875	3	3.38
2	2.000	2.375	10	4.38
3	3.000	3.500	10	8.68
4	4.000	4.500	10	11.35
5	5.000	5.500	10	14.0
6	6.000	6.625	10	21.1
8	7.981	8.625	10	27.7
10	10.020	10.75	10	41.6

* Pfaudler Co.

Plastic and Fiber-plastic Tubing

Haveg pipe and fittings are manufactured of acid-digested asbestos and either a phenol-formaldehyde resin or a furane resin. These materials have a low thermal and electrical conductivity and can be used to 130°C., or 265°F., with complete resistance to thermal shock. The material, termed Haveg "41," is highly resistant to most acidic chemicals, especially hydrochloric acid and chlorine, and the material Haveg "60" is very resistant to most acids and bases, interchangeably, as well as to many hydrocarbons, halogenated organic compounds, and organic acids. They are not resistant, however, to oxidizing acids such as nitric and

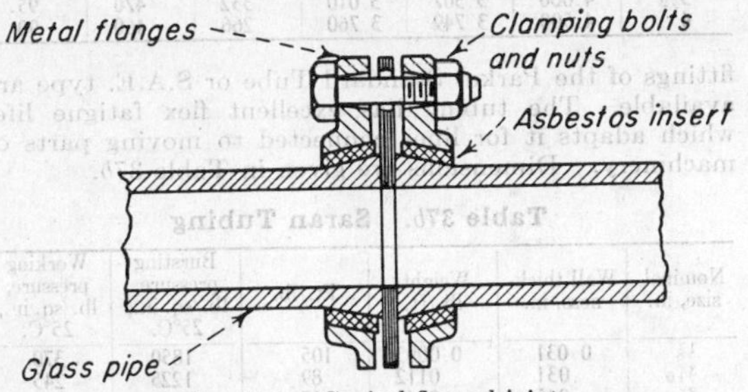

FIG. 69A. Conical flanged joint.

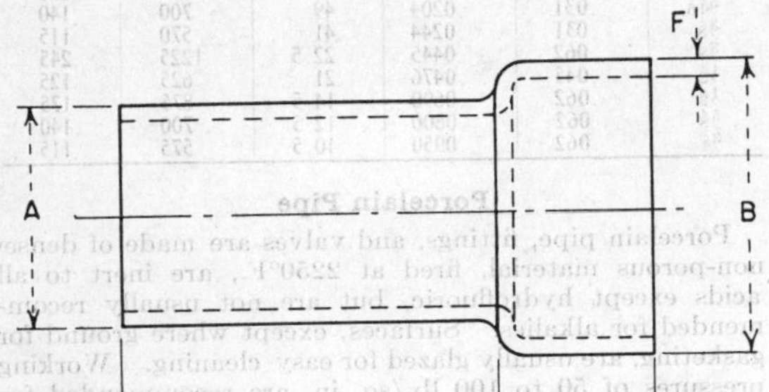

FIG. 69B. Bell-and-spigot joint.

strong sulfuric or to strong sodium hypochlorite. For use with hydrofluoric and related chemicals, Haveg "43" and Haveg "63," which have a carbonaceous filler instead of asbestos, are available. The pipe is made in two types, for liquids and for gases. Table 34 gives the dimensions.

The pipe and fittings for liquids are made with joints of several types, butt joints using split cast-iron flanges in tapered slots, and acme threaded sleeve connections. Hydraulic tests indicate the following pressures can be used safely at 100°C.: 1 to 2 in., 65 lb./sq. in.; 2½ to 5 in., 50 lb./sq. in.; 6 to 12 in., 35 lb./sq. in. Maximum lengths are 4 ft. for ½- and ¾-in. pipe, and 10 ft. for 1-in. pipe and larger. The outside diameter may vary +1⁄16 to +⅛ in.

The fume duct for gases is supplied with a butt joint and split flange connection or a bell-and-spigot joint, in lengths of 10 ft. maximum. The valves, pumps, fans, and related equipment can also be fabricated of Haveg.

Transparent plastics are available as tubing, one example being methyl methacrylate, or Lucite. Dimensions are given in Table 35. The standard length is 52 in.; all tubes are smooth on the inside and can be obtained polished on the outside at a higher price. The working pressures are about 15 per cent of those for ordinary wrought-iron pipe. The resin is resistant to

Table 34. Haveg Pipe*

Pipe for liquids			Fume duct			
Size, actual inside diam., in.	Outside diam., in.	Wall thickness, in.	Size, in.	Actual inside diam., in.	Outside diam., in.	Wall thickness, in.
½	1¼	⅜	2	2	2¾	⅜
¾	1½	⅜	3	3	3¾	⅜
1	1¾	⅜	4	4	3¾	⅜
1¼	2	⅜	5	5	6	½
1½	2½	½	6	6	7	½
2	3	½	8	8	9	½
2½	3½	½	10	10	11	½
3	4¼	⅝	12	12	13	½
3½	4¾	⅝	14	14½	15½	½
4	5¼	⅝	16	16½	17½	½
5	6½	¾	18	18¼	19¼	½
6	7½	¾	20	20½	21½	½
8	9¾	⅝	24	24	25¼	⅝
10	12	1	30	30¼	31½	⅝
12	14	1	36	36	37¼	⅝

* Haveg Corp.

hydrochloric acid and 50 per cent sulfuric acid at room temperature and to fruit juices, milk, vegetable extracts, sugars, gasoline (except benzol gasolines), naphtha, and neutral salt solutions. It is not resistant to concentrated sulfuric, nitric, hydrofluoric, and organic acids and to alcohols, esters, ethers, ketones, aromatic hydrocarbons, alkali, or strong ammonia.

Table 35. Lucite Tubing*

Outside diam., in.	Wall thickness, in.	Inside diam., in.	Weight per ft., lb.	Outside diam., in.	Wall thickness, in.	Inside diam., in.	Weight per ft., lb.
1½	⅛	1¼	0.28	3½	⅛	3¼	0.71
	3⁄16	1⅛	.40		3⁄16	3⅛	1.00
1⅝	⅛	1⅜	.30		¼	3	1.31
	3⁄16	1¼	.43		⅜	2¾	1.88
	¼	1⅛	.55	3¾	⅛	3½	0.75
1¾	⅛	1½	.33		3⁄16	3⅜	1.09
	3⁄16	1⅜	.46		¼	3¼	1.41
	¼	1¼	.60		⅜	3	2.04
2	⅛	1¾	.38	4	⅛	3¾	0.80
	3⁄16	1⅝	.55		3⁄16	3⅝	1.20
	¼	1½	.71		¼	3½	1.50
	⅜	1¼	.98		⅜	3¼	2.18
2¼	⅛	2	.43	4¼	⅛	4	0.83
	3⁄16	1⅞	.62		3⁄16	3⅞	1.23
	¼	1¾	.80		¼	3¾	1.60
	⅜	1½	1.13		⅜	3½	2.36
2½	⅛	2¼	0.48	4½	⅛	4¼	0.88
	3⁄16	2⅛	.71		3⁄16	4⅛	1.31
	¼	2	.91		¼	4	1.69
	⅜	1¾	1.28		⅜	3¾	2.50
2¾	⅛	2½	0.52	4¾	⅛	4½	0.94
	3⁄16	2⅜	.78		3⁄16	4⅜	1.65
	¼	2¼	1.00		¼	4¼	1.82
	⅜	2	1.19		⅜	4	2.67
2⅞	⅛	2⅝	0.55	5	⅛	4¾	0.98
	3⁄16	2½	.84		3⁄16	4⅝	1.45
	¼	2⅜	1.05		¼	4½	1.91
	⅜	2⅛	1.50		⅜	4¼	2.80
3	⅛	2¾	0.58	5½	⅛	5¼	1.08
	3⁄16	2⅝	.85		3⁄16	5⅛	1.60
	¼	2½	1.10		¼	5	2.11
	⅜	2¼	1.58		⅜	4¾	3.08
3⅛	⅛	2⅞	0.60	6	⅛	5¾	1.18
	3⁄16	2¾	.88		3⁄16	5⅝	1.77
	¼	2⅝	1.15		¼	5½	2.31
	⅜	2⅜	1.67		⅜	5¼	3.43
3¼	⅛	3	0.63	6 3⁄32	⅛	5 27⁄32	1.20
	3⁄16	2⅞	.92		3⁄16	5 23⁄32	1.82
	¼	2¾	1.21		¼	5 19⁄32	2.36
	⅜	2½	1.74		⅜	5 11⁄32	3.53

* E. I. duPont de Nemours & Co.

Laminated phenolic tubing is made in two types, rolled and molded. It is made by treating a base material such as paper or other fabric with a phenolic varnish and then winding on a mandrel. The rolled are oven-baked, and the molded are cured in molds under heat and pressure. The "Laminated Phenolic Products Standards," a publication of the National Electrical Manufacturers Association, New York, N.Y., defines nine main grades (individual manufacturers may offer special grades) and gives the dimensions shown in Table 36. Threaded pipe is made with outside diameters in "iron pipe sizes" (see Table 4) from 1/8 to 8 in. and larger. The inside diameters may be as much as 1/64 in. less than those for iron pipe of Schedules 80 and 160, since tubing mandrels (Table 36) are generally used; this results in the wall thickness being equal to, or greater than, that for iron pipe. The average fiber stress of the pipe and tubing is 8400 lb./sq. in., and the maximum advisable temperature is from 250° to 300°F. These materials are in general resistant to weak acids, salt solutions, oils, and organic solvents. They are attacked by strong oxidizing acids and alkalies. Any material that will attack paper or cloth will eventually cause the deterioration of paper- or cloth-base phenolic tubing. Fittings and plug cocks are available.

Table 36. Laminated Phenolic Tubing

Grade	Range of standard sizes, in.							
	Rolled				Molded			
	Inside diam.		Outside diam.		Inside diam.		Outside diam.	
	Min.	Max.	Min.	Max.	Min.	Max.	Min.	Max.
X	1/8	48	3/16	50	1/8	3 7/8	1/4	4
XX	1/8	48	3/16	50	1/8	3 7/8	1/4	4
XXX	....	..	....	..	1/8	3 7/8	1/4	4
C	3/8	48	1/2	50				
CE	....	..	....	..	1/4	3 7/8	3/8	4
L	....	..	....	..	1/8	3 7/8	3/16	4
LE	3/16	48	1/4	50	1/8	3 7/8	3/16	4
A	3/16	48	5/16	50	1/8	3 7/8	3/16	4
AA	1	48	1 1/4	50	3/8	3 3/4	5/8	4

Standard Steps in Diameters

Nominal inside and outside diameter, in.		
From	Up to and including	By steps of
1/8	1	1/32
1 1/16	3	1/16
3 1/8	6	1/8
6 1/4	8	1/4
8	25	1/2

Saran pipe is made from polyvinylidene chloride thermoplastic resin. It is most advantageous in applications utilizing its excellent chemical resistance. It is resistant to salts, strong acids, and strong alkalies with the exception of ammonium hydroxide. It is also resistant to organic oils and solvents with the exception of a few ethers such as cyclohexanone and dioxane.

The top temperature limit for Saran pipe is approximately 170°F. for continuous operation. At temperatures below 30° to 40°F. it should be protected from sharp impact because of its lowered impact strength at these temperatures. Saran pipe is classified as hydraulically smooth, the same as seamless drawn brass, lead glass, and similar materials.

The pipe may be threaded with conventional dies, and molded fittings including caps, couplings, elbows, tees, flanges, and reducers are available in sizes from 1/2 through 2 in. In addition, the pipe may be easily heat-welded by use of a hot iron or hot air. Therefore, welded fittings are also easily made. Nominal pipe sizes from 1/2 through 4 in. are available; see Table 37*a* for dimensions. Valves with all working parts made of Saran are in the developmental stage.

Saran is also available in the form of **tubing**, which is quite flexible and may be heat-formed easily. Its properties are the same as those of Saran pipe. Molded fittings of the Parker Standard Tube or S.A.E. type are available. The tubing has excellent flex fatigue life, which adapts it for lines connected to moving parts of machinery. Dimensions are given in Table 37*b*.

Table 37*a*. Saran Pipe

Nominal size, in.	Outside diam., in.	Inside diam., in.	Weight, lb./ft.	Ft./lb.	Bursting pressure lb./sq. in., 25°C.	Working pressure, lb./sq. in.
1/2	0.840	0.546	0.236	4.23	1300	260
3/4	1.050	.742	.320	3.12	1060	210
1	1.315	.957	.475	2.12	970	190
1 1/4	1.660	1.278	.650	1.550	820	160
1 1/2	1.900	1.500	.790	1.270	740	150
2	2.375	1.939	1.090	0.920	620	125
2 1/2	2.875	2.277	1.805	.554	570	115
3	3.500	2.842	2.480	.403	510	105
3 1/2	4.000	3.307	3.010	.332	470	95
4	4.500	3.749	3.760	.266	460	90

Table 37*b*. Saran Tubing

Nominal size, in.	Wall thickness, in.	Weight, lb./ft.	Ft./lb.	Bursting pressure, lb./sq. in., 25°C.	Working pressure, lb./sq. in., 25°C.
1/8	0.031	0.0095	105	1850	370
3/16	.031	.0112	89	1225	245
3/16	.045	.0149	67	1850	370
1/4	.031	.0156	64	875	175
1/4	.045	.0212	47	1350	270
5/16	.031	.0204	49	700	140
3/8	.031	.0244	41	570	115
3/8	.062	.0445	22.5	1225	245
1/2	.045	.0476	21	625	125
1/2	.062	.0690	14.5	875	175
5/8	.062	.0800	12.5	700	140
3/4	.062	.0950	10.5	575	115

Porcelain Pipe

Porcelain pipe, fittings, and valves are made of dense, non-porous material, fired at 2250°F., are inert to all acids except hydrofluoric, but are not usually recommended for alkalies. Surfaces, except where ground for gasketing, are usually glazed for easy cleaning. Working pressures of 50 to 100 lb./sq. in. are recommended for valves and piping. Temperatures of 400°F. or more can be used, but sudden thermal shocks must be avoided. Metal flanges (125 lb. A.S.M.E.) are permanently cemented to the outside of the porcelain (see Figs. 70*A* and *B* for pipe joint and valve). Table 38 lists standard pipe dimensions.

Table 38. Porcelain Pipe*

Inside diam., in.	Outside diam., in.	Wall thickness, in.	Weight per ft., lb.
1/2	1 1/2	1/2	2.10
1	2 3/16	19/32	3.75
1 1/2	2 13/16	21/32	5.50
2	3 3/8	11/16	7.35
3	4 1/2	3/4	11.4
4	5 5/8	13/16	16.4
6	7 7/8	15/16	24.4
8	10	1	36.0

Standard length is 60 in.
* Lapp Insulator Co., Inc.

Rubber and Rubber-lined Pipe

Hard-rubber pipe and fittings are made in two weights, standard for working pressures to 50 lb./sq. in., and extra heavy for working pressures up to 80 lb./sq. in. Dimensions and weights are given in Table 39. Standard lengths are 10 ft. The standard pipe is made screwed or flanged, and the extra heavy with extra heavy flanges. The limiting temperature is 120°F., or 50°C. The thermal expansion of hard-rubber pipe is 0.00004 in./in./°F. Hard rubber is resistant to many chemicals, including acetic acid, acetone, alum solutions, ammonium hydroxide (ammonia), chlorine solutions, copper sulfate, ferric chloride, ferric sulfate, formic acid, hydrobromic acid, hydrochloric acid, hydrofluoric acid, hydrofluosilicic

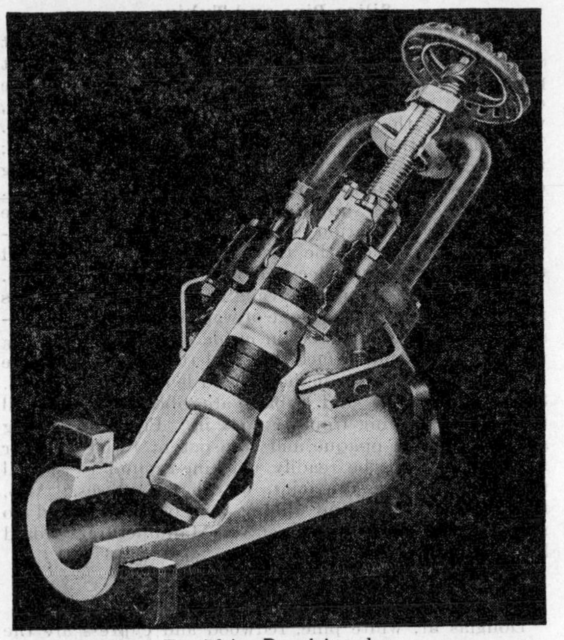

Fig. 70*A*. Porcelain valve.

acid, mercuric chloride, nickel sulfate, nitric acid up to 16°Bé., oxalic acid, phosphoric acid up to 75 per cent, potassium hydroxide, sodium acid sulfate, sodium borate, sodium chloride (brine), sodium hydroxide, sodium hypochlorite, sodium thiosulfate (hypo), sodium sulfide, stannic chloride, sulfurous acid, sulfuric acid up to 50° Bé., zinc chloride, and zinc sulfate. It is not resistant to such rubber solvents as ether, benzol, and aniline.

Table 39*a*. Hard-rubber Pipe*

Nominal diam., in.	Outside diam., in.	Standard			Extra heavy		
		Inside diam., in.	Wall thickness, in.	Weight per ft. lb.	Inside diam., in.	Wall thickness, in.	Weight per ft., lb.
¼	0.540	0.250	0.145	0.09			
⅜	.675	.375	.150	.14			
½	.840	.500	.170	.17			
¾	1.050	.687	.181	.26			
1	1.315	.937	.189	.34	0.750	0.282	0.54
1¼	1.660	1.250	.205	.54	1.000	.330	.82
1½	1.900	1.437	.231	.64	1.250	.325	.95
2	2.375	1.875	.250	.88	1.500	.437	1.58
2½	2.875	2.250	.312	1.13	1.875	.500	2.21
3	3.500	2.750	.375	1.99	2.250	.625	3.34
4	4.500	3.750	.375	2.88	3.000	.750	5.22
5	5.563	4.750	.406	3.90			
6	6.625	5.625	.500	5.20	4.875	.875	9.32
8	8.625	7.000	.812	10.80			

* American Hard Rubber Co.

Rubber-lined pipe is made with wrought steel, wrought iron, cast iron, and some types of spiral-welded steel pipe, using various types of natural and synthetic adhering rubber. The type of rubber is selected to provide the most suitable lining for the specific service. In general, soft rubber is used for abrasion resistance, semihard for general service, and hard for the more severe service conditions. A three-ply combination of soft-hard-soft (B. F. Goodrich Co. "Triflex") is available in sizes over 6 in. diameter for such severe service as is encountered in hot pickling of steel. Temperature limitations range from 150° to 185°F.

The **thickness** of lining ranges from ⅛ to ¼ in.

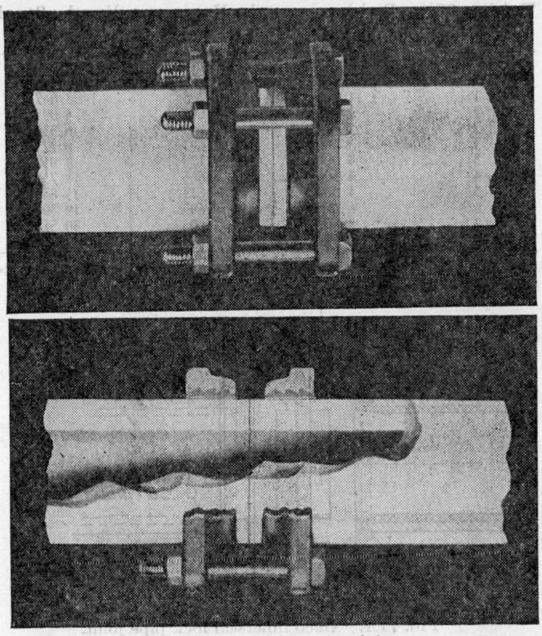

Fig. 70*B*. Porcelain pipe joints.

depending upon the service, the type of rubber, and the method of lining.

The maximum **lengths** of lined pipe, as reported by B. F. Goodrich Co., are 20 ft. for 1¼ in. and larger diameters, except 12 ft. for 3⁄16 and ¼ in. thickness in sizes under 3 in. Gates Engineering Co. reports standard steel pipe in sizes from 1½ to 10 in. lined with 3⁄16-in. hard rubber or Neoprene in lengths of 20 ft. (see Table 39*b*).

Joints are made in several ways. On flanged pipe, the lining is extended out over the flange face. On the Flexlock joint, the lining is extended out over the end of the pipe and carried back about 5 in. on the outside; the joint is then coupled with Flexlock rubber gaskets and rubber-lined steel or iron split sleeves, affording a flexible joint with respect to misalignment and to expansion and contraction.

A **new type of joint** is shown in Figs. 71*A*, *B*, and *C* for use on rubber-lined pipe of either hard or soft type (see Table 39*b*). The lined pipe is recessed at each end to receive a key ring (*B*) in two halves. The unlined flange (*A*) seats against the key ring. An insert of hard rubber (*CC*) and soft rubber (*C*) serves as the gasket or seal between the ends of the rubber lining in the pipe. The standard 20-ft. lengths of lined pipe are initially recessed for immediate joint assembly. Shorter lengths for final fit-up at point of use can be cut and recessed on the job, thus eliminating special factory fabrication from detailed piping drawings such as that required with rubber-lined flange and other types of joint for rubber-lined pipe. The joint is designed for 1200 lb./sq. in. hydrostatic pressure.

Fittings and valves for rubber-lined service are available in standard, specially cast, or fabricated flanged fittings or other types of joint.

Rubber-fabric Hose

Flexible rubber hose is made to resist internal or external pressure by reinforcing with fabric or metal. Many types and strengths are available; inside diameters range from ½ to 6 in. and in some types to 12 in.

Table 39b. Rubber- and Neoprene-lined Steel Pipe*

Standard steel pipe size, in.	Outside diam. steel pipe, in.	Inside diam. rubber lining, in.	Nominal weight per ft., lb.
1½	1.900	1.225	3.047
2	2.375	1.695	4.095
2½	2.875	2.095	6.336
3	3.500	2.695	8.275
3½	4.000	3.175	9.950
4	4.500	3.665	11.756
5	5.563	4.675	15.854
6	6.625	5.695	19.398
8	8.625	7.695	30.566
10	10.750	9.815	33.804

Standard lengths 20 ft., machined for use with "Gaco Inner-seal Lock Pipe Joint" (Fig. 71).
* Gates Engineering Co.

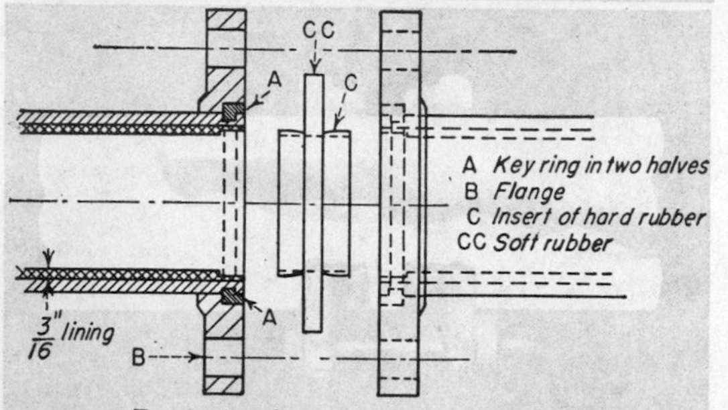

Fig. 71*A*. Gaco innerseal lock pipe joint.

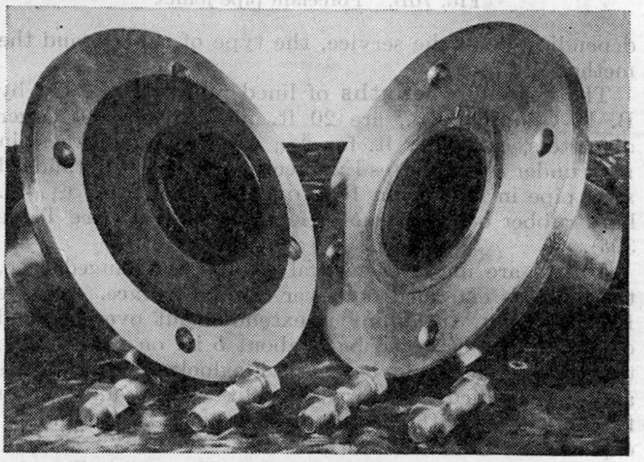

Fig. 71*B*. Gaco innerseal pipe joint.

Fig. 71*C*. Gaco innerseal joint.

Silica Pipe and Tubing

Fused silica or fused quartz, containing 99.8 per cent silicon dioxide, can be obtained as opaque or transparent pipe and tubing. Its melting point is 1710°C. The coefficient of expansion is 5.4×10^{-7} cm./cm./°C., which is ⅙ that of Pyrex or $\frac{1}{17}$ that of platinum. The tensile strength is approximately 7000 lb./sq. in.; the specific gravity is about 2.2. The pipe and tubing can be used continuously at temperatures up to 1000°C. and intermittently up to 1500°C. Its chief assets are the non-contamination of most chemicals in high-temperature service, the thermal-shock resistance, and the high-temperature electrical insulating characteristics.

Transparent tubing is available in inside diameters from 1 to 125 mm., in a range of wall thicknesses. Satin-surface tubing is available in inside diameters from $\frac{1}{16}$ to 2 in., and sand-surface pipe and tubing are available in ½ to 24 in. inside diameters and lengths up to 20 ft. Sand-surface pipe and tubing are obtainable in wall thicknesses varying from ⅛ to 1 in. Pipe and tubing sections in both opaque and transparent fused silica or fused quartz can be readily machine-ground to special tolerances for pressure joints or other purposes. Also, fused silica piping and tubing can be reprocessed to meet special-design requirements. Manufacturers should be consulted for specific details.

Wood Pipe

Douglas fir, white pine, redwood and cypress are the most common woods used for wood pipe. Dimensions are given in Table 40. The **wood-lined steel pipe** is suitable for temperatures up to 180°F., or 82°C., and for pressures from 200 lb./sq. in. for the 4-in. size, through 125 lb./sq. in. for the 10-in. size, to 100 lb./sq. in. for larger than 10 in. For **fume stacks** and similar uses the wood stave pipe with rods on 1-ft. centers is most satisfactory because it permits periodic tightening.

For underground installation, **machine-banded wood stave pipe** is generally suitable. The staves have double tongues and grooves on lateral edges and are wound with 1 in. wide steel strip of adequate gage and spacing to provide four pressure classes, 43, 86, 130, and 172 lb./sq. in., although it is recommended that the working pressure be limited to 60 per cent of the class pressure. The pipe is usually coated heavily with asphalt and sawdust. Lengths are random and up to 16 ft., and joints are mortise and tenon. Galvanized

Table 40. Wood Pipe*

Wood-lined steel pipe			Machine-banded wood stave pipe				Square wood pipe	
Inside diam., in.	Outside diam. in.	Usual metal wall thickness, B. & S. gage	Inside diam., in.	Outside† diam., in.	Wood wall thickness, in.	Weight per ft., lb.‡	Inside diam., in.	Outside dimensions, in.
4	5⅝	14	2	5	1½	8	2	4 × 4
5	6⅝	14	3	6	1½	10	2	5 × 5
6	7⅝	14	4	7¼	1⅝	13	3	6 × 6
8	10	14	6	9½	1¾	19	3	7 × 7
10	12	14	8	11½	1¾	22	3	8 × 8
12	14	14	10	13½	1¾	26	4	8 × 8
14	16	14	12	15½	1¾	32	4	9 × 9
16	18	12	14	17½	1¾	35	5	9 × 9
18	20½	12	16	19½	1¾	40	5	10 × 10
20	22¾	10	18	21½	1¾	44	6	10 × 10
21½	24¼	10	20	23½	1¾	50	6	12 × 12
23½	26¾	10	24	27½	1¾	55		
25½	29	10	24	29½	2¾	80		
			30	35½	2¾	100		
			36	41½	2¾	130		
			48	53½	2¾	175		

* Michigan Pipe Co.
† Dimension does not include banding and coating.
‡ Weights vary with types of wood.

or copper wire or stainless-steel bands can be supplied with or without the asphalt coating.

The **square wood pipe** is usually made of white pine with mortise and tenon joints. There is also steam-pipe casing for covering and insulating underground steam lines.

Fittings for wood pipe are made of cast iron, of wood, and of wood-lined steel.

PIPE JOINTS, FLANGES, AND FITTINGS

Pipe Joints

Screwed joints are the most common type of joint, and the pipe industry uses the American Pipe Threads Standards, A.S.A. Standard B2.1-1945. This joint is suitable for low and moderate pressures and at the lower temperatures. The maximum practical size is 4 in., although 6 and 8 in. are sometimes screwed in low-pressure use. Thread lubricants generally used are oils or oil-graphite mixtures. Materials used to ensure tightness are litharge and glycerin, red lead and glycerin, and phenol-formaldehyde resin varnishes. The screw thread necessitates pipe wall thicknesses greater than are required for pressure or other reasons; hence the use of other types of joint may make more economical use of the total thickness selected. It is not advisable to employ the screwed joint with stainless steels because of seizure and the damage done if disassembly and reassembly are required.

Flanged joints are widely used because they facilitate erection and dismantling and are readily joined to fittings and valves. There are many methods for attaching the flanges (see Fig. 72), the three main classes being screwed, lapped, and welded. **Screwed flanges** are of cast iron, brass, and steel, the latter in a wide variety of size and pressure classifications. The end of the pipe is sometimes faced after assembly.

Lap flange joints are fabricated to form a square corner on the pipe end, front and back faces being machined. The flanges are loose and can be rotated to suit. The advantages over the screwed type are: extra thickness for threads is eliminated, there are no threads to have potential leaks, and the pipe ends are held together without the flanges touching each other. This type is also available as welding nipples.

Welded flange joints are of several types, such as the threaded type back-welded, slip-on flanges front and back welded and refaced, and welding neck flange welded to the pipe end.

Welded joints have many advantages over the screwed and flanged types, provided that the welding

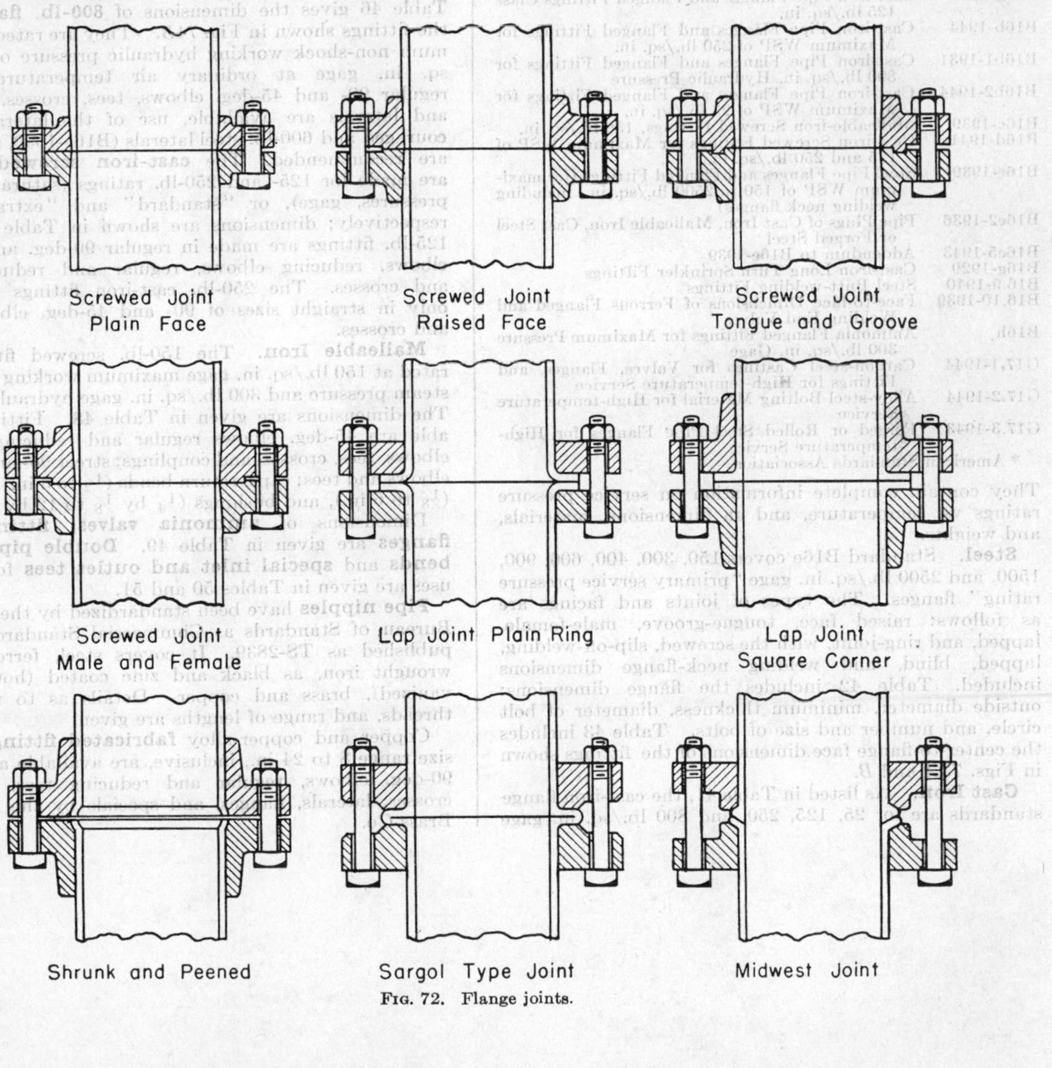

FIG. 72. Flange joints.

is done by qualified operators. Freedom from as many potential leaks, less weight, simplicity of insulating, and lower initial and maintenance costs are the principal factors. Disadvantages are that cheaper iron body valves cannot be welded in; the piping system cannot be readily dismantled; and, in some hazardous areas, welding is not permissible. When stress relieving is required, some flanges are required. For stainless steel in smaller sizes, welding can be simplified by using slip-on fittings and back welding.

Compression and **soldered** fittings are used widely with non-ferrous tubing, permitting the use of thinner walls. Compression fittings are used on some plastic tubing. There are numerous special joints, one of which is the Victaulic Co. coupling. Upset and machined ends are fitted with roll gaskets and clamped together with special couplers held with two bolts. The bell-and-spigot joint is described under cast-iron pipe.

Flanges and Fittings

The American Standards Association has issued the standards on pipe flanges and fittings listed in Table 41.

Table 41. Pipe Flanges and Fittings Standards*

A.S.A. No.	Title
B16a-1939	Cast-iron Pipe Flanges and Flanged Fittings Class 125 lb./sq. in.
B16b-1944	Cast-iron Pipe Flanges and Flanged Fittings for Maximum WSP of 250 lb./sq. in.
B16b1-1931	Cast-iron Pipe Flanges and Flanged Fittings for 800 lb./sq. in. Hydraulic Pressure
B16b2-1944	Cast-iron Pipe Flanges and Flanged Fittings for Maximum WSP of 25 lb./sq. in.
B16c-1939	Malleable-iron Screwed Fittings, 150 lb./sq. in.
B16d-1941	Cast-iron Screwed Fittings for Maximum WSP of 125 and 250 lb./sq. in.
B16e-1939	Steel Pipe Flanges and Flanged Fittings (for maximum WSP of 150 to 2500 lb./sq. in., including welding neck flanges)
B16e2-1936	Pipe Plugs of Cast Iron, Malleable Iron, Cast Steel or Forged Steel
B16e5-1943	Addendum to B16e-1939
B16g-1929	Cast-iron Long Turn Sprinkler Fittings
B16.9-1940	Steel Butt-welding Fittings
B16.10-1939	Face-to-face Dimensions of Ferrous Flanged and Welding End Valves
B16h	Ammonia Flanged Fittings for Maximum Pressure 300 lb./sq. in. Gage
G17.1-1944	Carbon-steel Castings for Valves, Flanges, and Fittings for High-temperature Service
G17.2-1944	Alloy-steel Bolting Material for High-temperature Service
G17.3-1943	Forged or Rolled Steel Pipe Flanges for High-temperature Service

* American Standards Association.

They contain complete information on service pressure ratings vs. temperature, and on dimensions, materials, and weights.

Steel. Standard B16e covers 150, 300, 400, 600, 900, 1500, and 2500 lb./sq. in. gage "primary service pressure rating" flanges. The types of joints and facings are as follows: raised face, tongue-groove, male-female, lapped, and ring-joint, with the screwed, slip-on-welding, lapped, blind, and welding neck-flange dimensions included. Table 42 includes the flange dimensions: outside diameter, minimum thickness, diameter of bolt circle, and number and size of bolts. Table 43 includes the center-to-flange face dimensions of the fittings shown in Figs. 73*A* and *B*.

Cast Iron. As listed in Table 41, the cast-iron flange standards are for 25, 125, 250, and 800 lb./sq. in. gage primary-service pressure ratings. Table 44 gives the dimensions of **25-lb. cast-iron flanges** and the center-to-flange—face dimensions of the fittings shown in Fig. 74*A*; sizes of fittings are from 4 to 72 in., inclusive. This class is for 25 lb./sq. in. gage maximum working saturated steam pressure and working gas pressure, and for 36 in. and smaller for maximum non-shock working hydraulic pressure of 43 lb./sq. in. gage at or near the ordinary range of air temperatures. Table 45 gives the dimensions of **125- and 250-lb. flanges** and the fittings shown in Fig. 74*B*. The 125-lb. flanges and fittings are rated for maximum saturated steam service pressures of 125 lb./sq. in. gage for 1- to 5-in. sizes, 100 lb. for 6- to 12-in. sizes, 80 lb. for 14- to 24-in. sizes, and 50 lb. for 30- to 48-in. sizes; also for water-service pressures of 175 lb./sq. in. gage for 1- to 12-in. sizes; and 150 lb. for 14- to 48-in. flanges only (not fittings). The 250-lb. flanges and fittings are rated as follows: all sizes for maximum working saturated steam pressure of 250 lb./sq. in. gage; sizes 10 in. and smaller for maximum non-shock fluids (other than steam) working hydraulic pressure of 325 lb./sq. in. gage at a temperature of 250°F., or 120°C.; sizes 10 in. and smaller for maximum non-shock working hydraulic pressure of 400 lb./sq. in. gage at or near the ordinary range of air temperatures. Table 46 gives the dimensions of **800-lb. flanges** and the fittings shown in Fig. 74*B*. They are rated for maximum non-shock working hydraulic pressure of 800 lb./sq. in. gage at ordinary air temperatures. While regular 90- and 45-deg. elbows, tees, crosses, reducers, and laterals are available, use of the laterals is discouraged and 600-lb. steel laterals (B16e-1939, see above) are recommended. The **cast-iron screwed fittings** are made for 125- and 250-lb. ratings (saturated steam pressures, gage), or "standard" and "extra heavy," respectively; dimensions are shown in Table 47. The 125-lb. fittings are made in regular 90-deg. and 45-deg. elbows, reducing elbows, regular and reducing tees, and crosses. The 250-lb. cast-iron fittings are made only in straight sizes of 90- and 45-deg. elbows, tees, and crosses.

Malleable Iron. The 150-lb. screwed fittings are rated at 150 lb./sq. in. gage maximum working saturated steam pressure and 300 lb./sq. in. gage hydraulic service. The dimensions are given in Table 48. Fittings available are 45-deg. elbows regular and reducing; 90-deg. elbows, tees, crosses, and couplings; street 90- and 45-deg. elbows and tees; caps, return bends (½ to 3 in.), locknuts (⅛ to 6 in.), and bushings (¼ by ⅛ to 12 by 6 in.).

Dimensions of **ammonia valves, fittings, and flanges** are given in Table 49. **Double pipe return bends** and **special inlet and outlet tees** for various uses are given in Tables 50 and 51.

Pipe nipples have been standardized by the National Bureau of Standards as Commercial Standard CS5-40, published as TS-2839. It covers steel, ferrous alloy, wrought iron, as black and zinc coated (hot-dip galvanized), brass and copper. Details as to materials, threads, and range of lengths are given.

Copper and copper-alloy **fabricated fittings** in the size range 8 to 24 in., inclusive, are available as 45- and 90-deg. elbows, regular and reducing tees, reducers, crosses, laterals, flanges, and specials by the American Brass Co.

Table 42. Steel Flanges*
All dimensions in inches

Nominal pipe size	150-lb. standard: Outside diam. flange	Thickness flange (minimum)	Diameter of bolt circle	Number of bolts	Size of bolts	300-lb. standard: Outside diam. flange	Thickness flange (minimum)	Diameter of bolt circle	Number of bolts	Size of bolts
½	3½	7/16	2⅜	4	½	3¾	9/16	2⅝	4	½
¾	3⅞	½	2¾	4	½	4⅝	⅝	3¼	4	⅝
1	4¼	9/16	3⅛	4	½	4⅞	11/16	3½	4	⅝
1¼	4⅝	⅝	3½	4	½	5¼	¾	3⅞	4	⅝
1½	5	11/16	3⅞	4	½	6⅛	13/16	4½	4	¾
2	6	¾	4¾	4	⅝	6½	⅞	5	8	⅝
2½	7	⅞	5½	4	⅝	7½	1	5⅞	8	¾
3	7½	15/16	6	4	⅝	8¼	1⅛	6⅝	8	¾
3½	8½	15/16	7	8	⅝	9	1 3/16	7¼	8	¾
4	9	15/16	7½	8	⅝	10	1¼	7⅞	8	¾
5	10	15/16	8½	8	¾	11	1⅜	9¼	8	¾
6	11	1	9½	8	¾	12½	1 7/16	10⅝	12	¾
8	13½	1⅛	11¾	8	¾	15	1⅝	13	12	⅞
10	16	1 3/16	14¼	12	⅞	17½	1⅞	15¼	16	1
12	19	1¼	17	12	⅞	20½	2	17¾	16	1⅛
14 o.d.	21	1⅜	18¾	12	1	23	2⅛	20¼	20	1⅛
16 o.d.	23½	1 7/16	21¼	16	1	25½	2¼	22½	20	1¼
18 o.d.	25	1 9/16	22¾	16	1⅛	28	2⅜	24¾	24	1¼
20 o.d.	27½	1 11/16	25	20	1⅛	30½	2½	27	24	1¼
24 o.d.	32	1⅞	29½	20	1¼	36	2¾	32	24	1½

Nominal pipe size	400-lb. standard†: Outside diam. flange	Thickness flange (minimum)	Diameter of bolt circle	Number of bolts	Size of bolts	600-lb. standard†: Outside diam. flange	Thickness flange (minimum)	Diameter of bolt circle	Number of bolts	Size of bolts
½	3¾	9/16	2⅝	4	½	3¾	9/16	2⅝	4	½
¾	4⅝	⅝	3¼	4	⅝	4⅝	⅝	3¼	4	⅝
1	4⅞	11/16	3½	4	⅝	4⅞	11/16	3½	4	⅝
1¼	5¼	13/16	3⅞	4	⅝	5¼	13/16	3⅞	4	⅝
1½	6⅛	⅞	4½	4	¾	6⅛	⅞	4½	4	¾
2	6½	1	5	8	⅝	6½	1	5	8	⅝
2½	7½	1⅛	5⅞	8	¾	7½	1⅛	5⅞	8	¾
3	8¼	1¼	6⅝	8	¾	8¼	1¼	6⅝	8	¾
3½	9	1⅜	7¼	8	⅞	9	1⅜	7¼	8	⅞
4	10	1⅜	7⅞	8	⅞	10¾	1½	8½	8	⅞
5	11	1½	9¼	8	⅞	13	1¾	10½	8	1
6	12½	1⅝	10⅝	12	⅞	14	1⅞	11½	12	1
8	15	1⅞	13	12	1	16½	2 3/16	13¾	12	1⅛
10	17½	2⅛	15¼	16	1⅛	20	2½	17	16	1¼
12	20½	2¼	17¾	16	1¼	22	2⅝	19¼	20	1¼
14 o.d.	23	2⅜	20¼	20	1¼	23¾	2¾	20¾	20	1⅜
16 o.d.	25½	2½	22½	20	1⅜	27	3	23¾	20	1½
18 o.d.	28	2⅝	24¾	24	1⅜	29¼	3¼	25¾	20	1⅝
20 o.d.	30½	2¾	27	24	1½	32	3½	28½	24	1⅝
24 o.d.	36	3	32	24	1¾	37	4	33	24	1⅞

Nominal pipe size	900-lb. standard‡: Outside diam. flange	Thickness flange (minimum)	Diameter of bolt circle	Number of bolts	Size of bolts	1500-lb. standard‡: Outside diam. flange	Thickness flange (minimum)	Diameter of bolt circle	Number of bolts	Size of bolts
½	4¾	⅞	3¼	4	¾	4¾	⅞	3¼	4	¾
¾	5⅛	1	3½	4	¾	5⅛	1	3½	4	¾
1	5⅞	1⅛	4	4	⅞	5⅞	1⅛	4	4	⅞
1¼	6¼	1⅛	4⅜	4	⅞	6¼	1⅛	4⅜	4	⅞
1½	7	1¼	4⅞	4	1	7	1¼	4⅞	4	1
2	8½	1½	6½	8	⅞	8½	1½	6½	8	⅞
2½	9⅝	1⅝	7½	8	1	9⅝	1⅝	7½	8	1
3	9½	1½	7½	8	⅞	10½	1⅞	8	8	1⅛
4	11½	1¾	9¼	8	1⅛	12¼	2⅛	9½	8	1¼
5	13¾	2	11	8	1¼	14¾	2⅞	11½	8	1½
6	15	2 3/16	12½	12	1⅛	15½	3¼	12½	12	1⅜
8	18½	2½	15½	12	1⅜	19	3⅝	15½	12	1⅝
10	21½	2¾	18½	16	1⅜	23	4¼	19	12	1⅞
12	24	3⅛	21	20	1⅜	26½	4⅞	22½	16	2
14 o.d.	25¼	3⅜	22	20	1½	29½	5¼	25	16	2¼
16 o.d.	27¾	3½	24¼	20	1⅝	32½	5¾	27¾	16	2½
18 o.d.	31	4	27	20	1⅞	36	6⅜	30½	16	2¾
20 o.d.	33¾	4¼	29½	20	2	38¾	7	32¾	16	3
24 o.d.	41	5½	35½	20	2½	46	8	39	16	3½

Nominal pipe size	2500-lb. standard: Outside diam. flange	Thickness flange (minimum)	Diameter of bolt circle	Number of bolts	Size of bolts
½	5¼	1 3/16	3½	4	¾
¾	5½	1¼	3¾	4	¾
1	6¼	1⅜	4¼	4	⅞
1¼	7¼	1½	5⅛	4	1
1½	8	1¾	5¾	4	1⅛
2	9¼	2	6¾	8	1
2½	10½	2¼	7¾	8	1⅛
3	12	2⅝	9	8	1¼
4	14	3	10¾	8	1½
5	16½	3⅝	12¾	8	1¾
6	19	4¼	14½	8	2
8	21¾	5	17¼	12	2
10	26½	6½	21¼	12	2½
12	30	7¼	24⅜	12	2¾

* A.S.A. Standards B16e-1939.
† The dimensions of 400- and 600-lb. flanges are identical for sizes ½ to 3½ in., inclusive.
‡ The dimensions of 900- and 1500-lb. flanges are identical for sizes ½ to 2½ in., inclusive.

Table 43. Center-to-flange Dimensions of Steel-flanged Fittings*

All dimensions in inches

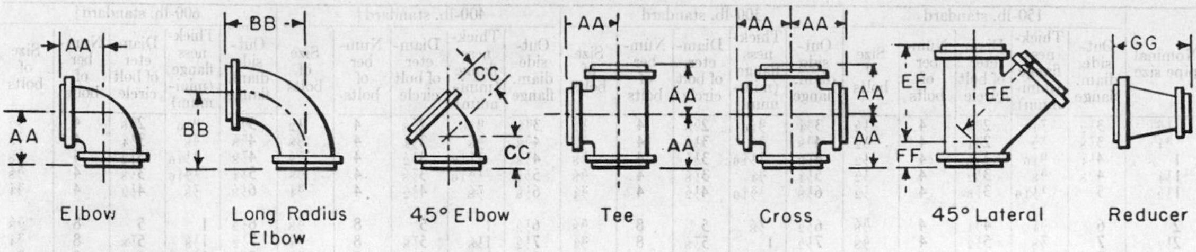

FIG. 73A.

Nominal pipe size	150-lb. standard† (see Fig. 73A) AA	BB	CC	EE	FF	GG	300-lb. standard† (see Fig. 73A) AA	BB	CC	EE	FF	GG
½												
¾												
1	3½	5	1¾	5¾	1¾		4	5	2¼	6½	2	4½
1¼	3¾	5½	2	6¼	1¾		4¼	5½	2½	7¼	2¼	4½
1½	4	6	2¼	7	2		4½	6	2¾	8½	2½	4½
2	4½	6½	2½	8	2½	5	5	6½	3	9	2½	5
2½	5	7	3	9½	2½	5½	5½	7	3½	10½	2½	5½
3	5½	7¾	3	10	3	6	6	7¾	3½	11	3	6
3½	6	8½	3½	11½	3	6½	6½	8½	4	12½	3	6½
4	6½	9	4	12	3	7	7	9	4½	13½	3	7
5	7½	10¼	4½	13½	3½	8	8	10¼	5	15	3½	8
6	8	11½	5	14½	3½	9	8½	11½	5½	17½	4	9
8	9	14	5½	17½	4½	11	10	14	6	20½	5	11
10	11	16½	6½	20½	5	12	11½	16½	7	24	5½	12
12	12	19	7½	24½	5½	14	13	19	8	27½	6	14
14 o.d.	14	21½	7½	27	6	16	15	21½	8½	31	6½	16
16 o.d.	15	24	8	30	6½	18	16½	24	9½	34½	7½	18
18 o.d.	16½	26½	8½	32	7	19	18	26½	10	37½	8	19
20 o.d.	18	29	9½	35	8	20	19½	29	10½	40½	8½	20
24 o.d.	22	34	11	40½	9	24	22½	34	12	47½	10	24

Nominal pipe size	400-lb. standard‡ (see Fig. 73A) A	C	E	F	G	600-lb. standard‡ (see Fig. 73A) A	C	E	F	G
½	3	1¾	5½	1½	4½	3	1¾	5½	1½	4½
¾	3½	2¼	6½	1¾	4½	3½	2¼	6½	1¾	4½
1	4	2¼	7	2	4½	4	2¼	7	2	4½
1¼	4¼	2½	7¾	2¼	4½	4¼	2½	7¾	2¼	4½
1½	4½	2¾	8¾	2½	4½	4½	2¾	8¾	2½	4½
2	5½	4	10	3¼	5½	5½	4	10	3¼	5½
2½	6¼	4¼	11¼	3¼	6¼	6¼	4¼	11¼	3¼	6¼
3	6¾	4¾	12½	3¾	6¾	6¾	4¾	12½	3¾	6¾
3½	7¼	5¼	13¾	4¼	7¼	7¼	5¼	13¾	4¼	7¼
4	7¾	5¼	15¾	4¼	7¾	8¼	5¾	16¼	4¼	8¼
5	8¾	5¾	16½	4¾	8¾	9¾	6¾	19¼	5¾	9¾
6	9½	6	18½	5	9½	10¾	7¼	20¾	6¼	10¾
8	11½	6½	22	5½	11½	12¾	8¼	24¼	6¾	12¾
10	13	7½	25½	6	13	15¼	9¼	29¼	7¾	15¼
12	14¾	8½	29½	6¼	14¾	16¼	9¾	31¼	8¼	16¼
14 o.d.	16	9	32½	6¾	16	17¼	10½	34	8¾	17¼
16 o.d.	17½	10	36	7¾	18	19¼	11½	38¼	9¾	19¼
18 o.d.	19	10½	39	8¼	19	21¼	12	41¾	10¼	21¼
20 o.d.	20½	11	42½	8¾	20½	23¼	12¾	45¼	10¾	23¼
24 o.d.	24	12½	50	10¼	24	27¼	14½	52¾	12¾	27¼

† A raised face of 1/16 in. is provided on the flange of each opening of these fittings and is included in the dimensions given.
‡ The raised face of ¼ in. is not included in the dimensions given. The dimensions are the same for the ½- to 3½-in. fittings, inclusive.

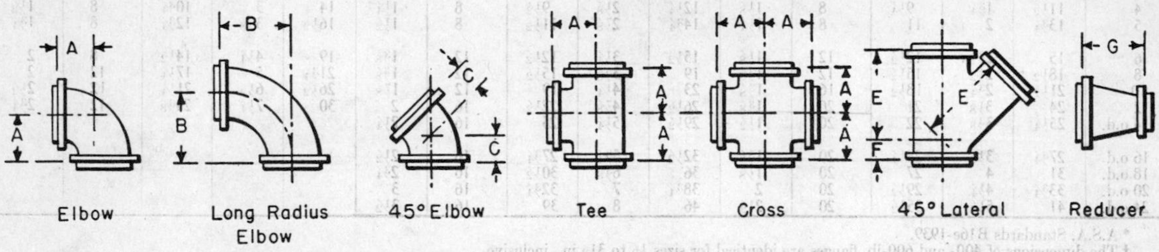

FIG. 73B.

Nominal pipe size	900-lb. standard§‖ (see Fig. 73B) A	C	E	F	G	1500-lb. standard§‖ (see Fig. 73B) A	C	E	F	G	2500-lb. standard§ (see Fig. 73B) A	C	E	F	G
½	4	2¾				4	2¾				4 15/16				
¾	4¼	3				4¼	3				5⅛				
1	4¾	3¼	8¾	2¼	4½	4¾	3¼	8¾	2¼	4½	5 13/16	3¾			
1¼	5¼	3¾	9¾	2¾	5¼	5¼	3¾	9¾	2¾	5¼	6⅝	4			
1½	5¾	4	10¾	3¼	5¾	5¾	4	10¾	3¼	5¾	7 5/16	4½			
2	7	4½	13	3¾	6¾	7	4½	13	3¾	6¾	8⅝	5½	15	5	9
2½	8	5	15	4¼	7¾	8	5	15	4¼	7¾	9¾	6	17	5½	10
3	7¼	5¼	14¼	4¼	7¼	9	5½	17	4¾	8¾	11⅛	7	19½	6½	11¼
4	8¾	6¼	17¼	5¼	8¾	10½	7	19	5¾	10¼	13	8¼	22¾	7½	13
5	10¾	7¼	20¾	6¼	10¾	13	8½	23	7¼	13¼	15⅜	9¾	27	9	15¼
6	11¾	7¾	22¼	6¼	11¾	13⅝	9⅛	24⅝	7⅞	14	17¾	11¼	31	10¼	17½
8	14¼	8¾	27¼	7¼	14¼	16⅛	10⅝	29⅝	8⅞	16½	19⅞	12½	35	11½	20
10	16¼	9¾	31¼	8¼	16¼	19¼	11¾	35¾	10	19¾	24¾	15¾	43	14½	25
12	18¾	10¾	34¼	8¾	17¼	22	13	40½	11¾	22½	27¾	17½	49	16	28½
14 o.d.	20	11¼	36¼	9¼	18½	24½	14	43¾	12¼	25¼					
16 o.d.	22	12¼	40½	10¼	20½	27	16	48	14½	27¾					
18 o.d.	23¾	13	45¼	11¾	24	30	17½	53	16¼	31					
20 o.d.	25¾	14¼	50	12¾	26	32½	18½	57½	17½	33½					
24 o.d.	30¼	17¾	59¾	15¼	30	38	20½	67	20¼	39¼					

* A.S.A. Standards B16e-1939.
§ A raised face of ¼ in. is provided on the flange of each opening of these fittings and is included in the dimensions given.
‖ The dimensions for the 900- and 1500-lb. fittings are identical for sizes ½ to 2½ in., inclusive.

Table 44. 25-lb. Cast-iron Flanges and Fittings*

All dimensions in inches

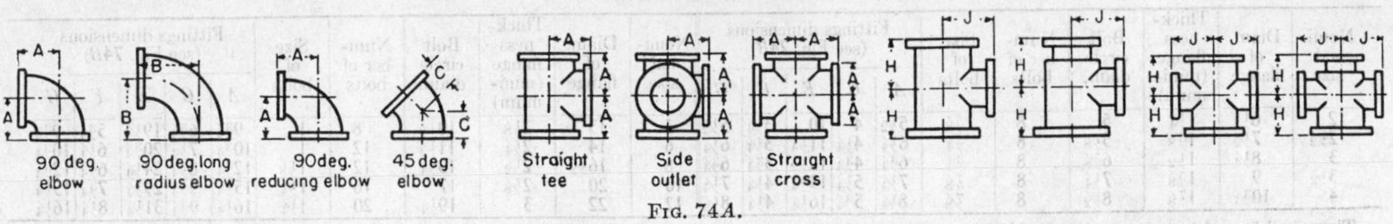

Fig. 74A.

Nominal size	Outside diam. flange	Thickness flange (minimum)	Diameter of bolt circle	Number of bolts	Size of bolts	Fitting dimensions (see Fig. 74A) A	B	C	H	J	Nominal size	Outside diam. flange	Thickness flange (minimum)	Diameter of bolt circle	Number of bolts	Size of bolts	Fitting dimensions (see Fig. 74A) A	B	C	H	J
4	9	¾	7½	8	⅝	6½	9	4			24	32	1⅜	29½	20	¾	22	34	11	15	19
5	10	¾	8½	8	⅝	7½	10¼	4½			30	38¾	1½	36	28	⅞	25	41½	15	18	23
6	11	¾	9½	8	⅝	8	11½	5			36	46	1⅝	42¾	32	⅞	28	49	18	20	26
8	13½	¾	11¾	8	⅝	9	14	5½			42	53	1¾	49½	36	1	31	56½	21	23	30
10	16	⅞	14¼	12	⅝	11	16½	6½			48	59½	2	56	44	1	34	64	24	26	34
12	19	1	17	12	⅝	12	19	7½			54	66¼	2¼	62¾	44	1	39	71½	27	29	37
14	21	1⅛	18¾	12	¾	14	21½	7½			60	73	2¼	69¼	52	1⅛	44	79	30	33	41
16	23½	1⅛	21¼	16	¾	15	24	8			72	86½	2½	81½	60	1⅛	53	94	36	40	48
18	25	1¼	22¾	16	¾	16½	26½	8½	13	15½	84	99¾	2¾	95½	64	1¼					
20	27½	1¼	25	20	¾	18	29	9½	14	17	96	113¼	3	108½	68	1¼					

* A.S.A. Standards B16b2-1944. Bolt holes are drilled ⅛ in. larger in diameter than the nominal diameter of the bolt. All 25-lb. cast-iron standard flanges have plain faces.

Screwed companion flanges should not be thinner than the "125-lb. American Standard" thickness on sizes 24 in. and smaller. Other types of flanges may have thicknesses as given in the table above.

Table 45. 125- and 250-lb. Cast-iron Flanges and Fittings*

All dimensions in inches

Nominal size	Diam. of flange	Thickness flange (minimum)	Bolt circle diam.	Number of bolts	Size of bolts	Fittings dimensions (see Fig. 74B) A	B	C	D	E	F	G	Diam. of flange	Thickness flange (minimum)	Bolt circle diam.	Number of bolts	Size of bolts	Fittings dimensions (see Fig. 74B) A	B	C	D	E	F	G
	125 lb. or "Standard"												250 lb. or "Extra Heavy"											
1	4¼	7/16	3⅛	4	½	3½	5	1¾	7½	5¾	1¾		4⅞	11/16	3½	4	⅝	4	5	2	8½	6½	2	
1¼	4⅝	½	3½	4	½	3¾	5½	2	8	6¼	1¾		5¼	¾	3⅞	4	⅝	4¼	5½	2½	9½	7¼	2¼	
1½	5	9/16	3⅞	4	½	4	6	2¼	9	7	2		6⅛	13/16	4½	4	¾	4½	6	2¾	11	8½	2½	
2	6	⅝	4¾	4	⅝	4½	6½	2½	10½	8	2½	5	6½	⅞	5	8	⅝	5	6½	3	11½	9	2½	5
2½	7	11/16	5½	4	⅝	5	7	3	12	9½	2½	5½	7½	1	5⅞	8	¾	5½	7	3½	13	10½	2½	5½
3	7½	¾	6	4	⅝	5½	7¾	3	13	10	3	6	8¼	1⅛	6⅝	8	¾	6	7¾	3½	14	11	3	6
3½	8½	13/16	7	8	⅝	6	8½	3½	14½	11½	3	6½	9	1 3/16	7¼	8	¾	6½	8½	4	15½	12½	3	6½
4	9	15/16	7½	8	⅝	6½	9	4	15	12	3	7	10	1¼	7⅞	8	¾	7	9	4½	16½	13½	3	7
5	10	15/16	8½	8	¾	7½	10¼	4½	17	13½	3½	8	11	1⅜	9¼	8	¾	8	10¼	5	18½	15	3½	8
6	11	1	9½	8	¾	8	11½	5	18	14½	3½	9	12½	1 7/16	10⅝	12	¾	8½	11½	5½	21½	17½	4	9
8	13½	1⅛	11¾	8	¾	9	14	5½	22	17½	4½	11	15	1⅝	13	12	⅞	10	14	6	25½	20½	5	11
10	16	1 3/16	14¼	12	⅞	11	16½	6½	25½	20½	5	12	17½	1⅞	15¼	16	1	11½	16½	7	29½	24	5½	12
12	19	1¼	17	12	⅞	12	19	7½	30	24½	5½	14	20½	2	17¾	16	1⅛	13	19	8	33½	27½	6	14
14	21	1⅜	18¾	12	1	14	21½	7½	33	27	6	16	23	2⅛	20¼	20	1⅛	15	21½	8½	37½	31	6½	16
16	23½	1 7/16	21¼	16	1	15	24	8	36½	30	6½	18	25½	2¼	22½	20	1¼	16½	24	9½	42	34½	7½	18
18	25	1 9/16	22¾	16	1⅛	16½	26½	8½	39	32	7	19	28	2⅜	24¾	24	1¼	18	26½	10	45½	37½	8	19
20	27½	1 11/16	25	20	1⅛	18	29	9½	43	35	8	20	30½	2½	27	24	1¼	19½	29	10½	49	40½	8½	20
24	32	1⅞	29½	20	1¼	22	34	11	49½	40½	9	24	36	2¾	32	24	1½	22½	34	12	57½	47½	10	24
30	38¾	2⅛	36	28	1¼	25	41½	15	59	49	10	30	43	3	39¼	28	1¾	27½	41½	15	....	....	....	30
36	46	2⅜	42¾	32	1½	28	49	18	....	....	....	36	50	3⅜	46	32	2							
42	53	2⅝	49½	36	1½	31	56½	21	....	....	....	42	57	3 11/16	52¾	36	2							
48	59½	2¾	56	44	1½	34	64	24	....	....	....	48	65	4	60¾	40	2							
54 o.d.†	66¼	3	62¾	44	1¾																			
60 o.d.†	73	3⅛	69¼	52	1¾																			
72 o.d.†	86½	3½	82½	60	1¾																			
84 o.d.†	99¾	3⅞	95½	64	2																			
96 o.d.†	113¼	4¼	108½	68	2¼																			

* A.S.A. Standards B16a-1939 and B16b-1944.

† These sizes are included for convenience where special fittings with larger flanges are required. They do not necessarily carry a definite rating.

All 250-lb. cast-iron flanges have a 1/16-in. raised face which is included in the dimensions given.

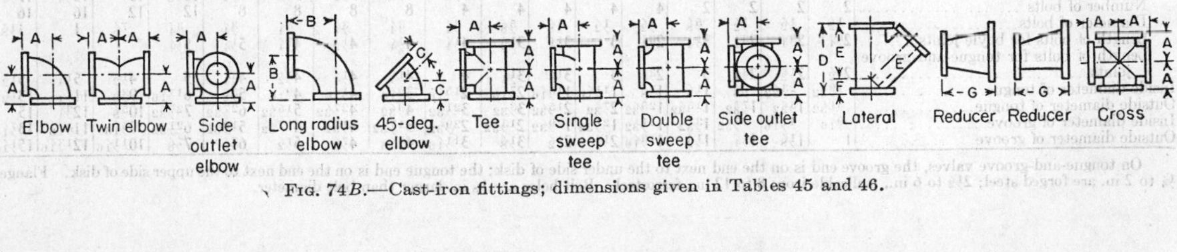

Fig. 74B.—Cast-iron fittings; dimensions given in Tables 45 and 46.

Table 46. 800-lb. Cast-iron Flanges and Fittings*

All dimensions in inches

Nominal size	Diam. of flange	Thickness flange (minimum)	Bolt circle diam.	Number of bolts	Size of bolts	Fittings dimensions (see Fig. 74B) A	C	E	F	G	Nominal size	Diam. of flange	Thickness flange (minimum)	Bolt circle diam.	Number of bolts	Size of bolts	Fittings dimensions (see Fig. 74B) A	C	E	F	G
2	6½	1¼	5	8	⅝	5½	4	10	3¼	5½	5	13	2⅛	10½	8	1	9¾	6¾	19¼	5¾	9¾
2½	7½	1⅜	5⅞	8	¾	6¼	4¼	11¼	3¼	6¼	6	14	2¼	11½	12	1	10¾	7¼	20¾	6¼	10¾
3	8¼	1½	6⅝	8	¾	6¾	4¾	12½	3¾	6¾	8	16½	2½	13¾	12	1⅛	12¾	8¼	24¼	6¾	12¾
3½	9	1⅝	7¼	8	⅞	7¼	5¼	13¾	4¼	7¼	10	20	2⅞	17	16	1¼	15¼	9¼	29¼	7¾	15¼
4	10¾	1⅞	8½	8	⅞	8¼	5¾	16¼	4¼	8¼	12	22	3	19¼	20	1¼	16¼	9¾	31¼	8¼	16¼

The raised face of ¼ in. is not included in "Thickness flange, minimum."
* A.S.A. Standards B16b1-1931.

Table 47. Cast-iron Screwed Fittings*

All dimensions in inches

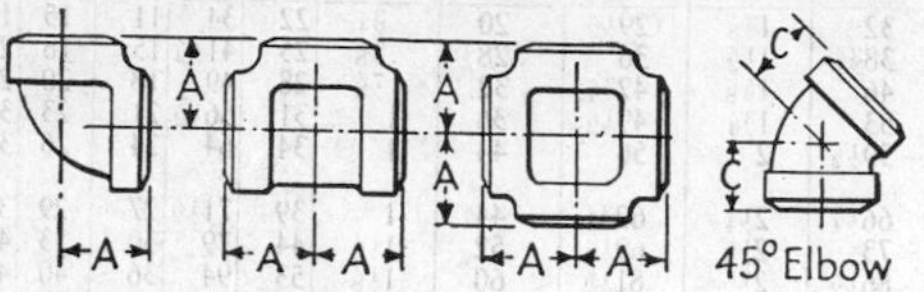

Fig. 75.

Nominal size	125 lb. A	125 lb. C	250 lb. A	250 lb. C
¼	0.81	0.73	0.94	0.81
⅜	.95	.80	1.06	.88
½	1.12	.88	1.25	1.00
¾	1.31	.98	1.44	1.13
1	1.50	1.12	1.63	1.31
1¼	1.75	1.29	1.94	1.50
1½	1.94	1.43	2.13	1.69
2	2.25	1.68	2.50	2.00
2½	2.70	1.95	2.94	2.25
3	3.08	2.17	3.38	2.50
3½	3.42	2.39	3.75	2.63
4	3.79	2.61	4.13	2.81
5	4.50	3.05	4.88	3.19
6	5.13	3.46	5.63	3.50
8	6.56	4.28	7.00	4.31
10	8.08	5.16	8.63	5.19
12	9.50	5.97	10.00	6.00
14 o.d.	10.40	....	11.00	
16 o.d.	11.82	....	12.50	

* A.S.A. Standards B16d-1941.

Table 48. Malleable-iron Screwed Fittings*

All dimensions in inches

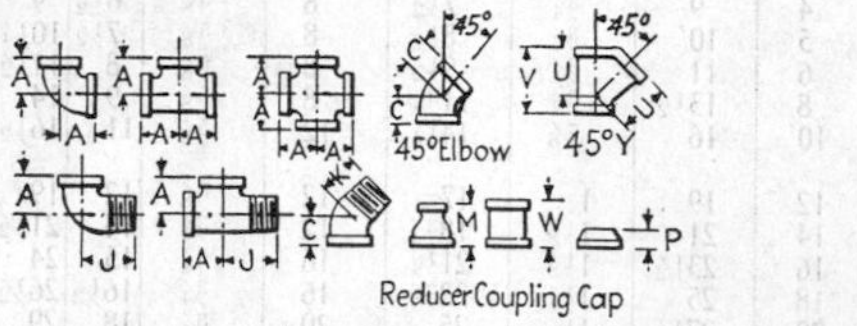

Fig. 76.

Nominal size	A	C	J	K	M	P	U	V	W
⅛	0.69	.....	1.00†	....	....	....	....	....	0.96
¼	.81	0.73	1.19	0.94	1.00	....	....	....	1.06
⅜	.95	.80	1.44	1.03	1.13	....	1.43	1.93	1.16
½	1.12	.88	1.63	1.15	1.25	0.87	1.71	2.32	1.34
¾	1.31	.98	1.89	1.29	1.44	.97	2.05	2.77	1.52
1	1.50	1.12	2.14	1.47	1.69	1.16	2.43	3.28	1.67
1¼	1.75	1.29	2.45	1.71	2.06	1.28	2.92	3.94	1.93
1½	1.94	1.43	2.69	1.88	2.31	1.33	3.28	4.38	2.15
2	2.25	1.68	3.26	2.22	2.81	1.45	3.93	5.17	2.53
2½	2.70	1.95	3.86	2.57	3.25	1.70	4.73	6.25	2.88
3	3.08	2.17	4.51	3.00	3.69	1.80	5.55	7.26	3.18
3½	3.42	2.49‡	5.09†	....	4.00	1.90	6.97	8.98	3.43
4	3.79	2.61	5.69	3.70	4.38	2.08	....	....	3.69
5	4.50	3.05‡	6.86†	....	....	2.32			
6	5.13	3.46‡	8.03†	....	....	2.55			

* A.S.A. Standards B16c-1939.
† Street tees are not made in these sizes.
‡ 45-deg. street elbows are not made in these sizes.

Table 49. Ammonia Valves, Fittings, and Flanges

All dimensions in inches

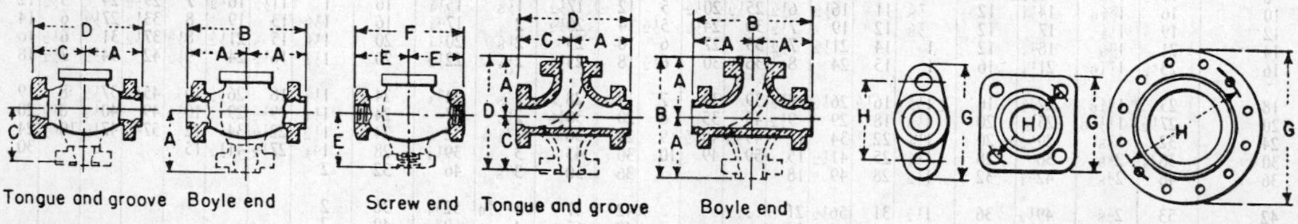

Fig. 77.

Style of flange	Oval				Square					Round							
Size of valve or fitting	¼	⅜	½	¾	1	1¼	1½	2	2½	3	3½	4	5	6	8	10	12
A. Center to face of tongue end	2⅞	3	3⅛	3⅜	3 9/16	4	4⅜	4⅞	5 9/16	6 13/16	7 5/16	7 11/16	8 11/16	9 11/16	10¼	11¾	13¼
B. Face to face of boyle end	5¾	6	6¼	6¾	7⅛	8	8¾	9¾	11⅛	13⅝	14⅝	15⅜	17⅜	19⅜			
C. Center to face of groove end	2 7/16	2 9/16	2 11/16	2 15/16	3⅛	3⅜	3¾	4¼	4¾	6	6½	6⅞	7¾	8¾	10	11½	13
D. Face to face of tongue and groove	5 5/16	5 9/16	5 13/16	6 5/16	6 11/16	7⅜	8⅛	9⅛	10 5/16	12 13/16	13 13/16	14 9/16	16 7/16	18 7/16	20¼	23¼	26¼
E. Center to end of plain screw valves	2¼	2⅜	2½	2¾	3⅛	3⅜	3¾	4¼									
F. End to end of plain screw valves	4½	4¾	5	5½	6¼	6¾	7½	8½									
G. Diamter of flange	3⅝	3 13/16	4	4¾	3 9/16	3 15/16	4 11/16	5 1/16	5 15/16	8¼	9	10	11	12½	15	17½	20½
H. Diameter of bolt circle	2⅜	2 9/16	2¾	3¼	3¼	3¾	4½	5	5⅞	6⅝	7¼	7⅞	9¼	10⅝	13	15¼	17¾
Number of bolts	2	2	2	2	4	4	4	4	4	8	8	8	8	12	12	16	16
Diameter of bolts	½	½	½	⅝	½	½	⅝	⅝	¾	¾	¾	¾	¾	¾	⅞	1	1⅛
Length of bolts for boyle joints	2¾	2¾	2¾	3¼	2¾	3	3¼	3½	4¼	4¼	4½	4¾	5¼	5¼			
Length of bolts for tongue-and-groove joints	2½	2½	2¾	3	2¾	3	3¼	3¼	4	4¼	4¼	4½	4¾	5	4½	5¼	5½
Inside diameter of tongue	19/32	23/32	⅞	1⅛	1⅜	1 11/16	1 15/16	2 7/16	2 15/16	3 9/16	4 1/16	4 9/16	5⅝	6 11/16	9⅜	11¼	13½
Outside diameter of tongue	31/32	1 3/32	1 7/32	1 15/32	1 25/32	2 7/32	2 15/32	3 3/32	3 21/32	4 9/32	4 27/32	5 15/32	6 27/32	7 27/32	10⅝	12¾	15
Inside diameter of groove	9/16	11/16	27/32	1 3/32	1 11/32	1 21/32	1 29/32	2 13/32	2 29/32	3 17/32	4 1/32	4 17/32	5 19/32	6 21/32	9 5/16	11 3/16	13 7/16
Outside diameter of groove	1	1⅛	1¼	1½	1 13/16	2¼	2½	3⅛	3 11/16	4 5/16	4⅞	5½	6⅞	7⅞	10 11/16	12 13/16	15 1/16

On tongue-and-groove valves, the groove end is on the end next to the under side of disk; the tongue end is on the end next to the upper side of disk. Flanges ¼ to 2 in. are forged steel; 2½ to 6 in., malleable iron; 8 to 12 in., ferrosteel. Bolt holes are ⅛ in. larger than bolt diameter.

Table 50. Double Pipe Return Bends

All dimensions in inches

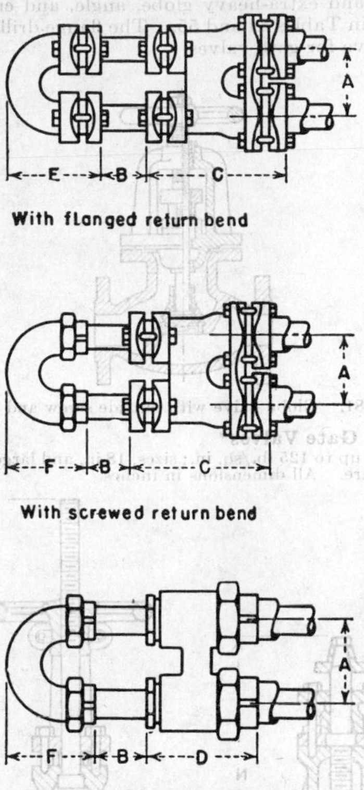

With flanged return bend

With screwed return bend

With screwed return bend

FIG. 78.

Size	2 × 1¼	2½ × 1¼	2½ × 2 × 1¼	3 × 2	4 × 3
A. Center to center	4⅝	4⅝	4⅝	6	8
B. Space for packing	3	3	3	3	3
C. Length of body	9⅜	9⅝	9⅝	10½	13¼
D. Length of body	6⅜				
E. Height of bend	6	6	6	7½	
F. Height of bend	5¼	5¼	5¼	6¾	9

Table 51. Special Inlet and Outlet Tees

For double pipe condensers, brine coolers, and water coolers

All dimensions in inches

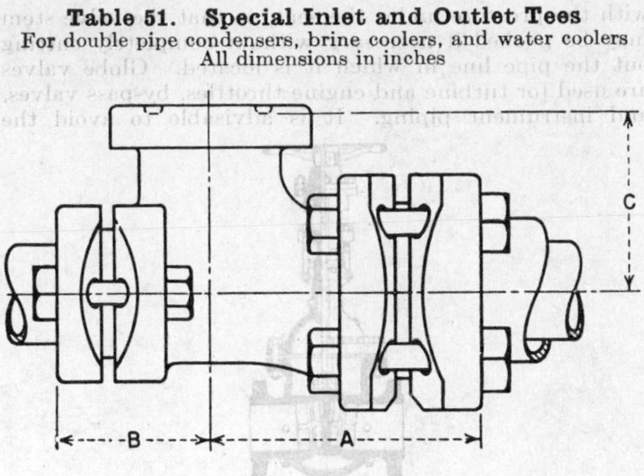

Tees

FIG. 79*A*.

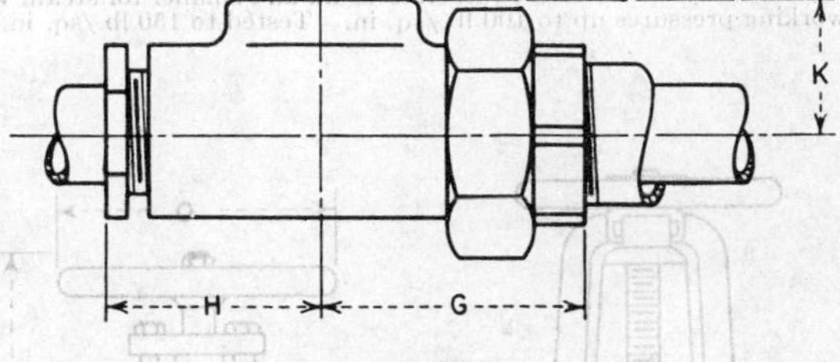

Water cooler tee

FIG. 79*B*.

Size	*A*	*B*	*C*	*G*	*H*	*K*
2 × 1¼ × ½	5¾	3⅝	3 7/16			
2 × 1¼ × ¾	5¾	3⅝	3 7/16			
2 × 1¼ × 1	5¾	3⅝	3⅞			
2 × 1¼ × 1¼	5¾	3⅝	3 11/16			
2 × 1¼ × 1½	5¾	3¼	3 13/16			
2 × 1¼ × 2	5¾	3⅝	3 11/16	4½	3⅝	2¼
2½ × 1¼ × 2	6⅜	3¼	3 13/16			
3 × 2 × ½	6¾	3⅞	4⅞			
3 × 2 × ¾	6⅛	4½	4⅜			
3 × 2 × 1¼	6¾	3⅞	4½			
3 × 2 × 1½	6¾	3⅞	4½			
3 × 2 × 2	6¾	3⅞	4⅜			

Valves

The satisfactory performance of any piping system depends to a large extent upon the proper selection and location of the valves that control and regulate the flow of the fluids to the connected equipment. Valves should be located so that they will be easy to operate and so that equipment can be made available for maintenance without interrupting operation of other connected units. A good valve will be designed so that strains due to temperature changes, pressure, and connected piping will not distort the seat; valve stem and gland should permit easy and efficient packing, and the seats and disks should be of such material and design that the valve will remain tight over a reasonable service period.

Gate valves (Fig. 80) are used with two types of seats. The wedge-shaped gate, which may be either solid or split, is most commonly used. The parallel-slide gate has the advantage that the disk cannot be jammed into the valve body. Gate valves may have either rising or non-rising valve stems. The advantage of the rising stem is that the position of the disk is indicated by the position of the stem. This type of valve, however, requires more room than the non-rising stem.

The gate valve is used whenever the pressure drop through the valve is a consideration and when its function is to stop the flow of fluid rather than to regulate it. Sectionalizing valves, stop valves in turbine and boiler leads, and guard valves are almost always of the gate type. General dimensions of standard, medium, and extra-heavy gate valves are given in Tables 52 and 53. The sizes of by-pass valves and the number of turns to open the main valve are given; data on by-passes are also given in A.S.A. Standards B31.1-1935. The flange drillings are identical with those of the appropriate steel flanges given in Table 43 and with those of the cast-iron flanges given in Tables 44 and 45.

Globe valves (Fig. 81) are classified as either inside screw or outside screw and yoke. Small valves generally are of the inside-screw type while in the larger sizes the outside screw and yoke is generally preferred. Globe valves invariably have rising stems. The pressure drop through this type of valve is much greater than for gate valves, and because of this they are seldom used in water lines. They are primarily used where the function of the valve is to regulate the flow of fluid. Ordinarily they are not used for working steam pressure of over 250 lb. except in the smaller sizes. They should be installed with the pressure under the seat, so that the valve stem may be packed if necessary without completely cutting out the pipe line in which it is located. Globe valves are used for turbine and engine throttles, by-pass valves, and instrument piping. It is advisable to avoid the use of globe valves when the fluid contains solid materials that might accumulate. General dimensions of standard, medium, and extra-heavy globe, angle, and cross valves are given in Tables 54 and 55. The flange drillings are as given above for gate valves.

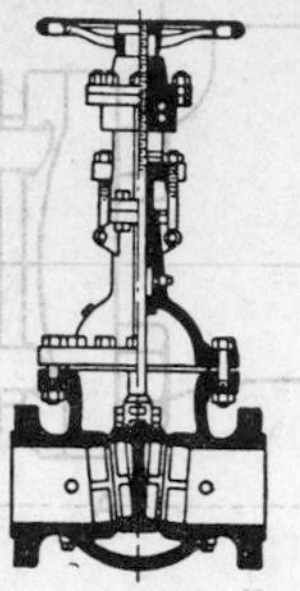

FIG. 80. High-pressure gate valve with wedge-type disk and seat.

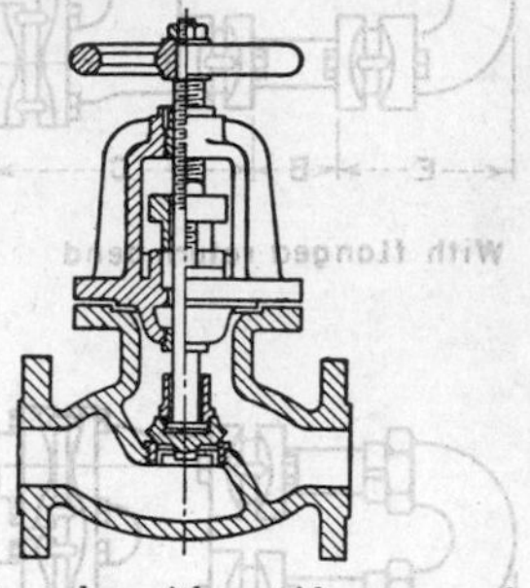

FIG. 81. Globe valve with outside screw and yoke.

Table 52. Dimensions of Standard Gate Valves*

Iron body with brass trim: sizes 16 in. and smaller for steam working pressures up to 125 lb./sq. in.; sizes, 18 in. and larger for steam working pressures up to 100 lb./sq. in. Tested to 150 lb./sq. in. hydraulic pressure. All dimensions in inches.

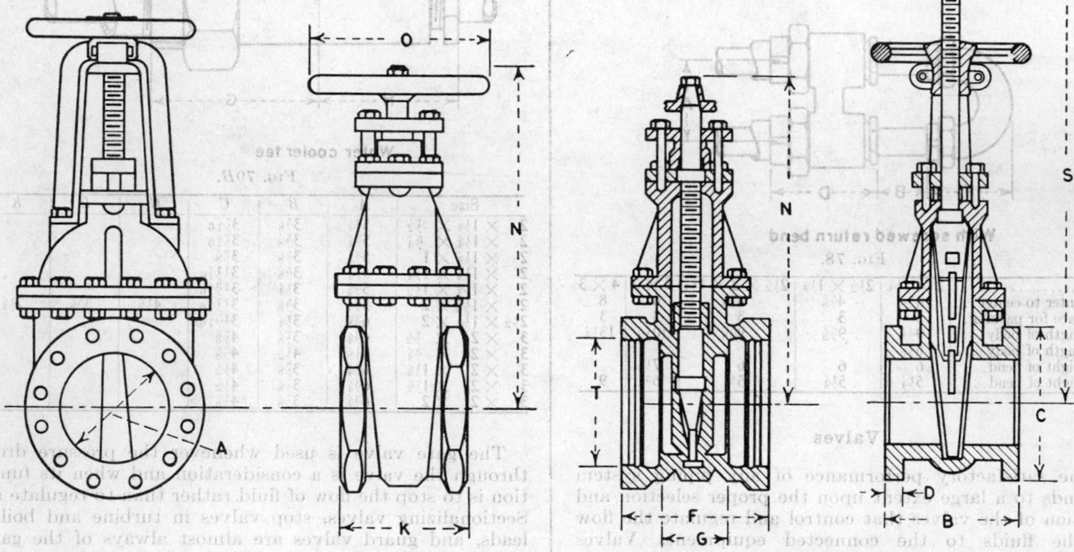

FIG. 82.

Size, *A*	*B*	*C*	*D*	*K*	*N*	*O*	*S*	*F*	*T*	*G*†	No. of turns to open
2	7	6	5/8	5 7/16	11 3/4	8	14 1/2	8 1/2	3 1/4	3	7
2 1/2	7 1/2	7	11/16	5 7/8	12 3/4	8	16	....	....	3 1/4	8
3	8	7 1/2	3/4	6 1/8	14 1/4	9	19	9	4 3/4	3 1/2	10 1/4
3 1/2	8 1/2	8 1/2	13/16	6 1/2	15 1/4	9	21 1/4	....	....	....	10 1/8
4	9	9	15/16	6 7/8	16 1/4	9	24	10 1/4	5 5/8	4 1/4	8 3/4
4 1/2	9 1/2	9 1/4	15/16	7 1/8	17 5/8	9	25 1/2	....	....	....	9
5	10	10	15/16	7 3/8	19	12	28 1/2	10 1/4	6 3/4	4 1/4	11
6	10 1/2	11	1	7 3/4	20 3/4	12	31 3/4	10 3/4	7 7/8	4 3/4	12 5/8
7	11	12 1/2	1 1/16	8 1/4	23	12	37 1/4	10 3/4	9	4 3/4	15 1/4
8	11 1/2	13 1/2	1 1/8	8 3/4	26	14	41	12	10	5	16
9	12	15	1 1/8	9 1/4	28	14	44 3/4	....	....	....	18 3/4
10	13	16	1 3/16	9 7/8	30 1/4	16	50	12 3/4	12 1/4	5 3/4	20 1/2
12	14	19	1 1/4	11 5/8	35 1/4	18	57 1/4	13 1/2	14 3/8	6 1/2	24 1/8
14	15	21	1 3/8	.....	39 1/4	20	66 3/4	13 3/4	16 1/2	6 3/4	28 1/4
15	15	22 1/4	1 3/8	.....	41 1/8	20	69 3/4	....	....	....	31 1/2
16	16	23 1/2	1 7/16	.....	44 1/4	22	75 1/4	16	18 3/4	8	33 1/4
18	17	25	1 9/16	.....	48 3/4	24	86	17	20 7/8	9	35 1/2
20	18	27 1/2	1 11/16	.....	52 1/2	24	91	17 1/2	23	9 1/2	42 1/4
22	19	29 1/2	1 13/16	.....	55 1/2	27	100	....	....	....	46
24	20	32	1 7/8	.....	63 1/2	30	109	19	27 1/4	11	50
26	23	34 1/4	2	.....	65 7/8	30	117 1/2	22	29 3/8	14	65
28	26	36 1/2	2 1/16	.....	70	36	125	26	31 1/2	17	80
30	24	38 3/4	2 1/8	.....	75 1/2	36	133	24	34	15	92 1/2
36	28	46	2 3/8	.....	83	..	158 1/2	27	40 1/2	18	108

* Crane Co.

† End to end of pipes in hub, without by-pass.

Hub end valves—12-in. and smaller—175 lb./sq. in. water working pressure.
14-in. and 16-in. —150 lb./sq. in. water working pressure.
18-in. and larger —120 lb./sq. in. water working pressure.

Crane Standard hub end valves are not rated for steam.

Table 53. Dimensions of Medium and Extra-heavy Gate Valves*

Medium valves: ferrosteel body with special brass seats, for steam working pressures, sizes 16 in. and smaller, up to 175 lb./sq. in.; sizes 18 in. and larger, up to 150 lb./sq. in. Tested to 500 lb./sq. in. hydraulic pressure. Extra-heavy valves: ferrosteel body with hard metal seats; for steam working pressure up to 250 lb./sq. in. Tested to 800 lb./sq. in. hydraulic pressure. All dimensions in inches.

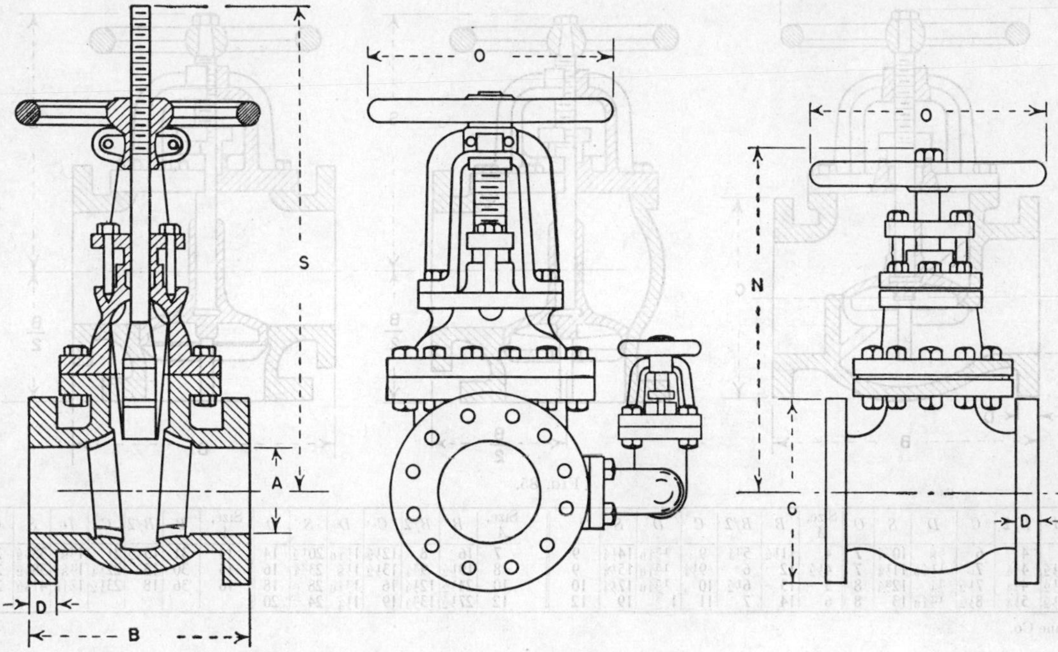

FIG. 83.

Size, A	Medium B	Medium B†	Medium C	Medium D	Medium N	Medium S	Medium T‡	Medium W§	Size, A	Extra heavy B	Extra heavy B†	Extra heavy C	Extra heavy D	Extra heavy N	Extra heavy S	Extra heavy O	Extra heavy W§
1¼	....	....	....	....	....	.....	...	.....	1¼	6½	5½	5¼	¾	8¾	10⅝	7	12
1½	....	....	....	....	....	.....	...		1½	7½	6¼	6⅛	13/16	9⅝	12¼	8	11
2	7½	5½	6½	⅞	11¼	14	...	6⅜	2	8½	7	6½	⅞	10½	13¾	8	14
2½	8	6	7½	1	12¼	15½	...	7⅞	2½	9½	8	7½	1	12⅞	16	9	15
3	9½	7¼	8¼	1⅛	14	18⅜	...	9¾	3	11⅛	9	8¼	1⅛	14⅝	19½	9	14
3½	10	7½	9	1 3/16	14¾	20½	...	11¼	3½	11⅞	10	9	1 3/16	15½	22	10	16
4	10½	7¾	10	1¼	16¼	23¾	...	9	4	12	11	10	1¼	17¾	24½	12	18
4½	11	8¼	10½	1 5/16	17	25	...	10	4½	13¼	12¼	10½	1 5/16	18¾	27	12	21
5	11½	8½	11	1⅜	19	28⅛	...	11	5	15	13½	11	1⅜	20¼	29¾	14	23
6	12	8¾	12½	1 7/16	21	31⅞	¾	25	6	15⅞	15⅞	12½	1 7/16	23	34⅛	16	28
7	12½	9¼	14	1½	22⅞	35¾	¾	30	7	16¼	16¼	14	1½	24¾	38	18	30
8	13½	10	15	1⅝	25⅞	40¾	¾	34	8	16½	16½	15	1⅝	28¾	42¾	20	34
9	14	10¾	16¼	1¾	27½	44¼	¾	38	9	17	17	16¼	1¾	30½	47	20	40
10	15	11½	17½	1⅞	30⅛	49¾	1	42	10	18	18	17½	1⅞	33¾	52¾	22	39
12	16	12½	20½	2	33¾	56⅞	1	50	12	19¾	....	20½	2	37¼	60	24	46
14	18	....	23	2⅛	38½	64½	1¼	59	14	22½	....	23	2⅛	42¾	67¾	24	52
15	18¾	....	24½	2 3/16	41	68¾	1¼	64	15	22½	....	24½	2 3/16	42¾	67¾	24	52
16	19½	....	25½	2¼	44¾	74½	1¼	67	16	24	....	25½	2¼	....	75¼	27	60
18	21	....	28	2⅜	47⅛	82¼	1½	76	18	26	....	28	2⅜	....	82¼	30	67
20	22½	....	30½	2½	52	92	1½	84	20	28	....	30½	2½	....	91½	30	74
22									22	29½	....	33	2⅝	....	101	36	82
24	25½	....	36	2¾	60	108	2	101	24	31	....	36	2¾	....	113	36	88

* Crane Co. † End to end, screwed, in. ‡ T = size of by-pass, in. § W = number of turns to open.

Check valves (Fig. 84) are represented by two types, the swinging check and the lift check, the former type being the one more commonly used. The lift-check valve must always be placed so that it lifts in a vertical direction, and the swing-check valve must be placed so that gravity will close the flapper.

Check valves are used to prevent reversal of flow but should not be counted upon to be absolutely tight. They are generally installed on pump and trap discharges before they join a common header and in feed lines close to the boiler to prevent water from blowing back in case of a feed-line rupture.

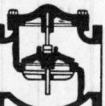

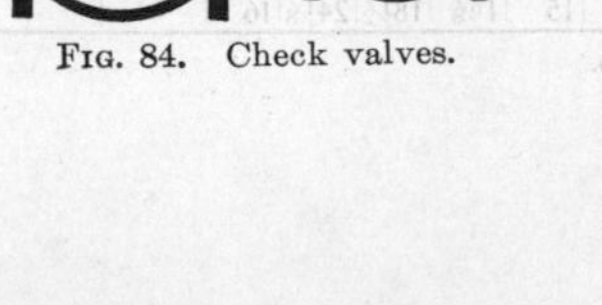

FIG. 84. Check valves.

Table 54. Dimensions of Standard Globe, Angle, and Cross Valves*

Iron-body valves with yokes and brass trim, for steam working pressures up to 125 lb./sq. in. Tested to 150 lb./sq. in. hydraulic pressure. All dimensions in inches.

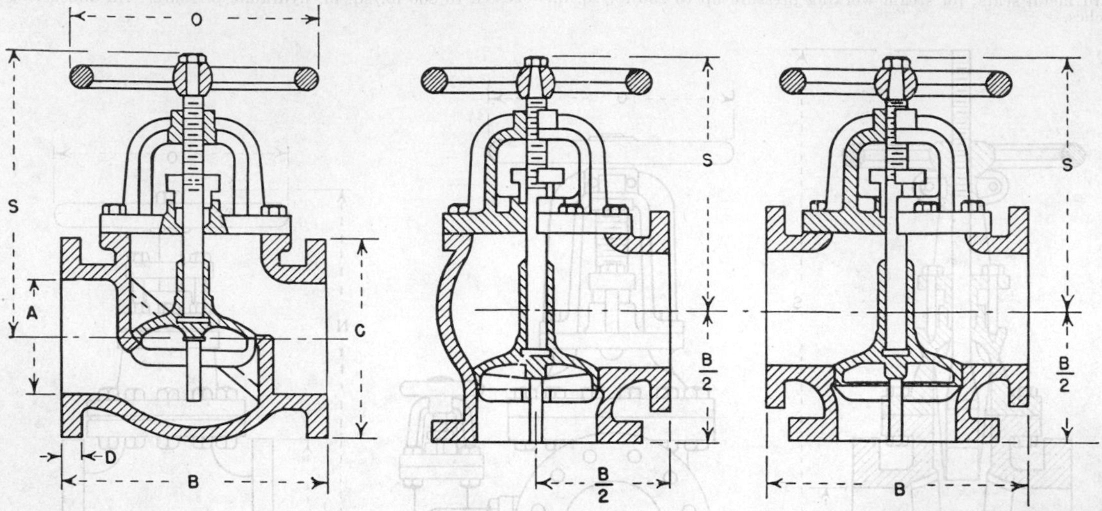

FIG. 85.

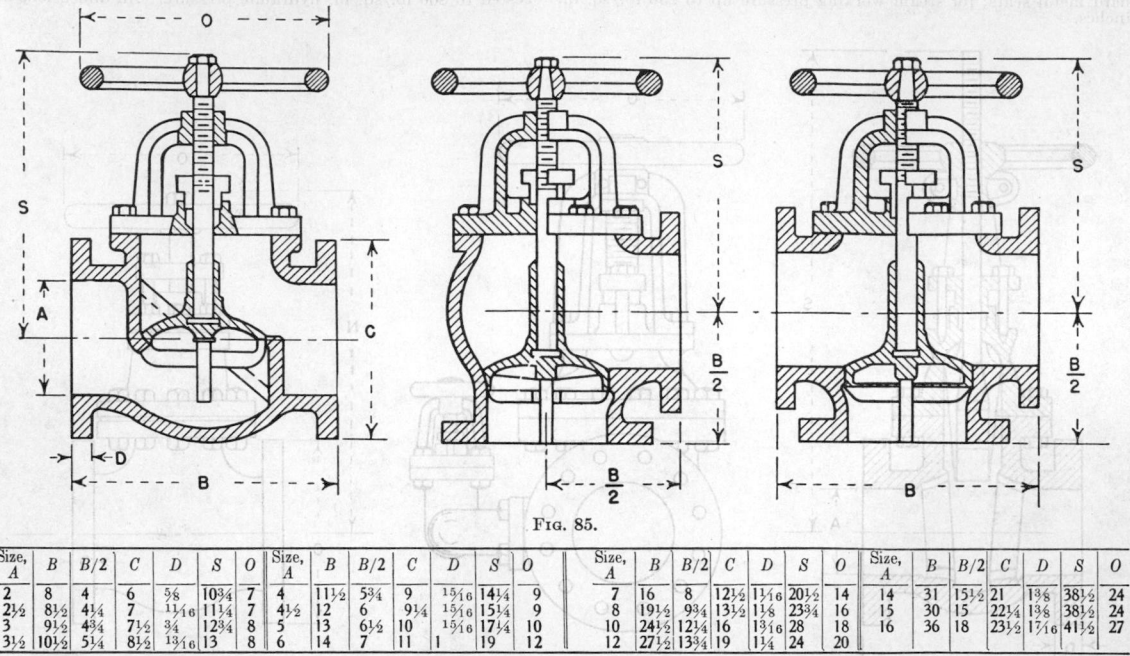

Size, *A*	*B*	*B*/2	*C*	*D*	*S*	*O*	Size, *A*	*B*	*B*/2	*C*	*D*	*S*	*O*	Size, *A*	*B*	*B*/2	*C*	*D*	*S*	*O*	Size, *A*	*B*	*B*/2	*C*	*D*	*S*	*O*
2	8	4	6	⅝	10¾	7	4	11½	5¾	9	15/16	14¼	9	7	16	8	12½	1 1/16	20½	14	14	31	15½	21	1⅜	38½	24
2½	8½	4¼	7	11/16	11¼	7	4½	12	6	9¼	15/16	15¼	9	8	19½	9¾	13½	1⅛	23¾	16	15	30	15	22¼	1⅜	38½	24
3	9½	4¾	7½	¾	12¾	8	5	13	6½	10	15/16	17¼	10	10	24½	12¼	16	1 3/16	28	18	16	36	18	23½	1 7/16	41½	27
3½	10½	5¼	8½	13/16	13	8	6	14	7	11	1	19	12	12	27½	13¾	19	1¼	24	20							

* Crane Co.

Table 55. Dimensions of Medium and Extra-heavy Globe, Angle, and Cross Valves*

Medium valves: ferrosteel body with special brass seats, for steam working pressures up to 175 lb./sq. in. Tested to 500 lb./sq. in. hydraulic pressure. Extra-heavy valves: ferrosteel body with hard metal seats, for steam working pressures up to 250 lb./sq. in. Tested to 800 lb./sq. in. hydraulic pressure. All dimensions in inches.

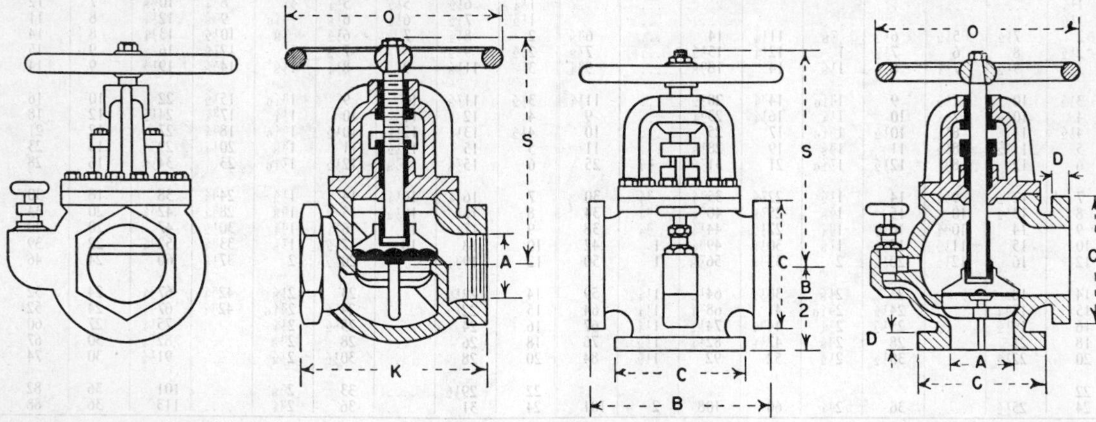

FIG. 86.

Size,	Medium						Size.	Medium						Extra heavy						Extra heavy					
A	*B*	*C*	*D*	*K*	*S*	*O*	*A*	*B*	*C*	*D*	*K*	*S*	*O*	*B*	*C*	*D*	*K*	*S*	*O*	*B*	*C*	*D*	*K*	*S*	*O*
2	9	6½	⅞	7¾	11¾	6½	10	22½	17½	1⅞	22½	28½	20	10½	6½	⅞	9½	13¾	9	24½	17½	1⅞	tinued in	33½	24
2½	10	7½	1	8	12⅝	7½	12	25½	20½	2	25½	31	20	11½	7½	1	10¾	14½	10	Extra heavy			screwed end		
3	11	8¼	1⅛	8¼	14¼	9								12½	8¼	1⅛	11¾	17½	10	globe, angle,			sizes larger		
3½	12	9	1 3/16	9½	15⅜	10	15							13¼	9	1 3/16	12¼	17½	10	and cross			than 6 in.		
4	13	10	1¼	10½	16⅜	10								14	10	1¼	13	19½	12	valves have					
4½	13½	10½	1 5/16	11¼	17⅜	12								15	10½	1 5/16	14	19½	12	been discon-					
5	14½	11	1⅜	12¼	18¼	12								15¾	11	1⅜	15	21½	14	tinued in sizes					
6	16	12½	1 7/16	14	20¼	14								17½	12½	1 7/16	These valves	25	16	larger than 10					
7	17½	14	1½	17	21¼	14								19¼	14	1½	have been	26¼	18	in. and are not					
8	20	15	1⅝	18½	24⅛	16								21	15	1⅝	discon-	29½	20	recommended.					

* Crane Co.

Materials Used for Valves. Valves for working pressures up to 250 lb./sq. in. are commonly made of cast iron. Valves of this material should not be used with temperatures exceeding about 450°F. For pressures up to 1350 lb./sq. in., valves are made with steel bodies, both cast and forged. Some manufacturers use cast chrome-nickel steel for valves used with high-temperature high-pressure steam. In power-plant practice, bronze or brass valves are used only in the small sizes and then mostly for instrument piping.

Various alloy steels are available for valve parts, such as seats, disks, and stems, which are subject to corrosion and erosion. The problem in seating materials is fivefold: (*a*) resistance to corrosion from salts in steam or water and from included air or CO_2; (*b*) resistance to erosion, especially for the flow of wet and dirty steam; (*c*) good seating-metal qualities, to avoid galling when seat and disk ride over each other; (*d*) maintenance of high strength at high temperature; (*e*) avoidance of distortion.

Accelerated tests on seats and disks, set 0.003 in. apart and blown with 250 lb./sq. in. steam for about 350 hr., showed that the following metals resisted erosion in about the order given, beginning with the most resistant:

1. Nickel-chromium alloy (20 per cent nickel, 75 per cent chromium).
2. Stainless steel (14 per cent chromium).
3. Monel and forged Everbright (30 per cent nickel).
4. Cast nickel; copper alloys (25 to 30 per cent nickel).
5. High-tin bronze (12 per cent tin).

Bronzes give satisfactory strength and resistance to corrosion for temperatures up to 550°F., especially the nickel bronze, but show high erosion and low strength for higher temperatures. Monel metal had for a number of years a monopoly for high-temperature work; but while it has good strength at high temperatures and excellent resistance to corrosion, it is not satisfactory in resistance to erosion, and it galls badly.

Distortion has much to do with seat tightness and durability. It is a function almost entirely of valve design, compensated by the fact that all present materials, monel, stainless steel, and nickel-chromium alloys, have coefficients of expansion exceeding those of cast or forged low-carbon steel by 25 to 45 per cent. Consequently some distortion is bound to occur with a valve that is assembled cold and heated to 600° to 700°F., aside from the additional distortion that may come from pressure effects on the valve body.

The Crane Co. gives the following data: Tensile strength of red brass, 30,000 lb./sq. in.; cast iron 22,500; ferrosteel, 33,500; hard metal (hard brass), 34,500; cast steel, 70,000. Comparative destructive hydraulic tests on cast-iron and ferrosteel extra-heavy gate valves showed the following results:

Size, in.	Bursting pressures, lb./sq. in.	
	Cast iron	Ferrosteel
4– 8	1600–1900	2450–2600
10–12	1350–1550	1750–1900
14–16	1100	1200–1350

It is observed from these tests that the factor of safety is apparently very high, but in a steam line pressure is not the only consideration. Allowance must be made for the strains of expansion and contraction, settling, weight of piping, water hammer, and the cutting effect of steam on the valve disk and seat.

The use of screwed valves larger than 6 in. is not recommended. On medium and extra-heavy valves it is desirable to have a by-pass on sizes larger than 8 in.

For determining the water working pressures for valves and fittings based on the steam working pressures, the following rule is recommended as conservative. For sizes 12 in. and smaller, add 40 per cent to the steam working pressure; for sizes 14 in. and larger, add 20 per cent. A much greater range may be safely used for very small sizes.

Cocks are generally designated under two headings, steam and gas, and are made with both brass and iron bodies. Brass-body cocks are regularly made in sizes from ¼ to 3 in., inclusive, and iron-body cocks in sizes from ½ to 6 in., inclusive. A variation from the conventional plug cock is the Reed Roller Bit Co. valve, in which the rotating cylinder does not form the seal as is usual but is fitted with two opposed disks which seat when the valve closes, the downstream disk being held against the seat by the fluid pressure. They are made in sizes from 1 to 8 in. and for use at 750°F. at pressures from 100 to 1500 lb./sq. in. A **porcelain** plug cock or Y valve is made by the Lapp Insulator Co. All parts in contact with the fluid are of dense porcelain; recommended working pressure is 40 lb./sq. in., and higher pressures can safely be handled by armoring the valve.

Gaskets

By H. H. Dunkle

The selection of a gasket will depend upon the mechanical features of the flanged assembly, the operating conditions, and the gasket characteristics. The interdependence of these three general classifications and their various subdivisions are illustrated in Fig. 87. A change in but a single condition may alter the gasket choice.

The **gasket characteristics** or loading constants are a measure of the ease or difficulty of sealing a gasket and of maintaining the seal under high pressure and are directly related to the hardness of the gasket construction. Loading constants for many standard gaskets have been determined and are published in the A.S.M.E. Unfired Pressure Vessel Code and elsewhere.

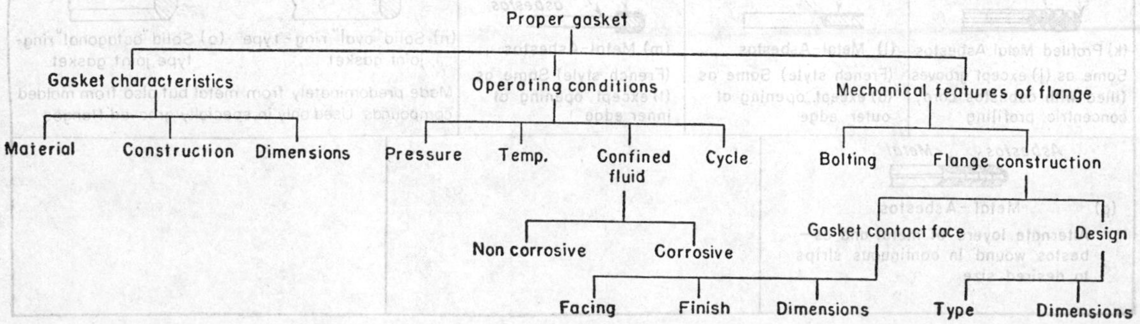

Fig. 87.

Of the **operating conditions** that influence gasket selection, the temperature and the corrosive nature of the application are the most important. The nonmetallic gaskets and those incorporating the low-melting-point metals are with few exceptions limited to service temperatures below 250°F. The semimetallic group comprised of gaskets made up of asbestos partly or completely clad with metal may be used at temperatures up to 850°F. Above 850°F. only all-metal gaskets are satisfactory. The corrosive nature of the service may limit the choice of gasket material (see Sec. 18) and thus have an important bearing on gasket construction and hardness.

Several **mechanical features** of the flanged assembly or closure have a direct bearing on gasket selection. The bolting will set the upper limit of gasket hardness by restricting the force available for seating the gasket or maintaining the seal at high pressures. The gasket dimensions will be defined by the size of the flange contact face and, of course, may be altered somewhat to raise or lower the stress on the gasket. Since all gasket designs cannot be fabricated in all widths, diameters, or shapes, the dimensions and shape of the flange contact face are further restrictions on gasket choice. Flange contact surface finish for most gaskets should be smooth and flat. For some soft gasket installations, serrated facings consisting of concentric V-shaped grooves spaced 1/32 in. are commonly used to provide radial reinforcement. Spirally grooved or so-called "phonograph" finished faces should never be used with metallic gaskets or for high pressures with any gasket.

Gasket Types. For ordinary service at low pressures and temperatures, leather, paper, cork compounds, rubber, plastics, compressed asbestos sheet gaskets (Fig. 88*a*) and plain or treated thick woven asbestos gaskets (Fig. 88*c*) are generally satisfactory, the woven asbestos type being particularly suitable for rough pitted or uneven surfaces and glass or glass-lined flanges.

Round or special shaped gaskets (Fig. 88*b*) molded from soft materials are useful in certain applications. Round-section metal gaskets have given excellent service in high-vacuum installations.

The plain corrugated metal gas et (Fig. 88*f*) generally installed with a coating of gasket paste and the asbestos-inlaid corrugated metal gasket (Fig. 88, *g* and *h*) are used for relatively low pressures, the latter types being particularly suitable for rough or pitted flanges.

The spirally wound metal-asbestos construction (Fig. 88*p*) is suitable for all flange conditions, requires relatively light bolting, and is the most resilient of all metallic gasket designs.

The metal-jacketed asbestos gaskets (Fig. 88*d*) are used for handhole openings and most small-diameter closures of narrow width.

Figure 88*e* is the standard double-jacketed metal-asbestos gasket for pipe flanges and heat exchangers, and Fig. 88*i* is a modification of this design having improved resilience and requiring less bolting.

The French-type construction (Fig. 88, *l* and *m*) is used for valve bonnets, cylinder heads, and vacuum service. A modification of this gasket using thick woven asbestos or rubber as a filler is finding extensive service in glass or glass-lined equipment.

The all-metal gaskets (Fig. 88, *j* and *k*) have to a great extent replaced plain flat metal gaskets because of their lower bolting requirements.

The ring-type joint gaskets (Fig. 88, *n* and *o*) of oval or octagonal cross section for use in ring-type joint-faced flanges have found almost universal acceptance in the petroleum industry for all temperatures and pressures.

Gasket Installation. Flange faces should be clean and as closely aligned as possible. Metal gaskets cannot compensate for much out-of-parallelism of flange faces.

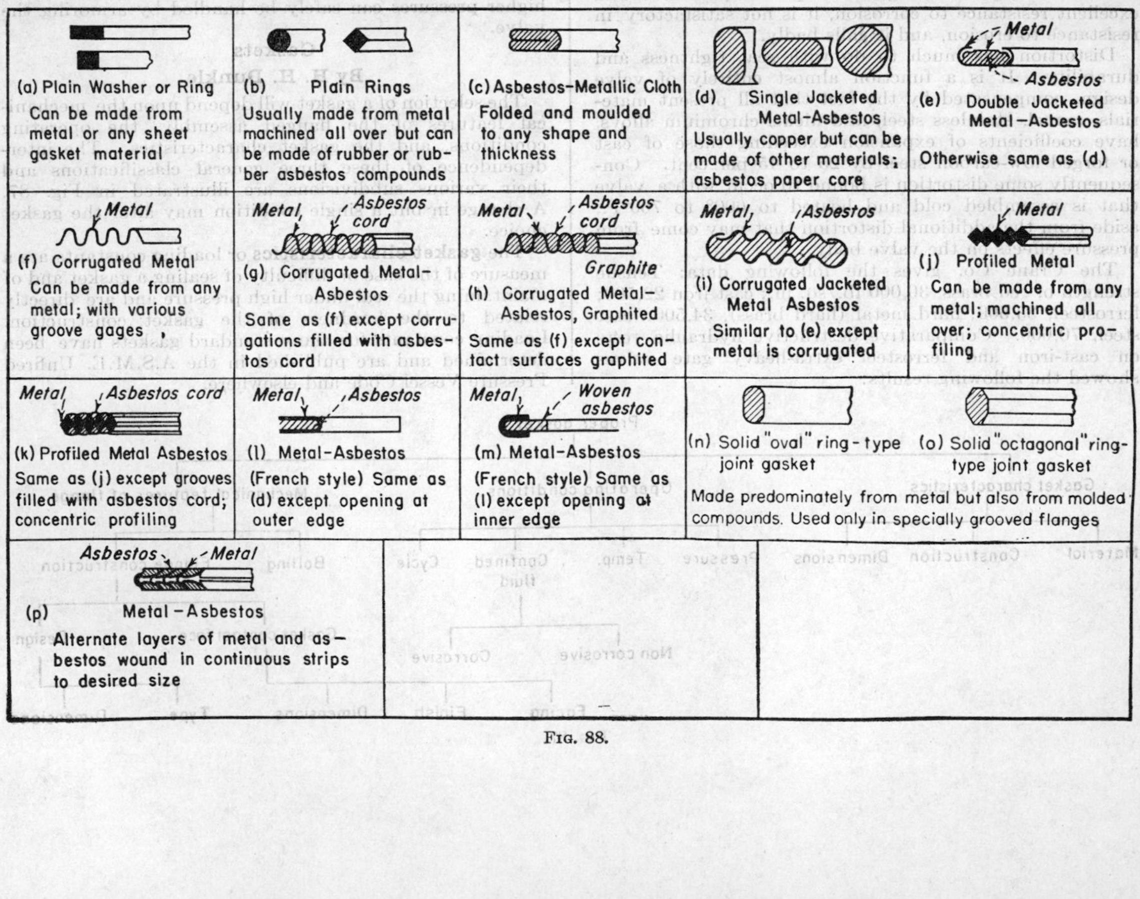

Fig. 88.

Non-metallic gaskets may be coated with graphite or oil and graphite to prevent sticking to the seats when the joint is broken. Care should be taken to center the gasket. Bolts should be tightened first by hand. Then, tightening two bolts 180 deg. apart, advance clockwise 90 deg. and repeat the operation, continuing until all the bolts are uniformly stressed.

The amount of bolt stress applied to the bolts at each step in the process of tightening should be small at the start in order to avoid cocking of the flange faces. Judicious use of a torque wrench will help a great deal in obtaining uniformly stressed bolts. It may be necessary to retighten the bolts in flanges operating at elevated temperatures after the equipment has been at heat for a few hours to overcome the effects of bolt, flange, and gasket relaxation.

Flange Facings. Flange faces are machined in various ways in order to retain the gaskets.

The **plain straight** face is standard for cast-iron flanges of the 25 and 125 lb./sq. in. ratings for low-pressure lines. Best results are obtained from non-metallic gaskets 1/16 in. thick or more. A full-faced gasket is preferred by some because it prevents excessive distortion of the flanges due to tightening. Full-faced gaskets are generally more difficult to install than gaskets dimensioned to fall entirely within the bolts.

In **raised-face flanges** the gasket contact face is 1/16 to 1/4 in. above the remainder of the flange. The 250 lb./sq. in. cast-iron and all-forged-steel flanges are provided with a raised face. Gaskets for these flanges are dimensioned so that their outer edge just touches the bolts to assist in centering. The raised-face flange is satisfactory for all pressures and is generally used.

The **tongue-and-groove** joint provides reinforcement for the gasket and is used for very high pressures. The **male-and-female** joint is similar to the tongue and groove, but the contact face is wider. Both the male-and-female and the tongue-and-groove joint are expensive to install and difficult to disassemble and clean or repair. The **ring-type joint** facing consists of a special shaped groove and requires the use of an all-metal gasket of special cross section (Fig. 88*n* or *o*).

Pipe Supports

A number of the more common methods of supporting pipes are illustrated in Fig. 89. In addition, pipe strap is commonly used to support pipes of small and medium sizes. For fragile pipe and lead, the use of angles and channels has met with success.

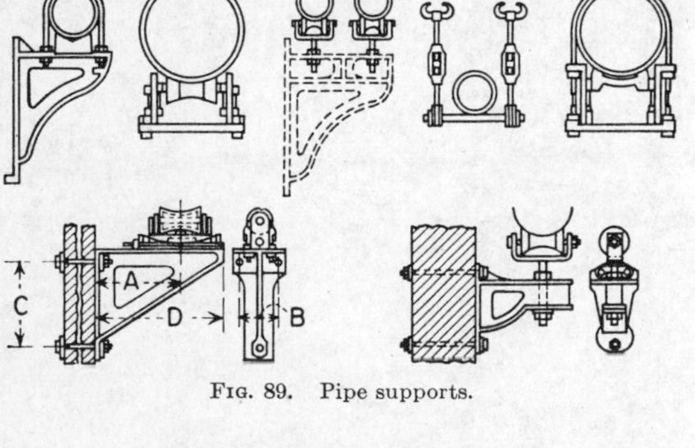

Fig. 89. Pipe supports.

Cements and Solders

Acid-resisting Cements. Cottrell ("Nitric Acid and Nitrates," Vol. 6 of Lunge's "Manufacture of Acids and Alkalies," Van Nostrand, New York, 1923) recommends the following cements:

1. *Silicate Cement* (for permanent joints). Coarse white asbestos powder made into a stiff dough with silicate of soda (sp. gr., 1.250).
2. *Soft Putty* (for HNO_3). 40 parts white asbestos powder; 8 parts blue asbestos powder; 10 parts China clay; 2.5 parts tallow; 21 parts boiled linseed oil.
3. *Rust Cement* (for iron to iron joints). 5 lb. filings; 1 oz. ammonium chloride; 2 oz. flowers of sulfur; and sufficient water to make damp.
4. Maurice A. Knight acidproof cement.

The **gas and oil industry** uses the following cements for practically all their requirements:

1. Red or white lead mixed with linseed oil.
2. Graphite and oil mixtures.
3. Shellac in methanol.
4. Litharge and glycerin.
5. Iron filings and ammonium chloride.
6. Soap, clay, and similar pasty materials.

Metal-joint Cements.

1. For Fe-Fe; Fe-Pb; Pb-Pb: 5 parts Fe_2O_3; 1 part boiled linseed oil.
2. Fe-Pb; Pb-Pb: 2.5 parts Fe_2O_3; 2.5 parts $BaSO_4$; 1 part boiled linseed oil.
3. For glass to metal: litharge and glycerin to a thick paste.
4. For glass to metal: DeKhotinsky cement.
5. For cracked castings: Smooth-On Manufacturing Co. product.

Brazing Solders. Alloys containing about 50 per cent each of copper and zinc. The S.A.E. specification calls for: copper 50 to 52 per cent; lead, maximum of 0.5 per cent; and zinc, remainder. This melts at 1550° to 1600°F.

Silver Solders. Compositions of eight grades of silver solder are given in Table 56 with melting and flow points. Compositions of six special silver solders are given in Table 57.

Table 56. Silver Solders*

H & H† letters	A.S.T.M. grade	Per cent‡ Ag	Cu	Zn	Cd	Melting point, °C.	Flow point, °C.	Color
	1	10	52	38	§	820	870	Yellow
	2	20	45	35	§	775	815	Yellow
ATT	3	20	45	30	5	775	815	Yellow
DE	4	45	30	25		675	745	Nearly white
ETX	5	50	34	16		695	775	Nearly white
Easy	6	65	20	15		695	720	White
Medium	7	70	20	10		725	755	White
IT	8	80	16	4		740	795	White

* A.S.T.M. Designation: B73-29.
† Handy and Harman Co. designations.
‡ Maximum impurities 0.15 per cent.
§ The addition of 0.5 per cent Cd to Grades 1 and 2 shall not be considered as a harmful impurity.

Table 57. Special Silver Solders*

No.	Per cent Ag	Cu	Zn	Cd	Pb	P	Melting point, °C.	Flow point, °C.
1	5.00			Remainder			338	393
2	5.00		16.60	78.40			249	316
3	5.50				Remainder		304	380
4†	2.50	0.25			97.25		298	353
5‡	50.00	15.50	16.50	18.00			627	635
6§	15.00	80.00				5	643	704

* A.S.M. "Metals Handbook," Cleveland, Ohio, 1939, p. 1212.
† Westinghouse Electric Corp.
‡ "Easy-Flo," Handy and Harman Co.
§ "Sil-Fos," Handy and Harman Co.

Soft Solders. (1) Tin 90 per cent and lead 10 per cent melts at 410°F. (2) tin 70 per cent and lead 30 per cent melts at 384°F. Plumber's solder is 2 of lead and 1 of tin. Ordinary solder is 50: 50, tin and lead.

Table 58. Melting and Solidifying Points of Fusible Alloys*

Alloy	Composition, in parts by weight				Melting point, °F.
	Tin	Lead	Bismuth	Cadmium	
Lipowitz's	4	8	15	3	150
Wood's	4	8	15	4	140–160
Darcet's	25	25	50	..	203
Cliché metal	2	2	5	..	221
Rose's	24.6	28.1	50	..	230
Bismuth solder	24.8	22.1	53.1	..	250

Alloys of Lead, Tin, and Bismuth

Element	Composition, %									
Lead	32.0	25.8	25.0	43.0	33.3	10.7	50.0	35.8	20.0	70.9
Tin	15.5	19.8	15.0	14.0	33.3	23.1	33.0	52.1	60.0	9.1
Bismuth	52.5	54.4	60.0	43.0	33.3	66.2	17.0	12.1	20.0	20.0
Solidifies at (°F.)	204.8	213.8	257.0	262.4	293.0	298.4	321.8	357.8	359.6	453.2

Low-melting-point Alloys

Element	Composition, %						
Cadmium	10.8	10.2	14.8	13.1	6.2	7.1	6.7
Tin	14.2	14.3	7.0	13.8	9.4		
Lead	24.9	25.1	26.0	24.3	34.4	39.7	43.4
Bismuth	50.1	50.4	52.2	48.8	50.0	53.2	49.9
Solidifies at (°F.)	150.0	153.5	155.3	155.3	169.7	194.0	203.0

* Marks, "Mechanical Engineers' Handbook," 4th ed., McGraw-Hill, New York, 1941.

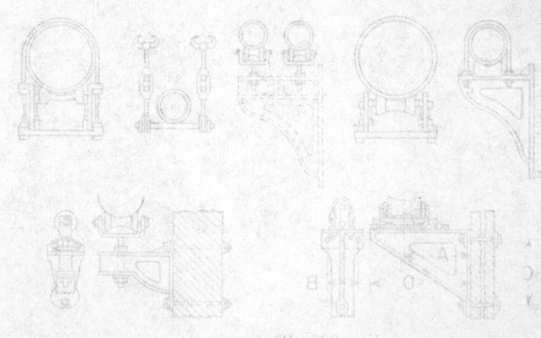

SECTION 6
HEAT TRANSMISSION

BY

William H. McAdams, Sc. D., Professor of Chemical Engineering, Massachusetts Institute of Technology; Member, American Institute of Chemical Engineers, American Chemical Society, American Society of Mechanical Engineers, American Society for Engineering Education, and American Academy of Arts and Sciences (Heat Transmission by Conduction and Convection)

Hoyt C. Hottel, B.A., S.M., Professor of Fuel Engineering and Director, Fuels Research Laboratory, Massachusetts Institute of Technology; Member, American Institute of Chemical Engineers, American Chemical Society, American Society of Mechanical Engineers, and American Academy of Arts and Sciences (Radiant-heat Transmission)

Allan P. Colburn, Ph. D., Assistant to the President and Professor of Chemical Engineering, University of Delaware; Member, American Institute of Chemical Engineers, American Society of Mechanical Engineers, and American Association for the Advancement of Science, American Chemical Society; Chairman, 1948, Heat Transfer Division American Society of Mechanical Engineers (Miscellaneous Over-all Coefficients of Heat Transfer)

Olaf P. Bergelin, Sc. D., Associate Professor of Chemical Engineering, University of Delaware; Member, American Institute of Chemical Engineers, American Chemical Society, American Society for Engineering Education. (Miscellaneous Over-all Coefficients of Heat Transfer)

CONTENTS

HEAT TRANSMISSION BY CONDUCTION AND CONVECTION

	Page
Conductivity Tables	456
Steady Conduction	456
Unsteady Conduction	462
Heat Transfer Coefficients	464
Dirt Deposit Factors	464
Mean Temperature Difference	464
Conversion Factors	466
Nomenclature Table	467
Fluids inside Tubes	407
Gases inside Tubes	467
Liquids inside Tubes	469
Gases outside Tubes	472
Liquids across Tubes	474
Natural Convection	474
Condensing Vapors	476
Effect of Non-condensable Gas	476
Boiling Liquids	478
Optimum Operating Conditions	479

MISCELLANEOUS OVER-ALL COEFFICIENTS

Miscellaneous Over-all Coefficients	480

RADIANT-HEAT TRANSMISSION

	Page
Nature of Thermal Radiation	483
Absorptivity; The Black Body; Kirchhoff's Law; Stefan-Boltzmann Law; Planck's Law; Emissivities of Surfaces	483
Radiation between Surfaces of Solids Separated by Non-absorbing Medium	484
Black Enclosure. All Temperatures Specified. The Factor F	484
Allowance for Refractory Surfaces in Equilibrium. The Factor $\bar{F}$	487
Allowance for Non-black Surfaces. The Factor $\mathfrak{F}$	488
Radiation from Non-luminous Gases	490
Carbon Dioxide and Water Vapor	490
Mean Beam Length	491
Sulfur Dioxide and Other Gases	492
Radiation from Clouds of Particles	493
Luminous Gas Flames	493
Powdered-coal Flames	495
The General Problem of Heat Transfer in a Combustion Chamber	495
Simplified Treatment of Combustion-chamber Performance	496
Billet-reheating Furnaces; Petroleum Heaters; Steam Boiler Furnaces	496

HEAT TRANSMISSION BY CONDUCTION AND CONVECTION*

BY WILLIAM H. McADAMS

REFERENCES: McAdams, "Heat Transmission," 2d ed., McGraw-Hill, New York, 1942. Brown and Marco, "Introduction to Heat Transfer," McGraw-Hill, New York, 1942. Jakob and Hawkins, "Elements of Heat Transfer and Insulation," Wiley, New York, 1942. Stoever, "Applied Heat Transmission," McGraw-Hill, New York, 1941. Schack, "Industrial Heat Transfer," translated by Goldschmidt and Partridge, Wiley, New York, 1933. McAdams, Some Recent Developments in Heat Transfer, *Purdue Univ. Eng. Bul.* **2**, No. 2, 104 (March, 1948).

There are three fundamental types of heat transfer: conduction, convection, and radiation.

1. Conduction is the transfer of heat from one part of a body to another part of the same body, or from one body to another in physical contact with it, without appreciable displacement of the particles of the body.

2. Convection is the transfer of heat from one point to another within a fluid, gas, or liquid, by the mixing of one portion of the fluid with another. In natural convection, the motion of the fluid is entirely the result of differences in density resulting from temperature differences; in forced convection, the motion is produced by mechanical means. When the forced velocity is relatively low, it should be realized that "free-convection" factors, such as density and temperature difference, may have an important influence.

3. Radiation is the transfer of heat from one body to another, not in contact with it, by means of wave motion through space.

All three types of heat transfer may occur at the same time, and it is advisable to consider the heat transfer by each type in any particular case.

Steady Conduction and Thermal Conductivity. Fourier's law is the fundamental differential equation for heat transfer by conduction:

$$\frac{dQ}{d\theta} = -kA\frac{dt}{dx} \qquad (1)$$

where $dQ/d\theta$ (quantity per unit time) is the rate of flow of heat, A is the area at **right angles** to the direction in which the heat flows, and $-dt/dx$ is the rate of change of temperature with the distance in the direction of the flow of heat, *i.e.*, the temperature gradient. The factor k is called the thermal conductivity, and is dependent upon the material through which the heat is flowing, and upon temperature.

Tables 1 to 11, inclusive, show thermal conductivities of various substances expressed as B.t.u./(hr.)(sq. ft.)(°F./ft.) numerically equal to p.c.u./(hr.)(sq. ft.)(°C./ft.). Conversion factors to other units are as follows:

Multiply by 12 to obtain B.t.u./(hr.)(sq. ft.)(°F./in.).

Multiply by 12 to obtain p.c.u.†/(hr.)(sq. ft.)(°C./in.).

Multiply by 0.00413 to obtain g.-cal./(sec.)(sq. cm.)(°C./cm.).

Multiply by 173 to obtain kilo-ergs/(sec.)(sq. cm.)(°C./cm.).

Multiply by 0.0173 to obtain watts/(sq. cm.)(°C./cm.).

The data are taken from Marks, "Mechanical Engineers'

* Helpful suggestions are acknowledged, particularly from T. H. Chilton and T. B. Drew. Alignment charts were kindly prepared by A. P. Colburn, T. H. Chilton, and H. C. Vernon, E. I. duPont de Nemours & Co.

† The p.c.u. is the pound-centigrade unit, sometimes known as pound-calorie unit or centigrade heat unit (c.h.u.). It is the quantity of heat necessary to raise 1 lb. water 1°C., or 1.8 B.t.u.

Table 1. Effect of Temperature upon Thermal Conductivity of Metals and Alloys*

Main body of table is k in B.t.u./(hr.)(sq. ft.)(°F./ft.)

t, °F. / t, °C.	32 / 0	212 / 100	392 / 200	572 / 300	752 / 400	932 / 500	1112 / 600	Melting point, °C.
Aluminum	117	119	124	133	144	155	...	660
Brass (70–30)	56	60	63	66	67	...	...	940
Cast iron	32	30	28	26	25		...	1275
Cast high silicon iron	30	...	...	...	...	...	...	1260
Copper (pure)	224	218	215	212	210	207	204	1083
Lead	20	19	18	18	...	...	...	327.5
Nickel	36	34	33	32	...	...	...	1452
Silver	242	238	...	...	...	...	...	960.5
Sodium	81	...	...	...	...	...	...	97.5
Steel (mild)	...	26	26	25	23	22	21	1375
Tantalum (at 18°C.)	32	...	...	...	...	...	...	2850
Tin	36	34	33	...	...	...	...	231.85
Wrought iron (Swedish)	...	32	30	28	26	23	...	1505
Zinc	65	64	62	59	54	...	...	419.4

* From "International Critical Tables," McGraw-Hill, New York, 1929, and other sources.

Table 2. Thermal Conductivities of Metals*

k = B.t.u./(hr.)(sq. ft.)(°F./ft.)

Substance	t, °F.	k
Metals		
Antimony	32	10.6
Antimony	212	9.7
Bismuth	64	4.7
Bismuth	212	3.9
Cadmium	64	53.7
Cadmium	212	52.2
Gold	64	169.0
Gold	212	170.0
Iron, pure	64	39.0
Iron, pure	212	36.6
Iron, wrought	64	34.9
Iron, wrought	212	34.6
Iron, cast	129	27.6
Iron, cast	216	26.8
Steel (1% C)	64	26.2
Steel (1% C)	212	25.9
Magnesium	32–212	92.0
Mercury	32	4.8
Nickel alloy (62 Ni, 12 Cr, 26 Fe)	68	7.8
Platinum	64	40.2
Platinum	212	41.9
Alloys		
Constantan (60 Cu, 40 Ni)	64	13.1
Constantan (60 Cu, 40 Ni)	212	15.5
Nickel silver	32	16.9
Nickel silver	212	21.5
Manganin {84 Cu, 4 Ni, 12 Mn}	64	12.8
	212	15.2
Platinoid (54 Cu, 25 Ni, 20 Zn)	64	14.5

* Marks, "Mechanical Engineers' Handbook," 4th ed., McGraw-Hill, New York, 1941.

Table 3. Thermal Conductivity of Chromium Alloys

k = B.t.u./(hr.)(sq. ft.)(°F./ft.)

American Iron and Steel Institute Type No.	k at 212°F.	k at 932°F.
301, 302, 302B, 303, 304, 316†	9.4	12.4
308	8.8	12.5
309, 310	8.0	10.8
321, 347	9.3	12.8
403, 406, 410, 414, 416†	14.4	16.6
430, 430F†	15.1	15.2
442	12.5	14.2
501, 502†	21.2	19.5

NOTE. Table 3 is based on recent information from manufacturers. See pp. 1543*ff.* for analysis corresponding to A.I.S.I. type numbers.

† Shelton and Swanger (National Bureau of Standards), *Trans. Am. Soc. Steel Treating*, **21**, 1061–1078, (1933).

Table 4. Thermal Conductivities of Some Building and Insulating Materials*

k = B.t.u./(hr.)(sq. ft.)(°F./ft.)

Material	Apparent density ρ, lb./cu. ft. at room temperature	t, °C.	k
Aerogel, silica, opacified	8.5	120	0.013
		290	.026
Asbestos-cement boards	120	20	.43
Asbestos sheets	55.5	51	.096
Asbestos slate	112	0	.087
	112	60	.114
Asbestos	29.3	−200	.043
	29.3	0	.090
	36	0	.087
	36	100	.111
	36	200	.120
	36	400	.129
	43.5	−200	.090
	43.5	0	.135
Aluminum foil (7 air spaces per 2.5 in.)	0.2	38	.025
		177	.038
Ashes, wood	....	0–100	.041
Asphalt	132	20	.43
Boiler scale (Note 1)			
Bricks:			
Alumina (92–99% Al_2O_3 by wt.) fused	....	427	1.8
Alumina (64–65% Al_2O_3 by wt.)	....	1315	2.7
(See also Bricks, fire clay)	115	800	0.62
	115	1100	.63
Building brick work	....	20	.4
Carbon	96.7	....	3.0
Chrome brick (32% Cr_2O_3 by wt.)	200	200	.67
	200	650	.85
	200	1315	1.0
Diatomaceous earth, natural, across strata (Note 2)	27.7	204	0.051
	27.7	871	.077
Diatomaceous, natural, parallel to strata (Note 2)	27.7	204	.081
	27.7	871	.106
Diatomaceous earth, molded and fired (Note 2)	38	204	.14
	38	871	.18
Diatomaceous earth and clay, molded and fired (Note 2)	42.3	204	.14
	42.3	871	.19
Diatomaceous earth, high burn, large pores (Note 3)	37	200	.13
	37	1000	.34
Fire clay (Missouri)	....	200	.58
		600	.85
		1000	.95
		1400	1.02
Kaolin insulating brick (Note 3)	27	500	0.15
	27	1150	.26
Kaolin insulating firebrick (Note 4)	19	200	.050
	19	760	.113
Magnesite (86.8% MgO, 6.3% Fe_2O_3, 3% CaO, 2.6% SiO_2 by wt.)	158	204	2.2
	158	650	1.6
	158	1200	1.1
Silicon carbide brick, recrystallized (Note 3)	129	600	10.7
	129	800	9.2
	129	1000	8.0
	129	1200	7.0
	129	1400	6.3
Calcium carbonate, natural	162	30	1.3
White marble	....	....	1.7
Chalk	96	....	0.4
Calcium sulfate ($4H_2O$), artificial	84.6	40	.22
plaster (artificial)	132	75	.43
(building)	77.9	25	.25
Cambric (varnished)	....	38	.091
Carbon, gas	....	0–100	2.0
Carbon stock	94	−184	0.55
		0	3.6
Cardboard, corrugated	....	....	0.037
Celluloid	87.3	30	.12
Charcoal flakes	11.9	80	.043
	15	80	.051
Clinker (granular)	....	0–700	.27
Coke, petroleum	....	100	3.4
		500	2.9
Coke, petroleum (20–100 mesh)	62	400	0.55
Coke (powdered)	....	0–100	.11
Concrete (cinder)	....	....	.20
(stone)	....	....	.54
(1:4 dry)	....	....	.44
Cotton wool	5	30	0.024
Cork board	10	30	.025
Cork (regranulated)	8.1	30	.026
(ground)	9.4	30	.025
Diatomaceous earth powder, coarse (Note 2)	20.0	38	.036
	20.0	871	.082
fine (Note 2)	17.2	204	.040
	17.2	871	.074
molded pipe covering (Note 2)	26.0	204	.051
	26.0	871	.088
4 vol. calcined earth and 1 vol. cement, poured and fired (Note 2)	61.8	204	.16
	61.8	871	.23
Dolomite	167	50	1.0
Ebonite	....	....	0.10
Enamel, silicate	38	....	0.5–0.75
Felt, wool	20.6	30	0.03
Fiber insulating board	14.8	21	.028
Fiber, red	80.5	20	.27
(with binder, baked)	....	20–97	.097
Gas carbon	....	0–100	2.0
Glass	....	....	0.2–0.73
Borosilicate type	139	30–75	0.63
Window glass	....	....	0.3–0.61
Soda glass	....	....	0.3–0.44
Granite	....	....	1.0–2.3
Graphite, longitudinal	....	20	95.
powdered, through 100 mesh	30	40	0.104
Gypsum (molded and dry)	78	20	.25
Hair felt (perpendicular to fibers)	17	30	.021
Ice	57.5	0	1.3
Infusorial earth, see diatomaceous earth			
Kapok	0.88	20	0.020
Lampblack	10	40	.038
Lava	....	....	.49
Leather, sole	62.4	....	.092
Limestone (15.3 vol. % H_2O)	103	24	.54
Linen	....	30	.05
Magnesia (powdered)	49.7	47	.35
Magnesia (light carbonate)	13	21	0.034
Magnesium oxide (compressed)	49.9	20	.32
Marble	....	....	1.2–1.7
Mica (perpendicular to planes)	....	50	0.25
Mill shavings	....	....	0.033–0.05
Mineral wool	9.4	30	0.0225
	19.7	30	.024
Paper	....	....	.075
Paraffin wax	....	0	.14
Petroleum coke	....	100	3.4
		500	2.9
Porcelain	....	200	0.88
Portland cement, see concrete	....	90	.17
Pumice stone	....	21–66	.14
Pyroxylin plastics	....	....	.075
Rubber (hard)	74.8	0	.087
(para)	....	21	.109
(soft)	....	21	0.075–0.092
Sand (dry)	94.6	20	0.19
Sandstone	140	40	1.06
Sawdust	12	21	0.03
Scale (Note 1)			
Silk	6.3	....	.026
varnished	....	38	.096
Slag, blast furnace	....	24–127	.064
Slag wool	12	30	.022
Slate	....	94	.86
Snow	34.7	0	.27
Sulfur (monoclinic)	....	100	0.09–0.097
(rhombic)	....	21	0.16
Wall board, insulating type	14.8	21	.028
Wall board, stiff paste board	43	30	.04
Wood shavings	8.8	30	.034
Wood (across grain):			
Balsa	7–8	30	0.025–0.03
Oak	51.5	15	0.12
Maple	44.7	50	.11
Pine, white	34.0	15	.087
Teak	40.0	15	.10
White fir	28.1	60	.062
Wood (parallel to grain):			
Pine	34.4	21	.20
Wool, animal	6.9	30	.021

* Marks, "Mechanical Engineers' Handbook," 4th ed., McGraw-Hill, New York, 1941, "International Critical Tables," McGraw-Hill, 1929, and other sources. For additional data, see pp. 458–459.

Note 1: B. Kamp [*Z. tech. Physik*, **12**, 30 (1931)] shows the effect of increased porosity in decreasing thermal conductivity of boiler scale. Partridge [University of Michigan, *Eng. Research Bull.* 15, 1930] has published a 170-page treatise on Formation and Properties of Boiler Scale.

Note 2: Townshend and Williams, *Chem. & Met.*, **39**, 219 (1932).

Note 3: Norton, "Refractories," 2d ed., McGraw-Hill, New York, 1942.

Note 4: Norton, private communication.

Table 5. Thermal Conductivities of Some Materials for Refrigeration and Building Insulation*

k = B.t.u./(hr.)(sq. ft.)(°F./ft.) at approximately room temperature

Material	Apparent density, lb./cu. ft. room temp.	k
Soft flexible materials in sheet form:		
Chemically treated wood fiber	2.2	0.023
Eel grass between paper	3.4–4.6	0.021–0.022
Felted cattle hair	11–13	0.022
Flax fibers between paper	4.9	.023
Hair and asbestos fibers, felted	7.8	.023
Insulating hair, and jute	6.1–6.3	0.022–0.023
Jute and asbestos fibers, felted	10.0	0.031
Loose materials:		
Cork, regranulated, fine particles	8–9	.025
Charcoal, 6 mesh	15.2	.031
Diatomaceous earth, powdered	10.6	.026
Glass wool, curled	4–10	.024
Gypsum in powdered form	26–34	0.043–0.05
Mineral wool, fibrous	6	0.0217
	10	.0225
	14	.0233
	18	.0242
Sawdust	12	.034
Wood shavings, from planer	8.8	.034
Semiflexible materials in sheet form:		
Flax fiber	13.0	.026
Semirigid materials in board form:		
Corkboard	7.0	.0225
	10.6	.025
Mineral wool, block, with binder	16.7	.031
Stiff fibrous materials in sheet form:		
Wood pulp	16.2–16.9	.028
Sugar-cane fiber	13.2–14.8	.028
Cellular gypsum	8	.029
	12	.037
	18	.049
	24	.064
	30	.083

* Abstracted from *U.S. Bur. Standards Letter Circ.* 227, Apr. 19, 1927. For additional data, see pp. 457–458.

Table 1 shows the effect of temperature upon k for certain metals. The presence of impurities, especially in metals, may cause variation in the thermal conductivity of 50 to 75 per cent. In using thermal conductivities it should be remembered that conduction is not the sole method of transferring heat and that, particularly with liquids and gases, radiation and convection may be much more important (see pp. 463–498).

Table 8. Thermal Conductivities of Insulating Materials at Low Temperatures (Gröber)*

k = B.t.u./(hr.)(sq. ft.)(°F./ft.)

Material	Weight, lb./cu. ft.	Temperatures, °F. 32	−50	−100	−200	−300
Asbestos	44.0	0.135	0.132	0.130	0.125	0.100
Asbestos	29.0	.0894	.0860	.0820	.0720	.0545
Cotton	5.0	.0325	.0302	.0276	.0235	.0198
Silk	6.3	.0290	.0256	.0235	.0196	.0155

* Marks, "Mechanical Engineers' Handbook," 4th ed., McGraw-Hill, New York, 1941.

For non-homogeneous solids the thermal conductivity at a given temperature is a function of the apparent or bulk density. Thus, at 32°F., k for asbestos wool is 0.052 when the bulk density is 24.9 lb./cu. ft. and is 0.111 for a density of 43.6 lb./cu. ft. Bridgman [*Proc. Am. Acad. Arts Sci.*, **59**, 141 (1923)] has shown that the thermal conductivity of liquids is increased only a few per cent under a pressure of 1000 atm. The effect of pressure on k for gases has not been determined, but calculations based on the equation of Eucken (Jeans, "Dynamical Theory of Gases," Cambridge, London, 1925) indicate an increase of approximately 30 per cent at 1000 atm. In determining the apparent thermal conductivity of granular solids, such as granulated cork or charcoal grains, Griffiths [*Spec. Rept.* 5, Food Investigation Board (1921), H.M. Stationery Office] finds that air circulates within the mass of granular solid. Under a certain set of conditions, the apparent thermal conductivity of a charcoal was 9 per cent greater when the test section was vertical than when horizontal. When the apparent conductivity of a mixture of cellular or porous non-homogeneous solid is determined, the observed temperature coefficient may be much larger than for the homogeneous solid alone, because heat is transferred not only by the mechanism of conduction, but also by convection in the gas pockets and by radiation from surface to surface of the individual particles. If internal radiation is an important factor, a plot of the apparent conductivity as ordinates vs. temperature should show a curve concave upward, since radiation increases with the fourth power of the absolute temperature. Griffiths notes that cork, slag wool, charcoal, and wood fibers, when of good quality and dry, have thermal conductivities of about 2.2 times that of still air, whereas a highly cellular form of rubber, 7 lb./cu. ft., had a thermal conductivity only 1.6 times that of still air. In measuring the apparent

Table 6. Thermal Conductivities of Insulating Materials at High Temperatures*

k = B.t.u./(hr.)(sq. ft.)(°F./ft.)

Material	For temperatures, °F. up to	Mean temperatures, °F. 100	200	300	400	500	600	800	1000	1500	2000
Laminated asbestos felt (approx. 40 laminations per in.)	700	0.033	0.037	0.040	0.044	0.048					
Laminated asbestos felt (approx. 20 laminations per in.)	500	.045	.050	.055	.060	.065					
Corrugated asbestos (4 plies per in.)	300	.050	.058	.069							
85% magnesia (density, 13 lb./cu. ft.)	600	.034	.036	.038	.040						
Diatomaceous earth, asbestos and bonding material	1600	.045	.047	.049	.050	.053	0.055	0.060	0.065		
Diatomaceous earth brick	1600	.054	.056	.058	.060	.063	.065	.069	.073		
Diatomaceous earth brick	2000	.127	.130	.133	.137	.140	.143	.150	.158	0.176	
Diatomaceous earth brick	2500	.128	.131	.135	.139	.143	.148	.155	.163	.183	0.203
Diatomaceous earth powder (density, 18 lb./cu. ft.)		.039	.042	.044	.048	.051	.054	.061	.068		
Rock wool		.030	.034	.039	.044	.050	.057				

Asbestos cement, 1.2; 85% magnesia cement, 0.05; asbestos and rock wool cement, 0.075 approx.

* Marks, "Mechanical Engineers' Handbook," 4th ed., McGraw-Hill, New York, 1941.

Table 7. Thermal Conductivities of Insulating Materials at Moderate Temperatures (Nusselt)*

k = B.t.u./(hr.)(sq. ft.)(°F./ft.)

Material	Weight, lb./cu. ft.	Temperatures, °F. 32	100	200	300	400	600	800
Asbestos	36.0	0.087	0.097	0.110	0.117	0.121	0.125	0.130
Burned infusorial earth for pipe coverings	12.5	.043	.046	.052	.057	.062	.073	.085
Insulating composition (loose)	25.0	.040	.046	.050	.053	.055		
Cotton	5.0	.032	.035	.039				
Silk hair	9.1	.026	.030	.034				
Silk	6.3	.025	.028	.034				
Wool	8.5	.022	.027	.033				
Pulverized cork	10.0	.021	.026	.032				
Infusorial earth (loose)	22.0	.035	.039	.045	.047	.050	.053	

* Marks, "Mechanical Engineers' Handbook," 4th ed., McGraw-Hill, New York, 1941.

Handbook" (4th ed., McGraw-Hill, 1941), the "International Critical Tables," vols. 2, 5 (McGraw-Hill, 1929), and other sources as shown by the headings to the tables.

Table 9. Thermal Conductivity of Liquids

k = B.t.u./(hr.)(sq. ft.)(°F./ft.)

A linear variation with temperature may be assumed. The extreme values given constitute also the temperature limits over which the data are recommended.

Liquid	t, °F.	k
Acetic acid 100%[5]	68	0.099
50%[5]	68	.20
Acetone[4]	86	.102
	167	.095
Allyl alcohol[11]	77–86	.104
Ammonia[8]	5–86	.29
Ammonia, aqueous 26%[5]	68	.261
	140	.29
Amyl acetate[7]	50	.083
alcohol (n-)[6]	86	.094
	212	.089
(iso-)[12]	86	.088
	167	.087
Aniline[9]	32–68	.100
Benzene[12]	86	.092
	140	.087
Bromobenzene[12]	86	.074
	212	.070
Butyl acetate (n-)[11]	77–86	.085
alcohol (n-)[4]	86	.097
	167	.095
(iso-)[4]	50	.091
Calcium chloride brine 30%[5]	86	.32
15%[5]	86	.34
Carbon disulfide[4]	86	.093
	167	.088
tetrachloride[10]	32	.107
	154	.094
Chlorobenzene[12]	50	.083
Chloroform[10]	86	.080
Cymene (para-)[12]	86	.078
	140	.079
Decane (n-)[12]	86	.085
	140	.083
Dichlorodifluoromethane[12]	20	.057
	60	.053
	100	.048
	140	.043
	180	.038
Dichloroethane[10]	122	.082
Dichloromethane[10]	5	.111
	86	.096
Ethyl acetate[7]	68	.101
alcohol 100%[2]	68	.105
80%	68	.137
60%	68	.176
40%	68	.224
20%	68	.281
100%[2]	122	.087
benzene[12]	86	.086
	140	.082
bromide[4]	68	.070
ether[4]	86	.080
	167	.078
iodide[4,5]	104	.064
	167	.063
Ethylene glycol[7]	32	.153
Gasoline[6,12]	86	.078
Glycerol 100%[1]	68	.164
80%	68	.189
60%	68	.220
40%	68	.259
20%	68	.278
100%[1]	212	.164
Heptane (n-)[12]	86	.081
	140	.079

Liquid	t, °F.	k
Hexane (n-)[12]	86	0.080
	140	.078
Heptyl alcohol (n-)[6]	86	.094
	167	.091
Hexyl alcohol (n-)[6]	86	.093
	167	.090
Kerosene[4]	68	.086
	167	.081
Mercury[7]	82	4.83
Methyl alcohol 100%[2]	68	0.124
80%	68	.154
60%	68	.190
40%	68	.234
20%	68	.284
100%	122	.114
chloride[8,10]	5	.111
	86	.089
Nitrobenzene[12]	86	.095
	212	.088
Nitromethane[12]	86	.125
	140	.120
Nonane (n-)[12]	86	.084
	140	.082
Octane (n-)[12]	86	.083
	140	.081
Oils[5,12*]	86	.079
Oils, castor[9]	68	.104
	212	.100
Oils, olive[9]	68	.097
	212	.095
Paraldehyde[12]	86	.084
	212	.078
Pentane (n-)[12]	86	.078
	167	.074
Perchloroethylene[10]	122	.092
Petroleum ether[4]	86	.075
	167	.073
Propyl alcohol (n-)[6]	86	.099
	167	.095
alcohol (iso-)[12]	86	.091
	140	.090
Sodium	212	49
	410	46
Sodium chloride brine 25.0%[5]	86	0.33
12.5%[5]	86	.34
Sulfuric acid 90%[5]	86	.21
60	86	.25
30	86	.30
Sulfur dioxide[8]	5	.128
	86	.111
Toluene[4,12]	86	.086
	167	.084
β-Trichloroethane[10]	122	.077
Trichloroethylene[10]	122	.080
Turpentine[7]	59	.074
Vaseline[7]	59	.106
Water[13]	32	.343
	100	.363
	200	.393
	300	.395
	420	.376
	620	.275
Xylene (ortho-)[7]	68	.090
(meta-)[7]	68	.090

* Thermal conductivity data for a number of oils are available from Reference 12. See also Table 10; for many oils an average value of 0.079 may be used.

1 Bates, *Ind. Eng. Chem.*, **28**, 494 (1936).
2 Bates, Hazzard, and Palmer, *Ind. Eng. Chem.*, **10**, 314 (1938).
3 Benning, A. F., private communication, 1940.
4 Bridgman, *Proc. Am. Acad. Arts Sci.*, **59**, 141 (1923).
5 Chilton and Genereaux, personal communication, 1939, based on data selected from the literature.
6 Daniloff, *J. Am. Chem. Soc.*, **54**, 1328 (1932).
7 "International Critical Tables," McGraw-Hill, New York, 1929.
8 Kardos, *Z. Ver. deut. Ing.* **77**, 1158 (1933); *Z. ges. Kälte-Ind.*, **41**, 1, 29 (1934).
9 Kaye and Higgins, *Proc. Roy. Soc. (London)*, **A117**, 459 (1928).
10 DuPont Chlorinated Hydrocarbons, *Tech. Bull.*, Electrochemicals Dept., du Pont, Buffalo, N. Y., 1938.
11 Shiba, *Sci. Papers Inst. Phys. Chem. Research (Tokyo)*, **16**, 205 (1931).
12 Smith, *Trans. Am. Soc. Mech. Engrs.*, **58**, 719 (1936).
13 Timrot and Vargaftik, *J. Tech. Phys. (U.S.S.R.)*, **10**, 1063 (1940).

thermal conductivity of diathermanous substances, such as quartz (especially when exposed to radiation emitted at high temperatures), it should be remembered that a part of the heat is transmitted by radiation.

Table 10. Thermal Conductivities of Petroleum Oils (*J. F. D. Smith*)
k = B.t.u./(hr.)(sq. ft.)(°F./ft.)

Designation of hydrocarbon oil	Aver. mol. wt.*	Viscosity, centipoises			k at 86°F.	k at 212°F.	Sp. gr. at 60°F.
		68°F.	140°F.	212°F.			
Light heat-transfer oil..	284	62.	9.5	3.2	0.0765	0.0748	0.925
Rabbeth spindle oil.....	303	24.5	5.7	2.37	.0825	.0805	.870
Velocite B oil..........	333	73.0	11.0	4.20	.0825	.0800	.897
Red oil.................	418	....	44.0	9.90	.0815	.0796	.928

* H. O. Forrest and L. W. Cummings, personal communication.

Steady Flow. For the steady flow of heat, the term $dQ/d\theta$ in Eq. (1) is constant and may be replaced by Q/θ or q. If k and A are independent of t and x, the equation may be expressed:

$$q = kA\frac{(t_1 - t_2)}{(x_2 - x_1)} = kA\frac{\Delta t}{x} \qquad (2)$$

wherein Δt represents the difference in temperatures. Usually, the thermal conductivity k is not constant but is a function of the temperature. In most cases, over the ranges of values used, the relation is linear. Integration of Eq. (1), with k linear in t, gives

$$q = k_{\text{avg}}A\frac{\Delta t}{x} \qquad (3)$$

where k_{avg} is the arithmetic average thermal conductivity between the temperatures t_1 and t_2. This average probably gives results which are correct within the precision of the data in the majority of cases, though a special integration can be made whenever k is known to be greatly different from linear in temperature.

In case the cross-sectional area A varies with the distance x, A may be expressed in terms of x, in order to integrate Eq. (1). This may, however, lead to complicated expressions, and it is customary to use Eq. (3), substituting the proper average value of A, as listed below.

1. A flat wall of constant area:

$$A_{\text{avg}} = A_1 = A_2 \qquad (4)$$

2. Area is proportional to first power of distance, as for long insulated pipes, L ft. long:

$$A_{\text{avg}} = \frac{(A_2 - A_1)}{2.3 \log_{10} (A_2/A_1)} \qquad (5)$$

where A_2 and A_1 are the larger and smaller areas, respectively. Whence,

$$q = \frac{2.73Lk(\Delta t)}{\log_{10} (D_2/D_1)} \qquad (5a)$$

3. Area is proportional to square of distance:

$$A_{\text{avg}} = \sqrt{A_2A_1} \qquad (6)$$

4. Rectangular bodies having walls of uniform thickness, the shortest inside dimension not exceeding twice the thickness. This is subdivided into several cases. [Langmuir, Adams, Meikle, *Trans. Am. Electrochem. Soc.*, **24**, 53 (1913).]

(a) **The lengths e of all the inside edges are between one-fifth and twice the thickness x of the walls:**

$$A_{\text{avg}} = A_1 + 0.54L\Sigma e + 1.2x^2 \qquad (7)$$

(b) **The length of one inside edge is less than one-fifth the thickness x of the walls.** In this case, the lengths of the four inside edges less than $x/5$ are neglected in determining Σe, and

$$A_{\text{avg}} = A_1 + 0.465x\Sigma e + 0.35x^2 \qquad (7a)$$

(c) **The length of two inside edges are each less than one-fifth the thickness of the walls, the longest inside dimension being E:**

$$A_{\text{avg}} = \frac{2.78Ex}{\log_{10} (A_2/A_1)} \qquad (7b)$$

(d) **All three interior dimensions are less than one-fifth the thickness of the walls:**

$$A_{\text{avg}} = 0.79\sqrt{A_1A_2} \qquad (7c)$$

Conduction through Several Bodies in Series. Figure 1 illustrates diagrammatically the temperature gradients accompanying the steady conduction of heat in series through three solids.

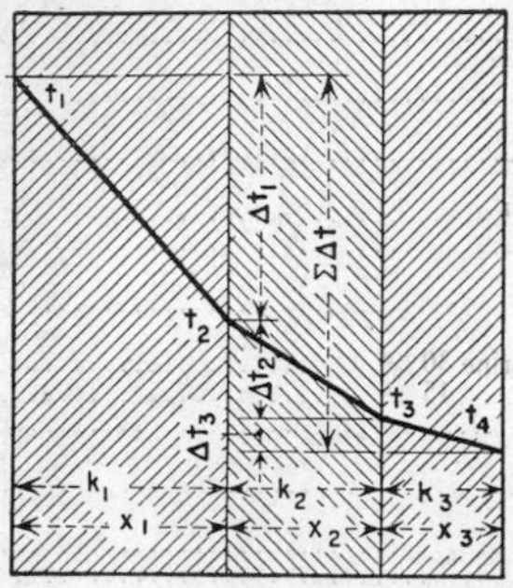

FIG. 1. Temperature gradients for steady heat conduction in series through three solids.

Since the heat flow through each of the three walls must be the same

$$q = \frac{k_1A_1\,\Delta t_1}{x_1} = \frac{k_2A_2\,\Delta t_2}{x_2} = \frac{k_3A_3\,\Delta t_3}{x_3} \qquad (8)$$

Since, by definition,

$$R = \frac{x}{kA} = \textbf{individual}\text{ thermal resistance} \qquad (9)$$

$$\Delta t_1 = qR_1, \qquad \Delta t_2 = qR_2, \qquad \Delta t_3 = qR_3 \qquad (10)$$

Adding the individual temperature drops, noting that q is uniform,

$$q(R_1 + R_2 + R_3) = \Delta t_1 + \Delta t_2 + \Delta t_3 = \Sigma\,\Delta t \qquad (11)$$

or

$$q = \frac{\Sigma\,\Delta t}{R_T} = \frac{t_1 - t_4}{R_T} \qquad (12)$$

where R_T is the **over-all** resistance and is the sum of the individual resistances in series, then

$$R_T = R_1 + R_2 + \cdots + R_n \qquad (12a)$$

When a wall is constructed of several layers of solids, the joints between adjacent layers may not perfectly exclude air spaces, and these additional resistances should not be overlooked. Consequently it is advisable to determine the over-all resistance of the composite wall. P. Nichols [*J. Am. Soc. Heating Ventilating Eng.* **30**, 35 (1924); *Ind. Eng. Chem.*, **16**, 490 (1924)] has described a "heat meter" which in practice can be used to determine the over-all resistance of a composite wall. In using this meter, care must be taken to secure data under steady conditions.

Conduction through Several Bodies in Parallel. For n resistances in parallel, the rates of heat flow are additive:

$$q = \frac{\Delta t}{R_1} + \frac{\Delta t}{R_2} + \cdots + \frac{\Delta t}{R_n} \qquad (13)$$

Table 11. Thermal Conductivities of Gases and Vapors

k = B.t.u./(hr.)(sq. ft.)(°F./ft.)

The extreme temperature values given constitute the experimental range. For extrapolation to other temperatures, it is suggested that the data given be plotted as log k vs. log T, or that use be made of the assumption that the ratio $c_p\mu/k$ is practically independent of temperature (or of pressure, within moderate limits).

Substance	t, °F.	k
Acetone[11]	32	0.0057
	115	.0074
	212	.0099
	363	.0147
Acetylene[3]	−103	.0068
	32	.0108
	122	.0140
	212	.0172
Air[1,11]	−148	.0095
	32	.0140
	212	.0183
	392	.0226
	572	.0265
Ammonia[3]	−76	.0095
	32	.0128
	122	.0157
	212	.0185
Benzene[11]	32	.0052
	115	.0073
	212	.0103
	363	.0152
	413	.0176
Butane (n-)[9]	32	.0078
	212	.0135
(iso-)[9]	32	.0080
	212	.0139
Carbon dioxide[12]	−58	.0068
	32	.0085
	212	.0133
	392	.0181
	572	.0228
disulfide[3]	32	.0040
	45	.0042
monoxide[1,3]	−312	.0041
	−294	.0046
	32	.0135
tetrachloride[11]	115	.0041
	212	.0052
	363	.0065
Chlorine[8]	32	.0043
Chloroform[11]	32	.0038
	115	.0046
	212	.0058
	363	.0077
Cyclohexane	216	.0095
Dichlorodifluoromethane	32	.0048
	122	.0064
	212	.0080
	302	.0097
Ethane[1,3]	−94	.0066
	−29	.0086
	32	.0106
	212	.0175
Ethyl acetate[11]	115	.0072
	212	.0096
	363	.0141
alcohol[11]	68	.0089
	212	.0124
chloride[11]	32	.0055
	212	.0095
	363	.0135
	413	.0152
ether[11]	32	.0077
	115	.0099
	212	.0131
	363	.0189
	413	.0209
Ethylene[3]	−96	.0064
	32	.0101
	122	.0131
	212	.0161
Heptane (n-)[11]	392	.0112
	212	.0103
Hexane (n-)[9]	32	.0072
	68	.0080

Substance	t, °F.	k
Hexene[11]	32	0.0061
	212	.0109
Hydrogen	−148	.065
	−58	.083
	32	.100
	122	.115
	212	.129
	572	.178
Hydrogen and carbon dioxide[7]	32	
0% H_2		.0083
20%		.0165
40%		.0270
60%		.0410
80%		.0620
100%		.10
Hydrogen and nitrogen[7]	32	
0% H_2		.0133
20%		.0212
40%		.0313
60%		.0438
80%		.0635
Hydrogen and nitrous oxide[7]	32	
0% H_2		.0092
20%		.0170
40%		.0270
60%		.0410
80%		.0650
Hydrogen sulfide[3]	32	.0076
Mercury[8]	392	.0197
Methane[1,3,9]	−148	.0100
	−58	.0145
	32	.0175
	122	.0215
Methyl alcohol[11]	32	.0083
	212	.0128
acetate[11]	32	.0059
	68	.0068
Methyl chloride[11]	32	.0053
	115	.0072
	212	.0094
	363	.0130
	413	.0148
Methylene chloride[11]	32	.0039
	115	.0049
	212	.0063
	413	.0095
Nitric oxide[3]	−94	.0103
	32	.0138
Nitrogen[2,3]	−148	.0095
	32	.0140
	122	.0160
	212	.0180
Nitrous oxide[2,3]	−98	.0067
	32	.0087
	212	.0128
Oxygen[1,2,6]	−148	.0095
	−58	.0119
	32	.0142
	122	.0164
	212	.0185
Pentane (n-)[9,11]	32	.0074
	68	.0083
(iso-)[11]	32	.0072
	212	.0127
Propane[9]	32	.0087
	212	.0151
Sulfur dioxide[2]	32	.0050
	212	.0069
Water vapor, zero pressure[10,14,*]	32	.0132
	200	.0159
	400	.0199
	600	.0256
	800	.0306
	1000	.0495

[1] Chilton and Genereaux, private communication, 1940.
[2] Dickens, *Proc. Roy. Soc. (London)*, **A143**, 517 (1934).
[3] Eucken, *Physik. Z.*, **12**, 1101 (1911); **14**, 324 (1913) (see footnote **17**).
[4] Gregory, *Proc. Roy. Soc. (London)*, **A149**, 324 (1935).
[5] Gregory and Archer, *Proc. Roy. Soc. (London)*, **A110**, 119 (1926).
[6] Gregory and Marshall, *Proc. Roy. Soc. (London)*, **A118**, 594 (1928).
[7] Ibbs and Hirst, *Proc. Roy. Soc. (London)*, **A123**, 134 (1929).
[8] "International Critical Tables," McGraw-Hill, New York, 1929.
[9] Mann and Dickens, *Proc. Roy. Soc. (London)*, **A134**, 77 (1931).
[10] Keenan and Keyes, "Thermodynamic Properties of Steam," Wiley, New York, 1944 (tenth impression).
[11] Moser, *Dissertation*, Berlin, 1913 (see footnote **17**).
[12] Sherrat and Griffiths, *Phil. Mag.*, **27**, 68 (1939).
[13] Spence and Dock, *Phil. Mag.*, **25**, 129 (1938).
[14] Varhaftik and Timrot, *J. Tech. Phys. (U.S.S.R.)*, 963 (1939).
[15] Varhaftik and Parquenov, *J. Exptl. Theoret. Phys. (U.S.S.R.)*, **8**, 189 (1938).
[16] Wüllner, *Ann. Physik*, **4**, 321 (1878).
[17] Data from Eucken and Moser are measurements relative to air. Data in this table from these sources are based on the thermal conductivity of air at 32°F. of 0.0140 B.t.u./(hr.)(sq. ft.)(°F./ft.).

* For saturated vapor (reference 10):

Lb./sq. in. abs.	250	500	1000	1500	2000
t, °F.	401	467	545	596	636
k	0.0248	0.0299	0.0395	0.0486	0.0578

$$q = \left(\frac{1}{R_1} + \frac{1}{R_2} + \cdots + \frac{1}{R_n}\right)\Delta t \qquad (13a)$$

$$q = (C_1 + C_2 + \cdots + C_n)\,\Delta t = \Sigma C\,\Delta t \qquad (13b)$$

where R_1 to R_n are the individual resistances and C_1 to C_n are the individual conductances; $C = kA/x$. Two-dimensional heat conduction can be predicted by the graphical method of Awbery and Schofield [*Proc. Intern. Congr. Refrig.*, 5th Congr., **3,** 591 (1929)], or by the relaxation procedure of Southwell, "Relaxation Method of Theory of Physics," Oxford Univ. Press, New York, 1946.

Heat Transfer in the Unsteady State (Heating and Cooling of Solids). In problems involving conduction of heat in the transient state, as in the warming or cooling of solid bodies, the temperature of the body varies with both time and the position of points in the body, and the mathematical relations are complicated. However, the basic differential equations for conduction have been integrated for various shapes and boundary conditions (see Ingersoll *et al.*, "Heat Conduction," McGraw-Hill, New York, 1948; Byerly, "Elementary Treatise on Fourier Series," Ginn, Boston, 1928; Carslaw, "Mathematical Theory of Heat," Macmillan, New York, 1921), and the results may be plotted as curves involving four ratios [Gurney and Lurie, *Ind. Eng. Chem.*, **15,** 1170 (1923)] defined as follows:

$$Y = \frac{t' - t}{t' - t_b} \qquad X = \frac{k\theta}{\rho c_p r^2{}_m} \qquad (14)$$

$$m = \frac{k}{h_T r_m} \qquad n = \frac{r}{r_m} \qquad (15)$$

Since each ratio is dimensionless, any consistent units may be employed in any ratio. The significance of the symbols is as follows: t', the temperature of the surroundings; t_b, the initial uniform temperature of the body; t,

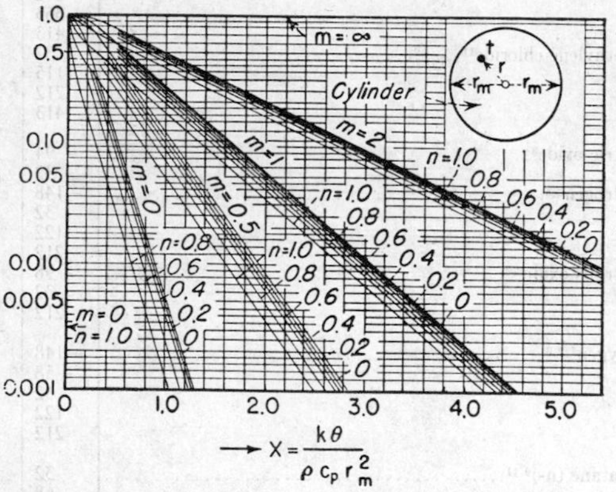

FIG. 2. Heating and cooling of a solid cylinder having infinite ratio of length to diameter.

the temperature at a given point in the body at the time θ measured from the start of the heating or cooling operations; k, the uniform thermal conductivity of the body; ρ, the uniform density of the body; c_p, the specific heat of the body; h_T, the coefficient of total heat transfer between the surroundings and the surface of the body (see p. 464), expressed as heat transferred per unit time per unit area of the **surface** per unit difference in temperature between surroundings and surface; r, the distance, in the direction of heat conduction, from the mid-point or mid-plane of the body to the point under consideration; r_m, the radius of a sphere or cylinder, one-half the thickness of a slab heated from **both** faces, the total thickness of a slab heated from one face and insulated perfectly at the other; x, the distance, in the direction of heat conduction, from the surface of a semi-infinite body (such as the surface of the earth) to the point under consideration. In making the integrations which lead to the curves shown, the following factors were assumed constant: c_p, h_T, k, r, r_m, t', x, and ρ.

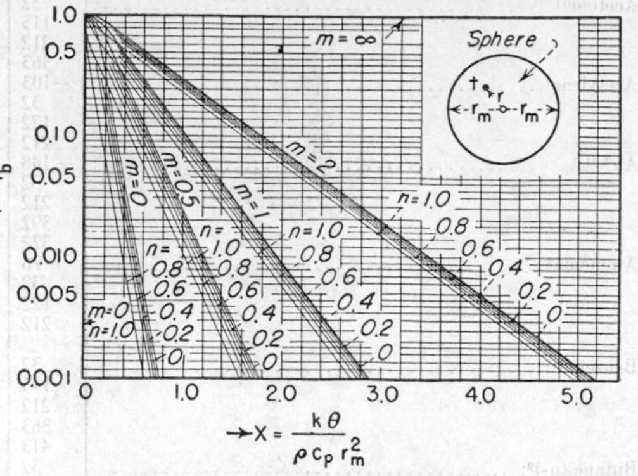

FIG. 3. Heating and cooling of a solid sphere.

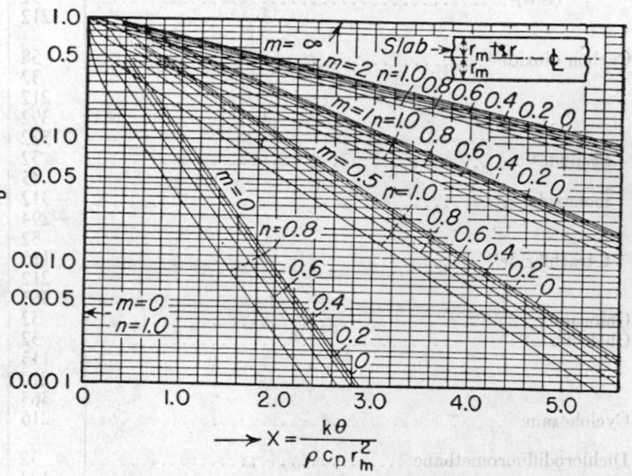

FIG. 4. Heating and cooling of a solid slab having a large face area relative to that of the edges.

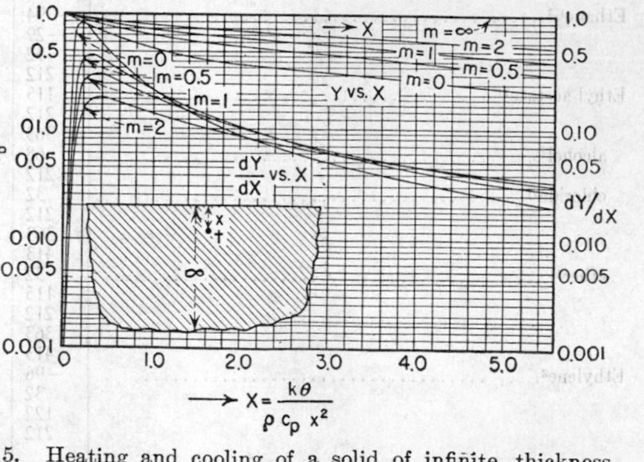

FIG. 5. Heating and cooling of a solid of infinite thickness, neglecting edge effects. (This may be used as an approximation in the zone near the surface of a body of finite thickness.)

The working curves are shown in Figs. 2 to 5 for the cylinders of infinite length, spheres, slabs of infinite faces, and semi-infinite solids, respectively, with Y plotted as ordinates on a logarithmic scale vs. X as abscissas to an arithmetic scale, for various values of the ratios m and n. To facilitate calculations involving instantaneous rates of cooling or heating of the semi-infinite body, Fig. 5 shows also a curve of dY/dX vs. X. Plots similar to Figs. 2–5, but to a larger scale, are given in McAdams, Brown and Marco, and Schack (see references p. 456).

Example. A flat slab of rubber, ½ in. thick, initially at 80°F., is to be placed between two electrically heated steel plates maintained at 287°F. The heating is to be discontinued when the temperature at the center line of the rubber slab reaches 270°F.

a. Calculate the length of the heating period.

b. At the end of the run, what would be the temperature of the rubber in a plane 0.1 in. from the center line?

Data. Using the ft.-lb.-hr.-°F.-B.t.u. system, for the rubber $k = 0.092$ and $k/\rho c_p = 0.0029$. Assume a contact coefficient h_T from metal to rubber of 1000.

Solution. All quantities will be expressed in the units mentioned above. Noting that the mid-plane distance r_m is $\frac{1}{48}$ = 0.0208 ft., $m = k/h_T r_m = 0.092/(1000)(0.0208) = 0.00442$. At the end of θ hr. of heating, $Y = (287 - 270)/(287 - 80) = 0.0821$. At the center line of the rubber slab, $n = r/r_m = 0$. From Fig. 4, for $Y = 0.0821$, $n = 0$, by interpolation to $m = 0.0044$, $X = 1.13 = k\theta/\rho c_p r_m^2 = 0.0029\theta/(0.0208)^2$, whence $\theta = 0.169$ hr., giving the answer to part (*a*). At this same time, $X = 1.13$, for the point 0.1 in. from the center line, $n = 0.1/0.25 = 0.4$, and, as before, $m = 0.0044$; from Fig. 4, $Y = 0.065 = (287 - t)/(287 - 80)$, whence $t = 273.5$°F., the answer to part (*b*). By similar procedure, temperatures at various positions and times may be predicted, thus making it possible to construct curves of t vs. θ for various positions, n.

With the infinite slab, in the early stages of the operation where Fig. 4 gives insufficient precision, Fig. 5 may be used for points near the surface. For a brick-shaped solid having the dimensions $2r_{m1}$, $2r_{m2}$, and $2r_{m3}$, the value of Y at a given time and position may be evaluated by the method of Newman [*Trans. Am. Inst. Chem. Engrs.*, **27**, 310 (1931)] as follows: Y equals the product $Y_1Y_2Y_3$, where Y_1 is evaluated from Fig. 4 at $X_1 = k\theta/\rho c_p r_{m1}^2$, $n_1 = r_1/r_{m1}$, and $m_1 = k/h_T r_{m1}$; similarly Y_2 and Y_3 are read for the same θ at X_2, n_2, and m_2, and at X_3, n_3, and m_3, corresponding to r_{m2} and r_{m3}.

When assumptions made for the analytical treatment need broadening (h_T varies with surface temperature, t_0 is non-uniform, etc.), close approximation for one-dimensional problems can be made by the ingenious graphical method of E. Schmidt ("Foppls Festschrift" Springer, Berlin), given in Sherwood and Reed ("Applied Mathematics in Chemical Engineering," p. 241, McGraw-Hill, New York, 1939), and the method of Dusinberre [*Trans. Am. Soc. Mech. Engrs.*, **67**, 703 (1945)]. For two- or three-dimensional cases the ingenious method of Southwell is applicable; see Emmons [*Trans. Am. Soc. Mech. Engrs.*, **65**, 607 (1943)] or Fowler [*Quart. Appl. Math.*, **3**, 361 (1946)].

Rubber. T. K. Sherwood [*Ind. Eng. Chem.*, **20**, 1181 (1928)] applies the theory of heating of solids to the rate of vulcanization of rubber.

Wood. MacLean [*Proc. Am. Wood Preserv. Assoc.*, p. 197 (1930)], in a study of the data of Wirka [*ibid.* (1924–1925)] on the steaming of green southern-pine timbers, found that the integrated equation (see Fig. 2 for the long cylinder) correlated the data on temperature gradients at various periods in the bath operation and obtained an average value of $k/\rho c_p = 0.00678$ sq. ft./hr. for the thermal diffusivity of the wood. The radii of the 33 specimens ranged from 3.37 to 6 in., the average moisture content was 63 per cent by weight, and the average density was 65.4 lb./cu. ft.

Blast-furnace Stoves and Heat Regenerators. When hot gases are intermittently turned into a chamber partly filled with cooler solids, or when cold air is allowed to flow intermittently over warmer solids, the mathematical relations become quite involved. See Hauser, *Z. angew. Math. Mech.*, **9**, 173 (1929); **11**, 105 (1931); *Tech. Mech. Therm.*, **1**, 219 (1930); Lubojatzky, *Metall Erz*, **28**, 205 (1931); Rummel, *J. Inst. Fuel*, **4**, 160 (1931); Schack. *Arch. Eisenhüttenw.*, **2**, 223, 481 (1929); Heiligenstaedt, "Regeneratoren, Rekuperatoren und Winderhitzer," Spamer, Leipzig, 1931.

An approximate result may be obtained from the equation of Rummel [*Arch. Eisenhuttenw.*, **4.**, 367 (1931)] as follows:

$$\frac{Q}{A} = \frac{(\Delta t_1 - \Delta t_2)/(\ln \Delta t_1/\Delta t_2)}{\dfrac{1}{h_A\theta_A} + \dfrac{1}{h_G\theta_G} + \dfrac{1}{2.5c_p\rho r'} + \dfrac{r'}{k(\theta_A + \theta_G)}} \tag{16}$$

in which Δt_1 and Δt_2 are the average differences in temperature between gas and air at the two ends of the regenerator; c_p, ρ, and k are, respectively, the specific heat, density, and thermal conductivity of the brick; r' is the ratio of the total volume to the exposed surface of the bricks; θ_G and θ_A are the length of the gas and air blows; and h_G and h_A are the individual heat-transfer coefficients from gas to bricks and from bricks to air, respectively, values of which are given by Rummel.

CONVECTION

Introduction to Convection. In many cases of heat transfer, involving either a liquid or a gas, convection is an important factor. In the majority of heat-transfer cases met in industrial practice, heat is being transferred from one fluid through a solid wall to another fluid. Assume a hot fluid, at a temperature t_1, flowing past one side of a metal wall, and a cold fluid, at t_7, flowing past the other side, to which a scale of thickness x_s adheres. In such a case, the conditions obtaining at a given section are illustrated diagrammatically in Fig. 6. In case of

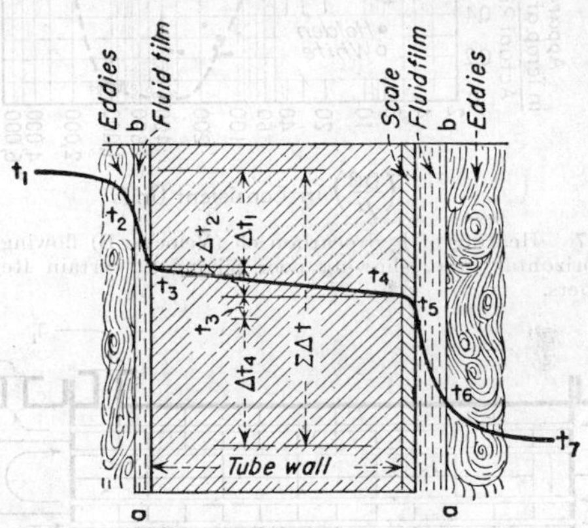

FIG. 6. Temperature gradients for steady flow of heat by conduction and convection from a warmer to a colder fluid separated by a solid wall.

turbulent flow of a fluid past a solid, it has long been known that, in the immediate neighborhood of the surface, there exists a relatively quiet zone of fluid, commonly called the "film," and that a considerable fraction of the total drop in temperature, between the main body of the fluid and the surface of the solid, occurs in the film. In more recent years it has been recognized that for **isothermal** turbulent flow, the motion in the film is laminar (streamline) in character, and the outer boundary ab of the film is now commonly defined as characterized by some critical value of the Reynolds dimensionless group, DG/μ (see p. 468). For convenience in visualization, it has often been assumed that the temperature gradient is wholly confined to the film flowing in laminar motion, although because of lack of perfect mixing in the main body of the fluid this assumption may be substantially in error.

Individual Coefficient of Heat Transfer. Since it is not convenient to measure the thickness of the fluid film or the temperature at the interface between the film and the main body of the fluid, and since both conduction and convection are involved, the differential rate of heat flow between fluid and solid is calculated from the equations

$$dq = h_i \cdot dA_i \cdot (t_1 - t_3) \quad (17)$$
$$dq = h_o \cdot dA_o(t_s - t_7) \quad (17a)$$

and the observed value of h is called the individual coefficient, "film coefficient," or surface coefficient, and includes the thermal resistances of the laminar film, "buffer" layer between film and core, and turbulent core. The coefficient h is determined by dividing the known rate of heat flow per unit surface of the wall by the difference between the temperatures of the fluid and surface. The temperature of the fluid is not uniform, because of the existence of the temperature gradient (Fig. 6), and it is therefore desirable to state at what point or area the fluid temperature was measured. The difference between the temperature at a point at the axis of the pipe and the average temperature is not serious for flow well in the turbulent region; but, for values of the Reynolds number DG/μ (see p. 468) below 2100 in consistent units, the tendency of the fluid to flow in streamlines may preclude good mixing. Figure 7 [Drew, Hogan, and

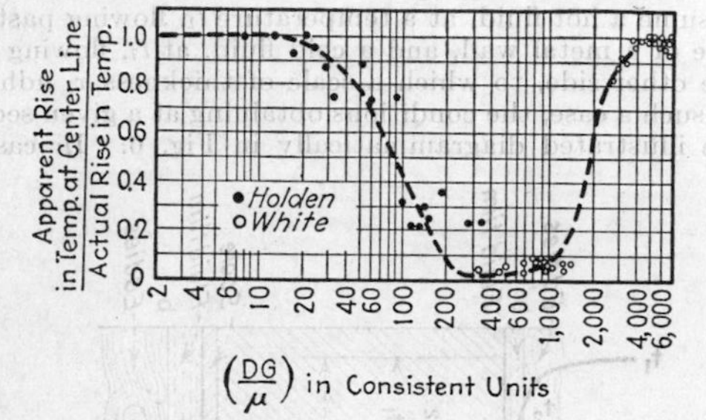

FIG. 7. Heating of hydrocarbon oil (Velocite B) flowing inside a horizontal tube, showing poor mixing at certain Reynolds numbers.

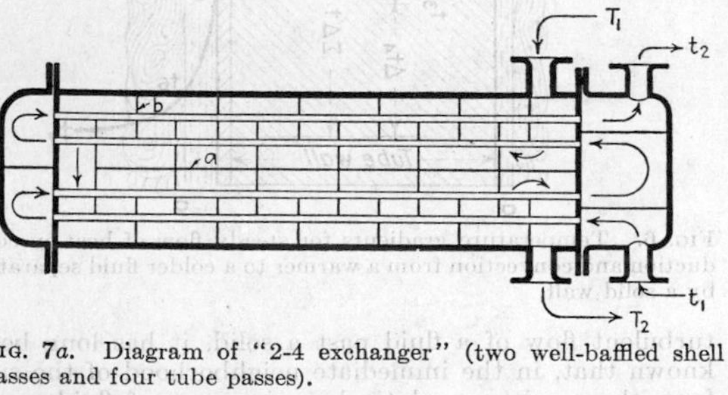

FIG. 7a. Diagram of "2-4 exchanger" (two well-baffled shell passes and four tube passes).

McAdams, *Trans. Am. Inst. Chem. Engrs.*, **26**, 81 (1931)], is based on data for the heating of Velocite B oil (see p. 460). When DG/μ was between 1500 and 300, the apparent temperature rise was not over one-tenth of the actual rise based on the temperature of the oil after mixing, but for values of DG/μ as low as 10 the rate of flow was so small that the entire body of fluid had been warmed to almost the temperature of the wall.

In determining the rate of heat flow from the weight of fluid flowing per unit time, its specific heat, and the apparent rise in temperature, the use of fluid temperature at the center line, in place of the average temperature obtained on mixing, is apt to introduce a far greater error in q than in the temperature difference. In experimental work it is highly desirable to provide two independent means of determining q, and these should be checked by a heat balance. In connection with the equations given herein one should employ the bulk temperature of the fluid, *i.e.*, the temperature obtained on mixing, in evaluating both q and the temperature difference.

Over-all Coefficient of Heat Transfer. In testing commercial heat-transfer equipment, it is not convenient to measure tube temperatures (t_3 or t_4 in Fig. 6) and hence the over-all performance is expressed as an over-all coefficient of heat transfer, U, based on a convenient area dA which may be dA_i, dA_o, or an average of dA_i and dA_o, and dA_o, say dA_{avg}. Whence, by definition,

$$dq = U \cdot dA \cdot (t_1 - t_7) \quad (18)$$

U is called the "over-all coefficient of heat transfer," or merely "over-all coefficient." The rate of conduction through the tube wall and scale deposit is given by

$$dq = \frac{k \cdot dA_{\text{avg}} \cdot (t_3 - t_4)}{x} = h_d \cdot dA \cdot (t_4 - t_5) \quad (19)$$

Upon eliminating t_3, t_4, and t_5 from Eqs. (17), (17*a*), (18), and (19), the complete expression for the steady rate of heat flow from one fluid through the wall and scale to a second fluid, as illustrated in Fig. 6, is

$$dq = \frac{t_1 - t_7}{\dfrac{1}{h_i \cdot dA_1} + \dfrac{x}{k \cdot dA_{\text{avg}}} + \dfrac{1}{h_d \cdot dA} + \dfrac{1}{h_o \cdot dA_o}} = U \cdot dA \cdot (t_1 - t_7) \quad (20)$$

Table 12. Heat Transfer Coefficients h_d for Scale Deposits from Water*

For use in Eq. (22)

Temperature of heating medium	Up to 240°F.		240–400°F.	
Temperature of water	125°F. or less		Above 125°F.	
Water velocity, ft./sec.	3 and less	Over 3	3 and less	Over 3
Distilled	2000	2000	2000	2000
Sea water	2000	2000	1000	1000
Treated boiler feed water	1000	2000	500	1000
Treated make-up for cooling tower	1000	1000	500	500
City, well, Great Lakes	1000	1000	500	500
Brackish, clean river water	500	1000	330	500
River water, muddy, silty†	330	500	250	330
Hard (over 15 g./gal.)	330	330	200	200
Chicago Sanitary Canal	130	170	100	130

Miscellaneous cases: refrigerating liquids, brine, clean petroleum distillates, organic vapors, 1000; refrigerant vapor, 500; vegetable oils, 330; fuel oil (topped crude), 200.

* From Standards of Tubular Exchanger Manufacturers Association, New York, N. Y., 1941.

† Delaware, East River (N.Y.), Mississippi, Schuylkill, and New York Bay.

Mean Temperature Difference. In a continuously operated heat exchanger the temperature difference between warmer and colder fluids varies, in general, throughout the length of the exchanger. To allow for this condition it is necessary to integrate the basic relation $dq = U \cdot dA \cdot \Delta t_0$, wherein Δt_0 denotes the **over-all** temperature difference between hot and cold fluids. The assumptions usually made are constant U, constant mass rates of flow, no changes in phase, constant specific heats, and negligible heat losses. For parallel or counterflow of fluids, the resulting equation is

$$q = U \cdot A \cdot \Delta t_{o,lm} = U \cdot A \left(\frac{\Delta t_{01} - \Delta t_{02}}{\ln \dfrac{\Delta t_{01}}{\Delta t_{02}}} \right) \quad (21)$$

in which the term in the parentheses is the **logarithmic** mean of the terminal temperature differences Δt_{01} and Δt_{02}. The value of UA is obtained from the resistance

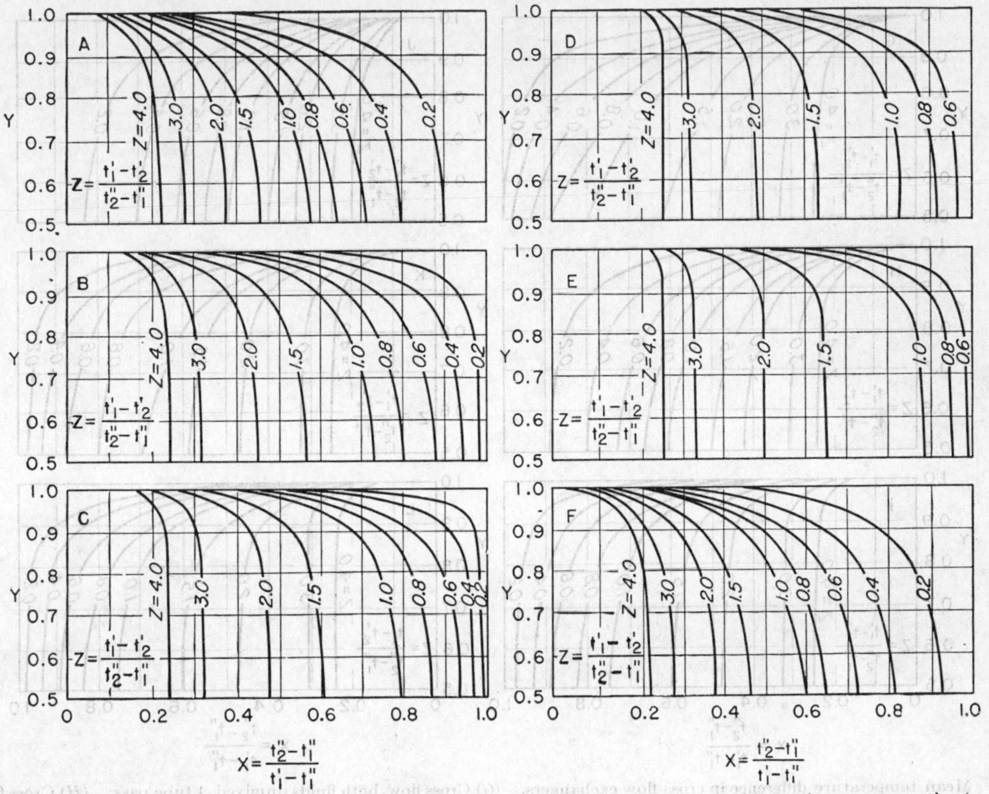

FIG. 7b. Mean temperature difference in reversed-current exchangers. (Shell side well mixed at a given cross section.) (A) One shell pass and 2, 4, 6, etc., tube passes. (B) Two shell passes and 4, 8, 12, etc., tube passes. (C) Three shell passes and 6, 12, 18, etc., tube passes. (D) Four shell passes and 8, 16, 24, etc., tube passes. (E) Six shell passes, and 12, 24, 36, etc., tube passes. (F) One shell pass and 3, 6, 9, etc., tube passes. (*Bowman, Mueller, and Nagle.*)

concept

$$\frac{1}{UA} = \frac{1}{h_iA_i} + \frac{1}{h_{di}A_i} + \frac{x}{kA_{\text{avg}}} + \frac{1}{h_{do}A_o} + \frac{1}{h_oA_o} \qquad (22)$$

For Z of 1 (p. 467), $\Delta t_{o,lm} = \Delta t_{o1} = \Delta t_{o2}$.

If U varies considerably with temperature, the apparatus should be visualized as divided into stages, in each of which variation of U with temperature or temperature difference is linear. Then for parallel or counterflow operation the following equation [Colburn, *Ind. Eng. Chem.*, **25**, 873 (1933)] may be applied to obtain the surface required in each stage:

$$q = A\left(\frac{\Delta t_{01}U_2 - \Delta t_{02}U_1}{\ln\dfrac{\Delta t_{01}U_2}{\Delta t_{02}U_1}}\right) \qquad (23)$$

Example. In a proposed design of a liquid-to-liquid cooler, Δt_{01} is 60° and Δt_{02} is 5°; the values of U_1 and U_2 are 305 and 141, respectively. Compare the values of q/A based on (*a*) the correct method, Eq. (23); (*b*) the arbitrarily chosen product of $(U_1 + U_2)/2$ and Eq. (21), and (*c*) the arbitrarily chosen product of $(U_1 + U_2)/2$ and $\Delta t_m = (\Delta t_{01} + \Delta t_{02})/2$.

a. By Eq. (23)

$$\frac{q}{A} = \frac{(141 \times 60) - (305 \times 5)}{2.3 \log_{10}\left(\dfrac{141 \times 60}{305 \times 5}\right)} = 4050 \text{ B.t.u./(hr.)(sq. ft.)}$$

$$b.\ \frac{q}{A} = \left(\frac{U_1 + U_2}{2}\right)\left[\frac{\Delta t_{01} - \Delta t_{02}}{2.3 \log\left(\dfrac{\Delta t_{01}}{\Delta t_{02}}\right)}\right] = (223)(22.2) = 4950 \text{ B.t.u./(hr.)(sq. ft.)}$$

$$Y = \Delta t_{om}/\Delta t_{o,lm}$$

$$c.\ \frac{q}{A} = \left(\frac{U_1 + U_2}{2}\right)\left(\frac{\Delta t_{01} + \Delta t_{02}}{2}\right) = \left(\frac{305 + 141}{2}\right)\left(\frac{60 + 5}{2}\right) = 7250 \text{ B.t.u./(hr.)(sq. ft.)}$$

Multipass and Cross-flow Exchangers. In these exchangers, where the flow is neither parallel nor countercurrent, the logarithmic mean temperature difference does not apply when the temperatures of both fluids change. Figure 7*a* shows an exchanger with two well-baffled passes in the shell and four passes in the tubes Making the same assumptions which led to Eq. (21), plus the assumption that the shell-side fluid is well mixed by suitable baffles, one employs the relation

$$q = U_m \cdot A \cdot \Delta t_{om} = U_m \cdot A \cdot Y \cdot \Delta t_{o,lm} \qquad (23a)$$

evaluating Y from Fig. 7*b* or 7*c*. [Bowman, Mueller, and Nagle, *Trans. Am. Soc. Mech. Engrs.*, **62**, 283 (1940).]

For example, assume an exchanger in which the hot fluid enters at 400° and leaves at 200°: the cold fluid enters at 100° and leaves at 200°. Assuming U independent of temperature, what will be the true mean-temperature difference from hot to cold fluid, (1) for counterflow and (2) for a reversed current apparatus with one well-baffled pass in the shell and two passes in the tubes?

1. With counterflow the terminal differences are 400 − 200 = 200, and 200 − 100 = 100; the logarithmic mean difference is 100/(2.3)(0.301) = 144°.

2. Z = (400 − 200)/(200 − 100) = 2, X = (200 − 100)/(400 − 100) = ⅓; from section A of Fig. 7*b*, Y = 0.81 = Δt_{om}/144; Δt_{om} = 117°.

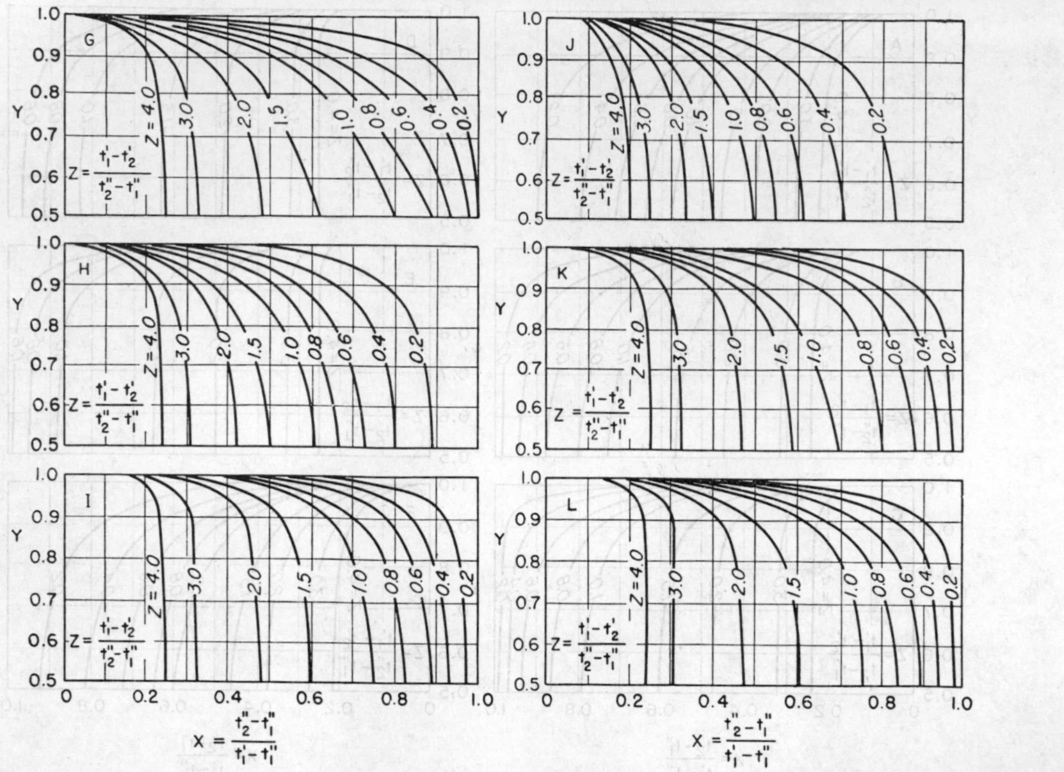

FIG. 7c. Mean temperature difference in cross-flow exchangers. (*G*) Cross flow, both fluids unmixed, 1 tube pass. (*H*) Cross flow, shell fluid mixed, 1 tube pass. (*I*) Cross flow, shell fluid mixed, 2 tube passes, shell fluid flows across second and first passes in series. (*J*) Cross flow, shell fluid mixed, 2 tube passes, shell fluid flows over first and second passes in series. (*K*) Cross flow (drip type), 2 horizontal passes with U-bend connections (trombone type). (*L*) Cross flow (drip type), helical coils with 2 turns. (*Bowman, Mueller, and Nagle.*)

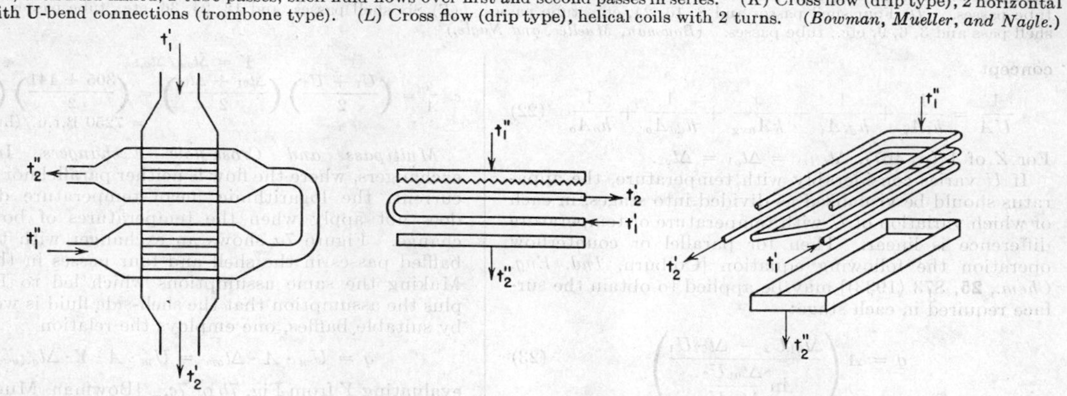

FIG. 7d.

If one of the temperatures remains constant, as in a condenser or in an evaporative cooler, Eq. (21) applies for parallel flow, counterflow, multipass, and cross flow.

Gardner treats unequal passes with 1:2 and 2:4 exchangers [*Ind. Eng. Chem.*, **33**, 1215 (1941)], an array of identical exchangers [*ibid.*, **34**, 1083 (1942)], and variable shell-side coefficient [*Trans. Am. Soc. Mech. Engrs.*, **67**, 33 (1945)].

Conversion Factors for Coefficients of Heat Transfer. Throughout this section, values of h and U are expressed in B.t.u./(hr.)(sq. ft.)(°F.). Conversion factors to other units are as follows:

Multiply by 1 to obtain p.c.u./(hr.)(sq. ft.)(°C.).

Multiply by 4.88 to obtain kg.-cal./(hr.)(sq. m.)(°C.).

Multiply by 0.0001355 to obtain g.-cal./(sec.)(sq. cm.)(°C.).

Multiply by 0.000568 to obtain watts/(sq. cm.)(°C.).

Multiply by 0.00204 to obtain watts/(sq. in.)(°F.).

Multiply by 0.000394 to obtain hp./(sq. ft.)(°F.).

The following section (pp. 467–479) deals with **individual coefficients** of heat transfer by **convection.** Heat transfer by radiation from solid to solid is treated on pp. 483–489, and heat transfer by radiation between solids and certain gases such as carbon dioxide and water vapor is treated on pp. 490–493. **Over-all coefficients** of heat transfer by convection are given on pp. 480–482, and also on pp. 474, 479, and 479–482.

Table 13. Nomenclature and Units

(The units are based on feet, pounds, hours, degrees Fahrenheit, and B.t.u. Any other consistent set may be used in the dimensionless relations given, but for the dimensional equations the units of this table must be used.)

A = area of heat-transfer surface, sq. ft.; A_i for inside, A_o for outside, A_m for average.
a = dimensionless constant in equations as specified.
b = empirical dimensional factor in Eq. (28c).
c_p = specific heat at constant pressure, B.t.u./(lb.)(°F.)
d = prefix, indicating differential.
D = diameter, ft.; D_c for helix; D_i for inside diameter; D_o for outside diameter; D_p for packing.
D' = diameter, in., D_i' for inside diameter; D_o' for outside diameter.
e = base of natural logarithms, 2.718.
f = friction factor, dimensionless (p. 381).
G = mass velocity, equals w/S, lb./(hr.)(sq. ft. of cross section occupied by fluid); G' = lb./(sec.)(sq. ft.).
G_{max} = mass velocity through minimum free area in a row of pipes normal to the fluid stream, lb./(hr.)(sq. ft.).
g_c = conversion factor 4.18×10^8, (lb. mass)(ft.)/(force lb.)(hr.)^2.
g_L = acceleration due to gravity, 4.18×10^8 ft./hr.^{2}
h = local individual coefficient of heat transfer, equals $dq/dA\ \Delta t$, B.t.u./(hr.)(sq. ft.)(°F. diff.).
$h_c + h_r$ = combined coefficient by conduction, convection, and radiation between surface and surroundings.
h_m = mean value of h for entire apparatus, based on Δt_m.
$h_{a.m.}$ = average h, arbitrarily based on arithmetic-mean temperature difference.
h_s = heat-transfer coefficient for deposit of dirt or scale on the surface.
j = dimensionless ordinate as specified in Fig. 8.
J = mechanical equivalent of heat, 778 ft.-lb./B.t.u.
k = thermal conductivity of fluid, B.t.u./(hr.)(sq. ft.)(unit temperature gradient, °F./ft.).

$$k_m = \frac{1}{(t_1 - t_2)} \int_2^1 k\, dt$$

k_f = k at the "film" temperature, $t_f = (t + t_w)/2$
L = length of heat-transfer surface, heated length, ft.; L' = length of horizontal tube, ft.
L_p = length of horizontal flat paddle; see page **474**.
ln = logarithm to base e; = 2.30 × logarithm to the base 10.
N = number of revolutions per hour.
n = number of rows in a vertical plane; exponent.
P = total pressure, atm.
Q = quantity of heat, B.t.u.
q = rate of heat flow, B.t.u./hr.
R = thermal resistance, $1/UA$.
r = radius, ft.
r_m = distance from surface to mid-plane, ft.
S = cross section, filled by fluid, in plane normal to direction of fluid flow, sq. ft.
T = temperature, °F. abs. = $t + 460°$.
T_1, T_2 = inlet and outlet bulk temperatures, respectively, of warmer fluid, °F.; t_1', t_2' in Fig. 7*b–d*.
t = bulk temperature (based on heat balance), °F.
t_v = saturation temperature of vapor.
t_w = wall temperature, °F.
t_1, t_2 = inlet and outlet bulk temperatures of colder fluid, °F.; t_1'', t_2'' in Fig. 7*b–d*.
t_∞ = temperature of stream of great depth, ambient temperature, °F.
t_i, t_o = temperatures of fluid inside and outside, °F.
t_f = $(t + t_w)/2$.
t_f' = $t_v - 0.75(\Delta t)_m$ for Eqs. (47) and (48).
U = over-all coefficient of heat transfer, B.t.u./(hr.)(sq. ft.)(°F.); U_o based on outside surface.
V = average velocity, ft./hr.
V_a = acoustic velocity, ft./hr.
V_s = average velocity, volumetric rate divided by cross section filled by fluid, ft./sec.
V_∞ = velocity of stream of great depth, ft./hr.
w = mass rate of flow from each tube, lb./(hr.)(tube).
x = length of conduction path, ft.
Y = dimensionless correction factor for Δt; see Figs. 7*b* and 7*c*; $Y = \Delta t_{om}/\Delta t_{o,lm}$.
y = dimensionless recovery factor, page 468.
Z = $(T_1 - T_2)/(t_2 - t_1)$.
z_p = twice the perimeter over which fluid flows in passing fin, ft.
α = a constant.
β = volumetric coefficient of thermal expansion, having units of reciprocal of Fahrenheit absolute temperature.
Γ = mass rate of flow from tube, lb./(hr.)(ft. of wetted periphery measured on a plane normal to direction of fluid flow); $w/\pi D$ for a vertical and $w/2L'$ for a horizontal tube.
Δt = temperature difference, °F., Δt_o for over-all, and Δt for individual.
$\Delta t_{a.m.}$, $\Delta t_{l.m.}$ = arithmetic and logarithmic means of terminal temperature differences, respectively, °F.
Δt_m = true average (length-mean) value of the terminal temperature differences, °F.
Δt_s = temperature difference between surface and surroundings, °F.
θ = time, hr.
λ = latent heat (enthalpy) of vaporization, B.t.u./lb.
μ = absolute viscosity at bulk temperature, lb. mass/(hr.)(ft.); equals 2.42 × centipoises = 105,800 × (lb. force)(sec.)/sq. ft.
μ_f = viscosity, lb. mass/(hr.)(ft.), at arithmetic mean of wall and fluid temperatures.
μ_w = viscosity at wall temperature, lb. mass/(hr.)(ft.).
π = 3.1416.
ρ = density, lb. mass/cu. ft.
ρ_∞ = density of stream of great depth, lb. mass/cu. ft.

Heating and Cooling of Fluids (No Change in Phase)

Figure 8 shows correlations of film coefficients for warming or cooling by forced convection for a number of shapes. In the **dimensionless** relations any self-consistent units may be employed, but those given in Table 13 (p. 467) are generally used. For the **dimensional** equations the units of Table 13 must be employed. Viscosities are given on p. 372, thermal conductivities on p. 459, and specific heats on p. 225*ff*. For gases at ordinary pressures the following values of the dimensionless Prandtl group $c_p\mu/k$ may be used, independent of temperature: air, oxygen, nitrogen, hydrogen, carbon monoxide, 0.74; methane, 0.78; carbon dioxide, 0.79; ethylene 0.81; a value of 0.78 may be used for steam at low pressures. The values of h do not include heat transfer by radiation, which at high temperature becomes important with gases (p. 490).

Gases in Long Straight Tubes. For the turbulent flow of gases in straight tubes, the following **dimensional** equation for forced convection is recommended for general use:

$$h = \frac{16.6 c_p (G')^{0.8}}{(D_i')^{0.2}} \tag{24}$$

where c_p is the specific heat of the gas at constant pressure, B.t.u./(lb.)(°F.) = g.-cal./(g.)(°C.) = p.c.u./(lb.)(°C.); G' is the mass velocity, expressed as lb. of gas/**sec.**/sq. ft. of cross section of the gas passage; and D_i' is the actual inside diameter in **inches**. Equation (24) is based on the data of Nusselt [*Z. Ver. deut. Ing.*, **53**, 1750, 1808 (1909)] who used air, carbon dioxide, and illuminating gas at temperatures around 120°F., and the data of a number of other investigators using air in tubes ranging from 0.17 to 3.77 in. diameter, and covers values of G' ranging from G_t' to 30. Below the transitional value G_t', free convection factors are important, and h should be predicted as shown on p. 472. Based on the data of Nusselt for air at approximately 120°F., flowing inside a horizontal tube having an inside diameter 0.868 in., and at temperature differences in the neighborhood of 100°F., A. P. Colburn (personal communication) finds that $G_t' = 0.46P^{0.645}$, where P ranges from 1 to 13 atm. abs. The effect of pressure up to 16 atm. for air is taken care of by the mass velocity term; at higher pressures, see Colburn, Drew, and Worthington, Heat Transfer in a 3:1 Hydrogen-nitrogen Mixture at High Pressure, *Ind. Eng. Chem.*, **39**, 958 (1947). Data on the flow of gases at high temperatures are rather inconclusive as to the effect of temperature on h; hence, at this time, it seems best not to introduce a temperature-correction term. For a similar reason, no term for the ratio of length to diameter is included, but the effect is probably small, except for ratios of length to diameter from 0 to 10. From the meager comparative data as to the effect of roughness or tube materials on h, one concludes that the effect is minor for metal tubes. The effect of a given scale deposit is far less serious for a gas than for water (see

p. 464), owing to the higher thermal resistance of the gas film relative to the water film. However, dust, or materials which sublime, such as sulfur, may seriously reduce heat transfer between gas and solid. The radiation of heat from certain gases becomes important at high temperatures (see pp. 490–493).

For mixtures of gases c_p should be a weighted mean on the weight fraction basis.

Heat Transfer to Gases Flowing at Very High Linear Velocities. When a temperature probe is placed in a stream of compressible fluid flowing at high velocity, the retardation of the stream near the probe, and consequent conversion of kinetic energy to enthalpy, substantially increases the temperature of the probe above that of the high-velocity stream. The stagnation temperature T_s for a perfect gas is defined as the temperature which would be obtained if the gas were adiabatically reduced to zero velocity ($T_s = T_m + V_m^2/2g_cJc_p$), where V_m is the temperature of the gas stream moving at velocity V_m. The actual temperature (T_{aw}) of the probe is related to T_s and T_m by the dimensionless recovery factor y:

$$y = \frac{T_{aw} - T_m}{T_s - T_m} = \frac{t_{aw} - t_m}{V_m^2/2g_cJc_p} \tag{25a}$$

Hottel and Kalitinsky [*J. Appl. Mech.*, **12**, A25 (March, 1945)] give recovery factors for subsonic flow of air past various probes.

When a gas flows at very high linear velocity through a heated tube a substantial fraction of the heat transferred may be consumed in accelerating the gas; under some conditions this fraction exceeds unity and the gas cools in flowing through the heated tube. With adiabatic flow of air through an ideally insulated tube the inner wall attains the adiabatic-wall temperature t_{aw} which lies between t_s and t_m, depending on the numerical value of the recovery factor [Eq. (25a)].

McAdams, Nicolai, and Keenan (*Nat. Advisory Comm. Aeronaut. Tech. Note* 985, Washington, D.C., June, 1945) find that subsonic flow of air at Mach numbers (V_m/V_a) ranging from 0.2 to 1.0, through a long straight tube, give a recovery factor of 0.86, according to the theoretical equation $y = \sqrt{c_p u/k} = \sqrt{0.74} = 0.86$. The heat transfer equation then becomes

$$\frac{dq}{dA} = h(t_{hw} - t_{aw}) \tag{25b}$$

where t_{hw} is the surface temperature of the heated wall,

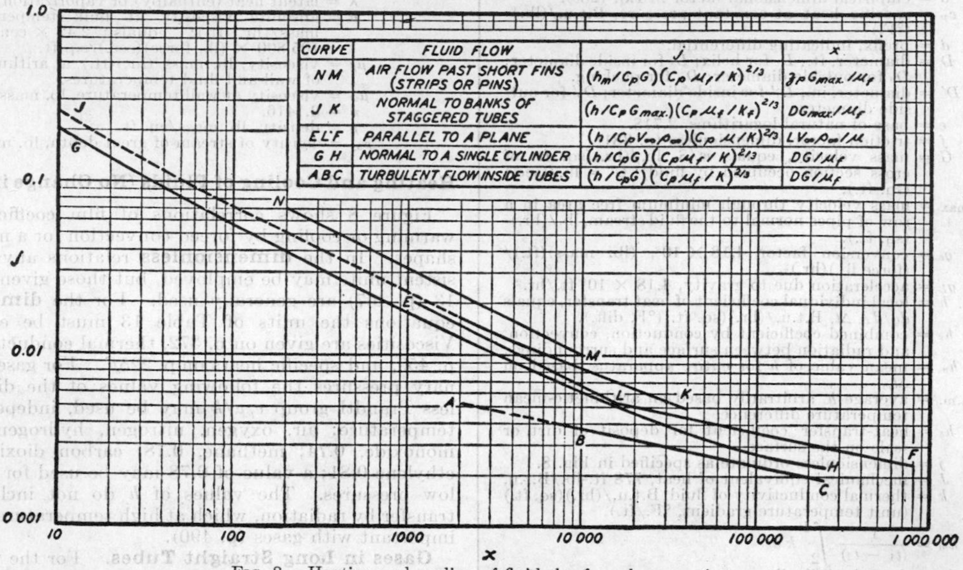

CURVE
FLUID FLOW
j
x
NM
AIR FLOW PAST SHORT FINS (STRIPS OR PINS)
$(h_m/C_pG)(C_p\mu_f/K)^{2/3}$
$\jmath\rho G_{max.}/\mu_f$
J K
NORMAL TO BANKS OF STAGGERED TUBES
$(h/C_pG_{max.})(C_p\mu_f/K_f)^{2/3}$
$DG_{max.}/\mu_f$
EL'F
PARALLEL TO A PLANE
$(h/C_pV_\infty\rho_\infty)(C_p\mu/K)^{2/3}$
$LV_\infty\rho_\infty/\mu$
G H
NORMAL TO A SINGLE CYLINDER
$(h/C_pG)(C_p\mu_f/K)^{2/3}$
DG/μ_f
ABC
TURBULENT FLOW INSIDE TUBES
$(h/C_pG)(C_p\mu_f/K)^{2/3}$
DG/μ_f
1.0
0.1
0.01
0.001
j
10
100
1000
10 000
100 000
1 000 000
x

FIG. 8. Heating and cooling of fluids by forced convection.

In designing a gas heater or cooler, instead of evaluating h, Eq. (24) may be combined with the usual heat balance:

$$q = 3600G'\left(\frac{\pi}{4}\right)\left(\frac{D_i'}{12}\right)^2 c_p(t_2 - t_1)$$
$$= \left(\frac{16.6c_p(G')^{0.8}}{(D_i')^{0.2}}\right)\left(\frac{\pi D_i'L}{12}\right)(\Delta t)_m$$

whence

$$\left(\frac{L}{D_i'}\right)\left[\frac{\Delta t_m}{(t_1 - t_2)}\right] = 4.52(D_i'G')^{0.2} \tag{25}$$

wherein t_1 is the lower temperature of the gas, t_2 is the higher, and Δt_m is the logarithmic-mean temperature difference between gas and tube surface, expressed in °F. The heated length L is expressed in feet. Where the tube temperature is substantially constant, as in a vapor-heated heater, the term $(t_2 - t_1)/\Delta t_m$ is equal to the natural logarithm of the ratio of the greater temperature difference to the lesser. Figure 9 is an alignment chart based on Eq. (25).

Example. To heat 600 lb./hr. of air from 44° to 210°F., with steam condensing at 220°F. outside the tubes.

Required. To calculate the tube length and the number of tubes in parallel.

Solution. The logarithmic mean of the initial and final differences is $(176 - 10)/2.3 \log 17.6 = 58.0$°F., and the ratio $\Delta t_m/(t_2 - t_1)$ is $58.0/166 = 0.349$. Assuming an optimum mass velocity G' of 1.81 (see p. 479) and tubes having an inside diameter of 1.30 in., align $D_i' = 1.3$ and $G' = 1.81$, reading 60 lb./hr. per tube on the right-hand scale of Fig. 9. Hence, 600/60 = 10 tubes in parallel are required. From the point where the straight line already drawn intersects the reference line, align with $\Delta t_m/(t_2 - t_1) = 0.349$, reading $L = 20$ ft.

For straight conduits of other than circular cross section, Eq. (24) applies with D_i' taken as four times the hydraulic radius in inches (clear cross section divided by the "wetted" perimeter).

and h is given by the dimensionless equation

$$\left(\frac{h}{c_pG}\right)\left(\frac{c_p\mu}{k}\right)^{2/3} = \frac{0.027}{(DG/\mu_m)^{0.23}} \quad (25c)$$

Gases in Coiled Pipes. Jeschke [*Z. Ver. deut. Ing.* **69**, *Ergänzungsheft*, 24, p. 1 (1925)] cooled air in helical coils of 1¼-in. seamless steel pipe, one helix 24.8 m. diameter with two turns, and the other 8 17 in. diameter with six turns. The flow, always turbulent, ranged up to values of DG/μ of 150,000 in consistent units (p. 467). From the two sets of data thus obtained, Jeschke proposed the dimensionless equation

$$\frac{hD}{k} = \left(0.039 + \frac{0.138D}{D_c}\right)\left(\frac{Dc_pG}{k}\right)^{0.76} \quad (26)$$

where D/D_c is the ratio of diameter of the pipe to that of the helix, consistent units being employed throughout Eq. (26). For ordinary use it is sufficient to multiply the value of h obtained from Eq. (24) by the term $1 + (3.54D/D_c)$.

Gases in Packed Tubes. Apparent coefficients of heat transfer for gases flowing through externally heated tubes filled with granular materials, based on the inside surface of the tube, were experimentally determined by Colburn [*Ind. Eng. Chem.*, **23**, 910 (1931)]. He passed air at velocities from 0.25 to 4.0 lb./(sec.)(sq. ft.) of gross section of the tube through both 1¼- and 3-in. tubes which were steam jacketed and filled with granular materials ranging from ⅛- to 1-in. particle size. The ratio of the observed apparent coefficient for the packed tube, to that for a 1-in. i.d. empty tube, both having the same mass velocity based on the gross cross section, depends on the ratio of the diameter of the packing to the inside diameter of the tube:

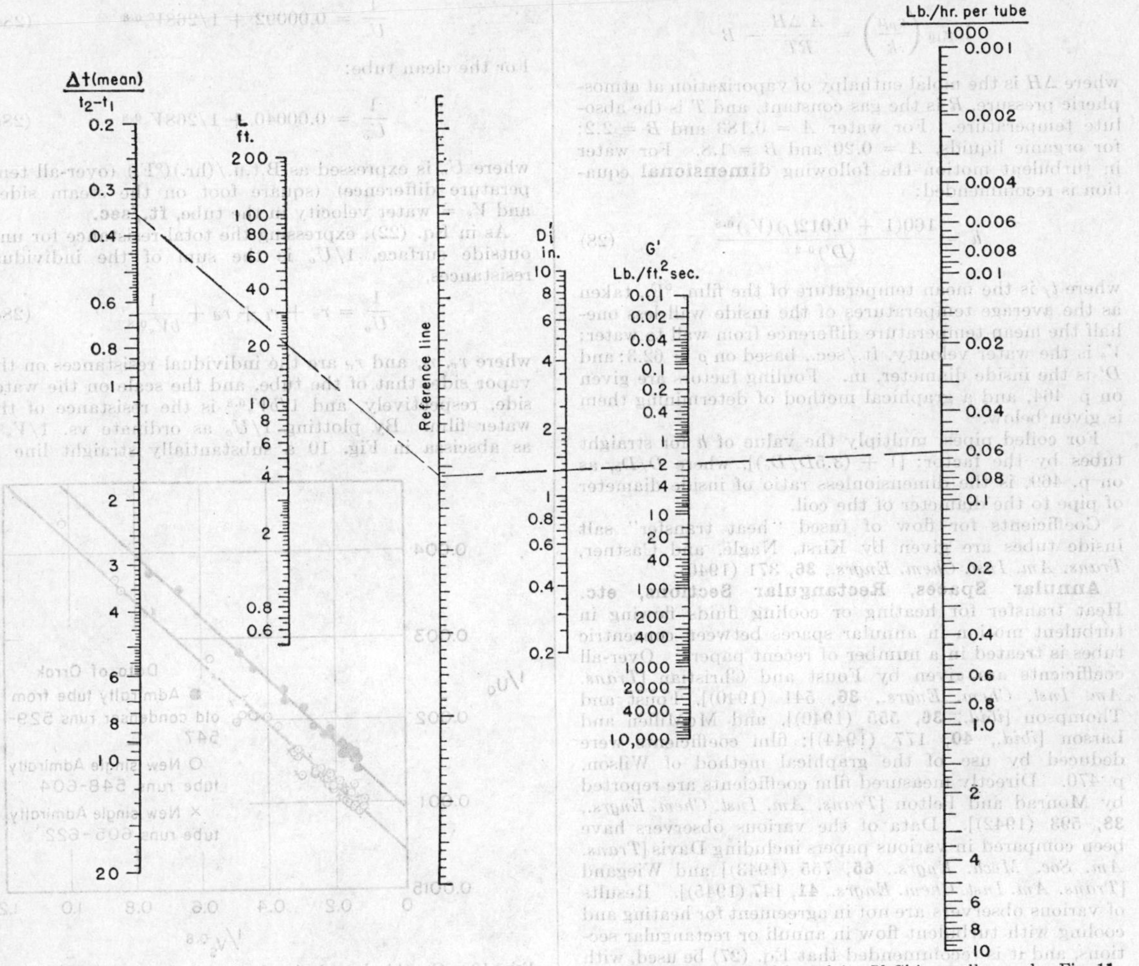

FIG. 9. Chart for designing tubular gas heaters and coolers based on Eq. (25). (*L* is heated length.) If G' is small, see also Fig. 11, p. 472.

D_p/D_i	0.05	0.10	0.2	0.3
Apparent h for packed/h for 1-in. empty	5.5	7.0	7.5	6.6

Coefficients of heat transfer between air and shallow beds of small cylindrical pellets are given by Wilkie and Hougen [*Trans. Am. Inst. Chem. Engrs.*, **41**, 445 (1945)] and by Hurt [*Ind. Eng. Chem.*, **35**, 522 (1943)].

Liquids of Low Viscosity inside Tubes. For water and liquids not more than twice as viscous, turbulent flow generally prevails ($DG/\mu > 2100$) and curve CA of Fig. 8 applies, which is fitted by the **dimensionless**

equation of Colburn [*Trans. Am. Inst. Chem. Engrs.*, **29**, 174–210 (1933)], in consistent units, as follows:

$$\frac{h}{c_p G}\left(\frac{c_p \mu_f}{k}\right)^{2/3} = 0.023\left(\frac{DG}{\mu_f}\right)^{-0.2} = \frac{f}{2} \qquad (27)$$

For liquids, if values of $c_p\mu/k$ are not available, these can be estimated from the dimensionless equation of Denbeigh [*J. Soc. Chem. Ind.*, **65**, 61–62 (1946)]

$$\log_{10}\left(\frac{c_p \mu}{k}\right) = \frac{A\,\Delta H}{RT} - B$$

where ΔH is the molal enthalpy of vaporization at atmospheric pressure, R is the gas constant, and T is the absolute temperature. For water $A = 0.183$ and $B = 2.2$; for organic liquids, $A = 0.20$ and $B = 1.8$. For water in turbulent motion the following **dimensional** equation is recommended:

$$h = \frac{160(1 + 0.012t_f)(V_s)^{0.8}}{(D')^{0.2}} \qquad (28)$$

where t_f is the mean temperature of the film, °F., taken as the average temperatures of the inside wall less one-half the mean temperature difference from wall to water; V_s is the water velocity, ft./sec., based on $\rho = 62.3$; and D' is the inside diameter, in. Fouling factors are given on p. 464, and a graphical method of determining them is given below.

For coiled pipes, multiply the value of h for straight tubes by the factor: $[1 + (3.5D/D_c)]$, where D/D_c, as on p. 469, is the dimensionless ratio of inside diameter of pipe to the diameter of the coil.

Coefficients for flow of fused "heat transfer" salt inside tubes are given by Kirst, Nagle, and Castner, *Trans. Am. Inst. Chem. Engrs.*, **36**, 371 (1940).

Annular Spaces, Rectangular Sections, etc. Heat transfer for heating or cooling fluids flowing in turbulent motion in annular spaces between concentric tubes is treated in a number of recent papers. Over-all coefficients are given by Foust and Christian [*Trans. Am. Inst. Chem. Engrs.*, **36**, 541 (1940)], Foust and Thompson [*ibid.*, **36**, 555 (1940)], and McMillen and Larson [*ibid.*, **40**, 177 (1944)]; film coefficients were deduced by use of the graphical method of Wilson, p. 470. Directly measured film coefficients are reported by Monrad and Pelton [*Trans. Am. Inst. Chem. Engrs.*, **38**, 593 (1942)]. Data of the various observers have been compared in various papers including Davis [*Trans. Am. Soc. Mech. Engrs.*, **65**, 755 (1943)] and Wiegand [*Trans. Am. Inst. Chem. Engrs.*, **41**, 147 (1945)]. Results of various observers are not in agreement for heating and cooling with turbulent flow in annuli or rectangular sections, and it is recommended that Eq. (27) be used, with D taken as the usual equivalent diameter D_e, equal to four times the ratio of free cross section to total wetted perimeter. For streamline flow (D_eG/μ below 2100) the same substitution is made in Eq. (29), p. 471 [Carpenter *et al.*, *Trans. Am. Inst. Chem. Engrs.*, **42**, 165 (1946)].

Mueller [*Trans. Am. Inst. Chem. Engrs.*, **38**, 613 (1942)] reports film coefficients for flow of air parallel to wires ranging from 0.0007 to 0.0030 in., for air velocities ranging from 7 to 50 ft./sec., and he obtained values of h ranging from 50 to 220 B.t.u./(hr.)(sq. ft.)(°F.).

Graphical Method of Interpreting Over-all Coefficients of Heat Transfer. Orrok [*Trans. Am. Soc. Mech. Engrs.*, **32**, 1773 (1910)] studied heat transfer from steam condensing under vacuum to cooling water in a single-tube surface condenser, obtaining results for both clean and fouled tubes. For each tube the over-all coefficient U was determined at a number of different water velocities. Wilson [*Trans. Am. Soc. Mech. Engrs.*, **37**, 47 (1915)] proposed a valuable graphical method of interpreting heat transfer in surface condensers. An application of Wilson's method to Orrok's data by McAdams, Sherwood, and Turner [*Trans. Am. Soc. Mech. Engrs.*, **48**, 1233 (1926)] showed the experimental results to be represented by the following empirical equations for turbulent motion:

For the fouled tube:

$$\frac{1}{U_o} = 0.00092 + 1/268V_s^{0.8} \qquad (28a)$$

For the clean tube:

$$\frac{1}{U_o} = 0.00040 + 1/268V_s^{0.8} \qquad (28b)$$

where U_o is expressed as B.t.u./(hr.)(°F.) (over-all temperature difference) (square foot on the steam side), and V_s = water velocity in the tube, **ft./sec.**

As in Eq. (22), expressing the total resistance for unit outside surface, $1/U_o$ is the sum of the individual resistances,

$$\frac{1}{U_o} = r_v + r_t + r_d + \frac{1}{bV_s^{0.8}} \qquad (28c)$$

where r_v, r_t, and r_d are the individual resistances on the vapor side, that of the tube, and the scale on the water side, respectively, and $1/bV_s^{0.8}$ is the resistance of the water film. By plotting $1/U_o$ as ordinate vs. $1/V_s^{0.8}$ as abscissa in Fig. 10 a substantially straight line is

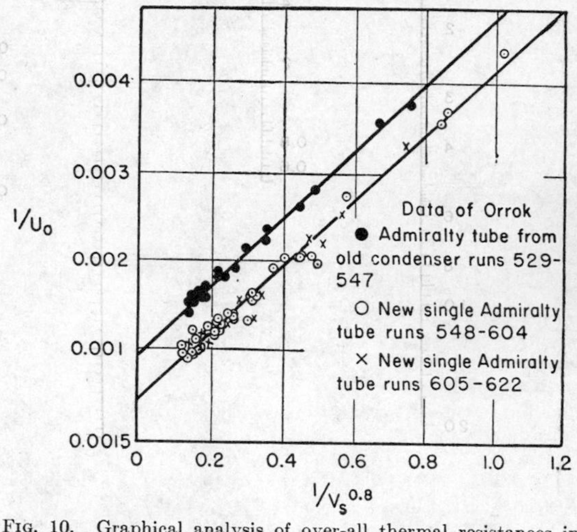

FIG. 10. Graphical analysis of over-all thermal resistances in surface condenser for steam.

obtained having a slope of $1/b$, where b is an empirical constant for the particular operating conditions. b is numerically equal to the water-side coefficient expressed per square foot on the steam side (268 in this case) for a water velocity of 1 ft./sec. The tube had outside and inside diameters of 1.00 and 0.902 in., respectively, and was made of Admiralty metal, $k = 63$. r_t is then 0.000068. If, for the new tube, r_d is assumed as zero, the intercept $0.00040 = r_t + r_v$, whence $r_v = 1/3010$, *i.e.*, the vapor-side coefficient is 3010 B.t.u./(hr.)(sq. ft.)(°F.). For the fouled tube, the intercept $0.00092 = r_v + r_t + r_d$; or the difference in the intercepts, $0.00092 - 0.00040 = r_d = 1/1920$, *i.e.*, the scale coefficient is 1920 based on the steam-side area. Since the inside diameter was 90.2 per cent of the outside diameter, the scale coefficient, based on the water-side area, is $1920/0.902 = 2130$.

The water-side coefficient expressed per square foot water side equals 268/0.902 = 297.

This graphical method, with appropriate modifications, may be applied when correlating over-all coefficients of heat transfer in liquid-to-liquid and gas-to-gas heat exchangers.

Streamline Flow. With liquids of high viscosity, such as the viscous oils flowing at values of DG/μ below 2100, streamline flow ensues, and the data are correlated by the dimensionless equation of Sieder and Tate [*Ind. Eng. Chem.*, **28**, 1429 (1936)]:

$$\frac{h_{a.m.}D}{k} = 2.0\left(\frac{wc_p}{kL}\right)^{1/3}\left(\frac{\mu}{\mu_s}\right)^{0.14} \quad (29)$$

As DG/μ increases from 2100 to 7000, the substantial effect of heated length upon h, shown by Eq. (29) for streamline flow, diminishes and becomes zero, as shown by Fig. 10*a*. Below DG/μ of 2100, h in Fig. 10*a* is based on

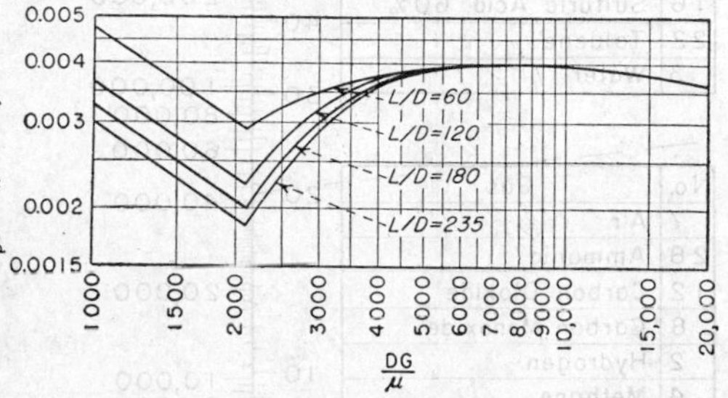

FIG. 10*a*. Heating and cooling of viscous oils flowing inside tubes. The curves for *Re* below 2100 are based on Eq. (29). *L* is length of each pass in feet.

the arithmetic mean of the terminal temperature differences; above DG/μ of 2100, the logarithmic mean is used. The curve in the streamline region of Fig. 10*a* is based only on data for runs where the tube-wall temperature was constant and single straight lengths of tubes were used; the curve lies within ±15 per cent of the data. Equation (29) may be rearranged to give

$$\frac{(t_2 - t_1)}{(\Delta t)_{a.m.}}\left(\frac{\mu_s}{\mu}\right)^{0.14} = 6.3\left(\frac{kL}{wc}\right)^{2/3} \quad (29a)$$

where w is the mass rate of flow per tube of heated length L.

When the Grashof number ($N_{Gr} = D^3\rho^2 g_L\beta\,\Delta t/\mu^2$, evaluated at the bulk temperature of the stream) is large, and the Reynolds number ($N_{Re} = DG/\mu$) is small, Kern and Othmer [*Trans. Am. Inst. Chem. Engrs.*, **39**, 517, 579 (1943)] find that the values of $h_{a.m.}$ for **horizontal** tubes may be considerably higher than those given by Eq. (29), and they recommend multiplying the right-hand side of Eq. (29) by the dimensionless term

$$\frac{2.3(1 + 0.01N_{Gr}^{1/3})}{\log_{10} N_{Re}} \quad (29b)$$

With a long tube and low rate of flow, equivalent to values of wc/kL of 10 or less, the liquid is exposed to a large surface for a very long time and hence will come practically to wall temperature, $(\Delta t)_2$ equals zero. Since, by definition, $q = wc(t_2 - t_1) = h_{a.m.}\cdot(\pi DL)\cdot\Delta t_{a.m.}$, and in this case t_2 equals t_s, the asymptotic value of h, in such a case, is

$$\frac{h_{a.m.}D}{k} = \left(\frac{2}{\pi}\right)\left(\frac{wc}{kL}\right) \quad (30)$$

In the range in which Eq. (30) applies, it is clear that the apparatus is overdesigned.

Liquids in Vertical Pipes. For usual cases of markedly turbulent flow of liquids in vertical pipes, the same procedure as for horizontal pipes should apply (p. 467). At velocities below the critical, and sometimes slightly above, free convection factors determine h_m. Colburn and Hougen [*Ind. Eng. Chem.*, **22**, 522 (1930)] give the following equation as representing their data on the flow of water in a 3-in. i.d. vertical pipe, at low velocities, up to 6 lb./(sec.)(sq. ft.), *i.e.*, up to approximately 0.1 ft./sec.:

$$h_m = 0.42t\sqrt[3]{\Delta t} \quad \text{for upward flow} \quad (31)$$
$$h_m = 0.49t\sqrt[3]{\Delta t} \quad \text{for downward flow} \quad (32)$$

where h_m is B.t.u./(hr.)(sq. ft.)(°F.); t is the average water temperature, °F.; and Δt is the average temperature difference between wall and fluid, °F.

They also correlated these same data on water by the following Nusselt-type dimensionless equation for fluids flowing upward at low velocities in vertical pipes:

$$\frac{h_m}{[k_f{}^2\rho_f{}^2c_p g_L\beta(\Delta t)/\mu_f]^{1/3}} = 0.13 \quad (33)$$

In this equation, k is the thermal conductivity; c_p, the specific heat at constant pressure; ρ, the density; β, the coefficient of expansion; Δt, the temperature difference between wall and fluid; g_L, the gravitational constant; and μ, the viscosity. All values are to be expressed in consistent units, and k, ρ, β, and μ are to be used at the average temperature of the **film**. Figure 11 is based on Eq. (33) and shows the relation between the substance involved, the temperature of the film, the temperature difference, and the coefficient h. For use with gases a pressure term appears on the chart, p. 472. For more accurate estimates, use the equations and graphs of Martinelli *et al.* [*Trans. Am. Inst. Chem. Engrs.*, **38**, 493 (1942)].

Liquid Films in Vertical Tubes. For the turbulent flow of water in layer form down the inside walls of vertical tubes, as on the water side of vertical ammonia condensers, the **dimensional** equation of McAdams, Drew, and Bays [*Trans. Am. Soc. Mech. Engrs.*, **62**, 627 (1940)] is recommended:

$$h_m = 120\Gamma^{1/3} \quad (34)$$

and is based on values of $\Gamma = w/\pi D$ ranging from 600 to 15,000 lb./(hr.)(ft.) of wetted perimeter.

T. B. Drew (personal communication, 1939) tentatively gives the following dimensional equations for any liquid flowing in layer form down vertical surfaces:

$$\text{For } \frac{4\Gamma}{\mu_f} > 2100: \qquad h_m = 7.5\left[\left(\frac{k^2\rho^2 c}{\mu_f}\right)\left(\frac{4\Gamma}{\mu_f}\right)\right]^{1/3} \quad (35)$$

$$\text{For } \frac{4\Gamma}{\mu_f} < 2100: \qquad h_{a.m.} = 55\left(\frac{k^2\rho^{4/3}c}{L\mu_f{}^{1/3}}\right)^{1/3}\left(\frac{4\Gamma}{\mu_f}\right)^{1/6} \quad (36)$$

For heating a petroleum oil, having viscosities of 610 lb./(hr.)(ft.) at 100°F. and 42 at 210°F. and a sp. gr. of 0.905 at 60°F., the following results were obtained by Bays (Mass. Inst. Tech. Thesis, 1936) for flow in layer form at $4\Gamma/\mu_f$ below 2100:

Γ, lb./(hr.)(ft.)	200	700	1400	2800
h_m for L = 2, Δt = 55°F.		40	48	53
h_m for L = 2, Δt = 100°F.	41	50	58	
h_m for L = 0.4, Δt = 100°F.	..	79	88	

Gases Flowing outside Tubes and over Plane Surfaces

Where a gas is flowing past two surfaces having unequal temperatures, it should be remembered that heat is transferred by radiation (see pp. 483–498) as well as by convection. Where forced circulation of gas over the

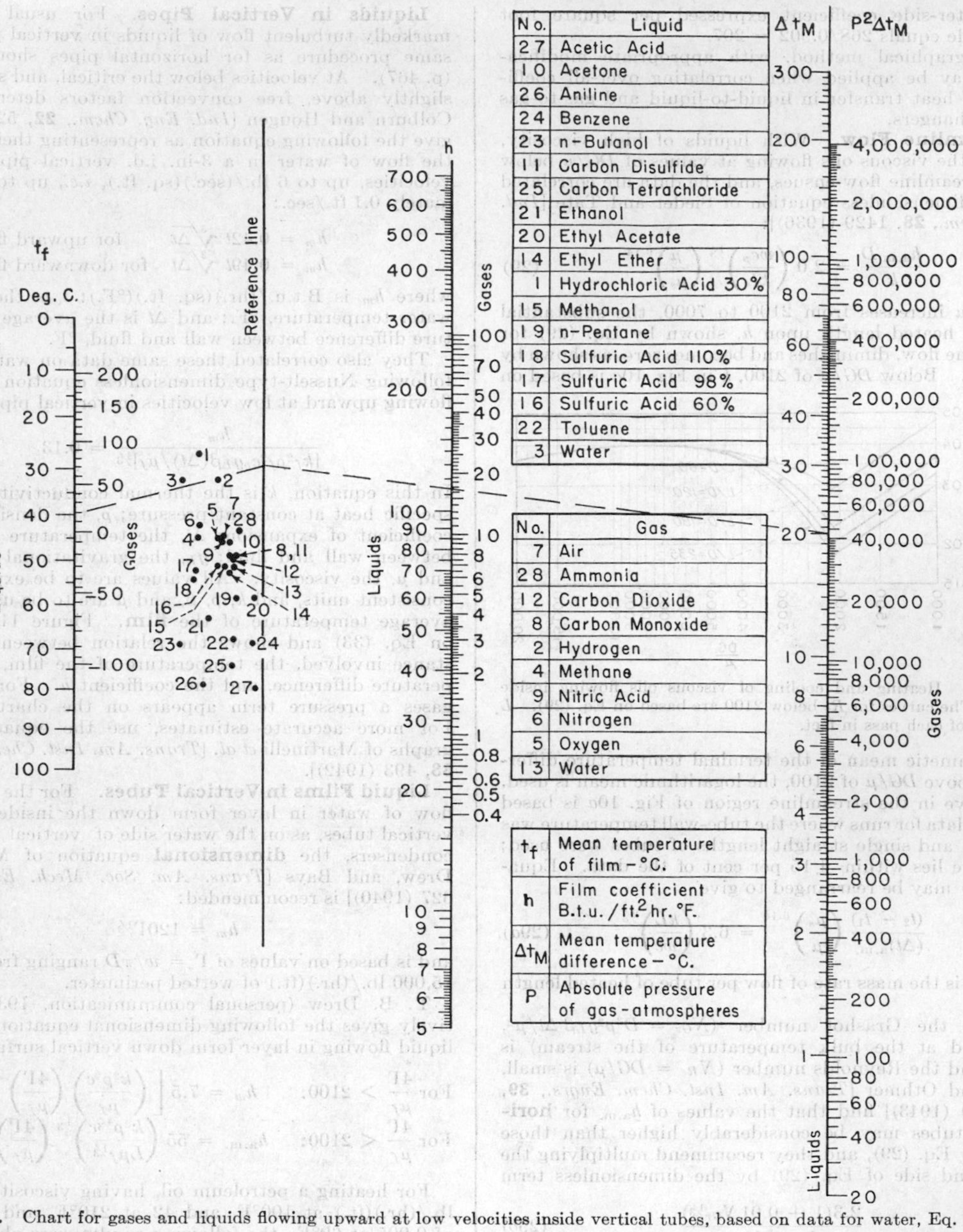

No. Liquid
27 Acetic Acid
10 Acetone
26 Aniline
24 Benzene
23 n-Butanol
11 Carbon Disulfide
25 Carbon Tetrachloride
21 Ethanol
20 Ethyl Acetate
14 Ethyl Ether
1 Hydrochloric Acid 30%
5 Methanol
19 n-Pentane
8 Sulfuric Acid 110%
7 Sulfuric Acid 98%
16 Sulfuric Acid 60%
22 Toluene
3 Water
No. Gas
7 Air
28 Ammonia
12 Carbon Dioxide
8 Carbon Monoxide
2 Hydrogen
4 Methane
9 Nitric Acid
6 Nitrogen
5 Oxygen
13 Water
t_f Mean temperature of film - °C.
h Film coefficient B.t.u./ft.^{2} hr. °F.
Δt_M Mean temperature difference - °C.
P Absolute pressure of gas-atmospheres
t_f
Deg. C.
h
Reference line
Liquids
Gases
Δt_M
$P^2\Delta t_M$

FIG. 11. Chart for gases and liquids flowing upward at low velocities inside vertical tubes, based on data for water, Eq. (33).

heating surface is not involved, a thermosiphon circulation develops because differences in density of the gas at different points give rise to conditions known as free convection (see p. 474).

Gases Flowing at Right Angles to a Bank of Tubes. When the pipes are staggered, the results are given by Curve JK of Fig. 8, p. 468. In the usual case, where D_oG/μ ranges from 3000 to 40,000, the following dimensionless equation is recommended:

$$\left(\frac{h_m}{c_pG_{\max.}}\right)\left(\frac{c_p\mu_f}{k}\right)^{\frac{2}{3}} = 0.33\left(\frac{D_oG_{\max.}}{\mu_f}\right)^{-0.4} \quad (37)$$

An approximate dimensional equation, based on a paper by Monrad [*Ind. Eng. Chem.*, **24**, 505–509 (1932)] is satisfactory for gases:

$$h_m = \frac{0.031c_pT^{0.3}G_{\max.}^{\frac{2}{3}}}{(D_o')^{\frac{1}{3}}} \quad (38)$$

wherein T is the mean gas temperature (°F. abs.), taken as the mean wall temperature less the mean temperature difference from wall to gas.

For tubes in line, deduct 25 per cent from the values given above. For flow in well-baffled exchangers, to allow for leakage around the baffles, assume no leakage and deduct 40 per cent from the values given by Eqs. (37) and (38) or Fig. 8. With mixtures of gases, use a mean c_p weighted according to the weight fractions present in the mixture. With tubes in line, both the surface coefficient of heat transfer and the friction factor are less than with staggered tubes; when these two arrangements are compared on a basis of equal power loss to offset friction, per unit surface, Colburn [*Purdue Univ. Eng. Bull.*, **26,** No. 1 (1942)] finds that the heat transfer coefficients for the in-line arrangement are slightly higher than with staggered tubes.

Gases Flowing at Right Angles to Single Tubes. Such data are valuable in calculating the true temperature of a gas based on the apparent temperature indicated by a thermocouple. Sometimes the thermocouple is placed directly in the gas stream and at other times it is protected by a pyrometer well; hence it is desirable to know the effect of diameter upon the convection coefficients between gas and a single cylinder for a large range of diameters. Data are available only for **air** at room temperature flowing at right angles to the axes of single cylinders ranging in diameter from 0.001 to 3.75 in.; the temperature of the cylinder ranged up to 1800°F. for the wires but extended only to 212°F. for the tubes. These data are correlated by curve *GH* of Fig. 8, p. 468, and by the dimensionless equations:

For D_oG/μ_f from 0.1 to 1000:

$$\frac{h_m D_o/k_f}{(c_p\mu_f/k_f)^{0.3}} = 0.35 + 0.47\left(\frac{D_oG}{\mu_f}\right)^{0.52} \qquad (39)$$

For D_oG/μ_f from 1000 to 50,000:

$$\frac{h_m D_o/k_f}{(c_p\mu_f/k_f)^{0.3}} = 0.26\left(\frac{D_oG}{\mu_f}\right)^{0.6} \qquad (39a)$$

For gases at moderate temperatures, with DG/μ_f ranging from 1000 to 50,000, a simplified dimensional equation is

$$h = \frac{0.3c_pG^{0.6}}{(D_o')^{0.4}} \qquad (40)$$

Calculation of True Temperature of a Gas. For calculating the true temperature of a gas from the reading of a thermocouple or pyrometer, placed in a gas stream and in sight of surrounding walls which may be at various temperatures, a heat balance per unit area of thermocouple gives the equation

$$q_{gr} + q_c = \Sigma q_r \qquad (41)$$

wherein q_{gr} is the rate of heat flow between gas and pyrometer by the mechanism of **gas radiation** (see pp. 490–493), q_c is the rate of heat flow between gas and pyrometer by **convection** and Σq_r is the sum of the various terms representing the radiant-heat interchange between the pyrometer and the various surfaces which it "sees," evaluated by the methods on pp. 483–498.

In the simple case of a gas stream having true temperature T_g and flowing through a duct of a diameter large compared with that of the pyrometer at temperature T_c, the inner surfaces of the walls having approximately constant temperature T_w, a heat balance expressed in B.t.u./(hr.)(sq. ft. of the pyrometer) gives the equation

$$q_{gr} + h_c(T_g - T_c) = (0.173)(\epsilon)\left[\left(\frac{T_c}{100}\right)^4 - \left(\frac{T_w}{100}\right)^4\right] \qquad (41a)$$

The value of ϵ, the emissivity of the surface of the pyrometer, is obtained from Table 1, p. 485; and T_c and T_w represent, respectively, the temperatures of the pyrometer and walls, expressed in °F. **absolute.** By definition, $q_r = \epsilon h_{ro}(t_1 - t_2)$, wherein h_{rb} is the appropriate radiation coefficient of heat transfer from solid to solid, plotted in Fig. 12 as a function of t_1 and t_2 expressed in °F. Hence, Eq. (41) may be written

$$q_{gr} + h_c(t_g - t_c) = \epsilon h_{rb}(t_c - t_w) \qquad (41b)$$

Under the conditions in which q_{gr} is negligible compared with q_c and q_r, Eq. (41b) simplifies to

$$t_g = t_c + \frac{(t_c - t_w)(\epsilon)(h_{rb})}{(h_c)} \qquad (41c)$$

Gases Flowing outside and Parallel to Axes of Tubes. Test data for this case are in rough agreement with the equations for gases inside pipes (pp. 467–469) when the equivalent diameter is taken as four times the ratio of free cross section to total "wetted" perimeter.

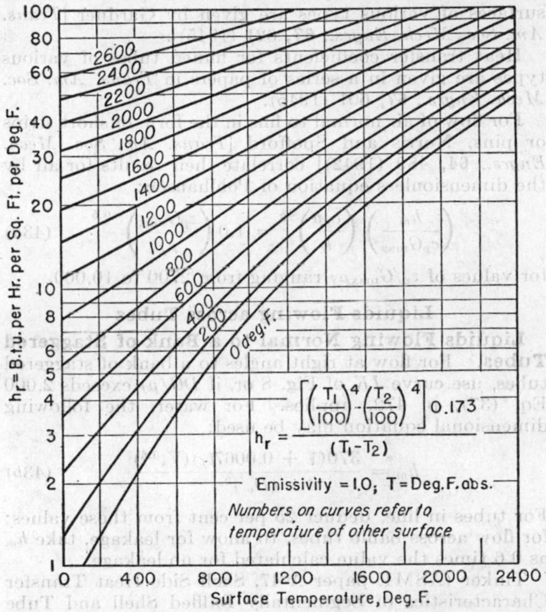

FIG. 12. Radiation coefficients of heat transfer h_r.

Gases Flowing Parallel to Planes. For forced flow of gas parallel to a **single** plane the data are correlated by curve *ELF* of Fig. 8, p. 468, and by the dimensionless equation of Colburn [*Trans. Am. Inst. Chem. Engrs.*, **29,** 174–210 (1933)]:

$$\left(\frac{h_m}{c_pV_\infty\rho_\infty}\right)\left(\frac{c_p\mu}{k}\right)^{2/3} = a\left(\frac{LV_\infty\rho_\infty}{\mu}\right)^{-n} \qquad (42)$$

where for $LV_\infty\rho_\infty/\mu$ above 20,000, a is 0.036 and n is 0.2; below $LV_\infty\rho_\infty/\mu$ of 20,000, a is 0.66 and n is 0.5.

For flow between parallel planes or in rectangular passages the case should be treated as flow inside tubes having the same hydraulic radius (see p. 470).

Finned Tubes. When the film coefficient on the outside of a metal tube is much lower than that on the inside, as when steam condensing in a pipe is being used to heat air, externally finned, or extended, heating surfaces are of value in increasing substantially the rate of heat transfer per unit length of tube. The data on extended heating surfaces, for the case of air flowing outside and at right angles to the axes of a bank of finned pipes, can be represented approximately by the dimensional equation

$$U = a(G')^n$$

The constants a and n are dependent on the nature and arrangement of the finned surface. For various commercial finned pipes, a ranges from 5.9 to 8.9 and n from

0.58 to 0.74. The *Aerofin Bulletin* (A.I.A. File 3004) gives the following equation for the high-pressure Aerofin heater:

$$\frac{1}{U} = 0.018 + \frac{1}{13(G_a')^{0.8}} \tag{43}$$

where G_a' is the mass velocity of the approaching air in lb./(sec.)(sq. ft.). This equation is given for a finned tube having approximately 4 sq. ft. of total surface exposed to air per foot of pipe length.

Fin efficiency is defined as the ratio of the mean temperature difference from surface to fluid divided by the temperature difference from fin to fluid at the base or root of the fin. Graphs of fin efficiency for extended surfaces of various types are given by Gardner [*Trans. Am. Soc. Mech. Engrs.*, **67,** 621 (1945)].

Heat transfer coefficients for finned tubes of various types are given in a series of papers in *Trans. Am. Soc. Mech. Engrs.*, **67,** 601 (1945).

For flow of air normal to fins in the form of short strips or pins, Norris and Spofford [*Trans. Am. Soc. Mech. Engrs.*, **64,** 489 (1942)] correlate their results for air by the dimensionless equation of Pohlhausen:

$$\left(\frac{h_m}{c_pG_{\max}}\right)\left(\frac{c_p\mu}{k}\right)^{2/3} = 1.0\left(\frac{z_pG_{\max}}{\mu_f}\right)^{-0.5} \tag{43a}$$

for values of $z_p/G_{\max}\mu_f$ ranging from 2700 to 10,000.

Liquids Flowing across Tubes

Liquids Flowing Normal to a Bank of Staggered Tubes. For flow at right angles to a bank of staggered tubes, use curve *JK* of Fig. 8 or, if DG/μ_f exceeds 2,000 Eq. (37), p. 472, applies. For water, the following dimensional equation may be used:

$$h_m = \frac{370(1 + 0.0067t_f)(V_s^{0.6})}{(D_o')^{0.4}} \tag{43b}$$

For tubes in line, deduct 25 per cent from these values; for flow across baffle tubes, to allow for leakage, take h_m as 0.6 times the value calculated for no leakage.

Tinker (ASME paper A-47, Shell Side Heat Transfer Characteristics of Segmentally Baffled Shell and Tube Heat Exchangers, Ross Heater & Manufacturing Co., Inc., 1947), gives a sound method of designing baffled heat exchangers.

Gravity Flow of Water Layers over Horizontal Tubes. For water flowing over a horizontal tube, data for several sizes of pipe are roughly correlated by the dimensional equation of McAdams, Drew, and Bays [*Trans. Am. Soc. Mech. Engrs.* **62,** 627 (1940)].

$$h_{\text{a.m.}} = 150\left(\frac{\Gamma}{D_o'}\right)^{1/3} \tag{44}$$

for Γ ranging from 100 to 1000 lb. water per hour per foot of pipe. Fouling factors are given on p. 464.

Stirred Kettles. For the heating or cooling water, oil, or glycerin in a jacketed kettle heated by a jacket and cooled by a helically wound coil (or cooled by the jacket and heated by the coil), and agitated by a horizontal flat paddle of length L_p attached to a central vertical shaft rotated at rate N, Chilton, Drew, and Jebens [*Ind. Eng. Chem.*, **36,** 510 (1944)] give the following dimensionless equation for $L^2N\rho/\mu$ ranging from 300 to 400,000:

$$\left(\frac{hD_j}{k}\right)\left(\frac{c\mu}{k}\right)^{-1/3}\left(\frac{\mu_s}{\mu}\right)^{0.14} = a\left(\frac{L_p^2N\rho}{\mu}\right)^n \tag{44a}$$

For h_j between jacket and fluid, $a = 0.37$, $n = 0.667$, μ_s at t_j.

For h_c between fluid and coil, $a = 0.87$ $n = 0.62$, μ_s at t_c.

Natural Convection

For the general case of free or natural convection of heat between a solid and a fluid, without change in state, Nusselt, Davis, Rice, and others suggest the dimensionless equation

$$\frac{h_cD}{k_f} = \frac{1}{\alpha}\left(\frac{D^3\rho_f^2g_L\beta\Delta t}{\mu_f^2}\right)^n\left(\frac{c_p\mu_f}{k_f}\right)^m \tag{45}$$

Fluids outside Single Horizontal Cylinder. For pipes having a large ratio of length to diameter, Rice gives $\alpha = 2.60$, $n = 0.27$, and $m = 0.25$; for approximate results, $\alpha = 2.12$, $n = m = 0.25$. Based on the approximate equation of Rice, Fig. 13 is an alignment chart based on the temperature of the film (arithmetic mean of the temperatures of the wall and fluid) expressed in degrees centigrade, the temperature difference between surface and fluid in degrees centigrade, the absolute pressure in atmospheres, and the diameter of the pipe in inches. While based on data for air, hydrogen, carbon dioxide, aniline, carbon tetrachloride, glycerin, olive oil, and toluene, reference points are shown in Fig. 13 to facilitate the extrapolation of these results to certain fluids other than those tested.

Fluids outside Various Shapes. For heat loss from surfaces by conduction and convection to air at atmospheric pressure and ordinary temperatures, the coefficients are given by dimensional equations, using the units of Table 13, p. 467:

Horizontal pipes:

$$h_c = 0.5(\Delta t_s/D_o')^{0.25} \tag{45a}$$

Long vertical pipes:

$$h_c = 0.5(\Delta t_s/D_o')^{0.25} \tag{45b}$$

Vertical plates less than two ft. high:

$$h_c = 0.28(\Delta t_s/H)^{0.25} \tag{45c}$$

Vertical plates more than three ft. high:

$$h_c = 0.27(\Delta t_s)^{0.25} \tag{45d}$$

Horizontal plates:
Facing upward:

$$h_c = 0.38\Delta t_s^{0.25} \tag{45e}$$

Facing downward:

$$h_c = 0.2\Delta t_s^{0.25} \tag{45f}$$

in which Δt_s is the temperature of the exposed surface less that of the ambient air in the room.

The experiments of Griffiths and Davis [*Spec. Rept.* 9, Food Investigation Board (Great Britain), 1922, rev. 1931] and of Schmidt [*Z. ges. Kälte-Ind.*, **35,** 213 (1928)] show that h_c for a vertical plate is not affected by the presence of a parallel plate more than 20 mm. away. Griffiths and Davis also determined heat transfer between a hot and a cold vertical plate separated by a completely enclosed air space, which was varied from ½- to 2-in. thickness and from 24.8 to 49.6 in. in height. Over this range the heat transfer by convection was found to depend only on the temperature difference and to be approximately 60 per cent of that found for a single vertical plate, not enclosed, when Δt is taken as the difference in temperature of the two plates, in °F.

Simultaneous Loss by Radiation. Radiation (pp. 483–498) is an important factor in the loss of heat from surfaces where the gas movement is due to free convection, except for small wires. The heat radiated from the solid to the surrounding walls may be expressed in terms of the coefficient of heat transfer by radiation, h_r (p. 473), and is obtained by multiplying the ordinate from Fig. 12 by the relative blackness or emissivity from

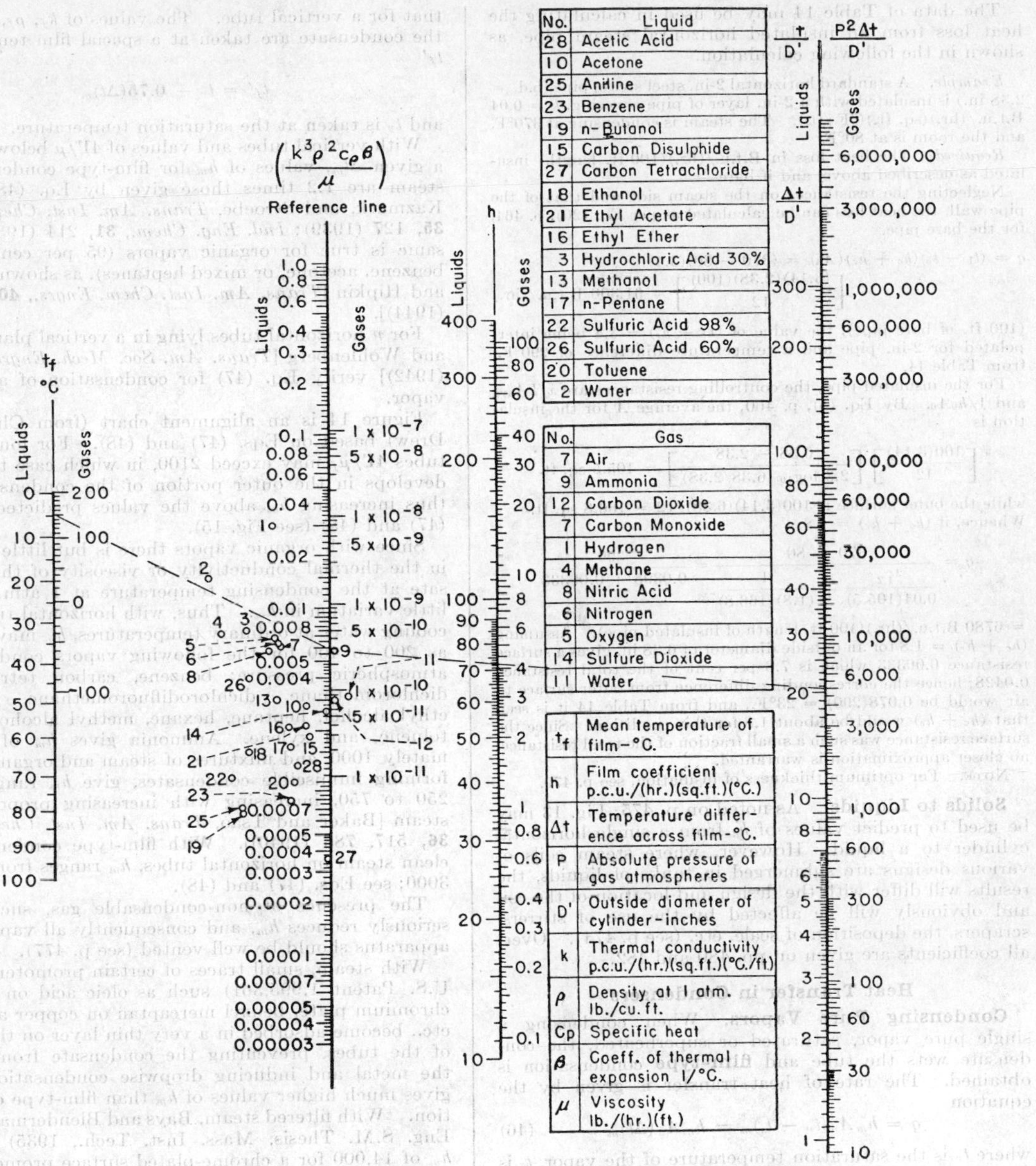

Reference line

$$\left(\frac{k^3 \rho^2 c_\rho \beta}{\mu}\right)$$

No.	Liquid
28	Acetic Acid
10	Acetone
25	Aniline
23	Benzene
19	n-Butanol
15	Carbon Disulphide
27	Carbon Tetrachloride
18	Ethanol
21	Ethyl Acetate
16	Ethyl Ether
3	Hydrochloric Acid 30%
13	Methanol
17	n-Pentane
22	Sulfuric Acid 98%
26	Sulfuric Acid 60%
20	Toluene
2	Water

No.	Gas
7	Air
9	Ammonia
12	Carbon Dioxide
7	Carbon Monoxide
1	Hydrogen
4	Methane
8	Nitric Acid
6	Nitrogen
5	Oxygen
14	Sulfure Dioxide
11	Water

t_f	Mean temperature of film—°C.
h	Film coefficient p.c.u./(hr.)(sq.ft.)(°C.)
Δt	Temperature difference across film—°C.
P	Absolute pressure of gas atmospheres
D'	Outside diameter of cylinder-inches
k	Thermal conductivity p.c.u./(hr.)(sq.ft.)(°C./ft)
ρ	Density at 1 atm. lb./cu. ft.
C_p	Specific heat
β	Coeff. of thermal expansion 1/°C.
μ	Viscosity lb./(hr.)(ft.)

FIG. 13. Heat transfer coefficients h_c for natural convection outside horizontal cylinders.

Table 1, p. 485. The value of h_r so obtained is added to h_c, giving the **combined coefficient for convection and radiation** $h_c + h_r$.

Table 14, based on data of Heilman and of McMillan, shows values of $(h_c + h_r)$ from single horizontal pipes of steel, with oxidized surfaces.

Table 14.   Values of $(h_c + h_r)$
B.t.u./(hr.)(sq. ft.) (°F. from pipe to room)
For horizontal bare standard steel pipe of various sizes in a room at 80°F.

Nominal pipe diam., in.	Temperature difference, °F.														
	30	50	100	150	200	250	300	350	400	450	500	550	600	650	700
1	2.16	2.26	2.50	2.73	3.00	3.29	3.60	3.95	4.34	4.73	5.16	5.60	6.05	6.51	6.98
3	1.97	2.05	2.25	2.47	2.73	3.00	3.31	3.69	4.03	4.43	4.85	5.26	5.71	6.19	6.66
5		1.95	2.15	2.36	2.61	2.90	3.20	3.54	3.90						
10	1.80	1.87	2.07	2.29	2.54	2.82	3.12	3.47	3.84						

Bailey and Lyell [*Engineering*, **147**, 60 (1939)] give values for $h_c + h_r$ up to Δts of 1000°F.

The data of Table 14 may be used in calculating the heat loss from an insulated horizontal steam pipe, as shown in the following calculation.

Example. A standard horizontal 2-in. steel steam pipe (o.d. = 2.38 in.) is insulated with a 2-in. layer of pipe covering, $k = 0.04$ B.t.u./(hr.)(sq. ft.)(°F./ft.). The steam is condensing at 370°F. and the room is at 80°F.

Required. The heat loss in B.t.u./(hr.)(100-ft. length) insulated as described above, and if bare.

Neglecting the resistances on the steam side and that of the pipe wall, the heat loss can be calculated as per Eq. (20), p. 464, for the bare pipe.

$$q = (t_1 - t_5)(h_c + h_r)(A_4) = (370 - 80)(3.4)\left[\frac{(3.14)(2.38)(100)}{12}\right] = 61{,}300 \text{ B.t.u./(hr.)}$$

(100 ft. of bare pipe), the value of $(h_c + h_r) = 3.4$ being interpolated for 2-in. pipe and a temperature difference of 290°F., from Table 14.

For the insulated pipe, the controlling resistances are x/kA_{avg} and $1/h_0A_0$. By Eq. (5), p. 460, the average A for the insulation is

$$\left[\frac{100(3.14)}{12}\right]\left[\frac{6.38 - 2.38}{2.3 \log_{10}(6.38/2.38)}\right] = 105.5 \text{ sq. ft.}$$

while the outer surface is $100(3.14)(6.38)/12 = 166.8$ sq. ft. Whence, if $(h_c + h_r) = 1.8$,

$$q = \frac{370 - 80}{\dfrac{2/12}{0.04(105.5)} + \dfrac{1}{(1.8)(166.8)}} = \frac{290}{0.0395 + 0.00333}$$

= 6780 B.t.u./(hr.)(100-ft. length of insulated pipe). Assuming $(h_c + h_r) = 1.8$ for an outside diameter of 6.38 in. gives a surface resistance 0.00333 which is 7.8 per cent of the total resistance 0.0428; hence the corresponding difference from outer surface to air would be $0.078(290) = 23$°F., and from Table 14 it is seen that $(h_c + h_r)$ would be about 1.8 for this condition. Since the surface resistance was such a small fraction of the total resistance, no closer approximation is warranted.

NOTE. For optimum thickness of insulation, see p. 479.

Solids to Liquids. As noted on p. 475, Fig. 13 may be used to predict values of h_c from a single horizontal cylinder to a liquid. However, where steam coils of various designs are submerged in tanks of liquids, the results will differ with the design and location of the coil and obviously will be affected by the use of stirrers, scrapers, the deposition of scale, etc. (see p. 474). Overall coefficients are given on pp. 480 and 482.

Heat Transfer in Condensers

Condensing Pure Vapors. When condensing a single pure vapor, saturated or superheated, the condensate wets the tube and **film-type** condensation is obtained. The rate of heat transfer is given by the equation

$$q = h_m A_w (t_v - t_s)_m = h_m A_w (\Delta t)_m \tag{46}$$

where t_v is the saturation temperature of the vapor, t_s is the surface temperature, and h_m is the average coefficient between vapor and wall. So long as the condensate flows in streamline motion ($4\Gamma/\mu_f < 2100$), the following dimensionless equations of Nusselt may be used:

For **horizontal** tubes:

$$\frac{h_m D}{k_f} = 0.73\left(\frac{D^3 \rho_f^2 g_L \lambda}{k_f \mu_f n (\Delta t)_m}\right)^{0.25} = 0.76\left(\frac{D^3 \rho_f^2 g_L}{\mu_f \Gamma}\right)^{1/3} \tag{47}$$

For **vertical** tubes:

$$\frac{h_m L}{k_f} = 0.94\left(\frac{L^3 \rho_f^2 g_L \lambda}{k_f \mu_f (\Delta t)_m}\right)^{0.25} = 0.93\left(\frac{L^3 \rho_f^2 g_L}{\mu_f \Gamma}\right)^{1/3} \tag{48}$$

These equations show that a tube of given dimensions, for the usual case where L/Dn exceeds 2.76, is more effective in a horizontal than in a vertical position. Thus for L/Dn of 100, a horizontal tube gives h_m 2.5 times that for a vertical tube. The values of k_f, ρ_f, and μ_f of the condensate are taken at a special film temperature, t_f'

$$t_f' = t_v - 0.75(\Delta t)_m \tag{49}$$

and t_v is taken at the saturation temperature.

With vertical tubes and values of $4\Gamma/\mu$ below 2100, for a given Δt_m, values of h_m for film-type condensation of steam are 1.2 times those given by Eq. (48) [Baker, Kazmark, and Stroebe, *Trans. Am. Inst. Chem. Engrs.*, **35**, 127 (1939); *Ind. Eng. Chem.*, **31**, 214 (1939)]. The same is true for organic vapors (95 per cent ethanol, benzene, acetone, or mixed heptanes), as shown by Baker and Hipkin [*Trans. Am. Inst. Chem. Engrs.*, **40**, 291–308 (1944)].

For n horizontal tubes lying in a vertical plane, Young and Wohlenberg [*Trans. Am. Soc. Mech. Engrs.*, **64**, 787 (1942)] verify Eq. (47) for condensation of an organic vapor.

Figure 14 is an alignment chart (from Chilton and Drew) based on Eqs. (47) and (48). For long vertical tubes $4\Gamma/\mu_f$ may exceed 2100, in which case turbulence develops in the outer portion of the condensate layer, thus increasing h_m above the values predicted by Eqs. (47) and (48) (see Fig. 15).

Since with organic vapors there is but little variation in the thermal conductivity or viscosity of the condensate at the condensing temperature at 1 atm., there is little variation in h_m. Thus, with horizontal tubes, with cooling water at ordinary temperatures, h_m may be taken as 200 to 400 for the following vapors condensing at atmospheric pressure: benzene, carbon tetrachloride, dichloromethane, dichlorodifluoromethane, diphenyl, ethyl alcohol, heptane, hexane, methyl alcohol, octane, toluene, and xylene. Ammonia gives h_m of approximately 1000, and mixtures of steam and organic vapors, forming immiscible condensates, give h_m ranging from 250 to 750, increasing with increasing proportion of steam [Baker and Tsao, *Trans. Am. Inst. Chem. Engrs.* **36**, 517, 783 (1940)]. With film-type condensation of clean steam on horizontal tubes, h_m ranges from 1000 to 3000; see Eqs. (47) and (48).

The presence of non-condensable gas, such as air, seriously reduces h_m, and consequently all vapor-heated apparatus should be well vented (see p. 477).

With steam, small traces of certain promoters (Nagle, U.S. Patent 1,995,361) such as oleic acid on nickel or chromium plate, benzyl mercaptan on copper and brass, etc., become adsorbed in a very thin layer on the surface of the tubes, preventing the condensate from wetting the metal and inducing dropwise condensation, which gives much higher values of h_m than film-type condensation. With filtered steam, Bays and Blenderman (Chem. Eng. S.M. Thesis, Mass. Inst. Tech., 1935) obtained h_m of 14,000 for a chrome-plated surface promoted with oleic acid, and Baum (Chem. Eng. S.M. Thesis, Mass. Inst. Tech., 1936) obtained 12,000 for copper promoted with benzyl mercaptan. Even higher values have been found with very short condensing surface [Shea and Krase, *Trans. Am. Inst. Chem. Engrs.*, **36**, 463 (1940)]. However, with dirty or corroded surfaces it is difficult to maintain dropwise condensation; with plant steam, not filtered, Fitzpatrick [*Trans. Am. Inst. Chem. Engrs.*, **35**, 97 (1939)] obtained h_m of 2000 to 4000 with brass treated with benzyl mercaptan. With filtered steam condensing on chrome plate promoted with oleic acid, Bartol and Hayes (S.M. Thesis, Naval Construction and Engineering, Mass. Inst. Tech., 1944) obtained h ranging from 12,000 to 36,000 with Δt of 60°F., giving q/A of substantially 2,000,000 B.t.u./(hr.)(sq. ft.); the water-side coefficients agreed with Eq. (27), extending its range to DG/μ_f of 500,000.

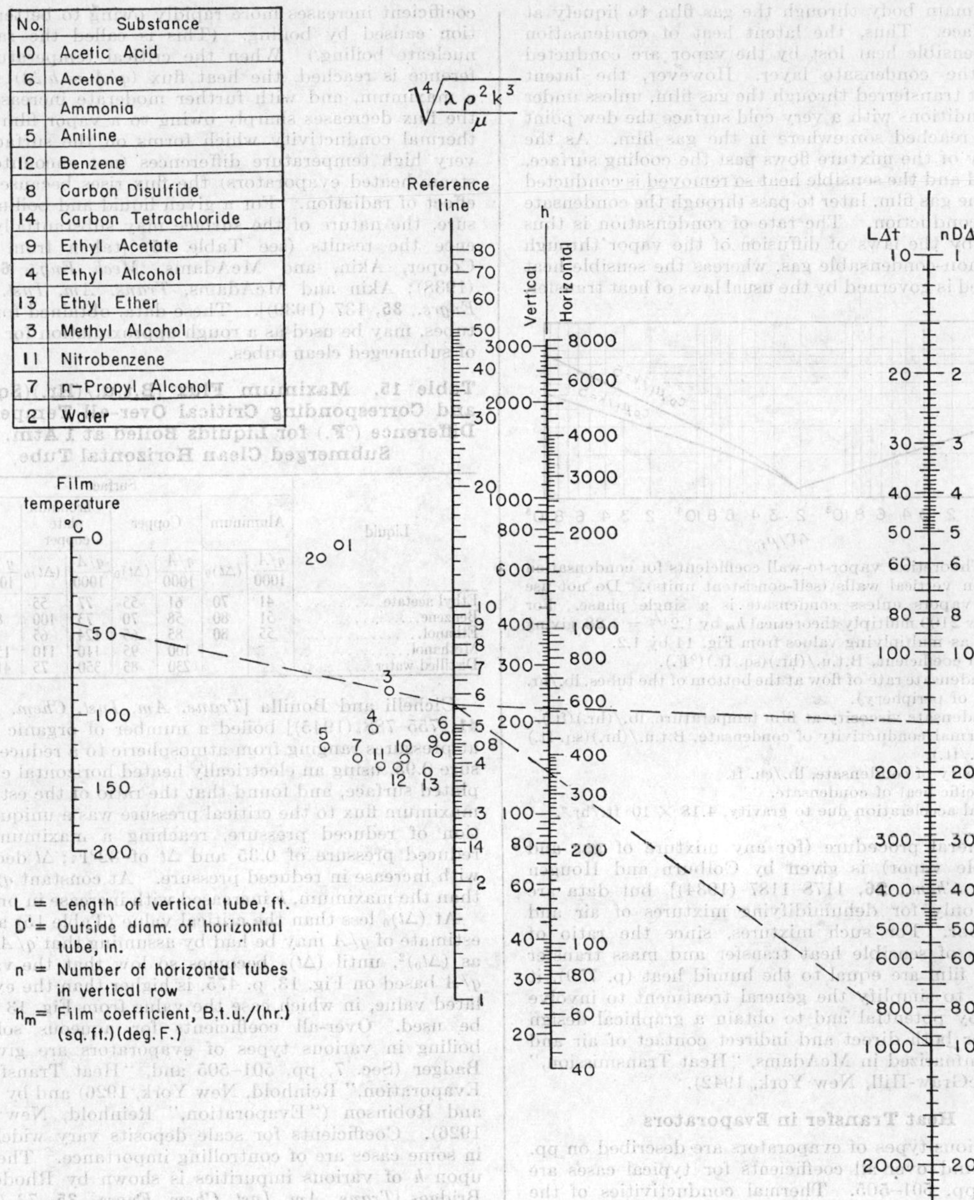
No. | Substance
10 | Acetic Acid
6 | Acetone
 | Ammonia
5 | Aniline
12 | Benzene
8 | Carbon Disulfide
14 | Carbon Tetrachloride
9 | Ethyl Acetate
4 | Ethyl Alcohol
13 | Ethyl Ether
3 | Methyl Alcohol
11 | Nitrobenzene
7 | n-Propyl Alcohol
2 | Water
$\sqrt[4]{\frac{\lambda \rho^2 k^3}{\mu}}$
Reference line
Film temperature °C
h
Vertical
Horizontal
LΔt
nD'Δt
V H
L = Length of vertical tube, ft.
D' = Outside diam. of horizontal tube, in.
n = Number of horizontal tubes in vertical row
h_m = Film coefficient, B.t.u./(hr.)(sq. ft.)(deg. F.)

FIG. 14. Chart for determining film coefficient h_m for film-type condensation of pure vapor, based on Eqs. (47) and (48). For vertical tubes multiply h_m by 1.2. If $4\Gamma/\mu_f$ exceeds 2100, use Fig. 15.

Effect of Superheat in Pure Vapor. If the temperature of the wall is above the saturation temperature at the prevailing pressure, there will be no condensation, and the case should be treated as cooling a gas (see p. 467).

Evaporation or Condensation of Vapor in Presence of Non-condensable Gas

When a mixture of a condensable vapor and a non-condensable gas is exposed to a surface colder than the dew point of the mixture, some condensation occurs. In the absence of dropwise condensation, a layer of condensate forms on the cooling surfaces and a film of a mixture of non-condensable gas and vapor collects next to the condensate layer, the concentration of vapor in the gas film being lower than in the main body of the mixture. As pointed out by Lewis [*Chem. & Met. Eng.*, **34**, 735 (1927)], owing to the differences in partial pressure of the vapor between the main body of the mixture and that at the interface between films, the vapor diffuses from the main body through the gas film to liquefy at the interface. Thus, the latent heat of condensation and the sensible heat lost by the vapor are conducted through the condensate layer. However, the latent heat is not transferred through the gas film, unless under special conditions with a very cold surface the dew point might be reached somewhere in the gas film. As the main body of the mixture flows past the cooling surface, it is cooled and the sensible heat so removed is conducted through the gas film, later to pass through the condensate layer by conduction. The rate of condensation is thus governed by the laws of diffusion of the vapor through a film of non-condensable gas, whereas the sensible heat transmitted is governed by the usual laws of heat transfer.

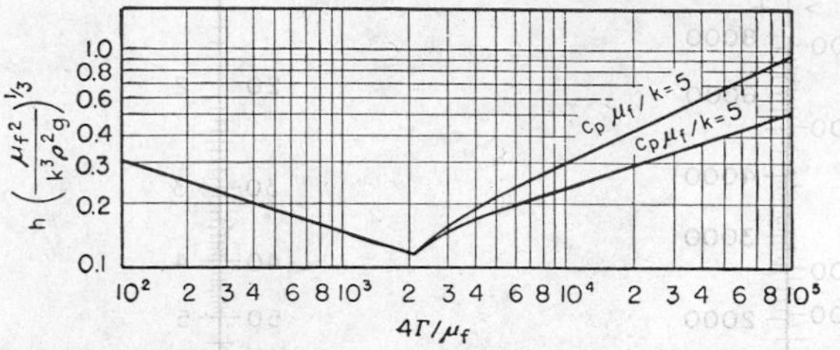

FIG. 15. Theoretical vapor-to-wall coefficients for condensation of vapors on vertical walls (self-consistent units). Do not use for mixed vapors unless condensate is a single phase. For $4\Gamma/\mu_f$ below 2100 multiply theoretical h_m by $1.2^{4/3}$ = 1.28, giving same result as multiplying values from Fig. 14 by 1.2.

h_m = film coefficient, B.t.u./(hr.)(sq. ft.)(°F.).

Γ = condensate rate of flow at the bottom of the tubes, lb./hr. (ft. of periphery).

μ_f = condensate viscosity at film temperature, lb./(hr.)(ft.).

k = thermal conductivity of condensate, B.t.u./(hr.)(sq. ft.)(°F./ft.).

ρ = density of condensate, lb./cu. ft.

c_p = specific heat of condensate.

g = local acceleration due to gravity, 4.18×10^8 ft./hr.².

The general procedure (for any mixture of gas and condensable vapor) is given by Colburn and Hougen [*Ind. Eng. Chem.*, **26**, 1178–1187 (1934)], but data are available only for dehumidifying mixtures of air and water vapor. For such mixtures, since the ratio of coefficients of sensible heat transfer and mass transfer of the gas film are equal to the humid heat (p. 759), it is possible to simplify the general treatment to involve an enthalpy potential and to obtain a graphical design method for both direct and indirect contact of air and water (summarized in McAdams, "Heat Transmission," 2d ed., McGraw-Hill, New York, 1942).

Heat Transfer in Evaporators

The various types of evaporators are described on pp. 499–522, and over-all coefficients for typical cases are cited on pp. 501–505. Thermal conductivities of the tubes are tabulated on p. 456, and steam-side coefficients are given on pp. 476–478. Where the use of a suitable promoter induces dropwise instead of film condensation on the steam side, the over-all coefficient increases substantially. Where the heating surface fouls rapidly, owing to deposition of scale, the over-all coefficient must be determined experimentally (see pp. 476–478).

Film Coefficients

Liquids Boiling outside Submerged Horizontal Tubes. In natural convection evaporators the heat may be supplied by submerged tubes, usually heated internally by condensing steam. At very small temperature differences the coefficients are of the order of those found when warming liquids (Fig. 13, p. 475), but, as the temperature difference is increased, the coefficient increases more rapidly owing to better agitation caused by boiling. (This is called the range of nucleate boiling.) When the critical temperature difference is reached, the heat flux ($q/A = h\,\Delta t$) reaches a maximum, and with further moderate increase in Δt the flux decreases sharply owing to a vapor film of low thermal conductivity which forms on the surface. At very high temperature differences (not encountered in steam-heated evaporators) the flux rises because of the effect of radiation. For a given liquid and boiling pressure, the nature of the surface may substantially influence the results (see Table 15) [taken from Sauer, Cooper, Akin, and McAdams, *Mech Eng.*, **60**, 669 (1938); Akin and McAdams, *Trans. Am. Inst. Chem. Engrs.*, **35**, 137 (1939)]. These data, obtained for single tubes, may be used as a rough approximation for a bank of submerged clean tubes.

Table 15. Maximum Flux [B.t.u./(Hr.)(Sq. Ft.)] and Corresponding Critical Over-all Temperature Difference (°F.) for Liquids Boiled at 1 Atm. with a Submerged Clean Horizontal Tube

Liquid	Surface: Aluminum		Copper		Chromium plate copper		Steel	
	q/A / 1000	$(\Delta t)_0$	q/A / 1000	$(\Delta t)_0$	q/A / 1000	$(\Delta t)_0$	q/A / 1000	$(\Delta t)_0$
Ethyl acetate	41	70	61	55	77	55		
Benzene	51	80	58	70	73	100	82	100
Ethanol	55	80	85	65	124	65		
Methanol	..	..	100	95	110	110	155	110
Distilled water	..	..	230	85	350	75	410	150

Cichelli and Bonilla [*Trans. Am. Inst. Chem. Engrs.*, **41**, 755–788 (1945)] boiled a number of organic liquids at pressures ranging from atmospheric to a reduced pressure 0.95, using an electrically heated horizontal chrome-plated surface, and found that the ratio of the estimated maximum flux to the critical pressure was a unique function of reduced pressure, reaching a maximum at a reduced pressure of 0.35 and Δt of 32°F.; Δt decreased with increase in reduced pressure. At constant q/A less than the maximum, h increased with increase in pressure.

At $(\Delta t)_0$ less than the critical value (Table 15) a rough estimate of q/A may be had by assuming that q/A varies as $(\Delta t_0)^2$, until $(\Delta t)_0$ becomes so low that the value of q/A based on Fig. 13, p. 475, is higher than the extrapolated value, in which case the value from Fig. 13 should be used. Over-all coefficients for aqueous solutions boiling in various types of evaporators are given by Badger (Sec. 7, pp. 501–505 and, "Heat Transfer and Evaporation," Reinhold, New York, 1926) and by Webre and Robinson ("Evaporation," Reinhold, New York, 1926). Coefficients for scale deposits vary widely and in some cases are of controlling importance. The effect upon h of various impurities is shown by Rhodes and Bridges [*Trans. Am. Inst. Chem. Engrs.*, **35**, 73 (1939); *Ind. Eng. Chem.*, **30**, 1401 (1938)]; the effect of the nature of the metal surface is shown by Bonilla and Perry [*Trans. Am. Soc. Mech. Engrs.*, **37**, 685–705 (1941)].

At a given Δt, in the range of nucleate boiling, the effect upon h of changing the boiling temperature is approximated by the equations of Cryder and Finalborgo [*Trans. Am. Inst. Chem. Engrs.*, **33**, 346 (1937)] as follows:

$$\log_{10} \frac{h}{h_n} = b(t - t_n) \tag{50}$$

where the subscript n refers to the normal boiling point at 1 atm. Expressing temperature in degrees Fahrenheit, the following values of b were obtained: 0.012 for carbon tetrachloride or kerosene, 0.014 for water or n-butanol, 0.015 for methanol or 26.3 per cent glycerol solution, 0.016 for 10.1 per cent sodium sulphate solution, and 0.017 for 24.2 per cent sodium chloride. The effect upon U of changing Δt is shown on p. 478.

Forced-circulation Evaporators. With vertical forced-circulation evaporators, the liquid enters the tube at substantially the saturation temperature in the overhead vapor-liquid separator but is at a pressure greater than in the separator, owing to hydrostatic head and to pressure drop caused by flow through the heated tube. The temperature rises as the fluid passes through the tube and finally reaches a section where ebullition starts, but with high entering velocities (5 to 15 ft./sec.) only a small fraction of the feed evaporates. For these conditions Boarts, Badger, and Meisenburg [*Trans. Am. Inst. Chem. Engrs.*, **33**, 363 (1937)] show that Eq. (27), used together with the length-mean Δt, correlates the data for water, and a similar finding was made earlier for sugar solutions by Logan, Fragen, and Badger [*Ind. Eng. Chem.*, **26**, 1044 (1934)]. At lower feed rates, N_{Re} from 30,000 to 60,000, the fraction evaporated increases, and, in the boiling section, h is approximately double that predicted by Eq. (27). Over-all coefficients for molasses and gelatin solutions, in a vertical forced circulation evaporator, are given by Coates and Badger [*Trans. Am. Inst. Chem. Engrs.*, **32**, 49 (1936)].

In a forced-circulation evaporator containing four horizontal passes, with water entering at velocities ranging from 0.3 to 0.9 ft./sec., McAdams, Woods, and Bryan [*Trans. Am. Soc. Mech. Engrs.*, **63**, 545 (1941)] showed that the local over-all coefficient reached a maximum of 2000 when 50 per cent by weight was vaporized, and thereafter decreased rapidly, approaching values obtained for superheating vapor [Eq. (24), p. 467]. A similar phenomenon was obtained with benzene. Hot wall vapor binding, and consequent decrease in heat flow per unit area, occurred with excessive Δt. McAdams, Woods, and Heroman [*ibid.*, **64**, 193–200 (1942)] give data for solutions of benzene in oil.

Natural-circulation Evaporators. In the long-tube vertical evaporator, where natural circulation is employed, much lower entering velocities are employed than in the forced-circulation vertical, and as a result a much larger fraction of the liquid evaporates. Data are given by Stroebe, Baker, and Badger [*Trans. Am. Inst. Chem. Engrs.*, **35**, 17 (1939); *Ind. Eng. Chem.*, **31**, 200 (1939)]; with water, film coefficients ranged from 1200 to 2600.

Optimum Operating Conditions

Optimum Velocity. In designing heat-transfer equipment involving the heating or cooling of a fluid flowing through tubes, usually the designer has some latitude in selecting the velocity of the fluid. In the final design of such apparatus, it is customary to calculate total costs for a number of velocities, selecting as the **optimum** velocity that which gives the minimum total cost. In preliminary designs where approximate results are satisfactory, this tedious procedure may be avoided by the use of equations or charts for estimating the optimum velocity.*

Optimum Amount of Cooling Water. In some cases the cooling water is under sufficient pressure in the plant mains to give the desired rate of flow, and the cost of water is directly proportional to the amount used. The optimum water rate is that which gives the minimum annual sum of cooling water costs and of fixed charges on the exchanger or condenser, and the corresponding optimum over-all temperature difference at the hot end ($\Delta t_2 = T_1 - t_2$) is given by the following dimensionless equation for counterflow:

* McAdams, "Heat Transmission," 2d ed., McGraw-Hill, New York, 1942.

$$\left[\frac{T_1 - T_2 - (\Delta t_2 - \Delta t_1)}{\Delta t_2 - \Delta t_1}\right]^2 \left[\left(\ln_e \frac{\Delta t_2}{\Delta t_1}\right) - \left(\frac{\Delta t_2 - \Delta t_1}{\Delta t_2}\right)\right] = \frac{Ue\theta}{bc} \qquad (51)$$

wherein b is the annual fixed charges on the heating surface, dollars per sq. ft.; e is the cost of cooling water, dollars per lb.; θ is the hours of operation per year, and the other symbols have the usual significance (p. 467). Figure 16, prepared by A. P. Colburn, is a graphical solution of the equation and eliminates trial-and-error calculations for the range of variables plotted. For values of z greater than 20, use Eq. (51) or the high-range chart of Douglass and Adams [*Ind. Eng. Chem.*, **33**, 1082–1083 (1941)].

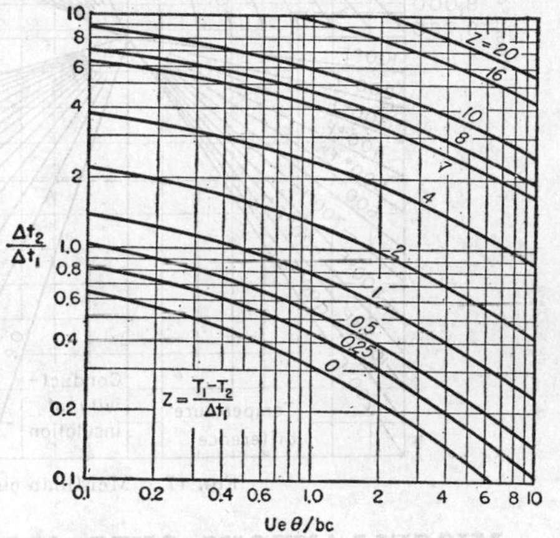

FIG. 16. Graphical solution of Eq. (51) for determining optimum Δt_2 in condensers and counterflow coolers.

T_1, T_2 = temperatures of the warmer fluid entering and leaving, respectively.

t_1, t_2 = temperatures of water entering and leaving exchanger, respectively.

Δt_1, Δt_2 = over-all temperature differences from the warmer fluid to the water at colder and warmer ends of exchanger, respectively; $\Delta t_1 = T_2 - t_1$; $\Delta t_2 = T_1 - t_2$.

Example. It is desired to design a counterflow gas cooler to cool a given gas ($c_p = 0.24$) from 200° to 90°F., using cooling water entering at 85°F. Water costs $0.20 per 1000 cu. ft., and annual fixed charges on the cooling surface are $0.50 per sq. ft. of inside surface, $D = 0.0875$ ft. It is planned to operate 8400 hr. per year. The power cost is such that the optimum mass velocity is 8600 lb./(hr.)(sq. ft.), and the corresponding U_i is 7.8. Calculate the optimum outlet temperature of the water, and the corresponding ratio of water to gas.

$$\frac{Ue\theta}{bc} = \frac{(7.8)(0.2/62300)}{(0.5/8400)(1.0)} = 0.422$$

$(T_1 - T_2)/\Delta t_1 = (200 - 90)/(90 - 85) = 22$, and from Eq. (51), $\Delta t_2/\Delta t_1 = 15.9 = \Delta t_2/5$, whence $\Delta t_2 = 79.5$°F. $= 200 - t_2$, and $t_2 = 120.5$°F. Per pound of gas cooled, $(200 - 95)(0.24) = 25.2$ B.t.u. are absorbed by the water in rising $120.5 - 85 = 35.5$°F.; hence $25.2/35.5 = 0.71$ lb. of water are required per pound of gas.

Economical Thickness of Insulation

The economical thickness of insulation can be readily determined for any given conditions of operation from Fig. 17 developed by McMillan [*Trans. Am. Soc. Mech. Engrs., Fuels Steam Power*, **51**, 349 (1929)]. Starting with the hours of operation per year on the upper left-hand margin, a line is drawn through the following successive intersections: Horizontal to the value of heat in dollars per 1,000,000 B.t.u., vertical to the temperature difference between the pipe and surrounding air, horizontal to the value of thermal conductivity of the insulation expressed in B.t.u./(hr.)(sq. ft.)(°F./in.), vertical to the cost of insulation in dollars per board foot (or as discount from standard list price for those materials having a standard list price of $0.30 per b.m.), horizontal to the rate of amortization in per cent per year, vertical to the nominal diameter of the pipe, and horizontal to the economical thickness of insulation in inches.

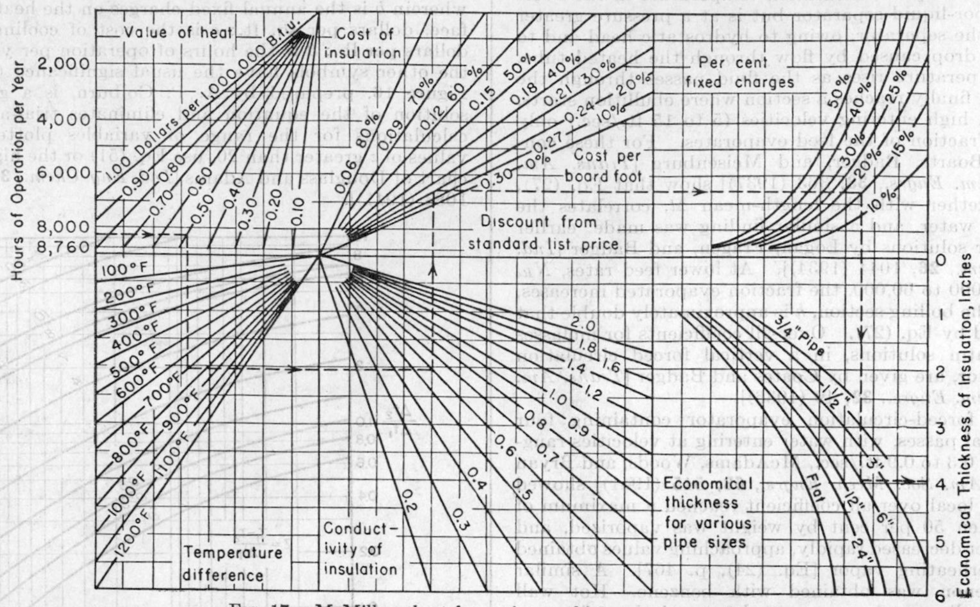

FIG. 17. McMillan chart for optimum thickness of insulation.

MISCELLANEOUS OVER-ALL COEFFICIENTS OF HEAT TRANSFER

BY OLAF P. BERGELIN AND ALLAN P. COLBURN

The following over-all coefficients either are the result of tests upon specific pieces of equipment or cover the range of values encountered in usual engineering practice. The values are approximate, since variation in factors such as the cleanliness of the heat transfer surface, the amount of non-condensable gases, or the viscosity of the fluids may cause a several-fold variation in coefficients. The over-all coefficients, therefore, are best used as a guide to preliminary design estimates or as a rough check upon design calculations.

Over-all Coefficients for Ammonia Condensers

For condensation of ammonia at 145 lb./sq. in. gage pressure and with cooling water at 60° to 70°F. Kratz, MacIntire, and Gould (*Univ. Ill. Eng. Exp. Sta. Bull.* 171, 186) report the values for U shown in Table 6. The ammonia condensed on the outside of the tubes in the vertical condenser with the cooling water flowing in a layer down the inside walls of the tubes. The horizontal condenser consisted of banks of horizontal 2-in.

Table 1. Over-all Coefficients for Heat Exchangers in Petroleum Service

Service of exchanger	Fluids: Tubes	Fluids: Shell	Velocity in tubes, ft./sec.	Δt_m, °F.	Over-all coefficient U, B.t.u./(hr.)(sq. ft.)(°F.)
Stabilizer reflux condensers	Water	Condensing vapors + residual gas	3.0	13.5	94
			5.0	22	145
		108°–118°A.P.I.	0.3–0.6	....	55–67*
			0.7	....	98–125*
Partial condensers	39°A.P.I. crude	58°A.P.I. gasoline	2.4	147	24
	55°A.P.I. crude	62°A.P.I. naphtha	4.4	80	37
	55°A.P.I. crude	62°A.P.I. naphtha	6.8	90	48
Stabilizer reboiler	Steam	58°A.P.I. naphtha		33.5	42
Absorber reboiler	Steam	37°A.P.I. oil	......	41.4	45
Stabilizer reboiler	Steam	67–74°A.P.I.	......	.....	43–183*
Oil preheater	42°A.P.I. oil	Steam	1.4	32	108
Exchangers	60°A.P.I.	58°A.P.I.	1.4	65	74
	63°A.P.I.	57°A.P.I.	4.6	69	139
	70–82°A.P.I.	67–74°A.P.I.	0.3–0.7	.....	18–37*
	70–82°A.P.I.	67–74°A.P.I.	0.3–0.7		35–45*
	43°A.P.I.	37°A.P.I.	1.6	59	33
	39°A.P.I. crude	13°A.P.I. residue	3.9	262	19
Coolers	Water	57°A.P.I.	......	40	52
	Water	44°A.P.I.		97	40
	Water	67–74°A.P.I.	0.2	.....	20*
	Water	67–74°A.P.I.	0.4–0.7	.....	51–53*

* McGiffin, *Trans. Am. Inst. Chem. Engrs.*, **38**, 761 (1942). All other data in Table 1 are from Higgins, "Heat Transfer," p. 56, a special publication of the A.S.M.E. (New York, 1936).

Table 2. Miscellaneous: A Range of Values of Miscellaneous Over-all Coefficients*

U, expressed in B.t.u./(hr.)(sq. ft.)(°F.) as found in practice. Under special conditions higher or lower values may be realized.

Type of heat exchanger	State of controlling resistance: Free convection, *U*	State of controlling resistance: Forced convection, *U*	Typical fluid	Typical apparatus
Liquid to liquid	25–60	150–300	Water	Liquid-to-liquid heat exchangers
Liquid to liquid	5–10	20–50	Oil	
Liquid to gas (atm. pressure)	1–3	2–10	....	Hot-water radiators
Liquid to boiling liquid	20–60	50–150	Water	Brine coolers
Liquid to boiling liquid	5–20	25–60	Oil	
Gas (atm. pressure) to liquid	1–3	2–10	....	Air coolers, economizers
Gas (atm. pressure) to gas	0.6–2	2–6	....	Steam superheaters
Gas (atm. pressure) to boiling liquid	1–3	2–10	....	Steam boilers
Condensing vapor to liquid	50–200	150–800	Steam-water	Liquid heaters and condensers
Condensing vapor to liquid	10–30	20–60	Steam-oil	
Condensing vapor to liquid	40–80	60–300	Organic vapor-water	
Condensing vapor to liquid	......	15–300	Steam-gas mixture	
Condensing vapor to gas (atm. pressure)	1–3	6–16	....	Steam pipes in air Air heaters
Condensing vapor to boiling liquid	40–100	......	....	Scale-forming evaporators
Condensing vapor to boiling liquid	300–800	......	Steam-water	
Condensing vapor to boiling liquid	50–150	......	Steam-oil	
Condensing vapor to boiling liquid	......	50–400	Steam-organic liquid	Steam-jacketed tubes

* Modified from Lucke, "Engineering Thermodynamics," p. 550, McGraw-Hill, New York, 1912.

Table 3. Coils Immersed in Liquids. Over-all Coefficients

U, expressed as B.t.u./(hr.)(sq. ft.)(°F.)

Substance inside coil	Substance outside coil	Coil material	Agitation	*U*	Reference
Steam	Water	Lead	Agitated	70	1
Steam	Sugar and molasses solutions	Copper	None	50–240	2
Steam	Boiling aqueous solution	....	....	600	3
Cold water	Dilute organic dye intermediate	Lead	Turbo-agitator at 95 r.p.m.	300	3
Cold water	Warm water	Wrought iron	Air bubbled into water surrounding coil	150–300	4
Cold water	Hot water	Lead	0.40 r.p.m., paddle stirrer	90–360	5
Brine	Amino acids	....	30 r.p.m.	100	3
Cold water	25% oleum at 60°C.	Wrought iron	Agitated	20	6
Water	Aqueous solution	Lead	500 r.p.m., sleeve propeller	250	1
Water	8% NaOH	....	22 r.p.m.	155	3
Steam	Fatty acid	Copper (pancake)	None	96–100	7
Milk	Water	....	Agitation	300	8
Cold water	Hot water	Copper	None	105–180	9
60°F. water	50% aqueous sugar solution	Lead	Mild	50–60	10
Steam and hydrogen at 1500 lb./sq. in.	60°F. water	Steel	....	100–165	10

NOTE. Chilton, Drew, and Jebens, *Ind. Eng. Chem.*, **36**, 510 (1944) give film coefficients for heating and cooling agitated fluids using a coil in a jacketed vessel.

References:

[1] Read, private communication.
[2] Stose and Whittemore, Thesis, Mass. Inst. Tech., 1922.
[3] Chambers and Steves, private communication.
[4] Chilton and Colburn, private communication.
[5] Pierce and Terry, *Chem. & Met. Eng.*, **30**, 872 (1924).
[6] Boertlein, private communication.
[7] Mills and Daniels, *Ind. Eng. Chem.*, **26**, 248–250 (1934).
[8] Feldmeier, *Adv. paper, Am. Soc. Mech. Engrs. Meeting*, Dec. 4, 1934; published in "Heat Transfer," 69–74, A.S.M.E., New York, 1936.
[9] Storrow, *J. Soc. Chem. Ind.*, **64**, 322 (1945).
[10] Private communication.

Table 4. Miscellaneous: Special Equipment and Materials

Type of equipment	Hot material	Cold material	*U*	Remarks	Reference
High-pressure boiler	Molten salt	Boiling water	100–150		1
Tubular exchanger	Molten salt	Oil	52–80		1
Steam superheater	Molten salt	Steam	70		1
Air heater	Molten salt	Air	6		1
Catalyst case	Gas	Molten salt	6	Fins on outside of tube	1
Double-pipe Karbate exchanger	Water	Water	300–500		2
Karbate trombone cooler	20°Bé. HCl	Water	300	Water Γ_1 = 1750	3
Karbate tube reboiler	Steam	20% HCl	136	Vertical thermosiphon reboiler	10
	Steam	35% HCl	472–575		10
Double-pipe pyrex glass ex. using heat exchanger tubing	Air-water vapor	Water	25–75	Cooling water in annulus	4
	Water	Water	80–110		4
	Condensing steam	Water	100–125		4
Glass trombone cooler	50% sugar solution	60°F. water	50–60	Sugar solution inside pipe	9
Glass pipe in trough	20°Bé. HCl	Water	25		3
Votator	Water	Water	520–1120	Rotor velocity = 300–1900 r.p.m.	5
Pebble heater	Solid pebbles	Air	4	Heating gases to 1900°F. using ½-in. pebbles	6
		Methane	9		6
		Hydrogen	22		6
Long-tube vertical evaporator	Condensing steam	Water	300–1200		7
Falling-film condenser	Condensing steam	Water	574–2300	Water Γ_2 = 400–21,000 inside tubes	8
Stainless-steel conveyor belt	Molten TNT	50°F. air	5–7	Air blowing under and over belt	9
Partial condenser	Hydrocarbons and chlorinated hydrocarbons	Boiling propane	55–76	Refrigerated condenser	10
Shell and tube reboiler	Hot water	Hydrocarbons	42–88	Hot water in tubes	10
Reboiler	Steam	Chlorinated hydrocarbons	67	Clean reboiler, Δt = 12°F.	10
			20	Same reboiler after several months service, Δt = 96°F.	

U = B.t.u./(hr.)(sq. ft.)(°F.).
Γ_1 = lb./(hr.)(ft.) of pipe length for each side of pipe.
Γ_2 = lb./(hr.)(ft.) of periphery.

[1] Newton and Shimp, *Trans. Am. Inst. Chem. Engrs.*, **41, 197 (1945).**
[2] Werking, *Trans. Am. Inst. Chem. Engrs.*, **35**, 489 (1939).
[3] Lippman, *Chem. & Met. Eng.*, **52**, No. 3, 112 (1945).
[4] Thompson and Foust, *Chem. & Met. Eng.*, **47**, 410 (1940).
[5] Houlton, *Ind. Eng. Chem.*, **36**, 522–528 (1944).
[6] Norton, *Chem. & Met. Eng.*, **53**, No. 7, 116 (1946).
[7] Cessna, Lientz, and Badger, *Trans. Am. Inst. Chem. Engrs.*, **36**, 759 (1940).
[8] McAdams, Drew, and Bays, *Trans. Am. Soc. Mech. Engrs.*, **62**, 627 (1940).
[9] Private communication.
[10] Breidenbach and O'Connell, *Trans. Am. Inst. Chem. Engrs.*, **42**, 761 (1946).

Table 5. Jacketed Vessels. Over-all Coefficients
U, expressed in B.t.u./(hr.)(sq. ft.)(°F.)

Fluid inside jacket	Fluid in vessel	Wall material	Agitation	U	Reference
Steam	Water	Enameled C. I.*	0–400 r.p.m.	96–120	1
Steam	Milk	Enameled C. I.	None	200	2
Steam	Milk	Enameled C. I.	Stirring	300	2
Steam	Milk boiling	Enameled C. I.	None	500	2
Steam	Milk	Enameled C. I.	200 r.p.m.	86	1
Steam	Fruit slurry	Enameled C. I.	None	33–90	1
Steam	Fruit slurry	Enameled C. I.	Stirring	154	1
Steam	Water	C. I. and loose lead lining	Agitated	4–9	3
Steam	Water	C. I. and loose lead lining	None	3	3
Steam	Boiling SO_2	Steel	None	60	3
Steam	Boiling water	Steel	None	187	3
Hot water	Warm water	Enameled C. I.	None	70	1
Cold water	Cold water	Enameled C. I.	None	43	1
Ice water	Cold water	Stoneware	Agitated	7	3
Ice water	Cold water	Stoneware	None	5	3
Brine, low velocity	Nitration slurry	....	35–58 r.p.m.	32–60	4
Water	Sodium alcoholate solution	"Frederking" (cast-in-coil)	Agitated, baffled	80	4
Steam	Evaporating water	Copper	....	381	5
Steam	Evaporating water	Enamelware	....	36.7	5
Steam	Water	Copper	None	148	6
Steam	Water	Copper	Simple stirring	244	6
Steam	Boiling water	Copper	None	250	6
Steam	Paraffin wax	Copper	None	27.4	7
Steam	Paraffin wax	Cast iron	Scraper	107	7
Water	Paraffin wax	Copper	None	24.4	7
Water	Paraffin wax	Cast iron	Scraper	72.3	7
Steam	Solution	Cast iron	Double scrapers	175–210	8
Steam	Slurry	Cast iron	Double scrapers	160–175	8
Steam	Paste	Cast iron	Double scrapers	125–140	8
Steam	Lumpy mass	Cast iron	Double scrapers	75–96	8
Steam	Powder (5% moisture)	Cast iron	Double scrapers	41–51	8

* C. I. = cast iron.

References:
[1] Poste, *Ind. Eng. Chem.*, **16**, 469 (1924).
[2] Bowen, *Agr. Eng.*, **11**. 27 (1930).
[3] Read, private communication.
[4] Chambers and Steves, private communication.
[5] Robson, *Australian Chem. Inst. J. & Proc.*, **3**, 47–54 (1936).
[6] Chemical Engineering Charts No. 4. *Ind. Chemist*, **82**, 374 (1931).
[7] Huggins, *Ind. Eng. Chem.*, **23**, 749–753 (1931).
[8] Laughlin, *Trans. Am. Inst. Chem. Engrs.*, **36**, 345 (1940).

tubes with cooling water flowing down over the outer surfaces. In the double-pipe condenser the ammonia condensed in the annular space between standard 2-in. and 1¼-in. steel pipes with water flowing in the inside pipe.

Table 6. Values of U for Ammonia Condensers

Type of condenser	Water rate	B.t.u./(hr.)(sq. ft.)(°F. over-all Δt) Δt_m = 1.5°F.	Δt_m = 3.5°F.	Δt_m = 7°F.	Δt_m = 2–6°F.
Vertical tube and shell......	Γ_v = 400	220	170	150	
	800	275	225	215	
	1200	310	270	260	
	1600	350	315	300	
	2000	...	390	340	
	2400	...	430	370	
Horizontal drip............	Γ_h = 400	...	...	...	250
	800	...	...	...	330
	1200	...	...	...	400
Double pipe............	V = 4	350	270	230	
	6	410	320	280	
	8	470	390	350	

Γ_v = lb. of water/hr./ft. of periphery.
Γ_h = lb. of water/hr./ft. of tube length for each side of tube.
V = ft./sec.

Cost Data for Heat Exchangers

The approximate cost of tubular heat exchangers can be estimated by using an exchanger of standard construction with carbon-steel shell and tubes as a basis and allowing for variations in design and materials by means of multiplying factors. The cost of steel exchangers, however, may vary from year to year as indicated in Table 7. Tables 8, 9, and 10 give multiplying factors for use with the cost of steel exchangers.

Table 7. Cost of All-steel Exchangers

Outside heat transfer area, sq. ft.	Cost in dollars/sq. ft. Happel (1946)*	Bliss (1941)†
100	......	\$6.00
200	\$6–8.00	4.00
300	5–6.00	3.00
500	4–4.50	2.50
1000	3–3.50	1.70
3000	2–2.50	1.00
4000	......	1.00

Table 8. Tube-side Material Factors‡

Material	Factor
Steel....................................	1.00
Aluminum 3S..............................	1.29
Copper...................................	1.33
Silicon bronzes (copper tubes)...........	1.35
70–30 cupronickel........................	1.51
Stainless 304............................	2.00
Monel....................................	2.09
Stainless 347............................	2.17
Nickel...................................	2.26
Stainless 316............................	2.34
Inconel..................................	2.34

Table 9. Pressure and Temperature Factors*

Pressure, Lb./Sq. In.	Factor
75	1.00
150	1.02–1.10
300	1.1 –1.2
400	1.2 –1.3
600	1.5 –1.6
Temp., °F.	
600–900	1.1 –2.0

Table 10. Construction Factors*

Type of Change	Factors
Bundle only..............................	0.3–0.6
Non-removable bundle.....................	0.5–0.7
Clamp-ring construction..................	0.9–1.0
Internal floating head...................	1.0
Expansion ring in shell..................	1.0–1.1
External floating head...................	1.1–1.2
Reboilers................................	1.1–1.2

* The 1946 cost data in Tables 7, 9, and 10 are taken from an article by Happel, Aries, and Borns, *Chem. & Met. Eng.*, **53**, No. 10, 99 (1946). The estimates are based upon an all carbon-steel exchanger meeting T.E.M.A. standards for 75 lb./sq. in. pressure and having a floating head.

† A detailed method of cost calculations, based upon tube and shell characteristics, is presented by Bliss in *Trans. Am. Inst. Chem. Engrs.*, **37**, 763 (1941). The 1941 cost figures in Table 7, taken from this reference, are for an all-steel exchanger, 150 lb. pressure, floating head construction with ¾-in., 16 B.W.G. tubes, an average number of baffles, and satisfying the A.P.I.-A.S.M.E. construction code.

‡ The relative cost of using various alloys in exchangers is indicated in Table 8, which specifically refers to an exchanger of fixed tube sheet design, having a 12¾-in. o.d. shell, 10-ft. tubes, and 126¾-in. 18 gage tubes rolled into tube sheets. The shell is carbon steel with expansion bellows, but all parts in contact with the tube side are of alloy. The shell is baffled on 6-in. centers with steel segment-cut baffles. The factor for alloy cost will increase with the size of the exchanger. The figures in Table 8 were furnished by courtesy of the Downingtown Iron Works.

RADIANT-HEAT TRANSMISSION

BY HOYT C. HOTTEL

The relative importance of the several mechanisms of the transfer of heat from one body to another differs greatly with the temperature level of the system. At very low temperatures the transfer is chiefly by conduction, the passing along, from one layer of molecules to another, of the kinetic energy of the molecules in excess of that of the adjacent layer—kinetic energy which the molecules have by virtue of their temperature. Superposed on this phenomenon, when the system is fluid, is that of convection, the transfer of energy by mass motion of a large portion of the fluid—large, that is, compared with molecular magnitudes. Even at moderate temperature levels, however, another phenomenon becomes appreciable. The molecules or atoms, because of some sort of excitation caused by temperature, give rise to radiant energy, emitted in an amount determined by the temperature level of the molecules and capable of passage with more or less absorption to a distant receiver of the radiation. If the phenomena of conduction and convection on the one hand are contrasted with thermal radiation on the other, it is found that the former are affected by temperature difference and very little by temperature level, whereas the latter increases rapidly with increase in temperature level. It follows that, at very low temperatures, conduction and convection are the major contributors to the total heat transfer; at very high temperatures, radiation is the controlling factor. The temperature at which radiation accounts for roughly one-half of the total heat transmission depends on such factors as the emissivity of the surface or the magnitude of the convection coefficient. For large pipes losing heat by free convection, this is room temperature; for fine wires of low emissivity, it is above a red heat.

Subject matter will be divided into (1) the nature of thermal radiation (pp. 483–484), (2) radiant-heat interchange between the surfaces of solids separated by a nonabsorbing medium (pp. 484–490), (3) radiation from non-luminous gases (pp. 490–493), (4) radiation from clouds of particles (pp. 493–495), and (5) the combined effect of all these mechanisms in the combustion chamber of a furnace (pp. 495–498). Nomenclature is summarized below.

Nomenclature

A = areas of surface, sq. ft. A_c for cold body (sink); A_R, A_S, A_T for refractory zones; A_1, A_2, . . . for source-sink type surfaces.

C_c = dimensionless factor to allow for effect of total pressure on CO_2 radiation.

C_w = dimensionless factor to allow for effects of total pressure and partial pressure on H_2O radiation.

$(C_p)_m$ = mean specific heat of combustion gas, B.t.u./(lb.)(°F.).

c_1, c_2 = dimensional constants in Planck's law [Eq. (2)].

F = dimensionless geometrical factor to allow for *direct* interchange between two surfaces, F_{12} from surface 1 to 2, based on A_1; F_{21} from surfaces 2 to 1, based on A_2 (see Figs. 1–4).

$\bar{F}$ = dimensionless geometrical factor to allow for net radiation between *black* surfaces, including the effect of refractory surfaces (see Figs. 2 and 4).

$\mathfrak{F}$ = dimensionless factor to allow for interchange between gray surfaces, defined by Eq. (15) and evaluated by Eq. (16). $\mathfrak{F}_{12}$ is based on A_1.

h = coefficient of heat transfer by convection, B.t.u./(hr.)(sq. ft.)(°F.).

i = enthalpy of entering fuel, air, and any recirculated flue gas, above a base temperature T_0 (water as vapor), B.t.u./hr.

KL = dimensionless factor, absorption strength (Fig. 10).

x_w/k = wall thickness/thermal conductivity, following Eq. (26) only.

L = beam length for gas radiation, ft. (see Table 2).

M = molecular weight.

N_B = intensity of radiation from a black surface, B.t.u./(sq. ft.)(hr.) per unit solid angle in direction normal to surface.

P = gas pressure, atm., P_c, P_w = partial pressure of CO_2, H_2O; P_T = total pressure.

q = rate of heat transfer, B.t.u./hr.; q_F, from flame; q_C to cold surface (sink); q_L, lost to surroundings.

r_{af} = weight ratio of air to fuel.

S = sulfur dioxide radiation, B.t.u./(sq. ft.)(hr.)

T = absolute temperature; Rankine, Fahrenheit absolute (460 + °F.); T_c, cold body (sink); T_F, flame; T_G, gas; T_g, T_r, green, red brightness temperature; T_S, surface; T_0, base temperature. In Figs. 9 and 10, T is in degrees Kelvin (273 + °C.).

t = thermometric temperature, °F.

U_R = over-all coefficient of heat transfer through refractory, B.t.u./(hr.)(sq. ft.)(°F.).

W = total emissive power, B.t.u./(sq. ft.)(hr.); W_B for black body; $W_{B\lambda}$ for monochromatic emissive power, B.t.u./(hr.)(sq. ft.)(cm.).

w_A = firing rate, defined after Eq. (31).

w_G = combustion-gas rate, lb./hr.

x, y, z = distances, ft., defined in Fig. 3.

Y, Z = dimension ratios, defined in Fig. 3.

α = absorptivity, dimensionless.

Δ = $T_g - T_r$ (see Fig. 9).

ϵ = emissivity, dimensionless.

ϵ' = "effective" emissivity, defined on p. 492.

θ = angle.

η = efficiency, dimensionless.

σ = Stefan-Boltzmann constant, p. 484.

THE NATURE OF THERMAL RADIATION

If two small bodies of areas A_1 and A_2 are placed in a large evacuated enclosure perfectly insulated externally, then, when the system has come to thermal equilibrium, the bodies will emit radiation at the rates A_1W_1 and A_2W_2, respectively, where W is the total emissive power,* energy per unit time per unit area of the surface [B.t.u./(sq. ft.)(hr.)] emitted throughout the hemisphere above each element of surface. Let the energy impinging on unit area of any small body in the enclosure, due to radiation from the walls of the latter, be I. If the bodies have *absorptivities* (fraction of incident radiation which is absorbed) of α_1 and α_2, then energy balances on the bodies will have the form

$$IA_1\alpha_1 = A_1W_1 \quad \text{and} \quad IA_2\alpha_2 = A_2W_2$$

from which $W_1/\alpha_1 = W_2/\alpha_2 (= W_x/\alpha_x$, where x is *any* body). This generalization, that at thermal equilibrium the ratio of the emissive power of a surface to its absorptivity is the same for all bodies, is known as *Kirchhoff's law*. Since α cannot exceed unity, Kirchhoff's law places an upper limit on W, called W_B; and any surface having this upper limiting emissive power is called a *perfect radiator*. Since such a surface must have an absorptivity of unity and therefore a reflectivity of zero, the perfect radiator is more commonly referred to as a *black body*. The ratio of the emissive power of an actual surface to that of a black body is called the *emissivity* ϵ of the surface. Kirchhoff's law restated is as follows: At thermal equilibrium the emissivity and absorptivity of a body are the same.

The emissive power of a black body depends on its temperature only, and the second law of thermodynamics may be used to prove a proportionality between emissive power and the fourth power of the absolute temperature.

* Sometimes called *emittance, total hemispherical intensity,* or *radiant flux density.*

The relation

$$W_B = \sigma T^4 \tag{1}$$

is known as the *Stefan-Boltzmann law;* and the proportionality constant σ is known as the Stefan-Boltzmann constant [0.173 $\times 10^{-8}$ B.t.u./(sq. ft.)(hr.)(°R.)^4; 5.71 $\times 10^{-5}$ ergs/(sq. cm.)(sec.)(°K.)^4; 4.92 $\times 10^{-8}$ kg.-cal./(sq. m.)(hr.)(°K.)^4].

Other properties of black-body radiation of interest in heat transmission are related to the nature of its distribution in the spectrum and the shift of that distribution with temperature. If $W_{B,\lambda}$ is the *monochromatic emissive power* at wave length λ such that $W_{B,\lambda} \cdot d\lambda$ is the energy emitted from a surface per unit area per unit time in the wave-length interval λ to $\lambda + d\lambda$, the relation among $W_{B,\lambda}$, λ, and T is given by *Planck's law*

$$W_{B,\lambda} = \frac{c_1\lambda^{-5}}{e^{\frac{c_2}{\lambda T}} - 1} \tag{2}$$

$c_1 = 1.176 \times 10^{-8}$ [B.t.u./(sq. ft.)(hr.)] (cm.)^4 or 0.885 $\times 10^{-12}$ (cal.)(sq. cm.)/(sec.); $c_2 = 2.58$ (cm.)(°R.) or 1.433 cm. °K. According to Planck's law the monochromatic emissive power at any temperature varies from 0 at $\lambda = 0$ through a maximum and back to 0 at $\lambda = \infty$; at any wave length it increases with temperature, but values at shorter wave lengths increase faster so that the maximum value shifts to shorter wave lengths as the temperature rises. The position of the maximum is inversely proportional to the absolute temperature (*Wien's displacement law*), derivable from Eq. (2). The relation is: $\lambda_{\max}T = 0.5193$ cm. °R. or 0.2885 cm. °K.

The emissivity ϵ of a surface (more properly the total hemispherical emissivity, to differentiate it from monochromatic emissivity ϵ_λ, the ratio of radiating powers at the wave length λ, and from directional emissivity ϵ_θ, the ratio of radiating powers in a direction making the angle θ with the normal to the surface) varies with its temperature, its degree of roughness, and, if a metal, its degree of oxidation. Table 1 gives the emissivities of various surfaces and emphasizes the large variation possible in a single material. Although the values in the table apply strictly to normal radiation from the surface (with few exceptions), they may be used with negligible error for hemispherical emissivity except in the case of well-polished metal surfaces, for which the hemispherical emissivity is 15 to 20 per cent higher than the normal value.

A few generalizations may be made concerning the emissivity of surfaces: (1) The emissivities of metallic conductors have been shown to be very low and proportional to the absolute temperature; and the proportionality constant for different metals varies as the square root of the electrical resistance at a standard base temperature. Unless extraordinary pains are taken to prevent any possibility of oxidation or imperfection of polish, however, a specimen may exhibit several times this theoretical minimum emissivity. (2) The emissivities of non-conductors are much higher and, in contrast to metals, generally decrease with increase in temperature. (3) The low temperature emissivity of most non-metals is above 0.8. (4) Iron and steel vary widely with the degree of oxidation and roughness, clean metallic surfaces having an emissivity of 0.05 to 0.45 at low temperatures to 0.4 to 0.7 at high temperatures; oxidized and/or rough surfaces, 0.6 to 0.95 at low temperatures to 0.9 to 0.95 at high temperatures.

The absorptivity α of a surface depends on the factors affecting emissivity and in addition on the quality of the incident radiation, measured by its distribution in the spectrum. One may assign two subscripts to α, the first to indicate the temperature of the receiver and the second that of the incident radiation It has already been seen that according to Kirchhoff's law, the emissivity of a surface at temperature T_1 is equal to the absorptivity $\alpha_{1,1}$ which the surface exhibits for black radiation from a source at the same temperature; *i.e.*, a surface of low radiating power is also a poor absorber (or good reflector or transmitter) of radiation from a source at its own temperature. If the monochromatic absorptivity α_λ varies considerably with wave length and much less with temperature (which is generally the case), it follows that the total absorptivity $\alpha_{1,2}$ will vary more with T_2 than with T_1. Data on $\alpha_{1,2}$ at $t_1 = 70$°F. for a large group of non-metals indicate a decrease with T_2 from 0.8–0.95 at 500°R. to 0.1–0.9 at 5000°R. The value of $\alpha_{1,2}$ for a metal is approximately its emissivity evaluated at $T = \sqrt{T_1 \cdot T_2}$.

If α_λ is a constant independent of λ, the surface is called *gray*, and its total absorptivity α will be independent of the spectral-energy distribution of the incident radiation; then $\alpha_{1,2} = \alpha_{1,1} \equiv \epsilon_1$, *i.e.*, emissivity ϵ may be used in substitution for α even though the temperatures of the incident radiation and the receiver are not the same.

Radiation between the Surfaces of Solids Separated by a Non-absorbing Medium

The net loss of energy by radiation from a body at temperature T_1 in *black* surroundings at T_2 is given by

$$q_{1,\text{net}} = 0.173A_1\left[\epsilon_1\left(\frac{T_1}{100}\right)^4 - \alpha_{1,2}\left(\frac{T_2}{100}\right)^4\right] \text{ B.t.u./hr.} \tag{3}$$

when A_1 is square feet and T is degrees Rankine.

When $\alpha_{1,2} = \epsilon_1$, *i.e.*, when the body is gray (see above), this simplifies to

$$q_{1,\text{net}} = 0.173A_1\epsilon_1\left[\left(\frac{T_1}{100}\right)^4 - \left(\frac{T_2}{100}\right)^4\right] \tag{4}$$

The more complicated but important case of radiation interchange in a system of several surfaces at different temperatures and emissivities involves the concept of a geometrical factor F. F_{12} is defined as the fraction of the radiation leaving surface A_1 in all directions which is intercepted by surface A_2. Evaluation of this factor is as follows: Visualize, on black surface A_1 of total emissive power W_{B1}, a small surface element dA_1 radiating in all directions from one side, and on black surface A_2 a small surface element dA_2 intercepting some of the radiation from dA_1. Let the straight line connecting dA_1 and dA_2 have length r, and let r make angles θ_1 and θ_2 with the normals to dA_1 and dA_2, respectively. The rate of radiation from dA_1 to dA_2, called $dq_{1\to 2}$, will be proportional to $dA_1 \cos\theta_1$, the apparent area of dA_1 viewed from dA_2; to $dA_2 \cos\theta_2$, the apparent area of dA_2 viewed from dA_1; and inversely proportional to the square of the distance separating the elements. Calling the proportionality constant N_{B1}, one may write

$$dq_{1\to 2} = N_{B1}\frac{dA_1 \cos\theta_1\, dA_2 \cos\theta_2}{r^2} \tag{5}$$

This equation defines N_B, the *intensity* of radiation from a black surface.

By integration of Eq. (5) over a receiving surface filling the field of view of dA_1, one obtains $W_{B1}\,dA_1$, the total rate of emission from dA_1 throughout the hemisphere. The integration gives $\pi N_{B1}\,dA_1$, from which one concludes that the emissive power W_B of a black surface is π times its intensity of radiation N_B. By integration of Eq. (5) over finite areas A_1 and A_2 to obtain the rate of radiation from one to the other and dividing the result by $A_1 . W_{B1}$, one obtains F_{12}, the

Table 1. The Normal Total Emissivity of Various Surfaces

A. METALS AND THEIR OXIDES

Surface	t, °F.*	Emissivity*	Reference
Aluminum			
Highly polished plate, 98.3% pure	440–1070	0.039–0.057	*SF*
Polished plate	73	0.040	*ES*
Rough plate	78	0.055	*ES*
Oxidized at 1110°F	390–1110	0.11–0.19	*RO*
Al-surfaced roofing	100	0.216	*RH*
Calorized surfaces, heated at 1110°F.			
Copper	390–1110	0.18–0.19	*RO*
Steel	390–1110	0.52–0.57	*RO*
Brass			
Highly polished:			
73.2% Cu, 26.7% Zn	476–674	0.028–0.031	*SF*
62.4% Cu, 36.8% Zn, 0.4% Pb, 0.3% Al	494–710	0.033–0.037	*SF*
82.9% Cu, 17.0% Zn	530	0.030	*SF*
Hard rolled, polished, but direction of polishing visible	70	0.038	*ES*
but somewhat attacked	73	0.043	*ES*
but traces of stearin from polish left on	75	0.053	*ES*
Polished	100–600	0.096	*RH*
Rolled plate, natural surface	72	0.06	*ES*
rubbed with coarse emery	72	0.20	*ES*
Dull plate	120–660	0.22	*Wam*
Oxidized by heating at 1110°F	390–1110	0.61–0.59	*RO*
Chromium; see Nickel Alloys for Ni-Cr steels	100–1000	0.08–0.26	
Copper			
Carefully polished electrolytic copper	176	0.018	*KH*
Comm'l, emeried, polished, but pits remaining	66	0.030	*ES*
Comm'l, scraped shiny but not mirror-like	72	0.072	*ES*
Polished	242	0.023	*WW*
Plate, heated long time, covered with thick oxide layer	77	0.78	*ES*
Plate heated at 1110°F	390–1110	0.57	*RO*
Cuprous oxide	1470–2010	0.66–0.54	*B*
Molten copper	1970–2330	0.16–0.13	*B*
Gold			
Pure, highly polished	440–1160	0.018–0.035	*SF*
Iron and steel			
Metallic surfaces (or very thin oxide layer):			
Electrolytic iron, highly polished	350–440	0.052–0.064	*SF*
Polished iron	800–1880	0.144–0.377	*VS*
Iron freshly emeried	68	0.242	*ES*
Cast iron, polished	392	0.21	*RO*
Wrought iron, highly polished	100–480	0.28	*Wam*
Cast iron, newly turned	72	0.435	*ES*
Polished steel casting	1420–1900	0.52–0.56	*P*
Ground sheet steel	1720–2010	0.55–0.61	*P*
Smooth sheet iron	1650–1900	0.55–0.60	*P*
Cast iron, turned on lathe	1620–1810	0.60–0.70	*P*
Oxidized surfaces:			
Iron plate, pickled, then rusted red	68	0.612	*ES*
completely rusted	67	0.685	*ES*
Rolled sheet steel	70	0.657	*ES*
Oxidized iron	212	0.736	*HN*
Cast iron, oxidized at 1100°F	390–1110	0.64–0.78	*RO*
Steel, oxidized at 1100°F	390–1110	0.79	*RO*
Smooth oxidized electrolytic iron	260–980	0.78–0.82	*SF*
Iron and Steel—(*Continued*)			
Oxidized surfaces—(*Continued*)			
Iron oxide	930–2190	0.85–0.89	*BF*2
Rough ingot iron	1700–2040	0.87–0.95	*P*
Sheet steel, strong rough oxide layer	75	0.80	*ES*
dense shiny oxide layer	75	0.82	*ES*
Cast plate, smooth	73	0.80	*ES*
rough	73	0.82	*ES*
Cast iron, rough, strongly oxidized	100–480	0.95	*Wam*
Wrought iron, dull oxidized	70–680	0.94	*Wam*
Steel plate, rough	100–700	0.94–0.97	*RH*
High temp. alloy steels (see Nickel Alloys).			
Molten metal			
cast iron	2370–2550	0.29	*T*
mild steel	2910–3270	0.28	*T*
Lead			
Pure (99.96%), unoxidized	260–440	0.057–0.075	*SF*
Gray oxidized	75	0.281	*ES*
Oxidized at 390°F	390	0.63	*RO*
Mercury	32–212	0.09–0.12	*C*
Molybdenum filament	1340–4700	0.096–0.292	*AW*
Monel metal, oxidized at 1110°F	390–1110	0.41–0.46	*RO*
Nickel			
Electroplated on polished iron, then polished	74	0.045	*ES*
Technically pure (98.9% Ni, + Mn), polished	440–710	0.07–0.087	*SF*
Electropolated on pickled iron, not polished	68	0.11	*ES*
Wire	368–1844	0.096–0.186	*VS*
Plate, oxidized by heating at 1110°F	390–1110	0.37–0.48	*RO*
Nickel oxide	1200–2290	0.59–0.86	*BF*-1
Nickel alloys			
Chromnickel	125–1894	0.64–0.76	*VS*
Nickelin (18–32 Ni; 55–68 Cu; 20 Zn), gray oxidized	70	0.262	*ES*
KA-2S alloy steel (8% Ni; 18% Cr), light silvery, rough, brown, after heating	420–914	0.44 0.36	*R*
after 42 hr. heating at 980°F	420–980	0.62–0.73	*R*
NCT-3 alloy (20% Ni; 25% Cr.). Brown, splotched, oxidized from service	420–980	0.90–0.97	*R*
NCT-6 alloy (60% Ni; 12% Cr). Smooth, black, firm adhesive oxide coat from service	520–1045	0.89–0.82	*R*
Platinum			
Pure, polished plate	440–1160	0.054–0.104	*SF*
Strip	1700–2960	0.12 –0.17	*F*
Filament	80–2240	0.036 0.192	*DW*
Wire	440–2510	0.073–0.182	*G*
Silver			
Polished, pure	440–1160	0.0198–0.0324	*SF*
Polished	100–700	0.0221–0.0312	*RH*
Steel, see Iron.			
Tantalum filament	2420–5430	0.194–0.31	*AW*
Tin—Bright tinned iron sheet	76	0.043 and 0.064	*ES*
Tungsten			
Filament, aged	80–6000	0.032–0.35	*FW*
Filament	6000	0.39	*Z*
Zinc			
Comm'l, 99.1% pure, polished	440–620	0.045–0.053	*SF*
Oxidized by heating at 750°F	750	0.11	*RO*
Galvanized sheet iron, fairly bright	82	0.228	*ES*
Galvanized sheet iron, gray oxidized	75	0.276	*ES*

B. REFRACTORIES, BUILDING MATERIALS, PAINTS, AND MISCELLANEOUS

Surface	t, °F.*	Emissivity*	Reference
Asbestos			
Board	74	0.96	*ES*
Paper	100–700	0.93–0.945	*RH*
Brick			
Red, rough, but no gross irregularities	70	0.93	*ES*
Silica, unglazed, rough	1832	0.80	*P*
Silica, glazed, rough	2012	0.85	*P*
Grog brick, glazed	2012	0.75	*P*
See Refractory Materials below.			
Carbon			
T-carbon (Gebr. Siemens) 0.9% ash	260–1160	0.81–0.79	*SF*
This started with emissivity at 260°F. of 0.72, but on heating changed to values given			
Carbon filament	1900–2560	0.526	*L*
Candle soot	206–520	0.952	*WW*
Lampblack-waterglass coating	209–362	0.959–0.947	*KH*
Carbon—(*Continued*)			
Same	260–440	0.957–0.952	*SF*
thin layer on iron plate	69	0.927	*ES*
thick coat	68	0.967	*ES*
Lampblack, 0.003 in. or thicker	100–700	0.945	*RH*
Enamel, white fused, on iron	66	0.897	*ES*
Glass, smooth	72	0.937	*ES*
Gypsum, 0.02 in. thick on smooth or blackened plate	70	0.903	*ES*
Marble, light gray, polished	72	0.931	*ES*
Oak, planed	70	0.895	*ES*
Oil layers on polished nickel (lub. oil)	68	....	*ES*
Polished surface, alone		0.045	
+0.001-in. oil		0.27	
+0.002-in. oil		0.46	
+0.005-in. oil		0.72	
∞ thick oil layer		0.82	

* When two temperatures and two emissivities are given, they correspond, first to first and second to second, and linear interpolation is permissible.

Table 1. The Normal Total Emissivity of Various Surfaces—(*Concluded*)

Surface	t, °F.*	Emissivity*	Reference
Oil layers on aluminum foil (linseed oil)	. . . . . .	. . . . . .	*HN*
Al foil	212	0.087†	
+1 coat oil	212	0.561	
+2 coats oil	212	0.574	
Paints, lacquers, varnishes			
Snowhite enamel varnish or rough iron plate	73	0.906	*ES*
Black shiny lacquer, sprayed on iron	76	0.875	*ES*
Black shiny shellac on tinned iron sheet	70	0.821	*ES*
Black matte shellac	170–295	0.91	*WW*
Black lacquer	100–200	0.80–0.95	*RH*
Flat black lacquer	100–200	0.96–0.98	*RH*
White lacquer	100–200	0.80–0.95	*RH*
Oil paints, sixteen different, all colors	212	0.92–0.96	*HN*
Aluminum paints and lacquers			
10% Al, 22% lacquer body, on rough or smooth surface	212	0.52	*HN*
26% Al, 27% lacquer body, on rough or smooth surface	212	0.3	*HN*
Other Al paints, varying age and Al-content	212	0.27–0.67	*HN*
Paint, lacquers, varnishes—(*Continued*)			
Al lacquer, varnish binder, on rough plate	70	0.39	*ES*
Al paint, after heating to 620°F	300–600	0.35	*SF*
Paper, thin			
Pasted on tinned iron plate	66	0.924	*ES*
rough iron plate	66	0.929	*ES*
black lacquered plate	66	0.944	*ES*
Plaster, rough lime	50–190	0.91	*Wam*
Porcelain, glazed	72	0.924	*ES*
Quartz, rough, fused	70	0.932	*ES*
Refractory materials, 40 different	1110–1830	. . . . . .	*KW*
poor radiators	. . . . . .	[0.65} –0.75	
	. . . . . .	0.70}	
good radiators	. . . . . .	0.80} – {0.85	
	. . . . . .	0.85} – {0.90]	
Roofing paper	69	0.91	*ES*
Rubber			
Hard, glossy plate	74	0.945	*ES*
Soft, gray, rough (reclaimed)	76	0.859	*ES*
Serpentine, polished	74	0.900	*ES*
Water	32-212	0.95-0.963	*H*

* When two temperatures and two emissivities are given, they correspond, first to first and second to second, and linear interpolation is permissible.

† Although this value is probably high, it is given for comparison with the data, by the same investigator, to show the effect of oil layers. See "Aluminum," part A of this table.

NOTE. The results of many investigators have been omitted because of obvious defects in experimental method. A comprehensive bibliography is given in reference *SF*, following this table.

C. REFERENCES IN TABLE OF EMISSIVITIES

Year	Author and source	Key
1907	K. Siegel, *Sitzungsber. Akad. Wien.*, **116**, 2A, 1203	*S*
1908	C. B. Thwing, *Phys. Rev.*, **26**, 190	*T*
1909	G. K. Burgess, *Bur. Standards Bull.* 6, *Sci. Paper* 121, p. 111	*B*
1911	F. Wamsler, *Z. Ver. deut. Ing.*, **55**, 599; *Forschungsarb. Ver. deut. Ing.*, **98**	*Wam*
1912	W. Westphal, *Verh. physik. Ges.* (2), **14**, 987; **15**, 897	*WW*
1913	Randolph and Overholzer, *Phys. Rev.*, **2**, 144	*RO*
1913	O. Lummer, *Electrotech. Z.*, **34**, 1428	*L*
1914	Burgess and Foote, *Bur. Standards Bull.* 11, *Sci. Paper* 224, pp. 41–64	*BFI*
1914	P. D. Foote, *Bur. Standards Bull.* 11, *Sci. Paper* 243, p. 607; *J. Wash. Acad.*, **5**, 1 (1915)	*F*
1915	Burgess and Foote, *Bur. Standards Bull.* 12, *Sci. Paper* 249 pp. 83–89	*BF2*
1915	V. A. Suydam, *Phys. Rev.*, (2), **5**, **497**	*VS*
1923	K. Hoffmann, *Z. Physik.*, **14**, 310	*KH*
1924	Davison and Weeks, *J. Optical Soc. Am.*, **8**, 581	*DW*
1925	Forsythe and Worthing, *Astrophys. J.*, **61**, **146**	*FW*
1925	W. Geiss, *Physica*, **5**, 203	*G*
1925	"Hütte," 25th ed., vol. 1, W. Ernst u. Sohn.	*Hü*
1925	C. Zwikker, *Arch. Néerb.*, IIIA, **9**, **207**	*Z*
1926	A. G. Worthing, *Phys. Rev.* (2), **28**, **190**	*AW*
1927	V. Polak, *Z. tech. Physik*, **8**, 307	*P*
1927	E. Schmidt, *Beih. Gesundh.-Ing.*, Beiheft 22, Reihe 1, p. 1.	*ES*
1927	M. Wenzl and F. Morawe, *Stahl und Eisen*, **47**, 867–871	*WM*
1927	K. Wetzler, Dissertation, Darmstadt	*KW*
1928	H. Schmidt and E. Furthmann, *Mitt. K.W.-Inst. Eisenforsch., Abhandl.*, **109**, **225**	*SF*
1928	Private communication from Standard Oil Development Co.	*HN*
1929	R. H. Heilman, *Trans. Am. Soc. Mech. Engrs.* F. St. Power Sec., Surface Heat Transmission	*RH*
1931	H. S. Rice, M.I.T. thesis in Fuel and Gas Engineering	*R*
1932	Calculated from formula of Foote	*C*
1932	Calculated from spectral data	*H*

desired fraction of the radiation leaving surface A_1 in all directions which is intercepted by surface A_2. Although the discussion has been restricted to black surfaces, it is apparent that for a non-black surface A_1 the emissivity of which is independent of angle of emission, F_{12} calculated by the method above will continue to represent the fractional radiation from A_1 intercepted by A_2 (though not necessarily absorbed unless A_2 is black).

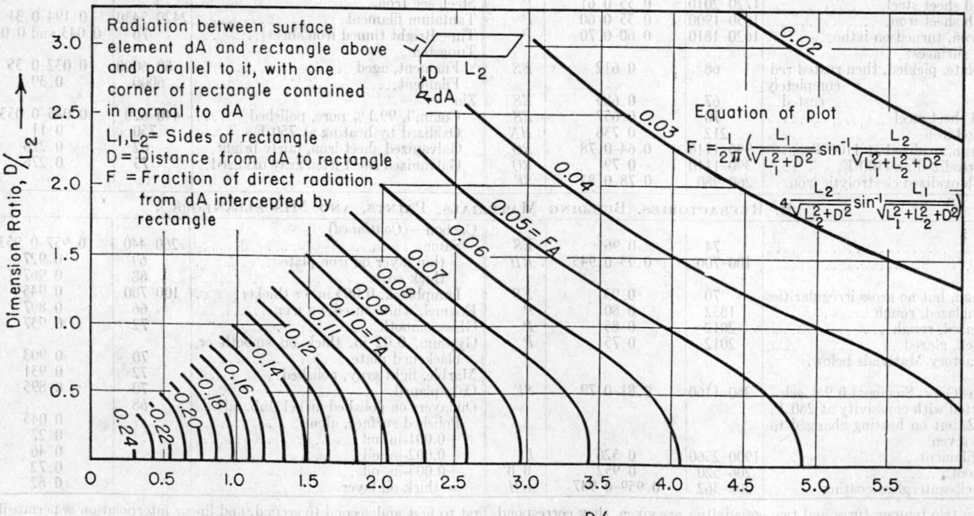

FIG. 18. Radiation from a plane element to a rectangle above it.

Values of F have been calculated for various surface arrangements on the assumption that emissivity ϵ_θ is constant, independent of θ (exact for black surfaces, quite good for most non-metallic or tarnished or rough metal surfaces). These values of F for a surface element dA and a rectangle in a parallel plane appear in Fig. 18; for opposed parallel rectangles and disks of equal size as lines 1 to 4 of Fig. 19; for adjacent rectangles in perpendicular planes in Fig. 20; for an infinite plane parallel to a system of parallel tubes as lines 1 and 3 of Fig. 21. Other cases are treated in the literature [Hottel, *Mech. Eng.*, **52**, 699 (1932)]. Important and useful concepts in evaluating F's are that

$$A_1F_{12} = A_2F_{21} \tag{6}$$

(since otherwise there would be a net heat flux between A_1 and A_2 when at the same temperature); that

$$F_{11} + F_{12} + F_{13} + \cdots = 1 \tag{7}$$

that, of course, $F_{11} = 0$ when A_1 can "see" no part of itself.

The rate of radiation from a black surface A_1 to black surface A_2 is now $A_1F_{12}\sigma T_1^4$; from A_2 to A_1, it is $A_2F_{21}\sigma T_2^4$; the net interchange is their difference, which may be written as either $A_1F_{12}\sigma(T_1^4 - T_2^4)$ or $A_2F_{21}\sigma(T_1^4 - T_2^4)$. One thus reaches the important conclusion that interchange may be obtained by evaluating the one-way radiation from either surface to the other, whichever is more convenient, and then replacing the emissive power by the difference of emissive powers of the two surfaces.

In an *enclosure of black surfaces* the net heat flux from A_1 is then given by

$$\begin{aligned} q_{1,\text{net}} &= (A_1F_{12}\sigma T_1^4 - A_2F_{21}\sigma T_2^4) \\ &\qquad + (A_1F_{13}\sigma T_1^4 - A_3F_{31}\sigma T_3^4) + \cdots \\ &\equiv A_1F_{12}\sigma(T_1^4 - T_2^4) + A_1F_{13}\sigma(T_1^4 - T_3^4) + \cdots \\ &\equiv A_1\sigma T_1^4 - (A_1F_{11}\sigma T_1^4 \\ &\qquad + A_2F_{21}\sigma T_2^4 + A_3F_{31}\sigma T_3^4 + \cdots) \end{aligned} \tag{8}$$

Allowance for Refractory Surfaces. The Factor $\bar{F}$. One of the commonest problems of radiant-heat transfer in industrial-furnace design is that in which a portion of the enclosure constitutes a heat source or heat sources (such as a fuel bed, a carborundum muffle, a row of electric resistors), another portion a heat sink or heat sinks (such as the surface of a row of billets, the tubes of a tube still or boiler furnace, etc.), and another

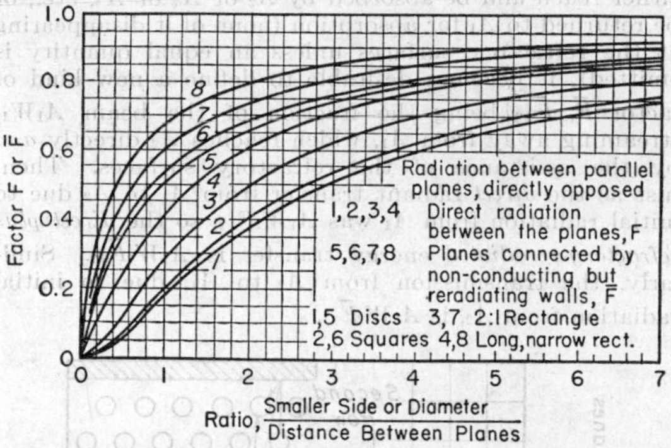

FIG. 19.

portion an intermediate refractory connecting-wall system which is a heat sink only to the extent that it loses heat by conduction through its walls to the furnace exterior. If the convection from gases on the inside of such refractory walls is approximately equal to the loss by conduction through the walls, then the net radiant heat interchange of the inside surface of the walls with the rest of the furnace interior is zero; and, since the radiation incident on the refractory walls is generally so enormous compared with the difference between gas convection and wall conduction, the assumption that the *net radiant*-heat transfer at the wall surface is zero

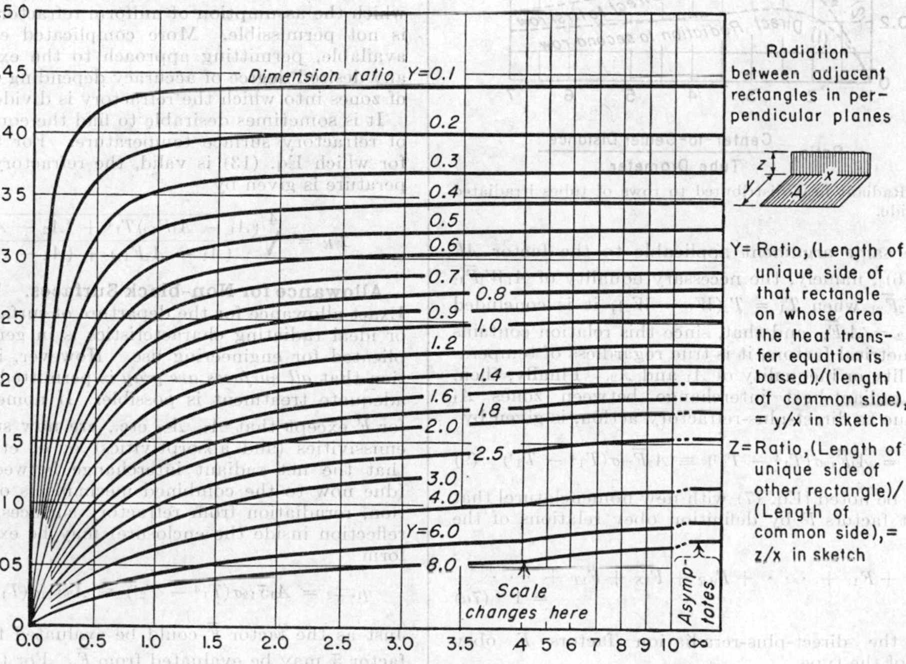

FIG. 20. Radiation between adjacent rectangles in perpendicular planes.

is an excellent one. It enormously simplifies the treatment of the problem of the transfer of heat from sources to sinks, and the effect of the refractory surfaces thereon.

Let the problem be restricted temporarily to source- or sink-type surfaces which are black, of areas A_1, A_2, etc., and to refractory surfaces A_R, A_S, A_T, etc., at which there is no net radiant flux. Since all the radiation A_1W_1 initially emitted by zone A_1 must ultimately either reach and be absorbed by A_2 or A_3 or A_4, etc., or be returned to A_1 for absorption (none of it disappearing at the refractory surfaces unless an equal quantity is emitted), it becomes desirable to define a new kind of factor $\bar{F}$, $\bar{F}_{12}$ being the fraction of the beam A_1W_1, streaming away from A_1, which reaches A_2 directly *and* by the assistance of the refractory surfaces. Then, just as the *direct* radiant transfer from A_1 to A_2 due to initial radiation from A_1 was $A_1W_1F_{12}$, so the *direct plus refractory-reradiated* energy transfer is $A_1W_1\bar{F}_{12}$. Similarly, the transmission from A_2 to A_1, due to initial radiation from A_2, is $A_2W_2\bar{F}_{21}$.

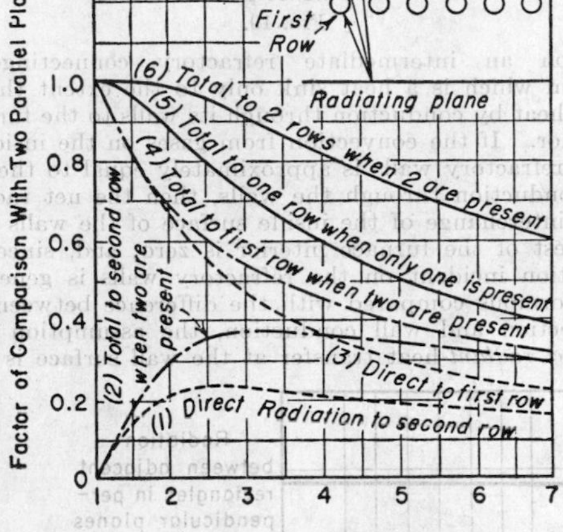

FIG. 21. Radiant heat distributed to rows of tubes irradiated from one side.

By the same argument applicable to the factor AF [see Eq. (6)], namely, the necessary equality of $A_1W_1\bar{F}_{12}$ and $A_2W_2\bar{F}_{21}$ when $T_1 = T_2(W_1 = W_2)$, it is concluded that $A_1\bar{F}_{12} = A_2\bar{F}_{21}$ and that, since this relation contains only geometrical factors, it is true regardless of temperature equality or inequality of A_1 and A_2. Finally, then, the net radiant-heat interchange between zones A_1 and A_2, due to direct-plus-refractory action, is given by

$$q_{1 \rightleftharpoons 2} = A_1\bar{F}_{12}\sigma(T_1^4 - T_2^4) \equiv A_2\bar{F}_{21}\sigma(T_1^4 - T_2^4) \tag{9}$$

It is to be noted [Eq. (7) with new nomenclature] that the direct factors F by definition obey relations of the type

$$F_{11} + F_{12} + F_{13} + \cdots + F_{1R} + F_{1S} + F_{1T} + \cdots = 1 \tag{7a}$$

whereas the direct-plus-reradiation factors $\bar{F}$ obey relations of the type

$$\bar{F}_{11} + \bar{F}_{12} + \bar{F}_{13} + \cdots = 1 \tag{10}$$

The factor $\bar{F}$ has been determined exactly for a few geometrically simple cases [Hottel and Keller, *Trans. Am. Soc. Mech. Engrs., Iron and Steel*, **55**, 39 (1933)] and may be approximated for others. If A_1 and A_2 are equal parallel disks, squares, or rectangles connected by non-conducting but reradiating refractory walls, then $\bar{F}$ is given by Fig. 19, lines 5 to 8. If A_2 represents an infinite plane and A_1 is one or two rows of infinite parallel tubes in a parallel plane and if the only other surface is a refractory surface behind the tubes, $\bar{F}_{21}$ is given by line 5 or 6 of Fig. 21. If an enclosure may be divided into several radiant-heat sources or sinks A_1, A_2, etc., and the rest of the enclosure (reradiating refractory surface) may be lumped together as A_R at a uniform temperature T_R, then the factor $\bar{F}_{12}$ is given in terms of the direct geometrical factors F by the expression

$$\bar{F}_{12} = F_{12} + \frac{F_{1R}F_{R2}}{1 - F_{RR}} \tag{11}$$

If there are but two source-sink-type surfaces, A_1 and A_2, the above expression reduces, by application of the principles expressed in Eqs. (6) and (7), to the more readily used

$$\bar{F}_{12} = F_{12} + \frac{1}{\dfrac{1}{F_{1R}} + \dfrac{A_1}{A_2}\,\dfrac{1}{F_{2R}}} \tag{12}$$

If this case is further simplified by considering that neither A_1 nor A_2 can "see" itself (*i.e.*, has no negative curvature), the above expression reduces to

$$\bar{F}_{12} = \frac{A_2 - A_1F_{12}^2}{A_1 + A_2 - 2A_1F_{12}} \tag{13}$$

which necessitates the evaluation of but one geometrical factor F. This case covers a major fraction of problems of radiant-heat interchange between source and sink in a furnace enclosure and is in error only to the extent to which the assumption of uniform refractory temperature is not permissible. More complicated expressions are available, permitting approach to the exact answer to any desired degree of accuracy depending on the number of zones into which the refractory is divided.

It is sometimes desirable to find the equilibrium value of refractory surface temperature. For the conditions for which Eq. (13) is valid, the refractory surface temperature is given by

$$T_R = \sqrt[4]{\frac{(A_1 - A_1F_{12})T_1^4 + (A_2 - A_1F_{12})T_2^4}{(A_1 - A_1F_{12}) + (A_2 - A_1F_{12})}} \tag{14}$$

Allowance for Non-black Surfaces. The Factor $\mathfrak{F}$. Exact allowance for the departure of surfaces from black or ideal radiating characteristics is in general too complicated for engineering use. However, if the assumption that *all surfaces are gray* is permitted, a simple and adequate treatment is possible. If nomenclature is as for $\bar{F}$ except that A_1, A_2, etc., are now surfaces having emissivities (and absorptivities) ϵ_1, ϵ_2, etc., it is found that the net radiant interchange between A_1 and A_2 (due now to the combined mechanisms of direct radiation, reradiation from refractory surfaces, and multiple reflection inside the enclosure) may be expressed in the form

$$q_{1 \rightleftharpoons 2} = A_1\mathfrak{F}_{12}\sigma(T_1^4 - T_2^4) \equiv A_2\mathfrak{F}_{21}\sigma(T_1^4 - T_2^4) \tag{15}$$

Just as the factor $\bar{F}$ could be evaluated from F, so the factor $\mathfrak{F}$ may be evaluated from $\bar{F}$. For the case of two non-refractory surfaces A_1 and A_2 and however many refractory zones,

$$\mathfrak{F}_{12} = \cfrac{1}{\cfrac{1}{\bar{F}_{12}} + \left(\cfrac{1}{\epsilon_1} - 1\right) + \cfrac{A_1}{A_2}\left(\cfrac{1}{\epsilon_2} - 1\right)} \qquad (16)$$

It is to be noted that the emissivity of the refractory surfaces forming the system is not a factor, *i.e.*, that whether a refractory surface maintains its equilibrium by complete absorption and black-body reradiation or by complete diffuse reflection and no radiation is immaterial.

The limitation of Eq. (16) to conditions for which the division of source- and sink-type surfaces into but two zones A_1 and A_2 must be remembered; it is valid only when all elements of surfaces on A_1 (or A_2) "see" substantially the same picture, *i.e.*, when $F_{dA_1 \rightarrow A_2}/F_{dA_1 \rightarrow A_R}$ is about the same for all points on A_1.

As in the case of $\bar{F}$, $\mathfrak{F}$ may be evaluated to any desired degree of accuracy by dividing the system into a sufficient number of zones; but most furnace problems do not justify going beyond the expression given above.

Recommended Procedure. The use of the preceding principles is best illustrated by some examples.

Example 1. What is the heat transfer by radiation between an oxidized nickel tube 4 in. outside diameter, at a temperature of 800°F., and an enclosing chamber of silica brick at 1800°F., the brick chamber being (*a*) very large relative to the tube diameter, and (*b*) 8 in. square inside?

a. If A_1 is the enclosed surface, $\bar{F}_{12} = F_{12} = 1$. Then, according to Eq. (16),

$$\mathfrak{F}_{12} = \cfrac{1}{\cfrac{1}{\bar{F}_{12}} + \left(\cfrac{1}{\epsilon_1} - 1\right)} = \epsilon_1$$

i.e., the interchange factor is independent of emissivity of surroundings when the latter are of great extent, and Eq. (3) applies. The emissivity of oxidized nickel at 800°F. is, by interpolation from Table 1, about 0.43; its absorptivity for radiation from a source at 1800° is approximately its emissivity at 1800°F., which by extrapolation is about 0.58. The tube area per foot length is $\pi 4/12 = 1.05$ sq. ft./ft. Then

$$q \text{ (per ft. length)} = 0.173 \times 1.05 \left[0.43\left(\frac{800 + 460}{100}\right)^4 - 0.58\left(\frac{1800 + 460}{100}\right)^4\right]$$
$$= 25{,}540 \text{ B.t.u./hr./ft. of tube}$$

The more usual procedure of using a single value for α and ϵ [Eq. (2)] would give, for $\epsilon = 0.58$, q per ft. = 24,840.

b. As before, $\bar{F}_{12} = 1$. When $\epsilon_1 = 0.58$ and $\epsilon_2 = 0.8$, Eq. (16) gives

$$\mathfrak{F}_{12} = \cfrac{1}{1 + \left(\cfrac{1}{0.58} - 1\right) + \cfrac{1.05}{2.67}\left(\cfrac{1}{0.8} - 1\right)} = 0.549$$

Therefore $q = 24{,}840 \times 0.549/0.58 = 23{,}500$.

If one wished to allow for the difference between ϵ and α, an approximation for this case would be to use

$$q_{\text{net}} = A_1\mathfrak{F}_{12}\sigma T_1^4 - A_1\mathfrak{F}_{12}\sigma T_2^4$$

and to evaluate $\mathfrak{F}_{12}$ in the first term using ϵ_1 and ϵ_2 at $T_1(\mathfrak{F}_{12} = 0.412)$ and in the second term using ϵ_1 and ϵ_2 at $T_2(\mathfrak{F}_{12} = 0.549)$.

$$q_{\text{net}} = 0.173 \times 1.05(0.412 \times 12.6^4 - 0.549 \times 22.6^4) = 24{,}140$$

Example 2. A muffle-type furnace in which the carborundum muffle forms a continuous floor of dimensions 15 by 20 ft. has its ultimate-heat-receiving surface in the form of a row of 4-in. tubes on 9-in. centers above and parallel to the muffle and backed by a well-insulated refractory roof; the distance from the muffle top to the row of tubes is 10 ft. The tubes fill the furnace top, of area equal to that of the carborundum floor. The average muffle-surface temperature is 2100°F.; the tubes are at 600°F. The side walls of the chamber are assumed substantially nonconducting but reradiating and are at some equilibrium temperature between 600° and 2100°F., such that they radiate just as much heat as they receive. The tubes are oxidized steel of emissivity 0.8; the carborundum has an emissivity of 0.7. Find the radiant-heat transmission between the carborundum floor and the tubes above, taking into account the reradiation from the side walls.

Call the area of the roof tubes A_1, that of the carborundum floor A_3, that of the refractory side walls of the furnace A_R. The problem must be broken up into two parts, first considering the roof with its refractory-backed tubes. To an imaginary plane A_2 of area 15 × 20 ft. located just below the tubes, the tubes emit radiation $A_1\mathfrak{F}_{12}T_1^4$, equal to $A_2\mathfrak{F}_{12}T_1^4$. To obtain $\mathfrak{F}_{21}$, one must first evaluate $\bar{F}_{21}$, which comes from Fig. 21, line 5, from which $\bar{F}_{21} = 0.84$. From Eq. (16)

$$\mathfrak{F}_{21} = \cfrac{1}{\cfrac{1}{0.84} + \left(\cfrac{1}{1} - 1\right) + \cfrac{9}{4\pi}\left(\cfrac{1}{0.8} - 1\right)} = 0.73^*$$

This amounts to saying that the system of refractory-backed tubes is equal in radiating power to a continuous plane A_2 replacing the tubes and refractory above them, having a temperature equal to the tubes and an equivalent or effective emissivity of 0.73.

The new simplified furnace now consists of an enclosure formed by a 15 × 20 ft. rectangle A_3 of emissivity 0.7, above and parallel to it a 15 × 20 ft. rectangle A_2 of temperature T_1 and emissivity 0.73, and refractory walls A_R to complete the enclosure. The desired heat transfer is $q_{2\rightleftharpoons 3}$.

$$q_{2\rightleftharpoons 3} = \sigma(T_1^4 - T_3^4)A_2\mathfrak{F}_{23}.$$

Normally to evaluate $\mathfrak{F}_{23}$, one would find F_{23} first, then evaluate $\bar{F}_{23}$ by Eq. (12)—an approximation to the extent that it assumes a constant side-wall temperature. For the present case, however, Fig. 19, line 6, presents an exact allowance for the continuous variation in side-wall temperature from top to bottom. The interchange factor between parallel 15 × 20 ft. rectangles separated by 10 ft. may be taken as the geometric mean of the factors for 15-ft. squares separated by 10 ft., and 20-ft. squares separated by 10 ft. Then, from Fig. 19, line 0, $\bar{F}_{23} = \sqrt{0.63 \times 0.69} = 0.66$. From Eq. (16),

$$\mathfrak{F}_{23} = \cfrac{1}{\cfrac{1}{0.66} + \left(\cfrac{1}{0.73} - 1\right) + 1 \cdot \left(\cfrac{1}{0.7} - 1\right)} = 0.433 = \mathfrak{F}_{32}$$

i.e., the floor and tubes interchange 43.3 per cent as much radiation as parallel black planes close together, each of area equal to the floor. The net interchange is

$$q_{\text{net}} = 0.173 \times (15 \times 20)(25.6^4 - 10.6^4)0.433 = 9{,}380{,}000 \text{ B.t.u./hr.}$$

Example 3. The distribution of radiant heat to the different rows of tubes in a tube nest irradiated from one side is desired, when the tubes are 4.0 in. outside diameter on 8-in. triangular centers. Let the area of the continuous plane below the tube nest be A_2 and the area of the tubes A_1. According to Fig. 21, curve 3, the first row of tubes will intercept directly 0.66 of the total. According to curve 1, the second row will intercept 0.21 of the total, leaving 1 − 0.66 − 0.21 = 0.13 to be intercepted by the remaining rows.

Suppose the tube nest replaced by a single row of tubes A_1 with refractory back wall A_R. Equation (11) gives $\bar{F}_{21}$. For the present case $F_{RR} = 0$ and $F_{2R} = 1 - F_{21}$ and $F_{R1} = F_{21}$; so Eq. (11) becomes

$$\bar{F}_{21} = F_{21} + (1 - F_{21})F_{21} = 0.66 + 0.34(0.66) = 0.88$$

a value which could have been read from Fig. 21, curve 5. A single tube and back wall will therefore be 88 per cent as effective a heat receiver as an infinite number of rows, so far as radiant-heat transmission is concerned.

Suppose the tube nest had been replaced by two rows of tubes with refractory back wall, instead of by a single row. According to Fig. 21, curves 4 and 2, the total radiation to the first row is 0.69, to the second 0.29, to both 0.69 + 0.29, or 0.98 as much as to an infinite number of rows (or to a continuous plane).

From Fig. 21, it is seen that only when the tubes are of small diameter relative to their distance apart is there any considerable quantity of radiant-heat penetration beyond the second

* The use of Eq. (16) was hardly justifiable here, since the "views" from spots on the top and the bottom of the tubes comprising the area A_1 are so different; but when A_1 is divided into two zones, the value of $\mathfrak{F}_{21}$ is raised only to 0.74.

row. The solution of a three- or four-row problem may be made readily by a graphical method [Hottel, *Trans. Am. Soc. Mech. Engrs., Fuel Steam Power*, **53,** 265 (1931)].

Radiation from Non-luminous Gases

If black-body radiation passes through a gas mass containing, *e.g.*, carbon dioxide, absorption occurs in certain regions of the infrared spectrum. Conversely,

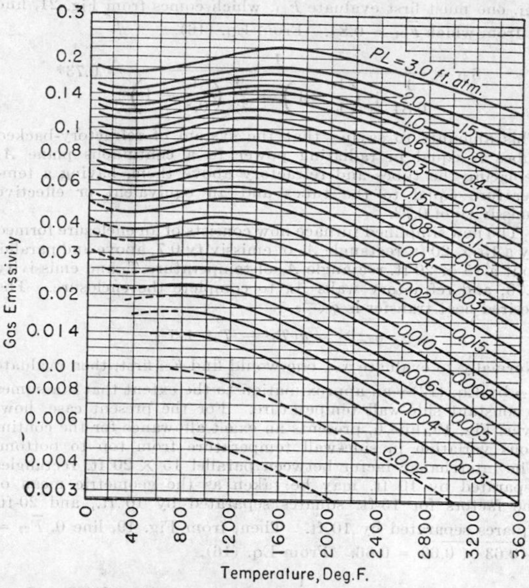

Fig. 22. Emissivity of carbon dioxide.

if the gas mass is heated it radiates in those same wave-length regions. This infrared spectrum of gases has its origin in simultaneous quantum changes in the energy levels of rotation and of interatomic vibration of the molecules; and, at the temperature levels reached in industrial furnaces, is of importance only in the case of

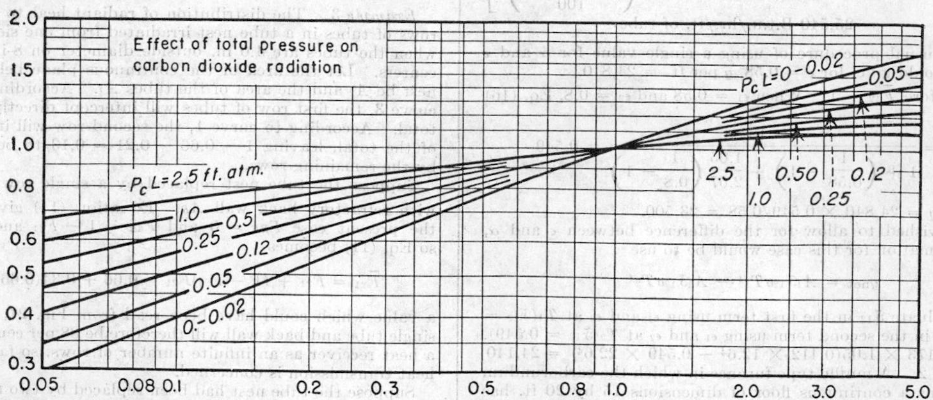

Fig. 22. Correction factor C_c for converting emissivity of carbon dioxide at 1 atm. total pressure to emissivity at P_T atm.

the heteropolar gases. *Of the gases encountered in heat transfer equipment, carbon monoxide, the hydrocarbons, water vapor, carbon dioxide, sulfur dioxide, ammonia, and hydrogen chloride are among those with emission bands of sufficient magnitude to merit consideration.* The gases with symmetrical molecules, hydrogen, oxygen, nitrogen, etc., have been found not to show absorption bands in those wave-length regions of importance in radiant-heat transmission at temperatures met in industrial practice.

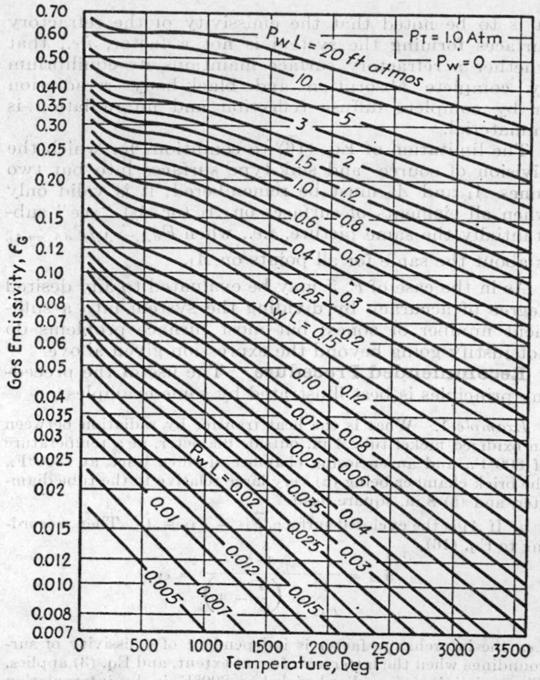

Fig. 23. Emissivity of water vapor at zero partial pressure in system with total pressure = 1 atm.

Consider a hemispherical gas mass of radius L containing carbon dioxide of partial pressure P_c, and let the problem be the evaluation of radiant-heat interchange between the gas at temperature T_G and a small element of surface at temperature T_s, located on the base of the hemisphere at its center. Per unit of surface the emission of the gas to the surface is $\sigma T_G^4 \cdot \epsilon_G$, where ϵ_G denotes gas emissivity. For carbon dioxide ϵ_G depends on T_G, the product term P_cL, and the total pressure P_T. Its value for P_T = 1 atm. is given in Fig. 22; an approximate correction factor C_c for a total pressure differing from 1 atm. is given in Fig. 22*a*. The absorption, by the gas, of radiation from the surface is $\sigma T_s^4 \cdot \alpha_G$, where α_G is the absorptivity of the gas for black-body radiation from the surface. Approximately α_G is obtained from the gas emissivity chart at the same value of P_cL as before but at the temperature T_s instead of T_G. Such an approximation is adequate if the gas is hotter than the surface and the absorption term consequently of secondary importance. If the reverse is the case, an accurate value of α_G may be obtained if one reads an emissivity from Fig. 22 at T_s as before, but at $P_cL(T_s/T_G)$ instead of P_cL, and then multiplies the result by $(T_G/T_s)^{0.65}$. The same correction factor C_c applies to absorptivity if the total pressure is not 1 atm.

The net radiant-heat interchange per unit of black bounding surface is then

$$(\sigma T_G^4 \epsilon_G - \sigma T_s^4 \alpha_G) C_c \tag{17}$$

In the case of water-vapor radiation the gas emissivity ϵ_G depends on T_G and P_wL, as before, and in addition somewhat on the partial pressure of water vapor P_w and the total pressure P_T. Correlation of the data of various experimenters is found possible by reducing all measured emissivities to values corresponding to an idealized case where $P_w = 0$ and $P_T = 1$, by the use of a factor depending on $(P_w + P_T)$ and on P_wL. The smoothed curves through the resulting corrected data appear in Fig. 23 as a plot of ϵ_G vs. T_G for the various values of P_wL, for the "ideal" system at zero partial pressure of water vapor and a total pressure of 1 atm. Allowance for departure from this "ideal" state is then made by multiplying ϵ_G as read from Fig. 23 by a factor C_w read from Fig. 23*a* as a function of $(P_w + P_T)$ and P_wL.

The absorptivity of water vapor for black-body radiation may be obtained like that of CO_2. Approximately, α_G is ϵ_G read at P_wL and T_s instead of T_G; more accurately, it is obtained by reading emissivity from Fig. 23 at T_S but at $P_wL(T_S/T_G)$ instead of P_wL, and then multiplying the result by $(T_G/T_S)^{0.45}$. The correction factor C_w still applies.

Table 2. Beam Lengths for Gas Radiation

Shape	Characterizing dimension, D	Factor by which D is multiplied to obtain mean beam length, L: When $PL = 0$	For average values of PL
Sphere	Diameter	⅔	0.60
Infinite cylinder	Diameter	1	0.90
Same, radiating to center of base	Diameter	......	0.90
Right circular cylinder, height = diameter, radiating to center of base	Diameter	......	0.77
Same, radiating to whole surface	Diameter	⅔	0.60
Infinite cylinder of half-circular cross section. Radiating to spot on middle of flat side	Radius	......	1.26
Space between infinite parallel planes	Distance between planes	2	1.8
Cube	Edge	⅔	0.60
1 × 2 × 6 rectangular parallelepiped, radiating to	Shortest edge		
2 × 6 face	....	1.18	1.06
1 × 6 face	....	1.24	
1 × 2 face	....	1.18	
All faces	....	1.20	
Space outside infinite bank of tubes with centers on equilateral triangles; tube diameter = clearance	Clearance	3.4	2.8
Same as preceding, except tube diameter = one-half clearance	Clearance	4.45	3.8
Same, except tube centers on squares; diameter = clearance	Clearance	4.1	3.5

When carbon dioxide and water vapor are present together, the total radiation due to both is somewhat less than the sum of the separately calculated effects, because each gas is somewhat opaque to radiation from the other. A correction for this effect may be read from Fig. 24, which gives the amount $\Delta\epsilon$ by which to reduce the sum of ϵ_G for CO_2 and ϵ_G for H_2O (each evaluated as if the other gas were absent) to obtain the ϵ_G due to the two together. The same type of correction applies in calculating α_G.

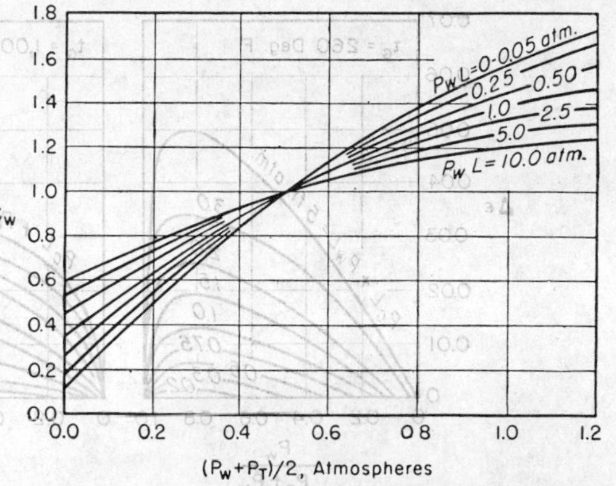

FIG. 23*a*. Correction factor C_w for converting emissivity of water to values of P_w and P_T other than 0 and 1 atm., respectively.

Discussion has so far been restricted to interchange between a gas and its bounding surface when the latter is black. If the surface is gray, with an emissivity (and absorptivity) equal to ϵ_s, multiplication by ϵ_s would make proper allowance for reduction in the primary beams from gas to surface and surface to gas, respectively; but some of the gas radiation initially reflected from the surface would have further opportunity for absorption at a surface, because the gas is but incompletely opaque to the reflected beam. Consequently, the factor to allow for surface emissivity lies between ϵ_s and 1; nearer the latter the more transparent the gas (*i.e.*, the lower P_cL and P_wL) and the more convoluted the surface. Rigorous treatment of the problem is tedious for engineering use. Fortunately, in the emissivity range of most industrial surfaces, 0.7 to 1.0, an adequate approximation consists in multiplying by an effective emissivity ϵ_s' lying halfway between the actual value of ϵ_s and unity.

The final formulation of radiant interchange between a gas and its bounding surface when the gas contains CO_2 and H_2O is now

$$\frac{q}{A} = \sigma\epsilon_s'(\epsilon_G T_G^4 - \alpha_G T_s^4)$$
$$= 0.173\epsilon_s' \left[\epsilon_G \left(\frac{T_G}{100}\right)^4 - \alpha_G \left(\frac{T_s}{100}\right)^4 \right] \tag{18}$$

To keep straight on nomenclature, a series of subscripts will be appended to the value of ϵ read from Fig. 22 or 23, the first representing the gas (whether CO_2 or H_2O), the second the temperature on the plot (whether T_G or T_s), the third the values of P_L at which ϵ is read. In this nomenclature, terms in Eq. (18) are defined as follows:

$$\epsilon_G = (\epsilon_{CO_2,T_G,P_cL})C_c + (\epsilon_{H_2O,T_G,P_wL})C_w - \Delta\epsilon_{T_G}$$
$$\alpha_G = \alpha_{CO_2} + \alpha_{H_2O} - \Delta\alpha$$
$$\alpha_{CO_2} = (\epsilon_{CO_2,T_s,P_cLT_s/T_G})\left(\frac{T_G}{T_s}\right)^{0.65} \cdot C_c$$
$$= \epsilon_{CO_2,T_s,P_cL} \cdot C_c \text{ approx.}$$

$$\alpha_{H_2O} = (\epsilon_{H_2O,T_s,P_wLT_s/T_G})\left(\frac{T_G}{T_s}\right)^{0.45} \cdot C_w$$
$$= \epsilon_{H_2O,T_s,P_wL} \cdot C_w \text{ approx.}$$

As previously pointed out, the error in q/A is negligible when α is evaluated as ϵ at T_s and P_cL or P_wL, if $T_s \ll T_G$. The maximum error so introduced is about 10 per cent when $T_s = 0.8T_G$.

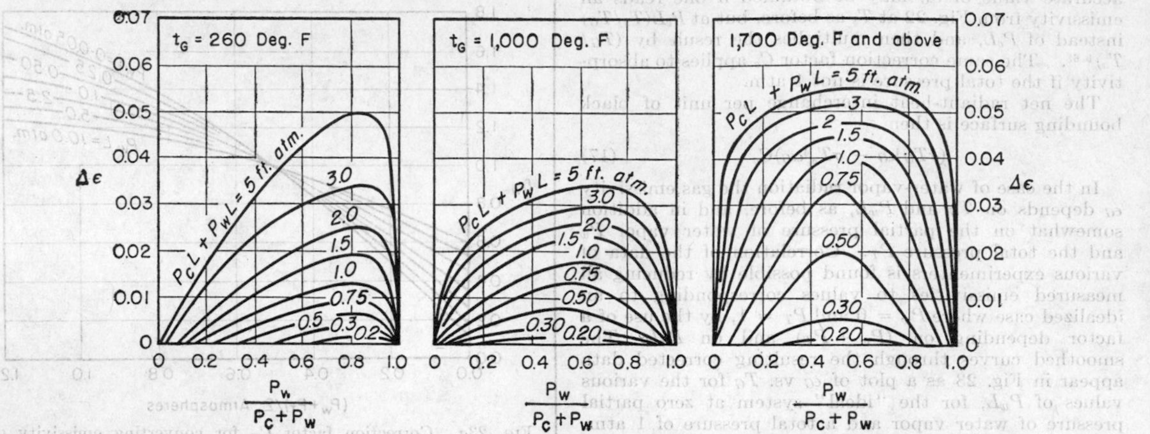

FIG. 24. Correction for superimposed radiation from carbon dioxide and water vapor.

The above expression was formulated for the case of interchange between a gas hemisphere and a spot on its base, *i.e.*, for the case in which the length of path L of the radiant beam is the same in all directions. For gas shapes of industrial importance it is found that any shape is approximately representable by an "equivalent" hemisphere of proper radius or that there is a mean beam length which can be used in evaluating gas emissivities and absorptivities from Figs. 22 and 23. As PL approaches zero, the mean beam length approaches as a limit the value, 4 × (ratio of gas volume to bounding area). For the range of PL encountered in practice, L is always less; 85 per cent of the limiting value is generally a satisfactory approximation (Port, Sc. D. Thesis, Mass. Inst. Tech., 1939). Table 2 summarizes the results of tedious graphical or analytical treatment of various shapes.

If gas radiation occurs in equipment in which there is a continuous change in temperature of the gas and the surface from one end to the other of the interchanger, exact allowance therefor can be made by conventional graphical integration. To a generally adequate degree of approximation, however, one may use a mean surface temperature equal to the arithmetic mean, and a mean gas temperature equal to the mean surface temperature plus the logarithmic mean of the temperature difference, gas to surface, at the two ends.

$$ts_{\text{avg}} = \frac{(ts_1 + ts_2)}{2} \quad (19)$$

$$tG_{\text{avg}} = ts_{\text{avg}} + \frac{(tG_1 - ts_1) - (tG_2 - ts_2)}{2.3 \log \dfrac{(tG_1 - ts_1)}{(tG_2 - ts_2)}} \quad (20)$$

Effect of Presence of Two Surfaces at Different Temperatures. When a radiating gas fills a chamber the walls of which consist of the ultimate heat-receiving surface and of an intermediate heat receiver and reradiator such as a refractory surface, the question arises as to how to evaluate the total heat interchange between gas and ultimate heat receiver by the combined mechanisms of direct radiation from the gas to the ultimate receiver, and radiation from the gas to the refractory surface, and thence to the ultimate receiver. This problem, in its general form involving heat balances, external heat losses from the furnace, and convection heat transfer inside the chamber, is treated in detail in the last part of the present section, p. 495. As an approximation, however, the total heat transfer to the ultimate receiver may be estimated by assuming that its effective area is that of itself plus a certain fraction x of that of the refractory, and that the only temperatures involved are those of the gas and the ultimate receiving surface. The fraction x, the effectiveness of the refractory surface, varies from zero when the ratio of refractory surface to ultimate receiving surface is very high, to unity when the ratio is very low and the value of ϵ_G is low. When the refractory-surface area and ultimate heat-receiving surface area are of the same order of magnitude, a value of 0.7 may be used for x, although for more exact calculations the method of the last section of this chapter should be used.

Radiation from Sulfur Dioxide. In the design of sulfur burners and of sulfur dioxide coolers the radiation from the gas may be a major factor in the evaluation of the total heat transferred. The data of Coblentz ("Investigations of Infra-red Spectra," Carnegie Institute, 1905) on the infrared absorption spectrum of sulfur dioxide, while hardly adequate as a basis for quantitative calculations, have been used for want of something better. The results are presented in Fig. 25, by S. A. Guerrieri, in a form similar to the water-vapor and carbon dioxide plots. The equation of radiant-heat transfer is

$$\frac{q}{A} = \epsilon_s'(S_g - S_s) \quad (21)$$

in which q/A is B.t.u./sq. ft. bounding surface per hr.; ϵ_s', the effective emissivity of the surface; S_g, the sulfur dioxide radiation, as read from Fig. 25 corresponding to the gas temperature; S_s the same, but corresponding to the surface temperature and representing, therefore, the amount of radiation from the surface which is absorbed by the gas.

Radiation from Other Gases. Measurements of total radiation from carbon monoxide have been made by Ullrich (Sc. D. Thesis, Mass. Inst. Tech., 1935) who found that the gas emissivity is a maximum at around

1600°F., that at $PL = 2$ its emission is about half that of carbon dioxide at all temperatures from 600° to 2500°F., and that at $PL = 0.01$ its emission varies from 40 to 90 per cent of that of carbon dioxide as the temperature varies from 600° to 2500°F.

Measurements of total radiation from ammonia have been made by Port (Sc. D. Thesis, Mass. Inst. Tech., 1939) who found that the gas emissivity is very high compared with carbon dioxide or water vapor, that it decreases continuously from room temperature up, that at $PL = 2$ it varies from one to two times that of water vapor, and that at $PL = 0.01$ it varies from 1.5 to 4 times that of water vapor over the range, room temperature to 2000°F.

For other gases of interest one must rely on evaluations similar to those on SO_2 above, based on the infrared absorption spectra of the gases in question. For the method of such calculation and for a more complete story on gas radiation see Schack [*Z. tech. Physik*, **5**, 266 (1924)], Hottel [*Ind. Eng. Chem.*, **19**, 888 (1927)], Schmidt [*Forsch. Gebiete Ingenieurw*, **3**, 57 (1932)], Hottel and Mangelsdorf [*Trans. Am. Inst. Chem. Engrs.*, **31**, 517 (1935)], Eckert [*V. deut. Ing. Forschungsheft*, 387 (1937)], Hottel and Egbert, [*Trans. Am. Soc. Mech. Engrs.*, **63** (1941); *Trans. Am. Inst. Chem. Engrs.*, **38**, 531 (1942)].

Example 4. Flue gas containing 6 per cent carbon dioxide and 11 per cent water vapor by volume (wet basis) flows through the convection bank of an oil tube still consisting of rows of 4-in. tubes on 8-in. centers, nine 25-ft. tubes in a row, the rows staggered to put the tubes on equilateral triangular centers. The flue gas enters at 1600° and leaves at 1000°F. The oil flows countercurrent to the gas and rises from 600° to 800°F. Tube surface emissivity is 0.8. What is the average heat input rate, due to gas radiation alone, per square foot of external tube area?

In addition to the direct radiation from gas to tubes, there will be some reradiation from the refractory walls bounding the chamber, the effect of which may be determined approximately by the method discussed on p. 492. With each row of tubes there is associated $\frac{8}{12} \times \sqrt{3}/2$ or 0.577 ft. of wall height, of area $(\frac{8}{12} \times 9 \times 2 + 25 \times 2) \times 0.577 = 35.8$ sq. ft. One row of tubes has an area of $\pi \times \frac{4}{12} \times 25 \times 9 = 235$ sq. ft. If the recommended factor of 0.7 on the refractory area is used, the effective area of the tubes is $\frac{235 + 0.7 \times 35.8}{235} = 1.11$ sq. ft./sq. ft. of actual area. The exact evaluation of outside tube temperature from the known oil temperature would involve a knowledge of oil-film coefficient, tube-wall resistance, and rate of heat flow into the tube, the evaluation usually involving trial and error. However, for the present purpose the temperature drop through the tube wall and oil film will be assumed 75°F., making the tube surface temperatures 675° and 875°F.; average 775°F. The radiating gas temperature is

$$t_g = 775 + \frac{(1600 - 875) - (1000 - 675)}{2.3 \log \frac{1600 - 875}{1000 - 675}} = 755 + 499 = 1274°F.$$

According to Table 2, $L = 2.8 \times$ the clearance between tubes, or $2.8 \times \frac{4}{12} = 0.935$ ft. $P_wL = 0.11 \times 0.935 = 0.102$; $P_cL = 0.06 \times 0.935 = 0.056$; $P_cL \cdot (T_S/T_G) = 0.056(775 + 460)/(1274 + 460) = 0.040$. From Fig. 22 for CO_2, ϵ_G(at $t_G = 1274$, $PL = 0.056$) = 0.064; α_G(at $t_s = 775$, $PL = 0.040$) $= 0.0535 \times (\frac{1734}{1235})^{0.65} = 0.067$. From Figs. 23 and 23*a* for H_2O, ϵ_G(at $t_G = 1274$, $PL = 0.102$, $(P_w + P_T)/2 = 1.11/2$) $= 0.064 \times 1.07 = 0.068$; at $t_s = 775$, $PL = 0.102$, $(P_w + P_T)/2 = 0.56$, $\alpha_G = 0.085 \times 1.07 = 0.091$. From Fig. 24, $\Delta\epsilon$ and $\Delta\alpha$ are both negligible. Substituting in Eq. (18), $q/A = 0.9[0.173 \times 17.34^4 (0.064 + 0.068) - 0.173 \times 12.35^4(0.067 + 0.091)] = 1275$ B.t.u./(sq. ft.)(hr)., exclusive of effect of refractory surfaces, or approximately $1275 \times 1.11 = 1415$ B.t.u. /sq. ft. tube area per hr. This is equivalent to a convection coefficient of $\frac{1415}{499}$ or 2.8, which is the order of magnitude expected of the convection coefficient itself.

Radiation from Clouds of Particles

The treatment of radiation from powdered-coal flames, from dust particles in flames, and from flames made luminous by the thermal decomposition of hydrocarbons to soot involves the evaluation of radiation from clouds of particles. Powdered-coal flames contain particles varying in size from 0.01 to 0.0 in., with an average size in the neighborhood of 0.001 in., and a composition varying from a high percentage of carbon to nearly pure ash. The suspended matter in luminous gas flames has its origin in the thermal decomposition of hydrocarbons in the flame due to incomplete mixing with air before being heated, consists of carbon and of very heavy hydrocarbons, and has an initial particle size of about 0.000012 in. The powdered-coal particles are sufficiently large to be substantially opaque to radiation incident on them, whereas the particles of a luminous flame* are so small as to act like semitransparent bodies with respect to thermal or long wave-length radiation. This difference in transparency of the individual particles justifies a separate treatment of the two types of flames.

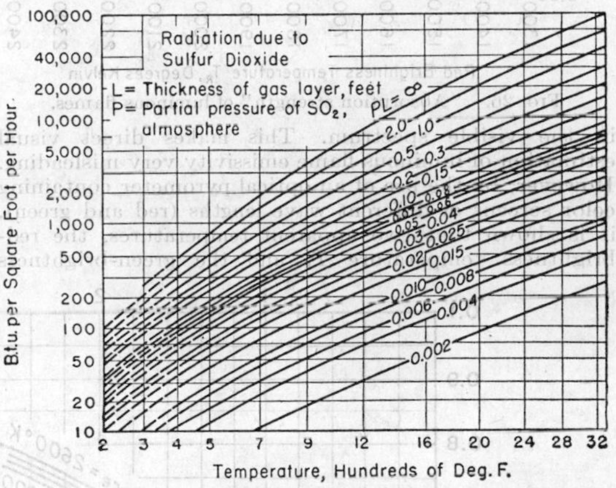

FIG. 25. Radiation from sulfur dioxide.

Luminous Flames. There are two methods of attacking the problem of developing a suitable method for predicting the radiation to be expected from a luminous flame. The first is to collect data on actual flames under varying conditions of aeration, fuel-gas composition, flame volume, etc., and to use the data as a basis for calculations. Unfortunately the published data of this sort are woefully inadequate, usually consisting of a measurement of total radiation from small laboratory flames, with no basis for determining the opacity of the flame or, consequently, the radiation from a larger flame of similar type. The changes in soot concentration attending changes in burner design, shape of combustion chamber, degree of primary and secondary aeration, fuel-gas composition, and draft regulation, all make the estimation of the luminous-flame radiation to be expected in a proposed installation exceedingly uncertain.

It is possible to show, however, how data may be obtained from a furnace with known conditions of combustion, and applied to a different size or shape of furnace in which the conditions of combustion are roughly the same. From a quantitative investigation of the variation, with wave length, of the monochromatic absorptivity of luminous flames it has been shown [Hottel and

* The term "luminous flame" as used in this article always refers to a flame made luminous by soot particles formed in the flame, not to the presence of macroscopic dust or powdered-coal particles or metal vapors, or to the bluish gas flames obtained with a high degree of primary aeration.

Broughton, The Determination of True Temperature and Total Radiation from Luminous Gas Flames, *Ind. Eng. Chem.*, anal. ed., **4**, 166 (1932)] that the absorptivity (and emissivity) decreases with increase in wave length and that the total emissivity is less than the emissivity

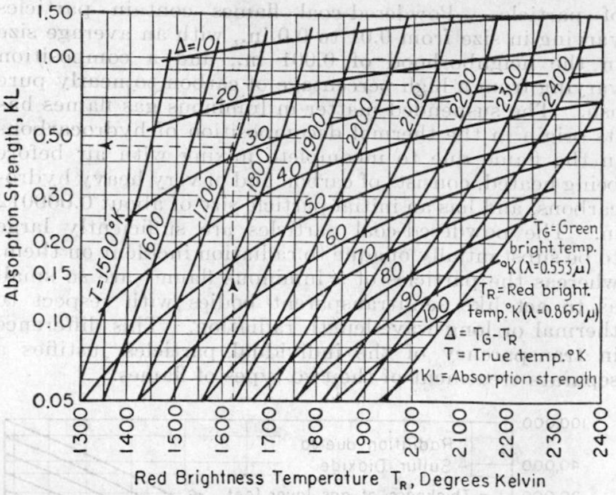

FIG. 26. "Absorption strength" of luminous flames.

in the visible spectrum. This makes direct visual estimation of luminous-flame emissivity very misleading. However, by the use of an optical pyrometer containing color screens of different wave lengths (red and green), it is shown that two apparent temperatures, the red-brightness temperature T_r and the green-brightness temperature T_g may be obtained, which permit a calculation of both true flame temperature and total flame emissivity. Figure 26 is a working plot (in °K.) from which the true temperature may be obtained, given T_r and $\Delta(= T_g - T_r)$. On the same plot one obtains the value of the absorption strength KL, in which K is a term measuring the soot concentration of the flame, and L is the thickness of flame through which the pyrometer is sighted. With absorption strength known, Fig. 27 may be used to determine the effective emissivity of the flame envelope. The transfer of heat from the flame envelope of area A and true flame temperature T_F to the confining walls of temperature T_s is given by

$$q = 0.172A\left[\left(\frac{T_F}{100}\right)^4 - \left(\frac{T_s}{100}\right)^4\right]\cdot \epsilon_F \cdot \epsilon_s' \qquad (22)$$

in which ϵ_F is the emissivity of the flame envelope as determined by Fig. 27, and ϵ_s' is the effective emissivity of the surroundings.

If an optical pyrometer with both red and green screens is not available, Fig. 26 may still be used to determine absorption strength: (*a*) if the red-brightness temperature T_r is determined with an ordinary optical pyrometer and the true temperature T_F with a high-velocity thermocouple; or (*b*) if a mirror is held behind the flame in the line of sight of the optical pyrometer. The first method is open to the serious objection that the absorption strength KL changes rapidly with a change in the usually small quantity $T_F - T_r$ representing the difference between the temperature readings of two entirely different kinds of instruments.

In using the two-color principle for determining the true temperature and total emissivity of a flame, it should be borne in mind that the pyrometer must not "see" anything but the flame itself; *i.e.*, the background of the flame should be an open peephole in the back wall of the furnace or a cold non-reflecting surface, never a

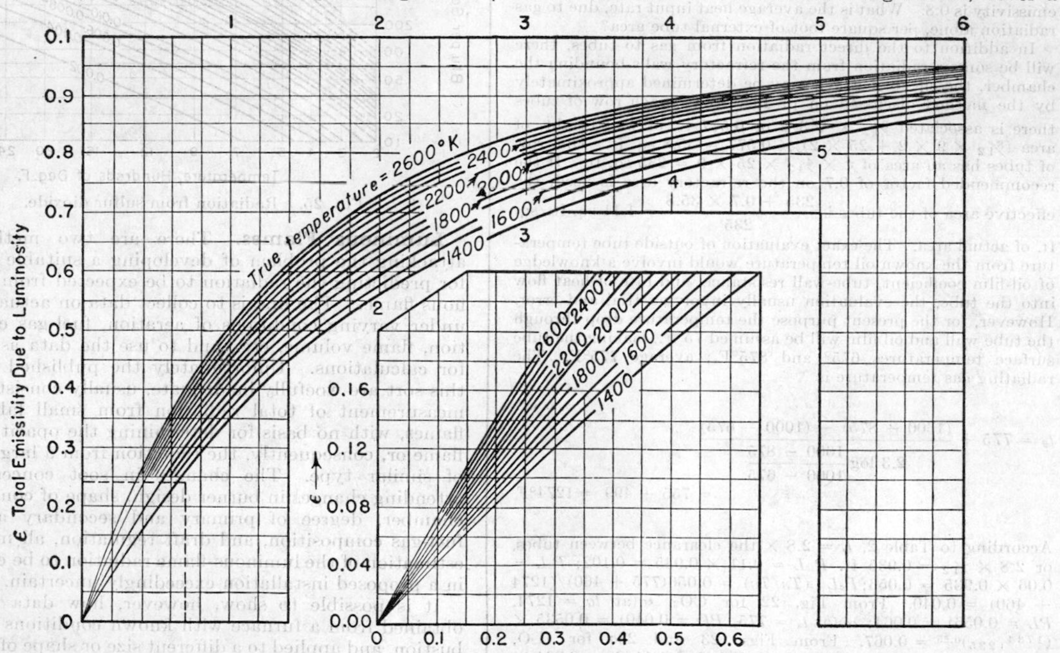

FIG. 27. Emissivity of luminous flames.

hot refractory surface. When temperature measurements have been made on an industrial flame of one size to determine its absorption strength KL, for the purpose of estimating the emissivity ϵ_F of a similar but larger flame, the absorption strength KL_1 determined from

Fig. 26 should be multiplied by the dimension ratio L_2/L_1 before Fig. 27 is used. In addition, the value of KL should correspond to the particular shape of flame under consideration, in accordance with the principles discussed in connection with the use of Table 2. An example will be found at the end of this article.

The data available on luminous flames in industrial furnaces indicate that radiation from the soot is frequently of a greater order of magnitude than non-luminous gas radiation. Lent [*Wärme*, **49,** 145 (1926)] has made a blast-furnace gas flame practically black by addition of benzene to form soot. Haslam and Boyer [*Ind. Eng. Chem.*, **19,** 4 (1927)] found that a luminous acetylene flame radiated roughly four times as much heat as when non-luminous, and the size of their experimental flame was such as to indicate that maximum blackness had not been obtained. Sherman [*Trans. Am. Soc. Mech. Engrs.*, **56,** 177 (1934)] has measured emissivities of luminous gas flames in an experimental furnace.

Powdered-coal Flames. The radiation from powdered-coal flames has been treated analytically by Wohlenberg and his associates [*Trans. Am. Soc. Mech. Engrs.*, **47,** 127 (1925); **48,** 849 (1926); *Fuel Steam Power*, **51,** 36 (1929)], by Haslam and Hottel [*Trans. Am. Soc. Mech. Engrs., Fuel Steam Power*, **50,** 3 (1928)], and by Lindmark *et al.* [*Ing. Vetenskaps Akad. Handl.*, No. 91 (1929), No. 109 (1931)]. Experimental studies have been conducted by Lindmark (*ibid.*) and by Sherman [*Trans. Am. Soc. Mech. Engrs., Fuel Steam Power*, **56,** 407 (1934)].

It may be shown that the emissivity of a cloud of opaque particles, based on the area of the envelope of the whole cloud, is of the form $(1 - \epsilon^{-x})$, in which x is the product term, (concentration of particles) (time-average cross section of a particle) (length of radiant beam through cloud), the last term being defined as under Gas Radiation, Table 2. By making suitable assumptions as to the laws of particle-size distribution in pulverized coal and the rate of combustion of individual particles, one may use the above exponential relation to calculate the emissivity of a pulverized-coal flame. Values so obtained, however, are almost invariably considerably lower than measured flame emissivities. The discrepancy is probably due to the contribution of cracked hydrocarbons producing luminosity as well as to residual ash particles not allowed for in the theoretical derivation. Fortunately, modern pulverized-coal installations involve such large flames that their emissivity is not far from unity.

Example 5. It is desired to determine approximately the radiation from a proposed luminous-flame burner installation, from measurements made on a similar combustion chamber all the dimensions of which are one-half those of the proposed installation. It is intended to keep aeration and mixing conditions as similar as possible in the two chambers. The flame in each case is roughly spherical in form. An optical pyrometer with red and green screens is sighted through the smaller flame at its diameter, the apparent temperatures obtained being T_r = 2699°F. (= 1755°K.), and T_g = 2740°F. (= 1778°K.). (*a*) What are the emissivity and true temperature of the smaller flame? (*b*) What will be the probable rate of heat transfer per square foot of flame envelope of the proposed larger installation if the flame temperature is the same and the surrounding walls are at 2600°F. and black?

a. From Fig. 26, when the red-brightness temperature T_r of the flame is 1755°K. and the difference between T_g and T_r is 1778 − 1755 = 23°K., the true-flame temperature is found to be 1811°K. and the absorption strength kL is 0.7. This value of KL, however, corresponds to length of radiant beam L equal to the diameter of the flame sphere. According to Table 2, the average value of L is 0.6 times the diameter when the radiating shape is spherical. Then the average absorption strength is 0.7 × 0.6 = 0.42. From Fig. 27, when KL = 0.42 and true-flame temperature = 1811°K., the flame emissivity is 0.20.

b. If the flame dimensions are doubled, other things being equal, the absorption strength KL will double. When KL = 2 × 0.42 = 0.84, Fig. 27 indicates that the flame emissivity will be 0.365 (not double the value 0.20). The net radiation per square foot of flame envelope will be

$$\frac{q}{A} = 0.173 \times 0.365 \times \left[\left(\frac{1811 \times 1.8}{100}\right)^4 - \left(\frac{2600 + 460}{100}\right)^4\right] = 16{,}100 \text{ B.t.u./(hr.)(sq. ft.)}$$

The General Problem of Heat Transfer in a Combustion Chamber

One of the most complex problems of heat transmission is the evaluation of the performance of a combustion chamber of a furnace, in which heat is being transmitted simultaneously by all or most of the mechanisms so far discussed. Two methods of treatment of this problem are possible: either (1) the theoretical one in which the attempt is made to consider the various individual factors, each acting in accordance with the principles discussed previously, and to combine them; or (2) the empirical one in which furnace test data are analyzed in the attempt to detect the effect of factors suspected of being of importance. These methods will be considered in order.

Allowance is to be made for the combined actions of direct radiation from the flame to the stock or heat sink; radiation from flame to refractory surfaces thence back through the flame (with partial absorption therein) to the sink; convection; and external losses. A solution of the problem is possible if the following assumptions are accepted: (1) external losses from refractory walls equal convection from flame to refractory; (2) the flame is gray and has an emissivity ϵ_F; (3) all refractory surfaces have a common average (but unknown) temperature; (4) a mean temperature T_F is assignable to the flame and combustion products in the chamber; (5) the heat sink or ultimate receiver has a uniform surface temperature T_C and is gray, with emissivity ϵ_C and area A_C. The solution of the problem, giving the net rate of heat transfer q_F from the flame by all mechanisms, is

$$q_F = \underbrace{\sigma(T_F{}^4 - T_C{}^4)A_C\mathfrak{F}_{CF}}_{\text{Radiation to sink}} + \underbrace{h_C A_C'(T_F - T_C)}_{\text{Convection to sink}} + \underbrace{U_R A_R(T_F - T_O)}_{\text{External loss (= convection to refractory)}} \quad (23)$$

in which

$$\mathfrak{F}_{CF} = \frac{1}{\dfrac{1}{F_{CF}} + \dfrac{1}{\epsilon_C} - 1} \quad (24)$$

$$F_{CF} = \epsilon_F\left(1 + \frac{A_R/A_C}{1 + \dfrac{\epsilon_F}{1 - \epsilon_F}\dfrac{1}{F_{RC}}}\right) \quad (25)$$

$$U_R = \frac{1}{\dfrac{1}{h_R} + \dfrac{x_w}{k} + \dfrac{1}{h_o}}$$

In these equations h_C, h_R, and h_o represent convection coefficients at the sink, inside refractory, and outside refractory surfaces, respectively; x_w and k are wall thickness and thermal conductivity of the refractory; T_O is outside air temperature; A_C' differs from A_C in excluding that cold surface or ultimate-receiver area which, though in view of the flame and receiving radiation, does not receive heat by convection from the gases until they leave the chamber.

It is to be noted that, as in the case of radiation in an enclosure containing no radiating or absorbing gas

(p. 489), $\mathfrak{F}$ is built up from $\bar{F}$ and ϵ_C, and $\bar{F}$ from F; but here the flame emissivity ϵ_F is in addition involved. Some simplification is possible if the geometrical factor F_{RC}—the fraction of the radiation leaving refractory surfaces which is directed toward the "cold" surface or heat sink—is replaced by $A_C/(A_R + A_C)$—a fair approximation when the refractory and cold surfaces are not completely segregated from each other. Then

$$q_F = \sigma(T_F^4 - T_C^4)A_C \left[\frac{1}{\frac{1}{\epsilon_C} + \frac{A_C}{A_C + A_R}\left(\frac{1}{\epsilon_F} - 1\right)}\right] + h_c A_C'(T_F - T_C) + U_R A_R(T_F - T_O) \quad (26)$$

From this simplified form much more readily than from the more general Eq. (23) it is to be noted that increasing the flame emissivity increases the heat transmission, but not proportionately; that decreasing surface emissivity (and absorptivity) from unity when the flame is very transparent produces almost no effect on the heat transmission; but that decreasing ϵ_C from unity when the flame is substantially opaque ($\epsilon_F = 1$) produces a proportional decrease in heat transmission.

The derivation of Eq. (23) [or Eq. (26)] was based on the assumption that A_C was composed of plane areas. Suppose instead that A_C is a row of tubes mounted in front of a refractory wall. A little consideration will show that the value of A_C to use in the radiation term of the above equations is the continuous plane A_p in which the tubes are located, multiplied by the proper factor $\bar{F}$ for tubes with a refractory background (see Fig. 21), and that the refractory surface A_R should be increased by the amount $(1 - \bar{F})A_p$.

Equation (23) [or Eq. (26)] expresses a relation between two unknowns T_F and q_F, and a second relation is necessary if a solution is to be obtained. The other relation is an energy balance. If one assumes turbulence to be so great that the mean flame temperature T_F used for calculation of radiation is the same as the temperature T_G of the gas leaving the chamber, then

$$q_F = i - w_G(C_p)_m(T_F - T_0) \quad (27)$$

where i represents the hourly enthalpy or heat content of the entering fuel, air, and recirculated flue gas, if any, above a base temperature T_0 (water as vapor); and $(C_p)_m$ represents the mean heat capacity (evaluated between T_F and T_0) of the gas leaving the chamber, at hourly mass rate w_G. Equations (23) [or (26)] and (27) may be solved by trial-and-error or by graphical methods involving superimposed plots.

The pair of equations just discussed apply strictly to one of two limiting furnace types—that one in which the assignment of a mean flame temperature equal to the temperature of the gases leaving is justifiable. For this to be the case, combustion must be relatively slow, delayed by retarded mixing of secondary air with the flame and progressing uniformly at all points in the chamber. Better agreement between predicted and experimental results is obtained on some furnaces when the assumption is made that flame temperature and exit gas temperature are not the same but differ by a constant amount. In a number of furnace tests the difference was about 300°F.

The other extreme in furnace types is that one in which combustion occurs substantially instantaneously at the burners (owing to complete premixing of fuel and air); the temperature attained is that generally known as theoretical flame temperature or adiabatic combustion temperature; and the temperature falls continuously as the gases flow from burner to outlet. When such a furnace is long compared with its cross section normal to the direction of gas flow, Eq. (26) [or Eq. (23)] may be considered as applying to a differential length of furnace, and the solution of the problem involves either a tedious but straightforward graphical integration or the use of a suitable mean of T_{F_1} and T_{F_2} in Eq. (26). Equation (27), of course, becomes $q_F = i - w_G(C_p)_m(T_{F_2} - T_0)$.

Simplified Treatment of Combustion-chamber Heat Transmission. Equation (23) or its equivalent has been used as a basis for deriving various simplified relations, easier to use but restricted in applicability in proportion to the degree of simplification. Several of these will be presented.

Billet-reheating Furnaces. For continuous billet-reheating furnaces, Eq. (23) has been modified as follows: (1) q is heat transferred to the stock, not from the flame; (2) convection terms have been omitted; (3) to compensate therefor and to allow for steadiness of furnace operation, $\mathfrak{F}_{CF}$ is evaluated using $1.2r_f \cdot \epsilon_F$ instead of ϵ_F, where r_f is the ratio of average billet-pushing rate over a period of several hours to pushing rate during periods of steady operation; (4) ϵ_F is flame emissivity due to CO_2 and H_2O only, as calculated in the previous example on gas radiation; (5) $F_{RC} = A_C F_{CR}/A_R = A_C/A_R$; (6) an average value of $(T_F^4 - T_C^4)$ is used, equal to the geometric mean of its value at the two ends of the furnace, and at the hot end T_F is taken as the calculated "theoretical" flame temperature, or adiabatic combustion temperature. The equation has been tested on reheating furnaces of various types and found satisfactory [Eberhardt and Hottel, *Trans. Am. Soc. Mech. Engrs.*, **58**, 185 (1936); *Heat Treatment Forging*, **22**, 144–149, 193–198, (1936)].

Petroleum Heaters. For cracking-coil and tube-still furnaces Eq. (23) has been modified as follows: (1) by omitting the last term, q becomes heat transferred to oil instead of heat lost by flame; (2) $h_C A_C'$ has for simplification been assigned an average value equal to $7A_C\mathfrak{F}_{CF}$ (the term is unimportant relative to the radiation term). The relation is then

$$q_C = [\sigma(T_F^4 - T_C^4) + 7(T_F - T_C)]A_C\mathfrak{F}_{CF} \quad (28)$$

Comments on p. 496 concerning the proper values of A_C and A_R for the case of tubes mounted on a wall apply. In evaluating $\mathfrak{F}_{CF}$, ϵ_F is calculated allowing for gas radiation only; $\epsilon_C = 0.9$. In applying Eq. (28) to data on 19 furnaces, Lobo and Evans [*Trans. Am. Inst. Chem. Engrs.*, **35**, 743 (1939)] found that F_{RC} was represented adequately by $A_C/(A_C + A_R)$ for values of A_R/A_C from 0 to 1, by A_C/A_R for values of A_R/A_C from 3 to 6.5. Since Eq. (28) involves heat received by oil rather than heat lost by the flame, when it is combined with the energy balance represented by Eq. (27), the latter must be modified. i is replaced by $i - q_L$, where q_L is the external heat loss from the combustion chamber. A simplified graphical treatment of the solution of Eqs. (28) and (27) is available in the reference given, together with a comparison on results with 85 tests on 19 furnaces of widely different types and excess air, burning fuel oil or refinery gas; the average deviation was 5.3 per cent; excluding tests almost certainly bad, the average deviation was less than 4 per cent (q_{exp} vs. q_{calc}).

A relation for petroleum heaters, somewhat simpler and quicker to use than Eq. (28) but not so safe, is obtainable by assuming certain terms in Eq. (23) constant, combining with Eq. (27) to eliminate T_F, and finding an expression different in form but numerically similar over the range of interest. The relation is

$$\eta = \frac{1}{1 + \dfrac{\sqrt{i/A_C\mathfrak{F}_{CF}}}{1.4\left[\dfrac{i/w_G(C_p)_m}{100}\right]^{1.6}}} \tag{29}$$

where η is the ratio of heat transferred to oil to the enthalpy of the entering air and fuel (net value). Other equations applicable in this field are available (Wilson, Lobo, and Hottel, *Ind. Eng. Chem.*, **24**, 486 (1932); Mekler, *Nat. Petroleum News*, July 27, 1938).

Steam Boiler Furnaces. For calculating heat transmission in the radiant sections of steam boiler furnace settings, many empirical relations are available. One of the simplest is the Orrok-Hudson equation

$$\eta = \frac{1}{1 + \dfrac{r_{af}\sqrt{w_A}}{27}} \tag{30}$$

in which r_{af} is the weight ratio of air to fuel; w_A is the firing rate expressed as pounds of equivalent good bituminous coal per hour per square foot of exposed tube area (complete circumference if not buried in wall).

Mullikin [*Trans. Am. Soc. Mech. Engrs.*, **57**, 517 (1935)] assumes that the flame emissivity ϵ_F is unity for large pulverized coal-, oil-, or gas-fired furnaces and that compensation for this somewhat too high value comes from use of the same value for gas temperature in Eqs. (23) and (27). When ϵ_F is unity, the term $A_C\mathfrak{F}_{CF}$ of Eq. (23) becomes simply $A_C\epsilon_C$ (though the remarks of p. 496 concerning proper evaluation of A_C apply). Mullikin introduces additional multiplying factors on A_C to allow for resistance of overlying slag or refractory facing on metal-block walls. These are 0.7 for bare-faced metal blocks on tubes and 0.35 for refractory-faced metal blocks on tubes. The simplification suggested is unsafe to use on small furnaces where ϵ_F is certainly not unity.

Wohlenberg [*Trans. Am. Soc. Mech. Engrs.*, **47**, 127 (1925); **57**, 531 (1935)] uses a relation intrinsically similar to Eq. (26), together with a heat balance involving the assumption of equality of flame and exit gas temperatures; he presents the relation for η in the form of the product of a number of quantities each making separate allowance for one of the variables under control.

Example 6. Natural gas is being burned for steam generation in a combustion chamber of which the back wall and floor are water-cooled. The gas passes through a tube nest directly above and covering the top of the combustion chamber. The chamber is 16 ft. wide by 16 ft. long by 20 ft. high. The gas, fired at the rate of 130,000 cu. ft./hr. (measured and fired at 60°F., 30 in. Hg, saturated) with 15 per cent excess air (saturated) has the equivalent composition $C_{1.25}H_{4.5}$ and a net heating value of 1070 B.t.u./cu. ft. The "cold" surfaces of the chamber have an average temperature of 350°F. What is the rate of heat input to the water-cooled walls, floor, and tubes above, exclusive of any convection to the roof tubes as the gas passes up through them? What percentage of the enthalpy of the entering fuel does this represent?

Derived Data. By stoichiometry, the products of combustion contain 8.60 per cent CO_2, 16.36 per cent H_2O, 2.44 per cent O_2, and 72.60 per cent N_2, wet basis; their total is 4911 lb.-moles. From specific heat charts the average molal heat capacity of the products between 2000° and 60°F. is 8.25; between 2500° and 60°F. it is 8.45.

Assumptions. The external loss from the refractory walls will be assumed equal to the convection to them on the inside. Convection coefficients inside the chamber = 2.0. Refractory wall conductance $k_w/x_w = 0.9$. The flame completely fills the chamber. To the emissivity of the flame due to non-luminous gases will be added 0.1 to allow for the luminosity due to cracked hydrocarbons in the flame (this varies enormously with burner type). The emissivity of the "cold" surfaces = 0.8, and absorptivity equals emissivity. The mean flame temperature is 100°F above the exit-gas temperature (these approach one another as firing rate increases).

Solution. Equations (23) and (27) are to be solved for q_F and T_F. $A_C = 16 \times 20 + 16 \times 16 \times 2 = 832$ sq. ft. (The effective area of the tube nest, for radiation reception, is that of a plane replacing the tubes.) $A_R = 16 \times 20 \times 3 = 960$ sq. ft. $A_C' = 16 \times 20 + 16 \times 6 = 576$ (plane of tube nest is excluded here). Evaluation of $\mathfrak{F}$ involves F_{RC} (or F_{CR}) and ϵ_F. In this problem F_{RC} and F_{CR} are equally tedious to evaluate; we shall choose the first. Since the three refractory rectangles do not all "see" the same arrangement of surfaces above them, it is necessary to determine the product A_RF_{RC} for each and to add them, then to divide by the total A_R. Consider first the front wall, 16 × 20 ft., which "sees" three cold faces, one directly opposite, one above, and one below. The fraction of its radiation intercepted by the wall opposite comes from Fig. 19, line 2. By the method of Example 2 of this section, $F = \sqrt{0.196 \times 0.26} = 0.225$, the fraction of the radiation from the front refractory wall intercepted by the rear water-cooled wall. To find the fraction intercepted by the water-cooled floor, reference is made to Fig. 20. From that figure, when $Y = {}^{20}\!/_{16}$ and $Z = {}^{16}\!/_{16}$, $F = 0.17$. Since the imaginary top plane replacing the tubes intercepts the same fraction as the water-cooled floor, the total fraction intercepted by cold surfaces is $0.17 \times 2 + 0.225 = 0.565$; and A_RF_{RC} for the front refractory wall is $16 \times 20 \times 0.565 = 181$ sq. ft. A similar procedure leads to the value $(0.17 \times 2 + 0.213)$ or 0.553 as the fraction of the radiation from either refractory side wall which is intercepted by the three cold faces. Then the final value of F_{RC} is

$$F_{RC} = \frac{(16)(20)(0.565) + (16)(20)(0.553)(2)}{16)(20) + (16)(20)(2)} = 0.56$$

Flame emissivity ϵ_F is next to be evaluated. The equivalent gray-body emissivity of a flame at T_G in interchange with cold surfaces at T_S is defined in the relation

$$q = \sigma(\epsilon_G T_G^4 - \alpha_G T_S^4)$$
$$\equiv \epsilon_F\sigma(T_G^4 - T_S^4)$$

from which

$$\epsilon_F = \frac{\epsilon_G - \alpha_G(T_S/T_G^4)}{1 - (T_S/T_G)^4}$$

One must first make a provisional guess as to the value of t_G and adjust later if necessary. Temporarily assume 2500°F. The effective beam length for gas radiation would be 0.6 times one side if the chamber were cubical (see Table 2); 0.6 times an average side of 18 ft., or 10.8 ft., may be used (a considerable error in this assumption will not materially affect the result). Then $P_CL = (0.086)(10.8) = 0.93$, and $P_WL = (0.1636)(10.8) = 1.77$. Because t_S is so low compared with t_G the approximate method of determining α_G for CO_2 and H_2O will be used (see p. 491). At $t_G = 2500$°F. and $t_S = 350$°, using Figs. 22, 22*a*, 23, 23*a*, and 24 and substituting into the above expression for ϵ_F, one obtains

$$\epsilon_F = \frac{(0.11 + 0.193 \times 1.08 - 0.05) - ({}^{810}\!/_{2960})^4(0.12 + 0.35 \times 1.08 - 0.028)}{1 - ({}^{810}\!/_{2960})^4}$$

due to gas radiation. (In this particular example ϵ_F could have been taken as the sum of the ϵ_G's with no allowance for the absorption terms.) Adding on an allowance for soot luminosity, $_F = 0.37$. From Eq. (25),

$$\overline{F}_{CF} = 0.37\left(1 + \frac{{}^{960}\!/_{832}}{1 + \dfrac{0.37}{0.63}\cdot\dfrac{1}{0.56}}\right) = 0.578$$

In using Eq. (24) to allow for the effect of receiver-surface emissivity, one should note that the radiation-receiving surfaces are of two kinds, plane surfaces in floor and back wall and a nest of tubes in the roof. The former will have an emissivity (or absorptivity) of 0.8. The tube nest will exhibit an effective absorptivity much higher because any beams penetrating up between tubes will have many chances for absorption after reflection. In the present example a mean value of 0.9 will be used on the whole of A_C. Then, by Eq. (24),

$$\mathfrak{F}_{CF} = \frac{1}{\dfrac{1}{0.578} + \dfrac{1}{0.9} - 1} = 0.544$$

This amounts to saying that the flame-wall system interchanges 54 per cent as much heat as a system of parallel black planes close together, having an area A_C and temperatures T_F and T_C. The over-all refractory-wall coefficient $= U = 1(\frac{1}{2} + 1/0.9 + \frac{1}{2}) = 0.47$. Substitution into Eq. (23) now gives

$$q_F = 0.173\left[\left(\frac{T_F}{100}\right)^4 - 8.1^4\right](832)(0.544) + (2)(576)(T_F - 810) + (0.47)(960)(T_F - 520)$$

An energy balance, Eq. (27) (with the gas-exit temperature assumed 100°F. below T_F), gives

$$q_F = (130{,}000)(1070) - (4911)(8.45)(T_F - 100 - 520)$$

Solution by trial and error of these two simultaneous equations gives $T_F = 2780(2320°\text{F.})$ and $q_F = 50{,}400{,}000$ B.t.u./hr. If the flame emissivity and heat capacity are adjusted to 2300° instead of 2500° and the solution of equations repeated, one obtains $t_F = 2290°$, and $q_F = 51{,}600{,}000$ B.t.u./hr., indicating that the final result is insensitive to the temperature at which ϵ_F and MC_p are evaluated. Not all the heat q_F goes to the water-cooled surfaces; the third term in the heat transfer equation represents loss through refractory walls. This is (0.47)(960)(2190) or 1,000,000 B.t.u./hr. Then, finally, the heat received by the water-cooled surfaces, exclusive of convection to the first tube row is 51,600,000 − 1,000,000 = 50,600,000 B.t.u./hr., or 50,600,000/(130,000)(1070) = 36.4 per cent of the enthalpy of the entering fuel.

SECTION 7

EVAPORATION

BY

W. L. Badger, M.S., Consulting Engineer. Formerly, Professor of Chemical Engineering, University of Michigan. Member, American Institute of Chemical Engineers, The American Chemical Society.

CONTENTS

	Page
Heat Transfer in Evaporators	500
Calculation of Heating Surface	500
Temperature Drop in Evaporators	500
Apparent Temperature Drops and Apparent Coefficients	500
Corrected Heat-transfer Coefficients	500
Quantity of Heat to Be Transferred	501
Heat-transfer Coefficients in Evaporators	501
General	501
Vertical-tube Evaporators	501
Horizontal-tube Evaporators	502
Inclined-tube Evaporators	502
Long-tube Evaporators	502
Miscellaneous Evaporator Types	503
Forced-circulation Evaporator	503
Effect of Liquor Level	505
Evaporator Construction	505
Evaporator Types—Classification	505
Apparatus Using Solar Heat	505
Apparatus Heated by Direct Fire	505
Porrion Evaporator	505
Jacketed Apparatus. Kettles	505
Horizontal-tube Evaporator	506
Standard Vertical-tube Evaporator	506
Basket-type Evaporator	506
Long-tube Evaporator	506
Forced-circulation Evaporator	507
Inclined-tube Evaporator	507
Coil Evaporator	507
Comparison of Evaporator Types	507
Cost of Evaporators	507
Calculations for Single-effect Evaporators	507
Principles of Multiple-effect Evaporation	508
General	508
Temperature Distribution in Multiple-effect Evaporators	509
Effect of Boiling-point Elevation on Multiple-effect Evaporators	509
Evaporator Operation	509
Steam Temperature	509
Optimum Vacuum	510
Methods of Feeding	510
Condensate Removal	511
Air Removal	511
Continuous vs. Batch Operation	512
Salt Removal	512
Scale Formation and Scale Removal	513
Salting	514
Foam and Entrainment	514
Materials of Construction	515
Multiple-effect Evaporator Calculations	515
General Principles	515
Symbols Used	515
Case I. Quadruple-effect Evaporator with Forward Feed	516
Case II. Quadruple-effect Evaporator with Backward Feed	516
Case III. Quadruple-effect Evaporator Fed in the Order II-III-IV-I	516
Economical Number of Effects	517
Extra Steam	519
Location of Operating Difficulties	519
Thermocompressors	519
Evaporator Accessories	520
Pumps	520
Condensers	520
Air Pumps	520
Evaporator Controls	520
Condenser Calculations	521
Water Required	521
Surface	521
Vapor Piping	521
Volume of Air Entering Condenser	521
Displacement of Vacuum Pump	521
Choice of Type and Size of Evaporator for a Given Problem	522

EVAPORATION

References: Badger, "Heat Transfer and Evaporation," Reinhold, New York, 1926. Badger and McCabe, "Elements of Chemical Engineering," 2d ed., McGraw-Hill, New York, 1936. Hausbrand-Hirsch, "Verdampfen, Kondensieren und Kühlen," Springer, Berlin, 1931. Walker, Lewis, McAdams, and Gilliland, "Principles of Chemical Engineering," 3d ed., McGraw-Hill, New York, 1937. Webre and Robinson, "Evaporation," Reinhold, New York, 1926.

HEAT TRANSFER IN EVAPORATORS

The rate of heat transfer in an evaporator (see pp. 464 and 478), like all other cases of heat transfer, is expressed by the general equation

$$q = UA\,\Delta T \tag{1}$$

In this equation, q is the rate of heat transfer, B.t.u./hr.; A is the heating surface, sq. ft.; ΔT is the mean temperature difference between steam and boiling liquid, °F.; U is the over-all coefficient in B.t.u./(hr.)(sq. ft. of heating surface)(°F. mean temperature difference).

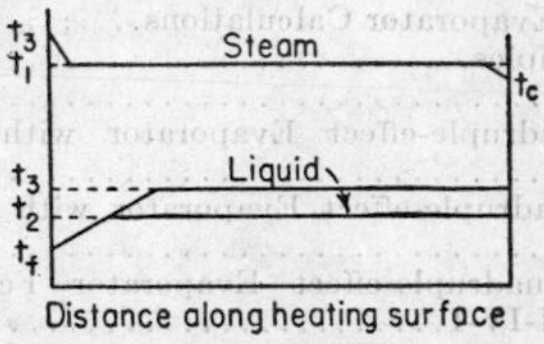

Fig. 1. Temperature drops in evaporators.

Calculation of Heating Surface. Where an over-all heat-transfer coefficient is the resultant of several individual thermal resistances, and these resistances have different areas in a direction perpendicular to the flow of heat, they may be calculated on the basis of any desired area by the expression

$$UA = \frac{1}{\dfrac{1}{h_1A_1} + \dfrac{1}{h_2A_2} + \dfrac{1}{h_3A_3}} \tag{2}$$

where U is the over-all heat-transfer coefficient; A is any desired area to which U is referred, and is the A to be used in the heat-transfer equation, Eq. (1); A_1 is the area of the resistance h_1; A_2 is the area of the resistance h_2, etc.

The over-all heat-transfer coefficient in an evaporator is usually made up of a steam film, the metal wall, and a water film. The thermal resistance of the metal wall can usually be neglected. Then h_1 is the steam-film coefficient. A_1 is the surface of the steam side of the tubes, h_3 is the liquid-film coefficient, and A_3 is the surface of the liquid side of the tubes. In practice, h_3 is usually small in comparison with h_1, and hence the term $1/(h_1A_1)$ is small in comparison with the term $1/(h_3A_3)$. The usual practice, therefore, is to use, instead of the above rigorous equation, the approximate one:

$$U = \frac{1}{\dfrac{1}{h_1} + \dfrac{1}{h_3}} \tag{3}$$

If A is taken as A_3 (the surface of the liquid side), the rigorous equation becomes sensibly equal to the approximate one. Hence the rule is, where the steam-film coefficient is large as compared with the liquid-film coefficient (and this is the usual case), calculate heating surfaces on the basis of the **liquid** side of the tubes. If the two film coefficients are of the same order of magnitude, use the full form of Eq. (2), in which the film coefficients are weighted in proportion to their respective areas.

Temperature Drop in Evaporators. Strictly speaking, ΔT should be the weighted mean temperature difference. It should take into consideration superheat in the steam, cooling of the condensate, elevation in boiling point of the liquid due to substances in solution and to hydrostatic head, and heating the feed. In practice, all these factors are usually neglected, except elevation in boiling point due to substances in solution. Superheat in the steam is usually neglected (for further discussion of superheat, see p. 477), and feed temperature is also neglected. This latter is a more questionable procedure than the former, but all evaporator heat-transfer coefficients have been calculated on this basis.

Apparent Temperature Drops and Apparent Coefficients. In ordinary evaporator practice, pressures may be determined more simply and accurately than temperatures. As a result, certain rather arbitrary units have become customary. In practice, the liquor temperature is calculated from the pressure in the vapor space by using the "Steam Tables,"* thus neglecting the boiling-point elevation; and the steam pressure is similarly calculated from the pressure in the steam space, thus neglecting the superheat. The **apparent temperature drop** is the difference between the steam and liquor temperatures so determined. The heat-transfer coefficient calculated by the use of such a temperature drop is called the **apparent heat-transfer coefficient.**

These considerations are shown in Fig. 1. Steam enters with some superheat at a temperature t_3, quickly drops to its saturation temperature t_1, and some heat is recovered by cooling the condensate to t_c. Liquid enters at t_f, and its boiling point calculated from the pressure in the vapor space is t_2. The apparent temperature drop then is $t_1 - t_2$.

Corrected Heat-transfer Coefficients. When the material being evaporated is a solution whose elevation in boiling point is known, the temperature calculated by the "Steam Tables" from the pressure in the vapor space may be corrected by adding the known elevation in boiling point. This gives the temperature t_3 in Fig. 1. When this is subtracted from the steam temperature, determined from the pressure in the steam space, this gives the temperature drop corrected for rise in boiling point $(t_1 - t_3)$. When this temperature drop has been used to calculate coefficients, the coefficients so obtained are called **coefficients corrected for rise in boiling point.**

In determining the boiling-point elevation of solutions, **Dühring's rule** (p. 294) is of value. The boiling points of a solution of constant composition, when plotted according to Dühring's rule, will give a straight line over the range of pressure ordinarily used in evaporator operation. Over very wide pressure ranges, however, the Dühring lines are not straight. Dühring's rule makes it possible to estimate the elevation in boiling point of a given solution at any desired pressure if the elevations are known at any two pressures.

* See p. 241 *et. seq.*

Quantity of Heat to Be Transferred. In the most general case, the heat that must be transferred to evaporate water consists of (1) sensible heat to bring the feed solution to its boiling point, (2) heat necessary to vaporize water at the boiling point, (3) heat of concentration of the solution, (4) heat of crystallization of products formed, and (5) radiation and other heat losses. In most cases, item 2 is by far the largest one, and sometimes it is the only one that need be considered. Data for items 3 and 4 are extremely scanty and unsatisfactory. As the elevation in boiling point of the solution increases, the relative magnitude of item 3 increases. The complete case, where all items in the above list must be determined, rarely arises. The only simple and workable solution of such a question is obtained from a total heat-concentration chart (sometimes known as the **Merkel chart**) [see Merkel, *Z. Ver. deut. Ing.*, **72**, 109–115 (1928)]. An actual Merkel chart exists for solutions of NaOH up to 78 per cent [Bertetti and McCabe, *Ind. Eng. Chem.*, **34**, 558 (1942)].

The amount of water to be evaporated is given by

$$E = 1000\left(1 - \frac{p}{P}\right) \tag{5}$$

where E = lb. water evaporated per 1000 lb. feed.
p = per cent solids in feed.
P = per cent solids in concentrated liquor.

Table 1 gives values calculated by this equation for various concentration ranges.

HEAT-TRANSFER COEFFICIENTS IN EVAPORATORS

General. The only rational way to discuss heat-transfer coefficients in evaporators would be to separate them into their component film coefficients. Some information on these film coefficients is given in Sec. 6, pp. 480–482. So little work has been done on the film coefficients for either the steam side or the liquor side, however, that most discussions of evaporator heat-transfer coefficients deal with over-all coefficients only.

Table 1. Pounds Water Evaporated per 1000 Lb. Feed

% Solids in thin liquor	% Solids in thick liquor																																	
	10	11	12	13	14	15	16	17	18	19	20	21	22	23	24	25	26	27	28	29	30	35	40	45	50	55	60	65	70	75	80	85	90	95	100
1	900	909	917	923	929	933	938	941	945	947	950	952	955	957	958	960	962	963	964	965	967	972	975	978	980	982	983	985	986	987	987	988	989	989	990
2	800	818	833	846	857	867	875	882	889	895	900	905	909	913	917	920	923	926	929	931	933	943	950	956	960	964	967	969	971	973	975	976	978	979	980
3	700	728	750	770	786	800	813	824	833	842	850	857	864	870	875	880	885	889	893	897	900	914	925	933	940	945	950	954	957	960	962	965	967	968	970
4	600	636	667	692	714	733	750	765	778	790	800	810	818	826	833	840	846	852	857	862	867	886	900	911	920	927	933	938	943	947	950	953	956	958	960
5	500	545	583	616	643	667	688	706	722	737	750	762	773	783	792	800	808	815	822	828	833	857	875	889	900	909	917	923	929	933	937	941	944	947	950
6	400	455	500	538	572	600	625	647	667	684	700	714	727	739	750	760	769	778	786	793	800	829	850	867	880	891	900	908	914	920	925	929	933	937	940
7	300	364	417	462	500	533	563	588	611	632	650	667	682	696	708	720	731	741	750	759	767	800	825	844	860	873	883	892	900	907	912	918	922	926	930
8	200	273	333	384	428	467	500	529	556	579	600	619	637	652	667	680	692	704	714	724	733	772	800	822	840	855	867	877	886	893	900	906	911	916	920
9	100	181	250	308	356	400	437	471	500	526	550	572	591	609	625	640	654	667	678	689	700	743	775	800	820	836	850	861	871	880	888	894	900	905	910
10	..	90	166	231	285	333	375	412	444	474	500	524	545	565	584	600	615	630	643	655	667	714	750	779	800	818	833	846	857	867	875	882	889	895	900
11	...	...	83	154	215	267	312	353	389	421	450	476	500	522	542	560	577	593	607	621	633	686	725	756	780	800	816	831	843	853	862	871	878	884	890
12	...	...	...	77	143	200	250	295	334	369	400	428	455	478	500	520	538	556	572	586	600	657	700	733	760	782	800	815	829	840	850	859	867	874	880
13	...	...	...	...	72	133	188	235	278	316	350	381	409	435	458	480	500	518	536	552	567	629	675	711	740	764	783	800	814	827	837	847	856	863	870
14	...	...	...	...	...	67	125	176	223	264	300	334	364	392	417	440	462	482	500	517	533	600	650	689	720	746	767	785	800	813	825	835	845	853	860
15	...	...	...	...	...	...	63	118	167	210	250	285	318	348	375	400	423	445	465	483	500	572	625	667	700	727	750	769	796	800	812	823	833	842	850
16	...	...	...	...	...	...	...	59	111	158	200	238	273	305	334	360	385	407	429	448	467	543	600	644	680	709	733	754	771	787	800	812	822	832	840
17	...	...	...	...	...	...	...	...	56	105	150	190	227	261	292	320	346	370	393	414	433	514	575	622	660	691	717	738	757	773	788	800	811	821	830
18	...	...	...	...	...	...	...	...	...	53	100	143	182	217	250	280	308	334	357	379	400	486	550	600	640	673	700	723	743	760	775	788	800	810	820
19	...	...	...	...	...	...	...	...	...	...	50	95	136	174	208	240	270	296	322	345	367	457	525	578	620	655	683	708	729	747	763	777	789	800	810
20	...	...	...	...	...	...	...	...	...	...	...	48	91	130	167	200	232	259	286	310	333	429	500	556	600	636	667	692	714	733	750	763	778	790	800
25	...	...	...	...	...	...	...	...	...	...	...	...	...	...	...	...	38	75	108	138	167	286	375	445	500	545	583	616	643	667	688	706	722	738	750
30	...	...	...	...	...	...	...	...	...	...	...	...	...	...	...	...	...	...	...	...	...	144	250	333	400	455	500	538	572	600	625	647	667	684	700
35	...	...	...	...	...	...	...	...	...	...	...	...	...	...	...	...	...	...	...	...	...	...	125	222	300	363	417	462	500	533	562	588	611	631	650
40	...	...	...	...	...	...	...	...	...	...	...	...	...	...	...	...	...	...	...	...	...	...	...	112	200	273	334	385	429	467	500	529	555	578	600
45	...	...	...	...	...	...	...	...	...	...	...	...	...	...	...	...	...	...	...	...	...	...	...	...	100	182	250	308	357	400	438	471	500	527	550
50	...	...	...	...	...	...	...	...	...	...	...	...	...	...	...	...	...	...	...	...	...	...	...	...	...	90	167	230	285	333	375	412	444	473	500

If the solution being concentrated has a very small rise in boiling point (up to 5°F.), the latent heat of vaporization (item 2) can usually be taken from the steam tables for the pressure in question. If the elevation in boiling point of the solution is considerable, the latent heat of vaporization can be calculated by the approximate formula:

$$Q = Q_w\left(\frac{T^2}{T_w{}^2}\right)\left(\frac{dT_w}{dT}\right) \tag{4}$$

where Q = the latent heat of vaporization of 1 lb. water at a temperature T, from a solution whose boiling point is T, °F.
Q_w = the latent heat of vaporization of water taken from the "Steam Tables" for a temperature of T_w.
T and T_w = boiling points of the solution and of water, respectively, at the same pressure, °F.
dT_w/dT = the slope of the Dühring line (p. 294).

The quantity of heat so calculated is the amount necessary to vaporize 1 lb. water from an infinite amount of solution, so that the concentration is not appreciably changed. It does not, therefore, involve item 3 in the above list.

Evaporator heat-transfer coefficients vary over wide ranges. The author has data, based on actual determinations, ranging from 4000 B.t.u./(sq. ft.)(°F.)(hr.) down to 2 in the same units. Even to attempt a preliminary design, or to get an approximate value for a coefficient, one must have some data from the type of machine in question, evaporating a liquor whose properties are like the one in question. Most of such information is in the private files of manufacturers of evaporators. Even so, for a given type of evaporator, a given temperature drop, a given boiling point, and a given solution of a given concentration, the heat-transfer coefficient may vary through considerable limits.

It cannot be stated too emphatically that the data in this section are included merely to give an idea of the order of magnitude of the values involved. These data are not offered for use in design, for actual coefficients may vary from 25 per cent above these figures to 75 per cent below, with apparently insignificant changes in conditions.

Heat-transfer Coefficients in Vertical-tube Evaporators. Figure 2 shows heat-transfer coefficients in a basket-type evaporator, 30 in. diameter, containing 24 tubes, 2 in. diameter by 48 in. long. The liquid evaporated was distilled water. These curves show that heat-transfer coefficients change widely with temperature drop and with boiling points, so that data that ignore these factors are very difficult to use. [Badger and Shepard, *Chem. & Met. Eng.*, **23**, 237, 281 (1920); *Trans. Am. Inst. Chem. Engrs.*, **13**, 101, Part I (1920).]

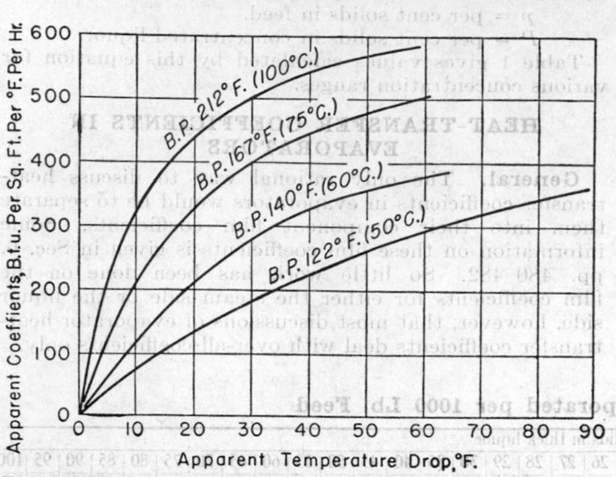

FIG. 2. Heat-transfer coefficients in a basket-type evaporator. (*Badger, "Heat Transfer and Evaporation,"* p. 131.)

Figure 3 gives a comparison between tubes 30 in. long and 48 in. long in this same evaporator, when the liquid is held at different levels. The liquid was distilled water, the steam temperature was 100°C., and the vapor temperature was 75°C. [Partly from Badger and Shepard, *Chem. & Met. Eng.*, **23**, 390 (1920); *Trans. Am. Inst. Chem. Engrs.*, **13**, 139, Part I (1920); partly from unpublished results.]

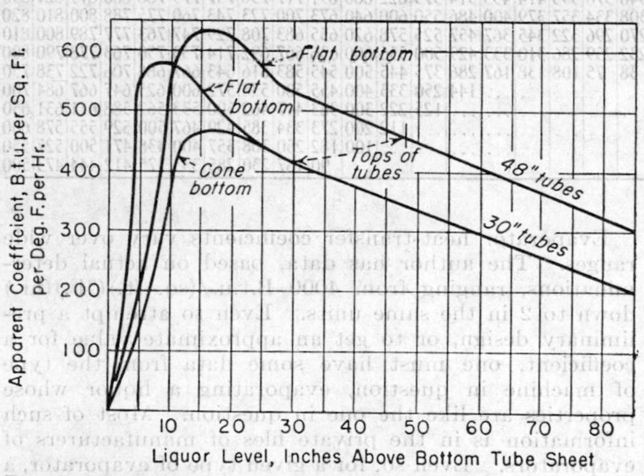

FIG. 3. Comparison of apparent heat-transfer coefficients at different liquid levels for tubes of 30- and 48-in. length. (*Badger, "Heat Transfer and Evaporation," p.* 112.)

Figure 4 shows tests on a high-dextrin malt sirup in the same evaporator but with tubes 5 ft. long. Steam temperature was 230°F., vapor temperature 130°F. (boiling-point elevation not determined). For additional tests on vertical-tube evaporators see Kerr's tests (Fig. 10, p. 504).

Foust, Baker, and Badger [*Ind. Eng. Chem.*, **31**, 206, 1939)] have published actual liquid velocities (1 to 4 ft./sec.) and over-all heat-transfer coefficients in the same experimental evaporator as above. The results cannot be expressed simply enough to reproduce here.

Heat-transfer Coefficients in Horizontal-tube Evaporators. Figure 5 shows heat-transfer coefficients in a horizontal-tube evaporator, 25 in. wide by 4 ft. long in plan, containing 156 iron tubes, 7/8 in. o.d. by 3/4 in. i.d. The liquid was distilled water, and the temperature drop and boiling points are given in the figure. [Badger, *Chem. & Met. Eng.*, **24**, 459 (1921); *Trans. Am. Inst. Chem. Engrs.*, **13**, 139, Part II (1920).]

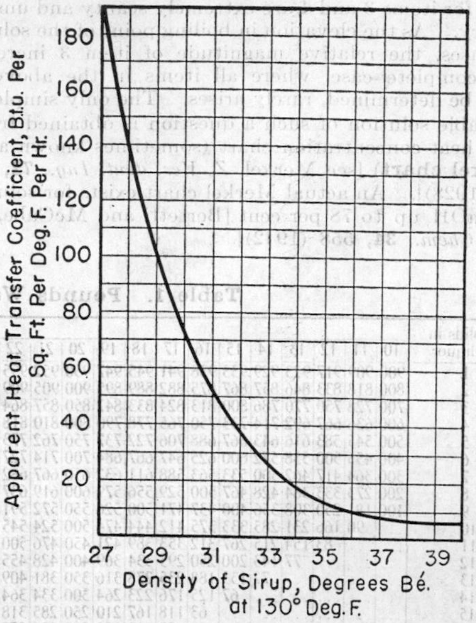

FIG. 4. Apparent heat-transfer coefficients in evaporating a high-dextrin malt sirup. Evaporator with 5-ft. tubes.

Heat-transfer Coefficients in Inclined-tube Evaporators. Figure 6 shows heat transfer in a Buflovac inclined-tube evaporator. The evaporator contained seven tubes, 3 in. o.d. by 4 ft. 10 in. long. The liquid was distilled water. [Van Marle, *Ind. Eng. Chem.*, **16**, 458 (1924).]

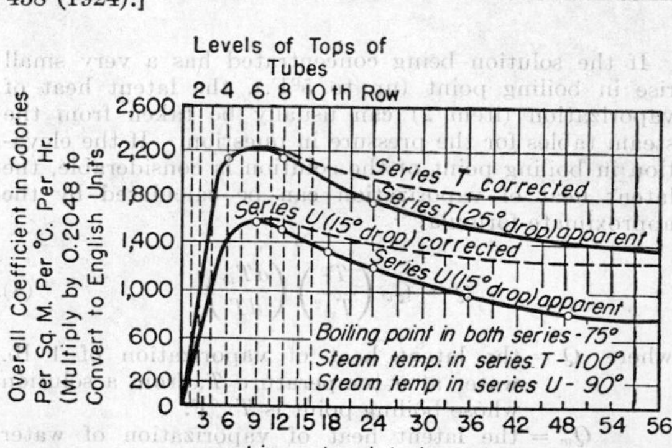

FIG. 5. Heat-transfer coefficients in a horizontal-tube evaporator. [*Badger, Trans. Am. Inst. Chem. Engrs.*, **13**, *Part* II, 148 (1920).]

Figure 7 shows results on a similar evaporator having one tube, 1-in. iron pipe size, 4.08 ft. long. The liquid was distilled water. [Linden and Montillon, *Ind. Eng. Chem.*, **22**, 708 (1930).]

Heat-transfer Coefficients in Long-tube Natural-circulation Evaporators. Kirschbaum, Kranz, and

Stark [*Forsch. Gebiete Ingenieurw.*, *Forschungsheft*, **375,** 1 (1935)] have studied an evaporator with tubes 40 mm. i.d. and 1.97 m. long, with and without recirculation. They concluded that no rational analysis was possible until the boiling and non-boiling sections of tube could be calculated separately. Brooks and Badger [*Ind. Eng. Chem.*, **29,** 918 (1937); *Trans. Am.*

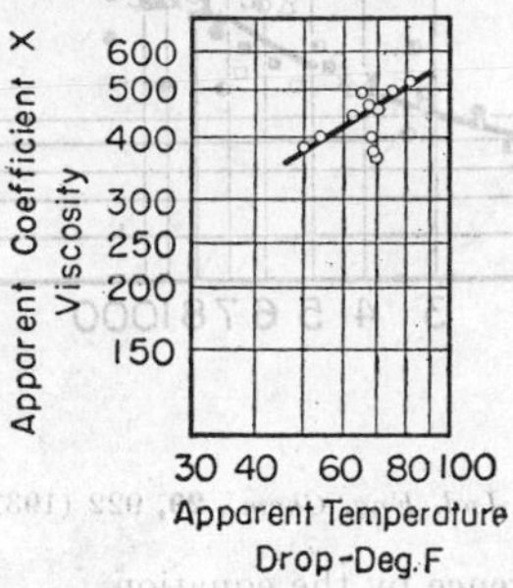

FIG. 6. Heat-transfer coefficients in a Buflovac inclined-tube evaporator. Seven tubes, 3 in. o.d. by 4 ft. 10 in. long. (Van Marle's results.) (*Badger, "Heat Transfer and Evaporation," p.* 144.)

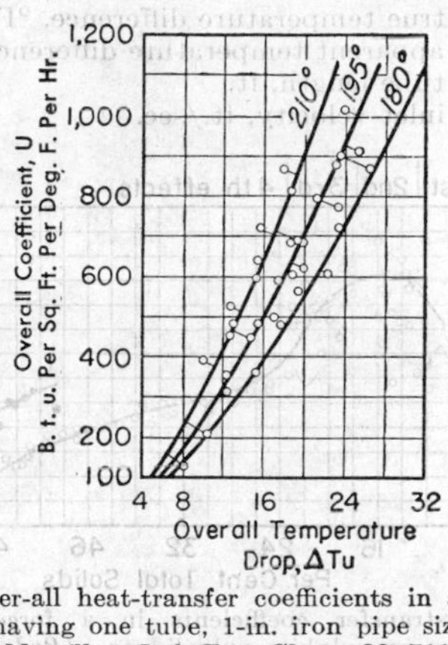

FIG. 7. Over-all heat-transfer coefficients in an inclined-tube evaporator having one tube, 1-in. iron pipe size, 4.08 ft. long. [*Linden and Montillon, Ind. Eng. Chem.*, **22,** 711 (1930).]

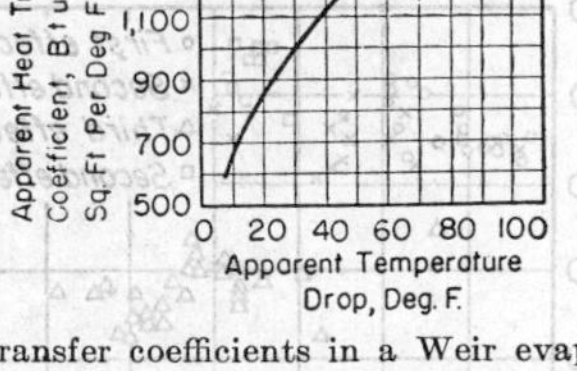

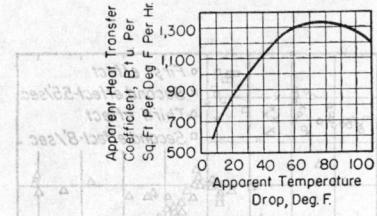

FIG. 8. Heat-transfer coefficients in a Weir evaporator acting on sea water. (*Royds, "Heat Transmission in Boilers, Condensers, and Evaporators," vol. 2, p.* 257.)

Inst. Chem. Engrs., **33,** 392 (1937)] made a separation by a vertical temperature traverse and found U_B, the over-all heat-transfer coefficient in the boiling zone calculated on true liquid temperature, to be given by $(U_B V^{0.27})/(W_f)^m = f\Delta t$, where V = lb. vapor produced per tube per hr., W_f = weight feed, lb. per tube per hr., m = exponent varying from 3.0 to 0.1 between $\Delta t = 5°$ and $\Delta t = 30°$, and f = a function taken from Fig. 9. Stroebe, Baker, and Badger [*Ind. Eng. Chem.*, **31,** 200 (1939)] made similar studies when the feed was preheated so that the whole tube was boiling. No one of the above studies has progressed far enough to be of much value in design.

Heat-transfer Coefficients in Miscellaneous Evaporator Types. Figure 8 shows heat-transfer coefficients in a Weir evaporator evaporating sea water. The heating surface was 12 U-shaped copper tubes, 1½ in. o.d., with a total heating surface of 38 sq. ft. Steam in the tubes varied from 248° to 360°F., and the temperature in the vapor space was correspondingly high. [Lang, *Trans. Soc. Eng. Shpb. Scot.*, **32**; Royds, "Heat Transmission in Boilers, Condensers and Evaporators," vol. 2, p. 257 (1921).]

Figure 10 gives a summary of Kerr's tests on full-size sugar evaporators in actual operation. Thirty-nine different evaporators, double-, triple-, and quadruple-effect, were tested. The evaporators were in various degrees of cleanliness and operated under different conditions of steam and vacuum. In plotting Fig. 10, Kerr says that certain results were eliminated; but, even so, the points scatter very badly, and it requires a great deal of judgment to decide the position of these curves. Curve A is for standard vertical-tube evaporators with a central downtake; curve B is for standard horizontal-tube evaporators, curve C is for the Lillie evaporator, and curve D is for the Kestner evaporator. It should be noted, for instance, that six tests on standard verticals with a boiling point at about atmospheric pressure gave coefficients varying from 180 to 550. This gives an idea of the reliability of the curves in Fig. 10. [Kerr, *Trans. Am. Soc. Mech. Engrs.*, **38,** 67 (1917).]

Table 2. Operating Data for the Evaporation of Sulphite Pulp Liquor*

	First effect	Second effect	Third effect	Fourth effect	
Total time of operation up to end of first series	7 hr. 10 min.	20 hr. 00 min.	26 hr. 55 min.	63 hr. 5 min.	
Total time of operation up to end of second series	73 hr. 55 min.	87 hr. 25 min.	100 hr. 20 min.	148 hr. 00 min.	
Number of test runs averaged	13	24	23	60†	35‡
Average % total solids	9.8	14.7	21.4	39.9	50.3
Average steam temperature, °F	230.79	214.05	192.59	175.47	246.76
Average vapor temperature, °F	210.67	194.10	167.25	130.94	210.92
Average temperature drop, °F	20.12	19.90	25.31	44.53	35.84
Average velocity, ft /sec	11.5	7.9	7.1	9.4	7.6
Average pressure drop, ft. H_2O	14.75	15.23	19.05	22.64	24.00
Average coefficients, B.t.u./(sq. ft)(hr.)(°F.)	735	989	723	346	337

* Badger, *Ind. Eng. Chem.*, **19,** 677 (1927).
† First series of runs, under fourth-effect conditions.
‡ Second series, finishing at atmospheric pressure.

Heat-transfer Coefficients in the Forced-circulation Evaporator. All the tests cited in this section were run on an experimental evaporator having eight tubes, 8 ft. long, ⅞ in. o.d. by ¾ in. i.d. Figure 11 shows the results when evaporating sulphate pulp liquor. The circulation velocity was 5 ft./sec., the vacuum was 26 in., the temperature drop was 40°F.

Figure 12 shows the results obtained when sulphite pulp liquor was evaporated. The operation was intended to duplicate the various effects of a multiple-effect evaporator, and the conditions were as shown in Table 2 above.

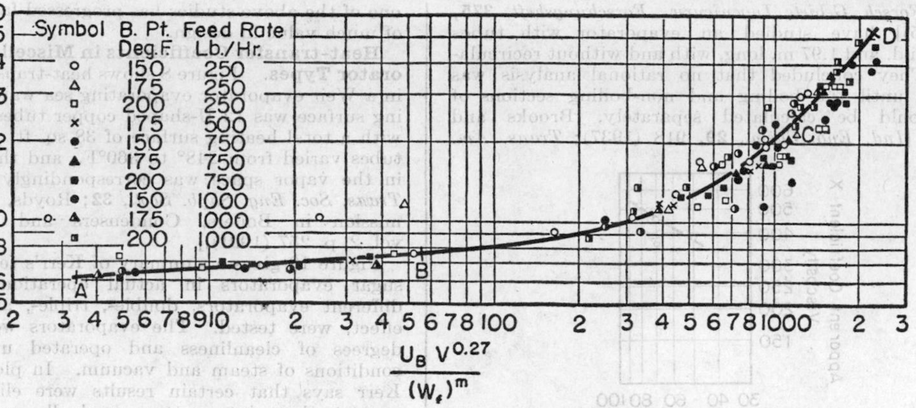

FIG. 9. Final correlation of boiling coefficient. [*Brooks and Badger, Ind. Eng. Chem.*, **29**, 922 (1937).]

Figure 13 shows heat-transfer coefficients in evaporating electrolytic caustic under conditions corresponding to triple-effect evaporation. The liquid was actual cell liquor. Average conditions were as given in Table 3.

Table 3*

Effect	% NaOH	Steam temp., °C.	Vapor temp., °C.	Liquor temp., °C.
1	12.8	120.9	97.1	109.4
2	18.4	115.9	89.9	104.2
3	26.9	84.5	44.1	59.5

* Badger, *Trans. Am. Inst. Chem. Engrs.*, **18**, 231 (1927).

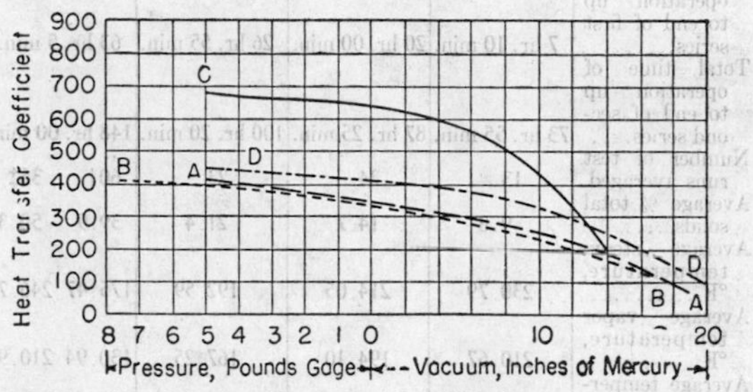

FIG. 10. Coefficients of heat transfer (Kerr's tests) with full-size sugar evaporators. [*Trans. Am. Soc. Mech. Engrs.*, **38**, 98 (1917).]

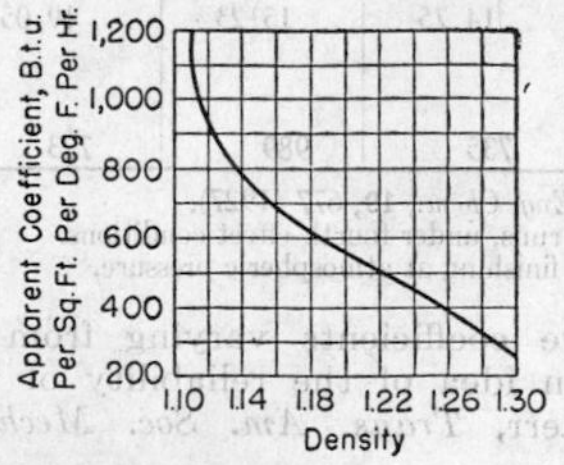

FIG. 11. Heat-transfer coefficients in a forced-circulation evaporator acting on sulphate pulp liquor. (*Badger, unpublished results.*)

Logan, Fragen, and Badger [*Ind. Eng. Chem.*, **26**, 1044 (1934)] and Boarts, Badger, and Meisenburg [*Ind. Eng. Chem.*, **29**, 912 (1937); *Trans. Am. Inst. Chem. Engrs.*, **33**, 363 (1937)] found that the liquid film coefficient for forced-circulation evaporators, at inlet velocities of 5 ft./sec. or more, calculated on true mean liquid temperatures, could be predicted by using the Colburn equation (see p. 465). The true mean temperature difference is related to the apparent temperature difference by the equation

$$\Delta T_T = \Delta T_A - \left(\frac{N}{2}\right)\frac{\Delta T_A}{9.5(u)^{0.73}}$$

where ΔT_T = true temperature difference, °F.
ΔT_A = apparent temperature difference, °F.
N = tube length, ft.
u = inlet velocity, ft./sec.

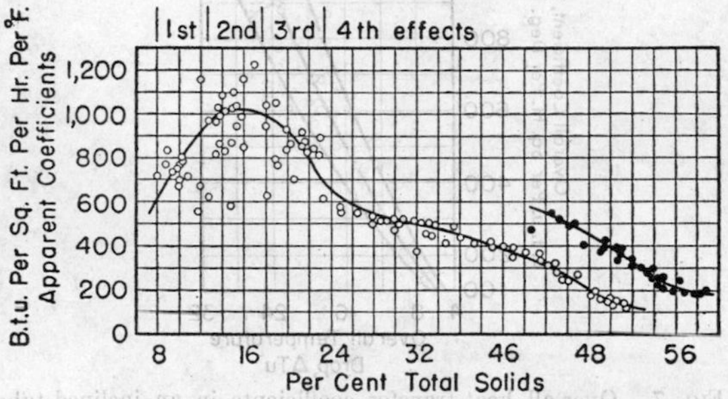

FIG. 12. Heat-transfer coefficients in a forced-circulation evaporator acting on sulphite pulp liquor. [*Badger, Ind. Eng. Chem.*, **19**, 679 (1927).]

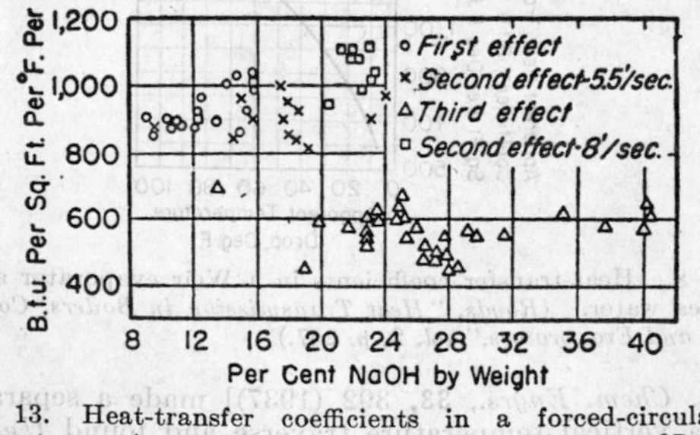

FIG. 13. Heat-transfer coefficients in a forced-circulation evaporator acting on electrolytic caustic solution. [*Badger, Trans. Am. Inst. Chem. Engrs.*, **18**, 237 (1926).]

Coates and Badger [*Trans. Am. Inst. Chem. Engrs.*, **32**, 49 (1936)] showed that in evaporating viscous liquids the over-all heat-transfer coefficient for any one inlet velocity is given by $U = K/\mu^{1.2}$ where K is a constant and μ is viscosity at mean film temperature. Fragen and Badger [*Ind. Eng. Chem.*, **28**, 534 (1936)] showed that for forced-circulation evaporators

$$U = \frac{435u^{0.45}}{\mu^{0.25}\Delta t^{0.1}}$$

where u = inlet velocity, ft./sec., μ = viscosity at entrance (English units), and Δt = over-all apparent temperature drop corrected for elevation of boiling point, °F.

It has been mentioned that heat-transfer coefficients in evaporators are extremely sensitive to minor changes in conditions. The most important factor affecting these coefficients is the velocity of the liquid past the heating surface. Since this velocity, in all types except the forced-circulation evaporator, is dependent on convection, it follows that any factor affecting the convection currents affects the heat-transfer coefficient. Thus, Fig. 3 shows how changing the shape of the bottom of the evaporator may change the coefficients.

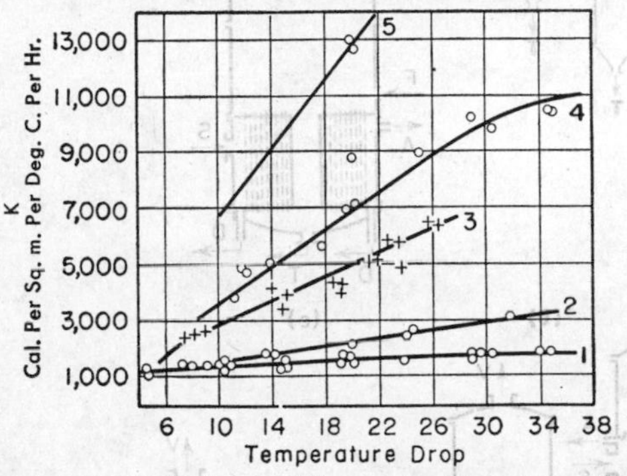

FIG. 14. Effect of tube conditions on heat-transfer coefficients. Coefficient K in metric units. Curve 1, rusty iron tubes; curve 2, clean, new iron tubes; curve 3, slightly dirty copper tubes; curve 4, polished copper tubes; curve 5, polished copper tubes slightly acid distilled water. [*Pridgeon and Badger, Ind. Eng. Chem.*, **16,** 475 (1924).]

Another important factor affecting heat-transfer coefficients is the condition of the surface. A layer of rust or solid deposit may be thin in itself but may serve to entangle a thicker stagnant film of water. Figure 14 shows the effect surface conditions may have on heat-transfer coefficients, though the highest coefficients here are due to a degree of cleanness probably impossible in commercial operation. This figure shows, however, how vast the difference may be between laboratory results and results obtained in large-scale plant operation.

Effect of Liquor Level. An inspection of Fig. 3 shows that for both 30- and 48-in. tubes the maximum rate of heat transfer came when the tubes were about one-fourth submerged, and was 140 per cent of the coefficient obtained when the liquid was even with the tops of the tubes. The shape of that part of the graph representing conditions below the top tube sheet is very sensitive to small changes in the construction of the evaporator body. The part of the graph representing conditions above the top tube sheet is much less sensitive. In practice, evaporators should be operated with a low liquor level, but the actual minimum level should be approached with care, as conditions here are unstable and a slight lowering below the optimum will result in dry tubes. Except in very special cases, operation at liquor levels appreciably above the top of the heating surface is indefensible, decreases capacity, and is used only by careless operators to minimize the attention they give the evaporator.

In most cases in practice, apparent heat-transfer coefficients will vary from 100 to 1000 B.t.u./(sq. ft.)(hr.)(°F.). In evaporators other than high-velocity or forced-circulation evaporators, on the one hand, and excluding evaporators operated at low boiling points, low temperature drops, or handling very viscous liquids, on the other hand, the average will be from 200 to 500 B.t.u./(sq. ft.)(hr.)(°F.). For further data and discussion on heat transfer in evaporation, see p. 478.

EVAPORATOR CONSTRUCTION

Evaporator Types—Classification

Evaporators may be classified as follows:

A. Apparatus using solar heat.
B. Apparatus heated by direct fire.
C. Apparatus with the heating medium in jackets, double walls, etc.
D. Steam-heated evaporators with tubular heating surfaces.
1. Tubes horizontal.
   a. Steam inside tubes (Fig. 15a).
   b. Steam outside tubes.
2. Tubes vertical.
   a. Standard type (Fig. 15c).
   b. Basket type (Fig. 15d).
   c. Long-tube type (Figs. 15b, 15e and 15g).
   d. Forced-circulation type (Fig. 15f).
3. Tubes inclined (Fig. 15h).
4. Tubes bent into special shapes such as hairpin tubes, etc. (Fig. 15i).

Apparatus Using Solar Heat. Solar evaporation is now confined to the evaporation of sea water for salt and, in a few instances, the evaporation of saline waters in deserts. The plant involves only shallow ponds, usually with earth floors and walls. Spray evaporators using atmospheric evaporation have not been successful for purposes within the scope of this chapter. The evaporation of 1 in. water corresponds to the evaporation of 113 tons water per acre of pond surface. Around San Francisco Bay evaporation exceeds rainfall by about 35 in. between Apr. 1 and Nov. 1. Solar-evaporation methods using ricks or piles of brush are obsolete.

Apparatus Heated by Direct Fire. This type of apparatus (except the steam boiler, which falls outside the scope of this article) has never been standardized. In designing evaporators to be heated by direct fire or flue gases, it must be remembered that, in the case of direct fire, radiation may be the most important factor (see p. 483, Sec. 6). In the case of apparatus not exposed to direct radiation but heated merely by hot gases, the rate of evaporation will be determined, not by the liquor-film coefficient, but by the gas-film coefficient. In some cases gas radiation may be important, but most installations of this type use waste gases at too low a temperature for radiation to be important (see p. 483, Sec. 6).

The Porrion evaporator is used only in the paper industry. It consists of circular disks of sheet steel assembled on a central shaft, and mounted over a trough containing the solution to be evaporated. Waste gases from an incinerator pass over the apparatus. As the shaft rotates, the disks carry films of liquid up into the gas where evaporation takes place.

Jacketed Apparatus. Kettles. Small batches may be evaporated in steam-jacketed kettles which show a wide variety of constructions. In the food industry, such kettles are usually open and may be made of sheet copper, sheet aluminum, or other metals. In some cases, the apparatus may be covered with a dome and operated under pressure or vacuum. Many kettles made of cast iron are in use, and they may be jacketed by fastening a sheet-metal outer shell to the underside of the kettle flange; or they may be cast so that the jacket and the kettle are integral. Figure 16 shows some details of kettle construction.

Olin *et al.* [*Chem. & Met. Eng.*, **31,** 116 (1924)] give tests on a 50-gal. cast-iron kettle showing heat-transfer coefficients of 220 to 280 when boiling water with steam in the jackets at 265° to 300°F. In another paper [*Chem. & Met. Eng.*, **32,** 370 (1925)], it is shown that a lining of 8-lb. lead reduced this coefficient to 50 to 60. A copper kettle gave coefficients ranging from 575 to 1575 according to the degree of cleanness of the surface and the temperature drop. These results (like those of Fig. 14 which they greatly resemble) represent surface conditions better than could be maintained in practice. All these coefficients would be greatly decreased when handling the viscous materials that are often heated in kettles. Huggins [*Ind. Eng. Chem.*, **23,** 749 (1931)] shows the effect of scraping the kettle walls when heating stiff masses.

Horizontal-tube Evaporator. Figure 15a shows a typical horizontal-tube evaporator. It has a vertical cylindrical shell to which are attached two steam chests. The tube sheet may be cast integral with one of the body castings, or it may be cast as part of a separate steam-chest casting and bolted to the body casting. Horizontal-tube evaporators invariably have packed tubes. If the body is rectangular in plan, the evaporator may be called a Wellner-Jelinek, though this term is fast disappearing. Tubes are usually ⅞ to 1¼ in. diameter, and may be from 4 to 12, or even 15 ft. long. It is difficult to design a horizontal-tube evaporator having over 5000 sq. ft. in one body. Horizontal-tube evaporators should be used only for non-viscous liquids and for cases where salt or scale is not formed.

Standard Vertical-tube Evaporator. This evaporator (Fig. 15c) consists of a vertical cylindrical shell with tube sheets across it horizontally. A downtake is usually provided in the center of the heating element. This type is the most widely used of all evaporators, and is **especially adapted for solutions that deposit scale or salt.** The tubes are 1¼ to 3 in. diameter and 30 in. to 6 ft. long. Single bodies may be built up to 20,000 sq. ft. per effect. The bottom may be a cone, as shown in Fig. 15d, or it may be a plain dished bottom as shown in Fig. 15c. There may be a single central downtake as shown, an eccentric downtake, several small downtakes (the Scott evaporator), or external downtakes. The central downtake is standard. It is sometimes called the "calandria" type, but the word "calandria" properly used refers to the heating element of any evaporator.

Basket-type Evaporator. This evaporator (Fig. 15d) is essentially the same as the standard evaporator of Fig. 15c, except that the heating element is a separate unit and the downtake is an annular ring between the shell and the heating element. Its proportions and field of usefulness are essentially the same as for the standard vertical, except that it cannot be built in quite such large units.

Long-tube Natural-circulation Evaporator. This type was first known as the Kestner, but the name Kestner is no longer commonly used. These evaporators are characterized by relatively long tubes with liquid inside. Figure 15e shows such an evaporator without a downtake; and 15g the same evaporator with a downtake. The evaporator without the downtake is more common. Tubes may be 1¼ to 2 in. diam., and 12 to 20 ft. long. The term "liquor level" has no significance in connection with these evaporators because the feed is introduced at a fixed rate and, in the evaporator without the downtake, is all discharged after passing

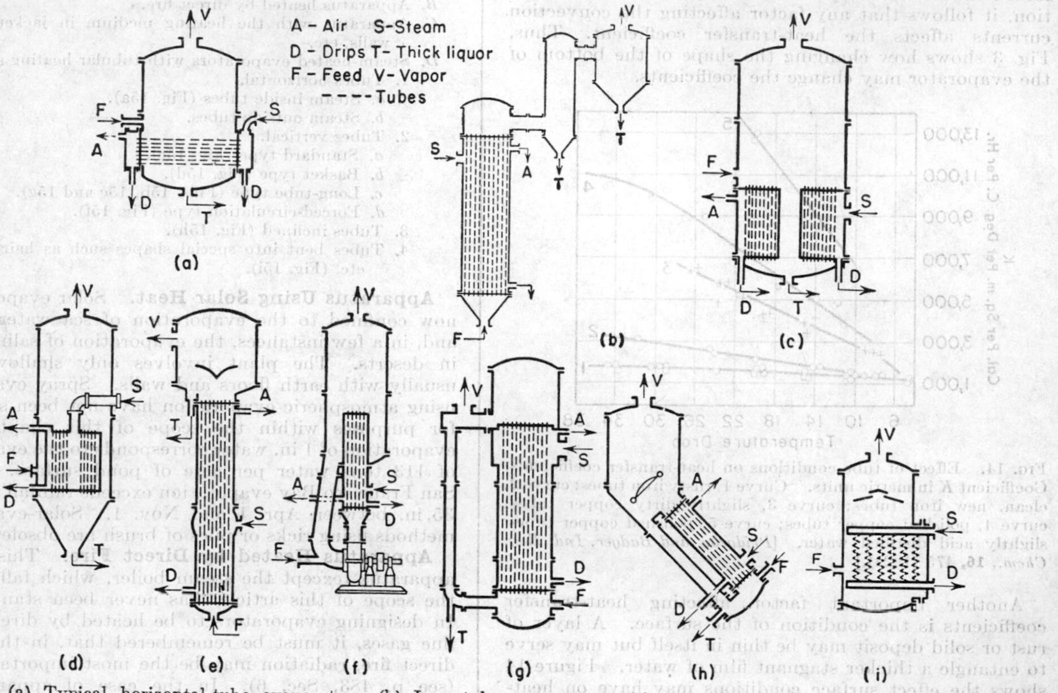

FIG. 15. (a) Typical horizontal-tube evaporator. (b) Long-tube evaporator without vapor head. (c) Standard vertical-tube evaporator. (d) Basket-type evaporator. (e) Long-tube natural-circulation evaporator without downtake. (f) Forced-circulation evaporator. (g) Long-tube natural-circulation evaporator with downtake. (h) Buflovac inclined-tube evaporator. (i) Griscom-Russell evaporator.

through the tubes. Figure 15b shows a recent modification in which the vapor head is replaced by primary and secondary vapor-liquid separators. This decreases the size of the body and makes the heating element more accessible for cleaning.

Forced-circulation Evaporator. Figure 15f shows the forced-circulation evaporator. In this case the liquid is pumped through the tubes with a positive velocity. The evaporator in this form may be used for concentrating clear liquids. If salt is separated, a salt receiver may be introduced in the circulation system. This evaporator gives very high heat-transfer coefficients. It is especially suitable where the liquid to be evaporated is viscous, where expensive materials must be used for the heating surface, where the available temperature drop is small, or where the operation separates salt that tends to grow on the tubes. The tubes are 7/8 in. diameter and 8 to 15 ft. long.

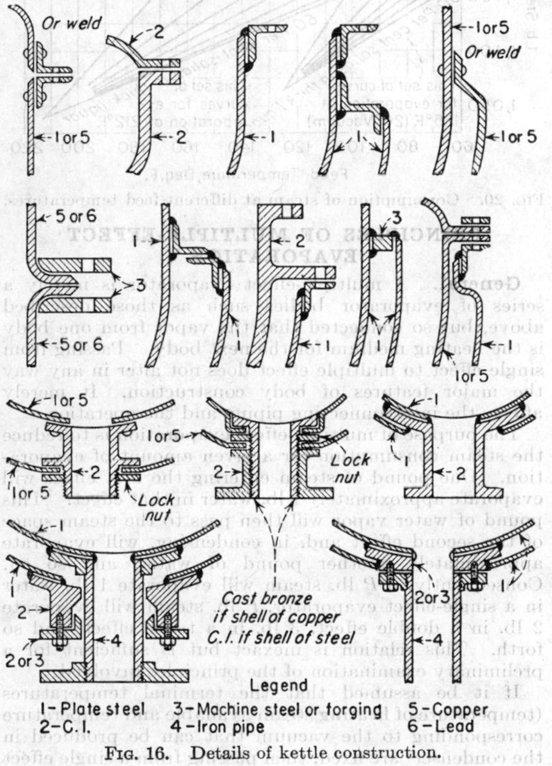

Fig. 16. Details of kettle construction.

Inclined-tube Evaporator. Figure 15h shows the Buflovac inclined-tube evaporator. This gives high velocity of circulation and therefore high coefficients. It is not suitable where the liquid will deposit salt or scale.

Coil Evaporator. Figure 15i shows the Griscom-Russell evaporator. Its use is practically confined to the production of distilled water for boiler-feed make-up in power plants; for if hard scale forms on the coils it may be cracked off by shutting off the steam and turning cold water into the coils.

Comparison of Evaporator Types. No hard and fast rules can be laid down as to the service for which a given type of evaporator is most suitable. In practically every field where evaporators are used, several different types will be found in exactly the same service and apparently giving equal satisfaction. The choice is often made on the basis of the tradition in a particular industry rather than for any definite technical reasons.

The horizontal-tube evaporator is best suited for non-foaming, non-viscous solutions that do not deposit scale or salt during evaporation. The vertical-tube evaporator has been in satisfactory operation in every conceivable kind of service and is probably the most versatile of all the types. It is primarily indicated where the liquid deposits salt or scale during evaporation, and it is often used to concentrate liquids to higher viscosities than are ordinarily obtained in the horizontal-tube evaporator. The coil evaporator is used only for making distilled water for boiler feed. The forced-circulation evaporator has been particularly successful on very viscous liquids, on foamy liquids, and on liquids which tend to deposit salt or scale during heating. It gives exceedingly low losses by entrainment. Its disadvantage is its high cost when applied to simple problems where one of the cheaper types would be satisfactory. The long-tube,

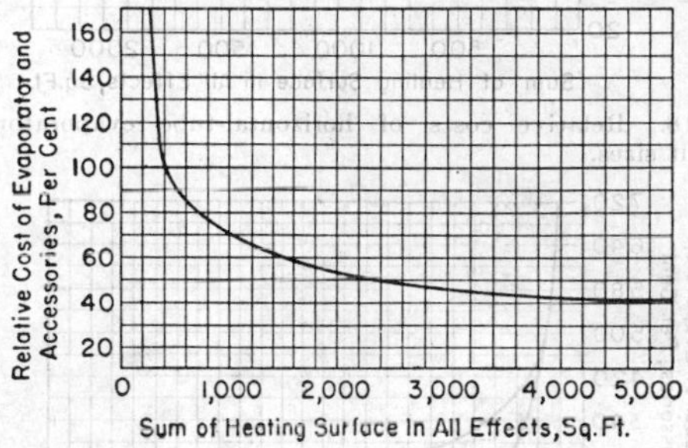

Fig. 17. Relative costs per square foot of evaporating surface of vertical-tube evaporators of different sizes.

natural-circulation evaporators have very high coefficients, give nearly as low entrainment losses as the forced-circulation evaporator, but cannot go to quite as high viscosities as the forced-circulation evaporator, and are not at all suited for salting liquids. All the above statements must be taken as indicating general tendencies only; and individual exceptions can be made to practically every one of them.

Because of the wide variety of conditions under which they may be used, the variations in capacity caused by these conditions, and the limited amount of data available, no general statements may be made regarding the relative capacities of different types of evaporators. It is probable that under most conditions the heat-transfer coefficients of short-tube vertical evaporators and horizontal-tube evaporators are of the same order of magnitude. The long-tube, natural-circulation evaporators and the forced-circulation evaporators usually have the highest coefficients of any type. However, Kerr's tests (Fig. 10) do not show high coefficients for the Kestner evaporator.

Cost of Evaporators. No builder of evaporators has ever been able to standardize his equipment sufficiently to have fixed prices for equipment of a given size. All evaporator installations are especially designed and vary greatly in such details as type and size of condenser, vacuum pumps, feed and condensate pumps, accessory piping, and special fittings. Figures 17, 18, and 19 give approximate curves showing the relation between the cost of evaporators with all their accessories and the total amount of heating surface involved. See Bliss, *Chem. Eng.*, **54**, No. 5, 134 (1947).

Calculations for Single-effect Evaporators. The calculations needed for single-effect evaporators are so simple that they do not need formulation. An illustrative example will suffice.

Example. A single-effect evaporator is to operate under 26-in. vacuum (referred to a 30-in. barometer) and is to be heated with steam at 5 lb. gage. It is to be fed with a 10 per cent solution at 70°F., which is to be concentrated to 50 per cent. It is to evaporate 1000 lb. water per hour. Assume: (1) no losses by radiation and convection, (2) no sensible heat recovered from condensate, (3) heating steam dry and saturated, (4) no heat of concentration or crystallization involved, (5) specific

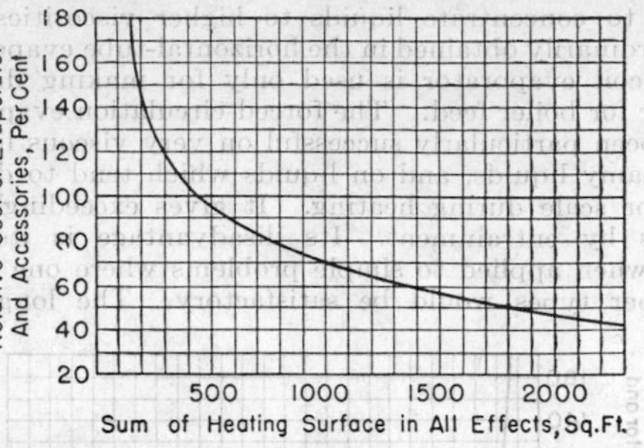

FIG. 18. Relative costs of horizontal-tube evaporators of different sizes.

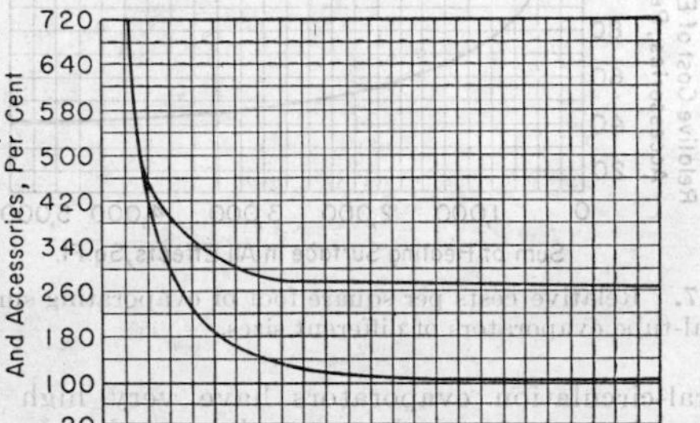

FIG. 19. Relative costs of forced-circulation evaporators of different sizes.

heat of feed and thick liquor to be 1.00, (6) coefficient of heat transfer to be 350. From Table 1, p. 501, it is seen that to concentrate 1000 lb. feed from 10 to 50 per cent calls for the evaporation of 800 lb. water. Hence, for an evaporation of 1000 lb., the feed will be 1250 lb./hr.

Heat needed to vaporize 1000 lb. water at 125°F.
= 1,022,000 B.t.u.
Heat needed to heat feed = 1250 × (125 − 70) = 69,000 B.t.u.
Total = 1,091,000 B.t.u.

One pound of heating steam gives up 960.4 B.t.u. at 227°F. Hence:

Total steam needed = 1,091,000/960.4 = 1136 lb./hr.
Heating surface needed = 1,091,000/(227 − 125) × 350
= 35.5 sq. ft.

If the feed were at 125°F., the steam consumption would be 1064 lb. and the surface needed would be 28.6 sq. ft. These figures are smaller than would be called for by any commercial installation, but obviously the steam consumption and heating surface for larger installations would be proportional to the amount to be evaporated.

If similar calculations were to be made for other feed temperatures, the curves for steam consumption vs. feed temperature, or heating surface vs. feed temperature, would be straight lines. Several such lines for different final concentrations, and for boiling points of 125° and 212°F., are shown in Fig. 20. For all cases the feed is supposed to be a 10 per cent solution. For each line the concentration of the thick liquor, boiling point, and temperature difference are constant. Figure 20 shows especially the importance of feed temperature. It would involve more space than is available to develop similar families of curves for a varying concentration and temperature of feed, varying concentration of thick liquor, varying boiling points, and varying temperature differences; but for any practical problem the values of some of these variables are fixed by the conditions of the problem so that all cases of interest can be expressed by relatively simple families of curves.

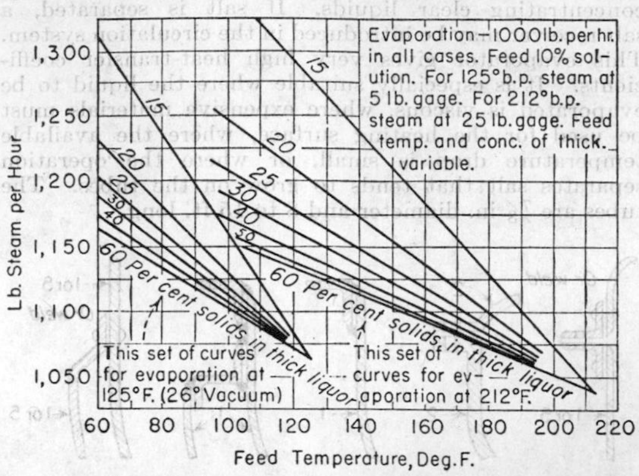

FIG. 20. Consumption of steam at different feed temperatures.

PRINCIPLES OF MULTIPLE-EFFECT EVAPORATION

General. A multiple-effect evaporator is merely a series of evaporator bodies such as those described above, but so connected that the vapor from one body is the heating medium for the next body. Passing from single effect to multiple effect does not alter in any way the major features of body construction. It merely affects the interconnecting piping and the operation.

The purpose of multiple-effect evaporation is to reduce the steam consumption for a given amount of evaporation. One pound of steam entering the first effect will evaporate approximately 1 lb. water in that effect. This pound of water vapor will then pass to the steam space of the second effect and, in condensing, will evaporate approximately another pound of water, and so on. Consequently, if P lb. steam will evaporate 1 lb. water in a single-effect evaporator, P lb. steam will evaporate 2 lb. in a double effect, 3 lb. in a triple effect, and so forth. This relation is inexact but is sufficient for a preliminary examination of the principles involved.

If it be assumed that the terminal temperatures (temperature of heating steam available and temperature corresponding to the vacuum that can be produced in the condenser) are fixed, then passing from a single effect to multiple effect does not increase the capacity of an evaporator. Under these assumptions there is a certain working temperature drop ΔT. If a single-effect evaporator is operating between these terminal conditions, assume that it will take A sq. ft. heating surface to accomplish the desired evaporation. If an N-effect evaporator is to be used between the same terminal conditions for the same weight of water evaporated, the temperature difference available for each effect will be decreased, so that it will take N bodies of A sq. ft. each to accomplish the same result. This statement is only approximate, and the precise relationship will have to be calculated for each particular case (see p. 515).

The above statements may be summarized by saying that **passing from single-effect to multiple-effect operation decreases steam cost but increases apparatus cost.**

Temperature Distribution in Multiple-effect Evaporators. If it be assumed that the temperature of the steam to the evaporator, and the pressure in the condenser (hence the saturation temperature of vapor leaving the last effect) are fixed, the evaporator will attain its own equilibrium and establish its own distribution of temperatures.

Let t_0 = saturation temperature of steam to the first effect.
t_1 = boiling point of liquid in first effect.
t_2 = boiling point of liquid in second effect.
t_n = boiling point of liquid in last effect.
t_v = saturation temperature of vapor going to condenser.

Consider a double effect, with liquid entering each effect at the boiling point in that effect and having no appreciable elevation in boiling point, no appreciable sensible heat recovered from the condensate and with negligible radiation losses. The heat transmitted in the first effect is

$$q_1 = U_1A_1(t_0 - t_1)$$

The heat transmitted in the second effect is (since $t_2 = t_v$ in this case)

$$q_2 = U_2A_2(t_1 - t_v)$$

But, under the assumptions made, all the heat passing through the heating surface of the first effect appears as latent heat in the vapor and must in turn pass through the heating surface of the second effect. Hence $q_1 = q_2$. It is usual (though not necessary) for all bodies of the same evaporator to have the same heating surface. Hence $A_1 = A_2$. Consequently,

$$U_1(t_0 - t_1) = U_2(t_1 - t_v) \tag{6}$$

and the temperature differences must be inversely proportional to the heat-transfer coefficients. If all the assumptions of this section are not true in any particular case, it merely means that the temperature differences would not be strictly proportional to the coefficients. In practice such quantities of heat as are used for heating feed or generated by "flashing" are small enough so that it still holds that the temperature differences across the individual effects are approximately inversely proportional to the heat-transfer coefficients. If a large amount of very cold feed is introduced into any one effect, that effect will deviate somewhat from this relationship.

In the same way it may be shown that for a triple effect that fulfills the assumptions here made

$$U_1A_1(t_0 - t_1) = U_2A_2(t_1 - t_2) = U_3A_3(t_2 - t_v)$$

and for n effects

$$U_1A_1(t_0 - t_1) = U_2A_2(t_1 - t_2) \cdot \cdot \cdot = U_nA_n(t_{n-1} - t_v) \tag{7}$$

If for any reason the coefficient in any effect should decrease, vapor would not condense as fast as before, the pressure in the next previous effect (and hence the temperature of that effect) would rise, and a new equilibrium would be established. In actual operation, the temperature distribution in a multiple-effect evaporator fluctuates more or less as conditions of operation fluctuate with corresponding effects on the heat-transfer coefficients.

Operators who do not understand this principle often try to control the temperatures in various effects to correspond to some preconceived idea as to what the distribution should be. If this is done by throttling vapor between effects, it merely introduces a pressure drop (and hence a temperature difference) between such effects and hence decreases capacity. It does not affect economy. If it is done by by-passing vapor around an effect, or introducing steam into effects other than the first, it merely means that such steam or vapor is used in less than n effects, if the evaporator is an n-effect evaporator, and hence economy is sacrificed though capacity may be increased.

Effect of Boiling-point Elevation on Multiple-effect Evaporators. Let E_1, E_2, E_3, E_n, = elevation of boiling point in 1st, 2d, 3d, and nth effect, respectively. Then, in the double effect discussed in the preceding section, the solution in the first effect boils at t_1, but the saturation temperature of the vapor is $(t_1 - E_1)$, and this is the temperature that is effective in the next effect. If the vacuum produced in the condenser is to be the same as in the case previously discussed, the **saturation** temperature of the vapor (t_v) must be the same in both

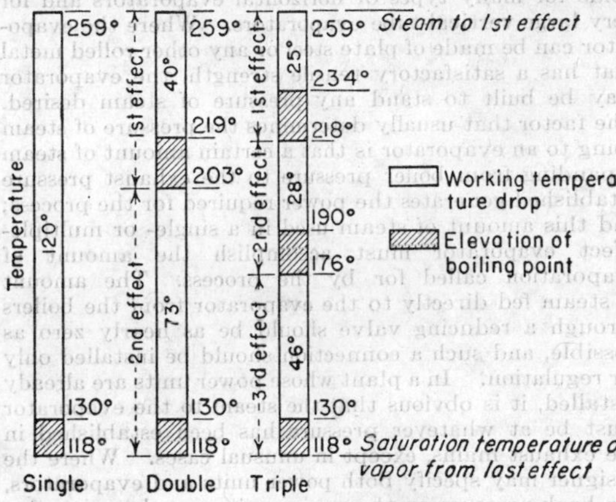

FIG. 21. Effect of elevation in boiling points of sodium chloride solutions on typical temperature differences for single-, double-, and triple-effect evaporators.

cases. In this case, however, if the vapor is to have a saturation temperature of t_v, the boiling point of the second effect is higher than before, because it is now equal, not to t_v, but to $(t_v + E_2)$. Hence Eq. (6) becomes

$$U_1(t_0 - t_1) = U_2[(t_1 - E_1) - (t_v + E_2)]$$

This means that, if U_1 and U_2 are the same as before, t_1 must be higher than before. The general equation is

$$U_1A_1(t_0 - t_1) = U_2A_2(t_1 - E_1 - t_2) = U_3A_3(t_2 - E_2 - t_3) = U_nA_n[(t_{n-1} - E_{n-1}) - (t_v + E_n)] \tag{8}$$

Hence the *elevation in boiling point is lost from the available temperature drop once for each effect in a multiple-effect evaporator*. This limits the total number of effects possible between given terminal temperatures to a smaller number for solutions with appreciable boiling-point elevations than for liquors with little or no elevation. Figure 21 shows the effect of the elevation in boiling point of sodium chloride solutions on fairly typical temperature differences for single-, double-, and triple-effect evaporators.

EVAPORATOR OPERATION

This section will discuss various features of evaporator operation as applied to both single and multiple effects and will also introduce certain data regarding evaporator construction.

Steam Temperature. The ideal plant has a steam flow sheet that shows a balance between steam used for power generation and steam used for heating and evaporating (usually called "process" steam). In a process that uses an evaporator of any appreciable size it will usually be found that the evaporator is the principal consumer of process steam and, therefore, the most important single factor in balancing the steam flow sheet.

Evaporators were largely developed when power was generated in reciprocating engines exhausting at atmospheric pressure or a few pounds above atmospheric. Consequently, evaporators are often considered as taking steam at from 5 to 20 lb./sq. in. gage. There is nothing in the principles of evaporator operation that requires adherence to this range. The pressure of the steam to a single effect, or to the first effect of a multiple-effect evaporator, may be from 24-in. vacuum to 250 lb./sq. in. gage or higher.

Many evaporators are constructed of cast iron and consequently cannot be easily designed for pressures in excess of 10 to 20 lb./sq. in. gage. This limit, therefore, holds for many types of horizontal evaporators and for very large vertical-tube evaporators. Where the evaporator can be made of plate steel or any other rolled metal that has a satisfactory tensile strength, the evaporator may be built to stand any pressure of steam desired. The factor that usually determines the pressure of steam going to an evaporator is that a certain amount of steam expanding from boiler pressure to the exhaust pressure established generates the power required for the process; and this amount of steam used in a single- or multiple-effect evaporator must accomplish the amount of evaporation called for by the process. The amount of steam fed directly to the evaporator from the boilers through a reducing valve should be as nearly zero as possible, and such a connection should be installed only for regulation. In a plant whose power units are already installed, it is obvious that the steam to the evaporator must be at whatever pressure has been established in the exhaust mains, except in unusual cases. Where the designer may specify both power units and evaporators, the back pressure on the power units may be any value that gives the desired balance.

Optimum Vacuum. The purposes of employing a vacuum are twofold: (1) to increase the available temperature difference; and (2) to protect liquids that would be damaged by high temperatures. An evaporator, either single or multiple effect, is not necessarily operated under a vacuum. The use of vacuum on evaporators is the result of the almost universal tendency to supply evaporators with steam at from 220° to 250°F. In order to obtain satisfactory working temperature differences (especially for multiple effects) it has been necessary to use a vacuum. If the desired temperature difference can be obtained, an evaporator may work under pressure above atmospheric. In the beet-sugar industry in Europe, triple-effect evaporators, operated with steam at not over 240°F., are operated with atmospheric pressure in the third effect. This means not only that the condenser and vacuum pump disappear, but the vapors from the third effect are available for process use. In one case in this country, an evaporator has been designed to take steam from the boilers at 250 lb./sq. in. gage and to deliver vapor at 80 lb./sq. in. gage for other process uses.

Where vacuum is employed, the temptation is to use the highest possible vacuum that may be produced by modern devices. As the vacuum increases, the boiling point decreases very rapidly and, therefore, the available temperature difference increases. This is the usual reasoning, but a reference to Fig. 2 will show that, as the boiling point decreases, the heat-transfer coefficients decrease very rapidly. It is possible to operate with so high a vacuum that the decrease in coefficients more than offsets the increased temperature difference. An evaporator operating with a 27.5-in. vacuum may have a lower capacity than when operated at a 27-in. vacuum, all other conditions being equal. The optimum vacuum varies with the type of evaporator, the type of liquid, and the cost of producing the vacuum. In some cases, such as the evaporation of fruit juices, gland extracts, gelatin, and other materials that are sensitive to heat, it may be desirable to employ the highest vacuum possible to preserve the quality of the product at the expense of decreased capacity.

In general, a 26-in. vacuum seems to be fairly well established as an average.

Methods of Feeding. Feeding a single-effect evaporator calls for little special comment. The feed inlet may usually be at any point that suits the convenience of the designer. It is well to have the feed introduced at such a point that the natural convection currents mix it with the liquid in the evaporator as completely as possible, but this is not necessary. In the case of salt evaporators with a salt-settling leg below the liquid space, the feed is sometimes introduced in the bottom of this leg to exert a washing and classifying action on the salt. In forced-circulation evaporators, the feed is preferably introduced at or near the suction of the circulation pump.

Multiple-effect evaporators may be fed in a variety of ways. The commonest method is **forward feeding.** This signifies introduction of the dilute liquor in the first effect, feeding from effect to effect by the difference in pressure between the effects, and withdrawing concentrated liquor from the last effect. **Backward feed** signifies introducing the dilute liquor in the last effect, advancing it from effect to effect by pumps, and withdrawing concentrated liquor from the first effect. **Parallel feed** signifies the introduction of dilute liquor into every effect. **Mixed feed** covers all other arrangements.

The advantages of forward feed are: simplest arrangement of equipment, simplest control, and only one pump whose suction is under vacuum. It gives a slightly higher evaporation in the last effect than in the first, because of "flash" as the liquor drops in temperature. The principal disadvantage of forward feed is that, if the feed is cold, it puts an undue load on the first effect and increases the steam consumption for a given amount of evaporation.

For cold feed, backward feed uses less steam for a given amount of evaporation than forward feed. Using the nomenclature of p.509 and letting t_f be the temperature of the feed, in an n-effect evaporator the feed is heated from t_f to t_n with vapor that has already evaporated roughly $(n - 1)$ times its weight of water, from t_n to t_{n-1} with vapor that has evaporated $(n - 2)$ times its weight of water, etc. In forward feed, all the heating from t_f to t_1 is done with steam direct from the mains.

Control of levels in the various bodies is apt to be troublesome in backward feed; there will be more than one pump with its suction under vacuum. This may be avoided when there is sufficient headroom for dropping the feed lines to form an inverted siphon between effects, with the feed pumps at the bottom of the siphon.

Backward feed should rarely be employed if the temperature of the feed is above the boiling point in the last effect. In this case, t_f is higher than t_n and the feed flashes down to t_n in the last effect. This not only sends unnecessary vapor to the condenser but means that the feed must be heated again from t_n to t_f in the evaporator. Even so, if the evaporator has a large total temperature difference and t_f is not too much above t_n, it may be more economical to heat from t_n to t_1 in backward feed than to heat from t_f to t_1 in forward feed. The question of whether backward feed or forward feed is more economical for any particular case where feed temperature is above the last-effect temperature can be determined only by making complete heat-balance calculations for the evaporator for both cases.

The average coefficient is likely to be higher in the case of backward feed because the concentrated liquor is boiled in the highest temperature effect so that its viscosity is lower than with forward feed. Backward feed calls for slightly less evaporation in the last effect than in the first. Hence, backward feed puts less load on the condenser than forward feed, unless the feed temperature is above the boiling point in the last effect.

Parallel feed is practically never used except in salt evaporators, or in any other case where the feed is a saturated solution and no thick liquor is withdrawn from the evaporator. It has no special advantages except convenience.

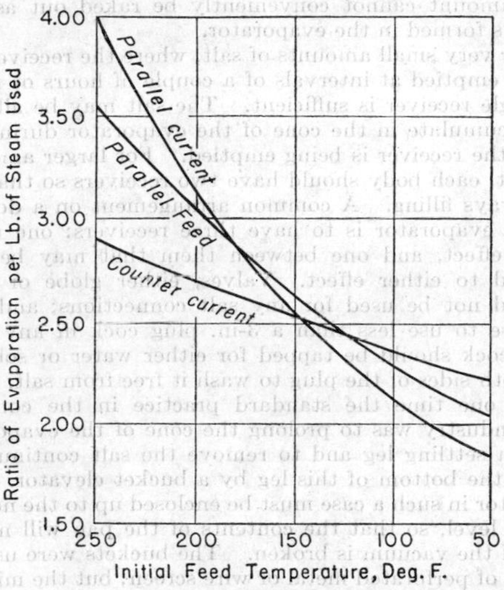

FIG. 22. Ratio of evaporation per pound of steam used for different feed temperatures and for different methods of feeding. (Parallel current = forward feed; countercurrent = backward feed.) [*Webre, Chem. & Met. Eng.*, **27,** 1078 (1922).]

Webre [*Chem. & Met. Eng.*, **27**, 1073 (1922)] has made calculations for an evaporator with a fixed total temperature range, a fixed ratio of feed to thick liquor, and a fixed set of heat-transfer coefficients. His curves, showing steam consumption per pound of water evaporated, for different feed temperatures and for different methods of feeding, are given in Fig. 22. These results are not general, as the precise relationships will vary if the factors mentioned above as fixed for this problem are varied.

Mixed feed is especially useful when a solution is to be concentrated through a wide range of densities and where the thick liquor is very viscous. In such cases, such a routing of liquor as II-III-IV-I is often used. This has the advantage of forward feed, requires only one pump to return the liquor from the last to the first effect, and evaporates the liquor of highest viscosity in the body with the highest boiling point. Many special cases of mixed feed arise, dictated by special circumstances; their characteristics can usually be determined only by carrying through a complete set of evaporator calculations.

Condensate Removal. The accessory apparatus used for this purpose is described on p. 520. The removal of condensate from single-effect evaporators requires no special comment. In very large evaporators it may be desirable to introduce a number of condensate connections rather than one large one. It is obvious that the condensate from any effect that takes steam from the mains should be returned as boiler feed, unless there is a chance of contamination.

In multiple effects, condensate from the first effect will always be returned to the boilers as hot as possible, unless the supply of steam to the first effect is from some source other than the boilers so that it may be contaminated. In case condensate from the second effect can also be kept clean, it may be used for boiler-feed make-up. All effects should have individual condensate-removal devices, and in no case should the condensate be passed from effect to effect. In the case of a multiple-effect evaporator where hot condensate is of no value in the process, condensate from all effects but the first and the last may be led into flash pots. Here the condensate is flashed down to the pressure of the steam space in the next effect, and the flashed vapor sent to the steam space of that effect. Care must be taken in such arrangements that the return of this flashed vapor does not interfere with the removal of air. It is often considered that the complications involved in this flash system, and the increased power to remove all the condensate from the vacuum existing in the steam space of the last effect, offset the possible value of the heat recovered by such a system.

Air Removal. The term air will be used in this discussion to include all non-condensed gases. In a single-effect evaporator, or in the first effect of a multiple-effect evaporator, air in the steam space comes only from the steam used for heating, unless the steam space is under a pressure less than atmospheric, in which case there may also be air from leaks. If the steam space is at a pressure greater than atmospheric, the air may be vented to the atmosphere. It is sometimes convenient to introduce the air into the vapor space of the same effect.

In multiple-effect evaporators, the air may be vented from the steam space directly into the vapor space of the corresponding effect, or all effects may be vented directly to the condenser. The former method is used where the amount of air to be expected is small, and is practiced in order to utilize the latent heat in the steam. In case there is much air, to avoid an undue concentration of air in the last effect with a corresponding decrease in the steam-film heat-transfer coefficient, all effects are vented directly to the condenser.

The control of air venting is usually accomplished by setting the vent valves purely by guess. In most cases it is considered more desirable to lose some steam through the vents than to decrease the heat-transfer coefficients by incomplete venting. It has been suggested that the vents might be controlled by inserting thermometers in the vent line. The vent valve would then be opened until this thermometer showed a temperature approaching the temperature of the steam space. The writer has never seen this method adopted in practice, but it has proved very useful in the laboratory.

The location of vent connections is not always easily determined. The general guide is that the air will collect at the end of the steam path, if the steam has a positive velocity in any particular direction. Thus, in horizontal-tube evaporators, air should be vented from the steam chest opposite the steam inlet. In evaporators like the Kestner, and others with a long cylindrical steam space, if steam is introduced at one end and air removed from the other, air removal will be fairly complete.

Steam in the standard vertical-tube evaporator has no positive velocity in any particular direction. Air vents are usually located around the outside edge of the steam space, and this is also true for the vertical basket type. Air vents should be provided on both the top and bottom tube sheets of these types. Air removal from the bottom of such evaporators can be accomplished by using oversized reciprocating condensate pumps. The fact that air is heavier than steam at the same pressure and temperature has no bearing on air removal, because circulation in the steam space is probably more than sufficient to keep the two gases well mixed, unless there is a positive velocity tending to force the air in a particular direction. Baffles, inside the steam space, for directing the steam flow or segregating the air at a particular point are not likely to be of value in most evaporators, because for over-all coefficients below 1000 the steam-film coefficient is so high that a considerable change in it will scarcely affect the over-all coefficient.

Continuous vs. Batch Operation. In general, continuous operation is always to be preferred wherever it is possible. If the finished material is very viscous, or for any other reason a very low coefficient exists in the effect containing concentrated liquor, and the differences in concentration between final and original liquors are considerable, batch operation may be desirable. With continuous operation, the effect containing the concentrated liquor is always working with a low coefficient. If the apparatus is operated in batches, this low coefficient prevails only a part of the time; and the increase in the average coefficient may offset the time lost in filling and emptying between batches.

An alternative to this is suggested by the fact that, if the change in concentration between initial and final liquors is large, concentrations sufficiently high to give liquors of objectionable viscosity do not appear until most of the water has been evaporated. Thus, in concentrating a 10 per cent solution to 50 per cent solids, evaporation of half the water to be removed gives a solution of 16.7 per cent solids, and evaporation of three-fourths of the total gives only 25 per cent solids. In such a case, if the scale of operations were sufficiently large to warrant the cost of the equipment, the material might be concentrated from 10 to 20 or 25 per cent in a continuously operated evaporator (single or multiple effect as conditions may dictate) and then finished from 20 or 25 to 50 per cent in a discontinuous single-effect evaporator, often called a "finishing pan." This is regular practice wherever the final solution has either very high viscosity or a very high elevation in boiling point. It is often possible to confine the finishing pan to concentration ranges that call for less than 10 per cent of the total evaporation.

Salt Removal. In this discussion the word salt will be used to cover any crystalline material deposited during evaporation and having a normal solubility curve; *i.e.*, substances whose solubility increases with increase in temperature. The salt may be the principal product desired from the operation, or it may be incidental to the main purpose to be accomplished by evaporation. Salt-removal methods may be classified as follows:

1. Dumping the entire charge.
2. Salt receivers or salt filters.
3. Bucket elevators.
4. Continuous pumping to an external settler.
5. Settlers in a closed-circulation system.

Dumping the entire charge is employed only where the mother liquor is very viscous, and the ratio of mother liquor to crystals is very small. The only place where this method is employed at present is in the "strike pans" in sugar manufacture.

Where small amounts of salt are to be removed, the evaporator may be provided with a cone bottom or hopper bottom closed by a plug cock. Below this cock are connected one or more receivers that can be equalized to the vapor space. During regular operation the cock is open, the receiver is equalized, and salt settles in the receiver as it is formed. When the receiver fills it is closed, cut off from the vapor space, and emptied. For very small amounts of salt a simple receiver may be used, and the entire contents may be pumped to a small filter or settling tank. For larger amounts of salt, the receiver may have a filter bottom built into it. In this case it will have a connection in the bottom for pumping out the liquor, connections in the top for air or steam to dry the salt, and a manhole in the side through which the dried charge can be raked out on to any desired form of conveyor. One ton per hour is the maximum amount of salt that should be handled by this method. More than this amount cannot conveniently be raked out as fast as it is formed in the evaporator.

For very small amounts of salt, where the receiver has to be emptied at intervals of a couple of hours or more, a single receiver is sufficient. The salt may be allowed to accumulate in the cone of the evaporator during the time the receiver is being emptied. For larger amounts of salt, each body should have two receivers so that one is always filling. A common arrangement on a double-effect evaporator is to have three receivers: one under each effect, and one between them that may be connected to either effect. Valves, either globe or gate, should not be used for any salt connections; and it is unwise to use less than a 3-in. plug cock in any case. The cock should be tapped for either water or solution on both sides of the plug to wash it free from salt.

At one time the standard practice in the common salt industry was to prolong the cone of the evaporator into a settling leg and to remove the salt continuously from the bottom of this leg by a bucket elevator. The elevator in such a case must be enclosed up to the normal brine level, so that the contents of the pan will not be lost if the vacuum is broken. The buckets were usually made of perforated metal or wire screen; but the mixture discharged from the elevator was usually very sloppy. The principal disadvantage of elevators is the space and height required, and the fact that they are extremely difficult to keep clean.

At present the most approved method of removing large quantities of salt is to prolong the cone of the evaporator into a settling leg and to remove continuously from the bottom of this leg a stream of salt suspended in solution. This is usually pumped into an overhead settler, from which the salt is drawn to centrifugals or filters and the liquid overflows to the feed tank. In the case of common salt, where the solution from all the effects has the same concentration, one settler may be provided for all effects, all the streams combined, and the overflow from the settler sent to the feed tank. In case the evaporator contains solution of a different concentration in each effect, there may be a separate settler for each effect, and the overflow solution will then be returned directly to the effect from which it came.

The average suspension pumped from the evaporator to the settlers is probably between 1 part salt to 5 parts solution, and 1 part salt to 10 parts solution, by weight. On some materials a 1:1 suspension can be pumped, but there is serious danger of stoppages in the lines when pumping such heavy suspensions. Lines for pumping such suspensions of salt in solution should be designed with rather high velocity (6 ft./sec. or more) and with long-radius bends.

In the case of forced-circulation evaporators, a salt settler may be incorporated in the circulation system itself. The quantities of liquid circulating, however, are so large that these settlers rarely give satisfactory separation. Most forced-circulation evaporators involve the use of a settler through which only a portion of the circulating liquid is diverted. This makes it possible to build settlers of reasonable size yet with sufficient cross section to give good separation. The settler is usually provided with a cone bottom, and the salt is pumped out continuously as in the method of the previous paragraphs.

Scale Formation and Scale Removal. The formation of scale is always due to the presence of a substance having an inverted solubility curve; *i.e.*, one whose solubility decreases with increasing temperature. Scale is not necessarily connected with the presence of materials of very low solubility. The most common scale-forming material is calcium sulfate. Other common scale-forming materials are calcium hydroxide, sodium sulfate, sodium carbonate, and the calcium salts of the organic acids found in sugar-beet juice and fruit juices. When dissolved in a solution in which the amount of CO_2 is constant, calcium carbonate has a normal solubility curve. However, when liquids containing calcium bicarbonate are boiled, as the CO_2 is removed, the calcium carbonate is precipitated in such a way as to form a true scale. When any solid particles are in suspension in the liquid and true scale is forming, such solids are usually included in the scale.

If scale-forming substances are present, there is no way of preventing scale formation. The **rate** of scale formation may be decreased by using high velocities or by introducing seed crystals of the scale-forming material in suspension. Scale cannot be entirely prevented by this method. Removal of scale-forming impurities is often impossible and often too expensive. Consequently the problem is usually the removal of scale or the decrease in the rate of scale formation rather than its prevention.

Scale may often be removed by boiling with suitable reagents. Calcium sulfate scale may be removed by boiling with soda or caustic followed by boiling with hydrochloric acid. Ordinary iron and steel apparatus can be boiled out with 0.5 per cent HCl for reasonable periods without injury. The possibility of corrosion is decreased by the use of acid inhibitors that have been recently developed. Some scales such as sodium sulfate and sodium carbonate can be removed by boiling with water. Sodium or ammonium fluoride may be used for scales high in silica. Phosphate scales or scales containing much Al_2O_3 are exceedingly difficult to remove.

Mechanical methods such as are used for removing scale from boilers may be used where the scale is inside the tubes. This is the case in the standard vertical and similar evaporators. The horizontal-tube evaporator is not suitable for evaporating liquids that form scale, because mechanical methods are not satisfactory for removing scale from the outside of tubes. If chemical methods of removal are satisfactory, the horizontal-tube evaporator may be employed for such liquids. In general, those evaporators having very long and narrow tubes are less suitable for scale removal by mechanical means than those having short and wide tubes.

When true scale is being formed, it may be shown [McCabe and Robinson, *Ind. Eng. Chem.*, **16**, 478 (1924)] that the decrease in the rate of heat transfer may be expressed by the equation

$$\frac{1}{U^2} = AT + B \tag{9}$$

where U is the rate of heat transfer at any particular time; T is the time since the beginning of the operation; A and B are the constants.

This is the equation of a straight line, and therefore the constants may be determined by two observations only, taken at different times; in practice, however, it is better to make several determinations. This equation is of value in predicting the amount of decrease to be expected in the capacity of the evaporator at different times after scale formation begins. Figure 23 shows the results of a test run on Na_2SO_4 plotted according to Eq. (9).

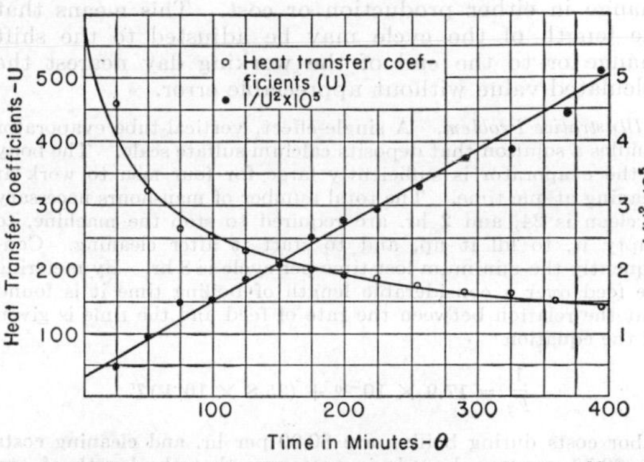

FIG. 23. Rate of scale formation, boiling sodium sulfate in basket-type evaporator. [*Badger and Othmer, Trans. Am. Inst. Chem. Engrs.*, **16**, *Part* II, 164 (1924).]

If it is not convenient to determine the heat-transfer coefficients in practice, any quantity that is proportional to the heat-transfer coefficient may be used. Thus, rate of feed, rate of salt production, or rate of condensate removal also follow such an equation as (9) if all other conditions except scale formation are constant. Hence

$$\frac{1}{F^2} = aT + b \tag{10}$$

where F = rate of feed, lb./hr. at any time; T = time since the beginning of the cycle, hr.; a and b = constants.

If the constants in Eq. (10) have been determined, the length of cycle to give the maximum capacity is given by the equation [Othmer and Badger, *Trans. Am. Inst. Chem. Engrs.*, **16**, 159, Part II (1924); McCabe, *Chem. & Met. Eng.*, **33**, 86 (1926)]

$$X_m = c + \frac{2}{a}\sqrt{abc} \tag{11}$$

where X_m = hr. boiling time per cycle for maximum production; c = time for emptying, cleaning, and refilling, hr.; a and b = constants (determined from test data) from Eq. (10).

Another useful equation is

$$P = \frac{48}{a}\left(\frac{\sqrt{aX + b} - \sqrt{b}}{X + c}\right) \tag{12}$$

where P = total lb. feed per 24 hr.; X = length of boiling time per cycle, hr.

The length of cycle to give minimum cost per unit of evaporation is given by the equation

$$X_n = \frac{M}{S} + \frac{2}{aS}\sqrt{abSM} \tag{13}$$

where X_n = hours of boiling for cycle of minimum cost; M = cost of one cleaning (labor and materials, dollars); S = labor cost during evaporation, dollars per hour.

It is not always convenient to fit the cycle calculated by Eq. (11) or (13) to production schedules. To determine the effect of a variation from the optimum, the following equation is useful:

$$T = \frac{aP}{48}\left(\frac{SX + M}{\sqrt{aX + b} - \sqrt{b}}\right) \tag{14}$$

where T = total operating costs, dollars per hour.

In most cases it will be found that a considerable variation may be made from the values given by Eq. (11) or (13) for the length of the cycle, without much change in either production or cost. This means that the length of the cycle may be adjusted to the shift change or to the end of the working day nearest the calculated value without appreciable error.

Illustrative Problem. A single-effect, vertical-tube evaporator handles a solution that deposits calcium sulfate scale. The body of the evaporator is sufficiently large for four men to work at cleaning at one time. The total number of man-hours necessary to clean is 24, and 2 hr. are required to stop the machine, to empty it, to fill it up, and to start it after cleaning. Consequently the minimum lost time per cycle is 8 hr. By metering the feed over a considerable length of boiling time it is found that the relation between the rate of feed and the time is given by the equation

$$\frac{1}{F_2} = 17.9 \times 10^{-10} + (35.8 \times 10^{-12})T$$

Labor costs during boiling are $0.90 per hr. and cleaning costs are $0.75 per man-hr. It is necessary that the length of one cycle be an integral number of days.

A. What cycle will give the maximum production?

B. If 380,000 lb. feed are to be handled per 24 hr., what cycle will give the minimum costs?

Solution A. Use Eq. (11). The values of the constants are

$$a = 35.8 \times 10^{-12}$$
$$b = 17.9 \times 10^{-10}$$
$$c = 8$$

Substituting in Eq. (11) gives

$$X_m = 48 \text{ hr.}$$

The total length of one cycle, therefore, is 48 + 8 or 56 hr. This is not an integral number of days. Accordingly a part of the curve showing the relation between production and time of boiling is drawn by substituting various values of X in Eq. (12) and solving for P. The resulting curve is shown in Fig. 24. From this it is seen that either 48 hr. or 72 hr. for the total cycle (40 or 64 hr. total boiling time) gives practically the same capacity as the calculated maximum.

Solution B. This is calculated by substituting the following values in Eq. (13)

$$a = 35.8 \times 10^{-12}$$
$$b = 17.9 \times 10^{-10}$$
$$M = 24 \times 0.75 = 18$$
$$S = 0.90$$

The results of the calculation give

$$X_m = 77 \text{ hr.}$$

Since Eq. (13) is independent of P, this does not tell whether or not the most economical cycle gives sufficient capacity for the required production. A reference to Fig. 24, however, shows that for 77 hr. boiling time (85 hr. total time per cycle) the production is well over the required 380,000 lb. To determine the effect on cost of altering the cycle to an even number of days, the curve for cost vs. boiling time is plotted from Eq. (14) with P = 380,000. This curve is also shown in Fig. 24. From this it is seen that a cycle of either 48, 72, or 96 hr. total time (40, 64, or 88 hr. actual boiling time) gives costs not greatly different from the minimum, while from the curve for maximum production the capacities for all these cycles are seen to be above the desired limit. The 40-hr. cycle shows a cost appreciably above the minimum, and the 88-hr. cycle shows that the maximum possible capacity of the machine is practically equal to the desired capacity. Consequently the cycle calling for 64 hr. boiling time, or 72 hr. total time, would be the proper one to adopt.

Salting. In concentrating solutions such as sodium chloride that do not have an inverted solubility curve but do have a solubility curve showing a very slight increase with temperature, the phenomenon known as **salting** occurs. This means a growth of crystals on the heating surface. The decrease in heat transfer caused by this growth does not follow Eq. (9) given on p. 513 for the rate of scale formation. Its cause is obscure. The rate of salting is decreased by high liquor velocities. It will usually be found that the beginning of salting occurs at a time when there has been some marked irregularity in the operation of the evaporator, such as a decrease in steam pressure, an increase in the boiling point, an increase in the liquor level, a stoppage in the circulation, or any other factor that causes a temporary decrease in the rate of boiling. After salting once starts, it usually progresses rather rapidly to the point where the evaporator must be shut down and boiled out.

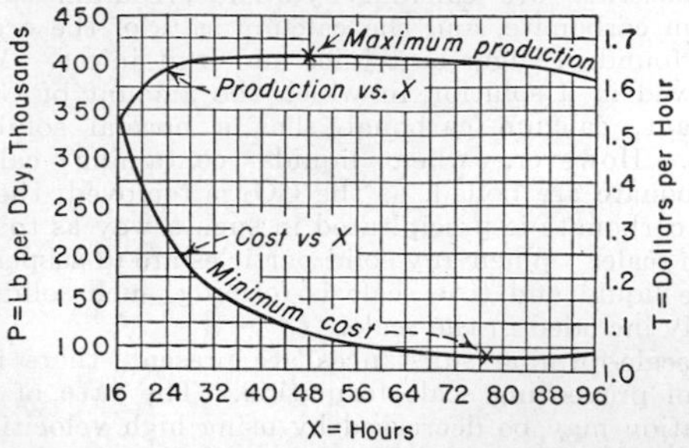

FIG. 24. Determination of optimum cycle for evaporation of material forming scale (solution of illustrative problem). [*McCabe, Chem. & Met. Eng.*, **33**, 87 (1927).]

Foam and Entrainment. These two words are often used together, but they signify two entirely different phenomena. **Entrainment** is the carrying over of drops of liquid from the evaporator, due to the fact that the vapor velocity is greater than the rate of settling of the drops. **Foam** implies some special property of the liquid that causes the formation of a mass of stable bubbles.

If a drop is ejected from the surface of a liquid into a body of vapor, it will rise to a certain height and then begin to fall. As it falls, it will finally reach a constant rate of fall given by Stokes's law. If the velocity of its ejection is so great that is does not begin to fall until it has risen to the vapor outlet, then it will be carried over. If the drop is so small that its terminal velocity of fall is less than the upward velocity of the vapor, it will be carried off with the vapor. Hence, increasing the vapor space above a boiling liquid decreases entrainment of the larger drops but has no effect on very small drops. It is *usually* considered that from 20 to 40 lb. evaporation per hr. per sq. ft. superficial liquid surface is as much as should be allowed if excessive entrainment is to be avoided. Power plant evaporators utilize rates much higher. Entrainment is probably connected with the surface tension of the liquid also. Vorkauf [*Forsch. Gebiete Ingenieurw., Forschungsheft,* **341** (1931)] has studied entrainment in steam boilers, but the experiments were carried out under such conditions that they are not directly applicable to evaporators. [See also Melhardt, *Chem. App.*, **18**, 165 (1931).]

Any ordinary type of steam separator may be used as an entrainment separator. Many of them are not especially effective. The essential characteristics of a successful entrainment separator are: (1) It should cause the vapor to turn a sharp corner at a high velocity, or impart a very rapid whirling motion to the vapor. In other words, successful entrainment separators are based on the fact that drops of liquid have a greater momentum than the vapor in which they are suspended.

(2) The separator must be so arranged that the liquid that has collected on a surface does not drop into a stream of high-velocity vapor. One entrainment separator that does not fulfill either of these requirements but is exceptionally effective is the umbrella-shaped baffle used in forced-circulation evaporators. When properly designed, the high-velocity curtain of liquid issuing from the edge of the baffle effectively scrubs the vapor free from entrainment. (3) It must cause a low pressure drop.

The cause of **foam** is obscure. It involves the presence of substances in solution that cause a change in surface tension, and it also involves the presence of suspended materials that stabilize the surface film. Pure liquids do not foam, even if finely divided solids are suspended in them. Solutions will not foam unless finely divided suspended material is present. Colloidal materials often fulfill both functions, and most solutions or suspensions of colloids will foam.

Obviously, foam may be prevented by removing all suspended solids and colloidal material. This is rarely practical. On the other hand, if a substance tends to foam, it requires very careful operation of the evaporator, and no single method is available that will surely break all foam. Operating with a low liquor level is often satisfactory. Steam-heated tubes in the upper part of the liquor space will sometimes break foam if it begins to form. Steam jets directed against the surface of a foam blanket will sometimes break the foam. The most successful evaporators for handling foam are those in which the liquid is made to impinge at very high velocities against a baffle plate, so that the foam is broken mechanically. The forced-circulation evaporator has been successful in this respect. Foam can also be controlled by injecting various oils. The fatty oils are better than petroleum oils, and free fatty acids are better than oils. Castor, cottonseed, and coconut oils have been used, and also the fatty acids prepared from these oils. Sulfonated castor oil is probably the most effective but is too expensive for most cases. The reason for the effect of such oils is unknown.

Materials of Construction. Evaporators have been made of practically every common material of construction. The greater proportion of them are of iron and steel. Cast iron is often preferred to plate steel because of its superior resistance to mild corrosion. Very large evaporators in cast iron are much more expensive than plate steel; hence in recent years all large evaporators for non-corrosive materials have been made of plate steel. Practically all the types described can be built of cast iron, or at least with those parts in contact with the liquid (except tubes) of cast iron. Tubes are usually iron or copper. No satisfactory form of cast-iron heating surface has ever been developed. Evaporators may also be built of copper, aluminum, lead, and bronze and similar alloys. Lead evaporators should have the body made of solid lead casting (not lead-lined). The tubes of such evaporators are usually lead-covered copper or Karbate. Nickel has been widely used for the heating surface of caustic evaporators, and nickel-lined bodies have been made for the same purpose. Enamel is not satisfactory, and no apparatus has been developed in this material, mainly because of the impossibility of making enameled tubes of small diameter. Satisfactory evaporators can be built with pyrex glass heating surfaces. In recent years there has been a rapidly growing tendency to use some of the various forms of chrome-nickel steels in evaporator construction. As more information is available regarding the properties of such steels and suitable methods of fabricating, this use will probably increase. Where no other material will serve the purpose, evaporator bodies may be lined with acidproof brick set in acidproof cement.

MULTIPLE-EFFECT EVAPORATOR CALCULATIONS

The calculations for single-effect evaporators have been discussed on p. 507.

General Principles. In this discussion of **multiple-effect calculations** the terminal temperatures will be assumed to be fixed, and the heat-transfer coefficient will be assumed to be known for each effect. In such a case the relative heating surfaces in the different effects and the temperature drops in the different effects are mutually interdependent. Fixing the ratios of the heating surfaces fixes the temperature distribution and vice versa.

The calculations are usually carried out on the basis of a fixed ratio of heating surfaces in the different effects. By far the commonest case is where all effects are to have the same heating surface. A set of equations is written involving a heat balance over each effect. A set of temperature differences is chosen at random, the heat-balance equations are solved on the basis of such temperatures, and the quantities of heat to be transmitted per effect are calculated. On the basis of these quantities of heat, the heat-transfer coefficients, the assumed temperature differences, and the heating surfaces are calculated. If these do not agree with the ratio originally assumed, a new set of temperatures is chosen and the process repeated till the required assumptions are fulfilled.

There are so many possible evaporator flow sheets that complete mathematical solutions for all cases cannot be given. A few typical sets will indicate the procedure by which the reader may develop his own solution for any particular case in hand.

Symbols Used.

Let I, II, III, IV = first, second, third, and fourth effect.

F = lb. thin liquor fed per hr.

E = lb. water evaporated per hr.

V = lb. steam to first effect per hr.

W, X, Y, Z = lb. water evaporated per hr. in I, II, III, IV.

A_1, A_2, A_3, A_4 = heating surface in I, II, III, IV, sq. ft.

t_0 = saturation temperature of steam in heating space of I, °F.

t_1, t_2, t_3, t_4 = saturation temperature of vapor from I, II, III, IV, °F.

$t_I, t_{II}, t_{III}, t_{IV}$ = boiling point of liquor in I, II, III, IV, °F.

t_f = temperature of feed, °F.

C_f = specific heat of feed, °F.

C_1, C_2, C_3, C_4 = specific heat of liquor in I, II, III, IV.

R_1, R_2, R_3, R_4 = B.t.u. lost per hr. by radiation from I, II, III, IV.

H_1, H_2, H_3, H_4 = heat of crystallization and concentration, B.t.u./hr. in I, II, III, IV.

L_0, L_1, L_2, L_3, L_4 = latent heat of vaporization of steam, B.t.u./lb. at t_0, t_1, t_2, t_3, t_4.

H_0 = heat present in heating steam as superheat, above t_0, B.t.u./lb.

$H_I, H_{II}, H_{III}, H_{IV}$ = heat present as superheat in vapor, B.t.u./lb., above t_1, t_2, t_3, t_4.

S_1, S_2, S_3, S_4 = lb. salt and liquor removed from I, II, III, IV.

U_1, U_2, U_3, U_4 = heat-transfer coefficients, B.t.u./(sq. ft.)(°F.)(hr.) in I, II, III, IV.

The following assumptions will be made:

1. Condensate leaves the steam chest at the saturation temperature of the steam.
2. Any salt withdrawn from any effect, together with mother liquor accompanying it, is at the same temperature as the boiling liquor in that effect.
3. Radiation is from liquor and vapor surfaces; *i.e.*, W, X, Y, and Z are net quantities of vapor actually condensed in the next effect.
4. Superheat in vapor does not affect the temperature difference in the next effect.
5. The sum of heat effects other than latent heat of vaporization (heat of crystallization and of concentration) is negative; *i.e.*, the net effect is an absorption of heat. If the reverse is true, the signs of the $H_1, \ldots H_4$ terms are to be reversed.
6. The coefficients used are corrected for elevation in boiling point but not for hydrostatic head.
7. Boiling-point elevations are known for the liquor in question at all concentrations and all pressures.

Case I. Quadruple-effect Evaporator with Forward Feed.

Heat-balance equations:

Across I:

$$V(H_0 + L_0) + FC_f(t_f - t_I) = W(L_1 + H_I) + H_1 + R_1 \tag{15}$$

Across II:

$$W(L_1 + H_I) + C_1(F - W - S_1)(t_I - t_{II}) = X(L_2 + H_{II}) + H_2 + R_2 \tag{16}$$

Across III:

$$X(L_2 + H_{II}) + C_2(F - W - X - S_1 - S_2)(t_{II} - t_{III}) = Y(L_3 + H_{III}) + H_3 + R_3 \tag{17}$$

Across IV:

$$Y(L_3 + H_{III}) + C_3(F - W - X - Y - S_1 - S_2 - S_3)(t_{III} - t_{IV}) = Z(L_4 + H_{IV}) + H_4 + R_4 \tag{18}$$

Material-balance equation:

$$E = W + X + Y + Z \tag{19}$$

The steps in the solution are as follows:

a. F, E, t_0, t_4, R_1, R_2, R_3, R_4, C_f, t_f, are usually given as basic data.

b. Values are assumed for t_1, t_2, and t_3.

c. Values are assumed for W, X, Y, and Z.

d. On the basis of steps *b* and *c*, values are calculated for t_I, t_{II}, t_{III}, t_{IV}, and C_1, C_2, C_3, C_4; S_1, S_2, S_3, S_4; H_1, H_2, H_3, H_4; L_1, L_2, L_3, L_4; H_I, H_{II}, H_{III}, H_{IV}.

e. Substituting these values from steps *b* and *d* in Eqs. (15) to (19) gives five equations with five unknowns. They may then be solved for V, W, X, Y, and Z.

f. If these values are greatly different from those assumed in step *c*, steps *c* to *e* are repeated till a check is secured. This rarely takes more than one adjustment.

g. Heating surfaces per effect are then calculated according to the equations:

$$A_1 = \frac{V(H_0 + L_0)}{U_1(t_0 - t_I)} \tag{20}$$

$$A_2 = \frac{W(L_1 + H_I)}{U_2(t_1 - t_{II})} \tag{21}$$

$$A_3 = \frac{X(L_2 + H_{II})}{U_3(t_2 - t_{III})} \tag{22}$$

$$A_4 = \frac{Y(L_3 + H_{III})}{U_4(t_3 - t_{IV})} \tag{23}$$

h. If the values so calculated for the heating surfaces do not correspond with the desired ratio, the temperatures t_I, t_{II}, t_{III}, and t_{IV} must be readjusted. This may usually be done by assuming that this readjustment does not greatly affect W, X, Y, and Z. Hence the readjustment may be made on the basis of Eqs. (20) to (23), until a set of values for t_I, t_{II}, t_{III}, and t_{IV} is found that give values for the A's satisfying the desired conditions.

j. The final values of t_I, t_{II}, t_{III}, and t_{IV} give corresponding values of t_1, t_2 and t_3. Steps *d* to *g* are then repeated on this basis. In most cases this will give values for the A's that satisfy the desired conditions. In exceptional cases the whole procedure may have to be repeated several times to secure the desired balance.

In most cases in practice one or more simplifying factors appear that reduce the complexity of the above calculations. There may be no salt separated; the heat of dilution may be negligible; a constant specific heat may be used; there may be so little elevation in boiling point that $t_I = t_1$, $t_{II} = t_2$, etc., and the terms for H_I, H_{II}, etc., disappear. A case in which all the terms involved above must actually be used is rare, though the concentration of electrolytic caustic soda solutions from cell liquor to high densities will involve almost the entire calculation outlined above. If all the data such as those shown in Fig. 2, p. 502, are available, the values for U must also be readjusted every time temperatures and temperature differences are readjusted. Such data are rarely available, and still more rarely may any evaporator calculation be carried out so accurately as to warrant such refinements.

Since the heating surfaces and the temperature differences are interdependent, it is not a necessary condition, in the readjusting outlined in steps *h* and *j*, that temperature differences be adjusted to give a required ratio of heating surfaces, though this is usually the problem. It is much cheaper to build a multiple-effect evaporator with all bodies alike than to introduce even small differences. In exceptional cases, however, the problem may involve adjusting the heating surfaces to give a desired set of temperature differences, or to give some definite temperature in some one effect. The method of making such adjustments is obviously similar to that outlined above.

Case II. Quadruple-effect Evaporator with Backward Feed.

Heat-balance equations:

Across IV:

$$Y(L_3 + H_{III}) = Z(L_4 + H_{IV}) + H_4 + R_4 + FC_f(t_{IV} - t_f) \tag{24}$$

Across III:

$$X(L_2 + H_{II}) = Y(L_3 + H_{III}) + H_3 + R_3 + C_4(F - Z - S_4)(t_{III} - t_{IV}) \tag{25}$$

Across II:

$$W(L_1 + H_I) = X(L_2 + H_{II}) + H_2 + R_2 + C_3(F - Z - Y - S_4 - S_3)(t_{II} - t_{III}) \tag{26}$$

Across I:

$$V(L_0 + H_0) = W(L_1 + H_I) + H_1 + R_1 + C_2(F - Z - Y - X - S_4 - S_3 - S_2)(t_I - t_{II}) \tag{27}$$

Material-balance equation:

$$E = W + X + Y + Z \tag{28}$$

The equations for the heating surfaces in the various effects are the same as before [Eqs. (20) to 23)].

Case III. Quadruple-effect Evaporator Fed in the Order II-III-IV-I.

Heat-balance equations:

Across I:

$$V(L_0 + H_0) = W(L_1 + H_I) + H_1 + R_1 + C_4(F - X - Y - Z - S_2 - S_3 - S_4)(t_I - t_{IV}) \tag{29}$$

Across II:

$$W(L_1 + H_I) = X(L_2 + H_{II}) + H_2 + R_2 + FC_f(t_{II} - t_f) \qquad (30)$$

Across III:

$$X(L_2 + H_{II}) + C_2(F - X - S_2)(t_{II} - t_{III}) = Y(L_3 + H_{III}) + H_3 + R_3 \qquad (31)$$

Across IV:

$$Y(L_3 + H_{III}) + C_3(F - X - Y - S_2 - S_3)(t_{III} - t_{IV}) = Z(L_4 + H_{IV}) + H_4 + R_4 \qquad (32)$$

Material-balance equation:

$$E = W + X + Y + Z \qquad (33)$$

Similar heat-balance equations may be written for any number of effects and for any arrangement.

Example. It is desired to concentrate a certain solution from 10 to 50 per cent solids in a triple-effect evaporator. Steam is available at 15 lb./sq. in. gage, and a vacuum of 26 in. referred to a 30-in. barometer is to be maintained in the third effect. The feed to the evaporator is 55,000 lb./hr. at 70°F. The solution has a negligible elevation in boiling point, and its specific heat may be taken as 1.00 at all concentrations. No salt or other solid separates on evaporation, and the heat of dilution is negligible. Condensate may be assumed to leave the steam chests at the saturation temperature of the steam. Radiation may be neglected. Heat-transfer coefficients may be assumed to be 550, 350, and 200 B.t.u./(sq. ft.)(hr.)(°F). All bodies are to be of the same size. Calculate the steam consumption, the heating surface of the evaporators, the temperature distribution, and the total heat going to the condenser when the evaporator is operated with forward feed.

Solution. The table on p. 501 shows that to concentrate 1000 lb. feed from 10 to 50 per cent solids requires an evaporation of 800 lb. water. Hence the evaporation in this case will be 55 × 800 or 44,000 lb.

Equations (15), (16), (17), and (19) cover this case. The terms S_1, S_2, S_3; H_I, H_{II}, H_{III}; H_1, H_2, H_3; R_1, R_2, R_3, and all terms involving z disappear. The terms C_1, C_2, C_3, all are equal to unity. Also, $t_1 = t_I$, $t_2 = t_{II}$, and $t_3 = t_{III}$. The equations become

$$VH_0 + F(t_f - t_1) = WL_1 \qquad (15a)$$

$$WL_1 + (F - W)(t_1 - t_2) = XL_2 \qquad (16a)$$

$$XL_2 + (F - W - X)(t_2 - t_3) = YL_3 \qquad (17a)$$

$$X + Y + W = 44{,}000 \qquad (19a)$$

The steps in the solution, given on p. 516, are as follows:

Step a. The data given are:

$$F = 55{,}000 \text{ lb.} \qquad t_0 = 249°\text{F.}$$
$$E = 44{,}000 \text{ lb.} \qquad t_3 = 125°\text{F.}$$
$$t_f = 70°\text{F.}$$

Step b. The values to be assumed for t_1 and t_2 may be chosen by noting that, except for flash or for heating feed, the temperature differences are inversely proportional to the heat-transfer coefficients. In this case, the first effect has a heavy heating load, so its temperature difference will be larger than that given by the above reasoning. The total available temperature difference is $t_0 - t_3 = 249 - 125 = 124°$F. Assume, as a first approximation, that the temperature differences in the three effects are 38°, 33°, and 53°, respectively. The conditions will then be:

Steam to first effect	249°	$= t_0$
Boiling point in first effect	211°	$= t_1$
Temperature difference in first effect	38°	$= \Delta t_1$
Vapor to second effect	211°	$= t_1$
Boiling point in second effect	178°	$= t_2$
Temperature difference in second effect	33°	$= \Delta t_2$
Vapor to third effect	178°	$= t_2$
Boiling point in third effect	125°	$= t_3$
Temperature difference in third effect	53°	$= \Delta t_3$

Step c. Since all properties of the liquor in this problem are independent of its concentration, this step and most of step *d* are unnecessary.

Step d. From the steam tables

$$L_0 = 946.0 \text{ B.t.u./lb.}$$
$$L_1 = 971.0 \text{ B.t.u./lb.}$$
$$L_2 = 991.0 \text{ B.t.u./lb.}$$
$$L_3 = 1021.6 \text{ B.t.u./lb.}$$

Step e. Substituting in Eqs. (15*a*) to (19*a*) gives

$$946V + 55{,}000(70 - 211) = 971W$$
$$971W + (55{,}000 - W)(211 - 178) = 991X$$
$$991X + (55{,}000 - W - X)(178 - 125) = 1021.6Y$$
$$W + X + Y = 44{,}000$$

Solving these equations gives

$$V = 22{,}150 \text{ lb.} \qquad W = 13{,}595 \text{ lb.}$$
$$X = 14{,}710 \text{ lb.} \qquad Y = 15{,}695 \text{ lb.}$$

Step f. This is unnecessary for the present problem.

Step g. Equations (20), (21), and (22) become

$$A_1 = \frac{VL_0}{U_1(t_0 - t_1)} \qquad (20a)$$

$$A_2 = \frac{WL_1}{U_2(t_1 - t_2)} \qquad (21a)$$

$$A_3 = \frac{XL_2}{U_3(t_2 - t_3)} \qquad (22a)$$

Substituting in these equations gives

$$A_1 = 1003 \text{ sq. ft.}$$
$$A_2 = 1141 \text{ sq. ft.}$$
$$A_3 = 1375 \text{ sq. ft.}$$

Step h. Since these heating surfaces do not conform to the condition that all three effects must be of equal size, the temperature differences must be readjusted. This may be done approximately by trying various values for Δt_1, Δt_2, and Δt_3 in Eqs. (20*a*) to (22*a*) till the heating surfaces so calculated are substantially equal. It is not necessary to work to fractions of a degree, for no evaporator calculation may be made with that degree of precision. After several trials it is found that temperature differences of 32°, 32°, and 60° are the nearest, thus:

$$A_1 = 1003 \times {}^{38}\!/_{32} = 1191 \text{ sq. ft.}$$
$$A_2 = 1141 \times {}^{33}\!/_{32} = 1176 \text{ sq. ft.}$$
$$A_3 = 1375 \times {}^{53}\!/_{60} = 1215 \text{ sq. ft.}$$

Step j. The revised conditions are:

Steam to first effect	249°	$= t_0$
Boiling point in first effect	217°	$= t_1$
Temperature difference in first effect	32°	$= \Delta t_1$
Vapor to second effect	217°	$= t_1$
Boiling point in second effect	185°	$= t_2$
Temperature difference in second effect	32°	$= \Delta t_2$
Vapor to third effect	185°	$= t_2$
Boiling point in third effect	125°	$= t_3$
Temperature difference in third effect	60°	$= \Delta t_3$

$$L_1 = 967.2$$
$$L_2 = 986.9$$

The revised equations are:

$$946V - 55{,}000(70 - 217) = 967.2W$$
$$967.2W + (55{,}000 - W)(217 - 185) = 986.9X$$
$$986.9X + (55{,}000 - W - X)(185 - 125) = 1021.6Y$$
$$W + X + Y = 44{,}000$$

From these and from Eqs. (20*a*) to (22*a*), the following values are obtained:

$$V = 22{,}540 \text{ lb.} \qquad W = 13{,}670 \text{ lb.}$$
$$X = 14{,}640 \text{ lb.} \qquad Y = 15{,}720 \text{ lb.}$$
$$A_1 = 1212 \text{ sq. ft.}$$
$$A_2 = 1182 \text{ sq. ft.}$$
$$A_3 = 1205 \text{ sq. ft.}$$

The heating surface per effect may be taken as the average of these values, or 1200 sq. ft. The heat going to the condenser is $1021.6Y$, or 16,060,000 B.t.u./hr.

Economical Number of Effects. This question is often decided by the temperature range that is available. If the pressure of the exhaust steam available for heating is fixed, and if the degree of vacuum is also fixed, there is a certain available temperature difference that cannot be altered by the design of the evaporator. From this, as mentioned in the paragraph on effect of boiling-point elevation (p. 509), the sum of all the elevations in boiling point must be subtracted. The amount of the net remaining temperature difference often limits the number of effects possible. A natural-circulation evaporator of any type has a certain minimum temperature difference below which it will not operate. If the temperature difference is lower than this, the evolution of steam bubbles is not sufficient to start a general circulation so that the liquid remains quiet with only a

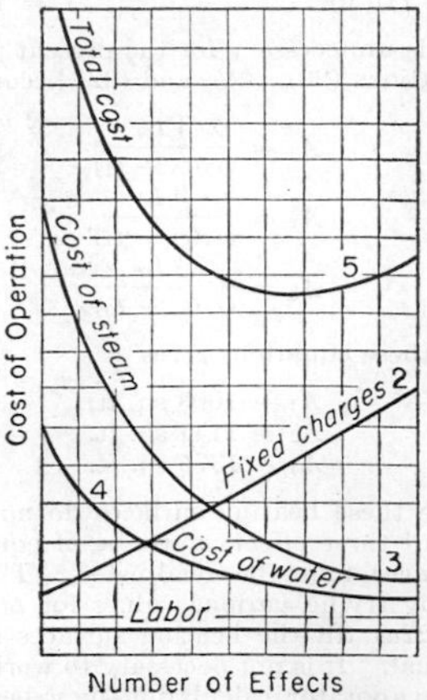

FIG. 25. Cost of operation of evaporators—determination of optimum number of effects.

slight simmering in the surface layers. This minimum temperature difference varies with the arrangement of the heating surfaces and the viscosity of the liquid. It is usually somewhere between 10° and 20°F. for ordinary liquids, and very much higher for more viscous liquids. The forced-circulation evaporator, on the other hand, is not dependent on convection and therefore will operate on temperature differences as low as may be desired.

Granting that there is sufficient available temperature difference or that operating conditions may be controlled by the designer, some particular number of effects will show the minimum cost per pound of water evaporated. It may be assumed that the labor cost is constant, irrespective of the number of effects. It may also be assumed that repairs, maintenance, and depreciation will be a constant percentage of the fixed cost, irrespective of the number of effects. In Fig. 25, curve 1 shows constant costs, such as labor. Curve 2 shows fixed charges. These are not proportional to the number of effects because, as shown in Figs. 17, 18, and 19, the cost per square foot of heating surface is decidedly greater for a single effect than for multiple effects; and this cost also decreases the more slowly, as the number of effects or the size of the unit increases. Curve 3 shows the cost of steam, which, it will be noted, obeys the law of diminishing returns; as does also curve 4, the cost of water used. The total cost per pound water evaporated is therefore the sum of curves 1, 2, 3, and 4, as shown in curve 5. This ordinarily determines a marked minimum cost for some one arrangement.

Figure 25 must obviously be based on the same method of heating feed in all cases. To determine the optimum arrangement, the calculations of Fig. 25 (which have all been made for, say, forward feed) must be repeated for all other appropriate feeding methods, and the various minimums so determined are then compared to find the optimum arrangement and the optimum number of effects.

Example. It is desired to concentrate a certain solution from 10 to 50 per cent solids in a standard vertical-tube evaporator. Steam is available at 15 lb./sq. in. gage, and a vacuum of 26 in. referred to a 30-in. barometer is to be maintained on the last effect. The feed is 55,000 lb./hr. at 70°F. The solution has a negligible elevation in boiling point, and its specific heat may be taken as 1.00 at all concentrations. No salt or other solid separates on evaporation, and the heat of dilution is negligible. Condensate may be assumed to leave the steam chests at the saturation temperature of the steam. Coefficients are to be taken from Fig. 2, and the costs from Fig. 17. All bodies are to be the same size. Fixed charges will be 25 per cent of the first cost per year, exhaust steam at 15 lb./sq. in. gage is worth 10¢ per 1000 lb., and water costs 2¢ per 1000 gal. The evaporator operates 24 hr. per day, 300 days per year. What is the optimum number of effects with forward or backward feed?

Solution. The problems are solved by the method given on p. 516. The final results show a slight variation in heating surface between the different cases, owing to the particular heat-transfer coefficients selected. These have been averaged, and all cases of forward feed are calculated on 1150 sq. ft. per effect, and all cases of backward feed on 1100 sq. ft. per effect. The single effect gives, of course, the same results in both cases A summary is given in Table 4, and the results are plotted in Fig. 26. This shows that for forward feed a quintuple effect is the optimum, but for backward feed a larger number of effects than five will be needed to give the minimum costs.

The problem has not been completed because it is not representative. The simplifying assumptions that have been made for the sake of shortening the calculations make this case deviate quite widely from results obtained in practice. In any actual case the optimum number of effects will be less than in this example, mainly because in practice almost any material that will be concentrated to a 50 per cent solution will have some rise in boiling point, and for most materials this rise will be considerable. This decreases the total temperature difference as the number of effects is increased, thus

Table 4. Calculation of Optimum Number of Effects

	Forward feed					Backward feed			
	Single	Double	Triple	Quadruple	Quintuple	Double	Triple	Quadruple	Quintuple
Steam used, lb./hr.	50,710	29,120	22,540	18,260	16,550	27,460	19,160	14,930	12,300
Lb. evaporated per lb. steam	0.869	1.51	1.95	2.41	2.66	1.60	2.30	2.95	3.58
Lb. evaporated per effect	0.869	0.730	0.650	0.602	0.530	0.800	0.767	0.762	0.716
Water, gal./hr.	134,950	69,620	48,200	38,950	31,630	61,430	35,660	22,120	13,810
Surface per effect, sq. ft.	1,150	1,150	1,150	1,150	1,150	1,100	1,100	1,100	1,100
Cost per sq. ft.	$6.70	$5.20	$4.20	$4.10	$4.00	$5.20	$4.20	$4.10	$4.00
Total surface, sq. ft.	1,150	2,300	3,450	4,600	5,750	2,200	3,300	4,400	5,500
First cost	$7,705	$11,960	$14,490	$18,860	$23,000	$11,440	$13,860	$18,040	$22,000
Fixed charges, per year	$1,926	$ 2,990	$ 3,622	$ 4,715	$ 5,750	$2,860	$ 3,465	$ 4,510	$ 5,550
Steam cost, per year	$36,511	$20,966	$16,229	$13,147	$11,916	$19,771	$13,795	$10,750	$ 8,856
Water cost, per year	$19,433	$10,022	$ 6,941	$ 5,609	$ 4,558	$ 8,846	$ 5,135	$ 3,186	$ 1,989
Total cost, per year	$57,870	$33,978	$26,792	$23,471	$22,224	$31,477	$22,395	$18,446	$16,345
Credit: Heat in thick liquor above 125°F., B.t.u.	......	......	......	......	......	781,000	979,000	1,089,000	1,155,000

decreasing both the working temperature difference and the heat-transfer coefficients for the whole evaporator and greatly increasing the surface needed per effect as the number of effects is increased.

It should be noted that the statement "1 lb. steam evaporates 1 lb. water in a single effect, 2 lb. in a double, etc.," is not even approximately true for such a case as this, with very cold feed. Here again this problem is not representative, for in practice a feed heater would greatly decrease apparatus costs, since a given surface in a heater is less expensive than in an evaporator. This problem shows how the ratio of evaporation to steam consumption varies with the method of feeding.

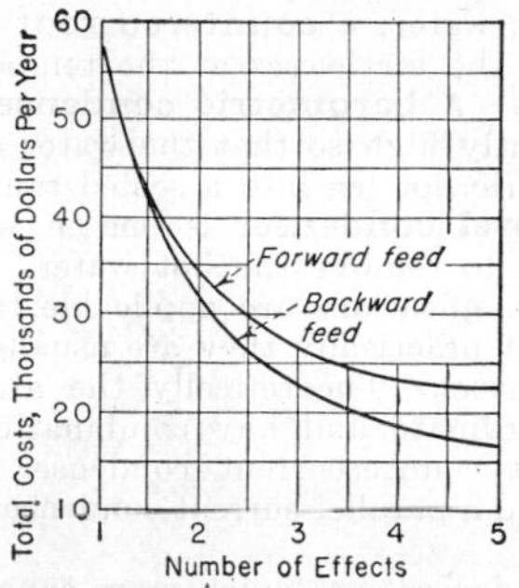

FIG. 26. Total costs of operation of evaporators for different number of effects and for forward and backward feeds.

In forward feed, the thick liquor always leaves at the temperature of the last effect, and hence at a constant temperature in all these cases. In backward feed, the thick liquor leaves at the temperature of the first effect. If it is to be used in another step in the process where this heat is of value, this constitutes a credit to the evaporator which has not been evaluated in this case. If, however, the thick liquor must be cooled after evaporation (as in crystallization processes), this extra heat in the thick liquor in backward feed may be a debit to the evaporator.

Extra Steam. Vapor from an evaporator is usually the most economical form of process steam. The savings accomplished by the use of such steam increases as it is drawn from later effects. The disadvantage of this method is that vapors from vacuum evaporators are often at too low a temperature to be useful. As evaporators are operated at higher pressures, this feature is of more value.

The steam so withdrawn is called **"extra steam"** and is easily taken into account in writing the heat-balance equations. By this means a calculation may be made of the disturbance caused by such vapor withdrawals. If moderate amounts (as compared with the total vapor generated in the effect in question) are withdrawn, the disturbances are slight, and no special provision need be made other than a connection to the vapor space. If the amounts of extra steam are large compared with the total amount of vapor generated, additional heating surface may have to be provided in the effect from which the vapor is withdrawn. Otherwise the result may be that other effects are robbed of temperature difference and their rate of evaporation too greatly decreased.

Location of Operating Difficulties. On p. 509 it was shown that (if minor factors are neglected) the heat-balance equations for a multiple-effect evaporator finally simplify into

$$U_1A_1\,\Delta t_1 = U_2A_2\,\Delta t_2 = U_3A_3\,\Delta t_3 = \cdots U_nA_n\,\Delta t_n \quad (7a)$$

Hence, for equal heating surfaces, the temperature differences are, roughly, inversely proportional to the heat-transfer coefficient. Any effect in which an undue amount of heating is done, or from which large amounts of extra steam are withdrawn, will deviate from Eq. (7*a*), but the extent and direction of such deviations can be calculated from the heat-balance equations or determined by observation when the evaporator is known to be working properly.

When a given evaporator is operating continuously and uniformly, with constant coefficients, the temperature distribution between effects will be nearly constant. If any pronounced change in the temperature distribution occurs, it is due to a change in the heat-transfer coefficients, and Eq. (7*a*) shows that any effect showing an unduly high temperature difference has an unduly low heat-transfer coefficient. Hence, when an evaporator shows diminished capacity, the temperature distribution at once gives an indication of where to look for the source of the difficulty.

Thermocompressors. The total heat content of 1 lb. vapor leaving an evaporator body is only slightly less than that of the steam used to heat that body; and only slightly lower in temperature in many cases. For over 80 years attempts have been made to devise systems in which the vapors could be compressed to a point where they might be used as heating steam. The advantage would be **multiple-effect economy in a single-effect evaporator.**

Compression may be accomplished by steam-jet ejectors, or by either reciprocating or centrifugal compressors. The former need very high pressure actuating steam to be efficient; the latter are very expensive in the sizes needed for evaporators. In both cases, the power (or steam) used for compression increases rapidly with an increase in the temperature range through which the vapor is compressed. This means that, if the evaporator is to be of reasonable size, the temperature difference is sufficiently large so that the economy of vapor recompression is lost. If the temperature difference is to be kept small so that multiple-effect economy is assured, the body is so large that the steam savings are negligible, owing to the increased first cost. If electric power to drive the centrifugal compressor is very cheap, such a system may show possibilities. In general, few plants in the United States will find thermocompression economical.

Two exceptions to this statement should be noticed. Consider a multiple-effect evaporator so arranged that the temperature difference across the first effect is small (about 10°F.). Suppose such an evaporator uses considerable amounts of high-pressure make-up steam, throttled at the evaporator. In such a case it would be advantageous to provide a steam-jet vapor compressor, allow the make-up steam to expand through the nozzle and entrain what vapor it would, instead of wasting the work of expansion in a throttle valve.

The second case is the evaporation of thermally sensitive liquids (such as fruit juices) that must be evaporated at the lowest possible temperature in order to avoid damage. This eliminates the possibility of a multiple-effect evaporator. Such liquids may be concentrated in a single-effect thermocompression evaporator, thus maintaining all the liquid at the desired low boiling point, but obtaining probably about double-effect economy.

Thermocompressor calculations are discussed by Pridgeon [*Chem. & Met. Eng.*, **28,** 1109 (1923)], and Badger [*Chem. & Met. Eng.*, **28,** 73 (1923); *Trans. Am. Inst. Chem. Engrs.*, **14,** 221 (1923)] has described thermocompression evaporators and given a bibliography up to 1923.

EVAPORATOR ACCESSORIES

Pumps. Evaporator feed pumps call for no special comment. Removal of material from an effect under vacuum, whether thick liquor or intermediate liquors, cannot be easily accomplished with ordinary centrifugal pumps. Even for withdrawing liquids from an effect under moderate vacuum, it is important that the packing be kept in first-class condition or the pumps will become air-bound. Self-priming centrifugal pumps are available for this service. For removing extremely heavy and viscous liquors from the finishing effect, centrifugal pumps are not suitable. Rotary positive-pressure pumps have been used for this service, but the preferred type is the so-called magma pump. This is a long-stroke slow-speed plunger pump with no suction valve. The thick liquid to be removed flows directly into the suction chamber, which must be equalized to the vapor space.

The removal of condensate is best accomplished by means of reciprocating pumps or centrifugals of the self-priming type. Steam traps may be used but are likely to be less positive in their action than pumps. Usually it is desirable to vent non-condensed gas through the condensate connection. When reciprocating pumps are used, this may be done by making the pump oversized. Centrifugal pumps of the self-priming type are not usually vented. Steam traps of the bucket or float type may easily be vented; but traps of the tilting-drum type call for a vent connection in the drum, and the movement of the drum makes vent connections inconvenient. Any steam space that is above atmospheric pressure may be vented directly to the air. When removing condensate from evaporators, the steam spaces of which are under vacuum, the vent lines may go to the vapor space of the body being drained if the amount of air to be removed is small. If considerable air is present (as, for instance, in beet-sugar evaporation), the vents should all be connected directly to the condenser. A simple equalizing line between the pump or trap and the steam space being drained is theoretically sufficient, but is not recommended in practice.

Single-effect evaporators and the first effect of multiple-effect evaporators should furnish uncontaminated condensate, and this should be returned as boiler feed. Except for losses due to leaks and blowdowns, this condensation should be equal to the weight of steam condensed in the evaporator. In any case, the return of such condensate will account for 90 per cent or more of the steam supplied to the evaporator. Condensate from the second effect, if free from contamination due to entrainment, may be used for the remainder of the make-up. In most evaporators of two effects or more, part of the condensate from the second effect and all the condensate from later effects will be wasted unless it is of value as hot process water. Systems for handling such drips may be classified as follows:

1. A pump or trap on each effect, discharging in parallel to a common hot well.
2. A pump or trap on each effect, discharging in series.
3. A common closed hot well into which all effects discharge directly, and from which drips are removed by a pump or return trap.
4. Pumps or non-return traps discharging from each effect to a closed hot well, from which condensate is withdrawn by a pump or return trap.

Method 1 is the commonest and is probably the best practice. Method 2 is undesirable because the drop in pressure between effects causes flash, which tends to vapor-bind the later pumps or traps. This may be avoided by flash pots, but these are frequently more trouble than they are worth. Method 3 is the simplest of all but requires at least 30 ft. between the top of the hot well and the bottom of the steam space of the last effect. Method 4 has the same disadvantage as method 3, is more complicated, and has no features to recommend it. In methods 3 and 4 the closed hot well must be vented to the condenser.

Condensers. Condensers may be classified as follows: A **surface condenser** is one in which the cold water and the vapor to be condensed are separated by metal surfaces; a **jet condenser** is one in which water and vapor are in direct contact. A **wet condenser** is one in which the same pump removes both air and hot water; a **dry condenser** is one in which a separate pump is used for air alone. A **parallel-current condenser** is one in which the air leaves at the temperature of the hottest water; a **countercurrent condenser** is one in which the air leaves at the temperature of the coldest water. A **barometric condenser** is one that is set sufficiently high so that the water drains from it by a barometric hot leg into a sealed tank or hot well; and a **low-level condenser** is one in which a pump must be used to remove the hot water. Theoretically these later classifications are applicable to any type of condenser, but practically they are usually applied only to jet condensers. Theoretically the above classifications are coordinate and any combination is possible. In practice a countercurrent condenser is practically always dry and a parallel-current condenser is practically always wet.

Surface condensers are much more expensive than jet condensers and use more water. They are never used on evaporator work unless the vapor to be condensed must be recovered separately from the cooling water. Barometric countercurrent dry condensers are the better practice in the larger sizes. Wet parallel-current low-level condensers are confined to extremely small installations.

Air Pumps. Air pumps may be classified as follows:

1. Wet reciprocating pumps.
2. Dry slide-valve reciprocating pumps.
3. Dry Corliss valve reciprocating pumps.
4. Steam-jet ejectors.

A wet reciprocating vacuum pump is usually nothing but an ordinary tank pump. The slide-valve reciprocating pumps are practically obsolete. Reciprocating vacuum pumps with Corliss-type valves (often called R.D.V. or rotary dry vacuum pumps) are common. Steam jets are much in favor and have the advantages of greater simplicity and no moving parts. They may be designed to produce any degree of vacuum desired down to 1 to 2 mm. Hg abs. For vacuums of 24 in. or less they are usually single-stage, from 24 to 28 in., two-stage, and for vacuums higher than 28 in., three-stage. They may have either surface or jet condensers between stages. The surface condensers are more expensive but do not load the nozzles with the air dissolved in the cold water. The jet condensers afford a cheaper installation but require more steam per pound air removed from the evaporator. At vacuums of 24 in. or less, the ejector is likely to use more steam than a reciprocating pump. For vacuums of 27 to 28 in. or more, the jet is more economical than the reciprocating pump. If the reciprocating pump is steam-driven, its exhaust may be sent to the evaporator and the pump merely becomes a reducing valve. Steam from the ejector cannot be so used, except when a feed-liquor heater may be used to carry part of the load of the intercondensers.

Evaporator Controls. Each body should be provided with a pressure-measuring device on the vapor space, and there should also be a gage on the steam space of the first effect. It is universal practice to use spring gages for this purpose, but spring gages for pressures less than 10 lb./sq. in. are quite certain to be inaccurate. The use of steam flowmeters, liquor flowmeters or measuring tanks for feed and thick liquor, and meters on water to the condenser is desirable, and these devices are not used as often as they should be. Because of removal of air through condensate lines, meters for condensate are seldom possible. There is no particular value in the use of recording thermometers on the various effects, but a record of feed temperature, hot-leg temperature, cooling-water temperature, and sometimes the boiling temperature of the last effect is valuable.

On many materials, the liquor feed valves may be operated manually and controlled by observation through ordinary sight glasses. Liquor-level regulators are available but are not satisfactory for viscous liquids or for those having appreciable salt in suspension. For the latter, there is no method of liquor-level regulation except manual regulation. For liquids that carry salt or that tend to freeze, the so-called J-type of level indicator is satisfactory. For viscous or dark-colored liquids, test cocks may be used for bodies under pressure greater than atmospheric; but a suitably designed hydrostatic gage is much better and will operate under both vacuum and pressure. Where possible, numerous sight glasses should be provided, for a good operator controls the evaporator by the appearance of the boiling liquid. Internal light sockets are not satisfactory. Sight glasses with lights directly opposite the operator's position are also unsatisfactory. The best illumination is obtained from a sight glass in the roof of the evaporator with a bulb and reflector sufficiently powerful to illuminate the entire interior. All sight glasses should be provided with water or thin liquor jets directed against the inner surface of the glass.

CONDENSER CALCULATIONS

(See pp. 476–477)

Water Required. The water needed to operate a condenser may be calculated by a simple heat balance.

Let
W = weight of steam to be condensed, lb./hr.
t_0 = saturation temperature of steam entering condenser, °F.
L = latent heat of vaporization at t_0, B.t.u./lb.
H_s = superheat in steam above t_0, B.t.u./lb.
H_w = total heat in exit condensate, B.t.u./lb.
H_{w_0} = total heat in liquid water at t_0, B.t.u./lb.
t_w = temperature of water at condenser inlet, °F.
t_e = temperature of water at condenser exit, °F.
Q = water used, lb./hr.
G = water used, gal./min.

The heat-balance equation is

$$W(L + H_s) + W(H_{w_0} - H_w) = Q(t_e - t_w)$$

Since the specific heat of water is nearly enough equal to 1.00 over the range involved in condenser practice, and since this discussion is concerned with jet condensers, H_w may be considered numerically equal to $(t_e - 32)$ and H_{w_0} equal to $(t_0 - 32)$. Hence,

$$Q = \frac{W(L + H_s) + W(t_0 - t_e)}{t_e - t_w} \quad (34)$$

$$G = \frac{Q}{60 \times 8.33} = 0.00201Q \quad (35)$$

Example. A jet condenser is to condense 2000 lb. steam per hr. at 24 in. vacuum referred to a 30-in. barometer. The steam is not superheated. The inlet water to the condenser is at 60°F. How much water will be used if the exit water temperature is (*a*) 120°F.; (*b*) 100°F.?

In Case *a*

$$W = 2000,\ t_0 = 125.4°,\ H_s = 0,\ t_e = 120°,\ t_w = 60°$$

$$Q = \frac{2000(1022.1 + 5.4)}{120 - 60} = 34{,}250 \text{ lb./hr.}$$

$$G = 0.00201 \times 34{,}250 = 68.8 \text{ gal./min.}$$

In Case *b*

$$Q = \frac{2000(1022.1 + 25.4)}{100 - 60} = 52{,}375 \text{ lb./hr.}$$

$$G = 105.3 \text{ gal./min.}$$

Surface. The necessary area in a surface condenser can be calculated from the heat to be removed from the vapor, mean temperature difference, and over-all coefficient of heat transfer. Data on water film and condensing vapor coefficients are given in Sec. 6, pp. 476–477.

Vapor Piping. Care should be taken to provide adequate connections from vacuum evaporators to their condensers. Because of the low density of steam under vacuum, the linear velocity of the vapor may be higher than would be allowable for steam lines. It should be remembered, however, that in the lower pressure ranges a very slight pressure loss due to friction in vapor pipes may mean an appreciable loss of total available temperature difference.

Volume of Air Entering Condenser. It is difficult to estimate the quantity of air entering an evaporator condenser. Such air will consist of

1. Air dissolved in the cooling water (assuming a jet condenser is used).
2. Air present in steam used for heating the first effect.
3. Air dissolved in the feed liquor.
4. Air entering from leaks.
5. Non-condensed gases liberated by reactions during evaporation.

Item 1 may be approximated by assuming that the cooling water is saturated with air, and that all this air is liberated in the condenser. This is not correct but results in an error in the safe direction. Ordinarily it may be assumed that the cooling water carries 2 per cent of its volume of dissolved air, if the air is measured at 60°F. and 30 in. Hg. Item 2 disappears when the first-effect steam space is under a pressure greater than atmospheric and is vented to the air. Items 3, 4, and 5 are exceedingly variable. One or two recorded tests indicate from 0.2 to 0.5 per cent air by volume in the vapors going to the condenser. A little carelessness in erection will multiply this figure many times.

Displacement of Vacuum Pump. The most important fact to remember in the calculation of air-pump displacement is that the air leaves saturated with water vapor. This fact results in certain considerations that lead to the basic distinctions between parallel-current and countercurrent condensers.

In discussing mixtures of gases, there are two possible points of view, both correct from a practical standpoint. Each gas may be considered to occupy the total volume at its particular partial pressure, or to occupy its particular fraction of the volume at the total pressure. Thus, 1 cu. ft. air at atmospheric pressure may be considered to consist of 1 cu. ft. oxygen at 0.21 atm. pressure and 1 cu. ft. nitrogen at 0.79 atm. pressure, or it may be considered, with equal correctness, as 0.21 cu. ft. oxygen at 1 atm. pressure and 0.79 cu. ft. nitrogen at 1 atm. pressure. In these calculations the latter point of view will be used.

Let W = weight of steam condensed, lb./hr.
t_0 = saturation temperature of steam entering condenser, °F.
V_s = specific volume of steam at t_0, cu. ft./lb.

t_{ac} = temperature of air leaving condenser in countercurrent operation, °F.
t_{ap} = temperature of air leaving condenser in parallel-current operation, °F.
t_e = temperature of water leaving condenser, °F.
t_w = temperature of water entering condenser, °F.
P = total pressure in condenser, mm. Hg.
p_{wc} = partial pressure of water vapor in air-water mixture leaving condenser in countercurrent operation, mm. Hg.
p_{wp} = partial pressure of water vapor in air-water mixture leaving condenser in parallel-current operation, mm. Hg.
p_{ac} = partial pressure of air in air-water mixture leaving condenser in countercurrent operation, mm. Hg.
p_{ap} = partial pressure of air in air-water mixture leaving condenser in parallel-current operation, mm. Hg.
a_e = air in vapor from evaporator, per cent by volume.
V_e = volume of air entering condenser from evaporators, measured at t_0 and P, cu. ft./hr.
V_w = volume of air entering condenser from cooling water, measured at 60°F. and 30 in. Hg, cu. ft./hr.
V_{pc} = air-pump displacement in cu. ft./min., countercurrent operation.
V_{pp} = air-pump displacement in cu. ft./min., parallel-current operation.

Then

$$V_e = \frac{a_e W V_s}{100} \tag{36}$$

$$P = p_{wc} + p_{ac} = p_{wp} + p_{ap} \tag{37}$$

$$V_{pc} = \frac{\left[V_e\left(\frac{460 + t_{ac}}{460 + t_0}\right) + V_w\left(\frac{460 + t_{ac}}{460 + 60}\right)\left(\frac{762}{P}\right)\right]\left(\frac{P}{p_{ac}}\right)}{60} \tag{38}$$

$$V_{pp} = \frac{\left[V_e\left(\frac{460 + t_{ap}}{460 + t_0}\right) + V_w\left(\frac{460 + t_{ap}}{460 + 60}\right)\left(\frac{762}{P}\right)\right]\left(\frac{P}{p_{ap}}\right)}{60} \tag{39}$$

In countercurrent operation, t_{ac} may be nearly the temperature of the inlet water, perhaps 5° to 10°F. higher. In parallel-current operation, t_{ap} is the temperature of the exit water.

Example. An evaporator sends 2000 lb. steam per hr. containing 0.3 per cent air, to a condenser in which the total pressure is 4 in. abs. Cooling water is available at 60°F. What is the air-pump displacement (*a*) in countercurrent operation with water leaving at 120°F. and air at 70°F.; (*b*) in parallel-current operation with water and air both leaving at 100°?

Solution. The temperature corresponding to 4 in. abs. is 125.4°. Strictly speaking, this should be calculated on the basis of the partial pressure of the water vapor in the mixture going to the condenser, not on the total pressure. Actually such a correction is seldom large enough to be comparable with the accuracy needed in these calculations. Then, for case *a*, $W = 2000$; $t_0 = 125.4°$; $V_s = 177.1$; $a_e = 0.2$.

From Eq. (36)

$$V_e = \frac{2000 \times 177.1 \times 0.2}{100} = 708 \text{ cu. ft.}$$

From the previous solution (p. 521) $Q = 34{,}250$ lb., and hence

$$V_w = \frac{34{,}250}{62.4} \times 0.02 = 10.9 \text{ cu. ft.}$$

At 70°F., the vapor pressure of water (p_{wc}) is 18.8 mm. P, which may be assumed constant throughout the condenser, is 101.4 mm. Hence, from Eq. (37), $P_{ac} = 101.4 - 18.8 = 82.6$ mm. Then since $t_{ac} = 70°$, from Eq. (38)

$$V_{pc} = \frac{\left(708 \times \frac{460 + 70}{460 + 125.4} + 10.9 \times \frac{460 + 70}{460 + 60} \times \frac{762}{101.4}\right)\left(\frac{101.4}{82.6}\right)}{60} = 14.8 \text{ cu. ft./min.}$$

In case (*b*), the previous solution shows $Q = 52{,}375$ lb. Hence $V_w = 16.8$; $t_{ap} = 100$; $p_{ap} = 101.4 - 48.9 = 52.5$.

From Eq. (39)

$$V_{pp} = \frac{\left(708 \times \frac{460 + 100}{460 + 125.4} + 16.8 \times \frac{460 + 100}{460 + 60} + \frac{762}{101.4}\right)\left(\frac{101.4}{52.5}\right)}{60} = 25.7 \text{ cu. ft./min.}$$

An inspection of the formulas and the sample problems shows that one advantage of a countercurrent condenser over a parallel-current condenser is that the air pump may be smaller for countercurrent operation. The real difference is in the last term in the numerator of Eqs. (38) and (39). As the amount of water fed to the condenser is reduced, t_e approaches t_0. This has no effect on countercurrent operation, for in Eq. (38), p_{ac} is dependent only on the *cold* water temperature and not at all on t_0. Hence in countercurrent operation the water may be reduced till t_e is nearly equal to t_0.

On the other hand, in Eq. (39), p_{ap} is dependent on the vapor pressure of water at the temperature t_e. As the water fed to a parallel-current condenser is decreased, t_e approaches t_0, and hence p_{wp} approaches P. Hence, p_{ap} approaches zero, and the numerator of Eq. (39) approaches infinity, so that for practical pump sizes the temperature of the exit water must be kept considerably below t_0. The real difference between parallel-current and countercurrent operation is, therefore, that parallel-current operation always takes more water for a given vacuum than countercurrent operation, other conditions being equal.

CHOICE OF TYPE AND SIZE OF EVAPORATOR FOR A GIVEN PROBLEM

It cannot be stated too emphatically that the factors affecting heat-transfer coefficients are so many and varied, their effects so little understood, and the published information so scanty, that the final working design of an evaporator should be attempted only by a specialist. Formerly it was possible to state, at least in certain lines, the type of evaporator to be used for a given service. At present practically all previously accepted standards have been discarded, and it seems that absolutely new standards are being adopted, so that in a short time what was the established practice up to 1929 will be considered obsolete. The change has largely been in the direction of applying the forced-circulation evaporator and the natural-circulation long-tube evaporator to a vastly wider range of problems. It is entirely possible that the standard horizontal, the standard vertical, and all the other older types of evaporators, will become entirely obsolete.

SECTION 8

GENERAL THEORY OF DIFFUSIONAL OPERATIONS

BY

Allan P. Colburn, Ph. D., Assistant to the President and Professor of Chemical Engineering, University of Delaware; Member, American Institute of Chemical Engineers, American Chemical Society, American Society of Mechanical Engineers, American Society for Engineering Education.

Robert L. Pigford, Ph. D., Professor and Head, Chemical Engineering Department, University of Delaware; Member, American Institute of Chemical Engineers, American Chemical Society, American Society for Engineering Education.

CONTENTS

INTRODUCTION

PAGE

Introduction 524

NOMENCLATURE

Nomenclature 524

EQUILIBRIUM RELATIONSHIPS

Raoult's Law 525
Binary Mixtures 525
Systems That Follow Raoult's Law 526
Principles of Non-ideal Liquid Mixtures 526
Liquid-activity Coefficients 526
Activity Coefficients as Functions of Composition 527
Binary Margules Equations 527
Binary Van Laar Equations 527
Application of Equations 527
Ternary Margules Equations 528
Direct Prediction of Values of Relative Volatility 529
Quaternary Equations 529
Effect of Temperature on Liquid-activity Coefficients 529
Utilization of Activity Coefficients 532
Prediction of Absorption Equilibrium 532
Henry's Law 532
Prediction of Azeotropes 532
Prediction of Heats of Solution 532
Prediction of Immiscibility 533
Prediction of Distribution in Liquid-liquid Extraction 533
Prediction of Activity Coefficients 533
Estimation from Azeotropic Data 533
Calculation from Boiling-point or Total-pressure Curves 534
Partly Miscible Solutions 534
Relationships to Physical Properties of Components 534
Calculation of Effect of Gas-law Deviations 535

RATE OF MASS TRANSFER

Diffusion in Gases 538
Diffusion in Liquids 540

PAGE

Applications of the Theory of Molecular Diffusion 540
Evaporation of Small Drops into a Quiet Atmosphere 540
Mixing of Gases in a Closed Cylindrical Container 541
Mass Transfer by Convection 541
Analogy among Mass Transfer, Heat Transfer, and Fluid Friction 541
Experimental Data for Surfaces Having a Definite Area 543
Inside Tubes 543
Viscous Flow inside Tubes 544
Annular Spaces 544
Plane Surfaces 544
Cylindrical Surfaces 545
Spherical Surfaces 546
Beds of Granular Solids 547

COUNTERCURRENT APPARATUS—NUMBER OF THEORETICAL PLATES OR TRANSFER UNITS REQUIRED TO CARRY OUT A SEPARATION

Material Balances—Operating and Equilibrium Lines 548
Mass Transfer between Phases 549
Use of Over-all Resistance to Transfer 549
Use of the Height Equivalent to a Transfer Unit (H.T.U.) 550
Use of Plate Efficiencies 550
Calculation of Number of Transfer Units 552
Calculation of Number of Theoretical Plates Required in Countercurrent Apparatus 554
Transient Behavior of Countercurrent Apparatus 555

SPECIAL METHODS FOR BATCH APPARATUS

Special Methods for Batch Apparatus 557

SIMULTANEOUS HEAT AND MASS TRANSFER

Wet- and Dry-bulb Thermometry 557
Simultaneous Heat and Mass Transfer for the System Liquid Water–air–water vapor 558
Simultaneous Heat and Mass Transfer under Conditions of Large-concentration Driving Forces 559

INTRODUCTION

REFERENCES: Benedict, Multistage Separation Processes, *Trans. Am. Inst. Chem. Engrs.*, **43**, 41 (1947). Carlson and Colburn, *Ind. Eng. Chem.*, **34**, 581 (1942). Chilton and Colburn, *Ind. Eng. Chem.*, **26**, 1183 (1934). Colburn, *Trans. Am. Inst. Chem. Engrs.*, **35**, 211 (1939); *Ind. Eng. Chem.*, **33**, 459 (1941). Dodge, "Chemical Engineering Thermodynamics," McGraw-Hill, New York, 1944. Hildebrand, "Solubilities of Non-Electrolytes," 2d ed., Reinhold, New York, 1936. Sherwood, "Absorption and Extraction," McGraw-Hill, New York, 1937. Walker, Lewis, McAdams, and Gilliland, "Principles of Chemical Engineering," 2d ed., McGraw-Hill, New York, 1937. Badger and McCabe, "Elements of Chemical Engineering," 2d ed., McGraw-Hill, New York, 1939.

The diffusional unit operations—distillation, solvent extraction, and gas absorption—have a common theoretical foundation. In each of these, mass is transferred from one fluid phase to another across an interface. The design of equipment for carrying out this function is based fundamentally on (1) allowable rates of fluid flow through the contacting device, (2) equilibrium relations between the concentrations of the diffusing substance in the phases, (3) rates of mass transfer between phases, and (4) stoichiometric calculations in which the rate data and the equilibrium data are combined to produce design specifications. The determination of flow rates is largely empirical, and the methods and data vary from one operation to another. Elements 2, 3, and 4 have a common theoretical foundation, whether applied in distillation, extraction, or absorption calculations; thus, the equilibrium relations are governed by basic thermodynamic principles.

The determination of the height of a column required for a specified separation is usually made in terms of the number of equilibrium stages, theoretical plates, or transfer units that are needed. The actual dimensions of the equipment are then determined from data on the value of the H.T.U. [height of a (mass) transfer unit], or the plate efficiency, and such data are largely characteristic of the operation involved. The stoichiometric calculation of the number of units required is made in the simpler cases by methods that are identical for all the diffusional operations.

Nomenclature

a = interfacial area, sq. ft./cu. ft.
B = second virial coefficient in equation of state, cu. ft.
B_F = film thickness, ft.
C = Sutherland constant.
c = concentration, lb.-moles/cu. ft.
c_p = specific heat, p.c.u./(lb.)(°C.).
c_s = humid heat, p.c.u./(lb.)(°C.).
D = diffusion coefficient, ft.²/hr. (occasionally cm.²/sec. in text).
d = diameter of surface, ft.
E = eddy kinematic viscosity, ft.²/hr.
E = over-all plate efficiency.
E_H = eddy thermometric conductivity, ft.²/hr.
E_M = eddy diffusivity, ft.²/hr.
E_{ML} = Murphree liquid efficiency.
E_{MV} = Murphree vapor efficiency of "wet" plate.
E_p = Murphree point efficiency.
f = friction factor.
G = mass velocity, lb./(hr.)(sq. ft.).
G' = flow rate of solute-free gas, lb./(hr.)(sq. ft.).
G_M = molar mass velocity, lb.-moles/(hr.)(sq. ft.).
g_c = gravitational conversion factor, 32.17 (lb. mass)(ft.)/(lb. force)(sec.²).
H = humidity, lb./lb.
H^* = humidity at saturation, lb./lb.
H_G = height of a transfer unit, based on gas-film resistance, ft.
H'_G = height of a gas-film transfer unit for heat transfer to beds of porous solids, ft.
H_L = height of a transfer unit, based on liquid-film resistance, ft.
H_{OG} = height of a transfer unit, based on over-all gas-phase resistance, ft.
H_{OL} = height of a transfer unit, based on over-all liquid-phase resistance, ft.
h = heat-transfer coefficient, p.c.u./(hr.)(sq. ft.)(°C.).
i = enthalpy of saturated air, B.t.u./lb.
J = mG_M/L_M, or in Table 15 also L_M/mG_M.
j_H = heat-transfer factor, $(h/c_jG)(c_p\mu/k)^{2/3}$.
j_M = mass-transfer factor, $(k_G/G_M)(\mu/\rho D)^{2/3}$.
K_G = over-all gas-phase mass-transfer coefficient, lb.-moles/(hr.)(sq. ft.)(mole fraction).
K_L = over-all liquid-phase mass-transfer coefficient, lb.-moles/(hr.)(sq. ft.)(mole fraction).
k' = mass-transfer coefficient, lb. transferred/(hr.)(sq. ft.)-(unit humidity difference, lb./lb.).
k_G = mass-transfer coefficient for equimolar counterdiffusion, lb.-moles/(hr.)(sq. ft.)(mole fraction).
k_G' = mass-transfer coefficient for diffusion of one component through a second, stationary component; same units as k_G; depends on value of $(1 - y)_f$, but the product, $k_G'(1 - y)_f$, is independent of concentration and is numerically equal to k_G for equimolar counterdiffusion.
k_L = mass-transfer coefficient for liquid film, lb.-moles/(hr.)(sq. ft.)(mole fraction).
L = length of plane surface, parallel to direction of flow, ft.
L_M = molar rate of liquid flow, lb.-moles/(hr.)(sq. ft.).
L' = flow rate of solute-free liquid, lb./(hr.)(sq. ft.).
L^0 = heat of solution at infinite dilution, p.c.u./lb.-mole.
l = radial distance from wall of cylindrical tube, ft.
M = molecular weight, lb./lb.-mole, or g./g.-mole.
m = $dy^*/dx = dy/dx^*$.
m = mass of droplet, lb.
m_c = Henry's law coefficient, or gas solubility.
N_A = lb.-moles of component A diffusing/(hr.)(sq. ft.).
N_{Re} = Reynolds number.
N_{Sc} = Schmidt number, $(\mu/\rho D)$.
N_{Sc}' = modified Schmidt number, $(E_M/E)(\mu/\rho D)$.
P = total pressure, atm.
P_R = reduced pressure, P/P_c.
p = partial pressure, atm.
P_c = critical pressure, atm.
Q = defined by Table 15.
R = universal gas constant, 82.06 (cc.)(atm.)/(g.-mole)(°K.), 1544 (ft.)(lb. force)/(lb.-mole)(°F.), 10.7 (cu. ft.)(lb./sq. in.)/(lb.-mole)(°R.), 1.99 cal./(g.-mole)(°K.).
r = radius of droplet, ft.
s = distance, ft.
s_x, s_y = slopes of lines used in graphical determination of number of transfer units.
T_c = critical temperature, °K.
T = temperature, °K.
T_R = reduced temperature, T/T_c.
t = temperature of bulk of gas, °C.
t_i = temperature of gas at interface, °C.
t_x, t_y = slopes of lines used in graphical determination of number of transfer units.
Ua = over-all heat-transfer coefficient, B.t.u./(hr.)(sq. ft.)-(°F.).
u = local fluid velocity, ft./sec.
u^+ = dimensionless friction velocity, $u/(\tau_0 g_c/\rho)^{1/2}$.
V = mean velocity of fluid, ft./sec.
V = molecular volumes, cc./g.-mole, or cu. ft./lb.-mole.
v = volume fraction.
w = weight fraction.
X = concentration in the liquid, weight ratio, lb./lb.; or mole ratio, lb.-mole/lb.-mole.

x = mole fraction of solute in liquid.
x_i = mole fraction of solute in liquid at interface.
x^* = mole fraction of solute in liquid in equilibrium with vapor having composition y.
Y = concentration in the gas, weight ratio, lb./lb., or mole ratio, lb.-mole/lb.-mole.
y = mole fraction of A in bulk of gas.
y_i = mole fraction of A in gas at interface.
y^+ = dimensionless position coordinate in cylindrical tube, $(l\rho/\mu)(\tau_0 g_c/\rho)^{1/2}$.
y^* = mole fraction of solute in gas in equilibrium with liquid having composition x.
Z = total height, ft.
z = height of tower, ft.
α = relative volatility.
γ = activity coefficient in liquid phase.
θ = time, hr.
θ_G = time of passage of vapor stream through one theoretical plate, or transfer unit, hr.
θ_L = time of passage of liquid stream through one theoretical plate, or transfer unit, hr.
λ = latent heat, p.c.u./lb.
μ = viscosity, lb. mass/(ft.)(hr.).
ν = kinematic viscosity, ft.²/sec.
ϵ = entrainment, lb.-mole liquid/lb.-mole vapor.
ρ = density, lb./cu. ft.
τ_0 = surface frictional stress, lb. force/sq. ft.

EQUILIBRIUM RELATIONSHIPS

In Secs. 9, 10, and 11 on Distillation, Absorption, and Extraction, the reliable equilibrium data are tabulated or referred to. It is the purpose of this section to present briefly the principles underlying such data and to indicate the means by which predictions for other conditions can be made.

RAOULT'S LAW

Vapor-liquid equilibrium is readily calculated for ideal solutions, *i.e.*, those which follow Raoult's law, and if the gas laws apply, the only data required are vapor pressures of the pure components. The treatment below will assume that the vapors follow the gas laws, a later section will consider deviations in these laws. For systems that follow Raoult's law, the partial pressure of any component in the vapor is equal to the product of its mole fraction in the liquid and the vapor pressure of the pure component at the same temperature. Thus, for systems of any number of components,

$$p_1 = P_1x_1 \qquad p_2 = P_2x_2, \text{ etc.} \tag{1}$$
$$Py_1 = P_1x_1 \qquad Py_2 = P_2x_2, \text{ etc.} \tag{2}$$

where p_1 = partial pressure of component 1 in the vapor, P_1 = vapor pressure of the pure component at the same temperature, P = total pressure, x_1 = mole fraction of component 1 in the liquid, and y_1 = mole fraction in the vapor.

The relative compositions of two components in vapor and liquid are

$$\frac{y_1}{y_2} = \frac{P_1x_1}{P_2x_2} = \alpha_{1\text{-}2}\frac{x_1}{x_2} \tag{3}$$

where $\alpha_{1\text{-}2}$ is the relative volatility of two ideal components; for ideal gases and liquids, $\alpha_{1\text{-}2} = P_1/P_2$.

Equations (1) to (3) are readily applied when the temperature is given and the total pressure is a variable. More often the total pressure and composition of one phase are given and the problem is to find the equilibrium temperature and the composition of the other phase. Equations (2) may be applied by cut and try. A temperature is assumed, vapor pressures are found, and values of y (or x) calculated. If the sum of the values of y (or x) is not unity, a new temperature is assumed and the calculation repeated.

Inasmuch as the ratios of vapor pressures are sometimes nearly constant over moderate temperature ranges, it is often more convenient to utilize the following relations:

$$y_1 = \frac{x_1P_1}{x_1P_1 + x_2P_2 + x_3P_3 + \cdots} = \frac{x_1P_1/P_i}{x_1P_1/P_i + x_2P_2/P_i + x_3P_3/P_i + \cdots} \tag{4a}$$

$$y_2 = \frac{x_2P_2/P_i}{x_1P_1/P_i + x_2P_2/P_i + x_3P_3/P_i + \cdots} \tag{4b}$$

where P_i = vapor pressure of any component chosen for reference (in multicomponent distillation, the i-component is usually chosen as the "heavy key"*). Note, similarly,

$$x_1 = \frac{y_1P_i/P_1}{y_1P_i/P_1 + y_2P_i/P_2 + y_3P_i/P_3 + \cdots} \tag{5a}$$

$$x_2 = \frac{y_2P_i/P_2}{y_1P_i/P_1 + y_2P_i/P_2 + y_3P_i/P_3 + \cdots} \tag{5b}$$

A rough value of temperature is usually satisfactory to evaluate the vapor-pressure ratios in these equations, a check is readily made afterward by using Eq. (2) to solve for P_1 and looking up the corresponding temperature.

Binary Mixtures. For binary mixtures, the pressure and one composition will fix the temperature, or the pressure and temperature will fix both compositions. At a given total pressure it is possible to choose a series of temperatures between the equilibrium temperatures of the two pure components and solve for the corresponding compositions as follows:

$$x_1 = \frac{P - P_2}{P_1 - P_2} \tag{6}$$

The value of y_1 can then be found by use of Eq. (2) or, if only values of y are desired,

$$y_1 = \frac{P_1(P - P_2)}{P(P_1 - P_2)} \tag{7}$$

One of the most convenient relations for binary ideal mixtures is obtained from Eq. (3) by noting that $y_1 = 1 - y_2$, $x_1 = 1 - x_2$, whence

$$y_1 = \frac{\alpha x_1}{1 + (\alpha - 1)x_1} \tag{8}$$

and

$$x_1 = \frac{y_1}{\alpha - (\alpha - 1)y_1} \tag{9}$$

where $\alpha = P_1/P_2$ (for ideal mixtures). Often the value of P_1/P_2 is nearly constant over the range of interest so that Eq. (8) or (9) is readily used. If a complete y-x curve is being computed, it is well to look up the values of vapor pressure at the boiling point of each pure component and to find the end values of the ratio, P_1/P_2. For small differences, interpolation is satisfactory.

A compilation of available data on ideal binary mixtures is given by Table 1. Where the boiling points of the two components are relatively close together, the limiting values of P_1/P_2 are seen to be not far different. For example, for benzene-toluene at 760 mm., at 80.1°C., $P_1/P_2 = 2.61$, and at 110.7°C., $P_1/P_2 = 2.315$. The calculation of a series of y-x values is carried out as shown in the table below, interpolating values of P_1/P_2 and employing Eq. (8).

* The "heavy key" is usually defined as the least volatile component the composition of which is specified in the overhead product.

x_1	P_1/P_2*	y_1
0	2.31	0
0.1	2.34	0.206
.2	2.37	.372
.4	2.43	.618
.6	2.49	.789
.8	2.55	.911
.9	2.58	.958
1.0	2.61	1.0

* Values of P_1/P_2 are interpolated according to x_1, which is not strictly correct but is close enough when the end values differ only by 20 per cent or ess.

Systems That Follow Raoult's Law. Raoult's law applies in limited cases as follows:

1. Members of homologous series and isomers, *e.g.*, benzene-toluene, propane-butane, methanol-ethanol, butane-isobutane.
2. Certain other mixtures such as benzene–ethylene dichloride and carbon tetrachloride–toluene. As classified according to hydrogen-bonding characteristics by Ewell, Harrison, and Berg [*Ind. Eng. Chem.*, **36,** 871 (1944)] certain mixtures are said to be "quasi-ideal," *i.e.*, either with slight positive deviations or ideal. These are classes III + III, III + IV, IV + IV, IV + V, and V + V, where the classes are defined in Table 5. A number of typical ideal systems are listed in Table 1.
3. For all mixtures, as any component approaches 100 per cent in composition, Raoult's law holds for that component. This condition is illustrated on p. 527 by the plots of activity coefficients (which are essentially deviation factors from Raoult's law). It will be seen that these curves become asymptotic with unity as the compositions approach unity. The extent to which this rule may be applied depends upon the degree of non-ideality of the system. Typical values of deviations at various compositions are given below for symmetrical systems:

Mole fraction	0	0.1	0.5	0.8	0.9	0.95	0.99
Deviation factor (activity coefficient)	2	1.75	1.19	1.03	1.007	1.0017	1.00007
	5	3.67	1.50	1.07	1.016	1.004	1.00016
	10	6.45	1.78	1.10	1.023	1.006	1.00023
	50	23.8	2.66	1.17	1.04	1.01	1.0004

Thus, for systems where the maximum deviation is of the order of 2, there would be a 3 per cent error or less to assume a deviation factor of unity for concentrations of 0.8 mole fraction and higher. Even if the maximum deviation is as high as 50 (as for a very slightly miscible system) Raoult's law is seen to apply at concentrations above 95 mole per cent with an error of 1 per cent or less.

Table 1. Relative Volatilities of Ideal Mixtures
Values of P_1/P_2 (approximately equal to α) are given at the boiling points of components 1 and 2, respectively

Mixture 1 and 2	B.p. of 1, °C.	P_1/P_2	B.p. of 2, °C.	P_1/P_2
Benzene–ethylene dichloride	80.1	1.113	83.48	1.109
Benzene-toluene	80.1	2.61	110.7	2.315
n-Butyl chloride-n-butyl bromide	77.5	2.08	101.6	1.87
Chloroform–carbon tetrachloride	61.1	1.71	76.6	1.60
Ethanol-isopropanol	78.3	1.18	82.3	1.17
Ethanol-propanol	78.3	2.18	97.2	2.03
Ethyl chloride–ethyl bromide	12.5	3.23	38.4	2.79
Ethyl ether–benzene	34.6	5.16	80.2	3.95
Ethylene dibromide–propylene dibromide	131.7	1.30	141.5	1.30
Ethylene dichloride–trichloroethane	83.5	2.52	113.7	2.33
n-Heptane–methylcyclohexane	98.4	1.058	100.3	1.056
n-Hexane-n-heptane	69.0	2.613	98.4	2.33
Methanol-ethanol	64.7	1.73	78.1	1.64
Methanol-isobutanol	64.6	6.1	107.5	4.4
Methanol-propanol	64.6	3.89	97.2	3.15
Methyl acetone–ethyl acetate	56.8	2.036	77.1	1.923
Phenol-o-cresol	181.2	1.30	190.6	1.275
Phenol-m-cresol	181.2	1.768	201.5	1.699
Phenol-p-cresol	181.2	1.793	202.2	1.728
Toluene–benzyl chloride	110.7	7.75	178.0	4.45
Toluene-chlorotoluene	110.7	4.76	162.0	3.65
Water–ethylene glycol	100.0	49.8	197.0	13.2
Water–ethylene glycol*	60.1	98	150.2	21
Water-glycerol†	38.1	76,400	202.0	244

* Pressure = 150 mm. Hg.
† Pressure = 50 mm. Hg.

PRINCIPLES OF NON-IDEAL LIQUID MIXTURES

An understanding of the physical principles underlying equilibrium data is desirable in order to provide a means of testing the reliability of available experimental data, to permit extension of data to different conditions, such as other temperatures, and to provide various means of estimating equilibrium data in the absence of complete experimental data. Furthermore the data can be utilized to obtain other physical properties, such as values of heat of solution, by means of thermodynamic relationships.

The treatment in this section will assume that the vapors follow the gas laws. For cases of appreciable deviations from ideal gases, methods are given later.

Liquid-activity Coefficients. Deviations from ideality in the liquid state are conveniently treated as multiplying factors, and these have the thermodynamic significance of activity coefficients, defined by the relations (for ideal gases):

$$Py_1 = \gamma_1 P_1 x_1 \qquad Py_2 = \gamma_2 P_2 x_2, \text{ etc.} \tag{10}$$

also

$$\frac{y_1 x_2}{y_2 x_1} = \alpha_{1\text{-}2} = \frac{\gamma_1 P_1}{\gamma_2 P_2} \tag{11}$$

and

$$y_1 = \frac{x_1\alpha_{1\text{-}i}}{x_1\alpha_{1\text{-}i} + x_2\alpha_{2\text{-}i} + x_3\alpha_{3\text{-}i} + \cdots} \tag{12a}$$

$$y_2 = \frac{x_2\alpha_{2\text{-}i}}{x_1\alpha_{1\text{-}i} + x_2\alpha_{2\text{-}i} + x_3\alpha_{3\text{-}i} + \cdots} \tag{12b}$$

$$x_1 = \frac{y_1/\alpha_{1\text{-}i}}{y_1/\alpha_{1\text{-}i} + y_2/\alpha_{2\text{-}i} + y_3/\alpha_{3\text{-}i} + \cdots} \tag{13a}$$

$$x_2 = \frac{y_2/\alpha_{2\text{-}i}}{y_1/\alpha_{1\text{-}i} + y_2/\alpha_{2\text{-}i} + y_3/\alpha_{3\text{-}i} + \cdots} \tag{13b}$$

where $\alpha_{1\text{-}i} = \gamma_1 P_1/\gamma_i P_i$, $\alpha_{2\text{-}i} = \gamma_2 P_2/\gamma_i P_i$, etc.

Note that any component can be picked for the reference or *i*-component (in multicomponent distillation the "heavy key" is usually so picked). For the component picked, α becomes unity, *e.g.*, if the second component is the *i*-component, $\alpha_{2\text{-}i} = \gamma_2 P_2/\gamma_2 P_2 = 1$.

Example. Find the vapor composition in equilibrium with the following mixture: $x_1 = 0.1$, $x_2 = 0.3$, $x_3 = 0.6$ under conditions where the activity coefficients are known to be $\gamma_1 = 3.4$, $\gamma_2 = 2.8$, $\gamma_3 = 1.1$. At a certain temperature the vapor pressures are $P_1 = 1240$, $P_2 = 810$, $P_3 = 505$ mm. Hg. Choosing component 2 as the reference component, $\alpha_{1\text{-}2} = \gamma_1 P_1/\gamma_2 P_2 = (3.4)(1240)/(2.8)(810) = 1.78$, and $\alpha_{3\text{-}2} = \gamma_3 P_3/\gamma_2 P_2 = (1.1)(505)/(2.8)(810) = 0.234$. Then by Eq. (12),

$$y_1 = \frac{(0.1)(1.78)}{(0.1)(1.78) + 0.3 + (0.6)(0.234)} = 0.288$$

$$y_2 = \frac{0.3}{(0.1)(1.78) + 0.3 + (0.6)(0.234)} = 0.485$$

$$y_3 = \frac{(0.6)(0.234)}{(0.1)(1.78) + 0.3 + (0.6)(0.234)} = 0.227$$

To check for the temperature of the equilibrium, use is made of Eq. (10) to solve for one of the values of vapor pressure, thus, $P_1 = Py_1/\gamma_1 x_1 = (760)(0.288)/(3.4)(0.1) = 641$ mm. Hg. One then refers to the vapor pressure–temperature relationship for this component to obtain the temperature. In this example the approximate temperature was chosen rather high, so that a second series of vapor pressures should be found at the new temperature and the calculations slightly adjusted if the new ratios of P_1/P_2 and P_3/P_2 are appreciably different.

The activity coefficients γ_1 and γ_2 are special functions of concentration and vary to different degrees with temperature as discussed later. The typical variation with concentration is shown by Figs. 1 and 2. In these figures, both γ_1 and γ_2 are plotted on logarithmic scales vs. x_1. In Fig. 1 the "deviations are positive," *i.e.*, the activity coefficients are greater than unity so that the logarithms of activity coefficients are positive numbers. In Fig. 2, the activity coefficients are fractional, and the logarithms are therefore negative. In both cases the activity coefficients approach unity as the concentration of the given component approaches unity as stated as condition 3 in the previous section.

The deviations are usually extreme as the concentration of the component approaches zero.

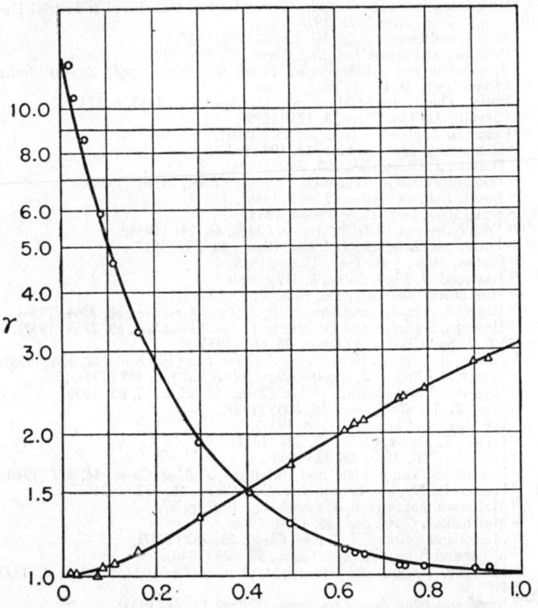

FIG. 1. Effect of concentration on activity coefficients for binary systems. n-Propyl alcohol in water. Curve calculated by Van Laar equation with $A = 1.13$ and $B = 0.49$. Points are the observed data of Gadwa (M.I.T. thesis, 1936) at atmospheric pressure.

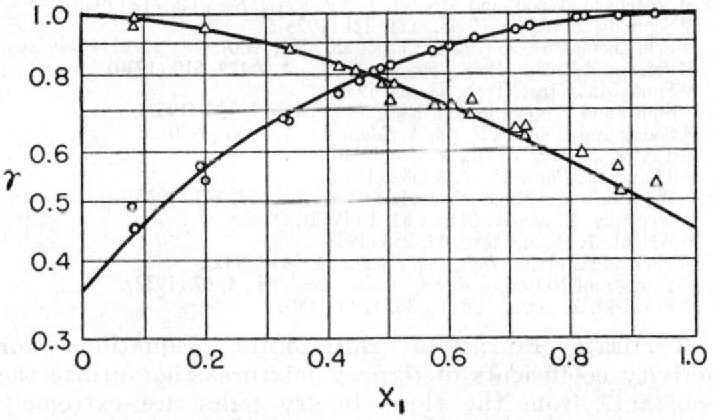

FIG. 2. Acetone in chloroform. Curve calculated by Van Laar equation with $A = -0.44$ and $B = -0.34$. Points are the observed data of Zawidski [*Z. physik. Chem.*, **35**, 129 (1900)] at 35.17°C.

Activity Coefficients as Functions of Composition. The fundamental thermodynamic equation relating activity coefficients and composition is the Gibbs-Duhem relation (Lewis and Randall, "Thermodynamics and the Free Energy of Chemical Substances," McGraw-Hill, New York, 1923)

$$x_1\left(\frac{\partial \ln \gamma_1}{\partial x_1}\right)_{T,P} + x_2\left(\frac{\partial \ln \gamma_2}{\partial x_1}\right)_{T,P} + \cdots = 0 \quad (14)$$

For binary mixtures, $\partial x_1 = -\partial x_2$, so that

$$x_1\left(\frac{\partial \ln \gamma_1}{\partial x_1}\right)_T = x_2\left(\frac{\partial \ln \gamma_2}{\partial x_2}\right)_T \quad (15)$$

These equations relate the slopes of the curves in Figs. 1 and 2 and provide a means of testing experimental data. It is much more convenient, however, to utilize integrated forms of these relations, and the most useful of these are given below in terms of constants which are also end values of the functions [Carlson and Colburn, *Ind. Eng. Chem.*, **34**, 581 (1942)]. A derivation and discussion of other more complicated equations for binary systems is given by Wohl [*Trans. Am. Inst. Chem. Engrs.*, **42**, 215 (1946)].

Binary Margules Equations. Margules (*cf.* Carlson and Colburn, *loc. cit.*) obtained a pair of exponential series of unlimited numbers of terms with the constants for one series obtained from those of the other by application of Eq. (15). Fortunately two terms are sufficient for most applications. The Margules two-term equations are conveniently expressed:

$$\log \gamma_1 = x_2^2[A_{1\text{-}2} + 2x_1(A_{2\text{-}1} - A_{1\text{-}2})] \quad (16a)$$
$$\log \gamma_2 = x_1^2[A_{2\text{-}1} + 2x_2(A_{1\text{-}2} - A_{2\text{-}1})] \quad (16b)$$

The constants $A_{1\text{-}2}$ and $A_{2\text{-}1}$ are the (maximum) end values of the activity coefficient curves, *i.e.*, when $x_1 = 0$, $\log \gamma_1 = A_{1\text{-}2}$, $\log \gamma_2 = 0$; when $x_2 = 0$, $\log \gamma_2 = A_{2\text{-}1}$, $\log \gamma_1 = 0$.

For symmetrical systems, $A_{1\text{-}2} = A_{2\text{-}1} = A$ and Eqs. (16) reduce to

$$\log \gamma_1 = Ax_2^2 \quad (17a)$$
$$\log \gamma_2 = Ax_1^2 \quad (17b)$$

Binary Van Laar Equations. These equations, as given by Carlson and Colburn (*loc. cit.*), are

$$\log \gamma_1 = \frac{A_{1\text{-}2}}{\left(1 + \dfrac{A_{1\text{-}2}x_1}{A_{2\text{-}1}x_2}\right)^2} \quad (18a)$$

$$\log \gamma_2 = \frac{A_{2\text{-}1}}{\left(1 + \dfrac{A_{2\text{-}1}x_2}{A_{1\text{-}2}x_1}\right)^2} \quad (18b)$$

similarly, as for the Margules equations, at $x_1 = 0$, $\log \gamma_1 = A_{1\text{-}2}$ and $\log \gamma_2 = 0$, at $x_2 = 0$, $\log \gamma_2 = A_{2\text{-}1}$, and $\log \gamma_1 = 0$. Also, for symmetrical systems, where $A_{1\text{-}2} = A_{2\text{-}1} = A$, Eqs. (18) reduce to Eqs. (17), indicating that the Van Laar and Margules equations are essentially the same when the constants $A_{1\text{-}2}$ and $A_{2\text{-}1}$ are similar. The equations differ more as the ratio of $A_{1\text{-}2}$ to $A_{2\text{-}1}$ departs increasingly from unity.

Application of Equations. These equations are exceedingly valuable as a means of testing experimental data and of extending incomplete information. Carlson and Colburn (*loc. cit.*) showed that data on most binary systems available agreed with the Van Laar equations with the following possible exceptions: (1) where the data covered a wide range of temperatures, and (2) where the systems were widely non-symmetrical, *i.e.*, where one A constant was more than double the other. Carlson and Colburn found that either the Margules or Van Laar equations worked well for systems approaching symmetry. For considerable non-symmetry, the Van Laar equations were preferable for mixtures of components having unequal molal volumes, such as mixtures of water and organic liquids, and the Margules for those of approximately equal molal volumes, *e.g.*, chloroform-ethanol.

Theoretically, one reliable equilibrium measurement on a binary system, including total pressure, temperature, and compositions of vapor and liquid phases, is sufficient to permit calculation of the two constants, from which the whole equilibrium diagram may be calculated. Actually several points are desirable. The calculation of the constants from experimental data can be done conveniently by use of the following equations obtained from Eqs. (16) and (18).

Margules:

$$A_{1\text{-}2} = \frac{\log \gamma_1}{x_2^2} + 2x_1\left(\frac{\log \gamma_2}{x_1^2} - \frac{\log \gamma_1}{x_2^2}\right) \qquad (19a)$$

$$A_{2\text{-}1} = \frac{\log \gamma_2}{x_1^2} + 2x_2\left(\frac{\log \gamma_1}{x_2^2} - \frac{\log \gamma_2}{x_1^2}\right) \qquad (19b)$$

Van Laar:

$$A_{1\text{-}2} = \log \gamma_1 \left(1 + \frac{x_2 \log \gamma_2}{x_1 \log \gamma_1}\right)^2 \qquad (20a)$$

$$A_{2\text{-}1} = \log \gamma_2 \left(1 + \frac{x_1 \log \gamma_1}{x_2 \log \gamma_2}\right)^2 \qquad (20b)$$

Typical reliable experimental data on vapor-liquid equilibrium have been tested with the Van Laar equations, and the resulting constants are listed in Table 2.

Table 2. Binary Constants for Equations for Activity Coefficients

NOTE. These constants were obtained for the Van Laar equations but can be utilized in the Margules equations where the ratio of $A_{1\text{-}2}$ to $A_{2\text{-}1}$ is within the range of about 0.75 to 1.3.

Where the data cover a rather wide range of temperature, the equations can be considered only approximate. The temperature and pressure ranges are indicated with the first and last numbers referring to the pure components; if there is a constant boiling mixture, its temperature is given between the other two.

The mixtures are given with the lower boiling component stated first, and the temperatures and constants apply in the same order. Pressure is 760 mm. Hg except where noted.

Mixture	Temp., °C.	$A_{1\text{-}2}$	$A_{2\text{-}1}$	Ref.
Acetaldehyde-ethanol	19.8–78.2	−0.10	−0.20	28
Acetaldehyde-water	19.8–100	.69	.78	6, 36, 41
Acetone-benzene	56.1–80.1	.176	.176	49, 51
Acetone-methanol	56.1–55.5–64.6	.243	.243	3, 18, 39
Acetone-water	56.1–100	.89	.65	3, 8, 18, 57
Acetone-water*	25	.82	.72	2, 52
Benzene-isopropanol	80.1–71.9–82.3	.591	.845	38
n-Butane-furfural†	37.8	1.10	1.26	34
	51.7	1.05	1.17	
	66.6	1.00	1.11	
	93.3	0.91	0.98	
Butanol-butyl acetate	117.7–116.6–126.1	.22	.24	5
Butene-1-furfural†	37.8	.84	1.03	34
	51.7	.80	0.99	
	66.6	.76	.95	
	93.3	.70	.90	
Carbon disulfide–acetone	46.3–39.5–56.1	.556	.778	19, 43, 59
Carbon disulfide–carbon tetrachloride	46.3–76.7	.10	.07	40, 42, 43
Carbon tetrachloride–benzene	76.4–80.2	.052	.046	45
Carbon tetrachloride–ethylene dichloride	76.4–74.5–83.5	.334	.258	25, 58
Ethanol-benzene	78.3–67.0–80.1	.845	.699	15, 29, 53
Ethanol-cyclohexane	78.3–66.3–80.8	.913	.751	37, 54
Ethanol-toluene	78.3–76.4–110.7	.763	.763	56
Ethanol-trichloroethylene	78.3–70.0–87.5	.845	.653	15
Ethanol-water‡	25	.67	.42	11, 48
Ethyl acetate–benzene	77.2–71.1–80.2	.50	.40	46
Ethyl acetate–ethanol	77.2–71.7–78.3	.389	.389	16, 24
Ethyl acetate–toluene	77.2–110.7	.04	.25	30
Ethyl ether–acetone	34.6–56.1	.322	.322	9, 18, 44
Ethyl ether–ethanol	34.6–78.3	.42	.55	10, 24, 31
n-Hexane–ethanol	68.9–59.3–78.3	.68	1.12	22
Isobutane-furfural†	37.8	1.14	1.31	34
	51.7	1.09	1.23	
	66.6	1.04	1.16	
	93.3	0.96	1.03	
Isopropanol-water	82.3–100	1.042	0.492	26
Isopropyl ether–isopropanol	68.5–66.1–82.3	0.42	.60	35
Methanol-benzene	56.1–55.5–64.6	.243	.243	14, 15, 27
Methanol–ethyl acetate	64.6–62.1–77.1	.505	.505	4
Methanol-trichloroethylene	64.6–59.8–87.5	.845	.845	15
Methanol-water	64.6–100	.36	.22	
Methanol-water§	25	.25	.20	7
Methyl acetate–methanol	57.2–53.7–64.6	.462	.462	3, 4
Methyl acetate–water	57.0–100	1.30	.82	32, 33
Methyl ethyl ketone–water	79.6–73.6–100	1.50	.75	33
Propanol-water	97.3–88.0–100	1.10	.492	7, 17, 55
Water-Cellosolve	100–134.5	0.26	.88	1, 12
Water-p-dioxane	100–87.7–101.3	.66	.87	20, 21, 50
Water-phenol	100–181	.36	1.40	47, 49
Water-pyridine	100–115.5	.38	0.62	13

* Pressure = 23.8 to 229.6 mm. Hg.
† Pressure = 5.9 to 13,260 mm. Hg.
‡ Pressure = 23.8 to 59 mm. Hg.
§ Pressure = 23.8 to 123.5 mm. Hg.

References:

[1] Baker, Hubbard, Huguet, and Michalowski, *Ind. Eng. Chem.*, **31**, 1260 (1939).
[2] Beare, McVicar, and Ferguson, *J. Phys. Chem.*, **34**, 1310 (1930).
[3] Bergstrom, from Hausbrand, "Principles and Practice of Industrial Distillation," Wiley, New York, 1926.
[4] Bredig, and Bayer, *Z. physik. Chem.*, **130** (1927).
[5] Brunjes and Furnas, *Ind. Eng. Chem.*, **27**, 396 (1935).
[6] Bushmakin and Kuchinskaya, *Trudy Gosudarst. Opyt. Zavoda Sintet. Kauchuka*, Litera B. IV (1935).
[7] Butler, Thomson, and Maclennan, *J. Chem. Soc.*, 1933, p. 674.
[8] Carveth, *J. Phys. Chem.*, **3**, 193 (1899).
[9] Cunaeus, *Z. physik. Chem.*, **36**, 232 (1901).
[10] Desmaroux, *Mém. poudres*, **23**, 198 (1928).
[11] Dobson, *J. Chem. Soc.*, **127**, 2866 (1925).
[12] Dominik and Wojciechowska, *Przemysl Chem.*, **23**, 61 (1939).
[13] Ewert, *Bull. soc. chim.*, **45**, 493 (1936).
[14] Fink, Mass. Inst. Tech. Thesis, 1933.
[15] Fritzweiler and Dietrich, *Angew. Chem.*, **46**, 241 (1933).
[16] Furnas and Leighton, *Ind. Eng. Chem.*, **29**, 709 (1937).
[17] Gadwa, Mass. Inst. Tech. Thesis, 1936.
[18] Haywood, *J. Phys. Chem.*, **3**, 317 (1899).
[19] Hirschberg, *Bull. soc. chim. Belg.*, **41**, 163 (1932).
[20] Hovorka, Schaefer, and Dreisbach, *J. Am. Chem. Soc.*, **58**, 2264 (1936).
[21] Hovorka, Schaefer, and Dreisbach, *J. Am. Chem. Soc.*, **59**, 2753 (1937).
[22] Isii, *J. Soc. Chem. Ind. Japan*, **38**, 661 (1935).
[23] Kireev, Klinov, and Grigorovich, *J. Chem. Ind. (U.S.S.R.)*, **12**, 936 (1935).
[24] Kireev and Popov, *J. Applied Chem. (U.S.S.R.)*, **7**, 489 (1934).
[25] Kireev and Skvortsova, *J. Phys. Chem. (U.S.S.R.)*, **7**, 63 (1936).
[26] Lebo, *J. Am. Chem. Soc.*, **43**, 1005 (1921).
[27] Lee, *J. Phys. Chem.*, **35**, 3558 (1931).
[28] Leeuw, *Z. physik. Chem.*, **77**, 284 (1911).
[29] Lehfeldt, *Phil. Mag.*, **46**, 42 (1898).
[30] Litkenhous, Van Arsdale, and Hutchison, *J. Phys. Chem.*, **44**, 377 (1940).
[31] Louder, Briggs, and Brown, *Ind. Eng. Chem.*, **16**, 932 (1924).
[32] McKeown and Stowell, *J. Chem. Soc.*, 1927, p. 97.
[33] Marshall, *J. Chem. Soc.*, **89**, 1350 (1906).
[34] Mertes and Colburn, *Ind. Eng. Chem.*, **39**, 787 (1947).
[35] Miller and Bliss, *Ind. Eng. Chem.*, **32**, 123 (1940).
[36] Morozov, Kagan, and Grossblyat, *J. Gen. Chem. (U.S.S.R.)*, **4**, 1322 (1934).
[37] Nagai and Ishii, *Proc. Imp. Acad. (Tokyo)* **11**, 23 (1935).
[38] Olsen and Eastburn, *J. Phys. Chem.*, **41**, 457 (1937).
[39] Othmer, *Ind. Eng. Chem.*, **20**, 743 (1928).
[40] Pahlavouni, *Bull. soc. chim. Belg.*, **36**, 533 (1927).
[41] Pascal, Dupuy, Ero, and Garnier, *Bull. soc. chim.* **29**, 9 (1921). Given also in International Critical Tables.
[42] Rosanoff, Bacon, and White, *J. Am. Chem. Soc.*, **36**, 1803 (1914).
[43] Rosanoff and Easley, *J. Am. Chem. Soc.*, **31**, 953 (1909).
[44] Sameshima, *J. Am. Chem. Soc.*, **40**, 1482 (1918).
[45] Scatchard, Wood, and Moehel, *J. Am. Chem. Soc.*, **62**, 712 (1940).
[46] Schmidt, *Z. physik. Chem.*, **121**, 221 (1926).
[47] Schreinemakers, *Z. physik. Chem.*, **35**, 459 (1900).
[48] Shaw and Butler, *Proc. Roy. Soc. (London)*, **A129**, 519 (1930).
[49] Sims, Mass. Inst. Tech. Thesis, 1933.
[50] Smith and Wojciechowski, *Roczniki Chem.*, **17**, 125 (1937).
[51] Soday and Bennett, *J. Chem. Education*, **7**, 1336 (1930).
[52] Taylor, *J. Phys. Chem.*, **4**, 355 (1900).
[53] Thayer, *J. Phys. Chem.*, **2**, 382 (1898).
[54] Washburn and Handorf, *J. Am. Chem. Soc.*, **57**, 441 (1935).
[55] Wrewsky, *Z. physik. Chem.*, **81**, 1 (1912).
[56] Wright, *J. Phys. Chem.*, **37**, 233 (1933).
[57] York and Holmes, *Ind. Eng. Chem.*, **34**, 345 (1942).
[58] Young and Nelson, *Ind. Eng. Chem.*, anal. ed., **4**, 67 (1932).
[59] Zawidski, *Z. physik. Chem.*, **35**, 129 (1900).

Ternary Margules Equations. Equations for activity coefficients of ternary mixtures that utilize the constants from the three binary pairs are extremely useful. Wohl (*loc. cit.*) has derived both Margules and Van Laar types of such equations, as well as more complicated formulas that will not be given here. The equations that Wohl calls three-suffix Margules are identical algebraically with the three-suffix equations used by Benedict, Johnson, Solomon, and Rubin [*Trans. Am. Inst. Chem. Engrs.*, **41**, 371–392 (1945)]. For the first component,

$$\begin{aligned}\log \gamma_1 = {} & x_2^2[A_{1\text{-}2} + 2x_1(A_{2\text{-}1} - A_{1\text{-}2})] \\ & + x_3^2[A_{1\text{-}3} + 2x_1(A_{3\text{-}1} - A_{1\text{-}3})] \\ & + x_2x_3[A_{2\text{-}1} + A_{1\text{-}3} - A_{3\text{-}2} + 2x_1(A_{3\text{-}1} - A_{1\text{-}3}) \\ & + 2x_3(A_{3\text{-}2} - A_{2\text{-}3}) - C(1 - 2x_1)] \qquad (21)\end{aligned}$$

Equations may be written for the second component by changing the subscripts from 1 to 2, from 2 to 3, and from 3 to 1; and in the same manner a relation for the third component can be obtained.

The great convenience of this form of Eq. (21) is that the constants $A_{1\text{-}2}$ and $A_{2\text{-}1}$ are end values of the logarithms of activity coefficients in the binary system.

1-2, etc. The constant C is evaluated from ternary data. When ternary equilibrium values are predicted from binary data only, the constant C is calculated from the relation:

$$C = \tfrac{1}{2}(A_{2\text{-}1} - A_{1\text{-}2} + A_{1\text{-}3} - A_{3\text{-}1} + A_{3\text{-}2} - A_{2\text{-}3})$$

according to Wohl (personal communication, 1947).

Ternary data on the system n-butane, butene-1, furfural are given by Gerster, Mertes, and Colburn [*Ind. Eng. Chem.*, **39**, 797 (1947)]. These data were found to be in good agreement with Eq. (21), in which the binary constants for hydrocarbon-furfural mixtures were evaluated from binary mixtures as given in Table 2, and the butane-butene mixture was assumed to be ideal, *i.e.*, $A_{1\text{-}2} = A_{2\text{-}1} = 0$. The ternary constant C was evaluated to be -0.20 at 100°F. and -0.17 at 200°F.

Direct Prediction of Values of Relative Volatility. For systems following the gas laws, the relative volatility is defined by Eq. (11)

$$\alpha_{1\text{-}2} = \frac{\gamma_1 P_1}{\gamma_2 P_2}$$

The ratio γ_1/γ_2 has been directly related to the constants in Eq. (21) by Wohl (personal communication), for ternary systems:

$$\log \frac{\gamma_1}{\gamma_2} = A_{2\text{-}1}(x_2 - x_1) + x_2(x_2 - 2x_1)(A_{1\text{-}2} - A_{2\text{-}1}) + x_3[A_{1\text{-}3} - A_{3\text{-}2} + 2x_1(A_{3\text{-}1} - A_{1\text{-}3}) - x_3(A_{2\text{-}3} - A_{3\text{-}2}) - C(x_2 - x_1)] \qquad (22)$$

In this relation, the constants $A_{2\text{-}1}$, etc., are the binary end constants, and C is the ternary constant. Equations may be written for $\log \gamma_2/\gamma_3$ and for $\log \gamma_3/\gamma_1$ by cyclic permutation of subscripts.

For systems where components 1 and 2 are ideal, like butane-butene-1-furfural, $A_{1\text{-}2} = A_{2\text{-}1} = 0$, and Eq. (22) simplifies to

$$\log \frac{\gamma_1}{\gamma_2} = x_3[A_{1\text{-}3} - A_{3\text{-}2} + 2x_1(A_{3\text{-}1} - A_{1\text{-}3}) - x_3(A_{2\text{-}3} - A_{3\text{-}2}) - C(x_2 - x_1)] \qquad (22a)$$

For the system n-butane-butene-1-furfural at 150°F., $A_{1\text{-}3} = 1.00$, $A_{3\text{-}2} = 0.95$, $A_{3\text{-}1} = 1.11$, $A_{2\text{-}3} = 0.76$, and $C = -0.19$, and Eq. (23) therefore becomes

$$\log \frac{\gamma_1}{\gamma_2} = x_3[1.00 - 0.95 + 2x_1(1.11 - 1.00) - x_3(0.76 - 0.95) + 0.1(x_2 - x_1)] = x_3(0.05 - 0.03x_1 + 0.19x_3 + 0.19x_2)$$

At $x_3 = 0.80$, $x_1 = 0.10$, $x_2 = 0.10$, $\log \gamma_1/\gamma_2 = 0.172$. In calculating values of α for this system, Gerster *et al.* (*loc. cit.*) found it necessary to make appreciable gas-law corrections by methods described later.

Quaternary Equations. Wohl (personal communication, 1946) has developed convenient expressions of the general three-suffix Margules equation for activity coefficients and for values of ratios of activity coefficients. Thus, for activity coefficients,

$$\begin{aligned}\log \gamma_1 = {} & x_2^2[A_{1\text{-}2} + 2x_1(A_{2\text{-}1} - A_{1\text{-}2})] \\ & + x_3^2[A_{1\text{-}3} + 2x_1(A_{3\text{-}1} - A_{1\text{-}3})] \\ & + x_4^2[A_{1\text{-}4} + 2x_1(A_{4\text{-}1} - A_{4\text{-}2})] \\ & + x_2x_3[A_{2\text{-}1} + A_{1\text{-}3} - A_{3\text{-}2} + 2x_1(A_{3\text{-}1} - A_{1\text{-}3}) \\ & + 2x_3(A_{3\text{-}2} - A_{2\text{-}3}) - (1 - 2x_1)C_{123}] \\ & + x_2x_4[A_{2\text{-}1} + A_{1\text{-}4} - A_{4\text{-}2} + 2x_1(A_{4\text{-}1} - A_{1\text{-}4}) \\ & + 2x_4(A_{4\text{-}2} - A_{2\text{-}4}) - (1 - 2x_1)C_{124}] \\ & + x_3x_4[A_{3\text{-}1} + A_{1\text{-}4} - A_{4\text{-}3} + 2x_1(A_{4\text{-}1} - A_{1\text{-}4}) \\ & + 2x_4(A_{4\text{-}3} - A_{3\text{-}4}) - (1 - 2x_1)C_{134}] \\ & + x_2x_3x_4{}^2(A_{4\text{-}2} - A_{2\text{-}4} + C_{234})\end{aligned} \qquad (23)$$

The equations for the other components are obtained by cyclic permutation of the subscripts, *e.g.*, writing 2 for 1, 3 for 2, 4 for 3, and 1 for 4. Note that there are four ternary constants, one for each ternary as in Eqs. (21) and (22), but no constant specifically for the quaternary is required in this equation.

Equations for ratios of two activity coefficients in quaternary mixtures are correspondingly simpler than for the single value. Thus

$$\begin{aligned}\log \frac{\gamma_1}{\gamma_2} = {} & A_{2\text{-}1}(x_2 - x_1) + x_2(A_{1\text{-}2} - A_{2\text{-}1})(x_2 - 2x_1) \\ & + x_3[A_{1\text{-}3} - A_{3\text{-}2} + 2x_1(A_{3\text{-}1} - A_{1\text{-}3}) - x_3(A_{2\text{-}3} - A_{3\text{-}2})] \\ & + x_4[A_{1\text{-}4} - A_{4\text{-}2} + 2x_1(A_{4\text{-}1} - A_{1\text{-}4}) - x_4(A_{2\text{-}4} - A_{4\text{-}2})] \\ & - (x_3C_{123} + x_4C_{124})(x_2 - x_1) - x_3x_4 \\ & [A_{1\text{-}3} - A_{3\text{-}1} + C_{134} - (A_{4\text{-}2} - A_{2\text{-}4} + C_{234})]\end{aligned} \qquad (23a)$$

Values of $\log \gamma_2/\gamma_3$, $\log \gamma_3/\gamma_4$, and $\log \gamma_4/\gamma_1$ are obtained by cyclic permutation of the subscripts. To obtain values of $\log \gamma_1/\gamma_3$

$$\begin{aligned}\log \frac{\gamma_1}{\gamma_3} = {} & A_{1\text{-}3}(x_3 - x_1) + x_1(A_{3\text{-}1} - A_{1\text{-}3})(2x_3 - x_1) \\ & + x_2[A_{2\text{-}1} - A_{3\text{-}2} + x_2(A_{1\text{-}2} - A_{2\text{-}1}) - 2x_3(A_{2\text{-}3} - A_{3\text{-}2})] \\ & + x_4[A_{1\text{-}4} - A_{4\text{-}3} + 2x_1(A_{4\text{-}1} - A_{1\text{-}4}) - x_4(A_{3\text{-}4} - A_{4\text{-}3})] \\ & - [x_2C_{123} + x_4(A_{1\text{-}3} - A_{3\text{-}1} + C_{134})](x_3 - x_1) - x_2x_4 \\ & [C_{124} - (A_{4\text{-}2} - A_{2\text{-}4} + C_{234})]\end{aligned} \qquad (23b)$$

The value of $\log \gamma_2/\gamma_4$ is obtained by cyclic permutation. It might be noted that these relations reduce to the ternary forms if the composition of one of the components becomes zero.

Effect of Temperature on Liquid-activity Coefficients. In general, values of activity coefficients tend toward unity with increasing temperature, with the rate of change more or less proportional to the magnitude of the activity coefficient. The quantitative effect is related to the relative partial molal enthalpy L by the thermodynamic relation

$$\frac{2.3d \log \gamma_1}{dT} = -\frac{L_1}{RT^2} \qquad (24)$$

where L_1 is the partial molal enthalpy of component 1 in solution minus the enthalpy of the pure liquid at the same temperature, T is the absolute temperature, and R is the gas-law constant in heat units [*i.e.*, 1.99 p.c.u./(lb.-mole)(°K.)]. When heat is evolved on mixing two liquids, L is negative and the activity coefficient rises with temperature. This is the case for most systems with negative deviations, *e.g.*, electrolytes; whereas, for systems with positive deviations, usually heat is absorbed on mixing, L is positive, and activity coefficients decrease with increasing temperature. These conditions support the generality stated above, since in both cases the activity coefficients tend toward unity. For some aqueous solutions the value of L changes in sign from negative to positive with increasing temperature, and the curve of activity coefficient vs. temperature must go through a maximum (*cf.* Fig. 3).

For systems where the activity coefficients are related by the Margules or Van Laar equations, which utilize the terminal values of $\log \gamma$ as constants, the effect of temperature on the constants is

$$\frac{dA_{1\text{-}2}}{dT} = \frac{-L_1^\circ}{2.3RT^2};\quad \frac{dA_{2\text{-}1}}{dT} = \frac{-L_2^\circ}{2.3RT^2} \qquad (25)$$

where L_1° and L_2° are the heats absorbed on solution at infinite dilution of 1 mole of component 1 and 2, respectively (as liquids). It is apparent that Eqs. (25) are particularly convenient in that the effect of temperature on activity coefficients at all compositions can be expressed in terms of the heats of solution at infinite dilution and of the Margules or Van Laar equations.

Equations (25) can also be written

$$\frac{dA_{1\text{-}2}}{d(1/T)} = \frac{L_1^\circ}{2.3R};\quad \frac{dA_{2\text{-}1}}{d(1/T)} = \frac{L_2^\circ}{2.3R} \qquad (26)$$

If the values of L° are nearly constant with temperature, Eqs. (26) are integrated to give the change between two temperatures as

$$(A_{1\text{-}2})_2 - (A_{1\text{-}2})_1 = \frac{(L_1^\circ)_{\text{avg}}}{2.3R}\left(\frac{1}{T_2} - \frac{1}{T_1}\right) \qquad (27a)$$

$$(A_{2\text{-}1})_2 - (A_{2\text{-}1})_1 = \frac{(L_2^\circ)_{\text{avg}}}{2.3R}\left(\frac{1}{T_2} - \frac{1}{T_1}\right) \qquad (27b)$$

Equations (27) are also the exact solutions for Eq. (26) in case the values of $L°$ are linear functions of $1/T$, (*i.e.*, $L_1° = a + b/T$), if the values of $(L°)_{avg}$ are the arithmetic averages of the values at T_1 and T_2. For example, data are plotted in Fig. 3 for the values of $L°/2.3R$ for methanol in water. Suppose the value of $A_{1\text{-}2}$ is given at 80°C. as 0.36 and one wishes the value at 25°C. The average ordinate over this temperature range is $L°/2.3R = -265$. By Eq. (27*a*),

$$(A_{1\text{-}2})_{25} - (A_{1\text{-}2})_{80} = -265(\tfrac{1}{298} - \tfrac{1}{353})$$

or

$$(A_{1\text{-}2})_{25} = 0.36 - 265(0.00336 - 0.00284) = 0.36 - 0.14 = 0.22.$$

According to Table 2, the experimental value of $A_{1\text{-}2}$ at 25°C. is 0.25; the correction is therefore approximately right.

Where the values of $L°$ vary with $(1/T)$ in a more complicated manner, solution of Eq. (26) is effected by graphical integration according to the formula

$$(A_{1\text{-}2})_2 - (A_{1\text{-}2})_1 = \frac{1}{2.3R}\int_{1/T_1}^{1/T_2} L_1° d\left(\frac{1}{T}\right) \qquad (28)$$

Plots are made of $L_1°$ and $L_2°$ vs. $1/T$, and the area under the curves between the respective temperatures permits calculation of the changes in the values of A.

Available data on heats of solution have been calculated in terms of the values at infinite dilution and are given in Tables 3 and 4. Unfortunately most of these data are limited to a single temperature, around 20°C. Where data have been determined over a range of temperatures, the values of L have been found, in case of aqueous solutions, to vary exceedingly with temperature, even going from positive to negative on occasion, as shown in Fig. 3. It is therefore not advisable to utilize values of L in temperature ranges beyond where the values have been determined, except qualitatively. If data on relative partial molal heat capacities are available, these can be used to predict the change of L with temperature,

$$\frac{dL_1°}{dT} = (MC)_1° - (MC)_1 \qquad (29)$$

where $(MC)_1°$ is the partial molal heat capacity at infinite dilution and $(MC)_1$ is the molal heat capacity of the pure component. These data are available on only a few systems, and they are also found to be quite dependent on temperature, at least in aqueous systems.

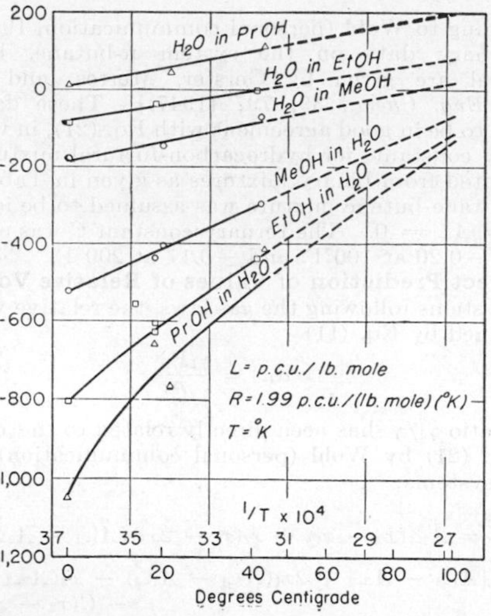

FIG. 3. Heats of solution at infinite dilution of alcohols and water. L = p.c.u./lb.-mole, R = 1.99 p.c.u./(lb.-mole)(°K.).

An approximate rule sometimes used for the effect of temperature on activity coefficients is given by $T \log \gamma_1$ = const., or $TA_{1\text{-}2}$ = const. This relation presumes that $L_1°/R = A_{1\text{-}2}T$, and holds only very qualitatively, *i.e.*, the more non-ideal the system, usually the larger the values of both A and L.

Table 3. Heats of Solution and Heat Capacities at Infinite Dilution of Aqueous Solutions

$MC°$ = partial molal heat capacity at infinite dilution
MC = molal heat capacity of pure component

System	t, °C.	$L_1°$ (p.c.u./lb.-mole)	$L_2°$ (p.c.u./lb.-mole)	Ref.	t, °C.	$(MC° - MC)_1$ (p.c.u./(lb.-mole)(°C.))	$(MC° - MC)_2$ (p.c.u./(lb.-mole)(°C.))	Ref.
Acetic acid–water	17	− 380	595	3, p. 1559	38	4	0.3	1, p. 115
	7	− 400	.....	1, p. 148				
	23	− 240	.....	1, p. 148				
	18.5	− 454	.....	1, p. 159				
	Room	− 122	347	1, p. 159				
Acetone-water	15	−2600	− 90	1, p. 157				
	17	−2780	.....	2, p. 852				
	Room	−2510	.....	1, p. 148				
Allyl alcohol–water	Room	−2000	.....	1, p. 148				
Aniline-water	16	179	.....	1, p. 149				
	24	550	.....	1, p. 149				
Butyric acid–water	9	120	1440	1, p. 159				
Ethanol-water	0	−3680	−405	1, p. 159	3	49.5	5.1	1, p. 116
	17.3	−2870	−180	1, p. 159	23	32.2	7.4	1, p. 116
	42.1	−2000	− 12	1, p. 160	41	34.3	4.5	1, p. 116
	17	−2760	.....	2, p. 852				
	13	−2540	.....	1, p. 148				
Formic acid–water	Room	− 296	−550	1, p. 159				
	7	− 79		1, p. 148				
Isoamyl alcohol–water	Room	−2800	.....	1, p. 149				
Isobutyl alcohol–water	Room	−2900	.....	1, p. 149				
Isopropanol-water	17	−4800	180	3, p. 1559				
	Room	−3760	.....	1, p. 148				
Methanol-water	0	−2390	−810	1, p. 159	5	28.85	2.6	1, p. 116
	19.7	−1840	−670	1, p. 159	20	16.4	2.4	1, p. 116
	42.4	−1340	−330	1, p. 159	40	12.4	4.8	1, p. 116
	Room	−2000	.....	1, p. 148				
Normal propanol–water	0	−4800	53	1, p. 160	5	59.9	6.2	1, p. 116
	21	−3500	200	1, p. 160	20	58.3	6.1	1, p. 116
	43.4	−1900	515	1, p. 160	40	41.2	12.1	1, p. 116
	Room	−3050	.....	1, p. 148				
Phenol-water	71	2100	530	3, p. 1559				
Propionic acid–water	8	239	2630	1, p. 159				
	Room	− 620	.....	1, p. 148				

References:
[1] "International Critical Tables," vol. 5.
[2] Landolt-Börnstein, "Physikalische-Chemische Tabellen," 1927.
[3] *Ibid.*, 1931.

Table 4. Heats of Solution and Heat Capacities at Infinite Dilution of Non-aqueous Solution

System	t, °C.	$L_1°$	$L_2°$	Ref.	t, °C.	$(MC° - MC)_1$	$(MC° - MC)_2$	Ref.
		p.c.u./lb.-mole				p.c.u./(lb.-mole)(°C.)		
Acetaldehyde-ethanol	Room	1800	2600	1, p. 156				
Acetone-benzene	Room	310	260	1, p. 153	10	− 4.1	−13.1	1, p. 127
					30	− 5.9	− 9.7	1, p. 127
					50	− 2.8	− 9.3	1, p. 127
Acetone-chloroform	14	−2200	−1375	1, p. 158	−40	43.2	− 9.2	1, p. 126
	25	−2010	−1200	1, p. 155	−10	30.6	18.8	1, p. 126
	Room	−1910	−1160	1, p. 151	20	13.3	25.7	1, p. 126
Acetone-ethanol	25	116	1250	3, p. 1560	35	5.1	21.8	1, p. 126
	Room	......	1220	1, p. 152				
-methanol	25	530	740	3, p. 1559				
	Room	500	......	1, p. 151				
-isopropanol	20	1400	1400	3, p. 1563				
Benzene-aniline	20	530	890	2, p. 857				
	Room	600	1160	1, p. 154				
-cyclohexane	Room	8250	7950	1, p. 157				
-ethylene bromide	Room	276	276	1, p. 156				
-normal hexane	Room	930	930	1, p. 157				
-nitrobenzene	20	300	200	2, p. 855	10	−39.5	−18.8	1, p. 128
					30	−30	−14.1	1, p. 128
					50	−22.6	−11.8	1, p. 128
					75	−11.7	− 8.6	1, p. 128
-normal propanol	15	515	3000	1, p. 158				
	Room	540	3500	1, p. 153				
n-Butane-furfural	40–100	1770	......	4				
Butene-1-furfural	40–100	1330	......	4				
-benzene	4	419	705	1, p. 158				
	14.5	415	700	1, p. 158				
	25	478	658	1, p. 155				
	Room	......	670	1, p. 151				
-carbon tetrachloride	25	280	370	3, p. 1563				
	Room	......	420	1, p. 150				
-chloroform	13	533	710	1, p. 157	−30	− 3.4	−18.8	1, p. 125
	25	473	575	1, p. 155	−10	− 3.1	−11.4	1, p. 125
	Room	......	580	1, p. 151	20	− 1.6	− 5.0	1, p. 125
Carbon disulfide-acetone	16	1060	1920	1, p. 158				
	Room	......	1790	1, p. 151				
-benzene	18	495	615	1, p. 158				
	0	420	709	1, p. 158				
-ethanol	0	565	540	1, p. 157				
	4.4	580	650	1, p. 157				
	15.5	630	770	1, p. 157				
	Room	......	1600	1, p. 151				
-methanol	20	830	3800	2, p. 853				
-toluene	18	268	417	1, p. 158				
Carbon tetrachloride-benzene	18	117	110	1, p. 157	20	−16.5	−18.4	1, p. 125
	25	116	116	1, p. 155	35	− 8.6	−13.2	1, p. 125
	Room	160	117	1, p. 151	50	− 5.3	− 7.3	1, p. 125
-chloroform	25	255	255	1, p. 155				
Chloroform-benzene	18	− 364	− 585	1, p. 158	6	8.6	0.3	1, p. 126
	25	− 396	520	1, p. 155	20	4.9	5.1	1, p. 126
	Room	− 239	− 430	1, p. 151	55	2.0	2.5	1, p. 126
-ethanol	25	−1350	2600	3, p. 1560				
	Room	−1440	......	1, p. 151				
-methanol	25	−1200	2000	3, p. 1559				
	Room	−1140	......	1, p. 151				
-normal propanol	25	−1250	2400	3, p. 1562				
Ethanol-benzene	0	770	540	1, p. 158	15	22.9	5.2	1, p. 126
	4.5	830	545	1, p. 158	20	4.4	4.0	1, p. 116
	15	1000	560	1, p. 158				
	Room	2480	300	1, p. 159				
	Room	1500	270	3, p. 1560				
	20	1800	400	3, p. 1560				
	Room	4000	360	1, p. 152				
-ortho-cresol	32.5	1030	2300	3, p. 1560				
-cyclohexane	52	1400	840	2, p. 854				
Ethyl acetate-benzene	17	216	156	1, p. 158				
	25	98	105	1, p. 156				
	Room	160	141	1, p. 153				
-ethanol	25	1160	1480	3, p. 1560				
	Room	1150	1800	1, p. 152				
Ethyl ether-acetone	25	435	615	2, p. 854; 3, p. 1562	−40	−37.9	− 9.3	1, p. 127
					−20	−31.9	− 8.3	1, p. 127
					0	−23.8	− 4.8	1, p. 127
					20	−14.5	− 2.8	1, p. 127
-aniline	20	− 310	− 860	1, p. 157	20	5.0	7.3	1, p. 127
-benzene	Room	100	......	1, p. 153	6	24	6.3	1, p. 116
	15	0	0	1, p. 158	6	35.3	6.6	1, p. 127
					20	19.7	6.3	1, p. 127
-bromoform	Room	−2050	−2300	2, p. 856				
-carbon disulfide	25	560	370	3, p. 1563	−30	− 5.5	− 4.2	1, p. 126
	Room	1000	......	1, p. 151	−10	− 4.3	− 2.4	1, p. 126
					20	− 1.8	− 0.6	1, p. 126
-chloroform	14	−1970	−2270	1, p. 158				
	25	−2250	−2180	1, p. 155	−50	85.4	83.4	1, p. 126
	Room	−2100	−2010	1, p. 151	−30	63.4	89.6	1, p. 126
	3	−2820	−2700	5	−10	45.2	44.6	1, p. 126
	15	−2300	−2300	5	0	36.2	35.6	1, p. 126
	24	−1950	−1950	5	20	13.2	14.6	1, p. 126
-ethanol	25	290	1400	3, p. 1560				
	Room	360	1100	3, p. 1560				
	Room	......	910	1, p. 152				

Table 4. Heats of Solution and Heat Capacities at Infinite Dilution of Non-aqueous Solution—(*Concluded*)

System	t, °C.	$L_1°$	$L_2°$	Ref.	t, °C.	$(MC° - MC)_1$	$(MC° - MC)_2$	Ref.
		p.c.u./lb.-mole				p.c.u./(lb.-mole)(°C.)		
-ethyl acetate	14	234	256	1, p. 158				
	25	256	268	1, p. 156				
-methanol	25	100	640	3, p. 1559				
	Room	......	600	1, p. 151				
-nitrobenzene	20	100	− 45	2, p. 855	20	0	0.6	1, p. 127
					5	−36.8	−34	1, p. 127
					20	−35.2	−34.6	1, p. 127
					35	−26.0	−21.6	1, p. 127
-normal propanol	25	460	1200	3, p. 1562				
Isobutane-furfural	40–100	1770	......	4				
Methanol-benzene	15	2550	327	1, p. 158				
	Room	2800	360	1, p. 151				
Methyl acetate–benzene	17	450	470	1, p. 158				

References:
[1] "International Critical Tables," vol. 5.
[2] Landolt-Börnstein, "Physikalisch-Chemische Tabellen," 1927.
[3] *Ibid.*, 1931.
[4] Mertes and Colburn, *Ind. Eng. Chem.*, **39**, 787 (1947).
[5] Macleod and Wilson, *Trans. Faraday Soc.*, **31**, 596 (1935).

UTILIZATION OF ACTIVITY COEFFICIENTS

Prediction of Absorption Equilibrium. In the absence of data for the conditions at hand, vapor-liquid equilibrium data for absorption can often be estimated from data at the boiling point, by use of activity coefficients. For the systems listed in Table 2, the values of $A_{1\text{-}2}$ and $A_{2\text{-}1}$ can be used under certain conditions.

Henry's Law. This relationship is often utilized for the solute in dilute solution and provides that, at a given temperature, the ratio of y_1/x_1 = const. The limitation in the use of this relation is found from the relation of Eq. (10), that $y_1/x_1 = \gamma_1 P_1/P$. For Henry's law to hold, γ_1 must be a constant. Inspection of Figs. 1 and 2 and of the Margules and Van Laar equations indicates that this relation holds only for very dilute solutions. Thus, for acetone-water at 25°C., the values of $A_{1\text{-}2}$ and $A_{2\text{-}1}$ from Table 1 are 0.82 and 0.72. The values of γ_1 are found from Eq. (16*a*) to be

x_1 =	0	0.01	0.02	0.05	0.10
γ_1 =	6.6	6.35	5.95	5.38	4.45

Thus, even at a concentration of only 1 mole per cent, the ratio of y_1/x_1 is about 4 per cent less and at 2 mole per cent concentration is 10 per cent less than at essentially zero concentration. This effect is greater the greater the values of A, and, of course, it is absent when $A = 0$, *i.e.*, when Raoult's law holds. In general it would appear to be unsafe to utilize Henry's law for concentrations greater than 1 mole per cent (equivalent to 0.556 mole of solute per liter in case of aqueous solutions).

When utilizing values in Table 2 to predict absorption equilibrium, and the values are for different temperatures, one can correct them by Eq. (27) if data on heats of solution are available. Otherwise, as a first approximation it can be assumed that the temperature effect on γ is negligible compared with the effect on the vapor pressure.

Prediction of Azeotropes. Inasmuch as an azeotrope (or constant-boiling mixture) occurs for miscible solutions when $\gamma_1 P_1/\gamma_2 P_2 = 1$, the existence of an azeotrope can be readily determined when data on vapor pressure and activity coefficients are available. A convenient procedure is to calculate the ratio γ_2/γ_1 and plot this vs. mole fraction of the low boiler, x_1. Another plot may be made of the ratio P_1/P_2 vs. temperature. An azeotrope occurs for conditions under which values of ordinates of the two curves are equal. If the vapor-pressure ratio is outside of the range of values of γ_2/γ_1, no azeotrope occurs.

For example, Fig. 4 shows curves of γ_1/γ_2 and P_2/P_1 vs. x, and temperature, respectively, for the system ethyl acetate–ethanol. At 80°C., $P_2/P_1 = 0.97$; for $\gamma_1/\gamma_2 = 0.97$, $x_1 = 0.51$. Again, at 20°C., $P_2/P_1 = 0.61$, for $\gamma_1/\gamma_2 = 0.61$, $x_1 = 0.78$. Thus the composition of the azeotrope changes from 0.51 mole fraction methyl acetate at 80°C. to 0.78 mole fraction acetate at 20°C. It should be noted that the ratios of γ_1/γ_2 used were obtained from data near 80°C. and the extrapolation to 20°C. involves some uncertainty.

To find the corresponding total pressures in the above examples, one recalls that, for a given component at an azeotrope, $\gamma_1 = P/P_1$.

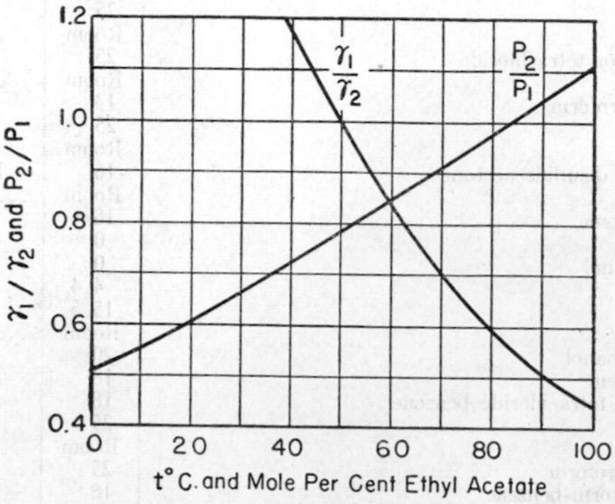

FIG. 4. Effect of temperature on the composition of the ethyl-acetate-ethyl alcohol azeotrope. At a given temperature the azeotropic composition is where $P_2/P_1 = \gamma_1/\gamma_2$.

Prediction of Heats of Solution. One of the main values of Eq. (26) is to permit the use of data on vapor-liquid equilibrium at various temperatures to obtain values of heats of solution useful in making exact thermal balances in distillation and similar problems. Values of heats of solution at infinite dilution are readily obtained by plotting values of A vs. $1/T$, and the slopes of the curves at any temperatures are equal to values of $L°/2.3R$. For values of relative partial molal heats of solution at other concentrations than infinite dilution, one may compute values of log γ_1 and log γ_2 at various temperatures with the Margules or Van Laar equations, plot these values vs. $1/T$, and from the slopes obtain values of $L/2.3R$. It is perhaps simpler to utilize the temperature relations of the terminal constants with the Margules equations as follows:

$$\frac{L_1}{2.3R} = \frac{d \log \gamma_1}{d(1/T)} = x_2^2 \left\{ \frac{dA_{1\text{-}2}}{d(1/T)} + 2x_1 \left[\frac{dA_{2\text{-}1}}{d(1/T)} - \frac{dA_{1\text{-}2}}{d(1/T)} \right] \right\} \tag{30a}$$

$$\frac{L_2}{2.3R} = x_1^2 \left\{ \frac{dA_{2\text{-}1}}{d(1/T)} + 2x_2 \left[\frac{dA_{1\text{-}2}}{d(1/T)} - \frac{dA_{2\text{-}1}}{d(1/T)} \right] \right\} \tag{30b}$$

where L_1 and L_2 are the relative partial molal heats absorbed on solution at concentrations x_1 and x_2, and $dA_{1\text{-}2}/d(1/T)$ and $dA_{2\text{-}1}/d(1/T)$ are the slopes of the curves of $A_{1\text{-}2}$ and $A_{2\text{-}1}$ vs. $1/T$. Values of integral heats of solution are then readily obtained from the usual equation:

$$H = L_1 x_1 + L_2 x_2 \tag{31}$$

where H = molal integral heat of solution, *i.e.*, the heat absorbed when 1 mole of solution is formed from the pure liquid components.

Prediction of Immiscibility. Given values of Van Laar constants, $A_{1\text{-}2}$ and $A_{2\text{-}1}$, it is possible to check if a region of immiscibility exists. For example, for symmetrical systems, *i.e.*, where $A_{1\text{-}2} = A_{2\text{-}1}$, immiscibility occurs for values of A greater than 0.87 (*cf.* Hildebrand, "Solubility of Non-electrolytes," Reinhold, New York, 1936). A convenient table relating the constants to degrees of solubility for symmetrical systems follows:

Van Laar or Margules Constants, $A_{1\text{-}2} = A_{2\text{-}1}$	Solubility Limits, Mole Fraction
0.87	0.5
.92	.3
1.0	.2
1.2	.1
1.43	.05
1.77	.02
2.03	.01

For non-symmetrical systems, immiscibility may occur where one Van Laar constant is less than 0.87 if the other is considerably greater. A test is readily made against the dashed curve on

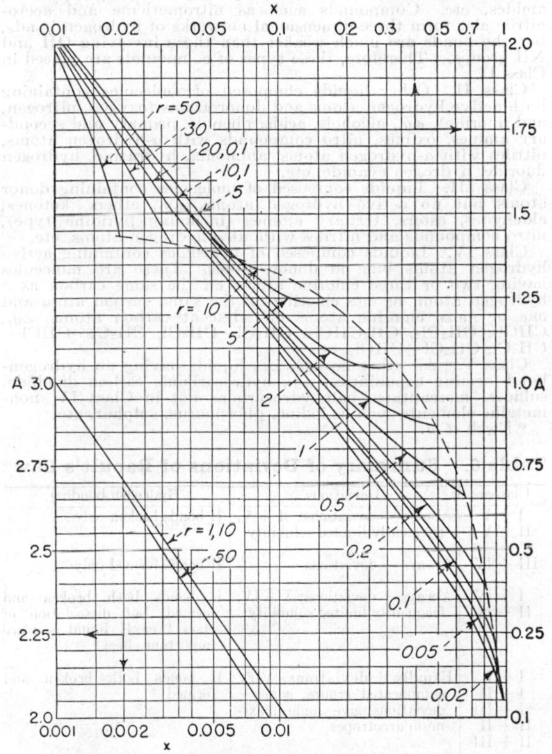

FIG. 5. Van Laar constants for binary immiscible systems.
A_{1-2}: $\gamma = x_2/x_1$; x_1
A_{2-1}: $\gamma = x_1/x_2$; x_2

Fig. 5, which represents the locus of critical solution points. Thus, using one constant, say $A_{1\text{-}2}$, a value of x_1 can be read from the curve. Subtracting this from unity gives the composition of the other component (at the critical solution point). From the second composition and the dashed curve, the corresponding value of $A_{2\text{-}1}$ is read for the critical point. If the given value of $A_{2\text{-}1}$ is greater, immiscibility occurs; if less, the solution is miscible throughout all compositions.

The solubility limits, in case of large values of $A_{1\text{-}2}$ and $A_{2\text{-}1}$, where non-symmetrical, can be found by trial and error, using Fig. 5.

Prediction of Distribution in Liquid-liquid Extraction. When a solute is distributed between two solvent liquids, at equilibrium its activity must be the same in both solvents, or

$$\gamma_{1\text{-}2}x_{1\text{-}2} = \gamma_{1\text{-}3}x_{1\text{-}3} \tag{32}$$

where $x_{1\text{-}2}$ and $x_{1\text{-}3}$ are the mole fractions and $\gamma_{1\text{-}2}$ and $\gamma_{1\text{-}3}$ the activity coefficients of component 1 in the liquid phases, composed mainly of components 2 and 3, respectively. The distribution ratio, $x_{1\text{-}2}/x_{1\text{-}3}$, then becomes

$$\frac{x_{1\text{-}2}}{x_{1\text{-}3}} = \frac{\gamma_{1\text{-}3}}{\gamma_{1\text{-}2}} \tag{33}$$

so that knowledge of the activity coefficients permits calculation of the ratio. If the components 2 and 3 are completely immiscible, even with the given amount of component 1 dissolved, then values of $\gamma_{1\text{-}2}$ and $\gamma_{1\text{-}3}$ are calculated as for binary systems. A convenient method of utilizing a series of such calculations, as shown by Hildebrand (*loc. cit.*) is to plot $\gamma_{1\text{-}2}x_{1\text{-}2}$ vs. $x_{1\text{-}2}$ and $\gamma_{1\text{-}3}x_{1\text{-}3}$ vs. $x_{1\text{-}3}$; from the same ordinate, equilibrium values of $x_{1\text{-}2}$ and $x_{1\text{-}3}$ are read from the two curves.

The Van Laar constants in Table 2 can be utilized as a first approximation, with the main problem being the fact that most of these values are for temperatures higher than normally employed in liquid-liquid extraction. As data on heats of solution become available, the constants can be corrected for temperature effect.

Where the two solvents are appreciably miscible, each of the solutions must be treated as a ternary mixture. In such cases values of $\gamma_{1\text{-}2}$ and $\gamma_{1\text{-}3}$ can be calculated by means of the ternary Margules or Van Laar equations, provided that the ternary solubility curve is known. A treatment of other methods of computing ternary values of γ is given by Treybal [*Ind. Eng. Chem.*, **36,** 875 (1944)].

When two fairly similar components are being separated by a solvent in which each is partly miscible, design calculations follow distillation methods and utilize the function, relative distribution ratio, as parallel to relative volatility. The relative distribution ratio β is defined:

$$\beta = \frac{x_{1\text{-}3}/x_{2\text{-}3}}{x_{1\text{-}2}/x_{2\text{-}2}} = \frac{\gamma_{1\text{-}2}/\gamma_{2\text{-}2}}{\gamma_{1\text{-}3}/\gamma_{2\text{-}3}} \tag{34}$$

where the subscripts 1-3 and 2-3 refer to compositions and activity coefficients in the solvent layer, and 1-2 and 2-2 refer to the layer composed mainly of components 1 and 2. Sometimes this ratio can be evaluated at the limiting concentrations of $x_1 = 0$ and $x_2 = 0$, based on values of activity coefficients for the binary systems calculated from the solubility data. An example of this procedure is given by Colburn and Schoenborn [*Trans. Am. Inst. Chem. Engrs.*, **41,** 421 (1945)].

PREDICTION OF ACTIVITY COEFFICIENTS

Estimation from Azeotropic Data. Given reliable measurements of composition, temperature, and total pressure of an azeotrope, one can readily calculate activity coefficients from the equations (since $y = x$ at an azeotrope):

$$\gamma_1 = \frac{P}{P_1} \qquad \gamma_2 = \frac{P}{P_2} \tag{35}$$

These values may be substituted in Eq. (19) or (20) and a complete equilibrium curve computed using the resulting constants $A_{1\text{-}2}$ and $A_{2\text{-}1}$. This procedure cannot be used satisfactorily to obtain both constants when one component is rather concentrated at the azeotrope, since the resulting activity coefficient will be close to unity so that the logarithm, which is approximately equal to $(\gamma - 1)/2.3$, cannot be known with any precision.

Example. An azeotrope in the water-dioxane system occurs at $x_1 = 0.53$ mole fraction water, with a boiling point of 87.8°C. at atmospheric pressure. At this temperature the vapor pressures of water and dioxane are 484 and 490 mm. Hg, respectively. Then, by Eq. (11), $\gamma_1 = {}^{760}\!/_{484} = 1.57$, and $\gamma_2 = {}^{760}\!/_{490} = 1.55$. Substituting in Eqs. (19) one obtains the

Margules constants $A_{1\text{-}2} = 0.664$ and $A_{2\text{-}1} = 0.875$, while in Eqs. (20), one obtains the Van Laar constants $A_{1\text{-}2} = 0.68$ and $A_{2\text{-}1} = 0.89$. These constants may be used with the respective Eq. (16) or (18) and with Eq. (10) to calculate a complete y-x curve.

Calculation from Boiling-point or Total-pressure Curves. The procedure suggested by Carlson and Colburn (*loc. cit.*) is to utilize the relations

$$\gamma_1 = \frac{P - \gamma_2 P_2 x_2}{P_1 x_1} \quad \text{and} \quad \gamma_2 = \frac{P - \gamma_1 P_1 x_1}{P_2 x_2} \tag{36}$$

By calculating values of γ_1 from data where x_1 is relatively dilute, values of γ_2 used in the computation can be taken as unity as a first approximation. The resulting values of γ_1 can be plotted as in Fig. 1 and extrapolated to $x_1 = 0$ in order to evaluate $A_{1\text{-}2}$. A similar procedure is used to find $A_{2\text{-}1}$. If desirable, a second approximation can be utilized in Eq. (36) based on the first values found for $A_{1\text{-}2}$ and $A_{2\text{-}1}$. This procedure is illustrated by Carlson and Colburn with an example.

This method is particularly useful for binary systems that are difficult to analyze. Measurements of total pressure as a function of liquid composition are easily made. If the binary system is composed of one volatile and one relatively non-volatile component, a curve of activity coefficient vs. composition can be established for the volatile component but not for the non-volatile one.

Partly Miscible Solutions. From knowledge of the solubility limits, activity coefficients can be computed for the miscible regions, and therefore the vapor-liquid equilibrium. Equations useful for solving for Margules constants are

$$\frac{A_{1\text{-}2}}{A_{2\text{-}1}} = \frac{(2 \log \bar{x}_2/x_2)(x_2^2 - \bar{x}_2^2 - x_2^3 + \bar{x}_2^3) + (\log \bar{x}_1/x_1)(x_1^2 - \bar{x}_1^2 - 2x_1^3 - 2\bar{x}_1^3)}{(2 \log \bar{x}_1/x_1)(x_1^2 - \bar{x}_1^2 - x_1^3 + \bar{x}_1^3) + (\log \bar{x}_2/x_2)(x_2^2 - \bar{x}_2^2 - 2x_2^3 + 2\bar{x}_2^3)} \tag{37}$$

$$A_{1\text{-}2} = \frac{\log \bar{x}_1/x_1}{(2A_{2\text{-}1}/A_{1\text{-}2} - 1)(x_2^2 - \bar{x}_2^2) - 2(A_{2\text{-}1}/A_{1\text{-}2} - 1)(x_2^3 - \bar{x}_2^3)} \tag{38a}$$

$$A_{2\text{-}1} = \frac{\log \bar{x}_2/x_2}{(2A_{1\text{-}2}/A_{2\text{-}1} - 1)(x_1^2 - \bar{x}_1^2) - 2(A_{1\text{-}2}/A_{2\text{-}1} - 1)(x_1^3 - \bar{x}_1^3)} \tag{38b}$$

where x_1 and x_2 are mole fractions in one phase, and $\bar{x}_1$ and $\bar{x}_2$ are mole fractions in the other.

Van Laar constants are similarly obtained from the equations

$$\frac{A_{1\text{-}2}}{A_{2\text{-}1}} = \frac{\left(\dfrac{x_1}{x_2} + \dfrac{\bar{x}_1}{\bar{x}_2}\right)\left(\dfrac{\log \bar{x}_1/x_1}{\log x_2/\bar{x}_2}\right) - 2}{\dfrac{x_1}{x_2}\dfrac{\bar{x}_1}{\bar{x}_2} - \dfrac{2x_1\bar{x}_1 \log \bar{x}_1/x_1}{x_2\bar{x}_2 \log x_2/\bar{x}_2}} \tag{39}$$

$$A_{1\text{-}2} = \frac{\log \bar{x}_1/x_1}{\dfrac{1}{\left(1 + \dfrac{A_{1\text{-}2}}{A_{2\text{-}1}}\dfrac{x_1}{x_2}\right)^2} - \dfrac{1}{\left(1 + \dfrac{A_{1\text{-}2}}{A_{2\text{-}1}}\dfrac{\bar{x}_1}{\bar{x}_2}\right)^2}} \tag{40}$$

A graphical solution of Eqs. (39) and (40) was developed by Colburn and Schoenborn (*loc. cit.*) and is given as Fig. 5.

Example. Consider the system methyl ethyl ketone–water boiling at atmospheric pressure, temperature = 73.8°C. The miscibility limits are $x_1 = 0.052$ and $\bar{x}_1 = 0.587$ mole fraction MEK. Using Fig. 5, for $x_1 = 0.052$, $r = 0.413/0.052 = 7.94$, $A_{1\text{-}2} = 1.43$. For $\bar{x}_2 = 0.413$, $r = 0.052/0.413 = 0.126$, $A_{2\text{-}1} = 0.62$. These results are only slightly lower than the values of $A_{1\text{-}2} = 1.50$ and $A_{2\text{-}1} = 0.75$, obtained from vapor-liquid equilibrium data of Marshall [*J. Chem. Soc.*, **89**, 1350 (1906)].

Relationships to Physical Properties of Components. Much effort has been made in attempts to develop quantitative and semiquantitative relations between activity coefficients and physical properties of the pure components. Hildebrand ("Solubility of Non-electrolytes," 2d ed., Reinhold, New York, 1936) reviewed the effects of association, polarity, etc., and concluded that the nearer the internal pressures of two components were, the less the deviation from Raoult's law.

Ewell, Harrison, and Berg [*Ind. Eng. Chem.*, **36**, 871 (1944)] explained deviations in terms of the hydrogen bonding characteristics of the components, and prepared the following tables:

Table 5. Classification of Liquids*

Liquids may be classified into the following groups based on their potentialities for forming hydrogen bonds:

Class I. Liquids capable of forming three-dimensional networks of strong hydrogen bonds, *e.g.*, water, glycol, glycerol, amino alcohols, hydroxylamine, hydroxy acids, polyphenols, amides, etc. Compounds such as nitromethane and acetonitrile also form three-dimensional networks of hydrogen bonds, but the bonds are much weaker than those involving OH and NH groups. Therefore, these types of compounds are placed in Class II.

Class II. Other liquids composed of molecules containing both active hydrogen atoms and donor atoms (oxygen, nitrogen, and fluorine), *e.g.*, alcohols, acids, phenols, primary and secondary amines, oximes, nitro compounds with α-hydrogen atoms, nitriles with α-hydrogen atoms, ammonia, hydrazine, hydrogen fluoride, hydrogen cyanide, etc.

Class III. Liquids composed of molecules containing donor atoms but no active hydrogen atoms, *e.g.*, ethers, ketones, aldehydes, esters, tertiary amines (including pyridine type), nitro compounds and nitriles without α-hydrogen atoms, etc.

Class IV. Liquids composed of molecules containing active hydrogen atoms but no donor atoms. These are molecules having two or three chlorine atoms on the same carbon as a hydrogen atom, or one chlorine on the same carbon atom and one or more chlorine atoms on adjacent carbon atoms, *e.g.*, $CHCl_3$, CH_2Cl_2, CH_3CHCl_2, $CH_2Cl—CH_2Cl$, $CH_2Cl—CHCl—CH_2Cl$, $CH_2Cl—CHCl_2$, etc.

Class V. All other liquids, *i.e.*, liquids having no hydrogen-bond-forming capabilities, *e.g.*, hydrocarbons, carbon disulfide, sulfides, mercaptans, halohydrocarbons not in Class IV, non-metallic elements such as iodine, phosphorus, sulphur, etc.

* Ewell *et al.*

Table 6. Summary of Deviations of Raoult's Law*

Classes	Deviations	Hydrogen bonding
I + V II + V	Always + deviations; I + V, frequently limited solubility	H bonds broken only
III + IV	Always − deviations	H bonds formed only
I + IV II + IV	Always + deviations; I + IV, frequently limited solubility	H bonds both broken and formed, but dissociation of class I or II liquid is more important effect
I + I I + II I + III II + II II + III	Usually + deviations, very complicated groups, some − deviations give some maximum azeotropes	H bonds both broken and formed
III + III III + V IV + IV IV + V V + V	Quasi-ideal systems, always + deviations or ideal; azeotropes, if any, will be mimina	No H bonds involved

* Ewell *et al.*

Note that positive deviations mean values of activity coefficients greater than 1, and negative deviations mean values less than 1.

A convenient method of studying systems with positive deviations from Raoult's law is investigation of the critical solution temperatures, as illustrated by Francis [*Ind. Eng. Chem.*, **36**, 764 (1944)] with series of hydrocarbons with various solvents. As brought out in an earlier discussion in this section, a decrease in mutual solubility corresponds to an increase in activity coefficients; the lower the critical solution temperature the less the solubility. Based on these studies, Francis also developed Fig. 6 [*Ind. Eng. Chem.*, **36**, 1096 (1944)]. Where one is dealing with components of the types given on this figure the degree of immiscibility and corresponding degree of deviation from Raoult's law are readily seen. This plot is particularly convenient in preliminary scouting for solvents for use in extractive and azeotropic distillation and liquid-liquid extraction.

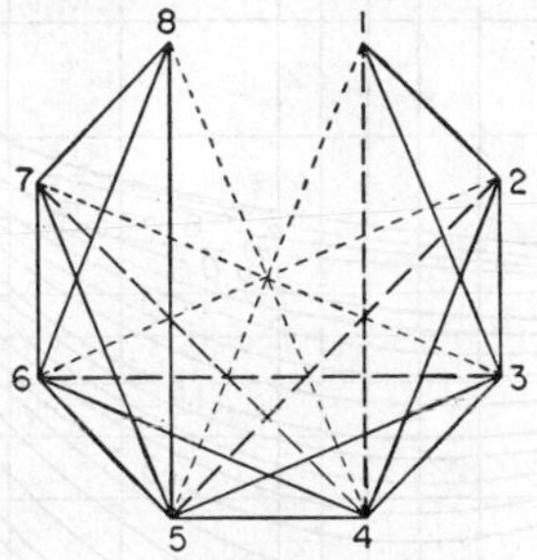

FIG. 6. Mutual solubility of liquids. ———— Complete miscibility; — — — — High solubility; - - - - - - Low solubility; no line, very low solubility. 1. Water; 2. Diethylene glycol; 3. Triethylene glycol; 4. Furfural; 5. Ethyl ether; 6. Benzene; 7. Cyclohexane; 8. n-Heptane.

CALCULATION OF EFFECT OF GAS-LAW DEVIATIONS

At elevated pressures gas-law deviations will affect vapor-liquid equilibrium in a marked manner, whether the system is ideal or not in the liquid. Even at atmospheric pressure there may be a slight effect which needs to be considered in careful work. In the range of moderate pressures, convenient generalizations are available that permit the prediction of the effect of compressibility for systems where experimental data on vapor-liquid equilibrium are lacking and that also provide a means of interpreting experimental results to evaluate their reliability. At pressures approaching the critical, the generalizations are not applicable, and experimental data on the actual systems are necessary; *cf.*, however, Smith and Watson, "High Pressure Vapor Liquid Equilibria Activity Coefficients for Ideal Systems," paper at Am. Inst. Chem. Engrs. meeting Sept. 15, 1948.

This treatment deals with three methods for determining the effect of compressibility on vapor-liquid equilibrium: (1) the use of fugacities from generalized correlations, (2) the use of a correction term involving a gas-law deviation term for the component in question, which may be obtained directly from compressibility data on the system or from generalized relationships, and (3) the use of direct experimental data. The former method has been employed largely for hydrocarbon systems up to fairly high pressures, the second for most other systems at moderate pressures, and the third for any systems near the critical.

1. *Use of Fugacities (for Ideal Liquid Solutions).* Correlation of fugacities of hydrocarbon vapors were early presented by Lewis and Luke [*Trans. Am. Soc. Mech. Engrs.*, **54**, 55 (1932)] and by Souders, Selheimer, and Brown [*Ind. Eng. Chem.*, **24**, 517 (1932)]. According to the latter, the vapor-liquid equilibrium ratio y_1/x_1 for any component may be written

$$\frac{y}{x} = \frac{f_L}{f_v} = K \tag{41}$$

where y = mole fraction of component in the vapor.
x = mole fraction of component in the liquid.
f_L = fugacity of pure component in liquid at temperature and pressure of the system.
f_v = fugacity of pure component in vapor at temperature and pressure of the systems.
K = equilibrium ratio, defined as equal to y/x.

The term K is usually assumed to be independent of the composition and a function, for any given component, of temperature and pressure only. Hence the term is used only for systems that are ideal in the liquid, *i.e.*, those which follow Raoult's law. At higher pressures, however, values of K for a given constituent may vary with the composition of the solution, as shown by Hanson, Ryasa, and Brown [*Ind. Eng. Chem.*, **37**, 1316 (1945)], for methane in different liquids.

The values of fugacity are evaluated as follows: The ratios of fugacity to pressure are correlated as functions of reduced temperature T_R and reduced pressure P_R. Two of the latest and best correlations are those of Gamson and Watson (*Nat. Petroleum News, Tech. Sec.* 36, R 554, Aug. 2, 1944); also Watson and Smith (*ibid.* 28, July 1, 1936) and of Kirkbride (*Petroleum Refiner* 24, December, 1945) the former of which is reproduced as Fig. 7. Thus the value of f_v is obtained by finding the ratio of f_v/P from the figure, where P = total pressure on the system. The value of f_L is correctly evaluated by finding f_1/P_1 from the figure and then utilizing the relation

$$\log \frac{f_L}{f_1} = \frac{V_L(P - P_1)}{2.3RT} \tag{42}$$

where P_1 = vapor pressure of pure component at temperature of system.
f_1 = fugacity of pure component in liquid phase at its vapor pressure which corresponds to the temperature of the system.
V_L = molal volume of component in liquid phase.
T = absolute temperature of system.
R = gas-law constant, whose units depend upon those used for V_L, P, and T.

At pressures below 500 lb./sq. in. the ratio of f_L/f_1 is essentially unity, and this correction for the fugacity of the liquid for the difference between the vapor pressure of the pure component and the total pressure can usually be ignored. The above relation is derived with the assumption of no change in liquid volume on mixing the components, which again limits this treatment to mixtures that are ideal in the liquid.

Data on the specific gravities of liquid hydrocarbons at temperature up to the critical are given in Fig. 8, as published by Kirkbride (*loc. cit.*). For densities of other liquids at elevated temperatures, the method of prediction proposed by Watson [*Ind. Eng. Chem.*, **35**, 398 (1943)] is very good, and requires a knowledge of only a single value of density at some lower temperature. Watson writes $\rho = (\rho_1/\omega_1)\omega$, or $v = (v_1\omega_1)/\omega$, where ρ_1, v_1, and ω, are the density, specific volume, and expansion factor, respectively, at a temperature where the liquid density is known, and the other values apply at the conditions of the problem. The values of expansion factor ω were found to be general functions of T_R and P_R for all the liquids studied. This function was given graphically, the coordinates being given in Table 7.

Table 7. Values of Expansion Factor ω*

T_R	$P_R = 0$	$P_R =$ 0.4	$P_R =$ 0.8	$P_R =$ 1.0	$P_R =$ 1.5	$P_R =$ 2	$P_R =$ 3	$P_R =$ 5
0.5	(0.1328)	......	......	0.1332	......	0.1338	......	0.1350
.6	.1242	......	......	.1250	......	.1258	......	.1275
.7	.1144	0.1150	......	.1158	......	.1170	0.1182	.1202
.8	.1028	.1042	0.1050	.1056	0.1070	.1077	.1098	.1125
.9	....	.0900	.0915	.0926	.0949	.0968	.1002	.1043
.95	....	......	.0810	.0831	.0872	.0902	.0943	.1000
1.0	....	......	......	.0440	.0764	.0818	.0875	.1954

* Watson, *Ind. Eng. Chem.*, **35**, 398 (1943).

More recently, Gamson and Watson (*Nat. Petroleum News, Tech. Sec.*, May 3, 1944), stated that, for values

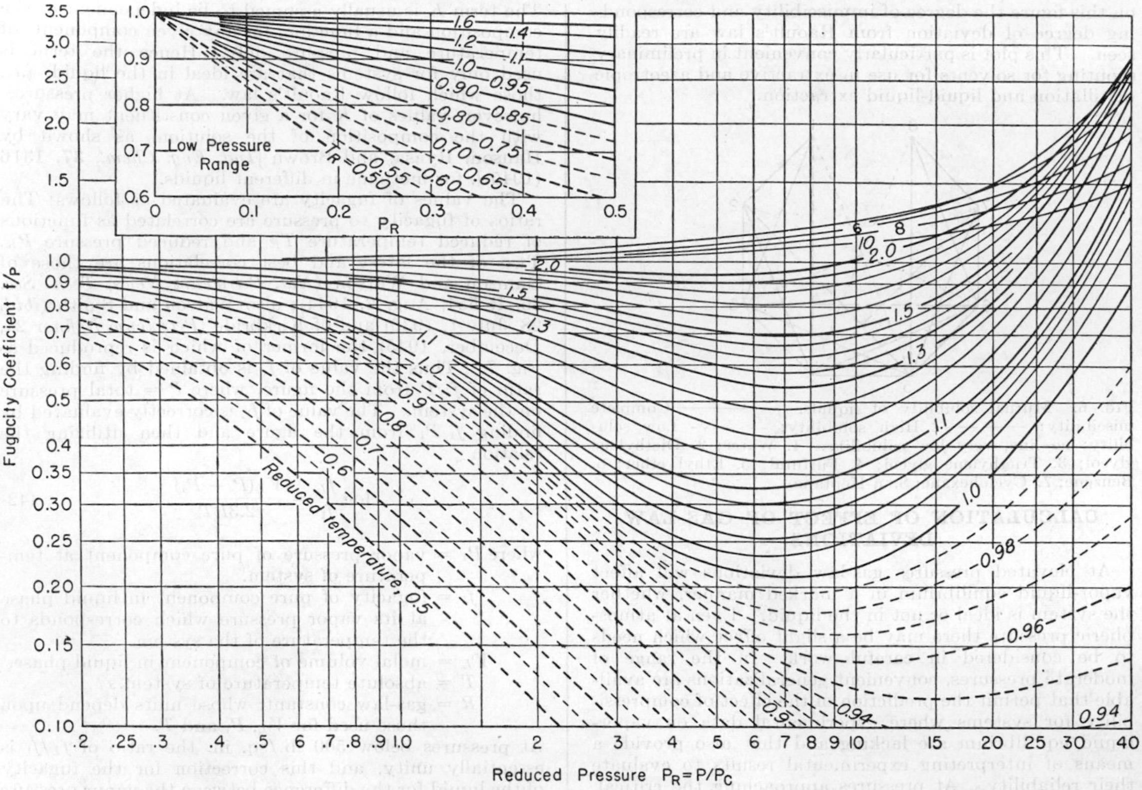

FIG. 7. Fugacity coefficients of gases and vapors. (*From Gamson and Watson.*)

of T_R below 0.65 and for pressures not exceeding 10 atm., the following relation holds:

$$\omega = 0.1745 - 0.0838T_R \tag{43}$$

A correlation of densities of saturated liquids with reduced temperature and pressure is given by Meissner and Paddison [*Ind. Eng. Chem.*, **33**, 1189 (1941)].

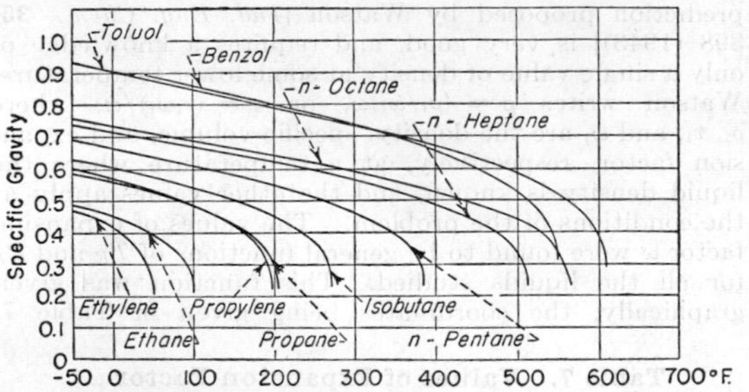

FIG. 8. Variation of specific gravity of liquid hydrocarbons with temperature.

Values of K for various hydrocarbons over extensive ranges of temperature and pressure are given in Sec. 9. Other plots of K for hydrocarbons are given with a greater scale by Brown and Souders ("The Science of Petroleum," edited by Dunstan *et al.*, vol. 2, Sec. 25, p. 1544, 1938) and by Gamson and Watson (*Nat. Petroleum News, Tech. Sec.* 36, R554, Aug. 2, 1944). Working-scale prints of the figures in the latter article are obtainable from K. M. Watson, University of Wisconsin, Madison, Wis.

Inasmuch as values of relative volatility, defined as $\alpha_{1-2} = y_1x_2/y_2x_1$, are equal to the ratios K_1/K_2 (assuming ideal liquid solutions), it is often preferable to provide correlations of α for various components, utilizing a given component as reference and providing complete K values for it. Thus Kirkbride [*Petroleum Refiner*, **24**, 11–24 (1945)] has prepared relative-volatility graphs consistent with Fig. 7 above, for hydrogen, methane, ethylene, propylene, propane, isobutane, butene-2, butene-1, isobutane, n-butane, isopentane, n-pentane, n-hexane, benzene, n-heptane, toluene, methylcyclohexane, and n-octane, all with reference to ethane, and a graph of K for ethane. (These charts are available in 11- by 17-in. size from Engineering Charts, College Station, Tex., or from *Petroleum Refiner*, Houston, Tex.)

As an example of the application of these methods, consider the prediction of the value of y/x for benzene in a mixture at 392°F. and 131 lb./sq. in. abs. Other data: P_1 = 206 lb./sq. in. abs., P_c = 704, T_c = 1011°R. Then $P_R = (131)/(704) = 0.186$, $T_R = (852)/(1011) = 0.843$. From Fig. 7, f_v/P is found to be 0.88, whence $f_v = (0.88)(131) = 115$. Now, $P_1/P_c = (206)/(704) = 0.292$; again from Fig. 7, f_1/P_1 is found to be 0.825, so $f_1 = (0.825)(206) = 170$. To utilize Eq. (42) to solve for f_L/f_1, V_L is found from Fig. 8 to be 1.9. In the units employed, $R = 10.7$, so that $\log_{10} f_L/f_1 = (1.9)(131 - 206)/(2.3)(10.7)(852) = -0.0068$. Thus $f_L/f_1 = 0.985$, $f_L = (0.985)(170) = 167$. Finally, $y/x = f_L/f_v = (167)/(115) = 1.45$. Now, the experimental value of y/x obtained by Griswold, Andres, and Klein [*Trans. Am. Inst. Chem. Engrs.*, **39**, 223 (1943)] was 1.44, and the value of 1.45 predicted by fugacities is therefore in excellent agreement and is far better than the value of P_1/P_2 of 1.57 which would be predicted by Raoult's and Dalton's laws.

2. *Correction Term for Moderate Pressures.* For fairly moderate pressures the relationship for vapor-liquid equilibrium of a given component may be written [*cf.* Scatchard and Raymond, *J. Am. Chem. Soc.*, **62**, 1278 (1944); Benedict, Johnson, Solomon, and Rubin, *Trans. Am. Inst. Chem. Engrs.*, **41**, 371 (1945); and Wohl, *ibid.*, **42**, 215 (1946)]

$$\log_{10}\left(\frac{y_1}{x_1}\right) = \log_{10}\frac{\gamma_1 P_1}{P} - \frac{(V_1 - B_1)(P_1 - P)}{2.3RT} \quad (44)$$

where y_1 = mole fraction of component in vapor.
x_1 = mole fraction of component in liquid.
P_1 = vapor pressure of pure component at temperature of system.
P = total pressure of system.
γ_1 = activity coefficient of component in liquid.
V_1 = molal volume of component in liquid state at temperature of system (the partial molal volume should properly be used here in case such data are available).
B_1 = virial coefficient of component 1 in the relation below for gas-law deviation.

$$V_G = (RT/P) + B \quad (45)$$

where V_G = molal volume of component in vapor phase. This simple equation of state is particularly useful when gas-law deviations are not large, hence at moderate pressures. The assumptions underlying the above equation for log y_1/x_1 are that no volume change occurs on mixing the pure liquids or vapors, or that the virial coefficients are the same for the pure component as for the portion in the mixture. Thus the relation can be used only as an approximation for non-ideal liquids.

It is of interest to apply the above relation to the direct solution for relative volatility, α. Thus

$$\log \alpha_{1\text{-}2} = \log\frac{y_1 x_2}{y_2 x_1} = \log\frac{\gamma_1 P_1}{\gamma_2 P_2} - \frac{(V_1 - B_1)(P_1 - P)}{2.3RT} + \frac{(V_2 - B_2)(P_2 - P)}{2.3RT} \quad (46)$$

The values of the virial coefficients of the pure components can be obtained readily from compressibility data such as those listed in Sec. 3 of this handbook, or from various correlations in terms of reduced temperatures and pressures. One of the best of these is the equation of Wohl [*Z. phys. Chem.*, **B2**, 77 (1929)].

$$B = \left(\frac{RT_c}{P_c}\right)\left(0.197 - 0.012T_R - \frac{0.4}{T_R} - \frac{0.146}{T_R^{3.27}}\right) \quad (47)$$

This relation is excellent in general but should not be used for the permanent gases such as H_2, He, and air, and is not too good for water, alcohol, and ammonia. It might also be noted that the term B is simply related to the compressibility factor Z, defined as $Z = PV/RT$:

$$B = \frac{(Z - 1)RT}{P} \quad (48)$$

Inasmuch as usual plots of Z vs. reduced temperature and reduced pressure are available, either values of B can be computed from the relation immediately above; or, in the previous relations for log y_1/x_1 and log α, substitutions may be made for B so that the equations are expressed in terms of Z.

This method of correcting values of y/x and α for gas-law deviations is very convenient and rather generally useful, even for non-ideal liquids, at moderate pressures. The method is not accurate if either the total or vapor pressure is higher than, roughly, 200 lb./sq. in. In those cases experimental values of vapor-liquid are essential, or the rigorous approach of Gamson and Watson (*loc. cit.*) may be satisfactory.

The application of this procedure is illustrated by the same example worked out under (1), *i.e.*, benzene at 392°F. and 131 lb./sq. in. abs. Solving first for the virial coefficient B,

$$B = \frac{(10.7)(1011)(-0.542)}{(704)} = -8.35$$

Table 8. Vapor-liquid Equilibrium of Mixtures*

System or compounds	Range		Ref.
	Temperature	Atm.	
$C_2H_5OH—H_2O$	To 345°C.	To 190	6
$C_2H_6—H_2O$	100° to 460°F.	To 680	17
n-Butane—H_2O	100° to 340°F.	To 680	15
Benzene-toluene	To 280°C.	To 34	5
Benzene	100° to 500°F.	To 41	3
Toluene	100° to 500°F.	To 41	3
Methyl cyclohexane	100° to 500°F.	To 41	3
$HCl—nC_4H_{10}$	70°, 120°, 180°F.	To 37	14
$H_2—nC_4H_{10}$†	24°, 82°, 116°C.	To 100	13
H_2-hexane	35°C.	To 150	20
H_2-cyclohexane	35°C.	To 150	20
H_2-benzene	35°C.	To 150	20
H_2-m-xylene	35°C.	To 150	20
H_2-isobutane	100° to 250°F.	34 to 200	2
H_2-2,2,4 trimethyl-pentane	100° to 302°F.	12 to 340	2
H_2-dodecanes	200° to 300°F.	34 to 340	2
$CH_4—C_2H_4$-isobutane	38°, 71°C.	34, 68	1
CH_4-decane	100 to 460°F.	To 680	18
C_2H_6-propylene	−30° to 90°C.	To 50	11
C_2H_4-propylene	54° to 59°C.	Critical region	21
$CO_2—CH_4$	100° to 460°F.	7 to 680	16
$He—CH_4$	90°, 106°K.	To 160	4
Propylene-isobutane	100° to 250°F.	To 600 lb./sq. in.	3
Propane-H_2S	125° to 225°F.	To 600 lb./sq. in.	3
Propylene-acetylene	....	....	12
Propane-acetylene	....	....	12
Natural gas − 95% diethylene glycol	100°F.	To 136	19
Natural gas − crude oil		To 560	22
CH_4—absorber oil	85°F.	To 200	7
C_2H_6—absorber oil	85°F.	To 200	7
C_3H_8—absorber oil	85°F.	To 200	7
nC_4H_{10}—absorber oil	85°F.	To 200	7
nC_5H_{12}—absorber oil	85°F.	To 200	7
Benzene—N_2‡	35° to 100°C.	1100	8
$CH_3OH—N_2$‡	0° to 75°C.	To 700	9
$CH_3OH—H_2$‡	0° to 75°C.	To 700	9
$CH_3OH—CH_4$	0° to 75°C.	To 700	9
$CH_3OH—CO_2$	0° to 75°C.	To 700	9
$NH_3—N_2$	90° to 148°C.	9650	10
$NH_3—CH_4$	45° to 100°C.	9650	10
$NH_3—N_2—H_2$	100°C.	5300	10

* Comings [*Ind. Eng. Chem.*, **29**, 948 (1947)].

† Two regions of isobaric retrograde condensation [Kay, *Chem. Rev.*, **29**, No. 3, 501 (1941)].

‡ These have minima in the curve of solubility vs. pressure. They represent solubility of a liquid in a compressed gas.

References:

1 Benedict, Solomon, and Rubin, *Ind. Eng. Chem.*, **37**, 55 (1945).
2 Dean and Tooke, *Ind. Eng. Chem.*, **38**, 389 (1946).
3 Drickamer and Bradford, *Ind. Eng. Chem.*, **36**, 1144 (1944).
4 Gonikberg and Fastovskii, *Foreign Petroleum Tech.*, **9**, 214 (1941).
5 Griswold, Andres, and Klein, *Trans. Am. Inst. Chem. Engrs.*, **39**, 223 (1943).
6 Griswold, Haney, and Klein, *Ind. Eng. Chem.*, **35**, 701 (1943).
7 Kirkbride and Bertetti, *Ind. Eng. Chem.*, **35**, 1242 (1943).
8 Kirchevskii and Gamburg, *Acta Physicochim. U.R.S.S.*, **16**, 362 (1945).
9 Krichevskii and Koroleva, *Acta Physicochim. U.R.S.S.*, **15**, 327 (1941).
10 Krichevskii and Tsiklis, *Acta Physicochim. U.R.S.S.*, **18**, 264 (1943).
11 Lu, Newitt, and Ruhemann, *Proc. Roy. Soc. (London)*, **A174**, 506 (1941).
12 McCurdy and Katz, *Oil Gas J.*, **43**, No. 44, 102 (1945).
13 Nelson and Bonnell, *Ind. Eng. Chem.*, **35**, 204 (1943).
14 Ottenweller, Hooloway, and Weinrich, *Ind. Eng. Chem.*, **35**, 207 (1943).
15 Reamer, Olds, Sage, and Lacey, *Ind. Eng. Chem.*, **36**, 381 (1944).
16 Reamer, Olds, Sage, and Lacey, *Ind. Eng. Chem.*, **36**, 88 (1944).
17 Reamer, Olds, Sage, and Lacey, *Ind. Eng. Chem.*, **35**, 790 (1943).
18 Reamer, Olds, Sage, and Lacey, *Ind. Eng. Chem.*, **34**, 1526 (1942).
19 Russell, Reid, Huntington, *Trans. Am. Inst. Chem. Engrs.*, **41**, 315 (1945).
20 Sattler, *Z. tech. Physik*, **21**, 410 (1940).
21 Schneider and Maass, *Can. J. Research*, **19B**, 231 (1941).
22 Standing and Katz, *Am. Inst. Min. Met. Eng.*, *Tech. Pub.* 1651 (1943).

Then, since $\gamma_1 = 1$ (for ideal liquids),

$$\log_{10}\frac{y_1}{x_1} = \log\frac{P_1}{P} - \frac{(1.9 + 8.35)(206 - 131)}{(2.3)(10.7)(852)}$$
$$= \log_{10} 1.57 - 0.0368 = 0.1594$$

whence $y_1/x_1 = 1.442$ This value is in almost exact agreement with the experimental value of Griswold *et al.* (*loc. cit.*) of 1.44; so that, in this instance, method 2 and method 1 give essentially the same result.

3. *Experimental Data.* A tabulation of experimental studies during the last 5 years of vapor-liquid equilibrium at elevated temperatures was recently presented by Comings [*Ind. Eng. Chem.*, **29**, 948 (1947)] and is included here as Table 8.

RATE OF MASS TRANSFER

When a homogeneous material, either gas, liquid, or solid, contains two or more components whose concentrations vary from point to point, there is a tendency for transfer of mass to take place in such a way as to cause the concentrations to become uniform. This phenomenon is associated with the thermal agitation of molecules; in a region where molecules of one kind are concentrated, there is a greater tendency for molecules of this kind to escape than to enter the region. The net rate of diffusion of material at a point is found from experiment as well as from theory to be proportional to the concentration gradient at the point.

$$N_A = -D\frac{\partial c}{\partial s} \qquad (49)$$

(for nomenclature, see p. 524). The rate of diffusion is rapid in gases and much slower in liquids. Values of diffusion coefficients may be estimated from kinetic theory. The treatment is quite different for gases and liquids.

DIFFUSION IN GASES

The study of diffusion, along with viscosity and thermal conduction in gases, is a part of the well-developed kinetic theory of gases. One of the most modern accounts of this theory is that of Chapman and Cowling ("The Mathematical Theory of Non-uniform Gases," Cambridge, London, 1939; *cf.* also Kennard, "Kinetic Theory of Gases," McGraw-Hill, New York, 1938), who give equations based on various assumptions regarding the nature of the interaction between molecules at collision, from which values of the diffusion coefficient for a binary mixture may be calculated. The most advanced treatment predicts that the diffusion coefficient should vary only slightly with composition, and this is confirmed by experiment.

Table 9 shows the experimentally observed variation of the diffusion coefficient with composition in the case of mixtures of hydrogen and carbon dioxide. This variation is greatest when the ratio of the masses of the molecules is large, but in no case does the maximum variation exceed 13 per cent.

Table 9. Diffusivity in Hydrogen–Carbon Dioxide Mixtures—Variation with Composition

% H_2	D, Cm.²/Sec.*
25	0.594
50.0	0.605
75	0.633

* Based on experiments at 15°C., 1 atm. by Lonius, *Ann. Physik*, **29** (4), 664 (1909).

These equations developed from the kinetic theory cannot be used in direct calculations of the diffusivity unless the diameters of the molecules are known. For this reason empirical equations such as that of Gilliland [*Ind. Eng. Chem.*, **26**, 681 (1934)] or of Arnold [*Ind. Eng. Chem.*, **22**, 1091 (1930)] are used when experimentally determined diffusivities are not available. According to Gilliland,

$$D = 0.0043\frac{T^{3/2}}{P(V_A^{1/3} + V_B^{1/3})^2}\sqrt{\frac{1}{M_1} + \frac{1}{M_2}} \qquad (50)$$

where D = diffusivity, cm.²/sec.
T = temperature, °K.
P = pressure, atm.
M_1, M_2 = molecular weights of gases.
V_A, V_B = molecular volumes at the normal boiling points, cc./g.-mole.

Values of V_A and V_B may be calculated from the molecular weight and the density at the normal boiling point if this is known. Otherwise the calculation may be made by adding together values of the atomic volumes listed in Table 10. If the units of D are to be ft.²/hr., the value of the constant should be 0.0166.

Equation (50) is based in part on the kinetic theory of a gas mixture composed of hard, smooth, spherical molecules, since this model leads to the effects of temperature and molecular weight given in the formula. Actually, the diffusivity is found to vary with a power of the absolute temperature which lies between 1.5 and 2.0 (usually 1.75) (*cf.* "International Critical Tables," vol. 5, p. 62, McGraw-Hill, New York, 1929). For this reason Eq. (50) can be expected to give results that are too low at high temperatures. There are some indications [*cf.* Klibanovr, Pomerantsev, and Frank-Kamenetsy, *J. Tech.*

Table 10. Atomic Volumes*

For use in calculating molal volumes at the normal boiling point, expressed in cc./g.-mole, as used in Eq. (50)

Element	Atomic Volume, Cc./g.-atom
Air	29.9
Antimony	34.2 (?)
Arsenic	30.5
Bismuth	48.0
Bromine	27.0
Carbon	14.8
Chlorine:	
Terminal, as in R—Cl	21.6
Medial, as in R—CHCl—R′	24.6
Chromium	27.4
Fluorine	8.7
Germanium	34.5
Hydrogen	3.7
In hydrogen molecule	7.15
Iodine	37.0
Lead	46.5–50.1
Mercury	19.0
Nitrogen	15.6
In primary amines	10.5
In secondary amines	12.0
Oxygen, doubly bound, as carbonyl oxygen	7.4
Coupled to two other elements:	
In aldehydes and ketones	7.4
In methyl esters	9.1
In methyl ethers	9.9
In higher ethers and esters	11.0
In acids	12.0
In union with S, P, N	8.3
Phosphorus	27.0
Silicon	32.0
Sulphur	25.6
Tin	42.3
Titanium	35.7
Vanadium	32.0
Water	18.8
Zinc	20.4
For three-membered ring, as in ethylene oxide, deduct	6
For four-membered ring, as in cyclobutane, deduct	8.5
For five-membered ring, as in furan, thiophene, deduct	11.5
For six-membered ring, as in benzene, cyclohexane, pyridine, deduct	15
For naphthalene ring, deduct	30
For anthracene ring, deduct	47.5

* LeBas, "The Molecular Volumes of Liquid Chemical Compounds," Longmans, London, 1915.

Phys. (*U.S.S.R.*), **12,** 14 (1942); Arnold, *Ind. Eng. Chem.*, **22,** 1091 (1930)] that the effect of temperature may be predicted from a formula due to Sutherland

$$\frac{D}{D_0} = \frac{C + T_0}{C + T}\left(\frac{T}{T_0}\right)^{3/2} \quad (51)$$

using values of the Sutherland constant of the mixture C. The Sutherland constant for the mixture is equal approximately to the geometric mean of the values for the pure components [Arnold, *Ind. Eng. Chem.*, **22,** 1091 (1930)]. If the Sutherland constant for a pure component is not available from viscosity data (*cf.* Sec. 5), it may be estimated from the empirical rule, $C = 1.47T_{b.p.}$, where $T_{b.p.}$ is the normal boiling point in °K.

Example. Calculate the molal volume of ethyl formate, $C_3H_6O_2$. The sum of the atomic volumes is

$$\begin{aligned}
3 \times C &= 3 \times 14.8 = 44.4\\
6 \times H &= 6 \times 3.7 = 22.2\\
1 \times O\ (\text{carboxyl}) &= 7.4\\
1 \times O\ (\text{ethyl ester}) &= \underline{11.0}\\
&\quad 85.0\ \text{cc./g.-mole (observed} = 84.6)
\end{aligned}$$

Example. Calculate the molal volume of nitroethane, $C_2H_5NO_2$. The sum of the atomic volumes is

$$\begin{aligned}
2 \times C &= 2 \times 14.8 = 29.6\\
5 \times H &= 5 \times 3.7 = 18.5\\
1 \times N &= 15.6\\
2 \times O\ (\text{doubly bound to N}) &= 2 \times 8.3 = \underline{16.6}\\
&\quad 80.3\ \text{cc./g.-mole (observed} = 80.4)
\end{aligned}$$

Actually, knowledge of the effect of temperature on the diffusivity alone is seldom required, since the diffusivity is usually used in combination with the density and viscosity as the dimensionless Schmidt group, $(\mu/\rho D)$. Both theory and experiment indicate that this quantity is independent of pressure and varies only slightly with the temperature, at least over moderate ranges. Typical experimental data are listed in Table 11.

Table 11. Diffusion Coefficients of Gases and Vapors in Air at 25°C., 1 Atm.

Substance	D, cm²/sec.	$(\mu/\rho D)$
Ammonia	0.28	0.78
Carbon dioxide	.164	.94
Hydrogen	.410	.22
Oxygen	.206	.75
Water	.256	.60
Carbon disulfide	.107	1.45
Ethyl ether	.093	1.66
Methanol	.159	0.97
Ethyl alcohol	.119	1.30
Propyl alcohol	.100	1.55
Butyl alcohol	.090	1.72
Amyl alcohol	.070	2.21
Hexyl alcohol	.059	2.60
Formic acid	.159	0.97
Acetic acid	.133	1.16
Propionic acid	.099	1.56
i-Butyric acid	.081	1.91
Valeric acid	.067	2.31
i-Caproic acid	.060	2.58
Diethyl amine	.105	1.47
Butyl amine	.101	1.53
Aniline	.072	2.14
Chloro benzene	.073	2.12
Chloro toluene	.065	2.38
Propyl bromide	.105	1.47
Propyl iodide	.096	1.61
Benzene	.088	1.76
Toluene	.084	1.84
Xylene	.071	2.18
Ethyl benzene	.077	2.01
Propyl benzene	.059	2.62
Diphenyl	.068	2.28
n-Octane	.060	2.58
Mesitylene	.067	2.31

References: "International Critical Tables," vol. 5, 1928; Landolt-Börnstein, "Physikalische-Chemische Tabellen," 1935.

Note: The group $(\mu/\rho D)$ in the above table is evaluated for mixtures composed largely of air.

In the application of the theory of diffusion in gaseous systems, it is frequently desirable to use integrated forms of Eq. (49) rather than the equation itself, which applies only at a single point.

Table 12. Diffusion Coefficients of Organic Esters in Air

No. of carbon atoms	D, cm.²/sec., at 25°C.	$\mu/\rho D$
2*	0.117	1.32
3†	.097	1.59
4‡	.086	1.79
5	.078	1.97
6	.069	2.23
7	.065	2.37
8	.057	2.70
9	.049	3.14

References: Pochettino, *Il. Nuovo Cimento*, **8,** Ser. 6, 5 (1914). Winkelmann, *Ann. Physik*, **22,** 1, 152 (1884); **23,** 203 (1884); **26,** 105 (1885); **33,** 445 (1888); **36,** 93 (1889).

Note: The group $(\mu/\rho D)$ in the above table is evaluated for mixtures composed largely of air.

* Methyl formate.

† Ethyl formate, methyl acetate.

‡ n-Propyl formate, i-propyl formate, ethyl acetate, methyl propionate.

1. *Equimolar, steady-state counterdiffusion* is the simplest case. Examples are the mixing of two gases in a cylindrical space at uniform pressure and the counterdiffusion of two components in binary distillation. In this case, Eq. (49) integrates to

$$N_A = \frac{D}{B_F}(c - c_i) = \frac{D}{RTB_F}(p - p_i) = \frac{DP}{RTB_F}(y - y_i) = k_G(y - y_i) \quad (52)$$

(For nomenclature, see p. 524.)

2. *Steady-state diffusion of one gas through a stationary layer of a second, non-diffusing gas* is slightly more complicated. Examples are the evaporation of a pure liquid through a film of an insoluble gas, and absorption of a soluble gas from a second insoluble gas. Diffusion of the soluble component A occurs under the influence of a partial pressure gradient, $P(y_1 - y_2)$. Since constant pressure is maintained in the gas mixture, a gradient in the concentrations of the second component, $\frac{d(1 - y)}{ds} = -\frac{dy}{ds}$ also exists, and this causes the second component B to diffuse in the opposite direction at a rate [*cf.* Colburn and Drew, *Trans. Am. Inst. Chem. Engrs.*, **33,** 197 (1937)] given by

$$N_B = \frac{DP}{RT}\frac{dy}{ds} \quad (53)$$

The concentration of this component is held uniform by a mass motion of both components simultaneously at a rate just sufficient to counterbalance N_B. Since this mass motion is in the same direction as N_A, the result is an apparent increase in N_A. The final value of N_A is

$$N_A = \frac{DP}{RT}\frac{dy}{ds} + \left(\frac{y}{1 - y}\right)\frac{DP}{RT}\frac{dy}{ds} = \frac{DP}{RT}\left(\frac{1}{1 - y}\right)\frac{dy}{ds} = -\frac{DP}{RT}\frac{d\ln(1 - y)}{ds} \quad (54)$$

The rate of diffusion across a layer of thickness B_F is then,

$$N_A = \frac{DP}{RTB_F}\frac{(y - y_i)}{(1 - y)_f} = k_G'(y - y_i) \quad (55)$$

where $(1 - y)_f$ represents the logarithmic mean of $1 - y$ and $1 - y_i$. (For nomenclature, see p. 524.)

Lewis and Chang [*Trans. Am. Inst. Chem. Engrs.*, **21,** 127 (1928)] and Sherwood ("Absorption and Extraction," p. 1, McGraw-Hill, New York, 1937) have derived integrated equations for this and other cases by the use of a kinetic theory suggested originally by Stefan [*Sitzber. Akad. Wiss. Wien.*, **63** (2), 63 (1871); **65** (2), 323 (1872)]. This theory is based on the assumption that the gradient in partial pressure of one component is proportional to the mean velocity of diffusion of molecules of that component relative to the second component. It leads to a diffusion coefficient which is indicated to be independent of concentration, but it gives no information concerning the way in which the coefficient depends on molecular properties, temperature, or pressure.

In case the second component is partly condensed or vaporized at the boundary, a more general relation, analogous to Eq. (55), should be used, as shown by Colburn and Drew [*loc. cit.*]. See p. 559 in this connection.

For the *unsteady-state* vaporization of a liquid into a long column of an insoluble gas held at constant pressure, Arnold [*Trans. Am. Inst. Chem. Engrs.*, **40**, 361 (1944)] has derived equations showing the effect of the continuously changing concentration of inert gas. His results are applicable also to the unsteady-state absorption of a soluble gas from a long column of inert gas.

DIFFUSION IN LIQUIDS

The theory of diffusion is not so well developed for liquids as it is for gases. There are few reliable experimental values of diffusion coefficients. Table 5 lists selected data for diffusion coefficients in water, ethanol, and benzene. Reference should be made to the "International Critical Tables" for further data, especially for diffusion of inorganic salts in water and for organic systems. If no experimental data are available, an estimate of the diffusion coefficient may be made from

Table 13. Diffusion Coefficients in Liquids at 20°C. Dilute solutions. For effect of temperature, see p. 540

Solute	Solvent	$D \times 10^5$ (cm.²/sec.)′ × 10⁵	$\left(\frac{\mu}{\rho D}\right)^*$	Ref.
O_2	Water	1.80	558	1
CO_2	Water	1 77	559	1
N_2O	Water	1.51	665	1
NH_3	Water	1.76	570	1
Cl_2	Water	1.22	824	1
Br_2	Water	1.2	840	2
H_2	Water	5.13	196	1
N_2	Water	1.64	613	1
HCl	Water	2.64†	381	2
H_2S	Water	1.41	712	1
H_2SO_4	Water	1.73	580	2
HNO_3	Water	2.6	390	2
Acetylene	Water	1.56	645	1
Acetic acid	Water	0.88	1140	1
Methanol	Water	1.28	785	1
Ethanol	Water	1.00	1005	1
Propanol	Water	0.87	1150	1
Butanol	Water	.77	1310	1
Allyl alcohol	Water	.93	1080	2
Phenol	Water	.84	1200	2
Glycerol	Water	.72	1400	1
Pyrogallol	Water	.70	1440	2
Hydroquinone	Water	.77	1300	2
Urea	Water	1.06	946	2
Resorcinol	Water	0.80	1260	2
Urethane	Water	.92	1090	2
Lactose	Water	.43	2340	2
Maltose	Water	.43	2340	2
Glucose	Water	.60	....	2
Mannitol	Water	.58	1730	2
Raffinose	Water	.37	2720	2
Sucrose	Water	.45	2230	2
Sodium chloride	Water	1.35	745	2
Sodium hydroxide	Water	1.51	665	2
CO_2	Ethanol	3.4	445	2
Phenol	Ethanol	0.8	1900	2
Chloroform	Ethanol	1.23	1230	2
Phenol	Benzene	1.54	479	1
Chloroform	Benzene	2.11	350	1
Acetic acid	Benzene	1.92	384	1
Ethylene dichloride	Benzene	2.45	301	1

References:
[1] Arnold, *J. Am. Chem. Soc.*, **52**, 3937 (1930).
[2] "International Critical Tables," vol. 5, p. 63.
* Based on μ/ρ = 0.01005 cm.²/sec. for water, 0.00737 for benzene, and 0.01511 for ethanol, all at 20°C.; applies only for dilute solutes.
† Extrapolated from another temperature, based on rules quoted by Arnold (*loc. cit.*), *i.e.*, for water solutions D increases 3 per cent per °C. temperature rise near 20°C., and increases about 2 per cent per °C. for other liquids.

an empirical equation due to Arnold [*J. Am. Chem. Soc.*, **52**, 3937 (1930)].

One of the best available theoretical treatments of diffusion in liquids (*cf.* Glasstone, Laidler, and Eyring, "The Theory of Rate Processes," pp. 516, McGraw-Hill, New York, 1941) indicates that the effect of temperature on D may be estimated by assuming that $D\mu/T$ is independent of temperature, where μ is the viscosity of the solution and T is the absolute temperature.

In concentrated solutions, the diffusivity varies with the concentration, especially if the solution is non-ideal. The fact that there is no net diffusion across an interface separating two liquid phases in equilibrium, in spite of an abrupt change in concentration, brings out the fact that the cause of diffusion is a difference of free energy or of activity rather than of concentration. Glasstone, Laidler, and Eyring (*loc. cit.*) show that the variation of the diffusion coefficient for a binary mixture with composition may be explained by assuming that $D\mu/[1 + (d \ln \gamma_1/d \ln x_1)]$ varies linearly with x_1 between the values of $D\mu$ for the pure components. In this expression γ_1 and x_1 are the activity coefficient and the mole fraction, respectively, of one component.

APPLICATIONS OF THE THEORY OF MOLECULAR DIFFUSION

Diffusion in gaseous and liquid systems is seldom static in the true sense because of convection currents caused by the differences in fluid density which almost invariably accompany differences in composition and temperature. Nevertheless, in a few instances, theories based on the assumption of perfectly static systems are approximately in agreement with fact. The mathematical solutions of the fundamental diffusion equation for such cases are analogous to those of the equation of heat conduction.

Evaporation of Small Drops into a Quiet Atmosphere. Such evaporation follows Eq. (49) if the droplet is small enough, as shown below. The instantaneous rate of evaporation is given by

$$-\frac{dm}{d\theta} = k_g 4\pi r^2 (y_i - y) \quad (56)$$

where

$$k_g = \frac{DP}{RTr} \quad (57)$$

In this equation, $-dm/d\theta$ is the instantaneous rate of evaporation expressed in lb.-moles/hr., r is the radius of the droplet, y_i and y are the mole fractions of the evaporating liquid in the gas at the interface and at a great distance, respectively, R is the universal gas constant and k_g is the mass-transfer coefficient, lb.-moles/(hr.)(sq. ft.)(mole fraction). Table 14 lists experimentally determined rates of evaporation and shows that, for sufficiently small particles, Eq. (57) is confirmed. Integration of the equations leads to the following expression [Eq. (58)] for the time θ_0 required for complete evaporation of a droplet

Table 14. Summary of Experimental Results on Evaporation of Small Drops of Liquid into Air

Material evaporated	Observed value of k_gRTr/DP	Observer
Iodine	1.03 ± 0.04	Topley and Whytlaw-Gray (1)
Benzophenone	1.2	Whytlaw-Gray and Patterson (2)
Water	1.0 ± 0.1	
Water	1.02	Houghton (3), according to Fuchs (4)
Water absorption by $CaCl_2$ solution	1.00 ± 0.05	Houghton and Radford (5)
Aniline	0.98 ± 0.02	Froessling (6)
Water	1.02 ± 0.03	

References:
[1] *Phil. Mag.*, **4**, 873 (1927).
[2] "Smoke," Arnold, London, 1932.
[3] *Physics*, **4**, 419 (1933).
[4] *Physik. Zeit. Sowjetunion*, **6**, 224 (1934).
[5] Papers in Physical Oceanography and Meteorology, M.I.T. and Woods Hole Oceanographic Institute, vol. 6, No. 3, 1938.
[6] *Gerlands Beitr. Geophys.*, **52**, 170 (1938).

having an initial radius r_0 into a volume of gas so large that the gas composition remains constant at y mole fraction.

$$\frac{\rho r_0{}^2 RT}{2MDP(y_i - y)} = \theta_0 \qquad (58)$$

Mixing of Gases in a Closed Cylindrical Container. This is the basis of one method used for the experimental determination of diffusion coefficients for permanent gases. The theory is based on the assumption that each half of the cylinder is filled initially with a gas uniform in composition and that a partition separating the ends is removed to allow diffusion to begin. To minimize convection currents in an experiment of this kind, the heavier gas is placed initially in the lower half of the cylinder, and diffusion is allowed to take place in a vertical direction. To obtain a uniform gas mixture in the shortest possible time by this method, natural convection should be encouraged by introducing the heavier gas above the lighter one.

The equation for this case has been derived by several authors (*cf.* Sherwood and Reed, "Applied Mathematics in Chemical Engineering," p. 225, McGraw-Hill, New York, 1939). Smith [*Ind. Eng. Chem.*, **26,** 1167 (1934)] has used the equation for the calculation of the time required for essentially complete mixing of equal volumes of two gases in a steel gas cylinder 125 cm. long. He finds, for example, that in order to form a 50-50 mixture of helium and methane in a cylinder at 25°C. and 100 atm. requires 280 hr. before the average composition in the upper and lower halves of the cylinder differ by only 1 per cent.

MASS TRANSFER BY CONVECTION

When mass is exchanged between a solid or a liquid surface and a moving fluid that carries the transferred material away from the surface, the transfer is said to take place by convection. This transfer process is called *forced* convection if the fluid motion is caused by an expenditure of external energy, as in the forced flow of a fluid through a tube or around an immersed object. The fluid motion that accompanies *free* or *natural* convection is caused by the action of gravity on portions of the fluid that have different densities. These differences in density may be due to differences in temperature accompanying heat transfer, to differences in concentration, or to both.

If the fluid motion is laminar, transfer of mass between adjacent layers of fluid takes place purely by molecular diffusion. If the velocity pattern of the flow is known, it is sometimes possible to calculate the over-all rate of mass transfer into the moving fluid by the use of the basic equations of molecular diffusion. If the flow is turbulent, however, such calculations are generally impossible, since the laws that govern the transport of matter by turbulent mixing of small volumes of fluid are very little understood. Prediction of mass-transfer rates under such conditions is frequently based on empirical methods.

Analogy among Mass Transfer, Heat Transfer, and Fluid Friction. The eddy motion that causes mass transfer or heat transfer in a moving fluid also causes fluid friction, owing to a transfer of momentum. The close similarity among the transfer of mass, heat, and momentum is brought out by the Reynolds analogy, which states that, when heat, mass, and momentum are supplied to the fluid in corresponding ways, the following ratios are equal:

$$\frac{\text{Rate at which mass is transferred from the solid surface}}{\begin{array}{c}\text{Total rate at which the component, in excess of the}\\ \text{interfacial concentration, flows past the surface}\end{array}}$$
$$= \frac{\text{rate of heat transfer from the solid surface}}{\begin{array}{c}\text{total rate at which heat, measured above the surface}\\ \text{temperature, flows past the surface}\end{array}}$$
$$= \frac{\text{rate of momentum loss due to friction}}{\text{total momentum of stream which flows past the surface}}$$

In terms of mathematical symbols, these statements may be written as

$$\frac{k_G(y - y_i)}{G_M(y - y_i)} = \frac{h(t - t_i)}{c_pG(t - t_i)} = \frac{g_c\tau_0}{\rho V^2} \qquad (59)$$

or

$$\frac{k_G}{G_M} = \frac{h}{c_pG} = \frac{f}{2} \qquad (60)$$

For nomenclature, see p. 524.

Experimental data for mass transfer into gas streams agree approximately with Eq. (60) when the value of the Schmidt number, $\mu/\rho D$, is near 1 and when the friction factor is calculated from the *skin* friction. For flow through a straight tube, or across a flat plate placed parallel to the direction of flow, the pressure drop is due entirely to skin friction against the surface. On the other hand, the frictional force exerted on an immersed body such as a sphere or a cylinder placed perpendicular to the direction of flow is due in part to fluid pressure exerted on the front face of the body which is not counterbalanced by equal and opposite pressure on the rear face. Equation (60) does not apply in such cases if the friction factor is calculated from the total drag, including the *form* drag.

The limited conditions under which the Reynolds analogy can be expected to hold may be seen from the equations that govern the rate of transfer through a turbulent fluid [von Karman, *Trans. Am. Soc. Mech. Engrs.*, **61,** 705 (1939)]. For mass transfer it is assumed that

$$N_A = N_m + N_t = -(D + E_m)\left(\frac{dc}{dx}\right) \qquad (61)$$

and for friction

$$\tau = \tau_m + \tau_t = -g_c(\mu + \rho E)\left(\frac{du}{dx}\right) \qquad (62)$$

where N_m = rate of mass transfer due to molecular diffusion, lb.-moles/(hr.)(sq. ft.).
N_t = rate of mass transfer due to turbulent mixing, lb.-moles/(hr.)(sq. ft.).
E_m = eddy diffusivity for mass transfer, ft.²/hr.
τ_m = shear stress due to molecular motion, lb. force/sq. ft.
τ_t = shear stress due to turbulent mixing, lb. force/sq. ft.
E = eddy kinematic viscosity, ft.²/hr.

In these equations the first term in the parentheses gives the rate of transfer of mass or of momentum due to molecular diffusion, and the second term gives the rate due to turbulent mixing. The Reynolds analogy follows from Eqs. (61) and (62) if it is assumed that (1) Either $\mu/\rho D = 1$, or both D and μ/ρ are much smaller than E_m and E, and (2) $E_m = E$ (3) N_A/τ is independent of position, x. Under these conditions the concentration and velocity fields are similar and, just as Reynolds assumed, mass and momentum are transferred in the same way.

Direct measurements of eddy diffusivity E_M have been made by Towle and Sherwood [*Ind. Eng. Chem.*, **31,** 457 (1939)] by observing the spread of a jet of pure hydrogen or carbon dioxide into a turbulent air stream flowing through a straight tube. The eddy diffusivity was independent of the density of the diffusing gas and was of the order of 100 times the molecular diffusivity, increasing approximately in proportion to the Reynolds number. The effect of a screen grid on the eddy diffusivity was investigated by Towle, Sherwood, and Seder [*Ind. Eng. Chem.*, **31,** 462 (1946)], who found that the diffusivity immediately following the grid was less than half the value obtained with no grid present. Over about 45 duct diameters downstream the diffusivity increased and approached an asymptotic value equal to that obtained at the same Reynolds number without a grid.

Simultaneous measurements of E_M and E were made by Sherwood and Woertz [*Ind. Eng. Chem.*, **31,** 1034 (1939)], who found that E_M was essentially constant over the central portion of a gas stream flowing turbulently through a narrow rectangular channel and was inversely proportional to the density of the gas at a constant Reynolds number. Determinations of eddy viscosity $E\rho$, made simultaneously by measuring pressure drops and observing velocity profiles in the center of the duct, showed that the ratio of E_M to E was independent of Reynolds number and of gas density and was equal to 1.6. This is about halfway between the value 2 predicted from Taylor's theory of the transport of vorticity [*Proc. Roy. Soc.* (*London*), **A151,** 494(1935); **A159,** 496(1937)] and Prandtl's theory for the transport of momentum (*cf.* Durand, "Aerodynamic Theory," vol. 3, pp. 162*ff.* Springer, Berlin, 1935). Taylor's theory assumes that vorticity is the quantity transported rather than momentum and thus takes account of the fact that the pressure fluctuations that occur in turbulent flow can cause a transfer of momentum although they cannot transfer heat or mass. Dryden [*Ind. Eng. Chem.*, **31,** 416 (1939)] discusses the theory of diffusion in turbulent flow, pointing out the effects of intensity and scale of turbulence on the rate of diffusion.

In the turbulent core of the stream flowing at a high velocity through a straight tube, the molecular diffusivity and viscosity are usually much smaller than the eddy diffusivity and viscosity. Within a thin layer near the tube wall, however, the reverse is true. Here turbulence is suppressed to such an extent that transfer occurs principally by molecular motion. Colburn [*Trans. Am. Inst. Chem. Engrs.*, **29,** 174 (1933)] and Chilton and Colburn [*Ind. Eng. Chem.*, **26,** 1183 (1934)] showed empirically that the resistance of this laminar film can be allowed for by the following modification of Reynolds analogy:

$$\frac{k_G}{G_M}\left(\frac{\mu}{\rho D}\right)^{2/3} = j_M = \frac{h}{c_pG}\left(\frac{c_p\mu}{k}\right)^{2/3} = j_H = \frac{f}{2} \qquad (63)$$

for turbulent flow through straight tubes and across plane surfaces, and

$$j_M = j_H < \frac{f}{2} \qquad (64)$$

for turbulent flow around cylinders. Experimental data show that Eqs. (63) and (64) are approximately valid for values of $(\mu/\rho D)$ near unity. Other dimensionless groups corresponding to (k_G/G_M) are (k'/G) and $(k_c\rho/G)$, where k' = mass transfer coefficient, lb./(hr.)(sq. ft.)(lb. solute/lb. dry gas). k_c = mass-transfer coefficient, lb.-mole/(hr.)(sq. ft.)(lb.-mole/cu. ft.).

For flow through smooth, cylindrical tubes these groups are equal also to $4d/H_G$, where d is the inside diameter of the tube and H_G is the height equivalent to a transfer unit, expressed in the same units as d.

In the case of fully developed turbulent flow through a straight cylindrical tube, more precise investigations of the relation between mass-transfer or heat-transfer rates and fluid friction have been carried out by a number of authors [Prandtl, *Physik. Zeit.*, **29,** 487 (1928); Taylor, *Brit. Adv. Comm. Aeronaut. Rept. and Memo.*, No. 272 (1917); Colburn and Hougen, *Bull. Univ. Wis. Eng. Exp. Sta.*, Ser. No. 70 (1930); Murphree, *Ind. Eng. Chem.*, **24,** 727 (1932); von Karman, *Trans. Am. Soc. Mech. Engrs.*, **61,** 705 (1939); Boelter, Martinelli, and Jonassen, *Trans. Am. Soc. Mech. Engrs.*, **63,** 447 (1941); Sherwood, *Trans. Am. Inst. Chem. Engrs.*, **36,** 817 (1940); Hofmann, *Forschungsheft*, **11,** 159 (1940); Mattioli, *Forschungsheft*, **11,** 149 (1940); Reichardt, *Z. angew. Math. Mech.*, **20,** 297 (1940)], all of whom base their theoretical developments on Nikuradse's measurements [*Ver. deut. Ing. Forschungsheft*, No. 356 (1932)] of the velocity distribution function for turbulent flow through straight smooth tubes.

One of the most logical of these theoretical analyses is that due to Reichardt (*loc. cit.*), who showed from Eqs. (61) and (62) that the relation between the concentration and velocity patterns is governed by the differential equation

$$\frac{dc}{du} = \frac{(N_A/\tau)(\mu/D)}{1 + (N_{Sc}' - 1)(\tau_t/\tau)} \qquad (65)$$

$$= \left(\frac{dc}{du}\right)_0 \frac{\left(\frac{N_A}{N_0}\frac{\tau_0}{\tau}\right)\left(\frac{\mu}{\mu_0}\frac{D_0}{D}\right)}{1 + (N_{Sc}' - 1)\left(\frac{\tau_t}{\tau}\right)} \qquad (66)$$

where the subscript zero denotes a value at the solid surface, and no subscript means a quantity in the fluid stream.

The ratio of the shear stress due to molecular motion to the total shear stress is related to the velocity distribution in the tube, which may be represented in a generalized form applicable to smooth tubes of any diameter by using the dimensionless velocity and distance variables

$$u^+ = \frac{u}{(\tau_0 g_c/\rho)^{1/2}}$$

the dimensionless velocity variable, and

$$y^+ = \left(\frac{l\rho}{\mu}\right)\left(\frac{\tau_0 g_c}{\rho}\right)^{1/2},$$

the dimensionless distance variable, with l measured from the tube wall. This distribution function, as determined from

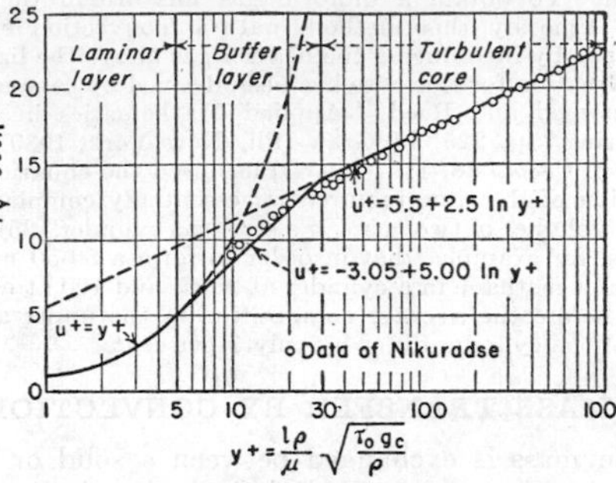

FIG. 9. Generalized velocity distribution for turbulent flow through smooth pipes. (*From McAdams, "Heat Transmission," 2d ed., McGraw-Hill, New York,* 1942.)

Nikuradse's data (*loc. cit.*), is shown in Fig. 9. Because of the definition of these variables, it is found that

$$\frac{du^+}{dy^+} = \frac{\mu\left(\frac{du}{dy}\right)}{\tau_0} = \frac{\tau_m}{\tau_0} \cong 1 - \left(\frac{\tau_t}{\tau}\right) \qquad (67)$$

By introducing a mathematical expression for the functional relationship between u^+ and y^+, the concentration profile may be found from Eq. (66). The slope of this concentration curve at the tube surface corresponds to a rate of mass transfer that may be expressed in terms of a mass-transfer coefficient by the equation

$$\frac{k_G/G_M}{f/2} = \frac{(\phi_m/\Delta c_m)(E_m/E)}{1 + b + \phi_m\sqrt{f/2}\,\psi(N_{Sc}')} \qquad (68)$$

where ϕ_m = mean fluid velocity/maximum velocity.
Δc_m = mean concentration difference/maximum concentration difference.
$N'_{Sc} = (\mu/\rho D)(E_M/E)$.

and b is a small, positive quantity that allows for the variation in the ratio N_A/τ across the tube. This quantity, which has been calculated by Reichardt (*loc. cit.*), depends on both the Reynolds number and the Schmidt number. The form of the function $\Psi(N_{Sc}')$ depends on the form that is assumed for the velocity distribution function in the laminar sublayer and the buffer layer near the solid surface. Reichardt obtains

$$\Psi(N_{Sc}') = 2(N_{Sc}' - 1) + 13.5\left(\frac{N_{Sc}'\ln(N_{Sc}')}{N_{Sc}' - 1} - 1\right) \qquad (69)$$

while Sherwood and von Karman give

$$\Psi(N_{Sc}') = 5(N_{Sc}' - 1) + 5\ln[1 + 5/6(N_{Sc}' - 1)] \qquad (70)$$

The corresponding heat-transfer problem does not differ mathematically from the mass-transfer problem outlined here. A similar result is obtained in which k_G/G_M is replaced by h/c_pG and $\mu/\rho D$ is replaced by $c_p\mu/k$, the Prandtl number. The eddy diffusivity E_M is replaced by the eddy thermometric conductivity E_H. See Martinelli, *Trans. Am. Soc. Mech. Engrs.*, **68,** 947 (1947).

Sherwood (*loc. cit.*) and von Karman (*loc. cit.*) assume that ϕ_m and Δc_m are equal to 1 and that b is zero. Boelter *et al.* include Δc_m but not ϕ_m, because of their method of carrying out the integration of Eq. (66). Their result is

$$\frac{k_G/G_M}{f/2} = \frac{(1/\Delta c_m)}{5\sqrt{\frac{f}{2}}\left[N_{Sc} + \ln(1+5\,Sc) + \frac{1}{2}\ln\left(\frac{N_{Re}}{60}\sqrt{\frac{f}{2}}\right)\right]} \qquad (71)$$

Figure 10 shows a plot of Eq. (68) as interpreted by several authors and compares the results of the theoretical analysis with the Reynolds analogy, represented by a straight horizontal line on the figure, and also with Colburn's empirical equation.

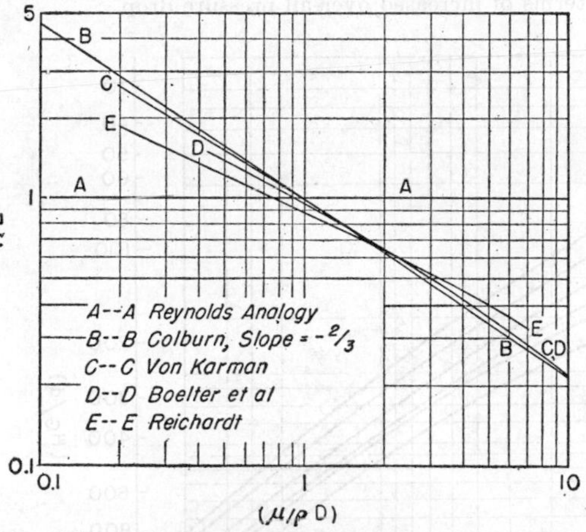

FIG. 10. Effect of Schmidt number on the analogy between mass transfer and fluid friction in smooth pipes.

It is evident from the figure that the use of the number ⅔ as an exponent on the Schmidt number gives too great an effect at values of $\mu/\rho D$ less than unity, though the agreement with the more exact theory is good for values of $\mu/\rho D$ or of $c_p\mu/k$ in the range 1 to 10.

Experimental Data for Surfaces Having a Definite Area. *Inside Tubes.* Several investigators [Gilliland and Sherwood, *Ind. Eng. Chem.*, **26**, 516 (1934); Chilton and Colburn, *Ind. Eng. Chem.*, **26**, 1183 (1934); Barnett and Kobe, *Ind. Eng. Chem.*, **33**, 436 (1941); Rommel, B. Ch. E. Thesis, Univ. of Del., 1943] have measured the rate of vaporization of pure liquids into air. The liquid stream is introduced at the top of the tube and flows as a film down the inside surface. Ordinarily the gas stream flows upward, countercurrently, but Gilliland and Sherwood (*loc. cit.*) showed that this reversal in direction had no effect on the rate of evaporation when the gas velocity relative to the tube wall was constant. Johnstone and Pigford [*Trans. Am. Inst. Chem. Engrs.*, **37**, 25 (1941)] distilled five different binary mixtures in a wetted-wall column. Krebs (Ph. D. Thesis, Univ. of Ill., 1937) absorbed SO_2 from air in water solutions of sodium acetate and acetic acid. The data of these various investigators are plotted in Fig. 11 to compare the results with the theoretical prediction based on the analogy between mass transfer and friction, as discussed above. This comparison shows that the experimental data from the several sources fall on a single line that is very nearly parallel to Boelter's theoretical line but lies about 30 per cent higher. If Sherwood and Woertz' value of 1.6 for E_M/E is assumed, the theory predicts the experimental data very closely.

A further check on the theory is afforded by measurements of the rate of heat transfer. A few of the most reliable data of this kind are also shown on the figure, indicating that excellent agreement is obtained for heat transfer to gases and to liquids by assuming that $E_M/E = 1$ rather than 1.6. It is perhaps surprising that different values of this ratio should be found necessary for the two processes, since it is usually assumed that heat and mass are transferred through the turbulent core by identical mechanisms.

When heat is transferred at the same time that mass is evaporated, however, the heat-transfer rate is invariably higher than when no mass is evaporated, as shown by the figure. When the two transfer processes occur simultaneously, as in the evaporation from a liquid surface at the wet-bulb temperature, heat transfer

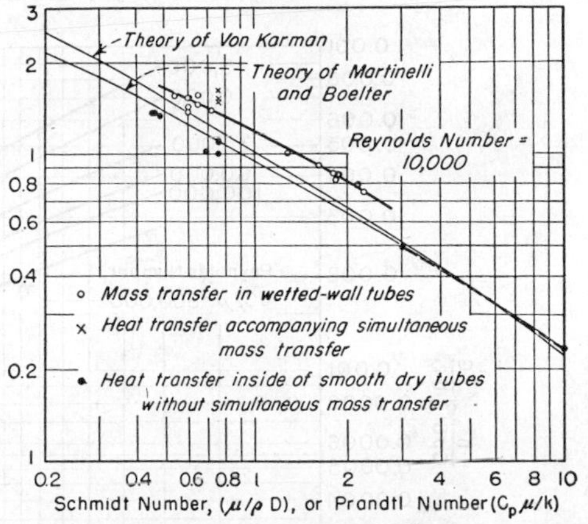

FIG. 11. Mass transfer and heat transfer inside wetted-wall tubes (turbulent flow). Data on vaporization of pure liquids into gases, distillation of binary liquid mixtures, absorption of soluble gases in liquids, heat transfer to gases and water.

occurs more rapidly than when the surface is dry. This may be due to the fact that the liquid surface inside a wetted-wall tower is slightly rough because of ripples. For this reason, the wet-bulb temperature obtained in the evaporation of water from the inside surface of a tube agrees essentially with that observed by means of a wet thermometer bulb, as determined by the dynamic equilibrium corresponding to $h/k' = c_s$, where c_s is the humid heat. Experimentally observed values of $h/k'c_s$ for wetted-wall columns do not agree with calculated values based on the theory, regardless of the assumption concerning the ratio of E_M and E_H to E. Experimental values of $h/k'c_s$ determined by Rommel (*loc. cit.*) and Barnett and Kobe (*loc. cit.*) are definitely higher than can be explained theoretically.

It is evident from Fig. 10 that even the most complete theories now available do not account fully for observed relations between heat transfer and mass transfer in wetted-wall towers. Use of any of the theories with the assumption of $E_M = E_H = E$ gives a result that is conservative from the point of view of equipment size and does not differ by more than 30 per cent from the best mass-transfer data. Conservative values of mass-transfer coefficients and H.T.U.'s may be estimated from Fig. 12, which is based on Boelter's equation.

Gilliland and Sherwood (*loc. cit.*) found no effect of total pressure on the group $k_G' \frac{(1-y)_f d}{D}$ $\left[\text{equal to } \frac{1}{4}\frac{d}{H_G}\left(\frac{dG}{\mu}\right)\left(\frac{\mu}{\rho D}\right)\right]$ where d is the tube diameter and the other terms are given in the table of nomenclature on p. 524], within the pressure range from 110 to 2330 mm. mercury, as shown in Fig. 13. It follows that there is no effect of total pressure on the value of H.T.U. and that k_g' varies inversely with the mean mole fraction of inert gas, $(1 - y)_f$.

All the data discussed thus far have been determined with apparatus containing a calming section preceding the test section. Abrupt changes in the distribution of velocities in the flowing stream were thus avoided. Greenewalt [*Ind. Eng. Chem.*, **18**, 1291 (1926)] measured the rate of absorption of H_2O vapor from air by 72 per cent sulfuric acid in a 2-in.-diameter wetted-wall tower into which the air stream was introduced by means of nozzles of different designs. The tower was 13.5 diameters high. Use of a small straight entrance nozzle gave transfer coefficients approximately twice as great as were observed when a gradually tapered venturi entrance nozzle was used. For all the entrance nozzles investigated, the transfer coefficient was found to be proportional to the 0.4 power of the pressure drop.

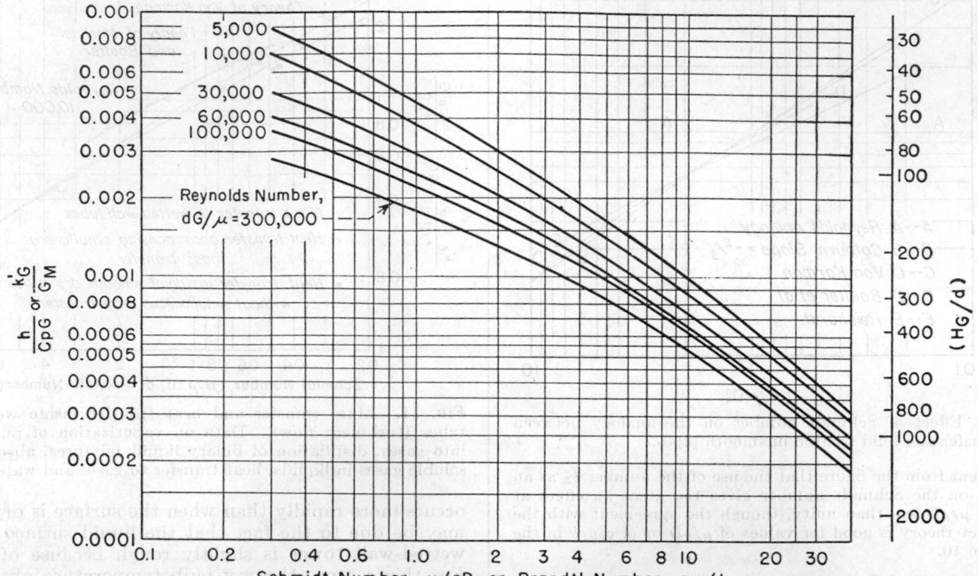

Fig. 12. Theoretical rates of mass transfer and heat transfer by convection in tubes. [*Boelter, Martinelli, and Jonassen, Trans. Am Soc. Mech. Engrs.*, **63**, 447 (1941).]

Kirschbaum and Keinzle [*Chem. Fabrik*, **14**, 171 (1941)] measured rates of vaporization of water and ethanol into air in a tube 10 diameters long, the inside surface of which was covered with wet filter paper. The absence of a calming section and the use of a mixing device just before the wetted surface caused the rate of mass and heat transfer to be about twice as high as would have been expected if a calming section had been provided. Surowiec and Furnas [*Trans. Am. Inst. Chem. Engrs.*, **38**, 53 (1942)] distilled solutions of ethanol and water in a wetted-wall tower 2.7 diameters high into which the vapor was introduced through a nozzle ½ as large as the tower. The observed values of H.T.U. were only about 40 per cent as great as though a calming section had been provided. It seems clear that the percentage increase in the rate of transfer which is obtained by introducing the gas into a wetted-wall tower as a jet must be smaller the shorter the wetted surface; and the advantage gained must be paid for in terms of increased over-all pressure drop.

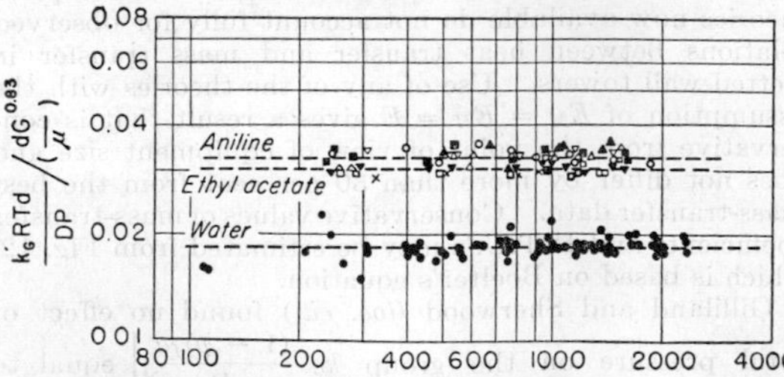

Fig. 13. Effect of total pressure on the rate of mass transfer. [*After Gilliland and Sherwood, Ind. Eng. Chem.*, **26**, 516 (1934).]

Viscous Flow inside Tubes. The data of Gilliland and Sherwood [*Ind. Eng. Chem.*, **26**, 516 (1934)] are available for the rate of vaporization of several different liquids into air from the inside surface of a vertical tube. These authors showed that their own data as well as those of Haslam, Hershey, and Kean [*Ind. Eng. Chem.*, **16**, 1224 (1924)] agree fairly well with a theory based on an integration of the fundamental equation of diffusion with the assumption that the gas flow is rodlike, *i.e.*, that the gas velocity is uniform across the tube. The fact that the theory based on a parabolic distribution of velocities does not agree with the data is explained by Boelter [*Trans. Am. Inst. Chem. Engrs.*, **39**, 557 (1943)], who showed that experimental data for vertical tubes could be predicted by allowing for natural convection currents due to the density differences in the air stream which accompany concentration gradients.

Annular Spaces. Weigand [*Trans. Am. Inst. Chem. Engrs.*, **38**, 569 (1942)] measured rates of vaporization of water into air flowing through the annular channels between a single outer tube and three different concentric inner tubes. The ratio of the inner diameter to the outer diameter was varied from 0 to 0.8. When vaporization took place from both surfaces simultaneously, the total rate of vaporization in the range of turbulent flow was found to follow the equation

$$j_M = 0.022 \left[\frac{(d_2 - d_1)G}{\mu}\right]^{-0.2} \qquad (72)$$

where d_2 and d_1 represent the larger and smaller diameters, respectively. The value of j_M was found to be independent of the diameter ratio. When vaporization occurred at the inner surface only, the mass-transfer coefficient agreed approximately with Eq. (72), but the value of j decreased as the ratio d_1/d_2 increased. Approximate agreement with Eq. (72) was also observed when water was vaporized at the outer surface only, but the variation with the d_1/d_2 was in the opposite direction and was smaller.

Plane Surfaces. The rates of evaporation of water and organic liquids into air from plane and cylindrical surfaces located parallel to the direction of flow have been measured by a number of investigators [Hinchley and Himus, *Trans. Inst. Chem. Engrs. (London)*, **2,** 57 (1924); Hine, *Phys. Rev.*, **24,** 79 (1924); Thiesenhusen, *Gesundh. Ing.*, **53,** 113 (1930); Powell and Griffiths, *Trans. Inst. Chem. Engrs. (London)*, **13,** 175 (1935); Powell, *Trans. Inst. Chem. Engrs. (London)*, **18,** 36 (1940); Shepherd, Hadlock, and Brewer, *Ind. Eng. Chem.*, **30,** 388 (1938); Pasquill, *Proc. Roy. Soc. (London)*, **A,** 75 (1943); Wade, *Trans. Inst. Chem. Engrs. (London)*, **20,** 1 (1942)]. Most of these data have been compared by Powell (*loc. cit.*) with his own data and those of Powell and Griffiths (*loc. cit.*) on wetted cloth surfaces ranging from 0.7 to 9.6 in. long in the direction of flow and air velocities up to 82 ft./sec. Figure 14 shows a comparison of some of these data with heat-transfer data, as correlated by Colburn [*Trans. Am. Inst. Chem. Engrs.*, **29,** 174 (1933); *cf.* also Jakob and Dow., *Trans. Am. Soc. Mech. Engrs.*, **68,** 123 (1946)].

The data for the laminar region below Reynolds numbers of about 20,000 agree fairly closely with a theoretical equation due to Pohlhausen [*Z. angew. Math. Mech.*, **1,** 252 (1921)],

$$j_M = \frac{f}{2} = 0.664 N_{Re}^{-1/2} \tag{73}$$

which is based on an exact integration of the equations of fluid flow and heat conduction. The equation has been confirmed by the heat-transfer measurements of Jakob and Dow (*loc. cit.*).

In the turbulent region most of the data agree with the equation

$$j_M = \frac{f}{2} = 0.036 Re^{-0.2} \tag{74}$$

which agrees with a theoretical result for the case $\mu/\rho D = c_p\mu/k = 1$ due to Latzko [*Z. angew. Math. Mech.*, **1,** 268 (1921); translated in *Nat. Advisory Comm. Aeronaut. Tech. Mem.* 1068 (1944)]. For evaporation of water into air at 25°C. and 1 atm., this equation reduces to

$$k_G b' = 0.136(b'V)^{0.8} \tag{75}$$

where b' is the length of the surface, in., V is the air velocity, ft./sec., and the units of k_G are lb.-moles of water/(hr.)(sq. ft.)(mole fraction).

The recent heat-transfer data of Jakob and Dow (*loc. cit.*) are 12 per cent lower than Eq. (74). These authors investigated the effect of unheated sections of the flat surface preceding the heated section and found that, for a constant Reynolds number based on the total length of the heated and unheated sections, the average coefficient of heat transfer was increased by the presence of the unheated section (for a constant air velocity, the coefficient would be reduced) by the factor

$$1 + 0.40\left(\frac{L_{st}}{L_{tot}}\right)^{2.75}$$

where L_{st} is the length of the unheated starting section and L_{tot} is the total length of the unheated and heated sections. In the absence of other data, this factor is recommended for use in mass-transfer problems for values of L_{st}/L_{tot} less than 0.6.

There is some doubt that the use of the two-thirds power of the Schmidt group can be justified for the case of turbulent flow, though there is adequate theoretical evidence for laminar conditions. Jakob and Dow (*loc. cit.*) assume that the thermal conductivity has no effect when the boundary layer of air covering the surface is fully turbulent. Pasquill (*loc. cit.*) compares his own measurements of the rate of evaporation of several different liquids from a flat plate located in the bottom surface of a wind tunnel with a theory of turbulent diffusion due to Sutton [*Proc. Roy. Soc. London*, **A146,** 701 (1934)], which is based on Taylor's concept of a turbulent eddy blending gradually with its changing surroundings [*Proc. London Math. Soc.*, **20,** 196 (1922)]. As applied by Pasquill [*Proc. Roy. Soc. London* **A182,** 75 (1943)] the theory indicates that the exponent on the Schmidt group should be 2/9 rather than 2/3. Pasquill's and Wade's data indicate that the correct value of the exponent lies between 2/9 and 2/3.

Powell's measurements (*loc. cit.*) of the rates of evaporation of water from circular disks facing an air stream are correlated by the equation

$$k_G d' = 0.68(Vd')^{0.65} \tag{76}$$

where d' = disk diameter, in.
V = air velocity, ft./sec.

This equation agrees with the data of Molstad, Farevaag, and Farrel [*Ind. Eng. Chem.*, **30,** 1131 (1938)]. Rates of evaporation from disks and rectangular plates facing

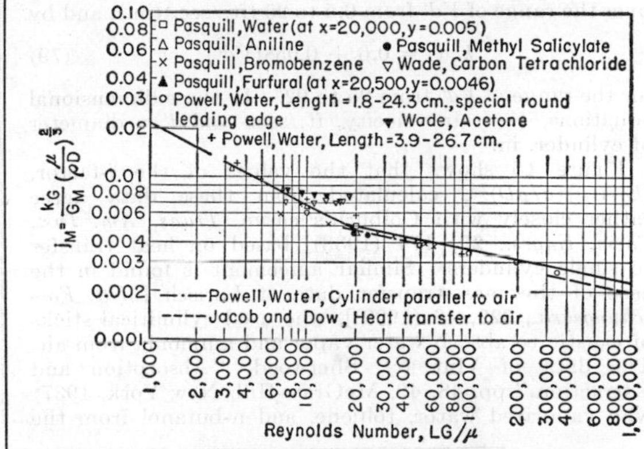

FIG. 14. Evaporation from plane surfaces parallel to wind.

downstream and tangentially were less than 40 per cent lower than for the same surface facing upstream. The effect of the angle of orientation of the surface with respect to the direction of air flow was marked only when the wetted surface faced downstream, the rate increasing to a maximum value as much as 65 per cent higher than for the parallel position for an angle of about 40°. This result was obtained for a plate 10.5 in. wide and 7.65 in. long in a direction parallel or perpendicular to the stream. Smaller increases at the minimum evaporation rate were observed with narrower plates at smaller angles.

Any obstruction that disturbs the gas flow in front of the evaporating surface causes a marked increase in the rate of transfer unless the ridge shields the wet surface from the air stream. In order to obtain the greatest rate of evaporation for a long surface, a number of ridges must be located at intervals along the surface in the direction of flow. Even with this arrangement, however, the rate of evaporation is less than if the surface were turned with its longest dimension perpendicular to the flow.

Rates of evaporation of water into quiet air at 1 atm.

by natural convection were studied by Boelter, Gordon, and Griffith [*Ind. Eng. Chem.*, **38**, 596 (1946)], who expressed their results for a circular pan 12 in. in diameter by the equations

$$w = 0.00138(p_w - p)^{1.20} \qquad (77)$$

where w = rate of evaporation, lb. water/(hr.)(sq. ft.).
p_w = partial pressure of water vapor in air at evaporating surface, mm. mercury.
p = partial pressure at great distance, mm. mercury.

For different experimental conditions, the rate of evaporation is dependent on the value of the Grashof number, $\frac{d^3 g_L \rho^2}{\mu^2}\left(\frac{MT_s}{M_s T_\infty} - 1\right)$ where M is the mean moelcular weight of the gas-vapor mixture and d is the surface diameter. This dependence is discussed by Boelter *et al.* (*loc. cit.*).

Cylindrical Surfaces. The measurements of Powell and Griffiths [*Trans. Inst. Chem. Engrs. (London)* **13,** 175 (1935)] and of Powell [*Trans. Inst. Chem. Engrs. (London)*, **18**, 36 (1940)] of the rates of evaporation of water from the surfaces of single cylinders about 1 ft. long and 0.063 to 14.7 in. in diameter mounted in a wind tunnel perpendicular to the direction of flow showed that the rate of vaporization at 1 atm. is given by

$$k_G d' = 0.286(Vd')^{0.6} \qquad (78)$$

over the range of Vd' from 0.5 to 25 (ft./sec.)(in.), and by

$$k_G d' = 0.6 + 0.058(Vd') \qquad (79)$$

for the range of Vd' from 25 to 92. In these dimensional equations, V = air velocity, ft./sec., and d' = diameter of cylinder, in.

Figure 15 shows that the values of the j-factor, $(k_G/G_M)(\mu/\rho D)^{2/3}$, calculated from these data agree rather closely with Colburn's curve [*Trans. Am. Inst. Chem. Engrs.*, **29**, 174 (1933)] based on heat transfer to single cylinders. Similar agreement is found in the case of the mass-transfer data of Lorisch [*Mitt. Forschungsarb.*, **322**, 46 (1929)], who used cylindrical sticks of caustic to absorb water vapor and ammonia from air. The data of Vint (*cf.* Sherwood, "Absorption and Extraction," pp. 48, 49, McGraw-Hill, New York, 1937) who vaporized water, toluene, and n-butanol from the surface of a 2-in.-diameter, cloth-covered cylinder into air, as well as the data of London, Nottage, and Boelter [*Ind. Eng. Chem.*, **33**, 467 (1941)], who evaporated water from heated 1-in. and ½-in.-diameter cylinders into air, are also in substantial agreement.

London *et al.* (*loc. cit.*), in their careful investigation, found the psychrometric ratio, $h/k'c_s$, for the evaporation of water from a cylinder to be 1.045.

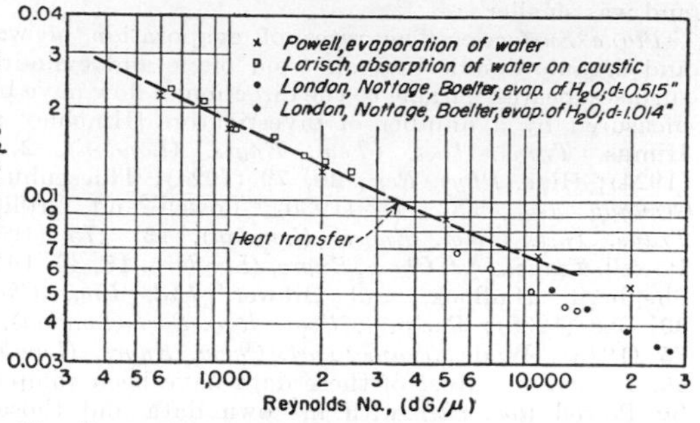

FIG. 15. Rates of mass transfer from the surface of circular cylinders perpendicular to an air stream.

The case of cylinders located parallel to the direction of the flow is considered in the section on plane surfaces.

The effect on the rate of transfer of vibrating the cylinder is considered by Martinelli and Boelter (*Proc. Intern. Congr. Appl. Mech.*, M.I.T., 1938, p. 578) for the analogous problem of heat transfer.

Powell [*Phil. Mag.*, (7), **29**, 274 (1940)] studied the rate of evaporation of water from a 0.92-cm.-diameter horizontal cylinder by natural convection. He found that the rate of evaporation reached a minimum value when the surface temperature was about 1°C. below the temperature of the surrounding air. On increasing the surface temperature, the movement of the surrounding air changed at this surface condition from a downward to an upward direction.

Spherical Surfaces. Rates of evaporation of small spheres into quiet air were shown above to be represented closely by the theoretical relation $k_G RTr/DP = 1$. This represents a minimum value for the rate of mass transfer. In the case of large spheres natural convection causes an increase in k_G.

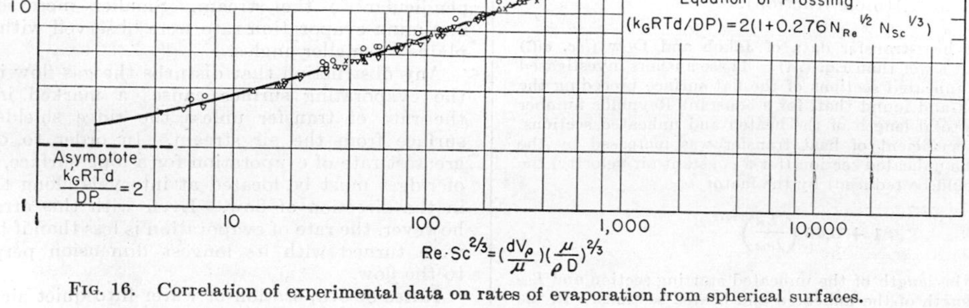

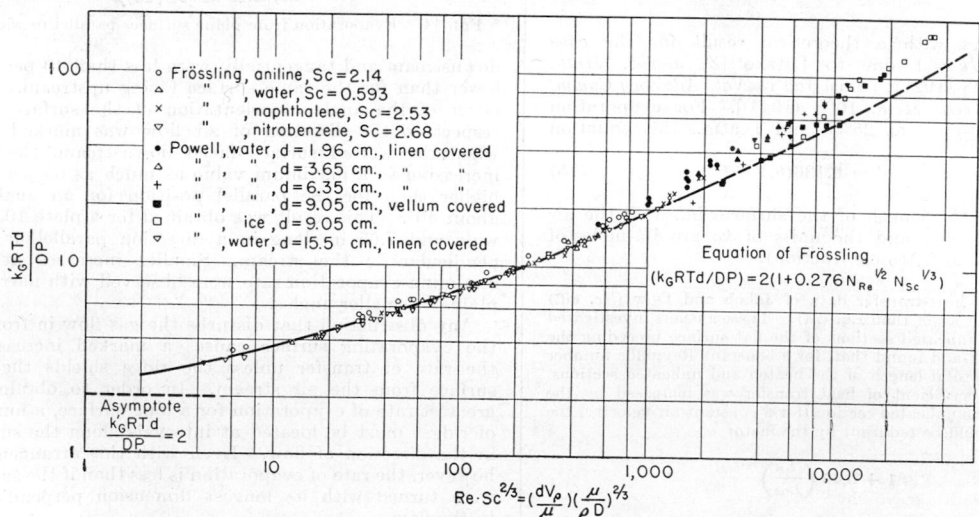

FIG. 16. Correlation of experimental data on rates of evaporation from spherical surfaces.

Rates of evaporation of single suspended spherical particles of aniline, naphthalene, nitrobenzene, and water into moving air have been measured by Frössling [Gerlands Beitr. Geophys., **52,** 170 (1938)]. Powell [*Trans. Inst. Chem. Engrs.* (*London*), **18,** 36 (1940)] has determined rates of evaporation of water from linen-covered and vellum-covered spheres ranging in diameter from 1.96 to 15.5 cm. and from an ice-coated sphere 9.05 cm. in diameter [*Proc. Brit. Assoc. Refrig.*, **36,** 61 (1939–1940)]. A comparison of Powell's and Frössling's and $c_p\mu/k$ for $\mu/\rho D$, indicating the similarity of the heat-transfer and mass-transfer processes outside spherical surfaces.

The rate of mass transfer into a drop of liquid from its surface is found to follow the laws of molecular diffusion if the drop is small or if it is composed of a viscous liquid. The surface contours of falling water drops larger than about 1 mm. in diameter are distorted by external gas pressure, and the resulting internal circulation of liquid causes mass transfer within large drops to take place

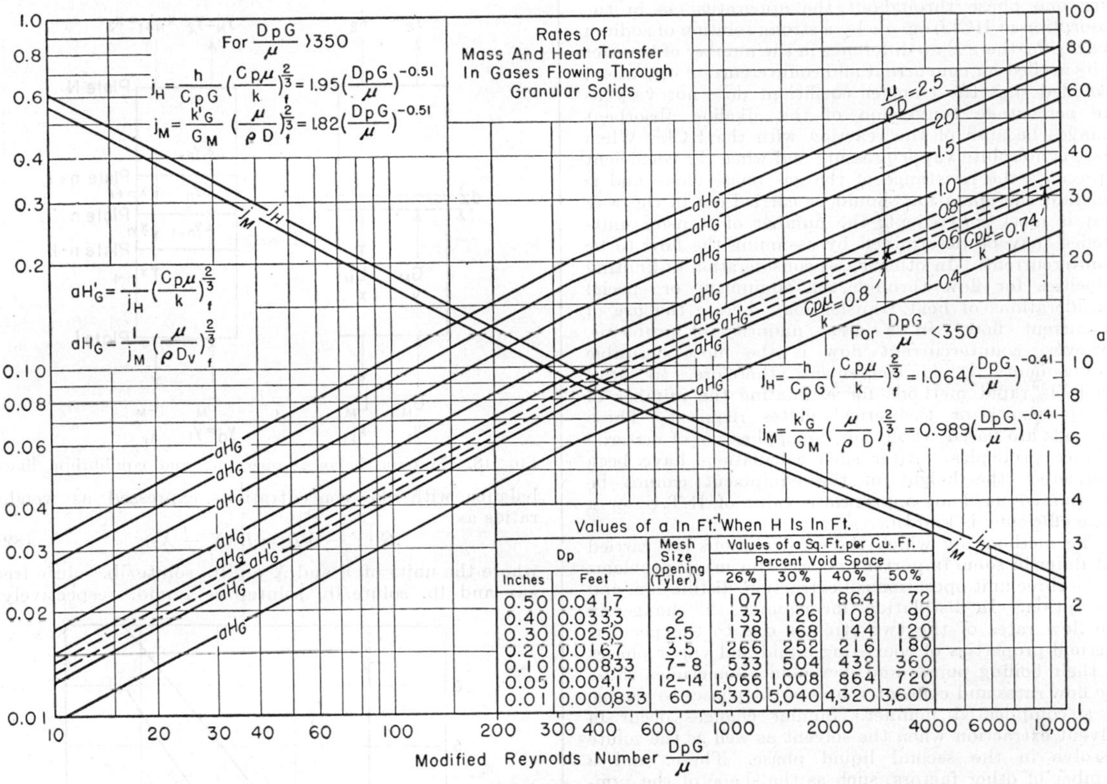

Fig. 17. Rates of mass transfer and heat transfer in gases flowing through granular solids.

data is shown in Fig. 16. Powell's data extend to higher Reynolds numbers and lie on the average about 25 per cent higher than Frössling's semi-empirical equation

$$\left(\frac{k_G}{G_M}\right)\left(\frac{dG}{\mu}\right)\left(\frac{\mu}{\rho D}\right) = \frac{k_GRTd}{DP}$$
$$= 2\left[1 + 0.276\left(\frac{dV\rho}{\mu}\right)^{1/2}\left(\frac{\mu}{\rho D}\right)^{1/3}\right] \quad (80)$$

where d is the diameter of the sphere. The data of Johnstone and Williams [*Ind. Eng. Chem.*, **31,** 993 (1939)] for the absorption of ammonia from air by drops of sulphuric acid, sulphur dioxide by sodium hydroxide, and hydrogen sulphide by sodium hydroxide–hydrogen peroxide solutions agree with the data of Fig. 16. The data of Hatta, Ueda, and Baba [*J. Soc. Chem. Ind. Japan*, **37,** 383 (1934)], Houghton [*Physics*, **2,** 467 (1932)], and Whitman, Long, and Wang [*Ind. Eng. Chem.*, **18,** 363 (1926)] for mass transfer to falling or suspended drops are also in agreement.

Equation (80) is also found to represent the available heat-transfer data if hd/k is substituted for k_GRTd/DP

more rapidly than would be expected from diffusion theory.

Powell (*loc. cit.*) shows that the rate of evaporation of water per unit surface area is greater for spherical surfaces than for other shapes.

Beds of Granular Solids. Figure 17 shows results obtained by Gamson, Thodos, and Hougen [*Trans. Am. Inst. Chem. Engrs.*, **39,** 1 (1943)] and Wilke and Hougen [*ibid.*, **41,** 441 (1945)] for rates of adiabatic evaporation of water and simultaneous heat transfer from packed beds of porous pellets of various sizes. The results are based on the assumption that the temperature of the evaporating surface is equal to the temperature of a wet-bulb thermometer, and for this reason the values of j_H and j_M are not independent. The results of Hurt [*Ind. Eng. Chem.*, **35,** 522 (1943)] are in substantial agreement with those of Wilke and Hougen. Hurt measured rates of mass transfer for the following systems: adsorption of water vapor from air by silica gel particles and by particles coated with P_2O_5, adiabatic humidification of air, and evaporation of naphthalene into air and into hydrogen.

COUNTERCURRENT APPARATUS—NUMBER OF THEORETICAL PLATES OR TRANSFER UNITS REQUIRED TO CARRY OUT A SEPARATION

When a specified separation is to be accomplished by contacting two streams of fluid, it often happens that the fewest number of transfer units or of theoretical plates is required if the streams flow through the apparatus countercurrently, since the greatest average driving force may be obtained in this way. In the special case where there is no vapor pressure of the solute material over one phase throughout the apparatus, as in the absorption of HCl from air by a strong solution of sodium hydroxide, there is no difference in the number of transfer units needed for concurrent and countercurrent operation, provided that the transfer coefficient does not vary as the percentage conversion of the alkaline absorbent changes because of the reaction with the HCl. When there is no solute vapor pressure and when the coefficient represents the resistance of the gas phase alone and is constant the operation should be carried out in the way that is most convenient; the number of transfer units needed may be determined by assuming the flow to be countercurrent. In other cases consideration of limiting velocities for flow through the equipment or special considerations of heat transfer may dictate the use of concurrent flow. In a great majority of problems, however, countercurrent flow is the most desirable arrangement. It is convenient, therefore, to have available rapid methods for estimating the number of transfer units or theoretical plates required. These methods are based wholly on the application of material balance principles. After such calculations have been completed, the height of the equipment cannot be determined until an appropriate value of H.T.U. or a plate efficiency is known.

The methods by which these calculations are carried out differ in some important respects for many problems in the three unit operations of extraction, distillation, and absorption. In distillation, for example, the changes in the flow rates of the two streams due to the peculiar thermal properties of coexisting liquid and vapor phases at their boiling points and dew points sometimes cause the flow rates and concentrations of the phases to change in a complicated manner. Similar changes occur in solvent extraction when the solvent as well as the solute dissolve in the second liquid phase. These, and a number of other factors, such as the shape of the symmetrical equilibrium curves that are encountered often in distillation but infrequently in the other operations, make it essential to explain special methods of calculation under each operation. (See Sec. 9 for the distillation of multicomponent mixtures, Sec. 11 for solvent extraction using solvents that are mutually soluble, and Sec. 10 for the absorption of several components from a rich gas.) Nevertheless, problems are often encountered in each of the diffusional operations which can be handled by simple procedures that can be described without reference to a particular field of application.

MATERIAL BALANCES—OPERATING AND EQUILIBRIUM LINES

A material balance on the solute component written around the top section of the packed, countercurrent diffusional operations equipment illustrated in Fig. 18 indicates that

$$G_M(y - y_2) = L_M(x - x_2) \tag{81}$$

in which G_M and L_M are assumed to be constant, as is the case when both streams are sufficiently dilute. In certain cases, such as the absorption of a gaseous material from an insoluble gas by a non-volatile liquid, the ratio of flow of the inert gas G' and of the non-volatile liquid L' remains constant through the apparatus, even when the streams are so concentrated that G_M and L_M change. In this case it may be better to write the material

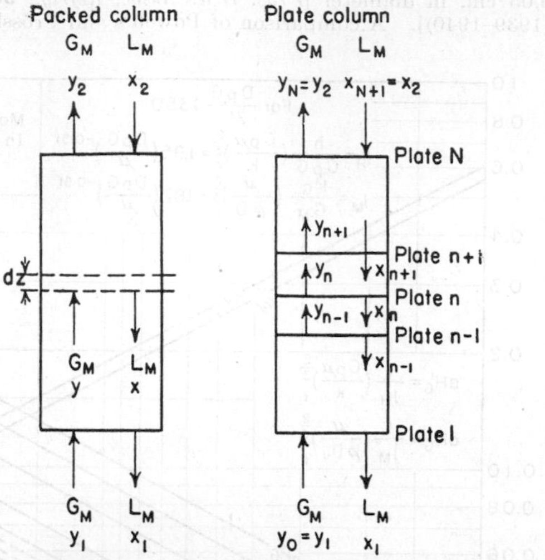

FIG. 18. Material balance—operating and equilibrium lines.

balance with the concentrations expressed as weight ratios as

$$G'(Y - Y_2) = L'(X - X_2) \tag{82}$$

where the units of Y and X are lb. solute/lb. solute-free gas and lb. solute/lb. solute-free liquid, respectively;

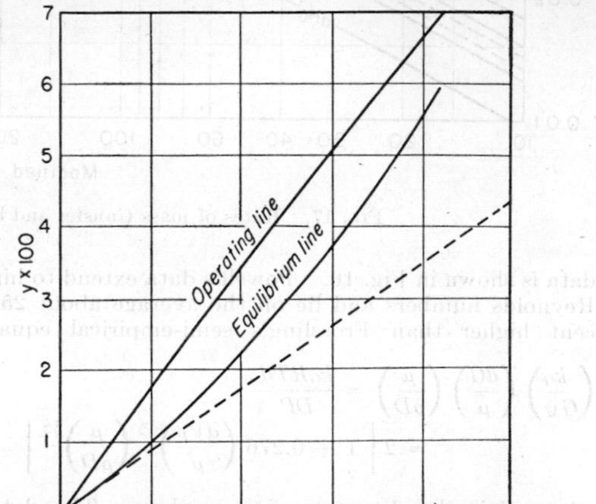

FIG. 19. Example of absorption of acetone, with a curved equilibrium line owing to heat of absorption. Dashed line is a tangent to the equilibrium curve at the origin.

or alternatively, Y and X may be expressed in lb.-moles of solute/lb.-moles of solute-free stream if G' and L' are in lb.-moles/hr. (solute-free). In these molar units, $Y = y/(1 - y)$ and $X = x/(1 - x)$, or $y = Y/(1 + Y)$ and $x = X/(1 + X)$. This equation gives the relation between the mean compositions of the two streams at each level in the apparatus, assumed to be operating under steady-state conditions. The line that represents this equation on a graph of y vs. x is called an operating line, as shown in Fig. 19.

The equilibrium curve may be plotted on the same graph by locating points that give the compositions of liquid and vapor phases at equilibrium with each other. Since the equilibrium relation depends in general on the temperature, heat effects must be allowed for in locating the equilibrium line. Comparison of the equilibrium and operating curves shows how far from equilibrium the two streams are at each point in the apparatus.

For a plate apparatus the material balance that leads to the operating line is applied between the top of the column and a section between the nth and the $(n + 1)$th plates (n is assumed to increase toward the top of the apparatus). The operating line equation is

$$G_M(y_n - y_2) = L_M(x_{n+1} - x_2) \tag{83}$$

when the subscript n denotes a stream flowing from the nth plate. Only a few discrete points on this operating line represent actual conditions within the apparatus; the operating line does not represent compositions on the plates, but only in the space between successive plates.

MASS TRANSFER BETWEEN PHASES

Use of Over-all Resistance to Transfer. When material is transferred from one phase to another across an interface that separates the two, the resistance to mass transfer in each phase causes a concentration gradient in each, as shown in Fig. 20. The concentra-

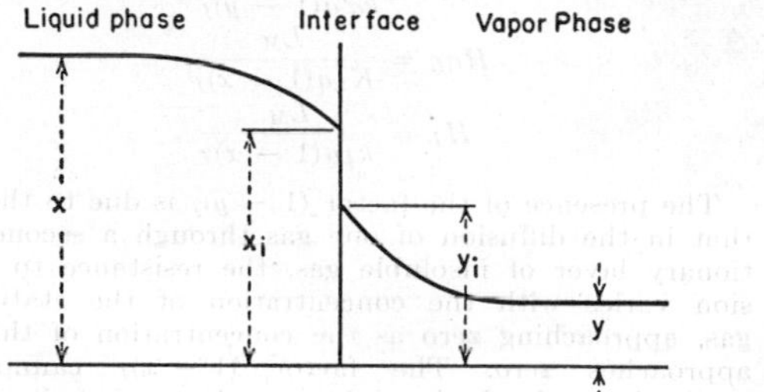

FIG. 20. Distribution of concentrations near an interface.

tions of the diffusing material in the two phases immediately adjacent to the interface are generally unequal, even if expressed in the same units, but are usually assumed to be related to each other by the laws of thermodynamic equilibrium, as discussed previously in this section.

In each phase the rate of transfer is proportional to the difference between the bulk concentration and the concentration at the interface. If, as is usually assumed [Whitman, *Chem. Met. Eng.*, **29**, 146 (1923)], the volume of the phases contained within the so-called "films," *i.e.*, the regions within which most of the variation in concentration occurs, is negligible, the rates of transfer in the two phases are equal. Thus

$$N_A = k_L(x - x_i) = k_G(y_i - y) \tag{84}$$

This equation may be used to find the interfacial concentrations corresponding to any set of values of x and y, provided that the ratio of the individual coefficients is known. Thus

$$\frac{y_i - y}{x - x_i} = \frac{k_L}{k_G} = \frac{L_M H_G}{G_M H_L} \tag{85}$$

Equation (85) may be solved graphically if a plot is made of the equilibrium vapor and liquid compositions, and a point is located representing the bulk concentrations x and y on this same diagram. A construction of this type is shown in Fig. 21.

In the design of equipment it is necessary to estimate the rate of mass transfer from known or predicted values of the transfer coefficients and the bulk concentrations. This may be done by solving Eq. (85) simultaneously with the equilibrium relation $y_i = F(x_i)$, to obtain y_i and x_i. The rate of transfer may then be calculated from Eq. (84).

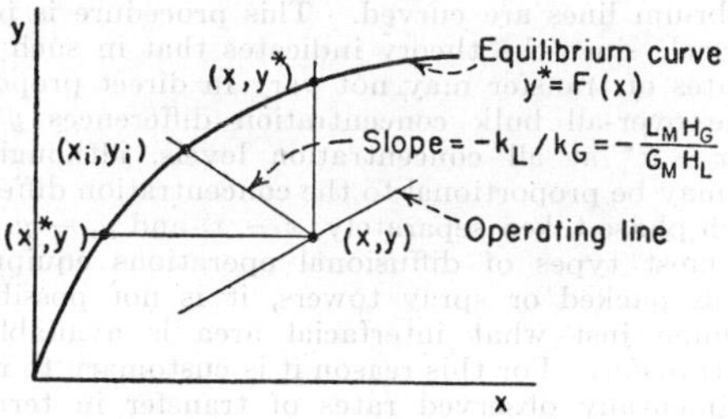

FIG. 21. Location of equilibrium concentrations for a point in a countercurrent tower.

If the equilibrium relation $y_i = F(x_i)$ is sufficiently simple, *i.e.*, if a plot of y_i vs. x_i is a straight line, not necessarily through the origin, the rate of transfer is proportional to the difference in the bulk concentration in one phase and the concentration in that same phase which would be in equilibrium with the bulk concentration in the other phase. One such difference is $y^* - y$, and another is $x - x^*$. In this case there is no need to solve for the interfacial compositions, as may be seen from the following derivation.

Since

$$N = k_G(y_i - y) = k_L(x - x_i) = K_G(y^* - y) \tag{86}$$

where y^* is the vapor composition in equilibrium with x,

$$\frac{1}{K_G} = \frac{1}{k_G}\left(\frac{y^* - y}{y_i - y}\right) = \frac{1}{k_G} + \frac{1}{k_G}\left(\frac{y^* - y_i}{y_i - y}\right)$$
$$= \frac{1}{k_G} + \frac{1}{k_L}\left(\frac{y^* - y_i}{x - x_i}\right) \tag{87}$$

in view of Eq. (85). If the equilibrium curve is a straight line, the term in parentheses is its slope, m. Thus

$$\frac{1}{K_G} = \frac{1}{k_G} + \frac{m}{k_L} \tag{88}$$

If the units of driving force on which k_L is based are liquid-phase concentrations expressed in lb.-moles of solute/cu. ft. of liquid and the units of k_G are partial pressures in atmospheres, Eq. (88) becomes

$$\frac{1}{K_G} = \frac{1}{k_G} + \frac{1}{m_c k_L} \tag{89}$$

where m_c is the Henry's law coefficient, (lb.-moles)/(cu. ft.)-(atm.).

When the equilibrium curve is not straight, there is no logical basis for the use of an over-all transfer coefficient. The calculation of the rate of transfer in such cases must be made by solving for the interfacial compositions, as described above. It is not possible to calculate a true average value of m for use in Eq. (88), since the value used must represent the relation between concentrations in equilibrium at the interface and depends, therefore, on the ratio k_L/k_G.

If it is desired to calculate the rate of transfer from the over-all concentration difference based on liquid compositions, $x^* - x$, the appropriate over-all coefficient K_L is related to the individual coefficients by the equations

$$\frac{1}{K_L} = \frac{1}{k_L} + \frac{1}{mk_G} \tag{90}$$

or

$$\frac{1}{K_L} = \frac{1}{k_L} + \frac{m_c}{k_G} \tag{91}$$

As in the case of Eq. (88), these equations apply only when the equilibrium line is straight.

Experimentally observed rates of mass transfer in diffusional operations equipment are often expressed in terms of over-all transfer coefficients, even when the equilibrium lines are curved. This procedure is purely empirical, since the theory indicates that in such cases the rates of transfer may not vary in direct proportion to the over-all bulk concentration differences $y^* - y$ and $x - x^*$ at all concentration levels, although the rates may be proportional to the concentration difference in each phase taken separately, $x - x_i$ and $y_i - y$.

In most types of diffusional operations equipment, such as packed or spray towers, it is not possible to determine just what interfacial area is available for mass transfer. For this reason it is customary to report experimentally observed rates of transfer in terms of transfer coefficients based on a unit volume of the apparatus rather than on a unit of interfacial area. Such volumetric coefficients are designated as K_Ga, k_La, etc., where a represents the interfacial area per unit of volume of the apparatus. Experimentally observed variations in the values of volumetric coefficients due to variations in flow rates, type of packing, etc., may be due as much to changes in the value of a as to changes in k. Calculation of the over-all coefficient from the individual coefficients is made by means of the equations

$$\frac{1}{K_Ga} = \frac{1}{k_Ga} + \frac{m}{k_La} \tag{92}$$

$$\frac{1}{K_Ga} = \frac{1}{k_Ga} + \frac{1}{m_ck_La} \tag{93}$$

$$\frac{1}{K_La} = \frac{1}{k_La} + \frac{1}{mk_Ga} \tag{94}$$

$$\frac{1}{K_La} = \frac{1}{k_La} + \frac{m_c}{k_Ga} \tag{95}$$

Because of the wide variation in the solubilities of gases in liquids, the variation in the value of m from one system to another sometimes has an important effect on the type of equipment that should be used for contacting. If, for example, it is desired to dissolve an insoluble gas, such as oxygen, in water, the large value of m for this system would cause the liquid-phase part of the over-all resistance to be extremely large in a spray tower, where the poor fluid mixing obtained in the liquid phase might result in a small value of k_L. On the other hand, this line of reasoning must be applied with caution, since gases with different solubilities are ordinarily absorbed under different conditions of operation; and the effect on the over-all resistance of changes in the solubility is therefore partly counterbalanced by changes in the specific resistances as the flow rates are changed.

Use of the Height Equivalent to a Transfer Unit (H.T.U.). Frequently the values of the individual coefficients of mass transfer vary so rapidly with the flow rates that the quantity obtained by dividing each coefficient by the flow rate of the phase to which it applies is more nearly constant than the coefficient itself. The quantity obtained by this division is called [Chilton and Colburn, *Ind. Eng. Chem.*, **27**, 255 (1935)] the height of one transfer unit, since it expresses in terms of a single length dimension the height of apparatus required to accomplish a separation of standard difficulty.

The number of over-all gas-phase transfer units required for changing the composition of the vapor stream from y_1 to y_2 is

$$N_{OG} = \int_{y_2}^{y_1} \frac{dy}{y^* - y} \tag{96a}$$

for equimolal diffusion, and

$$N_{OG} = \int_{y_2}^{y_1} \frac{(1 - y)_f\, dy}{(1 - y)(y^* - y)} \tag{96b}$$

for diffusion in one direction only. Convenient solutions of these equations are given later. (The number of transfer units required for a given separation is closely related to the number of theoretical plates or stages required to carry out the same separation in plate-type or stagewise apparatus.) In terms of H.T.U.'s the equations that express the addition of resistances become [Colburn, *Trans. Am. Inst. Chem. Engrs.*, **35**, 211 (1939)]

$$H_{OG} = H_G + H_L \left(\frac{mG_M}{L_M}\right) \frac{(1 - x)_f}{(1 - y)_f} \tag{97a}$$

and

$$H_{OL} = H_L + H_G \left(\frac{L_M}{mG_M}\right) \frac{(1 - y)_f}{(1 - x)_f} \tag{97b}$$

In these equations $(1 - y)_f$ is the logarithmic mean of $1 - y$ and $1 - y^*$, and $(1 - x)_f$ is the logarithmic mean of $1 - x$ and $1 - x^*$. These terms are omitted from these and the following equations for equimolal counterdiffusion. The following relations between the transfer coefficients and the values of (H.T.U.) apply:

$$H_{OG} = \frac{G_M}{K_G'a(1 - y)_f} \tag{98}$$

$$H_G = \frac{G_M}{k_G'a(1 - y)_f} \tag{99}$$

$$H_{OL} = \frac{L_M}{K_La(1 - x)_f} \tag{100}$$

$$H_L = \frac{L_M}{k_La(1 - x)_f} \tag{101}$$

The presence of the factor $(1 - y)_f$ is due to the fact that in the diffusion of one gas through a second stationary layer of insoluble gas, the resistance to diffusion varies with the concentration of the stationary gas, approaching zero as the concentration of this gas approaches zero. The factor $(1 - x)_f$ cannot be justified on the basis of kinetic theory for the liquid phase but is included in the equations on the basis of the assumption that diffusion through liquids is similar to that through gases. In binary distillation, where both components diffuse simultaneously, both these factors should be omitted (*cf.* p. 618).

The H. E. T. P. (height equivalent to one theoretical plate) is another quantity that is used occasionally to express the efficiency of a packing material for carrying out a separation. Experimental data should be reported as H.T.U.'s rather than H.E.T.P.'s, since the former quantity is theoretically correct for equipment, such as packed columns, in which mass is transferred by a differential rather than a stepwise action. If equilibrium and operating lines are parallel, *i.e.*, $mG_M/L_M = 1$, H.E.T.P.'s and H.T.U.'s are equal. If the equilibrium and operating lines are straight, but not parallel,

$$\frac{\text{H.E.T.P.}}{H_{OG}} = \frac{(mG_M/L_M) - 1}{\ln\,(mG_M/L_M)} \tag{102}$$

Use of Plate Efficiencies. The design of diffusional operations equipment of the stagewise countercurrent type involves the computation of the number of stages, or contacting steps, required for a specified separation. This calculation depends on a knowledge of the nearness of approach to complete equilibrium between phases on each plate, or stage. This degree of approach to theoretically perfect contacting is expressed conveniently in terms of the plate efficiency, which may be defined in a number of ways, as follows:

1. The *over-all plate efficiency* E is defined as the ratio of the number of theoretical plates required for a given separation to the number of actual, physical plates.

If E is known, the number of actual plates in a column may be found by calculating the number of theoretical plates, N_p, as described on p. 554, and then dividing N_p by E.

2. *The Murphree vapor efficiency* E_{MV} of a single plate is the ratio of the actual change in the average vapor composition accomplished by the plate to the change that would occur if the mixed vapor stream reached equilibrium with the exit liquid. Thus

$$E_{MV} = \frac{(y_n)_{\text{avg}} - (y_{n-1})_{\text{avg}}}{y_n^* - (y_{n-1})_{\text{avg}}} \tag{103}$$

If the plates are 100 per cent efficient, these two efficiencies are equal, and the same is true if the equilibrium line $y_n^* = F(x_n)$ and the operating line $(y_{n-1})_{\text{avg}} = F'(x_n)$ are straight and parallel. If these conditions do not exist, the two efficiencies are unequal. In the special case where the operating and equilibrium lines are straight (but not parallel), the relation between the two is

$$E = \frac{\ln [1 + E_{MV}(J - 1)]}{\ln J} \tag{104}$$

where

$$J = \frac{\text{slope of equilibrium line}}{\text{slope of operating line}} = \frac{mG_M}{L_M}$$

Figure 22 shows a plot of this equation. The difference between the two efficiencies is especially great when J is greatly

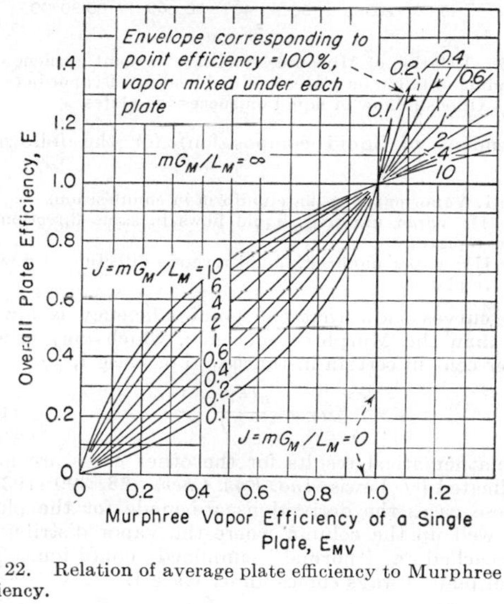

FIG. 22. Relation of average plate efficiency to Murphree vapor efficiency.

different from 1. For very small efficiencies, the equation reduces to

$$E = E_{MV}\frac{J - 1}{\ln J} \tag{105}$$

in which the factor involving J is equal to the ratio of the number of theoretical plates to the number of transfer units [*cf.* Eq. (102)].

If the value of the Murphree plate efficiency is known, the number of actual plates required for a specified separation may be found generally by plate-to-plate calculations, or, in the case of a binary mixture, by a slight modification of the stepwise graphical procedure used with the McCabe-Thiele diagram, as described below, p. 591. If the operating and equilibrium lines are straight, the Murphree efficiency may be converted to an equivalent average plate efficiency by means of Eq. (104) or Fig. 22. The number of actual plates may be calculated in this case as described above. If the Murphree efficiency is known but the over-all efficiency is not known, and either the operating or the equilibrium line is curved, a plate-to-plate determination must be used. In the case of a countercurrent operation with negligible vapor of the solute over the liquid phase, the concepts of theoretical plates and over-all plate efficiency have no significance. Determinations of required numbers of actual plates may be made by means of a stepwise calculation, using a value of the Murphree efficiency. Alternatively, if the operating line is straight, the equation

$$(N_p)_{\text{actual}} = \frac{\ln (y_1/y_2)}{\ln [1/(1 - E_{MV})]} \tag{106}$$

may be used.

If liquid is carried to the next plate above by entrainment in the vapor stream, the apparent efficiency of a single plate is lower than the over-all Murphree efficiency of the plate. An approximate relation between the two, as developed by Colburn [*Ind. Eng. Chem.*, **28**, 526 (1936)], is

$$E_a = \frac{E_{MV}}{1 + (\epsilon E_{MV}G_M/L_M)} \tag{107}$$

Figure 23 shows a plot of this relation.

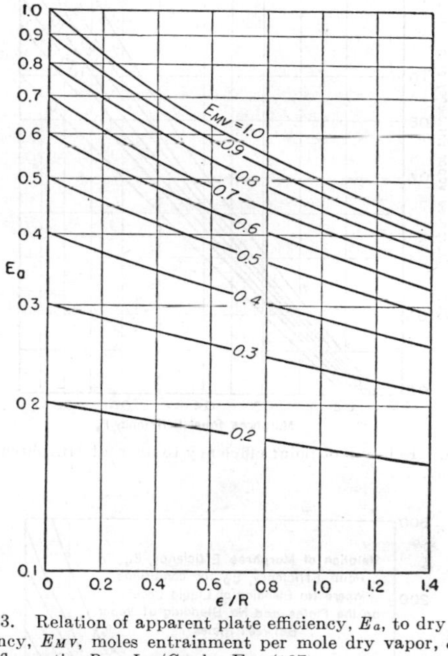

FIG. 23. Relation of apparent plate efficiency, E_a, to dry vapor efficiency, E_{MV}, moles entrainment per mole dry vapor, ϵ and the reflux ratio, $R = L_M/G_M$, by Eq. (107).

3. *The Murphree liquid efficiency* E_{ML} of a single plate is

$$E_{ML} = \frac{x_{n+1} - x_n}{x_{n+1} - x_n^*} \tag{108}$$

When the operating and equilibrium lines are straight, the two Murphree efficiencies are related by

$$E_{ML} = \frac{E_{MV}}{E_{MV} + \dfrac{1 - E_{MV}}{(mG_M/L_M)}} \tag{109}$$

4. *The point efficiency* E_p, expressed in terms of vapor compositions, describes the degree of approach to equilibrium between the vapor and the liquid at a single point on a plate. Especially on a wide plate, the liquid composition varies across the plate. The efficiency of contact at a single point is

$$E_p = \frac{y_n - y_{n-1}}{y_n^* - y_{n-1}} \tag{110}$$

The relation of this efficiency to the resistance to mass transfer between phases is brought out by the equation

$$E_p = 1 - e^{-N_{OG}} = 1 - \exp\left(-\frac{zK_Ga}{G_M}\right) \tag{111}$$

where N_{OG} is the number of transfer units available for mass transfer and z is the effective liquid depth. This equation is derived by assuming that the liquid has a constant composition in a vertical direction perpendicular to the plate. It shows that the departure of the point efficiency from 100 per cent is due to the resistance to mass transfer between phases.

It should be noted that the over-all mass transfer coefficient on a gas basis is related to the separate film coefficients and the slope of the equilibrium curve m (Eq. 92). The latter term becomes very important in many cases, such as absorption of CO_2 in water in bubble tray columns. Alternatively, N_{OG} can be replaced by z/H_{OG}, and then H_{OG} related to the separate values of H.T.U. by Eq. 97*a*.

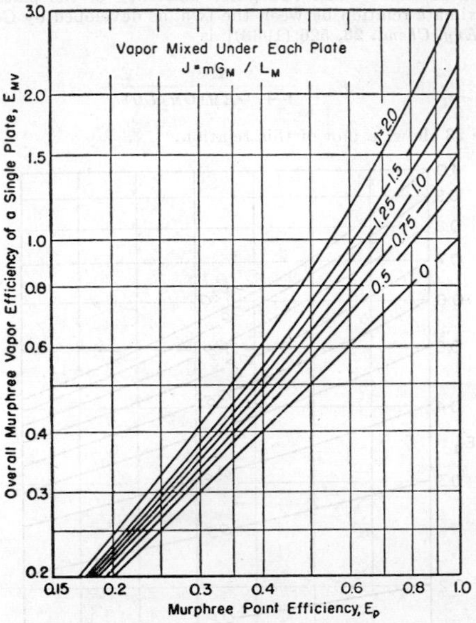

FIG. 24. Relation of point efficiency to over-all Murphree point efficiency.

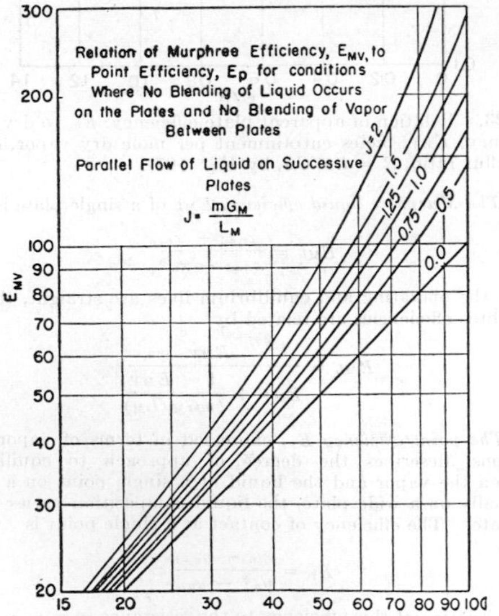

FIG. 25. Relation of Murphree efficiency to point efficiency for no blending of liquid on plates and no blending of vapor between plates. Parallel flow of liquid on successive plates.

The effect on the over-all Murphree efficiency of the variation in liquid composition on a plate is shown by Figs. 24, 25, and 26, which give relations between the Murphree efficiency and the point efficiency (assumed

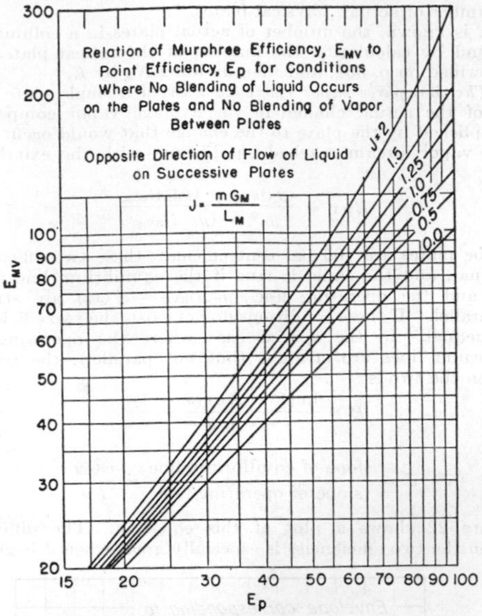

FIG. 26. Relation of Murphree efficiency to point efficiency for no blending of liquid on plates and no blending of vapor between plates. Opposite flow of liquid on successive plates.

independent of liquid composition) for the following cases:

Case I: Vapor entering plate uniform in composition.

Case II: Vapor unmixed, liquid flows in same direction on successive plates.

Case III: Vapor unmixed, liquid reverses direction of flow on successive plates.

These curves show that the point efficiency is always lower than the Murphree efficiency, which may exceed 100 per cent in certain instances. For Case I,

$$E_{MV} = \frac{e^{JE_p} - 1}{J} \tag{112}$$

The mathematical results for the other cases are more complicated [*cf.* Lewis, *Ind. Eng. Chem.*, **28**, 399 (1936)]. In these cases the derivations are made for the plates fairly well up the column where the vapor distribution has reached a "normal" unmixed condition. The bottom plate always comes under Case I.

CALCULATION OF NUMBER OF TRANSFER UNITS REQUIRED IN COUNTERCURRENT APPARATUS

In the case of apparatus such as a packed tower that operates in a truly countercurrent fashion, a material balance on a section of the tower having differential thickness dz, as shown in Fig. 18, shows that the ratio of the accomplished change in concentration to the driving force causing diffusion is simply related to the H.T.U. as follows:

$$\frac{dy}{y - y^*} = \frac{dz}{H_{OG}} \tag{113}$$

for distillation of a binary mixture, and

$$\frac{(1 - y)_f \, dy}{(y - y^*)(1 - y)} = \frac{dz}{H_{OG}} \tag{114}$$

for absorption of a gas from a second, insoluble gas by means of a non-volatile solvent. Comparison of Eqs. (113) and (114) with Eqs. (96a) and (96b) above shows that dz/H_{OG} represents an increment in the number of transfer units, since $N_{OG} = Z/H_{OG}$. The material balance for the absorption case is

$$d(G_M y) = G_M \frac{dy}{1 - y} = K_G'a(y^* - y)\,dz \qquad (115)$$

which may be rearranged to the form

$$\frac{(1 - y)_f\,dy}{(1 - y)(y^* - y)} = \frac{K_G'a(1 - y)_f\,dz}{G_M} = \frac{dz}{H_{OG}} \qquad (116)$$

The quantities on the right side of the equation are substantially independent of y, since K_Ga is proportional to a power of the total gas velocity G_M, which usually lies between 0.5 and 1 for systems in which most of the resistance is in the gas phase. The factor $(1 - y)_f$ is included on both sides of the equation to allow for the fact that the diffusional resistance is less the lower the concentration of the inert gas through which diffusion occurs.

The similar equations that apply to changes in the liquid-phase composition are, by analogy with those already given,

$$\frac{dx}{x^* - x} = \frac{dz}{H_{OL}} \qquad (117)$$

for distillation of a binary mixture, and

$$\frac{(1 - x)_f\,dx}{(x^* - x)(1 - x)} = \frac{dz}{H_{OL}} \qquad (118)$$

for absorption from an insoluble gas, as above. Since the ratio of the height of the apparatus to the value of the H.T.U. is the number of transfer units required, N_T, these differential equations may be integrated to give equations useful for calculating the number of units needed for a specified separation.

These calculations can be carried out by numerical or graphical integration for any problem with the aid of the operating line equations discussed above. Approximate graphical methods [White, *Trans. Am. Inst. Chem. Engrs.*, **36**, 359 (1940); Baker, *Ind. Eng. Chem.*, **27**, 977 (1935)] may be used for approximate calculations. In many practical problems these calculations can be avoided entirely and simplified analytical solutions for the above differential equations used instead.

White's method depends on the fact that, according to an approximate integration of Eq. (113) for a height equal to the value of H.T.U., *i.e.*, for one transfer unit, the increase in y is equal to the average value of the driving force, $y^* - y$. To step off one over-all gas-phase transfer unit on the x-y diagram shown in Fig. 27, a line is drawn from point (x_2,y_2) on the operating line toward the equilibrium line with a negative slope, s_y, equal to $-L_M/G_M$, where L_M/G_M is the slope of the operating line. At H, the point of intersection of this construction line with the equilibrium line, a second construction line is drawn with a positive slope t_y, equal to $3(L_M/G_M)$. The point of intersection of this line with the operating line has the coordinates (x_1,y_1), and the change in y is that corresponding to one over-all gas-phase transfer unit. This follows from the fact that $y_1 - y_2$ equals BH, the mean vertical distance between the operating and equilibrium lines. Once the slopes s_y and t_y have been established on the diagram, the second, third, and additional transfer units may be laid off quickly with the aid of two triangles or parallel rulers. If the construction is to be made to successively smaller values of y, the steeper line is laid off first, unless the operating line is below the equilibrium line, when the order of the construction is reversed.

Over-all liquid-phase transfer units may be constructed in a similar manner when most of the resistance to diffusion is in the liquid phase. If the operating line is below the equilibrium line and the construction is to be made to successively smaller values of x, the first line drawn has a slope s_x, equal to $\frac{1}{3}(L_M/G_M)$, and the second line, drawn from the equilibrium line to the operating line, has a slope t_x, equal to $-(L_M/G_M)$.

When the resistance to transfer is distributed between the two phases, the slopes of the construction lines should be

$$\begin{aligned} s_y &= \frac{-(L_M/G_M)}{1 + 2H_L/H_G} \\ s_x &= \frac{1}{3}\left(\frac{L_M}{G_M}\right)(1 - 2\,H_G/H_L) \\ t_y &= \frac{3(L_M/G_M)}{1 - 2H_L/H_G} \\ t_x &= -\left(\frac{L_M}{G_M}\right)(1 + 2H_G/H_L) \end{aligned} \qquad (119)$$

The shape of the operating line varies for many of the problems encountered in connection with the three main diffusional operations. If the problem deals with

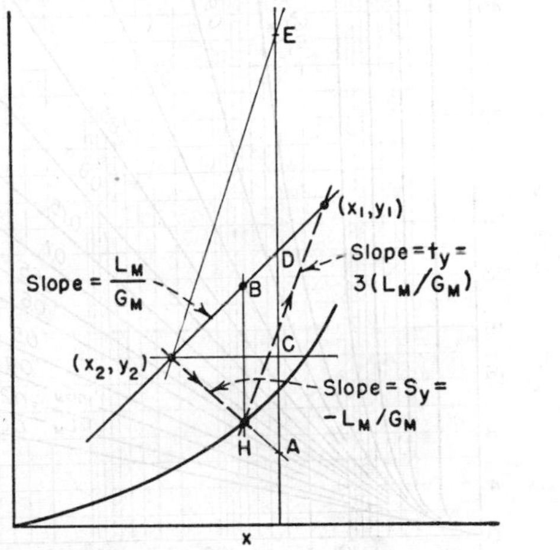

FIG. 27. Graphical determination of the number of gas-phase transfer units (N_{OG}) by method of White. To determine slopes s_y and t_y, draw any vertical line AE. Locate A and E such that $AC = CD$ and $CE = 3CD$.

the absorption of a solute component from its dilute mixture with an insoluble gas, or if it deals with distillation under the so-called "usual simplifying assumptions," or with the transfer of a single compound from a dilute liquid solution to a second immiscible-liquid phase, the operating line can usually be considered to be straight.

The assumption that $y^* = mx$ is usually allowable when dealing with dilute solutions, such as are encountered in the absorption of gases from dilute gas mixtures without appreciable heat of solution and distillation of materials of very low or very high purity. For applications involving dilute solutions, therefore, the number of transfer units becomes

$$N_T = \frac{\ln\,[(1 - J)Q + J]}{(1 - J)} \qquad (120)$$

where the symbols Q and J are identified by Table 15 for application to the different operations. If operating and equilibrium lines are straight, calculations made by means of this equation are equivalent to those in which a logarithmic-mean driving force is used. In some cases the equation gives results that are sufficiently exact even though the operating and equilibrium lines are curved slightly. These cases are those in which the operating and equilibrium lines approach each other very closely at one end of the apparatus, such as might occur in the absorption of essentially all of a material from a gas.

In such cases the value of mG_M/L_M used in the equation should be that corresponding to the conditions of the "pinch." Although the equation is simple to use by itself, it may also be solved by the use of Fig. 28.

The analytical solutions for the number of transfer units in cases where either the equilibrium line or the operating line, or both is curved slightly have been presented by Colburn [*Ind. Eng. Chem.*, **33**, 459 (1941)]. Slightly different results applicable to these cases were given by Othmer and Scheibel [*Trans. Am. Inst. Chem. Engrs.*, **38**, 339 (1942)].

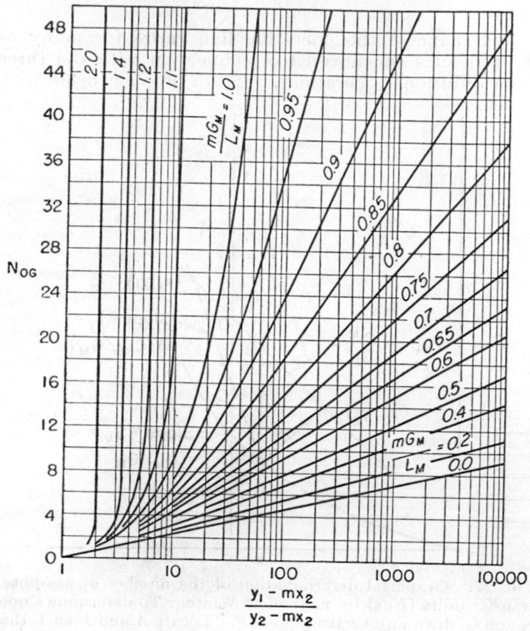

FIG. 28. Number of transfer units in an absorption column. Condition of constant $\frac{mG_M}{L_M}$, a plot of

$$N_{OG} = \frac{2.3}{\left(1 - \frac{mG_M}{L_M}\right)} \log \left[\left(1 - \frac{mG_M}{L_M}\right)\left(\frac{y_1 - mx_2}{y_2 - mx_2}\right) + \frac{mG_M}{L_M} \right]$$

N_{OG} = number of transfer units on an over-all gas concentration basis; m = slope of equilibrium curve; G_M/L_M = molal gas to liquor ratio; x and y are the mole fractions of solute in the liquid and gas, respectively; subscripts 1 and 2 refer to the concentrated and dilute ends, respectively.

As long as the treatment is restricted to dilute solutions and there are no heat effects, the operating and equilibrium lines are straight whether concentrations are expressed in terms of weight fractions, weight ratios, mole fractions, or mole ratios. The value of m must be modified, however, to be consistent with the units employed.

The number of transfer units required for an absorption operation is usually calculated in terms of gas compositions, and the number for a stripping operation is calculated by using liquid compositions. Thus the result of an absorption calculation is usually N_{OG} and for stripping N_{OL}. Each of these quantities may be read from Fig. 28, as indicated in Table 15. If it is desired to change this procedure, in order to calculate N_{OG} for a stripping operation, for example, the figure cannot be used directly. In order to calculate N_{OG} from the figure in this case, first calculate N_{OL}, as outlined in Table 15, and then multiply by L_M/mG_M.

CALCULATION OF THE NUMBER OF THEORETICAL PLATES REQUIRED IN COUNTERCURRENT APPARATUS

Because of the definition of the enrichment accomplished on each theoretical plate, the number of such plates may be determined numerically by means of a step-by-step calculation, or graphically by drawing in successive horizontal and vertical steps on a diagram of the type shown in Fig. 19. In many cases, these calculations may be carried out analytically by using the

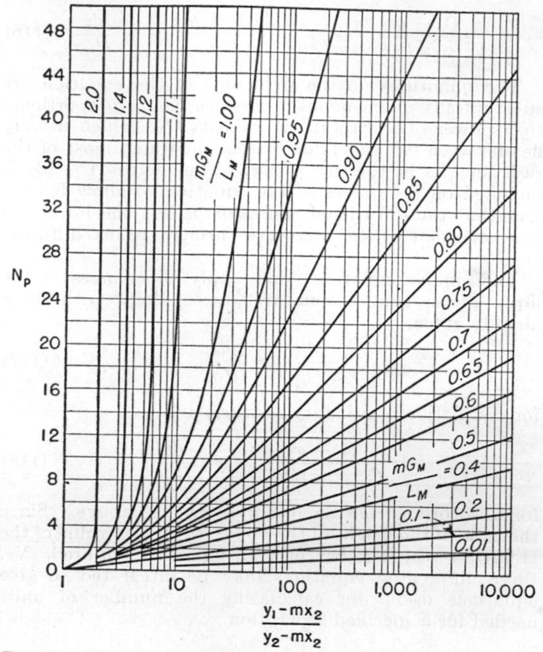

FIG. 29. Number of theoretical plates in an absorption column. Condition of constant $\frac{mG_M}{L_M}$, a plot of

$$N_p = \frac{\log \left[\left(1 - \frac{mG_M}{L_M}\right)\left(\frac{y_1 - mx_2}{y_2 - mx_2}\right) + \frac{mG_M}{L_M} \right]}{\log \frac{L_M}{mG_M}}$$

N_p = number of theoretical plates on a gas concentration basis; m = slope of equilibrium curve, G_M/L_M = molal gas to liquor ratio; x and y are the mole fractions of solute in the liquid and gas, respectively; and subscripts 1 and 2 refer to the concentrated and dilute ends, respectively.

calculus of finite differences, as described by Tiller and Tour [*Trans. Am. Inst. Chem. Engrs.*, **40**, 317 (1944)]. The result of such analytical calculations for the case of straight operating and equilibrium lines is

$$N_P = \frac{\ln [(1 - J)Q + J]}{\ln (1/J)} \tag{120a}$$

As before, the symbols J and Q are identified for the various cases in Table 15.

This equation is equivalent to one given by Kremser [*Nat. Petroleum News*, **22** (21), 42 (1930)] and others. Figure 29 allows solutions of this equation to be found quickly. When only a small number of stages is involved it may be better to use the equation directly, or to use an equivalent chart, such as the one published by Souders and Brown [*Ind. Eng. Chem.*, **24**, 519 (1932)] as well as by others.

Table 15. Summary of Equations

	Theoretical plates $N_P = \frac{\log[(1-J)Q+J]}{\log(1/J)}$	Transfer units $N_T = \frac{2.3 \log_{10}[(1-J)Q+J]}{(1-J)}$		
	N_P	N_T	J	Q
Absorption: Case I, constant $\frac{m}{R}$	N_P	N_{OG}	$\frac{mG_M}{L_M}$	$\frac{y_1 - mx_2}{y_2 - mx_2}$
Desorption Case I, constant $\frac{R}{m}$	N_P	N_{OL}	$\frac{L_M}{mG_M}$	$\frac{x_1 - y_2/m}{x_2 - y_2/m}$
Distillation, enriching*: Case I, constant $\frac{m}{R}$	N_P	N_{OG}	$\frac{mG_M}{L_M}$	$\frac{y_1 - mx_2}{y_2 - mx_2}$
Distillation, stripping, closed steam:† Case I, constant $\frac{R}{m}$	N_P	$N_{OL} + \frac{2.3 \log m/R}{1 - R/m}$	$\frac{L_M}{mG_M}$	$\frac{x_1 - x_2/m}{x_2 - x_2/m}$
Distillation, stripping, open steam:† Case I, constant $\frac{R}{m}$	N_P	N_{OL}	$\frac{L_M}{mG_M}$	$\frac{x_1}{x_2}$
Extraction, stripping: Case I, constant $\frac{R}{m}$	N_P	N_{OL}	$\frac{L_M}{mG_M}$	$\frac{w_1 - v_2/m}{w_2 - v_2/m}$
Extraction, enriching: Case I, constant $\frac{m}{R}$	N_P	N_{OG}	$\frac{mG_M}{L_M}$	$\frac{v_1 - mw_2}{v_2 - mw_2}$

Subscripts 1 and 2 refer to the concentrated and the dilute ends of the apparatus, respectively.

* Concentrations and m are based on high boiler or "heavy key."

† Concentrations and m are based on low boiler or "light key."

Example. Calculate the number of transfer units required in a packed column used to absorb acetone from air by water. The gas mixture fed to the column contains 7 mole per cent acetone and 93 mole per cent air nearly saturated with water vapor at 25°C. and 1 atm. The water fed to the top of the column is free of acetone, and its temperature is 25°C. The exit gas is to contain only 0.01 mole per cent acetone, and sufficient water is to be supplied to the column so that $J = mG_M/L_M = 0.5$ at the dilute end, where most of the transfer units are used. At 25°C. and 1 atm. total pressure, the vapor pressure of acetone over its dilute water solutions corresponds to $m = 1.75$. (This corresponds to an activity coefficient of 5.80 and a Van Laar constant of 0.764.)

1. Isothermal Operation Assumed: In this case the equilibrium line is straight throughout the column, and the operating line is substantially straight. Since there is no acetone in the inlet liquid, $x_2 = 0$. The abscissa for Fig. 28 is therefore $Q = y_1/y_2 = 7/0.01 = 700$. The figure shows that the number of transfer units on an over-all gas basis is 11.7.

The number of theoretical plates required for a plate-type tower is found similarly from Fig. 29 to be 8.45.

2. Non-isothermal Conditions Allowed For: The heat of solution of acetone in dilute solution is 10,000 p.c.u./lb.-mole of acetone. On the basis of 1 sq. ft. of tower cross section, the number of moles of acetone absorbed in 1 hr. is $G_{M_2}[(0.07/0.93) - 0.0001] = 0.0751\ G_{M_2}$. Since the specific heat of the liquid is 18 p.c.u./(lb.-mole)(°C.), the temperature rise of the water phase is

$$\frac{(10{,}000)(0.0751)(G_{M_2})}{(18)(L_{M_2})} = \frac{(10{,}000)(0.0751)(0.5)}{(18)(1.75)} = 11.9°\text{C.}$$

At this temperature, the value of y_1^*/x_1 corresponding to an activity coefficient of 5.8 is $5.8(^{363}/_{760}) = 2.77$, and the value of $(y_1^*/x_1)(G_M/L_M)$, is $0.5(2.77/1.75) = 0.792$.

Under these conditions the equilibrium line is curved upward at its upper end but is substantially straight below $y = 0.01$. The number of transfer units may be calculated in two parts. The number required between $y = 0.07$ and $y = 0.005$ may be found by using White's method or by graphical integration. Calculation shows that of the value of the integral

$$\int_{0.005}^{0.07} \frac{dy}{y^* - y} = 7.3$$

Between $y = 0.005$ and $y = 0.0001$, the number of transfer units required is found from Fig. 28 to be 6.5. The total number of transfer units is therefore 13.8.

Example. An aqueous solution containing 5 weight per cent methanol is fed at its boiling point to the top of a stripping column supplied at the bottom with open steam. The concentration of the liquid stream at the bottom of the column is to be 0.001 of the feed, and the reflux ratio L_M/G_M is to be 6. Calculate the number of transfer units required, based on liquid compositions.

To utilize the convenience of a straight operating line, a fictitious molecular weight of methanol of 42 is used (with this value, the molal latent heats of methanol and water are equal), and the feed strength becomes 2.2 fictitious mole per cent. A plot of the equilibrium data (see Sec. 9, p. 574) of Cornell and Montonna [*Ind. Eng. Chem.*, **25**, 1331 (1933)] shows that the equilibrium line is curved slightly, though it is approximated closely up to $x = 0.005$ by its tangent at the origin, $y^* = 7.6x$. The composition of the bottoms is to be 2.2×10^{-5} fictitious mole fraction. According to Table 15 and Fig. 28, the value of N_{OL} for the interval $x = 2.2 \times 10^{-5}$ to $x = 0.005$ corresponding to $J = L_M/mG_M = 6/7.6 = 0.790$, $Q = x_1/x_2 = 227$, is 18.5. Graphical computation of the integral $\int \frac{dx}{x^* - x}$ for the interval $x = 0.005$ to $x = 0.022$ gives $N_{OL} = 10.0$, making the total value of N_{OL} 28.5.

The number of theoretical plates required for carrying out this operation in a plate-type column is calculated by similar methods. Thus, between the limits $x = 2.2 \times 10^{-5}$ and $x = 0.005$ fictitious mole fraction, N_p is found from Fig. 29 to be 16.4. This corresponds, as before, to $J = 0.790$ and $Q = 227$. The graphical stepwise procedure is applied in the interval $x = 0.005$ to $x = 0.022$, yielding $N_p = 10.1$. The total number of theoretical plates required is therefore 26.5.

If closed steam had been used instead of open steam, the action of the still would supply one theoretical plate, and 25.5 theoretical plates would be required in the column. The composition of the vapor entering the column in this case is that in equilibrium with the dilute liquid flowing from the column, rather than pure steam, as in the case of open steam. If the column were packed, its height should correspond to $N_{OL} = 28.0$. This is calculated, as outlined in Table 15, by using $J = 0.790$, $Q = [0.005 - (0.000022/7.6)]/[0.000022 - (0.000022/7.6)] = 262$, giving $N_{OL} = 19.1$ for the interval $x = 0.000022$ to $x = 0.005$. For the interval $x = 0.005$ to $x = 0.022$, graphical integration yields $N_{OL} = 10.0$, as before. The number of over-all liquid-phase transfer units required for the packed section of the column is, according to Table 15, $19.1 + 10.0 - \frac{2.3 \log_{10}(1/0.790)}{1 - 0.790} = 28.0$.

TRANSIENT BEHAVIOR OF COUNTERCURRENT APPARATUS

It has been assumed in the development of equations for the number of transfer units or theoretical plates required in packed and plate-type countercurrent apparatus that the apparatus operates under steady-state conditions such that liquid and gas compositions throughout the equipment do not vary with time. In starting up such equipment, however, the steady state is approached only after a preliminary time during which compositions vary. The time required to reach steady conditions depends on the time of passage of both phases through the main apparatus and the auxiliary equipment and, to some extent, on the difficulty of the separation that is being carried out, *i.e.*, on the relative volatility.

Rate of Approach to Steady State for Distillation Column Operating Total Reflux. In starting up a distillation column, either a batch-type or a continuous column, it is customary to operate the equipment at total reflux for a time sufficient to allow the top-product composition to approach closely to the value corresponding to steady conditions before withdrawal of product is begun. The time that is required for this phase of the operation, exclusive of the time required for heating the liquid charge in the reboiler to the boiling point and the column to its operating temperature, may be estimated roughly from Fig. 30. These curves are based on the following assumptions: (1) the hold-up of liquid in the condenser and reflux return line is small compared with the hold-up in the column; (2) the vapor entering the base of the column has a constant composition; (in the case of batch rectification, this condition is met if the volume of the still charge is large compared with the column hold-up); (3) the equilibrium line is parallel to the 45 deg., total-reflux operating line. The ordinate on the figure is the number of times the total column hold-up must be displaced by the vapor flow. The degree of approach to steady conditions is defined as the difference between the instantaneous compositions at the top and bottom of the column divided by the same difference in the steady state. The time, expressed in terms of the hold-up, is seen to be proportional approximately to the number of transfer units or theoretical plates.

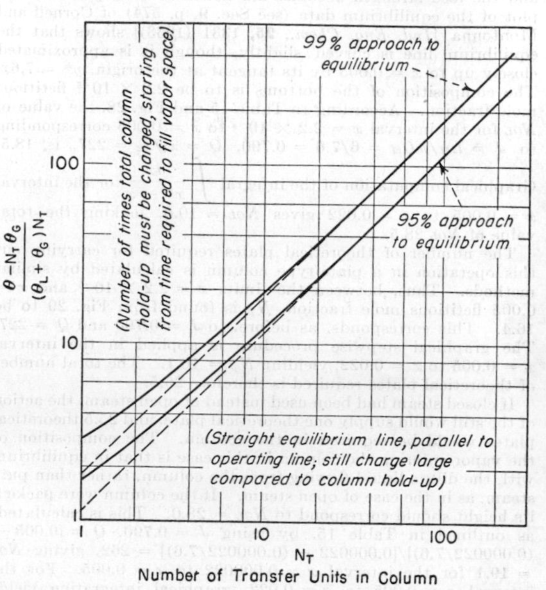

FIG. 30. Time required for a distillation column to reach equilibrium. θ = time, hr.; θ_L = time of contact of the liquid stream in a height of column equivalent to one gas-phase transfer unit, hr.; θ_G = time of contact of the vapor stream in a height of column equivalent to one gas-phase transfer unit, hr.

These same curves may be used for the case of a straight equilibrium line through the origin, or through the point $x = y = 1$, provided that the relative volatility is small, say less than 1.05. This case has been considered by Cohen [*J. Chem. Phys.*, **8**, 588 (1940)]. If the still charge is small compared with the column hold-up, the time is approximately one-fourth of that given by the figure.

Example. Estimate the time required for preliminary operation at total reflux in the distillation of a binary liquid mixture in a column having 10 theoretical plates, each with a liquid hold-up of 15 lb./sq. ft. of column cross section. Vapor of constant composition is supplied to the column. The superficial vapor velocity is 300 lb./(hr.)(sq. ft.). The time of contact of the liquid stream in the column, $N\theta_L$, is (15)(10)/300 = 0.5 hr. If a straight line, parallel to the equilibrium line, is assumed as a rough approximation for the equilibrium curve, and theoretical plates are assumed equivalent to transfer units, the time required for 95 per cent approach to steady conditions of operation is (12)(0.5) = 6.0 hr., according to Fig. 30.

Although no exact solution of the problem has been presented for the case where the equilibrium line is curved, Berg and James [*Chem. Eng. Progress*, **44**, 307 (1948)] have added certain approximations to the previous theoretical treatment of Cohen [*J. Chem. Phys.*, **8**, 588 (1940)] and have concluded that for total-reflux operation, using a mixture for which the relative volatility α is constant,

$$(t/N_p\theta_L) = \frac{\alpha^{N_p}}{(\alpha - 1)N_p \ln(\alpha)\phi} \ln(1/e)$$

$$\text{where } e = \frac{\left(\frac{x_P}{1 - x_P}\right)_{\text{s.s.}} - \left(\frac{x_P}{1 - x_P}\right)_t}{\left(\frac{x_P}{1 - x_P}\right)_{\text{s.s.}} - \left(\frac{x_W}{1 - x_W}\right)} \quad \textit{i.e.}, \text{ degree of approach to steady state}$$

x_P = mole fraction more volatile component in product at top of column.
x_W = mole fraction more volatile component in reboiler at bottom of column.
α = relative volatility.
ϕ = function of the product $N_p \ln \alpha$, *cf.* table.
N_p = number of theoretical plates in column.
$N_p\theta_L$ = time of passage of liquid stream through column.
"s.s." = steady-state conditions.
"t" = instantaneous composition at any time before steady state is reached.

$N_p \ln \alpha$	ϕ
0.4	19.5
0.6	10.9
0.8	7.10
1.0	4.63
1.2	3.64
1.4	2.89
1.6	2.48
1.8	2.17
2.0	1.92
2.6	1.49
3.2	1.32
4.0	1.16
5.0	1.07
infinity	1.00

Rate of Approach to Steady State for a Countercurrent Absorption or Extraction Column. The time required, expressed in terms of the contact time of the liquid on one theoretical plate θ_L, the contact time of the vapor on one theoretical plate θ_G, the number of theoretical plates N, and the ratio of the slopes of the equilibrium and operating lines mG_M/L_M are given by Fig. 31. The

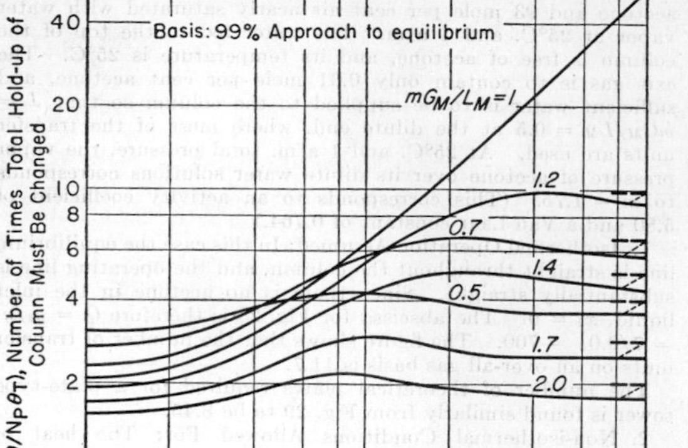

FIG. 31. Time required for a countercurrent plate absorber or extractor to reach equilibrium. θ = time, hr.; N_p = number of theoretical plates; $\theta_T = \theta_L + (mG_M/L_M)\theta_G$; θ_L = time of contact of the liquid phase on one theoretical plate, hr.; θ_G = time of contact of the gas phase on one theoretical plate, hr.

analysis leading to these results is based on the assumptions (1) that the compositions of the two streams fed to the column are constant, and (2) that the column is filled initially with liquid having a composition uniformly equal to that of the inlet liquid stream and with a vapor in equilibrium with this liquid throughout. The results given by the figure apply to a degree of approach to steady operation sufficient for the driving force at one end of the column to equal 99 per cent of its terminal value.

Example. Estimate the time of preliminary operation required for a countercurrent, sieve-plate extraction column having a hold-up per theoretical plate of 0.2 cu. ft./sq. ft. of column cross section of the "gas" phase and 1 cu. ft./sq. ft. of the "liquid" phase. The "gas" and "liquid" flow rates are 15 and 60 cu. ft./(hr.)(sq. ft.), respectively, and mG_M/L_M is 0.7. The apparatus contains 15 theoretical plates. According to Fig. 31, the value of $t/N\left(\theta_L + \frac{mG_M}{L_M}\theta_G\right)$ is 6.4. Since $\theta_L = 1/60 = 0.017$ hr. and $\theta_G = 0.2/15 = 0.013$ hr., the time required is calculated to be $(6.4)(15)[(0.017 + (0.7)(0.013)] = 2.5$ hr.

SPECIAL METHODS FOR BATCH APPARATUS

Drew, Hixson, and Knox (personal communication, 1948) have proposed an extension of the H.T.U. concept to the case of batchwise transfer of mass from one phase to another in an agitated vessel. This case covers transfer of a solute from one liquid to a second immiscible liquid, mutual solution of two pure liquids which have a limited solubility, or leaching a soluble compound from a solid. In any of these cases, the concentration of solute in each phase changes gradually with time, asymptotically approaching the equilibrium value in each phase.

Drew *et al.* show that an integral corresponding to the number of transfer units is related to the over-all mass-transfer coefficient on a volume basis, *Ka*. For the concentration in phase 1,

$$\int \frac{dy_1}{y_1 - y_1^*} = \frac{K_1at}{M_1} = \frac{t}{T_{o1}}$$

where y_1 represents the mole fraction of solute in phase 1, M_1 is the mass of phase 1 expressed in lb.-moles, K_1a is based on an over-all driving force in terms of phase 1 concentrations, t is the time, and T_{o1} is the over-all time of a phase 1 transfer unit. Similarly for phase 2,

$$\int \frac{dy_2}{y_2 - y_2^*} = \frac{K_2at}{M_2} = \frac{t}{T_{o2}}$$

Each over-all coefficient is related to the individual coefficients in the usual manner. Since the ratio of M to Ka is equal to the "time of a transfer unit" T, it follows that

$$T_{o1} = T_1 + T_2(mM_1/M_2)$$

where $T_{o1} = M_1/K_1a$, $T_1 = M_1/k_1a$, $T_2 = M_2/k_2a$, and $m = y_1^*/y_2$.

Drew *et al.* show further that according to experimental data on stirred vessels, the product of time of a transfer unit, vessel diameter, and stirring rate is approximately constant.

SIMULTANEOUS HEAT AND MASS TRANSFER

The general principles of diffusional operations have been discussed in this chapter as though diffusional operations equipment can be designed without regard for limitations that may be imposed by heat-transfer rates, just as though the heat-transfer and mass-transfer problems were separate. Actually mass transfer seldom occurs without simultaneous heat transfer. Almost invariably these two processes have a mutual effect on each other so that the rates are interdependent. In the design of apparatus for absorption of HCl vapor from air in water to form strong hydrochloric acid, for example, the removal of the heat of solution that is liberated at the liquid-vapor interface when HCl dissolves becomes such a difficult problem that design of such equipment is sometimes treated solely as a heat-transfer problem. Transfer of material from one phase to another is almost always accompanied by a heat effect that influences the temperatures of the phases and the temperature of the interface as well. Changes in the vapor-liquid equilibrium relation and in the fluid properties usually result.

Because of the analogy that exists between heat transfer and mass transfer across the same interface, it is sometimes possible to estimate heat-transfer rates from known mass-transfer rates, or vice versa. Thus it is possible, in theory at least, to calculate simultaneous temperature and concentration changes in the fluid streams that flow through a countercurrent apparatus by a series of incremental step-by-step calculations for a packed tower, or by plate-to-plate calculations for a plate tower.

WET- AND DRY-BULB THERMOMETRY*

One common application of the interrelation between heat transfer and mass transfer is in the use of wet- and dry-bulb thermometer readings to determine the vapor content of a gas stream. If the gas is unsaturated, evaporation of liquid from the wick that covers the wet bulb causes a heat effect that depresses the temperature of the wet bulb. This temperature difference causes a flow of sensible heat from the gas toward the bulb, and the bulb temperature is determined by the dynamic equilibrium that is established.

To develop an expression for the relation between wet-bulb depression and vapor content, we equate the sensible heat-flow rate to the heat effect caused by evaporation. Thus

$$h(t_g - t_{wb}) = \lambda k'(H_{wb} - H_g) \tag{121}$$

and

$$H_g = H_{wb} - \frac{h}{k'\lambda_{wb}}(t_g - t_{wb}) \tag{122}$$

where h = heat-transfer coefficient, B.t.u./(hr.)(sq. ft.)(°F.).
t_g = dry-bulb temperature, °F
t_{wb} = wet-bulb temperature, °F.
λ_{wb} = latent heat of vaporization at the wet-bulb temperature, B.t.u./lb.
k' = mass-transfer coefficient, lb. evaporated/(hr.)(sq. ft.)(lb. vapor/lb. dry gas).
H_g = humidity of gas stream, lb. vapor/lb. dry gas.
H_{wb} = humidity of saturated air at temperature of wet bulb, lb. vapor/lb. dry gas.

The ratio of the rate coefficient h/k' is best determined from experimentally observed wet-bulb depressions, although by methods outlined below prediction can be made from theoretical relations for heat and mass transfer rates by forced convection. Since the wet bulb is usually colder than its surroundings, heat is also transferred to it by radiation. The error from this source can be reduced by causing the air to flow past the wet bulb at a fairly high velocity, usually 10 ft./sec. or more, so that the rate of transfer by convection is enhanced. The radiation error may be particularly significant when the temperature level is high. The coefficient of total heat transfer h is the sum of h_c and h_r, where h_c represents the convection coefficient and h_r the radiation coefficient. Data on heat transfer to single cylinders (Sec. 6) may be used for estimating h_c, and h_r may be calculated from

$$h_r = 0.026\left(\frac{T_{\text{avg}}}{100}\right)^3 \quad \text{B.t.u./(hr.)(sq. ft.)(°F.)} \tag{123}$$

where the emissivity of the wick is assumed to be equal to 0.90, and T_{avg} is the arithmetic average of the wet-bulb temperature and the temperature of the surroundings, assumed equal to the dry-bulb temperature, expressed in °K. The fraction by which h/k' is increased due to radiation is $[1 + (h_r/h_c)]$.

When a gas contains the vapor of a volatile liquid as well as water vapor, the use of a thermometer wetted with the liquid gives unreliable measurements of the vapor content if the solubility of water in the liquid lowers its vapor pressure or if the temperature of the wet bulb is lower than the water dew point. Thus the concentration of acetone in humid air cannot be determined by this method. Neither can the method be used in the case of very volatile liquids in moist air because of the large values of the wet-bulb depression.

Experimental values of h/k' are plotted in Fig. 32, which shows that h/k' depends on the value of the Schmidt group, $(\mu/\rho D)$, as has been predicted theoretically [see Colburn, *Trans. Am.*

* Further discussion of this subject from a practical point of view is given in Secs. 12 and 13.

Inst. Chem. Engrs., **28**, 105 (1932)]. The data for organic liquids agree satisfactorily with a theory based on the assumption that $j_M = j_H$ and that h/k' varies as $(\mu/\rho D)^{2/3}$. The data for water do not agree with this assumption, as would be expected from experimental and theoretical studies of mass and heat transfer inside tubes (*cf.* p. 544), where the effect of $(\mu/\rho D)$ becomes progressively smaller as $(\mu/\rho D)$ itself becomes smaller. The best line drawn through the data agrees closely with a theory of the mass-transfer process suggested by Gilliland (Walker, Lewis, McAdams, and Gilliland, "Principles of Chemical Engineering," 3d ed., p. 590, McGraw-Hill, New York, 1937).

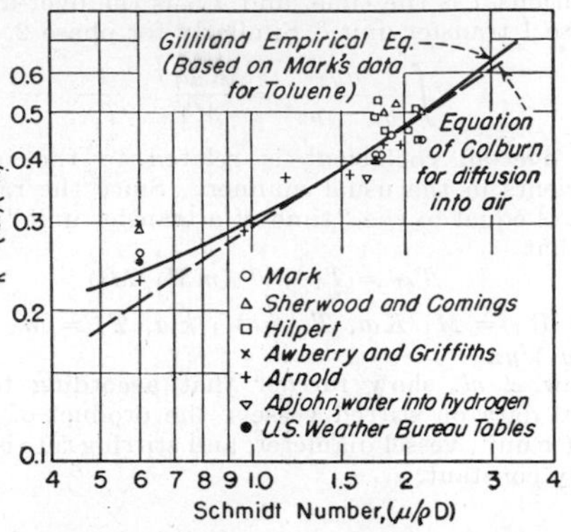

FIG. 32. Influence of physical properties on psychrometric ratio obtained with wet-bulb thermometer.

Most of the experimental data have been obtained under conditions of low vapor content and low temperature. [See, however, Awbery and Griffiths, *Proc. Phys. Soc.* (*London*), **44**, 132 (1932) for experimental data from 40° to 100°C.] Under these conditions, the assumption made in Eq. (121) that the driving force causing diffusion is a difference in absolute humidity, rather than the difference in partial pressure divided by the mean partial pressure of inert gas, is not significantly in error. Furthermore, no allowance is made in Eq. (121) for the sensible heat flow caused by diffusion of cold vapor molecules into the warmer gas surrounding the bulb (see p. 559 in this connection). The result of these effects is to cause the value of h/k' to decrease as the vapor content of the gas increases. This variation is sometimes assumed to be allowed for by assuming that $h/k'c_s$ is independent of vapor content, where c_s is the humid heat. It is advisable to use another method, such as chemical absorption, for determining the composition of a concentrated vapor in a gas.

SIMULTANEOUS HEAT AND MASS TRANSFER FOR THE SYSTEM LIQUID WATER–AIR–WATER VAPOR

The heat effects that accompany vaporization of water into air are put to use practically in the operation of atmospheric cooling towers (see Sec. 12). Because of the unusual relation between heat-transfer and mass-transfer rates for this system, simplified methods have been developed for calculating the capacity of equipment for transferring sensible and latent heat. The methods are not applicable when other vapors are present in the air at appreciable concentration, or when gases having physical properties different from those of air are present.

The psychrometric ratio h/k' for a thermometer bulb wet with water and immersed in an air stream is very nearly equal to the humid heat of water vapor–air mixtures, as noted above. Usually it may be assumed in the design of equipment, such as packed or spray-type towers, used for contacting air and water, that U/K' is also equal to the humid heat c_s, U is the over-all heat-transfer coefficient, based on the difference between the bulb temperatures of the water and the air, and K' is the mass-transfer coefficient based on the difference between the humidity of the air stream and the humidity of saturated air in equilibrium with liquid water at its bulk temperature. Merkel [*Forschungsarb.*, Heft 275 (1925)] was the first to show that, when the conditions of this assumption are fulfilled, the heat-transfer capacity of such equipment may be calculated relatively easily by use of a driving force based on the enthalpy of humid air. This same development has been described by McAdams ["Heat Transmission," 2d ed., p. 285, McGraw-Hill, New York, 1942] and also by others.

The increase in enthalpy of the humid air flowing through a section of a packed tower of height dz is given by the heat balance

$$G\,di = L\,dt_w = Ua(t_w - t_G)dz + \lambda K'a(H^* - H)dz \quad (124)$$

where G = air rate, (lb. dry air)/(hr.)(sq. ft.).
i = enthalpy of air stream, p.c.u./lb. dry air.†
L = water rate, assumed constant, lb./(hr.)(sq. ft.).
t_w = water temperature, °C.†
t_G = air temperature, °C.†
λ = latent heat of evaporation, p.c.u./lb.†
H^* = humidity of air in equilibrium with liquid water at t_w, lb. water vapor/lb. dry air.
H = humidity of air stream, lb. water vapor/lb. dry air.
Ua = over-all heat-transfer coefficient, p.c.u./(hr.)(cu. ft. of tower volume)(°C.).†
$K'a$ = over-all mass-transfer coefficient, lb. water vapor/(hr.)(cu. ft.)(unit difference in humidity).
z = tower height, ft.

Since the enthalpy is defined as

$$i = \lambda H + c_s t$$

Eq. (124) may be written as

$$\begin{aligned} G\,di = L\,dt_w &= \left(\frac{Ua}{c_s}\right)\left[\lambda H^* + \frac{Ua}{K'a}t_w - \lambda H - \frac{Ua}{K'a}t_G\right]dz \\ &= \left(\frac{Ua}{c_s}\right)\left[(\lambda H^* + c_s t_w) + c_s\left(\frac{Ua}{K'ac_s} - 1\right)t_w \right. \\ &\qquad \left. - (\lambda H + c_s t_G) - c_s\left(\frac{Ua}{K'ac_s} - 1\right)t_G\right]dz \\ &= \left(\frac{Ua}{c_s}\right)\left[i^* - i + c_s\left(\frac{U}{K'c_s} - 1\right)(t_w - t_G)\right]dz \quad (125) \end{aligned}$$

In view of the assumption that $U/K' = c_s$ this becomes

$$G\,di = L\,dt_w = \left(\frac{Ua}{c_s}\right)(i^* - i)\,dz \quad (126)$$

The height of apparatus required for a given duty may be calculated, therefore, from the equation

$$Z = \frac{Gc_s}{Ua}\int_{i_1}^{i_2}\frac{di}{i^* - i} = H_{iOG}\int_{i_1}^{i_2}\frac{di}{i^* - i} \quad (127)$$

Example. Calculate the number of transfer units Z/H_{iOG} required to cool water at 45° to 30°C. by means of air at 25°C. containing 0.01 lb. water/lb. dry air. The ratio of the air and water flows is assumed to be 1:1.

The enthalpy of the air entering the apparatus is

$$i_1 = 597.7(0.01) + 0.245(25) = 12.1 \text{ p.c.u./lb.}$$

based on liquid water and air bath at 0°C.
An over-all heat balance shows that

$$L(t_{w2} - t_{w1}) = G(i_2 - i_1)$$

or

$$45 - 30 = (1)(i_2 - 12.1), \qquad i_2 = 27.1 \text{ p.c.u./lb.}$$

In order to compute the value of the integral in Eq. (127), a plot may be made of the air enthalpy vs. the water temperature, as

† If i is expressed as p.c.u./lb., the temperature should be in °C. For i in B.t.u./lb., t is in °F.

shown in Fig. 33. At any point in the apparatus, the enthalpy driving force $(i^* - i)$ is equal to the vertical distance between the equilibrium and operating lines in Fig. 33. The area under a curve of $1/(i^* - i)$ vs. i between the limits $i_1 = 12.1$ and $i_2 = 27.1$ shows that $Z/(H_{iOG}) = 0.96$. A more rapid, approximate calculation, based on the use of a logarithmic-mean enthalpy difference, gives $\frac{Z}{(H_{iOG})} = \frac{i_2 - i_1}{(\Delta i)_{\text{l.m.}}} = \frac{27.1 - 12.1}{17.6} = 0.86$ where l.m. = log mean.

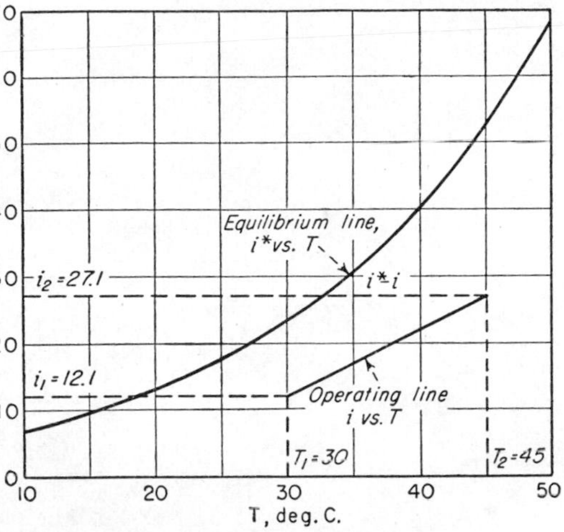

FIG. 33.

The humidity and dry-bulb temperature of the air leaving the tower are not given by this procedure. Usually they are not required, or the assumption that the exit air is saturated will give a satisfactory result.

The driving force in this calculation might have been taken as the average horizontal distance between the curves, equal to the mean difference between the water temperature and the wet-bulb temperature of the air. An infinitely tall tower would be required if the two lines touched, indicating that the wet-bulb temperature of the inlet air is the lowest temperature to which the water can be cooled.

SIMULTANEOUS HEAT AND MASS TRANSFER UNDER CONDITIONS OF LARGE-CONCENTRATION DRIVING FORCES

When a vapor diffuses through a gas in a region where the temperature varies, the sensible heat carried by the diffusing molecules must be taken into account in computing the heat-transfer rate if the rate of diffusion is large† [*cf.* Colburn and Drew, *Trans. Am. Inst. Chem. Engrs.*, **33**, 197 (1937); Ackermann, *Ver. deut. Ing. Forschungsheft*, No. 382, 1937]. The development is based on the assumptions that a stagnant or laminar film covers the surface to which heat and mass are being transferred and that the vapor does not condense until it reaches this surface. If h_g is the heat-transfer coefficient in the absence of simultaneous mass transfer, the coefficient that gives the total rate of transfer of sensible heat when multiplied by the temperature difference is

$$h_{\text{apparent}} = h_g \left(\frac{a}{1 - e^{-a}}\right) = h_g + \left(\frac{1}{1 - e^{-a}} - \frac{1}{a}\right)(wC_{pf}) \qquad (128)$$

where, for two diffusing components,

w = net rate of mass transfer of both components by diffusion, lb.-moles/(hr.)(sq. ft.), measured positively if diffusion takes place from a region of higher temperature toward a cooler region, *i.e.*, in the same direction as the flow of heat. The quantity w is calculated from

$$w = k_G \ln \left(\frac{z - y_i}{z - y_v}\right).$$

$C_{pf} = zC_{p_1} + (1 - z)C_{p_2}$, p.c.u./(lb.-mole)(°C.).

C_{p_1} = molal heat capacity of component 1, p.c.u./(lb.-mole)(°C.).

C_{p_2} = molal heat capacity of component 2, p.c.u./(lb.-mole)(°C.).

z = rate of diffusion of component 1/total net rate of diffusion.

$$a = \frac{wC_{pf}}{h_g} = \frac{(c_p\mu/k)^n}{(\mu/\rho D)^n}\frac{C_{pf}}{C_p} \ln \left(\frac{z - y_i}{z - y_v}\right)$$

where C_p = mean molal heat capacity of the gas mixture, p.c.u./(lb.-mole)(°C.).

y_i = mole fraction of diffusing component at surface or interface.

y = mole fraction of diffusing component in main body of gas stream.

The exponent n is usually taken as ⅔ but may be as small as ½ for values of $(\mu/\rho D)$ much less than 1. In the special case of equimolal counterdiffusion, $w = 0$, but wC_{pf} does not vanish, since $z = \infty$; instead $wC_{pf} = w_1(C_{p_1} - C_{p_2})$, where w_1 is the rate of diffusion of component 1. If component 1 diffuses through a stationary, non-diffusing film of component 2, $wC_{pf} = k_G C_{p_1} \ln \left(\frac{1 - y_i}{1 - y}\right)$.

The function of a in Eq. (128) assumes the following approximate forms for special values of a:

$$\text{For } a \sim 0: \quad h_{\text{apparent}} = h_g + \tfrac{1}{2}wC_{pf} \qquad (129)$$
$$\text{For } a \to +\infty: \quad h_{\text{apparent}} = h_g + wC_{pf} \qquad (130)$$
$$\text{For } a \to -\infty: \quad h_{\text{apparent}} = h_g e^{a} \qquad (131)$$

Example. What is the apparent coefficient of sensible heat transfer in the condensation of benzene from air saturated at 50°C. on a cold surface held at 20°C.? The total pressure is 1 atm. The rate of flow and the diameter of the condenser tubes are such that the heat-transfer coefficient in the absence of simultaneous mass transfer is expected to be 10 p.c.u./(hr.)(sq. ft.)(°C.).

The average material properties of benzene-air mixtures correspond to $c\mu/k = 0.74$, $\mu/\rho D = 1.76$. The vapor pressure of benzene is 269 mm. at 50°C. and 74.7 mm. at 20°C. The specific heat of benzene vapor is 0.27, and that of the air-benzene mixture is 0.246 p.c.u./(lb.-mole)(°C.). Thus $a = \left(\frac{0.74}{1.76}\right)^{2/3}\left(\frac{0.27}{0.246}\right) \ln \left(\frac{1 - 0.0984}{1 - 0.354}\right) = 0.859$, and $h_{\text{apparent}} = 10\frac{0.859}{1 - e^{-0.859}} = 14.9$ p.c.u./(hr.)(sq. ft.)(°C.). By the approximate equation, $h_{\text{apparent}} = 10 + (\tfrac{1}{2})(8.59) = 14.3$.

† Conversely, the rate of diffusion is affected by the presence of a temperature gradient, because of the phenomenon of thermal diffusion [*cf.* Chapman and Cowling, "The Mathematical Theory of Non-uniform Gases," Cambridge, London, 1939; Jones and Furry, *Rev. Modern Phys.*, **15**, 151 (1946)]. When diffusion of a lighter constituent and heat conduction take place in the same direction, the rate of diffusion is slower because of thermal diffusion, since the lighter molecules tend to concentrate in the hotter part of the gas. The effect is generally small, however. Thus, in the case of a mixture of equal volumes of hydrogen and nitrogen at 1 atm., diffusing between surfaces 1 cm. apart held at 100° and 300°C., the rate of thermal diffusion is about 1.5×10^{-6} g.-moles/(sec.)(sq. cm.). The rate of ordinary diffusion under the influence of a concentration gradient of 0.5 mole fraction/cm. is about 2.1×10^{-5} g.-moles/(sec.)(sq. cm.) at the same temperature. Even in this case, where the conditions are most favorable for thermal diffusion because of a large ratio of molecular weights, high concentrations of both components, and a large temperature gradient, the rate of mass transfer due to ordinary diffusion is fourteen times that due to thermal diffusion.

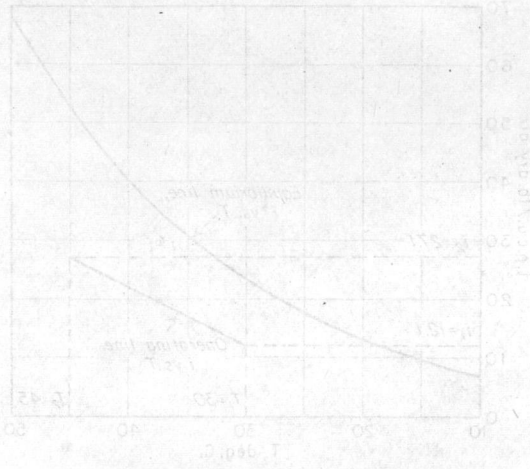

SECTION 9
DISTILLATION AND SUBLIMATION

BY

James S. Carey, Sc. D., Chemical Engineer, E. B. Badger & Sons Co.; Member, American Institute of Chemical Engineers, American Chemical Society; Licensed Professional Engineer, New York, Pennsylvania. (Distillation)

C. N. Collard, M. S., Chemical Engineer, E. B. Badger & Sons Co. (Plate Calculations for Multicomponent Mixtures)

Stanley B. Zdonik, S. B., S. M., Chemical Engineer, E. B. Badger & Sons Co.; Junior Member, American Institute of Chemical Engineers, Licensed Professional Engineer, Massachusetts. (Azeotropic and Extractive Distillations)

K. C. D. Hickman, Ph. D., Consulting Engineer, Member, American Chemical Society. (Molecular Distillation)

H. C. Vernon, M. S., Engineering Department, E. I. duPont de Nemours & Co.; Member, American Institute of Chemical Engineers. (Sublimation)

R. L. Pigford, Ph. D., Professor and Head, Chemical Engineering Department, University of Delaware; Member, American Institute of Chemical Engineers, American Chemical Society. (Plate Efficiencies for Distillation Columns, Packed Distillation Columns)

A. P. Colburn, Ph. D., Assistant to the President and Professor of Chemical Engineering, University of Delaware; Member, American Institute of Chemical Engineers, American Chemical Society, American Society for Mechanical Engineers, American Society for Engineering Education. (Plate Efficiencies for Distillation Columns, Packed Distillation Columns)

F. W. Woodfield, Jr., M. S., Chemical Engineer, General Electric Co.; Member, American Institute of Chemical Engineers, American Chemical Society.

CONTENTS

DISTILLATION

Definitions

PAGE
Distillation 563
Destructive Distillation 563
Rectification 563
Fractionation 563
Dephlegmation 563

Data and Generalizations Used in Distillation Calculations

Vapor Pressure 563
  Approximate Clapeyron-Clausius Equation 564
  Dühring's Rule 564
  Cox Chart 564
  Critical Pressures and Temperatures of Hydrocarbons 565
  Use of Vapor-pressure Plots 565
  Heat Contents and Other Properties 565
  Characterization Factor 565
The Perfect-gas Laws 567
Liquid-vapor Equilibrium Relations 568
Equilibrium Constants for Hydrocarbons 568
Effect of Pressure on Liquid-vapor Equilibrium Relations 569
Liquid-vapor Equilibrium Data for Binary Mixtures 573
Retrograde Phenomena 575

Use of Phase Rule in Distillation

Liquid-vapor Systems 576
Phase Diagrams for Binary Mixtures 577
Three-component Systems 579

Volatility and Relative Volatility

Volatility and Relative Volatility 579

Methods of Vaporization

Methods of Vaporization 580

Principles, Applications, and Methods of Calculation of Distillation Processes

Simple Batch Distillation 580
  Definition 580
  Applications 580
  Theory of Simple Batch Distillation 580
Applications of Simple Batch Distillation Equations 581
  Weathering of Natural Gasoline 581
Batch Steam Distillation 582
  Definition 582
  Applications 582
Theory of Batch Steam Distillation 582
  Effect of the Presence of Liquid Water in the Still 583
  Use of External Source of Heat 583
  Vaporization Efficiency in Steam Distillation 583
  Example of Calculation of Steam Consumption in Batch Steam Distillation 584
Continuous Equilibrium Vaporization 585
  Definition 585
  Applications 585
  Calculations of Continuous Equilibrium Vaporizations 585
  Continuous Equilibrium Vaporization of Petroleum Fractions 587
Continuous Rectification 588
  Principles of Continuous Rectification in Bubble Plate Towers 589
  Over-all Heat Balance 589
  General Material Balance Equations 589
  Sectional Heat Balances 590
  Simplifying Assumptions 590
  Use of Liquid-vapor Equilibrium Relationship 591
  McCabe-Thiele Graphical Method for Calculation of Number of Plates 591
  Effect of Thermal Condition of Feed 592
  Application of McCabe-Thiele Method 593
Batch Rectification 594

Types of Rectifying Columns or Towers

Types of Rectifying Columns or Towers 596

Design of Bubble-cap Plate Towers

Vapor Velocity in Plate Towers 597
Plate Spacings 598
Shape and Spacing of Caps 599
Types of Liquid Flow 599
Liquid-handling Capacities of Trays 599
Weirs and Down Spouts 600
Pressure Drop and Slot Opening 601

	Page
Cap-slot and Riser Areas	602
Reflux Supply	602
Reboil Heat	602

Petroleum Distillation

Criteria of Separation	603
Side Streams, or Withdrawing Liquid from Selected Plates of a Column	603
Estimation of Tower Temperatures	604
Use of Heat Balances in Design	604
Calculations	605
Discussion of Heat Balance	605

Laboratory Distillations

A.S.T.M. Distillation Test	606
Relation to Other Laboratory Distillations	607
True-boiling-point Distillations	607
True-boiling-point Curve Compared with A.S.T.M.	607
Applications of True-boiling-point Distillations	607
True-boiling-point Apparatus	608
Stedman Column Packing	608
Low-temperature Fractionations	609
Laboratory Distilling Columns of High Efficiency	609
Calibration of Columns	609

Plate Efficiencies for Distillation Columns

General	610

Packed Distillation Columns

Application	618
Packings	618
Design of Packed Columns	618
Diameter	618
Height	618
Values of H.T.U. and H.E.T.P.	619
Effect of Velocities	619
Effect of Temperature and Composition	619
Effect of Pressure	619
Effect of Packing and Column Dimensions	619
Effect of Reflux Ratio	619
Nomenclature	621

Plate Calculations for Multicomponent Mixtures

Introduction	622
Relationship between Theoretical Plates and Reflux Ratio	622
Minimum Plates at Total Reflux	623
Determination of Minimum Reflux Ratio	623
Location of the Optimum Feed Plate	624
Separations Involving Three or More Distributed Components	625
Prediction of the Performance of an Existing Tower	625
General Design Procedure	625
Sample Calculations	625
Material Balance	625
Determination of Condenser Outlet Temperature and Pressure	625
Top Tower Temperature	625
Reboiler Temperature	626
Equilibrium Vaporization of Feed	626
Estimation of Minimum Number of Plates at Total Reflux	626
Estimation of Minimum Reflux Ratio	626
Stepwise Plate Calculations	627
Stepwise Calculations from Reboiler to Feed Plate	628
Introduction of Components Lighter than Light Key below the Feed	628
Introduction of Components Heavier than Heavy Key above the Feed	629
Graphical Methods for Determining Number of Plates	629

Azeotropic and Extractive Distillations

General and Theoretical

The Occurrence of Azeotropes in Distillation	630
Formation of Homogeneous Azeotropes	630
Estimating Vapor-liquid Equilibrium from Azeotropic Data	631
Effect of Pressure on Composition of Homogeneous Azeotropes	631
Formation of Heterogeneous Azeotropes	631
Distillations Involving Heterogeneous Azeotropes	632
Equilibrium Data for Systems Forming Heterogeneous Azeotropes	633

Applications

Introduction and Definitions	634
Extractive Distillation	634
Choice of Separating Agent	643
Solvent Concentration	644
Novel Features of Extractive Distillation Columns	645
Design of an Extractive Distillation Column	645
Discussion	645
Illustrative Example	646
Azeotropic Distillation	651
Choice of Separating Agent	652
Solvent Concentration	652
Design of an Azeotropic Distillation Column	652
Illustrative Example	653
Comparison of Extractive and Azeotropic Distillation	655

Molecular Distillation

Historical	655
Definitions	655
The Simple Molecular Still	656
Mean Free Path Considerations	656
The Distilland Layer	657
Falling-film Stills	657
Centrifugal Stills	657
The Distilling Operation	657
Subsidiary Apparatus	658
Uses of the Molecular Still	659
Laboratory	659
Industrial	659
Comparative Properties of the Molecular Still	659
Fractionation	659

SUBLIMATION

Definitions	660
Uses	660
Limiting Factors	661
Methods—General	662
Simple Sublimation	662
Carrier Sublimation	662
Design Calculations	662
Physical Properties	662
The Operating Diagram	662
Capacity	662
Purification	663
Equipment	663
Plant for Simple Sublimation	664
Plant for Entrainer Sublimation	665

DISTILLATION

By James S. Carey

REFERENCES. *General Texts:* Hausbrand, "Principles and Practice of Industrial Distillation," 6th ed., trans. by Tripp, Wiley, New York, 1926. Young, "Distillation Principles and Practice," Macmillan, London, 1922. Sorel, "La Rectification de l'alcohol," Paris, 1893. Walker, Lewis, McAdams, and Gilliland, "Principles of Chemical Engineering," 3d ed., McGraw-Hill, New York, 1937. Robinson and Gilliland, "The Elements of Fractional Distillation," 3d ed., McGraw-Hill, New York, 1939. Nelson, "Petroleum Refinery Engineering," 2d ed., McGraw-Hill, New York, 1941.

LITERATURE. *General and Physical Properties:* Cox, *Ind. Eng. Chem.*, **15**, 592 (1923). Calingaert and Davis, *Ind. Eng. Chem.*, **17**, 1287 (1925). Watson and Nelson, *Ind. Eng. Chem.*, **25**, 880 (1933). Smith and Watson, *Ind. Eng. Chem.*, **29**, 1408 (1937). Kay, *Ind. Eng. Chem.*, **28**, 1014 (1936). Watson and Smith, *Nat. Petroleum News*, July 1, 1936. Lewis and Luke, *Trans. Am. Soc. Mech. Engrs.*, **54**, 17 (1932); *Ind. Eng. Chem.*, **25**, 725 (1933). Lewis and Kay, *Oil Gas J.*, Mar. 29, 1934. Cummings, *Ind. Eng. Chem.*, **23**, 900 (1931). Katz and Hachmuth, *Ind. Eng. Chem.*, **29**, 1072 (1937).

Simple Batch Distillation: Rayleigh, *Phil. Mag.* May 6, 1902, p. 521. Carey, Sc. D. Thesis, Mass. Inst. Tech., 1930.

Continuous Equilibrium Vaporization: Piroomov and Beiswenger, *Am. Petroleum Inst. Bull.* 10, Jan. 3, 1939, p. 52.

Continuous Rectification: McCabe and Thiele, *Ind. Eng. Chem.*, **17**, 605 (1925).

Batch Rectification: Fenske, *Ind. Eng. Chem.*, **24**, 482 (1932). Bogart, *Trans. Am. Inst. Chem. Engrs.*, **33**, 139 (1937). Smoker and Rose, *Trans. Am. Inst. Chem. Engrs.*, **36**, 285 (1940). Colburn and Stearns, *Trans. Am. Inst. Chem. Engrs.*, **37**, 291 (1940).

Tower Design: Sherwood and Jenny, *Ind. Eng. Chem.*, **27**, 265 (1935). Holbrook and Baker, *Ind. Eng. Chem.*, **26**, 1063 (1934). Pyott, Jackson, and Huntington, *Ind. Eng. Chem.*, **27**, 821 (1935). Peavy and Baker, *Ind. Eng. Chem.*, **29**, 1056 (1937). Rhodes, *Ind. Eng. Chem.*, **26**, 1333 (1934); **27**, 272 (1935). Colburn, *Ind. Eng. Chem.*, **28**, 526 (1936). Souders and Brown, *Ind. Eng. Chem.*, **26**, 98 (1934). Chute (Discussion), *Trans. Am. Inst. Chem. Engrs.*, **34**, 91 (1938). Souders *et al.*, *Ind. Eng. Chem.*, **30**, 86 (1938). Rogers and Thiele, *Ind. Eng. Chem.*, **26**, 524 (1934).

Laboratory Distillations: Peterkin and Ferris, *Ind. Eng. Chem.*, **17**, 1249 (1925). A.S.T.M. Standards, 1930, Part II, A.S.T.M. Designation D 86–30. Beiswenger and Child, *Ind. Eng. Chem.*, anal. ed., **2**, 284 (July 15, 1930). Peters and Baker, *Ind. Eng. Chem.*, **18**, 69 (1926). Podbielniak, *Ind. Eng. Chem.*, anal. ed., **3**, 177 (April 15, 1931).

DEFINITIONS

Distillation is the separation of the constituents of a liquid mixture by partial vaporization of the mixture and separate recovery of vapor and residue. The more volatile constituents of the original mixture are obtained in increased concentration in the vapor; the less volatile in greater concentration in the liquid residue. The completeness of separation depends upon certain properties of the components involved and upon the arrangement of the distillation process.

In general, *distillation* is the term applied to vaporization processes in which the vapor evolved is recovered, usually by condensation. *Evaporation* commonly refers to the removal of water from aqueous solutions of nonvolatile substances by vaporization. The vapor evolved, *i.e.*, the water, is discarded.

The majority of the applications of distillation are found in the separation of one or more of the components from mixtures of organic compounds, *e.g.*, the recovery of propane from mixtures of low-boiling hydrocarbons. Certain industrially important distillation processes in the purely inorganic field may be cited, however, *e.g.*, the separation of the constituents of air by liquefaction and distillation, and the recovery of ammonia from ammonia liquor.

Destructive distillation includes those operations in which the material under process first undergoes thermal decomposition and the volatile products formed are then withdrawn as vapors for recovery.

Rectification is a distillation carried out in such a way that the vapor rising from a still comes in contact with a condensed portion of vapor previously evolved from the same still. A transfer of material and an interchange of heat result from this contact, thereby securing a greater enrichment of the vapor in the more volatile components than could be secured with a single distillation operation using the same amount of heat. The condensed vapors, returned to accomplish this object, are termed **reflux.**

The generally used devices, in which vapors from a still on their way to a condenser can flow countercurrently to a portion of the condensate returned as reflux, are called **rectifying columns** or **towers.**

The term **reflux ratio** is commonly applied in two different ways in connection with rectification problems, as follows:

1. In plant usage, when referring to the operation of equipment, reflux ratio most frequently denotes the quantity of reflux per unit quantity of distillate removed from the process as a product.

2. For design purposes, reflux ratio is more logically taken as the ratio of liquid reflux to vapor at any given point in a rectifying column (see p. 589). Reflux ratio is used in both of the above senses in the literature, and, where, the term is not specifically defined, reference must be made to the accompanying text to determine the manner in which it has been used. It may be noted that reflux ratios in terms of definition 1 may vary from zero to infinity, while if definition 2 is used numerical values from zero to unity may be encountered.

"Fractionation" is synonymous with rectification and is the term commonly applied to rectification processes in the petroleum industry.

Dephlegmation, or **partial condensation,** is a term frequently applied to the cooling of a mixed vapor to a definite temperature thereby condensing a portion of the vapor richer in the higher boiling constituents than the original vapor. Dephlegmation usually is combined with fractionation or rectification. For example, a partial condenser or dephlegmator may be placed in the upper portion of a rectifying column to condense a portion of the rising vapors and return the condensate to the column as reflux.

DATA AND GENERALIZATIONS USED IN DISTILLATION CALCULATIONS

Vapor Pressure (see pp. 149 to 174, 293)

The vapor pressure of a liquid is the pressure of the vapor of that liquid at any given temperature at which the vapor and liquid phases of the substance can exist in equilibrium. If the temperature is held constant, and the vapor over the pure liquid is compressed, condensation will occur until no more vapor is left. Conversely, if the vapor space is expanded, evaporation ensues.

Vapor pressures are extensively employed to calculate the vapor and liquid compositions of mixtures when the vapor and liquid phases are in equilibrium. However,

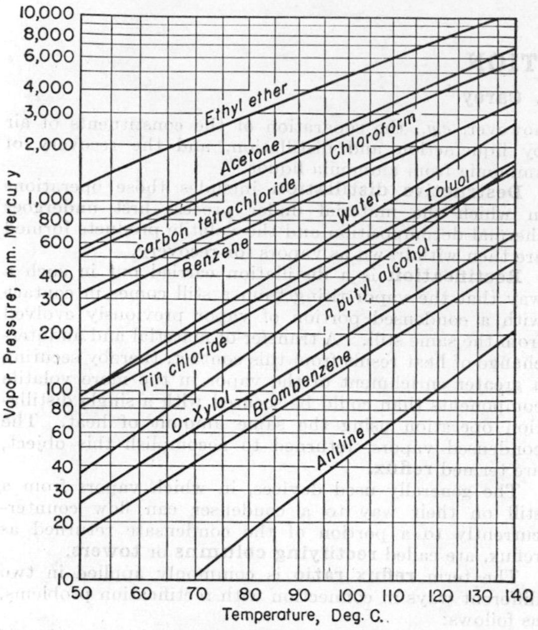

FIG. 1. Vapor-pressure curves of liquids.

the vapor pressure of a pure substance does not represent the vapor pressure of that substance when in solution as one of the components of a mixture. In distillation calculations, cognizance is taken of vapor-pressure lowering by applying the known laws of solutions or by utilizing available experimental data on partial vapor pressures of the respective components in a mixture (see Sec. 3, pp. 290 to 291, and Sec. 3, pp. 166 to 174).

Data on vapor pressures are fairly extensive in the literature. It frequently happens, however, that the vapor pressure vs. temperature relationship must be extrapolated or interpolated from meager or inconsistent data. For such cases certain useful thermodynamic relationships are summarized in the following sections. For ready reference, vapor-pressure data for some of the more common substances encountered in distillation problems are presented in plots, together with brief explanations of the methods used in construction of these plots.

The Approximate Clapeyron-Clausius Equation (see p. 293). A useful rule derived from the Clapeyron-Clausius equation for the correlation of vapor-pressure data is that a plot of logarithm (vapor pressure) against the reciprocal of the absolute temperature will, with most substances, yield a substantially straight line over narrow temperature ranges. Figure 1 illustrates the use of log p vs. $1/T$ plots.

Dühring's Rule. See pp. 294, 341.

Cox Chart. [Cox, *Ind. Eng. Chem.*, **15**, 592 (1923). Calingaert and Davis, *Ind. Eng. Chem.*, **17**, 1287 (1925).]

A method that has been found especially useful in plotting the vapor-pressure curves of petroleum hydrocarbons is illustrated by Fig. 2 and is developed as follows:

The desired range of vapor pressures is first laid off along the horizontal axis using logarithmic spacing. In Fig. 2 the horizontal scale covers pressures from 0.01 to 10,000 lb./sq. in. abs.

To determine the temperature scale, an arbitrarily placed straight line is drawn, sloping down toward the right, as shown in Fig. 2 by the broken line labeled "H_2O." This line is to represent the vapor-pressure curve of water; the temperature scale must therefore be properly adjusted to agree with the vapor-pressure scale. Thus at 100°F. the vapor pressure of water is 0.95 lb./sq. in. (see p. 277). Through a point on the water line,

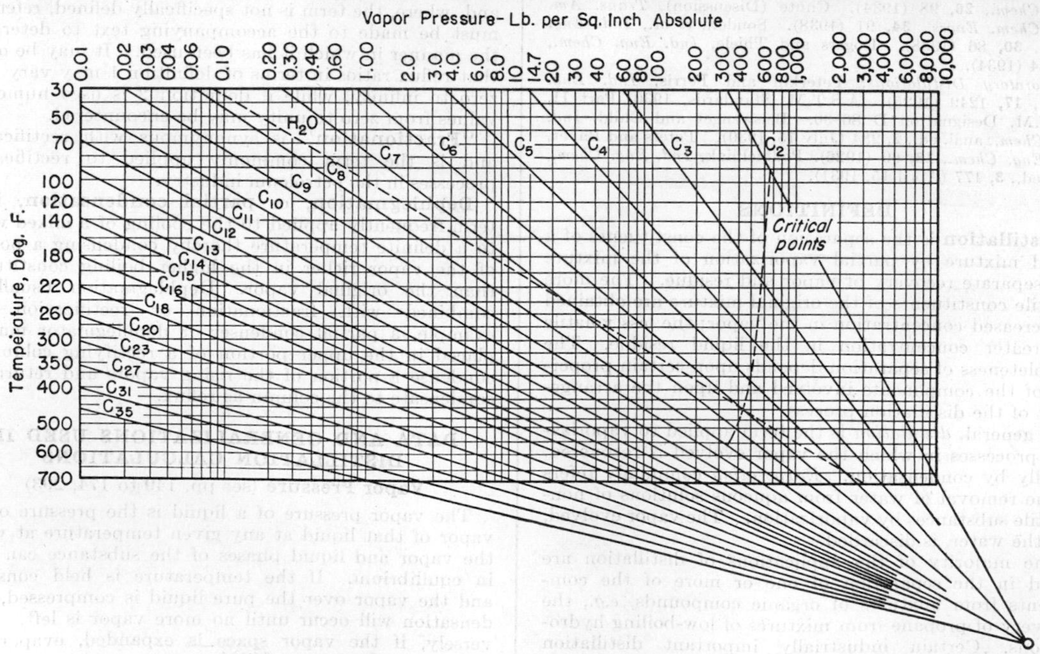

FIG. 2. Cox chart for vapor pressures of normal paraffin hydrocarbons.

corresponding to 0.95 lb./sq. in., an ordinate representing a temperature of 100°F. is drawn. The entire temperature scale is computed in this manner. The final position of the line representing the vapor-pressure curve for water is determined by trial to yield a plot with a well-proportioned temperature scale.

With the pressure and temperature scales determined in the manner outlined, the vapor-pressure curves of the normal paraffins and other hydrocarbons may be plotted. These are found to yield nearly straight lines having a common point of intersection, shown in Fig. 2 in the lower right-hand corner outside the range to which the temperature or pressure scales have been carried. The majority of the hydrocarbons occurring in petroleum have vapor-pressure curves closely paralleling those of the paraffins. With this chart one point on the vapor-pressure curve of a petroleum hydrocarbon, *e.g.*, its boiling point at atmospheric pressure, will permit an approximation of the entire vapor-pressure curve of the substance by drawing a line on the chart from the known point through the common point of intersection.

Critical Pressures and Temperatures of Hydrocarbons. A broken-line curve on the right-hand side of Fig. 2 is drawn through the points corresponding to the critical pressures and temperatures of paraffin hydrocarbons up to octane. Refer to p. 566 for the estimation of critical points of hydrocarbons and petroleum fractions.

Use of Vapor-pressure Plots. The vapor-pressure plots given in this section, because of their small size, cannot be read closely and should be used only for approximate work. The data of Sec. 3, pp. 149 to 174, should be used to construct full-sized plots for accurate calculations.

Heat Contents and Other Properties. In addition to vapor-liquid equilibrium relations (Sec. 4, pp. 293 to 294), properties required in distillation calculations are heat contents, molecular weights, and, to a more limited extent, critical properties. For mixtures of known chemical identities these properties can be determined directly from the data and correlations of Secs. 4, 8, and 10. When dealing with the distillation of petroleum and its fractions (whether natural or processed), as molecular weight increases, the multiplicity of compounds present within narrow boiling ranges increases rapidly. The bulk of commercial products from petroleum is represented by "fractions" that have appreciable boiling ranges as defined by the A.S.T.M. (see p. 606) or similar laboratory test. Therefore, average properties of these fractions are conventionally employed. Correlations for the estimation of these properties follow.

Characterization Factor. Watson and Nelson [*Ind. Eng. Chem.*, **25**, 880 (1933)] introduced the concept of a "characterization factor" to define numerically the chemical character of hydrocarbons and as a means for the correlation of the properties of mixtures. The Watson-Nelson factor is defined by Eq. (1):

$$K_{w-n} = \frac{(T_B)^{1/3}}{s} \tag{1}$$

where K_{w-n} = characterization factor.
T_B = atmospheric boiling point, °R.
s = specific gravity, 60°F./60°F.

In application to fractions, T_B is the cubic average boiling point as defined by the following paragraph and Fig. 3. (In addition to its relation to the correlations described herewith, characterization factor is widely used to define the properties of petroleum fractions ranging from naphthenic to paraffinic in general structure. Numerical values range from about 10.50 to 12.50; hence values are conventionally calculated to two figures following the decimal point.)

For the correlation of the various properties of petroleum fractions several differing average boiling points are used which are related to the A.S.T.M. distillation test by Fig. 3 of Smith and Watson [*Ind. Eng. Chem.*, **29**, 1408 (1937)]. In Fig. 3

$$\text{A.S.T.M. slope} = \frac{90 \text{ per cent temp.} - 10 \text{ per cent temp.}}{80}$$

and the volumetric average boiling point equals the arithmetic average of the 10, 30, 50, 70, and 90 per cent A.S.T.M. temperatures (for both slope and volume

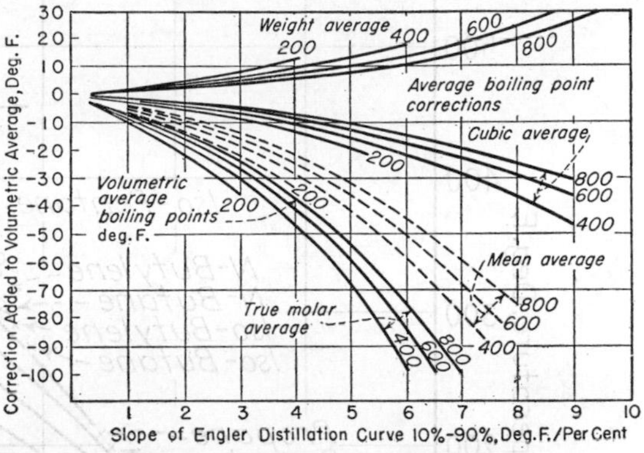

FIG. 3. Relationships of average boiling points.

average boiling point the percentage over is taken as distillate plus distillation loss). The correct boiling point for use in Eq. (1) is the cubic average boiling point as determined from the volume average boiling point and use of Fig. 3. The use of cubic average boiling point in Eq. (1) makes the characterization factor additive with the weight fraction for mixtures where changes in volume with mixing are negligible.

Figure 4 (Watson and Nelson, *loc. cit.*), correlates molecular weight against mean average boiling point, as determined from Fig. 3, and A.P.I.* gravity.

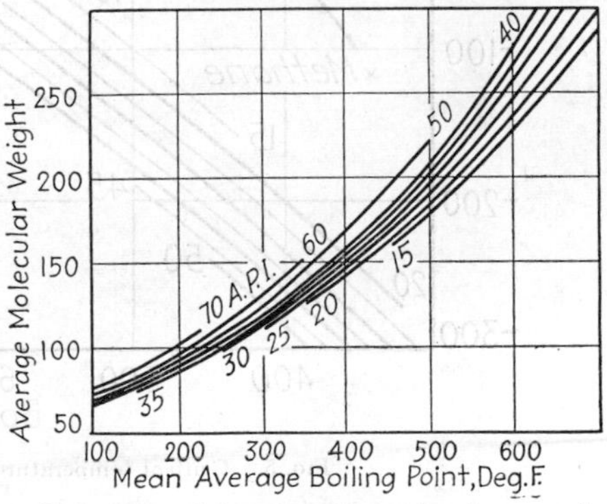

FIG. 4. Molecular weights of petroleum fractions as a function of boiling point and specific gravity.

Figure 5 (Smith and Watson, *loc. cit.*) correlates critical temperature against atmospheric boiling point and A.P.I. gravity. The weight average boiling point from Fig. 3 is used in conjunction with Fig. 5 to estimate the true critical temperature of a fraction; if the true molar average boiling point is used in conjunction with Fig. 5, the pseudocritical temperature is obtained (the pseudocritical properties of a hydrocarbon mixture

* See p. 604.

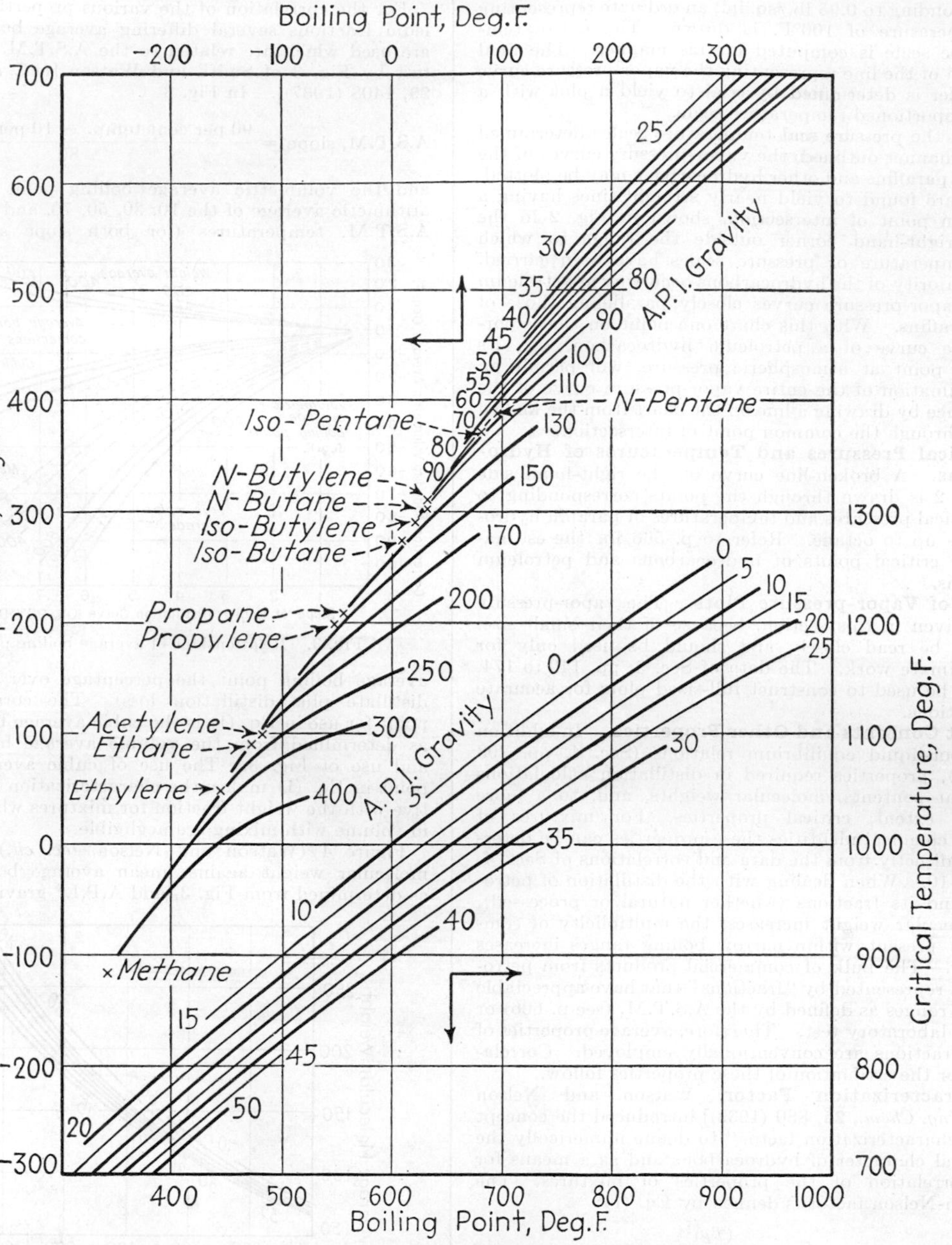

FIG. 5. Critical temperatures correlated against boiling point and gravity.

having an appreciable boiling range represent the critical points of a pure hydrocarbon having identical *P-V-T* relations in the superheated region as the mixture [see Kay, *Ind. Eng. Chem.*, **28**, 1014 (1936)].

Figure 6 (Smith and Watson, *loc. cit.*) correlates pseudocritical pressure against mean average boiling point and A.P.I. gravity, while Fig. 7 presents a similar correlation for light hydrocarbons in terms of molecular weight and A.P.I. gravity. In using Fig. 7 for the light hydrocarbons, the following extrapolated A.P.I. and specific gravities (60°F./60°F.) should be employed for calculation of mixture gravities:

	Component									
	CH_4	C_2H_2	C_2H_4	C_2H_6	C_3H_6	C_3H_8	$i\text{-}C_4H_8$	$n\text{-}C_4H_8$	$i\text{-}C_4H_{10}$	$n\text{-}C_4H_{10}$
°A.P.I.	440	167	213	213	138	145	104	99	114	110
Sp. gr.	0.247	0.473	0.41	0.41	0.526	0.511	0.600	0.613	0.576	0.585

Figure 8 (Smith and Watson, *loc. cit.*) gives a means of estimating the true critical pressure of a mixture by use of the ratio of true to pseudocritical temperature from Fig. 5 and the pseudocritical pressure from Fig. 6 or 7.

Figure 9 (Watson and Nelson, *loc. cit.*) correlates specific heats of liquid hydrocarbon fractions against temperature and A.P.I. gravity. A correction for characterization factor deviation from 11.8 is given.

Figure 10 (Watson and Nelson, *loc. cit.*) correlates specific heats of vapors at atmospheric pressure on a basis similar to Fig. 9.

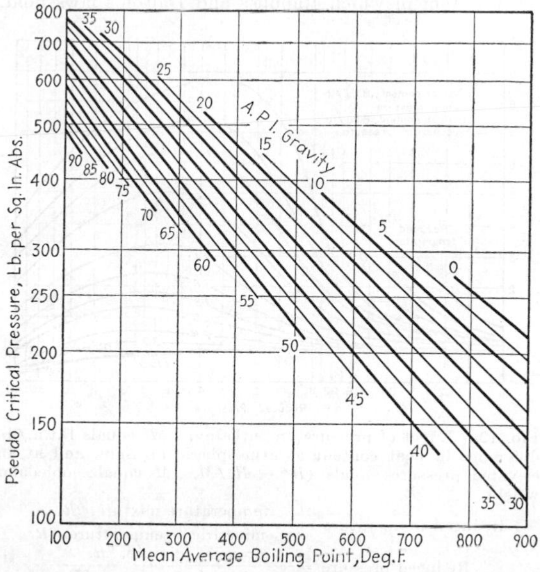

FIG. 6. Pseudocritical pressure vs. boiling point and gravity.

Figure 11 (Watson and Nelson, *loc. cit.*) correlates latent heats at atmospheric pressure against the true molar average boiling point and either characterization factor or molecular weight. In using Figs. 9, 10, and 11 in calculating heat content of a vapor at atmospheric pressure, the heat content of the liquid should be calculated from a reference temperature (*e.g.*, 32°F.) to the true molar average boiling temperature, the heat of

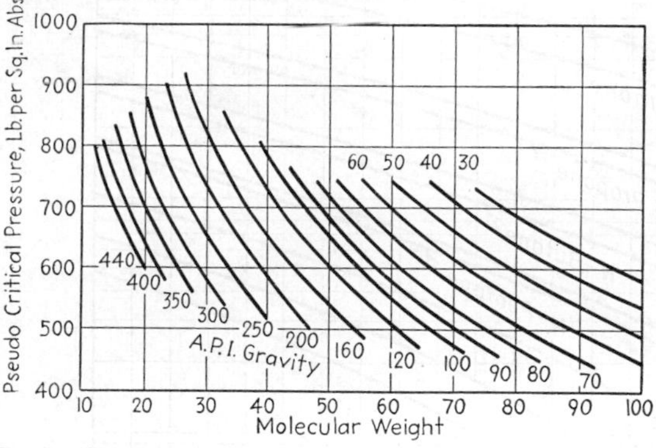

FIG. 7. Pseudocritical pressure vs. molecular weight and gravity.

vaporization added, and the additional heat content of the vapor from the vaporization temperature to final temperature determined.

Figure 12 (Watson and Smith, *Nat. Petroleum News*, July 1, 1936) gives a means for correction of heat content of vapors for pressure.

The Perfect-gas Laws (see pp. 289 to 293)

Two major applications of the gas laws are made in distillation calculations:

1. Calculation of partial pressures from Dalton's law.

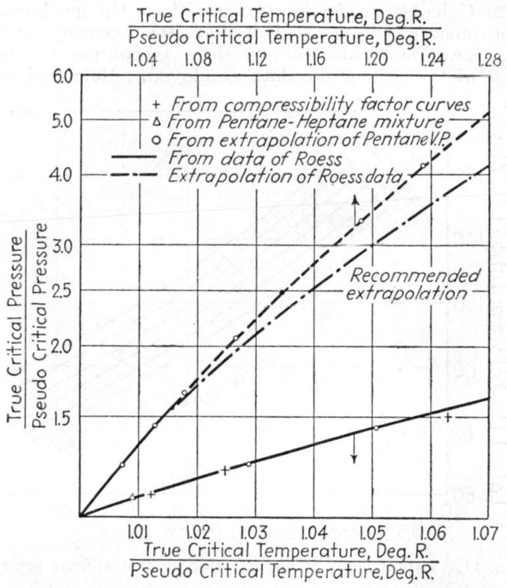

FIG. 8. Chart for estimation of true critical pressure.

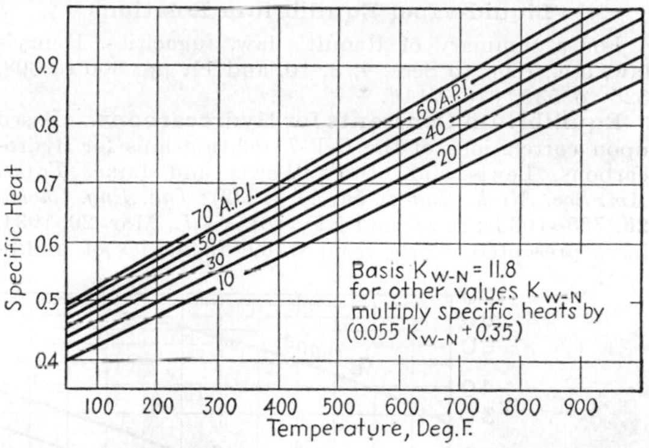

FIG. 9. Specific heats of liquid petroleum oils where $K = 11.8$ (mid-continent stocks). For other stocks multiply by $(0.055K + 0.35)$.

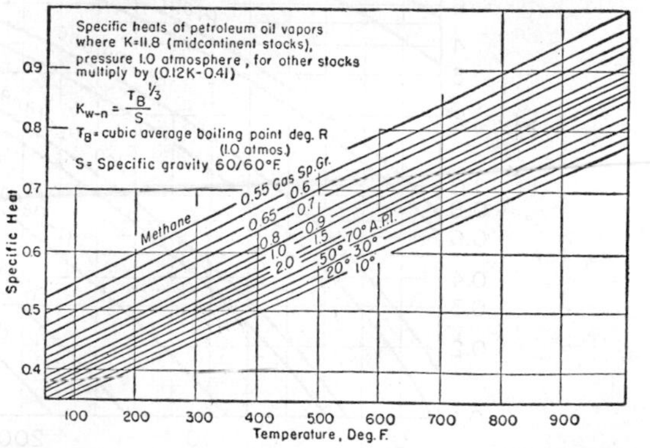

FIG. 10. Specific heats of paraffin gases and petroleum vapors at atmospheric pressure, where $K = 11.8$ (mid-continent stocks). For other stocks multiply by $(0.12K - 0.41)$.

2. Calculation of vapor volumes from the gas laws. For most distillation calculations the accuracy of the gas laws, at moderate pressures, is comparable with that of the rest of the data employed. Reference may

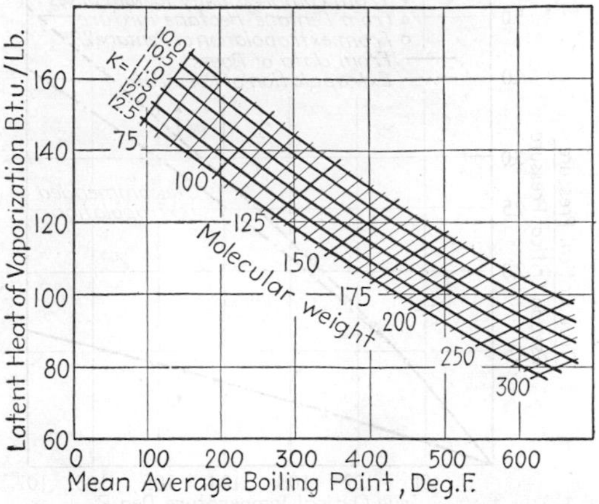

FIG. 11. Latent heats of vaporization at atmospheric pressure.

be made to Sec. 4, pp. 289 to 292, for illustrations of the use of the perfect-gas laws and other equations of state.

Liquid-vapor Equilibrium Relations

For a summary of Raoult's law, fugacities, Henry's law, etc., refer to Secs. 4, 8, 10, and 11, pp. 306 to 308, 525 to 532, 673, 714*ff*.

Equilibrium Constants for Hydrocarbons. Based upon correlations of the *P-V-T* relationships for hydrocarbons, Lewis and others [Lewis and Luke, *Trans. Am. Soc. Mech. Engrs.*, **54,** 17 (1932); *Ind. Eng. Chem.*, **25,** 725 (1933); Lewis and Kay, *Oil Gas J.*, Mar. 29, 1934] have presented charts from which fugacities of hydrocarbons may be calculated. For ready use the equilibrium relationship is conveniently expressed as

$$y = Kx \qquad (2)$$

where K is termed the equilibrium constant ($K = P_1/P$ for a system in which Raoult's and Dalton's laws hold.

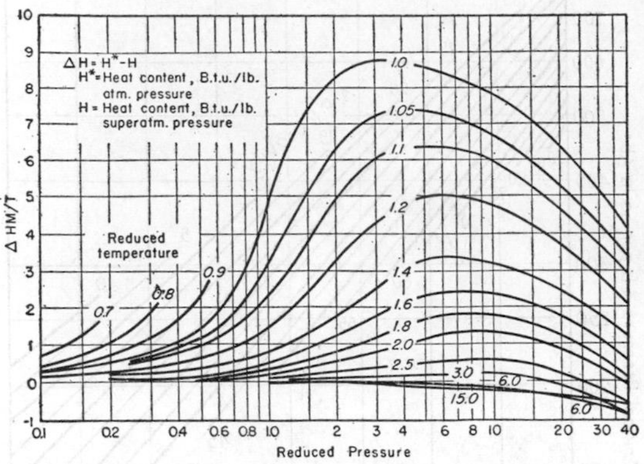

FIG. 12. Effect of pressure on enthalpy. ΔH equals B.t.u./lb. difference in heat content at atmospheric pressure, and at an elevated pressure equals $(H^* - H)/M$. M equals molecular weight.

$$\text{Reduced temperature} = \frac{\text{temperature mixture, °R.}}{\text{pseudocritical temperature, °R.}}$$

$$\text{Reduced pressure} = \frac{\text{pressure mixture, °R.}}{\text{pseudocritical pressure, °R.}}$$

See Sec. 8, p. 525*ff*). Taylor and Parker (unpublished data, Mass. Inst. Tech., 1935) calculated values of K for certain hydrocarbons between methane and octane using the Lewis *et. al.* (*loc. cit.*) fugacity correlations. Figures 13 to 20, inclusive, give plots of K vs. temperature for various pressures. Extrapolation of K values

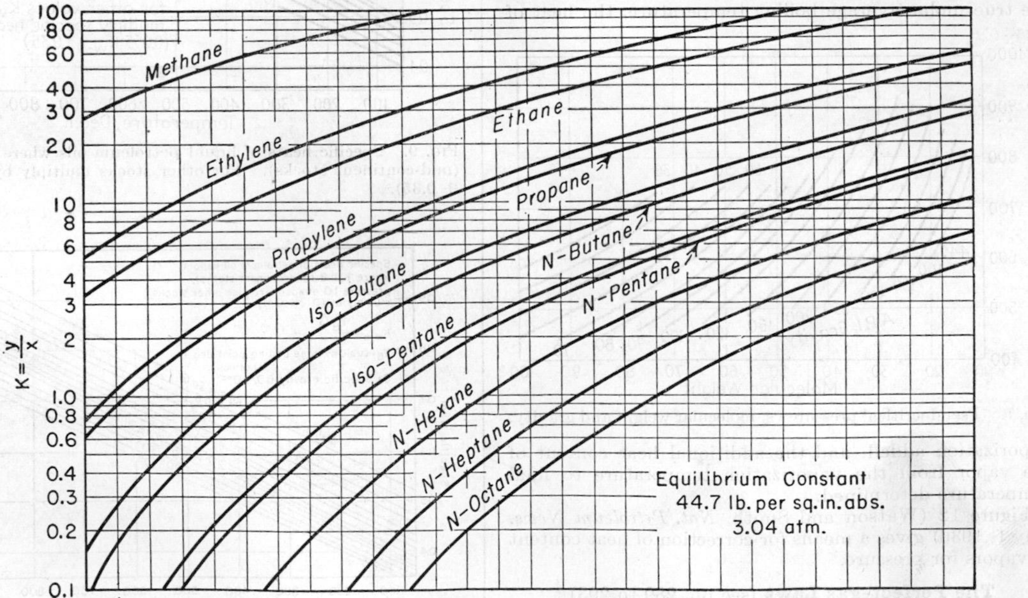

FIG. 13. Equilibrium constant K vs. temperature. 44.7 lb./sq. in. abs. or 3.04 atmospheres.

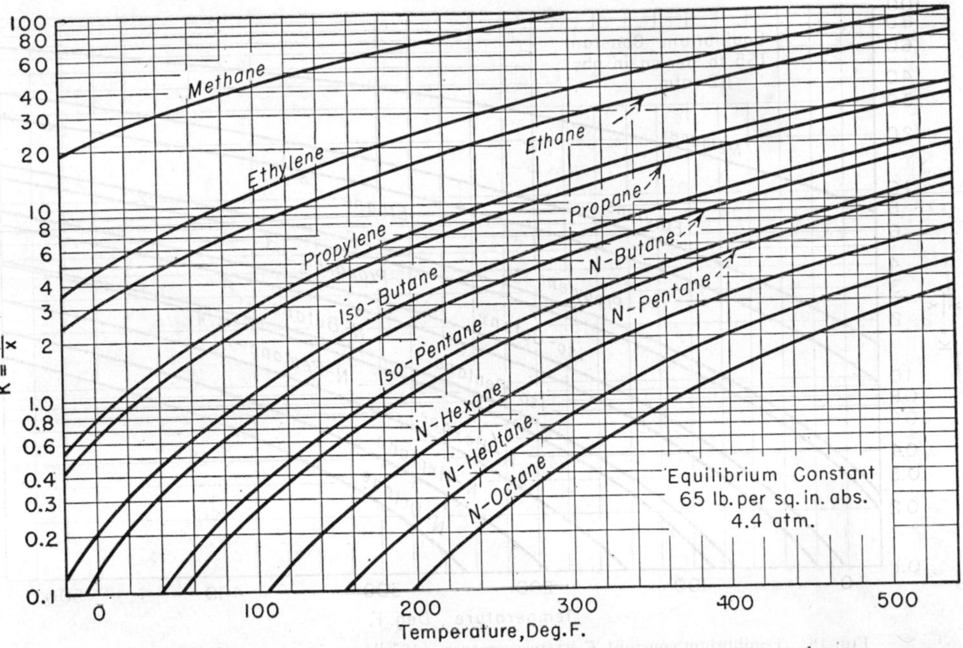

FIG. 14. Equilibrium constant K vs. temperature. 65 lb./sq. in. abs. or 4.4 atmospheres.

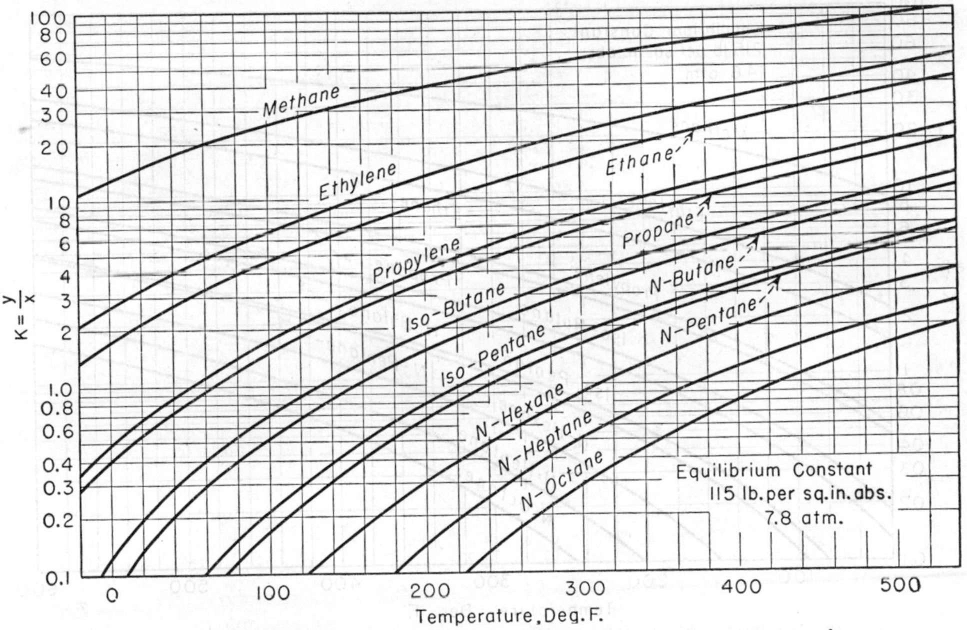

FIG. 15. Equilibrium constant K vs. temperature. 115 lb./sq. in. abs. or 7.8 atmospheres.

may be made by plotting log K vs. $1/T$ as approximate straight-line functions.

Table 1 gives the liquid-vapor equilibrium compositions for a number of binary mixtures. [From these data y vs. x diagrams at constant pressure may be plotted (see pp. 576–579).] The pressures at which the determinations were made and the equilibrium boiling temperatures are included.

Effect of Pressure on Liquid-vapor Equilibrium Relations. The effect of an increase in temperature is usually to decrease the relative differences in volatility (see definition, p. 579) between the components of a given mixture; conversely, a lowered vaporization temperature usually increases the volatility differences. Consequently the differential between vapor and liquid compositions is usually decreased by an increase in the pressure at which vaporization takes place, and this differential is increased by an appreciably lowered pressure. These considerations may be illustrated for the paraffin hydrocarbons by reference to the Cox chart

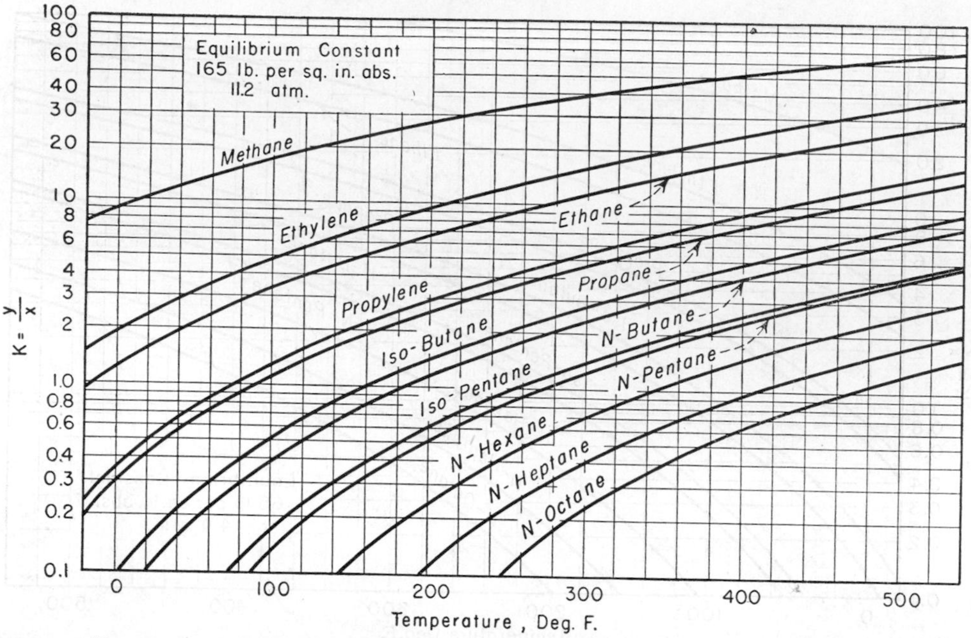

FIG. 16. Equilibrium constant K vs. temperature. 165 lb./sq. in. abs. or 11.2 atmospheres.

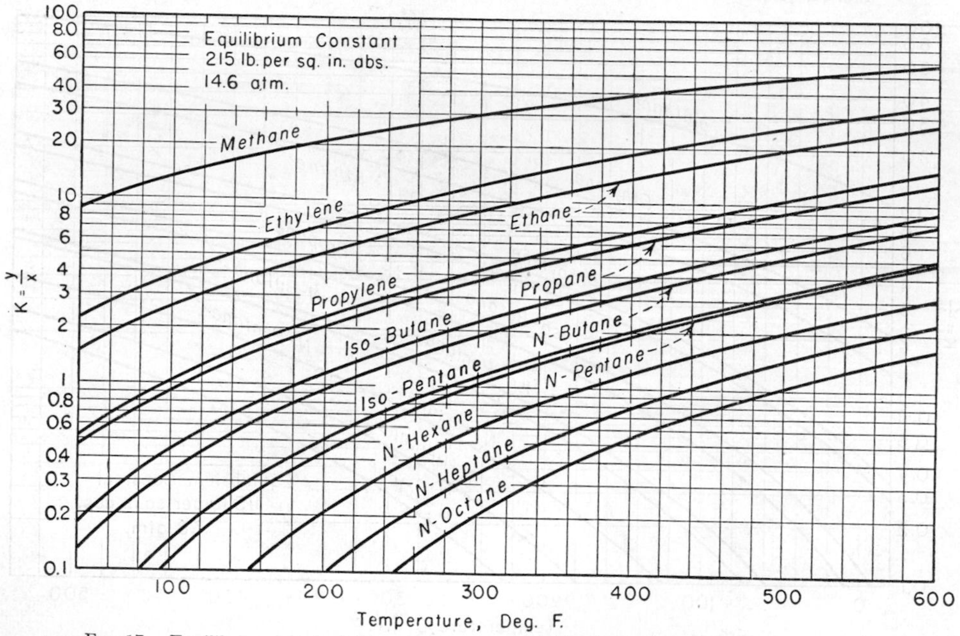

FIG. 17. Equilibrium constant K vs. temperature. 215 lb./sq. in. abs. or 14.6 atmospheres.

(Fig. 2). For mixtures of these hydrocarbons the volatilities of the components from a mixture are defined by the vapor pressure of the pure components. Inspection of the Cox chart shows that the relative differences in vapor pressures between the successive members of the paraffin series decrease with increase in temperature.

Rectification under high vacuum is suggested by these general considerations when mixtures consisting of components exhibiting small differences in volatility at ordinary temperatures must be separated.

Practical considerations dictate the use of pressure in certain distillation problems such as the separation of constituents that are in the gaseous state at ordinary temperatures and pressures. It is important, however, to recognize the limitations of pressure under which separation of given constituents must be accomplished in distillation processes. Cummings [*Ind. Eng. Chem.*, **23**, 900 (1931)] has investigated the effect of high pressure on the vapor-liquid equilibrium relations of certain binary mixtures in the region approaching and above

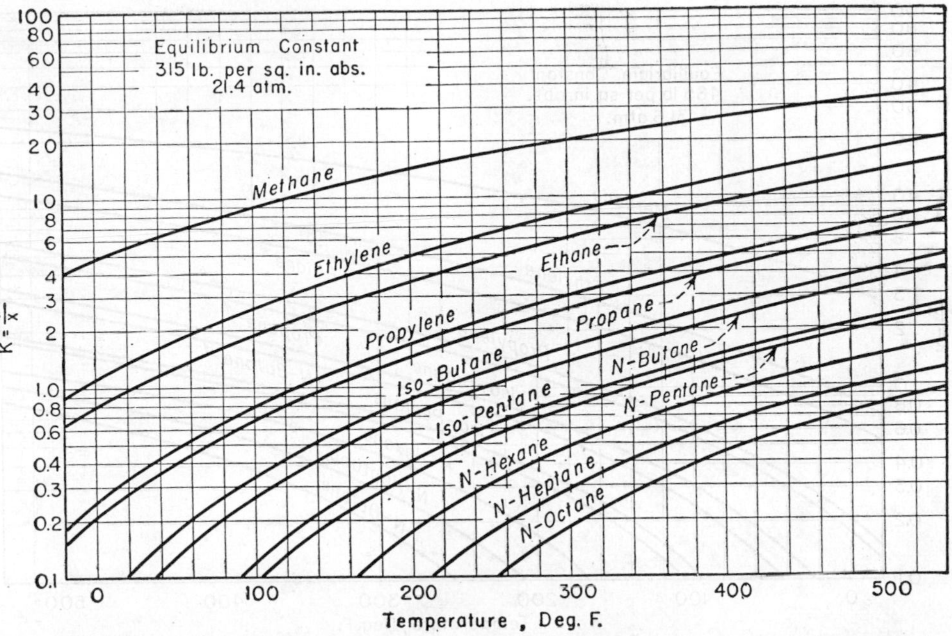

FIG. 18. Equilibrium constant K vs. temperature. 315 lb./sq. in. abs. or 21.4 atmospheres.

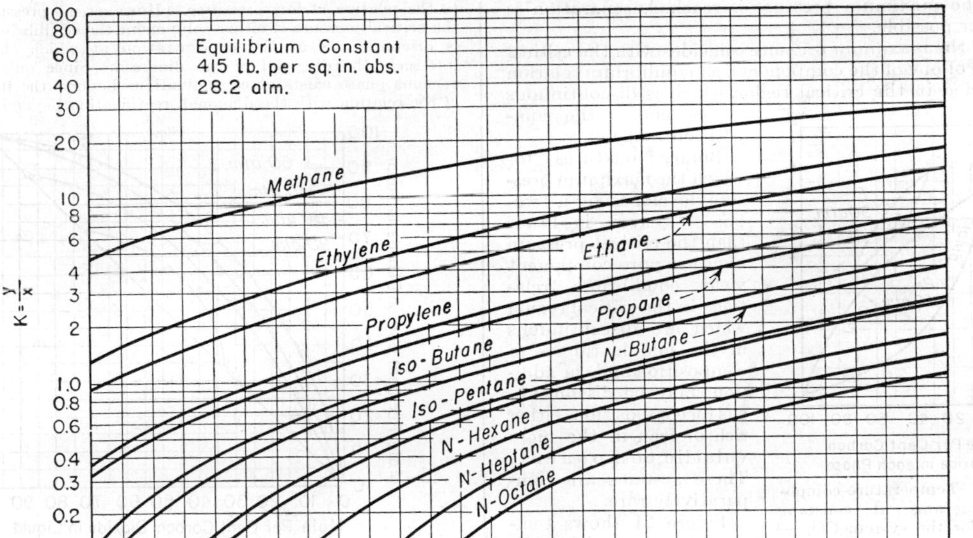

FIG. 19. Equilibrium constant K vs. temperature. 415 lb./sq. in. abs. or 28.2 atmospheres.

the critical points of the components. A brief summary of these results as applied to vapor-liquid equilibrium relations at constant pressure, together with quotations and illustrations from the original paper, will be given.

Cummings (*loc. cit.*) points out that a criterion of the behavior of systems under high-pressure conditions is the maximum pressure that a two-phase system of the binary mixture can exert.

Mixtures for which this maximum pressure is greater than the critical pressure of either pure component develop discontinuities in the vapor-liquid equilibrium relation at pressures above the critical pressure of either pure component. Between the critical pressure of either component and the maximum pressure, separation may be accomplished only in a limited range of compositions neither too poor nor too rich in the more volatile component, and this composition range becomes more restricted the closer the maximum pressure is approached.

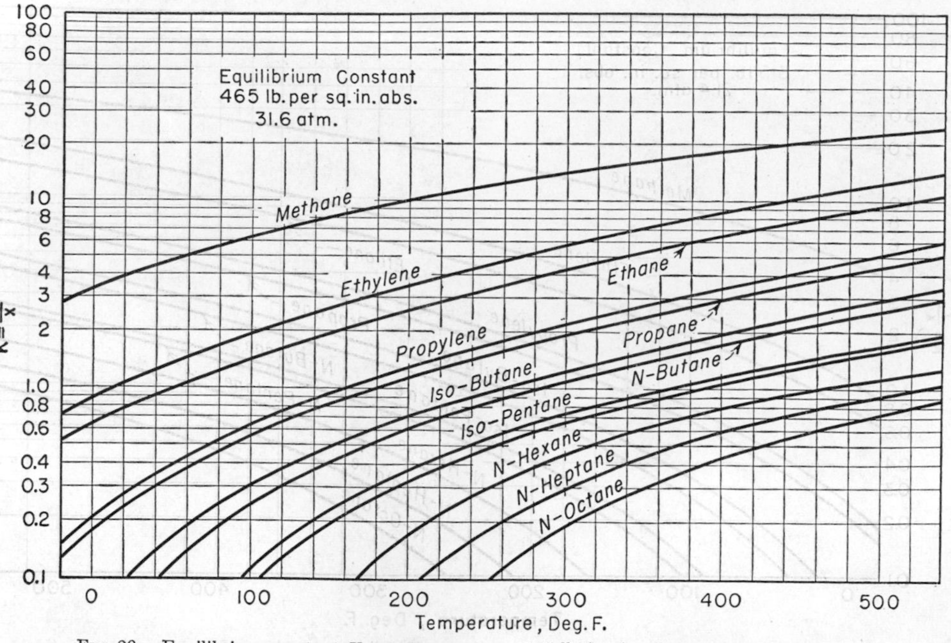

FIG. 20. Equilibrium constant K vs. temperature. 465 lb./sq. in. abs. or 31.6 atmospheres.

When the maximum pressure is reached, separation is no longer possible.

When the maximum pressure coincides with the critical pressure of one of the components, an equilibrium relation is obtained in the critical region which is discontinuous at one end of the composition range only.

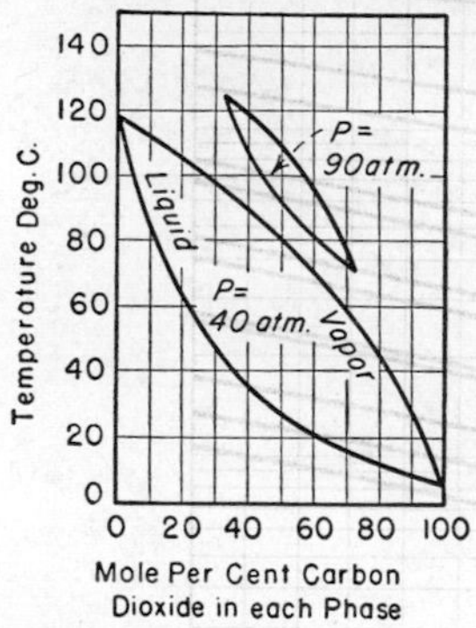

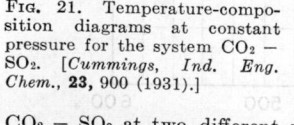

FIG. 21. Temperature-composition diagrams at constant pressure for the system $CO_2 - SO_2$. [*Cummings, Ind. Eng. Chem.*, **23**, 900 (1931).]

Binary mixtures for which the maximum pressure developed by a two-phase system is lower than the critical pressure of either pure component yield equilibrium relations in the critical region which are discontinuous in the middle ranges of composition and, in addition, may be discontinuous for compositions either rich or lean in the more volatile component. These systems are comparatively rare.

Figure 21 shows temperature-composition diagrams at constant pressure for the system $CO_2 - SO_2$ at two different pressures from Cummings' paper (*loc. cit.*). Forty atmospheres pressure is below the critical pressure of either component, and a continuous relationship is obtained. Ninety atmospheres is above the critical pressures of both CO_2 and SO_2 but below the maximum pressure of 95 atm. for this system.

Diagrams of vapor composition vs. liquid composition at constant pressure for the system $CO_2 - SO_2$ are given in Fig. 22. Figure 22 will be described by a direct quotation from Cummings as follows:

Below the critical pressure of both components the equilibrium relation is continuous through all concentrations, quite similar to the relation at low pressures. However, at pressures above the critical pressure of either component the equilibrium relation is discontinuous at the concentration at which the relation intersects the diagonal of the diagram. Since only a homogeneous phase exists at concentrations beyond the intersection of the relation with the diagonal, rectification beyond such con-

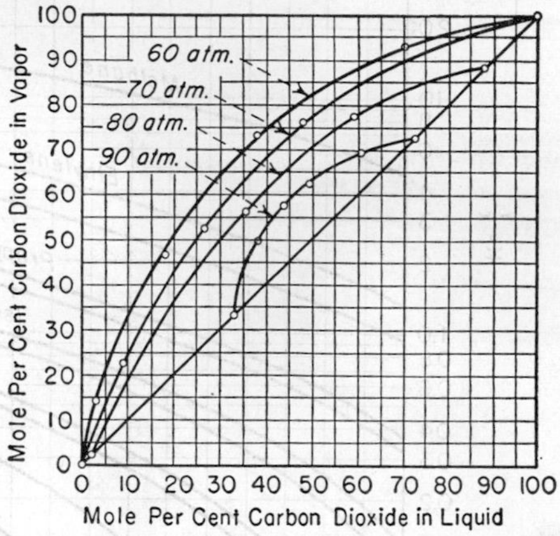

FIG. 22. Vapor-liquid equilibrium relations at constant total pressure for mixtures of CO_2 and SO_2. [*Cummings, Ind. Eng. Chem.*, **23**, 900 (1931).]

centrations cannot take place. At a total pressure of 74 atm. or below, it is possible to rectify all mixtures of carbon dioxide and sulphur dioxide, although the separation becomes more difficult at the higher pressures because the equilibrium relation approaches the diagonal as the pressure increases. Between 74 and 95 atm., rectification can be effected only within certain restricted concentrations. For example, at a total pressure of 90 atm. it is not possible to rectify mixtures containing less than 33 or more than 72.5 mole per cent of carbon dioxide. Above 95 atm. total pressure, which is the maximum pressure which the two-phase system exhibits, no rectification can take place.

Diagrams of vapor composition vs. liquid composition at constant pressure for oxygen-nitrogen mixtures are

Table 1. Constant-pressure Liquid-vapor Equilibrium Data for Binary Mixtures*

Component A	Component B	Mole % A in Liquid	Mole % A in Vapor	Temp., °C.	Total pressure, mm.	Ref.
Acetaldehyde	Water	0	0	100	760	15
		0.5	25	93.5		
		1	50	82.0		
		4	75	63.0		
		10	89	43.0		
		20	93	33.0		
		30	....	29.0		
		50	....	25.3		
		60	....	24.6		
		75	....	23.8		
Acetic acid	Benzene	0	0	80.2	760	14
		35.49	14.96	84.72		
		54.61	22.48	88.96		
		61.96	25.79	90.85		
		70.07	31.41	93.99		
		75.03	35.57	96.23		
		80.77	42.24	99.44		
		87.28	52.18	103.71		
		91.09	61.18	106.82		
		93.53	68.51	109.51		
		100	100	118.7		
Acetic acid	Water	100.0	100.0	118.1	760	2, 1, 8
(Molecular weight of		90.8	95.0	115.4		
acetic acid taken		83.3	90.0	113.8		
as 60.0)		69.8	80.0	110.1		
		57.5	70.0	107.5		
		47.0	60.0	105.8		
		37.4	50.0	104.4		
		28.4	40.0	103.2		
		20.5	30.0	102.1		
		13.6	20.0	101.3		
		7.0	10.0	100.6		
		3.7	5.0	100.3		
		0.0	0.0	100.0		
Acetone	Ethanol	0.0	0.0	78.3	760	4
		5.0	15.5	75.4		
		10.0	26.2	73.0		
		15.0	34.8	71.0		
		20.0	41.7	69.0		
		25.0	47.8	67.3		
		30.0	52.4	65.9		
		35.0	56.6	64.7		
		40.0	60.5	63.6		
		50.0	67.4	61.8		
		60.0	73.9	60.4		
		70.0	80.2	59.1		
		80.0	86.5	58.0		
		90.0	92.9	57.0		
		100.0	100.0	56.1		
Acetone	Methanol	0.0	0.0	64.5	760	1
		5.0	10.2	63.6		
		10.0	18.6	62.5		
		20.0	32.2	60.2		
		30.0	42.8	58.65		
		40.0	51.3	57.55		
		50.0	58.6	56.7		
		60.0	65.6	56.0		
		70.0	72.5	55.3		
		80.0	80.0	55.05		
		100.0	100.0	56.1		
Benzene	Ethanol	0	0	78.1	750	15
		6	20	74.4		
		11	30	72.4		
		20	40	70.1		
		39	50	68.3		
		57	56	67.8		
		72	60	68.3		
		89	70	70.8		
		96	85	75.2		
		100	100	79.7		
Benzene	Ethylene dichloride	0.0	0.0	83.48	760	10
		5.0	5.5	83.32		
		10.0	11.0	83.14		
		20.0	21.7	82.79		
		30.0	32.2	82.45		
		40.0	42.6	82.10		
		50.0	52.6	81.77		
		60.0	62.5	81.43		
		70.0	72.2	81.09		
		80.0	81.6	80.76		
		90.0	90.9	80.42		
		95.0	95.5	80.27		
		100.0	100.0	80.09		
Butanol (n-)	Water	0.1	1.9	99.4	760	19
		0.2	4.9	98.4		
		0.3	7.1	98.3		
		0.6	11.6	96.8		
		0.8	15.7	95.4		
		1.2	19.2	93.7		
		1.4	21.6	93.4		
		1.5	22.5	93.4		
		1.8	24.2	92.8		
		2.0	24.4	93.0		
		2.5	24.8	92.7		
		42.3	25.0	92.8		
		42.9	25.2	92.9		
		43.6	24.8	92.9		
		44.8	25.0	92.9		
		49.4	26.0	93.4		
		50.4	26.4	93.5		
		69.5	33.8	96.3		
		70.8	34.5	96.7		
		72.5	35.9	97.2		
		74.3	37.1	97.9		
		93.0	64.8	108.8		
		94.5	67.7	109.6		
		95.3	70.1	110.6		
		96.1	73.3	111.5		
Butanol (i-)	Water	0.2	4.3	98.9	760	19
		0.3	6.9	98.1		
		0.4	10.1	97.1		
		0.5	14.7	95.9		
		0.7	16.3	95.1		
		0.9	21.8	93.4		
		0.2	27.0	91.9		
		1.4	28.6	91.5		
		2.0	32.2	89.9		
		2.2	32.7	90.1		
		2.5	32.8	89.5		
		3.2	32.6	89.5		
		4.1	33.0	89.5		
		4.6	33.2	89.5		
		33.1	33.4	89.2		
		33.0	33.1	89.2		
		36.2	32.9	89.4		
		36.5	33.1	89.4		
		39.5	33.3	89.4		
		40.1	33.3	89.5		
		42.4	33.9	89.5		
		43.1	33.9	89.5		
		43.6	34.0	89.5		
		58.7	36.5	90.2		
		60.3	37.4	90.3		
		82.8	55.4	96.0		
		85.0	58.0	97.1		
		86.5	59.9	97.7		
Carbon tetrachloride	Benzene	0	0	80.0	760	15
		13.64	15.82	79.3		
		21.57	24.15	78.8		
		25.73	28.80	78.6		
		29.44	32.15	78.5		
		36.34	39.15	78.2		
		40.57	43.50	78.0		
		52.69	54.80	77.6		
		62.02	63.80	77.4		
		72.23	73.30	77.1		
Carbon tetrachloride	Ethyl acetate	0.0	0.0	74.1	685	9
		5.0	7.0	73.6		
		10.0	13.3	73.1		
		20.0	24.5	72.5		
		30.0	34.2	72.1		
		40.0	43.3	71.8		
		50.0	51.8	71.6		
		58.2	58.2	71.56		
		60.0	59.7	71.6		
		70.0	68.1	71.8		
		80.0	77.3	72.1		
		90.0	88.1	72.6		
		95.0	94.0	72.9		
		100.0	100.0	73.4		
Carbon tetrachloride	Toluene	0	0	110.4	762	20
		5.75	12.65			
		16.25	31.05			
		28.85	49.35			
		42.60	64.25			
		56.05	75.50			
		64.25	81.22			
		78.20	89.95			
		94.55	97.35	75.9		

* Arranged by H. C. Carlson and J. A. Lane, E. I. duPont de Nemours & Co., Wilmington, Del.

Table 1. Constant-pressure Liquid-vapor Equilibrium Data for Binary Mixtures*—(*Continued*)

Component A	Component B	Mole % A in Liquid	Mole % A in Vapor	Temp., °C.	Total pressure, mm.	Ref.
Carbon disulfide	Carbon tetrachloride	0	0	76.7	760	14
		2.96	8.23	74.9		
		6.15	15.55	73.1		
		11.06	26.60	70.3		
		14.35	33.25	68.6		
		25.85	49.50	63.8		
		39.08	63.40	59.3		
		53.18	74.70	55.3		
		66.30	82.90	52.3		
		75.74	87.80	50.4		
		86.04	93.20	48.5		
		100.0	100.0	46.3		
Carbon disulfide	Acetone	0	0	56.2	760	14
		1.90	8.32	54.0		
		4.76	18.50	51.4		
		13.40	35.10	46.6		
		18.58	44.30	44.0		
		29.12	52.75	41.4		
		37.98	57.40	40.3		
		44.77	59.80	39.8		
		53.60	62.70	39.3		
		65.30	66.10	39.1		
		78.94	70.50	39.3		
		80.23	72.30	39.6		
		87.99	76.00	40.5		
		96.83	88.60	43.5		
		100.0	100.0	46.3		
Chloroform	Acetone	0	0	56.2	760	14
		8.55	4.78	57.5		
		14.10	8.35	58.3		
		20.45	13.12	59.4		
		26.12	17.65	60.4		
		33.67	24.95	61.6		
		42.50	35.20	62.8		
		52.29	48.30	63.9		
		73.40	76.30	64.4		
		78.92	82.40	63.8		
		86.25	90.00	63.1		
		88.92	93.50	62.8		
		100.0	100.0	61.3		
Chloroform	Benzene	0	0	80.6	760	15
		8	10	79.8		
		15	20	79.0		
		22	30	78.2		
		29	40	77.3		
		36	50	76.4		
		44	60	75.3		
		54	70	74.0		
		66	80	71.9		
		79	90	68.9		
		100	100	61.4		
Chloroform	Methanol	0	0	64.9	757	15
		3.6	10.0	63.7		
		10.0	23.4	60.8		
		13.7	30.0	59.5		
		20.0	39.8	57.7		
		30.4	50.0	55.6		
		40.0	54.4	54.4		
		50.0	58.4	53.7		
		63.0	58.0	53.4		
		.....	68.0	53.6		
		71.0	......	53.7		
		100.0	100.0	61.4		
Ethanol	Water	0	0	100	760	16, 18
		1.90	17.00	95.5		
		7.21	38.91	89.0		
		9.66	43.75	86.7		
		12.38	47.04	85.3		
		16.61	50.89	84.1		
		23.37	54.45	82.7		
		26.08	55.80	82.3		
		32.73	58.26	81.5		
		39.65	61.22	80.7		
		50.79	65.64	79.8		
		51.98	65.99	79.7		
		57.32	68.41	79.3		
		67.63	73.85	78.74		
		74.72	78.15	78.41		
		89.43	89.43	78.15		
Ethyl acetate	Ethanol	0.0	0.0	78.3	760	5, 11
		5.0	10.2	76.6		
		10.0	18.7	75.5		
		20.0	30.5	73.9		
Ethyl acetate (*con't.*)	Ethanol	30.0	38.9	72.8		
		40.0	45.7	72.1		
		50.0	51.6	71.8		
		54.0	54.0	71.8		
		60.0	57.6	71.9		
		70.0	64.4	72.2		
		80.0	72.6	73.0		
		90.0	83.7	74.7		
		95.0	91.4	76.0		
		100.0	100.0	77.1		
Ethylene glycol	Water	100.0	100.0	160.6	228	12
		69.0	99.0	152.4		
		55.2	98.0	148.3		
		46.5	97.0	145.1		
		40.0	96.0	142.1		
		35.0	95.0	139.5		
		25.5	92.0	132.0		
		21.4	90.0	127.5		
		10.0	80.0	111.2		
		5.0	70.0	100.5		
		2.8	60.0	93.1		
		1.8	50.0	87.7		
		1.1	40.0	82.9		
		0.7	30.0	78.8		
		0.4	20.0	75.6		
		0.2	10.0	72.8		
		0.0	0.0	69.5		
Furfural	Water	0	0	100	760	15
		1	5.5	98.56		
		2	8.0	98.07		
		4	9.2	97.90		
		9.2	9.2	97.90		
		50	9.2	97.90		
		70	9.5	98.7		
		80	11	100.6		
		90	19	109.5		
		92	32	122.5		
		94	64	146.0		
		96	81	154.8		
		98	90	158.8		
		100	100	161.7		
Methanol	Water	0.0	0.0	100.0	760	2, 13, 3
		2.0	13.4	96.4		
		4.0	23.0	93.5		
		6.0	30.4	91.2		
		8.0	36.5	89.3		
		10.0	41.8	87.7		
		15.0	51.7	84.4		
		20.0	57.9	81.7		
		30.0	66.5	78.0		
		40.0	72.9	75.3		
		50.0	77.9	73.1		
		60.0	82.5	71.2		
		70.0	87.0	69.3		
		80.0	91.5	67.6		
		90.0	95.8	66.0		
		95.0	97.9	65.0		
		100.0	100.0	64.5		
Nitric acid	Water	8.36	0.627	106.5	760	15
		12.3	1.76	112.0		
		22.1	6.60	118.5		
		30.8	16.6	121.6		
		38.3	38.3	121.9		
		40.2	60.2	121.0		
		46.5	75.9	118.0		
		53.0	89.1	112.0		
		61.5	92.1	99.0		
Nitrogen	Oxygen	3.85	13.97	....	760	17
		8.02	26.10			
		12.40	36.60			
		17.05	46.00			
		22.20	54.20			
		27.73	61.60			
		33.8	67.95			
		40.47	73.74			
		47.83	78.95			
		56.62	84.35			
		66.65	88.95			
		78.40	93.50			
		91.90	97.70			
Nitrogen	Oxygen	4	9.0	....	3800	17
		12	27.0			
		22	42.0			
		33	56.0			

Table 1. Constant-pressure Liquid-vapor Equilibrium Data for Binary Mixtures*—(*Concluded*)

Component A	Component B	Mole % A in Liquid	Mole % A in Vapor	Temp., °C.	Total pressure, mm.	Ref.
Nitrogen (*cont.*)......	Oxygen	46	69.0			
		61	80.5			
		69	85.5			
		79	91.0			
		90	96.0			
Isopropyl ether.......	Isopropanol	0.0	0.0	82.3	760	7
		5.0	18.7	77.8		
		10.0	30.6	75.4		
		15.0	39.6	73.5		
		20.0	46.6	71.8		
		25.0	51.9	70.6		
		30.0	55.6	69.6		
		35.0	58.7	68.9		
		40.0	61.5	68.3		
		50.0	66.2	67.3		
		60.0	70.2	66.6		
		70.0	74.3	66.3		
		78.2	78.2	66.2		
		80.0	79.1	66.2		
		85.0	81.9	66.3		
		90.0	85.3	66.6		
		95.0	90.2	67.0		
		100.0	100.0	68.5		
Isopropanol..........	Water	0	0	100	760	15
		1	19	95.0		
		2	34	90.0		
Isopropanol (*cont.*)...	Water	3	43	86.7		
		6	50.5	83.5		
		15	56	81.5		
		30	58	81.0		
		50	63	80.7		
		70	70	80.5		
		80	77	81.0		
		90	83	82.3		
n-Propanol..........	Water	0.0	0.0	100.0	760	6, 3
		1.0	11.0	95.0		
		2.0	21.6	92.0		
		4.0	32.0	90.5		
		6.0	35.1	89.3		
		10.0	37.2	88.5		
		20.0	39.2	88.1		
		30.0	40.4	87.9		
		40.0	42.4	87.8		
		43.2	43.2	87.8		
		50.0	45.2	87.9		
		60.0	49.2	88.3		
		70.0	55.1	89.0		
		80.0	64.1	90.5		
		85.0	70.4	91.5		
		90.0	77.8	92.8		
		96.0	90.0	95.0		
		100.0	100.0	97.3		

1 Bergstrom, data from "Principles and Practice of Industrial Distillation" by Hausbrand, trans. by E. H. Tripp, Wiley, New York, 1926 (compositions below 26 mole per cent water and boiling points).
2 Cornell and Montonna, *Ind. Eng. Chem.*, **25**, 1331 (1933) (compositions of acetic acid-water above 26 mole per cent water. Compositions of methanol-water, which agree with Uchida and Kato's).
3 Doroszewsky and Polansky, *Z. physik. Chem.*, **73**, 192 (1910) (boiling points of aqueous mixtures).
4 Duffey, private communication to T. H. Chilton, 1935.
5 Furnas and Leighton, *Ind. Eng. Chem.*, **29**, 709 (1937).
6 Gadwa, Sc. D. Thesis in Chemical Engineering, Mass. Inst. Tech., 1936 (compositions of n-propyl alcohol-water).
7 Miller and Bliss, *Ind. Eng. Chem.*, **32**, 123 (1940).
8 Povarnina and Markova, *J. Russ. Phys.-Chem. Soc.*, **55**, 381 (1924) (boiling points of mixtures).
9 Schutz, *J. Am. Chem. Soc.*, **61**, 2691 (1939).
10 Smith and Matheson, *Bur. Standards J. Research*, **20**, 641 (1938) (data calculated from their vapor pressures using Raoult's law).
11 Stockhardt, private communication to T. H. Chilton, 1931.
12 Trimble and Potts *Ind. Eng. Chem.*, **27**, 66 (1935).
13 Uchida and Kato, *J. Soc. Chem. Ind., Japan*, **37**, 525 (1934) (compositions of mixtures).
14 Rosanoff and Easeley, *J. Am. Chem. Soc.*, **31**, 979 (1914).
15 "International Critical Tables," McGraw-Hill, New York.
16 Carey and Lewis, *Ind. Eng. Chem.*, **24**, 882 (1932).
17 Dodge and Dunbar, *J. Am. Chem. Soc.*, **44**, 608 (1927).
18 Noyes and Warfle, *J. Am. Chem. Soc.*, **23**, 463 (1901).
19 Stockhardt and Hull, *Ind. Eng. Chem.*, **23**, 1438 (1931).
20 Carey, Sc. D. Thesis, Mass. Inst. Tech., 1930.

given in Fig. 23 (also taken from Cummings). In the case of oxygen and nitrogen, the maximum pressure of the two-phase system and the critical pressure of the less volatile component, oxygen, are identical. In the

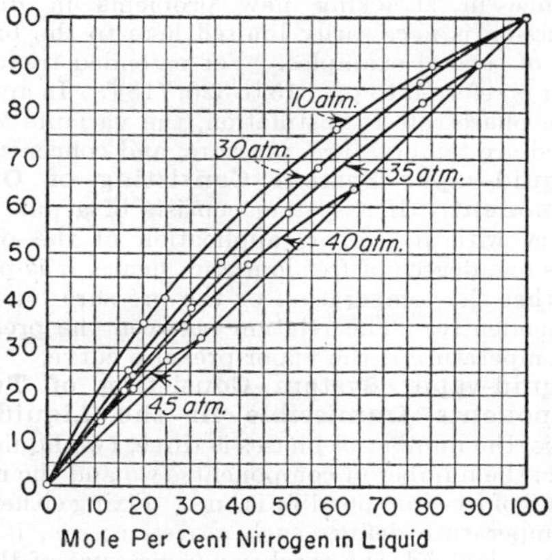

FIG. 23. Vapor-liquid equilibrium relations at constant total pressure for mixtures of oxygen and nitrogen. [*Cummings, Ind. Eng. Chem.*, **23**, 900 (1931).]

region between the critical pressure of nitrogen and the maximum pressure, an equilibrium relation is obtained which is discontinuous for mixtures rich in nitrogen only. In other respects, behavior is similar to that described for $CO_2 - SO_2$.

It is important to note that Cummings states that available data on mixtures of ethane and butane indicate that mixtures of normal paraffin hydrocarbons have a mixture pressure higher than the critical pressure of either component, and that their equilibrium relations in the critical region probably present limitations similar to those of the carbon dioxide-sulphur dioxide system.

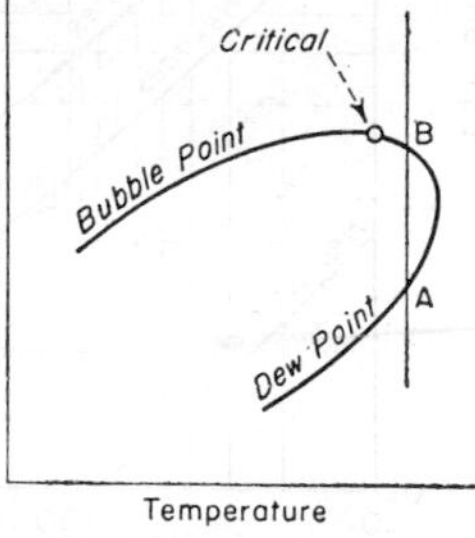

FIG. 24. Retrograde condensation in multicomponent hydrocarbon system.

Retrograde Phenomena. High-pressure oil and gas reservoirs are being increasingly exploited. In the various problems connected therewith, retrograde phenomena are encountered in these hydrocarbon systems, from the reservoir itself through subsequent processing. Retrograde condensation or retrograde vaporization occurs when one independant variable relating to the system as a whole is continuously changed in one direction while other variables sufficient to fix the path of the system are held constant. During this

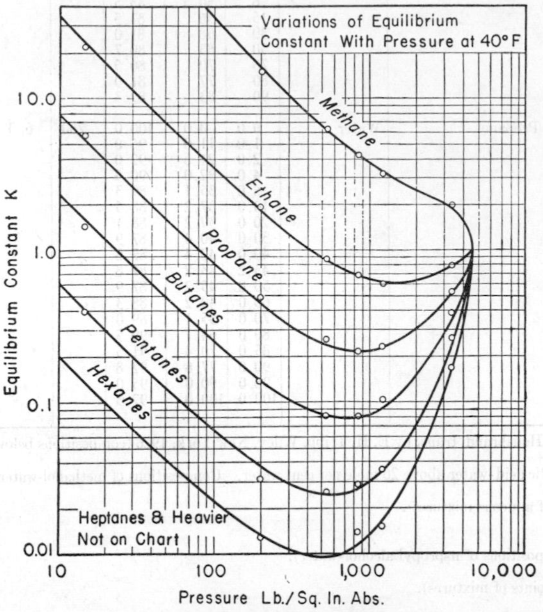

FIG. 25. Variations of equilibrium constant with pressure at 40°F.

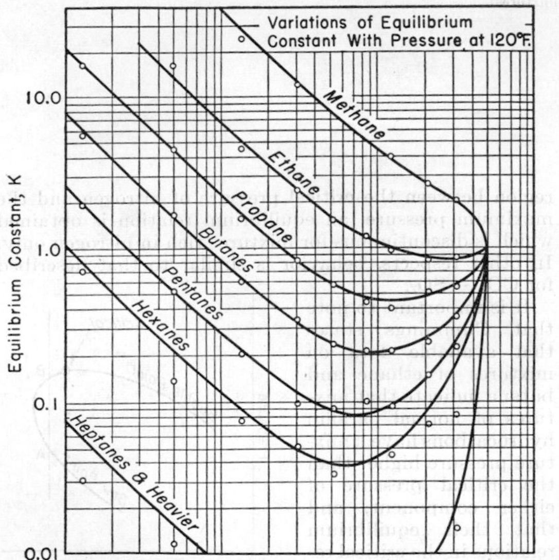

FIG. 26. Variations of equilibrium constant with pressure at 120°F.

continuous change in one variable, a phase appears, increases in quantity to a maximum, and then decreases to disappearance. Retrograde phenomena occur upon paths that cross either a dew-point curve or a bubble-point curve in two places because of the presence of a maximum or minimum in that curve with respect to the variable that is being held constant. The presence of such maxima or minima is usually found close to the critical state of the system.

In Fig. 24, which represents a multicomponent hydrocarbon system, the path from B to A represents isothermal retrograde condensation upon decrease in pressure.

Figures 25, 26, and 27 give equilibrium constants determined in admixture with crude oil at three respective temperatures [Katz and Hachmuth, *Ind. Eng. Chem.*, **29**, 1072 (1937)]. The minima in these curves as pressure is increased and their final convergence at the critical illustrate retrograde phenomena and furnish data for equilibrium calculations in the high-pressure region.

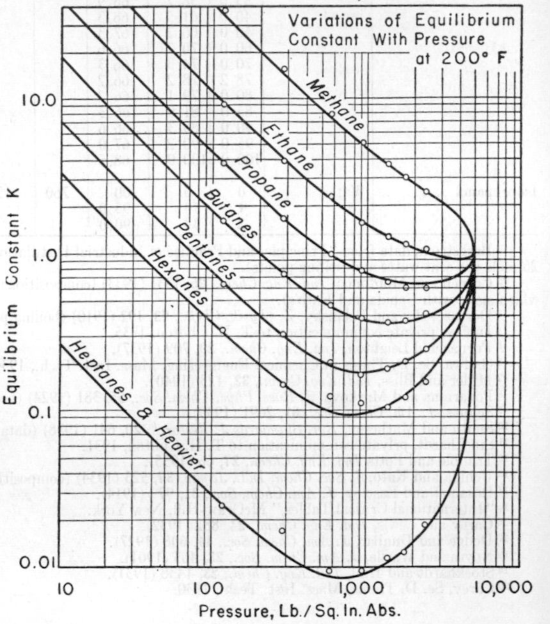

FIG. 27. Variations of equilibrium constant with pressure at 200°F.

USE OF PHASE RULE IN DISTILLATION

Application of the phase rule is frequently of convenience in attacking new problems in distillation. Discussion is necessarily limited here to the brief statement of several examples. For extended treatment the reader is referred to pp. 315 to 320, 718*ff*. In applications of the phase rule to distillation, the variants to be considered are temperature, pressure, and concentration.

Liquid-vapor System Consisting of One Pure Component. If a system consists of a pure liquid in contact with its vapor, application of the phase rule shows one degree of freedom and, hence, it is possible to fix either the temperature or the pressure but not both independently. The relation between the pressure and the temperature is the vapor-pressure curve.

Liquid-vapor System Consisting of Two Pure Components Immiscible in the Liquid Phase. In this, the number of phases is three, two liquid and one vapor; the number of components two and the number of degrees of freedom possible is one. Fixing either pressure or temperature defines such a system; *e.g.*, if the temperature is fixed, the total vapor pressure of the system is fixed as well as the vapor pressures of the constituents.

Liquid-vapor Systems Consisting of Two Completely Miscible Components. Here there are one liquid and one vapor phase while the number of components is two. The degrees of freedom are two. If two of the variants are fixed, the third is thereby fixed, *e.g.*, if the temperature and the molar composition of one component are fixed, the pressure is fixed.

Liquid-vapor Systems Consisting of Three Completely Miscible Components. With two phases and three components, the degrees of freedom are three. However, if the mole fractions of two of the components are stated, fixing either the pressure or the temperature will reduce the variance to zero and fix the system. In general, if there are n miscible components in a mixture, $n - 1$ relative compositions must be stated. If this be done, the treatment of all such systems becomes similar.

Liquid-vapor Systems Consisting of Three Miscible Components and One Non-miscible One. Two cases can be considered here: (1) the non-miscible component is present only in the vapor phase, and (2) the non-miscible component is present in both liquid and vapor phases.

In Case 1, there are four components and two phases which give four degrees of freedom. If the molar ratio of the non-miscible component and the molar ratios of two of the three miscible ones in the vapor phase are fixed, there remains one variant; hence fixing either temperature or pressure will fix the system. Thus, if steam is present in the vapor phase at such a partial pressure that it is superheated at the given temperature, the effect of the steam is to lower the vapor pressure the liquid mixture must exert to vaporize. The steam can be present in any ratio up to a partial pressure corresponding to saturation; the greater the molar ratio of the steam, the greater the lowering of the vapor pressure of the mixture.

In Case 2, if the non-miscible component is present in the liquid phase, the number of phases is three and the degrees of freedom three. Thus, if the molar ratios of two of the miscible components are fixed and one other variant, say the temperature, fixed, the variants are reduced to zero and the system is fixed. Thus, if water and three components mutually miscible but not miscible with water are in equilibrium with a vapor phase, fixing, say, the temperature and the molar concentration of two of the miscible components will fix the pressure and the molar ratio of water to the miscible components in the vapor evolved. These cases are of importance in considering steam-distillation problems.

Phase Diagrams for Binary Mixtures

The phase-rule variants in distillation problems are pressure, temperature, and composition. To present graphically all three variants would require a three-dimensional, or solid, diagram. If one of the variants is fixed, the relation between the remaining two may, for a given mixture, be presented on a plane diagram. Several types of such diagrams prove useful:

1. Partial-pressure-composition Diagrams at Constant Temperature

a. Completely Soluble Components. If both liquid and vapor are present, the number of phases is two and the number of components two, hence the degrees of freedom are two. If the temperature is fixed, one can plot the partial pressure of each component and the total vapor pressure against the composition. If the binary mixture follows Raoult's law in all concentrations, fixing the temperature fixes the vapor pressure of each component, and the partial vapor pressures are linear with the composition *e.g.*, for benzene and toluene at a temperature of 100°C.:

$$P_B = 1343 \text{ mm.} \qquad P_T = 560 \text{ mm.}$$

and

$$p_B = 1343x_B \qquad p_T = 560x_T = 560(1 - x_B)$$

This type of diagram is shown in Fig. 28 in which p_B represents the partial pressures of benzene, p_T the partial pressures of toluene, and P the total vapor pressure of the mixture.

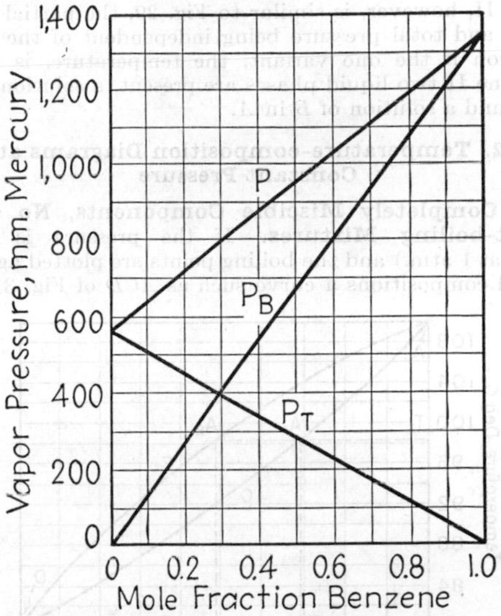

Fig. 28. Partial-pressure composition diagram at constant temperature (100°C.) for benzene-toluene mixtures.

If the partial pressures deviate from Raoult's law, the partial pressures and the total pressure will in general be represented by curves. These cases are more conveniently represented by other methods to be given.

b. Completely Insoluble Components. For this limiting case two liquid phases in addition to the vapor phase will be present; hence the system will be univariant. If the temperature is fixed, the system is fixed as is shown by Fig. 29.

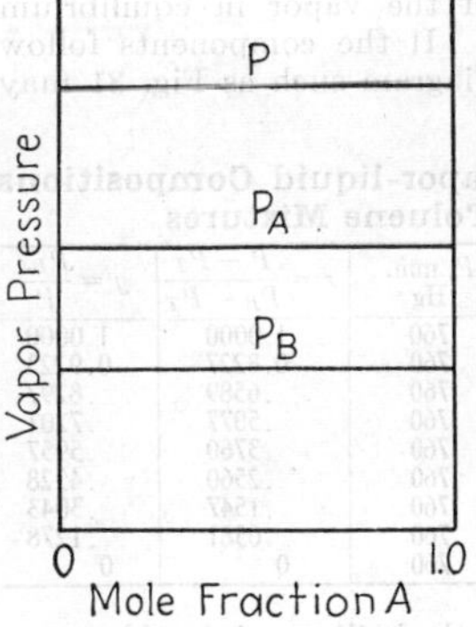

Fig. 29. Partial-pressure composition diagram at constant temperature for mixture of completely insoluble components A and B.

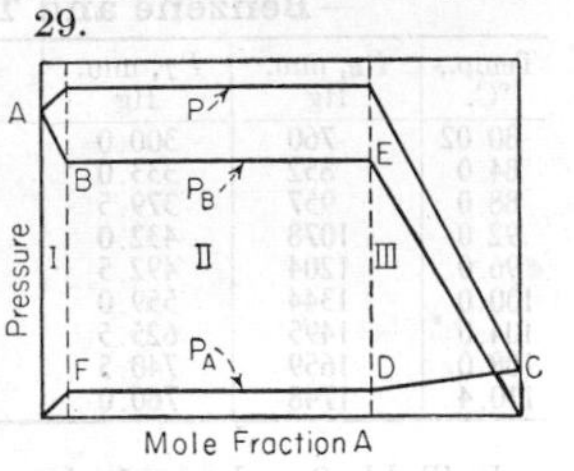

Fig. 30. Partial-pressure composition diagram at constant temperature for partially soluble components.

Figure 29 indicates that as long as both liquids are present their partial pressures and hence their molar vapor concentrations are fixed by fixing the temperature.

c. Partly Soluble Components. If a liquid A is soluble up to a certain concentration in a liquid B, liquid B will usually be found to be appreciably soluble in A. The extent of the mutual solubility is influenced by the temperature, and the partly soluble components frequently become completely miscible at higher temperatures.

Examining Fig. 30, three zones separated by the vertical dotted lines are marked off. In zones I and III, which cover the two solubility ranges, the components behave similarly to completely miscible liquids, *i.e.*, one liquid phase and one vapor phase are present, the system is bivariant and the partial pressures and total pressure vary with the composition at a fixed temperature. Zone II, however, is similar to Fig. 29, the partial pressures and total pressure being independent of the composition if the one variant, the temperature, is fixed. In zone II two liquid phases are present, a solution of A in B and a solution of B in A.

2. Temperature-composition Diagrams at Constant Pressure

a. Completely Miscible Components, No Constant-boiling Mixtures. If the pressure is fixed (*e.g.*, at 1 atm.) and the boiling points are plotted against liquid compositions a curve such as *ACD* of Fig. 31 will

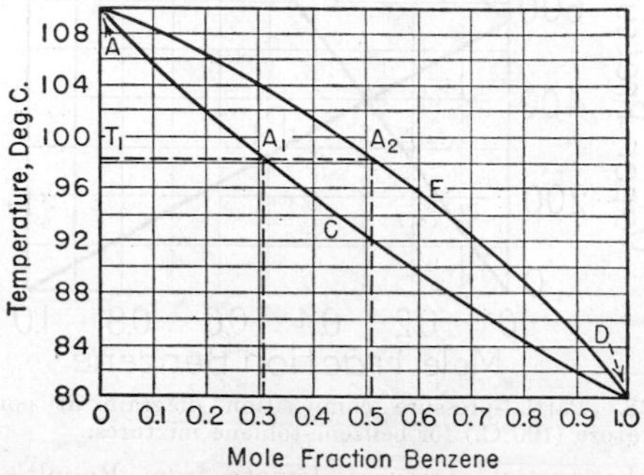

FIG. 31. Temperature-composition diagram at constant pressure (P = 760 mm.) for benzene-toluene mixtures.

be had. Vapor compositions corresponding to equilibrium with the liquid are plotted along *AED*. Thus in Fig. 31 a liquid of composition A_1 will boil at temperature T_1, and the composition of the vapor in equilibrium with the liquid will be A_2. If the components follow Raoult's law, points for a diagram such as Fig. 31 may be calculated as in Table 2.

Table 2. Equilibrium Vapor-liquid Compositions—Benzene and Toluene Mixtures

Temp., °C.	P_B, mm. Hg	P_T, mm. Hg	P, mm. Hg	$x = \frac{P - P_T}{P_B - P_T}$	$y = \frac{P_B x}{P}$
80.02	760	300.0	760	1.0000	1.0000
84.0	852	333.0	760	0.8227	0.9223
88.0	957	379.5	760	.6589	.8297
92.0	1078	432.0	760	.5077	.7201
96.0	1204	492.5	760	.3760	.5957
100.0	1344	559.0	760	.2560	.4528
104.0	1495	625.5	760	.1547	.3043
108.0	1659	740.5	760	.0581	.1278
110.4	1748	760.0	760	0	0

In Table 2, column 1 gives the boiling points of benzene and toluene mixtures at 760 mm., from that of pure benzene (80.02°C.) to pure toluene (110.4°C.). Column 2 gives the vapor pressures of pure benzene (P_B) corresponding to the temperatures of column 1, and column 3 gives the vapor pressures of pure toluene (P_T). Column 4 gives the total pressure of the vapor, 760 mm. in this illustration. Column 5 gives the calculated liquid compositions, in terms of mole fractions of benzene, boiling at the respective temperatures shown in column 1. Based on Raoult's and Dalton's laws (see pp. 525 and 290), a sample calculation of column 5, for 92.0°C. is

$$x = \frac{P - P_T}{P_B - P_T} = \frac{760 - 432}{1078 - 432} = 0.5077$$

Column 6 gives the vapor compositions, as mole fractions of benzene, in equilibrium with the liquid compositions of column 5. Column 6 is calculated from Raoult's law and, for 92.0°C. temperature, a sample calculation is

$$y = \frac{P_B x}{P} = \frac{(1078)(0.5077)}{760} = 0.7201$$

If the components do not follow Raoult's law the equilibrium vapor-liquid compositions and the boiling points must be obtained from experimental data.

b. Completely Miscible Components, Constant-boiling Mixtures Formed. In Fig. 31 the temperature-composition diagram is given for a mixture showing a variation in boiling point from pure B to pure A along the curve *ACD*. In certain cases minimum or maximum boiling points may be encountered in the complete composition range as shown by Fig. 32. In curve I of Fig. 32

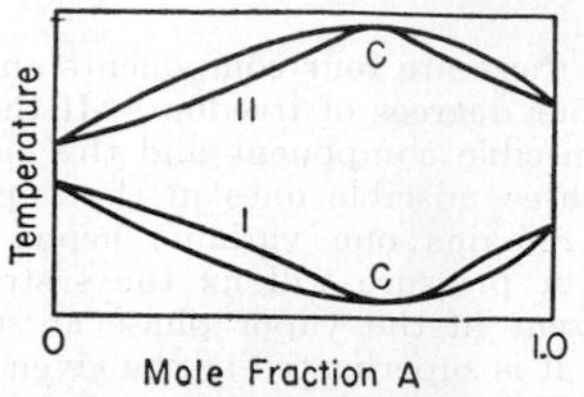

FIG. 32. Temperature-composition diagrams for liquids forming constant boiling binary mixtures.

in the region to the left of point C, behavior on vaporization is similar to the material of Fig. 31, the vapor given off being richer in component A than the liquid. At point C, the compositions of liquid and vapor are identical. To the right of point C, the volatility is reversed, *i.e.*, the vapors will be richer in B than the liquid. A mixture of the type represented by curve *I* is ethanol and water, which forms a minimum-boiling-point, constant-boiling mixture of approximately 95 weight per cent alcohol. Curve II represents a mixture with which a maximum boiling point is had at point C. To the left of point C vapors richer in component B than the liquid are evolved, while to the right of C vapors richer in component A are evolved. Representative of this type are mixtures of HCl and water, the well-known behavior of which is that, if a mixture of any composition is boiled at atmospheric pressure, a residual liquid containing approximately 20 per cent HCl is finally obtained. Materials that form either maximum- or minimum-boiling mixtures cannot be completely separated by a single distillation operation at a given pressure. However, the composition of the constant-boiling mixture may be altered by conducting the distillation at a different pressure, thus shifting the points C in Fig. 32. In this manner the constant-boiling mixture may be obtained as one of the products in one operation. A second vaporization at a different pressure will enable further separation of the original constant-boiling mixture. Another expedient resorted to is the addition of a third component to the constant-boiling mixture, which may so alter the volatility characteristics of the original mixture that separations are possible.

Acetone and methanol form a constant-boiling mixture and in addition have boiling points differing by but 9°C., at atmospheric pressure. A saturated aqueous solution of sodium thiosulphate added to a methanol-acetone mixture lowers the partial vapor pressure of the methanol from the mixture, so that the acetone is readily fractionated as a distillate from the methanol-sodium thiosulphate-water mixture.

c. Partly Miscible Components. A typical temperature-composition diagram for constituents of partial solubility is given in Fig. 33.

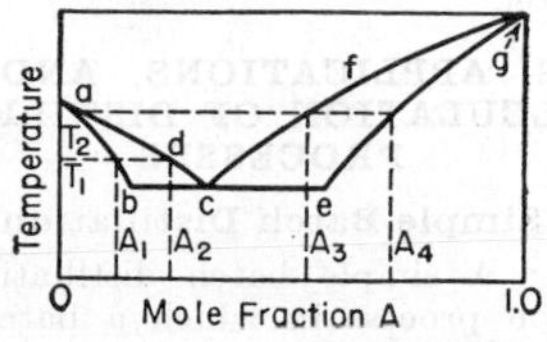

FIG. 33. Temperature-composition diagram for partially soluble binary mixture.

To the left of point c a liquid of composition A_1 has a boiling point of T_1, while the vapor evolved has a composition A_2. Similarly, to the right of point c, a liquid of composition A_4 will boil at T_2, the equilibrium vapor being represented by A_3. In the composition range b to e, however, vapors of the constant composition given by point c will be evolved at a fixed temperature.

3. Diagrams of Vapor Composition vs. Liquid Composition at Constant Pressure

Diagrams of this type are more useful in certain distillation calculations than the preceding types. In these diagrams the pressure is fixed and the concentration of the more volatile component in the vapor phase is plotted against the equilibrium concentration of the same component in the liquid phase. A separate boiling-point curve must be plotted if the boiling temperatures of the mixtures are wanted. The data calculated in Table 2 may be used to plot such a diagram for benzene-toluene. The several kinds of mixtures considered in connection with the previous diagrams are shown in Fig. 34.

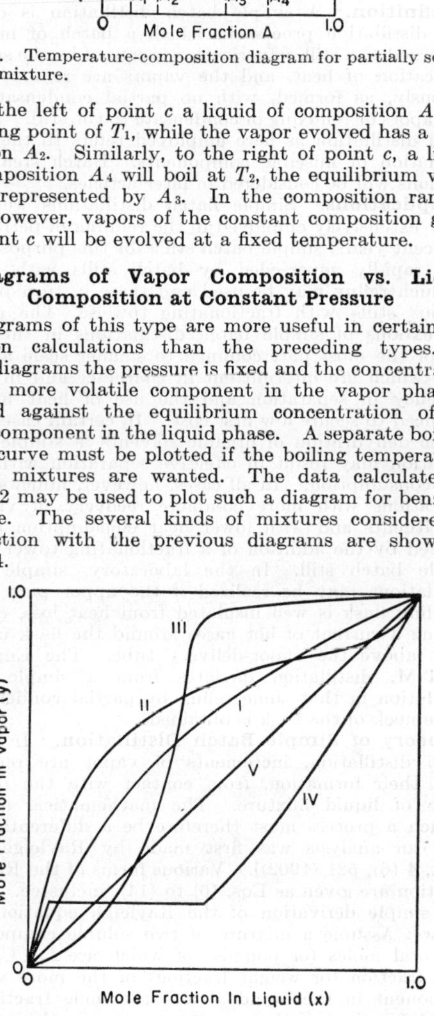

FIG. 34. Typical y vs. x diagrams at constant pressure.

I. Completely miscible, no constant-boiling mixture; *e.g.*, benzene-toluene.

II. Minimum-boiling-point, constant-boiling mixture; *e.g.*, ethanol-water.

III. Maximum-boiling-point, constant-boiling mixture; *e.g.*, HCl-water.

IV. Partly miscible; *e.g.*, aniline-water.

V. Equal compositions in vapor and liquid; *e.g.*, mixtures of stereoisomers.

In y vs. x diagrams the 45-deg. line of equality in vapor and liquid composition which also represents the yx curve for stereoisomers is usually included for reference. At the points where the curves cross this 45-deg. line, no enrichment of the vapor is obtained. It should be recalled that the points at which constant-boiling mixtures are formed may be shifted by changing the pressure of the distillation. The general effects of pressure changes on vapor-liquid equilibrium relations and the effects of pressures approaching the critical pressures of the components are discussed on p. 569.

Three-component Systems

For a three-component system a constant-temperature diagram may be constructed by representing the compositions of the three components along the sides of a triangle. If temperature is then represented by a perpendicular axis, contour lines may be plotted to show the various compositions at various temperatures.

Systems of three or more components become difficult to represent on diagrams. Fortunately, the systems containing the greatest numbers of components which are distilled on a large scale, *i.e.*, mixtures of petroleum hydrocarbons, approximate Raoult's law in their liquid-vapor equilibrium relations, or equilibrium data are available (see pp. 568 to 572). For other complex mixtures for which no data are available, experimental equilibrium vaporizations (see p. 585) must be made to determine the liquid-vapor equilibrium relationships.

VOLATILITY AND RELATIVE VOLATILITY

In general the term "volatility" is used to compare the vapor pressure of one pure substance with another, the substance having the higher vapor pressure at a given temperature being termed the more volatile. Vapor pressure alone does not define the ease of separation of the components from liquid mixtures as the vapor pressure of each is lowered by the presence of the others.

For a solution of components A, B, etc.,

$$v_A = p\frac{y_A}{x_A} \qquad v_B = p\frac{y_B}{x_B}$$

in which v_A and v_B are the activities of A and B, respectively. Relative volatility is defined as

$$\alpha = \frac{v_A}{v_B} = \frac{(y_A)(x_B)}{(y_B)(x_A)} \tag{3}$$

The activity coefficient γ is v/p for each component; hence

$$py_A = \gamma_A p_A x_A$$
$$py_B = \gamma_B p_B x_B$$

and

$$\alpha = \frac{(y_A)(x_B)}{(y_B)(x_A)} = \frac{\gamma_A p_A}{\gamma_B p_B} = \text{relative activity} = \frac{K_A}{K_B} \tag{4}$$

K_A and K_B are the equilibrium constants for A and B, respectively (see pp. 525*ff*).

If $\gamma_A = \gamma_B$,

$$\alpha = \frac{p_A}{p_B}$$

Relative volatility is a direct measure of the ease of separation of components by a distillation process; hence substances that are readily separated show large values of α. An α of unity means no separation is possible, and an $\alpha < 1$ means that the molar ratio of the constituents in the vapor will be α times their molar ratio in the liquid. Where a difference in volatilities exists the value of α will always be greater than 1, if the volatility of the more volatile be made the numerator in Eq. (4). Thus a relative volatility of 0.5 becomes $1/0.5 = 2.0$, if the ratio of volatility of the more volatile to the less volatile be used. The relative volatility of two components changes with the temperature. Viewed from the standpoint of ease of separation alone, the optimum temperature range in which to conduct a distillation is that in which the value of α is a maximum. The following relationship derived from Eq. (4) gives for a binary, the relationship between y and x for constant α

$$y_A = \frac{\alpha x_A}{1 + x_A(\alpha - 1)} \tag{5}$$

Figure 35 gives a plot of y vs. x for various constant values of α.

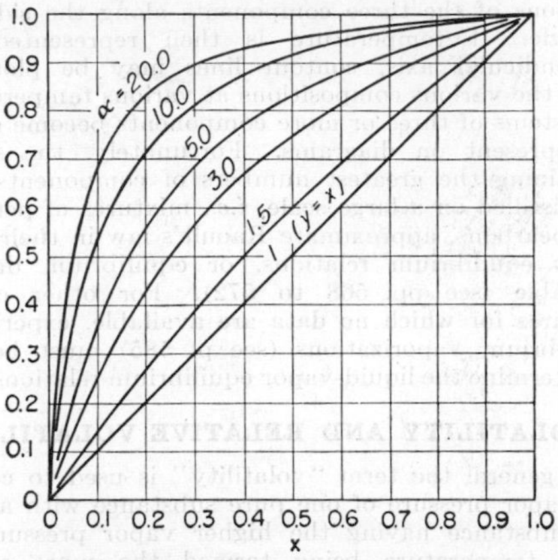

FIG. 35. Relationships of y vs. x for varying values α.

METHODS OF VAPORIZATION

A fundamental factor governing the method of attack of distillation calculations is the method of vapor formation. With the mechanism of vapor formation established, given sufficient data on the liquid-vapor equilibriums, the remainder of the calculations, from the distillation standpoint, may be nearly entirely handled by heat and material balances. In the design of complete distillation units, other problems must be solved in addition to those involving the separations of the materials, *e.g.*, adequate vapor and liquid passages must be calculated, the necessary heat-transfer surfaces must be designed, etc. Except where special considerations are involved, attention will be focused upon the distillation calculations in the following articles. References should be made to Secs. 5 and 6 (pp. 359 to 412 and 370 to 498), respectively, for information on the Flow of Fluids and the Heat Transmission.

Based upon the method of vapor formation, distillation operations may be classed as batch or continuous. In the batch method a given weight of material is charged into a suitable kettle or still and a portion of the charge distilled over. The vapors are continuously removed as formed. The more volatile components are found in greater concentration in the vapor than in the liquid; therefore the liquid grows poorer in the more volatile components as vaporization proceeds. Hence both the composition of liquid and the composition of the vapor evolved are changing during a batch process.

In a continuous-distillation process, the charge is pumped into the still or heater and vapor and liquid streams withdrawn at such rates that there is no accumulation or depletion of material in the system. The composition of the liquid in the still remains constant, as does the composition of the vapors evolved.

Either batch or continuous distillations may be carried out in conjunction with rectification, thereby securing better separations for the same expenditure of heat. In addition, either batch or continuous operations may be conducted with or without the use of open steam. In the following articles the fundamentals of the methods of calculation of these various combinations will be outlined, together with illustrations of the details of their application.

PRINCIPLES, APPLICATIONS, AND METHODS OF CALCULATION OF DISTILLATION PROCESSES

Simple Batch Distillation

Definition. A simple batch distillation is defined as a distillation process in which a batch of material is charged to a still, vaporization is caused by the suitable application of heat, and the vapors are removed continuously, as formed, with no partial condensation of the vapors or refluxing of condensate to the still. Simple batch distillation as here employed refers to the batch distillation of miscible components. Batch steam distillations will be considered in later articles.

Applications. Simple batch distillations were formerly extensively employed in the refining of petroleum. In recent years, simple batch stills for this purpose have been rapidly superseded by batch stills with towers in which reflux may be used and these in turn by continuous stills with fractionating towers. The present applications of simple batch distillations in the plant are for the most part confined to smaller scale applications which are intermittent in character and in which efficiency in separation and the use of heat may be sacrificed to secure low first cost. In certain cases, such as the dehydration of organic solvents, a simple batch operation may result in effective separation with small heat consumption. In all cases, however, much sharper separations with more complete recovery of valuable constituents and with lower heat consumption can be realized by the addition of a fractionating tower to the simple batch still. In the laboratory, simple batch distillations may be realized if the upper part of the distilling flask is well insulated from heat loss, *e.g.*, by passing a current of hot gases around the flask up to a point above the vapor-delivery tube. The Engler or A.S.T.M. distillation deviates from a simple batch distillation in that some reflux by partial condensation in the neck of the flask is obtained.

Theory of Simple Batch Distillation. In simple batch distillation, increments of vapor are removed, upon their formation, from contact with the residual batch of liquid mixture. The mathematical analysis of such a process must therefore be a differential one. Such an analysis was first made by Rayleigh [*Phil. Mag.*, **4** (6), 521 (1902)]. Various forms of the Rayleigh equation are given as Eqs. (6) to (14), inclusive.

A simple derivation of the Rayleigh equation is as follows: Assume a mixture of two soluble components, the total moles (or pounds) of which are L. Call the mole fraction (or weight fraction) of the more volatile component in the liquid x, and the mole fraction (or weight fraction) of the same component in the vapor in equilibrium with the original liquid y. Let dL moles (or lb.) be vaporized; the liquid will lose and the vapor will gain a differential quantity of the more volatile component. By a material balance,

$$(L - dL)(x - dx) + (y + dy)dL = Lx \tag{6}$$

Neglecting differentials of the second order and rearranging,

$$\frac{dL}{L} = \frac{dx}{y - x} \tag{7}$$

Integrating Eq. (7) between limits,

$$\ln \frac{L_1}{L_2} = \int_{x_2}^{x_1} \frac{dx}{y - x} \tag{8}$$

In Eq. (8) either consistent weight units or consistent molar units must be used, *e.g.*,

L_1 = moles (or pounds) of original charge.
L_2 = moles (or pounds) of residual charge after $L_1 - L_2$ have been distilled off.
x_1 = mole fraction (or weight fraction) of more volatile component in original charge L_1.
x_2 = mole fraction (or weight fraction) of more volatile component in residual charge L_2.

If experimental data giving the y vs. x relationship are available, the right-hand side of Eq. (8) may be integrated by plotting $1/(y - x)$ vs. x and measuring the area under the curve between the limits x_1 and x_2. If a mathematical relationship exists between y and x, the right-hand side of Eq. (8) may be directly integrated as in the following two cases:

a. $y - x$ Relationship from Relative Volatility. For a binary mixture the $y - x$ relationship in terms of relative volatility is

$$y = \frac{\alpha x}{1 + (\alpha - 1)x} \tag{9}$$

Substituting in Eq. (8) and integrating gives

$$\ln \frac{L_1}{L_2} = \frac{1}{\alpha - 1}\left(\ln \frac{x_1}{x_2} + \alpha \ln \frac{1 - x_2}{1 - x_1}\right) \tag{10}$$

During a simple batch distillation at constant pressure, the temperature rises as the residual liquid becomes poorer in the more volatile component. For mixtures of similar types of substances having low numerical values of α, α will be found to change very slowly with temperature; hence Eq. (10) should give trustworthy results. These conditions are met, for example, by adjacent members of a homologous series such as the paraffin hydrocarbons. Variation of α with temperature should be investigated, however, when applying Eq. (10).

b. Henry's Law Applies. If Henry's law applies, Eq. (11) may be obtained by integration of Eq. (8):

$$\ln \frac{L_1}{L_2} = \frac{1}{k - 1} \ln \frac{x_1}{x_2} \tag{11}$$

In Eq. (11), k is the constant in Henry's law, when written $y = kx$. $y = kx$ is an isothermal relationship, but cases may be found in which k changes slowly with temperature. Variation of k with temperature should be tested in applying Eq. (11).

Equations (8), (10), and (11) are expressed in terms of the total moles (or pounds) of the original charge and the residual charge left in the still after distillation, and the respective compositions of these liquids. These equations are therefore useful if one of these terms gives directly the answer sought, *e.g.*, if it is desired to calculate the percentage of charge remaining in the still to yield a residual liquid of composition x_2.

c. Application to Multicomponent Mixtures. In certain cases it is more convenient to deal with the total moles (or weights) of individual constituents in a mixture. For this purpose another form of the Rayleigh equation may be derived.

Let A, B, C, D, etc., represent the total moles of the respective components in a multicomponent mixture. If a differential amount of the mixture is vaporized,

$$\frac{-dA}{-dB} = \frac{y_A}{y_B} = \alpha \frac{x_A}{x_B} \tag{12}$$

$$x_A = \frac{A}{A + B + C + D} \quad \text{and} \quad x_B = \frac{B}{A + B + C + D}$$

Therefore

$$\frac{-dA}{-dB} = \alpha \frac{A}{B} \tag{13}$$

Integrating

$$\ln \frac{A_1}{A_2} = \alpha \ln \frac{B_1}{B_2} \tag{14}$$

where A_1 = total moles (or pounds) of component A in the original mixture.
A_2 = total moles (or pounds) of component A remaining in the residual liquid after the batch-distillation operation.
B_1 = total moles (or pounds) of component B in the original mixture.
B_2 = total moles (or pounds) of component B remaining in the residual liquid after the batch-distillation operation.

Equation (14) is most frequently used for mixtures of hydrocarbons for which $\alpha = \frac{K_A}{K_B}\left(\text{or, if Raoult's law holds, } \alpha = \frac{p_A}{p_B}\right)$. The variation of α under the conditions of the distillation should be investigated before applying Eq. (14).

Applications of Simple Batch Distillation Equations

Weathering of Natural Gasoline. If a mixture of volatile hydrocarbons is submitted to a simple-batch-distillation process, Eq. (14) furnishes a means for estimating the relative amounts of the several components remaining in the liquid at various stages of the distillation. Assume a mixture of propane, butane, and pentane of the following composition:

Component	Weight, %
C_3	8.0
C_4	65.6
C_5	26.4
Total	100.0

On the basis of 100 lb. of this original mixture, consider that approximately half of the butane, say 32.2 lb., is to be left in the residual liquid after a batch-distillation process. It is desired to calculate the amounts of propane and pentane, respectively, in the residual liquid.

General Assumption: Raoult's Law Applies:

a. Calculation of Initial Boiling Point.

Component	Wt., %	Molecular weight	Moles	Mole fraction	Vapor pressure at 21°F., mm.	Partial vapor pressure ($p = Px$), mm.
Propane	8.0	44	0.182	0.108	2950	319
Butane	65.6	58	1.131	0.674	620	417
Pentane	26.4	72	0.367	0.218	135	29
Total	100.0	..	1.680	1.000	....	765

The sum of the partial vapor pressures approximates atmospheric pressure sufficiently closely for the purposes of this calculation; hence 21°F. may be taken as the initial boiling point.

b. Calculation of Relative Volatilities. Assume the final temperature at the end of the distillation to be 45°F.

Component	Vapor pressure at 21°F., mm.	Vapor pressure at 45°F., mm
C_3	2950	4400
C_4	620	1010
C_5	135	250

For C_4/C_3,

$$\alpha_{21} = 620/2950 = 0.210$$
$$\alpha_{45} = 1010/4400 = 0.230$$
$$\alpha_{\text{avg}} = 0.220$$

For C_4/C_5,

$$\alpha_{21} = 620/135 = 4.59$$
$$\alpha_{45} = 1010/250 = 4.04$$
$$\alpha_{\text{avg}} = 4.32$$

c. Calculation of final weights of propane and pentane when the residual butane is 32.2 lb. on the basis of 100 lb. of original mixture. Using Eq. (14),

$$\log \frac{\text{initial butane}}{\text{final butane}} = 0.220 \log \frac{\text{initial propane}}{\text{final propane}}$$

$$\log \frac{65.6}{32.2} = 0.220 \log \frac{8.0}{\text{final propane}}$$

whence

Final propane = 0.316 lb.

$$\log \frac{\text{initial butane}}{\text{final butane}} = 4.32 \log \frac{\text{initial pentane}}{\text{final pentane}}$$

$$\log \frac{65.6}{32.2} = 4.32 \log \frac{26.4}{\text{final pentane}}$$

whence

Final pentane = 22.4 lb.

Component	Lb.	Weight, %
C_3	0.3	0.5
C_4	32.2	58.6
C_5	22.4	40.9
Total	54.9	100.0

d. Calculation of final boiling point to check the assumed final temperature of 45°F.

Component	Weight, %	Molecular weight	Moles	Mole fraction	Vapor pressure at 45°F., mm.	Partial vapor pressure ($p = Px$) mm.
C_3	0.5	44	0.011	0.007	4400	31
C_4	58.6	58	1.010	0.635	1010	641
C_5	40.9	72	0.568	0.358	250	90
Total	100.0	..	1.589	1.000	....	762

The final total vapor pressure corresponds to atmospheric pressure; therefore the assumed final temperature is satisfactory. Had the assumed final temperature been found to differ appreciably from that found by this check calculation, it would have been necessary to recalculate parts *b* and *c* using another value for the assumed final temperature.

e. Comparison with Actual Test Data. This illustration was based upon the method outlined by Robinson ("Elements of Fractional Distillation," McGraw-Hill, New York, 1930) and a portion of the test data quoted by Robinson is used in Table 3 to show the agreement in this case of the calculated with the observed results.

Table 3

Component	Initial weight, %	Final weight, % (calculated)	Final weight, % (observed)
C_3	8.0	0.5	0.4
C_4	65.6	58.6	58.0
C_5	26.4	40.9	41.6
Total	100.0	100.0	100.0

Discrepancies are due to analytical errors, deviations from Raoult's law, and the assumption that the average α's as calculated in part *b* may be used in the integrated Eq. (14) as constants.

When this method of calculation is applied to mixtures of more than three components, one component may be selected as the reference component and the relative volatilities of the several other components computed with respect to the reference substance, as in the foregoing example. It should be noted that, if the temperature range covered in the distillation is large, considerable change in relative volatilities of components with widely differing atmospheric boiling points may be expected.

Equation (14) lends itself readily to graphical solution. A plot of log A against log B is a straight line, the slope of which is the relative volatility α of A vs. B. If desired $\log \frac{A}{A_0}$ may be plotted against $\log \frac{B}{B_0}$, A_0 and B_0 representing, respectively, the initial quantities of the two components and A and B the quantities remaining in the liquid at various stages of the distillation.

Batch Steam Distillation

Definition. A batch steam distillation is a distillation in which vaporization of the volatile constituents of a batch of material is effected at a lowered temperature by introduction of steam directly into the charge. Steam used in this manner is termed "open steam."

The lowering of the partial pressure of the volatile constituents of the charge obtained by the use of open steam may be secured by the similar use of any chemically inert gas. The use of gases or vapors other than steam, however, in many cases introduces added problems in the condensation and recovery of the distillate as well as in the recovery of the gas. In most plants in which it is necessary to employ distillations of this nature, low-pressure exhaust steam is available at low cost. For these reasons steam is employed in the majority of cases.

Applications. Batch steam distillation in the laboratory purification of organic compounds is a valuable tool of the organic chemist. By means of steam distillations, volatile organic liquids may be separated from relatively non-volatile impurities at temperatures sufficiently low so that thermal decomposition does not occur. This type of operation, when carried over to plant scale, advantageously employs steam distillation. The purification of glycerin and of fatty acids are examples of plant-scale applications in this field.

Theory of Batch Steam Distillation (McAdams, Unpublished Notes on Distillation, Mass. Inst. Tech.). **Batch Steam Distillation of a Volatile Component from Its Solution in a Non-volatile Liquid.**

S = total moles of open steam required.
B = total moles volatile components present in the solution in the still at any time.
O = total moles non-volatile liquid in the still, a constant.
p_S = actual partial pressure of steam in the vapor.
p_B = actual partial pressure of component B in the vapor.
P_B = vapor pressure of pure component B.
E = vaporization efficiency of the steam distillation, *i.e.*, the ratio of p_B to the equilibrium pressure of B. If Raoult's law applies,

$$E = \frac{p_B}{P_B \dfrac{B}{B + O}} \tag{15}$$

P = total pressure.

(It will be assumed for the time that no liquid water is present in the still; hence $p_S <$ the saturation pressure of steam at the temperature in the still. The presence of liquid water and its effects will be discussed later.)

Case 1. *Steam Distillation of a Volatile Component from a Dissolved Non-volatile Component Present in Large Amount.*

$$\frac{+dS}{-dB} = \frac{p_S}{p_B} = \frac{P - p_B}{p_B} = \frac{P}{p_B} - 1 \tag{16}$$

But, from Eq. (15),

$$p_B = EP_B\left(\frac{B}{B + O}\right)$$

$$dS = -\left(\frac{P}{EP_B} - 1\right) dB - \frac{PO}{EP_B}\frac{dB}{B} \tag{17}$$

Assuming that the temperature in the still is maintained constant, P_B is constant. For constant total pressure, integration of Eq. (17) gives

$$S = \left(\frac{P}{EP_B} - 1\right)(B_1 - B_2) + \frac{PO}{EP_B} \ln \frac{B_1}{B_2} \qquad (18)$$

Case 2. Steam Distillation of a Volatile Component from a Relatively Small Amount of a Dissolved Non-volatile Component. Under these conditions O, as used in the preceding derivation, is small relative to B. Hence p_B can be considered constant, and, by Eq. (15),

$$p_B = EP_B$$

Therefore

$$\frac{+dS}{-dB} = \frac{p_S}{p_B} \qquad (19)$$

Integrating Eq. (19)

$$S = \frac{p_S}{p_B}(B_1 - B_2) \qquad (20)$$

or

$$\frac{S}{B_1 - B_2} = \frac{p_S}{p_B} \qquad (21)$$

If the weight ratio of steam to distillate is desired

$$\frac{W_S}{W_B} = \frac{p_S}{p_B}\frac{M_S}{M_B} \qquad (22)$$

where W_S/W_B = lb. steam/lb. distillate.
M_S = molecular weight of steam = 18.
M_B = molecular weight of distillate.

Effect of the Presence of Liquid Water in the Still. The effect of the presence of liquid water in the charge during a steam distillation can be analyzed by the application of the principles of the phase rule as illustrated in examples given on p. 577.

Assume that a volatile component immiscible with water, and containing a negligible amount of dissolved non-volatile impurity, is to be steam-distilled. If no liquid water is present there are two components and two phases and, therefore, two degrees of freedom. Both temperature and pressure can be independently varied. By reference to Eq. (21) it will be noted that the steam consumption of the process can be decreased by increasing the denominator and by reducing the numerator of the right-hand side of the equation. In the absence of liquid water, this can be effected by operating at the highest permissible temperature, reducing the operating pressure or a combination of both, since in this case p_S, the partial pressure of the steam, is equal to the difference between the total pressure and the partial pressure of the volatile component, or $p_S = P - p_B$.

If, on the other hand, water is present as a liquid phase, the number of phases is now three, which, with two components, indicates one degree of freedom. In this case fixing either the temperature or pressure fixes the state of the system. Referring again to Eq. (21), p_S is now fixed at the vapor pressure of water at the temperature of the distillation. If the total pressure is fixed at, say, atmospheric pressure, the temperature of the distillation is defined by that temperature at which $p_S + p_B = P$.

To avoid the accumulation of a liquid-water phase due to condensation of a portion of the open steam, the temperature of the operation should be so chosen that p_S $(= P - p_B)$ is less than the saturation pressure of steam for the given temperature.

If the non-volatile material is present in large amount, as considered in Eqs. (16) to (18) the partial pressure of steam rises during the progress of the batch distillation because of the depletion of the volatile component in the charge and the consequent decrease in p_B. Since it is usually desired to strip the solution down to a small residual concentration of component B, p_S will closely approach P toward the end of the batch process, hence the choice of an operating temperature such that the saturation pressure of steam $> P$ will ensure no condensation of open steam at any time during the process. For example, if the operation is at atmospheric pressure, the minimum temperature chosen to accomplish this purpose would be 212°F. or slightly above.

Use of External Source of Heat. The preceding discussion indicated the desirability of operating at the highest allowable temperature (the highest temperature at which no injury to the volatile material will be caused) and in the absence of a liquid-water phase to effect economy in the use of open steam. Saving in the use of open steam is not only desirable per se, but by decreasing the ratio of steam to distillate the total heat to be removed in the condenser is lowered and the condensing surface and cooling water requirements are decreased. For these reasons a means for supplying heat to the charge other than by condensation of a portion of the open steam is usually advisable. The external heat source may be supplied by steam condensing in a closed coil immersed in the charge. It should be noted that, if means of heating other than steam are contemplated, careful consideration should be given to avoidance of injury to the charge by too high film temperatures at the heat-transfer surface. In the estimation of the heat requirements and the design of the heating surfaces, consideration should be given to:

1. The time required to heat the cold charge to distillation temperature.
2. The heat required to supply the latent heat of vaporization during the batch process.

In considering 1 it is desirable to supply enough surface so the charge may be brought to temperature in a length of time that is well balanced with respect to the time of a complete cycle. Consideration 2 will be governed by the rate at which it is desired to distill the batch, *i.e.*, the time of a cycle.

Vaporization Efficiency in Steam Distillation. In a steam distillation, bubbles of steam rise through the liquid charge and escape from the surface of the liquid bearing a concentration of the volatile component governed by the partial vapor pressure of the volatile component from the charge. It has long been recognized that the partial pressure of the volatile material in the steam passing from the still is in most cases less than the partial pressure representing equilibrium with this constituent in the charge. Vaporization efficiency, as defined by Eq. (15), has been employed as an empirical factor of safety to make allowance for this departure from equilibrium conditions. Values of 0.6 to 0.7 have been commonly used in the past. The writer (Carey, Sc.D. Thesis, Mass. Inst. Tech., 1930) has shown that vaporization efficiency is related to the character of the substance distilled; the depth of liquid through which the steam bubbles pass; and the size of the individual steam bubbles, by the following expression:

$$E_v = \frac{p}{p^*} = 1 - e^{-K\frac{L}{D}} \qquad (23)$$

in which E_v = vaporization efficiency.
p = actual partial pressure of volatile material in steam.
p^* = partial pressure of volatile component which would be in equilibrium with the liquid charge.
e = base of Napierian logarithms.
L = depth of liquid charge through which steam rises.

D = diameter of steam bubbles.
K = constant characteristic of the substance distilled determined by its diffusion rate in the vapor state.

Certain laboratory data on the steam distillation of aniline indicated that the effect of depth of immersion of the steam spray is as given by Eq. (23). No data have been obtained to check the effect of bubble size in a steam-distillation operation.

To obtain high values of vaporization efficiency, the steam should be introduced to the charge through a large number of small orifices and at an appreciable depth below the surface of the liquid. With the holes in the steam spray ¼ in. in diameter and the depth of immersion 1 ft. or more it is probable that much higher values of vaporization efficiency may be safely employed than has been customary, *e.g.*, 0.9 to 0.95 instead of 0.6 to 0.7, when dealing with organic compounds of molecular weight less than 100. In the steam distillation of lubricating oils (mol. wt. > 250) calculations with data from refinery equipment show the vaporization efficiencies realized average about 0.5.

Example of Calculation of Steam Consumption in Batch Steam Distillation. It is proposed to debenzolize a batch of 10,000 lb. benzolized wash oil containing 10 per cent by weight of benzene by a batch steam distillation, simultaneously recovering the benzene removed. The proposed processing will utilize a batch steam still provided with a closed steam coil to heat and maintain the charge at the desired temperature and a suitably perforated steam spray to distribute the open steam through the liquid charge. The vapors from the still are to be condensed in a worm condenser, and the condensate will then flow to a settling tank where the benzene and water will separate into two layers by gravity. It is desired to calculate the pounds of open steam that must be blown through the charge to reduce the benzene content of the solution to 50 lb., under the two following conditions:

1. The contents of the still will be maintained at 350°F. during the distillation by means of high-pressure steam condensing in the closed coil. Atmospheric steam, superheated to 350°F., will be used in the steam spray.

2. The contents of the still will be maintained at 215°F. during the distillation by means of 15 lb./sq. in. exhaust steam condensing in the closed coil. Atmospheric steam at 215°F. will be used in the spray.

Data. The still is to be well insulated so that heat losses will be neglected in this preliminary calculation. Raoult's law applies to the solution of benzene in oil when the molecular weight of the oil is taken as 220. The vapor pressure of the oil is so low at distillation temperatures that its volatility may be neglected.

Vapor pressures of benzene:

350°F., 7100 mm.
215°F., 1380 mm.
Molecular weight of benzene = 78.

The distillation will be carried out at 760 mm. pressure. The immersion of the steam spray is such that the vaporization efficiency may be taken as 0.90.

Example 1. Distillation temperature = 350°F.

Component	Lb.	Molecular weight	Moles	Mole fraction (x)	Vapor pressure (P), mm.	Partial vapor pressure ($p = Px$) mm.
Benzene	1,000	78	12.81	0.239	7100	1700
Oil	9,000	220	40.85	0.761		
Total	10,000	. . .	53.66	1.000		

The partial pressure of benzene from this solution is noted to be greater than 760 mm. at 350°F. This indicates that a portion of benzene will vaporize during the heating of the charge to 350°F. and before the introduction of the process steam. The amount of benzene removed by this preliminary vaporization may be calculated as follows:

Let F = moles benzene vaporized. Mole fraction benzene in residual solution $= \dfrac{12.81 - F}{53.66 - F}$.

Partial-pressure benzene from residual solution

$$= \left(\frac{12.81 - F}{53.66 - F}\right) 7100 = 760 \text{ mm}$$

Solving, $F = 7.91$ moles benzene vaporized.
Moles benzene remaining = 12.81 − 7.91 = 4.90.
Using Eq. (18):

$$S = \left(\frac{P}{EP_B} - 1\right)(B_1 - B_2) + \frac{PO}{EP_B} \ln \frac{B_1}{B_2}$$

where $B_1 = 4.90$.
$B_2 = {}^{50}/_{78} = 0.64$.
$E = 0.90$.
$P_B = 7100$.
$O = 40.85$.
$P = 760$.

$$S = \left(\frac{760}{0.9 \times 7100} - 1\right)(4.90 - 0.64) + \left(\frac{760 \times 40.85}{0.9 \times 7100}\right)(2.3) \log \frac{4.90}{0.64}.$$

$S = 6.12$ moles steam.
Pounds steam = 6.12 × 18 = 110

Pounds steam per pound benzene stripped

$$= \frac{110}{(4.90 - 0.64)78} = 0.329$$

Pounds steam per pound total benzene recovered

$$= \frac{110}{(12.81 - 0.64)78} = 0.1$$

Example 2. Distillation temperature = 215°F.

Component	Lb.	Molecular weight	Moles	Mole fraction (x)	Vapor pressure (P), mm.	Partial vapor pressure ($p = Px$), mm.
Benzene	1,000	78	12.81	0.239	1380	330
Oil	9,000	220	40.85	0.761		
Total	10,000		53.66	1.000		

No benzene will be vaporized in warming up the charge in this case. Applying Eq. (18):

$$S = \left(\frac{P}{EP_B} - 1\right)(B_1 - B_2) + \frac{PO}{EP_B} \ln \frac{B_1}{B_2}$$

where $B_1 = 12.81$.
$B_2 = {}^{50}/_{78} = 0.64$.
$E = 0.90$.
$P_B = 1380$.
$O = 40.85$.
$P = 760$.

$$S = \left(\frac{760}{0.9 \times 1380} - 1\right)(12.81 - 0.64) + \left(\frac{760 \times 40.85}{0.9 \times 1380}\right)(2.3) \log \frac{12.81}{0.64}.$$

$S = 70.1$ moles steam.
Pounds steam = 70.1 × 18 = 1265

Pounds steam per pound benzene stripped and recovered

$$= \frac{1265}{(12.81 - 0.64)(78)} = 1.33$$

It should be noted that these calculations give the amount of steam actually blown through the charge.

Calculation will show that the external heating steam to heat the batch from, say, 70° to 350°F. is approximately double that required for heating to 215°F., and little saving in over-all steam consumption is secured by the higher temperature. In using a cheaper source of external heat, the higher temperature might be justified.

Continuous Equilibrium Vaporization

Definition. Continuous equilibrium vaporization is a distillation process in which a feed stock is partly vaporized under such conditions that equilibrium exists between all the vapor formed and all the remaining liquid. The requirements of continuity of operation presuppose a feed of constant composition supplied at a constant rate, from which vapor and liquid of constant compositions and amounts are formed and continuously withdrawn.

Applications. A simple illustration of this type of vaporization is a direct-fired shell still to which feed may be pumped at a constant rate. If firing is at a constant rate, a definite percentage of the charge will be vaporized and pass off as distillate. The residual liquid is withdrawn at a rate such that a predetermined level is maintained in the still. If thorough mixing of the contents of the still by the boiling process is assumed, the composition of the liquid in the still may be taken as that of the liquid effluent. The vapors from the still are of a composition corresponding to equilibrium with the liquid in the still.

Pipe stills comprised of tubing suitably arranged in a furnace setting have become widely used in recent years in petroleum distillation. The mechanism of vaporization in a pipe still is that of continuous equilibrium vaporization. Feed is pumped through the tubes and, when the necessary temperature is reached, vaporization sets in. All the vapor formed during the heating process remains in contact with the residual liquid until the outlet of the still is reached.

A considerable pressure drop, due to friction, may exist through the tubes of a pipe still, especially if a large amount of vaporization occurs. At the outlet of the tubular heater, when this pressure is removed, additional vaporization or "flashing" may occur, in which the latent heat of vaporization is supplied at the expense of a portion of the sensible heat of the liquid-vapor mixture. Whether vaporization takes place largely in the tubes of the heater, or chiefly by a flashing process, the vapor formed and the residual liquid should be in substantial equilibrium.

Since, in the primary distillation of petroleum, it is desired to avoid any cracking due to excessive temperatures, continuous equilibrium vaporizations have found wide application. As will later be shown numerically, continuous equilibrium vaporizations permit a greater percentage of material to be vaporized at a given temperature than simple batch distillations. Very effective distillation units have been evolved in which the vapors from a continuous equilibrium vaporization are fractionated into several products while the residual liquid is steam stripped of its light ends.

A **single flash distillation** is a continuous equilibrium vaporization in which all the vapor formed remains in contact with the residual liquid during the vaporization process.

Successive flash vaporization constitutes two or more continuous equilibrium vaporizations in which the vapors formed are separated from the residual liquid after each equilibrium vaporization. If the increments of vapor permitted to remain in contact with the residual liquid are made infinitely small and the vaporization is assumed to take place in an infinite number of steps, successive flash vaporization approaches simple batch distillation as a limiting case. In general, a greater percentage of the original material can be vaporized at a given temperature and pressure by a single flash distillation than by successive flash vaporizations.

Successive flash vaporizations are employed in petroleum refining under certain circumstances when it is desired to process the higher boiling fractions of the crude under vacuum. In this type of operation the lighter fractions of the crude are vaporized by heating in a pipe still and flashing in a fractionating tower. The vapors formed in this first vaporization are fractionated into the desired light products while the residual liquid, after steam stripping in the base of the tower, passes to a second tubular heater. The second heater discharges into a fractionating tower maintained at a reduced pressure in which the second flashing operation is completed, the vapors are fractionated, and the residual liquid is discharged from the base of the tower after steam stripping.

Calculations of Continuous Equilibrium Vaporizations. The method of calculation of the equilibrium vaporization of a binary mixture in which both components follow Raoult's law is illustrated in Table 2, p. 578, for benzene and toluene mixtures. The following examples will serve to illustrate the calculations possible with ternary or more complicated mixtures in which the components follow Raoult's law.

A mixture of n-propane, n-butane, and n-pentane of the same composition taken for the weathering of natural-gasoline illustration, will be used, *viz.*,

Component	Weight, %
C_3H_8	8.0
C_4H_{10}	65.6
C_5H_{12}	26.4
Total	100.0

1. Calculation of Vapor Composition in Equilibrium with Liquid at the Boiling Point of the Liquid under Atmospheric Pressure.

a. Calculation of Boiling Temperature.

Component	Weight, %	Molecular weight	Moles	Mole fraction	Vapor pressure at 21°F.	Partial vapor pressure ($p = Px$), mm.
C_3H_8	8.0	44	0.182	0.108	2950	319
C_4H_{10}	65.6	58	1.131	0.674	620	417
C_5H_{12}	26.4	72	0.367	0.218	135	29
Total	100.0	..	1.680	1.000	....	765

The temperature is obtained by successive approximations until the sum of the partial vapor pressures approximates 760 mm. within the precision with which the vapor-pressure charts can be read.

b. Calculation of Vapor Composition. By a combination of Dalton's and Raoult's laws

$$Py_1 = P_1x_1$$

where P = total pressure.
y_1 = mole fraction of any component in vapor.
x_1 = mole fraction of same component in liquid.
P_1 = vapor pressure of pure component.

Component	Px	P	$y = \frac{Px}{765}$
C_3H_8	319	765*	0.417
C_4H_{10}	417	....	0.545
C_5H_{12}	29	....	0.038
Total	765	....	1.000

* Where the sum of the partial pressures differs from the total pressure slightly (due to difficulty in reading any of the commonly used vapor-pressure charts with greater precision), the sum of the partial pressures should be used instead of the actual total pressure in calculating the vapor composition.

2. Calculation of Percentage Vaporized at a Given Temperature and the Compositions of the Vapor and the Residual Liquid. Basis: 100 moles of original liquid. By a material balance on total liquid and vapor

$$L + V = 100 \tag{24}$$

where L = total moles liquid.
V = total moles vapor.

By material balances on each of the three components:

$$x_3L + y_3V = 100x_{f3} \tag{25}$$
$$x_4L + y_4V = 100x_{f4} \tag{26}$$
$$x_5L + y_5V = 100x_{f5} \tag{27}$$

where x_3, x_4, x_5 = mole fractions propane, butane, and pentane, respectively, in residual liquid.
y_3, y_4, y_5 = mole fractions propane, butane, and pentane, respectively, in vapor in contact with residual liquid.
x_{f3}, x_{f4}, x_{f5} = mole fractions propane, butane, and pentane, respectively, in original liquid before vaporization.

By Dalton's and Raoult's laws:

$$y_3 = \frac{P_3x_3}{P} \tag{28}$$
$$y_4 = \frac{P_4x_4}{P} \tag{29}$$
$$y_5 = \frac{P_5x_5}{P} \tag{30}$$

where P_3, P_4, P_5 = vapor pressures of propane, butane, and pentane, respectively.
P = total pressure.

The sum of the x's of the residual liquid must equal unity:

$$x_3 + x_4 + x_5 = 1 \tag{31}$$

If the temperature and total pressure are fixed, the unknowns to be calculated are: L, V, x_3, x_4, x_5, y_3, y_4, y_5, or a total of eight unknowns. Eight independent equations are available; hence a direct solution is possible. However, to avoid unwieldy algebraic manipulations, it will be found simpler to solve by trial and error, by combining Eqs. (25) and (28), (26) and (29), (27) and (30), respectively, together with Eq. (24):

$$x_3 = \frac{100x_{f3}}{L + \frac{P_3}{P}(100 - L)} \tag{32}$$
$$x_4 = \frac{100x_{f4}}{L + \frac{P_4}{P}(100 - L)} \tag{33}$$
$$x_5 = \frac{100x_{f5}}{L + \frac{P_5}{P}(100 - L)} \tag{34}$$

By assigning a value to L and employing Eqs. (32), (33), and (34), x_3, x_4, and x_5 may be calculated. The correctness of the assumed value of L may then be tested by Eq. (31), the remaining independent equation.

Let the temperature to which the liquid of the preceding batch distillation example, p. 581, is heated in contact with the vapor formed be 45°F.

Component	Mole fraction in original liquid (x)	Vapor pressures at 45°F. (P), mm.
Propane	0.108	4400
Butane	0.674	1010
Pentane	0.218	250
Total	1.000	

Trial 1. Assume $L = 35$, $V = 65$. By Eqs. (32), (33), and (34),

$$x_3 = \frac{10.8}{35 + (\frac{4400}{760})(65)} = 0.026$$
$$x_4 = \frac{67.4}{35 + (\frac{1010}{760})(65)} = 0.555$$
$$x_5 = \frac{21.8}{35 + (\frac{250}{760})(65)} = 0.387$$

By Eq. (31):

$$x_3 + x_4 + x_5 = 0.026 + 0.555 + 0.387 = 0.968 \neq 1$$

Trial 2. Assume $L = 25$, $V = 75$. By a similar calculation,

$$x_3 = 0.024$$
$$x_4 = 0.540$$
$$x_5 = 0.439$$
$$x_3 + x_4 + x_5 = 1.003$$

Trial 3. Assume $L = 25.8$, $V = 74.2$.

$$x_3 = 0.024$$
$$x_4 = 0.542$$
$$x_5 = 0.434$$
$$x_3 + x_4 + x_5 = 1.000$$

By Trial 3, 74.2 mole per cent of the original mixture will be vaporized when heated to 45°F. The following comparison, Table 4, with the results of the batch weathering calculation will illustrate the differences in the two methods of vaporization.

Table 4. Comparison of Simple Batch and Continuous Equilibrium Distillations

Component	Mole fraction original mixture	Moles original mixture	Moles residual liquid by simple batch distillation to 45°F. final temperature	Mole fraction residual liquid, simple batch distillation	Moles residual liquid by continuous equilibrium vaporization at 45°F.	Mole fraction residual liquid by continuous equilibrium vaporization
Propane	0.108	10.8	0.4	0.008	0.6	0.024
Butane	.674	67.4	32.4	.632	14.0	.542
Pentane	.218	21.8	18.5	.360	11.2	.434
Total	1.000	100.0	51.3	1.000	25.8	1.000

By inspection of Table 4, it will be seen that 51.3 mole per cent remain unvaporized when distillation by simple batch is carried to a final temperature of 45°F., whereas 25.8 mole per cent are unvaporized at 45°F. by continuous equilibrium vaporization. This forcibly exemplifies the statement that a greater percentage of a given mixture can be vaporized at a given temperature by continuous equilibrium vaporization than by a simple batch distillation to the same final temperature.

3. Calculation of Dew-point of a Complex Vapor Mixture. Assume the hydrocarbon mixture of the composition used in the preceding examples, *viz.*, propane 0.108, butane 0.674, and pentane 0.218, all expressed in mole fractions, is completely vaporized. It is desired to calculate the temperature at which the first condensation of liquid will occur as well as the composition of the first drop formed.

If in Eqs. (32), (33), and (34), respectively, $L = 0$, *i.e.*, the material is 100 per cent vaporized, there results:

$$x_3 = \frac{Px_{f3}}{P_3} \tag{35}$$
$$x_4 = \frac{Px_{f4}}{P_4} \tag{36}$$
$$x_5 = \frac{Px_{f5}}{P_5} \tag{37}$$

Equations (35), (36), and (37), together with Eq. (36), make a direct solution possible. The calculations, most conveniently made by successive assumptions of temperatures until a solution is reached, are indicated in Table 5 showing a dew point of approximately 51.5°F.

Table 5. Calculation of Dew Point of Hydrocarbon Vapor Mixture

Component	x_f	P	Px_f	P_x at 51.5°F., mm.	$x = \frac{Px_f}{P_x}$
C_3H_8	0.108	760	82.1	4950	0.016
C_4H_{10}	.674	...	512.3	1170	.438
C_5H_{12}	.218	...	165.6	300	.552
Total	1.000	...	760.0	...	1.006

4. Calculation Using Equilibrium Constants. The preceding calculations of boiling point, percentage vaporization, and dew point for a multicomponent mixture utilized Raoult's law. If equilibrium constants are employed, the methods of calculation are identical except that the appropriate values of K are substituted for P_A/P.

In multicomponent rectification calculations the computation of actual mole fractions is frequently an unnecessary step. Boiling points, percentage vaporizations, and dew points may be calculated in terms of the total moles in the system as shown by the following equations:

a. Boiling Points (or Bubble Points).

Let

M = total moles of mixture.

$M_1, M_2, M_3, \ldots$ = moles components 1, 2, 3.

By trial and error the correct temperature at a given pressure (or pressure for a fixed temperature) is found so that

$$M_1K_1 + M_2K_2 + M_3K_3 + \cdots = M \tag{38}$$

b. Percentage Vaporized (or Condensed).

Let

M = total moles of liquid plus vapor.

$M_1, M_2, M_3, \ldots$ = moles components 1, 2, 3.

V = total moles vapor.

L = total moles liquid.

The trial and error is carried out by assuming a ratio V/L

Then

$$L = \frac{M_1}{1 + K_1\frac{V}{L}} + \frac{M_2}{1 + K_2\frac{V}{L}} + \frac{M_3}{1 + K_3\frac{V}{L}} \tag{39}$$

and $V = M - L$. The V/L ratio calculated must check that assumed. Or alternately assume L/V and calculate

$$V = \frac{M_1}{\frac{L}{VK_1} + 1} + \frac{M_2}{\frac{L}{VK_2} + 1} + \frac{M_3}{\frac{L}{VK_3} + 1} \tag{40}$$

$$L = M - V$$

The L/V ratio calculated must check that assumed.

c. Dew Point. By trial and error, using the same nomenclature

$$\frac{M_1}{K_1} + \frac{M_2}{K_2} + \frac{M_3}{K_3} + \cdots = M \tag{41}$$

Continuous Equilibrium Vaporization of Petroleum Fractions. The principles illustrated in the preceding examples are useful in calculations involving the lower boiling hydrocarbons found in natural gas and light gasolines. When dealing with higher boiling fractions or entire crudes, the multiplicity of compounds in each boiling range of very slightly differing boiling points render these calculations impractical. Recourse must be had to empirical correlations in terms of readily made laboratory distillation tests.

Of particular value in the design of oil-distillation equipment are so-called flash-distillation curves which give the relation between percentage of the oil vaporized and the temperature of the continuous equilibrium vaporization. Piroomov and Beiswenger (*Am. Petroleum Inst., Bull.* 10, p. 52, Jan. 3, 1929) have established a widely used correlation that makes it possible to estimate the flash curves of crudes from either the true boiling point (see p. 607) or the 10 per cent distillation (see below) of the crude or its fraction.

Figures 36 and 37 give Piroomov and Beiswenger's correlations. In the upper curve of Fig. 36, the slope of

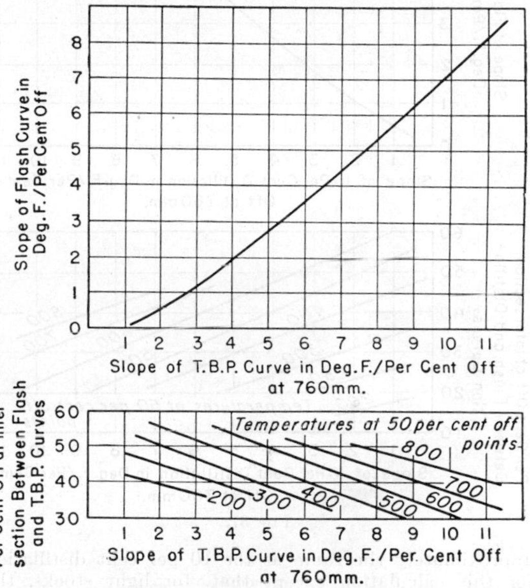

FIG. 36.

the equilibrium flash curve is plotted against the slope of the true-boiling-point curve (the slope taken as that of a straight line through the 10 and 70 per cent points). In the lower curve of Fig. 36, the per cent vaporized at the intersection between the flash and true-boiling-point curves is plotted against the true-boiling-point slope, a family of curves, each for a given 50 per cent true-boiling-point temperature being given. Figure 37 gives similar correlation for 10 per cent distillation data.

The use of Figs. 36 and 37 will be apparent from the following example. It will be assumed that the true-boiling-point curve of a crude has been obtained in an apparatus similar to that described on p. 607. The vapor temperatures are plotted as ordinates against liquid-volume per cent distilled as abscissas. Assume the following data to be read from the true-boiling-point curve (see Fig. 38):

Temperature at 70% vaporized..........	730°F.
Temperature at 10% vaporized..........	200°F.
Temperature at 50% vaporized..........	542°F.

Slope between 10 and 70% = (730 − 200)/(70 − 10).....8.84°F./per cent.

From Fig. 36, upper curve, slope flash curve = 6.15.

From Fig. 36, lower curve, the per cent off at intersection between true boiling point and flash = 32.

A point on the flash curve and its slope being established, the flash curve is drawn as a straight line. Figure 38 shows the true-boiling-point curve and the flash curve calculated in this illustration.

If a 10 per cent distillation rather than a true boiling point is available the method of calculation is similar using, however, Fig. 37. An A.S.T.M. (Engler) distillation (see p. 606) is more frequently available than the 10 per cent distillation and may be considered as

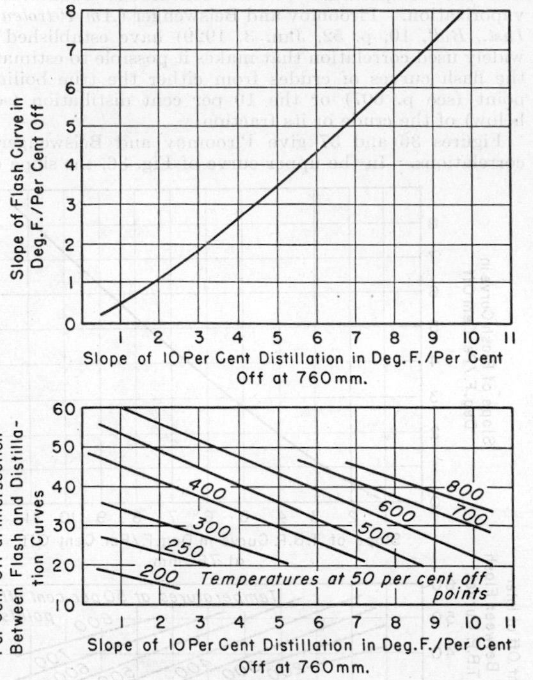

FIG. 37.

approximately the same as the 10 per cent distillation for this calculation except that, for light stocks, the A.S.T.M. temperatures are approximately 5°F. lower than the Saybolt 10 per cent temperatures.* The correlation, as noted by Piroomov and Beiswenger, is

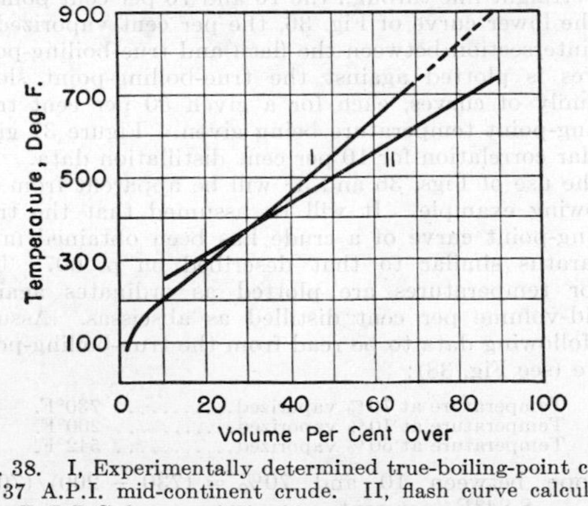

FIG. 38. I, Experimentally determined true-boiling-point curve for 37 A.P.I. mid-continent crude. II, flash curve calculated from T, B.P.C. by use of Fig. 36.

not so reliable for the 10 per cent distillation as for the true boiling point.

It should be noted that these correlations are based upon the following empirical facts:

* See original article for definition of Saybolt 10 per cent distillation.

1. The true-boiling-point and the 10 per cent distillation curves of many commonly encountered crudes and fractions are nearly straight lines between the 10 and 70 per cent points.
2. The flash curves of many crudes and fractions closely approximate straight lines.

Although these correlations are valuable tools, appreciable deviations may be expected in certain cases. An added precaution to be observed is that this method of calculation of the flash curve gives no evaluation of the pronounced "tails" at either end of the curve; hence the flash curves cannot be expected to give close approximations below the 10 per cent point or above the 80 and 90 per cent points.

The preceding discussion outlined the estimation of the flash curve at 760 mm. pressure from the true boiling point or the 10 per cent distillations as obtained at 760 mm. To correct the flash curve to other pressures Piroomov and Beiswenger recommend employing the temperature at the intersection of the true-boiling-point and the flash curve in conjunction with the Cox chart (Fig. 2).

When successive flashing is employed, a higher temperature is required to obtain a given total percentage vaporization than would be the case in a single flash vaporization. For the estimation of the flash curves of reduced crudes the reader is referred to Piroomov and Beiswenger's original article.

Continuous Rectification

Rectification (see Definition, p. 563) consists in the return of a portion of the condensed distillate from a still as reflux to interact with the vapor rising from the still. The reflux scrubs the higher boiling constituents from the vapor stream, at the same time being stripped of a portion of its content of low boiling material.

The **completeness of separation** of the low-boiling and the high-boiling components of a mixture by a rectification process depends upon

1. The volatilities of the respective components.
2. The ratio of reflux downflow to rising vapor.
3. The length of the path traveled countercurrently by reflux and vapor, *i.e.*, the tower height.
4. The efficiency of contact secured between liquid and vapor.

Item 1 is fixed by the nature of the components encountered in any specific problem. Item 2 must be properly balanced against cost of operation, height of tower required, and tower cross-sectional area; *i.e.*, as reflux ratio is increased, the necessary tower height is decreased, but the heat consumption and the cross-sectional area of the tower both are increased. Items 3 and 4 are determined for a given reflux ratio by the type of tower filling employed.

From the standpoint of internal construction, fractionating towers may be classified as

A. Bubble plate towers.
B. Sieve plate towers.
C. Packed towers.

Types *A* and *B* predominate in large-sized commercial applications, type *C* in smaller scale and laboratory apparatus. All three types are further discussed and compared in pp. 596 to 597.

Bubble plate towers are provided with a number of plates or trays, usually equally spaced in the tower shell. Reflux passes down the tower from tray to tray through suitable downflow pipes. Depths of liquid are maintained on each tray through which the rising vapors are made to bubble. From the rectification standpoint, bubble plate towers may be considered as attaining the desired separation in a number of steps. The vapors in passing through each tray undergo a change in composition fixed by the conditions of operation and the nature of the components. The tower height is determined by the number of these stepwise fractionations required and by the necessary spacing between the trays.

Useful methods are available for calculation of the number of trays required to effect a given separation in bubble plate towers. Information available on the heights of packed towers equivalent to one bubble tray is given on pp. 619 to 621.

Fractionating towers may be operated continuously or batch. Inasmuch as in continuous operation the flows and compositions of the liquid and vapor streams are constant at any given point in the apparatus, mathematical treatment is simpler for this case.

Principles of Continuous Rectification in Bubble Plate Towers. Assume a tower in continuous, balanced operation. Referring to Fig. 39 liquid feed F, consisting

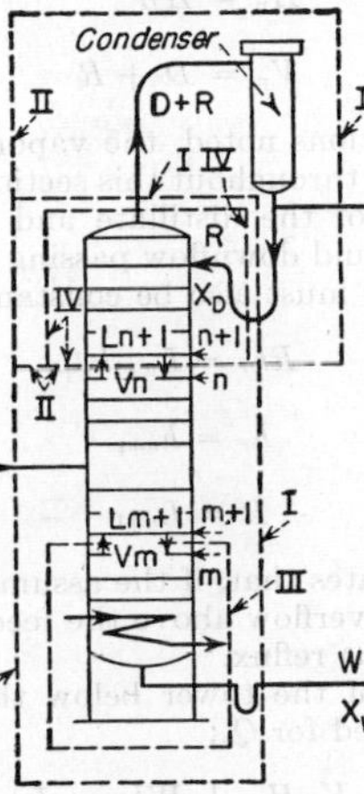

FIG. 39.

of a binary mixture, at its boiling point, is introduced on one of the trays in the mid-section of the column. The liquid feed together with reflux from the upper portion of the column flows down over the trays below the feed. In this illustration, heat to vaporize a portion of the feed and reflux is supplied by means of steam condensing in a closed coil in the base of the tower. The residue, or bottoms, W is withdrawn from the base of the tower. The vaporized feed and reflux pass up the tower undergoing rectification on each tray and, finally, to the total condenser. A portion of the total condensate is returned to the column as reflux R, while the rest is withdrawn as product or distillate P.

The change in composition of both liquid and vapor may be traced through a bubble plate tower by the use of three sets of relationships, applied to each tray of the tower in turn:

1. Material balances.
2. Heat balances.
3. The liquid-vapor equilibrium composition relationship, or the same corrected for failure to obtain complete equilibrium.

Under certain conditions (p. 590), the heat balances are not required, at least for preliminary calculations, which leads to considerable simplification of the calculations. Material balances and heat balances are outlined in following articles after which their application to actual problems in conjunction with the use of liquid and vapor equilibriums will be given.

Nomenclature.

x = mole fraction of more volatile component in liquid.
y = mole fraction of more volatile component in vapor.
D = moles of overhead distillate withdrawn per unit time.
x_D = mole fraction of more volatile component in overhead distillate.
n = the number of the plate under consideration, counting from the feed plate up.
m = the number of the plate under consideration, counting from the bottom of the tower up.
L_{n+1} = moles of liquid overflow from plate $n + 1$ to plate n, per unit time.
x_{n+1} = mole fraction of more volatile component in overflow, L_{n+1}.
V_n = moles of vapor passing from plate n to plate $n + 1$, per unit time.
y_n = mole fraction of more volatile component in vapor, V_n.
F = moles of mixture fed to the column per unit time.
x_F = mole fraction of more volatile component in feed, F.
W = moles of bottoms, per unit time.
x_w = mole fraction of more volatile component in bottoms.
R = moles of reflux, per unit time.
L'_{n+1} = theoretical minimum overflow from plate $n + 1$, per unit time.
E = plate efficiency defined as number of perfect plates divided by actual plates required.
h_f = sensible-heat content per mole of liquid feed.
h_r = sensible-heat content per mole of liquid reflux at point of return to the top plate of the column.
H_c = total heat content (sensible plus latent) per mole of vaporized reflux or distillate passing out the top of the tower.
h_w = sensible-heat content per mole of liquid bottoms.
Q_b = Heat required to heat and vaporize downflow reaching the bottom of the tower. Heat introduced at the base of a tower for this purpose is termed "reboil heat."

Over-all Heat Balance. For the tower of Fig. 39, equating total heat input to total heat output and neglecting heat losses from the tower give

$$(F \times h_f) + Q_b + (R \times h_r) = (R \times H_c) + (D \times H_c) + (W \times h_w) \quad (42)$$

Any convenient base temperature may be taken, above which to evaluate the heat contents.

For the conditions for which Eq. (42) is written, the feed F is assumed to enter at its boiling point. h_f therefore represents the molar heat of the liquid at the boiling temperature of the feed. In a general case the molar heat content of the feed may vary from nearly zero, if the feed is introduced cold, to the heat content of the completely vaporized feed.

The quantity $(H_c - h_r)R$ represents the heat that must be abstracted by the condenser to produce the reflux and is termed the "reflux duty." This reflux duty may be performed by returning condensate back to the top tray of the tower or by a partial condenser coil in the top of the tower.

If the feed rate F and its temperature, and the rates of withdrawal of overhead distillate D and bottoms W are fixed, the reflux duty and the heat introduced to the bottom of the tower Q_B are dependent upon each other, *i.e.*, the reflux cannot be increased without a corresponding increase in "reboil" heat duty.

General Material Balance Equations. Consider a section of the apparatus of Fig. 39 bounded by the dotted line II which includes the portion of the tower above the nth plate. A material balance on this section is

$$V_n = L_{n+1} + D \quad (43)$$

By a material balance on the more volatile component

$$V_n y_n = L_{n+1} x_{n+1} + D x_D \quad (44)$$

or

$$y_n = \left(\frac{L_{n+1}}{L_{n+1} + D}\right) x_{n+1} + \left(\frac{D}{L_{n+1} + D}\right) x_D \quad (45)$$

Similarly, for the section bounded by the dotted line III,

$$V_m + W = L_{m+1} \quad (46)$$
$$L_{m+1}x_{m+1} = V_m y_m + W x_w \quad (47)$$

or

$$y_m = \left(\frac{L_{m+1}}{L_{m+1} - W}\right) x_{m+1} - \left(\frac{W}{L_{m+1} - W}\right) x_w \quad (48)$$

Equations (43) to (48), inclusive, are based solely on material balances; hence they are valid under all conditions. By inspection of Eq. (45) it is found to contain three variables, L_{n+1}, y_n, and x_{n+1}. If one of these, say x_{n+1}, is known for any point in the column, another independent relationship will enable the remaining two to be calculated. A heat balance, properly drawn, will furnish this necessary equation. Similarly, Eq. (48) when combined with a heat balance permits the calculation of y_m and L_{m+1} if x_{m+1} is known for some point in the lower section of the column. In principle, these sectional heat balances are simple, but in actual application they may become involved in detail, or, in some cases, the necessary specific-heat and latent-heat data may be unavailable.

Sectional Heat Balances. For section IV of Fig. 39, assuming heat losses are negligible,

$$V_n H_n + R h_r = L_{n+1}(h_{n+1}) + (D + R) H_D \quad (49)$$

where H_n = total heat content per mole of vapor at nth tray.
h_r = sensible-heat content per mole of liquid reflux returned.
h_{n+1} = sensible-heat content per mole of liquid downflow from $n + 1$.
H_D = total heat content per mole of vaporized reflux and overhead distillate.

Similarly, for section III of Fig. 41, below the feed,

$$V_m H_m + W h_w = (L_{m+1})(h_{m+1}) + Q_b \quad (50)$$

where H_m = total heat content per mole of vapor at mth plate.
h_w = sensible-heat content per mole of liquid bottoms.
h_{m+1} = sensible-heat content per mole of overflow from $m + 1$.
Q_b = reboil heat at base of tower.

For fixed operating conditions Eq. (49) enables the relation between V_n and L_{n+1}, and in conjunction with Eq. (43) the ratio L_{n+1}/V_n to be established for any point above the feed plate. Similarly, Eq. (50) in conjunction with Eq. (46) evaluates L_{m+1}/V_m for any point below the feed plate. With the values of L_{n+1}/V_n and L_{m+1}/V_m determined, Eqs. (45) and (48), respectively, furnish relationships by which the vapor composition below any plate may be calculated from the liquid composition on the plate above and constants determined by the operating conditions.

In applying Eqs. (49) and (50), the heat contents of the various streams are influenced by the compositions of these streams, if the molar latent and specific heats of the components are unequal. However, a liquid composition on a plate is always a starting point in plate calculations. Consider that the composition on plate $n + 1$ is known. An approximation of the composition of the vapor entering the plate will permit a preliminary calculation of L_{n+1} and V_n, using Eqs. (43) and (49). Equation (45) will now permit a calculation of the vapor composition y_n. If this differs from the assumed value sufficiently to affect seriously the heat content of the vapor, one repetition of the calculations, using the preliminary value of y_n, to calculate the vapor-heat content will usually suffice.

Simplifying Assumptions. When certain simplifying assumptions can be made without introducing appreciable errors, the use of the sectional heat balances, Eqs. (49) and (50), in conjunction with the material balances, Eqs. (45) and (48), is unnecessary. Many common rectifications permit of this simplification. These assumptions are:

1. Sensible-heat changes through the tower are negligible in comparison with the latent heat. This condition will be more nearly fulfilled the less the temperature difference between the bottom and the top of the tower.
2. The molar latent heats of all components are equal.
3. The heat of mixing of the components is negligible.
4. Heat losses from the tower are negligible.

Applying assumptions 1, 2, 3, and 4 to Eq. (49), it is evident that, under these conditions, the heat content of the vapor passing any section above the feed plate is constant; therefore

$$V_n H_n = (D + R) H_D$$

also

$$H_n = H_D$$

whence

$$V_n = D + R \quad (51)$$

or, for the conditions noted, the vapor flow expressed in moles is constant throughout this section of the tower and equals the sum of the distillate and reflux. The heat content of the liquid downflow passing any section in this part of the tower must also be constant; hence

$$R h_r = L_{n+1} h_{n+1}$$

and

$$h_r = h_{n+1}$$

from which

$$R = L_{n+1} \quad (52)$$

Equation (52) states that, if the assumptions apply, on a molar basis the overflow above the feed plate is constant and is equal to the reflux.

For the part of the tower below the feed plate, Eq. (50) may be solved for Q_b:

$$Q_b = V_m H_m + W h_w - L_{m+1} h_{m+1}$$

Solving Eq. (42) for Q_b:

$$Q_b = R H_D + D H_D + W h_w - F h_F - R h_r$$

Combining:

$$V_m H_m - L_{m+1} h_{m+1} = R H_D + D H_D - F h_F - R h_r$$

The total heat of the vapor must be constant at all points under the assumptions 1, 2, and 3; hence,

$$V_m H_m = R H_D + D H_D$$

and

$$H_m = H_D$$

Therefore

$$V_m = D + R = V_n \quad (53)$$

or, for the assumption made, the molar vapor flow throughout the tower is constant and equals the sum of the distillate and reflux.

Similarly, the heat content of the liquid downflow must be constant, in this section,

$$L_{m+1} h_{m+1} = F h_F + R h_r$$

and

$$h_{m+1} = h_F = h_r$$

Therefore

$$L_{m+1} = F + R \quad (54)$$

The feed was assumed to enter the tower as liquid; therefore the overflow for plates below the feed equals the sum of the feed and reflux.

Equations (51), (52), (53), and (54) together with Eqs. (45) and (48) form the basis of simplified methods for calculation of the number of plates required in a tower.

Use of Liquid-vapor Equilibrium Relationship. The application of the material and heat balances discussed in the preceding articles enables the composition of vapor entering a plate to be calculated from the liquid composition on that plate under fixed operating conditions. If the composition of liquid on the plate below can be calculated from the vapor composition, the concentration changes from plate to plate may be computed. The relationship between vapor and liquid concentrations of any component at equilibrium furnishes this necessary step. In an actual tower, complete equilibrium is not attained on any of the plates. In one of the methods to be described for calculation of tower plates, complete equilibrium is assumed on every plate. The plates thus calculated are termed "theoretically perfect" plates. Correction for non-attainment of equilibrium is then made by dividing the number of theoretically perfect trays by a plate efficiency factor. For binary mixtures the equilibrium relationship may be conveniently represented on a plot of y vs. x (see p. 579).

McCabe-Thiele Graphical Method for Calculation of Number of Plates. [McCabe and Thiele, *Ind. Eng. Chem.*, **17**, 605 (1925).] Consider the tower of Fig. 37, assuming the feed to be a binary mixture. When the four assumptions noted on p. 590 may be made, linear relationships between the vapor composition entering any plate and the liquid composition on that plate are given by Eqs. (45) and (48) for plates above and below the feed plate, respectively. The compositions of the feed x_F, the overhead distillate x_D, and the bottoms x_w are dictated by the requirements of any particular problem. If the reflux is now fixed, the slopes of these lines may be calculated, and the determination of a single point on each line will determine the line.

Equations (45) and (48) may be plotted, using values of y as ordinates and values of x as abscissas. On the same plot, the y vs. x equilibrium curve for the given binary mixture is plotted. Such a diagram is shown in Fig. 40, to which a 45-deg. line passing through the origin has been added for reference. Equations (45) and (48) are represented by lines AC and CD, respectively, while the equilibrium curve is given by $OEFG$. The use of this diagram may be summarized by considering a plate on which the liquid composition is x_4. The composition of the vapor rising from this plate, y_4, is given by point L on the equilibrium curve if perfect equilibrium be assumed. The liquid composition on the plate above is x_3, represented by point R on the line CD, which constitutes the application of Eq. (45) graphically. The composition of the vapor entering plate 4, y_5, is given by point S. The changes in concentration through the column starting at the composition of the overhead distillate x_D may thus be traced by going stepwise from the line AC horizontally to the equilibrium curve and then, downward, vertically to the line AC as shown by the steps in Fig. 40. The computation can also be performed working "up" the column if desired.

It can be shown that the point C, the intersection of lines AC and CD, corresponds to the theoretical feed-plate composition. In tracing the compositions from plate to plate the point C will correspond exactly to a vertical step only by chance. In application, therefore, the feed plate may be taken as that plate corresponding to the first vertical step to the left of point C.

Total Reflux. Referring to Fig. 40, the equation of the line AC is given by Eq. (45)

$$y = \frac{L}{L + D}(x) + \frac{D}{L + D}x_D \tag{45}$$

Inspection of Fig. 40 will show that the slope of line AC, as drawn, is less than unity. This is confirmed by Eq. (45) in which the slope of AC is $\frac{L}{(L + D)}$. If D has a finite value, $\frac{L}{(L + D)}$ must be less than unity.

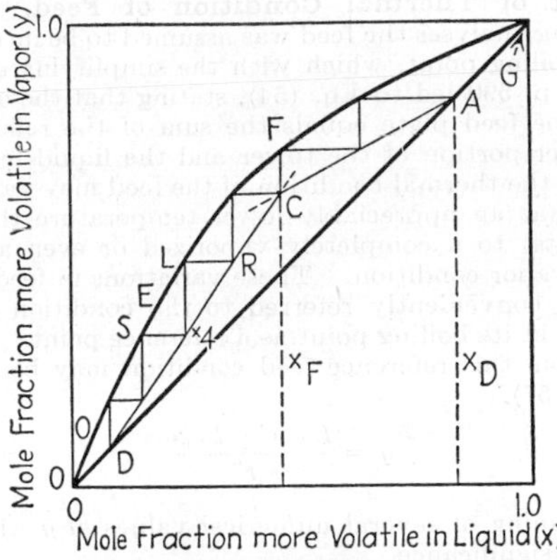

FIG. 40. McCabe-Thiele diagram.

Consideration of line CD, the equation of which is Eq. (48),

$$y = \frac{L}{L - W}x - \frac{W}{L - W}x_w$$

shows that, for a finite value of W, the ratio $L/(L - W)$, representing the slope of the line CD, must be greater than unity. If the lines AC and CD are made to approach the diagonal OG, the slope of AC increases, approaching unity; and the slope of CD decreases, approaching unity. For the slopes of AC and CD to equal unity, D must equal zero, and W must equal zero; hence F equals zero. This represents a tower to which no feed is introduced and from which no product is drawn. Under these conditions the minimum number of plates will be required to obtain given concentrations at the top and bottom of the tower, respectively. A corollary of the foregoing conditions is that a tower operating with an L/V ratio of unity (or reflux ratio R/D of infinity) has an infinite heat consumption per unit of product. It is evident that values of L/V approaching unity are impractical for commercial towers. Columns are sometimes operated at very high reflux ratios in order to effect difficult separations.

Theoretical Minimum Reflux Ratio. If the point C, Fig. 40, the intersection between the lines CD and AC, is raised until it reaches the equilibrium curve, an infinite number of steps will be required on either side of point C in making the stepwise calculation. An infinite number of plates will be required in such a tower; hence this condition represents the theoretical minimum reflux ratio for which the separation desired is possible. For liquids of normal volatilities the equilibrium curve will be concave downward at all points, as shown in Fig. 40. In this case the minimum ratio of downflow to vapor is given by

$$\frac{L'_{n+1}}{V_n} = \frac{x_D - y_F}{x_D - x_F} \tag{55}$$

If the equilibrium curve is concave upward in the region to the right of point C, the line AC will touch the equilibrium curve to the right of point C as point C is raised. In this case the minimum reflux ratio must be calculated by the equation

$$\frac{L'_{n+1}}{V_n} = \frac{x_D - y}{x_D - x} \qquad (56)$$

by reading a number of values of x and y from the equilibrium curve and inserting in Eq. (56). L'_{n+1}/V_n may be plotted against x and the minimum value of L'_{n+1}/V_n determined.

Effect of Thermal Condition of Feed. In the preceding analyses the feed was assumed to be introduced at its boiling point, which with the simplifying assumptions of p. 590 led to Eq. (54), stating that the overflow below the feed plate equals the sum of the reflux from the upper portion of the tower and the liquid feed. In practice the thermal condition of the feed may vary from "cold" (at an appreciably lower temperature than the feed plate) to a completely vaporized or even a superheated vapor condition. These variations in feed condition are conveniently referred to the condition of feed entering at its boiling point as a reference point. Deviations from this reference feed condition may be defined by Eq. (57).

$$q = \frac{L_{m+1} - L_{n+1}}{F} \qquad (57)$$

The meaning of several numerical values of q will illustrate its significance.

- $q = 1$: $F + L_{n+1} = L_{m+1}$, or the reference condition of feed entering as liquid at its boiling point.
- $q > 1$: $qF + L_{n+1} = L_{m+1}$, which corresponds to cold feed, or a condition in which the overflow below the feed equals the overflow from above plus the moles of liquid feed plus $(q - 1)$ moles of induced reflux per mole of feed.
- $q = 0$: $L_{n+1} = L_{m+1}$, representing entirely vaporized feed.
- $0 < q < 1$: partly vaporized feed.
- $q < 0$: (a negative quantity), corresponds to a lower downflow in the stripping section than in the rectifying section, or a superheated vapor feed.

q can be determined by a heat balance around the feed plate. For most purposes q can be regarded as the heat required to bring 1 mole of feed from inlet to feed-plate temperature and to vaporize it, divided by the molal latent heat of vaporization of the feed. The assumptions involved in this approximation are that the enthalpies of the liquid overflow from the feed plate and the plate above are equal and that the enthalpies of the vapor entering and leaving the feed plate are equal, all of which usually fall within the simplifying assumptions of p. 590.

Equations (44) and (47) (p. 589) are equations for the upper and lower operating lines, respectively, and these lines intersect for the condition of liquid feed preheated to its boiling point at an abscissa of $x = x_F$. To investigate the general case, Eqs. (44) and (47) can be rewritten using the values of y_i and x_i to denote the coordinates of intersection of the two operating lines.

$$V_n y_i = L_{n+1} x_i + D x_D \qquad (44a)$$
$$V_m y_i = L_{m+1} x_i - W x_w \qquad (47a)$$

Subtracting Eq. (44a) from Eq. (47a) and substituting, $F x_F = W x_w + D x_D$,

$$(V_m - V_n) y_i = (L_{m+1} - L_{n+1}) x_i - F x_F$$
$$\frac{(V_m - V_n) y_i}{F} = \frac{(L_{m+1} - L_{n+1})}{F} x_i - x_F$$
$$\frac{L_{m+1} - L_{n+1}}{F} = q$$

by Eq. (57) and

$$\frac{V_m - V_n}{F} = q - 1$$

by material balances; hence

$$(q - 1) y_i = q x_i - x_F \qquad (58)$$

or

$$y_i = \frac{q}{q - 1} x_i - \frac{x_F}{q - 1} \qquad (59)$$

Equation (59) is the locus of the points of intersection of the operating lines for all values of q and hence all thermal conditions of feed. This locus is a straight line passing through the diagonal at $y_i = x_i = x_F$. Equation (59) may thus be plotted on the diagram for any values of q and the operating lines laid out with a minimum of calculation. If desired, the actual points of intersection y_i and x_i may be calculated from the following equations obtained by combination of Eq. (59) with Eq. (44a) or (47a):

$$x_i = \frac{x_F + (q - 1)\left(1 - \dfrac{L}{V}\right) x_D}{q - (q - 1)\left(\dfrac{L}{V}\right)} \qquad (60)$$

or

$$x_i = \frac{\left(\dfrac{L}{D} + 1\right) x_F + (q - 1) x_D}{\dfrac{L}{D} + q} \qquad (61)$$

and

$$y_i = \frac{q x_F \left(1 - \dfrac{L}{V}\right) + \left(\dfrac{L}{V}\right) x_F}{q\left(1 - \dfrac{L}{V}\right) + \dfrac{L}{V}} \qquad (62)$$

or

$$y_i = \frac{\left(\dfrac{L}{D}\right) x_F + q x_D}{\dfrac{L}{D} + q} \qquad (63)$$

Figure 41 gives the generalized case of Fig. 40, showing the effects of various thermal conditions of the feed upon the intersections of the operating lines. In Fig. 41 a constant downflow to vapor (L/V) ratio in the upper portion of the column is employed, represented by the operating line AC_1 (C_2, C_3, C_4, C_5). The respective lines MC_1, MC_2, MC_3, MC_4, and MC_5 represent several values of q, or thermal condition of feed as follows:

MC_1: $q > 1$, or "cold" feed.
MC_2: $q = 1$, or feed introduced at its boiling point.
MC_3: q between 0 and 1, or partly vaporized feed.
MC_4: $q = 0$, or saturated vapor feed.
MC_5: $q < 0$, or superheated vapor feed.

It will be noted that, for the constant L/V taken for the upper portion of the column, the greater the heat content of the entering feed the higher the L/V ratio in the stripping section of the column. Thus, while column performances are usually compared in terms of reflux ratio in the upper portion of the column, complete design investigation requires evaluation for the particular separation of the relative fractionating duties of the stripping and enriching sections. In such investigation the following relations should not be forgotten:

1. A given value of L/V in the upper portion of the column represents a fixed heat abstraction in the condenser.

2. For a fixed heat abstraction in the condenser the column heat balance must be maintained by heat input to the reboiler and heat in the entering feed.
3. While "cold" feed may result in fewer total plates, the reboil heat requirement is thereby increased.
4. Heat input in the feed usually represents, in part at least, heat recovered by exchange, and heat introduced at the base of the tower represents external heat chargeable against the operation.
5. For the majority of separations some feed preheat is desirable, but overextension of heat input in the feed may result in unnecessary load on the upper portion of the column because of minimum stripping requirements.

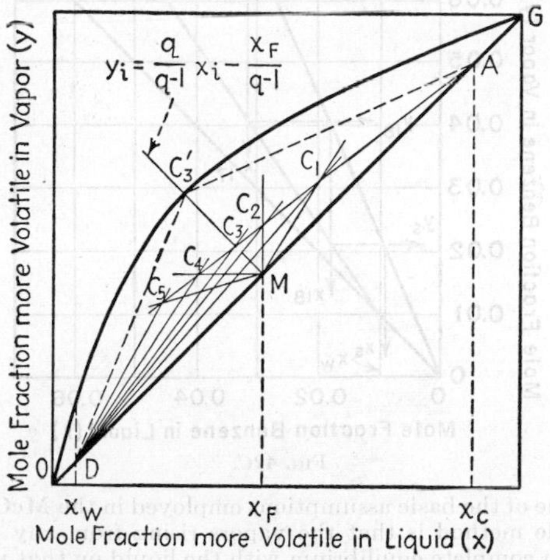

FIG. 41. The effect of thermal condition of the feed upon intersections of the operating lines and upon minimum reflux ratio.

Minimum Reflux Ratio Related to Thermal Condition of Feed. In Fig. 41 operating lines for a thermal condition of the feed corresponding to a value of q between 0 and 1 are given by DC_3 and C_3A, with the intersection at C_3. For a fixed thermal condition of the feed the locus of the intersections of the operating lines for any L/V ratio is given by Eq. (59). The minimum reflux ratio for this fixed value of q is found by raising C_3 along the line $y_i = \frac{q}{q-1} x_i - \frac{x_F}{q-1}$ until it reaches the equilibrium curve at C'_3. The minimum reflux ratio may be determined graphically by laying out the operating lines through C'_3, or it may be calculated by Eq. (64).

$$\frac{L'_{n+1}}{V_n} = \frac{x_D - y_f}{x_D - x_f} \tag{64}$$

Equation (64) differs from Eq. (55) only in that x_f designates actual feed-plate liquid composition rather than the composition of the feed x_F. As in the case of Eq. (55), Eq. (64) applies only to mixtures having equilibrium curves concave downward throughout. For curves concave upward in certain ranges (such as ethanol-water), the minimum value of L/V must be found by substituting several x,y values between the distillate and feed-plate compositions as read from the equilibrium curve to find the point of tangency of the operating line with the equilibrium curve.

Total Reflux. The condition of total reflux is approached by moving the intersection C_3, Fig. 41, along the locus toward M. Inasmuch as total reflux corresponds to zero feed input, the total reflux condition is independent of the thermal condition of the feed.

Selection of Reflux Ratio and Feed Preheat. The preceding discussion has indicated that the optimum design reflux ratio is only determined by economic balance studies. Reflux ratio is related to the thermal condition of the feed; hence a complete economic study involves investigation of reflux ratios for various thermal conditions of the feed. Reflux ratios may vary from the theoretical minimum corresponding to an infinite number of plates to total reflux corresponding to minimum plates. Determination of the minimum reflux ratio for binary mixtures has been previously discussed, while the determination of minimum plates at total reflux may be determined by a stepwise calculation between the equilibrium curve and the diagonal, or by the method described on p. 595.

Minimum plates at total reflux and the minimum reflux ratio give asymptotic values in the relation between reflux ratio and number of plates. The calculation of two points in the working range of reflux ratio will permit the entire curve to be approximated. It is desirable that the reflux ratios so selected fall within the usual working range. For separations not ordinarily requiring a large number of plates (*i.e.*, less than 40) the economic reflux (L/D) will usually be not more than 50 per cent greater than the theoretical minimum.

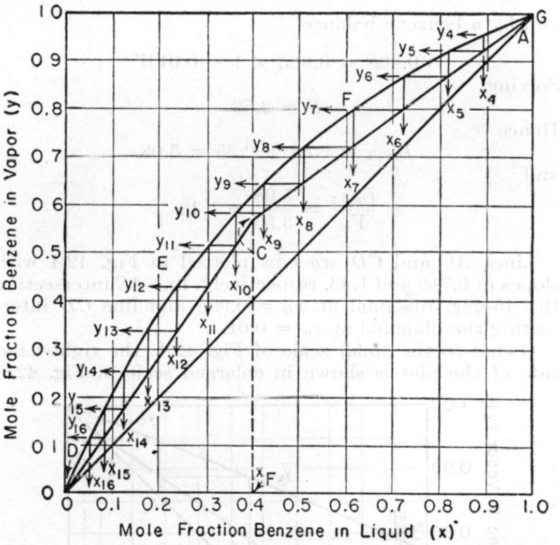

FIG. 42*A*.

Inasmuch as complete economic design investigation involves the thermal condition of the feed, a rough rule to assist in fixing this variable for preliminary calculations is desirable. The economic thermal condition of the feed will be found for many separations to lie between a condition of feed preheated approximately to the boiling point ($q = 1$) and partial vaporization (q between 0 and 1) such that the moles of feed vaporized slightly exceed the moles of top product.

Application of McCabe-Thiele Method. A mixture consisting of 40 mole per cent benzene and 60 mole per cent toluene is to be rectified continuously. It is desired to obtain an overhead distillate containing not less than 99.5 mole per cent benzene and a bottoms stream containing not more than 1 mole per cent benzene. An estimation of the number of plates necessary in the tower is required, assuming the plate efficiency is 60 per cent. Conditions and apparatus are similar to those of Fig. 39, with $q = 1$.

Solution. In Fig. 42*A*, the y vs. x curve for benzene-toluene mixtures has been plotted from the data of Table 5.

Calculation of Minimum Reflux Ratio. The benzene-toluene y vs. x curve is normal, hence the minimum reflux ratio may be calculated from Eq. (60); $\frac{L'_{n+1}}{V_n} = \frac{x_D - y_F}{x_D - x_F} = \frac{0.995 - 0.620}{0.995 - 0.400} = 0.630$ mole of overflow per mole of vapor. Since $V_n = R + D$ [Eq. (51)], and $R = L_{n+1}$ [Eq. (52)], if D is taken as unity

$$\frac{R'}{D} = 1.70 \text{ moles reflux per mole overhead distillate}$$

Assume 50 per cent more reflux than the theoretical minimum is used:

$$\frac{R}{D} = 1.70 \times 1.5 = 2.55$$

and

$$\frac{L_{n+1}}{V_n} = \frac{2.55}{3.55} = 0.718$$

By an over-all material balance

$$F = D + W = 1 + W$$

and by a benzene balance

$$0.40F = 0.995 \times 1 + 0.01W$$

Solving,

$$F = 2.53$$

Hence

$$L_{m+1} = 2.53 + 2.55 = 5.08$$

and

$$\frac{L_{m+1}}{V_m} = \frac{5.08}{3.55} = 1.43$$

Lines AC and CD are now plotted in Fig. 42A with slopes of 0.718 and 1.43, respectively, line AC intersecting the 45-deg. diagonal at $x_D = 0.995$ and line CD intersecting the diagonal at $x_w = 0.01$.

Owing to the small scale of Fig. 42A, the right-hand side of the plot is shown in enlarged scale in Fig. 42B.

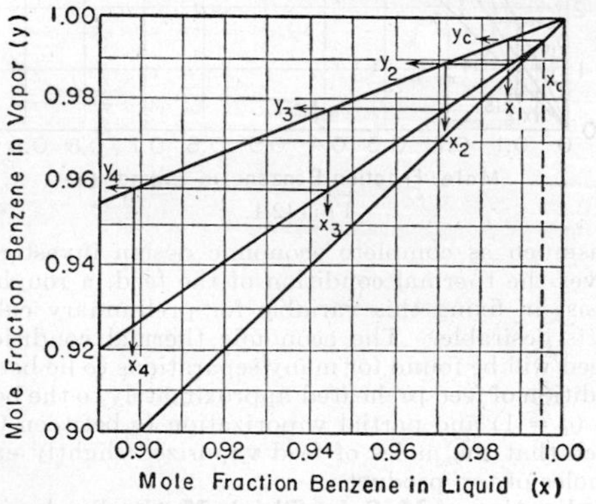

FIG. 42B.

Starting at $x_D = 0.995$, the vapor and liquid compositions are indicated by arrows for each step, each step representing a perfect plate. At x_4, the calculation is transferred to Fig. 42A and continued until x_{16} is reached when, for accuracy, it is continued on Fig. 42C, which gives the left side of the diagram on an enlarged scale.

The apparatus of Fig. 39 was provided with a still or reboiler at the base of the tower. Since the vapors from the reboiler will be in equilibrium with the liquid in the reboiler under actual operating conditions, the reboiler is always considered as a 100 per cent efficient plate. The number of perfect plates in the tower, as calculated, is 18 plus the reboiler. Inspection of Fig. 42A shows that the feed should be introduced on plate 10, giving nine perfect plates above the feed and nine below plus the reboiler.

Taking the plate efficiency as 60 per cent, the number of actual plates will be 15 plates above the feed and 15 below, or a total of 30 plates plus the reboiler.

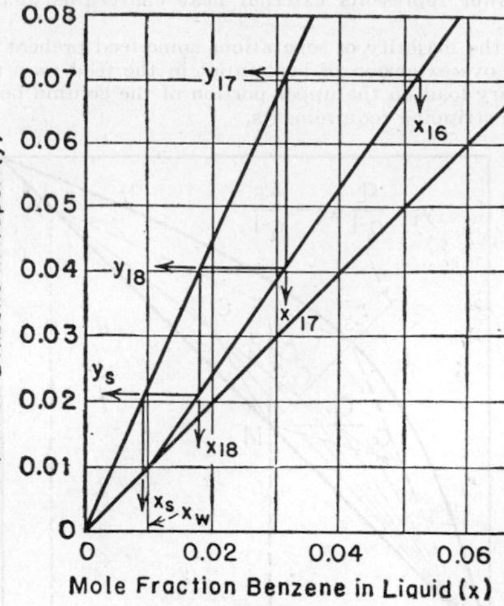

FIG. 42C.

One of the basic assumptions employed in the McCabe-Thiele method is that the vapors rising from any plate are in complete equilibrium with the liquid on that plate. Equilibrium is not attained in an actual tower, and the number of theoretically perfect plates is corrected by dividing by a plate efficiency to obtain the actual number of plates.

Batch Rectification

Differences between Batch and Continuous Rectifications. The preceding discussions have indicated that, for continuous rectification, if the compositions of the distillate and bottoms streams are fixed, for a given quantity and composition of feed, the design of the equipment resolves itself into a balance among reflux ratio, height of column (*i.e.*, number of plates or their equivalent), diameter of the column, and heat consumption. For a constant rate of delivery of distillate, increase in reflux ratio decreases the height of column required, but the diameter and heat consumption per unit of product increase.

In continuous operation the compositions and quantities of the vapor and liquid streams passing any given point in the system are constant. With batch operation, however, a batch of liquid is charged to the still and heated, and when vapors have reached the condensing system a portion of the condensed distillate is returned as reflux. (In actual operation when specification products are required it is advantageous to operate the column under total reflux conditions until a steady initial state is reached.) As top-product take-off is carried out, the material in the still becomes poorer in the more volatiles; hence as the distillation proceeds the reflux ratio must be increased to maintain a constant top-product quality, or a sacrifice in sharpness of separation must be made.

Engineering calculations on batch-rectification equipment are concerned with two general problems:

1. The design of new equipment to perform a given duty of processing a certain quantity of material for required separations. This requires a consideration of the turn-around time available for the processing of a given batch, including clean-out time,

charging and heating, and the actual production period, together with the sharpness of separation required on the one or more distillate fractions.

2. Adaptation of existing equipment to new uses. Batch-distillation equipment is frequently installed for all-purpose use. It is then necessary to estimate the utility of such equipment for new separations. In existing equipment the column size with respect to vapor and liquid capacities is fixed, as are the heating and condensing surfaces and the heat-transfer media.

Calculation methods are summarized herewith:

Determination of Plate Efficiency. When dealing with existing equipment details of internal construction may be difficult to determine in order to estimate plate efficiency. If the column can be operated with a given mixture at total reflux the Fenske equation [*Ind. Eng. Chem.*, **24**, 482 (1932)] can be used for determination of theoretical plates for a binary:

$$N + 1 = \frac{\log \left[\frac{x_p(1 - x_s)}{x_s(1 - x_p)} \right]}{\log \alpha} \tag{65}$$

where N = theoretical plates.
x_p = product concentration, mole fraction.
x_s = kettle concentration, mole fraction.
α = relative volatility.

Relation between Time of Batch Cycle and Vapor-handling Capacity of Column, Constant Composition of Top Product, Variable Reflux Ratio, Final Kettle Composition Fixed, Column Hold-up Negligible. Where a fixed specification of distillate must be maintained the reflux ratio must be increased during the batch rectification. For a fixed number of plates the relation between time of cycle and vapor-handling capacity, for a binary, is given by

$$\theta_T = \frac{S_1(x_p - x_{S_1})}{V} \int_{x_{S_2}}^{x_{S_1}} \frac{dx}{(1 - L/V)(x_p - x)^2} \tag{66}$$

where θ_T = time of batch cycle, hr.
S_1 = initial batch charge, moles.
S_2 = residue in kettle at end of cycle, moles.
x_p = fixed distillate composition, mole fraction.
x_{S_1} = initial still composition, mole fraction.
x_{S_2} = final still composition, mole fraction.
x = still composition, varying during cycle.
L = liquid downflow, moles/hr.
V = vapor flow, moles/hr.

[Bogart, *Trans. Am. Inst. Chem. Engrs.*, **33**, 139 (1937)].

Application:

1. Fix the required distillate composition and tolerance of more volatile in residue.
2. Assume the number of plates in the column, or fix as dictated by existing equipment.
3. If solving for vapor-handling capacity, fix θ_t as determined by turnaround time. If dealing with existing equipment, determine V.
4. On y-x diagram plot various enriching-section operating lines from required x_p value, and determine by stepwise procedure for the given number of plates the value of x in the still.
5. Integrate graphically, plotting x, still composition, against $[(1 - L/V)(x_p - x)^2]^{-1}$ between limits x_{S_1} and x_{S_2}. The calculations as required for the integration will give the values of reflux ratio as required during the progress of the batch cycle to maintain x_p constant.
6. For comparative purposes the same example as used for the McCabe-Thiele application to continuous rectification (see p. 593) will be employed, *i.e.*, the fractionation of a charge consisting of 40 mole per cent benzene and 60 mole per cent toluene into a distillate containing 99.5 mole per cent benzene and a bottoms containing 1.0 mole per cent benzene.

For comparative purposes the following basis will be employed: Still charge = S_1 = 100 moles. Time of batch distillation (exclusive of time for charging, heating, removing residue, and cleaning) = 1 hr. The still is equipped with 18 theoretical plates, as was determined for the continuous example.

With the distillation time and the number of plates fixed, the required vapor-handling capacity will be determined as a measure of column size, reboil heat, and condensing requirements.

Figure 43 shows the y-x diagram for benzene-toluene upon which the overhead product composition, x_p = 0.995, the initial still composition, x_{S_1} = 0.40, and the

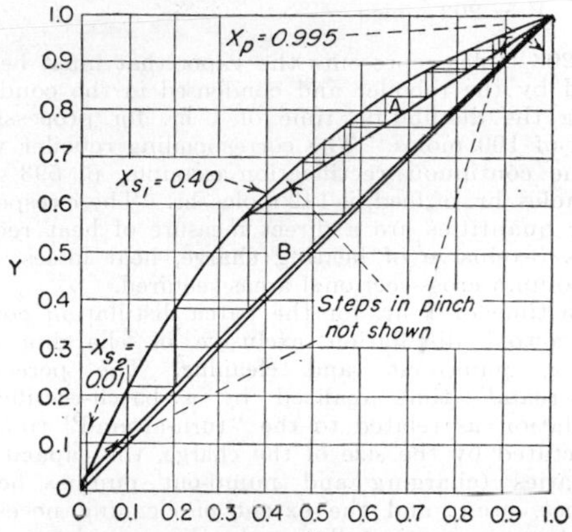

FIG. 43.

final still composition x_{S_2} = 0.01 have been designated. A series of enriching-section operating lines are drawn in from x_p on the $y = x$ line to intersect the equilibrium curve at such points that the 18 theoretical plates plus the still will give successive values of the still composition as the batch operation proceeds. (In Fig. 43, because of size limitations, only the initial and final lines designated as A and B, respectively, are shown.) From these lines values of L/V are calculated corresponding

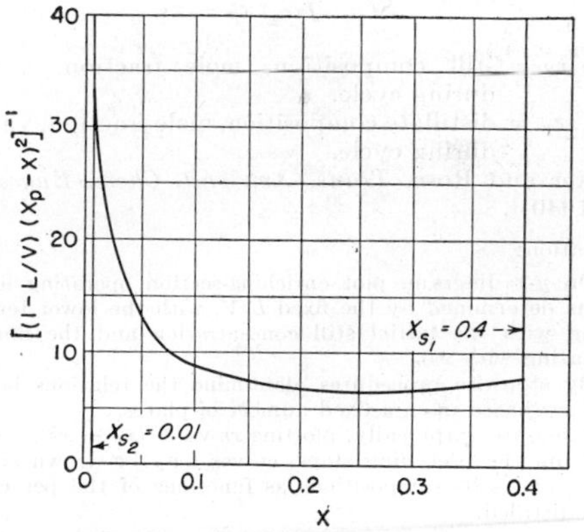

FIG. 44.

to the successive values of still composition. Equation (66) is solved by graphical integration, the following tabulation and Fig. 44 illustrating the steps.

Table 6

x_p (1)	x (2)	L/V (3)	$1 - L/V$ (4)	$(x_p - x)^2$ (5)	(4) × (5) (6)	1/(6) (7)
0.995	0.400	0.636	0.364	0.354	0.129	7.75
	.270	.713	.287	.526	.151	6.62
	.180	.782	.218	.664	.144	6.93
	.120	.840	.160	.765	.122	8.20
	.050	.915	.085	.893	.076	13.2
	.010	.969	.031	.970	.030	33.4

In Fig. 44, column (7) of Table 6 is plotted against column (2), the area giving 3.41 integration units. Applying Eq. (71)

$$\theta_T = \frac{S_1(x_p - x_{S1})}{V} \int_{x_{S_2}}^{x_{S_1}} \frac{dx}{(1 - L/V)(x_p - x)^2}$$

$$1 = \frac{(100)(0.595)(3.41)}{V}$$

$$V = 203 \text{ moles}$$

$V = 203$ moles represents the vapor that must be generated by the reboiler and condensed in the condenser during the distillation time of 1 hr. for processing a batch of 100 moles. The corresponding reboiler vapor for the continuous-rectification example, p. 593, basis 100 moles/hr. of feed, is 142 moles/hr. These respective vapor quantities are a direct measure of heat requirements (exclusive of heating charge, heat losses, etc.), and column cross-sectional areas required.

The time of 1 hr. for the batch distillation covered only actual distillation exclusive of charging time, heating, pump-out, and cleaning. The percentage "on-stream" time realized by a batch-rectification installation as related to the "turn-around" time will be dictated by the size of the charge, the capacities of auxiliaries (charging and pump-out pumps, heating surfaces, etc.), and the extent of cleaning necessary; but the average installation handling a clean charge probably cannot average over 80 per cent of the complete cycle time. Assuming an 80 per cent factor, the comparison between batch and continuous becomes

Type of Operation	Reboiler Vaporizing Capacity and Column Area
Batch	203/0.8 = 254 moles/hr.
Continuous	142 moles/hr.

Determination of Product Composition Variation during Batch Cycle at Fixed Reflux Ratio, Column Hold-up Negligible.

$$\ln \frac{S_1}{S_2} = \int_{x_{S_2}}^{x_{S_1}} \frac{dx_S}{x_p - x_S} \quad (67)$$

where x_S = still composition, mole fraction, variable during cycle.

x_p = distillate composition, mole fraction, variable during cycle.

[Smoker and Rose, *Trans. Am. Inst. Chem. Engrs.*, **36**, 285 (1940)].

Application:

1. On y-x diagram, plot enriching-section operating lines of slope as determined by the fixed L/V, with the lower terminus starting with the initial still concentration and the final line terminating with x_{S_2}.
2. By stepwise procedures, determine the relations between x_p and x_S, based upon a fixed number of plates.
3. Integrate graphically, plotting x_S vs. $1/(x_p - x_S)$.
4. From the preceding steps, curves may be drawn showing product and still compositions as functions of the per cent of charge distilled.

Effect of Column Hold-up in Batch Rectification. [Colburn and Stearns, *Trans. Am. Inst. Chem. Engrs.*, **37**, 291 (1940)].

In the event column hold-up is appreciable in relation to the still charge, Colburn and Stearns show that the following equations represent the course of the batch cycle:

$$\ln \frac{S_1}{S_2} = \int_{x_{S_2}}^{x_{S_1}} \frac{dx_S}{x_p - x_S - \dfrac{H\,dx_h}{dS}} \quad (68)$$

$$y_n = Rx_{n+1} + (1 - R)\left[x_p - \frac{N-n}{N}\left(\frac{H}{S_1}\frac{dx_h'}{dS_1/S_2}\right)\right] \quad (69)$$

where R = reflux, moles/hr.

N = total number of theoretical plates.

n = plates from given section to top.

H = column hold-up, moles.

x_h' = hold-up composition from n to top, mole fraction.

x' = over-all hold-up composition, mole fraction.

Equation (69) indicates that the operating line at any instant in the batch cycle is curved rather than straight like the conventional continuous lines. This curvature is such that the true operating line will in general lie below the line calculated for the given reflux ratio, and this will result in a sharper separation than that calculated on the basis of negligible hold-up. Equations (68) and (69) are illustrative only and cannot be directly applied except by trial-and-error procedure to determine x_h and x_h' relations at various periods in the cycle.

TYPES OF RECTIFYING COLUMNS OR TOWERS

Devices in which vapors from a still pass countercurrently to a portion of the condensate returned as reflux are termed "rectifying columns or towers" and are so called on account of their generally tall and upright form. The interchange of material between vapor and liquid phases which takes place in rectification is inherently a diffusional process, and the apparatus in which it takes place should provide for countercurrent flow of reflux and vapor and intimate mixing of the two streams throughout the countercurrent path.

Three types of rectifying columns have found widespread use:

1. Bubble-cap plate towers.
2. Sieve plate towers.
3. Packed towers.

Bubble plate towers are provided with a number of horizontal plates, or trays, which are usually equally spaced in the upright tower shell. The spacings used vary from 6 in. to 3 ft. Reflux passes down the tower from plate to plate by means of suitable downflow pipes. The entrances to the downflow pipes may constitute weirs over which the reflux must flow, and the height of these weirs determines the depth of liquid that is maintained on the tray. The lower ends of the downflow pipes are sealed in the liquid of the tray to which they deliver. The vapor rises through each of the trays in series. In passing through each tray the vapor is deflected beneath the surface of the liquid and is distributed throughout the liquid by means of bubble caps. Bubble caps present a wide variety in design, a common type consisting of a circular cap inverted over a vapor riser of smaller diameter than the cap. The periphery of the cap clears the plate by a small distance and is slotted to cause the vapor to pass into the liquid in the form of streams of small bubbles. With well-designed bubble trays intimate contact between vapor and liquid is obtained over wide ranges of vapor and liquid rates of flow, and the effectiveness of the contact does not change very rapidly with rates of flow of either vapor or liquid. Channeling of the liquid and vapor streams can be prevented, even in towers of large diameter; in fact there is practically no limit to tower size for which satisfactory bubble trays can be developed. Bubble-tray towers up to 32 ft. in diameter are in actual operation. Bubble-tray towers are the most widely used type of fractionating device in the larger scale installations, and various problems in their design are considered in some detail on pp. 597 to 602.

Sieve-plate columns were used by Coffey as early as 1832 and have been extensively employed ever since. In these columns a series of horizontal plates are spaced at distances of 6 in. or more in the column shell. The plates are perforated with a number of holes distributed over the plate, a convenient size being 3/16-in.-diameter holes spaced on ½-in. centers. The pressure and velocity of the vapor passing up through these holes must be sufficient to hold up a body of liquid reflux on each plate. To prevent too great a depth accumulating, each plate is provided with an overflow or downpipe, similar to those of the bubble-cap type, which rises about 1 in. above the plate and reaches down to about ½ in. from the plate below where it dips into a cup that forms a liquid seal against flow of vapor. Sieve plates of this kind cannot be operated at less than 50 per cent of the full vapor rate, or the liquid will drain through the perforations and the bubbling contact ceases which results in great loss in efficiency. For this reason sieve-plate columns are frequently provided with an automatic regulator on the heat supply, the control being actuated by the pressure at the base of the column. Sieve plates are also constructed of wire screens. Vapor velocities up to 3 ft./sec. are reported for sieve-plate columns, and plate efficiencies in rectification are said to be high when the plates are properly installed and operated at the designed rates of flow. Disadvantages of this type of plate are:

1. The plates operate properly only over a limited range of vapor and liquid rates of flow.
2. If the plates are not perfectly leveled, all the liquid will run through the low side, and the vapor will flow through the high side.
3. If the plates corrode, the holes become larger, and efficiency falls off.

(See p. 615 for data on performance of perforated plates.)

Packed columns consist of upright shells filled with loose pieces of solid material of uniform size thrown in at random in the interstices of which the reflux and vapor are distributed. Hempel introduced packed columns for laboratory use in 1881 [*Z. anal. Chem.*, **20**, 502 (1881)]. The filling was 4-mm.-diameter glass beads. Before 1890, Ilges employed porcelain balls 1 to 2 in. in diameter in plant columns (Maerker-Delbrück, "Spiritus Fabrikation," 1908 ed., p. 813). More recently, Raschig patented the use of cylindrical rings of height equal to the diameter. The size of the ones most used was 25 by 25 mm. The material of the rings may be glass, porcelain, copper, iron, or other material that will not be attacked or corroded by the material distilled. It is considered on theoretical grounds that the fractionating efficiency for ring packings varies inversely with the ring diameter so that if the rings are doubled in diameter the tower height must be doubled. Other varieties of tower packing have been introduced by Zeisberg, Prym, Lessing, and Brégeat. In the operation of packed towers, unless all the packing elements are thoroughly and continuously wetted with reflux downflow, channeling of the vapor is likely to occur, and the fractionating efficiency will be decreased. See pp. 616 to 621, for data on packed towers.

DESIGN OF BUBBLE-CAP PLATE TOWERS

The preceding articles have dealt with the number of plates necessary in towers, as governed by the separations required and the reflux used. In the design of the complete tower, adequate passages for the flow of the vapor and liquid streams, the distance between trays, the distribution of the reflux and its flow over the trays, and the type and arrangement of the bubble caps must be decided. The design of distillation equipment involves the application of scientific principles and empirical data in which engineering judgment based on past good practice is essential.

Vapor Velocity in Plate Towers. The classical treatment of plate-tower design from the vapor and liquid capacity standpoint has been to treat tolerances in each of the major factors as more or less separate identities. Within recent years it has become generally recognized that such independent consideration may lead to very unsatisfactory results and that an attempt must be made to evaluate each factor in relation to the others. This makes complete rationalization of this phase of column design extremely difficult and explains the fact that experience has continued to play a major role. Although separate discussions of the several principal factors follow, an attempt will be made to point out, at least qualitatively, their general relationships.

The safe vapor rate at which a column may be operated represents one of the primary factors in determining column size and economics. For many years it was considered that entrainment, or the mechanical carry-over from plate to plate of relatively small liquid droplets, defined the limitation in vapor velocity for any given conditions. The results of many quantitative determinations of actual entrainment obtained with various liquid-vapor systems and small-scale plate designs have been published within the past few years [Sherwood and Jenny, *Ind. Eng. Chem.*, **27**, 265 (1935). Holbrook and Baker, *Ind. Eng. Chem.*, **26**, 1063 (1934). Pyott, Jackson, and Huntington, *Ind. Eng. Chem.*, **27**, 821 (1935). Peavy and Baker, *Ind. Eng. Chem.*, **29**, 1056 (1937)]. These results, together with the mathematical analysis of the effect of entrainment per se on column performance [Rhodes, *Ind. Eng. Chem.*, **26**, 1333 (1934); **27**, 272 (1935). Colburn, *Ind. Eng. Chem.*, **28**, 526 (1936)], lead to the following conclusions:

1. Entrainment may reach a high ratio in terms of weight of entrained liquid per unit weight of vapor before fractionating efficiency is impaired. (In case where the elimination of minute traces of color or odor bodies is an accompanying problem, entrainment per se may become of more effect.)
2. Before vapor rates are reached at which entrainment interferes with fractionating efficiency, other factors usually become controlling. These include basic changes in the mechanism of vapor-liquid contact with vapor rate, froth or foam build-up on the tray, jetting or spouting action between caps, and weir and downflow performance.

In view of the above complexity of the factors limiting safe operating vapor rates, any attempts at generalized correlation must be regarded as empirical expressions of commercial practice.

For many years designers have employed various variations of Eq. (70) as a guide to vapor velocities.

$$u = K_v \sqrt{\frac{\rho_1 - \rho_2}{\rho_2}} \qquad (70)$$

where u = superficial vapor velocity (based on total column area), ft./sec.

ρ_1 = density of liquid downflow under column conditions of temperature.

ρ_2 = density of vapor under column conditions. [With the increased commercial necessity for high-pressure columns for certain purposes, the question of the vapor density to employ in Eq. (70) has frequently arisen. Most designers of column equipment appear to prefer the use of the perfect-gas-law vapor density in their correlations, rather than an actual or calculated corrected value. This discussion, therefore, assumes the gas-law vapor density with respect to ρ_2 and u.]

K_v = constant as defined by materials handled and tray-design variables.

Equation (70) can be derived upon a theoretical basis by equating the frictional upward pull of vapor upon a suspended liquid droplet to the downward gravitational effect on this same particle. Hence the original significance attached to Eq. (70) was as a means for the selection of a vapor rate that would substantially eliminate entrainment.

In view of the other factors recognized as of major importance, Eq. (70) can now be regarded as an empirical expression useful in approximating workable vapor rates and diameters of fractionating towers, which, upon subsequent investigation of liquid flows, weir arrangements, and cap spacings, will not require too great modification.

Table 7 represents an attempt to define semiquantitatively values of K_v in Eq. (70), based both upon the several published investigations cited and general commercial practice [*cf.* Souders and Brown, *Ind. Eng. Chem.*, **26**, 98 (1934)]. Referring to Table 7, it will be noted that tray spacing appears to lose appreciable significance in permitting higher vapor velocities for spacings above 30 in. Recognition of the effect of liquid seal (defined as height of liquid above the top of the cap slots based upon liquid density under flowing conditions but assuming absence of foam) is semiquantitative only down to plate spacings of approximately 16 in. The lower plate spacings are usually dictated by economic conditions extraneous to those of column design proper and are employed with special classes of materials. At the lower plate spacings the characteristics of the liquid downflow have greater effect on performance owing to height of foam or spray generated under operating conditions, and hence specific experience becomes of increasing importance. In general, Table 7 is considered to approximate conservative practice in petroleum and hydrocarbon distillation where tray spacings customarily range from 18 to 30 in., and operating pressures range from 50 mm. mercury to 400 lb./sq. in. abs. Important deviations from Table 7 are found in commercial hydrocarbon absorbers (absorption of C_3 to C_6 compounds with a heavier petroleum scrubbing oil), in which general practice appears to employ vapor velocities corresponding to 65 to 80 per cent of the K_v's of Table 7.

Table 7. Approximate Values of K_v in Eq. (70)

Plate spacing, in.	Liquid seal, in. 0.5	1	2	3
6	0.02–0.04			
12	0.09–0.11	0.07–0.09	0.05–0.07	
18	0.15	0.14	0.12	0.09
24	.185	.17	.16	.15
30	.195	.185	.18	.175
36	.205	.195	.19	.185

Examples of the use of Eq. (70) in approximating vapor rates for design of several oil fractionating towers are given herewith:

1. Atmospheric topping tower—gasoline and steam overhead.

Pressure	Atmospheric
Temperature	250°F.
Vapors	100 average molecular weight
Liquid	0.650 sp. gr. at 250°F.
Plate spacing	24 in.
Liquid seal	2 in.
K_v (from Table 7)	0.16

$$\rho_2 = \frac{100 \mid 492}{359 \mid 710} = 0.193$$

$$\rho_1 = 0.65 \times 62.4 = 40.5$$

$$u = 0.152\sqrt{\frac{40.5 - 0.193}{0.193}} = 2.3 \text{ ft./sec.}$$

2. Pressure distillate stabilizer, C_1 to C_4 overhead.

Pressure	200 lb./sq. in. abs.
Temperature	155°F.
Vapors	49.1 average molecular weight
Liquid	28.6 lb./cu. ft. at 155°F.
Plate spacing	24 in.
Liquid seal	1 in.
K_v (from Table 7)	0.17

$$\rho_2 = \frac{49.1 \mid 200 \mid 492}{359 \mid 14.7 \mid 615} = 1.49$$

$$u = 0.170\sqrt{\frac{28.6 - 1.49}{1.49}} = 0.725 \text{ ft./sec.}$$

3. Vacuum column, gas oil, and steam overhead.

Pressure	50 mm. Hg
Temperature	350°F.
Vapors	164 average molecular weight
Liquid	0.681 sp. gr. at 350°F.
Plate spacing	24 in.
Liquid seal	0.5 in.
K_v (from Table 7)	0.185

$$\rho_2 = \frac{164 \mid 50 \mid 492}{359 \mid 760 \mid 810} = 0.0183$$

$$\rho_1 = 0.681 \times 62.4 = 42.5$$

$$u = 0.180\sqrt{\frac{42.5 - 0.018}{0.0183}} = 8.9 \text{ ft./sec.}$$

4. Cracking plant tower (receiving reformer and light gas oil-cracking furnace discharges after tar elimination). Gasoline and gas overhead

Pressure	250 lb./sq. in. abs.
Temperature	450°F.
Vapors	75 average molecular weight
Liquid	0.626 sp. gr. at 450°F.
Plate spacing	24 in.
Liquid seal	1 in.
K_v (from Table 7)	0.170

$$\rho_2 = \frac{75 \mid 250 \mid 492}{359 \mid 14.7 \mid 910} = 1.92$$

$$\rho_1 = 0.626 \times 62.4 = 39.1$$

$$u = 0.170\sqrt{\frac{39.1 - 1.92}{1.92}} = 0.75 \text{ ft./sec.}$$

Vapor Velocity in Vapor Lines. The vapor lines leading from the top of a fractionating tower to the condensing equipment must be of sufficient size to prevent excessive pressure drop. This is particularly important in the design of vacuum towers in which it is desired to realize the lowest absolute pressure at the top of the tower for a given pressure maintained at the condenser outlet. Table 8 gives approximate ranges of vapor velocities which are employed for both atmospheric and vacuum conditions. In any case the pressure drop in the vapor lines may be calculated by the methods of Sec. 5 and a vapor velocity and the corresponding line size chosen to keep the pressure drop within the limit dictated by the conditions of design.

Table 8. Vapor Velocities in Vapor Lines

Absolute Operating Pressure	Vapor Velocity in Vapor Line, Ft./Sec.
Atmospheric	40– 60
100–50 mm. Hg	100–150
Below 50 mm. Hg	150–200

Plate Spacings. Plate spacings generally accepted as conventional in towers for specific services have been arrived at by the influence of factors related to performance and maintenance, together with the inevitable economic shakedowns that tend to standardize specialized equipment in general features. Columns that, for various reasons, are usually completely housed in buildings are frequently designed with tray spacings of 12 in. or less. These columns are often for sharp separations demanding a large number of plates; hence it is desirable to minimize height at the expense of cross section. By proper selection of materials, this class of columns usually requires little in the way of internal inspection and access. On the other hand, columns in the petroleum industry are not housed and present various degrees of internal cleaning, inspection, and maintenance demands. Columns for the distillation of crudes and for cracking-plant service usually are designed with plate spacings of 24 in. or more, permitting manholes between plates of sufficient size for removal of plate sections and permitting working space between plates. Plate spacings of 24 in. as compared with lower plate spacings for this type of tower do not impose economic burdens in tower cost, because the increased height is compensated for by the higher permissible vapor rates. The more rapid change of allowable vapor rates with plate spacing in the region of the lower spacings will lead to the conclusion that, considering column cost alone, larger plate spacings and smaller diameters would be more economical. Housing and other considerations frequently overbalance pure column design considerations for the lower plate spacings of 12 in. or less.

In general, as plate spacings are reduced, closer attention must be paid to cap submergences and foam or froth height upon the plate, since a greater proportion of the free disengaging area above the plate will be consumed by the presence of foam or an agitated liquid zone. To a lesser extent, pressure drops must be watched to ensure maintenance of the liquid downflow seal between plates.

Shape and Spacing of Caps. A wide variety of shapes and designs of bubble caps is to be found in commercial applications. In the larger capacity towers used in the petroleum industry, the principal types of caps used are the familiar round bell caps with diameters ranging from 4 to 7 in. and tunnel, or rectangular caps, of widths from 3 to 6 in. and lengths varying from 12 in. up. Both types are slotted around their lower periphery and are either anchored to the plate above the vapor risers with teeth resting on the plate or are suspended with adjustable leveling arrangements with the teeth clearing the plate. In general, the tunnel-type caps require a smaller total number for a given slot and riser area, and hence present fewer parts to handle and adjust in assembling a tray. This may represent simplification of maintenance where frequent cleaning of trays is necessary. The older style round bell caps appear, however, to give better fractionating performance and are readily adaptable to variations in tray layouts without necessitating caps of two or more different sizes in order to obtain the desired cap slot and riser area.

Spacing between caps is a balance between an effort to obtain a large proportion of vapor-riser and slot area and considerations of liquid flow on the plate. The statement appears frequently in the literature that caps should be located sufficiently close for impingement of vapor streams from adjacent caps to obtain greater agitation and turbulence. At the higher vapor rates now used, it is recognized that too close spacing of caps may result in jetting or spouting action between caps, particularly with tunnel-type caps that have lengths of slots facing one another. This jetting action may project liquid or coarse spray to a considerable height above the plate and thus induce carry-over.

Study of cap action in relation to liquid flow is frequently made using a full-scale section of a plate (representing the complete path of liquid flow from inlet to outlet weir) with air and water as the vapor and liquid media. By this means, the effects of cap spacing as related to vapor rate, cap submergence, and liquid flow can be appraised. As a result of the complicated relationships presented by cap spacing, it is possible to indicate only that the general range of spacings employed appears to be from 1 to 3 in.

Spacing of caps in relation to exit weirs has become recognized as of importance. With a large proportion of the caps on a tray adjacent to weirs, jetting or splashing over the weirs may materially reduce the liquid depth for which the tray was designed. If fairly stable froth or foam is produced on the tray under operating conditions, the flow over the exit weir may consist of this less dense material and result in a lowering of the effective depth on the tray. Hence in some cases it has been found desirable to provide a zone adjacent to the weirs in which no caps are placed and which permits subsidence of foam and prevents jetting over the weirs.

Types of Liquid Flow. The simplest type of liquid flow with respect to the mechanical problems of tray and down-spout arrangements is simple cross flow, in which the liquid travels from side to side in opposite directions on successive plates. By this arrangement all trays in a tower section can be of identical design, successive trays being installed in 180-deg. relationship. Downflows are segregated on opposite sides of the tower with no downflow piping manifolding consuming valuable disengaging space above the plates or interfering with access to the space between plates. Conditions that may limit this type of design are hydraulic gradients from inlet to outlet weir, or exit-weir lengths obtainable with the weir located at one side of the tower. The problem of weir lengths is frequently solved by utilizing central exit weirs on one tray and side exit weirs on the following tray. Hydraulic gradients, which may cause uneven vapor distribution between caps, may be allowed for by varying the cap heights across the tray or by "cascade" design, interspersing distributing weirs at several positions in the liquid flow.

Liquid-handling Capacities of Trays. Liquid flow across a bubble-cap tray requires a definite hydrostatic head between the liquid entrance and exit to overcome the resistance set up by the caps that the liquid must traverse. This hydraulic gradient can be such that no vapor passes through caps near the liquid entrance, forcing the vapor to pass through the remaining caps. The resultant overloading of the operating caps can cause entrainment, loss of efficiency, and danger of flooding. Good, Hutchinson, and Rousseau [*Ind. Eng. Chem.*, **34**, 1445 (1942)] have presented a quantitative analysis of the liquid-handling capacity of a tray which, though limited to the specific cap dimensions and spacings used in their tests, illustrates certain relationships.

Good *et al.* (*loc. cit.*) employ certain concepts and nomenclature as follow:

A stable tray is one on which all bubble caps are passing vapor. This does not mean that vapor distribution is uniform, since the closer the approach to instability the poorer the distribution.

An unstable tray is one in which certain caps are passing no vapor because of the hydraulic gradient set up from liquid inlet to outlet. In a large tower the area of inactive caps may shift from side to side; hence the term "unstable."

Liquid build-up is the term applied to the type of hydraulic gradient obtaining on a bubble tray.

S_m = minimum seal, difference between level of clear liquid at down pipe and level of top of slots, inches.

Δp_0 = pressure drop through plate at zero seal, *i.e.*, no liquid head above top of slots, in. water.

u = linear vapor velocity through entire free area of column, ft./sec.

ρ = vapor density, lb./cu. ft.

N = number of caps per sq. ft. of tower area upon which u is based ($N = 7.16$ for experimental plate).

Skirt clearance: distance from bottom of cap to tray.

Bubble-cap spacing: $4\frac{1}{4}$-in. equilateral triangular centers.

Bubble-cap dimensions: see Fig. 45.

Figure 46 shows the relationship of Good *et al.* between Δp_0 and $u(\rho)^{0.5}$. [$u(\rho)^{0.5}$ is a simplified form of the right-hand side of Eq. (70), p. 597.]

Figure 47 gives the generalized summary of the results of Good *et al.* The following example from the original paper illustrates the use of Fig. 47.

Example. An 8-ft.-diameter tower is designed to absorb ethyl alcohol from a dilute mixture of air and alcohol vapor, using water as absorbing medium. Each plate is to have 350

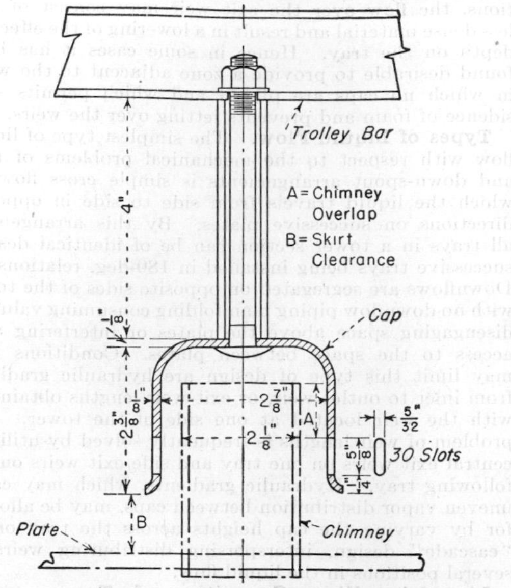

FIG. 45.

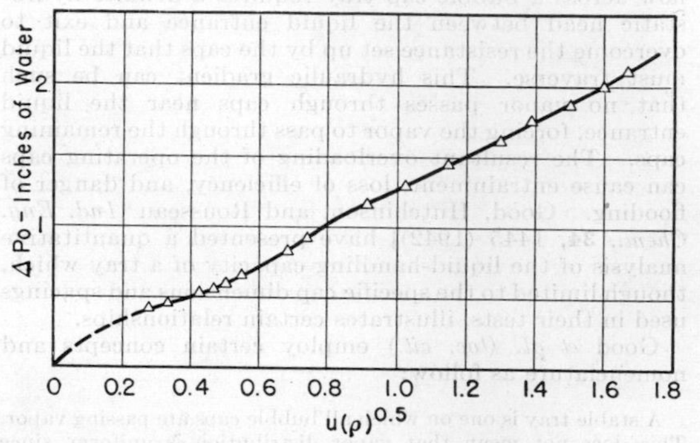

FIG. 46.

caps with dimensions as shown in Fig. 45. The caps are to be placed on 4¼-in. equilateral centers, providing staggered flow, and to have a skirt clearance of 1 in. The following characteristics apply:

Vapor rate, $u(\rho)^{0.5}$	0.65
Total liquid load, gal./hr.	7000
Plate width at short row of caps, ft.	6.5
Plate width at wide row of caps, ft.	8
Average plate width, ft.	7.25
Minimum seal, in.	1.0
Number of rows	19

$$\text{Actual liquid rate} = \frac{7000}{7.25} = 965 \text{ gal./hr./ft.}$$

$$N = \frac{350}{(8)^2 \times 0.785} = 6.95$$

$$u(\rho)^{0.5}\left(\frac{7.14}{N}\right) = 0.65\left(\frac{7.14}{6.95}\right) = 0.67$$

Capacity, from Fig. 47 for skirt clearance = 1 in., S_m = 1 in., and $u(\rho)^{0.5}\left(\frac{7.14}{N}\right) = 0.67$.

$$\frac{(\text{Gal./hr./ft.})(\text{number of rows})}{1000} = 19$$

$$\text{Allowable rate} = \frac{19 \times 1000}{19} = 1000 \text{ gal./hr./ft.}$$

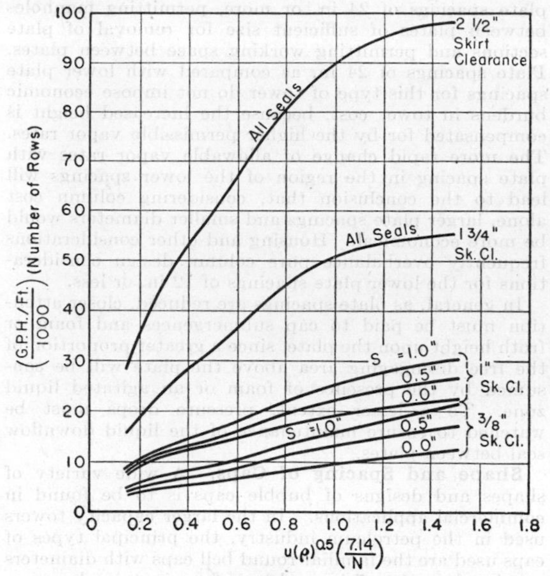

FIG. 47.

Since the actual liquid rate is 965 gal./hr./ft., the plate should be stable.

Weirs and Down Spouts. This portion of plate design has demanded greater attention with the present use of pressure operation, in which smaller tower cross sections are demanded by vapor flow, but in consequence of which liquid-handling capacity may become controlling.

Overflow weirs of various shapes have been employed. One of the simplest, which is frequently used on smaller scale columns, consists of a circular weir with the down spout a continuation of the pipe representing the weir. In larger towers straight or slightly curved weirs are more frequently employed.

The recognition of the probable presence of foam (rather than simple agitated liquid on trays) in many operations has led to various provisions for foam disengagement before the overflow enters the down pipe proper. A simple illustration lies in the provision of a straight exit weir as a chord across the tower at one side. Behind this weir one or more circular pipes, flush with the plate, may operate as submerged orifices. The circular segment behind the weir permits foam disengagement space before entrance of the liquid into the down spouts. These considerations of down-spout capacity as limited by foam disengagement are apparently old in the distillation art, as recently [Chute, (Discussion) *Trans. Am. Inst. Chem. Engrs.*, **34**, 91 (1938)] a patent disclosure of 1849 was noted in which the top of the down spout (and entrance weir) was a cup several times the diameter of the down spout.

In general, removal of liquid with low weir head prevents considerable variation in bubble-cap submergence with variations in liquid flow. Weir heads (calculated on the basis of foam-free liquid) up to 3 in. are frequently employed when the plate spacing is 18 in. or more. Simple cross-flow trays with weirs on opposite sides can usually be laid out with exit-weir lengths from 60 to 80 per cent of the tower diameter and with maintenance of vapor-riser and slot areas within recognized desirable minimum values.

For large vacuum towers in which liquid flow is low in relation to tower diameter, notched weirs are frequently employed to ensure liquid distribution and to eliminate uneven liquid flow due to small deviations in weir height. Notch widths may be varied in various stages across the weir to obtain desired liquid flow in relation to caps traversed.

The relation between volume of flow, weir perimeter, and liquid head on the weir may be estimated for rectangular weirs from the Francis weir formula:

$$V = 3.33LH^{1.50} \tag{71}$$

where V = flow over weir, cu. ft./sec.
L = length of weir, ft.
H = head on weir, ft.

Circular weirs are frequently used in towers and for these the following formula [Gourley, *Proc. Inst. Civil Engrs.*, **184**, 297 (1910–1911)] is recommended:

$$V = KLH^{1.42} \tag{72}$$

in which V and H are the same as in Eq. (71), L is the perimeter of the outside edge of the circular weir in feet and K varies from 2.93 to 3.03 as the diameter of the circular weir varies from 7 to 26 in.

The downflow passes from the weirs to downflow pipes which are sealed in the liquid on the tray beneath. When these downflows are continuations of circular weirs, their diameter may be considerably less than that of the weirs. In all cases their size should be carefully checked and an ample factor of safety allowed to preclude the possibility of backing up of liquid due to insufficient size to handle the liquid downflow.

Souders *et al.* [*Ind. Eng. Chem.*, **30**, 86 (1938)] showed that, for a circular weir forming the top of a down spout, three differing conditions of fluid head may apply:

1. At low heads ($H < D/4$) operation is as a weir, and the weir equations (71) or (72) can be employed.
2. At intermediate heads ($3D/2 > H > D/2$) operation is as a free-running orifice with sufficient vortex area to allow free separation of vapor from liquid. The orifice formula ($u = C\sqrt{2gh}$) with a $C = 0.5$ is recommended by Souders *et al.* (*ibid.*).
3. At high heads the orifice equation should still apply with a correction factor to allow for the decreased flow due to vapor separation within the orifice.

The results of Souders *et al.* (*ibid.*) were obtained with a 2-in.-diameter weir and down spout. This work, however, emphasizes the importance of considering froth or foam flow into the weir zones of plates and providing a froth subsidence zone before entrance into the down spout.

Pressure Drop and Slot Opening. Bubble caps may be divided into two types when considering pressure drops:

1. Those in which the slots extend to the bottom of the periphery in the form of open teeth, mounted so that the teeth clear the plate.
2. Caps in which the slots are closed rectangles with an unbroken band of metal forming the lower part of the periphery. Caps of type 1 are similar to those of type 2, if they are mounted so that the teeth rest on the plate.

Caps and mountings as in type 1 operate best at vapor rates such that the major portion of the slots are working, but without vapor escaping in the clearance between teeth and plate. For this type of cap, Rogers and Thiele [*Ind. Eng. Chem.*, **26**, 524 (1934)] give equations relating slot opening to vapor volume and physical dimensions of the slots for triangular and rectangular slots, respectively. These relations are as follows:

For triangular slots:

$$V = NC\left(\frac{B}{2A}\right)(0.001068)\sqrt{\frac{2g}{R}}\,h^{5/2} \tag{73}$$

where V = vapor volume per cap, cu. ft./sec.
N = number of slots per cap.
C = orifice coefficient = 0.51.
B = base of triangular slot, in.
A = altitude of triangular slot, in.
g = 32.2 ft./sec./sec.
$R = \dfrac{\text{density of vapor}}{(\text{density of liquid}) - (\text{density of vapor})}$
h = slot opening measured from top of slot, in.

For rectangular slots:

$$V = NCW(0.00134)\sqrt{\frac{2g}{R}}\,h^{3/2} \tag{74}$$

where V = vapor volume per cap, cu. ft./sec.
N = number of slots per cap.
C = orifice coefficient = 0.51.
W = width of slot, in.
h = slot opening measured from top of slot, in.

The pressure drop through the cap is then equal to the difference in liquid level between the outside and the inside of the cap, neglecting entrance loss, frictional losses through riser and annular space, and flow reversal, all of which are usually small. The liquid depth on the plate under flowing conditions is determined by the overflow weirs. The actual slot submergence or liquid head above the top of the slots may be calculated. The pressure drop then equals liquid head above the top of slots plus slot opening. In applying these equations, it should be noted that the head of liquid above the top of the slots in absolute pressure units may be less than that calculated from the weir flow due to the production of foam and the effect of this less dense material on weir action. However, if the liquid is assumed to be foam-free, pressure drops calculated in this manner will be conservative on the high side. Experimental data presented by Rogers and Thiele (*ibid.*) with several vapor-liquid combinations showed exponents differing somewhat from the respective theoretical values of $5/2$ and $3/2$ in certain cases, but the theoretical exponents as given are recommended for engineering purposes.

For rectangular slots, closed at the bottom and operated at vapor velocities such that vapor flows through all the slot opening, the orifice form of equation will give the pressure drop through the slots.

$$H_s = Cu_s^2\frac{\rho_2}{\rho_1} \tag{75}$$

where H_s = pressure drop, in. water.
C = coefficient = $\dfrac{12}{(C_0^2)(64.4)}$
u_s = slot velocity, ft./sec.
ρ_2 = vapor density.
ρ_1 = density water.

Souders *et al.* (*loc. cit.*) reported a value for C of 0.533 determined with a dry plate and slots 0.1875 in. wide by 1.0 in. deep. Rogers and Thiele's (*loc. cit.*) correlations indicated a value of 0.51 for C under flowing conditions.

In general, for atmospheric and pressure operation, pressure drops over plates are not a serious design limitation, provided that the general tolerances with respect to liquid depth and per cent of vapor-riser and slot area have been maintained. For vacuum operation, low pressure drop per tray is desirable, as pressure drop through the tower may result in a large percentage increase in bottom tower pressure over the low absolute pressure maintained at the condenser outlet. Hence lower slot submergences are employed for vacuum work, and bubble trays having individual pressure drops of from 1 to 2 mm. Hg per tray represent usual performance.

Cap-slot and Riser Areas. As previous discussion indicates, the vapor-riser and cap area must frequently be balanced against weir and down-spout design. This necessitates detailed study in the drafting room. In general, balances between weir lengths and riser area can be obtained with the riser area representing 10 to 20 per cent of the tower area for towers of 3 ft. and greater diameter, the greater riser areas being more readily obtainable with the larger diameters. Usual design practice maintains the vapor-riser area, the passage areas under the cap, and the area of slot openings approximately equal.

Reflux Supply. Reflux may be supplied to the tower by the following general methods:

1. Partial condensation of the overhead vapors and return of this partial condensate to the top tray as reflux.
2. Total condensation of the overhead vapors and return of a portion of the total condensate to the top tray. Reflux supplied in this manner is sometimes termed "wet reflux."
3. A combination of 1 and 2.
4. Removal of hot downflow liquid from the tower, cooling, and returning the cooled liquid to the tower.

1. Partial condensation produces a vapor enrichment in the lower boiling constituents by the partial removal of the higher boiling components and consequently has been frequently regarded as a portion of the fractionating system. It has been shown, both theoretically and practically, that the use of a partial condenser as part of the fractionating system is less efficient than when the rectification all takes place in the trays of the tower. The greater cost of condenser surface compared with other means of reflux supply, the inefficiency of such a device as a fractionator, and the lack of flexibility for control purposes have caused a definite trend away from the use of partial condensation in modern practice.

2. In the second method of refluxing, the overhead vapors pass directly to the final condensing system. If the temperature of the vapors from the top of the tower is sufficiently high, two condensers may be used in series: the first operating as a vapor heat exchanger to preheat the fresh feed, and the second as a water-cooled final condenser. In petroleum distillation the condensate from the final condenser is usually collected in a reflux drum from which the reflux supply is pumped and the remainder flows to storage as overhead distillate. In other fields it is usual to elevate the condensers and gravitate the reflux back to the column. If open steam is used in a tower distilling oil, a water separator may be added ahead of the reflux drum. Very effective reflux control may be obtained with this system, either by manual control of the reflux flow or by an automatic control set to maintain a constant temperature at the top of the tower by operating a valve in the reflux line. Control is thus much more direct than when a partial condenser is used. The condensate is usually cooled well below its boiling temperature as it passes to the reflux drum; in fact cooling is essential if low-boiling liquids are to be handled by pumping. Consequently the reflux pumped back to the top tray of the tower is in this case much cooler than the tower-top temperature. This causes a certain amount of condensation on the top tray of the tower, this tray serving to some extent as a partial condenser. The volume of the overflow from the top tray is thus greater than the volume of the cool reflux. This cooling effect on the top tray probably impairs the true fractionating action on this tray. The amount of space and cost of an additional tray are, however, usually much less than the equivalent dephlegmating surface required. Contact on a tray is so effective that evidence of this cooling effect is seldom found to extend more than two trays down the tower. It will be noted that, with cold reflux pump-back, the vapor volume beneath the top tray will be greater than that rising from the top tray. The tower diameter should be based on the higher vapor volume beneath the tray as estimated by a sectional heat balance.

3. Combinations of methods 1 and 2 are sometimes used. A dephlegmator of sufficient surface to handle the major portion of the reflux duty is employed, together with provision for returning a small amount of the final condensate as wet reflux. Control is effected by varying the external reflux.

4. The fourth method of refluxing is applied to special problems such as the production of additional refluxing at some point below the top tray of a tower.

Reboil Heat. Reference to the articles dealing with the theoretical methods of tower calculation will show the necessity of supplying heat in some manner to the base of the tower if true fractionating action is to occur on the trays below the point of introduction of the feed. In the simplest type of installation a direct-fired kettle or still may be surmounted by the fractionating column. In many cases the materials undergoing fractionation are liable to injury by too high temperatures in the film of liquid next to the heat-transfer surface. In such cases means must be taken to avoid high film temperatures. If the temperature level of the operation is sufficiently low, steam condensing in a closed coil affords a method of heating which definitely limits the maximum temperature reached to the condensing temperature of the steam under the pressure employed.

Liquid from the base of the tower may be pumped through a direct-fired tube still in which high velocity of the liquid in the tubes prevents excessive film temperatures, or a liquid, non-volatile at the temperatures used, may be circulated from a tube still through closed coils in the tower and returned to the tube still.

PETROLEUM DISTILLATION

The distillation operations employed in the general chemical industries usually have as their aim the isolation of fairly pure individual compounds from volatile liquid mixtures consisting of, at the most, several well-defined components. Unless the volatilities of two or more of the components from such a liquid mixture are nearly identical, or two or more of the components associate to form a constant-boiling mixture, these separations are readily performed and design of the fractionating equipment may be guided by well-developed theoretical methods as already given. The treatment of problems in the distillation of petroleum is modified by the nature of the raw material, the nature of the commercial products desired, and the scale of the unit operations. Although the underlying principles are the same as for the simpler distillations, the complexity of the problems necessitates the frequent use of empirical generalizations and recourse to the results obtained in past experience. The principal factors that introduce special considerations into the engineering of petroleum distillation equipment are:

1. The raw material is of exceedingly complex composition. The low-boiling constituents of many crudes are paraffin hydrocarbons up to pentane or hexane. As the boiling point increases, the character of the compounds changes and the paraffins are rapidly displaced in predominance by other types of hydrocarbons. The number of compounds in a given boiling range exhibiting very slight differences in volatility multiplies rapidly with rise in boiling point.
2. The major portion of the commercial products manu-

factured from petroleum are in themselves complex mixtures of boiling-point ranges defined by the specifications of the respective products.

3. The character and yields of the various fractions vary widely depending upon the source of the crude, and crudes from the same locality may exhibit marked variations. Distillation equipment must be designed with sufficient flexibility that it will process satisfactorily all the crudes that the refiner may desire to run and permit variation in the specifications of the various products to suit market conditions.

4. The temperature levels at which it is necessary to conduct many of the operations are much higher than in the majority of other applications of distillation. Materials having boiling points from less than 100°F. to the temperatures at which thermal decomposition of hydrocarbons is rapid must be recovered as overhead distillates. Special provisions to reduce the temperatures at which the heavier constituents will vaporize are employed.

5. The scale upon which the unit operations are conducted is much greater than in any other industry employing distillation.

Criteria of Separation. Specifications of the boiling range of the various fractionated petroleum products boiling up to 700° to 750°F. at atmospheric pressure are given in terms of the Engler or A.S.T.M. distillation test (see p. 606). The Engler distillation is a batch distillation carried out under definitely specified conditions. It deviates from simple batch distillation in that a certain amount of rectification is obtained owing to condensation and refluxing in the neck of the flask. Although of little theoretical significance, the results of this test are readily interpreted from the practical viewpoint, owing to its long use in the industry. This test is also used as an empirical measure of the degree of separation between the products obtained by the fractionation of oils.

A.S.T.M. distillations usually constitute one of the routine inspections of the lower boiling products fractionated from petroleum. If the distillation tests of the products from a distillation operation are compared, certain generalizations with regard to the degree of fractionation obtained may be applied. If the "end-point" A.S.T.M. temperature of any given product is lower than the "initial" A.S.T.M. temperature of the next higher boiling range material recovered in the same operation, excellent fractionation is regarded to have been obtained between the two products. The temperature interval between the end point of the lower boiling and the initial of the higher boiling is termed the "gap" between the products (see Fig. 48). If the end-point temperature of the lower boiling fraction exceeds the initial temperature of the higher boiling fraction,

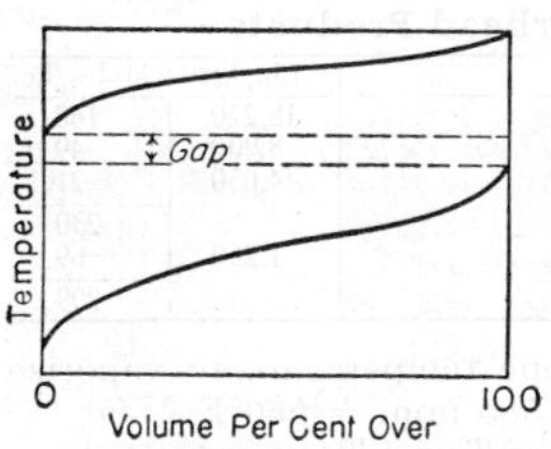

FIG. 48. Engler gap between fractions.

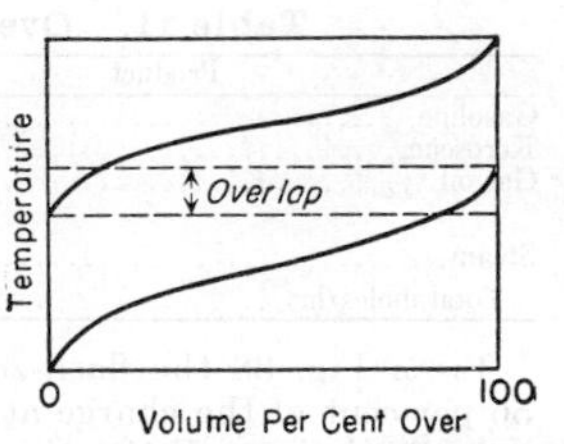

FIG. 49. Engler overlap between fractions.

the fractions are said to "overlap." The temperature interval between the initial of the higher boiling and the end point of the lower boiling is termed the "overlap" of the fractions (see Fig. 49).

The gap or overlap obtained between two streams is influenced by the shape of the distillation curve of the original feed stock, the fractionating equipment used, and the method of operation of the equipment, *i.e.*, the reflux ratio employed. The flatter the A.S.T.M. distillation curve of the feed stock, the more difficult it is to gap two products fractionated from it. Gaps of 50°F. or more may be obtained in commercial equipment with certain feed stocks. For the major portion of the commercial products, gaps of this magnitude are unusual. The degree of fractionation which should be employed is dependent upon an economic balance. Referring to the typical Engler distillation curves of Figs. 54 and 55, it is relatively easy to separate the main bodies of fractions as represented by, say, the 10 to 90 per cent ranges. On the higher boiling ends of the Engler curves the curvature turns sharply upward, and on the lower boiling ends pronounced downward curvatures are obtained. These "tails" represent small amounts of constituents of higher and lower boiling points, respectively, then represented by the main bodies of the fractions. The size of the gap between fractions is largely determined by the degree to which the tails are eliminated. The separation of these small amounts of lower or higher boiling constituents requires a disproportionate expenditure for equipment and operation. Consequently, in modern continuous distillation units, from which a number of fractionated products are obtained in one operation, a gap of zero, *i.e.*, the end point of the lower boiling not exceeding the initial of the higher boiling, is usually regarded as satisfactory fractionation, and even this degree of fractionation is frequently not economically justifiable.

A.S.T.M. distillations afford satisfactory empirical criteria of fractionation for the products that distill at temperatures less than 700° to 750°F. at atmospheric pressure. For higher boiling constituents the A.S.T.M. distillation is meaningless because cracking in the flask vitiates the results. For the determination of the boiling ranges and fractionation of the higher boiling gas-oil and lubricating-oil fractions, vacuum distillation tests are resorted to. The apparatus used may be similar to that employed in the Engler test, the distillation, however, being conducted at a greatly reduced pressure such as 10 mm. Hg. No standardized procedure similar to the A.S.T.M. distillation has as yet been generally adopted; hence vacuum distillation tests as performed by various laboratories may differ widely in results. The same concepts of "gaps" and "overlaps" are employed in judging the results of fractionations as is customary with the Engler test. A vacuum distillation apparatus that has found wide use is described by Peterkin and Ferris [*Ind. Eng. Chem.*, **17**, 1249 (1925)].

Side Streams, or Withdrawing Liquid from Selected Plates of a Column. In modern continuous units for the primary distillation of crudes, a number of fractions are commonly recovered from one fractionating tower. One of the fractions is recovered as an overhead vapor from the top of the tower. Side streams are withdrawn from various plates in the tower and a residue from the base of the tower. From the study of the simpler types of complex mixtures it has been demonstrated both theoretically and practically that the continuous separation of the components of a binary mixture in a fairly pure state may be effected in one tower, and the complete separation of a ternary mixture requires two towers, or in general the number of towers to effect the continuous complete separation of a complex mixture is equal to the number of components less one. It would appear from the foregoing that the prevalent practice of withdrawing several products from one tower is unsound. The justification for this procedure in current practice is that the major products distilled from oils are not pure compounds but complex mixtures of fairly wide boiling ranges and that the proper degree of separation between these fractions is a matter of economic balance.

In practice the removal of small amounts of contaminating lower boiling constituents from the side streams is effected by steam stripping, *i.e.*, the side streams are caused to flow down over the bubble trays in a separate small tower or section of the large tower where they are subjected to the stripping action of superheated steam.

Estimation of Tower Temperatures. An estimation of the temperature at which the tower top must be maintained to secure the desired distillate, and the temperatures at which the several side streams will be withdrawn, is necessary in order to draw up a heat balance as a basis for design of the tower. The following approximations will serve most practical purposes:

1. The temperature that must be maintained at the tower top (the outlet of the partial condenser if one is used) is approximately equal to the temperature at which 75 per cent of a sample of the condensed overhead vapor is distilled over in the A.S.T.M. distillation test, provided that the tower pressure is atmospheric and no open steam is used. If process steam is used in the tower the partial pressure of the overhead vapor must be calculated and the temperature as determined from the 75 per cent A.S.T.M. temperature corrected to the reduced partial pressure by use of the Cox chart or other vapor-pressure data. If the tower is operated at a reduced pressure, this reduced pressure is used in calculating the partial pressure of the oil vapor.

2. Side streams are withdrawn from plates the temperatures of which are approximately given by the temperatures at which 5 to 10 per cent of the respective samples are distilled over in the A.S.T.M. distillations, if the pressure in the tower is atmospheric. In the presence of appreciable amounts of process steam, or if the tower is operated under reduced pressure, the necessary correction by use of the Cox chart must be made. The 10 per cent A.S.T.M. temperature appears to represent a good average for the lower boiling fractions, *e.g.*, kerosene and gas oil. For the higher boiling fractions represented by the lubricating oil cuts, the 5 per cent A.S.T.M. temperature is recommended.

3. For the fractions of too high boiling range to secure an A.S.T.M. distillation test, a distillation test of the Engler type carried out under reduced pressure may be used in a similar manner.

The application of these approximations will be illustrated by the following heat balance.

Use of Heat Balances in Design. Modern continuous-distillation units for crudes are of the type shown diagrammatically in Fig. 50, consisting essentially of a

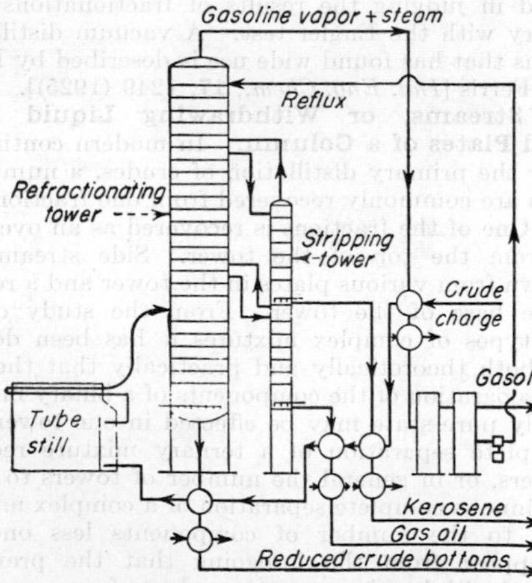

FIG. 50. Crude topping unit.

pipe still, a fractionating tower, and suitable condensing and cooling equipment. The use of a heat balance in the design of such equipment together with application of principles discussed in previous articles will be illustrated by a numerical example.

Table 9. A.S.T.M. Distillations

Gasoline	°F.	Kerosene	°F.	Gas oil	°F.
I.B.P.	100	I.B.P.	400	I.B.P.	510
10	180	10	425	10	565
20	210	20	435	20	570
30	230	30	450	30	575
40	248	40	460	40	579
50	262	50	470	50	582
60	280	60	482	60	585
70	300	70	495	70	588
80	320	80	510	80	592
90	350	90	530	90	595
E.P.	400	E.P.	560	E.P.	615
% Rec.	98	% Rec.	98	% Rec.	98

By use of these data Table 10 may be arranged:

Table 10. Summary of Data

Product	Liquid volume, %	Barrels per day	A.P.I.*	Lb./gal.	Lb./hr.	Mean boiling point†	Molecular weight‡
Gasoline + N.C. Gas (1.5%)	33.5	1,675	58	6.22	18,220	261	114
Kerosene	15.0	750	42	6.79	8,900	475	182
Gas oil	7.5	375	35	7.08	4,650	579	220
Reduced crude	44.0	2,200	24	7.58	29,200		
Crude	100.0	5,000	37.6	6.97	60,970		

$$* \text{Degrees A. P. I.} = \frac{141.5}{\text{sp. gr. } 60°\text{F.}/60°\text{F.}} - 131.5$$

† Mean boiling points (averages of I.B.P., E.P., and nine intermediate temperatures on A.S.T.M. of each cut): Gasoline = 261°F.; kerosene = 475°F.; gas oil = 579°F.

‡ Molecular weights are from Fig. 4, using the mean boiling points from the A.S.T.M. distillations.

For stripping purposes a total of 6.0 lb. of steam per barrel of oil charged will be used in the bottom of the tower and 800 lb./hr. for stripping of side streams in the side-stream strippers. Pounds steam per hour in bottom of tower $= \frac{5000}{24} \times 6 = 1250$. Total process steam = 1250 + 800 = 2050 lb./hr.

A superheater in the furnace setting will superheat low-pressure exhaust steam to 600°F. for use in stripping.

The tower will operate at substantially atmospheric pressure.

Example. Assume 5000 bbl. (42 gal.) of a Mid-Continent crude are to be topped continuously with the recovery of overhead distillates of gasoline, kerosene and gas oil, and a reduced-crude bottoms. The data available are:

1. A true-boiling-point work-up of the crude (see Fig. 38 and p. 607) or other laboratory or plant evaluation of the crude.
2. Approximate A.S.T.M. distillations of the products desired (Table 9).
3. Fundamental data, specific heats (Fig. 9), latent heats (Fig. 11), and molecular weights (Fig. 4).

Calculations. 1. *Estimation of Flash-zone Temperature.*

Partial-pressure oil vapors = $^{230}/_{299} \times 760 = 585$ mm.

Table 11. Overhead Products

Product	Lb./hr.	Moles/hr.
Gasoline	18,220	160
Kerosene	8,900	49
Gas oil	4,650	21
		230
Steam	1,250	69
Total moles/hr.	......	299

From Fig. 38 the flash-zone temperature to vaporize 56 per cent of the charge at 585 mm. = 560°F.

2. *Preliminary Estimation of Tower Temperatures.*

Tower top = 75 per cent on A.S.T.M. of gasoline = 310°F., to be corrected for partial pressure of oil vapors

Assume 1.4 lb. of reflux per pound of gasoline product.

Pounds of oil vapor per hour = 2.4 × 18,220 = 43,700

Moles oil vapor per hour = 43,700/114 = 394

Moles steam per hour = $^{2050}/_{18}$ = 114

508

Partial pressure oil vapors = $^{394}\!/_{508} \times 760 = 590$ mm.
On Cox chart (Fig. 2), 310°F. at 760 mm., corrected to 590 mm. = 290°F.
Kerosene draw-off: Assumed temperature = 400°F.
Gas-oil draw-off: Assumed temperature = 545°F.

Reduced-crude Bottoms: The reduced crude will be assumed to leave the bottom of the tower at 50° less than the flash-zone temperature, *e.g.*, 560 − 50 = 510°F.

3. *Heat Balance.* If the charge enters the tower from the still with a heat content just sufficient to vaporize the overhead products the cooling by means of reflux will be

a. Heat removal to cool the gasoline and steam vapor to 290°F.
b. Heat removal to cool the kerosene vapor to 400°F. and condense the kerosene.
c. Heat removal to cool the gas-oil vapor to 545°F. and condense the gas oil.
d. Cooling of bottoms to 510°F.

This reflux cooling may be calculated by a heat balance.

a. Sensible heat above 60°F. of liquid at tower draw-off temperature.
b. Latent heat at tower draw-off temperature.
c. Sensible heat of vapor from flash-zone temperature to tower draw-off temperature.
d. Sensible heat of liquid above 60°F. at flash-zone temperature.
e. Sensible heat of steam vapor above 60°F. (steam assumed superheated to 600°F.).
f. Sensible heat of liquid above 60°F. at bottoms draw-off temperature.
g. Sensible heat of steam vapor above 60°F. at tower-top temperature.

Table 12. Heat Balance. Basis: Flash-zone Temperature

Heat In, Above 60°F.				B.t.u./hr.
Gasoline + gas..	18,220 (290 − 60) (0.57)	(*a*)	=	2,390,000
	18,220 (110)	(*b*)	=	2,005,000
	18,220 (560 − 290) (0.56)	(*c*)	=	2,760,000
Kerosene........	8,900 (400 − 60) (0.57)	(*a*)	=	1,725,000
	8,900 (94)	(*b*)	=	836,000
	8,900 (560 − 400) (0.57)	(*c*)	=	812,000
Gas oil..........	4,650 (545 − 60) (0.60)	(*a*)	=	1,355,000
	4,650 (77)	(*b*)	=	358,000
	4,650 (560 − 545) (0.59)	(*c*)	=	41,000
Reduced crude...	29,200 (560 − 60) (0.57)	(*d*)	=	8,320,000
Steam..........	2,050 (600 − 60) (0.48)	(*e*)	=	530,000
Total..........			=	21,132,000
Heat Out, Above 60°F.				
Gasoline + gas...	18,220 (290 − 60) (0.57)	(*a*)	=	2,390,000
	18,220 (110)	(*b*)	=	2,005,000
Kerosene........	8,900 (400 − 60) (0.57)	(*a*)	=	1,725,000
Gas oil..........	4,650 (545 − 60) (0.60)	(*a*)	=	1,355,000
Reduced crude....	29,200 (510 − 60) (0.56)	(*f*)	=	7,350,000
Steam..........	2,050 (290 − 60) (0.48)	(*g*)	=	226,000
				15,051,000
Reflux duty..........				6,081,000
Total..........				21,132,000

4. *Calculations Based on Heat Balance.*

Total heat per pound required to heat reflux to tower-top temperature and vaporize:

$$(290 - 60)(0.57) + 110 = 241 \text{ B.t.u.}$$

Pounds per hour reflux pump-back at 60°F.	= 6,081,000/241 =	25,200
Pounds per hour hot reflux	= 6,081,000/110 =	55,200
Moles per hour hot reflux	= 55,200/114 =	485
Moles per hour gasoline	=	160
		645
Moles per hour steam		114
		759

Cubic feet per second of vapor below top tray

$$= \frac{759}{1} \left| \frac{359}{3{,}600} \right| \frac{750}{492} = 115$$

Taking allowable vapor velocity as 2 3 ft./sec.:

Tower area = 115/2.3 = 50.0 sq. ft.
Tower diameter = $\sqrt{50.0/0.785}$ = 8.0 ft.

5. *Check of Tower Temperatures Used in Heat Balance.*

Tower top:

Moles per hour cold reflux = 25,200/114	= 221
Moles per hour gasoline	= 160
	381
Moles per hour steam	= 114
	495

Partial-pressure oil vapors = $^{381}\!/_{495} \times 760 = 585$ mm. 585 mm. checks the originally estimated partial pressure of 590 mm. within the accuracy with which the usual vapor-pressure charts may be read; hence no temperature correction is required.

Kerosene draw-off: A heat balance may be drawn about the tower cutting above the kerosene draw-off. All items in the original heat balance will be identical except that the gasoline and steam will be cooled only to 400°F., the approximate kerosene-draw-off temperature.

Heat out above 60°F. with top product cooled to 290°F.	=	15,051,000
18,200 (400 − 290) (0.52)	=	1,040,000
2,050 (400 − 290) (0.48)	=	108,000
Heat out above kerosene draw-off	=	16,199,000
Equivalent reflux duty	=	4,933,000
	Total	21,132,000

$$\text{Pounds per hour hot reflux to kerosene plate} = \frac{4{,}933{,}000}{94} = 52{,}500$$

Applying a material balance to the section of the tower between the flash zone and the kerosene draw-off it is found that:

Vapor from kerosene tray = gasoline product + total reflux to kerosene tray

At kerosene tray:

Stream	Lb./hr.	Moles/hr.
Gasoline..........	18,220	160
Total reflux..........	52,500	290
		450
Steam..........	2,050	114
		564

Partial-pressure oil vapors = $^{450}\!/_{564} \times 760 = 605$ mm. Using the Cox chart, 425°F. at 760 mm. (see Table 9; 425°F. equals 10 per cent A.S.T.M. temperature), corrected to 605 mm. = 410°F. As compared with the originally estimated temperature of 400°F. this correction will not materially influence the original heat balance.

The gas-oil draw-off temperature may be corrected in a similar manner. With practice the temperature may be estimated sufficiently closely so that a revised heat balance need not be drawn.

Discussion of Heat Balance. The heat balance presented in the preceding illustration assumed a heat content of the oil from the furnace just sufficient to vaporize the overhead products. Consequently the reflux duty calculated represents that required to cool or condense and cool the several products. This reflux is a minimum for the operation. To increase the reflux ratio in the tower it is necessary to increase the heat content of the oil from the furnace or introduce heat in the base of the tower. The latter procedure is better from a theoretical standpoint, but practical difficulties encountered in transferring heat at the high temperature levels prevailing at the bottom of towers, and at the same time avoiding cracking of the bottoms material, have prevented this procedure from gaining widespread use.

LABORATORY DISTILLATIONS

A.S.T.M. Distillation Test. This test, originally known as the Engler distillation, has been used for many years in approximately its present form by the petroleum industry. For complete specifications of the apparatus and technique, as now standardized, the reader is referred to A.S.T.M. Standards, 1930, Part II, A.S.T.M. Designation D 86–30, p. 491, "Standard Method of Test for Distillation of Gasoline, Naphtha, Kerosene, and Similar Petroleum Products." The apparatus is shown in Figs. 51 and 52. The applications of this test to

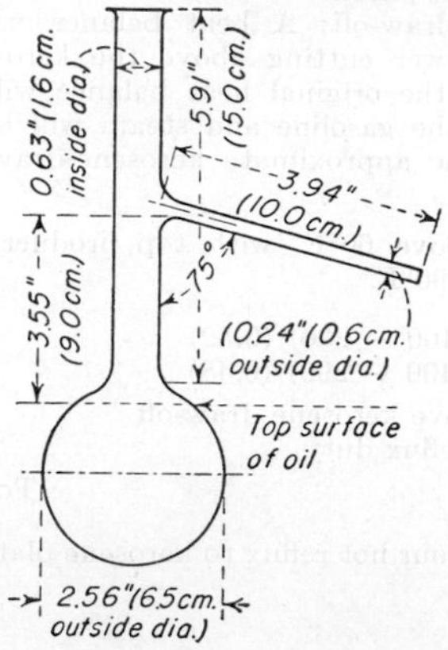

Fig. 51. A.S.T.M. distillation flask.

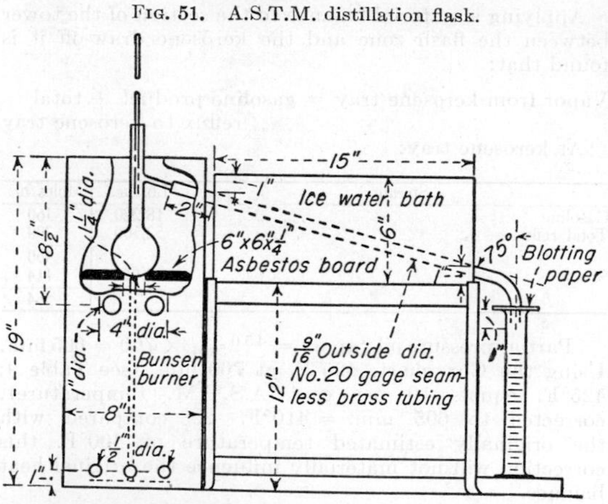

Fig. 52. A.S.T.M. distillation apparatus.

natural-gas gasoline and to gas oils are covered by A.S.T.M. Designation D 216–30 and A.S.T.M. Designation D 158–28, respectively. Equipment for these tests conforming to the A.S.T.M. specifications may be secured from any of the reliable manufacturers of laboratory apparatus.

The method of making the test for gasoline, naphtha, and kerosene is quoted from A.S.T.M. Standards, 1930, Part II, as follows:

Procedure. The condenser bath shall be filled with cracked ice, and enough water added to cover the condensed tube. The temperature shall be maintained between 32° and 40°F. (0° and 4.45°C.).

The condenser tube shall be swabbed to remove any liquid remaining from the previous test. A piece of soft cloth attached to a cord or copper wire may be used for this purpose.

One hundred milliliters of the product shall be measured in the 100-ml. graduated cylinder at 55° to 65°F. (12.8° to 18.3°C.) and transferred directly to the Engler flask. None of the liquid shall be permitted to flow into the vapor tube.

The thermometer provided with a cork shall be fitted tightly into the flask so that it will be in the middle of the neck and so that the lower end of the capillary tube is on a level with the inside of the bottom of the vapor outlet tube at its junction with the neck of the flask. The thermometer shall be approximately at room temperature when placed in the flask.

The charged flask shall be placed in the 1¼-in. (3.18-cm.) opening in the 6 by 6 in. (15.24 by 15.24 cm.) asbestos board with the vapor outlet tube inserted into the condenser tube. A tight connection may be made by means of a cork through which the vapor tube passes. The position of the flask shall be so adjusted that the vapor tube extends into the condenser tube not less than 1 in. (2.54 cm.) nor more than 2 in. (5.08 cm.).

The graduated cylinder used in measuring the charge shall be placed, without drying, at the outlet of the condenser tube in such a position that the condenser tube shall extend into the graduate at least 1 in. (2.54 cm.) but not below the 100-ml. mark. Unless the temperature is between 55° and 65°F. (12.8° and 18.3°C.), the receiving graduate shall be immersed up to the 100-ml. mark in a transparent bath maintained between these temperatures. The top of the graduate shall be covered closely during the distillation with a piece of blotting paper or its equivalent. cut so as to fit the condenser tube tightly.

Distillation. When everything is in readiness, heat shall be applied at a uniform rate, so regulated that the first drop of condensate falls from the condenser in not less than 5 nor more than 10 min. The distillation thermometer shall be read 2 min. after heat is applied and the indication recorded as the "correct temperature." This figure is of significance only in cases when there is a question as to the accuracy of the initial boiling point, as subsequently determined. When the first drop falls from the end of the condenser, the reading of the distillation thermometer shall be recorded as the **initial boiling point.** The receiving cylinder shall then be moved so that the end of the condenser tube shall touch the side of the cylinder. The heat shall then be so regulated that the distillation will proceed at a uniform rate of not less than 4 nor more than 5 ml./min. The volume of distillate collected in the cylinder shall be observed and recorded, to the nearest 0.5 ml., when the mercury of the thermometer reaches each point that is a multiple of 10°C., or the Fahrenheit equivalent of this point (30°C., 40°C., 50°C., 60°C., etc., or 86°F., 104°F., 122°F., 140°F., etc.). If preferred, the reading of the distillation thermometer may be observed and recorded when the level of the distillate reaches each 10-ml. mark on the graduate. In case a product is being tested to ascertain whether or not it conforms with a given specification, all necessary observations shall be made and recorded, whether or not they are included in the series ordinarily employed by the laboratory making the test.

No adjustment of the heat shall be made after the liquid residue in the flask is approximately 5 ml. unless the time required to bring over the last 5 ml. of distillate and reach the end point exceeds 5 min. The end point is the maximum temperature observed on the distillation thermometer and is usually reached after the bottom of the flask has become dry. If the bottom of the flask is not dry, the operator shall record this fact.

In case the time required to bring over the last 5 ml. of distillate and reach the end point exceeds 5 min., the test shall be repeated and the heat shall be adjusted when the liquid residue reaches 5 ml. This adjustment may be either an increase or a decrease but must accomplish the purpose of bringing the period required to vaporize the last 5 ml. of distillate and reach the end point within the limits of 3 and 5 min.

The total volume of the distillate collected in the receiving graduate shall be recorded as the **recovery.**

The cooled residue shall be poured from the flask into a small cylinder graduated in 0.1 ml., measured when cool, and the volume recorded as **residue.**

The difference between 100 ml. and the sum of the recovery and the residue shall be calculated and recorded as **distillation loss.**

Accuracy. With proper care and attention to detail, duplicate results obtained for initial boiling point and end point, respectively, should not differ from each other by more than 6°F. (3.3°C.). Duplicate readings of the volume of distillate collected in the cylinder when each of the prescribed temperature points is reached should not differ from each other by more than 2 ml. In case observations are made on the basis of prescribed percentage points, the differences in temperature readings should not exceed the amounts equivalent to 2 ml. of distillate at each point in question.

Relation to Other Laboratory Distillations. See next article dealing with true-boiling-point distillations.

True-boiling-point Distillations. True-boiling-point distillations have found wide application during the past several years in the evaluation of the yields of the several fractions that may be obtained from crudes (Beiswenger and Child, *Ind. Eng. Chem.*, anal. ed., **2**, 284, July 15, 1930) and also in the theoretical analysis of petroleum-fractionation problems (Lewis and Smoley, *Am. Petroleum Inst. Bull.* 11, p. 73, 1930). The true-boiling-point apparatus consists of a laboratory batch still with an efficient fractionating column. The column is operated at a high reflux ratio to secure as effective fractionation as possible.

Consider a mixture of, say, equal percentages of five pure components. If this mixture is batch-distilled in an apparatus in which efficient fractionation is obtained and the temperatures at the top of the column are read continuously, a plot of the overhead vapor temperatures vs. the per cent distilled, if perfectly sharp separation were obtained, would have the appearance of the dotted stepwise curve *A* in Fig. 53. In any actual apparatus

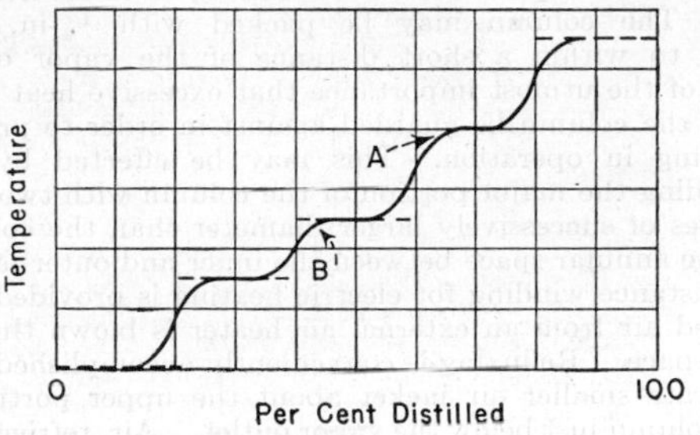

FIG. 53.

the steps are rounded to some extent as shown by the solid curves *B* of Fig. 53. The sharpness of the separation of the components is greater the higher the percentage of the individual constituents, the wider the differences in their boiling points, the more efficient the fractionation as determined by reflux ratio and tower height, and the less the "hold-up" or amount of liquid reflux required to fill the column. In view of these considerations, the distillation of a complex mixture as represented by a crude, or one of its fractions, yields a true-boiling-point curve that is generally smooth over its entire range, excepting in the case of low-boiling gasolines, as shown by Fig. 54.

True-boiling-point Curve Compared with A.S.-T.M. If the true-boiling-point curve of one of the lower boiling fractions of a crude is compared with the results of an A.S.T.M. distillation of the same fraction the following differences will be found:

1. The A.S.T.M. "initial" will be higher than the initial shown by the true-boiling-point curve.
2. The A.S.T.M. curve will cross the true-boiling-point curve at some point in the middle region of the distillation, *i.e.*, the slope of the main portion of the A.S.T.M. curve is less than the slope of the true-boiling-point curve.
3. The A.S.T.M. "end point" is lower than the true-boiling-point end point. These points are illustrated by Fig. 54. The differences are due to the fact that the A.S.T.M. distillation is a batch distillation in which only a slight amount of refluxing due to heat loss from the neck of the flask is obtained, while the true-boiling-point distillation at high reflux ratio results in very effective fractionation. The A.S.T.M. distillation approaches a simple batch distillation (see p. 580) in which the lower boiling constituents distill over, accompanied by appreciable amounts of the higher boiling.

If a quantity of charging stock, large in volume compared with the liquid hold-up in the column, is separated into two fractions during the true-boiling-point distillation, and the true-boiling-point curves of

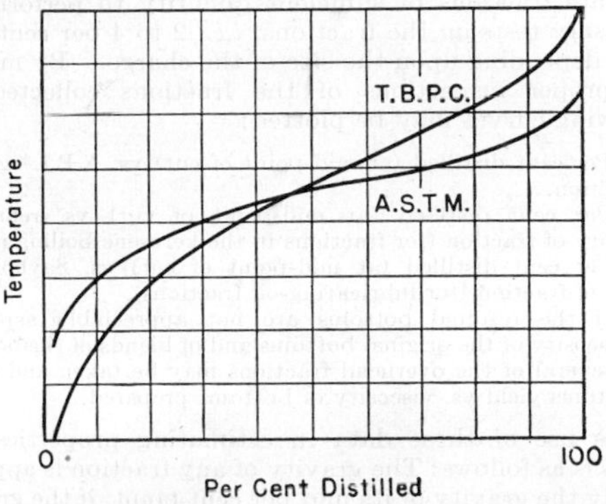

FIG. 54. True-boiling-point curve compared with A.S.T.M. distillation.

the original stock and the two fractions are compared, the results will appear as in Fig. 55, in which the solid curves represent the true-boiling-point curves. Examination of Fig. 55 shows that the end point of the lower boiling fraction on the true-boiling-point curve equals the initial of the higher boiling on its true-boiling-point curve, *i.e.* there is no overlap or gap. This will be the case if all three curves have been obtained in the same apparatus and with the same operating technique so that the same high degree of fractionation is obtained. The A.S.T.M. distillations of the two fractions are given by the dotted curves of Fig. 55, and show an appreciable gap, indicating good fractionation.

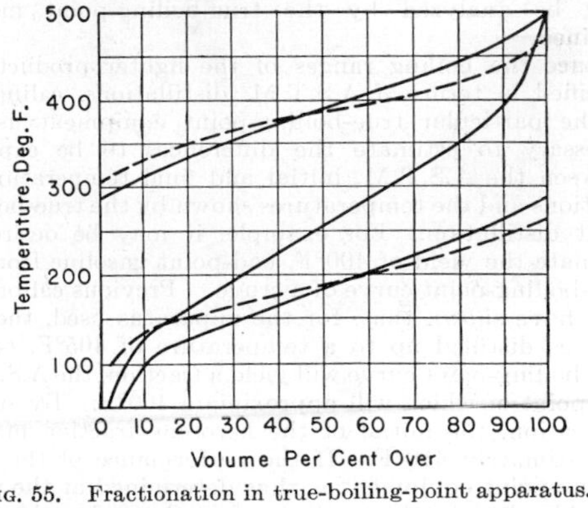

FIG. 55. Fractionation in true-boiling-point apparatus.

Applications of True-boiling-point Distillations. The applications of true-boiling-point distillations in petroleum technology are:

1. The analysis of crudes with respect to the yields of commercial products obtainable.
2. The theoretical analysis of problems in fractionation.

The analysis of crudes by the true-boiling-point method is given in the literature, to which reference should be made for all the details. In its essentials the analysis consists in the distillation of 1½ to 5 l. of the crude in a suitable true-boiling-point still. Distillation is conducted at atmospheric pressure up to a point where the temperature in the distilling flask is approximately 600°F. To avoid cracking in the flask, the apparatus is cooled and the pressure reduced by the use of a vacuum pump to some convenient reduced pressure, *e.g.*, 10 mm. Hg. Distillation is continued until the flask temperature again approaches 600°F. The distillate is collected in separate fractions of sufficient quantity to perform the necessary tests on the fractions, *i.e.*, 2 to 4 per cent fractions depending upon the size of the charge. By making the proper inspections of the fractions collected the following curves may be plotted:

1. Per cent distilled (at mid-point of cut) vs. A.P.I.* gravity of fraction.
2. Per cent distilled (at mid-point of cut) vs. refined-oil viscosity of fraction (for fractions in the kerosene boiling range).
3. Per cent distilled (at mid-point of cut) vs. Saybolt viscosity of fraction (for lubricating-oil fractions).
4. If the residual bottoms are not appreciably asphaltic, the viscosity of the original bottoms and of blends of the bottoms with several of the overhead fractions may be taken and a plot of bottoms yield vs. viscosity of bottoms prepared.

The use of these data in estimating properties and yields is as follows: The gravity of any fraction is approximately the gravity of its mid per cent point, if the gravity mid per cent curve does not exhibit too sharp curvature, *e.g.*, the gravity of the 30 to 40 per cent fraction of the curve would be estimated by reading the gravity at 35 per cent from curve 1.

The Saybolt viscosity of the lubricating-oil fractions is given by the viscosity at the mid per cent point of the fraction, provided that the cut is not too wide, *e.g.*, the viscosity of the 60 to 75 per cent fraction of the crude is given by the viscosity at 67.5 per cent as read from curve 3.

Curve 4 is directly read.

If the crude contains an appreciable asphaltic residuum, all the lighter materials, including the cylinder stock, may be separated from the asphalt by a preliminary continuous-flash distillation. The synthetic crude, represented by the overhead from this flash operation, may be analyzed by the true-boiling-point method outlined.

Since the boiling ranges of the lighter products are specified in terms of A.S.T.M. distillations, calibration of the particular true-boiling-point equipment used is necessary to estimate the differences to be expected between the A.S.T.M. initial and final temperatures of fractions and the temperatures shown by the true-boiling-point distillation. For example, it may be desired to estimate the yield of 400°F. end-point gasoline from the true-boiling-point curve of a crude. Previous calibration may have shown that, for the apparatus used, the percentage distilled up to a temperature of 405°F. on the true-boiling-point curve will yield a fraction, the A.S.T.M. end point of which will approximate 400°F. By similar calibration, the initial of the kerosene fraction may be approximately 410°F. If the performance of the true-boiling-point equipment is thus determined at the start, trial blending in the analysis of crudes is largely eliminated. A good true-boiling-point apparatus of the general type described in the following article will usually, but not necessarily, show better fractionation than plant equipment. Correlation of the various yields and properties from a true-boiling-point work-up against plant performance enables the refiner or designer of equipment to make empirical correlations where necessary to the true-boiling-point data to obtain close estimations of plant results.

* Degrees A.P.I. $= \dfrac{141.5}{\text{sp. gr. } 60°\text{F.}/60°\text{F.}} - 131.5.$

The theoretical analysis of petroleum-fractionation problems, including the calculation of fractionating trays required in towers, has been facilitated by the use of true-boiling-point data (Lewis and Smoley, *Am. Petroleum Inst. Bull.* 11, p. 73, 1930; Lewis and Wilde, *Trans. Am. Inst. Chem. Engrs.*, 1928). In these studies the charging stock was considered as being made up of a number of pure components characterized by small temperature intervals on the true-boiling-point curve. These components were traced up and down the tower by the application of the methods of pp. 589 to 591, and the results checked by actual data on large-scale equipment. The results indicated the soundness of these methods of calculation when applied to petroleum fractionation. The calculations are too time-consuming to permit of application to every design problem. The solution of a number of typical problems and their correlation with actual plant data will undoubtedly prove of value in the design of fractionating equipment.

True-boiling-point Apparatus [Peters and Baker, *Ind. Eng. Chem.*, **18,** 69 (1926); Beiswenger and Child, *Ind. Eng. Chem.*, anal. ed., **2,** 284 (1930)]. The major dimensions and general description of a true-boiling-point column that has been found to yield satisfactory results are as follows:

A round pyrex-glass flask of 5-l. capacity is sealed to a column approximately 4 ft. long and 1 in. inside diameter. The column is provided with a side vapor outlet near the top. The column may be packed with ¼-in. glass rings to within a short distance of the vapor outlet. It is of the utmost importance that excessive heat losses from the column be guarded against in order to prevent flooding in operation. This may be effected by surrounding the major portion of the column with two glass sleeves of successively larger diameter than the column. In the annular space between the inner and outer jackets a resistance winding for electric heating is provided, and heated air from an external air heater is blown through this space. Refluxing is conveniently accomplished by a separate smaller air jacket about the upper portion of the column just below the vapor outlet. Air, refrigerated if necessary, is blown through this jacket, and control of reflux is accomplished by control of the air supply. The apparatus is so arranged that it may be connected through the receiver and final condenser to a vacuum pump and a high vacuum maintained in the apparatus. When used with hydrocarbons, rubber connections should be avoided as far as possible and where used should be of the heavy-walled vacuum type of rubber tubing. Cork stoppers are used in preference to rubber ones. A clear nitrocellulose lacquer is useful for painting the cork stoppers and rubber connections to prevent vapor leaks and to render the apparatus vacuum-tight. Such a column is operated at such a reflux supply that it is just short of a flooded condition. The apparatus described, being entirely of glass, may be readily inspected for reflux flow. With this refluxing, distillate may be taken over at approximately 5 cc./min. under either atmospheric or vacuum operation.

Although described in conjunction with its application to petroleum technology, an apparatus of this type will be found useful for any laboratory work in which very effective fractionations are necessary.

Stedman Column Packing.* Stedman packing is a recently developed type of tower packing that is particularly applicable where low column height, low column weight, low pressure drop, and low hold-up, or combinations of them, may be required.

* By L. B. Bragg. Knolls Atomic Power Laboratory, General Electric Company.

The packing is fabricated of wire cloth which is punched, embossed, and welded to form a series of cells. The single-cell conical type, as shown in Fig. 56*A*, consists of a series of superposed, truncated conical disks. A semicircular hole is cut out of the side of each cone and

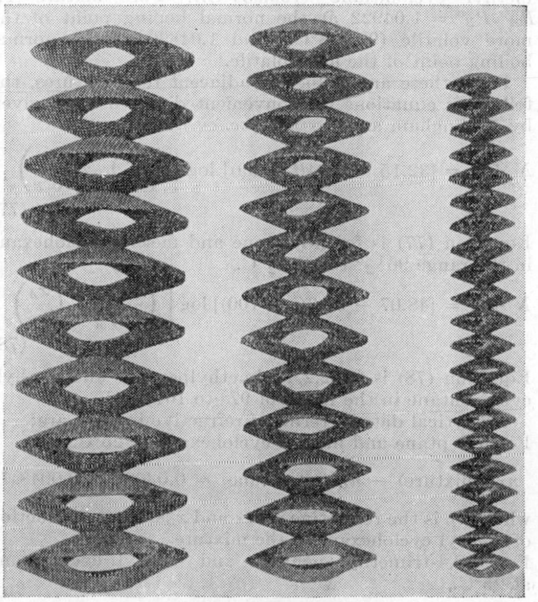

Fig. 56*A*.

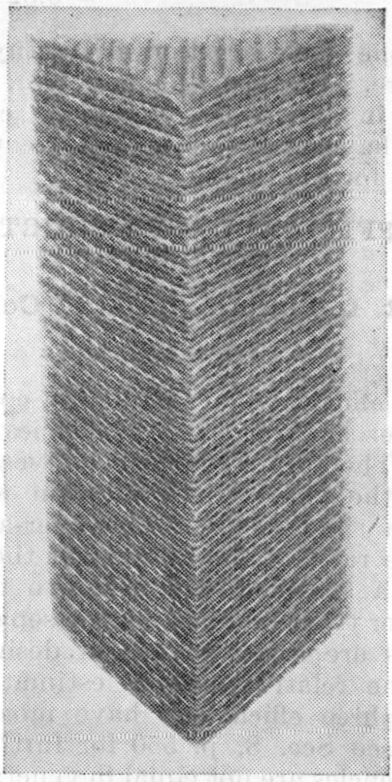

Fig. 56*B*.

extends about two-thirds of the distance from the edge of the cone to a flat at the center. The cones are welded together alternately flat to flat and edge to edge with the holes, which serve as vapor passageways, located on opposite sides of the cell thus formed.

The multiple-cell triangular-pyramid type of packing, as shown by Fig. 56*B*, is constructed of sheets having a regular pattern of raised triangular pyramids and an upturned lip around the perimeter. The pyramids are located on ⅜-in. equilateral triangular centers. The sheets are perforated with $\frac{3}{16}$-in.-diameter holes located between the pyramids. The pattern on the sheets is placed so that, by rotating adjacent sheets 120 deg. with respect to each other, the so-called valleys at the junctions of the bases of the pyramids of an upper sheet come immediately above the apexes of the pyramids of the next lower sheet. The sheets are welded together at these points of contact. This placing of adjacent sheets causes the holes of the lower sheets, which serve as vapor passageways, to be located below the apexes of the pyramids of the upper sheets.

With both types of packing, liquid flows downward along the wire cloth and seals the openings of the mesh. As the liquid flows from cell to cell, there is continual division and recombination of liquid streams which causes effective distribution and mixing. The vapor flows upward through the vapor openings and is likewise mixed by repeated division and recombination.

The packing is customarily fabricated of stainless steel or monel wire cloth of 60 by 40 meshes per inch, using wire 0.009 in. in diameter. In large commercial-sized columns, the packing is constructed with an equilateral triangular cross-sectional shape 6 in. along the side, and sections of packing are fitted against one another to fill the column completely. It is usually of a hexagonal cross-sectional shape. The packing can be adapted to fill completely circular cross sections.

The packing is available in sizes from ⅜ in. diameter up to 50 mm. diameter for the conical type, having H.E.T.P. values, respectively, from about 0.06 to 2.0 in. Sizes from 1½ to 6 in. in diameter for the triangular-pyramid type exhibit H.E.T.P. values, respectively, from about 1.5 to 2.0 in. In the larger columns, where the triangular-shaped packing has been used in columns from 12 in. to 11 ft. across the diagonal, H.E.T.P. values from about 2.5 to 3.3 in., respectively, have been obtained. At the same time the pressure drop and hold-up have been low.

Low-temperature Fractionations. Natural gas and gasoline consist principally of mixtures of the lower boiling paraffin hydrocarbons and their isomers from methane (CH_4) up to heptane (C_7H_{16}). Because of the attention given by refiners and producers to the recovery and fractionation of these hydrocarbons, an acute need for a practical and accurate method of analysis of such mixtures has been felt in recent years. While many references occur in the literature to the fractional distillation of low-boiling mixtures, the various forms of apparatus developed and described by Podbielniak [*Ind. Eng. Chem.*, anal ed., **3**, 177 (Apr. 15, 1931)] have found most widespread use for these purposes.

Podbielniak's apparatus consists of a well-insulated fractionating column in which condensation of the low-boiling constituents for reflux is accomplished by the use of liquid air or some other refrigerant. The overhead distillate is collected and measured in the vapor state under reduced pressure. Figure 57 shows one form of this apparatus.

Laboratory Distilling Columns of High Efficiency. A description of the assembly, testing, and operation of the laboratory distilling columns used at the National Bureau of Standards in the work of the American Petroleum Institute Research Project 6 on the analysis and purification of hydrocarbons is covered in Report on the Assembly, Testing and Operation of Laboratory Distilling Columns of High Efficiency by Willingham and Rossini (National Bureau of Standards, Dec. 31, 1945).

Calibration of Columns. The determination of the number of equivalent theoretical plates in a column is often essential in investigational work dealing with the more difficult separations. This is carried out by operation of the column at total reflux on a known binary test mixture until equilibrium conditions are attained, then sampling and analyzing distillate and bottoms. For columns having large numbers of theoretical plates the selection of the test mixture is of extreme importance, for the relative volatility must be low, the values of relative volatility must be accurately known, and accurate analytical data must be available.

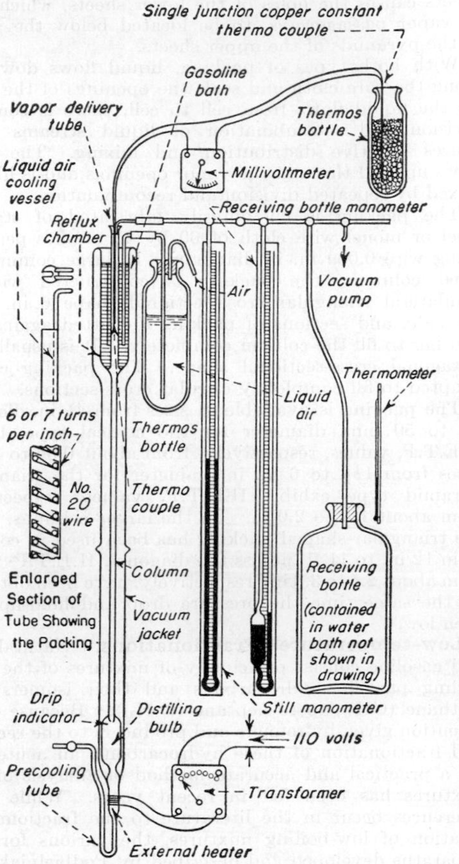

FIG. 57. Podbielniak's apparatus.

Willingham and Rossini (Report on the Assembly, Testing and Operation of Laboratory Distilling Columns of High Efficiency, A.P.I. Research Project 6, National Bureau of Standards, Dec. 31, 1945) have used mixtures as given herewith and give the essential data thereon. These mixtures were used in testing columns having on the order of 100 to 200 equivalent theoretical plates.

The Fenske equation [*Ind. Eng. Chem.*, **24**, 482 (1932)] in the following form is conveniently used as the basis for calculation:

$$N + 1 = \frac{1}{\log P_A°/P_B°}\left[\log\left(\frac{x_A}{x_B}\right)_p \Big/ \left(\frac{x_A}{x_B}\right)_s\right] \quad (76)$$

where N = number of theoretical plates.
$P°$ = vapor pressure of given component at the mean temperature.
A = more volatile component (subscript).
B = less volatile component (subscript).
x = mole fraction.
p = product (subscript).
s = still (subscript).

The mixtures used were (I) n-heptane and methyl cyclohexane and (II) 2,2,4-trimethyl pentane and methyl cyclohexane. For mixture I, $P_A°/P_B°$ has values of 1.07380 at the normal boiling point of n-heptane (98.426°C.) and 1.07584 at the normal boiling point of methyl cyclohexane (100.934°C.). For mixture II, $P_A°/P_B° = 1.04922$ at the normal boiling point of the more volatile (99.238°C.) and 1.04890 at the normal boiling point of the less volatile.

From these and values at adjacent temperatures, the following equations for convenient direct use are given by Willingham and Rossini (*loc. cit.*):

$$N + 1 = [32.15 - 0.34(t - 99)] \log\left[\left(\frac{x_A}{x_B}\right)_p \Big/ \left(\frac{x_A}{x_B}\right)_s\right] \quad (77)$$

Equation (77) is for n-heptane and methyl cyclohexane in the range 96½° to 101½°C.

$$N + 1 = [48.07 - 0.13(t - 100)] \log\left[\left(\frac{x_A}{x_B}\right)_p \Big/ \left(\frac{x_A}{x_B}\right)_s\right] \quad (78)$$

Equation (78) is for 2,2,4-trimethylpentane and methylcyclopentane in the range of 97° to 101½°C.

Analytical data in terms of refractive indexes are:

For n-heptane and methyl cyclohexane at 25°C.,

$$n_D \text{ (mixture)} - n_D \text{ (n-heptane)} = 0.0306x - 0.0048x^2$$

where n_D is the refractive index and x is the mole fraction of methyl cyclohexane in the mixture.

For 2,2,4-trimethyl pentane and methyl cyclohexane at 25°C.,

$$n_D \text{ (mixture)} - n_D \text{ (2,2,4-trimethyl pentane)} = 0.0244x - 0.0072x^2$$

where x is the mole fraction of methyl cyclohexane in the mixture. In Eqs. (97) and (98), t is the mean temperature in °C. The numerical coefficients are estimated to have uncertainties not greater than about +½ per cent for the ranges given.

PLATE EFFICIENCIES FOR DISTILLATION COLUMNS

By R. L. Pigford and A. P. Colburn

General

Contacting efficiencies of distillation column trays are reported either as Murphree efficiencies or as over-all efficiencies. The former definition gives the degree of approach of the vapor to equilibrium with the liquid running off a single plate. The over-all efficiency is defined as the ratio of the number of theoretical plates required for a specified separation to the number of actual plates or trays required for the separation. Over-all efficiencies are easier to use in design calculations because of the relative ease of estimating theoretical plates. Murphree efficiencies have more fundamental significance (see Sec. 8, p. 550 for further discussion). The two efficiencies are not equal in general, though they may be approximately equal unless operating and equilibrium lines are divergent (see Sec. 8, p. 551 for the conversion of these efficiencies when mG_M/L_M is constant).

The efficiency of contact between vapor and liquid at a single point on a plate is specified by means of Murphree point efficiency (*cf.* Sec. 8, p. 551). Values of this quantity are always less than 100 per cent, because of resistance to transfer of material from one phase to the other. As in packed columns or other contacting devices, this resistance is divided between liquid and vapor phases. It would be desirable to separate experimentally determined efficiencies into these component parts if this were possible, and some progress has been made along these lines [for example, Geddes, *Trans. Am. Inst. Chem. Engrs.*, **42,** 79 (1946); Walter and Sherwood, *Ind. Eng. Chem.*, **33,** 493 (1941)].

Experimentally determined Murphree or over-all plate efficiencies should be converted to point efficiencies whenever it is possible to make the necessary calculations with confidence. Methods available for converting to point efficiencies are based on the assumption that liquid flows across a tray without channeling and without lateral mixing (*cf.* Sec. 8, p. 552). This may be approximately true for large plates, but lateral mixing probably is so rapid on small plates having only one or two bubble caps that essentially no concentration gradients are present. For this reason Murphree and over-all plate efficiencies may be expected to be larger for larger plates, on which the vapor-liquid contact takes place under more nearly countercurrent conditions, than for smaller plates, on which the contact is substantially stagewise. Murphree plate and point efficiencies may be nearly equal for a plate having only one or two bubble caps.

Transfer of material between phases on a plate takes place by interaction between vapor bubbles formed at slots in the bubble caps, or at perforations in the plates, and the surrounding liquid; between liquid and vapor mixed into a froth or foam at the liquid surface; and from liquid spray thrown upward into the vapor space between the plates. The factors that affect the point efficiency can be partly clarified by studying the following relations of the fractional local efficiency E_{OG} to the vapor and liquid resistances. This can be done in terms of mass transfer coefficients, or, as recently suggested by Gerster, Colburn, Bonnet, and Carmody [*Trans. Am. Inst. Chem. Engrs.*, (1949)], in terms of numbers of transfer units. Expressed in terms of transfer coefficients,

$$\frac{1}{-2.3 \log (1 - E_{OG})} = \frac{G_M}{Z_v K_{OG}a} = \frac{G_M}{Z_v k_G a} + \frac{mG_M}{Z_v k_L a} \qquad (79)$$

where Z_v = effective depth of liquid, ft.
G_M = molar vapor velocity, lb. moles/(hr.)(sq. ft.).
$K_{OG}a$ = over-all mass transfer coefficient, lb. moles/(hr.)(cu. ft.)(unit Δy).
$k_G a$ = gas-film mass transfer coefficient, same units.
$k_L a$ = liquid film mass transfer coefficient, lb. moles/(hr.)(cu. ft.)(unit Δx).
m = dy^*/dx, or slope of equilibrium curve.

It is believed that the use of transfer units brings out the effect of variables better. Thus

$$\frac{1}{-2.3 \log (1 - E_{OG})} = \frac{1}{N_G} + \frac{mG_M/L_M}{N_L} \qquad (80)$$

where N_G = number of gas-film transfer units on the plate.
N_L = number of liquid-film transfer units.
L_M = molar liquid velocity through column (based on same cross-sectional area as G_M); hence L_M/G_M is the liquid-vapor ratio.

A convenient graphical solution of Eq. (80) is given by Fig. 58*A*. For values of E_{OG} less than 0.1, the following approximation can be used:

$$\frac{1}{E_{OG}} = \frac{1}{N_G} + \frac{mG_M/L_M}{N_L} \qquad (80a)$$

The virtue of Eqs. (80) and (80*a*) is that the quantity N_G varies little with velocity and other factors. Typical values for the system air-water are given in Fig. 58*B*. In the case of values of N_L, Gerster *et al.* recommend expressing data as

$$N_L = Z_H/H_L$$

where H_L = length of a liquid transfer unit in direction of liquid flow across plate, ft.
Z_H = distance of liquid travel, ft.

Typical values of H_L obtained by Gerster *et al.* for the system oxygen-water are given in Fig. 58*C*.

To convert values of N_G and N_L to other systems, the following relations can be used:

$$N_G(X) = N_G(\text{air-water})[N_{Sc}(\text{a-w})/N_{Sc}(X)]^{\frac{1}{2}}$$

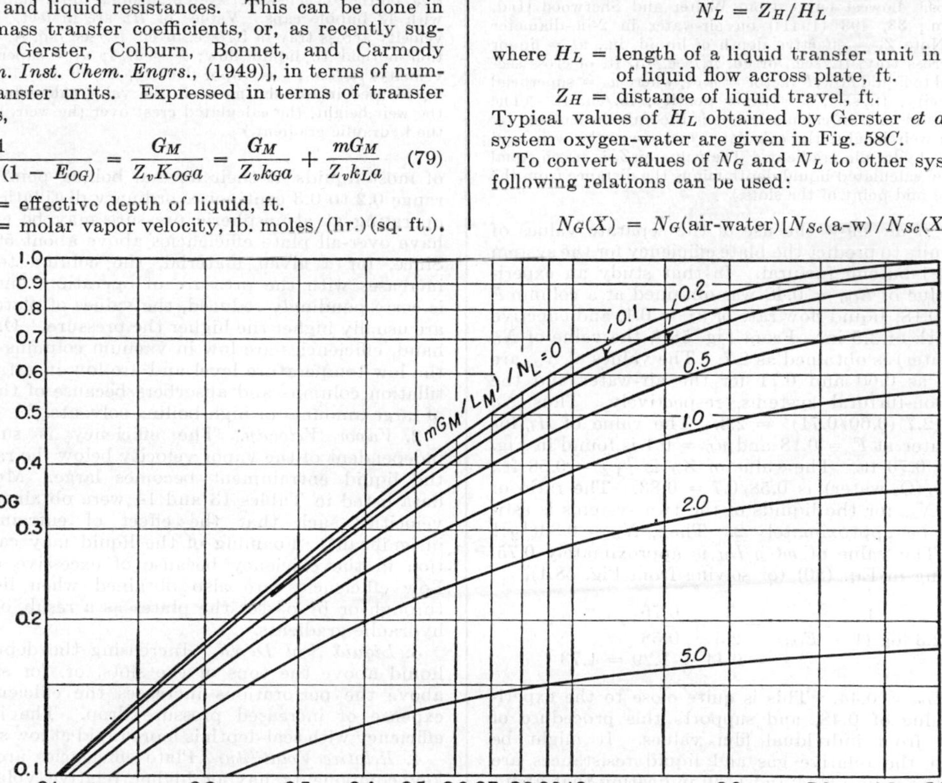

FIG. 58*A*. Relationship between E_{OG}, N_G, and N_L, according to Eq. (80). E_{OG} = local or transfer efficiency on a gas basis, N_G and N_L are number of gas and liquid transfer units, respectively, $m = dy^*/dx$, slope of equilibrium curve, and L_M/G_M = molar liquid-vapor ratio. [*Gerster, Colburn, Bonnet, and Carmody, Trans. Am. Inst. Chem. Engrs.*, Dec. (1949).]

where $N_{Sc} = \mu/\rho D_v$.
μ = vapor viscosity, lb./(hr.)(ft.).
ρ = vapor density, lb./cu. ft.
D_v = vapor diffusivity, ft.²/hr.

Also:

$$N_L(X) = N_L(O_2\text{-water})[N_{Sc}(O_2\text{-w})/N_{Sc}(X)]^{\frac{1}{2}}$$

where $N_{Sc} = \mu/\rho D_L$.
μ = liquid viscosity, lb./(hr.)(ft.).
ρ = liquid density, lb./cu. ft.
D_L = liquid diffusivity, ft.²/hr.

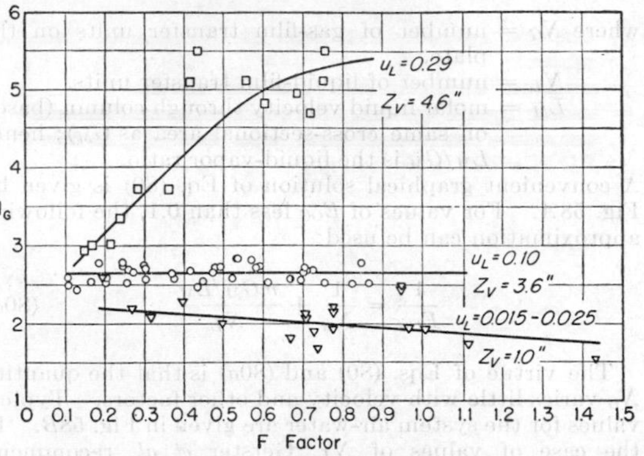

FIG. 58B. Typical values of N_G. Upper two curves from work of Gerster *et al.* on air-water on 13-in.-diameter tray with 13 bubble caps. Lowest curve from Walter and Sherwood [Ind. Eng. Chem., **33**, 493 (1941)] on air-water in 2-in.-diameter column. Note Z_v = effective depth of liquid, in., u_L = liquor velocity across tray, ft./sec., or cu. ft./sec./sq. ft. of cross section normal to liquid flow, F factor = $u_v\rho^{\frac{1}{2}}$, where u_v = superficial vapor velocity, ft./sec., ρ = vapor density, lb./cu. ft. (The depth of liquid used in calculating u_L is the value calculated by adding the weir height, the calculated crest over the weir, and one-half the hydraulic gradient. The value of Z_v is taken equal to the above calculated liquid depth minus the distance from the plate to the mid-point of the slots.)

Gerster *et al.* show the use of the separate values of transfer units to predict the plate efficiency for the system isobutane–1-butene–furfural. In that study an experimental value of $E_{OG} = 0.48$ was obtained at a column F factor of 0.18, liquid flow rate of $u_L = 0.1$, and effective liquid depth of 3.5 in. From Fig. 58B, the value of N_G (for air-water) is obtained as 2.7. The values of N_{Sc} are estimated as 0.60 and 0.71 for the air-water and the hydrocarbon-furfural systems, respectively. Thus N_G becomes $(2.7)(0.60/0.71)^{\frac{1}{2}} = 2.3$. The value of H_L for oxygen water at $F = 0.18$ and $u_L = 0.1$ is found in Fig. 58C to be 0.70 ft. The value of Z_H is $7/12 = 0.58$ ft., whence $N_L(O_2\text{-water})$ is $0.58/0.7 = 0.83$. The ratio of values of N_{Sc} for the liquids of the two systems is estimated to be approximately 2. Then, $N_L = 0.83(\frac{1}{2})^{\frac{1}{2}} = 0.58$. The value of mG_M/L_M is approximately 0.75. Substituting in Eq. (80) (or solving from Fig. 58A),

$$\frac{1}{-2.3 \log (1 - E_{OG})} = \frac{1}{2.3} + \frac{0.75}{0.58} = 0.44 + 1.29 = 1.73$$

Solving, $E_{OG} = 0.44$. This is quite close to the experimental value of 0.48, and supports this procedure of prediction from individual film values. It might be noted that the relative gas and liquid resistances are given by the values 0.44 and 1.29, indicating that about three-fourths of the resistance in this case is in the liquid. Note also that the predicted value of $E_{OG} = 0.44$ is well within the spread of experimental data on the isobutane–1-butene–furfural system as seen on Fig. 64.

The plate efficiency for an arbitrary set of conditions encountered in a design problem may be estimated from test data obtained from actual columns if the physical properties and the mechanical design features are similar. Reported values of plate efficiency for laboratory and commercial columns are given for sieve-plate columns in Table 13, and for bubble-cap columns in Table 14.

A study of the table shows that the principal variables that affect the plate efficiency are:

1. *Viscosity of the Liquid.* Increasing the viscosity decreases the efficiency, as expected. Since the viscosity

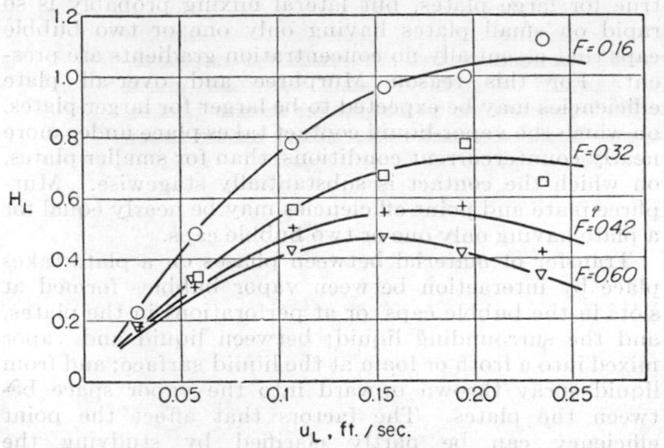

FIG. 58C. Typical values of H_L. Data from work of Gerster *et al.* on the oxygen-water system, using a 13-in.-diameter column with 13 bubble caps. Values of H_L are in feet. u_L = liquid velocity across tray in ft./sec. or cu. ft./sec./sq. ft. of cross section normal to liquid flow; $F = u_v\rho^{\frac{1}{2}}$; u_v = superficial vapor velocity, ft./sec.; ρ = vapor density, lb./cu. ft. (The depth of liquid used in calculating u_L is the value calculated by adding the weir height, the calculated crest over the weir, and one-half the hydraulic gradient.)

of most liquids at their normal boiling points is in the range 0.2 to 0.3 centipoise, ordinary distillation columns operating at atmospheric pressure may be expected to have over-all plate efficiencies above about 50 per cent. Since, for a given material, the column temperature increases with the pressure of operation, and viscosity is correspondingly reduced, the values of plate efficiency are usually higher the higher the pressure. On the other hand, efficiencies are low in vacuum columns because of the low temperature level and are low in extractive distillation columns and absorbers because of the presence of large amounts of high-boiling solvents.

2. *Vapor Velocity.* The efficiency is substantially independent of the vapor velocity below the range where the liquid entrainment becomes large. Most of the data listed in Tables 13 and 14 were obtained at vapor velocities such that the effect of entrainment was insignificant. Foaming of the liquid may cause reduction in the efficiency because of excessive carry-over. Low efficiencies are also obtained when liquid leaks through or by-passes the plates as a result of excessive hydraulic gradients.

3. *Liquid Seal Depth.* Increasing the depth of clear liquid above the tops of the slots, or, for sieve trays, above the perforations increases the efficiency at the expense of increased pressure drop. The increase in efficiency with seal depth is more rapid at low seal depths.

4. *Relative Volatility.* Plate efficiencies are lower for key components having higher relative volatilities, or for gases having low solubility in the liquid because of the increased importance of the liquid resistance, as indicated by Eq. (80). The effect is probably insignificant for relative volatilities less than about 5, however.

5. *Liquid Concentration.* Several investigators of plate efficiencies for the systems ethanol-water and i-propanol–water report low Murphree efficiencies near the composition of the azeotropes and in the region of low concentrations. These effects may be due to experimental errors involved in sampling and analysis, or they may be real, caused, for example, by the high volatility of alcohol in its dilute aqueous solution. Further investigation is needed before final conclusions can be drawn, but the variation with composition probably should be taken into account in the careful design of alcohol columns.

If comparable experimental data are not available in Tables 13 and 14, a conservative estimate of the over-all efficiency may be based on the work of Drickamer and Bradford [*Trans. Am. Inst. Chem. Engrs.*, **39**, 319 (1943)]. In this work, which applies principally to commercial hydrocarbon systems, the only physical property that was found to have an important effect was the liquid viscosity. Plate design was relatively unimportant within the range studied. The relation between over-all efficiency and viscosity is shown in Fig. 59. The viscosity used is the molal average viscosity of the feed—assumed 100 per cent liquid—taken at the arithmetic average of the top and bottom tower temperatures. In the case of absorbers, the viscosity is the calculated molal average viscosity of the rich oil at the temperature at which it leaves the absorber. Viscosities of hydrocarbons useful for computing plate efficiencies are given by Figs. 60*A* and 60*B*.

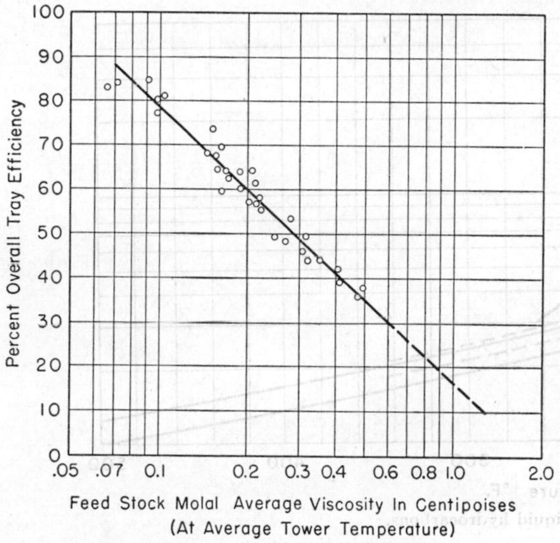

FIG. 59. Effect of liquid viscosity on plate efficiency for commercial hydrocarbon fractionators. [*Data of Drickamer and Bradford, Trans. Am. Inst. Chem. Engrs.*, **39**, 319 (1943).]

Figure 59 is based on plant tests of 54 refinery fractionating columns used for distillation or absorption of hydrocarbons. The efficiencies plotted are those for the key components, usually propane, butane, or pentane. Although the correlation developed for hydrocarbons agrees fairly well with a few of the data on commercial beer stills, acetic acid distillation towers, and alcohol towers, its use for the estimation of efficiencies for systems other than hydrocarbons of low relative volatility is questionable.

Substantially no effect of tray design was found by Drickamer and Bradford within the range studied, which was as follows:

Plate spacing: 18 to 30 in.
Length of liquid path across plate: about 2.5 ft. (liquid paths 4 to 5.5 ft. long gave slightly higher efficiencies).
Riser area: 7.2 to 11.5 per cent of tray area.
Weir length: 65 to 80 per cent of tower diameters.
Tower diameter: 4 to 7.5 ft.
Submergence (middle of slot to top of weir): ⅞ to 2⅛ in.
Bubble-cap design: Rectangular

Slot width = 3/32 to ¼ in.
Slot height = ¾ to 1¼ in.

A few of the test columns contained perforated trays.

Example. Estimate the over-all plate efficiency of a hydrocarbon distillation tower having a top temperature of 150°F.

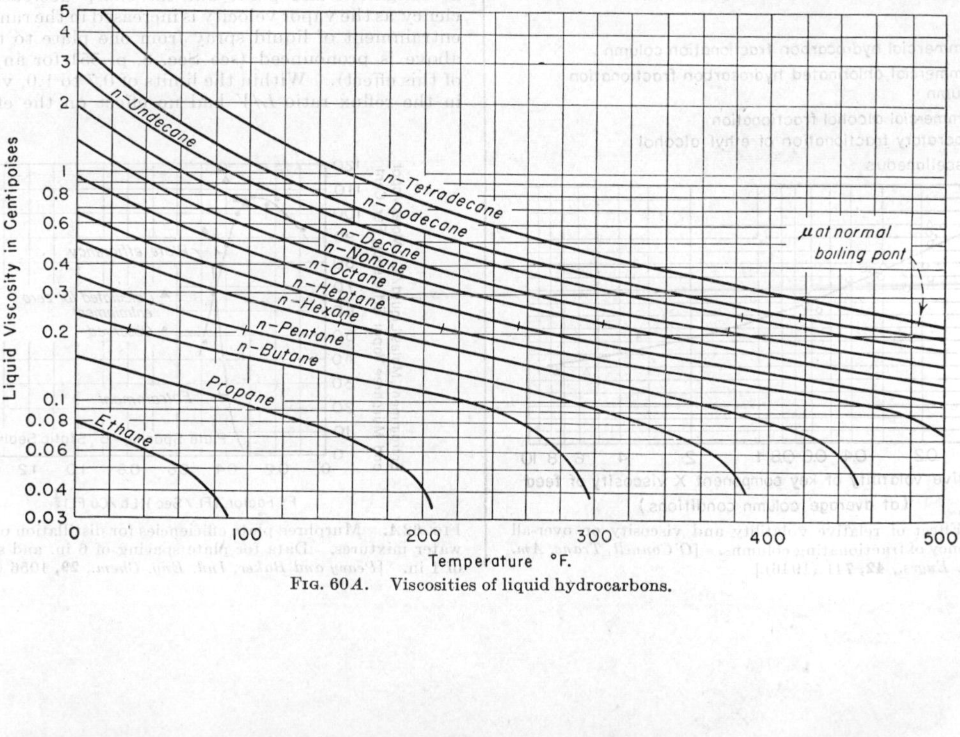

FIG. 60*A*. Viscosities of liquid hydrocarbons.

and bottom temperature of 230°F., and for the feed composition given below. The average tower temperature is 190°F.

Component	Mole fraction	Viscosity at 190°F., centipoises	(Mole fraction) × (viscosity)
C_3	0.2	0.048	0.0096
C_4	.3	.112	.0336
C_5	.2	.145	.0229
C_6	.3	.188	.0564

Molal average viscosity = 0.1286 centipoise.
From Fig. 59, the average plate efficiency is 73 per cent.

O'Connell [*Trans. Am. Inst. Chem. Engrs.*, **42**, 741 (1946)] states, however, that the plate efficiency predicted from Fig. 59 may be too high in the case of components having a very high relative volatility and that the plate efficiencies may be different for the various components in a multicomponent mixture. Figure 61 shows the correlation of over-all plate efficiencies developed by O'Connell. For absorbers, O'Connell utilizes a slightly different function—in place of relative volatility he employs gas solubility.

One of the most complete experimental investigations of plate efficiency is that of Peavy and Baker [*Ind. Eng. Chem.*, **29**, 1056 (1937)] for the system ethanol-water. The three-plate column used was 18 in. in diameter, and there were ten 3-in. caps located on 4.25-in. centers on each plate. The column was operated at atmospheric pressure, and the Murphree efficiencies reported were determined for the middle plate. The slot area was 23.75 sq. in. per plate, or 9.35 per cent of the free area of the column. Vertical baffle plates were used between the exit weir and the last row of caps to prevent liquid from splashing over the weir because of the violent plate action. Entrainment rates were determined colorimetrically.

Figures 62*A*, *B*, and *C* show some of the results of this investigation and point out the sharp decrease in efficiency as the vapor velocity is increased in the range where entrainment of liquid spray from one plate to the plate above is pronounced (see Sec. 8, p. 551 for an analysis of this effect). Within the limits of 0.7 to 1.0, variations in the reflux ratio L/V had no effect on the efficiency.

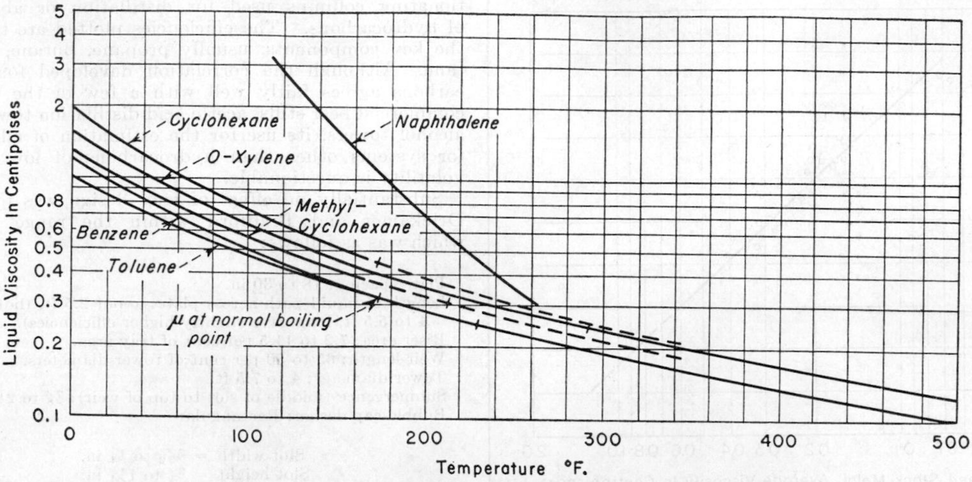

FIG. 60*B*. Viscosities of liquid hydrocarbons.

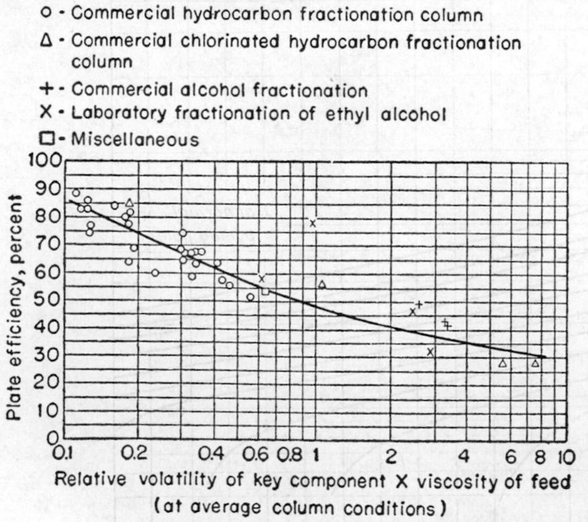

FIG. 61. Effect of relative volatility and viscosity on over-all plate efficiency of fractionating columns. [*O'Connell, Trans. Am. Inst. Chem. Engrs.*, **42**, 741 (1946).]

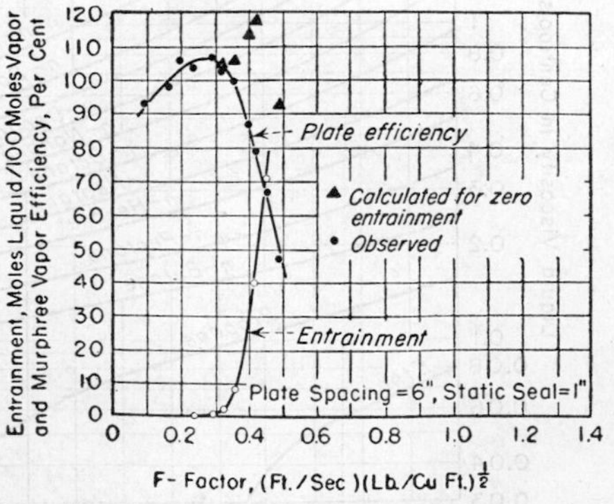

FIG. 62*A*. Murphree plate efficiencies for distillation of ethanol-water mixtures. Data for plate spacing of 6 in. and static seal of 1 in. [*Peavy and Baker, Ind. Eng. Chem.*, **29**, 1056 (1937).]

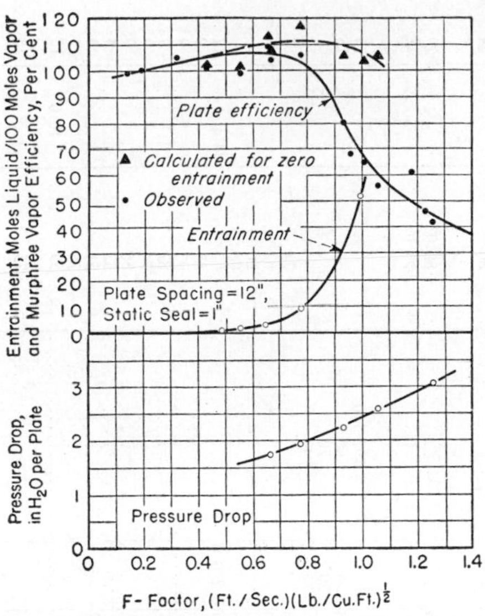

FIG. 62*B*. Murphree plate efficiencies for distillation of ethanol-water mixtures. Data for plate spacing of 12 in. and static seal of 1 in. (*Peavy and Baker, loc. cit.*)

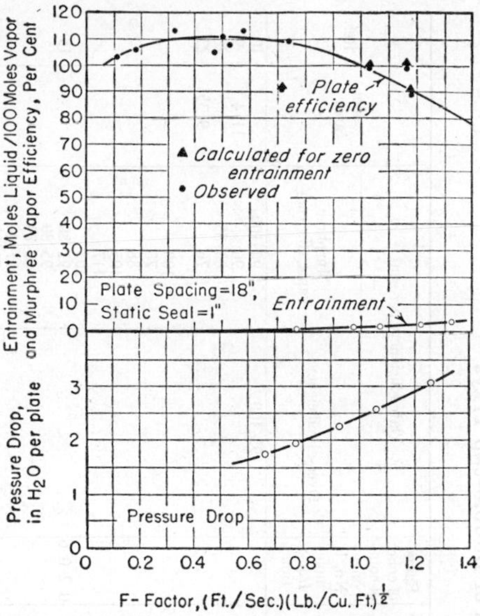

FIG. 62*C*. Murphree plate efficiencies for distillation of ethanol-water mixtures. Data for plate spacing of 18 in. and static seal of 1 in. (*Peavy and Baker, loc. cit.*)

Table 13. Plate Efficiencies of Sieve Trays

Mixture	Pressure, lb./sq. in. abs.	Approx. avg. temp., °C.	Viscosity of liquid feed at avg. column temp., cp.	Static seal depth, in.	Plate size, in.	Plate spacing, in.	Perforations: Diameter, in.	Perforations: Spacing, in.	No. trays	Range of F-factor (ft./sec.) (lb./cu. ft.)$^{1/2}$	% free area	Mole % of component 1	Plate efficiency: Over-all	Plate efficiency: Murphree efficiency	Ref.
C_3-C_6 + (Keys: C_4, C_5)	116	133–137	0.16–0.17	....	72	24	.....	.....	11 perforated out of 27	1.0–1.1	....	...	59–64		1
Ethanol-water.......	14.7	80	0.67	0.39	....	....	0.039	0.276	....	0.20	1.9	30		96*	5
			.67	...	....	....	.....	.....	..	.68	....	30		82*	
			.56	...	....	....	.....	.....	..	.21	....	75		84*	
			.56	...	....	....	.....	.....	..	.71	....	75		74*	
	14.7	80	.66	1.0	4.3	2.0	.079	.276		.09	7.4	50†		118‡	3
						3.9	.079	.276	..	.09	7.4	50†		128‡	
										.64	....	50†		89‡	
						7.9	.079	.276	..	.09	7.4	50†		127‡	
										.73	....	50†		86‡	
					15.7	7.9	.079	.276	..	.18§	7.4	50		110	
										.27§	....	50		119	
										.55§	....	50		102	
	17	100	.32	2.5	66	18	.50	1.125	..	0.54–0.61¶	14.5	0–2	40–43¶	20–50¶	2
Acetic acid–water...	14.7	101	.35	1.0	8	....	.2	.....	28	0.65	....	...	52		4

* Efficiency decreases with decreasing reflux ratio below $L/D = 6$. Pressure drop = 5.4 in. water per plate at $F = 0.32$, 1.2 in. water per plate at same F-factor for plate having 5.7 per cent free area. Efficiency varies slightly with free area, with maximum value at 7 per cent free area.

† Efficiency varies with concentrations. Value tabulated is near maximum.

‡ $L/D = 3$, efficiency is 8 to 13 per cent lower at $L/D = 80$.

§ Pressure drop = 0.5 in. H_2O at $F = 0.3$, 1.0 in. H_2O at $F = 0.5$, 1.5 in. H_2O at $F = 0.64$.

¶ Liquid flow = 100 gal./(min.)(ft. of plate width). Plates probably leaking. Normal period between cleanings = 170 days.

[1] Drickamer and Bradford, *Trans. Am. Inst. Chem. Engrs.*, **39**, 319 (1943).

[2] Gunness and Baker, *Ind. Eng. Chem.*, **30**, 1394 (1938).

[3] Kirschbaum, "Destillier—and Rektifiziertechnik," p. 197, Edwards Bros., Inc., Ann Arbor, Mich., 1943; *Z. Ver. deut. Ing., Beiheft Verfahrenstechnik*, No. 3, 1936. Kirschbaum and Andrews, *J. Inst. Petroleum Tech.*, **22**, 803 (1936).

[4] Peters, *Ind. Eng. Chem.*, **14**, 476 (1922).

[5] Volland, *Chem. Fabrik*, **1935**, 5.

Increasing the depth of unaerated liquid over the tops of the slots from 0 to 2 in. increased the efficiency from 78 to 118 per cent in the range where entrainment was small, on a plate with 12-in. spacing. This is shown by Fig. 63.

Analysis of their efficiency data by Peavy and Baker showed that, for columns operating with 10 to 15 per cent entrainment, the most economical plate spacing was 18 in., and that 12 in. was nearly as economical. The plate spacing is usually 24 to 30 in. for petroleum columns, which must be cleaned through manholes in the side of the column and are usually installed outdoors. According to Peavy and Baker's data, the optimum vapor velocity, at which the additional number of plates required because of lower plate efficiency just counterbalances the saving in cross-sectional area, was too high to be of practical interest (1.5, 3.3, and 76 ft./sec. for 6-, 12-, and 18-in. plate spacing, respectively). The velocity at which the entrainment was 10 to 15 per cent was considered to be more practical.

Extensive data on plate efficiencies during the extractive distillation of isobutane and 1-butene in the presence of dry and of aqueous furfural in a pilot-plant tower have recently been presented by Grohse, McCartney, Hauer,

Table 14. Summary of Experimentally Observed Efficiencies of Bubble-cap Trays
Non-hydrocarbon Systems

Mixture	Pressure, lb./sq. in., abs.	Approx. avg. temp., °C.	Viscosity of feed at avg. column temp. cp.	Static seal depth, in.	Plate size, in.	Plate spacing, in.	No. of caps per plate	Range of F-factor $u\rho^{1/2}$	Range of conc. of component No. 1, mole fraction	Plate efficiency, per cent: Over-all efficiency	Plate efficiency, per cent: Murphree vapor efficiency	% slot area	Ref.
Methanol-water	14.7	70			7.5 dia.		2	0.1 –0.3		95–100	90–100		10
Methanol-water	14.7	70		1.50	5 by 9	11	1	0.2 –0.5			96	11.4	9
Methanol-water	15.0	90		0.55	10.0	5	5	To 0.5			80–85		28
Ethanol-water	14.7	80		.00	18.0 dia.	6	10	0.1 –0.5			76	9.35	23
				.50									
				1.00				0.1 –0.35			100		
				0.00		12		0.2 –0.8			78		
				.50				0.2 –0.8			88		
				1.00				0.2 –0.8			105		
				2.00				0.3 –0.7			117		
				1.00		18		0.1 –0.9			105		
Ethanol-water	14.7	80		1.18	15.7 dia.	5.3	15	0.2 –0.4	0.2–0.6		92–80		15
						7.9		0.2 –0.6			95–80		
						10.6		0.2 –0.8			87–83		
						16.3		0.2 –1.1			97–84		
	4.85	55		1.18	15.7 dia.	5.3	15	0.2 –0.5			70–80		
						7.9		0.2 –0.7			75–90		
						10.6		0.2 –0.8			75–85		
	1.94	40		1.18	15.7 dia.	5.3	15	0.1 –0.5	0.5		70–90		
						7.9		0.1 –0.8			70–90		
						10.6		0.3 –0.9			70–100		
Ethanol-water	14.7	80		1.19	5 by 9	11.0	1	0.05–0.5	0.1–0.6		97–105		14
									0.6–0.8		74–86		
Ethanol-water	15.0	90		0.55	10.0	5.0	5	To 0.5			70		28
Ethanol-water	15.0	90			Oval cross section equiv. to 6.4 in. dia.		1				80		24
Ethanol-water	15.0	80		.37	9.0	6.8	3 rectangular caps	0.08–0.13			80–93		29
Ethanol-water	15.0	90			Square cross section equiv. to 6.4 in. dia.	6.0	7	To 0.25			54–69		27
Ethanol-water	15.0	80					7			92–110			25
Ethanol-water	15.0	80		.2	7.9	5.1	5	0.08–0.48	4%	60			16
									20%	70			
									72%	40			
Ethanol-water	14.7	80		.25	6.0		1 (2⅜ in.)	0.04–0.06			74		4
				.5	6.0		1 (2⅜ in.)	0.04–0.06			82		
				.75	6.0		1 (2⅜ in.)	0.04–0.06			88		
				1.25	6.0		1 (2⅜ in.)	0.04–0.06			96		
i-Propanol–water	14.7	85		1.19	5 by 9	11.0	1	0.16–0.25	0.2–0.6		92	11.4	17
n-Propanol–water	14.7	85		1.5	5 by 9	11.0	1	0.2 –0.9			85		9
i-Butanol–water	14.7	90		1.5	5 by 9	11.0	1	0.2 –0.5			95		9
Methanol–n-propanol	14.7	70		1.5	5 by 9	11.0	1	0.2 –1.0			87		9
Methanol–i-butanol	14.7	90		1.5	5 by 9	11.0	1	0.2 –1.0			74		9
Benzene–carbon tetrachloride	14.7	82		1.5	5 by 9	11.0	1	0.2 –0.8			86		9
Benzene-toluene	14.7	95			8.0	12.0	1 (4 in.)	0.3 –0.5			57		4
Benzene-toluene	14.7	95			6.0		1 (2⅜ in.)	0.09–0.12			51		4
Benzene-toluene	15.0	95					1				75		24
Benzene-toluene	14.7	95		0.5	8.0	16.0	1 tunnel cap perpendicular to flow	0.8			59	10.0	18
Benzene-toluene-xylene	14.7	105		0.5	8.0	16	1 tunnel cap perpendicular to flow				75		18

Benzene-toluene	15.0	95		.44	6.0	6	2	0.2 -0.7	65%	64	60		12
Carbon tetrachloride-toluene	14.7	93						0.3 -0.5			47		4
Trichloroethylene-toluene	15.0	87	0.30		8.0	7	2	0.17-0.60		52-56			26
Trichloroethylene-toluene, water	15.0	87	.30		8.0	7	2	0.50-0.97		51-55			26
Ethane-ethyl chloride-water-ethylene dichloride	27.5	105			30.0	18		0.45		29% for $H_2O-C_2H_4Cl_2$			22
										37% for $C_2H_5Cl-C_2H_4Cl_2$			
Trichloroethane-water	20.7	92	.37		24.0	21		.36		29% for $H_2O-C_2H_4Cl_2$			22
C_2, C_3, i-C_4, C_2H_5Cl	15.1	78	.11		30.0	18		.26		85% for i-C_4-C_2H_5Cl			22
H_2O-hexane	15.0	68			4.0 ft.			0.06-0.12			4-7.6		11
Liquid air	17.0	−185		.41	12.0	2	6*	0.2			84		20
								.5			77		
	17.0	−185		.28	12.0	1.5	6*	.1			60		
								.3			67		20
Hydrocarbon Systems													
C_1-C_6 + (Keys C_3-C_5)	347-368	154-159		0.07 -0.1	6.0 ft.	24-30		0.7- 1.0		77-84			7
C_3-C_6 + (Keys C_4, C_5)	116-117	127-137		0.15 -0.17	6.0 ft.	24		1.0 -1.1		59-64			7
					4.0 ft.	24		0.8 -1.0		68-73			7
C_3C_6 (Keys C_4C_5)	235	72		0.15 -0.17	4.5 ft.	20		0.4 -0.8		64-68			7
C_3C_6 + (Keys C_4C_5)	146-147	114-116		0.31 -0.32	6.0 ft.	30		0.9 -1.1		44-46			7
C_1C_6 + (Keys C_3, C_4)	362-365	92-109		0.220-0.222	6.5 ft. dia.	30		0.6 -0.7		55-58			7
C_6-C_{16} + steam	68-69	224-230		0.190-0.205	5.0 ft. dia.	30		0.2 -0.3		57-64			7
C_7-C_{18} + steam	58-59	250-264		0.25 -0.28	5.0 ft. dia.	30		0.03-0.07		48-52			7
C_3, i-C_4, n-C_4, C_5, etc	95	150°F.		0.25†	10.0 ft. dia.	27	115	0.74		140		13.9	6
	95	147°F.		.25‡	10.0 ft. dia.	27	136	0.70-0.73		75§		9.6	
C_2-C_7 (Keys i-C_4, N-C_4)	250	184°F.		.625	6.0 ft. dia.	30	34	0.70		75		11.0	2
	250	245°F.		1.0	6.0 ft. dia.	30	34	0.96		100			2
C_2, C_3, i-C_4, i-C_5, n-C_5, C_6	120	240		0.88	42.0	18	18	0.90-1.1		63		11.0	7, 21
	110	236						0.80-0.86		69			
	110	214						0.52-0.66		67			
	95	154						0.39-0.57		51			
C_1, C_2, C_3, i-C_4, C_5, C_7	285	92								108			3
Hydrocarbons boiling from 320° to 515°F. and steam	0	210			9.0 ft. dia.	20	111-115	0.60		65			19, 27
Methyl cyclohexane, toluene, phenol	16-20	150			7.0 ft. dia.	18				49-53			8
										61 (rectifying section)			8
										38 (stripping section)			8
Hydrocarbon mixture	215						30			80-90			
Naphtha	15	117		.5			1 tunnel cap perpendicular to flow				80	10.0	18
Naphtha + aniline	15	150									95		
Naphtha + pinene	15	147-151									85		
i-Butane, 1-butene, furfural**	60-115	60		2.16	13	24	13	0.13-0.81			48-60	11.7	12a
Toluene, paraffin, phenol	18	137			7	18				50			8a

* Tunnel caps with entrainment eliminator and baffles over tunnels.

† Weir height = 2.0 to 2.8 in., hydraulic gradient = 1.6 to 2.0 in., average depth of unaerated liquid above top of slots = 0.25 + (2.0 to 2.8) + (0.8 to 1.0) = 3.0 to 4.0 in.

‡ Weir height = 2.0 to 2.6 in., hydraulic gradient = 2.7 to 3.2 in., average depth of unaerated liquid above top of slots = 0.25 + (2.0 to 2.6) + (1.35 to 1.6) = 3.6 to 4.4 in.

§ Hydraulic gradient too great for stable operation. Liquid leaking through upstream rows of caps.

** Also with furfural containing up to 9 weight per cent water.

References:
[1] Brown and Holcomb, *Petroleum Engr.*, **12** (3), 107, 110 (1940).
[2] Brown and Lockhart, *Trans. Am. Inst. Chem. Engrs.*, **39**, 63 (1943).
[3] Brown, Souders, Nyland, and Hesler, *Ind. Eng. Chem.*, **27**, 383 (1935).
[4] Carey, Sc. D. Thesis, Mass. Inst. Tech., 1930.
[5] Carey, Griswold, Lewis, and McAdams, *Trans. Am. Inst. Chem. Engrs.*, **30**, 504 (1934).
[6] Cicalese, Davies, Harrington, Houghland, Hutchinson, and Walsh, *Petroleum Processing*, **1**, 296 (1946).
[7] Drickamer and Bradford, *Trans. Am. Inst. Chem. Engrs.*, **39**, 319 (1943).
[8] Drickamer, Brown, and White, *Trans. Am. Inst. Chem. Engrs.*, **41**, 555 (1945).
[8a] Drickamer and Hummel, *Trans. Am. Inst. Chem. Engrs.*, **41**, 607 (1945).
[9] Gadwa, Sc. D. Thesis, Mass. Inst. Tech., 1936.
[10] Gerster, Koffolt, and Withrow, *Trans. Am. Inst. Chem. Engrs.*, **39**, 37 (1943); **40**, 119 (1944); **41**, 393 (1945).
[11] Gester, *Chem. Eng. Progress*, **43**, 117 (1947).
[12] Griswold and Stewart, Paper Presented at Meeting of American Chemical Society, Pittsburgh, December, 1946.
[12a] Grohse, McCartney, Hauer, Gerster, and Colburn, Paper at A.I.C.E. Meeting, New York, Nov. 8 1948.
[13] Gunness, *Ind. Eng. Chem.*, **29**, 1092 (1937).
[14] Keyes and Byman, *Ill. Univ. Eng. Exp. Sta. Bull.* 328, May, 1941.
[15] Kirschbaum, *Z. Ver. deut. Ing.*, *Beiheft Verfahrenstechnik*, No. 5, 131 (1938); No. 3, 69 (1940).
[16] Kohrt, *Forsch. Gebiete Engenieurw.*, **4**, 286 (1933).
[17] Langdon and Keyes, *Ind. Eng. Chem.*, **35**, 464 (1943).
[18] Lewis and Smoley, *Am. Petroleum Inst.*, **65**, Serial No. 250 (1929).
[19] Lewis and Wilde, *Trans. Am. Inst. Chem. Engrs.*, **21**, 99 (1928).
[20] Lobo and Williams, *O.S.R.D. Rept.* 3768, Liquid Air Fractionation, June 13, 1944. Aston, *O.S.R.D. Rept.* 3699, Tests of Performance of Portable Units for Liquid Air Rectification, May 29, 1944.
[21] McGiffin, *Trans. Am. Inst. Chem. Engrs.*, **38**, 761 (1942).
[22] O'Connell, *Trans. Am. Inst. Chem. Engrs.*, **42**, 741 (1946).
[23] Peavy and Baker, *Ind. Eng. Chem.*, **29**, 1046 (1937).
[24] Rhodes and Slachman, *Ind. Eng. Chem.*, **29**, 51 (1937).
[25] Rumford, *J. Roy. Tech. Coll. (Glasgow)*, **4**, 239, 650 (1938).
[26] Schoenborn, Koffolt, and Withrow, *Trans. Am. Inst. Chem. Engrs.*, **37**, 997 (1941).
[27] Singer, Wilson, and Brown, *Ind. Eng. Chem.*, **28**, 824 (1936).
[28] Thomson, *Trans. Inst. Chem. Engrs. (London)*, **14**, 119 (1936).
[29] Uchida and Matsumoto, *J. Soc. Chem. Ind., Japan*, **39**, 224 (1936).
[30] Warden, *J. Soc. Chem. Ind., London*, **51**, 405T (1932).

Gerster, and Colburn [*Trans. Am. Inst. Chem. Engrs.*, Dec. (1949)]. A summary of their results, obtained under conditions simulating the stripping section of an extractive distillation column, are given in Fig. 64.

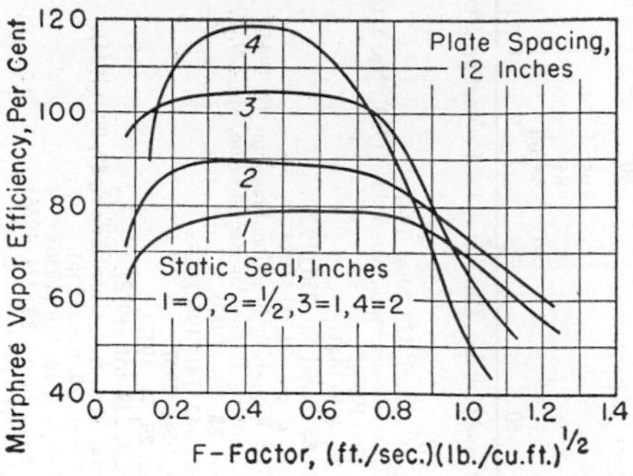

FIG. 63. Murphree plate efficiencies for distillation of ethanol-water mixtures. Data for plate spacing of 12 in. and static seals of 0, 0.5, 1, and 2 in. (*Peavy and Baker, loc. cit.*)

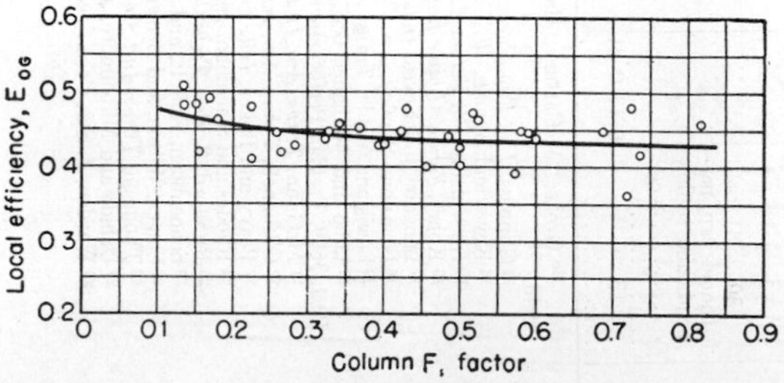

FIG. 64. Values of local plate efficiency for extractive distillation of isobutane and 1-butene in furfural. Conditions: total liquor rates, 9 to 38 gal./min.; total hydrocarbon in liquid, 12 to 27 mole per cent; concentration of water in furfural, 0 to 9 weight per cent; temperature, 110° to 150°F.; pressure, 40 to 100 lb./sq. in. gage; column diameter, 13 in.; number of caps per tray, 13; active slot submergence, 2.9 to 3.7 in. Note: F factor $= u\rho^{\frac{1}{2}}$, where u = superficial vapor velocity, ft./sec., and ρ = vapor density, lb./cu. ft. (*Grohse, et al., loc. cit.*)

Plate efficiencies for absorption columns are given in Sec. 10.

PACKED DISTILLATION COLUMNS

By R.L. Pigford and A.P. Colburn

Application. Packed towers are used in distillation in the following types of problems: separation of materials corrosive to metals (although ceramic, carbon, and glass plate towers are sometimes used in this application); where pressure drop must be maintained low as in vacuum distillation; where hold-up is desired low, as in batch distillation; for semiworks operation where the diameter of column is relatively small—usually less than 8 in., and where a very large number of plates or transfer units must be obtained in relatively low height on a small scale, as in analytical distillation. Although they are in general use for very small-scale distillation separations, packed columns have not found much employment on a large scale partly because of uncertainties of results and lack of reliable data on large packed columns, and partly because large bubble-cap columns are usually more economical.

Packings. Many of the tower packings described in the section on Absorption (p. 685) are useful in packed column distillation and reference is made to that section for data on types of packings. In general the smaller the size of the packing material the shorter the height of column but the lower the allowable vapor velocity and so the larger the diameter. As the particle size becomes quite small, the problem of maintaining satisfactory liquid distribution increases, and granular packings are therefore unsatisfactory. For very efficient packings as regards separating action, those promoting continuous film-type flow down extensive surface without drop formation appear best. Examples are the double spiral in Podbielniak columns, Fenske spirals, Stedman packing, and McMahon packing.

Design of Packed Columns

Diameter. The diameter of packed columns is determined by the actual vapor and liquid rates through the column, and the flooding velocity for the given packing. In general, design is made for a gas velocity 50 to 75 per cent of flooding in order to provide the possibility of increased capacity later. On the other hand, packed distillation columns usually give better separations if operated near flooding points, since the packing is then better irrigated and vapor distribution is better. Inasmuch as pressure drop is usually no consideration except for vacuum operation, there is good reason to operate packed distillation columns close to their flooding points. For large diameter packings this may be accomplished by maintaining the pressure drop at around 1 in. H_2O/ft. depth of packing.

Data on flooding velocity of various commercial packings are given on pp. 683 to 686. Data on pressure drop are given on pp. 680–683.

Adequate liquid distribution and an open packing support are important; these are also discussed in Sec. 10, p. 710.

Height. The determination of height of a packed column involves finding, first, the number of theoretical plates or transfer units required for the separation at hand and, second, the height of packing equivalent to a theoretical plate or transfer unit. The first step is an analytical problem; the second depends upon experimental data.

Separations carried out in packed columns are often characterized by the number of theoretical plates required and often by the number of transfer units. Fortunately, for tall columns these two concepts give results that are numerically nearly identical, and either method may therefore be used at will. This is because, in difficult separations, which require tall columns, the operating lines are nearly parallel to the equilibrium curve so that the term mV/L is near unity. As discussed on p. 550, when mV/L equals unity, the number of theoretical plates is exactly equal to the number of transfer units; the difference increases as mV/L departs more and more from unity but is insignificant for values between 0.9 and 1.1.

The number of theoretical plates for a given separation is determined in the usual manner, as treated elsewhere in this section. In case of a number of plates in the dilute region, a convenient analytical solution is available in Sec. 8, p. 553. This is especially useful where high purities are involved.

The number of transfer units may be obtained by graphically integrating the equation

$$N_{OG} = \int_{y_2}^{y_1} \frac{dy}{y^* - y} \qquad \text{or} \qquad N_{OL} = \int_{x_2}^{x_1} \frac{dx}{x - x^*}$$

(for nomenclature see p. 621), or by a special graphical method of White described in Sec. 8, p. 553, or, in case of distillation in the dilute region of one of the components, from Table 15 of Sec. 8, p. 555. The graphical integrations are carried out by first making a McCabe and Thiele diagram in order to locate values of y and y^*, then using values from this diagram to provide values of $1/(y^* - y)$ which are plotted vs. y; the area under the curve between limits of y_1 and y_2 is equal to N_{OG}. An example of a methanol stripping column is given in Sec. 8, p. 555. Other examples are shown by Chilton and Colburn [*Ind. Eng. Chem.*, **27**, 205 (1935)] and Colburn [*Ibid.*, **33**, 459 (1941)].

For cases of total reflux and mixtures of constant relative volatility α a convenient analytical solution for number of transfer units was given by Chilton and Colburn (*loc. cit.*) as follows:

$$N_{OG} = \frac{2.3}{\alpha - 1} \log \frac{y_1(1 - y_2)}{y_2(1 - y_1)} + 2.3 \log \frac{(1 - y_2)}{(1 - y_1)}$$

This equation is particularly useful in interpreting data on column testing with an ideal mixture. It gives results only slightly different from the Fenske-Underwood equation, p. 610, for number of theoretical plates for total reflux and constant α.

Based upon a continuous change in composition through the tower, the transfer-unit concept is more fundamental for this application than that of theoretical plates, though not usually so convenient. The use of theoretical-plate calculations can thus be considered as an approximation for transfer units, though limited to conditions where mV/L is near unity. For special cases where mV/L is considerably more or less than unity, transfer units must be employed for reliable results.

In distillation work, the number of transfer units is usually calculated as N_{OG}, *i.e.*, based on gas-composition changes, even though considerable transfer resistance is in the liquid stream. For a dilute volatile component in the stripping section, the analytical solution is preferably for N_{OL}, as given in Table 15, Sec. 8, p. 555. Additional discussion and examples of the transfer-unit concept in packed column distillation are given by Chilton and Colburn [*Ind. Eng. Chem.*, **27**, 255 (1935)], Colburn [*Trans. Am. Inst. Chem. Engrs.*, **35**, 211, 587 (1939)], and Colburn [*Ind. Eng. Chem.*, **33**, 459 (1941)].

Values of H.T.U. and H.E.T.P.

The available data on values of H.T.U. or H.E.T.P. are tabulated for convenience in Table 15. This table includes only data obtained during operation at total reflux, under which conditions the values of H.T.U. and H.E.T.P. are practically the same; and thus the values are given in a common column regardless of how they were calculated. Data at finite reflux are discussed later.

In extrapolating these data to other conditions, it is important to know the controlling variables. These have not been evaluated in a reliable manner for distillation; but, since the operation is analogous to absorption, conclusions may be drawn from that field. As discussed in Sec. 8, there are in general gas and liquid resistance to mass transfer, and these are expressed in terms of H.T.U. values as

$$H_{OG} = H_G + H_L\left(\frac{mV}{L}\right)$$

For difficult separations the term mV/L is near unity, and the effects of H_G and H_L are therefore of about equal weight.

Effect of Velocities. In absorption, values of H_G increase slightly with gas velocity but usually decrease with increasing liquor rate, since more contact area is provided. Values of H_L are essentially independent of gas velocity but increase somewhat with increasing liquor rate, at least for values above 1000 lb./(hr.)(sq. ft.). At lower rates the decreased contact area may reverse the effect. It is therefore believed that, regarding the over-all value of H_{OG}, liquid rate is an important variable and that, at constant values of mV/L, the gas rate can be ignored. Hence, Table 15 lists liquid rates only, and in general observed values of H_{OG} are seen to decrease with increasing liquid rate.

Effect of Temperature and Composition. The temperature has little if any effect on values of H_G, but increasing temperature causes a marked decrease in values of H_L, probably because of the increased value of diffusivity, and an even greater increase in value of the Schmidt modulus for the liquid, $Sc = \mu/\rho D$ (this modulus for gases is essentially independent of temperature and pressure).

The effect of composition and, for liquids, of temperature can presumably be estimated by assuming that the quotient (H.T.U.)/$(N_{Sc})^{1/2}$ is a constant. Typical values of gas and liquid diffusivities and the effect of temperature on the liquid value are given in Sec. 8, p. 538. Since temperature is thus an important factor, experimental values of temperature for the data in Table 15 may be estimated from the pressure of the distillation and boiling points of the constituents.

Effect of Pressure. The effect of pressure, per se, on gas and liquid values of H.T.U. is negligible. However, in distillation the pressure actually determines the temperature range encountered for a given system. Inasmuch as most liquids have a similar value of viscosity at their normal boiling points (about 0.3 centipoise), at a given pressure almost any liquid mixture of a narrow boiling range will have about the same viscosity and thus not too different values of the Schmidt number. Hence pressure actually sets the range of viscosity for close-boiling mixtures, and values are accordingly listed in Table 15.

Effect of Packing and Column Dimensions. The smaller the particle size of packing the greater the surface area; thus one would expect lower values of H.T.U. In general such is the case, although the effect is not direct. The through-put capacity for very small packings, however, is too low to make them very useful for large-scale operations.

There is considerable uncertainty regarding the effect of column diameter. It is generally believed that, owing to poorer liquid distribution, the values of H.T.U. become less favorable for a given packing the larger the diameter of column. The data in Table 15 for ½-in. rings and ¼-in. saddles bear out this belief.

In general, values of H.T.U. are found to be slightly less favorable for greater heights of packing, possibly because of progressive maldistribution effects. It is usually considered good practice to introduce redistributers every 10 ft. of packed height.

Effect of Reflux Ratio. There appears to be no reason why changing reflux ratio should affect the value of H_{OG} except possibly in changing the liquid flow rate and in changing the group mV/L and thereby the relative importance of gas and liquid resistances. Until the separate values of H_G and H_L are known, the latter effect cannot be predicted. On the other hand, Furnas and Taylor [*Trans. Am. Inst. Chem. Engrs.*, **36**, 135 (1939)] and Duncan, Koffolt, and Withrow [*Trans. Am. Inst. Chem. Engrs.*, **38**, 259 (1942)] both found unexpectedly large values of H_{OG} when testing at values of L/V around 0.6 and lower. It is believed that, under these special testing conditions, the liquor leaving the column must be very close to equilibrium with the entering vapor so that even mild maldistribution would affect the countercurrent action of the different streams and be reflected in high values of H.T.U. Such a type of

Table 15. Experimental Values of H.T.U. or H.E.T.P. of Packed Distillation Columns

Values are for atmospheric pressure and total reflux unless otherwise noted

Packing	Packing depth, ft.	Column diam., in.	Mixture	Liquor rate, lb./(hr.)(sq. ft.)	H_{OG} or H.E.T.P., ft.	Ref.
2-in. ceramic rings	10	12	EtOH-H_2O	816	2.2	9
1-in. ceramic rings	10	12	EtOH-H_2O	785	1.2	9
1-in. ceramic saddles	9	12	EtOH-H_2O	150	1.4	9
	9	12	EtOH-H_2O	960	1.1	9
½-in. ceramic saddles	10	12	EtOH-H_2O	190	1.5	9
	10	12	EtOH-H_2O	920	0.9	9
	8.5	2	n-Heptane–methyl cyclohexane	200	0.58	6
	8.5	2	n-Heptane–methyl cyclohexane	700	.56	6
	8.5	2	n-Heptane–methyl cyclohexane	1000	.52	6
½-in. aluminum saddles	8.5	2	n-Heptane–methyl cyclohexane	300	.50	6
	8.5	2	n-Heptane–methyl cyclohexane	700	.45	6
	8.5	2	n-Heptane–methyl cyclohexane	1200	.35	6
½-in. ceramic or carbon rings	9	5	EtOH-H_2O	550	1.0	5
	9	5	Trichloroethylene-toluene	200	1.45	5
	9	5	Trichloroethylene-toluene	700	1.25	5
	9	4.8	Trichloroethylene-toluene	400	1.3	13
½-in. ceramic or carbon rings	9	4.8	Trichloroethylene-toluene	1200	0.7	13
	5	4.8	Trichloroethylene-toluene	500	.87	13
	5	4.8	Trichloroethylene-toluene	1700	.7	13
	9	5	Carbon tetrachloride–benzene	250	2.00	5
	9	5	Carbon tetrachloride–benzene	1150	1.25	5
	4	6	Iso-PrOH-H_2O	150	0.3–0.6	4
	4	6	Iso-PrOH-H_2O	450	0.3–0.6	4
	8.5	2	n-Heptane–methyl cyclohexane	200	0.6	6
	8.5	2	n-Heptane–methyl cyclohexane	400	.7	6
	8.5	2	n-Heptane–methyl cyclohexane	700	.6	6
	8.5	2	n-Heptane–methyl cyclohexane	1000	.5	6
⅜-in. ceramic rings	8	12	EtOH-H_2O	320	1.3	9
	8	12	EtOH-H_2O	600	1.0	9
	8.5	2	n-Heptane–methyl cyclohexane	100	0.5	6
	8.5	2	n-Heptane–methyl cyclohexane	300	.6	6
	8.5	2	n-Heptane–methyl cyclohexane	800	.4	6
0.42-in. glass rings	3.3	1.36	EtOH-H_2O	50	.3	10
	3.3	1.36	EtOH-H_2O	700	.5	10
0.315-in. rings	3.5	4.33	EtOH-H_2O	120–240	.5	17
	7.2	4.33	EtOH-H_2O	120–240	.6	17
	13.2	4.33	EtOH-H_2O	120–204	.75	17
0.275-in. glass rings			Benzene-toluene, Comp.: 0.06–0.4	100–1000	.75	3
			Benzene-toluene, Comp.: 0.04–0.9	100–1000	.45	3
¼-in. carbon rings	8.5	2	n-Heptane–methyl cyclohexane	150	0.4–0.5	6
	8.5	2	n-Heptane–methyl cyclohexane	250	0.46	6
	8.5	2	n-Heptane–methyl cyclohexane	350	.4	6
¼-in. glass rings	8.5	2	n-Heptane–methyl cyclohexane	500	.5	6
	1.5	2	N_2-O_2	450	.26	11
	1.5	2	N_2-O_2	1370	.16	11
	0.83	2	N_2-O_2	450	.20	11
	0.83	2	N_2-O_2	1140	.13	11
¼-in. ceramic saddles	8.5	2	n-Heptane–methyl cyclohexane	200	.58	6
	8.5	2	n-Heptane–methyl cyclohexane	700	.56	6
	8.5	2	n-Heptane–methyl cyclohexane	1000	.52	6
	1.5	2	N_2-O_2	610	.39	11
	1.5	2	N_2-O_2	1800	.31	11
	4.8	8	N_2-O_2	930–1460	0.30–0.48	1
	1.5	2	N_2-O_2	500	0.28	11
¼-in. ceramic saddles	1.5	2	N_2-O_2	1400	.20	11
	1	10	N_2-O_2	400	.4	1
	1	10	N_2-O_2	670	.37	1
	4	8	N_2-O_2	400	.7	1
	4	8	N_2-O_2	670	.54	1
0.18-in. glass rings	3.3	1.36	EtOH-H_2O	50	.2	10
	3.3	1.36	EtOH-H_2O	700	.4	10
6 mesh carborundum	8.5	2	n-Heptane–methyl cyclohexane	270	0.15–0.3	6
4–6 mesh "Haydite"	4.8	8	N_2-O_2	1000	0.52	1
	4.8	8	N_2-O_2	1300	.40	1
4–8 mesh "Haydite"	1.5	2	N_2-O_2	400	.35	11
	1.5	2	N_2-O_2	1600	.37	11
1-turn glass helixes		0.39	n-Heptane–methyl cyclohexane	200	.12	16
	1.5	.39	n-Heptane–methyl cyclohexane	750	.12	16
0.125-in. diameter glass helixes	1.5	2	N_2-O_2	640	.31	11
	1.5	2	N_2-O_2	1020	.35	11
0.16-in. diameter No. 26 nickel wire helixes	1.5	1.26	n-Heptane–methyl cyclohexane	900	.45	16
	1.5	1.26	n-Heptane–methyl cyclohexane	1100	.34	16

Table 15. Experimental Values of H.T.U. or H.E.T.P. of Packed Distillation Columns—(*Concluded*)

Packing	Packing depth, ft.	Column diam., in.	Mixture	Liquor rate, lb./(hr.)(sq. ft.)	H_{OG} or H.E.T.P., ft.	Ref.
0.13-in. diameter No. 26 nickel wire helixes	1.5	1.26	n-Heptane–methyl cyclohexane	450	.45	8
	1.5	1.26	n-Heptane–methyl cyclohexane	830	.40	8
0.09-in. diameter No. 26 nickel wire helixes	1.5	0.79	n-Heptane–methyl cyclohexane	125	.09	16
	1.5	.79	n-Heptane–methyl cyclohexane	620	.12	16
0.094-in. diameter wire helixes	1.5	2	N_2-O_2	1040	.30	11
	1.5	2	N_2-O_2	1460	.53	11
Metal cloth, 17 lb./cu. ft.	1	2	N_2-O_2	1000	.49	11
	1	2	N_2-O_2	2660	1	11
"Fiberglas" 0.7 lb./cu. ft.	1	2	N_2-O_2	1970	.7	11
"Fiberglas" 10.7 lb./cu. ft.	1	2	N_2-O_2	1490	.6	11
"Fiberglas" 4.63 lb./cu. ft.	1.45	12	MeOH-H_2O	180	.6	12
	1.45	12	MeOH-H_2O	300	.9	12
	1.45	12	EtOH-H_2O	380	.7	12
"Fiberglas" 4.06 lb./cu. ft.	6.5	12	EtOH-H_2O Comp. 0.56–0.85	430	.85	12
	6.5	12	EtOH-H_2O Comp. 0.043–0.72	450	1.6	12
"Fiberglas" 4.06 lb./cu. ft.	6.5	12	Acetone-H_2O Comp. 0.18–0.89	830	1.8	12
	6.5	12	Acetone-H_2O Comp. 0.10–0.81	1040	2.8	12
Stedman No. 105		0.38	Benzene–ethylene dichloride	485	0.038	2
		.38	Benzene–ethylene dichloride	785	.075	2
Stedman No. 104		.75	Benzene–ethylene dichloride	108	.040	2
		.75	Benzene–ethylene dichloride	480	.097	2
Stedman No. 112		.98	Benzene–ethylene dichloride	63	.042	2
		.98	Benzene–ethylene dichloride	540	.097	2
Stedman No. 128		2.08	Benzene–ethylene dichloride	210	.076	2
		2.08	Benzene–ethylene dichloride	1760	.181	2
Stedman No. 107		3.08	Benzene–ethylene dichloride	105	.090	2
		3.08	Benzene–ethylene dichloride	1640	.192	2
Stedman No. 115		6.08	Benzene–ethylene dichloride	115	.097	2
		6.08	Benzene–ethylene dichloride	1730	.21	2
Stedman No. 116		12	Benzene–ethylene dichloride	200	.12	2
		12	Benzene–ethylene dichloride	1780	.19	2
Stedman 120 × 120 mesh	10.375	2.08	N_2-O_2	390	.15	11
	10.375	2.08	N_2-O_2	800	.17	11
Stedman 100 × 100 mesh	10.375	2.08	N_2-O_2	520–1300	0.12	11
Stedman 80 × 80 mesh	9.75	2.08	N_2-O_2	430	.09	11
	9.75	2.08	N_2-O_2	1400	.14	11
Stedman 60 × 40 mesh	10.75	2.08	N_2-O_2	470	.095	11
	10.75	2.08	N_2-O_2	1260	.11	11

[1] Aston, *O.S.R.D. Rept.* 3699, May 29, 1944, Tests of Performance of Portable Unit Columns for Air Rectification.
[2] Bragg, *Trans. Am. Inst. Chem. Engrs.*, **37**, 19 (1941); *Ind. Eng. Chem.*, anal. ed., **11**, 288 (1939); *Ind. Eng. Chem.*, **33**, 279 (1941); *Refiner*, **18**, 295 (1939).
[3] Carlson, M.S. Thesis, Mass. Inst. Tech., 1935.
[4] Deed, Schutz, and Drew, *Ind. Eng. Chem.*, **39**, 766 (1947).
[5] Duncan, Koffolt, and Withrow, *Trans. Am. Inst. Chem. Engrs.*, **38**, 259 (1942).
[6] Fenske, Larowski, and Tongberg, *Ind. Eng. Chem.*, **30**, 297 (1938).
[7] Fenske, Tongberg, and Quiggle, *Ind. Eng. Chem.*, **26**, 1169 (1934).
[8] Fenske, Tongberg, Quiggle, and Cryder, *Ind. Eng. Chem.*, **28**, 644 (1936).
[9] Furnas and Taylor, *Trans. Am. Inst. Chem. Engrs.*, **36**, 135 (1939).
[10] Jantzen, Dechema Monograph 5, No. 48, 142 pp., Verlag Chemie, 1932.
[11] Dodge, reported by Lobo and Williams, *O.S.R.D. Rept.* 3768, June 13, 1944, Liquid Air Fractionation.
[12] Minard, Koffolt, and Withrow, *Trans. Am. Inst. Chem. Engrs.*, **39**, 813 (1943).
[13] Schoenborn, Koffolt, and Withrow, *Trans. Am. Inst. Chem. Engrs.*, **37**, 997 (1941).
[14] Stedman, U.S. Patent 2,047,444 (July 14, 1936); *Can. J. Research*, **15-B**, 383 (1937).
[15] Tongberg, Larowski, and Fenske, *Ind. Eng. Chem.*, **29**, 957 (1937).
[16] Tongberg, Quiggle, and Fenske, *Ind. Eng. Chem.*, **26**, 1213 (1934).
[17] Wieman, *Beiheft Z. Ver. deut. Ing.*, No. 6, 15 pp. (1933).

operation is not general; in such cases reference should be made to the original papers for data.

Nomenclature.

H_{OG} = H.T.U. (height of transfer unit) on an over-all gas basis.
H_G = H.T.U. of gas film.
H_L = H.T.U. of liquid film.
H.E.T.P. = height equivalent to a theoretical plate.
m = dy^*/dx, or slope of equilibrium curve; in dilute region m is identical with the Henry law constant equal to y^*/x.
L = molar liquid rate passing down the column.
V = molar vapor rate passing up the column.
y = mole fraction of component in vapor.
y^* = vapor composition in equilibrium with x.
x = mole fraction of component in liquid.
x^* = liquid composition in equilibrium with y.
α = relative volatility = $\frac{y_A^* x_B}{y_B^* x_A}$
N_{Sc} = Schmidt modulus = $\mu/\rho D$, dimensionless.
μ = viscosity of vapor or liquid mixture.
ρ = density of vapor or liquid mixture.
D = diffusion coefficient of vapor or liquid mixture.

Subscripts 1 and 2 refer to exit and entrance.
Subscripts A and B refer to different components.

PLATE CALCULATIONS FOR MULTICOMPONENT MIXTURES

By C. N. Collard

REFERENCES: Lewis and Matheson, *Ind. Eng. Chem.*, **24**, 494 (1932). Brown and Martin, *Trans. Am. Inst. Chem. Engrs.*, **35**, 679 (1939). Gilliland, *Ind. Eng. Chem.*, **32**, 1220 (1940). Fenske, *Ind. Eng. Chem.*, **24**, 482 (1932). Underwood, *Trans. Inst. Chem. Engrs. (London)*, **10**, 112 (1932). Gilliland, *Ind. Eng. Chem.*, **32**, 1101 (1940). Colburn, *Trans. Am. Inst. Chem. Engrs.*, **37**, 805 (1941). Gilliland, *Ind. Eng. Chem.*, **32**, 918 (1940). Hummel, *Trans. Am. Inst. Chem. Engrs.*, **40**, 445 (1944). Thiele and Geddes, *Ind. Eng. Chem.*, **25**, 289 (1933). Jenny, *Trans. Am. Inst. Chem. Engrs.*, **35**, 635 (1939). Brown and Souders, *Trans. Am. Inst. Chem. Engrs.*, **30**, 438 (1934). Brown, Souders, Nyland, and Hesler, *Ind. Eng. Chem.*, **27**, 383 (1935). Smith, *Trans. Am. Inst. Chem. Engrs.*, **37**, 333 (1941). Edminster, *Trans. Am. Inst. Chem. Engrs.*, **42**, 15 (1946). Hengstebeck, *Trans. Am. Inst. Chem. Engrs.*, **42**, 309 (1946). Jenny and Cicalese, *Ind. Eng. Chem.*, **37**, 956 (1945). Scheibel, *Ind. Eng. Chem.*, **38**, 397 (1946). Underwood, *J. Inst. Petroleum*, **32**, 614 (1946).

Introduction. The basic principles of rectification previously discussed for binary mixtures also apply to separations involving three or more components. The application of these principles to a multicomponent mixture is more complex because of the increased number of variables in such a system. In a system composed of liquid and vapor in equilibrium and involving any number n of miscible components, the concentration of $(n - 1)$ components in either liquid or vapor, and either the pressure or temperature are required to define the system, in accordance with the phase rule (pp. 576 to 579). In a binary system, therefore, at constant pressure, the liquid and vapor compositions are fixed by the concentration of one component in either phase, and the relationship between composition of liquid and that of vapor may be plotted on a single y-x diagram, as a function of one component only.

In a multicomponent system, the relationship between the concentration of a given component in the liquid and that of a component in the vapor at constant pressure is variable depending upon the relative concentrations of the other components in the system. For this reason, compositions of liquid and vapor must be determined for each multicomponent system by calculation from equilibrium data.

Equilibrium data are usually presented in the form $K = y/x$, where the equilibrium constant K for each component present is a function of temperature and pressure (see pp. 568 to 576).

For systems where Raoult's law applies,

$$y_1 = \frac{P_1 x_1}{P}$$

The method of utilizing such data for calculating liquid-vapor relationships has been described (pp. 585 to 587).

A method for calculation of the number of theoretical plates required for a specified separation and reflux ratio as applied to a multicomponent mixture was presented by Lewis and Matheson [*Ind. Eng. Chem.*, **24**, 494 (1932)]. This method involves the use of equilibrium data combined with the material balance equations as developed for plate towers (p. 589). The composition and quantity of liquid and vapor on each plate in the rectifying section is found by starting with the desired overhead product composition and reflux ratio and calculating stepwise from plate to plate down the tower by means of material balance and equilibrium relationships until the feed plate is reached. Similarly, the compositions of liquid and vapor on the plates below the feed are found by starting at the base of the tower and calculating stepwise up the tower until the feed plate is reached.

The proper feed plate is that which requires the least number of total plates in the tower. The compositions on the rectifying plates near the feed must be adjusted by introducing the heavier components not present in the overhead distillate in any appreciable quantity; and, similarly, lighter components not present in the bottoms must be introduced on the plates just below the feed, until the feed-plate composition as calculated from the top of the tower matches its composition as calculated from the base of the tower. The procedure of matching at the feed plate is essentially trial and error.

If the simplifying assumptions (p. 590) do not hold, the method may be made rigorous by heat and material balances on each plate, or sectional heat balances may be made and the proper values of L and V used for each section, depending upon the degree of accuracy required.

Relationship between Theoretical Plates and Reflux Ratio. In selecting an economic reflux ratio, methods presented by Brown and Martin [*Trans. Am. Inst. Chem. Engrs.*, **35**, 679 (1939)], and by Gilliland [*Ind. Eng. Chem.*, **32**, 1220 (1940)], correlating reflux ratio and number of plates, and requiring a knowledge only of minimum reflux and minimum plates, may be used to advantage. The relationship developed by Gilliland is shown graphically in Fig. 65, where S = total number of theoretical steps from reboiler to overhead product, S_m = minimum number of steps at total reflux, L/D = molal ratio of reflux to overhead product in the section of the tower above the feed plate, and $(L/D)_m$ = minimum reflux ratio.

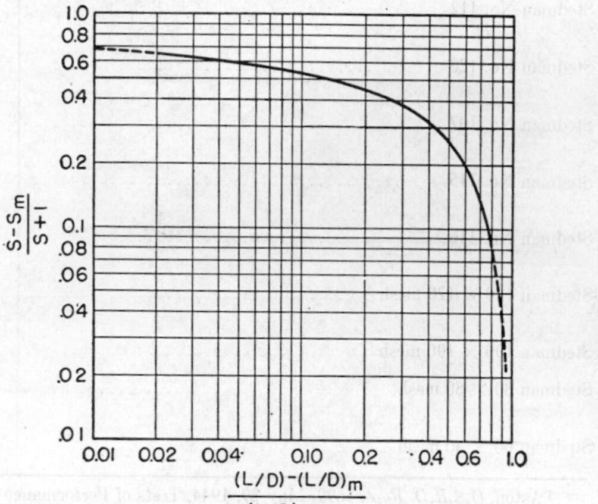

FIG. 65. Gilliland's correlation of theoretical plates as a function of reflux ratio.

For a tower employing a total condenser, so that the composition of the reflux is the same as that of the net overhead product,

$$S = N + 1$$

where N = number of theoretical plates in the tower.

If only a partial condenser is used and the overhead product is vented as a vapor, then

$$S = N + 2$$

When the simplifying assumptions (p. 590) may be made, the number of plates for a given reflux ratio may be predicted from this relationship with sufficient accuracy for most design work. The final design may be checked if necessary by plate-to-plate calculations, locating the proper feed plate during the procedure.

Minimum Plates at Total Reflux. The minimum number of plates required at total reflux may be approximated by the relationship developed by Fenske [*Ind. Eng. Chem.*, **24**, 482 (1932)] and Underwood [*Trans. Inst. Chem. Engrs.* (*London*), **10**, 112 (1932)]:

$$\left(\frac{x_{lk}}{x_{hk}}\right)_D = \left(\frac{x_{lk}}{x_{hk}}\right)_W (\alpha_{lk})^{S_m} \tag{81}$$

from which

$$S_m = \frac{\log (x_{lk}/x_{hk})_D(x_{hk}/x_{lk})_W}{\log \alpha_{lk}} \tag{82}$$

where x_{lk} = mole fraction of the light key component.
x_{hk} = mole fraction of the heavy key component, the subscripts D and W referring to the overhead and bottoms products, respectively.
α_{lk} = relative volatility of the light key component with respect to the heavy key component (see p. 579).

The "key" components referred to above are the two components with quantities fixed, respectively, in the distillate and bottoms by the sharpness of separation required. The "light key" component is generally defined as the most volatile component whose concentration is fixed in the bottoms, and the "heavy key" as the least volatile component whose concentration is fixed in the overhead.

If the value of α_{lk} varies through the tower, an average value may be computed from the α at the tower top and the reboiler such that $\alpha_{avg} = (\alpha_{top}\alpha_{reboiler})^{1/2}$.

Where the change in relative volatility through the tower is too great, Eq. (82) should not be used over the whole tower. In this case, stepwise calculations may be made down the tower from the top and up the tower from the bottom, using $L = V$, until the temperature gradient from tray to tray is approximately uniform; Eq. (82) may then be applied to the remaining section, using concentrations and relative volatilities at the top and bottom of this section in place of the terminal concentrations of the tower.

Determination of Minimum Reflux Ratio. A minimum reflux ratio exists for a specified separation of a multicomponent mixture. Whereas, for a binary mixture at minimum reflux, a region of infinitesimal change in concentration from plate to plate, or "pinch" section, occurs at the feed plate, in a multicomponent system there are two such regions, one above the feed plate and one below the feed. Concentration changes and temperature gradient are usually appreciable in the region of the feed plate because of rapid changes in concentration of components heavier than the heavy key immediately above the feed and lighter than the light key immediately below the feed. Methods for calculation of the minimum reflux ratio, based upon evaluation of compositions in the pinch sections above and below the feed plate, have been developed by Gilliland [*Ind. Eng. Chem.*, **32**, 1101 (1940)] and Colburn [*Trans. Am. Inst. Chem. Engrs.*, **37**, 805 (1941)]. Colburn's method consists in finding by trial and error the reflux ratio necessary to give the proper relationship between the compositions in the pinch sections. This relationship is expressed as a function of an empirical factor that was developed from an analysis of 37 examples. The average error in predicting the minimum reflux ratio for these examples was less than 1 per cent. The following equation is said to hold closely at minimum reflux:

$$\frac{r_m}{r_n} = \psi \tag{83}$$

where $\psi = \dfrac{1}{(1 - \Sigma C_m \alpha x_m)(1 - \Sigma C_n x_n)}$

r_m = ratio of concentration of light key to heavy key in stripping-section pinch.
r_n = ratio of concentration of light key to heavy key in enriching-section pinch.
$\Sigma C_m \alpha x_m$ = summation of values of $C_m \alpha x$ for all components heavier than the heavy key in stripping-section pinch.
$\Sigma C_n x_n$ = summation of values of $C_n x$ for all components lighter than the light key in the enriching-section pinch.
C_m = correction factor given by Fig. 66.
C_n = correction term given by Fig. 67.

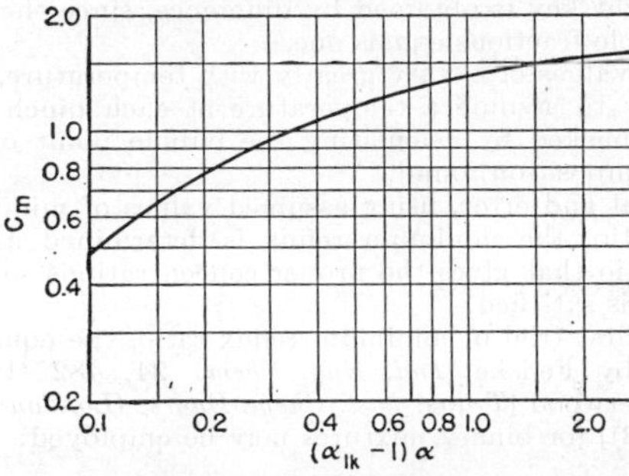

FIG. 66. Correction factor, C_m.

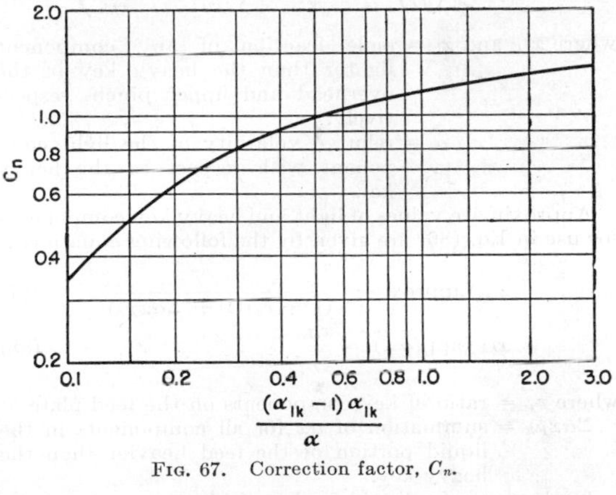

FIG. 67. Correction factor, C_n.

The concentrations of light components (components lighter than the heavy key) in the upper pinch are found by the equation

$$x_n = \frac{x_D}{(\alpha - 1)(L_n/D + \alpha x_{Dhk}/x_{hk})} \tag{84}$$

where x_D and x_n = mole fraction of a given light component in the overhead product and the liquid at the upper pinch, respectively.
x_{Dhk} and x_{hk} = mole fraction of the heavy key component in the overhead product and the liquid at the upper pinch, respectively.
α = relative volatility of the given component with reference to the heavy key component.

The concentrations of all light components may be calculated by means of Eq. (84), and the concentration of the heavy key is obtained by difference, since the sum of the mole fractions equals one.

Similarly, the concentrations of the heavy components (heavier than the light key) in the lower pinch may be found from the equation

$$x_m = \frac{\alpha_{lk} x_W}{(\alpha_{lk} - \alpha)(L_m/W) + \alpha x_{Wlk}/x_{lk}} \quad (85)$$

where x_W and x_m = mole fraction of a given heavy component in the bottoms and the lower pinch, respectively.

x_{Wlk} and x_{lk} = mole fraction of light key component in the bottoms and lower pinch, respectively.

Equation (85) is used to solve for the concentrations of each of the heavy components, and the concentration of the light key is obtained by difference, since the sum of the mole fractions equals one.

If the values of α vary greatly with temperature, it is necessary to assume a temperature at each pinch that can be checked by calculating the bubble point of the liquid composition found.

By trial and error, using assumed values of minimum reflux ratio, the minimum reflux is determined as the reflux ratio that gives the proper concentrations so that Eq. (83) is satisfied.

For a first trial of minimum reflux ratio, the equation derived by Fenske [*Ind. Eng. Chem.*, **24**, 482 (1932)] and Underwood [*Trans. Inst. Chem. Engrs. (London)*, **10**, 112 (1932)] for binary mixtures may be employed:

$$\text{Approx.}\ \left(\frac{L}{D}\right)_{\min} = \frac{1}{\alpha - 1}\left(\frac{x_{Dl}}{x_l} - \alpha\frac{x_{Dhk}}{x_{hk}}\right) \quad (86)$$

where x_{Dl} and x_l = mole fraction of any component lighter than the heavy key in the overhead and upper pinch, respectively.

α = relative volatility of the light component with respect to the heavy key.

Approximate values of light and heavy key components for use in Eq. (86) are given by the following equations:

$$x_{lk}\ (\text{approx.}) = \frac{r_f}{(1 + r_f)(1 + \Sigma\alpha x_{Fh})} \quad (87)$$

$$x_{hk}\ (\text{approx.}) = \frac{x_{lk}}{r_f} \quad (88)$$

where r_f = ratio of key components on the feed plate.

$\Sigma\alpha x_{Fh}$ = summation of αx for all components in the liquid portion of the feed heavier than the heavy key.

In the case of a liquid feed at the temperature of the feed plate, or of a partly vaporized feed, r_f may be taken as the ratio of key components in the liquid portion of the feed. For a cold feed or a superheated vapor feed, r_f may be estimated from Eq. (92) or (94).

The procedure indicated above for calculating the minimum reflux ratio may be summarized as follows:

1. Estimate r_f as described, and calculate $\Sigma\alpha x_F$. Using these values, calculate x_{lk} and x_{hk} by Eqs. (87) and (88), and estimate $(L/D)_{\min}$ from Eq. (86).

2. From the estimated value of $(L/D)_{\min}$, calculate L_m/W, the ratio of the liquid overflow in the stripping section to the bottoms. By Eqs. (84) and (85), the composition at the two pinches are calculated. The temperatures at the pinches for the first trial may be assumed to be at ⅓ and ⅔ of the temperature difference between the top and bottom of the tower.

3. Calculate the terms r_m/r_n and ψ. If the ratio r_m/r_n is greater than ψ, the estimated minimum reflux ratio was too high. The ratio r_m/r_n changes greatly with small changes in trial value of reflux ratio, but the factor ψ changes only slightly. By use of this fact, the true minimum reflux ratio may usually be predicted in a few trials.

In the case of a distributed component whose relative volatility lies between those of the key components, the concentrations of this component in the overhead and bottoms will be different from those at the operating reflux ratio. In order to use the method above for determining the minimum reflux ratio for this case, the quantities of distributed component going to the overhead and to the bottoms at minimum reflux must be estimated. The division is made such that the final value of x_n/x_m for the distributed component lies between the values of x_n/x_m for the key components.

The calculation of minimum reflux ratio may be made more simply for cases of ideal mixtures with constant relative volatilities by the method of Underwood [*J. Inst. Petroleum*, **32**, 614 (1946)]. The minimum reflux ratio is found by solving the equation:

$$\left(\frac{L}{D}\right)_{\min} + 1 = \frac{\alpha_a x_{Da}}{\alpha_a - \theta} + \frac{\alpha_b x_{Db}}{\alpha_b - \theta} + \frac{\alpha_c x_{Dc}}{\alpha_c - \theta} \cdots + \frac{\alpha_z x_{Dz}}{\alpha_z - \theta} \quad (89)$$

The proper value of the root θ is that which satisfies the equation:

$$\frac{\alpha_a z_{Fa}}{\alpha_a - \theta} + \frac{\alpha_b z_{Fb}}{\alpha_b - \theta} + \frac{\alpha_c z_{Fc}}{\alpha_c - \theta} \cdots + \frac{\alpha_z z_{Fz}}{\alpha_z - \theta} = 1 - q \quad (90)$$

In these equations subscripts $a, b, c, \ldots z$ refer to the individual components in order of decreasing volatility, subscripts D and F refer to overhead product and feed, respectively, and q is a term which defines the condition of the feed, *e.g.*, $q = 1$ for a liquid feed at the boiling point, and $q = 0$ for a vapor feed at the dew point (see p. 592). The terms $z_{Fa}, z_{Fb}, z_{Fc} \ldots z_{Fz}$ denote average mole fractions of components $a, b, c, \ldots z$, respectively, in the total feed. The correct value of θ lies between the values of α for the key components, although there are other values of θ which will satisfy equation (90).

Location of the Optimum Feed Plate. A criterion for the optimum feed-plate composition, as expressed by the ratio of the key components, has been developed by Gilliland [*Ind. Eng. Chem.*, **32**, 918 (1940)]. The ratio of key components at the intersection of the operating lines is given by the equation

$$\left(\frac{x_{lk}}{x_{hk}}\right)_i = \frac{F z_{Flk} + [(V_n/V_m) - 1]W x_{Wlk}}{F z_{Fhk} + [(V_n/V_m) - 1]W x_{Whk}} \quad (91)$$

where Fz_{Flk} = number of moles of light key component in the total feed.

Fz_{Fhk} = number of moles of heavy key component in the total feed.

Subscript i refers to the intersection of the operating lines.

The optimum ratio is given for three cases:

Case 1. The feed is either all liquid, or, if partly vaporized, the vapor portion is introduced in such a manner that it comes into intimate contact with the feed-plate liquid. For this condition,

$$\left(\frac{x_{lk}}{x_{hk}}\right)_f \leqq \left(\frac{x_{lk}}{x_{hk}}\right)_i \leqq \left(\frac{x_{lk}}{x_{hk}}\right)_{(f+1)} \quad (92)$$

where subscript f refers to the feed plate, subscript $(f + 1)$ refers to the plate above the feed, and subscript i refers to the point of intersection of the operating lines.

Case 2. The feed is partly vaporized, the liquid and vapor portions of the feed are in equilibrium, the liquid portion mixes with the feed-plate liquid, and the vapor portion mixes with the feed-plate vapor. For this case,

$$\left(\frac{x_{lk}}{x_{hk}}\right)_f \leqq \left(\frac{x_{lk}}{x_{hk}}\right)_F \leqq \left(\frac{x_{lk}}{x_{hk}}\right)_{f+1} \quad (93)$$

where subscript F refers to the liquid portion of the feed.

Equation (93) indicates that the ratio of key components in the liquid portion of the feed lies in a range between the ratios on the feed plate and the plate above.

Case 3. The feed is a superheated vapor that mixes with the vapor from the feed plate but does not come in contact with the feed-plate liquid. For this case,

$$\left(\frac{x_{lk}}{x_{hk}}\right)_{(f+1)} \leqq \left(\frac{x_{lk}}{x_{hk}}\right)_i \leqq \left(\frac{x_{lk}}{x_{hk}}\right)_{(f+2)} \qquad (94)$$

This states that the intersection ratio lies in a range between the ratios on the first and second plates above the feed plate.

Separations Involving Three or More Distributed Components. When the feed contains one or more components whose volatility lies between the key components, the distribution of these intermediate components between the overhead and bottoms with fixed terminal concentrations of key components varies with reflux ratio. Thus, at the minimum reflux ratio for the key components, the distribution is different from that at the operating reflux ratio.

By the stepwise method of calculation, it is necessary to estimate the distribution of the intermediate components based on relative volatilities, follow through the calculation to check the assumption, make a revised estimate if required, and repeat the procedure until a check is obtained.

Hummel [*Trans. Am. Inst. Chem. Engrs.*, **40**, 445 (1944)] has presented a method of plate-to-plate calculation which is a modification of methods presented by Thiele and Geddes [*Ind. Eng. Chem.*, **25**, 289 (1933)] and Jenny [*Trans. Am. Inst. Chem. Engrs.*, **35**, 635 (1939)], by which the trial-and-error procedure may be shortened. Temperatures and concentrations in the vicinity of the feed plate are estimated by Jenny's method, and simplified plate calculations for the distributed components only are made based upon the estimated temperatures of each plate.

Other methods of calculation, using absorption and stripping factors, have been presented by Brown and Souders [*Trans. Am. Inst. Chem. Engrs.*, **30**, 438 (1934)], Brown, Souders, Nyland, and Hesler [*Ind. Eng. Chem.*, **27**, 383 (1935)], and Smith [*Trans. Am. Inst. Chem. Engrs.*, **37**, 333 (1941)]. Edmister [*Trans. Am. Inst. Chem. Engrs.*, **42**, 15 (1946)] has presented a method of calculation based on the absorption and stripping factor concept together with the technique used by Jenny and by Hummel. Edmister's method is particularly convenient in handling a number of distributed components, though the solution involves trial and error as do the other methods.

For explanation of the above methods, the reader is referred to the original articles.

For determining the distribution between overhead and bottoms of components other than the keys, Hengstebeck [*Trans. Am. Inst. Chem. Engrs.*, **42**, 309 (1946)] has recommended the use of the relation, which is said to hold closely for all distributed components at operating reflux ratios,

$$\log \frac{Dx_D}{Wx_W} = C\alpha$$

where Dx_D = moles of a given component in the overhead.

Wx_W = moles of the given component in the bottoms.

α = relative volatility with reference to heavy key component.

C = a constant.

Thus a plot of log (Dx_D/Wx_W) vs. α is a straight line that may be drawn between the two points showing the proper relations for the key components, and the distribution of any other components may be approximated from the relation thereby indicated.

Prediction of the Performance of an Existing Tower. The separation possible in an existing tower where the quantity of reflux, number of plates, and feed-plate location are known involves substantially the same type of calculation previously described. In this case, however, it is necessary first to estimate the composition of the overhead and bottoms. The assumptions may then be checked by trial stepwise calculation, applying the technique used by Hummel or by the absorption-stripping-factor method of Edmister, as mentioned in a preceding paragraph.

General Design Procedure. For estimating the number of plates and reflux ratio required for a specified separation, the following procedure is recommended:

1. Calculate the minimum number of plates at total reflux.
2. Calculate the minimum reflux ratio.
3. By means of Fig. 65, make a plot of the number of plates vs. reflux ratio. From the information gained from this plot, and economic and practical considerations, choose a design reflux ratio.
4. Check the number of plates at the design reflux ratio by stepwise plate-to-plate calculations, and locate the proper feed plate by methods described above.

Sample Calculations. A mixture of hydrocarbons is to be fractionated so that the propane retained in the bottoms shall be not more than 1 mole per cent, and the overhead shall contain not more than 1.5 mole per cent butane. A partial condenser will be employed and the overhead delivered as a vapor.

Material Balance

Basis: 100 moles/hr. feed

Component	Feed	Overhead	Bottoms
C_2	3	3	
C_3	20	19.3	0.7
C_4	37	0.3	36.7
C_5	35	....	35
C_6	5	....	5
	100	22.6	77.4

In the calculations following, moles will be used instead of mole fractions where possible, thus eliminating the steps of conversion.

Determination of Condenser Outlet Temperature and Pressure. The available cooling-water temperature is assumed to be 85°F. Economic considerations govern condenser outlet temperature and operating pressure. Based on this cooling-water temperature, the economic condenser outlet temperature is estimated to be in the range of 115° to 120°F., which will require a pressure of approximately 250 lb./sq. in. gage, as shown by the following dew-point calculation (see p. 587).

Component	Moles d	250 lb./sq. in. gage 120°F. K	$\frac{d}{K}$
C_2	3	2.55	1.2
C_3	19.3	0.95	20.3
C_4	0.3	.38	0.8
	22.6	....	22.3

For C_3, $K = 22.3 \times 0.95/22.6 = 0.936$, and $t = 118$°F. The column d/K indicates the composition of the reflux.

Top Tower Temperature. When a partial condenser is used, the composition of the vapor leaving the tower will vary with reflux ratio; and the top tower temperature will likewise vary slightly. The top tower temperature will be calculated for $L/D = 3.5$.

By a dew-point calculation (see p. 587):

Component	Overhead d	Reflux l	Total vapor v_1	126°F. K	v_1/K
C_2	3.0	4.25	7.25	2.65	2.73
C_3	19.3	72.01	91.31	1.0	91.31
C_4	0.3	2.84	3.14	0.41	7.65
	22.6	79.1	101.7	....	101.69

$t = 126$°F. at 250 lb./sq. in. gage

Reboiler Temperature. By a bubble-point calculation (see p. 587):

Component	Bottoms w	250 lb./sq. in. gage 280°F. K	Kw
C_3..........	0.7	2.7	1.9
C_4..........	36.7	1.37	50.3
C_5..........	35.0	0.75	26.2
C_6..........	5.0	.435	2.2
	77.4	.....	80.6

For C_4, $K = 77.4 \times 1.37/80.6 = 1.31$, and $t = 274°F.$ at 250 lb./sq. in. gage.

Equilibrium Vaporization of Feed. The feed is preheated to 225°F. The composition and quantity of liquid and vapor portions entering the feed plate will be calculated by an equilibrium vaporization calculation (see p. 587).

Try $V/L = 0.2$.

	Moles feed f	250 lb./sq. in. gage 225°F. K	$K\frac{V}{L}$	$1 + K\frac{V}{L}$	$l = \frac{f}{1 + K\frac{V}{L}}$	$v = f - l$
C_2.....	3	5.0	1.0	2.0	1.5	1.5
C_3.....	20	2.0	0.4	1.4	14.3	5.7
C_4.....	37	0.97	.194	1.194	31.0	6.0
C_5.....	35	.48	.096	1.096	32.0	3.0
C_6.....	5	.25	.05	1.05	4.8	0.2
	100	....	.....	.....	83.6	16.4

$$\frac{V}{L} = \frac{16.4}{83.6} = 0.196 \qquad \text{check}$$

Estimation of Minimum Number of Plates at Total Reflux. C_3 and C_4 are the key components.

At 118°F.:

$$\alpha = \frac{K_{C_3}}{K_{C_4}} = \frac{0.936}{0.375} = 2.5$$

At 274°F.:

$$\alpha = \frac{2.64}{1.31} = 2.02$$

$$\alpha_{avg} = (2.02 \times 2.5)^{1/2} = 2.24$$

By Eq. (81), using moles instead of mole fractions,

$$\frac{19.3}{0.3} \times \frac{36.7}{0.7} = 3370 = (2.24)^{S_m}$$

$$S_m = 10.1$$

With partial condenser and reboiler

$$N = S_m - 2 = 8.1 \text{ theoretical plates}$$

Estimation of Minimum Reflux Ratio. The feed is partly vaporized so that, at the feed plate, $V_n = V_m + 16.4$ based on 100 moles of feed, and

$$r_f = \frac{14.3}{31.0} = 0.461$$

Estimated temperature at upper pinch:

$$t = 118 + \frac{(274 - 118)}{3} = 170°F.$$

Estimated temperature at lower pinch:

$$t = 118 + 2/3(274 - 118) = 222°F.$$

Relative volatilities and mole fractions:

Component	170°F. 250 lb./sq. in. gage		222°F. 250 lb./sq. in. gage		x_D	x_W	x_{Fh}
	K	α	K	α			
C_2..........	3.6	5.72	4.9	5.15	0.133		
C_3..........	1.4	2.22	1.95	2.05	.854	0.009	
C_4..........	0.63	1.0	0.95	1.0	.013	.474	
C_5..........	.275	0.436	.47	0.495	....	.452	0.382
C_6..........	.123	.195	.24	.253	....	.065	.057
					1.000	1.000	

For components heavier than the heavy key component in the liquid portion of the feed:

Component	x_{Fh}	α	αx_{Fh}
C_5..........	0.382	0.436	0.167
C_6..........	.057	.195	.011

$$\Sigma \alpha x_{Fh} = 0.178$$

By Eq. (87),

$$x_{lk} \text{ (approx.)} = \frac{0.461}{(1.461)(1.178)} = 0.268$$

By Eq. (88),

$$x_{hk} \text{ (approx.)} = \frac{0.268}{0.461} = 0.58$$

By Eq. (86),

$$\left(\frac{L}{D}\right)_{min} \text{(approx.)} = \frac{1}{2.22 - 1}\left(\frac{0.854}{0.268} - 2.22\,\frac{0.013}{0.58}\right) = 2.57$$

Try $\left(\frac{L}{D}\right)_{min.} = 2.57$

$$L_n = 22.6 \times 2.57 = 58.1$$
$$V_n = 22.6 + 58.1 = 80.7$$
$$V_m = 80.7 - 16.4 = 64.3$$
$$L_m = 64.3 + 77.4 = 141.7$$
$$\frac{L_m}{W} = \frac{141.7}{77.4} = 1.83$$

Composition in upper pinch by Eq. (84):

Since x_{Dhk} is small, the last term in the denominator will be neglected, so that $x_n = \dfrac{x_D}{(\alpha - 1)(L_n/D)}$

Component	$\alpha - 1$	$\frac{L_n}{D}$	x_D	$\frac{x_D}{\alpha - 1}$	x_n
C_2..........	4.72	2.57	0.133	0.0282	0.011
C_3..........	1.22	2.57	.854	.700	.272
					0.283

For C_4,

$$x_n = 1.00 - 0.283 = 0.717$$
$$r_n = \frac{0.272}{0.717} = 0.38$$

Composition in lower pinch by Eq. (85):

x_{Wlk} is small; hence the last term in the denominator will be neglected.

Component	α_{lk}	α	$\alpha_{lk} - \alpha$	x_W	$(\alpha_{lk} - \alpha)\frac{L_m}{W}$	$\alpha_{lk} x_W$	x_m
C_4....	2.05	1.0	1.05	0.474	1.92	0.97	0.505
C_5....	2.05	0.495	1.555	.452	2.84	.927	.326
C_6....	2.05	.253	1.797	.065	3.28	.133	.040
							0.871

For C_3,

$$x_m = 1.00 - 0.871 = 0.129$$
$$r_m = \frac{0.129}{0.505} = 0.255$$

Calculation of $\Sigma C_n x_n$ in upper pinch from Fig. 67:

For C_2,

$$\frac{(\alpha_{lk} - 1)}{\alpha}(\alpha_{lk}) = \frac{(2.22 - 1)2.22}{5.72} = 0.473$$
$$C_n = 0.97$$
$$\Sigma C_n x_n = 0.97 \times 0.011 = 0.0107$$

Calculation of $\Sigma C_m \alpha x_m$ in lower pinch:

Component	α_{lk}	α	$(\alpha_{lk} - 1)\alpha$	C_m	x_m	$C_m \alpha x_m$
C_5.......	2.05	0.495	0.52	1.2	0.326	0.193
C_6.......	2.05	.253	.266	0.9	.040	.009

$$\Sigma C_m \alpha x_m = 0.193 + 0.009 = 0.202$$

$$\psi = \frac{1}{(1 - 0.202)(1 - 0.0107)} = \frac{1}{0.798 \times 0.9893} = 1.27$$

$$\frac{r_m}{r_n} = \frac{0.255}{0.38} = 0.671$$

r_m/r_n is less than ψ; therefore estimated $(L/D)_{min}$ is too low. Try $(L/D)_{min} = 2.97$.

$$L_n = 22.6 \times 2.97 = 67.1$$
$$L_m = 67.1 + 83.6 = 150.7$$
$$\frac{L_m}{W} = \frac{150.7}{77.4} = 1.95$$

Composition in upper pinch by Eq. (84):

Component	$\frac{x_D}{\alpha - 1}$	$\frac{L_n}{D}$	x_n
C_2	0.0282	2.97	0.0095
C_3	.700	2.97	.235
			0.2445

For C_4,

$$x_n = 1.00 - 0.2445 = 0.7554$$
$$r_n = \frac{0.235}{0.7554} = 0.311$$

Composition in lower pinch by Eq. (85):

Component	$\frac{\alpha_{lk} x_W}{\alpha_{lk} - \alpha}$	$\frac{L_m}{W}$	x_m
C_4	0.923	1.95	0.473
C_5	.595	1.95	.305
C_6	.074	1.95	.038
			0.816

For C_3,

$$x_m = 1.00 - 0.816 = 0.184$$
$$r_m = \frac{0.184}{0.816} = 0.389$$
$$\frac{r_m}{r_n} = \frac{0.389}{0.311} = 1.25$$

In upper pinch:

$$\Sigma C_n x_n = 0.97 \times 0.0095 = 0.0092$$

Calculation of $\Sigma C_m \alpha x_m$ in lower pinch:

Component	C_m	α	x_m	$C_m \alpha x_m$
C_5	1.2	0.495	0.305	0.181
C_6	0.9	.253	.038	.0086

$$\Sigma C_m \alpha x_m = 0.1896$$
$$\psi = \frac{1}{(1 - 0.1896)(1 - 0.0092)} = 1.25$$

Thus Eq. (83) is satisfied and $(L/D)_{min} = 2.97$, provided that the assumed pinch temperatures are correct.

Calculate temperature at upper pinch:

Component	x_n	190°F. K	Kx_n
C_2	0.0095	4.05	0.038
C_3	.2350	1.6	.376
C_4	.7554	0.75	.566
	1.000		0.980

For C_4, $K = \frac{0.75}{0.98} = 0.765$, and $t = 193$°F.

Calculate temperature at lower pinch:

Component	x_m	220°F. K	Kx_m
C_3	0.184	1.95	0.369
C_4	.473	0.95	.449
C_5	.305	.47	.143
C_6	.038	.24	.009
			0.970

For C_4, $K = \frac{0.95}{0.97} = 0.98$, and $t = 226$°F.

Repeating the calculations above, using 190°F. instead of 170°F. at the upper pinch, gave a value of $(L/D)_{min} = 3.05$.

From the relationship shown by Fig. 65, the following tabulation was made showing number of plates as a function of reflux ratio:

$$\left(\frac{L}{D}\right)_m = 3.05 \qquad S_m = 10.1 \qquad N = S - 2$$

$\frac{L}{D}$	$\frac{L}{D} - \left(\frac{L}{D}\right)_m$	$\frac{(L/D) - (L/D)_m}{(L/D) + 1}$	$\frac{S - S_m}{S + 1}$	S	N
3.5	0.45	0.10	0.53	22.6	20.6
4.0	.95	.19	.46	19.5	17.5
5.0	1.95	.325	.361	16.37	14.37
6.0	2.95	.421	.30	14.86	12.86
7.0	3.95	.495	.26	14.0	12.0
8.0	4.95	.55	.225	13.3	11.3
10.0	6.95	.631	.18	12.5	10.5

These results are shown graphically in Fig. 68.

Stepwise Plate Calculations. A reflux ratio of $L/D = 5$ will be chosen for design, inasmuch as this value falls upon a part of the curve of Fig. 68 where further increases

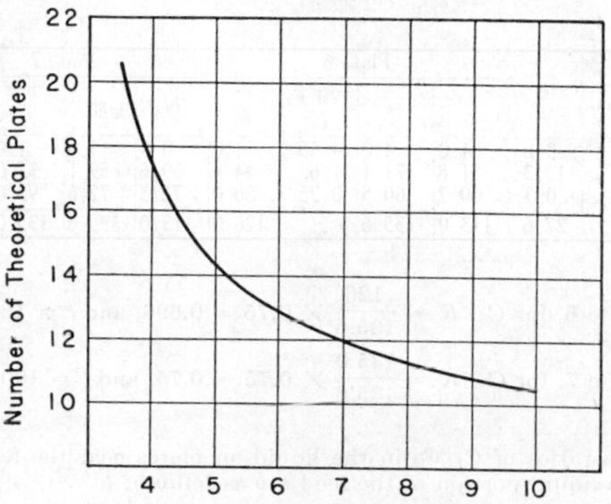

Fig. 68.

in L/D effect diminishing savings in plates. Moles will be used instead of mole fractions in illustrating the method of calculation, in order to reduce the amount or calculation required.

The feed plate will be located when Eq. (93) is satisfied.

Com-ponent	Moles d	120°F. 250 lb./sq. in. gage K	$\frac{d}{K}$	l_r	v_1	126°F. 250 lb./sq. in. gage K	$\frac{v_1}{K}$	l_1
C_2	3	2.55	1.2	6.0	9.0	2.65	3.4	2.8
C_3	19.3	0.95	20.3	103.0	122.3	1.0	122.3	101.5
C_4	0.3	.38	0.8	4.0	4.3	0.41	10.5	8.7
	22.6	....	22.3	113.0	135.6	....	136.2	113.0

Explanation:

$\frac{L}{D} = 5.$

$L = 5 \times 22.6 = 113.0.$

l_r = moles of a given component in the reflux.

v_1 = moles of a given component in the vapor from plate 1.

l_1 = moles of a given component in the liquid from plate 1.

$$l_r = \left[\frac{L}{\Sigma(d/K)}\right]\frac{d}{K} = \left(\frac{113.0}{22.3}\right)\left(\frac{d}{K}\right)$$
$$v_1 = d + l_r$$
$$l_1 = \left[\frac{L}{\Sigma(v_1/K)}\right]\left(\frac{v_1}{K}\right) = \left(\frac{113.0}{136.2}\right)\left(\frac{v_1}{K}\right)$$

Com-ponent	d	Plate 2 l_1	v_2	135°F. K	$\frac{v_2}{K}$	Plate 3 l_2	v_3	$\frac{v_3}{K}$
C_2	3.0	2.8	5.8	2.83	2.0	1.7	4.7	1.7
C_3	19.3	101.5	120.8	1.07	112.8	94.5	113.8	106.2
C_4	0.3	8.7	9.0	0.45	20.0	16.8	17.1	38.0
	22.6	113.0	135.6	....	134.8	113.0	135.6	145.9

Plate 2, for C_3, $K = \frac{134.8}{135.6} \times 1.07 = 1.063$, and $t = 134°F$.

Plate 3, for C_3, $K = \frac{145.9}{135.6} \times 1.07 = 1.15$, and $t = 145°F$.

Component	d	Plate 4				Plate 5		
		l_3	v_4	155°F. K	$\frac{v_4}{K}$	l_4	v_5	$\frac{v_5}{K}$
C_2	3.0	1.3	4.3	3.25	1.3	1.1	4.1	1.3
C_3	19.3	82.3	101.6	1.24	81.9	67.4	86.7	70.0
C_4	0.3	29.4	29.7	0.55	54.0	44.5	44.8	81.5
	22.6	113.0	135.6	....	137.2	113.0	135.6	152.8

Plate 4, for C_3, $K = \frac{137.2}{135.6} \times 1.24 = 1.26$, and $t = 156°F$.

Plate 5, for C_3, $K = \frac{152.8}{135.6} \times 1.24 = 1.4$, and $t = 170°F$.

Component	d	Plate 6				Plate 7			
		l_5	v_6	190°F. K	$\frac{v_6}{K}$	l_6	v_7	$\frac{v_7}{K}$	l_7
C_2	3.0	1.0	4.0	4.05	1.0	0.9	3.9	1.0	0.8
C_3	19.3	51.8	71.1	1.6	44.4	39.8	59.1	37.0	31.0
C_4	0.3	60.2	60.5	0.75	80.6	72.3	72.6	97.0	81.2
	22.6	113.0	135.6	....	126.0	113.0	135.6	135.0	113.0

Plate 6, for C_4, $K = \frac{126}{135.6} \times 0.75 = 0.696$, and $t = 181°F$.

Plate 7, for C_4, $K = \frac{135.0}{135.6} \times 0.75 = 0.75$, and $t = 190°F$.

The ratios of C_3/C_4 in the liquid on plates near the feed and in the liquid portion of the feed are as follows:

	Plate 6	Plate 7	Liquid feed
Ratio C_3/C_4	0.55	0.382	0.461

By Eq. (93), plate 6 is the plate above the feed.

Stepwise Calculations from Reboiler to Feed Plate. There are 16.4 moles of vapor in the feed; therefore, for an ideal system, per 100 moles of feed, $V_m = V_n - 16.4$. For $L/D = 5$, $V_n = 6D = 135.6$ for this case, and $V_m = 135.6 - 16.4 = 119.2$ moles.

By a heat balance around the tower and reboiler, it is found that at the reboiler $V_m = 131$ moles per 100 moles of feed, instead of 119.2 as at the feed plate. Therefore, the quantity of vapor will decrease as it rises through the tower from the reboiler to the feed plate. The quantity of vapor from each plate may be determined by a heat balance in conjunction with the stepwise calculations at each plate, if such accuracy is required. For this example, the average $V_m = 125.1$ will be used throughout.

Component	w	Reboiler			Plate 1B			
		280°F. K	Kw	v_r	l_{1B}	260°F. K	Kl_{1B}	v_{1B}
C_3	0.7	2.7	1.9	3.0	3.7	2.45	9.1	5.7
C_4	36.7	1.37	50.3	78.1	114.8	1.2	137.9	86.6
C_5	35.0	0.75	26.2	40.6	75.6	0.65	49.1	30.9
C_6	5.0	0.435	2.2	3.4	8.4	.36	3.0	1.9
	77.4	.....	80.6	125.1	202.5	....	199.1	125.1

Reboiler, for C_4, $K = \frac{77.4}{80.6} \times 1.37 = 1.31$, and $t = 274°F$.

Plate 1B, for C_4, $K = \frac{202.5}{199.1} \times 1.2 = 1.22$, and $t = 262°F$.

Explanation:

v_r = moles of component in vapor from reboiler.
l_{1B} = moles of component in liquid on plate 1B.

$$v_r = \left(\frac{V}{\Sigma Kw}\right) Kw = \left(\frac{125.1}{80.6}\right) Kw$$

$$l_{1B} = v_r + w$$

Component	w	Plate 2B				Plate 3B		
		v_{1B}	l_{2B}	250°F. K	Kl_{2B}	v_{2B}	l_{3B}	Kl_{3B}
C_3	0.7	5.7	6.4	2.3	14.7	9.4	10.1	23.2
C_4	36.7	86.6	123.3	1.13	139.3	89.0	125.7	142.0
C_5	35.0	30.9	65.9	0.60	39.5	25.3	60.3	36.2
C_6	5.0	1.9	6.9	.325	2.2	1.4	6.4	2.1
	77.4	125.1	202.5	.....	195.7	125.1	202.5	203.5

Component	w	Plate 4B				Plate 5B		
		v_{3B}	l_{4B}	240°F. K	Kl_{4B}	v_{4B}	l_{5B}	Kl_{5B}
C_3	0.7	14.3	15.0	2.19	32.9	20.7	21.4	46.9
C_4	36.7	87.3	124.0	1.07	133.0	83.5	120.2	128.8
C_5	35.0	22.2	57.2	0.55	31.5	19.8	54.8	30.1
C_6	5.0	1.3	6.3	.29	1.8	1.1	6.1	1.8
	77.4	125.1	202.5	....	199.2	125.1	202.5	207.6

Plate 4B, for C_4, $K = \frac{202.5}{199.2} \times 1.07 = 1.086$, and $t = 242°F$.

Plate 5B, for C_4, $K = 1.045$, and $t = 236°F$.

Component	w	Plate 6B				Plate 7B		
		v_{5B}	l_{6B}	230°F. K	Kl_{6B}	v_{6B}	l_{7B}	Kl_{7B}
C_3	0.7	28.3	29.0	2.05	59.5	36.8	37.5	77.0
C_4	36.7	77.5	114.2	1.0	114.2	70.7	107.4	107.4
C_5	35.0	18.2	53.2	0.505	26.9	16.6	51.6	26.1
C_6	5.0	1.1	6.1	.76	1.6	1.0	6.0	1.6
	77.4	125.1	202.5	.....	202.2	125.1	202.5	212.1

Plate 6B, for C_4, $K = 1.0$, and $t = 230°F$.
Plate 7B, for C_4, $K = 0.955$, and $t = 223°F$.

Component	w	Plate 8B				
		v_{7B}	l_{8B}	215°F. K	Kl_{8B}	v_{8B}
C_3	0.7	45.4	46.1	1.89	87.2	54.4
C_4	36.7	63.4	100.1	0.9	90.1	56.1
C_5	35.0	15.4	50.4	.44	22.2	13.8
C_6	5.0	0.9	5.9	.22	1.3	0.8
	77.4	125.1	202.5	....	200.8	125.1

Plate 8B, for C_4, $K = \frac{202.5}{200.8} \times 0.9 = 0.91$, and $t = 216°F$.

	Liquid plate 7B	Liquid plate 8B	Liquid feed
Ratio C_3/C_4	0.349	0.461	0.461

By Eq. (93), plate 8B is the feed plate.

The stepwise calculations indicate that 8 plates are required below the feed and 6 plates are required above the feed, giving a total of 14 plates at a reflux ratio $L/D = 5$. A value of 14.37 theoretical plates was obtained from the correlation of Fig. 65.

The calculations above are generally sufficient for determining the number of plates required and the feed-plate location. The error caused by omitting components lighter than the light key below the feed and heavier than the heavy key above the feed may be neglected in most cases.

If the compositions in the vicinity of the feed plate are required, where all components in the feed are present in significant quantities, they may be found by the procedure illustrated below.

Introduction of Components Lighter than Light Key below the Feed.

C_2 in liquid from plate 6	= 0.9 moles
C_2 in liquid feed	= 1.5 moles
C_2 in vapor from feed plate	= 2.4 moles

The temperature of the feed plate and plates below are estimated from temperatures calculated without components lighter than the light key, and the number of moles in the liquid and vapor from each plate is calculated from the relationships

$$l = v \frac{L}{VK}$$

and $v_m = l_{(m+1)} - w = l_{(m+1)}$ when $w = 0$.

For this example, $L/V = 1.619$, and $l = 1.619v/K$.

Plate	Est. temp., °F.	K_{C_2}	v	l
8*B*........	214	4.65	2.4	0.835
7*B*........	222	4.9	0.835	.276
6*B*........	230	5.1	.276	.0875
5*B*........	236	5.2	.0875	.0272

The first significant quantity of C_2 appears in the vapor from plate 5*B*. The stepwise plate calculations will be revised starting with the liquid on plate 6*B* and introducing the estimated quantity of C_2, as follows:

Component	w	Plate 6*B*, 230°F.				Plate 7*B*, 220°F.		
		l_{6B}	K	Kl_{6B}	v_{6B}	l_{7B}	K	Kl_{7B}
C_2.....	....	0.0875	5.1	0.445	0.275	0.275	4.8	1.3
C_3.....	0.7	29.0	2.05	59.5	36.7	37.4	1.92	71.9
C_4.....	36.7	114.2	1.0	114.2	70.5	107.2	0.94	100.9
C_5.....	35.0	53.2	0.505	26.9	16.6	51.6	.46	23.8
C_6.....	5.0	6.1	.26	1.6	1.0	6.0	.235	1.4
	77.4	202.59	.....	202.6	125.1	202.5	.....	199.3

Component	w	Plate 8*B*, 215°F.				
		v_{7B}	l_{8B}	K	Kl_{8B}	v_{8B}
C_2.........	....	0.82	0.82	4.7	3.85	2.4
C_3.........	0.7	45.1	45.8	1.89	86.7	53.1
C_4.........	36.7	63.4	100.1	0.9	90.1	55.3
C_5.........	35.0	14.9	49.9	.44	22.0	13.5
C_6.........	5.0	0.9	5.9	.22	1.3	0.8
	77.4	125.1	202.5	....	203.95	125.1

It is seen that the quantity of C_2 in the feed-plate vapor thus calculated checks the known quantity, as it should if the estimated temperatures agree closely with those calculated.

Introduction of Components Heavier than Heavy Key above the Feed.

	Moles C_5	Moles C_6
In vapor feed..............................	3.0	0.2
In vapor from feed plate..............................	13.5	0.8
In liquid from plate above feed..............................	16.5	1.0

The temperatures of the plates immediately above the feed are estimated from those calculated without components heavier than the heavy key, and the number of moles of heavy components in the liquid and vapor from each plate is calculated from the relationships

$$v = Kl\frac{V}{L}$$

and $l_n = v_{(n+1)} - d$, or $l_n - v_{(n+1)}$ when $d = 0$. For this example, $V/L = 1.2$; hence $v = 1.2\ Kl$.

Plate	Est. temp., °F.	C_5			C_6		
		K	l	v	K	l	v
6	195	0.36	16.5	7.13	0.173	1.0	0.208
5	175	.29	7.13	2.48	.131	0.208	.0326
4	157	.235	2.48	0.70			
3	145	.20	0.70	0.168			

Component	d	Plate 3, 145°F.				Plate 4, 155°F.		
		v_3	K	$\frac{v_3}{K}$	l_3	v_4	K	$\frac{v_4}{K}$
C_2............	3.0	4.7	3.05	1.54	1.28	4.28	3.25	1.32
C_3............	19.3	113.8	1.15	98.8	82.5	101.8	1.24	82.0
C_4............	0.3	17.	0.50	34.2	28.5	28.8	0.55	52.3
C_5............	....	0.168	.20	0.84	0.70	0.70	.23	3.04
	22.6	135.77	....	135.38	112.98	135.58	....	138.66

Plate 3, $t = 145°F$.

Plate 4: for C_3, $K = \frac{138.66}{135.6} \times 1.24 = 1.27$, and $t = 157°F$.

Component	d	Plate 5, 175°F.				Plate 6, 1.95°F.			
		l_4	v_5	K	$\frac{v_5}{K}$	l_5	v_6	K	$\frac{v_6}{K}$
C_2...	3.0	1.07	4.07	3.7	1.10	.92	3.92	4.2	0.93
C_3...	19.3	66.8	86.1	1.43	60.2	50.3	69.6	1.64	42.4
C_4...	0.3	42.6	42.9	0.66	65.0	54.4	54.7	0.775	70.5
C_5...	....	2.48	2.48	.29	8.55	7.15	7.15	.36	19.8
C_6...	....	0.0326	0.0326	.131	0.249	0.208	0.208	.172	1.21
	22.6	112.98	135.58	.....	135.10	112.98	135.58	.....	134.84

Plate 5, for C_4, $K = \frac{135.1}{135.6} \times 0.66 = 0.658$, and $t = 175°F$.

Plate 6, for C_4, $K = \frac{134.8}{135.6} \times 0.775 = 0.77$, and $t = 194°F$.

Component	l_6
C_2..............................	0.78
C_3..............................	35.5
C_4..............................	59.1
C_5..............................	16.6
C_6..............................	1.01
	112.99

The quantity of C_5 and C_6 in the liquid from plate 6 to the feed plate thus checks the known quantity within 1 per cent.

Graphical Methods for Determining Number of Plates. Graphical methods for solving multicomponent distillation problems, using a McCabe-Thiele type of diagram as applied to binary mixtures (p. 591), have been presented by Jenny [*Trans. Am. Inst. Chem. Engrs.*, **35**, 635 (1939)], Jenny and Cicalese [*Ind. Eng. Chem.*, **37**, 956 (1945)], Scheibel [*Ind. Eng. Chem.*, **38**, 397 (1946)], and Hengstebeck [*Trans. Am. Inst. Chem. Engrs.*, **42**, 309 (1946)]. The method of Hengstebeck gives a rapid and accurate solution for systems of constant molal overflow in the rectifying and stripping sections and for systems in which there is no component whose relative volatility lies between those of the key components. Distributed components lighter or heavier than the key components are handled satisfactorily by this method, and the feed-plate location is determined directly.

AZEOTROPIC AND EXTRACTIVE DISTILLATIONS

By Stanley B. Zdonik with F. W. Woodfield, Jr.

REFERENCES: Benedict and Rubin, *Trans. Am. Inst. Chem. Engrs.*, **41**, 353 (1945). Benedict, Johnson, Solomon, and Rubin, *ibid.*, **41**, 371 (1945). Carlson and Colburn, *Ind. Eng. Chem.*, **34**, 581 (1942). Colburn and Phillips, *Trans. Am. Inst. Chem. Engrs.*, **40**, 333 (1944). Colburn and Schoenborn, *ibid.*, **41**, 421 (1945). Dicks and Carlson, *ibid.*, **41**, 789 (1945). Drickamer, Brown, and White, *ibid.*, **41**, 555 (1945). Drickamer and Hummel, *ibid.*, **41**, 607 (1945). Dunn, Millar, Pierotti, Shiras, and Souders, *ibid.*, **41**, 631 (1945). Ewell, Harrison, and Berg, *Petroleum Engr.*, **255** (October, 1944); **263** (November, 1944); **219** (December, 1944); *Ind. Eng. Chem.*, **36**, 871 (1944). Ewell and Welch, *Ind. Eng. Chem.*, **37**, 1224 (1945). Guinot and Clark, *Trans. Inst. Chem. Engrs. (London)*, **16**, 187 (1938). Guinot, U.S. Patent 2,316,860 (1943). Happel, Cornell, Eastman, Fowle, Porter, and Schutte, *Trans. Am. Inst. Chem. Engrs.*, **42**, 189 (1946). Hartley, *Petroleum Refiner*, **12**, 131 (1945). Hildebrand, Solubility of Non-electrolytes, 2d ed., A.C.S. Monograph Series, Reinhold, New York, 1936. Horsley, *Anal. Chem.*, **19**, 508 (August, 1947). "International Critical Tables," Constant Boiling (Azeotropic) Mixtures, vol. 3, 318. Kireev, *Acta Physiochim. U.R.S.S.*, **14**, 371 (1941). Lake, *Trans. Am. Inst. Chem. Engrs.*, **41**, 327 (1945). Lange, "Handbook of Chemistry," 1208, Handbook Publishers, Inc., Sandusky, Ohio, 1941. Lecat, La Tension de vapeur des melanges de liquides, L'Azeotropisme, Bruxelles, 1918. Library Bulletin of Abstracts, U.O.P. Co., Chicago, Research Lab., No. 35, 139, Aug. 27, 1941. Marschner and Cropper, *Ind. Eng. Chem.*, **38**, 262 (1946). Matuszak and Frey, *Ind. Eng. Chem.*, anal. ed., 111 (Mar. 15, 1937). *Natl Petroleum News, Tech. Sec.*, **38**, R-283 (Apr. 3, 1946). Othmer, *Ind. Eng. Chem.*, **33**, 1106 (1941). Rossini, Mair, and Glasgow, *Oil Gas J.*, **158**, (Nov. 14, 1940). Scatchard and Hamer, *J. Am. Chem. Soc.*, **57**, 1805 (1935). Seyer and Todd, *Ind. Eng. Chem.*, **23**, 325 (1931). Sunier and Rosenblum, *Ind. Eng. Chem.*, anal. ed., 109 (Jan. 15, 1930). Swietoslawski, "Ebulliometry," Chemical Publishing Co., New York, 1937. Underwood, *Trans. Inst. Chem. Engrs. (London)*, **10**, 112 (1932). Updike, Langdon, and Keyes, *Trans. Am. Inst. Chem. Engrs.*, **41**, 717 (1945). Van Laar, *Z. physik. Chem.*, **72**, 723 (1910). White, *Trans. Am. Inst. Chem. Engrs.*, **41**, 539 (1945); *Natl. Petroleum News, Tech. Sec.*, **36**, R-731 (Nov. 1, 1944). Wohl, *Trans. Am. Inst. Chem. Engrs.*, **42**, 215 (1946). Young and Fortey, *J. Chem. Soc.*, **81**, 717, 739 (1902).

GENERAL AND THEORETICAL

The Occurrence of Azeotropes in Distillation

An azeotrope is a mixture of two or more liquid compounds whose boiling point does not change as vapor is generated and removed. The word "azeotrope" is synonymous with "constant-boiling mixture," sometimes written "C.B.M."

Azeotropes may be classified into two groups: those which exist as one liquid phase (homogeneous azeotropes), and those which exist as two or more liquid phases in equilibrium (heterogeneous azeotropes). The homogeneous group may be subclassified into minimum-boiling and maximum-boiling mixtures, as will be further discussed below. All reported heterogeneous azeotropes are minimum-boiling mixtures; and, as a corollary, all maximum-boiling azeotropes are homogeneous. About 3000 minimum azeotropes and about 250 maximum azeotropes have been recorded in the literature [Ewell, Harrison, and Berg, *Ind. Eng. Chem.*, **36**, 871 (1944)].

Tables 18 and 18*a* list data on some of the more common binary minimum- and maximum-boiling mixtures, and Table 18*b* lists a few reported ternary azeotropes. Table 18*c* reports a number of azeotropes involving hydrocarbons and may be useful as a guide in selecting azeotrope-forming agents for azeotropic distillations. Reference is made to the recent complete compilation of azeotropes and nonazeotropes by Horsley [*Anal. Chem.*, **19**, 508 (August, 1947)].

Occurrence of azeotropes is important because compounds that form an azeotrope cannot be completely separated by one simple fractionation at a given pressure. In the operation of azeotropic distillation, on the other hand (see p. 634), a separating agent is added deliberately to form an azeotrope and enhance the relative volatility of components to be separated.

Figure 69 shows a temperature-composition diagram for the binary system ethyl acetate–ethanol, illustrating a

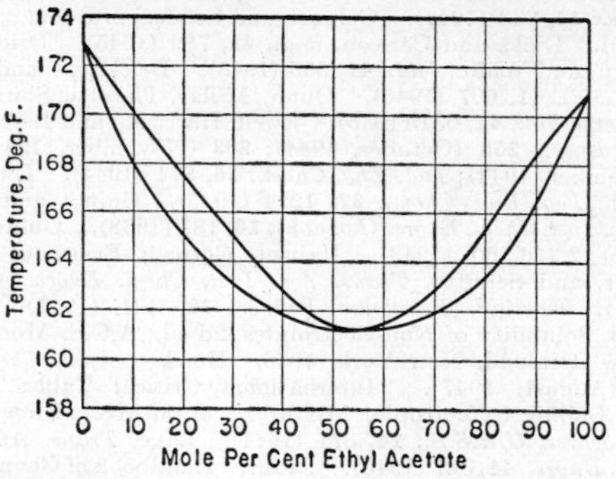

FIG. 69. Ethyl acetate–ethanol vapor-liquid equilibriums, 760 mm.

typical minimum-boiling homogeneous azeotrope containing 54 mole per cent ethyl acetate at atmospheric pressure. The azeotrope is represented by the point at which the vapor-dew-point (*e.g.*, upper) line and the liquid-boiling-point (*e.g.*, lower) line become tangent at the minimum boiling point for the system. Such a homogeneous minimum-boiling mixture is also called a "positive homoazeotrope" because the mixture exhibits positive deviations from Raoult's law. Starting with a feed containing 15 mole per cent ethyl acetate, simple fractionation is capable of producing substantially pure ethanol and the azeotrope. However, if the feed were 70 mole per cent ethyl acetate, an efficient fractionation would separate out substantially pure ethyl acetate and the azeotrope. In each case, the azeotrope would be taken overhead since it boils at the lower temperature. Simple fractionation, even with an infinite number of perfect plates, is incapable of carrying the separation beyond the azeotropic composition where vapor and liquid in equilibrium have an identical composition.

Figure 70 shows a temperature-composition diagram for the binary system chloroform-acetone, which forms a

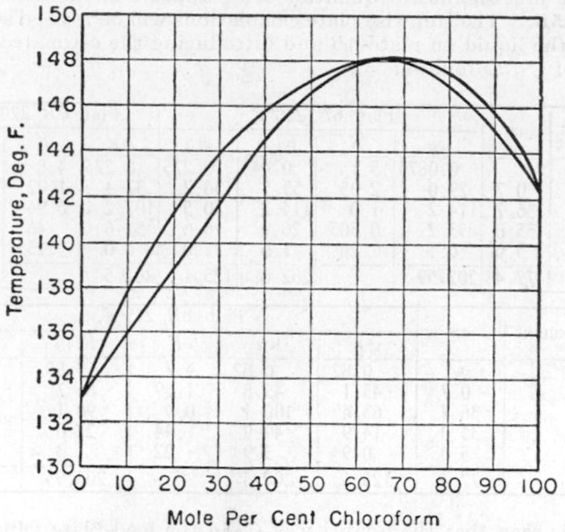

FIG. 70. Chloroform-acetone vapor-liquid equilibriums, 760mm.

maximum-boiling homogeneous azeotrope or "negative homoazeotrope." Only about one-tenth as many maximum azeotropes as minimum azeotropes have been reported. Efficient simple fractionations of chloroform-acetone mixtures at atmospheric pressure always should result in a bottoms stream approaching the azeotropic composition of approximately 65.5 mole per cent chloroform. Distillate should be substantially pure acetone or chloroform, respectively, depending on whether the feed is poorer or richer in chloroform than 65.5 mole per cent.

Formation of Homogeneous Azeotropes. Formation of a binary azeotrope depends upon (1) the magnitude of the deviations from Raoult's law and (2) the difference in boiling point between the two pure components. The smaller the deviations from Raoult's law for a pair of substances, the smaller the difference in boiling point must be before an azeotrope will form (Ewell, Harrison, and Berg, *loc. cit.*). These effects are illustrated graphically on Fig. 71 for systems with

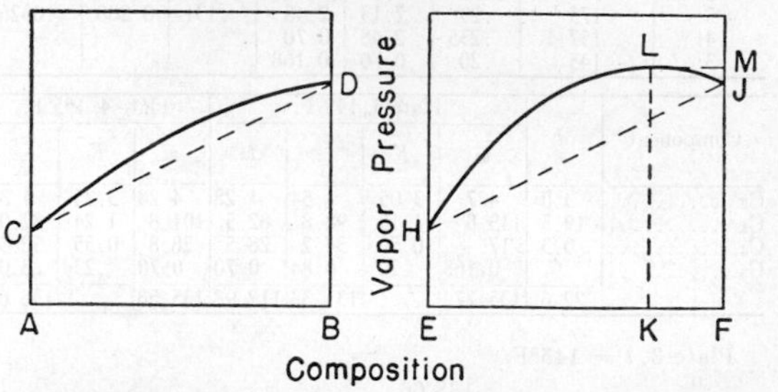

FIG. 71. Vapor-pressure curves. When deviations from ideal are insufficient to give an azeotrope, pure component *B* has the highest vapor pressure, *D*. When deviations are sufficient to give an azeotrope, some mixture of components *E* and *F*, such as *K*, will have the highest vapor pressure, *M*.

positive deviations from Raoult's law (*e.g.*, activity coefficient γ greater than unity). The dotted straight lines *C-D* and *H-J* represent vapor pressure vs. mole fraction for the binary mixtures, as calculated by Raoult's law. Positive deviations cause the measured vapor-pressure line to lie above the Raoult's-law line; and, if the vapor pressure of the mixture goes through a maximum value that exceeds the vapor pressure of the more volatile pure component, a minimum azeotrope is formed.

Negative deviations from Raoult's law (*e.g.*, γ less than unity) are represented by a measured vapor-pressure line located below the Raoult's-law line. If this measured vapor-pressure line goes through a minimum point that is lower than the vapor pressure of the less volatile pure component, a maximum-boiling azeotrope is formed.

When a homogeneous binary azeotrope occurs, $y_1 = x_1$, and by Raoult's law (including non-ideality activity coefficients),

$$\gamma_1 P_1 = \gamma_2 P_2 = P \tag{95}$$

where x_1 = mole fraction of component 1 in liquid.
y_1 = mole fraction of component 1 in vapor.
γ_1, γ_2 = activity coefficients of the components.
P_1, P_2 = vapor pressures of the pure components at the azeotropic boiling temperature.
P = total pressure.

Use of this relationship to predict the existence and composition of a homogeneous azeotrope when data on vapor pressures and activity coefficients are available is described in Sec. 8, p. 532.

A method of Ewell, Harrison, and Berg (*loc. cit.*) for predicting qualitatively when azeotropes may be expected is described in Sec. 8, p. 534. Their method is based on a classification of all liquids into five general groups according to their potentialities for forming hydrogen bonds.

Estimating Vapor-liquid Equilibrium from Azeotropic Data. In the absence of reliable measured vapor-liquid equilibrium data for binary systems that form homogeneous azeotropes, such data may be estimated by the method of Carlson and Colburn [*Ind. Eng. Chem.*, **34,** 581 (1942)]. Required data are vapor pressures of the pure compounds, plus composition and boiling point of the azeotrope at a known total pressure. Since $y = x$ at the azeotropic composition, activity coefficients γ may be calculated by the equations

$$\gamma_1 = \frac{P}{P_1} \qquad \gamma_2 = \frac{P}{P_2} \tag{96}$$

P_1 and P_2 are vapor pressures of the pure components at the boiling point of the azeotrope, and P is the total pressure. Extrapolation of these activity coefficients over the entire composition range, and utilization of the resulting values to calculate equilibrium data, are discussed further in Sec. 8, p. 533.

Data for a large number of binary azeotropes are listed in Tables 18, 18*a*, and 18*c*.

If the composition of a binary azeotrope is not known, this value may be estimated, usually with an accuracy better than ±10 per cent, from the boiling point of the azeotrope at a known pressure and vapor pressures of the pure compounds. Primary limitation of the method is assumption that the symmetrical two-suffix Margules equation (Sec. 8, p. 527) applies for calculating activity coefficients.

Taking logarithms of Eqs. (96) and equating these values to the appropriate two-suffix Margules terms,

$$\log \gamma_1 = \log \left(\frac{P}{P_1}\right) = (1 - x_1)^2 A \tag{97}$$

$$\log \gamma_2 = \log \left(\frac{P}{P_2}\right) = x_1^2 A \tag{98}$$

where x_1 is the mole fraction of the more volatile component 1 in the azeotrope, A is the logarithm of the end values of the activity coefficients at infinite dilution (being the same value for both components in a symmetrical system), and the other terms are as in Eqs. (96) If the square root of Eq. (97) is divided by the square root of Eq. (98), and the resulting expression is solved for x_1,

$$x_1 = \frac{1}{\left(\dfrac{\log P/P_1}{\log P/P_2}\right)^{1/2} + 1} \tag{99}$$

Equation (99) may be used to estimate the composition of a binary azeotrope when only the total pressure P and vapor pressures of the pure compounds P_1 and P_2 at the boiling point of the azeotrope are known. Accuracy is limited by the assumption of a symmetrical system, as stated above. An expression analogous to Eq. (99) was derived by Kireev [*Acta Physiochim. U.R.S.S.*, **14,** 371 (1941)].

Effect of Pressure on Composition of Homogeneous Azeotropes. A simple way of separating certain constant-boiling mixtures is by distillation under reduced pressure. The ethyl alcohol–water system is a good example, and data of Merriman [Young and Fortey, *J. Chem. Soc.*, **81,** 717, 739 (1902); Sunier and Rosenblum, *Ind. Eng. Chem.*, anal. ed., **109** (Jan. 15, 1930)] in Table 16 illustrate the possibilities.

Table 16. Composition of Ethyl Alcohol–Water Azeotrope at Different Pressures

Pressure, Mm.	Water in Azeotrope, Weight %
14,523.6	7.88
1,451.3	4.75
1,075.4	4.65
760.0	4.4
404.6	3.75
198.4	2.7
129.7	1.3
94.9	0.5
70.0	0.0

These data show the pronounced effect of pressure on azeotropic composition. At 70 mm., absolute alcohol boils at about 28°C.; and, if the data are correct, a complete separation of ethanol and water could theoretically be made by fractionation. It is reported, however, that complete alcohol-water separation under reduced pressure has been found difficult to achieve on a commercial scale.

Wrewski proposed a generalization on the effect of pressure upon azeotropic mixtures as follows (Swietoslawski, "Ebulliometry," Chemical Publishing Co., New York, 1937): When the boiling point of a positive azeotrope rises, its composition changes in favor of the component with the higher molecular latent heat of vaporization. The data tabulated above for ethyl alcohol and water follow this rule. Water has a higher latent heat of vaporization than ethyl alcohol, and the azeotrope of these two components progressively becomes richer in water as the pressure is increased. Conversely, in the case of negative azeotropes, when the boiling point rises, there is an increase in the concentration of the component with the lower latent heat of vaporization. This rule is not always followed, since some systems show a maximum azeotropic composition at a particular pressure.

The data of Karpinski (Swietoslawski, *loc. cit.*) for the benzene–ethyl alcohol–water ternary azeotrope are listed in Table 17 and show that the concentration of water in the azeotrope increases as the pressure increases.

Formation of Heterogeneous Azeotropes. A heterogeneous azeotrope is a constant-boiling mixture that exists as more than one (usually two) liquid phases in equilibrium. All reported heterogeneous azeotropes are minimum-boiling, and such systems are characterized by excessively large positive deviations from Raoult's law—deviations so large that immiscible liquid phases are present.

Table 17. Effect of Pressure on the Concentration of the Ternary Azeotrope Benzene–Ethyl Alcohol–Water

Pressure, atm.	Azeotrope temp., °C.	Weight % benzene	Weight % ethyl alcohol	Weight % water
1.0	64.85	74.1	18.5	7.4
3.41	101.06	68.1	22.4	9.4
6.13	122.08	63.7	24.9	11.4
9.69	140.95	61.5	25.6	12.9
10.29	143.27	61.5	25.5	13.0
19.02	171.35	60.0	24.4	15.6
19.12	171.85	60.3	24.5	15.2

Although most binary systems with immiscible regions at the boiling temperature form heterogeneous azeotropes, a few such systems are known that pass through the immiscible region and become a single liquid phase again before the azeotropic composition is reached. Two examples are methyl acetate–water, and methyl ethyl ketone–water, both at atmospheric pressure.

The binary systems benzoic acid–water and salicylic acid–water form heterogeneous mixtures that boil at a temperature greater than the lower boiling component and less than the higher boiling component. Though the condensation temperature of the vapor formed by these mixtures is cònstant, it is not equal to the boiling temperature of the liquid mixtures. The number of such mixtures seems to be limited (Swietoslawski, "Ebulliometry," Chemical Publishing Co., New York, 1937, p. 46).

Figure 72 illustrates the temperature-composition diagram at 1 atm. pressure for n-butanol–water. The

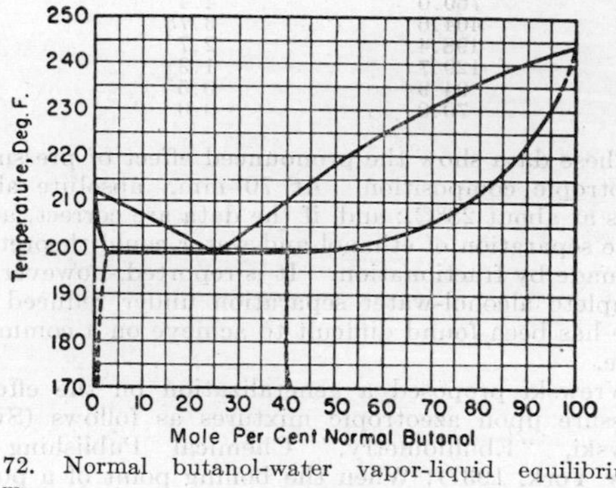

FIG. 72. Normal butanol-water vapor-liquid equilibriums, 760 mm.

upper V-shaped solid line represents dew-point temperature vs. composition, and the lower solid line represents boiling point vs. composition. The vertical dotted lines show solubility of the two liquids as a function of temperature, the immiscible liquid region lying between the solubility lines. Liquid in the immiscible region will separate into two layers with compositions of the layers determined by intersections of a horizontal temperature line with the solubility lines.

The diagram indicates that a heterogeneous azeotrope forms at 25.0 mole per cent n-butanol where the dew-point line becomes tangent to the boiling-point line at 199°F., the boiling point of the immiscible liquid phases. At 199°F., all liquids lying between the solubility lines will exist as two liquid layers, both of which are in equilibrium with the azeotropic vapor containing 25.0 mole per cent butanol. Only one liquid phase can exist in equilibrium with a vapor phase at temperatures above 199°F., and at atmospheric pressure.

Distillations Involving Heterogeneous Azeotropes. For feeds to a fractionating column containing greater than 25.0 mole per cent butanol, the maximum possible separation would produce the azeotropic vapor as an overhead product and pure butanol as a bottoms product. Feeds containing less than 25.0 mole per cent butanol would produce the azeotropic vapor overhead and pure water as a bottoms product.

As for homogeneous azeotropes, it is impossible in a single column to fractionate beyond the azeotropic vapor composition; because, whenever a liquid is obtained whose composition lies between the solubility lines, it separates into two liquid phases, and all vapor generated from the two liquid phases in equilibrium contact is of the azeotropic composition. However, if the azeotropic vapor is condensed and sent to a decanter, it separates into two liquid phases whose compositions depend upon the decanter temperature according to the solubility relations. One of these phases can be returned to the column as reflux, the other being fractionated in a separate column. Such a procedure may be utilized to separate a heterogeneous binary azeotrope in two columns without the addition of a third component as a separating agent.

Figure 73 represents a flow sheet for separating a feed containing 35 mole per cent n-butanol and 65 mole per

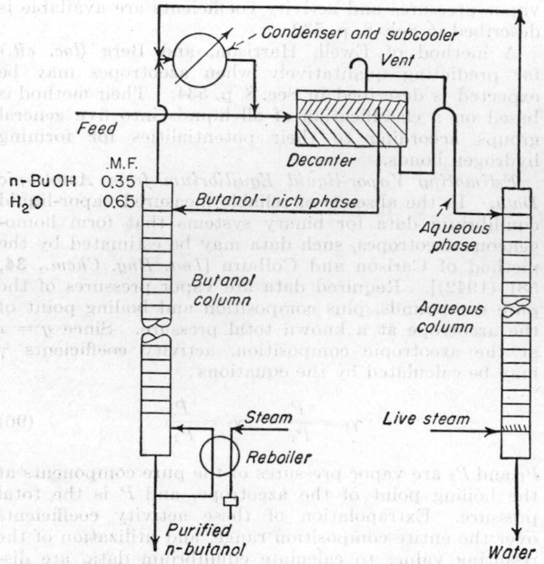

FIG. 73. Separation of n-butanol and water in two fractionating columns.

cent water into butanol and water streams of any desired purities and is representative of distillation processes involving two compounds that form a heterogeneous azeotrope. Since the fresh feed exists as two immiscible liquid phases, it is first sent to a gravity decanter where the n-butanol phase disengages from the aqueous phase and rises to the top. The butanol-rich phase is fed to the top of the butanol column, purified butanol being removed from the bottom, and vapors approaching the heterogeneous azeotropic composition being sent from the top of the column to the condenser. Aqueous phase from the decanter is fed to the aqueous column, practically pure water being removed as waste, and essentially all the butanol being taken overhead as vapor approaching the heterogeneous azeotropic composition. Vapor streams from the two columns are condensed and subcooled, in order to reduce mutual solubilities, and are combined with fresh feed to the decanter.

Although x-y equilibrium curves for typical immiscible

Table 18. Minimum-boiling-point Azeotropic Binary Mixtures*

Pressure 760 mm. Hg

System A	System B	Mole % A	Temp., °C.
Water	Ethanol	10.57	78.15
	Allyl alcohol	54.50	88.20
	Propionic acid	94.70	99.98
	Propyl alcohol (n)	56.83	87.72
	Isopropyl alcohol	31.46	80.37
	Methyl ethyl ketone	33.00	73.45
	Isobutyric acid	94.50	99.30
	Ethyl acetate (2 phase)	24.00	70.40
	Ethyl ether (2 phase)	5.00	34.15
	Butyl alcohol (n) (2 phase)	75.0	92.25
	Isobutyl alcohol	67.14	89.92
	Butyl alcohol (sec)	66.00	88.50
	Butyl alcohol (tert)	35.41	79.91
	Isoamyl alcohol (2 phase)	82.79	95.15
	Amyl alcohol (tert) (2 phase)	65.00	87.00
	Benzene (2 phase)	29.60	69.25
	Toluene (2 phase)	55.6	84.10
Carbon tetrachloride	Methanol	44.5	55.70
	Ethanol	61.3	64.95
	Allyl alcohol	73.0	72.32
	Propyl alcohol (n)	75.0	72.80
	Ethyl acetate	43.0	74.75
Carbon disulphide	Methanol	72.0	37.65
	Ethanol	86.0	42.40
	Acetone	61.0	39.25
	Methyl acetate	69.5	40.15
Chloroform	Methanol	65	53.5
	Ethanol	84	59.3
	Isopropyl alcohol	92	60.8
Butyl alcohol (n)	Cyclohexane	11	79.8
	Toluene	37	105.5
Isobutyl alcohol	Isoamyl bromide	60.0	103.80
	Benzene	10.0	79.84
	Toluene	50.0	101.15
	Pinene (α)	96.5	107.90
Amyl alcohol (n)	Amyl acetate (iso-)	96.4	131.3
	Butyl propionate (iso-)	85	130.5
Isoamyl alcohol	Chlorobenzene	42	124.3
	Xylene (o)	64	128
	Xylene (m)	58	127
	Xylene (p)	56	126.8
Nitrobenzene	Benzyl alcohol	39	204.3
	Borneol	60	207.75
	Menthol	60	207.9
Phenol	Bromotoluene (p)	58	176.2
	Carvene	49.5	169.0
	Pinene (α)	25	152.75
Aniline	Carvene	48	171.35
Benzyl alcohol	Guaiacol	38	204.4
	Naphthalene	64	204.3
Acetic acid	Chlorobenzene	72.5	114.65
	Benzene	2 5	80.05
	Toluene	62.7	105.4
	Xylene (m)	40	115.38
Ethyl alcohol	Methyl ethyl ketone	45	74.8
	Ethyl acetate	46	71.8
	Methyl propionate	67.5	73.2
	Propyl formate (n)	72	73.5
	Benzene	44.8	68.24
	Cyclohexane	44.5	64.9
Ethyl alcohol	Hexane (n)	33.2	58.68
	Toluene	81	76.65
	Heptane (n)	67	72
Allyl alcohol	Benzene	22.2	76.75
	Cyclohexane	26.6	74
	Hexane (n)	6.5	65.5
	Toluene	61.5	92.4
Acetone	Methyl acetate	61	56.1
	Isobutyl chloride	81	55.8
	Diethylamine	43.5	51.5
Propyl alcohol (n)	Ethyl propionate	64	93.4
	Benzene	20.9	77.12
	Hexane (n)	6	65.65
	Toluene	60	92.6
Isopropyl alcohol	Ethyl acetate	30.5	74.8
	Benzene	39.3	71.92
	Hexane (n)	29	61
	Toluene	77	80.6
Tetrachloroethylene	Ethanol	6	77.95
	Allyl alcohol	27	94.0
	Propionic acid	81	118.95
	Propyl alcohol (n)	24	94
	Isopropyl alcohol	8	81.7
	Butyl alcohol (n)	47	110
	Isobutyl alcohol	40	103.05
Trichloroethylene	Allyl alcohol	70	80.95
	Propyl alcohol (n)	69	81.75
	Isopropyl alcohol	54	74
	Isobutyl alcohol	86	85.4
	Butyl alcohol (tert)	74	75
	Amyl alcohol (tert)	83	84
Dichloroethylene	Allyl alcohol	76	79.6
		77	80
Chloral hydrate	Cyclohexane	13	76
Ethylene bromide	Acetic acid	20.7	114.35
	Propionic acid	65	127.75
	Isobutyl alcohol	22	106.2
	Isoamyl alcohol	52	123.2
	Ethyl benzene	83.5	131.1
Methanol	Trichloroethylene	70	60.2
	Acetonitrile	84.5	63.45
	Ethylene chloride	62	59.5
	1, 1-Dichlorethane	28.5	49.05
	Ethyl bromide	14	34.95
	Chloromethyl methyl ether	57.5	56
	Ethyl iodide	52.5	54.7
	Acetone	20	55.7
	Ethyl formate	30.5	50.95
	Methyl acetate	35	54.0
	Propyl bromide (n)	49	54.1
	Propyl iodide (n)	88	63.5
	Methylal	34.5	41.82
	Trimethyl borate	87	59
	Ethyl acetate	91.7	62.3
	Pentane (n)	13	31
	Pentane (iso-)	9	24.5
	Benzene	61.4	53.84
	Cyclohexene	63.0	55.9
	Cyclohexane	61.0	54.2
	Hexane (n)	51	50.6
	Heptane (n)	83	60.5
	Pinene (d)	98.5	64.5

* Abstracted from "International Critical Tables," McGraw-Hill.

systems often exhibit steep slopes so that at first analysis only a very few perfect plates appear to be required for nearly complete stripping of the bottoms streams, it is generally advantageous to use columns with on the order of 10 to 40 trays in order to minimize the quantity of material recycled through the condenser and decanter and back to the columns, thereby economizing on heating and condensing loads.

Over-all plate efficiencies for distillations involving removal of a heterogeneous azeotrope are often unusually low (see p. 613).

Equilibrium Data for Systems Forming Heterogeneous Azeotropes. Reliable measured vapor-liquid equilibrium data should always be preferred to calculated data.

If mutual solubilities in a binary system are exceedingly low, the mole fraction of each component in the azeotropic vapor may be approximated as the ratios of the vapor pressures (of the pure components) to the total pressure, the temperature being determined by trial and error until the sum of the vapor pressures equals the total pressure.

For systems exhibiting appreciable solubility, if

Table 18a. Maximum-boiling-point Azeotropic Binary Mixtures*

A	B	Mole % A	Temp., °C.	Pressure, mm.
Water	Hydrofluoric acid	65.4	120	760
	Hydrochloric acid	88.9	110	
	Perchloric acid	32.0	203	
	Hydrobromic acid	83.1	126	
	Hydriodic acid	84.3	127	
	Nitric acid	62.2	120.5	735
	Formic acid	43.3	107.1	760
Chloroform	Acetone	65.5	64.5	760
Formic acid	Diethyl ketone	48	105.4	
	Methyl propyl ketone	47	105.3	
Phenol	Cyclohexanol	90	182.45	
	Benzaldehyde	54	185.6	
	Benzyl alcohol	8	206.0	
Cresol (o)	Acetophenone	24	203.7	
	Phenyl acetate	42.5	198.6	
	Methyl hexyl ketone	97	191.5	
	Isoamyl butyrate	80	192.0	
Cresol (m)	Acetophenone	54	209.0	
	Isoamyl lactate	60	207.6	
Cresol (p)	Benzyl alcohol	38	207.0	
	Acetophenone	52	208.45	
	Camphor	38	213.15	

* From "International Critical Tables," McGraw-Hill.

Table 18b. Ternary Azeotropic Mixtures*
Pressure 760 mm. Hg.

A = 100 − (B + C)	B and C	Mole %	Temp., °C.
Water	Carbon tetrachloride	57.6	61.8
	Ethanol	23.0	2 phase
	Trichloroethylene	38.4	67.25
	Ethanol	41.2	2 phase
	Trichloroethylene	49.2	71.4
	Allyl alcohol	17.3	2 phase
	Trichloroethylene	51.1	71.55
	Propyl alcohol (n)	16.6	2 phase
	Ethanol	12.4	70.3
	Ethyl acetate	60.1	
	Ethanol	22.8	64.86
	Benzene	53.9	
	Allyl alcohol	9.5	68.3
	Benzene	62.2	
	Propyl alcohol (n)	8.9	68.48
	Benzene	62.8	
Carbon disulphide	Methanol	24.1	33.92
	Ethyl bromide	35.4	
Methyl formate	Ethyl bromide	23.8	16.95
	Isopentane	31.0	
	Ethyl ether	7.2	20.4
	Pentane (n)	48.2	
Propyl lactate (n)	Phenetol	35.2	163.0
	Menthene	34.1	

* From "International Critical Tables," McGraw-Hill.

mutual solubility data are available at the distillation temperature, activity coefficients may be calculated by methods described in Sec. 8, p. 534 (Partly Miscible Solutions), and the complete x-y curve may be estimated.

In the absence of solubility data, if Van Laar constants are known, the existence and extent of the immiscible region may be predicted by methods described in Sec. 8, p. 533 (Prediction of Immiscibility).

APPLICATIONS

Introduction and Definitions

Azeotropic and extractive distillations are fractionations that are facilitated by deliberately adding a new component to the system. The new component, being judiciously chosen to shift the vapor-liquid equilibrium in a favorable direction, is sometimes referred to as a "separating agent." These operations are most applicable in separating (1) those mixtures which are made up of similar compounds and whose boiling points are close together (relative volatility range of 1.05 to 1.20) and (2) those mixtures which are made up of dissimilar compounds that deviate from Raoult's law and cannot be separated at normal operating pressures by simple distillation even though their boiling points are not very close (azeotropes, relative volatility of 1).

The operation is called an *azeotropic distillation* if the separating agent forms one or more azeotropes with components in the system (usually minimum-boiling azeotropes) and by virtue of this fact is present in appreciable concentration on most of the plates in the column.

The operation is called an *extractive distillation* if the separating agent is relatively non-volatile compared with the components to be separated and therefore is charged continuously near the top of the distilling column in order to maintain an appreciable concentration of separating agent in the liquid on most of the plates in the column.

Both azeotropic and extractive distillations have been used in a few isolated industrial applications for a number of years, but probably most of the early applications were developed empirically.

These two processes are employed in the chemical and petroleum industries to separate into pure fractions components that form homogeneous azeotropes during ordinary distillations. Recent needs of the chemical industry for essentially pure compounds from petroleum fractions, such as propane, propylene, normal butane, isobutane, normal pentane, isopentane, mixed hexanes, benzene, toluene, and xylene have been fulfilled successfully by employing azeotropic or extractive distillations.

Both types of operation involve multicomponent distillation of non-ideal mixtures, with the result that even approximate calculations are often tedious. Basic principles and illustrative examples of the two operations are discussed on the ensuing pages. The reader is referred also to p. 630 for a discussion of the occurrence of azeotropes in distillation, and to Sec. 8, p. 526 to 532 for principles of non-ideal liquid mixtures.

Extractive Distillation

When a separating agent or solvent that is much less volatile than the feed components is used to facilitate a fractionation, the process is often called extractive distillation. The basis for extractive distillation is the change of volatility caused by introducing the solvent. The change is not the same for each component, and the resulting difference in volatility permits fractionating the feed components, usually with fewer plates and at a lower reflux ratio.

As an example of a typical extractive distillation, Fig. 74 shows a flow diagram for separating benzene from cyclohexane using phenol as the solvent. Benzene and cyclohexane have nearly the same boiling point at atmospheric pressure and form an azeotrope containing approximately 45 weight per cent cyclohexane (Table 18c, p. 643). However, on addition of relatively polar phenol, the ratio of cyclohexane to benzene in the equilibrium vapor becomes significantly greater than in the liquid [Dunn, Millar, Pierotti, Shiras, and Souders, *Trans. Am. Inst. Chem. Engrs.*, **41**, 631 (1945)]. The feed stream, composed of benzene and cyclohexane, is introduced continuously at point F near the middle of the extractive column, phenol being fed continuously at some point S between the benzene-cyclohexane feed

Table 18c. Hydrocarbon Azeotropes*

Paraffins

Name of entrainer	B.p. of entrainer, °C., 760 mm.	Azeotropic temp., °C., 760 mm.	Azeotropic composition, weight % H.C.†	Ref.
Ethane (−88.3°C.):				
Nitrous oxide	−89.5	Min.‡	......	4
Hydrochloric acid	−83.7	Min.	......	4
Carbon dioxide	−78.5	Min.	......	4
Sulfur dioxide (liq.)	−10.0	Min.	......	9
Methyl alcohol	64.8	....	......	4
Ethyl alcohol	78.3	....	......	4
Isopropyl alcohol	82.4	....	......	4
n-Propyl alcohol	97.3	....	......	4
Water	100.0	....	......	4
Butyl alcohol	117.7	....	......	4
Amyl alcohol	138	....	......	4
Propane (−42.2°C.):				
Sulfur dioxide (liq.) [at 7 kg./sq. cm. (7 atm. approx.)]	−10.0	Min.	78†	9
Methyl alcohol	64.8	Min.	......	4
Isobutane (−12.4°C.):				
Sulfur dioxide	−10.0	−24	......	10
Sulfur dioxide	−10.0	−27	......	13
Ethylene oxide	10.7	Min.	......	13
Methyl formate	31.5	Min.	......	13
n-Butane (−0.6°C.):				
Anhydrous ammonia (liq.)	−33.4	Min.	......	12
Sulfur dioxide	−10.0	−18	......	10
Sulfur dioxide	−10.0	−22	......	13
Methylamine	−6.5	Min.	......	12
Methyl bromide	4.5	−4.4	45.8	26
Ethylene oxide	10.7	Min.	......	13
Methyl formate	31.5	Min.	......	13
Isopentane (28.0°C.):				
Sulfur dioxide	−10.0	Min.	......	13
Ethylene oxide	10.7	Min	......	13
Ethyl chloride	13.3	12.0	5	1
Ethyl nitrite	17.4	16.5	15	1
Acetaldehyde	20.2	17	......	1
Methyl formate	31.9	17.0	53	1
Ethyl ether	34.6	No§	......	1
Ethyl mercaptan	36.2	No	......	1
Isopropyl chloride	36.3	24	......	1
Ethyl bromide	38.4	23.5	70	1
Isopropyl nitrite	40.0	27.5	94	1
Methylal	42.2	27.0	77	1
Carbon disulfide	46.2	27.9	98	2
Carbon disulfide	46.2	No	......	1
Propyl nitrite	47.8	No	......	1
n-Propylamine	50.0	25.5	>80	1
Ethyl formate	54.1	26.5	82	1
Ethyl formate	54.1	No	......	1
Acetone	56.2	<26.0	>80	1
Methyl alcohol	64.7	24.5	96	1
Ethyl alcohol	78.3	26.8	96.5	2
Ethyl alcohol	78.3	No	......	1
Isopropyl alcohol	82.4	27.8	95	2
Formic acid	100.7	27.2	96	1
n-Pentane (36.2°C.):				
Sulfur dioxide	−10.0	Min.	......	13
Ethylene oxide	10.7	Min.	......	13
Ethyl chloride	13.3	No	......	1
Ethyl nitrite	17.3	No	......	1
Acetaldehyde	20.2	No	......	1
Methyl formate	31.9	21.9	47	1
Ethyl ether	34.6	33.4	30	1
Ethyl mercaptan	36.2	32	50	1
Isopropyl bromide	36.3	32	48	1
Ethyl bromide	38.4	33	50	1
Methylpropyl ether	38.8	35.3	75	1
Isopropyl nitrite	40.0	34	60	1
Methylal	42.3	35.4	......	1
Methyl iodide	42.6	34	......	1
Carbon disulfide	46.2	35.7	90	2
n-Propyl nitrite	47.8	35.9	89	1
Ethyl formate	54.2	32.5	70	1
Diethylamine	56	35	85	1
Acetone	56.2	32	80	1
Acetone	56.2	No	......	1
Methyl acetate	57.0	No	......	1
Chloroform	61.2	No	......	1
Methyl alcohol	64.7	30.8	91	1
Isobutyl nitrite	67.1	No	......	1
Isobutylamine	68.5	No	......	1
Methyl borate	68.7	No	......	1
Ethyl alcohol	78.3	34.3	95	2
Isopropyl alcohol	82.4	35.5	94	2
n-Propyl alcohol	97.5	No	......	1
Formic acid	100.7	34.2	90	1

Paraffins—(*Continued*)

Name of entrainer	B.p. of entrainer, °C., 760 mm.	Azeotropic temp., °C., 760 mm.	Azeotropic composition, weight % H.C.†	Ref.
n-Pentane (36.2°C.):—(*Con't*).				
Nitrobenzene	210.9	No	......	1
n-Hexane (69.0°C.):				
Methyl formate	31.8	No	......	1
Ethyl ether	34.6	No	......	1
Isopropyl nitrite	40.0	No	......	1
Methylal	42.3	No	......	1
Carbon disulfide	46.3	No	......	1
Propyl nitrite	47.8	No	......	1
t-Butyl chloride	51.6	No	......	1
Ethyl formate	54.4	49.0	33	1
Acetone	56.2	49.8	41	2
Acetone	56.2	56	15	1
Methyl acetate	57.0	56.7	......	2
Methyl acetate	57.0	No	......	1
Chloromethylmethyl ether	59.5	58.8	10	1
Chloroform	61.2	59.9	28	1
Propylethyl ether	63.6	No	......	1
Methyl alcohol	64.7	50.6	72	2
Methyl borate	65	59	......	1
Isobutyl nitrite	67.1	65.0	44	1
Propyl mercaptan	67.5	<65	<45	1
Isobutylamine	68.5	61	50	1
Methyl borate	68.7	66.3	50	1
Isobutyl chloride	68.9	66.3	45	1
n-Propyl bromide	71	67.5	67	1
Ethyl iodide	72.2	68	24.5	3
t-Butyl bromide	73.5	68.7	......	1
Carbon tetrachloride	76.8	No	......	1
Ethyl acetate	77.2	66.0	58	1
Ethyl acetate	77.2	No	......	1
Butyl nitrite	77.8	68.0	82	1
Ethyl alcohol	78.3	56.7	79	2
Methylethyl ketone	79.0	64.2	63	2
Methyl propionate	79.9	67	88	1
n-Propyl formate	80.9	63	80	1
Isopropyl alcohol	82.3	61	78	1
Isopropyl alcohol	82.3	65.7	96	2
t-Butyl alcohol	82.6	63.7	74.8	1
Dichloroethylene	83.7	No	......	1
Ethyl nitrate	87.7	67.0	82	1
Methyl carbonate	90.4	No	......	1
Isopropyl acetate	90.8	No	......	1
Methyl isobutyrate	92.5	No	......	1
Allyl alcohol	97.0	65.2	95.5	2
n-Propyl alcohol	97.2	65.7	95.7	3
Isoamyl nitrite	97.2	No	......	1
Isobutyl formate	98.2	68.5	88	1
Ethyl propionate	99.2	No	......	1
sec-Butyl alcohol	99.5	67.2	92	2
Formic acid	100.7	60.5	72	1
Formic acid	100.7	67	92	2
Propyl acetate	101.6	No	......	1
Dimethylethyl carbinol	101.8	68.3	96	2
Isobutyl alcohol	107.9	68.3	97.5	2
n-Butyl alcohol	116.9	No	......	1
Acetic acid	118.5	No	......	1
Amyl alcohol	138	No	......	4
Nitrobenzene	210.9	No	......	4
n-Heptane (98.5°C.):				
Acetone	56.2	No	......	1
Methyl alcohol	64.8	59.1	48.5	2
Methyl alcohol	64.8	60.5	35.3	3
Ethyl acetate	77.2	No	......	1
Ethyl alcohol	78.3	70.9	51	2
Methylethyl ketone	79.6	77.0	30	1
Methyl propionate	79.9	No	......	1
Propyl formate	80.9	78.0	<30	1
Dichloroethylene	83.7	No	......	1
Trichloroethylene	86.9	No	......	1
Ethyl nitrate	87.7	82.5	32	1
Methyl carbonate	90.5	85.5	30	1
Isopropyl acetate	90.8	87.5	33	1
Isobutyl bromide	91.6	No	......	1
Methyl isobutyrate	92.5	89.5	30	1
Methylisopropyl ketone	95.4	90	50	1
Allyl alcohol	97.0	84.5	63	2
Isoamyl nitrite	97.2	94.5	45	1
n-Propyl alcohol	97.3	84.8	62	2
Isobutyl formate	97.9	92.0	50	1
Ethyl propionate	99.2	92.5	53	1
Formic acid	100.7	79.5	67	1
Propyl acetate	101.6	93.6	62	1
Diethyl ketone	102.2	93.5	65	1
Diethyl ketone	102.2	98.2	72	1
Methyl butyrate	102.7	94.9	66	1

* Abstracted from compilation by Ewell, Harrison, and Berg, *Petroleum Engr.*, **16**, 1,255 (October, 1944); **16**, 2,263 (November, 1944); **16**, 3,219 (December, 1944).

† H.C. = Hydrocarbon.

‡ "Min." means a minimum-boiling azeotrope was reported.

§ "No" means no azeotrope was found to exist at indicated conditions

Table 18c. Hydrocarbon Azeotropes*—(*Continued*)

Paraffins—(*Continued*)

Name of entrainer	B.p. of entrainer, °C., 760 mm.	Azeotropic temp., °C., 760 mm.	Azeotropic composition, weight % H.C.†	Ref.
n-Heptane (98.5°C.):—(*Con't*).				
Acetal	104.5	97.5	80	1
Pinacolin	106.2	<97	<85	1
Butyl formate	106.7	94.7	65	1
n-Butyl formate	106.9	90.7	60	1
Isobutyl alcohol	108	92.2		1
Ethyl isobutyrate	110.1	97.0	90	1
Methyl isovalerate	116.5	No		1
Isobutyl acetate	117.2	No		1
Acetic acid	118.5	91.9	67	2
Acetic acid	118.5	92.3	70	1
Ethyl borate	118.6	No	......	1
Ethyl butyrate	121.5	No		1
Isobutyl carbinol	131.3	97.7	93	2
Glycol	197.4	98.3	97	2
2,5-dimethylhexane (109.2°C.):				
Methyl alcohol	64.8	61.0	40	2
Ethyl alcohol	78.3	73.6	41	2
Isopropyl alcohol	82.4	79.0	38	2
t-Butyl alcohol	82.6	81.5	23	2
Ethyl nitrate	87.7	86.0	18	1
Chloral	97.8	<97.0	>10	1
Propyl acetate	101.6	98.5	30	1
Diethyl ketone	102.2	98	40	1
Methyl butyrate	102.7	100.0	>25	1
Acetal	103.6	103.0	25	1
Chloropicrin	111.9	<108.0	>10	1
Stannic chloride	113.9	108.0	>48	1
Epichlorohydrin	116.5	107.0	75	1
Chloroacetone	119.7	<108.0	>65	1
n-Octane (125.8°C.):				
Methyl alcohol	64.8	63.0	28	2
Ethyl alcohol	78.3	77	22	2
n-Propyl alcohol	97.2	95	22.7	3
Isoamyl nitrite	97.2	No		1
Formic acid	100.7	93.5	20	1
Propyl acetate	101.6	No		1
Methyl butyrate	102.7	No		1
Isobutyl alcohol	108	104		1
Epichlorohydrin	116.5	114.5	20	1
Isobutyl acetate	117.2	114.5	<30	1
Acetic acid	118.5	105.5	47.5	6
Chloroacetone	119.7	116.5	35	1
Ethyl butyrate	121.5	118.0	<40	1
Ethyl butyrate	121.5	No		1
Isobutyl ether	122.2	No		1
Propyl propionate	122.5	118.2	40	1
Isoamyl formate	123.6	<116.5	45	1
n-Butyl acetate	124.8	119.0	48	1
Mesityl oxide	129.5	123.0	70	1
Chloromethyl acetate	129.5	123.5	60	1
Chlorobenzene	131.8	No		1
Methyl lactate	143.8	<124.5	>80	1
Furfural	161.7	No		1
Acetamide	221.2	125.5	92	1
Isononane (135.2°C.):				
Acetic acid	118.5	108.8	......	7
n-Nonane (150.7°C.):				
Acetic acid	117.8	112.6	......	8
2,7-Dimethyloctane (160.2°C.):				
Formic acid	100.7	No	......	1
Acetic acid	118.5	No		1
Isobutyl carbinol	131.3	129.7	15	2
Propionic acid	140.9	138.3	30	1
Methylphenyl ether	153.9	No		1
Ethyl lactate	154.0	146.0	40	1
Isobutyric acid	154.4	147.6	45.0	3
Propyl isovalerate	155.5	<153.0	>30	1
Cyclohexanone	155.8	<153.0	>32	1
Bromobenzene	156.1	155.9	7	1
Chloracetal	157.4	155.5	35	1
Isoamyl propionate	160.3	155.0	50	1
Furfural	161.2	150.0	58	2
p-Chlorotoluene	161.3	158.5	50	1
n-Butyric acid	162.5	154.0	65	1
n-Butyric acid	162.5	155.8	70	1
Methyl oxalate	164.2	147.0	55	1
Isobutyl isovalerate	171.4	159.0	88	1
Isovaleric acid	176.5	158.0	82	1
Ethyl acetoacetate	180.4	159.0		1
Methyl malonate	181.5	<157	>70	1
Phenol	182.2	159.5	94	1
Phenol	182.2	No		1
Aniline	184.4	Min		14
Decane (173.3°C.):				
Methylheptanone	173.2	168.0	58	2
Aniline	184.4	Min.		14
n-Tridecane (234°C.):				
Pyrocatechol	245.9	229.7	70	1
Resorcinol	281.4	233.3	88	1

Naphthenes

Name of entrainer	B.p. of entrainer, °C., 760 mm.	Azeotropic temp., °C., 760 mm.	Azeotropic composition, weight % H.C.†	Ref.
Cyclohexane (80.8°C.):				
Carbon disulphide	46.3	No		1
Acetone	56.2	53.1	32.7	5
Methyl acetate	57.0	No		1
Methyl alcohol	64.7	54.2	62.6	3
Isobutyl nitrite	67.1	No		1
Methyl borate	68.7	No		1
Isobutyl chloride	68.9	No		1
Carbon tetrachloride	76.8	76.5		1
Ethyl acetate	77.2	72.8	46	1
Butyl nitrite	77.8	75.5	40	1
Ethyl alcohol	78.3	64.9	69.6	3
Methyl propionate	79.6	74	45	1
Methylethyl ketone	79.6	72	60	2
n-Propyl formate	80.9	75	52	1
Isopropyl alcohol	82.5	68.6	67	1
t-Butyl alcohol	82.6	71.8	63.1	3
Ethylene chloride	83.0	75		25
Dichloroethane	83.7	No		1
Trichloroethylene	86.9	No		1
Ethyl nitrate	87.7	74.5	62	1
Methyl carbonate	90.5	78	80	1
Propyl ether	90.6	No		1
Isopropyl acetate	90.8	78.9	75	1
Methyl isobutyrate	92.3	78.6	88	1
Methylisopropyl ketone	95.4	79.0	<88	1
Chloral hydrate	96	76	77.4	3
Propylene chloride	96.8	....	84	25
Allyl alcohol	97.0	74	80.0	3
n-Propyl alcohol	97.2	74.3	76.4	3
Isoamyl nitrite	97.2	No		1
Isobutyl formate	98.2	80	>80	1
Ethyl propionate	99.2	No		1
sec-Butyl alcohol	99.6	76		1
Formic acid	100.7	70.7	70	1
Propyl acetate	101.6	No		1
Dimethylethyl carbinol	101.7	78.5	85	2
Diethyl ketone	102.2	No		1
Pinacolin	106.2	No		1
Isobutyl alcohol	107.9	78.1	86.0	3
Diethyl carbinol	116.0	80.0	97	2
n-Butyl alcohol	117.8	79.8	90.3	3
Acetic acid	118.5	79.7	98	1
Methylcyclohexane (101.1°C.):				
Methyl alcohol	64.8	60	19	2
Methyl alcohol	64.8	60	30	1
Ethyl acetate	77.2	No		1
Ethyl alcohol	78.3	73	30	2
Ethyl alcohol	78.3	73	47	1
Methylethyl ketone	79.6	78.0	30	1
Methyl propionate	79.6	No		1
Ethyl nitrate	87.7	83.9	28	1
Isopropyl acetate	90.8	89.0	22	1
Methyl isobutyrate	92.5	91.0	25	1
Allyl alcohol	97.0	85.0	58	2
Isoamyl nitrite	97.2	95.5	18	1
n-Propyl alcohol	97.3	86.0	58.5	2
Isobutyl formate	97.9	92.4	43	1
Ethyl propionate	99.2	94.5	47	1
Formic acid	100.8	80.2	53.5	1
Nitromethane	101.2	<99		1
n-Propyl acetate	101.6	95.5	52	1
Methylpropyl ketone	101.7	100.6	59.7	3
Allyl iodide	102.0	99.0	30	1
Diethyl ketone	102.2	95.0	60	2
Diethyl ketone	102.2	100.5	60	1
Propyl iodide	102.4	99.5	>20	1
Methyl butyrate	102.7	97.0	55	1
Pinacolin	106.2	98	70	1
Ethyl isobutyrate	110.1	100.0	>80	1
Chloropicrin	111.9	100.7	71	2
Stannic chloride	113.9	No		1
Acetic acid	118.5	96.3	69	1
Ethyl borate	118.6	No		1
Ethyl butyrate	121.5	No	......	1
Isobutyl carbinol	131.3	100.0	94	2
Glycol	197.4	100.8	96	2
1,3-Dimethylcyclohexane (120.5°C.):				
n-Propyl alcohol	97.2	<94.0	<30	2
Chloroacetone	119.0	117.5		1
Isoamyl bromide	120.2	117.0	40	1
Tetrachloroethylene	120.8	118.0		1
Ethylcyclohexane (131.8°C.):				
Acetic acid	118.5	107.9		7
Nonanaphthene (136.7°C.):				
Acetic acid	118.5	109.6	......	7

Table 18c. Hydrocarbon Azeotropes*—(*Continued*)

Olefins

Name of entrainer	B.p. of entrainer, °C., 760 mm.	Azeotropic temp., °C., 760 mm.	Azeotropic composition, weight % H.C.†	Ref.
Ethylene (−103.9°C.):				
Sulphur dioxide (liq.)	−10.0	No	......	9
Propylene (−47.0°C.):				
Sulphur dioxide (liq.)	−10.0	No	......	9
Butene-1 (−6.5°C.):				
Anhydrous ammonia (liq.)	−33.4	Min.	......	12
Sulphur dioxide	−10.0	−16.0	......	10
Sulphur dioxide	−10.0	Min.	......	11
Sulphur dioxide	−10.0	−19.0	......	13
Methylamine	−6.5	Min.	......	12
Ethylene oxide	10.7	Min.	......	13
Methyl formate	31.5	Min.	......	13
Isobutylene (−6.7°C.):				
Sulphur dioxide	10.0	14.0	......	10
Sulphur dioxide	−10.0	Min.	......	11
Sulphur dioxide	−10.0	−17.5	......	13
Ethylene oxide	10.7	Min.	......	13
Methyl formate	31.5	Min.	......	13
Butene-2 trans (1.3°C.):				
Sulphur dioxide	−10.0	−14.0	......	10
Sulphur dioxide	−10.0	−16.0	......	13
Ethylene oxide	10.7	Min.	......	13
Methyl formate	31.5	Min.	......	13
Butene-2-cis. (3.7°C.):				
Sulphur dioxide	−10.0	−13.0	......	10
Sulphur dioxide	−10.0	−16.0	......	13
Ethylene oxide	10.7	Min.	......	13
Methyl formate	31.5	Min.	......	13
3-Methyl Butene-1 (22.5°C.) Isopropylethylene):				
Sulphur dioxide	−10.0	Min.	......	13
Ethylene oxide	10.7	Min.	......	13
Ethyl nitrite	17.3	<14.0	>30	1
Methyl formate	31.8	14.0	62	1
Methylal	42.3	No	......	1
Methyl alcohol	64.8	19.8	97	2
Ethyl alcohol	78.3	21.9	98	2
Formic acid	100.8	22.2	98	1
Pentene-1 (30.2°C.):				
Sulphur dioxide	−10.0	Min.	......	13
Ethylene oxide	10.7	Min.	......	13
Methyl formate	31.5	Min.	......	13
2-Methyl Butene-1 (32.0°C.) (methylethylethylene):				
Sulphur dioxide	−10.0	Min.	......	13
Ethylene oxide	10.7	Min.	......	13
Methyl formate	31.5	Min.	......	13
Isoprene (34.3°C.):				
Methyl formate	31.9	22.5	50	2
Ethyl ether	34.6	33.2	52	1
Ethyl bromide	38.4	32.0	>65	1
Dimethoxymethane	42.3	33.0	80	1
Methyl alcohol	64.7	29.5	......	2
Ethyl alcohol	78.3	32.7	97	2
Pentene-2 (35.8°C.):				
Sulphur dioxide	−10.0	Min.	......	13
Ethylene oxide	10.7	Min.	......	13
Methyl formate	31.5	Min.	......	13
2-Methyl Butene-2 (37.2°C.) (trimethylethylene):				
Sulphur dioxide	−10.0	Min.	......	13
Ethylene oxide	10.7	Min.	......	13
Methyl formate	31.8	24.3	46	1
Ethyl ether	34.6	34.2	12	1
Ethyl mercaptan	36.2	32.9	40	1
Ethyl bromide	38.4	35.2	40	1
Dichloromethane	41.5	36.9	>88	1
Methylal	42.3	35.3	27	1
Methyl iodide	42.6	No	......	1
Carbon disulphide	46.3	36.5	83	1
n-Propylamine	49.7	32.0	68	2
Ethyl formate	54.2	35.0	70	1
Ethyl formate	54.2	No	......	1
Methyl acetate	57.0	No	......	1
Chloroform	61.2	No	......	1
Methyl alcohol	64.8	31.8	93.3	3
Propyl mercaptan	67.5	No	......	1
Ethyl alcohol	78.3	35.3	96	2
Ethyl alcohol	78.3	No	......	1
Isopropyl alcohol	82.5	No	......	1
Formic acid	100.8	35.0	89.5	1
Isovaleric acid	176.5	No	......	1
3-Methyl-1,2-butadiene (40.8°C.):				
Methyl formate	31.9	26.5	32	2
Ethyl ether	34.6	No	......	1
Ethyl bromide	38.4	36.0	......	1
Methyl alcohol	64.7	35.0	90	2

Olefins—(*Continued*)

Name of entrainer	B.p. of entrainer, °C., 760 mm.	Azeotropic temp., °C., 760 mm.	Azeotropic composition, weight % H.C.†	Ref.
3-Methyl-1,2-butadiene (40.8°C.):—(*Continued*)				
Ethyl alcohol	78.3	39.0	......	2
1,5-Hexadiene (60.1°C.):				
Methyl formate	31.8	No	......	1
Ethyl ether	34.6	No	......	1
Ethyl bromide	38.4	No	......	1
t-Butyl chloride	51.6	No	......	1
Ethyl formate	54.2	45.2	42	2
Ethyl formate	54.2	47.7	40	1
Acetone	56.2	47.5	53	1
Acetone	56.2	54.5	40	1
Methyl acetate	57.0	51.0	40	2
Ethylidene chloride	57.5	56.5	33	1
Chloromethyl ether	59.2	55.5	45	1
Chloroform	61.2	55.0	......	2
Ethylpropyl ether	63.6	<60.0	<95	1
Methyl alcohol	64.7	47.1	77.5	2
Ethyl alcohol	78.3	53.5	87	1
Isopropyl alcohol	82.4	55.8	89	2
Formic acid	100.8	54.0	......	1
1,3-Hexadiene (80.8°C.):				
Formic acid	100.7	71.0	30	1
Acetic acid	118.5	No	......	1
Hexene (82.8°C.):				
Formic acid	100.7	71.5	31.5	1
Acetic acid	118.5	82.0	2	1
1,5-Hexadiene (85.4°C.):				
Methyl alcohol	64.8	47.1	75.9	3
Ethyl alcohol	78.3	53.5	86.4	3

Cyclic Olefins

Name of entrainer	B.p. of entrainer, °C., 760 mm.	Azeotropic temp., °C., 760 mm.	Azeotropic composition, weight % H.C.†	Ref.
1,3-Cyclohexadiene (80.5°C.):				
Acetone	56.3	56.1	......	2
Methyl alcohol	64.8	56.4	61.2	3
Methyl borate	68.7	No	......	1
Isobutyl chloride	68.9	No	......	1
Carbon tetrachloride	76.8	No	......	1
Ethyl acetate	77.2	72.8	46	1
Ethyl alcohol	78.3	66.7	65.8	3
Methylethyl ketone	79.6	73	60	1
Isopropyl alcohol	82.4	70.4	64.0	3
t-Butyl alcohol	82.6	73.4	61.3	3
Ethyl nitrate	87.7	76	>60	1
Allyl alcohol	97.0	75.9	79.2	3
n-Propyl alcohol	97.2	76.1	79	3
Formic acid	100.8	71	70	1
Dimethylethyl carbinol	102.0	79.7	85	2
Isobutyl alcohol	107.9	79.4	88.2	3
n-Butyl alcohol	116.9	No	......	1
Acetic acid	118.5	No	......	1
Cyclohexene (83.0°C.):				
Methyl acetate	57.0	No	......	1
Methyl alcohol	64.8	55.9	51.4	3
Carbon tetrachloride	76.8	No	......	1
Ethyl acetate	77.2	73.5	>15	1
Ethyl alcohol	78.3	66.7	64.6	3
Methyl propionate	79.9	75.5	......	1
Isopropyl alcohol	82.4	71	63.9	1
t-Butyl alcohol	82.6	73.7	61.9	3
Dichloroethylene	83.7	No	......	1
Thiophene	84.0	<82.5	<85	2
Trichloroethylene	86.9	No	......	1
Dichlorobromomethane	90.2	82.0	......	1
Propyl ether	90.6	No	......	1
Allyl alcohol	97.0	76.3	78.2	3
n-Propyl alcohol	97.2	76.6	77.8	3
Formic acid	100.8	71.5	69	1
Dimethylethyl carbinol	102.0	80.5	......	1
Isobutyl alcohol	107.9	80.5	85.7	3
n-Butyl alcohol	116.9	82.5	......	1
Acetic acid	118.5	81.8	93.5	1
α-Pinene (155.8°C.):				
Methanol	64.7	64.5	5	1
Ethanol	78.3	......	......	1
n-Propanol	97.2	97.1	1–2	1
Formic acid	100.7	118.2	1.5	1
Isobutanol	108.0	107.9	5	1
n-Butanol	117.8	117.4	12	2
Acetic acid	118.5	117.2	7	2
Isoamyl alcohol	131.8	130.0	20	1
n-Propyl isobutyrate	134.0	No	......	1

Table 18c. Hydrocarbon Azeotropes*—(Continued)

Cyclic Olefins—(*Continued*)

Name of entrainer	B.p. of entrainer, °C., 760 mm.	Azeotropic temp., °C., 760 mm.	Azeotropic composition, weight % H.C.†	Ref.
α-Pinene (155.8°C.):—(*Continued*)				
Propionic acid	140.7	136.2	41.5	1
Isoamyl acetate	142.1	No	......	1
Propyl butyrate	143.0	No	......	1
Ethyl chloroacetate	143.5	No	......	1
Methyl lactate	144.8	144.2	10	1
Butyl propionate	146.5	145.8	15	1
Isobutyl isobutyrate	147.3	No	......	1
Bromoform	148.3	146.0	30	1
Amyl acetate	149.0	148.0	25	1
Anisol	153.9	150.5	44	1
Isobutyric acid	154.4	146.7	35	1
Ethyl lactate	155.0	147.0	49	1
n-Propyl isovalerate	155.7	144.0	47	1
Cyclohexanone	155.8	149.8	60	1
Bromobenzene	156.1	153.4	50	1
Chloroacetal	156.8	151.0	45	1
Isobutyl butyrate	156.8	153.0	50	1
Trichlorohydrin	158.0	154.5	50	1
Ethyl bromoacetate	158.2	152.5	54	1
Isoamyl propionate	160.3	154.0	75	1
Cyclohexanol	160.7	149.9	64.5	1
Furfural	161.5	153.0	......	1
Furfural	161.5	143.4	62	2
Pentachloroethane	162.0	162.2(max.)	5	2
Methyl oxalate	163.3	144.1	61	1
Butyric acid	163.5	152.0	75	1
Ethyl silicate	165.0	149.0	65	1
Butyl butyrate	166.4	No	......	1
Ethyl caproate	167.8	No	......	1
Methyl acetoacetate	169.5	150.5	63	1
Phenetol	170.5	No	......	1
Isobutyl isovalerate	171.4	No	......	1
Methyl n-hexyl ketone	172.9	No	......	1
α-Dichlorohydrin	174.5	152.0	85	1
Isovaleric acid	176.5	154.2	89	1
2-Octanol	178.7	No	......	1
Ethyl acetoacetate	180.7	153.4	78	1
Methyl malonate	181.4	151.5	78	2
Phenol	181.5	152.8	81	1
α-Dichlorohydrin	183.0	No	......	1
Aniline	184.4	155.3	85	1
Ethyl oxalate	185.6	154.8	80	2
Monochloroacetic acid	189.4	152.0	......	2
Isobutyl carbonate	190.3	No	......	1
o-Cresol	190.8	No	......	1
Methyl succinate	195.5	155.5	90	1
Glycol	197.4	149.5	81.5	2
Ethyl malonate	199.2	No	......	1
Benzyl alcohol	205.5	No	......	1
Propyl oxalate	212.0	No	......	1
Acetamide	221.2	152.5	87	2
Camphene (158°C.):				
Formic acid	100.7	No	......	1
Acetic acid	118.5	118.2	3	2
Isobutyl carbinol	131.3	130.4	23	2
Amyl alcohol	131.8	No	......	1
Propionic acid	140.9	137.7	36	1
Isoamyl acetate	142.1	No	......	1
Bromoform	148.3	147.5	......	1
Anisol	153.9	152.7	23	1
Isobutyric acid	154.4	148.1	45	1
Ethyl lactate	155.0	149.0	35	1
n-Propyl isovalerate	155.7	145.0	35	2
Cyclohexanone	155.8	150.6	42	1
Bromobenzene	156.1	155.0	......	1
Isobutyl butyrate	156.8	153.0	37	1
Trichlorohydrin	158.0	156.0	......	1
Ethyl bromoacetate	158.2	154.0	......	1
Isoamyl propionate	160.3	155.5	50	1
Cyclohexanol	160.7	153.0	65	1
Furfural	161.5	146.8	60	2
n-Butyric acid	162.5	152.3	73	1
Methyl oxalate	163.3	145.0	60	1
Ethyl silicate	165.0	150.0	63	1
Ethyl caproate	167.8	159.0	85	1
Methyl n-hexyl ketone	172.9	158.0	90	1
α-Dichlorohydrin	174.5	154.0	80	1
Isovaleric acid	176.5	156.7	89	1
Butyl isovalerate	177.6	No	......	1
Isoamyl butyrate	178.5	No	......	1
Benzaldehyde	179.2	158.5	85	1
Ethyl acetoacetate	180.4	156.2	70	2
Methyl malonate	181.5	154.6	74	2
Camphene (158°C.):—(*Continued*)				
Phenol	181.5	155.0	78	1
Dichlorohydrin	183.0	156.0	75	2
Aniline	184.4	157.5	87	2
Ethyl oxalate	185.6	158.5	84	2
Monochloroacetic acid	189.4	154.7	85	2
Isobutyl carbonate	190.3	No	......	1
o-Cresol	191.1	No	......	1
Phenyl acetate	195.7	No	......	1
Methyl succinate	195.5	159.0	90	1
Ethyl malonate	199.2	No	......	1
Acetamide	221.2	156.0	85	2
Propionamide	222.1	156.4	90	2
β-Pinene (164°C.):				
Propionic acid	140.7	139.0	24	1
Chloracetal	157.4	156.2	23	1
p-Chloro toluene	161.3	160.2	......	1
Furfural	161.5	146.3	50	1
Pentachloroethane	162.0	166.0	58	2
Methyl oxalate	163.3	147.1	49	1
n-Butyric acid	163.5	158.0	<62	1
Butyl isovalerate	177.6	No	......	1
Benzaldehyde	179.2	162.0	75	1
Ethyl acetoacetate	180.7	159.5	65	1
Methyl malonate	181.5	158.0	72	1
Phenol	181.5	159.0	75	1
o-Cresol	190.8	No	......	1
Methene (170.8°C.):				
Cyclohexanol	160.7	157.5	38	1
Methyl oxalate	164.2	154.0	30	1
Methyl acetoacetate	169.5	160.0	48	1
Benzyl chloride	179.4	No	......	1
Methyl malonate	181.5	164.0	63	1
Phenol	181.5	164.0	67	1
α-Terpinene (p-menthadiene-1,3) (173.3°C.):				
Isobutyl butyrate	156.8	No	......	1
Chloracetal	157.4	No	......	1
Furfural	161.5	154.5	47	1
Methyl oxalate	164.2	155.0	32	1
Isobutyl isovalerate	171.4	170.5	35	1
Propyl lactate	171.7	164.0	50	1
Methyl malonate	181.5	167.0	55	1
Ethyl oxalate	185.7	170.5	70	1
Carvene (Limonene) 177.9°C.):				
Isoamyl alcohol	131.8	No	......	1
Propionic acid	140.9	No	......	1
Isobutyric acid	154.4	151.0	22	1
n-Hexanol	156.8	155.5	21	2
Cyclohexanol	160.7	159.3	26.5	1
Furfural	161.5	156.0	35	2
Methyl oxalate	163.3	156.7	25	1
n-Butyric acid	163.5	160.8	45	1
Butyl butyrate	166.4	No	......	1
Methyl acetoacetate	169.5	162.7	35	1
Phenetol	170.5	No	......	1
Isobutyl isovalerate	171.4	No	......	1
Propyl lactate	171.7	166.4	37	1
Isoamyl ether	172.7	No	......	1
Methyl n-hexyl ketone	172.9	170.5	45	1
α-Dichlorhydrin	174.5	165.8	57	1
o-Chlorophenol	175.5	175.0	......	1
Eucalyptol	176.3	No	......	1
Isovaleric acid	176.5	168.9	59	1
Butyl isovalerate	177.6	176.0	45	1
Ethyl α-bromoisobutyrate	178.0	174.0	55	1
Isoamyl butyrate	178.6	174.0	53	1
2-Octanol	178.7	174.4	65	1
Benzaldehyde	179.2	171.2	57	1
Benzyl chloride	179.4	174.8	54	1
Ethyl acetoacetate	180.7	169.0	57	1
Methyl malonate	181.5	169.0	56	1
Phenol	181.5	169.0	60	1
o-Bromotoluene	181.8	177.3	83	1
Isobutyl lactate	182.2	172.5	60	1
β-Dichlorhydrin	183.0	169.3	60	1
Aniline	184.4	171.4	61.2	1
Ethyl oxalate	185.0	172.2	59	1
Perchloroethane	185.0	No	......	1
p-Bromotoluene	185.0	No	......	1
Chloroacetic acid	186.5	167.8	66	1
Methyl sulphate	188.4	173.0	......	1
Isobutyl carbonate	190.3	No	......	1
o-Cresol	190.8	175.4	75	1

Table 18c. Hydrocarbon Azeotropes*—(*Continued*)

Cyclic Olefins—(*Continued*)

Name of entrainer	B.p. of entrainer, °C., 760 mm.	Azeotropic temp., °C., 760 mm.	Azeotropic composition, weight % H.C.†	Ref.
Carvene (Limonene) (177.9°C.): (*Continued*)				
Isoamyl isovalerate	192.7	No	......	1
Dimethyl aniline	194.1	174.0	73	1
Methyl succinate	195.0	No		1
Phenyl acetate	195.7	177.5	93	2
Trichloroacetic acid	196.0	171.0	......	1
Methyl aniline	196.1	174.5	87	1
Methyl succinate	196.4	175.5	74	2
Ethylene glycol	197.4	163.0	79	1
Ethyl malonate	198.6	177.5	90	2
Linalool	199.0	No	......	1
Methyl benzoate	199.5	No	......	1
p-Cresol	201.8	177.7	99	1
m-Cresol	202.2	No	......	1
Benzyl formate	202.3	No	......	1
Caproic acid	205.2	177.0	95	2
Benzyl alcohol	205.5	176.3	89	1
Propyl oxalate	212.0	No	......	1
Ethyl succinate	217.3	No	......	1
Acetamide	221.2	169.2	84	2
Glycerol	291.0	177.2	99	2
Thymene (179.7°C.):				
Isobutyric acid	154.4	154.0	......	2
Furfural	161.5	158.5	28	2
n-Butyric acid	162.5	160.5	32	1
Isovaleric acid	176.5	170.5	44	1
Methyl malonate	181.4	169.0	50	2
Phenol	182.2	172.3	60	1
Aniline	184.4	173.5	59	2
o-Cresol	191.9	178.5	74	1
Phenyl acetate	195.7	179.3	82	2
Methyl succinate	196.4	178.2	68	2
p-Cresol	201.7	No	......	1
m-Cresol	202.2	No	......	1
Benzyl alcohol	205.2	179.0	86	2
n-Caproic acid	205.2	179.0	97	2
Glycerol	291.0	179.6	99	2
γ-Terpinene (p-menthadiene-1,4) (179.9°C.):				
Cyclohexanol	160.7	159.8	......	1
Methyl oxalate	164.2	159.0	22	1
Phenetol	170.5	No	......	1
Methyl n-hexyl ketone	172.9	171.7	35	1
Isoamyl ether	173.4	No	......	1
Isoamyl butyrate	178.5	177.5	43	1
Benzaldehyde	179.2	173.0	52	1
Ethyl acetoacetate	180.7	171.0	49	1
Methyl malonate	181.5	168.0	52	1
Phenol	181.5	171.5	57	1
o-Bromotoluene	181.8	178.5	60	1
Ethyl oxalate	185.7	173.5	55	2
Monochloroacetic acid	189.4	170.0	......	2
o-Cresol	190.8	178.0	73	1
Isoamyl isovalerate	192.7	No	......	1
Methyl succinate	195.4	178.0	68	2
Phenyl acetate	195.7	180.3	85	2
Ethyl malonate	198.6	178.0	78	2
Methyl benzoate	199.5	No	......	1
Benzyl formate	202.3	No	......	1
Benzyl acetate	215.5	No	......	1
Ethyl succinate	217.3	No	......	1
Terpinolene [p-menthadiene-1,4(8)] (185.2°C.):				
Furfural	161.5	160.3	20	1
Methyl oxalate	164.2	160.0	10	1
Isobutyl isovalerate	171.4	No	......	1
Isoamyl butyrate	178.5	No	......	1
Methyl malonate	181.5	171.0	38	1
Ethyl oxalate	185.7	175.0	50	1
Methyl succinate	195.5	180.0	64	1

Aromatics

Name of entrainer	B.p. of entrainer, °C., 760 mm.	Azeotropic temp., °C., 760 mm.	Azeotropic composition, weight % H.C.†	Ref.
Benzene (80.2°C.):				
Ethyl ether	34.6	No	......	4
Ethyl bromide	38.4	No	......	4
Carbon disulphide	46.3	No	......	4
Ethyl formate	54.2	No	......	1
Acetone	56.2	No	......	4
Methyl acetate	57.0	No	......	4
Benzene (80.2°C.):—(*Continued*)				
Chloroform	61.2	No	......	4
Methyl alcohol	64.8	58.4	39.4	3
Methyl borate	68.7	No	......	1
Isobutyl chloride	68.9	No	......	1
Propyl bromide	71.0	No	......	1
Ethyl iodide	72.3	No	......	4
Butyraldehyde	75.7	No	......	15
Carbon tetrachloride	76.8	....	......	4
Ethyl acetate	77.2	77.0	6	1
Butyl nitrite	77.8	No	......	1
Ethyl alcohol	78.3	68.2	32.4	1
Methylethyl ketone	79.6	78.4	37.5	3
Methyl propionate	79.9	77.5	48	1
n-Propyl formate	80.8	72.0	52.1	3
Isopropyl alcohol	82.4	72.0	66.7	3
t-Butyl alcohol	82.6	74.0	63.6	3
Ethylene chloride	83.7	No	......	4
Dichloroethylene	83.7	No	......	1
Thiophene	84.0	No	......	1
Ethyl nitrate	87.7	80.1	85	1
Isobutyl mercaptan	88.0	No	......	1
Dichlorobromomethane	90.2	No	......	1
Methyl carbonate	90.4	77.5	74	1
Isopropyl acetate	90.8	No	......	1
Isobutyl bromide	91.6	No	......	1
Methyl isobutyrate	92.5	No	......	1
Methyl isopropyl ketone	95.4	No	......	1
Allyl alcohol	97.0	76.8	82.6	3
n-Propyl alcohol	97.2	77.1	83.1	3
Isoamyl nitrite	97.2	No	......	1
Isobutyl formate	98.2	No	......	1
sec-Butyl alcohol	99.6	79.0	......	1
Formic acid	100.7	71.1	69	1
Nitromethane	101.5	79.2	88.7	3
Propyl acetate	101.6	No	......	1
Dimethylethyl carbinol	102.0	No	......	1
Crotonaldehyde	102.2	No	......	1
Diethyl ketone	102.2	No	......	1
Acetal	104.5	No	......	1
Isobutyl alcohol	107.9	79.8	90.7	1
n-Butyl alcohol	116.9	No	......	1
Acetic acid	118.5	80.1	98	1
Isoamyl alcohol	130.5	No	......	4
Chlorobenzene	131.8	No	......	4
Propionic acid	140.7	No	......	1
Isobutyric acid	154.4	No	......	1
Bromobenzene	156.1	No	......	4
Cyclohexanol	160.7	No	......	1
Toluene (110.7°C.):				
Chloroform	61.2	No	......	4
Methyl alcohol	64.8	63.8	31	2
Carbon tetrachloride	76.8	No	......	4
Ethyl alcohol	78.3	76.7	31.9	3
Methylethyl ketone	79.6	No	......	1
Isopropyl alcohol	82.3	80.6	42.1	3
Isopropyl alcohol	82.3	No	......	4
Dichloroethylene	83.7	No	......	1
Ethyl nitrate	87.7	No	......	1
Allyl alcohol	96.5	92.4	49.8	3
n-Propyl alcohol	97.2	92.6	57.2	3
Isoamyl nitrite	97.2	No	......	1
Chloral	97.8	No	......	2
Isobutyl formate	98.2	No	......	1
Ethyl propionate	99.2	No	......	1
Formic acid	100.7	85.8	50	1
Nitromethane	101.0	95.0	......	16
Propyl acetate	101.6	No	......	1
Dimethylethyl carbinol	102.0	99.2	......	1
Crotonaldehyde	102.2	101.0	......	1
Diethyl ketone	102.2	No	......	1
Methyl n-propyl ketone	102.3	No	......	1
Propyl iodide	102.4	No	......	1
Methyl butyrate	102.7	No	......	1
Piperidine	105.7	No	......	1
Isobutyl alcohol	107.9	101.2	55.4	3
Ethyl isobutyrate	110.1	109.8	......	1
Chloropicrin	111.9	109.0	......	1
Stannic chloride	113.9	109.2	48	1
Pyridine	115.5	No	......	1
Epichlorohydrin	116.5	108.4	71	2
Methyl isovalerate	116.5	No	......	1
Isobutyl acetate	117.2	No	......	1
n-Butyl alcohol	117.7	105.5	67.9	3
Acetic acid	118.5	105.4	64	1
Ethyl borate	118.6	No	......	1

Table 18c. Hydrocarbon Azeotropes*—(Continued)

Aromatics—(Continued)

Name of entrainer	B.p. of entrainer, °C., 760 mm.	Azeotropic temp., °C., 760 mm.	Azeotropic composition, weight % H.C.†	Ref.
Toluene (110.7°C.):—(*Continued*)				
Isobutyl iodide	120.0	No		1
Isoamyl bromide	120.2	No		1
Ethyl butyrate	121.5	No		1
Propyl propionate	122.5	No		1
Isoamyl formate	123.8	No		1
Paraldehyde	124.0	No		1
Ethyl carbonate	125.9	No		1
Isoamyl alcohol	129.3	No		1
Mesityl oxide	129.5	No		1
Chloromethyl acetate	129.5	No		1
Amyl bromide	130.0	No		4
Isobutyl carbinol	131.3	110.1	86	2
Chlorobenzene	131.8	No		4
Acetylacetone	137.7	No		2
Propionic acid	140.7	No		1
Methyl lactate	143.8	110.4	82	2
Isobutyric acid	154.4	No		1
Bromobenzene	156.1	No		4
Cyclohexanol	160.5	No	.....	1
Butyric acid	163.5	No		1
Ethyl Benzene (136.1°C.):				
n-Propyl alcohol	97.2	No		1
Formic acid	100.7	94.0	32	1
Formic acid	100.7	99.5	3	1
Isobutyl alcohol	108.0	107.4	>20	2
Epichlorohydrin	116.5	No		1
n-Butyl alcohol	116.9	115.0		1
Acetic acid	118.5	114.7	34	1
Propyl propionate	122.5	No		1
Isoamyl formate	123.8	No		1
Paraldehyde	124.0	No		1
Butyl acetate	125.8	No		1
Ethyl carbonate	126.0	124.0	23	1
Ethyl carbonate	126.0	No		1
Isoamyl alcohol	130.5	125.9	50.7	3
Mesityl oxide	130.5	No		1
Ethylene bromide	131.5	131.1	10	1
Ethylene bromide	131.5	131.1	53.6	3
Chlorobenzene	131.8	No	.....	1
Ethyl isovalerate	134.5	No		1
Isobutyl propionate	136.9	133.0	52.1	3
Acetylacetone	137.7	135.0	65	1
Propionic acid	140.7	131.1	72	1
1,2-Dibromopropane	141.6	136.0	95.1	3
Isoamyl acetate	142.1	No		1
Butyl ether	142.2	No		1
n-Dipropyl ketone	143.6	No		1
Isobutyric acid	154.4	131.9	82	1
Isobutyric acid	154.4	134.3	88	1
Bromobenzene	156.1	No		1
n-Butyric acid	163.5	135.9	>97	1
Isovaleric acid	176.5	No		1
Acetamide	221.2	135.6	92	1
p-Xylene (138.3°C.):				
Formic acid	100.7	95.0	30	1
Formic acid	100.7	100.5	3	1
n-Butyl alcohol	117.8	115.7	32	1
Acetic acid	118.5	115.1	33	1
Paraldehyde	124.1	No		1
Isoamyl alcohol	130.5	126.8	49	1
Dibromoethylene	131.5	131.3		1
Propyl isobutyrate	134.0	No	.....	1
Isobutyl propionate	136.9	136.8	15	1
Propionic acid	140.7	132.0	64	1
1,2-Dibromopropane	141.7	137.5	88	1
Isoamyl acetate	142.1	No		1
Chloroethyl acetate	143.5	137.0	72	1
Isobutyric acid	154.4	133.5	80	1
Cyclohexanol	160.6	No		1
Furfural	161.5	138.0	95	1
n-Butyric acid	163.5	137.5	95	1
Isovaleric acid	176.5	No		1
Acetamide	221.2	137.5	91	1
m-Xylene (139.0°C.):				
Formic acid	100.7	94.2	29.8	1
Formic acid	100.7	No		1
Isobutyl alcohol	107.9	107.7	13	1
Isobutyl alcohol	107.9	No		4
n-Butyl alcohol	117.8	116.0	20	1
Acetic acid	118.5	115.4	27.5	1
Paraldehyde	124.1	No		1
Butyl acetate	125.8	No		1
Ethyl carbonate	126.0	No		1
Isoamyl alcohol	130.5	127.0	47	1
Ethylene bromide	131.5	No		1
Propyl siobutyrate	134.0	No	.....	1

Aromatics—(Continued)

Name of entrainer	B.p. of entrainer, °C., 760 mm.	Azeotropic temp., °C., 760 mm.	Azeotropic composition, weight % H.C.†	Ref.
m-Xylene (139.0°C.):—(*Continued*)				
Ethyl isovalerate	134.7	No		1
Isobutyl propionate	136.9	134.5		1
Isobutyl propionate	136.9	No		2
Isoamyl acetate	138.8	136.0	50	3
Allyl sulphide	139.0	137.0		1
Propionic acid	140.7	132.7	64.5	1
1,2-Dibromopropane	141.7	138.0	70	1
Isoamyl acetate	142.1	No		1
Butyl ether	142.2	No		1
Chloroethyl acetate	143.5	137.3	68	1
Propyl butyrate	143.5	138.7		1
Propyl butyrate	143.5	No	.....	1
Paraldehyde	144.8	134.0	70	1
Isoamyl nitrate	149.5	No		1
Isobutyric acid	154.4	134.0	79	1
Isobutyric acid	154.4	136.8	86	1
Cyclohexanol	160.7	138.9	94.9	3
Furfural	161.5	138.4	88	1
n-Butyric acid	163.5	138.1	94.5	3
Methylacetoacetate	169.5	No		1
Dichlorohydrin	174.5	No		1
Isovaleric acid	176.5	No	.....	1
Phenol	182.2	No	.....	1
Aniline	184.4	No		1
Propionamide	213.0	138.5		1
Acetamide	221.2	138.2	86	1
o-Xylene (143.6°C.):				
Formic acid	100.7	95.5	26	1
Formic acid	100.7	100.5	1	1
n-Butyl alcohol	117.8	116.8	25	1
Acetic acid	118.5	116.0	24	1
Morpholine	128.0	No	.....	17
Isoamyl alcohol	130.5	128.0	40.4	3
Isobutyl propionate	136.9	No		1
Propionic acid	140.9	135.0	58	1
1,2-Dibromopropane	141.7	139.2	30	1
Isoamyl acetate	142.1	No		1
Butyl ether	142.2	142.0		1
Propyl butyrate	143.5	143.2	45	1
n-Dipropyl ketone	143.6	143.0	58	1
Butyl propionate	146.5	No	.....	1
Amyl acetate	149.0	No		1
Anisol	153.9	No		1
Cyclohexanone	155.7	No		1
Furfural	161.5	140.5	87	1
n-Butyric acid	162.5	142.0	90	1
Phenol	182.2	No		1
Styrene (146°C.):				
Methyl alcohol	64.8	64.2		2
n-Propyl alcohol	97.2	97.0	8	2
Water	100.0	Min.		21
n-Butyl alcohol	117.8	116.5	21	2
Acetic acid	118.5	116.0	17	2
Isoamyl alcohol	130.5	128.5	34.7	3
Propyl isobutyrate	134.0	No		1
Chloroethyl acetate	143.5	140.2	40	1
Propyl butyrate	143.5	No		1
Methyl lactate	143.8	134.5	50	1
sym-Tetrachloroethane	146.3	143.5	45	1
Butyl propionate	146.5	145.5		1
Ethyl lactate	154.0	140.5	75	1
Furfural	161.5	141.0	85	2
Methyl oxalate	163.3	<142.5	88	2
Methyl acetoacetate	169.5	143.0	73	1
Dichlorohydrin	175.1	143.5	85	2
Ethyl acetoacetate	180.0	145.2	96.3	3
Phenol	181.5	No		1
Acetamide	221.2	144.0	88	2
Cumene (152.4°C.) (isopropylbenzene):				
Acetic acid	117.8	116.8	.....	8
n-Propyl Benzene (158.9°C.):				
Acetic acid	118.5	No	.....	1
Isobutyl carbinol	131.3	130.6	17	2
Isoamyl nitrate	149.5	No		1
Ethyl lactate	154.0	147.0	42	1
Propyl isovalerate	155.5	No	.....	1
Chloroacetal	157.4	<156.0	>25	1
Furfural	161.5	152.0	60	1
Methyl oxalate	164.2	152.5	58	1
Ethyl caproate	167.8	No		1
Methyl n-hexyl ketone	172.9	No		1
Methyl malonate	181.5	<159.0		1
Phenol	182.2	158.0	96	1
Monochloroacetic acid	189.5	156.0	.....	2

Table 18c. Hydrocarbon Azeotropes*—(Continued)

Aromatics—(Continued)

Name of entrainer	B.p. of entrainer, °C., 760 mm.	Azeotropic temp., °C., 760 mm.	Azeotropic composition, weight % H.C.†	Ref.
Mesitylene (164.6°C.) (1,3,5-trimethylbenzene):				
Isoamyl alcohol	131.8	No	.....	1
Propionic acid	140.7	139.3	23	1
Methyl lactate	143.8	142.0	<15	1
Isobutyric acid	154.4	148.5	48	1
Propyl isovalerate	155.5	No	.....	1
Bromobenzene	156.1	No	.....	1
Isobutyl butyrate	156.8	No	.....	1
Isoamyl propionate	160.3	No	.....	1
Cyclohexanol	160.7	156.3	44.8	3
p-Chlorotoluene	161.3	160.5	28	1
Furfural	161.5	155.2	40	1
Pentachloroethane	162.0	166.0 (max.)	60	2
Methyl oxalate	163.3	154.8	49.8	1
n-Butyric acid	163.5	158.5	33	3
Butyl butyrate	166.4	No	.....	1
Isobutyl isovalerate	168.7	163.0	.....	1
Methyl acetoacetate	169.5	160.5	57	1
Isobutyl isovalerate	171.4	No	.....	1
Phenetol	171.5	No	.....	1
Propyl lactate	171.7	160.0	75	1
Isoamyl ether	173.4	No	.....	1
Isovaleric acid	176.5	162.8	81	1
Ethyl acetoacetate	180.4	162.5	68	1
Methyl malonate	181.5	162.0	<90	1
Phenol	182.2	No	.....	1
Aniline	184.4	Min.	.....	14
Aniline	184.4	No	.....	1
Ethyl oxalate	185.7	No	.....	1
Monochloroacetic acid	189.4	162.0	83	2
o-Cresol	191.9	No	.....	1
Methyl succinate	195.5	No	.....	1
Glycol	197.4	155.0	88	2
Acetamide	221.2	160.0	85	1
Pseudocumene (168.2°C.) (1,2,4-trimethylbenzene):				
Isoamyl alcohol	131.8	No	.....	1
Propionic acid	141.1	140.0	90	2
Propionic acid	141.1	No	.....	1
Isobutyric acid	154.4	149.4	42	1
Isobutyric acid	154.4	152.3	37	1
Cyclohexanol	160.6	158.0	40	1
Furfural	161.5	156.0	40	1
n-Butyric acid	162.5	159.0	50	1
Methyl oxalate	163.3	157.0	35	2
Ethyl caproate	167.8	167.6	.....	1
Isobutyl isovalerate	168.7	166.5	49	1
Phenylethyl ether	170.5	168.1	>90	1
Isobutyl isovalerate	171.4	No	.....	1
Phenetol	171.5	168.9	.....	1
Isovaleric acid	176.5	163.8	77	1
Benzyl chloride	179.4	No	.....	1
Ethyl acetoacetate	180.4	166.5	63	1
Methyl malonate	181.5	<165.5	<80	1
Phenol	181.5	167.0	74	1
Aniline	184.4	167.5	>85	2
Ethyl oxalate	185.7	167.9	94	2
Caproic acid	205.2	No	.....	1
Cymene (176.7°C.) (methyl isopropyl benzene):				
Isoamyl alcohol	131.8	No	.....	1
Propionic acid	140.7	No	.....	1
Cyclohexanol	160.6	159.0	29	1
Furfural	161.5	157.8	32	2
Methyl oxalate	163.3	161.0	20	1
n-Butyric acid	163.5	160.5	43	1
Phenetol	170.5	No	.....	1
Isobutyl isovalerate	171.4	No	.....	1
Dichlorohydrin	174.5	165.0	52	1
o-Chlorophenol	175.5	173.5	50	1
Eucalyptol	176.4	176.0	.....	2
Isovaleric acid	176.5	168.0	63	1
p-Cresol methyl ether	177.1	No	.....	1
Isoamyl butyrate	178.6	<173.0	.....	1
sec-Octyl alcohol	178.7	172.5	.....	1
Benzaldehyde	179.2	171.0	72	1
Ethyl acetoacetate	180.0	170.5	.....	1
Ethyl acetoacetate	180.0	172.0	>55	1
Methyl malonate	181.5	169.0	60	1
Phenol	181.5	170.0	66	1
Isobutyl lactate	182.2	171.5	65	1
Aniline	184.4	170.0	70	2
Ethyl oxalate	185.7	174.0	85	1
Chloroacetic acid	186.5	166.0	65	1
o-Cresol	190.8	175.0	.....	1

Aromatics—(Continued)

Name of entrainer	B.p. of entrainer, °C., 760 mm	Azeotropic temp. °C., 760 mm.	Azeotropic composition, weight % H.C.†	Ref.
Cymene (176.7°C.) (methyl isopropyl benzene):—(Continued)				
Phenyl acetate	195.7	No	.....	1
Caproic acid	205.2	No	.....	1
Acetamide	221.2	170.5	81	2
Indene (182.4°C.):				
Formic acid	108.0	No	.....	1
Isobutyric acid	154.4	No	.....	1
Cyclohexanol	160.7	160.0	25	2
n-Butyric acid	162.5	161.0	.....	2
Isobutyl isovalerate	171.4	No	.....	1
Monoethanolamine	172.0	Min	.....	25
o-Chlorophenol	174.0	Min	.....	25
Isovaleric acid	176.5	173.5	45	1
Isoamyl butyrate	178.5	178.0	.....	1
Isobutyl lactate	182.2	177.0	48	1
Phenol	182.2	177.8	47	1
Aniline	184.4	176.8	55	2
Propylene glycol	187.4	Min.	.....	25
o-Cresol	191.1	182.3	91	1
Isoamyl isovalerate	192.7	No	.....	1
Diethylene glycol monomethyl ether	193.2	Min.	.....	25
Ethylene glycol	197.0	Min.	.....	25
Phorone	197.8	No	.....	1
p-Cresol	201.7	No	.....	1
m-Cresol	202.2	No	.....	1
Acetamide	221.2	177.6	82.5	2
n-Butyl Benzene (183.2°C.):				
Furfural	161.5	160.5	18	1
Methyl n-hexyl ketone	172.9	No	.....	1
Eucalyptol	176.4	No	.....	1
p-Cresol methyl ether	177.1	No	.....	1
sym-Triethyl Benzene (216°C.):				
Ethyl acetoacetate	180.7	No	.....	1
Phenol	181.5	No	.....	1
Aniline	184.4	No	.....	1
o-Cresol	191.1	No	.....	1
Methyl succinate	195.5	No	.....	1
Phenyl acetate	195.7	No	.....	1
p-Cresol	201.8	201.5	4	1
m-Cresol	202.2	No	.....	1
Benzyl formate	202.3	No	.....	1
Benzyl alcohol	205.5	200.2	45	1
Nitrobenzene	210.8	No	.....	1
Borneol	211.8	211.0	45	1
Propyl oxalate	212.0	210.0	30	1
Ethyl benzoate	212.6	No	.....	1
Benzyl acetate	215.5	214.5	50	1
p-Chlorophenol	217.0	214.7	82	1
o-Nitrophenol	217.3	214.3	55	1
Acetamide	221.2	198.0	73	2
Bornyl acetate	227.7	No	.....	1
Thymol	232.9	No	.....	1
n-Caproic acid	237.5	214.3	96	2
Pyrocatechol	245.9	214.7	.....	1
Pyrocatechol	245.9	214.7	.....	2
Resorcinol	281.4	No	.....	1
Naphthalene (218.1°C.):				
Ethanol	78.3	No	.....	1
n-Butyric acid	163.5	No	.....	1
Isovaleric acid	176.5	No	.....	1
Phenol	181.5	No	.....	1
β-Chlorohydrin	183.0	No	.....	1
Aniline	184.4	No	.....	1
Chloroacetic acid	186.5	184.0	15	1
o-Cresol	190.8	No	.....	1
Methyl succinate	195.5	No	.....	1
Phenyl acetate	195.7	No	.....	1
Trichloroacetic acid	196.0	190.0	.....	1
Methyl aniline	196.1	No	.....	1
Glycol	197.4	183.9	49	24
Ethyl malonate	198.9	No	.....	1
o-Toluidine	200.7	No	.....	1
p-Cresol	201.8	No	.....	1
Benzyl formate	202.3	No	.....	1
Caproic acid	205.2	202.0	30	1
Benzylidene chloride	205.2	No	.....	1
Benzyl alcohol	205.5	204.3	40	1
Ethyl aniline	206.0	205.0	10	1
Camphor	208.9	No	.....	1
Nitrobenzene	210.8	No	.....	1
Borneol	211.8	211.3	15	1
Propyl oxalate	212.0	No	.....	1
Ethyl benzoate	213.0	No	.....	1

Table 18c. Hydrocarbon Azeotropes*—(*Continued*)

Aromatics—(*Continued*)

Name of entrainer	B.p. of entrainer, °C., 760 mm.	Azeotropic temp., °C., 760 mm.	Azeotropic composition, weight % H.C.†	Ref.
Naphthalene (218.1°C.):—(*Continued*)				
Benzyl acetate	215.5	215.3	28	1
Methyl 1-ester of terpineol	216.0	No	.....	1
Diethyl aniline (N)	216.5	213.0	.....	1
Ethyl succinate	216.5	210.0	43	1
p-Chlorophenol	217.0	215.9	36.5	1
Tribromohydrin	220.0	216.5	.....	1
Acetamide	221.2	199.5	72.8	2
Propionamide	222.1	204.7	68.5	2
o-Nitrotoluene	222.3	No	.....	1
Citral	226.0	No	.....	1
Bornyl acetate	227.7	No	.....	1
Ethyl phenyl acetate	228.8	No	.....	1
Geraniol	229.5	No	.....	1
Propyl benzoate	231.2	No	.....	1
Isoamyl carbonate	232.2	No	.....	1
Thymol	232.9	No	.....	1
Safrol	235.9	No	.....	1
Caprylic acid	237.5	216.2	94	1
Pyrocatechol	245.9	217.5	88.5	1
Benzoic acid	249.0	No	.....	1
Benzoic acid	250.5	217.7	95	2
Propyl succinate	250.5	No	.....	1
Phenylacetic acid	266.5	No	.....	1
Resorcinol	281.4	No	.....	1
Glycerol	291.0	215.2	90	24
α-Methyl Naphthalene (245.1°C.):				
Ethyl succinate	217.3	No		1
Acetamide	221.2	209.8	56.2	2
Propionamide	222.1	215.0	48	2
Bornyl acetate	227.7	No	.....	1
Ethyl phenyl acetate	228.8	No	.....	1
Isoamyl carbonate	232.2	No	.....	1
o-Phenetidine	232.5	No	.....	1
Thymol	232.9	No	.....	1
Caprylic acid	237.5	233.5	48	1
Isobutyl benzoate	241.9	No	.....	1
Pyrocatechol	245.9	234.8	60	1
Butyl benzoate	249.5	No	.....	1
Propyl succinate	250.5	No	.....	1
Benzoic acid	250.5	239.6	73	1
Eugenol methyl ether	254.7	No	.....	1
Methyl cinnamate	261.9	No	.....	1
Phenylacetic acid	266.5	243.2	88	2
Resorcinol	281.5	243.1	85.5	2
α-Naphthol	288.0	No	.....	1
Diphenyl (255.9°C.):				
Glycol	197.4	192.0	36	24
Acetamide	221.2	213.0	49.5	2
Thymol	232.9	No	.....	1
Pyrocatechol	245.9	239.9	45	1
Butyl benzoate	249.5	No	.....	1
p-Phenetidine	249.9	249.0	30	1
Benzoic acid	250.7	246.1	50	2
Diphenyl (255.9°C.):—(*Continued*)				
β-Isosafrol	252.0	No	.....	1
Phenyl acrolein	253.5	250.0	60	1
Eugenol methyl ether	254.7	254.5	.....	1
Eugenol	254.8	243.2	50	1
Methyl cinnamate	261.9	No	.....	1
Isoamyl benzoate	262.0	No	.....	1
Phenylacetic acid	266.5	252.2	76.7	24
Isoamyl oxalate	268.0	No	.....	1
Resorcinol	281.4	252.2	79	2
Methyl phthalate	283.7	No	.....	1
α-Naphthol	288.0	No	.....	1
β-Naphthol	290.0	No	.....	1
Glycerol	291.0	243.8	45	24
Diphenyl Methane (265.6°C.):				
Acetamide	221.2	215.5	43.5	2
Propionamide	222.1	219.2	40	2
Pyrocatechol	245.9	242.8	35	1
p-Phenetidine	249.9	No	.....	1
Benzoic acid	250.5	249.0	17	1
Phenyl acrolein	253.5	No	.....	1
Methyl cinnamate	261.9	No	.....	1
Isoamyl benzoate	262.0	No	.....	1
Phenylacetic acid	266.5	258.7	65	1
Isoamyl oxalate	268.0	265.2	85	1
Ethyl cinnamate	272.5	No	.....	1
Resorcinol	281.4	259.0	74	1
Methyl phthalate	283.7	No	.....	1
α-Naphthol	288.0	265.0	90	1
β-Naphthol	290.0	No	.....	1
Acenaphthene (277.9°C.):				
Acetamide	221.2	217.0	35.8	2
Pyrocatechol	245.9	245.3	16	2
Benzoic acid	250.5	250.0	.....	2
Methyl cinnamate	261.9	No	.....	1
Isoamyl benzoate	262.0	No	.....	1
Phenylacetic acid	266.5	262.6	29	1
Isoamyl oxalate	268.0	No	.....	1
Isoeugenol methyl ether	270.5	No	.....	1
Ethyl cinnamate	272.5	No	.....	1
Resorcinol	281.4	266.2	59	2
Methyl phthalate	283.7	276.4	66.5	1
α-Naphthol	288.0	274.0	80	1
Glycerol	290.0	259.1	71	2
β-Naphthol	290.0	277.0	90	1
Dibenzyl (284°C.):				
Acetamide	221.2	218.2	32	2
Pyrocatechol	245.9	No	.....	1
Benzoic acid	250.5	No	.....	1
Phenylacetic acid	266.5	264.3	10	1
Isoamyl oxalate	268.0	No	.....	1
Ethyl cinnamate	272.5	No	.....	1
Resorcinol	281.9	269.7	53	2
Methyl phthalate	283.7	280.5	47	1
β-Naphthol	290.0	283.5	.....	1

Binary Azeotropes Where Both Components Are Hydrocarbons

Name of hydrocarbon No. 1	B.p. hydrocarbon No. 1, °C., 760 mm.	Name of hydrocarbon No. 2	B.p. hydrocarbon No. 2, °C., 760 mm.	Azeotropic temp. °C., 760 mm.	Azeotropic composition, weight % H.C.† No. 1	Ref.
Paraffin-paraffin:						
Ethane	−88.3	n-Butane	0.6	No	....	1
Isopentane	28.0	n-Pentane	36.2	No	....	1
n-Hexane	69.0	n-Octane	125.8	No	....	4
Paraffin-olefin:						
Ethane	−88.3	Acetylene	−88.5	Min.	....	1
Isopentane	28.0	3-Methylbutene-1	22.5	20.0	....	1
Isopentane	28.0	Isoprene	34.1	27.0	....	1
Isopentane	28.0	2-Methylbutene-2	37.2	?	....	1
n-Pentane	36.2	Isoprene	34.1	33.8	10	1
n-Pentane	36.2	2-Methylbutene-2	37.2	35.5	57	1
Paraffin-naphthene:						
n-Hexane	69.0	Cyclohexane	80.8	No	....	1
n-Hexane	69.0	Methylcyclopentane	71.8	Min.	....	18
n-Heptane	98.5	Methylcyclohexane	101.8	No	....	19
Paraffin-aromatic:						
n-Hexane	69.0	Benzene	80.2	68.9	19	1
n-Hexane	69.0	Benzene	80.2	No	....	1
n-Octane	125.8	Ethylbenzene	136.2	No	....	1
Paraffin-cyclic Olefin:						
n-Hexane	69.0	Cyclohexene	82.8	No	....	1
2,7-Dimethyloctane	160.3	Camphene	158.0	157.0	....	1

Table 18c. Hydrocarbon Azeotropes*—(*Concluded*)

Binary Azeotropes Where Both Components Are Hydrocarbons—(*Concluded*)

Name of hydrocarbon No. 1	B.p. hydrocarbon No. 1, °C., 760 mm.	Name of hydrocarbon No. 2	B.p. hydrocarbon No. 2, °C., 760 mm.	Azeotropic temp. °C., 760 mm.	Azeotropic composition weight % H.C.† No. 1	Ref.
Naphthene-aromatic:						
Methyl cyclopentane	71.8	Benzene	80.2	71.4	90.6	20
Cyclohexane	80.8	Benzene	80.2	77.5	45	1
Cyclohexane	80.8	Toluene	110.7	No	....	1
Naphthene-cyclic Olefin:						
Cyclohexane	80.8	1,3-Cyclohexadiene	80.8	79.2	52	1
Cyclohexane	80.8	Cyclohexene	82.8	No	....	1
Olefin-olefin:						
Ethylene	−103.9	Acetylene	−88.5	Min.	....	23
Isoprene	34.1	2-Methylbutene-2	37.2	No	....	1
2-Methylbutene-2	37.2	3-Methyl-1,2-Butadiene	40.8	No	....	1
Olefin-aromatic:						
2-Methylbutene-2	37.2	Benzene	80.2	Min.	....	1
Cyclic olefin-cyclic olefin:						
1,3-Cyclohexadiene	80.8	Cyclohexene	82.8	No	....	1
Cyclic olefin-aromatic:						
1,3-Cyclohexadiene	80.8	Benzene	80.2	No	....	1
Cyclohexene	82.8	Benzene	80.2	79.5	15	1
β-Pinene	163.8	Mesitylene	164.0	162.7	52	1
Menthene	170.8	Pseudocumene	169.0	168.0	20	1
Carvene	177.8	Cymene	175.3	174.4	25	1
Aromatic-aromatic:						
Benzene	80.2	Toluene	110.7	No	....	4
Benzene	80.2	Mesitylene	164.6	No	....	1
Toluene	110.7	Ethylbenzene	136.1	No	....	4
Ethyl benzene	136.2	p-Xylene	138.2	No	....	1
p-Xylene	138.2	m-Xylene	139.0	No	....	1
o-Xylene	142.6	Styrene	145.7	No	....	1
Mesitylene	164.0	Pseudocumene	169.0	No	....	1
Naphthalene	218.1	sym-Triethylbenzene	216.0	215.0	35	1

References:
[1] Lecat, La Tension de vapeur des melanges de liquides, L'Azeotropisme, Bruxelles, 1918.
[2] Timmermans, "Les Solutions concentrees," Masson et Cie, Paris, 1936.
[3] "International Critical Tables," Constant-boiling Mixtures, vol. 3, 318–324.
[4] Robinson and Gi liland, "The Elements of Fractional Distillation," 3d ed., 230, McGraw-Hill, New York, 1939.
[5] Field, Azeotropic Distillation, U.S. Patent 2,302,608 (1942).
[6] Shicktanz, *Bur. Standards J. Research*, **18**, 129 (1937) RP 967.
[7] White and Rose, *Bur. Standards J. Research*, **17**, 943 (1936) RP 955.
[8] White and Rose, *Bur. Standards J. Research*, **21**, 151 (1938) RP 1122.
[9] Guinot, Separation of Olefin-paraffin Mixtures, U.S. Patent 2,316,860 (1943).
[10] Schmitkons, Olefin Production, U.S. Patent 2,294,696 (1942).
[11] Britton, Nutting, and Horsley, Butene-1 and Isobutylene Separation, U.S. Patent 2,207,608 (1940).
[12] Deanesly, Increasing Olefin Concentration, U.S. Patent 1,866,800 (1932).
[13] Frey, Matuzak, and Snow, Separation of Olefin-paraffin Mixtures, U.S. Patent 2,186,524 (1940).
[14] Deanesly, Distillation Process, U.S. Patent 2,290,636 (1942).
[15] Greenburg, Benzene Separation, U.S. Patent 2,313,536 (1943).
[16] Geckler and Fragen, Toluene Production, U.S. Patent 2,316,126 (1943).
[17] Greenburg, Xylene Recovery, U.S. Patent 2,313,537 (1943).
[18] Bruun and Hicks-Bruun, *Bur. Standards J. Research*, **5**, 933 (1930) RP 239.
[19] Hicks-Bruun and Bruun, *Bur. Standards J. Research*, **8**, 525 (1932) RP 432.
[20] Griswold and Ludwig, *Ind. Eng. Chem.*, **35**, 117 (1943).
[21] Natta, Separation of Ethylbenzene from Styrene, U.S. Patent 2,308,229 (1943).
[22] Tongberg and Johnston, Vapor-liquid Equilibria for n-Hexane-benzene Mixtures, *Ind. Eng. Chem.*, **25**, 733 (1933).
[23] Churchill, Collamore, and Katz, Phase Behavior of the Ethylene-acetylene System, *Oil Gas J.*, **41** (13), 33–34, 36–37 (1942).
[24] Lange, "Handbook of Chemistry," 1208, Handbook Publishers, Inc., Sandusky, Ohio, 1941.
[25] Fisher, Separation of Solvents, U.S. Patent 2,341,433 (to Eastman Kodak Co.), 1944.
[26] Heldman, *J. Am. Chem. Soc.*, **66** (4), 661 (1944).

port and the top of the column. Phenol, being considerably less volatile than either benzene or cyclohexane, flows down the column and out the bottom. Operating conditions are adjusted to maintain a high concentration of phenol on all plates below S so that the benzene-cyclohexane azeotrope will not form and the volatility of cyclohexane will be appreciably greater than that of benzene. The section between F and the reboiler functions to strip cyclohexane from benzene and phenol. The section between F and S serves to absorb benzene from the vapor to the liquid stream such that vapor which passes S is substantially all cyclohexane and phenol. The column section above S is refluxed at the top with purified cyclohexane and accomplishes the separation between cyclohexane and phenol. Distillate is substantially pure cyclohexane. Bottoms from the extractive column contain all the benzene and phenol and are fed to a solvent recovery column that produces purified benzene as distillate and recovered phenol as bottoms.

Extractive distillation may be employed to separate two components A and B which form an azeotrope, if a suitable separating agent can be introduced to change the activity coefficients γ_A and γ_B of components A and B sufficiently so that γ_A/γ_B will not equal the ratio of vapor pressures P_B/P_A at any point within the working temperature range of the column. If this can be accomplished, the azeotrope will not form, and a substantially complete separation between A and B may be performed.

Choice of Separating Agent. The solvent chosen for extractive distillation should be non-corrosive to the equipment, non-reactive with the feed components, thermally stable, readily available, inexpensive, and non-toxic. It should have a boiling point sufficiently higher than the feed components so that separation of the solvent and the bottoms product is easily accomplished, but preferably not so high that the sensible heat requirement of the solvent cycle becomes a large fraction of the total heat input.

In order that the number of plates and the amount of reflux required to separate components A and B shall be reasonable, the solvent should cause a selectively greater increase in the volatility of one component than in that of the other. Highly selective solvents can usually be found if components A and B are of unlike kind, but it would probably be impossible to choose a solvent suitable for separating some stereoisomers, *e.g.*, the isomers of cyclohexane which boil at substantially the same temperature.

In general, where components *A* and *B* are of unlike kind, the greater the deviation from ideality produced by the solvent, the greater will be the increase in volatility ratio of *A* to *B*, but the greater will be the tendency for the solvent to form an insoluble liquid phase with one of the components. As a safe general rule, the solvent and the operating conditions should be chosen to avoid formation of immiscible liquid phases between the solvent and either of components *A* or *B* in an extractive distillation column. To illustrate, consider that *A* and *B* are being fractionated with addition of a selective solvent to enhance the relative volatility so that *A* may be distilled overhead and *B* may be removed from the bottom of the column with the solvent. Successful operation

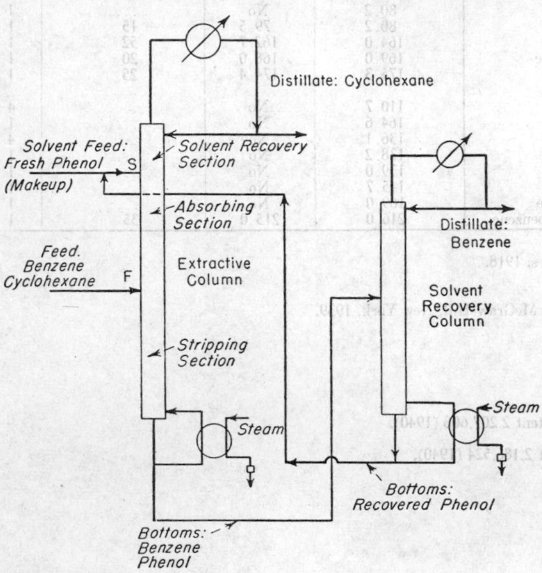

FIG. 74. Separation of benzene and cyclohexane by extractive distillation.

as an extractive distillation requires a high concentration of solvent (usually in the range 40 to 90 mole per cent solvent) in the liquid to increase the volatility of component *A*. However, if too selective a solvent is introduced so that immiscibility occurs, liquid in the column will be composed of an *A*-rich phase and a solvent-rich phase. Depending on solubilities, concentration of solvent in the *A*-rich phase will usually be insufficient to enhance the volatility of component *A* to the desired extent.

In plant columns separating C_4 hydrocarbons using furfural as the solvent, Happel *et al.* [*Trans. Am. Inst. Chem. Engrs.*, **42**, 189 (1946)] report that flooding occurs at relatively low rates of hydrocarbon feed if an immiscible liquid hydrocarbon phase is permitted to separate on trays below the hydrocarbon feed.

Table 19, based on the data of Souders and his associates [*Trans. Am. Inst. Chem. Engrs.*, **41**, 631 (1945)], illustrates the effect of various solvents on the relative volatility of a mixture of non-toluene and toluene contained in a narrow-boiling hydrocarbon cut. For convenience, volatility data were obtained on solutions containing 25 weight per cent toluene, 25 weight per cent non-toluene hydrocarbon, and 50 weight per cent solvent. Using the solvent-free concentrations in the liquid (x_1, x_2) and equilibrium vapor (y_1, y_2), relative volatilities were calculated as y_1x_2/x_1y_2. The "non-toluene" was a dearomatized straight-run saturated hydrocarbon fraction boiling between 210° and 235°F. ("true boiling point" 5 per cent and 95 per cent distillate points, respectively).

Table 19. Relative Volatility of Non-toluene to Toluene in Mixtures with Various Solvents

Solvent	B.p., °F	Weight % solvent	Relative volatility
Furfural	325	50	2.30
Nitrobenzene	412	50	2.16
Nitrotoluene	432	50	2.16
Phenol	360	50	2.10
Aniline	363	50	2.08
Acetophenone	396	50	1.95
Diethylene glycol mono-ethyl ether	396	50	1.85
Diacetone glycol	374	50	1.64

From the data, furfural appears the most selective, followed by nitrobenzene and nitrotoluene. However, furfural boils low enough to form azeotropes with the saturated hydrocarbons boiling near toluene, and therefore could not by simple fractionation be recovered for recirculating. Nitrobenzene, nitrotoluene, and aniline are toxic and of doubtful stability. Phenol is not so selective as furfural but is thermally stable, moderately inert, relatively high boiling, readily available, not expensive, and though toxic is not difficult to handle if proper care is practiced. Of the solvents listed, it would appear that phenol has the most desirable characteristics; and phenol is used in a commercial process for recovering toluene from petroleum.

Solvent Concentration. Effect of solvent concentration on the vapor-liquid equilibrium relations for mixtures of toluene and non-toluene is illustrated by Table 20, based on the data of Souders and his associates (*loc. cit.*).

Table 20. Effect of Solvent Concentration on Vapor-liquid Equilibrium

System: mixtures of toluene and non-toluene
Solvent: aniline or phenol
Pressure: 1 atm.

Weight % non-toluene on solvent-free basis

Liquid	Vapor			
	No solvent	25% solvent	50% solvent	75% solvent
10	13.5	17.0	20.4	23.9
20	25.8	30.7	35.7	41.2
30	36.9	42.4	47.9	
40	47.0	52.7	58.5	
50	56.3	62.0	67.5	75.2
60	65.0	70.2	75.4	
70	73.2	77.7	82.2	
80	81.7	85.1	88.5	92.2
85	86.0	88.6	91.6	
90	90.7	92.6	94.5	96.4
95	95.3	96.3	97.3	

Effect of the solvent increases approximately in proportion to its concentration in the liquid. In order to realize the maximum benefit from the solvent, it is usually added near the top of the column so that a high solvent concentration will be maintained in the liquid on all plates except the solvent-recovery section.

In a continuous column with a fixed feed rate of components *A* and *B*, and a fixed top reflux ratio, the solvent concentration in the liquid will be greater the greater the solvent feed rate. A high solvent feed rate will permit the separation to be accomplished with a column approaching the minimum number of plates, but the heat requirements become large as do the cross section of the extractive column and the size of the solvent stripping column. At the other extreme, a low solvent rate will require many more plates in the extractive column. Optimum design for each case will be dependent on economic balance considerations.

Since the separating agent employed for extractive distillation is usually of low volatility compared with the feed components, its concentration rapidly approaches nearly constant "pinch" values in the absorbing section and the stripping section. Thus in designing an extractive distillation column it is usually permissible to assume constant solvent concentrations in each of these two sections, and to use binary x-y data on a solvent-free basis for calculating fractionation of the feed components. These constant solvent concentrations may be calculated by the following equations:

For the absorbing section:

$$x_S = \frac{S}{(1 - \beta)L + \beta D/(1 - x_S)} \tag{100}$$

For the stripping section:

$$\bar{x}_S = \frac{S}{(1 - \beta)\bar{L} + \beta B/(1 - \bar{x}_S)} \tag{101}$$

where x_S, $\bar{x}_S$ = mole fraction solvent at the "pinch" composition in the absorbing section and stripping section, respectively.

β = average relative volatility of solvent to non-solvent.

S = moles pure solvent leaving tower per unit time.

L, $\bar{L}$ = moles total liquid per unit time (including solvent) flowing down the column in the absorbing and stripping sections, respectively.

D = moles distillate per unit time.

B = moles solvent-free bottoms per unit time.

It is noted that x_S and $\bar{x}_S$ appear also in the second term in the denominator, so that these solvent concentrations must be found by trial and error. However, the second term in the denominator is usually small and may become negligible for very non-volatile solvents when β becomes a small fraction of unity. Equations (100) and (101) are in the form used by Colburn (private communication) and are similar to equations derived by Benedict and Rubin [*Trans. Am. Inst. Chem. Engrs.*, **41**, 353 (1945)], who also demonstrate a graphical method for determining solvent concentration throughout the entire column.

In a typical extractive distillation column, the solvent concentration above the point of solvent introduction decreases very quickly to some low concentration and then decreases slowly until at the top of the column the concentration has been reduced to some practical value. Between the solvent nozzle and the feed plate the solvent concentration remains substantially constant at a value that may be estimated by Eq. (100). In this section the liquid on the plates consists of reflux and solvent. Solvent concentration in the stripping section remains practically constant to within a few plates from the bottom, at a value that may be estimated by Eq. (101). Near the reboiler, solvent concentration increases rapidly to a maximum value in the bottoms stream.

Referring to Eqs. (100) and (101), if the second terms in the denominators are negligible $x_S = \bar{x}_S$ if $L = \bar{L}$, a condition that is approximated for vapor feed at the dew point. For liquid feed F at the boiling point, $\bar{L} = (L + F)$, and $x_S > \bar{x}_S$. Thus, for vapor feed, solvent concentration remains nearly constant throughout most of the absorbing and stripping sections, and a single equilibrium curve based on data for this solvent concentration may be used to calculate fractionation of the feed components in both sections. For liquid feed, however, separate equilibrium curves should be calculated for the two sections based on the average solvent concentration in each section.

Novel Features of Extractive Distillation Columns. An important difference between simple fractionation and extractive distillation, besides the complicated solubility relationships existing between the solvent and feed components, is that a large portion of the heat requirement of an extractive distillation column is the sensible heat load of the solvent mixture flowing down the tower. Dropping the solvent temperature only a few degrees will result in a large internal condensation of vapors and will increase the reflux ratio and dilute the solvent. Such changes may be particularly critical if the column is operating close to an immiscible phase boundary and an undesirable phase separation takes place.

Another difference contrary to experience with simple fractionations is that, with given rates of feed and separating agent, an optimum reflux ratio exists. Increasing the non-solvent reflux beyond this optimum rate actually makes the separation more difficult because of dilution of solvent in the liquid. Souders *et al.* (*loc. cit.*) demonstrate, by working an example, that for each number of plates there is a particular reflux rate that requires the least accompanying solvent to accomplish the postulated separation.

These novel features about extractive distillation columns require consideration when automatic-control instruments are being selected. For example, if the solvent rate is large compared with the non-solvent reflux rate, it may be found impracticable to control top reflux by column temperature, in which case flow rate control of reflux may be preferable. Automatic control of the solvent feed rate and temperature is desirable because minor fluctuations in these values can upset steady-state operation of the extractive distillation column.

Design of an Extractive Distillation Column. Unusual features involved in designing an extractive distillation column result from the presence of the solvent stream and from the necessity of dealing with vapor-liquid equilibriums in a multicomponent non-ideal system. These conditions must be taken into account in estimating the number of perfect plates, the reflux ratio, and the solvent feed rate. Conventional design methods (see pp. 597 to 602 and 610 to 621) apply for establishing column diameter, plate construction (or choice of packing), plate efficiency (or H.T.U.), reboiler design, and condenser design. The number of perfect plates required for an extractive distillation at a given reflux ratio and solvent rate may be calculated with confidence by multicomponent plate-by-plate methods using heat and material balance equations, provided that enough data are available on vapor-liquid equilibriums, specific heats, heats of absorption, and latent heats of vaporization. Much shorter approximate methods are available, however, which result in designs of sufficient accuracy to be useful in a great many cases. Approximate designs may be suitable for order-of-magnitude cost estimates to compare the economics of alternative processes. The approximate methods are also useful for preliminary exploration of operating conditions prior to making one or two plate-by-plate calculations for final design.

The following illustrative example demonstrates an approximate method for estimating the number of perfect plates, the reflux ratio, and the solvent rate to separate isobutane and 1-butene by extractive distillation using anhydrous furfural as the solvent. Good binary and ternary data for this system have been published by Mertes and Colburn [*Ind. Eng. Chem.*, **39**, 787 (1947)] and by Gerster, Mertes, and Colburn [*ibid.*, **39**, 797 (1947)]. Although furfural containing approximately 4 to 6 weight per cent water is used commercially for this application [see Happel *et al.*, *Trans. Am. Inst. Chem. Engrs.*, **42**, 189 (1946)], the following example will be based on using dry furfural because of the better equilibrium data available. The principles of calculation will be demonstrated at least as well, since a large majority of extractive distillations use a single component for a separating agent. Water is added in the commercial processes supposedly to increase solvent selectivity, but also to reduce the maximum temperature in the reboiler and in the solvent-recovery equipment, thereby minimizing thermal decomposition of the furfural.

Conventional methods for designing fractionating columns apply for the solvent-recovery column, which is not discussed further here.

Heat and Material Balance Equations. On the basis of actual liquid compositions in the solvent–non-solvent solution, heat and material balance equations for an extractive distillation column are as follows (neglecting heat losses and assuming negligible non-solvent in the "solvent" feed):

For the absorbing section:

$$\frac{L_n}{V_{n+1}} = \left(\frac{y_{n+1} - rx_D}{x_n - rx_D}\right) = \left(\frac{H_{n+1} - rQ'_{SC}}{h_n - rQ'_{SC}}\right) \qquad (102)$$

For the stripping section:

$$\frac{L_{m+1}}{V_m} = \left(\frac{y_m - x_B}{x_{m+1} - x_B}\right) = \left(\frac{H_m - Q'_S}{h_{m+1} - Q'_S}\right) \qquad (103)$$

where L = liquid traffic including solvent, moles/hr.
V = vapor traffic, moles/hr.
y = mole fraction low boiler in the vapor.
x = mole fraction low boiler in the liquid, solvent included.
$r = D/(D - S)$.
D = distillate, moles/hr.
S = solvent flow, moles/hr.
H = molar heat content of vapor, B.t.u./mole.
h = molar heat content of liquid, B.t.u./mole.
$Q'_{SC} = h_D + Q_C/D - Sh_{SF}/D$.
Q_C = heat removed by condenser, B.t.u./hr.
$Q'_S = h_B - Q_S/B$.
B = bottoms stream, moles/hr.
Q_S = reboiler duty, B.t.u./hr.

Subscripts:

n refers to plate n in the absorbing section.
$n + 1$ refers to the next plate below n.
D refers to distillate.
m refers to plate m in the stripping section.
$m + 1$ refers to the next plate above m.
B refers to bottoms.
SF refers to solvent feed.

Equations (102) and (103) define true operating lines in the absorbing and stripping sections of an extractive distillation column. These equations may be used along with vapor-liquid equilibrium relationships to calculate the number of perfect plates required, by the plate-by-plate method.

Illustrative Example: Approximate Method for Number of Perfect Plates in an Extractive Distillation Column. It is desired to separate an equimolar mixture of isobutane and 1-butene into 99 mole per cent isobutane and 99 mole per cent butene by extractive distillation using anhydrous furfural as the separating agent. In the presence of furfural, the relative volatility of these components is approximately 2.0, compared with a normal relative volatility of 1.16 at 125°F.

Additional Data. The temperature of cooling water to be used is suitable for condensing isobutane at 95°F. (3400 mm. Hg vapor pressure). This will establish minimum operating temperature and pressure in the column. The hydrocarbon feed is to be 50 mole per cent vapor. The solvent feed is to be preheated to top column temperature.

Figures 75 and 76, from Mertes and Colburn (*loc. cit.*), give binary activity coefficient data for 1-butene in furfural and isobutane in furfural. Table 21 lists Margules binary constants. Data on heat of mixing and latent heat of vaporization are presented in Table 22. Table 23 tabulates vapor pressures. Colburn *et al.* also present gas-law corrections for isobutane, 1-butene, and furfural. Gas-law corrections will be neglected for this approximate solution. Vapor-liquid equilibrium data were measured in a continuous-flow system under isothermal conditions. This method has the advantage that variation of activity coefficients with composition may be determined at a series of constant temperatures, eliminating the simultaneous change of temperature with composition which occurs in the

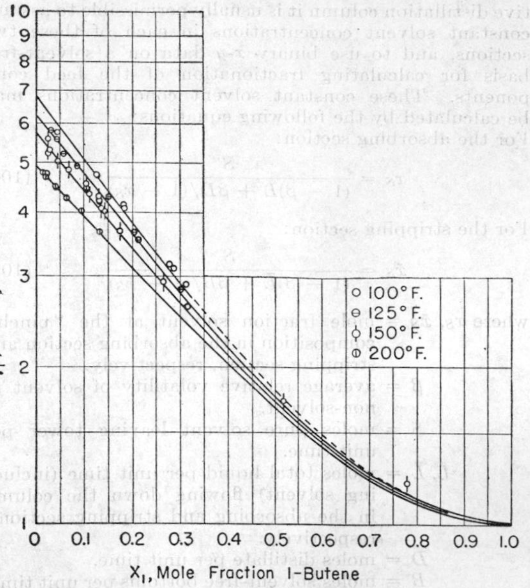

FIG. 75. Activity coefficients of 1-butene in furfural.

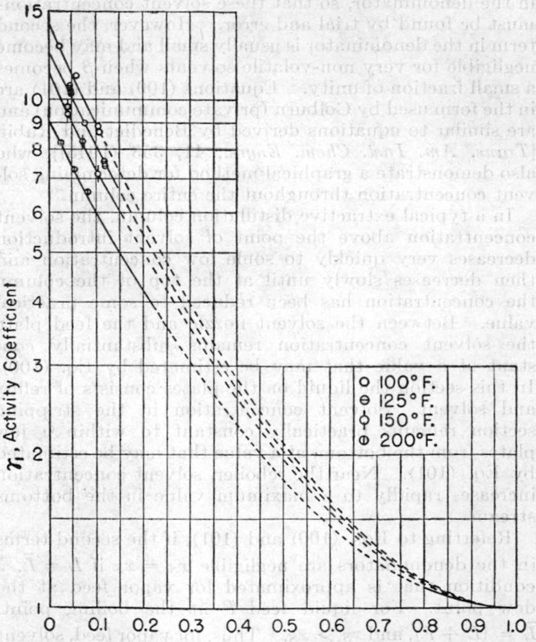

FIG. 76. Activity coefficients of isobutane in furfural.

constant-pressure equilibrium still. Measured vapor-liquid equilibrium data were correlated using the three-suffix binary and ternary Margules equations (see Sec. 8, p. 528). The binary and ternary constants were evaluated from equilibrium data, and the ternary equation was used to calculate the relative volatility of isobutane to 1-butene, as a function of temperature and ternary composition. Figure 77, taken from Colburn *et al.*, shows these relative volatilities. Parameter is mole per cent total isobutane plus 1-butene dissolved in furfural. Dashed, solid, and dotted lines are for 0, 50, and 100 mole per cent isobutane on a furfural-free basis. Lines were calculated by Gerster, Mertes, and Colburn [*Ind. Eng. Chem.*, **39**, 797 (1947)] using Margules ternary equations. The bottom dotted line (for 20

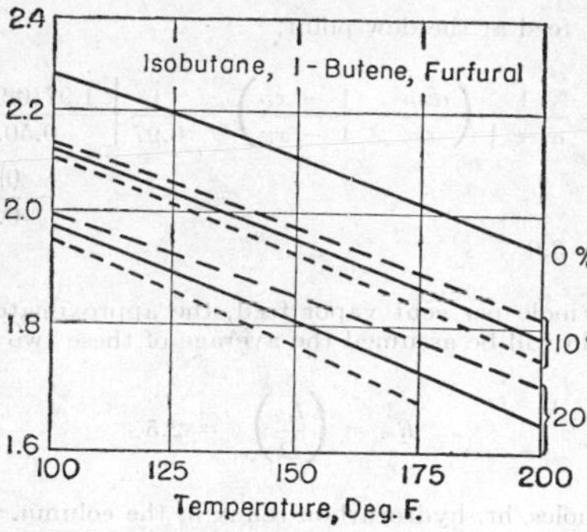

FIG. 77. Values of relative volatility, α, of isobutane to l-butene in furfural as a function of temperature.

mole per cent isobutane in furfural) is in error, being outside the solubility range for pure isobutane in furfural below approximately 180°F. (see Fig. 78).

Table 21. Margules Binary Constants for n-Butane, Isobutane, and 1-Butene with Furfural

Temp., °F.	n-Butane A_1	n-Butane A_2	Isobutane A_1	Isobutane A_2	1-Butene A_1	1-Butene A_2
100	1.096	1.257	1.142	1.310	0.842	1.029
125	1.045	1.171	1.090	1.231	.800	0.986
150	0.998	1.108	1.042	1.160	.763	.951
200	.908	0.975	0.955	1.030	.700	.900

Table 22. Heats Absorbed on Solution of Liquid n-Butane, Isobutane, and 1-Butene in Furfural and Latent Heats over Range 100° to 200°F.

	n-Butane	Isobutane	1-Butene
Compn., mole % C_4	Differential heats of solution, 100°–200°F., B.t.u./lb.-mole of C_4 dissolved		
0	3180	3180	2400
10	2820	2800	1920
20	2440	2430	1470
Compn., mole % C_4	Integral heats of solution, 100°–200°F., B.t.u./lb.-mole of C_4 dissolved		
0	3180	3180	2400
10	3000	2990	2160
20	2820	2800	1920
Temp., °F.	Latent heats of vaporization, B.t.u./lb.-mole		
100	8750	7890	
125	8310	7460	
150	7820	6940	
175	7290	6350	
200	6650		

Table 23. Vapor Pressures of Isobutane, 1-Butene, and Furfural at Various Temperatures*

Temp., °F.	Vapor pressures, mm. Hg Isobutane	1-Butene	Furfural
68	2,238	1,880	2.1
100	3,720	3,181	5.9
125	5,324	4,608	12.2
150	7,399	6,478	24.7
175	10,020	8,866	46.2
200	13,260	11,840	81.5
225	17,190	15,500	133
250	21,890	19,890	214

* Mertes and Colburn, *Ind. Eng. Chem.*, **39**, 787 (1947).

Although solubilities of isobutane and 1-butene in furfural were not listed in the publications referred to, binary constants in the Margules equation were presented. From these constants, solubilities have been calculated using Fig. 5 of Sec. 8, and are plotted in Fig. 78. Although Fig. 5 actually is based on the Van Laar equation, for systems that are nearly symmetrical (such as isobutane-furfural and 1-butene-furfural) the Van Laar and Margules constants are practically identical.

Figure 79 shows weight-balance data for the example, with the exception of furfural rate, which is to be determined.

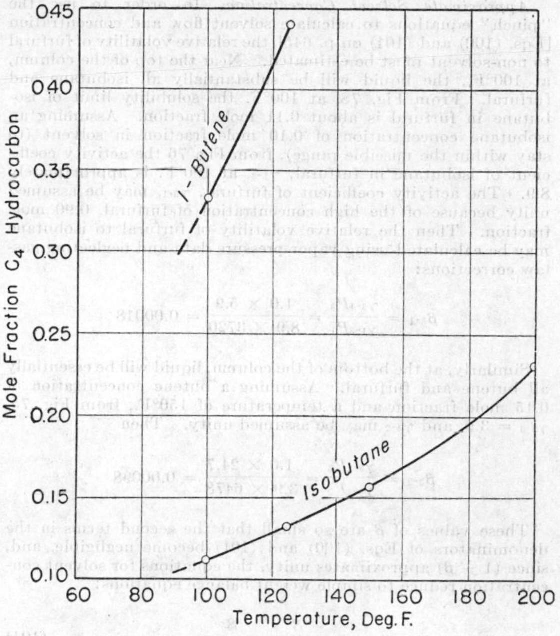

FIG. 78. Solubility of liquid hydrocarbons in furfural.

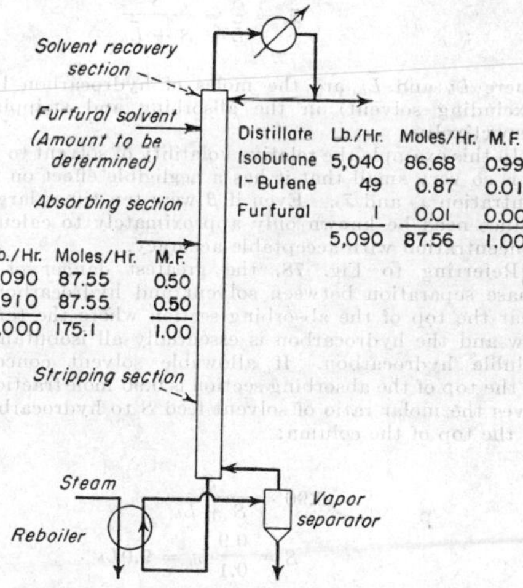

FIG. 79. Extractive distillation example. Separation of isobutane and 1-butene.

Solution. Top column pressure = 3400 mm. Hg abs., limited by cooling-water temperature.

Bottom column pressure = 3400 mm. + Δp, where Δp = pressure drop through the column. If the column is constructed of 100 bubble-cap plates with approximately 6 mm. pressure drop per plate, the approximate bottom column pressure = 4000 mm. Hg abs.

Top column temperature = 95°F.

Temperature will increase gradually down the column through the absorbing and stripping sections, as isobutane in the liquid is replaced by butene. Then, beginning several plates above the bottom, temperature will increase rapidly to a maximum value in the reboiler. Actual bottom temperature will vary with solvent concentration.

Approximate Solvent Concentration. In order to use the "pinch" equations to calculate solvent flow and concentration [Eqs. (100) and (101) on p. 645], the relative volatility of furfural to non-solvent must be estimated. Near the top of the column, at 100°F., the liquid will be substantially all isobutane and furfural. From Fig. 78, at 100°F. the solubility limit of isobutane in furfural is about 0.11 mole fraction. Assuming an isobutane concentration of 0.10 mole fraction in solvent (to stay within the miscible range), from Fig. 76 the activity coefficient of isobutane in furfural, $\gamma_{1\text{-}3}$, at 100°F. is approximately 8.9. The activity coefficient of furfural, $\gamma_{3\text{-}1}$, may be assumed unity because of the high concentration of furfural, 0.90 mole fraction. Then the relative volatility of furfural to isobutane may be calculated using vapor-pressure data and neglecting gas-law corrections:

$$\beta_{3\text{-}1} = \frac{\gamma_{3\text{-}1}P_3}{\gamma_{1\text{-}3}P_1} = \frac{1.0 \times 5.9}{8.9 \times 3720} = 0.00018$$

Similarly, at the bottom of the column, liquid will be essentially all butene and furfural. Assuming a butene concentration of 0.15 mole fraction and a temperature of 150°F., from Fig. 75, $\gamma_{2\text{-}3} = 3.9$, and $\gamma_{3\text{-}2}$ may be assumed unity. Then

$$\beta_{3\text{-}2} = \frac{\gamma_{3\text{-}2}P_3}{\gamma_{2\text{-}3}P_2} = \frac{1.0 \times 24.7}{3.9 \times 6478} = 0.00098$$

These values of β are so small that the second terms in the denominators of Eqs. (100) and (101) become negligible, and, since $(1 - \beta)$ approximates unity, the equations for solvent concentration reduce to simple weight balance equations:

$$x_s = \frac{S}{L} = \frac{S}{S + L_h} \tag{104}$$

$$\bar{x}_s = \frac{S}{\bar{L}} = \frac{S}{S + \bar{L}_h} \tag{105}$$

where L_h and $\bar{L}_h$ are the moles of hydrocarbon liquid traffic (excluding solvent) in the absorbing and stripping sections, respectively.

In this example the relative volatility of solvent to non-solvent, β, is so very small that it has a negligible effect on solvent concentration x_s and $\bar{x}_s$. Even if β were ten times larger, absolute values need be known only approximately to calculate solvent concentration with acceptable accuracy.

Referring to Fig. 78, the greatest danger of undesirable phase separation between solvent and hydrocarbon will occur near the top of the absorbing section where the temperature is low and the hydrocarbon is essentially all isobutane, the lesser soluble hydrocarbon. If allowable solvent concentration x_s in the top of the absorbing section is 0.90 mole fraction, Eq. (104) gives the molar ratio of solvent feed S to hydrocarbon reflux L_h at the top of the column:

$$0.90 = \frac{S}{S + L_h}$$

$$S = \frac{0.9}{0.1} L_h = 9.0 L_h \tag{106}$$

Note that L_h (moles/hr. hydrocarbon reflux in the absorbing section) increases down the column, since heat to raise the temperature of the large solvent flow must come from heat of absorption of the ascending hydrocarbon vapor in furfural solution. This is true even though solvent is preheated to the top column temperature.

Approximate Minimum Reflux Ratio. Assuming the "pinch" at minimum reflux occurs at a temperature intermediate between top and bottom column temperatures, say 120°F., and at equimolar isobutane-butene concentrations totaling approximately 0.15 mole fraction, the approximate relative volatility of isobutane to butene at the "pinch" is read from Fig. 77 as 1.97. The isobutane-butene system may be treated as a binary on a solvent-free basis to calculate minimum reflux using equations of Underwood [*Trans. Inst. Chem. Engrs.* (*London*), **10**, 112–52 (1932)]. For liquid feed at the boiling point,

$$R_m = \frac{1}{\alpha - 1}\left[\frac{x_D}{x_F} - \frac{\alpha(1 - x_D)}{1 - x_F}\right] = \frac{1}{0.97}\left[\frac{0.99}{0.50} - \frac{1.97(0.01)}{0.5}\right] = 2.0$$

For vapor feed at the dew point,

$$1 + R_m = \frac{1}{\alpha - 1}\left(\frac{\alpha x_D}{x_F} - \frac{1 - x_D}{1 - x_F}\right) = \frac{1}{0.97}\left[\frac{1.97(0.99)}{0.50} - \frac{0.01}{0.50}\right] = 4.0$$

$$R_m = 3.0$$

For 50 mole per cent vapor feed, the approximate minimum reflux ratio will be assumed the average of these two values, or

$$R_m = \left(\frac{L_h}{D}\right)_m = 2.5$$

L_h = moles/hr. hydrocarbon reflux in the column.
D = moles/hr. distillate.
Subscript m refers to "minimum."
Subscript D refers to distillate.
Subscript F refers to feed.

Operating Reflux Ratio and Solvent Feed Rate. As a first approximation, assume the hydrocarbon reflux ratio L_h/D at the top of the column is 1.3 times the minimum:

$$\frac{L_h}{D} = 1.3 \times 2.5 = 3.25$$

$$\frac{L_h}{V} = \frac{3.25}{4.25} = 0.765$$

(Internal reflux lower down in the absorbing section will be calculated later by a heat balance.) Since D = 87.56 moles/hr., L_h at the top of the column is

$$L_h = 3.25\,(87.56) = 285 \text{ moles/hr.}$$

Then, from Eq. (106),

S = 9.0 (285) = 2565 moles/hr. solvent feed rate, or 246,000 lb./hr. furfural

Solvent Concentration and Relative Volatility Data. The quantity of hydrocarbon vapor absorbed in the column to raise the temperature of the solvent must be estimated. Vapor absorbed into solution dilutes the solvent, thereby reducing the relative volatility of isobutane to butene. However, absorption increases the solvent-free ratios L_h/V and $\bar{L}_h/\bar{V}$ so that the number of plates calculated on the basis of L_h/V at the top of the column will usually be conservative for design purposes.

The total moles of hydrocarbon vapor absorbed in the column must be known in order to calculate the moles of vapor leaving the reboiler. This vapor flow affects both reboiler heat-transfer area and bottom-column diameter.

Approximate liquid temperature and solvent concentration x_S will be estimated for the plate above the hydrocarbon feed (designated plate n), and for the fifth plate above the bottom of the column. The trial-and-error procedure is as follows:

1. Assume a boiling temperature for liquid leaving the plate.
2. Calculate sensible heat gained by the solvent from solvent feed temperature to plate temperature.
3. Divide this heat quantity by the integral molar heat of absorption of hydrocarbon vapor to calculate moles of hydrocarbon absorbed. (Note that heat of absorption may be considerably different from latent heat of condensation unless heat of mixing of liquid hydrocarbon and furfural is zero.)
4. Add moles of hydrocarbon absorbed to moles of top column hydrocarbon reflux to calculate L_h, total moles of hydrocarbon liquid flowing across the plate.
5. Calculate solvent concentration from Eq. (104) or (105).
6. From the assumed temperature and the calculated liquid composition, calculate equilibrium partial pressures of isobutane, butene, and furfural. This requires vapor-pressure data and

activity coefficients for each component in solution. The sum of the three partial pressures is the calculated total pressure.

7. If the calculated total pressure does not equal the column operating pressure, the wrong initial temperature was assumed, and additional calculations may be made by the same procedure until the calculated pressure becomes a reasonable check of the operating pressure. The final approximation gives the solvent concentration.

Assumptions will be as follows:

1. Sensible heat to raise temperature of the hydrocarbon reflux is negligible compared with that required for the large flow of solvent.
2. Gas-law corrections are negligible.
3. Pressure at plate n equals 3700 mm. Hg (average of top and bottom pressure).
4. Isobutane concentration x_n on plate n is 0.50 mole fraction on a solvent-free basis.
5. Pressure at plate 5 is 4000 mm. Hg.
6. Butene concentration on plate 5 is substantially 1.0 mole fraction on a solvent-free basis.
7. Solvent feed enters at 100°F.

For solvent concentration on plate n, combining steps 2 through 5 mathematically gives the equation

$$x_{S_n} = \frac{1}{1 + (RD/S) + [(h_{St} - h_{SF})/\lambda_A]} \tag{107}$$

where x_{S_n} = mole fraction solvent on plate n.
R = top column reflux ratio, L/D.
D = moles distillate/hr.
S = moles solvent feed/hr.
h_{St}, h_{SF} = molar heat content of solvent at the temperature of plate n, and at the solvent feed temperature, respectively, B.t.u./(lb.-mole) above some base temperature.
λ_A = heat of absorption of hydrocarbon in furfural solution.

For this example,

$$R = 3.25$$
$$D = 87.56$$
$$S = 2565$$

At an assumed plate n temperature of 111°F.,

$(h_{St} - h_{SF}) = 96.1 \times 0.415 \times (111 - 100) = 438$
96.1 = molecular weight of furfural.
0.415 = specific heat of furfural [from Buell and Boatright, *Ind. Eng. Chem.*, **39**, 695 (1947)].
$\lambda_A = 7700 - 2400 = 5300$ B.t.u./mole.
7700 = latent heat of condensation of C_4 hydrocarbon at 111°F.
2400 = integral heat absorbed per mole of C_4 hydrocarbon on mixing equimolar isobutane and butene with furfural to produce a solution containing approximately 20 mole per cent hydrocarbon.

$$x_{S_n} = \frac{1}{1 + \frac{3.25(87.56)}{2565} + \frac{438}{5300}}$$

$$x_{S_n} = \frac{1}{1 + 0.111 + 0.083} = 0.836$$

Calculating total pressure at plate n by step 6, assuming equal mole fractions of isobutane and butene:

	x	Activity coefficient γ	Vapor pressure P, mm.	Partial pressure γPx
Isobutane	0.082	6.7*	4360	2400
Butene	0.082	4.1*	3700	1244
Furfural	0.836	1.1†	8.4	8
	1.000			3652

* Values from Figs. 75 and 76 read for $x = 0.164$ in each instance.

† Calculated by assuming mixture to be a binary with a hydrocarbon mole fraction of 0.164 and applying appropriate binary Margules equation (see Sec. 8, p. 527). The Margules constants from Table 21 for the systems isobutane-furfural and 1-butene-furfural were used to estimate the activity coefficient for furfural. An average of the two values determined was used.

The calculated total pressure of 3652 mm. Hg is reasonably close to the average column pressure of 3700 mm. Repeating the calculation for a slightly higher temperature would improve the agreement but would not change the value of x_{S_n} significantly. The accuracy of activity coefficient data and assumptions does not justify further calculation.

The estimated temperature and composition for plate n (plate above the hydrocarbon feed) are

t = 111°F.

	x
Isobutane	0.082
Butene	0.082
Furfural	0.836
	1.000

From Fig. 77, the relative volatility of isobutane to butene is

$$\alpha_{1\text{-}2} = 2.0$$

For solvent concentration on plate 5 (in the stripping section). Eq. (107) becomes

$$\bar{x}_{St} = \frac{1}{1 + (RD/S) + (qF/S) + [(h_{St} - h_{SF})/\lambda_A]} \tag{108}$$

where q = mole per cent of the feed that is liquid.
F = moles hydrocarbon feed per hour.

Note that, if the $(1 - \beta)$ term in Eqs. (100) and (101) is not substantially unity, the right-hand sides of Eqs (107) and (108) should be multiplied by $1/(1 - \beta)$.

For this example,

$$q = 0.50$$
$$F = 175.1$$

At an assumed plate 5 temperature of 130°F.,

$(h_{St} - h_{SF}) = 96.1 \times 0.415 \times (130 - 100) = 1200$
$\lambda_A = (7360 - 1800) = 5560$ B.t.u./mole

$$\bar{x}_{S5} = \frac{1}{1 + 0.111 + [0.50(175.1)/2565] + (1200/5560)}$$

$$\bar{x}_{S5} = \frac{1}{1 + 0.111 + 0.034 + 0.216} = 0.734$$

Calculating total pressure at plate 5 by step 6, assuming all hydrocarbon present is butene:

	x	Activity coefficient γ	Vapor pressure P, mm	Partial pressure γPx
Butene	0.266	3.05	4900	3970
Furfural	0.734	1.2	14	12
	1.000			3982

This calculated total pressure of 3982 mm. Hg is close to the approximate bottom column pressure of 4000 mm.; the assumed liquid temperature of 130°F. on plate 5 should therefore be a reasonably close approximation.

From Fig. 77, the relative volatility of isobutane to butene (for substantially zero mole fraction isobutane, 0.266 mole fraction butene, and 0.734 mole fraction furfural) is

$$\alpha_{1\text{-}2} = 1.83$$

Solvent-concentration and relative-volatility data may be summarized as follows:

Location	x_S	t, °F.	$\alpha_{1\text{-}2}$
Top plate in absorbing section	0.90	95–100	2.1
Plate above hydrocarbon feed	.836	111	2.0
Plate 5 from bottom	.734	130	1.83

Vapor-liquid Equilibrium Data. For the approximate determination of the number of perfect plates in the absorbing and stripping sections, it should be suitable to use a constant relative volatility in the absorbing section of 2.05, and another constant relative volatility in the stripping section of 1.85. Then, x-y data on a solvent-free basis may be calculated from the relative volatility equation

$$y^* = \frac{\alpha x}{1 + (\alpha - 1)x} \tag{109}$$

x	α	αx	$(\alpha - 1)x$	y^*
		Stripping section		
0	1.85	0	0	0
0.05	....	0.0925	0.0425	0.0887
.1	....	.185	.085	.1705
.2	....	.370	.170	.316
.3	....	.555	.255	.442
.4	....	.740	.340	.552
.5	....	.925	.425	.649
		Absorbing section		
0.4	2.05	0.820	0.420	0.577
.5	....	1.025	.525	.671
.6	....	1.230	.630	.754
.7	....	1.435	.735	.827
.8	....	1.640	.840	.890
.9	....	1.845	.945	.948
.95	....	1.947	.997	.975
1.00	....	2.050	1.050	1.000

These data were used to construct the equilibrium curve on Fig. 80. The method of using average relative volatility values for each of the two column sections results in a discontinuous equilibrium line, the discontinuity occurring above the hydrocarbon feed plate. Such a break actually does occur because liquid in the hydrocarbon feed dilutes the furfural solvent suddenly at the feed plate. The approximate method of calculation accentuates the break, however.

From Fig. 80, this line crosses the 45-deg. q-line (for 50 mole per cent vapor feed) at $x = 0.433$, $y = 0.564$. The slope of the stripping line is

$$\frac{\bar{L}_h}{\bar{V}} = \frac{0.564 - 0.010}{0.433 - 0.010} = \frac{0.554}{0.423} = 1.310$$

$$\bar{y} = 1.310\bar{x} - b$$
$$b = 0.010(0.310) = 0.0031$$
$$\bar{y} = 1.310\bar{x} - 0.0031$$

Number of Perfect Plates. By construction on Fig. 80, approximately 11 perfect plates are required in the absorbing section, and 17 perfect plates in the stripping section—based on a top column reflux ratio of 3.25. Since the hydrocarbon feed plate is the top plate in the stripping section, the change from the $\alpha = 2.05$ equilibrium curve to the $\alpha = 1.85$ equilibrium curve is made with the plate that crosses the intersection of the operating lines.

Based on an over-all plate efficiency of 25 per cent for extractive distillation of C_4 hydrocarbons with furfural (Happel, *loc. cit.*), the absorption section requires approximately 44 plates and the stripping section 68, or a total of 112 plates. No additional safety factor should be required, because the approximate calculation neglects increased hydrocarbon reflux in the column caused by absorption of vapor to heat the solvent.

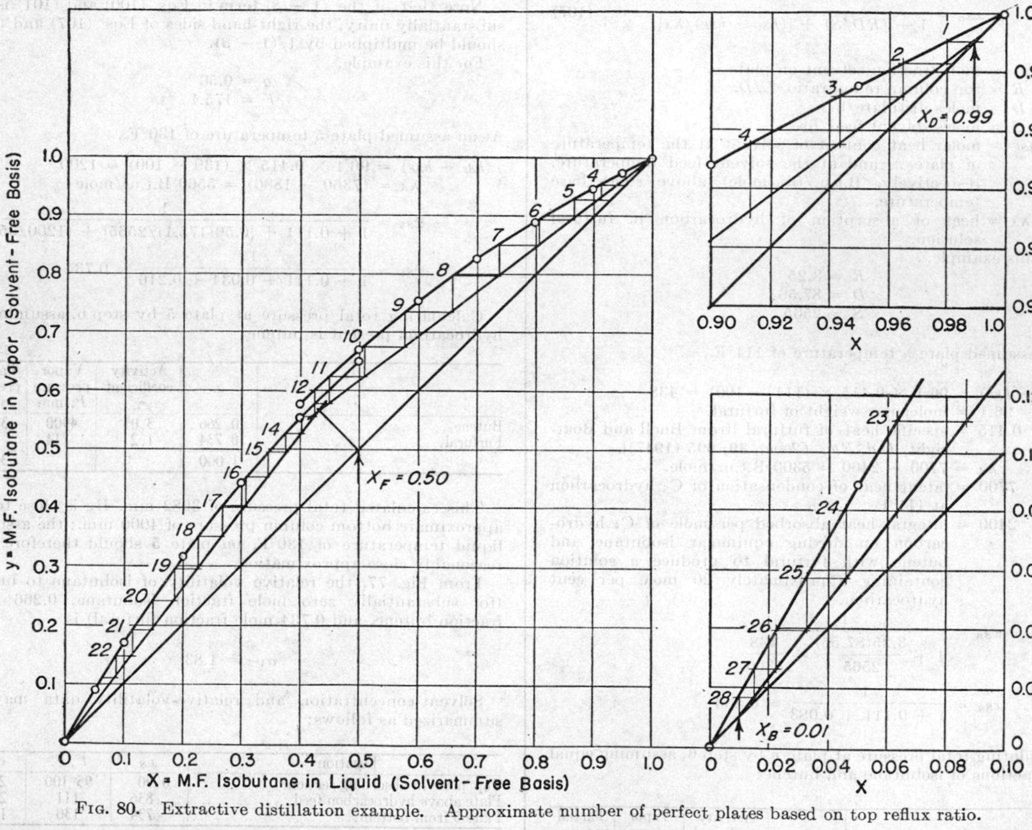

FIG. 80. Extractive distillation example. Approximate number of perfect plates based on top reflux ratio.

Operating Lines. For the approximate solution, it will usually be conservative to base the operating line in the absorbing section on the top column reflux ratio 3.25. Then the equation of the operating line in the absorbing section (on a solvent-free basis) is

$$y = ax + b$$
$$a = \frac{3.25}{4.25} = 0.765$$
$$b = 0.990(1 - 0.765) = 0.233$$
$$y = 0.765x + 0.233$$

If a higher top column reflux ratio were used, with increased solvent feed rate to maintain approximately the same solvent concentration, fewer perfect plates would be required in the absorbing and stripping sections. However, the larger solvent flow would require a correspondingly larger vapor flow to the bottom of the stripping section, requiring a larger reboiler surface and a larger diameter for the stripping section. Likewise, solvent-recovery equipment would be larger. Economic balance considerations determine optimum top column reflux ratio, and solvent-recovery facilities must be included in such a calculation.

Bottoms Composition and Boiling Point. Knowing the solvent flow rate, bottoms composition is established. Boiling temperature may be estimated by trial and error until the temperature is determined at which the sum of the partial pressures equals the bottom column pressure, say 4100 mm. Hg. Try 306°F. as the first trial.

Bottoms	Moles/hr.	Mole fraction x	Activity coefficient γ at 306°F.	Vapor pressure P, mm	Partial pressure γPx
Isobutane	0.87	0.0003	(5.8)*	34,800	60
Butene	86.68	.0325	(3.4)*	31,800	3510
Furfural	2565.	.967	1.0	560	540
	2652.55	0.9998			4110

* Extrapolated from Figs. 75 and 76.

Thus bottoms temperature is approximately 306°F.

Internal Liquid and Vapor Traffics. On the plate above the hydrocarbon feed,

$$L_h = S\frac{(1-xs)}{xs} = \frac{2565(0.164)}{0.836}$$

$$= 505 \text{ moles/hr. hydrocarbon liquid}$$

$$V = L_h + D = 505 + 88 = 593 \text{ moles/hr. vapor}$$

$$\frac{L_h}{V} = 0.85$$

On tray 5 from the bottom of the column,

$$\bar{L}_{h_5} = \frac{2565(0.266)}{0.734} = 930 \text{ moles/hr. hydrocarbon liquid}$$

$$\bar{V}_5 = \bar{L}_{h_5} - B = 930 - 88 = 842 \text{ moles/hr. vapor}$$

$$\left[\frac{\bar{L}_h}{\bar{V}}\right]_5 = 1.10$$

By comparison with slopes of the operating lines used in Fig. 80 for the approximate calculation, the above values indicate that the number of plates determined by the approximate method was conservative.

Reboiler Heat Load. By heat balance around the stripping section,

$$Q_B = V_{n+1}H_{n+1} + Bh_B - L_nh_n - [qh_F + (1-q)H_F]F$$

where Q_B = reboiler heat load, B.t.u./hr.
V_{n+1} = vapor from feed plate, moles/hr.
H_{n+1} = heat content of vapor from feed plate, B.t.u./mole.
B = bottoms stream, moles/hr.
h_B = heat content of bottoms stream, B.t.u./mole.
L_n = liquid from plate above feed = $(S + L_{hn})$, moles/hr.
h_n = heat content of liquid from plate above feed, B.t.u./mole.
q = mole per cent of feed as liquid.
F = feed stream, moles/hr.
h_F = heat content of liquid in feed, B.t.u./mole.
H_F = heat content of vapor in feed, B.t.u./mole.

V_{n+1} = 593
H_{n+1} = 10,500
B = (2565 + 88) = 2653
h_B = 11,400
L_n = (2565 + 505) = 3070
h_n = 3200
q = 0.5
F = 175.1
h_F = 2300
H_F = 10,300
Q_B = 6,250,000 + 30,300,000 − 9,850,000 − 1,100,000
Q_B = 25,600,000 B.t.u./hr.

Solvent-recovery Section. The solvent-recovery section at the top of the extractive distillation column serves to reduce solvent concentration in the distillate to 0.0001 mole fraction. Solvent concentration in vapor leaving the solvent feed plate is calculated below, neglecting gas-law corrections.

	x	γ	P	p	y
Isobutane	0.10	8.9	3720	3310	0.9984
Furfural	.90	1.0	5.9	5.3	.0016
				3315.3	1.0000

In the solvent-recovery section, the relative volatility of isobutane to solvent at 100°F. is

$$\alpha_{1\text{-}3} = \frac{\gamma_1 P_1}{\gamma_3 P_3} = \frac{1.0 \times 3720}{20.4 \times 5.9} = 30.8$$

The activity coefficient for furfural was obtained from Table 21 by setting the Margules constant A_2 at 100°F. equal to the log γ_3 and solving for γ_3 (*i.e.*, log γ_3 = 1.310). (See Sec. 8, p. 527 for theory.) Employing the equation of Underwood (*loc. cit.*) for the number of plates in the enriching section at finite reflux (the equation applies for a relatively pure product),

$$\left(\frac{L}{V}\alpha\right)^{n-1}\left[\frac{1}{1-x_n} - \frac{(L/V)(\alpha-1)}{(L/V)\alpha - 1}\right] = \frac{1}{\alpha(1-x_D)} \quad (110)$$

$$(0.765 \times 30.8)^{n-1}\left[\frac{1}{0.0016} - \frac{0.765(29.8)}{0.765(30.8) - 1}\right] = \frac{1}{30.8(0.0001)}$$

$$n - 1 = \frac{\log(0.521)}{\log 23.6} = -0.21$$

$$n = 0.79 \text{ perfect plate required.}$$

Thus two to five plates, depending on plate efficiency, should serve to reduce furfural concentration from 0.0016 mole fraction to less than 0.0001 mole fraction in the distillate.

Azeotropic Distillation

As defined on p. 634, azeotropic distillation depends on the formation of one or more azeotropes between the feed components and the separating agent. The separating agent for an azeotropic distillation may also be called an "azeotrope former," a "solvent," or an "entrainer," though the latter term should not be interpreted to imply a mechanism of mechanical entrainment. As in extractive distillation, the separating agent is added either to aid in separating a closely boiling pair or to separate an azeotrope.

Ewell, Harrison, and Berg [*Ind. Eng. Chem.*, **36**, 871 (1944)] summarize the alternative methods of using azeotrope formers as follows:

1. To separate a closely boiling pair or a maximum azeotrope:
   a. The entrainer forms a binary minimum azeotrope with only one component.
   b. The entrainer forms binary minimum azeotropes with each component, but one minimum is sufficiently lower than the other.
   c. The entrainer forms a ternary minimum azeotrope that is sufficiently lower than any binary azeotropes. The ratio of the original components in the ternary must be different from their ratio before the entrainer was added.
2. To separate a minimum azeotrope:
   a. The entrainer forms a binary minimum azeotrope with one component that is sufficiently lower than the original minimum azeotrope.
   b. The entrainer forms a ternary minimum azeotrope that is sufficiently lower than any binary minimum azeotrope and in which the ratio of the original components is different from their ratio in the binary minimum azeotrope.

As an example of an azeotropic distillation, ethyl alcohol may be used to separate a mixture of paraffins and aromatics that have about the same volatility, so that they cannot be fractionated by ordinary distillation. Ethyl alcohol forms one azeotrope with the paraffins and another with the aromatics, the boiling points of the two azeotropes differing sufficiently that a fractionation into two constant-boiling mixtures may be accomplished. Ethanol may be washed from the two mixtures with water, yielding the purified paraffin and aromatic products.

An entrainer that forms a heterogeneous azeotrope with one of the original components to be separated may be more practical than one that is completely miscible with both the original components, because recovery of the azeotrope former requires less equipment. Employment of a heterogeneous azeotrope former may be illustrated by a process for dehydrating ethanol-water mixtures using benzene. This procedure is of the type listed under 2*b* above, *viz.*, the benzene forms a ternary minimum azeotrope (with ethanol and water) which boils at a lower temperature and which contains a higher ratio of water to ethanol than the ethanol-water binary azeotrope.

Referring to Fig. 81, 96 per cent alcohol is fed to column *A*. The required quantity of benzene may be admitted with the feed or the reflux. The ternary azeotrope is taken overhead in this column, and absolute alcohol is obtained as bottoms product. Enough entrainer should be added to the system to effect rapid dehydration but not in such excess that it contaminates the absolute alcohol with more than 0.05 per cent benzene by volume [Guinot and Clark, *Trans. Inst. Chem. Engrs. (London)*, **16**, 187 (1938)].

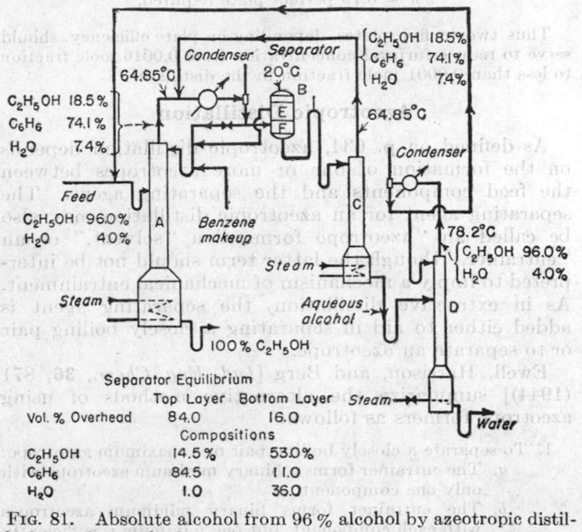

FIG. 81. Absolute alcohol from 96 % alcohol by azeotropic distillation. Atmospheric pressure operation.

Vapors of the ternary azeotrope are condensed and passed to decanter *B*, in which the mixture separates into two layers. The upper layer, rich in benzene, is returned to column *A* as reflux, and the lower layer is fed to column *C*, in which the benzene is recovered as the ternary azeotrope and returned to column *A*.

Aqueous alcohol from the base of column *C* is fed to column *D*, at the top of which 96 per cent alcohol is produced and returned to column *A* for dehydration. Practically pure water flows from the base of column *D*.

Since benzene is cycled continuously back to column *A*, the benzene serves indefinitely, any losses being made up by a small stream of fresh benzene added to the feed or reflux.

Azeotropic distillation is readily carried out batchwise by initially charging both feed and azeotrope former to the still, sufficient azeotrope former being added to maintain a portion in the still and column throughout the cycle. Laboratory batchwise distillations employing this principle do not require measurement of a continuous small flow of azeotrope former. Rossini and his associates at the National Bureau of Standards [*Oil Gas J.*, **158** (Nov. 14, 1940)] have prepared highly purified hydrocarbons from petroleum by distilling first without and then with an azeotrope former present in the still charge. Narrow-boiling fractions prepared by simple distillation were further separated into chemical constituents by azeotropic distillation.

Choice of Separating Agent. An ideal azeotrope former should be non-corrosive to the equipment, nonreactive with the feed components, thermally stable, readily obtainable, inexpensive, and non-toxic; and it should preferably have a low molar latent heat [Ewell, Harrison, and Berg, *loc. cit.*; White, *Natl. Petroleum News, Tech. Sec.*, **36**, R-731 (Nov. 1, 1944); Mair, Glasgow, and Rossini, *Bur. Standards J. Research*, **27**, 39 (1941)]. For separating close-boiling hydrocarbons, the following properties are desirable:

1. The azeotrope former should boil in the approximate range of 10° to 40°C. below the hydrocarbon mixture.
2. It should form, on mixing with the hydrocarbon, a large positive deviation from Raoult's law, to give a minimum azeotrope with one or more of the hydrocarbon types in the mixture.
3. It should be completely soluble in the hydrocarbon at temperatures existing in the column and reflux assembly.
4. It should be easily and cheaply separated from the hydrocarbons with which it forms azeotropes. The easiest separation occurs when the azeotrope former and hydrocarbon are immiscible at room temperature. When they are miscible, the azeotrope former is sometimes recovered by liquid-liquid extraction with water (or a suitable inexpensive solvent) from which the azeotrope former may be separated by distillation.
5. If the separating agent produces a minimum-boiling ternary mixture, it is desirable that the azeotrope former be completely miscible with one of the components to be separated and slightly miscible with the other. It should preferably not form any azeotrope with the miscible component, and the ternary mixture should separate into the phases over a large range of composition.

Final choice of a suitable separating agent for an azeotropic distillation requires some knowledge of vapor-liquid equilibrium data. However, the following generalizations may be useful in making preliminary selections. Nearly all polar organic molecules of the proper volatility form minimum constant-boiling mixtures with hydrocarbons, either paraffins, naphthenes, aromatics, or olefins. These polar compounds include those containing hydroxyl, carboxyl, cyanide, amine, nitro, and other groups. The relative lowering of the boiling point for the hydrocarbon–polar compound mixtures is greatest for paraffins and least for aromatics, the others being naphthenes, monoolefins, and diolefins in the order given.

The method of Ewell, Harrison, and Berg (*loc. cit.*) described in Sec. 8, p. 534, is useful in predicting qualitatively when azeotropes may be expected. Tables 18, 18*a*, 18*b*, and 18*c* on pp. 633 through 643 list data for a large number of azeotropes.

Solvent Concentration. In order to derive the maximum benefit from the separating agent, its concentration should be high (usually 50 to 80 mole per cent) on most of the plates in the column. Benedict and Rubin [*Trans. Am. Inst. Chem. Engrs.*, **41**, 353 (1945)] analyze the effect of introducing the separating agent in a continuous azeotropic distillation column (1) at the bottom of the column, (2) with the principal feed, and (3) at the top. They conclude that the optimum point for solvent addition will usually be at the top of the column. The bottom few plates in the column usually serve to reduce solvent concentration in the bottoms to a low value.

With solvent fed at the optimum point, control of solvent concentration on the plates in an azeotropic distillation column depends primarily on choosing an azeotrope former of suitable volatility. In practice, this limits the number of solvents suitable for any given azeotropic distillation—much more so than for extractive distillation where heat and material flows determine solvent concentration to a greater degree than do volatilities.

Design of an Azeotropic Distillation Column. Novel features involved in designing an azeotropic distillation column result from the necessity of dealing with vapor-liquid equilibriums in a non-ideal system. Once the choice of number of perfect plates, reflux ratio, and separating agent have been established, other design variables such as column diameter, plate construction (or choice of packing), plate efficiency (or H.T.U.), and heat loads are handled in conventional ways described on pp. 597 to 602 and 610 to 621.

The following illustrative example demonstrates the calculation of the number of theoretical plates for an azeotropic distillation using trichloroethylene as a separating agent to prepare anhydrous ethanol. The example employs an interpolation method for estimating ternary activity coefficients from binary activity coefficient data. Other procedures available for estimating ternary activity coefficients using various modifications and extensions of the Van Laar and Margules equations are discussed in Sec. 8, p. 528.

Illustrative Example: Calculation of the Number of Theoretical Plates Required for an Azeotropic Distillation Column. Anhydrous ethyl alcohol may be separated from water by distilling in the presence of trichloroethylene. The work of Colburn and Phillips [*Trans. Am. Inst. Chem. Engrs.*, **40**, 333 (1944)] illustrates the effectiveness of trichloroethylene as an "entrainer." The above authors obtained plate-to-plate compositions and temperatures in a 15-plate column from which activity coefficients were calculated and compared with activity coefficients predicted from known vapor-liquid equilibrium data for the binaries. Figure 82 gives the binary activity data, and Table 24

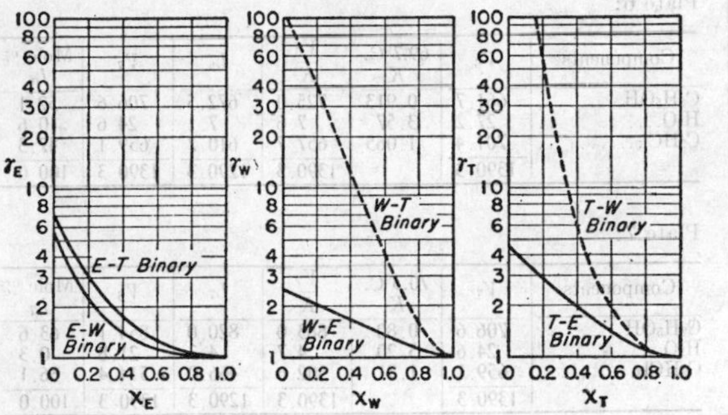

Fig. 82. Activity coefficients of the three components. Lines are indicated for the binary mixtures.

gives the results calculated by Colburn and Phillips for one of their runs.

From the experimental data of the above authors, it is possible to calculate either vapor-liquid equilibrium constants ($K = y/x$) or relative volatilities over the range of liquid compositions encountered in the experimental column and plot them as a function of temperature. The data shown in Fig. 83 make it possible to calculate plate by plate the number of theoretical plates required for a given separation.

Carlson and Colburn [*Ind. Eng. Chem.*, **34**, 581 (1942)] show how an approximate activity coefficient for any component in a ternary mixture may be predicted by interpolating (by logarithms) in terms of the mole fractions of the other two components present in the mixture. This is the method used by Colburn and Phillips in their work.

If experimental or calculated vapor-liquid equilibrium data are available for the binaries, curves similar to Fig. 82 may be constructed. In the absence of experimental ternary data, it is possible with the use of Fig. 82 and the method of Carlson and Colburn to estimate the ternary equilibrium data for ethyl alcohol, water, and trichloroethylene. The following example illustrates the calculation of one point on Fig. 83.

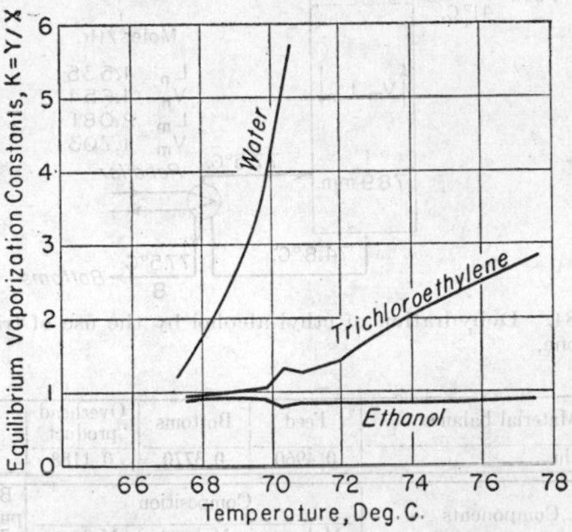

Fig. 83. Equilibrium constants for ethanol-water-trichloroethylene. Approximate total pressure, 760 mm.

For $x_E = 0.415$, $x_W = 0.070$, and $x_T = 0.515$ at $x_T = 0.515$ from Fig. 82, γ_T (water) = 2.77 and γ_T (ethanol) = 1.60.

$$\log \gamma_T = \frac{0.415 \log 1.60 + 0.070 \log 2.77}{0.415 + 0.070} = 0.238$$

$$\gamma_T = 1.73$$

Using the experimental value of γ_T of 1.69 and at a total pressure of 761 mm. Hg at 68.1°C., the vapor pressure of trichloroethylene is 420 mm. Hg.

$$y_T = \frac{x_T P_T \gamma_T}{P} = \frac{0.515 \times 420 \times 1.69}{761} = 0.481$$

$$K_T = \frac{y_T}{x_T} = \frac{0.481}{0.515} = 0.933$$

The following material balance and equilibrium data of Colburn and Phillips (run 5) will be used to illustrate a method for determining the number of theoretical plates required for a given separation in a non-ideal system (see Fig. 84 for flow sheet of the process).

Table 24. Calculated and Predicted Activity Coefficients for Run 5

Plate No.	Liquid mole fractions			Activity coefficients									Plate temp., °C.
	Alc.	H_2O	Tri.	Alc.			H_2O			Tri.			
				100%	70%	Pred.	100%	70%	Pred.	100%	70%	Pred.	
15	0.382	0.102	0.516	1.40	....	1.75	6.22	....	..	1.71	....	1.80	67.5
14	.415	.070	.515	1.43	1.38	1.65	6.76	7.2	16	1.69	1.69	1.73	68.1
13	.438	.046	.516	1.37	....	1.55	9.4	....	..	1.65	....	1.68	68.8
12	.462	.021	.517	1.38	1.36	1.5	12.9	14.0	17	1.64	1.64	1.65	69.2
11	.481	.019	.500	1.37	....	1.48	10.6	....	..	1.68	....	1.65	69.6
10	.512	.017	.471	1.32	1.29	1.4	11.4	11.5	14	1.65	1.70	1.70	69.8
9 *F*	.591	.014	.395	1.23	....	1.26	13.5	....	..	1.92	....	1.97	69.8
8	.634	.010	.356	1.14	1.09	1.20	5.5	6.2	8	2.24	2.34	2.12	70.3
7	.648	.005	.347	1.21	....	1.19	7.8	....	..	2.04	....	2.15	70.8
6	.666	.....	.334	1.18	1.16	1.17	.....	....	..	2.06	2.10	2.20	71.0
5	.698	.....	.302	1.17	....	1.13	.....	....	..	2.16	....	2.32	71.4
4	.733	.....	.267	1.18	1.12	1.10	.....	....	..	2.19	2.30	2.5	71.7
3	.786	.....	.214	1.15	....	1.06	.....	....	..	2.38	....	2.8	72.4
2	.846	.....	.154	1.13	1.08	1.04	.....	....	..	2.58	2.90	3.15	73.2
1	.900	.....	.100	1.09	....	1.02	.....	....	..	3.26	....	3.6	74.8
Still	.955	.....	.045	0.996	....	....	.....	....	..	3.41	....	....	77.5

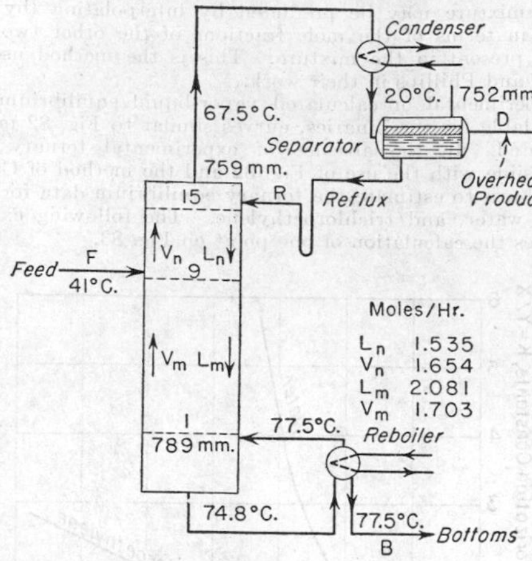

FIG. 84. Dehydration of ethyl alcohol by the use of trichloroethylene.

Material balance	Feed	Bottoms	Overhead product
Moles/hr	0.4960	0.3770	0.1188

Components	Composition			B.p., °C., pure comp. 760 mm.
	Mole %	Mole %	Mole %	
C_2H_5OH	80.2	95.6	36.7	78.4
H_2O	12.6	....	57.2	100.0
C_2HCl_3	7.2	4.4	6.1	87.2

Separator equilibrium at 20°C.	Total overhead	Water layer	Reflux
Moles/hr	1.4510	0.1188	1.3320

Components	Composition		
	Mole %	Mole %	Mole %
C_2H_5OH	34.1	36.7	33.9
H_2O	17.5	57.2	14.0
C_2HCl_3	48.4	6.1	52.1

The indicated overhead vapors are, according to Colburn and Phillips, fairly close to the composition of the ternary azeotrope.

Composition of binary azeotropes:

Azeotropes	Mole %			B.p., 760 mm., °C.
	C_2H_5OH	H_2O	C_2HCl_3	
$C_2H_5OH—H_2O$	89.4	10.6	....	78.15
$C_2H_5OH—C_2HCl_3$	52.5	....	47.5	70.8
$C_2HCl_3—H_2O$	....	35.0	65.0	73.0

Plate calculations:

$$\text{Rectifying section } \frac{Ln}{D} = 12.9$$

Plate 1:

Components	Mole % V_1	67.7°C. K	$\frac{V_1}{K}$	L_1	V_2	Mole % L_1
C_2H_5OH	34.1	0.910	37.5	484.3	518.4	37.5
H_2O	17.5	1.55	11.3	146.0	163.5	11.3
C_2HCl_3	48.4	0.945	51.2	660.0	708.4	51.2
	100.0		100.0	1290.3	1390.3	100.0

Plate 2:

Components	V_2	68.0°C. K	$\frac{V_2}{K}$	L_2	V_3	Mole % L_2
C_2H_5OH	518.4	0.930	558.0	518.0	552.1	40.2
H_2O	163.5	1.80	90.8	84.3	101.8	6.5
C_2HCl_3	708.4	0.955	741.5	688.0	736.4	53.3
	1390.3		1390.3	1290.3	1390.3	100.0

Plate 3:

Components	V_3	68.8°C. K	$\frac{V_3}{K}$	L_3	V_4	Mole % L_3
C_2H_5OH	552.1	0.935	591.0	548.0	582.1	42.5
H_2O	101.8	2.46	41.3	38.4	55.9	3.0
C_2HCl_3	736.4	0.970	758.0	703.9	752.3	54.5
	1390.3		1390.3	1290.3	1390.3	100.0

Plate 4:

Components	V_4	69.2°C. K	$\frac{V_4}{K}$	L_4	V_5	Mole % L_4
C_2H_5OH	582.1	0.931	625.7	581.1	615.2	45.0
H_2O	55.9	2.86	19.6	18.2	35.7	1.4
C_2HCl_3	752.3	1.01	745.0	691.0	739.4	53.6
	1390.3		1390.3	1290.3	1390.3	100.0

Plate 5:

Components	V_5	69.6°C. K	$\frac{V_5}{K}$	L_5	V_6	Mole % L_5
C_2H_5OH	615.2	0.911	676.0	627.6	661.7	48.6
H_2O	35.7	3.44	10.4	9.7	27.2	0.8
C_2HCl_3	739.4	1.05	703.9	653.0	701.4	50.6
	1390.3		1390.3	1290.3	1390.3	100.0

Plate 6:

Components	V_6	69.7°C. K	$\frac{V_6}{K}$	L_6	V_7	Mole % L_6
C_2H_5OH	661.7	0.913	725.0	672.5	706.6	52.1
H_2O	27.2	3.57	7.6	7.1	24.6	0.6
C_2HCl_3	701.4	1.065	657.7	610.7	659.1	47.3
	1390.3		1390.3	1290.3	1390.3	100.0

Plate 7:

Components	V_7	70.3°C. K	$\frac{V_7}{K}$	L_7	V_8	Mole % L_7
C_2H_5OH	706.6	0.80	883.0	820.0	854.1	63.6
H_2O	24.6	5.20	4.7	4.3	21.8	0.3
C_2HCl_3	659.1	1.31	502.6	466.0	514.4	36.1
	1390.3		1390.3	1290.3	1390.3	100.0

$$\text{Stripping section } \frac{Vm}{B} = 4.51$$

Still:

Components	Mole % B	77.5°C. K	KB	V_s	L_1	Mole % L_1
C_2H_5OH	95.5	0.91	87.2	393.4	488.9	88.7
C_2HCl_3	4.5	2.85	12.8	57.6	62.1	11.3
	100.0		100.0	451.0	551.0	100.0

Plate 1:

Components	L_1	74.1°C. K	KL_1	V_1	L_2	Mole % L_2
C_2H_5OH	488.9	0.86	420.5	344.0	439.5	79.8
C_2HCl_3	62.1	2.10	130.5	107.0	111.5	20.2
	551.0		551.0	451.0	551.0	100.0

Plate 2:

Components	L_2	72.5°C. K	KL_2	V_2	L_3	Mole % L_3
C_2H_5OH	439.5	0.84	368.5	301.7	397.2	72.1
C_2HCl_3	111.5	1.64	182.5	149.3	153.8	27.9
	551.0		551.0	451.0	551.0	100.0

Plate 3:

Components	L_3	72.0°C. K	KL_3	V_3	L_4	Mole % L_4
C_2H_5OH	397.2	0.84	334.0	273.5	369.0	67.0
C_2HCl_3	153.8	1.41	217.0	177.5	182.0	33.0
	551.0		551.0	451.0	551.0	100.0

Plate 4:

Components	L_4	71.2°C. K	KL_4	V_4	L_5	Mole % L_5
C_2H_5OH	369.0	0.84	310.0	254.0	349.5	63.4
C_2HCl_3	182.0	1.32	241.0	197.0	201.5	36.6
	551.0		551.0	451.0	551.0	100.0

Plate 5:

Components	L_5	70.4°C. K	KL_5	V_5	L_6	Mole % L_6
C_2H_5OH	349.5	0.828	289.0	236.5	332.0	60.2
C_2HCl_3	201.5	1.30	262.0	214.5	219.0	39.8
	551.0		551.0	451.0	551.0	100.0

It will be noted that the ratio of C_2H_5OH to C_2HCl_3 in the liquid on plate 5 in the stripping section is approximately the same as the ratio of C_2H_5OH to C_2HCl_3 in the liquid leaving plate 7 in the rectifying section with the exception that plate 5 liquid does not show any water present. Recalculate plate 4 in the stripping section by assuming that plate 4 liquid contains 0.1 mole per cent water in the liquid.

Recalculation of stripping section:

Plate 4:

Components	Mole % L_4	L_4	70.5°C. K	KL_4	V_4	L_5	Mole % L_5
C_2H_5OH	66.9	368.4	0.84	311.0	254.5	350.0	63.5
H_2O	0.1	0.6	5.70	3.4	2.8	2.8	0.5
C_2HCl_3	33.0	182.0	1.30	236.6	193.7	198.2	36.0
	100.0	551.0		551.0	451.0	551.0	100.0

Plate 5 liquid in the stripping section is now much closer to the analysis of plate 7 liquid in the rectifying section and must be a common plate, *i.e.*, the intersection of the stripping and rectifying sections or the feed plate. Therefore 11 theoretical plates and a still would be required for the given separation.

It is obvious from Table 24 that, by adding more plates to the stripping section, essentially pure ethyl alcohol could be produced.

Comparison of Extractive and Azeotropic Distillations

Extractive and azeotropic distillations are similar in that a solvent or separating agent is used to form non-ideal solutions with one or both of the feed components. The effect of the separating agent is to exaggerate the difference in volatility between the feed components. The two forms of distillation differ in the method used to maintain the required solvent concentration on the plates of the column. In extractive distillation, the required solvent concentration is maintained by introducing the solvent at a point near the top of the column, and because of its non-volatility it is found in high concentrations on all the plates. Azeotropic distillation may be compared with an extractive distillation where the solvent volatility is sufficiently high so that an azeotrope is formed with one or both of the feed components. In extractive distillation, the solvent is taken from the bottom of the main column. In azeotropic distillation most of the solvent is usually taken overhead in the main column, only a small fraction being removed with the bottoms.

Azeotropes involving the solvent and feed components must not occur in the absorbing section of the extractive column, although in some cases azeotropes may be present above the solvent inlet nozzle where the solvent concentration is low. In extractive distillation the volatility of the solvent must not match the volatility of the feed components, whereas for azeotropic distillation the solvent, in order to be effective, must boil within about 10° to 40°C. of some of the feed components so that an azeotrope can form. Consequently, a larger number of suitable separating agents can usually be found to effect a given separation by extractive distillation than by azeotropic distillation.

For a given separation by extractive distillation, a wide variety of towers may be designed because the solvent concentration in the tower may be controlled by heat and material inflow. Design of an azeotropic distillation column, on the other hand, is controlled by the composition of the azeotrope formed with the solvent.

The heat requirements in azeotropic distillation are usually greater than those in extractive distillation, since most of the solvent is vaporized and taken overhead. When the quantity of overhead is large and the ratio of solvent to overhead product in the azeotrope is high, this heat requirement becomes an important consideration in choosing between the two processes. On the other hand, where the overhead is a small fraction of the feed, the heat required to volatilize solvent may become insignificant compared with the total heat load; and, in such cases, azeotropic distillation may be the preferred process even for large-scale continuous operations.

Extractive distillation is not so readily adapted to batch distillation, because the solvent must be charged continuously to the top of the column. Consequently, azeotropic distillation is usually preferred for batch operation, either on a plant scale or in the laboratory.

MOLECULAR DISTILLATION*

By K. C. D. Hickman

Historical. The need for a better still for organic chemicals became evident to chemists in the period 1910 to 1925, when substances of ever-increasing molecular weight and boiling point were coming under study. The first molecular still appears to have been designed by Brönsted and Hevesy to separate the isotopes of mercury. Early high-vacuum stills for labile substances were constructed by H. I. Waterman, and the truly molecular still was first applied to organic substances by C. R. Burch. The history of these developments and those of Hickman, Washburn, and others is given in the bibliographies and reviews of Detwiler and Markely [*Oil & Soap*, **16**, 2 (1939)], Detwiler [*Oil & Soap*, **17**, 241 (1940); Abstracts of articles and patents on molecular or short-path distillation, U.S. Dept. Agr., Bur. Agr. Chemistry Eng. ACE-115, 98 pp., 1941; U.S. Regional Soybean Ind. Prod. Lab., Urbana, Ill.], Todd [*Oil & Soap*, **20**, 205 (1943)], and Hickman [*Chem. Rev.*, **34**, 51 (1944); *American Scientist*, **33**, 205 (1945)].

Definitions. The molecular still is the limiting type of high-vacuum still. Three broad classes of high-vacuum distillation can be discerned: (1) conventional or nearly conventional apparatus, generally consisting of a boiler, fractionating column, and condenser, operated under high vacuum; (2) unobstructed-path distillation operated under high or nearly high vacuum; and (3) molecular distillation where the vapor path is unobstructed and the condenser is separated from the evaporator by a distance less than the mean free path of the evaporating molecules. This article deals with categories 2 and 3. For this purpose a "high" vacuum is defined as a pressure of residual gas so low that further reduction does not change the performance of the apparatus. The high-vacuum range for the three classes is suggested very broadly in Table 25.

Table 25. High Vacuum Manometric Range

Class of still	Laboratory size, mm.	Industrial size, mm. Hg
1. Conventional stills	0.01–0.1	0.1 –5.0
2. Unobstructed-path stills	0.01–0.1	0.001 –0.01
3. Molecular stills	0.001	0.0001–0.001

The units of pressure adopted here are the millimeter of mercury and fractions thereof and the micron, which

* Communication 79 from the Laboratories of Distillation Products, Inc.

is 0.001 mm. Hg. The micron is thus $1\frac{1}{3}$ millionths of an atmosphere. Molecular and short-path distillations are generally conducted at pressures of 1 to 7 μ, rather less than 1 μ being readily available in the laboratory, whereas about 3 μ have proved the economical level in industry, where further reductions of pressure cost more than the benefits that accrue. It has become current practice to use micron terminology for relatively high pressures, *e.g.*, 500 μ instead of 0.5 mm. The practice is confusing, and it is preferable to use fractions of a millimeter for pressures above 100 μ. This is a useful distinction, because the short-path still and the pump next to the still that produces the high vacuum operate in the micron range, whereas the roughing pumps operate in the millimeter range. It will be appreciated that the difference between unobstructed path and molecular distillation is one of dimension and operating conditions; thus one piece of apparatus can occupy either category according to the rate of distillation and the pressure of residual gas.

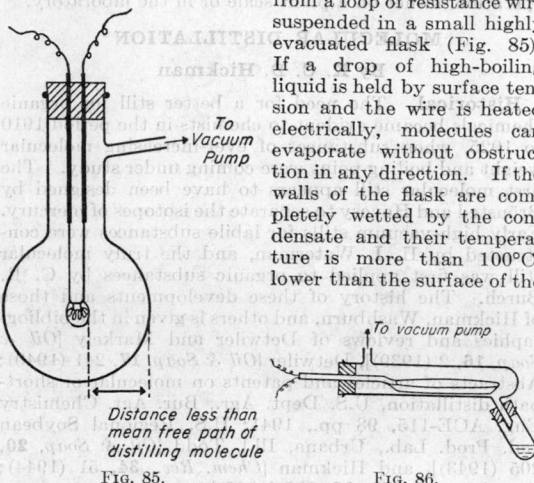

The Simple Molecular Still. The simplest and at the same time most perfect molecular still is formed from a loop of resistance wire suspended in a small highly evacuated flask (Fig. 85). If a drop of high-boiling liquid is held by surface tension and the wire is heated electrically, molecules can evaporate without obstruction in any direction. If the walls of the flask are completely wetted by the condensate and their temperature is more than 100°C. lower than the surface of the drop, all molecules will be caught and distillation will occur in a single unique act of evaporation.

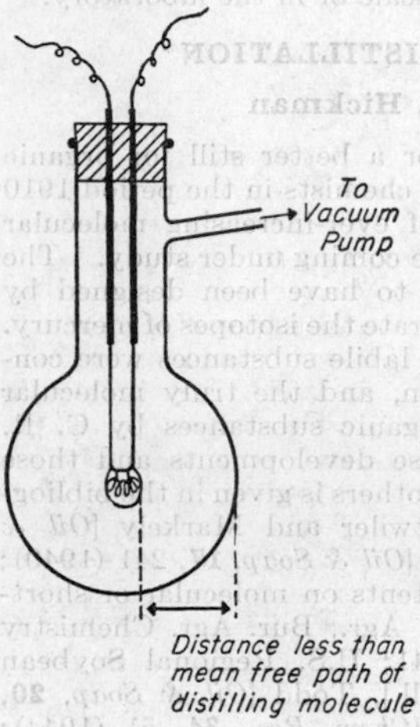

FIG. 85.

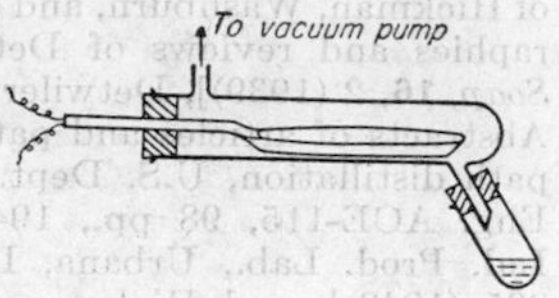

FIG. 86.

A simple and practical molecular still with an evaporative efficiency of about 0.5 is formed by an electrically heated tray suspended in an evacuated test tube (Fig. 86). An all-glass molecular pot still is shown in heavy outline superimposed on a conventional still and condenser in Fig. 87. Heat is applied beneath the still, and the ceiling is cooled by ice or a blast of air. Liquid condenses on the ceiling and drops into the receiver. Considering any typical element of liquid surface S, a molecule can evaporate in any direction embraced by a solid angle of 180 deg. Those molecules which proceed within the wide cone m, s, m' will condense on the ceiling and be collected; the rest will condense on the walls and return to the still. Contrast this with the conventional still where only those molecules within the cone d, s, d' will pass up the neck of the flask and only an exceedingly small fraction s, r, r' will wander into the receiver.

Ordinary laboratory "vacuum" distillations are done at 1 to 10 mm. in order to transfer the vapor rapidly enough to complete the distillation in reasonable time. The efficiency of transfer of the molecular pot still of Fig. 87 is 0.4 to 0.6; that of the conventional still is 0.001–0.0001. Stated another way, an evaporating molecule is likely to escape permanently at the first or second try in a molecular still, whereas it may return to the distilland a thousand times before finally reaching the exit of an ordinary vacuum still. The temperature of distillation and the hazard of decomposition are reduced in proportion to the lowered pressure of the molecular still.

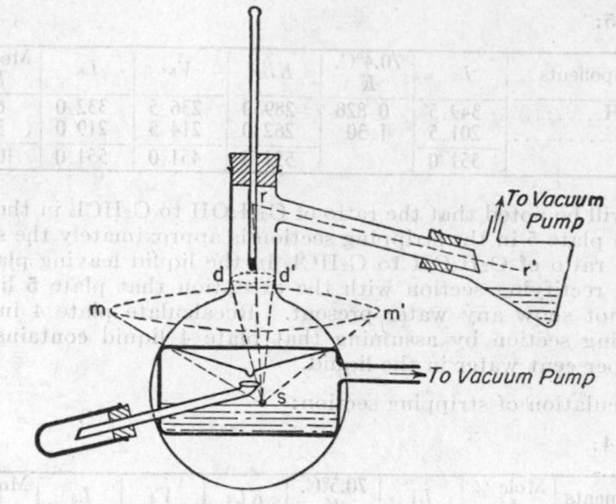

FIG. 87.

Mean Free Path Considerations. The more obvious deductions from the kinetic theory of gases have been in error with regard to the molecular still. The design of stills has proceeded faster under an empirical, purely observational approach.

The definition of molecular distillation as "within the mean free path of the emergent molecules" implies that, if the molecules suffer collision of any sort, particularly with molecules of residual gas, they will be delayed in reaching the condenser or may return to the distilland. Distillation will be slowed and will become progressively more equilibrant in kind. Burch showed how the chance that a molecule will reach the evaporator at one strike should vary with the distances, in multiples of the free path. Cox and Hickman measured the change in rate of distillation with change in pressure of residual gas and found that collisions mattered less than formerly supposed. The comparison of pressure, free path, and rate of distillation is given in Table 26. The rate is diminished by less than half even though not more than one molecule in 20,000 reaches the condenser without collision.

Table 26. Comparison of Pressure, Free Path, and Rate of Distillation

Pressure of residual air in 2-cm. gap, μ	2-cm. gap is approximate multiple m of free path	Comparative rate of distillation, %, with saturation pressures of 1μ	10μ	Number of molecules, %, reaching condenser without collision (Burch)
0.3	m = 0.2	100	100	80
4.0	= 3	77	89	
7	= 5	63	81	0.005
10	= 7	53	72	
25	= 17	35	42	
50	= 33	20	27	

As to the calculation of free paths, the physics textbooks devote much of their space to this one problem—Loeb ("Kinetic Theory of Gases," 2d ed., McGraw-Hill, New York, 1934) taking more than a hundred pages. There are three important calculations of the free path L:

Maxwell's:

$$L = \frac{1}{\frac{4}{3}\pi\sigma^2 N}$$

Clausius':

$$L = \frac{1}{\sqrt{2}\,\pi\sigma^2 N}$$

Tait's:

$$L_T = \frac{0.677}{\pi n \sigma^2}$$

where N is the number of molecules in unit volume and σ the diameter of the molecules. The mean free path L_{σ_2} of a large molecule σ_2, distilling into a permanent gas σ, is given by Loeb as

$$L_{\sigma_2} = \frac{1}{\sqrt{2}\,\pi N_2 \sigma_2^2 + \pi N_1 \sigma_1^2 \sqrt{\frac{\bar{C}_1^2 + \bar{C}_2^2}{\bar{C}_2}}}$$

in appropriate units. The measured paths of heavy molecules in a residual air vacuum of 1 μ and a distillation pressure of 1 μ are about 3 cm. for butyl phthalate and 2 cm. for olive oil (Jacobs, private communication).

The Distilland Layer. It is assumed in ordinary distillation that the surface of the distilland is at any moment a true sample of the liquid, that diffusion of volatile molecules to the surface of a liquid mixture is rapid compared with their rate of evaporation. This condition is demonstrably unfulfilled in the molecular still, where convection due to ebullition is absent and high viscosities and high molecular weights impede diffusion. Efficient molecular distillation requires mechanical renewal of the surface film. This can be done in three ways: (1) by vigorous agitation of the liquid in bulk, *i.e.*, the stirred pot still; (2) by gravitational flow in cascade or vertical falling film; or (3) by mechanical spreading in thin films. The last is accomplished by transferring the distilland to a heated roller or traveling band or by using centrifugal force to spread it on a rotating disk, cone, or cylinder. The heated rotating cone is currently the preferred form of evaporator.

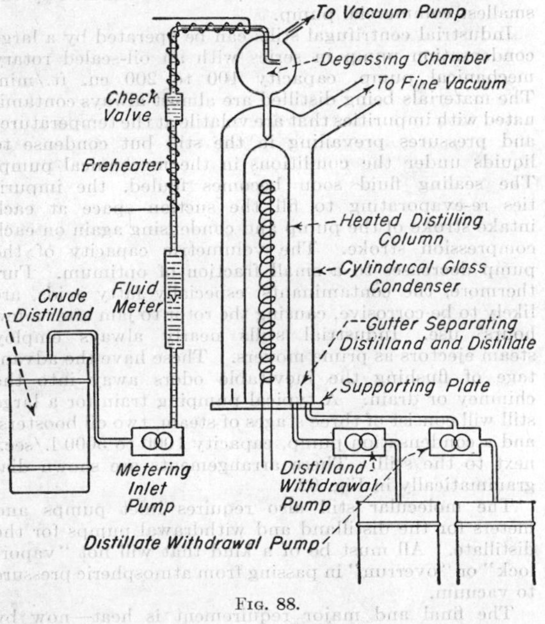

FIG. 88.

Falling-film stills have been made in single and multiple units, in sizes ranging from a few centimeters to 50 cm. diameter, 2 to 10 m. high with through-puts ranging from 1 to 60 l./hr. The elements of construction are shown in Fig. 88. The distilland is admitted through a metering device into the vacuum where it is *degassed* in one or more preliminary vessels and then allowed to pass on to the walls of a heated polished metal cylinder that is stationed within a concentric cooled condensing cylinder. The space between the two is maintained under high vacuum (1 to 5 μ) by fast pumps. Most falling-film stills have used chromium-plated evaporators, housed within tall glass cylinders.

Centrifugal stills, shown diagrammatically in Fig. 89, consist of a housing or base plate covered by a lid

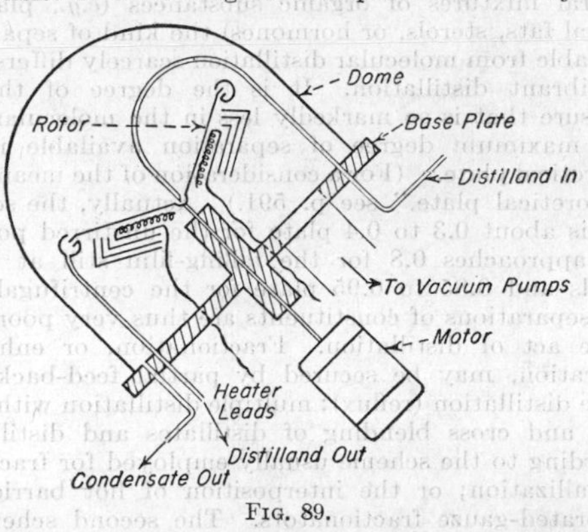

FIG. 89.

or dome. A rotor is supported on a shaft that passes through a bearing and stuffing box attached to the base plate. A radiant electrical heater warms the rotor, and radiation is conserved by baffles placed at the back and sides of the heater. The distilland is admitted from a preliminary degasser through a feed pipe to a depression at the center of the rotor, whence it is spun rapidly outward in an exceedingly thin and uniform layer. At the edge, the distilland, spent after evaporation of the volatile constituents, is picked up by a scoop (Fig. 90)

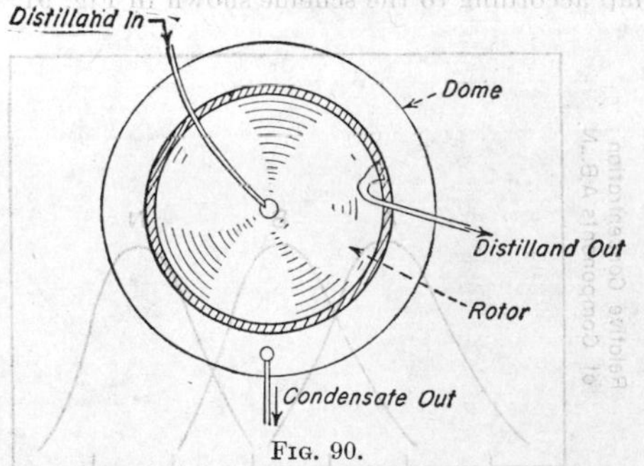

FIG. 90.

or collected by a concentric gutter. Since the distilland is held on the evaporator by centrifugal force, it is independent of gravity, and the rotor may be horizontal or vertical or even upside down. A convenient angle for small stills (12 to 36 in. in diameter) is with the evaporator facing upward at about 45 deg. The rotors of larger stills are generally cones instead of plates, and they spin on vertical shafts. The condenser hangs within the cone, and means are provided for pumping the condensates, collected separately from three concentric zones, over the rim. Stills of this type have an evaporating area of about 4.5 sq. m. and handle 200 to 700 l. of distilland per hour, collecting 2 to 400 l. of distillate per hour, according to the object of treatment. The consumption of power is about 100 kw., 60 per cent for heating the still, the remainder for pumps and subsidiary equipment.

The Distilling Operation. It is a safe generalization that all types of unobstructed-path stills provide non-equilibrant distillation. In ordinary distillation of a mixture, the quantities of constituents distilling are proportional to their partial pressures $P_1, P_2, \ldots P_n$, but, under molecular conditions, the quantities are $\frac{P_1}{\sqrt{M_1}}, \frac{P_2}{\sqrt{M_2}}, \ldots \frac{P_n}{\sqrt{M_n}}$. Since substances of like molecular weight distill at similar temperatures in natural mixtures of organic substances (*e.g.*, plant or animal fats, sterols, or hormones) the kind of separation available from molecular distillation scarcely differs from equilibrant distillation. It is the degree of thermal exposure that is so markedly less in the molecular still. The maximum degree of separation available is one theoretical plate. (For a consideration of the meaning of "theoretical plate," see p. 591.) Actually, the separation is about 0.3 to 0.4 plate for the unstirred pot still and approaches 0.8 for the falling-film still at useful speeds and 0.85 to 0.95 plate for the centrifugal still. The separations of constituents are thus very poor for a single act of distillation. Fractionation, or enhanced separation, may be secured by partial feed-back in a single distillation (reflux); multiple distillation with feed-back and cross blending of distillates and distillands, according to the scheme usually employed for fractional crystallization; or the interposition of hot barriers or perforated-gauze fractionators. The second scheme of multiple redistillation is by far the most effective.

The poor separations resulting from a simple act of molecular distillation mean that, in a series of distillates removed from a complex mixture, any given constituent will appear in *some concentration* in every fraction. If the fractions are withdrawn at equal time intervals and uniform increments of temperature, the variation of yield of a component A in the distillates follows a smooth probability curve, and this applies also to components B, C, and . . . N, so that the concentrations of each overlap according to the scheme shown in Fig. 91. The

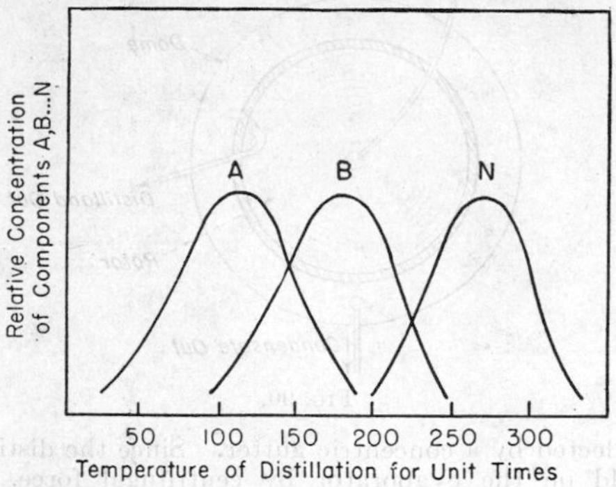

FIG. 91.

position of the maxima of these curves on the temperature axis, obtained under rigidly standardized conditions, provides the molecular substitute of a boiling point and one that can be made accurate, or at least reproducible within ±1°C.

The quantitative rate of distillation, under truly "projective" conditions of evaporation and complete non-return condensation, is equal to the quantity of material passing across an imaginary area in a saturated vapor of the distilland equal to the area of the evaporating surface and is given numerically by the Knudsen-Langmuir equation

$$w \text{ (grams)} = 5.83 \times 10^{-2}\, p \text{ (millimeters)} \sqrt{\frac{M}{T}}$$

where w = g./(sec.)(sq. m.), p = saturation pressure, mm. Hg, at the solution temperature T degrees K.; M = molecular weight. For the kind of substance handled in the molecular still, w = 0.5 g./(sec.)(sq. m.). Small though this figure may appear, it means that an area of less than 1 sq. mile could achieve the high-vacuum distillation of all substances of molecular weight 300 to 1000 as fast as produced chemurgically (*i.e.*, soybean, cottonseed, corn oils, animal fats, and waxes) in the world. In industry two stills 5 ft. in diameter can process a tank car of oil in 24 hr. A laboratory falling-film still can subject a sample of 200 cc. to 15 to 20 passes in half a working day.

Subsidiary Apparatus. The molecular still requires high-vacuum pumps in order to operate. Since the pressure in the still must be approximately one-millionth of an atmosphere and since no practical pump can achieve this reduction of pressure in one stage, it is usual to employ two or more pumps in series, each unit being designed especially for its place in the series. A solitary exception in the laboratory micro-pot still, which can be evacuated sufficiently well by a rotary oil pump (8 to 12 μ in the still).

The pump that is used for the ultimate vacuum is by general choice the Langmuir condensation pump. The laboratory falling-film still, handling 1 l./hr. of distilland, requires a condensation pump with a capacity of 100 l./sec. at 1 to 2 μ, which gas it will pass on to a mechanical *fore* pump at a pressure of 0.1 to 0.2 mm. The fore pump must have an effective volumetric capacity of at least 1 l./sec. If it is not desired to employ such a large mechanical pump, a vapor booster pump may be inserted between the condensation pump and the mechanical pump. The booster will take in gases at 0.2 mm. and 1 l. volume, compressing them to 0.5 mm. and 0.4 l. volume, which can be handled by the smallest commercial pump.

Industrial centrifugal stills can be operated by a large condensation pump in series with an oil-sealed rotary mechanical pump, capacity 100 to 200 cu. ft./min. The materials being distilled are almost always contaminated with impurities that are volatile at the temperatures and pressures prevailing in the still but condense to liquids under the conditions in the mechanical pump. The sealing fluid soon becomes fouled, the impurities re-evaporating to fill the suction space at each intake stroke of the pump and condensing again on each compression stroke. The volumetric capacity of the pump decreases to a small fraction of optimum. Furthermore, the contaminants, especially fatty acids, are likely to be corrosive, causing the rotor to jam after a few hours' use. Industrial stills nearly always employ steam ejectors as prime movers. These have the advantage of flushing the inevitable odors away into the chimney or drain. A typical pumping train for a large still will consist of three stages of steam, two oil boosters, and a condensation pump, capacity 1000 to 5000 l./sec., next to the still. These arrangements are shown diagrammatically in Fig. 92.

The molecular still also requires inlet pumps and meters for the distilland and withdrawal pumps for the distillate. All must be of a kind that will not "vapor-lock" or "overrun" in passing from atmospheric pressure to vacuum.

The final and major requirement is heat—now by common usage electrical heat. Heat is the only indispensable raw material of the molecular still. Because distillation is done at very high temperature (200° to 300°C.) and must often be repeated to secure proper separations, and because the hot evaporator loses heat by radiation to the cold condenser, less than 5 per cent of the input heat is actually used by distillation. When the heat required for the oil-vapor vacuum pump is also debited against the electrical consumption, the thermal efficiency may fall to 2 to 3 per cent. This extravagant use of electricity prompted inventors to experiment with other sources of heat and industrialists to shun molecular distillation altogether. Advances in design and reductions in cost of electricity, coupled with a truer appreciation of overhead and commercial charges, have placed the cost of heat in proper perspective, where it is seen to be a minor item in comparison

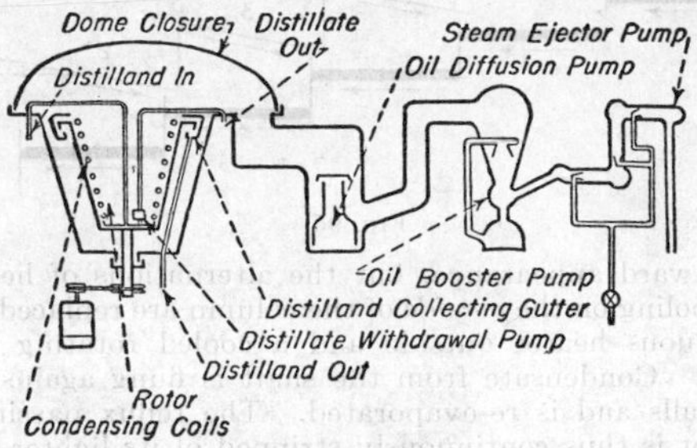

FIG. 92.

with investment, labor, and the usual commercial costs. The through-put of the modern molecular still is sufficiently high for the total operating cost to compare favorably with conventional distillation.

Item	Fraction of Total Cost, Per Cent
Capital charges	10
Labor	20
Administration, supervision, and overhead	40
Contingencies, repairs, and services other than power	20
Electricity, steam	10

Uses of the Molecular Still. *Laboratory.* The small demountable pot still (Fig. 93*A*) is a required intermediary step in the estimation of vitamin E in foods and oils. The same apparatus is useful for purifying small samples of drugs, dyes, sterols, and hormones. For the investigation of natural oils and waxes, the cyclic batch falling-film still is generally employed (Fig. 93*A*, *B*, and *C*). This instrument is designed for multiple repasses of the distilland under standard conditions of increasing temperature. The investigator can construct an *elimination curve* of materials under investigation which affords a precision method of analysis. The cyclic batch centrifugal still has recently been developed in small size, convenient for bench-top use (Fig. 93*C*).

Industrial. The falling-film still has largely been replaced by centrifugal units. These have evaporators approximately 1, 3, or 5 ft. in diameter and are generally grouped in blocks of three to seven to allow fractionation by multiple redistillation. Their chief uses are distillation of vitamin A esters from fish-liver oils, stripping of vitamins E (α-, β-, γ-, and δ-tocopherols) and sitosterols from vegetable oils, and the complete distillation of industrial high-boiling synthetics, plasticizers, fatty acid dimers, and the like. The quantities treated annually (1947) exceed 5,000,000 lb.

Comparative Properties of the Molecular Still. The modern high-vacuum still performs distillation at what are believed to be the lowest theoretically possible temperatures and accomplishes this in the shortest time at present available in distillation equipment. The combined reduction of thermal hazard is more than a thousand millionfold over a distillation lasting 1 hr. at atmospheric pressure. An industrial flash distillation done at 10 mm. involves a thermal exposure about 300,000 times higher than that caused by a high-speed molecular rotor. Against these advantages of the molecular still must be placed the poor separatory power of the unit act of distillation.

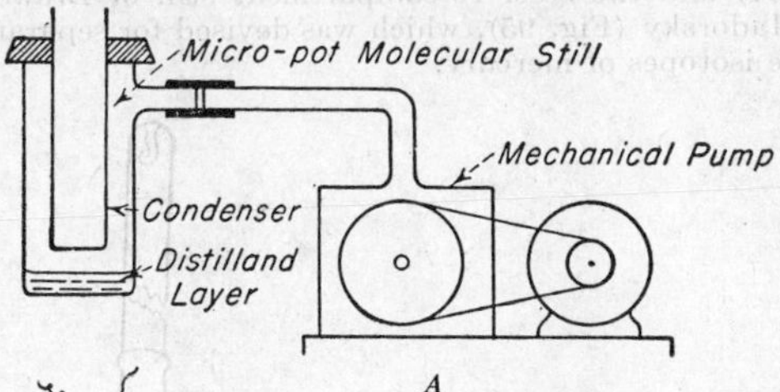

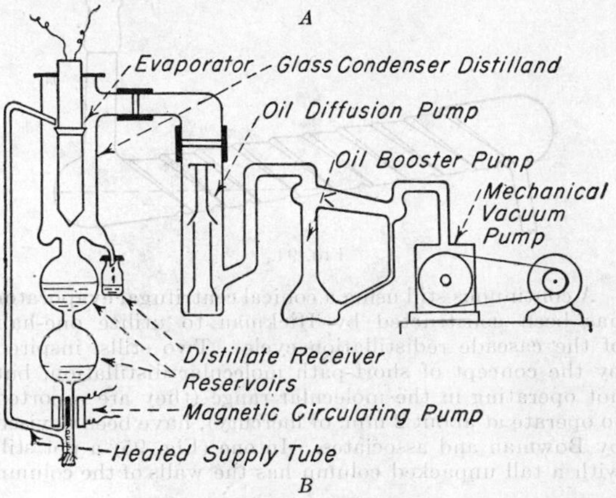

FIGS. 93*A* and *B*.

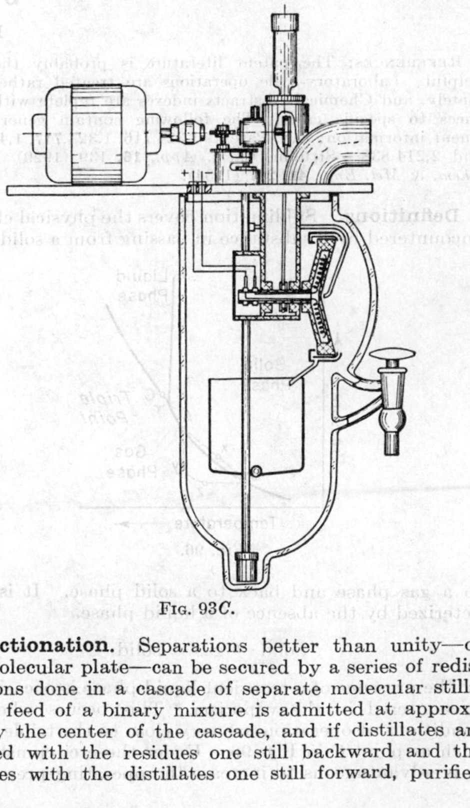

FIG. 93*C*.

Fractionation. Separations better than unity—or one molecular plate—can be secured by a series of redistillations done in a cascade of separate molecular stills. If the feed of a binary mixture is admitted at approximately the center of the cascade, and if distillates are blended with the residues one still backward and the residues with the distillates one still forward, purified components emerge at either end of the cascade. Two self-contained units have been described for doing this, the laboratory glass still of Wollner and associates (Fig. 94) and the steel 10-compartment still of Brewer and Madorsky (Fig. 95), which was devised for separating the isotopes of mercury.

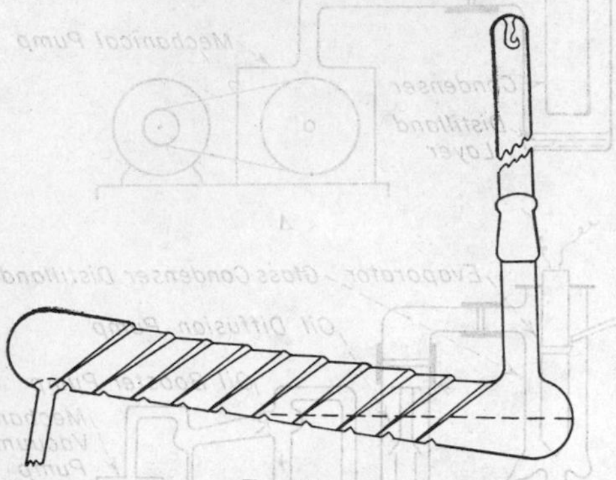

FIG. 94.

A continuous still using a conical centrifugal evaporator has been constructed by Hickman to utilize one-half of the cascade redistillation cycle. Two stills, inspired by the concept of short-path molecular distillation, but not operating in the molecular range (they are reported to operate at about 2 mm. of mercury), have been devised by Bowman and associates. In one (Fig. 95) a pot still with a tall unpacked column has the walls of the column alternately heated and cooled in concentric zones. Reflux returned at the top flows down the walls, where it is alternately re-evaporated and recondensed. This disturbs the laminar flow of vapor ascending from the pot and presents, in effect, a series of virtual baffles of negligible obstructive power. The other still is identical

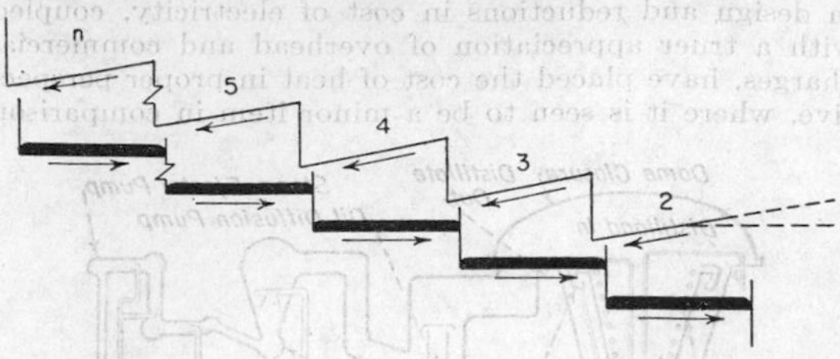

FIG. 95.

in outward appearance, but the alternations of heating and cooling on the *outside* of the column are replaced by a continuous heater outside and a cooled rotating shaft inside. Condensate from the shaft is flung against the hot walls and is re-evaporated. The reflux passing to the pot is thus continuously stripped of its lighter component, which ascends the column during its period as a vapor.

The thermal hazards to which the distilland is exposed in fractionating stills are in about the following increasingly harmful order: cascade redistillation in separate stills, centrifugal-cone fractionator, Wollner, Matchett, Levine, Brewer-Madorsky, and Bowman stills. In completeness of fractionation (number of plates) and ease of manipulation, the order is approximately reversed.

SUBLIMATION

By H. C. Vernon

REFERENCES: The patent literature is probably the most helpful. Laboratory-scale operations are treated rather completely, and Chemical Abstracts indexes are replete with references to specific cases. The following contain general pertinent information: U.S. Patents 1,324,716, 1,324,717, 1,464,844, and 2,214,838. Strubin, *Chem. App.*, **16**, 139 (1929). Bebie, *Chem. & Met. Eng.*, **41**, 247 (1934).

Definitions. Sublimation covers the physical changes encountered by a substance in passing from a solid phase

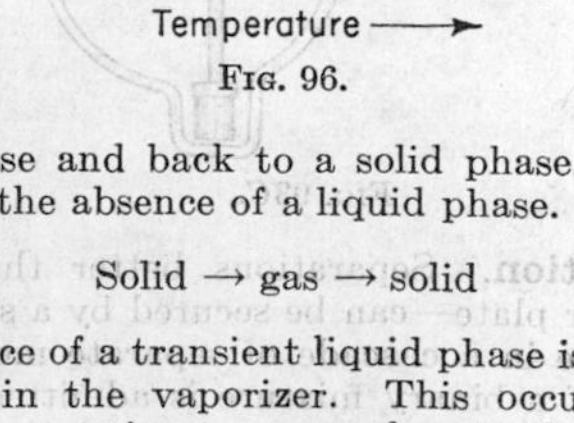

FIG. 96.

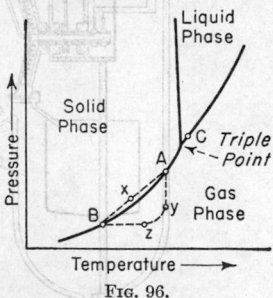

to a gas phase and back to a solid phase. It is characterized by the absence of a liquid phase.

Solid → gas → solid

The presence of a transient liquid phase is occasionally encountered in the vaporizer. This occurs when the conditions of operation are very close to the triple point, such as point *C* in Fig. 96. Under these circumstances, it is advantageous to increase the operating pressure to obtain a liquid phase in the vaporizer so as to increase the ease of transferring heat.*

That temperature at which the vapor pressure of the solid equals the total pressure of the gas phase in contact with it is defined as the **sublimation point.** (Note the direct analogy to boiling point as applied to a liquid phase.)

The **snow point** is defined as that temperature at which the partial pressure of the solid is equal to the partial pressure of the substance in the gas phase. (This is analogous to the dew point.)

When only the vaporization of the solid is considered, the term sublimation into the gas phase may be used.

Solid → gas

In some cases a compound is in the gas phase because of reaction or previous processing, and condensation to a solid phase is next involved. To describe the condensation to solid, the term **desublimation** has been used.

Gas → solid

Simple sublimation is the term applied when the gas phase is composed of the material being sublimed only.

Carrier sublimation is the term applied when the gas phase is composed of the material being sublimed plus other materials introduced either deliberately or unavoidably.

Uses. Sublimation is used for purification. It may also be used to form a particular crystal structure or

* A compound with a vapor pressure of 1 atm. but a few degrees above the melting point is ferric chloride. It is impractiable to attempt a plant process of vaporization at 1 atm. that is dependent on maintaining a liquid phase.

size. The fact that this operation may be carried on at low temperatures is useful when distillation would introduce problems of decomposition or corrosion. It is also useful as an alternate to crystallization from a solvent, when the solvents available introduce difficult problems.

Sublimation into the gas phase is a method useful for introduction of components into a gas-phase reaction. Desublimation is useful in removing products from a gas-phase reaction and simultaneously obtaining a pure finished product.

As a separation process, sublimation is directly analogous to a simple distillation. Under the most favorable circumstances, a two-stage separation can be performed. One stage takes place in the vaporizer and the other in a partial condenser. A simple distillation is rarely performed with deliberate introduction of an inert gas, since a reasonable vapor rate can be obtained because of good heat-transfer conditions in the boiler. A steam distillation is a general exception, and the steam is used for exactly the same reasons as the inert gas in a carrier sublimation.

Limiting Factors. The rate at which a sublimation may be carried out depends on the rate-limiting process in a series of operations:

1. Rate of heat flow to the solid.
2. Rate of change from solid to gas phase at constant temperature.
3. Rate of transport of mass from vaporizing zone to condensing zone.
4. Rate of change from gas to solid phase.
5. Rate of heat flow from the solid.

In any particular system one of these operations will determine the over-all rate. The rate-limiting process may change with time, principally because of changes in effective surfaces or areas in the equipment employed.

1. *Rate of Heat Flow to the Solid* (*Heat Transfer*). High rate of transfer of heat to a solid is difficult to achieve. This is particularly true when a heat-transfer surface is used and when the maximum temperature of this surface is limited by other considerations (decomposition, corrosion, etc.). Since this is due mainly to the relatively small effective areas of contact between the solid and the surface, means to promote contact should be used. The physical state of the feed material is most important; and, when process conditions will permit, the particle size should be reduced to as low a figure as practicable. The fine particles not only present a large surface-volume ratio but also permit a larger contact area between the solids and the heating surfaces. In addition, the greater mobility allows fresh particles to slip on the heated surfaces to replace those being used up. The generated vapor may give some slight aid by fluidizing action. The fluidized-solid techniques are most useful when it is necessary to heat by conduction through a wall. Fluidized-solids techniques are possible only when a gas (such as a carrier) is permissible, or when a high specific-vapor generation rate can be effected. Admixture with preheated inert solids is a method that provides close temperature control. Direct contact between the solids and a fluid (gas or liquid) heat-transfer medium will usually result in the maximum heat-transfer rates. For methods of calculation, see Sec. 6.

2. *Rate of Change from Solid to Gas* (*Phase Change*). This is of no practical significance. Those interested in theory may consult Alty, *Proc. Roy. Soc.* (*London*), **A161**, 68 (1937).

3. *Rate of Transport of Mass from Vaporizing Zone to Condensing Zone* (*Mass Flow*). This operation is a problem in fluid flow. Three basic mechanisms may be employed, and of these only two are of importance.

The mechanism of diffusional flow through a stagnant inert atmosphere is the slowest process and is to be avoided in all practical applications. The inadvertent introduction of this mechanism into some part of the system is often responsible for low over-all capacities.

When the gas phase contains no inert or non-condensable component, the pressure difference between the vaporizing zone and the condensing zone is determined by the effective temperatures of the vaporizing solid and of the condensing surface. (Flow rates may be calculated by the general methods given in Sec. 5.) This is merely another way of saying that the heat-transfer rates rather than the mass-transport rates are controlling. On the other hand, the maintenance of gas composition by avoiding introduction of air by leakage and entrapment in the solids feed is difficult of attainment. Gas evolution from the solid itself (from solution or by decomposition) can, on occasion, be a controlling factor. The presence of inert or non-condensable gas alters the mechanism either to the next case or to two processes in series, the one just discussed and diffusional flow through an inert layer that concentrates near the condensing surface.

The deliberate controlled introduction and removal of an inert or carrier gas provides a mechanism for sweeping the vapor from the vaporizing zone to the condensing zone. The total pressures of the system are fixed by design, and the total mass transport is determined by the inert gas flow.

Although it is difficult to get perfect contact between carrier gas and solid in the vaporizer and attain equilibrium conditions, the conditions in the condenser are such that equilibrium is essentially complete. Except for entrainment of dust, the loss from the condensing zone is as calculated from the vapor-pressure considerations. The percentage yield loss from the condenser consequently increases with low concentrations in the vaporizer, whether it is due to poor contacting or to low temperatures.

4. *Rate of Change from Gas to Solid Phase* (*Phase Change*). The physical properties of the crystal and the requirements for product form and size may determine the rates that may be attained per unit surface area. When crystal form and size are unimportant, this process will not control, and the maximum heat-transfer rates attainable will control the condensing-system capacity.*

The natural rate of crystal growth at the temperature of the solid must be greater than the rate of condensation if large crystals are required. The temperature gradients must be reduced for systems involving crystals that grow slowly. The mechanism of condensation must be reduction of gas phase to solid phase on the surface of the crystal. Since heat must be removed by conduction through a growing crystal, the rate of growth will decrease unless the temperature next to the cooling wall is progressively lowered. The way in which this lowering may be calculated can be found from the unsteady-state heat-transfer equations in Sec. 6. The natural rate of crystal growth must be determined experimentally. When very large crystals are required, batch operation with careful control is indicated. Low capacities will be experienced in the condensers.

5. *Rate of Heat Flow from the Solid* (*Heat Transfer*). There are three mechanisms for heat removal in the condensing zone. The capacity of the equipment and the physical form of the product will be determined by the mechanisms used.

Cooling by radiation to a cold wall will produce a very

* See Alty, *Proc. Roy. Soc.* (*London*), **A161**, 68 (1937).

fine crystal size. The size to which they grow will be determined by:

1. Natural crystal growth rate.
2. Concentration of the gas in contact with the crystals.
3. Time of contact.
4. Temperature gradients.

It is difficult to calculate the condensation rates involved only because of the lack of data on gas-emissivity constants (see Sec. 6).

Cooling by convection of the inert gas will also form small crystals. Included in this mechanism is that of admixture of a cold diluting gas. The final mixture temperature must be below the snow point.

Cooling by conduction to a crystal condensed on a cold surface is the mechanism for growing large crystals. Again, the natural rate of growth determines the size, but the insulating effect of the sublimate reduces heat-transfer rates progressively.* Small crystals will deposit on top of each other when the heat-transfer rate exceeds that corresponding to material crystal-growth rate.

Methods—General. *Simple Sublimation.* Unless the vapor pressure at the condensing temperature exceeds atmospheric (which is unusual), a vacuum system is necessary. The pressures in the system are the vapor pressures of the subliming materials at the temperatures of operation. Because of inleakage and introduction of entrapped gases, continuous pumping is used. Protection of the vacuum pump is necessary, and a cold trap followed by an air bleed or warming zone is needed to prevent fouling of the pumps. Production rates are high as long as the condensing surfaces are clean and residual gases are kept pumped out. Yields are very high.

Molecular distillation is a special case and is treated on pp. 655 to 660.

Carrier Sublimation. The introduction of a carrier gas to control the movement of the subliming molecules and avoid vacuum operation is the usual process. The carrier gas must be inert toward the products and have a vapor pressure in excess of the sublimate. The carrier may condense with the sublimate as long as it does not contaminate the product. Unless the carrier is recirculated, yield loss may be excessive. Recirculation of carrier gas will build up impurity concentrations. These may be reduced by a separate clean-up condensation at temperatures below the product condensation temperature, by scrubbing, or by bleeding the carrier to waste.

Design Calculations. *Physical Properties.* The physical properties required for process and equipment design are:

Vapor pressure–temperature relationships.†
Melting point.
Latent heat of sublimation (fusion + vaporization).
When crystal size control is required, specific crystal-growth rate.

Vapor pressure (see Sec. 5 and "International Critical Tables," Vol. 3) correlates with absolute temperature in most instances as

$$\log p = \frac{A}{T} + B \qquad \text{or} \qquad p = B'e^{\frac{A'}{T}}$$

where A, B = constants.*
p = vapor pressure of solid.
T = absolute temperature.

Dissociation in the gas phase leads to deviations. Theoretically and practically A is the latent heat of sublimation (see Sec. 5). Melting points (see Sec. 3) may be found in the literature or experimentally.

Latent heats of sublimation (see Sec. 3 and "International Critical Tables," vol. 5) can be estimated from the vapor pressure in accordance with the equation just given. Specific crystal-growth rate should be determined experimentally for the conditions expected in the sublimate condenser.

The Operating Diagram. The operating line is drawn on the phase diagram (see Fig. 96). The partial pressure of the solid in the vaporizer is shown at point A and in the condenser at point B. It is obvious that path A-x-B is not physically possible under equilibrium conditions. Some path such as A-y-z-B must be taken. That part of the path from A to y represents pressure reduction at constant temperature; from y to z reduction of both temperature and pressure, and from z to B reduction of temperature at constant pressure. The part A-y can be visualized as corresponding to the physical path from solid in the vaporizer through heated (or lagged) lines; the part y-z likewise corresponds to the entrance to the condenser; and the part z-B is in the condenser proper. A path such as from A to y can be accomplished by admixing additional carrier gas or by pressure reduction due to friction loss in equipment.

Capacity.

Table of Symbols

Any consistent set of units may be employed.

Let M = molecular weight.
p = partial pressure.
P = total pressure.
T = temperature, absolute.
V = volume-flow rate.
w = weight flow rate.
ρ = density.†

Subscript a refers to the inert.
Subscript b refers to the sublimate.
Subscript c refers to a third component.
Subscript i refers to a generalized component.
Subscript 0 refers to the pre-entrant zone.
Subscript 1 refers to the vaporizing zone.
Subscript 2 refers to the condensing zone.
Subscript 3 refers to the exit zone.

The following calculation assumes that all mechanisms are equalized, or mass transfer between zones is the rate-limiting process:

$$\frac{V_{a1}}{V_{b1}} = \frac{p_{a1}}{p_{b1}} \tag{1}$$

$$p_{a1} + p_{b1} = P_1 \tag{2}$$

$$p_{a1} = P_1 - p_{b1} \tag{2a}$$

$$V_{b1} = V_{a1}\frac{p_{b1}}{P_1 - p_{b1}} \tag{3}$$

or, in terms of mass (weight) of sublimate and volume of carrier gas,

$$w_b = \rho_{b1}V_{a1}\frac{p_{b1}}{P_1 - p_{b1}} \tag{4}$$

* The calculation of heat-transfer rates must be done by graphical integration, introducing the increasing resistance of the sublimate. A contact coefficient, the thermal conductivity of the crystal, and the time between cleaning operations are required. Continuous cleaning of the surface results in maximum capacity.

† Decomposition pressure is included for those cases in which the solid dissociates in the gas phase and recombines when condensed. A case in point is NH_4Cl.

* $B = \log B'$; $2.3A = A'$; when natural logarithms are used $A = A'$.

† The density may be calculated from the gas laws, taking the molecular weight of the gas. $\rho_i = \frac{M_i}{359} \times \frac{273}{T} \times p_i = 0.761 M_i \frac{p_i}{T}$ lb./cu. ft. (p in atm., T in °K.).

and the loss from the condenser becomes

$$w_{b2} = \rho_{b2} V_{a2} \frac{p_{b2}}{P_2 - p_{b2}} \tag{4a}$$

The concentrations in the gas phase at equilibrium can be calculated by the usual gas laws (due care must be given to association or dissociation in calculating M)

$$\text{Mole per cent } b = 100 \frac{p_b}{\Sigma p_i} = 100 \frac{p_b}{P}$$

$$\text{Weight per cent } b = 100 \frac{p_b/M_b}{\sum_i p_i/M_i}$$

Example. Sublimation of iodine by an air carrier stream.

P_2 = pressure in condensing zone = 1.0 atm. (760 mm.) (selected by design).

$P_1 - P_2$ = pressure drop through the system = 0.01 atm. (7.6 mm.) (assumed, subject to check calculation in final design).

Temperature in vaporizer = 105°C., selected in primary design.

p_{b1} = vapor pressure of solid = 0.079 atm. (60 mm.) (physical property).

P_1 = total pressure in vaporizer = 1.01 atm. (768 mm.) (sum of pressure drop and final total pressure).

V_{a0} = air rate (at 20°C., 1.0 atm.) = 10 cu. ft./min. (assumed; design rate will be in proportion to required capacity).

$$V_{a1} = 10 \times \frac{1.0}{1.0 - 0.079} \times \frac{378}{293} = 14 \text{ cu. ft./min.}$$

ρ_{b1} = density of iodine vapor = 0.0402 lb./cu. ft. (physical property).

$$w_{b1} = 0.0402 \times (14 \times 60) \frac{0.079}{1.01 - 0.079}$$
$$= 0.0402 \times 840 \times 0.0849 = 2.86 \text{ lb. of iodine processed per hr.}$$

The above is based on perfect contacting between the carrier air and the vaporizing solid. A fluidized-solid technique would approximate this very closely. Blowing the air over a pan of heated solids would not, and a contacting efficiency would have to be determined by experimental procedure. The detailed design of the vaporizer is the controlling variable; the condensing system must be designed for equal or greater capacity.

Temperature at outlet of condensing system 50°C.

p_{b2} = vapor pressure of solid = 0.00284 atm. (2.16 mm.).

p_{a2} = partial pressure of air = 1 − 0.00284 = 0.997 atm.

ρ_{b2} = density of iodine vapor = 0.00169 lb./cu. ft.

$$V_{a2} = 10 \times {}^{323}\!/_{293} = 11 \text{ cu. ft./min.}$$

$$w_{b2} = 0.00169 \times (11 \times 60) \frac{0.00284}{1.0 - 0.00284}$$
$$= 0.0169 \times 660 \times 0.00285 = 0.0318 \text{ lb. of iodine lost per hr.}$$

$$w_{b1} - w_{b2} = 2.86 - 0.03 = 2.83 \text{ lb. of iodine product per hr.}$$

$$\text{Yield} = 100 \times \frac{2.83}{2.86} = 98.9 \text{ per cent.}$$

The insertion of a tailing condenser to give an exit gas of 20°C. (or the improvement of the main condenser to give the same result) improves the yield thus:

p_{b3} = vapor pressure of solid 0.00062 atm. (0.471 mm.).

ρ_{b3} = density of iodine vapor 0.00108 lb./cu. ft.

V_{a3} = 10 cu. ft./min.

$$w_{b3} = 0.00108 \times (10 \times 60) \frac{0.00062}{1 - 0.00062}$$
$$= 0.00108 \times 600 \times 0.00062 = 0.000402 \text{ lb. of iodine lost per hr.}$$

$$\text{Yield loss} = 100 \times \frac{0.000402}{2.86} = 0.014 \text{ per cent.}$$

Purification. The composition of the gas leaving the vaporizer is given by the ratio of vapor pressures of the solids at the temperature of the vaporizer. This is true when there is sufficient impurity present in the feed to maintain this relation.* In other words, there will be less impurity in the gas leaving the vaporizer than in the feed when the ratio of vapor pressures of impurity to product is less than the mole ratio of impurity to product in the feed.

The composition of the solid in the condenser is given by the ratio of vapor pressures at the temperature of the condenser, provided that the partial pressure of the impurity in the gas phase is equal to or greater than the vapor pressure of the impurity at condenser temperature. Otherwise no impurity will condense at all.

A material balance on the condenser must be made to determine the purification. It is convenient to express this balance in terms of the carrier gas. Then, for 1 mole of carrier gas:

	Entering	Leaving
Moles of b	$\frac{p_{b1}}{p_{a1}}$	$\frac{p_{b2}}{p_{a2}}$
Moles of c	$\frac{p_{c1}}{p_{a1}}$	$\frac{p_{c2}}{p_{a2}}$

The value of p_{c2}/p_{a2} cannot exceed the value p_{c1}/p_{a1}. When the vapor-pressure relationship value of p_{c2} yields a greater value than p_{c1}/p_{a1}, no solid c can condense.

The carrier gas requires careful selection and adequate preheat temperature control. The gas must be inert toward the product to be sublimed, although it may react with the impurities under circumstances that produce no hazard or products that would cause harmful contamination. When carrier gas and subliming products are capable of forming explosive mixtures, care must be taken to maintain operating temperatures below those which yield a gas composition corresponding to the lower explosive limit. A carrier gas that enters at a temperature below the vaporization temperature will cause a loss in production rate by cooling the solids. A carrier gas that enters at a temperature above the vaporizer temperature will increase the heat transferred to the solid. This is permissible only if the increased solid temperature is not harmful. This may be used as a main source of heat supply, provided that the time of contact is sufficiently long to attain equilibrium. The most important consideration in vaporizer design is carrier-solid contact to ensure the maximum product-to-carrier ratio entering the condenser.

Equipment. Sublimation equipment has not been developed or standardized to any great extent. Figures 97 and 98 show a method of collecting fractions of varying composition by batch sublimation (Selden Co., U.S. Patents 1,324,716 and 1,324,717). The vaporization temperature is progressively raised to afford varying sublimate composition and to maintain yields. Preferential condensation may be obtained by a series flow through condensers each at a successively lower temperature. The relative slopes of the vapor-pressure curves will determine the efficacy of this approach. If only one grade of product is to be produced, two condensers in series are needed, and the material collected in the off-grade condenser must be recycled.

Pyrogallol is sublimed in equipment shown in Figs. 99 and 101. When large crystals are required, a vacuum

* An intimate mechanical mixture is implied. The presence of a solid solution, or chemical bonds of even a weak secondary nature, may alter vapor pressure–temperature relationships. A situation of this sort requires experimental determination of vapor-pressure data for the composition ranges involved. A laboratory-scale sublimation with composition and other conditions controlled to the anticipated plant values may be the best and simplest approach to a design.

is used. This is not always so low as required by true sublimation without carrier, and for efficient operation some means of recirculating the inerts present is needed. When "snow" is desired, atmospheric pressure is used and the carrier recirculated. Figure 100 is a diagram of equipment for salicylic acid sublimation. The sublimation of magnesium is carried out at 600°C. and 0.15 mm.

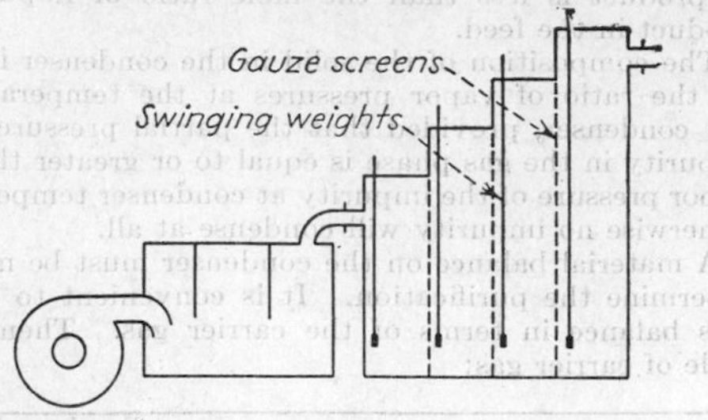

FIG. 97.

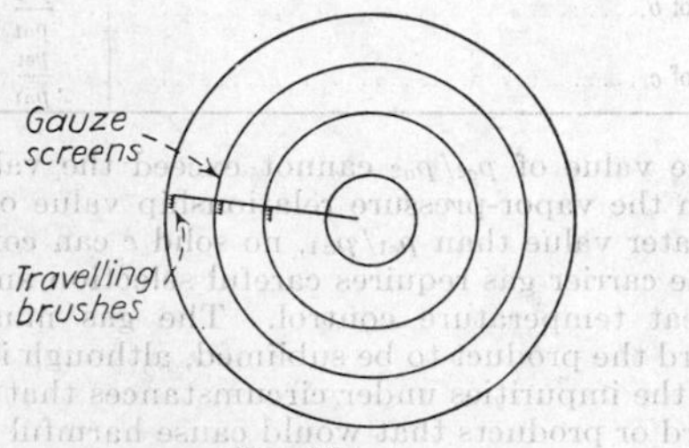

FIG. 98.

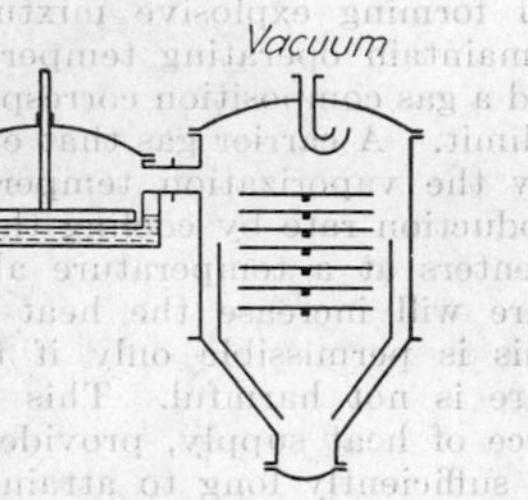

FIG. 99.

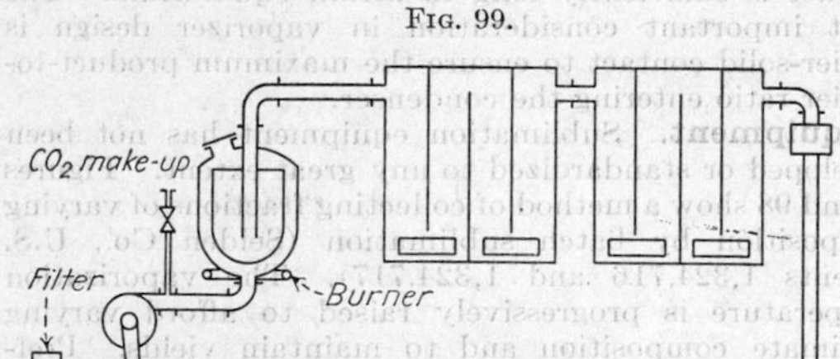

FIG. 100.

This is described by Bebie [*Chem. & Met. Eng.*, **41**, 247 (1934)].

All kinds of pots, kettles, retorts, and the like have been used for vaporizers. Close-clearance agitators are essential when jacketed pots are used. Heat transfer to solids is not rapid, and maximum agitation and wall cleaning are necessary for high rates. When the solid can be melted or when it can be mixed with a liquid (volatile or otherwise), more reasonable rates can be attained. If the liquid does not wet the solid, vortex entrainment by agitation improves contact. The most important consideration of a vaporizer is provision for intimate and long-time contact between the carrier gas and the hot solid (or liquid). Extreme precautions are necessary to avoid a drop in gas temperature between the vaporizing zone and the condensing zone. The tops of kettles and all piping must be externally heated to a temperature slightly (about 5°C.) above the vaporizing temperature. Failure to do so will cause plugging and lost production time. Piping should be large in cross section even when flow rates are low, if the provisions for heating them are not adequate.*

Continuous-vaporization equipment will afford maximum capacities with minimum investment and operating labor. No generalized designs have been published.

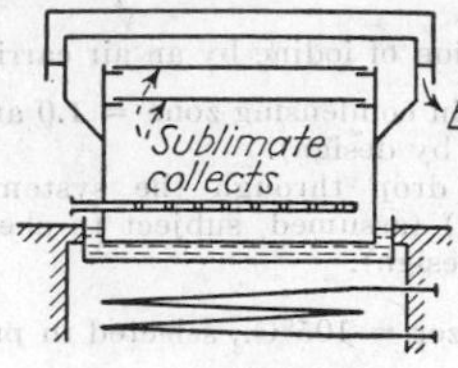

FIG. 101.

Condensing systems of many kinds have been used. Shell-and-tube condensers in parallel and operating on alternate cycles of condensing and cleaning give good yields in small space but take considerable operating labor. Large tubes, water- or air-cooled, with mechanical continuous scrapers are satisfactory when carefully installed. Large chambers using radiation cooling, liquid-spray cooling, or conduction-convection cooling to internal pipes or ducts can be used. Pervious cloth walls, metal screen walls, and chain walls, all with mechanical shaking, have been used. Recycled, cooled inert gas can be mixed with the gas from the vaporizer and the resulting "snow" settled out in large chambers or taken out in bag filters. The gas-cooling device will also collect solids and will require continuous (or cyclic) cleaning.

The wall temperature can be too cold, and a hard solid that is difficult to remove may form. Under these circumstances, a tailing condenser, easily cleaned, is advantageous to protect the yield.

Provided that the vapor-pressure relations of the substances in a mixture are well differentiated, much may be accomplished in the way of preferential vaporization and condensation by close temperature control. The temperature of the vaporizer may be raised progressively, and the sublimate collected in fractions richer and poorer in the desired component; this is well-known laboratory technique. Preferential condensation is carried out by passing the vapors through chambers maintained at steady but progressively lower temperatures; the chambers may be made concentric (Selden Co., U.S. Patents 1,324,716 and 1,324,717; *cf.* Figs. 97, 98). The normal method of obtaining a dry sublimate in a sublimation, when the vapor of water or other liquid is used as entrainer, by maintaining the condenser at a temperature just above the boiling point of the entrainer is an application of preferential condensation in a simple form.

Plant for Simple Sublimation. The high-vacuum sublimation of magnesium is representative [Bebie, *Chem. & Met. Eng.*, **41**, 247 (1934)]. The still is a closed vertical mild-steel cylinder; the lower third is enclosed in a furnace and the upper part in a flue, the temperature of which is controlled by dampers regulating the air flow.

* Condensation of crystals on the inner surface of connecting pipes acts as internal insulation. Since they continue to grow, sufficient cross section must be provided to give a reasonable time before plugging is complete.

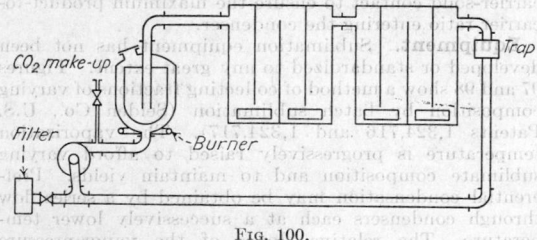

A split liner is fitted within the upper part, and on this the sublimed magnesium collects. The vapor pressure of magnesium at 300°C. is 0.001 mm., and at 651°, the melting point, 2.0 mm. Sublimation is found to proceed best at 600° and 0.15 mm.; this approaches the conditions for a true simple sublimation very closely. Welding vessels for these temperatures and pressures is a special technique; each layer of weld metal is peened vigorously before the next layer is applied.

Pyrogallol crystals are made by sublimation under reduced pressure. The pressure is not so low that the process is a true simple sublimation, but it is low enough to preclude the formation of snow-form condensate. A specimen plant is shown in Fig. 99; the crystals collect on the pivoted baffles and are tipped off at the end of the batch through the discharge door.

Plant for Entrainer Sublimation. *Supply of Entrainer.* The entrainer is usually air supplied by a blower. Steam is used for anthracene (Fig. 97) and for other materials insoluble in water. More costly entrainers, such as carbon dioxide or the 6 per cent carbon dioxide-air mixture used for salicylic acid, are collected after passing through the condenser and are recirculated (Fig. 100). The entrainer is preheated in a coil which can be conveniently placed in the furnace used for heating the vaporizer (Fig. 101) or in an outer jacket of the vaporizer itself (Fig. 100).

Vaporization. The vaporizer requires close attention. The entrainer must be properly distributed over the surface of the material or within its bulk. The heat transfer to the material must be good; the material is preferably kept molten; otherwise difficulties arise as the material shrinks away from the hot walls. Oil-bath heating, or heating by hot water in Frederking coils (Fig. 102), are useful as giving good temperature

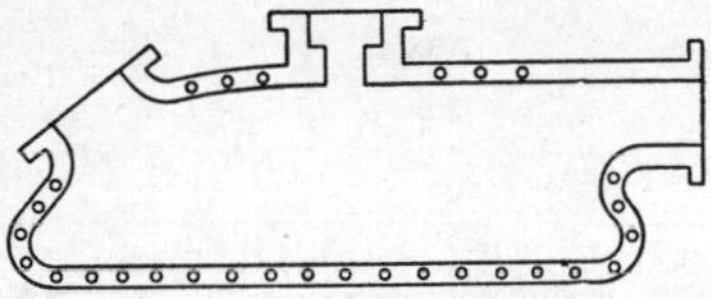

FIG. 102.

control. The duct to the condenser should be short, wide, and easily cleaned. It should be maintained at a temperature such that the sublimate is kept in the vapor phase either by flue-gas jacketing or by electrical heating. Provision for rodding through while the plant is in operation is a useful additional precaution.

Condensation. The vapor may be condensed directly by spraying the cooling medium into the condenser (Fig. 103). This gives rapid and efficient condensation and produces a loose powdery sublimate, but it may lead to trouble in recovering the product. A volatile or gaseous coolant may be used, and a solvent for the product may be used by recirculating the saturated solution.

Surface condensers are generally large air-cooled chambers. Water-cooled condensers are apt to produce too rapid condensation and to lead to the formation of hard deposits on the walls. Oil or glycerin at 100°C. is useful where water vapor at atmospheric pressure is the entrainer; this gives a dry sublimate directly. Baffles may be fitted to increase the condensing surface and to impede the direct passage of the vapors through the condenser. Arrangements should be made to keep surfaces free from deposits; swinging hammers and brushes,

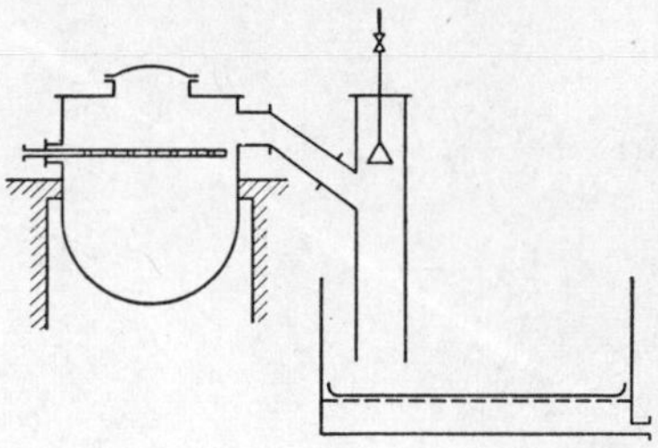

FIG. 103.

movable from without, are used. The discharge should be designed so that it is unnecessary for the workmen to enter the chamber to clean it thoroughly; as the sublimate is normally in the form of a light powder, it is the more easily raked out or discharged by gravity.

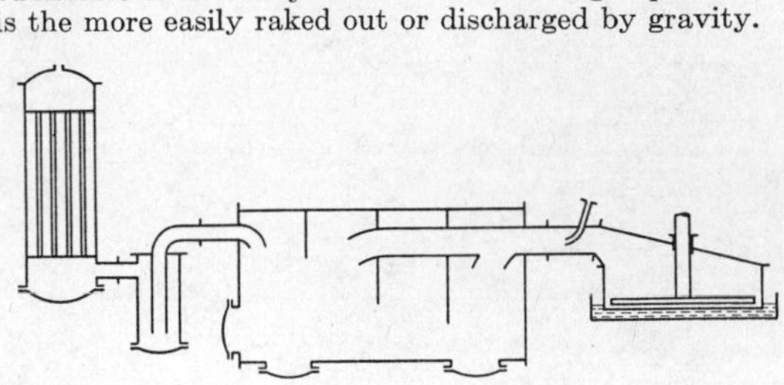

FIG. 104.

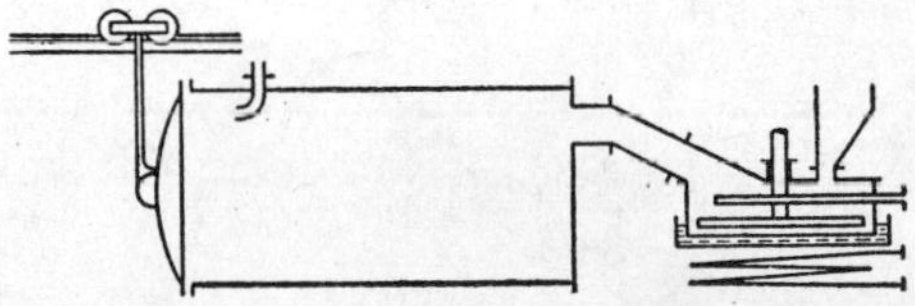

FIG. 105.

Auxiliary Equipment. Condensers and cold traps are used in vacuum sublimation to prevent traces of the sublimate reaching the vacuum pumps. In entrainer sublimation, when the exhaust is discharged to atmosphere, it is again important from both yield and health aspects that the very fine snow which escapes collection in the condenser should be trapped; cyclones are usually employed, but in severe cases scrubbing towers and cloth bags are necessary. The problem is not so acute when the entrainer is circulated in a closed system, but even here the escape of condensate from the condenser reduces the efficiency and may lead to trouble by blocked pipes and deterioration of quality by prolonging the exposure to heat.

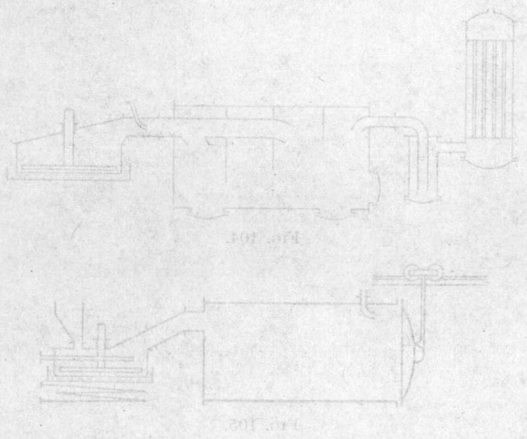

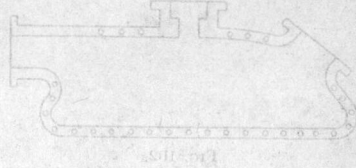

SECTION 10

GAS ABSORPTION

BY

Robert L. Pigford, Ph. D., Professor and Head of the Department of Chemical Engineering, University of Delaware; Member, American Institute of Chemical Engineers, American Chemical Society.

Allan P. Colburn, Ph. D., Assistant to the President and Professor of Chemical Engineering, University of Delaware; Member, American Institute of Chemical Engineers, American Chemical Society, American Society of Mechanical Engineers, American Society for Engineering Education.

CONTENTS

INTRODUCTION

	PAGE
Introduction	668
Outline of General Design Procedure	668

ILLUSTRATIVE EXAMPLE

Design of Column for Absorbing Acetone from Air by Water Scrubbing	669
Statement of Problem	669
Equilibrium Data	669
Liquor-gas Ratio	670
Choice of Packing and Column Diameter	670
H.T.U. of Packing	671
Optimum Exit-gas Strength	671
Number of Transfer Units	671
Column Height	672
Pressure Drop	672
Design Procedure for Multicomponent Systems	672

EQUILIBRIUM DATA

Solubility of Various Gases in Water	673
Solubility of Gases in Non-aqueous Pure Liquids and in Aqueous Solutions	676
Solubility Data for Hydrocarbons in Oil	676
Carbon Dioxide in Carbonate Solution	677
Sulfur Dioxide in Alkaline Solutions	679
Olefins in Cuprous Salt Solutions	679

LIMITING VELOCITIES AND PRESSURE DROPS FOR ABSORPTION COLUMNS

Packed Columns	680
Pressure Drop	680
Loading and Flooding Velocities	683
Liquid Hold-up	686
Plate Columns	686

RATE OF MASS TRANSFER IN ABSORPTION COLUMNS

Packed Columns—Gas-phase Resistance	687
Effect of Temperature and Pressure on *Hg*	687
Effect of Physical Properties of the Gas on *Hg*	687
Effect of Liquid Surface Tension on *Hg*	691

	PAGE
Packed Columns—Liquid-phase Resistance	691
Packed Columns—Over-all Resistance for Miscellaneous Systems	693
Ammonia in Water	693
Chlorine in Water	695
Carbon Dioxide in Water	695
Sulfur Dioxide in Water	696
Ketones in Water	696
Hydrogen Chloride in Water	697
Non-aqueous Systems	698
Plate Efficiencies of Bubble-tray Absorption Columns	698
Effect of Viscosity	698
Effect of Solubility	698
Effect of Solvent	698
Rates of Absorption in Spray Towers	699
Sulfur Dioxide Absorption	701
Rates of Absorption in Agitated Vessels	702
Absorption Accompanied by Chemical Reaction	702
Absorption of Carbon Dioxide in Alkaline Solutions	703
In Carbonate Solutions	703
In Ethanolamine Solutions	704
In Sodium or Potassium Hydroxide Solutions	704
Absorption of Ammonia in Acetic or Sulfuric Acids	706
Absorption of Sulfur Dioxide in Caustic	706
Absorption of Chlorine in Sodium Hydroxide	706
Absorption of Olefins and Acetylenes in Aqueous Copper Solutions	706
Absorption of Nitrogen Oxides in Water and Nitric Acid	706

ECONOMIC DESIGN OF ABSORPTION SYSTEMS

Packed Towers vs. Plate Towers	707
Column and Packing	707
Liquor-gas Ratio	707
Column Diameter (or Gas Velocity)	708
Column Height (Exit-gas Strength)	708
Costs of Packing Materials	709
Costs of Absorption Towers	709

DETAILS OF COLUMN CONSTRUCTION

Liquid Distribution in Packed Towers	710
Packing Supports	710

GENERAL NOMENCLATURE

General Nomenclature	710

GAS ABSORPTION

References: Sherwood, "Absorption and Extraction," McGraw-Hill, New York, 1937. Walker, Lewis, McAdams, and Gilliland, "Principles of Chemical Engineering," 3d ed., McGraw-Hill, New York, 1937. Badger and McCabe, "Elements of Chemical Engineering," 2d ed., McGraw-Hill, New York, 1936. Colburn, *Trans. Am. Inst. Chem. Engrs.*, **35**, 211 (1939). Colburn, *Ind. Eng. Chem.*, **33**, 459 (1941). Sherwood and Holloway, *Trans. Am. Inst. Chem. Engrs.*, **36**, 21, 39 (1940).

INTRODUCTION

Gas absorption is a unit operation in which a soluble component of a gas mixture is dissolved in a liquid. The inverse operation, called stripping or desorption, is employed when it is desired to transfer a volatile component from a liquid mixture into a gas. This section is concerned principally with the design of commercial equipment used for carrying out either of these operations continuously.

Usually the apparatus used for contacting a liquid and a gas stream continuously is a packed tower filled with irregular solid packing material, an empty tower into which the liquid is sprayed, or a plate-type unit containing a number of bubble-cap or sieve plates. Ordinarily, the gas and liquid streams are made to flow countercurrently past each other through the equipment in order that the greatest rate of absorption may be obtained.

The design procedure that is followed in order to specify the principal dimensions of such equipment is described in this section by means of an illustrative calculation for a problem that is typical of commercial practice. The experimental data that are required for making such calculations are summarized and presented in a form suitable for use.

There are three main steps in a design calculation for an absorption tower or a stripping tower, as follows:

1. Data on the vapor-liquid equilibrium relations for the system are used to determine the quantity of liquid needed to absorb the required amount of the soluble component from the gas or the quantity of gas needed to strip the required amount of the volatile component from the liquid.

2. Data on the liquid- and vapor-handling capacity of equipment of the type considered are used to determine the required cross section of the channels through which the liquid and vapor streams flow. Consideration of the economic factors involved may show that it is desirable to set the fluid velocities well below the maximum values that can be employed.

3. Equilibrium data and material balances are used to determine the number of equilibrium units (theoretical plates or transfer units) required for the separation desired. The difficulty of the separation depends on the degree of recovery that is economically most desirable. The required time of contact between the flowing streams, or the required height of the tower, can be calculated if data are available for the specific rate of transfer of material between the gas and liquid phases, expressed in terms of the plate efficiency or the height of one transfer unit (H.T.U.).

Experimental data applicable to each of these steps are summarized at the end of the illustrative calculation of an absorption tower.

OUTLINE OF GENERAL DESIGN PROCEDURE*

1. *Selection of Solvent.* Often a solvent is specified for a given absorption problem. Where choice is possible, consideration is given to liquids with high solubilities for the solute to reduce the amount to be circulated; sometimes a reversible chemical reaction will result in very high solubility and a minimum solvent rate. Data on the actual systems are necessary in the latter cases, and those available are given on pp. 673–680. Furthermore, the solvent should be relatively non-volatile, cheap, non-corrosive, stable, non-viscous, non-foaming, and preferably non-flammable. Since the exit gas usually leaves saturated with solvent, solvent loss may be costly; cheap solvents may therefore supplant ones of higher solubility. Water is generally used for gases fairly soluble in water, straw oil for hydrocarbons, and special chemical solvents for acid gases such as H_2S, CO_2, and SO_2.

2. *Selection of Vapor-liquid Equilibrium or Solubility Data.* Solubility values determine the liquid rate necessary for complete or economic solute recovery and so are essential to design. They may be obtained in one of the following ways:

a. From collection of data and references in this section beginning p. 673.

b. For ideal liquid solutions (similar chemical compounds, *e.g.*, series of hydrocarbons) use methods based on Raoult's law as discussed in Sec. 8, p. 535. For corrections due to high pressure, see p. 525.

c. For chemically dissimilar compounds not available under *a*, methods of extending and predicting data are given in Sec. 8, p. 532.

d. For mixtures for which no data are found or no predictions are possible under preceding paragraphs, reference is made to *Chemical Abstracts;* in the absence of the necessary data, actual experimental determinations may need to be performed.

3. *Calculation of Liquid-gas Ratio.* The minimum possible liquor rate is readily calculated from the entering-gas composition and solubility in the exit liquor, assuming saturation. It may be necessary to estimate the effect of the heat of solution of the gas on the exit-liquor temperature. Values of latent and specific heats are given in Sec. 3, and values of heat of solution (at infinite dilution) are given in Sec. 8, p. 530.

The actual liquor-gas ratio is greater than the minimum by from 25 to 100 per cent and may be arrived at by economic calculations, as shown by the example, p. 669, and discussed on p. 670.

4. *Selection of Column.* Usually packed columns are chosen for very corrosive materials, for very low pressure drop, for small-scale operation (say less than 2 ft. diameter), and for liquids that foam badly. Plate columns are preferred for large-scale operation (because they are cheaper), for very low liquor rates (where packing would be inadequately wetted), and where cooling is desired.

In packed towers the type of packing is chosen for reasons of mechanical strength, resistance to corrosion, and a balance of cost, capacity, and efficiency, as discussed on pp. 691 and 709. Packings found to be most economical and generally useful are 1- to 2-in. ceramic or

* Amplification of this outline will be found in the extensive example immediately following.

carbon rings (½-in. size for columns under 4 in. diameter), 1-in. saddles, 3-in. spiral or partition rings, drip-point tile, and wood grids. Cost data on packings and shells are given on p. 709*ff.*

5. *Calculation of Column Diameter.* Allowable vapor velocities in plate columns are sometimes chosen to limit entrainment to less than 10 per cent and at other times are calculated from established factors. Such information is given in Sec. 9. By using a value of *F*-factor within the range given in Table 36 of this section, safe design is ensured. For very high liquor-gas ratios, say over 5 to 1, liquor-flow capacity across the plate may determine the diameter; see p. 600.

For packed columns, flooding determines the minimum possible diameter, and usual design is for 50 to 75 per cent of the flooding velocity, data for which are given for common packings on p. 680*ff.* These safe operating velocities are normally rather close to the calculated economic velocities discussed on p. 709.

6. *Height of Column.* The height is dependent on the degree of removal of solute from the gases, which is usually an economic matter treated on p. 671 (for valuable solutes optimum recovery usually runs nearly complete, say 99+ per cent). To compute the economic recovery as well as the eventual height, it is necessary to have values of plate efficiency if a plate column, or of height of a transfer unit if a packed column. Data on plate efficiency of absorbers are given on p. 698. In case of packed columns, over-all values of H.T.U. given on pp. 693–698 are used if available for the conditions of the problem; otherwise, separate gas and liquid values of H.T.U., respectively, are estimated from data for the separate resistances, pp. 687–693, and combined as illustrated in the following example by Eq. (2).

The final calculation of height of a plate column requires finding the number of theoretical plates for the separation as described in Sec. 8, p. 548. For a packed column the number of transfer units is determined as illustrated in the example below, p. 671*ff.* and treated in Sec. 8, p. 552*ff.*

7. *Pressure Drop.* For plate columns methods of estimating pressure drop are given in Sec. 9, p. 601.

For packed columns data on typical packings are given on p. 681. The pressure drop at flooding for packings commonly used is around 2 in. of water per foot depth of packing. For operation at about 50 per cent of flooding the pressure drop is roughly ½ in. of water per foot depth. This value is a convenient one to keep in mind for operating purposes.

ILLUSTRATIVE EXAMPLE

Design of Column for Absorbing Acetone from Air by Water Scrubbing. The information that is required to begin a design of an absorber is (1) the total quantity of gas that must be handled per unit time, both under normal operating conditions and under conditions of peak loads; (2) the concentration of the soluble material as well as all other known materials in this gas, and its temperature and pressure; (3) the degree of recovery desired, or enough cost data to permit an estimation of this figure; (4) the solvent to be used, its temperature, and the concentration of the solute in the entering solvent; and (5) the pressure drop that is available for forcing the gas through the equipment. The corrosive properties of the materials must be known; they may limit the choice of the type of apparatus.

The following illustrative example is typical of design problems in which absorption occurs without chemical reaction.

1. *Statement of Problem.* A design is desired for a packed column for absorbing acetone vapor from air by water. Conditions given:

a. Normal quantity of total inlet gas = 100 lb.-moles/hr.
   Maximum overload possible = 50 per cent.
   Temperature = 35°C., pressure = 1 atm.
   Composition, by volume:
   acetone: 2 per cent.
   water: 4 per cent (about 70 per cent of saturation at 35°C.).
   air: 94 per cent.

b. Scrubbing liquor: water.
   Highest inlet temperature during year = 25°C.
   No acetone in inlet water.

c. Operation is based on 8400 hr./year.
   Cost of electricity = $0.008/kw.-hr.
   Water is clean, and gas is non-corrosive.
   Design is to be for optimum velocity and pressure drop.
   Acetone is to be recovered by distillation of the dilute water solution that flows from the absorber.

2. *Equilibrium Data.* A comparison of vapor-liquid equilibrium data for the acetone-water system is shown by Fig. 1, where the activity coefficient, or Raoult's-law

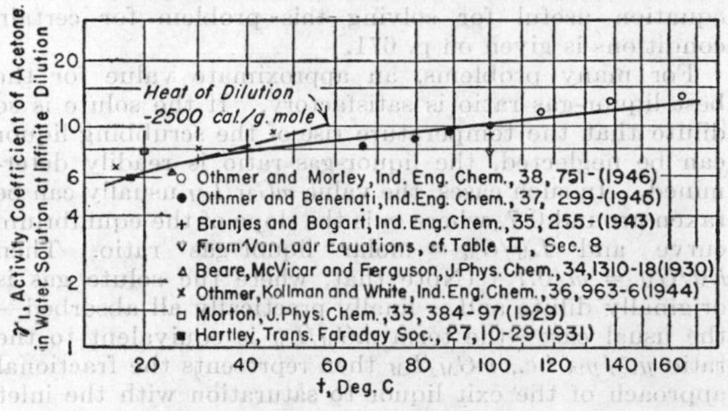

FIG. 1. Activity coefficient of acetone in dilute water solution.

deviation factor (cf. Sec. 8, pp. 526 to 535), extrapolated to very small concentrations of acetone in the liquid phase, is plotted against temperature. Below 50°C. most of the data have been obtained in experiments carried out at constant temperature with both air and acetone vapor in the vapor phase; above 50°C. the data are based on constant-pressure experiments with only acetone and water in the vapor. The dashed line shows the variation of the activity coefficient of acetone predicted from the partial molal heat of solution at 25°C. taken from Table 3, Sec. 8, p. 530.

A study of Fig. 1 shows that the vapor-liquid equilibrium data for this system at 30°C., the approximate temperature of the inert scrubbing liquid in this illustrative example, could not have been predicted exactly from high-temperature equilibriums, unless more extensive data on heats of solution had been available. Based on the data of Brunjes and Bogart [*Ind. Eng. Chem.*, **35**, 255 (1943)] obtained at the normal boiling point, the assumption that the activity coefficient is independent of temperature would have given an activity coefficient 30 per cent too large; and the assumption that the heat of solution, known to be −2500 cal./g.-mole at room temperature, is independent of temperature would have resulted in a value 40 per cent too small. Since the liquid-gas ratio required for the design is proportional to the activity coefficient in the dilute solution, the quantity of liquid required would have been in error by the same amounts. The effect on the calculated dimensions of the apparatus would be much less, however. Presumably the liquid rate could be adjusted after the equipment was placed in operation. This example illustrates the uncertainty that usually exists in the design of equipment of this type and points out the need for a careful analysis of available equilibrium data. It may often be advisable to determine the necessary equilibrium data experimentally.

The values of the solubility coefficient $K = y^*/x$,

derived from Fig. 1 are as follows (for nomenclature, see the end of the section):

t, °C.	25	30	35	40
γ_1, activity coefficient for acetone	6.7	7.1	7.5	7.8
P_1, vapor pressure of pure acetone, mm. Hg	229	283	346	421
$K = \frac{y^*}{x} = \gamma_1 \frac{P_1}{P}$	2.02	2.64	3.41	4.33

These values are to be used only when the value of x is small, less than 0.01, say. When the acetone concentration becomes larger its activity coefficient decreases, as shown in Sec. 8, p. 527.

3. *Liquor-gas Ratio.* The higher the liquor-gas ratio, the shorter the absorption column will be, but the weaker will be the liquor from the bottom. The best value of this ratio thus depends upon a balance between the cost of additional height of column and the cost of concentrating the solute in the scrubbing liquor. An equation useful for solving this problem for certain conditions is given on p. 671.

For many problems, an approximate value for the best liquor-gas ratio is satisfactory. If the solute is so dilute that the temperature rise of the scrubbing liquor can be neglected, the liquor-gas ratio is readily determined. In such cases, the value mG_M/L_M usually can be taken around 0.7, where m is the slope of the equilibrium curve and L_M/G_M = molar liquor-gas ratio. Then $L_M/G_M = m/0.7$. (Note that, where the solute gas is originally dilute and is finally practically all absorbed—the usual case—the term mG_M/L_M is equivalent to the ratio y_1^*/y_1; *i.e.*, mG_M/L_M then represents the fractional approach of the exit liquor to saturation with the inlet gas.)

Where the liquor temperature rises owing to heat of condensation of the solute, the value of m changes through the column, and a liquor-gas ratio must be chosen to give reasonable values of $K_1(G_M/L_M)_1$ and $m_2(G_M/L_M)_2$, where subscripts 1 and 2 refer to the bottom of the absorption column and the top, respectively. Where the temperature rise is considerable, the value of $m_2(G_M/L_M)_2$ should be taken somewhat less than 0.7 so that the value of $K_1(G_M/L_M)_1$ will not approach unity too closely.

To determine the value of K_1, the temperature changes are calculated. Let Q = heat loss by the gas, p.c.u./lb.-mole of entering gas mixture; this equals the heat of condensation and heat of solution of the solute, plus the sensible heat lost by the gas, less the heat of vaporization of any liquor added to the gas (which will leave nearly saturated with scrubbing liquor at the temperature of the inlet liquor). Then the rise in temperature of the liquor is found from

$$\text{Temperature rise} = \frac{QG_M}{C_L L_M} \qquad (1)$$

where C_L = molar heat capacity of the liquor.

The heat of solution of some organic compounds in water may amount to 2000 p.c.u./lb.-mole (*cf.* p. 530*ff.*, Sec. 8). For electrolytes like HCl, the heat of solution is of the same magnitude as a heat of vaporization (*cf.* p. 530). Data on heats of solution in dilute solutions are given in Tables 3 and 4 of Sec. 8, pp. 530 to 531. Latent heats are given in Sec. 3, pp. 210 to 218, and heat capacities on pp. 225 to 235.

In case of appreciable heat effects, consideration should be given to the possibility of installing cooling units in the lower part of the column to reduce the outlet temperature and thus make possible a greater exit strength. One procedure is to have several bubble-cap or sieve plates below the main packed section, since cooling coils can be conveniently installed on these plates. The plates will remove a large portion of the total solute and the resulting heat.

Another method for removing the heat of solution involves the use of an external cooler through which the lean liquid is circulated before it is returned to the column. It may be desirable, in a few instances, to return the liquid effluent directly to the top of the tower through a cooler, so that the liquid rate through the tower is several times greater than the rate at which fresh liquid is fed to the system. This is permissible only when the gas is very soluble. Otherwise the rate of absorption is limited by the higher average liquid concentration, and successive towers are required to obtain the driving forces corresponding to countercurrent conditions. Liquid recirculation is used also for batch absorption and when the net flow of liquid to the absorption system is too small to wet the packing evenly.

For the example previously outlined, the gas can be assumed to leave the tower at the temperature of the inlet water, 25°C., saturated with water vapor. The vapor pressure of water at 25°C. is 24 mm. Hg; the exit gas therefore contains 3.16 per cent vapor. The quantity of exit gas is then 94 moles/hr. of air and 3 moles/hr. of water vapor, making a total of 97 moles/hr. of exit gas, or 0.97 of the total quantity entering.

At around 25° to 30°C., the latent heat of pure acetone is 7500 p.c.u./lb.-mole, and that of water is 10,400 p.c.u./lb.-mole; the heat of dilution amounts to 2500 p.c.u./lb.-mole, making the total heat of condensation of acetone in dilute aqueous solution equal to 10,000 p.c.u./lb.-mole of acetone. Taking the molar heat capacity of air at 7 p.c.u./(lb.-mole)(°C.), the heat removed from the gas per mole of entering gas is:

Heat of condensation of acetone = (10,000)(0.02)
= 200 p.c.u./(hr.) (lb.-mole of entering gas)

Heat of condensation of water vapor (noting that $G_2 = 0.97G_1$)
= (10,400)[0.04 − (0.0316)(0.97)]
= 97 p.c.u./(hr.)(lb.-mole of entering gas)

Heat of cooling gas = 7(35 − 25)
= 70 p.c.u./(hr.)(lb.-mole of entering gas)

Heat lost, Q = 200 + 97 + 70
= 367 p.c.u./(hr.)(lb.-mole of entering gas)

Now the liquor-gas ratio is found by cut and try. For these dilute solutions, $K = m$. Taking $K_2(G_M/L_M)_2$ as, say, 0.65 gives, for m_2 = 2.02 (at 25°C.), $(L_M/G_M)_2$ = (2.02/0.65) = 3.10, and L_M = (3.10)(97) = 301 lb.-moles/hr. at the top of the column. For this liquor-gas ratio, noting that C_L = 18 p.c.u./(lb.-mole)(°C.) for water, the temperature rise of the liquor = (367)/(18)(3.10) = 6.6°C., and the exit-liquor temperature = 31.6°C. At the bottom of the column, L_M = 301 + 3 = 304, and $(L_M/G_M)_1 = 304/100 = 3.04$. At the exit-liquor temperature, K_1 = 2.85. Then $K_1(G_M/L_M)_1$ = 2.85/3.04 = 0.938. In view of the value of 0.65 taken for mG_M/L_M at the top of the tower, this value of KG_M/L_M can be regarded for the time being as not so great as to require an excessively tall tower.

Note that, on a weight basis, the approximate L/G = (3.1)(18)/(29) = 1.92 at the top of the column.

4. *Choice of Packing and Column Diameter.* In choosing a packing, one first considers factors such as corrosiveness of the gas or liquor, the availability of any packings in stock or obsolete apparatus, the necessity for cleaning in case of dirty scrubbing liquor, and physical properties such as friability. Then, from such packings as fill these requirements, the one is picked having the least cost for a given performance. The comparison of cost should be worked out as a comparison of annual cost of packing and column and of pressure drop when operated at the economic velocity or at a safe rate considering flooding. The cost factors that affect the choice of the packing are discussed on pp. 709 to 710.

The lower limit on column diameter is set by the velocity at which the column will flood at the given liquor-gas ratio. Usually from 50 to 75 per cent of this velocity is employed in design calculations to allow for the possibility that a future increase in capacity may be desired. Data on flooding and loading velocities are given on pp. 683 to 686.

Consideration is also given to the velocity that makes the sum of the costs a minimum. As shown on p. 708, this velocity may also be in the range of 50 to 75 per cent of the flooding velocity, unless the cost of the packing and the shell is large and the cost of power is low, in which case the calculated optimum velocity may exceed the flooding velocity.

One-inch stoneware rings are satisfactory from a construction-material viewpoint for this design; this packing is found to give a low annual cost. From Fig. 13, the flooding velocity is found, at $L/G = 1.92$, $\phi = 1$, to be $G = 1260$ lb./(hr.)(sq. ft.). The optimum velocity for this packing for a shell cost = \$9.50/cu. ft., packing cost = \$5.00/cu. ft., amortization rate = 0.2 [whence $C_1 = \$2.90/(\text{cu. ft.})(\text{yr.})$], cost of delivered energy = \$0.013/kw.-hr., 8400 hr./year operation, $b = 1.3$ in. water-pressure drop/ft. of packing is calculated from Eq. (45), p. 708, to be $G_{\text{opt}} = 735$ lb./(hr.)(sq. ft.). This value is 58 per cent of the flooding velocity, and hence a satisfactory value, since a 50 per cent overload is possible without flooding.

The gas quantity to the column is (taking the molecular weight roughly as 29) $w = (100)(29) = 2900$ lb./hr. Therefore the cross-sectional area required is $S = w/G = 2900/735 = 3.95$ sq. ft. This area is equivalent to a diameter of 2.24 ft., or very close to 27 in. diameter. Assume that the shell will be constructed from 30-in. standard steel pipe having an inside diameter of 29.38 in. Note that this diameter is equivalent to $G = 616$ lb./(hr.)(sq. ft.). The liquor rate is then $(1.92)(616) = 1181$ lb./(hr.)(sq. ft.).

5. *H.T.U. of Packing.* In the following design, the over-all value of H.T.U. based on the gas-phase driving force is utilized, including the resistances in both the gas and the liquor films. In the absence of reliable test values of H_{OG} for the conditions of this problem, the value is obtained as follows:

$$H_{OG} = H_G + H_L \left(\frac{mG_M}{L_M}\right)\left(\frac{1 - x_f}{1 - y_f}\right) \tag{2}$$

where H_G is the gas-film value of H.T.U., primarily a function of the gas and liquor rates and physical properties of the gas, and H_L is the liquor-film H.T.U., a function of liquor rate and temperature (see the end of the section for table of nomenclature). For derivation of this equation, see Sec. 8, p. 550. The values $(1 - x)_f$ and $(1 - y)_f$ can generally be taken as unity at this point unless the gas or liquid concentrations are large. The value m is equal to dy^*/dx. Where m varies through the column, the value at the dilute end of the column gives approximately the correct answer, since most of the transfer units are concentrated in the dilute end.

Data on H.T.U. for various packings are given on pp. 688–698. For 1-in. stoneware rings, the gas-film H.T.U. for ammonia-air for $G = 616$ lb./(hr.)(sq. ft.) and $L = 1181$ lb./(hr.)(sq. ft.) is found from Fig. 23 to be 1.2 ft. Now the value of $(\mu/\rho D)$ for the mixture acetone-air is found in Table 12, p. 539, Sec. 8, to be 1.59, whereas the value for ammonia-air is 0.78. Therefore, the gas film H.T.U. for the acetone-air mixture is $1.2\ (1.59/0.78)^{1/2} = 1.7$ ft.

Lacking a liquor-film H.T.U. for acetone-water on 1-in. rings, the H.T.U. will be estimated from Fig. 31. At a liquor rate of 1181 lb./(hr.)(sq. ft.), the H.T.U. for oxygen absorption is 0.77 ft. at 25°C., which is corrected to 35°C. by multiplying by 0.81, taken from the small temperature-correction plot on Fig. 31. Assuming a diffusion coefficient of 0.9×10^{-5} cm.²/sec. for acetone in water at 20°C., the H.T.U. for oxygen is multiplied by the ratio of the diffusion coefficient for oxygen to that of acetone raised to the 0.47 power, resulting in $H_L = 0.87$ ft.

Then, for $mG_M/L_M = 0.7$, say as an average,

$$H_{OG} = 1.7 + (0.87)(0.7) = 2.3 \text{ ft.}$$

6. *Optimum Exit-gas Strength.* The lower the exit-gas strength, the less the cost of solute wasted, but the higher will be the column required to exhaust the gas. A balancing of the cost of waste solute against the cost of column results in Eq. (48), p. 708, which gives the optimum mole fraction of solute in the exit gas:

$$(y_2 - mx_2)_{\text{opt}} = \frac{C_3 H_{OG}}{C_4 \theta G_{M_2}(1 - mG_M/L_M)_2} \tag{48}$$

where C_3 = annual cost of apparatus and power, \$/(cu. ft.)(year); C_4 = value of solute at its concentration in the exit liquor, \$/lb.-mole of solute; θ = number of hours per year operation; G_{M_2} = exit molar gas velocity, lb.-moles/(hr.)(sq. ft.). If there is no solute in the entering liquor, $x_2 = 0$.

If the gas velocity is the optimum value, $C_3 = 1.5\ C_1$; at other than the optimum velocity, $C_3 = C_1[(G_{\text{opt}}/G) + 0.5(G/G_{\text{opt}})^2]$; C_4 = value of 1 lb.-mole of pure solute less the cost of concentrating it (the concentration cost is usually small compared with the material value). The over-all H.T.U. is found for the liquor and gas rates used.

Explosive, toxic, or corrosive gases may require that the exit composition be lower than the optimum.

From paragraph 4 above, $C_1 = 2.90$; and, since the chosen value of G is practically the optimum, $C_3 = (1.5)(2.90) = \$4.35/(\text{cu. ft.})(\text{year})$. In paragraph 5, the value of H_{OG} was found to be 2.3 ft.

The value of acetone is taken as \$0.07/lb. recovered. Therefore, taking rectification cost as \$0.005/lb., $C_4 = (0.065)(58) = \$3.77/\text{lb.-mole}$.

Since $G = 616$ lb./(hr.)(sq. ft.), $G_M = 616/29 = 21.2$ lb.-moles/(hr.)(sq. ft.). The hours per year operation is $\theta = 8400$ hr./year.

Taking $m_2(G_M/L_M)_2$ from paragraph 3 as 0.65, the optimum mole fraction of acetone in the exit gas becomes

$$(y_2)_{\text{opt}} = \frac{(4.35)(2.3)}{(3.77)(8400)(21.2)(0.35)} = 0.000043$$

and

$$\frac{y_1}{y_2} = \frac{0.02}{0.000043} = 470$$

7. *Number of Transfer Units.* Case A. Constant mG_M/L_M. For the case of sufficiently dilute solutions so that m and L_M/G_M are considered constant through the column, the number of transfer units based on over-all changes in gas compositions is given by Eq. (120), Sec. 8, p. 554, or Fig. 28.

Case B. mG_M/L_M Not Constant. When the solute in the inlet gas is not dilute, the operating line will generally be curved, and the equilibrium values will form a curved line because of the rise in temperature of the liquid from the heat of absorption. An approximate equation can be derived by dividing the column into two sections, as shown by Colburn [*Ind. Eng. Chem.*, **33**, 459 (1941)]. In the dilute end, the above equation for constant (mG_M/L_M) applies. The equilibrium curve is assumed parabolic in the concentrated end, resulting in an equation for the number of transfer units in the whole column:

$$N_{OG} = \frac{2.3}{\left(1 - \frac{mG_M}{L_M}\right)_2} \times \log\left[\left(1 - \frac{mG_M}{L_M}\right)_2 \left(\frac{y_1 - mx_2}{y_2 - mx_2}\right) \frac{\left(1 - \frac{mG_M}{L_M}\right)_2}{\left(1 - \frac{KG_M}{L_M}\right)_1} + \left(\frac{mG_M}{L_M}\right)_2\right] \tag{3}$$

Values of $1 - [(mG_M/L_M)]_2$ and $1 - [(KG_M/L_M)]_1$ are to be calculated at the dilute and concentrated ends of the column, respectively.

Figure 28, Sec. 8, p. 554 may be used to solve this equation if the numbers on the abscissa are interpreted as values of

$$\frac{y_1 - mx_2}{y_2 - mx_2} \frac{1 - [(mG_M/L_M)]_2}{1 - [(KG_M/L_M)]_1}$$

and if the values of (mG_M/L_M) on the several curves are understood to refer to the dilute end of the column. This equation holds for the usual case where most of the solute is absorbed (say $y_1/y_2 > 30$). It applies with an error not exceeding a few per cent even when the solute concentration in the inlet gas is as high as 10 per cent and where the liquor temperature increases considerably in flowing down the column. The equation should not be used where values of y^* vary in an unusual manner with x, as in the case of concentrated HCl solutions or of SO_2 solutions in water.

Often where there is considerable change in mG_M/L_M, the logarithmic mean driving force can be used if the column is divided in two parts and the logarithmic mean is used over each part.

For special cases where the above methods do not apply and an exact design is desired, a graphical solution must be made, as described in Sec. 8, p. 553. When there is no equilibrium partial pressure from the solution, the number of transfer units is calculated by Eq. (37), below.

Returning now to the illustrative example of an acetone absorber, the number of transfer units is now calculated using Eq. (3), utilizing the following values, y_2 (optimum exit gas strength) = 0.000043, $y_1 = 0.02$, $m_2G_M/L_M = 0.65$, and $K_1G_M/L_M = 0.938$. The result is 19.5 transfer units.

8. *Column Height.* Given the number of transfer units and the H.T.U. of the packing, the height of the packed section then is

$$\text{Height} = (N)H_{OG} \tag{4}$$

To this height of packing must be added a few feet of shell at the top for liquor distribution and a few feet at the bottom for packing support and gas entrance.

For the example being considered, $N = 19.5$ and $H_{OG} = 2.3$ ft., whence height of packing = (19.5)(2.3) = 45 ft. Adding about 2 ft. on each end for support and liquor distribution gives a 49-ft. height of shell.

9. *Pressure Drop.* At the given gas and liquor velocities, the pressure drop is calculated from data on various packings as given by Figs. 7 to 9, pp. 680 to 682.

From Fig. 7 for 1-in. porcelain rings, at $G = 616$ lb./(hr.)(sq. ft.), $L = 1181$ lb./(hr.)(sq. ft.) and $\phi = 1$, $\Delta P = 0.33$ in water/ft. height. The over-all pressure drop is then (45)(0.33) = 15 in. water.

Design Procedure for Multicomponent Systems. If more than one soluble component is to be absorbed from an insoluble gas, the design conditions chosen (principally the liquor-gas ratio) should be determined by the volatility, or solubility, of the most insoluble constituent that it is economical to recover completely. Less volatile components will also be recovered completely; absorption of more volatile components will be incomplete, even though the effluent liquid becomes saturated with respect to these components. When the latter condition exists, even an infinite number of transfer units or plates gives only a limited, finite value of y_1/y_2, as shown by the vertical lines in Fig. 28, Sec. 8, for $mG_m/L_m > 1$. If there is no solute in the fresh liquid fed to the absorption column, the limiting value of y_1/y_2 is equal to $\lambda/(\lambda - 1)$, where $\lambda = mG_m/L_m > 1$. Figure 2 (from Sherwood, "Absorption and Extraction," p. 113, McGraw-Hill, New York, 1937) illustrates the significance of these considerations for a column used to absorb hydrocarbons in an oil. For this case, the liquor-gas ratio is chosen such that complete recovery of pentane and heavier components is obtained. Six theoretical plates are sufficient to give 77 per cent recovery of butane, but even an infinite number of plates would give only limited recovery of the lighter constituents.

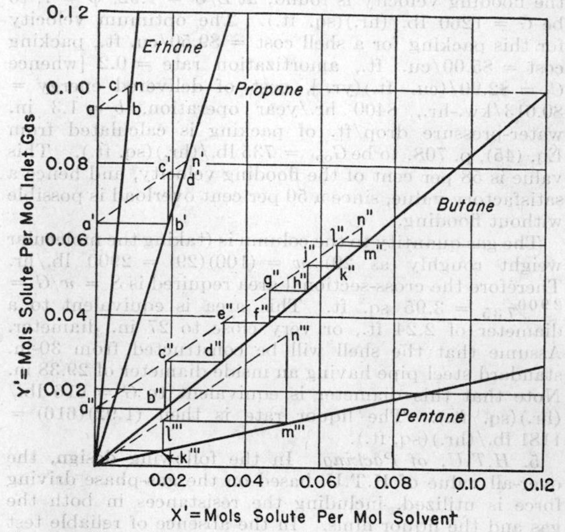

FIG. 2. Graphical computation of theoretical plates for multicomponent absorption. Points *a*, *c*, *e*, etc., show compositions of gas and liquid streams. Points marked *a* correspond to the bottom of the tower, points marked *n* to the top. (*After Sherwood, Absorption and Extraction, McGraw-Hill, New York,* 1937.)

When the gas stream is dilute, absorption of each constituent can be considered separately as though the other soluble components were absent. In concentrated gases, however, the change in gas- and liquor-flow rates within the column and the heat effects accompanying absorption of all components must be considered. Usually a trial-and-error calculation is involved. The degree of recovery of the light components is not known until mole fractions of the various components in the effluent liquid and the temperature of this liquid are known; these properties depend in turn on the degree of recovery. Examples involving concentrated gases are discussed by Sherwood (*op. cit.*, pp. 110–125) and by Sherwood and Jackson [*Trans. Am. Inst. Chem. Engrs.*, **37**, 959 (1941)].

Example. The air entering the tower in the example discussed above contains 1 per cent of acetaldehyde in addition to 2 per cent of acetone. (1) What will be the percentage recovery of acetaldehyde if the tower recovers the optimum amount of acetone? (2) What will be the percentage of acetone recovered if the tower is redesigned to recover the optimum amount of acetaldehyde?

1. The value of y^*/x for acetaldehyde is measured as 50 at the boiling point of a dilute solution, 93.5°C. Neglecting the heat of solution, y^*/x at 31.5°C. is equal to 50 (${}^{120}\!/_{7300}$) = 8.2, where the factor in parentheses is the ratio of pure acetaldehyde vapor

pressures at 31.5 and 93.5°C., respectively. Since $(L_m/G_m)_2 = 2.7$ in the acetone problem, $(mG_m/L_m)_2$ for the aldehyde is $8.2/3.1 = 2.64$, and y_1/y_2 for the aldehyde equals $2.64/1.64 = 1.61$, corresponding to only 38 per cent recovery.

2. When the liquor rate is set by the aldehyde equilibrium, $(L_m/G_m)_2 = 8.2/0.7 = 11.7$ approximately. Assuming $H_{OG} = 3.7$ ft. for acetaldehyde, the optimum exit-aldehyde strength is

$$y_2 = \frac{C_3H_{OG}}{C_4\theta G_m[1 - (mG_m/L_m)]_2} = \frac{(4.35)(3.7)}{(4.85)(8400)(21.2)(0.3)} = 6.2 \times 10^{-5} \text{ mole fraction}$$

based on C_3 = \$4.35/(cu. ft.)(year) for fixed charges and power, and C_4 = \$4.85/lb.-mole pure aldehyde. The number of transfer units required, for $y_1/y_2 = 1 \times 10^{-2}/6.2 \times 10^{-5} = 160$ and $mG_m/L_m = 0.7$, is 12.9 by Eq. (3). Since $m = 2.8$, $mG_m/L_m = 2.8/11.7 = 0.24$, $H_{OG} = 2.0$ ft., and $N_{OG} = 12.9(3.7/2.0) = 24$ for acetone, $y_1/y_2 = 1.0 \times 10^8$. The loss of acetone under these conditions is only about 0.000001 per cent of that entering.

EQUILIBRIUM DATA

Solubility of Various Gases in Water. In order to define the solubility factor of a gas in a liquid, it is generally necessary to state the temperature, the equilibrium partial pressure of the solute gas in the gas phase, and the concentration of the solute gas in the liquid phase. (Strictly speaking the total pressure on the system as well as the partial pressure of the solute gas should be stated, but where the total pressure is not more than a few, perhaps 5, atmospheres the solubility for a particular partial pressure of solute gas may be safely considered independent of total pressure.) The solubility of NH_3 (Table 3) at a temperature of 30°C. for a partial pressure of NH_3 of 260 mm. is given as 20 weights of NH_3 per 100 weights of H_2O. This method of stating temperature, partial pressure of solute gas in the gas phase, and concentration of solute in the liquid phase will be employed for systems where Henry's law does not hold.

If Henry's law holds, solubility is defined by giving the Henry's law constant H and the temperature where $H = p_A/x_A$ = atm./mole fraction of solute in solution. For quite a number of gases, Henry's law holds very well where the partial pressure of the solute gas does not exceed 1 atm. For partial pressures of solute gas greater than 1 atm., H is seldom independent of the partial pressure of the solute gas, and a given value of H can be used over only a narrow range of partial pressures. In defining gas solubility at these higher pressures, the partial pressure of the solute gas as well as the temperature and the value of H must be specified. In the following tables, if the partial pressure of the solute gas is not specified, the values of H may be safely used only for partial pressures of solute gas not greater than 1 atm. Where the partial pressure of the solute gas is specified, the given values of H may be used for partial pressures not more than perhaps an atmosphere higher or lower than the stated partial pressure. The use of Henry's law constants is illustrated by the examples given below.

Example 1. It is desired to find out how much hydrogen may be dissolved in 100 weights of water from a gas mixture when the total pressure is 760 mm., the partial pressure of H_2 is 200 mm., and the temperature is 20°C.

For partial pressures of H_2 up to 1 atm., the value of H is 6.83×10^4 at 20°C. (see Table 12).

$$x_A = \frac{p_A}{H}$$

$$p_A = \frac{200}{760} = 0.263 \text{ atm.}$$

$$x_A = \frac{0.263}{68{,}300} = 0.00000385$$

where x_A is the mole fraction of H_2 in the liquid phase. (Mole fraction is the ratio of the number of moles of a particular constituent contained in a given weight of the solution to the total moles of all constituents contained.) To calculate the units of weight of H_2 per 100 weights of H_2O, the following formula may be used:

$$\frac{x_A}{1 - x_A}\frac{M_A}{M_S}100 = \frac{0.00000385}{1 - 0.00000385}\frac{2.02}{18.02}100 = (0.0000431)$$

Thus, 0.0000431 weight of H_2 may be dissolved in 100 weights of H_2O at 20°C. from a gas mixture where the partial pressure of H_2 is 200 mm.

Example 2. Oxygen is dissolved in water to the extent of 0.03 weight of O_2 per 100 weights of H_2O. What equilibrium partial pressure of O_2 would this solution exert at 25°C.?

Take as a basis 100 weights of H_2O.

$$x_A = \frac{0.03/32}{0.03/32 + 100/18} = 0.0001688$$

$$p_A = Hx_A$$

If p_A is greater than 1 atm., the value of p_A should be known before the proper value of H can be selected. A trial-and-error solution is indicated. As a first approximation, assume that p_A will not exceed 1 atm. and select the value of H corresponding to 25°C. from Table 21.

$$H = 4.38 \times 10^4$$
$$p_A = 43{,}800 \times 0.0001688 = 7.39 \text{ atm.}$$

Select another value of H for a partial pressure of 7.39 atm. (5620 mm.) from Table 22, interpolating to obtain a value for 25°C.

$$H = 4.89 \times 10^4$$
$$p_A = 48{,}900 \times 0.0001688 = 8.25 \text{ atm.} = 6280 \text{ mm.}$$

A third approximation, using Table 22, assuming $p_A = 8.35$ atm., gives a value of p_A which is as accurate as the available values of H will permit.

$$H = 4.95 \times 10^4$$
$$p_A = 49{,}500 \times 0.0001688 = 8.35 \text{ atm.} = 6350 \text{ mm.}$$

Thus, 0.03 weight of O_2 dissolved in 100 weights of H_2O would exert a partial pressure of 6350 mm. at 25°C.

There may also be a sufficiently close, though less accurate, proportionality between the concentrations of the gas in the liquid and the gas phases when compositions are expressed in other units, particularly when comparatively dilute solutions are involved. Henry's law, though quite useful if it can be applied, must be checked experimentally in each instance to determine the accuracy with which it can be used. The following tables and charts give data on the solubility of some of the more **common gases in water.**

The solubility tables have been taken from the "International Critical Tables" and other reliable sources. In many instances, the tables here presented represent only a part of the solubility data given in the original. Where the data given are not sufficient for a particular problem, reference to the original is recommended. Markham and Kobe [*Chem. Rev.*, **28**, 519 (1941)] have summarized and critically reviewed the gas-solubility data that were available prior to 1941.

In the tables given below the various solute gases are considered in alphabetical order.

Table 1.[1] **Acetylene** (C_2H_2)

t, °C.	0	5	10	15	20	25	30
$10^{-3} \times H$*	0.72	0.84	0.96	1.08	1.21	1.33	1.46

"International Critical Tables," vol. 3, p. 260, McGraw-Hill, 1928.

[1] Superior numbers refer to table footnote references on p. 676.

* The H in these solubility tables is the proportionality constant for the expression of Henry's law, $p = Hx$, where x = mole fraction of the solute in the liquid phase; p = partial pressure of the solute in the gas phase, expressed in atmospheres; H = a proportionality constant and is in units of atmospheres of solute pressure in the gas phase per unit concentration of the solute in the liquid phase. (The unit of concentration of the solute in the liquid phase is moles solute per mole solution.)

Table 2.[2] Air

t, °C.	0	5	10	15	20	25	30	35
$10^{-4} \times H$*	4.32	4.88	5.49	6.07	6.64	7.20	7.71	8.23

t, °C.	40	45	50	60	70	80	90	100
$10^{-4} \times H$*	8.70	9.11	9.46	10.1	10.5	10.7	10.8	10.7

"International Critical Tables," vol. 3, p. 257.
* H is calculated from the absorption coefficients of O_2 and N_2, taking into consideration the correction for constant argon content.

Table 3.[3,4] Ammonia (NH_3)

Weight NH_3 per 100 weights H_2O	Partial pressure of NH_3, mm. Hg							
	0°C.	10°C.	20°C.	25°C.	30°C.	40°C.	50°C.	60°C.
100	947							
90	785							
80	636	987						
70	500	780						
60	380	600	945					
50	275	439	686					
40	190	301	470	.....	719			
30	119	190	298	.....	454	692		
25	89.5	144	227	.....	352	534	825	
20	64	103.5	166	.....	260	395	596	834
15	42.7	70.1	114	.....	179	273	405	583
10	25.1	41.8	69.6	....	110	167	247	361
7.5	17.7	29.9	50.0	....	79.7	120	179	261
5	11.2	19.1	31.7	....	51.0	76.5	115	165
4	.....	16.1	24.9	....	40.1	60.8	91.1	129.2
3	.....	11.3	18.2	23.5	29.6	45	67.1	94.3
2.5	.....	.....	15.0	19.4	24.4	(37.6)*	(55.7)	77.0
2	.....	.....	12.0	15.3	19.3	(30.0)	(44.5)	61.0
1.6	.....	.....	.....	12.0	15.3	(24.1)	(35.5)	48.7
1.2	.....	.....	.....	9.1	11.5	(18.3)	(26.7)	36.3
1.0	.....	.....	.....	7.4	.....	(15.4)	(22.2)	30.2
0.5	.....	.....	.....	3.4				

* Extrapolated values.

Table 4.[5,6] Bromine (Br_2)

t, °C.	0	5	10	15	20	25
$10^{-2} \times H$	0.213	0.275	0.366	0.466	0.593	0.737

, °C.	30	40	50	60	70	80
$10^{-2} \times H$	0.905	1.33	1.91	2.51	3.21	4.04

"International Critical Tables," vol. 3, p. 255.

Table 5.[7] Carbon Dioxide (CO_2)

t, °C.	0	5	10	15	20	25	30	35	40	45	50	60
$10^{-3} \times H$	0.728	0.876	1.04	1.22	1.42	1.64	1.86	2.09	2.33	2.57	2.83	3.41

"International Critical Tables," vol. 3, p. 260.

Table 6.[21] Carbon Dioxide (CO_2)

Total pressure, atm.	Weight of CO_2 per 100 weights of H_2O*								
	12°C.	18°C.	25°C.	31.04°C.	35°C.	40°C.	50°C.	75°C.	100°C.
25	....	3.86	....	2.80	2.56	2.30	1.92	1.35	1.06
50	7.03	6.33	5.38	4.77	4.39	4.02	3.41	2.49	2.01
75	7.18	6.69	6.17	5.80	5.51	5.10	4.45	3.37	2.82
100	7.27	6.72	6.28	5.97	5.76	5.50	5.07	4.07	3.49
150	7.59	7.07	....	6.25	6.03	5.81	5.47	4.86	4.49
200	....	....	....	6.48	6.29	6.28	5.76	5.27	5.08
300	7.86	7.35	....	....	....	....	6.20	5.83	5.84
400	8.12	7.77	7.54	7.27	7.06	6.89	6.58	6.30	6.40
500	....	....	....	7.65	7.51	7.26	....	....	....
700	....	....	....	....	....	....	7.58	7.43	7.61

* In the original, concentration is expressed in cubic centimeters of CO_2 (reduced to 0°C. and 1 atm.) dissolved in 1 g. of water.

Table 7.[2,9] Carbon Monoxide (CO)

t, °C.	0	5	10	15	20	25	30	35
$10^{-4} \times H$	3.52	3.96	4.42	4.89	5.36	5.80	6.20	6.59

t, °C.	40	45	50	60	70	80	90	100
$10^{-4} \times H$	6.96	7.29	7.61	8.21	8.45	8.45	8.46	8.46

"International Critical Tables," vol. 3, p. 260.

According to Whitney and Vivian [*Ind. Eng. Chem.*, **33**, 741 (1941)], the solubility of chlorine in water follows the equation

$$C = H'p + (K_e H'p)^{1/3} \quad \text{lb.-moles } Cl_2/\text{cu. ft.} \qquad (5)$$

obtained by assuming that the solubility of molecular chlorine follows Henry's law and that the equilibrium in

Table 8.[10] Carbon Monoxide (CO)

Partial pressure of CO, mm. Hg	$10^{-4} \times H$	
	17.7°C.	19.0°C.
900	4.77	4.88
2000	4.77	4.91
3000	4.77	4.93
4000	4.78	4.95
5000	4.80	4.97
6000	4.82	4.98
7000	4.86	5.02
8000	4.88	5.08

"International Critical Tables," vol. 3, p. 260.

Table 9.[22] Chlorine (Cl_2)

Partial pressure of Cl_2, mm. Hg	Solubility, g. of Cl_2 per liter					
	0°C.	10°C.	20°C.	30°C.	40°C.	50°C.
5	0.488	0.451	0.438	0.424	0.412	0.398
10	.679	.603	.575	.553	.532	.512
30	1.221	1.024	.937	.873	.821	.781
50	1.717	1.354	1.210	1.106	1.025	.962
100	2.79	2.08	1.773	1.573	1.424	1.313
150	3.81	2.73	2.27	1.966	1.754	1.599
200	4.78	3.35	2.74	2.34	2.05	1.856
250	5.71	3.95	3.19	2.69	2.34	2.09
300	.....	4.54	3.63	3.03	2.61	2.31
350	.....	5.13	4.06	3.35	2.86	2.53
400	.....	5.71	4.48	3.69	3.11	2.74
450	.....	6.26	4.88	3.98	3.36	2.94
500	.....	6.85	5.29	4.30	3.61	3.14
550	.....	7.39	5.71	4.60	3.84	3.33
600	.....	7.97	6.12	4.91	4.08	3.52
650	.....	8.52	6.52	5.21	4.32	3.71
700	.....	9.09	6.90	5.50	4.54	3.89
750	.....	9.65	7.29	5.80	4.77	4.07
800	.....	10.21	7.69	6.08	4.99	4.27
900	.....	......	8.46	6.68	5.44	4.62
1000			9.27	7.27	5.89	4.97
1200	$Cl_2 \cdot 8H_2O$ separates		10.84	8.42	6.81	5.67
1500	.....	......	13.23	10.14	8.05	6.70
2000	.....	......	17.07	13.02	10.22	8.38
2500	.....	......	21.0	15.84	12.32	10.03
3000	.....	......	......	18.73	14.47	11.70
3500	.....	......	......	21.7	16.62	13.38
4000	.....	......	......	24.7	18.84	15.04
4500	.....	......	......	27.7	20.7	16.75
5000	.....	......	......	30.8	23.3	18.46

Partial pressure of Cl_2, mm. Hg	Solubility, g. of Cl_2 per liter					
	60°C.	70°C.	80°C.	90°C.	100°C.	110°C.
5	0.383	0.369	0.351	0.339	0.326	0.316
10	.492	.470	.447	.431	.415	.402
30	.743	.704	.671	.642	.627	.598
50	.912	.863	.815	.781	.747	.722
100	1.228	1.149	1.085	1.034	.987	.950
150	1.482	1.382	1.294	1.227	1.174	1.137
200	1.706	1.580	1.479	1.396	1.333	1.276
250	1.914	1.764	1.642	1.553	1.480	1.413
300	2.10	1.932	1.793	1.700	1.610	1.542
350	2.28	2.10	1.940	1.831	1.736	1.661
400	2.47	2.25	2.08	1.965	1.854	1.773
450	2.64	2.41	2.22	2.09	1.972	1.880
500	2.80	2.55	2.35	2.21	2.08	1.986
550	2.97	2.69	2.47	2.32	2.19	2.09
600	3.13	2.83	2.59	2.43	2.29	2.19
650	3.29	2.97	2.72	2.55	2.41	2.28
700	3.44	3.10	2.84	2.66	2.50	2.37
750	3.59	3.23	2.96	2.76	2.60	2.47
800	3.75	3.37	3.08	2.87	2.69	2.56
900	4.04	3.63	3.30	3.08	2.89	2.74
1000	4.36	3.88	3.53	3.28	3.07	2.91
1200	4.92	4.37	3.95	3.67	3.43	3.25
1500	5.76	5.09	4.58	4.23	3.95	3.74
2000	7.14	6.26	5.63	5.17	4.78	4.49
2500	8.48	7.40	6.61	6.05	5.59	5.25
3000	9.83	8.52	7.54	6.92	6.38	5.97
3500	11.22	9.65	8.53	7.79	7.16	6.72
4000	12.54	10.76	9.52	8.65	7.94	7.42
4500	13.88	11.91	10.46	9.49	8.72	8.13
5000	15.26	13.01	11.42	10.35	9.48	8.84

the hydration reaction

$$Cl_2 + H_2O = HOCl + H^+ + Cl^-$$

is described by an equilibrium constant K_e. The following values of H' and K_e were observed:

Temp., °C.	Henry's law coefficient H', lb.-moles Cl_2/(cu. ft.)(atm.)	Equilibrium constant K_e, (lb.-moles/cu. ft.)^2
10	0.00707	7.10
15	.00584	8.55
20	.00469	10.7
25	.00390	12.8

Table 10.[2] Ethane (C_2H_6)

t, °C.	0	5	10	15	20	25	30	35
$10^{-4} \times H$	1.26	1.55	1.89	2.26	2.63	3.02	3.42	3.83

t, °C.	40	45	50	60	70	80	90	100
$10^{-4} \times H$	4.23	4.63	5.00	5.65	6.23	6.61	6.87	6.92

"International Critical Tables," vol. 3, p. 261.

Table 11.[1] Ethylene (C_2H_4)

t, °C.	0	5	10	15	20	25	30
$10^{-3} \times H$	5.52	6.53	7.68	8.95	10.2	11.4	12.7

"International Critical Tables," vol. 3, p. 260.

Table 12.[9,12,13] Hydrogen (H_2)

t, °C.	0	5	10	15	20	25	30	35
$10^{-4} \times H$	5.79	6.08	6.36	6.61	6.83	7.07	7.29	7.42

t, °C.	40	45	50	60	70	80	90	100
$10^{-4} \times H$	7.51	7.60	7.65	7.65	7.61	7.55	7.51	7.45

"International Critical Tables," vol. 3, p. 256.

Table 13.[10] Hydrogen (H_2)

Partial pressure H_2, mm. Hg	$10^{-4} \times H$ 19.5°C.	23°C.
900	7.42	....
1100	....	7.75
2000	7.42	7.76
3000	7.43	7.77
4000	7.47	7.81
5000	7.56	7.89
6000	7.70	8.00
7000	7.87	8.16
8200	....	8.41
8250	8.17	....

"International Critical Tables," vol. 3, p. 256.

Table 14.[14] Hydrogen Chloride (HCl)

Weights of HCl per 100 weights of H_2O	Partial pressure of HCl, mm. Hg 0°C.	10°C.	20°C.	30°C.
78.6	510	840		
66.7	130	233	399	627
56.3	29.0	56.4	105.5	188
47.0	5.7	11.8	23.5	44.5
38.9	1.0	2.27	4.90	9.90
31.6	0.175	0.43	1.00	2.17
25.0	.0316	.084	0.205	0.48
19.05	.0056	.016	.0428	.106
13.64	.00099	.00305	.0088	.0234
8.70	.000118	.000583	.00178	.00515
4.17	.000018	.000069	.00024	.00077
2.04	....	.0000117	.000044	.000151

Weights of HCl per 100 weights of H_2O	Partial pressure of HCl, mm. Hg 50°C.	80°C.	110°C.
78.6			
66.7			
56.3	535		
47.0	141	623	
38.9	35.7	188	760
31.6	8.9	54.5	253
25.0	2.21	15.6	83
19.05	0.55	4.66	28
13.64	.136	1.34	9.3
8.70	.0344	0.39	3.10
4.17	.0064	.095	0.93
2.04	.00140	.0245	.280

Table 15.[11] Hydrogen Sulfide (H_2S)

t, °C.	0	5	10	15	20	25	30	35
$10^{-2} \times H$	2.68	3.15	3.67	4.23	4.83	5.45	6.09	6.76

t, °C.	40	45	50	60	70	80	90	100
$10^{-2} \times H$	7.45	8.14	8.84	10.3	11.9	13.5	14.4	14.8

"International Critical Tables," vol. 3, p. 259.

Table 16.[2] Methane (CH_4)

t, °C.	0	5	10	15	20	25	30	35
$10^{-4} \times H$	2.24	2.59	2.97	3.37	3.76	4.13	4.49	4.86

t, °C.	40	45	50	60	70	80	90	100
$10^{-4} \times H$	5.20	5.51	5.77	6.26	6.66	6.82	6.92	7.01

"International Critical Tables," vol. 3, p. 260.

Table 17.[2,9] Nitric Oxide (NO)

t, °C.	0	5	10	15	20	25	30	35
$10^{-4} \times H$	1.69	1.93	2.18	2.42	2.64	2.87	3.10	3.31

t, °C.	40	45	50	60	70	80	90	100
$10^{-4} \times H$	3.52	3.72	3.90	4.18	4.38	4.48	4.52	4.54

"International Critical Tables," vol. 3, p. 259.

Table 18.[9,13,15] Nitrogen (N_2)*

t, °C.	0	5	10	15	20	25	30	35
$10^{-4} \times H$	5.29	5.97	6.68	7.38	8.04	8.65	9.24	9.85

t, °C.	40	45	50	60	70	80	90	100
$10^{-4} \times H$	10.4	10.9	11.3	12.0	12.5	12.6	12.6	12.6

"International Critical Tables," vol. 3, p. 256.
* Atmospheric nitrogen = 98.815 vol. % N_2 + 1.185 vol. % A.

Table 19.[10] Nitrogen (N_2)

Partial pressure of N_2, mm. Hg	$10^{-4} \times H$ 19.4°C.	24.9°C.
900	8.24	9.08
2000	8.32	9.15
3000	8.41	9.25
4000	8.49	9.38
5000	8.59	9.49
6000	8.74	9.62
7000	8.86	9.75
8100	9.04	
8200	....	9.91

See also Goodman and Krase [*Ind. Eng. Chem.*, **23**, 401(1931)] for values up to 169°C. and 300 atm.

Table 20.[16,17] Nitrous Oxide (N_2O)

t, °C.	0	5	10	15	20	25	30	35
$10^{-3} \times H$		1.17	1.41	1.66	1.98	2.25	2.59	3.02

"International Critical Tables," vol. 3, p. 259.

Table 21.[9,13,15,18,19] Oxygen (O_2)

t, °C.	0	5	10	15	20	25	30	35
$10^{-4} \times H$	2.55	2.91	3.27	3.64	4.01	4.38	4.75	5.07

t, °C.	40	45	50	60	70	80	90	100
$10^{-4} \times H$	5.35	5.63	5.88	6.29	6.63	6.87	6.99	7.01

"International Critical Tables," vol. 3, p. 257.

Table 22.[10] Oxygen (O_2)

Partial pressure of O_2, mm. Hg	$10^{-4} \times H$ 23°C.	25.9°C.
800	....	4.79
900	4.58	
2000	4.59	4.80
3000	4.60	4.83
4000	4.68	4.88
5000	4.73	4.92
6000	4.80	4.98
7000	4.88	5.05
8150	4.98	
8200	....	5.16

"International Critical Tables," vol. 3, p. 257.

Table 23.[20] **Propylene** (C_3H_6)

t, °C.	2	6	10	14	18
$10^{-3} \times H$	3.04	3.84	4.46	5.06	5.69

"International Critical Tables," vol. 3, p. 260.

Table 24.[3] **Sulfur Dioxide** (SO_2)

Weight of SO_2 per 100 weights of H_2O	Partial pressure of SO_2, mm. Hg							
	0°C.	7°C.	10°C.	15°C.	20°C.	30°C.	40°C.	50°C.
20	646	657						
15	474	637	726					
10	308	417	474	567	698			
7.5	228	307	349	419	517	688		
5.0	148	198	226	270	336	452	665	
2.5	69	92	105	127	161	216	322	458
1.5	38	51	59	71	92	125	186	266
1.0	23.3	31	37	44	59	79	121	172
0.7	15.2	20.6	23.6	28.0	39.0	52	87	116
.5	9.9	13.5	15.6	19.3	26.0	36	57	82.0
.3	5.1	6.9	7.9	10.0	14.1	19.7		
.2	2.8	3.7	4.6	5.7	8.5	11.8	.....	31.0
.15	1.9	2.6	3.1	3.8	5.8	8.1	12.9	20.0
.10	1.2	1.5	1.75	2.2	3.2	4.7	7.5	12.0
.05	0.6	0.7	0.75	0.8	1.2	1.7	2.8	4.7
.02	.25	.3	.3	.3	0.5	0.6	0.8	1.3

References for Tables 1 to 24
[1] Winkler, "Landolt-Börnstein Physikalisch-chemische Tabellen."
[2] Winkler, *Ber.*, **34**, 1408 (1901).
[3] Sherwood, *Ind. Eng. Chem.*, **17**, 745 (1925).
[4] Breitenbach, *Bull. Univ. Wis. Eng. Exp. Sta.*, *Ser.* **68.**
[5] Winkler, *Magyar Chemiai Folyóirat*, **4**, 33 (1898).
[6] Winkler, *Chem.-Ztg.*, **23**, 687 (1899).
[7] Bohr, *Ann. Physik*, **68**, 500 (1899).
[8] Sander, *Z. physik. Chem.*, **78**, 513 (1912).
[9] Winkler, *Z. physik. Chem.*, **9**, 171 (1892).
[10] Cassuto, *Physik. Z.*, **5**, 233 (1904).
[11] Winkler, *Mathematikai ès Termèszettudomanyi Ertesito*, Budapest, **25**, 86 (1907).
[12] Winkler, *Ber.*, **24**, 89 (1891).
[13] Winkler, *Math. naturw. Ber. Ungarn.* **9**, 195 (1892).
[14] Zeisberg, *Chem. & Met. Eng.*, **32**, 326 (1925).
[15] Winkler, *Ber.* **24**, 3602 (1891).
[16] Geffcken. *Z. physik. Chem.*, **49**, 257 (1904).
[17] Kunerth, *Phys. Rev.*, **19**, 512 (1922).
[18] Winkler, *Z. physik. Chem.*, **55**, 344 (1906).
[19] Winkler, *Ber.*, **22**, 1764 (1889).
[20] Than, *Liebigs Ann. Chem.*, **123**, 187 (1862).
[21] Wiebe and Gaddy, *J. Am. Chem. Soc.*, **61**, 315 (1939); **62**, 815 (1940).
[22] Adams and Edmonds, *Ind. Eng. Chem.*, **29**, 447 (1937).

Solubility of Gases in Non-aqueous Pure Liquids and in Aqueous Solutions. The solubility of a gas in a **non-aqueous pure liquid** is frequently of great interest to the chemical engineer. The possible combination of solutes and solvents that come under the above classification are very numerous. Solubility tables in the "International Critical Tables," "Landolt-Börnstein Physikalisch-chemische Tabellen," and Seidell, "Solubilities of Inorganic and Organic Compounds" present most of the available data. The list of solutes and solvents in Table 25 will give some idea of the type of systems for which solubility data are available. The list, which has been made up from tables appearing in the "International Critical Tables," vol. 3, pp. 261–270, does not include, by any means, all the solutes and solvents considered in the "International Critical Tables," but only some of those most commonly encountered.

Table 25

Solutes:	
Acetylene, C_2H_2	Hydrogen, H_2
Air	Hydrogen chloride, HCl
Ammonia, NH_3	Hydrogen sulfide, H_2S
Bromine, Br_2	Methane, CH_4
Carbon dioxide, CO_2	Methyl chloride, CH_3Cl
Carbon monoxide, CO	Nitric oxide, NO
Chlorine, Cl_2	Nitrogen, N_2
Ethane, C_2H_6	Nitrous oxide, N_2O
Ethylene, C_2H_4	Oxygen, O_2
	Sulfur dioxide, SO_2
	Etc.
Solvents:	
Acetic acid (glacial), $C_2H_4O_2$	Ethyl acetate, $C_4H_8O_2$
Acetic anhydride, $C_4H_6O_3$	Ethyl alcohol, C_2H_6O
Acetone, C_3H_6O	Ethylene chloride, C_2H_4Cl
Amyl alcohol, $C_5H_{12}O$	Ethyl ether, $C_4H_{10}O$
Aniline, C_6H_7N	Methyl acetate, $C_3H_6O_2$
Benzene, C_6H_6	Methyl alcohol, CH_4O
Bromobenzene, C_6H_5Br	Nitrobenzene, $C_6H_5NO_2$
Carbon disulfide, CS_2	Propyl alcohol, C_3H_8O
Carbon tetrachloride, CCl_4	Propylene, C_3H_6
Chlorobenzene, C_6H_5Cl	Toluene, C_7H_8
Chloroform, $CHCl_3$	Etc.

Solubilities of **gases in aqueous solutions** may be found in Seidell, Landolt-Börnstein and the "International Critical Tables," vol. 3, pp. 271–281. The solutes considered in the "International Critical Tables" include, in general, the same gases as listed in Table 25. The solvents are aqueous solutions containing various concentrations of both inorganic and organic compounds. Among the inorganic compounds are included many common acids, bases, and salts. Some of the organic compounds considered are: methyl alcohol, ethyl alcohol, glycerol, glucose, sucrose, chloral hydrate, and urea.

Solubilities of gases, particularly carbon dioxide, hydrogen, and nitrous oxide in certain **colloidal solutions** in water may be found in the "International Critical Tables," vol. 3, p. 281. Typical colloids considered are gelatin, starch, dextrin, egg albumen, serum albumen, glycogen, peptone, hemoglobin, arsenic trisulfide, ferric hydroxide, and silicic acid.

Data on solubility under pressures up to 200 atm. for N_2, H_2, O_2, CH_4, C_2H_4, C_3H_8, C_3H_6, and H_2S in water and a number of organic solvents are given by Frolich, Tauch, Hogan, and Peer [*Ind. Eng. Chem.*, **23**, 548 (1931)]. Goodman and Krase [*Ind. Eng. Chem.*, **23**, 401 (1931)] present experimental data on the solubility of N_2 in water at pressures from 100 to 300 atm. and temperatures from 0° to 170°C. Wiebe and Gaddy [*J. Am. Chem. Soc.*, **61**, 315 (1939)] give data up to 700 atm. and from 50° to 100°C.

Solubility data for certain systems that find important industrial application are of particular interest. The recovery of **ether vapor** is frequently accomplished by absorption **in** a liquid absorbent such as **sulfuric acid or meta-cresol.** Figures 3 and 4 give the solubility of ether in various solvents. All the solubility curves given with the exception of that for butyl alcohol are for 20°C. The butyl alcohol curve is for 15°C. (Robinson, "Recovery of Volatile Solvents," pp. 154–156, Reinhold, New York, 1922.)

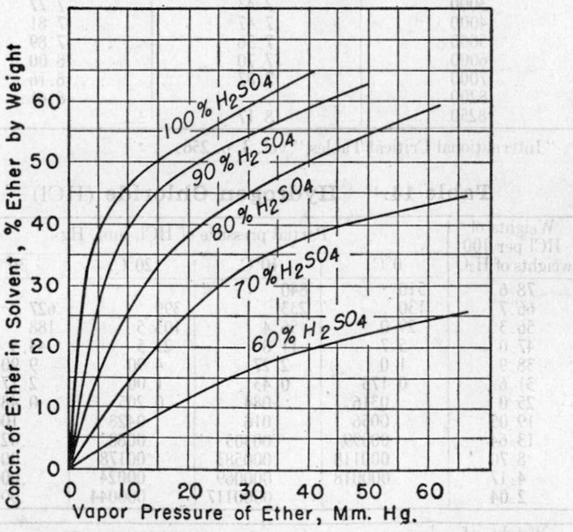

FIG. 3. Vapor pressure of ether at 20°C. from H_2SO_4 solutions of several concentrations of H_2SO_4 and H_2O.

Solubility data for hydrocarbons in oil are usually expressed in the form of an equilibrium constant $K = y^*/x$, where y^* = mole fraction of the solute in the gas phase, and x = mole fraction of the solute in the liquid

Table 26A. Equilibrium Data for Triethanolamine Solutions

Temp., °C.	Normality of amine	Partial pressure of CO_2, mm. Hg	Liquid concentration, moles CO_2/mole of amine
0.0	0.5	756.4	1.100
.0	.5	258.9	0.943
.0	.5	100.3	.805
.0	.5	45.6	.645
.0	.5	10.7	.378
25.0	.5	739.3	.921
25.0	.5	253.6	.715
25.0	.5	99.3	.512
25.0	.5	44.5	.375
25.0	.5	10.5	.191
50.0	.5	658.7	.623
50.0	.5	229.1	.408
50.0	.5	88.2	.262
50.0	.5	40.3	.162
50.0	.5	8.3	.0812
75.0	.5	474.7	.327
75.0	.5	129.9	.177
75.0	.5	50.1	.116
25.0	1.0	723.0	.805
25.0	1.0	259.0	.612
25.0	1.0	96.7	.424
25.0	1.0	43.4	.294
25.0	1.0	10.8	.161
25.0	1.0	1.4	.0587
0.0	2.0	752.0	.954
.0	2.0	259.3	.818
.0	2.0	100.4	.662
.0	2.0	45.8	.484
.0	2.0	10.7	.263
25.0	2.0	734.0	0.715
25.0	2.0	99.5	.316
25.0	2.0	45.4	.209
25.0	2.0	11.0	.0930
25.0	2.0	1.4	.0332
50.0	2.0	662.8	.382
50.0	2.0	230.3	.216
50.0	2.0	88.7	.130
50.0	2.0	40.4	.0791
50.0	2.0	9.4	.0346
75.0	2.0	485.9	.158
75.0	2.0	132.6	.0771
75.0	2.0	51.2	.0518
25.0	3.5	731.0	.595
25.0	3.5	420.0	.484
25.0	3.5	183.0	.312
25.0	3.5	46.6	.143
25.0	3.5	31.6	.114
25.0	3.5	10.0	.0620
25.0	5.0	738.6	.453
25.0	5.0	266.4	.258
25.0	5.0	98.7	.115
25.0	5.0	77.8	.108
25.0	5.0	44.5	.0729
25.0	5.0	10.6	.0292
50.0	5.0	678.8	.142
50.0	5.0	234.2	.0682
50.0	5.0	41.2	.0248
75.0	5.0	534.7	.0669
75.0	5.0	146.6	.0302
75.0	5.0	56.1	.0133

Table 26B. Equilibrium Data for Diethanolamine Solutions

Temp., °C.	Normality of amine	Partial pressure of CO_2, mm. Hg	Liquid concentration, moles CO_2/mole of amine
0.0	0.5	750.7	1.119
.0	.5	272.7	1.044
.0	.5	271.6	1.035
.0	.5	79.1	0.883
.0	.5	11.0	.741
25.0	.5	732.3	.987
25.0	.5	249.6	.912
25.0	.5	97.9	.797
25.0	.5	44.3	.714
25.0	.5	11.0	.551
50.0	.5	666.4	.883
50.0	.5	241.2	.778
50.0	.5	70.8	.588
50.0	.5	10.1	.336
75.0	.5	474.5	.630
75.0	.5	129.8	.456
75.0	.5	50.0	.355
0.0	2.0	751.0	.936
.0	2.0	272.1	.837
.0	2.0	80.1	.752
.0	2.0	11.5	.604
25.0	2.0	735.5	.753
25.0	2.0	729.0	.813
25.0	2.0	249.9	.717
25.0	2.0	99.3	.633
25.0	2.0	44.3	.553
25.0	2.0	10.5	.451
50.0	2.0	668.4	.680
50.0	2.0	242.3	.562
50.0	2.0	183.8	.548
50.0	2.0	71.0	.489
50.0	2.0	10.2	.302
75.0	2.0	488.6	.464
75.0	2.0	133.3	.356
75.0	2.0	51.1	.263
0.0	5.0	755.1	.762
.0	5.0	206.3	.683
.0	5.0	79.4	.638
.0	5.0	11.4	.526
25.0	5.0	741.1	.661
25.0	5.0	253.6	.589
25.0	5.0	44.8	.506
50.0	5.0	682.4	.562
50.0	5.0	246.1	.491
50.0	5.0	71.8	.414
50.0	5.0	10.4	.254
75.0	5.0	520.0	.403
75.0	5.0	142.6	.327
75.0	5.0	54.9	.242
25.0	8.0	744.0	0.582
25.0	8.0	268.4	.553
25.0	8.0	78.4	.480
50.0	8.0	703.5	.515
50.0	8.0	193.0	.458
50.0	8.0	74.5	.387
50.0	8.0	10.6	.250
75.0	8.0	574.0	.368
75.0	8.0	155.9	.302
75.0	8.0	58.9	.215

phase. The value of K varies with temperature, pressure, and composition (see Sec. 9, p. 568*ff.*). Values of K are given by Katz and Hachmuth [*Ind. Eng. Chem.*, **29,** 1072 (1937)]. See also Sherwood ("Absorption and Extraction," McGraw-Hill, New York, 1937) and a large series of articles by Sage *et al.* [*Ind. Eng. Chem.*, 1934 to date].

The absorption process for purifying **carbon dioxide** involves the absorption of carbon dioxide in an alkaline solution. This absorption is of a special type where the dissolved gas reacts with the absorbent to form a loose chemical compound. The absorbents most commonly employed for purifying carbon dioxide are solutions of sodium or potassium carbonate. Organic bases such as triethanolamine and diethanolamine are also used for absorption of carbon dioxide and hydrogen sulfide.

The vapor pressures of $\mathbf{CO_2}$ from **mono-, di-, and triethanolamine solutions** of various concentrations and at temperatures of 0° to 75°C. are given in Table 26 [Mason and Dodge, *Trans. Am. Inst. Chem. Engrs.*, **32,** 27 (1936)].

The solubilities of **hydrogen sulfide** in commercial **amine solutions** [Bottoms, *Ind. Eng. Chem.*, **23,** 501

Table 26C. Equilibrium Data for Monoethanolamine Solutions

Temp., °C.	Normality of amine	Partia pressure of CO_2, mm. Hg	Liquid concentration, moles CO_2 per mole amine
0.0	0.5	745.8	1.110
.0	.5	256.3	0.990
.0	.5	45.3	.817
.0	.5	10.6	.675
25.0	.5	735.7	1.004
25.0	.5	251.8	0.886
25.0	.5	99.6	.795
25.0	.5	44.2	.720
25.0	.5	10.8	.607
50.0	.5	661.3	.880
50.0	.5	228.3	.757
50.0	.5	40.1	.596
75.0	.5	475.8	.685
75.0	.5	130.3	.584
75.0	.5	50.0	.476
0.0	2.0	754.4	.900
.0	2.0	206.1	.776
.0	2.0	79.4	.718
.0	2.0	11.4	.601
25.0	2.0	736.4	.795
25.0	2.0	252.2	.697
25.0	2.0	98.6	.623
25.0	2.0	44.2	.589
25.0	2.0	10.6	.527
50.0	2.0	668.2	.698
50.0	2.0	183.1	.607
50.0	2.0	70.9	.556
50.0	2.0	10.1	.489
75.0	2.0	477.0	.560
75.0	2.0	130.6	.474
75.0	2.0	51.1	.430
0.0	5.0	751.5	.761
.0	5.0	272.2	.679
.0	5.0	206.2	.649
.0	5.0	80.1	.600
.0	5.0	11.5	.600
25.0	5.0	742.9	0.657
25.0	5.0	254.9	.601
25.0	5.0	98.7	.563
25.0	5.0	44.6	.539
25.0	5.0	10.6	.507
50.0	5.0	677.0	.574
50.0	5.0	245.3	.527
50.0	5.0	71.5	.505
50.0	5.0	10.4	.453
75.0	5.0	518.1	.493
75.0	5.0	142.6	.460
75.0	5.0	54.8	.418
0.0	9.5	752.4	.622
.0	9.5	272.2	.592
.0	9.5	79.2	.568
.0	9.5	11.4	.538
25.0	9.5	735.9	.588
25.0	9.5	252.2	.554
25.0	9.5	99.0	.532
25.0	9.5	44.8	.519
25.0	9.5	11.1	.495
50.0	9.5	701.3	.538
50.0	9.5	255.3	.522
50.0	9.5	74.3	.492
50.0	9.5	10.8	.443
75.0	9.5	559.7	.468
75.0	9.5	153.1	.458
75.0	9.5	56.7	.424
25.0	12.5	749.1	.548
25.0	12.5	256.3	.518
25.0	12.5	45.4	.521
50.0	12.5	716.2	.525
50.0	12.5	259.5	.501
50.0	12.5	196.0	.495
50.0	12.5	75.6	.483
50.0	12.5	10.9	.467
75.0	12.5	629.9	.479
75.0	12.5	168.1	.453
75.0	12.5	64.2	.395

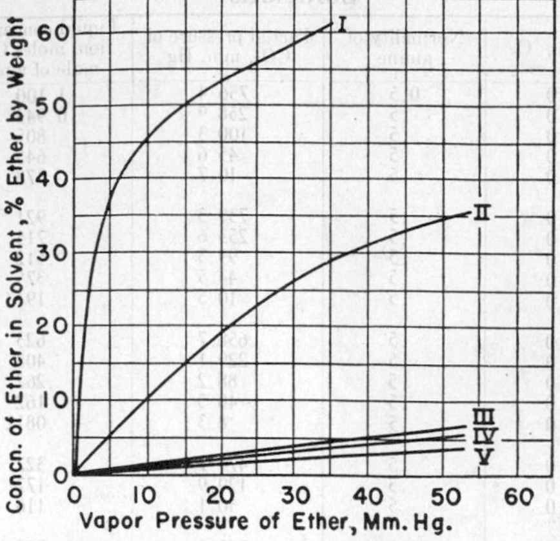

FIG. 4. Vapor pressure of ether from its solution in several solvents.

Curve	Solvent	Temp., °C.
I	100% H_2SO_4	20
II	*m*-Cresol	20
III	Amyl alcohol	20
IV	Butyl alcohol	15
V	Ethyl alcohol	20

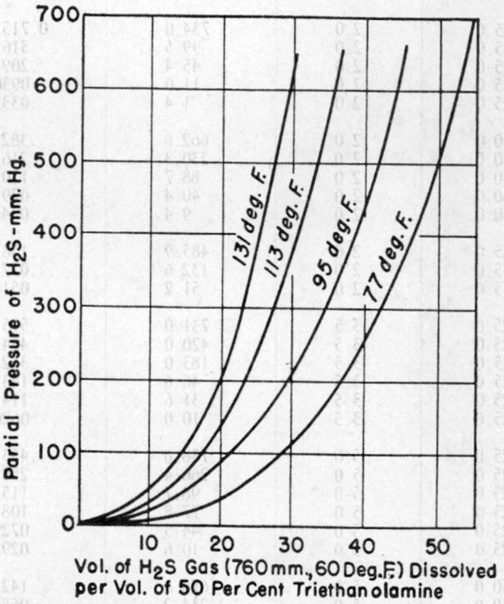

FIG. 5. Vapor pressure of H_2S from triethanolamine at 60°F.

(1931)] are given in the accompanying charts (Figs. 5 and 6).

Carbon Dioxide in Carbonate Solution. When carbon dioxide is dissolved in an aqueous solution of sodium carbonate (or potassium) the following reversible reaction occurs:

$$Na_2CO_3 + CO_2 + H_2O \rightleftarrows 2NaHCO_3$$

The solubility of CO_2 in such a solution depends on the ratio of carbonate to bicarbonate, the total amount of salt in the solution, the temperature, and the partial pressure of the carbon dioxide in the gas.

The relation between these variables was first worked out and formulated by McCoy [*J. Am. Chem. Soc.*, **29**, 437 (1903)]. Harte, Baker, and Purcell [*Ind. Eng. Chem.*, **25**, 528 (1933)] obtained more data and expressed

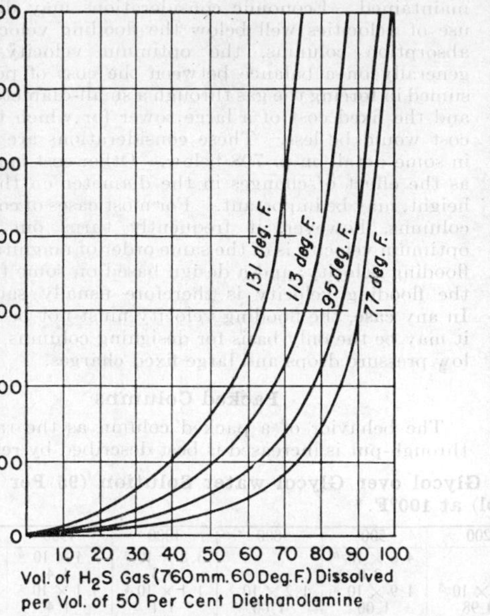

FIG. 6. Vapor pressure of H_2S from 50 per cent diethanolamine at 60°F.

the relation in an empirical formula, which reduces to

$$p_{CO_2} = \frac{137f^2N^{1.29}}{S(1-f)(365-t)} \qquad (6)$$

where p_{CO_2} = partial pressure of CO_2, mm. Hg.
f = fraction of total base present as bicarbonate.
N = sodium normality.
S = solubility of CO_2 in water under a pressure of 1 atm. of CO_2, g.-moles per liter. See Table 27.
t = temperature, °F.

Equation (6) has been tested only over the temperature range from 65° to 150°F. and sodium normalities from 0.5 to 2.0.

Sherwood ("Absorption and Extraction," p. 209, McGraw-Hill, New York, 1937) reports that the data of Sieverts and Fritzsche [*Z. anorg. allgem. Chem.*, **133**, 1 (1924)] on the *potassium* system may be approximated by the equation

$$p_{CO_2} = \frac{45f^2N^{1.29}}{S(1-f)(302-t)} \qquad (7)$$

for temperatures from 30° to 100°F., and *potassium* normalities from 1.0 to 2.0.

Table 27. Values of S for Eqs. (2) and (3)

Temperature, °F.	S = G.-moles CO_2 per L. at 1 Atm. CO_2
59	0.0455
77	.0336
95	.0262
113	.0215
131	.0175
145.4	.0151
167	.0120
185	.0090
212	.0065

Sulphur Dioxide in Alkaline Solutions. Johnstone [*Ind. Eng. Chem.*, **27**, 587 (1935); **30**, 101 (1938)] has determined solubilities of sulfur dioxide in aqueous solutions of ammonia, of sodium sulfite–bisulfite, and of methyl amine. The vapor pressure of SO_2 is given by the equation

$$p_{SO_2} = M\frac{(2S-C)^2}{C-S} \text{ mm. Hg} \qquad (8)$$

where S = total concentration of dissolved SO_2, moles/100 moles H_2O; C = total concentration of base, moles/100 moles H_2O, and

$$\log_{10} M = 4.519 - \left(\frac{1987}{T}\right) \text{ for sodium solutions} \qquad (9)$$

$$\log_{10} M = 5.390 - \left(\frac{2308}{T}\right) \text{ for methyl amine solutions} \qquad (10)$$

$$\log_{10} M = 5.865 - \left(\frac{2368}{T}\right) \text{ for ammonia solutions} \qquad (11)$$

The range of temperatures studied corresponds to 308° < T < 363°K. The vapor pressure of water over these solutions may be calculated from Raoult's law. The vapor pressure of ammonia over the solutions containing SO_2 follows the equation

$$p_{NH_3} = N\frac{C(C-S)}{2S-C} \text{ mm. Hg} \qquad (12)$$

where

$$N = 13.680 - \left(\frac{4987}{T}\right) \qquad (13)$$

The solutions investigated contained sodium and ammonium ions and methyl amine in the following concentration ranges: sodium, 4.0 to 7.8 g.-atoms/100 g.-moles H_2O; ammonia, 5.8 to 22.4 g.-moles/100 g.-moles H_2O; methyl amine, 7.3 to 22.0 g.-moles/100 g.-moles H_2O.

Olefins in Cuprous Salt Solutions. Gilliland and Seebold [*Ind. Eng. Chem.*, **33**, 1143 (1941)] determined the solubility of ethylene and propylene in aqueous solutions of cuprous chloride. The results are expressed by the following equations. For ethylene in a solution containing 1.90 g.-moles CuCl/l., 3.0 g.-moles NH_4Cl/l., and 2.52 g.-moles HCl/l.,

$$\log_{10}\left[\frac{X}{(1-X)f}\right] = \frac{2060}{T} - 8.20 \qquad (14)$$

in the range 288 < T < 330°K. For propylene in a solution containing 1.89 g.-moles CuCl/l., 3.0 g.-moles NH_4Cl/l., and 2.27 g.-moles HCl/l.,

$$\log_{10}\left[\frac{X}{(1-X)f}\right] = \frac{1520}{T} - 6.86 \qquad (15)$$

in the range 270 < T < 310°K. In Eqs. (14) and (15), f = fugacity of hydrocarbon, atm., and X = concentration of olefin in the liquid phase, g.-moles olefin/g.-atom copper.

The data of Morrell *et al.* [*Trans. Am. Inst. Chem. Engrs.*, **42**, 473 (1946)] on the solubility of unsaturated C_4 hydrocarbons in ammoniacal cuprous acetate solution containing 1.5 g.-moles/l. of Cu_2H ion are expressed by the equations

1-Butene:

$$\log_{10}\frac{(U.Cu_2H)}{[1.5-(U.Cu_2H)](pv)} = \frac{1860}{T} - 7.840 \qquad (16)$$

1-3-Butadiene:

$$\log_{10}\frac{(U.Cu_2H)}{[1.5-(U.Cu_2H)](pv)} = \frac{3053}{T} - 10.845 \qquad (17)$$

1-2-Butadiene:

$$\log_{10}\frac{(U.Cu_2H)}{[1.5-(U.Cu_2H)](pv)} = \frac{3157}{T} - 10.573 \qquad (18)$$

Trans-2-Butene:

$$\log_{10} \frac{[(2U).\mathrm{Cu_2H}]}{\{1.5 - [(2U).\mathrm{Cu_2H}]\}(p_U)^2} = \frac{2600}{T} - 11.013 \quad (19)$$

where (U.Cu$_2$H) and [(2U).Cu$_2$H] represent the concentrations of dissolved unsaturated hydrocarbon complex, g.-moles/l.; and p_U is the partial pressure of hydrocarbon over the solution, atm. Equations (17) and (18) may be used in the temperature range $0 < t < 40$°C. and Eqs. (16) and (19) within the range $0 < t < 10$°C.

Table 28. Natural Gas (94 Per Cent Methane) in Diethylene Glycol–water Solution (95 Per Cent by Weight Glycol) at 100°F.*

P, lb./sq. in. gage	200	700	1000	1500	2000
$(1/H')$, atm./(lb.-mole/cu. ft.)	4.1×10^3	5.2×10^3	4.9×10^3	5.4×10^3	5.8×10^3

* Russell, Reid, and Huntington, *Trans. Am. Inst. Chem. Engrs.*, **41**, 315 (1945).

maintained. Economic considerations may dictate the use of velocities well below the flooding velocity. For absorption columns, the optimum velocity depends generally on a balance between the cost of power consumed in forcing the gas through a small-diameter column and the fixed cost of a large tower for which the power cost would be less. These considerations are discussed in some detail on p. 708 below. Other cost factors, such as the effect of changes in the diameter on the column height, may be important. For most cases of commercial columns, however, it frequently turns out that the optimum velocity is of the same order of magnitude as the flooding velocity, and a design based on some fraction of the flooding velocity is therefore usually satisfactory. In any case, the flooding velocity must not be exceeded; it may be the only basis for designing columns that have low pressure drops and large fixed charges.

Packed Columns

The behavior of a packed column as the rate of gas through-put is increased is best described by reference to a plot of pressure drop vs. gas velocity, as shown in Fig. 7. For the dry packing, the pressure drop increases as the 1.8 power of the gas velocity, indicating that the flow through the packing material is turbulent. Approximately the same variation is observed when the packing is irrigated, up to the *loading velocity*, at which the curve bends sharply upward. A second break point occurs at the *flooding velocity*, and above this velocity the pressure-drop curve turns almost vertically upward. A lower break point in the curve sometimes is not observed in small packings.

Table 29. Vapor Pressure of Water and Diethylene Glycol over Glycol-water Solution (95 Per Cent by Weight Glycol) at 100°F.*

P, lb./sq. in. gage	0	200	500	800	1000	1500	2000
K_{glycol} $(= y^*/x)$	93×10^{-6}	....	4.6×10^{-6}	....	3.8×10^{-6}	3.4×10^{-6}	4.2×10^{-6}
γ_{glycol}†	1.2	....	2.0	....	3.3	4.4	7.2
K_{water} $(= y^*/x)$	5.4×10^{-2}	4.3×10^{-3}	1.9×10^{-3}	1.2×10^{-3}	1.1×10^{-3}	9.1×10^{-4}	1.2×10^{-3}
γ_{water}	0.84	0.98	1.00	1.00	1.1	1.4	2.5

* Russell, Reid, and Huntington, *Trans. Am. Inst. Chem. Engrs.*, **41**, 315 (1945).
† Based on vapor pressure of pure diethylene glycol = 0.06 mm. Hg at 100°F.

LIMITING VELOCITIES AND PRESSURE DROPS FOR ABSORPTION COLUMNS

The smallest diameter that can be used for a countercurrent absorption column designed to handle a specified total quantity of gas or liquid is determined by the greatest fluid velocity at which countercurrent flow can be

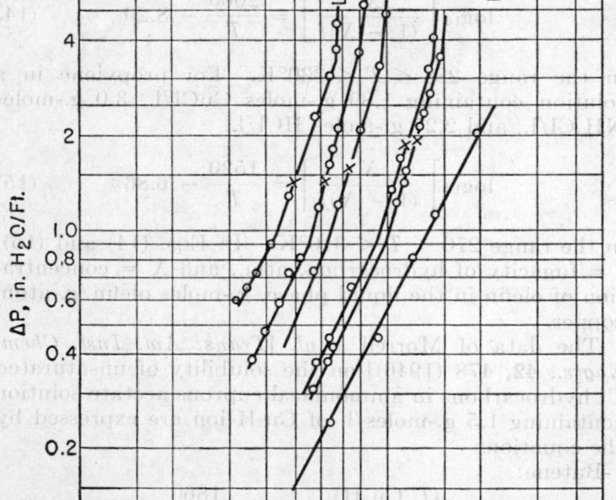

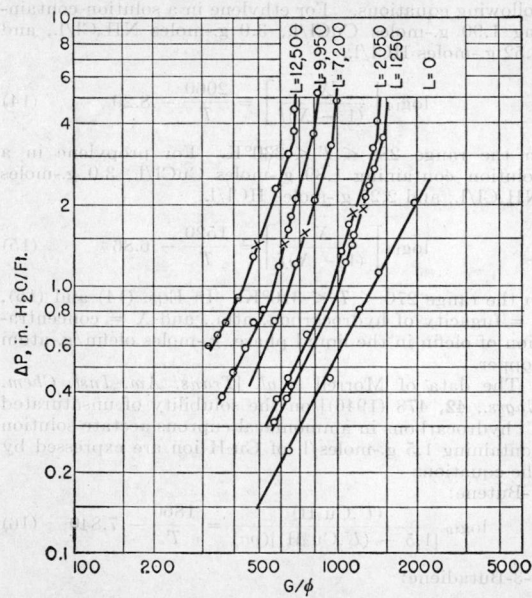

Fig. 7. Pressure drop data on 1-in. ceramic rings for water and air. Points marked X indicate visual build-up of water over the packing. [*Data of Sarchet, Trans. Am. Inst. Chem. Engrs.*, **38**, 283 (1942).]

L = water rate, lb./(hr.)(sq. ft.).
G = gas rate, lb./(hr.)(sq. ft.).
$\phi = (\rho/0.075)^{1/2}$.
ρ = gas density, lb./cu. ft.

Pressure Drop. Pressure drops are shown as functions of the gas velocity for several of the more common packing materials in Figs. 8 and 9, which are based on the experiments of Tillson (S.M. Thesis, Mass. Inst. Tech., 1939), using a 20-in.-diameter tower. The quantity ϕ which appears in the abscissa of these plots is defined by

$$\phi = \sqrt{\frac{\rho_g}{0.075}} \quad (20)$$

where ρ_g is the gas density, lb./cu. ft. This quantity allows the effects of variations in gas density to be taken into account, as shown by the work of Mach [*Dechema Monograph*, **6**, 38 (1933); *Z. Ver. deut. Ing. Forsch.*, 375 (1935)]. The group G/ϕ is used interchangeably with the "*F*-factor," $u\sqrt{\rho_g}$, and the relation between the two is

$$\frac{G}{\phi} = 985\, u \sqrt{\rho_g} \quad (21)$$

in which u = superficial gas velocity, based on cross-sectional area of the empty tower, ft./sec.

The pressure drop at the flooding point is found to lie between 1.5 and 4 in. H_2O/ft. for many packings when the liquid has the same viscosity as water. The pressure drop at flooding is smaller with more viscous liquids, as shown by Schoenborn and Dougherty [*Trans. Am. Inst. Chem. Engrs.*, **40**, 51 (1944)].

For grid packings, Johnstone and Singh [*Ind. Eng. Chem.*, **29**, 286 (1937)] propose the equations

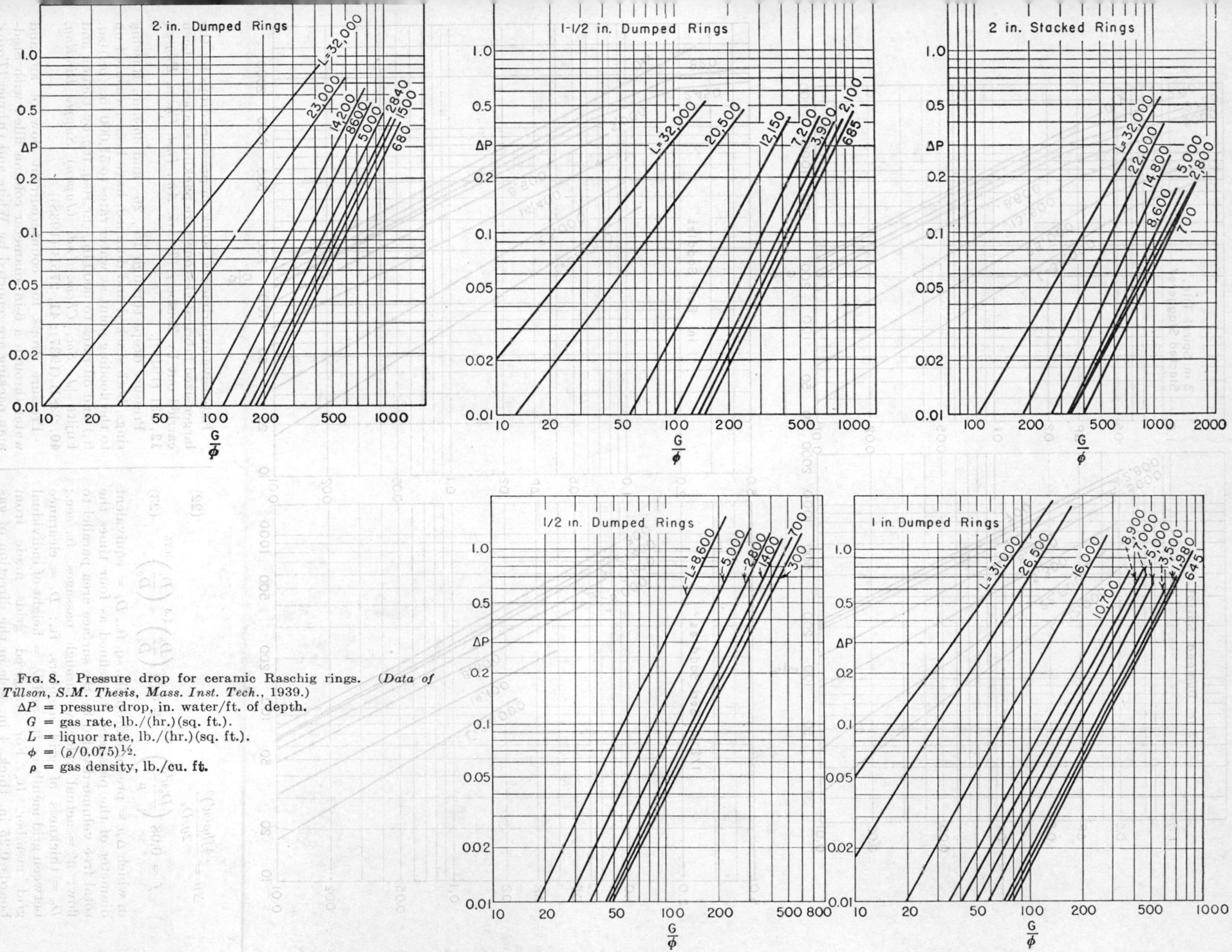

FIG. 8. Pressure drop for ceramic Raschig rings. (*Data of Tillson, S.M. Thesis, Mass. Inst. Tech.*, 1939.)

ΔP = pressure drop, in. water/ft. of depth.
G = gas rate, lb./(hr.)(sq. ft.).
L = liquor rate, lb./(hr.)(sq. ft.).
$\phi = (\rho/0.075)^{1/2}$.
ρ = gas density, lb./cu. ft.

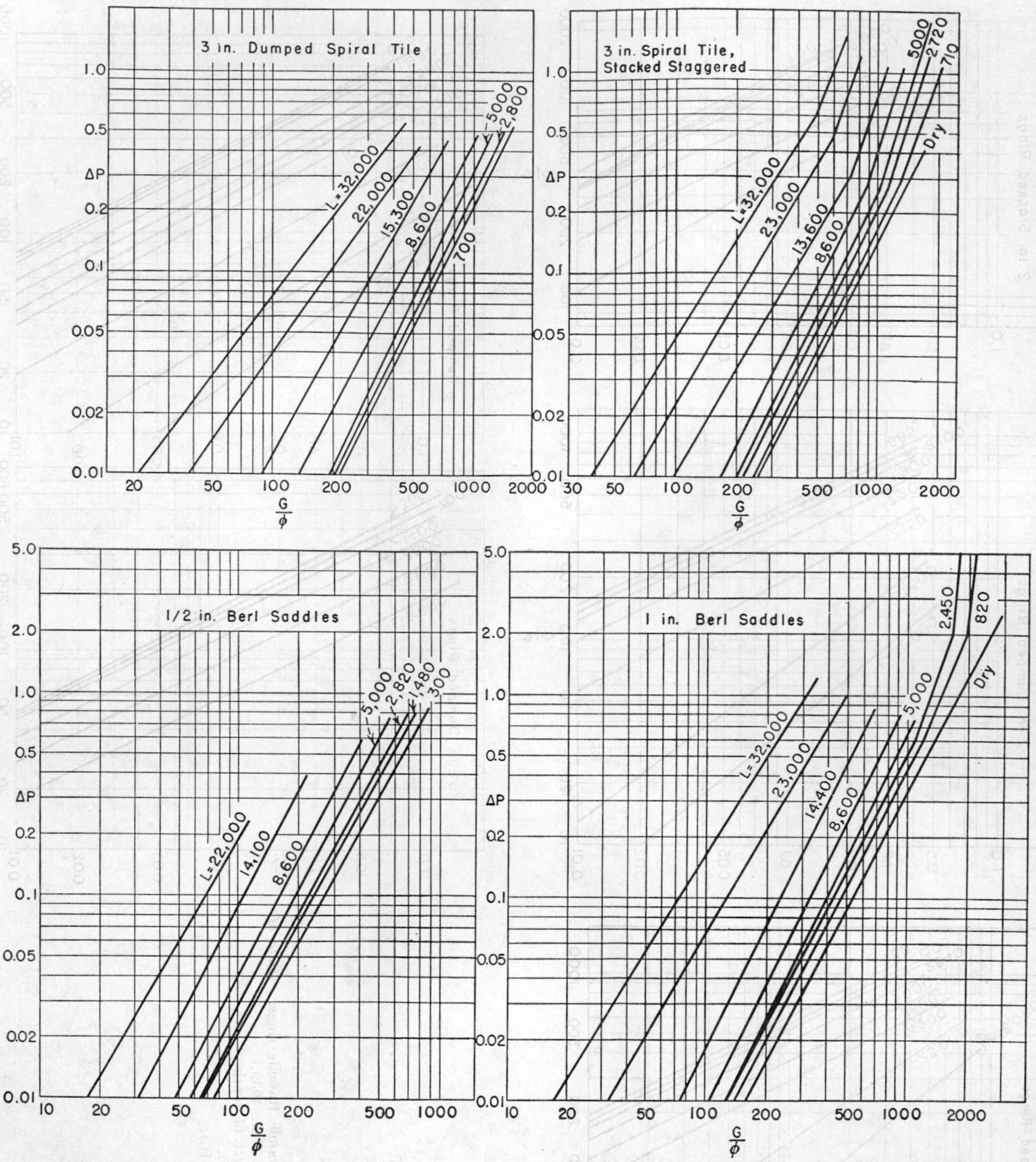

$$\Delta p = \frac{4fh\rho(u')^2}{2g_cD_e} \tag{22}$$

$$f = 0.08\left(\frac{D_eu'\rho}{\mu}\right)^{-0.2} + 0.52\left(\frac{D_b}{D_s}\right)^{1.5}\left(\frac{D_e}{D_g}\right)^{0.75} \tag{23}$$

in which Δp = pressure drop, lb./sq. ft., D_e = equivalent diameter of the passages, ft., defined as four times the total free volume divided by the surface area parallel to flow; u' = actual velocity through passages, ft./sec.; D_b = thickness of grid members, ft.; D_s = clearance between grid members, ft.; and D_g = height of individual grid member, ft. For staggered grids made from boards 0.25 in. thick, 4 in. wide in the direction of gas flow, spaced 1.25 in. apart, the pressure drop was 0.02 in. H_2O/ft. at G/ϕ = 2000 lb./(hr.)(sq. ft.).

Extensive experimental work on pressure drops for the larger sized fabricated stoneware packing shapes has been carried out by Zeisberg [*Trans. Am. Inst. Chem. Engrs.*, **12**, 231 (1919)].

Pressure drops through 15-, 26-, and 35-mm. Raschig rings, both stacked and dumped, have been measured up to the flooding point at water rates to 51,000 lb./(hr.)(sq. ft.) and oil rates to 6000 lb./(hr.)(sq. ft.) by Uchida and Fujita [*J. Soc. Chem. Ind. (Japan) Suppl. Binding*, **40**, 238B (1937); **41**, 275B (1938)].

Pressure drops for countercurrent flow of air and water through a 6-in.-diameter column filled with Stedman packing are reported by White and Othmer [*Trans. Am. Inst. Chem. Engrs.*, **38**, 1067 (1942)]. Values as high as 1 in. H_2O/ft. were observed at the flooding points

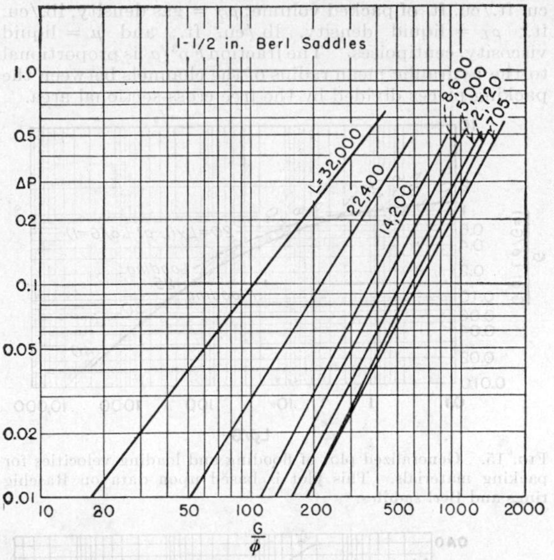

FIG. 9. Pressure drop for ceramic Berl saddles and spiral tile. (*Data of Tillson, S.M. Thesis, Mass. Inst. Tech.*, 1939) using air and water in a 20-in. tower, except for 1-in. saddles at liquor rates below 5000 lb./(hr.)(sq. ft.) from Mach [*Dechema Monograph*, **6**, 38 (1933); *Z. Ver. deut. Ing. Forsch.*, 375 (1935)].
ΔP = pressure drop, in. water/ft. of depth.
G = gas velocity, lb./(hr.)(sq. ft.).
L = liquid rate, lb./(hr.)(sq. ft.).
$\phi = (\rho/0.075)^{1/2}$.
ρ = gas density, lb./cu. ft.

(*cf.* Table 30). Similar data for McMahon packing—0.25-in. Berl saddles formed from 100-mesh wire screen—were reported by Forsythe. Stack, Wolf, and Conn (paper presented before Am. Chem. Soc., Div. of Ind. Eng. Chem., Pittsburgh, December, 1946). At G/ϕ = 700 lb./(hr.)(sq. ft.) and $L\phi/G = 1.6$, the pressure drop for this packing was 1.5 in. H_2O/ft.

For packing materials not included in Figs. 8 and 9, the pressure drop may be estimated roughly from the physical dimensions of the packing by methods described by Chilton and Colburn [*Trans. Am. Inst. Chem. Engrs.*, **26**, 178 (1931); *Ind. Eng. Chem.*, **23**, 913 (1931)]. See also Sec. 5, p. 393.

Loading and Flooding Velocities. The behavior of a packed column near the loading and flooding points is described in detail by Elgin and Weiss [*Ind. Eng. Chem.*, **31**, 435 (1939)]. As the gas velocity is increased near the flooding point, the hold-up of the liquid phase increases rapidly. When a ring-packed column is irrigated with water, a layer of liquid appears on top of the packing at the flooding point, and the gas bubbles through this liquid layer. When oil is used instead of water, the liquid head sometimes does not appear, but the liquid is entrained violently by the gas that flows from the packing. For Berl-saddle packings, the liquid phase may become continuous at a point just above the packing support, or slugs of foaming liquid may surge through the packing, as observed by Bain and Hougen [*Trans. Am. Inst. Chem. Engrs.*, **40**, 29, 389 (1944)]. The flooding point has been defined as the gas velocity at which a liquid layer builds up on top of the packing, as the second break point on a loglog plot of pressure drop vs. gas velocity, and also as the point at which the measured liquid hold-up increases abruptly. The visual flooding point is usually slightly less than that obtained from pressure-drop measurements.

Figures 10, 11, and 12 give loading velocities of commonly used Raschig-ring, Berl-saddle, spiral-tile, and grid-tile packings, as measured by P. Tillson (S.M. Thesis, Mass. Inst. Tech., 1939) using water and air in a 20-in. tower containing 3 ft. of packing, and by Molstad, Abbey, Thompson, and McKinney [*Trans. Am. Ints. Chem. Engrs.*, **38**, 387 (1942)]. These data agree substantially with the more limited data of Mach [*Dechema Monograph*, **6**, 38 (1933); *Z. Ver. deut. Ing. Forsch.*, 375 (1935)], White [*Trans. Am. Inst. Chem. Engrs.*, **31**, 1 (1935)], Uchida and Fujita [*J. Soc. Chem. Ind.* (*Japan*), **40**, 238B (1937)], Sarchet [*Trans. Am. Inst. Chem. Engrs.*, **38**, 283 (1942)], and Elgin and Weiss (*loc. cit.*).

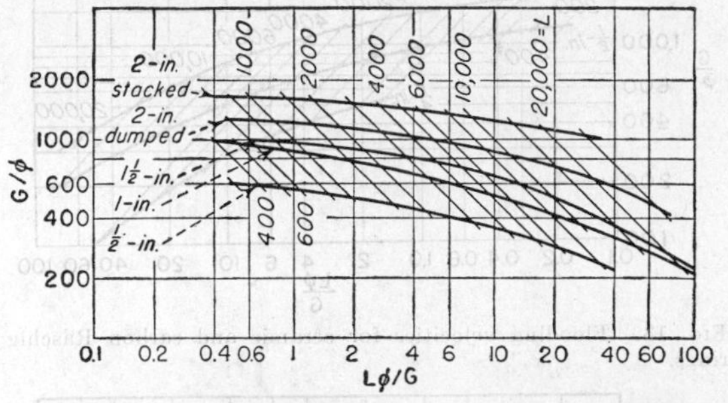

FIG. 10. Loading velocities for Raschig ring packing.

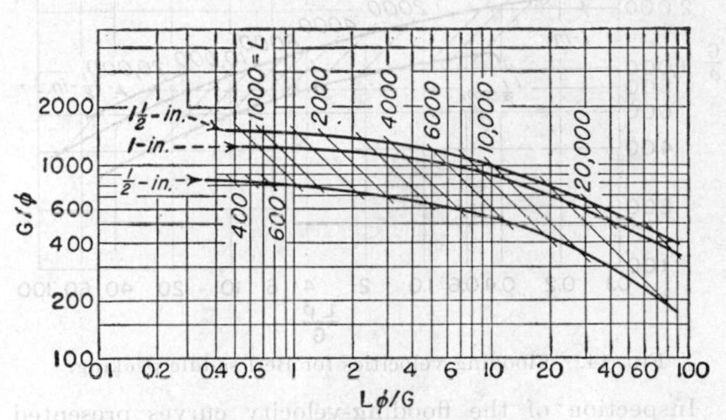

FIG. 11. Loading velocities for Berl saddle packing.

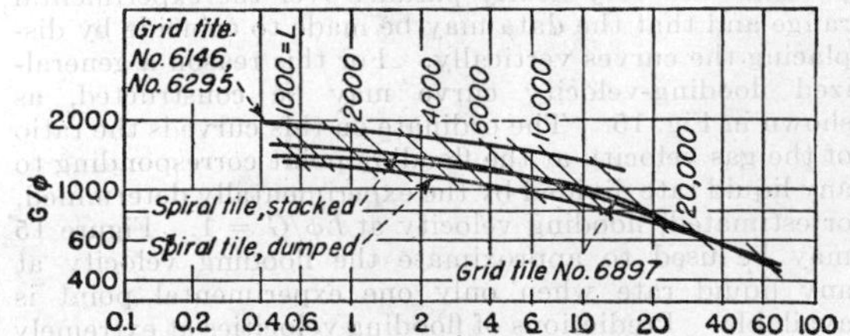

FIG. 12. Loading velocities for 3-in. single-spiral tile and drip-point grid tile.

Figures 13 and 14 summarize the available flooding data on water and air for Raschig rings and Berl saddles, as reported by Mach (*loc. cit.*), White (*loc. cit.*), Sarchet (*loc. cit.*), Bain and Hougen (*loc. cit.*), Sherwood, Shipley, and Holloway [*Ind. Eng. Chem.*, **30**, 765 (1938)], Uchida and Fujita (*loc. cit.*), and Schoenborn and Dougherty [*Trans. Am. Inst. Chem. Engrs.*, **40**, 51, 389 (1944)]. The effect of variations in gas density is allowed for by the inclusion of the ϕ factor, as indicated by the work of Sherwood, Shipley, and Holloway (*loc. cit.*) as well as by that of Bain and Hougen (*loc. cit.*). When the viscosity of the liquid differs from that of water (approximately 1 centistoke at 20°C.), the value of G read from the ordinate of the flooding-velocity curves must be multiplied by the factor ν^{-n}, where ν is the kinematic viscosity of the liquid, μ/ρ, expressed in centistokes, and the exponent n equals 0.15, 0.19, and 0.33 for 1-, ½-, and ¼-in. Raschig rings, respectively, and 0.12 for ½-in.

Berl saddles. A somewhat smaller effect of viscosity was found by Bain and Hougen (*loc. cit.*). Sherwood, Shipley, and Holloway (*loc. cit.*) found that variations in surface tension had only a slight effect. They also studied the effect of variations in liquid density from 49.9 to 74.1 lb./cu. ft. These results can be introduced into Figs. 13 and 14 if the abscissas are taken as $(L\phi/G) \times (62.3/\rho_L)^{1/2}$ and the ordinates as $(G/\phi)(62.3/\rho_L)^{1/2}$.

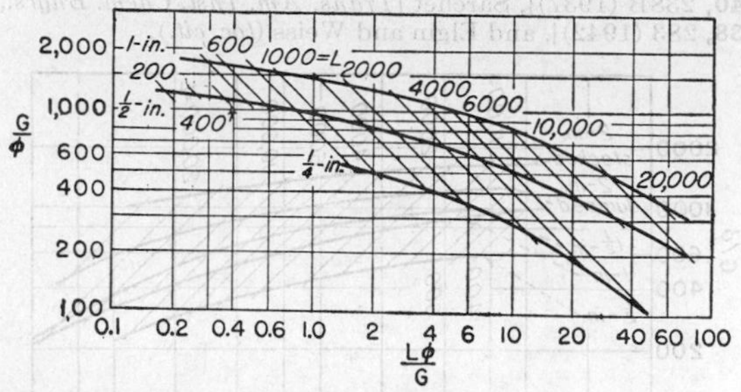

Fig. 13. Flooding velocities for ceramic and carbon Raschig rings.

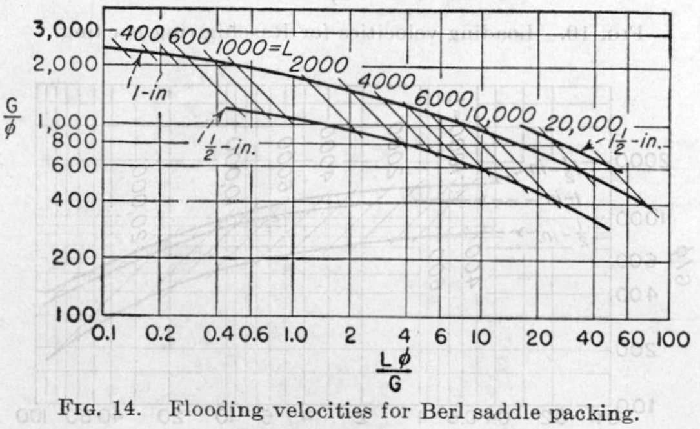

Fig. 14. Flooding velocities for Berl saddle packing.

Inspection of the flooding-velocity curves presented in Figs. 13 and 14 shows that the lines for the various packings are very nearly parallel over the experimental range and that the data may be made to coincide by displacing the curves vertically. For this reason a generalized flooding-velocity curve may be constructed, as shown in Fig. 15. The ordinate on this curve is the ratio of the gas velocity at the flooding point corresponding to any liquid rate divided by the experimentally determined, or estimated, flooding velocity at $L\phi/G = 1$. Figure 15 may be used to approximate the flooding velocity at any liquid rate when only one experimental point is available. Predictions of flooding velocities at extremely high liquid rates by means of Fig. 15 may be subject to some error, since the line on the figure indicates that the flooding velocity decreases almost to zero when the liquid rate is about forty times the rate corresponding to $L\phi/G = 1$. This may not be true for packings other than those which have been used as a basis for constructing the figure. Table 30 lists values of G/ϕ at $L\phi/G = 1$ for several packings not included in Figs. 13 and 14.

The effect of liquid rate on the flooding velocity indicated by Fig. 15 agrees closely with a curve presented by Sherwood, Shipley, and Holloway (*loc. cit.*), presented here as Fig. 16. If the physical properties of the packing are known, the flooding velocity at any liquid rate may be estimated from this figure. The ordinate on the figure is $(u_g^2 a'/g_c F_D^3)(\rho_G/\rho_L)(\mu^{0.2})$, where u_g = gas velocity at flooding, ft./sec., based on the empty tower; a' = total surface area of the packing, sq. ft./cu. ft. of packed volume; g_c = standard gravitational acceleration, F_D = fractional void space in the packing when dry, cu. ft./cu. ft. of packed volume; ρ_G = gas density, lb./cu. ft.; ρ_L = liquid density, lb./cu. ft.; and μ = liquid viscosity, centipoises. The fraction F_D^3/a' is proportional to the hydraulic mean radius of the channels between the packing pieces divided by the free cross-sectional area.

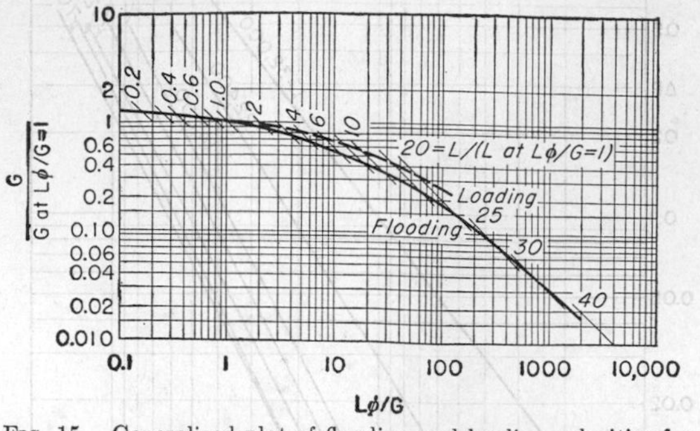

Fig. 15. Generalized plot of flooding and loading velocities for packing materials. This plot is based upon data on Raschig rings and Berl saddles.

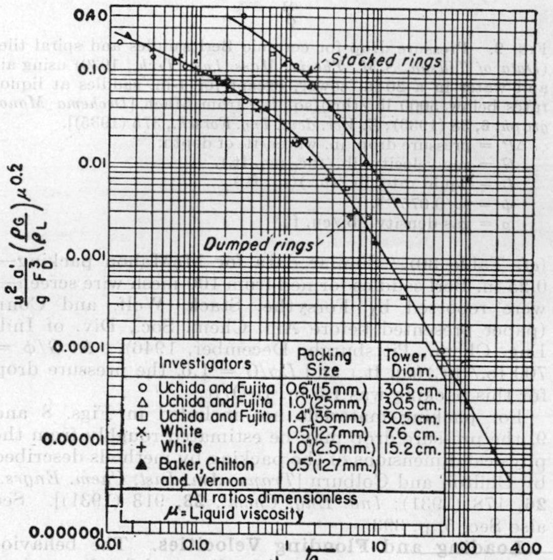

Fig. 16. Flooding velocities for ring packings. [*From Sherwood, Shipley, and Halloway, Ind. Eng. Chem.*, **30,** 768 (1938).]

a' = total surface area of packing, sq. ft./cu. ft. of packed tower volume.
F_D = fractional voids in dry packing, cu. ft./cu. ft. of tower volume.
G = gas rate, lb./(hr.)(sq. ft.).
g_c = standard gravitational acceleration, 32.2 (lb. mass) (ft.)/(lb. force)(sec.²)
L = liquid rate, lb./(hr.)(sq. ft.).
u = gas velocity in empty tower, ft./sec.
ρ_G and ρ_L = gas and liquid densities, respectively, lb./cu. ft.
μ = viscosity of liquid, centipoises.

Lobo, Friend, Hashmall, and Zenz [*Trans. Am. Inst. Chem. Engrs.*, **41,** 693 (1945)] have shown that Sherwood, Shipley, and Holloway's correlation of flooding velocities gives an average deviation of 11.5 per cent from a large volume of experimental data when the measured values of a'/F_D^3 are used rather than values reported by packing manufacturers. The value of a'/F_D^3 is smaller if the packing is dropped into a tower filled previously with liquid than if it is poured dry into an empty container. The latter procedure is generally used by packing

Table 30. Loading and Flooding Data for Packed Columns

Packing	Diam., in.	Col. diam., in.	G_1/ϕ, lb./(hr.)(sq. ft.)* Loading	Flooding	Ref.
Stacked:					
Rings	1.38	..	....	2900	8
	1.02	..	....	2300	8
	0.59	..	....	1800	8
Dumped:					
Rings	1.38	..	....	1650	8
	1.25	6	....	1800	9
	0.75	6	....	960	9
	.625	6	....	910	9
	.625	3	....	1250(?)	9
	.625	3	....	840	3
	.59	..	....	950	8
	.59	12	....	1020	6
	.375	3	....	700	9
	.315	12	....	780	6
Saddles	.59	12	870	1310	6
	.25	3	....	300	3
Broken solids	1.75	30	500	....	7
	1.38	..	....	860†	8
	1.0	..	....	750†	8
	0.59	..	....	600†	8
"Fiberglas" pads, horizontal	.....	12	....	1300	5
"Fiberglas" pads, vertical jackstraw arrangement, bulk density – 4.7 lb./cu. ft.	.....	12	1700–2050	2400–2800	11
Stedman packing·					
Conical	.....	1	310	370	2
Pyramidal	.....	3	760	1180	2
Pyramidal	.....	6	430	780	2
Pyramidal	.....	6	420–500	....	10
Hexagonal-pyramidal	.....	12	1060	1600	2
McMahon packing (0.25-in. saddles formed from wire gauze)	.....	6	700	....	4
0.5-in. wire helixes	.....	..	....	1000–1250	9

* $G_1/\phi = G/\phi$ corresponding to $L\phi/G = 1$; *cf.* Fig. 15.
† Visual flooding point; others determined graphically from ΔP vs. G curves.

References:

[1] Bain and Hougen, *Trans. Am. Inst. Chem. Engrs.*, **40,** 29, 389 (1944).
[2] Bragg, *Ind. Eng. Chem.*, **33,** 279 (1941); *ibid.*, anal. ed., **11,** 283 (1939); *Trans. Am. Inst. Chem. Engrs.*, **37,** 19 (1941).
[3] Elgin and Weiss, *Ind. Eng. Chem.*, **31,** 435 (1939).
[4] Forsythe, Stack, Wolf, and Conn, *Ind. Eng. Chem.*, **39,** 714 (1947).
[5] Herman and Kaiser, *Trans. Am. Inst. Chem. Engr*, **40,** 487 (1944); *cf.* also Lobo and Williams, *ibid.*, **41,** 143 (1945); Minard, Koffolt, and Withrow, *ibid.*, **39,** 813 (1943).
[6] Mach., *Forsch. Gebiete Ingenieurw.*, **6,** *Forschungsheft* 375, 9 (1935).
[7] Piret, Mann, and Wall, *Ind. Eng. Chem.*, **32,** 861 (1940).
[8] Uchida and Fujita, *J. Soc. Chem. Ind. (Japan)*, **39,** 886 (1936); **41,** 563 (1938); also **41,** 275B (1938), in English.
[9] White, *Trans. Am. Inst. Chem. Engrs.*, **31,** 390 (1935).
[10] White and Othmer, *Trans. Am. Inst. Chem. Engrs.*, **38,** 1067 (1942).
[11] Williams, Akell, and Talbott, *Chem. Eng. Progress*, **43,** 585 (1947).

manufacturers. If the packing is shaken after having been placed in a tower by the wet method, $a'/F_D{}^3$ may be higher than the manufacturer's figure, however. Tables 31, 32, 33, and 34 list typical values of packing constants. Lobo *et al.* (*loc. cit.*) give the following empirical equations for calculating F_D for Raschig rings:

For "dry-packed" method:

$$F_D = 1.046 - 0.658M \tag{24}$$

For "wet-packed, shaken" method:

$$F_D = 1.009 - 0.626M \tag{25}$$

For "wet-packed, unshaken" method:

$$F_D = 1.029 - 0.591M \tag{26}$$

here

$$M = \frac{1 - (C/B)^2}{(AB^2)^{0.017}} \tag{27}$$

where A = ring height, in.
B = outside diameter of ring, in.
C = inside diameter of ring, in.

Table 31. Comparative Data on Stoneware Tower Packings*

Packing	% Free space	Surface, sq. ft./cu. ft.	Weight, lb./cu. ft.	Number, units/cu. ft.
Berl saddles ½ in.	68	141	45	15,000
1 in.	69	79	42	2,300
1½ in.	70	50	42	650
Raschig rings ⅜ in.	53	148	65	26,000
½ in.	53	114	65	10,700
⅝ in.	54	80	64	5,800
¾ in.	67	72	46	3,000
1 in.	68	58	45	1,330
1¼ in.	73	44	38	670
1½ in.	68	36	45	380
2 in.	83	29	24	165
Spiral rings (in line) 3 × 3	59	42	58	64
4 × 3	60	31	56	27
6 × 6	67	20	47	8
Partition rings (staggered) 3 × 3	44	41	78	73
4 × 3	58	33	59	41
4 × 4	58	32	59	31
6 × 4	50	22	70	14
6 × 6	52	20	67	9

* Courtesy of Maurice A. Knight Co., Akron, Ohio.

Table 32. Metal Packing for Absorption Towers and Scrubbers

Packin	% Free space	Surface, sq. ft./cu. ft.	Weight, lb./cu. ft.	Number units/cu. ft.
Aluminum rings 2 × 2 in.	92	28	13.7	
Aluminum spirals	87	48.7	21.9	
Bregeat multiple spirals 45 × 30 mm.	93	82	....	320
30 × 30 mm.	90	130	....	700
Lessing rings 2 × 2 in.	95	37	....	160
1 × 1 in.	93	74	....	1,300
½ × ½ in.	91	130	....	9,000
⅜ × ⅜ in.	89	160	....	20,000
¼ × ¼ in.	87	250	....	72,000
Prym rings 9 × 9 mm.	88	190	....	28,000
6 × 6 mm.	86	240	....	80,000
Raschig rings 50 × 50 mm.	94.4	32	....	190
25 × 25 mm*	91.8	65		1,550
25 × 25 mm†	96.9			
15 × 15 mm	95.7	100	....	7,000
Steel rings 2 × 2 in.	92	28	38.7	
Steel spirals 2 × 1 in.	87	48.7	62.4	

* Rings made from sheet iron 0.8 mm. thick.
† Rings made from sheet iron 0.3 mm. thick.

Table 33. Carbon Raschig Rings*

Size, in.	*A*	*B*	*C*	*D*	*E*	*F*	*G*	*H*
Wall ¼ × 1⁄16	85,000	2.5	212	46	17	0.25	55	387
½ × 1⁄16	10,600	10.8	114	27	13	0.56	74	155
¾ × ⅛	3,200	22	70	34	42	1.5	66	106
1 × ⅛	1 350	43	58	27	44	2	73	79
1¼ × 3⁄16	651	67	43	29	89	4	70	61
1½ × ¼	400	96	38	34	138	6.4	67	56
2 × ¼	170	172	29	27	157	9	73	40
3 × ⅜	49	387	19	23	236	17.5	78	25

* Courtesy of National Carbon Co., Inc.

A = number of rings/cu. ft.
B = sq. ft. absorption surface/1000 rings.
C = sq. ft. absorption surface/cu. ft. tower space.
D = lb. of rings/cu. ft. tower space.
E = minimum crushing strength across diameter.
F = load on each bottom ring/10-ft. tower height.
G = % free gas space.
H = sq. ft. absorption surface/cu. ft. free gas space ($C \div G$).

These equations should not be used for $M < 0.2$ or for very thick-walled or solid cylinders.

Flooding velocities in vertical unpacked tubes have been determined by Holmes (private communication), who obtained the results shown in Figs. 17 and 18. Comparison of the two figures shows the advantage of tapering the lower end of the tube when flooding occurs at this end.

If gas approaches the tube from one side, the tapered end should be oriented so that the point faces the gas entrance. For systems other than water and air, the ordinate should be taken as G/ϕ' and the abscissa as $L\psi\phi'/G$, where $\phi' = [(\rho_G/0.075)(\rho_L/62.3)]^{1/2}$, $\psi = (73/\sigma)[\mu_L(62.3/\rho_L)^2]^{1/3}$, and σ is the liquid surface tension, dynes/cm.

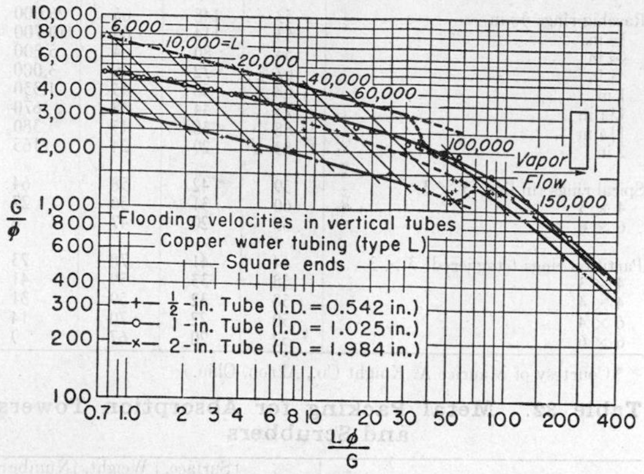

FIG. 17. Flooding velocities in empty vertical tubes. *(Data of Holmes, private communication, 1947.)*

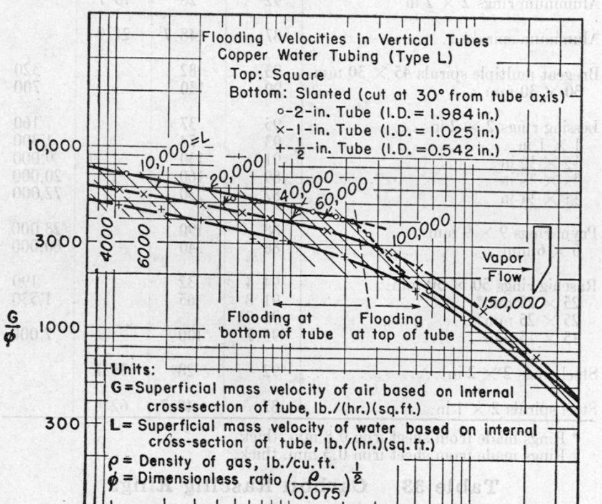

FIG. 18. Flooding velocities in empty vertical tubes. *(Data of Holmes, private communication, 1947.)*

Liquid Hold-up. Values of the liquid hold-up in packed towers have been measured by a number of investigators, notably Jesser and Elgin [*Trans. Am. Inst. Chem. Engrs.*, **39**, 277 (1943)] and Elgin and Weiss (*loc. cit.*). Figure 19, taken from Jesser and Elgin, summarizes the available data for water on various packings. The hold-up is substantially independent of the gas velocity, up to the flooding point. For liquids other than water, the hold-up is

$$h = h_w \mu^{0.1} \left(\frac{62.3}{\rho_L}\right)^{0.78} \left(\frac{73}{\sigma}\right)^n \tag{28}$$

where n varies from 0.42 at $L = 4000$ to 0.13 at $L = 24{,}000$ lb./(hr.)(sq. ft.); h_w is the hold-up for water, expressed in cu. ft. of water/cu. ft. of packing; ρ_L is the liquid density, lb./cu. ft.; and σ is the liquid surface tension, dynes/cm. Figure 20 shows values of hold-up as correlated by Cooper, Christl, and Peery [*Trans. Am. Inst. Chem. Engrs.*, **37**, 979 (1941)].

White and Othmer [*Trans. Am. Inst. Chem. Engrs.*, **38**, 1067 (1942)] have measured the hold-up of water in a 6-in. column filled with Stedman packing, obtaining values

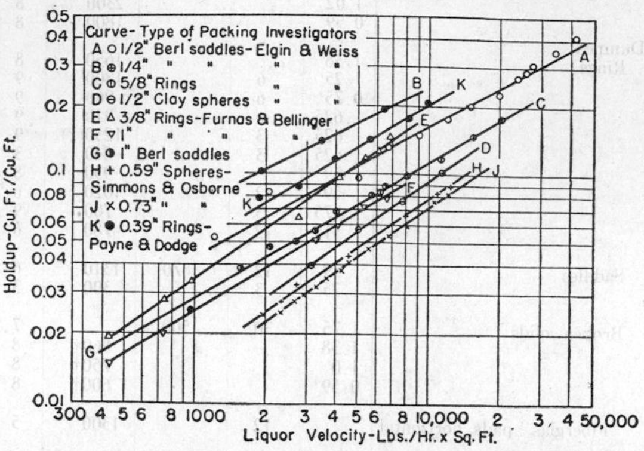

FIG. 19. Liquid hold-up of water for various packings. [*From Jesser and Elgin, Trans. Am. Inst. Chem. Engrs.*, **39**, 227 (1943).]

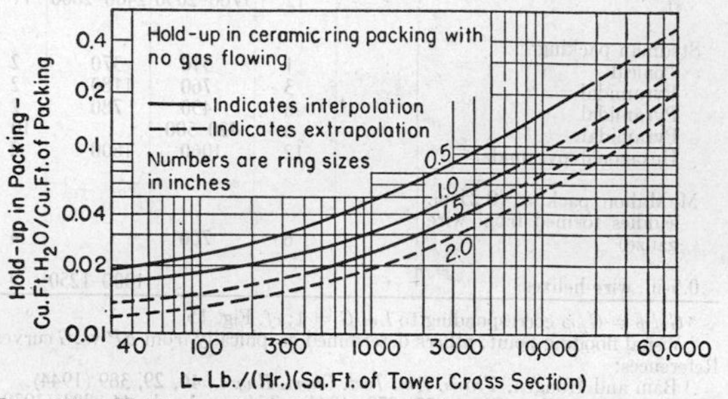

FIG. 20. Hold-up of water in ceramic ring packing. [*From Cooper, Christl, and Peery, Trans. Am. Inst. Chem. Engrs.*, **37**, 979 (1941).]

varying from 0.026 cu. ft./cu. ft. at zero liquid-flow rate to 0.18 cu. ft./cu. ft. at L = 5900 lb./(hr.)(sq. ft.)

Plate Columns

Pressure drops through bubble-cap and sieve plates, allowable vapor velocities, maximum liquid flows consistent with stable operation, and design of weirs and downspouts are discussed in Sec. 9, pp. 597 to 602. Plate designs used for absorbers are usually the same as those used for distilling columns.

RATES OF MASS TRANSFER IN PACKED ABSORPTION COLUMNS

There are two ways to estimate the H.T.U. for a packed absorption tower. (1) If reliable data on H_{OG} or H_{OL} are available for the same system of components under the desired conditions of operation, these should be used. Values of H_{OG} and H_{OL} that are available in the literature are summarized on pp. 693 to 698. Cases of chemical reaction require special treatment, as shown on p. 702*ff*. (2) When no over-all H.T.U. data are available for the system and packing under consideration, or when the experimental flow rates of gas or liquid are widely different from those required for economical absorption, a value of H_{OG} may be estimated from individual resistances through use of either of the equations

$$H_{OG} = H_G + H_L\left(\frac{mG_M}{L_M}\right) \qquad (29a)$$

$$H_{OL} = H_L + H_G\left(\frac{L_M}{mG_M}\right) \qquad (29b)$$

The derivation of these equations is discussed in Sec. 8, p. 550; the symbols are defined and their use is discussed on p. 524.

Recommended values of H_G are given by Figs. 21–29 on pp. 688–690. Values of H_L are given on p. 691*ff*.

When unusual conditions of operation are involved, such as unusual solutes or solvents, viscous liquids, unusual packings, or abnormally high or low flow rates, it may be necessary to determine H_{OG} from pilot-plant tests.

If values of the transfer coefficients k_Ga and k_La are desired, they may be calculated using the relations

$$k_Ga = \frac{G_M}{H_G(1-y)_f} \qquad (29c)$$

$$k_La = \frac{L_M}{H_L(1-x)_f} \qquad (29d)$$

(Nomenclature is given at the end of the section.) If the gas or liquid contains very little solute, say less than 2 mole per cent, the term $(1 - y)_f$ or $(1 - x)_f$ may be taken as unity.

Packed Columns—Gas-phase Resistance. As shown by Eq. (29), the liquid-film resistance is suppressed when (mG_M/L_M) is small. The system ammonia-air-water has been used frequently for measurements of H_G because the large solubility of this gas (small value of m) permits convenient operation under conditions such that mG_M/L_M is small. More soluble gases could be used, but the use of ammonia involves fewer experimental difficulties.

The most reliable and complete set of experimental data for this system are those of Fellinger (Sc. D. Thesis, Mass. Inst. Tech., 1941). Values of H_G have been calculated from these data by subtracting small values of the product $H_L(mG_M/L_M)$, using the H_L data of Sherwood and Holloway (see. p. 691). The resulting values of H_G are shown by dashed lines on Figs. 21 to 29 and are recommended for conservative use when H_{OG} or H_{OL} must be estimated from H_G and H_L. For solutes different from ammonia, the values of H_G read from Figs. 21 to 29 should be multiplied by the ratio $[(\mu/\rho D)/0.78]^{1/2}$, as discussed later.

In spite of a large volume of experimental work on the subject, no completely reliable data exist for the resistance to mass transfer in the gas phase for packed absorption towers. The methods that have been used for experimental investigations in this field have been (1) vaporization of pure liquids into an inert gas, (2) absorption of a gas in a liquid in which it is very soluble, and (3) absorption of a gas in a liquid from which no back pressure is exerted because of an irreversible, rapid chemical reaction. The most reliable data obtained directly by methods (1) and (3) are found to give values of H_G that are only one-third to one-half as great as the best values obtained indirectly by method (2). Estimates of equipment size based on H.T.U. data evaluated from rates of absorption of ammonia in water (method (2)) are likely to be very conservative, but the data that are available for this system are more abundant and more reliable and are suggested for use when values of H_G are required. A comparison of H.T.U. values measured by methods (1) and (3) on the one hand and by method (2) on the other is given in Table 34.

There appears to be no way in which the absence of liquid-phase resistance can be proved when a chemical reaction occurs. It seems likely, however, that, at least in a few such cases, as in the absorption of ammonia in acid or of sulfur dioxide in caustic, almost all the resistance is ordinarily in the gas phase. Experiments in which pure liquids are vaporized are difficult to carry out because the gas becomes nearly saturated with the evaporating liquid after passing through shallow beds of packing, for which end effects are large. Based on experiments in which water was evaporated into air using a 20-in.-diameter tower containing 8 in. of 1.5-in. Raschig-ring packing, and also on experiments in which sulfur dioxide was absorbed in caustic solution using 3-in. spiral tile [data of Johnstone and Singh, *Ind. Eng. Chem.*, **36,** 286 (1937)], Sherwood and Holloway [*Trans. Am. Inst. Chem. Engrs.*, **36,** 21, (1940)] proposed an equation equivalent to

$$H_G = \frac{1.01G^{0.31}}{L^{0.33}} \text{ ft.} \qquad (30)$$

This equation is said to hold for both 1.5-in. rings and 3-in. tile in the ranges $20 < G < 1830$ lb./(hr.)(sq. ft.) and $168 < L < 6100$ lb./(hr.)(sq. ft.). The comparison shown by Table 34 indicates that values of H_G estimated from this equation may be unsafe for design purposes, possibly because of large end effects in the experiments on which it was based.

Table 34. Comparison of Gas-phase Resistances to Mass Transfer Determined by Different Methods

Packing	Flow conditions	Height equivalent to one gas-phase transfer unit, H_G, ft.	
		From experiments on vaporization of H_2O into air, or absorption of SO_2 followed by rapid chemical reaction*	From experiments on absorption of NH_3 in water, corrected for liquid-phase resistance†
1.5-in. rings	$G = 500, L = 1500$	0.62	1.30
1.5-in. rings	$G = 1000, L = 1500$	.77	1.45
3-in. spiral tile.......	$G = 500, L = 1500$	.62	1.80
3-in. spiral tile.......	$G = 1000, L = 1500$	.77	2.40

* Calculated from Eq. (30).

† Data of Fellinger, Sc. D. Thesis, Mass. Inst. Tech., 1941, corrected for liquid-phase resistance by using values of H_L for desorption of oxygen from water.

A part of the discrepancy between values of H_G according to Eq. (30) and those based on ammonia-absorption measurements may be due to the reversible chemical reaction between NH_3 and H_2O, which occurs when ammonia is absorbed. The work of Vivian and Whitney, [*Chem. Eng. Progress*, **43,** 691 (1947)] suggests that the equilibrium between gaseous NH_3 and the various forms of ammonia that occur in the liquid at the interface may be different from the usual solubility relation, because the formation of the reacted non-volatile forms of ammonia does not take place instantaneously. Further experimental work may indicate that true values of H_G are smaller than those which must be recommended at this time, since the latter are calculated from ammonia-absorption data without taking account of the reaction between ammonia and water.

Effect of Temperature and Pressure on H_G. It is recommended that moderate temperature and pressure changes be assumed to have no effect on H_G. This is supported by the theory of mass transfer in wetted-wall towers (see Sec. 8, p. 543). A slight increase in H_G with temperature, ranging from 0.3 to 1.2 per cent/°C., has been reported by Dodge and Dwyer [*Ind. Eng. Chem.*, **33,** 485 (1941)], Molstad, McKinney, and Abbey [*Trans. Am. Inst. Chem. Engrs.*, **39,** 605 (1943)], and Kowalke, Hougen, and Watson (*Univ. Wis. Eng. Exp. Sta. Bull.* 68, June, 1925), each of whom measured rates of absorption of ammonia in water.

Effect of Physical Properties of the Gas on H_G. It is recommended that H_G be assumed proportional to $(\mu/\rho D)^{0.5}$, since the theory of mass transfer in wetted-wall tubes predicts approximately this effect for Schmidt numbers $(\mu/\rho D)$ within the range 0.5 to 3; see Sec. 8,

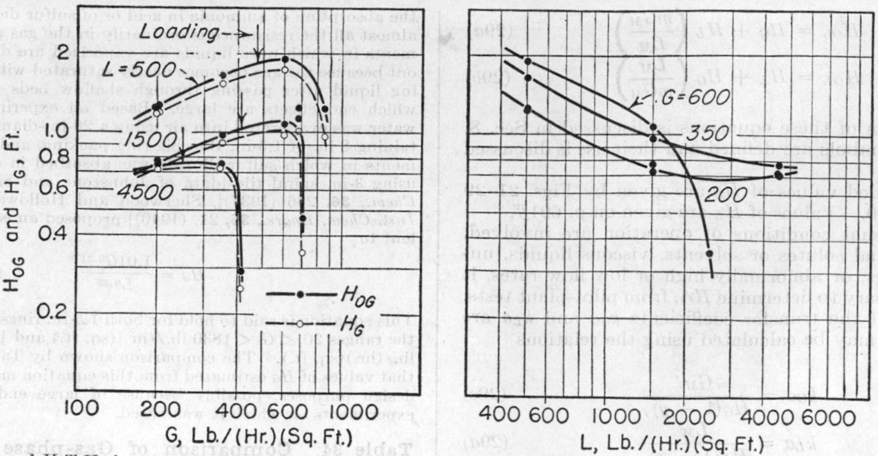

FIG. 21. Values of H.T.U. for absorption of ammonia in water—0.375-in. ceramic Raschig rings. (*Data of Fellinger, Sc.D. Thesis, Mass. Inst. Tech.*, 1941.)

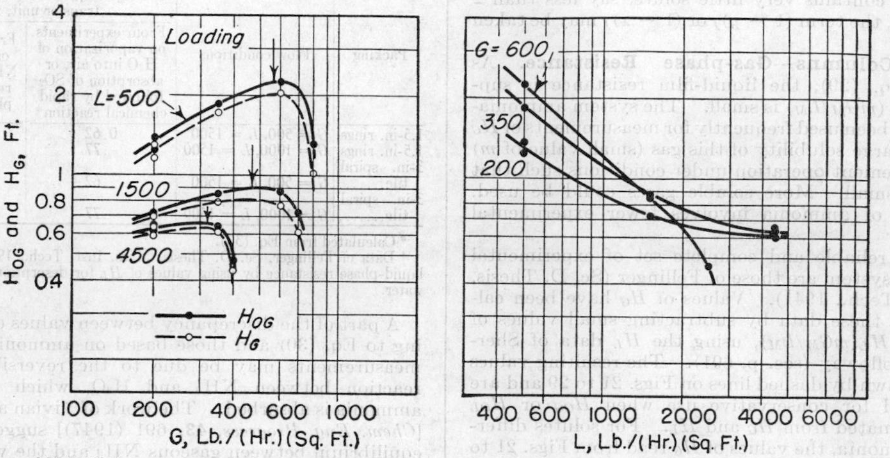

FIG. 22. Values of H.T.U. for absorption of ammonia in water—0.5-in. ceramic Raschig rings. (*Data of Fellinger, Sc.D. Thesis, Mass. Inst. Tech.*, 1941.)

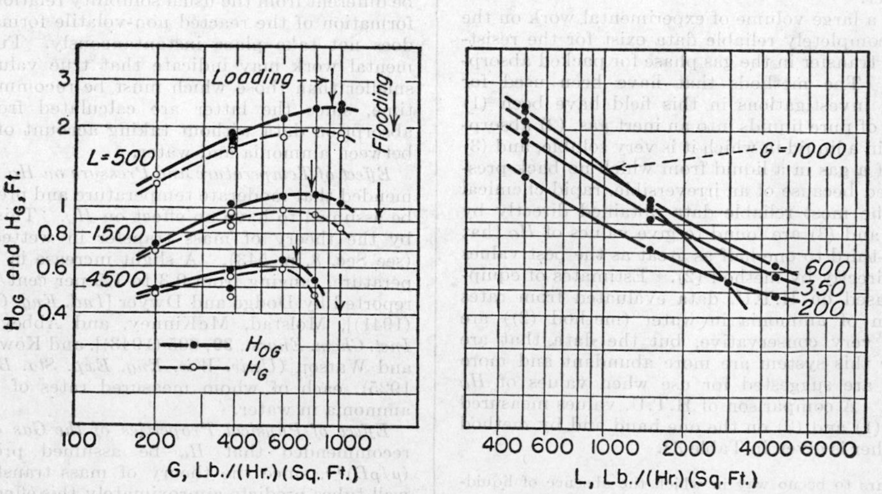

FIG. 23. Values of H.T.U. for absorption of ammonia in water—1-in. ceramic Raschig rings. (*Data of Fellinger, Sc.D. Thesis, Mass. Inst. Tech.*, 1941.)

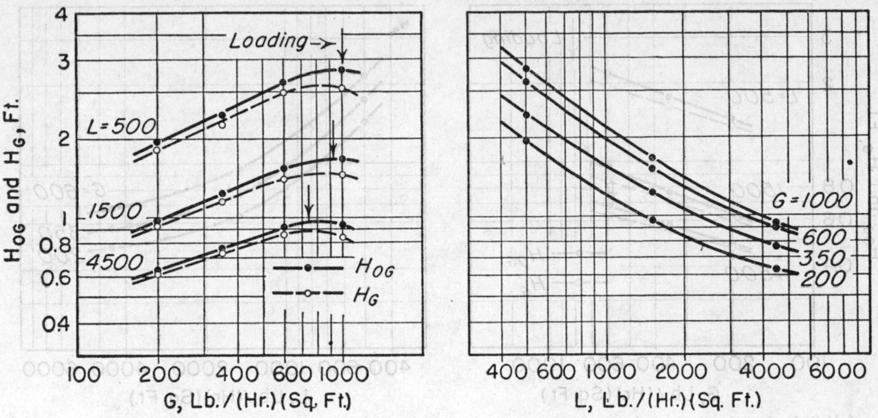

FIG. 24. Values of H.T.U. for absorption of ammonia in water—1.5-in. ceramic Raschig rings. (*Data of Fellinger, Sc.D. Thesis, Mass Inst. Tech.*, 1941.)

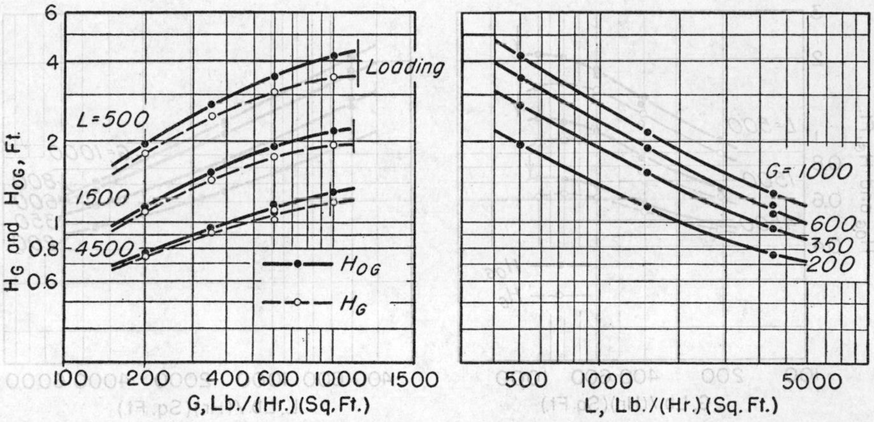

FIG. 25. Values of H.T.U. for absorption of ammonia in water—2-in. ceramic Raschig rings. (*Data of Fellinger, Sc.D. Thesis, Mass. Inst. Tech.*, 1941.)

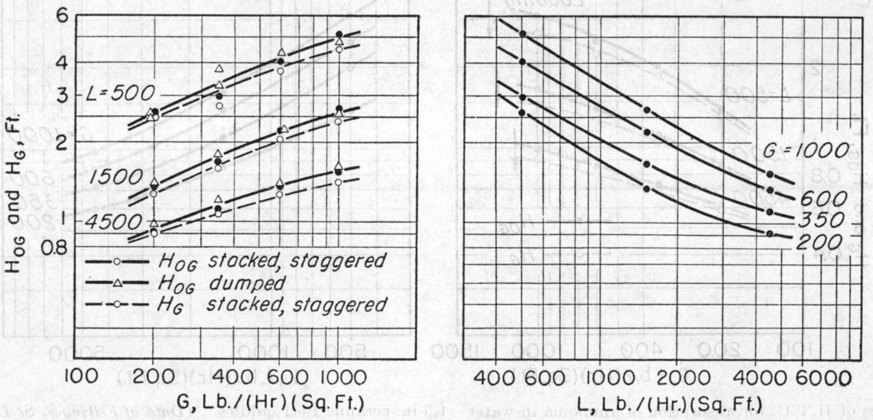

FIG. 26. Values of H.T.U. for absorption of ammonia in water—3-in. triple-spiral tile. (*Data of Fellinger, Sc.D. Thesis, Mass. Inst. Tech.*, 1941.)

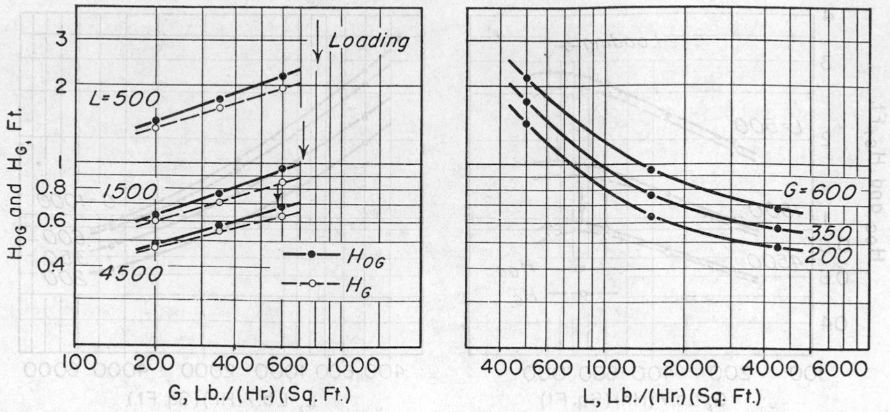

FIG. 27. Values of H.T.U. for absorption of ammonia in water—0.5-in. ceramic Berl saddles. (*Data of Fellinger, Sc.D. Thesis, Mass. Inst. Tech.*, 1941.)

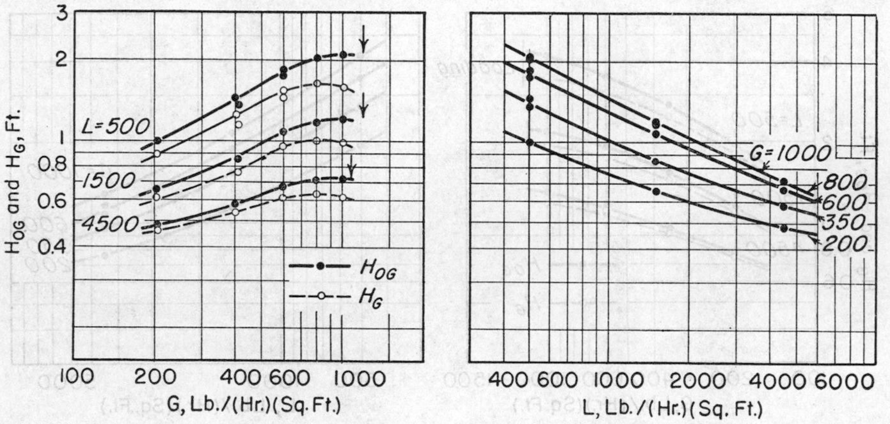

FIG. 28. Values of H.T.U. for absorption of ammonia in water—1-in. Berl saddles. (*Data of Fellinger, Sc.D. Thesis. Mass. Inst. Tech.*, 1941.)

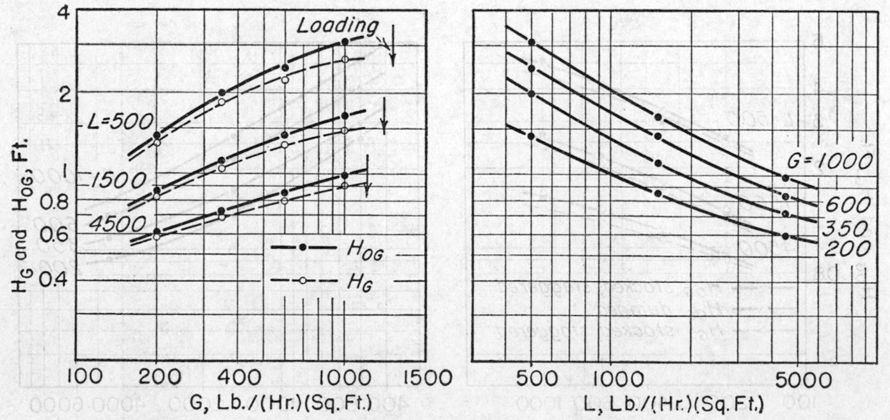

FIG. 29. Values of H.T.U. for absorption of ammonia in water—1.5-in. ceramic Berl saddles. (*Data of Fellinger, Sc.D. Thesis, Mass. Inst. Tech.*, 1941.)

p. 539, in this connection. Johnstone and Singh [*Ind. Eng. Chem.*, **29**, 286 (1937)] found that their data of evaporation of water, absorption of SO_2 in NaOH solution, and absorption of NH_3 in aqueous solutions of acetic acid were correlated on the basis of the ⅔ power of $\mu/\rho D$. On the other hand, Sherwood and Holloway [*Trans. Am. Inst. Chem. Engrs.*, **36**, 21 (1940)] report data obtained by Mehta and Parekh (S.M. Thesis in Chem. Eng., Mass. Inst. Tech., 1939) on the rates of vaporization of pure liquids in a 3.6-in.-diameter tower packed with 5 in. of 0.625-in. porcelain rings. These investigators found that H_G for toluene vaporaiztion into air was 12 per cent greater than H_G for water, corresponding to an exponent of 0.17 on the Schmidt group. The exponent probably depends on the packing size, but the variation has not yet been established. Numerical values of $\mu/\rho D$ are listed in Sec. 8, p. 539.

Effect of Liquid Surface Tension on H_G. Sherwood and Holloway (*loc. cit.*) summarize data from several sources on the effects obtained by adding small amounts of wetting agents on formaldehyde to water flowing through a packed tower. They conclude that a reduction in surface tension has no effect on the rate of mass transfer when the gas-phase resistance controls but that the liquid-phase resistance is increased by the presence of large organic molecules at the interface.

Packed Columns—Liquid-phase Resistance. The most reliable and extensive data on the resistance to mass transfer in the liquid phase in packed towers are those of Sherwood and Holloway [*Trans. Am. Inst. Chem. Engrs.*, **36**, 39 (1940)]. These investigators made a careful study of the rates of desorption of carbon dioxide, oxygen, and hydrogen from water in a 20-in.-diameter tower packed with 1.5-in. Raschig rings. Below the range of incipient flooding, the value of H_L was found to be independent of gas velocity, as shown by Fig. 30.

Figures 31 and 32 are plots of Sherwood and Holloway's original data for oxygen desorption from water at 20°C.

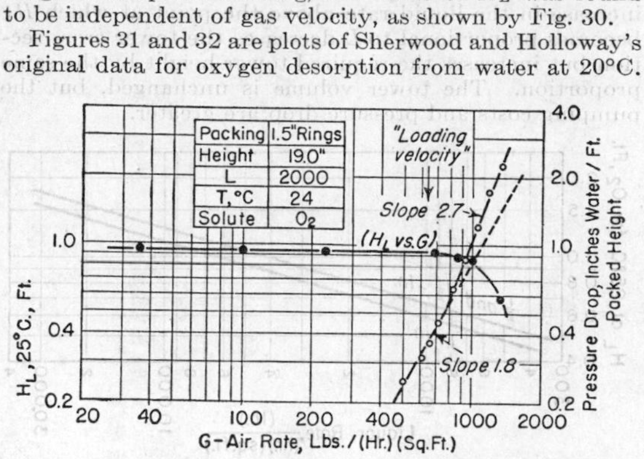

FIG. 30. Effect of gas velocity on H_L. [*Data of Sherwood and Holloway, Trans. Am. Inst. Chem. Engrs.*, **36**, 39 (1940).]

Using these curves, values of H_L may be estimated for other solutes by multiplying the H.T.U. value read from the chart by the ratio $(1.80 \times 10^{-5}/D, \text{cm}^2./\text{sec.})^{0.47}$. A number of experimentally determined values of D are listed in Table 13, Sec. 8, p. 540.

The liquid velocities at which the H_L data for Raschig rings break away from the straight lines in Fig. 31 are below the liquid rates at loading, indicating that at these

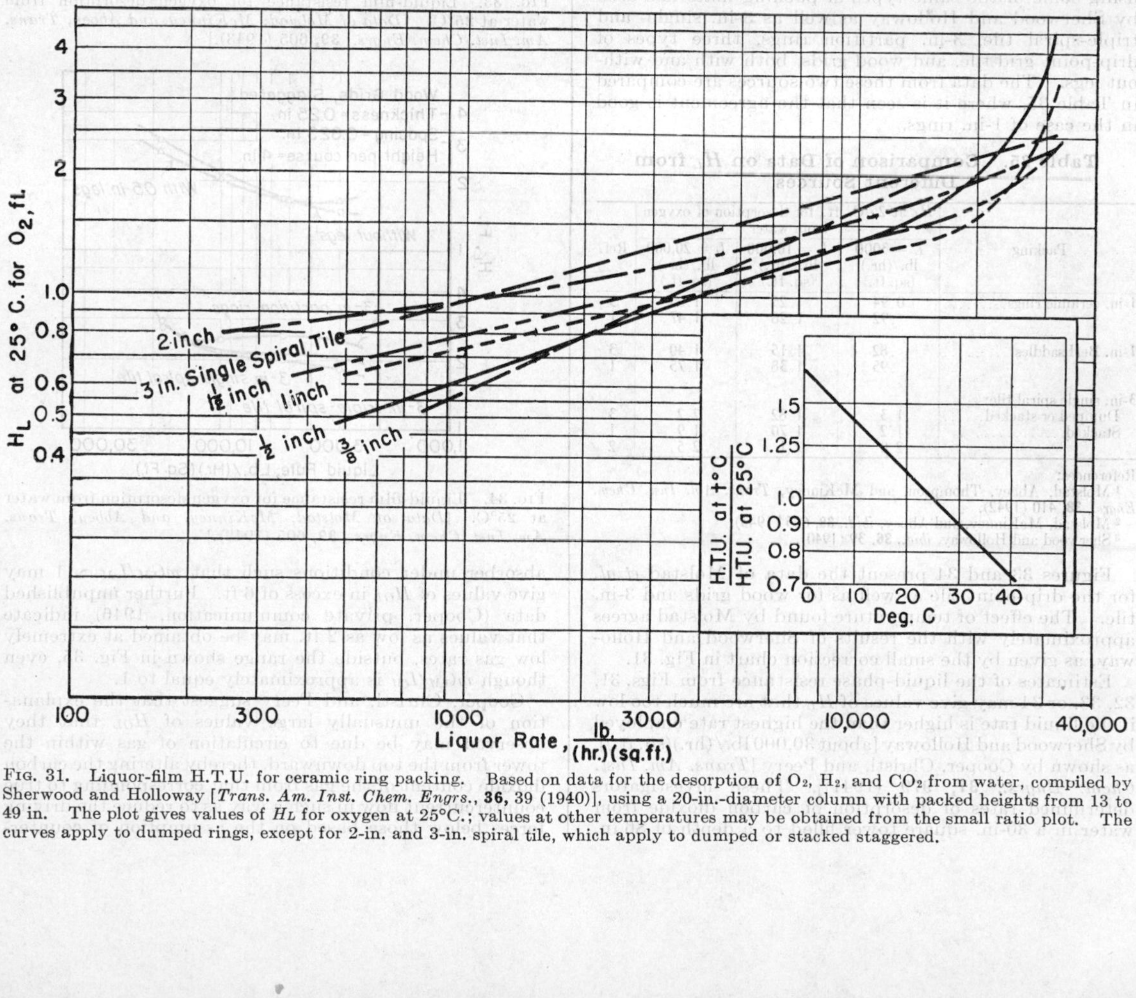

FIG. 31. Liquor-film H.T.U. for ceramic ring packing. Based on data for the desorption of O_2, H_2, and CO_2 from water, compiled by Sherwood and Holloway [*Trans. Am. Inst. Chem. Engrs.*, **36**, 39 (1940)], using a 20-in.-diameter column with packed heights from 13 to 49 in. The plot gives values of H_L for oxygen at 25°C.; values at other temperatures may be obtained from the small ratio plot. The curves apply to dumped rings, except for 2-in. and 3-in. spiral tile, which apply to dumped or stacked staggered.

points the liquid in the packing may begin to bridge over the spaces formerly filled with air. Operation at such high liquid rates is not economical, since an increase in the liquid rate above the point at which H_L becomes proportional to L decreases the tower cross section but increases the required tower height by the same proportion. The tower volume is unchanged, but the pumping costs and pressure drop are greater.

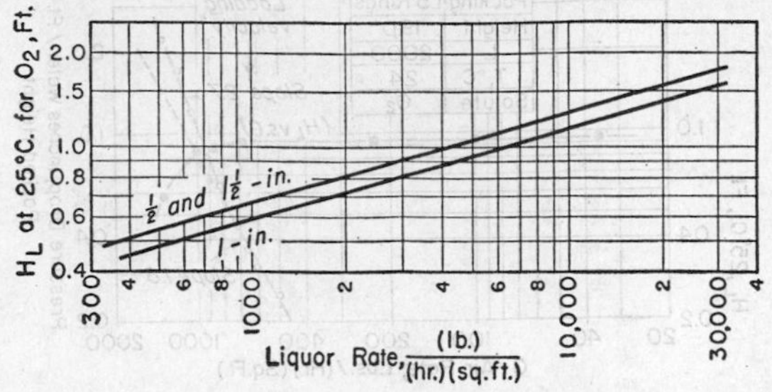

FIG. 32. Liquor-film H.T.U. for ceramic Berl saddles. Based on data for the desorption of O_2, H_2, and CO_2 from water, compiled by Sherwood and Holloway [*Trans. Am. Inst. Chem. Engrs.*, **36**, 49 (1940)], using a 20-in.-diameter column with packed heights from 15 to 22 in. The plot gives values of H_L for oxygen at 25°C.; values at other temperatures may be obtained from the ratio plot on Fig. 31.

Molstad, Abbey, Thompson, and McKinney [*Trans. Am. Inst. Chem. Engrs.*, **38**, 410 (1942)] and Molstad, McKinney, and Abbey [*ibid.*, **39**, 605 (1943)] have determined values of H_L for oxygen desorption from water using some of the same types of packing materials used by Sherwood and Holloway as well as 3-in. single- and triple-spiral tile, 3-in. partition rings, three types of drip-point grid tile, and wood grids, both with and without legs. The data from these two sources are compared in Table 35, where it is seen that the agreement is good in the case of 1-in. rings.

Table 35. Comparison of Data on H_L from Different Sources

Packing	H_L at 25°C., ft., for desorption of oxygen from water			Ref.
	L = 3000 lb./(hr.)(sq. ft.)	L = 10,000 lb./(hr.)(sq. ft.)	L = 20,000 lb./(hr.)(sq. ft.)	
1-in. ceramic rings..........	0.94	1.25	1.65	3
	.92	1.28	1.47	1
1-in. Berl saddles..........	.82	1.15	1.40	3
	.95	1.38	1.75	1
3-in. single spiral tile:				
Dumped or stacked......	1.3	1.82	2.2	3
Stacked................	1.2	1.70	1.9	1
	2.0	2.1	2.5	2

References:
[1] Molstad, Abbey, Thompson, and McKinney, *Trans. Am. Inst. Chem. Engrs.*, **38**, 410 (1942).
[2] Molstad, McKinney, and Abbey, *ibid.*, **39**, 605 (1943).
[3] Sherwood and Holloway, *ibid.*, **36**, 39 (1940).

Figures 33 and 34 present the data of Molstad *et al.* for the drip-point tile as well as for wood grids and 3-in. tile. The effect of temperature found by Molstad agrees approximately with the results of Sherwood and Holloway, as given by the small correction chart in Fig. 31.

Estimates of the liquid-phase resistance from Figs. 31, 32, 33, or 34 may give values of H_L that are much too low if the liquid rate is higher than the highest rate employed by Sherwood and Holloway [about 30,000 lb./(hr.)(sq. ft.)], as shown by Cooper, Christl, and Peery [*Trans. Am. Inst. Chem. Engrs.*, **37**, 979 (1941)]. These investigators determined rates of desorption of carbon dioxide from water in a 30-in. square tower filled to a depth of 86 in. with 2- by 2- by 1/16-in. steel Raschig rings. Water rates from 13,600 to 56,000 lb./(hr.)(sq. ft.) and air rates from 19 to 368 lb./(hr.)(sq. ft.) were employed. High liquor-gas flow ratios in this range are required for economical operation of apparatus used for absorption or stripping of insoluble gases. The results obtained by Cooper, Christl, and Peery are shown in Fig. 35, which indicates that operation of an atmospheric carbon dioxide

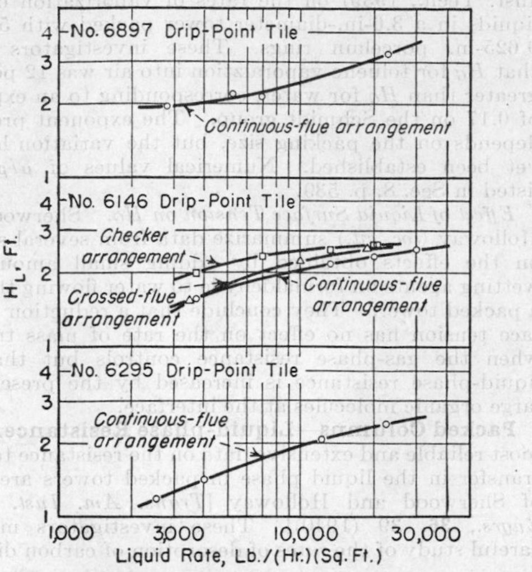

FIG. 33. Liquid-film resistance for oxygen desorption from water at 25°C. [*Data of Molstad, McKinney, and Abbey, Trans. Am. Inst. Chem. Engrs.*, **39**, 605 (1943).]

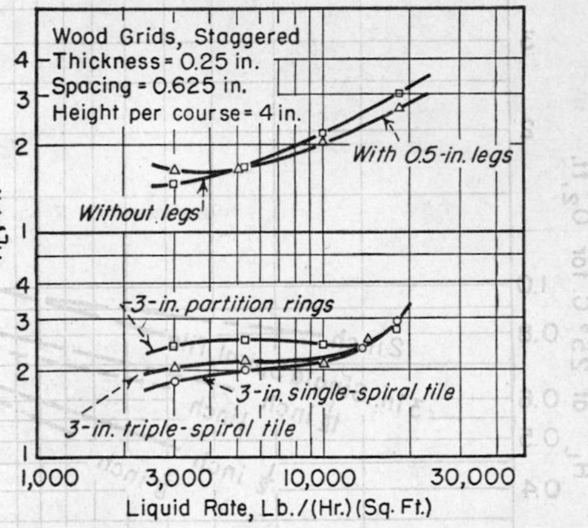

FIG. 34. Liquid-film resistance for oxygen desorption from water at 25°C. [*Data of Molstad, McKinney, and Abbey, Trans. Am. Inst. Chem. Engrs.*, **39**, 605 (1943).]

absorber under conditions such that $mG_M/L_M \sim 1$ may give values of H_{OL} in excess of 6 ft. Further unpublished data (Cooper, private communication, 1946) indicate that values as low as 2 ft. may be obtained at extremely low gas rates, outside the range shown in Fig. 35, even though mG_M/L_M is approximately equal to 1.

Cooper, Christl, and Peery suggest that the explanation of the unusually large values of H_{OL} that they observed may be due to circulation of gas within the tower from the top downward, thereby altering the carbon dioxide content of the gas from that corresponding to true countercurrent flow in such a way as to reduce the driving forces below those based on the assumption of countercurrent action. When the conditions are such that the ratio of the average linear liquid and gas velocities exceeds 1, the relatively fast moving liquid stream tends to sweep the gas downward, causing the H.T.U. values to exceed those observed by Sherwood and Holloway, as shown in Fig. 36. For this comparison the average linear water velocity is calculated by dividing the liquid hold-up, in cu. ft./ft. of packed height, into the liquor-flow rate, in cu. ft./sec.; and the linear gas velocity is that through the free space within the packing. Evidently the phenomenon is not caused by loading within the column, since the gas velocities employed were well below the loading velocities.

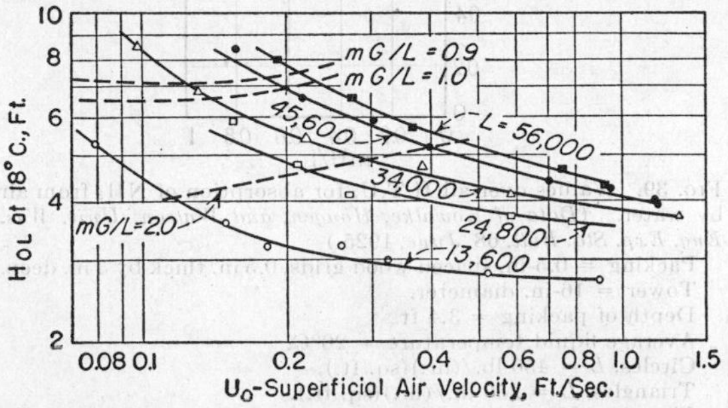

FIG. 35. Effect of gas velocity on H_{OL} at high liquor rates. From data on CO_2 absorption in water, using 2-in. steel Raschig rings. On this graph G and L are for molar rates. [*Cooper, Christl, and Peery, Trans. Am. Inst. Chem. Engrs.*, **37,** 979 (1941).]

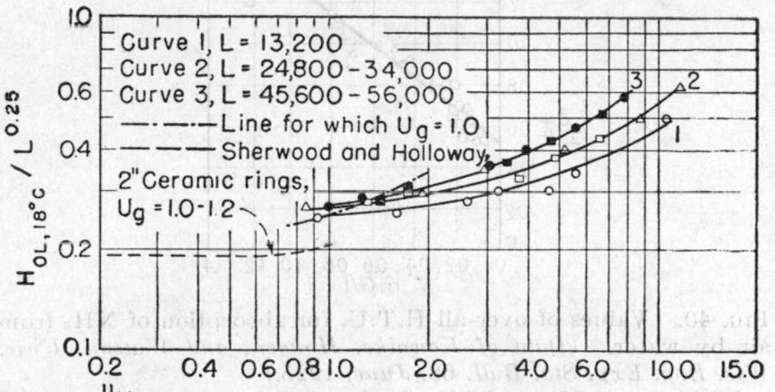

FIG. 36. Absorption of CO_2 in H_2O at high liquid rates. Correlation of data of Cooper, Christl, and Peery [*Trans. Am. Inst. Chem. Engrs.*, **37,** 979 (1941)] with those of Sherwood and Holloway [*Trans. Am. Inst. Chem. Engrs.*, **36,** 49 (1940)].

Packed Columns—Over-all Resistance for Miscellaneous Systems. *Ammonia in Water.* This system has been studied more frequently in packed towers than any other. The most reliable and complete set of experimental data are those of Fellinger, which are presented in Figs. 21 through 29. These data cover ceramic Raschig rings ranging in size from ⅜ to 2 in., Berl saddles from ½ to 1½ in., and 3-in. ceramic triple-spiral tile. Gas rates range from 200 lb./(hr.)(sq. ft.) to the flooding point, or to 1000 lb./(hr.)(sq. ft.), and liquid rates range from 500 to 4500 lb./(hr.)(sq. ft.). The effect of packing size is small.

The value of H_{OG} passes through a flat maximum near the gas velocity at loading. Similar behavior in this range of gas velocities near the flooding point has also been observed in the experiments on desorption of oxygen from water carried out by Holloway and Sherwood, as summarized above. Near the flooding point, the value of H_{OG} must increase sharply because of the extreme departure from countercurrent flow conditions.

Most of the early work on ammonia absorption was directed toward the determination of the gas-phase resistance to transfer; only recently it has been pointed out that, under economical operating conditions, a large part of the total resistance may be in the liquid phase [*cf.* Colburn, *Trans. Am. Inst. Chem. Engrs.*, **29,** 174 (1939)]. This is shown by Figs. 37 to 40 (Colburn, *loc. cit.*), where it is seen that H_{OG} is much greater than H_G, the intercept on the vertical axis, when mG_m/L_m is zero.

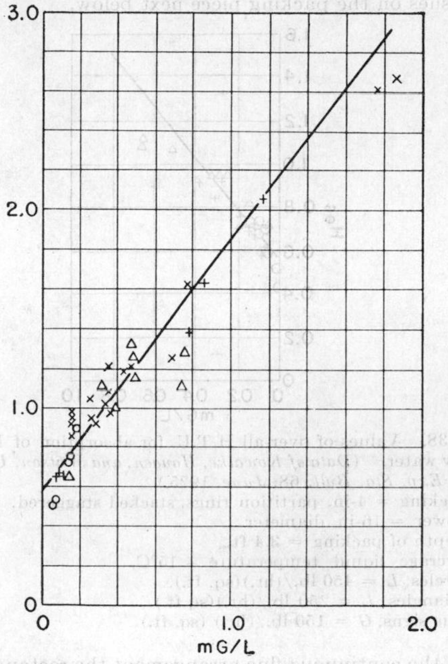

FIG. 37. Values of over-all H.T.U. for absorption of NH_3 from air by water. (*Data of Kowalke, Hougen, and Watson, Univ. Wis. Eng. Exp. Sta. Bull.* 68, *June,* 1925.)

Packing = 3-in. spiral rings, stacked staggered.
Tower = 16-in. diameter.
Depth of packing = 3.3 ft.
Average liquid temperature = 20°C.
Circles, L = 450 lb./(hr.)(sq. ft.).
Triangles, L = 250 lb./(hr.)(sq. ft.).
Crosses, G = 230 lb./(hr.)(sq. ft.).
Plus signs, G = 150 lb./(hr.)(sq. ft.).

The data of Molstad, McKinney, and Abbey [*Trans. Am. Inst. Chem. Engrs.*, **39,** 605 (1943)], Dodge and Dwyer [*Ind. Eng. Chem.*, **33,** 485 (1941)], as well as those of Borden and Squires, and of Dougherty and Johnson (*cf.* Sherwood and Holloway, *loc. cit.*), who used carbon rather than porcelain rings, agree approximately with those of Fellinger. Values of H_{OG} calculated from the experiments of Kowalke, Hougen, and Watson (*Univ. Wis. Eng. Exp. Sta. Bull.* 68, June, 1925) are considerably lower, probably because of the effect of spray sections above and below the packing.

Molstad *et al.* (*loc. cit.*) determined values of H_{OG} for absorption of ammonia from air by water using 3-in. triple-spiral tile, stacked staggered and also with the internal passages continuous; 3-in. single-spiral tile; and 3-in. partition rings. Their data for the single-spiral tile in the staggered arrangement agree closely with those of Fellinger, shown in Fig. 26. The values of H_{OG} were about 1.4 times as great for triple-spiral tiles and 2.8 times as great for partition rings, all at L = 3000 lb./(hr.)(sq. ft.). In the non-staggered stacked arrangement, the H_{OG} values for the triple-spiral tile were 1.2 to 4.3 times as large as in the staggered arrangement. The effect of variations in the liquid rate for triple-spiral tile was similar to that observed by Fellinger, but liquid rate had a greater effect for single-spiral tile and partition rings.

Molstad, McKinney, and Abbey (*loc. cit.*) report rates of absorption of ammonia from air by water in a 15.1-in.

square tower packed with three types of ceramic drip-point grid tile in the continuous-flue, crossed-flue, and checker arrangements. Some of their results are shown in Fig. 41. These packings are made in the form of blocks 7.5 and 5 in. square. Each block consists of 5, 8, or 9 slats separated by parallel channels for gas flow. The slats in one layer of packing are separated from those in the layers above and below by short legs. The liquid running off the slats collects on drip points and splashes on the packing piece next below.

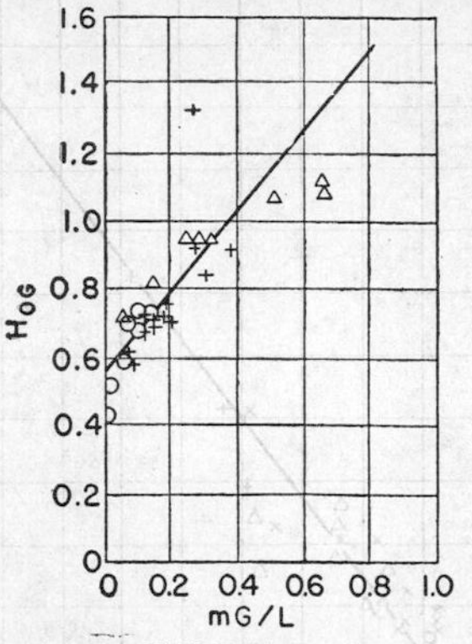

FIG. 38. Values of over-all H.T.U. for absorption of NH_3 from air by water. (*Data of Kowalke, Hougen, and Watson, Univ. Wis. Eng. Exp. Sta. Bull.* 68, *June*, 1925.)
Packing = 4-in. partition rings, stacked staggered.
Tower = 16-in. diameter.
Depth of packing = 3.4 ft.
Average liquid temperature = 15°C.
Circles, L = 450 lb./(hr.)(sq. ft.).
Triangles, L = 250 lb./(hr.)(sq ft.).
Plus signs, G = 150 lb./(hr.) (sq. ft.).

In the continuous-flue arrangement the rectangular gas passages form continuous channels; in the crossed-flue arrangement the long dimensions of the gas passages in successive layers of packing are oriented at right angles. Only 59 per cent as many packing pieces are used in the checker arrangement, since the packing pieces in each layer are arranged in a checkerboard pattern, each packing being surrounded by four blank spaces and four packing pieces, which it touches at the corners. With this arrangement, the values of H_{OG} were only about 30 per cent greater than with the other two arrangements under the worst conditions of the comparison, and the pressure drop was the lowest of any arrangement. The continuous-flue arrangement gave the same H.T.U. values as the crossed-flue arrangement with a pressure drop only 75 per cent as great.

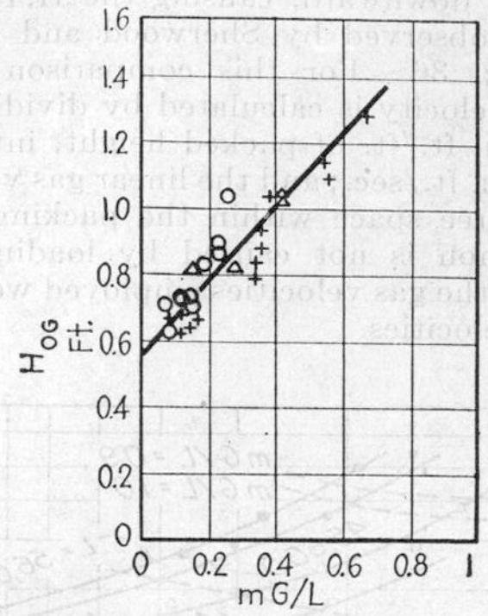

FIG. 39. Values over-all H.T.U. for absorption of NH_3 from air by water. (*Data of Kowalke, Hougen, and Watson, Univ. Wis. Eng. Exp. Sta. Bull.* 68, *June*, 1925.)
Packing = 0.5-in. spaced wood grids, 0.5 in. thick by 3 in. deep.
Tower = 16-in. diameter.
Depth of packing = 3.4 ft.
Average liquid temperature = 20°C.
Circles, L = 450 lb./(hr.)(sq. ft.).
Triangles, L = 250 lb./(hr.)(sq. ft.).
Plus signs, G = 250 lb./(hr.)(sq. ft.).

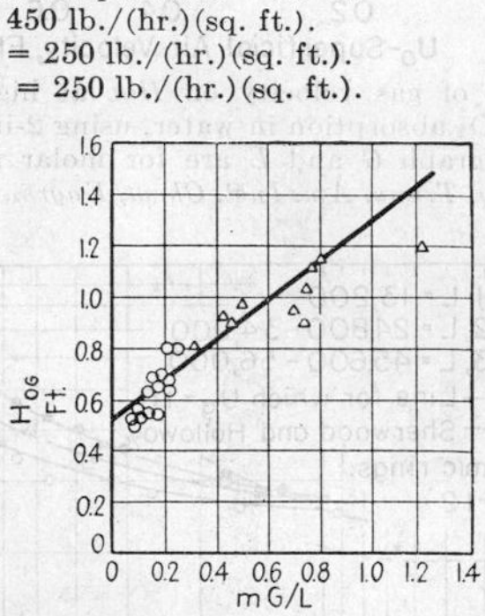

FIG. 40. Values of over-all H.T.U. for absorption of NH_3 from air by water. (*Data of Kowalke, Hougen, and Watson, Univ. Wis. Eng. Exp. Sta. Bull.* 68, *June*, 1925.)
Packing = 1.25 to 1.75 in., broken quartz.
Tower = 16-in. diameter.
Depth of packing = 3.4 ft.
Average liquid temperature = 22°C.
Circles, L = 450 lb./(hr.)(sq. ft.).
Triangles, L = 250 lb./(hr.)(sq. ft.).

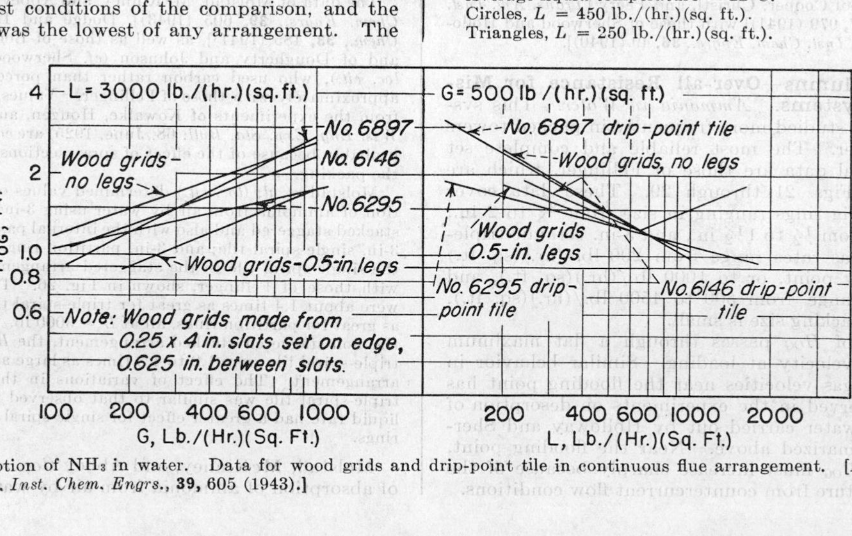

FIG. 41. Absorption of NH_3 in water. Data for wood grids and drip-point tile in continuous flue arrangement. [*Data of Molstad, et al., Trans. Am. Inst. Chem. Engrs.*, **39**, 605 (1943).]

Williams, Akell, and Talbott, [*Chem. Eng. Progress*, **43**, 585 (1947)] have measured rates of absorption of ammonia from air using a 6-in.-diameter tower packed in three 2-ft. sections with Fiberglas pads in the "vertical Jackstraw" arrangement. The density of the packing was 4.7 lb./cu. ft. Results of this work are shown by Fig. 42. This packing is used to best advantage when the liquid rate exceeds 2000 to 4000 lb./(hr.)(sq. ft.); the packing may be wetted insufficiently at lower liquid rates.

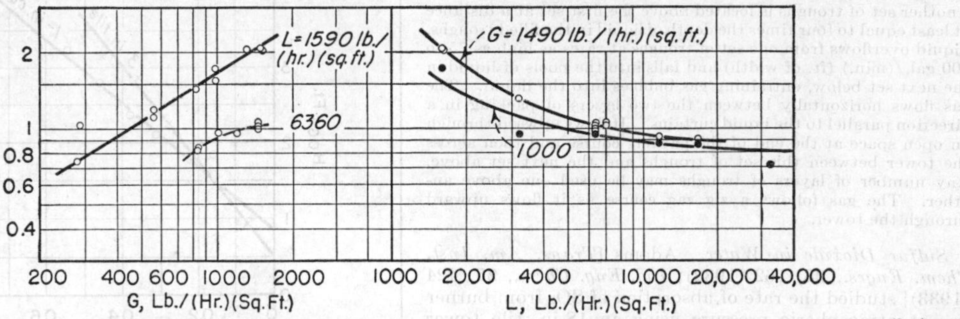

FIG. 42. Values of H.T.U. for absorption of NH_3 in water from air. Fiberglas pads, fibers vertical, bulk density of 4.7 lb./cu. ft. [*Data of Williams, Akell, and Talbott, Chem. Eng. Progress*, **43**, 585 (1947).]

Chlorine in Water. Rates of absorption of chlorine from air in water have been measured by Vivian and Whitney [*Chem. Eng. Progress*, **43**, 691 (1947)], Whitney and Vivian [*Paper Trade J.*, **110**, (20), 29 (1940); **113** (10), 31 (1941)], and Adams and Edmonds [*Ind. Eng. Chem.*, **29**, 447 (1937)]. The most extensive and reliable of these data are those of Vivian and Whitney, who used a 4-in.-diameter tower packed with 2 ft. of 1-in. tile Raschig rings and a 14-in.-diameter tower containing 8 ft. of 1-in. rings. Their results, shown in Fig. 43, agree approximately with those of the other investigators quoted. No effect of gas rate was found for the 4-in. tower within the range studied, 60 to 600 lb./(hr.)(sq. ft.); there was no effect of chlorine concentration in the gas. The effect of the temperature of the liquid was found to be essentially the same as that observed by Sherwood and Holloway in their experiments on oxygen stripping, as shown by the small correction chart in Fig. 31.

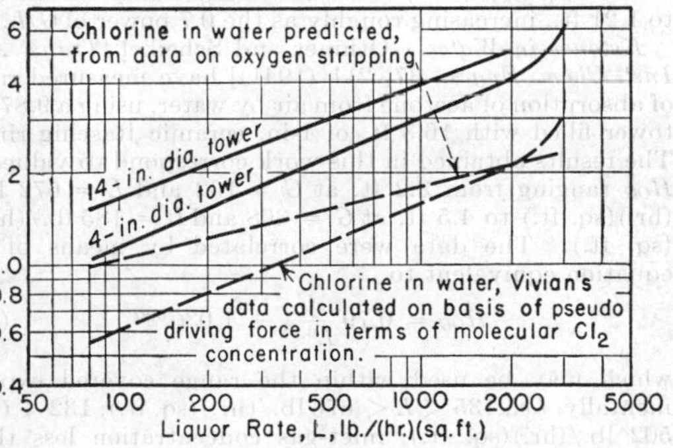

FIG. 43. Rates of absorption of chlorine from air in water. [*Data of Vivian and Whitney, Chem. Eng. Progress*, **43**, 691 (1947).]
Packing = 2 ft. of 1-in. Raschig rings.
Temperature = 70°F.
Gas velocity = 60–600 lb./(hr.)(sq. ft.) in 4 in.
Tower = 14-in., 75 lb./(hr.)(sq. ft.).

Figure 43 shows that Vivian's lowest values of H_{OL} are from 5 to 95 per cent larger than would have been expected from Sherwood and Holloway's liquid-film data, shown by the dotted line in the figure. Moreover, the effect of liquid rate on the H.T.U. for chlorine is greater than would have been expected from the oxygen data. This discrepancy is explained logically by Vivian, who points out that chlorine must enter the liquid phase at the interface in the form of molecular chlorine. In the body of the liquid phase, it exists partly in non-volatile form as hypochlorous acid and as chloride ion, having reacted with water according to the equation

$$Cl_2 + H_2O = HOCl + H^+_L + Cl^-$$

The driving force that causes diffusion into the liquid is the difference between the concentrations of dissolved chlorine in molecular form at the interface and in the main body of the liquid, rather than the larger difference between the total chlorine concentrations.

The dashed line in Fig. 43 shows Vivian's data obtained with his 4-in. tower, calculated by using molecular chlorine concentrations as the driving force. These concentrations are based on the experimental values of the Henry's-law constant for molecular chlorine listed on p. 675. It is to be noted that at high liquid rates this line intersects the line representing H.T.U. values for chlorine based on experiments with oxygen, which is inert toward water. At low liquid rates, however, the line representing the adjusted data lies much lower than the predicted line. At the highest liquid rates diffusion is so rapid compared with chemical reaction that most of the chlorine remains in molecular form until it reaches the bulk of the liquid. Under these conditions, the difference in molecular chlorine concentrations is the true driving force. At the lowest liquid rates, however, the reaction is relatively more rapid and is essentially complete at the interface. Under these conditions, the true driving force is nearly equal to the difference in the total chlorine concentrations. If this explanation is correct, rates of chlorine desorption from water should be greater than absorption rates, especially at high liquid rates.

Carbon Dioxide in Water. The most reliable data for this system are those of Sherwood and Holloway, and of Cooper, Christl, and Peery, given on pp. 691 to 693. Additional data have been obtained by Cantelo, Simmons, Giles, and Brill [*Ind. Eng. Chem.*, **19**, 989 (1927)] using a 3.5-in. tower packed with small glass rings, and by Simmons and Osborne [*Ind. Eng. Chem.*, **26**, 529 (1934)] using a small tower filled with 0.73-in. glass spheres. Values of H_{OL} observed by Cantelo *et al.* ranged from 3.4 ft. at L = 3000 lb./(hr.) (sq. ft.) to 8.0 ft. at 20,000 lb./(hr.)(sq. ft.), both at G = 10 lb./(hr)(sq. ft.) and at water temperature from 6° to 24°C.

A scrubbing apparatus particularly suited for applications, such as CO_2 absorption in water, in which the ratio of the liquid flow to the gas flow must be very large because of low solubility of the solute is described by Cooper [U.S. Patent 2,398,345 (1945)]. In water absorption of CO_2, when the ratio of the volumes of gas and liquid handled may be as low as 2 to 1, true countercurrent action is not obtained in a packed column because of the tendency for the large liquor flow to entrain gas downward within the packing. The cascade packing described by Cooper largely avoids this internal mixing.

The packing consists of troughs filled with liquid and arranged side by side to extend from one side of the tower to the other. Another set of troughs is located above the first set at a distance at least equal to four times the depth of liquid in the filled troughs. Liquid overflows from one set of troughs at rates as high as 75 to 300 gal./(min.) (ft. of width) and falls into the pools of liquid in the next set below, entraining gas bubbles into the liquid. The gas flows horizontally between the two layers of packing in a direction parallel to the liquid curtains. It flows upward through an open space at the end of the packing course and then across the tower between this set of troughs and the next set above. Any number of layers of troughs may be used one above another. The gas follows a zig zag course as it flows upward through the tower.

Sulfur Dioxide in Water. Adams [*Trans. Am. Inst. Chem. Engrs.*, **28,** 162 (1932); *Ind. Eng. Chem.*, **25,** 424 (1933)] studied the rate of absorption of SO_2 from burner gas at atmospheric pressure using an 18-in. tile tower packed with 3-in. spiral tile in the staggered stacked arrangement. The results of this investigation have been replotted by Colburn [*Trans. Am. Inst. Chem. Engrs.*, **35,** 211 (1939)], as shown in Fig. 44. The

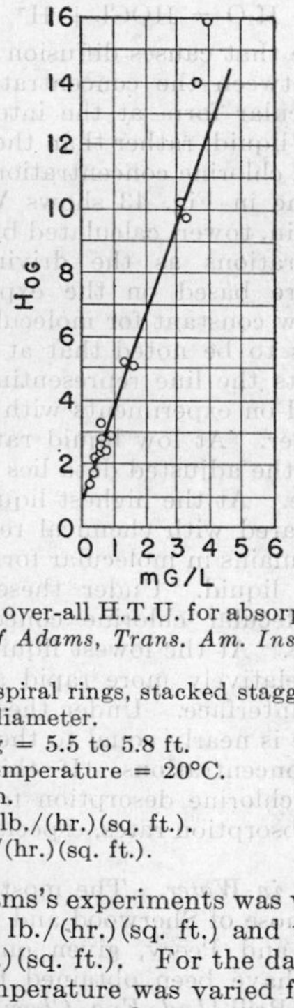

FIG. 44. Values of over-all H.T.U. for absorption of SO_2 from air by water. [*Data of Adams, Trans. Am. Inst. Chem. Engrs.*, **25,** 424 (1933).]
Packing = 3-in. spiral rings, stacked staggered.
Tower = 18-in. diameter.
Depth of packing = 5.5 to 5.8 ft.
Average liquid temperature = 20°C.
Pressure = 1 atm.
L = 150 to 3500 lb./(hr.)(sq. ft.).
G = 40 to 60 lb./(hr.)(sq. ft.).

water rate in Adams's experiments was varied from about 300 to over 3000 lb./(hr.)(sq. ft.) and the gas rate from 49 to 140 lb./(hr.)(sq. ft.). For the data plotted in Fig. 44, the water temperature was varied from about 12° to 28°C. When the water and gas rates were held constant, the value of H_{OG} was approximately halved when the water temperature was increased 25°C.

Haslam, Ryan, and Weber [*Trans. Am. Inst. Chem. Engrs.*, **15,** 177 (1923)] determined rates of SO_2 absorption from air by water in a 8-in. tower containing 30 in. of 3-in. spiral tile. The liquid rate was held constant in these experiments at 1930 lb./(hr.)(sq. ft.), and the temperature was 17°C. The gas rate was varied from 31 to 154 lb./(hr.)(sq. ft.), corresponding to a maximum value of mG_M/L_M of 0.46. The experimental data are shown in Fig. 45. Additional experiments were made, using the same tower filled with 1-in. coke packing. The values of H_{OG} are shown in Fig. 45.

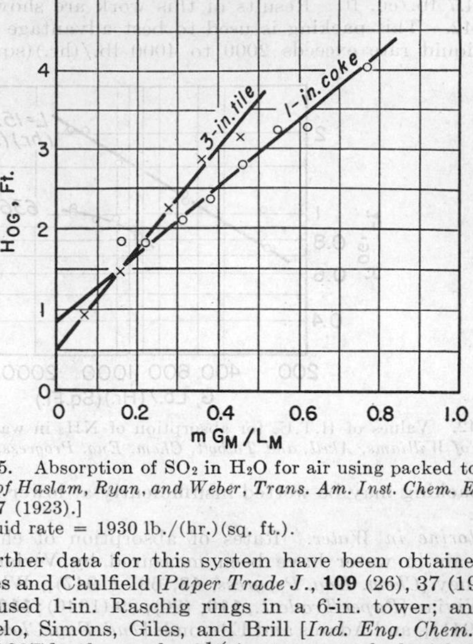

FIG. 45. Absorption of SO_2 in H_2O for air using packed towers. [*Data of Haslam, Ryan, and Weber, Trans. Am. Inst. Chem. Engrs.*, **15,** 177 (1923).]
Liquid rate = 1930 lb./(hr.)(sq. ft.).

Further data for this system have been obtained by Jennes and Caulfield [*Paper Trade J.*, **109** (26), 37 (1939)], who used 1-in. Raschig rings in a 6-in. tower; and by Cantelo, Simons, Giles, and Brill [*Ind. Eng. Chem.*, **19,** 989 (1927)], who used a 3½-in. tower packed with 33 in. of 0.25- and 0.4-in. glass rings. The SO_2 concentration entering varied from 10.9 to 18.9 per cent, the tower temperature from 10° to 16°C., the gas velocity from 0.037 to 0.30 ft./sec., and the liquid rate from 330 to 2600 lb./(hr.)(sq. ft.). Values of H_{OG} ranged from 0.32 to 1.21 ft., increasing roughly as the 0.7 power of G/L.

Ketones in Water. Othmer and Scheibel [*Trans. Am. Inst. Chem. Engrs.*, **37,** 211 (1941)] have measured rates of absorption of acetone from air by water, using a 9.87-in. tower filled with 16.3 ft. of 1-in. ceramic Raschig rings. The results obtained in this work correspond to values of H_{OG} ranging from 3.2 ft. at $G = 132$ and $L = 672$ lb./(hr.)(sq. ft.) to 4.5 ft. at $G = 308$ and $L = 185$ lb./(hr.)(sq. ft.). The data were correlated by means of an equation equivalent to

$$H_{OG} = 0.79\frac{G}{L^{0.95}} + 1.03G^{0.2} \tag{31}$$

which may be used within the range covered experimentally, *i.e.*, $135 < L < 676$ lb./(hr.)(sq. ft.), $132 < G < 502$ lb./(hr.)(sq. ft.), inlet-gas concentration less than 6 per cent by volume, and exit-liquid strengths less than 7 per cent acetone. The unusually high values of H_{OG} may have been due to heat effects accompanying acetone absorption and water vaporization, which were not allowed for by Othmer and Scheibel in their method of computing absorption coefficients.

Scheibel and Othmer [*Trans. Am. Inst. Chem. Engrs.*, **40,** 611 (1944)] carried out a similar investigation of rates of absorption and stripping of four different methyl ketones from air by water. The apparatus consisted of a 4-in.-diameter tower packed with 27 in. of 0.316-in. glass Raschig rings. The liquid rate was varied from 150 to 1800 lb./(hr.)(sq. ft.) and G from 120 to 660 lb./(hr.)(sq. ft.). The results were correlated by an equation equivalent to

$$H_{OG} = \frac{0.112G^{0.2}}{D_G} + \frac{0.00167G}{H'D_L L^{0.8}} \tag{32}$$

where D_G = diffusivity of ketone in air, ft.²/hr.; D_L = diffusivity of ketone in water, ft.²/hr.; H' = Henry's-law coefficient for ketone in water solution, lb.-moles/(atm.)(cu. ft.). Values of D_L used by Scheibel and Othmer are 3.17×10^{-5}, 2.90×10^{-5}, 2.66×10^{-5}, and 2.56×10^{-5} ft.²/hr. for acetone, methyl ethyl ketone, methyl isobutyl ketone, and methyl n-amyl ketone, all at 80°F. Solubility data suitable for calculating H' are given in the original paper. These data are equivalent to activity coefficients of 6.9, 51, 230, 2400 for acetone, methyl ethyl ketone, methyl isobutyl ketone, and methyl n-amyl ketone, respectively, at 70°F. Equation (32) should not be used for systems involving other solutes.

White and Othmer [*Trans. Am. Inst. Chem. Engrs.*, **38**, 1067 (1942)] measured rates of absorption of acetone from air in water using a 6-in. column filled with Stedman packing to a depth of 45 in. The values of H_{OG} obtained in these tests are shown in Fig. 46. Comparison of these

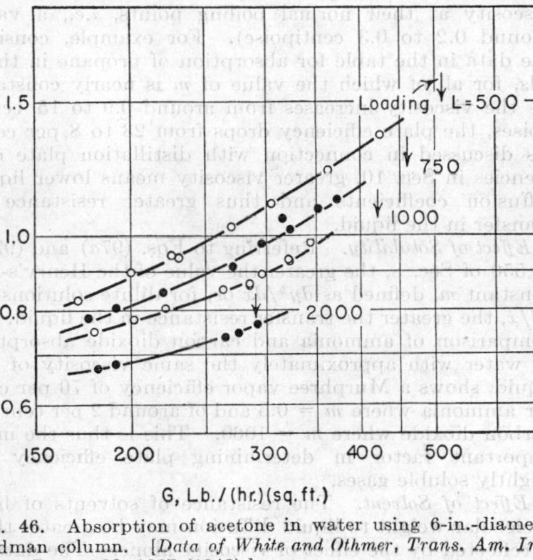

FIG. 46. Absorption of acetone in water using 6-in.-diameter Stedman column. [*Data of White and Othmer, Trans. Am. Inst. Chem. Engrs.*, **38**, 1067 (1942).]

data with the larger values of H_{OG} (2.5 to 5.0 ft.) measured in apparatus filled with 1-in. Raschig rings indicates the superior contacting efficiency of this packing, which is considerably more expensive, however.

Hydrogen Chloride in Water. The final step in the production of hydrochloric acid involves the absorption of HCl vapor from a gas containing a small quantity of insoluble constituents. This is accompanied by such a large heat effect that the principal problem is one of cooling the absorption liquor to dissipate the heat of absorption. Because of the extremely corrosive conditions encountered, equipment for HCl absorption is usually constructed from impregnated carbon or graphite, tantalum, quartz, Haveg, or glass.

No data are available for H.T.U. values for this system. When the acid strength does not exceed about 20 per cent by weight, the vapor pressure of HCl over the solution is so low that all the resistance to absorption can be assumed to be in the gas phase. For this case, use values of H_G read from Figs. 21 to 29.

Heat effects accompanying HCl absorption may be estimated from enthalpy data compiled by Van Nuys [*Trans. Am. Inst. Chem. Engrs.*, **39**, 663 (1943)]. Solution of gaseous HCl at 25°C. in a large volume of water at 25°C. is accompanied by the evolution of 17,880 cal./g.-mole of HCl. Dilution of HCl solutions is accompanied by the evolution of heat, as shown by the following values of $\phi_h - \phi_h°$, the heat evolved per mole of HCl when solutions of the compositions shown are diluted down to infinite dilution, the temperature being held constant throughout at 25°C. These data are taken from Rossini [*Bur. Standards J. Research*, **9**, 679 (1932)]. See also Wenner ("Thermochemical Calculations," p. 28, McGraw-Hill, New York, 1941).

Composition	$\phi_h - \phi_h°$, G.-cal./G.-mole HCl
HCl: 1600 H_2O	90
HCl: 400 H_2O	181
HCl: 200 H_2O	249
HCl: 100 H_2O	343
HCl: 50 H_2O	433
HCl: 25 H_2O	730
HCl: 20 H_2O	850
HCl: 15 H_2O	1050
HCl: 12 H_2O	1250
HCl: 10 H_2O	1460
HCl: 5 H_2O	2760
HCl: 3 H_2O	4470

For example, when 1 mole of HCl at 25°C. is dissolved in a solution containing 20 moles H_2O/mole HCl, the heat liberated amounts to 17,880 − 850 = 17,030 cal. if the resulting solution is cooled to 25°C.

Hatfield and Ford [*Trans. Am. Inst. Chem. Engrs.*, **42**, 121 (1946)] discuss the use of "Karbate" equipment for HCl absorption. Cooler absorbers constructed from this material can be used for contacting acid and HCl gas directly on a water-cooled surface that resists corrosion and gives high heat-transfer rates. These cooler absorbers are constructed with either horizontal or vertical tube bundles and, in the case of concentrated HCl gas, are preferred to the old system of a number of towers in series, each with its external cooler for the acid circulated over the tower. Cooler absorbers are characterized by compactness and low pressure drop. Hatfield and Ford state that a batch-type cooler absorber can be designed to produce absorption rates of about 10 lb. HCl/(hr.)(sq. ft. of exchanger surface area) from gases containing 90 per cent HCl or more if 20° to 22°Be. acid is produced and if cooling water is available at 80°F. Hydrochloric acid may be absorbed continuously from more dilute gases, containing 50 to 90 per cent HCl, inside a falling-film tower (Hatfield and Ford). The apparatus consists of a shell-and-tube heat exchanger constructed from Karbate. The acid film flows downward inside the vertical tubes in the same direction as the gas. Cooling water is circulated countercurrently in the shell. The weak exit gases are sent to a packed "tails" tower, where they are scrubbed with fresh water. One unit constructed from 72 0.875-in. inside-diameter Karbate tubes, each 9 ft. long, produced 22°Be. acid at 100°F. from gas containing 90 per cent HCl flowing into the tubes at a mass velocity of 7930 lb./(hr.)(sq. ft.).

Oldershaw, Simenson, Brown, and Radcliffe [*Chem. Eng. Progress*, **43**, 371 (1947)] discuss the operation of adiabatic and water-cooled packed HCl absorbers, showing that acid containing 32–33 per cent HCl can be produced commercially in adiabatic towers fed with gases containing 10 to 100 per cent HCl. Acid containing 35 per cent HCl, or more, can be produced in water-cooled units. Monolithic Karbate towers are usually cheaper than rubber- or brick-lined steel towers for small units, but the reverse is true for large installations. Adiabatic absorbers may be preferred to externally cooled units when dilute acid is being produced from gases containing chlorinated hydrocarbons and HCl since the exit gas temperature can be kept high enough to allow the hydrocarbons to escape with the vent gas.

The use of absorption towers constructed from tantalum metal in sizes up to a capacity of 500 lb./hr. of pure HCl feed gas is described by Hunter [*Trans. Am. Inst. Chem. Engrs.*, **37**, 741 (1941)]. When strong acid containing 30 per cent by weight HCl is made in the absorption system, water evaporates from the hot acid effluent, providing a cooling effect in the bottom part of the absorber. In the upper portions of the absorber, the low vapor pressure of HCl over the dilute water solution allows HCl to be absorbed even from a dilute gas mixture. Additional cooling due to water evaporation in this section can be obtained by introducing inert gas into the upper part of the absorber, as suggested by Hurt [U.S. Patent 2,220,570 (1940)].

Non-aqueous Systems. Simmons and Long [*Ind. Eng. Chem.*, **22**, 718 (1930)] and Simmons and Osborne [*Ind. Eng. Chem.*, **26**, 529 (1934)] have investigated the rate of absorption of benzene and ethylene dichloride in a light oil in small towers (2.8 to 3.6 in. inside diameter) with small-sized packing. Data from these sources have been recalculated and replotted by Sherwood ("Absorption and Extraction," McGraw-Hill, New York, 1937).

Gross and Simmons [*Trans. Am. Inst. Chem. Engrs.*, **40**, 121 (1944)] measured rates of absorption of benzene, trichlorethylene, and chloroform in kerosene in a 1-ft.-diameter column packed with 1-in. clay Berl saddles. Gas rates were below 80 lb./(hr.)(sq. ft.), and liquid rates ranged from 1000 to 3000 lb./(hr.)(sq. ft.). The value of H_{OG}, based on equilibrium data calculated from Raoult's law, ranged from 2.5 to 4.9 ft. Higher values were observed for chloroform and trichlorethylene than for benzene. The value of H_{OG} was substantially independent of the gas rate and the liquid rate, within the range studied.

Plate Efficiencies of Bubble-tray Absorption Columns. Available data on plate efficiencies of bubble-tray absorption columns are summarized in Table 36.

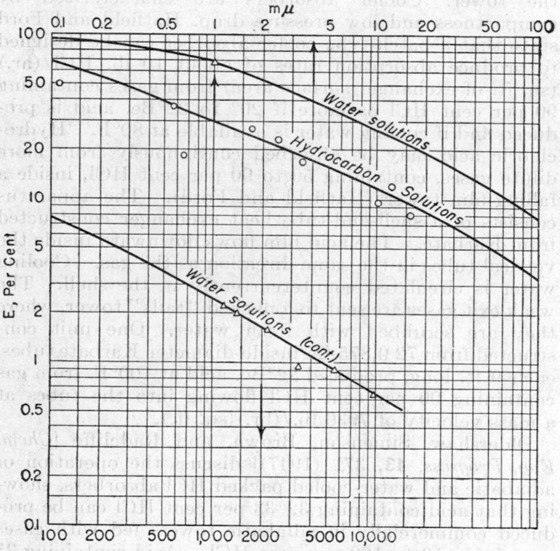

FIG. 47. Murphree plate efficiencies of absorbers as a function of gas solubility and liquid viscosity. (For values of mG_M/L_M near unity, these efficiency values may also be employed as over-all efficiencies; for other conditions over-all values may be calculated by methods of Sec. 8, p. 550.)

$m = dy^*/dx$ = slope of equilibrium curve.

μ = average liquid viscosity, centipoises.

No data are available on sieve-plate columns, and one therefore concludes that they are not often used in absorption. The table includes pertinent factors, some of which affect the values of efficiency, and others of which give some idea of the types of apparatus employed for this operation.

The values of plate efficiency are given as either over-all or Murphree values. The over-all efficiency is defined as the quotient of the number of theoretical plates required divided by the number of actually used. The Murphree efficiency is equal to the quotient of the change in vapor composition in passing through a given plate divided by the change that would have occurred if the vapor left in equilibrium with the liquid leaving. The latter concept is the more fundamental. The two values are related as discussed in Sec. 8, p. 550.

The process of mass transfer during absorption in a plate column is similar to distillation, and one would expect data on plate efficiencies during distillation, discussed in Sec. 9, p. 610, to be closely related to absorber efficiencies. In general it is seen that absorber efficiencies are lower than distillation values, probably owing to the greater importance of the liquid-film resistance, as discussed below.

Effect of Viscosity. Usually in absorption the scrubbing liquid is considerably below its boiling point, and the viscosity is therefore apt to be fairly high compared with its value in distillation (recalling the "rule of thumb" that most pure liquids have about the same viscosity at their normal boiling points, *i.e.*, a value around 0.2 to 0.3 centipoise). For example, consider the data in the table for absorption of propane in three oils, for all of which the value of m is nearly constant. As the viscosity increases from around 0.9 to 15 centipoises, the plate efficiency drops from 23 to 8 per cent. As discussed in connection with distillation plate efficiencies in Sec. 10, greater viscosity means lower liquid diffusion coefficients and thus greater resistance to transfer in the liquid.

Effect of Solubility. Referring to Eqs. (97*a*) and (97*b*), p. 550 of Sec. 8, the greater the value of the Henry's-law constant m, defined as dy^*/dx or, for dilute solutions, as y^*/x, the greater the transfer resistance in the liquid. A comparison of ammonia and carbon dioxide absorption in water with approximately the same viscosity of the liquid, shows a Murphree vapor efficiency of 70 per cent for ammonia where $m = 0.6$ and of around 2 per cent for carbon dioxide where $m = 1060$. This is thus the most important factor in determining plate efficiency for slightly soluble gases.

Effect of Solvent. The resistance of solvents of high molecular weight to liquid diffusion may be greater than is reflected by the effect of viscosity alone. The data in Table 36 appear to indicate higher plate efficiencies for absorption of gases in water than in petroleum oils.

A rough plot of the data in the table is given by Fig. 47, in which the values of plate efficiency are plotted vs. the product $(m)(\mu)$ and separate lines are drawn through the data for water and for oils as the scrubbing liquid. This plot should be convenient for estimation of rough values of plate efficiency for systems where no data are available. It should be realized that factors of plate construction, such as slot width and seal depth, no doubt affect the values, and further data are therefore needed for a reliable correlation. The choice of the factors of m and μ in the plot is from their effect on the liquid resistance to transfer as brought out by Eqs. (97*a*) and (111), p. 550 of Sec. 8. For values of $(m)(\mu)$ somewhat less than unity, the liquid resistance is small, since gas-film resistance controls. For large values of $(m)(\mu)$, say larger than 10, liquid resistance controls.

These effects have been brought out by Walter and Sherwood [*Ind. Eng. Chem.*, **33**, 493 (1941)], who derived the following equations from experimental data:

$$E = 1 - e^{-m} \tag{33}$$

where

$$m = \frac{h}{[2.50 + (0.370/H'P)]\mu^{0.68}w^{0.33}} \tag{34}$$

In these equations h = effective liquid depth, in. (taken to be the distance from the middle of the slots to the top

Table 36. Plate Efficiencies for Bubble-tray Absorbers

System	Average temp., °C.	Absolute pressure, lb./sq. in.	Average liquid viscosity, centipoises	$m = y^*/x$	L_M/G_M	Col. diam., in.	No. plates	Plate spacing., in.	No. caps per plate	Size of caps, in.	Slot area, % of tower cross section	Static seal depth, in.	Range of F-factor, (ft./sec.) (lb/cu. ft.)$^{1/2}$	Plate efficiency, % Over-all	Plate efficiency, % Murphree	Ref.
C_1—C_5 in 220 M.-W. oil.	38	78	0.81	0.68–0.93 for C_4	1.0	72	21	20	50	6.5	9.8	2	....	18	......	1
C_1—C_6 in 161 M.-W. oil	34	485	.42	0.27 for C_4	0.185	24	..	..	....	.....	.....	....	....	49 for n-C_4	......	2
	38	60	.42				10							50		
H_2S, C_1—C_5 in 185-M.W. oil.......	15	92	1.9	0.83–0.94 for C_3	1.16–1.24	108	19	30	92	6.31	8.04	0.75	0.18	26–27 for C_3	......	4
C_1—C_6 in 206 M.-W. oil.........	52	255	.40	....	0.05	48	24	18	..	.....	.....	.....	1.04	42	......	3
C_1—C_6 in 157 M.-W. oil.........	53	260	.41	....	0.25	60	16	30	..	.....	.....	.....	1.05	39	......	3
C_1—C_6 in 164 M.-W. oil.........	51	265	.50	....	0.25	60	16	30	..	.....	.....	.....	1.24	38	......	3
C_1—C_6 in 201 M.-W. oil.........	49	260	.48	....	0.08	48	24	18	..	.....	.....	.....	0.76	36	......	3
C_1—C_6 in 135 M.-W. oil.........	59	267	.22	....	0.4–0.5	60	16	30	..	.....	.....	.....	.46	56	......	3
C_1—C_5 in 135 M.-W. oil.........	55	254	.31	....	0.3–0.4	48	24	18	..	.....	.....	.....	.49	50	......	3
C_1—C_6 in 250 M.-W. oil.........	47	94	.41	....	0.5	48	24	18	..	.....	.....	.....	.88	10	......	3
i-C_4H_8 in heavy naphtha.......	26	66	.97	0.61	0.45	2	1	..	1	2	10	.75	.6	....	36	6
i-C_4H_8 in gas oil..	25–37	66	3.9–5.4	0.65–0.95	0.45	2	1	..	1	2	10	.75	.6	....	17	6
i-C_4H_8 in gas oil + lube oil	24	66	20.6	0.50	0.45	2	1	..	1	2	10	.75	.6	....	9–10	6
C_3H_6 in heavy naphtha........	18–43	46–66	0.74–1.10	1.95–3.29	0.47–0.73	2	1	..	1	2	10	.75	.6	....	22–24	6
C_3H_6 in gas oil.....	24–48	66	2.8–5.8	2.34–3.67	0.52	2	1	..	1	2	10	.75	.6	....	11–13	6
C_3H_6 in gas oil + lube oil.........	23–41	66	10.5–21.5	1.92–2.70	0.34	2	1	..	1	2	10	.75	.6	....	5–11	6
H_2O evaporation into air.........	20–31	14.7–55	0.8–1.0	....	....	2	1	..	1	2	10	.75	0.2–0.8	....	85–92	6
NH_3 in H_2O.......		14.7	....	....	....		1	..	5	3	.....	.....	....	....	65–85	5
NH_3 in H_2O......	11–17	14.7	....	0.52–0.69	2.9–16	18	1	..	7	4	10.6	.375	0.08–0.46	....	69	6
CO_2 in H_2O.......	10–12	14.7	....	1030–1100	2.2–16	18	1	..	7	4	10.6	.375	0.08–0.46	....	1.8–2.6	6
CO_2 in H_2O.......	13–59	14.7	1.20–0.48	1150–3330	6.3–41	5	4	11	1	3.5	15.2	1.5	0.2–1.5	....	1.5–3.5	6
CO_2 in H_2O-glycerol solution....	25	14.7	0.9	1640	14–22	5	4	11	1	3.5	15.2	1.5	0.33	....	2.0	6
			1.2	1840	....	..	..	..	..	....	.....	.....	....	....	1.6	6
			1.7	2080	....	..	..	..	..	....	.....	.....	....	....	0.96	6
			2.4	2340	....	..	..	..	..	....	.....	.....	....	....	0.96	6
			3.7	2720	....	..	..	..	..	....	.....	.....	....	....	0.65	6
CO_2 in sodium carbonate–bicarbonate solution; sodium normality = 1.7, %; conversion to bicarbonate = 35...	60	14.7	....	....	....		15		1						7	7

References:

[1] Atkins and Franklin, *Refiner Natural Gasoline Mfr.*, **15**, 30 (1936).
[2] Brown and Souders, *Oil Gas J.*, **32**, (45), 114 (1934).
[3] Drickamer and Bradford, *Trans. Am. Inst. Chem. Engrs.*, **39**, 319 (1943).
[4] Sherwood and Jackson, *Trans. Am. Inst. Chem. Engrs.*, **37**, 959 (1941).
[5] Reynolds and Saunders, Mass. Inst. Tech. Thesis, 1920, quoted in Ref. 6.
[6] Walter and Sherwood, *Ind. Eng. Chem.*, **33**, 493 (1941).
[7] Whitman and Davis, *Ind. Eng. Chem.*, **18**, 264 (1926).

of the liquid layer, including the crest of liquid over the weir); H' = Henry's-law coefficient, lb.-moles/(cu. ft.)-(atm.); P = total pressure, atm.; μ = viscosity of liquid, centipoises; w = slot width, in. This equation predicts efficiencies as low as 2 per cent for CO_2 absorption in water at 20°C.; under these conditions the resistance to transfer is probably within the liquid phase.

A recent correlation of O'Connell [*Trans. Am. Inst. Chem. Engrs.*, **42**, 741 (1946)] is similar to the plot given herewith but includes in the term for solubility the molecular weight of the scrubbing liquid.

It may be concluded from the data given here that the Murphree vapor efficiency, as well as the over-all plate efficiency in many multicomponent absorbers, such as natural gasoline absorbers, is dependent on the solubility of the component considered. In absorber design, however, only the key component need ordinarily be considered, since the oil leaves the tower essentially saturated with the components lighter than the key, and the gas leaves the absorber almost completely stripped of the components heavier than the key component.

Rates of Absorption in Spray Towers. The simplest spray absorption tower consists of an empty chamber into which liquid is sprayed at the top and gas is introduced at the bottom. Such equipment has the advantages of very low pressure drop through the spray chamber and simple inexpensive construction. Because of mixing of the gas within the spray chamber, entrainment of gas by the sprays, and entrainment of fine spray droplets within the tower, spray absorption units are not applicable where true countercurrent action and a large number of transfer units are required in order to obtain essentially complete absorption.

Experience with spray equipment indicates no definite maximum gas velocity corresponding to the flooding velocity in a packed tower. As the gas flow is increased the amount of entrained spray carried out by the exit gas increases gradually. Gas velocities as large as 800 lb./(hr.)(sq. ft.) have been employed in experimental equipment, and entrainment rates as low as 0.01 lb. liquid/lb. gas have been reported under these conditions, using small-sized commercial spray nozzles.

Gas rates as large as 2400 lb./(hr.)(sq. ft.) have been used in large spray installations of the wet-cyclone type described by Kleinschmidt and Anthony [*Trans. Am. Soc. Mech. Engrs.*, **63**, 349 (1941)] without excessive entrainment. In these centrifugal units the gas stream is intro-

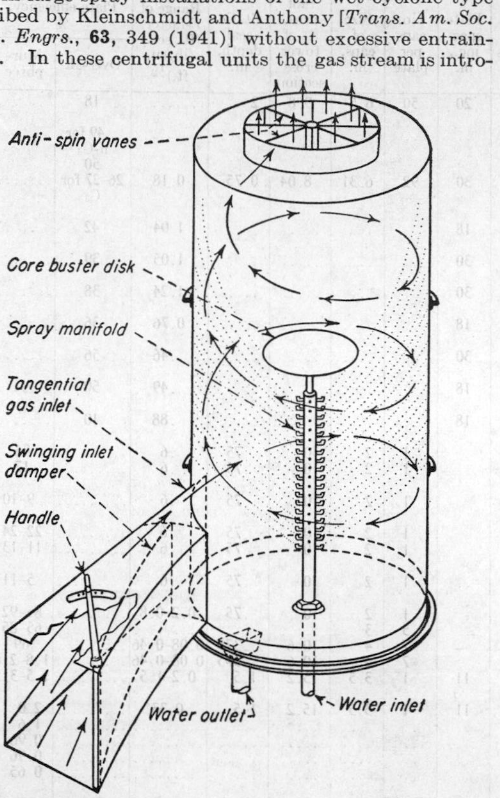

FIG. 48. Schematic view showing elements of cyclonic-spray scrubber. [*From Kleinschmidt and Anthony, Trans. Am. Soc. Mech. Engrs.*, **63**, 349 (1941).]

duced tangentially into the base of the tower and spirals upward, as illustrated in Fig. 48. The liquid is sprayed from nozzles attached to a manifold that is located along the tower axis. The spray droplets are thrown toward the tower walls by the centrifugal field. The cross-flow action of this equipment limits the rate of absorption to that corresponding to one theoretical plate. Dust particles can be collected simultaneously with absorption.

Recent data (Pigford and Pyle, unpublished data) showing the magnitude of the liquid-phase resistance to oxygen stripping from water in a spray tower are presented in Fig. 49. This information is based on experiments made with a tower 31.5 in. in diameter into which liquid was sprayed from six solid-cone nozzles spaced uniformly around a 12-in.-diameter circle at a point 52.5 in. above an air distributor in the base of the tower. Although the nozzles were inclined slightly toward the central axis of the tower to reduce the amount of spray striking the wall, the water running from the wall at the base of the spray chamber amounted to 60 to 80 per cent of the water sprayed. The values of H_{OL} given by the figure are based on analyses of the spray striking a separate collecting basin in the bottom of the tower. Although the values of H_{OL} are somewhat larger than would have been obtained by packing the tower, they are of the same order of magnitude. Thus at $L =$ 800 lb./(hr.)(sq. ft.) and $G = 450$ lb./(hr.)(sq. ft.) the values of H_{OL} at 30°C. are 1.2 ft. and 0.84 ft. for the spray tower and for 2-in. Raschig rings, respectively.

Further data based on experiments in which air was dehumidified in the same equipment used for oxygen desorption are also shown in Fig. 49. These data are presented in terms of H_{OG}, based on the specific enthalpy of the gas as a driving force (see pp. 557 to 559, Sec. 8 for an outline of related theory), and are considered to give the order of magnitude of the gas-phase resistance to mass transfer. Similar data for a spray chamber 7 ft. high and 2 by 3 ft. in cross section are presented by Niederman, Howe, Longwell, Seban, and Boelter [*Heating, Piping, Air Conditioning*, **13**, 591 (1941)]. Data on absorption of ammonia from air by water, employing the same equipment used by Pigford and Pyle for stripping oxygen and cooling water, are shown by Fig. 50.

Bosworth [*Australian Chem. Inst., J. & Proc.*, **13**, 53 (1946)] measured rates of absorption of CO_2 in a spray of viscous sugar solution containing lime. The spray was formed at the top of a 1-ft.-diameter tower by a shower head drilled with eighty-five $\frac{1}{16}$-in. holes. The central 9-in. core of the spray was collected after falling 5.5 to 96 in. through a gas stream flowing upward at 5.3 ft./sec. The results showed that the amount of CO_2 absorbed increased as the 0.25 power of the distance fallen, in agreement with a theory based on molecular diffusion of CO_2 through the stagnant interior of the drops. Other evidence indicates that there is circulation within drops formed from less viscous liquids, because of surface friction in a moving gas.

Kowalke, Hougen, and Watson (*Bull. Univ. Wis. Eng.*

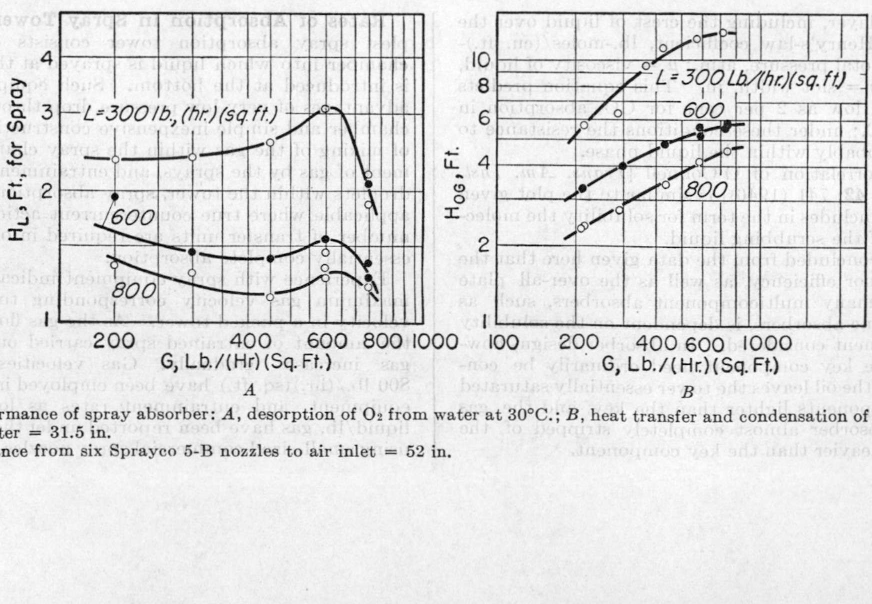

FIG. 49. Performance of spray absorber: *A*, desorption of O_2 from water at 30°C.; *B*, heat transfer and condensation of water from air. Tower diameter = 31.5 in.
Vertical distance from six Sprayco 5-B nozzles to air inlet = 52 in.

Exp. Sta., Ser. 68, June, 1925) obtained coefficients for absorbing ammonia in an 18-in. spray tower, 4 ft. high, in which water was introduced at the top by means of Vermorel sprays. The average gas temperature varied from 15° to 20°C., the water temperature from 7° to 15°C., and the tower was run at atmospheric pressure. These data have been plotted by Sherwood ("Absorption and Extraction," McGraw-Hill, New York, 1937).

diameter, 0.046 in.) which were distributed along a 6-ft. length of 1½-in. pipe located on the axis of the tower at its base. The gas entered the tower tangentially at the base through a rectangular entrance nozzle 13½ in. high and 4.5 to 12 in. wide.

The absorption that took place on the walls of the tower was determined in a separate series of experiments in which no sprays were used and the alkaline solution was

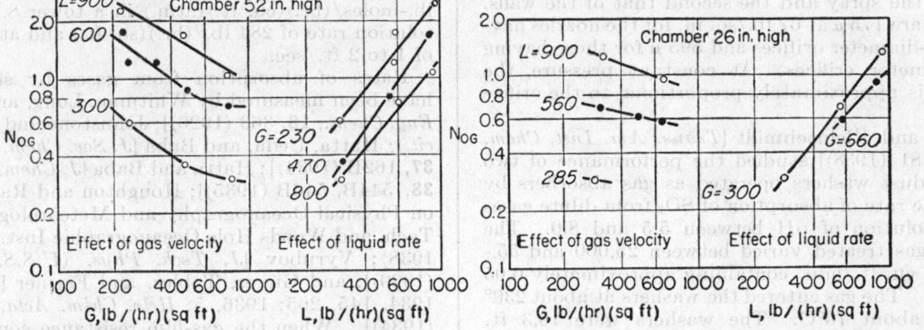

FIG. 50. Absorption of NH_3 in a spray-type column. Spray chamber 31.5 in. diameter, 52 in. or 26 in. from nozzles to base of chamber, six Sprayco 5-B nozzles on 12-in.-diameter circle at top of chamber. [*Data of Pigford and Pyle*, unpublished results (1947).]

The data of Hixson and Scott [*Ind. Eng. Chem.*, **27**, 307 (1935)] are shown, also, by Sherwood. These investigators used a small tower (2⅞ in. inside diameter) and formed the spray by means of thirty-seven 0.028-in. holes drilled through a brass plate. The calculated values of K_Ga apply to the central core of the spray. The liquid running down the walls was collected separately from that falling in the core of the spray.

Baker [*Chem. & Met. Eng.*, **22**, 173 (1920)], experimenting with a CECO spray on the absorption of NH_3 in water, obtained results that may be expressed in terms of the Murphree plate efficiency as follows:

Plate efficiency in per cent $= 100 - 0.09C$,
where C = cu. ft. of inert gas/min.

The spray consisted of a 12-in. disk driven at 1800 r.p.m., in a housing about 24 in. in diameter. Air was drawn in under the disk, while water led onto the disk was thrown outward in the form of drops. The temperature was about 20°C. The normal rating of the spray was $C =$ 300 cu. ft./min. The concentration of the solute in both the gas phase and liquid phase was low.

Sulfur Dioxide Absorption. Data on absorption of SO_2 were obtained by Haslam, Ryan, and Weber [*Trans. Am. Inst. Chem. Engrs.*, **15**, 177 (1923)], using an 8-in. spray tower 30 in. high, at atmospheric pressure and about 15°C., who obtained the equation

$$\frac{1}{K_Ga} = \frac{0.15}{u^{0.8}} + 1.17 \tag{35}$$

The absorption of sulfur dioxide from air by means of 0.6N sodium carbonate solution has been investigated by Johnstone and Silcox [*Ind. Eng. Chem.*, **39**, 808 (1947)], using a cyclone spray tower. The tower consisted of a chamber 14 ft. high which was tapered from 28.5 in. inside diameter at the bottom to 20.5 in. inside diameter at the top in order to represent a smoke stack. The apparatus was similar to a wet-cyclone scrubber of the Pease-Anthony type [*cf.* Fig. 48; Kleinschmidt and Anthony, *Trans. Am. Soc. Mech. Engrs.*, **63**, 349 (1941); Pease, U.S. Patent 1,992,762 (1935); Anthony, U.S. Patent 1,986,913 (1935)]. The alkaline solution was sprayed into it by means of 20 to 50 hollow-cone spray nozzles (orifice

distributed over the wall only. For the experiments in which the sprays were operating, the number of transfer units was calculated by using gas compositions and by assuming no back pressure of sulfur dioxide at the interface. The number of transfer units due to the sprays alone was calculated by subtracting the number due to the wall effect from the total number. For the experimental conditions, this correction amounted to 0.12 to 0.26 out of a total of 0.64 to 3.4 transfer units.

Figure 51 shows typical results of these experiments in terms of the number of transfer units for the spray as a

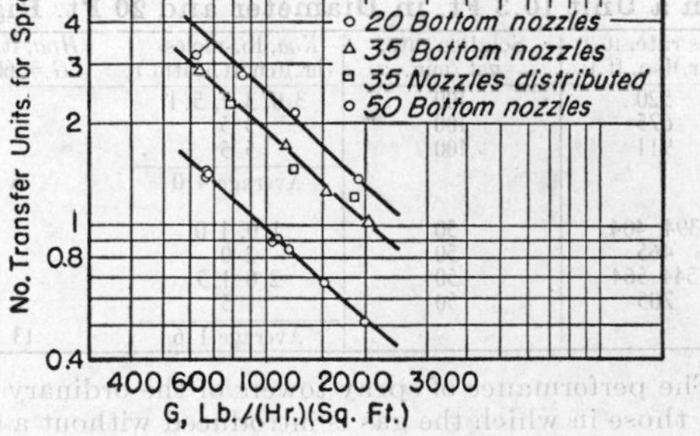

FIG. 51. Absorption of SO_2 from air in Na_2CO_3 solution using 28-in.-diameter cyclone spray tower. Liquid rate = 20 − 55 lb./hr.; 1 lb./min. per nozzle. [*Data of Johnstone and Silcox, Ind. Eng. Chem.*, **39**, 808 (1947).]

function of the gas rate based on the tower cross section. Different numbers of nozzles were used, and all the nozzles operated at 67 ± 2 lb./sq. in. pressure. The agreement between the data obtained when using 35 nozzles concentrated on the lower half of the manifold and those obtained when the same number of nozzles were distributed uniformly along the entire manifold suggests that there was negligible interference between adjacent sprays due to collisions between droplets. The data shown in Fig. 51, as well as data obtained when using different openings at the tangential gas entrance and also data obtained by Johnstone and Kleinschmidt (*loc. cit.*) using a 10.3-ft.-diameter tower, are correlated by the relation for the number of transfer units N_T:

$$N_T = 62.8 \frac{LD_tP}{G^{0.8}SD_dN_{Sc}^{2/3}} + 0.0064 \frac{A_wP}{G^{0.37}SN_{Sc}^{2/3}} \quad (36)$$

where A_w = wetted area of wall, sq. ft.; P = total pressure, atm.; G = mass velocity at entrance to tower, lb.-moles/(min.)(sq. ft.); S = area of tangential-gas entrance nozzle, sq. ft.; $N_{Sc} = \mu/\rho D$, Schmidt number of gas; L = total liquid rate, gal./min.; D_t = diameter of spray chamber, ft.; D_d = mass-median diameter of spray drops, microns. The first term in Eq. (36) gives the effect of the spray and the second that of the walls. Values of D_d are 175 μ at 67 lb./sq. in. for the nozzles having 0.046-in.-diameter orifices and 595 μ for those having 0.187-in.-diameter orifices. At constant pressure, the value of D_d is approximately proportional to the orifice diameter.

Johnstone and Kleinschmidt [*Trans. Am. Inst. Chem. Engrs.*, **34,** 181 (1938)] studied the performance of two wet-cyclone dust washers operated as gas absorbers by measuring the rate of absorption of SO_2 from dilute gases in Na_2CO_3 solution of pH between 5.5 and 8.0. The quantity of gas treated varied between 25,000 and 55,000 standard cu. ft./min. containing approximately 0.05 per cent SO_2. The gas entered the washers at about 230° and left at about 70°C. The washers were 10.3 ft. inside diameter and 20 ft. high. The cross section of the entrance was 13.5 sq. ft. Each washer was equipped with 80 lava spray nozzles of the Sturtevant type with 3/16-in. orifices. During the runs, 8 of the nozzles in one washer and 44 in the other were clogged so that the one washer operated with 100 gal. of solution/min. and the other on 50 gal./min. Approximately 95 and 85 per cent of the SO_2 was removed, respectively. The volumetric absorption coefficients and H.T.U.'s are summarized by Table 37. These values are applicable only to a scrubber 20 ft. high. The same number of transfer units would be obtained in taller or shorter towers if the same amount of liquid were sprayed.

Table 37. Performance of Cyclone Spray Absorber (Absorption of SO_2 from Air by Na_2CO_3 Solution in a Unit 10.3 Ft. in Diameter and 20 Ft. High)

Gas rate, lb./(hr.)(sq. ft.)	Solution rate, gal./min.	K_Ga, lb.-moles/(hr.)(cu. ft.)(atm.)	H_{OG}, ft. at G = 600
520	100	3.0, 3.1, 5.1	
675	100	3.5	
911	100	5.5	
		Average 4.0	5
394, 404	50	1.0, 1.0	
465	50	2.0	
544–564	50	2.0–1.5	
705	50	1.5	
		Average 1.6	13

The performance of spray towers of the ordinary type, *i.e.*, those in which the gas is introduced without a spiral motion and into which the liquid is sprayed from nozzles located at the top, has been analyzed theoretically by Johnstone and Williams [*Ind. Eng. Chem.*, **31,** 993 (1939)]. This analysis involves the use of experimental drop-sized distribution data for the calculation of the drop surface area exposed in the tower at any instant. Coalescence between drops that collide in the tower was allowed for, but the assumptions were made that all the droplets were ejected vertically downward along non-diverging paths originating at the top of the tower and that there was no entrainment of gas by the liquid spray. The interfacial areas calculated range from 0.1 to 0.2 sq. ft./cu. ft. of tower volume for a liquid rate of 500 lb./(hr.)(sq. ft.) and countercurrent gas velocities from 5 to 10 ft./sec. These calculated areas are surprisingly small, since an empty wetted-wall tube 20 ft. in diameter would expose 0.2 sq. ft./cu. ft., and smaller tubes would present larger areas.

Johnstone and Williams report experimentally determined absorption coefficients for a 42 in. inside diameter spray tower that agree well with values calculated from the estimated drop-surface areas and from transfer coefficients for single spheres. This spray tower was used to absorb SO_2 from flue gases in ammonium sulfite–bisulfite solution which was sprayed from 10 Sprayco No. 2-B hollow-cone nozzles arranged uniformly on a 21-in.-diameter circle. The data obtained at a nozzle pressure of 55 lb./sq. in. gave values of K_{Ga} ranging from 0.8 to 2.1 lb.-moles/(hr.)(cu. ft.)(atm.) in a tower 8.7 ft. tall at a solution rate of 284 lb./(hr.)(sq. ft.) and at gas velocities of 1 to 2 ft./sec.

Rates of absorption from gases by single droplets have been measured by Whitman, Long, and Wang [*Ind. Eng. Chem.*, **18,** 363 (1926)]; Johnstone and Williams (*loc. cit.*); Hatta, Ueda, and Baba [*J. Soc. Chem. Ind. (Japan)*, **37,** 162B (1934)]; Hatta and Baba [*J. Chem. Soc. (Japan)*, **38,** 544B, 546B (1935)]; Houghton and Radford (Papers on Physical Oceanography and Meteorology, Mass. Inst. Tech. and Woods Hole Oceanographic Inst., Vol. 6, p. 33, 1938); Vyrubov [*J. Tech. Phys. (U.S.S.R.)*, **9,** 1923 (1929)]; and Guyer, Tobler, and Farmer [*Chem. Fabrik*, 1934, 145, 265; 1936, 5; *Helv. Chim. Acta*, **17,** 257, 550 (1934)]. When the gas-film resistance controls the rate of absorption, the data agree with those of Fig. 16, Sec. 8, p. 546.

Rates of Absorption in Agitated Vessels. The data of Cooper, Fernstrom, and Miller [*Ind. Eng. Chem.*, **36,** 504 (1944)] describe rates of absorption of oxygen from air in an agitated vessel containing sodium sulfite solution are described. Carlson [*Ind. Eng. Chem.*, **38,** 14, 33 (1946)] has compared the absorption coefficients and the power requirements for oxygen absorption in an agitated vessel and in a packed tower used for oxygen transfer to water without chemical reaction (*cf.* p. 691).

In both cases the liquid-phase resistance is thought to control the absorption rate, in spite of the chemical reaction that occurs when sodium sulfite is present. The highest coefficient measured by Cooper, Fernstrom, and Miller was 5250 lb.-moles/(hr.)(cu. ft.) (mole fraction O_2 in the liquid) at a power input of 330 ft.-lb./min. to a vaned-disk impeller in a 9.5-in.-diameter jar. Based on a unit of liquid volume, this is equivalent to 3.4 hp./1000 gal. In addition to this power consumption, 150 ft.-lb./min. was required to force the gas through the liquid head, making the total power consumption 480 ft.-lb./min. In a packed tower to absorb oxygen at the same rate, the power consumption would be 1500 ft.-lb./min. for pumping the liquid and a negligible amount for gas resistance. The agitated tank has a slight advantage in size and power over the packed tower but is limited to gas velocities below 0.2 ft./sec., which is approximately the rate of rise of gas bubbles in a liquid.

Absorption Accompanied by Chemical Reaction. Absorption of difficultly soluble gases is frequently carried out by using a liquid absorbent that reacts chemically with the dissolved solute, forming a compound that is easily decomposed by heating the solution. Usually the solvent is recovered in a stripping operation carried out at a higher temperature (Example: Absorption of carbon dioxide in sodium carbonate–bicarbonate, or in ethanolamine solution), but the regeneration of the solvent may be accomplished by chemical means. (Example: Recovery of sulfur dioxide from flue gas by scrubbing with sodium sulfite–bisulfite solutions, followed by precipitation of zinc sulfite on addition of zinc oxide to the solution.) If the solvent contains an ingredient that reacts with only one component of a gas mixture, a separation of the gaseous solutes may be accomplished. (Example: Absorption of olefinic hydrocarbons in aqueous solutions of cuprous salts.) Gases are sometimes absorbed under conditions such that an irreversible reaction with the absorbent occurs. (Example: Absorption of nitrous gases or of carbon dioxide in sodium hydroxide solution.) An absorption tower may be used as a reaction vessel to produce a chemical product by means of a reaction between gases and the liquid circulated through the tower. (Example: Production of nitric acid by the reaction of nitric oxide, oxygen, and water.)

If an irreversible chemical reaction takes place in the liquid phase and if it occurs very rapidly, the rate of absorption may be governed only by the resistance to diffusion in the gas phase, as indicated by the theoretical work of Hatta [*Tech. Rept. Tôhoku Imp. Univ.*, **8,** 1 (1929); *ibid.*, **10**, 119 (1932); *cf.* Sherwood, "Absorption and Extraction," pp. 194, 200, McGraw-Hill, New York, 1937] for laboratory-scale batch absorption experiments. When there is no back pressure over the solvent and no liquid-phase resistance, the number of transfer units required for a packed column may be calculated from the equation Drew [*Trans. Am. Inst. Chem. Engrs.*, **35**, 681 (1939)]

$$N_G = \ln \left[\frac{\ln (1 - y_1)}{\ln (1 - y_2)}\right] \tag{37}$$

which is based on the assumption that the value of H.T.U. is essentially independent of gas velocity. For small values of the gas concentration y_1 and y_2, this is approximately equivalent to

$$N_G = \ln \left(\frac{y_1}{y_2}\right) \tag{38}$$

The development of gas absorption theory for cases in which chemical reaction occurs is incomplete. For this reason it is not possible at present to state conditions under which Eqs. (37) and (38) may be used. In each case the decision must be based on the analysis of experimental data. Cases that might be expected to come under this class, *e.g.*, absorption of CO_2 in strong NaOH solution, frequently give evidence that liquid-phase diffusion and reaction control the rate of absorption. At present, data for each system must be considered separately.

There is no sharp line dividing pure physical absorption from absorption accompanied by rapid chemical reaction. Most cases fall in the intermediate range. Absorption of oxygen or hydrogen in water appears to be a case of purely physical absorption, since these gases are inert toward water. In many cases that usually have been considered to fall in the class of purely physical processes, however, there is reaction between the solute and the solvent. Thus, in the absorption of ammonia, sulfur dioxide, or chlorine in water, a fraction of the dissolved molecules are present as ions resulting from a chemical reaction. Even in these intermediate cases, the equilibriums among the various forms of the diffusing substance set up by the reaction and the rates of these reactions may affect the rate of absorption.

Absorption of Carbon Dioxide in Alkaline Solutions. In many applications of equipment for carbon dioxide absorption the problem is one of recovery as well as removal of carbon dioxide. For this reason, absorbents such as potassium or sodium carbonate solutions and aqueous organic amine solutions which can be regenerated simply by heating are often employed. In a number of important cases, such as purification of gases for low-temperature operations, even the last traces of carbon dioxide must be removed. In these cases, solutions of sodium or potassium hydroxides are sometimes used, in spite of the fact that they react irreversibly with carbon dioxide and cannot be regenerated by heating.

Because of the variation of the absorption rate with the degree of conversion, and therefore with the amount of CO_2 absorbed, the tower height should be calculated in these cases from the equation

$$Z = G_M' \int_{y_2}^{y_1} \frac{dy}{K_G a(1 - y)^2(y - y^*)} \tag{39}$$

where Z = tower height, ft.; G_M' = superficial mass velocity of inert gas, lb.-moles/(hr.)(sq. ft.); y = mole fraction of CO_2 in gas (subscripts 1 and 2 refer to bottom and top, respectively); y^* = mole fraction of CO_2 in equilibrium with liquid; $K_G a$ = absorption coefficient, lb.-moles/(hr.)(cu. ft.)(unit Δy).

Absorption of Carbon Dioxide in Carbonate Solutions. Comstock and Dodge [*Ind. Eng. Chem.*, **29**, 520 (1937)] absorbed CO_2 in Na_2CO_3 and K_2CO_3 solutions in a 3-in.-diameter tower packed with 8-mm. inside diameter glass rings, 10 mm. long. The following ranges were covered: concentration of carbonate from 0.05 to 1.60*M*, temperature from 15° to 75°C., liquor flow from 0.74 to 19.7 gal./(min.)(sq. ft.), velocity from 0.12 to 0.67 ft./sec. for unpacked tower, CO_2 gas concentrations from 4.6 to 100 per cent, conversion of carbonate to bicarbonate from 0 to 100 per cent, and pH of the solution from 8.5 to 11.5.

The effect of temperature on the transfer coefficient may be found from Fig. 52, which, together with Fig. 53,

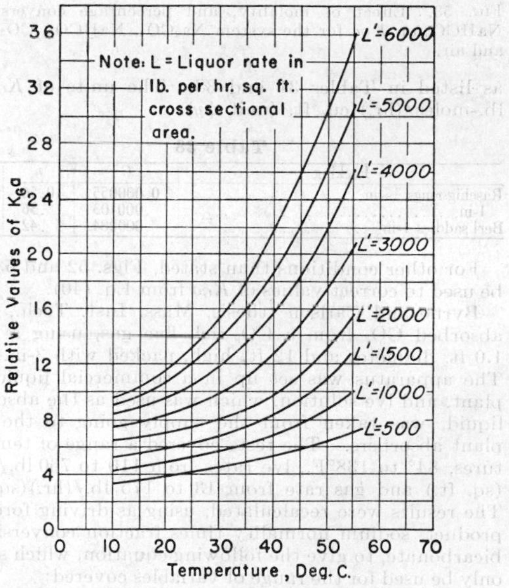

FIG. 52. Effect of temperature on the absorption coefficient system Na_2CO_3, $NaHCO_3$, H_2O, CO_2, and air.

was calculated from the data of Comstock and Dodge by Furnas and Bellinger, who used these data in their work.

Furnas and Bellinger [*Trans. Am. Inst. Chem. Engrs.*, **34**, 251 (1938)] determined over-all coefficients of absorption for CO_2 in Na_2CO_3 solutions in a 12-in. tower with three different packings: ⅜-in. Raschig rings, 1-in. Raschig rings, and 1-in. Berl saddles. The gas flow was varied from 0.15 to 1.11, 0.17 to 1.63, and 0.33 to 1.81 ft./sec.; liquor flow was from 750 to 6600, 1100 to 6900, and 1100 to 6900 lb./(hr.)(sq. ft.) for the three packings, respectively. These velocities are based on the cross-sectional area of the unpacked tower. The results are summarized for conditions at 25°C., Na_2CO_3 molality of 0.5, and 20 per cent of the sodium in the form of $NaHCO_3$, by the equation

$$K_G a = \frac{0.84 \times 10^{-8}}{C} L'^{(1-n)}(a')^n \qquad (40)$$

in which C and n are constants given in Table 38, and a' is the apparent surface area of the packing, sq. ft./cu. ft.,

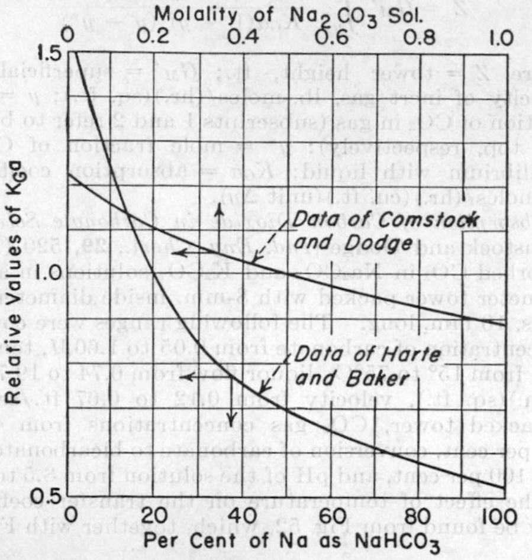

FIG. 53. Effect of molality, and percentage conversion to $NaHCO_3$ on K_Ga for the system Na_2CO_3, $NaHCO_3$, CO_2, H_2O, and air.

as listed in Tables 31 and 38. The units of K_Ga are lb.-moles/(hr.)(cu. ft.)(unit Δy).

Table 38

Packing	C	n	a'
Raschig rings ⅜-in.	0.000055	0.56	148
1-in.	.000103	.36	58
Berl saddles 1-in.	.000084	.42	79

For other conditions than stated, Figs. 52 and 53 may be used to correct values of K_Ga from Eq. (40).

Byrne and Carlson (thesis, Mass. Inst. Tech., 1921) absorbed CO_2 from a CO_2-rich flue gas, using a tower 1.0 ft. diameter and 12 ft. high, packed with 3-in. coke. The apparatus was set up in a commercial liquid CO_2 plant, and lye solution, which was used as the absorbing liquid, was taken from the supply going to the large plant absorbers. The tests covered a range of temperatures, 84° to 138°F., lye rates from 110 to 760 lb./(hr.)(sq. ft.) and gas rate from 13 to 175 lb./(hr.)(sq. ft.). The results were recalculated, using as driving force the product, sodium normality times fraction conversion to bicarbonate, to give the following equation, which should only be used for the range of variables covered:

$$K_L'a = 0.000074t + 0.0000048L' - 0.0055 \qquad (41)$$

where $K_L'a$ = capacity coefficient (lb.-moles CO_2)/(hr.)(cu. ft.)(unit driving force).
t = temperature, °F.
L' = lye rate, lb./(hr.)(sq. ft.).

No variation was observed with gas rate.

Whitman and Davis [*Ind. Eng. Chem.*, **18**, 265 (1926)] report a series of experiments in which a sodium carbonate lye was used to absorb carbon dioxide from flue gas. The absorption was carried out in a 15-plate column with one bubble cap per plate. See Table 39 for a summary of the results.

Absorption of Carbon Dioxide in Ethanolamine Solutions. Aqueous solutions of weak organic bases such as mono, di, or triethanolamines can be used to absorb acid gases, such as CO_2 and H_2S, and the alkaline absorbent

Table 39. Data on Absorption of CO_2 in Sodium Carbonate Lye in a 15-plate Bubble-cap Tower

t, °C	(Superficial gas velocity) × (gas density)$^{1/2}$, (ft./sec.)(lb./cu. ft.)$^{1/2}$	Lb.-moles of CO_2 absorbed/hr. per plate
27.8	0.04	0.00133
42.2	.07	.00193
50.0	.11	.00283
55.0	.09	.00261
60.0	.10	.00229
60.0	.11	.00301
60.5	.07	.00279
60.5	.05	.00182
63.5	.04	.00198
84.3	.10	.00153

can be regenerated by heating (to about 300°F. in the case of CO_2 and the monoamine). Industrial applications of this process are discussed by Storrs and Reed [*Trans. Am. Soc. Mech. Engrs.*, **64**, 299 (1942)] and by Reed and Wood [*Trans. Am. Inst. Chem. Engrs.*, **37**, 363 (1941)]. Vapor pressures of CO_2 and H_2S over these solutions are given on pp. 677*ff*.

Rates of absorption of CO_2 in diethanolamine solutions in a packed tower are reported by Cryder and Maloney [*Trans. Am. Inst. Chem. Engrs.*, **37**, 827 (1941)]. Typical values of K_Ga observed by these investigators for an 8-in. tower packed with ¾-in. Raschig rings are given by Fig. 54, which shows that the coefficient decreases as the percentage conversion increases. The latter quantity is equal to (2 × moles CO_2 absorbed/moles R_2NH), corresponding to the liquid-phase reactions

(1) $2R_2NH + CO_2 + H_2O \rightleftarrows (R_2NH)_2.H_2CO_3$
(2) $R_2NH + CO_2 + H_2O \rightleftarrows R_2NH.H_2CO_3$

In these reactions, R represents the group HOC_2H_4—.

Cryder and Maloney found that the rate of absorption was greatest for approximately $3N$ solutions, where normality is equal to g.-moles R_2NH/l. of solution, including both the unreacted amine and that combined with CO_2. The coefficient was substantially independent of gas rate in the range investigated, $160 < G < 330$ lb./(hr.)(sq. ft.), and varied only slightly with the temperature in the range $25 < t < 55$°C.

The data of Gregory and Scharmann [*Ind. Eng. Chem.*, **29**, 514 (1937)], who used monoethanolamine and diamino-isopropanol solutions in a 11.75-in. tower packed with 2-in. rings are analyzed by Cryder and Maloney. The values of K_Ga calculated from these data range from 0.15 to 0.25 lb.-mole CO_2/(hr.)(cu. ft.)(atm.), which are said by Cryder and Maloney to be somewhat higher than values for the diamine solutions, when the comparison is made for the same packing.

Absorption of Carbon Dioxide in Sodium or Potassium Hydroxide Solutions. Tepe and Dodge [*Trans. Am. Inst. Chem. Engrs.*, **39**, 255 (1943)] and Spector and Dodge [*Trans. Am. Inst. Chem. Engrs.*, **42**, 827 (1946)] have measured rates of absorption of CO_2 from air by sodium and potassium hydroxides in 6-in. and 12-in. towers packed with 0.5- and 0.75-in. Raschig rings and with 1-in. Berl saddles. Tepe and Dodge in their study of the sodium hydroxide system found the absorption coefficient to increase rapidly with increasing hydroxide concentration up to about $2N$ and to decrease with further increases in normality. The effect of the degree of conversion to carbonate (equal to moles Na_2CO_3/twice the atoms of total sodium) found by these investigators is shown by Fig. 55, which is based on G = 190 lb./(hr.)(sq. ft.), L = 1600 lb./(hr.)(sq. ft.), and liquid temperature = 78°F. The coefficient was found to be independent of the gas rate and to increase with the 0.28 power of the liquid rate within the range $800 < L < 10{,}000$ lb./(hr.)(sq. ft.) and the sixth power of the liquid temperature between 78° and 125°F.

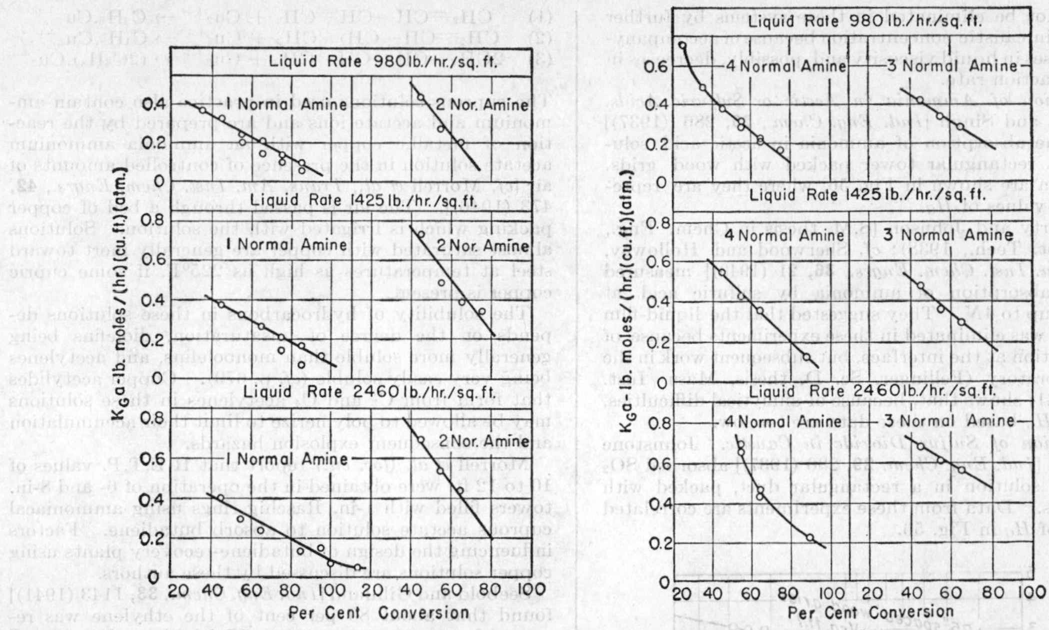

FIG. 54. Absorption of CO_2 by diethanolamine solutions at atmospheric pressure—absorption coefficient vs. per cent conversion of amine. [*From Cryder and Maloney, Trans. Am. Inst. Chem. Engrs.*, **37**, 827 (1941).]

Spector and Dodge (*loc. cit.*) measured rates of absorption of CO_2 at lower concentrations from air at pressures up to 100 lb./sq. in. gage using 0.75-in. rings and 1-in.

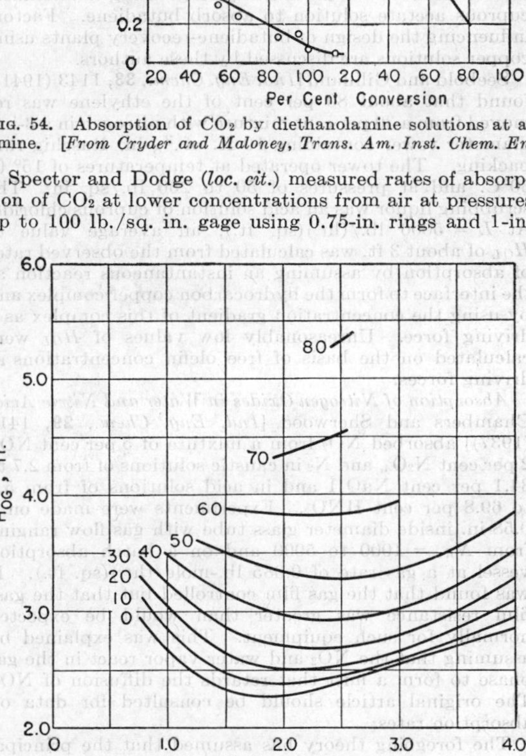

FIG. 55. Absorption of CO_2 in aqueous NaOH solutions. The parameter on the figure represents the per cent conversion of NaOH to Na_2CO_3. [*Data of Tepe and Dodge, Trans. Am. Inst. Chem. Engrs.*, **39**, 255 (1943).]

Packing = ½-in. Raschig rings.
Temperature = 298°K.; $H_{OG} \sim T^{-6}$.
Liquor rate = 1600 lb./(hr.)(sq. ft.); $H_{OG} \sim L^{-0.28}$.
Gas rate = 190 lb./(hr.)(sq. ft.); $H_{OG} \sim G$.

saddles. The observed variation of K_Ga with the 0.15 to 0.35 power of G was attributed to the range of gas concentrations, which was lower than that covered by Tepe and Dodge. Coefficients from 20 to 30 per cent greater were found for KOH than for NaOH. K_Ga decreased with the 0.5 power of the absolute tower pressure. The data obtained in these laboratory experiments were found to be 9.5 and 12 per cent lower than two coefficients reported for two 3-ft.-diameter by 9-ft. commercial towers used in series to reduce the CO_2 content of air to 3 p.p.m.

The absence of a large effect of gas velocity and the effects of liquor temperature and hydroxide concentration indicate that a part of the diffusional resistance is in the liquid phase even in this case of rapid chemical reaction. On the other hand, the variation with total pressure and the small effect of liquid rate indicate that at least a fraction of the resistance is in the gas phase.

Table 40 compares absorption coefficients for the absorption of CO_2 in different liquids. Apparently, the rate of the reaction between CO_2 and carbonate is so slow that almost all the resistance to transfer is in the liquid phase. Even for the most rapid reaction studied, the resistance is not entirely within the gas phase, as shown by the last two coefficients listed. Liquid-phase resist-

Table 40. Absorption Coefficients for CO_2 in Various Solutions

Basis: L = 2500 lb./(hr.)(sq. ft.); G = 300 lb./(hr.)(sq. ft.); T = 25°C. (liquid temperature)

Liquid	K_Ga, lb.-moles/(hr.)(cu. ft.)(unit Δy)	Ref.
Water	0.05	1
1N sodium carbonate, 20% of sodium as bicarbonate	.03	2
3N diethanol amine, 50% conversion to carbonate	.4	3
2N sodium hydroxide, 20% of sodium as carbonate	2.3	4
2N potassium hydroxide, 15% of sodium as carbonate	3.8	5
Approximate maximum value (gas-film coefficient)	24	6

References:
[1] Sherwood and Holloway, *Trans. Am. Inst. Chem. Engrs.*, **36**, 39 (1940).
[2] Furnas and Bellinger, *ibid.*, **34**, 251 (1938).
[3] Cryder and Maloney, *ibid.*, **37**, 827 (1941).
[4] Tepe and Dodge, *ibid.*, **39**, 255 (1943).
[5] Spector and Dodge, *ibid.*, **42**, 827 (1946).
[6] Sherwood and Holloway, *ibid.*, **36**, 21 (1940).

ance cannot be eliminated in these systems by further increases in caustic concentration because of accompanying increases in liquid viscosity and, possibly, decreases in specific reaction rate.

Absorption of Ammonia in Acetic or Sulfuric Acids. Johnstone and Singh [*Ind. Eng. Chem.*, **29**, 286 (1937)] studied the absorption of ammonia in acetic acid solutions in a rectangular tower packed with wood grids. These data are shown in Fig. 56, where they are represented by values of H_G.

Dougherty and Johnson [S.M. thesis in Chem. Eng., Mass. Inst. Tech., 1938); *cf.* Sherwood and Holloway, *Trans. Am. Inst. Chem. Engrs.*, **36**, 21 (1940)] measured rates of absorption of ammonia by sulfuric acid at strengths up to $4N$. They suggested that the liquid-film resistance was eliminated in these experiments because of rapid reaction at the interface, but subsequent work in the same laboratory (Fellinger, Sc. D. thesis, Mass. Inst. Tech., 1941) shows that, because of analytical difficulties, values of H_G based on these data are too low.

Absorption of Sulfur Dioxide in Caustic. Johnstone and Singh [*Ind. Eng. Chem.* **29**, 286 (1937)] absorbed SO_2 in NaOH solution in a rectangular duct, packed with wood grids. Data from these experiments are correlated in terms of H_G in Fig. 56.

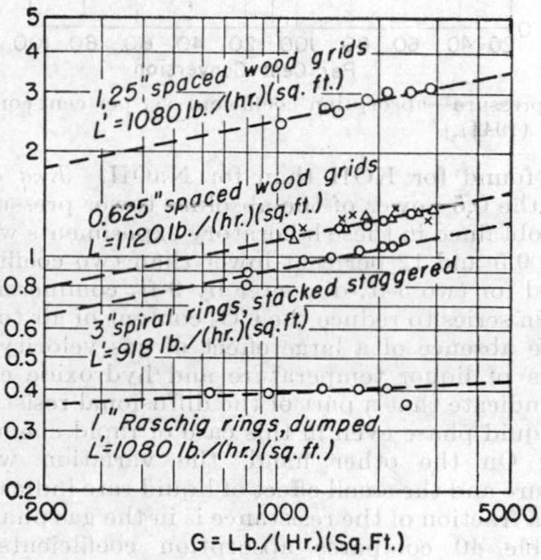

FIG. 56. Values of gas film H.T.U. Circles represent heat-transfer data between air and water. Triangles represent absorption of SO_2 by NaOH solution. Crosses represent absorption of NH_3 by acetic acid solutions. $\psi = (C\mu/k)^{2/3}$ for heat transfer and $(\mu/\rho D)^{2/3}$ for absorption. Column of rectangular cross section 5.25 by 15.25 in. Depth of packing = 1.25-in. spaced wood grids, 3.3 ft.; 0.625-in. spaced wood grids, 2 ft.; 3-in. spiral rings, 2.6 ft.; 1-in. rings, 0.5 ft. Dimensions of wood grids = 0.25 in. thick by 4 in. deep. [*Experimental data of Johnstone and Singh, Ind. Eng. Chem.*, **29**, 286 (1937).]

Absorption of Chlorine in Sodium Hydroxide. The data of Vivian and Whitney [*Chem. Eng. Progress*, **43**, 691 (1947)] on absorption of chlorine from air by sodium hydroxide solutions indicate that the resistance is entirely in the gas phase when the caustic concentration is greater than about $2N$. In this range, data obtained on a 4-in.-diameter tower filled with 2 ft. of 1-in. Raschig rings gave values of $(k_Ga)(p_{BM})$ ranging from 22 lo.-moles/(hr.)(cu. ft.) at G = 200 lb./(hr.)(sq. ft.) to 44 at G = 500, all at a caustic rate of 12,300 lb./(hr.)(sq. ft.) and a liquid temperature of 21° to 27°C.

Absorption of Olefins and Acetylenes in Aqueous Copper Solutions. Solutions of cuprous salts in water may be used for the selective absorption of unsaturated hydrocarbons that form loose chemical compounds with cuprous ions according to reactions such as the following:

(1) $CH_2{=}CH{-}CH{=}CH_2 + Cu_2^{++} \rightarrow C_4H_6.Cu_2^{++}$
(2) $CH_2{=}CH{-}CH_2{-}CH_3 + Cu_2^{++} \rightarrow C_4H_8.Cu_2^{++}$
(3) $2CH_3{-}CH{=}CH{-}CH_3 + Cu_2^{++} \rightarrow (2C_4H_8).Cu_2^{++}$

The cuprous solutions used in practice also contain ammonium and acetate ions and are prepared by the reaction of metallic copper with an ammonia–ammonium acetate solution in the presence of controlled amounts of air [*cf.* Morrell *et al.*, *Trans. Am. Inst. Chem. Engrs.*, **42**, 473 (1946)]. The air is passed through a bed of copper packing which is irrigated with the solution. Solutions almost saturated with copper are generally inert toward steel at temperatures as high as 225°F. if some cupric copper is present.

The solubility of hydrocarbons in these solutions depends on the degree of unsaturation, diolefins being generally more soluble than monoolefins, and acetylenes being very easily soluble (*cf.* p. 679). Copper acetylides that form from C_3 and C_4 acetylenes in these solutions may be allowed to polymerize to limit their accumulation and the consequent explosion hazards.

Morrell *et al.* (*loc. cit.*) report that H.E.T.P. values of 10 to 12 ft. were obtained in the operation of 6- and 8-in. towers filled with 1-in. Raschig rings using ammoniacal cuprous acetate solution to absorb butadiene. Factors influencing the design of butadiene-recovery plants using copper solutions are discussed by these authors.

Seebold and Gilliland [*Ind. Eng. Chem.*, **33**, 1143 (1941)] found that about 80 per cent of the ethylene was removed from a 30 per cent mixture with nitrogen in a 3-in.-diameter tower containing 10 ft. of ⅜-in. Raschig ring packing. The tower operated at temperatures of 15° to 30°C. and at pressures of 50 to 250 lb./sq. in. The scrubbing liquor was an acid solution of cuprous chloride. At L = 5000 lb./(hr.)(sq. ft.), an average value of H_{OL} of about 3 ft. was calculated from the observed rates of absorption by assuming an instantaneous reaction at the interface to form the hydrocarbon copper complex and by using the concentration gradient of this complex as a driving force. Unreasonably low values of H_{OL} were calculated on the basis of free olefin concentrations as driving forces.

Absorption of Nitrogen Oxides in Water and Nitric Acid. Chambers and Sherwood [*Ind. Eng. Chem.*, **29**, 1415 (1937)] absorbed NO_2 from a mixture of 5 per cent NO_2, 2 per cent N_2O_4, and N_2 in caustic solutions of from 2.7 to 34.1 per cent NaOH and in acid solutions of from 5.7 to 69.8 per cent HNO_3. Experiments were made on a 0.58-in. inside diameter glass tube with gas flow ranging from N_{Re} = 1000 to 5000 and on a batch absorption vessel at a gas rate of 0.085 lb.-mole/(hr.)(sq. ft.). It was found that the gas film controlled but that the gas-film resistance was greater than would be expected normally for such equipment. This was explained by assuming that the NO_2 and water vapor react in the gas phase to form a mist that retards the diffusion of NO_2. The original article should be consulted for data on absorption rates.

The foregoing theory has assumed that the principal resistance to gas absorption lies in either a gas or liquid film adjacent to the interface of the gas and liquid phases. In general this is true. The possibility exists, however, that a reaction which is essential to the absorption of the solute gas may take place in a single phase and at so slow a rate that this homogeneous reaction will determine in a large measure the rate at which gas absorption can proceed. The most notable case of this sort is the absorption of nitric oxide in water to form nitric acid. Taylor [*Ind. Eng. Chem.*, **19**, 1250 (1927); *cf.* also Partington and Parker, *J. Soc. Chem. Ind.*, **38**, 75T (1919); Burdick and Freed, *J. Am. Chem. Soc.*, **43**, 518 (1921); Toniolo, *Chem. & Met. Eng.*, **34**, 92 (1927); Sherwood,

"Absorption and Extraction," p. 222, McGraw-Hill, New York, 1937; and Wenner, "Thermo-chemical Calculations," McGraw-Hill, New York, 1941] presented a method for the calculation of the capacity of towers for nitric acid absorption. A large free volume is required for rapid absorption in order to permit time for the oxidation of NO to NO_2, as will be apparent from a study of the cycle of the reactions that are listed below:

(1) $3NO_2 + H_2O = 2HNO_3 + NO$
(2) $2NO + O_2 = 2NO_2$

The second reaction occurs in the gas phase, and the first is probably a liquid-phase reaction. Tower packing used for nitric acid towers should have not less than 60 per cent free volume and, preferably, as high as 80 per cent free volume.

Taylor's method is based on the use of the data of Burdick and Freed (*loc. cit.*) for the equilibrium vapor pressure of NO_2 consumed in the liquid phase according to reaction (1), and the data of Bodenstein [*Z. physik. Chem.*, **100**, 68 (1922)] and Briner, Pfeiffer, and Malet [*J. chim. phys.*, **21**, 25 (1926)] for the rate of the vapor-phase oxidation of the NO formed by reaction (1).

Taylor, Chilton, and Handforth [*Ind. Eng. Chem.*, **23**, 860 (1931)] have discussed the development by E. I. duPont de Nemours & Co. of the use of pressure in scrubbing towers in connection with the manufacture of nitric acid by the oxidation of ammonia. Reaction (1) above was shown to be the controlling reaction, and its rate was increased by using a series of graded acid strengths for absorption. This was performed with a bubble-cap column to avoid the necessity of a series of packed towers, since the operation of acid circulating pumps under pressure is difficult. Pressure operation is advantageous because the time for reaction (2) varies inversely as the square of the pressure, so that partial pressures of NO_2 are increased, giving a higher strength of acid as well as greater capacity.

A further factor is that reaction (1) is favored by low temperature, and cost studies showed that refrigeration would not pay but that use of available cooling water would. A method of cooling was developed for the individual plates of the column, and the improved cooling doubled the tower capacity with no increase in size.

The use of this development led to a column 5¼ ft. diameter and 40 ft. high with a capacity of over 25 tons of nitric acid per 24 hr.

ECONOMIC DESIGN OF ABSORPTION SYSTEMS

Although equations and experimental data for transfer coefficients or units are useful in calculating absorption problems once the various conditions involved are known, it is even more important to have a rational basis for choice of the operating variables. These variables include the type of column to be used, the packing to be chosen, the liquor-gas ratio, the diameter (or gas velocity), and the height (or exit-gas strength). All these factors involve economic balances, as shown by Colburn ("Absorption of Gases by Liquids," Collected Papers on the Teaching of Chemical Engineering, p. 269, Am. Inst. Chem. Engrs., New York, 1940).

Packed Towers vs. Plate Towers. The relative advantages and disadvantages of packed and plate towers may be summarized as follows:

1. Packed towers may be advantageous for vacuum operations because the pressure drop through a tower of this type can be less than for a plate tower.
2. Packed towers may be preferred in the case of liquids that foam.
3. The liquor hold-up is generally less in a packed tower.
4. Plate towers may be preferred when the operation involves liquids that deposit small amounts of solid material that must be removed periodically. A plate tower can be fitted with manholes, and the plates may be spaced vertically by 24 in. or more to facilitate the cleaning operation.
5. The total weight of a plate tower is usually less than for a packed tower designed for the same duty. The limited crushing strength of packing materials may make it impossible for one packing-support plate to bear the weight of a tall column of packing.
6. Plate towers may be more suitable when the absorption or distillation operation is carried out intermittently at temperatures either higher or lower than atmospheric temperature. Expansion and contraction of the shell under such circumstances may crush the packing.
7. Cooling coils are installed readily on plates, making plate towers more desirable when a large heat of solution must be removed.
8. Plate columns may be preferable for operations that require a large number of transfer units or theoretical plates because of the absence of channeling of the vapor or liquid streams and the consequent limitation in the amount of material that can be transferred.
9. Higher liquid rates usually can be handled in plate towers, provided that the distance the liquid must travel to cross each plate is not larger than a few feet.
10. Construction of packed towers is usually simpler and cheaper when acids or other corrosive substances must be handled.
11. Other things being equal, economical considerations usually show that packed columns are favored over plate columns for absorption when the column diameter is less than about 2 ft.

Column and Packing. Consideration is first made of factors such as corrosion and fouling, available apparatus, and life of the process. Then selection is made by comparison of costs of various types under the most favorable design conditions for each. Among packing materials 1-in. and ½-in. ceramic rings, 1-in. saddles, 3-in. spiral or partition rings, drip-point tile, and wood grids appear favorable from a cost viewpoint for absorption of soluble gases.

An interesting point in the selection of packed columns is that the cost of a shell varies approximately as the square of the diameter. This cost can therefore be expressed in terms of dollars per cubic foot of tower volume. This value can then be added to the cost per cubic foot of packing. It is readily apparent, then, that there is little advantage in searching for packings having a unit cost much less than that of the shell. For example, gravel costs only a fraction of a dollar per cubic foot compared with, say, $5 for 1-in. rings, but a shell of so much larger diameter is required when gravel is used that it is an uneconomic packing. Cost data are summarized on pp. 709 to 710.

Liquor-gas Ratio. The design factor of first importance is the value mG_M/L_M. This quantity is needed in determining the height of a transfer unit and the number of transfer units, and the liquor-gas ratio affects the column diameter. If the solute gas is dilute so that the term mG_M/L_M is apt to be constant through the column, the problem is simplified.

The choice of the magnitude of mG_M/L_M is often based on economic factors. The greater mG_M/L_M, the more concentrated will be the exit liquor and therefore the cheaper will be the operation of concentrating the solute in cases of solute recovery. On the other hand, the greater the value of mG_M/L_M, the higher and therefore more expensive the absorption column will be and also the more solute will be lost in the exit gases. A detailed example of the calculation of the optimum liquor-gas ratio in the absorption of acetone by water resulted in a value of mG_M/L_M of 0.7. For less valuable solutes a lower value of this ratio is employed, *e.g.*, in the absorption of refinery gases. In the design of stripping columns, the value of L_M/mG_M usually should be in the range 0.5 to 0.8.

Where the solute is not very dilute (*i.e.*, not less than

a few mole per cent in the exit liquor), the heat of solution of the solute will cause a temperature rise, and the value of m at the bottom of the column will be greater than at the top. Where, for this reason or for others, mG_M/L_M is not constant, the choice of the liquor-gas ratio is more difficult; however, the conditions at the dilute end are usually more important since, in case of nearly complete absorption, most of the transfer units are concentrated in the dilute region.

The equation resulting from a balance of the costs of absorption and of subsequent distillation to recover the solute is

$$\left(\frac{L_M}{m_2G_M} - 1\right)^2 = \frac{C_3H_{OG}[(y^*/x)_D - 1]B}{C_5\theta r G_M m_2} \tag{42}$$

where

$$B = \left[1 + n\left(\frac{L_M}{m_2G_M} - 1\right)\right] 2.3 \log_{10} \frac{y_1(1 - m_2G_M/L_M)^2}{y_2(1 - K_1G_M/L_M)} - \frac{(1 - m_2G_M/L_M) - 2(K_1/m_2 - 1)}{1 - K_1G_M/L_M}$$

$(y^*/x)_D$ = y^*/x at the boiling point of the feed to the stripping column.

m_2 = slope of equilibrium curve, y^*/x, at the temperature of the inlet liquid to the absorption column.

K_1 = y^*/x at the temperature of the rich absorption liquor as it runs from the absorption column.

C_3 = annual cost of apparatus and power for the absorption column, \$/(cu. ft.)(year), $= C_1\,[(G_{opt}/G) + 0.5\,(G/G_{opt})^2]$. (The latter terms are defined later.)

C_5 = product (cost of vapor for stripping, \$/lb.-mole) (total annual cost of distillation apparatus, cooling water, and steam/annual cost of steam).

y_1/y_2 = optimum ratio of solute mole fractions in gas stream flowing through absorber; *cf.* Column Height (Exit-gas Strength), pp. 708 to 709.

θ = hr. operation/year.

G_M = molal gas velocity through absorber, lb.-moles/(hr.)(sq. ft.).

r = ratio (actual reflux ratio in distillation column/minimum reflux ratio, defined as ratio of reflux to product).

n = exponent in the relation $H_{OG} \sim (G/L)^n$.

In deriving this equation, it is assumed that the distillation column produces essentially pure solute as an overhead product and that essentially no solute is withdrawn from the base of the distillation column and returned to the absorber with the stripped absorption liquid.

Example. Estimate the optimum value of mG_M/L_M for an absorber used to recover acetone vapor from air by water absorption followed by stripping with open steam. [In order to simplify the calculation for the purpose of illustration, neglect the temperature rise of the liquid in the absorption column, making $m_2 = K_1$ in Eq. (42).] The following data are assumed: C_3 = \$4.35/(cu. ft.)(year); C_5 = \$0.0108 per lb.-mole, *i.e.*, 30¢ per 1000 lb. for steam, times 2, the assumed ratio of the total annual distillation cost to the total annual steam cost; $r = 1.25$; G_M = 25.4 lb.-moles air/(hr.)(sq. ft.); θ = 8400 hr./year of operation; H_{OG} = 2.5 ft. and $n = 0.5$; $m_2 = K_1 = 2.7$; $(y^*/x)_D$ = 23; and $(y_1/y_2)_{opt} = 435$. Solution of Eq. (42) is made by trial and error, assuming $mG_M/L_M = 0.7$, as a first approximation. Substituting in Eq. (42),

$$B = [1 + (0.5)(0.43)]\ 2.3 \log_{10} [(0.3)(435)] - 1 = 4.91$$

$$\left(\frac{L_M}{mG_M} - 1\right)^2 = \frac{(4.35)(2.5)(23 - 1)(4.91)}{(0.0108)(8400)(1.25)(25.4)(2.7)} = 0.151$$

$$\frac{mG_M}{L_M} = 0.720$$

A second trial, assuming $mG_M/L_M = 0.72$ on the right of Eq. (42), gives $mG_M/L_M = 0.724$.

Column Diameter (or Gas Velocity). The gas velocity is selected by considering first the safe operating velocity with respect to flooding, and second the optimum velocity calculated by an economic balance between column cost and power cost. Data on flooding velocities are given on pp. 683 to 686. Design is usually for not greater than 50 to 60 per cent of flooding to allow for temporary fluctuations and for a possible increase in the capacity of the plant. Furthermore, since flooding of a column is serious and might lead to a plant shutdown, a liberal margin of safety is desirable.

After checking on a safe velocity with respect to flooding, it is of interest to see the result of an economic balance of column cost vs. power cost. Inasmuch as the column cross-sectional area varies inversely as the velocity whereas the power cost for blowing gas through the packing varies approximately as the cube of the velocity, there is a reasonably sharp optimum point.

Expressing the pressure drop through the column at constant liquor-gas ratio as

$$\Delta P = \frac{b'G^M}{\rho} \tag{43}$$

where ΔP = pressure drop per unit cross section and unit height, G = mass velocity, and ρ = gas density, the annual cost of power becomes, per unit weight of gas through-put and per unit height, $C_2'\theta b'G^M/\rho^2$. The annual cost of column per unit gas through-put and per unit height is C_1/G, where C_1 = annual cost of packing and shell, \$/(year)(cu. ft.), C_2' = cost of delivered energy, \$/ft.-lb., and θ = hr./year operation. Adding these values and solving for the velocity giving a minimum cost gives

$$G_{opt} = \left(\frac{C_1\rho^2}{MC_2'\theta b'}\right)^{\frac{1}{(M+1)}} \tag{44}$$

Often M is close to 2, under which conditions

$$G_{opt} = \left(\frac{C_1\rho^2}{2C_2'\theta b'}\right)^{1/3} \tag{45}$$

A more convenient expression of this equation is

$$G_{opt} = 2680\phi^{4/3}\left(\frac{C_1}{C_2\theta b}\right)^{1/3} \tag{46}$$

where $\phi = (\rho/0.075)^{1/2}$; C_2 = cost of delivered energy, \$/kw.-hr.; and b = pressure drop, in H_2O/ft. height at G/ϕ = 1000 lb./(hr.)(sq. ft.). Note that b may be an extrapolated value. Pressure-drop data are summarized on pp. 680 to 683.

If the column operates at the economic gas velocity, the total annual cost per transfer unit and per (lb./hr.) of gas through-put is

$$\text{Cost/(T.U.)(lb./hr.)} = \frac{1.5H_{OG}C_1^{2/3}(C_2\theta b)^{1/3}}{2680\phi^{4/3}} \tag{47}$$

which shows that the cost for a given operation varies directly as the value of H_{OG}, is slightly less sensitive to the unit investment charge, and varies only slightly as the annual cost of energy is varied.

It is of interest that both the flooding velocity and optimum velocity are related to the pressure drop. It is therefore not surprising that, for many packings, the two velocities are found to be of the same order.

At the economic velocity according to Eq. (44) or (45), the cost of energy per cubic foot turns out to be ½ of C_1, so that the total annual cost of column and energy in dollars per cubic foot C_3, becomes equal to $1.5C_1$.

Column Height (Exit-gas Strength). At a given value of mG_M/L_M, the required column height is dependent upon the value chosen for the exit-gas strength. The latter quantity may be determined by an economic balance between the cost of lost solute and the cost of additional column height. The annual cost of solute per unit cross-sectional area of column can be represented as $C_4\theta G_M y_2$, and the annual cost of column and pressure drop as $C_3H_{OG}N_{OG}$. Substituting for N_{OG} in terms of mG_M/L_M, y_1, and y_2 and solving for the value of y_2 at which the sum of the costs are a minimum there results, approximately,

$$y_2 - mx_2 = \frac{C_3H_{OG}}{C_4\theta G_M(1 - mG_M/L_M)} \tag{48}$$

for packed towers, where C_3 = annual cost of apparatus and energy for pressure drop, \$/(year)(cu. ft.); C_4 = value of solute at its concentration in the exit liquor, \$/lb.-mole of solute; and θ = hr./year operation.

A similar result for plate columns is

$$y_2 - mx_2 = \frac{C_6}{C_4 \theta G_M E (2.3 \log L_M/mG_M)} \tag{49}$$

where C_6 = annual cost of column and pressure drop, \$/(year)(plate)(sq. ft.), and E = over-all plate efficiency, fractional. Tiller [*Trans. Am. Inst. Chem. Engrs.*, **40**, 331 (1944)] gives an equivalent equation for the optimum number of plates in an absorber.

The optimum exit-liquor strength for a stripping column depends on a balance between the cost of lost solute in the exit liquor and the cost of additional tower height required for more stripping. Equations analogous to those for exit-gas strength in absorption are

$$\left(x_2 - \frac{y_2}{m}\right)_{\text{opt}} = \frac{C_3 H_{OL}}{C_4 L_M \theta (1 - L_M/mG_M)} \tag{50}$$

for a packed column, and

$$\left(x_2 - \frac{y_2}{m}\right)_{\text{opt}} = \frac{C_6}{C_4 L_M \theta E \ln (mG_M/L_M)} \tag{51}$$

for a plate column. In these equations, C_3 = annual cost of apparatus (amortization and depreciation of column and packing) and power, \$/(cu. ft.)(year); C_6 = annual cost of apparatus and power, \$/(year)(plate) (sq. ft. of cross section); C_4 = value of solute at concentration of exit gas, \$/lb.-mole of pure solute; E = over-all plate efficiency, fractional; H_{OL} = over-all H.T.U. based on liquid-phase driving force, ft.; θ = hr./year of operation; L_M and G_M = molar velocities, lb.-moles/(hr.)(sq. ft.); m = slope of equilibrium curve, y^*/x, at dilute end of stripper.

Costs of Packing Materials. Approximate costs of Raschig rings are shown by Fig. 57, which is based on

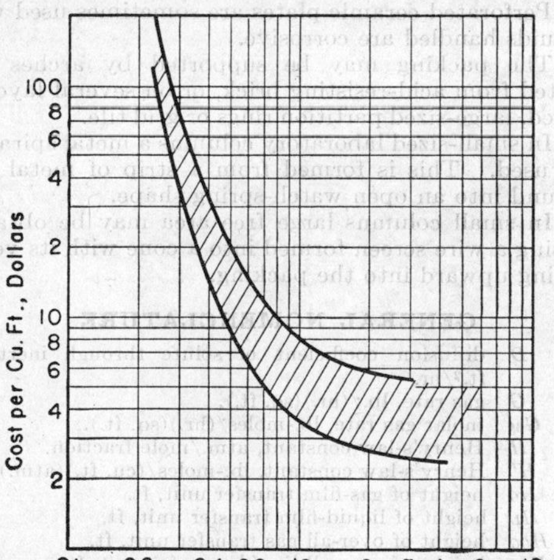

FIG. 57. Approximate costs of porcelain, stoneware, and metal Raschig rings and ceramic Berl saddles (1946).

1946 production costs. Shipping charges must be added to costs read from the figure; costs of installation should also be included in estimating total investment charges. The variations in price shown in Fig. 57 for rings of a given size are due to differences among different manufacturers as well as differences in costs of large and small orders. Rings constructed with thicker walls for greater resistance to crushing cost only slightly more than similar thin-walled rings made by the same manufacturer. The cost of stoneware rings ranges from 87 per cent of the cost of similar porcelain rings for the smaller sizes to 70 per cent for 2- and 3-in. rings. Costs of steel rings are approximately equal to the lowest values shown for ceramic rings. Metal rings made from stainless steel, aluminum, copper, or zinc are more expensive.

Ceramic Berl saddles are generally more expensive than Raschig rings of the same nominal size. Costs of these packings are equal approximately to the highest values read from Fig. 57.

Current (1946) costs of several packings of the larger sizes are given in Table 41.

Table 41. Approximate Costs of Spiral Tile, Partition Rings, and Ceramic Grids*

Type of packing	Size, in.	Cost	
		\$/1000 pieces	\$/cu. ft.
Single-spiral tile	3	40– 45	2.00–2.70
Triple-spiral tile	3¼	12– 27	0.71–1.90
Partition rings	3	54–111	2.60–7.40
Partition rings	4	90–275	2.80–8.40
Drip-point tile (solid-packed arrangement)	7½	......	1.40–1.70
Drip-point tile (checkered arrangement)	7½	......	1.00
Ceramic grids	...	......	1.00

* Based on information furnished in 1946 by the following manufacturers: General Ceramics and Steatite Corp., Custodis Construction Co., Inc., Lapp Insulator Co., Inc., General Refractories Co.

Fiberglas tower packing* is priced at about 38¢/lb. in the form of packs 20 by 20 by 1 in. and 20 by 25 by 1 in. To aid in shipping and handling during application, a water-soluble bonding material is used, amounting to about 5 per cent of the total weight. The density of the packing as ordinarily used is about 4 to 6 lb./cu. ft., making the unit cost of this packing \$1.90 to \$2.28 per cu. ft., exclusive of shipping and installation costs.

Costs of Absorption Towers. Approximate current costs of chemical porcelain towers of the bell-and-spigot type are listed in Table 42. The variation of the unit cost in dollars per cubic foot is due to the cost of the bottom section, the cover, the perforated plate for supporting the packing, and the liquor distributor. These items account for about half the unit cost of the 10-ft.-high towers. According to Bliss [*Trans. Am. Inst. Chem. Engrs.*, **37**, 763 (1941)], resin towers cost from 1.4 to 2.5 times as much as chemical stoneware towers of the larger and smaller sizes, respectively.

Table 42. Approximate Costs of Chemical Porcelain and Chemical Stoneware Towers*

Diameter, ft.	Cost, \$/cu. ft. of empty internal volume, for packed height equal to			
	10 ft.	20 ft.	30 ft.	∞
1	16.1	12.3	11.0	8.5
2	11.3	9.4	8.3	6.1
3	9.4	6.9	6.1	4.4
4	10.9	8.1	7.2	5.3
5	18.9	14.5	11.2	7.7

* Based on information furnished in 1946 by General Ceramics and Steatite Corp., U.S. Stoneware Co. for bell-and-spigot construction. Similar flanged towers cost from two to three times as much for the larger and smaller diameters, respectively. Costs of shipping and erection must be added to figures listed in this table.

The costs of metal tower shells are estimated by calculating the weight of metal required and multiplying this figure by the unit cost of steel fabricated in towers. These unit costs for steel amount to about \$0.15/lb. for weights of about 1000 lb.; \$0.11/lb. for 3000 lb.; and \$0.10/lb. for 8000 lb. The unit cost for aluminum construction is about \$1/lb.

For example, the costs of standard steel pipe per cubic foot of contents on the basis of \$0.15/lb. would be \$5.15/cu. ft. for 12-in. pipe and \$5.60/cu. ft. for 6-in. pipe.

* This information supplied through the courtesy of Owens-Corning Fiberglas Corp.

The cost of 9- by 4.5- by 2.5-in. acidproof brick used for lining towers exposed to corrosive acids is about $65/1000 bricks, exclusive of shipping and erection. This is equivalent to about 30¢/cu. ft. of tower volume for a 6-ft.-diameter tower.

Construction expense will be approximately one to two times the cost of a metal tower and will be at least 25 to 50 per cent of the costs of other types.

DETAILS OF COLUMN CONSTRUCTION

Liquid Distribution in Packed Towers. Estimates of packed-tower height for design purposes are generally based on the assumption that the liquid flows vertically downward through the packing at the same rate at every point throughout a tower cross section. Actually, all packing materials tend to spread the liquid stream radially toward the tower wall, and the liquid that reaches the wall may not be diverted back into the packing. Proper distribution of the liquid initially as it flows onto the packing is beneficial, but such uniform distribution may not persist throughout a tall tower. Oils have been observed to concentrate in a few streams within the packing. Redistribution of the liquid is sometimes provided for in tall towers by subdividing the packing into sections and providing additional distributor plates between the sections.

The detrimental effect of poor liquor or gas distribution is due to the fact that a stream that flows too slowly may become saturated quickly and remain essentially unchanged in composition as it flows through the remainder of the packing. This effect of poor liquid distribution is especially bad in operations that require essentially complete absorption in a liquid that exerts a back pressure opposing diffusion. Although there may be some effect of liquid distribution on the average transfer coefficient, the most important effect is that due to changes in the average driving force, caused by premature saturation of the liquid or gas in some parts of the tower.

An experimental study of liquor distribution within various packings has been carried out by Baker, Chilton, and Vernon [*Trans. Am. Inst. Chem. Engrs.*, **31**, 296 (1935); *cf.* also Kirschbaum, *Z. Ver. deut. Ing.*, **75**, 1212 (1931); Mayo, Hunter, and Nash, *J. Soc. Chem. Ind.*, **54**, 375T (1935); Uchida and Fujita, *J. Soc. Chem. Ind.* (*Japan*) **41**, 563 (1938); also, 275B (English); Tour and Lerman, *Trans. Am. Inst. Chem. Engrs.*, **35**, 709, 719 (1939); **40**, 79 (1944)]. A marked tendency for the liquor to concentrate at the tower walls was observed when the ratio of the tower diameter to the packing size was small. When this ratio was greater than 8 to 1, however, reasonably uniform liquor distribution was obtained when using either irregular packing materials or regularly shaped dumped packings. For example, in a 6-in. column into which water was fed through a single centrally located outlet over 0.5-in. broken-stone packing, the water assumed a uniform distribution at a point 4 ft. from the top, and the distribution did not change appreciably between points 4 and 15 ft. from the top. When the same packing was used in a 3-in. column, however, over 70 per cent of the water was found in the 25 per cent of the cross section adjoining the wall at the same elevation 4 ft. below the distribution point. In order to obtain good distribution in larger towers the number of initial distribution points should be equal to $(D/6)^2$, where D is the column diameter in inches.

The following methods have been used for distributing liquid in packed towers:

1. Perforated distributor plates can be fabricated from steel. A convenient design involves the use of a horizontal steel plate fitted with a number of gas risers made from short lengths of standard pipe and a number of liquid downcomers made from short lengths of smaller standard pipe. The upper end of each liquid tube should be located as nearly on the same horizontal plane as possible so that the weir height over each downcomer will be the same. The tops of the liquid tubes may be notched to minimize differences in liquor flow rate due to variations in the liquid depth across the plate.

Ceramic and stoneware distributor plates of this general type are available commercially for use where steel construction is unsatisfactory.

2. Spray nozzles may be used when the liquid contains no suspended solid particles. Spray-distribution systems should be inspected frequently for clogged nozzles.

3. A series of open troughs and weirs may be used to divide the liquid into a number of streams.

4. The liquid may be divided into a number of jets that are directed vertically downward at horizontal targets that distribute the liquid radially outward. This is a modified spray distributor that avoids trouble due to clogging.

5. The main liquid stream may be divided into a number of smaller streams by allowing it to flow into a central chamber to which a number of smaller tubes are connected. "Spider" distributors of this type are useful for experimental work where end effects must be avoided. For this purpose the legs of the spider can be made to discharge directly onto the packing without an intervening spray section.

Packing Supports. The packing support must have enough open area to avoid flooding the column at this point. The free space should be at least as great as the average free space within the packing. This amounts to 40 to 70 per cent for stacked packings, for which the free space can be calculated.

Several types of packing supports are used, as follows:

1. Steel subway grating combines good strength and large free area.

2. Perforated ceramic plates are sometimes used when the fluids handled are corrosive.

3. The packing may be supported by arches constructed from acid-resisting brick, or on several layers of stacked, large-sized partition rings or grid tile.

4. In small-sized laboratory columns a metal spiral has been used. This is formed from a strip of metal that is wound into an open watch-spring shape.

5. In small columns large free area may be obtained by using a wire screen formed into a cone with its vertex pointing upward into the packing.

GENERAL NOMENCLATURE

D	diffusion coefficient of solute through inert gas, ft.²/hr.
G	gas rate, lb./(hr.)(sq. ft.).
G_M	molar gas rate, lb.-moles/(hr.)(sq. ft.).
H	Henry's-law constant, atm./mole fraction.
H'	Henry's-law constant, lb.-moles/(cu. ft.)(atm.)
H_G	height of gas-film transfer unit, ft.
H_L	height of liquid-film transfer unit, ft.
H_{OG}	height of over-all gas transfer unit, ft.
H_{OL}	height of over-all liquid transfer unit, ft.
K_{OG}	over-all gas absorption coefficient, lb.-moles/(hr.)(sq. ft.) (Δy). [Note the driving force is often (atm.) instead of (Δy).]
K_{OL}	over-all liquid absorption coefficient (lb.-moles)/(hr.)(sq. ft.) (Δx).
L	liquid rate, lb./(hr.)(sq. ft.).
L_M	molar liquid rate, lb.-moles/(hr.)(sq. ft).
N_G	number of gas-film transfer units.
N_L	number of liquid-film transfer units.
N_{OG}	number of over-all gas transfer units.
N_{OL}	number of over-all liquid transfer units.
P	total pressure in the tower, atm.
R	gas constant, 0.729 (atm.)(cu. ft.)/(lb.-moles) °R.
S	cross-sectional area of the tower, sq. ft.
T	absolute temperature, °K.

X moles of solute per mole solvent in liquid phase.
Y moles of solute gas per mole inert gas.
a active interfacial area (of packing) per unit volume, sq. ft./cu. ft.
a' total area of packing per unit of tower volume, sq. ft./cu. ft.
g acceleration due to gravity, (ft.)/(sec.)(sec.).
K phase equilibrium constant, y^*/x.
k_G gas-film coefficient, (lb.-moles)/(hr.)(sq. ft.)(Δy).
k_L liquid-film coefficient, (lb.-moles)/(hr.)(sq. ft.)(Δx).
m slope of equilibrium line on x-y coordinates, dy^*/dx.
p partial pressure of solute gas, atm.
p_i partial pressure of solute gas at the interface, atm.
p^* partial pressure of solute gas which would be in equilibrium with liquid of the composition existing in the main body of the liquid phase, atm.
ps partial pressure of solvent vapor in the main body of the gas, atm.
Δp_{LM} logarithmic mean partial pressure of solute over the gas film, atm.
t temperature, °F.
u velocity of gas in empty tower, ft./sec.
x mole fraction of solute in body of liquid phase.
x^* mole fraction of solute in liquid in equilibrium with partial pressure of solute gas in body of gas phase.
x_i mole fraction of solute at liquid side of interface.
$(1 - x)_{LM}$ log mean of $(1 - x)$ and $(1 - x^*)$.
y mole fraction of solute in body of gas phase.
y_i mole fraction of solute at gas side of interface.
y^* mole fraction of solute in gas in equilibrium with liquid of composition existing in body of liquid phase.
ρ density, lb./cu. ft.
μ viscosity, (lb.)/(ft.)(hr.).
ϕ gas density correction factor, $\sqrt{\rho/0.075}$.

Subscript 1 refers to the "rich" end of the tower.
Subscript 2 refers to the "lean" end of the tower.

SECTION 11
SOLVENT EXTRACTION AND DIALYSIS

BY

Joseph C. Elgin, Ph. D., Professor of Chemical Engineering, Princeton University; Member, American Institute of Chemical Engineers, American Chemical Society, American Institute of Mining and Metallurgical Engineers, Society of Chemica Industry. (Solvent Extraction)

Raymond Wynkoop, Ph. D., Chemical Engineer, Standard Oil Company (Indiana); formerly Research Fellow in Chemical Engineering, Princeton University, Member, American Institute of Chemical Engineers, American Chemical Society. (Solvent Extraction)

James A. Lane, B.S., Chem., M.S., Ph. D., Chemical Engineer, Head of Design Department, Technical Division, Oak Ridge National Laboratory; Member, American Institute of Chemical Engineers. (Dialysis)

CONTENTS

SOLVENT EXTRACTION

PAGE
Introduction ... 714
Definition and Types of Solvent Extraction ... 714
  Liquid-liquid Extraction ... 714
  Leaching ... 715
  Washing ... 715
  Precipitative Extraction ... 715
  Application ... 715
  Typical Applications Tabulated ... 716
Elements of Extraction Operations ... 716
Systems of Operation ... 716
  Single Contact ... 716
  Simple Multistage (Cocurrent) Contact ... 716
  Countercurrent Multistage Contact ... 717
  Continuous Countercurrent Differential Contact ... 717
  Countercurrent Extraction with Reflux ... 717
  Batch (Pseudo) Countercurrent Multistage Extraction ... 718
  Application of Methods in Leaching ... 718
Phase Equilibriums in Condensed Systems ... 718
  Principles ... 718
  Binary Systems ... 718
    Liquid-liquid ... 718
    Liquid-solid ... 718
  Ternary Systems—No Pair Immiscible ... 719
    Graphical Representation ... 719
    Triangular Phase Diagram ... 719
    Phase Relationships on the Triangular Diagram ... 719
    Types of Phase Diagrams for Systems of Liquid Components ... 720
    Effect of Temperature ... 721
    Equilibrium-distribution Diagram ... 722
    Selectivity-selectivity Diagram ... 722
    Solvent Content–concentration Diagram ... 724
    Interpolation of Tie-line Data ... 724
    Triangular Phase Diagrams with Solid Components ... 725
    Data ... 725
    Aqueous and Non-aqueous Systems Tabulated ... 726
    Multicomponent Systems ... 728
  Ternary Systems—One Pair Immiscible ... 728
    Ideal-distribution Law ... 729
    Mathematical Expression of Equilibrium Distribution ... 729
Calculation and Design Methods ... 729
  Multistage Liquid-liquid Extraction Systems ... 729
    General Procedure (Multistage) ... 729
  Simple Multistage (Cocurrent) Extraction ... 729
    Calculation on the Triangular Diagram ... 729
    Calculation on the Solvent Content–concentration Diagram ... 730
  Countercurrent Multistage Extraction without Reflux ... 731
    Calculation on the Solvent Content–concentration Diagram ... 731
    Calculation on the Triangular Diagram ... 732
    Calculation on the Rectangular Equilibrium Distribution Diagram ... 733

PAGE
Countercurrent Liquid liquid Extraction with Reflux ... 733
  Solution on the Solvent Content–concentration Diagram ... 734
    Graphical Interpretation ... 735
    Calculation Procedure ... 736
    Reflux Limits ... 736
    Non-saturated Streams ... 736
    Non-isothermal Operation ... 736
    Illustrative Problem ... 736
  Solution on the Triangular Diagram ... 737
    Extraction Limits ... 739
    Algebraic Calculation of the Minimum Number of Stages ... 739
Special Cases Where Solvents Are Immiscible ... 739
  Ideal Distribution Law Invalid ... 739
    Simple Multistage Contact ... 740
    Countercurrent Multistage Contact ... 740
  Ideal Distribution Law Valid ... 740
Continuous Differential Extraction Systems ... 742
  Mechanism and Rate of Extraction ... 742
  Nomenclature for Continuous Differential Extraction ... 742
  Liquid-film Equations in Extraction ... 742
  Over-all Relationships in Extraction ... 743
  Calculation of Extraction-column Capacity ... 744
  More Useful Procedures Briefly Reviewed ... 744
    Use of Capacity Coefficient ... 744
    Use of Transfer Units ... 745
      Illustrative Example—Approximate Solution ... 745
    The H.E.T.S. ... 746
  Calculation of Leaching Operations ... 746
    Equipment Used in Leaching ... 746
Extraction Equipment ... 747
  General Factors in Design ... 747
  Phase Separation ... 747
  Liquid-liquid Equipment ... 747
  Mixers ... 747
  Wetted-wall Towers ... 748
  Spray and Packed Towers ... 748
    Some Generalizations Concerning Spray Towers ... 748
    Tabulated Rate Data for Packed and Spray Towers ... 748
  Bubble-cap and Sieve- or Perforated-plate Columns ... 752
    Collected Rate Data on Bubble Cap and Sieve or Perforated Plate Columns ... 753
  The H.E.T.S ... 752
    Tabulated Values of H.E.T.S ... 752
  Limiting Flows in Liquid-liquid Columns ... 752

DIALYSIS

Introduction ... 753
Theoretical Background ... 754
Calculation of Dialysis Coefficients ... 754
Applicability of Dialysis ... 755
Dialysis Equipment ... 756

SOLVENT EXTRACTION

REFERENCES. *Books:* Bowden, "The Phase Rule and Phase Reactions," Macmillan, London, 1938. Dunstan, "Science of Petroleum," vol. 3, Secs. 26, 27, 28, 29, Oxford. New York, 1938. Findlay, "Phase Rule," 7th ed., Longmans," New York, 1931. Hunter, "Science of Petroleum," edited by Dunstan, vol. 3, pp. 1779*ff*, Oxford, New York, 1938. Hildebrand, "Solubility of Non-electrolytes," Reinhold, New York, 1936. Jantzen, "Fractional Distillation and Extraction," Dechema Monograph, vol. 5, No. 48, Verlag Chemie, Berlin, 1932. Kalichevsky, "Modern Methods of Refining Lubricating Oils," Reinhold, New York, 1938. Mellan, "Industrial Solvents," Reinhold, New York, 1939. Roozeboom, "Die Heterogene Gleichgewichte," Brunswick, 2 vols., 1900, 1911. Seidell, "Solubilities of Inorganic and Organic Compounds," 3d. ed., vols. 1 and 2, Van Nostrand, New York, 1940, 1941. Sherwood, "Absorption and Extraction," McGraw-Hill, New York, 1937. Sherwood and Reed, "Applied Mathematics in Chemical Engineering," Chap. VIII, McGraw-Hill, New York, 1939. "International Critical Tables," vols. 3, 4, McGraw-Hill, New York, 1928. Technical data for solvents may be found in chemical manufacturers' publications.

Journals: Hunter and Nash, Application of Physical-chemical Principles to the Design of Liquid-liquid Contact Equipment, *J. Soc. Chem. Ind.*, **51,** 285T (1932); **53,** 95T (1934). Hunter and Nash, Contact Equipment for Extraction and Reaction in Two Phase Liquid Systems, *Ind. Chemist,* **9,** 245, 263, 313 (1933). Hunter and Nash, Study of the Principles of Solvent Extraction, Processes for Refining of Oil, *Proc. World Petroleum Congr.*, **2,** 340 (1933). Hunter and Nash, Liquid-liquid Extraction Systems, *Ind. Eng. Chem.*, **27,** 836 (1935). Saal and Van Dyck, *Proc. World Petroleum Congr., London,* **2,** 352, 1933. Evans, Extraction of Immiscible Solvents, *Ind. Eng. Chem.*, **26,** 439 (1934). Evans, Countercurrent and Multiple Extraction, *Ind. Eng. Chem.*, **26,** 860 (1934). Colburn, Simplified Calculation of Diffusional Processes, *Trans. Am. Inst. Chem. Engrs.*, **35,** 211 (1939). Colburn, Simplified Calculation of Diffusional Operations, *Ind. Eng. Chem.*, **33,** 459 (1941). Elgin, Design and Applications of Liquid-liquid Extraction, *Chem. & Met. Eng.*, **49,** 110 (1942). Ruthruff and Wilcock, Solvent Extraction of Vegetable Oils, *Trans. Am. Inst. Chem. Engrs.*, **37,** 649 (1941). Maloney and Schubert, Application of Rectangular Coordinate Methods to Solvent Extraction Design, *Trans. Am. Inst. Chem. Engrs.*, **36,** 741 (1940). Schiebel and Othmer, General Method for Calculating Diffusional Operations Such as Extraction, Distillation and Absorption, *Trans. Am. Inst. Chem. Engrs.*, **38,** 339, 383 (1942). Randall and Longtin, Separation Processes, *Ind. Eng. Chem.*, **30,** 1063, 1188, 1311 (1938). Smith and Funk, Extraction of Aromatics from Hydrocarbon Mixtures, *Trans. Am. Inst. Chem. Engrs.*, **40,** 211 (1944). Drew and Hixson, Solubility of High Molecular Weight Fatty Acids and Their Esters in Propane Near the Critical Temperature, *Trans. Am. Inst. Chem. Engrs.*, **40,** 75 (1944). Johnson and Bliss, Liquid-liquid Extraction in Spray Towers, *Trans. Am. Inst. Chem. Engrs.*, **42,** 331 (1946). Smith, Experimental Solvent Extraction Tower Unit, *Proc. Am. Petroleum Inst.*, **18,** (III) 67, 1937. Apparatus for Extraction of Liquids in Towers Filled with Raschig Rings, *J. Soc. Chem. Ind.*, **37,** 677A (1918). Othmer, Acetic Acid Dehydration, *Trans. Am. Inst. Chem. Engrs.*, **25,** 299 (1933–1934); *Chem. & Met. Eng.*, **41,** 81 (1934). Wilson, Keith, and Haglett, Use of Liquid Propane in Dewaxing, Deasphalting, and Refining Heavy Oils, *Trans. Am. Inst. Chem. Engrs.*, **32,** 364 (1936). Tuttle, Performance and Flexibility of the Duo-Sol Process, *Proc. Am. Petroleum Inst.*, **16,** III, 112 (1935). Sanders, *Chem. & Met. Eng.*, **39,** 161 (1932). Donald, Percolation Leaching in Theory and Practice, *Trans. Inst. Chem. Engrs., London,* **15,** 77 (1937). Poole and Wadsworth, *Oil Gas J.*, **32,** 49, 52 (1933); *Refiner Natural Gasoline Mfr.*, **12,** 412 and 452 (1933) (the application of solvent extraction methods for the improvement of lubricating oil quality). Miller, *Oil Gas J.*, **32,** 11 (1933); *Nat. Petroleum News,* **25,** 26 (1933) (use of two immiscible solvents for the manufacture of lubricating oils by the extraction process). Manley, McCarty, and Gross, *Oil Gas J.*, **32,** 78, 81 (1933); *Refiner Natural Gasoline Mfr.*, **12,** 420 (1933) (use of furfural as the extraction solvent to make motor oils having low viscosity index). Kalichevsky, Simpson, and Story, *Refiner Natural Gasoline Mfr.*, **16,** 250 (1937) (solvent refining of lubricating oils). Ferris and Peterkin, *Refiner Natural Gasoline Mfr.*, **12,** 435 (1933) (the nitrobenzene process, installation, and operating costs). Cottrell, *Oil Gas J.*, **32,** 64 (1933); *Refiner Natural Gasoline Mfr.*, **12,** 432 (1933) (the Edeleanu refining process may be applied to treat practically all petroleum products). Bahlke, Brown, and Diwoky, *Oil Gas J.*, **32,** 62 (1933); *Refiner Natural Gasoline Mfr.*, **12,** 445 (1933). Gloyer, Furans in Vegetable Oil Refining, *Ind. Eng. Chem.*, **40,** 228 (1948); Demmerle, Emersol Process, *Ind. Eng. Chem.*, **39,** 126 (1947); Marsel and Allen, Fatty Acid Processing, *Chem. Eng.*, **54,** No. 6, 104 (1947); Kenyon, Gloyer, and Georgian, Furfural Extraction of Vegetable Oils, *Ind. Eng. Chem.*, **40,** 1162 (1948); Kemp, *et al.*, Furfural as a Selective Solvent in Petroleum Refining, *Ind. Eng. Chem.*, **40,** 220 (1948); Kenyon, Kruse, and Clark, Solvent Extraction of Oil from Soybeans, *Ind. Eng. Chem.*, **40,** 186 (1948); Turehaus and Ehlers, Caustic Purification by Liquid-liquid Extraction, *Chem. Ind.*, **63,** 230 (1948); Skogan and Rogers, *Oil Gas J.*, **46,** 70 (1947).

Introduction. General theoretical principles common to the diffusional operations, of which solvent extraction is one, are presented in Sec. 8. The present section extends and applies those principles to solvent extraction and treats specifically the techniques, calculation methods, and equipment for operations involving only liquid, or liquid and solid phases. Liquid-liquid extraction is given principal attention. The principles and calculation procedures developed with minor modifications may be extended to liquid-solid extractions, and references to such applications are given. In addition, specific features, equipment, and applications of extraction, particularly for liquid-solid systems, are to be found in Sec. 15, pp. 927, 929, and Sec. 17, pp. 1196, 1217.

Definition and Types of Solvent Extraction. Solvent-extraction operations are defined as those in which the separation of mixtures of different substances is accomplished by treatment with a selective liquid solvent. At least one of the components of the mixture must be immiscible or partly miscible (soluble) with the treating solvent so that at least two phases are formed over the entire range of operating conditions used. In order for a separation to be effected, one or more of the components must be dissolved from the mixture by the solvent preferentially to the others. Solvent extraction that employs distribution between liquid and liquid or liquid and solid phases, *i.e.*, condensed phases, is analogous in basic principles and is a companion operation to absorption, distillation, evaporation, and adsorption in which separation is effected by distribution between vapor and liquid or vapor and solid phases. The separation of components difficult or impossible to separate by ordinary distillation may often be improved by vaporization in the presence of added solvents. Such operations, known as "extractive" or "azeotropic" distillation, have their basis in the alteration of the relative volatility of the original components in the presence of a third component or solvent.

Solvent-extraction operations are commonly classified into several types:

Liquid-liquid Extraction. In this case the mixture treated is liquid and the two phases (or layers) resulting from the solvent treatment are both liquids. The solvent-rich phase containing the preferentially dissolved component, or solute, is termed the "extract" layer; the residual phase formed by the undissolved component, or diluent, and usually containing some solvent is called the "raffinate" layer. Either layer may be the top, or light, layer or the bottom, or heavy, layer, depending upon their relative densities. The residual, or solvent-free, materials remaining from extract and raffinate layers after removing the solvent are often called the "extract" and "raffinate," respectively. The solvent may be partly miscible (non-consolute) with only one or with both components of the feed. In special cases, less frequently met in practice, the solvent is essentially immiscible with one of the feed components. The solvent is said to be *selective* for that feed component which is found in greater ratio to the other in the extract than in either the raffinate or the original feed.

Either or both liquid phases may contain dissolved solids. The distributed solute that is being extracted may be solid or liquid. Infrequently, more than two liquid phases may be formed simultaneously, and solid phases may also be involved. The extraction can involve only a physical process of selective solution, or in other cases chemical reactions between the extracted substance and the solvent or a solute present in the solvent may occur. Ordinarily the solvent is a single chemical individual. Advantages are sometimes gained, however, in the resulting phase-equilibrium relationships from the use of mixtures of liquids as solvents. Thus the selectivity of the primary solvent may be improved and the proportions of the two layers altered or partial miscibility obtained by the addition of a secondary solvent, or "antisolvent." A liquid-liquid extraction carried out by causing the components of a feed to distribute between two partly miscible solvents is called *fractional distribution.* In this case the two phases are produced by the solvents, one solvent selectively dissolving one feed component, the second the other.

Leaching. Extraction in which a solid mixture or phase is resolved into its components or a valuable component is removed and recovered from a solid mass by liquid treatment is called *leaching.* Broadly, operations involving the solvent treatment of solids may involve a number of different circumstances. The solid usually consists of a heterogeneous mixture of several constituents, one or more of which may be liquid or solid in solution, but it may be a homogeneous mixture such as a solid solution or a double salt. Solids to be leached are encountered in a variety of physical forms, and disintegration to afford a large contact surface with the solvent is often required. Where both the constituents between which separation is to be effected are soluble in the solvent to an important degree, the operation is properly classified as *fractional leaching.* Some leaching applications, such as the treatment of ores, involve chemical reaction with a constituent of the solvent.

Possibly the most important class of leaching operations is that in which removal of the soluble component from the interior of the inert solid must be effected by a slow process of diffusion through vegetable or animal membrane. Examples are the solvent recovery of oils from seeds, nuts, and similar vegetable materials; extraction of medicinals; tannins, turpentine, and rosin from wood by the solvent process; chlorophyl from alfalfa; and the extraction of sugar from the beet by water leaching.

Washing. The removal of soluble substances and impurities mechanically adhering to the surface of insoluble solids by solvent treatment is termed *washing.* In some cases the soluble substance is the valuable material; in others it is an undesired impurity. Washing usually involves the extraction of a solid precipitate with a portion of the same solvent from which the precipitation occurred. Solvent either pure or containing a smaller concentration of solute than is present in the solution adhering to the surface can be used. Ordinarily but not always water is the solvent concerned.

Precipitative Extraction. In many instances a liquid solution can be caused to split into two phases, liquid-liquid, or liquid-solid, by the addition of a third substance, which may be a liquid solvent or a solid and either wholly or partly miscible with the liquid components of the original mixture. Usually the distribution of the original feed components between these two phases is such as to effect their separation. If the conditions maintained are such that one of the phases separating is solid, the process might be called "solvent crystallization" depending upon whether the third substance or solvent is added before or after chilling. In separations of this type the two phases that result ordinarily, but not always, contain, respectively, the components of the original mixture in larger proportion, while the added substance is distributed between the two in a smaller proportion in each.

In the so-called "salting-out" process an organic substance is caused to separate from water solution by the addition of an inorganic substance such as sodium chloride. In other applications such as the propane deasphalting and the solvent dewaxing of lube oils and the separation of stearic from oleic acid in red oil, a greater separation of the solid component at a desired temperature is effected by the addition of a solvent such as liquid propane, ketones, methanol, etc., in which the solid is less soluble than in the original liquid. Often the addition of the solvent results in the added advantage of an easier and more complete mechanical separation of the two phases.

The possibility of improving the separation of two compounds forming a eutectic mixture on crystallization by the addition of a solvent, "extractive crystallization," merits investigation.

Application. As a chemical-engineering method, solvent extraction finds application in separating the components of condensed mixtures where vaporization methods (*e.g.*, distillation and evaporation) are impractical because the substances to be separated have comparable volatilities, they are relatively non-volatile, they are heat-sensitive, or one component is present in very small concentration. In general, solvent extraction is applied to the separation of molecules of different chemical types, whereas distillation finds use in separating molecules of different size or molecular weight. Solvent extraction is a valuable method for the removal of trace amounts of impurities and colored bodies, for the recovery and purification of a wide variety of organic compounds and pharmaceuticals, and for the treatment of liquors containing large quantities of gummy solids which would make difficult the heat transfer required by vaporization processes. If a proper solvent is available, solvent extraction can offer advantages and economies as a preliminary to distillation or evaporation even where the latter are feasible and readily performed. For example, where a valuable component is present in small amounts and is of lower volatility than the liquid in which it is dissolved, its extraction into a solvent may produce a more concentrated solution from which its recovery requires less vaporization. Further, the separation factors for distribution between two liquid phases are often considerably larger than between liquid and vapor. Since organic liquids usually have lower heats of vaporization than does water, with aqueous solutions extraction of the valuable component into an organic solvent and recovery from the latter by distillation or evaporation may consume less heat than direct vaporization of the aqueous solution.

In addition to examples already cited, some typical applications of solvent extraction are given in the following Table 1.

Table 1. Typical Applications of Solvent Extraction

Application	Typical Solvents
Production of improved lubricating oils—separation of paraffinic from naphthenic-aromatic components	Furfural, phenol, nitrobenzene, liquid propane, chlorex
Treatment of petroleum naphthas for purification and sulfur removal, especially RSH	Caustic soda and added organic solvents; liquid SO_2, copper solutions
Treatment of petroleum naphthas and diesel fuels to separate high and low antiknock fractions and cylics and paraffinics.	Liquid SO_2, furfural, aniline, and certain other amines, 97% and stronger acetic acid
Treatment of vegetable oils, *e.g.*, soybean, to separate drying (unsaturated) from non-drying (saturated) fractions	Furfural, nitroethane
Treatment of glyceride esters and fatty acid mixtures to separate components	Liquid propane, furfural, methanol
Treatment of tall oil to recover oleic and abietic acids	Liquid propane, furfural, nitroethane
Purification and improvement of vegetable and animal oils	Liquid propane
Concentration and recovery of dilute acetic acid	Isopropyl ether, ethyl acetate
Dewaxing of lubricating oils	Ketones, liquid propane
Recovery and purification of butadiene from other C_4 hydrocarbons	Copper ammonium acetate solutions, glycol plus methanol
Removal of polymerization inhibitors from butadiene and styrene	Aqueous caustic soda
Recovery and concentration of penicillin	Amyl acetate, cyclohexanone, chloroform
Purification of electrolytic caustic soda	Liquid ammonia
Purification and decolorization of crude wood rosin	Furfural-naphtha
Purification and decolorization of crude concentrated aqueous glycerin	Xylene
Recovery and concentration of vitamins, recovery of oils from vegetable beans and seeds	Hydrocarbons, chlorinated hydrocarbons

Many other industrial applications have been made, especially in the recovery, concentration, and purification of organic compounds in water solution, and numerous others have been proposed.

Elements of Extraction Operations. An extraction process consists of three operations, as follows: (1) mixing and bringing the material to be extracted into intimate contact with the solvent, (2) separation of the resulting phases or layers, (3) removal and recovery of the solvent from each of the phases for reuse. The mixing and settling operations constitute a unit or *extraction stage*. The system may consist of a single stage or a series of stages, the latter being the most frequent in industrial practice. A stage in which equilibrium between the two phases is attained is an *ideal* or *theoretical* extraction stage. The nearness with which equilibrium is approached is used as a measure of the "efficiency" of the stage.

The contacting operation and the degree of extraction and of separation obtained are the essential factors of an extraction process. In some cases the subsequent separation of the phases from each other and the removal and recovery of the solvent may be equally or more important and may determine the successful application of an extraction process. For example, a tendency toward emulsification of the two phases may introduce insurmountable difficulties in their separation after contacting. Approximation of the two phases to the same density offers similar difficulties.

In liquid-liquid extraction the removal and recovery of the solvent may be accomplished by distillation (see Sec. 9); evaporation (see Sec. 7); heating or cooling in order to diminish the solubility of the extract in the solvent or vice versa, and thus produce a separation into two phases; the addition of a third substance, *i.e.*, precipitative extraction or salting out; or by extraction of the primary solvent with a suitable secondary solvent and subsequent separation of primary and secondary solvents by distillation. Distillation is usually the most feasible and most frequently used method. The extraction of the primary solvent first into a secondary solvent sometimes offers advantages since the latter may frequently be selected so that its separation from the primary solvent by distillation is relatively much easier and less costly than that of the primary solvent from the extract.

Recovery of the solvent from both extract and raffinate layers is almost always necessary. The method employed is governed by the physical properties of the solvent and of the extract and raffinate. Whether the former or the latter is volatilized depends upon their relative volatilities from the two phases. This is an important factor and frequently controls the economics of an extraction process. Although a particular solvent may give excellent selectivity and separation between the two components of the feed, its application may be precluded in practice because of difficulty in separating it from the extract and raffinate, unless no other method of separating the feed is possible. In determining this, especially if the solvent is more volatile, the concentration of the extract in the extract or solvent layer as well as the heat of vaporization of the solvent is equally as important as its selectivity.

In *leaching operations*, where water is the most common solvent, further recovery of the solvent than that obtained in the phase-separating operation is usually not required, especially since the leached solids are often waste materials. When necessary, evaporation and drying are available methods. Liquors from leaching operations are often sufficiently concentrated, and either constitute the desired product or are further treated for the removal of the solvent in latter stages of the plant operation, separate from the extraction plant itself.

SYSTEMS OF OPERATION

Any of the operating methods by which interphase mass-transfer processes can be conducted may theoretically be used in the conduct of solvent extraction. In industrial practice continuous countercurrent multistage contact in a series of mixers and settlers or a tray-type tower, or countercurrent differential contact in a continuous tower of the packed, spray, or similar type are most commonly employed. In laboratory practice cocurrent batch operation by single or simple multistage contact is most common.

Single Contact. The simplest method and that most common to laboratory scale is to bring the entire quantities of solvent and feed to be extracted together in one contact and then recover the product and solvent without further extraction (see Fig. 1*a*). This method is the least effective and is rarely feasible on an industrial scale. As practiced, equilibrium is usually closely approached; hence the amount of solute extracted is fixed solely by the equilibrium relations and the quantity of solvent used. Recovery of extract is small and is limited by the amount of solvent employed. The concentration of the extract layer is low, and the degree of separation between components of the feed is poor. The operations may be either batch or continuous. This method can be regarded as analogous to simple continuous or to flash distillation.

Simple Multistage (Cocurrent) Contact. Figure 1*b* illustrates schematically this procedure (sometimes called "cocurrent multistage contact"). The total quantity of solvent used is divided into several portions. The feed undergoing extraction is then treated with each of these portions of fresh solvent in a series of successive steps or stages. That is, the raffinate from the first extraction step is treated with fresh solvent in a second stage, and so on. As the number of stages and the quantity of solvent are increased, the percentage recovery of the extracted component is increased.

If a sufficient number of stages and a sufficient quantity of solvent are used, the raffinate can be stripped of the extracted component to a high degree, but with any specified amount of solvent a finite removal of the extract is approached as the number of stages approaches infinity. The results are varied by the proportional division of the total quantity of solvent between the stages. If the ideal-distribution law holds, these are best with equal quantities of solvent in each stage. Unless the solvent in the feed is largely immiscible with the extracting solvent, separation between the components of the feed tends, however, to be poor. An excessively large quantity of solvent is required to obtain a high degree of extraction by this method, and the concentration of extract in the extract layer is low. The concentration of the extract layer becomes increasingly dilute in succeeding stages, and the combined layers are very dilute.

The method may be operated intermittently with a single mixer and settler unit or continuously with a series of such units.

The well-known Soxhlet extraction method corresponds to simple multistage contact with an infinite number of stages.

Countercurrent Multistage Contact. Figure 1*c* illustrates schematically the flow in this method. All the fresh solvent and the feed are sent to opposite terminals of a series of extraction stages. Extract and raffinate layers pass continuously and countercurrently from stage to stage through the system. The solvent feed to the nth stage is the extract layer from the $(n - 1)$th stage, and the raffinate layer leaving the nth stage becomes the feed to be extracted in the $(n - 1)$th stage. The finished extract and raffinate layers are withdrawn continuously from opposite ends of the system. Any number of stages may be employed, the more usual number being from three to six. The system may be composed of a series of mixers, each with its separate settler, or some form of continuous tray column may be used. The method corresponds to the method of operation in the plate distillation or absorption column, and, for a given amount of solvent and a fixed number of stages, its effectiveness exceeds that of simple multistage contact.

Continuous Countercurrent Differential Contact. Figure 1*d* illustrates this method. If one of the phases is subdivided and allowed to pass continuously and countercurrently through the other, which is not dispersed, continuous countercurrent differential operation is obtained. Theoretically, either of the phases may be subdivided or made discontinuous. Either the solvent or the liquid to be extracted may be subdivided and allowed to pass through the other. This may be conducted either with a packed column or by spraying one liquid into one end of an unpacked tower (spray tower) and allowing it to pass by gravity through a column of the other. The lighter liquid, if subdivided, enters at the bottom and passes upward due to difference in density or, if heavier, enters at the top and falls downward through the other.

In leaching, where a solid phase is being extracted, that phase is obviously, of necessity, the one that must be subdivided.

Theoretically, if equilibrium is approached, this operating system gives the maximum "efficiency." The actual results obtained are governed, however, by the rate of extraction and excellence of contact which can be obtained in the available equipment of the true continuous countercurrent type. Not only the relative size of equipment units but also their first costs and operating costs must be considered in judging the effectiveness of this method and the performance of continuous countercurrent equipment.

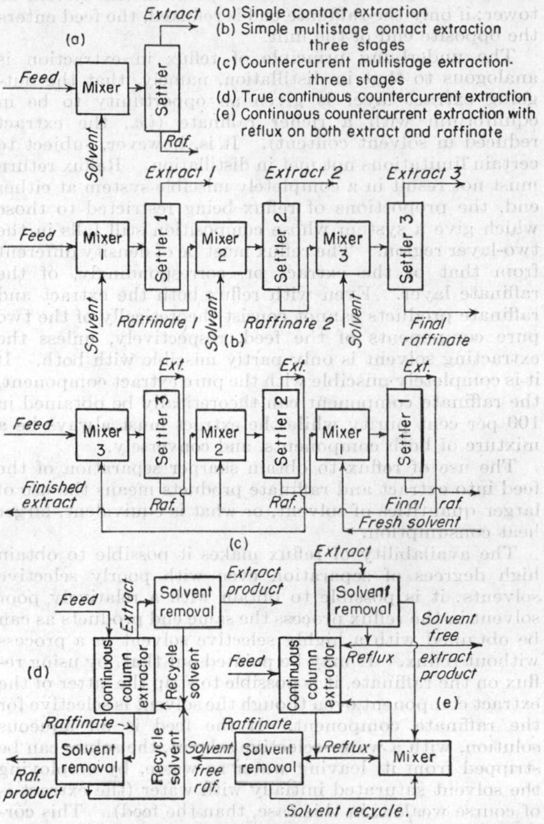

FIG. 1. Schematic flow diagrams illustrating various methods of operating extraction processes.

Countercurrent Extraction with Reflux. The flow scheme is illustrated in Fig. 1*e*. Unless one component of the feed is immiscible with the extracting solvent, neither true nor multistage continuous countercurrent extraction can completely, in general, separate the components of the feed in the extract and raffinate products. A theoretical limiting value of the extract product composition is attained which with infinite stages or tower height tends to approach equilibrium with the incoming feed.*

Reflux may be supplied to either one or both the extract and raffinate ends of the extraction system in order to increase the degree of separation between the components of the feed. Extract reflux is supplied by returning a portion of the extract layer from which the solvent has been wholly or partly removed (usually to just the point where the extract is saturated with the solvent). Similarly, for raffinate reflux a part of the raffinate layer is mixed with the incoming solvent (usually just to the point of saturation with the solvent). In general, with weaker feeds, extract reflux raises the con-

* If types of phase equilibrium for specific cases are examined, it will be seen that the highest concentration of the desired component in the extract does not always correspond to its highest concentration in the raffinate layer; in certain cases enhanced concentration in one layer accompanies diminishing concentration in the other.

centration of the extract component in the extract layer above that corresponding to equilibrium with the feed while raffinate reflux improves its degree of stripping from the raffinate. If the extract or both streams are refluxed, the feed enters an intermediate point in the system or tower; if only the raffinate end is refluxed, the feed enters the opposite end of the unit.

The underlying principle of reflux in extraction is analogous to that in distillation, namely, that the outgoing extract layer is given an opportunity to be in equilibrium with a richer raffinate (*i.e.*, the extract reduced in solvent content). It is, however, subject to certain limitations not met in distillation. Reflux return must not result in a completely miscible system at either end, the proportions of reflux being restricted to those which give a system whose composition still falls in the two-layer region. The reflux must be of density different from that of the extract or, correspondingly, of the raffinate layer. Even with reflux both the extract and raffinate products cannot consist theoretically of the two pure components of the feed, respectively, unless the extracting solvent is only partly miscible with both. If it is completely miscible with the pure extract component, the raffinate component can theoretically be obtained in 100 per cent purity while the extract must always be a mixture of both components, and conversely.

The use of reflux to obtain sharper separation of the feed into extract and raffinate products means the use of larger quantities of solvent, or what is equivalent, larger heat consumption.

The availability of reflux makes it possible to obtain high degrees of separation even with poorly selective solvents; it is possible to obtain with a relatively poor solvent in the reflux process the same end products as can be obtained with a highly selective solvent in a process without reflux. It may be pointed out that, by using reflux on the raffinate, it is possible to strip the latter of the extract component even though the solvent is selective for the raffinate component. If the feed is an aqueous solution, with a water-selective solvent, the solute can be stripped from it, leaving water as waste, by employing the solvent saturated initially with water (the extract is of course weaker, in this case, than the feed). This corresponds to the use of raffinate reflux. If the raffinate is water which is waste, fresh water may evidently replace a portion of the raffinate as reflux. Possibilities in the use of reflux must be determined in relation to the specific extraction system and its phase equilibriums.

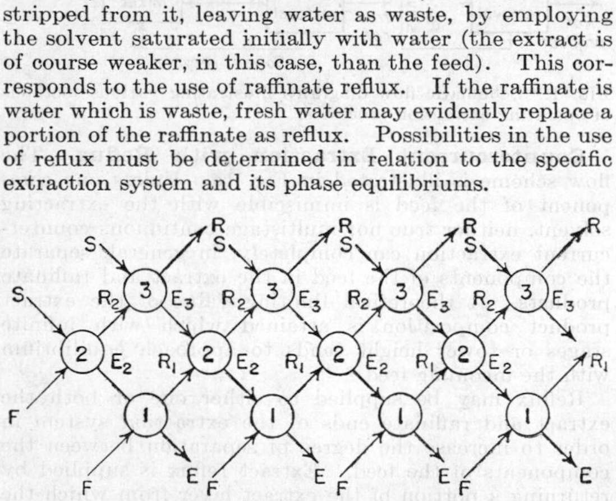

FIG. 2. Schematic flow diagram for a three-stage batch countercurrent multistage laboratory extraction.

Batch (Pseudo) Countercurrent Multistage Extraction. In order to simulate in the laboratory continuous multistage extraction, an operating scheme such as that illustrated in Fig. 2 can be used.

Application of Methods in Leaching. Although any of the preceding methods theoretically might be employed in leaching, restrictions are imposed because of difficulties in handling and contacting solids and liquids. The principle of countercurrent multistage contact is, however, widely used or closely simulated in leaching, *e.g.*, continuous countercurrent decanter systems, such as the Dorr, in washing operations.

PHASE EQUILIBRIUMS IN CONDENSED SYSTEMS

Quantitative phase-equilibrium (tie-line) and solubility data are essential for the rational engineering consideration and design of a solvent-extraction process. The thermodynamic principles involved, treatment, and prediction of such data are presented in Sec. 8. Generalizations, at least qualitatively useful in predictions, are also given by Hildebrand ("Solubility of Nonelectrolytes," Reinhold, New York, 1936); Francis [*Ind. Eng. Chem.*, **36,** 764, 1096 (1944)]; Ewell, Harrison, and Berg [*Ind. Eng. Chem.*, **36,** 871 (1944)]; Treybal [*Ind Eng. Chem.*, **36,** 875 (1944)]; Othmer and Tobias [*Ind. Eng. Chem.*, **34,** 693 (1942)]; and Carlson and Colburn [*Ind. Eng. Chem.*, **34,** 581 (1942)]. Solubility and phase-equilibrium behavior in multicomponent extraction systems are complex, and in general their experimental measurement for the particular solvent and system under consideration is necessary. The generalizations referred to serve to reduce considerably the experimental work and data required to define a given system accurately.

Binary Systems

Data for the mutual solubilities and the critical solution temperatures of the binary pairs between the feed components and the solvent are essential to the construction of the equilibrium relationships in the ternary system. They are also important in the location of potentially useful solvents.

Liquid-liquid. Mutual solubilities of liquids, the effect of temperature, and existent types of temperature-composition phase diagrams are treated in Sec. 4, pp. 315. The majority of the available binary solubility data will be found in the "International Critical Tables," in Seidell ("Solubilities of Inorganic and Organic Compounds," 3d ed., Vol. I, 1940, Vol. II, 1941, Van Nostrand, New York), and in the references of this section. Francis (*loc. cit.*) summarizes and gives data for the critical solution temperatures of typical hydrocarbons of the several series with a large number of organic compounds. Other solubility data for organic solvents with water and with other more common organic liquids not appearing in these sources are available in technical bulletins of solvent manufacturers. Reliable quantitative miscibility data are relatively few, and even qualitative data are still lacking for many potentially important cases.

Few reliable generalizations are possible. Ewell, Harrison, and Berg's (*loc. cit.*) qualitative classification of types will be found useful. A large fraction of the available measurements deal with organic liquids and water at room temperatures. Temperature materially alters the mutual solubility and may either increase or diminish it. Impurities in relatively small amounts radically alter miscibility, which may be increased or diminished. Inorganic salts usually diminish mutual solubilities, while many alkali metal fatty acid soaps and other high-molecular-weight organic salts appreciably increase it.

Liquid-solid. Solid-in-liquid solubility is treated in Sec. 4, pp. 319*ff*, and in Hildebrand. Data for the solubilities of inorganic substances are given in Sec. 3, Table 140. For other systems, see the "International Critical Tables" and Seidell.

Ternary Systems—No Pair Immiscible

Graphical Representation. For engineering purposes solubility and equilibrium data for ternary systems all of whose components are to some extent at least partly mutually soluble are most usefully represented graphically on plane diagrams at constant temperature and pressure. The equilateral triangular plot provides the most complete representation. Other plots on rectangular coordinates shown below and derivable from it are often more useful for design calculations and the comparison of potential extraction solvents. The influence of pressure can usually be neglected. The effect of temperature is best shown by a series of constant-temperature curves (isotherms) on the plane diagram. Representing temperature by a fourth coordinate giving a solid prism is inconvenient for practical purposes.

Triangular Phase Diagram. In this plot (Fig. 3) the vertexes represent the pure components, and the sides of the equilateral triangle are scaled to represent binary compositions of the three possible pairs, respectively. Ternary compositions are located within the triangle. Triangular coordinate paper (see p. 99) is available in several sizes, in some cases with a printed scale. Either weight or mole percentage can be employed. The former is ordinarily preferred for engineering calculations.

For example, in Fig. 3, the three pure components A, B, C lie at the vertexes, respectively. Point R on side AC

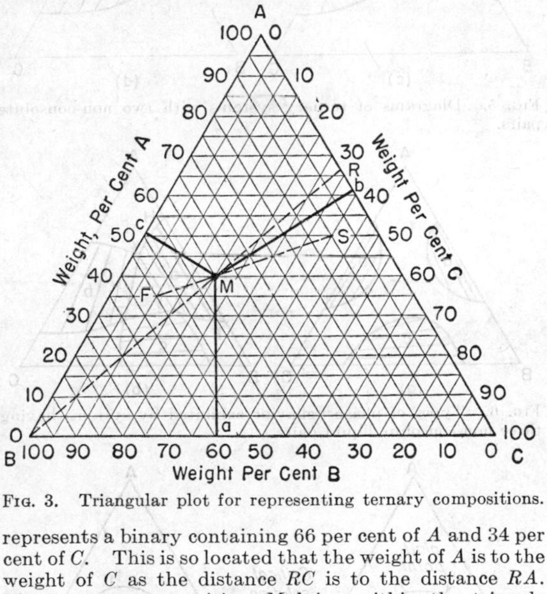

FIG. 3. Triangular plot for representing ternary compositions.

represents a binary containing 66 per cent of A and 34 per cent of C. This is so located that the weight of A is to the weight of C as the distance RC is to the distance RA. The ternary composition M lying within the triangle contains 40 per cent of A, 40 per cent of B, and 20 per cent of C. It is so located that the weight of A is to the weight of B is to the weight of C as the ratio of the perpendicular distances Ma, Mb, Mc from the point to the sides. The altitude of the triangle usually represents 100 per cent or unity. The sides are correspondingly divided into a number of equal parts to make up 100 per cent of the binary mixtures. The sum of the three perpendiculars to the sides, from any point within the triangle, equals the altitude. Percentages may be read off directly on a suitably drawn and scaled chart without needing to measure these lengths.

Other important geometrical properties of such diagrams are the following. If the third component B is added to the binary mixture R of A and C, all possible ternaries formed lie on the line BR, *e.g.*, mixture M, joining the vertex for pure B with the mixture composition R. Furthermore, the weight of B added to R to form M is to the weight of the binary mixture R as the length MR is to MB. Conversely, removal of B from M without removing any of A or C leaves as residue the mixture R. In general the over-all composition of any new mixture, *e.g.*, M, formed by mixing any two ternary mixtures, *e.g.*, F and S, must lie on the line joining the compositions of the two ternaries mixed, *i.e.*, line FS. The ratio of the distance MS to that of MF equals the ratio of the weights of the mixtures F and S which have been combined to form M. This is sometimes referred to as the "lever rule."

Phase Relationships on the Triangular Diagram. Consider a system composed of three components A, B, and C at some constant temperature. Assume that A is miscible with B and with C in all proportions whereas B and C are partly miscible. Refer to Fig. 4. The distance

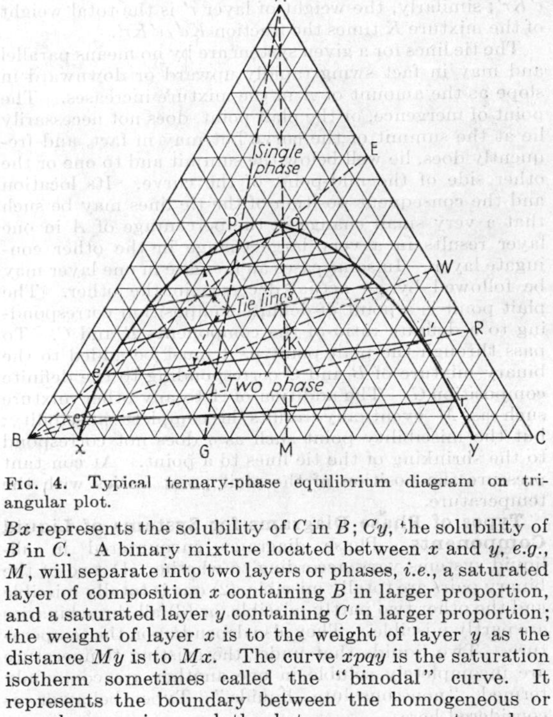

FIG. 4. Typical ternary-phase equilibrium diagram on triangular plot.

Bx represents the solubility of C in B; Cy, the solubility of B in C. A binary mixture located between x and y, *e.g.*, M, will separate into two layers or phases, *i.e.*, a saturated layer of composition x containing B in larger proportion, and a saturated layer y containing C in larger proportion; the weight of layer x is to the weight of layer y as the distance My is to Mx. The curve $xpqy$ is the saturation isotherm sometimes called the "binodal" curve. It represents the boundary between the homogeneous or one-phase region and the heterogeneous or two-phase region. The addition of A in any proportion to mixtures of B and C containing less than about 7 per cent C or more than about 89 per cent C gives a system of only one layer or phase. If A is added to any mixture of B and C lying between x and y, *e.g.*, M, in such a proportion as to form a mixture whose composition falls within the curve, *e.g.*, K or L, the resulting mixture separates at equilibrium into two layers or phases. The compositions of these two ternary layers which are in equilibrium, *conjugate layers*, lie on opposite sides of the saturation curve $xpqy$, *e.g.*, points e and r, and e' and r', respectively. Lines connecting the compositions of these two equilibrium or conjugate layers on the saturation curve are called "tie" lines. The component A may be regarded as distributing itself between the two conjugate layers.

As the amount of A in the mixture increases, the tie lines become shorter. The compositions of the conjugate layers approach each other and finally become identical. They merge at a definite point p called the *critical* or *plait point*. Above this point and outside the curve only one layer can exist.

An infinite number of tie lines exist, but for practical purposes it is necessary to know only a limited number spaced at conveniently frequent intervals over the desired range of working compositions. Others may then be interpolated. The compositions of the two equilibrium layers into which a mixture of any desired over-all composition will split are found by first locating the point representing the over-all mixture and then drawing through this a properly located tie line. Its extremities on the saturation curve give the desired compositions of the equilibrium layers. The amounts of these layers are read from the graph by measuring the length of the tie line and the segments into which it is divided by the point representing the over-all composition. Thus (Fig. 4), for the mixture K, the weight of the layer e' is found by multiplying the total weight of the mixture K by the ratio of the distance Kr' to the length of the tie line, $e'Kr'$; similarly, the weight of layer r' is the total weight of the mixture K times the fraction $Ke'/e'Kr'$.

The tie lines for a given system are by no means parallel and may in fact swing rapidly upward or downward in slope as the amount of A in the mixture increases. The point of mergence, or the plait point, does not necessarily lie at the summit of the curve but may in fact, and frequently does, lie well below the summit and to one or the other side of the mid-point of the curve. Its location and the consequent position of the tie lines may be such that a very small change in the percentage of A in one layer results in a very large change in the other conjugate layer. In some cases an increase in one layer may be followed by an actual decrease in the other. The plait point is a point of definite composition corresponding to a definite ratio of the components B and C. To pass through the plait point p, A must be added to the binary mixture of B and C corresponding to the definite composition G. The addition of A to any other mixture such as M eventually results in complete miscibility; but the miscibility point such as q does not correspond to the shrinking of the tie lines to a point. At constant pressure the position of the plait point varies with the temperature.

Types of Phase Diagrams for Systems of Liquid Components. Phase diagrams encountered in all-liquid systems vary according to whether (1) two of the binary pairs are totally miscible, (2) one is totally miscible and the other two partly miscible, or (3) all three binaries are partly miscible. They also depend upon the temperature. Two liquids that under the existing temperature are incompletely soluble in one another are commonly termed "non-consolute liquids." Type behavior is considered here.

1. *One Pair of Partly Miscible Liquids.* Figure 4 is typical of this case. By far the majority of commoner cases for which data are available fall into this class. The saturation curve and two-phase region vary from case to case in shape, position, and area. The position of the plait point and the slope of the tie lines vary widely. The systems chloroform–water–acetic acid; toluene–acetic acid–water; isopropyl ether–acetic acid–water; benzene–ethyl alcohol–water; and acetone–water–ethylene dichloride are examples of this class.

2. *Two Binary Pairs Are Partly Miscible.* Two hererogeneous regions surrounded by a homogeneous field may exist as in Fig. 5*a*; or the two-phase region may be a continuous band as in Fig. 5*b*. The system water–ethyl alcohol–ethyl cyanide, exhibits the first type of diagram. The system water–phenol–aniline at 50°C. shows the latter behavior. Other variations encountered are Figs. 5*c* and 5*d*.

3. *The Three Binary Pairs Are Partly Miscible.* Typical diagrams are Fig. 6*a*, in which three separate heterogeneous regions exist, and Fig. 6*b*, in which these regions overlap and a region of composition in which three liquid layers exist in equilibrium is formed. In the region bounded by xyz, any mixture separates into three layers

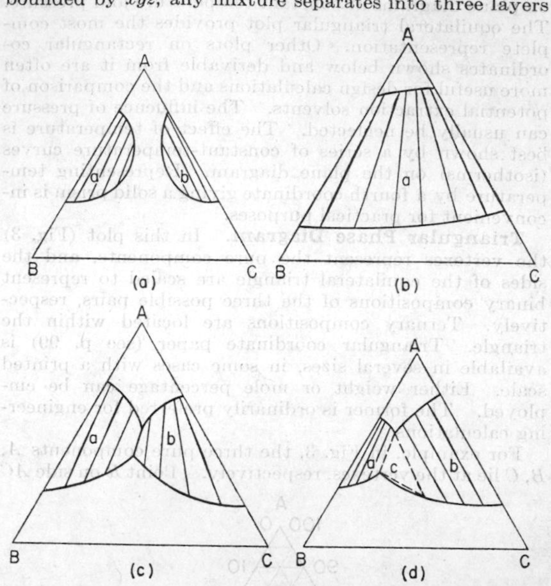

FIG. 5. Diagrams of ternary systems with two non-consolute pairs.

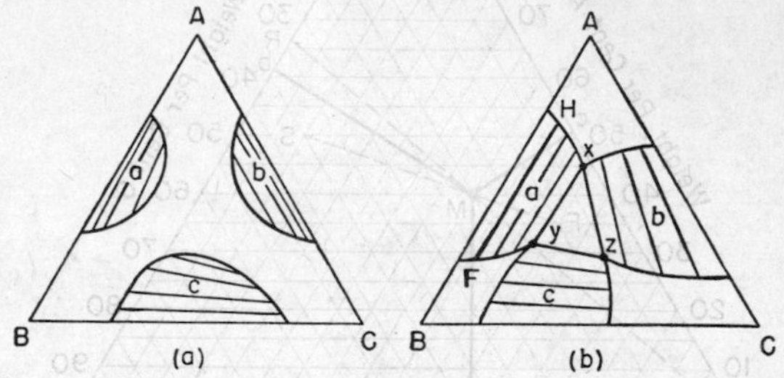

FIG. 6. Types of phase behavior exhibited by systems having three non-consolute liquid pairs.

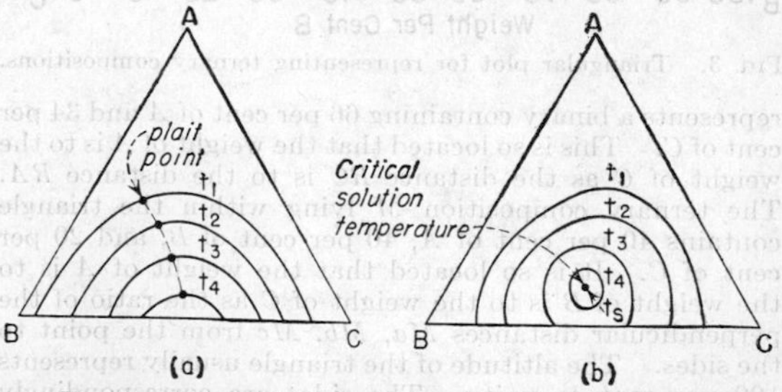

FIG. 7. Types of behavior with varying temperature.

whose compositions, respectively, regardless of the over-all composition of the original mixture, are given by the points x, y, and z. A mixture falling in the two-phase regions a, b, or c, immediately splits into two liquid layers whose compositions are given by the appropriate tie line, *e.g.*, HF.

Figures 8, 9, and 10 show triangular phase diagrams for typical ternary systems of the types commonly met in liquid-liquid extraction practice and illustrate the effect of temperature.

Effect of Temperature. Temperature has several possible effects on the behavior and type of diagram exhibited by a given system. A particular system theoretically may show several or all of these various types of behavior over a wide temperature range. It may be characterized by one type of diagram in one temperature range and a second type in another. Thus a

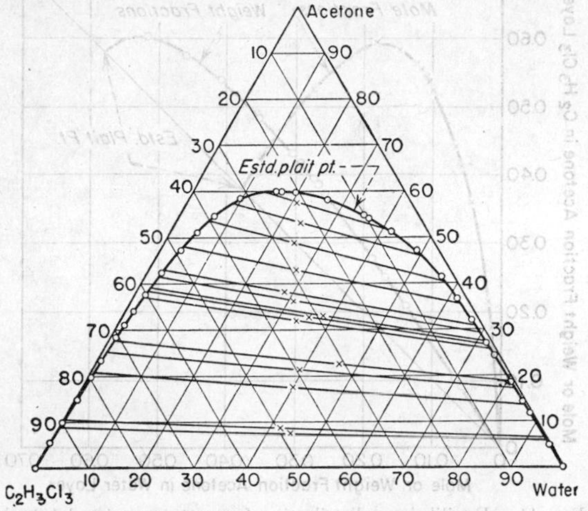

FIG. 8*a*. Liquid equilibriums of system acetone–water–1,1,2-trichloroethane at 25°C. [*Treybal, Weber, and Doley, Ind. Eng. Chem.*, **38**, 819 (1946).]

system having a diagram such as Fig. 4 at a given temperature may pass to a diagram such as Fig. 5*a* as temperature is lowered and subsequently to that of Fig. 6*a* on further temperature reduction. The behavior of Fig. 4 may exist at one temperature and may alter with temperature to a diagram such as Fig. 6*b* without passing through the type represented by Fig. 6*a*. A system with a 6*a* diagram may with changing temperatures pass to that of Fig. 5*b*.

The direction of the change in phase behavior produced in one system by lowering the temperature may, for

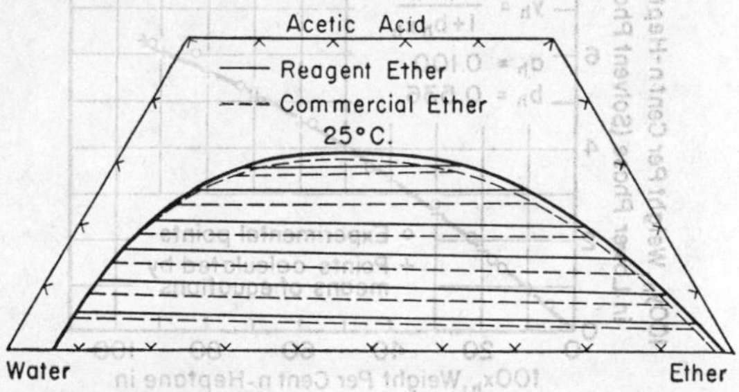

FIG. 8*b*. Mutual solubility curves and tie lines for acetic acid–ethyl ether–water system at 25°C. [*Major and Swenson, Ind. Eng. Chem.*, **38**, 834 (1936).]

a second system, result from an elevation of the temperature. Generally the two-phase regions of a diagram such as Fig. 5*a* may be made to expand by raising or lowering the temperature. If the juncture of the two-phase regions occurs so that the plait points meet, a continuous band results. If the juncture is effected in such manner that the plait points do not come in contact, a three-phase region will be shown over some temperature interval.

Systems that have one pair of partly miscible liquids such as Fig. 4, and that do not form a second pair as the temperature is altered before complete miscibility is reached, may show either one of two general types of behavior. No real ternary critical solution temperature may exist. Here as the temperature is raised (or lowered as the case may be), the two-phase region gradually diminishes in size and disappears as the two non-consolute liquids become completely miscible. Then lowering the temperature enlarges the heterogeneous region. This is

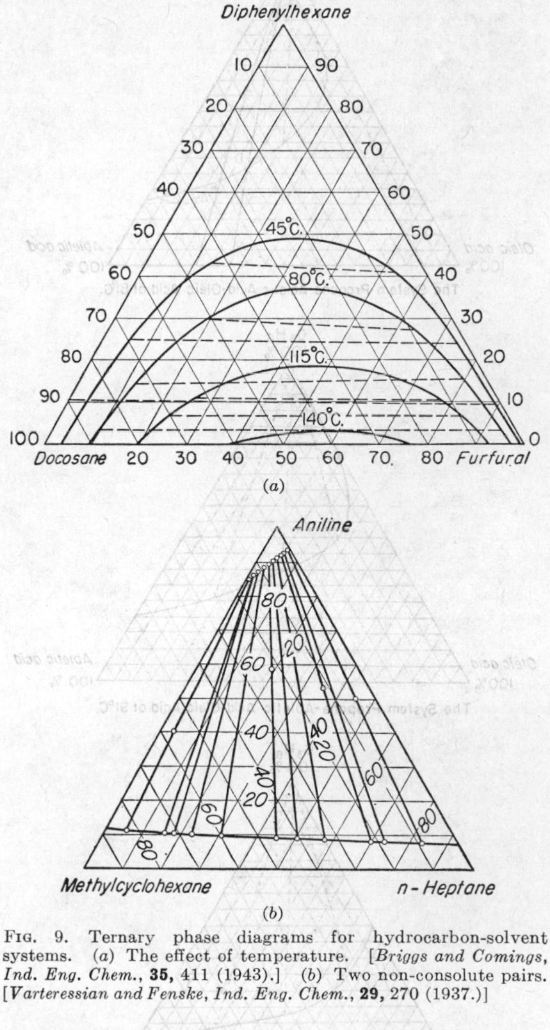

FIG. 9. Ternary phase diagrams for hydrocarbon-solvent systems. (*a*) The effect of temperature. [*Briggs and Comings, Ind. Eng. Chem.*, **35**, 411 (1943).] (*b*) Two non-consolute pairs. [*Varteressian and Fenske, Ind. Eng. Chem.*, **29**, 270 (1937).]

illustrated in Fig. 7*a*. Or a real ternary critical solution temperature shown by Fig. 7*b* may exist. In this case when the non-consolute pair becomes completely miscible there remains a heterogeneous region in which ternary mixtures of all three still split into two layers. This eventually disappears as the temperature is further changed.

Many such types of behavior with temperature are known, and in general it is not possible to predict the effect of temperature except by experimental determinations.

Temperature can have an important effect upon the conduct of extractions. In considering the effect of temperature, it is possible that the position of the tie lines also may be altered so as to affect the separation obtained.

Equilibrium-distribution Diagram. This is a simple plot on rectangular coordinates of the equilibrium percentages of the solute in the solvent-rich extract phase against the percentages of solute in the diluent raffinate phase both expressed on a solvent-containing

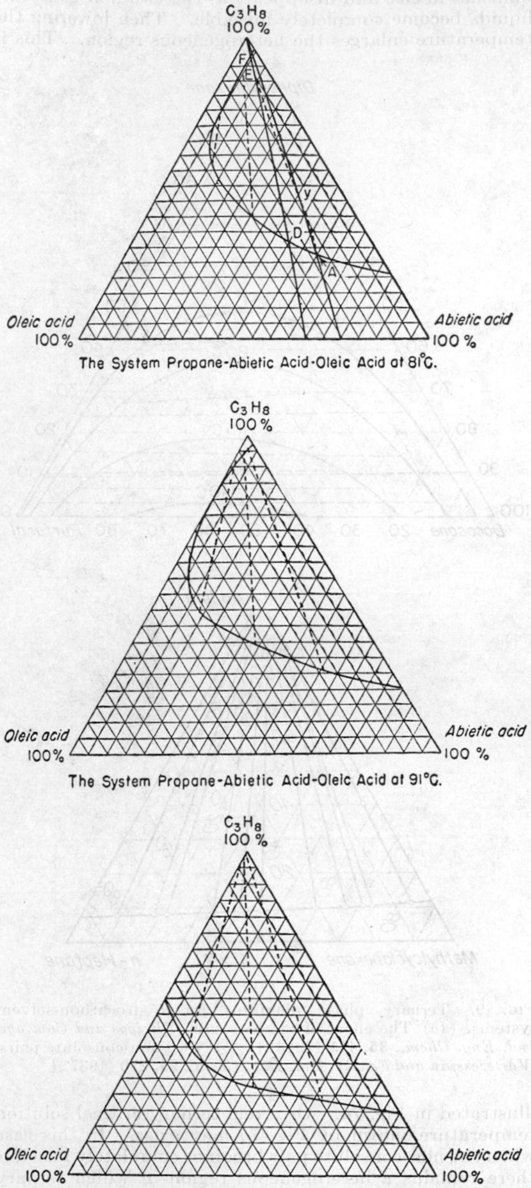

FIG. 10. Ternary phase diagrams for a fatty acid–liquid propane system showing effect of temperature. [*Hixson and Hixson, Trans. Am. Inst. Chem. Engrs.*, **37**, 941 (1941).]

basis. Either weight or mole percentages can be used, the former usually being preferred in engineering calculations. The percentages may be found directly from the extremities of the tie lines of the triangular diagram. The distribution curve and tie lines may be obtained directly one from the other. This diagram is useful in showing completely at all concentrations the distribution of solute between extract and raffinate phases and for the direct comparison of potential extracting solvents. It may be made the basis for calculating an extraction process, pp. 733 and 740. When the solvent is immiscible with the diluent it is the only diagram needed and is then analogous to the common *p-c* or *y-x* diagram

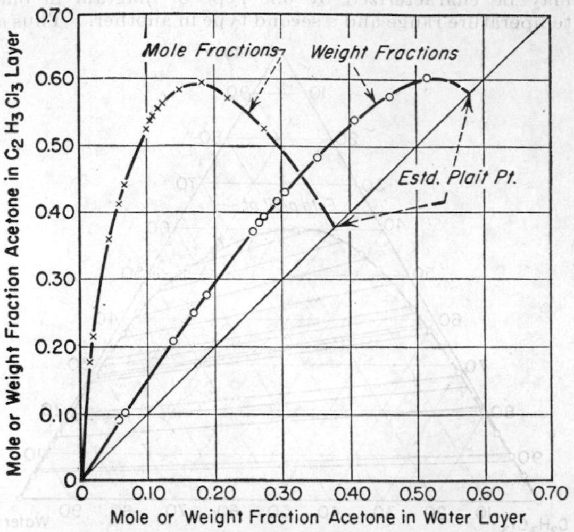

FIG. 11. Equilibrium distribution for acetone–water–1,1,2-trichloroethane system. [*Treybal, Weber, and Daley, Ind. Eng. Chem.*, **38**, 819 (1946).]

used for absorption calculations. Figure 11 is such a diagram for the system acetone–water–1,1,2-trichloroethane corresponding to the triangular diagram of Fig. 8*a* and showing the distribution of acetone between solvent and water layers. Figure 12 for the system of Fig. 9*b* shows the distribution of n-heptane between conjugate aniline and hydrocarbon phases. Figures 13, 14, and 15

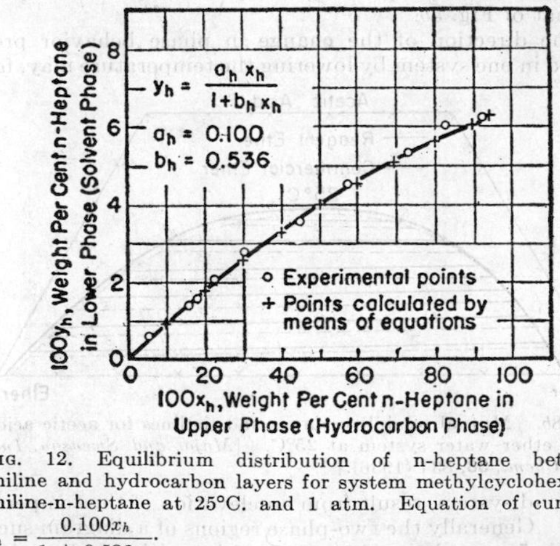

FIG. 12. Equilibrium distribution of n-heptane between aniline and hydrocarbon layers for system methylcyclohexane–aniline–n-heptane at 25°C. and 1 atm. Equation of curve is $y_h = \dfrac{0.100x_h}{1 + 0.536x_h}$.

show the distribution of acetic acid, acetone, and ethyl alcohol between water and a series of solvent phases.

Selectivity—Selectivity Diagram. Selectivity is defined on p. 715. If Fig. 4 is referred to and *B* is the solvent, the ratio of *A* to *C* will be higher in the extract than in the raffinate or feed when *B* is selective for *A*;

conversely, when for C. B may be non-selective, in which case the A/C ratio is the same in feed, extract, and raffinate. A second solvent may show the converse selectivity. The degree of selectivity may vary with the concentration range and to some degree with temperature. In some systems the same solvent may select A in one concentration range and C in another range;*

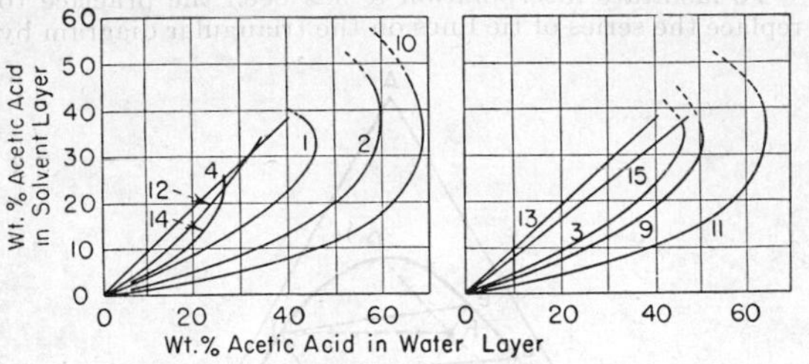

Fig. 13. Distribution of acetic acid between water and solvent. 1, isopropyl ether; 2, diisobutyl ketone; 4, methylcyclohexanone; 10, di-n-butyl ether; 12, methyl isobutyl ketone; 14, isoamyl acetate; 3, hexalin acetate; 9, fenchone; 11, octyl acetate; 13, isophorone; 15, di-isopropylcarbinol. [*Othmer, White, and Trueger, Ind. Eng. Chem.*, **33,** 1243 (1941).]

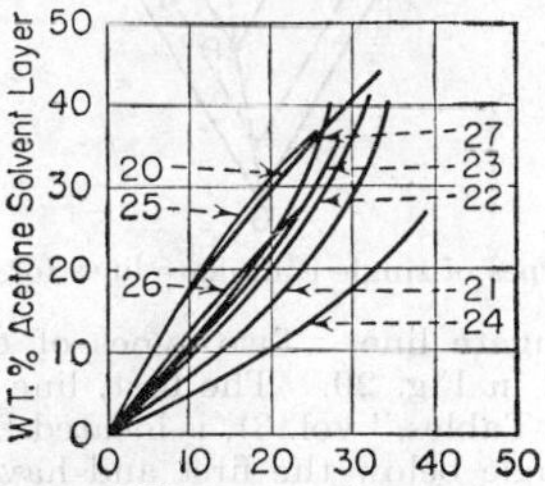

Fig. 14. Distribution of acetone between water and solvent. 20, tetrachloroethane; 21, xylene; 22, toluene; 23, monochlorobenzene; 24, di-n-butyl ether; 25, methyl isobutyl ketone; 26, furfural; 27, benzene. [*Othmer, White, and Trueger, Ind. Eng. Chem.*, **33,** 1245 (1941).]

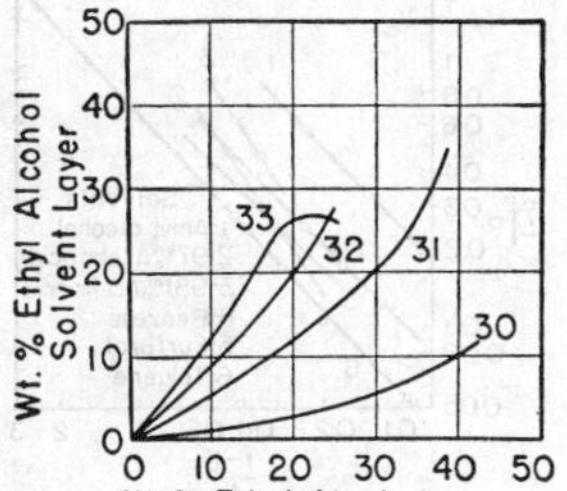

Fig. 15. Distribution of ethyl alcohol between water and solvent. 30, di-n-butyl ether; 31, di-n-propyl ketone; 32, n-amyl alcohol; 33, isoamyl alcohol. [*Othmer, White, and Trueger, Ind. Eng. Chem.* **33,** 1245 (1941).]

the selectivity reverses with concentration. Here there exists a particular composition for which the solvent is non-selective; above this it selects A, below it component C. This phenomenon is analogous to the azeotropic mixture in distillation and has similar implications. As expected, the composition of the mixture of non-selectivity varies with temperature. Experiment is the only certain method of predicting the selectivity of a solvent. The method of Francis [*Ind. Eng. Chem.*, **36,** 764, 1096 (1944)] based on the relative critical solution temperatures between proposed solvents and the pure components to be separated appears to be reliable and useful for hydrocarbon mixtures.

* This behavior has been observed by one of the authors in studying phase relationships for tert-amyl alcohol and mixtures of glycerol-water (unpublished experimental work). Previous evidence of this phenomenon is apparently rarely to be found in the literature.

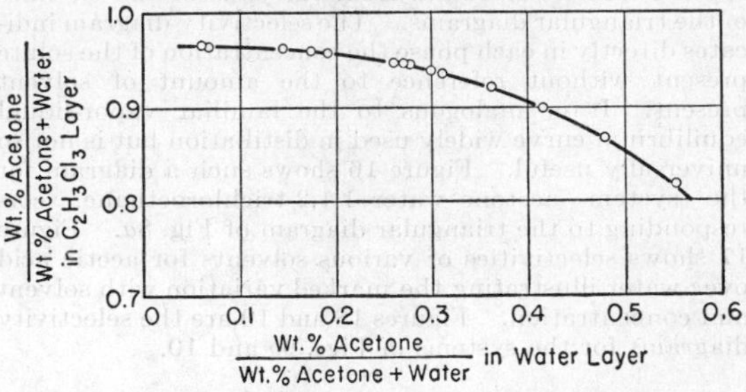

Fig. 16. Selectivity diagram for extraction of acetone from water by 1,1,2 trichloroethane at 25°C. [*Treybal, Weber, and Daley, Ind. Eng. Chem.*, **38,** 819 (1946).]

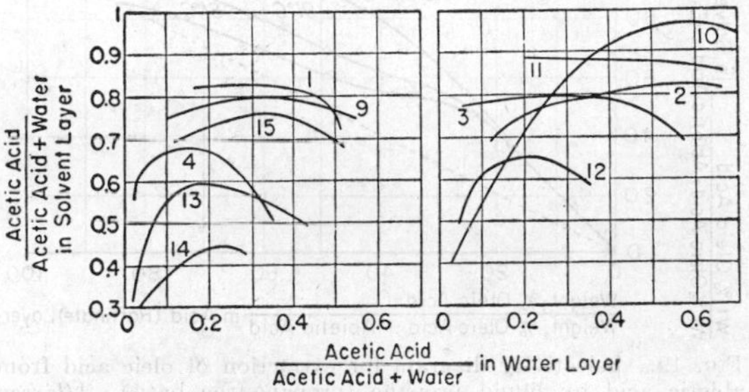

Fig. 17. Selectivity diagram for systems of acetic acid, water, and solvent. 1, isopropyl ether; 4, methylcyclohexanone; 9, fenchone; 13, isophorone; 14, isoamyl acetate; 15, diisopropylcarbinol; 2, diisobutyl ketone; 3, hexalin acetate; 10, di-n-butyl ether; 11, octyl acetate; 12, methyl isobutyl ketone. [*Othmer, White, and Trueger, Ind. Eng. Chem.*, **33,** 1247 (1941).]

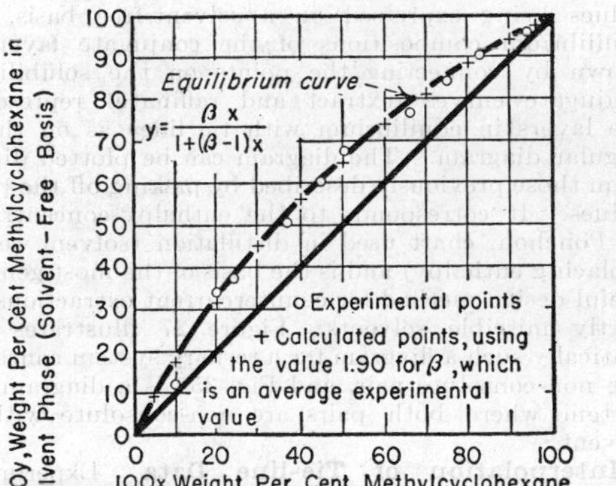

Fig. 18. Selectivity diagram for the system methylcyclohexane–aniline–n-heptane at 25°C. and 1 atm. with aniline as solvent (aniline-free basis). Equation of curve is $y = \dfrac{\beta x}{1 + (\beta - 1)x}$ where $\beta = 1.90$.

An *equilibrium-selectivity* or *effective-concentration* diagram provides an immediate quantitative index to degree of selectivity and its variation with concentration. It is useful in some cases for comparison of potential solvents and as a basis for design calculations. This diagram is made by plotting (per cent solute)/(per cent solute + per cent diluent) in the solvent phase against the same quantity in the diluent or raffinate phase, *i.e.*, on a solvent-free basis. These quantities are found by selecting the numerical values from the ends of the tie lines of the triangular diagrams. The selectivity diagram indicates directly in each phase the concentration of the solute present without reference to the amount of solvent present. It is analogous to the familiar vapor-liquid equilibrium curve widely used in distillation but is not so universally useful. Figure 16 shows such a diagram for the system acetone–water–1,1,2-trichloroethane, corresponding to the triangular diagram of Fig. 8*a*. Figure 17 shows selectivities of various solvents for acetic acid over water illustrating the marked variation with solvent and concentration. Figures 18 and 19 are the selectivity diagrams for the systems of Figs. 9*b* and 10.

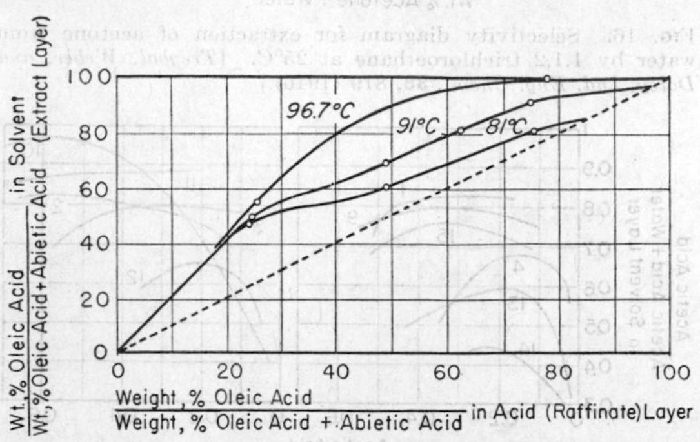

FIG. 19. Selectivity diagram for extraction of oleic acid from abietic acid by liquid propane (propane-free basis). [*Hixson and Hixson, Trans. Am. Inst. Chem. Engrs.*, **37**, 927 (1941).]

Solvent Content–concentration Diagram. This plot shows on rectangular coordinates the solvent content of the extract layer and of the raffinate layer against the percentage of the solute in the corresponding layer, both values being expressed on a solvent-free basis. The equilibrium compositions of the conjugate layers are shown by connecting the points on the solubility or conjugate curves (extract and raffinate) representing the layers in equilibrium with tie lines as on the triangular diagram. The diagram can be plotted directly from those previously described by picking off the proper values. It corresponds to the enthalpy-concentration, or Ponchon, chart used in distillation (solvent content replacing enthalpy) and is the basis of the most generally useful design method for countercurrent extractions with partly miscible solvents. Figure 27 illustrates schematically such a diagram for a ternary system containing one non-consolute pair, and Fig. 35*a* is a diagram for a system where both pairs are non-consolute with the solvent.

Interpolation of Tie-line Data. Experimental determinations of tie-line data are ordinarily laborious and time-consuming. Direct interpolation on the triangular diagram usually results in low accuracy. Unless the plait point is accurately known, extended extrapolation is often risky. A number of methods of interpolation and correlation have been proposed by Brancker, Hunter, and Nash [*Ind. Eng. Chem.*, anal. ed., **12**, 35 (1940)]; Bachman [*Ind. Eng. Chem.*, anal. ed., **12**, 38, (1940)]; Hand [*J. Phys. Chem.*, **34**, 1961 (1930)]; Othmer and Tobias [*Ind. Eng. Chem.*, **34**, 693 (1942)]; Treybal [*Ind. Eng. Chem.*, **36**, 875 (1944)]; and Campbell [*Ind. Eng. Chem.*, **36**, 1158 (1944)]. Of these, the procedures proposed by Othmer and Tobias or Treybal are recommended as likely to be most reliable. Extensions and variations of the original method of Hand, they depend upon the linearity of a loglog plot of selected concentrations or concentration ratios in the two conjugate phases and appear to have been tested for the larger variety of systems. All methods become more accurate for dilute concentrations and for solvents and diluents of greater immiscibility.

To facilitate interpolation it has been the practice to replace the series of tie lines on the triangular diagram by

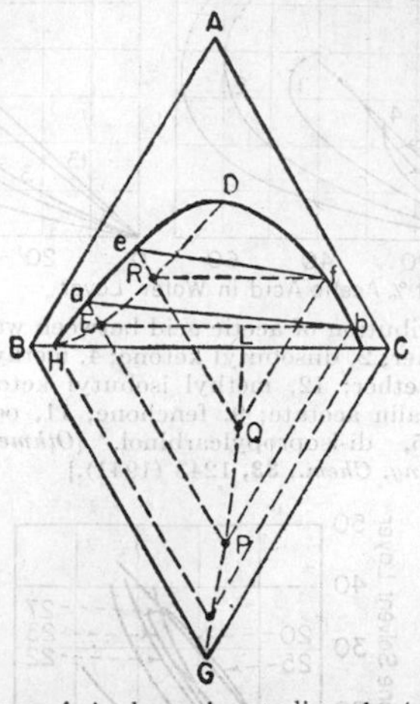

FIG. 20. Types of single conjugate lines for interpolation.

a single conjugate line. Two types of conjugate lines are illustrated in Fig. 20. The first, line *LG* ("International Critical Tables," vol. 3), is located by constructing a second triangle below the first and having a common base with it. By drawing from the opposite ends of a tie line parallels to opposite sides of the equilateral triangle, *e.g.*, *eQ* parallel to side *BG* and *fQ* parallel to *GC*, a triangle is obtained whose vertex is a point on the conjugate line *LG*, *i.e.*, point *Q*. If a similar procedure is followed

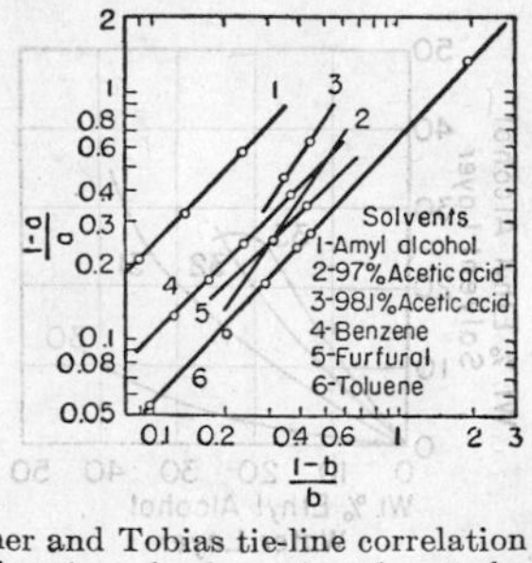

FIG. 21. Othmer and Tobias tie-line correlation plot for various systems. *a* is fraction of solvent in solvent phase and *b* is fraction of diluent in conjugate phase. Systems 1, 4, 5, and 6 are for acetaldehyde, water, and indicated solvents; systems 2 and 3 are for toluene and n-heptane with acetic acid of the given strengths (with water) as solvent.

for other tie lines, the conjugate line is located. The second type of conjugate line, *HD* (Sherwood, "Absorption and Extraction," McGraw-Hill, New York, 1937) is located by drawing from one end of the known tie line parallel to the base of the triangle, *e.g.*, from point *f*, line *fR* parallel to side *BC*. From the opposite end of the same tie line, a line is then drawn parallel to the side of the triangle opposite this point, *i.e.*, line *eR* parallel to side *AC*. The intersection of the two lines, *i.e.*, point *R*, locates a point on the conjugate line *HD*; other tie lines supply additional points. In either case interpolation is carried out from the known conjugate lines by reversing the procedure described for their location. Considerable tie-line data are required by either method since it is not required that the conjugate lines be straight.

Othmer and Tobias found tie-line data for a large number of systems, including many of appreciable miscibility, to yield approximately straight lines on a loglog plot of $(1 - a_1)/a_1$ against $(1 - b_2)/b_2$ where a_1 is the weight fraction of solvent in the solvent phase and b_2 is the weight fraction of diluent in the conjugate phase. Treybal proposed a rectilinear plot of log c_1/a_1 against log c_2/b_2 where c_1 is the fraction of solute in the solvent phase, c_2 is the fraction of solute in the diluent phase, a_1 is the fraction of solvent in the solvent phase, and b_2 is the fraction of diluent in the diluent phase, either weight or mole fractions being employed. Campbell showed a plot of log c_1 against log c_2 to correlate successfully the tie-line data for a number of systems. Othmer and Tobias plots are shown in Figs. 21 to 23, simultaneously presenting the tie-line data for these systems.

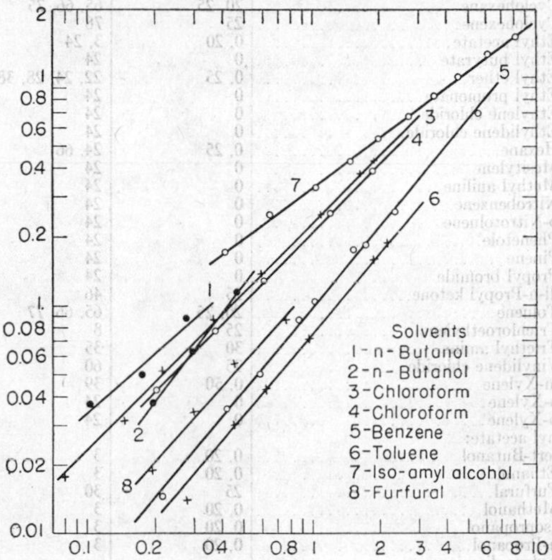

FIG. 22. Othmer and Tobias tie-line correlation plot for various systems.

Curve	System	Temp., °C.
1	Methanol, water, n-butanol	60
2	Methanol, water, n-butanol	15
3	Acetic acid, water, chloroform	60
4	Acetone, water, chloroform	25
5	Ethanol, water, benzene	25
6	Acetic acid, water, toluene	25
7	Ethanol, water, isoamyl alcohol	15
8	Acetone, water, furfural	25

Triangular Phase Diagrams with Solid Components. Ternary liquid-liquid extraction systems may involve a solid component. Two solids and one liquid are commonly concerned in leaching. For practical purposes their consideration is often simplified because the solubility of one solid in the solvent can be considered as negligible. When considering liquid-liquid extractions, the possibility of transforming an all-liquid system to one containing a solid component by altering the temperature should not be overlooked, and vice versa; the resulting change in phase relationships may affect to an important degree the conduct of the process.

Treatises on the phase rule (see references) should be consulted for extensive exemplification of the many varied types of phase relationships found with solid components and phases. Various solid and liquid phases can exist in equilibrium even in ternary systems, and compound formation between liquid and solid or two solid components may be encountered.

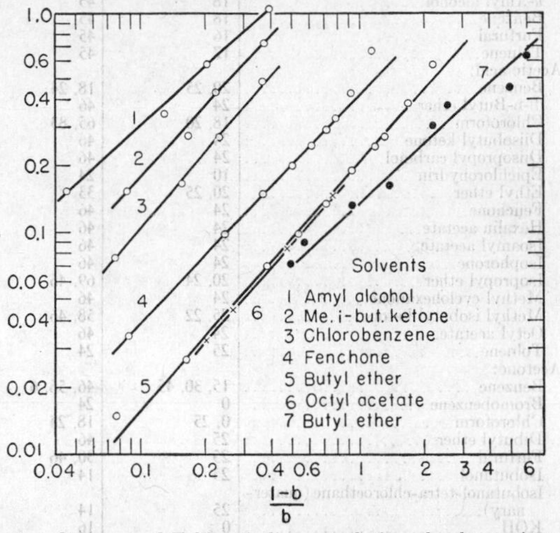

FIG. 23. Othmer and Tobias tie-line correlation plot for various systems.

Curve	System
1	Ethanol, water, n-amyl alcohol
2	Acetone, water, methyl isobutyl ketone
3	Acetone, water, monochlorobenzene
4	Acetic acid, water, fenchone
5	Acetone, water, dibutyl ether
6	Acetic acid, water, octyl acetate
7	Acetic acid, water, dibutyl ether

Data. Ternary condensed systems containing at least two liquid components for which mutual solubility and tie-line data are available are listed with source reference in Table 2. The list is substantially complete. The majority of phase-equilibrium data available are for systems containing water as one component. Data for a few typical systems are summarized in the charts of Figs. 8 to 19 and 21 to 23. Tie-line data for a number of systems are conveniently summarized in ref. 45, Table 2, and by Treybal (see below). Methods of interpolating tie-line data are given on p. 724. Theoretical methods for testing the reliability of and approximately estimating phase equilibrium in ternary systems from mutual solubilities, data for the binary systems involved, and vapor-liquid data are discussed in Sec. 8; see also Othmer and Tobias [*Ind. Eng. Chem.*, **34**, 690 (1942)]; Treybal [*Ind. Eng. Chem.*, **36**, 875 (1944)]; Scheibel and Friedland [*Ind. Eng. Chem.*, **39**, 1329 (1947)]; Scheibel [*Chem. Eng. Progress*, **44**, 681 (1948)]; and Brown [*Ind. Eng. Chem.*, **40**, 103 (1948)]. In addition to those listed in Table 2, mutual-solubility isotherms without tie lines have been determined for a number of systems; consult "International Critical Tables," Seidell ("Solubility of Inorganic and Organic Compounds," 3d ed., vols. I and II, 1940–1941, Van Nostrand, New York) Othmer, Trueger, and White [*Ind. Eng. Chem.*, **33**, 1240 (1939)]; Smith and Funk [*Trans. Am. Inst. Chem. Engrs.*, **40**, 211 (1944)]; Smith and Braun [*Ind. Eng. Chem.*, **37**, 1047 (1945)]; Elgin [*Ind. Eng. Chem.*, **38**, 26 (1946)]; and the references of Table 2.

Relative to the number of systems of potential use in industrial extraction operations, those for which the requisite equilibrium data for engineering calculation and design are available are few. In rare cases the equilibrium has been studied over the range of possible operating temperatures. Reliable generalizations or

Table 2. Ternary Condensed Systems for Which Phase-equilibrium Data Are Available

(*a*) Systems with water as one component

Components	Temp., °C.	Ref.
Acetaldehyde:		
n-Amyl alcohol	18	45
Benzene	18	45
Furfural	16	45
Toluene	17	45
Acetic acid:		
Benzene	20, 25	18, 24
di-n-Butyl ether	24	46
Chloroform	18, 20	65, 83
Diisobutyl ketone	24	46
Diisopropyl carbinol	24	46
Epichlorohydrin	10	24
Ethyl ether	20, 25	33
Fenchone	24	46
Hexalin acetate	24	46
Isoamyl acetate	24	46
Isophorone	24	46
Isopropyl ether	20, 24	69, 46
Methyl cyclohexanone	24	46
Methyl isobutyl ketone	25, 22	58, 46
Octyl acetate	24	46
Toluene	25	24
Acetone:		
Benzene	15, 30, 45	46, 56, 6
Bromobenzene	0	24
Chloroform	0, 25	18, 24
Dibutyl ether	25	46
Furfural	25	30, 46
Isobutanol	25	14
Isobutanol-tetra-chloroethane (quaternary)	25	14
KOH	0	16
Methyl isobutyl ketone	25	46
Monochlorobenzene	25	46
NaOH	0	16
Phenol	56, 5	30
Tetrachloroethane	25	46, 14
Toluene	25	46
Trichloroethane 1,1,2	25	70
Xylene	25	46
Allyl alcohol:		
Diallyl ether	22	13
Ammonia:		
1-Butene	25	49
Isoamyl alcohol:		
HBr	25	51
Aniline:		
Ethanol	0, 25	64
Formic acid	0, 20	50
Propionic acid	0, 20	1
Toluene	25	63
Benzene:		
Acetic acid	20, 25, 35	18, 24, 65
n-Butanol	25	81
Tert-butyl alcohol	25	59
Dioxane, 1,4,	25	5
Ethanol	20, 25	24, 39, 65, 67, 73, 79
Isopropanol	25	43
Pyridine	25	24
Isobutanol:		
Acetone	25	14
Ethyl ether	25	14
Tetrachlorethane	25	14
NaCl	25	14
NaOH	25	14
HBr	25	51
HCl	25	51
HI	25	51
n-Butanol:		
Methanol	0, 15, 30	42
Butylene glycol:		
Butanol	26, 50	44
Butyl acetate	25, 50	44
Butylene glycol diacetate	75, 50, 26	44
Methyl vinyl carbinol acetate	26, 50, 75	44
CS_2: acetic anhydride	0, 18	40
Carbon tetrachloride:		
Ethanol	0	24
Methanol	0	24
n-Propanol	0	24
Chloroform:		
Acetic acid	18, 20, 25	65, 83
Acetone	0, 25	18, 24
Ethanol	0	24
Methanol	0	24
Morpholine	25	7
Cyclohexanone:		
HCl	25	51
Ethanol:		
Isoamyl alcohol	15, 5, 28	24, 46
Isoamyl bromide	0	24
Isoamyl ether	0	24
Ethanol (*Continued*)		
n-Amyl alcohol	26	46
Benzaldehyde	0	24
Benzene	20, 25, 50	24, 39, 65, 67, 73, 79
Benzyl acetate	0	24
Benzyl alcohol	0	24
Benzyl ethyl ether	0	24
Bromobenzene	0	24
Bromotoluene	0	24
Carbon tetrachloride	0	24
di-n-Butyl ether	25	46
Isobutanol	0	24
Isobutyl bromide	0	24
Chloroform	0	24
Cottonseed oil	30	19
Cyclohexane	20, 25	65, 66, 75
Cyclohexene	25	78
Ethyl acetate	0, 20	3, 24
Ethyl butyrate	0	24
Ethyl ether	0, 25	22, 24, 28, 38
Ethyl propionate	0	24
Ethylene chloride	0	24
Ethylidene chloride	0	24
Hexane	0, 25	24, 66
Mesitylene	0	24
Methyl aniline	0	24
Nitrobenzene	0	24
p-Nitrotoluene	0	24
Phenetole	0	24
Pinene	0	24
Propyl bromide	0	24
di-n-Propyl ketone	25	46
Toluene	20, 25	65, 66, 77
Trichloroethylene	25	8
Triethyl amine	30	35
Vinylidene chloride	25	60
m-Xylene	0, 50	39
o-Xylene	0	24
p-Xylene	0	24
Ethyl acetate:		
tert-Butanol	0, 20	3
Ethanol	0, 20	3
Furfural	25	30
Methanol	0, 20	3
Isopropanol	0, 20	3
n-Propanol	0, 20	3
Ethylene diamine:		
NaOH	25	82
Ethylene glycol:		
n-Amyl alcohol	20	29
n-Hexyl alcohol	20	29
Ethyl ether:		
Ethanol	0, 25	20, 22, 24, 28
Isobutanol	25	14
Succinic nitrile		54
Triethyl amine	0, 12, 4, 30, 5	24
Furfural:		
Acetaldehyde	16	45
Acetone	25	30, 46
Isoamyl acetate	25	30
n-Butane	30.8–90	17
Ethyl acetate	25	30
Toluene	25	27
Glycerin:		
Acetone	25	84
tert-Amyl alcohol	7, 6, 25, 48, 6	48
Aniline	25, 75	78
Benzyl alcohol	25, 75	48
n-Butanol	25	48
Cyclohexanol	60	48
Methyl ethyl ketone	25	48
Lactic acid:		
n-Butanol	22	9
n-Butyl ether	22	9
Dichloroisopropyl ether	22	9
Hexone	22	9
Toluene	22	9
Methanol:		
Isoamyl alcohol	28	24
Bromobenzene	0	24
n-Butanol	0, 15, 30	42
Carbon tetrachloride	0	24
Chloroform	0	24
Cyclohexane	25	80
Cyclohezene	25	78
Ethyl acetate	0, 20	3
Ethyl bromide	0	24
Styrene	15	72
Toluene	25	31
Methyl isobutyl ketone:		
Benzoic acid	20, 7	26
Propionic acid	20, 7, 21	26

Table 2. Ternary Condensed Systems for Which Phase-equilibrium Data Are Available—*(Concluded)*

(a) Systems with water as one component—*(Concluded)*

Components	Temp., °C.	Ref.
Methyl ethyl ketone:		
Calcium chloride	23, 26	36
Gasoline	25	41
Trimethyl pentane 2,2,4	25	41
Morpholine-2-ethyl hexanol	25	7
Nitrobenzene:		
Ethanol	0	24
H_2SO_4	22	15
Phenol:		
Acetone	56, 5	24
Aniline		55
HCl	12	53
NaOH		37
KOH		37
Triethylamine	−2, 7, 10, 57, 75	24
Propionic acid:		
Aniline	0, 30	1
o-Toluidine	0, 20	1
Isopropanol:		
Cottonseed oil	30	19
Cyclohexane	15, 35	78
Ethyl acetate	0, 20	3
KNO_3	25 to 75	68
$(NH_4)NO_3$	25 to 75	68
Tetrachloroethylene	77	4
Toluene	25	77
n-Propanol:		
Isoamyl alcohol	25	10
Bromotoluene	0	24
Carbon tetrachloride	0	24
Ethyl acetate	0, 20	3
Toluene:		
Acetic acid	25	24
Ethanol	20, 25	39, 65
Methanol	25	34
Isopropanol	25	77

(b) Non-aqueous systems

Components	Temp., °C.	Ref.
Acetic acid (aqueous 97% and 98.1%):		
n-Heptane-toluene	23	45
Acetone-glycol:		
Benzene	27	71
Bromobenzene	27	71
Chlorobenzene	23	71
Nitrobenzene	22	71
Toluene	27	71
Xylene	25	71
Aniline-cetane:		
Benzene	25	23
Cyclohexane	25	23
n-Heptane	25	23
Aniline-cyclopentane:		
Neohexane	15, 25	57
Aniline-n-heptane:		
Cyclohexane	25	23
Methyl cyclohexane	25	74
Aniline-n-hexane:		
Methyl cyclopentane	25, 34.5, 45	11
Benzene-formic acid:		
Bromoform	25, 50, 70	2
Benzene-propylene glycol:		
Sodium oleate	20	47
Butadiene-isobutene-61.5% glycol + 26.4% methanol + 12.1% $Na_3Cu(CN)_4$	10	61
Butadiene-isobutene-57% triethanol amine + 43% methanol	0.65	61
Dicyanoethyl amine:		
Benzene–cyclohexane	25	7
Styrene–ethyl benzene	25	7
Toluene–troluoil	25	7
Docosane-diphenyhexane:		
Furfural	45, 80, 115	6
Ethanol-benzene-glycerol	25	31
Ethanol-carbon tetrachloride:		
Glycerol	25	32
Ethanol (96%)–olive oil:		
Oleic acid	18.5, 25, 96	52
Furfural-isobutene:		
Butadiene	0.65	61
Furfural-naphtha:		
Butadiene	0.65	61
Isobutene	0.65	61
Methanol–olive oil:		
Oleic acid	25, 96	52
Propane (liquid)–oleic acid–abietic acid	81, 91, 96.7	21
Propane (liquid)–oleic acid:		
Refined cottonseed oil	85, 98.5	20
Triolein	85	20
Propane (liquid)–stearic acid:		
Palmitic acid	95, 98	12
Toluene-Skellysolve B:		
70% methanol + 30% glycol	10.6, 21	62

References:

1 Angelescu, *Bull. soc. chim. Roomanie*, **10**, 160 (1929).
2 Avenarius and Tarasenkov, *J. Gen. Chem. (U.S.S.R.)*, **16**, 1777 (1946).
3 Beech and Glasstone, *J. Chem. Soc.*, 1938, 67.
4 Bergelin, Lockhart, and Brown, *Trans. Am. Inst. Chem. Engrs.*, **39**, 173 (1943).
5 Berndt and Lynch, *J. Am. Chem. Soc.*, **66**, 282 (1944).
6 Briggs and Comings, *Ind. Eng. Chem.*, **35**, 411 (1943).
7 Carbide and Carbon Chemicals Co.
8 Colburn and Phillips, *Trans. Am. Inst. Chem. Engrs.*, **40**, 333 (1944).
9 Congleton, Princeton Univ. Chem. Eng. Thesis, 1942.
10 Coull and Hope, *J. Phys. Chem.*, **39**, 967 (1935).
11 Darwent and Winkler, *J. Phys. Chem.*, **47**, 442 (1943).
12 Drew and Hixson, *Trans. Am. Inst. Chem. Engrs.*, **40**, 675 (1944).
13 Fairburn, *et al.*, *Chem. Eng. Progress*, **43**, 279 (1947).
14 Fritzsche and Stockton, *Ind. Eng. Chem.*, **38**, 737 (1946).
15 Gibby, *J. Chem Soc.*, 1932, p. 1540.
16 *Ibid.*, 1934, p. 9.
17 Griswold, *et al.*, *Chem. Eng. Progress*, **44** 839 (1948).
18 Hand, *J. Phys. Chem.*, **34**, 1961 (1930).
19 Harris, *et al.*, *J. Am. Oil Chemists' Soc.*, **24**, 370 (1947).
20 Hixson and Bockelmann, *Trans. Am. Inst. Chem. Engrs.*, **38**, 891 (1942).
21 Hixson and Hixson, *ibid.*, **37**, 927 (1941).
22 Horiba, *Mim. Coll. Eng. Kyoto Imp. Univ.*, **3**, 63 (1911).
23 Hunter and Brown, *Ind. Eng. Chem.*, **39**, 1343 (1947).
24 "International Critical Tables," vol. 3, pp. 398*ff.*, McGraw-Hill, New York, 1929.
25 Jackson, Princeton Univ. Chem. Eng. Thesis, 1937.
26 Johnson and Bliss, *Trans. Am. Inst. Chem. Engrs*, **42**, 311 (1946).
27 Knight, *ibid.*, **39**, 439 (1943).
28 Kono, *J. Chem. Soc. Japan*, **44**, 406 (1923).
29 Laddha and Smith, *Ind. Eng. Chem.*, **40**, 494 (1948).
30 Lloyd, Thompson, and Ferguson, *Can. J. Research*, **15B**, 98 (1937).
31 McDonald, *J. Am. Chem. Soc.*, **62**, 3183 (1940).
32 McDonald, Kluender, and Lane, *J. Phys. Chem.*, **46**, 946 (1942).
33 Major and Swenson, *Ind. Eng. Chem.*, **38**, 834 (1946).
34 Mason and Washburn, *J. Am. Chem. Soc*, **59**, 2076 (1937).
35 Meerburg, *Z. physik. Chem.*, **40**, 642 (1902).
36 Meissner and Stokes, *Ind. Eng. Chem.*, **36**, 816 (1944).
37 van Meurs, *Z. physik. Chem.* **91**, 313 (1916).
38 Miller and McPherson, *J. Phys. Chem.*, **12**, 706 (1908).
39 Mochalov, *Bull. inst. recherches biol. univ. Perm*, **11**, 25 (1937).
40 Mochalov, *J. Gen. Chem. (U.S.S.R.)*, **8**, 529 (1938).
41 Moulton and Walkey, *Trans. Am. Inst. Chem. Engrs.*, **40**, 695 (1944).
42 Mueller, Pugsley, and Ferguson, *J. Phys. Chem.*, **35**, 1313 (1931).
43 Olsen and Washburn, *J. Am. Chem. Soc.*, **57**, 303 (1935).
44 Othmer *et al.*, *Ind. Eng. Chem.*, **37**, 601 (1945).
45 Othmer and Tobias, *ibid.*, **34**, 690 (1942).
46 Othmer, White, and Trueger, *ibid.*, **33**, 1240 (1939).
47 Palit and McBain, *ibid.*, **38**, 741 (1946).
48 Plumb, Princeton Univ. Chem. Eng. Thesis, 1939.
49 Poffenberger *et al.*, *Trans. Am. Inst. Chem. Engrs.*, **42**, 815 (1946).
50 Pound and Wilson, *J. Phys. Chem.*, **39**, 709 (1935).
51 Reburn and Shearer, *J. Am. Chem. Soc.*, **55**, 1774 (1933).
52 Rius and Moreno, *Annales Fís y quím (Madrid)*, **42**, 123 (1947).
53 Schreinemakers and van den Bos, *Z. physik. Chem.*, **79**, 551 (1912).
54 Schreinemakers, *ibid.*, **25**, 543 (1898).
55 Schreinemakers, *ibid.*, **29**, 586 (1899).
56 Seidell, "Solubility of Inorganic and Organic Compounds," Van Nostrand, New York, 1928.
57 Serijan, Spurr, and Gibbons, *J. Am. Chem. Soc.*, **68**, 1763 (1946).
58 Sherwood, Evans, and Longcor, *Trans. Am. Inst. Chem. Engrs.*, **35**, 597 (1940).
59 Simonsen and Washburn, *J. Am. Chem. Soc.*, **68**, 235 (1946).
60 Skripach and Temkin, *J. Phys. Chem. (U.S.S.R.)*, **20**, 583 (1946).
61 Smith and Braun, *Ind. Eng. Chem.*, **37**, 1047 (1945).
62 Smith and Funk, *Trans. Am. Inst. Chem. Engrs.*, **40**, 211 (1944).
63 Smith and Drexel, *Ind. Eng. Chem.*, **37**, 601 (1945).
64 Tarasenkov and Avenarius, *J. Gen. Chem. (U.S.S.R.)*, **16**, 1577 (1946).
65 Tarasenkov and Paulsen, *Acta Physicochim. U.S.S.R.*, **11**, 75 (1939).
66 Tarasenkov and Paulsen, *J. Gen. Chem. (U.S.S.R.)*, **7**, 2143 (1937).
67 Tarasenkov and Paulsen, *ibid.*, **8**, 76 (1938).
68 Thompson and Molstad, *Ind. Eng. Chem.*, **37**, 1244 (1945).
69 Student Problem, *Trans. Am. Inst. Chem. Engrs.*, **36**, 594 (1940).
70 Treybal, Weber, and Daley, *Ind. Eng. Chem.*, **38**, 817 (1946).
71 Trimble and Frazer, *ibid.*, **21**, 1063 (1929).
72 Troyan, *Rubber Age*, **63**, 585 (1948).
73 Varteressian and Fenske, *ibid.*, **28**, 928 (1936).
74 Varteressian and Fenske, *ibid.*, **29**, 270 (1937).
75 Vold and Washburn, *J. Am. Chem. Soc.*, **54**, 4217 (1932).
76 Washburn and Beguin, *ibid.*, **62**, 579 (1940).
77 Washburn, Beguin, and Beckord, *ibid.*, **61**, 1694 (1939).
78 Washburn, Graham, Arnold, and Transue, *ibid.*, **62**, 1454 (1940).
79 Washburn, Hnizda, and Vold, *ibid.*, **53**, 3237 (1931).
80 Washburn and Spencer, *ibid.*, **56**, 361 (1934).
81 Washburn and Strandskov, *J. Phys. Chem.*, **48**, 241 (1944).
82 Wilson, *Ind. Eng. Chem.*, **27**, 867 (1935).
83 Wright, *Proc. Roy. Soc. (London)*, **49**, 174 (1891).
84 Young, Princeton Univ. Chem. Eng. Thesis, 1938.

predictions are as yet difficult if not impossible. Hence experimental measurement will frequently be necessary; for experimental methods, see Smith [*Ind. Eng. Chem.*, **34**, 234 (1942)]; Othmer, Trueger, and White [*Ind. Eng. Chem.*, **33**, 1240 (1939)]; and the references of Table 2.

Multicomponent Systems. When more than three components are involved, as in complex hydrocarbon oil- or vegetable oil-solvent systems, the representation of phase equilibriums is for practical purposes difficult and, where the number of components is large and these are unidentifiable, is impossible in terms of the pure components. Extraction processes for such materials usually have as their objective a separation into two fractions (an extract and a raffinate) characterized by different physical and chemical properties rather than the isolation of individual components. For engineering consideration and calculation, such cases may be treated as ternary systems by employing a physical property of the system (*e.g.*, density, refractive index, viscosity-gravity constant, viscosity index, iodine number, etc.) which is additive or assumed to be additive for the mixed solvent-free extract and raffinate and which characterizes the two fractions for the separation desired. Appropriate values of the selected physical property lying outside the limits affected by the extraction are then used to designate the extract and raffinate, respectively, and the mutual-solubility curve and tie lines are determined experimentally from a series of equilibrium laboratory extractions.

Data thus determined may be represented graphically on any of the types of diagram previously shown. For example, Hunter and Nash [*Proc. World Petroleum Congr., London*, **2**, 340 (1933); *Ind. Eng. Chem.*, **27**, 836 (1935)] thus represent the equilibrium in hydrocarbon oil-solvent systems on the triangular diagram in Fig. 24, using the

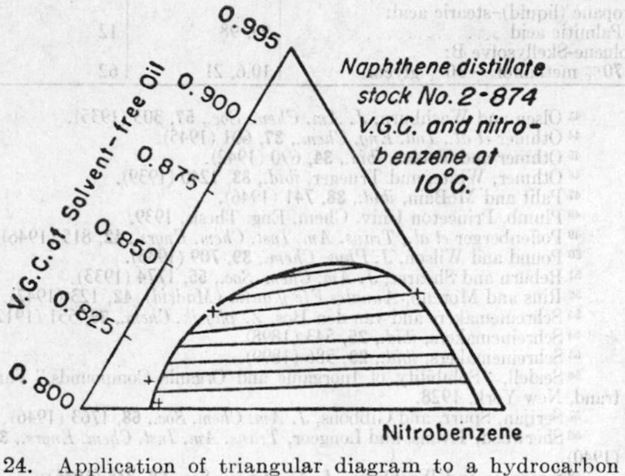

FIG. 24. Application of triangular diagram to a hydrocarbon oil–nitrobenzene system. [*Hunter and Nash, Ind. Eng. Chem.*, **27**, 841 (1935).]

viscosity-gravity constant [see Hill and Coats, *Ind. Eng. Chem.*, **20**, 641 (1928)]. The constant for a mixture of the two oils, extract and raffinate, resulting from an extraction is assumed to be a weighted mean of the two values of the constant for the two fractions. The vertexes of the triangle then represent the solvent, oil of low viscosity-gravity constant, and one of high viscosity-gravity constant (so selected that all oil fractions, extracts and raffinates, fall between them), respectively. The side opposite the solvent vertex is scaled to represent viscosity-gravity constants between the two extremes selected. Laboratory extraction data then locate the mutual-solubility curve and tie lines in terms of the constants of the solvent-free extracts and raffinates and the solvent content of each. Such a diagram can then be used for extraction calculations as are those for ordinary ternary systems.

The method rests on an experimental basis. Diagrams for a number of hydrocarbon oil–solvent systems may be found in the literature. Thompson (Dunstan, "Science of Petroleum," vol. 3, p. 1829, Oxford, New York, 1938) discusses the application of triangular graphs to multicomponent hydrocarbon oil systems and gives data and diagrams for a variety of solvents and oils. Kalichevsky [*Ind. Eng. Chem.*, **38**, 1009 (1946)] has correlated phase-equilibrium data at different temperatures for hydrocarbon oil–solvent systems by relatively simple mathematical equations that permit the estimation of the complete equilibrium diagram from four experimental determinations. Data for a variety of oil–solvent systems are also given.

Complex systems similarly may be represented on a solvent content-composition, or Ponchon (Maloney and Schubert), type of diagram. Ruthruff and Wilcock [*Trans. Am. Inst. Chem. Engrs.*, **37**, 649 (1941)] in the solvent-extraction separation of vegetable oils into drying (unsaturated) and non-drying (saturated) fractions thus represent (Fig. 25) the equilibrium for soybean oil with

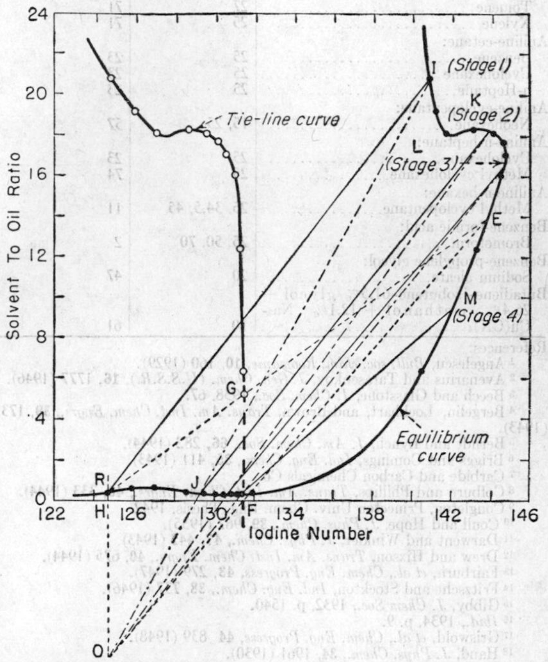

FIG. 25. Solvent content–concentration phase equilibrium diagram for soybean oil–furfural system. [*Ruthruff and Wilcock, Trans. Am. Inst. Chem. Engrs.*, **37**, 649 (1941).]

furfural in terms of the iodine number of the equilibrium extracts and raffinates.

Smith [*Ind. Eng. Chem.*, **36**, 68 (1944)], Hunter [*Ind. Eng. Chem.*, **34**, 963 (1942)], and Brown [*Ind. Eng. Chem.*, **40**, 103 (1948)] show methods for the graphical representation of tie lines and making calculations in quaternary liquid systems such as those involving extractions with a mixture of two solvents.

Ternary Systems—One Pair Immiscible

If the solvent and diluent are essentially immiscible over the solute concentration range involved, only the solute distributing itself between the two phases, the rectangular distribution chart (p. 722, illustrated schematically in Fig. 37) is sufficient to represent completely the phase equilibrium and to handle extraction calculations for the system. Unless the solute concentration is very dilute, this is rarely the case and is less frequently met in practice. In plotting the chart, concentrations may be expressed in weight per unit volume, mole fractions, or mole or weight ratio of solute in solvent or diluent layers, respectively. The latter is preferred for engineering calculations.

Ideal-distribution Law. In general the equilibrium-distribution curve is non-linear. The ideal-distribution law states that $y = mx$ where y and x are the mole fractions of the solute in the two phases, respectively, and m is the "distribution" coefficient which is constant independent of concentration. The law may also be expressed as $c_1 = Dc_2$ where c_1, c_2 are the respective solute concentrations in weight per unit volume. In the scientific literature, distribution coefficients D are usually reported in these units. If the law holds, as it rarely does in practical extractions, the equilibrium-distribution curve becomes a straight line through the origin of slope m (or D). This is analogous to Henry's law in gas absorption. Even though the law holds, it should be noted that, when weight or mole ratios are used, the relationship is still non-linear unless the phases are very dilute.

Mathematical Expression of Equilibrium Distribution. When mathematical expression of equilibrium distribution is possible, treatment of distribution data and extraction calculations are sometimes simplified. Where known changes such as dissociation, association, or chemical reaction are responsible for deviations, suitable corrections to the ideal law taking these into account will often lead to useful algebraic distribution equations. For the development of such equations, see, for example, Taylor ("Treatise on Physical Chemistry," 2d ed., vol. 1, pp. 479–485, Van Nostrand, New York, 1932) and Almquist [*J. Phys. Chem.*, **37**, 991 (1933)]. For algebraic treatments of special cases, see also Varteressian and Fenske [*Ind. Eng. Chem.*, **29**, 270 (1937)]; Happel and Robertson [*Ind. Eng. Chem.*, **27**, 941 (1935)]; and Rowley, Steiner, and Zimkin [*J. Soc. Chem. Ind.*, **65**, 237 (1946)]. The latter have developed the algebraic treatment of distribution in penicillin extraction. In some cases simple empirical equations are found to hold.

CALCULATION AND DESIGN METHODS

Multistage Liquid-liquid Extraction Systems

General Procedure. The calculation of multistage extraction systems is best based upon the concept of the "ideal" or "equilibrium" stage. The equipment may consist of a series of individual mixers and settlers or may be of the tray-column type. The number of actual non-equilibrium stages is then determined from the calculated ideal stages by use of either an individual or an average stage efficiency factor. As yet stage efficiencies must be experimentally determined, since no generally useful correlation has been presented. The average stage efficiency is ordinarily more useful. In basic principle, calculation methods for extraction are similar to those for distillation and absorption. Extraction operations, however, always involve in most practical cases at least a ternary system and partly miscible solvents. Consequently, simplifying assumptions often valid for distillation and absorption, such as constant molal overflow or constant gas-liquid ratio, are rarely so in extraction (*i.e.*, the operating line is curved), and the more complex general procedures are usually required.

The general material balance algebraic relationships for multistage extraction systems have been developed; see Varteressian and Fenske [*Ind. Eng. Chem.*, **28**, 1353 (1936)]; Hunter and Nash [*J. Soc. Chem. Ind.*, **51**, 285T (1932)]; and Underwood [*Ind. Chemist*, **10**, 129 (1934)]. Except in special cases where the solvents may be assumed immiscible and the ideal-distribution law holds, their solution is tedious, and graphical calculation methods are more convenient. Graphical calculation may be based upon (1) the rectangular solvent content–concentration (Ponchon) diagram, p. 724, (see Figs. 28, 33, and 35*a*, *b*, and *c*) [Maloney and Schubert, *Trans. Am. Inst. Chem. Engrs.*, **36**, 741 (1940); Randall and Longtin, *Ind. Eng. Chem.*, **30**, 1063, 1188, 1311 (1938); Thiele, *Ind. Eng. Chem.*, **27**, 392 (1935)]; (2) the triangular phase diagram (see Fig. 4) [Hunter and Nash, *J. Soc. Chem. Ind.*, **53**, 95T (1934); Saal and Van Dyck, *Proc. World Petroleum Cong., London*, **2**, 352 (1933); Varteressian and Fenske, *Ind. Eng. Chem.*, **28**, 1353 (1936); Skogan and Rogers, *Oil Gas J.*, **46**, No. 13, 10 (1947); Schiebel, *Chem. Eng. Progress*, **44**, 681 (1948); and Sherwood, "Absorption and Extraction," McGraw-Hill, New York, 1937]; or (3) the rectangular equilibrium-distribution diagram (see Fig. 11) (Varteressian and Fenske). The first is the most generally convenient for countercurrent operations. Illustrative graphical solutions of extraction problems will be found in the references.

Each of the above procedures may be adapted to the calculation of any of the possible operating methods, *i.e.*, countercurrent contact, countercurrent contact with reflux, or single or simple multistage contact. Often in extraction it is desired, for economic reasons, to prespecify the number of stages to be used rather than the quantity of solvent. Either may be done, but by any of these graphical methods trial-and-error solution is necessary if the number of stages is fixed in advance. True continuous countercurrent towers, *e.g.*, the packed or spray types, may be calculated in terms of theoretical stages by using the height equivalent to a theoretical stage, H.E.T.S., as is frequently done in distillation and absorption. The *extraction-coefficient* and *transfer-unit methods* of calculating such towers are treated on pp. 743 through 746 (see also Secs. 8 and 10). Practical application of these methods to systems of partly miscible solvents is difficult, and data for coefficients and H.T.U. are generally lacking. Hence the method of theoretical stages is frequently most useful.

Simple Multistage (Cocurrent) Extraction

This extraction method is illustrated in Fig. 1*b*, p. 717. Let it be desired to calculate the number of equilibrium or ideal contact stages required to recover component A from a quantity F of a feed mixture of A and B by treatment isothermally with a solvent S, totally miscible with component B for which it is selective and partly miscible with component A. The triangular equilibrium diagram is the type of Fig. 4. A quantity of solvent S_1, S_2, S_3, etc., will be used in each stage, respectively (usually equal amounts of solvent are employed in each stage). The allowable concentration of B in the raffinate product A is specified so that the final raffinate composition R_f is fixed.

Calculation on the Triangular Diagram. Figure 26 represents the usual triangular equilibrium diagram on which is shown the graphical stepwise construction for the solution of this method of operation. abc is the solubility isotherm for the system at the extraction temperature. The compositions of the extract layers from the successive stages E_1, E_2, E_3, . . . E_f, etc., lie on the solvent side of this curve ab. Those of the raffinate layers from the successive stages R_1, R_2, R_3, . . . R_f, etc., will all lie on the feed side bc of this saturation curve. The points M_1, M_2, M_3, . . . M_f, etc., represent the over-all compositions of the mixtures in the successive stages produced by adding a quantity of solvent S_2 to the raffinate R_1 from the preceding stage, etc. As shown, the incoming solvent is initially free of the feed components. If it contained any of these initially, the point representing its composition would be located within the diagram rather than at the solvent vertex S. F represents the solvent-free feed composition and R'_f the final solvent-free raffinate composition (A stripped substantially of B) desired. Allow the above symbols to represent also the quantities of each stream (weight or moles) as well as their composition.

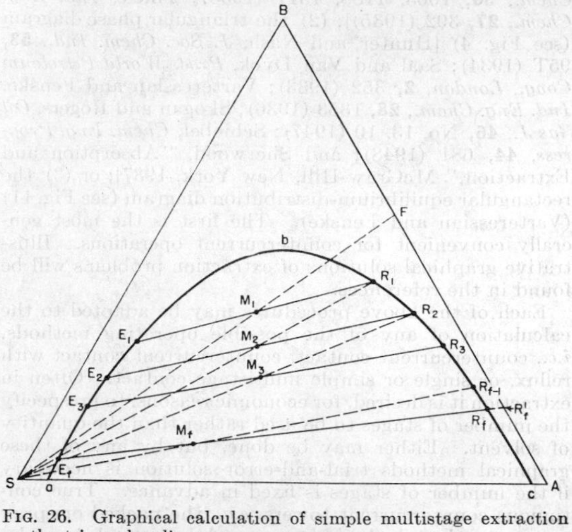

FIG. 26. Graphical calculation of simple multistage extraction on the triangular diagram.

Commencing at the feed end of the system, the over-all composition of the mixture in the first mixer M_1 (stage 1), formed by adding a weight of solvent S_1 to the weight of feed F, is found by dividing the line FS into segments proportional to the weights used, *i.e.*, M_1 is located on FS so that M_1F/M_1S = lb. S_1/lb. F (lever rule). The compositions of the extract and raffinate layers into which the mixture M_1 splits are then located at the extremities on abc of a tie line drawn through M_1, E_1, and R_1, respectively. M_1 divides this tie line into segments M_1R_1 and M_1E_1 proportional to the ratio of the weights of the two layers obtained, or M_1R_1/M_1E_1 = lb. E_1/lb. R_1. Since the total quantity $M_1 = F + S$, the amount of raffinate layer from this stage is lb. $R_1 = (M_1E_1/E_1R_1)(F + S)$, and similarly for the amount of extract. Similarly, the over-all composition in the second mixer (stage 2) M_2 is located on the line R_1S joining R_1 and the solvent vertex S by dividing it into segments so that M_2R_1/M_2S = weight S_2/weight R_1, *i.e.*, the ratio of solvent fed to this stage to the raffinate from stage 1. The extract and raffinate compositions from the second stage, E_2 and R_2, are located by the tie line through M_2, and the amounts of each are calculated as for the first stage, noting that lb. M_2 = lb. R_1 + lb. S_2. R_2 constitutes the raffinate feed to the third stage, and the construction and calculation are thus continued until the specified raffinate composition R_f (or R_f') is equaled or exceeded. The number of steps required is the number of ideal stages needed for the system. The sum of the solvent quantities used for all stages is the total solvent required, $S_1 + S_2 + S_3 + \cdots$ = total S.

The construction may also be commenced at the raffinate terminal, proceeding until the entering-feed composition is reached or exceeded. Various other combinations of conditions might have been specified in advance and the remaining ones similarly determined. The quantity of solvent required to strip the outgoing raffinate to a specified concentration depends not only on the number of contacts used but on the division of the total quantity of solvent to be used between the stages as well; or, conversely, the exhausting of the desired component from the raffinate, effected with a given number of stages and quantity of solvent, varies with the division of the latter between the stages. In addition to specifying in advance the total quantity of solvent to be used or the number of stages, either the concentration of the raffinate to be discharged from each intermediate stage (say, by stating that each is to effect an equal concentration decrease) or the division of solvent between stages (say, by using equal quantities in each) must be fixed. The raffinate compositions from intermediate stages vary with the specified division of the solvent, or, conversely, if the former are specified, the quantity of solvent necessary varies for each stage. If the division of the solvent is specified in advance, a trial-and-error solution is in general required. Thus in the present case both R_f and the number of stages to be used, together with the division of the solvent between the stages, might have been specified and the total quantity of solvent required calculated. A trial-and-error solution is then necessary.

Calculation of *single-contact* operation corresponds to that of the first stage in the simple multistage contact method.

Calculation on the Solvent Content–concentration Diagram. An alternate method of calculation that is sometimes more convenient employs the solvent content–concentration diagram (Fig. 27). For any stage, let

S' = amount of fresh solvent fed.
S = solvent content of extract layer on solvent-free basis, lb. S/lb. $(A + B)$.
s = solvent content of raffinate layer on solvent-free basis, lb. S/lb. $(A + B)$.
E = amount of extract layer, solvent-free.
R = amount of raffinate layer, solvent-free.
Y = fraction of B in the extract layer, solvent-free, lb. B/lb.$(A + B)$.
X = fraction of B in the raffinate layer, solvent-free, lb. B/lb.$(A + B)$.

Subscripts denote the particular stage, f representing the final stage from which stripped raffinate product is withdrawn and X_0 the initial B content of the feed. Material balances may be written around any individual stage, say stage 2, as follows:

Solvent balance per unit of raffinate product R, entering stage 2 from stage 1:

$$\frac{S_2'}{R_1} + s_1 = \frac{R_2}{R_1} s_2 + \frac{E_2}{R_1} S_2 = J_2 \qquad (1)$$

Define $J_2 = S_2'/R_1 + s_1$ by Eq. (1). Hence J_2 equals the total solvent content of stage 2 per unit of feed to this stage. Similarly, a B balance per unit of raffinate feed R_1 is

$$X_1 = X_2\frac{R_2}{R_1} + Y_2\frac{E_2}{R_1} \qquad (2)$$

and the corresponding $A + B$ balance is

$$R_1 = R_2 + E_2 \qquad (3)$$

From Eqs. (1) and (3),

$$S_2 - J_2 = \frac{R_2}{R_1}(S_2 - s_2) \qquad (4)$$

From Eqs. (2) and (3),

$$Y_2 - X_1 = \frac{R_2}{R_1}(Y_2 - X_2) \qquad (5)$$

and, eliminating R_2/R_1 from Eqs. (4) and (5),

$$\frac{S_2 - J_2}{Y_2 - X_1} = \frac{S_2 - s_2}{Y_2 - X_2} \tag{6a}$$

or

$$\frac{Y_2 - X_2}{Y_2 - X_1} = \frac{S_2 - s_2}{S_2 - J_2} \tag{6b}$$

Similar relations hold for any stage. The compositions and quantities of the extract and raffinate layers leaving any stage may be determined graphically on the diagram (Fig. 27) by the following construction. Thus, for stage

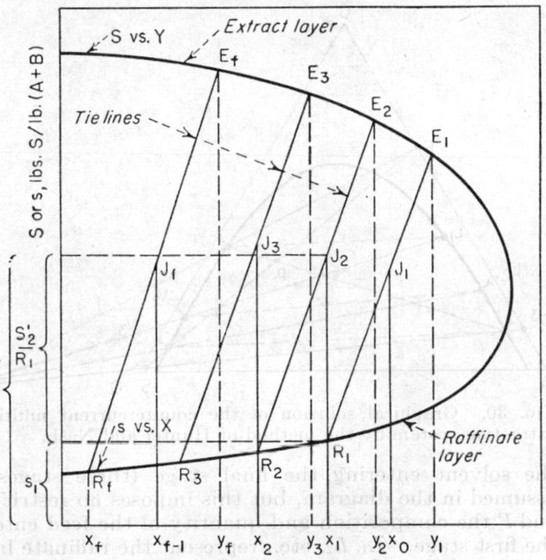

FIG. 27. Graphical calculation of simple multistage extraction on the solvent content–concentration diagram.

2, the height of the perpendicular at X_1, the concentration of the feed entering stage 2, is J_2 calculated from $S_2'/R_1 + s_1$ and representing the total solvent entering this stage. A tie line E_2R_2 through point J_2 locates at its terminals E_2 and R_2 on the solubi ity curve the extract composition Y_2 and the raffinate composition X_2 (both on a solvent-free basis) leaving stage 2. The quantity of raffinate R_2 can be calculated from either Eq. (4) or Eq. (5), the known value of R_1, and the values of S_2, s_2 or Y_2, X_2 given on the diagram by the terminals of the tie line.

If the above procedure is commenced at the feed entrance for stage 1 and repeated stepwise from stage to stage until the required final raffinate composition X_f (or R_f) is equaled or exceeded, the number of steps needed is the number of ideal stages for the operation, and the sum of the solvent feeds to each stage is the total solvent requirement. Thus, starting at the feed composition X_0, the perpendicular J_1X_0 is erected equal in height to $J_1 = S_1'/F$ (the solvent ratio for stage 1). The tie line E_1R_1 through J_1 locates E_1 and R_1, and the second stage is calculated by erecting at X_1 a perpendicular of length $J_2 = S_2'/R_1 + s_1$.

As before, if the number of stages and the division of solvent between stages is set in advance, trial-and-error solution is necessary. In both this and the preceding triangular method, inspection of the diagram readily shows the theoretical operating limits for the problem such as maximum solvent ratio for formation of two phases and maximum and minimum extract and raffinate compositions from any stage.

Countercurrent Multistage Extraction without Reflux

The general flow scheme is illustrated in Figs. 1c and 29. The equipment may consist of a series of individual mixers and settlers, or it may be of the tray-column type. Assume that F lb. (or moles)/hr. of a feed mixture consisting of A and B is to be stripped of B by countercurrent isothermal extraction with Q lb. (or moles)/hr. of solvent S. The solvent is totally miscible with component B, for which it is selective, and partly miscible with component A. Let Fig. 28 represent an extraction unit of M theoretical stages, it being required to calculate the necessary number of such stages to reduce the B content of the raffinate to a specified value X_r on a solvent-free basis. Three alternate methods of calculation are available, that based on the solvent content–concentration (Ponchon-type) diagram being usually the more convenient.

Calculation on the Solvent Content–concentration Diagram. This method represents the graphical solution on a Ponchon-type extraction diagram of the material balance and equilibrium tie-line relationships on a solvent-free basis. The following nomenclature is used, all flow quantities and compositions being *expressed on a solvent-free basis* unless otherwise specified.

F = amount of incoming feed, lb. $(A + B)$.
Q = amount of fresh solvent fed, lb. S.
V = amount of extract layer, lb. $(A + B)$.
L = amount of raffinate layer, lb. $(A + B)$.
L_r = final raffinate product, lb. $(A + B)$.
V_1 = final extract product, lb. $(A + B)$.
S = solvent content of extract layer, lb. S/lb. $(A + B)$.
s = solvent content of raffinate layer, lb. S/lb. $(A + B)$.
Y = fraction of B in the extract layer, lb. B/lb. $(A + B)$.
X = fraction of B in the raffinate layer, lb. B/lb. $(A + B)$.

subscripts 1, 2, $m - 1$, n, $m + 1$, etc., denote the stage referred to, f indicates the feed, r the final raffinate.

From the material-balance relationships per unit of raffinate product, in the manner developed subsequently for the raffinate section of a countercurrent system operating with reflux, pp. 733 through 735, the following equations may be derived:

$$\frac{Y_m - X_r}{S_m + N} = \frac{X_{m-1} - X_r}{s_{m-1} + N} = \frac{Y_m - X_{m-1}}{S_m - s_{m-1}} \tag{7a}$$

$$N = \frac{Q}{L_r} - s_r \tag{7b}$$

and

$$\frac{Y_m - X_r}{S_m - [s_r - (Q/L_r)]} = \frac{X_{m-1} - X_r}{s_{m-1} - [s_r - (Q/L_r)]} \tag{7c}$$

On the solvent content–concentration diagram, Eqs. (7a) and (7c) represent a series of straight lines radiating from a common operating point N (Fig. 28). N is located at an ordinate X_r, the solute B content of the raffinate product, and an abscissa $(Q/L_r) - s_r$. These lines intersect the extract and raffinate conjugate curves at Y_m, S_m and X_{m-1}, s_{m-1}, respectively. They relate, therefore, both the solvent and solute compositions of the raffinate stream entering any stage of the system with the corresponding compositions of the extract layer leaving that stage. Likewise, from material balances, the ratios of the two streams at any point are given by

$$\frac{L_{m-1}}{V_m} = \frac{N + S_m}{N + s_{m-1}} = \frac{S_m - [s_r - (Q/L_r)]}{S_{n-1} - [(s_r - Q/L_r)]} \tag{8}$$

Using the above equations, the number of equilibrium contact stages may be determined graphically on the solvent content–concentration diagram, as is illustrated in Fig. 28. Commencing at the raffinate terminal, the operating point N is located on a perpendicular at X_r, the final raffinate concentration, so that $N = (Q/L_r) - s_r$. Point J is located on a perpendicular at X_f, the feed composition, so that $J = Q/F$, the ratio of solvent to feed. A line from point $a(X_r, s_r)$ through J locates, at

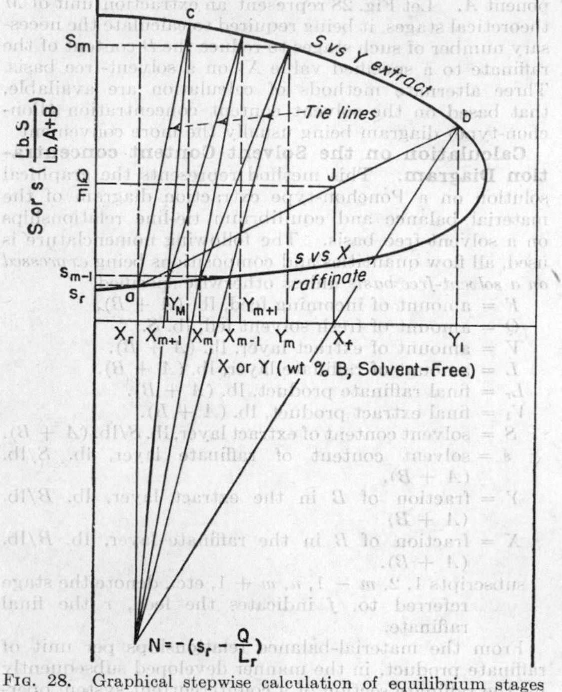

FIG. 28. Graphical stepwise calculation of equilibrium stages on the solvent content–concentration diagram for countercurrent multistage operation without reflux.

b, the final extract composition Y_1, S_1. N can thus also be located by drawing a line from b through X_f, the feed, and extending it to intersect, at N, the perpendicular erected at X_r. A tie line from a locates the extract composition (solvent and solute contents), at $c(Y_M, S_M)$, leaving the last stage M. Construction of an operating line; cN, intersects the conjugate curve at X_{m+1}, s_{m+1}, thus giving the raffinate compositions of the stage immediately preceding the last one. The tie line through

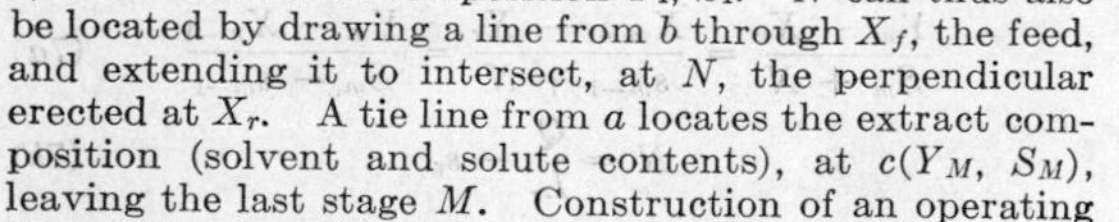

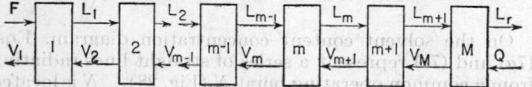

FIG. 29. Flow scheme for countercurrent multistage extraction without reflux.

X_{m+1}, s_{m+1} gives the equilibrium concentrations of extract leaving the $m + 1$ stage, Y_{m+1}, S_{m+1}. Repetition of this type of construction until the feed composition is equaled or exceeded fixes the number of stages required. The procedure is explained in more detail on pp. 735 through 737.

Calculation on the Triangular Diagram. Refer to Fig. 1c. Let Fig. 29 represent an extraction unit of M theoretical stages. F lb. per unit time of feed enter stage 1 at the left; feed passes successively as the raffinate stream through the unit and is withdrawn as final raffinate product from the final Mth stage. A fresh solvent feed Q is introduced at the Mth stage and flows countercurrently from right to left, leaving stage 1 as the final extract product.

The triangular phase diagram supplies the necessary relations between the composition and quantity of the extract layer leaving any stage and those of the raffinate layer leaving the same stage. Simultaneous graphical solution of the material balances on the equilibrium chart furnishes a graphical solution to a countercurrent extraction operation with ideal stages.

Refer to Fig. 30. Let Q represent the composition (with respect to all three components) and quantity of

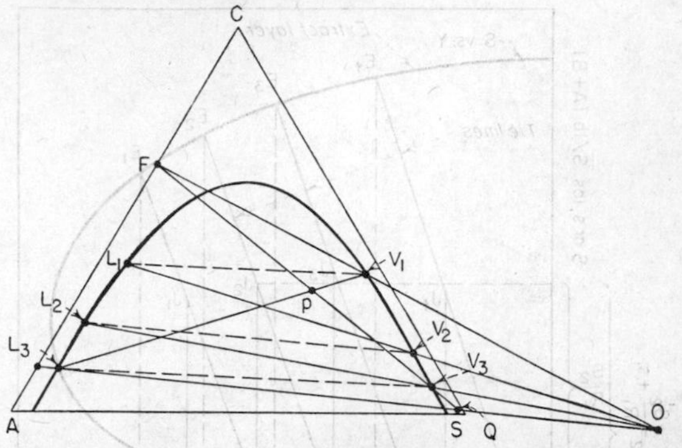

FIG. 30. Graphical solution of the countercurrent multistage extraction system by the method of Hunter and Nash.

the solvent entering the final stage (three stages are assumed in the diagram, but this imposes no restriction) and F the composition and quantity of the feed entering the first stage. L_1, L_2, etc., represent the raffinate layers leaving the first, second, etc., stages; V_1, V_2, etc., represent the extract layers leaving the first, second, etc., respectively. L_1 and V_1, L_2 and V_2, etc., are in equilibrium, and their compositions are related by the respective tie lines.

On the basis of over-all material balance equations, it may be shown that lines joining the incoming solvent and raffinate feed streams, and the leaving extract layer and raffinate layer terminal compositions, respectively, *i.e.*, lines (FQ) and V_1L_3, must intersect in a point, *i.e.*, p, so located that the ratio of the segments pF/pQ is equal to the weight of solvent Q divided by the weight of feed F. The ratio of the segments pL_3/pV_1 is equal to the weight of extract V_1 divided by the weight of final raffinate L_3. In other words, the same over-all composition p results whether the incoming feed and solvent are mixed or the leaving extract and raffinate layer are mixed. If the quantity of solvent is specified and the composition of the raffinate discharged from the final stage is likewise specified, the point p is located from the ratio of weights of solvent and feed. The composition of the terminal extract layer V_1 leaving the first stage is likewise determined by extending the line L_3p to intersect the saturation curve at V_1.

A line joining the composition of the raffinate layer entering any stage to the point representing the composition of the extract layer leaving that stage (or vice versa), *e.g.*, line L_2V_3, will intersect all the similar lines for each of the other stages in a common point, *i.e.*, O. This point ordinarily lies outside the triangle, but it may be located either within or outside the triangle. For convenience it is called the "operating point." This point has solely geometrical significance, representing the composition and amount of the hypothetical mixtures formed by mixing entering raffinate and leaving extract in positive or negative amounts. Its significance and geometrical location are discussed in detail by Hunter and Nash and by Evans (see references). The lines joining the extract and entering feed compositions at one terminal and the leaving raffinate and entering solvent compositions at the opposite terminal intersect in the same operating point O. Hence in practice this point O is readily located by first locating on the chart these compositions, drawing the lines connecting them, and then producing these lines until they intersect.

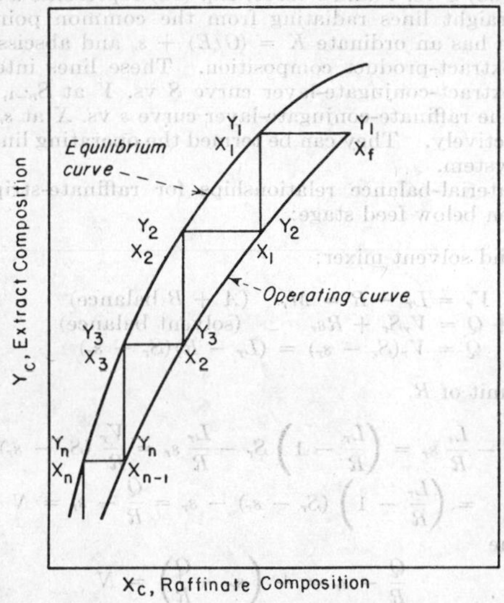

FIG. 31. Solution of countercurrent multistage extraction on equilibrium distribution chart (Varteressian and Fenske).

The terminal compositions and operating point having been determined graphically on the chart, the required number of stages is readily calculated graphically. Commencing at either terminal, *e.g.*, at stage 1 where the feed enters and the final extract V_1 leaves, a tie line V_1L_1 locates the composition L_1 of the raffinate layer leaving the first stage. An operating line drawn from L_1 to the operating point O intersects the saturation curve at L_2, *i.e.*, line L_1V_2O, locates at V_2 the composition of the extract layer leaving the second stage and entering the first. A tie line then locates L_2, the raffinate composition leaving stage 2, and so on. This process is continued until the raffinate composition found for a stage equals or exceeds that specified in advance for the final raffinate, and the number of steps that has been required is the number of theoretical stages necessary.

If the number of stages it is desired to use, rather than the quantity of solvent, is specified in advance, the procedure is the same. But in this case trial and error is necessary. This may be carried out by assuming a final extract composition or quantity of solvent and then solving graphically, using the given number of stages, until the terminal conditions thus found agree with those assumed.

Calculation on the Rectangular Equilibrium Distribution Diagram. This method employs the rectangular equilibrium distribution diagram (see p. 722) in conjunction with the triangular equilibrium chart, the final solution for the number of stages being carried out on the former. Solution of the material balance around the last stage and any other stage in the unit leads to an operating equation which is non-linear but for which particular solutions may be readily found by drawing operating lines at random on the triangular diagram from point O (see Fig. 30) and selecting the intersection points of these lines with the phase envelope as coordinates of the operating curve. Such a curve is shown in Fig. 31. The solution for the number of stages is carried out as a stepping procedure. See also p. 746, Fig. 42.

Smith [*Chem. Eng.*, **54,** No. 3, 123 (1947)] has recently proposed an unusual plot which is stated to retain both the utility of the rectangular coordinate system and the complete information of the triangular coordinate system. He plots S as ordinate vs. $\dfrac{A + 2S}{\sqrt{3}}$ or $\dfrac{B + 2S}{\sqrt{3}}$ as abscissa, and then uses the resulting diagram in the manner described for the triangular diagram. A, B, and S are percentages of these components, respectively.

Countercurrent Liquid-liquid Extraction with Reflux

A feed mixture of two completely miscible liquid components A and B, respectively, is to be separated into the components by extracting isothermally with a solvent S. The solvent is partly miscible with each component of the feed and preferentially selects component B. The triangular phase diagram is the type of Fig. 9*b*. Operation will be countercurrent multiple contact with reflux on both the extract and the raffinate products, the feed being fed at the proper intermediate stage. A flow diagram for the system is shown in Fig. 32.

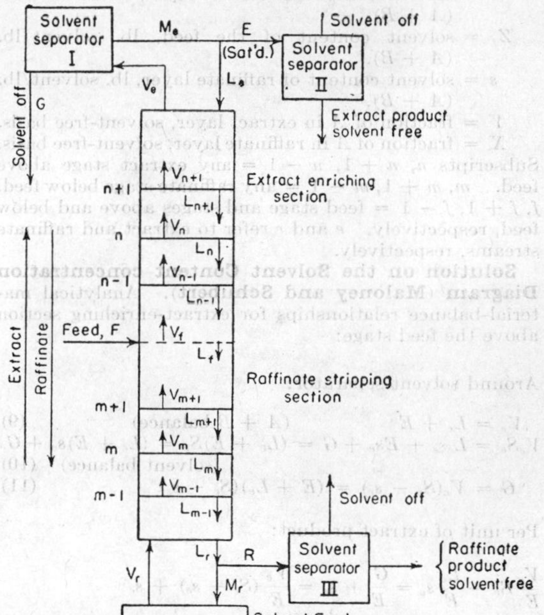

FIG. 32. Flow diagram for countercurrent multistage extraction with reflux.

Sufficient solvent is removed in the solvent separator I from the saturated extract layer to reduce this to a saturated raffinate layer. External extract reflux is furnished by returning a portion of this layer to the system, the remainder being withdrawn, after final removal of solvent in a supplementary solvent separator II, as extract product (A-rich). External raffinate reflux is supplied by mixing a portion of the raffinate layer leaving the system into the incoming solvent feed. The remaining raffinate is withdrawn, after its solvent is recovered in solvent separator III, as solvent-free raffinate product (B-rich). Equilibrium on each stage is assumed. The feed is assumed to enter just saturated with solvent.

Unless otherwise specified, the flow quantities of each stream are on a solvent-free basis and per unit time, *i.e.*, pounds or moles of $A + B$ (solvent-free) per hour. Symbols represent either weight fraction or mole fraction depending upon whether weight or mole basis is used. Let:

E = amount of extract product.
F = amount of feed.
V = extract-layer overflow from any stage.
L = raffinate-layer underflow from any stage.
G = amount of solvent removed in solvent separator, solute-free.
Q = amount of extracting solvent feed to raffinate end, solute-free.
R = amount of raffinate product.
L_r = raffinate layer (saturated) leaving last raffinate stage.
M_r = amount of raffinate reflux to mixer for return to system.
V_e = amount of extract layer leaving last extract stage to solvent separator.
M_e = amount of extract product from solvent separator.
L_e = amount of extract reflux return to extract end of system.
S = solvent content in extract layer, lb. solvent/lb. $(A + B)$.
Z_f = solvent content of the feed, lb. solvent/lb. $(A + B)$.
s = solvent content of raffinate layer, lb. solvent/lb. $(A + B)$.
Y = fraction of A in extract layer, solvent-free basis.
X = fraction of A in raffinate layer, solvent-free basis.

Subscripts n, $n + 1$, $n - 1$ = any extract stage above feed. $m, m + 1, m - 1$ = any raffinate stage below feed. $f, f + 1, f - 1$ = feed stage and stages above and below feed, respectively. e and r refer to extract and raffinate streams, respectively.

Solution on the Solvent Content–concentration Diagram (Maloney and Schubert). Analytical material-balance relationships for extract-enriching section above the feed stage:

Around solvent separator:

$$V_e = L_e + E \qquad (A + B \text{ balance}) \qquad (9)$$

$$V_eS_e = L_es_e + Es_e + G = (L_e + E)S_e = (L_e + E)s_e + G \qquad (\text{solvent balance}) \qquad (10)$$

$$G = V_e(S_e - s_e) = (E + L_e)(S_e - s_e) \qquad (11)$$

Per unit of extract product:

$$\frac{V_e}{E}S_e - \frac{L_e}{E}s_e = \frac{G}{E} + s_e = \frac{V_e}{E}(S_e - s_e) + s_e = \left(\frac{L_e}{E} + 1\right)(S_e - s_e) - s_e = K \qquad (12)$$

Define $\dfrac{G}{E} + s_e = K$

Around solvent separator and nth stage:

$$V_{n-1} = L_n + E \qquad (13)$$

$$V_{n-1}S_{n-1} = L_ns_n + Es_e + G \qquad (14)$$

$$\frac{V_{n-1}}{E}S_{n-1} - \frac{L_n}{E}s_n = \left(\frac{L_n}{E} + 1\right)S_{n-1} - \frac{L_n}{E}s_n = \frac{L_n}{E}(S_{n-1} - s_n) + S_{n-1} = \frac{G}{E} + s_e = K \qquad (15)$$

$$V_{n-1}Y_{n-1} = L_nX_n + EX_e \qquad (16)$$

$$\frac{V_{n-1}}{E}Y_{n-1} - \frac{L_n}{E}X_n = \left(\frac{L_n}{E} + 1\right)Y_{n-1} - \frac{L_n}{E}X_n = \frac{L_n}{E}(Y_{n-1} - X_n) + Y_{n-1} = X_e \qquad (A \text{ balance}) \qquad (17)$$

From Eqs. (15) and (17),

$$\frac{K - S_{n-1}}{X_e - Y_{n-1}} = \frac{S_{n-1} - s_n}{Y_{n-1} - X_n} = \frac{K - s_n}{X_e - X_n} \qquad (18)$$

On the solvent content–concentration extraction diagram (Fig. 33) S vs. Y and s vs. X, Eq. (18) represents a series of straight lines radiating from the common point K, which has an ordinate $K = (G/E) + s_e$ and abscissa X_e, the extract-product composition. These lines intersect the extract-conjugate-layer curve S vs. Y at S_{n-1}, Y_{n-1} and the raffinate-conjugate-layer curve s vs. X at s_n, X_n, respectively. They can be termed the operating lines for the system.

Material-balance relationships for raffinate-stripping section below feed stage:

Around solvent mixer:

$$V_r = L_r - R = M_r \qquad (A + B \text{ balance}) \qquad (19)$$

$$L_rs_r + Q = V_rS_r + Rs_r \qquad (\text{solvent balance}) \qquad (20)$$

$$Q = V_r(S_r - s_r) = (L_r - R)(S_r - s_r) \qquad (21)$$

Per unit of R,

$$\frac{V_r}{R}S_r - \frac{L_r}{R}s_r = \left(\frac{L_r}{R} - 1\right)S_r - \frac{L_r}{R}s_r = \frac{V_r}{R}(S_r - s_r) - s_r = \left(\frac{L_r}{R} - 1\right)(S_r - s_r) - s_r = \frac{Q}{R} - s_r = N \qquad (22)$$

Define

$$\frac{Q}{R} - s_r = -\left(s_r - \frac{Q}{R}\right) = N \qquad (22a)$$

Around solvent mixer and mth stage:

$$L_{m+1} = V_m + R \qquad (A + B \text{ balance}) \qquad (23)$$

$$L_{m+1}s_{m+1} + Q = V_mS_m + Rs_r \qquad (\text{solvent balance}) \qquad (24)$$

Per unit of R,

$$\frac{V_m}{R}S_m - \frac{L_{m+1}}{R}s_{m+1} = \left(\frac{L_{m+1}}{R} - 1\right)S_m - \frac{L_{m+1}}{R}s_{m+1} = \frac{L_{m+1}}{R}(S_m - s_{m+1}) - S_m = \frac{Q}{R} - s_r = N \qquad (25)$$

$$L_{m+1}X_{m+1} = V_mY_m + RX_r \qquad (A \text{ balance}) \qquad (26)$$

$$\frac{V_m}{R}Y_m - \frac{L_{m+1}}{R}X_{m+1} = \frac{L_{m+1}}{R}(Y_m - X_{m+1}) - Y_m = X_r \qquad (27)$$

From Eqs. (25) and (27),

$$\frac{Y_m - X_r}{S_m + N} = \frac{X_{m+1} - X_r}{s_{m+1} + N} = \frac{Y_m - X_{m+1}}{S_m - s_{m+1}} \qquad (28)$$

or

$$\frac{Y_m - X_r}{S_m - \left(s_r - \dfrac{Q}{R}\right)} = \frac{Y_{m+1} - X_r}{s_{m+1} - \left(s_r - \dfrac{Q}{R}\right)} \qquad (29)$$

On the Ponchon extraction diagram (Fig. 33) Eq. (29) represents a series of straight lines radiating from a common point N located at an ordinate X_r, the A content of the raffinate product, and abscissa $(Q/R) - s_r$. The lines intersect the extract- and raffinate-conjugate curves at Y_m, S_m and X_{m+1}, s_{m+1}, respectively. They locate the solvent and A composition of the raffinate stream entering a stage with the corresponding composition of the extract-layer upflow from that stage.

Internal and External Reflux Ratios.

Extract-enriching section: Rearranging and comparing

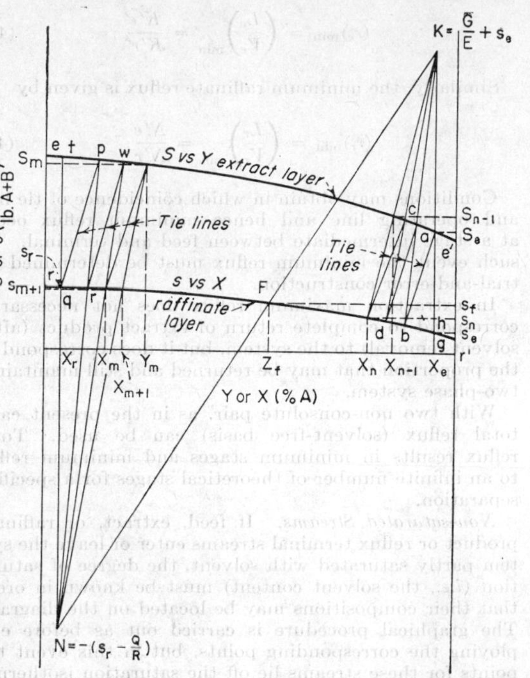

FIG. 33. Graphical stepwise calculation of equilibrium stages on the solvent content–concentration diagram for operation with reflux.

Eqs. (16) and (18) identifies the *internal reflux ratios as*

$$r_i = \frac{L_n}{V_{n-1}} = \frac{K - S_{n-1}}{K - s_n} \tag{30}$$

Similarly, the *external reflux ratio* at the extract terminal is

$$r_e = \frac{L_e}{V_e} = \frac{K - S_e}{K - s_e} = \frac{\left(\frac{G}{E} + s_e\right) - S_e}{G/E} \tag{31}$$

and, similarly, since

$$\frac{E}{V_e} = \frac{S_e - s_e}{K - s_e}$$

$$\frac{L_e}{E} = \frac{K - S_e}{S_e - s_e} \tag{32}$$

Raffinate-stripping section: The internal reflux ratio is

$$r_i = \frac{L_{m+1}}{V_m} = \frac{S_m + N}{s_{m+1} + N} = \frac{S_m - \left(s_r - \frac{Q}{R}\right)}{s_{m+1} - \left(s_r - \frac{Q}{R}\right)} \tag{33}$$

The corresponding external reflux at the raffinate terminal is

$$r_r = \frac{L_r}{V_r} = \frac{S_r + N}{s_r + N} = \frac{L_r}{M_r} = \frac{(S_r - s_r) + \frac{Q}{R}}{Q/R} \tag{34}$$

or, expressed in terms of the portion of raffinate returned with the solvent M_r to that withdrawn as product R,

$$\frac{M_r}{R} = \frac{Q/R}{S_r - s_r} \tag{35}$$

Location of the Feed Stage. From over-all material balances around the system, the following relations may be obtained:

For $A + B$ on solvent-free basis,

$$F = E + R \tag{36}$$

For solvent per unit of extract product,

$$\left(\frac{R}{E} + 1\right) Z_f = K - \frac{R}{E} N \tag{37}$$

For component A per unit of extract product,

$$\left(\frac{R}{E} + 1\right) X_f = \frac{R}{E} X_r + X_e \tag{38}$$

Combining and rearranging Eqs. (37) and (38),

$$\frac{R}{E} = \frac{K - Z_f}{Z_f + N} = \frac{X_e - X_f}{X_f - X_r} \tag{39}$$

and

$$\frac{F}{E} = \frac{K - N}{Z_f + N} = \frac{X_e - X_r}{X_f - X_r} \tag{40}$$

Equations (39) and (40) represent a single straight line on the Ponchon extraction diagram (Fig. 33) passing through

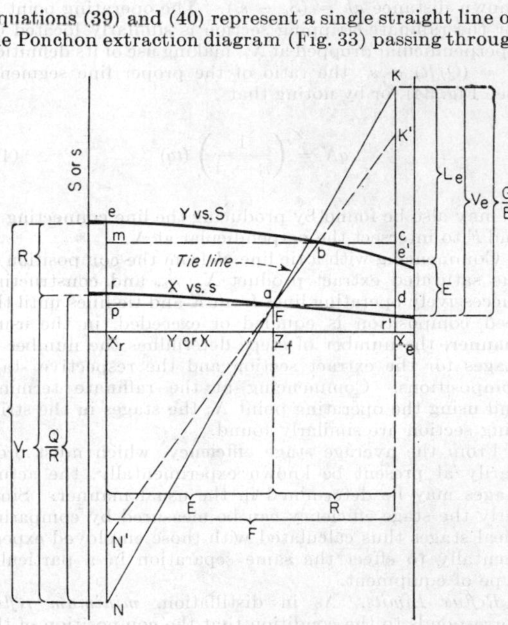

FIG. 34. Graphical representation of reflux and feed relations.

the operating points K and N and the feed composition F, Z_f. Hence the final operating lines for the extract-enriching and raffinate-stripping sections coincide to form a single straight line at the feed composition. Hence, in general, introduction of the feed at this point results in an optimum number of stages for a specified set of conditions.

Graphical Interpretation. The several terms of Eqs. (31), (32), (35), (39), and (40) are represented by line segments on the Ponchon extraction diagram, as shown in Fig. 34. The values of Q/R and G/E are absolute values read on the S, s scale (ordinate). All other segments are significant only as parts of ratios in the preceding equations. Thus the reflux ratio at the extract terminal is $r_e = L_e/V_e$ and is equal to the ratio of the segments indicated.

Calculation Procedure. Based on the preceding equations, the determination of the required number of theoretical extraction stages for fixed terminal conditions and reflux ratios is made graphically as follows: Refer to Fig. 33. The equilibrium-solvent-content curves vs. the percentage of the component A to be extracted on a solvent-free basis for extract and raffinate layers ee' and rr', respectively, are first plotted. The feed, extract product, and raffinate product composition Z_f, X_e and X_r, respectively, are located. The operating point K for the extract section and for the specified reflux ratio is located on a perpendicular at X_e by one of several methods: (1) so that the distance gK is equal to the value of $(G/E) + s_e$; or (2) so that the ratio of the segment aK to segment hK equals L_e/V_e; or (3) since

$$K - S_e = \left(\frac{r_e}{1 - r_e}\right)(S_e - s_e) \tag{41}$$

or

$$aK = \left(\frac{r_e}{1 - r_e}\right)(ah) \tag{41a}$$

K is located by producing the perpendicular beyond point $a(S_e, X_e)$ a distance equal to $r_e/(1 - r_e)$ times the known distance $ah = (S_e - s_e)$. The operating point N for the raffinate-stripping section is similarly located on a perpendicular dropped at X_r, making use of its definition $N = (Q/R) - s_r$, the ratio of the proper line segments (see Fig. 34), or by noting that

$$qN = \left(\frac{1}{r_r - 1}\right)(tq) \tag{42}$$

N may also be found by producing the line connecting K and F to intersect the perpendicular at N.

Commencing with a tie line ab from the composition of the saturated extract product X_e, S_e, and constructing successively operating lines from K and tie lines until the feed composition is equaled or exceeded, in the usual manner, the number of steps determines the number of stages for the extract section and the respective stage compositions. Commencing at the raffinate terminal and using the operating point N, the stages in the stripping section are similarly found.

From the average stage efficiency, which must ordinarily at present be known experimentally, the actual stages may be determined in the usual manner. Similarly the stage efficiency can be measured by comparing ideal stages thus calculated with those employed experimentally to effect the same separation in a particular type of equipment.

Reflux Limits. As in distillation, *minimum reflux* corresponds to the condition that the composition of the extract (solvent-free basis) on one stage is the same as that on the stage above and below, *i.e.*, $Y_n = Y_{n-1} = Y_{n+1}$, and the extract entering a stage has a composition corresponding to equilibrium with the raffinate layer on that stage. On the Ponchon extraction diagram (Fig. 33) this corresponds to coincidence of a tie line and an operating line. Ordinarily as the reflux is reduced such coincidence tends to occur at the feed stage so that $Y_{f+1} = Y_f = Y_{f-1}$. This corresponds, when operating without reflux, p. 731, to the situation that equilibrium between incoming feed and exit raffinate is attained at the feed end of the system. Hence the *minimum reflux ratio* for the extract section can be graphically determined by producing a tie line drawn from the saturated feed composition a to intersect the perpendicular at X_e, *i.e.*, on Fig. 34, the line abK'. The minimum reflux ratio is then given by the ratio of the corresponding line segments:

$$(r_e)_{\min} = \left(\frac{L_e}{V_e}\right)_{\min} = \frac{K'e'}{K'r'} \tag{43}$$

Similarly, the minimum raffinate reflux is given by

$$(r_r)_{\min} = \left(\frac{L_r}{V_r}\right)_{\min} = \frac{N'e}{N'r} \tag{44}$$

Conditions may obtain in which coincidence of tie line and operating line and hence minimum reflux occur at a stage intermediate between feed and terminal. In such event the minimum reflux must be determined by trial-and-error construction.

In extraction maximum reflux does not necessarily correspond to complete return of extract product (after solvent removal) to the system, but it does correspond to the proportion that may be returned and still maintain a two-phase system.

With two non-consolute pair, as in the present case, total reflux (solvent-free basis) can be used. Total reflux results in minimum stages and minimum reflux to an infinite number of theoretical stages for a specified separation.

Non-saturated Streams. If feed, extract, or raffinate product or reflux terminal streams enter or leave the system partly saturated with solvent, the degree of saturation (*i.e.*, the solvent content) must be known in order that their compositions may be located on the diagram. The graphical procedure is carried out as before employing the corresponding points, but in this event the points for these streams lie off the saturation isotherms, ee' and rr'.

If the feed enters solvent-free, the usual case, Eqs. (39) and (40) become, respectively,

$$\frac{R}{E} = \frac{K}{N} \tag{45}$$

and

$$\frac{F}{E} = \frac{K - N}{N} \tag{46}$$

Non-isothermal Operation. If the operating temperature varies from stage to stage, the same graphical procedure may be used. The temperature distribution must be known, and the solubility curves and tie lines for each stage temperature used must be plotted on the diagram and employed in the construction.

Illustrative Example (adapted from Maloney and Schubert, *loc. cit.*). It is proposed to separate a 50 weight per cent mixture of n-heptane and methyl cyclohexane into one product containing 90 weight per cent n-heptane and another product containing 90 weight per cent methyl cyclohexane. These are to be separated continuously at a rate of 100 lb. feed/hr. (solvent-free basis) in a countercurrent multistage extractor. It is desired to know the minimum reflux required for this separation, the minimum number of stages, and the number of stages required at an operating extract reflux (L_e/E) of 5 to 1.

a. Minimum Reflux. As discussed on p. 736, minimum reflux occurs when a tie line and an operating line coincide. In the ideal case this condition occurs at the feed point. On Figs. 35*a*, 35*b*, 35*c*, the equilibrium data of Varteressian and Fenske (see also Figs. 9 and 18) are reproduced on both the solvent content–composition diagram and in the conventional rectangular selectivity diagram. The loci of both the operating and equilibrium data are produced on the rectangular selectivity chart by projection of the terminal points of tie-line and operating-line data from the solvent content–composition diagram. The rectangular diagram is reproduced in order to show the relation between the conventional method (McCabe-Thiele) of stepping and the method used here. The feed is an unsaturated mixture, of the hydrocarbons, and the extended tie line that crosses this point intersects the verticals through the desired extract and raffinate concentrations (solvent-free) at D and M. By comparison with Fig. 34, the ratio DB to BC is seen to be $(L_e/E)_{min} = 3$ to 1. The conjunction of the operating lines and equilibrium curve, as in conventional distillation, will be noted on the rectangular projection.

b. Minimum Stages. The condition of minimum stages may not always correspond to total reflux as discussed on p. 736. In this case, however, two non-consolute pairs exist; and in general it is an easy matter to maintain two phases within the column, and thus total-reflux and minimum-stage operation are identical.

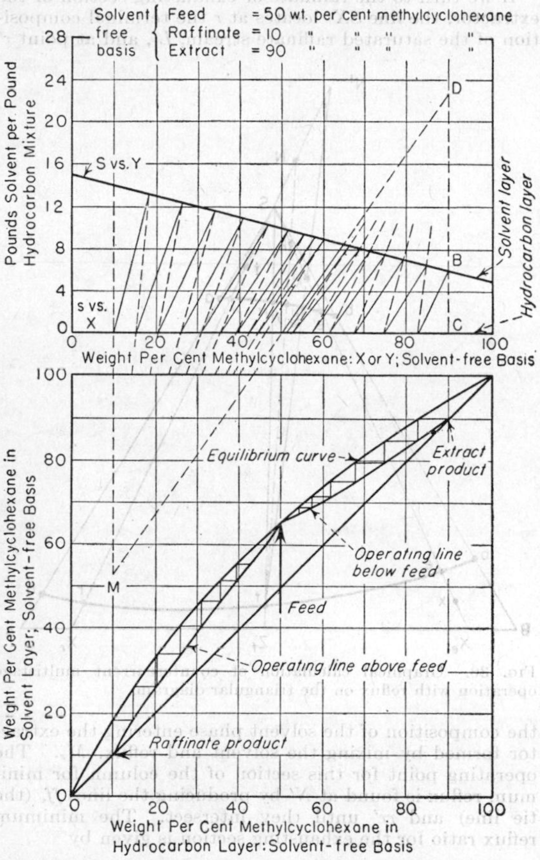

FIG. 35*a*. Graphical solution of illustrative problem.

In Fig. 35*b*, both solvent content–composition diagram and rectangular selectivity diagram are reproduced together. At total reflux the ratio (L_e/E) is infinite since E is zero. This corresponds to the existence of an operating point (*cf.* Fig. 34) at infinity, and thus all operating lines are vertical and parallel. Stepping on the solvent content–composition diagram shows the minimum number of stages to be 7. Projection onto the rectangular diagram confirms this figure and once again shows the comparison, exact in this case, to the McCabe-Thiele method.

c. Ideal Stages Required at $(L_e/E) = 5$ *to* 1. Operation at $(L_e/E) = 5$ is shown in Fig. 35*c*. The operating point D (*cf.* Fig. 34) has been placed so that $DB/BC = 5.0$. The number of stages, 11.5, has then been stepped off in the usual manner with D as the operating point until the feed tie line is crossed. At this juncture the stepping procedure is continued with point M as the operating point until the raffinate composition is reached. The absolute level of the operating point D is of further interest in that it represents the number of pounds of solvent that must pass through the system in order to produce 1 lb. of solvent-free extract of the desired composition. It should be noted in passing that the point C lies on the raffinate locus, and not on the zero-solvent-content ordinate as it might appear because of the low solubility of the solvent in the hydrocarbon in this case. The curvature of the McCabe-Thiele operating lines on the projected rectangular diagram serves to point out the inapplicability of the linear-operating-line method on partly miscible systems.

Solution on the Triangular Diagram (Varteressian and Fenske). Refer to Fig. 36 and the flow diagram in Fig. 32. The solvent-solute system used in this discussion is similar to that of the preceding illustrative example. The points r, f, and e located on the raffinate-saturation isotherm represent, respectively, the

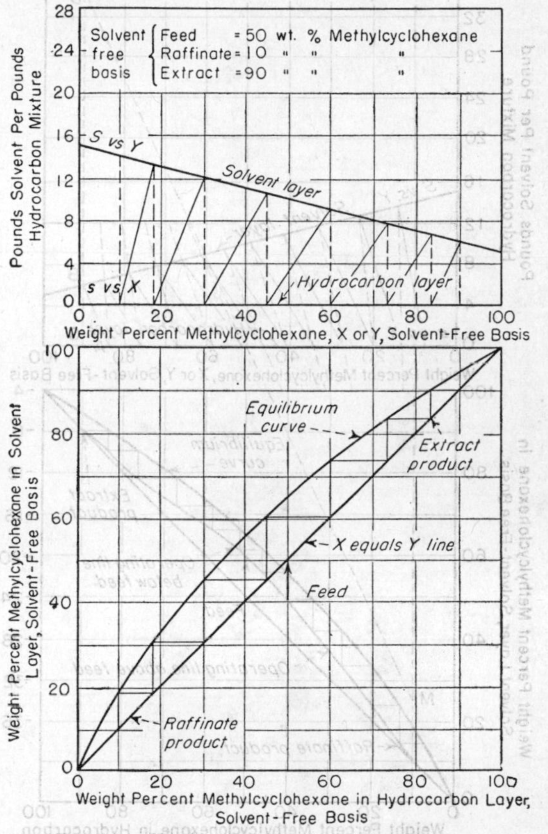

FIG. 35*b*. Graphical solution for minimum stages.

raffinate-product, the feed, and the extract-product compositions saturated with solvent. The corresponding points r', f', and e' located on the extract-layer-solubility curve represent, respectively, raffinate and feed layers which contain just sufficient solvent to convert them to saturated-extract or solvent layers, and the composition of the saturated-extract product. Note that points e and e', r and r', and f and f' lie on the lines joining the solvent-free extract composition X_e, the solvent-free raffinate composition X_r, and the solvent-free feed composition Z_f, respectively, to the solvent vertex S. f_e' represents the composition of the extract layer in equilibrium with the saturated feed located at the terminus of the tie line ff_e'.

The nomenclature previously employed, p. 734, will be used except that in this case, unless otherwise specified, *the quantities and compositions of all streams are on a solvent-containing basis.*

Minimum Reflux. Since at minimum reflux the solvent or extract layer at the feed level is in equilibrium with the entering feed, a tie line from point f locates at point f_e' the composition of the extract layer in equilibrium with the feed. If lines ee' and ff_e' are produced to intersect at point K', this locates at K' the operating point for the enriching section at minimum reflux. The composition of the solvent layer at positions in the extractor intermediately between the feed and the top is represented by points on the extract-phase locus lying between e' and f_e'.

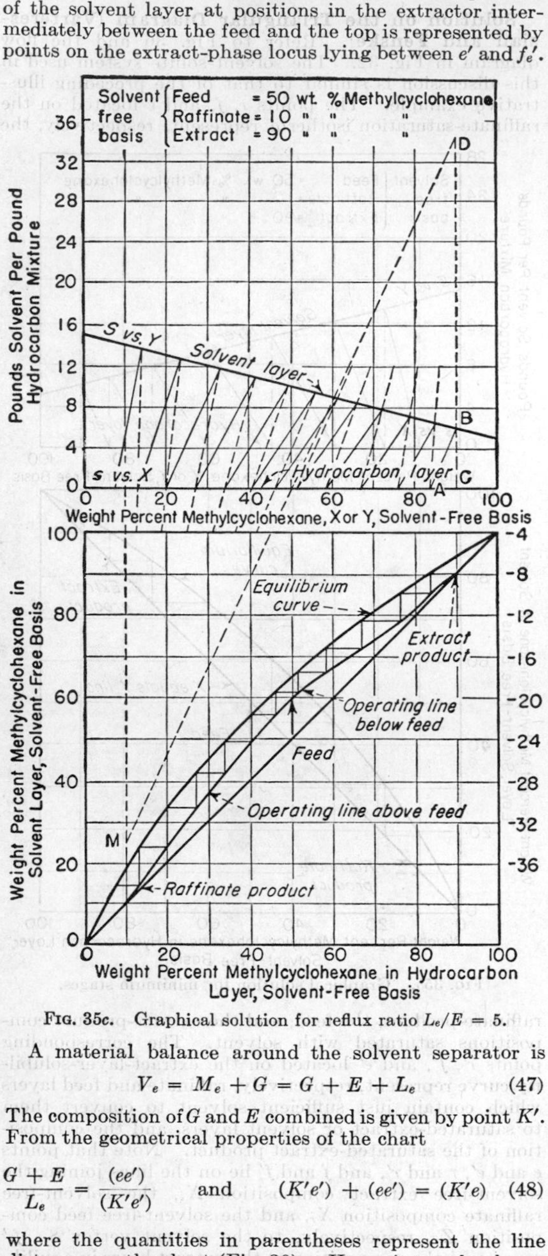

FIG. 35c. Graphical solution for reflux ratio $L_e/E = 5$.

A material balance around the solvent separator is

$$V_e = M_e + G = G + E + L_e \tag{47}$$

The composition of G and E combined is given by point K'. From the geometrical properties of the chart

$$\frac{G + E}{L_e} = \frac{(ee')}{(K'e')} \quad \text{and} \quad (K'e') + (ee') = (K'e) \tag{48}$$

where the quantities in parentheses represent the line distances on the chart (Fig. 36). Hence it may be shown from Eq. (47) that the minimum reflux ratio is

$$\left(\frac{L_e}{V_e}\right)_{\min} = \frac{(K'e')}{(K'e)} \tag{49}$$

or since $G/E = \dfrac{K'e'}{K'S}$

$$\left(\frac{L_e}{E}\right)_{\min} = \frac{(K'e')}{(ee')} \times \frac{(eS)}{(K'S)} \tag{50}$$

Similarly, the ratio of the amount of the raffinate overflow to that of the ascending solvent layer at the feed level is

$$\left(\frac{L_f}{V_{f-1}}\right) = \frac{(K'f_e)}{(K'f)} \tag{51}$$

$$\frac{L_f}{E} = \frac{(K'f_e)}{(ff_e)} \times \frac{(eS)}{(K'S)} \tag{52}$$

If we turn to the raffinate or exhausting section of the extractor, the line SX_r locates at r the terminal composition of the saturated raffinate stream, L_r, and at point r'

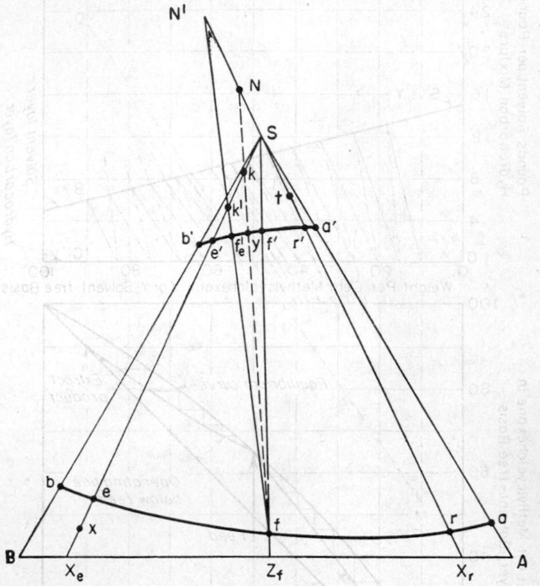

FIG. 36. Graphical calculation of countercurrent multistage operation with reflux on the triangular diagram.

the composition of the solvent phase entering the extractor formed by mixing the solvent and reflux, M_r. The operating point for this section of the column for minimum reflux is found at N' by producing the lines ff_e (the tie line) and rr' until they intersect. The minimum reflux ratio for the exhausting section is given by

$$\left(\frac{M_r}{V_r}\right)_{\min} = \frac{(N'r')}{(N'r)} \tag{53}$$

$$\frac{M_r}{R} = \frac{(N'r')}{(rr')} \times \frac{(rS)}{(N'S)} \tag{54}$$

Also

$$\frac{L_f + F}{V_f - 1} = \frac{(N'f_e)}{(N'e)} \tag{55}$$

By means of the above construction and Eqs. (49) and (53), or (50) and (54) the minimum reflux ratios for both sections of the column are readily determined. With minimum reflux the number of stages for the desired separation is infinite.

Total Reflux. The upper limit is complete reflux, and this corresponds to a minimum number of stages. In this case the operating point K' for the extract section and the operating point N' for the raffinate section coincide and fall on the solvent vertex at S. The composition of the solvent or extract phase at the feed level has B and A in the same ratio as the feed and is located at f' on the saturation curve $b'a'$ by a line drawn from f to the solvent vertex. Total reflux may not be attainable, as pointed out on p. 736.

Practical Reflux Ratios. For practical operating conditions the solvent layer at the feed level will have a composition represented by a point intermediate between f_e and f', say at y. Similarly, the operating point for the extract section will fall between K' (*i.e.*, minimum reflux) and S, *e.g.*, at point K. The operating point for the raffinate section will fall on line rr', produced, intermediate between N' and S, *e.g.*, at N.

Having specified a given feed, extract, and raffinate composition and the reflux ratio for the enriching and for the raffinate sections, the operating points K and N for each section may be readily determined from Eqs. (49) or (50) and (53) and (54) respectively, by following the described construction on the diagram. For example, if the specified reflux ratio (L_e/V_e) is $\frac{1}{4}$, the operating point K is located on the line $X_eee'S$ so that the distance Ke' is to the distance Ke as 1 is to 4; or, alternately, if L_e/E is specified, then K is so located that $\frac{(Ke')(eS)}{(ee')(KS)}$ is equal to the numerical value specified. A line from f through K to intersect rr', produced, then locates the operating point N for the exhausting section.

The solvent might if desired be removed from the extract layer in the solvent separator to such an extent that the extract reflux is less than saturated. Its composition would, however, remain on the line SX_e and at some point intermediate between e and X_e, *e.g.*, at x. In this case the reflux ratio would be determined as before, but the distance Kx would replace the distance Ke. In a similar manner, the quantity of raffinate reflux might have been such in proportion to the solvent that the composition of the solvent phase entering the extractor was less than saturated, or say at some point t.

Determination of the Number of Stages. With the operating points and terminal compositions located, the number of ideal stages required for the extract section and for the raffinate section are readily determined by either the method of Hunter and Nash or that of Varteressian and Fenske. The stages in each section are determined independently in the former method, using the appropriate operating point in each case. In the latter method an operating curve is plotted on the rectangular distribution diagram, as previously described. This is determined by drawing in operating lines at random in the enriching and in the exhausting sections, respectively, and plotting the compositions thus determined.

The reader is referred to the papers by Varteressian and Fenske for the solution of an actual example.

For methods of calculating fractional liquid extraction operations, see Scheibel, *Chem. Eng. Progress*, **44**, 771 (1948), and Bush and Densen, *Anal. Chem.*, **20**, 121 (1948).

Extraction Limits. In the general case for a fixed feed, unless the quantity of solvent used exceeds a definite value, it is impossible to secure 100 per cent extraction of solute from the raffinate even with an infinite number of stages. A definite finite value of the raffinate concentration is reached for each quantity of solvent even with infinite stages. The exact value depends upon the nature of the distribution relation and upon the quantity of solvent. Evans [*J. Chem. Education*, **14**, 408 (1937)] calculates that with *immiscible* solvents and the ideal distribution law valid about 94 per cent of the maximum removal possible with a given volume of solvent is obtained in five stages (no reflux) and concludes that it is usually hardly worth while to use more than five stages. He finds over 90 per cent of the maximum removal possible with infinite stages in five stages in the general case.

With partly miscible solvents, if an operating line and a tie line coincide, no further extraction is possible. In general for countercurrent extraction this occurs at one or the other terminals with an infinite number of stages. Depending upon the nature of the binodal curve, the slopes of the tie lines, the composition of the feed, and the quantity of solvent, coincidence of a tie line with an operating line may under certain circumstances occur at an intermediate point in the system. It is thus not always possible under every set of conditions to bring the outgoing extract into equilibrium with the feed with any given raffinate and solvent (see p. 736).

In other cases a better separation between the components of the feed may be obtained if the extract layer is brought out with a composition less than that corresponding to equilibrium with the feed. Varteressian and Fenske [*Ind. Eng. Chem.*, **28**, 1353 (1936); **29**, 270 (1937)] discuss the factors imposing limitations on the results of an extraction process and compare those obtained by different operating methods.

Algebraic Calculation of the Minimum Number of Stages. Varteressian and Fenske [*Ind. Eng. Chem.*, **29**, 270 (1937)] show that the minimum number of stages for a countercurrent extraction system with total reflux may be calculated algebraically in the specific case where the distribution of the two components between the phases at equilibrium is expressed by an equation of the type

$$\frac{Y_B}{Y_A} = \beta \frac{X_B}{X_A} \tag{56}$$

β is a constant analogous to the relative volatility.

Special Cases Where the Solvents Are Immiscible

If one of the components, A, of the feed is substantially completely immiscible with the solvent over the operating range desired, so that the solvent phase may be assumed with sufficient accuracy to contain only C and the raffinate phase to dissolve none of the solvent, extraction calculations are simplified. The process then involves only the distribution of C between the two phases and becomes almost completely analogous to the calculation procedures employed for absorption and distillation. Unfortunately, this situation is not the common one in extraction.

Ideal Distribution Law Invalid. The equilibrium data are best represented and used in terms of the equilibrium distribution plot, schematically represented by curve ON, Fig. 37. Compositions are conveniently expressed as the weight ratio of the extractable component to the solvent in each phase. Let x_0 represent the weight of extractable component C per pound of the solvent component A in the feed, x_1 the corresponding weight ratio in the raffinate, y_1 the weight ratio of C to the extracting solvent S in the extract in equilibrium with x_1, and y_0 the corresponding weight ratio of C to S in the entering fresh solvent; H is the pounds of solvent A fed and S the pounds of extracting solvent used. For a single-stage operation by a material balance the quantity of solvent is given by

$$S = H\left(\frac{x_0 - x_1}{y_1 - y_0}\right) \tag{57}$$

If the initial solvent S is pure, $y_0 = 0$. y_1 is related to x_1 by the equilibrium distribution diagram and calculations involve simply the solution of Eq. (57) in terms of this equilibrium value. The solution is conveniently made graphically on the distribution diagram Fig. 37. The coordinates of the point P are y_1, x_1, the equilibrium values in extract and raffinate. From Eq. (57) the slope of a line PQ connecting the concentration of the original feed x_0 (the solvent is assumed for simplicity to be pure) to point P is $-H/S$, the solvent ratio. The line is fixed by knowing either x_0 and x_1 or x_0 and H/S. If the entering solvent contains initially some of the solute, *e.g.*, composition y_0, a line with y_0 as ordinate is drawn parallel to the abscissa, *i.e.*, the line RK. This is employed in the construction instead of the abscissa. That is, a line from point K of slope $-H/S$ would be drawn to locate x_1, y_1.

Simple Multistage Contact. As Evans has shown [*Ind. Eng. Chem.*, **26**, 860 (1934)], the above procedure is simply extended to two or more stages for simple multistage calculations. Thus in Fig. 37 to locate

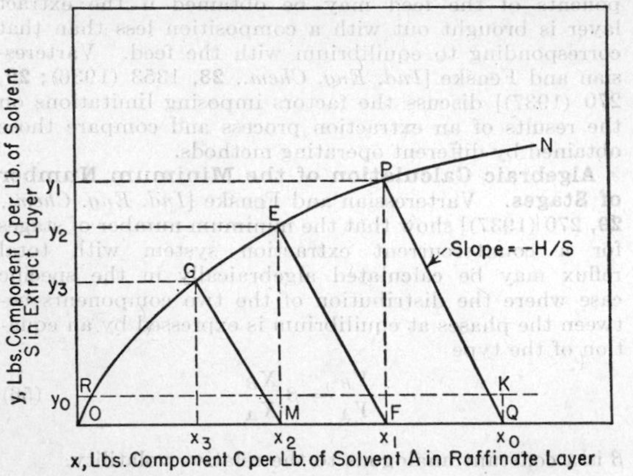

FIG. 37. Graphical calculation of simple multistage contact with immiscible solvents on equilibrium chart.

the composition of the raffinate from the second stage, the line EF is drawn from point x_1, representing the composition of the raffinate discharged from the first stage and having for its slope the ratio of the solvent H to the amount of solvent S_2 used in this stage, or $-H/S_2$. The composition of the raffinate from the third stage x_3 is similarly determined. The solvent ratio and hence the slope of the lines for each stage may have any specified value. If three stages are specified, x_3 is the final composition of the raffinate discharged from the operation. If the final raffinate is specified in advance, the construction is continued until the raffinate composition from a stage has this composition. The number of stages necessary is thus found or, in the case illustrated, three. The total quantity of solvent used is the sum of that used in each stage, or $S_1 + S_2 + S_3$. The average concentration of the total extract obtained is

$$y_{\text{avg}} = \frac{y_1S_1 + y_2S_2 + y_3S_3 + \cdots}{S_1 + S_2 + S_3 + \cdots} \tag{58}$$

Countercurrent Multistage Contact. For a system of N ideal stages a material balance on the component C around the first stage (*i.e.*, the terminal where the feed enters and the final extract leaves) and below the nth stage is

$$Hx_0 + Sy_{n+1} = Hx_n + Sy_1 \tag{59}$$

or

$$y_{n+1} = \frac{H}{S}x_n + \left(y_1 - \frac{H}{S}x_0\right) \tag{60}$$

where y_{n+1} is the composition of the extract entering the nth from the $(n + 1)$th stage, x_n the raffinate composition leaving the nth stage H the weight of solvent A in the feed, and S the weight of extracting solvent S fed to the system. In this case H/S is constant through the extraction system, and Eq. (60) is a straight line of slope H/S on the x-y distribution diagram, *i.e.*, the operating line OP on Fig. 38. Knowing the compositions at either one of the terminals, *i.e.*, x_0, y_1 or x_n, y_{n+1} and the solvent ratio H/S, or the compositions at both terminals, the operating line OP is readily constructed on the equilibrium distribution diagram, Fig. 38. The number of ideal stages is then found by stepping off the intervals from operating line to equilibrium curve and back to operating line, beginning at either end, as indicated on the diagram. The steps, PQR, RTG, etc., each represent a stage, and any point on the operating line has as its coordinates the composition of the raffinate leaving the stage and the extract entering this stage from the

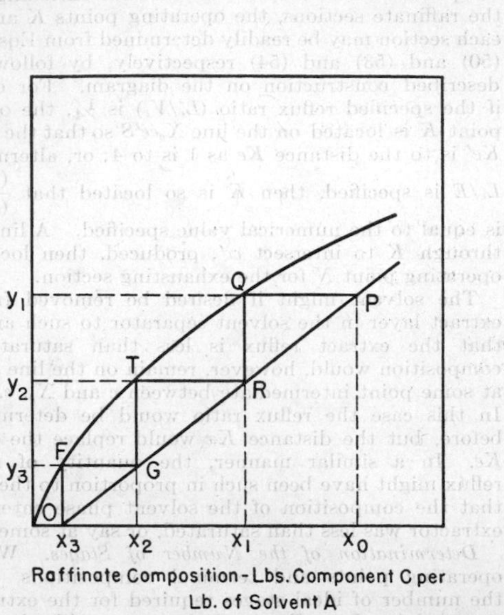

FIG. 38. Graphical solution of countercurrent multistage extraction on the distribution diagram; immiscible solvents.

stage preceding, *i.e.*, x_n, y_{n+1}. If the number of stages to be used is specified in advance, trial and error is required to locate an operating line which gives the required terminal compositions with this number of stages. When located, its slope gives the necessary quantity of solvent.

When the two solvents are immiscible, the separation between the two components of the feed in the extract is always complete.

Ideal Distribution Law Valid. If the ideal law (see p. 320) can be assumed, the equilibrium distribution curve, Fig. 37 or Fig. 38, becomes a straight line through the origin. This is the case commonly treated in textbooks of physical chemistry. Rather than employing the graphical method, convenient algebraic solutions of the extraction system are possible.

Single Contact. With no solute in the entering solvent Eq. (57) becomes

$$\frac{x_0 - x_1}{y_1} = \frac{S}{H} \tag{61}$$

and from the distribution law $(y_1 = Dx_1)$*

$$x_1 = \left(\frac{H}{H + SD}\right)x_0 \tag{62}$$

or

$$S = \frac{H}{D}\left(\frac{x_0 - x_1}{x_1}\right) \tag{63}$$

* As employed here D has the units (lb./lb. solvent A)/(lb./lb. solvent B). Its numerical value differs therefore from that frequently employed in the scientific literature when the units used are usually g. or g.-moles per unit volume.

The solute unextracted, or alternately, the quantity of solvent required for a specified fractional reduction in the concentration of the solute may thus be calculated algebraically.

Simple Multistage Contact. For equal quantities of fresh solvent containing no solute initially, the concentration of unextracted solute in the raffinate from the final contact is

$$x_N = \left(\frac{H}{H + SD}\right)^N x_0 \tag{64}$$

where N is the number of contacts. A nomographic solution of (64) has been prepared by Nord [*Ind. Eng. Chem.*, **38**, 560 (1946)].

For a fixed quantity of solvent ($S_f = NS$), x_N will be a minimum for a given value of N if the solvent is equally divided between the stages.

Equation (64) is solved conveniently for N in the form

$$\log \frac{x_N}{x_0} = N \log \left(\frac{H}{H + SD}\right) \tag{65}$$

or

$$\log Z_N + N \log \left(1 + \frac{SD}{H}\right) = 0 \tag{66}$$

where $Z_N = x_N/x_o$, the fraction unextracted after the Nth stage.

The chart of Fig. 39, in which Z_N is plotted vs. $(1 + SD/H)$ on logarithmic coordinates for various specified

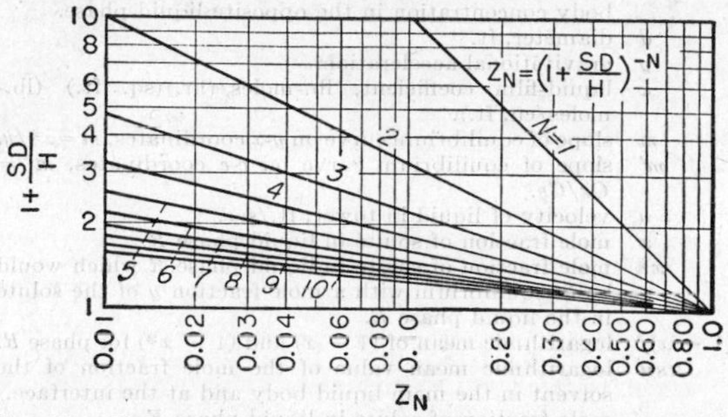

FIG. 39. Underwood's chart for graphical solution of Eq. (64). [*Underwood, Ind. Chemist*, **10**, 128 (1934).]

values of N [see Underwood, *Ind. Chemist*, **10**, 129 (1934)], provides a ready graphical solution of Eq. (64). The effect on Z_N of the number of contacts and of the quantity of solvent employed may be quickly seen from the chart.

If the extracting solvent contains initially some of the solute, Eq. (64) [Underwood, *op. cit.*] is modified to

$$x_N = C^N x_0 + \frac{C(1 - C^N)w}{1 - C} \tag{66a}$$

where w represents the weight of solute contained initially in the weight S of the solvent used per contact, and $C = \dfrac{H}{H + SD}$. Since $C < 1$, when $N = \infty$

$$x_N = \frac{Cw}{1 - C} = \frac{Hw}{SD} \tag{67}$$

which limits the removal of solute in this case.

If the total quantity of solvent S_f is fixed while N varies, since $S = S_f/N$, then

$$Z_N = \frac{x_N}{x_0} = \left(\frac{H}{H + \dfrac{S_f D}{N}}\right)^N = \left(1 + \frac{S_f D}{HN}\right)^{-N} \tag{68}$$

In Fig. 40 (see Underwood, *op. cit.*) Z_N is plotted as abscissa vs. $S_f D/H$ as ordinates on logarithmic coordinates for various values of N from 1 to ∞. From this chart x_N is readily determined graphically for any fixed quantity of solvent divided equally between any desired number of stages.

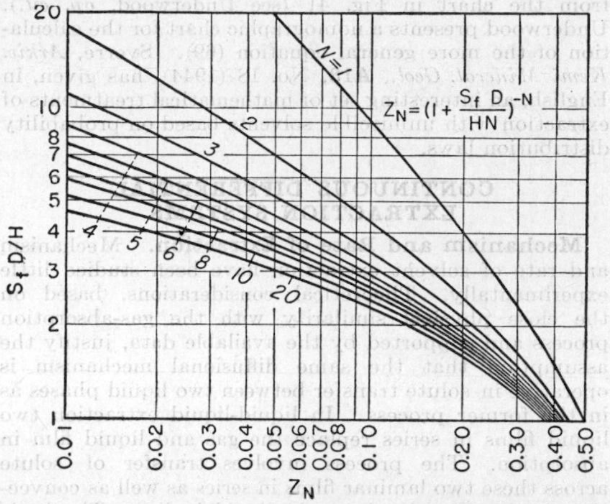

FIG. 40. Underwood's chart for graphical solution of Eq. (68). [*Underwood, Ind. Chemist*, **10**, 129 (1934).]

Countercurrent Multistage Contact. For continuous countercurrent multistage operation where the law $y = Dx^*$ (x^* is the equilibrium value of x corresponding to y) may be assumed, the following equation holds [see Underwood, *op. cit.*; and Hunter and Nash, *J. Soc. Chem. Ind.*, **51**, 285T (1932)].

$$Z_N = \frac{x_N}{x_0} = \frac{(E - 1) + r(E^N - 1)}{E^{N+1} - 1} \tag{69}$$

where $r = w/Hx_0 = Sy_0/Hx_0$

$E = SD/H$ and may be called the "extraction factor," (E is the ratio of the slope of the equilibrium curve to the slope of the operating line).

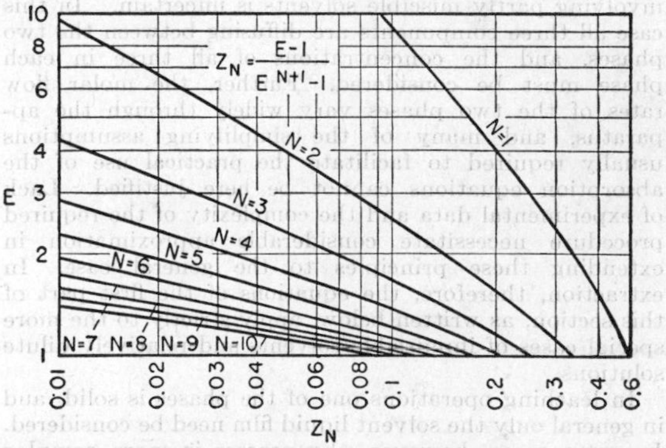

FIG. 41. Underwood's chart for graphical solution of Eq. (69*a*). [*Underwood, Ind. Chemist*, **10**, 129 (1934).]

In this case S and H represent, respectively, the weights of the solute-free solvents entering the system per unit time. If the extracting solvent contains initially no solute, $w = 0$, $r = 0$, and Eq. (69) reduces to

$$Z_N = \frac{E - 1}{E^{N+1} - 1} \tag{69a}$$

When E is specified, x_N, the exit composition for a fixed number of stages, or alternately, N, where x_N is specified, may be calculated algebraically from Eqs. (69) and (69*a*). The leaving extract composition results from a material balance around the terminals of the extractor. The solution of Eq. (69*a*) is quickly made graphically from the chart in Fig. 41 (see Underwood, *op. cit.*). Underwood presents a nomographic chart for the calculation of the more general equation (69). Sverre, *Arkiv. Kemi Mineral. Geol.*, **A18**, No. 18 (1944), has given, in English, an interesting set of mathematical treatments of extraction with unmiscible solvents based on probability distribution laws.

CONTINUOUS DIFFERENTIAL EXTRACTION SYSTEMS

Mechanism and Rate of Extraction. Mechanism and rate of solvent extraction have been studied little experimentally. Theoretical considerations, based on the close physical similarity with the gas-absorption process and supported by the available data, justify the assumption that the same diffusional mechanism is operative in solute transfer between two liquid phases as in the former process. In liquid-liquid extraction two liquid films in series replace the gas and liquid film in absorption. The process involves transfer of solute across these two laminar films in series as well as convective mechanisms in the main liquid bodies. The basic diffusional theory, equations, and methods of treating the rate of mass transfer across single liquid films and for two films in series have been developed in Sec. 8, pp. 538 to 547; 548 to 552, and 553 to 555. In general the pages noted deal with a gas phase and a liquid phase in contact. Identical equations and principles may be written for the case in which the gas phase is replaced by another liquid phase.

The equations and relationships developed for absorption in general consider the inert gas and the solvent liquid to be essentially immiscible. These equations are satisfactorily applied to extraction if the two liquid solvents can be considered immiscible (thus allowing for the interphase transfer of a single solute) and if the conditions assumed in their initial development are fulfilled by the particular extraction problem under consideration. Application to the more general case of extraction involving partly miscible solvents is uncertain. In this case all three components are diffusing between the two phases, and the concentrations of all three in each phase must be considered. Further, the molar flow rates of the two phases vary widely through the apparatus, and many of the simplifying assumptions usually required to facilitate the practical use of the absorption equations cannot be here justified. Lack of experimental data and the complexity of the required procedure necessitate considerable approximation in extending these principles to the general case. In extraction, therefore, the equations of the first part of this section, as written below, apply strictly to the more special cases of immiscible solvents and relatively dilute solutions.

In leaching operations one of the phases is solid, and in general only the solvent liquid film need be considered. In some cases, however, the process is more complex and the rate of removal of solute in, and from, the solid appears to be involved. Where the extracted substance must diffuse through an animal or vegetable membrane, the rate of this process may be of controlling importance. In other cases the rate of penetration of the solvent into the interior of the solid can be the important factor. Equations for the diffusion rates under certain of these conditions have been solved by Barrer "Diffusion in and through Solids," [Macmillan (Cambridge), New York, 1941], and the problem of the effect of structure on the diffusion theory has been examined by Osburn and Katz [*Trans. Am. Inst. Chem. Engrs.*, **40**, 511 (1944)]. Boucher, Brier, and Osburn [*ibid.*, **38**, 967 (1942)], and King, Katz, and Brier [*ibid.*, **40**, 533 (1944)] have applied these rate equations to soy oil extraction from soy meal and from porous plates. General application is lacking at the present time.

Nomenclature for Continuous Differential Extractions

A	total interfacial contact area, sq. ft. $= aV$.
B	effective thickness of a liquid film, ft.
D_L	diffusion coefficient of solute through the solvent, sq. ft./hr.
H.T.U.	height of transfer unit.
K	over-all liquid phase extraction coefficient, lb.-moles/(hr.)(sq. ft.) (lb.-moles/cu. ft.).
L'	total moles of liquid in unit time, solute plus solvent, lb.-moles/hr.
L	moles of liquid, solute free, per unit time.
N	moles of solute transferred per unit time.
N_t	number of transfer units.
Re	Reynolds number.
S	cross-sectional area of column, sq. ft.
V	effective volume of extraction column, cu. ft.
Z	height of column, ft.
a	effective interfacial contact area of column per unit volume, sq. ft./cu. ft.
c	concentration of solute, lb.-moles/cu. ft.
c^*	concentration of solute, lb.-moles/cu. ft., in one liquid phase which would be in equilibrium with the main-body concentration in the opposite liquid phase.
d	diameter, ft.
g	gravitational acceleration.
k	liquid-film coefficient, lb.-moles/(hr.)(sq. ft.) (lb.-moles/cu. ft.).
m	slope of equilibrium curve on y-x coordinates, $m = x^*/y$.
m'	slope of equilibrium curve on c-c coordinates, $m' = C_E/C_R$.
u	velocity of liquid in tower, ft./sec.
x	mole fraction of solute in liquid phase R.
x^*	mole fraction of solute in liquid phase R which would be in equilibrium with a mole fraction y of the solute in the liquid phase E.
$(1 - x)_f$	logarithmic mean of $(1 - x)$ and $(1 - x^*)$ for phase R.
x_{SM}	logarithmic mean value of the mole fraction of the solvent in the main liquid body and at the interface.
y	mole fraction of solute in liquid phase E.
y^*	mole fraction of solute in liquid phase E which would be in equilibrium with a mole fraction x of the solute in the main body of phase R.
$(1 - y)_f$	logarithmic mean of $(1 - y)$ and $(1 - y^*)$ for phase E.
μ	viscosity of liquid, lb./(hr.)(ft.).

Subscripts:

avg refers to average conditions.
c refers to continuous phase.
d refers to discontinuous phase.
E refers to extract phase.
i refers to interface value.
L refers to main body of a liquid phase.
lm refers to logarithmic mean.
O refers to over-all conditions.
R refers to raffinate phase.
S refers to solvent.
1 refers to rich end of column.
2 refers to lean end of column.

Liquid-film Equations in Extraction. The process here considered is one that involves the transfer of a solute from a raffinate phase R to an extracting phase E. The same relations may be written for the transfer of solute in the opposite direction by assigning the opposite algebraic sign to the concentration differences.

The following equations are adopted for extraction from the corresponding equations for absorption by considering the gas phase and gas film to have replaced a liquid phase and its corresponding liquid film. These absorption equations are discussed thoroughly in Sec. 8, pp. 549 through 550.

For solute transfer across a liquid film in extraction under steady-state conditions at any point in a continuously functioning apparatus [Eq. (84), p. 549, Sec. 8],

$$N = kA(c_L - c_i) = kAc_{\text{avg}}(x_L - x_i) \tag{70}$$

By analogy to Eq. (84), p. 549, Sec. 8, the film-extraction coefficient is given by

$$k = \frac{D_L}{Bx_{SM}} \tag{71}$$

B, the effective film thickness, is expected to be some function of the Reynolds number (*cf.* Sec. 8, p. 541 for effect in gaseous systems), and thus Eq. (71) supplies a qualitative indication of the factors that affect extraction rate under steady-state conditions.

All extraction processes, as previously noted, are carried out across two liquid films in series. The magnitudes of the individual film coefficients are of considerable interest, since they are of a more fundamental character than the usually measured over-all coefficient. Considerable effort has been expended in attempts to measure these rate constants and their dependence upon the physical properties of the fluids and upon the flow characteristics of the system. No extensive amount of data exists at the present time.

Fallah, Hunter, and Nash [*J. Soc. Chem. Ind.*, **54**, 49T (1935)] studied the extraction of phenol between water and kerosene in a wetted-wall column and found that the data for the kerosene film, when kerosene was the core liquid, could be correlated by the relation

$$\frac{kd}{D_L} = 0.94\left(\frac{du\rho}{\mu}\right)^{0.8}\left(\frac{\mu}{D_L\rho}\right)^{0.46} \tag{72}$$

This equation is comparable with one deduced by Gilliland for absorption [see also pp. 541–544].

Brinsmade and Bliss [*Trans. Am. Inst. Chem. Engrs.*, **39**, 679 (1943)], in studying the extraction of acetic acid from methyl isobutyl ketone with water in a wetted-wall tower were able to achieve a separation of film values. They report

For core fluid:

$$\frac{kd}{D_L} = 1.07(Re)^{0.67}\left(\frac{\mu}{D_L\rho}\right)^{0.62} \tag{73}$$

For wall fluid:

$$\left(\frac{\mu^2}{\rho^2 g}\right)^{1/3}\frac{k}{D_L} = 0.00135(Re)\left(\frac{\mu}{D_L\rho}\right)^{0.62} \tag{74}$$

where the various quantities are evaluated for the stream in which the coefficient is being computed.

Colburn and Welsh [*Trans. Am. Inst. Chem. Engrs.*, **38**, 179 (1942)], in a study on the binary system isobutanol-water in a spray tower, were able to separate what appear to be individual film resistances on a volume basis. They also suggest a method of plotting over-all data which may separate individual film values. The method has not been tested to any extent.

The analogy between absorption and extraction may be continued, and the transfer-unit concept of Colburn (*cf.* p. 550) can be introduced. Thus

$$dN_{tR} = \frac{x_{SM}\,dx}{(1 - x)(x - x_i)} = \frac{k_R x_{SM} c_{\text{avg}}\,dA}{L'_R} \tag{75}$$

which for dilute solutions becomes

$$dN_{tR} = \frac{dx}{(x - x_i)} = \frac{k_R x_{SM} c_{\text{avg}}\,dA}{L'_R} \tag{76}$$

Now the transfer unit is defined:

$$\text{H.T.U.}_R = \frac{Z}{N_{tR}} \tag{77}$$

So

$$\text{H.T.U.}_R = \frac{ZL'_R}{Ak_R x_{SM} c_{\text{avg}}} = \frac{L'_R/S}{k_R a x_{SM} c_{\text{avg}}} \tag{78}$$

Over-all Relationships in Extraction. When the two-film theory is applied to steady-state solute transfer across two liquid films in series in a continuous apparatus, the rate of transfer is given by

$$dN = k_E(c_{Ei} - c_{EL})dA = k_R(c_{RL} - c_{Ri})dA \tag{79}$$

$$dN = k_E c_{E\text{avg}}(y_{Ei} - y_{EL})dA = k_R c_{R\text{avg}}(x_{RL} - x_{Ri})dA \tag{80}$$

As in absorption, over-all coefficients may be defined

$$K_E = \frac{dN}{(c_E^* - c_{EL})dA} \tag{81}$$

$$K_R = \frac{dN}{(c_{RL} - c_R^*)dA} \tag{82}$$

In packed and spray columns the interfacial area is not measurable, and dA is replaced by $A\,dV$

$$K_E a = \frac{dN}{(c_E^* - c_{EL})dV} \tag{83}$$

$$K_R a = \frac{dN}{(c_{RL} - c_R^*)dV} \tag{84}$$

In using over-all coefficients, the phase offering the major resistance to transfer should be selected as a basis for calculation.

The over-all coefficients are related to the individual film coefficients (see Sec. 8, p. 549) in the following manner:

$$K_E a = \frac{1}{\dfrac{1}{k_E a} + \dfrac{m'}{k_R a}} \tag{85}$$

$$K_R a = \frac{1}{\dfrac{1}{k_R a} + \dfrac{1}{m' k_E a}} \tag{86}$$

Where $m' = c_{Ei}/c_{Ri} = c_{EL}/c_R^* = c_E^*/c_{RL}$ slope of equilibrium curve.

If the ideal-distribution law holds (m' = a constant), or if c_E is linear with c_R in the range under consideration, $K_E a$ and $K_R a$ will be constant and independent of concentration if $k_E a$ and $k_R a$ are constant. Otherwise the over-all coefficients vary with concentration even if the film coefficients are constant. This restricts the use of over-all coefficients. Except as an empirical approximation they should not be considered constants in the integration of the above equations, but they may be included under the integral sign by calculating their values at each concentration from Eq. (85) or (86). Alternatively the graphical procedure on p. 549, Sec. 8, may be used.

If the solute distribution largely favors phase E, *i.e.*, if m' is very large,

$$(c_{RL} - c_{Ri}) \simeq (c_{RL} - c_R^*)$$

and

$$k_R a \simeq K_R a$$

The phase R liquid film is then the major resistance to extraction, and changes in operating conditions affecting solely phase E do not greatly influence the rate of extraction. If m' is very small the converse relations are true. An intermediate case is possible in which neither film is entirely controlling and both must be considered. This two-film concept has had little experimental test in extraction as yet, and data are insufficient to classify actual extraction cases into any of the three possible cases. It should be noted that the relative magnitudes of the individual film coefficients also enter into the determination of the film offering the major resistance. The work of Comings and Briggs [*Trans. Am. Inst. Chem. Engrs.*, **38**, 179 (1942)] represents a series of rather ingenious attempts to observe individual film-resistance magnitudes by use of chemicals to reduce the transfer resistance of a particular phase. Few experimental data indicating relative film-coefficient magnitudes in particular extraction circumstances are available (*cf.* Brinsmade and Bliss and Colburn and Welsh, *loc. cit.*).

An over-all transfer unit [Colburn, *Trans. Am. Inst. Chem. Engrs.*, **35**, 211 (1939)] may be defined for extraction in terms of either liquid phase [see pp. 550 to 553, Eqs. (96*a*) to (118)].

Based on phase E:

$$N_{tOE} = \int_{y_2}^{y_1} \frac{(1-y)_f\,dy}{(1-y)(y^*-y)} = \int_{y_2}^{y_1} \frac{dy}{(1-y)\ln\dfrac{1-y}{1-y^*}} \tag{87}$$

where

$$(1-y)_f = \frac{y^*-y}{\ln\dfrac{1-y}{1-y^*}} \tag{87a}$$

Based on phase R:

$$N_{tOR} = \int_{x_2}^{x_1} \frac{(1-x)_f\,dx}{(1-x)(x-x^*)} = \int_{x_2}^{x_1} \frac{dx}{(1-x)\ln\dfrac{1-x^*}{1-x}} \tag{88}$$

where

$$(1-x)_f = \frac{x-x^*}{\ln\dfrac{1-x^*}{1-x}} \tag{88a}$$

For dilute solutions or where

$$\frac{(1-y)}{(1-y^*)} \simeq 1.0$$

and

$$\frac{(1-x^*)}{(1-x)} \simeq 1.0$$

(87) and (88) reduce to

$$N_{tOE} = \int_{y_2}^{y_1} \frac{dy}{y^*-y} \tag{89}$$

$$N_{tOR} = \int_{x_2}^{x_1} \frac{dx}{x-x^*} \tag{90}$$

The over-all H.T.U.'s are defined by

$$\text{H.T.U.}_{OE} = \frac{Z}{N_{tOE}} = \frac{L'_E}{K_E a c_{E\text{avg}}(1-y)_f S} \tag{91}$$

$$\text{H.T.U.}_{OR} = \frac{Z}{N_{tOR}} = \frac{L'_R}{K_R a c_{R\text{avg}}(1-x)_f S} \tag{92}$$

The relationships between the over-all and individual H.T.U. values may be written

$$\text{H.T.U.}_{OE} = \text{H.T.U.}_E + \text{H.T.U.}_R \frac{L'_E}{mL'_R}\frac{(1-y)_f}{(1-x)_f} \tag{93}$$

$$\text{H.T.U.}_{OR} = \text{H.T.U.}_R + \text{H.T.U.}_E \frac{mL'_R}{L'_E}\frac{(1-x)_f}{(1-y)_f} \tag{94}$$

These equations assume a linear equilibrium curve. The retention of the $(1-y)_f$ and $(1-x)_f$ terms cannot be justified on any firm basis as in gas absorption. For the dilute case,

$$\text{H.T.U.}_{OE} = \text{H.T.U.}_E + \text{H.T.U.}_R \frac{L'_E}{mL'_R} \tag{95}$$

$$\text{H.T.U.}_{OR} = \text{H.T.U.}_R + \text{H.T.U.}_E \frac{mL'_R}{L'_E} \tag{96}$$

The utilization and significance of the H.T.U.'s in column calculations and the significance of the terms included are fully discussed on pp. 550, 555. It may be pointed out here that, on the customary graphical extraction diagram for immiscible solvents (see Fig. 38), L'_R/L'_E is the slope of the operating line and m the slope of the equilibrium-distribution curve at any point. The term mL'_R/L'_E is therefore the ratio of the two slopes. If the operating and equilibrium curves are straight this term will be constant, and the over-all H.T.U.'s will be constant if the film H.T.U.'s are constant. This is more likely to be the case with dilute solutions where L'_E and L'_R do not change appreciably through the column and where the ideal-distribution law holds. Where this is not true, over-all H.T.U.'s vary through the column, and their use requires, rigorously, a graphical integration for column height. Under these conditions the use of the over-all H.T.U. is likely to be more accurate when one or the other film offers the major resistance and the over-all H.T.U. value is substantially equal to the film value.

Calculation of Extraction-column Capacity. The use of the preceding point relations in column calculations requires their integration over the column between the desired terminal compositions. In the general case graphical integration of the rigorous relations is possible with allowance for the variation of over-all H.T.U. with column height if neither film controls. Since the latter requires a knowledge of the individual film H.T.U. values and their variation with conditions, the general case is of little use at the present time. The situation is the same if the use of capacity coefficients replaces the H.T.U. method.

For methods of integrating and approximation where capacity coefficients are employed and various simplifications are justified, see Walker, Lewis, McAdams, and Gilliland ("Principles of Chemical Engineering," 3d ed., Chap. XV, McGraw-Hill, New York, 1937); and Sherwood ("Absorption and Extraction," Chap. III, McGraw-Hill, New York, 1937).

For methods of integrating and approximation for the transfer-unit method, see pp. 552, 555, and also Table 15, p. 555, Sec. 8 or the original paper from which this table is abstracted, Colburn [*Ind. Eng. Chem.*, **33**, 459 (1941)]. Also see Colburn [*Trans. Am. Inst. Chem. Engrs.*, **35**, 211 (1939)]; Wiegand and Drew [*ibid.*, **36**, 679 (1940)]; and Scheibel and Othmer [*ibid.*, **38**, 339 (1942)].

More Useful Procedures Briefly Reviewed. *Use of the Capacity Coefficient.* In dilute solutions where the ideal-distribution law holds,

$$V = ZS = \int \frac{dN}{K_E a(c_E^* - c_{EL})} = \frac{N}{K_E a\,\Delta c_{E_{lm}}} \tag{97}$$

where

$$\Delta c_{E_{lm}} = \frac{(c_E^* - c_{EL})_1 - (c_E^* - c_{EL})_2}{\ln\dfrac{(c_E^* - c_{EL})_1}{(c_E^* - c_{EL})_2}} \tag{98}$$

and subscripts 1 and 2 refer to the rich and lean ends of the column, respectively. Further, the moles of solute transferred per unit time, N, may be given by

$$N = L_R\left(\frac{x_1}{1 - x_1} - \frac{x_2}{1 - x_2}\right) = L_E\left(\frac{y_1}{1 - y_1} - \frac{y_2}{1 - y_2}\right) \quad (99)$$

$$N = L'_R(x_1 - x_2) = L'_E(y_1 - y_2) \quad (100)$$

or by any equivalent expression involving the concentration and rate of flow.

Similarly, we may also use the other phase as a basis for calculation:

$$V = ZS = \int \frac{dN}{K_Ra(c_{RL} - c_R^*)} = \frac{N}{K_Ra\,\Delta c_{Rlm}} \quad (101)$$

Further, if an individual film is known to be controlling, the over-all coefficient may replace the individual-film coefficient.

For the case in which the system is dilute and one phase is known to be controlling but the ideal distribution law is not valid (non-linear-equilibrium curve), the equations take the following form (written here for phase R controlling):

$$\frac{ZSK_Rac_{R\text{avg}}}{L_R} = \int_{x_2}^{x_1} \frac{dx}{x - x^*} \quad (102)$$

where

$$c_{R\text{avg}} = \frac{(c_{R\text{avg}})_1 + (c_{R\text{avg}})_2}{2} \quad (103)$$

The right-hand term may be graphically integrated by reference to a plot on which the equilibrium curve and operating line are drawn.

Use of Transfer Units. If the problem is considered in terms of H.T.U.$_{OE}$ where the solutions are dilute and the ratio of the slope of the equilibrium curve to the operating line is constant, L'_R and L'_E are also very nearly constant. Accordingly the following may be written:

Linear equilibrium:

$$x^* = my \quad (104)$$

Dilute operation:

$$(x - x_2) = (y - y_2)\frac{L'_E}{L'_R} \quad (105)$$

Combining (104) and (105),

$$x^* = \frac{mL'_R}{L'_E}(x - x_2) + my_2 \quad (106)$$

and, with the aid of (106), the equation for the number of transfer units is integrated:

$$N_{tOR} = \int_{x_2}^{x_1} \frac{dx}{x - x^*} = \frac{1}{1 - \dfrac{mL'_R}{L'_E}} \ln\left[\left(1 - \frac{mL'_R}{L'_E}\right)\left(\frac{x_1 - my_2}{x_2 - my_2}\right) + \frac{mL'_R}{L'_E}\right] \quad (107)$$

and the height of the required column is computed by means of the relation expressed in Eq. (77). A similar expression may be written for the other phase. When the entering solvent contains no solute, $y_2 = 0$, and the terms in (107) involving y_2 disappear. Equation (107) is solved graphically in Fig. 28 p. 554, Sec. 8. The values of y and x appearing in this figure should be interchanged in accordance with the nomenclature used here.

In the general case neither $\dfrac{mL'_R}{L'_E}$ nor the over-all H.T.U. is constant. Graphical integration is possible if individual film H.T.U. values are known, but this is seldom the case. Colburn [*Ind. Eng. Chem.*, **33**, 459 (1941)] and Scheibel and Othmer [*Trans. Am. Inst. Chem. Engrs.*, **38**, 339 (1942)] have developed methods by which a satisfactory analytical approximation to the exact graphical method may be obtained. These methods are particularly successful when the number of transfer units required in the lean end of the column is large; and, since this condition corresponds to operation at high recovery, the methods are generally applicable to industrial installations.

In the Colburn method the column is divided into two parts, and the transfer units are calculated separately for each part and added. The calculations are based upon the existence of a linear variation of the operating and equilibrium lines in the dilute region and upon a second-degree variation in the concentrated region. The division between the two regions is made at x_a, where $x_a = \sqrt{x_1x_2}$. Integration is accomplished for the dilute region as indicated in Eqs. (104) to (107). Integration for the concentrated region is accomplished by the use of certain approximations. The sum of the two integrals turns out to be

$$N_{tOR} = \frac{1}{1 - \dfrac{mL'_R}{L'_E}} \ln\left[\left(1 - \frac{mL'_R}{L'_E}\right)\left(\frac{x_1 - my_2}{x_2 - my_2}\right)\left(\frac{1 - \dfrac{mL'_R}{L'_E}}{1 - \dfrac{x_1^*}{x_1}}\right) + \frac{mL'_R}{L'_E}\right] \quad (108)$$

where m is the slope in the dilute region and L'_R/L'_E is evaluated at the lean end of the column. Equation (108) can be solved by Fig. 28, p. 554, Sec. 8 if

$$\left(\frac{x_1 - my_2}{x_2 - my_2}\right)\left[\frac{1 - (mL'_R/L'_E)}{1 - (x_1^*/x_1)}\right]$$

is used as the abscissa. See also Table 15, p. 555, Sec. 8, where similar relations are summarized for use with Fig. 28. To the values obtained by this method, the value from Wiegand's approximation [*Trans. Am. Inst. Chem. Engrs.*, **36**, 679 (1940)] should be added. This method does not allow for variation of H.T.U. with mL'_R/L'_E; and, in calculating column height, H.T.U.$_{OE}$ at the lean end of the column should be used.

The method proposed by Scheibel and Othmer assumes that the equilibrium line may be approximated by the assumption that x^*/y varies linearly with y. Their equations have been solved graphically, and reference should be made to Fig. 3 of their paper. When the ratio mL'_R/L'_E is greater than unity, Eq. (108) fails, whereas the equations of Scheibel and Othmer are still valid.

Illustrative Example. It is desired to extract an aqueous 25 weight per cent solution of acetic acid with isopropyl ether at 20°C. so that the exit water is to contain 0.01 weight per cent of acetic acid. Prior economic considerations dictate the use of approximately 3.8 lb. of acid-free ether per pound of aqueous acid solution fed to the extractor. How many over-all transfer units, based on the extract phase, in a continuous differential contactor will be required?

Operating and Equilibrium Lines. The data for this system have been given by Nichols *et al.* [*Trans. Am. Inst. Chem. Engrs.*, **36**, 601, 609 (1940)] and are shown on Fig. 42.

In accordance with the material balance criteria discussed on p. 738 and in Fig. 36, the line FQ is drawn on Fig. 42, and the point P is placed such that $FP/PQ = 3.8/1.0$. This is in effect an entering stream material balance, and the exit streams must also satisfy the same balance. The line L_nV is accordingly also drawn through P. The lines L_nQ and FV_1 when extended must intersect at the operating point O. Points may now be determined at random for plotting on a rectangular distribution

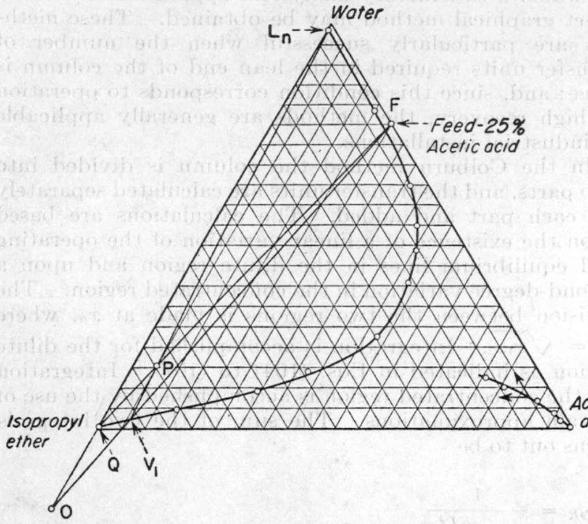

FIG. 42. Graphical solution of illustrative example.

chart by drawing operating lines that intersect the acetic acid–water base line between 100 per cent water and F. The points where these lines cross the phase envelope are plotted on Fig. 43 as the operating line. It will be noted that considerable curvature exists because of the large volume change involved. Tie-line data are also taken from Fig. 42 and plotted on Fig. 43 as the equilibrium line. The slope of the operating line at the dilute end of the system is found to be 6.0; that of the equilibrium line is 4.2. The dimensionless ratio mL'_R/L'_E is thus found to be $4.2/6.0 = 0.70$.

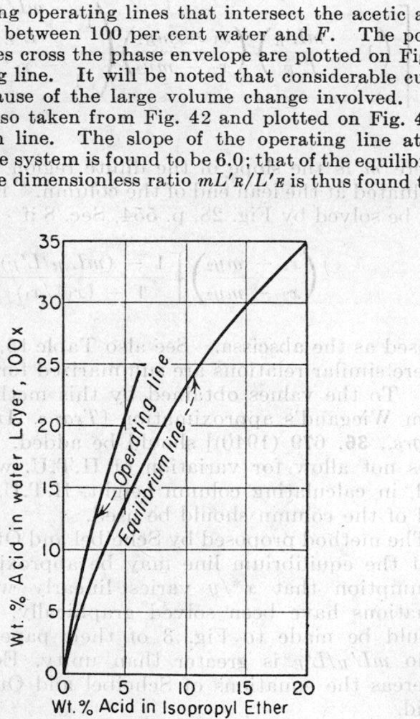

FIG. 43. Graphical solution of illustrative example.

Number of Transfer Units. The number of transfer units may now be estimated for this case by means of Eq. (108), which, although derived on a mole-fraction basis, is usable in this case because only ratios are involved. For $y_2 = 0$, (108) becomes

$$N_{tOR} = \frac{1}{1 - (mL'_R/L'_E)} \ln \left[\left(\frac{x_1}{x_2}\right) \frac{(1 - (mL'_R/L'_E)^2}{\left(1 - \frac{x_1^*}{x_1}\right)} + \frac{mL'_R}{L'_E} \right]$$

Evaluating from the statement of the problem and from Fig. 43,

$$x_1 = 0.25$$
$$x_2 = 0.0001$$
$$x_1^* = 0.16$$
$$\frac{mL'_R}{L'_E} = 0.70$$

Solving Eq. (108) yields

$$N_{tOE} = 21 \text{ approx.}$$

Alternatively, the quantity

$$\left(\frac{x_1}{x_2}\right)\left[\frac{1 - (mL'_R/L'_E)}{1 - (x_1^*/x_1)}\right]$$

could have been computed and used as ordinate on Fig. 28, p. 554, Sec. 8, in order to obtain a rapid graphical solution. The foregoing calculations are subject to the approximations noted on pp. 743 through 744. The approximation of Wiegand [*Trans. Am. Inst. Chem. Engrs.*, **36**, 679 (1940)] could also have been included. Exact graphical integration yields 20.7 transfer units. See also Table 15, Sec. 8, p. 555.

The H.E.T.S. Continuous extraction columns may be calculated in terms of the H.E.T.S., height equivalent to a theoretical stage. The number of theoretical stages may be evaluated by stepping methodsfor both dilute and concentrated or miscible systems, as on pp. 733, 737, 738. For dilute systems, Colburn has produced approximate analytical equations which are solvable by means of Table 15, p. 555, Sec. 8, and Fig. 29, p. 554, Sec. 8. The height of the column required for a specific operation is obtained by multiplying the number of stages required by the H.E.T.S. This empirical procedure is probably the best one for use at the present time, particularly for partly miscible solvents; but unfortunately the H.E.T.S. in other than dilute systems is dependent upon a large number of factors concerning which there is little quantitative information at present. In the mixer-settler type of operation where equilibrium is nearly assured in each stage, the advantages of the method are obvious.

Calculation of Leaching Operations. For methods of calculating leaching operations, see Baker [*Trans. Am. Inst. Chem. Engrs.*, **32**, 62 (1936); *Chem. & Met. Eng.*, **42**, 669 (1935)]; Elgin [*Trans. Am. Inst. Chem. Engrs.*, **32**, 451 (1936)]; Hawley [*Ind. Eng. Chem.*, **9**, 866 (1917); **12**, 482 (1920)]; Kammermeyer [*Ind. Eng. Chem.*, **34**, 1228 (1942)]; and Ravenscroft [*Ind. Eng. Chem.*, **28**, 851 (1936)]. These methods handle the calculation in terms of the number of theoretical contacts. Ruth [*Chem. Eng. Progress*, **44**, 71 (1948)] has extended the method of Kammermeyer by an algebraic solution.

Equipment Used in Leaching. The equipment used in leaching operations varies widely with the type and quantity of carrier material being extracted. Thus, for example, the following classification [Callaham, *Chem. & Met. Eng.*, **49**, 120 (1942)] serves to divide rather sharply the types of equipment used.

For treating coarse solids:
- Open tanks and vats.
- Diffusion batteries.
- Single-deck rake classifiers.
- Multideck rake classifiers.
- Screw-fed countercurrent contactors.

For treating fine materials:
- Agitators.
  - Simple.
  - Pachuca tank (cone and draft tube)
  - Dorr agitators.
- Thickeners.
- Continuous centrifuge.

Open tanks and vats and diffusion batteries have been used for leaching of materials that are difficult to transfer, such as waxed paper. The rake classifiers are used principally in extraction of coarse mineral materials, and the screw-fed countercurrent contactor has recently [*Chem. & Met. Eng.*, **48**, 128 (1941); see also Kenyon, Kruse, and Clark, *Ind. Eng. Chem.*, **40**, 186 (1948); and Bilbe, *Oil Mill Gaz.*, **53**, No. 1, 39 (1948)] been used to contact soy flakes continuously with hot hexane under pressure.

In the treatment of the finer materials the most successful methods have used various adaptations of the agitated or unagitated settler. Riegel ("Chemical Machinery," Reinhold, New York, 1944) has given an excellent review of the variations of this unit that are commercially available. These units may be single- or multistage in design, and several units may be used together. Typical applications are in the extraction of bauxite with sulfuric acid; causticization; and flocculation.

EXTRACTION EQUIPMENT

General Factors in Design. Extraction equipment has not been standardized. Widely varying equipment is possible and has been found useful.

Mixing and contacting the two phases and their separation from each other are carried out either as separate and distinct steps in different equipment units or simultaneously and continuously in the same or different units. A single-stage extraction plant may, for example, consist of a single vessel in which solvent and feed material are first mixed and then, after the agitation is stopped, the two phases are allowed to separate by settling in the same vessel. Two vessels may be employed, one for mixing, the second for settling; the first is then used for a second batch while the first batch is settling in the second vessel. Or the contents of the first vessel may be mixed continuously by means of a pump or analogous device and the mixture may be passed to a second vessel where it settles continuously.

Extraction equipment should be designed to have the contact area and the intimacy of contact as high as possible with the expenditure of a minimum amount of energy. This usually implies a fine degree of subdivision of one phase in the other, which is continuous. A practical limit in this direction is set by the fact that increasing dispersion and subdivision increases the difficulty of subsequently settling and separating the two phases. Securing sufficient subdivision to provide a feasible contact area is, however, usually the more difficult problem, since the interfacial tension between two liquids is commonly high compared with, for example, that between gas and liquid. For this reason it is often difficult to secure sufficient subdivision with many common solvents in the common types of equipment available for extraction. This frequently accounts for the relatively poor efficiency of many common types of contact equipment, *e.g.*, the bubble-cap column, when used for extraction.

High interfacial tension likewise increases the power consumption required to secure subdivision and hence the operating costs. In selecting extraction equipment both the intimacy of contact and rate of extraction obtained and the corresponding power consumption required should be considered and properly balanced. The densities and viscosities of the two phases, as well as interfacial tension, also affect the degree of subdivision that is attained. Olney and Carlson [*Chem. Eng. Progress*, **43**, 473 (1947)] present experimental data on mixer power consumption in two-phase liquid systems.

The fundamentals of mixing and contacting liquids and liquids and liquids and solids are discussed in Sec. 17, pp. 1216, 1217. In leaching it is frequently desirable to reduce the size of the solid to be extracted by cutting or grinding prior to contact with the solvent if this is feasible.

Phase Separation. Gravity settling and centrifuging are the available means of separating two liquid phases. The former is more usual. In addition, where solids are being extracted, the extracted solid residue may be removed from the extract by filtration (see Sec. 15). In some leaching methods the solid is stationary in the extraction vessel while the solvent flows through or over it. This is commonly referred to as *percolation.* A method of separation of intimately mixed organic and aqueous phases has been reported in which porous alundum thimbles are introduced into the agitated vessel. Half of the thimbles are treated with a hydrophilic substance, and half with a hydrophobic substance so that the two pure phases may be withdrawn continuously by pumping through the thimbles. Fine materials such as Fiberglas and excelsior are used to coalesce mixtures of oil and brine droplets in the extractive desalting of crude oil. Packing is also used to coalesce fine dispersions in the Scheibel extractor. *Cf.* "Mixers." Centrifugal separation of liquids from liquids and solid from liquid is treated in Sec. 15, pp. 992 to 1013. Settling theory is also treated in Sec. 15, pp. 937*ff.*

Liquid-liquid Equipment. Equipment for mixing liquid phases in liquid-liquid extraction may be roughly classified as follows:

1. Mixers:
   a. Air agitators.
   b. Mechanical mixers.
   c. Flow mixers.
   d. Column mixers.
   e. Pumps.
2. Countercurrent columns operating continuously:
   a. Wetted-wall.
   b. Spray.
   c. Packed.
   d. Sieve or perforated plate.
   e. Modified bubble plate.
   f. Baffled columns.
   g. Columns with internal agitators.

Mixers. The various types of mixers are described and compared in Sec. 17. Data on relative effectiveness and on power consumptions are also given. There are essentially no published data on the efficiency of mixers as extracting devices, although in general it may be said that, by suitable mechanical arrangement and expenditure of power, stage efficiencies of 100 per cent may be closely approached. Mixer-settler equipment has been widely used in the petroleum field. Hunter (Dunstan, "Science of Petroleum," vol. 3, p. 1786, Oxford, New York, 1938) summarizes data for a few miscellaneous cases. (This reference also contains a discussion of factors affecting the efficiencies of other types of liquid-liquid contacting equipment and a summary or data.) See also Ferris, Meyers, and Peterkin [*Proc. Am. Petroleum Inst.* **14** (III), 55 (1933)] for a description of the nitrobenzene–lube oil extraction process; and Bahlke, Brown, and Diwoky [*ibid.*, **14** (III), 77 (1933)] for a description of the Chlorex–lube oil process.

Earlier installations of the Furfural process [Manley, McCarty, and Gross, *Proc. Am. Petroleum Inst.*, **14** (III), 55 (1933)] employed mixers and settlers. Later installations are using Raschig ring-packed countercurrent columns. The plant columns are 7 ft. in diameter, packed with 1½-in. stoneware rings and charge from 10 bbl. for medium- or low-viscosity oils per square foot of tower area per 24 hr. For this process a 12-ft.-high semiplant column packed with 1-in. rings is reported to be equivalent to six or seven theoretical stages.

In the phenol process (Stratford in Dunstan, "Science of Petroleum," vol. 3, p. 1914, Oxford, New York, 1938) a seven-stage countercurrent extractor is used, each stage consisting of a Leaver-type mixer and a settling drum.

In the Duosol process (Sheldon and Story, Dunstan, *ibid.*, p. 1926) the components of the oil are distributed between two solvents, cresylic acid and propane, which pass countercurrently through the extractor. The latter consists of two horizontal cylindrical tanks each 8 ft. in diameter by 75 ft. long. Each is divided into compartments from 15 to 25 ft. long, four compartments in the first, five in the second. Each compartment is provided with an individual pump to circulate and regulate the flow of the liquid phases. The two layers are mixed by specially designed nozzles as they enter each compartment. Settling takes place in each compartment, and the lower layer passes in one direction to the next compartment through its mixing nozzle, while the top layer passes to the next compartment in the reverse direction. In refining kerosenes with liquid sulfur dioxide (Hall, Dunstan, *ibid.*, p. 1890) ring-packed towers 30 ft. high by 3 ft. in diameter are used for contacting the two liquids.

The Podbielniak centrifugal contactor combines the features of a mixer and centrifugal separator in one small unit. The device contains about 300 ft. of steel ribbon wound and spaced spirally inside a steel casing, which is then rotated rapidly. Two liquid phases may be pumped countercurrently through the spiral channel under a centrifugal force much greater than ordinary gravitational force. In some models the ribbon is perforated, thus allowing a certain degree of short circuiting of the spiral path, which, because of the creation of more interfacial area, increases the efficiency of the unit. About five theoretical stages may be obtained in a single unit. The physical size is small—about the size of a 5-hp. motor, and the allowable through-put is large.

Scheibel, *Chem. Eng. Progress*, **44**, 681,771 (1948) has described a novel countercurrent column, which is in reality a series of mixer-settlers, or more properly mixer-coalescers. Each stage consists of a column section in which there is a small turbine agitator (a shaft runs concentrically through the length of the column). Above and below the mixing chamber section is a short packed section which coalesces the emulsion formed by the turbine. Agitated and packed sections alternate throughout the column. Stage efficiencies of 85 to 110 per cent and H.E.T.S. values as low as 2.2 in. are reported for small-diameter towers of this design.

Wetted-wall Towers. Wetted-wall towers and modifications of the wetted-wall idea in which the tower is operated horizontally have been used extensively on an experimental scale because they offer an opportunity for determining the interfacial area of contact. Commercially these towers have been so far reported as being of little importance because of their low contacting efficiency and relatively critical operation.

The methods of Fallah, Hunter, and Nash, and of Brinsmade and Bliss (see p. 743) and Treybal and Work [*Trans. Am. Inst. Chem. Engrs.*, **38**, 203 (1943)] are recommended as examples of the means by which the problem of individual resistance determination may be attacked; and the apparatus of Treybal and Work is cited as typical of the equipment used in this type of tower.

In a similar attempt to attack the problem of correlation of physical and flow properties with extraction rates, Bergelin, Lockhardt, and Brown [*Trans. Am. Inst. Chem. Engrs.*, **39**, 173 (1943)], noting that in a conventional wetted-wall column a continuous wall film was difficult to maintain unless the average wall fluid velocity were kept between five and twenty times that of the core fluid, studied extraction rates in a horizontal-duct type of extractor which permitted much greater range in the setting of relative flow rates.

Because of the present relative industrial unimportance of this type of unit, no attempt has been made here to tabulate or correlate the available performance data in this type of equipment.

Spray and Packed Towers. The spray and packed tower remain the most widely used forms of the continuous tower. They differ mainly in that the former depends upon solid packing to obtain sufficient contact area while the latter depends upon a suitable entrance device or nozzle to give a sufficiently fine subdivision which is largely maintained through the column. In either, the lighter liquid enters the bottom and passes upward under gravity, while the heavier is fed to the top and passes countercurrently downward. Either may be broken up and made discontinuous while the other is continuous. If the lighter is discontinuous, the interface between the two is held at the top of the column, and conversely. Separation of the two phases takes place continuously at the top or bottom of the column in suitably provided settling sections. The interface level is maintained by conventional liquid-level-control devices such as the non-siphoning loop in atmospheric columns, or the float- or differential-type controller in high-pressure columns.

Coalescence of liquid droplets in passing through such a column with consequent reduction in contact area in long columns is a possibility. Tendency to coalesce is more pronounced where interfacial tension between the two liquids is high and, with high flow rates, where the number of drops of the discontinuous phase per unit volume is large. It has been the authors' experience that difficulties due to coalescence in such columns, especially the spray type, have been overexaggerated if the entrances are suitably designed. These should be of the diffusion type, gradually expanding from the column diameter, rather than of the abrupt-orifice type. Constriction of the column cross section at the entrance by the introduction of the inlet nozzle is particularly to be avoided. With properly designed entrances with the spray column, the high countercurrent velocity of the continuous phase between the droplets of the discontinuous phase apparently acts to prevent their coalescence, which is in accord with hydrodynamic theory.

With a packed column in absorption, the liquid preferentially wets the packing and is always the discontinuous phase. In extraction either liquid may preferentially wet the packing. If the liquid preferentially wetting the packing is made discontinuous, it will flow over the packing in continuous layers much the same as in absorption. If, however, the other liquid is made discontinuous, it will be found to pass the packing in discrete drops or globs similarly to the spray column except that the path is tortuous and the passage velocity slower because of the packing. This is the case, for example, with toluene and water and a clay packing where the latter preferentially wets. This factor can have a pronounced effect on the efficiency of the extraction column under otherwise comparable conditions.

Further, the rate coefficients may change markedly with physical condition of the packing. For example, Zipser (M.S. thesis, Worcester Polytech. Inst., 1942) notes, in studying the extraction of aniline from water with nitrobenzene in a 2.94-in.-diameter tower packed with ⅜-in. earthenware rings, that a progressive increase in rate of extraction was noted with time. Cleaning of the packing with gasoline reduced the coefficients to the original values, but fouling soon reappeared with accompanying increase in rate. Zipser believes that the better wetting of the packing was promoted by accumulated aniline degeneration products.

In those cases in which filming or wetting of the packing does not seem to be an important factor, the choice of the dispersed and continuous phases is often dictated by the comparatively slow rates of diffusion within the main body of liquids, *i.e.*, the phase having the greater volume is made the dispersed phase, thus making the interfacial area a maximum. This latter basis for choice is also frequently used in considering spray columns. Some available data indicate a serious loss of efficiency in passing from small to larger diameter packed and spray towers. The factors involved are as yet unknown.

Elgin and Browning [*Trans. Am. Inst. Chem. Engrs.*, **31,** 439 (1935)] have analyzed the factors determining the area of contact and the rate of extraction in a spray column. They point out that the contact area per unit volume a is not constant but varies greatly with drop size and the rates of flow of both phases (increasing as

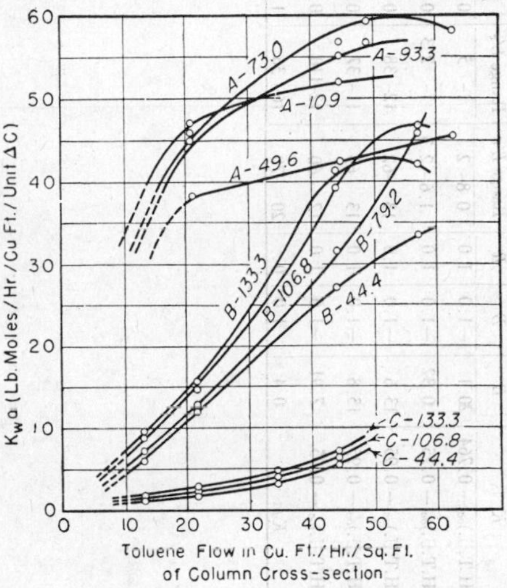

FIG. 44. Effect of varying flow of the dispersed phase on the capacity coefficient of a spray column extracting benzoic acid from toluene with water, and the toluene dispersed. Letters indicate the entrance nozzle corresponding to approximate drop diameters of 0.06 to 0.12 in. for A, 0.2 in. for B, 0.4 in. for C. Numbers on curves show the rate of water flow.

the former decreases and increasing with the latter). The performance of such a column depends greatly on the drop size that is obtained and the flow conditions. The effect of these factors and the contact area obtained with varying conditions may have more influence on the performance of the spray column than the effect of operating variables on the extraction coefficient itself. The above authors give a theoretical relation for the contact area in terms of the rates of flow of the two phases, drop diameter, and the velocity of the drops through a static column of the continuous phase.

The performance of packed and spray extraction columns may be measured either in terms of the extraction capacity coefficient Ka, the H.T.U., or the H.E.T.S.

Tables 3 and 4 contain most of the available literature data on extraction in packed and spray columns. The mode of presentation has been chosen not because of any theoretical significance but rather because the form of the correlation equation used seems to give a fair representation of the data over the ranges chosen. Equations like (95) or (96) might have also been chosen and the effect of average concentration on the slope of the equilibrium curve probably thus included. However, (95) and (96) are without real theoretical significance in extraction and thus reduce to empirical relations. The method used in Tables 3 and 4 seems to give a better fit in most cases, and in some is directly reducible to (95) and (96).

Observations have been made of rate data in both packed and spray towers that do not fit the empirical relation of Tables 3 and 4. These data are recorded in

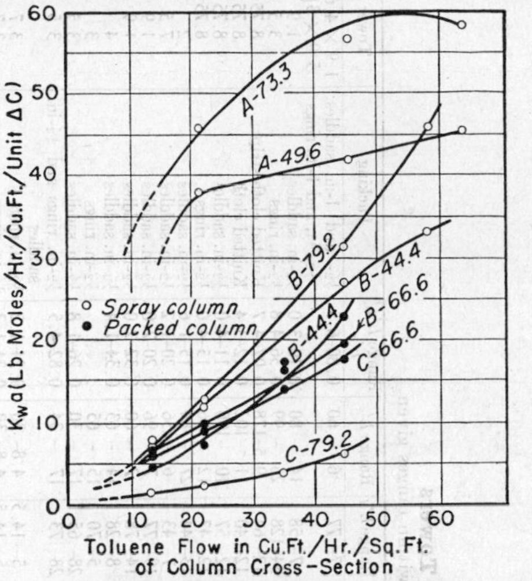

FIG. 45. Comparison of the capacity of spray and saddle-packed columns with different drop sizes when extracting benzoic acid from toluene with water, toluene dispersed. Letters designate different drop sizes (see Fig. 44) and numerals the water rate.

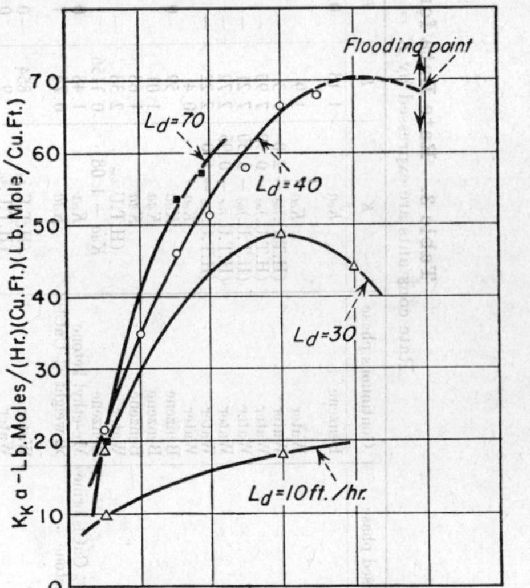

FIG. 46. Capacity coefficients based on dispersed (ketone) phase, in extraction of acetic acid from water with methyl isobutyl ketone. Packing, ½″ rings.

Figs. 44 and 45, which show the effect of drop size and flow rates on the system benzoic acid, toluene, water, in a spray tower, and in a tower packed with ½-in. saddles [Appel and Elgin, *Ind. Eng. Chem.*, **29,** 451 (1937)]. Figures 46 and 47 show that the data of Sherwood, Longcor, and Evans [*Ind. Eng. Chem.*, **31,** 1144 (1939)] in which sharp maxima were obtained.

These same authors have also made an interesting

Table 3. Rate Data for Packed Towers

Rate constants are expressed by $K = BL'^{R}_{d}L'^{S}_{c}$ within ranges given

Solute	Dispersed phase	Continuous phase	K	B	R	S	Range L'_d	Range L'_c	Range L'_d/L'_c	Packing	Tower $d \times Z$, in.	Ref.*
Acetic acid	Water	Benzene	K_{ca}	1.18	0.5	0.25	7– 77	6 – 60	0.80–6.5	½- and 1-in. saddles; ½, ¾, and 1-in. rings	1.9 × 4.6; 7.5 × 4.6; 5.9 × 31	5
Benzoic acid	Benzene	Water	K_{ca}	1.9	.5	.25	7– 58	14 – 80	0.10–3.0	½-in. saddles	1.9 × 4.6	5
Benzoic acid	Kerosene	Water	$(H.T.U.)_{oc}$ – 2.0	9.3	–1.0	1.0	34–128	26 – 99	0.36–4.8	½-in. rings	3.5 × 48	1
Benzoic acid	Toluene	Water	$(H.T.U.)_{oc}$ – 0.92	7.95	–1.0	1.0	12– 64	9.5–178	0.07–4.7	Knitted cloth strips	8.75 × 69	9
Benzoic acid	Toluene	Water	$(H.T.U.)_{oc}$ – 3.50	7.21	–1.0	1.0	12– 40	11 –102	0.12–3.4	Knitted cloth	8.75 × 69	9
Benzoic acid	Toluene	Water	$(H.T.U.)_{oc}$ – 0.95	2.72	–1.0	1.0	12– 57	10 –100	0.11–4.0	½-in. saddles	8.75 × 69	9
Benzoic acid	Toluene	Water	$(H.T.U.)_{oc}$ – 1.0	2.22	–1.0	1.0	12– 47	12 – 86	0.15–4.0	½-in. rings	8.75 × 69	9
Benzoic acid	Toluene	Water	K_{ca}	0.4	1.0	0	13– 44	27 – 84	0.15–1.6	½-in. saddles	2 × 60	2
Benzoic acid	Water	Benzene	K_{da}	.59	0.5	0.25	7– 43	6 – 36	0.20–1.2	½-in. saddles	7.5 × 4.6	5
Benzoic acid	Water	Benzene	K_{da}	1.01	.5	.25	19– 77	7 – 96	0.20–7.0	½-in. saddles	1.9 × 35.6	5
Benzoic acid	Water	Benzene	K_{da}	1.68	.5	.25	14– 78	7 – 60	0.32–2.0	½-in. saddles	1.9 × 4.6	5
Furfural	Toluene	Water	$(H.T.U.)_{oc}$	2.38	–1.15	1.15	8– 28	14 – 63	0.34–1.4	½-in. saddles	4 × 13.6	7
Phenol	Water	Benzene	K_{da} – 1.08	0.1156	1.0	0	5– 70	15 – 65	....	½-in. rings	3 × 88	3
Water	50 weight % $CaCl_2$ brine	Me-ethyl ketone	K_{ca}	1.48	0.34	0.40	28– 65	17 – 30	0.26–1.8	½-in. saddles	3.5 × 24	8
Water	Me-ethyl ketone	50 weight % $CaCl_2$	K_{da}	0.74	1.0	0	28– 73	17 – 34	0.82–4.5	½-in. rings and ½-in. saddles	3.5 × 24	8
Water or isobutanol	Isobutanol	Water	$(H.T.U.)_c$	.854	–0.75	0.75	5– 14.5	4.8– 30	0.24–1.2	½-in. rings	3.7 × 21	4
Water or isobutanol	Isobutanol	Water	$(H.T.U.)_d$	.9	0	0	5– 14.5	4.8– 30	0.24–1.2	½-in. rings	3.7 × 21	4
Water or isobutanol	Water	Isobutanol	$(H.T.U.)_c$	1.71	–0.75	0.75	8– 16	4 – 40	0.24–1.2	½-in. rings	3.7 × 21	4
Water or isobutanol	Water	Isobutanol	$(H.T.U.)_d$	0.6	0	0	8– 16	4 – 40	0.24–1.2	½-in. rings	3.7 × 21	4

Table 4. Rate Data for Spray Towers

Rate constants are expressed by $K = BL'^{R}_{d}L'^{S}_{c}$

Solute	Dispersed phase	Continuous phase	K	B	B	R	Range L'_d	Range L'_c	Range L'_d/L'_c	Notes	Tower $d \times Z$, in.	Ref.
Acetic acid	Isopropyl ether	Water	$(H.T.U.)_{oc}$ – 0.764	20.1	–1.0	1.0	0.8– 2.1	1– 5	0.3 –1.6	Entrance nozzle diameter = 0.0104 ft.	2.03 × 48.75	6
Acetic acid	Water	Isopropyl ether	$(H.T.U.)_{oc}$ – 0.25	0.32	–1.0	1.0	1.6– 2.3	1– 2.3	0.87–2.3	Entrance nozzle diameter = 0.0104 ft.	2.03 × 48.75	6
Benzoic acid	Toluene	Water	$(H.T.U.)_{oc}$ – 0.95	13.6	–1.0	1.0	13 –62	12– 38	0.35–5.5	3/32-in. diameter distributor holes	8.75 × 69	9
Benzoic acid	Toluene	Water	$(H.T.U.)_{oc}$ – 0.95	13.8	–1.0	1.0	13 –63	11– 37	0.33–5.5	1/8-in. diameter distributor holes	8.75 × 69	9
Benzoic acid	Toluene	Water	$(H.T.U.)_{oc}$ – 0.95	7.91	–1.0	1.0	12 –60	12–110	0.08–4.9	1/16-in. diameter distributor holes	8.75 × 69	9
Water	Me-ethyl ketone	50 weight % $CaCl_2$ brine	K_{da}	0.4	1.0	0	20 –70	16– 36	1.4 –3.7		3.5 × 65	8

References:
[1] Allerton, Strom, and Treybal, *Trans. Am. Inst. Chem. Engrs.*, **39**, 361 (1943).
[2] Appel and Elgin, *Ind. Eng. Chem.*, **29**, 451 (1937).
[3] Bretton, M. S. thesis, Worcester Polytech. Inst., 1943.
[4] Colburn and Welsh, *Trans. Am Inst. Chem. Engrs.*, **39**, 179 (1942).
[5] Comings and Briggs, *Trans. Am. Inst. Chem. Engrs.*, **38**, 143 (1942).
[6] Elgin and Browning, *Trans Am. Inst. Chem. Engrs.*, **31**, 439 (1935); **32**, 105 (1936).
[7] Knight, *Trans. Am. Inst. Chem. Engrs.*, **39**, 439 (1943).
[8] Meissner, Stokes, Hunter, and Morrow, *Ind. Eng. Chem.*, **36**, 917 (1944).
[9] Row, Koffolt, and Withrow, *Trans. Am. Inst. Chem. Engrs.*, **37**, 559 (1941)

comparison of the effects of packing type on the capacity coefficients for the two systems above. These results are reproduced in Figs. 48 and 49.

Figures 48 and 49 may be interpreted as showing that packed towers in general have higher volumetric-rate coefficients than spray towers. However, spray towers

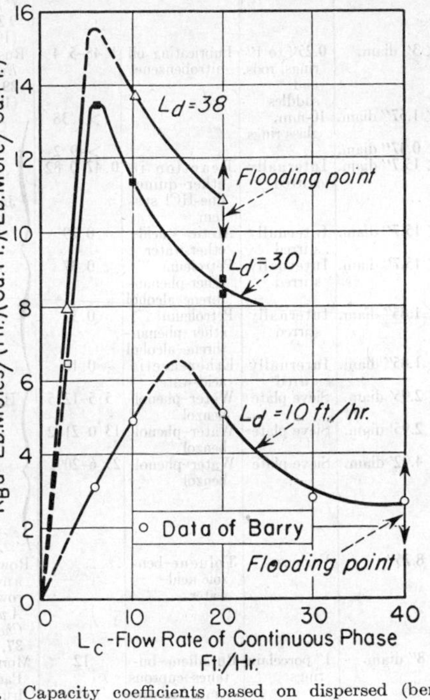

FIG. 47. Capacity coefficients based on dispersed (benzene) phase, in extraction of acetic acid from water with benzene. Packing, ½" rings.

are capable of handling much higher flow rates than packed towers, and thus for certain applications spray towers may show decided advantages in saving of total tower volume.

Johnson and Bliss [*Trans. Am. Inst. Chem. Engrs.*, **42**, 331 (1946)] have made a rather complete study of a small spray tower 1.8-in. in diameter and 23 to 43 in.

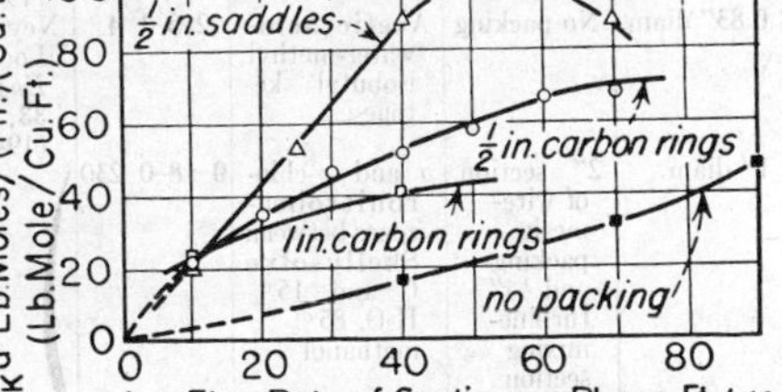

FIG. 48. Comparison of capacity coefficients based on dispersed (ketone) phase for various packings in extraction of acetic acid from water with methyl isobutyl ketone. Ketone flow rate, 40 ft./hr.

long. The systems studied were methyl isobutyl ketone, acetic acid, water; methyl isobutyl ketone, propionic acid, water; and methyl isobutyl ketone, benzoic acid, water. Factors studied were effect of distributor, drop geometry, flooding, hold-up, effect of operating variables on rate of extraction, effect of tower length, effect of solute, effect of solvent.

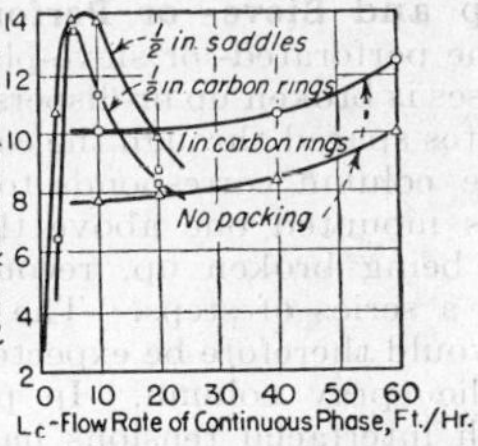

FIG. 49. Comparison of capacity coefficients for various packings extracting acetic acid from water with benzene, with the benzene dispersed at a flow of 30 ft./hr.

Certain conclusions reached are of general interest.

1. A satisfactory distributor for the dispersed phase in a spray tower is a group of tubes. The optimum tube diameter is about 0.1 in. The number of such tubes required may be estimated by

$$\text{No. of tubes} = \frac{VdS}{0.078} \tag{109}$$

2. Diffuser-type entrances permit high through-puts.
3. The effect of dispersed-phase flow rate on the coefficients is nearly linear.
4. The effect of continuous-phase flow rate is small but not negligible.
5. The effect of solute concentration is small.
6. It is desirable to disperse the stream having the higher flow rate, unless the flows are comparable, in which case it appears desirable to disperse the phase receiving the solute. The latter procedure seems to minimize coalescence. It is assumed that the individual resistances involved are of the same order of magnitude.

In addition, Johnson and Bliss suggest that, for spray towers, the following equation may be used to translate from one *solute* to another under otherwise similar flow and configuration situations if the individual resistances are approximately equal:

$$\frac{Ka_2}{Ka_1} = \frac{1 + m_1}{1 + m_2} \tag{110}$$

where Ka is based on the non-aqueous phase and m is the ratio of the concentration in the non-aqueous phase to the aqueous phase at equilibrium. The subscripts 1 and 2 refer to the two solutes being considered.

For change in both *solute* and *solvent* it is suggested that the change in coefficient is due to change in drop size, as well as distribution coefficient.

$$\frac{Ka_2}{Ka_1} = \left(\frac{\Delta\rho_2}{\Delta\rho_1} \times \frac{\gamma_1}{\gamma_2}\right)^n \left(\frac{1 + m_1}{1 + m_2}\right) \tag{111}$$

where $\Delta\rho$ = difference in density, between aqueous and non-aqueous phases.
γ_1 = interfacial tension.
n = an empirical constant, believed by Johnson and Bliss to be about 0.87.

Descriptions of the operation and performance of commercial-scale Raschig ring packed extraction towers in the furfural extraction of vegetable oils, of petroleum oils, recovery and purification of butadiene from refrigerated C_4 hydrocarbon cuts by extraction with copper ammonium acetate solutions, and in recovering ammonia from butene by water extraction are given, respectively, by Gloyer [*Ind. Eng. Chem.*, **40**, 228 (1948)], Kemp, *et al.* [*Ind. Eng. Chem.*, **40**, 220 (1948)], Morrell, *et al.* [*Trans. Am. Inst. Chem. Engrs.*, **42**, 473 (1946)], and Poffenberger, *et al.* [*Trans. Am. Inst. Chem. Engrs.*, **42**, 815 (1946)]. Turehaus and Ehlers [*Chem. Inds.*, **63**, 230 (1948)] describe a spray tower in the purification of electrolytic caustic soda by extraction with liquid ammonia.

Bubble-cap and Sieve- or Perforated-plate Columns. In the perforated- or sieve-plate column one of the liquid phases is broken up or dispersed at intervals by perforated plates spaced through the column. In a sense the sieve-plate column corresponds to a series of short spray columns mounted one above the other, the dispersed liquid being broken up, recombined, and then redispersed in a series of steps. The factors governing performance would therefore be expected to be analogous to those in the spray column. In particular, systems with very high interfacial tensions have been found to give low efficiencies, while partly miscible systems with low interfacial tensions have been found to have quite respectable efficiencies. Sieve-plate columns have found favor in conducting liquid-liquid extractions on an industrial scale. They are reported to be particularly effective, but no quantitative data have been published.

One of the authors has found that as much as 20 to 70 per cent of the total extraction taking place in a 4-ft.-high spray column, depending on conditions, may occur within 1 in. or more of the entrance nozzle (Hardy, M.S. thesis, Princeton Univ., 1943). Similar conclusions have recently been reached by Nandri and Viswanathan [*Current Science*, **15**, 162 (1939)] in the study of the extraction of acetic acid between nitrobenzene and water. This effect must also be bound up with distributor design, but it should be pointed out that other investigators [Johnson and Bliss, *Trans. Am. Inst. Chem. Engrs.*, **42**, 331 (1946)] do not observe so profound an effect. It should be noted, however, that dispersion of a liquid in another consumes energy, and hence the energy consumption of a sieve-plate column will be considerably greater than that of a spray column of equivalent height.

Rogers and Thiele [*Ind. Eng. Chem.*, **29**, 529 (1937)] studied the system chlorex–lube oil distillate in a small bubble-cap column and noted stage efficiencies of about 30 per cent. The low efficiency was attributed to poor mixing in the viscous system. Moulton and Walkey [*Trans. Am. Inst. Chem. Engrs.*, **40**, 695 (1944)] studied the extraction of methyl ethyl ketone from gasoline with water in a perforated-plate column and obtained stage efficiencies between 3.6 and 8.6 per cent. They noted that, as the tray spacing was increased, the efficiency dropped, indicating that turbulence and redistribution were important factors.

The remaining literature data on bubble-cap and sieve-tray columns are presented graphically in Fig. 50.

The H.E.T.S. Many operations of industrial importance involve the treatment of systems of high mutual solubility in which large concentration changes are frequently encountered. The methods of computing the number of stages required in these cases have been covered on pp. 729 through 739. No simple or fundamental method now exists for predicting or correlating extraction-rate data for such systems. The use of the concept of height equivalent to a theoretical stage frequently permits full-scale design from moderate-sized pilot equipment. Table 5 contains observed values of H.E.T.S. for various systems reported in the literature.

Limiting Flows in Liquid-liquid Columns. Maximum through-put for extraction towers is limited by a flooding point above which the discontinuous liquid commences to be ejected back out of the tower along with the continuous liquid. This results in unsatisfactory operation. The larger the flow of one phase, the smaller is that permissible to the other before flooding is reached. In general, the greater the difference in densities between the liquids and the larger the drop size in the spray tower or the packing size in the packed portion (flooding velocity varies with the type of packing, saddles, for example, permitting higher through-puts than the same size rings), the higher is the possible

Table 5. Miscellaneous Data on Efficiency of Extraction Columns on H.E.T.S. Basis

Size of column	Packing	System	H.E.T.S., ft.	Ref.
9.84′ × 0.55″ diam.	4-mm. metal rings	Alcohol-benzene-water	2.46–4.7	Varteressian and Fenske, *Ind. Eng. Chem.*, **28**, 928, 1353 (1936)
6′ × 3″ diam.	0.25″ to 1″ rings, rods, and saddles	Lubricating oil nitrobenzene	1.48–5.4	Rushton, *Ind. Eng. Chem.*, **29**, 309 (1937)
3.28′ × 1.57″ diam.	10-mm. glass rings		>3.38	
49.2′ × 0.47″ diam.			>49.2	
1.64′ × 15.7″ diam.	Internally stirred	Reaction in ether–quinoline-HCl system	0.47–0.82	
1.64′ × 15.7″ diam.	Internally stirred	Acetic acid–ether-water	0.29	Jantzen, Dechema Monograph, vol. **5**, No. 48, p. 81 (1932)
1.64′ × 15.7″ diam.	Internally stirred	Petroleum ether-phenanthrene–alcohol	0.37	
1.64′ × 1.85″ diam.	Internally stirred	Petroleum ether–phenanthrene–alcohol	0.30	
1.64′ × 1.85″ diam.	Internally stirred	Ether–acetic acid-water	0.16	
13.1′ × 2.95′ diam.	Sieve plate	Water–phenol-benzol	5.5–12.5	Hunter and Nash, *Proc. World Power Conf. Chem. Eng. Congr.*, 1936
13.1′ × 2.95′ diam.	Sieve plate	Water–phenol-benzol	13.0–21.2	
46′ × 4.92′ diam.	Sieve plate	Water–phenol-benzol	23.6–209	
8′ × 8.75″	Various	Toluene–benzoic acid–water	....	Row, Koffolt, and Withrow, *Trans. Am. Inst. Chem. Engrs.*, **37**, (1941)
30′ × 8″ diam.	1″ porcelain rings	Butadiene–butenes–cuprous ammonium acetate solution	12	Morrell, Paltz, Packie, Asbury, and Brown, *Trans. Am. Inst. Chem. Engrs.*, **42**, 473 (1946)
2.5′ × 0.83″ diam.	14″ Berl saddles	Acetic acid–water-methyl isobutyl ketone	1.1–0.6	Ney and Lochte. *Ind. Eng. Chem.*, **33**, 825 (1941)
2.5′ × 0.83″ diam.	14- to 18-mm. spinning cylinder	Acetic acid–water-methyl isobutyl ketone	0.7–0.55	Ney and Lochte, *Ind. Eng. Chem.*, **33**, 825 (1941)
2.5′ × 0.83″ diam.	No packing	Acetic acid–water-methyl isobutyl ketone	2.5–1.4	Ney and Lochte, *Ind. Eng. Chem.*, **33**, 825 (1941)
42″ × 1″ diam.	2″ section of wire-mesh packing and ½″ turbine-mixing section	*o* and *p*-chloronitrobenzenes between Skellysolve *C* and 15% H_2O, 85% methanol	0.18–0.230	
42″ × 1″ diam.	2″ section of wire-mesh packing and ½″ turbine-mixing section	Acetic acid between water and methyl isobutyl ketone	0.20–0.283	Scheibel, *Chem. Eng. Progress*, **44**, 681, 771 (1948)
42″ × 1″ diam.	2″ section of wire-mesh packing and ½″ turbine-mixing section	Distribution of 1-propanol and ethanol between water and benzene	0.39–0.46	

through-put. Generally, allowable liquid through-puts are considerable larger for spray than for packed towers, the former having two to six times larger allowable

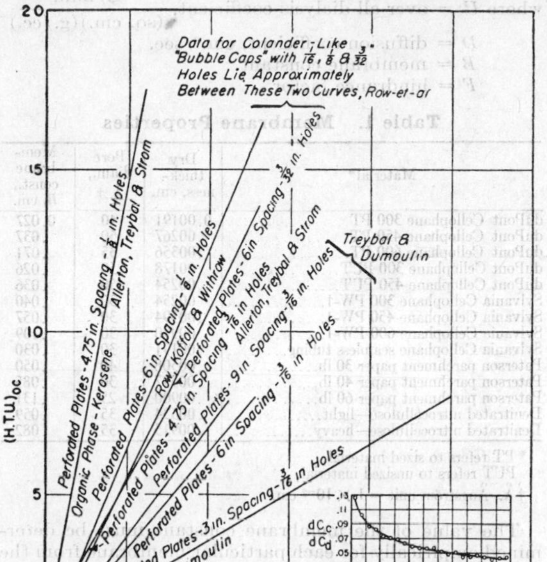

FIG. 50. Extraction rates for sieve-plate and modified bubble-plate columns. System benzoic acid, water, toluene except where noted. [*Allerton, Strom, and Treybal, Trans. Am. Inst. Chem. Engrs.*, **39**, 361 (1943); *Row, Koffolt, and Withrow, Ibid.*, **37**, 559 (1941); *Treybal and Dumoulin, Ind. Eng. Chem.*, **34**, 709 (1942).]

rates depending upon particular conditions and tower design.

In a recent study, Blanding and Elgin [*Trans. Am. Inst. Chem. Engrs.*, **38**, 305–338 (1942)] described in detail the flooding characteristics of spray and packed towers and reported data for two liquid pairs with different drop sizes and type of 0.5-in. packing. They show that entrance construction has a great influence on flooding. If the entrance is designed so that the velocity of the continuous liquid at its entrance, and more especially at its exit where the discontinuous fluid enters, is less than the main tower contacting section, higher through-puts are possible. This state of affairs can be arranged by the use of a conical entrance section as in Fig. 51.

In discussion of the work of Blanding and Elgin (*loc. cit.*), Colburn has pointed out that a fairly good correlation is obtained for flooding rates in packed extractors when the volumetric-flow rate of the discontinuous phase divided by the specific-gravity difference of the two phases is plotted vs. the ratio of the volumetric rate of the continuous phase to the discontinuous phase (see Fig. 52).

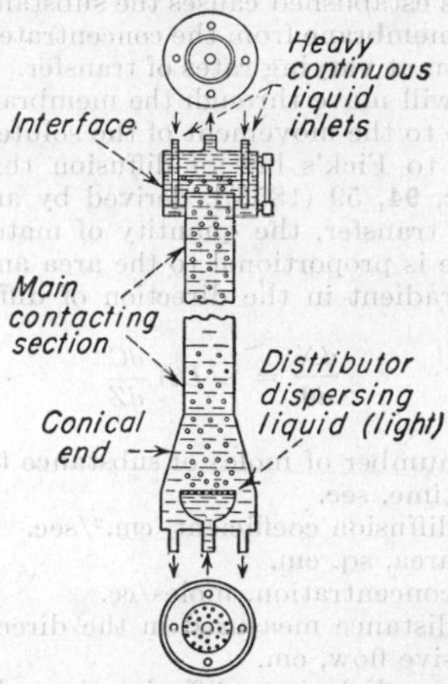

FIG. 51. Entrance design for a spray-type extraction column.

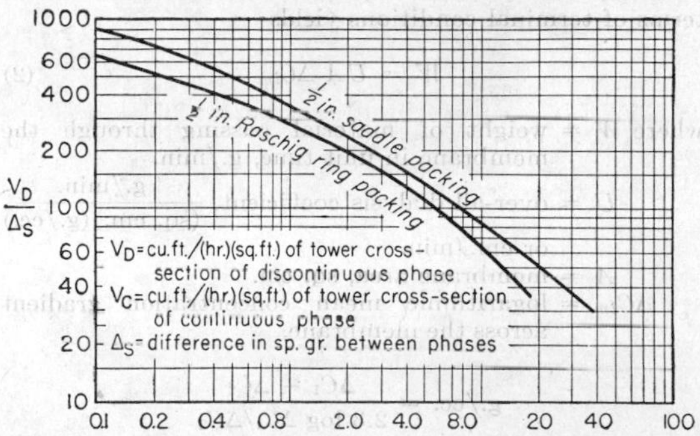

FIG. 52. Colburn correlation of flooding data for packed extraction columns.

This correlation certainly does not include all the variables involved in the rather complex process of extractor flooding, but it does provide the best generalized representation of the problem to date. Comparison of unpublished data with the curves originally presented by Colburn indicates that the position of the curves is good, and the Colburn plot is accordingly presented here with the suggestion that a generous safety factor, say 1.75, be used.

DIALYSIS

Introduction. Various substances in solution having widely different molecular weights may be separated by diffusion through a semipermeable membrane. Such a process of separation, termed "dialysis," is used commercially to recover caustic soda from a mixture of caustic and hemicellulose and in the refining of beet sugar. Other applications of dialysis are limited to laboratory or research scale.

Dialysis, which depends on the relative diffusion rates of two solutes, is most efficiently applied to separate low-molecular-weight substances in solution from very high-molecular-weight substances.

Complete separation is possible only if the latter substances are too large to pass through the membrane pores. The degree of separation or the extent of recovery of the low-molecular-weight solute may determine the capacity of the dialysis equipment. In most cases this capacity is very low because of the slow rate of transfer in dialytic cells. In the recovery of caustic, for example, one commercial dialyzer of 850 sq. ft. membrane area can handle 36 gal./hr. of caustic at 90 per cent recovery.

Theoretical Background. If an aqueous solution containing substances of different molecular weights is kept separated from a more dilute solution of these substances by means of a semipermeable membrane such as cellophane or parchmentized paper, the concentration gradient thus established causes the substances to diffuse through the membrane from the concentrated to the more dilute solution at varying rates of transfer. At the same time, water will move through the membrane in a direction opposite to the movement of the solutes.

According to Fick's law of diffusion through liquids [*Ann. Physik*, **94**, 59 (1855)], derived by analogy to the laws of heat transfer, the quantity of material diffusing per unit time is proportional to the area and to the concentration gradient in the direction of diffusion. Thus

$$\frac{dN}{d\theta} = -DA\frac{dC}{dZ} \tag{1}$$

where N = number of moles of substance transferred.
θ = time, sec.
D = diffusion coefficient, cm.²/sec.
A = area, sq. cm.
C = concentration, moles/cc.
Z = distance measured in the direction of diffusive flow, cm.

Similarly, for dialysis or diffusion through membranes, an integration of Fick's equation for continuous flow in terms of terminal conditions yields

$$W = UA\,\Delta C_{lm} \tag{2}$$

where W = weight of material passing through the membrane in unit time, g./min.
U = over-all dialysis coefficient, $\frac{\text{g./min.}}{\text{(sq. cm.)(g./cc.)}}$ or cm./min.
A = membrane area, sq. cm.
ΔC_{lm} = logarithmic mean concentration gradient across the membrane,

$$\text{g./cc.} = \frac{\Delta C_1 - \Delta C_2}{2.3 \log \Delta C_1/\Delta C_2}$$

ΔC_1, ΔC_2 = inlet and outlet concentration differences, respectively.

For batch dialysis a similar equation applies except that ΔC_1 and ΔC_2 refer to initial and final concentration differences.

Calculation of Dialysis Coefficients. In order to apply Eq. (2) to the design of dialysis equipment, a knowledge of the over-all dialysis coefficient of the solute is necessary. When experimental data are not available, the over-all dialysis coefficient may be estimated from the diffusion coefficient, related to the properties of the solution, and from the properties of the membrane being used. Analogous to heat transfer, the resistance to transfer through a membrane is given by a liquid film on each side of the membrane as well as the membrane itself. Thus

$$\frac{1}{U} = \frac{1}{U_1} + \frac{1}{U_2} \tag{3}$$

where $1/U$ = over-all resistance, min./cm.
$1/U_1$ = combined film resistance, min./cm.
$1/U_2$ = membrane resistance, min./cm.

Substituting in Eq. (3) the diffusion coefficient of the solute and an average value for the combined thickness of the liquid films gives

$$U = \frac{60D}{0.05 + B/F} \tag{4}$$

where U = over-all dialysis coefficient, $\frac{\text{g./min.}}{\text{(sq. cm.)(g./cc.)}}$
D = diffusion coefficient, cm.²/sec.
B = membrane constant.
F = hindrance factor.

Table 1. Membrane Properties

Material*	Dry thickness, cm.	Pore diam., Å.†	Membrane const., *B*, cm.
duPont Cellophane 300 PT	0.00191	40	0.027
duPont Cellophane 450 PT	.00267	50	.037
duPont Cellophane 600 PT	.00356	35	.071
duPont Cellophane 300 PUT	.00178	50	.026
duPont Cellophane 450 PUT	.00254	50	.036
Sylvania Cellophane 300 PW-1	.00254	30	.040
Sylvania Cellophane 450 PW-1	.00394	30	.057
Sylvania Cellophane 600 PW-1	.00990	30	.109
Sylvania Cellophane seamless tubing	.00191	30	.030
Paterson parchment paper 30 lb.	.00508	40	.050
Paterson parchment paper 40 lb.	.00597	30	.085
Paterson parchment paper 60 lb.	.00900	25	.131
Denitrated nitrocellulose—light	.00534	35	.059
Denitrated nitrocellulose—heavy	.00915	35	.082

* PT refers to sized material.
PUT refers to unsized material.
† Å., Ångström unit = 1×10^{-8} cm.

The value of the membrane constant must be determined empirically for each particular membrane from the ratio of the dialysis and diffusion coefficients. Furthermore, the dialysis must be carried out under controlled conditions where the film resistances and hindrance factor are either negligible or calculable. The average membrane pore diameter must also be determined experimentally from the dialysis coefficient of a large molecule of known molecular diameter, the membrane constant, and the Faxen equation. The membrane constant is dependent on the degree of swelling of the membrane, which, in turn, is sometimes affected by the solute. This is true in the case of dialyzing caustic. Values given in Table 1 are for aqueous solutions without hydroxide ions present.

Values of B are given in Table 1 for various commercial membranes. The hindrance factor F is calculated from the Faxen equation [Bacon, *J. Franklin Inst.*, **221**, 251–258 (1936)] as follows:

$$F = 1 - 2.104S + 2.09S^3 - 0.95S^5 \tag{5}$$

where S = ratio of particle diameter to pore diameter. The value of S may be best obtained from measured values of F and a plot of S vs. F.

Values of diffusion coefficients necessary for calculating dialysis coefficients from Eq. (4) are given on p. 540. When necessary data are lacking, diffusion coefficients may be estimated from the molecular weight of the solute by various equations given in the literature. The most notable of these are the Stokes-Einstein equation [von Wogan, *Ber. deut. physik. Ges.*, **6**, 542 (1908)], the Arnold equation [Arnold, *J. Am. Chem. Soc.*, **52**, 3937 (1930)], and, more recently, that proposed by Powell, Roseveare, and Eyring [*Ind. Eng. Chem.*, **33**, 430 (1941)].

In general, diffusion coefficients may be estimated sufficiently accurately for dialysis design from the data given in Table 2 obtained by plotting diffusion coefficients against molecular weight. Values of the average diameter of the solute molecules are also given.

In dealing with the diffusion of electrolytes, it should be pointed out that the effective diameter of the molecule depends on the extent of hydration of the electrolyte, and the molecular weight of the hydrated salt should be used.

Table 2. Diffusion Coefficients and Molecular Diameters of Non-electrolytes

Molecular wt.	D, cm.²/sec. × 10⁶	Molecular diam., Å.
10	2.20	2.9
100	0.70	6.2
1,000	.25	13.2
10,000	.11	28.5
100,000	.05	62.0
1,000,000	.025	132

Applicability of Dialysis. Two factors are involved in the determination of the applicability of dialysis to

Fig. 53. Assembled dialyzer unit.

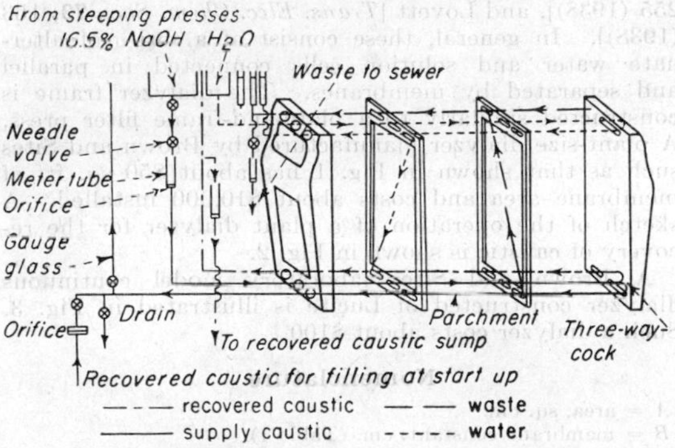

Fig. 54. Diagrammatic drawing of dialyzer.

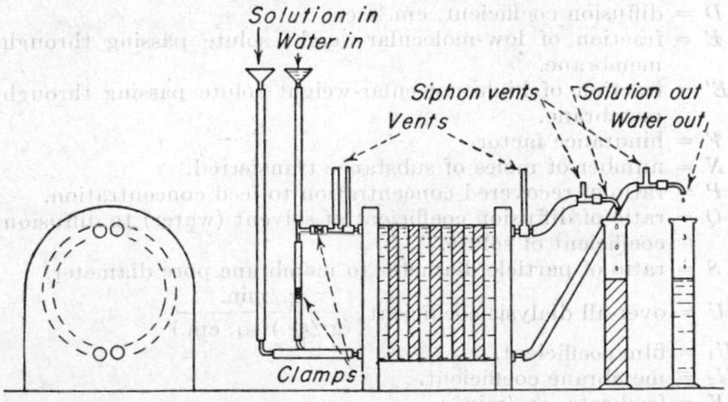

Fig. 55. Continuous countercurrent dialyzer.

a particular problem of separation. The first is the size of equipment required and the second the degree of separation that may be achieved. The method usually employed for dialyzer design is first to estimate the size of equipment or, in other words, the membrane area required and then to use this figure to calculate the degree of separation of low- and high-molecular-weight substances in the solution. An adaptation of Fick's equation in which the concentration gradient is corrected for the osmotic effect or counterdiffusion of solvent permits calculation of the membrane area. Substituting flow rates, the solute concentration, and the fraction of solute passing through the membrane in Eq. (2) gives

$$\frac{V}{UA} = \frac{1 - P - r(1 - E)}{2.3E \log [1 - P/r(1 - E)]} \tag{6}$$

where V = feed rate, cc./min.

U = over-all dialysis coefficients, $\dfrac{\text{g./min.}}{(\text{sq. cm.})(\text{g./cc.})}$.

A = membrane area, sq. cm.

P = ratio of recovered concentration to feed concentration.

E = fraction of low-molecular-weight solute passing through the membrane.

r = ratio of inlet and outlet flow rates on the solution (or feed) side of the membrane.

The value of r moreover may be calculated from the relative diffusion rates of the solute and solvent (in most cases water).

$$r = \frac{1}{QEC_F + 1} \tag{7}$$

where Q = ratio of diffusion coefficient of solvent to solute, ($D_{H_2O} = 4.0 \times 10^{-5}$ cm.²/sec.).

C_F = concentration of solute in the feed.

E = fraction of solute through membrane.

Equation (7) assumes that the diffusion of solute through the membrane and the counterdiffusion of water are independent of head or that the difference in heads of the solution and water side is small. For practical purposes and for the usual operation of commercial dialyzers, it is believed that this assumption is justified, at least for design calculations.

In cases where the solutions being dialyzed are concentrated, values of the required membrane area calculated from Eq. (6) will be too low due to the fact that Fick's equation does not hold for large transfer of solute and solvent through the membrane. For such cases, a general dialysis equation based on weight concentrations and weight flows has been derived by the writer by means of which the true membrane area may be calculated. For countercurrent dialysis of aqueous solutions,

$$\frac{UA}{X_{si} - X_{so}} = \left(\frac{W_{wo} - W_{si}}{G - H}\right)^2 \left\{ \frac{2.3 \log \dfrac{\Delta C\, X_1}{\Delta C\, X_2}}{\Delta C\, X_1 - \Delta C\, X_2} - \frac{Q - 1}{2}\left[\frac{G(W_{wo} + W_{wi}) - H(W_{so} + W_{si})}{(W_{wo} - W_{si})^2}\right] \right\} \tag{8}$$

and

$$G = QX_{wi} + Y_{wi} = QX_{wo} + Y_{wo}$$
$$H = QX_{si} + Y_{si} = QX_{so} + Y_{so}$$

Q = ratio of diffusion coefficient of water to solute

$$\Delta C\, X_1 = \frac{X_{si}}{QX_{si} + Y_{si}} - \frac{X_{wo}}{QX_{wo} + Y_{wo}} \tag{8a}$$
$$\Delta C\, X_2 = \frac{X_{so}}{QX_{so} + Y_{so}} - \frac{X_{wi}}{QX_{wi} + Y_{wi}}$$

where X and Y refer to the solute and water, respectively, and the subscripts s and w refer to the terminal conditions on the solution and water sides of the membrane. Also

W_{si}, W_{so} = inlet and outlet total weights of material on solution side.

W_{wi}, W_{wo} = inlet and outlet total weights of material on water side.

The following examples will serve to illustrate the use of the equations and to compare values of membrane area calculated by Eqs. (6) and (8).

1. It is desired to recover the sugar from an aqueous molasses solution containing 200 g. sugar and 10 g. of a high-molecular-weight impurity per liter of solution. (*a*) Assuming that 80 per cent of the sugar will be recovered and that the concentration of the recovered solution will be one-half of the feed concentration, calculate the membrane area required to handle 10 gal. of feed per hour using duPont Cellophane 450 PT membranes. (*b*) If the molecular weight of the impurity is 10,000, calculate the concentration of this impurity in the recovered sugar solution.

a. The over-all dialysis coefficient of sugar is first calculated from Eq. (4).

$$U = \frac{60D}{0.05 + B/F}$$

$$D_{\text{sucrose}} = 0.45 \times 10^{-5} \text{ cm.}^2/\text{sec.}$$

$$B \text{ (from Table 1)} = 0.037$$

$$\begin{aligned} F &= 1 - 2.104(9/50) + 2.09(9/50)^3 - 0.95(9/50)^5 \\ &= 1 - 2.104(0.18) + 2.09(0.0058) - 0.00002 \\ &= 0.635 \end{aligned}$$

$$U = \frac{60(0.45)10^{-5}}{0.05 + 0.058} = \frac{2.7 \times 10^{-4}}{1.08 \times 10^{-1}} = 2.5 \times 10^{-3} \text{ cm./min.}$$

Now

$$Q = \frac{4.0 \times 10^{-5}}{0.45 \times 10^{-5}} = 8.9$$

From Eq. (7),

$$r = \frac{1}{(8.9)(0.8)(0.2) + 1} = \frac{1}{2.42}$$

and Eq. (6),

$$\begin{aligned} \frac{V}{UA} &= \frac{(1 - 0.5) - (1 - 0.8)/2.42}{2.3(0.8) \log [1 - 0.5/(1 - 0.8)/2.42]} \\ &= 0.29 \\ V &= \frac{(10)3785}{60} = 631 \text{ cc./min.} \\ U &= 2.5 \times 10^{-3} \text{ cm./min.} \\ A &= \frac{631}{(0.29)(2.5 \times 10^{-3})} = 8.71 \times 10^5 \text{ sq. cm} \\ &= 936 \text{ sq. ft.} \end{aligned}$$

b. The diffusion coefficient of the impurity is 0.11×10^{-5} cm.^{2}/sec., and the molecular diameter is 28.5 Å. Thus

$$F = 1 - 2.104\left(\frac{28.5}{50}\right) + 2.09\left(\frac{28.5}{50}\right)^3 - 0.95\left(\frac{28.5}{50}\right)^5 = 0.14$$

$$U = \frac{60(0.11)10^{-5}}{0.05 + (0.037/0.14)} = \frac{6.6 \times 10^{-5}}{0.314} = 2.1 \times 10^{-4} \text{ cm./min.}$$

From previous calculation, $r = 1/2.42$; $P = 0.5$

$$\frac{V}{UA} = \frac{631}{(8.71)(10^5)(2.1)(10^{-4})} = 3.45$$

Equation (6) gives

$$3.45 = \frac{0.5 - (1 - E')/2.42}{2.3E' \log [0.5/(1 - E')/2.42]}$$

where E' = fraction of impurity passing through the membrane. By trial and error,

$$E' = 0.125$$

The recovered solution contains 100 g. of sugar and 1.25 g. of impurity per liter of solution.

2. *Calculation of Membrane Area by General Equation*. The conditions are the same as those of Example 1. Assume the density of the feed solution is 1.0. For 1000 g. of feed,

$$X_{si} = 200 \qquad \text{and} \qquad Y_{si} = 800$$

$$\begin{aligned} W_x &= X_{si} - X_{so} \\ W_x &= 200(0.8) = 160 = X_{so} \end{aligned}$$

$$\begin{aligned} Y_{wo} &= \frac{160}{0.1}(0.9) = 1440 \\ W_y &= Y_{wi} - Y_{wo} \\ W_y &= 160(Q) = 160(8.9) = 1425 \\ X_{so} &= 200 - 160 = 40 \\ Y_{so} &= 1425 + 800 = 2225 \\ X_{wi} &= 0 \\ Y_{wi} &= 1425 + 1440 = 2865 \\ G &= 8.9(0) + 2865 = 2865 \\ H &= 8.9(200) + 800 = 2580 \end{aligned}$$

$$\Delta C\,X_1 = \frac{200}{2580} - \frac{160}{2865} = 0.0217$$

$$\Delta C\,X_2 = \frac{40}{2580} - \frac{0}{2865} = 0.0155$$

$$\begin{aligned} \frac{UA}{X_{si} - X_{so}} &= \left(\frac{600}{285}\right)^2 \left[\frac{2.3 \log 0.0217/0.0155}{0.0217 - 0.0153} - \frac{7.9}{2}(12.1)\right] \\ &= 4.44(54.1 - 47.8) \\ &= 28.0 \end{aligned}$$

$$A = \frac{(28)160}{2.5 \times 10^{-3}} = 1.79 \times 10^6 \text{ sq. cm. for 1000 g./min.}$$

For 631 g./min.

$$A = 11.3 \times 10^5 \text{ sq. cm.} = 1200 \text{ sq. ft.}$$

which compares with the value of 930 sq. ft. from Eq. (6). This indicates that Eq. (6) is sufficiently accurate for order-of-magnitude calculations.

Dialysis Equipment. Commercial dialysis is best carried out by means of continuous countercurrent dialyzers, which are discussed by Volbrath [*Chem. & Met. Eng.*, **43**, 303–306 (1936)], Eynon [*J. Soc. Chem. Ind.*, **52**, 1731 (1939)], Bassett [*Chem. & Met. Eng.*, **45**, 254–255 (1938)], and Lovett [*Trans. Elec. Chem. Soc.*, **73**, 163 (1938)]. In general, these consist of a series of alternate water and solution cells connected in parallel and separated by membranes. The dialyzer frame is constructed similarly to a plate-and-frame filter press. A plant-size dialyzer manufactured by Brown and Sites such as that shown in Fig. 1 has about 850 sq. ft. of membrane area and costs about $10,000 installed. A sketch of the operation of a plant dialyzer for the recovery of caustic is shown in Fig. 2.

A Brown and Sites laboratory model continuous dialyzer constructed of Lucite is illustrated in Fig. 3. Such a dialyzer costs about $100.

Nomenclature

A = area, sq. cm.
B = membrane constant, cm. (Table 1).
C = concentration of solute, moles/cc.
C_F = concentration of solute in the feed, g./cc.
D = diffusion coefficient, cm.^{2}/sec.
E = fraction of low-molecular-weight solute passing through membrane.
E' = fraction of high-molecular-weight solute passing through membrane.
F = hindrance factor.
N = number of moles of substance transferred.
P = ratio of recovered concentration to feed concentration.
Q = ratio of diffusion coefficient of solvent (water) to diffusion coefficient of solute.
S = ratio of particle diameter to membrane pore diameter.
U = over-all dialysis coefficient, $\dfrac{\text{g./min.}}{(\text{g./cc.})(\text{sq. cm.})}$.
U_1 = film coefficient.
U_2 = membrane coefficient.
V = feed rate, cc./min.
W = weight of solute passing through membrane, g./min.
Z = distance measured in direction of diffusive flow, cm.
θ = time, sec.

SECTION 12

HUMIDIFICATION, DEHUMIDIFICATION, AND COOLING TOWERS AND SPRAY PONDS

BY

W. G. Hillen, M. E., Director, Sales Training, Carrier Corporation. (Humidification, Dehumidification, and Spray Ponds)

W. H. Carrier, M. E., D. E., Chairman Emeritus of the Board of Carrier Corporation. (Humidification, Dehumidification, and Spray Ponds)

James G. DeFlon, B. S., Technical Advisor of Sales Department, Fluor Corporation, Ltd. Member: American Society of Heating and Ventilating Engineers, A.S.M.E. Power Test Code Committee on Water Cooling Equipment. (Cooling Towers)

CONTENTS

	Page
Physical Principles	758
Dehumidification	758
Humidification	758
Definitions	759
Humidity	759
Percentage Humidity	759
Percentage Relative Humidity	759
Dew Point	759
Humid Heat	759
Humid Volume	759
Saturated Volume	759
Wet-bulb Temperature	759
Psychometric Chart	759
Examples of Problems	767, 775
Determination of Humidity	776
Dew-point Method	776
Wet-bulb Method	777
Thermal-conductivity Method	777
Hair-hygrometer Method	777
Calculation of Non-aqueous Vapors	777
Air Conditioning	777
Indirect Humidifiers	779
Direct Humidifiers	782
Comparison of Direct and Indirect Systems	783
Systems of Dehumidification	783
Unit Air Conditioners	784
Spray-type Unit Air Conditioners	784
Coil-type Unit Air Conditioners	784
Adsorption Unit	786
Adsorption Cycle	786
Activation Cycle	786
Moisture Removal	786
Humidity-controlling Devices	787
Thermostats	787
Hygrostats	787
Control	789
Data on Performance and Cost of Air-conditioning Equipment	789
Cooling Towers and Spray Ponds	790
Principles	790
General Considerations	790
Mechanical-draft Towers	790
Atmospheric Cooling Towers	793
Cooling Tower Spray Nozzles	796
Spray Ponds	796
Maintenance of Constant Humidity by Reagents	798

11
12

HUMIDIFICATION, DEHUMIDIFICATION, AND COOLING TOWERS AND SPRAY PONDS

REFERENCES: *Heating, Piping and Air Conditioning Magazine*, November, 1929: The Control of Humidity and Temperature as Applied to Manufacturing Processes and Human Comfort, by Willis H. Carrier assisted by a committee appointed by the American Society of Heating and Ventilating Engineers as follows: Willard, Fleisher, Lewis, Still, Armspach, Gant, Yaglou, Ellis, Houghten, and Harding. *A.S.H. & V.E. Guide*, 1948. Marks, "Mechanical Engineers' Handbook," 4th ed., McGraw-Hill, New York, 1941. Kent, "Mechanical Engineers' Handbook," 10th ed., John Wiley & Sons, New York, 1938. Cooling Tower Company, Inc., New York. Carrier Corporation, Syracuse, N. Y. Carrier and Lindsay, Temperatures of Evaporation of Water into Air, *Bull.*, Carrier Engineering Corporation. Walker, Lewis, McAdams, and Gilliland, "Principles of Chemical Engineering," 3d ed., McGraw-Hill, New York, 1937. Badger and McCabe, "Elements of Chemical Engineering," 2d ed., McGraw-Hill, New York, 1936. Schütte & Koerting Company, Philadelphia. Foster-Wheeler Corporation, New York. Spraco, Inc., Boston. Carrier and Busey, "Air Conditioning Apparatus," presented at 1911 meeting, A.S.M.E. F. Kastner, "Luftbefeuchtungsanlagen," Oldenbourg, Munich and Berlin, 1931. Goff and Gratch, "Thermodynamic Properties of Moist Air." Palmetier and Wile, "A New Psychrometric Chart," Carrier Corp.

The moisture and temperature changes resulting when a liquid is brought into contact with a gas form the basis for many processes. Human comfort depends on cooling of the body surface, and this takes place partly by evaporation. The moisture content of the surrounding air is therefore important to comfort and health. Many materials such as textiles, leather, paper, and rayon are best worked under certain conditions of moisture content in the air which in turn affects their moisture content. The field of air conditioning has therefore arisen for the purpose of processing air by adjusting its moisture content. Where dry air is to have its moisture content increased, humidifiers are used, while dehumidifiers are employed to reduce the moisture content.

Since evaporation of a liquid requires an absorption of heat, the process of humidification will usually tend to cool the liquid. Thus, by means of cooling towers and cooling ponds, this principle has been utilized to cool water.

It is the purpose of this section to present the basic principles underlying humidification and dehumidification, methods of determining humidity, descriptions of air-conditioning equipment, and the available methods and data for design.

PHYSICAL PRINCIPLES*

Humidification is the process of evaporation of a liquid into a gas and is essentially the transfer to the main body of gas (by diffusion and convection) of vapor molecules from the gas layer in contact with, and having a partial pressure of vapor practically the same as that of, the liquid. **Dehumidification** is the process of condensation of a vapor from a mixture with a non-condensing gas and is the reverse of humidification as it depends on the transfer of vapor from the main body of gas to the layer of gas next to the liquid because of a partial pressure gradient. At the same time, there are temperature changes of the liquid and gas because of heat exchange between the fluids and the latent heat supplied to, or removed from, the vapor transferred.

* For a complete discussion of the principles of absorption, of which humidification is a special case, the reader is referred to Sec. 8.

Dehumidification. In dehumidification the liquid surface is colder than the dew point of the gas-vapor mixture, and therefore, both the latent heat of the vapor transferred and the sensible heat transferred must be taken up by the water surface. This heat either passes through a film of water to a colder surface, as in tubular dehumidifiers; or goes to raise the temperature of the water itself, as in spray dehumidifiers. The rate of vapor transfer from the gas depends on the mean difference in partial pressure of vapor between the gas and liquid, while the rate of sensible-heat transfer from the gas depends on the mean temperature difference between the gas and liquid. Thus the cooling of the gas stream is calculated from the sensible-heat transfer and is independent of the vapor transfer. The fundamental equations for dehumidification are therefore:

$$w = KA\,\Delta p_m \tag{1}$$
$$q_s = hA\,\Delta t_m \tag{2}$$
$$q_T = q_s + wr \tag{3}$$

where w = rate of vapor transfer, lb./hr.
q_s = rate of sensible-heat transfer, B.t.u./hr.
q_T = rate of total heat transfer, B.t.u./hr.
K = mass-transfer coefficient, lb./(hr.)(sq. ft.)(in. Hg) difference in partial pressure.
h = heat-transfer coefficient, B.t.u./(hr.)(sq. ft.)(°F.) difference in temperature.
A = surface area, sq. ft.
Δp_m = mean difference in partial pressure, in. Hg.
Δt_m = mean difference in temperature, °F.
r = latent heat of vaporization at the liquid temperature, B.t.u./lb.

Humidification. In humidification, vapor is transferred from the liquid to the gas because of a partial

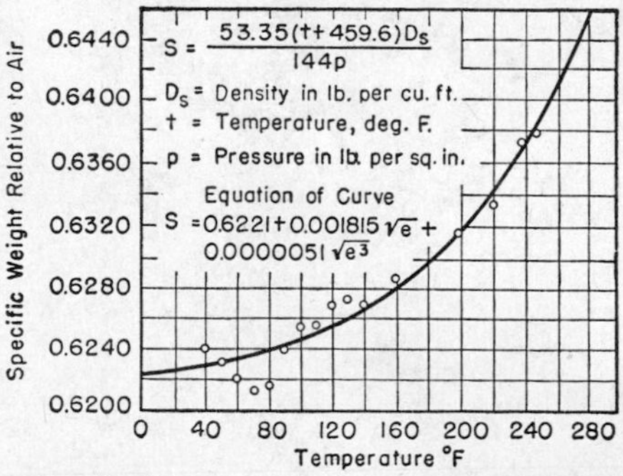

FIG. 1. Specific weight of water vapor. e is partial pressure of water vapor, in. Hg. (*Heating, Piping and Air Conditioning*, p. 538, *November*, 1929.)

pressure gradient, but the gas may be warmer or colder than the liquid so that sensible heat may flow in either direction. The basic equations for vapor and sensible-heat transfer are the same as for dehumidification given by Eqs. (1) and (2). When the liquid is warmer than the gas, Eq. (3) applies and the total heat transferred comes from the liquid. When the gas is warmer than the liquid, sensible heat is transferred to the liquid while latent heat is removed from it. The resultant heat flow through the liquid is then the difference of the two terms in Eq. (3).

The special case of "adiabatic humidification" results when the gas is just sufficiently warmer than the liquid so that the latent heat is just supplied by the sensible heat transferred, and therefore there is no heat transferred through the liquid. Then

$$q_s = wr$$

or

$$hA\,\Delta t_m = KA\,\Delta p_m\,r \qquad (4)$$

This case is of special importance and will be discussed later in showing the basis for the psychrometric charts.

DEFINITIONS

The system of units proposed by Grosvenor and used by Walker, Lewis, McAdams, and Gilliland ("Principles of Chemical Engineering," McGraw-Hill) and by Badger and McCabe ("Elements of Chemical Engineering," McGraw-Hill) is given here.

Humidity (H) is the number of pounds of water carried by 1 lb. dry air (sometimes called "absolute humidity"). This

Percentage relative humidity is the quotient of the partial pressure of water vapor at any temperature divided by the vapor pressure of water at the same temperature.

Dew point is the temperature at which a given mixture of air and water vapor is saturated with water vapor.

Humid heat (s) is the number of B.t.u. necessary to raise the temperature of 1 lb. dry air and the water vapor it contains 1°F. Assuming the specific heats of dry air and water vapor to be constant over the range of temperature usually involved and equal to 0.24 and 0.45, respectively, then

$$s = 0.24 + 0.45H$$

Humid volume is the volume, in cubic feet, of 1 lb. dry air and the water vapor it contains.

Saturated volume is the volume, in cubic feet, of 1 lb. dry air and the water vapor it contains when saturated.

Wet-bulb temperature is the dynamic equilibrium temperature attained by a water surface when exposed to air in a manner such that the sensible heat transferred from the gas to the liquid is equal to the latent heat carried away by evaporation of water vapor into the gas.

PSYCHROMETRIC CHART*

Three charts are given in Figs. 2 to 4 for low-, medium-, and high-temperature ranges. In addition to the temperature-humidity scales and the adiabatic saturation lines, other scales and lines are included for convenience

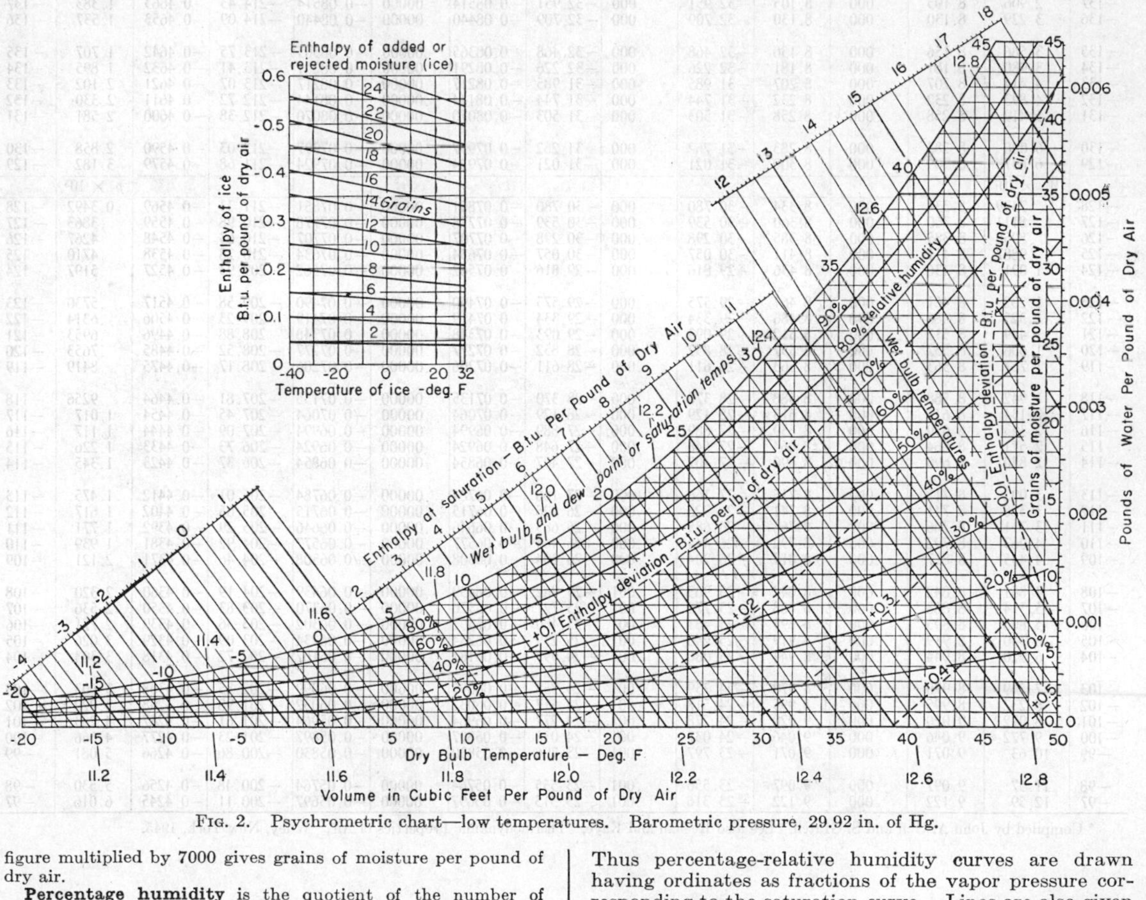

FIG. 2. Psychrometric chart—low temperatures. Barometric pressure, 29.92 in. of Hg.

figure multiplied by 7000 gives grains of moisture per pound of dry air.

Percentage humidity is the quotient of the number of pounds of water vapor carried by 1 lb. dry air divided by the number of pounds of vapor which 1 lb. dry air would carry, were it completely saturated at the same temperature multiplied by 100.

Thus percentage-relative humidity curves are drawn having ordinates as fractions of the vapor pressure corresponding to the saturation curve. Lines are also given for estimating specific volume of the mixture per pound of

* See Sec. 13 for psychrometric charts involving organic vapors mixed with air.

Table 1. Thermodynamic Properties of Moist Air* (Standard Atmospheric Pressure, 29.921 in. Hg)

Temp. t, °F.	Humidity ratio $W_s \times 10^8$	Volume, cu. ft./lb. dry air			Enthalpy, B.t.u./lb. dry air			Entropy, B.t.u./(°F.) (lb. dry air)			Condensed water			Temp. t, °F.
		v_a	v_{as}	v_s	h_a	h_{as}	h_s	s_a	s_{as}	s_s	Enthalpy, B.t.u./lb. h_w	Entropy, B.t.u./(lb.)(°F.) s_w	Vapor press., in. Hg $p_s \times 10^6$	
−160	0.2120	7.520	0.000	7.520	−38.504	0.000	−38.504	−0.10300	0.00000	−0.10300	−222.00	−0.4907	0.1009	−160
−159	.2394	7.545	.000	7.545	−38.262	.000	−38.262	−0.10219	.00000	−0.10219	−221.68	−0.4896	.1139	−159
−158	.2703	7.571	.000	7.571	−38.021	.000	−38.021	−0.10139	.00000	−0.10139	−221.36	−0.4885	.1286	−158
−157	.3049	7.596	.000	7.596	−37.779	.000	−37.779	−0.10059	.00000	−0.10059	−221.04	−0.4874	.1450	−157
−156	.3435	7.622	.000	7.622	−37.538	.000	−37.538	−0.09980	.00000	−0.09980	−220.72	−0.4864	.1635	−156
−155	.3869	7.647	.000	7.647	−37.296	.000	−37.296	−0.09901	.00000	−0.09901	−220.40	−0.4853	.1842	−155
−154	.4354	7.673	.000	7.673	−37.055	.000	−37.055	−0.09822	.00000	−0.09822	−220.07	−0.4843	.2073	−154
−153	.4897	7.699	.000	7.699	−36.813	.000	−36.813	−0.09743	.00000	−0.09743	−219.75	−0.4832	.2331	−153
−152	.5502	7.724	.000	7.724	−36.572	.000	−36.572	−0.09664	.00000	−0.09664	−219.42	−0.4822	.2620	−152
−151	.6178	7.750	.000	7.750	−36.330	.000	−36.330	−0.09586	.00000	−0.09586	−219.10	−0.4811	.2942	−151
−150	.6932	7.775	.000	7.775	−36.088	.000	−36.088	−0.09508	.00000	−0.09508	−218.77	−0.4800	.3301	−150
−149	.7772	7.801	.000	7.801	−35.847	.000	−35.847	−0.09430	.00000	−0.09430	−218.44	−0.4789	.3701	−149
−148	.8709	7.826	.000	7.826	−35.606	.000	−35.606	−0.09352	.00000	−0.09352	−218.11	−0.4779	.4146	−148
−147	.9750	7.851	.000	7.851	−35.364	.000	−35.364	−0.09274	.00000	−0.09274	−217.78	−0.4768	.4641	−147
−146	1.091	7.876	.000	7.876	−35.123	.000	−35.123	−0.09198	.00000	−0.09198	−217.45	−0.4758	.5194	−146
−145	1.219	7.902	.000	7.902	−34.881	.000	−34.881	−0.09121	.00000	−0.09121	−217.12	−0.4747	.5807	−145
−144	1.362	7.927	.000	7.927	−34.640	.000	−34.640	−0.09044	.00000	−0.09044	−216.78	−0.4737	.6488	−144
−143	1.521	7.953	.000	7.953	−34.399	.000	−34.399	−0.08967	.00000	−0.08967	−216.45	−0.4726	.7243	−143
−142	1.698	7.978	.000	7.978	−34.157	.000	−34.157	−0.08892	.00000	−0.08892	−216.11	−0.4716	.8082	−142
−141	1.893	8.004	.000	8.004	−33.916	.000	−33.916	−0.08816	.00000	−0.08816	−215.78	−0.4705	.9011	−141
−140	2.109	8.029	.000	8.029	−33.674	.000	−33.674	−0.08740	.00000	−0.08740	−215.44	−0.4695	1.004	−140
−139	2.348	8.054	.000	8.054	−33.433	.000	−33.433	−0.08664	.00000	−0.08664	−215.11	−0.4684	1.118	−139
−138	2.613	8.079	.000	8.079	−33.192	.000	−33.192	−0.08589	.00000	−0.08589	−214.77	−0.4674	1.244	−138
−137	2.906	8.105	.000	8.105	−32.951	.000	−32.951	−0.08514	.00000	−0.08514	−214.43	−0.4663	1.383	−137
−136	3.229	8.130	.000	8.130	−32.709	.000	−32.709	−0.08440	.00000	−0.08440	−214.09	−0.4653	1.537	−136
−135	3.586	8.156	.000	8.156	−32.468	.000	−32.468	−0.08365	.00000	−0.08365	−213.75	−0.4642	1.707	−135
−134	3.980	8.181	.000	8.181	−32.226	.000	−32.226	−0.08291	.00000	−0.08291	−213.41	−0.4632	1.895	−134
−133	4.414	8.207	.000	8.207	−31.985	.000	−31.985	−0.08217	.00000	−0.08217	−213.07	−0.4621	2.102	−133
−132	4.893	8.232	.000	8.232	−31.744	.000	−31.744	−0.08144	.00000	−0.08144	−212.72	−0.4611	2.330	−132
−131	5.419	8.258	.000	8.258	−31.503	.000	−31.503	−0.08070	.00000	−0.08070	−212.38	−0.4600	2.581	−131
−130	6.000	8.283	.000	8.283	−31.262	.000	−31.262	−0.07997	.00000	−0.07997	−212.03	−0.4590	2.858	−130
−129	6.637	8.309	.000	8.309	−31.021	.000	−31.021	−0.07924	.00000	−0.07924	−211.68	−0.4579	3.182	−129
	$W_s \times 10^7$												$p_s \times 10^5$	
−128	.7339	8.334	.000	8.334	−30.780	.000	−30.780	−0.07851	.00000	−0.07851	−211.33	−0.4569	0.3492	−128
−127	.8111	8.360	.000	8.360	−30.539	.000	−30.539	−0.07778	.00000	−0.07778	−210.98	−0.4559	.3863	−127
−126	.8958	8.385	.000	8.385	−30.298	.000	−30.298	−0.07707	.00000	−0.07707	−210.63	−0.4548	.4267	−126
−125	.9887	8.411	.000	8.411	−30.057	.000	−30.057	−0.07634	.00000	−0.07634	−210.28	−0.4538	.4710	−125
−124	1.091	8.436	.000	8.436	−29.816	.000	−29.816	−0.07562	.00000	−0.07562	−209.93	−0.4527	.5197	−124
−123	1.202	8.461	.000	8.461	−29.575	.000	−29.575	−0.07490	.00000	−0.07490	−209.58	−0.4517	.5730	−123
−122	1.325	8.486	.000	8.486	−29.334	.000	−29.334	−0.07419	.00000	−0.07419	−209.23	−0.4506	.6314	−122
−121	1.459	8.512	.000	8.512	−29.093	.000	−29.093	−0.07348	.00000	−0.07348	−208.88	−0.4496	.6953	−121
−120	1.606	8.537	.000	8.537	−28.852	.000	−28.852	−0.07277	.00000	−0.07277	−208.52	−0.4485	.7653	−120
−119	1.767	8.563	.000	8.563	−28.611	.000	−28.611	−0.07206	.00000	−0.07206	−208.17	−0.4475	.8419	−119
−118	1.942	8.588	.000	8.588	−28.370	.000	−28.370	−0.07135	.00000	−0.07135	−207.81	−0.4464	.9256	−118
−117	2.134	8.613	.000	8.613	−28.129	.000	−28.129	−0.07064	.00000	−0.07064	−207.45	−0.4454	1.017	−117
−116	2.343	8.639	.000	8.639	−27.889	.000	−27.889	−0.06994	.00000	−0.06994	−207.09	−0.4444	1.117	−116
−115	2.571	8.664	.000	8.664	−27.648	.000	−27.648	−0.06924	.00000	−0.06924	−206.73	−0.4433	1.226	−115
−114	2.820	8.690	.000	8.690	−27.407	.000	−27.407	−0.06854	.00000	−0.06854	−206.37	−0.4423	1.345	−114
−113	3.092	8.715	.000	8.715	−27.166	.000	−27.166	−0.06784	.00000	−0.06784	−206.01	−0.4412	1.475	−113
−112	3.388	8.741	.000	8.741	−26.926	.000	−26.926	−0.06715	.00000	−0.06715	−205.65	−0.4402	1.617	−112
−111	3.711	8.766	.000	8.766	−26.685	.000	−26.685	−0.06646	.00000	−0.06646	−205.29	−0.4392	1.771	−111
−110	4.063	8.792	.000	8.792	−26.444	.000	−26.444	−0.06577	.00000	−0.06577	−204.92	−0.4381	1.939	−110
−109	4.445	8.817	.000	8.817	−26.204	.000	−26.204	−0.06508	.00000	−0.06508	−204.46	−0.4371	2.121	−109
−108	4.861	8.842	.000	8.842	−25.963	.001	−25.962	−0.06439	.00000	−0.06439	−204.19	−0.4360	2.320	−108
−107	5.314	8.868	.000	8.868	−25.772	.001	−25.721	−0.06370	.00000	−0.06370	−203.83	−0.4350	2.536	−107
−106	5.806	8.893	.000	8.893	−25.481	.001	−25.480	−0.06302	.00000	−0.06302	−203.46	−0.4339	2.771	−106
−105	6.340	8.919	.000	8.919	−25.240	.001	−25.239	−0.06234	.00000	−0.06234	−203.09	−0.4329	3.026	−105
−104	6.920	8.944	.000	8.944	−25.000	.001	−24.999	−0.06167	.00000	−0.06167	−202.72	−0.4318	3.303	−104
−103	7.549	8.970	.000	8.970	−24.759	.001	−24.758	−0.06099	.00000	−0.06099	−202.35	−0.4308	3.603	−103
−102	8.232	8.995	.000	8.995	−24.518	.001	−24.517	−0.06032	.00000	−0.06032	−201.98	−0.4298	3.929	−102
−101	8.972	9.020	.000	9.020	−24.278	.001	−24.277	−0.05964	.00000	−0.05964	−201.61	−0.4287	4.283	−101
−100	9.772	9.046	.000	9.046	−24.037	.001	−24.036	−0.05897	.00000	−0.05897	−201.23	−0.4277	4.666	−100
−99	10.63	9.071	.000	9.071	−23.797	.001	−23.796	−0.05830	.00000	−0.05830	−200.86	−0.4266	5.081	−99
−98	11.57	9.097	.000	9.097	−23.556	.001	−23.555	−0.05764	.00000	−0.05764	−200.48	−0.4256	5.530	−98
−97	12.59	9.122	.000	9.122	−23.316	.001	−23.315	−0.05697	.00000	−0.05697	−200.11	−0.4245	6.016	−97

* Compiled by John A. Goff and S. Gratch. See also Keenan and Kaye, "Thermodynamic Properties of Air," Wiley, New York, 1945.

Table 1. Thermodynamic Properties of Moist Air (Standard Atmospheric Pressure, 29.921 in. Hg)— *(Continued)*

Temp. t, °F.	Humidity ratio $W_s \times 10^6$	Volume, cu. ft./lb. dry air v_a	v_{as}	v_s	Enthalpy, B.t.u./lb. dry air h_a	h_{as}	h_s	Entropy, B.t.u./(°F.) (lb. dry air) s_a	s_{as}	s_s	Condensed water: Enthalpy, B.t.u./lb. h_w	Entropy, B.t.u./(lb.)(°F.) s_w	Vapor press., in. Hg $p_s \times 10^4$	Temp. t, °F.
−96	1.369	9.147	0.000	9.147	−23.075	0.001	−23.074	−0.05631	0.00000	−0.05631	−199.73	−0.4235	0.6542	−96
−95	1.489	9.173	.000	9.173	−22.835	.002	−22.833	−0.05565	.00000	−0.05565	−199.35	−0.4225	.7111	−95
−94	1.617	9.198	.000	9.198	−22.594	.002	−22.592	−0.05500	.00000	−0.05500	−198.97	−0.4214	.7725	−94
−93	1.756	9.224	.000	9.224	−22.353	.002	−22.341	−0.05434	.00000	−0.05434	−198.59	−0.4204	.8388	−93
−92	1.906	9.249	.000	9.249	−22.113	.002	−22.111	−0.05369	.00001	−0.05368	−198.21	−0.4193	.9105	−92
−91	2.068	9.275	.000	9.275	−21.872	.002	−21.870	−0.05303	.00001	−0.05302	−197.83	−0.4183	.9879	−91
−90	2.242	9.300	.000	9.300	−21.631	.002	−21.629	−0.05237	.00001	−0.05236	−197.44	−0.4173	1.071	−90
−89	2.430	9.325	.000	9.325	−21.391	.003	−21.388	−0.05172	.00001	−0.05171	−197.06	−0.4162	1.161	−89
−88	2.634	9.351	.000	9.351	−21.150	.003	−21.147	−0.05107	.00001	−0.05106	−196.67	−0.4152	1.259	−88
−87	2.852	9.376	.000	9.376	−20.909	.003	−20.906	−0.05042	.00001	−0.05041	−196.29	−0.4142	1.363	−87
−86	3.089	9.401	.000	9.401	−20.669	.003	−20.666	−0.04978	.00001	−0.04977	−195.90	−0.4131	1.476	−86
−85	3.342	9.426	.000	9.426	−20.428	.003	−20.425	−0.04913	.00001	−0.04912	−195.51	−0.4121	1.597	−85
−84	3.615	9.451	.000	9.451	−20.188	.004	−20.184	−0.04849	.00001	−0.04848	−195.12	−0.4110	1.728	−84
−83	3.909	9.477	.000	9.477	−19.947	.004	−19.943	−0.04785	.00001	−0.04784	−194.73	−0.4100	1.868	−83
−82	4.225	9.502	.000	9.502	−19.706	.004	−19.702	−0.04721	.00001	−0.04720	−194.34	−0.4090	2.019	−82
−81	4.564	9.527	.000	9.527	−19.466	.005	−19.461	−0.04658	.00001	−0.04657	−193.95	−0.4079	2.181	−81
−80	4.930	9.553	.000	9.553	−19.225	.005	−19.220	−0.04595	.00001	−0.04594	−193.55	−0.4069	2.356	−80
−79	5.322	9.578	.000	9.578	−18.984	.005	−18.979	−0.04532	.00002	−0.04530	−193.16	−0.4059	2.543	−79
−78	5.742	9.604	.000	9.604	−18.744	.006	−18.738	−0.04469	.00002	−0.04467	−192.76	−0.4048	2.744	−78
−77	6.193	9.629	.000	9.629	−18.503	.006	−18.497	−0.04406	.00002	−0.04404	−192.37	−0.4038	2.960	−77
−76	6.677	9.654	.000	9.654	−18.263	.007	−18.256	−0.04343	.00002	−0.04341	−191.97	−0.4027	3.192	−76
−75	7.196	9.680	.000	9.680	−18.022	.007	−18.015	−0.04280	.00002	−0.04278	−191.57	−0.4017	3.441	−75
−74	7.753	9.705	.000	9.705	−17.782	.008	−17.774	−0.04218	.00003	−0.04215	−191.17	−0.4007	3.707	−74
−73	8.349	9.730	.000	9.730	−17.541	.008	−17.533	−0.04155	.00003	−0.04152	−190.77	−0.3996	3.992	−73
−72	8.990	9.756	.000	9.756	−17.301	.009	−17.292	−0.04093	.00003	−0.04090	−190.37	−0.3986	4.298	−74
−71	9.675	9.781	.000	9.781	−17.060	.010	−17.050	−0.04031	.00003	−0.04028	−189.97	−0.3975	4.625	−71
−70	10.40	9.806	.000	9.806	−16.820	.011	−16.809	−0.03969	.00003	−0.03966	−189.56	−0.3965	4.976	−70
−69	11.19	9.831	.000	9.831	−16.579	.011	−16.568	−0.03908	.00004	−0.03904	−189.16	−0.3954	5.351	−69
−68	12.03	9.856	.000	9.856	−16.339	.012	−16.327	−0.03846	.00004	−0.03842	−188.75	−0.3944	5.752	−68
−67	12.92	9.882	.000	9.882	−16.098	.013	−16.085	−0.03785	.00004	−0.03781	−188.35	−0.3934	6.181	−67
−66	13.88	9.907	.000	9.907	−15.858	.014	−15.844	−0.03724	.00004	−0.03720	−187.94	−0.3924	6.640	−66
−65	14.91	9.932	.000	9.932	−15.617	.015	−15.602	−0.03663	.00005	−0.03658	−187.53	−0.3913	7.130	−65
	$W_s \times 10^5$												$p_s \times 10^3$	
−64	1.601	9.958	.000	9.958	−15.377	.016	−15.361	−0.03602	.00005	−0.03597	−187.12	−0.3903	0.7654	−64
−63	1.718	9.983	.000	9.983	−15.137	.018	−15.119	−0.03541	.00005	−0.03536	−186.71	−0.3893	.8213	−63
−62	1.843	10.009	.000	10.009	−14.896	.019	−14.877	−0.03481	.00006	−0.03475	−186.30	−0.3882	.8810	−62
−61	1.976	10.034	.000	10.034	−14.656	.020	−14.636	−0.03420	.00006	−0.03414	−185.89	−0.3872	.9447	−61
−60	2.118	10.059	.000	10.059	−14.416	.022	−14.394	−0.03360	.00006	−0.03354	−185.47	−0.3861	1.0127	−60
−59	2.269	10.085	.000	10.085	−14.175	.023	−14.152	−0.03300	.00007	−0.03293	−185.06	−0.3851	1.0852	−59
−58	2.431	10.110	.000	10.110	−13.935	.025	−13.910	−0.03240	.00007	−0.03233	−184.64	−0.3841	1.1624	−58
−57	2.603	10.135	.000	10.135	−13.695	.027	−13.668	−0.03180	.00008	−0.03172	−184.23	−0.3830	1.2447	−57
−56	2.786	10.161	.000	10.161	−13.454	.029	−13.425	−0.03120	.00008	−0.03112	−183.81	−0.3820	1.3324	−56
−55	2.982	10.186	.000	10.186	−13.214	.031	−13.183	−0.03061	.00009	−0.03052	−183.39	−0.3810	1.4258	−55
−54	3.190	10.211	.000	10.211	−12.974	.033	−12.941	−0.03002	.00009	−0.02993	−182.97	−0.3799	1.5253	−54
−53	3.411	10.237	.000	10.237	−12.733	.035	−12.698	−0.02943	.00010	−0.02933	−182.55	−0.3789	1.6312	−53
−52	3.646	10.262	.001	10.263	−12.493	.038	−12.455	−0.02884	.00011	−0.02873	−182.13	−0.3779	1.7438	−52
−51	3.897	10.288	.001	10.289	−12.253	.041	−12.212	−0.02825	.00011	−0.02814	−181.71	−0.3769	1.8635	−51
−50	4.163	10.313	.001	10.314	−12.012	.043	−11.969	−0.02766	.00012	−0.02754	−181.29	−0.3758	1.9910	−50
−49	4.446	10.338	.001	10.339	−11.772	.046	−11.726	−0.02707	.00012	−0.02695	−180.87	−0.3748	2.1264	−49
−48	4.747	10.364	.001	10.365	−11.532	.049	−11.483	−0.02649	.00013	−0.02636	−180.44	−0.3738	2.2702	−48
−47	5.066	10.389	.001	10.390	−11.292	.053	−11.239	−0.02590	.00013	−0.02577	−180.02	−0.3728	2.4230	−47
−46	5.406	10.414	.001	10.415	−11.051	.056	−10.995	−0.02532	.00014	−0.02518	−179.59	−0.3717	2.5854	−46
−45	5.766	10.440	.001	10.441	−10.811	.060	−10.751	−0.02474	.00015	−0.02459	−179.16	−0.3707	2.7578	−45
−44	6.149	10.465	.001	10.466	−10.571	.064	−10.507	−0.02416	.00016	−0.02400	−178.73	−0.3696	2.9408	−44
−43	6.555	10.490	.001	10.491	−10.330	.068	−10.262	−0.02358	.00017	−0.02341	−178.30	−0.3686	3.1349	−43
−42	6.985	10.516	.001	10.517	−10.090	.073	−10.017	−0.02301	.00019	−0.02282	−177.87	−0.3676	3.3408	−42
−41	7.441	10.541	.001	10.542	−9.850	.078	−9.772	−0.02243	.00020	−0.02223	−177.44	−0.3665	3.5591	−41
−40	7.925	10.566	.001	10.567	−9.609	.083	−9.526	−0.02186	.00021	−0.02165	−177.01	−0.3655	3.7906	−40
−39	8.437	10.592	.001	10.593	−9.369	.089	−9.280	−0.02129	.00022	−0.02107	−176.58	−0.3645	4.0359	−39
−38	8.980	10.617	.002	10.619	−9.129	.094	−9.035	−0.02072	.00024	−0.02048	−176.14	−0.3634	4.2958	−38
−37	9.556	10.642	.002	10.644	−8.889	.100	−8.789	−0.02015	.00025	−0.01990	−175.71	−0.3624	4.5711	−37
−36	10.16	10.668	.002	10.670	−8.648	.106	−8.542	−0.01958	.00026	−0.01932	−175.27	−0.3614	4.8626	−36
−35	10.81	10.693	.002	10.695	−8.408	.113	−8.295	−0.01902	.00028	−0.01874	−174.84	−0.3604	5.1713	−35
−34	11.49	10.718	.002	10.720	−8.168	.121	−8.047	−0.01845	.00030	−0.01815	−174.40	−0.3593	5.4980	−34
−33	12.21	10.744	.002	10.746	−7.927	.128	−7.799	−0.01789	.00032	−0.01757	−173.96	−0.3583	5.8437	−33

Table 1. Thermodynamic Properties of Moist Air (Standard Atmospheric Pressure, 29.921 in. Hg)—(*Continued*)

Temp. t, °F.	Humidity ratio $W_s \times 10^4$	Volume, cu. ft./lb. dry air v_a	v_{as}	v_s	Enthalpy, B.t.u./lb. dry air h_a	h_{as}	h_s	Entropy, B.t.u./(°F.) (lb. dry air) s_a	s_{as}	s_s	Condensed water: Enthalpy, B.t.u./lb. h_w	Entropy, B.t.u./(lb.)(°F.) s_w	Vapor press., in. Hg $p_s \times 10^2$	Temp. t, °F.
−32	1.298	10.769	0.002	10.771	−7.687	0.136	−7.551	−0.01733	0.00034	−0.01699	−173.52	−0.3573	0.62093	−32
−31	1.378	10.794	.002	10.796	−7.447	.145	−7.302	−0.01677	.00036	−0.01641	−173.08	−0.3563	.65979	−31
−30	1.464	10.820	.002	10.822	−7.207	.154	−7.053	−0.01621	.00038	−0.01583	−172.64	−0.3552	.70046	−30
−29	1.554	10.845	.003	10.848	−6.966	.163	−6.803	−0.01565	.00040	−0.01525	−172.20	−0.3542	.74365	−29
−28	1.649	10.870	.003	10.873	−6.726	.173	−6.553	−0.01509	.00043	−0.01466	−171.75	−0.3531	.78928	−28
−27	1.750	10.896	.003	10.899	−6.486	.184	−6.302	−0.01453	.00045	−0.01408	−171.31	−0.3521	.83748	−27
−26	1.856	10.921	.003	10.924	−6.246	.196	−6.050	−0.01398	.00048	−0.01350	−170.86	−0.3511	.88838	−26
−25	1.969	10.946	.004	10.950	−6.005	.207	−5.798	−0.01342	.00051	−0.01291	−170.42	−0.3500	.94212	−25
−24	2.087	10.972	.004	10.976	−5.765	.219	−5.546	−0.01287	.00054	−0.01233	−169.97	−0.3490	.99885	−24
−23	2.212	10.997	.004	11.001	−5.525	.232	−5.293	−0.01232	.00057	−0.01175	−169.52	−0.3480	1.0587	−23
−22	2.344	11.022	.004	11.026	−5.285	.246	−5.039	−0.01177	.00061	−0.01116	−169.07	−0.3469	1.1219	−22
−21	2.483	11.048	.004	11.052	−5.044	.261	−4.783	−0.01122	.00064	−0.01058	−168.62	−0.3459	1.1885	−21
−20	2.630	11.073	.005	11.078	−4.804	.277	−4.527	−0.01067	.00068	−0.00999	−168.17	−0.3449	1.2587	−20
−19	2.785	11.098	.005	11.103	−4.564	.293	−4.271	−0.01012	.00072	−0.00940	−167.72	−0.3439	1.3327	−19
−18	2.948	11.124	.005	11.129	−4.324	.310	−4.014	−0.00958	.00076	−0.00882	−167.26	−0.3428	1.4107	−18
−17	3.120	11.149	.006	11.155	−4.083	.328	−3.755	−0.00904	.00080	−0.00824	−166.81	−0.3418	1.4929	−17
−16	3.301	11.174	.006	11.180	−3.843	.348	−3.495	−0.00850	.00084	−0.00766	−166.35	−0.3408	1.5795	−16
−15	3.491	11.200	.006	11.206	−3.603	.368	−3.235	−0.00796	.00089	−0.00707	−165.90	−0.3398	1.6706	−15
−14	3.692	11.225	.007	11.232	−3.363	.389	−2.974	−0.00743	.00094	−0.00649	−165.44	−0.3387	1.7666	−14
−13	3.903	11.250	.007	11.257	−3.123	.412	−2.711	−0.00689	.00099	−0.00590	−164.98	−0.3377	1.8677	−13
−12	4.125	11.275	.008	11.283	−2.882	.436	−2.446	−0.00636	.00104	−0.00532	−164.52	−0.3367	1.9740	−12
−11	4.359	11.301	.008	11.309	−2.642	.461	−2.181	−0.00582	.00109	−0.00473	−164.06	−0.3357	2.0859	−11
−10	4.606	11.326	.008	11.334	−2.402	.487	−1.915	−0.00529	.00115	−0.00414	−163.60	−0.3346	2.2035	−10
−9	4.865	11.351	.008	11.359	−2.162	.514	−1.648	−0.00475	.00121	−0.00354	−163.14	−0.3336	2.3272	−9
−8	5.137	11.376	.009	11.385	−1.922	.543	−1.379	−0.00422	.00128	−0.00294	−162.67	−0.3326	2.4573	−8
−7	5.423	11.401	.010	11.411	−1.681	.574	−1.107	−0.00369	.00135	−0.00234	−162.21	−0.3316	2.5940	−7
−6	5.724	11.427	.010	11.437	−1.441	.606	−0.835	−0.00316	.00142	−0.00174	−161.74	−0.3305	2.7377	−6
−5	6.040	11.452	.011	11.463	−1.201	.639	−0.562	−0.00263	.00149	−0.00114	−161.28	−0.3295	2.8886	−5
−4	6.371	11.477	.012	11.489	−0.961	.675	−0.286	−0.00210	.00157	−0.00053	−160.81	−0.3285	3.0472	−4
−3	6.720	11.502	.013	11.515	−0.721	.712	−0.009	−0.00157	.00165	0.00008	−160.34	−0.3275	3.2137	−3
−2	7.085	11.528	.013	11.541	−0.480	.751	0.271	−0.00105	.00174	.00069	−159.87	−0.3264	3.3885	−2
−1	7.469	11.553	.014	11.567	−0.240	.792	0.552	−0.00052	.00183	.00131	−159.40	−0.3254	3.5720	−1
	$W_s \times 10^3$													
0	0.7872	11.578	0.015	11.593	0.000	0.835	0.835	0.00000	0.00192	0.00192	−158.93	−0.3244	3.7645	0
1	.8295	11.604	.015	11.619	.240	.880	1.120	.00052	.00202	.00254	−158.46	−0.3234	3.9666	1
2	.8739	11.629	.016	11.645	.480	.928	1.408	.00104	.00212	.00316	−157.99	−0.3223	4.1785	2
3	.9204	11.654	.017	11.671	.721	.977	1.698	.00156	.00223	.00379	−157.52	−0.3213	4.4007	3
4	.9692	11.679	.018	11.697	.961	1.030	1.991	.00208	.00234	.00442	−157.04	−0.3203	4.6337	4
5	1.020	11.705	.019	11.724	1.201	1.085	2.286	.00260	.00246	.00506	−156.57	−0.3193	4.8779	5
6	1.074	11.730	.020	11.750	1.441	1.142	2.583	.00312	.00258	.00570	−156.09	−0.3182	5.1339	6
7	1.130	11.756	.021	11.777	1.681	1.202	2.883	.00364	.00271	.00635	−155.61	−0.3172	5.4022	7
8	1.189	11.781	.022	11.803	1.922	1.266	3.188	.00415	.00285	.00700	−155.13	−0.3162	5.6832	8
9	1.251	11.806	.024	11.830	2.162	1.332	3.494	.00467	.00299	.00766	−154.65	−0.3152	5.9776	9
10	1.315	11.831	.025	11.856	2.402	1.401	3.803	.00518	.00314	.00832	−154.17	−0.3141	6.2858	10
11	1.383	11.857	.026	11.883	2.642	1.474	4.116	.00569	.00330	.00899	−153.69	−0.3131	6.6085	11
12	1.454	11.882	.028	11.910	2.882	1.550	4.432	.00620	.00346	.00966	−153.21	−0.3121	6.9462	12
13	1.528	11.907	.029	11.936	3.123	1.630	4.753	.00671	.00363	.01034	−152.73	−0.3111	7.2997	13
14	1.606	11.933	.030	11.963	3.363	1.713	5.076	.00721	.00380	.01101	−152.24	−0.3100	7.6696	14
15	1.687	11.958	.032	11.990	3.603	1.800	5.403	.00772	.00399	.01171	−151.76	−0.3090	8.0565	15
16	1.772	11.983	.034	12.017	3.843	1.892	5.735	.00822	.00418	.01240	−151.27	−0.3080	8.4612	16
17	1.861	12.009	.035	12.044	4.083	1.988	6.071	.00873	.00438	.01311	−150.78	−0.3070	8.8843	17
18	1.953	12.034	.038	12.072	4.324	2.088	6.412	.00923	.00459	.01382	−150.29	−0.3059	9.3267	18
19	2.051	12.059	.040	12.099	4.564	2.192	6.756	.00973	.00481	.01454	−149.80	−0.3049	9.7889	19
20	2.152	12.084	.042	12.126	4.804	2.302	7.106	.01023	.00504	.01527	−149.31	−0.3039	10.272	20
21	2.258	12.110	.044	12.154	5.044	2.416	7.460	.01073	.00528	.01601	−148.82	−0.3029	10.777	21
22	2.369	12.135	.046	12.181	5.284	2.536	7.820	.01123	.00553	.01676	−148.33	−0.3018	11.305	22
23	2.485	12.160	.049	12.209	5.525	2.661	8.186	.01173	.00579	.01752	−147.84	−0.3008	11.856	23
24	2.606	12.186	.051	12.237	5.765	2.792	8.557	.01223	.00607	.01830	−147.34	−0.2998	12.431	24
25	2.733	12.211	.054	12.265	6.005	2.929	8.934	.01273	.00635	.01908	−146.85	−0.2988	13.032	25
26	2.865	12.236	.057	12.293	6.245	3.072	9.317	.01322	.00665	.01987	−146.35	−0.2977	13.659	26
27	3.003	12.262	.059	12.321	6.485	3.221	9.706	.01372	.00696	.02068	−145.85	−0.2967	14.313	27
28	3.147	12.287	.062	12.349	6.726	3.377	10.103	.01421	.00728	.02149	−145.36	−0.2957	14.996	28
29	3.297	12.312	.065	12.377	6.966	3.540	10.506	.01470	.00761	.02231	−144.86	−0.2947	15.709	29
30	3.454	12.338	.068	12.406	7.206	3.709	10.915	.01519	.00796	.02315	−144.36	−0.2936	16.452	30
31	3.617	12.363	.071	12.434	7.446	3.887	11.333	.01568	.00832	.02400	−143.86	−0.2926	17.227	31
32	3.788	12.388	.075	12.463	7.686	4.072	11.758	.01617	.00870	.02487	−143.36	−0.2916	18.035	32
32†	3.788	12.388	.075	12.463	7.686	4.072	11.758	.01617	.00870	.02487	0.04	0.0000	18.037	32†
33	3.944	12.413	.079	12.492	7.927	4.242	12.169	.01666	.00904	.02570	1.05	.0020	18.778	33
34	4.107	12.438	.082	12.520	8.167	4.418	12.585	.01715	.00940	.02655	2.06	.0041	19.546	34

† Extrapolated to represent metastable equilibrium with undercooled liquid.

Table 1. Thermodynamic Properties of Moist Air (Standard Atmospheric Pressure, 29.921 in. Hg)— *(Continued)*

Temp. t, °F.	Humidity ratio $W_s \times 10^3$	Volume, cu. ft./lb. dry air			Enthalpy, B.t.u./lb. dry air			Entropy, B.t.u./(°F.) (lb. dry air)			Condensed water			Temp. t, °F.
		v_a	v_{as}	v_s	h_a	h_{as}	h_s	s_a	s_{as}	s_s	Enthalpy, B.t.u./lb. h_w	Entropy, B.t.u./(lb.)(°F.) s_w	Vapor press., in. Hg p_s	
35	4.275	12.464	0.085	12.549	8.407	4.601	13.008	0.01764	0.00977	0.02741	3.06	0.0061	0.20342	35
36	4.450	12.489	.089	12.578	8.647	4.791	13.438	.01812	.01016	.02828	4.07	.0081	.21166	36
37	4.631	12.514	.093	12.607	8.887	4.987	13.874	.01861	.01056	.02917	5.07	.0102	.22020	37
38	4.818	12.540	.097	12.637	9.128	5.191	14.319	.01909	.01097	.03006	6.08	.0122	.22904	38
39	5.012	12.565	.101	12.666	9.368	5.403	14.771	.01957	.01139	.03096	7.08	.0142	.23819	39
40	5.213	12.590	.105	12.695	9.608	5.622	15.230	.02005	.01183	.03188	8.09	.0162	.24767	40
41	5.421	12.616	.109	12.725	9.848	5.849	15.697	.02053	.01228	.03281	9.09	.0182	.25748	41
42	5.638	12.641	.114	12.755	10.088	6.084	16.172	.02101	.01275	.03376	10.09	.0202	.26763	42
43	5.860	12.666	.119	12.785	10.329	6.328	16.657	.02149	.01323	.03472	11.10	.0222	.27813	43
44	6.091	12.691	.124	12.815	10.569	6.580	17.149	.02197	.01373	.03570	12.10	.0242	.28899	44
45	6.331	12.717	.129	12.846	10.809	6.841	17.650	.02245	.01425	.03670	13.10	.0262	.30023	45
46	6.578	12.742	.134	12.876	11.049	7.112	18.161	.02293	.01478	.03771	14.10	.0282	.31185	46
47	6.835	12.767	.140	12.907	11.289	7.391	18.680	.02340	.01534	.03874	15.11	.0302	.32386	47
48	7.100	12.792	.146	12.938	11.530	7.681	19.211	.02387	.01591	.03978	16.11	.0321	.33629	48
49	7.374	12.818	.151	12.969	11.770	7.981	19.751	.02434	.01650	.04084	17.11	.0341	.34913	49
50	7.658	12.843	.158	13.001	12.010	8.291	20.301	.02481	.01711	.04192	18.11	.0361	.36240	50
51	7.952	12.868	.164	13.032	12.250	8.612	20.862	.02528	.01774	.04302	19.11	.0381	.37611	51
52	8.256	12.894	.170	13.064	12.491	8.945	21.436	.02575	.01839	.04414	20.11	.0400	.39028	52
53	8.569	12.919	.178	13.097	12.731	9.289	22.020	.02622	.01906	.04528	21.12	.0420	.40492	53
54	8.894	12.944	.185	13.129	12.971	9.644	22.615	.02669	.01976	.04645	22.12	.0439	.42004	54
55	9.229	12.970	.192	13.162	13.211	10.01	23.22	.02716	.02047	.04763	23.12	.0459	.43565	55
56	9.575	12.995	.200	13.195	13.452	10.39	23.84	.02762	.02121	.04883	24.12	.0478	.45176	56
57	9.934	13.020	.208	13.228	13.692	10.79	24.48	.02809	.02197	.05006	25.12	.0497	.46840	57
58	10.30	13.045	.216	13.261	13.932	11.19	25.12	.02855	.02276	.05131	26.12	.0517	.48558	58
59	10.69	13.071	.224	13.295	14.172	11.61	25.78	.02902	.02357	.05259	27.12	.0536	.50330	59
60	11.08	13.096	.233	13.329	14.413	12.05	26.46	.02948	.02441	.05389	28.12	.0555	.52159	60
61	11.49	13.121	.242	13.363	14.653	12.50	27.15	.02994	.02527	.05521	29.12	.0574	.54047	61
62	11.91	13.147	.251	13.398	14.893	12.96	27.85	.03040	.02616	.05656	30.12	.0594	.55994	62
63	12.35	13.172	.261	13.433	15.134	13.44	28.57	.03086	.02708	.05794	31.12	.0613	.58002	63
64	12.80	13.197	.271	13.468	15.374	13.94	29.31	.03132	.02803	.05935	32.12	.0632	.60073	64
65	13.26	13.222	.282	13.504	15.614	14.45	30.06	.03177	.02901	.06078	33.11	.0651	.62209	65
66	13.74	13.247	.292	13.539	15.855	14.98	30.83	.03223	.03002	.06225	34.11	.0670	.64411	66
67	14.24	13.273	.303	13.576	16.095	15.53	31.62	.03269	.03106	.06375	35.11	.0689	.66681	67
68	14.75	13.298	.315	13.613	16.335	16.09	32.42	.03314	.03213	.06527	36.11	.0708	.69019	68
69	15.28	13.323	.327	13.650	16.576	16.67	33.25	.03360	.03323	.06683	37.11	.0727	.71430	69
	$W_s \times 10^2$													
70	1.582	13.348	.339	13.687	16.816	17.27	34.09	.03405	.03437	.06842	38.11	.0746	.73915	70
71	1.639	13.373	.351	13.724	17.056	17.89	34.95	.03450	.03554	.07004	39.11	.0765	.76475	71
72	1.697	13.398	.364	13.762	17.297	18.53	35.83	.03495	.03675	.07170	40.11	.0784	.79112	72
73	1.757	13.424	.377	13.801	17.537	19.20	36.74	.03540	.03800	.07340	41.11	.0803	.81828	73
74	1.819	13.449	.392	13.841	17.778	19.88	37.66	.03585	.03928	.07513	42.10	.0821	.84624	74
75	1.882	13.474	.407	13.881	18.018	20.59	38.61	.03630	.04060	.07690	43.10	.0840	.87504	75
76	1.948	13.499	.422	13.921	18.259	21.31	39.57	.03675	.04197	.07872	44.10	.0859	.90470	76
77	2.016	13.525	.437	13.962	18.499	22.07	40.57	.03720	.04337	.08057	45.10	.0877	.93523	77
78	2.086	13.550	.453	14.003	18.740	22.84	41.58	.03765	.04482	.08247	46.10	.0896	.96665	78
79	2.158	13.575	.470	14.045	18.980	23.64	42.62	.03810	.04631	.08441	47.10	.0914	.99899	79
80	2.233	13.601	.486	14.087	19.221	24.47	43.69	.03854	.04784	.08638	48.10	.0933	1.0323	80
81	2.310	13.626	.504	14.130	19.461	25.32	44.78	.03899	.04942	.08841	49.09	.0952	1.0665	81
82	2.389	13.651	.523	14.174	19.702	26.20	45.90	.03943	.05105	.09048	50.09	.0970	1.1017	82
83	2.471	13.676	.542	14.218	19.942	27.10	47.04	.03987	.05273	.09260	51.09	.0989	1.1379	83
84	2.555	13.702	.560	14.262	20.183	28.04	48.22	.04031	.05446	.09477	52.09	.1007	1.1752	84
85	2.642	13.727	.581	14.308	20.423	29.01	49.43	.04075	.05624	.09699	53.09	.1025	1.2135	85
86	2.731	13.752	.602	14.354	20.663	30.00	50.66	.04119	.05807	.09926	54.08	.1043	1.2529	86
87	2.824	13.777	.624	14.401	20.904	31.03	51.93	.04163	.05995	.10158	55.08	.1062	1.2934	87
88	2.919	13.803	.645	14.448	21.144	32.09	53.23	.04207	.06189	.10396	56.08	.1080	1.3351	88
89	3.017	13.828	.668	14.496	21.385	33.18	54.56	.04251	.06389	.10640	57.08	.1098	1.3779	89
90	3.118	13.853	.692	14.545	21.625	34.31	55.93	.04295	.06596	.10890	58.08	.1116	1.4219	90
91	3.223	13.879	.716	14.595	21.865	35.47	57.33	.04339	.06807	.11146	59.07	.1135	1.4671	91
92	3.330	13.904	.741	14.645	22.106	36.67	58.78	.04382	.07025	.11407	60.07	.1153	1.5135	92
93	3.441	13.929	.768	14.697	22.346	37.90	60.25	.04426	.07249	.11675	61.07	.1171	1.5612	93
94	3.556	13.954	.795	14.749	22.587	39.18	61.77	.04469	.07480	.11949	62.07	.1188	1.6102	94
95	3.673	13.980	.822	14.802	22.827	40.49	63.32	.04513	.07718	.12231	63.07	.1206	1.6606	95
96	3.795	14.005	.851	14.856	23.068	41.85	64.92	.04556	.07963	.12519	64.06	.1224	1.7123	96
97	3.920	14.030	.881	14.911	23.308	43.24	66.55	.04600	.08215	.12815	65.06	.1242	1.7654	97
98	4.049	14.056	.911	14.967	23.548	44.68	68.23	.04643	.08474	.13117	66.06	.1260	1.8199	98
99	4.182	14.081	.942	15.023	23.789	46.17	69.96	.04686	.08741	.13427	67.06	.1278	1.8759	99
100	4.319	14.106	.975	15.081	24.029	47.70	71.73	.04729	.09016	.13745	68.06	.1296	1.9333	100
101	4.460	14.131	1.009	15.140	24.270	49.28	73.55	.04772	.09299	.14071	69.05	.1314	1.9923	101
102	4.606	14.157	1.043	15.200	24.510	50.91	75.42	.04815	.09591	.14406	70.05	.1332	2.0528	102
103	4.756	14.182	1.079	15.261	24.751	52.59	77.34	.04858	.09891	.14749	71.05	.1350	2.1149	103
104	4.911	14.207	1.117	15.324	24.991	54.32	79.31	.04900	.1020	.1510	72.05	.1367	2.1786	104

Table 1. Thermodynamic Properties of Moist Air (Standard Atmospheric Pressure, 29.921 in. Hg)—(*Continued*)

Temp. t, °F.	Humidity ratio $W_s \times 10$	Volume, cu. ft./lb. dry air			Enthalpy, B.t.u./lb. dry air			Entropy, B.t.u./(°F.) (lb. dry air)			Condensed water			Temp. t, °F.
		v_a	v_{as}	v_s	h_a	h_{as}	h_s	s_a	s_{as}	s_s	Enthalpy, B.t.u./lb. h_w	Entropy, B.t.u./(lb.)(°F.) s_w	Vapor press., in. Hg p_s	
105	0.5070	14.232	1.155	15.387	25.232	56.11	81.34	0.04943	0.1052	0.1546	73.04	0.1385	2.2439	105
106	.5234	14.258	1.194	15.452	25.472	57.95	83.42	.04985	.1085	.1584	74.04	.1403	2.3109	106
107	.5404	14.283	1.235	15.518	25.713	59.85	85.56	.05028	.1118	.1621	75.04	.1421	2.3797	107
108	.5578	14.308	1.278	15.586	25.953	61.80	87.76	.05070	.1153	.1660	76.04	.1438	2.4502	108
109	.5758	14.333	1.321	15.654	26.194	63.82	90.03	.05113	.1189	.1700	77.04	.1456	2.5225	109
110	.5944	14.359	1.365	15.724	26.434	65.91	92.34	.05155	.1226	.1742	78.03	.1472	2.5966	110
111	.6135	14.384	1.412	15.796	26.675	68.05	94.72	.05197	.1264	.1784	79.03	.1491	2.6726	111
112	.6333	14.409	1.460	15.869	26.915	70.27	97.18	.05239	.1302	.1826	80.03	.1508	2.7505	112
113	.6536	14.435	1.509	15.944	27.156	72.55	99.71	.05281	.1342	.1870	81.03	.1525	2.8304	113
114	.6746	14.460	1.560	16.020	27.397	74.91	102.31	.05323	.1384	.1916	82.03	.1543	2.9123	114
115	.6962	14.485	1.613	16.098	27.637	77.34	104.98	.05365	.1426	.1963	83.02	.1560	2.9962	115
116	.7185	14.510	1.668	16.178	27.878	79.85	107.73	.05407	.1470	.2011	84.02	.1577	3.0821	116
117	.7415	14.536	1.723	16.259	28.119	82.43	110.55	.05449	.1515	.2060	85.02	.1595	3.1701	117
118	.7652	14.561	1.782	16.343	28.359	85.10	113.46	.05490	.1562	.2111	86.02	.1612	3.2603	118
119	.7897	14.586	1.842	16.428	28.600	87.86	116.46	.05532	.1610	.2163	87.02	.1629	3.3527	119
120	.8149	14.611	1.905	16.516	28.841	90.70	119.54	.05573	.1659	.2216	88.01	.1646	3.4474	120
121	.8410	14.637	1.968	16.605	29.082	93.64	122.72	.05615	.1710	.2272	89.01	.1664	3.5443	121
122	.8678	14.662	2.034	16.696	29.322	96.66	125.98	.05656	.1763	.2329	90.01	.1681	3.6436	122
123	.8955	14.687	2.103	16.790	29.563	99.79	129.35	.05698	.1817	.2387	91.01	.1698	3.7452	123
124	.9242	14.712	2.174	16.886	29.804	103.0	132.8	.05739	.1872	.2446	92.01	.1715	3.8493	124
125	.9537	14.738	2.247	16.985	30.044	106.4	136.4	.05780	.1930	.2508	93.01	.1732	3.9558	125
126	.9841	14.763	2.323	17.086	30.285	109.8	140.1	.05821	.1989	.2571	94.01	.1749	4.0649	126
127	1.016	14.788	3.401	17.189	30.526	113.4	143.9	.05862	.2050	.2636	95.00	.1766	4.1765	127
128	1.048	14.813	2.482	17.295	30.766	117.0	147.8	.05903	.2113	.2703	96.00	.1783	4.2907	128
129	1.082	14.839	2.565	17.404	31.007	120.8	151.8	.05944	.2178	.2772	97.00	.1800	4.4076	129
130	1.116	14.864	2.652	17.516	31.248	124.7	155.9	.05985	.2245	.2844	98.00	.1817	4.5272	130
131	1.152	14.889	2.742	17.631	31.489	128.8	160.3	.06026	.2314	.2917	99.00	.1834	4.6495	131
132	1.189	14.915	2.834	17.749	31.729	133.0	164.7	.06067	.2386	.2993	100.00	.1851	4.7747	132
133	1.227	14.940	2.930	17.870	31.970	137.3	169.3	.06108	.2459	.3070	101.00	.1868	4.9028	133
134	1.267	14.965	3.029	17.994	32.211	141.8	174.0	.06148	.2536	.3151	102.00	.1885	5.0337	134
135	1.308	14.990	3.132	18.122	32.452	146.4	178.9	.06189	.2614	.3233	103.00	.1902	5.1676	135
136	1.350	15.016	3.237	18.253	32.692	151.2	183.9	.06229	.2695	.3318	104.00	.1918	5.3046	136
137	1.393	15.041	3.348	18.389	32.933	156.1	189.0	.06270	.2778	.3405	105.00	.1935	5.4446	137
138	1.439	15.066	3.462	18.528	33.174	161.2	194.4	.06310	.2865	.3496	106.00	.1952	5.5878	138
139	1.485	15.091	3.580	18.671	33.414	166.5	199.9	.06350	.2954	.3589	107.00	.1969	5.7342	139
	W_s													
140	0.1534	15.117	3.702	18.819	33.655	172.0	205.7	.06390	.3047	.3686	107.99	.1985	5.8838	140
141	.1584	15.142	3.829	18.971	33.896	177.7	211.6	.06430	.3142	.3785	108.99	.2002	6.0367	141
142	.1636	15.167	3.961	19.128	34.136	183.6	217.7	.06470	.3241	.3888	109.99	.2018	6.1930	142
143	.1689	15.192	4.098	19.290	34.377	189.7	224.1	.06510	.3343	.3994	110.99	.2035	6.3527	143
144	.1745	15.218	4.239	19.457	34.618	196.0	230.6	.06549	.3449	.4104	111.99	.2051	6.5160	144
145	.1803	15.243	4.386	19.629	34.859	202.5	237.4	.06589	.3559	.4218	112.99	.2068	6.6828	145
146	.1862	15.268	4.539	19.807	35.099	209.3	244.4	.06629	.3672	.4335	113.99	.2084	6.8532	146
147	.1924	15.293	4.698	19.991	35.340	216.4	251.7	.06669	.3790	.4457	114.99	.2101	7.0273	147
148	.1989	15.319	4.862	20.181	35.581	223.7	259.3	.06708	.3912	.4583	115.99	.2117	7.2051	148
149	.2055	15.344	5.033	20.377	35.822	231.3	267.1	.06748	.4038	.4713	116.99	.2134	7.3867	149
150	.2125	15.369	5.211	20.580	36.063	239.2	275.3	.06787	.4169	.4848	117.99	.2150	7.5722	150
151	.2197	15.394	5.396	20.790	36.304	247.3	283.6	.06827	.4304	.4987	118.99	.2167	7.7616	151
152	.2271	15.420	5.587	21.007	36.545	255.9	292.4	.06866	.4445	.5132	119.99	.2183	7.9550	152
153	.2049	15.445	5.788	21.233	36.785	264.7	301.5	.06906	.4591	.5282	120.99	.2200	8.1525	153
154	.2430	15.470	5.996	21.466	37.026	273.9	310.9	.06945	.4743	.5438	121.99	.2216	8.3541	154
155	.2514	15.496	6.213	21.709	37.267	283.5	320.8	.06984	.4901	.5599	122.99	.2232	8.5599	155
156	.2602	15.521	6.439	21.960	37.508	293.5	331.0	.07023	.5066	.5768	123.99	.2248	8.7701	156
157	.2693	15.546	6.675	22.221	37.749	303.9	341.7	.07062	.5237	.5943	124.99	.2265	8.9846	157
158	.2788	15.571	6.922	22.493	37.990	314.7	352.7	.07101	.5415	.6125	125.99	.2281	9.2036	158
159	.2887	15.597	7.178	22.775	38.231	326.0	364.2	.07140	.5600	.6314	127.00	.2297	9.4271	159
160	.2990	15.622	7.446	23.068	38.472	337.8	376.3	.07179	.5793	.6511	128.00	.2313	9.6556	160
161	.3098	15.647	7.727	23.374	38.713	350.1	388.8	.07218	.5994	.6716	129.00	.2329	9.8876	161
162	.3211	15.672	8.020	23.692	38.954	363.0	402.0	.07257	.6204	.6930	130.00	.2345	10.125	162
163	.3329	15.698	8.326	24.024	39.195	376.5	415.7	.07296	.6423	.7153	131.00	.2361	10.367	163
164	.3452	15.723	8.648	24.371	39.436	390.5	429.9	.07334	.6652	.7385	132.00	.2377	10.614	164
165	.3581	15.748	8.985	24.733	39.677	405.3	445.0	.07373	.6892	.7629	133.00	.2393	10.866	165
166	.3716	15.773	9.339	25.112	39.918	420.8	460.7	.07411	.7142	.7883	134.00	.2409	11.123	166
167	.3858	15.799	9.708	25.507	40.159	437.0	447.2	.07450	.7405	.8150	135.01	.2426	11.385	167
168	.4007	15.824	10.098	25.922	40.400	454.0	494.4	.07488	.7680	.8429	136.01	.2441	11.652	168
169	.4163	15.849	10.508	26.357	40.641	471.8	512.4	.07527	.7969	.8722	137.01	.2457	11.925	169
170	.4327	15.874	10.938	26.812	40.882	490.6	531.5	.07565	.8273	.9030	138.01	.2473	12.203	170
171	.4500	15.900	11.391	27.291	41.123	510.4	551.5	.07603	.8592	.9352	139.01	.2489	12.486	171
172	.4682	15.925	11.870	27.795	41.364	531.3	572.7	.07641	.8927	.9691	140.01	.2505	12.775	172
173	.4875	15.950	12.376	28.326	41.605	553.3	594.9	.07680	.9281	1.0049	141.01	.2521	13.069	173
174	.5078	15.975	12.911	28.886	41.846	576.5	618.3	.07718	.9654	1.0426	142.02	.2537	13.369	174

Table 1. Thermodynamic Properties of Moist Air (Standard Atmospheric Pressure, 29.921 in. Hg)— *(Concluded)*

Temp. t, °F.	Humidity ratio W_s	Volume, cu. ft./lb. dry air v_a	v_{as}	v_s	Enthalpy, B.t.u./lb. dry air h_a	h_{as}	h_s	Entropy, B.t.u./(°F.) (lb. dry air) s_a	s_{as}	s_s	Condensed water: Enthalpy, B.t.u./lb. h_w	Condensed water: Entropy, B.t.u./(lb.)(°F.) s_w	Condensed water: Vapor press., in. Hg p_s	Temp. t, °F.
175	0.5292	16.001	13.475	29.476	42.087	601.1	643.2	0.07756	1.005	1.083	143.02	0.2553	13.675	175
176	.5519	16.026	14.074	30.100	42.328	627.1	669.4	.07794	1.047	1.125	144.02	.2568	13.987	176
177	.5760	16.051	14.710	30.761	42.569	654.7	697.3	.07832	1.091	1.169	145.02	.2584	14.304	177
178	.6016	16.076	15.386	31.462	42.810	684.1	726.9	.07870	1.137	1.216	146.03	.2600	14.628	178
179	.6288	16.102	16.104	32.206	43.051	715.2	758.3	.07908	1.187	1.266	147.03	.2616	14.958	179
180	.6578	16.127	16.870	32.997	43.292	748.5	791.8	.07946	1.240	1.319	148.03	.2631	15.294	180
181	.6887	16.152	17.689	33.841	43.534	783.9	827.4	.07984	1.296	1.376	149.03	.2647	15.636	181
182	.7218	16.177	18.565	34.742	43.775	821.9	865.7	.08021	1.357	1.437	150.04	.2662	15.985	182
183	.7572	16.203	19.504	35.707	44.016	862.5	906.5	.08059	1.421	1.502	151.04	.2678	16.340	183
184	.7953	16.228	20.513	36.741	44.257	906.2	950.5	.08096	1.490	1.571	152.04	.2693	16.702	184
185	.8363	16.253	21.601	37.854	44.498	953.2	997.7	.08134	1.565	1.646	153.05	.2709	17.071	185
186	.8805	16.278	22.775	39.053	44.740	1004	1049	.08171	1.645	1.727	154.05	.2724	17.446	186
187	.9283	16.304	24.047	40.351	44.981	1059	1104	.08208	1.731	1.813	155.05	.2740	17.828	187
188	.9802	16.329	25.427	41.756	45.222	1119	1164	.08245	1.825	1.907	156.06	.2755	18.217	188
189	1.037	16.354	26.934	43.288	45.463	1184	1229	.08283	1.928	2.011	157.06	.2771	18.614	189
190	1.099	16.379	28.580	44.959	45.704	1255	1301	.08320	2.039	2.122	158.07	.2786	19.017	190
191	1.166	16.405	30.385	46.790	45.946	1332	1378	.08357	2.161	2.245	159.07	.2802	19.427	191
192	1.241	16.430	32.375	48.805	46.187	1418	1464	.08394	2.296	2.380	160.07	.2817	19.845	192
193	1.324	16.455	34.581	51.036	46.428	1513	1559	.08431	2.444	2.528	161.08	.2833	20.271	193
194	1.416	16.480	37.036	53.516	46.670	1619	1666	.08468	2.609	2.694	162.08	.2848	20.704	194
195	1.519	16.506	39.785	56.291	46.911	1737	1784	.08505	2.794	2.879	163.09	.2864	21.145	195
196	1.635	16.531	42.885	59.416	47.153	1871	1918	.08542	3.002	3.087	164.09	.2879	21.594	196
197	1.767	16.556	46.402	62.958	47.394	2022	2069	.08579	3.238	3.324	165.10	.2895	22.050	197
198	1.917	16.581	50.426	67.007	47.636	2195	2243	.03616	3.507	3.593	166.10	.2910	22.514	198
199	2.091	16.607	55.074	71.681	47.877	2395	2443	.08653	3.817	3.904	167.11	.2925	22.987	199
200	2.295	16.632	60.510	77.142	48.119	2629	2677	.08689	4.179	4.266	168.11	.2940	23.468	200

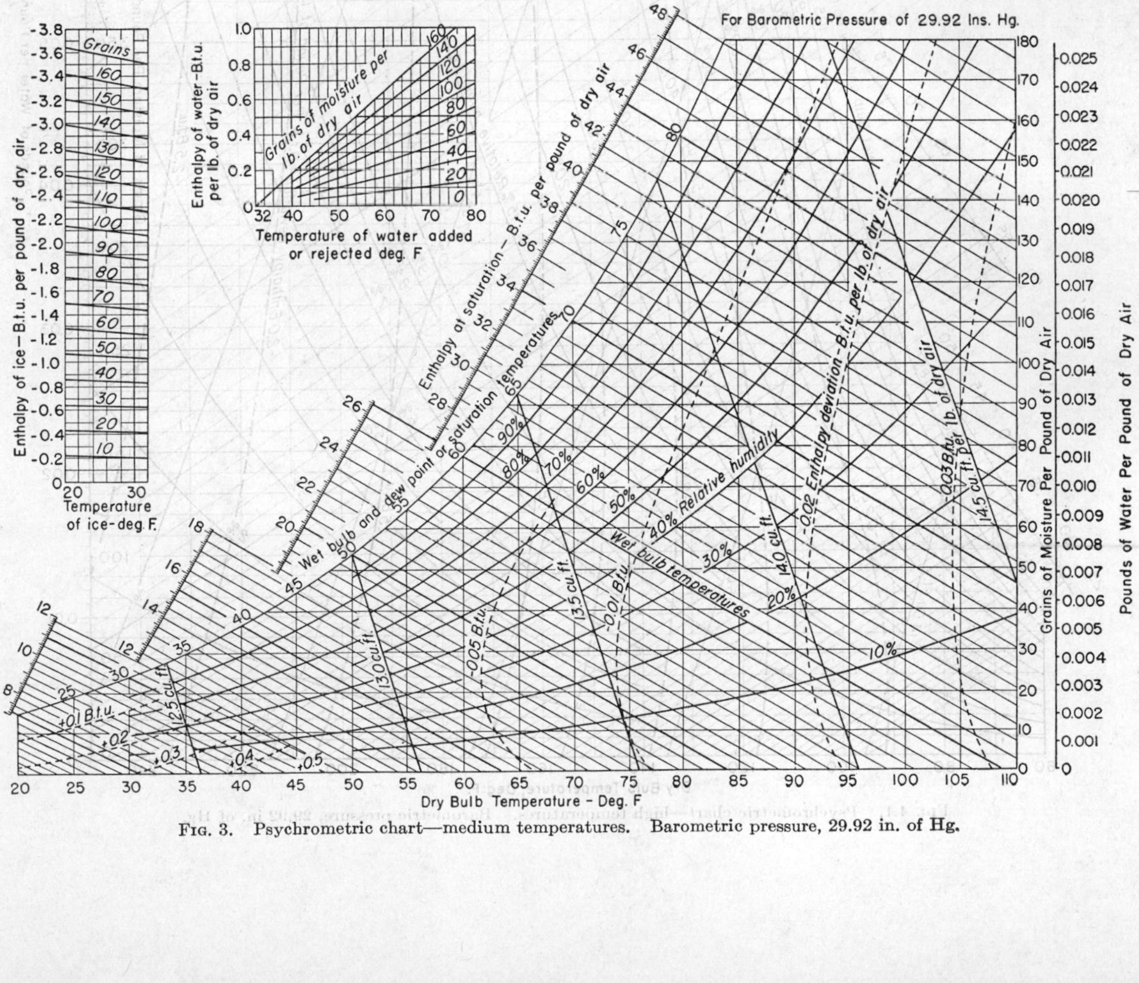

FIG. 3. Psychrometric chart—medium temperatures. Barometric pressure, 29.92 in. of Hg.

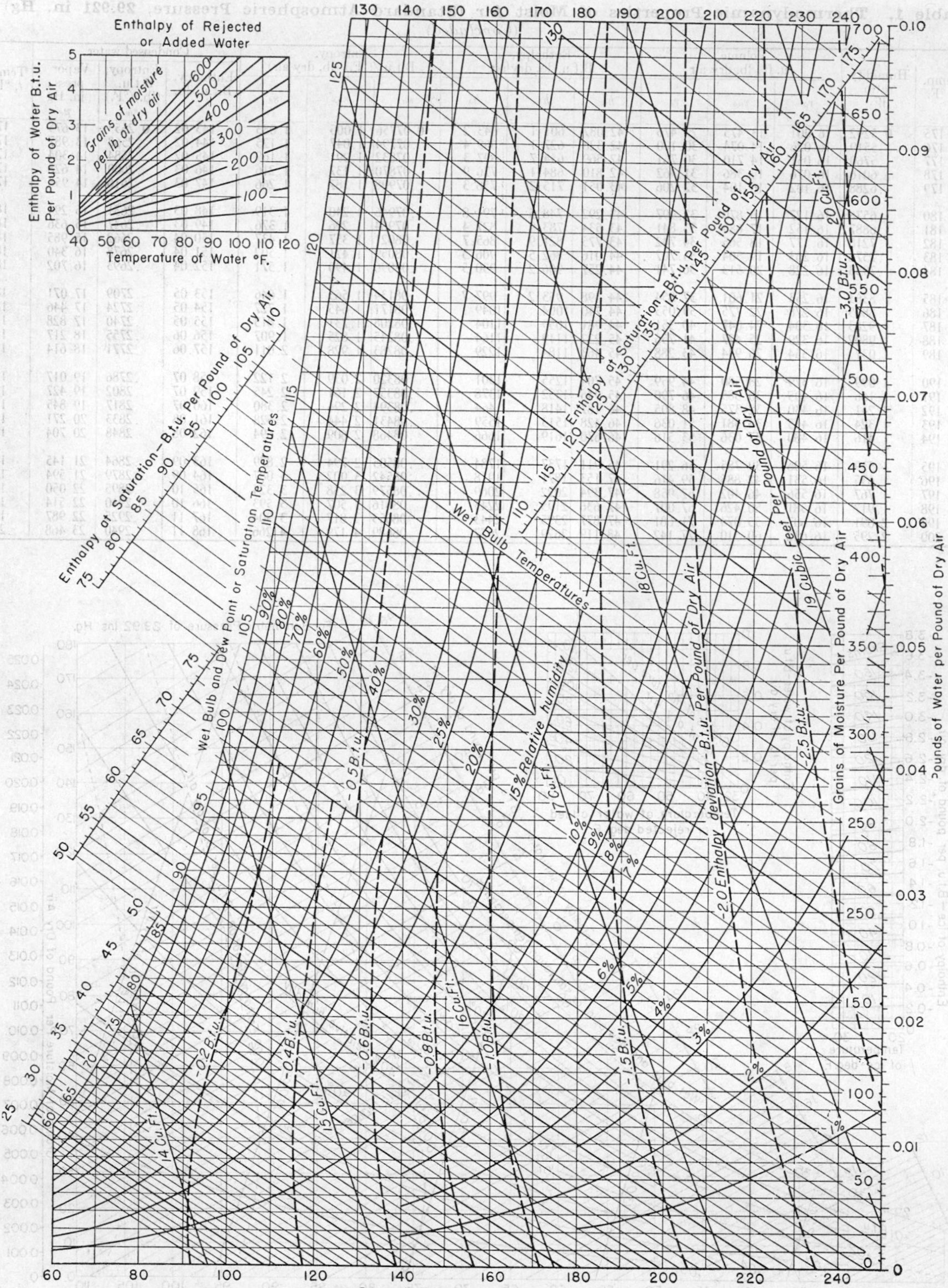

FIG. 4A. Psychrometric chart—high temperatures. Barometric pressure, 29.92 in. of Hg.

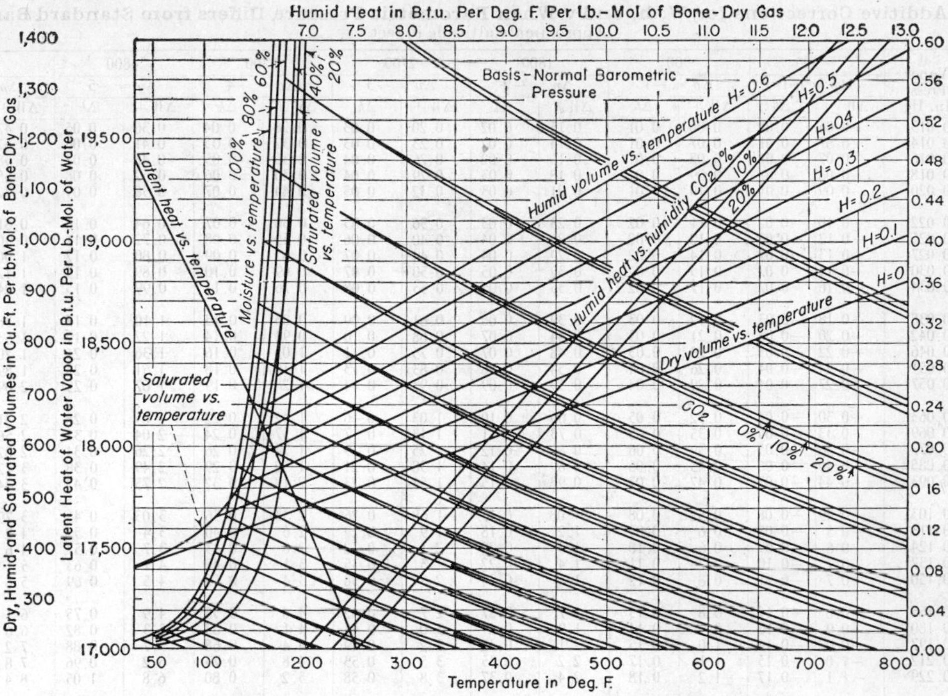

FIG. 4B. Revised form of high-temperature psychrometric chart for air and combustion gases, based on pound-moles of water vapor and dry gases. *(Courtesy of Sirozi Hatta, Chem. & Met. Eng., p. 164, March, 1930.)*

dry air, or humid volume. The wet-bulb temperature curves are spoken of as adiabatic saturation curves, or lines of constant total heat.

These charts are for normal barometric pressure of 29.92 in. Hg. A chart in Grosvenor's units is given in Sec. 13, p. 811.

Examples of Problems Solved by Use of the Psychrometric Chart

The following problems include both an exact and an approximate determination for change in enthalpy. The approximate method ignores the enthalpy of water added to or rejected from the system and the deviation of enthalpy along the wet-bulb lines for conditions other than saturation. In the comfort air-conditioning range, the approximate method will give results that are generally within 1.5 per cent of the correct value and, fortunately, on the high side so that a factor of safety is provided when estimating a cooling or heating load.

With the new type of chart an exact solution can be obtained with very little additional effort.

t = dry-bulb temperature.
t' = wet-bulb temperature.
t'' = dew-point temperature.
w = moisture content per lb. dry air.
W = moisture added to or rejected from the air stream per lb. dry air.
h' = enthalpy at saturation, B.t.u./lb. dry air.
D = enthalpy deviation, B.t.u./lb. dry air.
$h = h' + D$ = true enthalpy, B.t.u./lb. dry air.
h_w = enthalpy of the water added to or rejected from the system, B.t.u./lb. dry air.
H = heat added to the system, B.t.u./lb. dry air.
R = heat removed from the system, B.t.u./lb. dry air.

Subscripts 1, 2 3, etc., indicate entering and progressive state points (t_1, t_2, etc.).

Example 1. Find the properties of moist air when the dry bulb is 80°F. and the wet bulb is 67°F.

Solution (read directly from psychrometric chart, Fig. 4A).

Moisture content w = 78.2 grain/lb. dry air
Enthalpy at saturation h' = 31.62 B.t.u./lb. dry air
Enthalpy deviation D = 0.1 B.t.u./lb. dry air
True enthalpy h = 31 52 B.t.u./lb. dry air
Specific volume V = 13.8 cu. ft./lb dry air
Relative humidity = 50 per cent
Dew point t'' = 60.3°F.

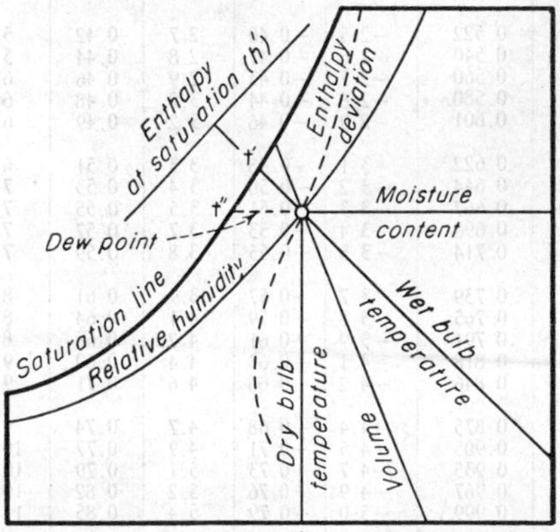

FIG 5. Diagram of psychrometric chart showing the properties of moist air.

Example 2. Air is heated by a steam coil from 30°F. dry bulb and 80 per cent relative humidity to 75°F. dry bulb. Find the relative humidity, wet-bulb temperature, and dew point of the heated air.

Table 2. Additive Corrections for W, h, and v When Barometric Pressure Differs from Standard Barometer

Approximate altitude in feet

Wet Bulb Temp. t'	Sat. Vapor Press. In. Hg	−900		900		1800		2700		3700		4800		5900	
		$\Delta p = +1$		$\Delta p = -1$		$\Delta p = -2$		$\Delta p = -3$		$\Delta p = -4$		$\Delta p = -5$		$\Delta p = -6$	
		ΔW_s[1]	Δh	ΔW_s[1]	Δh	ΔW_s[1]	Δh	ΔW_s[1]	Δh	ΔW_s[1]	Δh	ΔW_s[1]	Δh	ΔW_s[1]	Δh
−20	0.013	−0.06	−0.01	0.06	0.01	0.13	0.02	0.20	0.03	0.28	0.04	0.36	0.05	0.47	0.07
−18	0.014	−0.07	−0.01	0.07	0.01	0.14	0.02	0.23	0.03	0.32	0.05	0.41	0.06	0.52	0.08
−16	0.016	−0.07	−0.01	0.08	0.01	0.16	0.02	0.26	0.04	0.36	0.05	0.46	0.07	0.58	0.09
−14	0.018	−0.08	−0.01	0.09	0.01	0.18	0.03	0.29	0.04	0.40	0.06	0.52	0.08	0.65	0.10
−12	0.020	−0.09	−0.01	0.10	0.01	0.21	0.03	0.32	0.05	0.44	0.07	0.58	0.09	0.72	0.11
−10	0.022	−0.10	−0.02	0.11	0.02	0.23	0.03	0.36	0.05	0.50	0.07	0.64	0.10	0.81	0.12
−8	0.025	−0.12	−0.02	0.12	0.02	0.26	0.04	0.40	0.06	0.55	0.08	0.72	0.11	0.90	0.13
−6	0.027	−0.13	−0.02	0.14	0.02	0.29	0.04	0.44	0.07	0.62	0.09	0.80	0.12	1.00	0.15
−4	0.030	−0.14	0.02	0.15	0.02	0.32	0.05	0.50	0.07	0.69	0.10	0.89	0.13	1.12	0.17
−2	0.034	−0.16	−0.02	0.17	0.02	0.35	0.05	0.55	0.08	0.76	0.11	0.99	0.15	1.24	0.19
0	0.038	−0.18	−0.03	0.19	0.03	0.39	0.06	0.61	0.09	0.85	0.13	1.10	0.17	1.38	0.21
2	0.042	−0.20	−0.03	0.21	0.03	0.44	0.07	0.68	0.10	0.94	0.14	1.22	0.19	1.53	0.23
4	0.046	−0.22	−0.03	0.23	0.03	0.48	0.07	0.75	0.11	1.05	0.16	1.36	0.21	1.70	0.26
6	0.051	−0.24	−0.04	0.26	0.04	0.54	0.08	0.83	0.13	1.16	0.18	1.51	0.23	1.89	0.29
8	0.057	−0.27	−0.04	0.29	0.04	0.59	0.09	0.93	0.14	1.28	0.19	1.67	0.25	2.09	0.32
10	0.063	−0.30	−0.04	0.32	0.05	0.66	0.10	1.03	0.16	1.42	0.22	1.85	0.28	2.31	0.35
12	0.069	−0.33	−0.05	0.35	0.05	0.73	0.11	1.13	0.17	1.57	0.24	2.04	0.31	2.56	0.39
14	0.077	−0.36	−0.05	0.39	0.06	0.81	0.12	1.25	0.19	1.74	0.26	2.26	0.34	2.82	0.43
16	0.085	−0.40	−0.06	0.43	0.06	0.89	0.14	1.38	0.21	1.92	0.29	2.49	0.38	3.12	0.48
18	0.093	−0.44	−0.07	0.47	0.07	0.98	0.15	1.53	0.23	2.12	0.32	2.75	0.42	3.44	0.53
20	0.103	−0.49	−0.08	0.52	0.08	1.08	0.17	1.68	0.26	2.33	0.36	3.03	0.46	3.79	0.58
22	0.113	−0.5	−0.08	0.6	0.09	1.2	0.18	1.9	0.29	2.6	0.40	3.4	0.52	4.2	0.64
24	0.124	−0.6	−0.09	0.6	0.10	1.3	0.20	2.1	0.32	2.8	0.43	3.7	0.57	4.6	0.71
26	0.137	−0.7	−0.10	0.7	0.11	1.4	0.22	2.3	0.35	3.1	0.48	4.1	0.63	5.1	0.78
28	0.150	−0.7	−0.11	0.8	0.12	1.6	0.24	2.5	0.38	3.4	0.52	4.5	0.69	5.6	0.86
30	0.165	−0.8	−0.12	0.8	0.13	1.7	0.27	2.7	0.42	3.8	0.58	4.9	0.75	6.1	0.92
32	0.180	−0.9	−0.13	0.9	0.14	1.9	0.29	3.0	0.45	4.1	0.63	5.3	0.82	6.6	1.01
34	0.197	−0.9	−0.14	1.0	0.15	2.1	0.32	3.2	0.49	4.4	0.68	5.7	0.88	7.2	1.11
36	0.212	−1.0	−0.15	1.1	0.17	2.2	0.35	3.5	0.53	4.8	0.74	6.2	0.96	7.8	1.20
38	0.229	−1.1	−0.17	1.2	0.18	2.4	0.37	3.8	0.58	5.2	0.80	6.8	1.05	8.4	1.30
40	0.248	−1.2	−0.18	1.3	0.20	2.6	0.41	4.1	0.63	5.7	0.88	7.4	1.14	9.2	1.42
42	0.268	−1.3	−0.20	1.4	0.21	2.8	0.44	4.4	0.69	6.1	0.94	8.0	1.23	10.0	1.54
44	0.289	−1.4	−0.22	1.5	0.23	3.1	0.47	4.8	0.74	6.7	1.04	8.7	1.34	10.8	1.67
46	0.312	−1.5	−0.23	1.6	0.25	3.3	0.51	5.2	0.80	7.2	1.11	9.4	1.45	11.7	1.81
48	0.336	−1.6	−0.25	1.8	0.27	3.6	0.56	5.6	0.87	7.8	1.21	10.2	1.58	12.6	1.95
49	0.3491	−1.7	−0.26	1.8	0.28	3.7	0.58	5.3	0.90	8.1	1.25	10.5	1.63	13.1	2.03
50	0.3624	−1.7	−0.27	1.9	0.29	3.9	0.60	6.1	0.94	8.4	1.30	10.9	1.69	13.6	2.11
51	0.3761	−1.8	−0.28	2.0	0.30	4.0	0.63	6.3	0.97	8.7	1.35	11.3	1.75	14.1	2.18
52	0.3903	−1.9	−0.29	2.0	0.32	4.2	0.65	6.5	1.01	9.0	1.40	11.8	1.83	14.7	2.28
53	0.4049	−1.9	−0.30	2.1	0.33	4.4	0.68	6.7	1.05	9.3	1.44	12.2	1.89	15.2	2.36
54	0.4200	−2.0	−0.31	2.2	0.34	4.5	0.70	7.0	1.09	9.7	1.50	12.7	1.97	15.8	2.45
55	0.4356	−2.1	−0.32	2.3	0.35	4.7	0.73	7.3	1.13	10.1	1.57	13.2	2.05	16.4	2.54
56	0.4518	−2.2	−0.34	2.4	0.37	4.9	0.76	7.6	1.18	10.5	1.63	13.7	2.13	17.1	2.66
57	0.4684	−2.3	−0.35	2.4	0.37	5.1	0.79	7.9	1.22	10.9	1.69	14.2	2.21	17.7	2.75
58	0.4856	−2.3	−0.37	2.5	0.39	5.3	0.82	8.2	1.27	11.3	1.76	14.7	2.28	18.4	2.86
59	0.5033	−2.4	−0.38	2.6	0.41	5.4	0.85	8.5	1.32	11.7	1.82	15.3	2.38	19.1	2.97
60	0.522	−2.5	−0.40	2.7	0.42	5.7	0.88	8.8	1.37	12.2	1.90	15.9	2.47	19.9	3.09
61	0.540	−2.6	−0.41	2.8	0.44	5.9	0.91	9.2	1.43	12.7	1.98	16.5	2.57	20.7	3.22
62	0.560	−2.7	−0.43	2.9	0.46	6.1	0.95	9.5	1.48	13.2	2.05	17.1	2.66	21.4	3.33
63	0.580	−2.8	−0.44	3.0	0.48	6.3	0.98	9.9	1.54	13.7	2.13	17.7	2.76	22.3	3.47
64	0.601	−2.9	−0.46	3.2	0.49	6.5	1.02	10.2	1.59	14.2	2.21	18.4	2.87	23.1	3.60
65	0.622	−3.1	−0.48	3.3	0.51	6.8	1.06	10.6	1.65	14.7	2.29	19.1	2.98	23.9	3.73
66	0.644	−3.2	−0.50	3.4	0.53	7.1	1.10	11.0	1.72	15.3	2.38	19.8	3.09	24.8	3.87
67	0.667	−3.3	−0.51	3.5	0.55	7.3	1.14	11.4	1.78	15.8	2.47	20.5	3.20	25.7	4.01
68	0.690	−3.4	−0.53	3.7	0.57	7.6	1.18	11.8	1.84	16.4	2.56	21.3	3.32	26.7	4.16
69	0.714	−3.5	−0.55	3.8	0.59	7.9	1.23	12.2	1.90	17.0	2.65	22.1	3.45	27.7	4.32
70	0.739	−3.7	−0.57	3.9	0.61	8.1	1.27	12.7	1.98	17.6	2.75	22.9	3.58	28.7	4.48
71	0.765	−3.8	−0.59	4.1	0.64	8.4	1.32	13.1	2.05	18.2	2.84	23.7	3.70	29.7	4.64
72	0.791	−3.9	−0.61	4.2	0.66	8.7	1.36	13.6	2.13	18.8	2.94	24.6	3.84	30.9	4.82
73	0.818	−4.1	−0.63	4.4	0.69	9.0	1.41	14.1	2.20	19.5	3.05	25.5	3.99	31.9	4.99
74	0.846	−4.2	−0.66	4.6	0.71	9.4	1.46	14.6	2.28	20.2	3.16	26.4	4.14	33.1	5.18
75	0.875	−4.4	−0.68	4.7	0.74	9.7	1.52	15.1	2.36	20.9	3.27	27.4	4.28	34.3	5.37
76	0.905	−4.5	−0.71	4.9	0.77	10.0	1.57	15.7	2.46	21.7	3.39	28.3	4.42	35.5	5.56
77	0.935	−4.7	−0.73	5.1	0.79	10.4	1.63	16.3	2.55	22.5	3.52	29.4	4.61	36.9	5.77
78	0.967	−4.9	−0.76	5.2	0.82	10.8	1.69	16.9	2.65	23.3	3.65	30.5	4.77	38.2	5.98
79	0.999	−5.0	−0.79	5.4	0.85	11.2	1.75	17.5	2.74	24.2	3.79	31.6	4.95	39.6	6.20
80	1.032	−5.2	−0.82	5.6	0.88	11.6	1.82	18.1	2.84	25.1	3.93	32.7	5.13	41.0	6.43
81	1.067	−5.4	−0.85	5.8	0.91	12.0	1.88	18.8	2.95	26.0	4.08	33.9	5.32	42.5	6.66
82	1.102	−5.6	−0.88	6.0	0.94	12.5	1.96	19.5	3.06	27.0	4.24	35.1	5.51	44.0	6.90
83	1.138	−5.8	−0.91	6.2	0.97	12.9	2.02	20.2	3.17	28.0	4.39	36.4	5.71	45.6	7.15
84	1.175	−6.0	−0.94	6.4	1.00	13.3	2.10	20.9	3.28	28.9	4.54	37.7	5.92	47.2	7.41
85	1.214	−6.2	−0.97	6.7	1.05	13.8	2.17	21.6	3.39	29.9	4.69	39.0	6.12	48.9	7.67
86	1.253	−6.4	−1.00	6.9	1.08	14.3	2.24	22.3	3.50	30.9	4.85	40.4	6.34	50.6	7.94
87	1.294	−6.7	−1.05	7.1	1.11	14.8	2.32	23.1	3.63	32.0	5.02	41.8	6.56	52.3	8.21
88	1.335	−6.9	−1.08	7.4	1.16	15.3	2.40	23.9	3.75	33.1	5.20	43.2	6.79	54.2	8.51
89	1.378	−7.1	−1.12	7.7	1.21	15.9	2.50	24.8	3.90	34.3	5.39	44.8	7.04	56.2	8.83

Table 2. Additive Corrections for W, h, and v When Barometric Pressure Differs from Standard Barometer
(Continued)

Wet Bulb Temp. t'	Sat. Vapor Press. In. Hg	−900 $\Delta p = +1$ ΔW_s^1	Δh	900 $\Delta p = -1$ ΔW_s^1	Δh	1800 $\Delta p = -2$ ΔW_s^1	Δh	2700 $\Delta p = -3$ ΔW_s^1	Δh	3700 $\Delta p = -4$ ΔW_s^1	Δh	4800 $\Delta p = -5$ ΔW_s^1	Δh	5900 $\Delta p = -6$ ΔW_s^1	Δh
90	1.422	−7.4	−1.16	7.9	1.24	16.5	2.59	25.7	4.04	35.6	5.60	46.4	7.29	58.2	9.15
91	1.467	−7.6	−1.20	8.2	1.29	17.0	2.67	26.6	4.18	36.9	5.80	48.1	7.56	60.3	9.48
92	1.514	−7.9	−1.24	8.5	1.34	17.6	2.77	27.5	4.33	38.2	6.01	49.8	7.83	62.5	9.83
93	1.561	−8.2	−1.29	8.8	1.39	18.3	2.88	28.5	4.49	39.6	6.23	51.5	8.11	64.7	10.18
94	1.610	−8.5	−1.34	9.1	1.43	18.9	2.98	29.5	4.64	41.0	6.46	53.4	8.41	67.0	10.55
95	1.661	−8.8	−1.39	9.4	1.48	19.6	3.09	30.5	4.80	42.4	6.68	55.2	8.69	69.3	10.92
96	1.712	−9.1	−1.43	9.8	1.54	20.2	3.18	31.5	4.96	43.8	6.90	57.2	9.01	71.7	11.30
97	1.766	−9.4	−1.48	10.1	1.59	20.9	3.29	32.6	5.14	45.3	7.14	59.2	9.33	74.2	11.70
98	1.820	−9.7	−1.53	10.4	1.64	21.7	3.42	33.8	5.33	47.0	7.41	61.3	9.67	76.8	12.11
99	1.876	−10.1	−1.59	10.8	1.70	22.4	3.53	35.0	5.52	48.6	7.67	63.5	10.02	79.6	12.56
100	1.933	−10.4	−1.64	11.2	1.77	23.2	3.66	36.3	5.73	50.4	7.95	65.7	10.37	82.5	13.02
101	1.992	−10.8	−1.71	11.6	1.83	24.0	3.79	37.6	5.94	52.2	8.24	68.0	10.74	85.4	13.48
102	2.053	−11.1	−1.75	12.0	1.90	24.8	3.92	38.9	6.14	54.1	8.54	70.5	11.13	88.5	13.98
103	2.115	−11.5	−1.82	12.4	1.96	25.7	4.06	40.3	6.37	56.0	8.85	72.9	11.52	91.6	14.47
104	2.179	−11.9	−1.88	12.8	2.02	26.6	4.20	41.6	6.58	57.9	9.15	75.5	11.93	94.8	14.98
105	2.244	−12.4	−1.96	13.3	2.10	27.6	4.36	43.1	6.82	59.9	9.47	78.2	12.37	98.2	15.53
106	2.311	−12.8	−2.02	13.7	2.17	28.6	4.52	44.6	7.06	62.1	9.82	81.1	12.83	101.7	16.09
107	2.381	−13.2	−2.09	14.2	2.25	29.6	4.68	46.2	7.31	64.3	10.17	84.1	13.31	105.3	16.66
108	2.450	−13.7	−2.17	14.7	2.33	30.6	4.84	47.7	7.55	66.5	10.53	87.0	13.77	109.1	17.27
109	2.523	−14.2	−2.25	15.3	2.42	31.7	5.02	49.4	7.82	68.9	10.91	90.0	14.25	113.0	17.90
110	2.597	−14.7	−2.33	15.8	2.50	32.8	5.20	51.3	8.13	71.3	11.30	93.1	14.75	117.0	18.54
111	2.673	−15.2	−2.41	16.3	2.58	34.0	5.39	53.1	8.42	73.8	11.70	96.4	15.28	121.4	19.24
112	2.751	−15.7	−2.49	16.9	2.68	35.2	5.58	55.0	8.72	76.4	12.11	99.9	15.84	125.9	19.96
113	2.831	−16.3	−2.58	17.5	2.78	36.4	5.77	56.9	9.03	79.2	12.56	103.5	16.42	130.4	20.68
114	2.913	−16.9	−2.68	18.1	2.87	37.7	5.98	58.9	9.50	82.0	13.01	107.3	17.03	135.0	21.42
115	2.996	−17.4	−2.76	18.8	2.98	39.1	6.21	61.0	9.68	85.0	13.49	111.1	17.64	139.7	22.18
116	3.082	−18.0	−2.86	19.4	3.08	40.4	6.42	63.2	10.03	88.0	13.97	115.1	18.28	144.7	22.98
117	3.170	−18.7	−2.97	20.1	3.19	41.8	6.64	65.4	10.38	91.2	14.49	119.2	18.94	150.0	23.83
118	3.260	−19.3	−3.07	20.8	3.31	43.3	6.88	67.8	10.77	94.4	15.00	123.5	19.63	155.4	24.73
119	3.353	−20.0	−3.18	21.6	3.43	44.9	7.14	70.3	11.18	97.9	15.56	128.0	20.35	161.1	25.61
120	3.448	−20.7	−3.29	22.4	3.56	46.6	7.41	72.8	11.58	101.4	16.13	132.7	21.10	167.1	26.58
121	3.545	−21.4	−3.40	23.2	3.77	48.3	7.68	75.5	12.01	105.1	16.72	137.6	21.89	173.2	27.56
122	3.644	−22.2	−3.53	24.0	3.82	50.0	7.96	78.2	12.45	109.0	17.35	142.6	22.70	179.6	28.58
123	3.746	−23.0	−3.66	24.9	3.96	51.8	8.25	81.1	12.91	112.9	17.98	147.9	23.55	186.3	29.66
124	3.850	−23.8	−3.79	25.8	4.11	53.7	8.55	84.0	13.38	117.1	18.65	153.3	24.42	193.2	30.77
125	3.956	−24.7	−3.94	26.7	4.25	55.6	8.86	87.1	13.88	121.4	19.34	159.1	25.35	200.5	31.95
126	4.065	−25.6	−4.08	27.6	4.40	57.7	9.20	90.3	14.39	125.9	20.07	165.0	26.30	208.0	33.15
127	4.177	−26.5	−4.23	28.6	4.56	59.8	9.54	93.6	14.93	130.6	20.83	171.2	27.30	215.8	34.41
128	4.291	−27.5	−4.39	29.7	4.74	62.0	9.89	97.1	15.49	135.5	21.61	177.6	28.33	224.0	35.73
129	4.408	−28.5	−4.55	30.8	4.92	64.3	10.26	100.7	16.07	140.6	22.44	184.3	29.41	232.6	37.12
130	4.527	−29.5	−4.71	32.0	5.11	66.7	10.64	104.5	16.68	145.9	23.29	191.4	30.55	241.5	38.55
131	4.650	−30.6	−4.89	33.2	5.30	69.2	11.05	108.5	17.33	151.4	24.18	198.7	31.73	250.3	40.05
132	4.775	−31.8	−5.08	34.4	5.50	71.8	11.47	112.6	17.99	157.2	25.11	206.3	32.96	260.6	41.63
133	4.903	−32.9	−5.26	35.8	5.72	74.5	11.91	116.9	18.68	163.3	26.10	214.3	34.25	270.8	43.28
134	5.034	−34.2	−5.47	37.1	5.93	77.4	12.37	121.4	19.41	169.6	27.12	222.7	35.60	281.4	44.99
135	5.168	−35.5	−5.68	38.5	6.16	80.3	12.84	126.0	20.15	176.2	28.18	231.4	37.01	292.6	46.80
136	5.305	−36.8	−5.89	40.0	6.40	83.4	13.34	130.9	20.94	183.1	29.30	240.5	38.48	304.2	48.67
137	5.445	−38.2	−6.11	41.5	6.64	86.6	13.86	136.0	21.77	190.3	30.46	250.1	40.03	316.4	50.64
138	5.588	−39.7	−6.36	43.2	6.92	90.0	14.41	141.4	22.64	197.8	31.67	260.1	41.65	329.3	52.73
139	5.735	−41.2	−6.60	44.8	7.18	93.6	14.99	147.0	23.55	205.7	32.95	270.6	43.34	342.7	54.89
140	5.884	−42.8	−6.86	46.5	8.45	97.3	15.59	152.8	24.48	214.0	34.29	281.6	45.12	356.8	57.17

t = Dry bulb temperature (°F).
t' = Wet bulb temperature (°F).
p = Barometric pressure (in. of Hg).
Δp = Pressure difference from standard barometer (in. of Hg).
W = Moisture content of air (gr. per lb. of dry air).
W_s^1 = Moisture content of air saturated at wet bulb temperature t' (gr. per lb. of dry air).
ΔW = Moisture content correction of air when barometric pressure differs from standard barometer (gr. per lb. of dry air).
ΔW_s^1 = Moisture content correction of air saturated at wet bulb temperature when barometric pressure differs from standard barometer (gr. per lb of dry air).
NOTE: To obtain ΔW reduce value of ΔW_s^1 by 1% where $t - t' = 24$°F and correct proportionally when $t - t'$ is not 24°F.
h = Enthalpy of moist air (B.t.u. per lb. of dry air).
Δh = Enthalpy correction when barometer pressure differs from standard barometer, for saturated or unsaturated air. (B.t.u. per lb. of dry air).
v = Volume of moist air (cu. ft. per lb. of dry air).

$$= \frac{.754\,(t + 459.8)}{p}\left[1 + \frac{W}{4360}\right]$$

Example: At a barometric pressure of 25.92 with 220°F DB and 100°F WB, determine W, h, and $v\Delta$. $p = -4$ and from table $\Delta W_s^1 = 50.4$. From note above,

$$\Delta W = \Delta W_s^1 - \left(\frac{120}{24} \times .01 \times 50.4\right) = 50.4 - 2.5 = 47.9$$

Therefore $W = 102$ (from chart) $+ 47.9 = 149.9$ gr. per lb of dry air. From table $\Delta h = 7.95$. Therefore h = saturation enthalpy from chart + deviation $+ 7.95 = 71.7 - 2.0 + 7.95 = 77.65$ B.t.u. per lb. of dry air. From equation above

$$v = \frac{.754\,(220 + 459.7)}{25.92}\left[1 + \frac{149.9}{4360}\right] = 20.43 \text{ cu. ft. per lb. of dry air}$$

Table 3. Thermodynamic Properties of Water at Saturation*

Temp. t, °F.	Absolute press. $p_s \times 10^5$ Lb./sq. in.	In. Hg	Specific volume, cu. ft./lb. Sat. solid v_i	Evap. $v_{ig} \times 10^{-6}$	Sat. vapor $v_g \times 10^{-6}$	Enthalpy, B.t.u./lb. Sat. solid h_i	Evap. h_{ig}	Sat. vapor h_g	Entropy, B.t.u./(lb.) (°F). Sat. solid s_i	Evap. s_{ig}	Sat. vapor s_g	Temp. t, °F.
−160	0.4949	1.008	0.01722	36.07	36.07	−222.05	1212.43	990.38	−0.4907	4.0456	3.5549	−160
−159	.5592	1.138	.01722	32.03	32.03	−221.73	1212.55	990.82	−0.4896	4.0325	3.5429	−159
−158	.6312	1.285	.01723	28.47	28.47	−221.41	1112.67	991.26	−0.4886	4.0196	3.5310	−158
−157	.7121	1.450	.01723	25.32	25.32	−221.09	1212.79	991.70	−0.4875	4.0067	3.5192	−157
−156	.8026	1.634	.01723	22.54	22.54	−220.76	1212.90	992.14	−0.4864	3.9939	3.5075	−156
−155	.9040	1.840	.01723	20.08	20.08	−220.44	1213.02	992.58	−0.4854	3.9812	3.4958	−155
−154	1.017	2.072	.01723	17.90	17.90	−220.12	1213.15	993.03	−0.4843	3.9685	3.4842	−154
−153	1.144	2.329	.01723	15.97	15.97	−219.79	1213.26	993.47	−0.4832	3.9559	3.4727	−153
−152	1.286	2.618	.01723	14.26	14.26	−219.47	1213.38	993.91	−0.4822	3.9435	3.4613	−152
−151	1.444	2.939	.01723	12.74	12.74	−219.14	1213.49	994.35	−0.4811	3.9311	3.4500	−151
−150	1.620	3.298	.01723	11.39	11.39	−218.82	1213.62	994.80	−0.4801	3.9188	3.4387	−150
−149	1.816	3.698	.01723	10.19	10.19	−218.49	1213.73	995.24	−0.4790	3.9065	3.4275	−149
−148	2.035	4.143	.01724	9.123	9.123	−218.16	1131.84	995.68	−0.4780	3.8944	3.4164	−148
−147	2.278	4.639	.01724	8.174	8.174	−217.83	1213.95	996.12	−0.4769	3.8823	3.4054	−147
−146	2.549	5.190	.01724	7.330	7.330	−217.50	1214.06	996.56	−0.4758	3.8702	3.3944	−146
−145	2.850	5.803	.01724	6.577	6.577	−217.17	1214.17	997.00	−0.4748	3.8583	3.3835	−145
−144	3.184	6.483	.01724	5.905	5.905	−216.88	1214.28	997.45	−0.4737	3.8464	3.3727	−144
−143	3.556	7.240	.01724	5.305	5.305	−216.50	1214.39	997.89	−0.4727	3.8346	3.3619	−143
−142	3.967	8.076	.01724	4.770	4.770	−216.16	1214.49	998.33	−0.4716	3.8229	3.3513	−142
−141	4.423	9.005	.01724	4.292	4.292	−215.83	1214.60	998.77	−0.4706	3.8112	3.3406	−141
−140	4.928	10.03	.01724	3.864	3.864	−215.49	1214.70	999.21	−0.4695	3.7996	3.3301	−140
−139	5.487	11.17	.01724	3.481	3.481	−215.16	1214.82	999.66	−0.4685	3.7881	3.3196	−139
−138	6.106	12.43	.01724	3.138	3.138	−214.82	1214.92	1000.10	−0.4674	3.7766	3.3092	−138
−137	6.790	13.82	.01724	2.831	2.831	−214.48	1215.02	1000.54	−0.4664	3.7653	3.2989	−137
−136	7.546	15.36	.01725	2.555	2.555	−214.14	1215.12	1030.98	−0.4653	3.7540	3.2887	−136
−135	8.380	17.06	.01725	2.308	2.308	−213.80	1215.22	1001.42	−0.4643	3.7428	3.2785	−135
−134	9.301	18.94	.01725	2.086	2.086	−213.46	1215.32	1001.86	−0.4632	3.7315	3.2683	−134
−133	10.32	21.01	.01725	1.886	1.886	−213.11	1215.42	1002.31	−0.4622	3.7205	3.2583	−133
−132	11.44	23.28	.01725	1.707	1.707	−212.77	1215.52	1002.75	−0.4611	3.7093	3.2482	−132
−131	12.67	25.79	.01725	1.545	1.545	−212.43	1215.62	1003.19	−0.4601	3.6984	3.2383	−131
−130	14.03	28.56	.01725	1.400	1.400	−212.08	1215.71	1003.63	−0.4590	3.6874	3.2284	−130
−129	15.52	31.60	.01725	1.269	1.269	−211.73	1215.80	1004.07	−0.4580	3.6766	3.2186	−129
	$p_s = 10^6$			$v_{ig} = 10^{-7}$	$v_g \times 10^{-7}$							
−128	1.716	3.494	.01726	11.51	11.51	−211.39	1215.91	1004.52	−0.4569	3.6658	3.2089	−128
−127	1.897	3.862	.01726	10.45	10.45	−211.04	1216.00	1004.96	−0.4559	3.6551	3.1992	−127
−126	2.095	4.265	.01726	9.489	9.489	−210.69	1216.09	1005.40	−0.4548	3.6444	3.1896	−126
−125	2.312	4.708	.01726	8.622	8.622	−210.34	1216.18	1005.84	−0.4538	3.6338	3.1800	−125
−124	2.551	5.194	.01726	7.839	7.839	−209.99	1216.27	1006.28	−0.4527	3.6232	3.1705	−124
−123	2.812	5.726	.01726	7.131	7.131	−209.64	1216.37	1006.73	−0.4517	3.6127	3.1610	−123
−122	3.099	6.310	.01726	6.491	6.491	−209.28	1216.45	1007.17	−0.4506	3.6022	3.1516	−122
−121	3.413	6.949	.01726	5.911	5.911	−208.93	1216.54	1007.61	−0.4496	3.5919	3.1423	−121
−120	3.757	7.649	.01726	5.386	5.386	−208.58	1216.63	1008.05	−0.4485	3.5815	3.1330	−120
−119	4.133	8.414	.01726	4.911	4.911	−208.22	1216.71	1008.49	−0.4475	3.5713	3.1238	−119
−118	4.544	9.251	.01727	4.480	4.480	−207.86	1216.80	1008.94	−0.4464	3.5611	3.1147	−118
−117	4.993	10.17	.01727	4.088	4.088	−207.51	1216.89	1009.38	−0.4454	3.5510	3.1056	−117
−116	5.484	11.16	.01727	3.733	3.733	−207.15	1216.97	1009.82	−0.4444	3.5409	3.0965	−116
−115	6.019	12.26	.01727	3.411	3.411	−206.79	1217.05	1010.26	−0.4433	3.5308	3.0875	−115
−114	6.604	13.44	.01727	3.118	3.118	−206.43	1217.13	1010.70	−0.4423	3.5209	3.0786	−114
−113	7.241	14.74	.01727	2.852	2.852	−206.07	1217.21	1011.14	−0.4412	3.5109	3.0697	−113
−112	7.936	16.16	.01727	2.610	2.610	−205.70	1217.29	1011.59	−0.4402	3.5011	3.0609	−112
−111	8.693	17.70	.01727	2.389	2.389	−205.34	1217.37	1012.03	−0.4391	3.4912	3.0521	−111
−110	9.517	19.38	.01728	2.189	2.189	−204.98	1217.45	1012.47	−0.4381	3.4815	3.0434	−110
−109	10.41	21.20	.01728	2.006	2.006	−204.61	1217.52	1012.91	−0.4370	3.4718	3.0348	−109
−108	11.39	23.19	.01728	1.839	1.839	−204.24	1217.60	1013.36	−0.4360	3.4621	3.0261	−108
−107	12.45	25.35	.01728	1.687	1.687	−203.88	1217.68	1013.80	−0.4350	3.4526	3.0176	−107
−106	13.60	27.70	.01728	1.549	1.549	−203.51	1217.75	1014.24	−0.4339	3.4430	3.0091	−106
−105	14.86	30.25	.01728	1.422	1.422	−203.14	1217.82	1014.68	−0.4329	3.4335	3.0006	−105
−104	16.22	33.01	.01728	1.307	1.307	−202.77	1217.89	1015.12	−0.4318	3.4240	2.9922	−104
−103	17.69	36.02	.01728	1.201	1.201	−202.40	1217.96	1015.56	−0.4308	3.4146	2.9838	−103
−102	19.29	39.28	.01728	1.104	1.104	−202.03	1218.04	1016.01	−0.4298	3.4053	2.9755	−102
−101	21.03	42.81	.01729	1.016	1.016	−201.65	1218.10	1016.45	−0.4287	3.3960	2.9673	−101
−100	22.91	46.64	.01729	0.9352	0.9352	−201.28	1218.17	1016.89	−0.4277	3.3868	2.9591	−100
−99	24.94	50.79	.01729	.8613	.8613	−200.90	1218.23	1017.33	−0.4266	3.3775	2.9509	−99
−98	27.15	55.28	.01729	.7936	.7936	−200.53	1218.30	1017.77	−0.4256	3.3684	2.9428	−98
−97	29.54	60.14	.01729	.7314	.7314	−200.15	1218.37	1018.22	−0.4246	3.3593	2.9347	−97
	$p_s \times 10^5$			$v_{ig} \times 10^{-6}$	$v_g = 10^{-6}$							
−96	3.212	6.539	0.01729	6.745	6.745	−199.78	1218.44	1018.66	−0.4235	3.3502	2.9267	−96
−95	3.491	7.108	.01729	6.223	6.223	−199.40	1218.50	1019.10	−0.4225	3.3412	2.9187	−95
−94	3.793	7.722	.01729	5.743	5.743	−199.02	1218.56	1019.54	−0.4214	3.3322	2.9108	−94
−93	4.118	8.385	.01730	5.303	5.303	−198.64	1218.62	1019.98	−0.4204	3.3233	2.9029	−93
−92	4.470	9.102	.01730	4.899	4.899	−198.25	1218.68	1020.43	−0.4194	3.3145	2.8951	−92
−91	4.850	9.876	.01730	4.528	4.528	−197.87	1218.74	1020.87	−0.4183	3.3056	2.8873	−91
−90	5.260	10.71	.01730	4.186	4.186	−197.49	1218.80	1021.31	−0.4173	3.2969	2.8796	−90
−89	5.702	11.61	.01730	3.872	3.872	−197.10	1218.85	1012.75	−0.4162	3.2880	2.8718	−89
−88	6.179	12.58	.01730	3.583	3.583	−196.72	1218.92	1022.20	−0.4152	3.2794	2.8642	−88
−87	6.692	13.63	.01730	3.317	3.317	−196.33	1218.97	1022.64	−0.4142	3.2708	2.8566	−87
−86	7.245	14.75	.01730	3.072	3.072	−195.94	1219.02	1023.08	−0.4131	3.2621	2.8490	−86
−85	7.841	15.96	.01730	2.846	2.846	−195.56	1219.03	1023.52	−0.4121	3.2536	2.8415	−85
−84	8.482	17.27	.01731	2.638	2.638	−195.16	1219.12	1023.96	−0.4110	3.2450	2.8340	−84
−83	9.171	18.67	.01731	2.446	2.446	−194.78	1219.18	1024.40	−0.4100	3.2366	2.8266	−83
−82	9.913	20.18	.01731	2.270	2.270	−194.38	1219.23	1024.85	−0.4090	3.2282	2.8192	−82

Table 3. Thermodynamic Properties of Water at Saturation—(*Continued*)

Temp. t, °F.	Absolute press. $p_s \times 10^5$		Specific volume, cu. ft./lb.			Enthalpy, B.t.u./lb.			Entropy, B.t.u./(lb.) (°F).			Temp. t, °F.
	Lb./sq. in.	In. Hg	Sat. solid v_i	Evap. $v_{ig} \times 10^{-6}$	Sat. vapor $v_g \times 10^{-6}$	Sat. solid h_i	Evap. h_{ig}	Sat. vapor h_g	Sat. solid s_i	Evap. s_{ig}	Sat. vapor s_g	
−81	10.71	21.81	.01731	2.106	2.106	−193.99	1219.28	1025.29	−0.4079	3.2197	2.8118	−81
−80	11.57	23.55	.01731	1.955	1.955	−193.60	1219.33	1025.73	−0.4069	3.2114	2.8045	−80
−79	12.49	25.42	.01731	1.816	1.816	−193.20	1219.37	1026.17	−0.4058	3.2030	2.7972	−79
−78	13.47	27.43	.01731	1.687	1.687	−192.81	1219.43	1026.62	−0.4048	3.1948	2.7900	−78
−77	14.53	29.59	.01732	1.568	1.568	−192.41	1219.47	1027.06	−0.4038	3.1866	2.7828	−77
−76	15.67	31.91	.01732	1.458	1.458	−192.01	1219.51	1027.50	−0.4028	3.1784	2.7756	−76
−75	16.89	34.39	.01732	1.356	1.356	−191.62	1219.56	1027.94	−0.4017	3.1702	2.7685	−75
−74	18.20	37.05	.01732	1.262	1.262	−191.22	1219.60	1028.38	−0.4007	3.1622	2.7615	−74
−73	19.60	39.91	.01732	1.175	1.175	−190.82	1219.65	1028.83	−0.3996	3.1540	2.7544	−73
−72	21.10	42.96	.01732	1.094	1.094	−190.41	1219.68	1029.27	−0.3986	3.1460	2.7474	−72
−71	22.71	46.24	.01732	1.020	1.020	−190.01	1219.72	1029.71	−0.3976	3.1381	2.7405	−71
−70	24.43	49.74	.07132	0.9501	0.9501	−189.61	1219.76	1030.15	−0.3965	3.1301	2.7336	−70
−69	26.27	53.49	.01732	.8858	.8858	−189.20	1219.80	1030.60	−0.3955	3.1222	2.7267	−69
−68	28.24	57.51	.01733	.8261	.8261	−188.80	1219.84	1031.04	−0.3945	3.1143	2.7198	−68
−67	30.35	64.80	.01733	.7707	.7707	−188.39	1219.87	1031.48	−0.3934	3.1064	2.7130	−67
−66	32.60	66.38	.01733	.7193	.7193	−187.98	1219.90	1031.92	−0.3924	3.0987	2.7063	−66
−65	35.01	71.28	.01733	.6715	.6715	−187.58	1219.94	1032.36	−0.3914	3.0910	2.6996	−65
	$p_s \times 10^3$			$v_{ig} \times 10^{-5}$	$v_g \times 10^{-5}$							
−64	0.3758	0.7652	0.01733	6.272	6.272	−187.17	1219.98	1032.81	−0.3903	3.0832	2.6929	−64
−63	.4033	.8211	.01733	5.859	5.859	−186.76	1220.01	1033.25	−0.3893	3.0755	2.6862	−63
−62	.4326	.8808	.01734	5.476	5.476	−186.35	1220.04	1033.69	−0.3882	3.0678	2.6796	−62
−61	.4639	.9444	.01734	5.120	5.120	−185.93	1220.06	1034.13	−0.3872	3.0602	2.6730	−61
−60	.4972	1.012	.01734	4.788	4.788	−185.52	1220.10	1034.58	−0.3862	3.0526	2.6664	−60
−59	.5328	1.085	.01734	4.479	4.479	−185.11	1220.13	1035.02	−0.3851	3.0450	2.6599	−59
−58	.5708	1.162	.01734	4.192	4.192	−184.69	1220.15	1035.46	−0.3841	3.0376	2.6535	−58
−57	.6112	1.244	.01734	3.925	3.925	−184.28	1220.18	1035.90	−0.3831	3.0301	2.6470	−57
−56	.6543	1.332	.01734	3.675	3.675	−183.86	1220.20	1036.34	−0.3820	3.0226	2.6406	−56
−55	.7001	1.426	.01734	3.443	3.443	−183.44	1220.23	1036.79	−0.3810	3.0152	2.6342	−55
−54	.7489	1.525	.01735	3.226	3.226	−183.02	1220.25	1037.23	−0.3800	3.0079	2.6279	−54
−53	.8009	1.631	.01735	3.024	3.024	−182.60	1220.27	1037.67	−0.3789	3.0005	2.6216	−53
−52	.8562	1.743	.01735	2.836	2.836	−182.18	1220.29	1038.11	−0.3779	2.9932	2.6153	−52
−51	.9151	1.863	.01735	2.660	2.660	−181.76	1220.31	1038.55	−0.3769	2.9860	2.6091	−51
−50	.9776	1.990	.01735	2.496	2.496	−181.34	1220.34	1039.00	−0.3758	2.9786	2.6028	−50
−49	1.004	2.126	.01735	2.343	2.343	−180.92	1220.36	1039.44	−0.3748	2.9715	2.5967	−49
−48	1.115	2.270	.01736	2.200	2.200	−180.49	1220.37	1039.88	−0.3738	2.9643	2.5905	−48
−47	1.190	2.422	.01736	2.066	2.066	−189.06	1220.38	1040.32	−0.3727	2.9571	2.5844	−47
−46	1.270	2.585	.01736	1.941	1.941	−179.64	1220.40	1049.76	−0.3717	2.9501	2.5784	−46
−45	1.354	2.757	.01736	1.824	1.824	−179.21	1220.42	1041.21	−0.3707	2.9430	2.5723	−45
−44	1.444	2.940	.01736	1.715	1.715	−178.78	1220.43	1041.65	−0.3696	2.9359	2.5663	−44
−43	1.539	3.134	.01736	1.612	1.612	−178.35	1220.44	1042.09	−0.3686	2.9289	2.5603	−43
−42	1.641	3.340	.01736	1.516	1.516	−177.92	1220.45	1042.53	−0.3676	2.9219	2.5543	−42
−41	1.748	3.559	.01736	1.427	1.427	−177.49	1220.47	1042.98	−0.3665	2.9149	2.5484	−41
−40	1.861	3.790	.01737	1.343	1.343	−177.06	1220.48	1043.42	−0.3655	2.9080	2.5425	−40
−39	1.982	4.035	.01737	1.264	1.264	−176.63	1220.49	1043.86	−0.3645	2.9012	2.5367	−39
−38	2.110	4.295	.01737	1.191	1.191	−176.19	1220.49	1044.30	−0.3634	2.8942	2.5308	−38
−37	2.245	4.570	.01737	1.122	1.122	−175.76	1220.50	1044.74	−0.3624	2.8874	2.5250	−37
−36	2.388	4.862	.01737	1.057	1.057	−175.32	1220.51	1045.19	−0.3614	2.8807	2.5193	−36
−35	2.540	5.170	.01737	0.9961	0.9961	−174.88	1220.51	1045.63	−0.3604	2.8739	2.5135	−35
−34	2.700	5.497	.01737	.9391	.9391	−174.45	1220.52	1046.07	−0.3593	2.8671	2.5078	−34
−33	2.870	5.843	.01738	.8857	.8857	−174.01	1220.52	1046.51	−0.3583	2.8604	2.5021	−33
	$p_s \times 10^2$			$v_{ig} \times 10^{-4}$	$v_g \times 10^{-4}$							
−32	0.3049	0.6208	0.01738	8.355	8.355	−175.57	1220.52	1046.95	−0.3573	2.8538	2.4965	−32
−31	.3239	.6595	.01738	7.883	7.883	−173.13	1220.52	1047.40	−0.3562	2.8470	2.4908	−31
−30	.3440	.7003	.01738	7.441	7.441	−172.68	1220.52	1047.84	−0.3552	2.8405	2.4853	−30
−29	.3652	.7435	.01738	7.025	7.025	−172.24	1220.52	1048.28	−0.3542	2.8339	2.4797	−29
−28	.3876	.7891	.01738	6.634	6.634	−171.80	1220.52	1048.72	−0.3532	2.8274	2.4742	−28
−27	.4113	.8373	.01738	6.267	6.267	−171.35	1220.51	1049.16	−0.3521	2.8207	2.4686	−27
−26	.4363	.8882	.01738	5.921	5.921	−170.91	1220.51	1049.60	−0.3511	2.8143	2.4632	−26
−25	.4627	.9420	.01739	5.596	5.596	−170.46	1220.51	1050.05	−0.3501	2.8078	2.4577	−25
−24	.4905	.9987	.01739	5.290	5.290	−170.01	1220.50	1050.49	−0.3490	2.8013	2.4523	−24
−23	.5199	1.059	.01739	5.003	5.003	−169.56	1220.49	1050.93	−0.3480	2.7949	2.4469	−23
−22	.5509	1.122	.01739	4.732	4.732	−169.12	1220.49	1051.37	−0.3470	2.7885	2.4415	−22
−21	.5836	1.188	.01739	4.477	4.477	−168.66	1220.48	1051.82	−0.3460	2.7822	2.4362	−21
−20	.6181	1.259	.01739	4.237	4.237	−168.21	1220.47	1052.26	−0.3449	2.7757	2.4308	−20
−19	.6545	1.333	.01739	4.011	4.011	−167.76	1220.46	1052.70	−0.3439	2.7695	2.4256	−19
−18	.6928	1.410	.01740	3.797	3.797	−167.31	1220.45	1053.14	−0.3429	2.7632	2.4203	−18
−17	.7332	1.493	.01740	3.596	3.596	−166.85	1220.43	1053.58	−0.3418	2.7568	2.4150	−17
−16	.7757	1.579	.01740	3.407	3.407	−166.40	1220.42	1054.02	−0.3408	2.7506	2.4098	−16
−15	.8204	1.670	.01740	3.228	2.228	−165.94	1220.41	1054.47	−0.3398	2.7444	2.4046	−15
−14	.8676	1.766	.01740	3.060	3.060	−165.48	1220.39	1054.91	−0.3388	2.7383	2.3995	−14
−13	.9172	1.867	.01740	2.901	2.901	−165.03	1220.38	1055.35	−0.3377	2.7320	2.3943	−13
−12	.9694	1.974	.01740	2.750	2.750	−164.57	1220.36	1055.79	−0.3367	2.7259	2.3892	−12
−11	1.024	2.086	.01740	2.609	2.609	−164.11	1220.34	1056.23	−0.3357	2.7198	2.3841	−11
−10	1.082	2.203	.01741	2.475	2.475	−163.65	1220.32	1056.67	−0.3347	2.7138	2.3791	−10
−9	1.143	2.327	.01741	2.349	2.349	−163.18	1220.30	1057.12	−0.3336	2.7076	2.3740	−9
−8	1.207	2.457	.01741	2.229	2.229	−162.72	1220.28	1057.56	−0.3326	2.7016	2.3690	−8

Table 3. Thermodynamic Properties of Water at Saturation—(*Continued*)

Temp. t, °F.	Absolute press. $p_s \times 10^2$		Specific volume, cu ft./lb.			Enthalpy, B.t.u./lb.			Entropy, B.t.u./(lb.)(°F.)			Temp. t, °F.
	Lb./sq. in.	In. Hg.	Sat. liquid v_i	Evap. $v_{ig} \times 10^{-4}$	Sat. vapor $v_g \times 10^{-4}$	Sat. liquid h_i	Evap. h_{ig}	Sat. vapor h_g	Sat. liquid s_i	Evap. s_{ig}	Sat. vapor s_g	
−7	1.274	2.594	.01741	2.116	2.116	−162.26	1220.26	1058.00	−0.3316	2.6956	2.3640	−7
−6	1.344	2.737	.01741	2.010	2.010	−161.79	1220.23	1058.44	−0.3306	2.6896	2.3590	−6
−5	1.419	2.888	.01741	1.909	1.909	−161.33	1220.21	1058.88	−0.3295	2.6836	2.3541	−5
−4	1.496	3.047	.01742	1.814	1.814	−160.86	1220.18	1059.32	−0.3285	2.6777	2.3492	−4
−3	1.578	3.213	.01742	1.723	1.723	−160.39	1220.15	1059.76	−0.3275	2.6718	2.3443	−3
−2	1.664	3.388	.01742	1.638	1.638	−159.92	1220.13	1060.21	−0.3264	2.6658	2.3394	−2
−1	1.754	3.572	.01742	1.557	1.557	−159.45	1220.10	1060.65	−0.3254	2.6600	2.3346	−1
	p_s			$v_{ig} \times 10^{-3}$	$v_g \times 10^{-3}$							
0	0.01849	0.03764	.01742	14.81	14.81	−158.98	1220.07	1061.09	−0.3244	2.6541	2.3297	0
1	.01948	.03966	.01742	14.08	14.08	−158.51	1220.04	1061.53	−0.3234	2.6483	2.3249	1
2	.02052	.04178	.01742	13.40	13.40	−158.04	1220.01	1061.97	−0.3224	2.6425	2.3201	2
3	.02616	.04400	.01743	12.75	12.75	−157.56	1219.97	1062.41	−0.3213	2.6367	2.3154	3
4	.02276	.04633	.01743	12.14	12.14	−157.09	1219.94	1062.85	−0.3203	2.6309	2.3106	4
5	.02396	.04878	.01743	11.55	11.55	−156.61	1219.90	1063.29	−0.3193	2.6252	2.3059	5
6	.02521	.05134	.01743	11.00	11.00	−156.14	1219.88	1063.74	−0.3182	2.6194	2.3012	6
7	.02653	.05402	.01743	10.48	10.48	−155.66	1219.84	1064.18	−0.3172	2.6138	2.2966	7
8	.02791	.05683	.01743	9.979	9.979	−155.18	1219.80	1064.62	−0.3162	2.6081	2.2919	8
9	.02936	.05977	.01744	9.507	9.507	−154.70	1219.76	1065.06	−0.3152	2.6025	2.2873	9
10	.03087	.06286	.01744	9.060	9.060	−154.22	1219.72	1065.50	−0.3142	2.5969	2.2827	10
11	.03246	.06608	.01744	8.636	8.636	−153.74	1219.68	1065.94	−0.3131	2.5912	2.2781	11
12	.03412	.06946	.01744	8.234	8.234	−153.26	1219.64	1066.38	−0.3121	2.5857	2.2736	12
13	.03585	.07300	.01744	7.851	7.851	−152.77	1219.59	1066.82	−0.3111	2.5801	2.2690	13
14	.03767	.07669	.01744	7.489	7.489	−152.29	1219.55	1067.26	−0.3101	2.5746	2.2645	14
15	.03957	.08056	.01744	7.144	7.144	−151.80	1219.50	1067.70	−0.3090	2.5690	2.2600	15
16	.04156	.08461	.01745	6.817	6.817	−151.32	1219.46	1068.14	−0.3080	2.5635	2.2555	16
17	.04363	.08884	.01745	6.505	6.505	−150.83	1219.41	1068.58	−0.3070	2.5581	2.2511	17
18	.04581	.09326	.01745	6.210	6.210	−150.34	1219.36	1069.02	−0.3060	2.5526	2.2466	18
19	.04808	.09789	.01745	5.929	5.929	−149.85	1219.31	1069.46	−0.3049	2.5471	2.2422	19
20	.05045	.1027	.01745	5.662	5.662	−149.36	1219.26	1069.90	−0.3039	2.5417	2.2378	20
21	.05293	.1078	.01745	5.408	5.408	−148.87	1219.21	1070.34	−0.3029	2.5364	2.2335	21
22	.05552	.1130	.01746	5.166	5.166	−148.38	1219.16	1070.78	−0.3019	2.5310	2.2291	22
23	.05823	.1186	.01746	4.936	4.936	−147.88	1219.10	1071.22	−0.3008	2.5256	2.2248	23
24	.06105	.1243	.01746	4.717	4.717	−147.39	1219.05	1071.66	−0.2998	2.5203	2.2205	24
25	.06400	.1303	.01746	4.509	4.509	−146.89	1218.98	1072.09	−0.2988	2.5150	2.2162	25
26	.06708	.1366	.01746	4.311	4.311	−146.40	1218.93	1072.53	−0.2978	2.5097	2.2119	26
27	.07030	.1431	.01746	4.122	4.122	−145.90	1218.87	1072.97	−0.2968	2.5045	2.2077	27
28	.07365	.1500	.01746	3.943	3.943	−145.40	1218.81	1073.41	−0.2957	2.4991	2.2034	28
29	.07715	.1571	.01747	3.771	3.771	−144.90	1218.75	1073.85	−0.2947	2.4939	2.1992	29
30	.08080	.1645	.01747	3.608	3.608	−144.40	1218.69	1074.29	−0.2937	2.4887	2.1950	30
31	.08461	.1723	.01747	3.453	3.453	−143.90	1218.63	1074.73	−0.2927	2.4835	2.1908	31
32	.08858	.1803	.01747	3.305	3.305	−143.40	1218.56	1075.16	−0.2916	2.4783	2.1867	32
	p_s			v_{ig}	v_g							
32†	0.088586	0.18036	0.01602	3304.6	3304.6	0.00	1075.16	1075.16	0.00000	2.1867	2.1867	32†
33	.092227	.18778	.01602	3180.5	3180.5	1.01	1074.59	1075.60	.00205	2.1811	2.1831	33
34	.095999	.19546	.01602	3061.7	3061.7	2.01	1074.03	1076.04	.00409	2.1755	2.1796	34
35	.099908	.20342	.01602	2947.8	2947.8	3.02	1073.46	1076.48	.00612	2.1700	2.1761	35
36	.10396	.21166	.01602	2838.7	2838.7	4.02	1072.90	1076.92	.00815	2.1644	2.1726	36
37	.10815	.22020	.01602	2734.1	2734.1	5.03	1072.33	1077.36	.01018	2.1589	2.1691	37
38	.11249	.22904	.01602	2633.8	2633.8	6.03	1071.77	1077.80	.01220	2.1535	2.1657	38
39	.11699	.23819	.01602	2537.6	2537.6	7.04	1071.20	1078.24	.01422	2.1480	2.1622	39
40	.12164	.24767	.01602	2445.4	2445.4	8.04	1070.64	1078.68	.01623	2.1426	2.1588	40
41	.12646	.25748	.01602	2356.9	2356.9	9.05	1070.06	1079.11	.01824	2.1372	2.1554	41
42	.13145	.26763	.01602	2272.0	2272.0	10.05	1069.50	1079.55	.02024	2.1318	2.1520	42
43	.13660	.27813	.01602	2190.5	2190.5	11.05	1068.94	1079.99	.02224	2.1265	2.1487	43
44	.14194	.28899	.01602	2112.3	2112.3	12.06	1068.37	1080.43	.02423	2.1211	2.1453	44
45	.14746	.30023	.01602	2037.3	2037.3	13.06	1067.81	1080.87	.02622	2.1158	2.1420	45
46	.15317	.31185	.01602	1965.2	1965.2	14.06	1067.24	1081.30	.02820	2.1005	2.1387	46
47	.15907	.32387	.01602	1896.0	1896.0	15.06	1066.68	1081.74	.03018	2.1052	2.1354	47
48	.16517	.33629	.01602	1829.5	1829.5	16.07	1066.11	1082.18	.03216	2.0999	2.1321	48
49	.17148	.34913	.01602	1765.7	1765.7	17.07	1065.55	1082.62	.03413	2.0947	2.1288	49
50	.17799	.36240	.01602	1704.3	1704.3	18.07	1064.99	1083.06	.03610	2.0895	2.1256	50
51	.18473	.37611	.01602	1645.4	1645.4	19.07	1064.42	1083.49	.03806	2.0842	2.1223	51
52	.19169	.39028	.01602	1588.7	1588.7	20.07	1063.86	1083.93	.04002	2.0791	2.1191	52
53	.19888	.40492	.01603	1534.3	1534.3	21.07	1063.30	1084.37	.04197	2.0739	2.1159	53
54	.20630	.42003	.01603	1481.9	1481.9	22.08	1062.72	1084.80	.04392	2.0688	2.1127	54
55	.21397	.43564	.01603	1431.5	1431.5	23.08	1062.16	1085.24	.04587	2.0637	2.1096	55
56	.22188	.45176	.01603	1383.1	1383.1	24.08	1061.60	1085.68	.04781	2.0586	2.1064	56
57	.23006	.46840	.01603	1336.5	1336.5	25.08	1061.04	1086.12	.04975	2.0535	2.1033	57
58	.23849	.48558	.01603	1291.7	1291.7	26.08	1060.47	1086.55	.05168	2.0485	2.1002	58
59	.24720	.50330	.01603	1248.6	1248.6	27.08	1059.91	1086.99	.05361	2.0434	2.0970	59
60	.25618	.52160	.01603	1207.1	1207.1	28.08	1059.34	1087.42	.05553	2.0385	2.0940	60
61	.26545	.54047	.01604	1167.2	1167.2	29.08	1058.78	1087.86	.05746	2.0334	2.0909	61

† Extrapolated to represent metastable equilibrium with undercooled liquid.

Table 3. Thermodynamic Properties of Water at Saturation—(*Continued*)

Temp. t, °F.	Absolute press. p_s		Specific volume, cu. ft./lb.			Enthalpy, B.t.u./lb.			Entropy, B.t.u./(lb.)(°F).			Temp. t, °F.
	Lb./sq. in.	In. Hg	Sat. liquid v_l	Evapo. v_{lg}	Sat. vapor v_g	Sat. liquid h_l	Evap. h_{lg}	Sat. vapor h_g	Sat. liquid s_l	Evapo. s_{lg}	Sat. vapor s_g	
62	.27502	.55994	.01604	1128.7	1128.7	30.08	1058.22	1088.30	.05937	2.0284	2.0878	62
63	.28488	.58002	.01604	1091.7	1091.7	31.08	1057.65	1088.73	.06129	2.0235	2.0848	63
64	.29505	.60073	.01604	1056.1	1056.1	32.08	1057.09	1089.17	.06320	2.0186	2.0818	64
65	.30554	.62209	.01604	1021.7	1021.7	33.08	1056.52	1089.60	.06510	2.0136	2.0787	65
66	.31636	.64411	.01604	988.63	988.65	34.07	1055.97	1090.04	.06700	2.0087	2.0757	66
67	.32750	.66681	.01605	956.76	956.78	35.07	1055.40	1090.47	.06890	2.0039	2.0728	67
68	.33900	.69021	.01605	926.06	926.08	36.07	1054.84	1090.91	.07080	1.9990	2.0698	68
69	.35084	.71432	.01605	896.47	896.49	37.07	1054.27	1091.34	.07269	1.9941	2.0668	69
70	.36304	.73916	.01605	867.95	867.97	38.07	1053.71	1091.78	.07458	1.9893	2.0639	70
71	.37561	.76476	.01605	840.45	840.47	39.07	1053.14	1092.21	.07646	1.9845	2.0610	71
72	.38856	.79113	.01606	813.95	813.97	40.07	1052.58	1092.65	.07834	1.9797	2.0580	72
73	.40190	.81829	.01606	788.38	788.40	41.07	1052.01	1093.08	.08022	1.9749	2.0551	73
74	.41564	.84626	.01606	763.73	763.75	42.06	1051.46	1093.52	.08209	1.9701	2.0522	74
75	.42979	.87506	.01606	739.95	739.97	43.06	1050.89	1093.95	.08396	1.9654	2.0494	75
76	.44435	.90472	.01606	717.01	717.03	44.06	1050.32	1094.38	.08582	1.9607	2.0465	76
77	.45935	.93524	.01607	694.88	694.90	45.06	1049.76	1094.82	.08769	1.9560	2.0437	77
78	.47478	.96666	.01607	673.52	673.54	46.06	1049.19	1095.25	.08954	1.9513	2.0408	78
79	.49066	.99900	.01607	652.91	652.93	47.06	1048.62	1095.68	.09140	1.9466	2.0880	79
80	.50701	1.0323	.01607	633.01	633.03	48.05	1048.07	1096.12	.09325	1.9419	2.0352	80
81	.52382	1.0665	.01608	613.80	613.82	49.05	1047.50	1096.55	.09510	1.9373	2.0324	81
82	.54112	1.1017	.01608	595.25	595.27	50.05	1046.93	1096.98	.09694	1.9328	2.0297	82
83	.55892	1.1380	.01608	577.34	577.36	51.05	1046.37	1097.42	.09878	1.9281	2.0269	83
84	.57722	1.1752	.01608	560.04	560.06	52.05	1045.80	1097.85	.10062	1.9236	2.0242	84
85	.59604	1.2136	.01609	543.33	543.35	53.05	1045.23	1098.28	.10246	1.9189	2.0214	85
86	.61540	1.2530	.01609	527.19	527.21	54.04	1044.67	1098.71	.10429	1.9144	2.0187	86
87	.63530	1.2935	.01609	511.60	511.62	55.04	1044.10	1099.14	.10611	1.9099	2.0160	87
88	.65575	1.3351	.01610	496.52	496.54	56.04	1043.54	1099.58	.10794	1.9054	2.0133	88
89	.67678	1.3779	.01610	481.96	481.98	57.04	1042.97	1100.01	.10976	1.9008	2.0106	89
90	.69838	1.4219	.01610	467.88	467.90	58.04	1042.40	1100.44	.11158	1.8963	2.0079	90
91	.72059	1.4671	.01610	454.26	454.28	59.03	1041.84	1100.87	.11339	1.8919	2.0053	91
92	.74340	1.5136	.01611	441.10	441.12	60.03	1041.27	1101.30	.11520	1.8874	2.0026	92
93	.76684	1.5613	.01611	428.38	428.40	61.03	1040.70	1101.73	.11701	1.8830	2.0000	93
94	.79091	1.6103	.01611	416.07	416.09	62.03	1040.13	1102.16	.11881	1.8786	1.9974	94
95	.81564	1.6607	.01612	404.17	404.19	63.03	1039.56	1102.59	.12061	1.8741	1.9947	95
96	.84103	1.7124	.01612	392.65	392.67	64.02	1039.00	1103.02	.12241	1.8698	1.9922	96
97	.86711	1.7655	.01612	381.51	381.53	65.02	1038.43	1103.45	.12420	1.8654	1.9896	97
98	.89388	1.8200	.01612	370.73	370.75	66.02	1037.86	1103.88	.12600	1.8610	1.9870	98
99	.92137	1.8759	.01613	360.30	360.32	67.02	1037.29	1104.31	.12778	1.8566	1.9844	99
100	.94959	1.9334	.01613	350.20	350.22	68.02	1036.72	1104.74	.12957	1.8523	1.9819	100
101	.97854	1.9923	.01614	340.42	340.44	69.01	1036.16	1105.17	.13135	1.8480	1.9793	101
102	1.0083	2.0529	.01614	330.96	330.98	70.01	1035.58	1105.59	.13313	1.8437	1.9768	102
103	1.0388	2.1149	.01614	321.80	321.82	71.01	1035.01	1106.02	.13490	1.8394	1.9743	103
104	1.0700	2.1786	.01614	312.93	312.95	72.01	1034.44	1106.45	.13667	1.8351	1.9718	104
105	1.1021	2.2440	.01615	304.34	304.36	73.01	1033.87	1106.88	.13844	1.8309	1.9693	105
106	1.1351	2.3110	.01615	296.02	296.04	74.01	1033.29	1107.30	.14021	1.8266	1.9668	106
107	1.1688	2.3798	.01616	287.96	287.98	75.00	1032.73	1107.73	.14197	1.8224	1.9644	107
108	1.2035	2.4503	.01616	280.14	280.16	76.00	1032.16	1108.16	.14373	1.8182	1.9619	108
109	1.2390	2.5226	.01616	272.58	272.60	77.00	1031.58	1108.58	.14549	1.8140	1.9595	109
110	1.2754	2.5968	.01617	265.24	265.26	78.00	1031.01	1109.01	.14724	1.8098	1.9570	110
111	1.3128	2.6728	.01617	258.14	258.16	79.00	1030.44	1109.44	.14899	1.8056	1.9546	111
112	1.3510	2.7507	.01617	251.25	251.27	80.00	1029.86	1109.86	.15074	1.8015	1.9522	112
113	1.3902	2.8306	.01618	244.57	244.59	80.99	1029.30	1110.29	.15248	1.7973	1.9498	113
114	1.4305	2.9125	.01618	238.10	238.12	81.99	1028.72	1110.71	.15423	1.7932	1.9474	114
115	1.4717	2.9963	.01618	231.82	231.84	82.99	1028.15	1111.14	.15596	1.7890	1.9450	115
116	1.5139	3.0823	.01619	225.73	225.75	83.99	1027.57	1111.56	.15770	1.7849	1.9426	116
117	1.5571	3.1703	.01619	219.83	219.85	84.99	1026.99	1111.98	.15943	1.7809	1.9403	117
118	1.6014	3.2606	.01620	214.10	214.12	85.99	1026.42	1112.41	.16116	1.7767	1.9379	118
119	1.6468	3.3530	.01620	208.54	208.56	86.98	1025.85	1112.83	.16289	1.7727	1.9356	119
120	1.6933	3.4477	.01620	203.16	203.18	87.98	1025.28	1113.26	.16461	1.7687	1.9333	120
121	1.7409	3.5446	.01621	197.93	197.95	88.98	1024.70	1113.68	.16634	1.7647	1.9310	121
122	1.7897	3.6439	.01621	192.85	192.87	89.98	1024.12	1114.10	.16805	1.7606	1.9286	122
123	1.8396	3.7455	.01622	187.93	187.95	90.98	1023.54	1114.52	.16977	1.7566	1.9264	123
124	1.8907	3.8496	.01622	183.15	183.17	91.98	1022.96	1114.94	.17148	1.7526	1.9241	124
125	1.9430	3.9561	.01622	178.51	178.53	92.98	1022.39	1115.37	.17319	1.7486	1.9218	125
126	1.9966	4.0651	.01623	174.00	174.02	93.98	1021.81	1115.79	.17490	1.7446	1.9195	126
127	2.0514	4.1768	.01623	169.63	169.65	94.97	1021.24	1116.21	.17660	1.7407	1.9173	127
128	2.1075	4.2910	.01624	165.38	165.40	95.97	1020.66	1116.63	.17830	1.7367	1.9150	128
129	2.1649	4.4078	.01624	161.26	161.28	96.97	1020.08	1117.05	.18000	1.7328	1.9128	129
130	2.2237	4.5274	.01625	157.25	157.27	97.97	1019.50	1117.47	.18170	1.7289	1.9106	130
131	2.2838	4.6498	.01625	153.36	153.38	98.97	1018.92	1117.89	.18339	1.7250	1.9084	131
132	2.3452	4.7750	.01626	149.58	149.60	99.97	1018.34	1118.31	.18508	1.7211	1.9062	132
133	2.4081	4.9030	.01626	145.91	145.93	100.97	1017.76	1118.73	.18676	1.7172	1.9040	133
134	2.4725	5.0340	.01626	142.34	142.36	101.97	1017.18	1119.15	.18845	1.7134	1.9018	134
135	2.5382	5.1679	.01627	138.87	138.89	102.97	1016.59	1119.56	.19013	1.7095	1.8996	135
136	2.6055	5.3049	.01627	135.50	135.52	103.97	1016.01	1119.98	.19181	1.7056	1.8974	136

Table 3. Thermodynamic Properties of Water at Saturation—(*Continued*)

Temp. t, °F.	Absolute press. p_s		Specific volume, cu. ft./lb.			Enthalpy, B.t.u./lb.			Entropy, B.t.u./(lb.)(°F).			Temp. t, °F.
	Lb./sq. in.	In. Hg	Sat. liquid v_i	Evap. v_{ig}	Sat. vapor v_g	Sat. liquid h_i	Evap. h_{ig}	Sat. vapor h_g	Sat. liquid s_i	Evap. s_{ig}	Sat. vapor s_g	
137	2.6743	5.4450	.01628	132.22	132.24	104.97	1015.43	1120.40	.19348	1.7018	1.8953	137
138	2.7446	5.5881	.01628	129.04	129.06	105.97	1014.85	1120.82	.19516	1.6979	1.8931	138
139	2.8165	5.7345	.01629	125.94	125.96	106.97	1014.26	1121.23	.19683	1.6942	1.8910	139
140	2.8900	5.8842	.01629	122.94	122.96	107.96	1013.69	1121.65	.19850	1.6903	1.8888	140
141	2.9651	6.0371	.01630	120.01	120.03	108.96	1013.11	1122.07	.20016	1.6865	1.8867	141
142	3.0419	6.1934	.01630	117.16	117.18	109.96	1012.52	1122.48	.20182	1.6828	1.8846	142
143	3.1204	6.3532	.01631	114.40	114.42	110.96	1011.94	1122.90	.20348	1.6790	1.8825	143
144	3.2006	6.5164	.01631	111.70	111.72	111.96	1011.35	1123.31	.20514	1.6753	1.8804	144
145	3.2825	6.6832	.01632	109.09	109.11	112.96	1010.77	1123.73	.20679	1.6715	1.8783	145
146	3.3662	6.8536	.01632	106.54	106.56	113.96	1010.18	1124.14	.20845	1.6678	1.8763	146
147	3.4517	7.0277	.01633	104.06	104.08	114.96	1009.59	1124.55	.21010	1.6641	1.8742	147
148	3.5390	7.2056	.01633	101.65	101.67	115.96	1009.01	1124.97	.21174	1.6604	1.8721	148
149	3.6282	7.3872	.01634	99.306	99.322	116.96	1008.42	1125.38	.21339	1.6567	1.8701	149
150	3.7194	7.5727	.01634	97.022	97.038	117.96	1007.83	1125.79	.21503	1.6530	1.8680	150
151	3.8124	7.7622	.01635	94.799	94.815	118.96	1007.24	1126.20	.21667	1.6493	1.8660	151
152	3.9074	7.9556	0.01635	92.635	92.651	119.96	1006.66	1126.62	0.21830	1.6457	1.8640	152
153	4.0044	8.1532	.01636	90.528	90.544	120.97	1006.06	1127.03	.21994	1.6421	1.8620	153
154	4.1035	8.3548	.01636	86.477	88.493	121.97	1005.47	1127.44	.22157	1.6384	1.8600	154
155	4.2046	8.5607	.01637	86.480	86.496	122.97	1004.88	1127.85	.22320	1.6348	1.8580	155
156	4.3078	8.7708	.01637	84.536	84.552	123.97	1004.29	1128.26	.22482	1.6312	1.8560	156
157	4.4132	8.9853	.01638	82.642	82.658	124.97	1003.70	1128.67	.22645	1.6276	1.8540	157
158	4.5207	9.2042	.01638	80.798	80.814	125.97	1003.11	1129.08	.22807	1.6239	1.8520	158
159	4.6304	9.4276	.01639	79.001	79.017	126.97	1002.51	1129.48	.22969	1.6204	1.8501	159
160	4.7424	9.6556	.01639	77.251	77.267	127.97	1001.92	1129.89	.23130	1.6168	1.8481	160
161	4.8566	9.8882	.01640	75.546	75.562	128.97	1001.33	1130.30	.23292	1.6133	1.8462	161
162	4.9732	10.126	.01640	73.885	73.901	129.97	1000.74	1130.71	.23453	1.6097	1.8442	162
163	5.0921	10.368	.01641	72.267	72.283	130.98	1000.13	1131.11	.23614	1.6062	1.8423	163
164	5.2134	10.615	.01642	70.690	70.706	131.98	999.54	1131.52	.23774	1.6027	1.8404	164
165	5.3372	10.867	.01642	69.153	69.169	132.98	998.94	1131.92	.23935	1.5990	1.8384	165
166	5.4634	11.124	.01643	67.654	67.670	133.98	998.35	1132.33	.24095	1.5956	1.8365	166
167	5.5921	11.386	.01643	66.194	66.210	134.98	997.75	1132.73	.24255	1.5920	1.8346	167
168	5.7233	11.653	.01644	64.770	64.786	135.98	997.16	1133.14	.24414	1.5887	1.8328	168
169	5.8572	11.925	.01644	63.382	63.398	136.99	996.55	1133.54	.24574	1.5852	1.8309	169
170	5.9936	12.203	.01645	62.029	62.045	137.99	995.95	1133.94	.24733	1.5817	1.8290	170
171	6.1328	12.487	.01645	60.710	60.726	138.99	995.36	1134.35	.24892	1.5782	1.8271	171
172	6.2746	12.775	.01646	59.423	59.439	139.99	994.76	1134.75	.25051	1.5748	1.8253	172
173	6.4192	13.070	.01647	58.168	58.184	141.00	994.15	1135.15	.25209	1.5713	1.8234	173
174	6.5666	13.370	.01647	56.944	56.960	142.00	993.55	1135.55	.25367	1.5679	1.8216	174
175	6.7168	13.676	.01648	55.750	55.766	143.00	992.95	1135.95	.25525	1.5644	1.8197	175
176	6.8699	13.987	.01648	54.586	54.602	144.00	992.35	1136.35	.25683	1.5611	1.8179	176
177	7.0259	14.305	.01649	53.450	53.466	145.00	991.75	1136.75	.25841	1.5577	1.8161	177
178	7.1849	14.629	.01650	52.341	52.357	146.01	991.14	1137.15	.25998	1.5543	1.8143	178
179	7.3469	14.959	.01650	51.260	51.276	147.01	990.54	1137.55	.26155	1.5508	1.8124	179
180	7.5119	15.295	.01651	50.203	50.220	148.01	989.93	1137.94	.26312	1.5475	1.8106	180
181	7.6801	15.637	.01651	49.173	49.190	149.02	989.32	1138.34	.26468	1.5442	1.8089	181
182	7.8514	15.986	.01652	48.168	48.185	150.02	988.72	1138.74	.26625	1.5408	1.8071	182
183	8.0258	16.341	.01652	47.187	47.204	151.02	988.12	1139.14	.26781	1.5375	1.8053	183
184	8.2035	16.703	.01653	46.229	46.246	152.03	987.50	1139.53	.26937	1.5341	1.8035	184
185	8.3845	17.071	.01654	45.294	45.311	153.03	986.89	1139.92	.27093	1.5308	1.8017	185
186	8.5688	17.446	.01654	44.381	44.398	154.04	986.28	1140.32	.27248	1.5275	1.8000	186
187	8.7565	17.829	.01655	43.489	43.506	155.04	985.67	1140.71	.27404	1.5242	1.7982	187
188	8.9476	18.218	.01656	42.619	42.636	156.04	985.07	1141.11	.27559	1.5209	1.7965	188
189	9.1422	18.614	.01656	41.769	41.786	157.05	984.45	1141.50	.27713	1.5176	1.7947	189
190	9.3403	19.017	.01657	40.939	40.956	158.05	983.84	1141.89	.27868	1.5143	1.7930	190
191	9.5420	19.428	.01658	40.128	40.145	159.06	983.22	1142.28	.28022	1.5111	1.7913	191
192	9.7473	19.846	.01658	39.337	39.354	160.06	982.61	1142.67	.28176	1.5078	1.7896	192
193	9.9563	20.271	.01659	38.563	38.580	161.06	982.00	1143.06	.28330	1.5045	1.7878	193
194	10.169	20.704	.01659	37.807	37.824	162.07	981.38	1143.45	.28484	1.5013	1.7861	194
195	10.386	21.145	.01660	37.069	37.086	163.08	980.76	1143.84	.28638	1.4980	1.7844	195
196	10.606	21.594	.01661	36.348	36.365	164.08	980.15	1144.23	.28791	1.4949	1.7828	196
197	10.830	22.050	.01661	35.643	35.660	165.08	979.54	1144.62	.28944	1.4917	1.7811	197
198	11.058	22.515	.01662	34.954	34.971	166.09	978.91	1145.00	.29097	1.4884	1.7794	198
199	11.290	22.987	.01663	34.281	34.298	167.10	978.29	1145.39	.29250	1.4852	1.7777	199
200	11.526	23.468	.01663	33.623	33.640	168.10	977.68	1145.78	.29402	1.4820	1.7760	200
201	11.767	23.957	.01664	32.980	32.997	169.11	977.05	1146.16	.29554	1.4789	1.7744	201
202	12.011	24.455	.01665	32.351	32.368	170.11	976.43	1146.54	.29706	1.4756	1.7727	202
203	12.260	24.961	.01665	31.737	31.754	171.12	975.81	1146.93	.29858	1.4725	1.7711	203
204	12.513	25.476	.01666	31.136	31.153	172.12	975.19	1147.31	.30010	1.4693	1.7694	204
205	12.770	26.000	.01667	30.549	30.566	173.13	974.56	1147.69	.30161	1.4662	1.7678	205
206	13.031	26.532	.01667	29.974	29.991	174.14	973.94	1148.08	.30312	1.4631	1.7662	206
207	13.297	27.074	.01668	29.413	29.430	175.14	973.32	1148.46	.30463	1.4600	1.7646	207
208	13.568	27.625	.01669	28.863	28.880	176.15	972.69	1148.84	.30614	1.4568	1.7629	208
209	13.843	28.185	.01669	28.326	28.343	177.16	972.06	1149.22	.30765	1.4536	1.7613	209
210	14.123	28.754	.01670	27.801	27.818	178.17	971.43	1149.60	.30915	1.4506	1.7597	210
211	14.407	29.333	.01671	27.287	27.304	179.17	970.81	1149.98	.31065	1.4474	1.7581	211
212	14.696	29.921	.01671	26.784	26.801	180.18	970.17	1150.35	.31215	1.4444	1.7565	212

Solution (read directly from psychrometric chart).

$$\text{Dew point } t'' = 25.2°\text{F.}$$
$$\text{Wet-bulb temperature of heated air } t' = 51.5°\text{F.}$$
$$\text{Relative humidity} = 15 \text{ per cent}$$

Example 2a. Find the heat added per pound of dry air in Example 2.

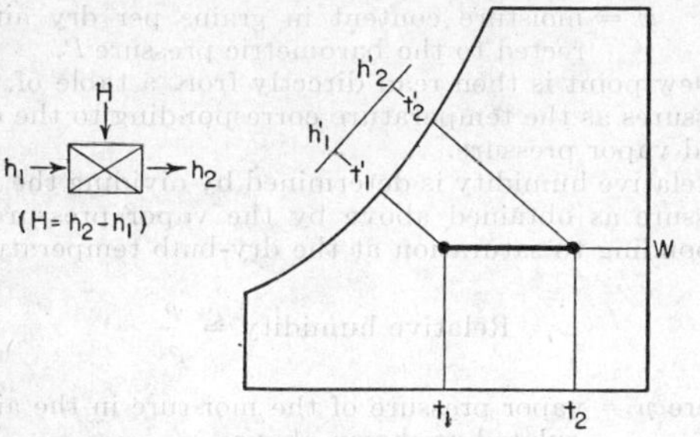

FIG. 6. Heating process.

Approximate Solution.

$h_1' = 10.1$ and $h_2' = 21.1$; $\Delta h = h_2' - h_1' = 11$ B.t.u.

Note that this result is 1.5 per cent higher than the following exact solution.

Exact Solution.

$$h_1 = h_1' + D_2 = 10.1 + 0.06 = 10.16 \text{ B.t.u.}$$
$$h_2 = h_1' + D_2 = 21.1 - 0.1 = 21.0 \text{ B.t.u.}$$
$$h = h_2 - h_1 = 10.84 \text{ B.t.u.}$$

Example 3. Air at 95°F. dry bulb and 70°F. wet bulb passes through a water spray where its relative humidity is increased to 90 per cent. The spray water is recirculated, and the make-up water enters at 70°F. Determine the leaving dry-bulb temperature, wet-bulb temperature, change in enthalpy of the air, and the amount of moisture added per pound of dry air.

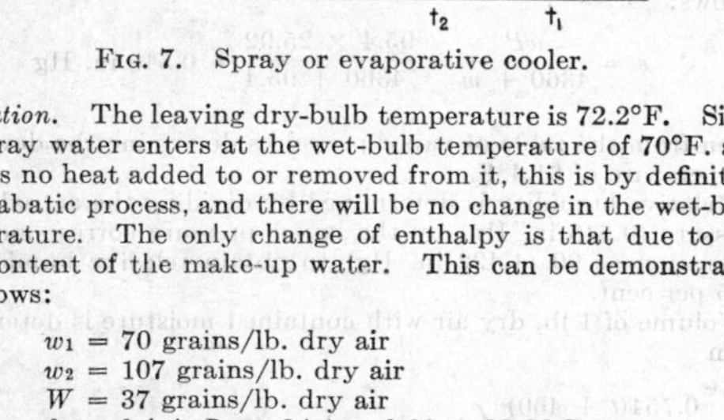

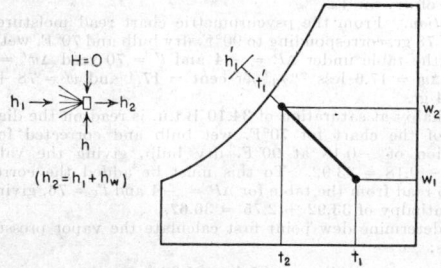

FIG. 7. Spray or evaporative cooler.

Solution. The leaving dry-bulb temperature is 72.2°F. Since the spray water enters at the wet-bulb temperature of 70°F. and there is no heat added to or removed from it, this is by definition an adiabatic process, and there will be no change in the wet-bulb temperature. The only change of enthalpy is that due to the heat content of the make-up water. This can be demonstrated as follows:

$$w_1 = 70 \text{ grains/lb. dry air}$$
$$w_2 = 107 \text{ grains/lb. dry air}$$
$$W = 37 \text{ grains/lb. dry air}$$
$$h_1 = h_1' + D_1 = 34.1 - 0.22 = 33.88 \text{ B.t.u.}$$
$$h_2 = h_2' + D_2 = 34.1 - 0.02 = 34.08 \text{ B.t.u.}$$
$$h_w \text{ (from small diagram, 37 grains at 70°F.)} = 0.2 \text{ B.t.u.}$$
$$H = h_1 - h_2 + h_w = 33.88 - 34.08 + 0.2 = 0$$

Example 4. Find the cooling effect per pound of dry air where the air enters a cooling coil at 83°F. dry bulb, 69°F. wet bulb and leaves at 53°F. dry bulb, 52°F. wet bulb and the condensate is rejected at 52°F.

Approximate Solution.

$$h_1' = 33.24 \text{ B.t.u.}; \quad h_2' = 21.42 \text{ B.t.u.};$$
$$R' = h_1' - h_2' = 11.82 \text{ B.t.u.}$$

Note that this value is 1.6 per cent greater than the following exact solution.

Exact Solution.

$$h_1 = h_1' + D_1 = 33.24 - 0.12 = 33.12 \text{ B.t.u.}$$
$$h_2 = h_2' + D_2 = 21.42 - 0.01 = 21.41 \text{ B.t.u.}$$
$$W = w_1 - w_2 = 84 - 56 = 28 \text{ grains}$$
$$h_w \text{ (from small diagram, 28 grains at 52°F.)} = 0.08$$
$$R = h_1 - h_2 - h_w = 33.12 - 21.41 - 0.08 = 11.63$$

Example 4a (Refer to Fig. 4). Find the cooling load per pound of dry air due to infiltration of room air at 80°F. dry bulb, 67°F. wet bulb into a cooler maintained at 30°F. dry bulb, 28°F. wet bulb where moisture freezes on the coil which is maintained at 20°F.

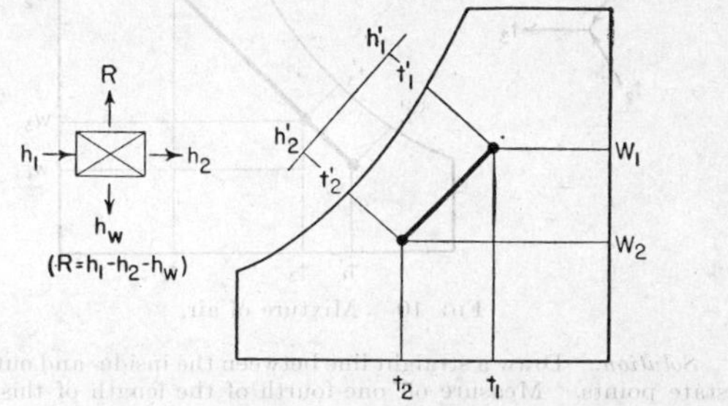

FIG. 8. Cooling and dehumidifying process.

Approximate Solution.

$$h_1' = 31.62 \text{ B.t.u.}; \quad h_2' = 10.10 \text{ B.t.u.};$$
$$R' = h_1' - h_2' = 21.52 \text{ B.t.u.}$$

Note that this result is 5 per cent lower than the following exact solution.

Exact Solution.

$$h_1 = h_1' + D_1 = 31.62 - 0.1 = 31.52 \text{ B.t.u.}$$
$$h_2 = h_2' + D_2 = 10.10 + 0.06 = 10.16 \text{ B.t.u.}$$
$$W = w_1 - w_2 = 78 - 19 = 59 \text{ grains}$$
$$h_w = \text{(from diagram, 59 grains at 20°F.)} = -1.26 \text{ B.t.u.}$$
$$R = h_1 - h_2 - h_w = 31.52 - 10.16 + 1.26 = 22.62$$

Example 5. Determine the water consumption and the amount of heat dissipated per 1000 cu. ft. of entering air at 90°F. dry bulb and 70°F. wet bulb when the air leaves at 110°F. saturated and make-up water is at 75°F.

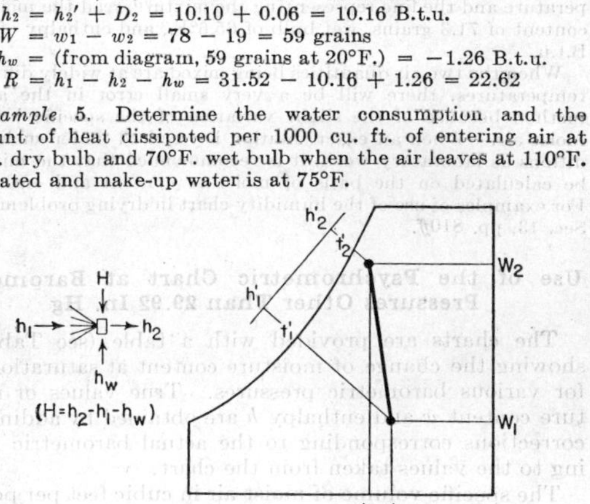

FIG. 9. Cooling tower.

Approximate Solution (high-temperature chart).

$$W = w_1 - w_2 = 416 - 78 = 338 \text{ grains}$$
$$H' = h_2' - h_1' = 92.34 - 34.1 = 58.24 \text{ B.t.u.}$$
$$\text{Volume of entering air} = 14.1 \text{ cu. ft./lb. dry air}$$
$$\text{Heat dissipated per 1000 cu. ft./min.} = (58.24 \times 1000)/14.1 = 4130 \text{ B.t.u.}$$

Note that this result is 3.4 per cent greater than the following exact solution.

Exact Solution.

$$W = w_1 - w_2 = 416 - 78 = 338 \text{ grains}$$
$$h_w = {}^{338}\!/_{7000} \times (75 - 32) = 2.08 \text{ B.t.u.}$$
$$h_1 = h_1' + D_1 = 34.1 - 0.18 = 33.92$$

$h_2 = h_2' = 92.34$
$H = h_2 - h_1 - h_w = 92.34 - 33.92 - 2.08 = 56.34$
Specific volume of entering air = 14.1 cu. ft./lb. dry air

$$\text{Heat dissipated per 1000 cu. ft./min.} = \frac{(56.34 \times 1000)}{14.1} = 3990$$

Example 6 Inside air at 75°F. dry bulb and 62°F. wet bulb is mixed with outside air at 95°F. dry bulb and 75°F. wet bulb in the proportion of 1 part outside air to 3 parts inside air. Find the properties of the resulting mixture

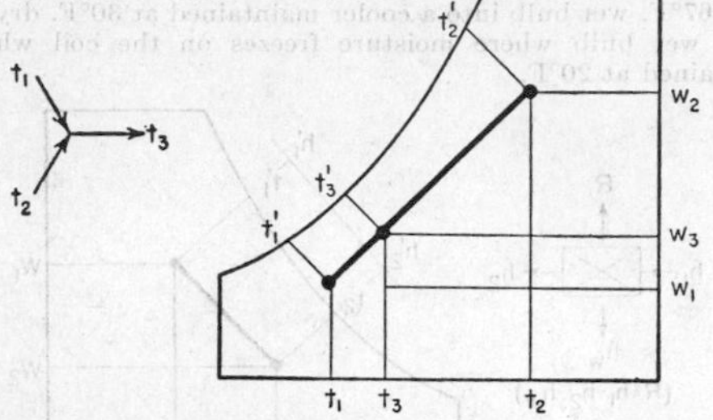

FIG. 10 Mixture of air.

Solution. Draw a straight line between the inside- and outside-state points. Measure off one-fourth of the length of this line, starting from the inside state. This will establish a point representing the mixture of 1 part outside air with 3 parts inside air. Properties of the mixed air are then determined as explained in Example 1.

It is often convenient to use the dry-bulb temperature scale or the moisture-content scale as the means of dividing the mixture line into the desired proportions In this example, for instance, one-fourth of the difference between the inside and outside dry-bulb temperatures is 5°, and thus the mixture of 1 part outside air and 3 parts inside air will be at a temperature of 75 + 5 = 80°F. At the intersection of 80°F. dry-bulb temperature and the line representing the mixture, read the moisture content of 71.3 grains, wet bulb of 65.6°F., and enthalpy of 30.6 B.t.u.

When the two air quantities being mixed are at widely different temperatures, there will be a very small error in the above method because of the slight variation of the specific heat of moist air. When an exact solution is required under such conditions as laboratory work, it is recommended that the mixture be calculated on the basis of moisture content and enthalpy. For examples of use of the humidity chart in drying problems, see Sec. 13, pp. 810*ff.*

Use of the Psychrometric Chart at Barometric Pressures Other Than 29.92 In. Hg

The charts are provided with a table (see Table 2) showing the change of moisture content at saturation W' for various barometric pressures. True values of moisture content w and enthalpy h are obtained by adding the corrections corresponding to the actual barometric reading to the values taken from the chart.

The specific volume of moist air in cubic feet per pound of dry air can be determined for any barometric pressures by use of the formula

$$V = \frac{0.754(t + 460)}{P}\left(1 + \frac{w}{4360}\right)$$

where V = specific volume per lb. dry air, cu. ft./lb.
t = dry-bulb temperature, °F.
P = barometric pressure, in. of Hg.
w = grains moisture content per lb. dry air corrected for the actual barometric pressure.

Relative humidity and dew point can be determined for other than standard barometric pressures by the use of a table of vapor pressures and a simple calculation. To determine either relative humidity or dew point, it is necessary to calculate the vapor pressure of the moisture in the air by:

$$p = \frac{wP}{4360 + w}$$

where P = barometric pressure.
p = vapor pressure.
w = moisture content in grains per dry air corrected to the barometric pressure P.

Dew point is then read directly from a table of vapor pressures as the temperature corresponding to the calculated vapor pressure.

Relative humidity is determined by dividing the vapor pressure as obtained above by the vapor pressure corresponding to saturation at the dry-bulb temperature.

$$\text{Relative humidity} = \frac{p}{p_s}$$

where p = vapor pressure of the moisture in the air calculated as shown above.
p_s = saturated vapor pressure corresponding to the dry-bulb temperature.

Note that the factor 4360 is obtained by multiplying the specific weight of water vapor referred to air by 7000, the number of grains per pound. The specific weight of water vapor varies slightly with temperature, and the above factor should be modified in accordance with the following table:

Up to 90°F. use 4360
90° to 150°F. use 4380
150° to 200°F. use 4400

Example 7. For a barometric pressure of 25.92 in. Hg ($\Delta P = -4$) and a reading of 90°F. dry bulb and 70°F. wet bulb temperature, determine the following: moisture content w, enthalpy h, dew point t'', relative humidity, and volume per pound of dry air V.

Solution. From the psychrometric chart read moisture content of 78 gr. corresponding to 90°F. dry bulb and 70°F. wet bulb. From the table under $\Delta P = -4$ and $t' = 70$ read $\Delta w' = 17.6$. Then $\Delta w = 17.6$ less $^{20}\!/_{23}$ per cent = 17.4 and $w = 78 + 17.4 = 95.4$ gr.

Enthalpy at saturation of 34.10 B.t.u. is read on the diagonal scale of the chart for 70°F. wet bulb and corrected for the deviation of −0.18 at 90°F. dry bulb, giving the value of 34.10 − 0.18 = 33.92. To this must be added the correction of 2.75 read from the table for $\Delta P = -4$ and $t' = 76$, giving the true enthalpy of 33.92 + 2.75 = 36.67.

To determine dew point first calculate the vapor pressure as follows:

$$e = \frac{wP}{4360 + w} = \frac{95.4 \times 25.92}{4360 + 95.4} = 0.548 \text{ in. Hg}$$

Then from the table of vapor pressures determine the dew-point temperature of 61.4°F.

Relative humidity is determined by dividing the actual vapor pressure (0.548 in. Hg) by the vapor pressure corresponding to saturation at 90 (1.422 in. Hg) to obtain relative humidity of 38.5 per cent.

Volume of 1 lb. dry air with contained moisture is determined from

$$V = \frac{0.754(t + 460)}{P}\left(1 + \frac{w}{4360}\right)$$

$$= \frac{0.754(90 + 460)}{25.92}\left(1 + \frac{95.4}{4360}\right) = 16.4 \text{ cu. ft./lb. dry air}$$

DETERMINATION OF HUMIDITY

Dew-point Method. The dew point of air is measured directly by observing the temperature at which moisture begins to form on an artificially cooled mirror surface. Such surfaces are usually cooled either by evaporation of a low-boiling solvent such as ether or by a temperature-regulated stream of water. To be accurate, this method requires that there be no temperature difference between the mirror surface and the liquid surrounding the thermometer bulb; and, for cases of high temperatures and low humidities, the accuracy especially of the ether container, seems questionable.

Wet-bulb Method. The wet-bulb temperature is measured by passing the air over the bulb of a thermometer that is covered by a cloth or wick saturated with water. If no other heat is supplied to or removed from the wick except by adiabatic evaporation, the wick and the thermometer bulb rapidly attain the wet-bulb temperature. A comparison of wet- and dry-bulb temperatures readily gives the humidity by use of the psychrometric charts given by Figs. 2 to 4. The difficulties of this method are (1) heat is supplied to the wick by radiation so that the process is not quite adiabatic and (2) it is sometimes difficult to keep the wick completely wet. Both tend to make the wet-bulb reading high with a correspondingly higher humidity reading than the true value. The first difficulty is minimized by use of a high velocity of gas flow across the bulb, which makes the rate of heat transfer by radiation negligible compared with that by convection. Carrier and Lindsay [*Mech. Eng.*, **47**, 327 (1925)] have shown that, with a velocity of 17 ft./sec. across the surface of the bulb, the error in the wet-bulb depression or difference between dry- and wet-bulb readings is about 1.3 per cent at about 50°F. wet bulb, and about 0.8 per cent at 80°F. wet bulb; with a velocity of 0.4 ft./sec., which is a reasonable natural convection velocity over a stationary bulb, the error is about 12 per cent at 50°F. wet bulb and about 7 per cent at 80°F. wet bulb. See also Carrier and Mackey [*Trans. Am. Soc. Mech. Eng.*, **59**, 33 (1937)] and Dropkin (*Cornell Univ., Eng. Exp. Sta. Bull.* 23). The difficulty of keeping the wick wet is a mechanical one and is greater the higher the velocity of gas flow across the bulb and the higher the temperature because of the greater rate of evaporation. Incorrect readings of the wet-bulb temperature will be obtained if there is solvent present which will condense on the wick.

For measuring the wet-bulb temperature in a room, the sling psychrometer shown by Fig. 11 has been widely

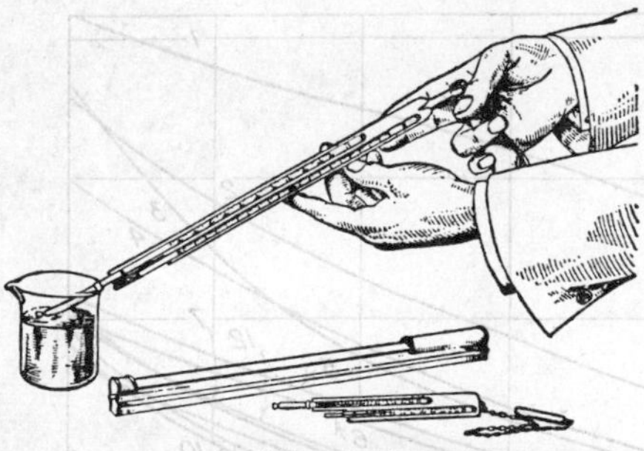

FIG. 11. Sling psychrometer.

used. This is composed of wet- and dry-bulb thermometers mounted in a sling which can be whirled manually to give the desired gas velocity across the bulbs.

Where the sling psychometer cannot be used, air samples are forced, or drawn, over stationary wet- and dry-bulb thermometers at a velocity of about 15 ft./sec., and precautions are taken to keep the wick over the wet-bulb thermometer wet, and to shield both bulbs from radiation from warm or cold surfaces.

Thermal-conductivity Method. Because of the difference in thermal conductivity between air and water vapor, this property may be accurately measured for a mixture and the humidity obtained from a comparison of the measured conductivity with previous calibrations. This method is very reliable provided the air contains no vapors other than water but requires elaborate apparatus such as that described by Hamilton [*Ind. Eng. Chem.*, anal. ed., **2**, 234 (1930)].

Hair-hygrometer Method. This instrument is based on the change in length of hygroscopic materials, such as hair, with change in relative humidity.

Calculation of Non-aqueous Vapors. (See Sec. 13.) For mixtures of dry air and other vapors, the following formulas may be used for the purpose of calculation:

$$\text{Lb. vapor/lb. dry air} = w_v = \left(\frac{p_v}{P - p_v}\right)\left(\frac{M_v}{M_a}\right)$$

$$\text{Per cent humidity} = 100\left(\frac{p_v}{P - p_v}\right)\left(\frac{P - p_s}{p_s}\right)$$

$$\text{Per cent relative humidity} = 100\,\frac{p_v}{p_s}$$

where w_v = vapor content, lb./lb. dry air.
p_v = partial pressure of vapor in air, in. of Hg.
P = total pressure, in. of Hg.
p_s = vapor pressure of pure vapor at the dry-bulb temperature, in. of Hg.
M_v = molecular weight of vapor.
M_a = molecular weight of air.

Equations for the determination of humidity from wet- and dry-bulb temperature when the wet bulb is covered with a saturated salt solution have been developed by G. C. Williams and R. O. Schmitt, *Ind. Eng. Chem.*, **38**, 967 (1946).

For further discussion of non-aqueous vapors, see Sec. 13, pp. 812*ff*.

AIR CONDITIONING

The control of atmospheric conditions within an enclosure with reference to temperature, humidity, air motion, and cleanliness is termed "air conditioning." In industrial applications, control of temperature and humidity is frequently necessary to maintain material in process in a uniform and desirable condition for use. Materials such as textile fibers are hygroscopic, *i.e.*, they take up or give off moisture to the atmosphere to an extent governed by the temperature and humidity. Figure 12 shows, for instance, the hygroscopic moisture of paper as a function of the percentage relative humidity; Fig. 13 gives the variation in dimensions of the papers of Fig. 12 as a function of the relative humidity; Fig. 14 gives the hygroscopic moisture of natural fiber textile materials as a function of the water content in per cent of dry material; Fig. 15 gives the hygroscopic moisture of artificial textile fibers compared with the crude constituents and natural silk; Fig. 16 gives the hygroscopic moisture of various fibrous materials prepared for electrical insulation; Fig. 17 gives similar data for leather and rubber; Fig. 18 gives similar data for cereal foods; Fig. 19 gives similar data for several inorganic substances; Fig. 20 shows the effect of varying temperature on the equilibrium water content at a constant relative humidity of 50 per cent; Fig. 21 gives the hygroscopic moisture data of several organic substances; Fig. 22 gives similar data for various carbon products. It may be necessary to maintain the humidity below that at which an inorganic compound is in equilibrium with its saturated solution, to prevent caking of crystals, etc. (see p. 798).

Air conditioning is much used for comfort of workers in industrial establishments, as well as in railway cars, office buildings, theaters, and auditoriums. Figure 23 shows a series of **equal-comfort** lines superimposed on the psychrometric chart, which may be called **effective temperature lines.**

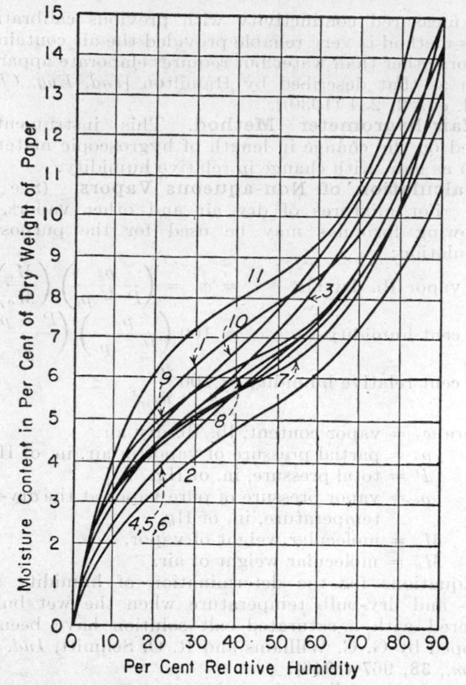

FIG. 12. Hygroscopic moisture of various papers. 1. Sulfite cellulose pulp. 2. News print. 3. Writing. 4. Fine white writing. 5. White bond. 6. Fine white bond. 7. Commercial ledger. 8. White ledger. 9. Index bristol. 10. Kraft wrapping. 11. Rope manila. ("*International Critical Tables*," *vol.* 2, *p.* 322.)

The numerical designation of the effective temperature line is where it crosses the 100 per cent relative humidity line. In other words, 71°F. effective temperature line is that line which starts at the 71°F. dewpoint saturated condition indicated on the chart.

Any combination of wet- and dry-bulb temperatures which falls along the same effective temperature line gives the same sensation of comfort. For instance, one has the same feeling of comfort at 80°F. dry bulb and 58°F. wet bulb as at 74°F. dry bulb and 67°F. wet bulb.

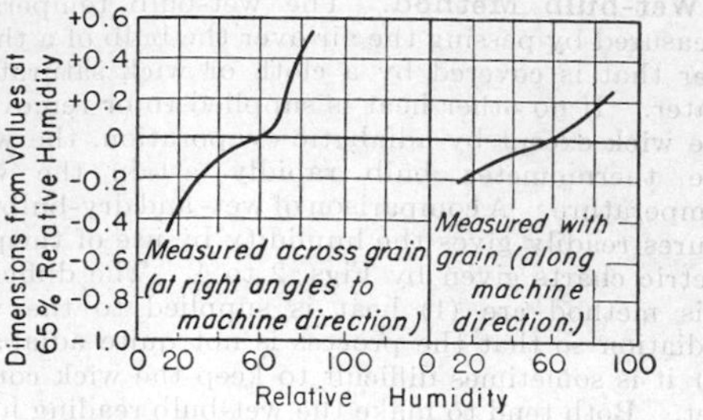

FIG. 13. Variation in dimensions with variation in relative humidity. Composite curves for all papers of Fig. 3. ("*International Critical Tables*," *vol.* 2, *p.* 322.)

At the high dry-bulb temperature, the body loses less heat by radiation and convection than at the lower temperatures. As a result, at the higher temperatures, the body must lose more heat by evaporation than at the lower temperatures. Since lower relative humidities are more conducive to evaporation than high humidities, it will be found that combinations of higher temperatures and low relative humidities give the same feeling of comfort as do lower temperatures and higher relative humidities.

The general zone in which the majority of people are comfortable for summer and winter conditions is shown.

The optimum condition or where the greatest number of people are comfortable is indicated by the 66°F. effective temperature line for winter, and the 71°F. effective temperature line for summer. This difference between the winter and summer conditions is caused by the human body becoming more acclimated to the higher temperature in summertime and also due, to some extent, to the clothing worn.

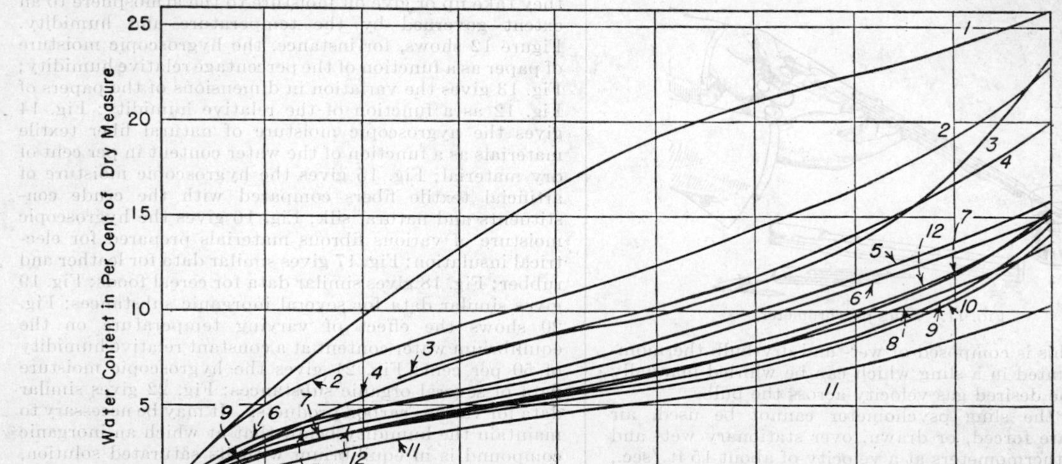

FIG. 14. Hygroscopic moisture (25°C.) of natural fiber textile materials. 1. Absorbent cotton. 2. Wool, worsted. 3. Silk, new yellow. 4. Jute. 5. Manila hemp. 6. Sisal hemp. 7. Indian cotton. 8. Cotton cloth. 9. Egyptian cotton. 10. American cotton. 11. Linen. 12. Flax. ("*International Critical Tables*," *vol.* 2, *p.* 323.)

Favorable Conditions of Temperature and Humidity Artificially Created and Maintained in Manufacturing Processes*

Industry and product	Process	t °C.	Relative humidity, %
Cotton	Carding	20–23	50
	Combing	20–23	60–65
	Roving	20–23	50–60
	Spinning	20–23	60–65
	Spooling, twisting	20–23	65
	Warping	20–23	65
	Weaving	20–23	75–80
Wool	Carding	23–25	65–70
	Spinning	23–25	55–60
	Weaving	20–23	50–55
	Storage for shipping	20–23	55–60
Silk	Dressing	21–25	60–65
	Spinning	21–25	65–70
	Throwing	21–25	65–70
	Weaving	21–25	60–70
Confectionery	Chocolate covering	18	≯55
	Hard-candy making	21	≯50
	Storage	−1	≯70
		+15†	≯55
Tobacco	Softening	29	85
	Cigar and cigarette making	21–23	55–70
Printing	Lithographing	21	45
	Relief and offset	25	45
	Folding	25	65
	Binding	21	45
Baking	Dough fermentation	27	65
	Proofing	32–35	80–90
	Loaf cooling	21	65
Electrical cable	Winding insulation	≯40	≯ 5
Cellulose lacquers	Application	24	≯20
Munitions	Fuse loading	21	55
Cereals	Seal packing prepared, crisp cereals	23	45–50

* "International Critical Tables," vol. 2, p. 322, McGraw-Hill
† Divergence in practice.

Air-conditioning Equipment.

1. Apparatus for the addition to, or removal from, the room being conditioned, of heat and moisture.
2. Automatic controls for the apparatus, so that the temperature and humidity may be controlled within the desired limits.

Apparatus for accomplishing the necessary changes in moisture content are generally known as humidifiers and dehumidifiers.

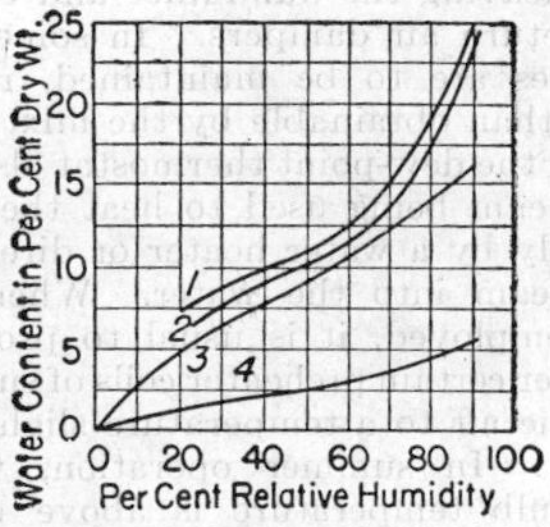

FIG. 15. Hygroscopic moisture (25°C.) of artificial textile fibers compared with crude constituents and natural silk. 1. Viscose rayon (artificial silk). 2. Natural silk, new yellow. 3. Nitrocellulose. 4. Cellulose acetate. (*"International Critical Tables," vol. 2, p. 323.*)

Humidifiers may be divided into the following general types, depending upon the method of operation:

1. **Indirect** system which introduces moistened air into the room.
2. **Direct** system which sprays water directly into the room.
3. **Combined** system which is a combination of the first two.

Indirect Humidifiers. These humidifiers are similar in operation to spray-type air washers except that the water is sprayed directly against the incoming air. Such humidifiers comprise a chamber, usually from 6 to 8 ft. in length, through which the air is drawn at a velocity

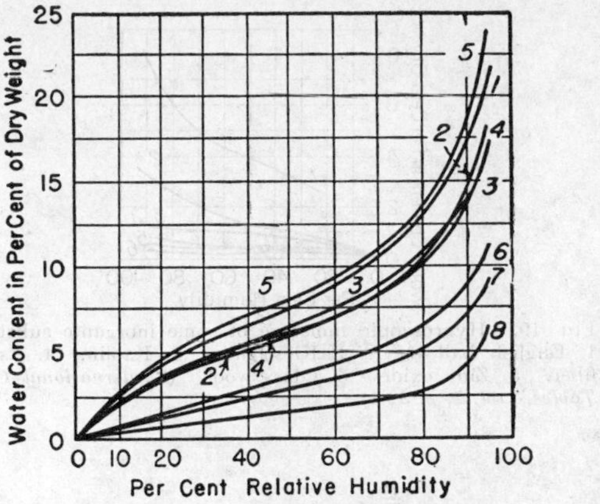

FIG. 16. Hygroscopic moisture of various fibrous materials prepared for electrical insulation. 1. Manila paper. 2. Red rope paper. 3. Pressboard. 4. Leatheroid paper. 5. Silk. 6. Red rope paper (varnished). 7. Empire cloth. 8. Asbestos paper. (*"International Critical Tables," vol. 2, p. 323.*)

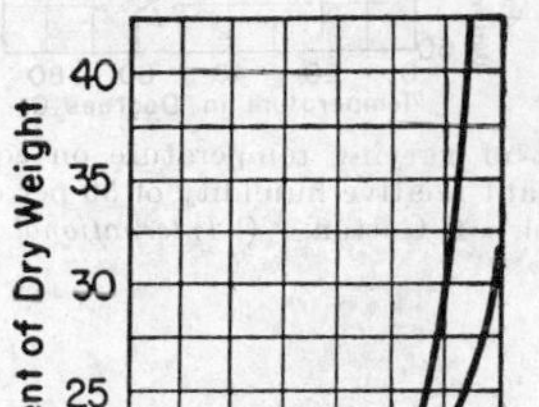

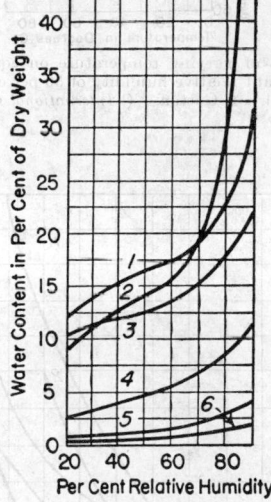

FIG. 17. Hygroscopic moisture of leather and rubber. 1. Leather (sole oak tanned). 2. Sheepskin. 3. Gold's beater skin. 4. Latex, dipped cord. 5. Reclaimed rubber. 6. Smoked, crepe sheet. (*"International Critical Tables," vol. 2, p. 324.*)

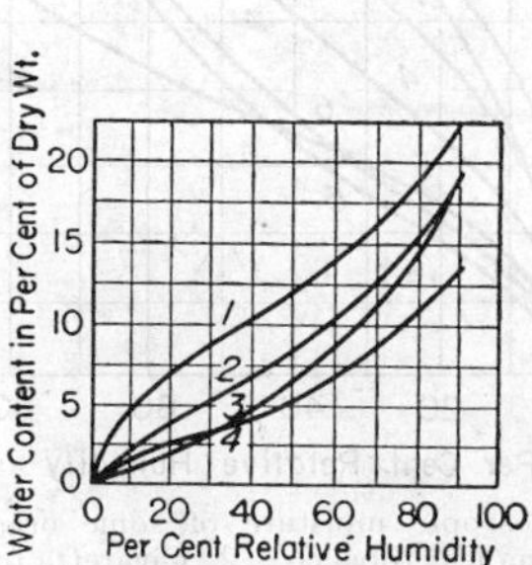

FIG. 18. Hygroscopic moisture of cereal foods. 1. Macaroni. 2. Flour (patent). 3. Bread. 4. Crackers. (*"International Critical Tables," vol. 2, p. 324.*)

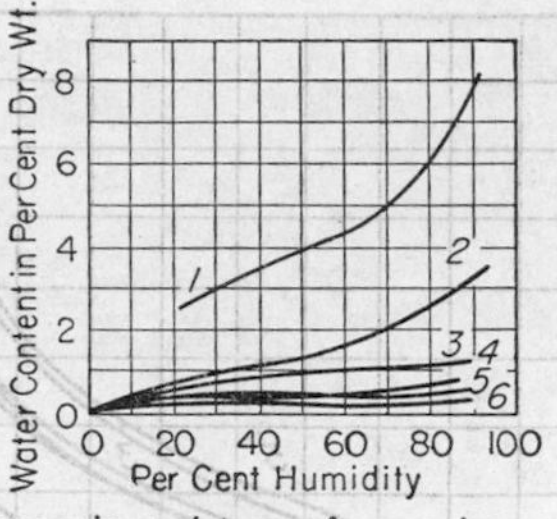

FIG. 19. Hygroscopic moisture of some inorganic substances. 1. English ball clay. 2. Kieselguhr. 3. Kaolin. 4. Asbestos fiber. 5. Zinc oxide. 6. Glass wool. (*"International Critical Tables," vol.* 2, *p.* 324.)

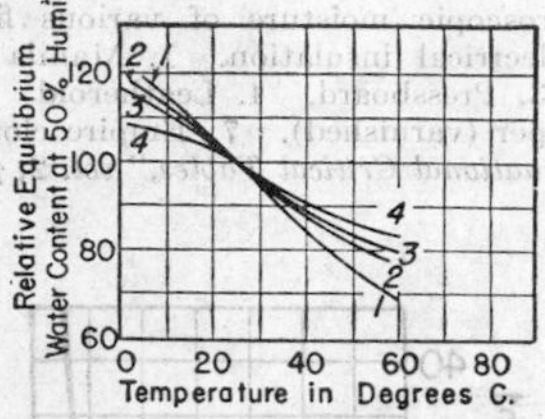

FIG. 20. Effect of varying temperature on equilibrium water content at constant relative humidity of 50 per cent. 1. Wood 2. Silk. 3. Wool. 4. Cotton. (*"International Critical Tables," vol.* 2, *p.* 324.)

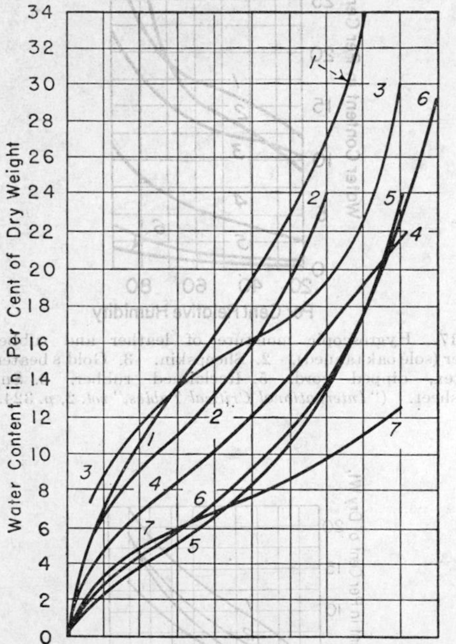

FIG. 21. Hygroscopic moisture of some organic substances. 1. North Carolina leaf tobacco. 2. Cigarette tobacco (Fatima). 3. Sole leather (oak tanned). 4. Catgut. 5. Soap (Ivory). 6. Lumber. 7. Glue (hide, first grade). (*"International Critical Tables," vol.* 2, *p.* 325.)

from 500 to 700 ft./min. Inside the chamber are placed one or more banks of spray nozzles distributed uniformly over the cross-sectional area of the chamber. These nozzles create a finely divided spray through centrifugal action and require water pressures for effective humidification of from 35 to 45 lb./sq. in. At the intake of the humidifying chamber, a set of baffles distributes the air and prevents the spray from escaping from the chamber, and at the outlet of the humidifier chamber an eliminator is provided. This eliminator, consisting of a series of metal baffles, is so designed as to separate all the free, unevaporated moisture from the humidified air. Thus the air leaving the humidifier is completely saturated but without any entrainment or unevaporated water particles. The general construction of this type of humidifier is shown in Fig 24.

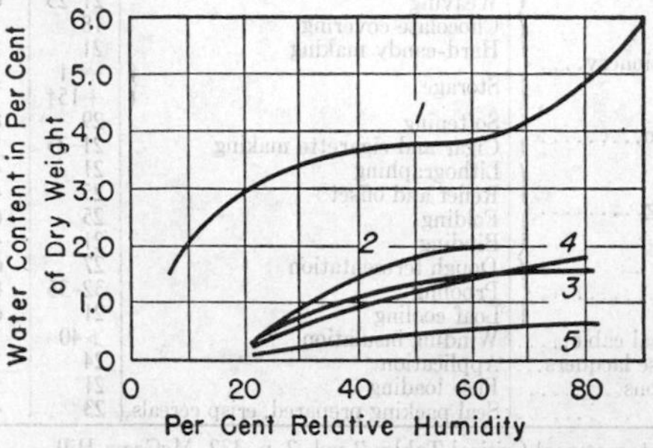

FIG. 22. Hygroscopic moisture of carbon products. 1. Carbon black, for rubber trade 2. By-product furnace coke (Franklin Co., Illinois coal) 3. By-product coke, domestic size (Pittsburgh bed coal) 4 By-product coke (domestic size). 5 Connellsville, 72-hr. beehive foundry coke. (*"International Critical Tables." vol.* 2, *p.* 325.)

Such humidifiers are used only in connection with ventilating systems. They are practically always placed on the inlet side of the ventilating fan. In the winter, provision is made for maintaining a constant dew point by means of a thermostat placed in the path of the saturated air leaving the humidifier and controlling the outside and return air dampers. In some cases, where high humidities are to be maintained, requiring dew points higher than obtainable by the mixture of outside and return air, the dew-point thermostat also controls the quantity of steam being used to heat the spray water, either indirectly by a water heater or directly by introducing the steam into the water. Where all outside air must be employed, it is usual to provide in front of the air washer certain preheater coils of sufficient capacity to bring the air to a temperature slightly above the freezing point. In summer operation, whenever the outside wet-bulb temperature is above the minimum dew point desired in the building, all outside air is taken and, it will be remembered from the psychrometric principles previously discussed, that the air issuing from the humidifier will have been cooled to the wet-bulb temperature of the entering air, provided of course that the water is neither heated nor cooled during recirculation, which is usually the case in such installations.

In order to calculate the quantity of air that must be supplied with this fan-type system, it is necessary to compute all the heat that will be dissipated in the space to be conditioned by such sources as the sun; transmission of heat through walls; heat given up by people; and other sources such as lights, motors, etc The cubic feet per minute then required may be obtained from

$$\text{Cu. ft /min.} = \frac{\text{B.t.u.} \times 56}{\text{temperature rise}}$$

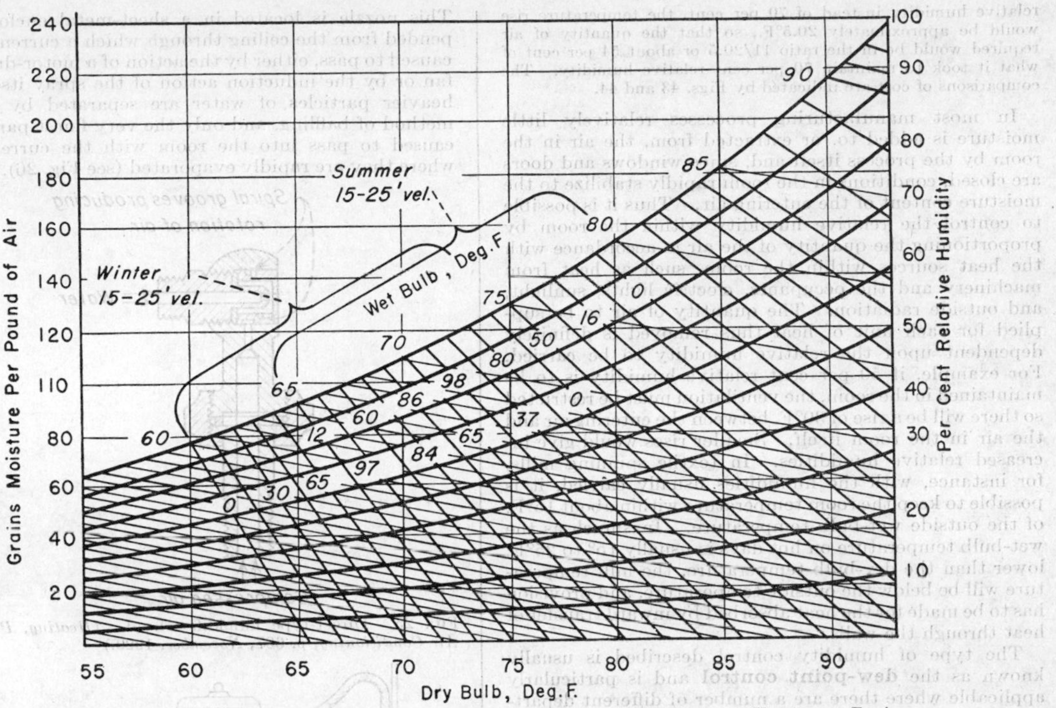

FIG. 23. The comfort chart. (*Research Laboratory, American Society of Heating and Ventilating Engineers.*)

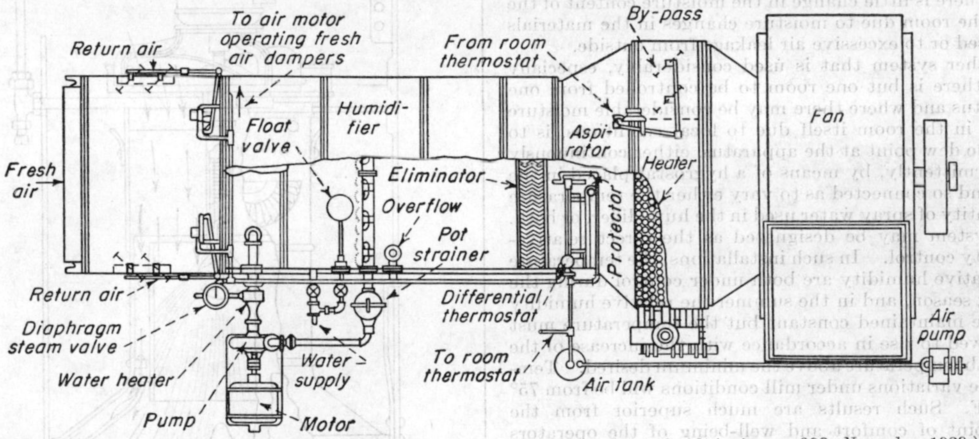

FIG. 24. Humidifier with humidity control—plan. (*Heating, Piping and Air Conditioning, p. 608, November, 1929.*)

The factor 56 is the approximate number of cubic feet of air raised 1°F. by 1 B.t.u. The temperature rise is from the dew point to the dry bulb, since the air after leaving the humidifier saturated at the dew-point temperature must rise to the temperature in the room, which will give the required relative humidity. It is when rising from the dew point to the dry-bulb temperature that the heat given up in the space being conditioned is absorbed. This temperature rise may be found from the psychrometric chart and is dependent upon the conditions to be maintained.

Example. Assume that there is an outside temperature of 85°F. dry bulb and 64°F. wet bulb. If this air is drawn into the humidifier it will saturate itself adiabatically at its wet-bulb temperature of 64°F. This saturation temperature of 64°F. is now the dew point of the air going into the fan and being distributed into the space being conditioned. This air must then rise from 64° to 75°F. in order to give 70 per cent relative humidity. In following this on the psychrometric chart, it will be observed that the air cools along the wet-bulb line to 64°F. and 100 per cent relative humidity, because the wet-bulb line is the line of adiabatic change. When the air is discharged into the room and begins to absorb heat, it heats up along the 64°F. dew-point horizontal line. This is a line of constant moisture content. The distance along the horizontal from the saturation line to 70 per cent relative humidity corresponds to about 11°F. for the ordinary range of temperature.

It can therefore be seen that assuming the heat to be absorbed as constant, which would be true for a particular problem, the cubic feet per minute required is dependent on the dew-point depression, which is the difference in temperature between the dry bulb and dew point. This dew point is dependent upon the relative humidity. If we were to maintain only 50 per cent relative humidity instead of 70 per cent, the temperature rise would be approximately 20.5°F., so that the quantity of air required would be in the ratio 11/20.5 or about 54 per cent of what it took to maintain 70 per cent relative humidity. The comparisons of cost are indicated by Figs. 43 and 44.

In most manufacturing processes relatively little moisture is added to, or extracted from, the air in the room by the process itself and, since windows and doors are closed, conditions in the room rapidly stabilize to the moisture content of the entering air. Thus it is possible to control the relative humidity within the room by proportioning the quantity of the air in accordance with the heat sources within the room, such as heat from machinery and the occupants, electric lights, sunlight, and outside radiation. The quantity of air to be supplied for each unit of heat thus removed is definitely dependent upon the relative humidity to be carried. For example, if 50 per cent relative humidity is to be maintained in the room, the ventilation must be restricted so there will be a rise of 30°F. between the entering air and the air in the room itself. Smaller rises would give increased relative humidities. In textile spinning mills, for instance, with the humidities usually carried, it is possible to keep the room temperature within about 15°F. of the outside wet-bulb temperature. Inasmuch as the wet-bulb temperature on hot days is usually 15° to 25°F. lower than the dry-bulb temperature, the mill temperature will be below the outside temperature, and provision has to be made for the heat absorbed by inward transfer of heat through the walls.

The type of humidity control described is usually known as the **dew-point control** and is particularly applicable where there are a number of different departments to be served from one central equipment, and where there is little change in the moisture content of the air in the room due to moisture changes in the materials processed or to excessive air leakage from outside.

Another system that is used considerably, especially where there is but one room to be controlled from one apparatus and where there may be considerable moisture change in the room itself due to local conditions, is to vary the dew point at the apparatus, either continuously or intermittently, by means of a hygrostat placed in the room and so connected as to vary either the temperature or quantity of spray water used in the humidifier, or both. This system may be designated as the direct relative-humidity control. In such installations, the temperature and relative humidity are both under control during the heating season, and in the summer the relative humidity must be maintained constant but the temperature must be allowed to rise in accordance with the increase of the wet-bulb temperature above the minimum desired. Temperature variations under mill conditions will be from 75° to 90°F. Such results are much superior from the standpoint of comfort and well-being of the operators than those previously experienced. This has resulted in greatly increasing the effectiveness of the workers during this period, but the amount of ventilation provided has also greatly improved conditions in mills as regards dust, lint, and odors.

Direct Humidifiers. Before the advent of modern air conditioning, the direct-type humidifier was largely used for increasing the humidity in cotton mills and similar places. Great improvement has been made in this type of humidifier, which still finds wide application, especially in the textile industry.

One of these types uses atomizer heads operated by compressed air in which the water is drawn into the nozzle and then finally atomized by a small jet of air under high pressure (see Fig. 25).

In another type, water is supplied at high pressure (usually about 200 lb./sq. in.) to an atomizing nozzle. This nozzle is located in a sheet-metal enclosure suspended from the ceiling through which a current of air is caused to pass, either by the action of a motor-driven disk fan or by the induction action of the spray itself. The heavier particles of water are separated by a simple method of baffling, and only the very finest particles are caused to pass into the room with the current of air where they are rapidly evaporated (see Fig. 26).

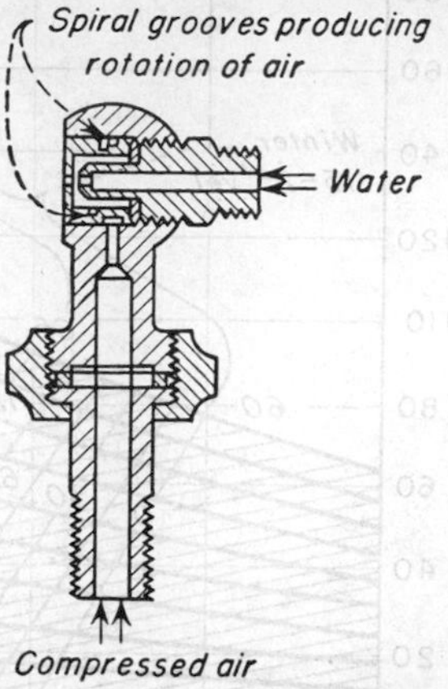

FIG. 25. Direct-type humidifier head. (*Heating, Piping and Air Conditioning, p.* 607, *November,* 1929.)

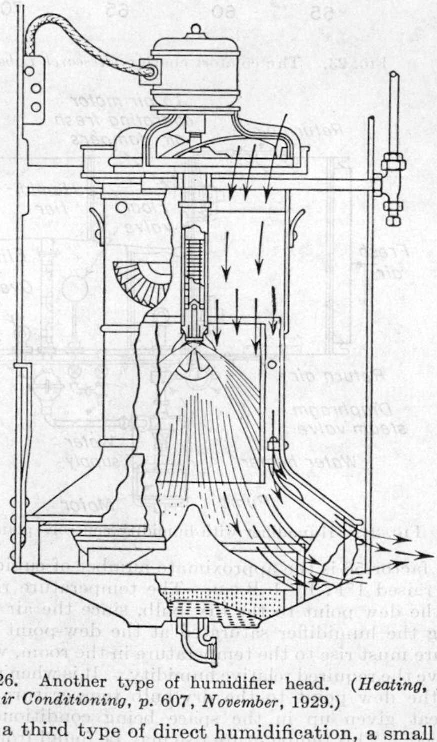

FIG. 26. Another type of humidifier head. (*Heating, Piping and Air Conditioning, p.* 607, *November,* 1929.)

In a third type of direct humidification, a small jet of water is thrown upon a disk rotating at high velocity within a stationary circular comb which consists of fine metal strips or teeth against which the water thrown from the disk impinges. The finely atomized particles that are thus produced are carried out with the current of air created by a disk fan operated from the motor shaft (see Fig. 27).

Comparison of Direct and Indirect Systems. The three types of direct humidifiers described are suitable only for application in certain types of industrial plants and do not provide either for controlled ventilation or for dehumidification, which is essential in many industries as well as for applications providing for human comfort.

Where large quantities of power are generated in a limited space, and where higher than 60 per cent relative humidity is required, it is feasible and economical to use a combination of direct and indirect humidification. The indirect humidification is for the purpose of securing the desired quantity of ventilation and cooling, and the additional direct humidification is placed in the rooms to increase the humidity and, at the same time, secure additional cooling effect. In general, it may be stated that direct humidification is most satisfactory where higher humidities, with but little cooling or ventilation, are required.

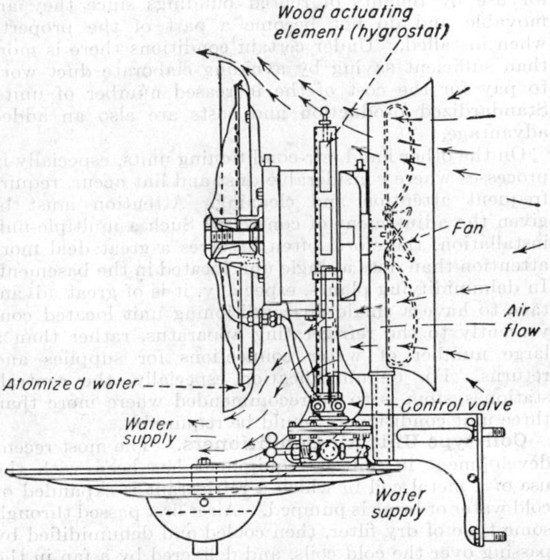

FIG. 27. In this type of humidifier a small jet of water is thrown on a rotating disk. (*Heating, Piping, and Air Conditioning*, p. 608, *November*, 1929.)

Therefore, the indirect system is best where lower relative humidities with maximum cooling and ventilating effects are desired, while for conditions where there are large quantities of heat to be absorbed by ventilation and high humidities are needed, the combination system of direct and indirect humidification is more desirable.

Inasmuch as the cooling and humidification requirements in summer are much more severe than those in winter, an excess of humidifying and cooling capacity must be provided to meet extreme conditions, and this requires, for the best results, careful automatic control. Such applications of humidity control may be considered standard practice today in the United States. The same type of equipment as used in cotton mills is largely applicable to other textile industries, including rayon, and also to tobacco factories, paper mills, and the like.

Systems of Dehumidification. In many industries, as already pointed out, it is as important to control the temperature in summer as in winter; at the same time, the relative humidity must be controlled both summer and winter, the air being warmed and the humidity increased in winter, and in summer the air must be cooled and dehumidified, so that both temperature and humidity may be held at a definite point regardless of outside weather conditions or conditions within the plant itself.

The design of air-distributing equipment and the external appearance of such equipment are the same as those used for systems of humidification. The main differences are to be found in the internal construction of the dehumidifier, in the use of refrigeration as well as heat for controlling the water temperature, and difference in the general methods of control.

Dehumidifiers are of three general types: (1) the spray type in which the water is cooled outside of the spray chamber and then introduced; (2) the type in which the refrigerating coils are placed directly in the spray chamber and the water is sprayed over this surface, air coming in contact with both the wetted coils and the spray; and (3) the dry-coil type in which air passes over a coil in which refrigeration is expanded or through which cold water or brine passes. No water is sprayed on the outside of the coil, although it does become wet because of moisture condensed from the air. The first and second types of dehumidifiers have air-distributing baffles at the inlet and eliminators at the outlet. The dehumidifier is usually considerably longer than the humidifier, and in the spray type there are two sets of sprays, one or both of which are opposed to the air flow.

The individual spray nozzles are considerably larger in capacity than the spray nozzles used in the humidifier and operated at a much lower pressure, usually from 20 to 25 lb./sq. in., the object being to get a coarser spray and, at the same time, a much larger quantity of spray per unit of air to be treated than in the humidifier. An exceedingly finely divided, or atomized, spray is not so effective in dehumidification, since heat capacity as well as surface in the spray water must be provided. The usual allowable rise in spray-water temperature entering and leaving the dehumidifier is from 6° to 10°F., and in well-designed dehumidifiers of the spray type the final temperature of the air is substantially identical with the final temperature of the water leaving. This is made possible by the counterflow effect employed in such apparatus. The velocities through the dehumidifier are usually lower than those employed in the humidifier, normal velocities being from 400 to 550 ft./min. on the cross-sectional area.

Greater care must also be taken with the elimination of free water than with the humidifier, since an entrainment beyond the eliminator defeats the purpose of the equipment.

Dehumidifier systems are usually provided with two essential control elements, a dew-point control at the apparatus and a room control. The former maintains a uniform temperature of saturated air leaving the dehumidifier at all times of the year. In summer, in industrial work, the dew-point control is constructed so as to use nearly all return air, thus reducing the refrigerating load required; while, when the wet-bulb temperature outside is at or below the dew point required within the room, the dew point is controlled by a mixture of outside and return air as required under control of the dew-point thermostat. By the second element, the temperature in the room is independently controlled at a definite point above the dew-point or saturation temperature, by either a thermostat or a hygrostat, depending upon whether the more important factor is the temperature or the relative humidity. Such a control either operates upon volume dampers controlling the quantity of air introduced, or upon heaters that supply the necessary heat for temperature regulation. Usually both the volume and the heating of the air are controlled.

In auditorium installations the requirements are quite different, since there must be a considerable volume of air circulated for ventilation and cooling purposes and for effective distribution. The moisture content must be relatively constant, but its temperature varies according to the requirements of the auditorium. Satisfactory solution of this problem has proved difficult, but special systems of control are now in general use which permit the temperature and humidity to be controlled independently in the auditoriums and, at the same time, to be immediately responsive to changes in occupancy; for the most part, this is accomplished without the use of any external heat.

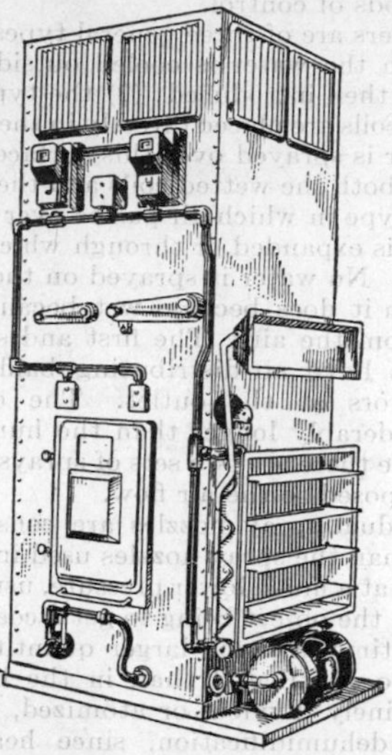

FIG. 28. Unit air conditioner (5000 cu. ft./min.)

Spray-type Unit Air Conditioners. The indirect humidifiers and the dehumidifiers just described are what is known as the central-station type; *i.e.*, they consist of separate pieces of equipment, such as humidifiers, pumps, fans, distributing ducts, heaters, and controls, all assembled at the point of erection. They usually consist of but few sets of apparatus, more often only one set of equipment for the entire installation.

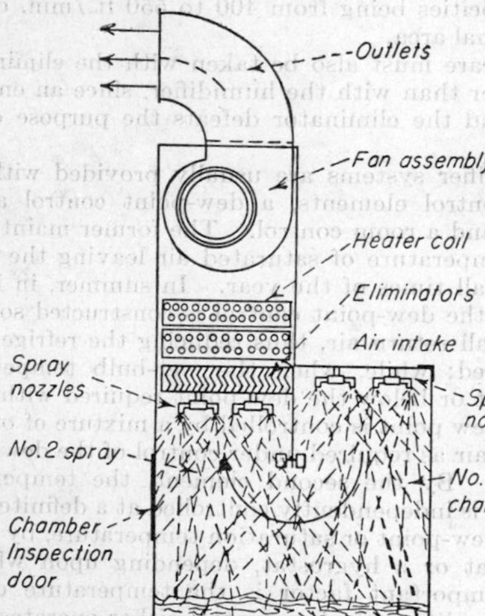

FIG. 29. Section of a unit conditioner. (*Heating, Piping and Air Conditioning*, *p.* 610, *November*, 1929.)

There is now being introduced quite extensively what is known as the unit type of air conditioner (Figs. 28, 29). With this equipment, the requirements of a given room are obtained by installing the necessary number of standard complete air-conditioning units, usually without any distributing duct work, the necessary air distribution being accomplished by the location of the units themselves. Each unit is a complete air-conditioning plant in itself including spray-air containers (humidifier and dehumidifier), air reheater, fan, spray pump, automatic control, etc. These units are usually built in graduated standard sizes having capacities from 1000 to 10,000 cu. ft./min. each.

There are several advantages of such equipment, as well as some disadvantages. The principal advantages are that the desirable results of the indirect humidification system can be applied to old buildings without installing duct work. Where manufacturing processes are changed and departures are subdivided or enlarged, the units are easily moved or increased in number to suit the new requirements. They are particularly adapted for use by tenants of rented buildings since they are movable and do not become a part of the property when installed. Under certain conditions there is more than sufficient saving by avoiding elaborate duct work to pay for the cost of the increased number of units. Standardized production and costs are also an added advantage.

On the other hand, air-conditioning units, especially in processes where considerable dust and lint occur, require frequent attention and cleaning. Attention must be given the adjustment of controls. Such a multiple-unit installation, therefore, often involves a great deal more attention than does a single unit located in the basement. In dehumidifying plants, especially, it is of great advantage to have a single air-conditioning unit located conveniently to the refrigerating apparatus, rather than a large number of water connections for supplies and returns. For dehumidification especially, the central-station system is to be recommended where more than three unit conditioners would be required.

Coil-type Unit Air Conditioners. The most recent development in unitary equipment has been with the use of a metal coil in which a refrigerant is expanded or cold water or brine is pumped. Air is first passed through some type of dry filter, then cooled and dehumidified by passing over the cold coils, and delivered by a fan in the unit either directly into the space conditioned or indirectly by means of duct work.

Winter heating and humidifying are also available by means of a heating coil and humidifier built into the unit. This type of unit may be floor-mounted or hung from the ceiling.

Figure 30 shows a typical unit for air-conditioning service, available in ratings from 6 to 60 tons of refrigeration and 2000 to 8000 cu. ft./min.

Where the capacity required is larger than available with the standard unit, cooling-coil assemblies are available of large capacity to be used in connection with a centrifugal fan and duct system. Either direct-expansion refrigerant or cold water may be used for cooling and dehumidification. For heating, a steam coil is used, and for humidification, a spray.

This type of equipment is shown in Fig. 31, which represents a surface dehumidifier 4000 to 32,400 cu. ft./min. air handling capacity, 6 to 100 tons refrigeration.

This type of unit consists of an encased bank of cooling coils, eliminator, and drip pans made as one unit. The coils may be sprayed and the characteristic performance of a spray-type dehumidifier obtained, securing compactness for duty. Units are furnished in a number of sizes capable of being stacked or grouped to yield a variety of capacity and dimensional combinations.

Another type of unitary equipment, called "self-contained units," in which the refrigeration machine is

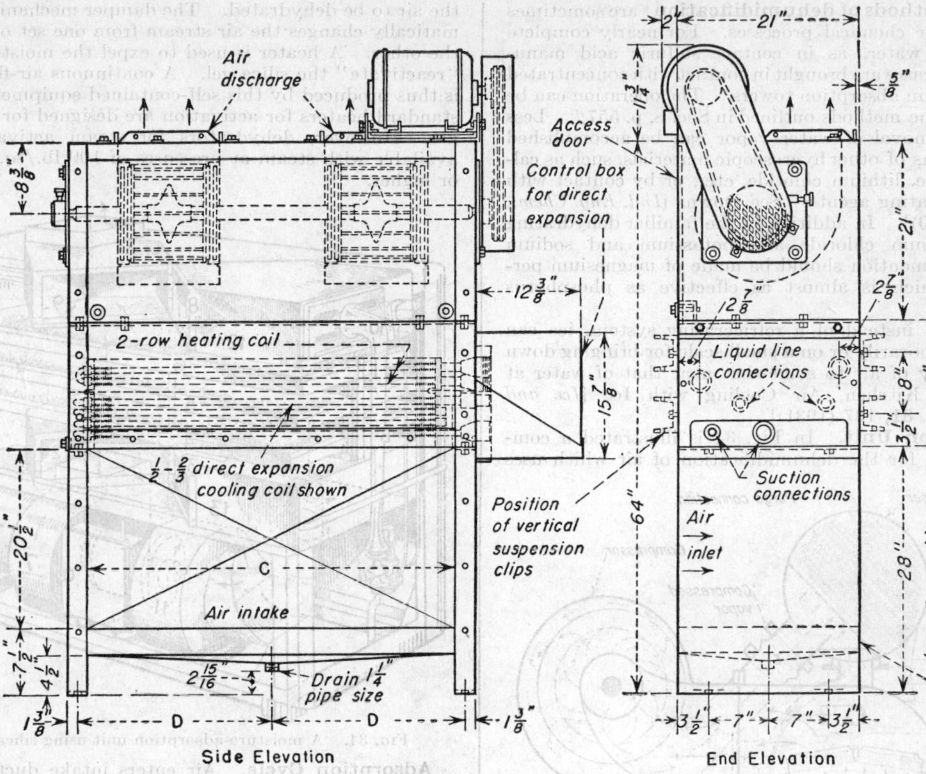

FIG. 30. Coil-type unit air conditioner.

built into the cabinet with the fans, coils, etc., provides a unit completely assembled in the factory which can be installed very simply, requiring only outside air connections and electrical connections to standard room socket.

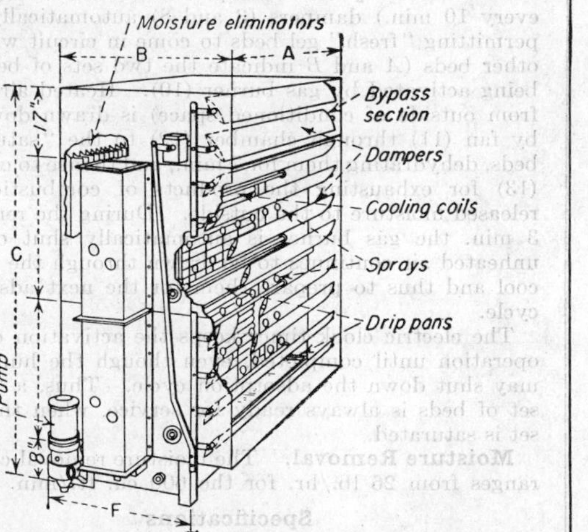

FIG. 31. Single-unit air conditioner—floor mounted.

Figure 32 shows such a unit with a $\frac{3}{4}$-hp., refrigeration compressor motor complete with fan, cooling coils, filters, etc. Heating is available with electric heaters. This unit is available with air-cooled and water-cooled refrigeration condensers. This unit is 32 in. wide, 16½ in. deep, and 31 in. high and is intended for installation in the space conditioned.

Refrigeration used with all air-conditioning systems whether central station or unitary is of many types. Reciprocating compressors are used with ammonia, carbon dioxide, methyl chloride, and "Freon-12" and are available in many sizes. Centrifugal refrigeration as

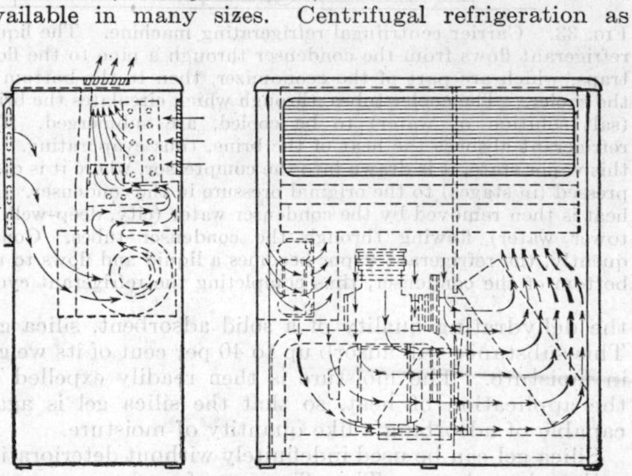

FIG. 32. A typical coil-type air conditioner using a small reciprocating refrigeration machine.

shown in Fig. 33 is widely used with trichlorofluoromethane ($CFCl_3$), "Freon-11." These machines are available in sizes from 100 to 1200 tons refrigeration capacity, at normal air-conditioning temperature operation. Steam refrigeration is also used in which a vacuum is maintained in the cooler by means of steam jets (see Sec. 25).

Other methods of dehumidification* are sometimes employed for chemical processes. For nearly complete removal of water, as in contact sulfuric acid manufacture, the gases are brought in contact with concentrated sulfuric acid in absorption towers. The operation can be treated by the methods outlined in Sec. 8, p. 557*ff*. Less complete removal of water vapor can be accomplished with solutions of other hygroscopic materials, such as calcium chloride, lithium chloride, etc., or by contact with solid dehydrating agents. See Downs [*Ind. Eng. Chem.*, **31**, 134 (1939)]. In addition to the familiar dehydrating agents, calcium chloride and potassium and sodium hydroxides, mention should be made of magnesium perchlorate, which is almost as effective as phosphorus pentoxide.

Moreover, instead of a refrigerating system, ice can be used temporarily or on a small scale for bringing down the humidity of air to somewhat near that of water at 32°F. See Kitchen, Air Cooling with Ice [*Ice and Refrigeration*, **81**, 137 (1931)].

Adsorption Unit. In Fig. 34 is illustrated a commercial unit for the dehumidification of air which uses

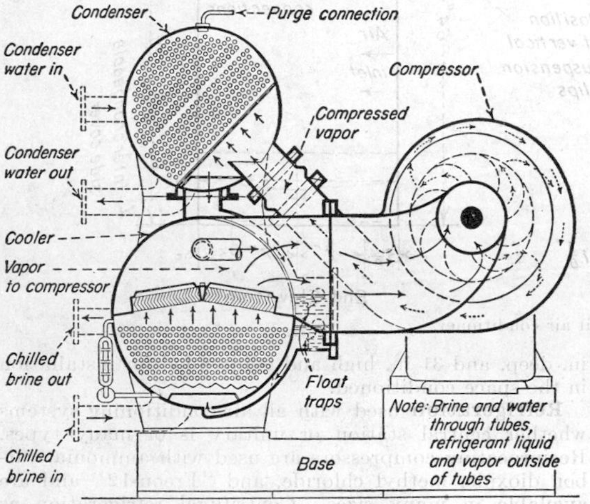

FIG. 33. Carrier centrifugal refrigerating machine. The liquid refrigerant flows from the condenser through a pipe to the float traps, which are part of the economizer, then to the bottom of the cooler. The cooler tubes, through which circulates the brine (salt solution or water) to be cooled, are submerged. The refrigerant absorbs the heat of the brine, thus evaporating. In this vapor state, it is drawn into the compressor, where it is compressed (in stages) to the original pressure in the condenser. Its heat is then removed by the condenser water (city, deep-well, or tower water) flowing through the condenser tubes. Consequently, the refrigerant vapor becomes a liquid and flows to the bottom of the condenser, thus completing the refrigerant cycle.

the dehydrating quality of a solid adsorbent, silica gel. This substance will adsorb up to 40 per cent of its weight in moisture. The moisture is then readily expelled by the application of heat, so that the silica gel is again capable of adsorbing a like quantity of moisture.

Silica gel can be used indefinitely without deterioration or loss in volume. The efficiency of moisture removal increases as the inlet air becomes drier. There is virtually no limit to the dryness of air which may be secured, and thus, by selecting the proper size of unit, moisture levels from practically anhydrous air to any higher level may be secured.

The equipment used to dehydrate air by this principle is fully automatic. It consists of two series of silica gel trays or beds, together with motor-driven fans to propel the air to be dehydrated. The damper mechanism automatically changes the air stream from one set of beds to the other. A heater is used to expel the moisture or to "reactivate" the silica gel. A continuous air-flow cycle is thus produced by this self-contained equipment. The standard heaters for activation are designed for use with gas. However, dehydrators for steam activation are available with steam at pressures of 100 lb./sq. in. gage or higher.

* See Sec. 13 for the drying of gases.

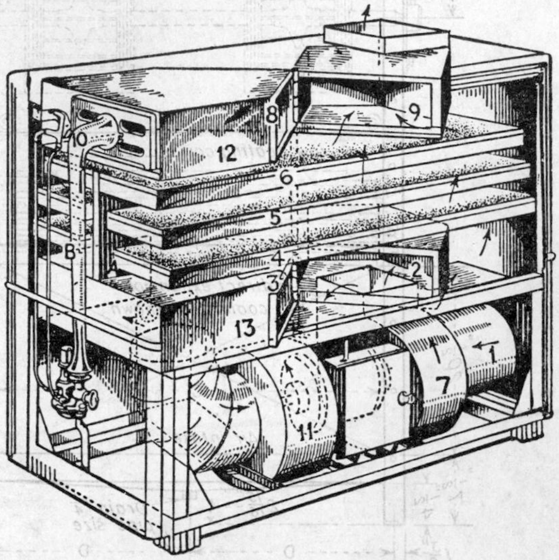

FIG. 34. A moisture-adsorption unit using silica gel.

Adsorption Cycle. Air enters intake duct (1) and passes into the adsorption inlet chamber (2). The position of damper (3) guides it *up* through the previously activated gel beds (4, 5, and 6) by fan (7). Here it gives up a large percentage of its moisture. Now guided by damper (8) it must pass out supply duct (9) to be cooled as required.

Activation Cycle. At a predetermined time (usually every 10 min.) dampers (3 and 8) automatically shift, permitting "fresh" gel beds to come in circuit while the other beds (*A* and *B* indicate the two sets of beds) are being activated by gas burner (10). Heated air (taken from outside of conditioned space) is drawn downward by fan (11) through chamber (12) to the "saturated" beds, dehydrating them for 7 min., and thence to chamber (13) for exhausting the products of combustion and released moisture to the outside. During the remaining 3 min. the gas burner is automatically shut off, and unheated air continues to be drawn through the beds to cool and thus to prepare them for the next adsorption cycle.

The electric clock timer keeps the activation cycle in operation until completed, even though the humidistat may shut down the adsorption cycle. Thus, a "fresh" set of beds is always ready for service, when the other set is saturated.

Moisture Removal. The moisture removal capacity ranges from 26 lb./hr. for the 600 cu. ft./min. unit to

Specifications

Air capacity, cu. ft./min.	Gas consumption, B.t.u./hr.	No. of Motors and hp.	Dimensions, in.		
			Length	Width	Height
500	65,000	Two, ¼	41	43	29
800	95,000	One, ¾	53	47	41
1300	145,000	One, 1	82	51	60
2700	262,000	One, 3	94	51	86
5000	525,000	One, 5	120	67	118

218 lb./hr. for the 5000 cu. ft./min. unit, based on an entering air condition of 9 gr./cu. ft. The units may be used in multiple to obtain a wide range of dehydrating capacities.

HUMIDITY-CONTROLLING DEVICES*

Thermostats. Devices used in temperature control in air-conditioning equipment are largely common forms that have long been in use and are well known. Three general forms of these thermostats are in use:

First, the most common, is the bimetal thermostat which operates on the principle of the differential thermal

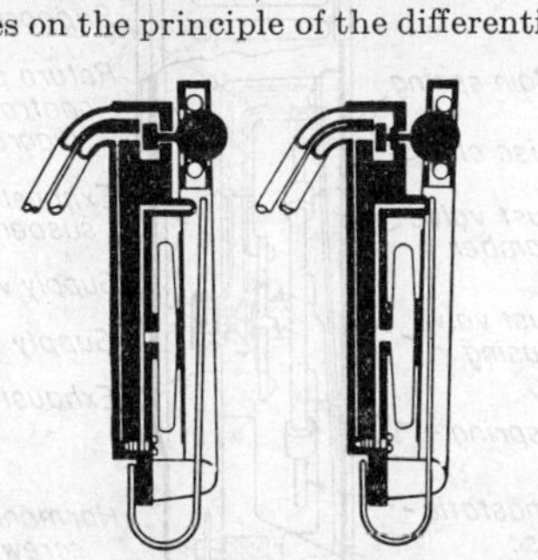

FIG. 35. A bimetal thermostat. (*Heating, Piping and Air Conditioning*, p. 611, *November*, 1929.)

expansion of two metals having different coefficients of expansion, these metals being brazed together in a curved strip. The result of a change of temperature is to cause a change in the curvature of this strip, thus actuating a contact arm, or lever, a pneumatic valve or an electric contact which provides the motive power for the actual operation of the controlling valves or dampers (Fig. 35).

The valve is carried directly upon the inner non-expansive element. The outer expansion element and the inner element carry the pneumatic valve which operates directly upon an adjustable seat connected to the expan-

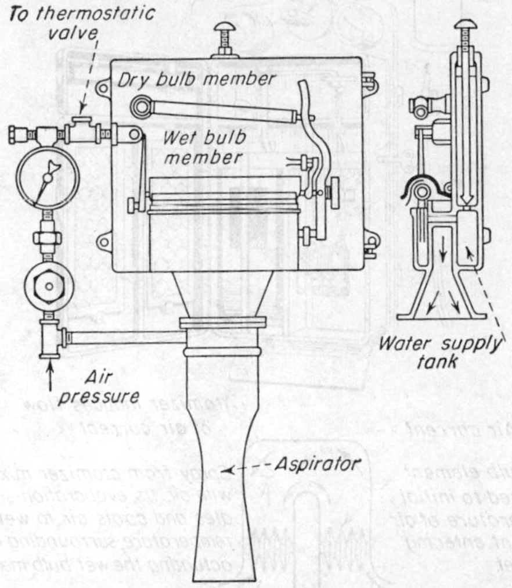

FIG. 37. Differential hygrostat. (*Heating, Piping and Air Conditioning*, p. 613, *November*, 1929.)

sive element. This is also provided with a constant thermal leak. The action of this air-supply valve in connection with the constant leak is such as to increase or decrease the pressure upon the operating diaphragms of the valve, or damper air motors. This type of thermo-

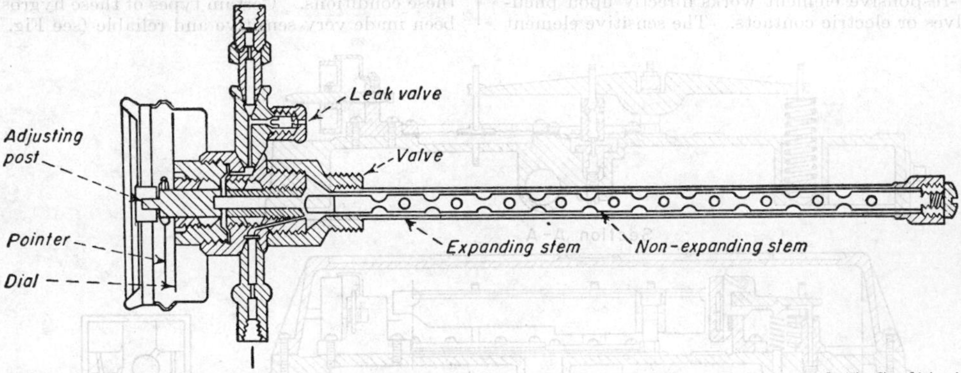

FIG. 36. The sensitive element of this type of thermostat is inserted in the duct. (*Heating, Piping and Air Conditioning*, p. 612, *November*, 1929.)

The **second** type employs a metal diaphragm or bellows filled with a volatile liquid, the vapor pressure of which changes with change of temperature and actuates directly a pneumatic valve through which it controls the governing valves or dampers in the air-conditioning apparatus.

The **third** type of thermostat is the duct thermostat in which the sensitive element is inserted in the duct, while the setting dial and air-pressure connections are on the outside. This type of thermostat consists of an outside tubular thermoresponsive casing of brass or aluminum, or other highly thermoresponsive metal, and an inner rod or tube of special nickel steel or other metal, which is relatively unaffected by temperature changes.

* For additional information on the measurement and control of humidity see Sec. 19, pp. 1298–1300.

stat is very rugged and responds to slight temperature changes with proportionate graduated effect. The construction of this type of thermostat is shown in Fig. 36.

Hygrostats. These are of two general types: The first type is, in effect, a form of differential thermostat in which one expansive element is affected by the wet-bulb temperature, which temperature is suitably maintained either by a wick or by a spray of water. These thermostatic elements are so interconnected that they will not be affected by changes of temperature alone, but only by changes in the relation between the wet- and dry-bulb temperatures which would result in changing the relative humidity (Fig. 37). Such a correlation is made possible by the fact that there is practically a straight-line relationship between the relative temperature changes of wet and dry bulb for any definite relative humidity. As in the thermostat, these elements may be of either the solid-expansion or vapor-pressure type. A common form of this apparatus is shown in Fig. 38.

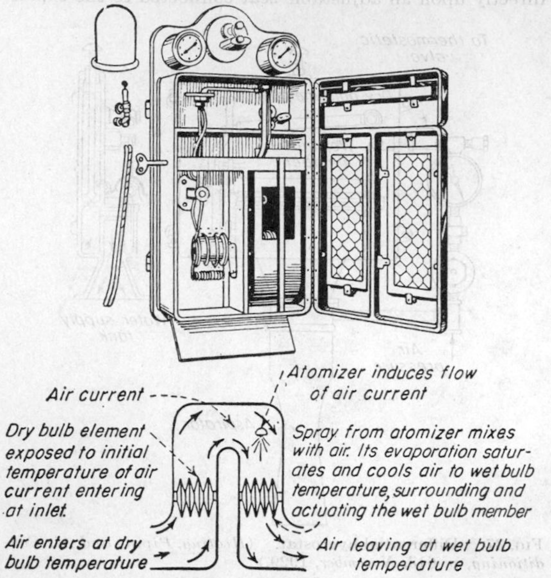

FIG. 38. A common form of hygrostat. (*Heating, Piping and Air Conditioning*, p. 614, *November*, 1929.)

The second form of hygrostat is that depending upon the dimensional changes of hygroscopic materials due to changes in moisture content. In these instruments the humidity-responsive element works directly upon pneumatic valves or electric contacts. The sensitive element responds to relative humidity and affects the actuating mechanism precisely as the thermostat responds to temperature. It is found that suitable hygroscopic materials show no material change in extension with

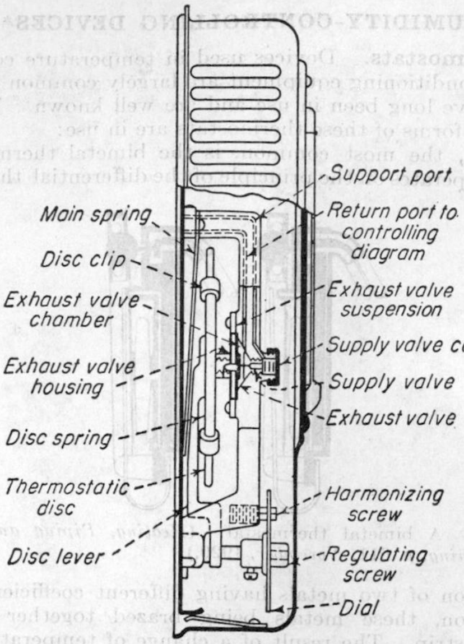

FIG. 40. Application of thermostats. (*Heating, Piping and Air Conditioning*, p. 615, *November*, 1929.)

changes of temperature as long as the relative humidity is constant, although the actual moisture of the hygroscopic material itself undergoes slight changes under these conditions. Certain types of these hygrostats have been made very sensitive and reliable (see Fig. 39).

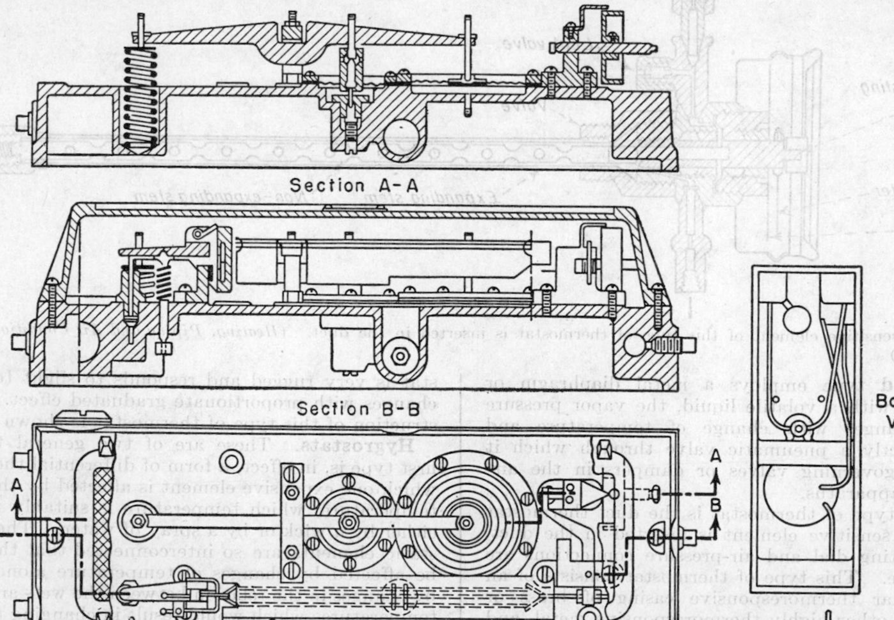

FIG. 39. Four views of a hygrostat. (*Heating, Piping and Air Conditioning*, p. 615, *November*, 1929.)

Control. Temperature- and humidity-controlling devices usually act upon the large control valves or dampers in the air-conditioning apparatus through the medium of compressed air or electricity. In either case, the actuation is accomplished by suitable motors, either pneumatic or electric. For all the larger installations the pneumatic control is practically universally used because of simplicity, reliability, and also because of the fact that it permits graduation of dampers or valves with temperature or humidity changes. In smaller

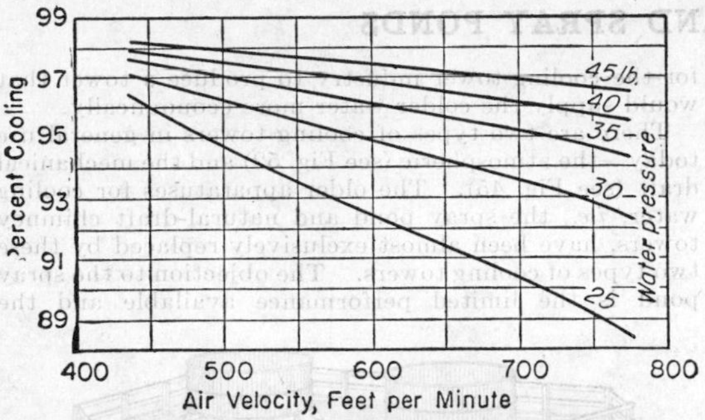

FIG. 41. Percentage of cooling of the wet-bulb depression accomplished by standard commercial-type humidifier for various water pressures on the nozzles at varying air velocities through the humidifier.

humidifying devices, electric controls are very serviceable because of simplicity of the circuits and the fact that there are no auxiliaries, such as air compressors, air tanks, air filters, and relief valves required. In the larger systems, such accessories are warranted by the increased size of equipment, and the increased number of controls operated from the same compressed-air source. In the unit types of equipment, direct-acting thermostats and hygrostats are sometimes used in which the respon-

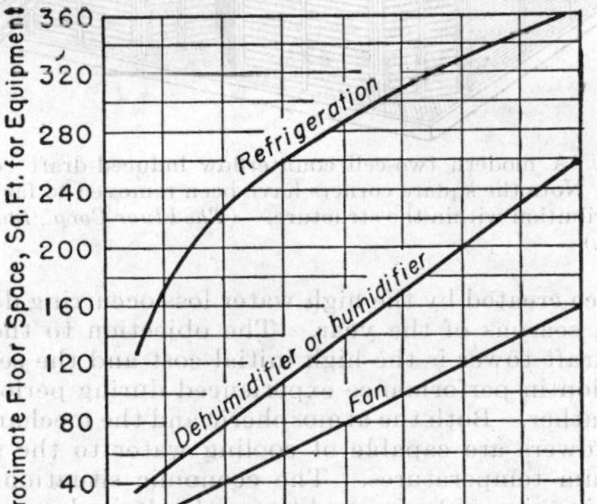

FIG. 42. Floor space per 1000 cu. ft./min.

sive element works directly upon the valves controlled without the intermediary use of either compressed air or electric current. Such applications are illustrated for thermostats in Fig. 40, and are shown for hygrostats in Fig. 27.

Data on Performance and Cost of Air-conditioning Equipment. Figure 41 shows the percentage of cooling of the wet-bulb depression accomplished by the standard commercial-type humidifier for various water pressures on the nozzles at varying air velocities through

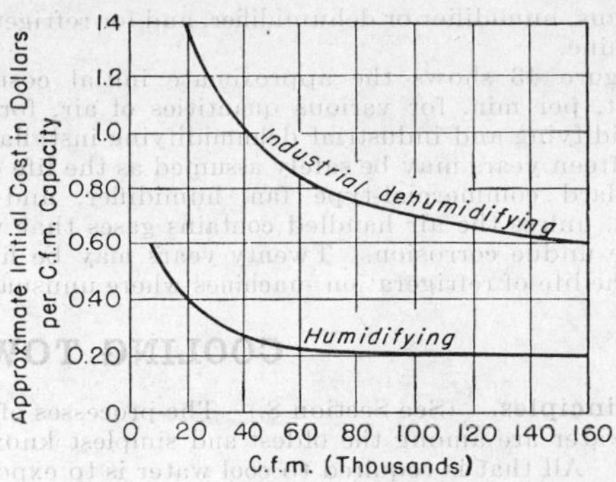

FIG. 43. Initial cost per 1000 cu. ft./min.

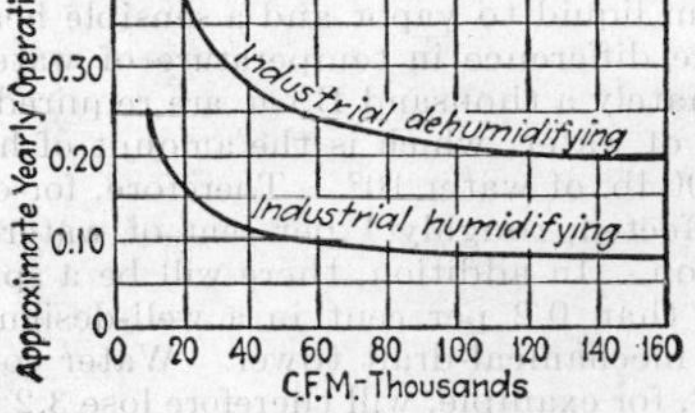

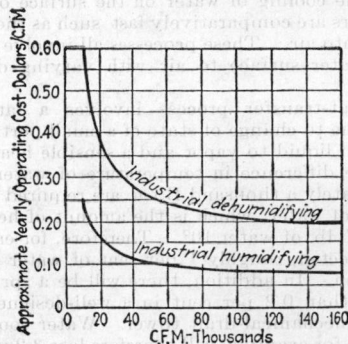

FIG. 44. Total operating cost.

the humidifier. These standard humidifiers are approximately 10 ft. long with a velocity of air through them of 700 ft./min. Two banks of sprays are used facing each other and from 1.5 to 3 spray nozzles per sq. ft. per bank of nozzles.

Figure 42 indicates the approximate floor area required

Summer and Winter Loads for Ventilating and Air Conditioning

Type of building	Initial cost per sq. ft.	Annual over-all cost per sq. ft.	Unit costs (based on annual over-all costs)	Annual power costs (rate in parenthesis)
Department store.... 160,000 sq. ft. 6,250 people	\$1.50	\$0.27	27¢/yr./sq. ft. of sales area	\$13,000 (3¢/kw.-hr.)
Office building....... 325,000 sq. ft. 2,540 people	1.25	0.19	1¢/man-hr.	13,896 (1.2¢/kw.-hr.)
Plant office building.. 42,560 sq. ft. 300 people	2.00	0.31	3¢/man-hr.	2,694 (3¢/kw.-hr.)
Restaurant and cafeteria.......... 12,644 sq. ft. 760 people	3.60*	1.16*	1¢/meal	4,748* (3¢/kw.-hr.)
Hotel (guest rooms).. 120,000 sq. ft. 600 people	1.50	0.50†	20¢/room per day	13,600 (3¢/kw.-hr.)

NOTE: Both initial and annual costs are based on heating, ventilating, and air conditioning. In most cases, heating and ventilating are normal expenses that must be paid anyway. The actual added cost of adding air conditioning, which also provides for heating and ventilating, is on the average one-third of the figures used above.

* Higher initial and operating costs are due to a larger required capacity to provide for unusual heat load, more frequent air change, and larger number of people.

† Higher operating cost due to longer period of daily operation.

for fans, humidifier or dehumidifier, and for refrigeration machine.

Figure 43 shows the approximate initial costs per cu. ft. per min. for various quantities of air, for both humidifying and industrial dehumidifying installations.

Fifteen years may be safely assumed as the life of the standard commercial-type fan, humidifier, and duct work, unless the air handled contains gases that would cause undue corrosion. Twenty years may be figured for the life of refrigeration machines where unusual care and maintenance are used. Figure 44 gives the approximate yearly depreciation and interest, maintenance, cost of steam, labor, water and power for humidifying and industrial dehumidifying installations. The steam figured was only that required to heat from the dew point to the dry bulb and does not take care of the heat losses from the building. This yearly operating cost may be approximately divided into 25 per cent depreciation and interest, 5 per cent for maintenance, 70 per cent steam, labor, water, power, etc.

COOLING TOWERS AND SPRAY PONDS

Principles. (See Section 8.) The processes of cooling water are among the oldest and simplest known to man. All that is required to cool water is to expose its surface to air. Some of these cooling processes are slow, such as the cooling of water on the surface of a pond; while others are comparatively fast such as the spraying of water into air. These processes all involve the exposure of water surface to air with varying degrees of efficiency.

The heat-transfer process involves a latent heat transfer due to change of state of a small portion of the water from liquid to vapor and a sensible heat transfer due to the difference in temperature of water and air. Approximately a thousand B.t.u. are required to evaporate 1 lb. of water, which is the amount of heat lost in cooling 100 lb. of water 10°. Therefore, for each 10° of cooling effected, roughly 1 per cent of water is lost by evaporation. In addition, there will be a spray loss of not more than 0.2 per cent in a well-designed atmospheric or mechanical-draft tower. Water cooling from 120 to 90°, for example, will therefore lose 3.2 per cent of its weight (3 plus 0.2 per cent) with each passage through the tower.

In cooling towers in which the water is warmer than the air, the heat removed from the water and transferred to the air is the sum of the sensible heat and the latent heat of evaporation. The sensible heat qs is small in comparison with the latent heat transferred wr.

$$qT = qs + wr$$
$$w = KA\,\Delta p_m$$

The factors that influence the performance of a cooling tower are therefore those which affect the expression $KA\,\Delta p_m$. The value KA depends on the construction of the equipment, the extent of water surface exposed, and the velocity of the air. Δp_m depends only on the temperature of the water and the temperature and humidity of the air. Actually, Δp_m represents very nearly the difference in vapor pressure between the water, at its temperature, and the water if it were at the wet-bulb temperature of the air.

It is evident that the water cannot be cooled below the wet-bulb temperature of the entering air. The wet-bulb temperature, or, to be more precise, the adiabatic saturation temperature, represents the minimum temperature that the water would reach with infinite time of contact between water and air in a cooling tower. This must be kept in mind when a plant is designed to operate on cooling-tower water. Design wet-bulb temperatures for the warmest months of the year, in different parts of the world, are given in a table on weather conditions, p. 797.

General Considerations. As process refinements have appeared, industry has demanded closer and closer control of its production units, resulting in more exacting demands being made upon cooling towers. In many plants, each additional degree of cooling has meant hundreds and even thousands of dollars per day in increased production. It has therefore been necessary for the cooling-tower industry to produce a tower that would supply the colder water more economically.

There are two types of cooling towers in general use today—the atmospheric (see Fig. 52) and the mechanical draft (see Fig. 45). The older apparatuses for cooling water, *i.e.*, the spray pond and natural-draft chimney towers, have been almost exclusively replaced by these two types of cooling towers. The objection to the spray pond is the limited performance available and the nuisance created by the high water loss occurring during certain seasons of the year. The objection to the natural-draft tower is the high initial cost and the serious reduction in performance experienced during periods of hot weather. Both the atmospheric and the mechanical-draft towers are capable of cooling water to the same minimum temperatures. The economic situation, the prevailing atmospheric conditions, the desired approach to the wet-bulb temperature, and the amount of space available, will indicate which type to select.

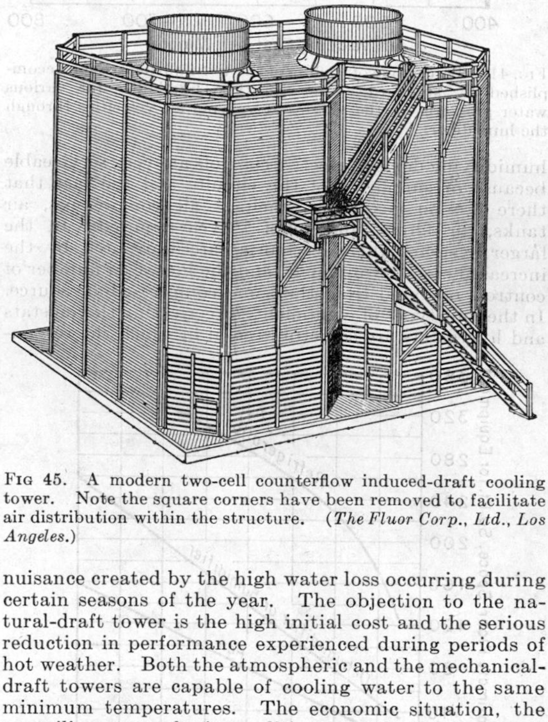

FIG 45. A modern two-cell counterflow induced-draft cooling tower. Note the square corners have been removed to facilitate air distribution within the structure. (*The Fluor Corp., Ltd., Los Angeles.*)

Mechanical-draft Towers. Two types of mechanical-draft towers are in use today—the forced draft and the induced draft. In the forced-draft tower, the fan is mounted at its base and the air is forced in the bottom and discharged through the top at low velocity. In the induced-draft tower, the fan is mounted on the roof of the structure and air is pulled upward and discharged at a high velocity.

The forced-draft type is fast losing because it is more often subjected to the recirculation of the hot humid exhaust vapors back into the air intakes than is the induced-draft type. This occurs under certain atmospheric conditions because the velocity of the humid exhaust air is so low that the suction created by the fans tends to draw it back into the tower. Since the wet-bulb temperature of the exhaust air is considerably above the wet-bulb temperature of the ambient air, there is a decrease in performance evidenced by an increase in cold-water temperature.

Except for the location of the fans, the structural and operational features of the two types of mechanical-draft towers are essentially the same. A cross-sectional view of the induced-draft tower with the various parts labeled is shown in Fig. 46. The entrained moisture is removed

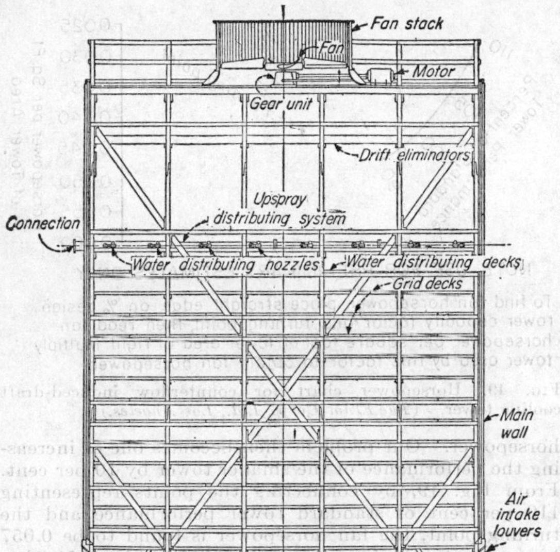

FIG. 46. Cross-sectional view of a counterflow induced-draft cooling tower. With the upspray distributing system, it is necessary to allow sufficient space between the drift eliminators and the nozzles for a spray chamber. (*The Fluor Corp., Ltd., Los Angeles.*)

from the exhaust air by the drift eliminator, which is placed just above the spray chamber and below the fan. The water is pumped to the main header located in the top of the tower where it is distributed to the various nozzles. This water is sprayed up in a manner similar to that used in a spray pond and is intimately mixed with the exhaust air before dropping to the decks below. (In performance, the upspray distributing system represents the equivalent of adding 8 or 9 ft. to the height of the cooling tower over that of the gravity-type system.) The fall of the water is interrupted by the slat-type grids as it flows countercurrently to the air. In flowing countercurrently, the coldest water contacts the dryest air and the warmest water contacts the most humid air. Maximum performance is thus obtained, since the temperature of all the cold water is approaching the wet-bulb temperature of the entering dry air. This was not true of the older cross-flow and parallel-flow types of cooling towers.

The performance of a given type of cooling tower is governed by the ratio of the weights of air to water and the time of contact between water and air. In commercial practice, the variation in the ratio of air to water is first obtained by keeping the air velocity constant at about 350 ft./min./sq. ft. of active tower area and varying the water concentration (gal./min./sq. ft. of tower area). As a secondary operation, the air velocity is varied to make the tower accommodate the cooling requirement. The time of contact between water and air is governed largely by the time required for the water to discharge from the nozzles and fall through the tower to the basin. The time of contact is therefore obtained in a given type of unit by varying the height of the tower. Should the time of contact be insufficient, no amount of increase in the ratio of air to water will produce the desired cooling. It is therefore necessary that a certain

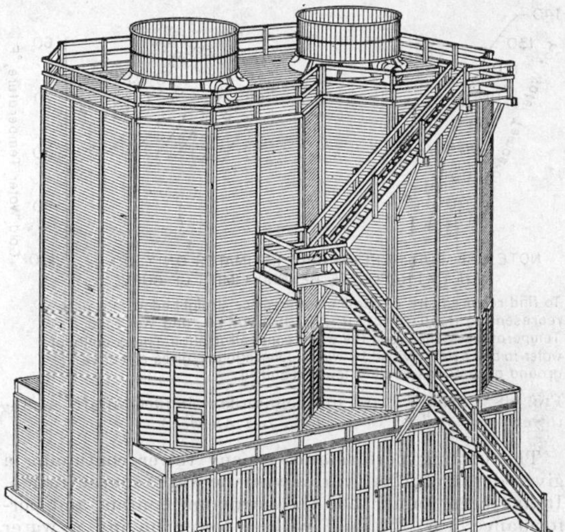

FIG. 47. Counterflow induced-draft cooling tower placed over a coil shed. The tower is used to cool water which, in turn, is used to cool other water, gases, or liquids circulated through the coils installed in the lower part of the tower. (*The Fluor Corp., Ltd., Los Angeles.*)

minimum height of cooling tower be maintained. Where a wide approach* of 15° to 20°F. to the wet-bulb temperature and a 25° to 35°F. cooling range† is required, a relatively low cooling tower will suffice. A tower in which the water travels 15 to 20 ft. from the distributing system to the basin is sufficient. Where a moderate approach of 8° to 15°F. and a cooling range of 25° to 35°F. is required, a tower in which the water travels 25 to 30 ft. is adequate. Where a close approach of 4° to 8°F. with a 25° to 35°F. cooling range is required, a tower in which the water travels from 35 to 40 ft. is required. It is usually not economical to design a cooling tower with an approach of less than 4°F., but it can be accomplished satisfactorily with a tower in which the water travels 35 to 40 ft.

Figure 48 shows the relationship of the hot-water, cold-water, and wet-bulb temperatures to the water concentration.‡ From this, the minimum area required for a given performance of a well-designed counterflow

* The approach is the difference between the cold-water temperature and the wet-bulb temperature of the inlet air.

† The cooling range is the difference between the hot-water temperature and the cold-water temperature.

‡ See also London, Mason, and Boelter, *Trans. Am. Soc. Mech. Engrs.*, **62**, 41 (1940). Lichtenstein, *Trans. Am. Soc. Mech. Engrs.*, **65**, 779 (1943). Simpson and Sherwood, *J. Am. Soc. Refrigerating Engrs.*, **52**, 535, 574 (December, 1946). Simons, *Chem. & Met. Eng.*, **49** (5), 138 (1942); **49** (6), 83 (1942); **46**, 208 (1939). Hutchinson and Spivey, *Trans. Inst. Chem. Engrs. (London)*, **20**, 14 (1942).

induced-draft cooling tower can be obtained. The air-water ratio can be computed from a heat balance. Figure 49 gives the horsepower per square foot of tower area required for a given performance. These curves do not apply to parallel or cross-flow cooling, since these processes are not so efficient as the counterflow process. Also they do not apply where the approach to the cold-water temperature is less than 5°. These charts should be considered approximate and for preliminary estimates only. Many factors not shown in the graphs must be included in the computation, and hence the manufacturer should be consulted for final design recommendations.

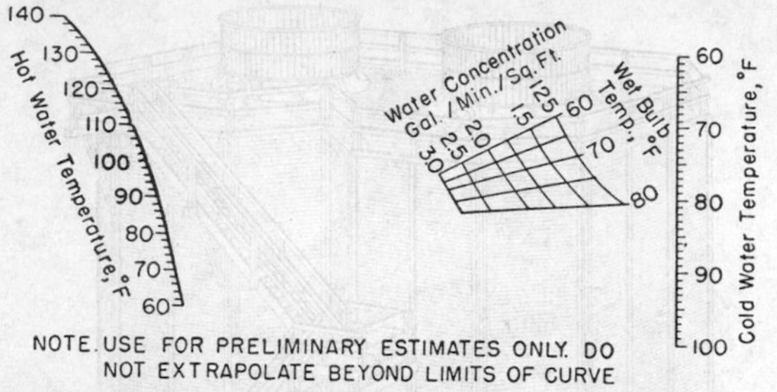

FIG. 48. Sizing chart for counterflow induced-draft cooling tower. (*The Fluor Corp., Ltd., Los Angeles.*)

The cooling performance of any tower containing a given depth of filling varies with the water concentration. It has been found that the maximum contact and performance are obtained with a tower having a water concentration of 2 to 3 gal. water per min. per sq. ft. of ground area. Thus the problem of calculating the size of a cooling tower becomes one of determining the proper concentration of water required to obtain the desired results. A higher tower will be required if the water concentration falls below 1.6 gal./sq. ft. Should the water concentration exceed 3 gal./sq. ft., a lower cooling tower may be used. Once the necessary water concentration is obtained, the tower area can be calculated by dividing the gal./min. circulated by the water concentration in gal./(min.)/(sq. ft.). The required tower size then is a function of the following:

1. Cooling range (hot-water temperature minus cold-water temperature).
2. Approach to wet-bulb temperature (cold-water temperature minus wet-bulb temperature).
3. Quantity of water to be cooled.
4. Wet-bulb temperature.
5. Air velocity through the cell.
6. Tower height.

To illustrate the use of the charts, let us assume that we have the following cooling conditions:

Hot-water temperature = 102°F.
Cold-water temperature = 78°F.
Wet-bulb temperature (T_{wb}) = 70°F.
Gal./min. = 2000

Laying a straightedge across Fig. 48 and connecting the points representing the design water and wet-bulb temperatures, we find that a water concentration of 2 gal./sq. ft. is required. Dividing the quantity of water circulated by the water concentration, we find that the theoretical area of the tower is 1000 sq. ft.

To obtain the theoretical fan horsepower, we use Fig. 49. Connecting the points representing the 100 per cent of standard tower performance with the turning point, we find that it will require 0.041 hp./sq. ft. of actual effective tower area. Multiplying by the tower area of 1000 sq. ft., we find that it requires 41.0 fan hp. to perform the necessary cooling.

Suppose that the commercial tower size is such that the actual tower area is 910 sq. ft. We can still obtain the cooling equivalent to 1000 sq. ft. of standard tower area by increasing the air velocity through the tower. Within reasonable limits, the shortage of actual area can be compensated for by an increase in air velocity through the tower, which, in turn, requires a higher fan

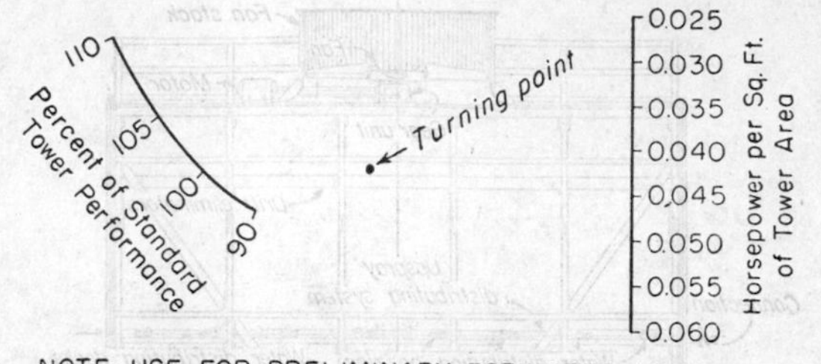

FIG. 49. Horsepower chart for counterflow induced-draft cooling tower. (*The Fluor Corp., Ltd., Los Angeles.*)

horsepower. Our problem then becomes one of increasing the performance of the smaller tower by 10 per cent. From Fig. 49, by connecting the points representing 110 per cent of standard tower performance and the turning point, the fan horsepower is found to be 0.057 hp./sq. ft. of actual tower area, or $0.057 \times 910 = 51.9$ hp.

On the other hand, suppose the commercial tower size is such that the actual tower is 1110 sq. ft., the cooling equivalent to 1000 sq. ft. of standard tower area can be accomplished with less air and less fan horsepower. By the use of Fig. 49, the theoretical fan horsepower for a tower doing only 90 per cent of standard performance is found to be 0.031 hp./sq. ft. of actual tower area, or 34.5 hp.

This illustrates how sensitive the fan horsepower is to small changes in tower area. The importance of designing a tower that is slightly oversize in ground area and in providing plenty of blower capacity become immediately apparent.

Assume that we have the same cooling range and approach as used in the first example, except that the wet-bulb temperature is lower. The design conditions would then be

Gal./min. = 2000
Range = 24°F.
Approach = 8°F.
T_1 = 92°F.
T_2 = 68°F.
Wet-bulb temperature (T_{wb}) = 60°F.

From Fig. 48, we find the water concentration required to perform the cooling is 1.75 gal./(min.)(sq. ft.), giving a theoretical tower area of 1145 sq. ft. as compared with 1000 sq. ft. for a 70° wet-bulb temperature. This shows that, the lower the wet-bulb temperature for the same cooling range and approach, the larger the area of the tower required and therefore the more difficult the cooling job.

The problem of estimating the performance of an existing tower at other than design conditions is often encountered by the plant operator. For example, suppose we have a tower that was designed for the following conditions:

Gal./min. =	1000
Range =	30°F.
Approach =	10°F.
T_1 =	110°F.
T_2 =	80°F.
Wet-bulb temperature =	70°F.

What will the cold-water temperature T_2 be when the wet-bulb temperature T_{wb} drops to 60°, provided of course that the heat load and water quantity remain constant? From Fig. 48, we find that the water concentration is 2.0 gal./sq. ft. at design conditions. This water concentration does not change, since the volume of water and the tower area remain constant. With the water concentration at 2.0 and the wet-bulb temperature at 60°F. by adjusting the angle of the straightedge on Fig. 48 until

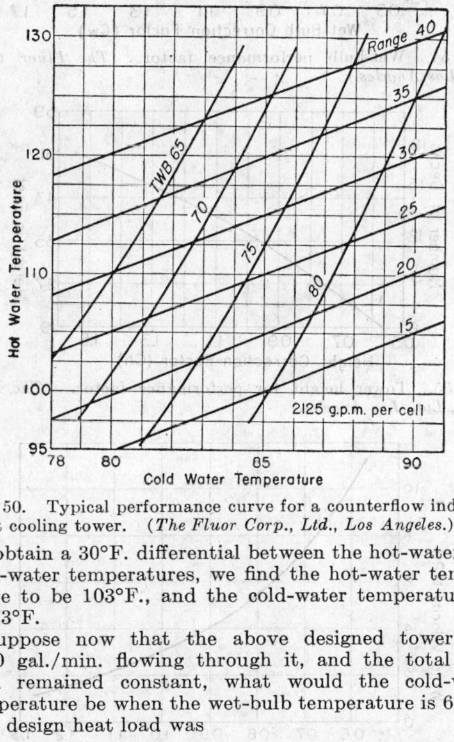

FIG. 50. Typical performance curve for a counterflow induced-draft cooling tower. (*The Fluor Corp., Ltd., Los Angeles.*)

we obtain a 30°F. differential between the hot-water and cold-water temperatures, we find the hot-water temperature to be 103°F., and the cold-water temperature to be 73°F.

Suppose now that the above designed tower had 1500 gal./min. flowing through it, and the total heat load remained constant, what would the cold-water temperature be when the wet-bulb temperature is 65°F.? The design heat load was

$$1000 \times 8.33 \times 30 = 250,000 \text{ B.t.u./min.}$$

The new cooling range (heat load remaining constant) when circulating 1500 gal./min. over the tower would be

$$\frac{250,000}{1500 \times 8.33} = 20.0°\text{F.}$$

Theoretically, the design area of the tower from Fig. 48 was 500 sq. ft. [1000 gal./min. ÷ 2.0 gal./(min.) (sq. ft.) = 500 sq. ft.]. The water concentration when circulating 1500 gal./min. is

$$1500/500 = 3.0 \text{ gal./sq. ft.}$$

Now, referring to Fig. 48, with a water concentration of 3.0 gal./sq. ft. and 65°F. wet-bulb temperature, adjust the straightedge until a difference of 20°F. exists between the hot-water and cold-water temperatures. This shows the hot-water temperature to be 100°F. and the cold-water temperature to be 80°F.

This indicates that the possibility of a lower cold-water temperature obtained by the lower existing wet-bulb temperature was lost because of the adverse effect of the increased water quantity.

Figure 50 shows the type of performance curve furnished by the cooling-tower manufacturer. This shows the variation in performance with change in wet-bulb and hot-water temperature, while the water quantity is maintained constant.

Atmospheric Cooling Towers. A cross-sectional view of a typical atmospheric cooling tower is shown in Fig. 51. In atmospheric towers, the water is pumped

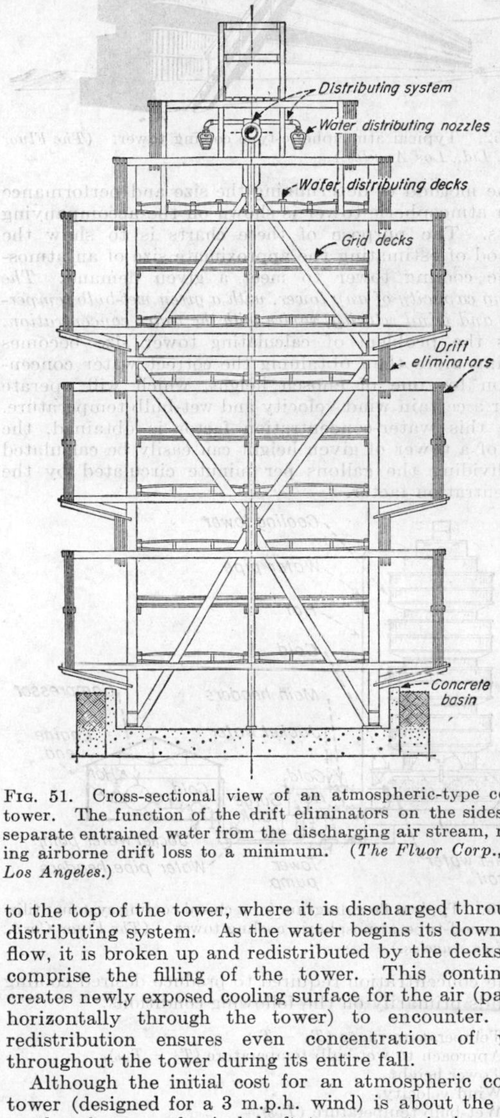

FIG. 51. Cross-sectional view of an atmospheric-type cooling tower. The function of the drift eliminators on the sides is to separate entrained water from the discharging air stream, reducing airborne drift loss to a minimum. (*The Fluor Corp., Ltd., Los Angeles.*)

to the top of the tower, where it is discharged through a distributing system. As the water begins its downward flow, it is broken up and redistributed by the decks that comprise the filling of the tower. This continually creates newly exposed cooling surface for the air (passing horizontally through the tower) to encounter. The redistribution ensures even concentration of water throughout the tower during its entire fall.

Although the initial cost for an atmospheric cooling tower (designed for a 3 m.p.h. wind) is about the same as that for a mechanical-draft tower, certain important limitations govern its performance. It must be located broadside to the prevailing wind in an exposed area. Any surrounding structures, hills, or other barriers would tend to block off the wind.

Originally the main objection to atmospheric towers was the excessive spray loss occurring during periods of high winds. This high loss was caused by the lack of a method of separating the entrained water from the air in the conventional louver-type tower. As a result, most manufacturers have now solved this problem by incorporating drift eliminators in the louvers. A typical example of this type of tower is shown in Fig. 52.

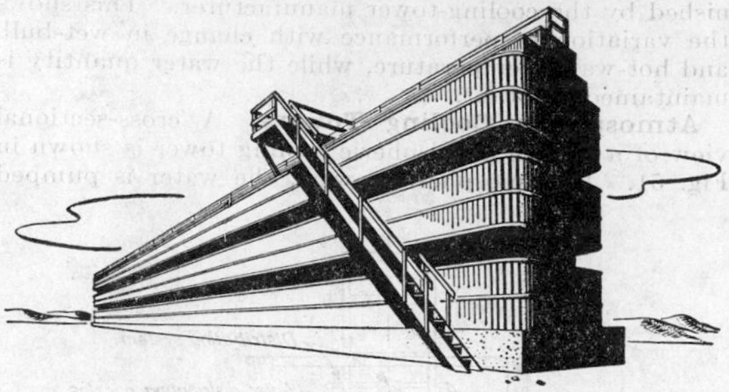

FIG. 52. Typical atmospheric-type cooling tower. (*The Fluor Corp., Ltd., Los Angeles.*)

The method of determining the size and performance of an atmospheric tower is shown on the accompanying charts. The purpose of these charts is to show the method of estimating the approximate size of an atmospheric cooling tower to meet a given demand. *The cooling capacity of any tower, with a given wet-bulb temperature and wind velocity, varies with the water concentration.* Thus the problem of calculating tower size becomes nothing more than obtaining the correct water concentration for one of chosen height, which will operate under a certain wind velocity and wet-bulb temperature. Once this water-concentration factor is obtained, the area of a tower of given height can easily be calculated by dividing the gallons per minute circulated by the concentration factor.

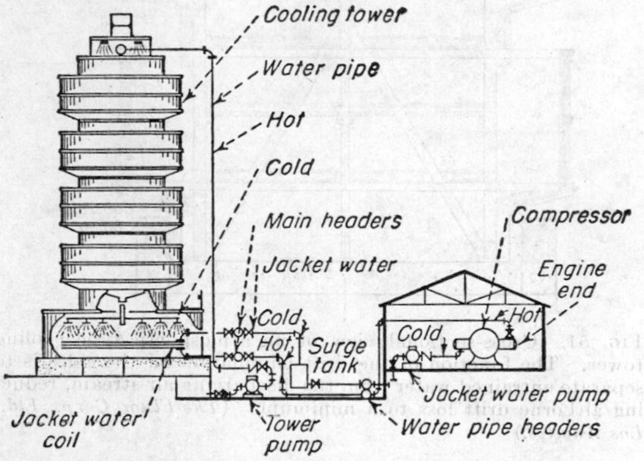

FIG. 53. Typical flow diagram for a closed cooling system utilizing the coil-type atmospheric cooling tower. (*The Fluor Corp., Ltd., Los Angeles.*)

The concentration required to produce desired cooling depends primarily on the following conditions:

1. Temperature range ($T_1 - T_2$).
2. Approach to wet-bulb temperature ($T_2 - T_{wb}$).
3. Tower height.
4. Wind velocity.
5. Wet-bulb temperature (T_{wb}).

It is easily seen that, because of the infinite number of possible combinations of these values, it is impractical if not impossible to have one curve from which to obtain the correct concentration factor. Figure 58 gives the required concentration for cooling water through a certain range and with a certain approach to the wet-bulb temperature; but this curve assumes a wet-bulb temperature of 70°F., a tower height of 35 ft., and a wind velocity of 3 m.p.h. Should any of these three values change, the concentration would have to be corrected for the new conditions by the use of one or more of the correction factors shown in Figs. 54 to 56. [See also Hutchinson and Spivey, *Trans. Inst. Chem. Engrs. (London)*, **20**, 14 (1942).]

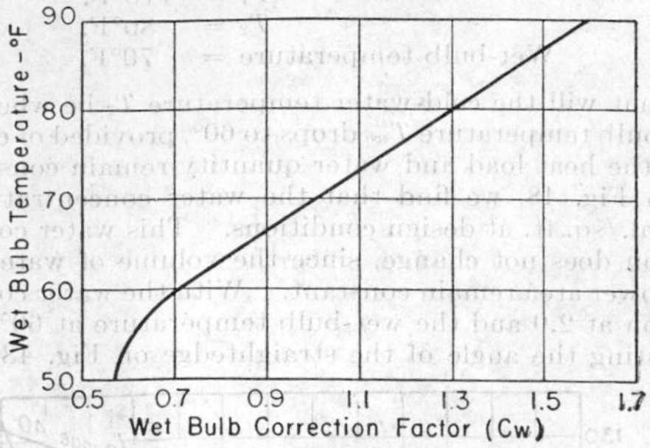

FIG. 54. Wet-bulb performance factor. (*The Fluor Corp., Ltd., Los Angeles.*)

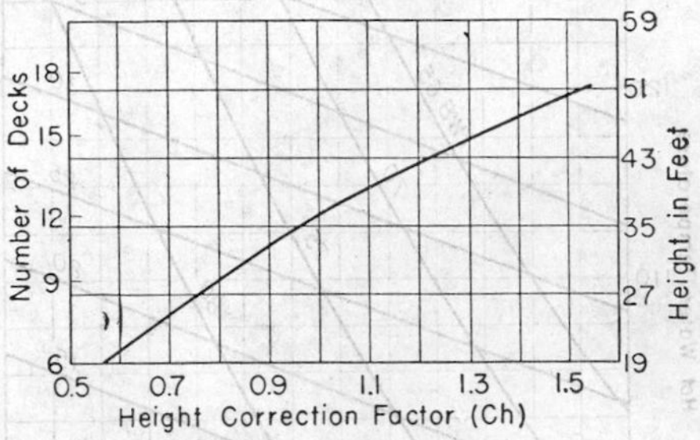

FIG. 55. Tower height for performance factor. (*The Fluor Corp., Ltd., Los Angeles.*)

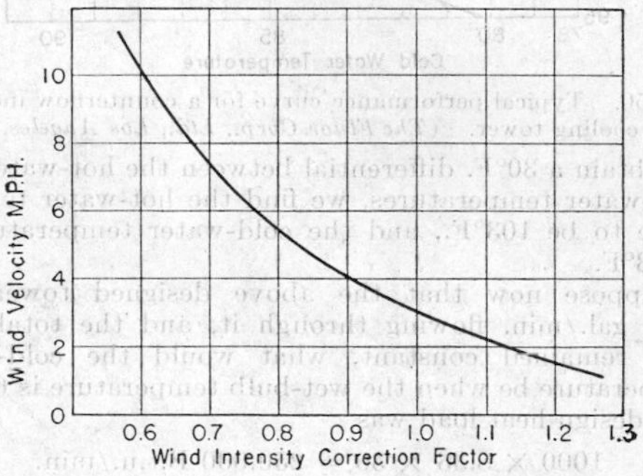

FIG. 56. Wind-intensity performance factor. (*The Fluor Corp., Ltd., Los Angeles.*)

Let us see how a variation in any of the three aforementioned conditions would affect the concentration of water.

1. *Wind Velocity.* The higher the wind velocity the greater the amount of air that goes through the tower. This results in greater cooling. Therefore, when the wind velocity is higher, the concentration can be greater and still obtain equal cooling.

2. *Tower Height.* In general, it is found in atmospheric as well as mechanical-draft towers that the greater the cooling range and the closer the approach to the wet-bulb temperature, the higher will be the tower required to give sufficient time of contact between water and air to accomplish the desired cooling. In atmospheric towers, the performance is limited by both maxi-

mum and minimum water concentrations. Should the water concentration fall below 1 gal./min./sq. ft. of tower area, it will be necessary to employ the next size higher tower. This is because, in extremely light water concentrations, the water will not be uniformly distributed and the performance as predicted by the accompanying curves will not be achieved. Should the water concentration exceed 3 gal./min./sq. ft. of tower area, it will then be necessary to choose the next size lower tower. The higher water concentrations tend to blanket

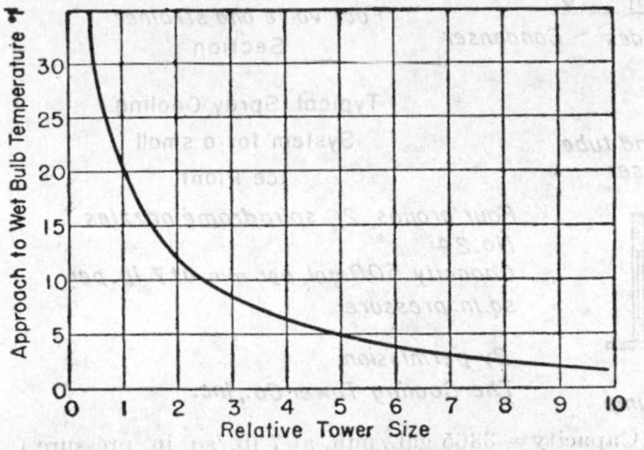

FIG. 57. Variation of tower size with approach to wet-bulb temperature. (*The Fluor Corp., Ltd., Los Angeles.*)

the tower and will not allow sufficient air to pass through to accomplish the cooling shown on the curves.

3. *Wet-bulb Temperature.* Theoretically, a cooling tower cannot cool water to a temperature lower than the prevailing wet-bulb temperature. Because of this limitation, the economical approach of the cold-water temperature to the wet-bulb temperature becomes an important factor. Air has a greater capacity for absorbing heat at the higher wet-bulb temperatures. At the lower wet-bulb temperature, air, in passing through the

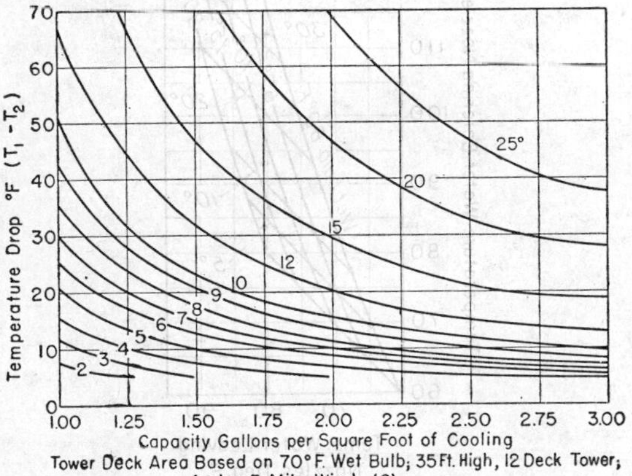

FIG. 58. Capacity curves—approach to wet-bulb temperature. (*The Fluor Corp., Ltd., Los Angeles.*)

tower, must have a greater temperature rise to accomplish the same cooling. Therefore, to obtain the same approach at the lower wet-bulb temperatures, it is necessary to reduce the concentration.

To calculate the size of an atmospheric cooling tower with effective width of 12 ft., the following general formula may be used:

$$L = \frac{\text{gal./min.} \times W}{C \times 12 \times Cw \times Ch}$$

where L = length of tower, ft.
gal./min. = quantity of water, gal./min.
W = wind correction factor.
C = concentration of water/sq. ft. of cooling-tower area.
Cw = wet-bulb correction factor.
Ch = tower-height correction factor.
T_1 = inlet temperature.
T_2 = outlet temperature.
$(T_1 - T_2)$ = temperature range.
T_{wb} = wet-bulb temperature.
$(T_2 - T_{wb})$ = approach to wet-bulb temperature.

Problem 1. Determine the length of a 35-ft.-high tower required to cool 1500 gal./min. from 90° to 75°F. with a 70°F. wet-bulb temperature and a 3 m.p.h. wind. From these conditions we know that the approach $(T_2 - T_{wb})$ = 5°F. and the range $(T_1 - T_2)$ = 15°F. From Fig. 58, it is found that these two values require a concentration of C = 1.17. The correction factors Ch, Cw, and W will be equal to 1, as shown on their respective correction-factor curves, because Fig. 58 was based upon values equal to those given in the problem. Substituting the corresponding values in the above formula, we find that a tower 107 ft. long is required.

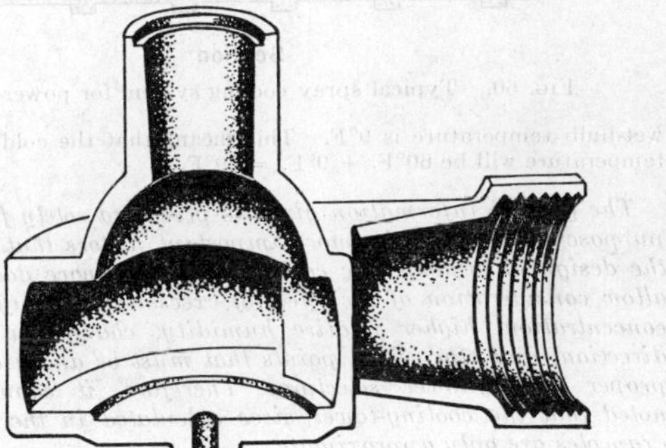

FIG. 59. This cutaway view of an upspray nozzle shows the side entrance which starts the spiral action upward to the discharge orifice. The removable screw plate can be drilled as shown for automatic drainage to prevent winter freezing. (*The Fluor Corp., Ltd., Los Angeles.*)

Problem 2. Let us take the same conditions that exist in Problem 1 and see what the effect would be when the wind velocity is increased to 5 m.p.h. From Fig. 56, when the wind velocity is 5 m.p.h., W = 0.80. Substituting this value for W in the formula, we find that an 85.3-ft.-long tower is required.

Problem 3. Again assume the same conditions of Problem 1 except that, because of space considerations it is necessary to have a 51-ft.-high tower. From Fig. 55, when tower height is 51 ft., Ch = 1.63. Substituting the new Ch in the formula, we find that a 65.5-ft.-long tower is required.

Problem 4. Assume the conditions of this problem are identical to those of Problem 1 except that water has to be cooled from 85° to 70°F. and the wet-bulb temperature is 65°F. From Fig. 54, when the wet-bulb temperature is 65°F., Cw = 0.86. Substituting this factor in the general formula, we find tower length to be 124 ft.

Problem 5. Once a tower is installed, the problem often arises of determining what cold-water temperature one can expect under conditions differing from those for which the tower was designed. For example, let us take the tower calculated for Problem 1. This tower is 35 ft. high, 12 ft. wide, and 107 ft. long. What cold-water temperature T_2 can be expected when the wet-bulb temperature is 60°F., wind velocity 4 m.p.h., cooling range 15°F., and water circulation 2000 gal./min.?

The wet-bulb-temperature correction factor Cw from Fig. 54 is 0.71. The wind-velocity correction factor W from Fig. 56 is 0.875. By substituting these values in the general formula and solving for water concentration, we find

$$C = \frac{2000 \times 0.875}{107 \times 12 \times 0.71 \times 1} = 1.92 \text{ gal./sq. ft.}$$

Turning to Fig. 58, we find that, when the cooling range is 15°F. and the water concentration is 1.9, the approach to the

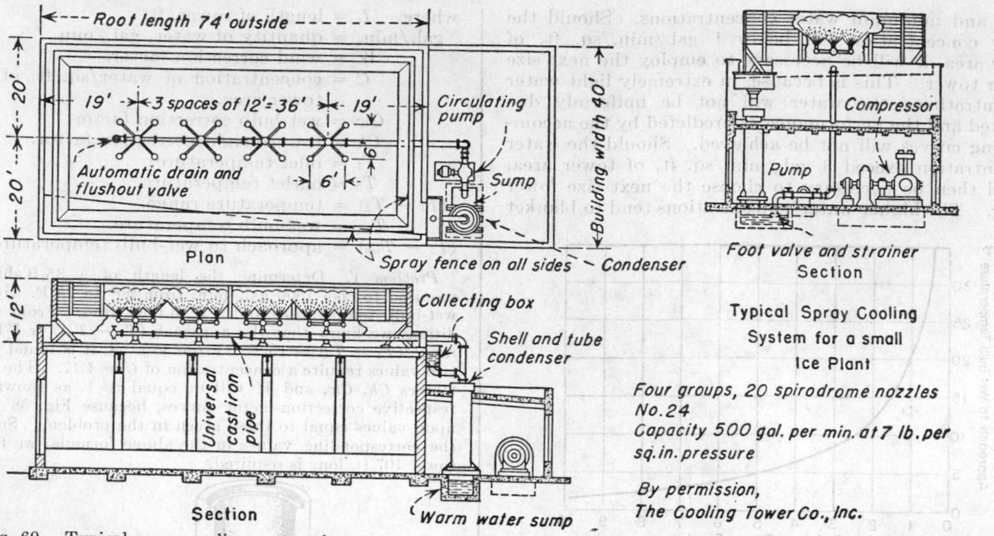

FIG. 60. Typical spray cooling system for power-plant service. (Capacity = 3365 gal./min. at 7 lb./sq. in. pressure.)

wet-bulb temperature is 9°F. This means that the cold-water temperature will be 60°F. + 9°F. = 69°F.

The general information given is presented solely for the purpose of showing the more important factors that affect the design of atmospheric cooling towers. Space does not allow consideration of the effect of excessive or insufficient concentration, higher relative humidity, change in wind direction, and other finer points that must be analyzed for proper cooling-tower selection. Therefore, it should be noted that the cooling-tower sizes calculated in the above examples are only approximate.

Cooling Tower Spray Nozzles. The types of nozzle used in upspray cooling-tower distributing systems and spray ponds is shown in Fig. 59. The table shown below gives the performance of a particular family of nozzles. In upspray distributing systems, a pressure of 7 lb./sq. in. is common practice; however, 5 lb./sq. in. is adequate.

Nozzle Capacity
U.S. gallons per minute

Nozzle number	Pipe size, in.	Orifice size, in.	Pressure, lb./sq. in. 5	6	7	8	9	10	12	15	20	25	30
2.4	3/8	3/16	....	....	1.0	1.1	1.2	1.3	1.4	1.6	1.8	2.0	2.2
2.4	3/8	5/16	1.8	1.9	2.1	2.2	2.3	2.3	2.7	3.0	3.4	3.8	4.2
4SB*	3/4	11/32	4.5	6.4	5.3	5.7	6.0	6.3	6.8	7.6	8.8	9.8	10.7
7LB	1	7/16	6.5	7.1	7.7	8.2	8.7	9.2	10.1	11.3	13.1	14.6	16.0
7SB*	1	7/16	7.0	7.7	8.3	8.9	9.5	10.0	11.0	12.2	14.2	16.0	17.5
14SB*	1	5/8	10.2	11.0	12.0	12.7	13.5	14.2	15.5	17.4	19.7	22.0	24.1
14LB	1	5/8	11.5	12.6	13.6	14.5	15.4	16.2	17.7	19.8	22.7	25.3	27.8
19	1½	7/8	16.8	18.3	19.8	21.1	22.4	23.5	25.8	28.7	32.9	36.5	40.0
24	1½	1	20.0	22.0	23.8	25.3	26.9	28.3	31.0	34.7	40.0	44.8	49.0
39	2	1¼	33.0	36.2	39.0	41.8	44.3	46.7	51.0	57.0	65.8	73.3	80.0
49	2½	1 25/32	39.5	43.0	46.5	49.5	52.5	55.0	60.5	67.0	78.0	86.0	94.0

* SB (short barrel), downspray only; LB (long barrel), for upspray only.

The nozzle shown in the table is of the non-clogging type. It does not depend on small orifices to obtain minimum drop size but depends rather on centrifugal force. The water is given a spiral action by its tangential entrance into the spiral chamber. The dome-shaped approach to the discharge orifice increases this spiral action as it approaches the discharge orifice. This whirling action, which generates the necessary velocity for fine break-up, assures uniform drop size and efficient water distribution over a maximum area. This formation of uniformly small drops assures maximum contact with air, resulting in high cooling efficiency.

Spray Ponds. Where space is available, circulating water may be cooled by spraying into the air. The contact of the sprayed water with the air before reaching the surface of the pond is comparatively limited, and therefore cooling is possible only through a small range. The nozzle layout must provide that spray clouds from one nozzle do not interfere with spray clouds from

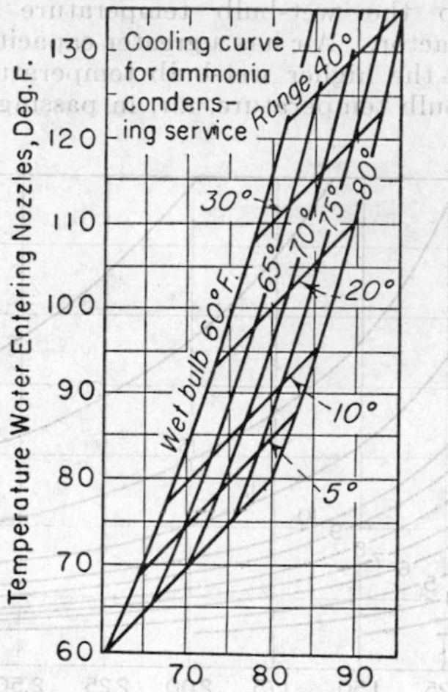

FIG. 61. Spray ponds—cooling curves for ammonia-condensing service.

another nozzle. This allows the humid air to escape from the pond, and dry air to take its place. Pipe lines are ordinarily supported 2 ft. above the water level, and the pond itself should be about 3 ft. deep. The pond should be large enough to take care of driftage, or else louvers should be provided.

Figure 60 shows a typical installation at a power plant.

Performance curves are given by Figs. 61 and 62 for rates as used in ammonia- and in steam-condensing service. Other uses include supplying cooling water for

Weather Conditions for United States and Foreign Countries*

Thirty-year average from data taken during the hottest month of the year

State	City	Dry bulb, °F.	Wet bulb, °F.	% Relative humidity	M.p.h.	Direction of wind
	Data below taken in July					
Alabama	Montgomery	81.0	73.5	70.0	5.0	S. W.
	Mobile	80.5	75.5	79.0	6.0	S.
	Birmingham	79.8	73.3	73.0	5.0	S. W.
Arizona	Yuma	91.0	71.0	36.0	....	W. & S.
	Phoenix	90.5	68.5	31.0	....	E.
Arkansas	Little Rock	80.0	73.0	72.0	5.0	S. & S. W.
	Ft. Smith	80.5	72.5	68.0	5.0	E.
California	San Francisco	57.3	54.3	82.0	14.0	W.
	Sacramento	72.4	60.6	51.0	9.7	S.
	Los Angeles	67.4	61.6	72.0	4.5	W.
	Red Bluff	82.1	62.0	30.0	5.7	S. E.
	San Diego	66.9	62.4	78.0	5.4	S. W.
Colorado	Denver	71.8	57.8	45.0	7.5	S.
	Colorado Springs	67.9	56.4	51.0		
	Pueblo	72.6	58.0	44.0	7.0	
Connecticut	New Haven	71.9	65.9	73.0	8.0	S.
Dist. Columbia	Washington	76.8	69.3	69.0	5.3	S.
Florida	Jacksonville	80.9	75.0	75.0	8.0	S. W.
	Key West	83.7	76.7	73.0	8.0	E.
	Pensacola	81.3	75.8	77.0	8.0	S. W.
	Tampa	79.9	75.0	79.0	6.0	N. W.
Georgia	Atlanta	77.6	69.6	67.0	8.6	E. & N. E.
	Augusta	80.5	74.0	74.0	5.0	S. & S. E.
	Savannah	80.5	75.5	79.0	6.4	S. W.
Illinois	Cairo	78.6	72.4	75.0	6.2	S.
	Chicago	72.3	65.8	71.0	15.1	S. W.
	Springfield	76.1	68.0	64.0	6.6	S. W.
Indiana	Indianapolis	76.4	68.0	65.0	8.2	S. W.
Iowa	Davenport	75.4	67.2	65.0	7.4	S. W.
	Des Moines	75.0	67.0	66.0	7.1	S. W.
	Dubuque	74.7	66.7	66.0	5.5	N. W.
	Keokuk	77.0	69.0	67.0	6.9	S.
Kansas	Concordia	77.7	67.2	58.0	....	S.
	Dodge City	77.7	66.7	58.0	....	S. E.
	Wichita	78.3	69.5	65.0	....	S.
Kentucky	Lexington	76.0	68.0	64.0	8.0	S. W.
	Louisville	78.6	69.6	64.0	6.1	S. W.
Louisiana	New Orleans	81.3	75.3	76.0	6.5	S. W.
	Shreveport	82.1	75.0	72.0	5.0	S. & S. E.
Maine	Eastport	59.8	56.3	81.0	....	S. & S. W.
	Portland	68.0	62.0	71.0	....	S.
Maryland	Baltimore	77.3	69.6	70.0	6.6	S. W.
Massachusetts	Boston	71.3	64.8	70.0	9.3	S. W.
	Nantucket	67.5	64.5	85.0	....	S. W.
Michigan	Alpena	65.8	60.3	78.0	8.0	S. E. & W.
	Detroit	72.1	65.1	69.0	9.0	S. W.
	Grand Haven	69.7	63.2	70.0	9.0	S. W.
	Escanaba	66.7	61.2	73.0	8.0	S.
	Marquette	64.9	59.0	70.0	9.0	N. W.
	Port Huron	69.0	63.0	72.0	9.0	N. E.
	Grand Rapids	72.6	65.4	68.0	9.0	S. W.
Minnesota	St. Paul	72.1	64.5	67.0	7.2	N.
	Duluth	66.0	60.0	71.0	11.0	N. E.
Mississippi	Vicksburg	80.4	74.4	75.0	6.0	S. W.
Missouri	Kansas City	76.9	68.9	67.0	7.5	S.
	St. Louis	79.1	70.6	66.0	8.2	S. W.
	Springfield	75.7	68.5	70.0	7.7	S.
Montana	Havre	68.5	57.0	50.0	....	S. W.
	Helena	66.9	53.4	43.0	....	W.
	Miles City	73.0	62.0	55.0	....	W. & N. W.
Nebraska	North Platte	73.9	64.9	61.0	9.0	S. E.
	Omaha	76.5	66.5	59.0	7.0	S.
	Valentine	73.1	62.5	57.0	10.0	S.
New Jersey	Atlantic City	72.5	69.0	84.0	8.0	S. W.
New Mexico	Santa Fe	68.7	55.2	46.0	6.1	N. E.
New York	Albany	72.0	65.5	71.0	7.7	S.
	Buffalo	70.2	64.0	71.0	10.9	S. W.
	New York	73.5	67.0	71.0	9.1	S. & S. W.
	Rochester	70.4	63.2	67.0	7.1	S. W.
North Carolina	Charlotte	77.8	70.3	69.0	5.0	S. W.
	Raleigh	77.8	71.3	73.0	....	S. W.
	Wilmington	78.7	73.7	79.0	....	S. E.
	Asheville	72.2	....	....	4.7	S. W.
North Dakota	Bismarck	70.2	61.2	61.0	9.0	N. W.
	Williston	69.2	59.2	56.0	9.0	N. W.
Ohio	Cincinnati	77.7	68.7	63.0	6.6	S. W.
	Cleveland	72.5	65.8	70.0	11.7	S. E.
	Columbus	75.0	66.5	64.0	8.7	S. W.
	Toledo	73.7	65.7	65.0	8.0	S. E.

State	City	Dry bulb, °F.	Wet bulb, °F.	% Relative humidity	M.p.h.	Direction of wind
	Data below taken in July					
Oklahoma	Oklahoma City	79.0	70.0	64.0	9.0	S.
Oregon	Portland	66.3	58.0	60.0	7.9	N. W.
	Roseburg	66.1	57.0	57.0	4.0	N. W.
Pennsylvania	Erie	71.8	64.8	69.0	9.0	W.
	Philadelphia	75.8	67.8	66.0	9.4	S. W.
	Pittsburgh	74.6	67.1	68.0	5.2	S. W.
	Scranton	71.8	64.5	68.0	6.1	S. W.
	Harrisburg	74.5	67.0	67.0	5.7	W.
South Carolina	Charleston	81.3	75.3	76.0	9.8	S. W.
	Columbia	81.1	74.1	72.2	7.0	S. W.
	Aiken	81.3	72.5	65.5	....	
South Dakota	Huron	71.6	63.6	65.0	10.4	S. E.
	Pierre	75.0	63.0	51.0	9.6	S. E.
	Rapid City	71.9	59.5	47.0	7.5	W.
	Yankton	74.6	66.8	66.0	6.5	S.
Tennessee	Chattanooga	77.8	70.8	71.0	5.2	S. W.
	Knoxville	76.2	70.2	74.0	5.0	S. W.
	Memphis	80.7	72.7	68.0	7.4	S. W.
	Nashville	79.4	71.4	68.0	5.0	S. W.
Texas	Abilene	82.2	69.2	52.0	9.0	S.
	Corpus Christi	81.9	76.9	80.0	12.0	S. E.
	El Paso	80.5	64.5	44.0	9.7	W. & E.
	Galveston	83.0	77.0	76.0	10.0	S. W.
	San Antonio	82.4	73.4	65.0	7.0	S. E.
	Ft. Worth	82.5	69.5	51.0	10.0	S.
Utah	Salt Lake City	76.2	58.2	34.0	6.3	S. E.
Vermont	Northfield	66.1	61.6	78.0	....	S.
	Burlington	68.2	63.2	76.0	....	S.
Virginia	Lynchburg	77.3	69.8	69.0	....	S. W.
	Norfolk	78.4	72.6	74.0	....	S. W.
	Richmond	79.2	72.0	70.0	....	S.
Washington	Seattle	63.3	56.3	64.0	5.5	W.
	Spokane	68.8	55.0	41.0	....	S. W.
	Walla Walla	74.3	59.0	40.0	....	S.
West Virginia	Parkersburg	74.9	67.5	68.0	4.0	W.
Wisconsin	Green Bay	69.6	62.6	68.0	8.0	N. & S.
	La Crosse	72.6	66.0	71.0	6.0	S.
	Milwaukee	69.7	64.7	77.0	9.8	N.
Wyoming	Cheyenne	67.4	54.6	46.0	8.0	N. W. & S.
	Lander	66.6	52.6	43.0	4.0	S. W.
Bermuda	Hamilton	78.7	74.7	84.0	9.5	S. E.
Canada	Calgary	60.6	53.6	65.6	7.5	N. W.
	Montreal	69.5	65.0	79.0	11.3	S. W.
	Toronto	68.5	62.3	70.5	7.9	N. W.
	Victoria	60.0	54.0	69.0	9.5	S. W.
	Winnipeg	66.0	61.0	77.5	11.3	N. W.
China	Hongkong	81.9	77.7	82.0	5.1	S. E.
	Macão	79.0	75.3	83.7	9.9	S. E.
Cuba	Havana	80.5	76.6	81.5	10.0	E.
	Santiago	80.2	73.9	74.0	5.4	N.
England	Durham	59.5	56.0	79.0	3.5	S. W.
	Sheffield	60.7	57.0	79.0	3.5	W.
	Birmingham	60.4	56.0	75.0	3.5	W.
	London	62.8	57.0	71.0	3.5	S. W.
France	Marseilles	70.7	62.8	63.3	5.0	{S. W. W. / N. W.
	Nice	71.8	65.9	73.3	....	E. S. E.
Ireland	Dublin	60.5	56.0	77.0	3.5	S.
	Armagh	58.4	....	....	3.5	
Italy	Turin	73.0	64.0	61.0	....	N. N. E. by E.
Jamaica	Kingston	80.3	73.0	70.0	6.3	N.
Porto Rico	San Juan	79.9	75.2	80.0	10.0	E.
Russia	Kharkov	69.3	61.8	66.0	4.0	N. E. E. and S. W. W.
	Taganrog	74.1	....	....	5.3	N. E. E. and S. W. W.
Scotland	Glasgow	58.0	54.0	77.0	3.5	W.
	Data below taken in June					
India	Bareilly	89.8	78.0	58.0	3.1	S. E.
	Bhavnagar	89.3	81.3	71.0	11.5	S. W.
	Bombay	82.4	78.4	84.0	11.1	W. S. W.
	Calcutta	84.8	79.8	80.0	4.3	S.
	Delhi	93.2	76.5	45.0	3.6	S. E.
	Data below taken in December					
Chile	Valparaiso	65.1	58.6	68.0	5.0	S. W.

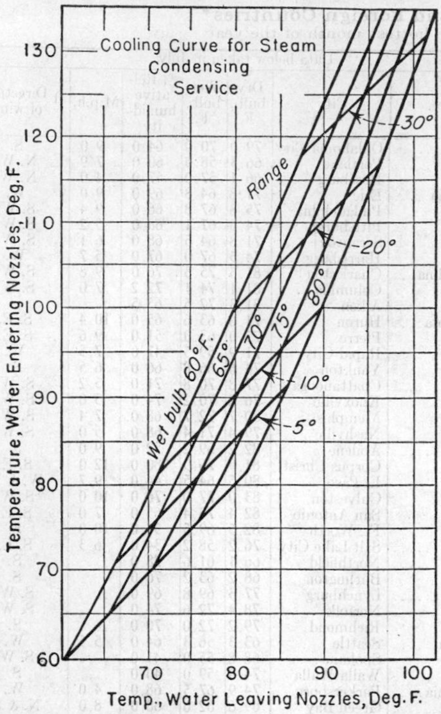

FIG. 62. Spray ponds—cooling curves for steam-condensing service.

Diesel engines, electrical transformers, oil refining, furnace doors, heat-treatment of metals, and a variety of special uses.

There has been considerable improvement in spray nozzles, largely in the development of non-clogging nozzles.

Maintenance of Constant Humidity by Reagents

The following systems, rearranged in the order of increasing humidities, are taken from "International Critical Tables," vol. 1, pp. 67–68. A saturated aqueous solution of a salt in contact with an excess of a definite solid phase and at a definite temperature will maintain a constant humidity within an enclosed space.

Solid phase	Max. temp., °C.	% humidity
$H_3PO_4.\frac{1}{2}H_2O$	24.5	9
$ZnCl_2.\frac{1}{2}H_2O$	20	10
$KC_2H_3O_2$	168	13
$LiCl.H_2O$	20	15
$KC_2H_3O_2$	20	20
KF	100	22.9
NaBr	100	22.9
$CaCl_2.6H_2O$	24.5	31
$CaCl_2.6H_2O$	20	32.3
$CaCl_2.6H_2O$	18.5	35
CrO_3	20	35
$CaCl_2.6H_2O$	10	38
$CaCl_2.6H_2O$	5	39.8
$K_2CO_3.2H_2O$	24.5	43
$K_2CO_3.2H_2O$	18.5	44
$Ca(NO_3)_2.4H_2O$	24.5	51
$NaHSO_4.H_2O$	20	52
$Mg(NO_3)_2.6H_2O$	24.5	52
$Ca(NO_3)_2.4H_2O$	18.5	56
$NaClO_3$	100	54
$Mg(NO_3)_2.6H_2O$	18.5	56
$NaBr.2H_2O$	20	58
$Mg(C_2H_3O_2).4H_2O$	20	65
$NaNO_2$	20	66
$(NH_4)_2SO_4$	108.2	75
$(NH_4)_2SO_4$	20	81
$NaC_2H_3O_2.3H_2O$	20	76
$Na_2S_2O_3.5H_2O$	20	78
NH_4Cl	20	79.2
NH_4Cl	25	79.3
NH_4Cl	30	79.5
KBr	20	84
Tl_2SO_4	104.7	84.8
$KHSO_4$	20	86
$Na_2CO_3.10H_2O$	24.5	87
K_2CrO_4	20	88
$NaBrO_3$	20	92
$Na_2CO_3.10H_2O$	18.5	92
$Na_2SO_4.10H_2O$	20	93
$Na_2HPO_4.12H_2O$	20	95
NaF	100	96.6
$Pb(NO_3)_2$	20	98
$TlNO_3$	100.3	98.7
TlCl	100.1	99.7

For a more complete list of salts, and for references to the literature see "International Critical Tables," vol. 1, p. 68.

SECTION 13

DRYING

BY

W. R. Marshall, Jr., Ph. D., Associate Professor, Chemical Engineering Department, University of Wisconsin; Member, American Institute of Chemical Engineers.

Samuel J. Friedman, M. S., Chemical Engineer, E. I. du Pont de Nemours & Co.; Member, American Institute of Chemical Engineers, American Chemical Society, American Society of Mechanical Engineers.

CONTENTS

THE DRYING OF SOLIDS

PAGE
Introduction 800
Theory and Fundamental Concepts 801
Internal vs. External Conditions 801
Internal Mechanism of Liquid Flow 801
The Periods of Drying 802
The Constant-rate Period 802
Effect of Air Velocity 803
Determination of the True Surface Temperature 803
Estimation of the Constant Rate 804
The Constant-rate Period in Through-circulation Drying 805
Evaporation from Liquid Drops 805
Drying at Air Temperatures above the Boiling Point of the Liquid 806
Constant-rate Period When Heat Transfer Depends on Conduction and Radiation 806
The Falling-rate Period 806
The Zone of Unsaturated Surface Drying 806
Zone Where Internal Liquid Flow Controls 806
Critical Moisture Content 807
Approximate Equations for Estimating Drying Time 808
Methods of Correlating Drying Data 808
Tests on Plant Dryers 809
Equilibrium Moisture Content 810
Application of Psychrometry to Drying 811
Humidity Charts for Other Solvent Vapors 812
Classification of Dryers 813
Direct Dryers 813
Indirect Dryers 814
Types of Dryers 815
Tray and Compartment Dryers 815
Batch Through-circulation Dryers 820
Tunnel Dryers and Vertical Turbodryers 821
Continuous Through-circulation Dryers 823
Direct Rotary Dryers 828
Pneumatic Conveying Dryers 834
Spray Dryers 838
Direct Continuous-sheeting Dryers 848
Vacuum Shelf Dryers 853
Vacuum Freeze Dryers 854
Agitated-pan Dryers 856
Vacuum Rotary Dryers 858
Indirect Rotary Dryers 859
Screw-conveyor Dryers 862
Vibrating-conveyor Dryers 863
Drum Dryers 863
Cylinder Dryers 866
Infrared Dryers 868
Dielectric or High-frequency Dryers 870
Selection of Drying Equipment 874
Drying with Solvent Recovery 875

THE DRYING OF GASES

Introduction 877
Methods of Drying Gases 877
Drying Gases with Desiccants 878
Liquid or Soluble Desiccants 878
Solid Desiccants 880
Adsorption Rates on Solid Desiccants 882
Drying Gases by Compression 884
Drying Gases by Refrigeration 884

GENERAL REFERENCES: Badger and McCabe, "Elements of Chemical Engineering," 2d ed., McGraw-Hill, New York, 1936. Ceaglske and Hougen, *Trans. Am. Inst. Chem. Engrs.*, **33**, 283 (1937). Hirsch, "Die Trockentechnik," 2d ed., Springer, Berlin, 1932. Hougen, McCauley, and Marshall, *Trans. Am. Inst. Chem. Engrs.*, **36**, 183 (1940). Krischer *et al.*, *Z. Ver. deut. Ing.*, **82**, 373 (1938); *Verfahrenstechnik* (4), 104 (1938); (5), 140 (1938); (1), 17 (1940). McCready and McCabe, *Trans. Am. Inst. Chem. Engrs.*, **29**, 131 (1933). Macey, *Trans. Brit. Ceramic Soc.* **41** (4), 73 (1942). Marshall, *Heating, Piping and Air Conditioning*, **17**, 472 (1945); **18**, 71 (May, 1946). Shepherd, Hadlock, and Brewer, *Ind. Eng. Chem.*, **30**, 388 (1938). Sherwood, *Ind. Eng. Chem.*, **21**, 976 (1929); **25**, 311 (1933); **26**, 1096 (1934). Symposium on Drying, *Ind. Eng. Chem.*, **30** (1938). Tiemann, "The Kiln Drying of Lumber," Lippincott, Philadelphia, 1917. Walker, Lewis, McAdams, and Gilliland, "Principles of Chemical Engineering," 3d ed., McGraw-Hill, New York, 1937.

THE DRYING OF SOLIDS

INTRODUCTION

Drying means the removal of a liquid from a solid by **thermal means.** This definition distinguishes drying from mechanical methods of removing liquids from solids, but it does not differentiate drying from evaporation in which heat is used to evaporate large amounts of water from solutions or slurries. Drying and evaporation can be differentiated on the basis of the equipment involved and by the fact that evaporation processes generally remove much larger quantities of liquid per hour than do drying processes.

The term *dehydration* has also been applied to drying processes, but its use is restricted almost entirely to the drying of foods. This term also refers to the removal of chemically bound water from inorganic salts, and to the simultaneous removal of hydrogen plus hydroxyl groups from organic compounds.

Drying operations play a significant role in most chemical processes. **Investment costs** of dryers range from \$1000 to \$200,000, depending on the capacities involved. The *cost* of drying 1 lb. of solid ranges from \$0.0005 to \$0.025, depending on the drying method and the capacity. Continuous dryers, in general, should evaporate water at a cost of around 0.1¢/lb. of water removed.

The **reasons for drying** are manifold, but they ordinarily fall into one of the following categories: (1) to facilitate handling in further processing, (2) to permit satisfactory utilization of the final product, (3) to reduce shipping costs, (4) to increase the capacity of other equipment in the process, (5) to preserve a product during storage and shipment, and (6) to enhance the value and usefulness of waste or by-products.

Drying maintains almost a fixed relationship to other operations in a process. Thus it frequently follows a filtration or centrifuging operation and precedes a grinding or packaging step [see Marshall, *Heating, Piping and Air Conditioning*, **17**, 472 (1945)]. In this respect, drying is frequently regarded as a "finishing operation" because in most cases it occurs near the end of a process just prior to preparation of the product for shipment. Drying is frequently preceded by a mechanical dewatering process, since it is much less expensive and frequently easier to remove a liquid from a solid by mechanical methods than by thermal methods, *i.e.*, by drying.

The generally accepted definitions, peculiar to drying, are given alphabetically below:

Bound moisture is that liquid held by a solid which exerts a vapor pressure less than that of the pure liquid at the same temperature. Liquid may become *bound* by retention in small capillaries, solution in cell or fiber walls, homogeneous solution throughout the solid, and by chemical or physical adsorption on solid surfaces. Bound moisture can be removed from a solid only under specific conditions of humidity in the external surroundings.

Capillary flow is the flow of liquid through the interstices and over the surface of a solid, caused by liquid-solid molecular attraction.

Commercial dry basis expresses moisture content as pounds of water per pound of solid as it leaves the dryer.

Constant-rate period is that drying period during which the rate of water removal per unit of drying surface is constant.

Critical moisture content is that obtaining when the constant-rate period ends.

Dry-weight basis indicates the moisture content of wet solid as pounds of water per pound of bone-dry solid. The advantage of using this basis is that the moisture loss is obtained by subtraction of the moisture contents before and after drying. See *wet-weight basis* definition.

Dryer efficiency is that fraction of the total heat supplied by fuel used to evaporate water. "Over-all efficiency" is sometimes used to distinguish "total system efficiency" from "efficiency of the drying space."

Equilibrium moisture content is that to which a given material can be dried under specific conditions of air temperature and humidity. See discussion under Theory and Fundamental Concepts, p. 810.

Falling-rate period is that drying period during which the instantaneous drying rate continually decreases. See Theory and Fundamental Concepts, pp. 806–808.

Fiber saturation point is the moisture content of cellular materials (wood, etc.) at which the cell walls are completely saturated while the cavities are liquid-free. It may be defined as the equilibrium moisture content as the humidity of the surrounding atmosphere approaches saturation.

Free moisture content is that liquid which is removable at a given temperature and humidity. It may include bound and unbound moisture.

Funicular state is that condition in drying a porous body when capillary suction results in air being sucked into the pores.

Humidity denotes the amount of water vapor actually present in a gas. See Theory and Fundamental Concepts, p. 811, for methods of expression.

Hygroscopic material is one that may contain bound moisture.

Initial moisture distribution refers to the moisture distribution throughout a solid at the start of drying.

Internal diffusion. Diffusion is a single-phase phenomenon; internal diffusion must therefore occur as solid through solid, liquid through liquid, or gas through gas. Internal diffusion occurs when the moving phase obeys the fundamental laws of diffusion.

Moisture content of a solid is usually expressed as moisture quantity per unit weight or volume of the dry or wet solid. A weight (dry or wet) basis is preferred.

Moisture gradient refers to the distribution of water in a solid at a given moment in the drying process, the nature of which depends on the characteristics of the solid involved.

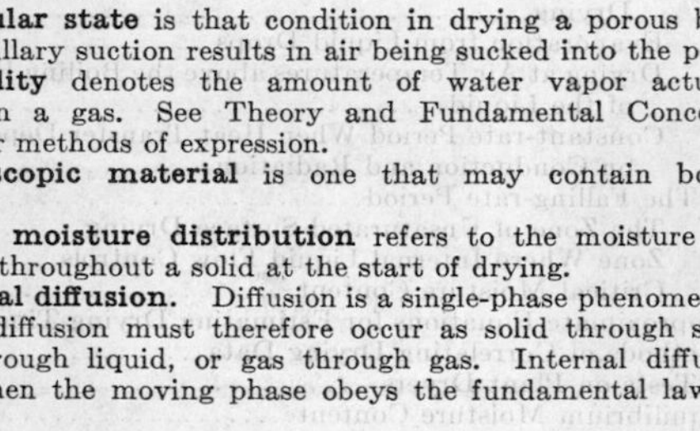

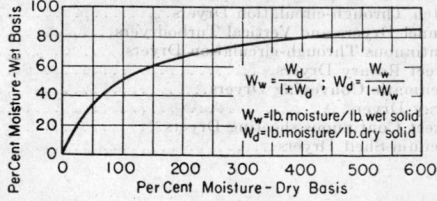

FIG. 1. Relationship between wet-weight and dry-weight bases.

Non-hygroscopic material is one that can contain no bound moisture.

Pendular state is that state of a liquid in a porous solid when a continuous film of liquid no longer exists around and between discrete particles so that flow by capillarity cannot occur. This state succeeds the funicular state.

Unaccomplished moisture change is the ratio of the free moisture present at any time to that initially present.

Unbound moisture in a hygroscopic material is that moisture in excess of the equilibrium moisture content corresponding to saturation humidity. All water in a non-hygroscopic material is unbound water.

Wet-weight basis expresses the moisture in a material as a percentage of the weight of the *wet* solid. This basis is less satisfactory than the dry-weight basis with which the percentage change of moisture is constant for all moisture contents. See Fig. 1 for the relationship between the dry- and wet-weight bases. When the wet-weight basis is used to express moisture content, a 2 or 3 per cent change at high moisture contents (above 70 per cent) actually represents a 15 to 20 per cent change in evaporative load. This might tax the capacity of a dryer.

THEORY AND FUNDAMENTAL CONCEPTS

When a solid dries, two fundamental and simultaneous processes occur: (1) transfer of heat to evaporate liquid and (2) transfer of mass as internal moisture and evaporated liquid. The factors governing the rate of each process determine the rate of the drying process.

Commercial drying operations may utilize heat transfer by convection, conduction, radiation, or a combination of any of these mechanisms. Industrial dryers differ fundamentally by the methods of heat transfer employed. See Classification of Dryers, p. 813. However, regardless of the heat-transfer mechanism, the heat must flow first to the outer surface and thence to the interior of the solid. The single exception is drying with high-frequency electricity, which generates heat internally and hence produces a higher temperature within the solid than on its surface; this in turn results in a heat flow from the interior to the external surface.

Mass is transferred in drying as (1) liquid and/or vapor within the solid, and (2) vapor from wet surfaces. The liquid-concentration gradient depends on the liquid-flow mechanism within the solid.

Internal vs. External Conditions

A study of how a solid dries may be based on the **internal mechanism** of liquid flow or on the effect of the **external conditions** of temperature, humidity, air flow, state of subdivision, etc., on the drying rate of the solid. The former procedure represents a fundamental study of the internal conditions. The latter procedure, although less fundamental, is more generally used because the results have greater immediate application.

Internal Mechanism of Liquid Flow. Internal liquid flow may occur by several mechanisms, depending on the structure of the solid. Some of the possible mechanisms are as follows: (1) *diffusion* in continuous homogeneous solids, (2) *capillary flow* in granular and porous solids, (3) flow caused by *shrinkage* and *pressure* gradients; (4) flow caused by *gravity*, and (5) flow caused by a *vaporization-condensation* sequence.

In general, one mechanism predominates at a given time in a solid during drying, and it is not uncommon to find different mechanisms predominating at different times during the drying cycle.

The particular mechanism that occurs during the drying of a given solid can be determined by a study of **internal moisture gradients.** The determination of such gradients is a difficult experimental problem. The usual technique involves cutting specially prepared specimens that have been dried for different time intervals into segments and determining the moisture content in each segment. Objections to this technique are that the moisture gradient may be disrupted during cutting, moisture loss may occur from the edges of the specimen, and only rigid or semirigid samples can be easily handled. McCready and McCabe [*Trans. Am. Inst. Chem. Engrs.*, **29**, 131 (1933)] determined moisture gradients in drying paper and paper pulp. Ceaglske and Hougen [*Trans. Am. Inst. Chem. Engrs.*, **33**, 283 (1937)] made similar determinations for sand. Bateman, Hohf, and Stamm [*Ind. Eng. Chem.*, **31**, 1150 (1939)] determined moisture gradients in wood. Macey [*Trans. Brit. Ceramic Soc.*, **41** (4), 73 (1942)] studied the distribution of moisture in clays caused by pressure gradients.

Hougen, McCauley, and Marshall [*Trans. Am. Inst. Chem. Engrs.*, **36**, 183 (1940)] discussed the conditions under which capillary flow and diffusional flow may be expected to occur in a solid, and a summary of published experimental moisture gradients for the two cases was presented and analyzed. Their curves indicated that capillary flow is typified by a moisture gradient involving a double curvature and point of inflection (Fig. 2*a*), while diffusional flow is a smooth curve, concave downward (Fig. 2*b*), as would be predicted from the integrated diffusion equations. They also showed that for diffusional flow the *liquid diffusivity*, usually assumed constant, is *not constant* but decreases as the moisture content decreases. The dashed curve in Fig. 2*b* is for a constant diffusivity, and the solid curve is experimental, indicating variable diffusivity. Hence the integrated diffusion equations for constant diffusivity do not strictly apply to drying, even when liquid diffusion does occur. Hougen, McCauley, and

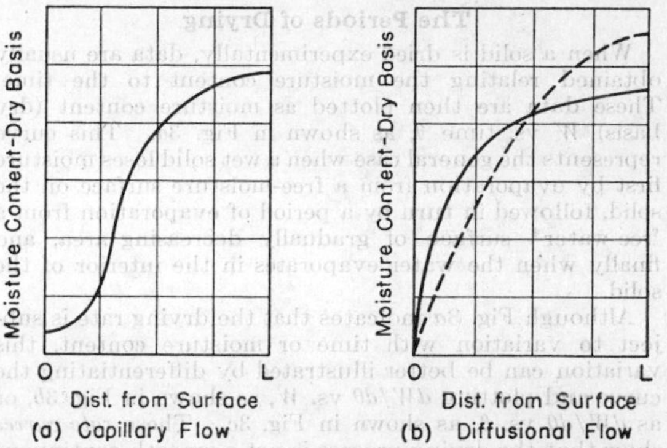

FIG. 2. Two types of internal moisture gradients obtained in drying solids.

Marshall classified solids on the basis of capillary and diffusional flow as follows:

Capillary Flow. Moisture held in the interstices of solids as liquid covering the surface and as free moisture in cell cavities is subject to movement by gravity and capillarity, provided that passageways for continuous flow are present. In drying, liquid flow due to capillarity applies to liquids not held in solution and to all moisture above the fiber-saturation point, as in textiles, paper, leather, and to all moisture above the equilibrium moisture content at atmospheric saturation, as in fine powders and granular solids, such as paint pigments, minerals, clays, soil, and sand.

Vapor Diffusion. Vapor may move by vapor diffusion through the solid, provided that a temperature gradient is established by heating, which creates a vapor-pressure gradient. Vaporization and vapor diffusion may occur in any solid where heating takes place at one surface and drying from the other, and where liquid is isolated between granules of solid.

Liquid Diffusion. The movement of liquids by diffusion in solids is restricted to the equilibrium moisture content below the point of atmospheric saturation, and to single-phase solid systems in which moisture and solid are mutually soluble. The first class applies to the last stages in the drying of clays, starches, flour, textiles, paper, and wood; and the second class includes the drying of soaps, glues, gelatins, and pastes.

Methods of predicting moisture-gradient curves from capillary-suction data have been suggested for capillary flow by Ceaglske and Hougen [*Trans. Am. Inst. Chem. Engrs.*, **33**, 283 (1937)]; for diffusional flow by Sherwood [*Ind. Eng. Chem.*, **21**, 12; 976 (1929)], Newman [*Trans. Am. Inst. Chem. Engrs.*, **27**, 203 (1931)], Gilliland and Sherwood [*Ind. Eng. Chem.*, **25**, 1134 (1933)], and Tuttle [J. Franklin Inst., **200**, 609 (1925)]; and for flow due to pressure gradients by Macey [*Trans. Brit. Ceramic Soc.*, **41** (4), 73 (1942)].

Knowledge of the internal mechanism of liquid flow in a solid during drying is of particular value in analyzing the operation of dryers from the standpoint of improving **performance** and increasing **capacity.** It is also important in the development of new drying techniques.

Extensive studies of the internal mechanism of drying have been made and reported by Krischer *et al.* [*Z. Ver. deut. Ing.*, **82** (13), 373 (1938); *Z. Ver. deut. Ing., Verfahrenstechnik* (4), 104 (1938); *Z. Ver. deut. Ing., Verfahrenstechnik* (5), 140 (1938); *Z. Ver. deut. Ing., Verfahrenstechnik* (1), 17 (1940); *Forsch. Gebiete Ingenieurw. Forschungsheft* 402, **11**, 1 (1940); *Forschungsheft* 415, **13**, 1 (1942)].

External Variables. A study of drying based on the effects of the external variables is the most common method used to investigate the drying characteristics of solids. This is because the results so obtained are usually directly applicable to the design and operation of dryers.

The principal external variables involved in any drying study are temperature, humidity, air flow, state of subdivision of the solid, agitation of the solid, method of supporting the solid, and the contact between hot surfaces and wet solid. All these variables will not necessarily occur together in one problem.

The Periods of Drying

When a solid is dried experimentally, data are usually obtained relating the moisture content to the time. These data are then plotted as moisture content (dry basis) W vs. time θ, as shown in Fig. 3*a*. This curve represents the general case when a wet solid loses moisture first by evaporation from a free-moisture surface on the solid, followed in turn by a period of evaporation from a free-water* surface of gradually decreasing area, and finally when the water evaporates in the interior of the solid.

Although Fig. 3*a* indicates that the drying rate is subject to variation with time or moisture content, this variation can be better illustrated by differentiating the curve and plotting $dW/d\theta$ vs. W, as shown in Fig. 3*b*, or as $dW/d\theta$ vs. θ, as shown in Fig. 3*c*. These *rate curves* show that the drying process is not a smooth continuous one in which a single mechanism controls throughout. The rate curve in Fig. 3*c* has the advantage of showing how long each drying period predominates.

Section BC on each curve represents the **constant-rate period.** In Fig. 3*a*, it is shown by a straight line of constant slope $dW/d\theta$, which becomes a horizontal line on the rate curves in Figs. 3*b* and 3*c*.

The curved portion CD of Fig. 3*a* is termed the falling-rate period, and, as shown in Figs. 3*b* and 3*c*, it is typified by a continuously changing rate throughout the remainder of the drying cycle. Point C, where the constant rate ends and the drying rate begins to fall, is termed the **critical moisture content.** The portion designated by AB represents a warming-up period, and it may or may not be a significant item.

The Constant-rate Period. During this period drying is typified by evaporation from a free-water surface on the surface of the solid. The rate of evaporation is essentially independent of the solid, and the rate of drying is essentially equivalent to the rate of evaporation for the same external conditions from the surface of a water layer with no solid present. For such a case, the rate of drying is determined by the rate of diffusion of water vapor through the air film at the surface of the solid out into the main body of the air stream. A constant rate of evaporation on the surface of the solid tends to maintain the surface at a constant temperature, which, in the absence of other heat effects, is nearly the wet-bulb temperature. However, when heat arrives at the surface of evaporation by radiation and/or conduction, in addition to convection, this constant temperature will lie somewhere between the air temperature and the wet-bulb temperature and in turn will produce a higher constant rate.

When heat is transferred to a wet solid by conduction through hot surfaces, and heat transfer by convection is not a factor, the boiling-point temperature rather than the wet-bulb temperature is significant. In such cases, the drying rate will be appreciably higher than by convection drying with air at the same temperature as the heating surfaces. This is utilized in indirect dryers (see Classification of Dryers, pp. 813*ff.*) in which the material is made to contact hot surfaces, frequently with vigorous agitation.

* The term *water* is used throughout for convenience with the understanding that the discussion applies equally well to other liquids.

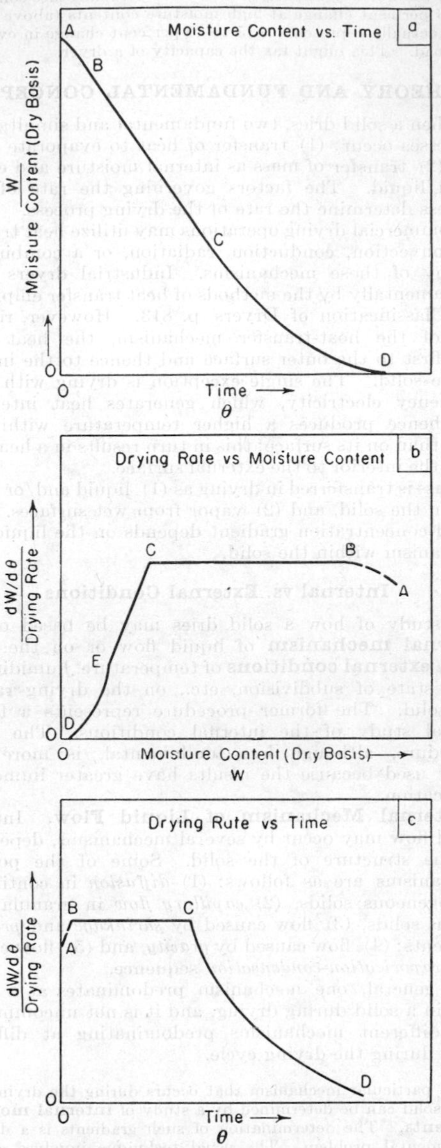

FIG. 3. The periods of drying.

Radiation is effective in increasing the constant rate by raising the surface temperature above the wet-bulb temperature. Radiation is usually only secondary to convection or conduction, although in some cases it is a primary mechanism, as in infrared drying.

When the heat for evaporation in the constant-rate period is supplied by a hot gas, a dynamic equilibrium is established between the rate of heat transfer to the material and the rate of vapor removal from the surface. This equilibrium between heat- and mass-transfer rates can be expressed as follows:

$$\frac{dw}{d\theta} = \frac{h_t A \, \Delta t}{\lambda} = k_g A \, \Delta p \qquad (1)$$

where $dw/d\theta$ = drying rate, lb. water/hr.; h_t = total heat-transfer coefficient, B.t.u./(hr.)(sq. ft.)(°F.); A = area of heat transfer and evaporation, sq. ft.; λ = latent heat of evaporation at t_s, B.t.u./lb.; k_g = mass-transfer coefficient, lb./(hr.)(sq. ft.)(atm.); $\Delta t = (t_a - t_s)$, where t_a = air temperature, °F., and t_s = temperature of surface of evaporation, °F.; $\Delta p = (p_s - p_a)$, where p_s = vapor pressure of water at t_s, surface temperature, atm.; and p_a = partial pressure of water vapor in the air, atm. When h_t is the coefficient of heat transfer by convection only, then t_s under equilibrium conditions is the wet-bulb temperature of the air, and p_s is the vapor pressure at this temperature. If heat is also supplied by radiation, then h_t is the sum $(h_c + h_r)$, where h_r is the radiation coefficient and h_c is the convection coefficient, and t_s becomes higher than the wet-bulb temperature. A similar result occurs when heat reaches the surface of evaporation by convection and conduction. When the surface is at the wet-bulb temperature, the value of Δp in mm. Hg is almost exactly one-half the wet-bulb depression in °C.

It is evident from Eq. (1) that the magnitude of the constant rate depends upon three factors: (1) the heat- or mass-transfer coefficient, (2) the area exposed to the drying medium, and (3) the difference in temperature or humidity between the air stream and the wet surface of the solid. All these factors are the external variables, as noted above. The internal mechanism of liquid flow does not affect the constant rate.

Effect of Air Velocity. The chief effect of air velocity in the constant-rate period is on h_c and k_g, since the rate of transfer of heat and mass in this period depends chiefly on the diffusion of heat and vapor through the air film at the surface of the solid. Air velocity is the principal factor in varying the thickness of this film, and its influence on the rate of drying can be expressed in terms of its influence on h_c and k_g. Shepherd, Hadlock, and Brewer [*Ind. Eng. Chem.*, **30**, 388 (1938)] showed that h_c may be expressed in terms of the mass velocity for flow parallel to plane surfaces as follows:

$$h_c = 0.0128G^{0.8} \tag{2}$$

where h_c = convection heat-transfer coefficient, B.t.u./(hr.)(sq. ft.)(°F.), and G = mass velocity of dry air, lb./(hr.)(sq. ft.). The recommended expression for the constant rate in drying from plane surfaces with air flow parallel to the surface of evaporation when there are no radiation or conduction effects is given by the following heat-transfer expression:

$$\frac{dw}{d\theta} = \frac{0.0128G^{0.8}A}{\lambda}(t_a - t_w) \tag{3}$$

where t_w = wet-bulb temperature of the drying air, °F. *Heat-transfer coefficients rather than mass-transfer coefficients should be used to estimate drying rates, since heat-transfer coefficients for drying are generally more reliable*, and, unless the temperature of the drying surface is measured, it must be calculated by means of heat-transfer considerations before mass-transfer coefficients can be applied for drying-rate predictions. The assumption that the surface of drying is at the wet-bulb temperature of the air introduces a more serious error in the computation of mass transfer than in the computation of heat transfer.

If Eq. (3) is expressed on a weight basis instead of an area basis, the following equation is obtained:

$$\frac{dW}{d\theta} = \frac{0.0128G^{0.8}(t_a - t_w)}{\rho_s L \lambda} \tag{3a}$$

where $dW/d\theta$ = drying rate, lb. water/(hr.)(lb. dry stock); ρ_s = density of the dry solid, lb./cu. ft.; L = thickness of the solid being dried, ft., assumed constant.

When the air is blown perpendicularly to the drying surface, Molstad, Farevaag, and Farrell [*Ind. Eng. Chem.*, **30**, 1131 (1938)] found that

$$h_c = 0.37G^{0.37} \tag{4}$$

Equation (4) is valid when the source of air is close to the surface. If the air impinges upon the material from slots, nozzles, or perforated plates, the coefficients will be higher than predicted by Eq. (4).

Determination of the True Surface Temperature. Frequently, radiation and conduction cause the temperature of evaporation to exceed the wet-bulb temperature of the air. When this occurs it is necessary to estimate the true surface temperature in order to calculate the constant rate. This may be done from heat-transfer considerations.

When radiation raises the surface temperature above the wet-bulb temperature, this increased surface temperature must be calculated from a heat balance equating the rate of convection and radiation heat transfer to the rate of evaporation. The latter item, for the problem, is more conveniently expressed by modifying Eq. (1) in terms of a humidity difference rather than a vapor-pressure difference as follows:

$$k_g(p_s - p_a) = k_g'(H_s - H_a) \tag{5}$$

where $k_g' = Pk_g(M_a/M_v)$ is an approximation at low humidities; M_a = molecular weight of air; M_v = molecular weight of the diffusing vapor; p_s and p_a = vapor pressure and partial pressure of the liquid on the drying surface and in the air, respectively, atm.; H_s = saturation humidity of the air at the temperature of the drying surface, lb./lb. dry air; H_a = humidity of the drying air, lb./lb. dry air; and P = total pressure, atm. For air–water vapor mixtures, $k_g' = 1.6k_g$, approximately.

A rate balance between evaporation and heat transfer when radiation occurs may now be written as follows:

$$k_g'A\lambda(H_s - H_a) = h_cA(t_a - t_s) + h_r\epsilon A(t_r - t_s) \tag{6}$$

where λ = latent heat of evaporation, B.t.u./lb. at t_s.
A = area of both heat and mass transfer, sq. ft.
h_c = heat-transfer coefficient by convection, B.t.u./(hr.)(sq. ft.)(°F.), Eq. (2) or Eq. (4).
h_r = heat-transfer coefficient by radiation, B.t.u./(hr.)(sq. ft.)(°F.), as defined in Sec. 6, Fig. 12.
t_s = temperature of the wet surface, °F.
t_r = temperature of source radiating heat to the wet surface, °F.
ϵ = emissivity of surface receiving radiation.

Equation (6) may be modified by means of the empirical relationship for air–water vapor mixtures, $h_c/k_g' = c_s$, where c_s = heat capacity of humid air, B.t.u./(lb. dry air)(°F.), as defined on p. 811. Thus Eq. (6) becomes

$$\frac{\lambda}{c_s}(H_s - H_a) = (t_a - t_s) + \epsilon\frac{h_r}{h_c}(t_r - t_s) \tag{6a}$$

Equation (6a) may be solved by trial and error, or graphically as indicated in Example 1, to estimate the true values of H_s and t_s, and hence the actual drying rate. The proper values of λ and h_r depend on the value of t_s but generally do not vary over a wide range with the temperatures usually encountered in air drying. The application of Eq. (6a) is illustrated as follows:

Example 1. A wet material is drying in a tray exposed to air at 300°F. and a humidity of 0.02 lb. water/lb. dry air. The air velocity is 400 ft./min. Determine the true surface temperature.

Solution. If no radiation occurs, the wet surface will assume a temperature of 113°F., with a corresponding value of H_s = 0.0647. If a metal tray directly above the wet material attains the air temperature of 300°F., however, then t_s will be above 113°F. For this case, let ϵ = 0.9. From Eq. (2), h_c = 3.7, and from Sec. 6, Fig. 12, h_r is estimated to be 1.5, λ will be about 1020, c_s = 0.25. Substituting in Eq. (6a),

$$\frac{1020}{0.25}(H_s - 0.02) = (300 - t_s)\left(1 + \frac{0.9 \times 1.5}{3.7}\right)$$

or

$$(H_s - 0.02) = 0.000334(300 - t_s)$$

The value of t_s and H_s may now be obtained from the humidity chart (Fig. 8) by drawing a line through the point H = 0.02, t = 300 with slope 0.000334, and reading at the intersection with the saturated humidity curve (*cf.* Fig. 5) the values of H_s = 0.08, and t_s = 120°F., which check in the above equation. Thus the constant rate is increased by (0.08 − 0.02)/(0.0647 − 0.02) = 1.34, or by 34 per cent.

In the case where t_r does not equal t_a, additional manipulation is required to obtain the required form for the graphical solution illustrated above.

Estimation of the Constant Rate. The constant rate of evaporation from a plane surface at the wet-bulb temperature may be calculated from Eq. (3) or may be estimated from Fig. 4. This figure is based on an air

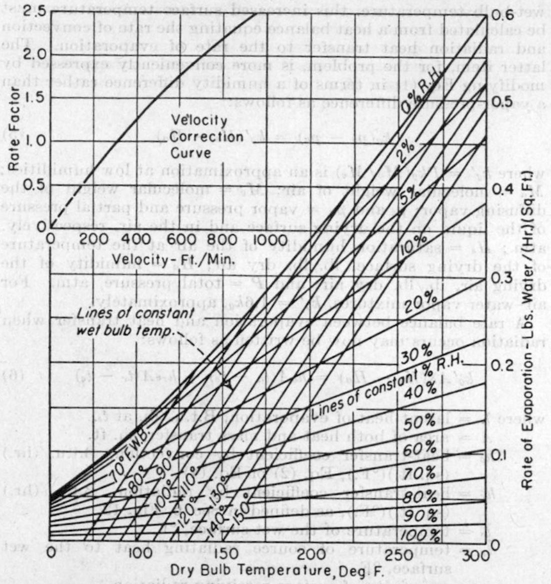

FIG. 4. Chart for estimating the rate of drying in the constant-rate period. [*Shepherd, Hadlock, and Brewer, Ind. Eng. Chem.*, **30**, 388 (1938).]

velocity of 300 ft./min. at 150°F. and 30 per cent relative humidity with the surface of evaporation at the wet-bulb temperature. For these conditions, a heat-transfer coefficient of 3.1 B.t.u./(hr.)(sq. ft.)(°F.) was used. This was purposely selected about 10 per cent less than Eq. (2) would predict for this air rate. A correction curve for other air rates is given on the chart. The drying rate given by this chart does not include the effects of radiation or conduction through unwetted surfaces. When such effects are important, the rate determined from Fig. 4 must be corrected, not only for radiation and conduction, but also for the resulting increased surface temperature, *i.e.*, decreased temperature difference. When conduction is negligible a radiation correction can be applied to the rate determined from Fig. 4 by multiplying this rate by the factor $(h_c + h_r)/h_c$. If drying takes place from one surface and there are no radiation effects, the over-all coefficient for total heat transfer is given by

$$h_t = h_c\left(1 + \frac{A_u}{1 + Lh_c/k}\right) \tag{7}$$

where A_u = ratio of outside unwetted surface to wetted surface; L = depth of material in the tray, ft.; and k = thermal conductivity of the wet material B.t.u./(hr.)(sq. ft.)(°F./ft.). In Eq. (7), the factor in parentheses can be used in conjunction with a temperature correction to adjust the drying rate read from Fig. 4 for conduction through uninsulated trays, the usual case in commercial dryers.

When both radiation and conduction are significant, the over-all coefficient for total heat transfer to the wetted surface is given by

$$h_t = (h_c + h_r)\left[1 + \frac{A_u}{1 + L(h_c + h_r)/k}\right] \tag{8}$$

The constant rate must finally be corrected by the ratio $(t_a - t_s)/(t_a - t_w)$. A general relationship for estimating t_s based on the use of Eq. (8) to determine h_t is as follows:

$$(H_s - H_a) = \frac{h_t c_s}{\lambda h_c}(t_a - t_s) \tag{9}$$

Figure 5 indicates how H_s and t_s may be determined graphically on a humidity chart by the point of intersection on the saturation-humidity curve of a straight line of slope $h_t c_s / h_c \lambda$ and passing through (H_a, t_a).

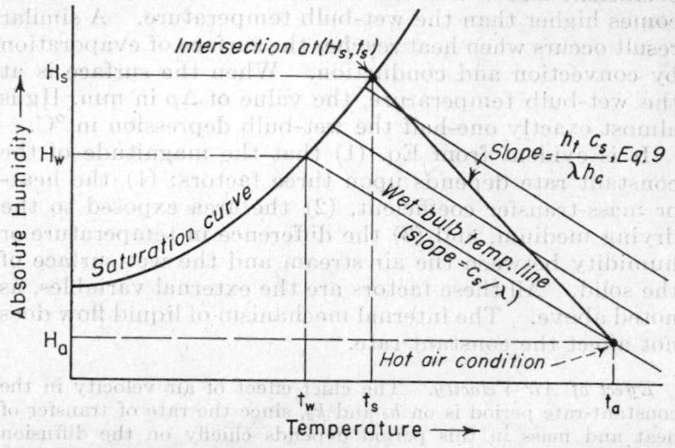

FIG. 5. Graphical estimation of the true surface temperature during the constant-rate period.

Example 2. An inorganic pigment is being dried in a tray dryer that consists of 2 tiers of 44 stainless-steel trays, 1.25 in. deep and spaced 1.5 in. apart. The trays are 26 in. square. The air velocity across the trays averages 300 ft./min., and the average air temperature is 225°F. The average relative humidity across the trays is 12 per cent. Estimate the constant drying rate by means of Fig. 4. Determine how much increase in rate can be effected by doubling the air rate and lowering the relative humidity to 5 per cent.

Solution. From Fig. 4, the constant rate corresponding to the given conditions is 0.27 lb./(hr.)(sq. ft.). This must be corrected for conduction through unwetted surfaces and for radiation. For the trays of the dimensions specified, $A_u = 1$. Also, $L = 0.104$ ft., $h_c = 3.1$, and $h_r = 1.5$ by Fig. 12, Sec. 6. The value of k, however, is not so easy to determine. Shepherd, Hadlock, and Brewer [*Ind. Eng. Chem.*, **30**, 388 (1938)] found for wet sand that $k = 2.0$, although k for sand alone is 0.2 and for water alone is about 0.4. For this problem, let $k = 1.0$. Then, from Eq. (8),

$$h_t = (3.1 + 1.5)\left[1 + \frac{1}{1 + 0.104(3.1 + 1.5)/1.0}\right] = 7.7$$

The correction factor for radiation and conduction is then $7.7/3.1 = 2.5$.

The true surface temperature is obtained from Eq. (9) by substituting the appropriate values as given in the problem. Thus

$$(H_s - 0.114) = \frac{7.7 \times 0.295}{3.1 \times 1010}(225 - t_s)$$

A straight line through $H_a = 0.114$ and $t_a = 225$°F., with slope = 0.000725, intersects the saturation curve at $H_s = 0.174$ and $t_s = 144$. Hence the temperature correction is (225 − 144)/(225 − 137) = 0.92, and the final correct constant drying rate becomes 0.27(2.5)(0.92) = 0.62 lb./(hr.)(sq. ft.). The rate from Fig. 4 could also have been corrected by multiplying it by $(H_s - H_a)/(H_w - H_a)$, which gives in this problem 0.27(0.174 − 0.114)/(0.140 − 0.114) = 0.62 lb./(hr.)(sq. ft.). The latter correction factor must be determined carefully, because differences between nearly equal numbers are involved.

If an air rate of 600 ft./min. is used, at 225°F. and 5 per cent relative humidity, the rate from Fig. 4 is 0.33 lb./(hr.)(sq. ft.); while the actual constant drying rate is (0.33)(1.76)(1.85) = 1.1

lb./(hr.)(sq. ft.), where 1.76 is the velocity correction, and 1.85 represents the total conduction, radiation, and surface-temperature correction as shown in the first part of the problem.

The Constant-rate Period in Through-circulation Drying. The methods developed for estimating the rate of evaporation when air flows across a free-water surface must be modified for the case of air flow through a permeable bed of solids. Marshall and Hougen [*Trans. Am. Inst. Chem. Engrs.*, **38**, 91 (1942)] showed that the constant rate in through-circulation drying depends on the air rate, the air temperature, the air humidity, the particle size in the permeable bed, and the method by which the material is formed to produce a permeable bed. Of four methods of preforming (see Types of Dryers, p. 823) studied, the highest rate is obtained by granulating the wet feed prior to drying; the next highest rates, in order, are obtained by extruding the wet material through round holes, by forming sticks on a heated finned drum, and by pelleting. The magnitude of the constant rate attributed to each method is due, for the most part, to the amount of drying surface created.

The following general expression for the constant rate in through-circulation drying for the system water and air was developed by Gamson, Thodos, and Hougen [*Trans. Am. Inst. Chem. Engrs.*, **39**, 1 (1943)] from experiments on the rate of evaporation of water from the surface of wet granular solids:

$$\frac{dW}{d\theta} = \frac{0.42aG^{0.59}(\Delta H)_m}{\rho_s D_p^{0.41}} = \frac{0.37c_s aG^{0.59}\Delta t_m}{\rho_s \lambda D_p^{0.41}} \tag{10}$$

where $dW/d\theta$ = constant drying rate, lb. water/(hr.)(lb. dry stock); a = drying area, sq. ft./(cu. ft. bed volume); G = superficial mass velocity, lb. dry air/(hr.)(sq. ft.); $(\Delta H)_m$ = logarithmic mean of inlet and outlet humidity driving forces across the air film adjacent to the particle, through which the water vapor diffuses, lb./lb. (the surface humidity being taken as the humidity corresponding to the wet-bulb temperature of the drying air); ρ_s = bulk density of dry granular bed, lb./cu. ft.; D_p = average diameter of the particle, ft.; Δt_m = logarithmic mean difference between the temperature entering and leaving the bed and the wet-bulb temperature, °F.; c_s = humid heat, B.t.u./(lb. dry air)(°F.); and λ = latent heat of evaporation, B.t.u./lb. From Eq. (10), a close approximation to the height of a mass- or heat-transfer unit through the bed is

$$H_t = \frac{2.5}{a}(D_pG)^{0.41} \tag{11}$$

For the more general case of through-circulation drying, when liquids other than water vaporize into gases other than air, the following equations from Gamson, Thodos, and Hougen [*Trans. Am. Inst. Chem. Engrs.*, **39**, 1 (1943)] are recommended:

$$\frac{dW}{d\theta} = \frac{MGa}{p_{gf}M_m\rho_s}\left(\frac{D_pG}{\mu}\right)^{-0.41}\left(\frac{\mu}{\rho_g D}\right)^{-0.67}(\Delta p)_m$$
$$= \frac{1.06c_pGa}{\lambda\rho_s}\left(\frac{D_pG}{\mu}\right)^{-0.41}\left(\frac{c_p\mu}{k}\right)^{-0.67}(\Delta t)_m \tag{12}$$

where G = mass velocity, lb. total gas/(hr.)(sq. ft.); μ = viscosity of gas, lb./(hr.)(ft.); ρ_g = density of gas, lb./cu. ft.; D_v = diffusivity of evaporated liquid, sq. ft./hr.; p_{gf} = mean partial pressure of inert gas across gas film at solid surface, atm.; $(\Delta p)_m = (p_s - p_g)_m$, where p_s = vapor pressure of liquid at the wet-bulb temperature of the gas, atm., and p_g = actual vapor pressure of the liquid in the gas stream, atm.; c_p = average specific heat of gas mixture, B.t.u./(lb.)(°F.); k = thermal conductivity of gas mixture, B.t.u./(hr.)(sq. ft.)(°F./ft.); M = molecular weight of evaporating liquid; M_m = mean molecular weight of gas stream.

Equations (10) and (12) apply only when $(D_pG/\mu) > 350$. For $(D_pG/\mu) < 350$, Wilke and Hougen [*Trans. Am. Inst. Chem. Engrs.*, **41**, 445 (1945)] recommend the following equations:

$$\frac{dW}{d\theta} = \frac{1.82MGa}{p_{gf}M_m\rho_s}\left(\frac{D_pG}{\mu}\right)^{-0.51}\left(\frac{\mu}{\rho_g D}\right)^{-0.67}(\Delta p)_m$$
$$= \frac{1.95c_pGa}{\lambda\rho_s}\left(\frac{D_pG}{\mu}\right)^{-0.51}\left(\frac{c_p\mu}{k}\right)^{-0.67}(\Delta t)_m \tag{12a}$$

where the symbols have the same meaning as for Eq. (12).

In order to use Eq. (10), (12), or (12*a*), values of a and ρ_s must be known. The value of a is not always easy to estimate without experimental data. However, if the void fraction is known, a can sometimes be estimated from one of the following relations:
For spherical particles:

$$a = \frac{6(1 - F)}{(D_p)_m} \tag{13}$$

For uniform cylindrical particles:

$$a = \frac{4(D/2 + Z)(1 - F)}{DZ} \tag{14}$$

where F = void fraction; $(D_p)_m$ = harmonic mean diameter of spherical particles, ft.; D = diameter of cylinder, ft.; and Z = height of cylinder, ft. Equation (13) should be used where the granular bed can be assumed to consist of spherical particles, while Eq. (14) applies to cylindrical particles, such as extrusions. When the extrusions are long in comparison with their diameter, Eq. (14) may be simplified to

$$a = \frac{4(1 - F)}{D} \tag{15}$$

Example 3. Estimate the constant drying rate in the through-circulation drying of extruded titanium dioxide. The average diameter of the extrusions is ¼ in., and their length is large compared with their diameter. The superficial air flow is 250 ft./min. The inlet air temperature is 275°F. at 3 per cent relative humidity. The depth of bed is 3 in., and the dry-bulk density of the layer is estimated to be 60 lb./cu. ft. The void fraction is estimated to be 50 per cent.

Solution. Referring to Eq. (10), G = 750 lb./(hr.)(sq. ft.), ρ_s = 60 lb./cu. ft., D_p = 0.0208 ft., and $a = 4(1 - 0.5)/0.0208$ = 96 sq. ft./cu. ft. To determine $(\Delta H)_m$, ΔH_2, the driving force of the air leaving the bed, must be determined. To find this, observe that the number of transfer units (see Sec. 8), N_t, is given by $N_t = (H_2 - H_1)/(\Delta H)_m$. For the constant-rate period $(\Delta H)_m$ = logarithmic mean = $(H_2 - H_1)/\ln(\Delta H_1/\Delta H_2)$, since the surface humidity is constant and assumed equal to the wet-bulb temperature humidity. Hence, $N_t = \ln(\Delta H_1/\Delta H_2) = L/H_t$, where L = bed depth and H_t is given by Eq. (11). Thus $\Delta H_2 = \Delta H_1 e^{-L/H_t}$. By Eq. (11), $H_t = (2.5)(15.6)^{0.41}/96$ = 0.082 ft. or 1.0 in. Hence N_t = 3/1.0 = 3.0, and $\Delta H_2 = \Delta H_1 e^{-3.0}$, $\Delta H_1 = H_w - H_1$, and H_1 = 0.0638 lb./lb., H_w = 0.102 lb./lb. obtained from a humidity chart for 275° F. air at 3 per cent relative humidity. Thus $\Delta H_2 = 0.0382\ e^{-3.0} = 0.00189$, and $H_2 = 0.102 - 0.00189 = 0.100$. Then $(\Delta H)_m = (H_2 - H_1)/\ln(\Delta H_1/\Delta H_2)$ = (0.100 − 0.0638)/3.0 = 0.0121. Substituting in Eq. (10), the constant rate is

$$\frac{dW}{d\theta} = \frac{0.42 \times 96 \times (750)^{0.59} \times 0.0121}{60 \times (0.0208)^{0.41}} = 2.0 \text{ lb./(hr.)(lb. dry stock)}$$

Evaporation from Liquid Drops. The general theory covering this subject may be found in Sec. 8, p. 546. For the important problem of spray drying, evaporation of water drops may be estimated by means of Fig. 16, Sec. 8. Below a Reynolds number of 20 for spherical particles, the heat-transfer coefficient across the gas film surrounding the drop is approximately

$$h = \frac{2k_f}{D_p} \tag{16}$$

where h = heat-transfer coefficient based on area of drop, B.t.u./(hr.)(sq. ft.)(°F.); k_f = thermal conductivity of gas film, B.t.u./(hr.)(sq. ft.)(°F./ft.); and D_p = particle diameter, ft.

Equation (16) is applicable to spray dryers when the Reynolds number involved is less than 20. Drop diameters are almost always less than 500 μ and are usually in the range of 20 to 150 μ.

The rate of evaporation of drops may be expressed in terms of heat or mass transfer. In terms of heat transfer, the evaporation rate is given by the equation

$$\frac{dw}{d\theta} = \frac{2\pi k_f D_p}{\lambda}(t_a - t_s) \tag{17}$$

where $dw/d\theta$ = evaporation rate, lb./hr.; λ = latent heat of liquid at t_s, B.t.u./lb.; t_a = gas temperature, °F.; and t_s = particle temperature, °F. A similar expression based on mass transfer is

$$\frac{dw}{d\theta} = \frac{2\pi M D_v D_p}{RT}(p_s - p_a) \tag{18}$$

where D_v = the diffusivity of the vapor, sq. ft./hr.; T = absolute temperature of the gas, °R.; R = gas constant, (cu. ft.)(atm.)/(°R.); M = molecular weight of diffusing vapor; p_s = vapor pressure at the particle surface corresponding to the liquid temperature, atm.; p_a = vapor pressure of liquid in the drying medium, atm.

Both Eq. (17) and Eq. (18) are based on the assumption that Eq. (16) holds. If Eq. (17) is integrated for a constant drop diameter, *i.e.*, if it is assumed that the solid in the liquid drop forms a structure that becomes rigid at a fixed D_p, and evaporation proceeds as from a pure liquid drop, an expression for the time of evaporation is obtained as follows:

$$\theta = \frac{\lambda W \rho_s D_p^2}{12k_f(t_a - t_s)} \tag{19}$$

where θ = time, hr.; λ = latent heat of evaporation, B.t.u./lb.; W = water content, dry basis, of the drop as it enters the drying chamber, lb./lb. dry solid; and ρ_s = density of dry particle, lb./cu. ft. The temperature difference between drop and gas $(t_a - t_s)$ is essentially constant for a single drop evaporating in a large mass of gas. *However, in spray dryers this is not true, and an over-all average temperature difference must be used.*

When the drop diameter varies as evaporation proceeds, the expression for the time of evaporation becomes

$$\theta = \frac{\lambda \rho_L[(D_{p_1})^2 - (D_{p_2})^2]}{8k_f(t_a - t_s)} \tag{20}$$

where ρ_L = density of the evaporating liquid, lb./cu. ft.; D_{p_1} = drop diameter at the start of evaporation, ft.; and D_{p_2} = drop diameter of dry particle, ft. Equation (20) assumes that the drop density is essentially that of the liquid.

Drying at Air Temperatures above the Boiling Point of the Liquid. When the temperature of the drying air is maintained above the boiling point of the evaporating liquid, the usual equations expressing evaporation rate as a function of the vapor-pressure difference lose significance, since the vapor-pressure driving force can apparently approach zero while the drying rate is still appreciable. Such cases can be treated conveniently on a basis of heat transfer, since a temperature difference must always exist in order for drying to proceed. Victor [*Chem. & Met. Eng.*, **52**, 105 (1945)] has discussed the psychrometry of this case, but he did not consider it from the standpoint of the drying rate. Above a dry-bulb temperature of 260°F. the amount of moisture the air can pick up apparently increases with an increase in dew point.

The effect of high humidities on the drying rate at temperatures above the boiling point can be interpreted by Eq. (9). If t_a = 300°F., H_a = 0.03, and $h_c = h_t$, a humidity chart shows that $(H_s - H_a) = 0.046$, which is proportional to the rate. Now, if H_a is increased to 0.07 at 300°F., the value of $(H_s - H_a)$ is still 0.046, which indicates no adverse effect on the drying rate. In fact the rate may actually increase. This fact has been verified in commercial dryers, operating at air temperatures above the boiling point, by maintaining as high an air recirculation as the physical construction of the dryer will permit with no reduction in drying time.

Constant-rate Period When Heat Transfer Depends on Conduction and Radiation. In indirect dryers where heat transfer and drying do not depend on convection from heated gases, the drying rate depends on heat conduction through retaining walls to wet material in contact with such surfaces, or on radiation, or on both. This applies to drum dryers, indirect continuous-sheeting dryers, steam-tube rotary dryers, agitated-pan dryers, vacuum rotary and vacuum tray dryers, and infrared dryers (see Types of Dryers, p. 815*ff*.).

The chief difference between indirect and direct drying is that the material is usually at a higher temperature than the surrounding air, so that heat is actually transferred to the air instead of from the air.

For the constant rate when drying occurs in contact with hot surfaces, the rate of heat transfer, and hence the rate of drying, is given by the general expression

$$\frac{Q}{A\theta} = \frac{(t_h - t_s)}{(1/h_w) + (1/h_m) + (1/h_p)} \tag{21}$$

where Q = total heat transferred, B.t.u.; θ = time, hr.; A = total heat-transfer surface exposed to wet material, sq. ft.; t_h = temperature of heating medium, °F.; t_s = temperature of solid being dried, °F.; and h_m, h_w, and h_p = heat-transfer coefficients of heating medium (usually condensate), metal wall, and wall to wet product, respectively, B.t.u./(hr.)(sq. ft.)(°F.).

Generally, the individual heat-transfer coefficients in Eq. (21) are difficult to determine and estimate, and over-all coefficients are therefore used according to the equation

$$q = UA(t_h - t_s) \tag{22}$$

where q = rate of heat transfer, B.t.u./hr.; U = over-all heat-transfer coefficient based on the temperature difference between the heating medium and the product, B.t.u./(hr.)(sq. ft.)(°F.). The over-all coefficient is a function of dryer type. Thus, in agitated-pan dryers, U depends on the degree of agitation, which affects primarily h_p in Eq. (21), temperature of the surface, the physical properties of the wet material, etc. Values of U and their variation with operating conditions are considered in the discussion of the types of indirect dryers, p. 859.

As long as the temperature difference in Eq. (22) remains constant, a constant drying rate usually will be maintained. As drying proceeds, however, the material temperature will begin to increase after a certain moisture content, so that a falling-rate period is also encountered with indirect dryers. U is frequently defined for the entire drying period on the basis of an over-all mean temperature difference; thus

$$q = UA(\Delta t)_m \tag{23}$$

where $(\Delta t)_m$ = average temperature difference over the entire drying period, °F.

The Falling-rate Period. In the discussion of the periods of drying, p. 802, it was shown that the drying process is discontinuous, consisting of a period of a constant rate of evaporation and a period in which the rate continuously decreases (see Fig. 3, line *CD*). The latter period is usually designated as the falling-rate period, and it begins when the constant-rate period ends at the critical moisture content. If the critical moisture content is less than the required final moisture content, the constant-rate period will constitute the whole drying process. On the other hand, if the initial moisture content is less than the critical moisture content, as in the case of some slow-drying materials, such as soap and wood, the whole drying process will be in the falling-rate period. This period, in the most general case, can be divided into two zones: (1) the zone of unsaturated surface drying and (2) the zone where internal liquid flow controls.

The Zone of Unsaturated Surface Drying. This period follows immediately after the critical point; the decrease in the rate of drying in this zone is caused by a decrease in the wetted surface of the material. The surface is no longer completely wetted, and dry portions of the solid protrude into the air film, reducing the rate of evaporation per unit of total surface. The effective wetted surface in this zone is frequently a linear function of the water content, so that the rate of drying varies linearly with the average water content of the solid, as shown by line *CE* in Fig. 3*b*. The factors discussed above as influencing the rate of drying in the constant-rate period have similar effects on the rate of drying in this zone since the mechanism of drying is essentially the same.

Zone Where Internal Liquid Flow Controls. During the second zone of the falling-rate period, the rate of internal liquid flow controls the drying rate; and, in drying to low moisture contents, this period predominates in determining the drying time.

Studies of internal moisture flow have indicated the possibility of several controlling mechanisms, the more significant ones having been postulated previously as diffusion, capillarity, and pressure gradients due to shrinkage. Of these mechanisms, internal moisture movement by diffusion has been treated extensively, and capillary flow and flow caused by shrinkage and pressure gradients have received preliminary consideration.

The *limitations* of the diffusion equations in drying have been summarized, and their restriction to certain classes of materials and certain times of the drying cycle noted [Hougen, McCauley, and Marshall, *Trans. Am. Inst. Chem. Engrs.*, **36**, 183 (1940)]. When liquid diffusion controls in the falling-rate period, it obeys the same fundamental laws that apply to the diffusion of heat. Sherwood [*Ind. Eng. Chem.*, **21**, 12 (1929)] adapted the diffusion equations to the falling-rate period for the case where the surface is dry or at its equilibrium moisture content and the solid has a *uniform* initial moisture distribution. The following expression was obtained:

$$\frac{W - W_e}{W_o - W_e} = \frac{8}{\pi^2}\left[e^{-D\theta\left(\frac{\pi}{2L}\right)^2} + \frac{1}{9}e^{-9D\theta\left(\frac{\pi}{2L}\right)^2} + \frac{1}{25}e^{-25D\theta\left(\frac{\pi}{2L}\right)^2} + \cdots\right] \quad (24)$$

where W, W_o, W_e = average-moisture contents (dry basis) at any time θ, at the start of the diffusional flow period, and in equilibrium with the external conditions, respectively, lb./lb.; D = liquid diffusivity, sq. ft./hr.; and L = one-half the thickness of the solid layer through which the liquid is diffusing, ft. This equation assumes that evaporation is occurring from two opposite faces of the solid. When evaporation occurs from only one surface, L = total thickness of solid layer, ft.

Equation (24) was derived on the assumption that D is constant. This is rarely true, however, and D has been shown to vary with moisture content, temperature, and humidity [Hougen, McCauley, and Marshall, *Trans. Am. Inst. Chem. Engrs.*, **36**, 183 (1940); Bateman, Hohf, and Stamm, *Ind. Eng. Chem.*, **31**, 1150 (1939)]. Thus, in using Eq. (24), the basis for the value of diffusivity used should be known. Van Arsdel [*Trans. Am. Inst. Chem. Engrs.*, **43**, 13 (1947)] has developed a graphical method for taking into account the variation of D with moisture content in drying hydrophilic solids to low moisture contents.

When the time becomes large, a limiting form of Eq. (24) is obtained as follows:

$$\frac{W - W_e}{W_o - W_e} = \frac{8}{\pi^2}e^{-D\theta\left(\frac{\pi}{2L}\right)^2} \quad (25)$$

From Eq. (25), an expression for the rate of drying may be derived to give

$$\frac{dW}{d\theta} = -\frac{\pi^2 D}{4L^2}(W - W_e) \quad (26)$$

where $dW/d\theta$ = drying rate, lb./(hr.)(lb. dry material). Equation (26) states that, when internal diffusion controls for long times, the rate of drying is directly proportional to the free moisture content ($W - W_e$) and the liquid diffusivity D, and that the drying time varies as the square of the material thickness. However, Eq. (26) holds only when $(W - W_e)/(W_o - W_e) < 0.6$. When this term exceeds 0.6, the rate of drying vs. moisture content gives a curve that is concave upward.

Equations (24), (25), and (26) hold only for a slab-shaped solid the length of which is large compared with its thickness. For other shapes, reference should be made to Newman [*Trans. Am. Inst. Chem. Engrs.*, **27**, 310 (1931)].

The falling rate frequently can be expressed with fair accuracy over the required range of moisture content by an equation similar to Eq. (26); thus

$$\left(\frac{dW}{d\theta}\right)_f = -K(W - W_e) \quad (27)$$

where K is a function of the constant rate as follows:

$$K = -\frac{(dW/d\theta)_c}{(W_c - W_e)} \quad (28)$$

where $(dW/d\theta)_c$ = constant drying rate, lb./(hr.)(lb. dry material); and W_c = critical moisture content, lb./lb. dry material. Substituting the expression for $(dW/d\theta)_c$ as given by Eqs. (1) and (3*a*), the value of K becomes

$$K = \frac{h_t(t_a - t_s)}{\rho_s L\lambda(W_c - W_e)} \quad (29)$$

and hence the falling rate for this case is given by

$$\left(\frac{dW}{d\theta}\right)_f = -\frac{h_t(t_a - t_s)(W - W_e)}{\rho_s L\lambda(W_c - W_e)} \quad (30)$$

For materials obeying Eq. (30), the drying time varies directly as the thickness. When the surface temperature in the constant-rate period is at the wet-bulb temperature, t_w can be substituted for t_s and $0.0128G^{0.8}$ can be substituted for h_t in Eqs. (29) and (30).

The drying time for each case of the falling-rate period may be obtained by integration of Eqs. (26) and (30), respectively, to give:

1. *Diffusion law:*

$$\theta_f = \frac{4L^2}{D\pi^2}\ln\left(\frac{W_c - W_e}{W - W_e}\right) \quad (31)$$

2. *Proportional-to-thickness law:*

$$\theta_f = \frac{\rho_s L\lambda(W_c - W_e)}{h_t(t_a - t_s)}\ln\left(\frac{W_c - W_e}{W - W_e}\right) \quad (32)$$

The following table gives an approximate classification of materials that obey Eqs. (31) and (32).

Table 1

Materials Obeying Eq. (31)	Materials Obeying Eq. (32)
1. Single-phase solid systems, such as soap, gelatin, glue	1. Coarse granular solids, such as sand, paint pigments, minerals, etc.
2. Wood and similar solids below the fiber-saturation point	2. Materials in which moisture flow occurs at concentrations above the equilibrium moisture content at atmospheric saturation, or above the fiber-saturation point
3. Last stages of drying starches, textiles, paper, clay, hydrophilic solids, and other materials when bound water is being removed	

Equations (30) and (32) hold for cross-circulation drying. When through-circulation drying is involved, the appropriate constant-rate expression given by Eq. (10) must be used to determine K in Eq. (28). Thus, for through-circulation drying in the falling-rate period when Eq. (27) holds, the rate is given by

$$\left(\frac{dW}{d\theta}\right)_f = -\frac{0.37c_{sa}G^{0.59}(\Delta t)_m}{\rho_s\lambda D_p^{0.41}(W_c - W_e)}(W - W_e) \quad (33)$$

where the symbols have been defined for Eqs. (10), (24), and (28).

Critical Moisture Content. To use the above equations for estimating the drying time in the falling-rate period, it is necessary to know values of the critical moisture content. Such values are difficult to obtain without making actual drying tests, which in themselves would give the required drying time and thereby obviate the necessity of the calculations. However, in those cases where drying tests are not feasible, some estimate of the critical moisture content must be made. Broughton [*Ind. Eng. Chem.*, **37**, 1184 (1945)] correlated critical moisture contents for the drying of kaolin and china clay by cross circulation. His correlation, however, applies only to those solids in which liquid diffusion is the internal mechanism of moisture flow.

Values of critical moisture content for some representative materials are given in Table 2 for drying by cross circulation and in Table 6 for drying by through circulation. The values tabulated are approximate at best, since the critical moisture content varies with the rate of drying and the thickness of the layer being dried.

It appears that the constant-rate period ends when the moisture content at the surface reaches some specific value. If the rate of drying is great, the moisture gradients within the solid will be steep and the average moisture content considerably greater than that at the surface. For this reason the critical moisture content (average through the material) increases with an increase in the rate of drying and with an increase in the thickness of the layer being dried.

Table 2. Approximate Critical Moisture Contents Obtained on the Air Drying of Various Materials, Expressed as Percentage Water on the Dry Basis

Material	Thickness, in.	Critical moisture, % water, dry basis
Barium nitrate crystals, on trays	1.0	7
Beaverboard	0.17	Above 120
Brick clay	.62	14
Carbon pigment	1	40
Celotex	0.44	160
Chrome leather	.04	125
Copper carbonate (on trays)	1–1.5	60
English china clay	1	16
Flint clay refractory brick mix	2.0	13
Gelatin, initially 400% water	0.1–0.2 (wet)	300
Iron blue pigment (on trays)	0.25–0.75	110
Kaolin		14
Lithol red	1	50
Lithopone press cake (in trays)	0.25	6.4
	.50	8.0
	.75	12.0
	1.0	16.0
Niter cake fines, on trays		Above 16
Paper, white eggshell	0.0075	41
Fine book	.005	33
Coated	.004	34
Newsprint		60–70
Plastic clay brick mix	2.0	19
Poplar wood	0.165	120
Prussian blue		40
Pulp lead, initially 140% water		Below 15
Rock salt (in trays)	1.0	7
Sand, 50–150 mesh	2.0	5
Sand, 200–325 mesh	2.0	10
Sand, through 325 mesh	2.0	21
Sea sand (on trays)	0.25	3
	.5	4.7
	.75	5.5
	1.0	5.9
	2.0	6.0
Silica brick mix	2.0	8
Sole leather	0.25	Above 90
Stannic tetrachloride sludge	1	180
Subsoil, clay fraction 55.4%		21
Subsoil, much higher clay content		35
Sulfite pulp	0.25–0.75	60–80
Sulfite pump (pulp lap)	0.039	110
White lead		11
Whiting	0.25–1.5	6–9
Wool fabric, worsted		31
Wool, undyed serge		8

Approximate Equations for Estimating Drying Time

An estimate of the over-all drying time for a given drying problem involves an estimate of both the constant-rate drying time and the falling-rate drying time. An approximate equation for the over-all drying time applicable to the tray drying of material of the type listed in Table 1 as obeying Eq. (32) may be obtained from Eqs. (3*a*) and (32) as follows:

$$\theta_t = \theta_c + \theta_f = \frac{(W_o - W_c)\lambda L \rho_s}{h_t(t_a - t_s)} + \frac{\rho_s L \lambda (W_c - W_e)}{h_t(t_a - t_s)} \ln \frac{W_c - W_e}{W - W_e}$$
$$= B\left(\frac{W_o - W_c}{W_c - W_e} + \ln \frac{W_c - W_e}{W - W_e}\right) \qquad (34)$$

where $B = \dfrac{\rho_s L \lambda (W_c - W_e)}{h_t(t_a - t_s)} = \dfrac{1}{K}$

θ_t = total drying time, hr.
θ_c = drying time for constant-rate period, hr.
θ_f = drying time for falling-rate period, hr.
W_o = average initial moisture content, lb./lb. dry solid.
W_c = average critical moisture content, lb./lb. dry solid.
W_e = average equilibrium moisture content, lb./lb. dry solid.
W = average moisture content at time θ_t, lb./lb.
h_t = total over-all heat-transfer coefficient given by Eq. (8), B.t.u./(hr.)(sq. ft.)(°F.).
t_a = air temperature, °F.
t_s = temperature of the surface of wet material during the constant-rate period obtainable from Eq. (9), °F.
L = depth of material in tray, ft.
λ = latent heat of evaporation at t_s, B.t.u./lb.
ρ_s = density of dry solid, lb./cu. ft.

Equation (34) applies to those materials satisfying Eq. (32) when they are not dried to very low moisture contents.

For through-circulation drying, an expression identical to Eq. (34) is obtained except for the coefficient. Thus the total drying time for through-circulation drying is given by

$$\theta_t = B'\left(\frac{W_o - W_c}{W_c - W_e} + \ln \frac{W_c - W_e}{W - W_e}\right) \qquad (35)$$

where $B' = \dfrac{2.7 \rho_s \lambda D_p^{0.41} (W_c - W_e)}{c_s a G^{0.59} (\Delta t)_m}$.

Example 4. In Example 2, if the initial moisture content is 1.0 lb. water/lb. dry stock, estimate the time required to dry the pigment to 0.01 lb./lb. under the two conditions of the problem. Assume a critical moisture content of 0.30 lb./lb. for the first case, and 0.50 for the second case. The dry density is 60 lb./cu. ft.

Solution. For 300 ft./min., 225°F., and 12 per cent humidity, B in Eq. (34) becomes

$$\frac{\rho_s L \lambda (W_c - W_e)}{h_t(t_a - t_s)} = \frac{(60)(0.104)(1010)(0.3 - 0)}{(7.7)(225 - 144)} = 3.04.$$

Then

$$\theta_t = \theta_c + \theta_f = \frac{3.04(1.0 - 0.3)}{(0.3 - 0)} + 3.04 \ln \left(\frac{0.3}{0.01}\right) = 7.1 + 10.3 = 17.4 \text{ hr.}$$

The ratio of constant-rate time to falling-rate time is 7.1/10.3 = 0.69.

For 600 ft./min., 225°F., and 5 per cent relative humidity, the value of B is $\dfrac{\rho_s L \lambda (W_c - W_e)}{h_t(t_a - t_s)} = \dfrac{(60)(0.104)(1020)(0.50 - 0)}{(11.2)(225 - 125)}$ = 2.9. Then the total time becomes 2.9(1.0 − 0.5)/(0.5 − 0) + 2.9 ln (0.5/0.01) = 2.9 + 11.3 = 14.2 and the ratio of times for the constant- and falling-rate periods is 2.9/11.3 = 0.257. Notice that B changed very little in the two cases because of the increased critical moisture content resulting from the higher constant rate.

Methods of Correlating Drying Data

Drying tests may be made to select a suitable dryer and secure design data, to study efficiency or capacity potential on existing plant dryers, to study the effects of operating variables on product quality, or to study the fundamentals of liquid flow.

When tests are made to obtain design data and to select the proper dryer for a given problem, data on the effects of the various external variables are usually required. These tests should be conducted in an experimental unit that properly simulates the large-scale dryer. Data should be obtained relating the moisture content to the time, and should be plotted as moisture content, dry basis, as ordinate against time as abscissa, shown by curve *A* in Fig. 6. If the moisture content is on a wet basis, the time curve will resemble curve *B* in Fig. 6.

To analyze the characteristics of the drying curve so obtained, the latter, plotted on a dry basis, should be differentiated graphically, and the drying rate so obtained should be plotted to determine the nature and extent of the periods in the cycle. It is customary to plot the drying rate vs. moisture content on arithmetic coordinates as in Fig. 3*b*. However, such a plot gives no clue to the duration of each period. The duration of the periods is shown by plotting the drying rate vs. time, as shown in Fig. 3*c*. Rate curves can also be plotted on loglog paper to permit easy reading at low moisture contents or long times. A fourth method of plotting consists of a semilogarithmic plot of moisture content (dry basis) vs. time to determine if a simple relationship exists in the falling-rate period. Thus a straight line on such a plot (curve *B* in Fig. 7) indicates that Eq. (26) or (27) is valid. If Eq. (27) holds, the slope of the straight line should be a function of the constant rate; and, if Eq. (26) holds, the drying time should vary as the square of the material thickness, curve *A* in Fig. 7.

A modification, shown in Fig. 7, which has advantage, is to plot the logarithm of the **unaccomplished moisture change,** defined as the ratio of the free water still to be removed at time θ to the total free water initially present, $(W - W_e)/(W_o - W_e)$, as a function of time, instead of log W vs. time. This reduces a series of different tests, in which the initial moisture content shows some variation, to a common basis, so that all points on the ordinate start at a value of 1.0 and go down. This method is also

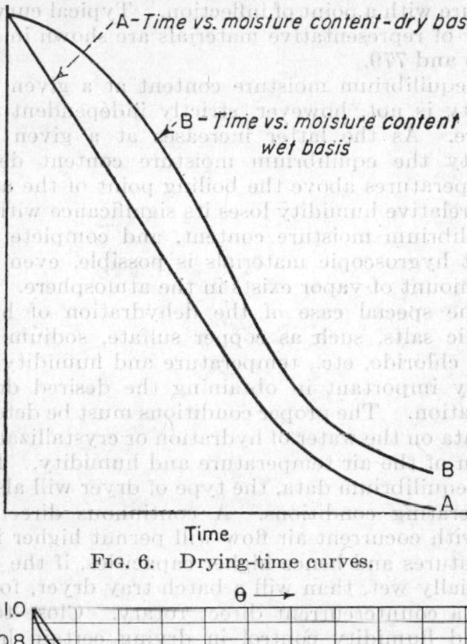

FIG. 6. Drying-time curves.

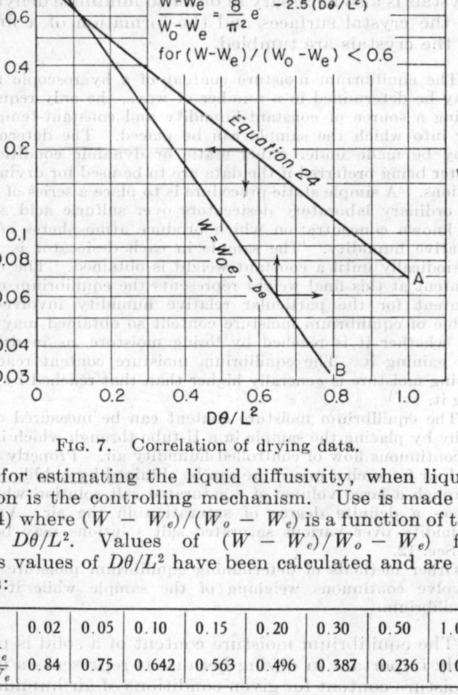

FIG. 7. Correlation of drying data.

useful for estimating the liquid diffusivity, when liquid diffusion is the controlling mechanism. Use is made of Eq. (24) where $(W - W_e)/(W_o - W_e)$ is a function of the group $D\theta/L^2$. Values of $(W - W_e)/W_o - W_e)$ for various values of $D\theta/L^2$ have been calculated and are as follows:

$\frac{D\theta}{L^2}$	0.02	0.05	0.10	0.15	0.20	0.30	0.50	1.00
$\frac{W - W_e}{W_o - W_e}$	0.84	0.75	0.642	0.563	0.496	0.387	0.236	0.069

A plot of these values is shown in Fig. 7. Other values may be calculated from Eq. (26). Similar values for other shapes are given by Newman [*Trans. Am. Inst. Chem. Engrs.*, **27**, 310 (1931)]. The above figures are used as follows: If a straight line is obtained on Fig. 7 for experimental data, and if it has also been established that the drying time varies as the square of the thickness, then the **average liquid diffusivity** can be obtained by equating an experimental value of $D\theta/L^2$ to a theoretical value given above for the same value of $(W - W_e)/(W_o - W_e)$. Thus

$$D_{\text{avg}} = \frac{(D\theta/L^2)_{\text{theo}}}{(\theta/L^2)_{\text{exp}}} \tag{36}$$

where D_{avg} = the experimental average value of liquid diffusivity, sq. ft./hr. The value of diffusivity calculated from Eq. (36) must be recognized as an average value over the entire range of drying from $(W - W_e)/(W_o - W_e) = 1$ to the value of $(W - W_e)/(W_o - W_e)$ at which θ/L^2 was evaluated. Further, Eq. (36) is an approximation since it assumes that the theoretical curve goes through the point (0,1). This is not true, however, and a more accurate value of D_{avg} can be obtained either by adding a value of 0.09 to $(D\theta/L^2)_{\text{theo}}$ in Eq. (36), or by taking a ratio of the slopes of the curves in Fig. 7. Thus, the ratio of the slope of the experimental curve of unaccomplished moisture change vs. drying time on a semilogarithmic plot to the slope, at the same unaccomplished moisture change, of Eq. (24) on a semilogarithmic plot (Fig. 7) equals the quantity D/L^2. If L is known, D can be evaluated.

Example 5. Assume that a material dries to give the unaccomplished moisture change vs. time relationship shown by line B in Fig. 7 with the abscissa having the unit of hours. Assume this material is 0.2 ft. in thickness and is drying from both sides. Determine the diffusivity by Eq. (36) at unaccomplished changes of 0.8, 0.3, and 0.07, and the diffusivity from the ratio of slopes.

Solution. The average diffusivities are calculated by Eq. (36). $L = 0.2/2 - 0.1$ ft. At unaccomplished moisture changes of 0.8, 0.2, and 0.07, $(D\theta/L^2)_{\text{theo}}$ = 0.03, 0.57, and 0.99, respectively, and θ = 0.055, 0.36, and 0.61, respectively, from curves A and B in Fig. 7. Thus, at an unaccomplished moisture change of 0.8, $D_{\text{avg}} = (0.03)(0.1)^2/(0.055) = 0.00546$ sq. ft./hr. At an unaccomplished change of 0.3, $D_{\text{avg}} = (0.57)(0.1)^2/(0.36) = 0.0158$ sq. ft./hr. Finally, at an unaccomplished change of 0.07, $D_{\text{avg}} = (0.99)(0.1)^2/(0.61) = 0.0162$ sq. ft./hr. The value of D of 0.00546 is in error because the unaccomplished change exceeded 0.6.

More precise values of diffusivities are obtained by the method of slopes outlined above. At unaccomplished moisture changes of 0.8, 0.3, 0.07, the slope of line A in Fig. 7 is -1.74, -1.065, and -1.065, respectively, and the slope of line B is -1.90 for all values. Then $D_{\text{inst}} = (-1.90)(0.1)^2/(-1.74) = 0.0109$ sq. ft./hr. at an unaccomplished change of 0.8 and equals $(-1.90)(0.1)^2/(-1.065) = 0.0178$ sq. ft./hr. for unaccomplished changes of 0.3 and 0.07. The value of the diffusivity is thus constant, as should be expected. The value of $D = 0.0109$ is not acceptable because more than one term of Eq. (24) is involved at this point.

Tests on Plant Dryers. The objectives of tests on existing plant dryers are usually different from those of design tests and generally are made to check present performance and to determine the capacity potential of the dryer. Generally, the former objective is an adjunct of the latter, inasmuch as present performance must be determined before performance under other operating conditions can be predicted. Data from plant tests are usually used to make over-all heat and material balances and to estimate drying rates or heat-transfer coefficients. The data to be taken in plant tests on continuous dryers to make the above calculations are:

1. Inlet and outlet moisture contents.
2. Feed rates.
3. Inlet- and outlet-gas temperatures.
4. Gas rates.
5. Inlet and outlet humidities.
6. Inlet and outlet material temperatures.
7. Material temperatures at various points within the dryer, when possible.

8. Retention time or time of passage through dryer.
9. Fuel consumption.

From the drying rate, heat-transfer coefficients can be obtained for application to other conditions of operation. For a good analysis of the operation, two sets of data for different conditions of operation, preferably different feed rates, are desirable.

Correlation of plant data may be made on the basis of heat-transfer coefficients or the **length of a transfer unit** (see Sec. 8). The former is generally applicable to all types of dryers, and the latter usually applies only in the case of continuous direct dryers. The relationship between the heat-transfer coefficient and length of a transfer unit is as follows:

$$L_t = \frac{Gc_p}{Ua} \tag{37}$$

where L_t = length of transfer unit, ft.; G = mass velocity of the gas, lb./(hr.)(sq. ft.); c_p = specific heat of the gas, B.t.u./(lb.)(°F.); and Ua = volumetric heat-transfer coefficient, B.t.u./(hr.)(cu. ft.)(°F.).

The number of transfer units in any direct dryer is given by the equation

$$N_t = \frac{t_1 - t_2}{(\Delta t)_m} \tag{38}$$

where N_t = number of transfer units, t_1 = inlet-gas temperature, t_2 = temperature of gas after heat transfer for drying but not for heat losses, and $(\Delta t)_m$ = mean temperature difference between gas and solid throughout the dryer. When gas and solid temperatures throughout the dryer cannot be measured, the terminal Δt's must be averaged, either arithmetically or logarithmically.

The transfer-unit concept is valuable in analyzing dryer performance, since good operation requires that continuous direct dryers should furnish a certain minimum number of transfer units. The concept has application to rotary dryers, tunnel dryers, spray dryers, pneumatic conveying dryers, etc.

Example 6. Data obtained on a plant rotary dryer 5 by 30 ft. are as follows: inlet dry-bulb temperature = 250°F., wet-bulb temperature = 110°F., outlet-air dry-bulb temperature = 150°F., and wet-bulb temperature = 109°F. Determine the number of transfer units and the length of a transfer unit for this dryer.

Solution. Because of the slight change in wet-bulb temperature throughout the dryer, an average material temperature throughout will be assumed to be about 115°F., which is selected slightly high to compensate for any radiation and conduction effects. By Eq. (38), the number of transfer units is $N_t = (250 - 150)/(\Delta t)_m$, where in this case $(\Delta t)_m = (250 - 150)/\ln(^{135}\!/_{35})$. Hence $N_t = \ln(^{135}\!/_{35}) = 1.35$, and the length of a transfer unit is $L_t = 30/1.35 = 22.2$ ft. These results show that the dryer may be operating slightly under capacity, since good practice for rotary dryers indicates that N_t should be between 1.5 and 2.0.

In any capacity test of a plant dryer, the effect of the following variables should be studied:

1. *The effect of increased temperature.* This is often the greatest single potential source of increased capacity, and plant dryers frequently can be operated at temperatures higher than their design temperatures.
2. *The effect of increased final moisture.* This, too, can be a great source of increased capacity if overdrying occurs. Generally, it is uneconomical to dry a material much below its normal regain moisture in atmospheric air. Because of the marked increase in drying time required to dry to low percentages, the permissible maximum final moisture content should always be established.
3. *The effect of increased air velocity* should be estimated. Frequently higher air rates are necessary simply to provide the required additional heat at higher capacities.
4. *Uniformity of air flow should be established.* Non-uniform air distribution can seriously reduce dryer capacity and efficiency.
5. Possible benefits from *air recirculation* should be considered.

Equilibrium Moisture Content

In the drying of solids it is important to distinguish between hygroscopic and non-hygroscopic materials. A hygroscopic material adsorbs and retains a definite percentage of moisture under definite conditions of air humidity. The moisture so retained is termed the **equilibrium moisture content** because it is in a state of equilibrium with the water vapor in the surrounding air. Equilibrium moisture may be adsorbed surface films or condensed in fine capillaries at reduced vapor pressure, and its concentration varies with the temperature and humidity of the surrounding air. However, at low temperatures, *e.g.*, 60° to 120°F., a plot of equilibrium moisture content vs. per cent relative humidity, expressed as $100(p/p_s)$, is essentially independent of temperature. Such a plot usually results in a curve of double curvature with a point of inflection. Typical curves for a number of representative materials are shown in Sec. 12, pp. 778 and 779.

The equilibrium moisture content at a given relative humidity is not, however, strictly independent of temperature. As the latter increases at a given relative humidity the equilibrium moisture content decreases. At temperatures above the boiling point of the adsorbed liquid, relative humidity loses its significance with regard to equilibrium moisture content, and complete dryness of most hygroscopic materials is possible, even when a large amount of vapor exists in the atmosphere.

In the special case of the dehydration of hydrated inorganic salts, such as copper sulfate, sodium sulfate, barium chloride, etc., temperature and humidity control are very important in obtaining the desired degree of dehydration. The proper conditions must be determined from data on the water of hydration or crystallization as a function of the air temperature and humidity. In addition to equilibrium data, the type of dryer will also affect the operating conditions. A continuous direct rotary dryer with cocurrent air flow will permit higher inlet air temperatures and hence higher capacities, if the crystals are initially wet, than will a batch tray dryer, for example, or a countercurrent direct rotary. Close temperature and humidity control in drying certain hydrated crystals is also necessary in order to minimize dehydration of the crystal surfaces and the formation of a fine dust as the crystals are tumbled.

The equilibrium moisture content of a hygroscopic material may be determined in a number of ways, the only requirement being a source of constant-humidity and constant-temperature air into which the sample can be placed. The determination may be made under either static or dynamic conditions, the latter being preferred if the data are to be used for drying calculations. A simple static procedure is to place a series of samples in ordinary laboratory desiccators over sulfuric acid solutions of known concentration which produce atmospheres of known relative humidity. The sample in each desiccator is weighed periodically until a constant weight is obtained. The moisture content at this final weight represents the equilibrium moisture content for the particular relative humidity involved. The value of equilibrium moisture content so obtained may depend on whether it is reached by losing moisture, as in drying, or by gaining it. The equilibrium moisture content reached by losing moisture is generally higher than that reached by adsorbing it.

The equilibrium moisture content can be measured dynamically by placing the sample in a U-tube through which is drawn a continuous flow of controlled-humidity air. Properly humidified air for such a procedure can be obtained by bubbling dry air through a large volume of a saturated salt solution which produces a definite degree of saturation in the air. Values of humidity over various saturated salt solutions may be found in Sec. 12.

Other methods of determining equilibrium moisture content involve continuous weighing of the sample while it reaches equilibrium.

The equilibrium moisture content of a solid is particularly important in drying because it represents a limiting moisture content for given conditions of air humidity and temperature. Drying costs can be unnecessarily high if a material is dried to a moisture content less than that which it normally possesses in equilibrium with atmospheric air. For example, if a dryer dries to 1 per cent moisture a material which on standing under normal atmospheric humidities regains moisture to 5 per cent, the material is overdried, and in all probability the dryer would be capable of a considerably higher capacity and efficiency with a uniform 5 per cent final moisture content.

Application of Psychrometry to Drying

Drying a solid by hot air or hot gases may be divided into two separate processes: (1) transfer of heat to evaporate the water, and (2) removal of the vapor by the air or gas stream. Likewise, two processes are involved in the design and operation of direct dryers: (1) the estimation of the drying rate or, conversely, drying time, and the effect of the external variables on the drying rate, and (2) the calculation of the heat and air quantities required. The first estimates have been considered previously. The background required for the second calculations will be presented here and will be based on the use of the humidity chart.

The form of humidity chart suitable for most drying calculations is that suggested by Grosvenor [*Trans. Am. Inst. Chem. Engrs.*, **1**, 184 (1908)] in which the absolute humidity, lb. vapor/lb. dry air, is related to the atmospheric temperature. It is usually based on a pressure of 1 atm. Figure 8 shows a humidity chart for air-water vapor mixtures. The various curves are defined as follows:

1. The **saturated humidity curve** gives the maximum weight of water vapor that 1 lb. of dry air can contain when the partial pressure of water vapor in the air is equal to the vapor pressure of water at that temperature. It is defined by

$$H_s = \frac{p_s}{P - p_s}\left(\frac{18}{28.9}\right) \tag{39}$$

where H_s = saturated humidity, lb./lb. of dry air; p_s = vapor pressure of water at temperature t_s; P = absolute pressure; and (18/28.9) = ratio of the molecular weight of water to that of air. Similarly, the humidity at any condition less than saturation is given by

$$H = \frac{p}{P - p}\left(\frac{18}{28.9}\right) \tag{40}$$

2. The **dew point** is the temperature at which the air becomes saturated with water vapor. The saturated humidity curve is sometimes referred to as the dew-point curve.

3. The **per cent relative humidity** curves, which are plotted against temperature, are defined by

$$H_R = \frac{100p}{p_s} \tag{41}$$

where H_R = per cent relative humidity; p = partial pressure of water vapor in the air at temperature t; and p_s = vapor pressure of water at the same temperature.

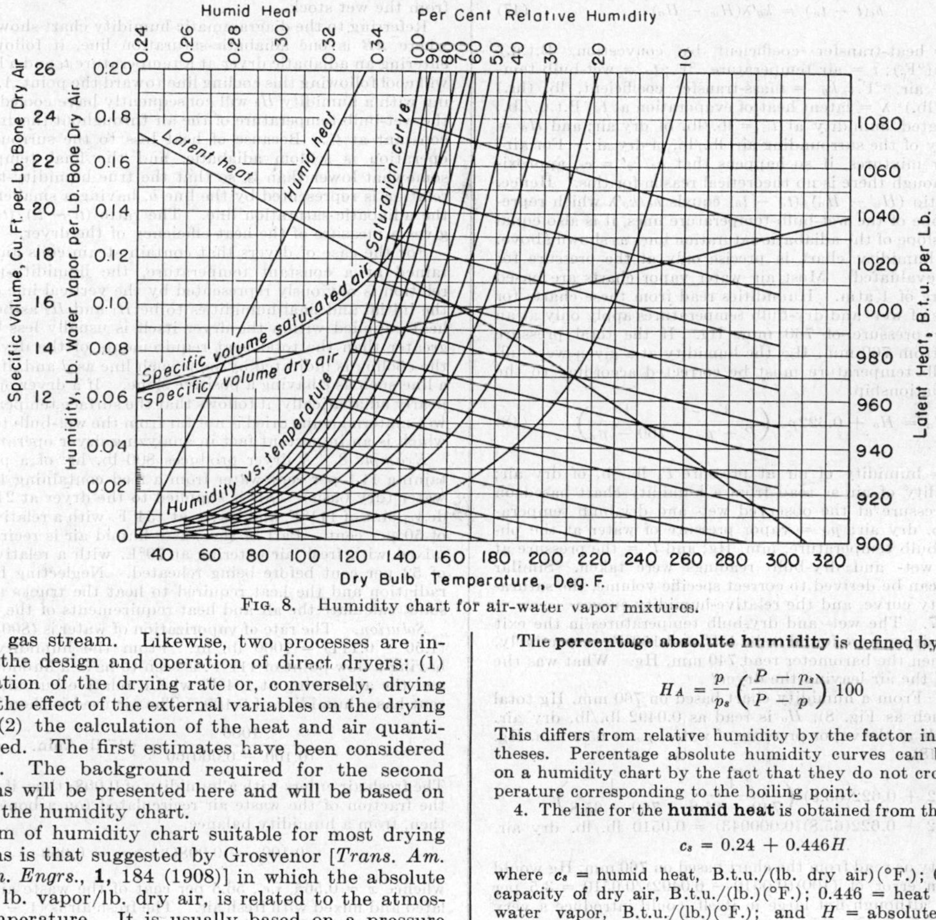

FIG. 8. Humidity chart for air-water vapor mixtures.

The **percentage absolute humidity** is defined by

$$H_A = \frac{p}{p_s}\left(\frac{P - p_s}{P - p}\right)100 \tag{42}$$

This differs from relative humidity by the factor in the parentheses. Percentage absolute humidity curves can be detected on a humidity chart by the fact that they do not cross the temperature corresponding to the boiling point.

4. The line for the **humid heat** is obtained from the following:

$$c_s = 0.24 + 0.446H \tag{43}$$

where c_s = humid heat, B.t.u./(lb. dry air)(°F.); 0.24 = heat capacity of dry air, B.t.u./(lb.)(°F.); 0.446 = heat capacity of water vapor, B.t.u./(lb.)(°F.); and H = absolute humidity, lb./lb. dry air.

5. The line of **specific volume** of *dry* air is given by

$$V_d = 11.57 + 0.025t \tag{44}$$

and for *saturated* air by

$$V_s = 0.0405(460 + t)(0.622 + H_s) \tag{45}$$

where V_d and V_s = specific volumes of dry and saturated air, cu. ft./lb. of dry air; t = temperature of air, °F.; and H_s = saturation humidity, lb./lb. dry air.

6. The **latent heat** as a function of temperature is obtainable from steam tables.

7. The series of lines of negative slope, starting from the saturation curve, represent both **adiabatic-saturation** and **wet-bulb-temperature** lines for air-water vapor only. They are based on the relationship

$$(H_s - H) = \frac{c_s}{\lambda}(t - t_s) \tag{46}$$

where H_s and t_s = adiabatic saturation humidity and temperature, respectively, corresponding to the air conditions represented by H and t, and c_s = humid heat for humidity H. The slope of the adiabatic-saturation curves is c_s/λ, where λ = the latent heat of evaporation at t_s. These lines show the relationship between the temperature and humidity of air passing through a continuous dryer operating adiabatically.

The wet-bulb temperature is established by a dynamic equilibrium between heat and mass transfer when liquid evaporates from a small mass, such as the wet bulb of a thermometer, into a very large mass of gas such that the latter undergoes no temperature or humidity change. It is expressed by the relationship

$$h_c(t - t_w) = k_g'\lambda(H_w - H_a) \tag{47}$$

where h_c = heat-transfer coefficient by convection, B.t.u./(hr.)(sq. ft.)(°F.); t = air temperature, °F.; t_w = wet-bulb temperature of air, °F.; k_g = mass-transfer coefficient, lb./(hr.)(sq. ft.)(lb./lb.); λ = latent heat of evaporation at t_w, B.t.u./lb.; H_w = saturated humidity at t_w = lb./lb. of dry air; and H_a = the humidity of the surrounding air, lb./lb. of dry air. For air–water vapor mixtures, it so happens that $h_c/k_g' = c_s$, approximately, although there is no theoretical reason for this. Hence, since the ratio $(H_w - H_a)/(t_w - t_a)$ equals $h_c/k_g'\lambda$ which represents the slope of the wet-bulb-temperature lines, it is also equal to c_s/λ, the slope of the adiabatic saturation lines as shown above.

A given humidity chart is precise only at the pressure for which it is evaluated. Most air–water vapor charts are based on a pressure of 1 atm. Humidities read from these charts for given values of wet- and dry-bulb temperatures apply only at an atmospheric pressure of 760 mm. Hg. If the total pressure is different from 760 mm. Hg, the humidity at a given wet-bulb and dry-bulb temperature must be corrected according to the following relationship:

$$H_a = H_o + 0.622p_w\left(\frac{1}{P - p_w} - \frac{1}{760 - p_w}\right) \tag{48}$$

where H_a = humidity of air at pressure P, lb./lb. of dry air; H_o = humidity of air as read from a humidity chart based on 760 mm. pressure at the observed wet- and dry-bulb temperatures, lb./lb. dry air; p_w = vapor pressure of water at the observed wet-bulb temperature, mm. Hg; and P = the pressure at which the wet- and dry-bulb readings were taken. Similar corrections can be derived to correct specific volume, the saturation-humidity curve, and the relative-humidity curves.

Example 7. The wet- and dry-bulb temperatures in the exit duct from a dryer were found to be 110° and 150°F., respectively, on a day when the barometer read 740 mm. Hg. What was the humidity of the air leaving the dryer?

Solution. From a humidity chart based on 760 mm. Hg total pressure (such as Fig. 8), H_o is read as 0.0492 lb./lb. dry air. From a table of the vapor pressure of water, p_w = 65.8 mm. Hg. From Eq. (48),

$$H_a = 0.0492 + 0.622(65.8)\left(\frac{1}{740 - 65.8} - \frac{1}{760 - 65.8}\right)$$
$$= 0.0492 + 0.622(65.8)(0.000043) = 0.0510 \text{ lb./lb. dry air.}$$

The humidity as read from the chart based on 760 mm. Hg would represent an error of (100)(0.0510 − 0.0492)/0.0510 = 3.5 per cent which, although not large in itself, could introduce a very large error if *humidity differences* were involved.

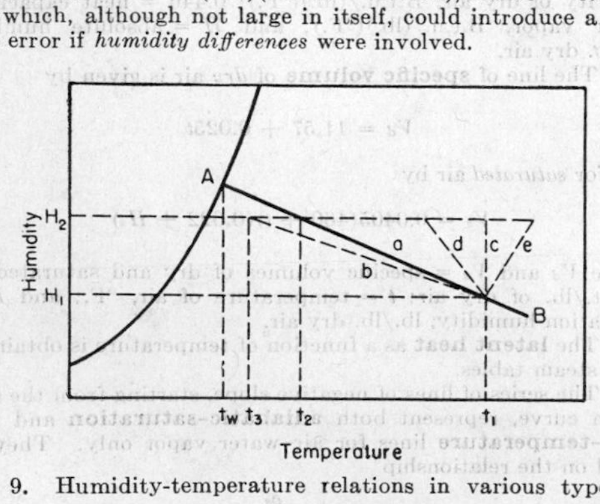

FIG. 9. Humidity-temperature relations in various types of dryers.

Methods of using the humidity chart for air conditioning have been demonstrated in Sec. 12, p. 767. In drying problems, the humidity chart finds its greatest utility in analyzing the operation of existing dryers, in making design calculations, and in checking calculations of air quantities. The application of the humidity chart in estimating the true surface temperature during the constant-rate period has been illustrated in Examples 1 and 2, pp. 803 and 804. It is equally useful in interpreting the humidity-temperature relations within the dryer. As pointed out above, the adiabatic cooling lines on the humidity chart indicate the relation between the temperature and the humidity of air passing through a truly adiabatic dryer, *i.e.*, one in which all the sensible heat given up by the air is used to evaporate water from the wet stock.

Referring to the diagrammatic humidity chart shown in Fig. 9, where AB is one adiabatic-saturation line, it follows that air entering an adiabatic dryer at a temperature t_1 and a humidity H will cool following this cooling line toward the point A. Air leaving with a humidity H_2 will consequently have cooled to t_2, while the wet-bulb temperature of the air throughout the dryer remains constant at t_w. Because of heat loss to the surroundings, the operation is seldom adiabatic, and the final temperature is somewhat lower than t_2, so that the true humidity-temperature relation is represented by the line b, having a smaller slope than the adiabatic-saturation line. The ratio $(t_1 - t_2)/(t_1 - t_3)$ then gives a measure of the heat efficiency of the dryer.

For the case of dryers that contain steam coils and are maintained at a constant temperature, the humidity-temperature relation is obviously represented by the vertical line c, assuming the initial and final humidities to be H_1 and H_2 as before. The heat supplied within the dryer itself is usually less but may be greater than the total heat requirements of the dryer. If less, the cooling is indicated by some such line as d and, if greater, by a line such as e having a positive slope. If a dryer operates very nearly adiabatically, it follows that the surface temperature of the wet material being dried is not far from the wet-bulb temperature, which is an important fact in analyzing dryer operation.

Example 8. A dryer produces 800 lb./hr. of a product containing 11.1 per cent water from a feed containing 150 per cent water (dry basis). Air is supplied to the dryer at 212°F. with a dew point of 111°F. and leaves at 154°F. with a relative humidity of 50 per cent. Part of the waste humid air is recirculated and mixed with fresh air entering at 70°F. with a relative humidity of 52 per cent before being reheated. Neglecting heat loss by radiation and the heat required to heat the trucks and the dry solid, calculate the air and heat requirements of the dryer.

Solution. The rate of vaporization of water is (800)(1.0/1.111)(1.50 − 0.111) = 1000 lb./hr. From the humidity chart, air having a dew point of 111°F. is seen to have a humidity of 0.060, and the air leaving at 154°F. with a relative humidity of 50 per cent has a humidity of 0.100. The air used is therefore

$$\frac{1000}{(0.100 - 0.060)60} = 417 \text{ lb./min.}$$

The fresh air enters with a humidity of 0.008; thus, if x represents the fraction of the waste air recirculated (on a bone-dry basis), then, from a humidity balance,

$$0.100x + 0.008(1 - x) = 0.060$$

whence x = 0.565, *i.e.*, 56.5 per cent of the waste air is recirculated and mixed with fresh air. The fresh air is (1 − 0.565)(416) = 181 lb./min.; and, since the specific volume of the fresh air is 13.4 cu. ft./lb., the volume of fresh air is (181)(13.4) = 2425 cu. ft./min. at 70°F. The total heat requirements of the dryer are most easily calculated as the sum of the heats required to heat the water and form vapor at 154°F., plus the heat to heat the fresh air from 70° to 154°F., *i.e.*,

$$1000(84 + 1005) + 0.243(181)(60)(154 - 70)$$
$$= 1{,}311{,}000 \text{ B.t.u./hr.}$$

or 1311 B.t.u./lb. water evaporated.

Other applications of the humidity chart will be indicated in examples under Types of Dryers.

Humidity Charts for Other Solvent Vapors

Humidity charts for other solvent vapors may be prepared in a manner analogous to Fig. 8. There is one important difference involved, however, in that the wet-bulb temperature differs considerably from the adiabatic-saturation temperature for vapors other than water.

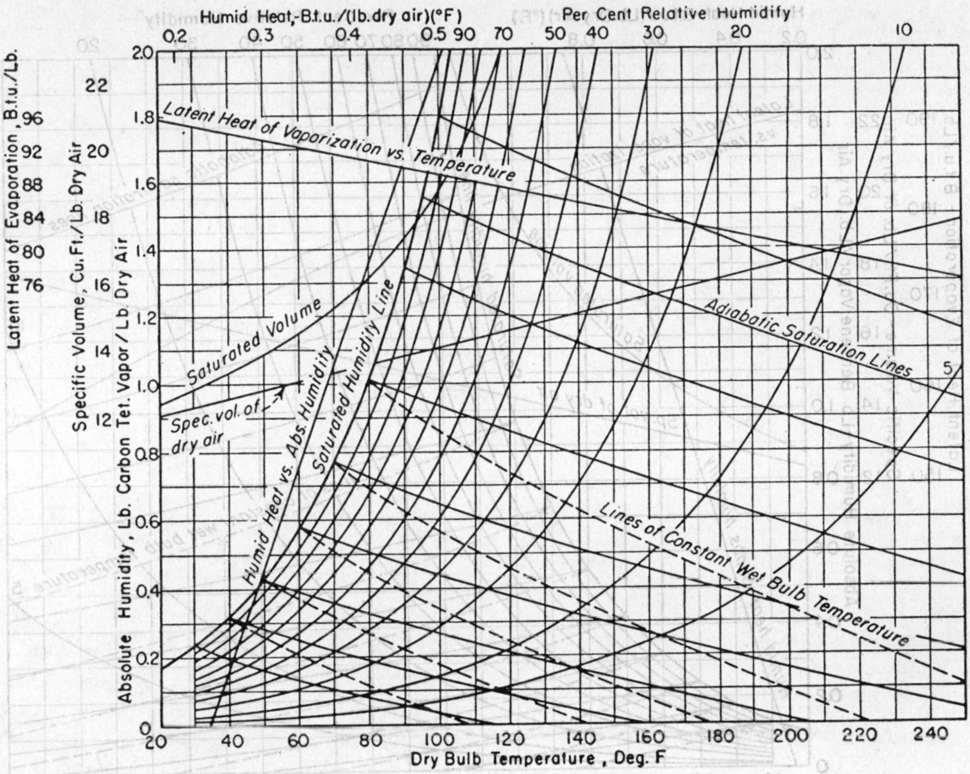

FIG. 10. Humidity chart for air-carbon tetrachloride vapor mixtures.

Figures 10 to 13 show humidity charts for carbon tetrachloride, benzene, toluene, and o-xylene. The lines on these charts have been calculated in the manner outlined for air–water vapor, except for the wet-bulb-temperature lines. The determination of these lines depends on data for the psychrometric ratio h_c/k_g', as indicated by Eq. (47). For the charts shown, the wet-bulb-temperature lines are based on the following equation:

$$(H_w - H) = \frac{\alpha h_c}{\lambda_w k_g'} (t - t_w) \tag{49}$$

where α = radiation correction factor, a value of 1.06 having been used for these charts. Values of h_c/k_g', obtained from values of $h_c/k_g'c_s$ as presented by Walker, Lewis, McAdams, and Gilliland ("Principles of Chemical Engineering," 3d ed., McGraw-Hill, New York, 1937), where c_s = the humid heat of air with respect to the vapor involved, are as follows:

Material	Carbon tetrachloride	Benzene	Toluene	o-Xylene
$h_c/k_g'c_s$	0.51	0.54	0.47	0.49

A discussion of the theory of the relationship between h_c and k_g' may be found in Sec. 8. Because both theoretical and experimental values of h_c/k_g' apply only to dilute gas mixtures, the wet-bulb lines at high concentrations have been omitted from Figs. 10 to 13. For a discussion of the precautions to be taken in making psychrometric determinations of solvent vapors at low solvent wet-bulb temperatures in the presence of water vapor, see the paper by Sherwood and Comings [*Trans. Am. Inst. Chem. Engrs.*, **28,** 88 (1932)].

CLASSIFICATION OF DRYERS

Two methods can be used to classify dryers. The first is based on the handling characteristics and physical properties of the wet material. It is better suited as a guide for selecting a group of dryers for preliminary consideration in a given drying problem. The second method of classification is based on the method of transferring heat to the wet solid and reveals differences in dryer design and operation.

A classification chart based on heat transfer is shown in Fig. 14, by Marshall [*Heating, Piping Air Conditioning*, **18,** 71 (May, 1946)]. This chart classifies dryers as **direct** or **indirect,** with subclasses of continuous and batch. **Direct dryers** utilize hot gases in direct contact with the wet solid, to supply heat and carry away the vaporized liquid. **Indirect dryers** accomplish drying by transferring heat through a retaining wall to the wet solid, the vaporized liquid being removed independently of the heating medium.

Direct Dryers

The general operating characteristics of direct dryers follow:

1. Drying depends on heat transfer to the wet solid from a hot gas, the latter removing the vaporized liquid.
2. The hot gases may be steam-heated air, combustion products, an inert gas, or a superheated vapor.
3. Drying temperatures may range up to 1400°F., the limiting temperature for most common metals of construction. At the higher temperatures, radiation becomes an important source of heat.
4. At gas temperatures below the boiling point, the vapor content of gas influences the rate of drying and the final moisture content of the solid. With gas temperatures above the boiling point throughout, the vapor content of the gas has only a slight retarding effect on the drying rate and final moisture content. Thus superheated vapors of the liquid being removed can be used for drying.
5. For low-temperature drying, dehumidification of the drying air may be required when atmospheric humidities are excessively high (see Drying of Gases, p. 877).
6. A direct dryer consumes more fuel per pound of water evaporated the lower the final moisture content. Likewise, the investment cost increases markedly.

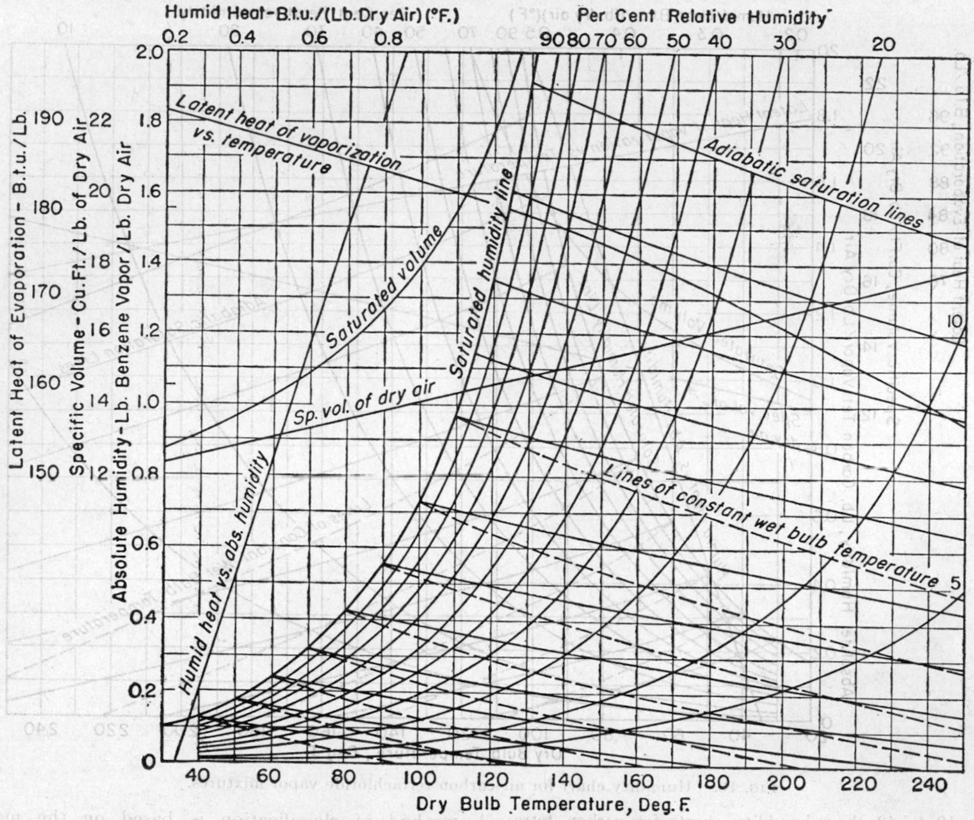

FIG. 11. Humidity chart for air-benzene vapor mixtures.

7. Efficiency increases with an increase in the inlet-gas temperature for a constant exhaust temperature.

Direct continuous dryers usually handle above 100 lb./hr. of dry product. They may handle as low as 50 lb./hr. when initial moisture contents are in excess of 2 lb. water/lb. dry solid.

The **over-all thermal efficiency** of direct continuous dryers in which the air is not reheated between inlet and outlet is, in terms of terminal temperatures,

$$\eta = \frac{(t_1 - t_2)100}{(t_1 - t_a)} \quad (50)$$

where η = per cent over-all thermal efficiency; t_1 = inlet-gas temperature; t_2 = reduced gas temperature caused by evaporation only; and t_a = temperature of air entering heaters. This equation compares the evaporation actually obtained with the total heat supplied.

The **evaporative efficiency** in a direct continuous dryer with no reheating is given by

$$\eta_e = \frac{(t_1 - t_2)100}{(t_1 - t_s)} \quad (51)$$

where η_e = evaporative efficiency; and t_s = adiabatic-saturation temperature of the entering gas. This expression compares the evaporation actually obtained with that which is theoretically possible.

Equations (50) and (51) are applicable to direct rotary dryers, tunnel dryers with no reheat sections, spray dryers, pneumatic conveying dryers, direct continuous-sheeting dryers, and other direct continuous types in which air passes through without being reheated. These equations assume the humid heat remains constant.

The following observations apply to Eqs. (50) and (51):

1. Increasing t_1 with t_2 constant increases efficiency and reduces the amount of air required.

2. Increasing t_a or t_s, by recirculation of exhaust gases, for example, increases efficiency. Benefits of recirculation must be balanced against larger dryer sizes.

Over-all **operating costs** for continuous direct dryers expressed as cost per pound of product are usually lower than for batch direct dryers because of lower labor and fuel costs, and higher production rates. They may range from \$0.0005 to \$0.025/lb. of dry product for labor, power, fuel, and maintenance.

Direct batch dryers are used for low production rates and for special handling of high-cost product. They are characterized by long drying times (6 to 40 hr.), and unsteady-state operation with air temperature, humidity, material temperature, and moisture content changing continually with time at a given position in the dryer. Direct batch dryers do not dry uniformly unless carefully designed from the standpoint of tray spacing and uniformity of air flow.

High fuel and labor costs for these dryers result in high over-all operating costs per pound of product. Fuel consumption in some cases may run as high as 6 to 8 lb. steam/lb. water evaporated. This ratio is seldom less than 2.5 and increases as the final moisture content is decreased.

Indirect Dryers

Indirect dryers differ from direct dryers with respect to heat transfer and vapor removal. Their general operating characteristics follow:

1. Heat is transferred to the wet material by conduction through a solid retaining wall, usually metallic. The source of heat may be condensing steam, hot water, gases of combustion, molten heat-transfer salts, hot oil, electricity, etc.

2. Surface temperatures may range from below freezing in the case of frozen state dryers, p. 854 to 1000°F. in the case of indirect rotary dryers heated by direct combustion.

3. Indirect dryers are suited to drying under reduced pressures and inert atmospheres to permit the recovery of solvents and

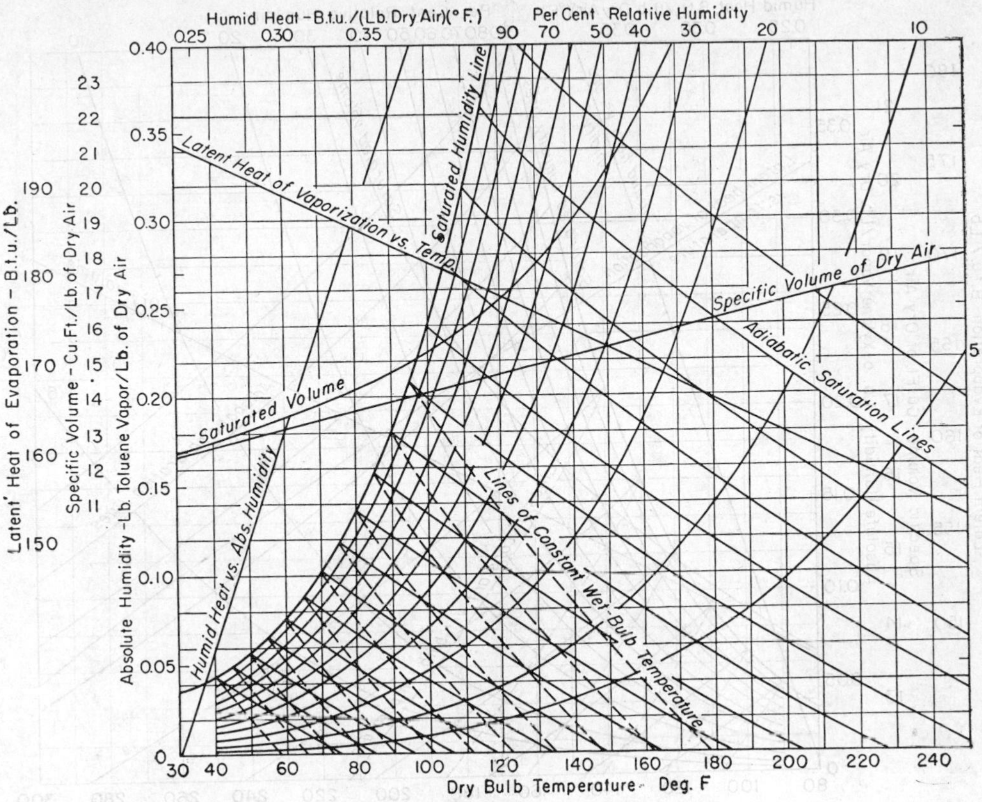

FIG. 12. Humidity chart for air-toluene vapor mixtures.

to prevent the occurrence of explosive mixtures or oxidation of easily decomposed materials.

4. Indirect dryers using condensing steam are generally economical from the standpoint of heat consumption, since they furnish heat only in accordance with the demand made by the material being dried. However, their efficiency decreases as the final moisture content decreases.

5. Dust recovery and dusty materials can be handled most satisfactorily in indirect dryers.

6. Indirect dryers may utilize some method of agitation to ensure good contact at the hot metal surface and to eliminate moisture gradients in the charge.

Indirect continuous dryers are usually more economical to operate than direct dryers.

Indirect continuous dryers can sometimes operate at pressures less than atmospheric. With good seals at the charging and discharging points, negative pressures of 27 to 28 in. Hg can be maintained during continuous operation. This feature permits **continuous drying and solvent recovery** (see Drying and Solvent Recovery, p. 875).

Indirect batch dryers are suited for evaporating and drying solutions or slurries, for drying pastes and granular solids, and for drying under high vacuum. They may be divided into two groups, in which: (1) the solid remains stationary throughout the cycle, and (2) the solid is agitated during the drying cycle. The former is typified by the vacuum shelf dryer, and the latter by the agitated-pan dryer, which may or may not operate under vacuum.

The major cost in the operation of batch indirect dryers is labor for charging and discharging the dryers and for clean-up. Fuel requirements lie in the range of 1.5 to 3.0 lb. steam/lb. water evaporated, depending on the desired degree of dryness. Power costs depend on the degree of agitation required, the nature of the material, and, if vacuum is used, the degree of evacuation.

Infrared and Dielectric Dryers. **Infrared** dryers depend on the transfer of radiant energy to evaporate moisture. This energy may be generated electrically or by incandescent refractories heated by gas. The latter method has the added advantage of convection heating. Infrared heating has not found wide application in the chemical industries for the removal of moisture. Its principal use is in baking or drying paint films and in heating thin layers of materials. Power costs for this method will be two to four times the fuel costs for the dryers described above.

Dielectric dryers have not yet found a wide field of application. Their fundamental characteristic of generating heat within a solid indicates potentialities for drying large massive objects such as wood, sponge-rubber objects, ceramics, etc. Power costs may range to ten times the fuel costs for more conventional methods.

TYPES OF DRYERS

This section describes those dryers with industrial application, their fields of application, theory and design data, auxiliary equipment commonly used, and, where possible, reliable performance and cost data. The investment cost data represent estimates for 1946, suitable only for preliminary process evaluations before drying tests are feasible. Firm cost figures should be obtained from the dryer manufacturers. Final dryer designs should always be based on experimental tests in dryers that simulate the large-scale type to be used in the plant.

Tray and Compartment Dryers

Description. In tray and compartment dryers, heated air circulates over the wet material until the latter reaches the required final moisture content. The method of supporting the wet solid depends on its physical form. Lumber, ceramics, and similar materials are stacked in piles or on racks; rayon skeins, painted objects, hides, etc.,

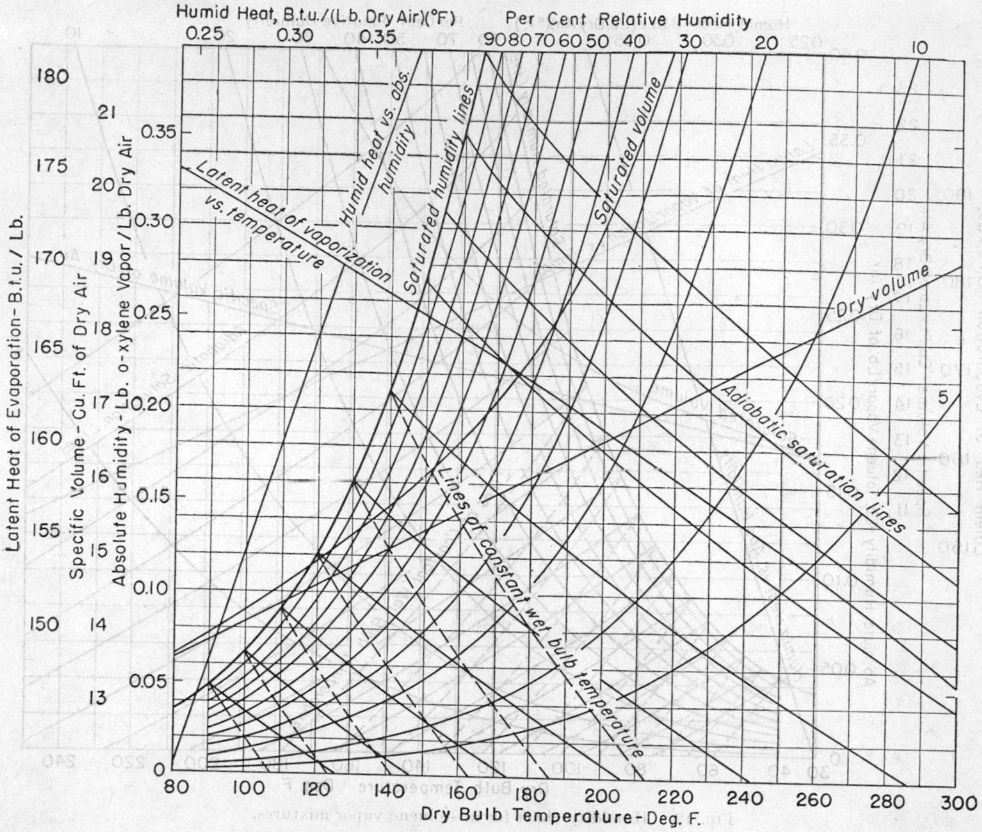

FIG. 13. Humidity chart for air–o-xylene vapor mixtures.

are suspended on hangers; granular materials, pastes, slurries, and liquids are placed in trays, which may be supported on stationary or movable racks.

Modern tray and compartment dryers usually consist of a well-insulated enclosure, fans and heating coils integrally installed, and suitable supports for the material. These dryers are more efficient and permit closer control of the drying operation than the early loft dryers in which air circulated over the material by natural convection. Satisfactory operation of tray-type dryers depends on maintaining a constant temperature and a uniform air velocity over all the material being dried.

In tray dryers, circulation of air at velocities of 400 to 1000 ft./min. is desirable to improve the surface heat-transfer coefficient and to eliminate stagnant air pockets. Proper air flow in tray dryers depends on sufficient fan capacity, on the design of ductwork to modify sudden changes in direction, and on properly placed baffles. *Non-uniform air flow is one of the most serious problems in the operation of tray dryers.*

Tray dryers may be of the tray-truck or stationary-tray type. In the former, the trays are loaded on trucks which are pushed into the dryer; and, in the latter, the trays are loaded directly onto stationary racks within the dryer. Trucks may be fitted with flanged wheels to run on tracks, or with flat swivel wheels for travel anywhere in the plant. They may also be suspended from and moved on a monorail. Trucks usually contain two tiers of trays, with 18 to 48 trays per tier, depending upon the tray dimensions.

Trays may be square or rectangular, with 4 to 8 sq. ft. per tray, and may be fabricated from any material compatible with the corrosion and temperature conditions. When the trays are stacked in the truck, there should be a clearance of not less than 1½ in. between the material in one tray and the bottom of the tray immediately above. Where material characteristics and handling permit, the trays should have screen bottoms for additional drying area. Metal trays are preferable to non-metallic trays, since they conduct heat more readily and withstand rougher handling. Tray loadings should be such as to give cycle times compatible with ease of loading and the over-all process schedule. Cycle times may be 6 to 48 hr.

Steam is the usual heating medium. The usual heater arrangement consists of a main heater before the fan and auxiliary coils between trucks in the dryer to reheat the air after it passes across the trays. When steam is not available or the drying load is small, electrical heat can be used. For temperatures above 300°F., products of combustion from coal, oil, or gas can be used if direct contact between these gases and the product is permissible. Fans may be belt- or direct-driven. With high temperatures, the fan bearings may require cooling. It is not uncommon to provide controllers to reverse the direction of air flow at definite times in the cycle.

Examples of commercial tray and tray-truck dryers are shown in Figs. 15 to 18.

Figure 15 shows a Proctor & Schwartz three-truck dryer. Heated air is forced through several trucks in series with auxiliary heating coils between the trucks. Louvers at the side of the housing enclosing the trucks are intended to provide uniform air distribution across the trays. However, the louvers as shown are not wholly effective. At air rates below 300 ft./min., velocity distribution is poor in this dryer. In operation, a portion of the air is exhausted to maintain the humidity in the

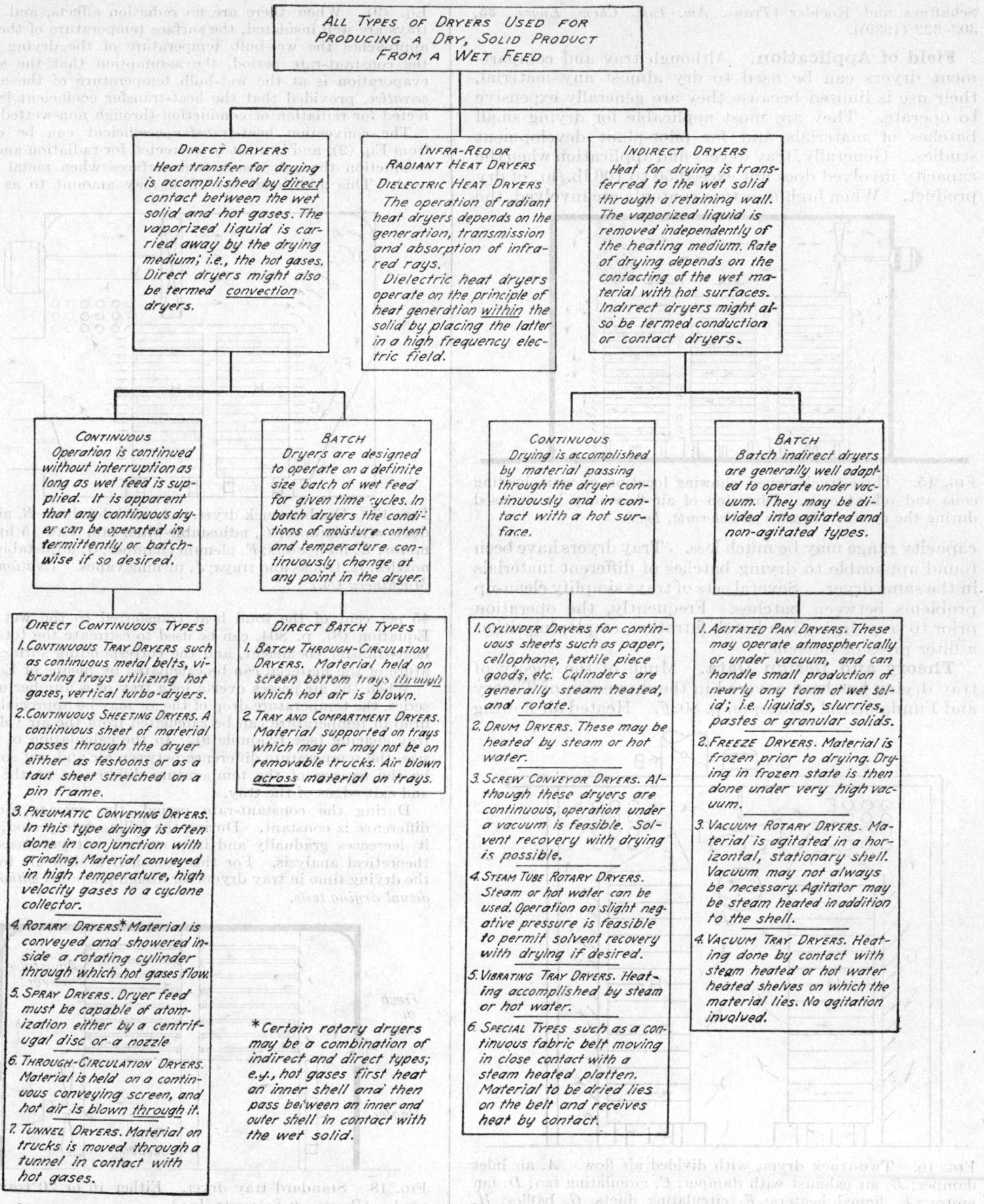
ALL TYPES OF DRYERS USED FOR PRODUCING A DRY, SOLID PRODUCT FROM A WET FEED
DIRECT DRYERS
Heat transfer for drying is accomplished by direct contact between the wet solid and hot gases. The vaporized liquid is carried away by the drying medium; i.e., the hot gases. Direct dryers might also be termed convection dryers.
INFRA-RED OR RADIANT HEAT DRYERS
DIELECTRIC HEAT DRYERS
The operation of radiant heat dryers depends on the generation, transmission and absorption of infra-red rays.
Dielectric heat dryers operate on the principle of heat generation within the solid by placing the latter in a high frequency electric field.
INDIRECT DRYERS
Heat for drying is transferred to the wet solid through a retaining wall. The vaporized liquid is removed independently of the heating medium. Rate of drying depends on the contacting of the wet material with hot surfaces. Indirect dryers might also be termed conduction or contact dryers.
CONTINUOUS
Operation is continued without interruption as long as wet feed is supplied. It is apparent that any continuous dryer can be operated intermittently or batchwise if so desired.
BATCH
Dryers are designed to operate on a definite size batch of wet feed for given time cycles. In batch dryers the conditions of moisture content and temperature continuously change at any point in the dryer.
CONTINUOUS
Drying is accomplished by material passing through the dryer continuously and in contact with a hot surface.
BATCH
Batch indirect dryers are generally well adapted to operate under vacuum. They may be divided into agitated and non-agitated types.
DIRECT CONTINUOUS TYPES
1. CONTINUOUS TRAY DRYERS such as continuous metal belts, vibrating trays utilizing hot gases, vertical turbo-dryers.
2. CONTINUOUS SHEETING DRYERS. A continuous sheet of material passes through the dryer either as festoons or as a taut sheet stretched on a pin frame.
3. PNEUMATIC CONVEYING DRYERS. In this type drying is often done in conjunction with grinding. Material conveyed in high temperature, high velocity gases to a cyclone collector.
4. ROTARY DRYERS.* Material is conveyed and showered inside a rotating cylinder through which hot gases flow.
5. SPRAY DRYERS. Dryer feed must be capable of atomization either by a centrifugal disc or a nozzle
6. THROUGH-CIRCULATION DRYERS. Material is held on a continuous conveying screen, and hot air is blown through it.
7. TUNNEL DRYERS. Material on trucks is moved through a tunnel in contact with hot gases.
DIRECT BATCH TYPES
1. BATCH THROUGH-CIRCULATION DRYERS. Material held on screen bottom trays through which hot air is blown.
2. TRAY AND COMPARTMENT DRYERS. Material supported on trays which may or may not be on removable trucks. Air blown across material on trays.
*Certain rotary dryers may be a combination of indirect and direct types; e.g., hot gases first heat an inner shell and then pass between an inner and outer shell in contact with the wet solid.
1. CYLINDER DRYERS for continuous sheets such as paper, cellophane, textile piece goods, etc. Cylinders are generally steam heated, and rotate.
2. DRUM DRYERS. These may be heated by steam or hot water.
3. SCREW CONVEYOR DRYERS. Although these dryers are continuous, operation under a vacuum is feasible. Solvent recovery with drying is possible.
4. STEAM TUBE ROTARY DRYERS. Steam or hot water can be used. Operation on slight negative pressure is feasible to permit solvent recovery with drying if desired.
5. VIBRATING TRAY DRYERS. Heating accomplished by steam or hot water.
6. SPECIAL TYPES such as a continuous fabric belt moving in close contact with a steam heated platten. Material to be dried lies on the belt and receives heat by contact.
1. AGITATED PAN DRYERS. These may operate atmospherically or under vacuum, and can handle small production of nearly any form of wet solid; i.e. liquids, slurries, pastes or granular solids.
2. FREEZE DRYERS. Material is frozen prior to drying. Drying in frozen state is then done under very high vacuum.
3. VACUUM ROTARY DRYERS. Material is agitated in a horizontal, stationary shell. Vacuum may not always be necessary. Agitator may be steam heated in addition to the shell.
4. VACUUM TRAY DRYERS. Heating done by contact with steam heated or hot water heated shelves on which the material lies. No agitation involved.

FIG. 14. Classification of dryers, based on methods of heat transfer. [*Marshall, Heating, Piping, Air Conditioning*, **18**, 71 (1946).]

dryer low enough not to retard the drying rate but high enough to maintain economical operation. Air flow may be in either direction and may be reversed, if required, several times during the drying cycle. In this dryer, 10,000 to 15,000 cu. ft. of air/min. will be circulated, depending on whether a 3- or 5-hp. fan is used.

Figure 16 shows a modified Proctor & Schwartz two-truck dryer with divided air flow. The particular advantage of this arrangement is that air is passed over only one truck and then reheated. Propeller-type fans are generally used.

A National Drying Machinery Company double-truck dryer is shown in Fig. 17. An adjustable-pitch fan, capable of developing 1 to 15 hp., circulates the air. In order to obtain uniform air distribution, turning vanes are used to direct the air into the plenum chamber, which has adjustable air orifices to direct the air across the trays. The unit is made symmetrical to permit conversion to a tunnel dryer with temperature zoning and reversal of air flow and velocity control.

Figure 18 illustrates a standard type Proctor & Schwartz tray dryer for drying pharmaceuticals, pilot-plant development work, research, and university laboratories. The trays are held on stationary racks and must be removed individually. It is easily adapted to changes that may be required in experimental work. The performance of this type of dryer has been reported by

Schaffner and Koehler [*Trans. Am. Inst. Chem. Engrs.*, **35**, 303–322 (1939)].

Field of Application. Although tray and compartment dryers can be used to dry almost any material, their use is limited because they are generally expensive to operate. They are most applicable for drying small batches of materials and for pilot-plant development studies. Generally, tray dryers find application when the capacity involved does not exceed 50 to 100 lb./hr. of dry product. When high moisture contents are involved, the

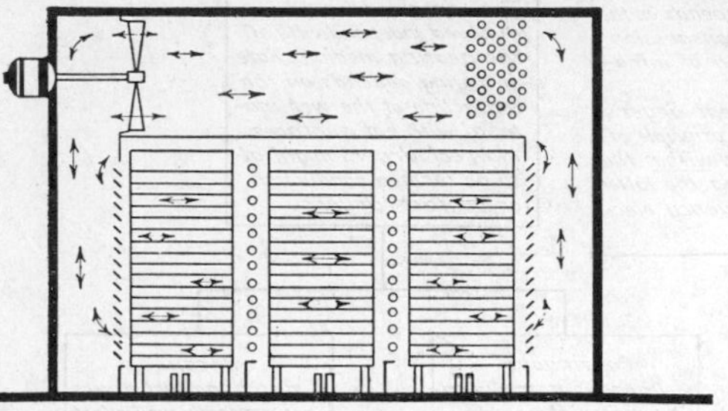

FIG. 15. Three-truck dryer, showing location of main heating coils and reheat coils. Direction of air flow may be reversed during the cycle. (*Proctor & Schwartz, Inc.*)

capacity range may be much less. Tray dryers have been found applicable to drying batches of different materials in the same dryer. Several sets of trays simplify clean-up problems between batches. Frequently, the operation prior to tray drying is a batch filtration, employing either a filter press or nutsche.

Theory and Design Data. Much of the theory of tray drying has been covered in the discussion of Theory and Fundamental Concepts, p. 802*ff*. Heated air flowing

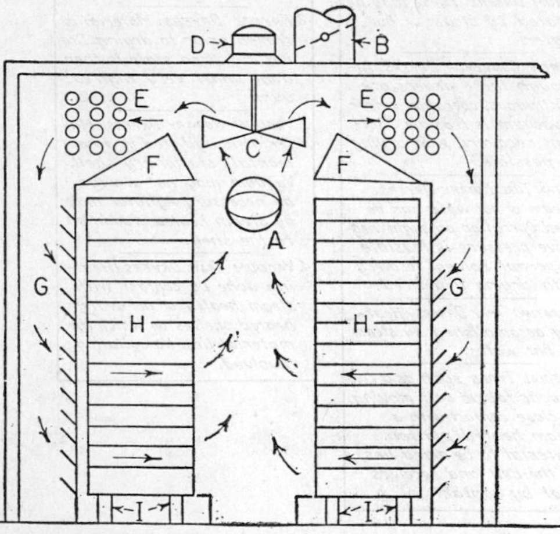

FIG. 16. Two-truck dryer, with divided air flow. *A*, air inlet damper; *B*, air exhaust with damper; *C*, circulating fan; *D*, fan motor; *E*, finned heaters; *F*, circulating ducts; *G*, baffles; *H*, trucks and trays; *I*, truck wheels. (*Proctor & Schwartz, Inc.*)

across the exposed surfaces of the wet material causes heat transfer and consequently drying. The constant drying rate of materials in trays is determined, principally, by the rate of heat transfer by convection from the air to the wet material according to Eq. (1), p. 802. In tray dryers when the wetted area is constant, the drying rate depends on the *total* heat-transfer coefficient and the difference between the air and surface temperatures.

When evaporation takes place from a completely wetted surface, the true surface temperature can be evaluated by Eq. (9). When there are no radiation effects, and when the trays are well insulated, the surface temperature of the material approaches the wet-bulb temperature of the drying air. For the constant-rate period, the assumption that the surface of evaporation is at the wet-bulb temperature of the air is *conservative*, provided that the heat-transfer coefficient is not corrected for radiation or conduction through non-wetted surfaces.

The convection heat-transfer coefficient can be calculated from Eq. (2), and it must be corrected for radiation and for heat conduction through non-wetted surfaces when metal trays are used. This additional heat flow may amount to as much as

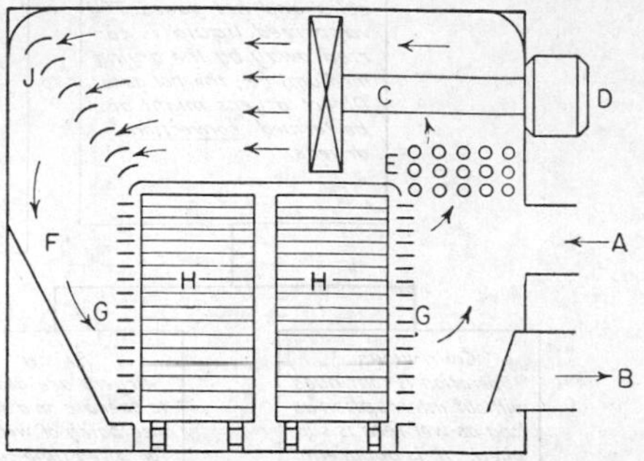

FIG. 17. Double-truck dryer. *A*, air inlet duct; *B*, air exhaust duct with damper; *C*, adjustable-pitch fan, 1 to 15 hp.; *D*, fan motor; *E*, fin heaters; *F*, plenum chamber; *G*, adjustable air-blast nozzles; *H*, trucks and trays; *J*, turning vanes. (*National Drying Machinery Co.*)

45 per cent of the total heat transferred to the wet material. Equation (8), p. 804, can be used to estimate the total over-all heat-transfer coefficient, and the corresponding surface temperature of the material can be calculated by means of Eq. (9).

When the air passes over a long tray or a number of trays in series, the temperature drop of the air may be appreciable. This drop in temperature can be calculated from the air velocity and tray spacing (see Example 9). In the calculation of the dryer size, the temperature difference between air and wet solid should be the *average* of the temperature differences at the entrance and exit edges of the tray.

During the constant-rate period, the *average* temperature difference is constant. During the falling-rate period, however, it decreases gradually and in a manner not yet susceptible to theoretical analysis. For this reason it is difficult to estimate the drying time in tray dryers for design purposes *without making actual drying tests*.

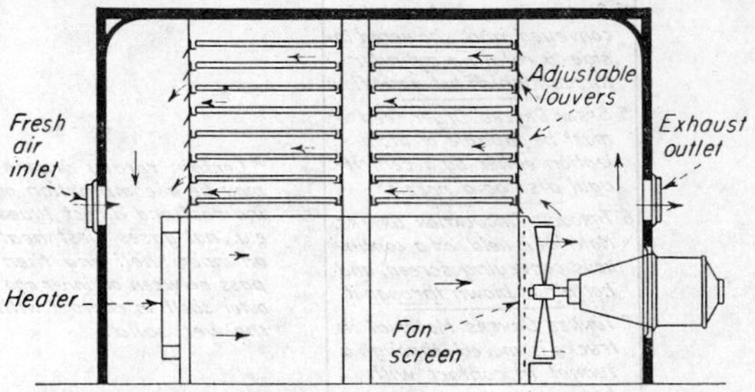

FIG. 18. Standard tray dryer. Either 16 or 20 trays may be used. (*Proctor & Schwartz, Inc.*)

The **effect of depth of loading** on the constant-rate period is given by Eq. (3*a*). In the falling-rate period, if diffusion controls the movement of moisture in the material, the drying time is proportional to the *square* of the depth of loading [Eq. (26)]. However, other forces such as capillarity and shrinkage may be active during this period with the result that the drying time is often a linear function of the depth of loading. For accurate estimates of this effect, experimental tests should be made.

When Eq. (34), p. 808, applies, an estimate of the drying time can be made if the initial moisture content, the desired final moisture content, the bulk density of the material, and the critical moisture content of the material are known. In addition, the desired air rate, temperatures, and humidities must be specified.

Example 9. It is desired to produce 4500 lb. dry product/24 hr.-day from a granular, filter-press cake. The initial moisture content is 0.80 lb. water/lb. dry solid, and the final moisture content is to be 0.01 lb./lb. The critical moisture content is about 0.40 lb./lb. The dry density is 40 lb./cu. ft. An air temperature of 220°F. is to be used, and the air velocity should not be less than 400 ft./min. across the trays. It is desired to design one or more two-truck dryers with 30-in.-square trays, 1 in. deep, to furnish the required production. Assume the free space for air flow between trays will be about 1½ in. Also, assume that the inlet humidity of the air is 0.04 lb./lb. dry air, and the atmospheric humidity is $H_a = 0.015$ lb./lb.

Solution. First estimate the total drying time on the assumption that Eq. (34) applies. In order to determine B in this equation, h_t and t_s must be estimated. h_t is found to be 9.8 B.t.u./(hr.)(sq. ft.)(°F.) from Eq. (8) for $L = 0.0832$ ft., $A_u = 1$, $k = 1.0$ B.t.u./(hr.)(ft.)(°F.), $h_c = 4.05$ from Eq. (2), $h_r = 1.8$ from Sec. 6, Fig. 12. Then, from Eq. (9), $t_s = 126$°F., and $H_s = 0.098$ lb./lb. Now the air in passing across two 30-in. trays suffers a drop in temperature so that an average temperature difference $(t_a - t_s)_m$ must be used in estimating the constant-rate time. The temperature of the air leaving the trays is given by $t_2 = t_s + (t_1 - t_s)e^{-N_t}$, where t_1 = entering temperature, °F.; $N_t = h_t L_T / G c_p b$ = number of transfer units; L_T = tray length, ft.; and b = free space between trays, ft. Hence $t_2 = 126 + (220 - 126)e^{-1.37} = 150$°F. This represents a large temperature drop, which is almost too great for good operation. The logarithmic mean temperature difference is now 51°F. The time for the constant-rate period now becomes

$$\theta_c = \frac{\rho_s \lambda L(W_o - W_c)}{h_t(t_a - t_s)_m} = \frac{(40)(1025)(0.083)(0.8 - 0.4)}{(9.8)(51)} = 2.75 \text{ hr.}$$

For the falling-rate period, the drying time is $B \ln (W_c/W) = 2.75 \ln (0.4/0.01) = 10.2$ hr. Hence **the total drying time = 12.95 hr.** Use a 14-hr. cycle to allow for loading and unloading. The required hold-up in the dryer for the desired production is (4500)(14.0)/(24) = 2625 lb. The tray loading is (40)(0.083) = 3.32 lb./sq. ft., and hence the number of square feet of tray area required is 2630/3.42 = 770 sq. ft. The area per tray is 6.25 sq. ft., so that the number of trays required = 770/6.25 = 123. If a truck 6.5 ft. high with two tiers of trays of 31 trays per tier is used, the required number of trucks is two, and hence only one two-truck dryer is required. The total air flow is determined from the total free space and air velocity, which in this case is (5)(0.125)(31)(400) = 7750 cu. ft./min. However, a design value of 10,000 cu. ft./min. should be used to eliminate any decrease in air rate caused by air leakage above and around the trucks. The size of the heater should be such that the heat requirements in the constant-rate period will be satisfied. This, of course, means that the heater will be oversize for the falling-rate period, an inherent shortcoming of this type of dryer. The percentage recirculation required to maintain an inlet humidity of 0.04 is $100x/w = 100(H_1 - H_a)/(H_2 - H_a) = 100(0.04 - 0.015)/(0.061 - 0.015) = 54.3$ per cent, where H_1, H_2, H_a are absolute humidities of air entering trays, air leaving trays, and atmosphere, respectively.

Air is circulated by propeller or centrifugal fans. If the fan is mounted within the dryer itself, the fan bearings can be operated up to temperatures of 400°F. without water cooling. Above this temperature, either water-cooled bearings must be used or they should be located outside the hot zone. The pressure drop in tray and compartment dryers generally does not exceed 1 to 2 in. of water.

Recirculation of air in this type of dryer is usually between 80 and 95 per cent. For some materials which dry mainly in the falling-rate period, recirculation can be as high as the physical construction of the dryer permits. This is achieved in practice by allowing natural leakage from and into the dryer and ductwork to provide for exhausting water vapor and for admitting an equal amount of fresh air. For further discussion of air recirculation, see articles by Victor [*Chem. & Met. Eng.*, **52**, 105 (1945)] and Marshall [*Heating, Piping Air Conditioning*, **15**, 567 (1943)].

Any of the standard extended-surface air heaters can be used to heat the air. Finned-tube units are desirable because of their compactness. Data on such heaters may be found in Sec. 6.

If noxious gases, fumes, or dust are given off during the drying operation, dust- or fume-recovery systems should be installed following the air-exhaust duct.

In order to minimize heat losses and to prevent condensation on dryer walls at high humidities, it is essential that the dryer be suitably insulated. It may be necessary to control the amount of exhaust air in the early stages of drying to maintain a sufficiently low dew point.

Dryer doors should be as tight as possible to prevent leakage of hot air, resulting in heat losses, working discomfort, and safety hazards if toxic materials are being handled.

It is good practice to provide an **extra set of trucks and trays** for each dryer to permit loading and unloading while the dryer is in operation.

Controls for tray or compartment dryers may consist of a recording temperature controller on the air and a humidity controller to control the amount of air exhausted from the dryer or to inject a controlled amount of steam to maintain a given humidity. Cycle controllers are sometimes used to change the drying temperature at definite times during the cycle. The cycle may start at a high temperature during the constant-rate period, when the material is kept cool by surface evaporation, and finish at lower temperatures throughout the falling rate period.

Auxiliary Equipment. Very little auxiliary equipment is needed with a self-contained tray or compartment dryer. In some instances, tray loaders are used to ensure uniform tray loading. **Uniform tray loading is essential** to obtaining a minimum drying time and uniform air flow across the trays.

Special conveyors are sometimes employed to transport filter-press cakes to the dryers.

Auxiliary air-drying equipment is sometimes required for low-temperature drying operations (see Drying of Gases, p. 877).

Table 3. Manufacturer's Performance Data for Tray and Tray-truck Dryers*

Material	Color	Chrome yellow	Toluidine red	Half-finished Titone	Color
Type of dryer	2 truck, Fig. 16	16-tray dryer, Fig. 18	16-tray dryer, Fig. 18	3 truck, Fig. 15	2 truck
Capacity, lb. product/hr	24.6	35.5	4.1	125	10.5
Number of trays	80	16	16	180	120
Tray spacing, in.	4	4	4	3	3½
Tray size, in.	24 × 30 × 1½	26 × 38 × ⅞	26 × 38 × ⅞	24 × 27 × 1½	24 × 26½ × 1
Depth of loading, in.	1–2	1⅛–1¼	1⅛–1½	1⅛	
Initial moisture, % dry basis	207	46	220	223	116
Final moisture, % dry basis	4.5	0.25	0.1	25	0.5
Air temp., °F	185–165	212	122	200	80–210
Loading, lb. product/sq. ft.	2.05	6.9	1.6	3.05	1.9
Drying time, hr	33	21	41	20	96
Air velocity, ft./min.	175	450	450	590	500
Air volume, cu. ft./min.	8000	....	....	10,000	
Over-all drying rate, lb. water/(hr.)(sq. ft.)	0.12	0.134	0.084	0.24	0.023
Steam consumption, lb./lb. water evaporated	2.5	3.0	....	2.75	
Total installed power, hp	2.0	1.0	1.0	3.00	2.0
Approx. operating cost, cents/lb. water evaporated	1.4	4.6	14.0	0.4	4.3
Approx. purchase cost	$3000	$1400	$1400	$3000	$2500

NOTE: See text for bases of cost figures.

* Courtesy of Proctor & Schwartz, Inc.

Standard air-drying units using solid desiccants can be obtained for such purposes.

Performance and Cost Data. Performance data on some typical tray and compartment dryers are tabulated in Table 3. These indicate that an **over-all rate of evaporation** of 0.03 to 0.30 lb. water/(hr.)(sq. ft. of tray area) may be expected from tray and tray-truck dryers.

The **operating costs** shown in Table 3 include labor at \$1.50/man-hr., fuel, power, maintenance at 5 per cent, and depreciation at 6 per cent. The **purchase costs** include fans, motors, and heaters but not trays and are based on current costs of September, 1946.

Thermal efficiencies of this type of dryer usually vary from 20 to 50 per cent, depending on the drying temperature used and the humidity of the exhaust air. In drying to very low moisture contents under temperature restrictions, however, the thermal efficiency may be on the order of 10 per cent.

Purchase costs of tray dryers vary with the size of the dryer and the materials of construction and range from \$3 to \$12/cu. ft. of total volume.

The major **operating cost** of a tray dryer is the labor involved in loading and unloading the trays. About 2 man-hr. are required to load and unload a standard two-truck tray dryer. In addition to this, about one-third to one-fifth of a man's time is required to supervise the dryer during the drying period.

Steam consumption will range from 2.5 to 4 lb. steam/lb. water evaporated. In exceptional cases, steam consumptions as high as 10 lb./lb. water evaporated may be encountered.

Power for tray and compartment dryers will be in the neighborhood of 1 hp./truck in the dryer.

Maintenance on tray and compartment dryers will run from 3 to 5 per cent of the installed cost per year.

Batch Through-circulation Dryers

Description. In one type of batch through-circulation dryer, heated air passes **through** a stationary **permeable** bed of the wet material placed on removable screen-bottom trays suitably supported in the dryer (Fig. 19). This type is similar to a standard tray dryer except that hot air passes **through** the wet solid instead of across it. The pressure drop through the bed of material does not usually exceed about 1 in. of water in types shown in Fig. 19.

In another type, deep perforated-bottom trays are placed on top of plenum chambers in a closed circuit, hot-air circulating system. Gunpowder is dried this way. In some food-dehydration and grain-drying plants the material is placed in *finishing bins* with perforated bottoms; heated air passes up through the material and is removed from the top of the bin, reheated, and recirculated. The latter types involve pressure drops through the bed of material of 2 to 18 in. of water at relatively low air rates. Finishing bins usually require a dehumidification system to produce dry air at low temperatures.

Steam is the usual heating medium employed with these dryers, although any other convenient source of heated air can be used.

Field of Application. Batch through-circulation dryers are used where production requirements are higher than can be obtained with cross-circulation tray dryers but are not high enough for continuous drying. They are restricted to granular materials that permit *through circulation of air*. Drying times are usually much shorter than in cross-circulation tray dryers. The drying chamber may also be the storage chamber in the case of grains and dehydrated foods.

Theory and Design Data. In this type of dryer, heat and mass transfer occur between the individual particles and the heated air passing through the granular bed. Drying rates during the constant-rate period can be calculated from Eq. (10), if the particle size of the material is known. There is no wholly satisfactory method for calculating the falling-rate period. The assumption that the rate during this period is proportional to the free moisture content is the only approximation available. In any case, it is necessary to know the critical moisture content. Since the air drops in temperature and increases in humidity in passing through the bed of material, it is essential to use an **average temperature or humidity driving force** when calculating rates (see Example 3, p. 805). As in tray drying, drying tests are required for accurate design.

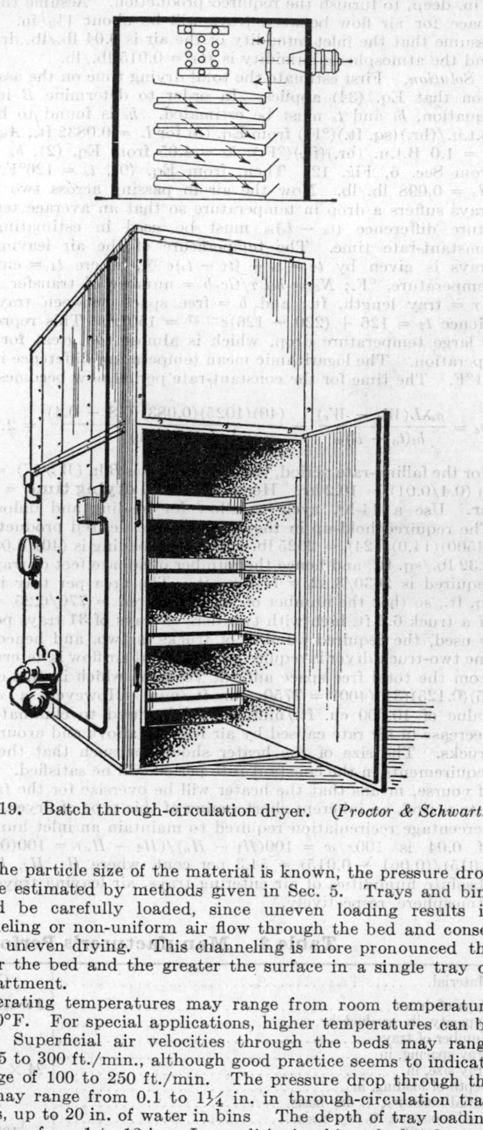

FIG. 19. Batch through-circulation dryer. (*Proctor & Schwartz, Inc.*)

If the particle size of the material is known, the pressure drop can be estimated by methods given in Sec. 5. Trays and bins should be carefully loaded, since uneven loading results in channeling or non-uniform air flow through the bed and consequent uneven drying. This channeling is more pronounced the deeper the bed and the greater the surface in a single tray or compartment.

Operating temperatures may range from room temperature to 300°F. For special applications, higher temperatures can be used. Superficial air velocities through the beds may range from 5 to 300 ft./min., although good practice seems to indicate a range of 100 to 250 ft./min. The pressure drop through the bed may range from 0.1 to 1¼ in. in through-circulation tray dryers, up to 20 in. of water in bins The depth of tray loading can range from 1 to 12 in. In conditioning bins, the height may be several feet. Air recirculation should range from 70 to 80 per cent.

Auxiliary Equipment. Auxiliary equipment for this type of dryer is essentially the same as that for tray and compartment dryers.

Performance and Cost Data. Few data are available on the performance or cost of this type of dryer. Drying rates ranging from 0.02 to 2.5 lb. water/(hr.)(sq. ft. of tray area) may be expected. Table 4 gives some typical performance and cost data for tray-type through-circulation dryers.

Table 4. Performance and Cost Data for Batch Through-circulation Dryers*

Kind of material	Granular polymer	Vegetable	Vegetable seeds
Capacity, lb. product/hr	270	93.8	61
Number of trays	16	24	24
Tray spacing, in.	17	17	17
Tray size, in. × in.	36 × 41	36 × 41	33.5 × 38.5
Depth of loading, in.	2.5	2.4	1.5
Physical form of product	Crumbs	0.25-in. diced cubes	Washed seeds
Initial moisture content, % dry basis	11.1	669.0	100.0
Final moisture content, % dry basis	0.1	5.0	9.9
Air temp., °F	190	170 dry-bulb, 120 wet-bulb to 145 dry-bulb	97
Air velocity, superficial, ft./min.	180	120 to 190	200
Tray loading, lb. product/sq. ft.	3.30	1.06	1.38
Drying time, hr.	2.0	8.5	5.5
Over-all drying rate, lb. water evaporated/(hr.)(sq. ft.)	0.182	2.43	0.233
Steam consumption, lb./lb. water evaporated	4.0	2.42	6.8
Installed power, hp	10.0	25.0	25.5
Approximate operating cost, ¢/lb. water evaporated†	4.1	0.35	3.21
Approximate purchase cost‡	\$5000–\$8000	\$5000–\$8000	\$5000–\$8000

* Courtesy of Proctor & Schwartz, Inc.
† Includes steam, electricity, labor, and repairs.
‡ Includes fans, motors, heaters, temperature controls, trays.

Thermal efficiency, steam consumption, labor costs, and maintenance costs are comparable with those encountered in tray dryers, and the power for the fans may be twice as much.

Tunnel Dryers and Vertical Turbodryers

Description. Tunnel dryers are essentially several batch truck dryers in series. In operation, wet material is placed on trucks which are moved progressively through the tunnel in contact with hot gases. Usually, each truck occupies successively all positions in the tunnel for a given period of time, resulting in semi-continuous operation. In some cases chain drives fastened to the bottom of the trucks move them slowly, but continuously, through the tunnel. Schematic diagrams of some typical tunnel dryers are shown in Fig. 20.

Air may flow in one pass through the tunnel in the line of truck movement without reheating, **adiabatic operation**, or it may pass through each truck perpendicular to the line of truck movement with reheating of the air after passing through each truck, **constant-temperature operation.**

The material to be dried may be loaded on the trays of the trucks (pottery and ceramic pieces), placed on racks in trucks (rayon cakes), or draped over rods carried by continuous conveyors (rayon skeins).

Tunnel dryers may operate at steam-pressure temperatures, or with hot gases from gas or oil burners.

Air flow may be parallel (Fig. 20*b*) or countercurrent (Fig. 20*a*) to the material flow, or a combination of both may be used as in the so-called center-exhaust tunnel dryer (Fig. 20*c*). In the latter type, the first portion of the tunnel operates with parallel air flow and the latter portion with countercurrent flow, the

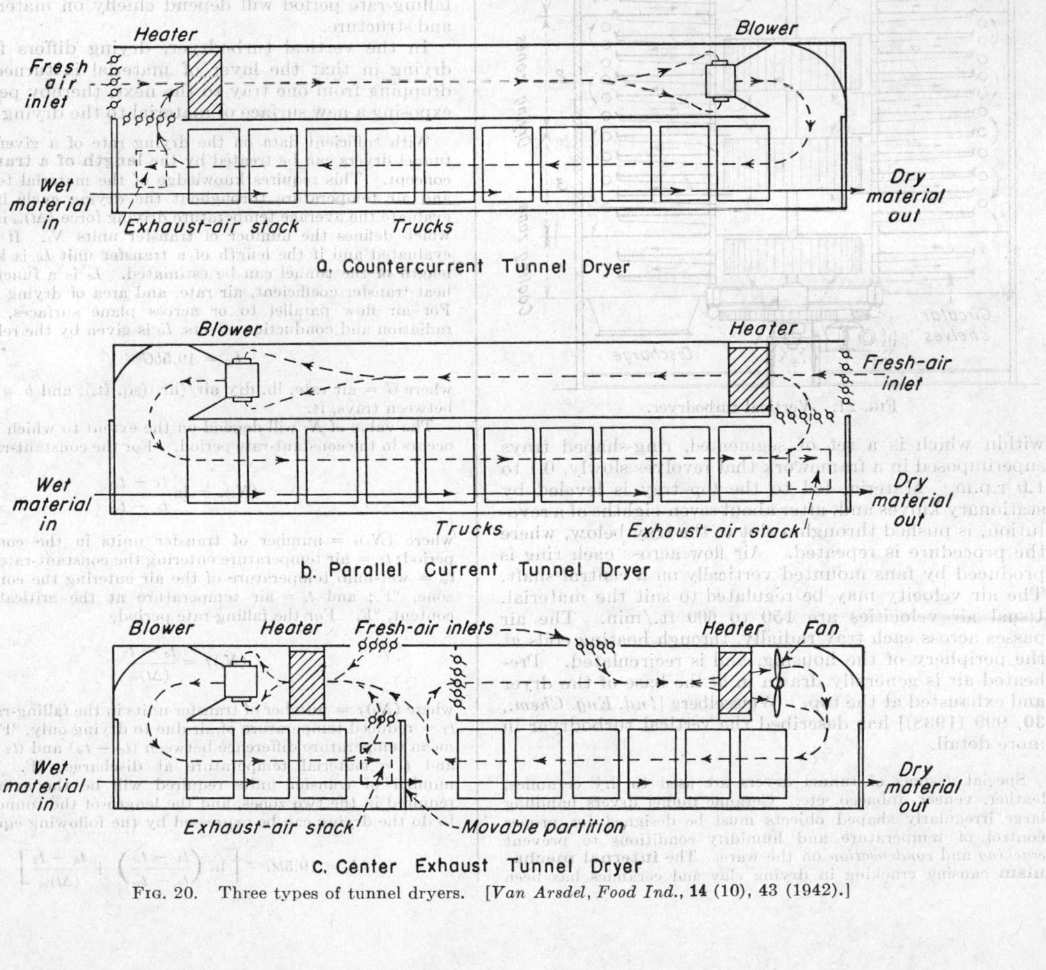

FIG. 20. Three types of tunnel dryers. [*Van Arsdel, Food Ind.*, **14** (10), 43 (1942).]

exhaust gases leaving at the point where the change from parallel to counterflow operation is effected. If cross flow of air at right angles to the line of trucks is used, the temperature and humidity of the air along the length of the tunnel can be effectively controlled. This is often advantageous when temperature- or humidity-sensitive materials are being dried. It has also been used successfully to dry silica gel for desiccant use. Van Arsdel [*Food Industries*, **14** (10), 43; (11), 47; (12), 47 (1942)] has described the operation of tunnel dryers used in the drying of vegetables.

One type of **batch tunnel dryer** employs several trucks which are placed in a series of connected compartments and left there throughout the cycle, the air for drying entering at one end and passing through all trucks before returning for reheating. This type of tunnel dryer possesses all the inherent disadvantages of a batch dryer, plus the added disadvantage of an inordinately long passage of the air across a series of stationary trays. This results in low-temperature humid air contacting the last truck which consequently dries at a greatly retarded rate until the other trucks are nearly dry. Large quantities of air must be circulated to minimize this dsadvantage.

The European-developed **vertical turbodryer** can be classified as a **continuous** tray dryer. It consists of a cylindrical or hexagonal vertical housing (Fig. 21),

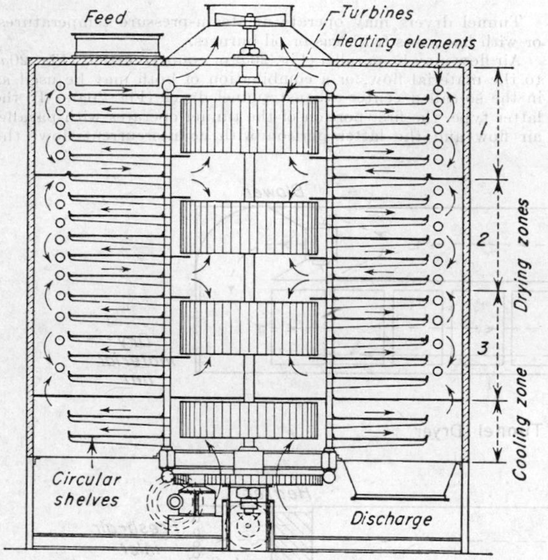

FIG. 21. Vertical turbodryer.

within which is a set of segmented, ring-shaped trays superimposed in a framework that revolves slowly, 0.1 to 1.0 r.p.m. Material fed to the top tray is leveled by stationary knives and, after about seven-eighths of a revolution, is pushed through a slot to the tray below, where the procedure is repeated. Air flow across each ring is produced by fans mounted vertically on a central shaft. The air velocity may be regulated to suit the material. Usual air velocities are 150 to 600 ft./min. The air passes across each tray radially, through heating coils at the periphery of the housing, and is recirculated. Preheated air is generally drawn in at the base of the dryer and exhausted at the top. Weisselberg [*Ind. Eng. Chem.*, **30**, 999 (1938)] has described the vertical turbodryer in more detail.

Special designs of tunnel dryers are used to dry ceramics, leather, veneer, tobacco, etc. Ceramic tunnel dryers handling large irregularly shaped objects must be designed for proper control of temperature and humidity conditions to prevent *cracking* and *condensation* on the ware. The **internal mechanism** causing cracking in drying clay and ceramics has been studied by Macey [*Trans. Brit. Ceramic Soc.*, **41** (4), 73 (1942)]. The problems of design for ceramic dryers have been outlined in part by Carruthers [*J. Am. Ceramic Soc.*, **14**, 8 (1931); *Trans. Am. Soc. Mech. Engrs.*, **57**, 439 (1935)], Garve [*J. Am. Ceramic Soc.*, **16**, 118 (1933)], and Hummel and Twells [*Bull. Am. Ceramic Soc.*, **19**, 434 (1940)]. In leather dryers, the hide may be held under tension on a toggle frame to prevent shrinkage, or it may be pasted wet on a large glass plate that imparts special surface properties to the leather. The frames or plates are then moved through suitably designed tunnels under proper conditions of temperature and humidity. The calculation of a tunnel dryer for chrome leather hung in sheets for both constant-temperature and adiabatic operation together with a consideration of optimum air recirculation has been presented by Hougen [*Chem. & Met. Eng.*, **47** (3), 160 (1940)]. Veneer and tobacco are dried on conveying belts with heated air passing across the material.

Since tunnel dryers expose considerable surface, they must be thoroughly insulated to prevent major heat losses as well as condensation on the inside of the dryer housing.

Applications. Tunnel dryers may be used for all types of materials that can be dried in tray and truck dryers. They are commonly used for handling large volumes of materials and have found wide application in drying foods, especially fruits, and in drying rayon cakes.

The vertical turbodryer is usually used for materials that are relatively free flowing throughout the drying cycle, although it has been claimed to operate successfully on starch and glue.

Theory and Design Data. Both the tunnel dryer and the vertical turbo may be treated by the theory discussed under Tray and Compartment Dryers. The drying cycle consists of a constant-rate and a falling-rate period. Thus air velocity, temperature, and humidity will control during the constant-rate period; and the falling-rate period will depend chiefly on material depth and structure.

In the vertical turbodryer, drying differs from tray drying in that the layer of material is turned over in dropping from one tray to the next, thereby periodically exposing a new surface of material to the drying air.

With sufficient data on the drying rate of a given material, tunnel dryers can be treated by the **length of a transfer unit** concept. This requires knowledge of the material temperature and air temperature throughout the drying cycle in order to evaluate the average temperature driving force $(\Delta t)_m$ in Eq. (38), which defines the number of transfer units N_t. If N_t can be evaluated and if the length of a transfer unit L_t is known, the length of the tunnel can be estimated. L_t is a function of the heat-transfer coefficient, air rate, and area of drying [Eq. (37)]. For air flow parallel to or across plane surfaces, neglecting radiation and conduction effects, L_t is given by the relation

$$L_t = 19.5bG^{0.2} \qquad (52)$$

where G = air rate, lb. dry air/(hr.)(sq. ft.), and b = free space between trays, ft.

The value of N_t will depend on the extent to which the drying occurs in the constant-rate period. For the constant-rate period,

$$(N_t)_c = \ln \frac{t_1 - t_w}{t_c - t_w} \qquad (53)$$

where $(N_t)_c$ = number of transfer units in the constant-rate period; t_1 = air temperature entering the constant-rate zone, °F.; t_w = wet-bulb temperature of the air entering the constant-rate zone, °F.; and t_c = air temperature at the critical moisture content, °F. For the falling-rate period,

$$(N_t)_f = \frac{t_c - t_2}{(\Delta t)_m} \qquad (54)$$

where $(N_t)_f$ = number of transfer units in the falling-rate period; t_2 = reduced temperature of air due to drying only, °F.; $(\Delta t)_m$ = mean temperature difference between $(t_c - t_w)$ and $(t_2 - t_s)$, °F.; and t_s = material temperature at discharge, °F. The total number of transfer units required will be the sum of those required in the two zones, and the length of the tunnel required to do the drying can be expressed by the following equation:

$$L = 19.5bG^{0.2}\left[\ln\left(\frac{t_1 - t_w}{t_c - t_w}\right) + \frac{t_c - t_2}{(\Delta t)_m}\right] \qquad (55)$$

The values of t_c and t_2 must be determined by means of heat and material balances, which involve the tunnel cross section and air rate. Equation (55) applies to parallel-flow tunnels as written, and with a negative sign to counterflow tunnels.

As in all dryer designs, the final design of tunnel and vertical turbodryers should be based on **actual drying tests in equipment simulating each type.**

Example 10. It is required to *estimate* the size of a tunnel dryer for producing 500 lb./hr. (dry basis) of a product with 1 per cent moisture. The wet feed consists of a press cake at 60°F. containing 1.5 lb. water/lb. dry product. The product is essentially non-hygroscopic. The dry-bulk density is 35 lb./cu. ft. Shrinkage during drying is negligible, and tests have indicated that the critical moisture content of the material is about 0.40 lb. water/lb. dry stock. Counterflow operation is to be used, and the inlet-air temperature will be 300°F. The specific heat of the dry material is 0.3 B.t.u./(lb.)(°F.). Fresh air will be at 60°F. and a humidity of 0.01 lb./lb. dry air. The maximum air mass velocity that can be used without serious dusting is 2000 lb./(hr.)(sq. ft.).

Solution. Assuming that the material will leave the dryer at 290°F. and air will be exhausted at 140°F., the heat requirements are approximately as follows:

Total heat to evaporate moisture =

$$(500)(1.50 - 0.01)[(212 - 60) + 970 + 0.45(140 - 212)] = 815{,}000 \text{ B.t.u./hr.}$$

Heat to raise material temperature =

$$0.3(500)(290 - 60) = 34{,}500 \text{ B.t.u./hr.}$$

$$\text{Total} = 849{,}500 \text{ B.t.u./hr.}$$

say 850,000

The entering-air humidity will be assumed to be 0.03 lb./lb. to permit some recirculation. This corresponds to 1 per cent relative humidity at 300°F.

The quantity of air required is (850,000)/(0.254)(300 − 140) = 20,936 lb./hr., where 0.254 = humid heat at 300°F. and 1 per cent relative humidity. The rate of water removal is 500(1.50 − 0.01) = 745 lb./hr. The humidity of the air leaving the dryer is 0.03 + 745/20,900 = 0.0656, which corresponds to 48 per cent relative humidity at 140°F. The wet-bulb temperature corresponding to these conditions is 116.0°F. The pounds of air required to recirculate to give the desired entering humidity are (20,900)(0.03 − 0.01)/(0.0662 − 0.01) = 7500 lb./hr. The heat requirement for the constant-rate section is 610,000 B.t.u./hr. Thus the air temperature at the point where surface drying stops will be 140 + (300 − 140)(610,000)/(850,000) = 255°F.

The number of transfer units in the constant-rate section from Eq. (53) is $(N_t)_c = \ln[(255 - 116.0)/(140 - 116.0)] = 1.75$. If the assumptions upon which Eq. (34) is based are assumed to hold, then the correct $(\Delta t)_m$ to use in Eq. (54) to obtain the number of transfer units in the falling-rate section is the logarithmic mean Δt. $\Delta t_1 = 255 - 116.0 = 139$°F.; $\Delta t_2 = 300 - 290 = 10$°F.; and $(\Delta t)_m = (139 - 10)/\ln(139/10) = 49.0$°F. By Eq. (54), $(N_t)_f = (300 - 255)/49.0 = 0.92$. The total number of transfer units required will then be 1.75 + 0.92 = 2.67.

The cross-sectional area needed for air flow is determined by dividing the total air requirements by the allowable air velocity, 20,900/2000 = 10.45 sq. ft. Twenty-eight trays, 3 ft. square, 1 in. deep, with a free space between trays of 1.5 in. on 6-ft.-high trucks will give the required cross-sectional area for air flow.

From Eq. (52) and the tray spacing of 0.125 ft., $L_t = 19.5(0.125)(2000)^{0.2} = 11.1$ ft. The total tunnel length required is then (2.67)(11.1) = 29.6 ft. The tunnel length will be specified as 30 ft. so that it will accommodate five trucks, each loaded with two tiers of 3-ft.-square trays, 28 trays high.

Each tray has 9 sq. ft. of area and will hold (9)(35)/(12) = 26.3 lb. of dry material. The total hold-up of material in the tunnel will be (2)(28)(5)(26.3) = 7364 lb. The retention time in the tunnel will be 7364/500 = 14.7 hr. Trucks should be added to the dryer every 14.7/5 = 2.9 hr. The above calculations do not include heat losses through the tunnel walls and the heat to raise the trucks to tunnel temperature. This should be included and the air flow increased in proportion. The increased air rate will not greatly change the above results.

The pressure drop through the tunnel and trucks may be estimated by the methods presented in Sec. 5.

Performance and Cost Data. There are not many published data on the performance of either tunnel or vertical turbodryers. Air rates, **power costs,** and **fuel consumption** for tunnel dryers will be roughly comparable with tray dryers. **Labor costs** will be slightly less than for batch tray dryers, since one man can tend a greater volume of material from a tunnel dryer. **Maintenance** should not exceed 10 per cent of the installed cost.

Steam consumption for a vertical turbodryer will be about 1.7 to 3.0 lb. steam/lb. water removed. **Power costs** will be about 50 per cent of the power cost for a tray dryer having the same tray area.

Tunnel dryers should operate with high air volumes to ensure uniform distribution and to compensate for any by-passing around the sides, bottom, and top of the trucks.

Continuous Through-circulation Dryers

Description. Continuous through-circulation dryers operate on the principle of blowing hot air through a permeable bed of wet material passing continuously through the dryer. Drying rates are high because of the large area exposed and the short distance of travel for the internal moisture.

A widely used type is the **horizontal conveying-screen dryer** in which wet material is conveyed as a layer, 1 to 6 in. deep, on a horizontal screen or apron, while heated air is blown either **upward or downward** through the bed of material. A dryer of this type, shown diagrammatically in Fig. 22, is described by Hurxthal [*Ind. Eng. Chem.*, **30,** 1004 (1938)]. Its drying characteristics were studied by Marshall and Hougen [*Trans. Am. Inst. Chem. Engrs.*, **38,** 91 (1942)]. This dryer usually consists of a number of individual units, complete with fan and heating coils, arranged in series to form a housing or tunnel through which the conveying screen travels. As shown in the cross-sectional view in Fig. 22, the air in this dryer circulates around through the wet material and is reheated before entering the bed. It is customary to circulate the hot air or gas **upward** in the wet end, and **downward** in the dry end. A portion of the air from each unit is exhausted continuously by one or two exhaust fans, not shown in the sketch, which handle air from several units. Since each unit can be operated individually, extremely flexible operation is possible, with high temperatures usually at the wet end, followed by lower temperatures, and in some cases a unit with specially humidified air for final conditioning.

Through-circulation drying requires the wet material to be in such a state of subdivision that hot air may be readily blown through it. Many materials meet this requirement without special preparation. Many other materials require special and often elaborate pretreatment to render them suitable for through-circulation drying. The process of putting a wet solid into a form suitable for through circulation of air is called **preforming,** and often the success or failure of this drying method depends on the preforming step.

Fibrous, flaky, and coarse granular materials are usually amenable to through-circulation drying without preforming. They can be loaded directly onto the conveying screen by suitable spreading feeders of the oscillating-belt or vibrating type, by hand, or by spiked drums or belts feeding from bins. When materials must be preformed, several methods are available, depending on the physical state of the wet solid.

Relatively dry materials such as centrifuge cakes can sometimes be **granulated** to give a suitably porous bed on the conveying screen. A typical granulator is shown in Sec. 16, Figs. 95 and 96.

Pasty materials can often be preformed by **extrusion** to form spaghetti-like pieces, about ¼ in. in diameter and several inches long.

Wet pastes that cannot be granulated or extruded may be **predried** and preformed on a steam-heated **finned drum.** Preforming on a finned drum differs basically by the fact that predrying is done (see Drum Dryers, p. 864).

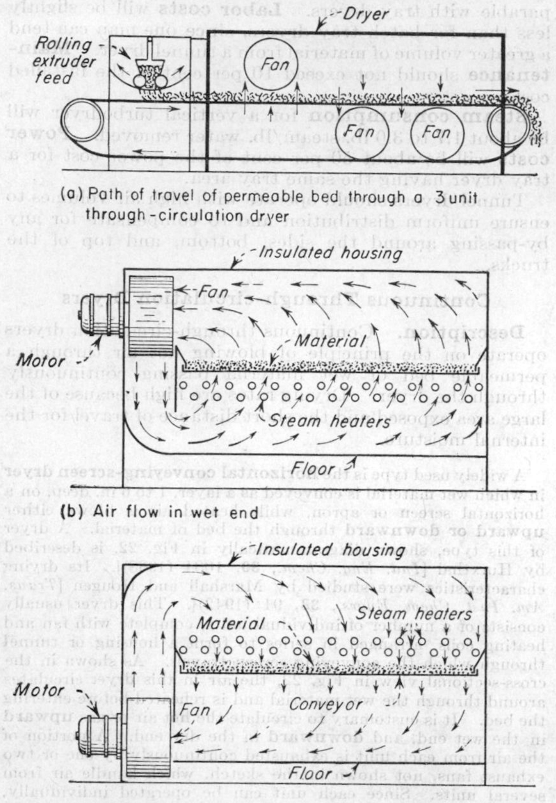

FIG. 22. Diagrammatic sketches of a typical through-circulation dryer for pastes, filter cakes, etc. (*Proctor & Schwartz, Inc.*)

Thixotropic filter cakes from rotary vacuum filters that cannot be preformed by any of the above methods can often be **scored,** by knives, on the filter, the scored cake discharging in pieces suitable for through-circulation drying. This method is especially suitable for starch.

Materials that shrink markedly on drying, such as vegetables and other high-moisture-content solids, are often reloaded during the drying cycle to two to six times the original loading at the point where shrinkage has destroyed the effectiveness of contact between the air and solid. Such a system is shown in Fig. 23.

In a few cases, powders have been **pelleted** or formed in **briquettes** to eliminate dustiness and permit drying by through circulation. Table 5 gives a classification of materials based on various methods of preforming for through-circulation drying.

Steam-heated air is the usual heat-transfer medium used with these dryers, although gases of combustion have been used. Temperatures above 600°F. are not feasible because of the problems of lubricating conveyors, chain and roller drives, etc. **Recirculation** of air is in the range of 60 to 90 per cent and may be adjusted separately for each unit by proper damper settings.

Conveyors may be made of wire-mesh screen or perforated-steel plate. The minimum practical screen size is about 30 mesh. Conveyor widths may range from 2 to 9 ft. Two or more conveyors may be built into a single dryer. Dryer lengths may range from 1 to 25 units, or from 17 to 160 ft.

A special form of the conveying-screen dryer is the D-L-O (Dwight-Lloyd-Oliver) dryer in which high-temperature flue gases are drawn down through the bed of material by means of a single large fan [Irvin, *Ind. Eng. Chem.*, **30**, 1002 (1938)]. The design of this dryer is such that the material can be mechanically dewatered by the gases as well as dried. It has special application in the metallurgical industry.

Another type of combined filter dryer is the **Oliver crystal filter dryer,** which is essentially a continuous, rotary, top-feed filter enclosed to permit steam-heated air or hot gases to be drawn into the filter chamber and through the bed of material to the center of the drum. This equipment can be used for dewatering and washing in addition to drying.

Through-circulation dryers may also employ a vibrating screen or grate to convey the material while hot air or flue gases are passed upward or downward through the bed of material. Examples of this type are the **Jeffrey-Taylor vibrating dryer,** the **McNally-Pittsburg "Vissac" coal dryer,** and the **Link-Belt "shaking-screen" coal dryer.** Mechanical dewatering also can be accomplished in these types.

Two dryers usually classified as rotary dryers are basically through-circulation dryers. These are the **Roto-louvre dryer,** and the **Perkins dryer.** In the former, hot air is blown through louvers in the wall of a horizontal cylinder and **up through** the bed of granular material which moves continuously as the cylinder

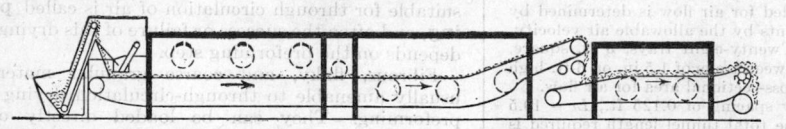

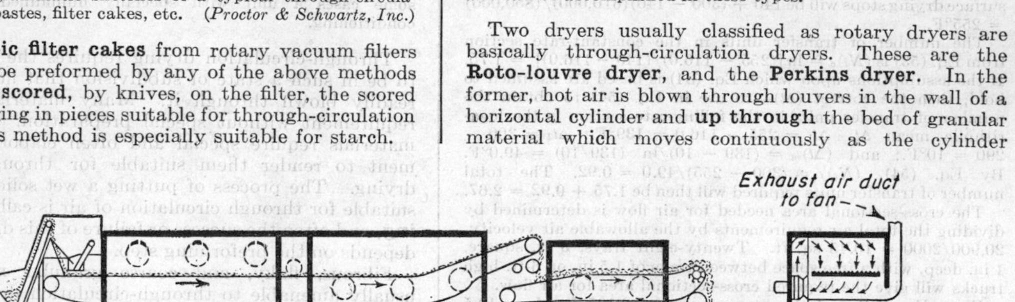

FIG. 23. Through-circulation dryer for vegetables and similar high-moisture-content materials in which excessive shrinkage occurs. (*Proctor & Schwartz, Inc.*)

Table 5. Methods of Preforming Some Materials for Through-circulation Drying

No preforming required	Scored on filter	Granulation	Extrusion	Finned drum	Flaking on chilled drum	Briquetting and squeezing
Cellulose acetate	Starch	Kaolin	Calcium carbonate	Lithopone	Soap flakes	Soda ash
Silica gel	Aluminum hydrate	Cryolite	White lead	Zinc yellow		Cornstarch
Scoured wool		Lead arsenate	Lithopone	Calcium carbonate		Synthetic rubber
Sawdust		Cornstarch	Titanium dioxide	Magnesium carbonate		
Rayon waste		Cellulose acetate	Magnesium carbonate			
Fluorspar		Dye intermediates	Aluminum stearate			
Tapioca			Zinc stearate			
Breakfast food						
Asbestos fiber						
Cotton linters						
Rayon staple						

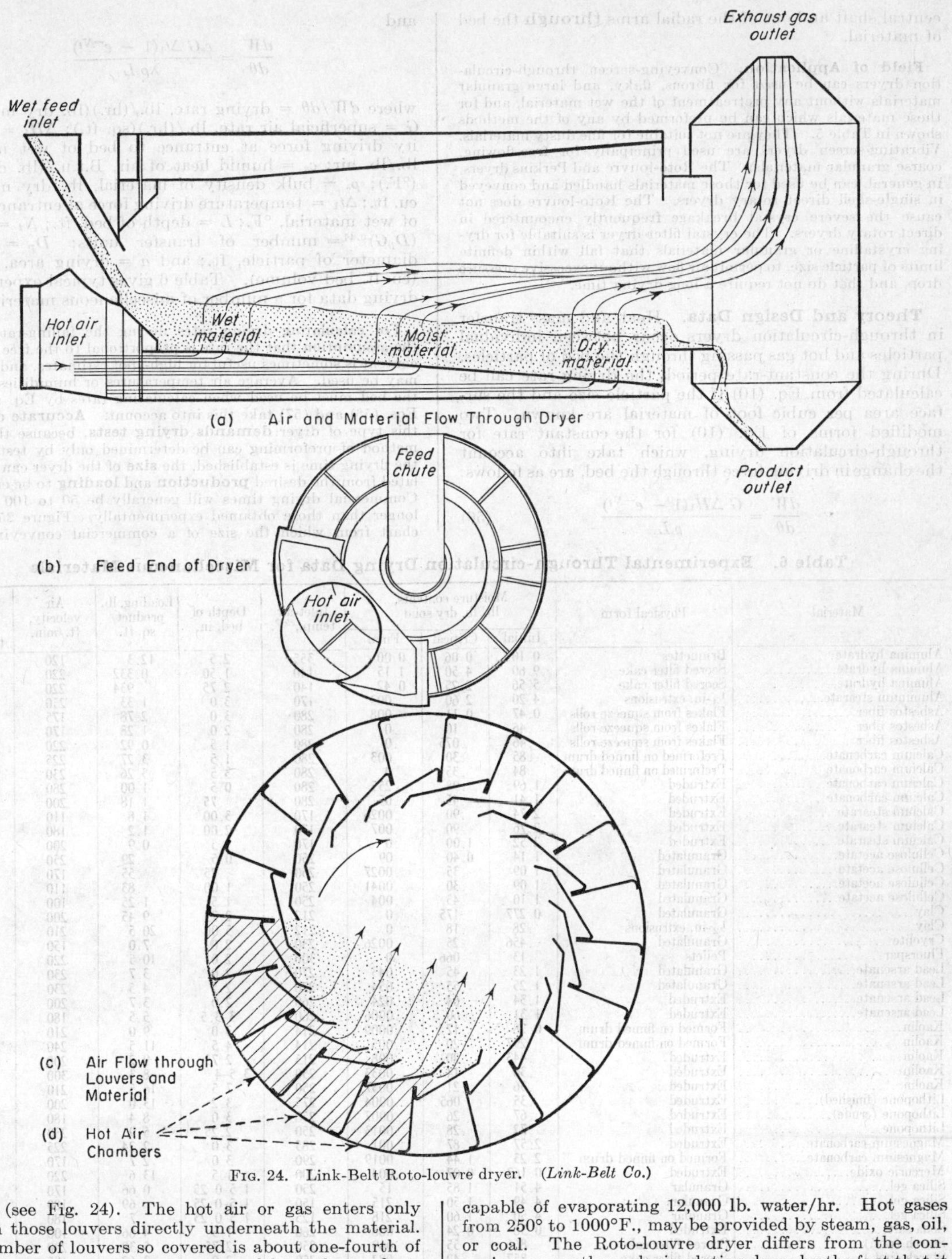

FIG. 24. Link-Belt Roto-louvre dryer. (*Link-Belt Co.*)

rotates (see Fig. 24). The hot air or gas enters only through those louvers directly underneath the material. The number of louvers so covered is about one-fourth of the total. Exhaust gases are removed from the product-discharge end of the rotating cylinder. The rotation of the cylinder, usually 2 to 3 r.p.m., keeps the material in continuous agitation and conveys it through the dryer [Erisman, *Ind. Eng. Chem.*, **30**, 996 (1938)].

Roto-louvre dryers range in size from 2½ to 11½ ft. in diameter, and 8 to 35 ft. long. The largest unit is capable of evaporating 12,000 lb. water/hr. Hot gases from 250° to 1000°F., may be provided by steam, gas, oil, or coal. The Roto-louvre dryer differs from the conveying-screen through-circulation dryer by the fact that it constantly **mixes** the material by its rotary operation, whereas in the latter the material remains stationary.

The Perkins dryer consists of a single-shell rotary dryer fitted with a hollow central shaft having radial arms dipping into the bed of material, which rolls on the bottom of the rotating shell. Hot air or gas is forced through the central shaft and out of the radial arms **through** the bed of material.

Field of Application. Conveying-screen through-circulation dryers can be used for fibrous, flaky, and large granular materials without any pretreatment of the wet material, and for those materials which can be preformed by any of the methods shown in Table 5. They are not suitable for fine dusty materials. Vibrating-screen dryers are used principally for free-flowing, coarse granular materials. The Roto-louvre and Perkins dryers, in general, can be used for those materials handled and conveyed in single-shell direct rotary dryers. The Roto-louvre does not cause the severe crystal breakage frequently encountered in direct rotary dryers. The crystal filter dryer is suitable for drying crystalline or granular materials that fall within definite limits of particle size, to permit air flow without excessive pressure drop, and that do not require a long drying time.

Theory and Design Data. Heat and mass transfer in through-circulation dryers occur between individual particles and hot gas passing through the bed of particles. During the constant-rate period, the drying rate can be calculated from Eq. (10) if the particle size and the surface area per cubic foot of material are known. Two modified forms of Eq. (10) for the constant rate for through-circulation drying, which take into account the change in driving force through the bed, are as follows:

$$\frac{dW}{d\theta} = \frac{G\,\Delta H_1(1 - e^{-N_t})}{\rho_s L} \tag{56}$$

and

$$\frac{dW}{d\theta} = \frac{c_s G\,\Delta t_1(1 - e^{-N_t})}{\lambda \rho_s L} \tag{57}$$

where $dW/d\theta$ = drying rate, lb./(hr.)(lb. dry material); G = superficial air rate, lb./(hr.)(sq. ft.); ΔH_1 = humidity driving force at entrance to bed of wet material, lb./lb. air; c_s = humid heat of air, B.t.u./(lb. dry air)-(°F.); ρ_s = bulk density of material, lb. dry material/cu. ft.; Δt_1 = temperature driving force at entrance to bed of wet material, °F.; L = depth of bed, ft.; $N_t = 0.4aL/(D_pG)^{0.41}$ = number of transfer units; D_p = average diameter of particle, ft.; and a = drying area, sq. ft./(cu. ft. bed volume). Table 6 gives typical experimental drying data for a number of miscellaneous materials.

For calculating drying rates during the falling-rate period, the assumption that the rate is proportional to the free moisture content is sometimes useful for high-spot estimates, and Eq. (35) may be used. Average air temperatures or humidities through the bed must be used when calculating rates by Eq. (35), but Eqs. (56) and (57) take this into account. **Accurate design** of this type of dryer **demands drying tests,** because the proper method of preforming can be determined only by tests. Once the drying time is established, the **size** of the dryer can be calculated from the desired **production** and **loading** to be employed. Commercial drying times will generally be 50 to 100 per cent longer than those obtained experimentally. Figure 25 gives a chart from which the size of a commercial conveying-screen

Table 6. Experimental Through-circulation Drying Data for Miscellaneous Materials

Material	Physical form	Moisture contents, lb./lb. dry solid			Inlet-air temp., °F.	Depth of bed, in.	Loading, lb. product/sq. ft.	Air velocity, ft./min.	Experimental drying time, min.
		Initial	Critical	Final					
Alumina hydrate	Briquettes	0.105	0.06	0.00	355	2.5	12.3	120	30
Alumina hydrate	Scored filter cake	9.60	4.50	1.15	140	1.50	0.332	220	150
Alumina hydrate	Scored filter cake	5.56	2.25	0.42	140	2.75	.934	220	180
Aluminum stearate	¼-in. extrusions	4.20	2.60	.003	170	3.0	1.33	250	60
Asbestos fiber	Flakes from squeeze rolls	0.47	0.11	.008	280	3.0	2.78	175	9.3
Asbestos fiber	Flakes from squeeze rolls	.46	.10	.0	280	2.0	1.28	170	6.0
Asbestos fiber	Flakes from squeeze rolls	.46	.075	.0	280	1.5	0.92	220	4.5
Calcium carbonate	Preformed on finned drum	.85	.30	.003	280	1.5	3.27	225	20
Calcium carbonate	Preformed on finned drum	.84	.35	.0	280	3.5	5.26	230	30
Calcium carbonate	Extruded	1.69	.98	.255	280	0.5	1.00	280	15
Calcium carbonate	Extruded	1.41	.45	.05	280	.75	1.18	200	20
Calcium stearate	Extruded	2.74	.90	.0026	170	3.00	1.8	110	95
Calcium stearate	Extruded	2.76	.90	.007	170	2.00	1.2	180	70
Calcium stearate	Extruded	2.52	1.00	.0	170	1.5	0.9	200	40
Cellulose acetate	Granulated	1.14	0.40	.09	250	0.5	.29	250	3
Cellulose acetate	Granulated	1.09	.35	.0027	250	.75	.55	170	12
Cellulose acetate	Granulated	1.09	.30	.0041	250	1.00	.83	110	18
Cellulose acetate	Granulated	1.10	.45	.004	250	1.5	1.25	100	30
Clay	Granulated	0.277	.175	.0	212	2.75	9.45	200	32
Clay	½-in. extrusions	.28	.18	.0	212	5.00	20.5	210	73
Cryolite	Granulated	.456	.25	.0026	230	2.0	7.0	150	40
Fluorspar	Pellets	.13	.066	.0	300	2.0	10.5	220	13
Lead arsenate	Granulated	1.23	.45	.043	270	2.0	3.7	230	30
Lead arsenate	Granulated	1.25	.55	.054	270	2.5	4.5	230	40
Lead arsenate	Extruded	1.34	.64	.024	260	2.0	3.7	200	60
Lead arsenate	Extruded	1.31	.60	.0006	260	3–3.5	5.5	180	70
Kaolin	Formed on finned drum	0.28	.17	.0009	214	3.0	9.0	210	35
Kaolin	Formed on finned drum	.297	.20	.005	214	4.5	11.5	240	25
Kaolin	Extruded	.443	.20	.008	215	2.75	9.2	200	30
Kaolin	Extruded	.36	.14	.0033	250	3.5–4	8.3	300	20
Kaolin	Extruded	.36	.21	.0037	250	7.5	16.5	210	50
Lithopone (finished)	Extruded	.35	.065	.0004	275	3.2	13.0	200	30
Lithopone (crude)	Extruded	.67	.26	.0007	250	3.0	8.4	180	85
Lithopone	Extruded	.72	.28	.0013	250	2.25	5.9	230	30
Magnesium carbonate	Extruded	2.57	.87	.001	285	3.0	2.24	225	29
Magnesium carbonate	Formed on finned drum	2.23	1.44	.0019	290	3.0	2.7	170	40
Mercuric oxide	Extruded	0.163	0.07	.004	200	1.5	13.6	220	40
Silica gel	Granular	4.51	1.85	.15	250	1.5–0.25	0.66	170	25
Silica gel	Granular	4.49	1.50	.215	150	1.5–0.25	.69	180	105
Silica gel	Granular	4.50	1.60	.218	125	1.5–0.25	.7	180	110
Soda salt	Extruded	0.36	0.24	.008	280	1.5	4.66	100	85
Starch (pot.)	Scored filter cake	.866	.55	.069	250	2.75	5.38	200	45
Starch (pot.)	Scored filter cake	.857	.42	.082	250	2.0	3.62	185	25
Starch (corn)	Scored filter cake	.776	.48	.084	160	2.75	5.4	146	90
Starch (corn)	Scored filter cake	.78	.56	.098	225	2.75	5.6	150	40
Starch (corn)	Scored filter cake	.76	.30	.10	160	0.75	1.57	131	25
Titanium dioxide	Extruded	1.02	.60	.10	310	1.5	1.38	270	10.5
Titanium dioxide	Extruded	1.07	.65	.29	310	3.2	3.28	170	10
White lead	Formed on finned drum	0.238	.07	.001	180	2.5	15.7	220	50
White lead	Extruded	.49	.17	.0	200	1.5	6.9	200	45
Zinc stearate	Extruded	4.63	1.50	.005	190	1.75	0.85	170	60

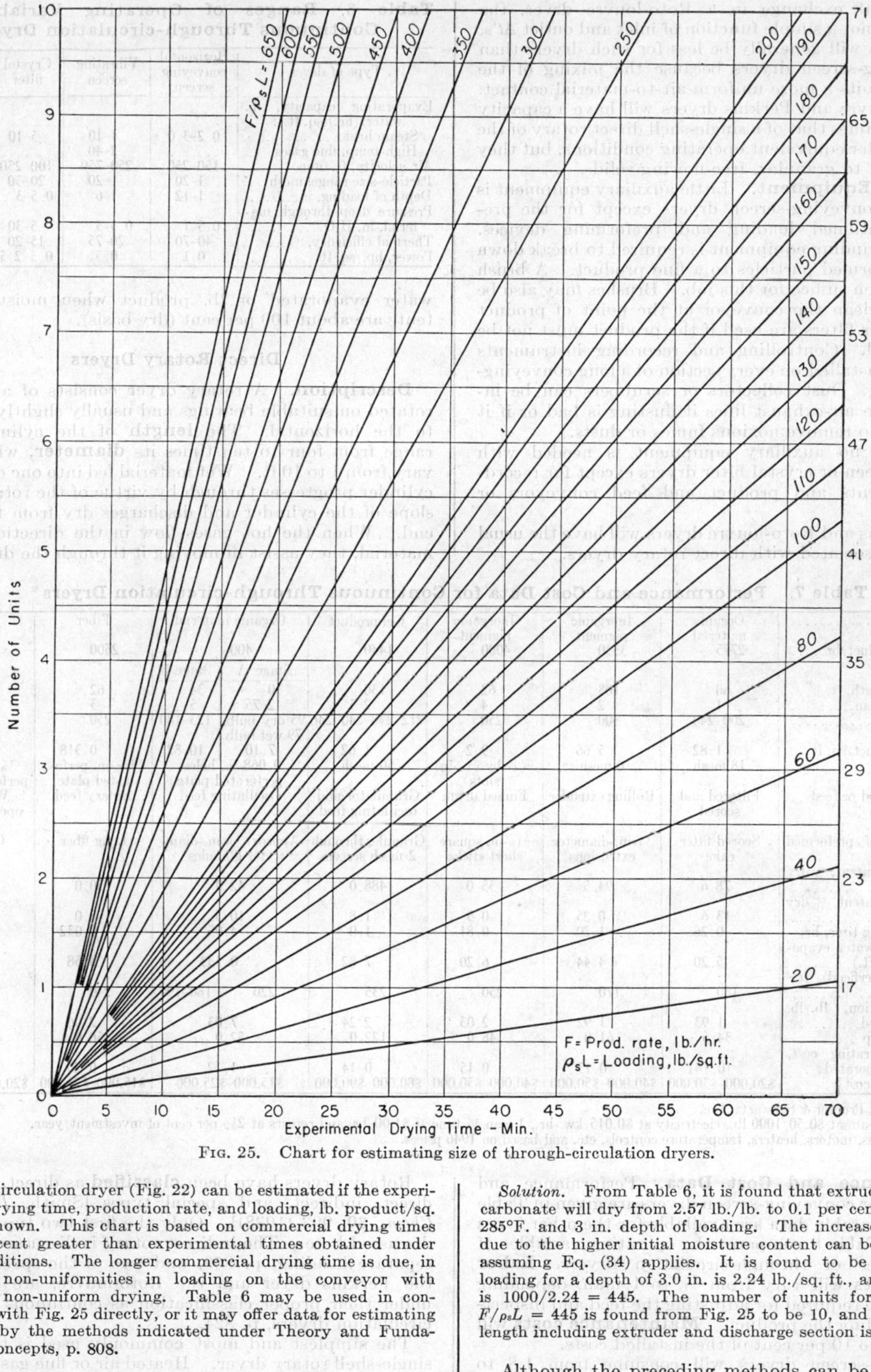

FIG. 25. Chart for estimating size of through-circulation dryers.

through-circulation dryer (Fig. 22) can be estimated if the experimental drying time, production rate, and loading, lb. product/sq. ft., are known. This chart is based on commercial drying times 100 per cent greater than experimental times obtained under ideal conditions. The longer commercial drying time is due, in part, to non-uniformities in loading on the conveyor with resultant non-uniform drying. Table 6 may be used in conjunction with Fig. 25 directly, or it may offer data for estimating the time by the methods indicated under Theory and Fundamental Concepts, p. 808.

Example 11. It is desired to estimate the size of a conveying-screen through-circulation dryer required to produce 1000 lb./hr. of magnesium carbonate. The initial moisture content is 3.0 lb. water/lb. dry solid, and the final moisture content is to be 0.1 per cent.

Solution. From Table 6, it is found that extruded magnesium carbonate will dry from 2.57 lb./lb. to 0.1 per cent in 29 min. at 285°F. and 3 in. depth of loading. The increased drying time due to the higher initial moisture content can be estimated by assuming Eq. (34) applies. It is found to be 31 min. The loading for a depth of 3.0 in. is 2.24 lb./sq. ft., and hence $F/\rho_s L$ is 1000/2.24 = 445. The number of units for 31 min. and $F/\rho_s L = 445$ is found from Fig. 25 to be 10, and the total dryer length including extruder and discharge section is 71 ft.

Although the preceding methods can be used to estimate drying times for conveying-screen dryers, and possibly for a crystal filter dryer, they cannot be used to estimate drying times for vibrating-screen dryers or for the Roto-louvre and Perkins dryers. Because of the crossflow heat exchange in a Roto-louvre dryer, the average Δt is not a simple function of inlet and outlet Δt's. Drying times will generally be less for such dryers than for conveying-screen dryers because the mixing of the material permits a more uniform air-to-material contact. The Roto-louvre and Perkins dryers will have a capacity of about 1.5 times that of a single-shell direct rotary of the same size under equivalent operating conditions, but they are restricted to granular, free-flowing solids.

Auxiliary Equipment. Little auxiliary equipment is needed for conveying-screen dryers except for the previously mentioned loading and preforming devices. Sometimes grinding equipment is required to break down the dry preformed particles to a fine product. A brush sifter will often suffice for this job. Brushes may also be required to clean the conveyor at the point of product removal. Air filters are used if the product must not be contaminated. Controlling and recording instruments are usually installed on every section of a long conveying-screen dryer. Dust collectors or scrubbers can be installed on the air-exhaust lines if dusting is bad or if it is necessary to remove noxious fumes or dusts.

Relatively no auxiliary equipment is needed with vibrating-screen or crystal filter dryers except for recording instruments and product and feed conveyors or hoppers.

The Perkins and Roto-louvre dryers will have the usual auxiliaries associated with direct rotary dryers.

Table 8. Ranges of Operating Variables for Continuous Through-circulation Dryers

Type of dryer	Horizontal conveying screen	Vibrating screen	Crystal filter	Roto-louvre
Evaporating capacity, lb. water/(hr.)(sq. ft.):				
Steam heat	0.2–3.0	1–10	5–10	2–10
High-temp. flue gases	....	2–40	....	5–40
Air velocity, ft./min	150–250	250–750	100–250	100–350
Particle-size range, mesh	1–20	1–20	20–30	1–150
Depth of loading, in	1–12	1–6	0.5–3	4–16
Pressure drop through material, in. H_2O	0.5–1	0.5–5	5–30	2–6
Thermal efficiency, %	40–70	20–75	15–20	35–70
Power, hp./sq. ft	0.1	0.5	0.5–2.5	0.1–0.3

water evaporated or lb. product when moisture contents are about 100 per cent (dry basis).

Direct Rotary Dryers

Description. A rotary dryer consists of a cylinder rotated on suitable bearings and usually slightly inclined to the horizontal. The **length** of the cylinder may range from four to ten times its **diameter**, which may vary from 1 to 10 ft. Wet material fed into one end of the cylinder progresses through by virtue of the rotation and slope of the cylinder and discharges dry from the other end. When the hot gases flow in the direction of the material, they assist in moving it through the dryer.

Table 7. Performance and Cost Data for Continuous Through-circulation Dryers*

Kind of material	Organic material	Inorganic pigment	Inorganic pigment	Gel product	Organic material		Fiber	Fiber
Capacity, lb. product/hr	2775	3330	4000	1440	400		2500	1500
					Stage A	Stage B		
Approx. dryer length, ft	60	108	62	130	30	34	62	62
Depth of loading, in	1	2	4	2.5	2.75	5	3	2
Air temp., °F	200–240	300	250	212–238–240–246	95 dry bulb, 79 wet bulb	125–180	250	240
Loading, lb. product/sq. ft	1.82	5.66	9.2	1.62	7.10	10.82	0.318	0.170
Type of conveyor	18 mesh	6 mesh	3/64- by 3/16-in. slots	50 mesh	0.068-in. holes, perforated plate		3/16-in. perforated plate	1/8-in. holes, perforated plate
Preforming method or feed	Filtered and scored	Rolling extruder	Finned drum	Granulator and oscillating feed	Oscillating feed		Rotary feed	Wet stock opener, rotary feed
Type and size of preformed particle.	Scored filter cake	1/4-in.-diameter extrusions	5/16-in. square short sticks	Granules through 2-mesh screens	Approx. 1/8-in.-diameter globules		Long fiber	Cut fiber
Initial moisture content, % dry basis	78.6	94.5	55.0	488.0	42.9		50.0	100.0
Final moisture content, % dry basis	13.6	0.25	0.5	1.8	10.5		8.0	5.5
Commercial drying time, hr	0.26	1.20	0.81	1.0	4.95		0.032	0.043
Drying rate, lb. water evaporated/(hr.)(sq. ft.)	5.20	4.44	6.20	7.82	0.333		3.58	4.95
Air velocity (superficial), ft./min	180	170	250	235	220	180–190	240	240
Steam consumption, lb./lb. water evaporated	1.93	1.92	2.03	2.24	7.03		2.06	2.01
Installed power, hp	34.5	61.0	48.0	122.0	52.0		56.0	62.75
Approximate operating cost, ¢/lb. water evaporated†	0.141	0.13	0.15	0.14	1.12		0.20	0.18
Approx. purchase cost‡	\$20,000–\$30,000	\$40,000–\$50,000	\$40,000–\$50,000	\$80,000–\$90,000	\$15,000–\$25,000		\$15,000–\$25,000	\$20,000–\$30,000

* Courtesy of Proctor & Schwartz, Inc.
† Includes steam at \$0.50/1000 lb., electricity at \$0.015/kw.-hr., 1 man 1/4 time at \$1.00/hr. and repairs at 2½ per cent of investment/year.
‡ Includes fans, motors, heaters, temperature controls, etc. and based on 1946 prices.

Performance and Cost Data. Performance and cost figures for conveying-screen dryers are given in Table 7. No comparable data are available for the other types of dryers. Table 8 compares the operating variables of various types of through-circulation dryers. **Labor requirements** will vary from 1/6 to 1 man, depending upon the time required for adjusting the feed and inspecting and handling the product. **Maintenance costs** will vary from 5 to 10 per cent of the installed costs.

Conveying-screen dryers will consume from 1.8 to 2.2 lb. steam/lb. water evaporated, depending on the final moisture content and the per cent recirculation of air in the dryer. The **direct operating costs** of such dryers should be in the range of \$0.001 to \$0.0025/lb.

Rotary dryers have been **classified** as direct, indirect-direct, indirect, and special types [Smith, *Ind. Eng. Chem.*, **30**, 993 (1938)]. Only the first two types will be discussed here. The indirect rotary is discussed under a separate heading, p. 859, and two of the special types, the Perkins dryer and the Roto-louvre, are discussed under their proper classification as continuous through-circulation dryers, p. 824.

The simplest and most commonly used is the direct single-shell rotary dryer. Heated air or flue gas is passed either counter-current to or parallel with the flow of material. Its elemental form is the rotary kiln in which the material travels as a solid bed *on the bottom* of the rotating cylinder. Rotary kilns, which are brick-lined for ex-

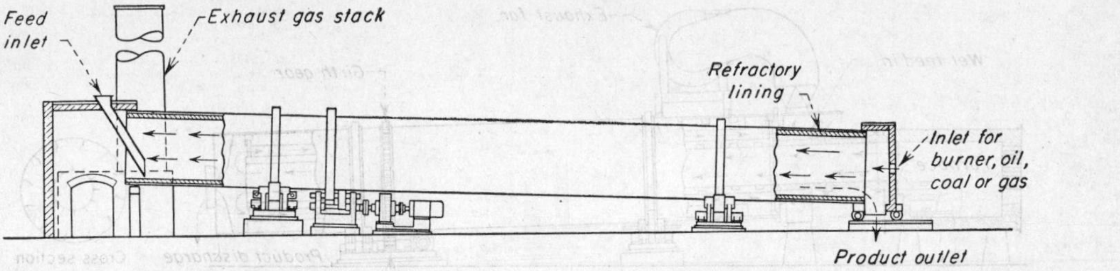

FIG. 26. Rotary kiln, refractory-lined, for roasting and calcining. In special cases, drying is also accomplished. (*Hardinge Co.*)

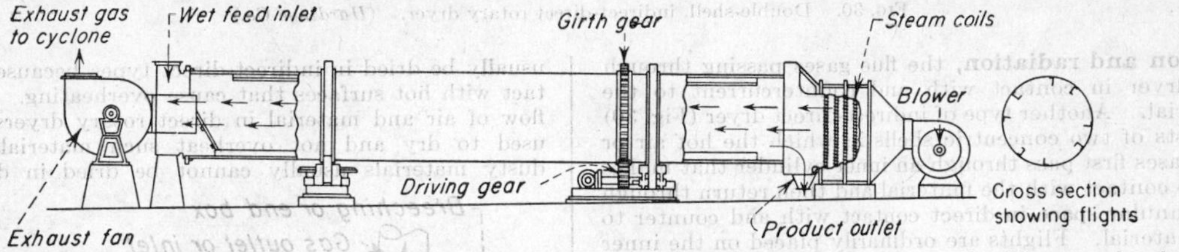

FIG. 27. Single-shell, direct rotary dryer using steam-heated air and balanced pressure by means of a blower and exhauster. (*Hardinge Co.*)

tremely high-temperature operation, are usually used for calcining and roasting, and drying is usually secondary. A typical high-temperature kiln is shown in Fig. 26.

The **direct rotary dryer** is usually equipped with **flights** on the interior surface of the shell for **lifting** and

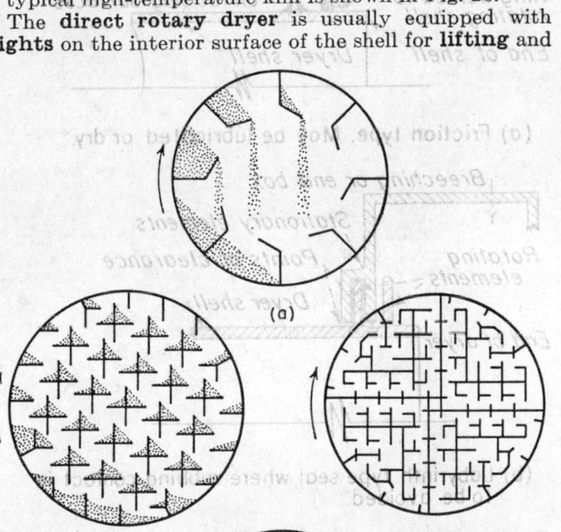

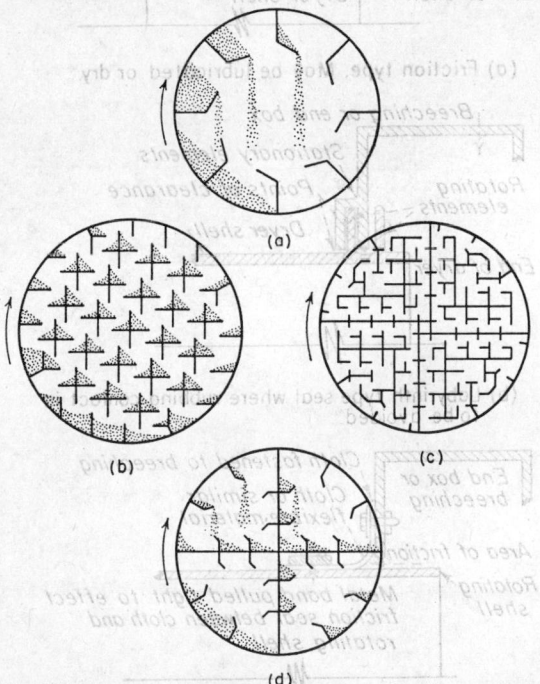

FIG. 28. Various flight designs and internal arrangements used in single-shell, direct rotary dryers.

showering the wet material through the hot gases during passage through the cylinder (Fig. 27). These flights may extend continuously the entire length of the dryer, or they may be offset every 2 to 6 ft. The shape of the flights depends upon the handling characteristics of the material. For free-flowing granular materials, a radial flight with a 90-deg. lip is used. For sticky materials, a straight radial flight without any lip is used. Many intermediate types have been designed to give **maximum showering action** during rotation. When materials change characteristics during drying, the flight design is often changed along the dryer length. The Ruggles-Coles Class XH dryer is equipped with special lifting and agitating flights for handling flotation concentrates that would ordinarily ball up and stick to the shell of a dryer having conventional flights. **Spiral flights** are usually used for the first few feet at the feed end of the dryer to accentuate a forward motion of the material into the dryer before normal flight action begins.

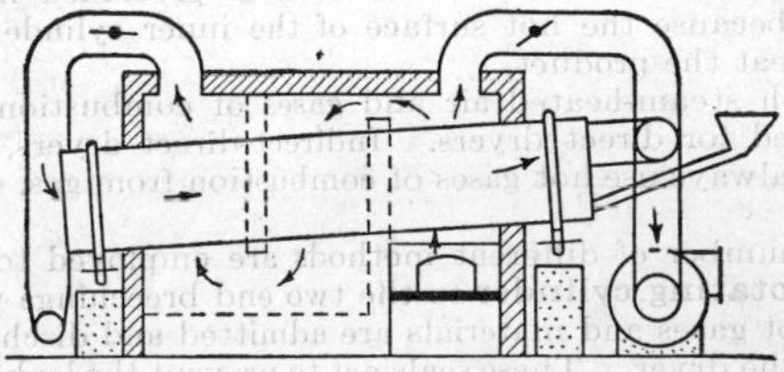

FIG. 29. Single-shell, indirect-direct rotary dryer.

When parallel air flow is used, flights are often left out of the last few feet of the dryer to prevent excessive dust carry-over in the exhaust gases. Some dryers and many kilns have lengths of chain attached to the inside of the shells. The free ends of the chains tumble over the wall of the dryer or kiln during rotation, removing material that would normally adhere to the wall. In kilns, the chains contribute markedly to the heat transfer. Their use increases maintenance costs, especially when flights are present. In dryers of large cross section, internal elements or partitions are sometimes used to increase the effectiveness of the material distribution and air distribution and to reduce dusting and grinding action on the material. Some examples are shown in Fig. 28. Use of such internal members increases the difficulty of cleaning the dryer and increases maintenance costs when abrasive materials are handled.

In **indirect-direct rotary dryers,** heat is transferred from the hot gases to the material by conduction through metal surfaces and by direct contact. One type (Fig. 29) is simply a single-shell rotary dryer encased in a brick or steel housing. Gas, oil, or powdered coal fired in the housing transmits heat to the shell of the dryer by **con-**

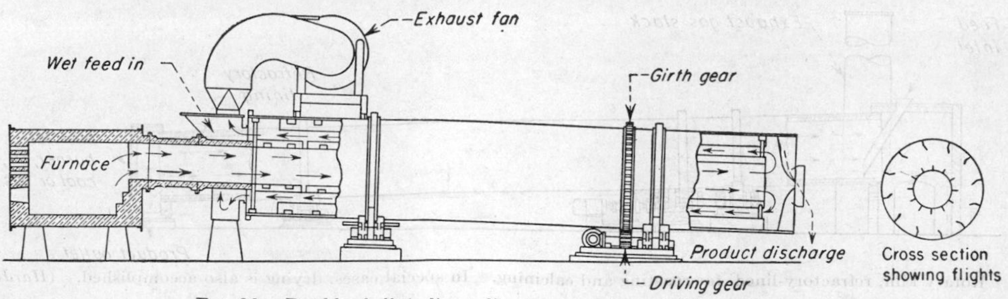

FIG. 30. Double-shell, indirect-direct rotary dryer. (*Hardinge Co.*)

vection and radiation, the flue gases passing through the dryer in contact with and countercurrent to the material. Another type of indirect-direct dryer (Fig. 30) consists of two concentric shells in which the hot air or flue gases first pass through an inner cylinder that makes direct contact with the material and then return through the annular space in direct contact with and counter to the material. Flights are ordinarily placed on the inner surface of the outer cylinder and on the outer surface of the inner cylinder. Variations of this principle have been incorporated in other types. Indirect-direct rotary dryers, although more efficient than the direct rotary, are often more **difficult to clean** and will have high maintenance costs when abrasive materials are handled.

In the direct type, the gas may flow **parallel or countercurrent** to the flow of material. Countercurrent flow gives greater efficiency with a given inlet-gas temperature, but parallel flow can be used to dry heat-sensitive materials at higher inlet-gas temperatures than can be employed with counterflow. The indirect-direct type (Fig. 30) almost always employs countercurrent flow, during direct drying, since this is more efficient. The drying of heat-sensitive materials is precluded in this type because the hot surface of the inner cylinder will overheat the product.

Both steam-heated air and gases of combustion may be used for direct dryers. Indirect-direct dryers, however, always use hot gases of combustion from gas, oil, or coal.

A number of different methods are employed to **seal the rotating cylinder** to the two end breechings where the hot gases and materials are admitted and discharged from the dryer. These seals act to prevent the leakage of air into the cylinder or out of the breeching as well as to prevent leakage of material out of the dryer. Three common types are shown in Fig. 31. Many variations of these methods are possible, depending on the requirements of the particular problem involved.

Hot gases are forced through the drying cylinder by either a blower or an exhauster or a combination of the two. With the latter arrangement, it is possible to run the dryer at the same pressure as the room surrounding it. This is often required. An exhauster alone tends to draw in cold air through discharge ports, feed hoppers, and the seals, thus decreasing the effective drying temperature. A blower alone tends to blow hot air and dry dusty product out of the aforementioned openings, causing unpleasant working conditions and heat and yield losses.

Field of Application. Rotary dryers are applicable to the drying of relatively *free-flowing granular materials.* Materials that are not completely free-flowing in the wet condition are sometimes handled by special devices, such as premixing with a portion of the dry product before feeding to the dryer, by using a special flight construction, by knockers on the shell, or by attaching chains to the inside of the dryer. Heat-sensitive materials cannot usually be dried in indirect-direct types because of contact with hot surfaces that cause overheating. Parallel flow of air and material in direct rotary dryers can be used to dry and not overheat such materials. Fine dusty materials usually cannot be dried in direct or indirect-direct types if they tend to float in the air stream while being showered and to be carried out as dust. Fine materials that **flocculate** may be dried without appreciable dust carry-over. For fine dusty heat-stable materials, a totally indirect rotary dryer is best suited, p. 859.

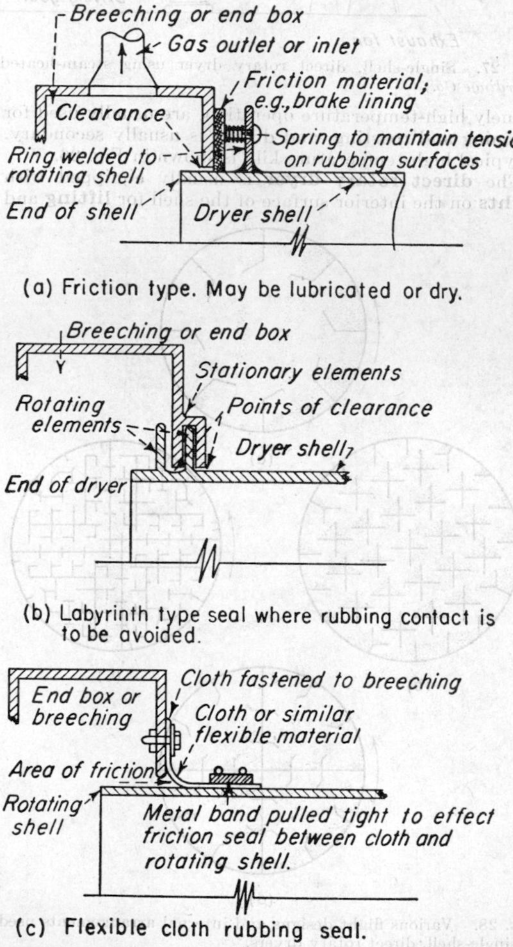

FIG. 31. Three types of seals used on rotary dryers.

Solids that must meet exacting crystal size and shape specifications are usually not dried in direct rotary dryers because of breakage and abrasion produced by showering the material. This breakage may be severe in dryers of large diameter.

Theory and Design Data. Drying in a direct single-shell rotary dryer is best expressed as a heat-transfer mechanism as follows:

$$q_t = UaV(\Delta t)_m \tag{58}$$

where q_t = total heat transferred, B.t.u./hr.; Ua = volumetric heat-transfer coefficient, B.t.u./(hr.)(cu. ft. dryer volume)(°F.); V = dryer volume, cu. ft.; and $(\Delta t)_m$ = true mean temperature difference between the hot gases and the material, °F.

Drying in a continuous dryer can be divided into three zones: (1) heating the wet material to a constant surface temperature that tends to approach the air wet-bulb temperature, (2) drying at this constant temperature, and (3) removing last traces of moisture during a period of rising material temperature. The mean over-all temperature difference for the three zones can be calculated from heat balances according to the following formula:

$$\frac{1}{(\Delta t)_m} = \frac{q_p}{q_t(\Delta t)_p} + \frac{q_v}{q_t(\Delta t)_v} + \frac{q_s}{q_t(\Delta t)_s} \tag{59}$$

where q_p = heat transferred to the wet material while heating it to the air wet-bulb temperature, B.t.u./hr.
q_v = latent heat transferred to the material while moisture is being evaporated at constant temperature, B.t.u./hr.
q_s = sensible heat transferred to the dry material while heating it to the discharge temperature, B.t.u./hr.
q_t = total heat transferred in the dryer, B.t.u./hr.
$(\Delta t)_p$ = mean temperature difference between the air and the material while heating to the wet-bulb air temperature, °F.
$(\Delta t)_v$ = mean temperature difference between the air and the material at constant temperature, °F.
$(\Delta t)_s$ = mean temperature difference between the air and material while heating to its discharge temperature, °F.
$(\Delta t)_m$ = over-all mean temperature difference in the dryer, °F.

Equation (59) holds for **all continuous direct dryers.** When considerable moisture is removed from a material, q_v will almost equal q_t. For this case, a good approximation to $(\Delta t)_m$ is the logarithmic mean between the wet-bulb depressions of the drying air at the inlet and exit of the dryer.

Data for evaluating Ua are given by Friedman and Marshall [*Chem. Eng. Progesss,* **45,** 482, 573 (1949)], who reported tests on a small rotary dryer. Although they showed that Ua is a complex function of the feed rate, air rate, and physical properties of the material being handled, a conservative design can be based on the following equation

$$Ua = 10G^{0.16}/D \tag{60}$$

where G = air mass velocity, lb./(hr.)(sq. ft. of dryer cross section), and D = dryer diameter, ft.

The volume of the dryer V can be calculated by Eq. (58) if q_t is known and Eqs. (59) and (60) are used to evaluate $(\Delta t)_m$ and Ua. Unless material characteristics limit the gas temperature, the inlet-temperature difference is usually fixed by the heating medium employed; *i.e.,* 250° to 300°F. for steam, or 1000° to 1200°F. for gas-, coal-, or oil-fired burners. The proper exit-gas temperature is actually an economic function. Its value is determined on the basis of good practice as follows: Equation (60) combined with a heat balance on the gas gives the following:

$$N_t = \frac{t_1 - t_2}{(\Delta t)_m} = \frac{10L}{c_pG^{0.84}D} \tag{61}$$

where N_t = **number of heat-transfer units** based on the gas; t_1 = initial gas temperature, °F.; t_2 = exit gas temperature corrected for heat losses, °F.; L = length of the dryer, ft.; and c_p = specific heat of the heating gases, B.t.u./(lb.)(°F.). Equation (61) can be used to select an exit-gas temperature, since it has been found (empirically) that rotary dryers are most economically operated when N_t = 1.5 to 2. Equation (61) may be modified to

$$L_t = 0.1c_pG^{0.84}D \tag{62}$$

where L_t = **length of one transfer unit.**

The air mass velocity G normally employed is the highest that will not cause excessive dusting. Once G is established and the exit-air temperature selected, the diameter D of the dryer can be calculated from a heat balance. Knowing D, t_2, t_1, G, and c_p, the length of the dryer, when found from Eqs. (61) and (62), should be such that L/D is from 4 to 10. If it is not, another value of N_t should be selected, which will give an L/D ratio within this range.

In rotary kilns, the material is not showered through the air stream but is retained in the lower part of the cylinder and mixed by its rotation. Hence the **effective heat-transfer area** is a function of the inner kiln surface. For kilns, the following semi-empirical equation is recommended for the evaluation of the volumetric heat-transfer coefficient:

$$Ua = \frac{0.12G^{0.46}}{D} \tag{63}$$

Neither the above equation nor the ones for rotary dryers account for the effect of radiation at high temperatures when the dryer or kiln is fired directly. Capacities will be somewhat greater under this condition. An analysis of the heat transfer occurring in direct-fired rotary kilns is given by Gilbert [*Cement,* **5,** 417 (1932); **6,** 79, 189, 262, 327, 369 (1933); **7,** 1, 123 (1934)]. Equations for predicting the capacity of lime kilns are given by Gibbs [*Chem. & Met. Eng.,* **53** (4), 99 (1946); **53** (5), 139 (1946)].

Heat transfer in **indirect-direct rotary dryers** is complex, occurring by a combination of conduction, radiation, and convection. A double-shell indirect-direct rotary dryer gives approximately 35 per cent more over-all heat transfer than a single-shell direct rotary of the same volumetric capacity operating at comparable temperatures.

Rotary dryers and kilns usually run with **3 to 12 per cent of their volume** filled with material. Under these conditions, the dryer usually can be made to hold a material long enough to accomplish the removal of **internal moisture.** If the **hold-up** in the dryer is not great enough, the time of passage through the dryer may be too short to achieve the desired degree of removal of internal moisture, and its capacity will be less than that predicted by the previous equations. Dryer performance at low hold-up will be erratic.

The **time of passage** in **rotary kilns** (from which hold-up can be calculated) can be predicted by the following formula based on time-of-passage studies of Sullivan, Maier, and Ralston [*U.S. Bur. Mines Tech. Paper* 384 (1927)]:

$$\theta = \frac{0.19L}{NDS} \tag{64}$$

where θ = time of passage in the kiln, min.; L = kiln length, ft.; N = rate of rotation, r.p.m.; and S = slope of the kiln, ft./ft. Other equations for estimating the time of passage when dams are used within the kiln and at the kiln discharge are presented by Bayard [*Chem. & Met. Eng.*, **52** (3), 100 (1945) in the form of nomographs.

Time of passage is defined as hold-up divided by feed rate. It can be measured indirectly in rotary dryers if the hold-up and feed rate can be measured directly. Hold-up cannot always be measured conveniently on large plant dryers, however, unless a period of shut-down occurs when the dryer can be discharged and its contents weighed directly. Direct methods have been resorted to, one of which consists of adding a pound or two of an inert detectable solid to the feed, and analyzing for it in the discharged product. The time required for the maximum concentration to occur represents the **average time of passage**. This method requires continuous sampling and analyses to ascertain the peak concentration, and it is restricted to those cases where an easily detected material can be added to the product without fear of contamination.

The time of passage in rotary dryers can be predicted by the relationships developed by Friedman and Marshall [*Chem. Eng. Progress*, **45**, 482, 573 (1949), (paper presented at Am. Inst. Chem. Engrs. meeting, New York, Nov. 10, 1948)] as given below:

$$\theta = \frac{0.23L}{SN^{0.9}D} \pm 0.6\frac{\beta LG}{F} \qquad (65a)$$

$$\beta = 5(D_p)^{-0.5} \qquad (65b)$$

where β = a constant depending upon the material being handled and approximately defined by Eq. (65*b*); D_p = weight average particle size of material being handled, microns; F = feed rate to dryer, lb. dry material/(hr.)(sq. ft. of dryer cross section); θ = time of passage, min.; and G = air mass velocity, lb./(hr.)(sq. ft.). The plus sign refers to countercurrent flow and the negative sign to parallel flow.

Air mass velocities in rotary dryers range from 200 to 10,000 lb./(hr.)(sq. ft.). It is customary to employ the highest air velocity possible without serious dusting. The **amount of dusting** occurring during operation is a complex function of the material being dried, its physical state, the air velocity employed, the hold-up in the dryer, the number of flights, the rate of rotation, and the construction of the breeching at the end of the dryer. It can be estimated only by experimental tests. An air rate at 1000 lb./(hr.)(sq. ft.) can usually be safely used with 35-mesh solids. Information on the dusting of a number of materials in a 1-ft. by 6-ft. rotary dryer has been presented by Friedman and Marshall (*ibid.*).

Rotary dryers operate at **peripheral speeds** of 30 to 150 ft./min., *i.e.*, ND = 25 to 35; while for rotary kilns ND = 0.7 to 5.0, where N = r.p.m. and D = diameter, ft.

Slopes of rotary dryer shells vary from 0 to 0.08 ft./ft. The slope is usually adjusted to give a hold-up of from 3 to 12 per cent after the diameter, length, and rate of rotation have been fixed by the preceding equations. In some cases of parallel-flow operation, negative slopes have been used.

The **radial flight height** in a single-shell direct dryer will range from one-twelfth to one-eighth of the dryer diameter. The **number of flights** will range from $2D$ to $3.5D$ for dryers larger than 2 ft. in diameter and should be designed to carry and shower all the hold-up and minimize any kiln action, *i.e.*, the tendency of the material simply to roll over on the bottom of the dryer without showering, as in a kiln.

Inlet-air temperatures will range from 100° to 350°F. for dryers employing steam heat. With flue gases the materials of construction limit the inlet-air temperature. For standard steel construction, temperatures as high as 1400°F. may be employed with direct rotary dryers and as high as 1200°F. with indirect-direct dryers. Kilns with refractory linings operate with inlet-air temperatures as high as 3000°F.

Auxiliary Equipment. A combustion chamber is required for high temperatures, and steam coils, usually finned, are used for low temperatures. If contamination of the product is undesirable, direct-fired units may not be suitable.

The **method of feeding** rotary dryers depends upon the material, and the location and type of prior processing equipment. When the feed comes from above the dryer, a chute extending into the dryer cylinder can be employed. Screw or vibrating feeders are used if gravity feed is not possible. With parallel flow, a cold air or water jacket on the feed chute prevents overheating of heat-sensitive materials. If the material to be dried is too wet or sticky to handle satisfactorily in the dryer, it may be mixed with a portion of the dry product in a suitable solids blender to make it free-flowing. Any type of solids conveyor can be used to **recycle** the dry product. This conveyor should be insulated to eliminate heat loss from the hot dry product. 50 to 60 per cent recirculation can be used economically. One method of feeding utilizes the exhaust gases to convey the wet feed. The latter is added to the exhaust gases at high velocity from the dryer. The wet feed plus dust from the dryer separates from the exhaust gases in a cyclone and drop into the feed end of the drying cylinder. This method is a combination of pneumatic conveying (p. 834) and rotary drying. The increased efficiency of this system is due to the two parallel-flow stages operating countercurrently.

The product from rotary dryers can be discharged directly into storage or packaging bins or conveyed to some other part of the plant for processing. Pneumatic conveyors are sometimes used both to cool and to convey the product. Other cooler-conveyors that can be used are cooled screw conveyors and rotary coolers similar to the rotary dryer.

It is customary to use cyclone collectors (see Sec. 15, p. 1023) to remove dust from exhaust gases. The product from the cyclones may be discharged back into the dryer or collected separately. For expensive products, bag collectors can be used following the cyclones. When toxic fumes or solids are present in the gases, wet scrubbers are used following the cyclone. Cyclones and bag collectors should be insulated and may have to be steam-chased to prevent condensation of the moisture in the exit gases. The pressure drop in the dust-collection system of a rotary dryer will be 50 to 90 per cent of the total pressure drop in the unit.

Rotary dryers operating above steam temperatures should be insulated to prevent undue heat losses. Steam-heated dryers of less than 3 ft. in diameter should also be insulated. **Insulation** is particularly necessary for parallel-flow dryers. It is not unusual for the product to cool in the last 10 to 50 per cent of the length of parallel-flow dryers if they are not well insulated.

Sticking of material to the dryer shell can sometimes be alleviated by knockers on the outside of the shell.

When only a blower or an exhauster is used on a rotary dryer, it will operate under either a positive or negative pressure, and it is usually advisable to place rotary valves in the feed and product chutes to prevent undue leakage of air in or out of the dryer.

It is difficult to control very closely the feed rate to a rotary dryer. For best operation, however, the feed should be as uniform as possible, with regard to both moisture content and rate. The inlet-air temperature is usually controlled in direct rotary dryers. Ideally, control should be by the temperature of the product, but this temperature is often inconvenient or difficult to measure accurately and consistently. It is customary to have instruments recording the inlet- and exit-air temperatures, and sometimes the product temperature. For oil- or gas-fired dryers, a totalizing meter is usually placed in the fuel line.

Performance and Cost Data. Performance data on rotary dryers are given by Miller, Smith, and Schuette [*Trans. Am. Inst. Chem. Engrs.*, **38**, 841 (1942)], Alliot [*J. Soc. Chem. Ind.*, **38**, 173T (1919)], and Horgan [*Trans. Inst. Chem. Engrs. (London)*, **6**, 131 (1928)]. Gas-, oil-, or coal-fired direct rotary dryers will evaporate from 2 to 7 lb. water/cu. ft. dryer volume; if steam-heated air is used, the capacity will range from 0.2 to 2 lb. water/cu. ft. Indirect-direct types will evaporate 4 to 9 lb. water/cu. ft.

Evaporation capacity increases to some extent with increasing moisture content of the feed and with increasing allowable moisture content of the product. Evaporative capacities of kilns will be about 20 to 35 per cent of that for high-temperature direct rotary dryers of the same size. Table 9 gives typical performance and cost data for three types of rotary dryers.

Table 9. Manufacturer's Data on Performance and Cost of Rotary Dryers*

Data	Direct single-shell rotary dryer with countercurrent air flow			Direct single-shell rotary dryer with parallel air flow		Double-shell indirect-direct rotary dryer		
Material handled	Manganese ore	Stone	Ammonium sulfate	Shale and stone	Pebble phosphate	Silica sand	Zinc slimes	Anthracite coal
Dryer diameter, ft	5.00	7.5	4.00	8.67	7.5	5.83	5.83	7.5
Dryer length, ft	40	75	25	80	55	35	35	55
Moisture in feed, % wet basis	8.0	5.4	1.75	10.0	13.8	5.0	43.0	17.5
Moisture in product, % wet basis	0.9	0.1	0.1	1.0	1.3	0.1	22.0	2.9
Production rate, lb./hr	30,000	100,000	5,300	180,000	110,000	30,000	11,000	35,600
Evaporation rate, lb. water/hr	2,250	5,400	87.5	18,000	16,000	1,550	4,050	6,500
Capacity, lb. water/(cu. ft. of dryer volume)(hr.)	2.9	1.6	0.28	3.8	6.6	1.7	4.3	2.7
Type of fuel	Oil	Coal	Steam	Oil	Oil	Oil	Oil	Coal
Fuel consumed, gal./hr. or lb./hr	34	670	290	180	193	16	40.2	800
Calorific value of fuel, B.t.u./gal. or B.t.u./lb	150,000	13,500	945	150,000	150,000	150,000	151,000	11,300
Efficiency, B.t.u. supplied/lb. water evaporated	2,220	1,550	3,050	1,500	1,810	1,500	1,520	1,360
Total power, hp	22	70	9	145	85	25	19	61
Approx. cost of dryer and auxiliaries, f.o.b. factory†	$8,000	$24,000	$7,200	$34,000	$23,500	$14,000	$14,000	$27,000
Approx. erection and installation cost	$1,000	$ 3,200	$ 800	$ 4,800	$ 3,200	$ 1,800	$ 1,800	$ 3,200
Drying cost‡ per ton of product	$ 0.15	$ 0.09	$ 0.08	$ 0.14	$ 0.23	$ 0.08	$ 0.48	$ 0.24
Drying cost per ton of water removed	$ 1.88	$ 1.49	$ 4.45	$ 1.36	$ 1.56	$ 1.52	$ 1.32	$ 1.26

* Courtesy of the Hardinge Co.

† Standard dryer construction throughout. Stainless steel or special alloys will increase the cost 150 to 300 per cent.

‡ Fuel at $0.40/million B.t.u., power at $0.01/hp.-hr. No other operating costs included.

Labor: ½ man or less for oil- or gas-fired units; 1 man for coal-fired units.

Maintenance labor and materials: $0.01 to $0.03/ton of product for large dryers; $0.05 to $0.15/ton of product for small dryers.

Thermal efficiencies of a high-temperature direct rotary dryer will range from 55 to 75 per cent, and with steam-heated air from 30 to 55 per cent. The indirect-direct rotary will have a thermal efficiency of 75 to 85 per cent at high temperatures.

The **purchase cost** in dollars of a direct single-shell rotary dryer in ordinary steel will be about $35(DL + 90)$. where D = diameter, ft., and L = length, ft. The **total installed cost** with auxiliaries will range from 150 to 250 per cent of the purchase cost. Stainless-steel dryers may cost two to three times as much as ordinary steel dryers. Indirect-direct rotary dryers will cost from 1.5 to 2.0 times as much as single-shell rotary dryers of the same volume. These figures refer to prices as of September, 1946.

Operating costs are usually lower for rotary dryers than for other types of dryers. From 20 to 35 per cent of one man's time is required. **Total horsepower** for fans, dryer drive, and feed and product conveyors will be in the range $0.5D^2$ to D^2, where D = diameter, ft. Yearly maintenance costs may range from 5 to 10 per cent of the total installed cost.

Example 12. A preliminary design and cost estimate are desired for a direct rotary dryer to dry 1000 lb./hr. of a crystalline material (65-mesh, dry bulk density = 40 lb./cu. ft.) from 0.15 to 0.005 lb. water/lb. dry stock. The material is discharged continuously from a solid-bowl centrifuge at 70°F., minimum temperature. The specific heat of the dry material is 0.2 B.t.u./(lb.)(°F.). Almost all the water on the material is surface moisture. The material cannot be heated above 300°F. without danger of rapid decomposition and fire. Preliminary tests have indicated that the material will handle satisfactorily in a rotary dryer. The minimum temperature of the air in the building in which the dryer is to be located is 60°F.

Solution. A parallel-flow dryer employing high-temperature air cannot be used without considerable risk, since some of the material may stick to the walls or feed flights and become heated to a dangerously high temperature. Therefore a countercurrent, steam-heated, hot-air, direct rotary dryer is indicated. Countercurrent flow is necessary to reach the low final moisture content. A single-shell dryer is preferred because of ease of cleaning.

Assume an inlet-air temperature of 290°F. and a material-discharge temperature of 250°F. Since the majority of heat required is heat of vaporization of the water, the mean temperature difference can be assumed to be equivalent to the logarithmic mean average of the wet-bulb depressions of the inlet and exit air. When this is the case, Eq. (61) can be modified to

$$N_t = \ln \frac{t_1 - t_w}{t_2 - t_w} \tag{66}$$

For air at 290°F. and $H = 0.025$, $t_w = 114$°F. from the humidity chart in Fig. 8. Assume that two transfer units can be obtained in the dryer. Substituting in Eq. (66), $2 = \ln [(290 - 114)/(t_2 - 114)]$ and solving for the exit-air temperature, t_2 is found to be 138°F.

The heat load q_t for the dryer is

$$\begin{aligned}
&\text{Heat to solid material} = \\
&\qquad 0.2(1000)(250 - 70) = 36{,}000 \text{ B.t.u./hr.} \\
&\text{Heat to remove moisture} = \\
&\quad (0.150 - 0.005)(1000)[(114 - 70) \\
&\qquad + 1028 + 0.45(138 - 114)] = \underline{157{,}000} \text{ B.t.u./hr.} \\
&\qquad q_t = 193{,}000 \text{ B.t.u./hr.}
\end{aligned}$$

The total quantity of air required for drying is then (193,000)/(0.24)(290 − 138) = 5300 lb./hr. This quantity of air will be increased by about 10 per cent to allow for heat losses from the dryer shell, so that the total air required will be 6000 lb./hr.

An air velocity of 500 lb./(hr.)(sq. ft.) can probably be tolerated, so that the diameter of the dryer becomes

$$D = \sqrt{\frac{(6000)}{(0.785)(500)}} = 3.92 \text{ ft.}$$

The length of a transfer unit from Eq. (62) is $L_t = 0.1\ (0.24)(500)^{0.84}(4) = 18.5$ ft. Since the number of transfer units was assumed to be 2, the required length of the dryer is 2(18.5) = 37 ft. The closest standard size of 4 by 40 ft. should be used.

The rate of rotation to use is $^{30}\!/_4 = 7.5$ r.p.m. and the number of flights, 3(4) = 12. The flight height should be 1.5(4) = 6 in. The total power required will be about $0.8(4)^2 = 12.8$ hp. The slope is obtained by assuming that the dryer will operate with 7 per cent hold-up. Since the total dryer volume is 502 cu. ft., the hold-up will be 35.1 cu. ft. The time of passage will then be (35.1)(40)/(1000) = 1.40 hr. = 84 min., and the slope is calculated from Eqs. (65*a*) and (65*b*) as follows:

$$\beta = 5/(208)^{0.5} = 0.347$$

$$S = \frac{0.23(40)}{(7.5)^{0.9}(4)\left[84 - \dfrac{0.6(347)(40)(500)}{83.3}\right]} = 0.0128 \text{ ft./ft.}$$

The heat load on the steam coils will be 0.24(6000)(290 − 60) = 331,000 B.t.u./hr. Assuming 125 lb./sq. in. gage steam to be used in the heating coils, the steam requirements will be 331,000/867 = 382 lb./hr. This corresponds to $^{382}\!/_{145} = 2.63$ lb. steam/lb. water removed.

The purchase cost of this dryer in standard steel construction should be about 35[(4)(40) + 90] = $8750. The installed cost with auxiliaries should be about $15,000. The drying cost per 100 lb. of product can then be estimated as follows, based on 7200 hr./year operation:

Depreciation at 6 per cent	=	$ 900
Power at $0.005/kw.-hr. = (7200)(13)(0.746)(0.005)	=	349
Labor at $1.50/hr. = (7200)(0.25)(1.50)	=	2700
Steam at $0.25/1000 lb. = (7200)(0.382)(0.25)	=	689
Maintenance at 5 per cent	=	750
Total operating cost/year	=	$ 5388

$$\text{Operating cost/100 lb. product} = \frac{5388(100)}{(7200)(1000)} = \$0.0748$$

Pneumatic Conveying Dryers

Description. In this type of dryer, moisture removal is accomplished by **dispersing the material** to be dried in a hot gas zone followed by conveying at high velocities. The dryer is essentially a device for dispersing a wet solid in hot gases, a duct through which these gases convey the dispersed particles, and a collection system for removing the dry product from the air stream. Drying, disintegration, and/or grinding, pneumatic conveying, and classification can be accomplished simultaneously in such equipment by proper design of the component parts.

roller or hammer mill pulverizes the wet material in contact with hot air which conveys away the fine dry product to cyclones and bag collectors (Fig. 34). This system has the advantage of producing a fine dry product in a single step, eliminating the additional handling required when grinding and drying are accomplished separately. The system shown in Fig. 34 is suitable for drying and pulverizing clay, copper sulfate, steamed bone, aluminum stearate, synthetic resins, etc.

The conveying portion of the system may be a square or circular vertical duct in which the solid is conveyed by the hot gas stream. To increase the time of contact with the heated gases, the conveying duct, in one modification, consists of an inverted cone in which the gas velocity decreases upward so that only the material that has become sufficiently dry can be carried out to the collector. For satisfactory operation of this system, the particle size of the material should not cover a too wide range. The **Trump "Vertex"** dryer is an example of this type.

Cyclones are usually the primary collectors in these dryers. They may be so arranged that the product is classified into small and large particle fractions during the collection. Where very fine materials are handled, it

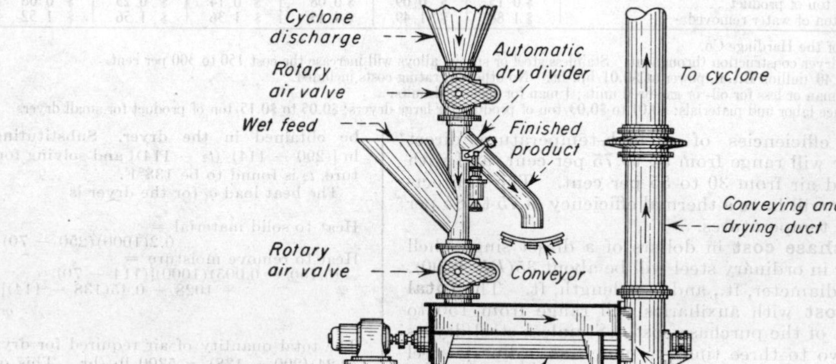

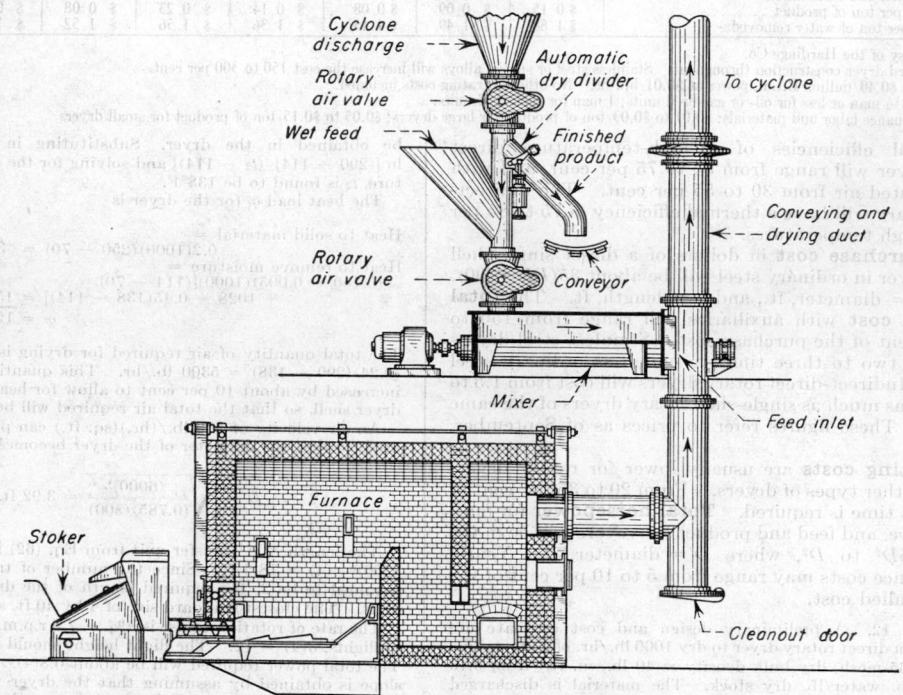

FIG. 32. Raymond Flash dryer without disintegrator but with dry-product return. (*Combustion Engineering Co., Flash Dryer Division.*)

Granular materials that are relatively free flowing in the wet state, such as coal, sodium chloride, potassium persulfate, etc., can be **dispersed** by merely dropping them into the heated air stream, as shown in Fig. 32.

Sludges, filter-press cakes, and similar non-granular materials, however, must be **disintegrated** by a cage mill or similar disintegrator through which heated air is circulated (Fig. 33). In some cases, pasty materials must be mixed with the dry product to permit suitable disintegration, as shown in Figs. 32 and 33. The system in Fig. 33 can be used for distiller's spent grain, sewage sludge, corn gluten, gypsum, clay, diatomite, pigments, calcium carbonate, etc.

When **fine grinding and drying** are required, a ring

may be necessary to provide bag collectors, wet scrubbers, or Cottrell precipitators after the cyclones to prevent high yield losses. In some arrangements, a portion of the air from the collector is recirculated. When this is done, any fines coming from the cyclone are recirculated through the system.

Any coarse wet material collected in the cyclone must be recirculated through the drying system. The product may also be recirculated, even though dry, if it does not meet particle-size specifications from the grinding operation.

Another dryer which may be included in this classification is a combination dryer and fine grinder developed by Stephanoff (U.S. Patent 2297726). It consists of

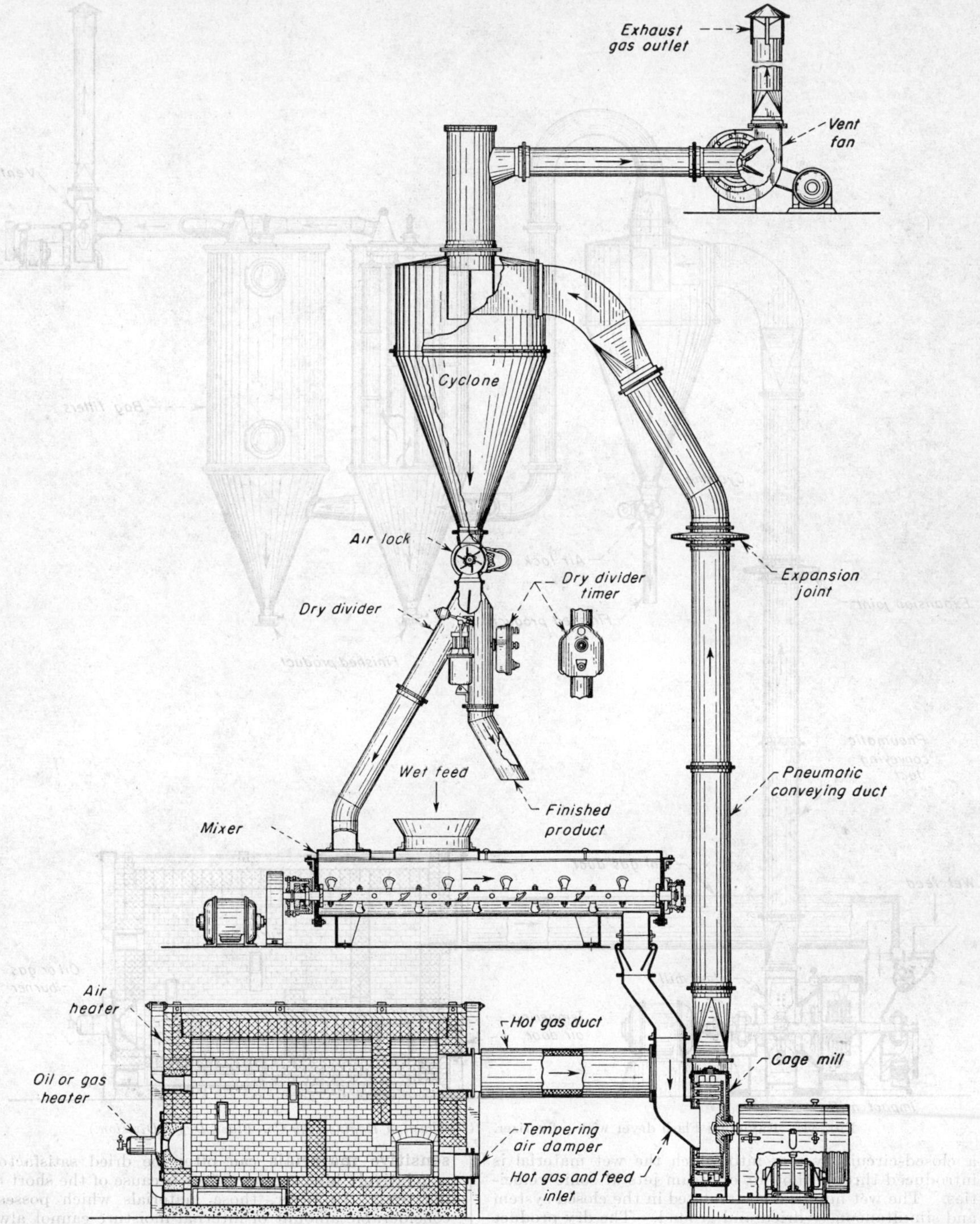

FIG. 33. Raymond Flash dryer with disintegrator and dry-product recycling system. (*Combustion Engineering Co., Flash Dryer Division.*)

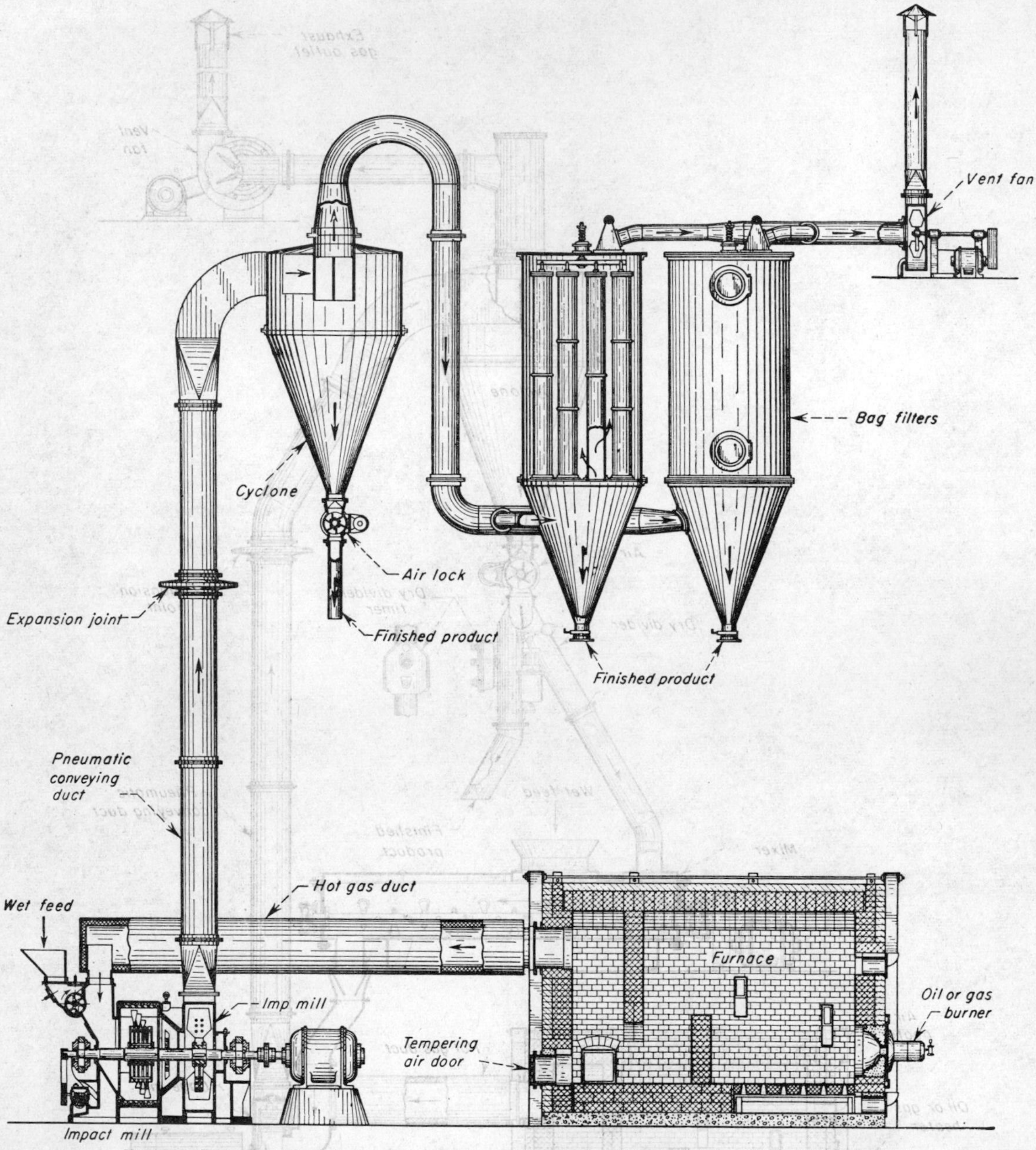

FIG. 34. Raymond Flash dryer with pulverizer. (*Combustion Engineering Co., Flash Dryer Division.*)

a closed-circuit system into which the wet material is introduced through hot air or steam jets at sonic velocities. The wet material is circulated in the closed system and simultaneously dried and ground. The dry product is removed by classification when it reaches a desired degree of dryness and a specified particle size.

Field of Application. Pneumatic conveying dryers are suitable for all materials that can be *suitably dispersed* in a gas stream and conveyed. Very abrasive materials, or materials that should not be broken up or scarred, are not suitable, nor are those materials which tend to stick to the walls of ducts. **Temperature-sensitive materials** can often be dried satisfactorily at relatively high temperatures because of the short drying time. However, those materials which possess a considerable amount of internal moisture cannot always be dried in one pass and may require a two-stage operation.

Dryers of this type are particularly applicable to the handling of **large capacities** and are usually about the cheapest type of equipment for evaporation capacities above 2000 lb. of water per hour. Some typical applications are given in Table 10.

Theory and Design Data. There is no published information on the fundamental theory of this method of

Table 10. Typical Products Dried in Pneumatic Conveying Dryers

Material	Initial moisture, lb./lb. dry solid	Final moisture, lb./lb. dry solid	Production rate, lb./hr.	Remarks
Potassium persulfate	0.05	0.002	1000	Dried in system of Fig. 32. No disintegration required
Synthetic casein....	4.0	0.11	250	Product between 40 and 80 mesh
Chrome green.......	0.33	0.001	500	Feed from filter press
Acid-treated clay....	1.50	0.176	5000	
Copper sulfate......	$CuSO_4.5H_2O$	$CuSO_4.H_2O$	1500	Fine grinding used. 95–99% through 325 mesh. 110 hp.-hr./ton prod.
Fruit pulp..........	5.66	0.11	300	Pulverizing and drying
Gypsum............	0.33	0.05	500	Feed of ¾-in. particles pulverized
Sewage sludge.......	4.00	0.11	25,000	Product may be sold or incinerated
Starch by-product....	....	0.03	150–200	Feed from filter press
Whey..............	0.54	0.025	500	Steam-heated air
Yeast filter cake....	2.34	0.05	200	Closed circuit, oversize returned to dryer

* Manufacturer's data, courtesy of Combustion Engineering Co., Flash Dryer Division.

drying. If the particle size of the material is known and is such that the heat-transfer coefficient is independent of the conveying velocity, the heat-transfer rate to the individual particles in the conveying gases can be estimated from Eq. (16). The inlet-air temperature is usually fixed by the heat sensitivity of the material. The exit-air temperature is such that the dryer will usually have two to three transfer units.

If only surface moisture is present on the wet material, its temperature during drying will approach the wet-bulb temperature of the drying air. From these temperatures, the mean temperature difference between the hot air and the material can be evaluated and the heat-transfer rate can be calculated from the equation $q = hA(\Delta t)_m$, where A is the area of the particle. The total heat to be transferred to the largest particle divided by the rate of heat transfer as calculated above will give the required drying time. The conveying duct must be made sufficiently long to keep this particle in the heated air stream for the calculated time.

The total heat to be transferred is given by

$$Q = \rho_s V \lambda (W_1 - W_2) \tag{67}$$

where Q = heat load, B.t.u.; ρ_s = particle density, lb./cu. ft.; V = volume of particle, cu. ft.; W_1 = initial moisture content, lb./lb.; W_2 = final moisture content, lb./lb.; and λ = latent heat of evaporation, B.t.u./lb. For spherical particles, $V = \pi(D_p)^3/6$, where D_p = particle diameter, ft. The drying time is then given by

$$\theta = \frac{\rho_s \pi (D_p)^3 \lambda (W_1 - W_2)}{6hA(\Delta t)_m} \tag{68}$$

where θ = drying time, hr.; h = surface heat-transfer coefficient, B.t.u./(hr.)(sq. ft.)(°F.); A = area of particle, sq. ft.; and $(\Delta t)_m$ = average temperature difference between inlet and outlet conditions, °F. If h is obtained from Eq. (16) and A is the area of a spherical particle, the time is given by

$$\theta = \frac{\rho_s (D_p)^2 \lambda (W_1 - W_2)}{12k_f(\Delta t)_m} \tag{69}$$

where k_f = thermal conductivity of air film surrounding the particle, B.t.u./(hr.)(sq. ft.)(°F./ft.). The length of the conveying duct is then given by

$$L = \theta v \tag{70}$$

where L = length of duct, ft.; and v = gas velocity, ft./hr.

This procedure applies only to the case where the wet feed is dispersed by the hot gas stream and does not pass through a dispersing device. When dispersing systems are used, most of the drying is accomplished prior to conveying, thus invalidating the preceding method for such systems.

Previous experience and drying tests must be relied upon for the proper design of this type of dryer. Since grinding and classification are often important, the systems are frequently tailor-made to the individual material being handled.

Operating temperatures for these dryers may range from 300° to 1300°F. The lower temperatures are employed with some sacrifice in heat economy. The initial moisture content may range from 3 to 900 per cent (dry basis) or 3 to 90 per cent (wet basis).

Air quantities required are determined by the amount of moisture to be removed, the available temperature drop of the air, and the quantity of solid material to be conveyed. The **ratio of solids to conveying gas** normally ranges from 0.05 to 1.0 lb. solid/lb. gas, although slightly higher loadings are possible in carefully designed systems. The **gas velocities** in the conveying duct must be high enough to transport the largest particle in the system. This velocity can be calculated by methods given in Sec. 5. In actual practice, the velocity should be about 100 per cent greater than the free-falling velocity of the largest particle and may range from 50 to 200 ft./sec. in these systems. A good **average velocity for high-spot design estimates** in a majority of cases is 75 ft./sec., based on the exhaust-gas temperature. Since in those systems using a dispersing device most of the drying occurs prior to conveying, the gas velocity will not change much from the point of disintegration to collection. In systems of the type shown in Fig. 32, however, a large velocity variation may occur.

Auxiliary Equipment. Air circulation is provided by high-pressure blowers or fans since the pressure drop is usually in the range of 8 to 15 in. water. Fan bearings must be protected from high temperatures, and fan impellers from both high temperatures and abrasive action of the solids.

Heated air is usually furnished from furnaces fired by gas, oil, or coal. Where it is undesirable to contact the material with flue gases, indirect fired heaters must be employed. For low-temperature operation, extended-surface heaters using steam may be used.

When the feed is too wet to feed directly into the dryer, a portion of the dry material is often mixed with it to bring it to a consistency that can be handled. This mixing is usually done in a pug mill, or double-paddle mixer, although a screw conveyor can sometimes be used. These mixers should do a thorough job of blending the wet with the dry material. The proper proportion of dry product returned to the mixer can be automatically apportioned by a **dry divider** (Fig. 33) consisting of a timer-actuated thruster which actuates a splitter damper. The latter held in one position discharges all the flow from the cyclone to the mixer for a portion of a predetermined cycle, and in a second position discharges the flow to the product conveyor for the remainder of the cycle.

Since the pressures in different parts of the system vary, rotary valves or air locks are required for discharging the product to a hopper or a premixer. These valves should give a positive seal, since leakage into the bottom of a cyclone severely impairs its collection efficiency. Vibrators above these valves are used to prevent bridging of the material in the duct.

To prevent excessive heat losses, the drying system should be thoroughly insulated. Insulation thickness generally ranges up to 4 in.

Bag collectors following the cyclones should be of the continuous-cleaning type, so that the unit does not have to be shut down.

The dryer is usually controlled by maintaining the exhaust-gas temperature constant by adjusting the inlet-gas temperature to meet varying demands on the system. Feed rate to the system is kept as constant as a rotary or disk feeder, or premixer will permit.

Performance and Cost Data. Few generalizations can be made concerning performance of this type of equipment. A well-designed system has two to three transfer units, as indicated by typical performance data for Raymond Flash Drying Systems shown in Table 11.

Table 11. Manufacturer's Performance Data for Pneumatic Conveying Dryers*

Data	Sewage sludge filter cake	Feed (distillers and corn products)	Clay (dry and grind)	Coal ¼ × 0
System involved	Fig. 33	Fig. 33	Fig. 34	Fig. 32
Dryer size, evaporative capacity, lb./hr.	3500	10,000	2060	7,480 actual 9,680 equivalent†
Inlet moisture, % (wet basis)	80	62	27	9 (surface)
Outlet moisture, % (wet basis)	10	12	5	3 (surface)
Inlet-gas temperature, °F.	1300	1,000	980	1,200
Outlet-gas temperature, °F.	250	225	165	175
Air rate, cu. ft./min., vented	7600	28,000	5000	32,000
Production rate, lb./hr.	1000	7,600	6770	113,300
Ratio of solids to air in conveying system	0.044	0.09	0.34	1.0
Material temperature in, °F	60	70	60	60
Material temperature out, °F	160	150	120	135
Number of transfer units‡	2.4	2.3	2.8	3.0
Type of fuel	Digestion gas	Coal	Oil	Coal
Fuel consumption, B.t.u./lb. water evaporated (heat input to air heater or furnace)	1725	1,750	1700	1,600 (equivalent evaporation)
Power consumption, kw.-hr./lb. water evaporated	0.012	0.010	0.037	0.010 (equivalent evaporation)

* Courtesy of Combustion Engineering Co., Flash Dryer Division.
† Equivalent evaporation includes sensible heat added to coal.
‡ Number of transfer units based on logarithmic mean Δt between inlet and outlet Δt's.

Thermal efficiencies may be as high as 65 to 75 per cent for well-insulated systems operating above 800°F., and as low as 25 to 30 per cent when indirect or steam-heated air is employed without recirculation.

Approximate **installed costs** and space requirements for Raymond Flash dryers using three different methods for dispersing the wet feed are given in Table 12. To use this table, determine the evaporative capacity required, the inlet temperature, and the system required, *i.e.*, method of disintegration, if any. Refer to Table 12 and use the lower cost values for high temperatures, and vice versa. Generally, the installed cost of a pneumatic conveying dryer will be less than that of a rotary dryer for the same evaporative capacity. This is particularly true at high evaporation loads. The costs in Table 12 are based on prices current in 1946.

The usual **operating costs** are labor, power, fuel, and maintenance. For labor, it can be figured that one operator can supervise one or more dryers regardless of size, the number depending on the wet-feed material-handling system. The same operator can also bag the finished product from small dryers but not from the larger sizes. The **range of other operating costs** are as follows:

System	Drying without disintegration	Drying with disintegration	Drying and pulverizing
Power, kw.-hr./lb. water evaporated	0.008–0.015	0.01–0.02	0.015–0.05
Fuel, B.t.u./lb. water evaporated	1600–3750	1600–2800	1600–2800
Maintenance, ¢/ton product	1.0–3.0	2.0–5.0	3.0–10.0

When other operations, such as conveying, grinding, and classifying are performed, the investment and operating costs chargeable to the drying operation may be only 40 to 50 per cent of the above figures, indicating that under these conditions pneumatic conveying drying may be particularly attractive.

Table 12. Installed Cost of Raymond Flash Drying Systems

Mild steel construction* (motors, drives, and air heater or furnace **included**) [dewatering, material handling, secondary dust collectors (if required) and building **not included**]

Evaporative capacity, lb. water/hr.	Approx. building space, width, ft. × length, ft. × height, ft.	*a.* Drying without disintegration, inlet temp. 700°–1300°F.	*b.* Drying with disintegration, inlet temp. 700°–1300°F.	*c.* Drying and pulverizing, inlet temp. 1000°–1300°F.
500	15 × 15 × 25	$ 4,500–$ 7,000	$ 7,500–$ 8,500	$10,500–$12,000
1,000	15 × 20 × 25	5,000– 7,500	9,000– 10,500	12,000– 14,500
1,500	20 × 25 × 30	7,000– 9,000	10,500– 14,000	15,000– 17,000
2,000	25 × 25 × 35	10,000– 13,000	15,000– 20,000	18,000– 22,000
5,000	30 × 30 × 40	18,000– 24,000	22,000– 30,000	
8,000	36 × 40 × 50	27,000– 37,000	32,000– 45,000	
10,000	40 × 40 × 50	34,000– 45,000	40,000– 55,000	
15,000	45 × 45 × 55	49,000– 66,000	59,000– 79,000	
20,000	48 × 50 × 60	65,000– 87,000	79,000–106,000	

NOTE: Although the lowest inlet temperature for use with the table above is 700°F., flash drying is widely applied using inlet temperatures as low as 300°F., particularly for heat-sensitive materials.
* The above figures may be 60 to 100 per cent higher for special alloy-steel construction. Costs apply as of September, 1946. (Courtesy of Combustion Engineering Co., Flash Dryer Division.

Spray Dryers

Description. A spray dryer creates a **highly dispersed liquid state** in a high-temperature gas zone. The operation of a spray dryer involves three fundamental aspects: (1) atomization, (2) spray-gas mixing, and (3) drying of liquid drops Atomization is accomplished by any one of several atomizing devices. The principal ones are: (1) high-pressure nozzles, (2) two-fluid nozzles, and (3) high-speed rotating disks. With such atomizers a pumpable liquid, whether clear solution or thick slurry, can be dispersed into droplets as small as 2 μ with thin solutions. Larger drop sizes do not usually exceed 500 μ, or about 30 mesh. Because of the large total drying surface and small drop diameters created, the actual drying time in a spray dryer is measured in fractions of a second, while the over-all time in the dryer will not usually exceed 30 sec. Lewis [*Ind. Chemist*, **10**, 439, 499 (1934); **11**, 71 (1935)] has presented a review of spray-drying techniques and design problems. A more recent review by Marshall and Seltzer (Am. Inst. Chem. Engrs., meeting, New York, 1948) considered the above-mentioned fundamental aspects in detail, as well as the design and operating characteristics of modern spray dryers.

Spray dryers consist essentially of a drying chamber, a source of hot gases, a means of atomizing the liquid feed, and a method of separating the dry product from the exhaust gases. Special designs may provide for cooling air to enter around the chamber, closed systems to recover volatile solvents, wet dust collectors to recover excessively fine dust, air sweepers or mechanical rakes to remove dry product from within the chamber, etc. Spray dryers may operate with countercurrent or cocurrent gas flow and may be heated by oil-, gas-, or coal-fired furnaces, indirect steam heaters, indirect coal-fired heaters, or by waste gases from plant boiler houses. **Inlet-gas temperatures** may range from 200° to 1400°F. with special materials of construction for the latter temperature. A light distillate, such as a No. 2 oil, is usually suitable when the product must not be contaminated. When CO_2 must be kept out of the drying atmosphere, indirect heaters are used at an increase in investment and operating costs. When oxygen must be eliminated, an inert gas system can be used. Superheated steam or other vapors also can be used as heating mediums in specially designed closed systems.

Inlet gases to a spray dryer flow down or up around the atomizer when the latter is a centrifugal disk, combined, in some designs, with flow into the chamber from side inlets. Countercurrent flow is difficult to use with this type of atomizer because it produces a flat or horizontal spray pattern. With nozzles, the direction of flow may be counter to or parallel with the spray, or a combination of both. The more nearly vertical particle trajectories from nozzles are more amenable to counterflow operation.

The centrifugal-disk atomizer has found favor in the chemical industries and nozzles in the soap and food industries. However, the superiority of one method over another for a given problem can be answered only on the basis of all the factors involved, and finally by actual experimental tests. The principal characteristics of three common methods of atomization are noted as follows:

1. **Centrifugal disks** atomize liquids by extending them in thin sheets which are discharged at high speed from the periphery of the rapidly rotating specially designed disk. The principal objectives in disk design are to ensure bringing the liquid up to

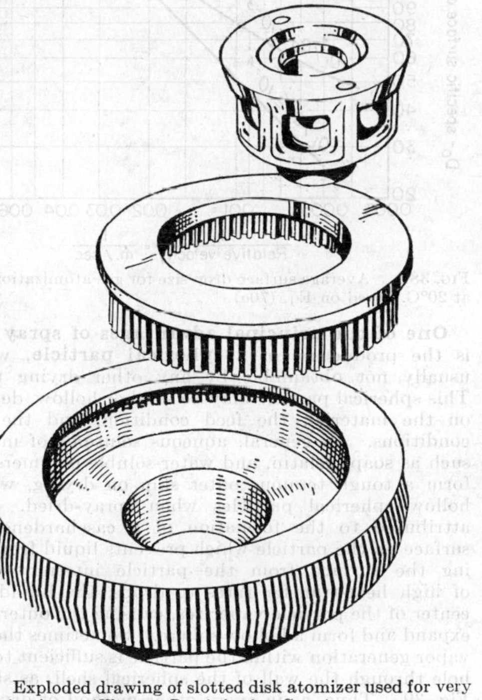

Fig. 35. Exploded drawing of slotted disk atomizer used for very fine atomization. (*Western Precipitation Corp.*)

disk speed and to obtain a uniform drop-size distribution in the atomized liquid. Disk diameters may range from 2 in. on small laboratory models to 12 or 14 in. for plant-size dryers. In one design a 32-in. wheel is used with the liquid flowing through the spokes of the wheel to the periphery. Disk speeds may range from 3000 to 50,000 r.p.m. The high rate is usually used in small-diameter dryers. Usual speeds on plant-size dryers range from 3600 to 12,000 r.p.m., depending on disk diameter and the degree of atomization desired. The degree of atomization as a function of disk speed is believed to be affected by the product of disk diameter and speed, *i.e.*, by **peripheral speed,** as opposed to angular speed. Thus a 5-in. disk operating at 30,000 r.p.m. would be expected to atomize more finely than a 2-in. disk of the same design running at 50,000 r.p.m. Examples of different disk designs are shown in Figs. 35 to 38.

Centrifugal-disk atomization is particularly advantageous for atomizing suspensions and pastes that would erode and plug nozzles. Very thick pastes can be handled if positive-pressure pumps are used to feed them to the disk. Disks are capable of operating over a rather wide range of feed rates and disk speeds without producing too variable a product. Nozzles, however, must operate at fixed conditions to give a specified product at a given capacity.

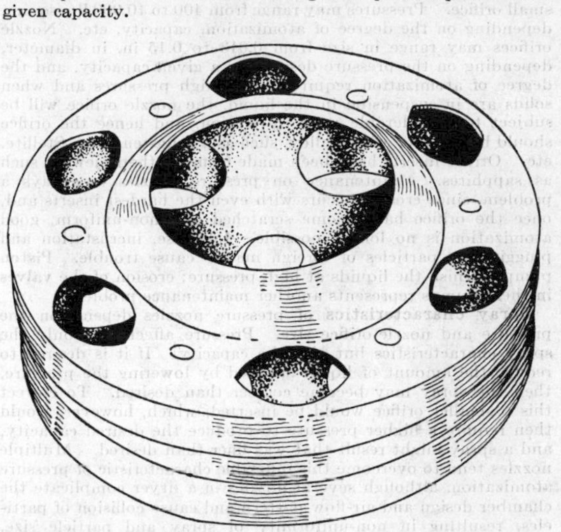

Fig. 36. Double-discharge disk atomizer. Liquid impacts against the edge of the bowl, flowing up and down to discharge from holes at the top and bottom. (*Instant Drying Co.*)

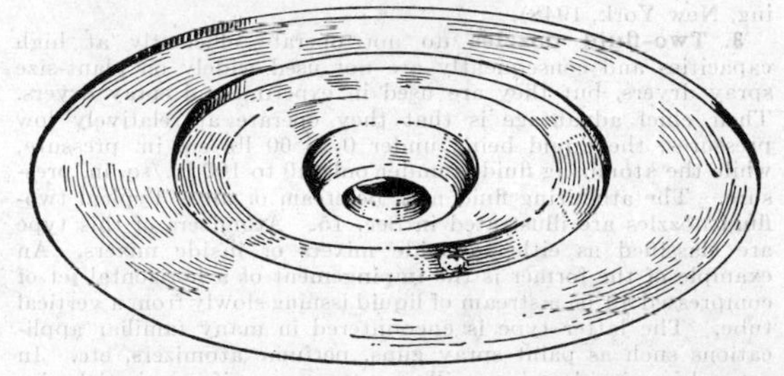

Fig. 37. Smooth, saucer-shaped disk with sharp, vertical edge. (*Bowen Engineering, Inc.*)

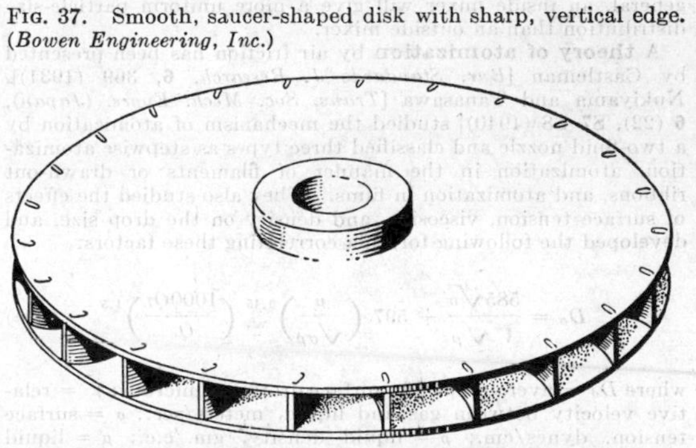

Fig. 38. Vaned disk for fine atomization. Liquid travel to periphery is lengthened by S-shaped vanes. (*Bowen Engineering, Inc.*)

Centrifugal disks may be belt-driven, direct-driven by a high-speed electric motor powered by a frequency changer, or driven by a steam turbine. Air turbines have been used on laboratory models. Direct drive by an electric motor appears to have advantages where very high speeds are required and where closely controlled speed variations are necessary. However, direct-driven disks require specially designed motor assemblies to operate in hot-gas zones as well as to permit feeding the liquid onto the disk. Positive lubrication is required with such motors, with special systems of oil removal from the bearings to prevent product contamination.

2. Pressure nozzles effect atomization by forcing the liquid under high pressure and with a high degree of spin through a small orifice. Pressures may range from 400 to 10,000 lb./sq. in., depending on the degree of atomization, capacity, etc. Nozzle orifices may range in size from 0.013 to 0.15 in. in diameter, depending on the pressure desired for a given capacity, and the degree of atomization required. For high pressures and when solids are in suspension in the liquid, the nozzle orifice will be subject to considerable wear by erosion, and hence the orifice should be made of a hard alloy, such as a tungsten alloy, Stellite, etc. Orifice inserts have been made from synthetic jewels, such as sapphires. Maintenance on pressure nozzles is always a problem, since erosion occurs with even the hardest inserts and, once the orifice has become scratched and non-uniform, good atomization is no longer possible. Likewise, incrustation and plugging by particles of foreign matter cause trouble. Piston pumps furnish the liquids at high pressure; erosion of the valves in these pumps represents another maintenance problem.

Spray characteristics of pressure nozzles depend on the pressure and nozzle-orifice size. Pressure affects not only the spray characteristics but also the capacity. If it is desired to reduce the amount of liquid sprayed by lowering the pressure, then the spray may become coarser than desired. To correct this a smaller orifice would be inserted, which, however, would then require a higher pressure to produce the desired capacity, and a spray might result that was finer than desired. Multiple nozzles tend to overcome this inflexible characteristic of pressure atomization, although several nozzles in a dryer complicate the chamber design and air-flow pattern and cause collision of particles, resulting in non-uniformity of spray and particle size. Fogler and Kleinschmidt [*Ind. Eng. Chem.*, **30**, 1372 (1938)] reported a study of atomization by pressure nozzles. The factors contributing to atomization by pressure nozzle have been summarized by Marshall and Seltzer (Am. Inst. Chem. Engrs., meeting, New York, 1948).

3. Two-fluid nozzles do not operate efficiently at high capacities and consequently are not used widely on plant-size spray dryers, but they are used in experimental spray dryers. Their chief advantage is that they operate at relatively low pressures, the liquid being under 0 to 60 lb./sq. in. pressure, while the atomizing fluid is under only 10 to 100 lb./sq. in. pressure. The atomizing fluid may be steam or air. Typical two-fluid nozzles are illustrated in Sec. 15. Atomizers of this type are classified as either outside mixers or inside mixers. An example of the former is the impingement of a horizontal jet of compressed air on a stream of liquid issuing slowly from a vertical tube. The latter type is encountered in many familiar applications such as paint spray guns, perfume atomizers, etc. In general, an inside mixer will give a more uniform particle-size distribution than an outside mixer.

A **theory of atomization** by air friction has been presented by Castleman [*Bur. Standards J. Research*, **6**, 369 (1931)]. Nukiyama and Tanasawa [*Trans. Soc. Mech. Engrs. (Japan)*, **6** (22), S7–S8 (1940)] studied the mechanism of atomization by a two-fluid nozzle and classified three types as stepwise atomization, atomization in the manner of filaments or drawn-out ribbons, and atomization in films. They also studied the effects of surface tension, viscosity, and density on the drop size, and developed the following formula correlating these factors:

$$D_o = \frac{585\sqrt{\sigma}}{V\sqrt{\rho}} + 597\left(\frac{\mu}{\sqrt{\sigma\rho}}\right)^{0.45}\left(\frac{1000Q_L}{Q_a}\right)^{1.5} \quad (70a)$$

where D_o = average specific surface drop size, microns; V = relative velocity between gas and liquid, meters/sec.; σ = surface tension. dynes/cm.; ρ = liquid density, gm./c.c.; μ = liquid viscosity, poises; Q_L = volumetric liquid flow rate; Q_a = volumetric gas rate. Equation (70*a*) has been plotted in Fig. 38*a* for the atomization of water at 20°C. It shows the effect on average drop size of increasing the ratio of air rate to liquid rate.

The **particle-size distribution** obtained by any one of the above three methods of atomization depends on a number of factors. There are no published data comparing the three methods for a given set of conditions, although it is generally believed that centrifugal-disk atomizers give the most uniform particle size, the pressure nozzle ranking second and the two fluid nozzle third. In general, the size distribution will depend on atomizer design, liquid properties, and the degree of atomization or particle sizes involved. If the finest atomization possible is attempted, a limiting condition is approached, and the particle-size range, regardless of the method of atomization, will be narrow. This is true of pressure nozzles, in which uniformity of size increases with pressure. On the other hand, in the production of a coarse product with a high percentage of large particles, the method of atomization will have a large effect on the particle-size distribution. Production of uniform coarse particles from centrifugal disks frequently can be obtained by proper disk design.

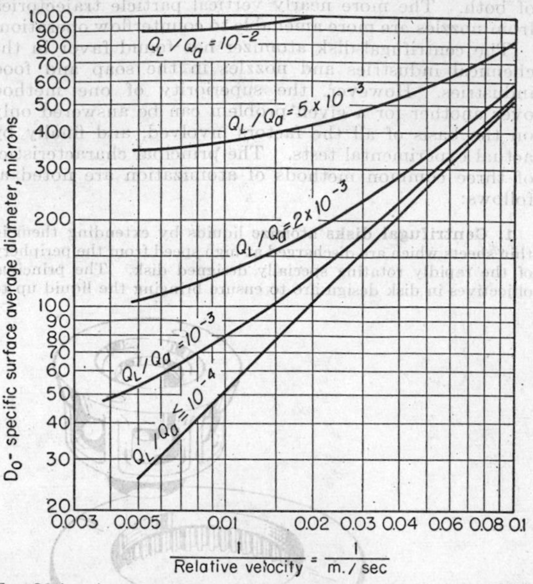

FIG. 38*A*. Average surface drop size for gas atomization of H_2O at 20°C. based on Eq. (70*a*).

One of the principal advantages of spray drying is the production of a **spherical particle,** which is usually not obtainable by any other drying method. This spherical particle may be solid or hollow, depending on the material, the feed condition, and the drying conditions. In general, aqueous solutions of materials, such as soap, gelatin, and water-soluble polymers, which form a tough tenuous outer skin on drying, will form hollow spherical particles when spray-dried. This is attributed to the formation of a casehardened outer surface on the particle which prevents liquid from reaching the surface from the particle interior. Because of high heat-transfer rates to drops, the liquid at the center of the particle vaporizes, causing the outer shell to expand and form a hollow sphere. Sometimes the rate of vapor generation within the particle is sufficient to blow a hole through the wall of the spherical shell, as shown in Fig. 39. Thus, as in every other method of drying, the properties of the solid play an important role in controlling the drying process. Spherical particles may be obtained from true solutions or slurries and may be produced by any of the above-described atomizers.

The **bulk physical properties** of spray-dried materials are subject to considerable variations, depending on the direction of flow of the inlet gas and its temperature, the degree and uniformity of atomization, the solids content of the feed, the temperature of the feed, and the degree of aeration of the feed. The properties of the product usually of greatest interest are (1) particle size, (2) bulk density, and (3) dustiness. Generally, the required final moisture is more or less easily obtained. The particle size varies inversely with disk speed or nozzle pressure, but the exact relationship among these varia-

FIG. 39. Spherical, hollow particles obtained by spray drying, showing vapor holes through the particle shell. Magnified 24 times.

bles must be determined. It has been suggested that maximum speeds and pressures exist above which additional energy expenditure is useless (Sauter, *Forsch. Gebiete Ingenieurw.*, No. 313, 1928). The particle size is also a function of the solids content, liquid viscosity, liquid density, and feed rate or capacity. In general, particle size increases with solids content, viscosity, density, and feed rate.

The **bulk density of spray-dried materials** is often a critical property, subject to considerable fluctuation if any of the above conditions are varied. The importance of bulk density in packaging costs becomes extremely significant if, for example, some change in the liquid feed or drying conditions reduces the product density by one-half. The bulk density of spray-dried materials frequently may be increased as follows: (1) by increasing the solids content of the liquid feed; (2) by decreasing the inlet-gas temperature; (3) by countercurrent gas flow instead of cocurrent flow; (4) by deaerating the liquid feed; (5) by crushing hollow particles; (6) by effecting a wide range in size distribution; (7) by fine atomization; or (8) by using certain additives, where possible. In general, the converse of these statements applies to decreasing the bulk density of a spray-dried material.

Some spray dryers and spray-drying systems are illustrated in Figs. 40 to 44. Figure 40 illustrates a typical Swenson spray-drying system originally developed for the production of dried skim milk. It has been widely used to dry cream, whole milk, ice-cream mix, eggs, and other foods in a liquid form, such as *purées*, coffee extract, etc. All these materials are dried at 280° to 300°F. This dryer has been modified recently for use in the chemical industry and has been used to dry distillery by-products, organic and inorganic dyestuffs, intermediates, and inorganic pigments and salts at temperatures up to 1000°F.

The basic Swenson design includes a vertical cyclonic drying chamber, which acts as a primary separator for the product; a collector which may be either wet or dry; a source of heat; a high-pressure nozzle atomizer; and auxiliaries such as a pneumatic conveyor for the product, a product collector, fans, pumps, etc. The heated air is introduced into the drying chamber through a number of tangential openings located in the upper section of the dryer. The spray-nozzle assembly is located in the center of the chamber and at approximately the same elevation as the top of the cone. The dried product is removed at the bottom of the cone while the air leaves from the top of the chamber. The hot gases enter the chamber through side inlets and spiral in an ever-tightening vortex to the bottom of the inverted cone from whence they pass upward around the spray nozzle and out the top of the dryer. This system of air flow produces very effective mixing of spray and gases and tends to spread the cone angle of the spray. A chain sweeper driven by the rotating air flow keeps the walls free from adhering product. Because of the grinding action of this sweeper and the very high cyclonic velocity of the gases, it is impossible to produce a coarse product with this dryer. The collector may operate dry, as an ordinary dust separator (Sec. 15); and, when it is so used, *e.g.*, in the drying of eggs, the over-all process yield may be better than 99.5 per cent, by weight, of the solids through-put.

In many processes it is advantageous to operate the dust separator as a wet collector (Sec. 15) in which the dilute feed to the dryer is first circulated through the collector for two reasons: (1) washing the gases of any entrained solids and thereby recovering them, and (2) preconcentrating the feed by humidification of the dryer exhaust gases, which leave the system within a relatively few degrees of the dew point. The capacity can be increased by the installation of a steam-heated tubular heater through which the liquid in the wet collector system is circulated. Over-all process losses by entrainment when using a wet collector are negligible.

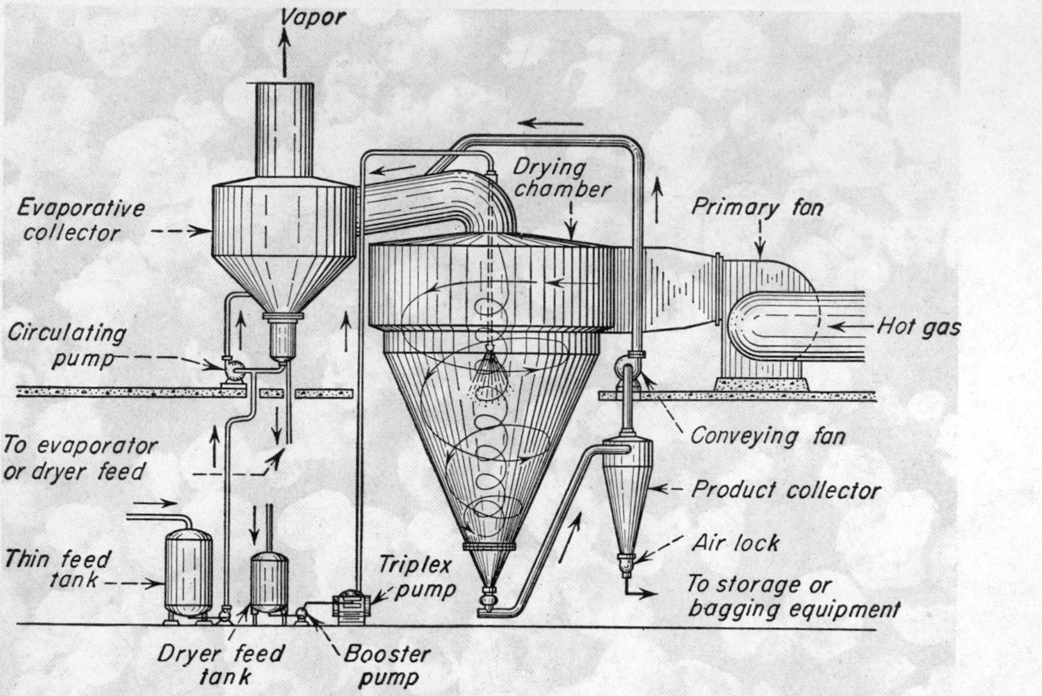

Fig. 40. Swenson spray dryer showing cyclonic air flow causing product separation in chamber and combined cocurrent-countercurrent air flow. (*Swenson Evaporator Co.*)

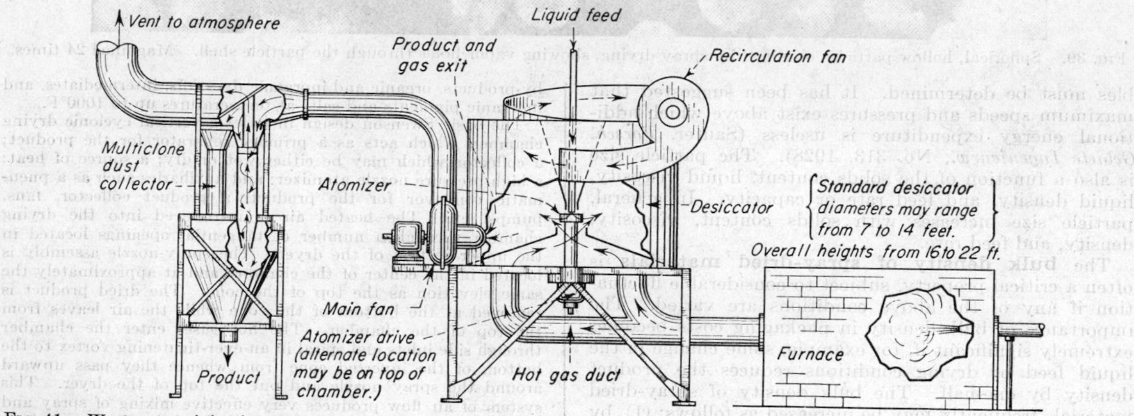

Fig. 41. Western precipitation spray dryer with upward cocurrent air flow, and product removal with exhaust gas. (*Western Precipitation Corp.*)

The pressure atomizing system includes a horizontal single-acting triplex plunger pump using either ball- or wing-type valves, a small-diameter high-pressure pipe line, and a nozzle body made up of a whizzer and an orifice plate. The whizzer translates linear velocities of the fluid being pumped into rotational velocity, and the orifice serves to form the fine spray required for drying. The design of the whizzer varies with the material being handled. Nozzle orifices may vary from $\frac{1}{32}$ to $\frac{1}{4}$ in. in diameter, and again the design is empirical, dictated by the materials being handled. When producing dehydrated food products, the operating pressure of the atomizing system usually ranges between 5000 and 6000 lb./sq. in. Operating pressures for most chemical applications range between 1200 and 2500 lb./sq. in.

Figure 41 illustrates the principle of the Western Precipitation "Turbulaire" spray dryer, showing the furnace location, drying chamber, hot-gas inlet, recirculating fan, main fan, and multiclone dust collector. A powder conveying and cooling system may be used to discharge the product from the dust collectors. Spray dryers of this type utilize centrifugal-disk atomization (see Fig. 35 for disk type) and depend on fine atomization to permit drying in relatively small diameter chambers. It should be observed that all the dry product passes through the main fan, which has a grinding action on the product. The atomizer speed may range from 9000 to 15,000 r.p.m., and the disk may be direct-driven or belt-driven. The motor and atomizer may be mounted on either the bottom or the top of the chamber. The atomizer size ranges from 8 to $9\frac{1}{2}$ in.

The "Turbulaire" spray dryer has been designed to operate in **closed systems** to effect the recovery of solvents (Fig. 69). It has also been designed for closed-circuit operation as a single-effect flash evaporator to remove impurities from a liquid by spray-vaporizing the latter to throw out the solid impurities in the spray chamber, the vaporized liquid being condensed as a

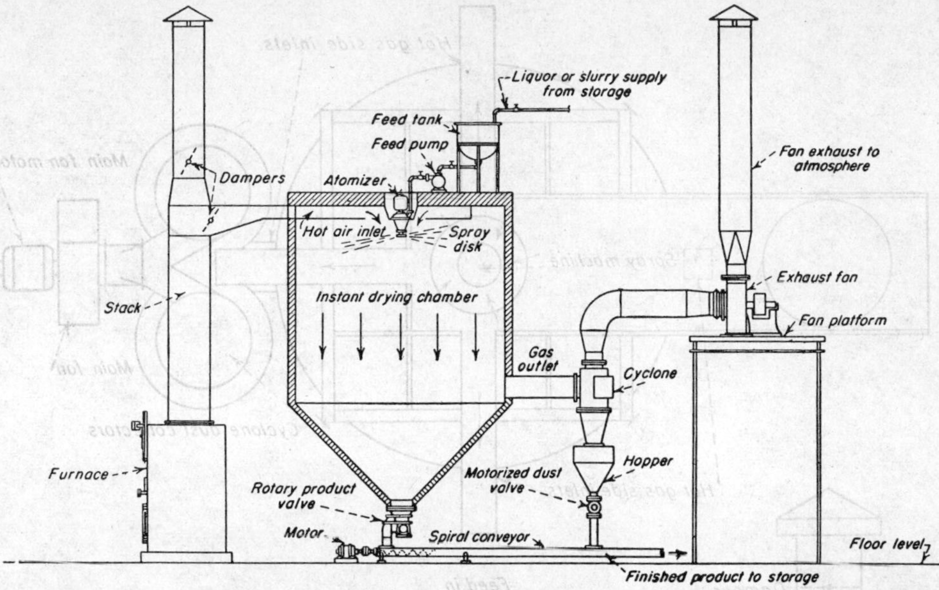

Fig. 42. Instant spray dryer with cocurrent air flow and primary separation of product in chamber. (*Instant Drying Co.*)

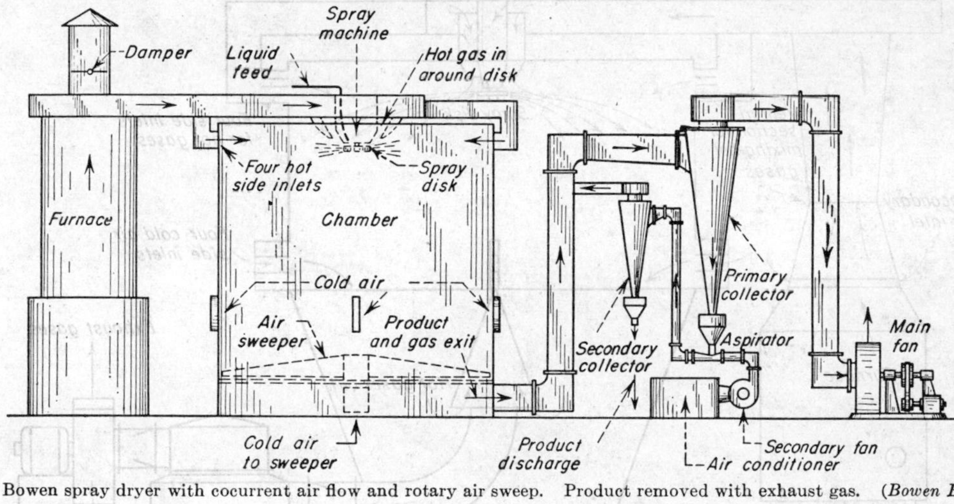

Fig. 43. Bowen spray dryer with cocurrent air flow and rotary air sweep. Product removed with exhaust gas. (*Bowen Engineering, Inc.*)

fog and removed by a Cottrell precipitator, and the carrier gas passing back through a direct, oil-fired air heater with careful combustion control to maintain zero per cent oxygen. "Turbulaire" dryers have also been designed with multiple-spray disks in a single chamber.

Fuel consumption for "Turbulaire" spray dryers varies from 1500 to 2000 B.t.u./lb. of water evaporated. All the product is removed from the dryer by the spent drying gases and recovered in either double- or single-stage mechanical dust collectors that are at times followed by bag filters.

Figure 42 illustrates the Instant spray-drying system. This dryer utilizes a centrifugal-disk atomizer direct-connected to an electric motor capable of speeds up to 10,000 r.p.m. The motor shaft is hung from one bearing and maintained in position at its lower end by spring-loaded carbon brushes, which yield position during any period of unbalance. Hot gases enter the chamber at the top and circulate down around the disk with a swirling motion imparted by a specially designed scroll-shaped duct with adjustable louvers. The exhaust gases leave the chamber through the bottom of the cone. The product is generally discharged with the air stream and collected in cyclones. An alternative design utilizes a flat bottom with a rotating rake to discharge the product, while the gases leave from four outlets around the chamber. Chambers are constructed of steel or concrete and range up to 24 ft. in diameter.

A typical Bowen type spray dryer (Fig. 43) utilizes a centrifugal-disk atomizer. Hot gases entering around the rotating disk are given a swirling motion by fixed vanes, which aids in producing intimate mixing between the hot gases and spray. Hot gases are also introduced through side inlets. Introduction of gas at these points is claimed to increase the capacity of a spray dryer by keeping the spray from hitting the chamber walls and permitting a higher feed rate for a given chamber diameter. Cold air may also be introduced through side inlets near the bottom. The hot gases convey the dry product from the dryer to a cyclone collector. An air sweeper, driven by the reaction of high-velocity air jets issuing from its trailing edge, keeps the chamber floor clean by blowing any product that accumulates on the floor into the exhaust gas. The entire system is under suction. Dryers of this type may be constructed of steel, concrete, or tile, and have been built in diameters up to 40 ft.

A modified Bowen dryer for materials where a separation of the

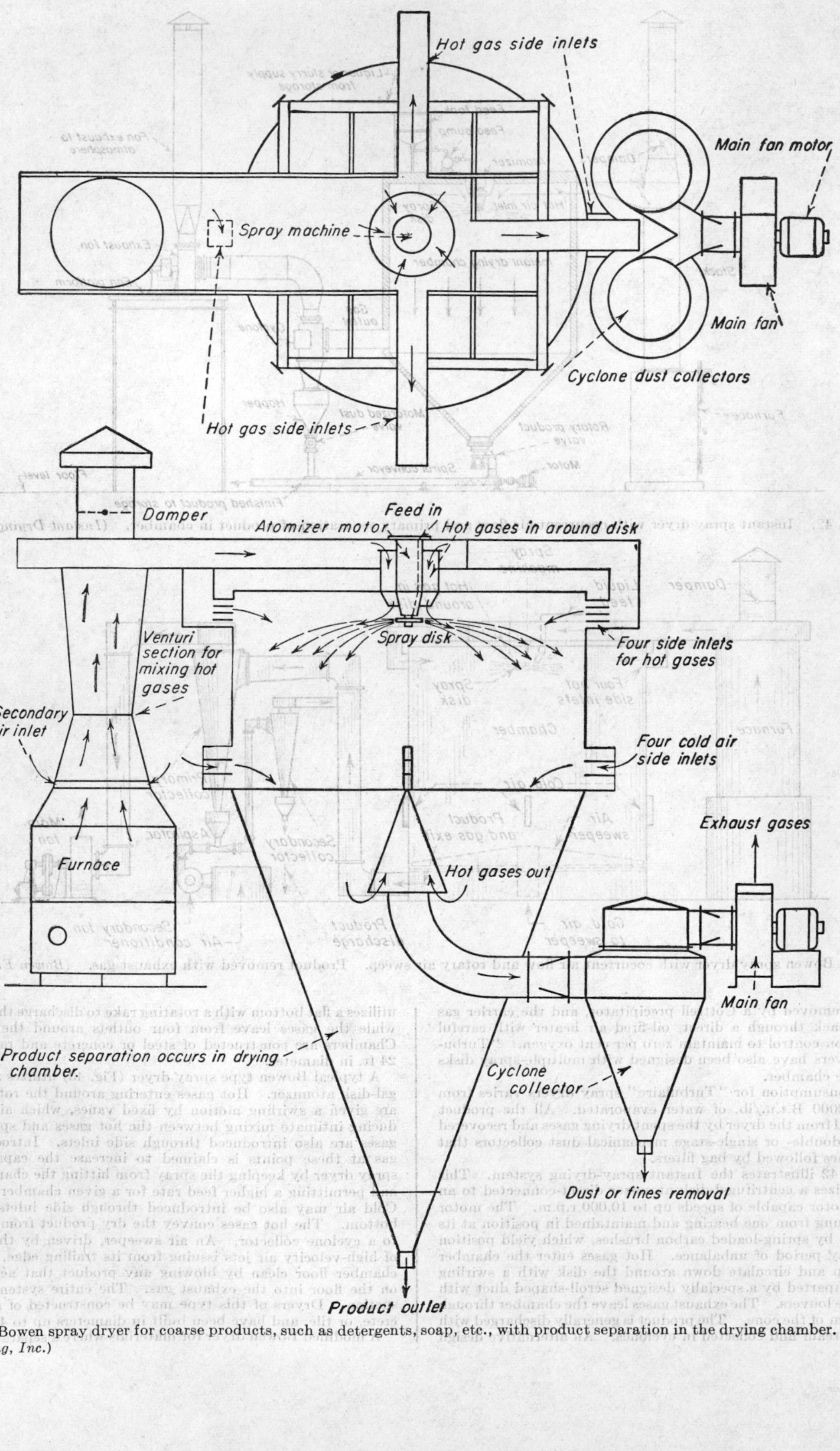

FIG. 44. Bowen spray dryer for coarse products, such as detergents, soap, etc., with product separation in the drying chamber. (*Bowen Engineering, Inc.*)

gases and product in the chamber is desirable is shown in Fig. 44. This differs from the design shown in Fig. 43 by having a conical bottom with a central internal gas offtake. It is especially suited to the production of coarse granular products.

Bowen spray disks may have a variety of designs, depending on the properties of the material to be dried and the particle size required. Smooth, inverted-saucer-shaped disks (Fig. 37) are used for producing coarse products, and vaned (Fig. 38) or slotted disks may be used to obtain fine atomization. Disk speeds may range from 6000 to 20,000 r.p.m. in commercial Bowen dryers, from 15,000 to 30,000 on pilot-plant models, and up to 50,000 on table or laboratory models using 2-in. disks. Up to 100 hp. may be installed on the atomizer motor.

Application. Spray dryers are used to dry solutions, slurries, and pastes. In general, if a material can be pumped, it can be spray-dried. Pastes of press-cake consistency have been pumped onto spray disks and atomized and dried successfully. **Viscous liquids** may require preheating before atomizing. In some cases, high-pressure nozzles are the only feasible means of atomizing viscous liquids that show a tendency to string.

Spray drying is particularly applicable to the drying of **heat-sensitive materials** such as pharmaceuticals, foods, etc. This is due to the **short contact times** involved as well as to the fact that the liquid drops remain cool when evaporation is taking place in the hot zone of the dryer; whereas, at the point where they are no longer wet, the surrounding gases have been cooled by the evaporation process, thereby eliminating further danger of overheating. Bullock and Lightbown [*Ind. Chemist*, **19** (223), 455 (1943)] spray-dried a number of pharmaceuticals that ordinarily suffer decomposition if heated to the drying temperature used in the tests. They showed that negligible decomposition was obtained in all cases.

Determination of the spray-drying characteristics of a given material is almost entirely empirical; and, although much valuable qualitative information can be determined by small-scale equipment, it is advisable whenever possible to confirm such data by tests in a commercial-scale dryer. Table 13 indicates a variety of materials that have been successfully spray-dried in a plant-size dryer. Spray dryers, when used as **spray coolers or crystallizers**, do not require a source of heat, since a molten anhydrous material requires only cooling and removal of latent heat for solidification. Spray cooling has found wide application to a number of processes such as the cooling of sodium bisulfate, ammonium nitrate, molten soap, phenothiazine, etc.

Table 13. Some Materials That Have Been Successfully Spray-dried in an 18-ft.-diameter by 18-ft.-high Chamber with a Centrifugal-disk Atomizer*

Material	Air temp., °F. In	Air temp., °F. Out	% water in feed	Evaporation rate, lb./hr.
Blood, animal	330	160	65	780
Yeast	440	140	86	1080
Zinc sulfate	620	230	55	1320
Lignin	400	195	63	910
Aluminum hydroxide	600	130	93	2560
Silica gel	600	170	95	2225
Magnesium carbonate	600	120	92	2400
Tanning extract	330	150	46	680
Coffee extract *A*	300	180	70	500
Coffee extract *B*	500	240	47	735
Magnesium chloride	810	305	53	1140 (to dihydrate)
Detergent *A*	450	250	50	660
Detergent *B*	460	240	63	820
Detergent *C*	450	250	40	340
Manganese sulfate	600	290	50	720
Aluminum sulfate	290	170	70	230
Urea resin *A*	500	180	60	505
Urea resin *B*	450	190	70	250
Sodium sulfide	440	150	50	270
Pigment	470	140	73	1750

NOTE 1: The fan on this dryer handles about 11,000 cu. ft./min. at outlet conditions.

NOTE 2: The outlet-air temperature includes cold air in-leakage, and the true temperature drop caused by evaporation must therefore be estimated from a heat balance.

* Courtesy of Bowen Engineering, Inc.

Theory and Design Data. There is little published information on design procedures for spray dryers. Theoretically, it should be possible to estimate the time required to evaporate moisture from a drop of a given size by Eq. (19) or (20) and then to estimate a chamber size based on particle-trajectory calculations. However, in a spray-drying process, many complicating factors due to collisions among particles, size distribution, and air currents produced by atomization destroy the validity of estimates made for single isolated particles. Furthermore, the constant-rate period and the falling-rate period must be considered. However, in spray drying, material advantage is taken of the limiting condition of very short drying times obtained by the use of high temperatures normally impractical with heat-sensitive materials dried by methods requiring much longer drying times. This may be illustrated by the following table giving the theoretical time required to evaporate a given quantity of water from a drop of constant diameter according to Eq. (19):

Table 14. Time to Evaporate Drops with Δt_m of 275°F.

Time, sec.	0.0042	0.017	0.067	0.15
Drop size, μ	10	20	40	60

Table 14 is based on Eq. (19), in which $W = 3.0$, $\rho_s = 30$, $\lambda = 1000$, logarithmic mean $\Delta t = 275°$ (750°F. inlet and 250°F. outlet), and $k_f = 0.025$. For other values of these quantities other times may be obtained by direct ratios.

In order to estimate the size of a spray dryer, data must be obtained on the following items: (1) capacity required, lb./hr., (2) initial moisture content, (3) final moisture content, (4) allowable inlet-gas temperature, (5) outlet-gas temperature, (6) degree of atomization required, (7) method of atomization. Obviously, many more items are involved, such as materials of construction, type of fuel to be used, required product properties, etc. However, a preliminary high-spot size estimate requires only the first items mentioned. The **evaporative load, drying rate,** and the **gas-temperature drop through the dryer** determine the volume of gases required, and thus the chamber size. The chamber design involves three primary items: (1) atomization, (2) mixing of the spray and hot gases, and (3) relative direction of flow of spray and gases.

"Turbulaire" spray dryers may be estimated,* roughly, for a condition of 750°F. inlet temperature and 250°F. outlet; *i.e.*, 500°F. drop, by the relation

$$V = aw \tag{71}$$

where V = chamber volume, cu. ft.; w = water evaporation rate, lb./hr.; and a ranges from 0.33 to 0.67. When a temperature drop of 500°F. is not possible because of physical or chemical limitations of the material, the capacity of a given unit will be roughly proportional to the temperature drop of the gases, with power, heat demand, and investment and operating costs remaining essentially constant. Under the same conditions for which Eq. (71) holds, the volumetric gas rate will be given by

$$Q = bw \tag{72}$$

where Q = volume of gases, cu. ft./min.; w = rate of evaporation, lb./hr.; and b = 5 to 7. From Eqs. (71) and (72), it is evident that the average contact time in "Turbulaire" spray dryers is on the order of 4 to 7 sec. This requires fine atomization.

When a **coarse product** is desired, with particle sizes in the range of 40 to 150 mesh, longer contact times in the dryer are

* Courtesy of Western Precipitation Corp.

Table 15. Performance Data for Swenson Spray Dryers*

Product	Dryer size, diam., ft.	Total solids in feed, %	Evaporation rate, lb./hr.	Product rate, lb./hr.	Inlet-gas temp., °F.	Source of heat	Fuel requirements	Horsepower: Connected	Horsepower: Brake	Direct operating costs, ¢/lb. dry product
Dried skim milk	10	9.0	2110†	225	260–270	Steam	1225 B.t.u./lb. H_2O evaporated	39.5	33	1.75
Dried skim milk	14	9.0	5275†	525	260–270	Steam	1225 B.t.u./lb. evaporated	82	68.5	1.25
Dried skim milk	18	9.0	7285†	725	260–270	Steam	1225 B.t.u./lb. evaporated	103.5	84.7	1.0
Powdered eggs	14	26.5	1430‡	525	300–310	Steam	3600 lb./hr. steam at 110 lb. gage	56.5	53.5	1.1
Distillery by-product	14	25.0	2660§	900	450	Flue gas	....	120.5	8.9	0.3
Distillery by-product	18	25.0	4720§	1600	450	Flue gas	....	205.5	164	.3
Organic chemical	14	22.0	1380‡	400	400	Oil	25 gal./hr. Bunker C oil	69.5	58	.9
Inorganic salt	18	25.0	7800¶	2700	900	Producer gas	60,000 cu. ft./hr.	146	105.5	.2
Starch	18	42.0	3000‡	2175	340	Steam	12,500 lb./hr. steam at 140 lb. gage	....	....	.5

* Courtesy of Swenson Evaporator Co.,
† Evaporative collector plus heat exchanger.
‡ Dry collector.
§ Evaporative collector as pre-evaporator.
¶ Evaporative collector.

required, and hence larger chambers. The concept of contact time in spray dryers is fundamentally related to the rate of heat and mass transfer, and the drying characteristics of the solid in the drop. Its usefulness for high-spot design estimates is indicated by the following formula:

$$V = \theta Q \tag{73}$$

where V = chamber volume, cu. ft.; Q = average volumetric gas rate, cu. ft./min.; and θ = over-all average contact time, min. Q, of course, depends on the air-temperature drop and the capacity involved. One application of Eq. (73) can be illustrated by the following example:

Example 13. It is desired to spray-dry a detergent at a rate of 2000 lb./hr. from a slurry containing 50 per cent solids. It is desired to produce a coarse product, 40 to 100 mesh, with non-dusting characteristics. The inlet-gas temperature will be 450°F. and the outlet 200°F. Estimate the volume and dimensions of the spray chamber if the chamber height and the diameter are to be equal. This specification on height and diameter implies the use of a disk atomizer.

Solution. For the specified production rate, an evaporation load of 2000 lb./hr. of water must be handled, or about 2,000,000 B.t.u./hr. must be supplied by the hot gases. If a chamber efficiency of 67 per cent is assumed, the weight of gases to supply this heat is given by $W = 3{,}000{,}000/(450 - 200)0.25 = 48{,}000$ lb./hr. The average volumetric gas flow between inlet and outlet conditions is given by Wv_{avg}, where v_{avg} = the average of the inlet- and outlet-gas specific volumes, cu. ft./lb. The inlet specific volume is about 25 cu. ft./lb., and the outlet is 18.3 cu. ft./lb., taking into account the change in humidity. Hence $v_{avg} = 22$ cu. ft., and $Q = (48{,}000)(22)/60 = 17{,}600$ cu. ft./min.

To determine the chamber volume, some estimate of the over-all average contact time must be available. Estimates of actual evaporation times given by Eq. (19) result in such short times that any volume estimated on this basis is physically impractical. Hence an average contact time must necessarily involve other factors. Since no correlation or published data are available on contact times, experience with other materials is the only means for estimating this quantity. For this problem, 20 sec. is known to be a suitable value, and hence the chamber volume is given by $17{,}700(\frac{1}{3}) = 5900$, say 6000 cu. ft. From this result, the proper chamber dimensions must be determined. This should be done by taking into account particle trajectories in a manner similar to the method of Lapple and Shepherd [*Ind. Eng. Chem.*, **32**, 605 (1940)]. Even their work is not strictly applicable, however, because it applies to single particles, a condition not realized in spray dryers. If a centrifugal-disk atomizer is used, it will be evident that a flat horizontal trajectory is involved. This means that the chamber diameter must be at least equal to the height, and possibly larger. For the case of equal height and diameter, the chamber dimensions become 20 by 20 ft. The chamber size used for purposes of high-spot cost estimates and preliminary layouts, consequently, would not be less than 20 by 20 ft., and possibly greater, depending on the design of the dryer, atomizer speed, etc.

The above example is suitable only for very preliminary estimating and cannot be used at the present time for final designs.

With regard to contact times required in spray dryers, the following generalizations should be noted: Contact time, and hence chamber size, increases as the inlet-gas temperature decreases, as the particle size increases, and as the initial moisture content increases. Contact time decreases as the inlet-gas temperature increases, as the particle size decreases, and as the inlet moisture content decreases. Probably the greatest effect on the time is particle size, which, according to Eq. (19), affects the drying time as the square of the particle diameter.

Table 16. Estimated Cost of Producing 13,200 Lb. of Dried Skim Milk from 150,000 Lb. of Sweet Fluid Skim Milk on a Swenson Spray-process Milk-drying Plant during a Day of 20 Hr. Continuous Operation

	Per Day	
Direct manufacturing costs:		
Electricity for driving motors, 1600 kw. at \$0.015/kw.-hr.	\$ 24.00	
Fuel for firing the boilers, 20,750 lb. coal at \$6.00/ton	62.25	
Labor:		
3 men to operate drying plant, 8 hr. each = 24 hr. at \$1.00/hr.	24.00	
2 men to clean plant and truck barrels, handle shipping, etc., 8 hr. each = 16 hr. at \$0.80/hr.	12.80	
Water for cleaning, estimated cost of pumping	1.00	
Miscellaneous—packing, oil, grease, scrubbing brushes, cleaning compounds, etc., estimated	5.00	
Containers—58⅔ barrels with double paraffined liners at \$1.85 each	108.52	
Total daily direct manufacturing costs, estimated		\$237.57
Cost per pound of dried skim milk produced, estimated		\$0.018
Estimated overhead expense:		
Depreciation on building and machinery at standard income-tax allowable rate, estimated	\$ 14.60	
Insurance, estimated	1.25	
Taxes, estimated	1.10	
Interest at 5 per cent on investment, estimated	10.10	
Office expense, chargeable to drying operation, estimated	5.00	
Total daily overhead expense, estimated		\$ 32.05
Total daily manufacturing costs, estimated		\$269.62
Cost per pound of dried skim milk produced, estimated		\$0.0204

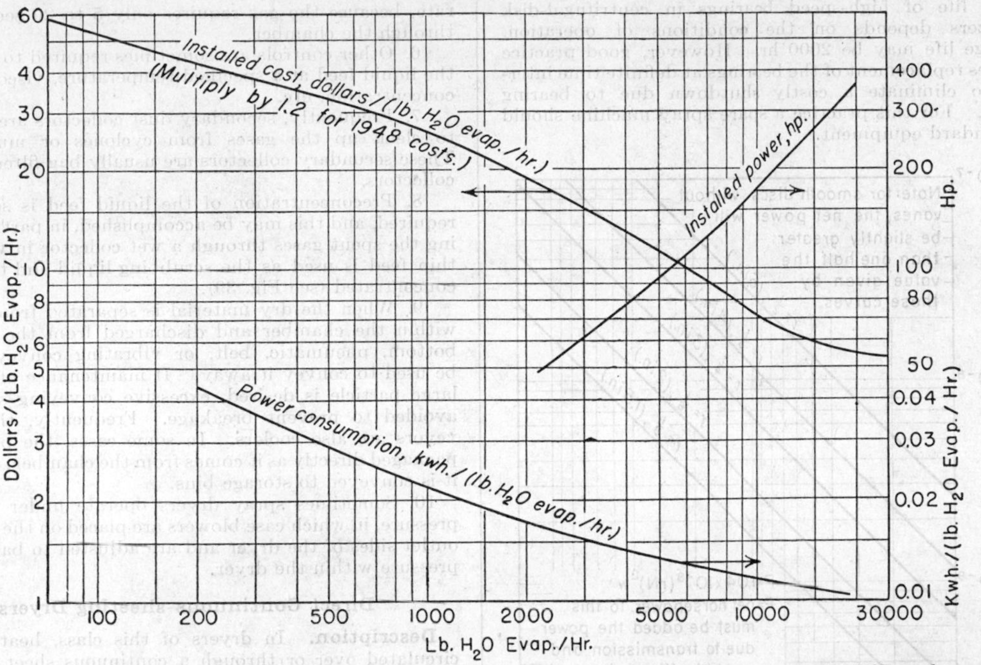

FIG. 45. Cost and power data for Turbulaire spray dryers. All data based on 750°F. inlet temperature and 250°F. outlet. (*Western Precipitation Corp.*)

Material properties, of course, influence the drying rate in the same way as for other dryers, and materials that tend to caseharden and cause high internal resistance to liquid flow will require a longer contact time.

Performance and Cost Data. Typical performance data for several Swenson spray dryers are given in Table 15. These indicate the marked advantages of a high inlet-gas temperature on capacity and operating costs. Although the data in this table will serve as a guide for preliminary size estimates of this type of spray dryer, the final analysis and selection of the proper dryer size and range of operating conditions should be made by the manufacturer of such equipment.

The **installed costs** of Swenson spray dryers will generally lie in the following ranges: 18-ft.-diameter unit, $65,000 to $90,000; 14-ft.-diameter unit, $45,000 to $70,000; 10-ft.-diameter unit, $30,000 to $50,000. These cost figures are only approximations. Firm estimates can be obtained only from the manufacturer. The above ranges are functions of the materials of construction, auxiliaries, etc.

Table 16 gives the manufacturing costs and overhead costs for a skim-milk spray-drying plant. Of special significance in the direct manufacturing costs is the cost of containers. This cost item reflects the low bulk density of the product.

Figure 45 shows a correlation of the **installed cost,** $/(lb. water evaporated/hr.) for "Turbulaire" spray dryers. The data in this figure are of July, 1946, and are based on a 500°F. gas-temperature drop through the dryer, a direct gas-fired air heater, and black iron construction throughout. If indirect heat is used or closed-circuit operation is desired, the costs given in Fig. 45 will be from 60 to 100 per cent higher. These ratios also apply if alloy steels or other special materials are used extensively in the construction.

Figure 46 shows the approximate installed costs of spray dryers for various drying temperatures plotted as $/(lb. water evaporated/hr.) vs. evaporative capacity, lb./hr. These curves are based on average evaporation loads between readily spray-dried and difficultly spray-dried materials. Thus each curve for a given problem may be displaced up or down depending on the drying properties of the material. Lines of constant chamber diameter indicate how temperature affects the capacity of a given size of spray dryer. The cost data of Fig. 46

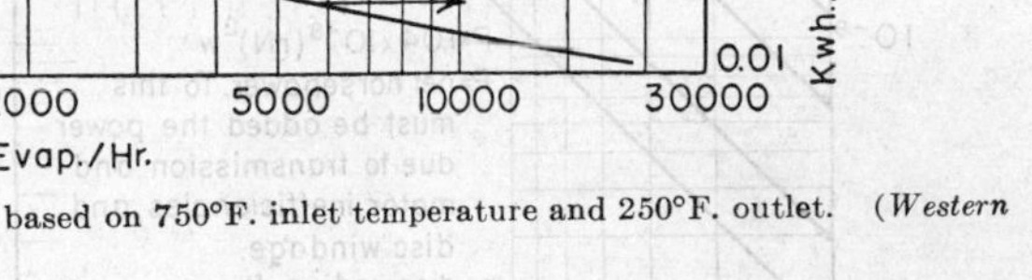

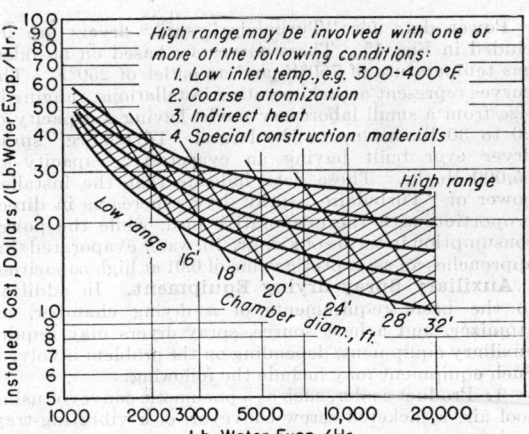

FIG. 46. Range of investment costs for spray dryers for a wide variation of operating conditions.

are installed costs for dryers constructed of ordinary steel; include furnace, dust collectors, atomizer and spray machine, fans and motors; and are based on prices current for September, 1946.

The **power requirements of centrifugal-disk atomizers** will vary with speed and capacity. Figure 47 shows how the power for vaned-disk atomizers varies with speed, diameter, and feed rate.

The life of high-speed bearings in centrifugal-disk atomizers depends on the conditions of operation. Average life may be 2000 hr. However, good practice dictates replacement of the bearings at definite time intervals to eliminate a costly shutdown due to bearing failure. For this practice a spare spray machine should be standard equipment.

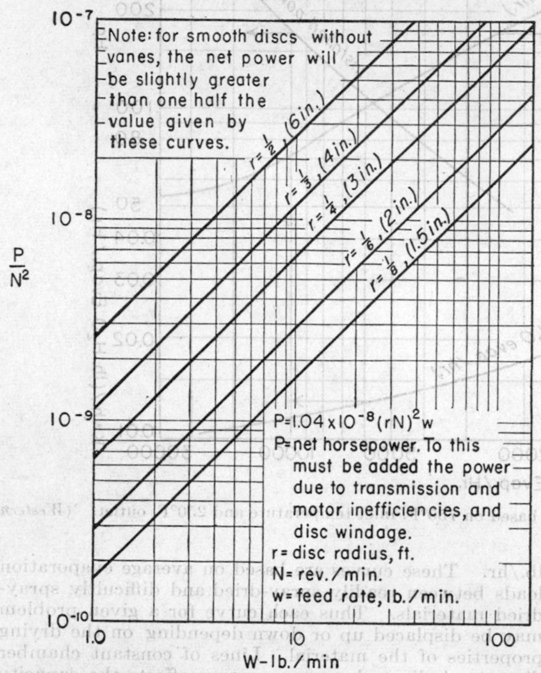

FIG. 47. Theoretical power consumption for centrifugal disk atomizers.

Power data for "Turbulaire" spray dryers are included in Fig. 45. These curves are based on an inlet-gas temperature of 750°F. and an outlet of 250°F. The curves represent actual operating installations, ranging in size from a small laboratory model having a capacity of 20 to 30 lb./hr. up to the largest "Turbulaire" spray dryer ever built having an evaporative capacity of 15,000 lb./hr. These data indicate that the installed power of "Turbulaire" spray dryers increases in direct proportion with the evaporative load, while the power consumption in terms of kw.-hr./lb. water evaporated/hr. approaches an asymptotic value of 0.01 at high capacities.

Auxiliary Spray-drying Equipment. In addition to the basic requirements of a drying chamber, an atomizer, and a heat source, spray dryers may require auxiliary equipment, depending on the problem involved. Such equipment may include the following:

1. Product cooler, such as a pneumatic conveyor using cool air, a jacketed screw conveyor, or a vibrating-tray cooler.

2. Condensers or solvent-recovery units for closed-circuit operation.

3. Specially designed sweepers or rakes for product removal from the drying chamber.

4. Special methods of maintaining the walls of the drying chamber cold.

5. Control systems on the dryer feed. Spray dryers can be **effectively controlled** by controlling the rate of feed to produce a definite outlet-gas temperature. This temperature is exceedingly sensitive to variations in feed rate, because the gas requires only 5 to 20 sec. to pass through the chamber.

6. Other controls are sometimes required to maintain the liquid feed at a specified temperature, viscosity, pH, concentration, etc.

7. Frequently, secondary dust collectors are required to clean up the gases from cyclones or multiclones. These secondary collectors are usually bag filters, or wet collectors.

8. Preconcentration of the liquid feed is sometimes required, and this may be accomplished, in part, by passing the spent gases through a wet collector in which the thin feed is used as the scrubbing liquid and thereby is concentrated (see Fig. 39).

9. When the dry material is separated from the gas within the chamber and discharged from the chamber bottom, pneumatic, belt, or vibrating conveyors may be used to convey it away. If maintenance of a coarse large particle is desired, excessive conveying should be avoided to prevent breakage. Frequently, these conveyors are also coolers. In some cases, the product is packaged directly as it comes from the chamber; in others, it is conveyed to storage bins.

10. Sometimes spray dryers operate under balanced pressure, in which case blowers are placed on the inlet and outlet sides of the dryer and are adjusted to balance the pressure within the dryer.

Direct Continuous-sheeting Dryers

Description. In dryers of this class, heated air is circulated over or through a continuous sheet material which is suitably suspended. This is contrasted with drying of continuous sheets in contact with hot surfaces, as described under cylinder dryers. A wide variety of types operating under a wide variety of conditions have been built to dry printed fabrics, specially treated paper, coated fabrics, rugs, etc.

In the **festoon or loop dryer,** material is festooned on rolls or rods fastened at their ends to endless chains which carry them through the dryer. The sheet material is sometimes made to move continuously over the rolls to prevent sustained contact with the supports. However, in some cases, where sustained contact is unobjectionable, loose rolls or sticks are placed under the material at the conveyor inlet end of the dryer.

Figure 48 shows a typical festoon dryer for drying dyed or finished piece goods without tension. Curing is frequently done in the last half of the dryer. This particular dryer can handle one or more strands of goods up to the width of the dryer, and it is so constructed that, when handling two strands, for example, either strand may be stopped without interfering with the processing of the other. The goods are automatically draped on the rolls to give the desired loop length. These loops progress through the dryer where they are subjected to a downward blast or flow of heated air which is controlled by adjustable louvers. At predetermined intervals, a ratchet rotates the rolls to prevent sustained contact with the same portion of the goods through the dryer. Simultaneously with the rolling action, a patented feature causes the roll to expand in length, producing a slight tension across the cloth width to prevent creases from forming when the pole turns. The air heaters, placed underneath a false floor, radiate some heat to the cloth. The dry material is automatically removed from the rolls at the discharge end of the dryer. These dryers are built up from standard 39-in. units, each individually equipped with fan and heating coils so that temperature and velocity zoning can be maintained. In the dryer shown in Fig. 48, variable-pitch propeller-type fans are used. They provide about 8000 to 12,000 cu. ft./min. in each unit.

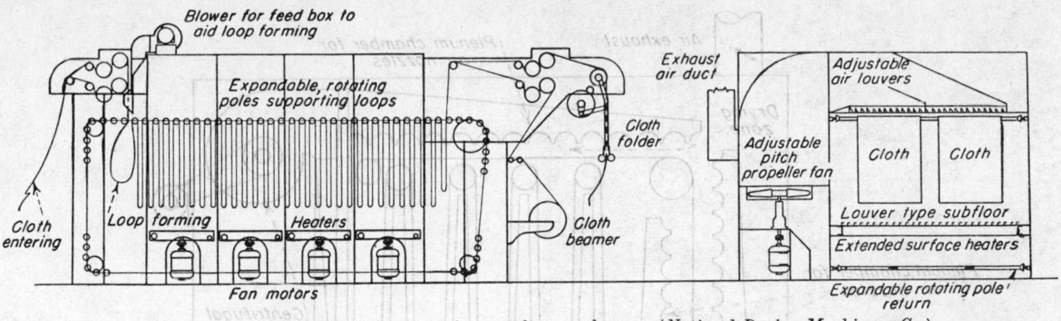

FIG. 48. Standard National four-unit loop or festoon dryer. (*National Drying Machinery Co.*)

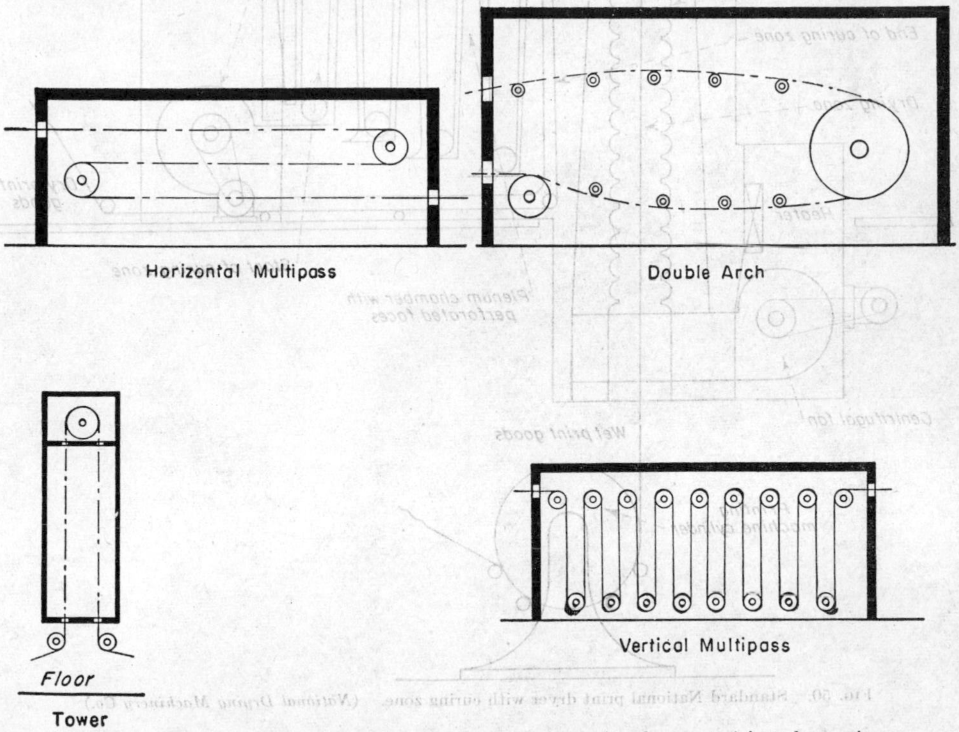

FIG. 49. Schematic diagrams of various arrangements for drying sheet materials under tension.

Another common type employs fixed rolls within the heated enclosure which support the material under tension so that it makes one or more horizontal or vertical passes within the drying chamber. Variations of this type are shown schematically in Fig. 49. Heated air is blown across the sheet. Figure 50 illustrates a dryer of this class that is used to dry and cure printed and coated fabrics. Hot air from nozzles impinges on the goods at nozzle velocities of 4000 to 5000 ft./min. The dryer is designed to dry the treated surface sufficiently to permit its contact with the rolls in the curing section. In this section, heated air flows onto the material from perforated tapered stacks inserted between two moving sheets of goods.

Figure 51 shows the so-called **"Multipass Air-lay" dryer,** in which air is blown through slots perpendicular to the sheeting as it moves through the dryer in a number of vertical passes. The air may actually penetrate the sheeting if it is porous. The heated air holds the sheeting against the conveyor by virtue of its impact pressure.

Tenter dryers are used for drying sheets of material in a stretched condition if the sheet has strength when wet. The material is stretched in a horizontal position between two endless conveying chains equipped with clips that automatically grip the edge of the sheeting as it enters the dryer. Heated air is usually blown perpendicularly from slots or nozzles against both sides of the sheet. Another method of holding the material is by fastening it to a pin frame, similar to a curtain stretcher. Figure 52 shows a single section of a standard tenter dryer. Large machines will range up to 10 of such sections, each section having individual fans and motors. Tenter dryers may also be of multipass construction, as shown in Fig. 53, by using pin frames to keep the material under tension.

Some dryers for continuous sheeting employ a combination of steam-heated cylinders and/or infrared lamps in conjunction with forced circulation of heated air.

Field of Application. Continuous-sheeting dryers are suitable for sheet materials that are sufficiently strong mechanically in the wet and dry states to support their

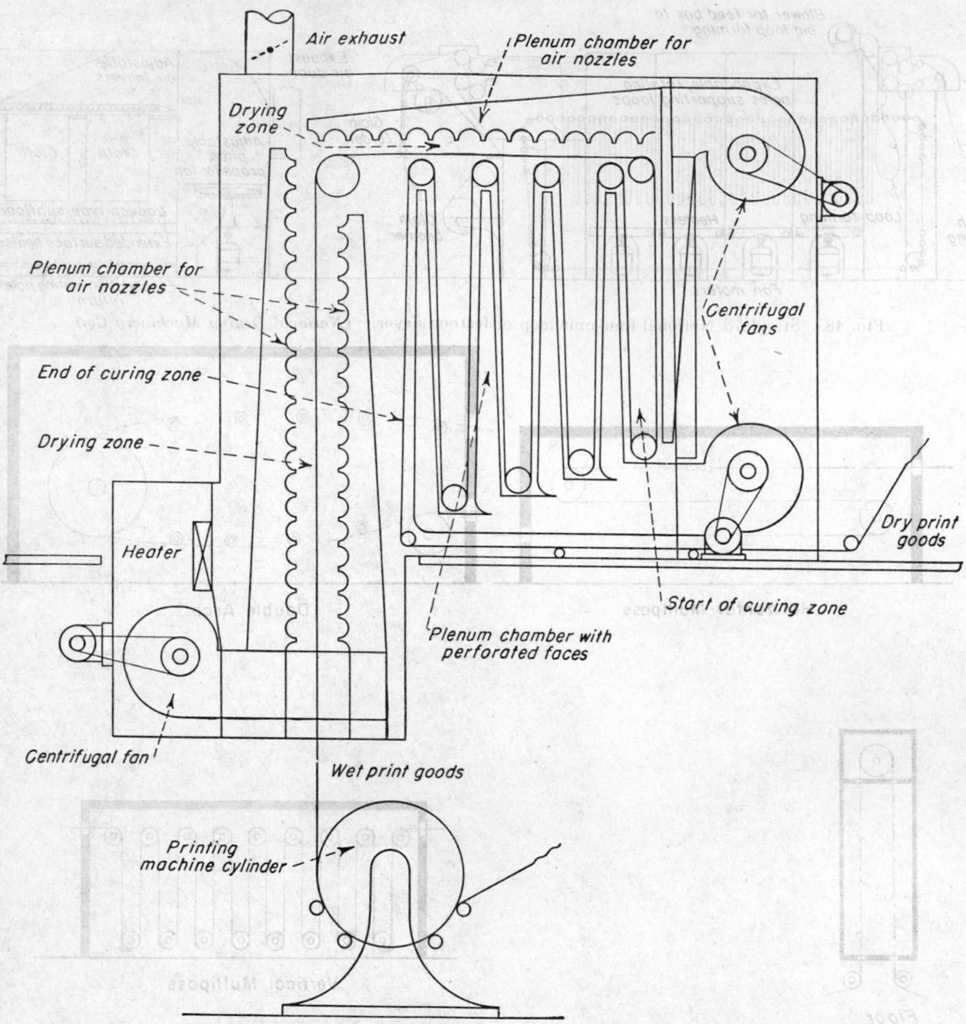

FIG. 50. Standard National print dryer with curing zone. (*National Drying Machinery Co.*)

own weight or withstand any tension required. Their widest use is in the textile fields, in the manufacture of coated and impregnated fabrics, and in the preparation of films and coated papers.

Festoon or loop dryers are used to dry, under no tension, materials such as rayon draping fabric, rayon piece goods, worsted wool, tubular-knit cotton goods, dyed fabrics, etc. "Multipass Air-lay" dryers are used to dry materials such as rubber-backed cotton, rayon fabric, cotton crepe, Neoprene sheet, plastic sheeting, etc. Tenter dryers dry under tension such materials as wool piece goods, cotton gauze, rayon, fabrics of cotton and rayon mixtures, etc.

Theory and Design Data. The drying of sheet materials follows the general principles of drying in that both a constant-rate and a falling-rate period may occur. During the constant-rate period, Eqs. (1) and (2) can be used to predict the drying rate when the air flow is parallel to the wet material. When the air flow is perpendicular to the sheet material, Eqs. (1) and (4) can be used during the constant-rate period, provided that the air is not impinging upon the material from jets, and if the material is not porous. If jets of air are used, the drying rate depends upon the size, spacing, and position of the jets and may be as much as two to twenty times higher than the rates predicted by Eq. (4). If the material is porous, there is no method of predicting drying rates.

In drying wet piece goods, such as cottons, the drying will probably occur almost entirely in the constant-rate period. In drying porous fabrics, the drying rate may actually increase when the liquid film ceases to close the pores and the air begins to flow through the goods.

The drying times for continuous-sheeting dryers are short, ranging from 2 to 15 min. for loop or festoon dryers, and from 6 to 120 sec. for high-speed tenter dryers. There is no published method for predicting drying rates in the falling-rate period, although, if the critical moisture content is known, a reasonable estimate can be obtained by assuming the drying rate to be proportional to the free moisture content. Since most materials dried by this method have high equilibrium moisture contents, it is essential that the residual moisture content be expressed as the free moisture content (see Definitions, p. 800).

Drying often follows a treating bath in which the

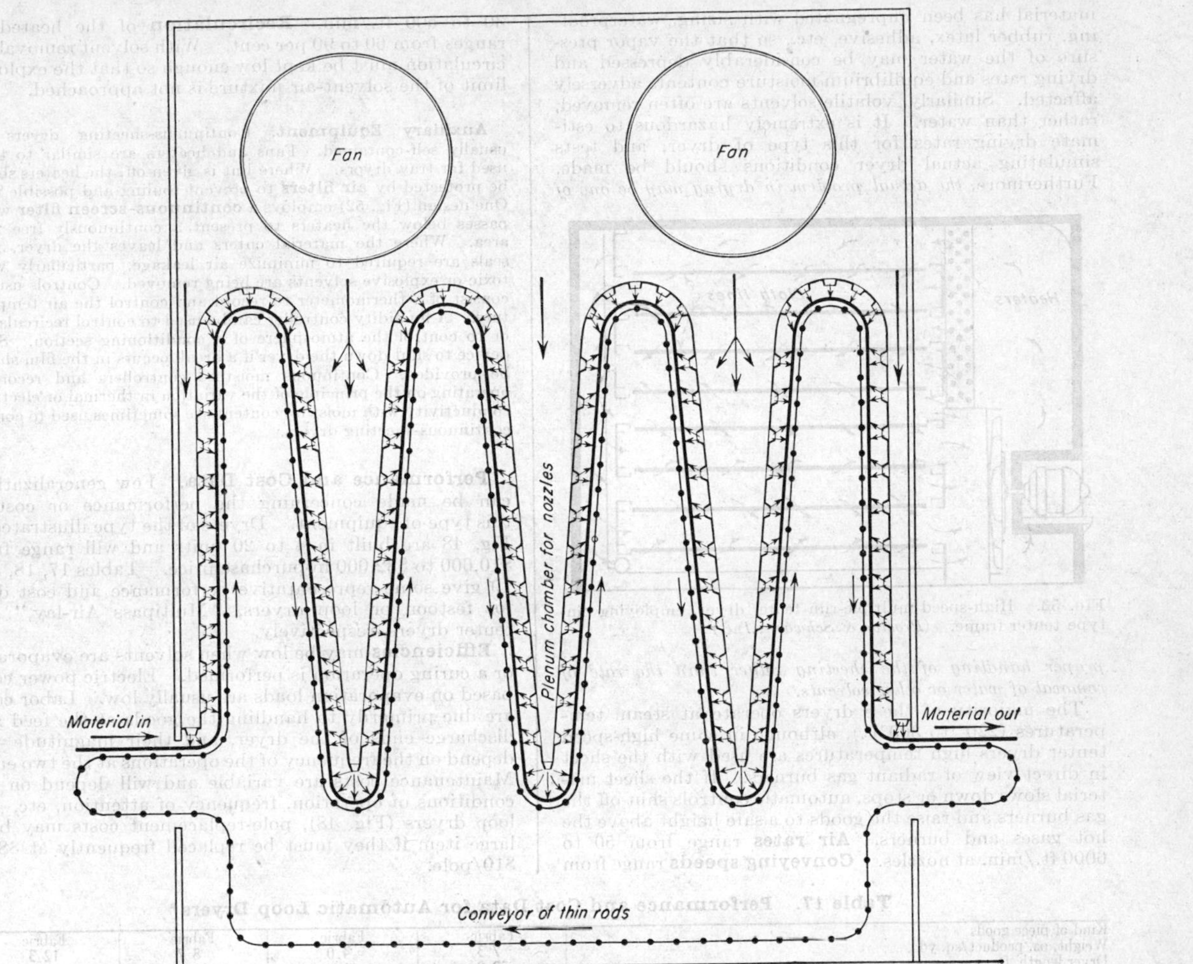

FIG. 51. Multipass Air-lay dryer for drying sheet material with absence of tension. (*Proctor & Schwartz, Inc.*)

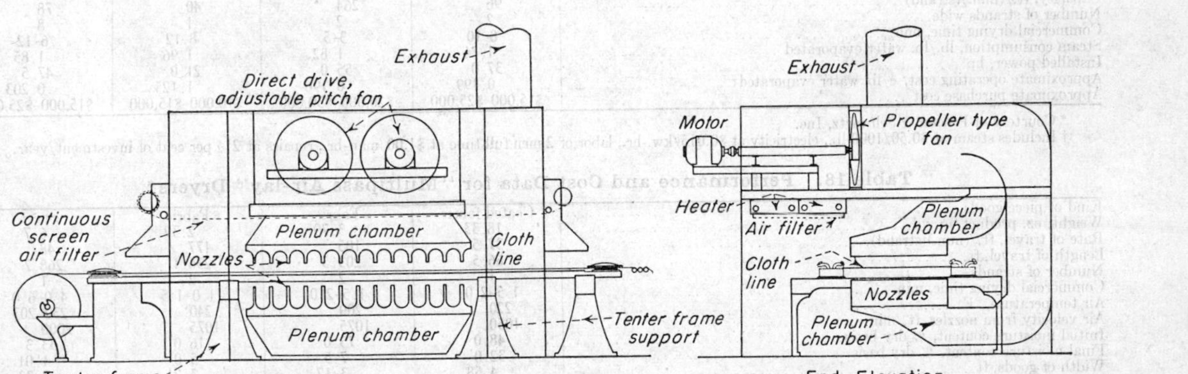

FIG. 52. Single-unit, high-speed tenter dryer. (*National Drying Machinery Co.*)

material has been impregnated with sizing, waterproofing, rubber latex, adhesive, etc., so that the vapor pressure of the water may be considerably depressed and drying rates and equilibrium moisture contents adversely affected. Similarly, volatile solvents are often removed, rather than water. It is extremely hazardous to estimate drying rates for this type of dryer, and tests simulating actual dryer conditions should be made. Furthermore, *the actual problem in drying may be one of proper handling of the sheeting rather than the rate of removal of water or other solvents.*

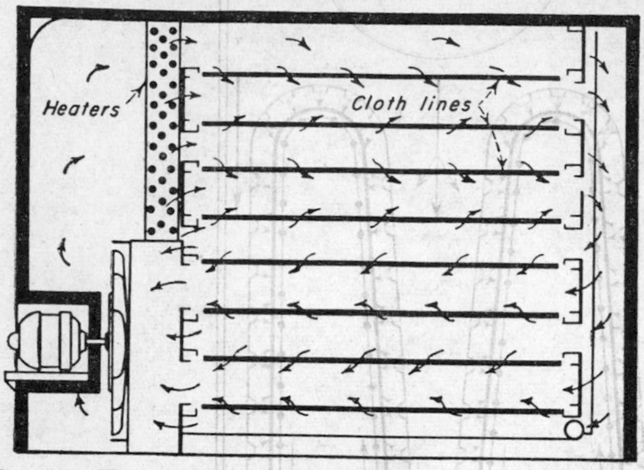

Fig. 53. High-speed multiple-run tenter dryer, employing pin-type tenter frame. (*Proctor & Schwartz, Inc.*)

The majority of these dryers operate at steam temperatures (220° to 300°F.), although in some high-speed tenter dryers high temperatures are used with the sheet in direct view of radiant gas burners. If the sheet material slows down or stops, automatic controls shut off the gas burners and raise the goods to a safe height above the hot gases and burners. **Air rates** range from 50 to 6000 ft./min. at nozzles. **Conveying speeds** range from 30 to 500 ft./min. **Recirculation** of the heated air ranges from 60 to 90 per cent. With solvent removal, recirculation must be kept low enough so that the explosive limit of the solvent-air mixture is not approached.

Auxiliary Equipment. Continuous-sheeting dryers are usually self-contained. Fans and heaters are similar to those used for tray dryers. Where lint is given off, the heaters should be protected by **air filters** to prevent fouling and possible fires. One design (Fig. 52) employs a **continuous-screen filter** which passes below the heaters to present a continuously free filter area. Where the material enters and leaves the dryer, good seals are required to minimize air leakage, particularly when toxic or explosive solvents are being removed. Controls usually consist of a thermometer to record and control the air temperature. A humidity controller can be used to control recirculation or to control the atmosphere of a conditioning section. Some device to shut down the dryer if a break occurs in the film should be provided. Continuous moisture controllers and recorders operating on the principle of the variation in thermal or electrical conductivity with moisture content are sometimes used to control continuous-sheeting dryers.

Performance and Cost Data. Few generalizations can be made concerning the performance or cost of this type of equipment. Dryers of the type illustrated in Fig. 48 are built in 4 to 20 units and will range from $10,000 to $32,000 in purchase price. Tables 17, 18, and 19 give some representative performance and cost data for festoon or loop dryers, "Multipass Air-lay," and tenter dryers, respectively.

Efficiencies may be low when solvents are evaporated or a curing operation is performed. Electric power costs based on evaporation loads are usually low. Labor costs are due primarily to handling the goods at the feed and discharge ends of the dryer, and their magnitude will depend on the frequency of the operations at the two ends. Maintenance costs are variable and will depend on the conditions of operation, frequency of attention, etc. In loop dryers (Fig. 48), pole-replacement costs may be a large item if they must be replaced frequently at $8 to $10/pole.

Table 17. Performance and Cost Data for Automatic Loop Dryers*

Kind of piece goods	Fabric	Fabric	Fabric	Fabric
Weight, oz. product/sq. yd	7.3	4.0	8.7	12.3
Dryer length, ft	52.0	61	32	52
Air temperature, °F	290	280	200	280
Air velocity across material, ft./min	500	500	444	500
Initial moisture content, % dry basis	100.0	100.0	40.0	85.0
Final moisture content, % dry basis	5.0	5.0	7.0	2.0
Width of goods, ft	4.5	3.33	5.0	0.918
Capacity, ft./(min.)(strand)	96	264	40	78
Number of strands wide	2	2	1	8
Commercial drying time, min	8–10	3–5	8–12	6–12
Steam consumption, lb./lb. water evaporated	1.86	1.82	1.96	1.85
Installed power, hp	37.0	55.0	21.0	47.5
Approximate operating cost, ¢/lb. water evaporated†	0.199	0.196	1.125	0.203
Approximate purchase cost	$15,000–$25,000	$20,000–$30,000	$10,000–$15,000	$15,000–$25,000

* Courtesy of Proctor & Schwartz, Inc.
† Includes steam at $0.50/1000 lb., electricity at $0.015/kw.-hr., labor of 2 men full time at $1.00/man-hr , repairs at 2½ per cent of investment/year.

Table 18. Performance and Cost Data for "Multipass Air-lay" Dryers*

Kind of piece goods	Coated fabric	Fabric	Fabric	Sheet
Weight, oz. product/sq. yd	18.33	3.79	2.39	6.7
Rate of travel, ft./(min.)(strand)	37.5	105	177	48
Length of travel, ft	66.5	203.5	236.1	268.7
Number of strands	1	2	2	1
Commercial drying time, min	1.5–2.0	1.5–2.0	1.0–1.5	4.0–6.0
Air temperature, °F	220	200	240	230–203
Air velocity from nozzles, ft./min	1050	1075	1075	2000
Initial moisture content, % dry basis	48.0	75.0	116.0	33.3
Final moisture content, % dry basis	32.0	7.5	5.0	1.01
Width of goods, ft	4.58	3.17	3.21	8.33
Steam consumption, lb./lb. water evaporated	2.385	1.97	2.06	5.14
Installed power, hp	7.0	43.0	48.5	87.0
Approx. operating cost, ¢/lb. water evaporated†	1.43	0.290	0.322	1.13
Approx. purchase cost	$8,000–$10,000	$10,000–$15,000	$20,000–$30,000	$40,000–$50,000

* Courtesy of Proctor & Schwartz, Inc.
† Includes steam at $0.50/1000 lb., electricity at $0.015/kw.-hr., labor of 2 men full time at $1.00/man-hr., repairs at 2½ per cent of investment/year.

Table 19. Performance and Cost Data for Tenter Dryers*

Type of dryer	Multiple-run	Single-pass	Single-pass	Single-pass
Kind of piece goods	Fabric	Fabric	Fabric	Fabric
Weight, oz. product/sq. yd.	9.6	1.545	2.62	3.49
Rate of travel, ft./min	72	390	150	390
Length of travel, ft	135	70	52	31.0
Air temperature, °F	290	280	260	290
Air velocity from nozzles, ft./min	95 (no nozzles)	4000	3800	4600
Initial moisture content, % dry basis	70.0	130.0	100.0	70.0
Final moisture content, % dry basis	8.0	3.0	10.0	40.0
Commercial drying time, sec	90–120	9–12	18–24	3.5–5.0
Width of goods, ft	5.0	4.0	3.67	3.33
Steam consumption, lb./lb. water evaporated	1.91	1.90	2.17	2.29
Installed power, hp	48.0	52.0	27.0	15.0
Approx. operating cost, ¢/lb. water evaporated†	0.41	0.30	0.58	0.66
Approx. purchase cost	\$15,000–\$25,000	\$8,000–\$12,000	\$5,000–\$8,000	\$4,000–\$6,000

* Courtesy of Proctor & Schwartz, Inc.
† Includes steam at \$0.50/1000 lb., electricity at \$0.015/kw.-hr., labor of 2 men full time at \$1.00/man-hr. and repairs at 2½ per cent investment/year.

Vacuum Shelf Dryers

Description. Vacuum shelf dryers are indirect batch dryers consisting of a vacuum-tight chamber usually constructed of cast-iron or steel plate, heated supporting shelves within the chamber, a vacuum source, and usually a condenser. Cast-iron chambers are generally square, and steel chambers can be either square or cylindrical. One or two doors are provided, depending on the size of the chamber. The doors are fitted with a pliable gasket of rubber or similar material.

Hollow shelves of flat steel plate are fastened permanently inside the vacuum chamber and are connected in parallel to inlet and outlet headers. The heating medium entering through one header and passing through the hollow shelves to the exit header is generally steam ranging in pressure from 100 lb./sq. in. gage, to subatmospheric pressures for low-temperature operation. Low temperatures can be provided by circulating hot water, and high temperatures can be obtained by circulating hot oil or Dowtherm. Some experimental dryers employ electrically heated shelves.

The material to be dried is placed in pans or trays which, in turn, are placed on the heated shelves. The trays are generally of metal with solid bottoms to ensure good heat transfer between the shelf and the tray.

Vacuum shelf dryers may vary in size from 1 to 20 shelves, the larger chambers having over-all dimensions of 9 ft. wide, 18 ft. long, and 12 ft. high.

Vacuum is applied to the chamber and vapor is removed through a large pipe which is connected to the chamber in a manner such that, if the vacuum is suddenly broken, the inrushing air will not greatly disturb the material being dried. This line usually leads to a condenser where the moisture or solvent that has been vaporized is condensed. The uncondensed exhaust goes to the vacuum source which may be a wet or dry vacuum pump or a steam jet ejector.

Applications. Vacuum shelf dryers are used extensively for drying pharmaceuticals, temperature-sensitive or easily oxidizable materials that would otherwise be handled in other types of dryers, and materials that are so valuable labor cost is insignificant. It is particularly useful for handling small batches of materials wet with toxic or valuable solvents. Recovery of the solvent is easily accomplished without danger of passing through the explosive range. Dusty materials may be dried with negligible dust loss. Hygroscopic materials may be completely dried at temperatures below that required in atmospheric dryers.

From the standpoint of high production rates, vacuum shelf dryers are unsatisfactory. However, their flexibility may outweigh, in many cases, the disadvantages of high labor costs, high investment cost for vacuum chamber, condenser, and vacuum pump, and the high costs of vacuum and condenser operation.

Theory and Design Data. In vacuum shelf dryers, heat is transferred to the wet material by conduction through the shelf and then through the bottom of the tray on the shelf, and by radiation from the shelf above. The radiant heat transfer may be nearly equal to the heat transfer by conduction.

The drying cycle usually consists of two periods: (1) a constant-rate period when the temperature of the material is approximately at the boiling temperature of the liquid at the absolute pressure prevailing in the drying chamber, and (2) a falling-rate period during which the temperature of the material approaches that of the heating medium. The critical moisture content will not necessarily be the same as for atmospheric tray drying [Ernst, Ridgway, and Tiller, *Ind. Eng. Chem.*, **30**, 1122 (1938)]. During the constant-rate period, moisture is rapidly removed. Often 50 per cent of the moisture will evaporate in the first hour of a 6- to 8-hr. cycle.

The drying time has been found to be proportional to between the first and second power of the depth of loading. When the time of exposure of a material to the drying temperature is a factor, it may be found feasible and profitable to reduce the drying time by reducing the loading, even at the expense of more frequent handling with higher labor costs.

The magnitude of the over-all heat-transfer coefficient obtained in vacuum shelf dryers is on the order of 1 B.t.u./(hr.)(sq. ft. tray surface)(°F. temperature difference between heating medium and solid). This coefficient is equivalent to an evaporation rate of 0.03 to 0.2 lb. water/(hr.)(sq. ft.). Harcourt (paper presented at Am. Soc. Mech. Engrs. meeting, Niagara Falls, Sept. 17–18, 1936) reported values of evaporation rates as high as 0.6 lb./(hr.)(sq. ft.) (see Table 20). High rates apply to high-moisture-content materials and low rates to low-moisture materials carried to complete dryness. See Ernst *et al.* [*Ind. Eng. Chem.*, **30**, 1119, (1938)].

Table 20. Performance Data of Vacuum Shelf Dryers

Material	Sulfur black	Calcium carbonate	Calcium phosphate
Loading, lb. dry material/sq. ft.	5.25	4.1	6.7
Steam pressure, lb./sq. in. gage	60	60	30
Vacuum, in. Hg	27–28	27–28	27–28
Initial moisture content, % (wet basis)	50	50.3	30.6
Final moisture content, % (wet basis)	1	1.15	4.3
Drying time, hr	8	7	6
Evaporation rate, lb./(hr.)(sq. ft.)	0.66	0.58	0.49

This type of equipment will employ from 20 to 29.5 mm. Hg vacuum. Shelf temperatures will range from less than 32°F. to more than 200°F. Depth of loading in trays will generally be 1 in. or less. Minimum spacing between shelves in these dryers is 2 in. and maximum spacing about 6 in.

Reliable estimates for cost and size must be based on

actual tests in vacuum dryers. The above data simply permit high-spot estimating.

Auxiliary Equipment. Vacuum shelf dryers require a source of vacuum, which can be either a steam-jet ejector or a vacuum pump. A condenser is usually placed in the line before the vacuum source to reduce the load on the pump or the ejector, as well as to recover valuable solvent if required.

The pump or ejector should be designed large enough to evacuate the chamber to the desired vacuum in a suitably short time. Too small a vacuum pump may not be able to handle the air leakage around the door, which is almost certain to exist after the dryers have been in use for a period of time. Similarly, the condenser must be large enough to accommodate the high rate of vapor flow at the beginning of the cycle.

Trays should be designed and maintained as flat as possible to obtain maximum area of contact with the heated shelves. For the same reason, the shelves should be kept free from scale and rust. Air vents should be installed on the steam-heated shelves to vent off non-condensable gases.

The heating medium should not be applied to the shelves until after the vacuum has been produced in the dryer in order to prevent the possibility of the material overheating or boiling over at the start of drying. **Case-hardening** can sometimes be avoided by retarding the rate of drying sufficiently in the first part of the cycle.

Control of these dryers can be accomplished by observing the rate of condensation or by following the temperature of the discharged vapors, which increases toward the temperature of the heating medium near the end of the cycle.

Performance and Cost Data. The thermal efficiency of vacuum shelf dryers is usually on the order of 60 to 80 per cent. For high-moisture-content materials, the fuel requirements vary from 1.3 to 2.0 lb. steam/lb. water evaporated. Table 20 gives operating data for one organic color and two inorganic compounds. The evaporation rates given are somewhat higher than those usually encountered in industrial operation.

Operating costs for vacuum shelf dryers including the vacuum system may range from $0.001 to $0.02/lb. product. Labor may constitute 50 per cent of these operating costs, and maintenance 20 per cent. Absolute **maintenance costs** amount to 5 to 10 per cent of the total installed cost. Actual **labor costs** will depend on drying time, facilities for loading and unloading trays, etc. The power required for these dryers is only that for the vacuum system, and for vacuums of 27 to 29 in. Hg the power requirements are on the order of 0.007 to 0.015 hp./sq. ft. tray surface. When a steam jet vacuum system is used, electric power is negligible.

Purchase costs for this type of dryer in standard steel construction, including condensers, motors, and vacuum pumps will range from $50/sq. ft. of tray surface for a dryer containing 20 sq. ft. of surface to $10/sq. ft. of tray surface for a dryer containing 1150 sq. ft. of surface.

Vacuum Freeze Dryers

Description. In low-temperature vacuum drying, a solvent (usually water) is removed from a material at a very low temperature under vacuum. Frequently, the solvent will be in a **frozen state** so that **drying will occur by sublimation.**

Equipment for vacuum freeze drying processes consists of three primary items: (1) an *evaporator* in which heat is supplied and solvent vaporized, (2) a *piping system* to transport vapor to the vapor-removal system, and (3) the *vapor-removal system* including the condensable and non-condensable gas-exhaust system.

Representative types of evaporators are: 1. Vacuum shelf dryers for materials held in trays or bottles. 2. Dryers in which material is sprayed or spread upon the interior vertical walls of a cylindrical vessel. Heat is introduced through the wall by circulating warm water in a shell. Material may be introduced and removed by the use of air locks without breaking the vacuum. 3. Horizontal rotary vacuum dryers in which the material tumbles and slides over the rotating surface. This type may be operated continuously by admitting the feed and discharging the product through air locks.

The vapor-piping system must be designed to transport the required amount of vapor, at the existing pressures, under a pressure drop no greater than the maximum allowable dryer pressure minus the lower pressure attainable in the pumping system. The large specific volumes of gases at very low pressures necessitate the use of large-diameter piping.

The vapor-removal system handles non-condensables as well as condensable vapors. Water vapor is condensed wherever possible to reduce pumping requirements. Non-condensables are removed by diffusion pumps, mechanical vacuum pumps, or steam ejectors.

When operating at vapor pressures below 500 to 1000 μ Hg, water vapor is condensed as ice on surfaces refrigerated with solid CO_2, Freon, or ammonia, adsorbed on solid desiccants, or absorbed by low-vapor-pressure solutions. Static surface condensers require large condensing areas because of the decrease in heat-transfer rate with increasing thickness of ice deposit. Continuous rotary condensers employing scrapers to keep the condensing surface clean require considerably less area. The scraped ice may be removed from the vacuum system batchwise or continuously by means of ice ejectors. Figure 54 shows a diagrammatic sketch of a low-temperature, continuous rotary condenser.

Below $-40°C.$, the rate of decrease in vapor pressure of ice becomes progressively slower. Few products are dried with temperatures lower than $-30°C.$

Desiccants (see Drying of Gases, p. 877) may be used in place of surface condensers for removal of water vapor. Their cost is generally higher than refrigeration, but they are sometimes used for laboratory dryers.

Direct pumping of the water vapor without condensation can be used. However, the same pressures must be produced which are obtained with condensers and desiccants. Either ejector pumps or oil-sealed rotary pumps may be used. A multistage steam ejector with interstage condensers is suitable and is widely used in large-scale operation. Combinations of these or of other accessory means, such as oil-diffusion or oil-ejector pumps, may be used. In any event, the volume occupied by the water vapor under expanded conditions of high vacuum demands pumps of high volumetric capacity. The water vapor is usually condensed on the high-pressure side of a rotary pump, resulting in operation against a back pressure that is much below atmospheric. This approximates a second stage for the pump, so that under actual operating conditions a single-stage oil-sealed rotary pump will have its efficiency maintained close to 100 per cent at 100 or 200 μ. However, the capacity of any mechanical pump within practical limits is such that in large-scale operation a jet-type ejector is preferred.

Application. This method of drying is useful for drying substances that are chemically or physically altered by the temperature and/or gases encountered in the usual atmospheric dryers. Dilute solutions and suspensions may be dried almost completely at temperatures well *below the freezing temperature* even when in proximity to heating surfaces as warm as 200°F.

Some of the principal characteristics* of vacuum freeze drying are:

* E. W. Flosdorf, private communication. See also Flosdorf. *Chem. Eng. Progress*, **43** (7), 343 (1947).

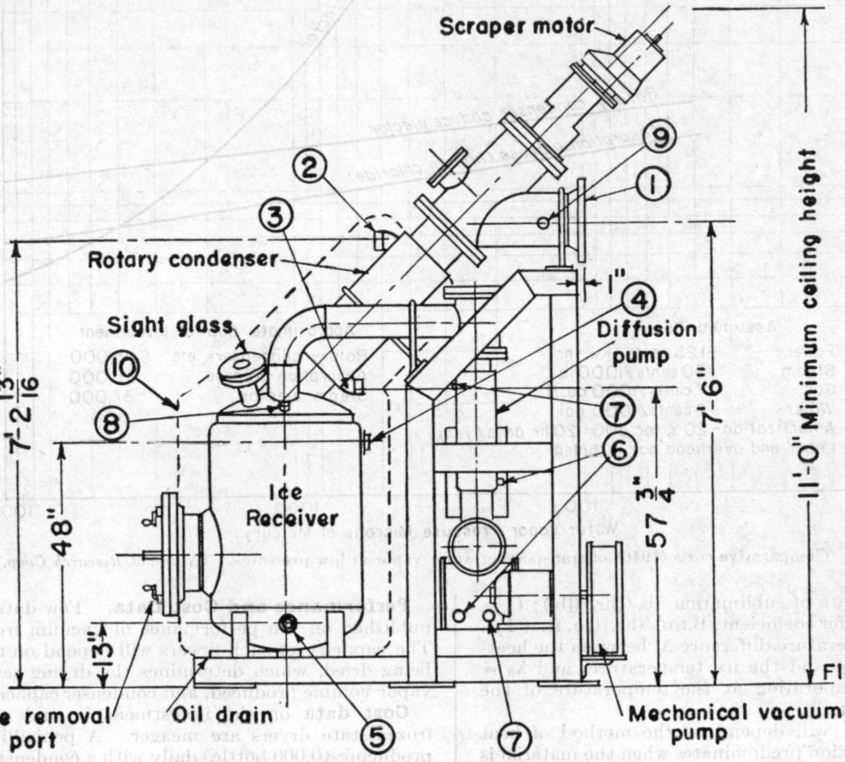

FIG. 54. Low-temperature vacuum scraper condenser, rated at 10 lb. of water/hr. at 100 microns. (1) vapor inlet; (2) refrigerant outlet; (3) refrigerant inlet; (4) refrigerant outlet; (5) refrigerant inlet; (6) water inlet; (7) water outlet; (8) air inlet; (9) McLeod gage connection; (10) cork insulation. (*National Research Corp.*)

1. The temperature is below that at which many labile substances undergo chemical change. This applies to labile components in blood, to viruses and most forms of microorganisms, and to other biologicals and pharmaceuticals.

2. Because of the low temperature, the loss of volatile constituents is minimized. This is particularly important in application to many foods like orange juice and pineapple juice.

3. Since the product is frozen, there is no bubbling or foaming, which, in the case of drying some substances, would result in changes due to surface action, such as the surface denaturation of proteins, which occurs in drying their solutions even at low liquid temperatures under vacuum.

4. The solute remains evenly dispersed and distributed without undergoing concentration as the frozen solvent sublimes, and the remaining dry residue emerges as a highly porous, solid framework. It occupies essentially the same total space as the original solution did, and the final residue is not a fine powder so often obtained with other drying methods. It consists of a friable, interlocking, spongelike structure. As a result, solubility is extremely rapid and complete.

5. During drying, the surface of the evaporating ice layer gradually recedes to leave more and more of the highly porous residue of solute exposed. As a result, casehardening never occurs. A far lower content of moisture may be obtained in the final product without use of excessively high final temperature. Because of this lower moisture content, a greater degree of stability results than is the case after any other method of drying.

6. Bacteriological growth and enzymatic changes cannot take place under the frozen conditions of drying and are resisted in the dry product. This is important for foods, as well as in medical products used in parenteral injection.

7. Because of the high vacuum used, in contrast with the vacuum used in ordinary low-temperature liquid evaporation, the amount of oxygen present is so extremely small that even the most readily oxidizable constituents are protected. For a detailed history of the development of freeze drying with an extensive bibliography, see Flosdorf *et al.* [*J. Immunol.*, **29**, 389 (1935); **50**, 21 (1945)].

Theory and Design Data. In drying a frozen solid under very low vacuum, the usual constant-rate and falling-rate periods are obtained. In the first period, a frozen liquid evaporates at a constant maximum rate which is a function of the area exposed and the temperature difference between the solid and heat source. The rate may be expressed by the general heat-transfer expression

$$\frac{dW}{d\theta} = \frac{UA\,\Delta t}{\lambda_0} \tag{74}$$

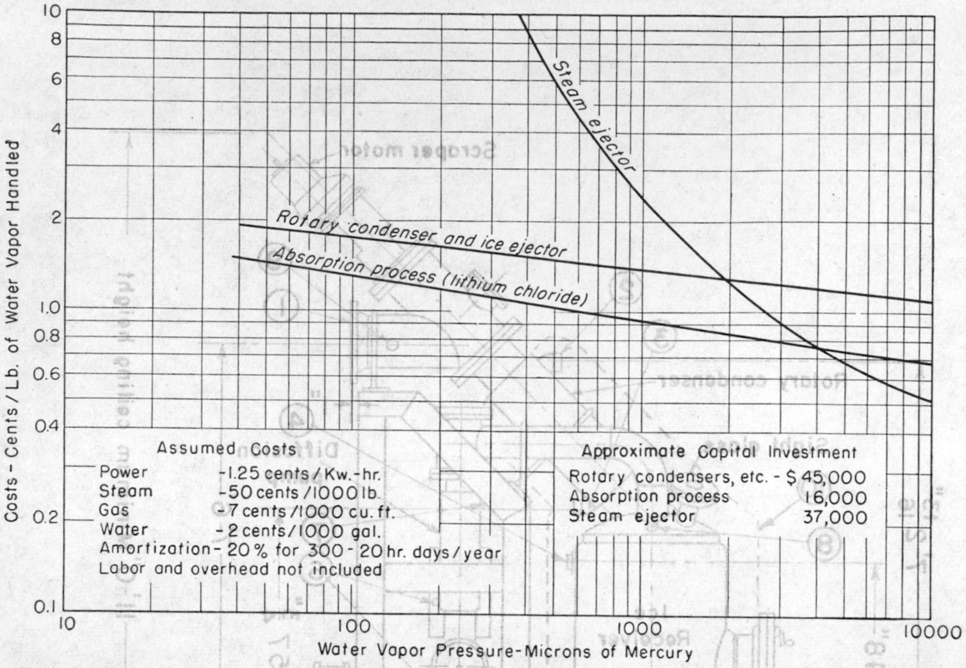

FIG. 55. Comparative costs (1946) of transporting water vapor at low pressures. (*National Research Corp.*)

where $dW/d\theta$ = rate of sublimation, lb./(hr.)(lb.); U = over-all heat-transfer coefficient, B.t.u./(hr.)(sq. ft.)(°F.), based on the temperature difference Δt between the heat-source temperature and the ice temperature; and λ_0 = latent heat of evaporation at the temperature of the frozen liquid, B.t.u./lb.

The value of U will depend on the method of heat transfer. Conduction predominates when the material is in direct contact with the heated surface, and radiation predominates when the material is held in containers because of the small contact area resulting from surface irregularities.

A value of U = 1.0 B.t.u./(hr.)(sq. ft.)(°F.) for conduction has been reported for the constant-rate period. Likewise, for radiant heat transfer between oxidized or darkened steel shelves at 180° to 200°F. and glass or steel containers at 30°F., a combined angle and emissivity factor F_{AE} of 0.6 is proposed* based on the area receiving radiation.

For a given product, the actual desiccating conditions must be determined by experimental tests. This is necessary because, as in most types of dryers, the period of a falling rate can at present be studied only by actual tests.

Auxiliary Equipment. Auxiliaries may include prefreezing equipment as well as concentrating and deaerating equipment. Prefreezers are of two general types: (1) shell freezers in which the charge is frozen on the sides of bottles rotating in contact with a refrigerated brine; and (2) tray freezers in which the charge is frozen in a stationary container contacting a refrigerated fluid which may be liquid air or high-velocity cold air. Preconcentration can be effected by (1) vacuum evaporation; (2) selective freezing and separation of ice by centrifuging; (3) selective solvent action; and (4) dialysis. Deaeration is a desirable preliminary to liquid-phase vacuum operation to reduce foaming and bubbling and to reduce the load on the noncondensable gas-pumping system.

* Private communication from Vacuum Engineering Division of National Research Corp.

Performance and Cost Data. Few data have been published on the performance of vacuum freeze dryers. The capacity of such dryers will depend on the material being dried, which determines the drying temperatures, vapor volume produced, and condenser capacity required.

Cost data on the investment in and operation of frozen-state dryers are meager. A penicillin plant for producing 10,000 bottles daily with a condensing capacity of only 10 to 15 liters may cost on the order of $50,000, whereas an orange-juice plant requiring a condensing capacity of 50 tons of water vapor in producing 1,000,000 lb. of dried orange juice per season may cost several hundred thousand dollars.

Figure 55 shows a comparison of the cost of removing and condensing 200 lb./hr. of water vapor over a range of pressure for three different systems. These data are based upon an amortization rate of 20 per cent per year of 300 20-hr. days but do not include dryer cost or labor and overhead costs. They are meant to show, principally, the relative orders of magnitude of the costs of the processes mentioned over a range of pressures. Sufficient operating experience is not available to evaluate their accuracy.

The total cost to produce dry orange-juice crystals by sublimation drying has been given by Flosdorf [*Food Industries*, **17**, 22 (1945)] as $1.067/lb. This is based on a plant producing 4800 lb./day and includes total drying costs, amortization, raw-material costs, all other processes, packaging, interest, overhead, and royalty. For sublimation drying of meat, the cost including labor, power, amortization, interest, overhead, and royalty is about $0.22/lb. dry meat, or about $0.04/lb. fresh meat.

Agitated-pan Dryers

Description. Agitated-pan dryers, which may operate atmospherically or under vacuum, usually consist of a shallow circular pan, jacketed on the bottom and partway up the sides for steam or other heating medium. A central vertical shaft supports an agitator to stir the material in the pan and bring fresh material in contact with the hot surfaces. The agitator can be designed to scrape the inside surface or set to a very close clearance. The agitator shaft may enter the pan from above through the cover or from below through a hub. Atmospheric pan dryers, if not open to the atmosphere, may be closed with covers containing suitable manholes and sight glasses with an outlet through which the heated vapors escape by natural draft. Vacuum pans usually have dome covers fitted with manholes, sight glasses, and an appropriate vacuum connection. Both types have manholes in the side, flush with the bottom of the pan, through which the dry product is discharged. In some cases, metal fingers just clearing the agitator extend into the pan from either the top or the side. These prevent the charge from rotating in the pan as a single mass. Typical atmospheric and vacuum pan dryers are shown in Fig. 56. The atmospheric pan dryer is sometimes called a **graining bowl.** One modification of this type of dryer incorporates a screen filter in the bottom of the pan for dewatering prior to drying.

The Dopp kettle, which consists of a dish-bottom kettle containing a double-motion agitator, possessing spring-loaded scraper blades to give positive scraping action, can sometimes be used as a dryer. This kettle can be jacketed, may operate under pressure or vacuum, and is frequently used when the material being dried undergoes a change in consistency from a thin fluid state to a pasty viscous mass before drying to a dry crumbly mass.

Field of Application. Agitated-pan dryers are used to handle small batches of materials that must be agitated during drying. They also can be used to handle pastes and slurries if a continuous dryer cannot be justified. If solvents are to be removed and recovered, small batches of slurries can be handled more economically in a vacuum pan dryer than on a vacuum drum dryer. This type of dryer is not suitable for materials that cannot suffer particle-size degradation during drying. Pan dryers are accessible for cleaning and are therefore adaptable to handling small batches of a number of different materials.

Theory and Design Data. The factors governing the operation of these dryers are primarily rate of heat transfer and degree of agitation. The over-all heat-transfer coefficient [Eq. (23)] will range from 5 to 35 B.t.u./(hr.)(sq. ft. contact surface)(°F.). Slightly higher coefficients, 20 to 75 B.t.u./(hr.)(sq. ft.)(°F.), may be expected when a positive scraping action occurs, as in the Dopp kettle, or when liquids are being handled. The value of the coefficient depends on the density and moisture content of the material, and on the degree of agitation. The actual drying time depends on the area of heating surface per volume of charge, and the drying time may vary approximately as the inverse of this ratio.

Auxiliary Equipment. No auxiliary equipment is required for an atmospheric pan dryer. The auxiliary equipment for the vacuum pan dryer will be similar to that described under vacuum rotary dryers. Except for a pressure gage or thermometer in the jacket, these dryers are seldom instrumented, since the appearance of the material is often the criterion for its dryness. The best method of determining the progress of the drying would be to measure the material temperature.

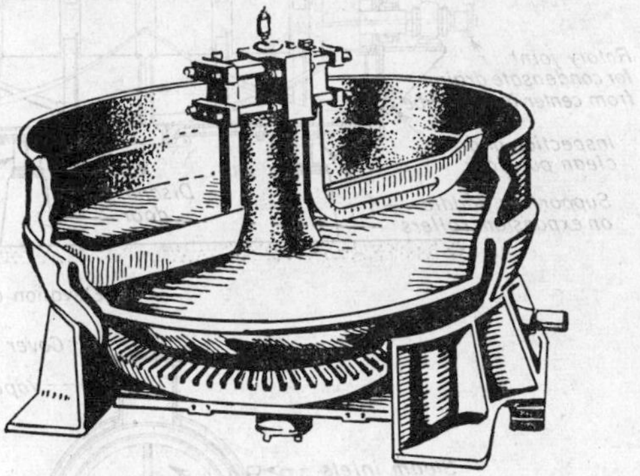

(a)

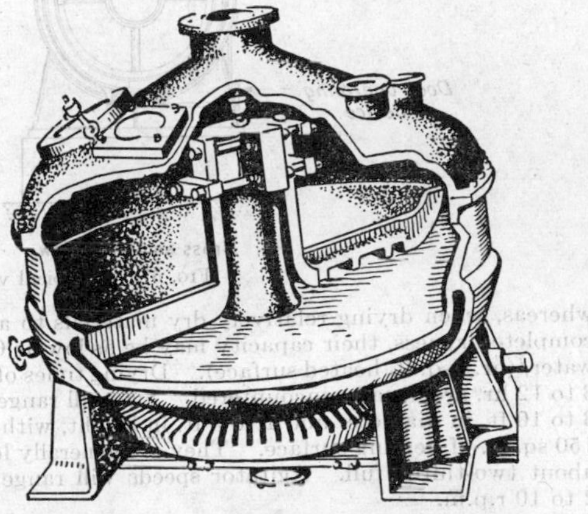

(b)

FIG. 56. Agitated pan dryers. (a) Atmospheric pan dryer. (b) Vacuum pan dryer. (*Buflovak Equipment Division.*)

Performance and Cost Data. Table 21, which was presented by Harcourt (Am. Soc. Mech. Engrs. meeting, Niagara Falls, Sept. 17–23, 1936), gives typical performance data for atmospheric and vacuum pan dryers. Performance data on a Dopp kettle used for drying are given by Laughlin [*Trans. Am. Inst. Chem. Engrs.*, **36**, 345 (1940)]. These dryers will have a capacity of 1 to 3 lb. water/(hr.)(sq. ft. heated surface) when drying relatively wet materials to high final moisture content;

Table 21. Performance Data for Atmospheric Pan Dryers

Material	Pan diam., ft.	Initial moisture, %	Steam pressure, lb./sq. in.	Agitator speed, r.p.m.	Evaporation, lb./(hr.)(sq. ft.)	Final moisture, %	Time, hr.
Sodium sulfate	3	45	40–50	2	1.5	0.5	11
Sodium carbonate	3	74	40	2	2.98	14	6
Sodium carbonate	3	83.5	40	2	2.96	0.1	
Calcium carbonate	3	37.2	60	3	3.26	0.1	4

Performance Data for Vacuum Pan Dryers

Material	Pan diam., ft.	Initial moisture, %	Steam pressure, lb./sq. in.	Agitator speed, r.p.m.	Evaporation, lb./(hr.)(sq. ft.)	Final moisture, %	Vacuum, in. Hg	Time, hr.
Sodium sulfate	3	57	55	2	4.3	7.5	27.5	5.5
Sodium chloride	3	81.5	30	2	8.3	0.6	27.5	10
Calcium phosphate	3	46.8	40	2.5	3.5	2.6	28.3	3

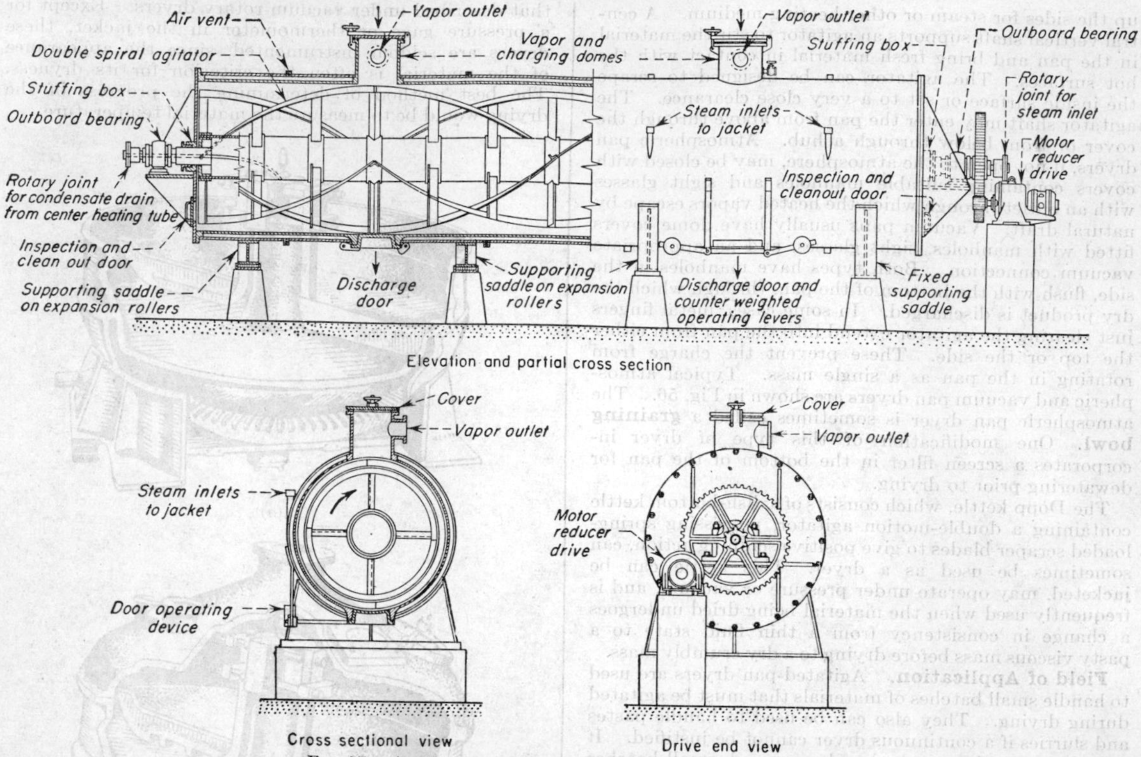

FIG. 57. A typical vacuum rotary dryer. (*Blaw-Knox Co.*)

whereas, when drying relatively dry materials to almost complete dryness, their capacity may be as low as 0.3 lb. water/(hr.)(sq. ft. heated surface). Drying times of from 3 to 12 hr. are usually encountered. Size will range from 3 to 10 ft. in diameter and 1 to 3 ft. in height, with 15 to 150 sq. ft. of heating surface. They are generally loaded about two-thirds full. Agitator speeds will range from 2 to 10 r.p.m.

Thermal efficiencies of 70 to 80 per cent and 65 to 75 per cent are common with vacuum pan and atmospheric pan dryers, respectively.

Purchase costs for these dryers will range from $30 to $200/sq. ft. of heating surface for standard iron construction, the high figure corresponding to a dryer with 8 sq. ft. of heating surface, and the low figure to 150 sq. ft.

The power requirement (exclusive of vacuum equipment) expressed in horsepower will be about one to two times the pan diameter. Maintenance costs will be about 5 to 10 per cent of the total installed cost. Labor costs will be high, since considerable attention is involved in charging and discharging. About 2 man-hr. per batch will be required to load and unload a dryer, and about one-half of a man's time when the dryer is in operation.

Vacuum Rotary Dryers

Description. Vacuum rotary dryers are batch dryers available in two different forms. The more common type (Fig. 57) consists of a stationary cylindrical shell, mounted horizontally, in which a set of agitator blades mounted on a revolving central shaft stir and agitate the material being dried. Heat is furnished by circulating a suitable heating medium (hot water, steam, or Dowtherm) through a jacket around the shell and, in larger dryers, through the hollow central shaft. The agitator is usually of the discontinuous-spiral or double-spiral type with the blades set as close to the cylinder wall as possible without scraping. The dryer is charged through a manhole at the top of the shell and discharged through one or more manholes at the bottom of the dryer. If material is to be removed from one end of the dryer only, special agitator blades and drive are provided which, when reversed, drive material to the discharge end. Vacuum is applied and maintained by any of the conventional methods (steam jets, vacuum pumps, etc.). Another type of vacuum rotary dryer consists of a rotating cylindrical shell, suitably jacketed. Vacuum is applied to this type of unit through hollow trunnions, with suitable packing glands, supporting the dryer. Rotating glands must also be used for admitting and removing the heating medium from the rotating shell. The inside of the shell may have lifting flights to help agitate the material.

Field of Application. Vacuum rotary dryers dry large batches of materials which must be kept out of contact with air or from which the solvents are to be recovered. Materials ranging from sludges to coarse crystals can be handled in this type of dryer if they do not cake severely to the agitator or flights or to the shell. Some grinding will occur with the rotating central agitator; and thus, when breakage or grinding is not permissible, the rotating-shell type of unit is used. Vacuum rotary dryers are not usually recommended unless they are to dry the same material over a long continuous period.

Theory and Design Data. The rate of heat transfer from the heating medium through the dryer wall to the wet material can be expressed by $q = UA\,\Delta t_m$, Eq. (23). Drying can be divided into distinct periods with the material, during the constant-rate period, at substantially the boiling point of the evaporating liquid under the pressure conditions existing in the dryer and approaching the wall temperature during the falling-rate period. The over-all heat-transfer coefficient is almost entirely dependent upon the coefficient between the material and dryer wall and varies with the type of material being handled and the total weight bearing on the drying surfaces, increasing with increasing bulk density, particle size, and total load on the drying surfaces. Over-all coefficients will range from 5 to 35 B.t.u./(hr.)(sq. ft. contact surface)(°F.) if the dryer walls are kept reasonably clean. Coefficients as low as 1 or 2 may be encountered if caking on the walls occurs.

When solvents are to be recovered, the vacuum applied to these dryers is usually limited, since the pressure in the system must be high enough so that the boiling point of the solvent is above the cooling-water temperature; otherwise condensation cannot occur. If solvent is not to be recovered or if water is being removed, vacuums of 28 to 29 in. Hg are usually employed. Air in-leakage through gasketed surfaces and at the bearings will be in the range of 0.15 lb. air/(hr.) (lin. ft. gasketed surface). These dryers are usually charged from 50 to 60 per cent full; with smaller charges, 40 per cent, for very dense materials; and larger charges, 85 per cent, for light fluffy materials. Agitator speeds usually range from 3 to 5 r.p.m. Faster speeds consume greater power but improve the heat transfer slightly. Power requirements are usually approximately equal to $0.2\ LD$, where L = dryer length, ft., and D = diameter, ft.

Auxiliary Equipment. Vacuum rotary dryers require a source of vacuum, which can be a steam jet, a vacuum pump, or a condenser and barometric leg or any combination thereof. If a vacuum pump is used, a condenser should precede it to prevent condensation of vapors in the pump. If a solvent is being removed, it should be recovered by a condenser in the vacuum line. If the solvent is extremely valuable, a vacuum pump should be used and a condenser should be placed on the pressure side of the pump as well as on the vacuum side.

Dust-collection equipment is frequently required with dusty products or if the rotary agitator produces fines. Dust can be recovered in bag filters, cyclones, or wet scrubbers (see Sec. 15). Condensation of the solvent on filter bags may cause trouble unless provision is made to heat the bags or the exhaust vapors. A cyclone on top of the dryer, feeding dust back into the dryer continuously, combined with a wet scrubber is commonly used. Dust should be removed from the vapors before they pass to the condenser; otherwise the condenser surface will foul up rapidly. If a liquid heating medium is used, it is desirable to baffle the jacket or to use nozzles to ensure a high liquid velocity in the jacket and to prevent by-passing and formation of stagnant pockets. Spray nozzles are sometimes installed to facilitate cleaning of the shell if it is to be used for a number of different materials.

The ideal method of controlling vacuum rotary dryers is by the temperature of the material being dried. However, it is difficult to install a thermometer in conjunction with the agitator. Control is usually accomplished by measuring the exit-vapor temperature, which increases rapidly as the drying nears completion.

Performance and Cost Data. Typical performance data (Harcourt, paper presented at Am. Soc. Mech. Engrs. meeting, Niagara Falls, Sept. 17, 1936) for vacuum rotary dryers are given in Table 22. Drying cycles will range from 4 to 48 hr. depending upon the material. Short cycles involve high labor costs for loading and unloading, and long cycles involve inordinately high investment costs. Sizes range from 1.5 ft. diameter by 3.5 ft. long to 6.5 ft. diameter by 36 ft. long, with the heating surface ranging from 17 to 1000 sq. ft. The length is limited by the span that the central shaft can stand without excessive deflection. Drying rates range from 2.5 lb./(hr.)(sq. ft. heated surface) for high moisture materials not carried to complete dryness to 0.1 lb./(hr.)(sq. ft. heated surface) for materials dried to almost complete dryness.

Table 22. Performance Data of Vacuum Rotary Dryers

Material	Diam. × length, ft.	Initial moisture, %	Steam pressure, lb./sq. in.	Agitator speed, r.p.m.	Batch dry weight, lb.	Final moisture, %	Vacuum, in. Hg	Time, hr.	Evaporation, lb. H_2O/(hr.)(sq. ft.)
Cellulose acetate	5 × 30	87.5	14	5.25	1350	8	26.5–27	7	0.3
Starch	5 × 30	45–48	15	4	8000	12	26 –27	4.75	1.2
Sulfur black	5 × 30	50	30	4	7000	1	27	6	0.9

For materials of relatively high moisture content, steam requirements (exclusive of steam for vacuum equipment) will range from 1.3 to 1.8 lb./lb. water evaporated. When drying to almost complete dryness, steam requirements may exceed 5 lb./lb. water evaporated. The thermal efficiency of a well-insulated dryer may exceed 90 per cent. Steam or electrical power requirements for vacuum equipment can be calculated from the amount of gases (primarily leakage if a condenser is used) that must be handled. Water requirements of the condenser can be calculated from the heat loads anticipated.

Representative purchase costs for vacuum rotary dryers, including dry and wet dust collectors, vapor piping, barometric condenser, ejector, speed reducer, drive and motor, in both steel and stainless-steel construction are given in Fig. 58. The installed cost may range from 1.5 to 3 times the purchase cost in ordinary steel. Maintenance costs (including vacuum equipment) will range from 5 to 10 per cent of the installed cost per year. This usually consists of replacing packings and gaskets and realigning or replacing scraper blades if an abrasive material is being handled.

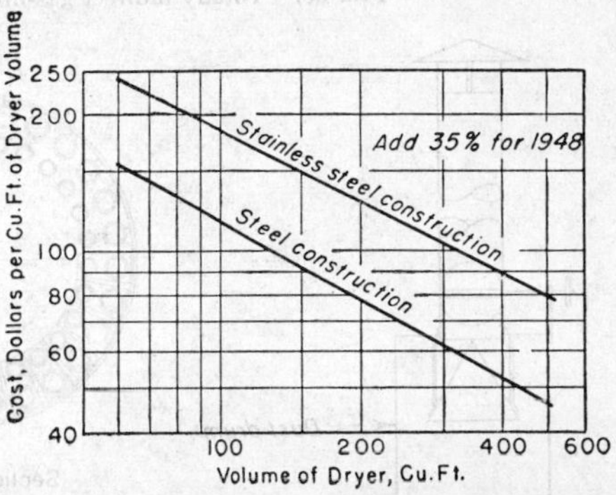

FIG. 58. Cost of vacuum rotary dryers, including dry and wet dust collectors, vapor piping, barometric condenser, ejector, speed reducer drive, and motor. (*Buflovak Equipment Division, Blaw-Knox Co.*)

Vacuum rotary dryer labor costs may vary from 4 to 24 man-hr. per day depending upon the size and number of dryers, facilities for feeding and discharging the dryer, and the drying cycle.

Indirect Rotary Dryers

Description. The indirect rotary dryer is similar mechanically to the direct and indirect-direct rotary dryers, p. 828. It consists of a rotating cylinder inclined to the horizontal, with material fed to one end and removed from the opposite end. Drying, however, is accomplished entirely indirectly, with heat being conducted to the material through the metal shell or tubes, or through cylinders placed within the rotating shell.

The simplest type of indirect rotary dryer consists of a rotating cylinder encased in a brick or steel housing that is also the combustion chamber. The combustion gases may go directly to a stack or pass through a concentric rotating cylinder within the dryer shell for additional heat utilization. Flights are used on the main drying cylinder and on the outside of the inner hot-air duct. Moisture vapor usually leaves at the feed end of the dryer through a stack or an exhauster.

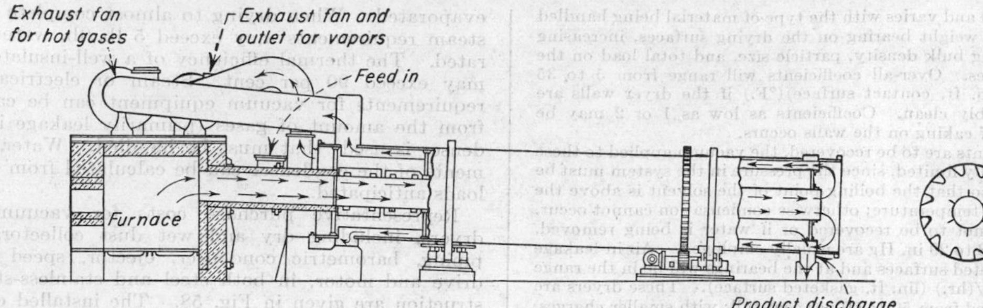

FIG. 59. Totally indirect gas-fired rotary dryer with heated flights. (*Hardinge Co.*)

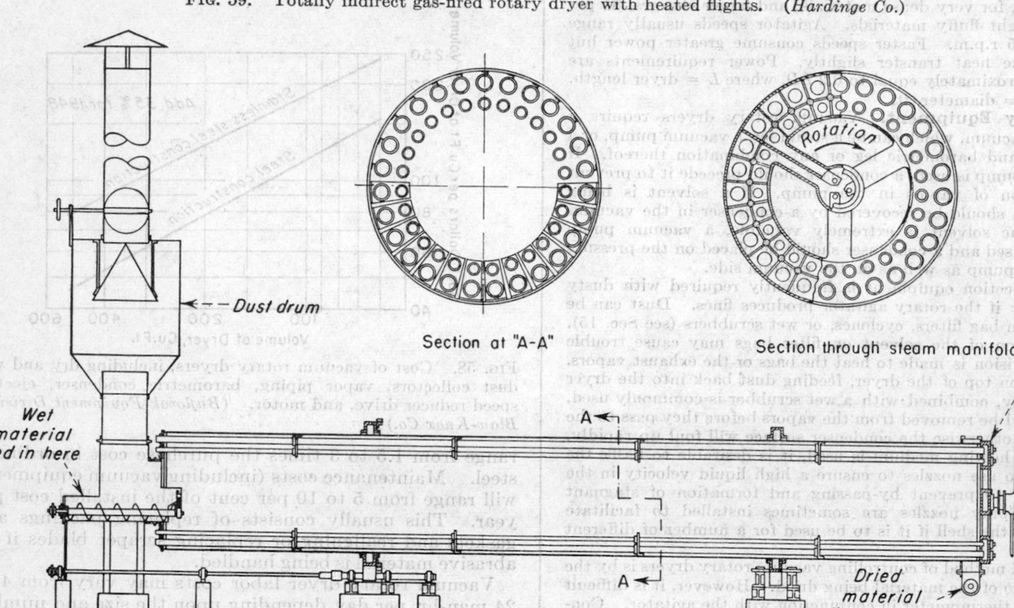

FIG. 60. Steam-tube rotary dryer. (*General American Process Equipment Div., Louisville Drying Machinery Co.*)

In another type, hot gases pass through an inner concentric fluted cylinder and then back through hollow flights attached to the outer cylinder (Fig. 59).

In another type, hot gases pass through *stationary* tubes extending the entire length of the rotating dryer cylinder. Wet material showered by flights attached to the inside of the rotating cylinder falls over the tube. Material and hot gases usually flow countercurrently.

The most common type of indirect rotary dryer is the steam-tube dryer [Bill, *Ind. Eng. Chem.*, **30**, 997 (1938)], Fig. 60. Steam-heated tubes running the full length of the dryer are fastened symmetrically in one, two, or three concentric circles to the inside of the cylinder and rotate with it. With sticky materials, only one row of tubes is used, and the tubes are shielded at the feed end of the dryer to prevent material from caking behind them. Flights are sometimes used to promote agitation of the material.

Wet feed may enter the dryer through a screw conveyor or gravity chute. It passes through the dryer by virtue of the latter's slope and rotation. The product discharges from the dryer through peripheral openings in the shell, since the steam manifold, located at the discharge end of the dryer, prevents removal through the end. To retain a deep bed of material within the dryer, the discharge openings can be equipped with chutes extending radially into the dryer. These can be made removable to allow complete discharge of the dryer. Steam is admitted to the tubes through a revolving packing gland to a large rotating pipe leading to the manifold. Condensate is also removed through this manifold and rotating joint. The tubes are fastened rigidly to the shell at the manifold end of the dryer and are supported simply by a close-fitting annular plate at the other end to permit expansion. The tubes are continuously vented at the far end to prevent the accumulation of non-condensables. Vapors from drying are removed at the feed end of the dryer and vented to the atmosphere by a natural-draft stack or an exhauster.

Field of Application. Indirect rotary dryers are used for the continuous drying of granular or powdery materials that can be dried at high temperatures but cannot stand contamination from products of combustion. They are especially suited to materials that are too dusty to be handled in a direct rotary dryer. The steam-tube dryer is particularly applicable to the handling of dusty materials that must be dried at steam temperatures; and whenever it can be used satisfactorily, it usually proves to be the cheapest and the most economical to operate of all continuous dryers. Indirect rotary dryers cannot be used for materials that tend to stick to hot surfaces, since a coating on the heat-transfer surface rapidly decreases the equipment capacity and may produce an unsatisfactory lumpy product. Rotary steam-tube dryers have also found application to continuous drying and solvent recovery when low pressures are not required.

Theory and Design Data. Heat-transfer coefficients in indirect rotary dryers range from about 2 to 10 B.t.u./(hr.)(sq. ft. heating surface)(°F.) at steam temperatures. Coefficients will increase with increasing temperature of operation because of increased heat transfer by radiation. In steam-tube dryers employing steam at 300°F., the value of $U\,\Delta t$ (based on the total heat-transfer surface in the dryer) in the usual heat-transfer equation will vary from 300 for fine light powdery materials to 2000 for coarse dense granular materials like sand.

The **time of passage,** or hold-up of material, in these dryers can be estimated from Eq. (64), if the rate of rotation and dryer slope are known. The dryer hold-up will generally be 5 to 15 per cent, whereas the direct types of similar design operate with 2 to 10 per cent. Rates of rotation of a steam-tube dryer range from one-third to one-fifth of those for other types of indirect or direct rotary dryers, and dryer slopes will be in the same range as the direct rotary dryers.

Length-to-diameter ratios of these dryers will range from 4 to 10. The amount of heating surface available with the hot-gas indirect rotary dryer can be calculated from a knowledge of the dryer dimensions. Table 23 gives the heating surface available per foot of length for Louisville steam-tube dryers of various sizes and numbers of concentric tube circles.

Table 23. Heating Surface of Steam-tube Dryers*

Dryer size,† diam. × length, ft. × ft.	Number and diam. of tubes, No.—in.	Total heating surface, sq. ft.	Heating surface, sq. ft./ft. length	Free cross-sectional area, sq. ft.
3.17 × 15	(14—4½)	230	16.5	6.3
3.17 × 25	(14—4½)	395	16.5	6.3
3.17 × 35	(14—4½)	560	16.5	6.3
4.5 × 30	(18—4½)(18—2½)	955	33.0	13.3
4.5 × 40	(18—4½)(18—2½)	1285	33.0	13.3
6 × 25	(27—4½)(27—3)	1270	53.0	24.0
6 × 35	(27—4½)(27—3)	1800	53.0	24.0
6 × 45	(27—4½)(27—3)	2335	53.0	24.0
6 × 55	(27—4½)(27—3)	2865	53.0	24.0
8 × 40	(36—4½)(54—3)	3300	84.6	43.0
8 × 60	(36—4½)(54—3)	5000	84.6	43.0

* Courtesy of General American Process Equipment, Louisville Drying Machinery Co.

† Dryer lengths available by 5-ft. increments.

The maximum operating temperature of indirect rotary dryers is about 1200°F., unless special alloy steels are used. Steam-tube dryers seldom operate at temperatures in excess of 350°F. and may operate at temperatures below 212°F. with hot water or vacuum steam. The air velocity through these dryers is very low and is often maintained by a natural-draft stack. When dust collectors are used, however, exhausters remove the moist air at dew points ranging from 80° to 140°F.

Auxiliary Equipment. Auxiliary equipment for these dryers is similar to that for direct rotary dryers. The problem of sealing indirect rotary dryers is usually not as difficult as for direct rotary dryers, since high air velocities and pressure differentials within the cylinder are not involved. **Recirculation of dry product** to permit handling of a sticky or pasty feed may frequently be feasible with this type of dryer, which requires auxiliary equipment.

For dust collection, wet-type collectors are usually used because of the high humidity (80 to 90 per cent) of the exit vapors. Knockers are frequently employed to alleviate sticking to the dryer shell and hot surfaces. Control of the operation of indirect dryers should be based on the product temperature. This method is used to control the feed rate in steam-tube rotary dryers, whereas either the inlet-gas temperature, inlet-gas rate, or the feed rate is controlled in gas-heated indirect dryers.

Performance and Cost Data. An indication of the performance of indirect rotary dryers can be obtained from Table 24, which compares relative capacities of various types of direct and indirect rotary dryers based on equal dryer volume and similar operating temperatures:

Table 24. Relative Capacities of Various Types of Rotary Dryers

Type	Relative Capacity
Direct rotary	1.0
Indirect-direct (double shell)	1.35
Indirect (hot gases)	0.7
Indirect (steam-tube)	3.0

Typical performance and cost data for indirect rotary dryers are given in Tables 25 and 26.

Table 25. Manufacturer's Data on Performance and Cost of Indirect Rotary Dryers*

Data	Double-shell indirect rotary dryer, Fig. 59		Steam-tube rotary dryer	
Material handled	Chalk	Kaolin	Scrap rubber	Corn gluten
Dryer diameter, ft	5	7.5	4	5.83
Dryer length, ft	35	55	30	30
Moisture in feed, % dry basis	38.9	29.9	61.3	100
Moisture in product, % dry basis	0.1	2.1	9.9	13.6
Production rate, lb./hr	3,000	15,000	1,000	1,350
Evaporation rate, lb. water/hr	1,160	4,100	466	1,030
Capacity, lb. water/(cu. ft. dryer volume)(hr.)	1.69	1.68	1.24	1.29
Type of fuel	Oil	Coal	Steam	Steam
Fuel consumed, gal./hr. or lb./hr	19.5	705	805	2,340
Calorific value of fuel, B.t.u./gal. or B.t.u./lb	140,000	13,600	895	880
Efficiency, B.t.u. supplied/lb. water evaporated	2,340	2,400	1,600	2,050
Total power, hp	18	65	6	10
Approx. cost of dryer and auxiliaries, f.o.b. factory†	$12,000	$32,500	$7,200	$12,000
Approx. erection and installation cost	$ 1,300	$ 3,200	$ 600	$ 1,000
Drying cost per ton product‡	$ 0.88	$ 0.60	$ 0.70	$ 1.38
Drying cost per ton water removed	$ 2.30	$ 2.18	$ 1.56	$ 1.84

* Courtesy of the Hardinge Co., July, 1946.

† Standard dryer construction throughout; stainless steel or special alloys will increase the cost 150 to 300 per cent.

‡ Fuel at $0.40/million B.t.u., power at $0.01/hp.-hr. No other operating costs included. Labor: ½ man or less for oil- or gas-fired units; 1 man for coal-fired units. Maintenance labor and materials: $0.01 to 0.03/ton of product for large dryers; $0.05 to $0.15/ton of product for small dryers.

The thermal efficiency of a hot-gas indirect dryer will range from 40 to 65 per cent, and that of a steam-tube dryer will be from 75 to 90 per cent. **Purchase costs** of an indirect rotary dryer will be from 1.5 to 2 times that of a direct single-shell rotary dryer of equal volume. Figure 61 gives data for high-spot estimates of steam-tube rotary dryers. The installed cost may be about double the purchase cost.

Operating costs for these dryers are low, the lowest being for the steam-tube rotary. From 20 to 35 per cent of a man's time is required for operation. Total power, including that for all auxiliaries, will be $0.5D^2$ to D^2 for the hot-gas indirect dryer and $0.3D^2$ to $0.7D^2$ for the steam-tube dryer, where D = dryer diameter, ft. Yearly maintenance should not exceed 10 per cent of the installed cost.

Example 14. It is desired to dry 1500 lb./hr. of a powdery free-flowing organic material coming from a continuous centrifuge at 0.15 lb. water/lb. material and 60°F. to 0.005 lb. water/lb. material. The material cannot be heated above 325°F. without some decomposition occurring. The bulk density of the material is 45 lb./cu. ft., and it has a specific heat of 0.3 B.t.u./(lb.)(°F.). Prepare a high-spot size estimate for a steam-tube rotary dryer to handle this material.

Table 26. Steam-tube Dryer Performance and Cost Data*

	Class 1	Class 2	Class 3
Class of materials handled.	High moisture organic, distillers' grains, brewers' grains, citrus pulp	Pigment filter cakes, blanc fixe, barium carbonate, precipitated chalk	Finely divided inorganic solids, water-ground mica, water-ground silica, flotation concentrates
Description of class.......	Wet feed is granular and damp but not sticky or muddy and dries to granular meal	Wet feed is pasty, muddy or sloppy. Product is mostly hard pellets	Wet feed is crumbly and friable. Product is powder with very few lumps
Normal moisture content of wet feed, % dry basis	233	100	54
Normal moisture content of product, % dry basis	11	0.15	0.5
Normal temp. wet feed, °F.	100–120	50–70	50–70
Normal temp. product, °F.	175–185	225–275	200–225
Evaporation per lb. product, lb..................	2	1	0.53
Heat load per lb. product, B.t.u..................	2250	1190	625
Steam pressure normally used, lb./sq. in. gage....	125	125	125
Heating surface required per lb. product, sq. ft...	1.67	2.00	0.35
Steam consumption per lb. product, lb...........	3.33	1.72	0.85
Power required, kw.:			
Small dryer...........	2	3	3
Medium dryer........	4	5	5
Large dryer..........	6	8	8

* Courtesy of General American Process Equipment Div., Louisville Drying Machinery Co.

Solution. The heat that must be supplied to the material, assuming an exit-material temperature of 250°F. and exit-vapor temperature of 212°F., is

Heat to evaporate water at 212°F. =

$$1500(0.150 - 0.005)[(212 - 60) + 960] = 241{,}860 \text{ B.t.u./hr.}$$

Sensible heat to material =

$$1500(0.3)(212 - 60) = 68{,}400 \text{ B.t.u./hr.}$$

$$\text{Total} = 310{,}260 \text{ B.t.u./hr.}$$

The use of 100 lb./sq. in. steam in the tubes is assumed. From previous experience with similar materials, it is known that a value of $U\,\Delta t$ = 1200 B.t.u./(hr.)(sq. ft.) can be expected from

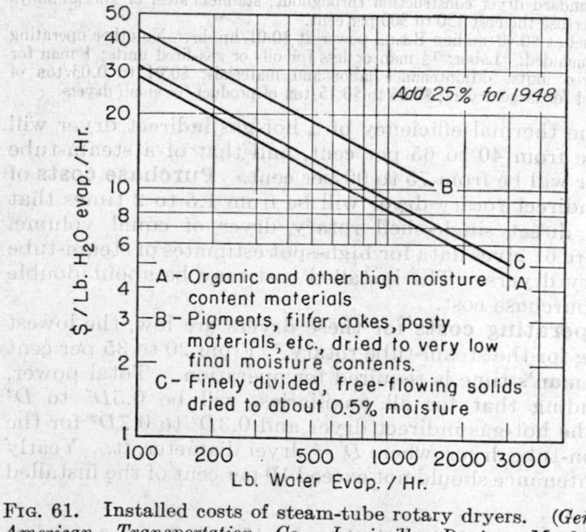

FIG. 61. Installed costs of steam-tube rotary dryers. (*General American Transportation Co., Louisville Drying Machinery Division.*)

a steam-tube dryer. The area required in the dryer is then 310,400/1200 = 259 sq. ft. Table 23 indicates that the nearest size is a 38-in. by 20-ft. steam-tube dryer employing fourteen 4.5-in. tubes with 315 sq. ft. of heating surface. This size is adequate for the job, and its cost may be estimated from Fig. 61.

Screw-conveyor Dryers

Description and Field of Application. The screw-conveyor dryer is essentially a jacketed conveyor in which material is heated and dried as it is conveyed. In one type, the jacket may extend only to the top of the conveyor, which is left open to the atmosphere. This is commonly termed a **trough dryer.** When the jacket encloses the conveyor completely, a slight negative pressure is required to sweep out the evaporated moisture. The heating medium depends upon the temperature sensitivity of the material being dried and may be hot water, steam, or a high-temperature heat-transfer medium such as Dowtherm. Frequently the screw conveyor is also heated. Drying may be accomplished in one or more passes. With a multipass unit, the conveying sections are usually mounted horizontally one above the other with material dropping from one pass to the next lower pass and traveling in a direction opposite to the pass above. Each pass may be individually controlled, and it is not uncommon to operate the last pass as a product cooler.

Screw-conveyor dryers are usually used for granular free-flowing materials. Although this type of dryer has not been widely used, indications are that it should be one of the cheapest methods of continuously drying a granular material. It is convenient to use when the wet material must be conveyed a considerable distance to the next point of processing. It should not be used where crystal structure or size specifications are important unless actual tests indicate that the size degradation occurring during drying is not objectionable.

Dryers of this type have a particular advantage in that they are adaptable to **continuous operation under vacuum.** By careful design, a vacuum as low as 27 to 28 in. Hg is feasible. Because of this feature, screw-conveyor dryers are particularly suited to continuous drying and solvent recovery. For vacuum operation, specially designed charging and discharging ports must be used.

A screw-conveyor dryer may be used as an auxiliary to another type of dryer, such as a direct rotary, to effect an increase in capacity. It has the added advantage of acting as a product conveyor from the rotary while it performs as a dryer.

Auxiliary equipment depends on the particular problem. The feed rate should be kept fairly constant. If the material comes from a batch process, a special feeder may be required. The product from rotary filters, continuous centrifuges, etc., may be fed directly to the screw conveyor. Dust collectors are not usually necessary. It is essential to minimize heat losses by proper insulation. When cooling is required, a coolant must be available, and vacuum operation requires suitable vacuum equipment. In drying pasty materials, recycling of the dry product into the feed may be required to permit suitable handling in the dryer. In such cases, premixing equipment must be used.

Theory and Design Data. The theory involved in drying with this equipment is the same as for vacuum rotary or agitated-pan dryers. For conveyor speeds of 2 to 20 r.p.m., the heat-transfer coefficient will be on the order of 2 to 12 B.t.u./(hr.)(sq. ft.)(°F.). For conveyor speeds up to 200 r.p.m., coefficients as high as 25 to 30 B.t.u./(hr.)(sq. ft.)(°F.) may be obtained. Screw-conveyor dryers may operate from 30 to 70 per cent full, depending upon the loading that can be handled without spilling or binding.

Performance and Cost Data. Screw-conveyor dryers will evaporate from 1 to 3 lb. water/(hr.)(sq. ft.) from materials of relatively high moisture content but will remove only 0.1 to 0.5 lb. water/(hr.)(sq. ft.) when drying only slightly wet materials to complete dryness. Thermal efficiencies range between 50 to 70 per cent for well-insulated dryers. Dryer diameters may be 6 to 24 in., and any desired length may be used if internal bearings can be tolerated. In the large-diameter units, the conveyor is usually heated. Meager data indicate that these dryers in standard iron construction may cost from $7 to $40/sq. ft. of heating surface, depending upon the type of construction employed. Installed costs may be from two to three times the purchase cost. Power for this type of dryer can be calculated from data on screw conveyors in Sec. 20. Each pass may be individually powered, if desired. Labor costs will be low, with one-third to one-fifth of a man's time required. Maintenance will run from 10 to 15 per cent of the installed cost.

Vibrating-conveyor Dryers

This type of dryer consists of a vibrating conveyor with a heated solid deck over which the wet solid moves. This deck can be heated by hot gases, steam, methanol vapors, or Dowtherm vapors which pass through a jacket fastened to the deck upon which the material is conveyed. Direct gas flames can also be used as the source of heat. A hood equipped with an exhaust fan and placed over the deck removes the evaporated liquid. Infrared lamps may be mounted above the deck to increase drying rates.

This type of dryer can be used only for granular or powdery free-flowing materials that can be conveyed without sticking on a vibrating conveyor. It is particularly applicable for continuous drying of materials that must not suffer degradation and that require a high degree of uniformity in the finished product. It will handle fine powders without excessive dusting. Slow-drying materials usually require excessively large equipment.

These dryers may be built in stages with several jackets that can be individually heated and controlled. In some cases, several conveyors may be built in a tier, the product passing from one conveyor to the next below. Such an arrangement is required for materials requiring a long contact time.

Few auxiliaries are needed with this type of dryer. A spreading feed is required to load the material on the vibrating deck. Water-cooled sections are often provided immediately after the drying sections to cool the material to a temperature suitable for packaging.

Heat-transfer coefficients obtained with this equipment should be slightly better than those obtained with screw-conveyor dryers. The thermal efficiency will be 40 to 60 per cent when hot air is used as a heating medium, which can be recirculated 100 per cent through the jackets, and 60 to 90 per cent when using condensing vapors. Labor involved should not exceed one-third of a man's time.

Drum Dryers

Description and Application. Drum drying consists of applying a liquid material, solution, slurry, or paste, to a revolving heated metal drum which **conducts** heat to the wet film to evaporate the water during a partial revolution of the drum. The dry material is scraped from the drum by a stationary knife. Drum dryers, may be designated as: (1) the atmospheric double-drum dryer, (2) the atmospheric single-drum dryer, (3) the atmospheric twin-drum dryer, (4) the vacuum single-drum dryer, and (5) the vacuum double-drum dryer.

Selecting a drum dryer involves choosing the type best suited to the problem, and the best method of applying the feed liquid. The primary bases for the selection are the quality and physical appearance of the dry product, and the capacity, lb. dry product/(hr.)(sq. ft. drum surface), which involves the rate of evaporation and heat transfer.

In drum drying, the product is exposed to heat for only short periods of time. This has the advantage that, although the product may approach the temperature of the drum surface, there usually is no adverse effect from overheating.

The **atmospheric double-drum dryer** has found wide application because of its reasonable cost, flexibility of operation, automatic control of the thickness of the film on the drums, and lack of residue of feed liquid at the end of the operating period.

A typical atmospheric double-drum dryer with a **perforated-pipe feed** is shown in Fig. 62. The direction of rotation of the

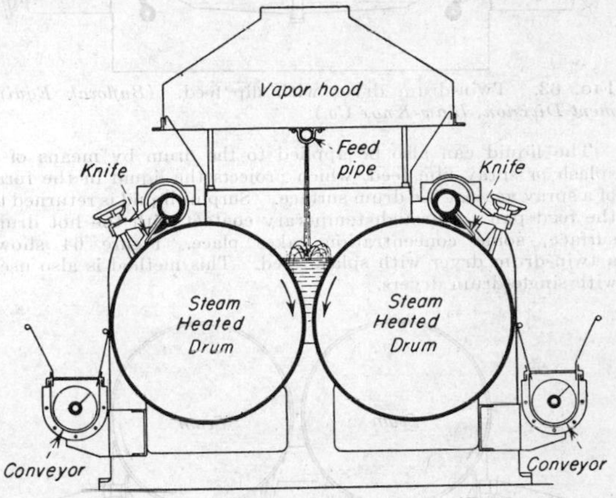

FIG. 62. Double-drum dryer with pipe feed. *(Buflovak Equipment Division, Blaw-Knox Co.)*

double-drum dryer is downward, toward the nip of the rolls. **End boards**, faced with a material suitable for the liquid being dried, confine the liquid between the drums. The facing must have good wear- and heat-resisting properties, and materials such as canvas, bakelite, wood, and metals are used, mounted on a rigid backing of galvanized steel. A single adjustment is commonly used to hold the end boards against the ends of the drums, although a three-point adjustment is sometimes used.

The reservoir of liquid between the drums is heated to the boiling point and partly concentrated by the hot drum surface. Since the film of liquid adhering to the drum surfaces must pass through the narrow clearance between the drums, the liquid is evenly distributed. The thickness of the film is regulated by adjustment of this clearance.

Liquid is distributed between the drums by a perforated pipe, by a trough with serrated edges, or by a swinging pipe suspended in the vapor hood. Control of the feed is obtained by means of a rotary or gear pump with variable-speed drive or by a centrifugal pump with an automatic feed valve controlled by the level between the drums. **Level control** stabilizes dryer operation, maintains more uniform concentration between the drums, and reduces fluctuations in the moisture content of the dry material, thus decreasing wear of the knives.

The atmospheric double-drum dryer is particularly suited to drying dilute solutions, solutions near the saturation point if the crystals that separate out are not abrasive, concentrated solutions of highly soluble products, and fairly heavy sludges that can be pumped. It is less adaptable to handling solutions of inorganic salts of limited solubility, or slurries that tend to settle out and cause undue pressure between the drums.

The atmospheric **single-drum dryer** and the **twin-drum dryer**, the latter having two drums turning in a direction opposite to the double-drum dryer, operate on the same principle. The feed liquid is contained in a separate reservoir located below or on the side of the drum or drums. The simplest feed method for a twin-drum dryer is the dip feed (Fig. 63) in which the drum takes up a liquid film by dipping into a reservoir of the liquid under the drum. The liquid in the reservoir is concentrated to a considerable extent, resulting in separation of the solid component if solubility is limited. To minimize such concentration, either a single or double transfer roll may be used to transfer the liquid to the drum surface. Such a roll also may be placed near the top of the drum to eliminate the feed pan and still permit complete drying of the liquid, the same as with a double-drum dryer. These rolls introduce a number of complications, require careful adjustment, and frequently become coated.

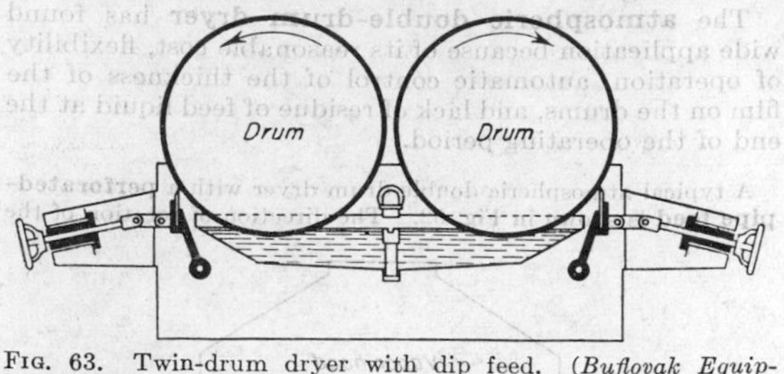

FIG. 63. Twin-drum dryer with dip feed. (*Buflovak Equipment Division, Blaw-Knox Co.*)

The liquid can also be applied to the drum by means of a splash or spray film feed, which projects the liquid in the form of a spray against the drum surface. Surplus liquid is returned to the feed pan. Through temporary contact with the hot drum surface, some concentration takes place. Figure 64 shows a twin-drum dryer with splash feed. This method is also used with single-drum dryers.

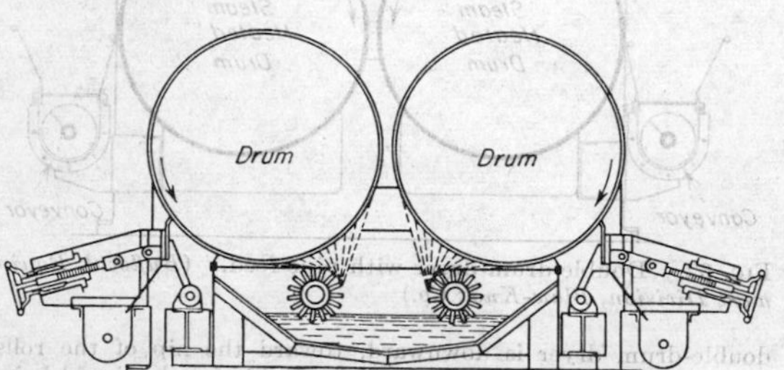

FIG. 64. Twin-drum dryer with splash feed. (*Buflovak Equipment Division, Blaw-Knox Co.*)

In all these various feed methods the principal objective is to apply a uniform film. This is frequently difficult, and special devices such as spreader bars and rolls, which create additional problems, are sometimes used to overcome this difficulty. The advantage of the twin-drum dryer over the single-drum dryer is that the two drums assist in confining the spray and that the drums can be set close enough together to make the coating more uniform. With a top feed, as used for a double-drum dryer (Fig. 62) a heavy film obtains, which is rarely uniform because of boiling and bubbling of the liquid.

Single- and twin-drum dryers find their greatest application in drying slurries and concentrated inorganic salt solutions which are sufficiently fluid. In drying slurries, these dryers preserve the particle size of the suspended solid better than the double-drum dryer.

Vacuum drum dryers operate on the same principle as the corresponding atmospheric dryers. They are used for heat-sensitive materials, for products subject to oxidation, or where solvents have to be recovered. Such dryers are enclosed in a vacuum-tight casing. Figure 65 shows a vacuum single-drum dryer with a patented spray film feed. The dried material, removed from the drum by a knife, falls into a screw conveyor which delivers it into receivers. Large dryers are supplied with two receivers so that the vacuum in one may be broken and its contents discharged while the other is being filled under vacuum.

The operation of vacuum drum dryers is complicated by the vacuum condition. The adjustment of knives, feed, spreaders, etc., inside the dryer must be made from the outside by means of continuous connections passing through vacuum-tight stuffing boxes. Observation glasses permit a view of the operation in the vacuum chamber.

A special type of drum dryer suitable for drying pastes is a **finned** single-drum dryer. In this dryer the drum is grooved around the circumference by machining on a lathe. The fins thus formed are from ¼ to ⅜ in. apart and about 5⁄16 in. deep. and substantially increase the heating surface. The wet material is fed into the grooves by means of a pasting roll, and the dried material is removed by finger scrapers that fit between the fins. The final product is usually in the form of short sticks rather than a fine powder. This type of dryer is usually used as a preforming machine for a continuous through-circulation dryer (see p. 823) and is seldom used for complete drying.

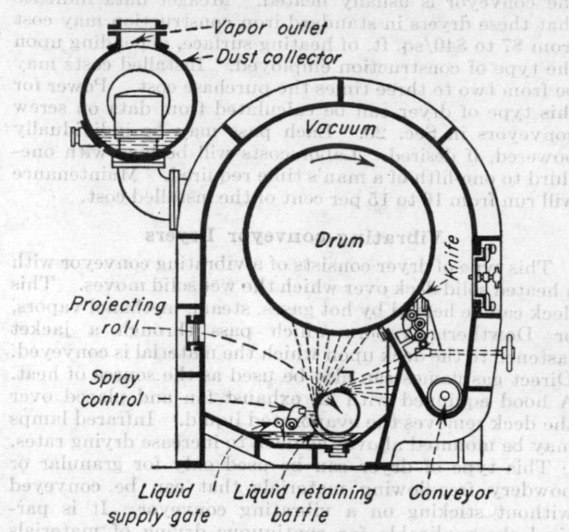

FIG. 65. Vacuum single-drum dryer showing patented spray film feed. (*Buflovak Equipment Division, Blaw-Knox Co.*)

The following table gives a partial list of materials to which drum drying is applicable:

Double-drum dryers	Single-drum dryers	Twin-drum dryers
Aluminum hydrate	Barium sulfate	Barium carbonate
Buttermilk	Calcium acetate	Barium sulfate
Cereals	Calcium arsenate	Calcium acetate
Colloidal clay	Calcium carbonate	Calcium carbonate
Colors	Calcium hydrate	Calcium phosphate
Dextrines	Calcium phosphate	Clay
Distillery slop	Carbonate sludge	Disodium phosphate
Dye pastes	Caustic mud	Glue
Intermediates	Clay	Kaolin
Lithopone	Ferrous hydrate	Lime sludge
Milk	Glue	Mineral salts
Potatoes	Iron oxide	Sodium benzene sulfonate
Sodium acetate	Kaolin	Sodium carbonate
Sodium sulfate	Lead arsenate	Sodium chloride
Sodium sulfonates	Lime sludge	Sodium sulfate
Yeast	Ocher	Trisodium phosphate
	Resins	
	Sulfite waste	
	Vegetable glue	
	Zinc oxide	

Theory and Design Data. Drum drying involves heat transfer from condensing steam through a metal wall of considerable thickness to a thin layer of material from which liquid is vaporized at or near its boiling point. The over-all rate of heat transfer varies greatly, largely because of varying resistance to the removal of liquid as the film becomes dry. The heat-transfer rate also varies greatly from one material to another. The principal resistances to heat transfer in drum drying are indicated by Eq. (21) under Theory and Fundamental Concepts.

The maximum rate of evaporation is obtained with dilute solutions that evaporate readily. Published data on film coefficients for drum dryers are meager. Very favorable conditions may give rise to an evaporation rate as high as 18.5 lb. water/(hr.)(sq. ft. drum surface). In general, the temperature and the heat-transfer coefficient around the circumference of the drum are not uniform.

Roeser and Mueller (*Bur. Standards Research Paper* 231, 1930) measured the material temperature for the drum drying of milk and found that the temperature of the milk film near the point of removal was within 3°F. of the metal temperature.

Only at very high rates of evaporation does the steam-side coefficient become important. In such cases, the removal of condensate and noncondensable gases must receive special attention. Contrary to similar heat-transfer problems, the resistance to heat flow through the metal wall is a relatively significant factor.

Heat-transfer coefficients are difficult to determine because it is practically impossible to establish an accurate temperature difference. The temperature of both the film and the drum surface varies around the circumference of the drum, and their difference reaches a minimum near the knife that removes the film.

Van Marle [*Ind. Eng. Chem.*, **30**, 1006 (1938)] reported that, under optimum conditions, the *over-all* heat-transfer coefficient on a double-drum dryer (Fig. 62) is about 360 B.t.u./(hr.)(sq. ft.)(°F.) in the area between the drums and about 220 between the point of closest approach of the drums and knives. Under other conditions, the heat-transfer coefficient may fall as low as 30.

In order to determine the size of a drum dryer for a given production, it is essential to conduct actual drum drying tests. Depending on the nature of the product and the concentration of the liquid, capacities usually vary between 1 and 10 lb. product/(hr.)(sq. ft. drying surface). Capacity, with very few exceptions, is proportional to the active drum area. Commercial dryers generally operate at somewhat higher capacity than small experimental dryers because of steady operation over longer periods. Short tests on laboratory-size drum dryers give a fairly good indication of the capacity that may be expected and also provide an opportunity to determine the physical characteristics of the dry product.

Since the success of drum drying depends on the application of a *uniform film of maximum thickness* to the drum surface, concentration of the feed liquid should be as high as possible as long as it does not interfere with film uniformity.

The temperature of the heating medium and the drum speed affect the thickness of the film. In most cases, thickness increases with temperature up to the point where the hot drum begins to repel the liquid. This increase in thickness in most cases is more than would be expected from the increase in temperature difference. Higher temperatures also permit an increase in drum speed, which provides a further increase in capacity.

Greater drum speed has a less pronounced effect because the film is applied at a higher rate, tending to reduce its thickness. At very low speeds the increase in capacity is practically proportional to the increase in drum speed, while at intermediate speeds it is much less. Doubling the speed may increase the capacity only about 50 per cent. At higher speeds the capacity hardly changes.

Auxiliary Equipment. In addition to the standard equipment required with all drum dryers, certain items of auxiliary equipment may frequently be required as follows:

1. Variable-speed drives to regulate and vary the drum speed.
2. Adjustable knife for various products.
3. Controls to maintain a constant liquid level, either between the rolls or in the dip tank.
4. Air circulation over the drums, in addition to vapor hoods, to ensure positive removal of vapors.
5. Dust collectors following hoods for those cases in which the product loses a large percentage of fines in the exhausted vapors.
6. Sometimes a twin-drum dryer is designed to operate as a double-drum dryer by providing a double set of knives and provision for reversing the direction of rotation.
7. Two-stage drum drying, as developed by Buflovak, consists of partial drying on a twin-drum dryer which discharges the paste or slurry to a double-drum dryer located directly underneath it.

Performance and Cost Data. Table 27, due to Harcourt (Am. Soc. Mech. Engrs., meeting, Niagara Falls, Sept. 17–23, 1936) presents performance data for double-drum, twin-drum, single-drum, and vacuum single-drum dryers. This table indicates that the capacity of a drum dryer is affected greatly by the final moisture content of the product. It is frequently difficult to obtain excessively low moisture contents from drum dryers, and to do so may require installation of greatly increased amounts of drying surface in proportion to the slightly increased drying obtained. In such cases, drum dryers have been used in series with some other type; *e.g.*, a rotary steam-tube dryer, which is more suited for drying to low moisture contents. One such arrangement has tripled the capacity of a drum dryer.

The power consumption for drum dryers depends on the dryer size, drum speed, and product handled which involves the tension on the knives and the end boards. A safe estimate of power for atmospheric double-drum dryers is 0.67 hp./(r.p.m.)(100 sq. ft. drum surface).*

The heat requirements for drum dryers involve latent heat of evaporation, sensible heat to the dry product and liquid, and in some instances heats of solution and dehydration of crystals. Heat losses depend on dryer size, drying temperature, and the percentage of bare drum between knife and point of feed, and they will be large if the initial and final moisture contents are low. In general, however, drum dryers are more efficient than direct dryers. Steam consumption should seldom exceed 2.0 lb. steam/lb. water evaporated and will usually be about 1.5 lb./lb. water evaporated.

The approximate **cost of atmospheric double-drum dryers** of the type shown in Fig. 62 is shown graphically in Fig. 66. Curve A is based on cast-iron drums and

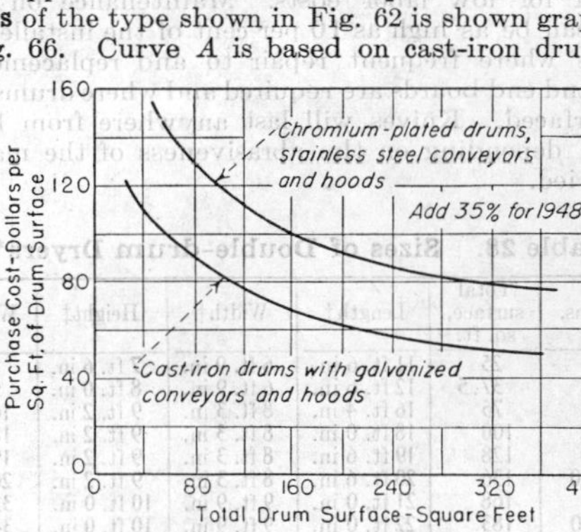

FIG. 66. Approximate purchase costs for atmospheric double-drum dryers. (*Buflovak Equipment Division, Blaw-Knox Co.*)

galvanized-iron conveyors and vapor hoods, and curve B is for chromium-plated drums and stainless-steel conveyors and vapor hood. Dryer drums are seldom fabricated of stainless steel.

In order to estimate double-drum dryers to the nearest standard size, Table 28 gives sizes of Buflovak double-drum dryers for use in conjunction with Fig. 66 on costs.

The operating costs of drum dryers consist of fuel, power, labor, and cooling water for vacuum operation. Fuel can be estimated from the heat requirements, as noted above. Power consumption for drum dryers, although varying for different products, does not represent

* D. J. Van Marle, private communication.

Table 27. Performance Data for Drum Dryers

Material	Method of feed	Moisture content, % wet basis		Steam pressure, lb./sq. in.	Drum speed, r.p.m.	Feed temp., °F.	Capacity, lb. product/(hr.)(sq. ft.)	Vacuum, in. Hg
		Feed	Product					
Double-drum dryer:								
Sodium sulfonate	Trough	53.6	6.4	63	8½	164	7.75	
Sodium sulfate	Trough	76.0	0.06	56	7	150	3.08	
Sodium phosphate	Trough	57.0	.9	90	9	180	8.23	
Sodium acetate	Trough	39.5	.44	70	3	205	1.51	
Sodium acetate	Trough	40.5	10.03	67	8	200	5.16	
Sodium acetate	Trough	63.5	9.53	67	8	170	3.26	
Single-drum dryer:								
Chromium sulfate	Spray film	48.5	5.47	50	5	...	3.69	
Chromium sulfate	Dip	48.0	8.06	50	4	...	1.30	
Chromium sulfate	Pan	59.5	5.26	24	2½	158	1.53	
Chromium sulfate	Splash	59.5	4.93	55	1¾	150	2.31	
Chromium sulfate	Splash	59.5	5.35	53	4¾	154	3.76	
Chromium sulfate	Dip	59.5	4.57	53	5¾	153	3.36	
Vegetable glue	Pan	60–70	10–12	20–30	6–7	...	1–1.6	
Calcium arsenate	Slurry	75–77	0.5–1.0	45–50	3–4	...	2–3	
Calcium carbonate	Slurry	70	0.5	45	2–3	...	1.5–3	
Twin-drum dryer:								
Sodium sulfate	Dip	76	0.85	55	7	110	3.54	
Sodium sulfate	Top	69	.14	60	9½	162	4.27	
Sodium sulfate	Top	69	5.47	32	9½	116	3.56	
Sodium sulfate	Splash	71	0.10	60	6	130	4.30	
Sodium sulfate	Splash	71.5	.17	60	12	140	5.35	
Sodium sulfate	Splash	71.5	.09	60	10	145	5.33	
Sodium phosphate	Splash	52.5	.59	58	5½	208	8.69	
Sodium phosphate	Dip	55	.77	60	5½	200	6.05	
Sodium sulfonate	Top	53.5	8–10	63	8½	172	10.43	
Vacuum single-drum dryer:								
Extract	Pan	59	7.75	35	8	...	4.76	27.9
Extract	Pan	59	2.76	35	6	...	1.92	27.7
Extract	Pan	59	2.09	36	4	...	1.01	Atm.
Extract	Pan	56.5	1.95	35	7½	...	3.19	27.5
Extract	Pan	56.5	1.16	50	2½	...	0.75	Atm.
Skim milk	Pan	65	2–3	10–20	4–5		2.5–3.2	
Malted milk	Pan	60	2	30–35	4–5		2.6	
Coffee	Pan	65	2–3	5–10	1–1½	...	1.6–2.1	
Malt extract	Spray film	65	3–4	3–5	0.5–1.0	...	1.3–1.6	
Tanning extract	Pan	50–55	8–10	30–35	8–10	...	5.3–6.4	
Vegetable glue	Pan	60–70	10–12	15–30	5–7		2–4	

a major item. The approximation given above should be adequate. Labor requirements for drum dryers should not be more than ½ man per dryer, if proper controls are provided. In particular, liquid-level controllers are essential for low labor costs. Maintenance on drum dryers can be as high as 10 per cent of the installed cost in cases where frequent repair to and replacement of knives and end boards are required and where drums must be resurfaced. Knives will last anywhere from 1 to 6 months, depending on the abrasiveness of the material being dried.

Table 28. Sizes of Double-drum Dryers*

Drum dimensions, in.	Total surface, sq. ft.	Length†	Width†	Height‡	Weight§
24 × 24	25	11 ft. 6 in.	6 ft. 9 in.	7 ft. 6 in.	8,500
24 × 36	37.5	12 ft. 6 in.	6 ft. 9 in.	8 ft. 0 in.	9,200
32 × 52	75	16 ft. 4 in.	8 ft. 3 in.	9 ft. 2 in.	16,800
32 × 72	100	18 ft. 0 in.	8 ft. 3 in.	9 ft. 2 in.	18,400
32 × 90	128	19 ft. 6 in.	8 ft. 3 in.	9 ft. 2 in.	19,600
32 × 100	136	20 ft. 6 in.	8 ft. 3 in.	9 ft. 2 in.	20,500
42 × 90	168	21 ft. 0 in.	9 ft. 9 in.	10 ft. 0 in.	32,500
42 × 100	183	22 ft. 0 in.	9 ft. 9 in.	10 ft. 0 in.	34,000
60 × 144	377	26 ft. 0 in.	13 ft. 6 in.	14 ft. 0 in.	60,000

* Courtesy of Buflovak Division of Blaw-Knox Co.
† Approximate, does not include working space.
‡ To top of vapor hood.
§ Approximate pounds.

Cylinder Dryers

Description. Cylinder dryers, sometimes called "can" dryers or drying rolls, are differentiated from drum dryers in that they are used for materials in **continuous-sheet** form, whereas drum dryers are used for materials in a liquid or paste form. Cylinder dryers may consist of one large cylindrical drum, such as the so-called Yankee dryer, but more often they comprise a number of drums arranged so that a continuous sheet of material may pass over them in series. Typical of this arrangement are Fourdrinier-paper-machine dryers, cellophane dryers, slashers for textile piece goods and fibers, etc.

The construction of the individual cylinders or drums is similar in most respects to that of drum dryers. Special designs are used to obtain uniform distribution of steam within large drums where uniform heating across the drum surface is critical.

Multiple cylinders are arranged in various ways. Generally they are staggered in two horizontal rows, one above the other, such that the upper rolls are placed above the spaces between the lower. In any one row, the cylinders are placed close together. The wet sheet material contacts the undersurface of the lower rolls and passes over the upper rolls, contacting over half the total amount of drum surface. The cylinders may also be arranged in a single horizontal row, in more than two horizontal rows, or in one or more vertical rows. When it is desired to contact only one side of the sheet with the drying cylinder, unheated guide rolls are used to conduct the sheeting from one cylinder to the next. For sheet materials that shrink on drying, it is frequently necessary to drive the cylinders at progressively slower speeds through the dryer. This requires elaborate individual electric drives on each cylinder.

Cylinder dryers usually dry under atmospheric pressure. However, the Minton paper dryer [*Tech. Assoc. Pulp Paper Ind.*, **7**, 1 (1924)] is designed for operation under vacuum. The drying cylinders are usually heated by steam, but occasionally single cylinders may be gas-heated as in the case of the Pease blueprinting machine. Humid air around the surface of the sheeting is generally removed by natural draft furnished by hoods placed over the dryer cylinders. In some cases, however, provision is made for blowing atmospheric or heated air across a sheet to increase the drying rate. In pulp and paper dryers the sheet is frequently held in contact with the rolls by an endless belt that absorbs the moisture from the sheet and is itself dried.

Single-cylinder dryers are used for drying paperboard and similar sheet materials. The Yankee dryer is a large single-cylinder dryer designed for drying paper to give it a machine-glazed finish.

Application. Cylinder dryers are applicable to almost any form of sheet material that is not injuriously affected by contact with steam-heated metal surfaces. They are used chiefly where the sheet possesses certain properties, such as a tendency to shrink, or lacks the mechanical strength necessary for most types of continuous-sheeting air dryers. Specifically, cylinder dryers are used to dry films of various sorts, paper pulp in sheet form, paper sheets, paperboard, textile piece goods and fibers, etc. They are also used in some cases to impart a special finish to the surface of the sheet.

Theory and Design Data. Upon contacting the drying cylinders, the sheet material is first heated to an equilibrium temperature somewhere between the wet-bulb temperature of the surrounding air and the boiling point of the liquid under the prevailing total pressure. The various resistances to heat transfer in this type of dryer are given by Eq. (21). The resistance of the vapor layer between the sheet and the dryer may be quite significant. The greater the relative movement of air across the sheeting the nearer will the equilibrium temperature be to the wet-bulb temperature of the air with a resultant increase in drying rate. After the sheet is dried below the critical moisture content and internal moisture flow controls, the temperature of the sheet rises.

For drying a paper stock, 120 lb. weight, the **critical moisture content** was shown to be between 0.25 to 0.30 lb. water/lb. dry stock, the initial being 1.50 [Baxter, *Paper Ind. and Paper World*, **21**, 236 (1939)]. For a 70-lb. book paper, the critical moisture was 0.10 lb. water/lb. dry stock with 1.50 initially. In the first case, a constant rate existed over 12 out of 16 cylinders, and, for the second, the rate was constant for 19 out of 27 cylinders. Burstein [*Paper Trade J.*, **122** (10), 35 (1946)] compared critical moisture contents obtained in drying various papers under oven and machine conditions.

High air velocity across the sheet is effective in reducing the drying time and may be obtained by the use of air jets directed at the surface of the sheet. A high sheet speed also results in a high relative velocity between the sheet and the air. However, the effect of this relative air velocity is not so great on sheet materials as it is in the drying of materials in trays.

The value of the coefficient of heat transfer from steam to sheet is determined by the conditions prevailing on the inside and on the surface of the dryers. Low coefficients may be caused by: (1) poor removal of air or other non-condensables from the steam in the cylinders, (2) poor removal of condensate, (3) accumulation of oil or rust on the interior of the drums, and the accumulation of a fiber lint on the outer surface of the drums. In a test reported by Lewis *et al.* (*Pulp Paper Mag. Can.*, February, 1927, p. 22) on a sulfite-paper dryer, in which the actual sheet temperatures were measured, a value of 33 B.t.u./(hr.)(sq. ft.)(°F.) was obtained for the coefficient of heat flow between the steam and the paper sheet. Burstein [*Paper Trade J.*, **122** (10), 35 (1946)] showed that the rate of drying paper on a machine increases linearly with surface temperature up to 212°F., above which further temperature increase results in a lesser increase in rate.

Performance and Design Data. The satisfactory performance of a cylinder dryer requires a study of both the drying conditions and the mechanical operation. The principal drying conditions to be considered include the temperature of the rolls, air temperature and velocity, the sheet speed, the type of material, and the moisture contents of feed and product.

Production of a uniformly dried sheet is one of the most difficult problems in drying on cylinder dryers, since a thin sheet is particularly sensitive to drying conditions and to any **non-uniform temperatures** on the rolls because of its small mass per unit area exposed. Uneven drying generally results in a poor quality of product. Another problem is encountered when uneven initial moisture distribution occurs, since such non-uniformity generally persists throughout the dryer and thus affects adversely the quality of the dry sheet. This effect is eliminated, to some extent, in paper dryers by employing a felt to hold the sheet material against the rolls and to absorb excess surface moisture from the material.

The **size of commercial cylinder dryers** covers a wide range. The individual rolls may vary in diameter from 2 to 6 ft. and up to 20 ft. in width. In some cases, the width of rolls decreases throughout the dryer in order to conform to the shrinkage of the sheet. A single-cylinder dryer, such as the Yankee dryer, generally has a diameter between 9 and 15 ft. Cylinder dryers for paper and pulp generally contain from 10 to 40 rolls.

The **capacity** of cylinder dryers is not easy to estimate without a knowledge of the sheet temperature, which, in turn, is difficult to predict. For a number of tests on

Table 29. Test Data on Newsprint and Bond-paper Dryers*

Data	Test								
	A	*B*	*C*	*D*	*E*	*F*	*G*	*H*	*I*
Paper	News	News	News	News	News	Bond	Bond	Bond	Bond
Cylinder diameter, ft.	6	4	4	4	3	3	3	......	3
Number of dryers	37	34	28	30	19	20	27	......	20
Effective surface, sq. ft. (approx.)	5,760	3,700	2,410	3,210	1,100 (?)	625	1,080	......	1,050 (?)
Sheet speed, ft./min.	1,000	690	623	505	286	155	100	60	
Absolute humidity, room air, lb./lb. dry air	0.0074	0.0147	0.0118	0.0108	0.0163	0.0100	0.0114	0.013	0.005
Absolute humidity, exhaust air, lb./lb. dry air	.114	.0536	.080	.0429	.0401	.0304	.035	.022	.012
Room-air temp., °F.	77.1	84.1	84	75	93	82	89	87	75
Exhaust-air temp., °F.	132	109.3	126	105	113	113	113	100	89
Feed moisture content, lb. water/lb. dry paper	2.375	2.52	2.645	2.66	1.895	1.940	1.652	2.00	2.027
Product moisture content, lb. water/lb. dry paper	0.095	0.081	0.092	0.093	0.086	0.063	0.037	0.025	0.031
Production rate, lb./hr., dry paper	6,810	5,010	3,457	3,270	1,037	606	1,017	926	1,028
Evaporation rate, lb. water evaporated/hr.	15,520	12,200	8,830	8,400	1,875	1,137	1,650	1,830	2,050
Air rate, lb. dry air/hr. (calculated)	145,600	316,000	129,300	259,000	79,000	56,000	70,000	204,000	292,000
Steam consumption, lb./hr.	29,030	25,214	11,850	12,140	2,793	1,422	2,630	2,783	5,055
Efficiency, lb. steam/lb. water evaporated	1.87	2.07	1.34	1.45	1.49	1.27	1.59	1.52	2.46
Steam temp., °F.	223	250	341	230	341	233	240.5	187	252
Unit air rate, lb. dry air/lb. water evaporated (calculated)	9.4	25.7	14.7	30.9	42	49.2	42.4	111	143
Sensible heat of exhaust air above room temp., B.t.u./lb. water evaporated	134	168	160	240	218	394	264	375	520
Sensible heat of room air above 50°F., B.t.u./lb. water evaporated	66	228	130	201	469	410	429	1,070	930
Average rate of drying, lb./(hr.)(sq. ft.) (effective surface)	4.83	3.3	3.66	2.61	1.7 (?)	1.75	1.53		1.95 (?)

* *Tech. Assoc. Pulp Paper Ind.*, **5**, 32 (1922); **6**, 112 (1923); **7**, 182 (1924).

paper dryers, the over-all average drying rate expressed as lb. product/(hr.)(sq. ft. effective surface) is shown in Table 29 together with other performance data. In another correlation of capacity data from a large number of plant installations of paper dryers, it was shown that steam temperature is the largest single factor affecting capacity (Table 30). These over-all evaporation rates are based on the total surface area of the dryers and cover a range of sheet speeds between 500 and 1250 ft./min. In general, evaporative rates above 2.2 lb. water/(hr.)(sq. ft.) occurred in installations equipped with some special system for circulating air around the cylinders.

Table 30. Effect of Steam Temperature on Capacity of Cylinder Dryers

Type of sheet	Steam temp., °F.	Over-all capacity, lb. water evaporated/(hr.)(sq. ft. total surface)
Newsprint	220	1.5
	260	2.7
Kraft paper	220	0.9
	280	2.7
Paperboard	220	0.7
	300	2.4

The capacity of a cylinder dryer may be increased appreciably by the use of air jets directed at the face of the sheet across its width. High air velocities tend to reduce the air-film thickness on the surface of the sheet; but at the same time, unless hot air is used, the temperature of the sheet will be reduced, nullifying to some extent the benefit of reduced film thickness. Such air jets should be directed only at the wet sheet and not at the bare surface of the cylinder, which would cause undue heat losses. The results of a survey of paper drying published by the Technical Association of the Pulp and Paper Industry include performance data from a number of mills on newsprint and writing paper [*Tech. Assoc. Pulp Paper Ind.*, **17**, 210 (1934); **19**, 199 (1936)], and miscellaneous papers [*Paper Trade J.*, **103**, (3) 26 (1936)]. Data on the performance of board and kraft papers have been assembled by Stamm [*Tech. Assoc. Pulp Paper Ind.*, **15**, 208 (1932)] and by Montgomery [*Paper Trade J.*, **96**, 11, 32 (1933); **108**, 11, 44 (1939)].

In general, about 60 to 70 per cent of the total cylinder surface is in effective contact with the sheet. For paper dryers, about 50 to 70 per cent of the total length of the sheet contacts the cylinders.

The sheet speed ranges from 100 to 1300 ft./min. on dryers handling paper and pulp sheeting. The speed of rotation of dryer rolls is varied as shrinkage takes place to eliminate undue strain on the sheet. No data are available to show the exact effect of sheet speed on capacity.

The thermal efficiency of cylinder dryers is about 55 to 70 per cent for normal atmospheric conditions; or, as shown by Table 29, steam consumption may range from 1.34 to 2.5 lb./lb. water evaporated. The steam consumption will vary with the atmospheric air temperature, being higher in the winter months when low outside temperatures occur. The figures for steam consumption in Table 29 are for winter months.

Operating cost data for these dryers are meager. Power costs may be estimated by assuming 1 hp./cylinder for diameters of 4 to 6 ft. Data on labor and maintenance costs are also lacking.

Several miscellaneous operating and design items should be noted as follows: (1) air vents and adequate condensate removal should be provided for on each drum; (2) air vents should be closed before shutdown to prevent air being sucked in as the rolls cool, which would otherwise have to be expelled and would delay start-up; (3) close temperature control of the heating medium in the cylinders should be provided; and (4) the surface of the rolls should be maintained clean.

Infrared Dryers

Description and Application. Infrared dryers operate on the principle of radiant-heat transfer. This heat may be generated electrically by means of incandescent lamps or by gas-heated incandescent refractories. In general, the electrically heated radiating source will be at color temperatures around 2500°K., which distinguishes such radiation from the low-wave-length infrared rays obtained from steam-heated surfaces or other surfaces at less than 800°K. Infrared dryers utilizing electric lamps are usually arranged like a tunnel, formed by banks of lamps, through which a conveyor carries the material to be dried.

The shapes of such tunnels depend on the shapes of the objects being dried and the energy concentrations required. Frequently a tunnel must be made adaptable for drying articles of several different sizes and shapes. This can be done with special lamp holders or "trees," which can be adapted as required. Some tunnels resemble inverted U's in cross section. Others consist of two flat banks facing one another. Such arrangements are designed to prevent liquid from dripping on the lamps and reflectors.

In **cylindrical tunnels**, the highest energy concentration is obtained, and considerable flexibility is possible. Their principal use is in drying or heating massive irregular objects. One or more sides of a tunnel should be movable for accessibility and maintenance. By suitable baffling, convection currents that cause heat losses are minimized and stray radiations are redirected toward the work. However, heat losses will occur unless tunnels are suitably insulated, and present practice points toward well-insulated tunnels. Frequently, exhaust systems are required to remove organic vapors. The material to be dried may be conveyed through the tunnels by overhead trolleys, belt or wire conveyors, chain conveyors, vibrating conveyors, etc.

A typical lamp arrangement for drying discontinuous sheets or other flat materials is a flat horizontal bank mounted directly above a belt or vibrating conveyor. Such an arrangement on a small scale is also suitable for experimental studies. For such test purposes, the smallest recommended bank is twelve of the proposed lamps. With this number, the work under the center unit will receive approximately the same concentration of energy as will be obtained from a large bank of such lamps on the same centers.

In infrared drying, the temperature attained by the work is considerably above any wet-bulb temperature and depends on (1) the intensity of radiation concentrated on the surface of the material; (2) the percentage of energy absorbed by the material; (3) the heat capacity of the material; and (4) heat losses due to conduction, convection, and radiation to low-temperature surroundings.

Infrared lamps generally employ tungsten filaments that operate at color temperatures on the order of 2500°K., compared with 2960°K. for ordinary illuminating lamps. Carbon filaments, although more efficient in *radiating heat* when operating at 2200°K., become less efficient than tungsten filaments after only 100 hr. of operation.

Reflectors used with lamps may be electrolytically gold-plated, Alzak-finished aluminum, or vaporized aluminum. Gold-plated reflectors are the most efficient reflectors for infrared, the Alzak finish about 10 per cent less efficient, and vaporized aluminum reflectors fall between the two. The reflectors are generally parabolic in shape. In some types the reflector is built into the lamp.

Infrared drying has been used successfully to dry photographic film, glue used in sealing cartons, plastic powders prior to molding, glue adhesives in plywood manufacture, ground enamel frit prior to firing, blueprints, coconut shreds, latex on cotton gauze, knit goods, banknote paper, and washed metal surfaces.

The use of infrared radiation for removing moisture from solids is restricted to the drying of **thin film or sheet material.** It is not generally applicable to drying large massive objects or materials in which internal moisture flow is the controlling factor in the drying process. Probably the greatest field of application for this type of equipment is in drying and baking paint films and in preheating thin objects or surfaces. These applications differ somewhat from drying problems that involve large heat requirements for evaporation.

The warm-up period for infrared dryers depends on the time required for the air temperature to reach its equilibrium value. Lamps may be turned off and on as required. Thus the material passing through a tunnel may be used to intercept photocell light beams to turn the lamps on at each position. When material is no longer passing, lamps will automatically go out.

Theory and Design Data. The fundamentals of infrared heating have been discussed by Tiller and Garber [*Ind. Eng. Chem.*, **34**, 773 (1942)], who considered the temperature-time relationships in the heating of thin metal sheets. No application was made to drying. They presented the following equation for determining the degree of approach to the maximum temperature rise in heating a thin sheet of material by infrared radiation:

$$\frac{t_m - t}{t_m - t_0} = e^{-Fh_c\theta/c\rho L} \qquad (75)$$

where t = temperature of the sheet at time θ, °F.; t_0 = initial temperature of the sheet, °F.; h_c = convection heat-transfer coefficient during heating, B.t.u./(hr.)(sq. ft.)(°F.); c, ρ, and L = respectively, specific heat of the sheet, B.t.u./(lb.)(°F.), density of sheet, lb./cu. ft., and thickness of the sheet, ft.; θ = time, hr.; F = ratio of the area from which convectional heat transfer occurs to area under direct radiation; t_m = the maximum temperature obtainable by the sheet under the existing conditions of energy input, absorptivity of the stock, and surrounding air temperature, expressed by $t_m = t_a + \frac{3.41\alpha I}{Fh_c}$, where t_a = temperature of the surrounding atmosphere, °F.; α = absorptivity of the sheet receiving radiation, ratio of heat absorbed to heat received; and I = intensity of radiant energy on the surface, watts/sq. ft.

When moisture is removed, the rate of drying during the constant-rate period can be expressed as follows:

$$\frac{dw}{A\,d\theta} = \frac{3.41\alpha I - h_c(t_s - t_a)}{\lambda} \qquad (76)$$

where $dw/A\,d\theta$ = lb. evaporated/(hr.)(sq. ft.); t_s = temperature of surface of evaporation, °F., and λ = latent heat of evaporation at t_s, B.t.u./lb. Equation (76) assumes that the areas receiving radiation, losing moisture, and losing heat by convection are all equal.

Penetration of the radiant energy into the solid will depend on the wave lengths involved and on the absorption characteristics of the solid.

Drying times for the removal of moisture by infrared radiation are usually considerably higher than for other infrared applications. The drying rate depends primarily on the absorptivity of the surface receiving radiation, and the heat losses to the surrounding atmosphere. The removal of water from washed metal parts requires 2 to 3 min. with a 250-watt lamp in a 12-in. reflector, and photographic prints under the same conditions require 6 to 9 min. Drying time for textiles* run from 15 to 60 sec. in a continuous oven with intensities between 1450 and 2100 watts/(sq. ft.) on each side, 2900 to 4200 watts/sq. ft. total. Stout, Caplan, and Baird [*Trans. Am. Inst. Chem. Engrs.*, **41**, 283 (1945)] compared infrared with air and vacuum drying and concluded that infrared drying produced no significant changes in the mechanism of drying.

The design of infrared dryers depends principally on determination of the energy concentration required, expressed as watts per unit area of receiving surface. Some typical concentrations and drying times for several infrared applications are given in Table 31, from a report on radiant energy heating by the Utilities Research Commission, Chicago, Ill.

In general, as the drying lamps are moved away from the work, the energy distribution on the surface will become more uniform, although the average energy concentration will at the same time decrease. The total energy required for a given evaporation load will depend on the evaporative rate required, the lamp-bank efficiency, and the percentage of incident energy absorbed, *i.e.*, the absorptivity. Lamp-bank efficiencies may range from 40 to 80 per cent, and absorptivities from 10 to 90 per cent.

Table 31. Typical Drying Times and Energy Concentrations for Infrared Dryers

Material	Description	Distance of lamp from work, in.	No. and wattage of lamps	Watts/sq. in.	Time under lamps, min.
Photographic emulsion......	Coated on glass	24–36	6—260	0.25	2
Glue on carton flaps..........	Cardboard boxes	8	10—250	2.75	About 1
Glue..........	Gluing silk linings	16	6—260	1.10	4–6
Plastic granules..	Drying prior to molding	12–36	24—260	1.33–0.66	15–45
Plywood sheets...	Moisture reduced from 25 to 15%	8	16—260	3.2	7.5
Enamel frit......	Water removal from ground enamel	10	24—260	2.6	9
Aluminum oxide grains.........	Abrasive	10–12	32—260	3.7	20
Blueprints.......	Drying after complete immersion	18	....	2	2.5–6
Coconut.........	Shredded	16	....	6.0	4
Graphite........	Water wet	16	....	3.0	6
Latex..........	Dipped on gauze	4	....	3.0	6
Knit goods......	33% moisture removal	4	....	6.0	3
Silver nitrate....	Mirror coating	16	....	5.7	2
Paper..........	Banknote grade	6	....	4.9	0.5

Example 15. Estimate the electrical energy, kilowatts, required to evaporate 55 lb./hr. of water with infrared radiation if the lamp-bank efficiency is 65 per cent and the absorptivity is 50 per cent.

Solution. The heat required to preheat and evaporate water at a rate of 55 lb./hr. is 61,350 B.t.u./hr. The combined efficiency for the lamp bank and absorption is 32.5 per cent. Hence the heat input must be 188,000 B.t.u./hr., which is equivalent to 55 kw. If sensible heat is a large item, this must be added to the heat load.

The **over-all efficiency** of an infrared system depends on four factors: (1) the efficiency of the lamp bank in delivering input energy to the receiving surface, which may range from 40 to 80 per cent; (2) the absorptivity of the work, which in practice will lie between 0.5 and 1.0; (3) the space factor, which is the ratio of the energy intercepted by the work to the energy reaching the work plane, which can be increased by redirecting the non-

* F. M. Tiller, private communication.

intercepted energy; and (4) heat losses due to convection and the lower surrounding temperatures. On the basis of these various factors, it is evident that an average over-all efficiency for infrared systems will range from less than 30 per cent up to 60 per cent for well-designed systems.

When radiant heat is derived from gas, the **added heat transfer by convection** from the combustion gases may be appreciable.

Performance and Cost Data. In general, energy costs for infrared dryers may range as high as ten times the cost for steam-heated dryers. The method is restricted to thin sheets of material or to use as an **auxiliary drying means** to augment the capacity of other types of dryers, especially continuous-sheeting dryers.

In general, the investment costs for infrared dryers are low. However, this is largely offset not only by the high power costs but also by lamp-replacement costs. The problem of lamp breakage with resultant contamination of the product with glass is an added disadvantage. Lamp costs will range from $1 to $8 for 125- and 1000-watt sizes, respectively. Modern lamps usually employ a tungsten filament and built-in reflectors, although large-wattage lamps use external reflectors, which necessitate periodic cleaning and additional maintenance.

Dielectric or High-frequency Dryers

Description. In dielectric drying, the material to be dried is heated by placing it in a strong electrostatic field produced by high-frequency voltage. In its simplest form, such a dryer might consist of two flat metallic plates between which lies the material to be dried, as illustrated in Fig. 67. Such a combination represents a

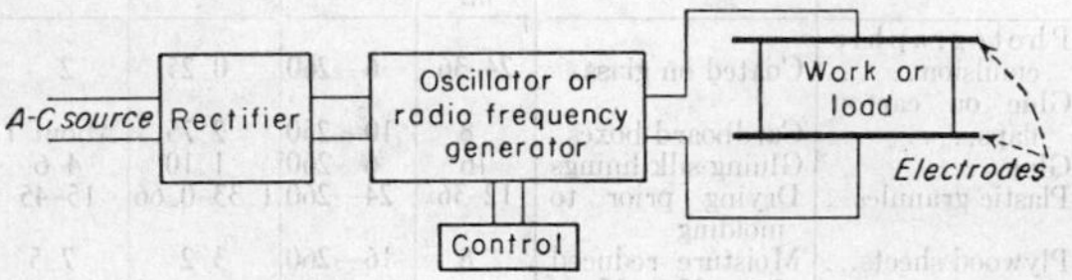

FIG. 67. Simplified diagram of dielectric or high-frequency dryer showing location of material to be dried between two electrodes.

capacitor, the plates of which are connected to the terminals of a high-frequency electronic generator. During one portion of the electrical cycle, the upper plate is charged positively and the lower negatively, thereby creating a stress in one direction on the molecules of the wet material, which acts as the dielectric of the capacitor. A half-cycle later, the polarities of the plates have been reversed, thereby reversing the atomic or molecular stress in the dielectric or wet material. This reversal of polarity occurs with extreme rapidity, corresponding to the frequency of the electronic generator, which may be in the range of 2 to 100 megacycles (2×10^6 to 100×10^6 cycles/sec.).

This rapidly oscillating electric field generates heat in the intervening wet material or dielectric. One explanation of this heating is that the continued rapid application of alternating stresses on the molecules causes a phenomenon similar to frictional stresses that create heat. Because the electric field creating the stress is uniform through the thickness of the dielectric, the heating is likewise very nearly uniform, which represents *the outstanding characteristic of dielectric drying.* It permits heating a material at its center, even when wet, as rapidly as it heats at its surface, a mechanism directly contrary to the other methods of drying, in which heat must be conducted into the interior of the solid only as it becomes dry. Heat can be generated so rapidly in some thick pieces of material that steam is actually generated in the interior, causing rupture of the solid.

For drying objects of irregular shape, it is sometimes necessary to design specially shaped electrodes to conform more closely with the surface contours of the material to be dried. This is necessary to provide an even electrical field within the material and to eliminate possibilities of local spots of overheating.

Dielectric dryers are rated according to their kilowatt output and may range in size from 1 to 100 kw.

Application. Few industrial applications of dielectric drying have been reported. Numerous experimental studies have been made of this drying method to establish its field of application. Studies have been made on large massive objects such as wood, ceramic materials, and foundry molds that normally take many hours or days to dry. Rayon cakes also have been the subject of study by dielectric drying. The patent literature has suggested other applications, such as the drying of large blocks of cellulose sponge and sponge rubber and the drying or setting of glues in the manufacture of shoes.

Dielectric heating has found considerable application in the plywood industry in bonding wood layers with thermosetting resins, the latter being set by dielectric heating.

An early experimental study of drying clay and ceramic pieces dielectrically was reported by Vaughn *et al.* (*Va. Polytechnic Inst., Eng. Exp. Sta. Bull.* 42, p. 34, 1940). A description of the dielectric drying of wood was given by Berkness [*Wood Prod.*, **45** (9), 12 (1940)]. Bosomworth [*Can. Chem. Process Ind.*, **30** (7), 28 (1946)] has described the application of high-frequency heating in the rubber and plastics industry. Nonken [*Tech. Assoc. Papers*, **27**, 625 (1944)] has discussed the possible application to the paper industry, and Rusca [*Textile World*, **96** (5), 118 (1946)] has discussed its application to textiles.

Dielectric heating may be used for the initial evaporation of water from heat-sensitive organic solutions and slurries. Commercial application has been made of this in the drying of penicillin, Brown et al. [*Proc. Inst. Radio Engrs.*, **34** (2), 58 (1946)].

Theory and Design Data. Little can be presented on the theory and design of dielectric dryers. In general, any design procedure will involve a calculation of the power required for a given problem, after which the problem resolves into an electrical and mechanical design of equipment.

Power requirements can be estimated from the following formulas expressing the sensible and latent heat requirements:

1. *Sensible-heat requirement:*

$$P_s = 2.93 \times 10^{-4} c_p M (t_2 - t_1) \qquad (77)$$

where P_s = power, kw.-hr.; M = weight of dry material to be heated, lb.; c_p = specific heat of dry material, B.t.u./(lb.)(°F.); $(t_2 - t_1)$ = temperature rise of the mass M, °F. The same relation holds for heating the water to its evaporation temperature. By dividing the power P_s by the time of the cycle, the result gives the required kilowatt output rating of the generator for heating only. Heat losses must be added to get the actual power required.

2. *Latent-heat requirement:*

$$P_e = 0.3W \qquad (78)$$

where P_e = power, kw.-hr.; W = evaporation load, lb.; 0.3 = coefficient when evaporating water. Hence the total power P_t, kw.-hr., required for drying is

$$P_t = P_s + P_e + P_L \qquad (79)$$

where P_L = power required to compensate for losses due to radiation, convection, and conduction.

It is important to know the amount of power that can actually be introduced into the material to be heated by dielectric heating. An equation giving this information is as follows:

$$E = 1.4 \left(\frac{V}{d}\right)^2 fe'' \quad (80)$$

where E = rate of heat generation, watts/cu. in.; V/d = voltage gradient, kv./in.; d = thickness of material, in.; f = frequency, megacycles/sec.; e'' = loss factor of the material. The loss factor e'' is a function of the material and is defined as the product of power factor and dielectric constant. For a given material it may vary over wide limits with frequency, temperature, moisture, etc.; hence it must be measured over ranges of these variables to determine the expected variation for the process. Sometimes an optimum frequency occurs for the loss factor for a given material. Generally, this optimum covers a wide band of frequencies (5 to 10 megacycles).

Approximate loss factors for a number of common materials together with their specific heats are given in Table 32.

Table 32. Dielectric Loss Factors e'' for Miscellaneous Materials

Frequency, megacycles	1	10	20	30	Specific heat
Paperboard (soft)	....	0.024	0.026	0.027	
Paperboard (hard)	....	.11	.12	.11	0.4
Fiber	0.25				
Porcelain	.04	.044	.....	.06	.26
Oak, dry	.09				
Birch, dry	.32	....	.....	.....	.40
Hard rubber	.03	.017	.....	.017	.33
Phenol-formaldehyde	....	0.20–0.50	.....	.....	.30
Urea-formaldehyde	....	0.16–0.21	.....	.....	.40
Cellulose acetate	....	0.03–0.45	.....	.....	.35
Cellulose nitrate	....	0.43–0.62	.....	.....	

Performance and Cost Data. Only generalities can be presented concerning the performance of dielectric dryers. The frequency selected for a given application is influenced by several factors, such as the area to be heated, how the energy is introduced, practicability of oscillator or generator design, power required, etc. Although the frequency should be as high as practical to keep the voltage within reasonable limits, it may be limited by voltage distribution and oscillator design. For a given electrode size, the higher the frequency, the greater the probability of **unequal voltage distribution.** This can be minimized by making the longest dimension of the electrode a small fraction of the wave length, or by scanning the work. However, the generator design requires proper operation of vacuum tubes, and cannot be changed so simply.

At present, it appears undesirable to use frequencies above 20 megacycles when power above 10 kw. is involved. However, satisfactory performance can be obtained in many applications with frequencies on the order of 2 to 15 megacycles. This range is likewise suitable for the size of electrodes required for most applications.

Ordinarily the voltage across the work should be kept under 15,000 volts. It is not impossible to use voltages above this, but the additional precautions necessary to avoid corona and arc-overs are often more expensive than some other compromise of the engineering factors. The permissible voltage gradient varies widely with the type of material. The radio-frequency voltage that will puncture a given dielectric material is generally much lower than the voltage it will stand at 60 cycles Hence it is desirable to keep the voltage gradient below 2000 volts/in for porous materials, and below 5000 volts/in. on essentially nonporous materials, whenever possible.

In general, it costs considerably more to evaporate 1 lb. of water by dielectric heat than by steam, oil, gas, or radiant heat. For comparison, if water can be removed from a solid by steam heat for 0.1¢/lb., removal with radiant heat will cost about 1¢/lb., and removal with dielectric heat will cost 1.5 to 2.5¢/lb. Figure 68 shows the power required for various evaporation rates, and the cost per pound of water evaporated for various generator sizes. The high cost of dielectric heat reflects the inefficiencies involved in producing and applying high-frequency currents. The first cost of high-voltage and high-frequency apparatus and suitable conveying means for the material is very high.

Because of the type and skill of the help required to operate dielectric dryers, labor costs considerably more than for other types of dryers.

A major item of maintenance in dielectric dryers is the tubes used to generate high-frequency currents. Such tubes cost $200 or more apiece, and their life is only about 2000 hr.

An approximate idea of performance and cost data for this type of drying can be obtained from the following applications:*

1. The setting of glue in plywood manufacture involves heating the glue to 300°F. in a 4-ft. by 8-ft. by 4½-in. plywood sheet accompanied by a reduction in water content from 10 to 5 per cent. Two pieces of plywood of this size are heated simul-

* Data by courtesy of the Westinghouse Electric Corp.

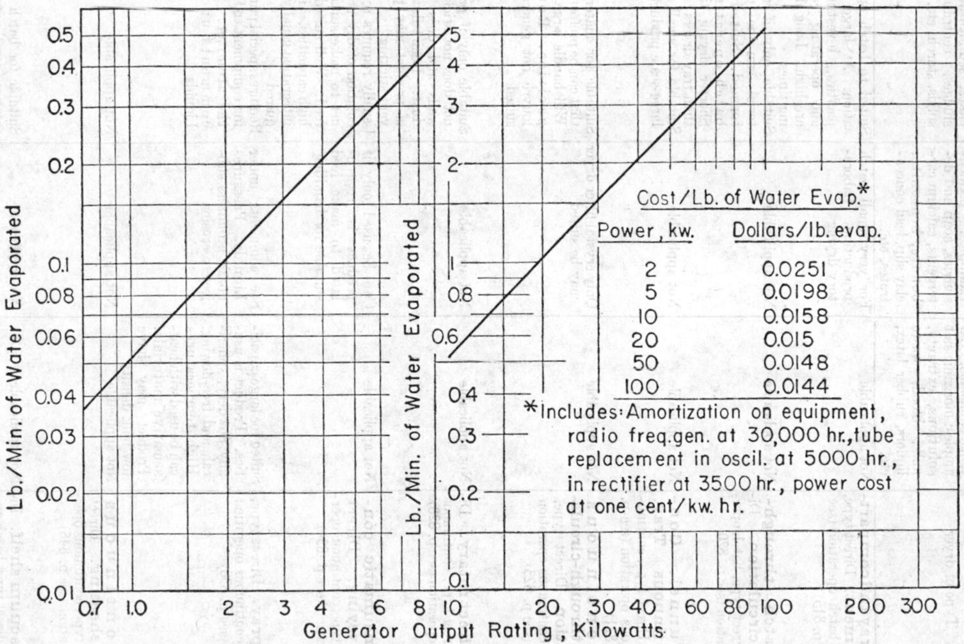

FIG. 68. Rate of evaporation vs. power into the work (output rating of generator).

Table 33. Classification of Commercial Dryers Based on Materials Handled

Type of dryer	Liquids	Slurries	Pastes and sludges	Free-flowing powders	Granular, crystalline, or fibrous solids	Large solids, special forms and shapes	Continuous sheets	Discontinuous sheets
	True and colloidal solutions; emulsions. Examples: inorganic salt solutions, extracts, milk, blood, waste liquors, rubber latex, etc.	Pumpable suspensions. Examples: pigment slurries, soap and detergents, calcium carbonate, bentonite, clay slip, lead concentrates, etc.	Examples: filter-press cakes, sedimentation sludges, centrifuged solids, starch, etc.	100 mesh or less. Relatively free flowing in wet state. Dusty when dry. Examples: centrifuged precipitates, pigments, clay, cement.	Larger than 100 mesh. Examples: rayon staple, salt crystals, sand, ores, potato strips, synthetic rubber.	Examples: pottery, brick, rayon cakes, shotgun shells, hats, painted objects, rayon skeins, lumber.	Examples: paper, impregnated fabrics, cloth, cellophane, plastic sheets.	Examples: veneer, wallboard, photograph prints, leather, foam rubber sheets.
Tray and compartment. Direct type, batch operation (see p. 815)	Not applicable	For very small batch production. Laboratory drying	Suited to batch operation. At large capacities, investment and operating costs are high. Long drying times	Dusting may be a problem. See comments under Pastes and Sludges	Suited to batch operation. At large capacities, investment and operating costs are high. Long drying times	See comments under Granular solids	Not applicable	See comments under Granular solids
Batch through-circulation. Direct type, batch operation (see p. 820)	Not applicable	Not applicable	Suitable only if material can be preformed. Suited to batch operation. Shorter drying time than tray dryers	Not applicable	Usually not suited for materials smaller than 30 mesh. Suited to small capacities and batch operation	Primarily useful for small objects	Not applicable	Not applicable
Tunnel. Continuous Tray. Direct type, continuous operation (see p. 821)	Not applicable	Not applicable	Suitable for small and large-scale production.	See comments under Pastes and Sludges. Vertical-turbo applicable	Essentially large-scale, semicontinuous tray **drying.**	Suited to a wide variety of shapes and forms. Operation can be made continuous. Widely used	Not applicable	Suited for leather, wallboard, veneer.
Continuous through-circulation. Direct type, continuous operation (see p. 823)	Not applicable	Only crystal filter dryer may be suited	Suitable for materials that can be preformed. Will handle large capacities. Roto-louvre not generally suited	Not generally applicable, except Roto-louvre in certain cases	Usually not suited for materials smaller than 30 mesh. Material does not tumble, except in Roto-louvre dryer. Latter operates at higher temperatures	Suited to smaller objects that can be loaded on each other. Can be used to convey materials through heated zones. Roto-louvre not suited.	Not applicable	Special designs are required. Suited to veneers. Roto-louvre not applicable
Direct rotary. Direct type, continuous operation (see p. 828)	Not applicable	Not applicable	Suitable only if product does not stick to walls and does not dust. Recirculation of product may prevent sticking	Suitable for most materials and capacities, provided that dusting is not too severe	Suitable for most materials at most capacities. Dusting or crystal abrasion will limit its use	Not applicable	Not applicable	Not applicable
Pneumatic conveying. Direct type, continuous operation (see p. 834)	Not applicable	Can be used only if product is recirculated to make feed suitable for handling	Usually requires recirculation of dry product to make suitable feed. Well suited to high capacities. Disintegration usually required	Suitable for materials that are easily suspended in a gas stream and lose moisture readily. Well suited to high capacities	Suitable for materials that are easily suspended in a gas stream. Well suited to high capacities. Product may suffer physical degradation	Not applicable	Not applicable	Not applicable
Spray. Direct type, continuous operation (see p. 838)	Suited for large capacities. Product is usually powdery, spherical, and free-flowing. High temperatures can be used with heat-sensitive materials. Product may have low bulk density	See comments under Liquids. Pressure-nozzle atomizers subject to erosion	Requires special pumping equipment to feed the atomizer. See comments under Liquids	Not applicable	Not applicable	Not applicable	Not applicable	Not applicable
Continuous sheeting. Direct type, continuous operation (see p. 848)	Not applicable	Not applicable	Not applicable	Not applicable	Not applicable	Not applicable	Different types are available for different requirements. Suitable for drying without contacting hot surfaces	Not applicable
Vacuum shelf. Indirect type, batch operation (see p. 853)	Not applicable	Not applicable	Suitable for batch operation, small capacities. Useful for heat-sensitive or readily oxidizable materials. Solvents can be recovered	See comments under Pastes and Sludges	Suitable for batch operation, small capacities. Useful for heat-sensitive or readily oxidizable materials. Solvents can be recovered	See comments under Granular solids	Not applicable	See comments under Granular solids

Vacuum freeze. Indirect type, batch or continuous operation (see p. 854)	Usually used only for pharmaceuticals such as penicillin and blood plasma. Expensive. Used on heat-sensitive and readily oxidized materials	See comments under Liquids	See comments under Liquids	See comments under Liquids	Expensive. Usually used on pharmaceuticals and related products which cannot be dried successfully by other means. Applicable to live chemicals.	See comments under Granular solids	Applicable in special cases such as emulsion-coated films.	See comments under Granular solids
Pan. Indirect type, batch operation (see p. 856)	Atmospheric or vacuum. Suitable for small batches. Easily cleaned. Solvents can be recovered. Material agitated while dried	See comments under Liquids	See comments under Liquids	See comments under Liquids	Suitable for small batches. Easily cleaned. Material is agitated during drying, causing some degradation	Not applicable	Not applicable	Not applicable
Vacuum rotary. Indirect type, batch operation (see p. 858)	Not applicable	May have application in special cases	Use is questionable. Material usually cakes to dryer walls and agitator. Solvents can be recovered	Suitable for non-sticking materials. Useful for large batches of heat-sensitive materials and for solvent recovery	Useful for large batches of heat-sensitive materials or where solvent is to be recovered. Product will suffer some grinding action. Dust collectors may be required	Not applicable	Not applicable	Not applicable
Indirect rotary. Indirect type, continuous operation (see p. 859)	Not applicable	Not applicable	Generally requires recirculation of dry product. Little dusting occurs	Chief advantage is low dust loss. Well suited to most materials and capacities, particularly those requiring drying at steam temperature	Low dust loss. Inexpensive at almost all capacities. Material must not stick or be temperature-sensitive	Not applicable	Not applicable	Not applicable
Screw conveyor. Indirect type, continuous operation (see p. 862)	Not applicable	Not applicable	Can only be used if material does not stick or cake	Suitable for materials that will not stick. Usually used with low moisture content materials. Can be used to convey the material	Usually used with low-moisture-content materials. Can be used as conveyor and auxiliary dryer	Not applicable	Not applicable	Not applicable
Vibrating tray. Indirect type, continuous operation (see p. 863)	Not applicable	Not applicable	Not applicable	Suitable for free-flowing materials that are dusty. Usually operated at steam temperatures	Suitable for free-flowing materials that can be conveyed on a vibrating tray. Usually operated at steam temperatures, but higher temperatures are feasible	Not applicable	Not applicable	Not applicable
Drum. Indirect type, continuous operation (see p. 863)	Single, double, or twin. Atm. or vacuum operation. Product flaky and usually dusty. Maintenance costs may be high	See comments under Liquids. Twin-drum dryers are widely used	Can be used only when paste or sludge can be made to flow. See comments under Liquids	Not applicable	Not applicable	Not applicable	Not applicable	Not applicable
Cylinder. Indirect type, continuous operation (see p. 866)	Not applicable	Not applicable	Not applicable	Not applicable	Not applicable	Not applicable	Suitable for thin or mechanically weak sheets which can be dried in contact with a heated surface. Special surface effects obtainable	Suitable for materials which need not be dried flat and which will not be injured by contact with hot drum
Infrared. Batch or continuous operation (see p. 868)	Only experimental use has been reported. Expensive	See comments under Liquids	See comments under Liquids	Expensive. Has been used on expensive products. Danger of overheating	Primarily suited to drying surface moisture. Not suited for thick layers. Expensive	Specially suited for drying and baking paint and enamels	Usually used in conjunction with other methods. Useful when there are space limitations	Useful for laboratory work or in conjunction with other methods
Dielectric. Batch or continuous operation (see p. 870)	Very expensive. Used commercially only on penicillin	See comments under Liquids	See comments under Liquids	Very expensive. No commercial use reported	Very expensive. No commercial uses reported	Rapid drying of large objects suited to this method. Power costs excessive	Applications not developed	Successful on foam rubber. Not fully developed on other materials

taneously with a high-voltage plate separating them. Platens, electrically heated to 300°F., hold the work under a pressure of 200 lb./sq. in. A frequency of 6.8 megacycles is used, with a radio-frequency generator rated at 100 kw. The temperature at the center of the work is 300°F. and at the edges is 250°F. The power requirements are calculated to be 36 kw. for sensible heat, 21.8 kw. for latent heat and 4 kw. for heat losses. The heating time is about 28 min.

2. A water-mixed glue was set at a temperature not greater than 170°F., to eliminate bubble formation with a consequent weakening of the bond. Laminated glue strips were glued to a wooden block under 100 to 300 lb./sq. in. pressure. The glue lines were perpendicular to the electrode, permitting concentration of the heat in the glue. A frequency of 30 megacycles from a power source rated at 2 kw. performed the heating in 1 min.

3. Foam sponge rubber has been dried from 37 to 1 per cent moisture by dielectric drying. Rubber mattresses weighing 50 lb. are dried at an evaporative rate of 300 lb./hr. of water requiring a 100-kw. generator. The total drying time is 1 hr., which represents a combination of dielectric heating for 45 sec., followed by hot-air drying for 1 hr. The previous drying time with hot air alone was 16 hr. The cost of power for such a drying operation, however, will be about 2¢/lb. water evaporated.

SELECTION OF DRYING EQUIPMENT

A careful consideration of many factors is necessary in the final selection of the most suitable type of dryer for a given application. Such a selection is complicated by the large number of different types of dryers available on the market. Commercial dryers are not usually sufficiently flexible to compensate for design inaccuracies or for unconsidered problems in the physical handling of the material. For this reason it is particularly important that all pertinent facts are considered and that experimental tests are made before a dryer is finally selected for a given problem.

The following procedure is recommended for **selecting the best dryer** to perform a given operation most economically:

1. Select those dryers which appear best suited to handling the wet material and dry product, which fit into the continuity of the process as a whole, and which will produce a product of the desired physical properties. This preliminary selection can be made with the aid of Table 33, which classifies the various types of dryers on the basis of the materials handled.

2. The dryers so selected should be evaluated, approximately, from available cost and performance data, many of which have been previously listed under the various types of dryers. From this evaluation, those dryers which appear to be far out of line costwise or performancewise should be eliminated from immediate consideration.

3. Drying tests should be conducted in those dryers still subject for consideration. Such tests will determine the optimum operating conditions and the product quality and will form the basis upon which vendors of equipment can furnish firm quotations.

4. From the results of the drying tests and price quotations, a final selection of the most suitable dryer can be based on accurate quality and cost data.

Preliminary Dryer Selection. The important factors to consider in the preliminary selection of a dryer are the following:

1. The properties of the material being handled.
   - *a.* Physical characteristics when wet.
   - *b.* Physical characteristics when dry.
   - *c.* Corrosiveness.
   - *d.* Toxicity.
   - *e.* Inflammability.
   - *f.* Particle size.
   - *g.* Abrasiveness.
2. The drying characteristics of the material.
   - *a.* Type of moisture (bound or unbound, or both).
   - *b.* Initial moisture content.
   - *c.* Final moisture content (maximum).
   - *d.* Permissible drying temperature.
   - *e.* Probable drying time for different dryers.
3. The flow of material to and from the dryer.
   - *a.* Quantity to be handled per hour.
   - *b.* Continuous or batch operation.
   - *c.* Process prior to drying.
   - *d.* Process subsequent to drying.
4. Product qualities.
   - *a.* Shrinkage.
   - *b.* Contamination.
   - *c.* Uniformity of final moisture content.
   - *d.* Decomposition of product.
   - *e.* Overdrying.
   - *f.* State of subdivision.
   - *g.* Product temperature.
   - *h.* Bulk density.
5. Recovery problems.
   - *a.* Dust recovery.
   - *b.* Solvent recovery.
6. Facilities available at site of proposed installation.
   - *a.* Space.
   - *b.* Temperature, humidity, and cleanliness of air.
   - *c.* Available fuels.
   - *d.* Available electric power.
   - *e.* Permissible noise, vibration, dust, or heat losses.
   - *f.* Source of wet feed.
   - *g.* Exhaust-gas outlets.

The physical nature of the material to be handled is the primary item for consideration. A slurry will demand a different type of dryer from that required by a coarse crystalline material, which, in turn, will be different from that required by a sheet material. Table 33 lists the various types of dryers best suited for handling eight different classes of feed material, and it facilitates the preliminary selection of a group of dryers on the basis of material handling.

Initial Comparison of Dryers. After the preliminary selection of the suitable types of dryers, a high-spot evaluation of the size and cost should be made of these types to eliminate those far out of line costwise. The information necessary to make such an evaluation is contained, for the most part, under the discussion of each dryer type as given previously. Where the data are insufficient, preliminary cost and performance data can usually be obtained from the appropriate manufacturer. In comparing dryer performance, proper weight should be given to the factors listed above that affect design and operation. The possibility of eliminating or simplifying other unit operations, such as filtration, grinding, or conveying, should be carefully considered. The physical appearance and properties of the product may be of great importance.

Drying Tests. The critical comparisons suggested usually eliminate all but three or four types of dryers from further consideration. Final evaluation of the remaining dryers requires tests in experimental dryers that simulate each type. Such tests should establish optimum operating conditions, product quality, and dryer size. The principal manufacturers of drying equipment are usually prepared to perform drying tests on dryers simulating their equipment.

Since plant-scale dryers seldom cost less than $5000 installed and often cost more than $25,000, extensive test work is justified in establishing the correct type of dryer and the correct operating conditions. Once a given type and size of dryer is installed, the product characteristics and capacity of the unit can be changed only within rather narrow limits. It is much cheaper, therefore, and far more satisfactory to experiment in small-scale units than on the dryer that is finally installed.

Final Selection. Based upon the results of the drying tests that establish size and operating conditions, formal quotations and guarantees should be obtained from dryer manufacturers. First costs, installation costs, operating costs, product quality, and dryer operation and flexibility can now be given consideration in the **final evaluation** and selection.

Example of Dryer Selection. Assume that equipment is desired to dry 1000 lb./hr. of a pigment coming from a continuous filter, and the following hypothetical conditions are imposed: The initial moisture content is 1.0 lb./lb. dry material, and this must be reduced to 0.02 lb./lb. dry material. The product, which has an ultimate particle size of 1μ or less, should be free-flowing and readily dispersible in water. It is known that exposure of this material to temperatures greater than 180°F. for longer than 5 min. produces changes in color. Ordinary steel construction will be satisfactory. The filter cake can be pumped in a gear pump.

From Table 33, the following types of dryers are selected as having possible application:

1. Spray dryer.
2. Pneumatic conveying dryer.
3. Drum dryer.
4. Continuous-conveying screen-type through-circulation dryer.
5. Steam-tube rotary dryer.
6. Screw-conveyor dryer.

Batch dryers were not considered because the filtration operation is continuous and because the production rate is higher than that which can be handled economically by batch operation. Infrared and dielectric dryers were eliminated on a cost basis. Direct rotary dryers were eliminated on the basis that the material would cake on the inside of the dryer.

The feasibility of a spray dryer and a drum dryer depends on the pumpability of the filter cake. The use of a continuous through-circulation dryer is contingent upon preforming the material to form a satisfactory permeable bed. Successful operation of the pneumatic conveying, steam-tube rotary, and screw-conveyor dryers may require partial recirculation of the dry product.

High-spot dryer sizes, investments, and operating costs may be estimated from information available under Types of Dryers with the following results:

Type of dryer	Estimated installed cost	Estimated operating cost per 100 lb. product
Steam-tube rotary	$ 25,000	$0.20
Pneumatic conveying	26,500	.30
Continuous through-circulation with a finned drum	40,000	.30
Drum	55,000	.30
Spray	66,500	.40
Screw-conveyor	160,000	.72

Operating costs include labor, power, fuel, maintenance, and depreciation.

From this high-spot cost comparison, the screw-conveyor dryer can be eliminated from further consideration. The remaining five dryers offer advantages and disadvantages that can be determined only by actual tests in dryers simulating each type.

Hypothetical results of such tests on small-scale dryers simulating steam-tube rotary, pneumatic conveying, through-circulation, drum, and spray dryers are given below:

Type of Dryer	Results
Steam-tube rotary	Material sticks to hot steam tubes, even with 80% recirculation of dry product. Product lumpy. Low rates of heat transfer obtained. Dryer unsuitable
Pneumatic conveying	50% recirculation of dry product required for successful handling of feed. Inlet air temperature of 800°F. can be used without hurting the product. A cage mill is required to suspend product in air stream. Fine free-flowing powder obtained
Through-circulation	The only satisfactory preforming method found was by means of a finned drum, but this method caused grit formation and product degradation. Not suitable
Drum	Difficult to obtain film of uniform thickness on dryer drum. Product tended to stick to drum, making removal with knives difficult. Formed grit. Very low heat-transfer rates. Dryer unsuitable
Spray	In order to obtain a slurry of suitable atomizing properties, dilution to 33% solids was required. Inlet-air temperature of 800°F. can be used without hurting the product. Fine free-flowing powder obtained

From the above results, only two dryers are left for further consideration. Color strength and grit tests on the products from the pneumatic conveying and spray dryers are identical.

The product from the pneumatic conveying dryer has a slightly higher bulk density, and the spray-dried product is less dusty and disperses in water a little more rapidly. Firm cost estimates were obtained for spray and pneumatic conveying dryers, and installation costs were estimated by the company purchasing the dryer. A summary of the final data facilitates the final dryer selection.

The following tabulation indicates that both the capital investment and the operating cost for the pneumatic conveying dryer will be less than for the spray dryer. The dispersion rate and decreased dustiness of the spray-dried product are not sufficiently superior to outweigh the cost advantage. On this basis the pneumatic conveying dryer should receive first consideration for selection.

Type of dryer	Spray	Pneumatic conveying
Size of dryer	12 ft. diam. by 12 ft. high	18 × 12 × 20 ft.
Purchase cost, including auxiliaries	$32,000	$10,000
Installation cost, excluding building but including storage facilities	15,000	5,000
Total installed cost	$47,000	$15,000
Operating cost (including depreciation, labor, power, fuel, and maintenance) per 100 lb.	$0.60	$0.40
Packaging cost	0.11	0.09
Total operating cost per 100 lb.	$0.71	$0.49
Product characteristics:		
Bulk density, lb./cu. ft.	45	55
Color	Standard	Standard
Dustiness	Slight	Some
Dispersion rate, sec.	10	12

DRYING WITH SOLVENT RECOVERY

When a material wet with a liquid other than water is dried, it is frequently necessary to recover the solvent vapors to ensure the lowest possible operating cost, or to eliminate a safety hazard. Thus, if 500 lb./hr. of a material containing 15 per cent methanol is to be dried, the value of the vaporized methanol amounts to about $27,000 per year, so that a certain expenditure for the recovery of this solvent appears justified. In other instances, toxic or flammable solvents must be recovered from a safety rather than an economic standpoint.

Several of the dryers previously discussed are particularly well suited for drying when solvent vapors are to be recovered. Of the indirect batch dryers, the vacuum rotary, vacuum pan, and vacuum shelf dryers are widely used for this purpose. For small quantities of a number of different materials that do not have to be agitated or cannot be agitated during drying, the vacuum shelf dryer is ordinarily used. For small to medium-size batches that can be agitated during drying, the vacuum pan dryer is used. Large batches are usually handled in vacuum rotary dryers. Vacuum drum dryers are used to dry liquids and slurries continuously under vacuum over a wide range of capacities. The foregoing dryers are always operated batchwise, with the exception of the drum dryer, which can be operated continuously by removing the dried material from the vacuum chamber through a screw conveyor with converging pitch. A limited class of materials, such as soap, can be sufficiently packed in the conveyor by this method to form an adequate vacuum seal. Feed from the storage tank to the drum dryer enters through a closed piping system. Depending upon the air leakage and the type of recovery system, solvent recoveries may range from 90 to 99 per cent with the preceding dryers.

With indirect dryers the solvent is removed under slight or high vacuum with only a slight quantity of inert gas which is originally present at the start or which leaks into the system during operation. The solvent vapors from the dryer pass through a dust collector, if much dust occurs in the vapors, to prevent yield losses and fouling of condenser surfaces. The dust collector may be either a bag collector, cyclone, or wet scrubber. The latter is frequently preferred, since condensation of vapors may occur with either of the first two. Sometimes both a cyclone and a wet scrubber are used if considerable dusting occurs. Discharge from the wet scrubber can be pumped back to the filtration operation and the dust reprocessed in the dryer.

The vapors leaving the dust collector usually pass through a surface condenser, where solvent is condensed and collected by a barometric leg or run into a tank that is kept at a reduced pressure. The vapors leaving the condenser consist of inert gases that leak into the system plus a certain amount of solvent corresponding roughly to equilibrium conditions for the solvent-inert gas mixture at the temperature and pressure of the gases leaving the condenser, although the vapors actually may leave at less than saturation.

Vacuum sources for these systems consist of steam jets (one or more stages), vacuum pumps, or a combination thereof. With steam jets, care must be taken to keep the pressure well above the vapor pressure of the solvent at the condenser temperature. If this is not done, considerable solvent may be lost into the jets; and, if the pressure is allowed to become less than the solvent vapor pressure at the condenser temperature, all the solvent will be lost in the steam-jet exhaust. If the condensing temperature of the solvent is lower than the available water temperature under the desired vacuum conditions, brine-cooled condensers should be used. However, these will be expensive to operate, and, where possible, the pressure in the system should be kept at levels where water-cooled condensers can be used. If a vacuum pump is used, a small condenser can be placed on the discharge side of the vacuum pump to condense vapors passing through the pump. Care must be exercised in the selection of a pump, since condensation may occur during compression. Since the size of either a vacuum pump or a steam-jet system depends upon the volume of gases handled, it is desirable to condense as much of the solvent as possible before the vacuum source.

Batch direct dryers using air or inert gases as the heating medium can sometimes be used for drying and solvent recovery. These dryers, which can be of the tray, compartment, or batch through-circulation type [see Dew and Mackey, U.S. Patent 2,201,560] consist primarily of a drying chamber, a condenser, heating coils, and fan arranged in a closed system as shown in Fig. 69. Heated

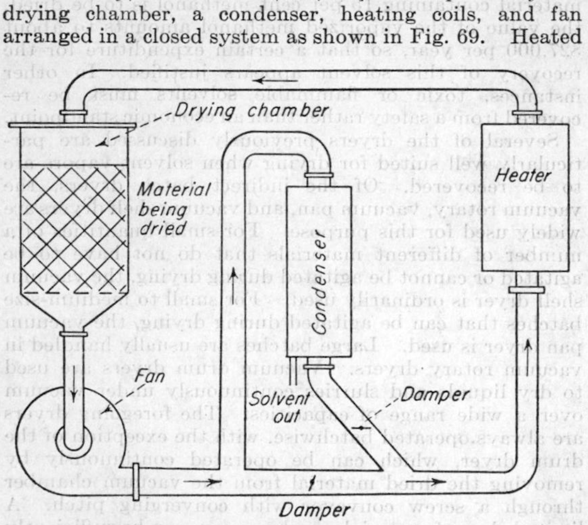

FIG. 69. Typical batch direct dryer with solvent-recovery system.

gas passing over or through the solvent-wet material vaporizes solvent and passes through the condenser where the solvent vapor is condensed. The gas from the condenser is then recirculated through the heater back through the drying chamber. Although such systems avoid vacuum equipment, heat economy is poor, since a considerable portion of the heat is used to reheat the gases after solvent removal. If air is used as the circulating gas, care must be exercised that the solvent-air mixture within the dryer is not within any explosive limits. If inert gases are used, the cost of inert-gas make-up in the system must be considered.

Instead of a condenser, an activated-carbon bed or other adsorbent can sometimes be used to remove solvent from the gas stream. If the dryer is used continually, at least two adsorbers must be installed so that one can be regenerated while the other is in the circulating system. Cost of regeneration of the adsorbent bed plus distillation costs if the solvent is steamed out must be considered with this system. Activated-carbon adsorbers are usually used where the concentration of the solvent in the gas stream is low. Solvent recoveries of over 90 per cent can be expected from either of these two systems if they are carefully designed.

It is frequently desirable to perform drying and solvent recovery continuously. However, continuous drying and solvent recovery have received little attention. The chief problems are charging and discharging the product to maintain a high vacuum if such is required, and eliminating excessive quantities of inert gases in atmospheric operation to reduce the size and cost of vapor-recovery equipment.

Only two types of continuous dryers employing solvent recovery are in extensive use. One of these, a **cylinder dryer** from which solvents from the drying of printing inks are adsorbed in an activated-carbon system, is highly specialized. The other type is the **screw-conveyor dryer**, which has been used to remove solvents from extraction residues such as soybeans. When employed with solvent recovery, the screw-conveyor dryer is kept under a slight vacuum (1 to 10 in. water), which is generally sufficient to ensure vapor removal without drawing excessive quantities of air into the recovery system. The solvent vapors are recovered in a condenser operating at almost atmospheric pressure. Vacuums on the order of 26 to 28 in. Hg are possible with screw-conveyor dryers when charging and discharging of material are not obstacles. Indirect continuous dryers such as the **vibrating tray** and the **steam-tube rotary dryer** are suited for continuous drying with solvent recovery, because they can be operated with small amounts of inert gases under atmospheric pressures. Continuous vacuum dryers have been offered on the market, but continuous feeding and discharging of these dryers is difficult and usually unsatisfactory for all but a few limited types of materials.

Direct continuous dryers can be used to dry solvents from materials when the solvents are discarded. Solvent recovery with these dryers is costly and difficult, however, since separation must occur from a high percentage of non-condensable gases. If air is used, an explosion hazard may exist. If combustion gases are used, water may contaminate the solvent. The cost of make-up gas makes the use of inert gases unattractive. However, where the particular method of drying yields a highly desirable product or one of sufficiently better quality to offset the high cost of recovery, continuous direct dryers such as a spray dryer, for example, can be adapted to continuous drying and solvent recovery. Figure 70 shows a diagrammatic sketch of a **solvent-recovery closed system involving a spray dryer.**

Calculation procedures and theory outlined previously are the same for the removal of solvents as for the removal of water. All the heat-transfer equations are valid. However, the psychrometry is different in that the psychrometric ratio for solvent vapor–air mixtures is not numerically equal to the humid heat of the mixture as in the purely fortuitous case for water vapor–air mixtures. Consequently, in humidity charts for solvent vapor–air mixtures, the adiabatic-saturation lines will not coincide with the wet-bulb-temperature lines. During the constant-rate period, the driving force for heat transfer will be the difference between the air temperature and the wet-bulb temperature of

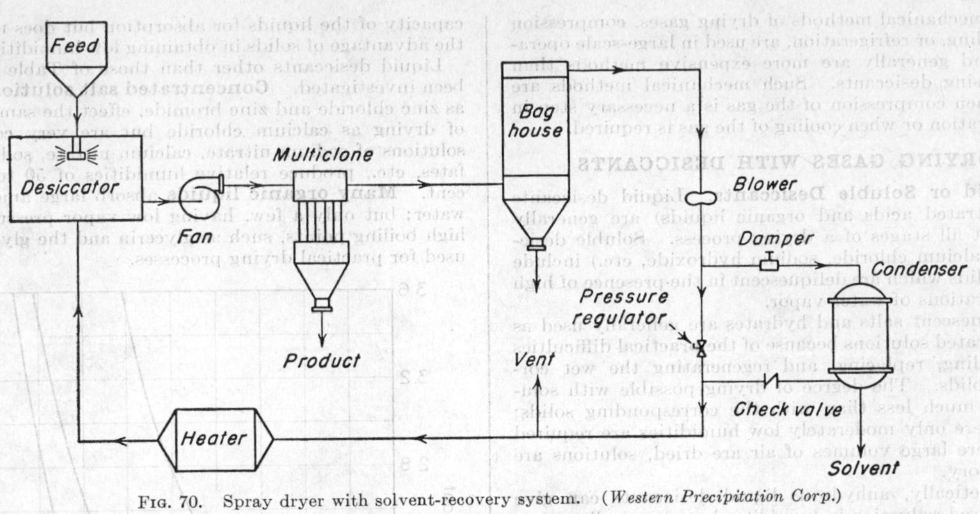

FIG. 70. Spray dryer with solvent-recovery system. (*Western Precipitation Corp.*)

the solvent-wet stock as read from an appropriate humidity chart or as calculated by Eq. (47) under Theory.

When countercurrent or cocurrent adiabatic drying during the constant-rate period takes place in continuous direct dryers, the humidity in the dryer will follow an adiabatic-saturation line, as in the case of water removal. However, since this line is not coincident with the wet-bulb line, the temperature of the stock will decrease with increasing humidity. This factor must be taken into account in the calculation of drying rates for solvent removal in continuous direct dryers that operate countercurrently or cocurrently. Humidity charts for four different air-solvent mixtures have been given in Figs. 10 to 13. Methods for calculating humidity charts for air-solvent mixtures are discussed by Barta and Garber [*Chem. & Met. Eng.*, **47,** 287 (1940)] and outlined briefly on p. 812.

The recovery of solvents in a surface condenser depends upon the amount of non-condensable gas present, and the pressure and temperature at which the condensation occurs. When the above factors are set, the percentage removal in the condenser will be $100(H_2 - H_1)/H_2$, where H_2 = humidity of the solvent-gas mixture entering the condenser, lb. solvent/lb. dry gas; and H_1 = humidity leaving the condenser. H_2 is obtained from a knowledge of the drying rate and the amount of non-condensables entering the system. H_1 is the saturation humidity at the conditions of the condenser outlet and can be obtained from Eq. (39) or from a humidity chart of the particular air–solvent vapor mixture. When the non-condensables are completely recirculated, the case for direct dryers, the over-all recovery should be 100 per cent. The solvent concentration entering the condenser can then be calculated from the following equation:

$$H_2 = H_1 + \frac{M(dW/d\theta)}{w} \qquad (81)$$

where M = weight of dry material in dryer, lb.; $dW/d\theta$ = drying rate, lb. solvent/(hr.)(lb. dry material); and w = weight rate of flow of non-condensables, lb./hr. It should be noted that Eq. (81) indicates a method for determining experimentally the drying rate in closed systems by measuring the inlet and outlet solvent concentrations to and from the condenser, and the flow rate. The determination of H_2 and H_1 may require knowledge of the psychrometric ratio, which is discussed in Sec. 8.

THE DRYING OF GASES*

INTRODUCTION

The removal of 95 to 100 per cent of the water vapor in air or other gases is frequently necessary. Gases having a dew point of −40°F. are considered "commercially dry." The more important reasons for the removal of water vapor from air are: (1) comfort, as in air conditioning; (2) control of the humidity of manufacturing atmospheres; (3) protection of electrical equipment against corrosion, short circuits, electrostatic discharges; (4) requirement of dry air for use in chemical processes where moisture present in air adversely affects the economy of the process; (5) prevention of water adsorption in pneumatic conveying; and (6) as a prerequisite to liquefaction.

METHODS OF DRYING GASES

Gases may be dried by the following processes:

1. **Absorption.** (*a*) Use of spray chambers with such organic liquids as glycerin, or aqueous solutions of salts such as lithium chloride, etc. (*b*) Use of packed columns with countercurrent flow of sulfuric acid, phosphoric acid, or organic liquids.
2. **Adsorption.** Use of solid adsorbents such as activated alumina and silica gel.
3. **Compression.** Compression to a partial pressure of water vapor greater than the saturation pressure to effect condensation of liquid water.
4. **Cooling.** Cooling below dew point of the gas with surface condensers or cold-water sprays (see Sec. 12).
5. **Compression and cooling.**

As indicated above, both solid and liquid desiccants are used in the drying of gases. **Liquid desiccants** are used in continuous processes in spray chambers and packed towers. **Solid desiccants** are generally used in intermittent operation that requires periodic interruption for regeneration of the spent desiccant.

Desiccants are classified as follows:

1. Solid adsorbents, which remove water vapor by the phenomena of surface adsorption and capillary condensation, such as silica gel and activated alumina.
2. Solid absorbents, which remove water vapor by chemical reaction, such as fused anhydrous calcium sulfate, lime, and magnesium perchlorate.
3. Deliquescent absorbents, which remove water vapor by chemical reaction and dissolution, such as calcium chloride and potassium hydroxide.
4. Liquid absorbents, which remove water vapor by absorption, such as sulfuric acid, lithium chloride solutions, and ethylene glycol.

* The material in this section is based largely on a report by O. A. Hougen and F. W. Dodge on the "The Drying of Gases," 1941, released in 1946 through the Office of Technical Service, U.S. Dept. of Commerce.

The mechanical methods of drying gases, compression and cooling, or refrigeration, are used in large-scale operations and generally are more expensive methods than those using desiccants. Such mechanical methods are used when compression of the gas is a necessary step in the operation or when cooling of the gas is required.

DRYING GASES WITH DESICCANTS

Liquid or Soluble Desiccants. Liquid desiccants (concentrated acids and organic liquids) are generally liquid at all stages of a drying process. Soluble desiccants (calcium chloride, sodium hydroxide, etc.) include those solids which are deliquescent in the presence of high concentrations of water vapor.

Deliquescent salts and hydrates are generally used as concentrated solutions because of the practical difficulties in handling, replacing, and regenerating the wet corrosive solids. The degree of drying possible with solutions is much less than with the corresponding solids; but, where only moderately low humidities are required and where large volumes of air are dried, solutions are satisfactory.

Theoretically, anhydrous liquid desiccants can give any desired reduction in humidity, but practically excessive quantities of reagent are required and regeneration to the anhydrous state may be difficult and expensive.

The advantages and disadvantages of liquid dessicants are:

Advantages

1. Readily circulated, thus permitting continuous operation
2. Readily dispersed by sprays or tower packings to ensure large surface exposure and low pressure drop
3. Readily pumped, thus making a central regenerator feasible to serve a number of drying units
4. Desired degree of drying effected with close control by maintaining desiccant at proper concentration and temperature

Disadvantages

1. Relative humidities below 10 to 30 per cent unobtainable in practice
2. For extreme drying, acids and organic liquids must be nearly anhydrous. This causes loss of desiccant and contamination of dried product due to high vapor pressure of desiccant
3. Unless desiccant is so cheap that part of spent solution may be discarded, reconcentration must be effected by evaporation or heating in an air stream, which may involve decomposition
4. Corrosion is a problem at some regeneration temperatures
5. Some liquids vaporize during regeneration and must be recovered

The **general requirements for liquid desiccants** are: (1) vapor pressure of water in liquid or solution must be low; (2) it must maintain its drying ability over wide concentration range; (3) vapor pressure of desiccant itself must be low to avoid loss and contamination; (4) aqueous solutions as used must be non-corrosive; (5) viscosity must not be excessive; (6) for salt solutions, salt solubility must be great enough to avoid solidification and excessive crystallization at temperatures used; (7) reaction or solution heat must be low; (8) liquid must be chemically stable, non-hydrolyzing, and thermally stable; (9) regeneration must require no excessive heat loads or special treatment; (10) desiccants must be nontoxic; and (11) desiccants must be low cost for economical installation and make-up cost. Properties of some of the more common liquid and soluble desiccants are given in Table 34. Of these, lithium chloride and the glycols have the widest applications because of ease of handling, non-toxicity, non-corrosiveness, and ease of regeneration. The acids and bases, however, are widely used in certain chemical industries despite handling hazards.

The equilibrium moisture contents of various liquid desiccants as a function of percentage relative humidity are given in Fig. 71. Comparison of these curves with those of Fig. 72, for solid desiccants, indicates the greater capacity of the liquids for absorption but does not show the advantage of solids in obtaining low humidities.

Liquid desiccants other than those of Table 34 have been investigated. **Concentrated salt solutions,** such as zinc chloride and zinc bromide, effect the same degree of drying as calcium chloride but are very corrosive; solutions of sodium nitrate, calcium nitrate, sodium sulfates, etc., produce relative humidities of 50 to 70 per cent. **Many organic liquids** absorb large amounts of water; but only a few, having low vapor pressures and high boiling points, such as glycerin and the glycols, are used for practical drying processes.

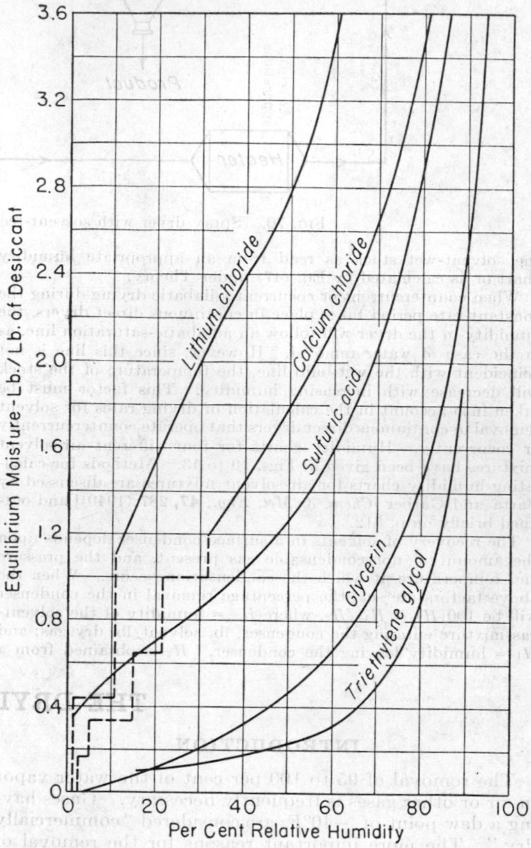

FIG. 71. Equilibrium moisture of liquid and solid desiccants.

The design of gas-drying systems using liquid desiccants involves:

1. Determination of degree of moisture removal necessary.
2. Determination if system being considered can effect the desired moisture reduction. Here, performance data or mass-transfer coefficients for the device involved are necessary. See Sec. 8 for methods of predicting such data and applying them to design.
3. Determination of the quantity of desiccant required from performance data and desired temperature rise of desiccant in the device involved. The temperature rise results from sensible-heat transfer from gas to liquid and the heat of dilution as water is absorbed by the desiccant.
4. Estimate the heat requirements for regeneration and the cooling requirements for the regenerated desiccant. For illustrations of design methods for spray and packed towers, see Sec. 10. A typical diagrammatic flow sheet for a lithium chloride system with precooling is shown in Fig. 73 together with the various heat requirements involved [Hougen and Dodge, "Drying of Gases," PB-23201, Office of Technical Services,

Table 34. Properties of Liquid and Soluble Desiccants

Liquid and soluble desiccants	Relative humidity, %, at room temp. economically obtainable	Temp. range, °F., for practical operation	Concentration range, %	Approx. cost per lb. dry desiccant	Toxicity	Corrosiveness	Stability	Principal applications	Remarks
Calcium chloride (aqueous solution).	20–25	90–120	40–50 as solution. Sometimes used as a dissolving solid	$0.09 for 50-lb. lots. $0.02 for ton lots	Non-toxic	Non-corrosive	Stable	Drying city gas	Same as for $LiCl_2$ solutions
Diethylene glycol.......	5–10	60–110	70–95	$0.15	....	Non-corrosive	Stable	Dehydration of natural gas	High boiling point: 473°F. eliminates need of elaborate refluxing equipment. Regeneration at 300°F. removes all but last traces of water without loss of desiccant
Glycerol..............	30–40 with 70–80 conc., 5 with anhydrous	70–100	70–80	$0.14	Non-toxic	Non-corrosive	Oxidizes and decomposes at high temperatures	Drying manufactured gas	Regenerated with vacuum evaporation. Low heat requirements. At concentrations as low as 50 to 60%, glycerol can still absorb considerable water
Lithium chloride (aqueous solution).	10–20	70–100	Usual: 40–45. However, 30 has good capacity	$1.00	Non-toxic	Non-corrosive except on Mg alloys. Elect. corrosion at regen. temp.	Stable at boiling temperatures	Air conditioning. Industrial processes. Low-temp. vacuum drying operations	Aftercoolers to remove heat of condensation and solution. Solution cooler to control inlet solution temperature. Hydrometer control to regulate concentration during regeneration
Phosphoric acid........	5–20	60–130	80–95	$0.12	Toxic	Corrosive	Stable	Used as laboratory desiccant, especially the anhydride	Corrosive, toxic nature makes it impractical for an industrial drying agent; regeneration does not have a fuming problem
Sodium and potassium hydroxides.	10–20	85–120	Sat. sol.; also used as dissolving solids	NaOH = $0.035 for 76% flake, carload lots. KOH = $0.083 for 90% flake, carload load	Toxic	Corrosive	Stable	Liquid air manufacture; compressed gases; gas plants	High heats of solution and corrosiveness preclude use in ordinary installations. Frequently used to remove CO_2 and H_2O simultaneously
Sulfuric acid...........	5–20	70–120	60–70	$0.05 for 50-lb. lots. $0.01 for ton lots	Toxic	Corrosive	Stable	Miscellaneous gas-drying problems in chemical processes	Is being displaced by other desiccants because of handling hazards. Where its use is feasible, it is highly efficient
Triethylene glycol......	5–10	60–110	70–95	$0.25	....	Non-corrosive	Stable	Drying natural gas	Boiling point = 550°F. Regeneration without loss of desiccant

U.S. Department of Commerce, 1946]. The design of a bubble cap tower for drying natural gas under 1000 lb./sq. in. gage has been presented in detail by Senatoroff [*Oil Gas J.*, **44** (32), 98 (1945)].

Published cost data for liquid-desiccant processes are meager. Tupholme [*Gas Age-Record*, **63**, 311 (1939)] gives data for a plant using glycerin to dry city gas: Dew point reduced from 62.5° to 27°F. at a cost of 0.3¢/1000 cu. ft. gas treated for a plant handling 1,000,000 cu. ft. gas/24 hr. with depreciation at $1.30/day and steam, cooling water, and maintenance at $1.95/day.

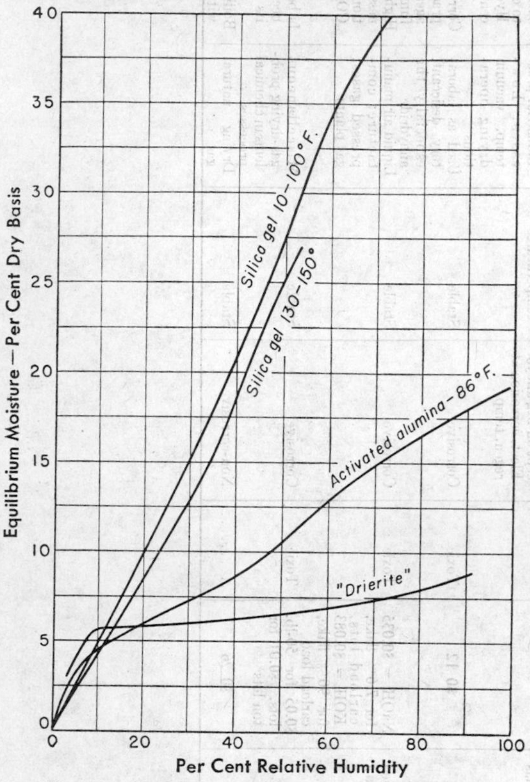

FIG. 72. Equilibrium moisture for solid desiccants.

Solid Desiccants. Water vapor sorbed by solid desiccants may be retained by one or more of the following mechanisms: (1) absorption by chemical reactions as in the formation of hydrates or hydroxides; (2) adsorption in a monomolecular layer on the desiccant surface; the water so retained is small unless the solid surface is greatly extended; (3) adsorption by capillary condensation where the desiccant is highly porous, of high capillarity, and wetted by water.

The general requirements for solid desiccants are:

1. Water vapor should be sorbed rapidly; *i.e.*, the desiccant should have high "activity."
2. Drying efficiency of desiccant should be high, *i.e.*, a large fraction of the water vapor should be removed from the gas. Bower [*Bur. Standards J. Research*, **12**, 241 (1934)] compares the drying efficiencies of several solid desiccants. His data for silica gel were obtained on low-grade gel and should be disregarded.
3. Desiccant should have a large drying capacity or equilibrium moisture content. Data for several solid desiccants are given in Fig. 72.
4 Drying efficiency should be maintained over a wide range of temperature, humidity, and gas rate.
5. Where desiccant regeneration is involved, the heat of adsorption should not exceed too greatly the heats of condensation and wetting. However, a high adsorption heat indicates high drying capacity.
6. Desiccant should permit economical regeneration.
7. Solid desiccants should not become sticky or liquid during adsorption.
8. Resistance to gas flow should be low.
9. Desiccant should have good mechanical strength to avoid crushing and dust formation.
10. Desiccant should not change volume during adsorption, should have a high bulk density and be chemically inert and non-corrosive, and should be non-toxic and cheap.

Activated carbon, silica gel, and alumina gel are among the solid desiccants that owe their water-adsorption properties to surface adsorption and capillary condensation. Porous calcium sulfate, magnesium and barium perchlorates, and calcium and barium oxides are among those which react chemically and still maintain a rigid solid structure. The properties of various solid desiccants follow:

1. Silica gel possesses a high adsorptive power because of its extreme capillarity, the capillary pores occupying approximately 50 per cent of its specific volume. The capillaries are probably spine-shaped, and the average pore diameter has been estimated to be 4×10^{-7} cm., which is only about ten times the diameter of a molecule of adsorbate. The drying efficiency of silica gel depends upon the concentration of water in the gas mixture, the temperature of the gel and gas, the properties of the condensed liquid, its wettability, and the state of the gel itself. It preferentially adsorbs water vapor in the presence of other vapors (see Sec. 14 for further description of this and other properties). It is readily capable of drying air to a dew point below −94°F., or 0.002 mm. Hg.

Other important properties of silica gel are:

a. 3- to 8- or 6- to 16-mesh gel is usually used for gas drying.
b. Specific heat is 0.2 B.t.u./(lb.)(°F.).
c. Bulk density is 40 to 45 lb./cu. ft., depending on mesh size.
d. The apparent density is between 0.65 and 0.73.
e. It is chemically, physically, and thermally stable.
f. It is reactivated at about 300°F. and can stand an unlimited number of reactivations.
g. When reactivated at temperatures above 500°F., it loses adsorptive capacity.
h. It possesses a residual water content of 5 per cent after reactivation. Removal of this water requires temperatures above 600°F., which destroy adsorptive capacity. Practical moisture contents are based on the commercially dry silica gel, *i.e.*, gel with 5 per cent moisture.
i. To remove 1 lb. adsorbed water from silica gel requires in commercial practice about 2000 to 2500 B.t.u.
j. Silica gel maintains good drying efficiency until it has adsorbed about 20 per cent of its weight of water.
k. The pressure drop through a silica gel bed is given by the empirical equation

$$\frac{\Delta p}{L} = b'V + c'V^2 \qquad (82)$$

where Δp = pressure drop, in. water; L = bed depth, in.; V = air velocity, cu. ft./(min.)(sq. ft. total cross section); and b' and c' are constants depending on mesh size as follows:

Mesh size	b'	c'
4–8	0.002	2.2×10^{-5}
6–16	.0053	5.9×10^{-5}
12–28	.0114	2.5×10^{-4}

2. Activated Alumina is prepared from alumina trihydrate. It is a granular porous adsorbent with properties similar to those of silica gel. Alumina gel has many applications in addition to gas drying. It is used to adsorb gases and vapors from gaseous mixtures, to dry liquids, etc. Its principal physical properties are given in Sec. 14.

3. Anhydrous calcium sulfate, or "Drierite," is prepared from high-grade gypsum, $CaSO_4.2H_2O$, which is dried, crushed, sized, and heated to 450° to 500°F. for 2 hr. This leaves a granular porous form of anhydrous calcium sulfate, with sufficient mechanical strength to support its own weight.

Anhydrous calcium sulfate has the following properties as a desiccant:

a. In the absence of liquid water, Drierite takes on water in moist air to form the hemihydrate, $CaSO_4.\frac{1}{2}H_2O$, which represents only 6.6 per cent water by weight. This low moisture adsorption is a serious limitation compared with other desiccants. However, its porous structure permits some *adsorption* to bring the total capacity to 12 to 14 per cent in saturated air.

b. The drying rate compares favorably with the best desiccants. Dew points down to −80°F. can be obtained.

c. It maintains its drying efficiency until nearly exhausted, even at temperatures up to 150°F.

d. It is an excellent drying agent for organic liquids and vapors, where the boiling points are below 220°F.

e. Regeneration is effected by heating at 400° to 425°F. with a stream of hot air. Drierite will withstand 200 or more regenerations.

f. The total heat of adsorption on Drierite is considerably higher than on silica gel and alumina gel; and more cooling during adsorption, as well as more heat during activation, is therefore required. In addition to the heat of condensation, 540 B.t.u./(lb. water adsorbed) are evolved from the chemical reaction involved.

g. The equilibrium moisture content of Drierite is shown in Fig. 72.

The chief advantages of Drierite are its high activity, drying power, and constant efficiency until nearly exhausted over a wide temperature range. Its disadvantages are low capacity and a limited number of regenerations. The cost is moderate.

4. Magnesium perchlorate, "Anhydrone," is the equal of any desiccant from the standpoint of drying efficiency. The adsorption rate is rapid, and the first hydrate does not lose water until 275°F. is reached, thereby permitting its use for drying at higher temperatures than most commercial desiccants. It passes through three hydrate stages, namely, di-, tri-, and hexahydrate. The latter, $Mg(ClO_4)_2.6H_2O$, represents saturation after adsorption of 48.6 per cent of the dry weight of water, a high capacity. Although drying efficiency tends to decrease after each hydrate formation, even the trihydrate is a superior drying agent to solid NaOH and $CaCl_2$.

The regeneration problems of magnesium perchlorate are the major disadvantages of this desiccant. Temperatures of 400° to 500°F. are required, and a high vacuum (not over a few millimeters of Hg) is necessary to prevent fusing of the crystals, which destroys their drying ability. In practice, the temperature is increased gradually as the hydrates are successively decomposed. The heat of hydration of magnesium perchlorate hexahydrate is 264 B.t.u./lb.

Another disadvantage of magnesium perchlorate is its high cost. Its chief use at present is in the final drying of high-pressure and liquefied gases.

Perchlorates are unsafe in drying organic vapors because of explosion hazards.

5. Other solid desiccants that have been used or studied for gas drying are oxides, such as barium and calcium oxide, and activated carbon. **Barium oxide** maintains a high drying activity up to 1000°F., a condition of red heat. This is due to the formation of barium hydroxide, which is stable at these temperatures. This fact implies, however, that regeneration by ordinary means is almost impractical. Barium hydroxide has 10.5 per cent water, anhydrous basis. A significant expansion occurs with the hydroxide formation, which makes it unsuited to tower operation. Its specific gravity is 5.72. Barium oxide would appear to have marked possibilities for the *drying of gases at high temperatures.*

Calcium oxide has long been used as a desiccant because of its low cost. Although it is capable of high drying efficiency, its capacity is low because of the formation of carbonates on its surface.

Activated carbon is old historically and probably has been as intensively studied as any single adsorbent. Although capable of adsorbing large amounts of water vapor, activated carbon finds its major use in solvent recovery, in odor and taste removal, and as a catalyst and catalyst carrier. Instead of water vapor being adsorbed preferentially, as in the case of alumina and silica gel, organic vapors tend to displace any water present on the carbon. This displacement can be used to maintain **isothermal operation** in carbon beds that have been regenerated by steam stripping by allowing them to remain wet with water when placed into the adsorption line, the water so remaining absorbing the heat of adsorption to evaporate and keep the bed at a

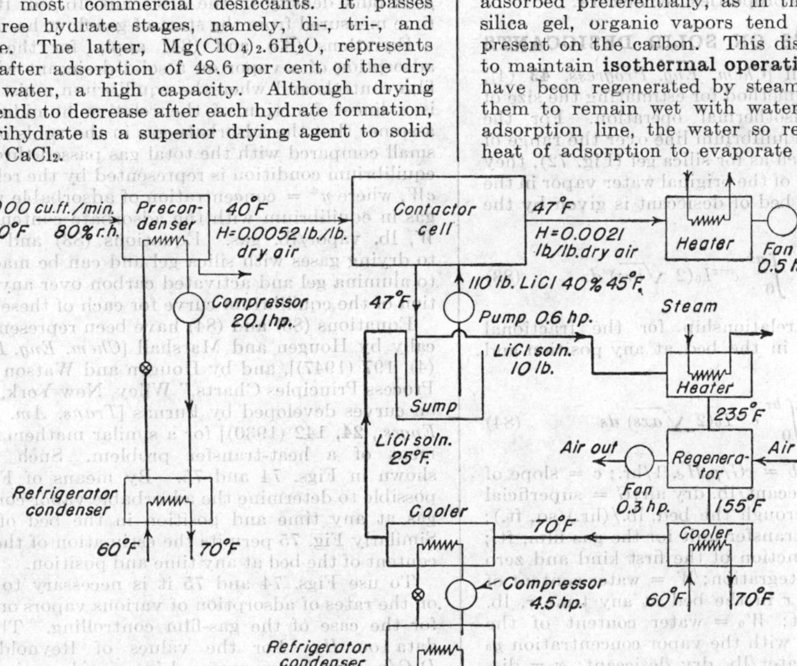

Fig. 73. Precondenser and lithium chloride system.

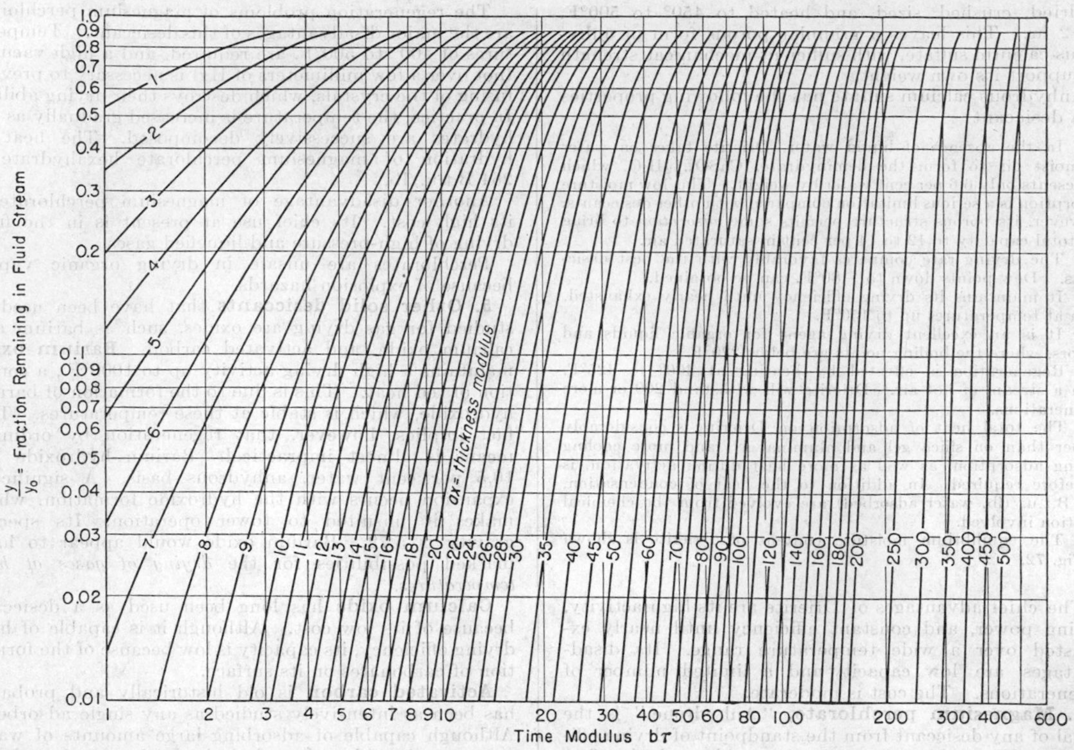

FIG. 74. Fractional nonremoval of a component from a fluid stream by a stationary granular bed. Transfer in fluid phase controlling.

constant temperature. For a complete discussion of activated carbon and its properties, see Sec. 14.

ADSORPTION RATES ON SOLID DESICCANTS

Hougen and Marshall [*Chem. Eng. Progress*, **43** (4), 197 (1947)] developed a method for estimating the size of gas-drying units for isothermal operation. For the special case of a linear equilibrium line over the range of operating conditions, such as for silica gel (Fig. 72), they showed that the fraction of the original water vapor in the air leaving a stationary bed of desiccant is given by the following equation:

$$\frac{y}{y_0} = 1 - e^{-b\tau} \int_0^{ax} e^{-z} I_0(2\sqrt{b\tau z})\, dz \qquad (83)$$

and a corresponding relationship for the fractional approach to saturation in the bed at any position and time was found to be

$$\frac{W}{W_0} = e^{-ax} \int_0^{b\tau} e^{-s} I_0(2\sqrt{axs})\, ds \qquad (84)$$

where $a = 1/H_d$, 1/ft.; $b = cG/\rho_s H_d$, 1/hr.; c = slope of equilibrium line, lb. desiccant/lb. dry air; G = superficial mass velocity of gas through the bed, lb./(hr.)(sq. ft.); H_d = height of a mass-transfer unit for the gas film, ft.; I_0 = modified Bessel function of the first kind and zero order; s = variable of integration; W = water content of desiccant at any point x in the bed at any time τ, lb. water/lb. dry desiccant; W_0 = water content of the desiccant in equilibrium with the vapor concentration y_0 (*i.e.*, $W_0 = y_0/c$), lb. water/lb. dry desiccant; x = distance from gas-inlet end of bed, parallel to gas flow, ft.: y = humidity of gas at any time τ and any position x, lb. water/lb. dry gas; y_0 = humidity of the entering gas, lb. water/lb. dry gas; z = variable of integration; ρ_s = bulk density of the desiccant, lb./cu. ft.; and τ = time measured from the start of gas flow, hr.

Equations (83) and (84) apply for the isothermal adsorption of a vapor on a solid desiccant when the gas film controls and when the equilibrium moisture content is a linear function of the relative humidity. It also assumes that the adsorbate gas in the voids of the bed is small compared with the total gas passed through. The equilibrium condition is represented by the relation $y^* = cW$, where y^* = concentration of adsorbable vapor in the gas in equilibrium with the adsorbate content of the gel W, lb. vapor/lb. gas. Equations (83) and (84) apply to drying gases with silica gel and can be made to apply to alumina gel and activated carbon over any linear portion of the equilibrium curve for each of these desiccants.

Equations (83) and (84) have been represented graphically by Hougen and Marshall [*Chem. Eng. Progress*, **43** (4), 197 (1947)], and by Hougen and Watson "Chemical Process Principles Charts," Wiley, New York, 1946 based on curves developed by Furnas [*Trans. Am. Inst. Chem. Engrs.*, **24**, 142 (1930)] for a similar mathematical treatment of a heat-transfer problem. Such curves are shown in Figs. 74 and 75. By means of Fig. 74 it is possible to determine the adsorbable vapor content of the gas at any time and position in the bed of desiccant. Similarly Fig. 75 permits the estimation of the adsorbate content of the bed at any time and position.

To use Figs. 74 and 75 it is necessary to have data on the rates of adsorption of various vapors on desiccants for the case of the gas-film controlling. This requires data on H_d. For the values of Reynolds number, D_pG/μ, usually encountered in gas adsorption, Gamson, Thodos, and Hougen [*Trans. Am. Inst. Chem. Engrs.*, **39**, 1 (1943)] showed that the gas-film height of a mass-transfer unit is given by

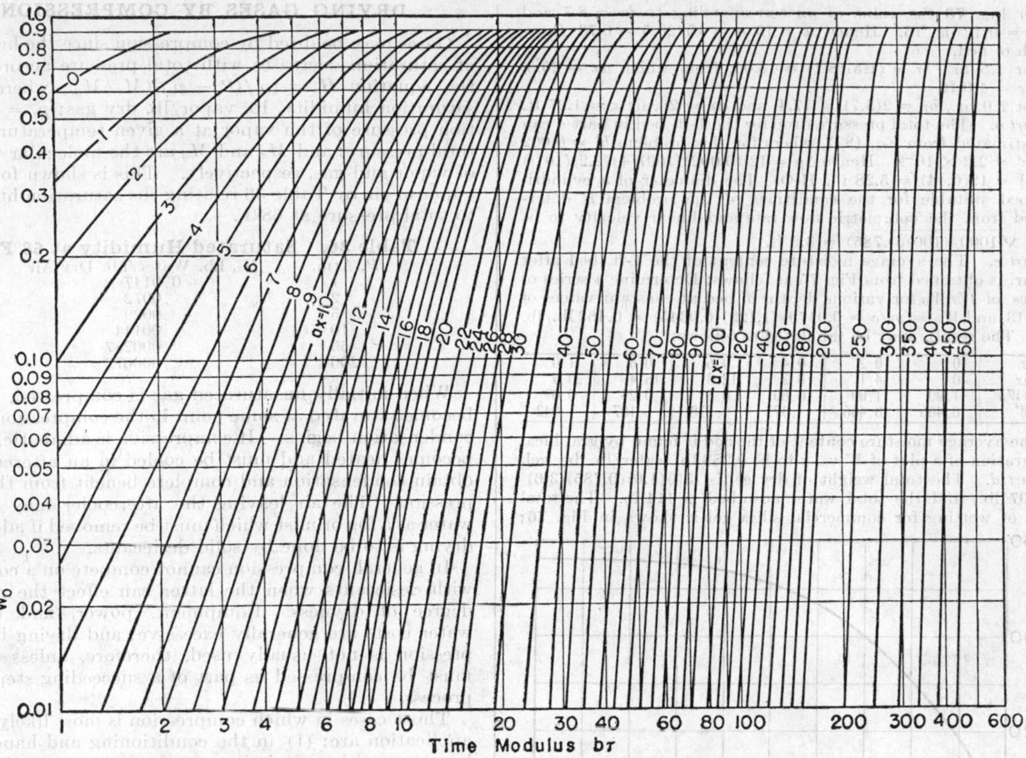

FIG. 75. Fraction of adsorbent saturation at any time and position in the bed.

$$H_d = \frac{0.55}{a_v}\left(\frac{D_pG}{\mu}\right)^{0.51}\left(\frac{\mu}{\rho D_v}\right)^{0.67} \tag{85}$$

where H_d = height of a mass-transfer unit, ft.; a_v = external area of solid particles, sq. ft./(cu. ft.); D_p = diameter of sphere having the same surface area as the particle, ft.; G = mass velocity of dry gas, lb./(hr.)(sq. ft.); μ = gas viscosity, lb./(ft.)(hr.); ρ = gas density, lb./(cu. ft.); D_v = diffusivity of vapor in the carrier gas, sq. ft./hr. Values of D_p and a_v for silica gel are given in Table 35.

Table 35. Properties of Silica Gel Particles

Mesh size	D_p, ft.	a_v, sq. ft./cu. ft.
2– 4	0.0220	117
4– 6	.0128	202
6– 8	.00909	284
8–10	.00634	407
10–12	.00576	448
10–14	.00446	576
14–20	.00320	805
20–28	.00226	1140

The equilibrium moisture content for a typical silica gel based on Fig. 72 is given by

$$y^* = cW = 1.122p_sW \tag{86}$$

where $c = 1.122\ p_s$, and p_s = vapor pressure of water at the temperature of adsorption, atm.

Experimental data of Ahlberg [*Ind. Eng. Chem.*, **31**, 988 (1939)] on the adsorption of water vapor from air by silica gel were correlated by Hougen and Marshall [*Chem. Eng. Progress*, **43** (4), 197 (1947)] to give the following expression for the height of an over-all mass-transfer unit for water-vapor adsorption:

$$H_{do} = \frac{1.42}{a_v}\left(\frac{D_pG}{\mu}\right)^{0.51} \tag{87}$$

where H_{do} = over-all height of a mass-transfer unit, ft. Equation (87) gives values somewhat higher than would be predicted by Eq. (85). This may be accounted for, to some extent, by solid-phase resistance.

For the case of isothermal adsorption of water vapor on silica gel, the values of the constants for Eqs. (83) and (84) may be written as follows:

$$a = \frac{1}{H_d} = 0.703a_v\left(\frac{D_pG}{\mu}\right)^{-0.51} \tag{88}$$

$$c = 1.122p_s$$

$$b = \frac{Gc}{\rho_s H_d} = 0.79\frac{Gp_sa_v}{\rho_s}\left(\frac{D_pG}{\mu}\right)^{-0.51} \tag{89}$$

The application of Figs. 74 and 75 to the case of water-vapor adsorption on silica gel is illustrated in the following example:

Example 16. Air at 80°F. and 80 per cent relative humidity (y_0 = 0.0179) is to be dried to 10 per cent relative humidity (y = 0.00214) at a rate of 1000 cu. ft./min. The gel is 6 to 8 mesh with a bulk density of 39 lb./cu. ft. Isothermal conditions will be maintained in the bed by means of cooling coils. Determine the following: (*a*) The depth of bed required for a drying time of 1, 1.5, and 2.0 hr. (*b*) The total pressure drop through the bed and the cross-sectional area of the bed for an air rate of 100 ft./min. to be used in part *a*. (*c*) The *average* moisture content of the bed corresponding to 1.5 hr. drying time. (*d*) The heat required to regenerate the bed if reactivation occurs at 350°F.

Data for Calculations. ρ_s, density of gel = 39 lb./cu. ft.; ρ_0, density of air entering = 0.0715 lb./cu. ft.; G = (100)(60)(0.0715) = 429 lb./(sq. ft.)(hr.); μ = 0.0447 lb./(ft.)(hr.); D_p for 6 to 8 mesh (Table 35) = 0.009 ft.; a_v = 284 sq. ft./cu. ft.; D_pG/μ = 86.4; $(D_pG/\mu)^{-0.51}$ = 0.103.

At 80°F., p_s = 0.0345 atm. Hence $a = 0.704a_v(D_pG/\mu)^{-0.51}$ = 0.704(284)(0.103) = 20.5; and $b = 0.789(Gp_sa_v/\rho_s)(D_pG/\mu)^{-0.51}$ = 0.789(429)(0.0345)(284)(0.103)/39 = 8.7.

Part a. For 1 hr., $b\tau$ = 8.7; y/y_0 = 0.00214/0.0179 = 0.12.

From Fig. 73 the value of ax corresponding to $b\tau = 8.7$ and $y/y_0 = 0.12$ is 15. Hence $x = 15/a = 15/20.5 = 0.73$ ft. = depth of bed.

For 1.5 hr., $b\tau = (1.5)(8.7) = 13.05$, from which $ax = 20.5$, and $x = 1.0$ ft.

For 2.0 hr., $b\tau = 2(8.7) = 17.4$, and $ax = 26$, or $x = 1.27$ ft.

Part b. The total pressure drop for the 1-ft. bed in part *a* can be estimated from Eq. (82), where for this problem $b' = 0.002$ and $c' = 2.2 \times 10^{-5}$. Hence $\Delta p = 12[(0.002)(100) + (2.2)(10^{-5})(10^4)] = 12(0.44) = 5.28$ in. H_2O. The diameter of a cylindrical bed suitable for the conditions of the problem is determined from the volumetric flow rate and linear velocity to be $D = \sqrt{1000/(100)(0.785)} = 3.6$ ft.

Part c. The average moisture content of the 1-ft. bed after 1.5 hr. is obtained from Fig. 75 as follows: Determine a series of values of W/W_0 for various depths of bed at constant values of $b = 13$, and $W_0 = y_0/c = 0.0179/(1.122)(0.0345) = 0.462$ lb./lb. gel. The results may be tabulated as follows:

x	0	0.2	0.4	0.6	0.8	1.0
ax	0	4.1	8.2	12.3	16.4	20.5
W/W_0	1.00	1.00	0.80	0.51	0.23	0.09
W	0.466	0.466	.373	.238	.107	.042

The average moisture content of the bed is found by graphical integration of a plot of W vs. x to be 0.288 lb. water/lb. dry gel.

Part d. The total weight of dry gel is $(39)(1.0)(0.785)(3.6)^2 = 397$ lb., and the total water adsorbed is 114 lb. The total heat of wetting for commercial silica gel is shown in Fig. 76;

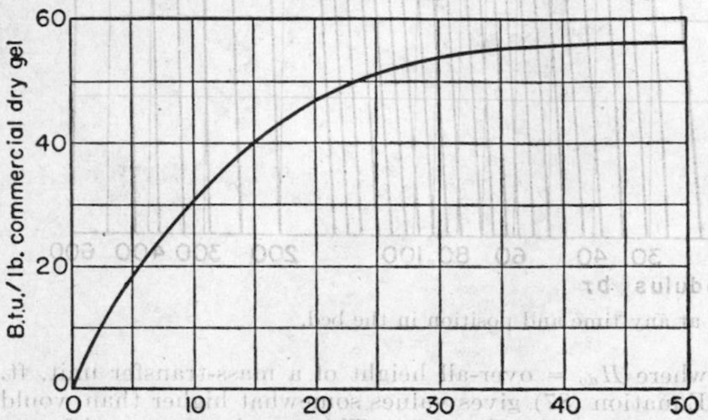

FIG. 76. Heat of wetting of silica gel. [*Ewing and Bauer, J. Am. Chem. Soc.*, **59**, 1548 (1937).]

and, for the moisture content of the bed in this problem, the total heat of unwetting is 23 B.t.u./lb. gel. Sensible heat to the gel is $397(350 - 80)(0.2) = 13{,}500$ B.t.u. Heat of unwetting is $397(53) = 21{,}000$ B.t.u. Heat of evaporation and superheating is $114(1172) = 134{,}000$ B.t.u. Total heat required = 168,500 B.t.u. If efficiency of reactivation is 50 per cent, the total heat to be supplied is 337,000 B.t.u.

For the case of isothermal adsorption when the equilibrium relationship is **non-linear,** Hougen and Marshall [*Chem. Eng. Progress*, **43** (4), 197 (1947)] illustrated a graphical procedure for determining the adsorbate-concentration curves in the gas and desiccant for any time and position. Another procedure could be followed for this case when the equilibrium curve can be approximated by a series of straight, line segments and when the conditions in the bed do not change from one line to another. Each linear segment can be represented by a linear equation of the form $y^* = cW + \beta$, where β, is the intercept on the y-axis. In the linear case treated above, $\beta = 0$. For that particular portion of the equilibrium curve involved in the adsorption process, the appropriate value of β can be employed and use made of the curves in Figs. 73 and 74 as illustrated in the above example. The use of a linear relationship involving the intercept β requires only slight modifications of the left-hand sides of Eqs. (83) and (84) as follows:

For Eq. (83) the left-hand side becomes $(y - \beta)/(y_0 - \beta)$, and for the Eq. (84), $(W - \beta)/(W_0 - \beta)$, and these ratios may be used in conjunction with Figs. 74 and 75 in place of y/y_0 and W/W_0 as illustrated above.

DRYING GASES BY COMPRESSION

Gases may be dried by compression since the humidity at saturation decreases with total pressure according to the equation $H_s = [p_s/(P - p_s)](M_v/M_G)$, where H_s = saturation humidity, lb. vapor/lb. dry gas; p_s = saturation pressure of the vapor at a given temperature; P = total pressure; and M_v and M_g are the molecular weights of vapor and gas, respectively. This is shown for water vapor in air by Table 36 relating the saturation humidity to total pressure at 68°F.

Table 36. Saturated Humidity at 68°F.

P, Atm.	H_s, Lb. Water/Lb. Dry Air
1	0.0147
2	.0072
5	.0029
10	.00144
50	.000287
200	.000072

When initially unsaturated gas is compressed, it must become saturated at some point in the compression before condensation begins. If compression is adiabatic, the air becomes heated and must be cooled in an aftercooler to obtain condensation and complete benefit from the compression. The air leaving the aftercooler may contain water as a fog or mist which must be removed if additional drying is to be done by solid desiccants.

In general, compression cannot compete on a cost basis with desiccants when the latter can effect the required degree of dryness. Equipment, power, and cooling-water costs are generally excessive; and drying by compression is not usually used, therefore, unless the gas must be compressed as part of a succeeding step in the process.

Those cases in which compression is most likely to find application are: (1) in the conditioning and handling of large quantities of air in manufacturing operations; (2) in drying air used for pneumatic conveying; (3) in drying special gases that are to be compressed into cylinders, in which case last traces of moisture are removed by desiccants; and (4) in the transporting of individual gases long distances.

DRYING GASES BY REFRIGERATION

When air is cooled well below its dew point by such means as brine, ammonia, or Freon expansion coils, or solid CO_2, it is possible to condense much of the water vapor and greatly reduce the air humidity. Saturated air at 11°F. is only 10 per cent of saturation at 68°F., and at −20°F. only 2 per cent. It is usual practice in cooling gases to low temperatures to contact them with refrigerated surfaces such as pipe or finned tubes. However, the condensation and freezing that occur in cooling to low humidities require frequent shutdown of sections of the coils to remove the ice. Some types of coolers permit continuous removal of this ice, one example having been illustrated under vacuum freeze dryers (Fig. 54).

A gas leaving cooling coils may be somewhat less than saturated at its dry-bulb temperature because its dew point tends to approach the surface temperature of the coils as a result of the heat and mass transfer that occur at the boundary between the gas and cooling coil. The degree to which the leaving gas is less than saturated depends on the flow rate and on the properties of the gas film as given by the Prandtl and Schmidt numbers, $(c\mu/k)$ and $(\mu/\rho D_v)$, where μ is the gas viscosity, k and c its thermal conductivity and heat capacity, respectively, and ρ and D_v its density and diffusivity, respectively (see Sec. 8).

Large plants have been built for the purpose of drying air for blast furnaces by passage over refrigerated brine coils (Gayley dry blast). Refrigeration is also used to cool the water in spray dehumidifiers.

SECTION 14

ADSORPTION

BY

C. L. Mantell, Ph. D., Consulting Chemical Engineer, New York, Member, American Institute of Mining and Metallurgical Engineers, American Institute of Chemical Engineers.

CONTENTS

	Page
Theoretical Aspects of Adsorption	886
Adsorption as a Unit Operation	887
Industrial Adsorbents	888
Fuller's Earth	889
Percolation Plants	892
Revivification of Fuller's Earth	893
Acid-treated Clays	896
Bauxite	896
Bone Char	897
Decolorizing Carbons	898
Applications	901
Water Purification	902
Gas-adsorbent Carbons	904
Recovery of Gasoline from Natural Gas	906
Elimination of Odors	907
Alcohol Recovery	908
Gas Masks	908
Metal-adsorbent Chars	909
Medicinal Carbons	909
Activated Alumina	909
Properties and Applications	910
Silica Gel	911
CO_2 Purification	913
Gas Dehydration	913
Air Conditioning	913
Liquid-phase Adsorption	914
Petroleum Refining	914
Catalysis	915
Miscellaneous Applications	915
Ion Exchangers	915
Magnesia	916
Properties and Applications	916

ADSORPTION

REFERENCES: *Theory:* Brunauer, "The Adsorption of Gases and Vapors—Physical Adsorption," Princeton University Press. Princeton, N.J., 1943.

Practice: Mantell, "Adsorption," McGraw-Hill, New York, 1945; "Industrial Carbon, Its Elemental, Adsorptive and Manufactured Forms," 2d ed., Van Nostrand, New York, 1946.

Pressure Drop in Beds of Granular Solids: Leva, *Chem. Eng. Progress*, **43,** 549 (1947); *Ind. Eng. Chem.*, **39,** 857 (1947). Weintraub and Leva, *Chem. Eng. Progress*, **44,** 801 (1948).

General: Hougen and Marshall, *Chem. Eng. Progress*, **43,** 197 (1947). Amero, Moore, and Capell, *Chem. Eng. Progress*, **43,** 349 (1947). Hougen and Dodge, "Drying of Gases," PB-23201, Office of Technical Services, U.S. Dept. Commerce, 1946. Berg, *Trans. Am. Inst. Chem. Engrs.*, **42,** 665 (1946).

Silica Gel: Taylor, *Ind. Eng. Chem.*, **37,** 649 (1945). Mairs and Forziati, *J. Res. Bureau of Standards* (April, 1944), **34,** 435 (1945).

Ion Exchange: Kunin, *Ind. Eng. Chem.*, **40,** 41 (1948). Roberts and Thompson, *Chem. Eng. Progress*, **43,** 97 (1947).

Theoretical Aspects of Adsorption. The phenomenon of adsorption is difficult to define or limit. In some cases adsorption is similar to chemical reaction, but in extreme cases it is very different. Adsorption often takes place where any chemical reaction could hardly be expected. Chemical compounds have never been found containing the rare gases such as helium and argon, yet they are markedly adsorbed by charcoal. In similar ways, inactive substances are found to be adsorbed by the noble metals such as platinum and palladium.

The forces involved in adsorption phenomena are different from those encountered in chemical reactions and ordinary chemical compounds. Various theories have been advanced at different times to account for adsorption. It might be assumed that, at the boundary or surface of the adsorbent, part of the valences holding the atoms together are free, existing as partial or secondary valences. Adsorption may be due to the action of these partial-valence forces. Such a viewpoint would make possible a distinction between chemical and adsorption compounds. When we are able to calculate the force between the electrons and the positive nuclei in the atoms, the difference between chemical compounds and adsorption compounds will probably be less marked than it is now. It might be possible to account for all physical and chemical phenomena by calculating the forces between the molecules, electrons, and atoms. If these forces are all electrical in nature, the forces now considered as primary and secondary valences would be of the same kind but differing in degree.

If the relation between the adsorbed quantity and the concentration in the solution is studied, and the quantity of substance taken from the solution or the amount adsorbed on the surface of a solid adsorbent is plotted against the concentration of the solution, curves concave to the concentration axis (Fig. 1) are obtained. Schmidt [*Z. phys. Chem.*, **77,** 644 (1911); **78,** 667 (1912)] developed a more complicated relation

$$xe^{\frac{A(s-x)}{s}} = kcs$$

which agrees with experimental evidence, and where e is the base of the natural logarithm and A is a constant. The expression is purely empirical.

Langmuir [*J. Am. Chem. Soc.*, **38,** 2267 (1916); **39,** 1883 (1917); **40,** 1361 (1918); *cf.* Dushman, "High Vacuum," p. 205, 1922] proposed a theory which assumes that forces acting in adsorption are similar in kind to those involved in chemical combination. The forces acting in ordinary chemical combination result from strong deviations of the orbits of the outer electrons in the atomic structure, while the forces acting in adsorption are small deviations. When Langmuir's formula is applied to absorption of gases at high pressures, divergence is found between the experimental and calculated values. Langmuir assumes that elementary spaces or points of residual

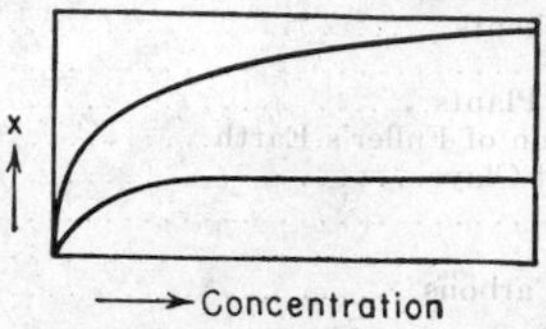

FIG. 1. Adsorption curves.

valency exist on the surface of a crystal. The adsorptive forces are concentrated at these points, and adsorption consists of the fixing of the adsorbed atoms in the elementary spaces for a certain time. He assumes that these spaces or points of residual valency can hold only one atom or molecule; therefore the adsorbed layer can be only one molecule thick. Thus the force between the atoms of the surface of the adsorbent and the atoms of the adsorbed substance decreases rapidly with the distance between the atoms. Langmuir proposes a relation between the concentration in the solution and the adsorbed quantity on the assumption that equilibrium is established between the rate of adsorption and the rate of desorption. In the case of gases, the equilibrium is between the rate of condensation on the surface and the rate of evaporation.

Langmuir obtained experimental verification of his theory in a series of determinations on adsorption of gases at relatively low pressures on plane surfaces of mica, glass, and platinum. When experimental evidence for adsorption layers more than one molecule thick is obtained, it is claimed that the adsorbing surface is not quite even, and roughness may increase its actual area considerably.

Brunauer ("The Adsorption of Gases and Vapors," Princeton University Press, Princeton, N.J., 1943) critically reviewed the adsorption theories and their supporting data and presented a multimolecular layer theory, according to which the physical adsorption of the first layer is determined by the net heat of adsorption and the surface of the adsorbent.

Mantell ("Adsorption," McGraw-Hill, New York, 1945) states that, except as background, the engineer designing adsorption systems and the operator are little interested in the theoretical phases or the theories of adsorption. They know the operation works. They are much more interested in factors like the distribution of the adsorbent; its settling rate or ability to be brought into contact with the material to be treated; its cost; its advantages and disadvantages in dealing with adsorption from liquids; or its hardness, crushing strength, mesh size, pressure drop, ability to stay in place, physical permanence, adsorptive capacity per unit of weight, or volume in terms of capacity of the treated product—none of which is closely related to the theoretical aspects of adsorption or the mathematical expressions of the proposers of theories of adsorption. To the engineer or designer, particularly in the gas-phase adsorption, the adsorbent is a physical part of the system, to be considered in the same way as a fan or blower, a tower or tank, a length of pipe or a fitting that generates pressure energy or consumes it in the course of passage of gas through the system in the process of treatment. An adsorbent of high equilibrium capacity would be inferior to one of lower capacity if the first blew out of the system and clogged it and the second stayed in place. The first would fail, and the second would work. The engineering properties of the adsorbent are therefore very important.

Adsorption as a Unit Operation. In its industrial applications the unit operation of adsorption might be considered as complementary and supplementary to other unit operations and processes, each of which might have definite limitations with respect to small concentrations of substances to be removed or eliminated.

Processing of liquids concerns itself with maintaining composition within definite limits. Raw material is subjected to stage treatment to obtain products of satisfactory purity. Each step effects increased concentration of material desired in the product and decreased concentration of undesirable constituents. These concentration changes involve chemical, physical, or mechanical separations. Chemical methods are effective when compounds of definite composition in true or colloidal solution are to be separated; otherwise physical or mechanical methods with chemical treatment as a preparatory step are favored. Each method has limitations with respect to small concentrations of substances usually classed as impurities.

The most commonly used chemical methods are

1. **Precipitation** by bringing about reaction with the substance to be separated, rendering it insoluble.
2. **Volatilization** by reaction forming a gas.
3. **Precipitation of colloids or suspensoids** by addition of a suitable coagulating agent.

The physical or mechanical methods include for separation of

1. **Suspended solids from liquids:** filtration, settling, thickening, centrifuging.
2. **Miscible liquids of different boiling points:** distillation, extraction.
3. **Immiscible liquids of different specific gravities:** centrifuging, settling.
4. **Dissolved solids from liquids:** evaporation followed by crystallization.

Often chemical action on substances in liquids is not sufficiently quantitative to produce the desired results. Further refinement in the chemical process would entail expense disproportionate to the results obtained. Impurities of organic origin are often more cheaply removed by a solid adsorbent.

When colloidal substances are precipitated from organic liquids or solutions, chemical reagents often fail to give complete elimination. These reactions depend on closely defined conditions of temperature, pH, etc., uncontrollable slight variations in which alter the equilibrium point in many cases in the opposite direction to that desired.

Filtration, settling, and centrifuging are effective in removing undissolved solids from liquids only when the particles are of sufficient size. The diameter of the pores in the filter medium determines the completeness of separation. Colloids and many ultrafine suspensions cannot be filtered, settled, or centrifuged out. Moreover, the rate of flow of a given liquid is a direct function of the permeability or porosity of the filter medium; hence separation of particles in an extremely fine state of division can be accomplished only at the expense of capacity.

Adsorbents remove solids by surface attraction and allow easy filtration.

Liquids to be distilled may, for practical purposes, be divided into two classes: (1) those consisting of two or more substances whose boiling points are constant, and (2) those exhibiting a more or less continuous variation of boiling point with quantity distilled.

In the former case, when the boiling points of any of the constituents are nearly the same, repeated distillation with a rectifying column will effect a separation, but last traces are exceedingly difficult to remove. This applies particularly to liquids whose taste or odor is markedly affected by the presence of very slight quantities of another liquid. Adsorbents can often readily remove impurities, leaving the distillates colorless and of satisfactory taste and odor.

Distillation will serve only to define or limit the boiling range. Distillation is useless for separation of any substance whose boiling point falls within the boiling range of the liquid itself. Such substances may be removed by adsorbents.

Many impurities emulsoidally dispersed, which do not settle or yield to centrifuging, can be separated by adsorption.

Soluble crystalline substances can be purified to any desired extent by dissolving in a suitable solvent, evaporating, and recrystallizing several times; but this procedure is costly and slow, and the yield is poor. Probably the most far-reaching effect of adsorbents is on solutions of substances to be crystallized. They may be either organic or inorganic, and the solvent may be aqueous or otherwise. The purity of the crystals is a function of the purity of the solution. If color bodies are present, there is always a tendency to occlude them, making the crystals colored when the product should be colorless. Treatment of the solution with adsorbents prior to crystallization will frequently increase the purity of the solution, eliminate colored bodies, and give high-purity crystals. Foreign matter in the solution also interferes with the shape and size of the crystals in many cases. This adversely affects the ease with which they are freed from mother liquor and slows up drying. The appearance of the finished product suffers from lack of brilliance and sparkle. Suitable adsorbents eliminate such impurities. Last, but not least in importance, is the greatly lowered yield caused by impurities in the solution. Even in very small concentrations, they may inhibit crystallization to a marked degree. Adsorbents increase the yield by removing impurities exerting a depressant effect on crystallization.

Adsorption from Gases. Solids of an essentially porous nature, each with a decided affinity for the adsorption of certain vapors, have been developed for industrial use in the recovery of solvents, in fractionation of mixed gases, as well as in other applications. The commercial materials include a variety of clays, chars, activated carbons, gels, alumina, and silicates. With most of them a selective preference is shown for the adsorption of vapors. They are more or less granular in form and are supported in beds or columns of suitable thickness, through which the gas from which vapor is to be adsorbed may be passed. Inasmuch as adsorption may be made practically complete even with very low vapor content, the procedure lends itself readily to recovery operations. Bulkeley [*Chem. & Met. Eng.*, **45**, 300 (1938)] reviewed the use of adsorbents:

Depending on a variety of factors, adsorbents are generally able to take up from 8 to 25 per cent of their weight of vapors.

The depth of bed required may vary from a few inches to several feet, which, of course, affects the power consumption of the system. The effect of the adsorbent on the solvent is to accomplish condensation at a temperature much above that of saturation. In the process, the latent heat of the solvent vapor is given up to the adsorbent bed and to the unadsorbed gas or air (mostly to the latter, since the specific heat of adsorbents is relatively low) so that it may in some cases be necessary to cool the gas. In addition, it should be noted that both the capacity and efficiency of adsorption fall off with rising temperature of the bed. An important point in favor of carbon is that, since it is not selectively adsorbent of water vapor, little latent heat is released through such action and, with a moist gas, less heating of both the exit gas and adsorbent bed occurs. When steam is used for regeneration of a carbon bed, the moisture remaining in the carbon is immediately evaporated owing to its selective displacement by solvent vapor when adsorption is resumed, and the resultant cooling effect often makes a precooling period unnecessary.

Recovery is, of course, not completed with adsorption. To release the adsorbed solvent the adsorption bed must be regenerated, usually by heating the bed above the boiling temperature of the solvent at atmospheric pressure, by submerged heating elements or by circulating hot air, combustion gases, or steam through the bed. The volatilized solvent is then condensed in a surface or other type of condenser and the gas returned to the adsorbers for the stripping of any uncondensed solvent. The recovered solvent from the condenser can be further purified by rectification and the regenerated adsorber cooled, if necessary, and returned to service. In some instances it is practicable to accomplish regeneration at lower temperatures by application of a vacuum, in which case the need for a heat-carrying gas is eliminated. The volatilized solvent can then be condensed in vacuo. Sometimes adsorption can be accomplished at pressures higher than atmospheric.

Three types of arrangement of adsorbent units are in use, the type depending both on the performance ability of the adsorbent with relation to a particular solvent, and the requirements of the process. Generally two units suffice for continuous operation, for in most cases a single bed or series of beds of adsorbent is capable of trapping all of the solvent. In this case one unit will be in operation while the other is being regenerated. In discontinuous operations where there is a shutdown period, say every 20 hr., of sufficient duration to accomplish regeneration, a single unit may be sufficient provided that its capacity is adequate to recover all of the solvent. But for fully continuous operation where a second-stage adsorber is necessary to remove the solvent completely, the so-called three-step cycle must be used. This consists of a group of three interconnected adsorbers, two of which are operated on the recovery line in series, while the third is being regenerated.

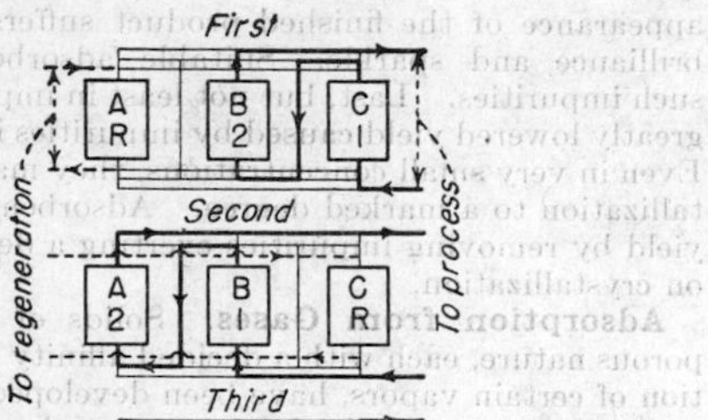

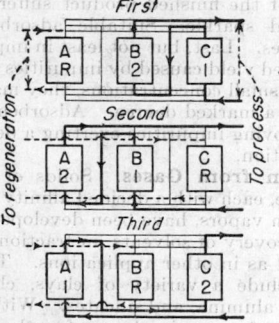

FIG. 2. Diagram illustrating operation of three-step adsorber cycle.

Figure 2 suggests diagrammatically the connections used in such a setup, showing how the connections are altered for the three steps of the cycle. In any one step the first adsorber in the line is the partially charged unit which was second in the line in the preceding step, while the second adsorber is the unit which was being regenerated in the preceding step. Meanwhile the third adsorber is undergoing regeneration (and cooling if necessary). By this method each unit is rotated in turn through the three steps of regeneration, second stage and first stage adsorption, and since the exit gas from the final adsorber should be practically free of solvent, it can be permitted to go to waste. The use of two adsorption stages permits utilizing the ultimate adsorption capacity of each unit, since as the first unit on the line approaches capacity and rises in temperature, any solvent escaping will be caught by the second stage fresh unit.

Generally speaking, an adsorption recovery system is relatively high in first cost, but usually pays a handsome return on the investment, with operating and maintenance costs that are low in comparison with the value of the recovered solvent. Small systems can be manually operated with efficiency equal to that of the larger systems, which should be fully automatically controlled.

INDUSTRIAL ADSORBENTS

The **commercially important solid adsorbents** are in the order of their tonnages consumed:

Adsorbent	Important Industrial Uses
Fuller's earth	Refining of petroleum fractions, vegetable and animal oils and fats, and waxes
Bauxite	Percolation treatment of petroleum fractions
Acid-treated clays	Contact filtration of petroleum fractions
Bone char or bone black	Sugar refining; ash removal from solutions
Decolorizing carbons and water carbons	Sugar refining; refining of vegetable and animal oils and fats, and of waxes; miscellaneous decolorizing of inorganic and organic substances; water purification; purification of dry-cleaning fluids; purification of food products
Gas-adsorbent carbon	Solvent recovery; recovery of gasoline from natural gas; elimination of industrial odors; purification of CO_2 and industrial gases; gas masks
Alumina	Dehydration of air, gases, and liquids
Silica gel	Dehydration and purification of air and industrial gases; air conditioning; refining of petroleum distillates; gas masks
Ion-exchange materials	Water treatment, chemical processing
Magnesia	Treatment of gasoline and regeneration of dry-cleaning solvents
Medicinal carbons	Elimination of bacteria and toxic poisons; an addition to animal foods
Metal-adsorbent chars	Recovery of precious metals

The properties, characteristics, and industrial applications of these adsorbents are individually discussed below. In only a few cases are the different adsorbents competitive in nature, because each is specific in its properties, applications, and capabilities. Rather than competitive, in some uses they are complementary, each doing that part of the work for which it is best fitted.

The adsorbent is industrially applied by two different methods termed **percolation** and **contact filtration**. In the **percolation method**, the adsorbent is held in place and the liquid or solution to be treated is caused to flow over, by, through, and around it. In the percolation method, the adsorbent is usually granular in form and must have certain mechanical and physical requirements to hold its shape and be retained in place. In **contact filtration**, the finely divided adsorbent is mixed with the liquid or solution to be treated, agitated, and the adsorbent removed from the treated liquid by filtration or by combination methods of settling and filtration. An example of percolation "bleaching" or treatment is found in the bone-char process for sugar refining, or in the treatment of petroleum oils by granular fuller's earth in cylindrical vertical towers. Examples of contact filtration are found in the Suchar or Norit process of sugar refining, or the ordinary application of finely powdered decolorizing chars, or the contact methods employing either finely divided earth or acid-treated clays in the clarification and filtration of petroleum stocks in the manufacture of lubricating oils. The equipment, types, and operation employed in the percolation method of adsorption are similar whether applied to sugar, petroleum oil, or organic solvents, while the same holds true for contact treatment whether applied to inorganic liquids, solutions of inorganic salts, glycerol, petroleum oil, sugar, or a host of other substances.

Fuller's Earth

From a tonnage viewpoint, the adsorbent most widely used is fuller's earth. The Florida grade is accepted as being of highest quality.

Fuller's earth is obtained as a clay, semiplastic when wet, rocklike when dry, and, as mined, it usually contains 40 to 60 per cent of free moisture. A distinctive feature of the Florida earth is its laminated structure, probably accounted for by its marine origin.

After mining, the earth is hauled to the mill where it is dried in rotary kilns to remove nearly all the free, and part of the combined, moisture and is then ground on roller mills, separated on silk screens into the various commercial grades representing meshes of approximately 16 to 30, 30 to 60, 60 to 100, through 100 and through 200, respectively. A chemical analysis is found in Table 2, and a typical flow sheet for the manufacture of percolation-grade fuller's earth is given in Fig. 3. Table 1 gives physical data on representative clays.

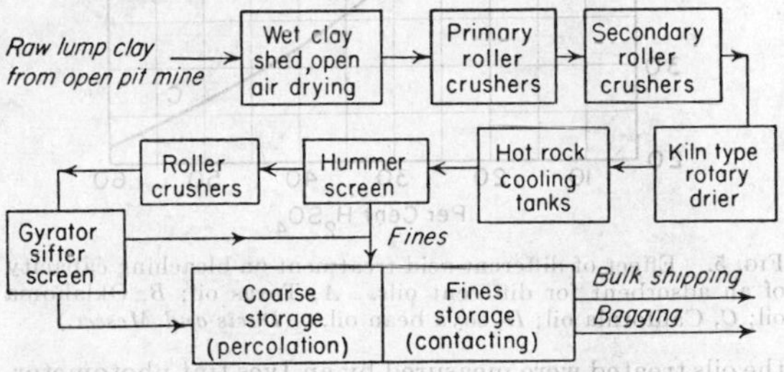

FIG. 3. Flow sheet for manufacture of fuller's earth (percolation grade).

Volatile matter is determined as per cent loss in weight after a 10-min. ignition at 1800°F. Moisture content is determined by drying to constant weight at 220°F. The volume occupied by a given clay after 5 min. mechanical tamping time is calculated in pounds per cubic foot. The acidity of a bleaching clay is the number of milligrams of KOH required to neutralize the distilled water extract from 1 g. of clay with phenolphthalein as the indicator. Oil retention for contact grade clays is the increase in weight of clay after contact with oil and blowing the cake with air at 40 lb./sq. in. pressure at 375°F. Water-content correction is made. Decolorizing tests are run by comparison with so-called standards.

Table 1. Typical Tests on Representative Clays

Dry natural clay	Fuller's earth, Florida-Georgia type		Natural bleaching clay southwest type fine grade	Artificially activated clay domestic type fine grade
	Fine grade (contacting)	Coarse grade (percolation)		
Volatile matter, %	16.0	16.5	16.0	21.0
Density, lb./cu. ft.	31.0	35.0	53.0	36.0
Acidity, mg. KOH/g	Neutral	Neutral	Neutral	3.5
Screen test, % through 200 mesh	95.0	....	70.0	88.0
Mineral-oil decolorization value (efficiency $a = 100\%$), %	100	....	40–110	150–500

Fuller's earth is highly hygroscopic, picking up or giving off moisture readily, which causes the free-moisture content to vary markedly. This, however, averages about 6 to 8 per cent when it comes from the mill. In the newer practice of clay processing, the material after mining is broken up in smooth-roll crushers. If insufficient water is present, more is added and the mixture pugged. The product is then extruded through multiple-orifice dies through screw-type machines. Extrusion reduces the density and breaks up laminations. Fines are recycled. Clay burning or calcination is carried to 800°F.

Fuller's earth is used for bleaching, clarifying, or neutralizing (sometimes combining decolorizing and neutralizing) mineral, vegetable, and animal oils, fats, and greases. Petroleum products so treated are bright stocks, cylinder stocks, long-residuum or blending stocks, neutral and spindle oils, transformer and cable oils, paraffin wax, petrolatum, petroleum jelly, kerosene, gasoline, and refined oils. Lubricating oil that has been treated with H_2SO_4 is often treated with fuller's earth to neutralize, decolorize, and improve demulsibility. An important use is in the vapor-phase cracking of gasoline, where by its use it is claimed that the acid-treating, neutralization, and redistillation processes are eliminated and a water-white, gum-free product which retains a large proportion of its antiknock value is obtained.

Fuller's earth removes gum or potential gum from petroleum fractions and increases the oxidation stability. Gray towers in practice are operated until the product no longer meets the gum specification. Kerosene is treated for the removal of water by passage through fuller's earth, while white oils are decolorized and made more stable with particular reference to acid stability tests. There is some application of fuller's earth for the stabilizing of adsorber oils employed in the purification of coke oven and natural gas in bubble tower and similar absorbers. Acid-treated oils are always handled by the contact method, whereas the percolation procedure commonly necessitates neutralization in advance of treatment.

Table 2. Analyses of Clays*
In per cent

Constituents	A	B	C	D	E	F	G	H
SiO_2	72.95	58.10	58.72	57.95	58.04	47.38	59.30	56.53
Al_2O_3	12.65	15.43	16.01	1.57	15.81	15.38	9.53	11.57
Fe_2O_3	3.56	4.95	2.12	0.85	3.10	2.57	1.70	
FeO	0.47	0.30	.....	.....	.....	.....	.....	3.32
MgO	.57	2.44	3.30	19.71	4.12	4.24	3.20	6.29
CaO	1.00	1.75	1.05	4 17	1.19	2.25	1.13	3.06
Na_2O	0.20	0.27	2.11	1.84	1.62	0.59	0.40 }	1.28
K_2O	.68	66	1.50	0.43	0.75	.....	..... }	
CO_2	.....	.84	.....	1.22	.....	None	None	
(Loss below 105°C.)	5.77	4.59	6.21	4.82	4.92	20.50	15.10	
(Loss above 105°C.)	1.25	9.45	8.61	8.11	10.03	7.10	8.79	
Water (in composition)	.....	.....	.....	.....	.....	.....	.....	17.95

* Davis and Messer, *Trans. Am. Inst. Mining Met. Eng.*, 1929, p. 293.

A. Commercial fuller's earth from Florida. Analyst, H. K. Shearer (Bauxite and Fuller's Earth of the Coastal Plain of Georgia, *Geol. Survey Georgia, Bull.* 31, 1917).

B. Commercial fuller's earth from Florida. Analyst, H. K. Shearer (*ibid.*).

C. Clay from Nevada with no bleaching action. Analysts, Davis and Messer (*Trans. Am. Inst. Mining Met. Eng.*, 1929, p. 288).

D. Commercial high-magnesia earth from Nevada. Analysts, Davis and Messer (*ibid.*).

E. Earth from which commercial acid-treated adsorbents are prepared. Analysts, Davis and Messer (*ibid.*).

F. Earth from which commercial acid-treated adsorbents are prepared. Analyst, Salmi ("The Chemically Prepared Adsorptive Clays and Their Application in the Purification of Oils, Fats, and Waxes," Los Angeles, 1926).

G. Commercial adsorbent prepared from earth *F*. Analyst, V. Salmi (*ibid.*).

H. Commercial fuller's earth from Florida (Floridin Co. analysis, not from Davis and Messer paper).

Among the animal oils treated with fuller's earth are: tallow, lard, bone oil, whale, fish, cod, seal, neat's-foot, etc. Vegetable oils treated include: linseed, cottonseed, rape, coconut, corn oil, palm, sesame, poppy seed, peanut, wood oils, turpentine, olive, sunflower, mustard, castor, etc.

More than 95 per cent of the annual fuller's earth production in the United States is used by the petroleum industry, and from 2 to 3 per cent for vegetable oils and fats.

Certain types of fuller's earth have the ability to withstand repeated burning or roasting for revivification, while maintaining a high degree of efficiency.

In refinery practice it is customary to use the coarser grades of earth (16 to 30, 30 to 60, or 60 to 100) in the percolating method from three to twenty or more times, depending upon the needs of the particular plant and the efficiency of the burning or revivifying process.

The name fuller's earth was derived from its early use in "fulling," the operation of removing grease from woolen goods. It is used as a general term to designate mineral matter containing hydrous aluminum silicates of the clay group, which has a high capacity for the removal of colors from oils, whether the oil is of the animal, vegetable, or mineral variety. These earths are usually dried and broken up after mining. They may be "tempered" by heating or treated with reagents such as water and various mineral acids to improve their efficiency. Most fuller's-earth deposits are of sedimentary nature, while others are the result of alterations of basic dikes, glacial silts, volcanic ash, or other decomposition products of igneous rocks.

There appears to be no relation between the adsorptive power of fuller's earth and clays and their chemical analysis. Davis and Messer (*Trans. Am. Inst. Mining Met. Eng.*, 1929, p. 293) give the analytical results on various clays (Table 2). *A* and *B* are good bleaching materials but differ markedly in their chemical analysis. *B* and *C* approximate each other in their chemical analysis. but *C* has no bleaching action. A high silica-to-alumina ratio appears to be characteristic of the earths having good bleaching powers. Earths that may be converted into good adsorbents by acid treatment have the property of swelling in water, but not all earths that swell make good oil bleachers on treatment with acid. When fuller's earths that contain no soluble salts have a high apparent acidity in water, they show good bleaching action. Many English clays as well as those from the coastal plain of the Southeastern United States, particularly Georgia and Florida, require from 10 to 150 cc. of 0.1*N* NaOH per 100 g. of earth to neutralize the apparent acidity. On the other hand, some earths are highly efficient bleaching agents but do not show acidity.

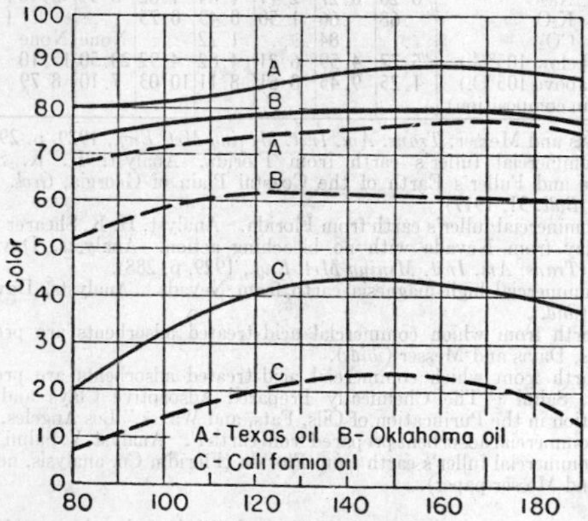

FIG. 4. Effect of contacting temperature on the color of oils. Solid line indicates red, broken line indicates yellow. (*Davis and Messer.*)

The adsorbing power of fuller's earth has been estimated as being so great that a cubic foot would adsorb about 0.6 lb. lime and thus be equivalent to the acidity of a 2 per cent H_2SO_4 solution.

The action of earths from different localities is often quite specific. Oils from different oil fields also require specific treatment, bleaching well with some clays and poorly with others. The same clay may be very effective with one oil and less efficient with another oil from a different source. The variables to be considered are similar to those encountered in decolorization by carbons. The most important variables are temperature, time, pH, concentrations of color and adsorbent, as well as the types of colors to be removed. In Figs. 4, 5, and 6, the colors of

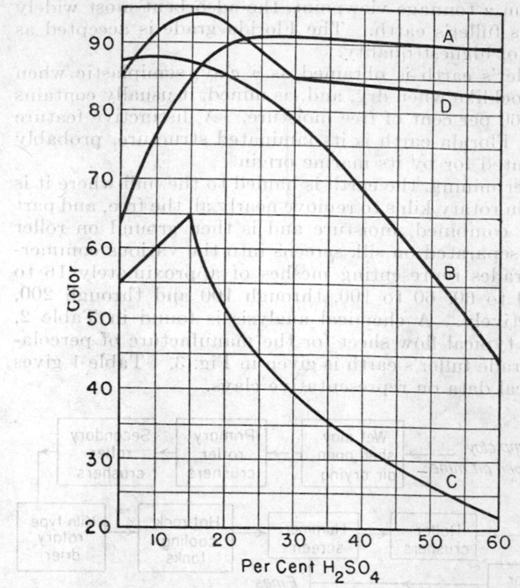

FIG. 5. Effect of different acid treatment on bleaching capacity of an adsorbent for different oils. *A*, Texas oil; *B*, Oklahoma oil; *C*, California oil; *D*, soya bean oil. (*Davis and Messer.*)

the oils treated were measured by an Ives tint photometer, the instrument being graduated from 0 to 100 in such a way that the lighter the color of the oil the higher the reading. The effect of temperature on color removal is shown in Fig. 4. The effect of different acid treatments

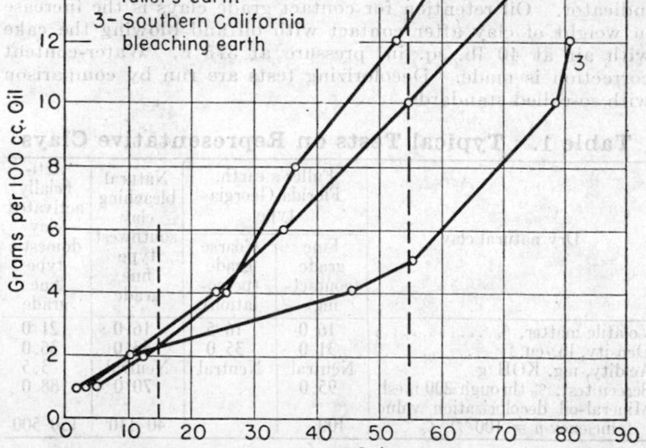

FIG. 6. Effect of different concentrations of adsorbents on California motor oil. (*Davis and Messer.*)

of the earth before it is used as a bleaching agent is shown in Fig. 5. The effect of different concentrations of the adsorbent is given in Fig. 6.

Rogers, Grimm, and Lemmon [*Ind. Eng. Chem.*, **18**, 164 (1926)] have correlated much of the earlier adsorption

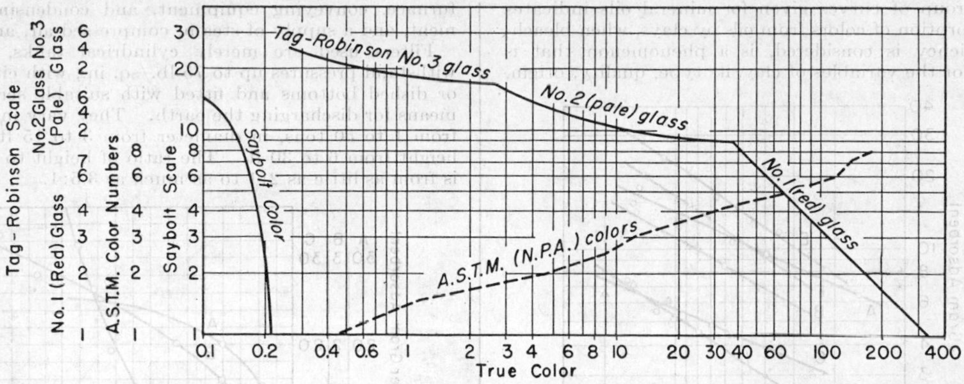

FIG. 7. Relation of various color systems and true color scale. (*Rogers, Grimm, and Lemmon.*)

studies on the decolorization of mineral oils. Various experimenters have used a wide range of testing methods and different types of colorimeters. Figure 7 gives the relation among the various color systems. The preferred method of color measurement employs the optical density color system [Ferris and MacIlvaine, *Ind. Eng. Chem.*, anal. ed., **6**, 23 (1934)]. Color conversions vary from oil to oil, inasmuch as the "bloom" enters. The A.S.T.M. curves differ for resulting colors in oils that have been

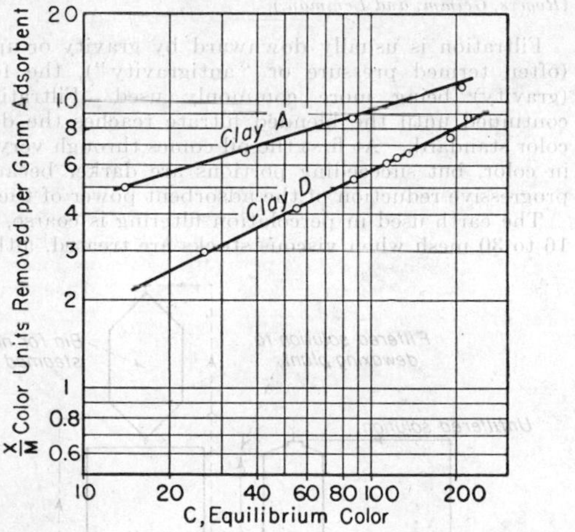

FIG. 8. Application of exponential adsorption equation to decolorization of mineral oil. Curves *A* and *D* represent clays from different geographical locations. (*Rogers, Grimm, and Lemmon.*)

treated with fuller's earth or bauxite. Rogers *et al.* plotted the results of their adsorption studies employing the equation

$$\log \frac{x}{m} = \log K + \frac{1}{n} \log C$$

plotting log x/m against log C to obtain straight lines with the slope of $1/n$. The results are shown in Fig. 8. Similar results are obtained for the decolorization of kerosene (Fig. 9), cylinder stock (Fig. 10), and paraffin wax (Fig. 11). The slopes $(1/n)$ of the adsorption curves increase in the regular way as one passes from darker to lighter oils. The decolorizing action of the adsorbents is quite specific for different oils and the types of colors that need to be removed. This is shown in Fig. 12. In general, only the results of adsorption of the colors from a single oil by different amounts of adsorbent can be grouped together. The conditions are not those of a simple solution of a solute in a single solvent. Results of tests of an oil diluted with a colorless naphtha are not comparable with tests made on the original oil, even if the amount of adsorbent is expressed correctly in relation to the amount of original oil. Decolorization of a partly decolorized oil gives results widely different from those with the original oil. Oils of the same kind produced by

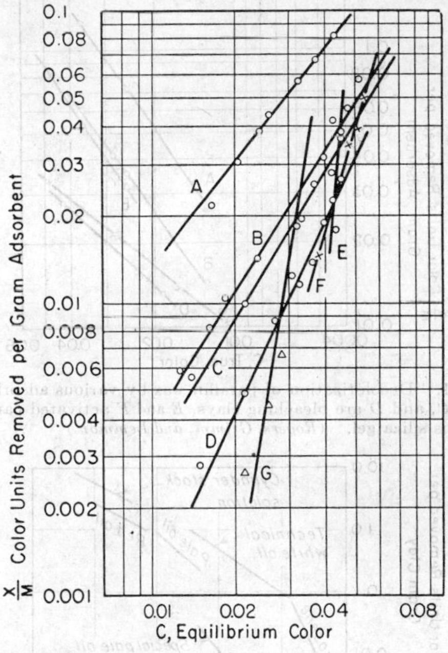

FIG. 9. Decolorization of kerosene by various adsorbents. *A*, *B*, *C*, and *D* are bleaching clays, *E* and *F* activated carbons, and *G* is silica gel. (*Rogers, Grimm, and Lemmon.*)

the same refinery operations will give different adsorption curves if they differ considerably in color. The age of the oil, as the result of the development of oxidized substances, also affects adsorption values.

Comparative curves for the adsorption of iodine from toluene by charcoal, which is one of the methods for the testing of activated carbons, the adsorption of coloring matter from cylinder stock by clay, and the adsorption of coloring matter from kerosene by clay, given in Fig. 13, are phenomena all of the same type conforming to the same general laws.

The group of curves given for mineral oils indicates that adsorption of colors from oils by clays, when bleaching efficiency is considered, is a phenomenon that is specific for the variables of clay, its type, quality, origin,

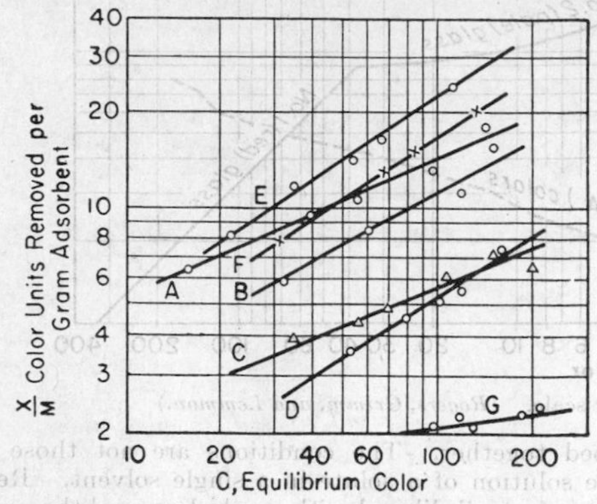

FIG. 10. Decolorization of cylinder stock solutions by various adsorbents. *A, B, C,* and *D* are bleaching clays, *E* and *F* activated carbons, and *G* is silica gel. (*Rogers, Grimm, and Lemmon.*)

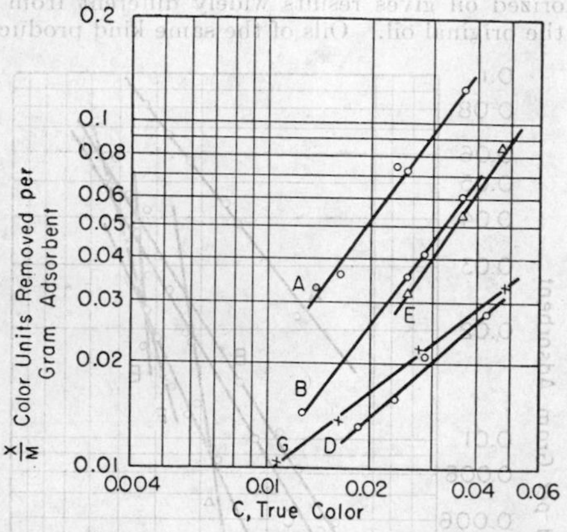

FIG. 11. Decolorization of paraffin wax by various adsorbents. *A, B, C,* and *D* are bleaching clays, *E* and *F* activated carbons, and *G* is silica gel. (*Rogers, Grimm, and Lemmon.*)

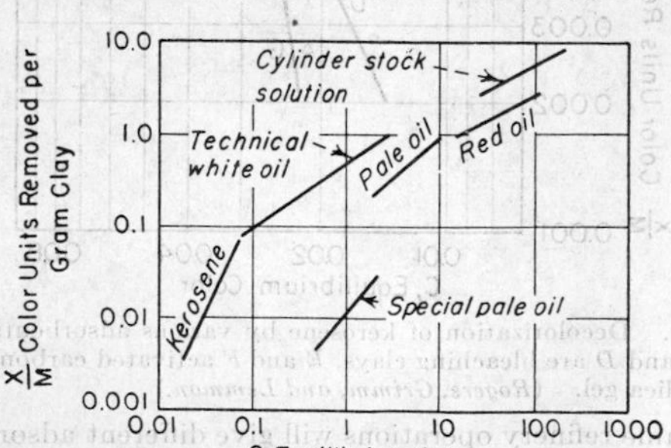

FIG. 12. Comparison of decolorization curves for various petroleum oils. (*Rogers, Grimm, and Lemmon.*)

and treatment; of the oil, its type, color, acidity, temperature, kind of color, and source of oil, so that each particular usage of a particular adsorbent on a particular oil must in plant practice be studied separately.

Percolation Plants. The essential equipment for a percolation plant includes tanks, pumps, filters, earth furnace, conveying equipment, and condensing equipment, and a supply of steam, compressed air, and water.

Filters used are merely cylindrical tanks, built to withstand pressures up to 75 lb./sq. in., with either cone or dished bottoms and fitted with suitable screens and means for discharging the earth. They vary in capacity from 1 to 50 tons, in diameter from 3 to 15 ft., and in height from 6 to 30 ft. The ratio of height to diameter is from as little as 2:1 to as much as 3.5:1.

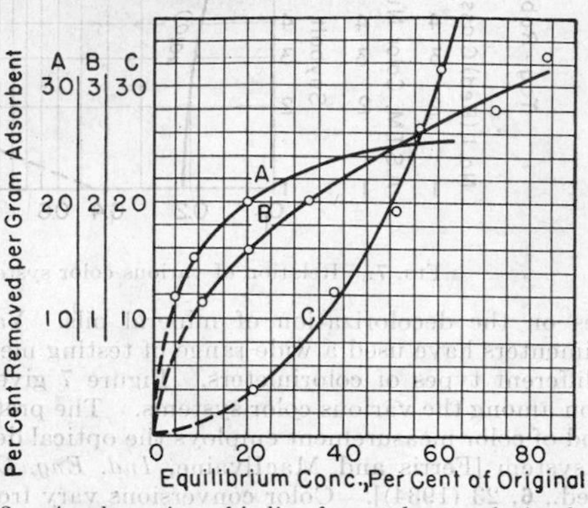

FIG. 13. *A*, adsorption of iodine from toluene solution by charcoal; *B*, adsorption of coloring matter from cylinder stock solution by clay; *C*, adsorption of coloring matter from kerosene by clay. (*Rogers, Grimm, and Lemmon.*)

Filtration is usually downward by gravity or upward (often termed pressure or "antigravity"), the former (gravity) being more commonly used. Filtration is continued until the blended filtrate reaches the desired color standard. At first the oil comes through very light in color, but succeeding portions are darker because of progressive reduction of the adsorbent power of the clay.

The earth used in percolation filtering is coarse, being 16 to 30 mesh when viscous stocks are treated, although

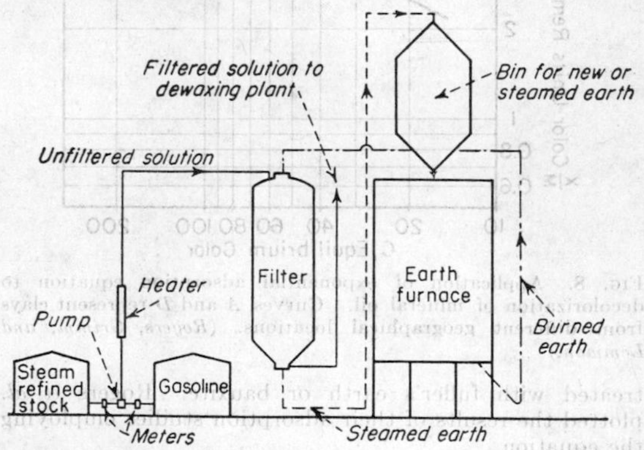

FIG. 14. Typical flow chart of filter plant running bright stock solution. (*Floridin Co.*)

in the average or normal practice 30- to 60-mesh material is employed. When the oils are diluted with gasoline or naphtha, finer mesh material may be employed. A typical flow chart is shown in Fig. 14.

The variables of percolation filtration as determined from plant operation, all other conditions being equal, are:

1. The higher the column of earth (depth of filter) the better the bleach (light color). The diameter of the filter may not affect the color, but soaking rates and yield are related to the diameter. In other words, height of column gives bleach; diameter of column gives quantity.

2. The finer the earth the shorter the column necessary.
3. The higher the viscosity at a given temperature the longer the time required.
4. In order to obtain maximum efficiency, the viscosity of the oil to be treated should be reduced, either by heating or by dilution, to a point where it will penetrate the minute pores of the earth.

Typical Filter Cycle

No. 7 burn fuller's earth.
16-ton filter, approximately 18 ft. high by 8 ft. diameter.
Pennsylvania residual stock (150 viscosity at 210°F.) in naphtha solution, 40 % oil, 60 % naphtha.
Filter temperature, 130°F.
Head pressure, 30 lb./sq. in.
Charge with oil, 4 hr.
Fill from rerun, 2 hr.
Shut in, allow air release, 2 hr.
Run to No. 1 true color, 72 hr. }
Run to rerun, 10 hr. } running time.
Run to cycle, 20 hr. }
Pumping off, 2 hr. (draining).
Soak, naphtha, 4 hr.
Wash, 8 hr.
Steam, 13 hr.
Dumping, 3 hr.
Total time, 140 hr.

Running Rates

Type of stock	Rate, bbl./ton/hr.	Temp., °F.
Pennsylvania residual stock in 60/40 naphtha solution	0.2–1	150
Pennsylvania bright stock, straight	0.1–0.5	200
Pennsylvania 180 neutral, straight	0.5–1.0	90–120
Pennsylvania 225 neutral, straight	0.3–0.6	150–160
Paraffin wax (1½ A.S.T.M.)	0.7–1.2	130–160
Petrolatums, Pennsylvania 123°F. melting point	0.3–0.8	180–220
Mid-continent 165° melting point amorphous wax	0.2–0.4	210

The filters are charged with oil dissolved in naphtha and are allowed to stand long enough to release entrained air. The discharge valves of the filters are then opened and the oil is permitted to drain out. The mechanically held or entrained oil in the adsorbent earth is recovered by washing with naphtha. The filter is then steamed to recover the wash naphtha. After steaming, the fuller's earth is dumped from the filter and sent on to kilns for burning. The adsorbent is thus revivified for another cycle. After burning, the adsorbent is again loaded into the filter shell and another purification cycle started. A typical filter cycle is outlined in the table shown on p. 893. Running rates for different oils are given in barrels of oil per ton of fuller's earth.

The filtration of petroleum through a finely porous adsorbent gives separations into fractions of various gravities, viscosities, color, and other characteristics. Table 3 gives the data [Kauffman, *Chem. & Met. Eng.*, **30**, 156 (1924)] on antigravity filtration of a steam-refined lubricating stock which has been diluted with naphtha to aid filtration and decolorization. The oil characteristics after reducing (*i.e.*, after the gasoline dilution had been distilled off) are given. The separation of the oil into fractions is clearly seen.

After the earth is saturated and the filtrate runs too dark, the operation is stopped and the surplus oil washed out of the earth by naphtha. The filter is unloaded and the earth transferred to kilns or furnaces for revivification. It is ignited and heated to 1000°–1400°F., although in the case of some earths lower temperatures are employed.

To illustrate the other method of use of fuller's earth and adsorbent clays in petroleum technology, a typical flow sheet of contact filtration will be discussed.

In the distillation of crude petroleum, the gasoline, kerosene, gas oil, and part of the wax-distillate fraction are separated as overhead distillates. The residue left in the still is used as a base for the manufacture of either long-residuum or bright stock, depending upon the extent to which the crude oil has been treated in the still and whether or not the flash point is relatively high or low.

According to the particular refinery's practice, long residuums as low as 390° to 400°F. flash test and 60 to 65 sec. viscosity at 212°F., and bright stocks as high as 540° to 550°F. flash and 180 to 190″/210°F.* viscosity are made. The flow sheet in question (Fig. 15) [Kauffman, *Chem. & Met. Eng.*, **34**, 155 (1927)] deals with a long residuum of 440° to 450°F. flash and 80 to 85″/210°F.† viscosity. This long residuum is first treated with H_2SO_4 at the rate of 40 to 45 lb. of 66° acid per barrel of oil at 140° to 150°F. After treatment, the sludge is settled at the bottom of the agitator and drawn off. The oil with a small amount of clay is filtered to remove emulsified H_2SO_4, after which it is mixed with clay in mixing agitators and pumped to a pipe still or other heating unit, then follows through the processing as shown in the flow sheet. After filtration and cooling, the oil is diluted with naphtha and dewaxed by chilling and centrifuging. The amount of bleaching clay used is 0.5 lb./gal. oil. Figures 16 and 17 show the layout of the treatment plant. The amount of earth needed to produce a desired color in the finished lubricating oil is a function of the temperature, as shown in Fig. 18.

Revivification of Fuller's Earth. Four types of furnaces, or burners as they are more often called, are in general use for the revivification of earth. The first is the **gravity-drop** burner, of which the Paris, Brockway, Keubler, and Mohr are typical. In this the earth falls by gravity through the gases of combustion and over a series of fire-clay baffles or slides, set at an angle of about 45 deg., the material being subject to heat for less than a minute. Capacities range from 10 to 24 tons per 24-hr. day.

The second type is the **rotary kiln,** of which the Bonnot is a representative; a steel cylinder lined with fire brick and slightly inclined from the horizontal, which revolves slowly, being fired from one end and fed with earth at the other. The earth remains in the kiln from 15 to

* 190 sec. at 210°F. Saybolt viscosity.
† 85 sec. at 210°F. Saybolt viscosity.

Table 3. Antigravity Filter Fractionation Data on "Cut-back" Steam-refined Lubricating Stock

Description	Original stock	First through clay	Stream after 10 bbl. yield	Stream after 23 bbl. yield	Stream after 75 bbl. yield	Stream after 148.4 bbl. yield	Stream after 294.6 bbl. yield	Stream after 500.4 bbl. yield	Stream after 589.4 bbl. yield
Temperature of filter stream, °F.	......	106	136	118	106	84	68	68	68
Gravity	41.6	49.0	48.3	46.2	43.2	42.4	41.9	41.8	41.5
Sulfur, %	0.134	0.011	0.020	0.045	0.080	0.128	0.125	0.133	0.129
After Reducing									
Gravity	23.0	30.7	29.7	27.7	25.4	23.9	24.1	22.9	23.3
Flash	430	460	480	440	420	435	470	440	440
Fire	540	535	540	540	525	540	545	540	535
Viscosity at 100°F	3122	819	892	1148	1937	2291	2662	2722	2798
Viscosity at 210°F	156	90	87	95	125	135	140	140	140
Pour	60	55	55	55	65	65	65	65	65
Iodine value	18.0	0.61	1.2	4.3	12.1	14.8	16.4	17.4	18.1
A.S.T.M. carbon residue, %	2.25	0.008	0.027	0.115	0.546	0.932	1.49	1.61	1.79

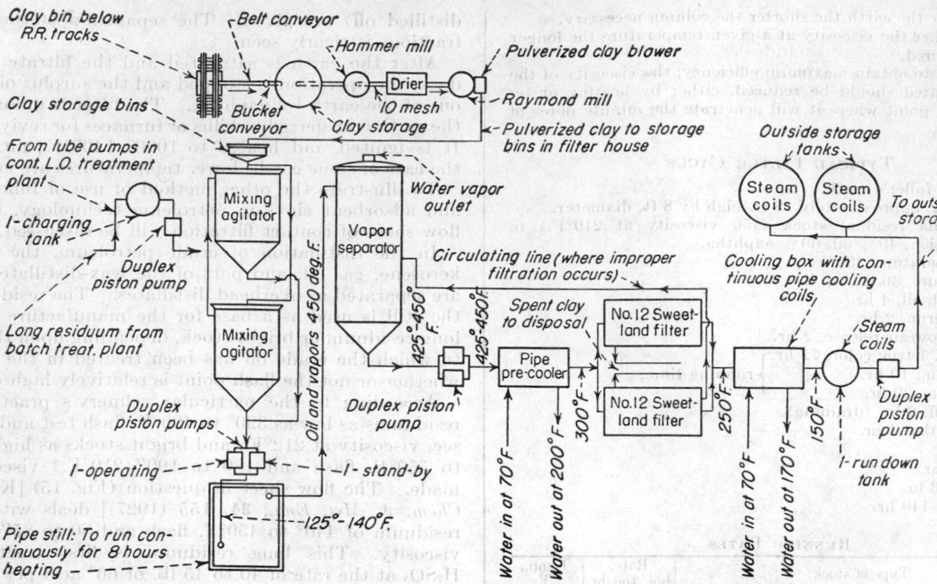

FIG. 15. Diagrammatic flow sheet for the operation of a contact filter plant.

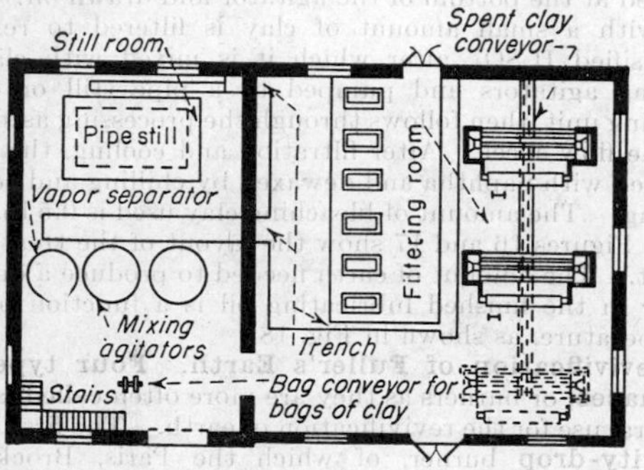

FIG. 16. Layout of principal equipment in contact filter plant.

30 min. Sizes range from 2.5 to 6 ft. in diameter and 24 to 60 ft. in length, and usual capacities are from 24 to 48 tons per day.

The third type is the **multiple-hearth** furnace, adapted from those used for roasting ores, of which the Nichols-Herreshoff and Wedge are examples. In this type the earth is slowly rabbled across each hearth, dropping from one to the other, being retained in the furnace from 1 to 2 hr.

Furnaces in general use range in number of hearths from 7 to 12 and in outside diameter from 16 to 25 ft. Capacities range from 24 to more than 100 tons per day. Fuel consumption, when natural gas or refinery gas is used, is from 750 to 1250 cu. ft. or more per ton.

At least two bins for steamed earth and one for burned earth are usually provided. New earth is stored in bins or in bags as received. Means of conveying earth must be provided. Belt and bucket elevators and belt conveyors are preferred. Chain and bucket elevators, screw conveyors, and flight conveyors are also used. Pneumatic conveyors are not to be recommended except under very favorable conditions where the vertical lift is short and the changes in direction few. Industrial cars and platform elevators are used to a certain extent, where tonnage handled is small. When filters are blown out

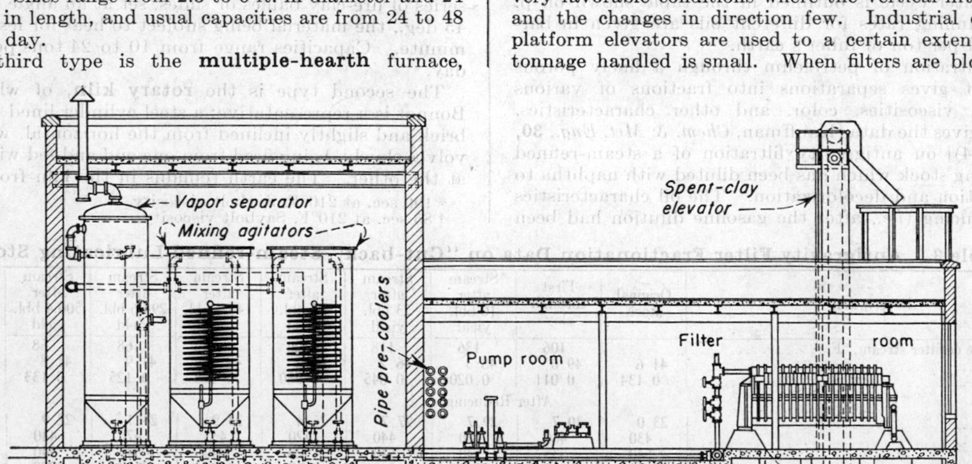

FIG. 17. Sectional elevation of filter house showing arrangement of vapor separator, mixing agitators, cooling coils, pumps, and filter presses.

with steam, this steam is either piped directly to a condenser or used in the redistillation of filter wash.

Auxiliary equipment includes the piping system, which in some plants is quite elaborate, allowing the filtrate from any filter to be diverted to any other filter in a battery; temperature and pressure regulators for the unfiltered-oil lines; reducing valves and water traps for the steam lines; relief valves and pressure gages, recorders, pyrometers for earth furnaces, heat exchangers, etc.

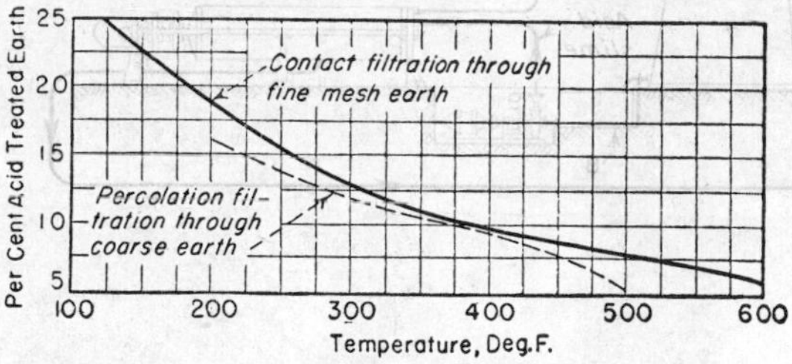

FIG. 18. Temperature vs. per cent acid-treated bleaching clay necessary to obtain type color on 90 seconds viscosity at 210°F. Pennsylvania cylinder stock.

Following are some installation costs (courtesy of the Floridin Co., Warren, Pa.): A typical percolation filter plant, with brick buildings and having a filter capacity of 400 tons of earth, would cost about $300,000 complete. This cost includes 16- to 25-ton filters; 2 to 10 hearths, 16-ft.-diameter Nichols-Herreshoff furnaces; conveyors; elevators; pumps; and all necessary equipment.

Gravity slide retorts cost approximately $4000/ton/hr. capacity, multiple-hearth furnaces complete cost approximately $15,000, and rotary kilns $18,000 to $20,000/ton/hr. capacity.

The comparative efficiencies of Florida fuller's earth after revivifying are shown in Fig. 19. In multiple-hearth furnaces of the Herreshoff or Wedge type, the earth may be revivified from twenty-five to thirty times before its efficiency has been reduced to so low a figure that it must be discarded. Most plants find it more economical to discard earth at or before the tenth burn or revivification. On the other hand, an earth burned thirty times in a multiple-hearth furnace may be better than the seventh burn of earth handled in a gravity-slide burner.

The fourth is the **Thermofor** kiln [Simpson and Payne, *Natl. Petroleum News*, **31**, R-486; *Refiner Natural Gasoline Mfr.*, **18**, 438, 455; *Oil Gas J.*, **38**, 27, 147, 150, 152, 154, 156, 159 (1939)], consisting essentially of a stationary

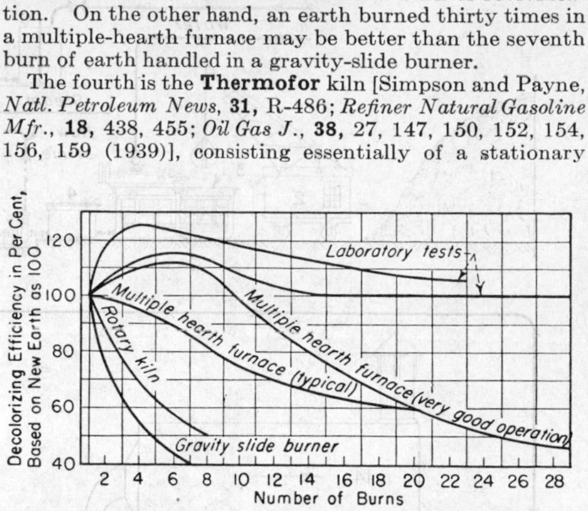

FIG. 19. Comparative efficiencies of Florida fuller's earth for decolorizing Pennsylvania steam-refined stock, when reburned in various burners or furnaces. *(Floridin Co.)*

chamber containing vertical heat-transfer tubes through which a heat-transfer medium flows. The heating agent is molten metal or salt at 850° to 1150°F. The heat-transfer tubes are surrounded by an interlaced structure of angle irons which are stacked in such a manner that the tube bundle represents a huge honeycomb.

A 50 tons per day capacity kiln weighs approximately 30,000 lb. and is stated to have an equivalent hearth area of 1800 sq. ft. Heat in the kiln is obtained from burning the adsorbed carbonaceous material in the fuller's earth or clay undergoing revivification. The kilns are mounted vertically, the clay feed being at the top and the discharge at the bottom, while the heat-transfer medium is cir-

Table 4*

A. Effect of Mesh Size and Filter Rate on Efficiency of Fuller's Earth and Bauxite

	Fuller's earth						Bauxite					
Screen analysis (Tyler standards)	0/16- 0.8 16/20- 7.8 20/24-52.6 24/32-38.2 §T/32- 0.6	0/16- 3.6 16/20-25.0 20/28-52.3 28/42-14.8 42/60- 3.7 T/60- 0.6	0/24- 1.3 24/28- 7.0 28/35-37.8 35/60-46.8 60/65- 4.1 65/80- 0.4 T/80- 2.6			0/60- 6.2 60/65-44.9 65/80-32.6 80/100-15.3 T/100- 1.0	0/16- 0.8 16/20- 9.2 20/24-40.0 24/32-49.0 T/32- 1.0	0/24- 6.5 24/28- 9.4 28/35-45.9 35/60-33.8 60/65- 2.5 65/80- 1.0 T/80- 0.9	0/24- 1.0 24/28- 1.5 28/35-40.9 35/60-50.2 60/65- 4.9 65/80- 1.0 T/80- 0.5			0/60-31.7 60/65-31.3 65/80-31.2 80/100- 4.9 T/100- 0.9
Percolation evaluations:†												
Burning temp., °F	900	900	900			900	1200	1200	1200			1200
Burned density, no. per cu. ft	31.0	31.0	30.9			29.7	54.9	54.8	54.5			53.7
Rate, cc./min	15	15	5	10	15	9	15	15	5	10	15	11
Yield to 6 A.S.T.M. (solution, color), bbl. viscosity oil per ton	12.4	13.1	15.5	15.2	15.0	18.3	7.35	8.7	10.4	9.9	9.6	12.9
Percolation efficiency:												
Weight basis	83	87	103	101	100	122	49	58	69	66	64	78
Volume basis	83	87	103	101	100	117	87	103	122	116	113	136

B. Relation of N.P.A. to O.D. Colors‡

Fuller's earth				Bauxite			
Solution (60% naphtha)		Viscosity oil		Solution (60% naphtha)		Viscosity oil	
N.P.A.	O.D.	N.P.A.	O.D.	N.P.A.	O.D.	N.P.A.	O.D.
4+	52	5	150	4	48	5	140
4¾	93	6	265	4¼+	87	6	250
5½−	152	7	425	4¾−	130	7	365
6	236	8	640	5−	175	8	485

Note: Solutions, upon reduction, gave oils of the colors indicated.

* Attapulgus Clay Co.

† Based on filtration of Pennsylvania "A" cylinder stock solution (60% naphtha) to 6 A.S.T.M. solution color at 135°F., using 2 by 25¾-in. column or approximately 570 g. of 30/60 fuller's earth and an equivalent volume of 30/60 bauxite. Efficiencies based on 30/60 mesh fuller's earth yield at 15 cc./min.

‡ N.P.A. refers to National Petroleum Assoc., A.S.T.M. O.D. means oil depth, analogous to Lovibond type system.

§ T = through, so that T/32 means through 32 or below 32.

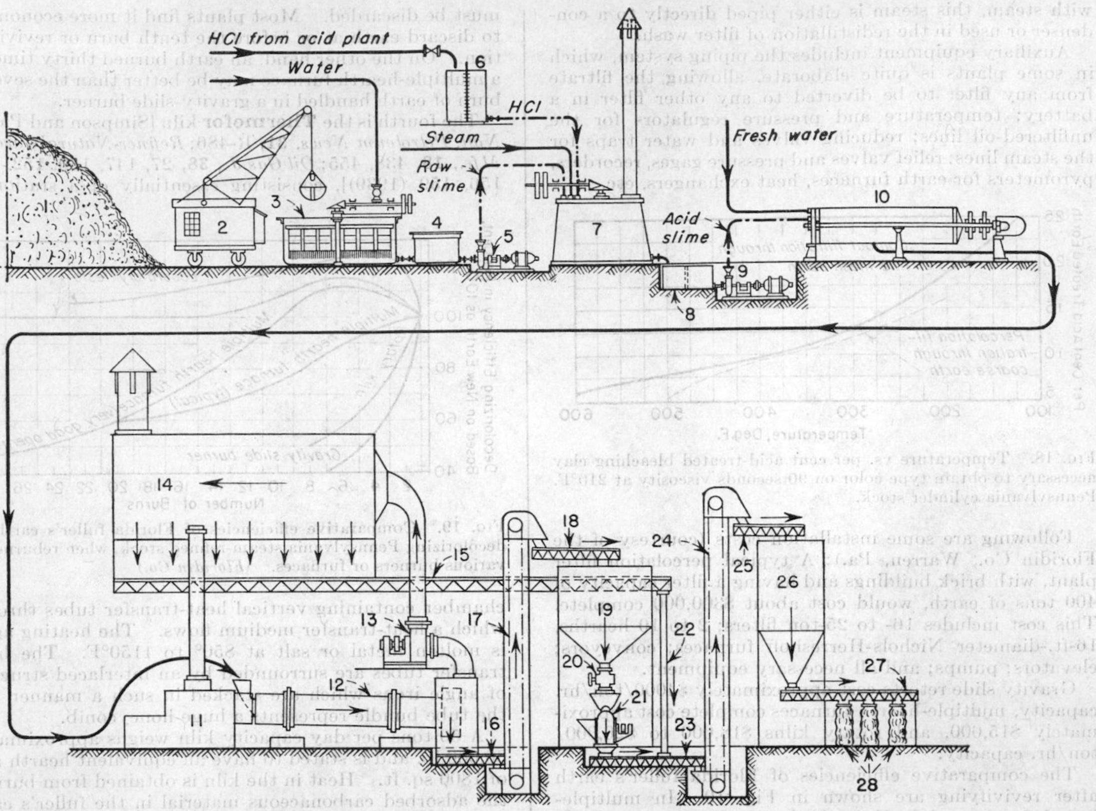

FIG. 20. Schematic arrangement for the production of highly active bleaching clays.

(1) Raw material.
(2) Crane with gripper.
(3) Slime apparatus.
(4) Screen.
(5) Slime pump.
(6) Acid vessel.
(7) Reaction vessel.
(8) Sieve.
(9) Filter press pump.
(10) Filter press.
(11) Furnace.
(12) Drying kiln.
(13) Fan.
(14) Electrical precipitator.
(15) Dust screw conveyor.
(16) Exhaust screw conveyor.
(17) Elevator.
(18) Dry material screw conveyor.
(19) Dry material storage.
(20) Discharge mechanism.
(21) Mill.
(22) Dust downpipe.
(23,25) Power screw conveyor.
(24) Powder elevator.
(26) Powder storage.
(27) Packing screw.
(28) Finished product.

culated through the vertical tubes. Temperature control and uniformity are stated to be features of the kiln.

Acid-treated Clays. Acid-treated clays of the proper type are much more efficient adsorbing agents than most varieties of fuller's earth or bleaching clays. Their manufacture has been highly developed in Germany and the Western part of the United States where the available clays did not have high bleaching powers. A flow sheet for the manufacture of acid-treated clays as described by Burghardt [*Ind. Eng. Chem.*, **23**, 800 (1931)] is given in Fig. 20.

The clay is converted into a thick slime and screened, after which it is pumped to a reaction vessel where it is treated with a mineral acid, the mixture being heated by live steam at 2 to 3 atm. Burghardt says that HCl is preferable to H_2SO_4 because the former gives clays that have better filtering properties. In the Western part of the United States, however, H_2SO_4 is commonly used as the activating agent. With HCl, the activation at 105°C. takes 2 to 3 hr., while twice that time is required with H_2SO_4. In the manufacture of a German clay, 28 to 30 per cent of HCl calculated as dry HCl on the weight of the dry clay is used. This amounts to a ton of technical HCl (19° to 21°Be.) per ton of finished clay. When decomposition of the clay is complete, the acid sludge is pumped to a filter press, the clay filtered out and washed until dissolved salts and free acid are removed, after which the clay is dropped from the press and sent to dryers. After drying, it is disintegrated and sent to storage or packed. The acid-treated clays are stated to be from three to five times as efficient as a high-grade Florida or Georgia clay.

Bauxite

As an industrial adsorbent, bauxite finds application in the same type of equipment and under the same sort of conditions as does fuller's earth, but in general the effect of different variables on the operating characteristics is much greater than in the case of fuller's earth. Hubbell and Ferguson (*Refiner Natural Gasoline Mfr.*, March, 1938) state:

> The lubricating oil decolorizing characteristics of bauxite show this material to possess the qualities of (*a*) revivification without substantial degradation in efficiency, (*b*) high decolorizing value of light colors, (*c*) a solvent-like effect on oil quality somewhat greater than that of other adsorbents.

These qualities are influenced by the type of stock, burning temperature, and temperature of filtration.

A comparison of fuller's earth and bauxite in the effect of mesh size and filter rate is given in Table 4.

Filtering temperatures for petroleum oils are higher for bauxite than for fuller's earth. Inasmuch as bauxite does not retain adsorbed materials as strongly as clay does, naphtha requirements for washing are smaller. For some waxes and petrolatum, bauxite is several times as effective as fuller's earth. The material shows greater selectivity than do clays. Only those bauxites which after burning are completely inert to the action of steam are suitable for adsorbent use for lubricating oils. As in the case of clays, there is no relation between chemical analysis and adsorbing ability or efficiency.

Jones (Floridin Co.) states that in general the efficiency of new bauxite to new commercial Florida Georgia fuller's earth is 70 to 85 per cent on steam refined stocks, 50 to 80 per cent on neutral oils and 150 to 400 per cent on white petrolatums.

Bone Char

Bone black or bone char is used in very large tonnages in the decolorization and refining of sugar. It is the carbonaceous residue obtained as the result of the destructive distillation of bones. Fresh hard bones free from flesh, fat, and oil are used. Skeletons of marine mammals and fish bones produce soft unsatisfactory chars. Products deficient in carbon are obtained when bones that have been exposed to atmospheric action and partly decomposed are used. The principal constituents of bones are $Ca_3(PO_4)_2$, $Mg_3(PO_4)_2$, $CaCO_3$, alkaline salts, and fatty and cartilaginous matter.

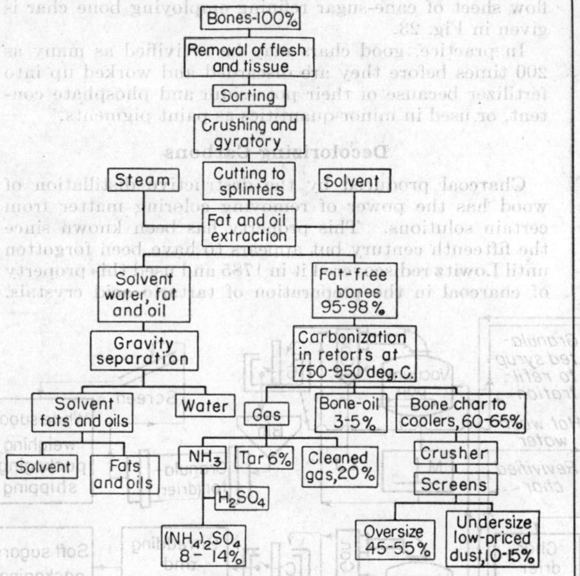

FIG. 21. Flow sheet of bone-char manufacture.

Figure 21 is the flow sheet of the manufacture of bone char. The bones are sorted, crushed, and ground, and the fat and oily matter removed by solvent extraction. The residue is then carbonized. Operating details are in Table 5.

The char always retains a certain percentage of nitrogen resulting from the destruction of the bone cartilage. Table 6 gives analyses of char. Column (1) represents new char, column (2) char that has been used, and column (3) is a recommended specification. Bone char should be dull black in color, have a firm physical structure and a high degree of porosity, yield a uniform white or cream-tinted ash upon ignition, and have a somewhat metallic sound when struck. It should be uniform in its grain.

Table 5. Manufacture of Bone Char

Type plant	Vertical retorts	Horizontal retorts
Charges, lb.	250	500
Carbonization:		
Time, hr	6–8	8–10
Temperature, °C	750–950	750–950

Yields, per cent:	
Bone oil	3–5
Ammonia as $(NH_4)_2SO_4$	8–14
Gases, weight	20
Tar	6
Bone char	60–65
Apparent density, lb./cu. ft.:	
New char	40–46
After 50 revivifications	60–75
Average life in plant	2 years
Number of revivification cycles	150–200

Table 6. Analyses of New Bone Char

Constituents	(1)	(2)	(3)
Carbon	9.30	11.50	9 min., 11 max.
Sand, etc.	0.42	0.75	0.5 max.
Tricalcium phosphate	75.00	82.00	70–75
Calcium carbonate	6.23	2.70	
Calcium sulfate	0.08	0.65	0.2 max.
Calcium sulfide	.01	.11	0.1 max.
Ferric oxide	.23	.47	0.15 max.
Water	.....	.....	8 max.

In use, the carbon, calcium phosphate, iron oxide, and calcium sulfate generally tend to increase; the calcium sulfide should be kept down as low as possible by proper regeneration; the calcium carbonate decreases.

The cellular structure of the bones from which it is made is found in the final bone char. The most desirable size is 16 to 20 mesh, as this wears down more slowly, and appreciable amounts of materials of smaller size retard filtration. As with other adsorbents, the size of the grain of the char, the temperature at which it is used, the concentration of the sugar solution treated, and the types of color, etc., are all variables that determine the relative efficiency of the char. Bone char consists of a skeleton of calcium phosphates and carbonates, cellular in structure, with a great number of minute tubes and channels. The skeleton is entirely coated or lined by carbon in a state of very fine subdivision and high activity. The adsorptive power of bone char resides in its activated carbon content, for if the carbon is burned off, the calcium phosphate skeleton has little or no decolorizing power, although it will adsorb dissolved salts.

In sugar refineries the char is contained in large vertical cylinders or filters made of steel plate, being generally 20 to 22 ft. high and 6 to 14 ft. in diameter. The dimensions vary of course with the size of the refinery, the filtration rate, the desired time of contact between the sugar solution and the char, and the age of the char. The bottom of the filter is conical in shape. The char is supported in the filter by porous plates covered with a coarse blanket on which is superposed a blanket of closer weave. The cylindrical tank is loaded at the top, and discharge of the adsorbent when necessary for revivification is possible through a number of manholes near the bottom. Generally, downward percolation of the sugar liquors through the char is practiced.

Char is loaded in the filters at a temperature not higher than 130°F., and, in refining, the sugar solutions are at 160° to 170°F. With newly charged filters, the highest grade sugar solutions are sent through first, and these are followed without interruption by grades lower in purity. If the same grade of sugar solution be filtered exclusively, the bed reaches an equilibrium with the color bodies in that particular sirup. As a result, each successive batch leaving the filter has a higher residual concentration of color. The unit, however, is still capable, as a result of its "reserve power," of bringing a still darker liquor down to an intermediate color. This is merely a specific example of the general case of adsorption indicated in the curves in the theoretical treatment.

The ratio of char to raw sugar to be purified is about 1 lb. char/lb. raw sugar but varies with the capacity of the char, the purity of the raw sugar, the rate of filtration, the

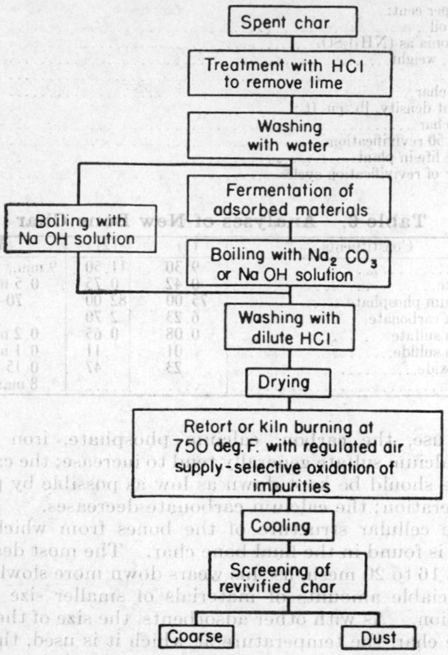

FIG. 22. Revivification of bone char.

type color, as well as other factors. When high-grade centrifugal sugars are worked, the char-to-sugar ratio may be as low as ¾ lb. to 1, but with poor quality it may be as high as 1½ lb. of char to 1 of sugar. Normally, beet-sugar refining requires less char than cane sugar. For the production of cube sugars, practically complete decolorization is required. With normal cane sugar which has been treated by the affination and mechanical-filtration process, from 175 to 225 per cent of char taken as the percentage of the dissolved solids in the liquor may be necessary. In normal operation, 200 per cent of char removes about 97 per cent of the coloring matter and 45 to 55 per cent of the mineral salts. The percentage of char employed in any particular instance is determined by the ratio of the amount of liquor passed through the filter to a constant quantity of charcoal in the filter. With second-refined liquors which will yield granulated sugar, 30 to 40 per cent of char is needed. This removes 80 to 90 per cent of the color and 25 to 35 per cent of the ash. The total quantity of char used for liquors and mother sirups, expressed as the percentage of the raw sugar entering the refinery, varies with the quality of the raw sugar and the percentage of white sugar produced. A refinery making a small proportion of brown sugars must pass more mother sirups over the char than a refinery producing a larger proportion of brown sugars. Generally, with cane sugar, the percentage varies from 70 to 100 per cent.

When the char has become so charged with impurities that it will no longer decolorize efficiently, the sugar solutions in the filter are displaced with water to dissolve the sugars from the char and the char is washed and unloaded from the filter for revivification. A flow sheet of normal revivification practice is given in Fig. 22, and the flow sheet of cane-sugar refining employing bone char is given in Fig. 23.

In practice, good chars may be revivified as many as 200 times before they are discarded and worked up into fertilizer because of their potassium and phosphate content, or used in minor quantities as paint pigments.

Decolorizing Carbons

Charcoal produced by the destructive distillation of wood has the power of removing coloring matter from certain solutions. This property has been known since the fifteenth century but appears to have been forgotten until Lowitz rediscovered it in 1785 and used this property of charcoal in the preparation of tartaric acid crystals.

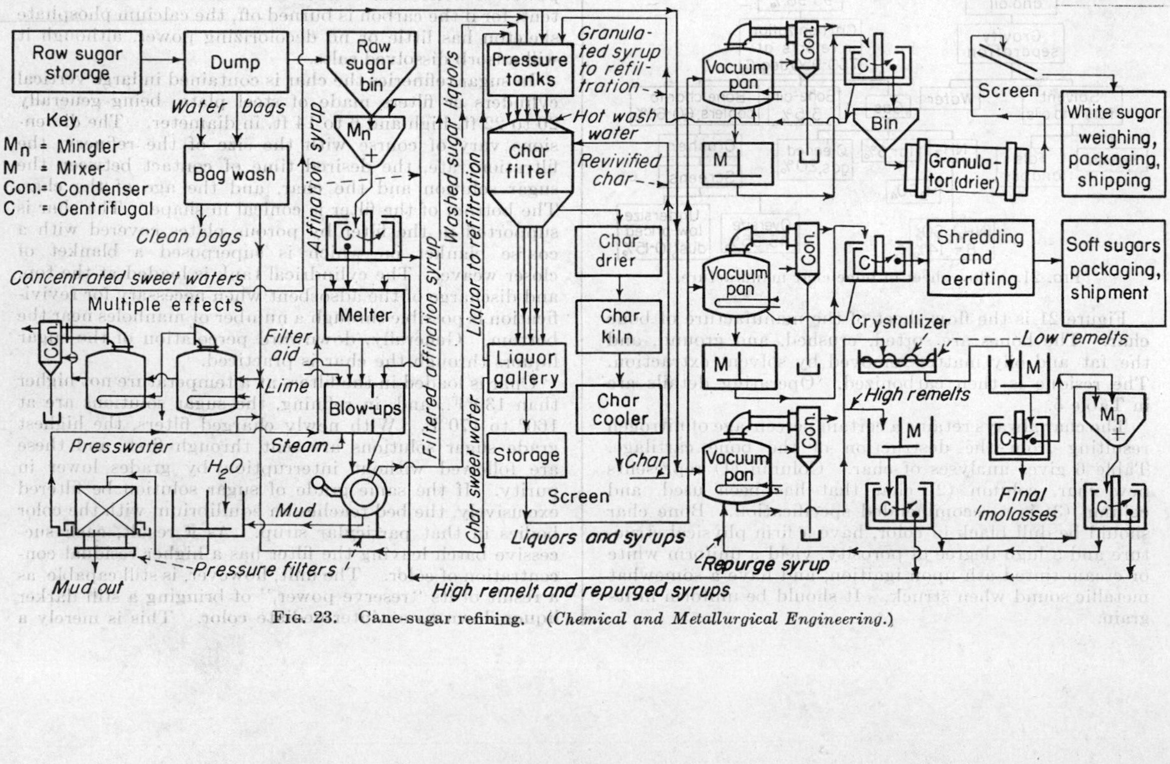

FIG. 23. Cane-sugar refining. (*Chemical and Metallurgical Engineering.*)

Records show that an English refinery used wood charcoal in the clarification of raw sugar as early as 1794. Since that time, many processes for the preparation of decolorizing carbons have been brought forward. Almost any vegetable substance can be destructively distilled to produce a decolorizing char. Mantell ("Industrial Carbon," p. 117, Van Nostrand, New York, 1946) states that **manufacturing methods for the preparation of decolorizing carbons** may be divided into several classes.

Class 1. Carbon may be deposited as a layer on a porous inorganic base. Carbonizable vegetable materials such as sawdust, seaweed, peat, and molasses, as well as a large number of other materials, may be mixed with porous substances such as infusorial earth, pumice stone, insoluble salts, and many other materials with or without the addition of a liquid binding medium. The mixture is strongly heated, whereupon the carbon in the vegetable matter is deposited throughout the porous base. Sometimes a natural high-ash vegetable product such as rice hulls or husks containing an appreciable proportion of silica is used as a base material. The object of the Class 1 process is to produce a material somewhat similar to bone char in that it has a porous structure with carbon distributed over a large area, as well as appreciable mechanical strength.

Class 2. Carbon may be deposited on an inorganic base which is afterward separated from the carbon by chemical means. Vegetable materials are mixed with chemical reagents such as lime, chalk, H_2SO_4, $CaCl_2$, $ZnCl_2$, $MgCl_2$, H_3PO_4, etc., and after carbonization the inorganic matter is dissolved out, leaving the resulting carbon. The $ZnCl_2$, H_3PO_4, etc., function as activating agents. Materials are usually carbonized at relatively low temperatures. After preparation, the inorganic acids, bases, and salts are removed by chemical treatment or by leaching.

Class 3. A number of important decolorizing chars are made by carbonizing materials, such as lignite, waste-pulp liquors, black-ash residues, sawdust, woods, and similar material, in carbonizing retorts under controlled conditions of temperature and atmosphere. The desired porosity, compactness, and mechanical strength of the resulting carbon will vary widely, depending upon the conditions under which it is initially carbonized. In some cases the material is given a second carbonization when in the form of carbon particles from which most of the volatile matter has been eliminated. After preparation, the carbon is activated by air, oxides of carbon, chlorine, superheated steam, or mixtures of steam and air. When gaseous compounds of carbon are used, carbon may be deposited therefrom in an active form on the material undergoing activation, especially if relatively low temperatures are maintained. The bulk density of the carbon particles is reduced somewhat during activation.

The market forms and characteristics of the commercial decolorizing carbons vary widely, some of them being neutral, some acid, and some alkaline. Generally they are relatively soft, black, glistening powders. Attempts to evaluate them for commercial practice have caused a large number of testing methods to be developed. Mantell ("Industrial Carbon, Its Elemental, Adsorptive, and Manufactured Forms," p. 119, Van Nostrand, New York, 1946) sums up the present state of testing methods in the comment:

> Data regarding the action of carbons on one solution cannot be applied to a different solution, but each one to be decolorized must be tested separately. The so-called standard methods of estimation have little practical value. The adsorption of color by carbon has been repeatedly shown to be a definite equilibrium reaction in which other materials than coloring matters assist or take part in the determination of the equilibrium.
>
> When different adsorbents (see "Laboratory Testing Manual," Darco Sales Corp., New York) such as carbons are to be compared on a practical basis, they may be evaluated by plotting adsorption isotherms on the particular samples or materials in connection with which the adsorbent is to be used. In this manner, the common error of assuming that the ratio of colors removed with the same weight of adsorbent is equivalent to the ratio of weights required to remove equal color can be avoided. Monetary ratios, which of course include the price of the adsorbent, will vary with different degrees of decolorization when isotherms are not parallel. In this case it can be seen that false conclusions will be arrived at when one-point comparisons are made. Isotherms permit calibration of the value of a series of carbon samples at any and all degrees of decolorization.

Commercial carbons of the activated type employed as adsorbents may be grouped, on the basis of their physical structure, properties, and applications, into four classes: decolorizing, gas-adsorbent, metal-adsorbent, and medicinal carbons. No one type of carbon can be universally used or effective for all purposes. Granular, activated carbons which are mechanically strong, relatively dense, and highly active are required for industrial gas and vapor adsorption, for which purpose the soft, finely pulverized, highly porous, and generally effective active decolorizing carbons are practically worthless. Two activated carbons may contain the same percentage of active carbon; one may be extremely valuable as a vapor adsorbent while the other is valuable as a sugar decolorizer, but both become almost useless when an attempt is made to interchange their applications.

Although there is a voluminous literature on the decolorization of sugar and other solutions by carbon, theoretical discussions of the properties of the adsorbents have failed to bring forth any broad generalization of a satisfactory nature.

When a liquid containing impurities is brought into contact with a carbon, the attraction of the carbon for the impurities is greater than the attraction of the liquid for the impurities. The carbon therefore adsorbs the impurities, such as coloring matter, odor, flavor, until equilibrium is reached, after which the carbon will no longer remove these substances from that particular solution. If at this point the carbon and liquid are separated, and the carbon introduced into a further quantity of the original (or a darker) liquid, more impurities from the liquid would be taken up by the carbon until a second equilibrium was reached. A smaller quantity of impurities will be taken up in the second use of the carbon than in the first, and, if a third use be made, still smaller quantities are taken up. Theoretically, this practice of using the "reserve power" of the carbon can be repeated a number of times, but in practice the economic, mechanical, and operating factors determine the manner of employing the carbon and the number of times that it may be used. By suitably decreasing the amounts of liquid with which the carbon comes in contact in each successive treatment—which is in effect successively increasing the proportion of carbon employed—the percentage of impurities removed becomes the same. In this way, three separate lots of liquid, each smaller than the preceding one, are all decolorized to the same degree.

In use, the rate of removal of impurities from a solution by carbon is very rapid during the first interval of contact and gradually reaches a point where increased time of contact gives no further decolorization. Color removal is usually greater at higher temperatures. With time of contact and temperature both fixed, the percentage of total impurities removed varies directly but logarithmically with the percentage of carbon used. If two solutions, one containing a much greater amount of impurities than the other, are treated similarly as regards time of contact, temperature, and amount of carbon per unit of solution, the more dilute solution will have a greater percentage of its total impurities removed.

The majority of colored materials encountered in industry are negatively charged, and ordinarily carbons will give greater decolorization with increase in acidity of the solution. Both the carbon and the impurity to be removed carry electrical charges, and results show that in general the adsorption efficiency of carbons is dependent largely upon the difference in electrical charges between the carbon and the particle, colloid, or color (or ion) to be adsorbed. The generalized curves in Fig. 24 show the relation between acidity, neutrality, and alka-

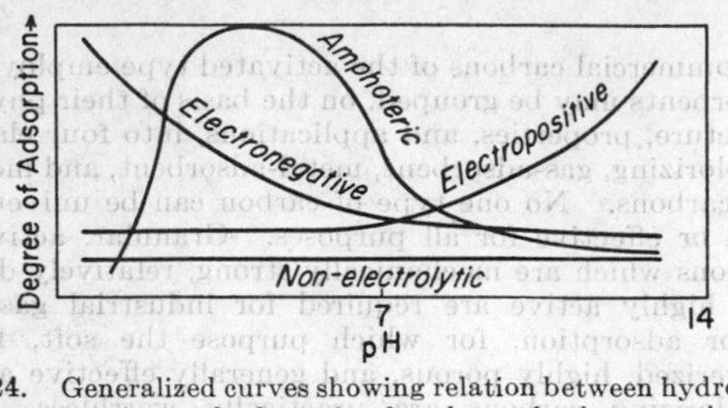

FIG. 24. Generalized curves showing relation between hydrogen-ion concentration and adsorption by chars of substances having various electrical properties.

linity of the solution and adsorption by chars of substances having various electrical properties.

It will be noted that the adsorption of non-electrolytic materials, such as sugar, is a function of the carbon and is not affected by acidity or alkalinity. Electropositive materials (colors like ponceau red) are taken up more effectively in alkaline solution, while electronegative materials (colors like methylene blue, and most colored impurities) are removed by the carbon most effectively in acid solutions. Amphoteric substances such as colloids, proteins, some natural colors, etc., which, depending upon the pH value of the solution, may act either as

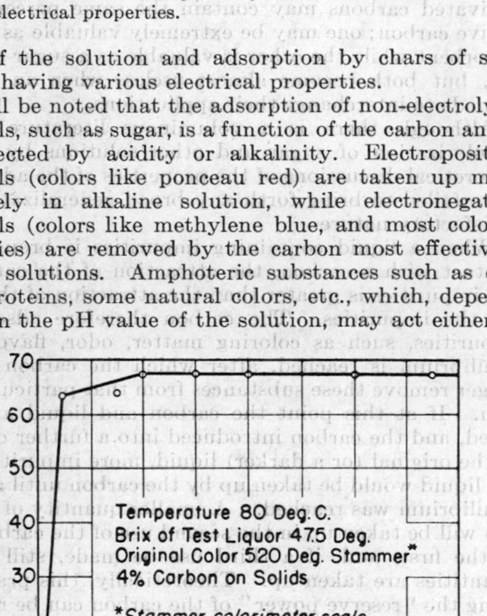

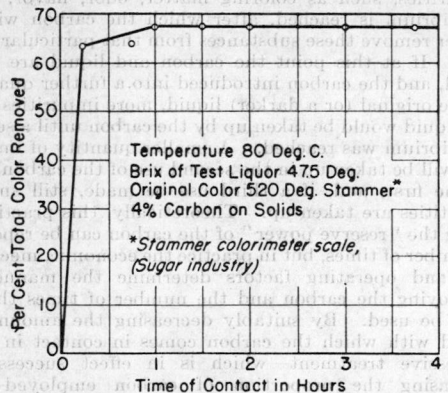

FIG. 25. Rate of decolorization by carbon. (*Blowski and Bon.*)

acids or as bases, are adsorbed most effectively near the isoelectric point, where they show neither acid nor basic properties.

An effect of the pH on the adsorption of color bodies is frequently on the solubility of these materials. In the case of amphoteric substances, the isoelectric point is also the point of minimum solubility. The pH of the solution will affect the degree of dissociation of the color body. The pH of the adsorbent itself is an important factor, as this may affect the pH of the liquid. Frequently color changes in indicators, which color changes are functions of pH, are mistaken for adsorption effects.

In commercial practice, the optimum conditions for adsorption frequently cannot be obtained, as undesirable factors interfere. For example, decolorization of sugar solutions is more effective in acid solutions, but the acidity would cause inversion losses of sugar.

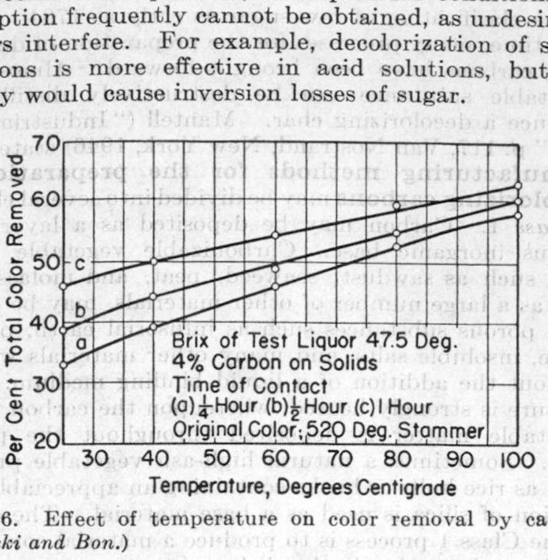

FIG. 26. Effect of temperature on color removal by carbon. (*Blowski and Bon.*)

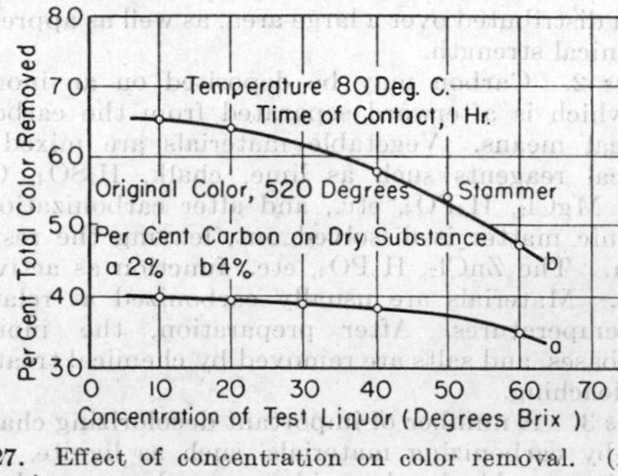

FIG. 27. Effect of concentration on color removal. (*Blowski and Bon.*)

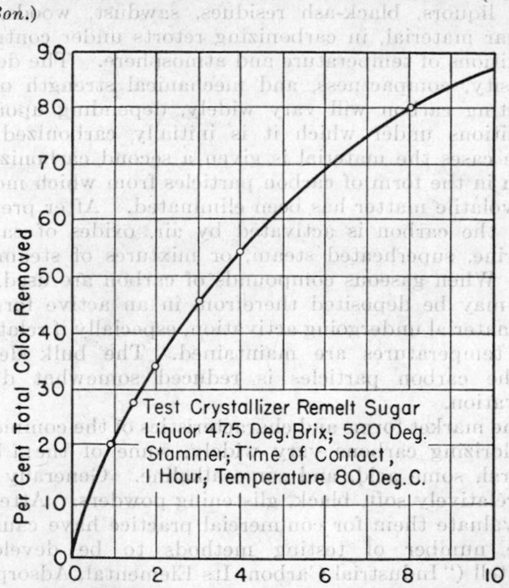

FIG. 28. Effect of amount of carbon on removal of color. (*Blowski and Bon.*)

For specific cases of decolorization in sugar solutions, the relation between time of contact and color removal is given in Fig. 25, the effect of temperature on color removal in Fig. 26, the effect of concentration of the sugar solution on color removal in Fig. 27, the effect of the amount of carbon on the percentage of total color removed in Fig. 28, and the relation between pH of the liquor and percentage of total color removed in Fig. 29.

Different types of colors occurring at different stages of sugar refining show different isotherms with any one given carbon. The influence of the type of color on the relation between the per cent total color removed and the amount of carbon used is shown in Fig. 30. In the removal of any one color, the effect of the percentage carbon and of the various types of carbons is given for different sugar liquors in Figs. 31 and 32.

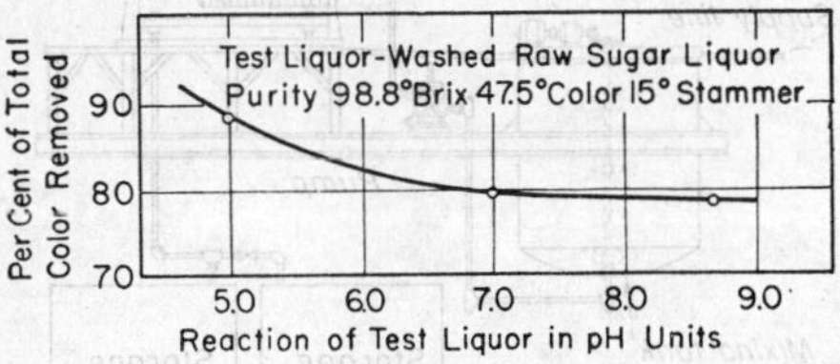

FIG. 29. Effect of pH on decolorizing. (*Blowski and Bon.*)

Applications of Decolorizing Carbons. Some of the more important applications of decolorizing carbons in industry are given in Table 7.

The manner of using activated carbons is a simple one (Fig. 33). The liquor to be treated is mixed thoroughly and brought into intimate contact with an amount of carbon determined previously by a laboratory test, the mass brought up to the most suitable temperature, and agitated for 15 to 30 min. The carbon, now holding

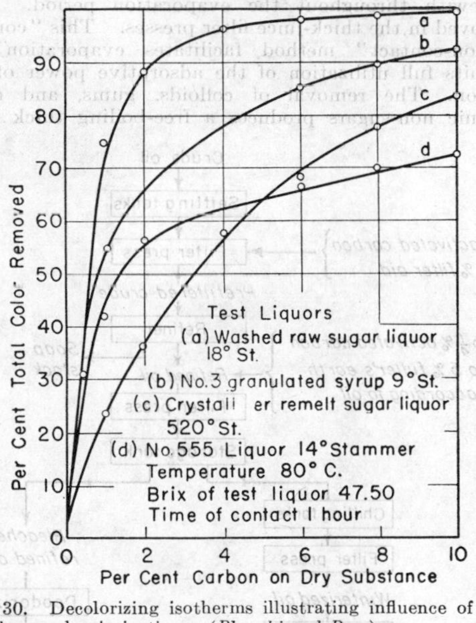

FIG. 30. Decolorizing isotherms illustrating influence of type of color on decolorization. (*Blowski and Bon.*)

the impurities adsorbed on its enormous surface, is removed from the liquor by settling or filtration. Decolorizing, purifying, and deodorizing results are greatly affected by the degree of contact, which in turn is a function of the fineness of the carbon, its effective surface, and the manner and continuity of agitation.

A flow sheet showing the use of activated carbons on cottonseed oil is shown in Fig. 34.

The characteristics of activated carbons are given in Table 8.

Sugar refineries using the activated-carbon contact process are in operation in most of the sugar-producing

Table 7. Applications of Decolorizing-type Carbons

Product	Material removed	Remarks
Inorganic salts:		
Borax, $MgSO_4.7H_2O$, $Na_2SO_4.10H_2O$, $PbCl_2$	Color	
$ZnSO_4$	Color and odor	
Inorganic acids: H_3BO_3, H_3PO_4	Color	
Tartaric acid, Citric acid	Color and colloids	Aids crystallization
Iodine		Iodine from sea water and brines
NaCl	Br	Br recovery
Gallic acid	Color and colloids	Aids crystallization
Cider	Color and odor	Gives better flavor
Wines	Color, odor, and bacteria	Deodorizing, clarification and "aging" of vinous beverages
Glycerin	Color and colloids	Eliminates distillation, prevents foaming when glycerin is used as anti-freeze
Beet sugar	Color and colloids	Aids in refining
Cane sugar	Color, colloids, and ash	Replaces bone char
Glucose	Color, odor, and ash	Replaces bone char
Corn sirup, Corn sugar	Color, odor, and ash	Purification of liquors before concentration by evaporation
Maple sirup	Color, odor	Production of uniform product throughout season
Sorghum sirup	Color, colloids, and gums	Uniform product made
Used preserving sirup	Color, flavor, and odor	Rendered available for reuse
Gelatin	Color, odor, and flavor	Product of improved appearance
Drugs:		
Salicylic acid, salicylates, quinine and salts, acetanilid, caffein and theine, alkaloids.	Color and miscellaneous impurities; colloids and gums	Purified products
Photographic chemicals	Color, etc.	Makes crystallization easier
Phenol	Color, odor, and thiophens	
Organic liquids:		
Alcohols, acetone oils, cologne spirits.	Color and extraneous odor	Improved product
Dry-cleaning fluids:		
Naphtha, gasoline, CCl_4, Stoddard solvent.	Color, odor, grease, colloids	Allows reuse of solvent, purification permits continuous systems
Essential oils	Color and extraneous odor	Improved product
Agar-agar	Color and odor	Allows production of edible product
Potable water	Color, odor, and taste	Particularly useful for removal of taste due to minute amounts of impurities
Vegetable oils	Color and odor	Carbon usually employed in conjunction with fuller's earth or bleaching clays in vegetable-oil refining
Crude vegetable oils	Colloids, resins, phosphetids, etc.	Carbon usually employed with diatomaceous earth; gives oil of better keeping quality and lower free fatty-acid content
Off-grade oils	Color and colloids	Recovery of deteriorated products
Lard	Color and odor	Carbon used with diatomaceous earth
Fish oils	Color and odor	
Medicinal oils	Color, odor, and taste	Makes product more palatable

Table 8. Characteristics of Adsorbent Carbons

Name of carbon	% carbon	% ash	% HCl soluble	pH water extract
Carboraffin	84–87	3–4	0.5	2–3
Cliffchar	89–94	3.5–6	1.6–3	9–10
Darco S-51	74–76	23–26	0.5	4.5–6
Darco G-60	93–96	3–6	0.2	5–6
Klearit	95.6	0.2	1.0	8.7
Norit	94–98	3–8	2–4	4–8.5
Nuchar	97+	2.5–4.5	2–3	6.8–7.4
Suchar	95–96	3–5	2.5–3.5	6.7–7.4

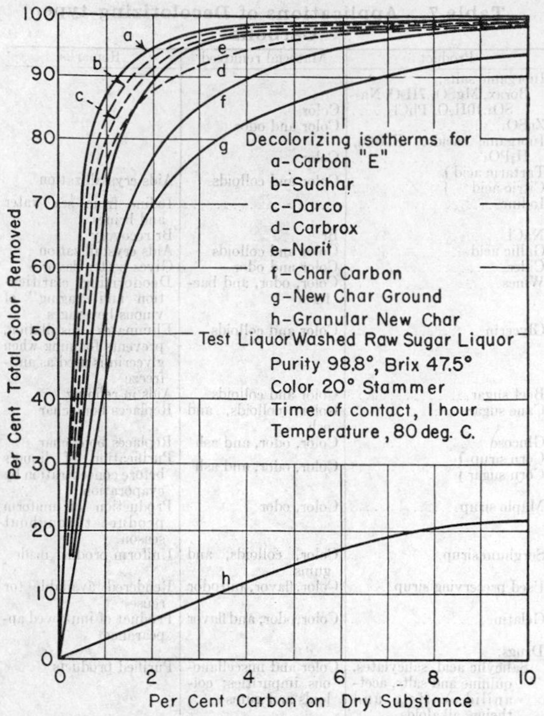

FIG. 31. Comparison of effect of various vegetable carbons on washed raw sugar liquor. (*Blowski and Bon.*)

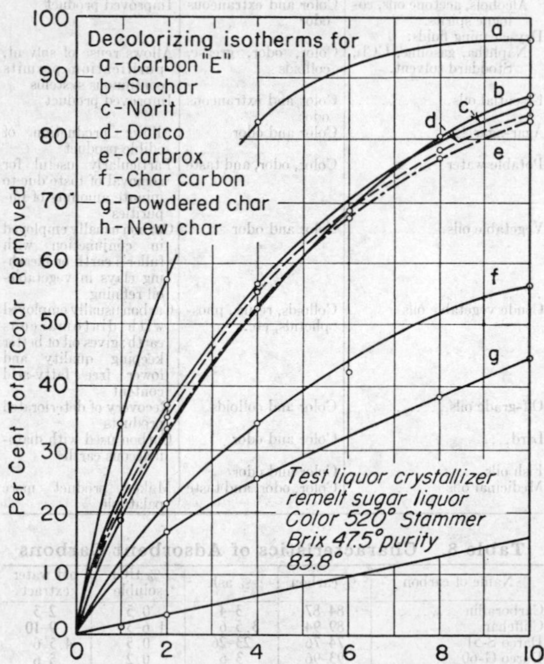

FIG. 32. Comparison of effects of various vegetable carbons on crystallizer remelt sugar liquor. (*Blowski and Bon.*)

areas of the world. The percentage of carbon on the sugar is 1.3 to 2.5 per cent, varying with the quality of the raw sugar and the nature of the desired product. The carbons on a weight basis may be considered from thirty to forty times as active as bone char. A typical flow sheet for the Suchar process is given in Fig. 35.

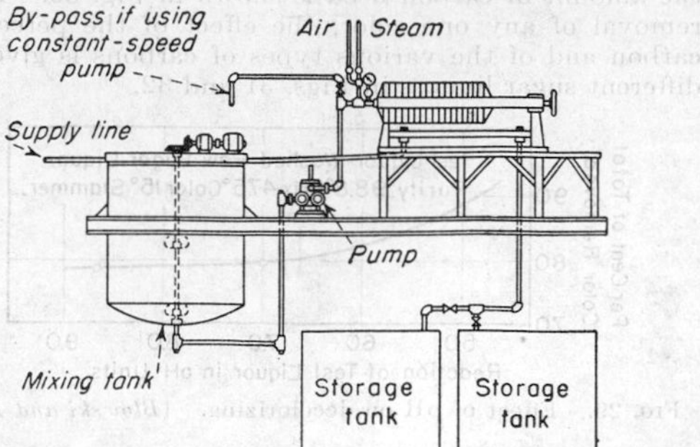

FIG. 33. Typical plant for use of carbon adsorbents ("contact filtration").

In the manufacture of beet sugar, activated carbon is introduced into the carbonated and "sulfured" thin juice as it enters the evaporators, remaining in contact therewith throughout the evaporation period. It is removed in the thick-juice filter presses. This "concentration-contact" method facilitates evaporation but permits full utilization of the adsorptive power of the carbon. The removal of colloids, gums, and other organic non-sugars produces a free-boiling thick juice

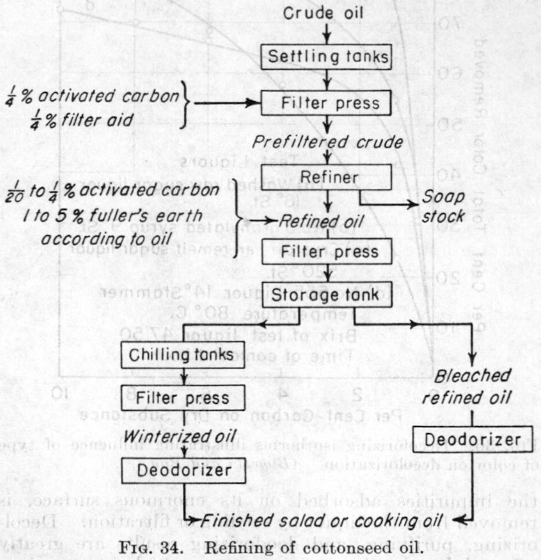

FIG. 34. Refining of cottonseed oil.

in the vacuum pans, shortens the time of boiling, and gives whiter crystals. The proportion used varies between 0.05 and 0.20 per cent, based on sugar produced.

Water Purification. Bone char and, in their early stages, some of the activated chars have been used to a limited extent in household or other small filters for a number of years. Activated carbons are of value in the removal of odors or flavors from potable waters.

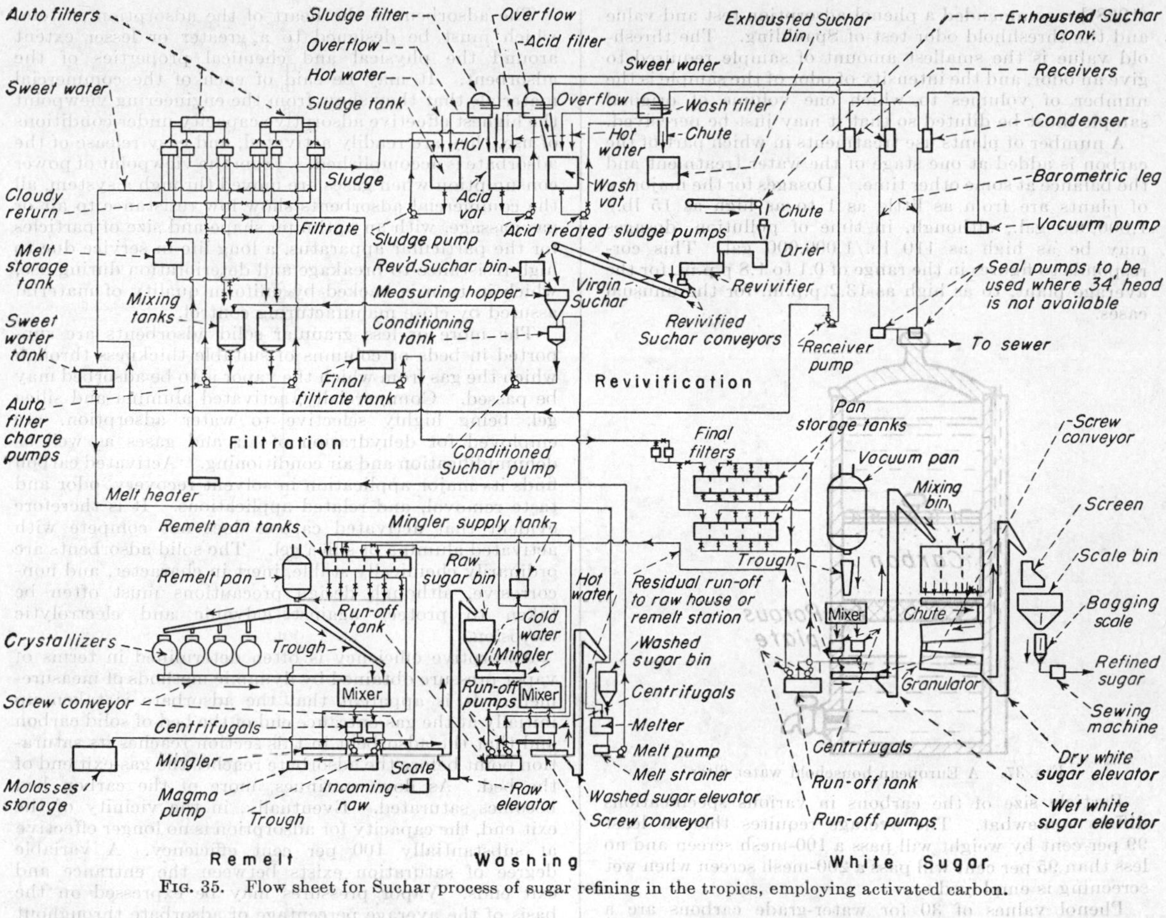

FIG. 35. Flow sheet for Suchar process of sugar refining in the tropics, employing activated carbon.

Many waters are poluted with such a variety of compounds that it may be more advantageous to "overchlorinate" to break down such compounds as may be attacked by chlorine and, then, to use activated carbon for removal of the excess chlorine and certain compounds not attacked by chlorine. This method of treatment is suitable for some cases. In others, chlorination may produce odors that are difficult to remove with activated carbon. In these cases it is desirable to make the activated-carbon treatment before chlorination. Each individual water supply, as the result of the wide variety of contaminating influences, must be treated as an individual case. The result is the production of a potable water of excellent taste. Carbon has been used for this purpose in England for a number of years. Excess of chlorine added to water results in the conversion of some of the taste-producing compounds to others that have little or no taste. Chemical methods of removal of excess chlorine, such as the addition of SO_2, probably do nothing more than remove the excess of chlorine, converting it to HCl. Such treatments remove little or nothing from the water but convert part of the offensive to inoffensive compounds. Activated carbons, on the other hand, actually take out these undesirable constituents.

Owing to differences in their physical properties, the commercial carbons are employed either in powder form, in which case they are added in small quantities to the water and later allowed to settle out or are filtered out, or in granular form, when the carbon is employed in stationary filters through which the water passes. Such an installation is illustrated in Fig. 36. In Europe there is considerable application of activated carbon in small household water filters, the construction of which is shown in Fig. 37.

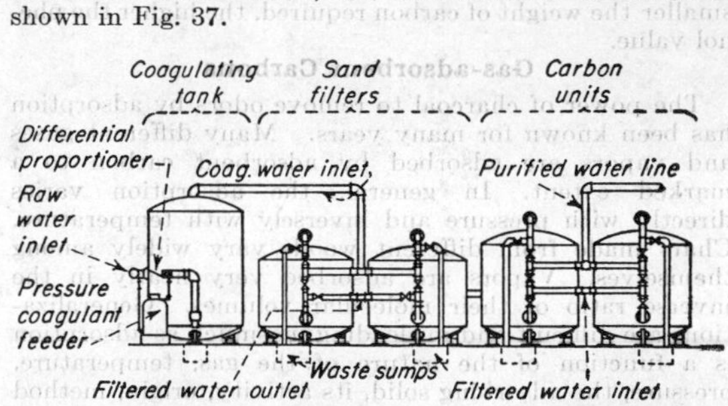

FIG. 36. An activated carbon installation for filtration of water, consisting of pressure coagulating tank, pressure filters, and carbon purifiers.

Baylis ("Elimination of Taste and Odor in Water," McGraw-Hill, New York, 1935) described the use of activated carbon in the powdered and granular forms and the mechanical devices employed therewith. The Subcommittee on Activated Carbon of the American Water Works Association [*J. Am. Water Works Assn.*, **30,** 1133

(1938)] recommended a phenol-adsorption test and value and the threshhold odor test of Spaulding. The threshold value is the smallest amount of sample required to give an odor, and the intensity of odor of the sample is the number of volumes to which one volume of original sample must be diluted so that it may just be perceived.

A number of plants use treatments in which part of the carbon is added at one stage of the water treatment and the balance at some other time. Dosages for the majority of plants are from as little as 1 to as high as 15 lb./1,000,000 gal., although, in time of pollution, dosages may be as high as 110 lb./1,000,000 gal. This corresponds to figures in the range of 0.1 to 1.8 p.p.m. for the average plant, to as high as 13.2 p.p.m. for the unusual cases.

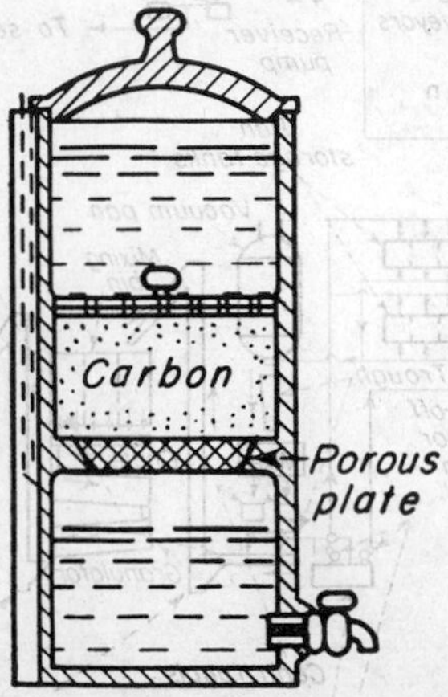

FIG. 37. A European household water filter.

Particle size of the carbons in various specifications differs somewhat. The average requires that at least 99 per cent by weight will pass a 100-mesh screen and no less than 95 per cent will pass a 200-mesh screen when wet screening is employed.

Phenol values of 30 for water-grade carbons are a recommended standard for some specifications, while others call for differing figures. The phenol value is determined by an exacting method in terms of the unit weight of carbon required to adsorb 90 per cent of the phenol in a sample containing 100 parts per billion. The smaller the weight of carbon required, the higher the phenol value.

Gas-adsorbent Carbons

The power of charcoal to remove odors by adsorption has been known for many years. Many different gases and vapors are adsorbed by adsorbent carbon to a marked extent. In general, the adsorption varies directly with pressure and inversely with temperature. Chars made from different woods vary widely among themselves. Vapors are adsorbed very nearly in the inverse ratio of their molecular volumes. Generalizations are difficult and misleading inasmuch as adsorption is a function of the nature of the gas, temperature, pressure, the adsorbing solid, its activity, origin, method of preparation, and previous history. Commercial activated gas-adsorbent carbons may vary widely in porosity, density, strength, hardness, and adsorbing ability.

The data in Table 9 on activated carbon apply specifically to the Columbia brand of the National Carbon Co. These products are more or less granular in form, although in recent years extruded and preformed blocks or rods of activated carbon, first used in European practice, have found favor in the United States.

The adsorbent is the heart of the adsorption system, which must be designed to a greater or lesser extent around the physical and chemical properties of the adsorbent. It may be said of each of the commercial materials that they show from the engineering viewpoint the highest effective adsorptive capacity under conditions of use, they are readily activated, and easy release of the adsorbate is accomplished. From the viewpoint of power consumption when gases are moved through a system, all the commercial adsorbents show low resistance to air or gas passage, with an optimum shape and size of particles for the particular apparatus, a long life in service due to high resistance to breakage and deterioration during use, which in turn is backed by uniform quality of material assured by close manufacturing control.

The more or less granular solid adsorbents are supported in beds or columns of suitable thickness through which the gas from which the vapor is to be adsorbed may be passed. Commercially, activated alumina and silica gel, being highly selective to water adsorption, are employed for dehydration of air and gases as well as dehumidification and air conditioning. Activated carbon finds its major application in solvent recovery, odor and taste removal, and related applications. It is therefore evident that activated carbon does not compete with activated alumina or silica gel. The solid adsorbents are ordinarily chemically stable, inert in character, and noncorrosive, although design precautions must often be taken to protect against galvanic and electrolytic corrosion.

Adsorptive efficiency is often determined in terms of vapor pressure obtained by dynamic methods of measurement. It is apparent that the adsorbate is taken up initially at the gas-entrance end of the bed of solid carbon and that the adsorbent in this section reaches its saturation point before the adsorbate reaches the gas-exit end of the bed. As flow continues, more of the carbon bed becomes saturated. Eventually, in the vicinity of the exit end, the capacity for adsorption is no longer effective at substantially 100 per cent efficiency. A variable degree of saturation exists between the entrance and exit ends. Vapor pressures may be expressed on the basis of the average percentage of adsorbate throughout the bed and the vapor pressure of the adsorbate in the exit gas.

As adsorbate is taken up by the adsorbent, heat is liberated. This may be removed by various cooling methods so that the adsorbent reaches a normal equilibrium point or "break." Adsorption continues, however, but at decreasing efficiency until saturation is reached. If the heat of adsorption is not removed, a hot zone forms at the area of maximum adsorption. This hot zone follows layer by layer the course of adsorption from the entrance to the exit end of the apparatus. The quantity of heat liberated is dependent upon the concentration of the adsorbate, the heat content of the gas, the rate of flow, frictional effects, and other factors. The heat liberated during adsorption from a practical viewpoint may be considered as about equal to the thermal requirements necessary to vaporize an equal amount of adsorbate during activation.

For the same adsorbate, the total weight taken up at 100 per cent efficiency in a given apparatus is nearly independent of the initial concentration of adsorbate in the gas.

In the case of activated carbon for gas masks, the "service life against chlorpicrin" is determined on a volume basis and is a measure of the ability of the carbon to protect against substantially all organic vapors that are not catalytically decomposed. The high density of the carbons enables them to give maximum protection per unit of volume. The "heat of wetting" is believed by

Table 9. Properties and Applications of Solid Adsorbents*

	Activated carbon†	Activated Alumina‡	Silica gel§
Apparent density	0.7–0.9	1.6	0.7
Specific gravity	1.75–2.1	3.25–3.35	2.1–2.3
Average weight, lb./cu. ft.	28–34	50	38–40
Average porosity, %	50–55	51	50–65
Thermal conductivity, B.t.u./sq. ft./hr./°F./in.		1.45 (200°F.) 1.0 (100°F.)	1
Specific heat, B.t.u./lb./°F.	0.2	0.24	0.22
Reactivation temp. range, °F.	Average 220–240 Special up to 575	350–600	300–350
Activation method:			
Live steam, direct contact	X		
Preheated gas or air		X	X
Waste gas or flue gas from gas or oil		X	X
Embedded electrical heaters	X	X	X
Embedded steam coils	X	X	X
Specific surface, sq. cm./g.			6×10^6
Average pore diameter, cm.			4×10^{-7}
Hardness, resistance to breakage (U.S.C.W.S.)	80–98		
Adsorptive capacity, CCl_4 from air at 25°C., sat. at 0°C., weight %	50–90		15–35
Retentive capacity, CCl_4 retained at 25°C. after 6 hr. in dry air stream, weight %	25–40		2
Service life (accel. chlorpicrin test, U.S.C.W.S.), min.	40–65		
Heat of wetting with benzene, cal./g.	20–26		
Commercial flow rates for complete drying of gases, cu. ft./hr./lb.		25–50	25–75
Commercial gas velocities, ft./min.	20–120	25–100	25–100
Effective bed temp., °F.	40–120	32–80	40–90
Weight adsorption of water at break point, %, dry basis		12–14	10–20
Weight of water at saturation, %, dry basis		20–25	40
Applications:			
Drying air		X	X
Drying non-reactive gases		X	X
Drying reactive gases			X
Gas and air conditioning		X	X
Gas concentrating (SO_2)			X
Petroleum vapor desulfurizing			X
Gaseous fuel drying		X	X
Oil vapor adsorption (filtration)	X	X	X
Solvent recovery	X		
Gasoline from natural gas	X		
Benzene from manufactured gas	X		
Odor, taste removal from CO_2	X		
Fractionation of gases (He from natural gas)	X		
Stench abatement	X		
Odor removal, air conditioning	X		
H_2S adsorption	X	X	X
Alcohol from fermenter gases	X		
Commercial sizes, screen mesh	4–8, 6–8 4–14, 6–14, 8–14 10–24, 14–35 through 200	2–1 in., 1–2, 2–4 4–8, 8–14, 14–20 20–40, 40–80 through 80	Various, also through 200
Pressure drop for 8–14 mesh bed 1 ft. thick, in. water:			
At 75 f.p.m. lin. vel.	8	4.5	
At 50 f.p.m. lin. vel.	5	2.7	
At 25 f.p.m. lin. vel.	1.7	1.0	
Common adsorbent arrangement:			
Vertical columns	X	X	X
Trays			X
Horizontal adsorbers	X		
Screens	X	X	X

* Mantell, *Chem. & Met. Eng.*, **47**, 305–307 (1940).
† Typical analysis: H_2O, 0.5 to 2.5; ash, 0.5 to 4.0 per cent; balance C.
‡ Typical analysis: Al_2O_3, 92; Na_2O <7; SiO_2 <0.1; Fe_2O_3 <0.1; TiO_2 <0.01 per cent.
§ 99.5+ per cent SiO_2.

military authorities to measure the specific adsorptive power of the carbon and therefore to measure the ability of the carbon to adsorb efficiently and retain organic vapors even when these vapors are present in extremely low concentration and where water vapor is present. The "hardness" of the carbons is a measure of their resistance to breakage or dusting in transportation and use.

Activated carbons are used in relatively thick beds through which the vapor-laden air is passed. The larger particle sizes are preferred in order to keep the resistance to a minimum. The adsorptive capacity data given in the tabulation measure the ability of the carbons to adsorb relatively concentrated vapors whereas the retentive capacity data measure the specific adsorptive power or ability of the carbons to adsorb and retain at normal temperatures vapors present in low concentration and in the presence of water vapor. Both characteristics must be considered in determining the effective adsorptive capacity of carbons under specific recovery conditions. The adsorbed vapors which are retained by the carbon at normal temperatures are readily released when low-pressure steam is introduced, approximately 3 lb. of steam being usually required to remove 1 lb. of adsorbed solvent. Relatively high "apparent density" is important because it affects the size of carbon containers and the water-adsorbing ability of the carbon. The hardness is important because it means greater resistance to breakage during use, and consequent long life.

In commercial plants employing adsorption processes with solid adsorbents, the depth of bed required may vary from a few inches to several feet which, of course, affects the power consumption of the system. Figures 38, 39 and 40 give the pressure drop at different linear velocities and different bed densities in pounds per square foot of bed area for activated carbons of different meshes. Bed thicknesses may be calculated by dividing the bed densities, which are in pounds per square foot, by the weight per cubic foot of the adsorbent.

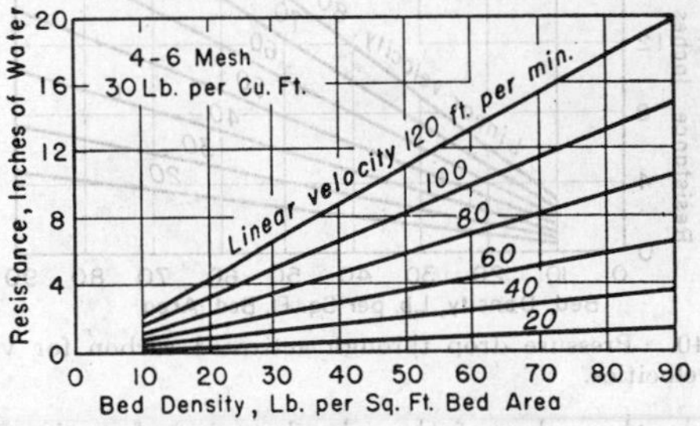

FIG. 38. Pressure drop through activated carbon for various meshes.

The effect of the adsorbent on the adsorbate is to accomplish condensation at a temperature above that of saturation. The latent heat of the solvent vapor is given up to the adsorbent and to the unadsorbed gas or air. It may be necessary to cool the gas. Both the capacity and efficiency of adsorption fall off with rising temperature of the bed. Carbon does not selectively adsorb water vapor, so that little latent heat is released through such action. With a moist gas, less heating of both the exit gas and adsorbent bed occur. When steam is used for regeneration, the moisture remaining in the adsorbent is evaporated owing to its selective displacement by solvent vapor when adsorption is resumed. The cooling effect often makes a precooling period unnecessary.

The adsorption of gases and vapors by activated carbon and other adsorbents is dependent upon two factors, one the retentivity or specific adsorptive capacity and the other the capillary adsorption action, the sum of these two being the total adsorptive capacity or saturation value. The effect of differences in adsorptive character-

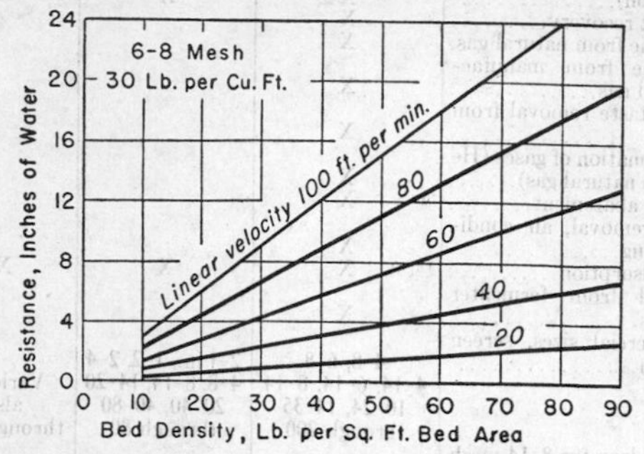

FIG. 39. Pressure drop through activated carbon for various bed densities.

istics of different adsorbents may be shown. Up to a break point, the carbon and gel are equally efficient for the removal of benzol, but they differ beyond that point. The relative values of adsorbents for vapors from gases may be determined by comparison of their adsorptive capacities at the break points. The break point is defined as the point beyond which the efficiency of removal rapidly decreases, and up to which the adsorbent has shown 100 per cent efficiency. These values will be affected somewhat by size of particles, depth of bed, rate of passage of gas, concentration of vapors, etc., but are always lower than total saturation values. When the concentration of the gases or vapors to be adsorbed

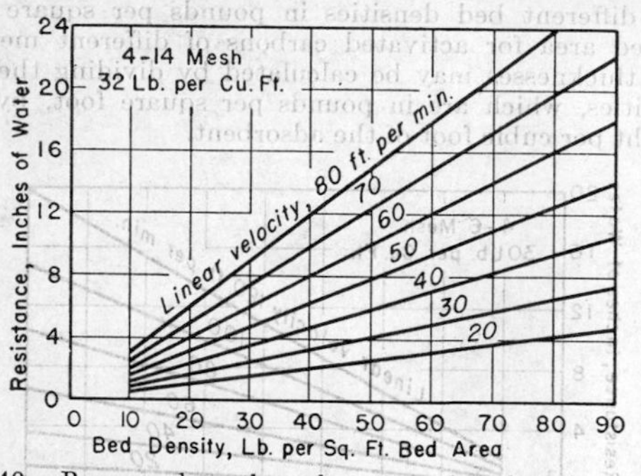

FIG. 40. Pressure drop through activated carbon for various flow velocities.

is high, the value of the adsorbent is a function of its saturation point; but when the concentration of the gases or vapors to be adsorbed is low, the value of the adsorbent is determined by its retentivity.

Activated carbon specifically adsorbs hydrocarbon vapors in preference to water vapor, while silica gel specifically adsorbs water in preference to hydrocarbons. A wet activated carbon will still adsorb benzol, while a wet gel will not.

The adsorbed vapors can be readily recovered from the carbon by heating at 100° to 150°C., at which point the adsorptive power is sufficiently decreased to allow the

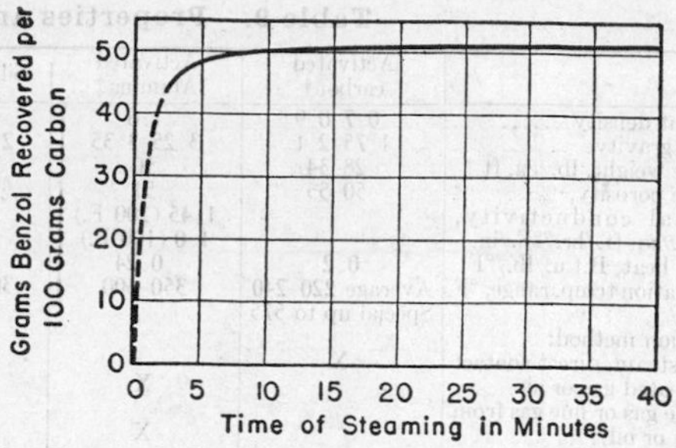

FIG. 41. Removal of adsorbed benzol from activated carbon by wet steam. (*Ray.*)

easy removal of the vapors and gases. This is shown in Fig. 41 and Table 10.

The more important industrial applications of gas-adsorbent carbon are listed in Table 11. A number of these are discussed individually.

Table 10. Weight Adsorption of Various Vapors by Activated Carbon

Substance	Adsorption, weight, %	Retentivity, weight, %
Benzene	45–55	5.9 (steam)
Carbon tetrachloride	80–110	27–30
Gasoline	10–20	2–3
Methyl alcohol	50	1.2 (1 hr. at 150°C.)
Ethyl alcohol	50	1.05 (1 hr. at 150°C.)
Isopropyl alcohol	50	1.15 (1 hr. at 150°C.)
Ethyl acetate	57.5	4.87 (1 hr. at 150°C.)
Acetone	51	3.0 (1 hr. at 150°C.)
Acetic acid	70	2.5 (1 hr. at 150°C.)

Table 11. Applications of Gas-adsorbent Carbons

Product	Material removed or adsorbed	Remarks
CO_2	Odor and taste	Effectively purifies fermentation gas so that it can meet the high standards of CO_2 for beverages
Gasoline	Gasoline	Recovery from natural gas, still, and refinery gases
Industrial gases	H_2S	Purification and sulfur recovery
Manufactured gas	Benzol	Benzol recovery
Manufactured gas	H_2S, odor	Deodorization and sulfur removal
Helium	Gases other than helium	Recovery from natural gas
Volatile solvents	Volatile solvents	Recovery from air or from gases in ventilating systems
Air and gaseous products from industrial operations.	Odors and smells	Abatement of industrial stenches

Recovery of Gasoline from Natural Gas. In the adsorption of gasoline from natural gas, the adsorbent charcoal is placed in two, three, or more adsorbers. These are cylindrical towers or chambers which serve to hold the charcoal. Material of 8- to 14-mesh size is supported on screens or packed in such a way that the passage of gas is not interfered with. Natural gas is passed through the charcoal until the latter is saturated with gasoline. This corresponds to 10 to 20 per cent of the weight of the charcoal. The gas flow is converted by means of appropriate valves to another adsorber and saturated steam is forced through the charcoal in the first adsorber to drive out the gasoline. The gasoline vapor and steam are condensed in water-cooled condensers and the water trapped from the gasoline. In order to cool and dry the previously steamed charcoal in preparation for a subsequent adsorption, residual gas which had been stripped of its gasoline is passed through the bed of carbon. The adsorbers are thus operated in cycles of adsorbing, gasoline removal or "distillation," and cooling

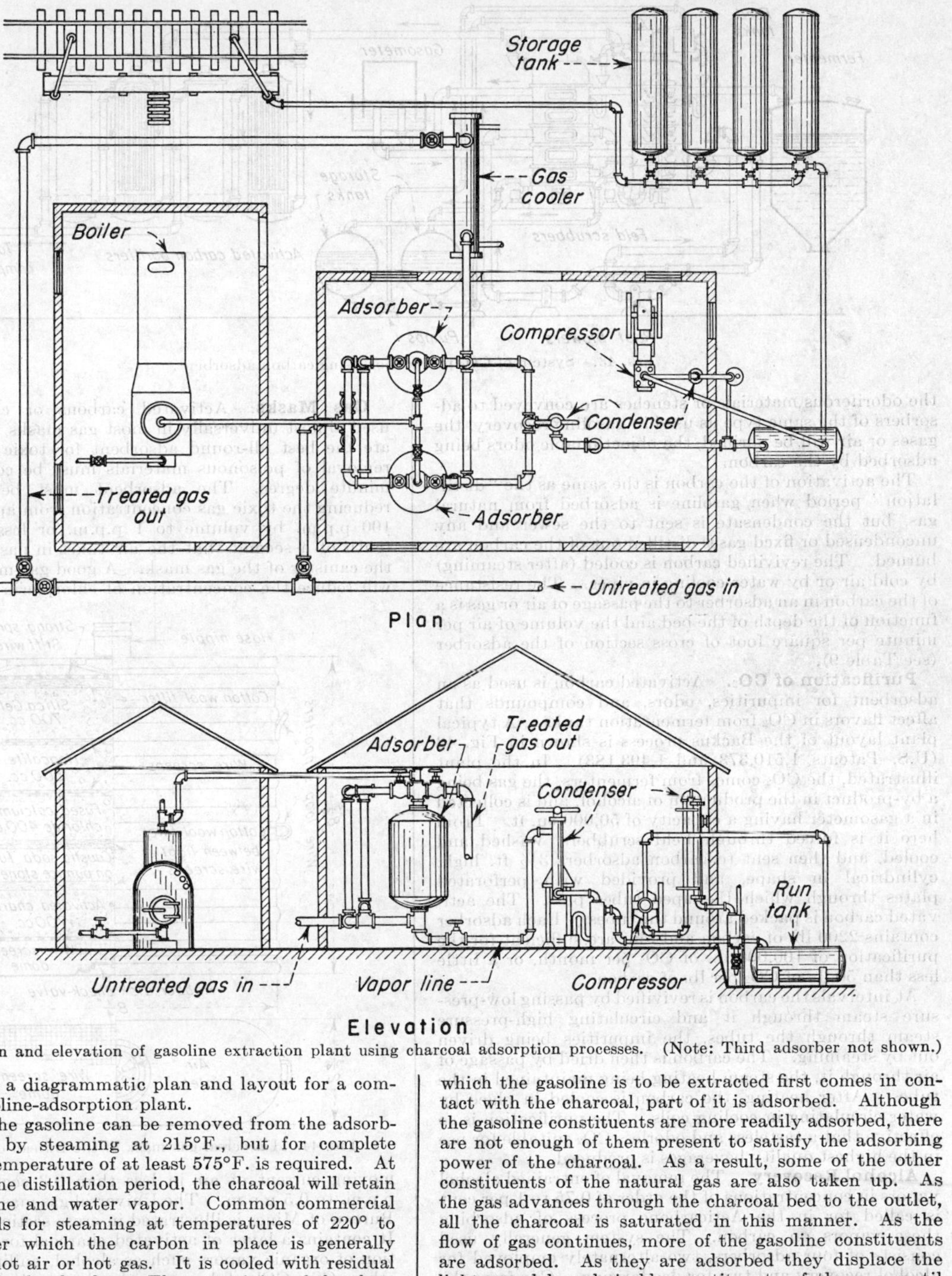

FIG. 42. Plan and elevation of gasoline extraction plant using charcoal adsorption processes. (Note: Third adsorber not shown.)

Figure 42 is a diagrammatic plan and layout for a commercial gasoline-adsorption plant.

Most of the gasoline can be removed from the adsorbent carbon by steaming at 215°F., but for complete removal a temperature of at least 575°F. is required. At the end of the distillation period, the charcoal will retain some gasoline and water vapor. Common commercial practice calls for steaming at temperatures of 220° to 240°F., after which the carbon in place is generally dried with hot air or hot gas. It is cooled with residual gas from a fresh adsorber. The retentivity of the charcoal will cut down its adsorption capacity for gasoline in all subsequent adsorptions after its first-time use.

Natural gasoline is essentially a mixture of paraffin hydrocarbons, the first of which—methane—is the least readily adsorbed by charcoal. The next member—ethane—is more readily adsorbed than methane, propane more readily than ethane, etc. When natural gas from which the gasoline is to be extracted first comes in contact with the charcoal, part of it is adsorbed. Although the gasoline constituents are more readily adsorbed, there are not enough of them present to satisfy the adsorbing power of the charcoal. As a result, some of the other constituents of the natural gas are also taken up. As the gas advances through the charcoal bed to the outlet, all the charcoal is saturated in this manner. As the flow of gas continues, more of the gasoline constituents are adsorbed. As they are adsorbed they displace the lighter and less desirable constituents of the gas, until finally all the gasoline constituents that can be retained are held within the pores of the charcoal until the latter is saturated.

Elimination of Odors. Many industrial processes, of which rendering of slaughterhouse waste and waste fats for the recovery of tallow are examples, give off objectionable odors and vapors. When the gases or air containing

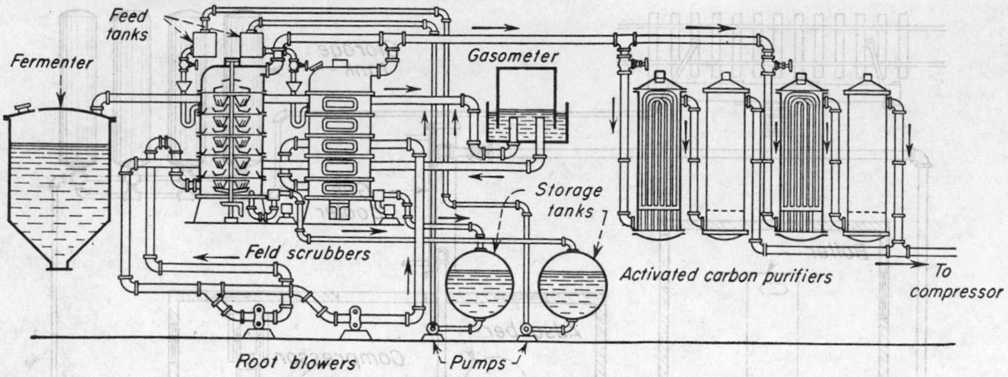

FIG. 43. System of CO_2 purification using carbon adsorbent.

the odoriferous materials or stenches are conveyed to adsorbers of the same type as used in gasoline recovery, the gases or air will be purified, the objectionable odors being adsorbed by the carbon.

The activation of the carbon is the same as the "distillation" period when gasoline is adsorbed from natural gas, but the condensate is sent to the sewer and any uncondensed or fixed gases distilled out of the carbon are burned. The revivified carbon is cooled (after steaming) by cold air or by water cooling in pipes. The resistance of the carbon in an adsorber to the passage of air or gas is a function of the depth of the bed and the volume of air per minute per square foot of cross section of the adsorber (see Table 9).

Purification of CO_2. Activated carbon is used as an adsorbent for impurities, odors, and compounds that affect flavors in CO_2 from fermentation tanks. A typical plant layout of the Backus process is shown in Fig. 43 (U.S. Patents 1,510,373 and 1,493,183). In the plant illustrated, the CO_2 comes from fermenters, the gas being a by-product in the production of alcohol, and is collected in a gasometer having a capacity of 50,000 cu. ft. From here it is forced through Feld scrubbers, washed and cooled, and then sent to carbon adsorbers 3½ ft. high, cylindrical in shape, and provided with perforated plates through which U-shaped tubes pass. The activated carbon is packed around the tubes. Each adsorber contains 2200 lb. of carbon and four are sufficient for the purification of 100,000 lb. of CO_2 per month, or a little less than 50 lb. of CO_2 per lb. of carbon.

At intervals the carbon is revivified by passing low-pressure steam through it and circulating high-pressure steam through the tubes, the impurities being driven out by steaming. The carbon is then dried by passage of air through it, the steam heating being maintained in the tubes. After heating, the carbon is cooled in place by water circulating in cooling coils. The purification is so effective that a tasteless and odorless CO_2 suitable for use in the highest quality beverages is produced.

Alcohol Recovery. The removal of entrained alcohol vapors in concentrations of the order of 0.75 to 1 per cent is called for in the Acticarbone process of adsorbing these vapors on carbon. The system generally used consists of four adsorbers, two alternately employed for alcohol recovery and two for deodorizing. Gas from the fermenters, after passage through a separator to remove entrained water, is led to the adsorbers operated in cycle. When saturated, the adsorber is steamed to remove the alcohol, the distillate is condensed and sent to a beer still as a solution of from 10 to 12 per cent alcohol, the adsorber is dried with warm air and cooled to at least 90°F. before it is returned to cycle.

Gas Masks. Activated carbons or charcoals are used almost universally in most gas masks because they are the best all-round adsorbent for toxic gases. The removal of poisonous materials must be complete to a minute degree. The adsorbent must be capable of reducing the toxic gas concentration from approximately 100 p.p.m. by volume to 1 p.p.m. or less during the tenth of a second that the air takes in passing through the canister of the gas mask. A good gas-mask charcoal will reduce the concentration of chlorpicrin in a rapidly

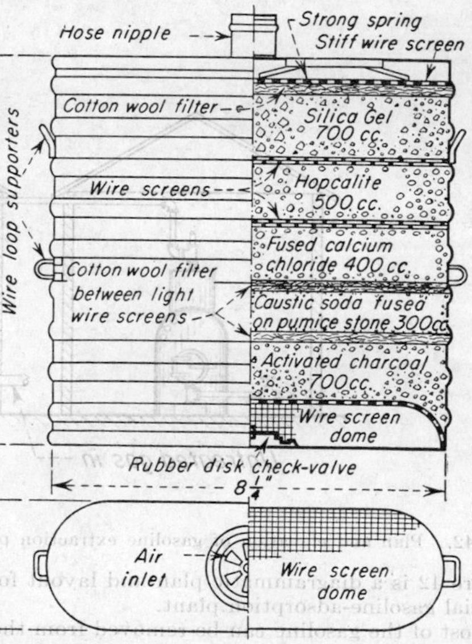

FIG. 44. Universal gas mask of the Bureau of Mines.

moving current of air in less than 0.3 sec. from 7000 p.p.m. to 0.5 p.p.m. The Universal gas mask of the U.S. Bureau of Mines is illustrated in cross section in Fig. 44. It contains a layer of activated charcoal for the adsorption of organic vapors such as alcohol, aniline, benzene, ether, CS_2, CCl_4, toluene, and other toxic materials; a filter of cotton wool to take out suspended solids including smoke, mist, and dust; a layer of caustic soda on granules of pumice stone to adsorb acid gases such as CO_2, chlorine, formic acid, HCN, HCl, oxides of nitrogen, and SO_2, as well as to take up water vapor; a layer of $CaCl_2$ to take up water vapor and particles that have passed through the caustic soda; a synthetic material (hopcalite) to remove CO by catalytic oxidation from the dry gas; and, finally, a top layer of silica gel to adsorb ammonia and other organic vapors which have passed through the previous layers of adsorbents, as well as to guard against moisture reaching the hopcalite from above. In operation, this mask has a life in excess of 6 hr., but it is not efficient in protecting the wearer against methane or any atmospheres deficient in oxygen. It is not recommended for use in atmospheres containing toxic gases having a concentration of more than 1 to 2 per cent.

Metal-adsorbent Chars

Alkaline-activated chars that can be used for removing metals such as gold and silver from their solutions are, according to McKee and Horton [*Chem. & Met. Eng.*, **32**, 13, 56, 164 (1925)], structurally identical with ordinary decolorizing carbons and can easily be converted into them by acid treatment. This change is not mere neutralization, for the original properties cannot be restored by alkaline washing. A drastic treatment with alkali at about 850°C. is required. Like colloids, the alkaline char remains suspended in water, possesses a negative electric charge, and at a definite pH (3.8) flocculates and settles rapidly. At this pH it becomes electrically neutral, loses its metal-adsorbing power, and becomes an ordinary decolorizing char.

The action of various chars under the influence of an electric field is interesting. The alkaline chars, when suspended in pure water, move toward the positive pole, whereas the acid-treated chars are practically stationary. This indicates that the alkaline chars are negatively charged and that the acid-treated chars are neutral or at their **isoelectric point**. A char made with soluble alkaline impregnating agents, such as Na_2CO_3, and then thoroughly washed will be practically inert in a neutral or only slightly acid solution. If this char is previously treated with acid, washed, and even ignited to red heat, it behaves as a normal decolorizing char. This change in the nature of the char is fundamental and has nothing to do with the question of the reaction of the medium to be decolorized. In other words, the char has been conditioned to act in a specific manner. The original alkaline char is very effective as a metal adsorbent; but the property is practically lost on treatment with acid and is not regained by treating the char with an aqueous solution of an alkali.

Early in the nineteenth century Graham [*Dinglers polytech. J.*, **40**, 443 (1831)], obtained metallic precipitates on charcoal from solutions. Moore and Edmands (Australian Patent 566, Feb. 2, 1917; U.S. Patent 1368520, 1921) developed and put in practice a method of precipitation of gold on charcoal. In 1916 the announcement was made of the use of finely divided charcoal as a precipitant in place of zinc at the Yuanmi mine, in western Australia.

Many observers have stated that there is an apparent saturation point to the capacity of the charcoal for metals deposited from cyanide solution. There seems to be no similar limit for metallic precipitation. In U.S. Bureau of Mines tests (Gross and Scott, *U.S. Bur. Mines Tech. Paper* 378, 1927), loads of 16,700 oz. gold per ton of charcoal were obtained from a chloride solution, and a load of 3700 oz. silver per ton from a nitrate solution. Both metals were plainly visible as metallic precipitates. From cyanide solutions, however, the load limit on pine charcoal was about 2000 oz. gold per ton and 1000 oz. silver per ton.

Medicinal Carbons

The use of charcoal in medicine and pharmacy has grown rapidly in recent years. Many of the commercial synthetic vegetable carbons find application, in small tonnages to be sure, in the manufacture of various types of pills, digestive tablets, and the like. The adsorbent action of charcoal for alkaloids, pathogenic bacteria, enzymes, toxins, and poisons of various sorts is well known (Mantell, "Industrial Carbon," Van Nostrand, 1946; Kausch, "Die aktive Kohle," Verlag von Wilhelm Knapp, Halle, 1928).

Considerable quantities of carbon are used in hog, poultry, and cattle food to protect the animals from certain diseases as the result of the adsorption by the charcoal of disease-producing organisms and other toxins.

Activated Alumina

REFERENCES: Derr, *Ind. Eng. Chem.*, **30**, 384 (1938). Housley, *Trans., Am. Inst. Elec. Engrs.*, **38** (April, 1939). Derr and Willmore, *Ind. Eng. Chem.*, **31**, 866 (1939). Mantell, "Adsorption," McGraw-Hill, New York, 1945.

Table 12. Properties of Activated Alumina

Volume-weight Relations

Size of Activated Alumina	Weight of packed material, lb./cu. ft.	Voids, %	Pores, %	Total air space, %
2- 1 in.	50	50	25	75
1- 2 mesh	50	50	25	75
2- 4 mesh	51	49	26	75
4- 8 mesh	51	49	26	75
8-14 mesh	51	49	26	75
14-20 mesh	54	46	28	74
20 40 mesh	54	46	28	74
40-80 mesh	59	41	30	71
Minus 80 mesh	64	36	33	69

Resistance to Crushing under Static Pressure

Depth of bed 4 in.

Size	Pressure, lb./sq. in. 10-100	300	500
	Percentage through the finer screen		
8-14 mesh	2.8	3.4	16.5
4- 8 mesh	10.8	18.9	42.6
2- 4 mesh	12.5	25.6	48.1
1- 2 mesh	33.2	50.8	57.7
	Percentage through 14-mesh screen		
8-14 mesh	2.8	3.4	16.5
4- 8 mesh	1.7	4.9	13.6
2- 4 mesh	1.4	4.3	12.2
1- 2 mesh	2.7	6.7	9.5

Resistance to Shock

Particles dropped 20 ft. on to steel plate through 6-in. pipe

Size	Percentage through the finer screen	Percentage through 14-mesh screen
8-14 mesh	2.2	2.2
4- 8 mesh	5.8	0.9
2- 4 mesh	6.8	0.7
1- 2 mesh	15.0	0.9

Resistance to Repeated Shock

500 g. 8-14 mesh in steel container

Nature of shock	Percentage through 14-mesh screen
25 drops of 3 ft. each	25.8
50 drops of 3 ft. each	30.5
25 drops of 5 ft. each	27.7
50 drops of 5 ft. each	31.0

Resistance to Abrasion

500 g. 8-14 mesh in rotating steel drum

Tumbling at 80 r.p.m., min.	Percentage through 14-mesh screen
15	20.0
30	22.3

Activated Alumina* is a granular adsorbent consisting essentially of aluminum trihydrate rendered highly porous and adsorbent. A typical composition of the commercial grade A material is Al_2O_3, 92 per cent; moisture or loss on ignition, 7 per cent; Na_2O, less than 1 per cent; SiO_2, less than 0.1 per cent; Fe_2O_3, less than 0.1 per cent; and TiO_2, less than 0.01 per cent. Substantially all the soda is combined with the silica and the

* Registered trade-mark of Aluminum Ore Co.

alumina as an insoluble compound. Standard mesh sizes range from powder up to 1½-in. lumps. The substance is chemically inert, non-corrosive, non-toxic, insoluble in water and neutral liquids, of high resistance to abrasion, and of satisfactory mechanical strength as a packing in towers or columns.

There are three standard grades. The regularly processed material, grade A, is employed in most gas-drying procedures for the dehydration of liquids and as a desiccant in miscellaneous applications. For many desiccant applications grade C impregnated with calcium chloride is advantageous because it has about twice the moisture-adsorptive capacity of grade A. It is less

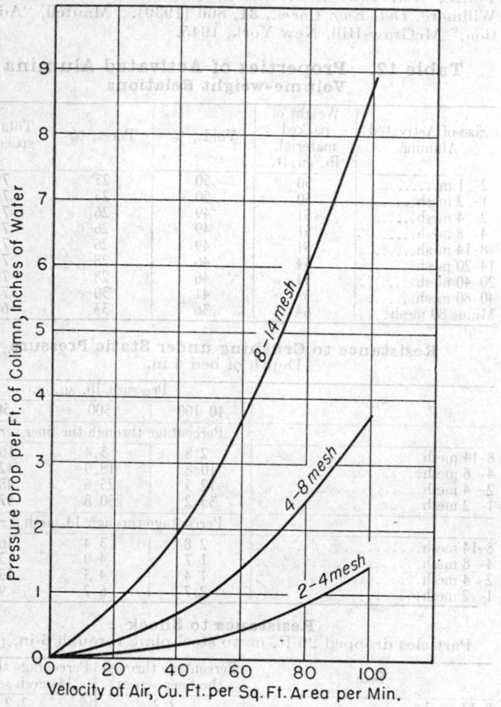

Fig. 45. Resistance to gas flow by Activated Alumina in columns.

stable and should not be reactivated at temperatures above 350°F.

Grade E is impregnated with a cobalt salt which is blue when the alumina is dry and turns pink to white as moisture is adsorbed.

Properties. The properties are given in Tables 9 and 12 while the resistance to gas flow by various meshes in towers or columns is indicated in Fig. 45.

Figure 46 shows the relation of vapor pressure and per cent water adsorbed by Activated Alumina as determined by the dynamic method. Activated Alumina is capable of removing substantially 100 per cent of the moisture in a non-reacting gas until it has taken up water equivalent to 12 to 14 per cent of its own weight. Additional moisture is progressively adsorbed with continued flow but at rapidly decreasing efficiencies. The adsorbed moisture is retained in the pore structure and is released by the application of controlled heat. Reactivation is normally accomplished by passing air or gas, heated within the temperature range of 350° to 600°F. through the adsorbent. When cooled, the material is again useful as an adsorbent.

It is to be noted in Fig. 46 that the ordinate is the vapor pressure in the air leaving the alumina.

Adiabatic dehydration of a gas is illustrated in Fig. 47 wherein the temperatures existing in an uncooled bed at several cross sections are plotted against time in hours. Heat is stored in the bed and maximum temperatures occur at the cross section where adsorption at high efficiency is taking place. The temperature of the exit air remains relatively low for a considerable period

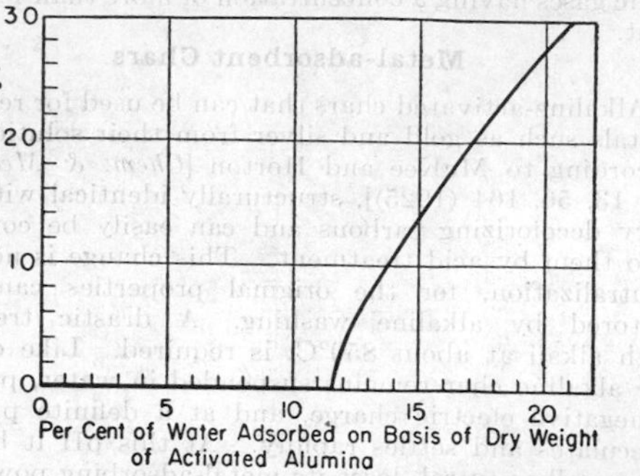

Fig. 46. The variation of vapor pressure, with percentage adsorption from air at 30°C. and 90 per cent relative humidity for 8- to 14-mesh Activated Alumina by the dynamic method.

and then rises. Dry air is produced until the temperature of the exit air falls below its maximum.

Isothermal dehydration such as is obtainable commercially is illustrated in Fig. 48. This graph shows the moisture adsorbed from air of several moisture concentrations and at several space velocities. Equal quantities of water are not taken up by an adsorbent at different relative humidities. Approximately equivalent quantities are adsorbed because less heat is developed in gas of low moisture concentration while the same cooling surface

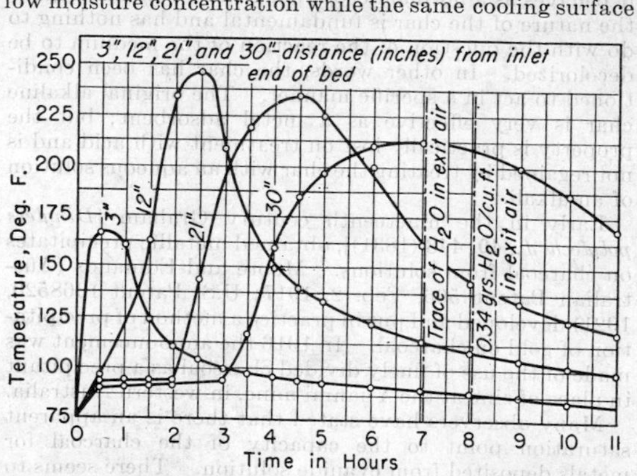

Fig. 47. Temperatures existing at various cross sections of an uncooled Activated Alumina bed (32 in. high, 1 ft. diameter) when drying air of 75°F. dry-bulb temperature containing 9 gr. of moisture per cu. ft. and flow of 5.2 cu. ft./hr./lb. of alumina. [*Derr, Ind. Eng. Chem.*, **30**, 384 (1938).]

maintains lower temperatures than when gas of high humidity is dried.

Strictly, the operations discussed are not isothermal. Theoretically, temperature rises due to adsorption in runs 2 and 3 in Fig. 48 are about 48° and 15°F., respectively. If the temperature of the exit end of the bed at the break point were 100°F. in run 2, it might be 80° to 85°F. in run 3, thus accounting for the fact that the bed held almost the same amount of water in both runs.

Gas may be dried with adsorbents to 0.0008 mg. of moisture per liter. A gas with a dew point of −40°F. is considered commercially dry. In large-scale, properly designed equipment, dew points of dehydrated gases of −76°C. (−105°F.) may be reached. Where partial drying or dehumidification is required, shallow beds of adsorbent are employed with air at high space velocities. Several types of commercial equipment are available for this purpose.

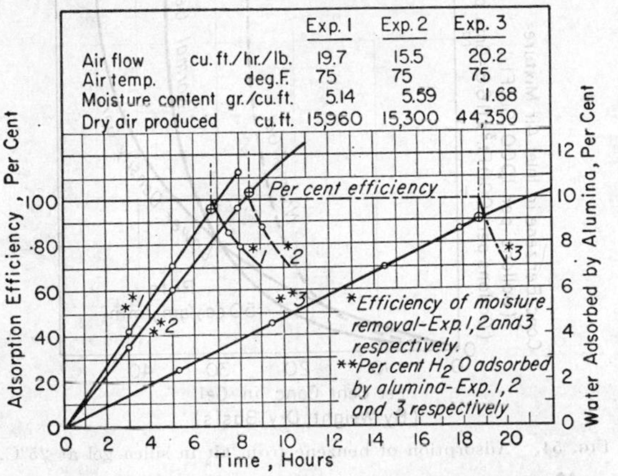

FIG. 48. Adsorption of moisture by 120 lb. of Activated Alumina in a bed 42 in. deep by 0.78 sq. ft. cross section (bed temperature, 90°–100°F.). [*Derr, Ind. Eng. Chem.*, **30**, 384 (1938).]

Dehydration of organic liquids is shown in Figs. 49 and 50, illustrating the removal of dissolved water from butyl acetate and gasoline, respectively. Dehydration is accomplished by passing the organic liquids through beds of adsorbent at room temperature. Reactivation is accomplished by applying a vacuum and heating with internal steam coils. Another satisfactory method of reactivation is the recirculation of an inert gas through a heater, the alumina bed, and a cooler to trap out moisture and retained solvent.

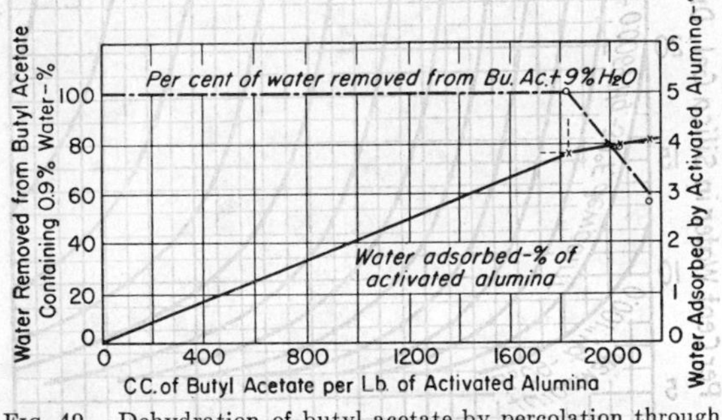

FIG. 49. Dehydration of butyl acetate by percolation through 8- to 14-mesh Activated Alumina. [*Derr and Willmore, Ind. Eng. Chem.*, **31**, 866 (1939).]

Other applications involve the drying of compressed inert and combustible gases and dehydrating refrigerants in the liquid phase. Gases at high pressure prevent an appreciable rise in temperature due to the heat of adsorption.

Favorable results in maintaining the essential characteristics of electrical insulating oils are obtained by employing adsorbents in thermosyphons in perforated containers in direct contact with the oil. The development of acidity and sludge in the oil is prevented or

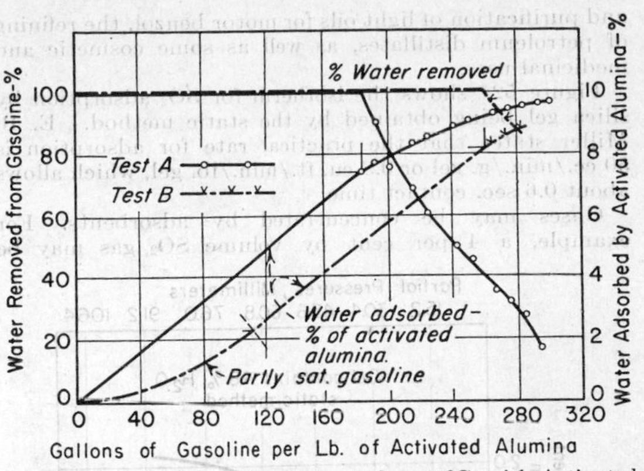

FIG. 50. Drying of gasoline saturated at 26°C. with Activated Alumina. Saturation 0.0048 per cent water by volume; flow, 80 gal./hr./lb. [*Derr and Willmore, Ind. Eng. Chem.*, **31**, 866 (1939).]

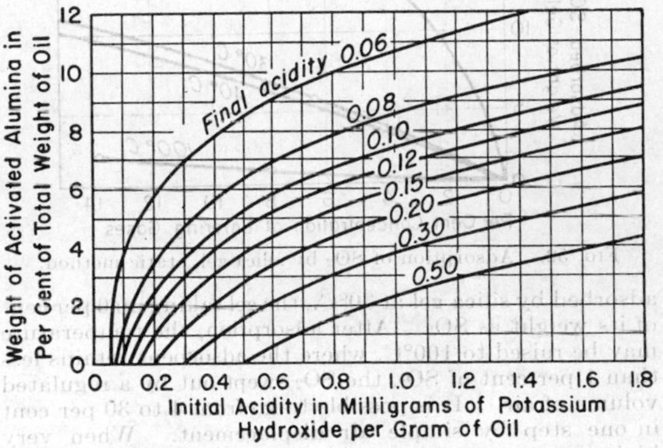

FIG. 51. Reduction of transformer oil acidity by Activated Alumina.

removed. The quantity needed to remove acid from a given transformer oil is shown in Fig. 51.

Silica Gel

Silica gel is an adsorbent prepared from the coagulation of a colloidal solution of silicic acid. The term "gel" merely indicates the condition of the material at one stage of its manufacture. As used, it is a hard glassy substance, similar in appearance to clear quartz sand, having the composition represented by the formula $SiO_2 \times H_2O$. It is a highly porous structure, characterized by uniformity of the arrangement of the pores and their size. It is produced by interraction under controlled conditions and proportions of liquid sodium silicates and H_2SO_4. The mixture is coagulated into a hydrogel which is washed to remove the Na_2SO_4 formed in the reaction, the hydrogel containing 90 per cent water and 10 per cent silica. It is dried to produce the hard glassy silica gel of commerce. The average diameter of pores of silica gel is estimated as 4×10^{-7} cm., which is roughly about ten times as large as the diameter of the average molecule. The moisture content of the silica gel varies between 4.5 and 7 per cent. The properties of silica gel are given in Table 9.

Commercial applications of silica gel include the dehydration and purification of industrial gases such as CO_2, H_2, O_2, N_2, Cl_2, air conditioning, dehydration of air for combustion, refrigeration where the silica gel is an adsorbent for a refrigerant such as SO_2, desulfurization and purification of light oils for motor benzol, the refining of petroleum distillates, as well as some cosmetic and medicinal uses.

Figure 52* shows the isotherm for SO_2 adsorption by silica gel being obtained by the static method. E. B. Miller states that the practical rate for adsorption is 50 cc./min./g. gel or 0.8 cu. ft./min./lb. gel, which allows about 0.6 sec. contact time.

Gases may be concentrated by adsorbents. For example, a 4 per cent by volume SO_2 gas may be

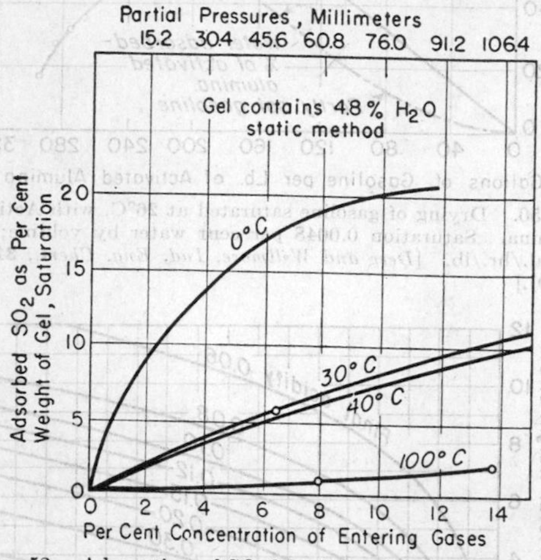

FIG. 52. Adsorption of SO_2 by silica gel, static method.

adsorbed by silica gel at 30°C., the gel taking up 6 per cent of its weight as SO_2. After adsorption, the temperature may be raised to 100°C. where the adsorbent retains less than 1 per cent of SO_2, the SO_2 swept out by a regulated volume of air. It is possible to go from 4 to 30 per cent in one step by simple air displacement. When very high percentage gases are required, the adsorbed gas can be liberated at 100°C. or above by evacuation. By proper regulation of the temperature, pressure, and volume of air factors, almost any desired concentration of gas may be obtained.

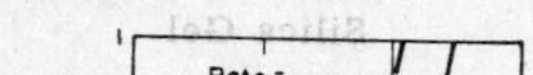

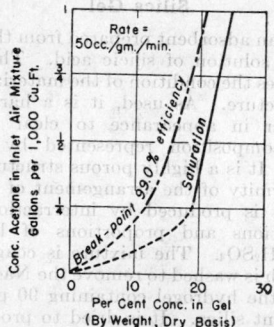

FIG. 53. Adsorption of acetone from air in silica gel at 25°C.

Silica gel adsorbs vapors of organic solvents and can be applied for the recovery of these solvents in the same manner as that previously described for other adsorbents. Curves for the adsorption of acetone and benzene are shown in Figs. 53 and 54. The break point on the curves is defined as the per cent concentration of vapor in the dry weight of gel when the adsorption efficiency falls to 99 per cent, while saturation designates the per cent

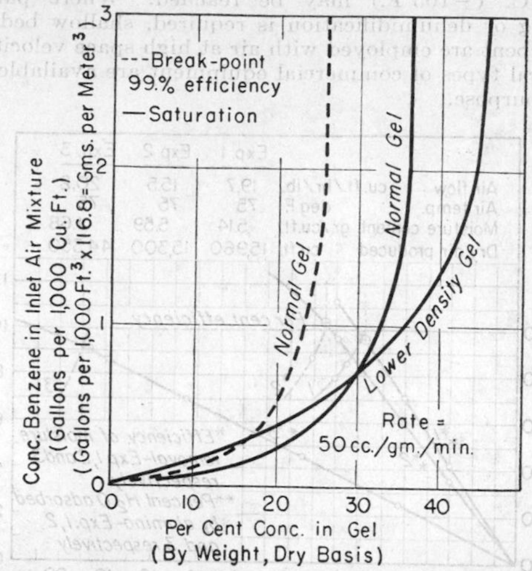

FIG. 54. Adsorption of benzene from air in silica gel at 25°C.

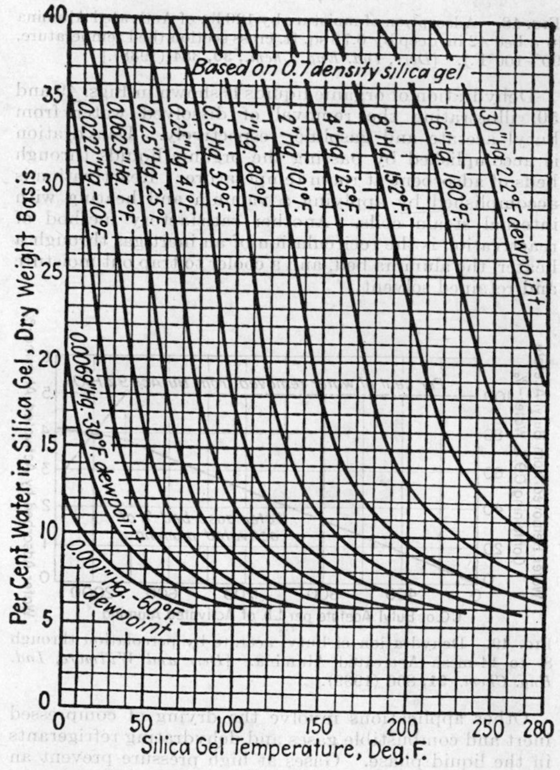

FIG. 55.

concentration of vapor by weight in the dry gel when the adsorption efficiency is zero—*i.e.*, when the partial pressure of the vapor in the outlet gas stream is identical with that in the inlet gas stream.

Silica gel shows specific selective adsorption for water. Isotherms are given in Fig. 55. The percentage removal

* Figures 52 to 54 and 56 to 60, inclusive, are by courtesy of The Silica Gel Corp. and have been taken from a chapter on Silica Gel by E. B. Miller in Alexander's "Colloid Chemistry," vol. 3, p. 119, Chemical Catalog Company, Inc., New York, by permission of the authors and publishers. Also, see article by J. B. Ledman, *Proc. Am. Gas Assoc.*, 1934.

of water at various air velocities is shown in Fig. 56. The rate of removal is mainly dependent upon the air velocity and increases only slightly when the temperature is increased from 125° to 200°C.

CO_2 Purification. Reich [*Chem. & Met. Eng.*, **38**, 136 (1931)] has described CO_2 purification by silica gel. The system consists of two pressure-type purifiers, an activator, and a silica gel trap. This equipment is placed next to the compressor between the first and second stages and is operated under a pressure of about 80 lb./sq. in. The purifiers are connected in parallel

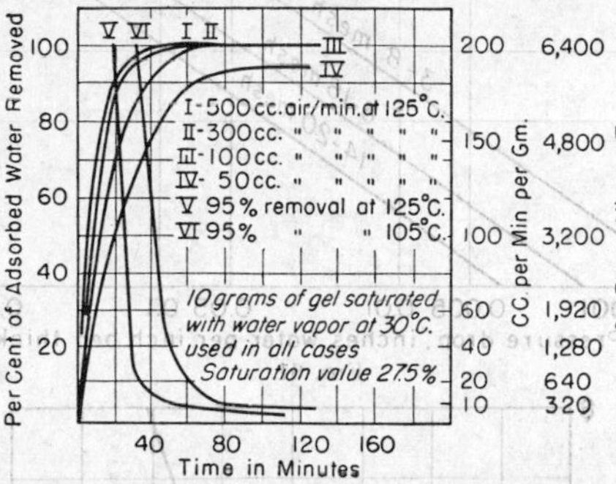

FIG. 56. Removal of adsorbed water by air at normal pressure.

and are provided with a perforated bottom on which the silica gel is resting. The CO_2 passes through this and gives up its impurities, also moisture, during its passage. The adsorbers are operated in cycle, in a manner similar to the method described under activated carbon. A typical flow sheet is shown in Fig. 58.

Gas Dehydration. Silica gel will adsorb water vapor up to 40 per cent of its weight. This water vapor may be removed by heating, and the silica gel will again be capable of adsorbing a like quantity. Its retentivity is

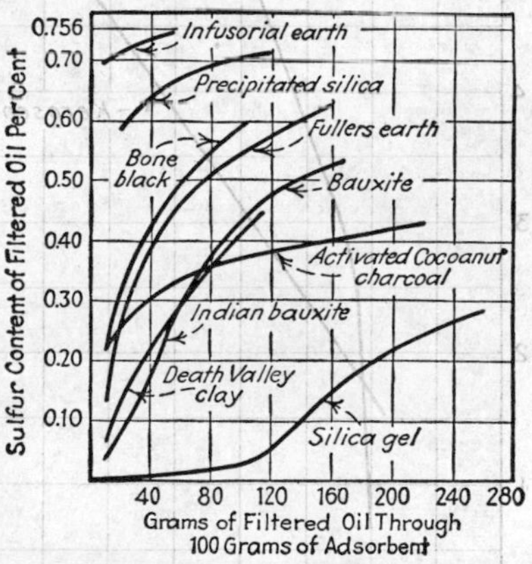

FIG. 57. Desulfurization of Mexican unrefined kerosene distillate, sulfur 0.756 per cent, by percolation.

about 5 per cent, *i.e.*, it will retain 5 per cent of its weight as water not readily removed by temperatures reached in steam heating. A typical system is shown in Fig. 59. In operation, air to be dehydrated passes through a dirt-and-dust filter to a precooler and then to a fan delivering the air to adsorbers which are operated in cycles of adsorption, revivification by heating, and cooling in preparation for a further cycle. From the adsorber the dry air or gas passes to an aftercooler and then to the air outlet. Activation of the adsorbers is by means of heated air. Air when being dehydrated is under pressure from the fan in its passage through the adsorber, while heated activating air is sucked through the adsorbers by the activation and cooling fan. The same system can be employed for dehydration of industrial gases.

Air and other gases at atmospheric pressure are dried to extremely low dew points (−80° to −85°F.) by means of silica gel. The useful water capacity of the gel in the case varies from 10 to 30 per cent of the initial weight, de-

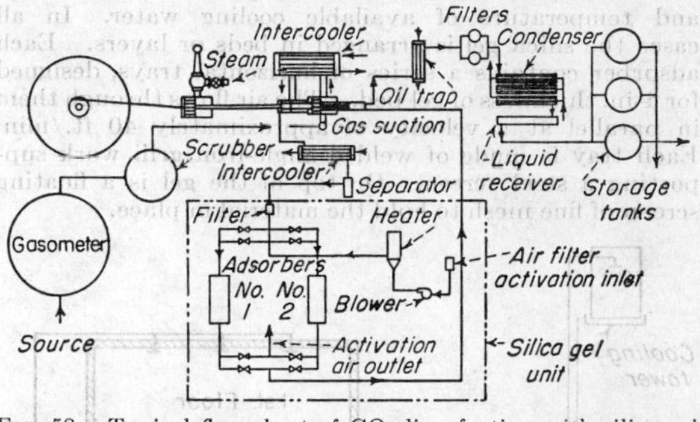

FIG. 58. Typical flow sheet of CO_2 liquefaction with silica-gel purification and dehydration unit.

pending on equipment design and operating conditions. Still lower dew points are obtained if the gas is dried at elevated pressure and then expanded. Typical values are −100° to −108°F. for the gas at atmospheric pressure after drying at 100 lb./sq. in. Additional data are given by Bower [J. J. Bower, *J. Research Nat. Bur. Standards*, **33**, 199 (1944)].

Air Conditioning. A modification of this system of air dehydration is found in the silica gel air-conditioning

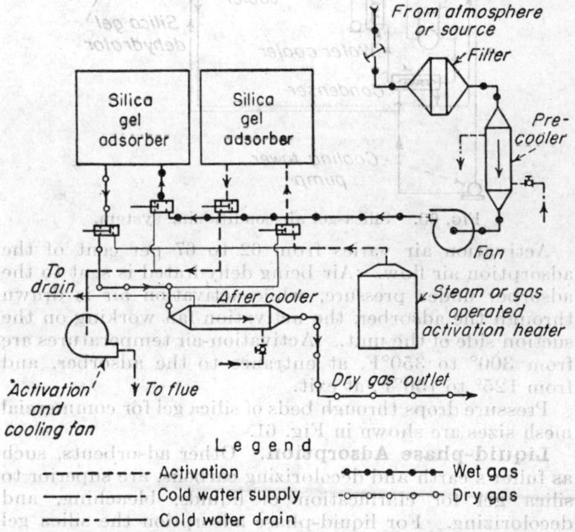

FIG. 59. Typical system for dehydration of gases by silica gel.

system, as illustrated in Fig. 60. Fresh air is drawn by fans from outdoors through a filter for the removal of dust and dirt, passes through beds of silica gel where it is dehumidified, then passes to circulating fans, air coolers, and is delivered to the rooms where it is to be used. The rate of passage of the air over the silica gel and the air-moisture-silica gel-air velocity ratios and factors are controlled to give the desired moisture adsorption and resulting humidity in the air.

Lead-covered telephone cable for toll circuits must have the last trace of moisture removed from their internal insulations. Practically all toll cables in North America are manufactured by companies employing silica gel for dehydration of air to reduce the humidity in the rooms in which the cables are manufactured to as low as 1 per cent.

The mechanical features of air-conditioning systems using adsorbents vary, depending upon the size of the installation, the control, adaptability, locality, humidity, and temperature of available cooling water. In all cases the silica gel is arranged in beds or layers. Each adsorber contains a series of horizontal trays, designed for 4-in. thickness of gel bed. The air flows through them in parallel at a velocity of approximately 40 ft./min. Each tray is made of welded angle-iron grill work supporting a steel screen. On top of the gel is a floating screen of fine mesh to hold the material in place.

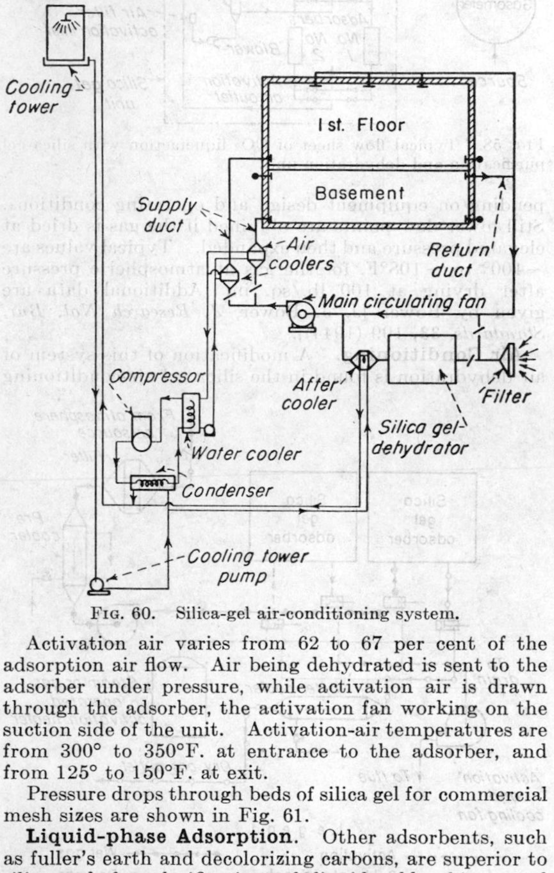

Fig. 60. Silica-gel air-conditioning system.

Activation air varies from 62 to 67 per cent of the adsorption air flow. Air being dehydrated is sent to the adsorber under pressure, while activation air is drawn through the adsorber, the activation fan working on the suction side of the unit. Activation-air temperatures are from 300° to 350°F. at entrance to the adsorber, and from 125° to 150°F. at exit.

Pressure drops through beds of silica gel for commercial mesh sizes are shown in Fig. 61.

Liquid-phase Adsorption. Other adsorbents, such as fuller's earth and decolorizing carbons, are superior to silica gel for clarification of liquids, bleaching, and decolorizing. For liquid-phase adsorption the silica gel must be ground to about 200 mesh. The desulfurization of kerosene distillates by various adsorbents is shown in Fig. 57.

Silica gel is effective in the removal of water from such substances as hydrocarbons and the halogen-containing refrigerants. Figure 62 shows that both Freon-12 and a typical hydrocarbon (kerosene) can be dried to 10 p.p.m. residual water with a ratio of silica gel to liquid low enough to be well within the practical range. Additional data have been reported by Veltman and Waring [*Refrig. Eng.*, **54,** 550 (1947)].

Petroleum Refining. The original silica-gel process for oil refining was primarily for the purpose of removing sulfur, but it is now applied principally for the elimination of "gum." Crude oil is slightly acidulated with 0.2 to 0.4 per cent by weight of H_2SO_4 and forced through silica

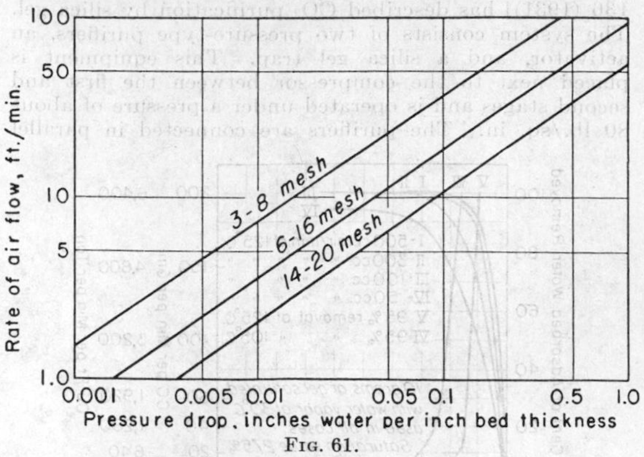

Fig. 61.

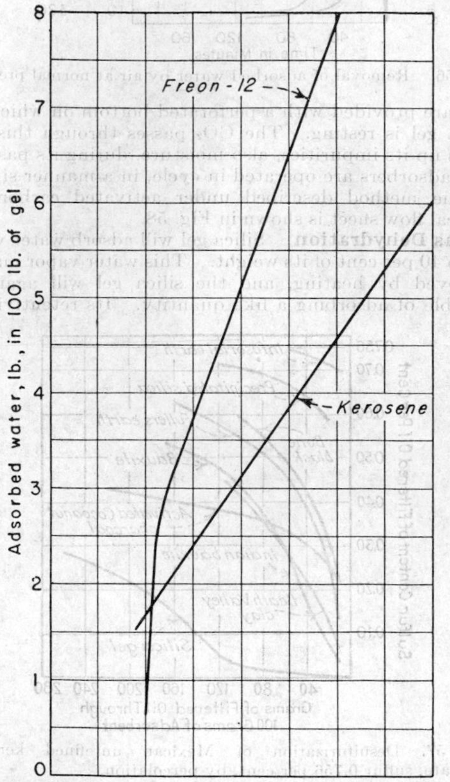

Fig. 62.

gel at temperatures of approximately 275°F. under sufficient pressure to keep it in the liquid phase. The small amount of acid is used to start the formation of gum. The process is primarily a catalytic one in which the gel serves as a carrier for the gum, the latter being the active catalyst. The surface of the gel and gum as the catalyst makes for the accumulation of more gum, the action continuing until the pores of the gel are completely sealed and it can adsorb no more gum. The gel is then removed and reconditioned for reuse.

Catalysis. Silica gel has a very large internal surface, some types having a specific area as great as 8600 sq. ft./g. It has thus found extensive use as a catalyst carrier, especially in the form of "plural gels" in petroleum cracking reactions.

Miscellaneous Applications. Very fine air-floated powders ("flowers of silica gel") show 98 per cent of the adsorptive capacity of the material from which the powder is made. It finds medicinal use as a dusting material for the human skin, keeping it dry and sweet under plaster casts applied over fractured portions of the body. As an adsorbent for odors and moisture, it finds application in toilet powders of various sorts.

ION EXCHANGERS

Adsorption of electrolytes from solution is so slight that it is difficult to make accurate measurements.

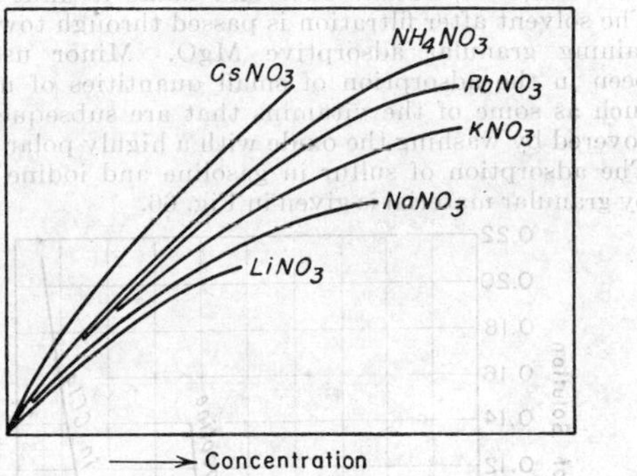

FIG. 63. Adsorption curves for different cations.

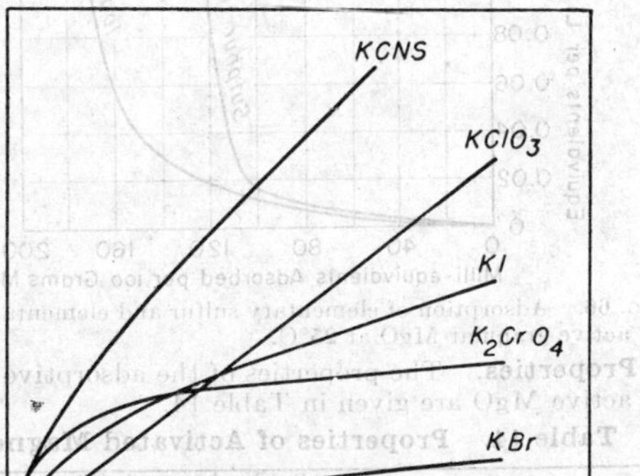

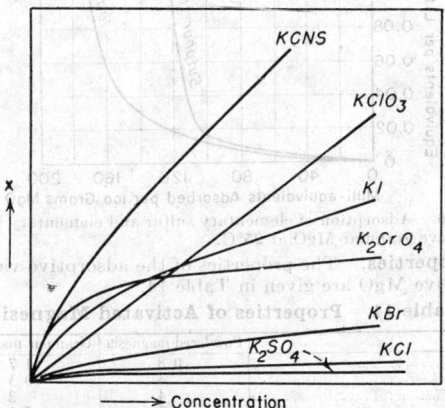

FIG. 64. Adsorption curves for different anions.

Osaka (*Mem. Coll. Sci., Kyoto Imp. Univ.*, vol. 1, No. 6, 1915) and Odén [Odén and Andersson, *J. Phys. Chem.*, **25**, 311 (1911); Odén and Langelius, *J. Phys. Chem.*, **25**, 385 (1921)] have shown that the differences in adsorption of different salts and ions, while small, are sufficiently marked to be measurable. The alkali ions are plotted in Fig. 63, different anions in Fig. 64, and the results of numerous investigators on the adsorption of different anions and cations are tabulated in Table 13.

In operation, the ion exchangers give up an ion from their structure for another in the solution that is being treated. Until 1935, the only products available for ion exchange were the natural or synthetic zeolites. The

Table 13. Order of Adsorption of Ions

Anions			Cations	
1	2	3	4	5
OH	OH		H	
CNS	CNS		Al	
I	I	I		
ClO_3			Cu	
NO_3	NO_3	NO_3	Zn	
CrO_4			Mg	
Br	Br	Br	Ca	
Cl	Cl	Cl		Cs
	HPO_4		NH_4	NH_4
SO_4	SO_4	SO_4	Rb	
			K	K
			Na	Na
			Li	

1 Odén and Langelius, *J. Phys. Chem.*, **25**, 385 (1921).
2 Rona and Michaelis, *Biochem. Z.*, **94**, 240 (1919).
3 Osaka, *Mem. Coll. Sci., Kyoto Imp. Univ.*, **1**, 257 (1915).
4 Rona and Michaelis, *loc. cit.*
5 Odén and Andersson, *J. Phys. Chem.*, **25**, 311 (1921).

main function of these base-exchanging zeolites was to reduce the hardness of water in water treatment by exchanging the sodium base of the zeolite for the calcium and magnesium ions in the water. The zeolites are double silicates capable of undergoing reversible base-exchange reactions. The designation has come to include all granular cation-exchange materials operating on the "sodium cycle" for removing hardness from a water supply and employing common salt as a regenerant, as illustrated in the following well-known reactions:

Let Z = zeolite.

$$\left.\begin{matrix}Ca\\Mg\end{matrix}\right\}\text{salts} + Na_2Z = \left.\begin{matrix}Ca\\Mg\end{matrix}\right\}Z + Na_2\ \text{salts}$$

The reversible regeneration reaction is

$$\left.\begin{matrix}Ca\\Mg\end{matrix}\right\}Z + 2NaCl = Na_2Z + \left.\begin{matrix}Ca\\Mg\end{matrix}\right\}Cl_2$$

More recently, organic cation exchangers called carbonaceous zeolites were commercially introduced. While retaining the simplicity and hardness-removal characteristics of zeolite softeners, the carbonaceous zeolites are unusual in that, in addition to operating on a sodium cycle, they may also be regenerated with acid. In this case they will exchange hydrogen ion for sodium, calcium, and magnesium ions, thus operating on the so-called "hydrogen cycle." The carbonaceous exchangers reduce alkalinity and total solids by converting all the carbonates and bicarbonates to carbonic acid, which is then removed by mechanical degasification. This is the only known method other than distillation for removing sodium bicarbonate from solution. Because the carbonaceous exchangers are non-siliceous, they can be used to soften water low in silica without the danger of silica pick-up from the zeolite itself. Therefore, they have an advantage over the older zeolites in boiler feed-water conditioning, where even minute increases in silica content caused by dissolved silica from siliceous zeolites have been troublesome. In 1934, Adams and Holmes [(*J. Soc. Chem. Ind.*, **54**, 1 (1935)] showed that specific resins of the phenol-formaldehyde type could be used as cation exchangers similar to the zeolite adsorbents. This, along with the development of anion adsorbents, opened the way for full application of the principles of ion exchange to numerous industrial processes. The anion adsorbents, coupled with the cation exchangers, have made practicable a two-step ion-exchange process of removing salts from water and solutions ("demineralizing"), thereby producing an effluent comparable in quality with distilled water.

Exchange adsorbents may be classified:

A. Cation-exchange adsorbents.
 1. Inorganic (siliceous).
  a. Natural (*e.g.*, modified green sand, clays).
  b. Synthetic (*e.g.*, synthetic gel zeolites).
 2. Organic.
  a. Natural (peat, lignite).
  b. Sulfonated coals, wood, etc. (carbonaceous).
  c. Synthetic (*e.g.*, tannin-formaldehyde resins).

B. Anion-exchange adsorbents.
 1. Inorganic.
  a. Natural (*e.g.*, dolomite).
  b. Synthetic (*e.g.*, heavy-metal silicates).
 2. Organic.
  a. Synthetic (*e.g.*, amine-formaldehyde resins).

and according to functional groups as:

Types of Synthetic-resin Ion Exchangers

Functional Group	Principal Region of Application
Acid Resins	
—SO_3H (nuclear)	Very low pH
—CH_2SO_3H	Low pH
—COOH	Neutral solutions
—OH (phenolic)	High pH
—CH_2OH } —CH_2SH }	Not yet investigated
Basic Resins	
—NH_2 (aromatic)	Acid solutions
—NH_2 (aliphatic)	Acid and neutral solutions
=NH (aromatic and aliphatic)	Not fully investigated
≡N (aromatic and aliphatic)	Not fully investigated

The synthetic-resin exchangers function as cation exchangers in either a sodium cycle or a hydrogen cycle and may be regenerated with sodium chloride or sulfuric acid depending on the cycle used. The anion adsorbents are regenerated with sodium hydroxide or sodium carbonate or other alkaline solutions.

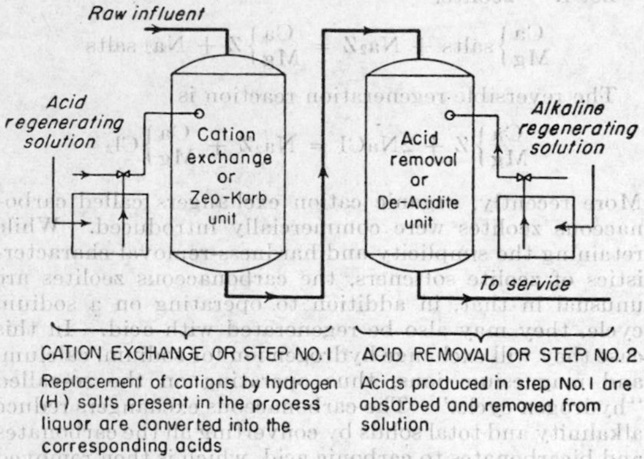

FIG. 65. Two-step process for demineralizing water.

The reactions involved in two-step demineralizing processes may be written as follows:

$$CaCl_2 + H_2Z \rightleftharpoons CaZ + 2HCl$$
$$HCl + R_3N \rightleftharpoons R_3N.HCl$$

where Z = cation exchanger.

R_3N = anion exchanger.

An illustration of the usual two-step process for demineralizing water is shown in Fig. 65. Step 1 involves cation exchange wherein the salts present in the process liquor are converted into the corresponding acids. In step 2 the acids produced in step 1 are adsorbed and removed from solution by the process of anion exchange. Demineralizing does not remove silica, carbon dioxide, or the anions of other very weak acids. The carbon dioxide is removed by a degasifying operation. The final effluent is comparable in quality with distilled water, as shown by the following average analysis:

Expressed as p.p.m. of $CaCO_3$:	
Total hardness	0– 2
Alkalinity to methyl orange	1– 6
Chlorides	0– 4
Sulfates	0– 3
Free CO_2, p.p.m.	5–10
Color (A.P.H.A.)	Below 5

The quantity of cation exchanger necessary for demineralizing a particular water is partly dependent on the type of anions present. To ensure maximum removal of metallic cations, a larger cation-exchanger regenerant dosage is employed.

Magnesia

Active magnesium oxide has found application as an adsorbent in the treatment of gasoline and the regeneration of used dry-cleaning solvent. In the first application it removes H_2S and converts mercaptans to H_2S and organic sulfides, removing the H_2S formed. Its use in the regeneration of dry-cleaning solvent rests upon the fact that it adsorbs most polar compounds and so removes color, acids, and other materials that tend to give the solvent an unpleasant odor and make it unfit for use. The solvent after filtration is passed through towers containing granular adsorptive MgO. Minor uses have been in the adsorption of small quantities of materials such as some of the vitamins that are subsequently recovered by washing the oxide with a highly polar solvent. The adsorption of sulfur in gasoline and iodine in CCl_4 by granular material is given in Fig. 66.

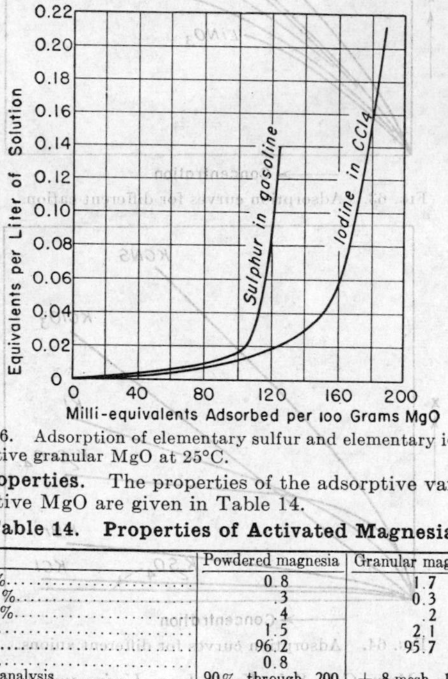

FIG. 66. Adsorption of elementary sulfur and elementary iodine by active granular MgO at 25°C.

Properties. The properties of the adsorptive variety of active MgO are given in Table 14.

Table 14. Properties of Activated Magnesia

	Powdered magnesia	Granular magnesia
SiO_2, %	0.8	1.7
Fe_2O_3, %	.3	0.3
Al_2O_3, %	.4	.2
CaO	1.5	2.1
MgO	96.2	95.7
SO_3	0.8	
Screen analysis	90% through 200 mesh	+ 8 mesh 1.0% +28 mesh 96.0% +48 mesh 99.3%
Loss on ignition, %	7–8	7–8
Bulk density, lb./cu. ft	25	
Surface, sq. cm./g	2×10^8	

Regeneration of Active Magnesium Oxide. Magnesium oxide that has been used for adsorption of coloring matter from non-aqueous solvents can be regenerated by treating with clean solvent containing 1 to 5 per cent of ethanol, methanol, ethyl ether, or other strongly polar non-aqueous substances, which later is removed by evaporation in dry air at a temperature not exceeding 100°C.

SECTION 15

MECHANICAL SEPARATIONS

BY

Robert Ammon, B. S., Chief Metallurgist, American Zinc, Lead and Smelting Co., St. Louis, Mo.; Member, American Chemical Society, American Institute of Mining and Metallurgical Engineers, American Society for Testing Materials, Institute of Metals (London). (Concentration of Ores by Sink-and-float Methods)

Anthony Anable, B. S., Engineer, The Dorr Co., Inc., New York; Member, American Institute of Chemical Engineers, American Institute of Mining and Metallurgical Engineers. (Classification, Sedimentation)

Hugh W. Bellas, B. S., Grasselli Chemicals Dept., E. I. duPont de Nemours & Co., Wilmington, Del.; Member, American Institute of Chemical Engineers, American Chemical Society. (Filtration)

C. E. Berry, B. S., Chemical Engineer, E. I. duPont de Nemours & Co., Inc., Wilmington, Del.; Member, American Chemical Society, American Institute of Chemical Engineers, American Institute of Mining and Metallurgical Engineers. (Editor: Miscellaneous Methods of Mechanical Separation and Concentration)

C. P. Cabell, B. S., Senior Industrial Engineer, Engineering Division. General Electric Co., Richland, Wash.; Member, American Institute of Chemical Engineers. (Screening)

W. S. Calcott, Ch. E., Ll. D., Assistant Director (Development) Chemical Division, Organic Chemicals Department, E. I. duPont de Nemours & Co., Inc., Wilmington, Del.; Member, American Chemical Society, American Institute of Chemical Engineers, Society of Chemical Industry, American Association for the Advancement of Science, Chemical Society (London), Army Ordnance Association, Institute of Chemists. (Sweating)

Fred D. DeVaney, E. M., M. S., Research Metallurgist, Pickands, Mather & Co., Hibbing, Minn.; Member, American Institute of Mining and Metallurgical Engineers. (Jigging, Tabling, Elutriation, Froth Flotation, and Magnetic Separation)

A. E. Flowers, M. E. in E. E., M. M. E., Ph. D., Late Engineer-in-charge of Development, The De Laval Separator Co., Poughkeepsie, N.Y.; Member, American Society of Mechanical Engineers, American Institute of Electrical Engineers, American Society of Testing Materials, American Physical Society. (Centrifuges)

J. L. Gillson, Sc. D., Geologist, E. I. duPont de Nemours & Co., Inc., Wilmington, Del.; Member, American Institute of Mining and Metallurgical Engineers, Society of Economic Geologists; Fellow, Geological Society of America, Mineralogical Society of America. (Electrostatic Methods of Concentration)

C. Fred Gurnham, B. S., M. Ch. E., D. Eng. Sc., Consulting Engineer; Professor and Head of Department of Chemical Engineering, Tufts College, Medford, Mass.; Member, American Chemical Society, American Electroplater's Society, Electrochemical Society, National Society of Professional Engineers, American Institute of Cnemical Engineers. (Expression)

Selden H. Hall, M. E., Consulting Engineer and Registered Patent Agent, Poughkeepsie, N.Y. (Centrifuges)

Donald F. Irvin, B. S., Engineer, Oliver United Filters, Inc., New York; Member, American Institute of Mining and Metallurgical Engineers. (Filtration)

C. E. Lapple, B. S., Ch. E., Chemical Engineer, Engineering Department, E. I. duPont de Nemours & Co.; Member, American Institute of Chemical Engineers, American Chemical Society. (Dust and Mist Collection)

Warren L. McCabe, Ph. D., Director of Research, The Flintkote Co.; Member, American Institute of Chemical Engineers, American Chemical Society. (Crystallization)

J. W. Stillman, Ph. D., E. I. duPont de Nemours & Co.; Member, American Chemical Society. (Sampling)

CONTENTS

CLASSIFICATION

	Page
Draining	922
Definition	922
Purpose	922
Equipment Used	922
Drag Conveyor	922
The Grainer	922
The Dorr Classifier	923
Adjustments	924
Capacity	924
Chip-removing Device	924
Suction Box	924
Materials of Construction	924
The Dorr Bowl Classifier	924
Sizes	925
Power Consumptions	925
Costs	925
The Akins Classifier	926
Hardinge Contercurrent Classifier	926
Washing (Granular Material)	927
Definition	927
Purpose	927
Equipment Used	927
Operation	928
Adjustments	928
Capacity	928
Accessories	928
Materials of Construction	928
Washing Calculations and Practical Examples	928

	Page
Leaching	929
Definition	929
Purpose	929
Equipment Used	929
Operation	929
Leaching Calculations and Practical Examples	929
Closed-circuit Grinding	930
Definitions	930
Purpose	930
Theory	931
Work Units	931
Open- vs. Closed-circuit Grinding	931
Closed-circuit Grinding Equipment	931
Circulating Load Limits	932
Size Control	932
Fineness of Grinding	932
Savings	933
Automatic Control	933
Metallurgical Operating Data	933
Closed Circuiting Increases Mill Capacity	933
Classifier Determines Capacity, Not Mill	933
Closed Circuiting Reduces Unit Power Cost	933
Closed Circuiting Reduces Wear on Liners and Balls	934
The Higher the Circulating Load the Lower the Unit Grinding Cost	934
Operating Costs	934
Cement-slurry Grinding	934
Power Requirements	936
Whiting, Lithopone, and Abrasive Grinding	936
Whiting Classification	936
Lithopone Classification	936
Abrasives Classification	937

SEDIMENTATION

Continuous Thickening	937
Definition	937
Mechanics of Clarification	937
Clarification Capacity	938
Thickening Capacity	939
Mechanics of Thickening	939
Destabilization of Colloids	939
Equipment Used. Intermittent Settling Tanks	939
Operation	939
Non-mechanical Continuous Thickeners	940
Equipment Used	940
The Allen Cone	940
The Callow Cone	940
Mechanical Continuous Thickeners	941
The Single-compartment Dorr Thickener	941
The Dorr Torq Thickener	941
The Dorr Tray Thickener	942
The Dorr Traction Thickener	942
The Dorr Clarifier	943
The Dorr Sifeed Clarifier	943
The Dorr Squarex Clarifier	943
Equipment Costs	944
The Hardinge Auto-Raise Thickener	944
Center Pier-type Hardinge Auto-Raise Thickener	945
Hardinge Diaphragm Pump	945
Dorr Thickener Types, Sizes, and Drive Ratings	945
Dorrco Pumps	945
Selection of Type of Continuous Mechanical Thickener	947
Thickening Costs	947
Filter Thickeners	947
The Oliver-Borden Filter Thickener	948
Hardinge Sand-filter Clarifier	948
Hardinge Automatic Backwash Sand Filter	948
Briggs Filter Thickener	949
Continuous Countercurrent Decantation (C.C.D.)	950
Definition	950
Purpose	950
Theory	950
Use of C.C.D. in Chemical Processing	951
C.C.D. Calculations	951
Various Types of C.C.D. Flow Sheets	951
Economic Advantages of C.C.D	952
Typical Operating Figures from C.C.D. Practice	952
Accessories	953
Complete C.C.D. System in a Single Unit	953
Wash-water Feeders	954
Mixing between Thickeners	954

SCREENING

	Page
Definitions	955
Screening	955
Sizing	955
Scalping	955
A Sieve Scale	955
Screen Cloth	955
Mesh	955
Open Area	955
Aperture or Screen Size	955
Particle-size Distribution	955
Purposes	955
Equipment	955
Grizzly	955
Shaking Screens	956
Vibrating Screens	956
High-amplitude Normal Vibration	956
Low-amplitude Normal Vibration	957
Oscillating Screens	957
Reciprocating Screens	957
Sifters	958
Riddles	958
Screening Surfaces	958
Punched Metal Plates	958
Bar Screens	958
Wire Screens	959
Comparative Openings for Silk and Wire Cloths	960
Screen Performance	960
Factors Affecting Efficiency and Capacity	960
Variables in Material Being Screened	962
Bulk Specific Gravity	962
Size of Separation	962
Stickiness of Material	962
Shape of Material	962
Particle-size Distribution	962
Mechanical Variables	962
Screen Cloth, Percentage of Open Area	962
Vibration Amplitude and Frequency	962
Angle of Inclination	962
Direction of Vibration	962
Feed Rate	962
Arrangement of Screens	962
Typical Capacity Data	962
Sieve Testing	962
Tyler Standard Sieve Scale	964
Mechanical Shakers	964
Interpretation	964

FILTRATION

Introduction	964
Theory of Filtration	965
Constant-pressure Equations	965
Constant-rate Equations	965
Practical Significance	965
Application of Filtration Theory to Interpretation of Data	966
Filter Mediums	967
Cotton Ducks	967
Cotton Twills	967
Cotton Chain	967
Wool Cloths	968
Glass Fabrics	968
Metal Fabrics	968
Rubber Mediums	968
Miscellaneous	968
Vinyon	968
Filtration Leaf Tests	968
Filter Aids	969
Types of Filters	970
Gravity Filters	970
Nutsche Filter	970
Pressure Filters	970
Life of Wood Used in Plate-and-frame Presses	970
Intermittent Vacuum Filters	970
Continuous Vacuum Filters	971
Makes of Filters	971
Plate-and-frame Presses	971
Solid Filling	971
Center Filling	971
Discharge	971
Open and Closed Filtrate Discharge	971

	Page
Merrill Press	972
Merrill Precipitation Press	972
Kelly Filter	972
Sweetland Filter	973
Vallez Filter	974
Burt Filter	974
Intermittent Vacuum Filters	975
Moore Filter	975
Butters Filter	976
Continuous Vacuum Filters	976
Oliver Filter	976
Oliver Continuous Precoat Filter	978
Feinc Filter	979
Dorrco Filter	979
Double-drum Filter	980
Oliver Top-feed Filter	980
Oliver Horizontal Filter	981
American Filter	981
Lurgi Filter	981
Bird-Young Filter	983
Oliver Pipe-line Strainers	984
Streamline Filter	984
Corrosion-resisting Construction	984
Pulp Agitation	985
Range in Operating Vacuum	985
Industrial Applications	985
Cyanide Pulp	985
Metallurgical Flotation Concentrates	985
Metallurgical Gravity Concentrates and Sands	986
Cane-sugar Sirups	986
Paper Pulp	986
Sewage Sludge	986
Cement Slurry	986
Salt and Crystals	986
Beet-sugar Plant Filtration	987
Petroleum Products	987
Settled Cane-mud Filtration	987
Food Products	988
Sparkler Filter	988
Niagara Filter	988
Republic Filter	988
Oliver Pressure Filter	988
Caustic Soda	988
Phosphoric Acid	988
Pigments	988
Power Used on Filters	989
Unit Costs	989
Floor Space Available	989
Cost of Equipment	989
Corrosion	989
Filter Auxiliaries	990
Pressure Pumps	990
Vacuum Pumps	990
Receivers or Separators	990
Moisture Traps and Condensers	991
Filtrate Pumps	991
Barometric Legs	991
Blowers	991
Flappers	991
Mechanical Vibrators	991
Compression Belts	991
Compression Rolls	992
Filter Operation	992

CENTRIFUGES

Definitions	992
Theory	992
Liquid-solid Separation (Clarification)	994
Settling Rate—Large Particles; Turbulent Motion	994
Settling Rate—Small Particles; Viscous Resistance	995
Settling Rate—Microscopic Particles—Hindered Motion	996
Liquid-liquid Separation	996
Liquid-liquid-solid Separation	997
General Design	997
Hollow Bowls	997
Bowls with Shallow Settling Spaces	997
Apparatus	997
Liquid-solid Separators	997
Dryers	997
Basket Centrifuges	997
Centrifuges for Continuous Discharge of Dried Solids	998
Hollow-bowl Clarifiers	999
Disk-bowl Clarifiers	999
Concentrator Bowls with Nozzle Discharge of the Solids	999
Valve Bowls	999
Liquid-liquid Separators	1000
Concentrators	1000
True Separator Bowls	1000
Liquid-liquid-solid Separators	1001
Concentrating Separators with Nozzle Discharges	1001
Separator Bowls with Sludge Pockets	1001
Traveling-cushion Bowls	1001
Special Types of Centrifuges	1001
Test Apparatus	1001
Centrifugal Filters	1002
Closed-cover Feed, and Discharge Machines	1002
Liquid-sealed Discharges	1002
Bearing-type Sealed Passages	1002
Clarifiers with Both Inlet and Outlet Passages Sealed	1002
Hermetic Separators and Clarifiers	1002
Accessory Apparatus	1002
Pipes	1002
Pumps	1002
Presettling Tanks	1003
Heaters or Coolers	1003
Radiant Electric Heaters	1003
Air-separating Traps	1003
Feed Regulators	1003
Processing and Preparation	1003
Settling and Settling Combined with Decanting	1003
Coagulating	1004
Partial Centrifuging	1004
Adsorbent Treatments	1004
Washing and Chemical Treatments	1004
Temperature Effects	1004
Emulsion Breaking	1004
Machine-operating Features	1004
Starting Devices and Driving Mechanisms	1004
Lubrication	1005
Balance and Vibration	1005
Wear and Corrosion	1005
Energy and Power Requirements	1005
Stored Energy	1005
Power Lost in Friction	1005
Power Lost in Accelerating the Through-put	1006
Relation of Size, Driving Power, and Costs	1006
Cream Separators	1006
Oil Purifiers	1006
Basket Centrifuges	1007
Costs of Centrifuging	1007
Applications	1007
The Dairy Industry	1007
Petroleum Refineries	1008
Vegetable Oil Refining	1008
Wool-grease Recovery	1008
Lubricating-oil Purification	1009
Steam Turbine Oil Purification	1009
Diesel-engine Lubricating-oil Purification	1009
Rolling-mill Oil Purification	1009
Gasoline-engine Crankcase Oil Purification	1009
Block-test Lubricating-oil Purification	1010
Diesel-fuel Purification	1010
Transformer-oil Purification	1010
Electric Switch Oil Purification	1010
Cable Oil Purification	1010
Varnish and Lacquer Clarification	1011
Pigment Lacquer Clarification	1011
Dry Cleaners' Solvent Clarifier	1011
Dry Cleaners' Dryers	1011
Laundry Drying	1011
Sugar Refineries	1011
Breweries	1011
Rubber Latex	1012
Chemical Plant Dryers, Chips, Cutting-oil Reclamation	1012
Grinding Coolant Clarification	1012
Activated Sewage Clarification	1012
Miscellaneous Applications	1012

DUST AND MIST COLLECTION

References	1013
Nomenclature	1014

PAGE
Purpose of Dust and Mist Collection 1016
Properties of Particle Dispersoids 1016
Particle Size 1016
Particle Classification 1016
Explosion Hazards 1016
Health Hazards 1017
Miscellaneous Properties 1017
Particle Measurements 1017
Atmospheric-pollution Measurements 1017
Process-gas Sampling 1017
Particle-size Analysis 1017
Particle-size Representation 1017
Particle Dynamics 1017
Collection Equipment 1021
Gravity Settling Chambers 1021
Impingement Separators 1022
Cyclone Separators 1023
Fields of Application 1024
Flow Pattern 1024
Pressure Drop 1024
Collection Efficiency 1026
Cyclone Design Factors 1027
Commercial Equipment 1028
Cyclone Costs 1028
Entrainment Separation 1028
Mechanical Centrifugal Separators 1028
Miscellaneous Inertial Separators 1028
Packed-bed Separators 1029
Cloth Collectors 1029
Scrubbers 1034
Chamber Scrubbers 1035
Cyclone Scrubbers 1036
Inertial Scrubbers 1038
Mechanical Scrubbers 1038
Packed Scrubbers 1038
Film Scrubbers 1039
Miscellaneous Scrubbers 1039
Electrical Precipitators 1039
Field Strength 1039
Potential and Ionization 1039
Current Flow 1039
Electric Wind 1040
Charging of Particles 1040
Particle Mobility 1040
Collection Efficiency 1040
Application 1041
Single-stage Precipitators 1041
Two-stage Precipitators 1044
Alternating-current Precipitators 1045
Air Filters 1045
Viscous Filters 1046
Dry Filters 1046
Automatic Filters 1046
General Design and Performance of Air Filters 1047
Miscellaneous Air-filter Equipment 1049
Miscellaneous Collectors 1049

CRYSTALLIZATION

Commercial Importance of Crystal Size and Shape 1050
Theory of Crystallization 1050
Types of Crystals 1050
Crystal Forms. Law of Haüy 1050
Isomorphism 1050
Crystallographic Systems 1050
Liquid Crystals 1051
Yield of Crystallization Process 1051
Purity of Product 1052
Heat Effects in a Crystallization Process 1052
Heat of Crystallization 1052
Enthalpy-concentration Chart 1052
Fractional Crystallization 1053
Crystal Formation 1054
Supersaturation 1054
Seeded vs. Unseeded Crystallization 1054
Miers' Theory 1055
Methods of Forming Crystals in Solutions 1055
Effects of Impurities 1057
Kinetics of Nucleation 1057
Geometry of Crystal Growth 1057
Parallel Displacement of Faces. Translation Velocities 1057
Crystal Habit 1058
Invariant Crystals 1058
Overlapping Principle 1058
Variation of Translation Velocities 1058
Crystal Growth 1058
Rate of Solution 1058
The Berthoud-Valeton Theory 1058
Effect of Impurities on Crystal Growth 1059
The ΔL Law 1059
Calculation of Screen Analysis of Product from That of Seeds 1059
Limitations of ΔL Law 1060
Simultaneous Growth and Formation of Crystals 1061
Crystallization Apparatus 1061
Tank Crystallization 1062
The Wulff-Bock Crystallizers 1062
Agitated Batch Crystallizers 1062
The Howard Crystallizer 1063
The Double-pipe Crystallizer 1063
The Swenson-Walker Crystallizer 1063
Crystallizing Evaporators 1065
Vacuum Crystallizer 1065
The Krystal Classifying Crystallizer 1068
Comparison of Vacuum and Liquid-cooled Crystallizers 1069
Prilling 1069
Cost of Crystallization Equipment 1070
Crystallization Costs 1070
Engineering and Cost Data 1070
Caking of Crystals 1070
Critical Humidity 1070
Prevention of Caking 1071

MISCELLANEOUS METHODS OF MECHANICAL SEPARATION AND CONCENTRATION

Miscellaneous Methods of Mechanical Separation 1072
Concentration of Materials 1072
Expression 1072
Definition 1072
Purpose 1072
Expression Equipment 1072
Batch Presses 1072
Box 1072
Platen 1073
Pot 1073
Curb 1073
Cage 1073
Continuous Presses 1073
Screw Press 1073
Roller Mills 1074
Auxiliary Equipment 1074
Expression Theory 1074
Equilibrium Conditions 1074
Empirical Equations 1074
Sweating 1075
Equipment 1075
Operation 1075
Application 1075
Jigging 1075
Types of Jigs 1076
Harz 1076
Jeffrey (Baum) 1076
Conset 1076
Hancock 1076
Capacity 1076
Cooley 1076
Jeffrey (Baum) 1076
Conset 1076
Hancock 1076
Harz 1076
Power Requirements 1077
Water Consumption 1077
Harz 1077
Hancock 1077
Jig Feed 1077
Jigs for Placer Gold 1077
The Denver Mineral Jig 1077
Cost 1078
Tabling 1078
Wet Tables 1078
Spiral Concentrators 1079
Dry Tables 1079
Agglomeration Tabling 1080

PAGE
Concentration of Ores by Sink-and-float Methods........ 1080
Scope of Heavy-media Separation Processes............ 1081
General Field of Application........................ 1081
Range of Sizes Amenable to Separation............. 1081
Range of Separating Gravities Available............ 1081
Characteristics of Magnetic Media................. 1081
Preparation of Feed............................. 1081
Heavy-media Separations.......................... 1081
Removal of Medium from the Separation Products... 1081
Reclamation and Cleaning of the Medium for Reuse 1081
Application of Heavy-media Separation of Fine Sizes... 1082
Separatory Cones.............................. 1082
Reagents...................................... 1082
The Dutch State Mines Cyclone Separator Processes..... 1083
Construction and Operation of Cyclone Separator...... 1083
Medium Used in the Cyclone Separator.............. 1083
Methods of Cleaning Medium...................... 1083
Capacity of the Cyclone Separator................... 1083
Elutriation...................................... 1084
Wet Elutriation................................ 1084
Dry Elutriation................................ 1084
Haultain Infrasizer............................. 1085
Froth Flotation.................................. 1085
References.................................... 1085
General Principles.............................. 1088
Collectors.................................... 1088
Cationic Reagents.............................. 1088
Frothers...................................... 1088
Depressors.................................... 1088
Deflocculating Agents........................... 1088
Activation and Miscellaneous Reagents............ 1088
Flotation Machines.............................. 1089
General Operations............................. 1089
Steffensen Flotation Machine..................... 1089
Denver Sub-A (Fahrenwald) Machine................ 1091
Mixing and Aeration Zone...................... 1091
Separation Zone............................... 1091
Concentrate Zone.............................. 1091
Magnetic Separation.............................. 1091
Removal of Ferromagnetic Materials................ 1091
Strongly Magnetic Material Forms Small Part of Feed 1091
Dry Methods................................ 1091
Wet Methods................................ 1091
Strongly Magnetic Material Forms Large Part of Feed 1091
Dry Methods................................ 1092
Wet Methods................................ 1092
Removal of Moderately Magnetic Materials........... 1092
Dry Methods................................. 1092
Wet Methods................................. 1092
Removal of Extremely Weak Paramagnetic or "Diamagnetic" Materials.............................. 1092

PAGE
Special Magnetic Machines and Processes............ 1093
Magnetic Roasting.............................. 1093
Electrostatic Methods of Concentration................ 1093
Principle...................................... 1093
Types of Equipment.............................. 1093
Rotor Type................................... 1093
Toboggan Type................................ 1094
Table Type................................... 1094
Commercial Machines............................ 1095
Ritter Products............................... 1095
Huff... 1095
Sutton, Steele and Steele...................... 1095
Feldspathic................................... 1095
Effect of Humidity.............................. 1095
Power Generation............................... 1095
Application of Electrostatic Methods............... 1095
Capacity of Electrostatic Machines................ 1095
Patent Art..................................... 1095

SAMPLING

References.. 1095
Methods of Sampling................................ 1097
Grab Sampling.................................. 1097
Car Sampling.................................. 1097
Boat Sampling................................. 1097
Coning and Quartering........................... 1097
Shovel Sampling................................ 1100
Pipe Sampling.................................. 1100
Mechanical Sampling............................ 1100
Bottle or Beaker Sampling........................ 1100
Continuous Liquid Sampling...................... 1100
Dipper Sampling.............................. 1100
Thief Sampling............................... 1100
Practical Methods Employed in Different Industries..... 1100
Coal.. 1100
Iron Ore....................................... 1101
Pig Iron....................................... 1101
Steel.. 1101
Metals and Alloys............................... 1101
Cement.. 1101
Soft Solids..................................... 1101
Lumpy Solids................................... 1101
Petroleum Products.............................. 1101
Lacquer Solvents............................... 1102
Water... 1102
Molten Metal................................... 1102
Gases... 1102
Flue Gas..................................... 1102
Air or Gas in Confined Space.................. 1102
Dust- or Fume-laden Air....................... 1102
General Rules for Sampling.......................... 1102

CLASSIFICATION

BY ANTHONY ANABLE

REFERENCES: Taggart, "Handbook of Mineral Dressing," Wiley, New York, 1945. Richards and Locke, "Textbook of Ore Dressing," 3d ed., McGraw-Hill, New York, 1940. Gaudin, "Principles of Mineral Dressing," McGraw-Hill, New York, 1939.

Classification may be defined as the separation of solids into fractions from their suspension in a fluid medium, usually water. The general wet classification methods are: (1) mechanical classification, (2) non-mechanical classification, and (3) hydraulic classification. Of these, only mechanical classification is discussed in this section. See Sec. 16, pp. 1118–1119, for a discussion of the use of air classifiers for dry classification.

The largest use of classification is in the dressing of metallic and non-metallic ores. Other important uses of classification methods and machines are: (1) draining, in which particles are separated from suspension together as one fraction, without regard to size or specific gravity; (2) washing, in which the main object is the removal of already dissolved substances from particles collected from suspension in one fraction without regard to size or specific gravity; (3) leaching, in which the main object is dissolving and removal by washing of substances from particles, collected from suspension in one fraction; and (4) classification for closed-circuit grinding, in which the main object is the separation of partly ground material into finished and unfinished fractions in order to obtain increased capacity and savings in grinding costs. The last operation is probably the largest and most important application of classification today. It is especially valuable in ore dressing, but it is also applicable with advantage to practically any grinding operation. Also see p. 1135 ff. of Sec. 16 for dry-grinding applications of air classifiers.

The above four uses of classification are discussed in detail in this section.

DRAINING

Definition. Draining may be defined as the dewatering action taking place when moist particles of a granular substance are placed on, or advanced over, a flat or inclined surface. Dewatering by centrifugal machines and by filters accomplishes the same purpose but is discussed separately under Centrifuges and Filtration.

Purpose. It is the purpose of draining to separate, insofar as is possible by gravitational force alone, the two constituents of the mixture—the solid portion and the liquid portion—so that each may subsequently be subjected to such further more complete and generally more expensive separation treatment as may be required, such as, in the case of the solids, centrifugation, filtration, or drying and, in the case of the liquid, clarification or filtration.

Draining, by itself, is seldom a complete treatment, as it does not result in the production of a crystal-clear solution and a bone-dry solid. Because of its relative cheapness, however, it is an approved preliminary unit operation, preparing the two constituents of the mixture so as to meet the requirements of the additional unit operations that follow.

Equipment Used. a. The Drag Conveyor. The simplest type of mechanical apparatus for draining is the drag conveyor shown in Fig. 1. This device consists of a shallow, inclined, rectangular box, equipped with an endless belt or chain to which scrapers are attached. The belt or chain is carried on two sets of pulleys or sprockets, located respectively at the upper and lower ends of the inclined box. The shafts upon which the pulleys or sprockets are mounted lie with their axes at right angles to the long axis of the box and at such an elevation above the inclined bottom that the scrapers just clear the bottom as they are drawn along from the lower to the upper end.

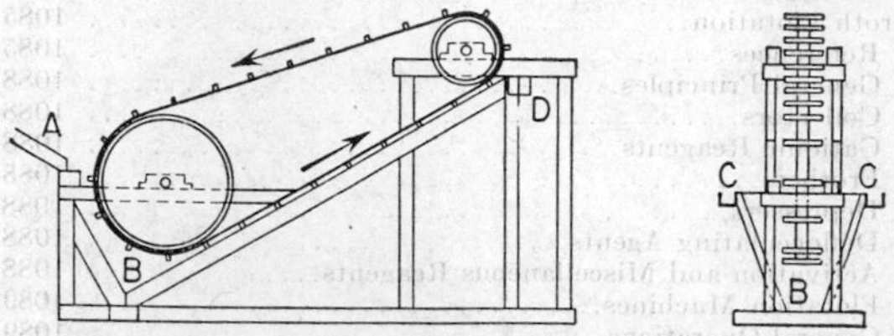

FIG. 1. Drag conveyor.

The mixture of solids and liquids enters through a trough at the lower end. The heavy granular solids settle to the bottom, are picked up by the scrapers, and are dragged up the incline, out of the solution, over the draining deck and discharged from the upper end of the box. Solution entrained with the solids partly drains in the process and flows back into the lower portion of the box.

The solution from which the heavy solid particles have settled overflows across a weir at the lower end of the box or through a pipe. Finely divided solids, contained in the feed mixture and settling more slowly than the granular portion, remain suspended in the solution and pass off with it as an overflow product.

Drag classifiers are generally about 3 ft. to 5 ft. 6 in. wide, set at slopes ranging from 2 to 6 in./ft., the steeper slope corresponding to the coarser separations. The belts are equipped with wood or steel blades on 12- to 18-in. centers and are driven at 12 to 40 ft./min.

Capacity is roughly proportional to width and belt speed and roughly inversely proportional to the blade spacing. Data from commercial installations indicate that, when making coarse separations at from 28 to 48 mesh, a capacity of 5 to 7 tons/24 hr. may be secured per foot of tank width per 1 ft./min. belt speed. When making finer separations around 100 to 200 mesh, the capacity may be as low as 1.5 to 2.5 tons/24 hr./ft. of tank width/1 ft./min. belt speed.

b. The Grainer. (See p. 933.) The salt grainer, a special type of surface crystallizer widely used in the manufacture of coarse salt crystals from brine, is shown in Fig. 2. This machine collects the crystals formed in a steam-heated bath of saturated solution and drains them before discharging them to stock piles or to rotary dryers.

The grainer described is essentially a surface crystallizer, and the draining obtained is merely incidental to mechanical discharge. Rake speed is adjusted to suit crystal formation independent of drainage conditions, and subsequent drainage in stock piles or by filters or centrifuges is used.

The grainer tank, constructed generally of concrete, is

100 to 150 ft. long, about 12 to 18 ft. wide, and about 2 ft. deep. It is filled with brine and equipped with steam coils that warm the contents slightly below the boiling point to the point of surface crystallization.

A reciprocating mechanism actuated by a steam or water piston travels slowly back and forth over the bottom of the tank and over the drainage deck at the

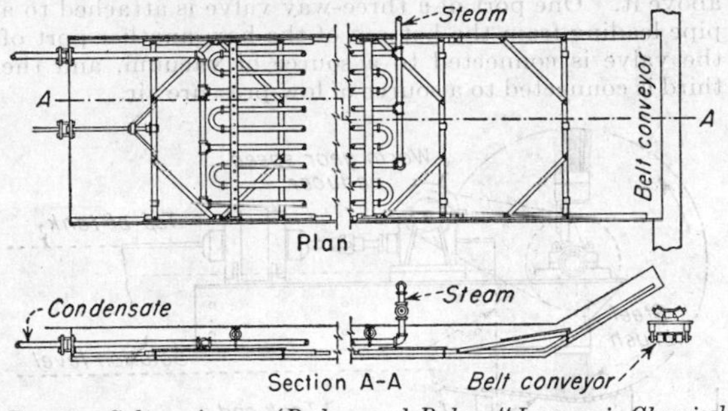

FIG. 2. Salt grainer. (*Badger and Baker, "Inorganic Chemical Technology," McGraw-Hill.*)

discharge end. Raking blades are hinged to the mechanism at regular intervals, being arranged in such a manner as to swing on the hinges with a feathering motion. The blades lie parallel to the tank bottom and to the drainage deck on the return stroke and then immediately take a position at right angles to the bottom on the forward or crystal-advancing stroke.

The salt crystals are pushed along the tank bottom at regular intervals, eventually emerging from their bath of mother liquor and being subjected to a period of draining on the inclined deck. At intervals of a few weeks the grainer is shut down and cleaned, as scale forms rapidly on the coils and impurities collect in the liquor to a degree endangering the purity of the finished salt.

The capacity of a grainer 150 ft. long by 18 ft. wide by 22 in. deep is 11.6 to 14.5 tons/day (Badger and Baker, "Inorganic Chemical Technology," 2d ed., p. 15, McGraw-Hill, New York, 1941).

c. The Dorr Classifier. The Dorr classifier is shown in Fig. 3. It consists of a settling box of wood, concrete, or steel, in the form of an inclined trough, with or without lining of rubber, lead, or special metals, with the upper end open, in which are placed mechanically operated rakes or scrapers which carry the quick-settling granular material to the point of discharge at the open end. Each rake is carried by two hangers, one at the discharge end and the other at the overflow end. Special covering or special metals may be used with corrosive solutions.

The rakes are lifted and lowered vertically by the hangers by the action of the head motion, transmitted through eccentrics and cranks. The horizontal motion is obtained directly from the head-motion crank. The rakes can be raised several inches at the lower end by a lifting device and operated in that, or any intermediate, position. This allows the classifier to be started readily when nearly filled with solids after an unexpected shutdown.

The feed enters continuously from a distributing trough near the overflow end. The heavy, quick-settling particles sink to the bottom of the tank and are advanced up the inclined deck by the reciprocating rakes. At the upper end, this crystalline or granular product emerges from the liquid, the excess solution drains off and the product is discharged with a low moisture content

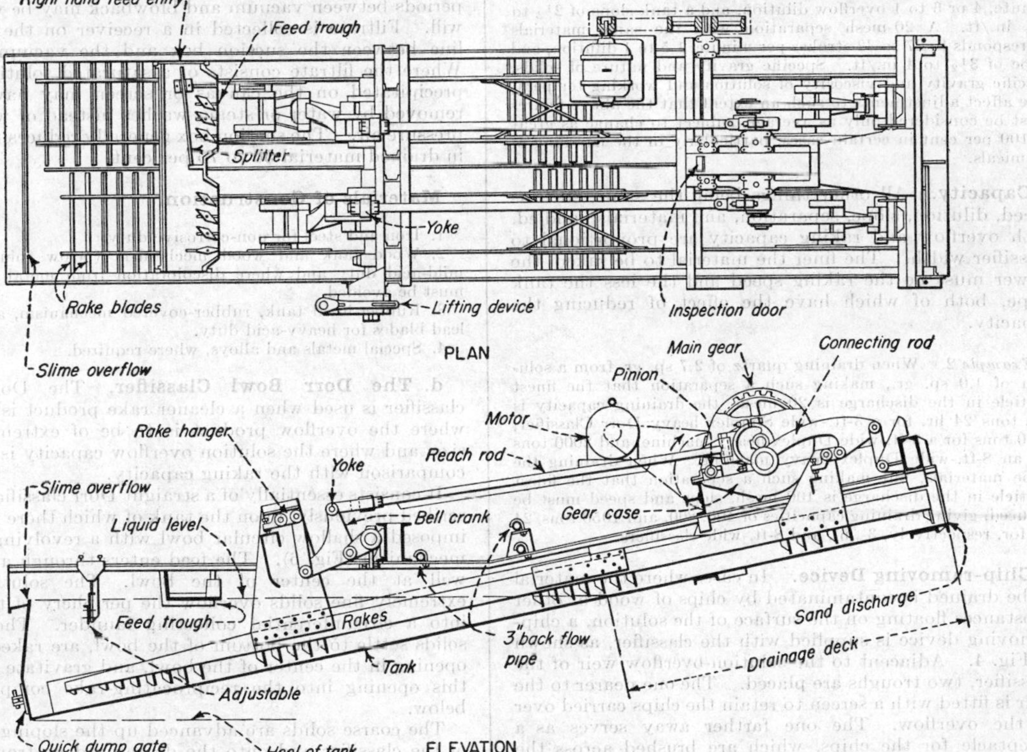

FIG. 3. The Dorr classifier.

Table 1. Dorr Classifiers

Type of mechanism	Width	Max. length	Min. length	Aver. conn. hp.*
The Dorr FR—light duty	1′6″	18′4″	12′0″	1½
For small open-circuit operation where	2′0″	23′4″	15′0″	2
the sand is small compared with the	3′0″	23′4″	15′0″	2
overflow	4′0″	23′4″	15′0″	3
The Dorr FH—intermediate duty	4′0″	30′0″	18′4″	3
For small closed-circuit operations with	5′0″	30′0″	18′4″	5
relatively light circulating loads				
The Dorr F—normal duty	6′0″	30′0″	18′4″	7½
For the average or normal closed-circuit	8′0″	30′0″	18′4″	7½
operation—the general run-of-mill	12′0″	30′0″	18′4″	10
classification job	16′0″	30′0″	18′4″	15
The Dorr FX—heavy duty	5′0″	31′6″	24′0″	10
For the heavy and really tough closed-	6′0″	31′6″	24′0″	10
circuit operation, where the circulating	7′0″	31′6″	24′0″	15
load is high, the sands are heavy, and	8′0″	31′6″	24′0″	15
the service is exacting	12′0″	31′6″	24′0″	20
	14′0″	31′6″	24′0″	25
	16′0″	31′6″	24′0″	25

* Average connected horsepower is the power rating of the motor supplied for direct drive and is 25 to 50 per cent in excess of the actual power required to drive the classifier on a job of average severity.

The agitation near the bottom of the tank, caused by the reciprocating motion of the rakes, throws the fine material into suspension, and this is carried off with the overflow product at the lower end of the tank.

Adjustments. All other things being the same, the mesh of separation between discharge and overflow solids is determined by rake speed, overflow dilution and slope, of tank bottom. The greater the rake speed, the lower the dilution; and the steeper the slope, the coarser is the separation.

Example 1. With quartz of 2.7 sp. gr., suspended in water, a 100-mesh separation corresponds to 16 to 20 raking strokes per minute, 4 or 6 to 1 overflow dilution, and a tank slope of 2½ to 2¾ in./ft. A 20-mesh separation with the same materials corresponds to 27 to 32 strokes per minute, 1.5 to 1 dilution, and slope of 3½ to 4 in./ft. Specific gravity and nature of solids, specific gravity and viscosity of solution and working temperature affect adjustments to such an extent that the above figures must be considered only as average, subject to change as great as 100 per cent in certain cases, particularly in the handling of chemicals.

Capacity. All other things being the same, such as speed, dilution, slope, separation, and materials handled, both overflow and raking capacity are proportional to classifier width. The finer the material to be raked, the slower must be the raking speed and the less the tank slope, both of which have the effect of reducing the capacity.

Example 2. When draining quartz of 2.7 sp. gr. from a solution of 1.0 sp. gr., making such a separation that the finest particle in the discharge is 20 mesh, the draining capacity is 750 tons/24 hr. for a 3-ft.-wide Simplex heavy Dorr Classifier, 1500 tons for a 6-ft.-wide Duplex heavy machine, and 2000 tons for an 8-ft.-wide Duplex heavy machine. When draining the same materials, but making such a separation that the finest particle in the discharge is 100 mesh, slope and speed must be reduced, giving draining capacities of 400, 800, and 1050 tons/24 hr. for, respectively, 3-, 6-, and 8-ft.-wide classifiers.

Chip-removing Device. In cases where the material to be drained is contaminated by chips of wood or other substances floating on the surface of the solution, a chip-removing device is supplied with the classifier, as shown in Fig. 4. Adjacent to the solution-overflow weir of the classifier, two troughs are placed. The one nearer to the weir is fitted with a screen to retain the chips carried over in the overflow. The one farther away serves as a receptacle for the chips, which are brushed across the screen by a scraper.

Suction Box. The suction box, shown in Fig. 5, may be supplied with the classifier in order to reduce the moisture of the material below that point which is possible with ordinary gravitational draining.

It consists of a metal box attached to the underside of the drainage deck and equipped with a suitable screen or canvas to replace the portion of tank bottom directly above it. One port of a three-way valve is attached to a pipe leading from the bottom of the box, another port of the valve is connected to a source of vacuum, and the third is connected to a source of low-pressure air.

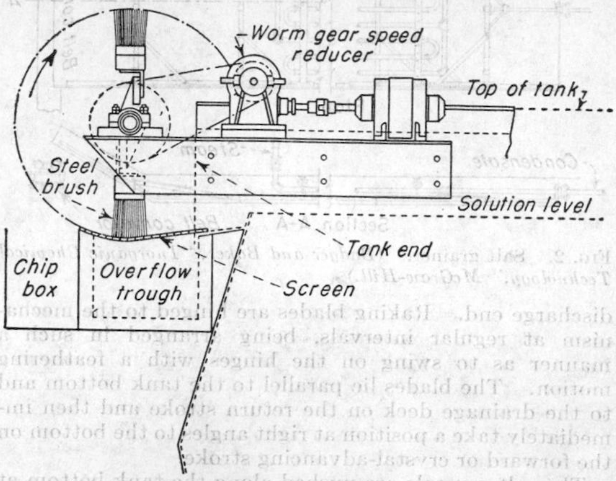

FIG. 4. Chip-removing device on Dorr classifier.

This valve is operated mechanically from the classifier mechanism. Adjustments are provided so that the periods between vacuum and blowback may be varied at will. Filtrate is collected in a receiver on the vacuum line between the suction box and the vacuum pump. Where the filtrate consists of a saturated solution, salts precipitated on the canvas or screen may have to be removed by water or steam washes instead of with low-pressure air. The suction box generally reduces moisture in drained material 50 to 75 per cent.

Materials of Construction.

1. Iron and steel for non-corrosive duty.
2. Wood tank and wood mechanism (below solution) for mild-acid duty and where discoloration from metal corrosion must be avoided.
3. Rubber-lined tank, rubber-covered mechanism, and hard-lead blades for heavy-acid duty.
4. Special metals and alloys, where required.

d. The Dorr Bowl Classifier. The Dorr bowl classifier is used when a cleaner rake product is desired, where the overflow product is to be of extremely fine size, and where the solution overflow capacity is large in comparison with the raking capacity.

It consists essentially of a straight Dorr classifier, as described previously, upon the tank of which there is superimposed a shallow circular bowl with a revolving raking mechanism (Fig. 6). The feed enters through a loading well at the center of the bowl. The solution and extremely fine solids overflow the periphery of the bowl into a circumferential collecting launder. The coarse solids settle to the bottom of the bowl, are raked to the opening in the center of the bowl, and gravitate through this opening into the reciprocating-rake compartment below.

The coarse solids are advanced up the sloping bottom of the classifier and onto the draining deck from which they are presently discharged by the reciprocating rakes.

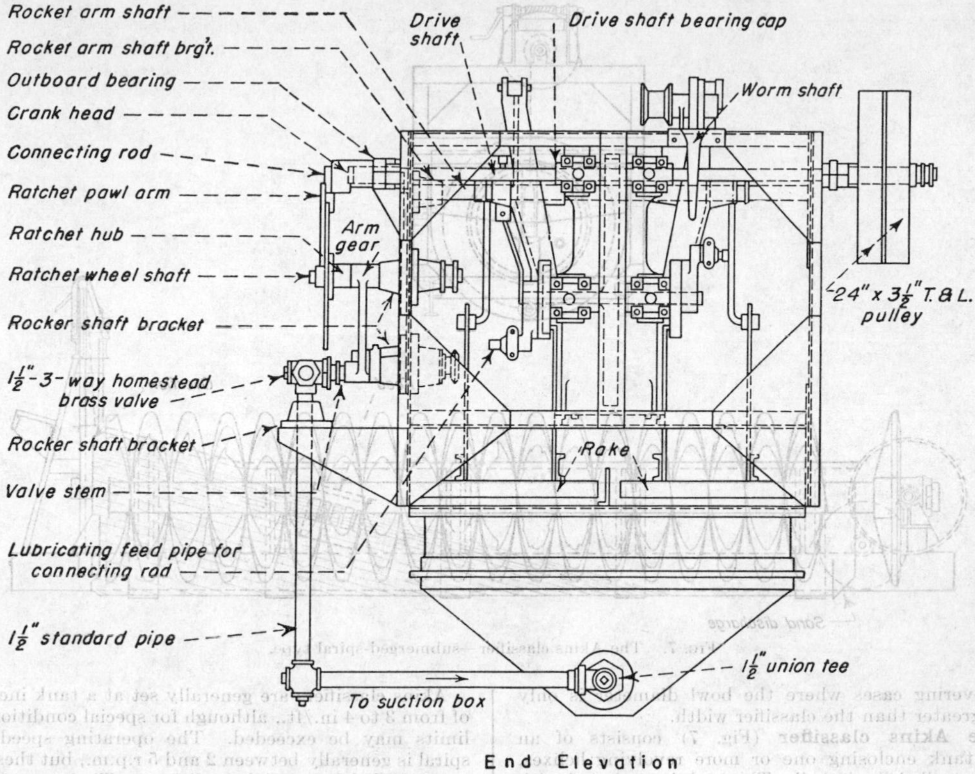

FIG. 5. Suction box on Dorr classifier.

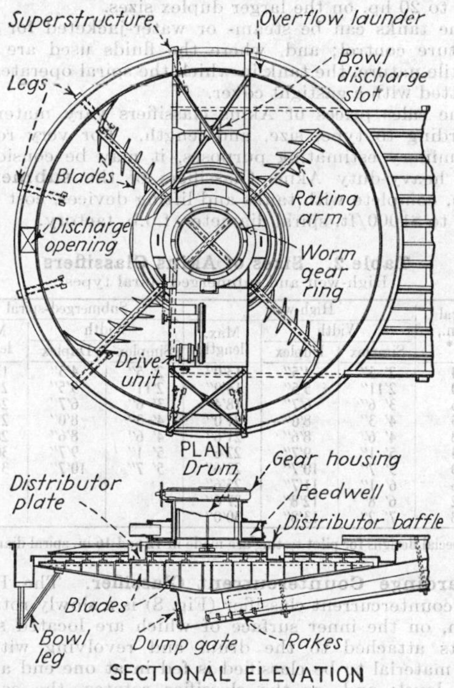

FIG. 6. The Dorr bowl classifier.

Wash water, introduced near the center of the reciprocating-rake compartment, flows countercurrently with respect to the coarse solids and, after passing through the opening in the bowl bottom, leaves the classifier as a portion of the overflow product.

Sizes. Bowls may be used with all types and sizes of Dorr classifiers, as listed previously. Bowls range in diameter from 3 ft. for use with a 15-in.-wide classifier to 28 ft. for use with the larger units.

Power consumptions for the bowl alone, which must be added to that of the classifier with which it is used, range from less than ¼ hp. for the small units to 2 to 3 hp. for the largest.

Costs. Owing to the various types and sizes of Dorr classifiers furnished, as well as the variety of materials of construction which may be used for acid-resistant duty, sales prices per unit of sand-raking capacity vary widely. For rough and very preliminary estimating purposes only, the net sales price of a light-duty Dorr classifier of iron and steel construction and including a steel tank, all f.o.b. factory, may be taken at $600 to $850/ft. of width. For the heavy-duty machine on the same basis, $750 to $1600/ft. of width may be used. Acid-resistant construction, such as wood, lead, rubber-covered steel, hard-lead or special alloys, materially increases the price.

The sales prices of Dorr bowl classifiers vary even more than those of the regular, single-stage Dorr classifiers, since a wide range of bowl sizes may be used with each size and type of classifier. Accordingly, for rough and preliminary estimating purposes, it may be considered that a Dorr bowl classifier may cost from 50 to 200 per cent in excess of the price of the regular classifier to which the bowl is attached, the higher figure corresponding to the use of large bowls with narrow classifiers, and the lower

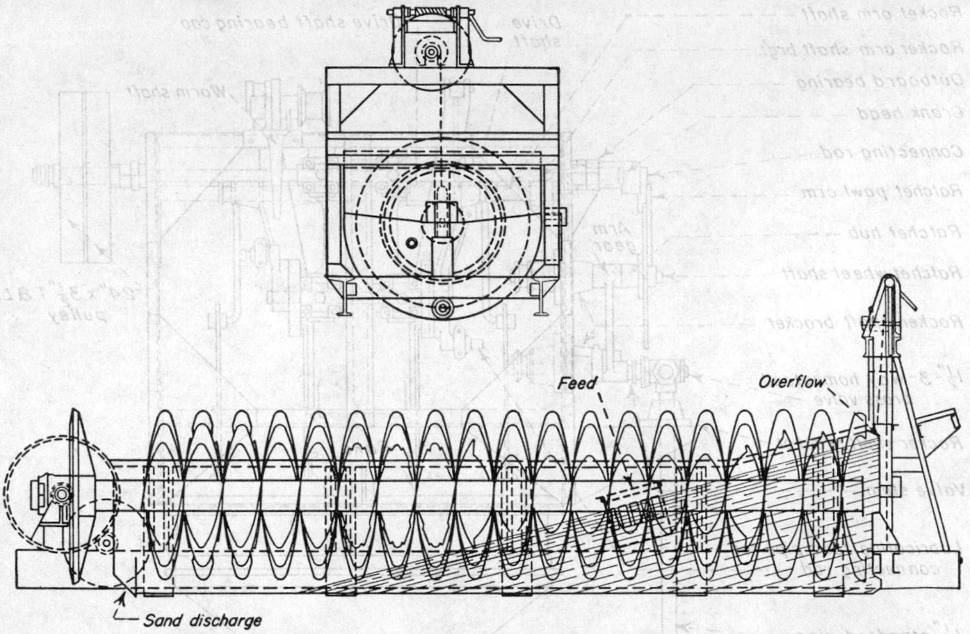

FIG. 7. The Akins classifier—submerged-spiral type.

figure covering cases where the bowl diameter is only slightly greater than the classifier width.

e. The Akins classifier (Fig. 7) consists of an inclined tank enclosing one or more revolving helixes, commonly called "spirals." The tank is equipped with an overflow weir and an overflow box for collecting the overflow product which generally consists of fine solids and water, although under certain conditions very nearly clear water may constitute the overflow. The settled product, commonly termed "sands," is discharged at the upper end of the tank by the revolving spiral.

The sand-raking mechanism is made up of double or single spirals mounted on a heavy hollow shaft, spirals being continuous from the overflow weir to a point above the feed entrance. The double- and single-spiral constructions are termed double-pitch and single-pitch, respectively. In a simplex machine, with one spiral assembly, the sand load is conveyed up one side of the tank, leaving a drainage channel between the spiral and the tank on the opposite side.

The feed enters through a feed opening at one, or, in some cases, both sides of the tank, and the overflow is discharged at the lower end over an adjustable weir. In some cases the feed and overflow points are reversed. The heavier, or coarser, solids settle to the bottom and in the case of simplex machines are advanced by the action of the spiral along the bottom and one side of the tank out of the settling pool to the point of discharge. In the case of duplex machines, the spirals are operated toward each other so that the sands are conveyed up the center of the tank between the spirals, causing the sands to be carried considerably higher on the spirals.

The lower end of the spiral is equipped with a lifting mechanism, and the spiral drive is so arranged that the spiral may rotate and be lifted or lowered simultaneously.

Akins classifiers are now furnished in two general types, *i.e.*, the **high-weir type** in which the spiral is not completely submerged at the lower end, and the **submerged-spiral** type in which the spiral is completely submerged to provide a larger settling pool.

Akins classifiers are generally set at a tank inclination of from 3 to 4 in./ft., although for special conditions these limits may be exceeded. The operating speed of the spiral is generally between 2 and 5 r.p.m., but these limits are exceeded for special conditions. The power required will depend upon the size of the classifier and the sand load. It varies from less than 1 hp. on the small simplex sizes to 20 hp. on the larger duplex sizes.

The tanks can be steam- or water-jacketed for temperature control; and, where the fluids used are of a volatile nature, the tank in which the spiral operates can be fitted with a gastight cover.

The sales prices of Akins classifiers vary materially according to type, size, and length. For very rough, preliminary estimating purposes, it may be considered that heavy-duty Akins Classifiers of the **submerged type**, complete with tanks and lifting devices, cost from $700 to $1000/ft. spiral diameter, f.o.b. factory.

Table 2. Sizes of Akins Classifiers
High-weir and submerged-spiral types

Spiral diam., in.*	High-weir Width Simplex	High-weir Width Duplex	High-weir Max. length	Submerged-spiral Width Simplex	Submerged-spiral Width Duplex	Submerged-spiral Max. length
24	2′ 5″	4′5″	13′0″	2′ 5″	4′5″	17′0″
30	2′11″	5′5″	14′9″	2′11″	5′5″	21′0″
36	3′ 6″	6′7″	18′4″	3′ 6″	6′7″	24′0″
45	4′ 3″	8′0″	21′0″	4′ 3″	8′0″	25′3″
48	4′ 6″	8′6″	21′6″	4′ 6″	8′6″	28′9″
54	5′ 1″	9′7″	22′6″	5′ 1″	9′7″	30′0″
60	5′ 7″	10′7″	23′6″	5′ 7″	10′7″	31′0″
66	6′ 1″	11′7″	25′6″			
72	6′ 8″	12′8″	27′6″			
78	7′ 2″	13′8″	30′0″			

* Special designs for pilot testing are made in 12 and 16 in. spiral diameters.

Hardinge Countercurrent Classifier. The Hardinge countercurrent classifier (Fig. 8) is a slowly rotating drum, on the inner surface of which are located spiral flights attached to the drum and revolving with it. The material to be classified is fed in at one end at the pulp level; and, as the classifier rotates, the coarser particles are settled out, moved forward by the spiral flights, and repeatedly turned over in a forward motion, releasing any fines mixed with them. The fines, with any wash water added, are discharged as an overflow product through an opening at the lower end of the classifier. The sands or oversize are dewatered and elevated by buckets to an elevation, whereby the classifier may be operated in closed circuit with ball or pebble mills without the use of auxiliary oversize return equipment. The oversize settled solids are continually being

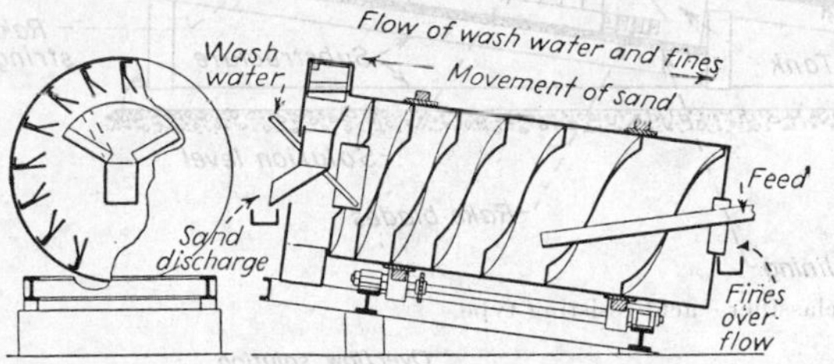

FIG. 8. Principle of operation of the Hardinge countercurrent classifier.

turned over and may be washed by the countercurrent action of the wash water, which is added at the oversize discharge end.

Two steel tires, near its end, support the rotating drum on four rollers, mounted on two parallel shafts, which in turn are driven by sprockets and chain from a driving motor. On large units, a pinion and a circumferential spur gear on the drum are used to revolve it.

Some of the uses of this classifier are for closed-circuit grinding of various materials, dewatering sand, and washing and scrubbing various granular materials.

It is generally supplied of standard iron and steel construction. However, it has also been supplied in the chemical industry of stainless-steel construction; also with rubber lining and with lead lining for classification and washing operations where corrosive mixtures must be handled.

The classifier is supplied in sizes varying from 18 to 10 in. diameter and from 4 to 30 ft. in length.

The exact classification desired is obtained by varying the slope of the classifier, by varying the speed or the size of the overflow opening.

WASHING (GRANULAR MATERIAL) *

Definition. Washing of granular substances may be defined as the displacement of dissolved substances, adhering in solution form to a solid substance, by one or more applications of water or other suitable displacing agent, each application being followed by a dewatering step to remove the bulk of the dissolved substances thus displaced.

Purpose. The purpose of washing is to recover the greatest possible amount of either the dissolved substance or the solid substance, or both. Its object is also to recover each substance, contaminated to the least possible extent by the other substance and in as pure a form as possbile. For example, in electrolytic caustic manufacture, it is the purpose of salt washing not only to recover the salt in as nearly a caustic-free condition as possible so that it may be marketed as such or reused to make up cell liquor, but also to recover the caustic soda solution diluted by wash solution to the least extent in order to promote good evaporator economy.

Equipment Used. The Dorr multideck classifier is the only standard, self-contained unit utilizing the alternate administrations of a displacing solution and of a draining process for the washing of soluble substances from insoluble granular substances. Centrifugation is applicable to the washing of such substances; but, since it is not classed under draining operations, it is discussed elsewhere (p. 992).

The Dorr multideck classifier consists of a series of two or more Dorr classifier mechanisms connected together and driven from a common driving mechanism. A single tank is used, divided into from two to six individual washing compartments and drainage decks. As shown in Fig. 9, the various compartments are connected by backflow launders, located on the outside of the tank, so that wash solution may flow continuously through the series from the discharge to the feed end of the tank. The solid substance is advanced mechanically by the reciprocating rakes from the feed to the discharge end of the tank.

As shown in Fig. 10, two or more Akins classifiers may be arranged in tandem for countercurrent washing. The direction of flow of wash solution and solids is the same as in the Dorr washing classifier. As each washing operation is carried out in a separate classifier instead of (as in

* See article on Countercurrent Decantation (p. 928 and 950) and on Filtration (p. 964) for washing of finely divided materials.

Data on Various Hardinge Countercurrent Classifiers Operations

Material	Porphyry gold ore	Porphyry gold ore	Limestone	Heavy pyrite zinc ore	Quartz and schist	Porphyry ore	Porphyry gold ore	Limestone
Separation mesh size	28	35	48	48	65	100	200	325
Specific gravity	2.7	2.7	2.4	4.0	2.7	2.7	2.7	2.6
Classifier size	6 × 14	6 × 14	3 × 8	8 × 20	5 × 12	5 × 12	6 × 14	1.5 × 4
Slope, in./ft.	1.25	1.25	0.75	1.25	1	1	0.75	0.5
Speed, r.p.m.	2.3	1.7	2.5	1.5	2	1.5	0.5	2
Feed, tons/24 hr. (dry basis)	1745	630	53	3000	715	942	170	6
Overflow, tons/24 hr.	520	210*	21	1400	400	192	90	1.2
Return, tons/24 hr.	1225	420	1.32	1600	315	750	78	5.0
Feed % solids	73	66	65	75	70	78	35	55
Overflow % solids	45	43	35	51	42	30	14	11
Return % solids	80+	80+	76	85	76.1	81	70	72
Feed size	75.0% − 28 mesh	61.7% − 35 mesh	75.3% − 48 mesh	36% + 65 mesh	61.5% + 100 mesh	90% − 10 mesh	57.8% − 200 mesh	5% + 48 mesh
Overflow	96.5% − 28 mesh	98.0% − 35 mesh	100% − 48 mesh	91.6% − 65 mesh	82.2% − 100 mesh	98.4% − 100 mesh	97.0% − 200 mesh	95% − 325 mesh
Return	48.3% − 28 mesh	38.1% − 35 mesh	59.0% − 48 mesh	52.3% + 65 mesh	14% − 100 mesh	7.3% − 200 mesh	15.2% − 200 mesh	4.8% − 325 mesh
Remarks	C.C. with 2–8 × 30 Hardinge ball mills	....	C.C. with 4½ × 13 Hardinge mill classifier underload	6.5 hp. required	2 hp. required	C.C. with 7′ × 36″ Hardinge mill	Clean oversize secured	Note clean oversize

* Not the maximum capacity of the classifier.

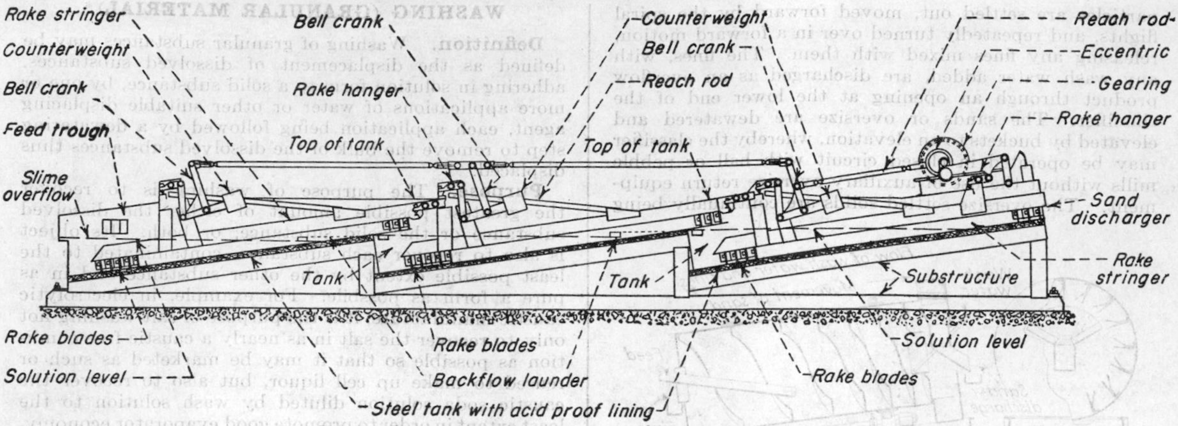

FIG. 9. Dorr multideck washing classifier—acid-resisting type.

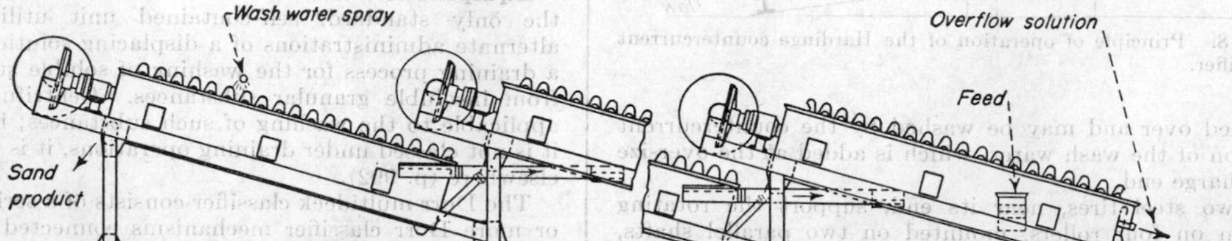

FIG. 10. Akins classifiers arranged for multiple-stage washing.

the case of the Dorr) in individual compartments of a single machine, external pipes or troughs are used for transferring the product between the different units.

Operation. The mixture of a solid substance and a solution is introduced at the overflow end of the tank, the washed solids are drained and discharged at the other end, and wash water or solution is introduced in the last compartment. The wash solution flows in a direction countercurrent to the solids and thus becomes progressively enriched in soluble salts until it finally overflows from the first compartment in relatively concentrated form.

In a similar manner the solids, being advanced toward the last compartment, come in contact with wash solutions progressively less concentrated. The mild agitation caused by the reciprocating rakes in each compartment assures efficient displacement of the soluble constituents in the solids which have just been discharged from the preceding compartment, and the draining on the inclined deck assures the removal of the bulk of the solution before the solids are remixed with the next weaker solution in the following compartment.

Adjustments are as given for single-stage classifiers except that the slope is not variable. Rake speed determines the time the solids are retained in each washing compartment, and hence the capacity is generally determined by the time required for displacement of dissolved material rather than by the mesh at which the separation is to be made.

The greater the volume of the wash solution, the more complete is the displacement of dissolved material and the less concentrated is the solution overflowing the first compartment. When both high purity of washed solids and high concentration of solution are required, a larger number of compartments are required in order to increase the number of washes which may be given, by the permissible volume of wash solution, which is necessarily limited by the desired strength of the solution.

Capacity. Same as given for single-stage classifier of same type and width. Capacity may vary up to 100 per cent from figures given, which are quartz (2.7 sp. gr.) in water, because of variations in specific gravity, viscosity, temperature, and time required for displacement of dissolved substances. High speeds (greater than 12 r.p.m.) are seldom used because of the time element in the displacement washing.

Accessories. Chip-removing devices and suction boxes may be supplied, these being identical with those discussed above. Steam coils may be placed in the compartments or in the solution launders between compartments to maintain proper temperature.

Materials of Construction. These are the same as for the single-stage Dorr classifier discussed on p. 924.

Washing Calculations and Practical Examples. This method of washing crystalline or granular substances in a Dorr multideck classifier uses the principles of countercurrent decantation (abbreviated C.C.D.), but strictly speaking it is countercurrent washing, not decantation. The washing principles used are the same also as those utilized in large-scale vat leaching, *e.g.*, copper leaching, using "advancing washes," although, in this multideck classifier washing, both solids and solutions move countercurrently as in true C.C.D., while in vat leaching only the solutions move. The over-all washing efficiency, concentration of solutions in all compartments, and purity of final product may be determined mathematically, if data are available on feed characteristics, moisture content to which the solids will drain, wash water permissible, etc. The method of calculation is best illustrated by an actual example on a typical set of conditions.

Example 3. In the electrolysis of brine, the conversion of NaCl to NaOH is generally incomplete. In the subsequent evaporation of the solution, the sodium chloride crystallizes out, is recovered in salt catchers operated in conjunction with the evaporator stages, and is discharged suspended in a solution of caustic soda. The problem is to separate the salt from the caustic and to wash it as completely as possible in a four-compartment classifier with the amount of saturated brine available.

Given. 75 tons NaCl crystals per day. 100 tons 10 per cent NaOH solution per day. 25 tons of water (containing 8 tons dissolved NaCl) available for washing. One four-compartment washing classifier. One suction box mounted on last deck. Salt drains to 25 per cent moisture without vacuum. Salt drains to 15 per cent moisture with vacuum.

Problem. To find concentrations of solution in all compartments, washing efficiency, and percentage NaOH in washed salt.

Procedure. Draw diagrammatic flow sheet (Fig. 11 below) letting W, X, Y, and Z represent pounds of NaOH per ton water in each compartment and letting T represent tons of water in circulation at each point.

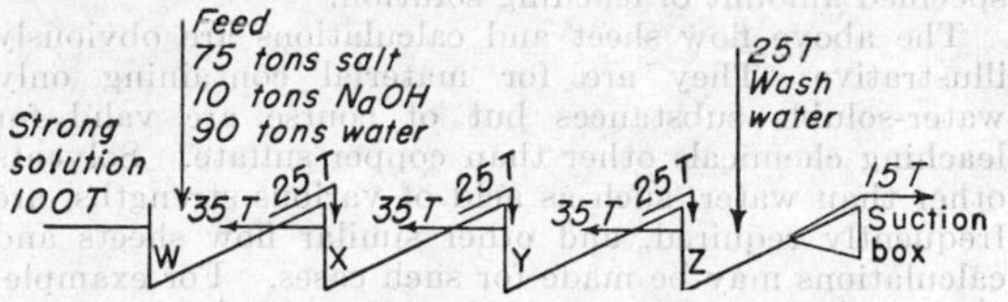

FIG. 11. Washing-classifier flow sheet.

Equating pounds of dissolved NaOH out of and into each compartment, set up the following simultaneous equations:

$$100W + 25W = 35X + 10 \times 2000 \quad (1)$$
$$35X + 25X = 25W + 35Y \quad (2)$$
$$35Y + 25Y = 25X + 35Z \quad (3)$$
$$35Z + 15Z = 25Y + 25 \times 0 \quad (4)$$

Solving: By substitution:

$W = 194.51$ lb. NaOH per ton water
$X = 123.39$ lb. NaOH per ton water
$Y = 72.58$ lb. NaOH per ton water
$Z = 36.29$ lb. NaOH per ton water

Summarizing.

Compartment	Solution Strength, % NaOH*
1	8.86
2	5.81
3	3.50
4	1.78

* Salt in saturated brine-wash solution neglected.

$$\text{Recovery of NaOH} = \frac{100 \times 194.51}{10 \times 2000} \times 100 = 97.26 \text{ per cent}$$

$$\text{NaOH in washed salt} = \frac{15 \times 36.29}{(75 \times 2000) + (15 \times 36.29)} \times 100 = 0.36 \text{ per cent}$$

From the preceding example, it should be obvious that the same method of calculation may be applied for finding other unknowns, *viz.:*

a. To find the minimum amount of wash solution which will give a washed product containing not more than a specified percentage of soluble impurities, when given the number of washing compartments to be used and the moisture content to which product will drain with and without suction box.

b. To find the minimum number of washing compartments required to give a specified percentage of soluble impurities in the washed product, when given the maximum permissible amount of wash solution and the moisture content to which the product will drain with and without suction box.

Other Applications. *a.* The washing of sodium phosphate crystals free from mother liquor.

b. The removal of finely divided clay or bond from artificially prepared abrasives such as carborundum, aloxite, etc.

c. The washing of phosphate rock to remove clay and fine sands.

LEACHING

Definition. Leaching may be defined as a process of removing, by the application of a solvent, that constituent of the substance being treated which is readily soluble in the solvent applied. In this article, the leaching of coarse, granular substances is alone considered.

Purpose. It is the purpose of leaching to recover the greatest possible amount of the soluble constituent in a solution of a concentration suitable for the succeeding process steps, such as precipitation of the soluble materials or evaporation and crystallization.

Equipment Used. As is true in the case of washing, the Dorr multideck classifier is the only standard self-contained unit applying the principle of continuous-countercurrent flow of the substance to be leached and the solvent, to the leaching of granular substances. Discontinuous methods include the use of percolation vats, filter-bottom tanks, etc., but, since these are not classed as draining operations in this article, they are not discussed here. The leaching of finely divided substances is discussed in the article on Counter-current Decantation (p. 927 and on p. 928).

The Dorr multideck classifier used for leaching is described completely in the preceding section on Washing.

Operation. The substance to be leached is introduced into the first compartment at the overflow end of the tank, the residue remaining after leaching is discharged from the last compartment, solvent is introduced into the last compartment, and concentrated solution containing the soluble constituent passes off across a weir in the first compartment. The solvent flows in a direction countercurrent to the solids and thus becomes progressively enriched in the soluble constituent of the treated substance until it finally overflows from the first compartment in a relatively concentrated form.

In a similar manner, the substance to be leached being advanced toward the last compartment comes in contact with weaker and weaker solutions and is thus progressively impoverished in soluble constituents. The mild agitation set up in each compartment assures efficient penetration of solvent and good leaching of the substance discharged from the preceding compartment. The draining on the inclined deck assures the removal of the bulk of the enriched solvent before the substance is retreated with the next weaker solvent in the ensuing compartment.

Leaching Calculations and Practical Examples. The method of calculating the extraction, washing efficiency, concentration of solutions, etc., in a leaching or dissolving operation is similar to that used in washing calculations. The unknowns may be calculated from data available on feed characteristics, moisture content to which the residue will drain, the amount of lixiviant or solvent that may be used, etc.

Example 4. A calcine containing water-soluble copper is to be leached. The rate at which the copper dissolves having been determined experimentally and a multideck classifier having been selected to give the required time of contact in each compartment, it is desired to know the probable extraction of copper, the concentration of the solutions in the various compartments, and the percentage of copper remaining in the residue after treatment.

Given. One hundred tons of calcine containing 5 per cent (10,000 lb.) of soluble copper mixed with 100 tons of water; one six-compartment multideck classifier equipped with suction box; 50 tons of available wash water; and a residue that drains to 25 per cent moisture or to 15 per cent moisture with the aid of a suction box.

The following rates of dissolution of copper have been determined experimentally:

Pounds dissolved before reaching classifier	2000
Pounds dissolved in first compartment	4000
Pounds dissolved in second compartment	3000
Pounds dissolved in third compartment	1000
Pounds dissolved in fourth, fifth, and sixth compartments	0

Problem. To find the concentration of the solutions in all six compartments, the over-all extraction efficiency, and copper content of the final residue.

Procedure. Calculate the amount of solids raked in each compartment, allowing for the dissolution of soluble copper that takes place in each. Calculate the amount of water advanced with the raked solids in each case, allowing for 25 per cent moisture in the first five compartments and 15 per cent in the last one.

Compartment *A*

$$\text{Solids raked} = 100 - \frac{(2000 + 4000)}{2000} = 97 \text{ tons}$$

Water in raked solids = 32.33 tons

Compartment *B*

$$\text{Solids raked} = 97 - \frac{3000}{2000} = 95.5 \text{ tons}$$

Water in raked solids = 31.83 tons

Compartments *C*, *D*, and *E*

$$\text{Solids raked} = 95.5 - \frac{1000}{2000} = 95 \text{ tons}$$

Water in raked solids = 31.66 tons

Compartment *F*

Solids raked = 95 tons
Water in raked solids = 16.76 tons

Draw the diagrammatic flow sheet shown in Fig. 12 and assign to each flow line figures to show the amount of water (not solids) in circulation at that point. Water enters at two points: 100 tons with the calcine in the first compartment, and 50 tons as wash water in the last compartment. The balance of the water in circulation is either overflow from compartments or moisture contained in the raked residues.

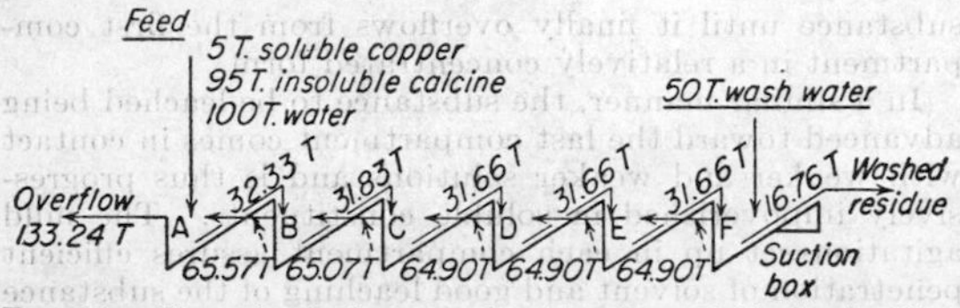

FIG. 12. Flow sheet of copper leaching in multideck classifier.

Let *A*, *B*, *C*, *D*, *E*, and *F* represent the pounds of dissolved copper per ton of water in the respective compartments of the classifier. Then set up the following six simultaneous equations by equating the pounds of copper entering and leaving each compartment.

Equate pounds of dissolved copper entering and leaving each compartment as follows:

Compartment *A* $133.24A + 32.33A = 65.57B + 6000$
Compartment *B* $65.57B + 31.83B = 32.33A + 65.07C + 3000$
Compartment *C* $65.07C + 31.66C = 31.83B + 64.90D + 1000$
Compartment *D* $64.90D + 31.66D = 31.66C + 64.90E$
Compartment *E* $64.90E + 31.66E = 31.66D + 64.90F$
Compartment *F* $64.90F + 16.76F = 31.66E$

Simplifying,

$$165.57A = 65.57B + 6000$$
$$97.40B = 32.33A + 65.07C + 3000$$
$$96.73C = 31.83B + 64.90D + 1000$$
$$96.56D = 31.66C + 64.90E$$
$$96.56E = 31.66D + 64.90F$$
$$81.66F = 31.66E$$

Solving by substitution,

A = 74.25 lb. copper per ton water
B = 96.10 lb. copper per ton water
C = 61.05 lb. copper per ton water
D = 28.60 lb. copper per ton water
E = 12.70 lb. copper per ton water
F = 4.92 lb. copper per ton water

Checking calculations,

Total copper in classifier overflow	= 133.24 × 74.25	= 9890 lb.
Total copper in classifier discard	= 16.76 × 4.92	= 82.5
		9972.5 lb.
Error due to neglected decimals in slide-rule computations		= 27.5
Total copper in feed to classifier		= 10,000 lb.

$$\text{Recovery of copper} = 100 - \frac{16.76 \times 4.92}{133.24 \times 74.25 + 16.76 \times 4.92} \times 100$$
$$= 100 - \frac{82.5}{9972.5} \times 100 = 99.173\%$$

Copper content of leached residue

$$= \frac{16.76 \times 4.92}{95 \times 2000 + 16.76 \times 4.92} \times 100 = 0.043\%$$

The same method of calculation may be used to find the minimum amount of water required for leaching to give a specified extraction or to find the minimum number of compartments to give a specified extraction with a specified amount of leaching solution.

The above flow sheet and calculations are obviously illustrative. They are for material containing only water-soluble substances but of course are valid for leaching chemicals other than copper sulfate. Solvents other than water, such as acid of various strengths, are frequently required, and other similar flow sheets and calculations may be made for such cases. For example, in the flow sheet shown, acid may be added as a part of the feed if required for dissolving acid-soluble copper.

The permissible amount of water for washing in a balanced system is limited mainly by two factors: (1) the amount of moisture discharged with the tailings and (2) the amount of evaporation, the allowable water-wash volume being the sum of these two. Thus, in the above illustration, assuming this to be a balanced cyclic system, the amount of evaporation required will be the difference between the overflow volume and the volume of water in the return feed, or 33.24 tons. Efficient leaching and washing without other than natural evaporation is practicable, *e.g.*, large-scale copper leaching.

CLOSED-CIRCUIT GRINDING*

Closed-circuit wet grinding or classified grinding, introduced and perfected in the metallurgical industry, has been successfully applied to such chemical engineering operations as the grinding of lithopone, the grinding of phosphate rock with weak acid in phosphoric acid manufacture, the preparation of water-floated whiting, the wet grinding of cement slurry, abrasives, etc. A brief résumé of this subject may therefore properly be included in a discussion of classification and mechanical classifiers.

Definitions. [Dorr and Marriott, Importance of Classification in Fine Grinding, *Trans. Am. Inst. Mining Met. Engrs.*, Milling, pp. 109–154, 1930.] **Open-circuit grinding** is a method of comminution aiming to secure the desired reduction in particle size by a single passage of the material through the mill. **Closed-circuit grinding** is a method of comminution in which a partly finished mill discharge is separated by a classifier into a finished overflow product and an unfinished rake product which is returned to the mill for further grinding. **Overflow** is the comparatively finer, more slowly settling portion of the mill discharge which is carried over the tail board, or lip, of the classifier by the flow of water. **Rake product** is the comparatively coarser, more rapidly settling portion of the mill discharge which is discharged from the classifier by the mechanical action of the rakes. In closed-circuit grinding, the rake product is frequently referred to as *the circulating load* in that it travels in the mill-classifier circuit until reduced to overflow fineness. **Mill** is the generic term used to describe grinding mills, whether ball, pebble, or rod mills.

Purpose. It is the purpose of closed-circuit grinding to center the sizing of the ground product at one point, the classifier, so that the mill, now responsible only for grinding, may be fed at such a rate and loaded in such a

* See pp. 1117 to 1118.

manner that it may operate at maximum efficiency. As will be shown later, a judicious loading of the mill, unlimited by size specifications for the mill discharge, permits large savings in power, less wear on liners, and less wear on grinding mediums. These three items largely determine the unit cost of grinding.

Theory. Theory and practice concur in the hypothesis that the work done in a rotary mill increases with rate of feed. Work done is measured by the actual reduction in particle size from feed to discharge and may be measured by the tons of material of a specific size actually produced in a given time or by the more refined methods of Kick and of Rittinger, described later (see p. 931).

Table 3. Relation between Rate of Feed to Ball Mill, Work Done, and Unit Power Consumption*

Feed rate, lb./hr. of 3/8-in. limestone	Finished product in discharge, lb./hr. 65-mesh limestone	% finished material in mill discharge	Rittinger's work units (S.U.)	Kick's work units (E.U.)	Kw.-hr./ton 65-mesh limestone
1000	600	60	40,000,000	900,000	13.3
2000	970	48.5	60,000,000	1,500,000	8.25
3000	1200	40	90,000,000	2,100,000	6.67
4000	1400	35	100,000,000	2,250,000	5.70
5000	1650	33	110,000,000	2,350,000	4.85

* Courtesy of the University of Minnesota.

This relationship is brought out in Table 3 and the accompanying graph (Fig. 13) based upon studies conducted at the Mines Experimental Station of the University of Minnesota, under the direction of E. W. Davis. It will be seen that the work done in the experimental

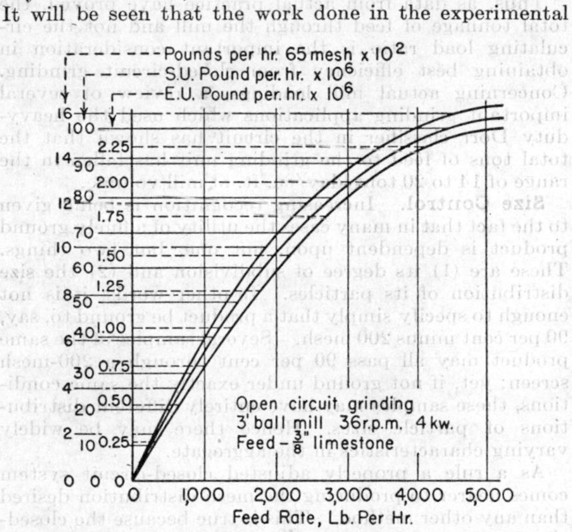

FIG. 13. Relation between rate of feed and work done in a rotary mill.

mill increases as the feed is increased progressively to five times its original value, regardless of which of these methods is used to compute work, and in practice it has been found that the work curve is still increasing even when the rate of feed is increased to more than ten times the open-circuit capacity of the mill.

Work Units. Authorities appear divided as proponents of the Kick and the Rittinger methods of computing the work done in cylindrical mills. Rittinger's method is based upon the hypothesis that the work done in crushing is proportional to the increased surface produced, a function of the square of the diameter of the particle, and the Rittinger unit of measurement is called a "surface unit" (S.U.). Kick, on the other hand, claims that the work done is inversely proportional to the change in volume of the particle, a function of the cube of the diameter, and the Kick unit of measurement is known as an "energy unit" (E.U.). Practical plant operators prefer to measure work by the amount of material of a specific size actually produced, since this may readily be determined from screen analyses of feed and of discharge and the rate of new feed to the mill.

Open- vs. Closed-circuit Grinding. The fact that capacity increases without a corresponding increase in power may be attributed to the more rapid elimination of fines, the reduction of uneconomical overgrinding, and the increased amount of material that may be exposed to the cascading action of the balls at one time. Theory indicates that uneconomical, open-circuit grinding is a

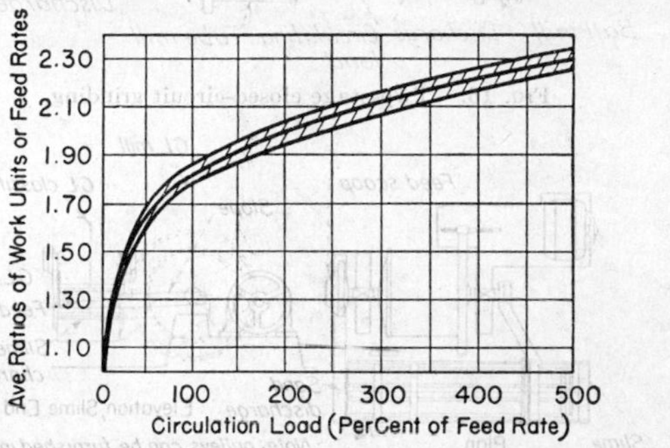

FIG. 14. Relation between circulating load and work done in cylindrical mill.

result of hampering the work of the mill by the imposition of a specification for the fineness of discharge, which results in overgrinding of the bulk of the product in order that all may pass a given sieve size and in a tendency for the accumulated fines to act as a cushion, damping the effective impact of the balls upon the unfinished material.

In closed-circuit grinding, the classifier builds up the feed to the mill to the optimum value by returning unfinished oversize to the mill for further comminution. Furthermore, it grades the mill discharge so that only material of finished size may escape from the circuit as an overflow product. As shown in Fig. 14, the work done in a cylindrical mill increases with the amount of the circulating load or rake product.

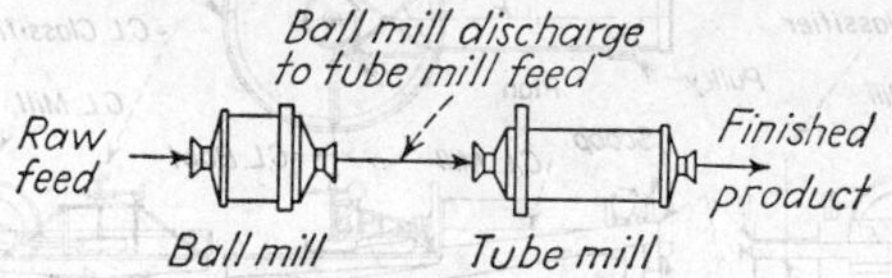

FIG. 15. Two-stage open-circuit grinding.

Closed-circuit Grinding Equipment. Figure 15 is a diagram representing a two-stage open-circuit grinding installation, and Fig. 16 represents a two-stage closed-circuit grinding installation. In the case of the latter, it is to be noted that the amount of the circulating load in both ball- and tube-mill circuits is several times the amount of the new feed and finished product.

The closed-circuit grinding equipment consists of a cylindrical wet-grinding mill, a mechanical classifier, and two launders or troughs, the one conveying the mill discharge to the feed end of the classifier and the other conveying the oversize from the classifier to the feed box of the mill. The mill is equipped with a spiral scoop feeder, which picks up the new feed and the classifier discharge from its feed box and introduces them into the mill through the hollow feed trunnion. Water is added in the feed box to give the optimum consistency for grinding while additional water is added at the classifier to give the dilution corresponding to the desired separation and for washing the rake product free from adhering fines.

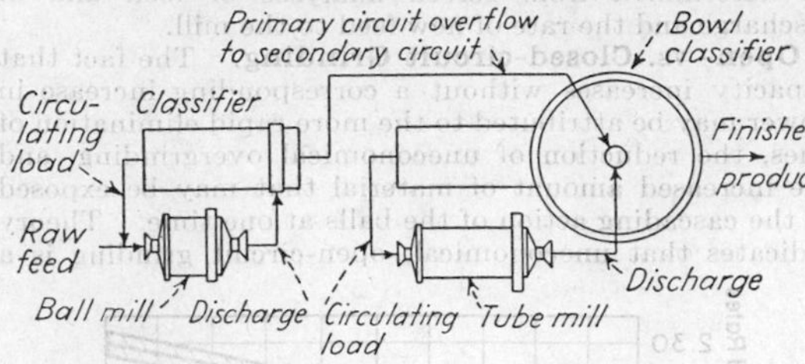

FIG. 16. Two-stage closed-circuit grinding.

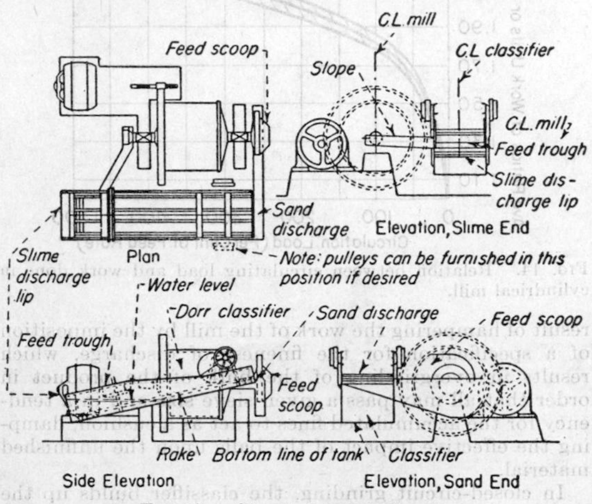

FIG. 17. Closed-circuit ball mill.

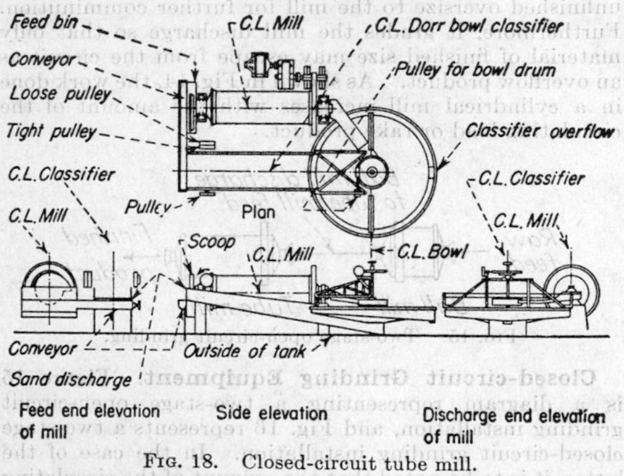

FIG. 18. Closed-circuit tube mill.

Two stages of grinding are becoming increasingly common while in certain instances three and four stages have been adopted profitably. In general, coarse grinding to, say, 35 to 48 mesh is carried out in ball mills (relatively large in diameter and short in length) loaded with balls ranging in size from 5 to 2 in. in diameter. Single-stage classifiers are generally used with ball mills.

Fine grinding to 100 to 325 mesh is generally accomplished in tube mills (relatively small in diameter and great in length) loaded with balls ranging in size from 2 to ¾ in. Bowl classifiers are generally used with tube mills.

The general arrangement of closed-circuit grinding equipment is shown in Fig. 17, representing a primary closed-circuited ball mill for relatively coarse grinding, 35 to 48 mesh, and Fig. 18, representing a secondary closed-circuit tube mill for relatively fine separations, 100 to 325 mesh.

CIRCULATING LOAD LIMITS

It has already been pointed out that the circulating load may profitably be built up to several times the new feed to the mill. That being the case, the question is frequently asked, "What circulating load ratio should be carried?" This is really an ambiguous question and one that will lead to great difficulty if not properly considered. What is actually required for good grinding is a proper loading of the mill. To bring this about, there should always be added to the new feed the correct amount of circulating load from the classifier to give the optimum tonnage through the mill. When the problem is considered in this manner, it takes on a new aspect, as the following will show:

In every grinding unit there is available an effective volume through which feed will pass, which, if properly utilized, will lead to maximum capacity. However, if the tonnage of total feed becomes excessive for the total volume, then the mill will "choke." Changing the dilution of the pulp in the mill will not remedy the congestion, for the available mill volume has been taxed beyond capacity, and the stoppage is due to overloading.

Thus, as data from actual practice have proved, the total tonnage of feed through the mill and not the circulating load ratio is the important consideration in obtaining best efficiency from closed-circuit grinding. Concerning actual mill loadings, a review of several important grinding applications which used the heavy-duty Dorr classifier in the circuit has shown that the total tons of feed to the grinding unit has fallen in the range of 14 to 20 tons/day/cu. ft. of mill volume.

Size Control. Increasing recognition is being given to the fact that in many cases the utility of a finely ground product is dependent upon, not one, but two things. These are (1) its degree of subdivision and (2) the size distribution of its particles. In other words, it is not enough to specify simply that a product be ground to, say, 90 per cent minus 200 mesh. Several samples of the same product may all pass 90 per cent through a 200-mesh screen; yet, if not ground under exactly the same conditions, these samples may have entirely different distributions of particle sizes. Hence there may be widely varying characteristics in the aggregate.

As a rule a properly adjusted closed-circuit system comes nearer to producing the mesh distribution desired than any other method. This is true because the closed-circuit system allows one to "bunch" the grind. For example, if we wish to grind a ¼-in. feed to 97 per cent minus 200, open-circuit grinding will produce a much greater quantity of this as minus 325-mesh superfines than will closed-circuit grinding. At the same time there will certainly be considerably more plus 150-mesh material in the open-circuit product.

Fineness of Grinding. The operating data tabulated below are from the Portland cement industry and give a comparison of size distributions and fineness of grinding obtained by grinding in open and closed circuit. At the plant where these results were obtained, a year's operation with open-circuit grinding gave an average fineness of 89 per cent minus 200 mesh and 5 per cent plus 100 mesh. Under closed-circuit conditions the average fineness has been increased to 97.5 per cent minus 200 mesh and nothing on 100 mesh, while at times, when handling only 50 tons/hr. per mill, the fineness has been held at 99.9 per cent minus 200 mesh and 95 per cent minus 325 mesh.

The table below shows the distribution of sizes as determined by screen analyses, elutriation tests, and microscopic examinations of particle sizes.

Table 4. Comparison of Size Distribution

	Mesh	Open circuit, % +	Open circuit, % cum.	Closed circuit, % +	Closed circuit, % cum.
Sieve No.:					
1	20	1.0	1.0		
2	28	0.25	1.25		
3	35	0.50	1.75		
4	48	0.50	2.25		
5	65	0.75	3.00		
6	100	2.00	5.00		
7	200	6.00	11.00	2.6	2.6
8	325	(Not recorded)		8.4	11.0
Elutriation jar:					
1	340*	10.7	21.7	16.6	27.6
2	540*	7.8	29.5	5.9	33.5
3	820*	3.4	32.9	6.7	40.2
4	1400*	8.4	41.3	8.0	48.2
Overflow........	.....	58.7	.....	51.8	

* Average particle size determined by microscopic examination.

A study of these particle-size determinations indicates (1) an elimination of stray oversize, coarser than critical size, (2) a substantial reduction in plus 200-mesh material, and (3) less superfine as indicated by final elutriator overflow.

Another size analysis that shows how the substitution of closed circuit for open circuit results in a reduction in plus 150-mesh material and also in the superfines, is shown in Table 5. This size analysis was made by means of an Oden sedimentation balance, a device for making very accurate size determinations in the micron range. See pp. 1111 to 1113 for other particle-size-measurement methods. The open-circuit analysis is of a month's composite sample of cement slurry ground by the open-circuit method at a well-known cement mill. The closed-circuit sample was taken from closed-circuit grinding operations at the same mill.

Table 5. Oden Size Analysis of Cement

Diameter, microns	Equivalent sieve mesh	Open circuit, % cum.	Closed circuit, % cum.
295	48	0.32	
208	65	.68	
147	100	2.18	0.08
104	150	4.10	.38
74	200	7.85	3.75
53	325 dry	13.8	15.6
42	325 wet	15.9	19.15
40	......	19.3	23.3
30	......	23.5	31.0
25	......	27.6	34.7
20	......	33.1	39.4
15	......	39.9	44.9
10	......	48.5	51.7
7.5	......	53.4	57.0
6.0	......	57.6	60.5

Savings. At the cement plant from which the above data were obtained the following savings in grinding power and grinding mediums consumption were recorded:

Table 6. Savings in Power and Grinding Medium

Consumption per bbl.	Open circuit	Closed circuit	Saving, %
Power, kw.-hr....................	5.5	3.01	45.3
Primary mediums, lb.............	0.086	0.057	33.7
Secondary mediums, lb..........	.321	.112	65.1

Automatic Control. A point worth noting is that in closed-circuit grinding the classifier gives a control that automatically takes care of such small fluctuations in character and rate of feed as would otherwise cause variations in the finished product. Harder feed, for example, simply increases the circulating load.

Metallurgical Operating Data

The operating data given below, taken from metallurgical practice, are arranged to emphasize certain important facts relating to closed-circuit grinding.

EXAMPLES OF CLOSED-CIRCUIT GRINDING IN MINING INDUSTRIES*

Closed Circuiting Increases Mill Capacity. When a substance is being ground so that all of it shall pass a screen of a given size, the capacity of the mill may be increased by closed circuiting it with a classifier.

Example 1. Mill = 6 by 20 ft.

Feed = −6 mesh.
Product = 8 per cent + 100 mesh.
Capacity, open circuit = 144 tons, 24 hr.
Capacity, closed circuit = 240 tons, 24 hr.
Capacity increase = 96 tons, 24 hr.
Percentage increase in capacity = 66.
(Taggart, "Handbook of Ore Dressing," p. 455, Wiley, New York, 1927.)

Example 2. Wright-Hargreaves Gold Mines, Kirkland Lake, Ont.:

Product = −200 mesh.
Capacity, open circuit = 190 tons, 24 hr.
Capacity, closed circuit = 275 tons, 24 hr.
Capacity increase = 85 tons, 24 hr.
Percentage increase in capacity = 44.7.

Example 3. Phelps-Dodge Corp., Morenci, N. M.:

Product = −65 mesh.
Capacity, open circuit = 89 tons, 24 hr.
Capacity, closed circuit = 174 tons, 24 hr.
Capacity increase = 85 tons, 24 hr.
Percentage increase in capacity = 95.5.

Classifier Determines Capacity, Not Mill. The capacity of the closed circuit is frequently, if not always, determined by the classifier, not the mill; *i.e.*, additional classifiers increase the capacity of the circuit without any increase in size of the mill.

Example 1. Tough-Oakes Gold Mines, Ltd, Kirkland Lake, Ont.: By using two classifiers in closed circuit with a given mill, the capacity of the circuit was 28 per cent greater than the capacity of the mill in closed circuit with a single classifier.

Example 2. Consolidated Mining and Smelting Co. of Canada, Kimberly, B.C.: Same as Example 1, only the capacity was increased 35 per cent.

Example 3. Nevada Consolidated Copper Co., Hurley, N. M.:

Product = −65 mesh.
One mill in closed circuit with one classifier:
Capacity = 150 tons, 24 hr.
One mill in closed circuit with six classifiers:
Capacity = 800 tons, 24 hr.
Capacity increase = 650 tons, 24 hr.
Percentage increase in capacity = 433.

Closed Circuiting Reduces Unit Power Cost. The power required to drive a given mill remains practically constant regardless of tonnage fed; and, accordingly, increased capacity, due to closed circuiting, results in diminished power costs per ton finished product.

Example 1. Lake Shore Mines, Kirkland Lake, Ont.:

Percentage closed-circuiting increased capacity = 44.70.
Power consumption was reduced 10 per cent (due, no doubt, to better balance with heavy feed).
Percentage reduction in power per ton of finished product = 37.

* Anable, Closed Circuit Fine Grinding and What It Should Accomplish in the Cement Industry, *Rock Products*, Jan. 5, 1929.

Example 2. Lucky Tiger Mine:

Mill = 5 by 14-ft. tube.
Feed = 11 to 20 per cent + 20 mesh.
Finished product = −100 mesh.
Power was 47 hp. throughout tests.
Capacity, open circuit = 22 tons, 24 hr.
Capacity, closed circuit = 37 tons, 24 hr.
Tone of −100 mesh, per hp.-hr., open circuit = 0.016.
Tons of −100 mesh, per hp.-hr., closed circuit = 0.055.
Percentage unit power cost reduced = 71.
(Taggart, "Handbook of Ore Dressing," p. 456, Wiley, New York, 1927.)

Example 3. Mill = 6- by 20-ft. tube.

Feed = −6 mesh.
Finished product = 100 per cent + 100 mesh.
Capacity, open circuit = 144 tons, 24 hr.
Power consumption, open circuit = 75 kw.
Tons of 10 per cent + 100 mesh per hp.-hr. = 0.0565.
Capacity, closed circuit = 240 tons, 24 hr.
Power consumption, closed circuit = 65 kw.
Tons of 10 per cent + 100 mesh per hp.-hr. = 0.1087.
Percentage unit power cost reduced = 48.
(Taggart, "Handbook of Ore Dressing," p. 455, Wiley, New York, 1927.)

Closed Circuiting Reduces Wear on Liners and Balls. Through a better loading of the mill with coarse classifier oversize, the abrasion of metal liners and the consumption of grinding mediums are greatly reduced.

Example 1. Lake Shore Mines, Kirkland Lake, Ont.:

Consumption of steel per ton finished product:
Open circuit = 6.5 lb./ton.
Closed circuit = 3.2 lb./ton.
Percentage reduction in steel loss = 51.

Example 2. Chino Copper Company, Hurley, N. M.:

One mill and one classifier in closed circuit:
New feed = 150 tons, 24 hr.
Ball and liner wear = 3.2 lb./ton finished product.
One mill and six classifiers in closed circuit:
New feed = 240 tons, 24 hr.
Ball and liner wear = 1.5 lb./ton of finished product.
Percentage reduction in steel loss = 53.

The Higher the Circulating Load the Lower the Unit Grinding Cost. Increasing the circulating load of a closed-circuit mill has the same effect on the efficiency of grinding as increasing the new feed to an open-circuit mill. Capacity increases and all unit costs decrease throughout the entire range of circulating loads from 0 to 1100 per cent of the new feed, and the upper limit of this relationship has never been reached.

Example 1. Quoting from Oughtred's paper (*Trans. Can. Inst. Mining Met.*, 1928, p. 310) on the Sullivan concentrator of the Consolidated Mining and Smelting Company, Kimberly, B. C.: "An abnormally high circulating load is maintained, consistent with the mechanical limitation of the machines. Normal circulating load at the present time ranges from 1000 to 1100 per cent, or an equivalent of 3000 tons of sand per standard classifier."

Example 2. Quoting from the report of high-circulating-load tests at the Nevada Consolidated Copper Co., Hurley, N. M.: "Only the structural limitations of Section 7 prevented us from obtaining the ultimate capacity of a ball mill in these tests, but we learned enough to discover that we could reduce the cost of producing −65-mesh flotation feed from around 20¢ to about 5¢/ton by using all the classifiers on a single mill. We definitely learned that 3500 tons per day is not too great a feed for the above conditions."

(NOTE. The ball mill referred to was 7 ft. in diameter by 10 ft. long.)

Summing up, it may be stated that, in the field of metallurgy, both theory and practice agree on the desirability of closed-circuiting wet-grinding mills with classifiers, since closed-circuit operation increases capacity, reduces unit power cost of grinding, cuts down the wear on liners and grinding mediums and in general permits the mill to be loaded to its maximum output of material of a specified size. The above advantages have the cumulative effect of reducing the unit cost of grinding, controlling accurately the size of the maximum particle in the finished product, and limiting the tendency to overgrind a large portion to semicolloidal size.

Operating Costs. The unit power consumption for fine closed-circuit grinding at metallurgical plants varies somewhat, depending upon the degree to which the ore is crushed before grinding and the fineness of the overflow from the classifiers.

Studies made at eight plants have yielded the data presented in the table below. All these mills crushed their ore to 1/4 to 1/2 in. before grinding. Expressing the power consumption in terms of tons of 100-mesh and tons of 200-mesh material actually produced eliminates any error that otherwise might be introduced by different classifier settings. Most of these plants are making separations at from 48 to 65 mesh.

Table 7. Unit Power Consumption for Fine Grinding in Western Copper Plants*

Plant	Kw.-hr./ton, 100 mesh	Kw.-hr./ton, 200 mesh
A	7.90	10.80
B	8.53	9.73
C	8.95	11.66
D	9.20	11.69
E	10.10	13.10
F	10.61	14.56
G	10.74	14.80
H	11.93	11.69
Average	9.74	12.25

* Anable, Cement Industry Investigates Metallurgical Grinding Methods, *Eng. Mining J.*, **129** (4), 188 (Feb. 24, 1930).

The gradual wearing out of mill liners and steel balls represents an appreciable item of the operating cost. Table 8 below gives some practical operating data on this point.

Table 8. Consumption of Liners and Grinding Mediums at Closed-circuit-grinding Installations in Metallurgy

	Primary circuit	Secondary circuit	Total, lb./ton
Lake Shore Mines, Kirkland Lake, Ont.	.....	.....	3.2
Chino Copper Co., Hurley, N. M.	.....	.....	3.2–1.5
Cananea Consolidated Copper Co., Sonora, Mex.	0.459	1.283	1.742
Phelps-Dodge Corp., Bisbee, Ariz.	.....	.....	3.72

The cost of replacing mediums and liners amounts to 80 to 90 per cent of the maintenance cost of cyclindrical mills. **Grinding mediums,** such as **balls, rock,** or **flint pebbles,** wear much more rapidly than the steel-mill liners, probably in the ratio of about 3.5 to 4.5:1. **Forged-steel balls** cost from 4 to 5¢/lb., the higher price being for the smaller sizes. **High-carbon steel rods** sell at 0.25¢/lb. over the base price for merchant bars which usually range from 4 to 4½¢/lb. **Manganese-steel liners** cost from 13 to 16¢/lb., depending upon the size and weight of the individual castings, the higher price being for the small, light castings.

Cement-slurry Grinding

In this industry, open-circuit two-stage grinding has been standard practice for many years. Combination or compartment mills are generally used. The mill is divided into two portions by a vertical, slotted grid, the primary compartment being loaded with 2- to 5-in. balls and the secondary with 3/4- to 1-in. balls. Recently, however, many wet-process cement mills have adopted two-stage closed-circuit grinding, either using the existing combinations or compartment mills or preferably separate ball mills. Four mills of this type are in operation and several more under consideration. The following description relates to the mill of Universal Atlas Cement Co. at Leeds, Ala., as reported by Counselman in *Tech. Pub.* 1096, *Am. Inst. Mining Met. Engrs.*, 1939. See pp. 1152 to 1154 for discussion of dry grinding of cement materials.

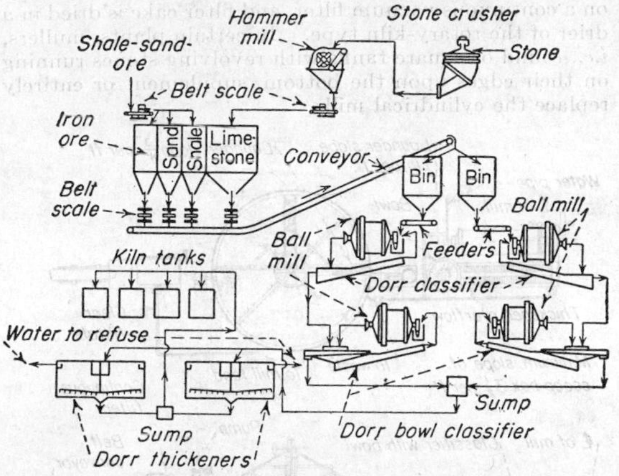

FIG. 19. Two-stage closed-circuit cement grinding.

CLOSED-CIRCUIT GRINDING CEMENT RAW MATERIALS AT LEEDS

After several years study, the Universal Atlas Cement Co. decided in 1937 to rebuild its plant at Leeds, Ala. The entire old plant, which was to operate during the new construction, was then to be scrapped with the exception of two kilns. The quarry was to be mechanized, new and larger kilns installed, with the most modern type of equipment all through the plant.

Two-stage, closed-circuit grinding (Fig. 19) was decided upon to reduce 1-in. material to 200 mesh. In selecting the ball mills, it was felt desirable to have the primary and secondary mills of exactly the same size, so that all parts would be interchangeable. After study of the grindability tests, mills were specified to be 9 ft. diameter inside the liners by 8 ft. long from the head liner to the grid plate. Throat and discharge trunnions are 22 in. and 24 in. clear diameter. The feed scoops are 4 ft. 0 in. radius. The discharge grates of the primary mills have slots 7/8 in. wide inside, tapering to 1 in. outside, and the plates are 1½ in. thick. On the secondary mills the tapered slots are 3/8 in. wide inside by ½ in. wide outside, same plate thickness. These slots extend inward for 13 in. from the inside of the liners toward the axis of the mill, the central portion being blanked off. All the material passing through the mill must therefore pass through these slots near the periphery, so that these are low-level mills. Internal radial lifters inside the discharge head raise the mill product to the discharge trunnion and are so arranged that there is no spillback through the slots. The mills have water-cooled babbitted bearings.

The primary mills rotate at 19 r.p.m. and are fed heat-treated steel balls 4 in. in diameter, of about 500 Brinell. The secondary mills run at 17 r.p.m., and 2-in. balls are fed as make-up. Mixed charges were put in the mills at the start. Motors are 350 hp., 180 r.p.m. synchronous, 2200 volt, capable of 100 per cent overload at 40°C. temperature rise.

The discharges from the primary mills flow by gravity at a slope of 13/16 in./ft. to 8 ft. by 25 ft. 6 in. classifiers with 20-hp. motors. Actual power consumption is about 12 hp. Sands are conveyed to the primary ball-mill scoop boxes by vibrating launders having 7½-hp. motors.

The overflows of the primary classifiers go by gravity, in a launder having a slope of 2 in./ft., to the bowl classifiers, which are 16 ft. wide by 39 ft. long by 24 ft. in diameter, and any finished material overflows without getting into the secondary ball mills. The rakes of these machines are driven by 20-hp. constant-speed motors, actually taking 13 hp. with normal load. The bowl is driven by a 10-hp. motor with a variable-speed reducer to provide adjustment for helping control the fineness. Actual power consumption is about 6½ hp.

The rake products of these classifiers are elevated by 12-ft.-diameter sand wheels, rotating at 9½ r.p.m., driven by 10-hp. motors, and flow down a slope of 4½ in./ft. to the scoop boxes of the secondary ball mills described above. The ball-mill discharges flow to the bowl classifiers, the launder slopes being 15/8 in./ft.

The overflows of the two bowl classifiers flow by gravity to a sump, and from there are pumped by one of two rubber-lined pumps, with 75-hp. motors at 2500 gal./min., through a 12-in.

Table 9. Typical Results—Closed-circuit Cement Grinding

Opening	New feed	Primary mill discharge	Primary classifier sands	Primary classifier overflow	Secondary mill discharge	Bowl classifier sands	Bowl classifier overflow	Primary circulating load	Secondary circulating load
Unit No. 1.* 7/8-in. Wide Slots in Primary Mill Grates									
Inch:									
+1 ..	2.6								
+3/4 ..	5.4	0.3	1.2						
+½ ..	10.7	3.7	3.5						
+3/8 ..	23.0	6.8	7.1						
Mesh:									
+ 4 .	45.7	13.2	16.1						
+ 5 .	48.3	14.5	18.2						
+ 8 .	62.5	21.9	32.1	....	....	....	....	214	
+ 14 .	74.0	37.2	58.7	1.6	....	0.4	....	166	
+ 28 .	84.5	61.2	86.5	18.6	2.1	7.4	....	168	351
+ 48 .	89.0	75.0	94.0	37.4	15.1	26.1	....	198	340
+100 .	92.2	81.6	96.2	52.1	48.8	64.0	....	202	342
+200 .	94.0	85.4	97.0	61.0	78.9	92.6	13.0	210	351
+325 .	....	....	....	....	....	....	22.4		
Specific surface.	....	....	....	....	....	....	3792	Best	Best
Tons/hr..	54	164	110	54	187	187	54	avg.	avg.
H_2O, per cent...	2.3	23.8	24.5	50.1	35.6	23.4	83.9	203	346
Unit No. 2. 3/8-in. Wide Slots in Primary Mill Grates									
Inch:									
+1 ...	2.6								
+3/4 ..	5.4								
+½ ..	10.7	....	0.2						
+3/8 ..	23.0	0.6	0.9						
Mesh:									
+ 4 .	45.7	3.5	7.8						
+ 5 .	48.3	4.0	8.9						
+ 8 .	62.5	9.2	18.9	....	....	....	....	95	
+ 14 .	74.0	20.6	41.4	0.6	....	....	....	96	
+ 28 .	84.5	45.0	76.3	13.6	1.0	4.9	....	100	
+ 48 .	89.0	60.5	89.1	32.6	8.0	22.6	....	98	223
+100 .	92.2	71.3	93.3	48.4	37.4	62.1	....	104	196
+200 .	94.0	76.6	95.2	58.9	69.2	91.5	10.1	95	210
+325 .	....	....	....	....	....	....	19.2		
Specific surface.	....	....	....	....	....	....	4093	Best	Best
Tons/hr..	50	100	50	50	102	102	50	avg.	avg.
H_2O, per cent...	2.3	32.8	21.6	55.2	34.2	30.5	84.5	100	203

* Approximately 6000 lb. less ball load than No. 2 unit.

Table 10. Sedimentation Analyses of Bowl-classifier Overflows by Wagner Turbidimeter Method*

Particle size		Unit No. 1	Unit No. 2
Mesh:			
+200		13.0	10.1
+325		22.4	19.2
Microns:			
−60	+55	23.4	20.1
	+50	25.5	22.6
	+45	28.8	25.6
	+40	31.2	28.3
	+35	33.2	31.2
	+30	37.3	33.2
	+25	40.9	37.4
	+20	44.8	40.7
	+15	49.6	45.7
	+10	56.6	52.4
	+ 7.5	61.5	58.0

* It is assumed, for specific surface calculations, that the average diameter of the minus 7.5 μ particles is 2.6 μ.

Table 11. Tonnages and Power Consumptions

	Test data, tons/hr.		Power readings, kw. Unit No. 1	Power readings, kw. Unit No. 2
Belt-scale settings:		Primary ball mills	290	310
Limestone	91.0	Secondary ball mills	305	305
Shale	15.0	Primary classifiers	9	9.5
Sandstone	2.2	Bowl-classifier rakes	10	9.5
		Bowl-classifier bowls	3.8	3.8
		Thickener, rotation	1.5 for both units	
		Kw-hr. per barrel	3.58	4.0

Table 12. Power Consumptions per Barrel* of Cement

Equipment	Kw.	Kw.-hr./bbl.
Primary mill	290	3.44
Secondary mill	305	
Primary classifier	9	0.14
Bowl classifier:		
Rakes	10	
Bowl	3.8	
Thickener	1.5	
Total power		3.58

* 625 lb. of raw materials—limestone, shale, and sandstone—are required to make a barrel of finished cement.

line to the distribution box between the two thickeners, the underflows of which are pumped to feed tanks ahead of the kilns. Five tables of operating data, shown above, give typical closed-circuit grinding results as of Nov. 19, 1938.

POWER REQUIREMENTS

In addition to more uniform grinding, one of the principal advantages of closed-circuit grinding is saving in power. At most plants having open-circuit wet grinding in a two-compartment mill, the power required for the mills alone for raw-grinding enough material to make a barrel of cement (the relative grindability of the mix being almost the same as at Leeds) amounts to about 7.5 kw.-hr. At Leeds, when grinding in one unit 54 tons/hr., equivalent to 173 bbl., the power readings were 3.58 kw.-hr.—a saving of 3.92 kw.-hr./bbl., or, on a 4000 bbl. a day basis, a saving of 15,700 kw.-hr./day.

The power consumption of 3.58 kw.-hr./bbl., for reduction from 1 in. to 200 mesh, is equivalent to 11.45 kw.-hr./ton. This can be broken down to a consumption of 5.55 kw.-hr./ton, for reduction from 1 in. to 14 mesh, and 5.90 kw.-hr./ton for reduction from 14 to 200 mesh. This compares very favorably with metallurgical practice.

Whiting, Lithopone, and Abrasive Grinding

Whiting (French, Cuban, English, or domestic chalk), lithopone (chemically precipitated $BaSO_4.ZnS$) and abrasives (Aloxite, Carborundum, etc.) used to be ground and sized by relatively crude intermittent methods. The substance was first wet ground to a fineness approaching that desired for the final product. The charge was withdrawn from the mill in the form of a thin paste or slurry, and the suspended solids were then sized by diluting the paste or slurry with water and allowing it to flow through a series of tanks of various sizes, the first in the series being the smallest and the last the largest. In certain cases, especially in the abrasive industry, as many as a dozen or more tanks are frequently used. The material overflows from each tank into the next succeeding one, and, owing to different settling periods provided, the coarsest and most quickly settling material settles in the first and smallest tank, the finest and most slowly settling material in the last and largest tank, and the intermediate sizes are collected in the tanks between the two extremes.

In certain cases, the materials are dry ground and sized pneumatically. Dry treatment, however, is outside the scope of this section.

Whiting Classification. The tendency toward the production of but a single grade or fineness of product in preference to variously sized products led to the development of the arrangement shown in Fig. 20 for the preparation of −300-mesh, water-floated whiting. Crushed chalk is ground in a tube mill in closed circuit with a bowl classifier. Classifier overflow is thickened in a continuous thickener, thickener discharge is dewatered on a continuous vacuum filter, and filter cake is dried in a drier of the rotary-kiln type. In certain plants, mullers, *i.e.*, round or square tanks with revolving stones running on their edges upon the bottom, supplement or entirely replace the cylindrical mills.

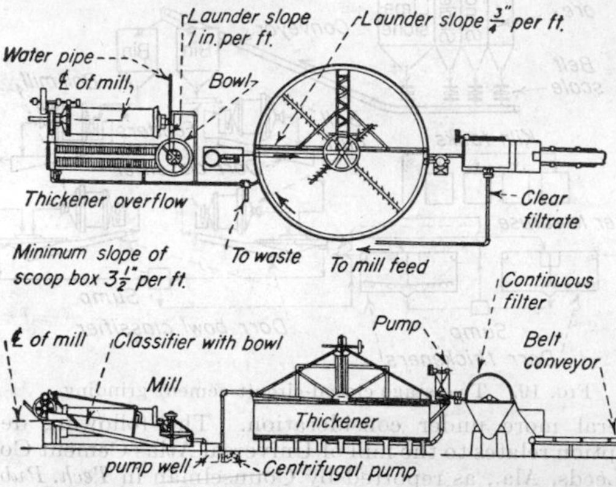

FIG. 20. Wet closed-circuit grinding of whiting.

Tables 13 and 14 give certain operating data from two installations of this general type.

Table 13. Plant A. 31 Tons Whiting per Day

	Dilution water to solid	Screen analysis—cumulative % +65	+100	+150	+200	+325	−325
Bowl-classifier feed	6.1:1	0.9	1.7	3.5	5.0	7.8	92.2
Bowl-classifier overflow	10.1:1	....	....	....	0.5	2.5	97.5
Bowl-classifier rake product	1:1	11.7	22.4	41.6	58.4	73.3	26.8
Thickener discharge	1.07:1	....	....	....	0.5	2.5	97.5
Vacuum-filter discharge	0.33:1	....	....	....	.5	2.5	97.5

Table 14. Plant B. 40 Tons Whiting per Day

	Dilution water to solid	Screen analysis—cumulative % +65	+100	+150	+200	+325	−325
Bowl-classifier feed	12.6:1	..	1.4	1.8	2.3	4.8	89.7
Bowl-classifier overflow	16.7:1	..	....	....	0.6	4.1	95.9
Bowl-classifier rake product	......	11	24.2	50.3	70.9	86.4	13.6
Thickener discharge	1.1:1	..	....	....	0.6	4.1	95.9

Lithopone Classification. Finished lithopone, after calcination and quenching, is generally reduced to the desired fineness by wet grinding in a tube mill in closed circuit with a classifier and a hydroseparator. The principle of operation is the same as that used in the closed-circuit grinding of metallurgical ores, cement slurry, and whiting, but the product is ground to a much finer degree.

Since discoloration of the pigment must be avoided at all cost, the classifying equipment must be constructed of wood or such special alloys as aluminum-zinc. Instead of using a bowl classifier, it is the practice to use a standard classifier of wood construction in direct-closed circuit with the tube mill and a hydroseparator (an undersized thickener) also constructed of wood for reclassifying the classifier overflow. This arrangement is equivalent, insofar as results are concerned, to a tube-mill bowl-classifier layout and may more easily be provided in acid-resisting construction.

The tube mill, lined with silex blocks and loaded with flint pebbles, received the quenched lithopone and discharges to the classifier. The classifier overflow is pumped to the hydroseparator for reclassification, and the classifier rake product is returned to the mill for further grinding in the conventional manner.

The hydroseparator, a shallow thickener insufficiently large to settle all the lithopone, overflows the finest material and discharges a pulp containing intermediate-sized lithopone, which is pumped back to the tube mill for regrinding with the classifier returns. The fine hydroseparator overflow is dewatered in a thickener, filtered on vacuum filters or presses, dried, and packed for shipment. The pulp and solutions are generally maintained at a temperature of about 140° to 150°F. in order to facilitate the making of the fine separations required.

The following data were obtained at one of the plants grinding lithopone in this manner:

Table 15. Making 350-mesh Lithopone

	Dilution water to solids	% + 350 mesh
Feed to tube mill	1.5:1	7.0
Tube-mill discharge	3.5:1	4.5
Classifier overflow	10 :1	2.6
Hydroseparator overflow	20 :1	0.25

Abrasives Classification. In the preparation of finely divided abrasives, such as Carborundum, Aloxite, Alundum, etc., wet grinding and classification are generally employed. Experience has shown that hydraulic classification into a variety of different sizes is greatly simplified if the feed to the series of hydraulic cones has been classified so as to remove, first, the very coarse grains and, second, the very fine—almost colloidal—grains. Apparently the presence of these end sizes interferes with the proper regulation of the hydraulic cones.

To meet this condition, a unique closed-circuit-grinding layout has been adopted by four of the large abrasive companies. The crushed abrasive is ground in a ball mill, operating in closed circuit with a bowl classifier. All oversize material is eliminated at this point and returned to the ball mill for further grinding. The overflow from this first bowl is pumped to a second bowl classifier which overflows to waste those superfine grains which are injurious to the operation of the hydraulic cones. The rake product of this second classifier is the finished product and contains a range of sizes of grain which may later be easily separated into the desired commercial grades.

These abrasive installations are unique in that the second bowl classifier makes a separation around the equivalent of 600 to 700 mesh. So delicate is the adjustment that sometimes a dispersing agent, sodium silicate, for example, is used, especially if the plant's water supply contains a trace of acid or other flocculating agent.

The following operating data are typical of results secured at a fine abrasive plant producing as finished product grains ranging in size from 0.10 to 0.005 mm.:

Table 16. Moisture Determination and Microscopic Analyses of Products in Classification Circuit

Product sampled	% solids	Dry lb./hr.	Approximate distribution of grain sizes, mm.*: 0.06 and coarse	0.06–0.05	0.05–0.025	0.025–0.005	Finer than 0.005
No. 1 bowl classifier:			%	%	%	%	%
Feed	41.5	870	25	35	20	10	10
Overflow	6.9	317	3	47	30	15	5
Rake product	70.0	553	80	20			
No. 2 bowl classifier:							
Overflow	1.16	144	..	..	Trace	Trace	Balance
Rake product†	70.0	173	15	50	20	10	

* Made by microscopic examination.
† Finished product.

REFERENCE: Dorr and Lasseter, Solid-liquid Separation, from Alexander, "Colloid Chemistry," vol. 6, 1946.

SEDIMENTATION

BY ANTHONY ANABLE

CONTINUOUS THICKENING

Definition. The term "sedimentation" or "thickening" generally implies gravitational settling of solid particles that are suspended in a liquid. It may be divided into two general classes, sedimentation of sandy material and sedimentation of slimes. Usually the term sedimentation or thickening includes the removal of the bulk of the liquor or water from the slime after settling.

Mechanics of Clarification (Deane, Settling Problems, *Trans. Am. Electrochem. Soc.*, p. 659, 1920). In all clarification problems, dilution ratio, or the weight ratio of liquids to solids, has a most important bearing. If, for example, a thin pulp, say a greatly diluted metallurgical slime, is poured into a glass cylinder and is allowed to settle, the following is observed: First, we see that a classification takes place, in which the coarsest particles settle to the bottom at a comparatively rapid rate, while the finest particles, settling at a slower rate, remain on top, with gradations in size ranging between these limits. All the particles have free movement and, excepting those of colloidal size, settle at a constant velocity, which is expressed mathematically by the formula of Stokes.

Stokes's formula for spherical grains is

$$u = \frac{gD^2(\rho_s - \rho)}{18\eta}$$

where u = terminal velocity; D = diameter of the sphere; g = acceleration due to gravity, or 981 cm./sec.; ρ_s = density of sphere; ρ = density of the fluid; η = coefficient of viscosity of the fluid; the quantities all being expressed in c.g.s. units.

For water at 20°C.,

$$\rho_1 = 1 \quad \text{and} \quad \eta = 0.010$$

Expressing u in millimeters per second and D in millimeters, the Stokes formula becomes

$$\frac{u}{10} = \frac{981}{18}\frac{(\rho_s - 1)}{0.010}\left(\frac{D}{10}\right)^2$$

Simplifying,

$$u = 545(\rho_s - 1)D^2$$

Later we see that a gradual clarification takes place, relatively slow in the last stages if very fine particles are present, and that there is an absence of any line of demarcation between the settling solids and the supernatant liquid.

Now let us gradually increase the density of the pulp by adding amounts of solids and note what happens after each addition after the resulting pulp has been well mixed and allowed to settle. We observe, first, that a dilution is reached in which the fastest settling particles form into a zone and settle from then on collectively and at a retarded rate; second, that this zone commences to form at progressively earlier periods until eventually a point is reached where the initial subsidence of solids is in mass, no independent particle movement being discernible, and proceeds with a sharp line of separation between it and the supernatant liquid; third, from this point, where the subsidence takes place more or less at a constant rate, a point in concentration is reached where there is a marked decrease of the rate of settling. This is called the **point of compression** and marks the dividing line between the zone of **clarification** and the zone of **thickening.** Accordingly, pulps may be classified as follows:

Table 1. Character of Subsidence of Different Types of Pulps

	Type of pulp	Character of subsidence	Description	Examples
Clarification free settling zone	Dilute	Independent particle subsidence	Particles or flocs settle independently. No definite line of subsidence. Settling unhindered and dependent upon size of particle or floc	Turbid water, sewage, and trade wastes
	Intermediate	Phase subsidence	Upper zone of independent particle subsidence. Lower zone of collective subsidence. Line of demarcation not sharp	Chemical and metallurgical pulps
Point of compression	Concentrated	Collective subsidence	Definite line of subsidence. Settling rate decreases with increasing concentrations of solids. Settling rate retarded by particle or floc interference	Chemical and metallurgical pulps
Thickening compression zone	Compact	Compact subsidence	Flocs and particles in intimate contact. Subsidence due to compression	All pulps by sedimentation pass into this zone

The following is an abstract from Coe and Clevenger's paper, Methods for Determining the Capacities of Slime-thickening Tanks (*Trans. Am. Inst. Mining Met. Engrs.*, pp. 356–384, 1916):

If a thin pulp, of a dilution of, say 10 to 1, is placed in a 1000-cc. cylinder, after thorough mixture, at least momentarily, it forms a homogeneous mass as shown in Fig. 21*E*. After a short time, however, it assumes a flocculent structure which, after settling a brief period of time, forms four distinct zones, *A*, *B*, *C*, and *D*, as in Fig. 21*F*.

The first particles that reach the bottom of the cylinder are the coarser granular sands which may be present in the pulp. Immediately following this and somewhat contemporaneously with the settling of the sand, the slime flocs nearest the bottom settle, filling the interstitial spaces between the sand particles, and build up, one upon another, in a zone of increasing depth. This we term zone *D*, which may be defined as that portion of the pulp wherein the flocs, considered as integral bodies, have settled to a point where they rest directly one upon another. After the pulp enters zone *D*, further separation of liquid must come through liquid pressed out of the flocs and out of the interstitial spaces between the flocs.

Immediately above zone *D* is a transition zone *C*. The pulp in zone *C* decreases in percentage solids from the bottom, where the flocs enter zone *D*, to the top, where the consistency of the flocculated pulp is the same as that of the original pulp. In speaking of flocculated pulp, it is intended to eliminate from consideration the coarser portion of the contained sand which falls directly through the overlying zones into zone *D*.

Above *C* is zone *B*, of constant consistency of flocculated pulp and of the same consistency as the flocculated pulp in the feed pulp. Zone *A*, overlying zone *B*, is clear water or solution. In the case of a very rapidly settling slime, zone *A* in the earlier stages may be turbid, due to finely divided matter remaining in suspension. Later this very fine material settles and the liquid becomes clear, although there are cases, especially when the liquid contains very little electrolyte, where it remains turbid for a long time.

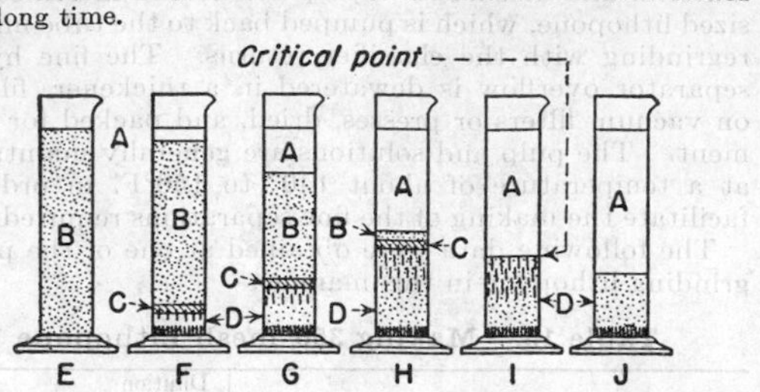

Fig. 21. Six phases of sedimentation illustrating intermittent thickening.

Figure 21*E* shows a cylinder freshly filled with pulp of a consistency of about 10 parts water to 1 part ore. In this illustration zone *B* occupies the total depth. *F*, *G*, and *H* of Fig. 21 show progressive stages of settling in which zones *A* and *D* are growing deeper, zone *B* is decreasing in depth, and zone *C* remains constant, a feature of this particular pulp. Figure 21*I* shows the condition when all the pulp has entered zone *D* and compression of the slime flocs is going on. Figure 21*J* shows the final stage of settling, beyond which the pulp will not thicken further.

With intermittent operation, any one of the stages described may represent the condition in the thickener depending upon the length of time that the pulp has been allowed to settle. In the operation of continuous thickeners, the feed of the thin pulp at the center of the tank, the overflow of clear liquid at the periphery of the tank, and the discharge of the thickened pulp at the bottom are generally continuous. In a continuous thickener, the four zones previously described in discussing intermittent settling are generally present as shown in Fig. 22. At the top there is a zone of clear water, *A*. Beneath this is a zone *B*, consisting of flocculated pulp of uniform consistency. Directly beneath this is a transition zone *C*, and at the bottom a zone *D* of pulp which is undergoing compression. In making tests, the settling rates of thin pulps are determined by readings taken at the juncture of zones *A* and *B*, *i.e.*, where the pulp surface joins the liquid.

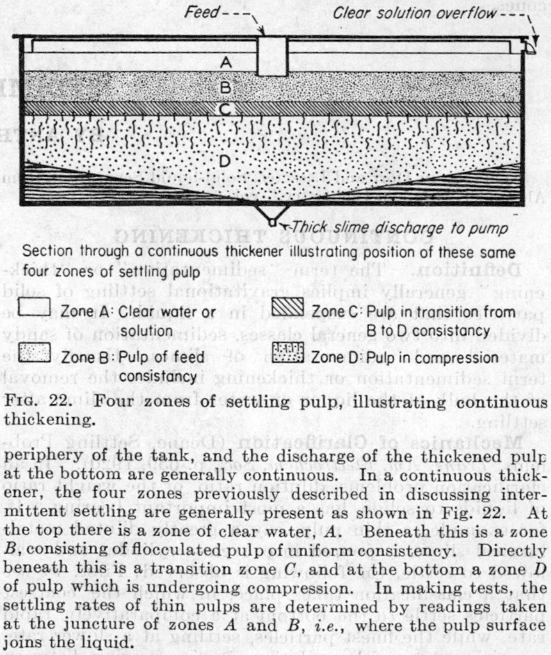

Fig. 22. Four zones of settling pulp, illustrating continuous thickening.

Clarification Capacity. The relationship between the settling rates of particles at their various dilutions in terms of thickener area required may be expressed by the following formula:

$$A = \frac{1.333(F - D)}{R \times \text{sp. gr.}}$$

where A = sq. ft./ton dry solids/24 hr.; R = settling rate, ft./hr., of a feed with F dilution; sp. gr. = specific gravity of liquid; F = weight ratio of liquid to solids for the rate R; and D = weight ratio of liquid to solids in discharge.

By applying this formula to pulps of different densities, ranging in dilution from feed to discharge density, the zone requiring the greatest unit area is found, and it is this zone which determines the area that must be provided for the pulp being tested.

Thickening Capacity. The volume provided in a tank in the thickening zone depends directly upon the period of detention required for the sludge to reach the desired density and may be determined by the following formula:

$$V = \frac{4T(G - \text{sp. gr.})}{3G(S - \text{sp. gr.})}$$

where V = volume, cu. ft., required for thickening per ton of solids per 24 hr.; S = average specific gravity of thickened pulp during compression period; sp. gr. = specific gravity of clear solution; G = specific gravity of solids in pulp; and T = period of detention, hr.

Mechanics of Thickening. There is a remarkable difference in the thickening properties of various pulps. Some consolidate within a few hours to a dense sludge that will barely flow through a pipe, whereas the moisture content of others after settling for days will not be reduced below 95 per cent. These effects are largely due to differences in the specific gravity of the solids and the physical differences in the character and structure of the flocs.

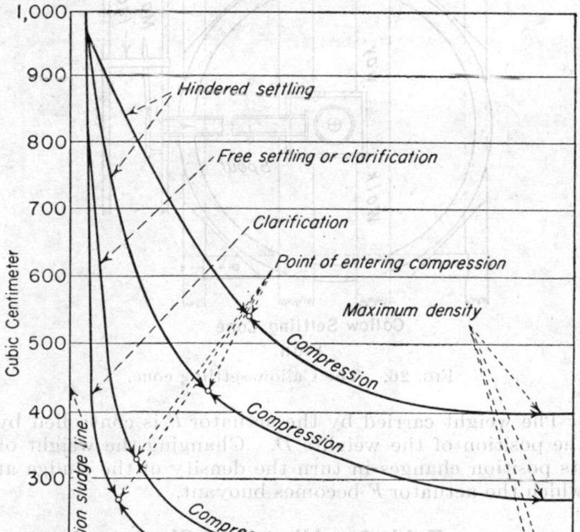

FIG. 23. Settling behavior of different types of pulps.

Figure 23 illustrates the phases of the mechanics of clarification and thickening of different types of pulp.

Destabilization of Colloids. In all clarification-thickening problems, the aim is to obtain the lowest operating area (in terms of square feet per ton of solids treated) permissible with satisfactory operation, including a safe allowance for variations in feed conditions, such as dilution, temperature, degree of flocculation, particle size, etc. From the formula for area it is seen that the rate of settlement or clarification is one of the controlling factors. The area of a thickener varies inversely with the settling or clarification rate, other conditions remaining constant.

Any agency that will effect an increase in settling rate will bring about a corresponding reduction in tank area required. Since the rate of subsidence is the settling rate of the most slowly settling particles, the control of the settling rates of the finest particles is extremely important. As the particles become finer, the effect of gravity is gradually reduced and is eventually overbalanced by the forces of surface energy and Brownian movement, and a colloidal state is reached where the dispersed particles remain in permanent suspension in the liquid. From a state of coarse suspension, through finer suspensions and colloidal solutions to true solutions, there exists a perfect continuity in change of particle size, and the change from one state to another is not sharply defined.

In general metallurgical practice **sands** are considered to be particles coarser than 200 mesh, 0.074 mm.; and **slimes** any material finer than 200 mesh. According to Zsigmondy's classification of solid particles, the following ranges of sizes are given:

Suspensions. Particles coarser than 0.0001 mm. in mean diameter.

Colloidal solutions. Particles under 0.0001 mm. and over 0.000001 mm. in mean diameter.

True solutions. Particles under 0.000001 mm. in mean diameter.

The metallurgical and chemical pulps encountered in practice vary from fine to coarser suspensions, with relatively small amounts of colloidal material. Probably the largest content of colloidal material is found in certain clays and in sewage and trade wastes.

The most widely accepted theory in explanation of the stability of the colloid is based on the assumption of an electric charge carried by the particles, a surface phenomenon derived by preferential adsorption of either positive or negative ions from dissociation of compounds. Particles, charged with like signs, all positive or all negative, will be mutually repelled and remain dispersed through the liquid medium in permanent suspension.

In order to destroy this condition of stability and induce clarification, it is necessary to neutralize the charge on the particle by introducing a charge of an opposite sign, by the addition of either an electrolyte or another colloid. Accordingly, negatively charged colloids are precipitated by positive ions, and positively charged colloids by negative ions, and the neutralized particles agglomerate into flocs, producing the condition known as **flocculation.** The use of lime in cyaniding is a good illustration of such an effect of flocculation.

Equipment Used. Intermittent Settling Tanks. This is the simplest and oldest device for thickening.

The intermittent settling tank is of any convenient shape or size, rectangular tanks, however, being more common than cylindrical. A discharge valve or gate is placed in the bottom for the removal of the thickened material. The clarified solution is withdrawn either by a swing siphon as shown in Fig. 24 or through draw-off connections, located at suitable intervals along the side. This is a homemade device on which few or no operating data are available.

Operation. The tank is filled with the pulp to be thickened, which is then allowed to stand undisturbed for the period of time that experience has shown necessary for settlement of the solids and their compacting as a heavy dense sludge in the lower part of the tank (Fig. 24).

The supernatant layer of clarified solution is thereupon removed, by either gradually lowering the swing siphon or opening the drawoff connections one by one, starting with the uppermost one and working downward. When the decanted liquid begins to show appreciable amounts of sludge, or the sludge level is exposed, the decantation is stopped because the sludge line has been reached and further separation by settlement cannot be obtained.

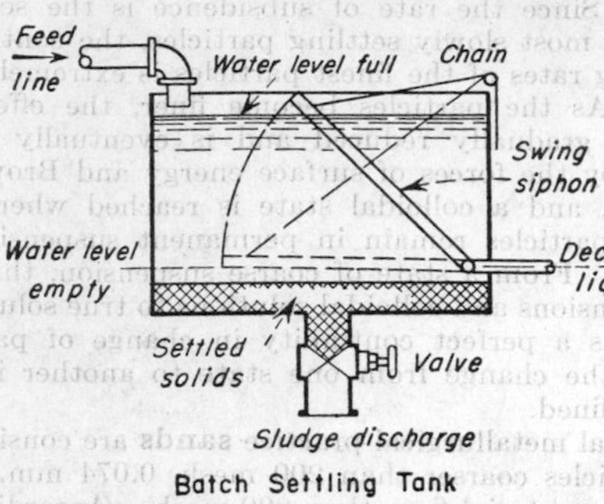

FIG. 24. Intermittent settling tank.

The sludge-discharge valve or gate is now opened, and the compacted mass flows out, aided possibly by a stream of water from a hose or by manual shoveling or sweeping. The valves are then closed, the tank refilled, and, after a suitable settling period, the cycle of operations is repeated. Intermittent settling tanks are usually operated in groups of a half dozen or so, one or more being filled or emptied while settlement is taking place in others.

Non-mechanical Continuous Thickeners. The settling cone is a non-mechanical settler. Continuous operation can be secured, since it is equipped with

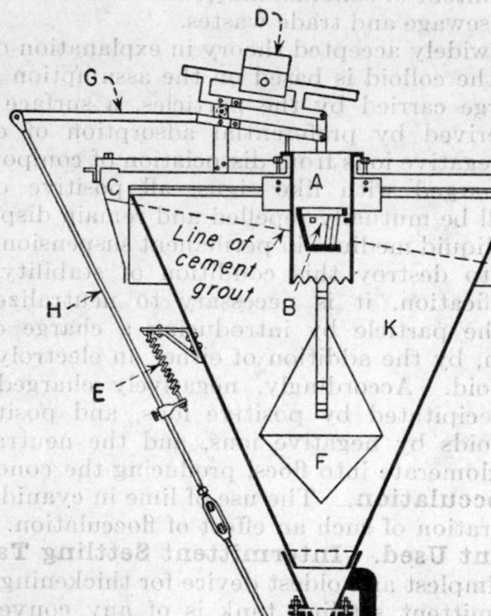

FIG. 25. The Allen settling cone.

facilities for the continuous overflow of clarified solution and the continuous discharge of thickened sludge.

The cone, as its name implies, consists of a conical tank, the angle at the apex of which is 45 to 60 deg. The conical tank is mounted with its apex pointing directly downward, being provided with a manually or automatically controlled sludge-discharge valve at the bottom and a centrally located loading well and peripheral overflow trough at the top.

Equipment Used. a. The Allen Cone. This type is shown in Fig. 25. The feed enters through the central loading well. *A*, the clarified solution, overflows into the externally located peripheral trough *C*, and the solids settle in the base of the cone *K*. The baffle *B* in the loading well prevents undue agitation.

The solids settling in the cone *K* increase the density of its contents until the buoyant power of the sludge causes the actuator *F* to rise. This motion of the actuator is transmitted by means of the connecting parts *G*, *H*, and *I* to the ball valve *J*, which is thereupon unseated from the orifice in the apex of the cone, thus allowing the thickened material to be discharged.

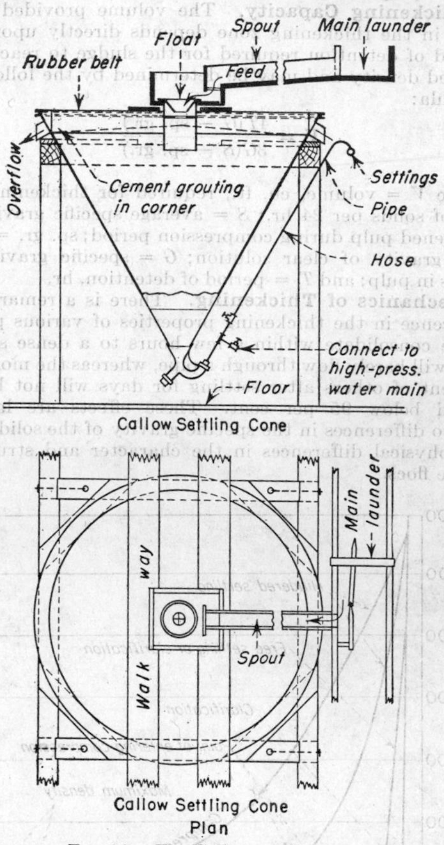

FIG. 26. The Callow settling cone.

The weight carried by the actuator *F* is controlled by the position of the weights *D*. Changing the weight or its position changes in turn the density of the sludge at which the actuator *F* becomes buoyant.

Table 2. Allen Cone Sizes

Diameter	Depth, Inlet to Spigot
3′6″	5′2″
4′6″	6′2″
6′0″	7′8″
8′0″	9′11″

b. The Callow Cone. The apex angle of the Callow cone is 60 deg. As shown in Fig. 26, the feed enters through a semisubmerged well at the top center of the tank. The peripheral trough for the collection of clarified solution is formed by a metal strip, attached to the inside of the tank near the top, and fitted with an adjusta-

Table 3. Callow-cone Sizes

Outside diam. (diam. tank top)	Inside diam. (diam. overflow weir)	Depth (tank top to bottom spigot)	Overflow capacities, gal./min.
2′ 9″	2′ 0″	3′ 4″	
3′ 3″	2′ 6″	3′ 9″	
3′ 9″	3′ 0″	4′ 2″	
4′ 9″	3′ 0″ and 4′ 0″	5′ 0½″	6–8
5′ 9″	4′ 7″ and 5′ 0″	6′ 0′	10–12
6′ 9″	5′ 5″ and 6′ 0″	6′ 10¼″	14–18
8′ 9″	7′ 5″ and 8′ 0″	8′ 7″	25-30

ble strip of belting to ensure uniform overflow across the entire periphery.

The thickened material is discharged from the bottom of the tank by a bushing, a plug valve, or by an adjustable gooseneck siphon. The purpose of the gooseneck siphon is to control the discharge density. The greater the elevation of the siphon discharge above the apex of the cone, the greater is the density of the discharge, and vice versa.

Mechanical Continuous Thickeners. The Dorr thickener, of which there are several different types, consists essentially of a shallow, cylindrical settling tank, equipped with a central feed well, peripheral overflow-collection trough, pump-regulated sludge-discharge outlet at the bottom and a slowly revolving, centrally located shaft, equipped with radial arms and plow blades for moving the settled sludge gently to the central sludge outlet. Distinctive features are the use of shallow cylindrical tanks, mechanical methods for the collection of the settled solids, and volumetric regulation of discharge density by means of a diaphragm pump with variable displacement.

The Single-compartment Dorr Thickener. This type is shown in Fig. 27. The tank is cylindrical and flat-bottomed, constructed of steel, concrete, or wood, and if steel, may be rubber-lined, or, if wood, lead-lined for acid-resisting duty. An overflow-collection trough, annular in plan view and rectangular in section, is provided around the inside top of the tank and is equipped with a leveling strip to ensure uniform overflow of clarified solution across its entire length. A discharge casting, constructed of cast iron or hard lead (for acid-resisting duty), is secured in the center of the tank bottom. A discharge line extends from this cone to the suction side of a Dorrco diaphragm pump.

A structural-steel superstructure spans the top of the tank, supported by the tank sides in relatively small units and by steel or concrete columns in the case of the larger machines. In addition to supporting the thickener mechanism, provision may be made for supporting the Dorrco pump, motor drive unit, and drive details such as pulleys, shafting, and speed reducer.

The thickener mechanism consists of a central vertical shaft carried by bearings and a supporting bracket on the superstructure. The worm gear, splined to the vertical shaft, is driven by a worm on a tangential shaft mounted on the supporting bracket. The worm shaft is provided with a pulley, sprocket, or gear for driving by belt, chain, or directly connected motor. At the lower end of the vertical shaft, a spider is secured from which there extend four radial arms, two long and two short. These arms are inclined to the horizontal plane, with the result that a slightly conical bottom of settled solids is built up under the plow blades which are attached to the undersides of these arms. Where the settled solids are too valuable to be used in forming the conical bottom, a cement or dirt fill may be used. These plow blades are mounted at an angle to a radial line and are so arranged that the entire area of the bottom is swept by them each revolution and solids deposited thereon moved gently toward the central discharge outlet. A manually operated lifting device, mounted on top of the superstructure and engaging a cap, turned on the upper end of the thickener shaft, makes it possible to lift the mechanism vertically a foot or so to relieve the load when starting up after a shutdown or operating interruption. The thrust of the worm shaft is borne by a spring-loaded thrust bearing, the displacement of which actuates a visual, overload-indicating pointer which calls to the

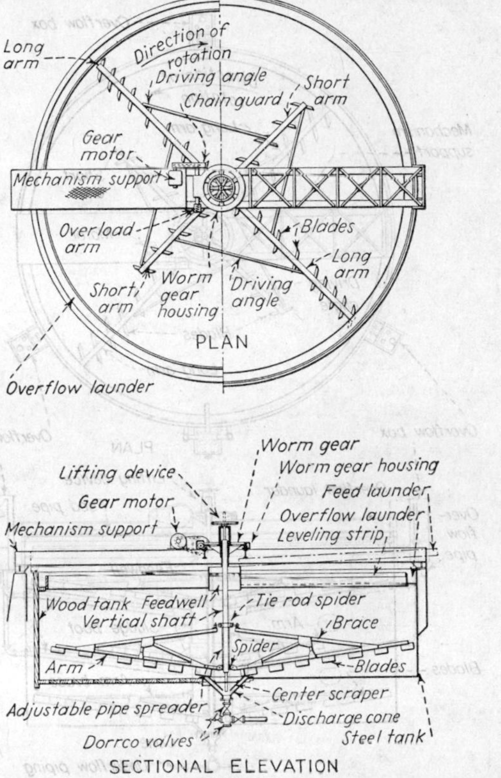

FIG. 27. The Dorr thickener.

attention of the operator the presence and degree of overload. Should the overload increase to the danger point threatening a mechanical breakage, an electrical contact is made which shuts down the drive motor and rings a warning bell.

The Dorr Torq Thickener. This type of thickener, shown in Fig. 28, is also a single-compartment thickener but differs from the unit described in the foregoing by

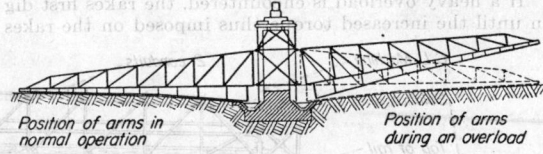

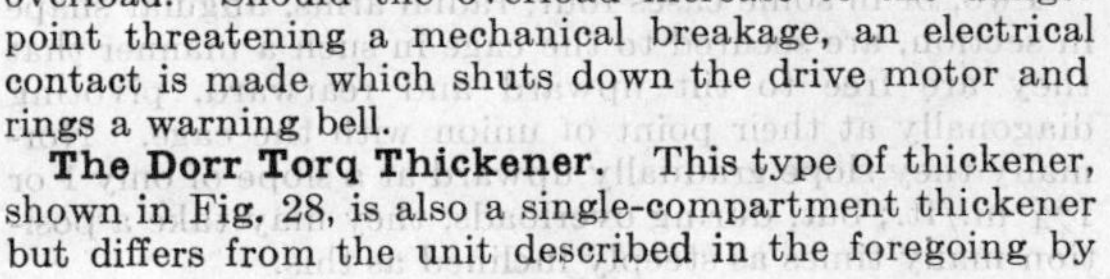

FIG. 28. The Dorr Torq thickener.

reason of its special antistalling arm construction. The tank is cylindrical and flat-bottomed, constructed of steel or concrete and infrequently of wood. Feed enters centrally, overflow is collected peripherally, and sludge is removed from a central annular depression by a diaphragm pump.

The unique feature from which this machine derives its name is a torque-actuated automatic lifting construction that causes the rake arms to raise when an overload is encountered and to lower them again to the normal operating position after the overload has been passed.

At the center of the tank there is placed a stationary column or pier, constructed of steel or concrete. A compact drive unit is mounted directly on top of the center column a foot or so above the water level. This drives a ball-bearing-mounted turntable from which there is hung a central revolving cage, concentric with the column.

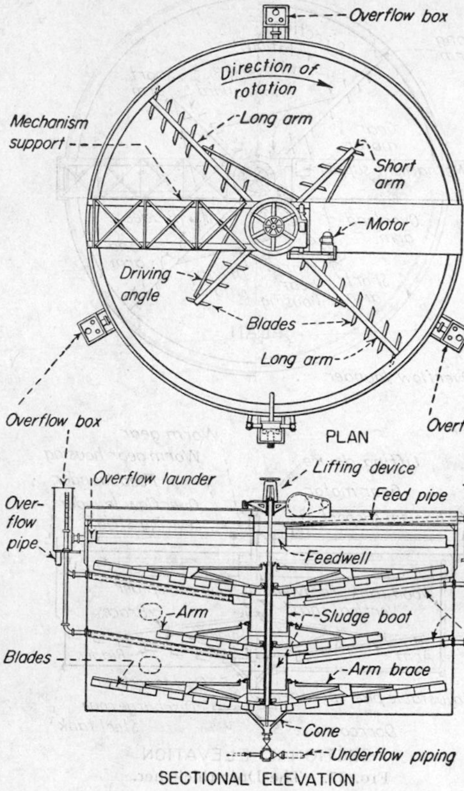

FIG. 29. The Dorr tray thickener.

Two, or in some cases four, radial arms, angular shape in section, are secured to the cage in such a manner that they are free to tilt upward and rearward, pivoting diagonally at their point of union with the cage. Normally they slope gradually upward at a slope of only 1 or $1\frac{3}{4}$ in./ft.; but, during overloads, they may take a position many times as steeply inclined as this.

If a heavy overload is encountered, the rakes first dig in until the increased torque thus imposed on the rakes reaches a safe predetermined point, well within the structural limits of the machine. Then, as the torque increases above this predetermined point, this increased torque is utilized to cause the rake arms to swing gradually upward, pivoting near the tank center. The greater the torsional load, the higher swing the arms until the rakes completely clear the obstruction or reduce it to a lower torsional equivalent.

As the overload is reduced by the continuous raking action, the torque decreases and the rake arms drop lower and lower. Finally, when the torque has decreased to a value less than that predetermined in the design of the machine, the rakes resume their normal operating position.

The Dorr Tray Thickener. As shown in Fig. 29, the tank of the Dorr tray thickener is subdivided into a multiplicity of shallow superimposed settling compartments by a number of slightly conical steel trays which are self-supporting and attached to the sides of the tank by rim angles at their peripheries. Each compartment has an individual connection for feed pulp and for clarified overflow solution. The solids settled in each compartment are discharged by gravity into the lower portion of the compartment directly below through a centrally located downcast seal ring which is concentric with an upcast ring attached to the mechanism of the compartment below. The bed of sludge effectively prevents the intermingling of sludge and solution.

A feed box, located slightly above the top of the tank, is provided with as many V-notch meters as there are settling compartments, thus assuring that each compartment receives its portion of the total feed. The overflow pipes from all compartments terminate in an overflow-collection box near the top of the tank. Adjustable rings on the ends of these pipes permit close regulation of the volume of solution clarified in each compartment and assist in the maintenance of the correct depth of sludge bed in each. All sludge eventually reaches the lowermost compartment from which it is continuously removed at the proper rate by a Dorrco diaphragm pump.

The tray-type thickener, similar in principle and in major structural details to the single-compartment thickener, provides the maximum capacity per unit of floor space. Since capacity is proportional to settling area and since each compartment operates substantially as an individual thickener, each compartment added increases the capacity of the original single-compartment tank approximately 100 per cent. Several other types of Dorr tray thickeners are furnished which differ from the above only in the construction of the tray and the feed, discharge, and overflow features, which latter are affected to a certain degree by the character of the pulp handled.

The Dorr Traction Thickener. As shown in Fig. 30, this is essentially a single-compartment thickener, the

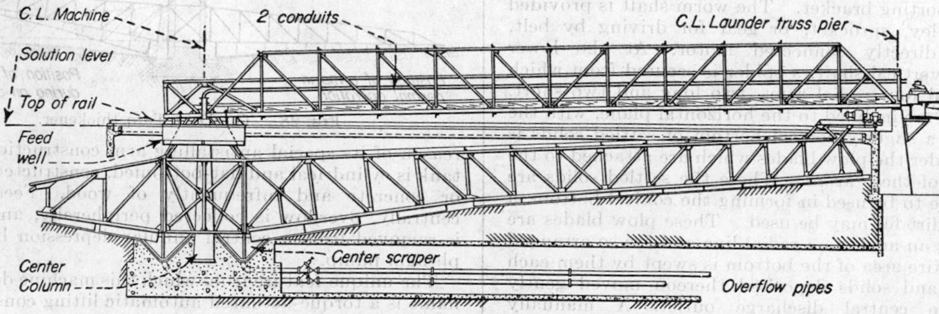

FIG. 30. The Dorr traction thickener.

distinctive feature of which is the application of the driving power to the end of a radial truss by means of a motor-driven carriage, running on the top of the tank sides. Feed enters through a loading well at the center, overflow is collected in a circumferential trough at the periphery, and sludge is continuously removed from the center by a Dorrco diaphragm pump.

The driving truss extends from a stationary vertical column at the center of the tank, on which its central bearing is mounted, to the periphery of the tank where its driving unit is attached. It also extends in the opposite direction from the center of the tank to a distance approximately one-third of the tank radius. Plow blades, secured to the lower chord of the truss, sweep settled solids into an annular depression, concentric with the central column, from which the sludge-discharge line extends.

Connected to a source of electrical power, the central column supports slip rings which are in contact with brushes on the revolving truss, electrically connected to the driving motor. A stationary bridge-type truss, extending from the tank periphery to the center, supports the feed trough and electrical conduits and serves as a walkway for operators.

An under-speed alarm rings a bell when an overload occurs, and, if the operator fails to correct this condition at once, an automatic device shuts off the feed of the unit by opening a by-pass in the feed trough.

The traction thickener is very rugged structurally and especially adapted to large tonnages and handling of severe raking loads. It may be furnished with corrosion-resisting construction.

The Dorr Clarifier. The Dorr clarifier is especially adapted to the handling of light finely divided solids such as trade wastes, water purification sludges, and domestic sewage. It is a modification of single-compartment thickeners, described previously. It is furnished in two types—circular and square.

The two main types of Dorr clarifiers have certain common and essential features—a shallow symmetrical concrete tank; provision for introducing the feed, overflowing the clarified liquor, and discharging the thickened sludge; and a motor-driven revolving mechanism for sweeping the settled solids to a central discharge hopper in the bottom of the tank. Positive removal of sludge is effected by a diaphragm or plunger pump. Skimming devices may be furnished, if desired, for continuously removing scum and other light material that tends to float on the surface.

The Dorr Sifeed clarifier is installed in circular tanks. Its name "Sifeed," is a contraction for the words "siphon feed," which is one of its distinguishing features.

Feed enters centrally below the water level through suitable connections, terminating in a slotted cylindrical diffuser. The central column, supporting the revolving mechanism, and the central drum form a conduit for the influent (shown in Fig. 31).

The feed leaves the central drum in a radial direction through slots. The head of water above the slots and the circular baffle have the effect of tapering the velocity of flow and giving quiescent feed conditions.

Distribution continues throughout the tank on the same radial diverging lines. The rate of diffusion is gradually decelerated as the circle of propagation increases, so that the velocity reaches the absolute minimum as the flow approaches the side of the tank.

A continuous annular trough with a continuous weir on its inboard side extends around the complete periphery of the tank. This gives maximum trough and weir length for any tank of equivalent capacity and assures minimum velocity of flow at the point of take-off. A circular scum baffle may be provided just inside the weir.

The clarifier mechanism consists of two radial trussed arms, driven by a motor on the stationary central column, and equipped with plow blades that just clear the bottom and sweep settled solids to the central discharge hopper in the bottom. The rake arms are attached to a central drum, concentric with the central column.

Sludge is removed continuously by a diaphragm pump. Where scum and floating solids tend to accumulate on the surface of the tank, positive mechanical means are provided for its removal.

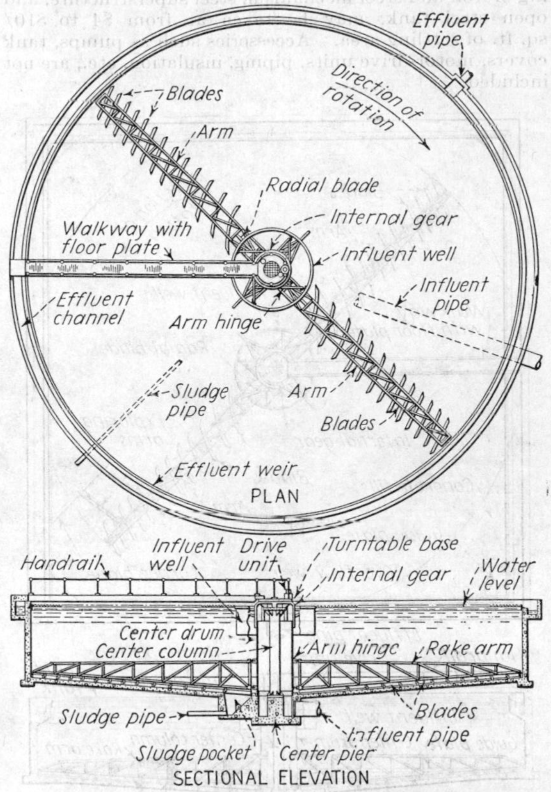

FIG. 31. The Dorr Sifeed clarifier.

The Dorr Squarex clarifier is installed in square sedimentation tanks. It follows closely the arrangement of the Dorr Sifeed clarifier with the exception that the tank is square, not round, and one of the rigid arms of the clarifier mechanism is equipped with a special corner blade that reaches out into the four corners of the tank and moves the settled sludge in to the point where it may be picked up by the regular mechanism. The action of the corner blade is positive and controlled automatically. Every square foot of the tank bottom is swept at each revolution of the mechanism.

Feed enters centrally through suitable connections, is distributed radially by a submerged diffuser, and is collected peripherally across a continuous weir extending around the four sides of the tank. Two radial arms with plow blades attached are secured to a central revolving drum and are driven by a gear motor mounted on top of the center pier. These revolving rake arms sweep the area of a circle inscribed within the square bottom of the tank.

A diaphragm or plunger pump is used for sludge removal and for control purposes. Scum-skimming devices of several different types are supplied to meet different conditions (Fig. 32).

Equipment Costs. The three chief types of Dorr thickeners—the single-compartment central drive, the traction, and the tray—are priced substantially the same per unit of settling area provided. The cost of construction, however, is such that the cost per unit of area varies materially with different sizes, the unit cost being much greater for relatively small thickeners than for the larger ones.

For rough and very preliminary estimating purposes only, the cost of a Dorr thickener, f.o.b. factory, consisting of iron and steel mechanism, steel superstructure, and open steel tank, may be taken at from \$4 to \$10/sq. ft. of settling area. Accessories such as pumps, tank covers, motor-drive units, piping, insulation, etc., are not included.

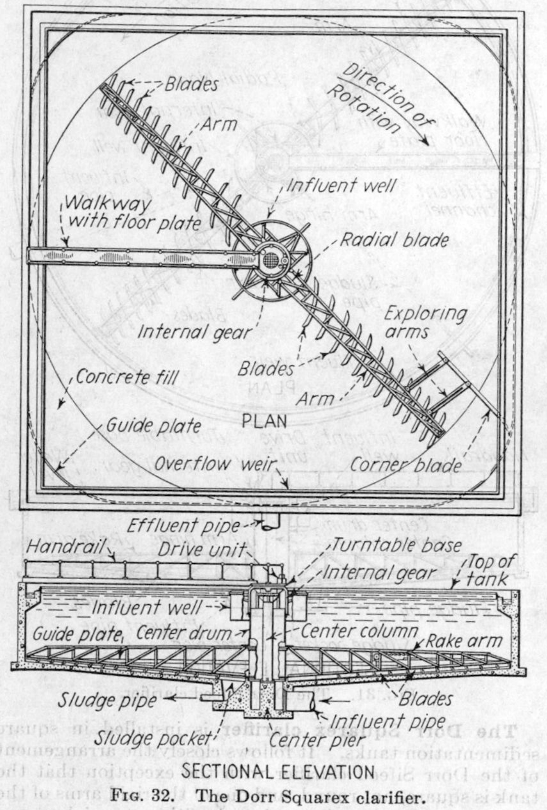

FIG. 32. The Dorr Squarex clarifier.

Special materials of construction, such as lead- or rubber-covered steel, wood, hard lead, and special alloys, alter the unit cost to such a wide extent that no reliable estimating figures may be given for thickeners of such special construction.

The Hardinge Auto-Raise Thickener. The outstanding and interesting feature of the Hardinge thickener is the Auto-Raise device included as part of the drive mechanism. With it, possible breakage due to overloads and manual raising of the mechanism when they occur is eliminated, and maintenance and attention are reduced to a minimum.

The Auto-Raise mechanism includes two concentric torque tubes, the outer and shorter one being entirely above the thickener liquid level. A yoke at the top of the inner torque tube has extended rollers which normally rest at the bottom of two diagonally opposite sloping slots in the outer torque tube. When the scraper encounters an obstruction, the abnormal resistance created causes the aforementioned rollers to move along and up the sloping slots with a telescoping effect of the two torque tubes and a shortening of their total length.

When the overload or resistance is decreased, the scrapers automatically lower, by the effect of their own

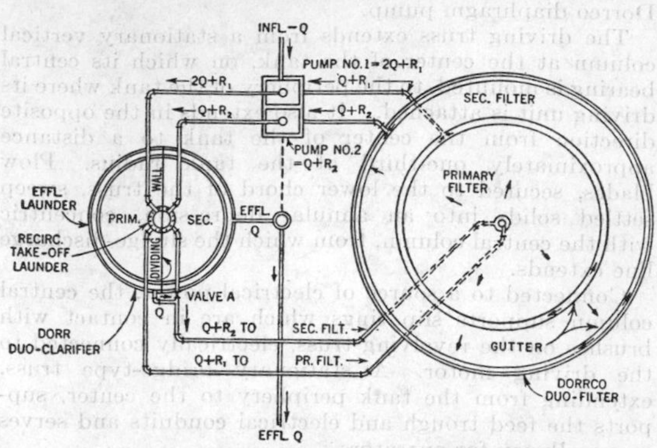

FIG. 32*A*. Clarifier-filter combination. Dorr Duo-Clarifier (left) and Dorrco Duo-Filter (right) provide a compact, two-stage biofiltration plant for the treatment of milk and biological wastes.

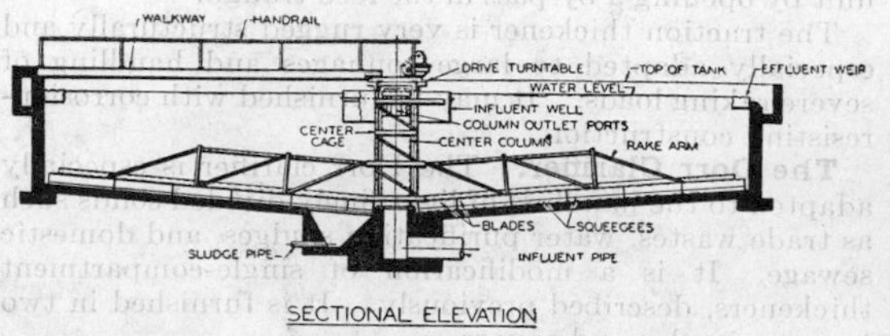

FIG. 32*B*. Sectional elevation of Dorr S-7 Clarifier. Diameter, 30 to 100 ft.; novel features are trussed-frame rake arms, longer and fewer rake blades, double overload protection.

weight, to their normal operating position. If the over, load increases, the scrapers raise near their maximum distance, sound an alarm, and cut off the driving motor.

The general arrangement of the Auto-Raise thickener is shown in Fig. 33. A support structure of I-beam or low-truss construction spans the top of the tank and supports the rotating mechanism.

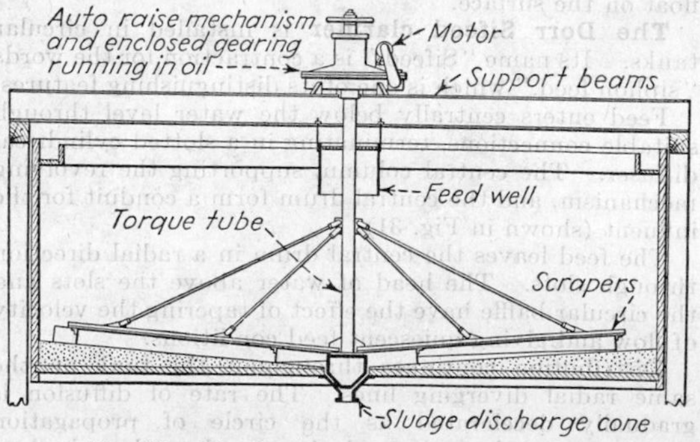

FIG. 33. The Hardinge Auto-Raise thickener.

The driving mechanism is compact and includes fully enclosed oil-lubricated gears and overload protection devices. The rotating mechanism is supported on a ring-type ball bearing which is designed to prevent any sway of the mechanism and which operates in an oil bath. The thickener is supplied in sizes from 6 to 100 ft. diameter, with full double spiral scrapers, with segmental scrapers and with scrapers of wood, stainless steel, or rubber-covered steel construction. Power requirements are low, a 1-hp. motor being ample for a 40-ft.-diameter machine.

Center Pier-type Hardinge Auto-Raise Thickener. For large thickening or settling tanks, the Hardinge Auto-Raise Thickener is arranged so that it is supported on the top of a center pier or stationary center column, as shown in Fig. 34. The rotating thickening mechanism and scrapers are supported from the top of the pier on a ring-type ball bearing. A bridge and walkway connect the side of the tank to the top of the pier and carry the thickener feed launder. The drive unit, located on top of the pier, consists of a spur pinion, meshing with a spur gear, supported on ball bearings, in an oil-filled housing, and the Auto-Raise functions by means of sloping drive brackets attached to the drive gear and operating against rollers at the top of the drive cylinder and torque frame.

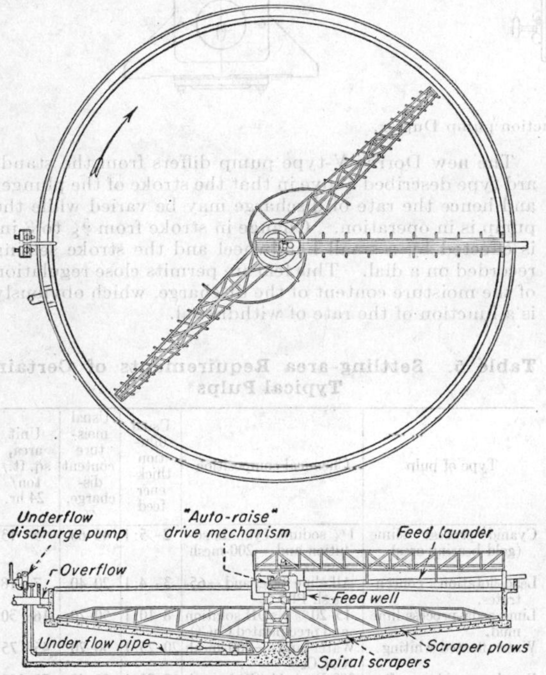

FIG. 34. The Hardinge thickener.

Large Hardinge 110-ft.-diameter Thickeners use spiral scrapers in the central area, approximately 50 ft. in diameter only. Ordinary plows are used to cover the area from the end of the double spirals to the end of the plow arm.

Hardinge Diaphragm Pump. Hardinge diaphragm pumps are used for the removal and control of sludge or pulp as underflow from Hardinge thickeners. The general arrangement of the pump is shown in Fig. 35.

The outstanding feature of the Hardinge diaphragm pump is the easy control attachment by which the stroke and capacity can be varied without stopping the pump. This easy control feature is a valuable one where it is desirable to closely regulate the moisture content of sludge coming from the thickeners, particularly in countercurrent decantation washing plants where a small decrease in the moisture content of the sludge discharge from each thickener adds substantially to the plant efficiency.

The constant-speed eccentric supplies the primary motion, and the variation in the movement of the diaphragm piston is obtained by moving the connecting rod attached to the eccentric along a lever arm.

The pump is supplied in 3- and 4-in. sizes, in simplex, duplex, and triplex arrangements, and arranged for belt or direct motor drive. It is also supplied in special materials of construction for the handling of corrosive mixtures. Usual speed is 50 r.p.m.

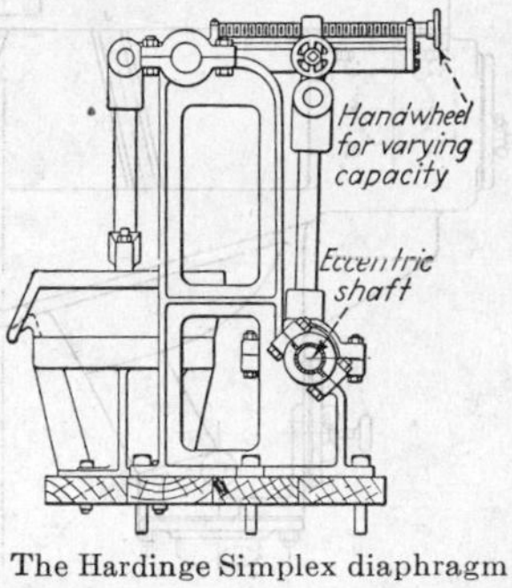

FIG. 35. The Hardinge Simplex diaphragm pump.

Dorr Thickener Types, Sizes, and Drive Ratings. Dorr traction thickeners are furnished in sizes ranging from 6 ft. in diameter to 375 ft. with motor-drive unit ratings ranging from ½ hp. for the smallest to 7½ hp. for the largest.

Dorr tray thickeners range in size from 10 to 75 ft. in diameter with any number of trays up to seven. The smallest units are driven by a ½-hp. motor, while the largest multitray thickeners require a 5-hp. motor.

Unit-type central-drive Dorr thickeners range in size from 6 to 200 ft. in diameter with motor-drive units ranging from ½ to 5 hp.

Dorrco Pumps. This pump is a thickener accessory which is virtually essential for satisfactory continuous operation. Two types are furnished. the suction type (Fig. 36) capable of lifting sludge against a head equivalent to 14 ft. of water, and the pressure type (Fig. 37), capable of operating against a pressure head equivalent to 45 ft. of water.

Table 4. Dorrco Pumps
Type VM

Size	Hp.	Capacity, cu. ft./min. Max.	Capacity, cu. ft./min. Min.	Weight	Stroke, in. Max.	Stroke, in. Min.
0	1/20	0.55	0.16	100	5/8	3/16
1	¼	1.70	0.41	250	1⅛	5/16
2	½	4.50	1.17	540	1¾	½
3	1	7.50	2.10	900	2¼	⅝
4	1½	12.20	3.30	1400	2¾	¾
6	3	30.00	9.20	3000	3¾	15/16
8	7½	75.00	12.00	5700	5½	⅞

Dorrco Pressure Pumps

Size and type	Width	Length	Height	Rating, motor-drive unit, hp.*	Stroke, in. Max.	Stroke, in. Max. recommended	Displacement,† cu. ft. water/min.
2 simplex.....	2′ 3″	2′ 4″	4′ 5¼″	2	2	1½	2.03
2 duplex......	2′ 3″	3′ 10″	4′ 5¼″	5	2	1½	4.06
2 triplex......	2′ 3″	5′ 4″	4′ 5¼″	5	2½	1½	6.09
3 simplex.....	2′ 9″	2′ 8″	5′ 3¼″	3	2½	2	3.60
3 duplex......	2′ 9″	4′ 4″	5′ 3¼″	5	2½	2	7.20
3 triplex......	2′ 9″	6′ 0″	5′ 3¾″	7½	3	2	10.80
4 simplex.....	2′ 9″	2′ 8″	5′ 3¾″	3	3	2½	5.72
4 duplex......	2′ 9″	4′ 4″	5′ 8¾″	5	3	2½	11.54
4 triplex......	2′ 9″	6′ 0″	5′ 8¾″	7½	3	2½	17.16
4 quadruplex..	2′ 9″	8′ 0″	5′ 8¾″	10	3	2½	22.88
4 quintuplex..	2′ 9″	10′ 0″	5′ 8¾″	10	3	2½	28.60

Size refers to inside diameter of pipes leading to and from each pump bowl.
* Actual power consumption about 50 per cent of motor rating.
† Discount displacements 20 per cent for 50 per cent moisture in sludge, and 30 per cent for 40 per cent moisture.

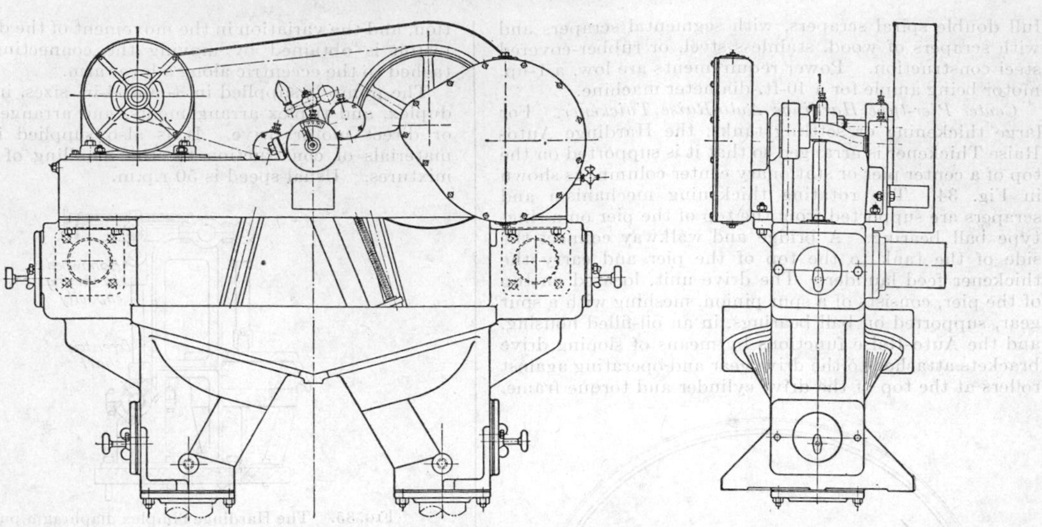

FIG. 36. The Dorrco suction pump Duplex.

Both suction and pressure pumps are of the diaphragm type, the diaphragm, of rubber-cord construction, being clamped rigidly around its periphery to the pump bowl by means of a metal retaining ring. Intake and discharge valves are of the ball type, constructed of rubber with a lead slug in the center to give the correct weight.

A connecting rod, driven from the pump shaft by means of an adjustable eccentric, actuates the diaphragm, the central portion of which oscillates while the periphery remains stationary. The volume of sludge displaced at each stroke may be varied at will from zero to the maximum for which the pump is designed, adjustment being made by varying the eccentricity by a handwheel on the eccentric.

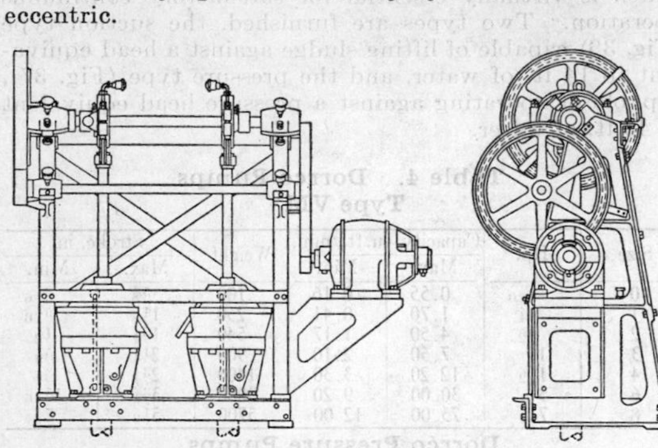

FIG. 37. The Dorrco suction pump.

Once adjusted to average conditions, the pump tends to maintain constant sludge density, since the more dilute the discharge becomes the smaller is the amount of solids removed so that the tendency is to bring the dilution back to normal. The final, fine adjustment of capacity is obtained by permitting a small amount of air to enter the pump bowl through a small needle valve connected thereto.

Dorrco pumps are supplied in four sizes, 1, 2, 3, and 4 in. and in five types, simplex, one pump body; duplex, tow pump bodies; triplex; quadruplex; and quintuplex. Pumps are arranged with tight and loose pulleys for belt drive or equipped with individual drive motors and double-reduction gears, or silent chains and sprockets for giving the required reduction in speed.

The new Dorrco V-type pump differs from the standard type described above in that the stroke of the plunger and hence the rate of discharge may be varied while the pump is in operation. Change in stroke from 5/8 to 3 in. is effected by a small handwheel and the stroke setting recorded on a dial. This feature permits close regulation of the moisture content of the discharge, which obviously is a function of the rate of withdrawal.

Table 5. Settling-area Requirements of Certain Typical Pulps*

Type of pulp	Chemical composition	Usual dilution thickener feed	Usual moisture content discharge, %	Unit area, sq. ft./ton/24 hr.
Cyanide-process slime (gold-bearing ores).	1% sodium cyanide solution and −200-mesh quartz	2– 5:1	45–60	5– 13
Lead-flotation concentrates.	Alkaline water and −65-mesh PbS	3– 4:1	20–40	7– 18
Lime-soda-process lime mud.	15–20% NaOH solution and precipitated $CaCO_3$	8–10:1	50–70	16– 30
Water-floated whiting..	Water and −300-mesh $CaCO_3$	20–25:1	50–70	45– 75
Bauxite residue, after H_2SO_4 digestion.	30° Be. $Al_2(SO_4)_3$ and fine silica	10–20:1	20–40	75–150
Water-floated clay.....	Water and 300–325-mesh clay	30–60:1	12–60	50–225

* The figures above are general averages for illustrative purposes only, since each material must be checked by tests before selection of size of machine required.

The unit settling area required for thickening pulp varies greatly, not only between pulps of different composition but also between pulps of seemingly identical composition. In general, pulps prepared from metal-bearing ores by wet grinding to a moderate mesh, 65 to 100, exhibit the most rapid settling rates and require the smallest unit areas. Pulps consisting of chemical precipitates suspended in a solution exhibit medium settling rates and medium unit areas. Pulps prepared from non-metallic minerals, ground to a fine mesh, 250 to 325, exhibit generally the slowest settling rates and require the largest unit area.

The variation in settling rate and unit-area requirement is even more striking in the case of seemingly identical pulps. In cyanide pulps, unit areas vary from as low as 2 sq. ft./ton/24 hr. for extremely granular solids to as high as 15 for claylike slimy solids.

Lime mud pulps ($CaCO_3$), precipitated in the lime-soda process of caustic soda manufacture, vary widely with respect to the unit areas required for thickening. Variations from 2 to 40 sq. ft./ton/24 hr. are common. The settling characteristics are determined not so much by the precipitate itself as by the physical conditions during causticizing, including time, temperature, speed of agitator, and strength of solution. In clay pulps, the unit areas vary from 5 to 225 sq. ft. dependent upon the type of clay, its physical character, the dilution of the pulp, and, finally and most importantly, the natural flocculating or deflocculating characteristics exhibited.

Selection of Type of Continuous Mechanical Thickener

The selection of the type of thickener for handling a given pulp is generally based on the following considerations:

1. Floor Space Occupied. The tray type gives the greatest capacity, settling area, and volume, per unit of floor space occupied.

2. Conservation of Heat. The tray type, easily covered and insulated, gives the least temperature drop between feed and overflow.

3. Large Raking Capacity and Structural Ruggedness. The traction type has the greatest raking capacity per unit of area and is especially adapted to handling difficult raking problems, since power is applied at the end of a long arm and deeper and larger plow blades are used than on other types.

4. Handling Corrosive Materials. The single-compartment (central-drive) type is best suited to handling acid and corrosive solutions, since the portion of the mechanism below the solution level may be constructed of any one of several materials, the efficiency of which has long been established for acid-resisting duty: *e.g.*, wood, lead-covered steel, hard lead, rubber-covered steel, or such alloys as Duriron, Pioneer metal, bronze, stainless steel, etc. The traction type may be furnished with certain corrosion-resisting materials.

5. Periodic Overloads. The torque type, with automatic self-raising and lowering arms, adjusts itself to the severity of the raking load.

Thickening Costs

1. Erection. The following average figures are suitable for preliminary estimates:

Erection of thickener mechanisms	$150/ton
Erection of thickener superstructure	$100/ton
Erection of thickener tanks (steel)	$100/ton
Erection of thickener tanks (wood)	$150/ton
Foundations and concrete tanks:	
Excavation	$1.25/cu. yd.
Concrete in place	$45/cu. yd.
Beams and joists (wood) in place	$75/1000 ft. b.m.
Columns and beams (steel) in place	$150/ton

2. Labor. This is a very small item as attention is generally confined to pump adjustment starting and stopping (not over once or twice per shift), and lubrication, requiring about 10 min. per day per thickener.

At the Phelps-Dodge Corp., Morenci branch, one laborer devoted 3 hr./24 hr. to the operation of one 200-ft. thickener, handling 2000 tons of tailings per day. At the Inspiration Copper Co., one man per shift at a wage of $4.40 operated eight 60-ft., three 80-ft., three 100-ft., and one 200-ft. thickeners (1924–1926).

At the average chemical plant operating four to six thickeners, one man and a helper operate the thickeners, as well as other equipment such as agitators, filters, etc., and, in addition, carry out routine analyses for chemical control.

3. Power. See figures given on p. 945 under Dorr Thickener Types, Sizes, and Drive Ratings. These figures refer to horsepower ratings of motor-drive units and should be discounted 50 per cent to give approximate power consumption during continuous operation.

4. Repairs and Supplies. Owing to slow speed of rotation and location of all wearing parts above solution level, repairs and supplies are virtually negligible. Whatever breakages do occur are generally due to faulty or negligent operation, resulting in severe overloads.

5. Maintenance on 12 Thickeners at the Inspiration Copper Company.

	Labor	Material	Total
1924	$25.49	$15.16	$ 40.65
1925	$81.55	$77.96	$159.51
1926 (6 mo.)	......	$ 8.40	$ 8.40
Total (2½ years)	......	......	$208.56

Table 6. Typical Thickening Costs from Practice

Plants	Daily tonnage	Thickeners installed	Cost per thickener per ton
Tonapah Extension Mining Co.	350	Four 30 × 10 ft.	$0.02
South American Development Co.	250	Six 39 × 12 ft.	$0.0115
Tom Reed Gold Mines	250	Ten 30 × 10 ft. Five 40 × 12 ft.	$0.0086
McIntyre Porcupine Gold Mines, Ltd.	1600	Four 50 × 10 ft. Ten 30 × 10 ft.	$0.0050
Inspiration Copper Co. (copper tailings).	9000	Seven 60 ft. One 80 ft. Three 100 ft. One 200 ft.	$0.0014
Inspiration Copper Co. (flotation concentrates).	400	Two 80 ft. One 60 ft.	$0.0374

Dorr Type S Clarifiers have shown extremely low repair and replacement costs. A recent survey of a considerable number of installations of all sizes in various services, mainly sewage- and water treatment plants over 10 years gave average repair and replacement costs of less than ½ of 1 per cent of the original purchase price f.o.b. factory.

The preceding description has been of thickening only. Addition of reagents and flocculation before thickening are required in many problems. Earlier practice employed equipment for flocculation as a separate step ahead of clarification and thickening. Among apparatus for this purpose are the Dorrco flocculator, the Jeffrey flocculator, and others. Recent developments are combination units which, in a single tank, provide for flocculation in separate cells preceding the clarification and thickening steps. Machines of this type are the Dorrco Clariflocculator and the Dorr multifeed clarifier, this type being widely used in the clarification of beet and sugar-cane juices.

Recent practice in water softening for large-scale use involves treatment so that the softening reactions take place in a sludge "blanket," resulting presumably in the development of fewer nuclei and the promotion of relatively dense, fast-settling particles. Machines using this principle are the Dorrco Hydro-Treator, the Spaulding Precipitator, and the Infilco Accelator.

Numbers of other machines have been developed for special clarification and settling problems, *e.g.*, those combining flotation and sedimentation principles. For further details the reference at the head of this article (Dorr and Lasseter), from which the above has been abstracted, should be consulted.

Filter Thickeners

The filter thickener, as the name implies, operates on a combination of the thickener and the filter principles. It consists essentially of a cylindrical or rectangular tank, equipped with a connection for the introduction of feed; a connection, generally in the tank bottom, for the discharge of sludge; and one or more submerged filter elements.

The solution is drawn through a filtration medium by vacuum or by gravity. The filtration medium, in some cases, consists of a multiplicity of fabric-filter elements, immersed in the pulp contained in the tank, and in other cases of a layer of sand or other granular substance laid upon a false bottom.

The submerged cake, forming upon the filtration medium, is periodically removed either by the application of low-pressure air or water on the reverse side of the medium or else by a mechanical scraping device, traveling over the medium.

The solid material in the pulp is in all cases discharged as a thick sludge, rather than as a filter cake. This sludge collects in the bottom of the tank and is discharged either by a spigot or by some type of pump.

The Oliver-Borden Filter Thickener. This thickener (Fig. 38) consists of a two-compartment steel tank 1; of a rectangular horizontal cross section, each compartment having a V-shaped bottom 2; fitted with a multi-bladed impeller mixer.

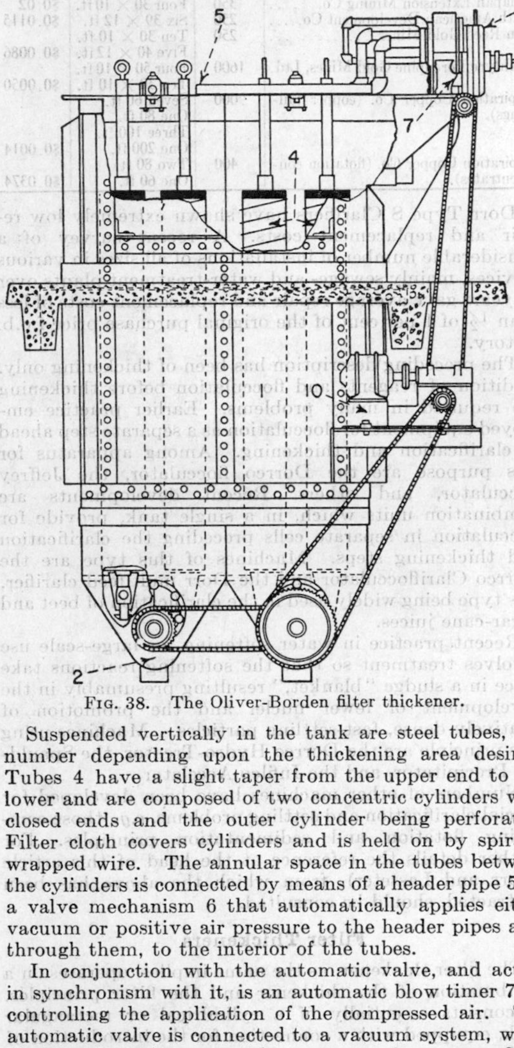

FIG. 38. The Oliver-Borden filter thickener.

Suspended vertically in the tank are steel tubes, the number depending upon the thickening area desired. Tubes 4 have a slight taper from the upper end to the lower and are composed of two concentric cylinders with closed ends and the outer cylinder being perforated. Filter cloth covers cylinders and is held on by spirally wrapped wire. The annular space in the tubes between the cylinders is connected by means of a header pipe 5, to a valve mechanism 6 that automatically applies either vacuum or positive air pressure to the header pipes and, through them, to the interior of the tubes.

In conjunction with the automatic valve, and acting in synchronism with it, is an automatic blow timer 7 for controlling the application of the compressed air. The automatic valve is connected to a vacuum system, while the blow timer is connected to a source of compressed air, both the vacuum and the air pressure being controlled by regulators.

Solution samplers are provided so that clarity of filtrate from each group of tubes can be readily determined.

The V-shaped bottoms of the thickener tank, carrying the impeller mixers, are each fitted with a gooseneck pipe outlet leading from the center of the bottoms and controlled by means of a throttle valve.

The automatic valve 6, the blow timer 7, and the impeller mixer are all driven through a roller-chain drive, by a motor 10, connected to a speed reducer.

To facilitate the placing of the tubes of the tank and their removal therefrom when necessary, an overhead track and trolley with light chain block are used.

Hardinge Sand-filter Clarifier. The Hardinge sand filter is used for clarification operations where a crystal-clear liquid product is desired. In many cases it is used for the clarification of overflows from settling tanks or thickeners where the desired clarification cannot be economically accomplished by sedimentation methods.

The general arrangement of the Hardinge sand filter is shown in Fig. 39. The sand-filter bed in the bottom of a round wood, steel, or concrete tank is supported on a wooden false bottom or a gravel drainage bottom. A steel truss carries the sand-cleaning mechanism, which consists of a central vertical shaft suspended and operated from the truss and having spiral-shaped scrapers attached to its lower end. The threaded shaft is supported from a threaded ratchet, which can be used for lowering the scrapers, which are revolved by a worm-driving gear keyed to the shaft below the ratchet. As the scraper revolves, it moves the material caught on the surface of the sand bed to the sludge-discharge well; and, if it is lowered an infinitesimal amount, a thin layer of the sand bed is scraped to the center with the mud. The sludge is drawn off through a spigot on the end of the sludge line.

The capacity varies: $\frac{1}{30}$ gal./(sq. ft.)(min.) is obtained on 50 Be. caustic liquor; $\frac{1}{3}$ gal. on brine; and 1 gal. on gasoline.

Operating Principles. The four factors controlling the filter rate, in the order of their importance, are as follows:

1. The quality of suspended solids.
2. Quantity of suspended solids.
3. The viscosity of liquid.
4. Hydrostatic head or vacuum used.

If the suspended solids are semicolloidal, the surface of the sand may become coated so rapidly by an impermeable layer that the rate will decrease very quickly and the scrapers may have to be operated continuously to keep the surface of the sand bed clear. On the other hand, if the suspended solids are crystal, then a fairly high continuous filtration rate can be maintained for many hours, and a scraper may be operated and the filter bed cleaned once every 8 or 24 hr.

The velocity of the liquid through the filter must be retarded sufficiently to prevent dragging the suspended particles into or through the voids in the filter bed, and the filter rate must be controlled so that the suspended solids will be caught at the upper opening of the voids and the solids will be removed by the cleaning scrapers. In some operations vacuum is supplied underneath the filter bed by a centrifugal pump, which serves to remove the filtrate as well as create the vacuum.

Hardinge Automatic Backwash Sand Filter. In conjunction with various sedimentation operations, the settling tank effluent may be passed through a Hardinge Automatic Backwash Sand Filter to obtain improved clarification. This unit is used particularly in conjunction with the treatment of waste waters and the treatment of industrial-water supplies.

The general arrangement of this filter is shown in Fig. 40. The filter consists of a compartmented sand bed with under-drains and with a suitable backwash cleaning mechanism traveling on tracks on the sidewalls of the filter tank. It is supplied in sizes varying from 3 to 16 ft in width and in length up to 100 ft.

The filter bed is divided into a series of narrow compartments, each connected to a separate effluent and backwash port. The cleaning mechanism consists of a

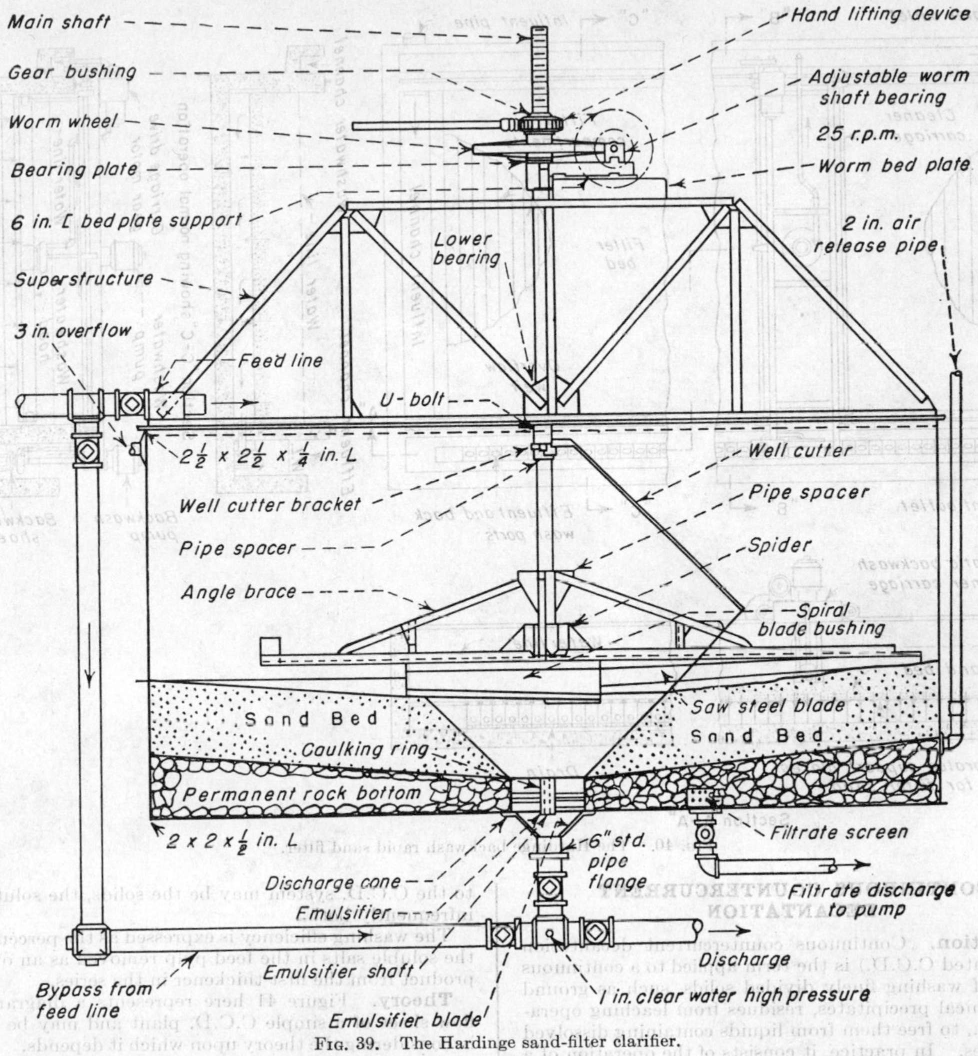

FIG. 39. The Hardinge sand-filter clarifier.

motor-driven carriage, which is equipped with two pumps; a backwash pump, which takes the filtrate from the effluent channel and pumps it through the backwash port and up through the sand bed; and a wash-water pump, for removing the dirty wash water which is collected inside of a hood over the compartment being washed. Filter heads vary from 6 to 10 in. and as the head in the filter tank reaches a predetermined level, an electrical contact starts the washing carriage and pumps. The wash water is delivered into a wash-water launder alongside the filter tank.

One installation on industrial waste water reduces the suspended matter in the overflow from the sedimentation unit from 190 p.p.m. to 17 p.p.m. A similar unit, treating an industrial water supply, removes 97.4 per cent of the suspended matter from a feed carrying 11 p.p.m.

Briggs Filter Thickener. The Briggs thickener is a tube-type machine but differs from similar thickeners in that the tubes depend from a rotated header and are moved about on a circular path through the tank containing the slurry to be thickened. A manifold is connected to the source of vacuum during most of the cycle and is then automatically switched to a source of backwash liquid. The motion of the tubes is regarded as the principal feature of this thickener and is that mainly claimed as the superior feature. This is because, it is claimed, the motion permits the cake to slough off the tubes after it has been loosened at the seam by the backwash.

A minimum amount of dilution is claimed and a complete removal of the solids. The speed of rotation of the tubes is critical in its operation; it must be great enough to give the sloughing but not enough to stir up the thickened solids at the bottom of the tank. The tank has a conical bottom from which the accumulated solids may be pumped continuously or discontinuously.

A paddle agitator can be installed near the cone end if desired. The thickener usually operates at a vacuum of about 22 to 25 in. A low-capacity pump can be used, since there is a very little air leakage: 0.5 cu. ft./(sq. ft.)(min.); 0.1 gal. filtrate is required per sq. ft. of tube area, and this amount of filtrate is backwashed over an interval of about 1 sec. Sizes and prices are available from the Briggs Filtration Co.

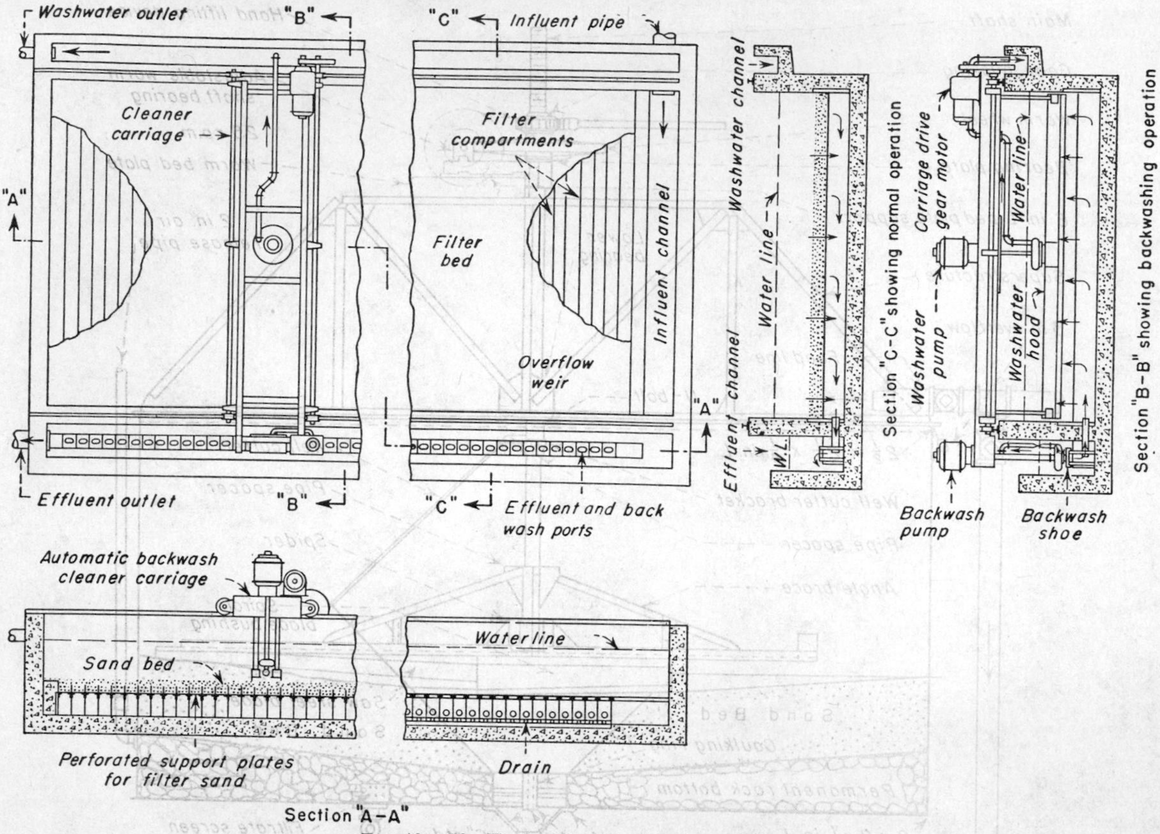

FIG. 40. The Hardinge backwash rapid sand filter.

CONTINUOUS COUNTERCURRENT DECANTATION

Definition. Continuous countercurrent decantation (abbreviated C.C.D.) is the term applied to a continuous system of washing finely divided solids, such as ground ore, chemical precipitates, residues from leaching operations, etc., to free them from liquids containing dissolved substances. In practice, it consists of the operation of a series of continuous thickeners so that the solids to be

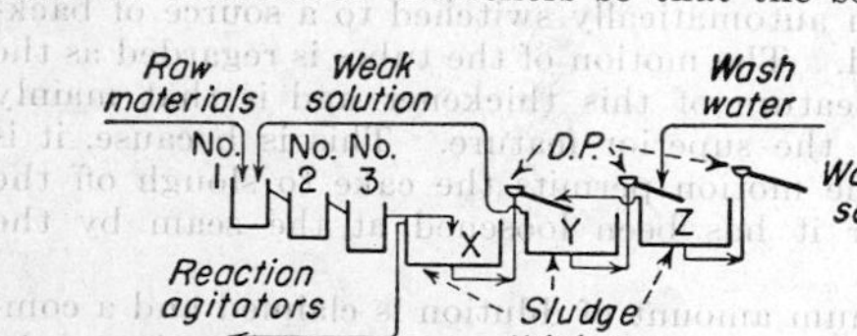

FIG. 41. Simple C.C.D. flow sheet.

washed pass through them in series, being diluted after each settling by a weaker liquid overflowing from subsequent thickeners in the system and flowing in the opposite direction.

Purpose. It is the purpose of C.C.D. to attain a high washing efficiency (separation of soluble materials from insoluble materials) with a minimum number of decantations and with the use of minimum amount of wash liquid. In practice, the desirable portion of the pulp fed to the C.C.D. system may be the solids, the solution, or infrequently both.

The washing efficiency is expressed as the percentage of the soluble salts in the feed pulp removed as an overflow product from the first thickener in the series.

Theory. Figure 41 here represents a diagrammatic flow sheet of a simple C.C.D. plant and may be helpful in considering the theory upon which it depends.

The chemical reaction taking place in the three continuous agitators results in the formation of a pulp consisting of a concentrated solution and insoluble solids. It is desired to recover both solution and solids in as pure form as possible and with the least possible reduction in the concentration of the solution.

This pulp is sent to the first of the three continuous thickeners, X, Y, and Z, arranged in series as shown. The clear concentrated solution overflows to further treatment, while solids settle on the bottom in the form of a thick sludge and are removed by a diaphragm pump.

The sludge pumped from thickener X is repulped with weak wash solution overflowing thickener Z, and the pulp so formed is rethickened in thickener Y. The overflow from thickener Y, weaker than the original solution in thickener X but more concentrated than that in thickener Z, is used in place of fresh water in the agitation step, thus conserving the soluble salts contained in it. The sludge pumped from the thickener Y is repulped with fresh water, and the pulp so formed resettled in thickener Z. The weak solution overflowing from thickener Z flows to the trough feeding thickener Y, being there used for repulping the sludge from thickener X. The sludge discharged from thickener Z is finished washed solids. The overflow from thickener X is the finished concentrated solution.

In C.C.D. operation the solids move in a direction countercurrent to the wash solution and are progressively impoverished in soluble-salt content by coming in contact with progressively weaker wash solutions and finally with fresh water. Similarly, the wash water flowing in a direction opposite to that of the solids becomes progressively enriched in soluble salts by coming in contact with solids containing greater and greater amounts of soluble salts.

The efficiency of a C.C.D. system is dependent, first, upon removing the settled solids from each thickener with the minimum amount of solution, *i.e.*, at greatest or final density; and second, upon obtaining thorough repulping or remixing of the wash solution and sludge between thickeners and prior to resettling. When these two conditions are satisfied in plant operation, the washing efficiency will be that predicted from C.C.D. calculations, discussed later.

Use of C.C.D. in Chemical Processing. This method is applicable to all problems in chemical-engineering practice wherein a pulp, consisting of finely divided, "settleable" solids and a solution, is, first, to be separated into its two constituents; and, second, the solids are to be washed for the purpose of either recovering the solution entrained by them or for the purpose of purifying the solids by the removal of the contaminating solution. The objective in any case is the maximum recovery of solution and solids, each contaminated to the least possible extent by the other.

Examples. Aluminum Sulfate. Bauxite ore, containing Al_2O_3, digested with sulfuric acid, yields a solution of aluminum sulfate [$Al_2(SO_4)_3$], in which there is suspended insoluble rock residue, chiefly silica. C.C.D. treatment of this pulp yields a strong aluminum sulfate solution which is concentrated and crystallized to form commercial "alum," and a washed silica residue which may be discarded virtually free from the valuable aluminum salt.

Caustic Soda. Soda ash (Na_2CO_3) solution, causticized with lime, yields a solution of caustic soda in which there is suspended precipitated calcium carbonate ($CaCO_3$). C.C.D. treatment of this pulp yields a strong caustic soda solution and a washed calcium carbonate.

Phosphoric Acid. Phosphate rock, containing P_2O_5, digested with sulfuric acid, yields a solution of phosphoric acid in which there is suspended finely divided calcium sulfate ($CaSO_4.2H_2O$) precipitate. Both the acid and the synthetic gypsum being of value, continuous countercurrent decantation gives two products: first, a strong phosphoric acid and, second, a washed gypsum sludge suitable for the manufacture of building materials.

C.C.D. Calculations. The washing efficiency, purity of washed solids, and concentrations of solutions may be

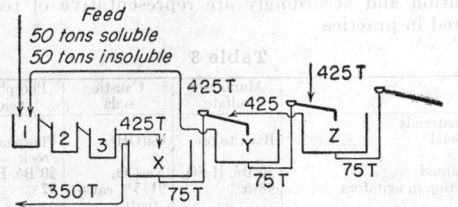

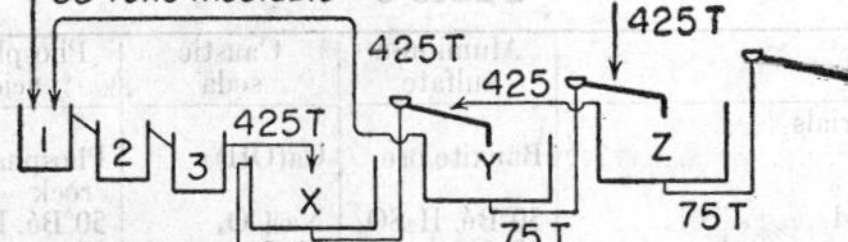

FIG. 42. Flow-sheet setup for C.C.D. calculations.

determined by simple calculations, if reliable data are available on the character of the pulp to be handled, the amount of wash water available, and the number of thickeners to be used. The diagrammatic flow sheet (Fig. 42) and the accompanying calculations will serve to illustrate the method of procedure.

Given. (1) 100 tons per day of 50 per cent soluble material to be leached with 425 tons of water; (2) insoluble residue settles to 60 per cent moisture.

To Find. (1) Solution concentration in all thickeners; (2) washing efficiency with three thickeners (C.C.D. system); (3) percentage soluble in washed residue.

Calculation for Flow-sheet Tonnages.

Solids with sludge from each thickener = 50 tons (given)

Water with sludge from each thickener

$$= 50 \times {}^{60}\!/_{40} = 75 \text{ tons}$$

Overflow 3d thickener

$$= 425 + 75 - 75 = 425 \text{ tons } H_2O$$

Overflow 2d thickener

$$= 425 + 75 - 75 = 425 \text{ tons } H_2O$$

Feed 1st thickener = 425 tons

Overflow 1st thickener

$$= 425 - 75 = 350 \text{ tons } H_2O$$

Calculations for Solution Concentrations.

a. Let X, Y, and Z represent pounds of soluble salts per ton of water in the respective thickeners.

b. Then equating pounds of dissolved salts into and out of each thickener, set up the following simultaneous equations:

Thickener 1	$350X + 75X = 425Y + 100{,}000$ lb.
Thickener 2	$425Y + 75Y = 425Z + 75X$
Thickener 3	$425Z + 75Z = 425 \times 0 + 75Y$

c. Solving by substitution:

$$X = 282.80 \text{ lb. soluble salts/ton}$$
$$Y = 47.83 \text{ lb. soluble salts/ton}$$
$$Z = 7.17 \text{ lb. soluble salts/ton}$$

d. Solution strength:

Thickener 1 =	282.80 lb./2000 lb. water =	12.4 per cent
Thickener 2 =	47.83 lb./2000 lb. water =	2.3 per cent
Thickener 3 =	7.17 lb./2000 lb. water =	0.36 per cent

e. Washing efficiency:

Washing efficiency

$$= \frac{\text{salt recovered}}{\text{salt in feed}} \times 100$$

$$= \frac{350 \times 282.80}{100{,}000} \times 100 = 98.98 \text{ per cent}$$

f. Percentage of soluble material in washed sludge:

Water with sludge

$$= 75 \text{ tons}$$

Soluble salts with sludge

$$= 75 \times 7.17 = 537.75 \text{ lb.}$$

Solids with sludge

$$= 50 \times 2000 = 100{,}000 \text{ lb.}$$

Soluble salt content

$$= \frac{537.75}{100{,}000 + 537.75} \times 100 = 0.535 \text{ per cent}$$

Various Types of C.C.D. Flow Sheets. Although the flow sheet described above is the one most generally used in C.C.D. operations, various types of related flow sheets have been developed for special cases.

1. The Double-reaction Flow Sheet (Fig. 43). With this arrangement, reactions such as leaching, digestion, or causticizing take place in two stages with an intermediate stage of thickening to effect a change of solution. It facilitates the rapid removal of finished solution and subjects the incompletely treated solids to additional agitation under carefully controlled conditions.

It is widely used in the cyanidation of gold-bearing ores, the first stage of leaching taking place in the wet-grinding mills which grind the ore in cyanide solution. When used in double-acid digestion, the acid added in the first stage is less than enough to complete the extraction and in the second stage more than enough to complete the extraction of the remaining soluble materials, thus assuring (1) a basic final solution and (2) a high recovery of insoluble material. In the double

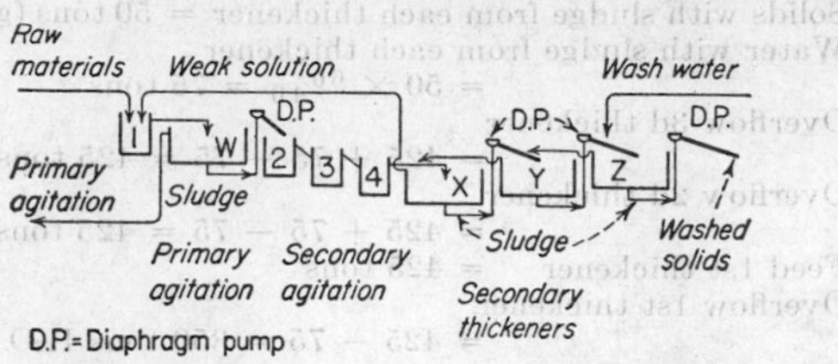

FIG. 43. Double-stage reaction flow sheet.

causticization of sodium carbonate solution, an excess of lime is used in the first stage and an excess of sodium carbonate in the second. This results in (1) a rapid and relatively complete reaction in the first stage due to the presence of excess lime and (2) a relatively complete utilization of this excess lime through the subsequent reagitation of the sludge with more than sufficient sodium carbonate to satisfy the reaction requirements.

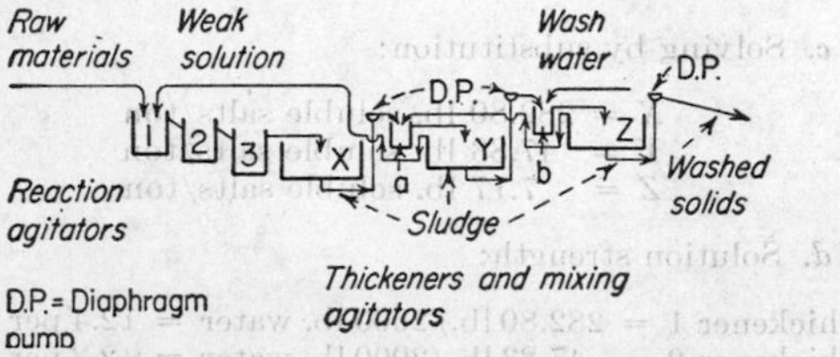

FIG. 44. Intermediate-agitation flow sheet.

2. The Intermediate-agitation Flow Sheet (Fig. 44). Certain residues from reactions exhibit adsorptive powers to such an extent that the mixing of sludges and wash solutions, as generally carried out in repulping troughs, does not result in the usual displacement of soluble materials. In such cases, small agitators are placed between all thickeners so that longer and more violent repulping may release the soluble materials adsorbed by the solid constituent of the sludge being washed.

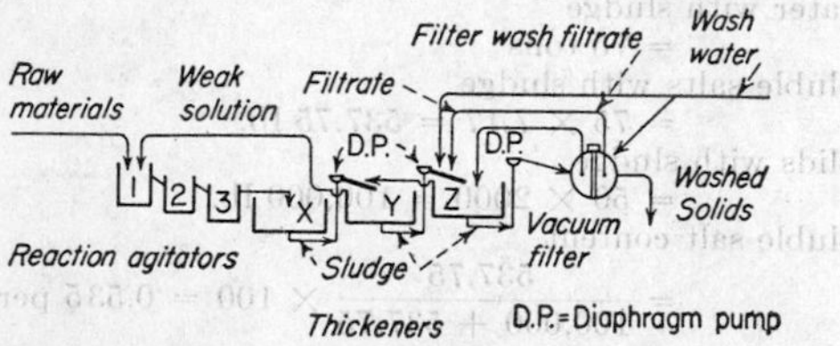

FIG. 45. Flow sheet with filter at end.

3. Continuous-countercurrent-decantation Flow Sheet with Filter at End (Fig. 45). When the solids are of value and are to be delivered in as moisture-free condition as possible, a continuous vacuum filter may be placed at the end of the series of thickeners. If desired, the filter may be arranged for cake washing in which case all or a portion of the water used in the C.C.D. series may be applied through the filter.

In such a case the filtration cycle is divided into two parts: (1) straight dewatering of the thickened sludge and (2) washing of the filter cake. The filtrates are kept separate because of their difference in soluble-salt content and are disposed of in different ways for this same reason. The first filtrate, being of the same concentration as the last thickener overflow, is returned to the last thickener loading well, eventually to overflow with the balance of the solution in this thickener and be used as a dilute wash solution at a preceding stage. The second filtrate, consisting of the filter wash water, containing only traces of soluble materials displaced from the already well-washed sludge, is utilized as a weak wash solution and is applied to the sludge from the next to the last thickener before the last stage of thickening.

Economic Advantages of C.C.D. The use of a series of continuous agitators and thickeners instead of a number of intermittent agitation and settling tanks has made possible certain distinct operating economies not possible otherwise. The advantages may be subdivided into two classifications, tangible and intangible.

1. Tangible Advantages:

a. Less operating labor.
b. Less steam for heating.
c. Higher washing efficiency, *i.e.*, extraction and recovery.
d. Higher temperature of finished liquor.
e. Higher concentration finished liquor.

Example. The following actual example is taken from an "alum" plant where C.C.D. replaced batch digestion of rock and washing of residue with savings of the following order:

Table 7

	C.C.D. plant	Batch plant	Annual saving
Labor	1 man per shift	3 men per shift	$ 7,776
Steam in reaction			5,760
Over-all recovery, available Al_2O_3	97.5%	90%	13,406
Temperature of finished liquor and saving of evaporator steam	90°C.	30°C.	6,660
Concentration of finished liquor and saving evaporator steam	35°Be.	30°Bé.	9,000
Total			$42,602

Data used in above comparison were as follows:

(1) Capacity, 100 tons "alum" (17 per cent Al_2O_3) per day.
(2) Labor, 48¢/hr.
(3) Bauxite, 50 per cent available Al_2O_3 at $14.25/ton.
(4) Steam at 37¢/1000 lb.

Intangible Advantages:

a. Less tanks and less floor area occupied.
b. Reduction of human factor in efficient operation.
c. Simple chemical control through routine analyses of first thickener overflow and last thickener discharge.
d. Reduction of "unaccounted losses" since finished product can leave system only at one point.

Typical Operating Figures from C.C.D. Practice. The data presented below are averaged from good plant operation and accordingly are representative of results secured in practice.

Table 8

	Aluminum sulfate	Caustic soda	Phosphoric acid
New materials			
a. Solid	Bauxite ore	$Ca(OH)_2$	Phosphate rock
b. Liquid	50°Bé. H_2SO_4	Na_2CO_3	50°Bé. H_2SO_4
Extraction in agitators	98.5%	91.5% causticity	97%
Strength of finished solution	35°Bé., hot	15.8°Bé., hot	30°Bé. (22% P_2O_5)
Temperature of finished solution	90°C.	74°C.	
Number of agitators	4	3	3
Number of thickeners	3	3	6
Washing efficiency	99%	99.3%	99%
Over-all recovery	97.5% of available Al_2O_3	....	96% of available P_2O_5

Complete C.C.D. System in a Single Unit. The Dorr washing type of tray thickener (Fig. 44) differs from tray thickeners previously described in that in this case the several superimposed settling compartments operate in series, just as do the thickeners in a C.C.D. system, and not in parallel, as do the compartments of the other types of tray thickeners. The feed pulp is introduced into the first or uppermost compartment, and strong solution overflows from this compartment, while wash water is introduced into the last (lowest) compartment.

In this thickener, all five tray compartments operate in series to give five stages of countercurrent washing. This unit is equivalent in capacity and washing to five separate thickeners series-connected in the conventional manner. In Fig. 46 the heavily shaded area represents

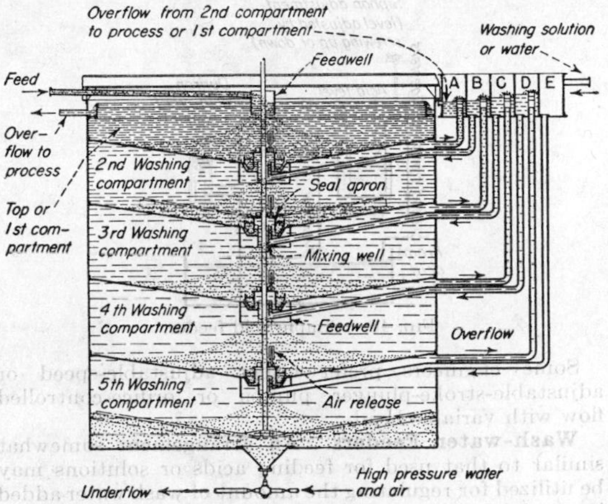

FIG. 46. The Dorr washing-tray thickener.

settled sludge in the various compartments. The more lightly shaded areas denote the solutions; the heavier the shading the stronger the solution, and vice versa.

Feed enters the top compartment via a central semi-submerged feed well. Strong solution overflows into a peripheral launder. Solids settle to the bottom, are raked to the center, and flow into the top of the second compartment, directly below, via an inverted cup and seal.

Integral with the sludge seal is a downcast boot which projects into the second compartment and serves as a mixing well. Wash solution enters here to repulp and dilute the settled sludge just prior to its being thickened again in the second compartment.

The wash solution used in the second compartment is the overflow from the third compartment and is controlled in section *B* of the overflow box. The overflow from the second compartment is controlled in section *A* of the overflow box and is generally returned to the agitator or other processing steps and sometimes to the first compartment. Settled sludge passes from the second to the third compartment as before, being repulped and diluted on the way with a weak overflow solution from the fourth compartment.

This operation is repeated twice more, first in the fourth compartment and then in the fifth, or bottom, compartment. In each case, as before, the sludge is repulped with a weak wash solution overflowing from a later stage of decantation. In each case, after thickening, the sludge flows through the seal to the next stage of washing, and the overflow is collected and utilized for washing purposes in an earlier stage of decantation.

Solids pass downward, compartment by compartment, against a counterflow of wash solution and finally of fresh water. Thus, in accordance with the basic C.C.D. principle, the solids transfer their dissolved values to the wash solution and finally are removed by a diaphragm pump from the bottom compartment virtually free from values. Similarly, the wash water, as it flows through the system, becomes increasingly enriched in dissolved values until it overflows from compartment 2 and returns to process.

The washing type of thickener is especially adapted to relatively small C.C.D. operations where floor space is limited for a multithickener washing series. It is well adapted to the handling of hot solutions since it is easily insulated against temperature drop. It is regularly supplied in diameters up to 60 ft. and depths up to 40 ft., this depth corresponding to a machine with five compartments.

Economic Advantages of Washing-tray Thickener Compared with a Multithickener C.C.D. Series. A comparison between a five-thickener C.C.D. plant and its equivalent in the form of a single, five-compartment washing-tray thickener shows the following points of advantage in favor of the latter:

	C.C.D. plant	Tray thickener	Saving, per cent
Floor area	8,439 sq. ft.	2,615 sq. ft.	69
Exposed radiating surface	21,128 sq. ft.	12,367 sq. ft.	41
Building volume	270,929 cu. ft.	122,126 cu. ft.	55
Diaphragm pumps	5	2	60
Foundations	5	1	80
Power consumption	10	3	70

Typical Washing-tray Thickener Operating Data. At Cactus Mines Co., Mojave, Calif., there are two washing plants operating on identical feeds and discharging to identical further processing. The feeds are gold flotation tailings cyanided in a continuous agitation series. One washing system, receiving 40 per cent of the flow, consists of four single-compartment thickeners—24 ft. in diameter by 10 ft. deep, arranged in the conventional C.C.D. manner. The other system, receiving 60 per cent of the flow, is a four-compartment washing-tray thickener, 35 ft. in diameter by 27 ft. deep. Flows to each are proportional to the settling areas provided, which in turn determine capacity.

The two tables that follow are from Johnson, *Am. Inst. Mining Eng., Tech. Pub.* 1082, 1939. They show that the solution strengths are approximately the same in the corresponding steps of the two systems and that the over-all recovery of dissolved gold, which is really the washing efficiency, is only ½ per cent less in the tray unit than in the individual thickener unit. The installed cost of the tray unit per ton of capacity was 25 per cent less than for the individual thickener unit—a more than sufficient saving to offset the slightly lower recovery secured.

Table 9. Comparative Solution and Residue Values
Assays, oz. Au per ton

	Thickener unit	Tray unit
Overflow solution 1	0.0344	0.0340
Overflow solution 2	.0147	.0146
Overflow solution 3	.0084	.0077
Overflow solution 4	.0033	.0036
Underflow residue 4	.0196	.0200
Underflow solution 4	.0037	.0046

Accessories. Successful operation of a C.C.D. plant depends to a large extent on the continuous feeding of measured volumes of raw materials to the agitators and wash water to the last thickener in the series and to the intimate mechanical mixing of solutions and sludges in the repulping troughs between thickeners. Certain accessory devices have been developed to satisfy these essential requirements.

Table 10. Comparative Efficiencies of the Two Units

C.C.D. Thickeners and Repulpers	
Washed tails No. 4 thickener..	0.0196 oz. = $0.686/ton ore
Underflow solution No. 4 thickener..........	0.0037 oz. = 0.1295
Soluble loss per ton of ore.....	1.26 × 0.0037 = 0.163
Total gold dissolved per ton:	× $35
Agitator No. 1 heads.....	0.16 oz.
Agitator No. 3 residue....	0.023 oz.
Dissolved in agitators.....	0.137 oz. = $4.795
Gold dissolved in C.C.D. (0.023–0.0196) × $35	= 0.119
Total dissolved.........	$4.914
Soluble loss............	0.163
Total dissolved gold recovered per ton..........	$4.751
Percentage recovered.........	$\frac{\$4.751}{4.914} \times 100 = 96.6$ per cent
Washing Tray Thickener	
Washed tails No. 4 compartment..........	0.020 oz. = $0.700
Underflow solution No. 4 compartment........	0.0046 oz. = 0.161
Soluble loss per ton of ore....	1.19 × 0.0046 = 0.1916
Total gold dissolved per ton:	× $35
Agitators.............	$4.795
Dissolved in washing thickener (0.023–0.02) × $35.......	= 0.105
Total dissolved.........	$4.900
Soluble loss..........	0.1916
Total dissolved gold recovered per ton..........	$4.7084
Percentage recovered.........	$\frac{\$4.7084}{4.9000} \times 100 = 96.1$ per cent

1. Feeders for Rock or Other Solid Substances. Feeders must be accurate to at least 1 per cent, adjustable to deliver a moderate range of tonnages, able to start easily under load, and easily operated and adjusted by common labor.

The belt-type weigh meter (Fig. 47) has been found very satisfactory in C.C.D. practice. The material is

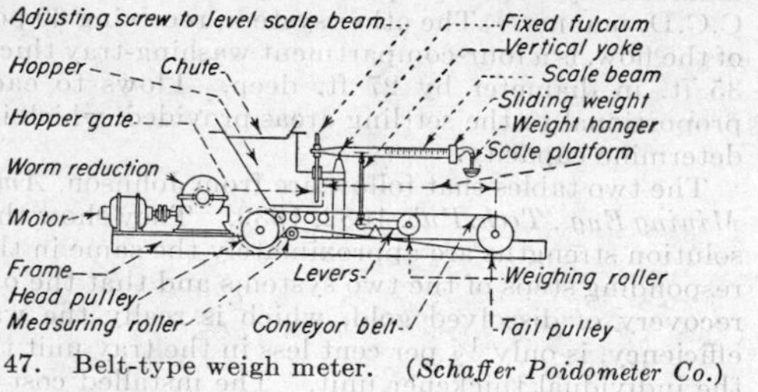

FIG. 47. Belt-type weigh meter. (*Schaffer Poidometer Co.*)

drawn from the feed hopper in the form of a continuous ribbon by a short endless belt located at the bottom of the hopper. The drive pulley is connected to a constant-speed motor. The belt passes over rollers balanced by an adjustable weighing element, which in turn actuates the sliding gate at the hopper opening through which the loaded belt issues. Once the weights on the beam of the weighing element have been set at a certain point, the machine continues to deliver the tonnage or poundage desired. The more closely the material is sized the greater is the accuracy, and vice versa. In the case of −100-mesh, dry, pneumatically sized bauxite, the machine is said to be accurate to within 0.25 per cent.

Other feeders of reliability and fair accuracy are the belt feeder, apron (or pan) feeder, rotary-drum feeder, revolving-disk feeder, reciprocating-plunger feeder, etc.

2. Solution Feeders. There are various feeders on the market for this purpose, namely, those of the adjustable displacement plunger-pump type, revolving bucket-wheel type, and weir (or V-notch) type many of which may be supplied in materials resistant to corrosive solutions.

The arrangement shown in Fig. 48 is simple in construction and operation and is sufficiently accurate for feeding acid or other chemicals to a C.C.D. plant.

Referring to Fig. 46, acid, stored in a lead-lined tank, is continuously kept in circulation by an acid-resisting plunger pump which delivers the acid to a small lead feed tank with a gravity-return line to maintain a constant acid level therein. A lead siphon, adjusted by a handwheel, feeds the required amount of acid to the agitators, the rate of feed for different positions of the siphon being determined experimentally and the results transferred to a calibration scale for the use of the operator.

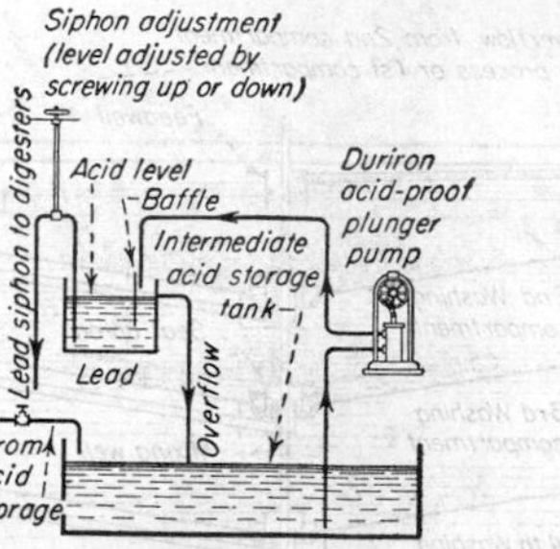

FIG. 48. Simple acid feeder.

Some engineers prefer either adjustable-speed or adjustable-stroke-plunger pumps or orifice-controlled flow with variable head.

Wash-water Feeders. An arrangement somewhat similar to that used for feeding acids or solutions may be utilized for regulating the amount of wash water added in the last thickener of the C.C.D. series. In this case, the intermediate storage tank is supplied with water by a float-controlled valve so that virtually no attention is required.

V-notch meters are sometimes used, these being mounted in boxes attached to, and connected with, the small tank. The overflow-return pipe in the feed tank is vertical and passes through the tank bottom. Removable adjusting rings of various widths may be placed on the top of the vertical overflow pipe to change the water level and accordingly the amount of water passing the V-notch per unit of time.

Heat-exchange coils may be submerged in the feed tank where it is desirable to maintain the contents of the agitators and thickeners at an elevated temperature. Condensate or low-pressure exhaust steam is generally used as the heating medium.

Modern developments of recording and controlling meters have rendered the old V-notch system obsolete for chemical work. Most of these instruments work on the principle of the differential pressure on the two sides of an orifice and will do anything from recording and controlling of flow including compensation for variable head to the proportionate mixing of flows from variable head sources.

Mixing between Thickeners. Complete mixing of the thickened sludge from one thickener with wash solution from another is essential before resettlement of the mixture so formed. Incomplete mixing takes place in the inclined feed troughs, even if baffles are used, and accordingly mechanical agitation is needed.

The Dorrco repulper (Fig. 49) consists of a trough of square or rectangular cross section, located with one end close to a sludge pump in the C.C.D. series and extending in the general direction of the thickener into which the repulped sludge is to be fed. On top of this trough, and

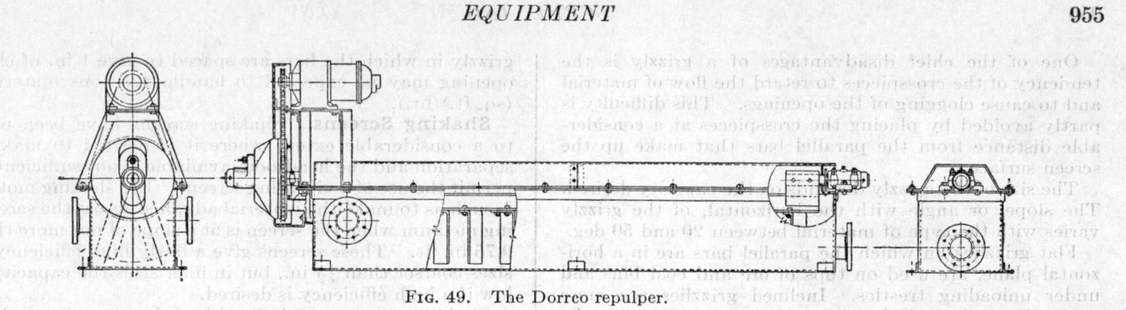

FIG. 49. The Dorrco repulper.

running parallel with it, is a shaft to which paddles are attached. Through a crank at one end of the shaft an oscillating motion is imparted to the paddles. The paddles are set at an angle and staggered so that, in addition to providing thorough mixing, they convey the pulp toward the discharge end of the repulper.

The repulper is driven with an individual motor-drive unit. The trough may be set at a slope of ¼ in./ft. instead of the ¾ in./ft. required with the non-agitated trough, thus saving a great deal of headroom in a multi thickener C.C.D. plant. It is supplied in standard and acid-resisting construction in lengths up to 50 ft.

SCREENING

BY C. P. CABELL

REFERENCES: Anon., Slurry Plant Uses Vibrating Screen for Dewatering, *Coal Age*, **50**, 95–96 (March, 1945). Atkinson, Black Heat Speeds Clay Screen, *Elect. West.*, **87**, 56 (1941). Banning and Lamb, Vibrating Screen Surface for the Removal of Flat and Elongated Pieces from Crushed Stone, *U.S. Bur. Mines R. I.* 3781, 1944; Directions for Laboratory Mineral Sizing, *U.S. Bur. Mines I. C.* 7224, 1942. Fowle, Better Feeding Means Better Screening, *Rock Products*, **43**, 29–32 (March, 1940). Jarman, The Overstrom Mud Screen for Ceramic Products, *Bull. Am. Ceramic Soc.*, **19**, 251–253 (1940). Kohn, Graphical Methods of Interpreting Sieve Analyses, *Eng. Mining J.*, **143**, 60–61, 94 (1942); *Food Industries*, **13**, 48–50 (January, 1941). Nettleton, How to Study Screen Efficiency, *Rock Products*, **43**, 44–45 (May, 1940); Fair Tolerances for Screening, *Rock Products*, **43**, 29–30 (December, 1940); Sand Separation on Vibrating Screens, *Rock Products*, **48**, 75–77 (May, 1945); 86–87 (June, 1945); 59–61 (July, 1945); 86–87 (August, 1945); 66–67 (September, 1945); 112 (October, 1945). Taggart, "Handbook of Mineral Dressing," Wiley, New York, 1945.

Definitions. *Screening* is the separation by means of a screen of a mixture of various sizes of grains into two or more portions; the separation is effected in such a manner that the grains of any portion are of more uniform size than those of the original mixture.

Material that stays on a given screen is the oversize or plus (+) of the screen; that passing is the undersize or minus (−).

Sizing is the operation of screening a feed containing about 60 per cent undersize.

Scalping is the operation of screening in which 85 to 95 per cent of the feed material is considerably smaller than the screen mesh. *By-passing* is a screening operation in which there are only 5 to 20 per cent of throughs in the feed.

A sieve scale is a series of test sieves having successively larger or smaller openings. For example, each sieve in the *Tyler standard sieve scale* has $\sqrt{2}$ times larger openings than the next smaller sieve.

Screen cloth is a mesh surface, metallic or fabric, used for screening. The term "screen cloth" (as opposed to a "screen") usually is reserved for surfaces having an opening of 1 in. or smaller, and made of relatively fine wire.

Mesh. In the coarser sizes of screens, the term "mesh" means either the distance between adjacent wires or rods or the distance between centers of adjacent wires or rods. In the finer sizes of screen cloths the mesh means the number of openings per linear inch.

Open Area. The open area of a screen or screen cloth is the combined area of all the openings presented, and the percentage open area is equal to 100 × open area divided by total screen surface.

Aperture or Screen Size. Aperture or screen size is defined as the minimum clear space between the edges of the opening in the screen. The aperture A, mesh M, and wire diameter D are interrelated by the following equation in which P is the percentage of opening:

$$P = \frac{A^2}{(A + D)^2} = (1 - MD)^2$$

Particle-size Distribution. This term may be defined as the relative percentage by weight of grains of each of the different size fractions represented in the sample. It is one of the most important factors in a screening operation.

Purposes. The purposes of screening are: (1) to scalp off the coarse end of a long-range product, usually for further reduction; (2) to cut off the fine end from crusher feeds and thus save power and prevent overgrinding; (3) to grade products into commercial sizes; and (4) to perform a step in a concentration process.

EQUIPMENT

Screens may be divided into four main classes: grizzlies, shaking screens, vibrating screens, and oscillating screens. Crizzlies and shaking screens are customarily used for separations above 1 in., but vibrating screens are now competing in this field. The chief interest of the chemical engineer is in separations from about 4 mesh to 325 mesh, a field that is almost exclusively served by vibrating and oscillating screens.

Grizzly. A grizzly consists of a set of parallel bars held apart by spacers at some predetermined opening. Bars are frequently made of manganese steel (12 per cent Mn) to reduce wear. Figure 50 shows a typical grizzly and the shape of bar most commonly used in its construction. A grizzly is frequently used before a primary crusher in rock- or ore-crushing plants to remove the fines before the ore or rock enters the crusher.

In Table 1 are shown some of the sizes of grizzlies available.

Table 1. Various Sizes of Grizzlies*

Size, ft.		Openings, in.	Weight, lb.
Width	Length		
3	8	1	1125
3	8	1½	955
3	8	1¾	865
4	8	½	1950
4	8	1	1505
4	8	1½	1235
4	10	1	1835
4	10	1¼	1645
4	10	1½	1540
4	10	2	1365

* From Catalog 1, Allis-Chalmers Co.

One of the chief disadvantages of a grizzly is the tendency of the crosspieces to retard the flow of material and to cause clogging of the openings. This difficulty is partly avoided by placing the crosspieces at a considerable distance from the parallel bars that make up the screen surface.

The size of the grizzly depends on the capacity desired. The slope, or angle with the horizontal, of the grizzly varies with the type of material between 20 and 50 deg.

Flat grizzlies, in which the parallel bars are in a horizontal plane, are used on tops of ore and coal bins and under unloading trestles. Inclined grizzlies are more generally used, and the slope of these varies with the angle of repose of the material and the velocity at which the material strikes the grizzly. In general it may be said that moist material requires a greater slope of grizzly bar than dry material and that the larger the size of the material fed to the grizzly, the greater the slope.

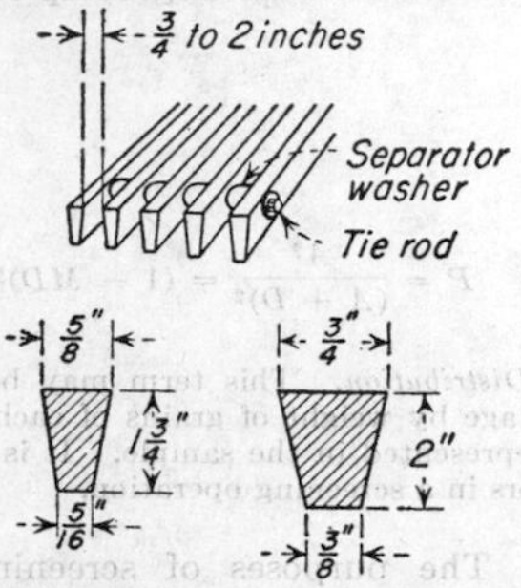

FIG. 50. Grizzly.

Although the simple horizontal or sloping grizzly is still the most generally used, a number of improved types provided with mechanical features are now available and are used in the metallurgical and mineral aggregate fields. A self-cleaning grizzly has been developed, in which the bars are bent to form one-quarter of the circumference of a circle. Heavy metal arms attached to a shaft revolve between the bars, preventing oversize from clogging. In another type, known as the moving-bar grizzly, the bars are mounted at one end on eccentrics, these eccentrics being 180 deg. apart for adjacent bars. When the eccentric shaft is driven at a speed of about 50 r.p.m., the material is moved gently across the screening surface, and a better opportunity is given for fine material to pass through to a hopper below.

In the chain type of grizzly, endless chains passing over sheaves replace the grizzly bars. Adjacent chains are mounted on sheaves of slightly different diameter. This tends to break up lumps and to prevent blinding of the open area. The traveling-bar grizzly consists of a multiplicity of short lengths of bar running transversely between two slowly moving sprocket chains. The undersize passes through the upper run of bars and is diverted into hoppers by chutes directly under the upper run and directly above the lower. The distinctive feature of the shaking grizzly is the pivoting of the bars on a horizontal shaft at the upper end and the rapid raising and lowering of the lower ends of the bars by a connecting member actuated by an overhead eccentric. A vibrating-bar grizzly particularly useful for wet sticky materials has recently been placed on the market.

The capacity of a grizzly varies with the efficiency desired and the spacing of the bars as well as its slope. The lower the efficiency desired, the greater the distance between bars; and the greater the slope, the greater the capacity of the grizzly. Taggart's "Handbook of Mineral Dressing" is authority for the statement that a grizzly in which the bars are spaced to give 1 in. of clear opening may be expected to handle 125 tons material/(sq. ft.)(hr.).

Shaking Screens. Shaking screens have been used to a considerable extent where it is desired to make a separation and the headroom available is not sufficient to permit the use of a vibrating screen. The shaking motion is such as to make the material advance across the screening medium when the screen is at a slope of not more than 0.75 in./ft. These screens give a fairly high efficiency in sizes coarser than $\frac{1}{2}$ in., but in finer sizes the capacity is low if a high efficiency is desired.

Shaking screens may be used both for screening and for conveying. For example, in the anthracite-coal industry, innumerable "shaking troughs" with perforated bottoms are used both for grading the fine anthracite particles and for conveying them to subsequent treatment steps.

Shaking screens require from 0.05 to 0.10 hp./sq. ft. screening surface, and speeds range from 60 strokes of 9-in. amplitude/min. to 800 strokes of $\frac{3}{4}$-in. amplitude. These screens are generally inclined at an angle of 18 to 20 deg., with an average of 14 deg. Capacity ranges from 2 to 8 tons/(sq. ft.) (24 hr.)(mm. aperture). The chief disadvantage of shaking screens is the high cost of maintenance of the screen and supporting structure.

Vibrating Screens. Where large capacity and high efficiency are desired, the use of vibrating and oscillating screens is standard practice. The capacity, especially in the finer sizes, is so much greater than any of the other screens that they have practically replaced all other types where the efficiency of the screen is an important factor. Another advantage of these screens is that the vibrations of the screen cloth reduce the blinding effect to a minimum. A vibrating screen consists essentially of a flat or slightly convex screening surface to which is applied a rapid vibration normal or nearly normal to the surface. The vibrating means may be eccentric shafts, an unbalanced flywheel, a cam and tappet arrangement, or an electromagnet. Many vibrating screens are on the market, but they can be classified as (1) high-amplitude normal vibration screens and (2) low-amplitude electrically vibrated screens.

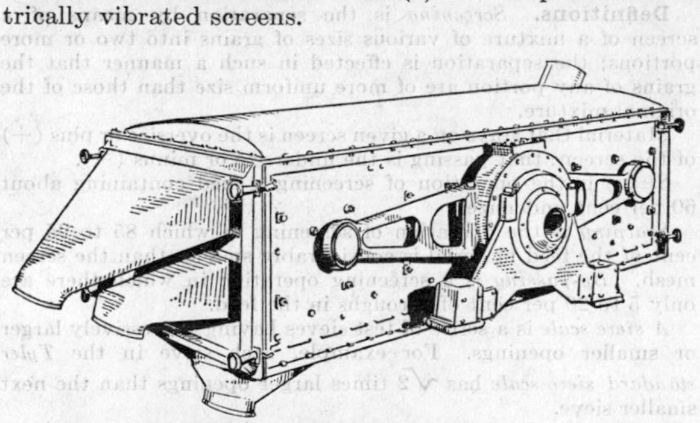

FIG. 51. The Selectro Screen. (*Productive Equipment Corp.*)

High-amplitude Normal Vibration. In this type of screen a high-amplitude (about $\frac{9}{64}$-in.) vibration is applied by a mechanical vibration source (eccentric shafts, unbalanced flywheels, or cam and tappet) operating at about 1200 to 1800 r.p.m.

One of the better known screens employing high-amplitude normal vibration is the Selectro, manufactured by the Productive Equipment Corp. of Chicago (see Fig. 51). Vibration is supplied by a horizontal adjustable positive eccentric. The stroke of the Selectro screen can be adjusted from 0 to $\frac{3}{8}$-in., which permits efficient service on operations ranging from coarse coal or gravel down to 225 mesh. The Selectro also has a trunnion mounting that permits changing the slope while in operation so that the rate of flow of the material over the surface can be closely regulated.

Two popular screens, the Niagara (Fig. 52) and the Ty-Rock (Fig. 53), are made by the W. S. Tyler Co., Cleveland, Ohio. These are positive eccentric machines with the amplitude of vibration set at the factory. Improvements embodied in the Ty-Rock screen are said to give it longer life for heavy work.

The high-amplitude vibrating screen is the only vibrating screen used for heavy work (above about 1-in. opening).

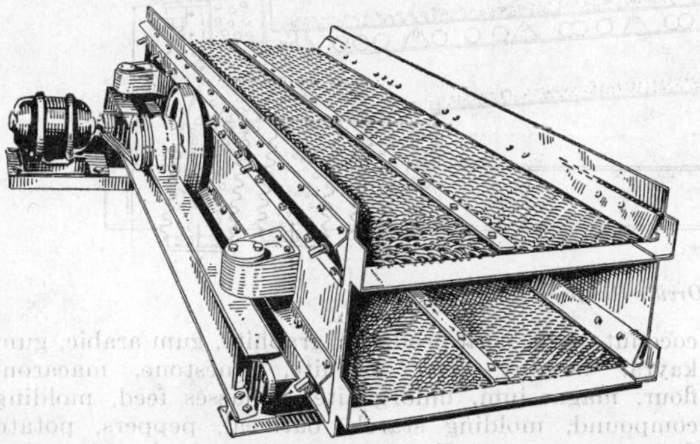

Fig. 52. The Niagara screen. (*W. S. Tyler Company.*)

In dry screening in chemical work, these screens are normally used for screening from 1 in. down to about 35 mesh. The screen usually operates at an angle of about 20 deg. with feed entering at the upper end and traveling down. For operations with little headroom, low-head screens (Allis-Chalmers) are available.

For wet screening, vibrating screens are very popular. A low angle (5 to 10 deg.) is used, and feed normally enters at the top of the screen. However, in many operations, it is found desirable to feed the slurry at the lower end of the screen and have the dried material discharge at the top. The Selectro screen has been successfully used in this manner for TNT crystals, coal, synthetic rubber, and tomato refuse.

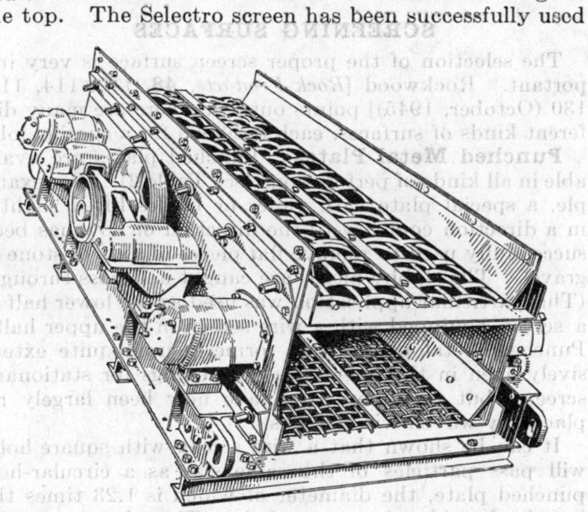

Fig. 53. The Ty-Rock screen. (*W. S. Tyler Company.*)

For the removal of small quantities of oversize from slurries, vibrating screens are widely used. The Overstrom mud screen (Separations Engineering Corp., New York) is well known in the oil and ceramic industries. G. W. Jarman, Jr. [The Overstrom Mud Screen, *Bull. Am. Ceramic Soc.*, **19**, 251–253 (1940)], reports that a 5-ft. by 54-in. screen with 20 sq. ft. of free-screening area has a capacity of 55,000 to 60,000 gal. oil-well mud/hr. through a 30- by 20-mesh screen at 1750 r.p.m. with $\frac{1}{16}$-in. amplitude vibration. For ceramic slip, the capacity is 60,000 gal./hr. through a 120-mesh screen.

The dewatering of coal and other minerals is an important use of vibrating screens. For example, a Robins screen [*Coal Age*, **50**, 3, 95–96 (March, 1945)], made by Hewitt-Robins, Inc., Passaic, N.J., 4 by 16 ft. in a typical operation handles 25 to 60 tons/hr. of $\frac{1}{8}$-in. coal, removing about 25,000 gal. water/hr. The screen cloth is approximately 60 mesh. In this type of operation the screen inclination is often reversed, the material working uphill toward the discharge.

Low-amplitude Normal Vibration. This type is the electric screen. A low-amplitude vibration is applied normal to the screen cloth by electromagnets at a frequency of 1800 to 7200 vibrations/min., the low frequency for coarser screening (8 mesh) and the high for fine screening below about 80 mesh. A typical screen is the Hum-mer (see Fig. 54) made by the W. S. Tyler Co.

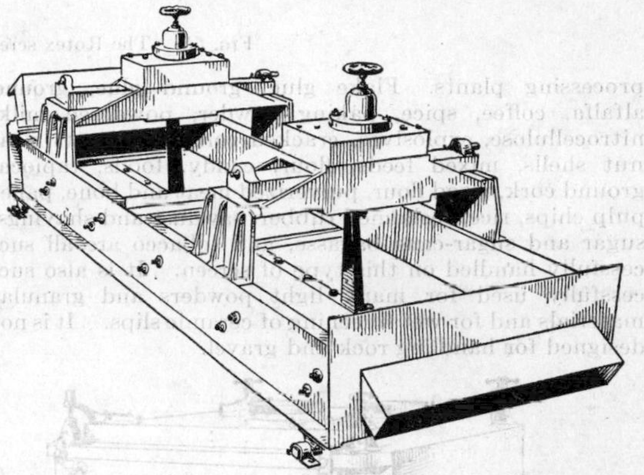

Fig. 54. The Hum-mer screen. (*W. S. Tyler Company.*)

These screens are not normally used for screening any coarser than about 8 mesh. They are, however, used for separations at 100 mesh and finer. For such fine work two or more small screens are used in series.

These screens have been very successful in wet screening to remove 1 to 2 per cent of a slurry. They are not recommended for heavy materials having high separable-solids content.

Oscillating Screens. This type of screen is characterized by low-speed (300 to 400 r.p.m.) oscillation in a plane essentially parallel to the screen cloth.

Screens in this group are usually used from 0.5 in. to 60 mesh. Some light free-flowing materials, however, can be separated at 200 to 300 mesh. Silk cloths are often used.

Reciprocating Screens. This is an important screen for chemical work. An eccentric under the screen supplies oscillation, ranging from gyratory (about 2 in. diameter) at the feed end to reciprocating motion at the discharge. Frequency is 500 to 600 r.p.m.; and, since the screen is inclined about 5 deg., a secondary high-amplitude normal vibration of about $\frac{1}{10}$ in. is also set up. Further vibration is caused by balls bouncing against the lower surface of the screen cloth.

Reciprocating screens are very popular in this country and England. They are used for a variety of chemicals, usually dry, and have been employed for very fine separations (300 mesh) when screening some light materials such as aluminum powder. A typical machine is the Rotex, made by the Orville-Simpson Co., Cincinnati, Ohio (see Fig. 55).

For screening light or bulky materials the oscillating screen is standard equipment in many chemical and

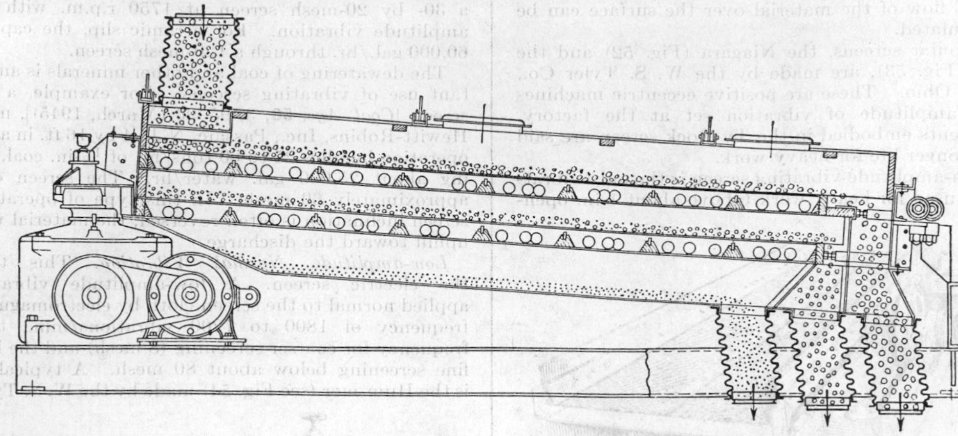

FIG. 55. The Rotex screen. (*Orville-Simpson Company.*)

processing plants. Flake glue, ground glue, ground alfalfa, coffee, spice, baking powder, powdered milk, nitrocellulose, explosives, cracked corn, meal, oats, peanut shells, mixed feeds, flour, candy, foods, tapioca, ground cork, wood flour, pulverized meat and bone, paper pulp chips, rice, reclaimed rubber, sawdust and shavings, sugar and sugar-cane bagasse, and tobacco are all successfully handled on this type of screen. It is also successfully used for many light powders and granular materials and for wet screening of ceramic slips. It is not designed for handling rock and gravel.

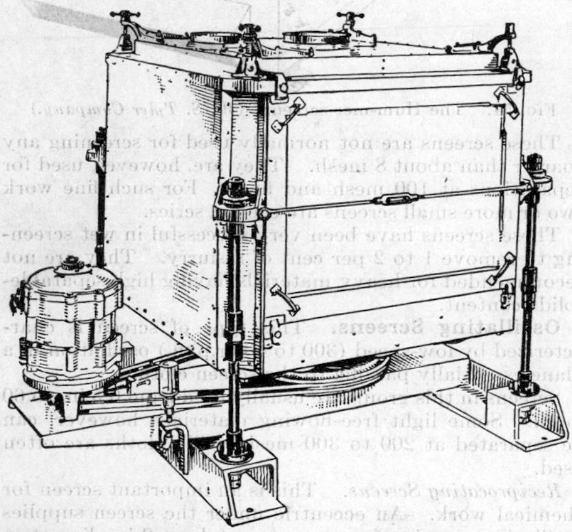

FIG. 56. The Gyratory sifter. (*Allis-Chalmers Mfg. Co.*)

Sifters. A sifter is a boxlike machine, with a series of screen cloths nested atop one another. Oscillation, supplied by eccentrics or counterweights, is in a circular or near-circular orbit. In some machines a supplementary whipping action is set up. Most sifters have an auxiliary vibration caused by balls bouncing against the lower surface of the screen cloth.

The sifter is a relatively new device which is chiefly used for dry-screening light materials. Sifters have been successfully used for aceto-acet anilid, alfalfa meal, antimony sulfide, arsenic, barium nitrate, barium sulfate, barley flour, bentonite, bone black, bran, brick, bronze, calcinated diatomaceous earth, calcium phosphate, carbon black, casine, charcoal, cinnamon and ginger, clay, coconut, cork, corn products, graphite, gum arabic, gum kayra, hominy feed, kyanite, limestone, macaroni flour, magnesium, milorganite, molasses feed, molding compound, molding starch, oatmeal, peppers, potato flour, quartz, resin Formvar, resinous powder, rice, rubber, rye flour, sawdust, silica gel, sodium phosphate, soybean flakes, spices, stone, sugar, sugar-beet pulp, sulfathiazole, tile dust, turmeric, vermiculite, wheat flour, yeast, and zinc concentrate. Typical machines are made by the Allis-Chalmers Co. (see Fig. 56).

Riddles. A riddle is a screen driven in an oscillating path by a mechanism attached to the sole support of the screen, a vertical bar extending from the top of the screen bar. The riddle is the cheapest screen on the market. It is intended normally for batch screening, wet or dry. One supplier is the Great Western Manufacturing Co., Leavenworth, Kans.

SCREENING SURFACES

The selection of the proper screen surface is very important. Rockwood [*Rock Products*, **48**, 112, 114, 116, 130 (October, 1945)] points out that there are many different kinds of surfaces, each designed for a specific job.

Punched Metal Plates. Punched plates are available in all kinds of perforations (see Table 2). For example, a special plate with vanes or baffles lifted slightly in a direction contrary to the direction of flow has been successfully used to remove flat pieces of slate or stone or gravel. The slabby pieces are caught and pass through. (This particular application was used as the lower half of a screen equipped with a wire screen in the upper half.) Punched metal plates were formerly used quite extensively even in the finer sizes of openings for stationary screens, but the finer sizes have now been largely replaced by woven-wire screens.

It can be shown that a wire screen with square holes will pass particles of the same size as a circular-hole punched plate, the diameter of which is 1.23 times the length of a side of a square hole. The advantages of a punched plate may be summarized as strength, long life and ruggedness, while the disadvantages are chiefly great weight per unit of area and difficulty of handling. Punched-plate screens are chiefly used in metal- and coal-mining operations and in crushed-stone and gravel sizing, where the openings are ½ in. in diameter or greater. Woven-wire screens are preferred where the openings are less than ½ in.

Bar Screens. Bar screens have been utilized for the removal of fines from cement. Bar screens have been used to solve one of the serious problems in the crushed-stone industry, the removal of flat and elongated pieces from crushed stone. Banning and Lamb (*U.S. Bur. Mines* R. I. 3781, 1944) found that this separation could be made by use of a special V-opening surface on a vibrating screen. The V-opening surface was made of ¼-in. pipes set parallel to the length of the screen. The pipes at the feed end were set on a horizontal plane, but at the discharge end every other pipe was raised above the original horizontal plane. Thus the spaces between the pipes had a V-shape that became increasingly large toward the discharge end.

Table 2. Punching Steel Plates*†

Diameter of hole			U.S. Standard		
Mm.	In.	Decimal of an inch	No.	Thickness, in.	Weight/sq. ft.
¾		0.02952	26	0.018	0.765
1		.03937	24	.025	1.02
	3/64	.04687	22	.031	1.275
1¼		.04921	20	.037	1.53
1½		.05906	18	.050	2.04
	1/16	.06250	18	050	2.04
	5/64	.07812	16	.0625	2.55
2		.07874	16	.0625	2.55
2¼		.08858	16	.0625	2.55
	3/32	.09375	16	.0625	2.55
2½		.09843	16	.0625	2.55
3		.11811	14	.078	3.187
	⅛	.125	14	.078	3.187
3¼		.12795	14	.078	3.187
3½		.1378	12	.109	4.46
	9/64	.14062	12	.109	4.46
4		.15748	12	.109	4.46
4½		.17717	10	.1406	5.737
	3/16	.18750	10	.1406	5.737
5		.19685	10	.1406	5.737
5½		.21654	10	.1406	5.737
6		.23622	10	.1406	5.737
	¼	.25	8	.1718	7.012
6½		.25591	8	.1718	7.012
7		.27559	3/16	.1875	7.65
	9/32	.28125	3/16	.1875	7.65
	5/16	.3125	3/16	.1875	7.65
8		.31496	3/16	.1875	7.65
9		.35433	3/16	.1875	7.65
	⅜	.375	3/16	.1875	7.65
10		.3937	¼	.25	10.2
11		.43307	¼	.25	10.2
	7/16	.4375	¼	.25	10.2
12		.47244	¼	.25	10.2
	½	.5	¼	.25	10.2
13		.51181	¼	.25	10.2
4		.55518	¼	.25	10.2
15		.59055	¼	.25	10.2
	19/32	.59375	¼	.25	10.2
	⅝	.625	¼	.25	10.2
19		.74803	¼	.25	10.2
	¾	.75	¼	.25	10.2
22		.86614	¼	.25	10.2
	⅞	.875	¼	.25	10.2
25		.98425	¼	.25	10.2
	1		5/16	.312	12.75

* *Allis-Chalmers Co. Bull.*

† This table gives the greatest thickness of steel in which it is advisable to punch round or square holes of given diameters or sizes. Spacing, strain upon the plate, wear of dies, and other considerations determine what is advisable. While the table is offered as a convenient guide in ordering, thinner plates will generally answer every requirement and cost less.

This construction was developed as a means of preventing blinding by the flat pieces. The design was entirely satisfactory; the flat pieces passed through as undersize, and the cubical pieces were the oversize.

Wire Screens. The simplest wire screen is the "piano wire" screen, which is made of parallel wires stretched lengthwise of the screen and supported at intervals by crossties with any spacing desired.

Square mesh cloth is the conventional type of screen cloth, but there are many types of cloth with an oblong weave. This construction provides greater open area and, in addition, makes it possible to use stronger wire for the same size of screen opening and for the same percentage of open area.

For shaking and vibrating screens, the rectangular opening has considerable advantage over the square mesh in that the percentage of open area is larger and in that grains which should just pass through the screen cannot touch more than three sides and many will not touch more than two sides of the opening; whereas, with the square mesh, these grains are almost sure to touch three sides and may touch four sides of the opening. The screen cloth with a rectangular opening also does not tend to blind with wet sticky materials to the same extent as do square-mesh screens.

Figure 57 shows six screens with the same opening but different sizes of wire. The top screen has 14 meshes to

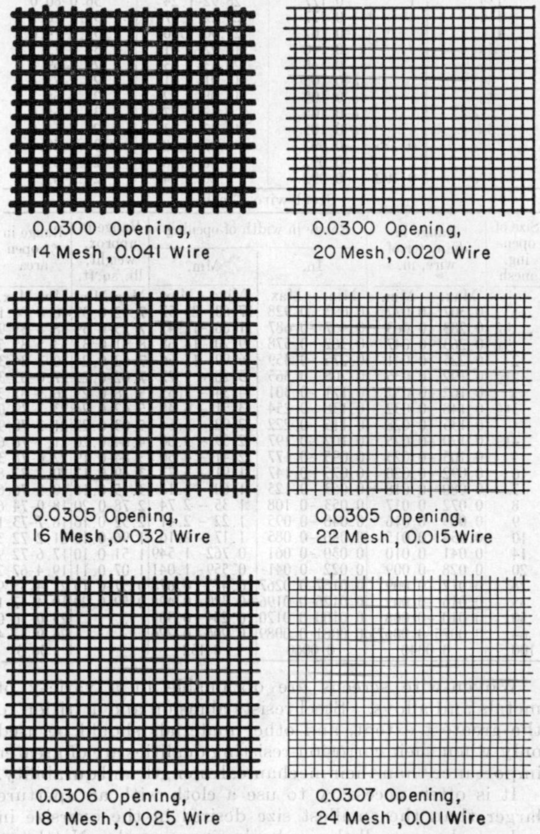

FIG. 57. Effect of wire size on per cent open area for fixed aperture.

the linear inch, whereas the bottom screen has 24 meshes to the linear inch; therefore the screen with the coarser wire has less than 40 per cent of the open area of the screen with the finer wire, and the capacity of the screen with the coarser wire will be less than 40 per cent of that with the finer wire.

It must be remembered that the life of the screen with the finer wire is much less than that of one with heavier wire. If the material to be screened contains a large amount of abrasive, it may not be practical to use a screen with a fine wire; whereas, if the material is damp and sticky, the life of the screen may be of only secondary importance.

Table 3 shows the largest and smallest diameter of rod or wire for some of the various woven screens manufactured by the W. S. Tyler Co., Cleveland, Ohio (see their catalogue for the intermediate sizes of screens as well as rods or wires).

Table 3. Steel-wire Screens and Cloths

Heavy Steel-wire Screens

Size of opening, in.	Range in diameter of wire, in.		Range in approx. weight, lb./sq. ft.	Range in % open area
	Max.	Min.	Max. Min.	Min. Max.
4	1	– 3/8	13.70–2.16	64.0–83.6
3¾	1	– 5/16	14.44–1.61	62.3–85.2
3½	1	– 5/16	15.29–1.72	60.5–84.3
3¼	1	– 5/16	16.23–1.74	58.5–83.2
3	1	– ¼	17.35–1.29	56.3–85.2
2¾	1	– ¼	18.49–1.40	53.7–84.0
2½	1	–0.225	19.93–1.16	51.0–84.1
2¼	1	–0.207	21.67–1.17	47.9–83.8
2	1	–0.192	23.69–1.14	44.4–83.2
1¾	1	–0.192	26.02–1.30	40.5–81.2
1½	1	–0.177	28.92–1.24	36.0–80.0
1¼	¾	–0.177	20.22–1.47	39.1–76.7
1	¾	–0.162	23.40–1.50	32.6–74.0
7/8	5/8	–0.148	18.97–1.41	34.0–73.5
¾	5/8	–0.148	20.53–1.63	29.7–69.8
5/8	9/16	–0.105	17.91–1.01	27.7–73.4
½	7/16	–0.105	14.08–1.22	28.4–68.3
7/16	5/16	–0.105	9.48–1.37	34.0–65.0
3/8	5/16	–0.105	10.42–1.33	29.7–61.0
5/16	0.225	–0.092	6.87–1.42	33.8–59.6
¼	¼	–0.092	8.96–1.66	25.0–53.4
3/16	0.192	–0.092	7.60–2.04	24.4–45.1

Steel-wire Cloth

Size of opening, mesh	Range in diameter of wire, in.		Range in width of opening, In.		Range in width of opening, Mm.		Range in approx. weight, lb./sq. ft.	Range in % open area
	Max.	Min.	Min.	Max.	Min.	Max.	Max. Min.	Min. Max.
1	0.307	–0.072	0.693	–0.928	17.60	–23.57	7.52–0.39	48.0–86.1
5/8	0.283	–0.063	0.467	–0.687	11.86	–17.45	7.75–0.39	38.8–83.9
¾	0.263	–0.047	0.362	–0.578	9.20	–14.68	8.51–0.31	33.5–85.5
2	0.225	–0.041	0.275	–0.459	6.99	–11.66	7.62–0.24	30.3–84.3
2½	0.192	–0.035	0.208	–0.365	5.28	– 9.27	7.02–0.22	27.0–83.3
3	0.162	–0.032	0.171	–0.301	4.34	– 7.65	5.78–0.23	26.3–81.5
3½	0.148	–0.032	0.138	–0.254	3.51	– 6.45	5.62–0.24	23.3–79.0
4	0.135	–0.028	0.115	–0.222	2.92	– 5.64	5.03–0.22	21.2–78.9
4½	0.120	–0.025	0.102	–0.197	2.59	– 5.00	4.64–0.20	21.1–78.6
5	0.105	–0.023	0.095	–0.177	2.41	– 4.50	3.84–0.19	22.6–78.3
6	0.092	–0.020	0.075	–0.147	1.91	– 3.73	3.49–0.17	20.2–77.8
7	0.080	–0.018	0.063	–0.125	1.60	– 3.18	3.15–0.16	19.5–76.6
8	0.072	–0.017	0.053	–0.108	1.35	– 2.74	2.78–0.20	18.0–74.6
9	0.063	–0.016	0.048	–0.095	1.22	– 2.41	2.52–0.16	18.7–73.1
10	0.054	–0.015	0.046	–0.085	1.17	– 2.16	1.96–0.16	21.2–72.3
14	0.041	–0.010	0.030	–0.061	0.762	– 1.549	1.51–0.10	17.6–72.9
20	0.028	–0.009	0.022	–0.041	0.559	– 1.041	1.07–0.11	19.4–67.2
28	0.017	–0.009	0.0187	–0.0267	0.475	– 0.678	0.57–0.15	27.4–55.9
35	0.016	–0.009	0.0126	–0.0196	0.320	– 0.498	0.59–0.20	19.4–47.1
50	0.010	–0.008	0.0100	–0.0120	0.254	– 0.305	....	25.0–36.0
64	0.0075	–0.00675	0.0081	–0.0089	0.206	– 0.226	....	26.9–32.4
100	0.0040		0.0060		0.152		....	36.0

Woven-wire screens are obtainable in a variety of metals and alloys. Steel resists abrasion and attrition to the greatest extent, and other materials should be used only when their corrosion-resisting qualities are of greater importance than their mechanical strength and durability.

It is often necessary to use a cloth with an aperture larger than the smallest size desired in the oversize in order to be sure all the undersize is removed. Nettleton [*Rock Products*, **43**, (12), 29–30 (December, 1940)] discusses the practical aspects of this point.

Comparative Openings for Silk and Wire Cloths. In screening any material, the size of the finished product is determined by the actual mesh opening and not by the number of meshes per linear inch.

Table 4 (supplied by the Orville-Simpson Co.) shows the relationship between the various meshes and grades of silk and wire cloth according to the actual opening in each mesh. The left-hand column of the table shows the mesh opening of each screen listed, arranged in numerical order, with the coarsest at the top and the finest at the bottom. The silk and wire cloths are arranged in the table with the lightest weights to the left and the progressively heavier cloths to the right. The extreme right-hand column shows the Tyler testing screen series, which is for reference only, since these screens are not ordinarily used for anything but screen analysis.

The best commercial cloth will vary from the listed figures. Tolerances of 1.5 to 7.5 per cent in average mesh openings, and from 10 to 20 per cent in wire diameters, depending on the fineness of the screen, are permitted by the National Bureau of Standards in their specifications of wire screen for use in testing sieves.

This table will be found convenient in the selection of a screen with a certain opening or a heavier or lighter screen of approximately the same opening.

SCREEN PERFORMANCE

The efficiency of a screen can be defined only in terms of the specifications the screening operation is required to meet. The most common method of definition is to express efficiency as the ratio of the undersize obtained to the undersize in the feed. However, when the object of the screening operation is to remove undesirable undersize from the feed, efficiency may be calculated as the ratio of the true oversize to the oversize obtained.

The W. S. Tyler Co. gives the following formula for screen efficiency when R = fractional recovery of fines in screening, E = percentage of fines recovered in screening, a = percentage of coarse in feed to screen, b = percentage of fines in feed to screen, and c = percentage of coarse in oversize after screening:

$$R = \frac{100(c - a)}{(bc)}$$

$$E = 100R$$

Certain other formulas for the derivation of screen efficiency are used. Taggart, in his "Handbook of Mineral Dressing," suggests the formula

$$E = 100 \times \frac{100(u - o)}{u(100 - o)}$$

where E is equal to the efficiency of screen, u is the percentage of undersize in the feed, and o is the percentage of undersize in the screen oversize.

A formula having somewhat wider applicability is given by Newton [*Rock Products*, **35**, 26 (1932)] as follows:

$$E = \frac{(o - b)(u - a)}{ou(1 - a - b)}$$

where E = screen efficiency (as a decimal).
o = per cent oversize in feed (as a decimal).
u = per cent undersize in feed (as a decimal).
a = per cent undersize in tailings (as a decimal).
b = per cent oversize in fines (as a decimal).

Graphical methods of evaluating efficiency, using sieve analyses, are given by Kohn [*Eng. Mining J.*, **143** (7), 60–61, 94 (1942); *Food Industries*, **13** (1), 48–50 (January, 1941)]. For any serious research on screening graphical methods of this general type are recommended.

FACTORS AFFECTING EFFICIENCY AND CAPACITY

Screening is an operation where relatively inexpensive equipment is used to accomplish many different results. The variables encountered are so numerous and so complex that no satisfactory method of correlating them has been developed.

Therefore, no specific recommendations can be made as to type of screen or cloth to be selected, or capacity and efficiency to be obtained. It is suggested that the reader review the "equipment" section, select therefrom the type of machine that appears likely to answer his requirements, and then submit samples to a manufacturer for testing.

However, many variables in screening are susceptible to change in the field, and the practical operating engineer

Table 4. Comparative Openings, Silk and Wire*

Opening between meshes, inches	Double extra silk		Grit gauze silk		Treble extra grit gauze silk		Tuf-Tex light wire bolting cloth		Tinned steel mill screen		Market grade bronze cloth		Tyler standard testing screen series	
	No.	Mesh/in.	No.	Mesh/in.	No.	Mesh/in.	Mesh/in.	Diam. wire, in.	Mesh/in.	Diam. wire, in.	Mesh/in.	Diam. wire, in.	Mesh/in.	Diam. wire, in.
0.0750					14 xxxGG	13.5					10	0.025		
.0689			14 GG	13.5										
.0600					16 xxxGG	15.5					12	.023		
.0582			16 GG	15.5										
.0528					18 xxxGG	17.5								
.0519			18 GG	17.5										
.0510	0000xx	18									14	.020		
.0456			20 GG	19	20 xxxGG	19								
.0420			22 GG	21										
.0411					22 xxxGG	21								
.0386	000xx	23									18	.017		
.0376			24 GG	23										
.0367					24 xxxGG	23								
.0331					26 xxxGG	25								
.0332			26 GG	25										
.0305			28 GG	27							22	.015		
.0296					28 xxxGG	27								
.0287	00xx	29	30 GG	29					24	0.0130				
.0278					30 xxxGG	29								
.0260			32 GG	31										
.0257					32 xxxGG	31			28	.0100				
.0242			34 GG	33										
.0233					34 xxxGG	33								
.0224			36 GG	35	36 xxxGG	35								
.0215			38 GG	37										
.0210					38 xxxGG	37								
.0206	0xx	38												
.0197			40 GG	39	40 xxxGG	39								
.0188			42 GG	40.5					36	.0090				
.0184			44 GG	42.5	42 xxxGG	40.5								
.0170			46 GG	44.5	44 xxxGG	42.5								
.0164					46 xxxGG	44.5							35	0.0122
.0157	1xx	48	48 GG	46.5	48 xxxGG	46.5								
.0148			50 GG	48.5										
.0144			52 GG	50.5										
.0138	2xx	54	54 GG	52.5	50 xxxGG	48.5							42	.0100
.0135					52 xxxGG	50.5								
.0131			56 GG	54.5	54 xxxGG	52.5		0.0045						
.0127			58 GG	56.5			58				45	.0095		
.0125	3xx	58			56 xxxGG	54.5			50	.0075				
.0118			60 GG	58	58 xxxGG	56.5								
.0115			62 GG	60	60 xxxGG	58								
.0112					62 xxxGG	60		.0045	55	.0070				
.0111	4xx	62	64 GG	62			64							
.0109					64 xxxGG	62								
.0105	5xx	66												
.0102			66 GG	64					60	.0065				
.0097			68 GG	66	66 xxxGG	64							60	.0070
.0094			70 GG	68										
.0092	6xx	74			68 xxxGG	66	76	.0040						
.0088			72 GG	72			78	.0040						
.0085					70 xxxGG	68	80	.0040						
.0083					72 xxxGG	72								
					Fine treble extra silk									
.0082	7xx	82			8xxx	82								
.0076	8xx	86					86	.0040					65	.0072
.0074					9xxx	86	88	.0040						
.0062					10xxx	97								
.0059	9xx	97									90	0.00525		
.0055	10xx	109									100	.00450		
											Extra fine bronze cloth			
.0049	11xx	116												
.0046													115	.0038
.0043					12xxx	116					120	.0037		
.0042	12xx	125			13xxx	125					130	.0034		
.0039	13xx	129									140	.0029		
.0038					14xxx	129					160	.0025		
.0037	14xx	139			15xxx	139								
.0035	15xx	150									170	.0024	170	.0024
.0034					16xxx	150								
.0031					17xxx	157					190	.0022		
.0030	16xx	157												
.0029					18xxx	163					200	.0021	200	.0021
.0028											220	.0017		
	Standard silk													
.0026	21	178									240	.0016		
.0024	25	200									250	.0016	250	.0016

* Courtesy of Orville-Simpson Co.

can do much to improve many existing installations and to modify present installations to make them usable for new products or processes.

Capacity and efficiency in screening operations are closely related. If a low efficiency is not objectionable, the capacity may be large. Usually, as the tonnage of the feed to the screen is increased, the efficiency is decreased.

Efficiency in commercial operations is about 60 per cent, and 75 per cent is unusually good. Therefore, the problem in screening is to select the correct combination of variables that will give about 70 per cent efficiency and maximum capacity.

Variables in Material Being Screened. *Bulk Specific Gravity.* Often this factor determines the type of screen motion to be used. Light bulky materials are very successfully screened on oscillating screens.

Size of Separation. For sizes above about 1.5 in., grizzlies are required. Below 1.5 in., vibrating or oscillating screens are more satisfactory. Below about 35 mesh, capacity falls off rapidly with decrease in screen mesh used. For screening at fine meshes, oscillating or high-frequency vibrating screens are used.

Stickiness of Material. Sticky materials are hard to screen. A strong high-amplitude vibration should be used to break up the mass, and rectangular screen cloth is often effective in reducing blinding of openings. Exceedingly sticky materials must be screened wet—at not more than 33 per cent solids—and subsequently dried. (It is better to add water to the material beforehand than to spray the screen surface. For separations at fine meshes, sprays above the screen cloth actually increase blinding; sprays from below are better.) Both vibrating and oscillating screens may be used for wet screening. Capacity for efficient wet screening is much greater than for dry screening.

Shape of Material. Irregularly shaped particles tend to blind a screen cloth. High-amplitude vibration should be used, and a rectangular cloth is recommended to reduce blinding.

Particle-size Distribution. One of the most important variables in screening is the proportion of particles in the feed that are nearly the same as the aperture of the cloth. This factor makes most "tables of capacity" meaningless, since it alone can change capacity 300 per cent or more. As Taggart points out, undersize much smaller than the aperture (less than 0.7 times aperture) passes through very rapidly, and no normal variation in the amount of such material has any effect on capacity. Likewise, oversize much larger (1.5 times) than the aperture slides freely over the screen surface and readily allows undersize material to pass through it to the screen surface. But, when the proportion of particles of a size near the screen aperture is large, such particles, which have relatively small interstitial spaces, hold back the fine material from the screen. The particles only slightly larger than the screen apertures wedge in and cause blinding, and those only slightly smaller than the aperture pass through with difficulty. These statements have been borne out by tests and practical experience on many types of material.

One way to overcome this trouble is to use a screen cloth of much larger aperture than the size desired in the final product and to make a relatively inefficient separation.

Mechanical Variables. *Screen Cloth Percentage of Open Area.* As the percentage of open area of a screen cloth decreases, capacity decreases but screen life increases. The use of rectangular openings and various special types of construction has been previously mentioned.

Vibration Amplitude and Frequency. Speed and amplitude of vibration must be adjusted to convey the material and prevent blinding of the cloth. After these requirements are met, adjustments should be made so that proper stratification of the feed takes place at the feed rate required.

It is generally agreed that optimum speed and amplitude of vibration depend on the median size of the material. As the size decreases, optimum amplitude decreases and speed increases. It is difficult to be more specific, but it may be said in general that optimum conditions are reached when the feed is vibrated violently but not thrown so far in the air that particles jump down the screen.

Efficiency can often be changed markedly by changing speed and/or amplitude.

Angle of Inclination. The angle of inclination is probably not so important a variable as speed or stroke. With vibrating screens, optimum angle to prevent blinding usually increases as feed size decreases, but this does not apply to oscillating screens. Generally speaking, angle of inclination must be coordinated with speed and stroke for best results.

Direction of Vibration. Usually somewhat greater efficiency can be obtained by counterflow, *i.e.*, having the material move down the screen against the vibration. This is accomplished at a loss in capacity.

Feed Rate. Maximum capacity and efficiency cannot be obtained unless the screen is fed properly. As pointed out by Fowle [*Rock Products*, **43**, 3, 29–32 (March, 1940)], it is necessary to distribute the feed evenly and to prevent the material from striking the screen surface with too great velocity in the direction of screen travel. Best practice calls for arranging the feed chute so that the feed material comes in countercurrent to the material going over the screen and thus must stop and reverse its direction of flow.

Arrangement of Screens. It is generally agreed that the most efficient screening results when a series of single-deck screens is used. This is because lower decks of multiple-deck screens are not fed so that their entire area is used and because different sizes of material require different amplitudes of vibration.

Typical Capacity Data. The figures given below are quite meaningless for design purposes, since the particle-size distribution of the feed was not recorded and since these are not necessarily maximum figures. However, they do serve to illustrate the wide range of capacity that may be expected in screening work.

Table 5. Capacity of Vibrating Screens for Sizing

Material	Wire cloth	Capacity, tons feed/(hr.)(sq. ft. screening surface)
Powdered soap	7 mesh	0.1
Trisodium phosphate	20 mesh	.065
Clay (dry)	10 mesh	.625
Clay (damp)	4 mesh	.169
Coal	1¼ in.	5.0
Coal	⅛ in.	2.0
Coke	1 in.	1.5
Coke	⅛ in.	0.5
Crushed stone	3 mesh	2.5
Sand	7/16 in.	2.0
Sugar	12 mesh	0.5

SIEVE TESTING

Many specifications now call for definite sizes of material, and the use of test sieves is required. Test sieves are also generally used to determine the efficiency of screening devices and the work of crushing and grinding machinery.

It is essential that standard sieves, with standard size of opening, be used for sieve analysis. The time of screening and the method of agitating the material on the

Table 6. The Tyler Standard Screen-scale Sieves

1	2	3	4	5	6	7	8	9
Tyler standard screen scale $\sqrt{2}$ or 1.414 openings, in.	Every other sieve from 0.0029 to 0.742 in. ratio of 2 to 1	Every other sieve from 0.0041 to 1.050 in. ratio of 2 to 1	Every 4th sieve from 0.0029 to 0.742 in. ratio of 4 to 1	For closer sizing sieves from 0.0029 to 1.050 in. ratio $\sqrt[4]{2}$ or 1.189	Openings, mm.	Openings in fractions of an inch (approximate)	Mesh	Diameter of wire, in.
1.050	......	1.050	......	1.050	26.67	1	.....	0.148
				0.883	22.43	7/8	.....	.135
0.742	0.742	......	0.742	.742	18.85	3/4	.....	.135
				.624	15.85	5/8	.....	.120
.525	......	0.525	......	.525	13.33	1/2	.....	.105
				.441	11.20	7/16	.....	.105
.371	.371	......	......	.371	9.423	3/8	.....	.092
				.312	7.925	5/16	2½	.088
.263	......	.263	......	.263	6.680	1/4	3	.070
				.221	5.613	7/32	3½	.065
.185	.185	......	.185	.185	4.699	3/16	4	.065
				.156	3.962	5/32	5	.044
.131	......	.131	......	.131	3.327	1/8	6	.036
				.110	2.794	7/64	7	.0328
.093	.093	......	......	.093	2.362	3/32	8	.032
				.078	1.981	5/64	9	.033
.065	......	.065	......	.065	1.651	1/16	10	.035
				.055	1.397	....	12	.028
.046	.046	......	.046	.046	1.168	3/64	14	.025
				.0390	0.991	....	16	.0235
.0328	......	.0328	......	.0328	.833	1/32	20	.0172
				.0276	.701	....	24	.0141
.0232	.0232	......	......	.0232	.589	....	28	.0125
				.0195	.495	....	32	.0118
.0164	......	.0164	......	.0164	.417	1/64	35	.0122
				.0138	.351	....	42	.0100
.0116	.0116	......	.0116	.0116	.295	....	48	.0092
				.0097	.246	....	60	.0070
.0082	......	.0082	......	.0082	.208	....	65	.0072
				.0069	.175	....	80	.0056
.0058	.0058	......	......	.0058	.147	....	100	.0042
				.0049	.124	....	115	.0038
.0041	......	.0041	......	.0041	.104	....	150	.0026
				.0035	.088	....	170	.0024
.0029	.0029	......	.0029	.0029	.074	....	200	.0021
For Coarser Sizing—3 to 1½-in. Opening								
						3	....	0.807
						2	....	.192
						1½	....	.148

Table 7. The I. M. M. Series*

Mesh	Diameter wire, in.	Aperture, in.	Aperture, mm.	Ratio of each to one below	Screening area, %
5	0.1	0.1	2.540	1.61	25.00
8	.063	.062	1.574	1.24	24.60
10	.05	.05	1.270	1.20	25.00
12	.0417	.0416	1.056	1.33	24.92
16	.0313	.0312	0.792	1.25	24.92
20	.025	.025	.635	1.25	25.00
25	.02	.02	.508	1.205	25.00
30	.0167	.0166	.421	1.17	24.80
35	.0143	.0142	.416	1.135	24.70
40	.0125	.0125	.317	1.25	25.00
50	.01	.01	.254	1.20	25.00
60	.0083	.0083	.211	1.17	24.80
70	.0071	.0071	.180	1.145	24.70
80	.0063	.0062	.157	1.24	24.60
90	.0055	.0055	.139	1.10	24.50
100	.005	.005	.127	1.51	25.00
120	.0041	.0042	.107	1.273	25.40
150	.0033	.0033	.084	1.32	24.50
200	.0025	.0025	.063	....	25.00

* The Institute of Mining and Metallurgy, England.

sieve should also be standard, and in many industries the practice of specifying the size of opening (or a certain standard sieve) and the time and the method of screening are followed. An excellent elementary treatise for beginners in sieve testing is given by Dasher [Directions for Laboratory Mineral Sizing, *U.S. Bur. Mines I. C.* 7224, 1942].

When it is desired to obtain exceedingly accurate results, a wet analysis is made. The sample to be tested should be washed through the finest sieve to be used in the test. The oversize from this washing should be dried and then screened through the various test sieves including the one the sample was washed through, as the dried sample may contain a considerable amount of material that will pass through the sieve used for washing. Results from a wet sieve analysis usually show a much

Table 8. U.S. Sieve Series

Meshes/lin. in.	Sieve No.	Sieve opening, in.	Sieve opening, mm.	Wire diameter, in.	Wire diameter, mm.
2.58	2½	0.315	8.00	0.073	1.85
3.03	3	.265	6.73	.065	1.65
3.57	3½	.223	5.66	.057	1.45
4.22	4	.187	4.76	.050	1.27
4.98	5	.157	4.00	.044	1.12
5.81	6	.132	3.36	.040	1.02
6.80	7	.111	2.83	.036	0.92
7.89	8	.0937	2.38	.0331	.84
9.21	10	.0787	2.00	.0299	.76
10.72	12	.0661	1.68	.0272	.69
12.58	14	.0555	1.41	.0240	.61
14.66	16	.0469	1.19	.0213	.54
17.15	18	.0394	1.00	.0189	.48
20.16	20	.0331	0.84	.0165	.42
23.47	25	.0280	.71	.0146	.37
27.62	30	.0232	.59	.0130	.33
32.15	35	.0197	.50	.0114	.29
38.02	40	.0165	.42	.0098	.25
44.44	45	.0138	.35	.0087	.22
52.36	50	.0117	.297	.0074	.188
61.93	60	.0098	.250	.0064	.162
72.46	70	.0083	.210	.0055	.140
85.47	80	.0070	.177	.0047	.119
101.01	100	.0059	.149	.0040	.102
120.48	120	.0049	.125	.0034	.086
142.86	140	.0041	.105	.0029	.074
166.67	170	.0035	.088	.0025	.063
200	200	.0029	.074	.0021	.053
238.10	230	.0024	.062	.0018	.046
270.26	270	.0021	.053	.0016	.041
323	325	.0017	.044	.0014	.036

Table 9. Comparison of Tyler and U.S. Sieve Series

Tyler standard sieve series				U. S. series approx. equivalent No.
Opening, in.	Opening, mm.	Tyler mesh	Diam. of wire, in.	
3	......	.....	0.207	
2	......	.....	.192	
1½	......	.....	.148	
1.050	26.67	.....	.148	
0.883	22.43	.....	.135	
.742	18.85	.....	.135	
.624	15.85	.....	.120	
.525	13.33	.....	.105	
.441	11.20	...	.105	
.371	9.423	.....	.092	
.312	7.925	2½	.088	
.263	6.680	3	.070	
.221	5.613	3½	.065	
.185	4.699	4	.065	4
.156	3.962	5	.044	5
.131	3.327	6	.036	6
.110	2.794	7	.0328	7
.093	2.362	8	.032	8
.078	1.981	9	.033	10
.065	1.651	10	.035	12
.055	1.397	12	.028	14
.046	1.168	14	.025	16
.0390	0.991	16	.0235	18
.0328	.833	20	.0172	20
.0276	.701	24	.0141	25
.0232	.589	28	.0125	30
.0195	.495	32	.0118	35
.0164	.417	35	.0122	40
.0138	.351	42	.0100	45
.0116	.295	48	.0092	50
.0097	.246	60	.0070	60
.0082	.208	65	.0072	70
.0069	.175	80	.0056	80
.0058	.147	100	.0042	100
.0049	.124	115	.0038	120
.0041	.104	150	.0026	140
.0035	.089	170	.0024	170
.0029	.074	200	.0021	200
.0024	.061	250	.0016	230
.0021	.053	270	.0016	270
.0017	.043	325	.0014	325
.0015	.038	400	.0010	

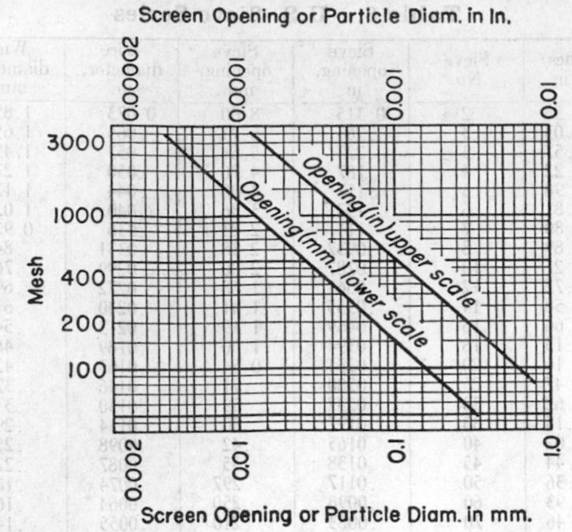

FIG. 58. Screen opening and mesh for testing sieves, Tyler series.

greater proportion of fine material than is indicated by dry analysis.

The Tyler Standard Sieve Scale. This series has as its base a 200-mesh screen in which the opening is 0.0029 in. and the wire diameter 0.0021 in. This screen has been adopted by the U.S. Bureau of Standards as the standard 200-mesh screen. In the Tyler standard, the width of opening for each succeeding screen is increased by the square root of 2, or 1.414; thus the area of each succeeding opening is twice that of the preceding screen (see Fig. 58).

Table 6 shows the Tyler standard screen-scale sieves (Catalogue 48, W. S. Tyler Co., Cleveland, Ohio). The I.M.M. series is shown in Table 7 and the U.S. sieve series in Table 8, and a comparison of the Tyler and U.S. series is shown in Table 9.

Mechanical Shakers. The Ro-Tap shaker manufactured by the W. S. Tyler Co. is the standard machine for automatically carrying out sieve-test procedure with accuracy and dependability. This device consists of a receptacle to hold a series of superimposed standard sieves and a motor-driven shaker that imparts to the sieves both a circular and a tapping motion.

The Ro-Tap is equipped to handle from 1 to 13 sieves at a time and is equipped with a timing element that automatically terminates the test after any predetermined time.

Another mechanical shaker is the Newark End-shake, made by the Newark Wire Cloth Co., Newark, N.J. Sieves used are Newark test sieves, made to conform with the U.S. standard series.

Interpretation (see Sec. 16, p. 1111*ff*). Sieve analyses are most readily interpreted by plotting them on special types of paper that give simple curves. Logarithmic-probability paper is often used. A number of methods are described by Austin [*Ind. Eng. Chem.*, anal. ed., **11**, 334 (1939)]. To his list should be added the methods of Weinig [*Colo. School Mines Quart.*, **28** (3) (1933)], Roller [*J. Franklin Inst.*, **223** (1937)], and Kohn [*Eng. Mining J.*, **143** (7), 60–61, 94 (1942); *Food Industries*, **13**, 48–50 (January, 1941)].

Discussions of methods of particle-size measurement are given by Heywood [*Engineering*, **146**, 192 (1938)] and Work [*Chem. & Met. Eng.*, **45**, 247 (1938)]. Data for testing screens and sieve sizes used for coal appear on pp. 1640 and 1641.

FILTRATION

BY DONALD F. IRVIN

INTRODUCTION

REFERENCES: *Theory.* Carman, *Trans. Inst. Chem. Engrs.* (*London*), **16**, 168–188 (1938); *Ind. Eng. Chem.*, **31**, 1047 (1939). Hermans and Bredee, *Rec. trav. chim.*, **54**, 680 (1935); *J. Soc. Chem. Ind.*, **55T**, 1 (1936). Ruth, *Ind. Eng. Chem.*, **27**, 708, 806 (1935); *ibid.*, **38**, 564 (1946). Ruth, Montillon, and Montonna, *Ind. Eng. Chem.*, **25**, 76, 153 (1933).

Application. Bonilla, *Trans. Am. Inst. Chem. Engrs.*, **34**, 243 (1938). Carman, *Trans. Inst. Chem. Engrs.* (*London*), **12**, 229 (1934). Larian, *Trans. Am. Inst. Chem. Engrs.*, **35**, 623 (1939). McMillen and Webber, *Trans. Am. Inst. Chem. Engrs.*, **34**, 213 (1938). Sperry, *Ind. Eng. Chem.*, **36**, 323 (1944). Walas, *Trans. Am. Inst. Chem. Engrs.* **42**, 783 (1946).

Equipment. General equipment. Anon., *Chem. & Met. Eng.*, **51** (1), 117 (January, 1944). Barnebl, *Trans. Am. Inst. Chem. Engrs.*, **37**, 707 (1941). Clarifying filters. Field, *Product Eng.*, **16**, 409 (1945). Filter media. Prior and Walker, *Chem. & Met. Eng.*, **45**, 250 (1938).

Definition. Filtration is defined here as the separation of solids from a liquid and is effected by passing the liquid through a porous medium. The solids are retained upon the surface of the medium in the form of a cake.

Purpose of Filtration. The purpose of filtration in industrial work is to separate the liquid from solids suspended in it, either one or both being valuable.

For instance, in causticizing plants, "lime mud" (chiefly calcium carbonate) is filtered from a solution of sodium hydroxide. In this case, the object of filtration is to remove caustic solution from the lime mud as completely as possible, obtaining high percentage separation of caustic and at the same time producing lime-mud cake nearly free from soda so that the cake may be calcined and converted into quicklime, which is again used in the further production of caustic soda.

The petroleum refiner filters wax from paraffin-base oil which he is processing, both products being valuable.

The metallurgist filters cyanide-slime pulp, obtaining valuable gold- and silver-bearing solutions and discards the solids as unprofitable for further treatment; or he filters concentrates obtained by flotation and discards the water.

In many of these processes, filtration is preceded by thickening the pulp, enabling high capacity to be obtained from the filter and separating an initial portion of clear solution. This is dealt with under Sedimentation (p. 937).

Usually when the liquid contains a very small proportion of solids in suspension, clarification by settlement should precede filtration. There are exceptions to this statement, but it is of quite general application.

When the solids are present in small proportion, the process is usually spoken of as **clarification.** In this article it is not intended to deal with the clarification of potable water or of municipal water supplies or boiler-feed water. In most cases, however, there is a comparatively large volume of solids to be removed from the liquid. Between the two extremes we have all proportions that might be found in a single industry.

The broad application of industrial filtration is realized by a survey of industries in which filtration plays an important part: pulp and paper, metallurgy, oil refining, chemical manufacturing, beet- and cane-sugar milling, sugar refining, sewage disposal, cement manufacture, etc.

THEORY OF FILTRATION

By Hugh W. Bellas

Filtration has been developed as a practical art rather than as a science, but the theory of filtration has received more and more attention in industry during the past years.

Filtration theory, although seldom used in the actual design of a filter for a given operation, is valuable in interpreting laboratory tests, in seeking the optimum conditions for filtration, and in predicting effects of changes in operating conditions. The use of filtration theory is limited by the fact that the filtering characteristics must always be determined on the actual slurry in question, data obtained on one slurry being inapplicable to another.

Filtration usually results in the formation of a layer (or cake) of solid particles on the surface of the porous body, frequently a textile fabric, that forms the filtering medium. Once this layer has formed, its surface acts as the filter medium, solids being deposited and adding to the thickness of the cake while the clear liquor passes through. The cake is therefore composed of a bulky mass of particles of irregular shape, among which run small capillaries. The flow of liquor through the capillaries is always streamline and may therefore be represented by Poiseuille's equation, which may be adapted in the following form:

$$\frac{dV}{A\ d\theta} = \frac{P}{\mu[\alpha(W/A) + r]} \tag{1}$$

[Carman, *Trans. Inst. Chem. Engrs. (London)*, **16**, 174 (1938); also, Walker, Lewis, McAdams, and Gilliland, "Principles of Chemical Engineering," McGraw-Hill, New York, 1937], expressing the differential or instantaneous rate of filtration per unit area as the ratio of a driving force, pressure, to the product of viscosity by the sum of cake resistance and filter medium resistance.

The rate of filtration can usually be expressed in terms of volume of filtrate collected V, area of filtering surface A, and time θ. The pressure P is the total drop through the filter medium and the cake upon it. The viscosity μ is that of the filtrate. (Any convenient units may be used, inconsistencies being absorbed in the cake and cloth resistances.)

W is the weight of dry-cake solids, which may be replaced by one of several equivalent terms, since

$$W = wV = \left(\frac{\rho c}{1 - mc}\right) V$$

where w is the weight of dry-cake solids per unit volume of filtrate, ρ is the density of the filtrate, c is the weight fraction of cake solids in the solute-free slurry, and m is the weight ratio of washed wet cake to washed dry cake.

The symbol α represents the average specific cake resistance, which is a constant for the slurry in its immediate condition. In the usual range of operating conditions it is related to the pressure by the expression

$$\alpha = \alpha' P^s$$

where α' is a constant determined largely by the size of the particles forming the cake; s is the cake compressibility, varying from 0 for rigid incompressible cakes, such as fine sand and kieselguhr, to 1.0 for very highly compressible cakes. For most industrial slurries, s lies between 0.1 and 0.8. The symbol r represents the resistance of unit area of filter cloth, as well as pressure drop in lines, etc.

Equation (1) can be integrated as follows for constant-pressure filtration, giving the relationship between the total time and filtrate measurements:

$$\frac{\theta}{(V/A)} = \frac{\mu\alpha}{2P}\left(\frac{W}{A}\right) + \frac{\mu r}{P} \tag{2}$$

$$\frac{\theta}{(V/A)} = \frac{\mu\alpha w}{2P}\left(\frac{V}{A}\right) + \frac{\mu r}{P} \tag{2a}$$

For a given constant-pressure filtration, these may be simplified to

$$\frac{\theta}{(V/A)} = K_p\left(\frac{W}{A}\right) + C = K_p'\left(\frac{V}{A}\right) + C \tag{2b}$$

where K_p, K_p', and C are constants for the conditions employed.

Equation (1) may be integrated for constant rate of filtrate flow (or cake deposition) to give the following equation, in which filter-medium resistance is treated as a constant pressure to be deducted from the rising total pressure [Ruth, *Ind. Eng. Chem.*, **27**, 717 (1935)]:

$$\frac{\theta}{(V/A)} = \frac{1}{(\text{rate per unit area})} = \frac{\mu\alpha}{(P - P_1)}\left(\frac{W}{A}\right) \tag{3}$$

which may also be written

$$\frac{\theta}{(V/A)} = \frac{1}{(\text{rate per unit area})} = \frac{\mu\alpha w}{(P - P_1)}\left(\frac{V}{A}\right) \tag{3a}$$

In these equations P_1 is the pressure drop through the filter medium.

$$P_1 = \mu r\left(\frac{V}{A\theta}\right)$$

For a given constant rate run, the equations may be simplified to

$$\frac{V}{A\theta} = \text{rate per unit area} = \frac{P}{K_r} + C' \tag{3b}$$

where K_r and C' are constants for the given conditions.

In the filtration of small amounts of fine particles from liquids by means of bulky filter mediums (absorbent cotton, felt, etc.), it has been found that the above equations based upon the resistance of a cake of solids do not hold, since no cake is formed. For these cases, where filtration takes place in the capillaries of a thick medium, Hermans and Bredee [*J. Soc. Chem. Ind.*, **55T**, 1–4 (1936)] have developed equations which they have found applicable to the constant-pressure filtration of viscose, sugar solutions, etc.

Practical Significance of the Filtration Equations. The differential form [Eq. (1)] of the filtration equation yields interesting information on the mutual effects of the operating variables.

When the cake is composed of hard granular particles that make it rigid and incompressible, an increase in pressure results in no deformation of the particles or their interstices, whereby $s = 0$, and, neglecting filter-medium resistance, Eq. (1) becomes

$$\frac{dV}{d\theta} = \frac{AP}{\mu\alpha'(W/A)}$$

For incompressible cakes, therefore, the flow rate is directly proportional to the area and pressure and inversely to the viscosity, to the total amount of cake (or filtrate), and to α'.

When the cake consists of extremely soft, easily deformed particles, such as ferric and other metal hydroxides, s approaches 1.0, whereby Eq. (1), again neglecting the filter medium, reduces to

$$\frac{dV}{d\theta} = \frac{A}{\mu\alpha'(W/A)}$$

For very compressible cakes, therefore, the rate is independent of pressure.

The *effect of pressure* shown above is modified in most industrial filtrations, where the cake compressibility usually lies between 0.1 and 0.8. Furthermore the resistance of the filter medium reduces the effects of the respective variables. It has been found true, however, that in the filtration of granular or crystalline solids an increase in pressure causes a nearly proportionate increase in flow rate. Flocculent or slimy precipitates have their filtration rates increased only slightly by an increase in pressure. Some materials have a critical pressure above which a further increase results in an actual decrease in flow rate.

In the filtration of certain non-homogeneous sludges, such as those of slimy solids to which filter aids have been added, it has been found that a constant flow rate during filtration is more satisfactory than a constant pressure, which latter results in poor initial clarity of the filtrate and a rapid build-up of cake resistance. As a matter of fact, filtration of any but the most incompressible sludges is more satisfactory when a low pressure is used at the beginning of the run This is especially important in filtering slurries of low solid content.

Since most pressure filters are fed by centrifugal pumps, their operation is seldom either constant pressure or constant rate but, in accordance with the characteristic of the pump, is essentially constant rate during its early stages and constant pressure during much of the latter part of the cycle. Pumps having steep head-discharge characteristics do not operate at either constant rate or constant pressure during any part of the cycle, but always under intermediate conditions of increasing pressure and decreasing flow rate.

Cake thickness is an important factor in determining the capacity and design of a filter, and upon it the cycle of operation depends. Filtration theory shows that, cloth resistance neglected, the average flow rate during a filtration is inversely proportional to the amount of cake deposited.

If the cake has a high resistance relative to that of the filter medium, therefore, the highest capacity of a given filter is reached with zero cake thickness. Consideration of the fact that a thin cake does not usually discharge easily, however, together with the important factor of time required to clean the filter, leads to the selection of an appreciable cake thickness. Filter capacity is often measured in terms of dry solids handled per unit of filtering area.

If the cake has a low resistance compared with that of the filter medium, the economic cake thickness will be increased.

In washing filter cakes it is usually found that there is a definite cake thickness at which a given ratio of wash water to cake solids will produce a minimum soluble salts content of cake. Conversely, the ratio of wash water to cake solids which is found necessary to produce a given soluble content of the cake is a minimum at this cake thickness. In many cases, however, the effect of cake thickness on washing efficiency is not marked. Minimum volume of wash water is desirable since excessive volumes may derange plant procedure.

The *effect of temperature* upon the filtration rate of most incompressible cakes is evident through its effect on viscosity. A temperature rise lowers the viscosity of the filtrate and causes the flow rate to change in inverse proportion to the viscosity.

Many compressible sludges are affected in other ways by temperature change, although the general effect is toward an increase in flow rate with temperature.

The *effect of particle size* on cake and cloth resistances is marked. Even small changes in particle size affect the coefficient α' in the equation for cake resistance, $\alpha = \alpha' P^s$, and larger changes affect the compressibility s. Decreased particle size results in lower filtration rates and higher moisture content of the cake but sometimes in better washing efficiency. It is important, therefore, that close control be kept of the particle size in the feed to the filter. Agglomeration of particles by coagulation is often an important aid in filtering difficultly filterable materials.

It has occasionally been found possible to increase the filtering rate of a slurry by adding larger non-compressible particles to it. Where there is a very wide range in the size of particles in the slurry, however, care must be taken to avoid excessive settling in the filter.

The *effect of the type of filter medium* is often not fully recognized. In selecting the medium for a given filtration, a balance must be struck between as open a weave as possible in order to reduce plugging and as tight a weave as is necessary to prevent excessive "bleeding" of fine particles. After a small thickness of cake has formed on the medium, bleeding often stops, fine particles being caught in the cake.

Of the weaves of filter cloths described under a following section, the number duck weaves have the greatest ability to retain fine solids, followed in decreasing ability by chains (broken twills), twills, and hose ducks. The tendency to plug, however, is in the reverse order. Thick, stiff cloths tend to plug more readily than thin, pliable ones. The effect of cloth plugging on filtration rate is so appreciable that it will ultimately be the cause of replacement of the cloth. It also results in a need for using a safety factor in predicting filter capacities.

The *effect of solid content* of the slurry on the rate of filtration is shown in Eqs. (2*a*) and (3*a*), where it is expressed as w, the weight of cake-forming solids per unit volume of filtrate. These equations show that, filter-medium resistance neglected, the rate of filtrate flow is inversely proportional to the ratio of solids to filtrate but that the rate of cake deposition is directly proportional to this ratio. If a slurry is thickened before filtration, time required for its filtration on a given filter area will be reduced in direct proportion to the decrease in ratio of liquid to solids in the slurry.

Application of Filtration Theory to the Interpretation of Data

The filtration equations are useful in predicting the effect of a change in any variable if the constants are determined from data taken on the slurry in question. For example, vacuum test data can be extrapolated to show the approximate filtering rates that could be obtained if the slurry were filtered under pressure. Another problem often of interest is the effect of cake thickness or time cycle on over-all filtration rate.

If a *constant-pressure test* is run on a slurry, care being taken that not only the pressure but also the temperature and the solid content remain constant throughout the run and that time readings begin at the exact start of filtration, one can observe values of filtrate volume or weight and time. With the use of the known filtering area, values of $\theta/(V/A)$ can be calculated for various values of (V/A) which, when plotted with $\theta/(V/A)$ as the ordinate and (V/A) as the abscissa (Fig. 59), result in a straight line having the slope $\mu\alpha w/2P$, and an intercept on the vertical axis of $\mu r/P$. Since μ, w, and P are known, α and r can be calculated from

$$\alpha = \frac{2P}{\mu w} \times \text{(slope)}$$

and

$$r = \frac{P}{\mu} \times \text{(vertical intercept)}$$

The effect of a change in any variable except P (which affects α) may now be estimated.

To determine the effect of a change in pressure, it is necessary to run a test under one or more other pressures, and to calculate α and r at those pressures in the same way. By plotting α and r against P on loglog paper (or by plotting log α and log r vs. log P), straight lines result from which it is possible to determine α and r at any reasonable pressure (Fig. 60). In many cases it has been found that r does not vary appreciably with pressure, in which case an average value can be used at all pressures. This is often due to the fact that a low filtering pressure is used at the start of filtration.

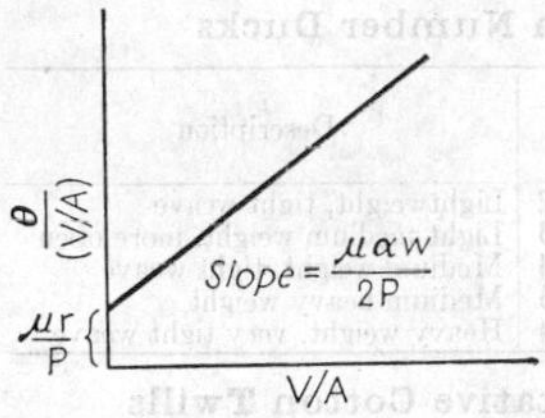

FIG. 59.

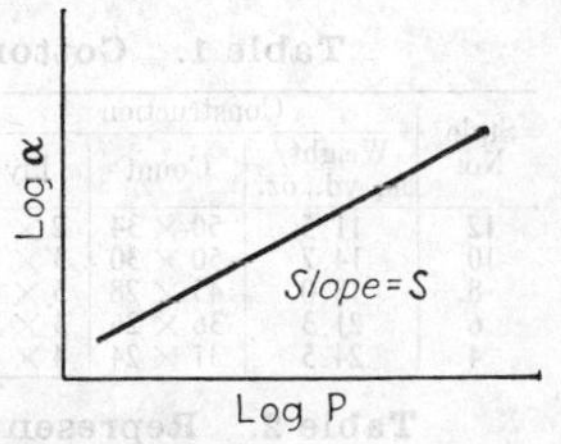

FIG. 60.

FIGS. 59 and 60. Typical plots of filtration data.

When a low pressure is used for any appreciable time at the start of filtration, the beginning of time and filtrate readings should be delayed until the constant pressure is reached, in which case r is the resistance of the filter medium plus that of the cake deposited at low pressure. When the weight of the dry cake is measured vs. time, as is usually done in vacuum leaf tests, two or three tests are sufficient to permit plotting the straight-line function of θ/V vs. W. The ability to interpolate or extrapolate on this line eliminates the need for a large number of tests.

In *constant-rate filtration* it is suggested that the method of Bonilla [*Trans. Am. Inst. Chem. Engrs.*, **34**, 243 (1938)] be used, involving the determination of P_1, and α_0 in the equation

$$\alpha = \alpha_0 + \alpha'(P - P_1)^s$$

and plotting $(\alpha - \alpha_0)$ vs. $(P - P_1)$ to determine α' and s.

FILTER MEDIUMS

By Hugh W. Bellas

The choice of the filter medium is often the most important consideration in assuring efficient operation of a filter. This is true in spite of the fact that in most filtrations the medium does not do the actual filtering but merely acts as a support for the cake of solids that is deposited and on which the separating process takes place.

The filter medium should be selected primarily for its ability to retain the solids without plugging and without undue bleeding of particles at the start of filtration. However, all the following attributes of a good filter medium should be sought in varying degrees depending on the specific problem:

1. Ability to bridge solids across its pores within a reasonable time after beginning to feed.
2. Minimum resistance to flow of filtrate.
3. Avoidance of wedging particles into its pores, thereby greatly increasing resistance to flow.
4. Sufficient strength to withstand the filtering pressure and mechanical wear.
5. Resistance to chemical attack.
6. Smooth surface for easy discharge of cake.

Filter mediums are manufactured from cotton, wool, linen, jute, silk, glass fiber, nitrated cotton, porous carbon and other solids, metals, rayon and other synthetics, and miscellaneous materials such as porous rubber.

Cotton fabrics are by far the most common type of medium, primarily because of their low first cost and availability in a wide variety of weaves. Cotton is attacked by all mineral acids and by many organic acids that can crystallize at operating temperatures, but it is not usually affected by volatile organic acids. Strong alkalies, acidic salts, and metallic ammonium salts also attack cotton. Operating temperatures should always be kept below 200°F. In describing cotton fabrics, reference is made to (1) weave, (2) style number, (3) weight, (4) count, (5) ply, and (6) yarn number. Of these, only the *style number* is completely definitive, but unfortunately it is a purely arbitrary number assigned by each manufacturer, except in the case of the so-called "number ducks." *Yarn number*, defining the weight of the original twisted filaments, is set by the weight of the fabric and its other characteristics and so is seldom a factor for the user to consider. *Ply* is defined as the number of small yarns twisted together to form the final thread, whereas *count* expresses the number of threads per inch in each direction. *Weight* is best given in terms of ounces per square yard. Fabrics of heavy weight and low count, in multi-ply construction, make the strongest cloths but in general show more tendency either to plug readily or to retain coarse solids. A wide variety of weaves is available, of which the simplest is the *plain weave*, in which the cross threads (filling) are woven over and under the long threads (warp) alternately, giving a somewhat square appearance.

Cotton ducks are the most common of the plain weaves, the term covering a wide range of constructions. They are low in first cost, have good mechanical strength and resistance to wear, and discharge cakes readily; but, when sufficiently tight to retain fine solids, they have high resistance to flow and plug rather quickly. The "number ducks" listed in Table 1 are frequently used in filter presses. Also of interest are other plain-weave cloths such as "ounce ducks," *e.g.*, 10-oz. duck, having a loose open weave useful only for coarse solids; "hose ducks," rather heavy and still more open in weave; cider-press cloths of about ⅛-in. thread spacing, useful as a backing cloth; sheetings, lightweight (2 to 8 oz./sq. yd.) open weaves for mild service; and specially finished weaves for straining "dopes," etc., such as cambrics, nainsooks, combed lawns, balloon cloths, and voiles.

Cotton twills are characterized by a diagonal weave resulting from interlacing the warp and filling yarns with a progression of one at the point of interlacing. They have less resistance to flow and less tendency to plug than ducks but consequently are more likely to pass fine particles at the start of filtration. Table 2 lists the six most common weaves, which serve for the average filtering problem. In the selection of a twill, the weight is important for strength and mechanical wear, while increasing count improves the ability to retain fine solids. Heavy cloths are necessary in large filter presses to provide satisfactory gasketing, also. *Drills* are a lightweight variation of the twill weave, while *canton flannel* has a nap on one side, making it useful in clarifying operations not requiring easy cake removal.

Cotton chain weaves, or broken twills, are woven with a 1-2-4-3 interlacing of warp and filling threads.

They are usually intermediate between twills and number ducks in tightness, plugging tendencies, and other properties. At moderate or low operating pressures, they are superior to number ducks from most standpoints. Here again, high-count weaves are tighter than low-count fabrics of equal weight.

Wool cloths are sometimes used in filtering dilute acid solutions and in clarifying viscous liquids. They are characterized by severe plugging tendencies and rapid attack by alkalies, however. *Jute* cloths have been used extensively for filter pressing of coarse solids. *Nitrated-cotton* fabrics and *human-hair* cloths have found use in filtering sulfuric acid solutions up to about 30 per cent in strength, and other treatments of cotton have been applied for improving resistance to mildew, weak acids, and alkalies. *Fibrous-glass* fabrics have been used for severe acid or temperature conditions where physical strength and wear resistance are not required, as in leaf filters. The flexing strength of glass fabrics has been improved by admixture of asbestos. *Silk* has been used in a very light, plain weave on string-discharge filters, where its free-filtering properties are utilized under mild mechanical conditions.

In the field of **synthetic fibers**, *nylon* has found use as a silk substitute and for resistance to alkalies and weak acids. *Vinyon* has been extensively employed for its resistance to all concentrations of acids and alkalies, including sulfuric acid up to about 70 per cent strength. The maximum temperature limitation of 130°F. can be exceeded with a new fiber, Vinyon-N, which is claimed to be satisfactory up to 180°F. Exposure of Vinyon to higher temperatures results in softening of the fiber and severe shrinkage. Table 4 lists some typical Vinyon constructions.

Metal fabrics are available in steel, KA2S, KA2S-Mo, monel, nickel, copper, brass, bronze, aluminum, and Everdur, and in several types of weave. In the plain weave, 400 mesh is the closest wire spacing available, thus limiting the use of this weave to coarse crystalline slurries, pulps, etc. The so-called "Dutch weaves," employing straight warp wires and crimped filling, can be woven much closer, providing a good medium for filtering fine crystals and pulps. However, this type of weave tends to plug readily when soft amorphous particles are filtered, making the use of filter aid desirable. The long life of the proper wire cloth in corrosive and high-temperature filtrations makes it desirable to install them more or less permanently.

Two types of **rubber** medium were available that resist dilute acids and alkalies at temperatures up to 160°F. *Multipore* is a perforated sheet suitable for filtering relatively coarse particles, and *Microporous rubber* can remove very fine solids. Their manufacture has now been discontinued.

Porous-block mediums are made from a number of materials. *Carbon and graphite* in the form of plates, tubes, and special shapes are available in a wide range of porosities and are resistant to all acids and alkalies under non-oxidizing conditions. *Aloxite, alundum, silica,* and *porcelain* are similarly utilized and are resistant to all acids except hydrofluoric, but not to strong alkalies. The medium can be mounted in press plates, filter leaves, nutsches, or other filter types, although this type is most commonly used in removal of relatively small amounts of solids from corrosive slurries, or in "polishing" of beverages, filter aids often being employed.

Cotton batting finds extensive use in filtering gelatinous particles from paints, spinning solutions, and other viscous liquids, and for removal of dirt from milk, etc., the batting being discarded after use. Filtration takes place by deposition of the particles on the fibers throughout the mat, which is usually supported on both sides by gauze or light sheeting. Filter cotton is available in various stages of refinement and purification, from soft absorbent cotton to wool-like material with low filtering resistance. *Cotton table felt* is also used in clarifying liquids.

Wool felts and cotton-wool mixtures are available in both pressed and woven types. The former are less expensive, but the latter can be reused after cleaning and/or washing.

Filter papers and pulps are often used for the retention of very fine solids and for the clarification of liquids containing small amounts of solids. They are available in various degrees of permeability, thickness, and strength, and some are resistant to strong acids and alkalies up to 30 per cent concentration. They must be well supported in the filter; and, if a cake is to be removed without destroying the paper, they must be covered with a sheeting.

Granular beds, such as sand and coal filters, are widely used for filtration of water and chemical solutions to remove small quantities of easily coagulated solids. They are cleaned by backwashing at a rate sufficient to disrupt the bed and carry out the fine solids.

Table 1. Cotton Number Ducks

Style No.	Construction			Description
	Weight/sq. yd., oz.	Count	Ply	
12	11.5	50 × 34	2 × 2	Lightweight, tight weave
10	14.7	50 × 30	3 × 3	Light medium weight, more open
8	18.0	45 × 28	3 × 4	Medium weight, tight weave
6	21.3	36 × 26	3 × 3	Medium heavy weight
4	24.5	31 × 24	4 × 4	Heavy weight, very tight weave

Table 2. Representative Cotton Twills

Classification		Construction (approx.)			Uses
Weight	Count	Weight/sq. yd., oz.	Count	Ply	
Light	Low	15.5	38 × 28	4 × 4	For light service, *e.g.*, vacuum filters
Light	High	15.5	66 × 44	2 × 2	For light service, *e.g.*, vacuum filters
Medium	Low	17.5	36 × 25	3 × 3	For general use, including leaf filters
Medium	High	18.0	67 × 36	2 × 4	For general use, including leaf filters
Heavy	Low	22	34 × 24	4 × 4	For severe service, *e.g.*, filter presses
Heavy	High	20	58 × 42	3 × 4	For severe service, *e.g.*, filter presses

Table 3. Representative Cotton Chains (Broken Twills)

Classification		Construction (approx.)			Uses
Weight	Count	Weight/sq. yd., oz.	Count	Ply	
Very light	High	12	56 × 50	2 × 2	Vacuum filters, etc.
Light	High	15	68 × 42	2 × 4	Vacuum and low-pressure work
Medium	High	18	57 × 37	3 × 3	For general use
Heavy	Low	22	34 × 30	3 × 4	For filter-press service, etc.
Heavy	High	20	67 × 38	3 × 5	For filter-press service, etc.

Table 4. Representative Vinyon Weaves

Style No.	Weave	Construction			Porosity ranking*
		Weight/sq. yd., oz.	Count	Ply	
1	Plain	13.66	54 × 34	3 × 5	B
2	Chain	14.28	54 × 42.5	3 × 5	D
3	Twill	26.33	36 × 25.5	10 × 10	F
5	Twill	29.25	42 × 26	10 × 10	E
7	Chain	23.68	63.5 × 36	3 × 10	C
13	Chain	26.40	36 × 25	10 × 10	E
17	Plain	22.23	67 × 23.5	5 × 10	A

* A is the tightest weave, F is the most open.

FILTRATION LEAF TESTS

By Donald F. Irvin

It is unusual to be able to forecast what may be accomplished in the filtration of an untested product, and even the results obtained upon known products vary greatly with the conditions of filtration. Therefore, unless exact data have already been established, preliminary tests should be made to determine the filter requirements for a given filtration problem. Such tests are easy to make and require very simple, small-scale test equipment. Whether vacuum or pressure filtration is to be used is generally known beforehand. Occasionally tests are made for comparison.

Vacuum Tests. The leaf shown in Fig. 61 is connected to a filtrate receiver equipped with a vacuum gage. The receiver is connected to an aspirator. Different filter mediums may be used on this leaf for comparative tests.

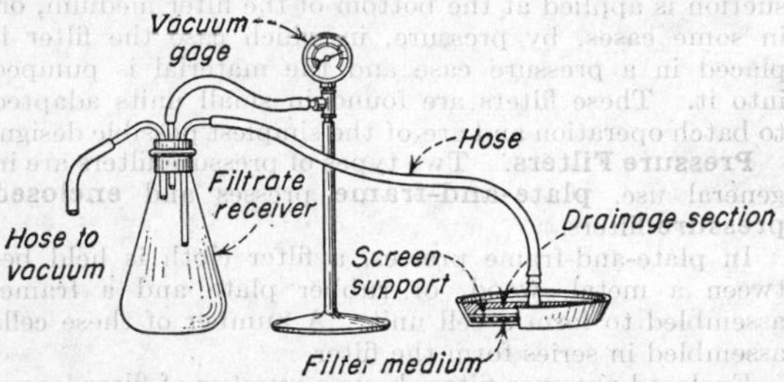

FIG. 61. Small-scale vacuum filtration testing unit. *(Oliver United Filters, Inc.)*

In making leaf tests, the operation of a continuous vacuum filter should be kept in mind. The cycle is divided into three periods, **cake formation** (or "pickup"), **drying**, and **discharge**. Sometimes pickup is followed by a period of displacement washing, and the cake may also be subjected to compression during drying. These things should be considered, and a plan of the cycle or cycles to be tested should be formed.

If the object of filtration is simply the removal of solids from the liquor, the cycle may be: one-third pickup, one-third drying, and one-third discharge and reentry time. While under vacuum, the test leaf is submerged for the pickup period in the material to be tested. The leaf is then removed and held with the drain pipe down for the drying time allotted. Observations should be made during the test such as vacuum readings during pickup and drying; time at which cracks in the cake appear; temperature of the material; percentage of cake-forming solids present; acidity or alkalinity.

Usually a few preliminary tests will indicate the time range. Careful tests may then be made and, in these, variations in temperature, dilution, conditioning agents, etc., should be tried, and capacities and clarity of filtrate noted.

Pressure Tests. For plate-and-frame press work, tests are best made with a laboratory-size model. This will give representative "cake packing," etc. The apparatus shown in Fig. 62 is used for tests to obtain data for operation with a shell-type pressure filter.

Operation of the commercial unit should be kept in mind and the cycle arranged accordingly. After determining the cake-building or filling time, displacement washing and drying the cake with compressed air should be tried. For wet discharge it is advisable to open the cell and experiment upon washing the cake away with a jet of water. For dry discharge the effect of a gentle air blast in the test leaf should be tried.

In both vacuum and pressure tests the daily filter capacity is determined by the dry weight of cake per unit area of test leaf multiplied by cycles per 24 hr. and multiplied by the filter area. Capacity in solids is usually expressed in pounds per square foot per day and filtrate in gallons per square foot per minute or per day.

The material tested should be representative, and samples should be tested immediately after they are taken. In some instances, samples stored for several days have given results very different from those obtained when tested immediately because of changes that occur upon standing. All tests should be made under conditions that represent large-scale operations so far as possible.

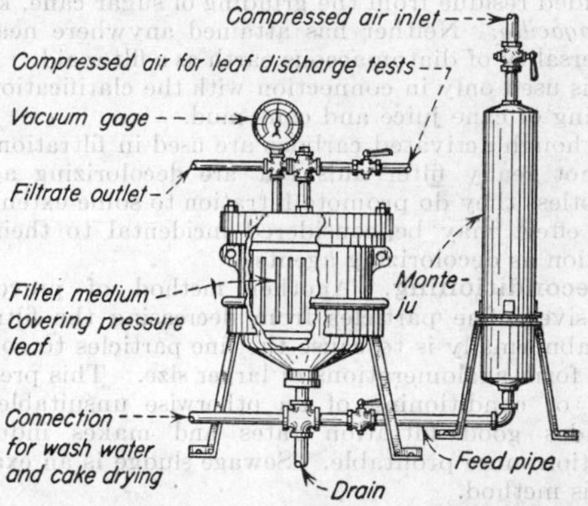

FIG. 62. Small-scale pressure-filtration testing unit. *(Oliver United Filters, Inc.)*

Results obtained by leaf tests for capacity are irregular with extremely free-filtering materials, such as crystals in mother liquor. In such cases it is better to employ small-scale equipment.

Before undertaking test work it is advisable to consult a manufacturer of filtration equipment, giving as many data on the materials as possible, together with the objects of filtration.

FILTER AIDS*

Filter aids are useful when handling finely divided solids and colloidal materials. A filter aid should be of low specific gravity so that when mixed in the liquid to be filtered it will remain in suspension. It should be porous rather than dense and must be chemically inert to the liquid being filtered.

Kieselguhr or diatomaceous earth having a high silica content is little affected by solutions, is free filtering, and is of light gravity. It is the most widely used of all filter aids in the filtration of sugar juices, vegetable oils, petroleum products, fruit juices, beverages, etc. Paper pulp is also used in the clarification of wine and beer.

The amounts of filter aid added are comparatively small, and the expense is more than counterbalanced by increased filter efficiency. Both paper pulp and kieselguhr can be washed and revivified so that they may be reused several times. Fuller's earth, charcoal, asbestos, sawdust, magnesia, salt, and gypsum are used as filter aids in special cases.

Decolorizing carbons and earths, such as Darco, Carbrox, Suchar, Norit, Filtrol, Palex, and activated clays act both as decolorizers and as filter aids for oils, fats, etc. In many cases, a coating of the filter aid is applied to the filter medium to act as a clarifying agent and to prevent blinding of the filter medium.

The most commonly used filter aid is diatomaceous earth prepared by various manufacturers. This material, being skeletal remains of diatoms, has a very high filter rate, does not fill the pores of filter mediums, and is used either as a precoating of the filter medium itself, or as a pulp mixture which must be filtered. Various degrees of purification used on the raw diatomaceous

* See p. 885.

earth provide a material with varying filtration properties. These different types are given various trade names identifying them for certain work, as Filter-cel, Dicalite, etc.

For specific amounts of filter aid, the producers of various grades give their own recommendations.

Other materials have been employed as filter aids, among them macerated paper pulp, and the finely shredded residue from the grinding of sugar cane, known as *Bagacillo*. Neither has attained anywhere near the universality of diatomaceous earth as a filter aid. Bagacillo is used only in connection with the clarification and filtering of cane juice and cane mud.

Although activated carbons are used in filtration they are not really filter aids but are decolorizing agents. Doubtless they do promote filtration to some extent, but such effect may be considered incidental to their real function as decolorizing agents.

Preconditioning. Another method of preventing excessively fine particles from decreasing the filtration rate abnormally is to cause the fine particles to coalesce or to form agglomerations of larger size. This pretreatment or conditioning of an otherwise unsuitable feed provides good filtration rates and makes industrial filtration more profitable. Sewage sludge is an example of this method.

Coagulation of sewage sludge prior to filtration is effected by the use of such reagents as alum, ferric chloride, or other chemicals.

Effect of Temperature and Viscosity. By increase of temperature, water decreases in viscosity and gives rates of flow proportional to the following data:

Temperatures	0°C.	20°C.	40°C.	60°C.
Rate of flow	1.0	1.8	2.7	3.7

Thus the rate of flow is doubled by raising the temperature from 20° to 60°C. In practice there are many cases where the filter capacity is materially increased by heating the filter feed, as when filtering cement slurry, clays, some flotation concentrates, sirups, oils, etc. The economy of heating a given filter feed may be determined by tests and computations.

When a concentrated solution is filtered, its viscosity may be high and give a low filtration rate. By diluting the filter feed with fresh water, the viscosity of the solution is reduced and gives a higher rate of filtration. The total filtration time for increased bulk of filter feed as diluted may be much less than for the smaller bulk of the more viscous strong solution.

Whether dilution can be adopted depends upon the purpose of filtration. If the solution is required at high concentration for subsequent treatment, this may preclude dilution, but if it is to be discharged afterward or if it may be reconcentrated by evaporation, then dilution may be allowable.

TYPES OF FILTERS

Filters may be conveniently grouped under four heads:

1. Gravity filters.
2. Pressure filters.
3. Intermittent vacuum filters.
4. Continuous vacuum filters.

Centrifuges and centrifuging are discussed in the next article, p. 992.

Gravity Filters. A gravity filter generally consists of a tank with a false floor covered by a filter medium. Leaching tanks used in cyanide plants have a cloth-covered filter bottom and may be termed "gravity filters." By far the largest number of gravity filters employ a bed of sand as the filter medium and are used for clarifying solutions or water. To be effective such a filter medium must be comparatively thick and the amount of solution large compared with the amount of removable solids. Strainers and sand and charcoal filter beds, used for water purification, are good illustrations of gravity filters. Gravity filters are often useful for small-batch operations in chemical industries where corrosion is excessive.

The Nutsche Filter. This is the simplest form of gravity filter and is made by only one U.S. manufacturer, being usually built by the plant in which it is to be used.

In most cases it consists of a simple support for the filter medium chosen and a vessel in which this filter medium is placed. It may be operated either by gravity drainage through the filter medium, by vacuum in which suction is applied at the bottom of the filter medium, or, in some cases, by pressure, in which case the filter is placed in a pressure case and the material is pumped into it. These filters are found in small units adapted to batch operation and are of the simplest possible design.

Pressure Filters. Two types of pressure filters are in general use, **plate-and-frame** presses and **enclosed pressure** filters.

In plate-and-frame presses, a filter cloth is held between a metal, wood, or rubber plate and a frame, assembled to form a cell unit. A number of these cells assembled in series form the filter.

Enclosed pressure filters have a number of filter leaves suspended inside a shell into which the material to be filtered is charged under pressure. The leaves may be parallel or perpendicular to the horizontal axis of the filter shell. In most instances the leaves are stationary, but in some cases they can be rotated during the filtration cycle. In one instance the filter medium is installed as an inner lining of the shell and rotates with the shell.

The principle of operation of all pressure filters is the same. A filtering medium is stretched over a frame provided with channels for the collection and drainage of solution, and the material to be filtered is forced under pressure into the space between the filter medium and outer housing or frame.

Plate-and-frame presses must be taken apart and cleaned by hand at the end of each cycle. Pressure filters having leaves enclosed in a shell are opened and closed by mechanical means, and manual labor is reduced. In both types, a sluicing mechanism may be employed for discharging the cake, resulting in a material saving in time and decreased maintenance cost.

A disadvantage common to all pressure filters is their intermittent operation. The advantages are that high pressures may be used to dry the cake. Higher costs for labor and renewals of filter medium are characteristic of pressure filters.

Life of Wood Used in Plate-and-frame Presses under Acid-filtrate Conditions.* The Independent Filter Press Co., Inc., of Brooklyn, New York, advises that the life of various types of lumber under acid filtration differs considerably, and it is not possible to give a definite figure thereon.

They do state that, after trying various kinds of lumber, they chose longleaf yellow pine to be the most durable. Even so, they state that the degree of acidity is the most important factor. Two cases are cited wherein longleaf yellow-pine plates lasted, in one instance, under normal acid filtration, for 2 to 3 years; on the other hand, in a particular case where the conditions were trying, a set of longleaf yellow-pine plates and frames lasted about 2 months. No record is available by this company as to the relative life of longleaf yellow pine compared with maple or cypress in this work.

Intermittent Vacuum Filters. The intermittent type generally consists of a series of frames or leaves over

* See p. 1552.

which the filter medium is stretched, the leaves being provided with channels for the drainage of liquid.

Several leaves are connected to a common header, which is in turn connected to a vacuum line by flexible hose. By completely submerging the leaves in a tank of material to be filtered and applying vacuum, the cake is formed. As soon as sufficient cake has been formed, washing and discharging of the cake are performed by hoisting the leaves out of the tank and, while still under vacuum, transporting them to a tank containing wash water or to the point where discharge is to be made. Discharge is accomplished by release of vacuum and inflating the leaves by compressed air.

The intermittent vacuum filter gained a wide use for some years in cyanide-plant operation, in separation of gold- and silver-bearing solutions, but has been supplanted generally by continuous vacuum filters.

Continuous Vacuum Filters. Continuous vacuum filters are of two types: the **rotary drum** and the **rotary disk**. The drum type is a cylinder whose periphery forms the filtering surface; this surface may be either external or internal and is divided into separate compartments. Each compartment is separately connected to an automatic control valve which regulates the period under vacuum for forming, and compressed air for discharging, the filter cake. The internal drum filter receives its feed inside the drum, but the external drum filter is mounted in a feed tank.

The top feed filter is an external drum filter that has no tank. Feed is supplied near the top of the drum, and a hopper below the drum receives the cake when discharged.

The rotary-disk type has its filtering medium shaped in segments that are assembled to make up a disk. Each segment is connected, through a central axis or shaft, to an automatic valve, similar to that used on the rotary-drum filter. A number of disks may be assembled on the common shaft.

MAKES OF FILTERS

Plate-and-frame Presses. A plate-and-frame press consists of a series of solid vertical plates and hollow frames cast with side lugs, so that they may be mounted on two parallel horizontal bars and clamped together. Figure 63 illustrates a typical press. Each plate and

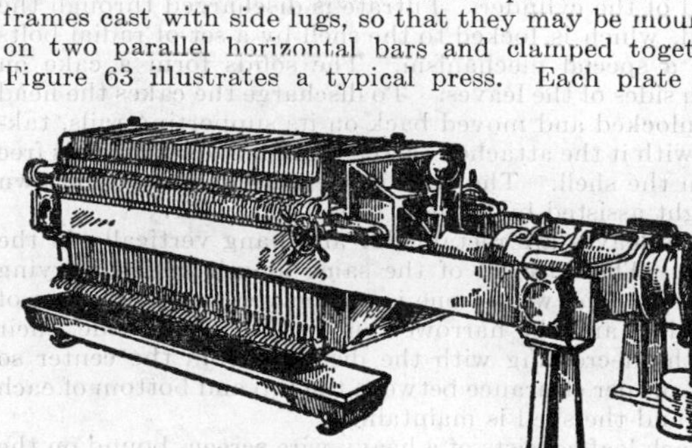

Fig. 63. Shriver plate-and-frame filter press.

frame is accurately machined to give a tight joint when clamped together with a cloth between them, and the faces of the plate are ribbed and channeled or have a pyramid surface to provide for drainage of the filtrate (Fig. 64). A filter cloth is laid over each plate to cover both faces, and a frame is set on each side of it so that, by alternating plates and frames and clamping them together, a series of filter cells is formed. Each cell consists of the empty frame bounded on both sides by filter cloth behind which are the plates.

One end of the series is closed by the head of the press and the other end by the final plate against which a capstan screw is tightened, thereby clamping the series together and enabling the whole to be operated under pressure. A ratchet gear and pinion or hydraulic closing device may be used to obtain greater force. The feed channel is formed by a hole in each plate and frame, these holes registering together. In each frame there is an opening from this channel that admits feed into the frame, and at the bottom of each plate there is an outlet for the filtrate. If the filtrate from a plate becomes turbid, that particular plate is removed from service by closing the outlet cock. The filter cloths are examined after each run, and any defective ones are replaced.

When it is necessary to wash the filter cake, two methods are used, depending upon the type of plates. In one of these, the wash water is forced through the feed channel and follows the filtrate. In the other method, wash is supplied from a channel that passes through the plates and frames similarly to the feed channel. From this wash channel there is an inlet in one corner of each alternate plate. The former construction is used for the method of operation called center filling and the latter for solid filling.

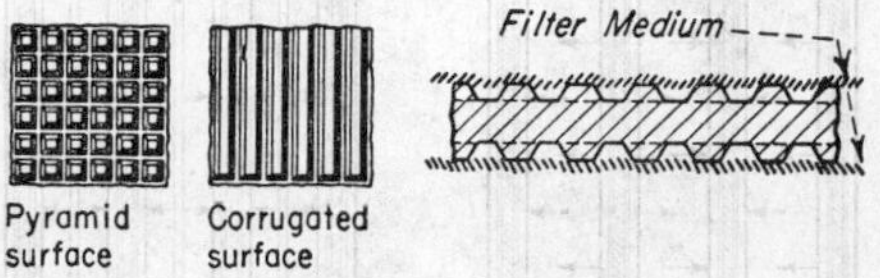

Fig. 64. Filter plate.

Solid Filling. When the solids quickly form a thick cake at low pressure (25 to 50 lb./sq. in.), feed is continued until the frames are completely filled, as indicated by the filtrate ceasing to flow from the drain cock at the bottom. Filling may be followed by an air blow to expel solution before wash is applied; and, as the cell is completely filled with solids, no displacement of cake can occur, though there may be cracking during this air blow. Washing is then performed by admitting wash solution or water under pressure behind the filter cloth of each alternate plate. The wash is forced through the cake and the cloth on the opposite side and issues through the drain cock shown in Fig. 65. Air blow may again be given to displace the wash and to obtain as dry a cake as possible. Dryness of cake is an advantage of plate-and-frame presses.

Center Filling. In this method the feed is stopped while there is still a space down the center of each frame, thus dividing the cake into two halves (Fig. 65). Wash solution is forced in behind the feed without allowing the pressure to drop so that there is no interval during which the cake can dry and crack or slough off. Washing proceeds from the center out toward both sides of the frame so that less washing time is required than with solid filling. This is an advantage when dealing with slimy materials that are slow in cake building and washing. However, there is the disadvantage of diffusion of wash with the last portion of filtrate.

Discharge. When filling, washing, and air blow are finished, the cake is discharged. The labor involved in this discharging is the chief drawback of presses. The press screw is released, the plates and frames separated, and the frames are emptied into a tray or into a conveyor below the press.

Open and Closed Filtrate Discharge. An open press is one having a control cock to each plate outlet, all outlets delivering to a launder. A closed press collects its filtrate in a channel inside the press similar to the feed channel.

The advantages of plate-and-frame presses are simple construction, low cost, simplicity of operation, and dryness of cake. They may be used for high pressures, and for acid filtrates the plates and frames may be made of wood, stainless steel, rubber-covered steel, vulcanized rubber, and the like.

The disadvantages are high labor charges when handling large tonnages and high cloth consumption due to damage at the joints when opening and closing the press. Washing is imperfect, and the percentage of idle time in the cycle is high because of the time required for opening and closing.

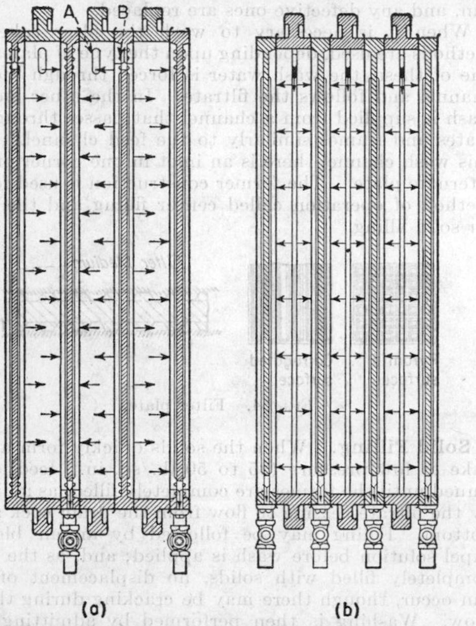

FIG. 65. Cross sections of plate-and-frame filter presses, showing operation of solid (*a*) and center (*b*) filling.

Failure at the edges of a cloth where it serves as a gasket between plate and frame cannot be remedied without dismantling the press and installing a new cloth; so, for the balance of the cycle, the feed leaks at the point of defect, resulting in losses and a dirty filtering plant.

The plate-and-frame press was the pioneer in filtration and still finds frequent use.

Merrill Press. The Merrill press is essentially a plate-and-frame press with an automatic sluicing device that enables the cake to be discharged without opening the press.

Along the median line at the bottom of the press, and passing through the plates and frames, is a continuous channel, within which is a sluicing pipe with nozzles, one projecting into each compartment. This pipe is slowly rotated back and forth through an arc of approximately 180 deg., by means of a rack and pinion, driven from a line shaft or small individual motor. Simultaneously water under pressure issues from the nozzles and plays upon the slime cake in the chambers, washing it down into the annular space around the sluicing pipe from which it leaves the press through a number of discharge cocks. The press finds application in cases where discharge in the wet stage is permissible. It was originally developed to handle a large daily tonnage of slime in which the treatment with cyanide took place (at least partly) inside the press. Its success marked an advance in slime filtration by plate-and-frame processes, since manual labor was reduced and cloth life lengthened. However, pressure filters have now been generally replaced by continuous vacuum filters for this work.

Merrill Precipitation Press. The collection of gold and silver precipitated from cyanide solution by zinc dust was formerly effected by using a plate-and-frame filter press of triangular cross section. This duty is now performed by an ordinary press of rectangular cross section or by a series of cylindrical cloth bags. The bags are fixed on a header connected to the discharge of a centrifugal pump and are at all times submerged in barren solution which passes outward through the cloth bags. To clean up the bags, the solution is drained from the tank, the bags are blown with compressed air, and the inner bag liners are then removed for fluxing and melting.

Kelly Filter. The Kelly filter consists primarily of a steel cylindrical tank enclosing a number of rectangular filter leaves. The axis of the tank is set approximately horizontal (Fig. 66). The leaves are

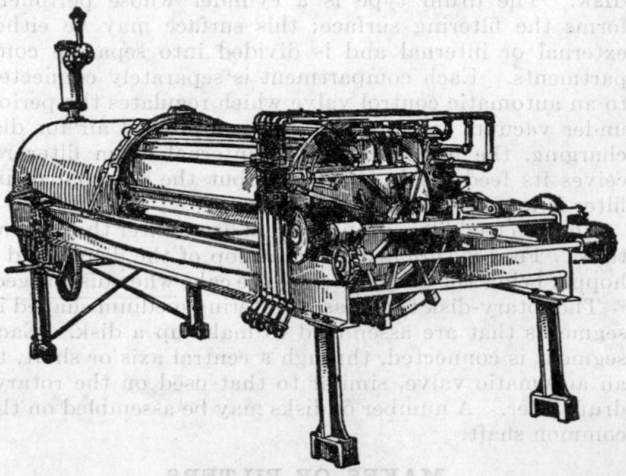

FIG. 66. Kelly filter.

supported by a carriage attached to the movable head of the cylinder. Filtrate is discharged through the head, which is locked to the shell by a set of radial bolts and a special mechanism. The solids form a cake on both sides of the leaves. To discharge the cakes the head is unlocked and moved back on its supporting rails, taking with it the attached carriage and leaves until all is free from the shell. The cake is then discharged by its own weight assisted by a slight back blast of air.

The leaves are rectangular and hang vertically in the shell. They are all of the same length but of varying widths. The widest one is on the vertical center line of the shell and the narrower ones are on either side, their widths decreasing with the distance from the center so that proper clearance between the top and bottom of each leaf and the shell is maintained.

Each leaf consists of a heavy wire screen, bound on the four edges with a rolled-steel shape similar in cross section to a slotted flattened pipe. The screen forms the drainage element, and the steel shape provides rigidity, protects the filter cloth from the edges of the screen, and serves as a channel for the filtrate. The upper corners of the leaves next to the head are connected to the head by nipples and unions forming outlets for the filtrate. The filtrate passes through these fittings and then through passages in the head to a trough or filtrate header outside.

The leaves are enclosed in bags of filter cloth. These bags are made with the front end open; they are slipped over the leaves, and the open ends are sewed by hand. Metallic filter cloths may be used.

The largest Kelly filters are of the twin type, *i.e.*, there are two shells mounted on the same beams with the head

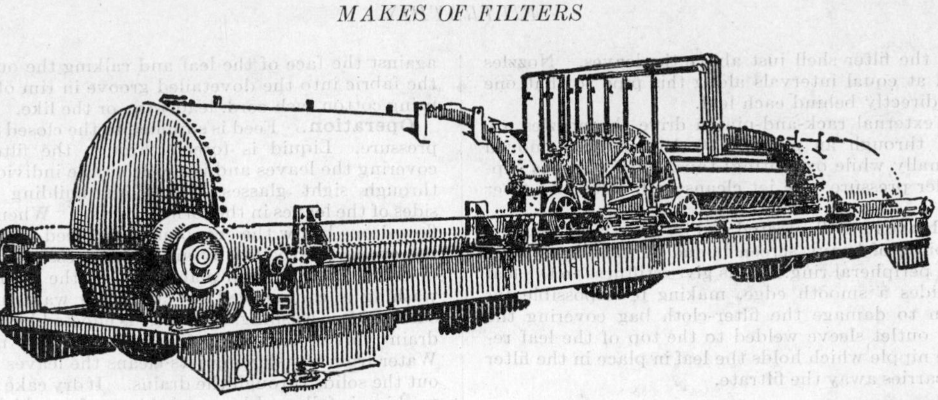

FIG. 67. Kelly filter (twin type).

ends facing one another and using a space common to both for running out the carriages. This arrangement (Fig. 67) gives a greater filter area per unit of floor space and therefore reduces the cost of the filtration process.

The two halves of the unit are opened and closed automatically, means being provided for releasing the chain opening device (which is common to both shells) from one half while opening and closing the other half. The maximum travel for the carriage of either half is fixed by a spring bumper.

Kelly filters of standard design operate at pressures up to 60 lb./sq. in. Where higher pressures are used, greater strength is secured by heavier steel plates for the shell and by making the closing head of cast steel instead of cast iron. They are operated at pressures as high as 250 lb./sq. in. and are well adapted to high as well as subzero temperatures, because of the ease of insulation.

Operation. The material to be filtered is charged into the shell under pressure. The filtrate is forced through the cloth and through outlets from each frame into a launder, while the cake is being formed on both sides of the leaves.

Filtration is stopped when the desired cake thickness is obtained. The excess feed is drained off, and, as the level drops in the shell, the float of an automatic air regulator drops and opens an air valve. This admits air at a pressure of 3 or 4 lb./sq. in., which holds the cake upon the leaves and forces the excess liquor more rapidly from the shell.

After draining, the wash liquid is forced into the shell, following the same path as the filtrate. Excess wash is drained off, and if a dry-cake discharge is desired the drain valve is closed and an air blast is used.

For discharging, the air is turned off, the release valve is opened, the head is unlocked, the carriage rolled out, and a slight back blast of air causes the cake to discharge.

For wet discharge, the excess feed is drained out, the filter is opened, the carriage run out, and the cake sluiced off.

The Kelly filter competed with the intermittent vacuum filters in early cyanide-slime operations and later was adopted in beet-sugar manufacture.

It is now widely used in petroleum refining, for handling lubricating oils, for pressure-still sludges, and for dewaxing operations, for which duties its design makes it especially suitable.

Sweetland Filter. The Sweetland filter (Fig. 68) consists of a series of circular filter disks suspended inside a cylindrical cast-iron shell transversely to its axis. The shell is divided along the horizontal center line into two halves, hinged together along the back. The upper half is rigidly fastened to supports, so that the lower half may swing open, thus exposing the interior of the filter for cleaning. The lower half (Fig. 69) is counterweighted to facilitate opening and closing, and a special locking mechanism makes it possible to open or close the filter within a fraction of a minute. The edges of the two halves of the cast-iron shell are accurately machined and grooved to hold a composition gasket which forms a tight joint when the filter is closed.

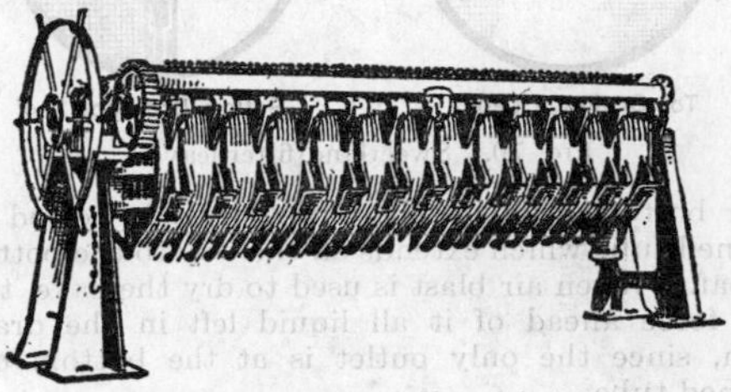

FIG. 68. Sweetland filter (closed).

A "boss" is cast along the top half of the filter body, and holes are drilled through to receive the filter-leaf outlet nipples. Each hole is counterbored on the inside to receive a filter-leaf rubber washer and, on the outside, to receive the cap nut and the lead washer.

Inside the upper half, leaf spacers are placed along the front and back sides of the filter to keep the leaves in alignment.

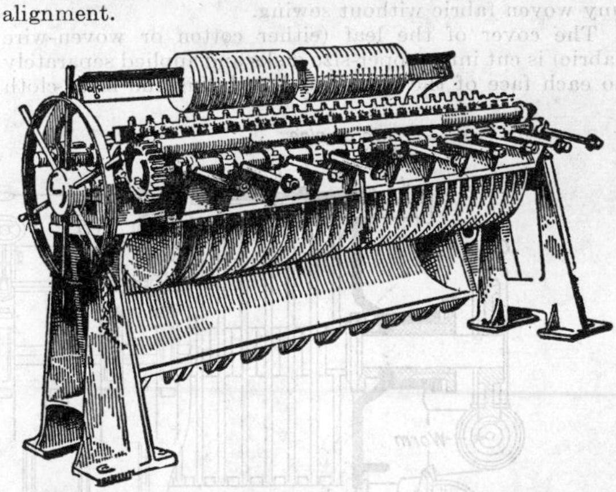

FIG. 69. Sweetland filter (open).

Special uses have been found for the Sweetland filter in many industrial chemical plants as well as in gasoline refining and in some phases of cane-sugar and beet-sugar refining.

Automatic sluicing mechanism is provided consisting of a manifold pipe passing through the entire length of the filter shell just above the leaves. Nozzles are fitted at equal intervals along this pipe so that one nozzle is directly behind each leaf.

By an external rack-and-pinion drive the nozzles are oscillated through an arc of 110 deg. and are moved longitudinally while oscillating; thus, when water is supplied under pressure, one jet cleans both sides of a filter leaf.

Filter leaves are illustrated in Fig. 70. Each leaf is a circular piece of heavy screen bound at the edge with a U-shaped peripheral ring. This gives stiffness to the leaf and provides a smooth edge, making it impossible for the screen to damage the filter-cloth bag covering the leaf. An outlet sleeve welded to the top of the leaf receives the nipple which holds the leaf in place in the filter and also carries away the filtrate.

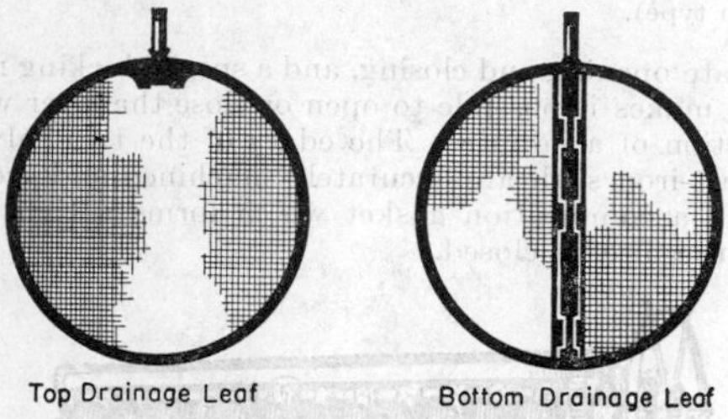

FIG. 70. Sweetland filter leaves.

For bottom drainage the outlet sleeve is welded into a flattened tube which extends all the way to the bottom of the leaf. When air blast is used to dry the cake, the air must force ahead of it all liquid left in the drainage screen, since the only outlet is at the bottom of the flattened tube.

Hence the liquid or wash water is displaced and dry-cake discharge secured.

An improved type of leaf construction has lately been devised, which gives a better means of cloth attachment. This is provided by a special peripheral member (in place of the conventional U-shape mentioned above), which has a dovetailed groove in each outlet side of the U-shaped ring. Such design makes it possible to apply any woven fabric without sewing.

The cover of the leaf (either cotton or woven-wire fabric) is cut into proper-size disks and applied separately to each face of the filter leaf by placing the filter cloth against the face of the leaf and calking the outer edge of the fabric into the dovetailed groove in rim of U-shaped, using cotton sash cord, lead wire, or the like.

Operation. Feed is supplied to the closed shell under pressure. Liquid is forced through the filter medium covering the leaves and issues from the individual outlets through sight glasses, the solids building upon both sides of the leaves in the form of a cake. When the rate of flow drops below the economic limit, feed is shut off, and the excess is expelled by air under pressure sufficient to hold the filter cake in place. When the shell is cleared, washing is commenced by admitting water under pressure. After washing, discharge is effected by opening the drain valves and operating the sluicing mechanism. Water from the sluicing jets cleans the leaves and sluices out the solids through the drains. If dry cake is required, washing is followed by an air blast, after which the lower half of the shell is swung open and a low air pressure turned into the leaves to discharge the cake, which falls into a hopper or conveyor below.

Vallez Filter. In the Vallez filter (Figs. 71 and 72), the leaves rotate inside a cylindrical pressure shell. The leaves are annular screen disks covered with filter medium and assembled in parallel at regular intervals upon a horizontal hollow shaft which rests in bearings at either end of the shell. This shaft serves as a filtrate channel and is rotated by worm-gear drive. The shell is divided into halves at the horizontal center line, and the upper half is provided with inspection doors opposite the disks.

To replace a disk, it is necessary to unbolt the upper half of the shell and remove it with an overhead crane; and, since the disks are not sectored, the shaft is next raised and the disks removed until the defective one can be taken off in its turn.

Feed is forced in from a manifold pipe having openings into the bottom of the shell. For discharge, a sluicing pipe is located at the extreme inside top of the shell with holes drilled so that jets impinge at an angle on each side of every leaf. At the extreme bottom of the shell is a trough equipped with a revolving scroll to remove the sluiced solids through an opening in the bottom of the shell.

The Vallez filter was designed for certain phases of sugar-refining work, and a variant of the Vallez type has been used in oil-refinery operations.

Burt Filter. The Burt filter (Fig. 73) is a steel cylinder which is rotated like a cement kiln. It has a hollow trunnion at the feed end, and a tire and rollers toward the

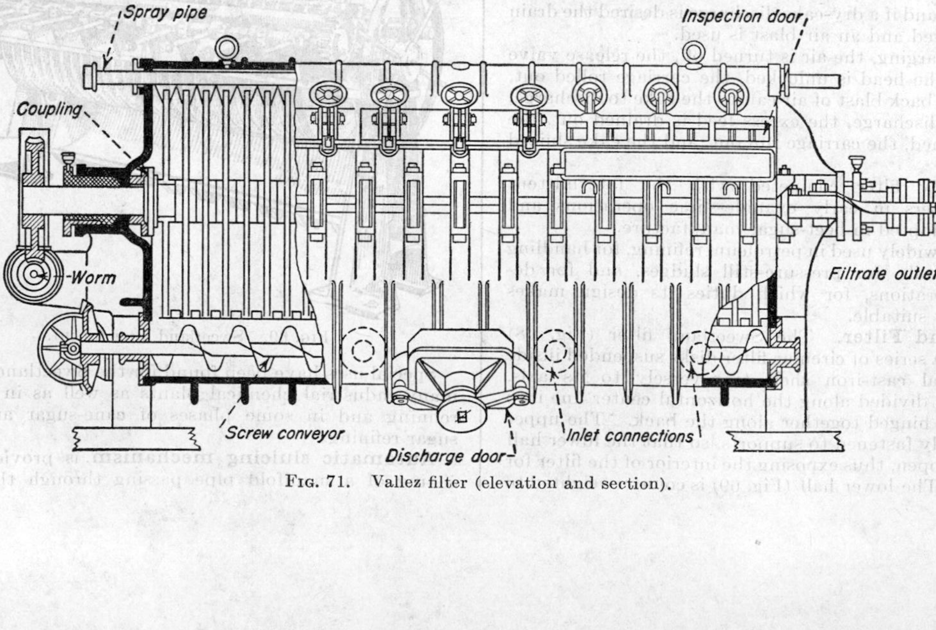

FIG. 71. Vallez filter (elevation and section).

other end. It is revolved by a pinion drive, and its speed is varied according to the nature of the material being filtered. The cylinder is lined on its inner periphery with drainage panels covered by filter cloth. Each panel has one or more outlet nipples passing through the cylinder shell, and external stationary launders receive the filtrate dropping from each circle of nipples. The rear end is

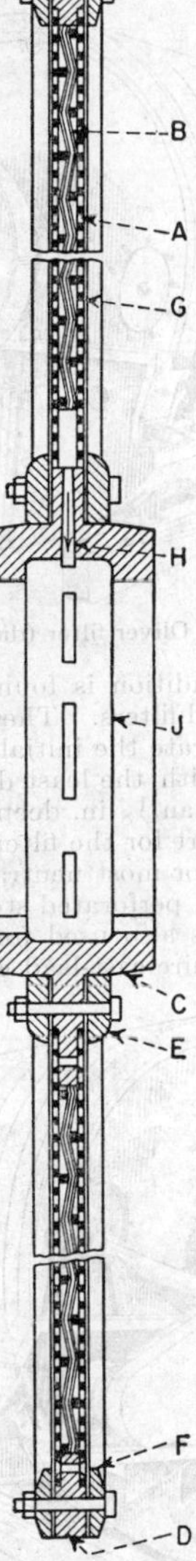

FIG. 72. Vallez filter (section of leaf).

closed by a cast-iron door equipped with a quick-opening outlet for discharge.

Operation. The material to be filtered is fed through the hollow trunnion at the head end, as the filter revolves. When the required charge has been introduced, the feed inlet is closed, air is admitted under pressure, and this pressure is maintained in order to force filtrate through the filter while forming and holding the cake in place. During rotation the filter cake forms while the filter medium is submerged by the feed. The cake is homogeneous because of rotation. Any cracks that form are sealed when the cake reenters the feed. When the flow of filtrate ceases, air pressure is released, and wash water is admitted. Rotation continues filling the cake with the water. Compressed air is then admitted and forces the wash through the cake. By using a muddy wash, all cracks or pit holes are sealed and a uniform cake washing secured. For discharge, the air pressure is cut off, water is admitted, the discharge ports are opened, and rotation is continued.

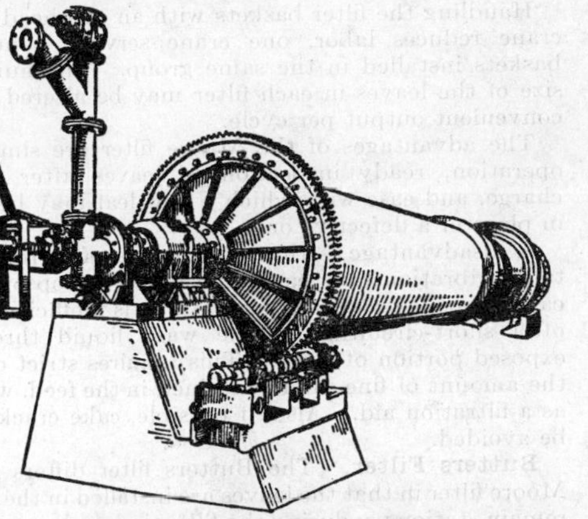

FIG. 73. Burt filter.

The Burt filter had a limited use in cyanidation some years ago, since which time it has been used to some extent in handling zinc sulfate liquors in electrolytic zinc refineries.

Intermittent Vacuum Filters

Moore Filter. The Moore filter (Fig. 74) was the first to use vacuum filtration on a commercial scale. The leaf consists of a frame over which a bag is stretched to form the filter. The frame is rectangular and is made of perforated pipe connected to a vacuum system. Collapse of the bag is prevented and drainage is provided for, by wooden slats sewed vertically into the bag. When the

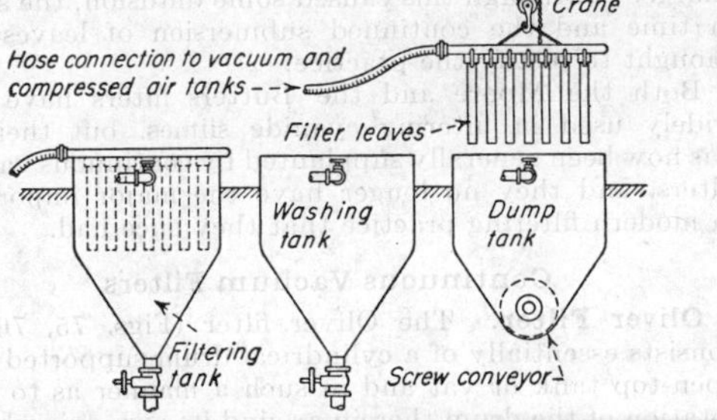

FIG. 74. Moore filter.

leaf is submerged in the material to be filtered and vacuum is applied, the cake forms on the outside of the leaf, while the filtrate is drawn through it. A number of leaves suspended side by side constitute a *basket*, the capacity of which is directly proportional to the number of leaves multiplied by the area of the two sides of a leaf.

When a sufficiently thick cake has formed, the basket is raised out of the feed tank and transported to another tank containing wash water, vacuum being continued meanwhile by means of a flexible hose connection. After sufficient washing, the basket is again raised and transported to the discharge point, where the vacuum is shut off and air pressure is applied. The pressure distends the bags and discharges the cake. The moisture that remained on the inside walls of the filter cloth is blown back and wets the outer side. It is lubricating action of this water, coupled with the weight of cake and the flexing of the cloth, that causes the cake to slide off. As the leaves hang vertically from the frame, they are in the best position for a complete discharge to be made.

Handling the filter baskets with an overhead traveling crane reduces labor, one crane serving a number of baskets installed in the same group. The number and size of the leaves in each filter may be altered to give a convenient output per cycle.

The advantages of the Moore filter are simplicity of operation, ready inspection of leaves after cake discharge, and ease with which a new leaf may be installed in place of a defective one.

A disadvantage is that, during transfer from tank to tank, vibration may cause the cake to drop off. If the cake drops from a leaf, the washing is inefficient because of a short-circuiting of the wash liquid through the exposed portion of cloth. This requires strict control of the amount of fine sand contained in the feed, which acts as a filtration aid. Also, if possible, cake cracking must be avoided.

Butters Filter. The Butters filter differs from the Moore filter in that the leaves are installed in the tank and remain stationary during the filtration cycle.

Feed is admitted until the leaves are submerged, after which vacuum is applied. When the required load of cake has formed, the remaining feed is pumped into a stock tank, the discharge valve is closed, and wash solution or wash water is run in until the leaves are again submerged. Vacuum is maintained in the meantime to prevent the cake from falling off. After washing, the solution or water is pumped out. The cake is discharged by opening the discharge valve of the tank, shutting off the vacuum, and applying air pressure, when the cake falls off and is sluiced away.

Small-scale operations require only one filter and one tank, whereas the Moore arrangement requires two or three tanks. For large-scale work, several tanks and baskets are needed in each method.

At times the wash solution has been pumped onto the surface of the charge in the filter while drawing off excess charge. Although this caused some diffusion, the saving in time and the continued submersion of leaves were thought to justify the practice.

Both the Moore and the Butters filters have been widely used in filtering cyanide slimes, but their use has now been generally supplanted by continuous vacuum filters, and they no longer have the major importance in modern filtering practice that they once had.

Continuous Vacuum Filters

Oliver Filter. The Oliver filter (Figs. 75, 76, 77) consists essentially of a cylindrical drum supported in an open-top tank or vat and in such a manner as to allow rotation of the drum therein around its own axis which is in a horizontal plane. The position of the drum in the tank is such that its lower portion is confined within the tank walls, while the upper portion is exposed above.

The ends of the drum are either open spiders or closed heads which carry the two main trunnions by means of which the drum is supported. The drum shell is composed of a number of shallow compartments over which is secured a covering of filter cloth. The cloth is supported by a drainage grid and is held in place by a spiral winding of wire uniformly spaced.

Since the pulps handled on the Oliver filter differ widely in the percentage of liquid content and hence in the filtering rate also, the nature of the drainage grid is determined by the use of the filter.

For the simpler types of minerals and chemical products, the screen grid is $\frac{7}{8}$ in. deep, handling moderate amounts of filtrate. If built for various free-filtering material (sulfite pulp as an example), the filtrate passages must be ample and drainage grids are therefore deeper, *i.e.*, $1\frac{1}{2}$ to $1\frac{3}{4}$ in.

Fig. 75. Oliver filter (discharge side).

The opposite condition is found in the very shallow grid in the cane-mud filters. These filters handle smaller flows and also separate the initial cloudy filtrate; hence, sharp separation with the least dilution occurs with the shallow grid (less than $\frac{1}{8}$ in. deep).

The screen support for the filter medium is a specially milled cedar grille for most neutral or acid filtrates, but, with caustic filtrate, perforated steel sheet or cast iron is used. The latter is also used for salt solutions. Concentrated acids require cast-lead grids.

Fig. 76. Oliver filter (agitator side).

The interior of each compartment communicates through a separate conduit (28), to a valve mechanism (31) which, during operation, automatically applies either suction or positive air pressure to the several conduits in rotation and through them in turn to the interior of the compartments. The automatic valve (31) is connected to a vacuum system and to a source of compressed air.

As the automatic valve is an indispensable control,

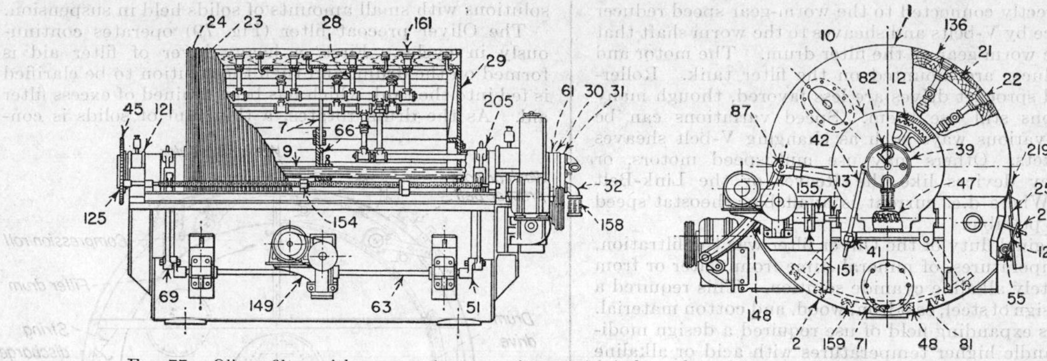

FIG. 77. Oliver filter with concentric-type agitator with independent drives for filter drum and agitator.

PARTS LIST

1. Filter drum
2. Filter tank
7. Drum arms
9. Drum shaft
10. Housed bearing
12. Worm-drive gear
13. Worm shaft
21. Wood staves
22. Division strips
23. Filter medium
24. Wire winding
25. Scraper blade
26. Scraper adjuster
28. Drum piping
29. Closed drum head
30. Wear plate
31. Automatic valve
32. Vacuum connection
39. Felt washer
41. Valve-adjusting pivot
42. Valve-adjusting rod
45. Wiring sprocket
47. Worm
48. Agitator rakes
51. Agitator crank
55. Scraper bearing
61. Pipe plate
63. Crank shaft
66. Center spider
71. Handhole cover
81. Agitator arc
82. Automatic valve flange
87. Connecting-rod pin
121. Rear bearing
123. Scraper plate
125. Feed-screw sprocket
136. Filter cover support
148. Type A drum drive
149. Type A agitator drive
151. Shaft coupling link
154. Wiring center dolly box
155. Shaft coupling
158. Diaphragm vacuum connections
159. Saddle clips
161. Drum nipples
205. Oscillating spider
219. Scraper tip

it has been specially developed in the Oliver filter. The earliest designs of automatic valves for Oliver filters made provision for separation of wash solution from filtrate. This was made by placing a bridge or stop in the valve interposing a barrier between the filtrate outlet in the lower half of the valve and the wash solution coming from the upper half. In many cases no separation of initial filtrate from wash solution is required. For such cases and for simple dewatering operations no bridge in the valve is required, and only one outlet from the valve is used.

Two outlets are usually enough, but in some cases a "cloudy filtrate" port or outlet is provided when using "open," or wire-mesh, filter covers, which diverts a cloudy filtrate produced in some uses of the filter just as the cake begins to form. This small amount of filtrate is returned to the filter feed, and thus the remaining filtrate is kept quite clear.

The usual bridges or stops in the valve are set before starting operation of the filter. In some cases they are arranged to be set by a handwheel on the outside of the filter valve, the handwheel allowing exact adjustment without removal of the valve for setting the bridges. This is a patented feature.

Other valve bridges, also externally controllable, permit modifying the amount of vacuum without altering the peripheral location of the bridges.

For high rates of filtrate flow the vacuum passages or conduits leading to the valve from the surface of the filter drum are made very capacious to avoid frictional losses. In extreme cases the passages convert the entire interior of the filter drum into manifold wedge-shaped compartments, which take the place of the usual tubular conduits.

Under the drum, barely clearing the bottom of the tank, is suspended a framework (81) supporting horizontal rakes (48) which, during operation, slowly oscillate, thus agitating the feed.

The filter cake is usually discharged from the drum surface by a scraper blade which is set in a vertical position, and in the present design the blade does not touch the wire winding of the drum. On the edge of the scraper is affixed a detachable rubber tip. The scraper itself is mounted upon the edge of the filter tank. The low-discharge point of the Oliver filter cake permits cake discharge by pressure reversal only, in most cases, the scraper serving as a diversion plate only.

There are important exceptions to this among, which are the various fibrous materials formed in paper making. For these fibrous sheets, so called, vacuum alone does not reduce the moisture content sufficiently low in some cases, nor does the scraper always give the best discharge.

By applying pressure upon the wet sheet of fiber with heavy steel or cast-iron rolls, much additional moisture is expressed, and the same sheet is suitably removed from the drum of the Oliver filter by couch rolls or lead rolls in contact with the sheet; the effect is by adhesion to the couch roll which rolls upon the sheet surface. The sheet is removed from the couch roll by a knife or "doctor."

The lead roll provides a lifting effect on the sheet which passes over the lead roll without adhesion.

A widely used means for removal of fibrous sheet is obtained by the effect of vacuum applied externally to the sheet at point of discharge. Another discharge method uses jets of air and/or water introduced beneath the fibrous sheet externally.

Above the exposed portion of the drum, and connected to the filter tank, is a steel framework supporting a number of horizontal water headers and fitted with spray nozzles, the whole being enclosed by a sectionalized housing. The entire filter mechanism is driven by a motor and speed changer, or alternative mechanical devices.

The simplest type of drive for the Oliver filter is by a belt and pulley.

Various mechanical drives are possible for the Oliver filter. The one most favored now is a combination of motor directly connected to the worm-gear speed reducer and thence by V-belts and sheaves to the worm shaft that drives the worm gear on the filter drum. The motor and speed reducer are mounted on the filter tank. Roller-chain and sprocket drives are less favored, though many installations still use them. Speed variations can be made in various ways such as changing V-belt sheaves or sprockets. Others may use multispeed motors, or cone-pulley devices like the Reeves or the Link-Belt drives. Where d.c. current is available, rheostat speed control is possible.

The original duty of the Oliver filter was the filtration, at air temperatures, of mineral slime from water or from a moderately alkaline cyanide solution. This required a simple design of steel, cast iron, wood, and cotton material.

Soon its expanding field of use required a design modified to handle higher temperatures with acid or alkaline solutions. The results were such that the Oliver filter now uses, besides the original type of steel, wood, cast iron, and cotton, filters of all wood, others of all steel, stainless steel, or monel metal.

Certain duties require an all cast-lead filter, and there are also all cast-iron filters. Lead-lined steel filter tanks are built, and in some cases a sheathing of brass or copper is used over the wood and steel portion. Any practical material of construction may be used in building the Oliver filter, and filter covers are provided in a wide variety of textiles, woven-wire cloth, as well as finely perforated metal sheets.

Operation. During operation, the drum rotates slowly while the tank is supplied with the material to be filtered and the level is maintained to ensure a constant depth of submergence of the lower portion of the filter drum. In some types this depth may be set between limits ranging from zero to almost complete submersion of the drum. Once chosen, the valve is set for the given conditions.

Through the action of the automatic valve, vacuum is applied to those compartments of the drum passing through the sludge. The vacuum created within the compartments causes a flow of filtrate through the filter medium, conduits, and automatic valve, and a layer of cake solids is deposited upon the filter medium covering the submerged portion of the drum.

As the drum revolves, the vacuum in the compartments is maintained, and the layer of cake solids emerges and passes through the arc included by the upper or exposed portion of the drum. It is subjected to washing by water from the spray nozzles; the wash water permeates the cake and displaces the liquid contained.

During this washing operation, the replaced liquid, together with some of the wash water, flows through the filter medium and conduits and is discharged from the automatic valve in the same manner as the liquid from the cake-forming operation. Wash liquor may be discharged separately from the original filtrate at the automatic valve.

Following the washing period, as each successive compartment reaches the scraper, the vacuum is cut off by the action of the automatic valve, and air at low pressure is applied. This operation, with or without the action of the scraper, effects the discharge of the cake, after which the cleaned filter surface again rotates into the tank and the cycle outlined above is repeated.

Oliver Continuous Precoat Filter. This filter is a major modification of the standard Oliver continuous drum-type filter. Its unique principle of filtration and method of cake discharge have opened up many opportunities for the continuous filtration or clarification of products that have hitherto been difficult to handle. Precoat filtration is particularly adapted to handling solutions with pasty, gummy, or colloidal substances, or solutions with small amounts of solids held in suspension.

The Oliver precoat filter (Fig. 79) operates continuously in cycles. First, a heavy layer of filter aid is formed on the drum, and then the solution to be clarified is fed into the tank which has been drained of excess filter aid. As the drum rotates, a thin film of solids is con-

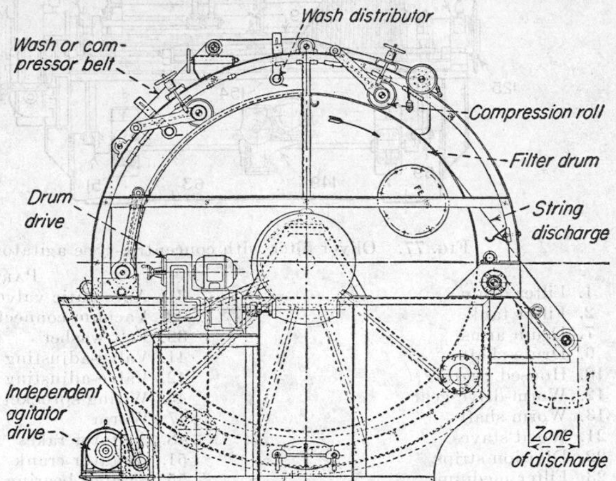

Fig. 78. End view of FE chemical washing filter showing compact assembly. (*Filtration Engineers Incorporated.*)

tinuously formed on the surface of the filter aid and is rotated through the washing and drying zone to the discharge point. Here, an advancing knife-edge shaves off the film of solids and usually some filter aid. The cleaned surface of filter aid rotates on into the tank for

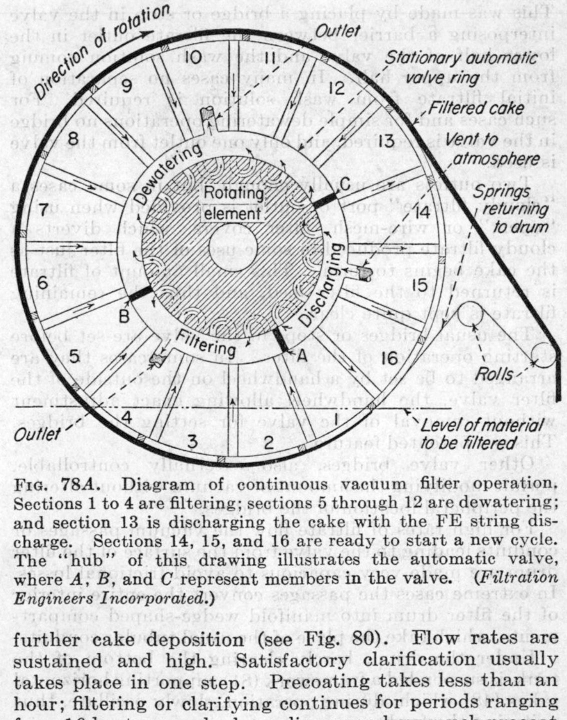

Fig. 78*A*. Diagram of continuous vacuum filter operation. Sections 1 to 4 are filtering; sections 5 through 12 are dewatering; and section 13 is discharging the cake with the FE string discharge. Sections 14, 15, and 16 are ready to start a new cycle. The "hub" of this drawing illustrates the automatic valve, where *A*, *B*, and *C* represent members in the valve. (*Filtration Engineers Incorporated.*)

further cake deposition (see Fig. 80). Flow rates are sustained and high. Satisfactory clarification usually takes place in one step. Precoating takes less than an hour; filtering or clarifying continues for periods ranging from 16 hr. to a week, depending upon how much precoat is removed with the cake.

The Oliver precoat filter is made for both continuous-vacuum or continuous-pressure operation. Enclosed units are made for handling products giving off noxious or volatile gases or where insulation must be provided.

Feinc Filter. This filter (Figs. 78 and 78*A*) is a continuous, rotary-drum vacuum filter similar to the Oliver already described but differing essentially as follows:

The discharge of cake is effected by a system of endless strings passing around the drum and over a roll for lifting the cake from the drum.

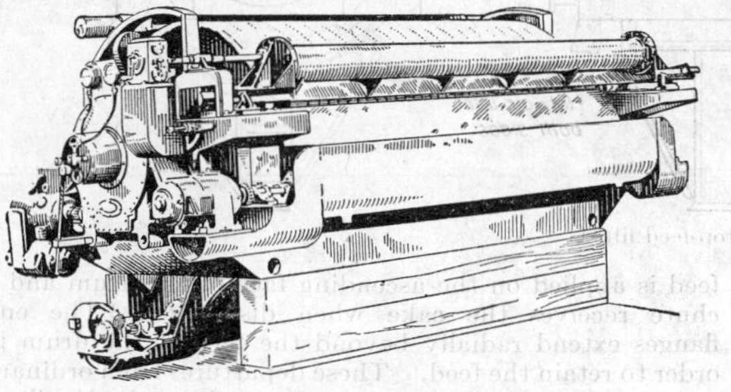

Fig. 79. Oliver precoat filter.

Outlet pipes from the drum compartments terminate in ports on the periphery of the rotating hub instead of at the face or end of the hub.

The valve is annular in shape enclosing the hub and has ports that break the vacuum on those compartments where the strings lift the cake from the filter drum.

No back blow of air is used.

The drainage members in each compartment are formed of spirally wound wire mats cut in sections to fit.

Operation. The feed is supplied to the tank in which the drum revolves and vacuum is applied, causing cake to form and to embed the endless strings on the submerged sections.

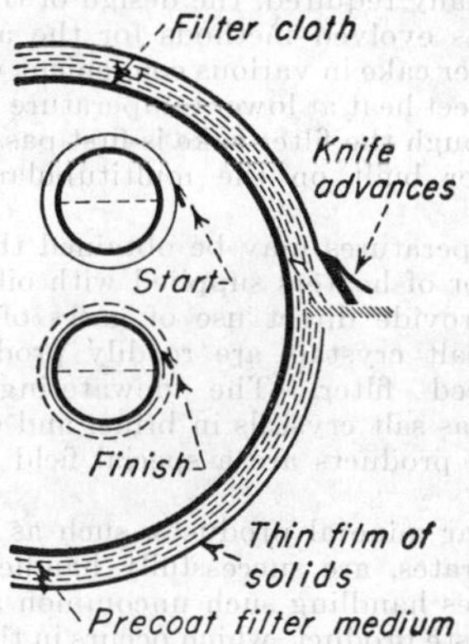

Fig. 80. Operating method of Oliver precoat filter.

As the drum revolves, the cake emerges and travels around with the drum to the point of discharge, which is normally on the descending side. Here the vacuum is broken by the automatic valve and the strings leave, lifting the cake from the drum. They pass over a discharge roll, as shown in Fig. 78*A*, the flexure causing the cake to fall from the strings. As the strings return to the drum they pass through a comb to keep them in alignment and to remove any adhering cake.

Dorrco Filter (Fig. 81). This is a vacuum filter of the rotary-drum type with the filter medium placed on the inner surface of the drum as a series of panels parallel to the drum axis. The drum serves also as the container for the pulp, and no supplementary tank is used. Except for the annular retaining ring which creates the bath, the drum is open at one end for convenient inspection.

The drum is supported by a tire and riding rolls at each end. It is driven by a motor and speed reduction connected to the riding-roll shaft by chain and sprocket. The automatic filter valve is the same as the one used on the Oliver filter, although formerly it was of annular type. Bridges used in this valve are the same as those in the Oliver filter, but additional bridges permit alternating suction and pressure in the cake-discharge section.

A troughed belt conveyor runs through the machine at one side of the center line collecting the cake as it is discharged and delivering it through the end of the machine at the center-line elevation. This conveyor is driven through the opening in the annular valve, and the drive arrangement is such that the speed of the filter drum can be adjusted independently of the speed of the conveyor. In some cases cake is removed by a spiral scroll in a trough.

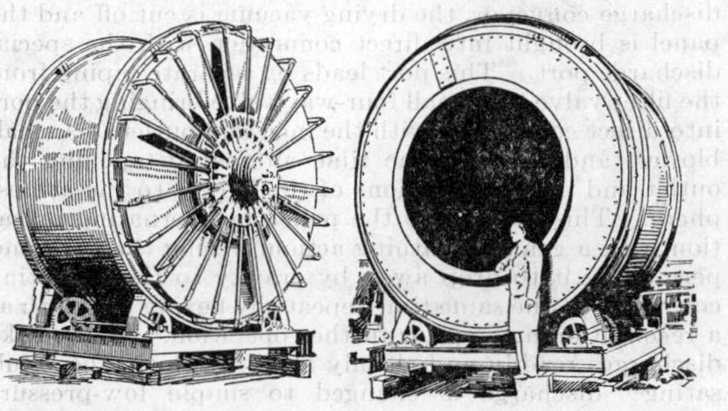

Fig. 81. Dorrco filter.

The feed is arranged to enter at either end of the drum and is distributed uniformly on a line, the full length of the drum, by means of an inclined riffled launder or some other simple feed distributor.

The inner face of the drum is composed of a series of individual filter panels. In the case of a fabric filtering medium, the cloth is loosely stretched over the panel surface and conveniently held by packing rope on the four sides of the panel. This packing rope is pressed down into narrow grooves, leaving a smooth and practically continuous exposure of active filtering surface. These panel coverings are supported on a punched plate which is rigidly spaced away from the outer shell to allow passage for air and filtrate, and the filtering medium is in turn held free from the punched plate by an intermediate layer of backing cloth. The cloth can be applied as a one-piece blanket for the filter, or, by simply cutting the cloth in rectangular strips, the individual panels may be separately reclothed. The panel compartments communicate through the end of the drum with externally located piping, leading to individual parts in the replaceable rotating wearing plate of the automatic head valve, so that the vacuum, pressure, steam, etc., can be applied in proper sequence during the cycle.

Operation. As the filter revolves, the cloth passes down into the pulp bath underneath the line of feed and the cake-forming vacuum is applied automatically as adjusted by the setting of bridges in the head valve, permitting, where desired, the precoating of the cloth with the coarsest, most rapidly segregating material in the feed. As the cloth emerges from the bath, the cake is drained out and washing sprays may be brought to bear upon it while it is still resting in an inclined position against the drum. Following this early application of washing sprays, the cake is dried by vacuum until discharge.

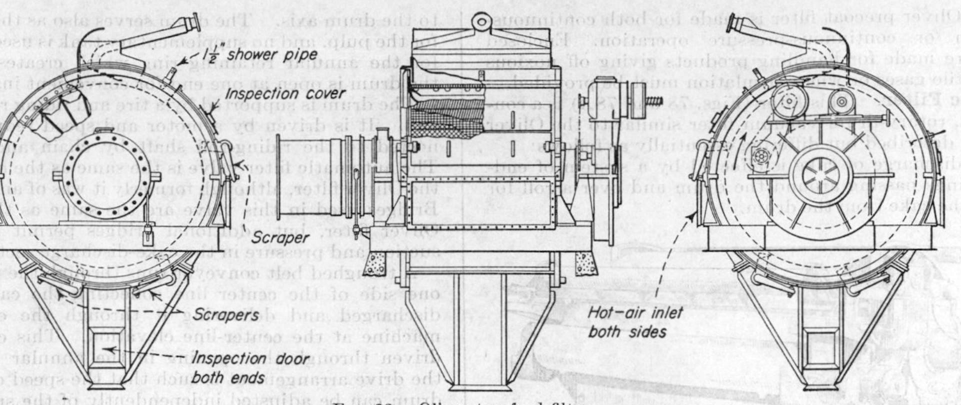

FIG. 82. Oliver top-feed filter.

As the cake passes over the hopper which guards the discharge conveyor, the drying vacuum is cut off and the panel is brought into direct connection with the special discharge port. This port leads by separate piping from the filter valve to a small four-way valve bringing the port into direct connection with the inlet and outlet of a small blower, and at the same time alternately opening the outlet and inlet connections of the blower to the atmosphere. This imparts to the panel an alternating pulsation with a gentle breathing action, freeing the cake and permitting it to drop away by gravity to the collecting conveyor, at the same time repeatedly flexing the cloth as a reconditioning step for further operation. When cake discharges readily and cleanly from the cloth, the "pulsating" discharge is changed to simple low-pressure "blow" discharge.

Following the discharging of the cake, the cake cloth is exposed and, if desired, the pulsation may be continued and spraying or steaming of the cloth accomplished with the separate division of the valve available for this use.

No agitation of the pulp bath is used because any settlement or segregation that takes place is in the direction of cake formation, and neither the pulp in the bath nor the cake on the cloth comes in contact with any mechanical device or stationary surface.

In the case of free-filtering materials, permitting high speeds of filter operation and handling of large tonnages on comparatively small drums, the face of the filter is short in proportion to its diameter and a simple chute is substituted for the belt conveyor. The short-faced filter lends itself particularly well to the distribution of extremely heterogeneous feeds where any appreciable depth of bath is difficult to maintain, the coarse material being merely sluiced onto the narrow path of filtering surface revolving under it. When operating conditions prevent the use of cotton filter cloth, a special type Dorrco filter has been designed to use woven-wire cloth. The separate panels in this design are inwardly convex, instead of uniformly concave like those in the standard Dorrco. These panels have special clamping bars and plates to hold the wire cloth in place.

Double-drum Filter. This scheme has two conventional filter-drums mounted with axes parallel and surfaces almost tangent. The drums rotate away from each other, and filter-feed is poured into the space above the tangent point. Two end pieces prevent fluid feed from running off the drums. Cake forms on the two drums, discharging at the usual point. This scheme suggests two top-feed filters and is known as the Peterson Synchro-drum filter.

Oliver Top-feed Filter. On this filter (Fig. 82) the feed is applied on the ascending face of the drum and a chute receives the cake when discharged. The end flanges extend radially beyond the face of the drum in order to retain the feed. These departures from ordinary rotary vacuum-filter construction are shown by the illustration. The cake is carried around the filter nearly to the point of cake formation before discharging, or about 315 deg. travel on the circumference.

In cases where minimum moisture concentrations are needed, a hood is provided in which the incoming air is heated before it is drawn through the cake, thus obtaining a dry product by direct evaporation with high thermal efficiency.

Large volumes of air are drawn through the cake at low vacuum (2 to 4 in. Hg), and the power consumption is reduced to a minimum by designing the filter with the least possible internal resistance—no piping being used in the vacuum conduits of the filter.

Since very low moisture concentration in a filtered product is usually required, the design of the Oliver top-feed filters has evolved methods for the application of heat to the filter cake in various convenient ways.

Where indirect heat at lower temperature is needed, the air drawn through the filter cake is first passed through a heat exchanger built on the multitubular plan, using steam.

Higher temperatures may be obtained through use of direct steam or of heaters supplied with oil, gas, or coal fuel, which provide direct use of gases of combustion. "Bone-dry" salt crystals are readily produced on the Oliver top-feed filter. The dewatering of crystal magma, such as salt crystals in brine, and other crystalline chemical products are a special field of the Oliver top-feed filter.

Other similar mineral products, such as metallurgical table concentrates, are successfully handled by it also. Its use includes handling such uncommon substances as cellophane-waste product, which occurs in tiny micaceous flakes.

A special economic advantage of the Oliver top-feed salt filter is obtained by using vacuum exhausters that are direct-driven from steam turbines, the exhaust from which is used in vacuum-pan evaporators.

The tonnage that can be handled on the Oliver top-feed filter of a given area depends chiefly upon the fineness of the material and the final degree of moisture content required. On materials for which it is particularly adapted, the capacities obtained are from 5 to more than 35 tons/sq. ft. filter area/24 hr.

A recent modification of the Oliver top-feed salt filter is known as the Oliver-Robison filter. In this design, the trunnion of the filter is not provided with the usual filter valve, but instead the trunnion extends directly into the vacuum filtrate receiver. Special drum construction also provides the means of obtaining greater drying efficiency.

In special cases the drum face may be divided into a series of hoppers by inserting partitions of suitable height between the extended drum flanges. These partitions are set above the division strips so that each hopper corresponds to an individual section of the drum. This construction makes possible the handling of large volumes of loosely packed materials as it prevents the slippage of the cake while dewatering is being effected.

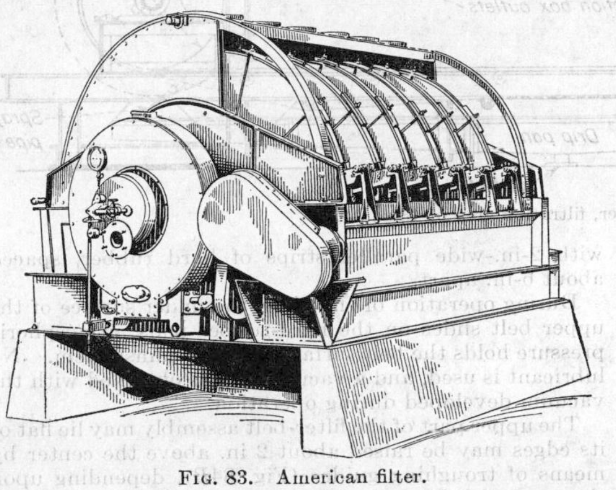

FIG. 83. American filter.

Oliver Horizontal Filter. This unit is a continuous vacuum filter whose rigid filter surface is rotated in a horizontal plane. This surface is annular in shape and divided into separate segments, each of which is connected to the central automatic valve (see Fig. 84). The width of the annular surface varies with the size of the unit, which is 15 ft. diameter in largest size.

The segments have separately removable and attachable filter covers. The annular surface has walls 6 in. high around the inner and outer rims which restrain flow of feed. After cake forming, washing, and drying, the cake of solids is removed by a spiral-scroll cake discharger. The shaft of the scroll rotates parallel to the surface of the filter. This filter has been applied successfully both to dewatering and washing problems and to products from +28 mesh to −200 mesh. It has special merit in dewatering (and washing, if that is required), materials that are granular and not slimy and that readily form suction-filter cakes. Separate filtrates of consistent character are readily obtained by the four-outlet automatic filter valve, using separate wash liquids.

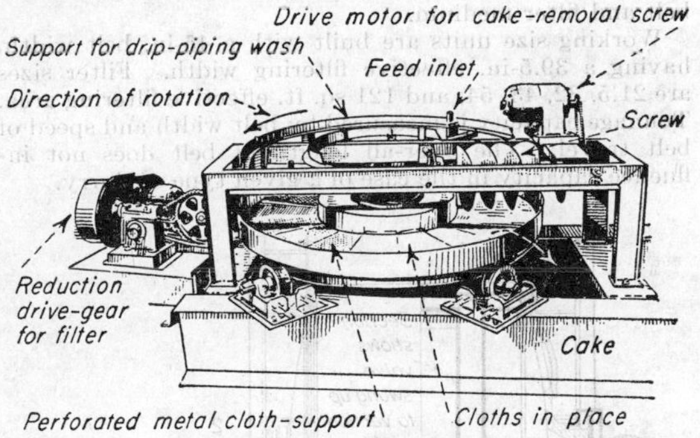

FIG. 84. Oliver horizontal filter.

About ⅛ in. of solids remain on the filter surface after cake removal; by slight back-pressure air under the filter surface, a lively ebullition occurs in the oncoming feed, with which the residual ⅛ in. of solids is quickly incorporated.

American Filter. The American filter (Fig. 83) is a continuous rotary vacuum filter consisting essentially of a number of filter disks mounted at regular intervals around a hollow cast-iron center shaft, as illustrated in Fig. 83. Rotation is by a gear drive. Each disk consists of a sector of wood, iron, or bronze, ribbed on both sides to support the filter cloth and to provide drainage.

Some of the recent American filter units are provided with a built-up steel central shaft in place of the hollow cast-iron shaft.

Each sector has an outlet nipple that passes through an opening in the cast-iron center shaft joining a conduit running the entire length of the shaft and terminating in a port at the automatic valve. Each serves as a filtrate channel for all sectors along the shaft on that line. The automatic valve is similar to those used for other continuous rotary vacuum filters. Sectors are held in place by radial rods, each rod having a clamp and nut on the outer end that holds two adjacent sectors in place. Any sector can be replaced without disturbing the others, and at slow speeds it is not necessary to stop the filter to make the change.

Filter covers are in the form of bags slipped over the sectors, and the outer edges are folded under the clamps. At the filtrate nipple, a cord is tied round the neck of the bag and a rubber washer makes a tight joint between the nipple and the center shaft. The assembly of filter disks on the center shaft is mounted in a feed tank so that the sectors are completely submerged during the cake-building portion of the cycle. On the discharge side, the filter tank is crenelated to accommodate the disks. The space between these divisions is utilized for cake discharge; scrapers or tapered discharge rolls for each disk are mounted at the top of the tank. In some cases discharge is effected by fine water jets under pressure.

Operation. During operation, the feed is supplied at the bottom of the tank through a manifold pipe having one supply nozzle under each disk. A homogeneous mixture is maintained by forcing a steady stream of feed through these nozzles, the excess pulp returning to the supply tank through an overflow in the filter tank. The disks rotate slowly, and, as soon as the sectors are submerged, vacuum is applied by the action of the automatic valve. A layer of cake solids forms upon the cloth on both sides of the sectors, and the filtrate passes from the sector through the conduit in the center shaft and out through the automatic valve. Vacuum is still maintained when the sectors emerge and are exposed to the air, and wash is applied if required. As each sector reaches the scraper, or discharge, roll, vacuum is cut off and a gentle air blast is applied. This causes the filter bag to inflate, since it is not held fast to the sector by any grid or wiring. Contact of the bags with scrapers or with rotating discharge rolls causes the cake to drop between the tank divisions. In some uses, the feed enters the American filter from a launder placed along the rim of the filter tank.

Lurgi Filter. The Lurgi filter differs from other continuous vacuum filters by filtering and washing on a horizontal endless rubber belt. This carries the filter medium over suction-boxes and around an end pulley, at which cake discharge is accomplished by gravity, a scraper, or by sluicing sprays.

Design and Construction. Figure 84*A* shows the design to be a flanged, rubber-covered driving drum and a corresponding driven drum with a horizontal axis. These drums are 48 in. wide with 40 in. diameter. An endless

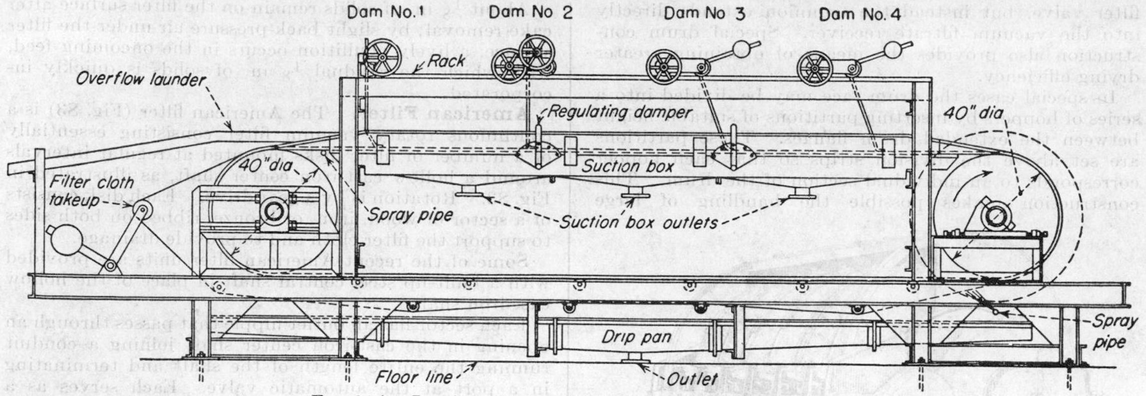

FIG. 84*A*. Lurgi continuous filter, filtration area of 43 sq. ft.

belt encircles the two drums, which are about 20 ft. between diameters; this belt (Fig. 84*B*) is formed of two elements.

The section in contact with the pulley is of fabric reinforced with rubber, having a channel cross section, and a series of ½- by 2-in. drainage perforations along the center line. The flanges on this section of the belt are about 2 in. wide, and a slot is provided in each, into which

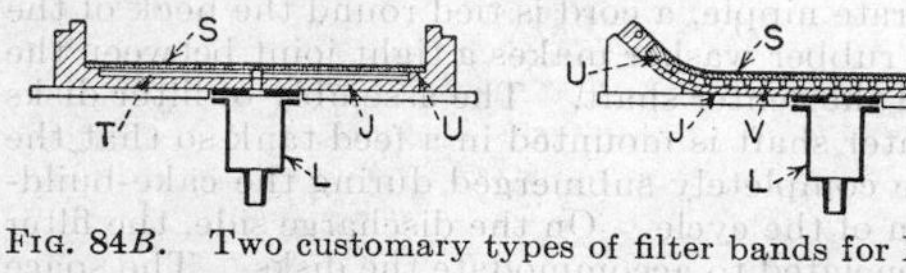

FIG. 84*B*. Two customary types of filter bands for Lurgi filter.
J. Filter band (rubber band with cloth insertions)
L. Suction box
S. Filter cloth
T. Rubber ribs of band sections
U. Perforated rubber band
V. Spacers on the perforated rubber band

the edges of the filter cloth are calked. The interior section of the filter belt lies between the flanges of the main belt and consists of a perforated rubber sheet (with spacers attached to the surface in contact with the main belt) which supplies drainage area. Several filtrate boxes are located beneath the main belt and these are connected to the usual vacuum receivers and to the source of vacuum. The upper edges of these boxes are supplied with 2-in.-wide parallel strips of hard rubber, spaced about 6-in. apart.

During operation of the filter, the under surface of the upper belt slides on the hard-rubber pads; atmospheric pressure holds the belt surface tightly against them. No lubricant is used, and a vacuum seal is obtained with the vacuum developed during operation.

The upper part of the filter-belt assembly may lie flat or its edges may be raised about 2 in. above the center by means of troughing guides (Fig. 84*B*), depending upon the type of belt used. The filter cloth is a continuous strip, having its ends lapped and sewed together, the edges generally calked into grooves in belt flanges. Cross dams of rubber and steel separate the filter pulp from the wash liquid. The lower edges of these dams slide upon the cake surface or upon the filter cloth. Adjustable closures are supplied for regulation of the flow of filtrate entering a given box, and a suction box extending beyond the last cross-dam provides a dewatering zone, ahead of cake discharge. The lower, or "return," side of the filter belt is supported by idler rolls. At this point, washing sprays are supplied for cleaning both the belt and filter medium.

Working size units are built with a 45-in. belt width, having a 39.5-in. effective filtering width. Filter sizes are 21.5, 32, 43, 54, and 121 sq. ft. effective filtering area. Tonnage capacity is measured by belt width and speed of belt travel. The over-all length of belt does not influence capacity in the case of a given type of slurry.

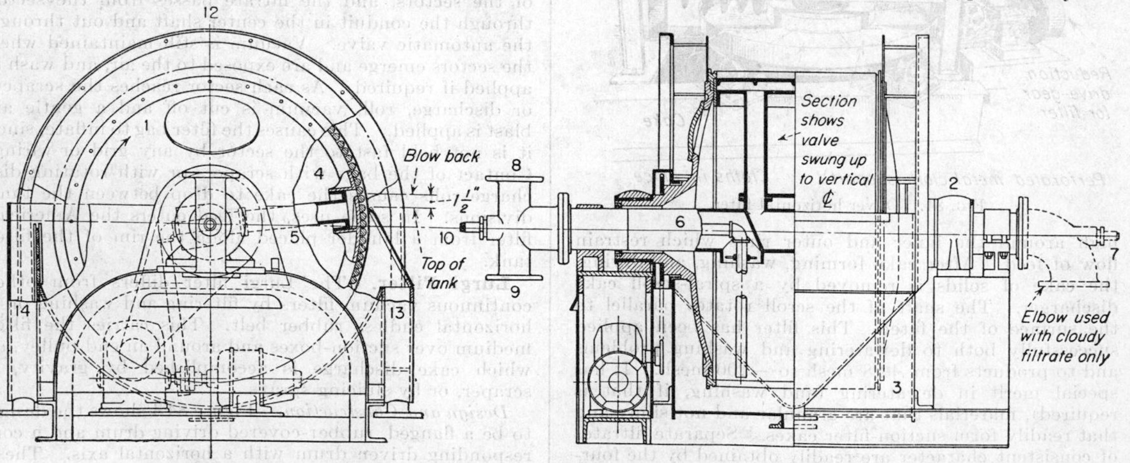

FIG. 84*C*. Bird-Young continuous vacuum filter, this model 4 ft. diameter by 2 ft.

Operation. The feed slurry is run onto the belt between cross dams 1 and 2 the excess overflowing dam 1 passing down an overflow. Cake is formed on the belt, the filtrate being drawn through the cloth into the suction boxes beneath the belt. Belt speed may vary from 0.3 to 15 ft./min., depending upon the inherent ease of filtration of the slurry. Movement of the belt carries the cake through the wash liquid behind each cross dam, where the cake is washed. After passing the last cross dam, the belt moves over a dewatering suction box and then down around the end drum, where cake is discharged. On the return of the belt, the cloth and belt are washed by sprays.

About 25 per cent of the belt length has effective filtering surface. The life of the rubber belt may be 1½ to 2 years, but it is much affected by conditions of filtration. Tests for the performance of the Lurgi filter should be made on a suction leaf covered with the actual filter cloth intended for use on the filter.

This filter is offered usually for freely filtering slurries formed by granular particles suspended in a solution. Its chief application has been in filtering and washing gypsum solids in a slurry of phosphoric acid. Advantages of this filter type are the countercurrent washing obtainable, the control of operating variables, and the ability to scrub the filter medium during the course of operation. Chief disadvantages are the relatively high initial cost per unit area of effective filtration surface, costly belt replacements, and the temperature and corrosive limitations that accompany the use of natural or synthetic rubber.

Bird-Young Filter. This filter has neither internal drum piping nor automatic rotary valve; the entire inside of the drum is subject to vacuum, and cake discharge is done by a pulsating-air blowback, without help of either scraper or strings. This filter's special purpose is handling slow-filtering slurries, because of its ability to operate with very thin cakes.

Design. The filter is shown in Fig. 84C, in combined elevation and in section. The drum is made to bear atmospheric pressure and is supported, with axis horizontal, by trunnions at the drum ends.

The filter-feed bowl (3) is mounted so that from 5 to 50 per cent of the drum surface will be submerged in the slurry. Mounted inside the drum at the point desired for cake discharge is a shoe arrangement (4) which supplies compressed-air blowback to the surface sections of the drum as they pass the cake discharge point. This shoe is supported within a few thousandths of an inch from the inner drum surface by a radial supporting pipe (5) which is in turn supported by a stationary axial pipe (6) which passes through packing rings at the trunnions. The axial pipe acts as the channel for blowback air (7), the channel for evacuated air (8), the channel for filtrate removal (9) except at low submergence, when the siphon pipe (10) is used for filtrate removal, and the channel for wash filtrate (11).

The drainage surface of the metal drum shell is divided into 50 to 100 longitudinal surface sections by slotted division strips. The filter cover is applied as a single piece, being held in place by crosswires in each cross-division strip. This gives the effect of separate sections of surface. Filtrate from each section drops to the bottom of the drum and flows out via the axial pipe, or by pumping out through the siphon pipe.

If washing is carried on, the wash product is caught upon the internal sector fans, flowing from them out through separate piping within the axial pipe header. Wash piping may be mounted upon the supporting rings.

A variable-speed drum drive is used to provide 10 to 1 variation, depending in turn upon the nature of the slurry itself, and a range of 6 min./revolution to 2 sec./revolution, which are claimed to be practical limits.

Construction. Sizes of this filter range from 1 to 140 sq. ft. of filtering area, with operations arranged for a fixed point of submergence. Materials used may be plain steel, stainless steel, or rubber-covered steel. Wood is unsuited to withstand the required atmospheric pressure.

Sizes range from 4 ft. diameter by 2 ft. face to 5½ ft. diameter by 8 ft. face, with areas of 25 sq. ft. and 140 sq. ft., respectively. There is also a test size, with 1 ft. diameter by 4 in. face, having 1 sq. ft. area. Filter prices on this design have been reported in 1948 to vary for the above range of sizes from $100 to $900/sq. ft. of filter area for plain-steel construction. For stainless steel or rubber-covered steel the price is about 1.75 to 2.0 times the amounts quoted above.

Operation. Slurry feed is maintained through the bowl and this flow is relied upon to keep solids in suspension. A dry-vacuum pump is preferred since its pulsating air discharge is used as a source of blowback air for cake discharge.

The interior of the filter drum serves as a vacuum receiver. At 50 per cent submergence, air and filtrate flow out of drum through the axial pipe. At low submergence, filtrate is removed through the siphon pipe by a self-priming pump. As drum-rotation continues, filtrate flowing from cake drops into the bottom of drum. Later (in case cake washing is wanted) the wash filtrate falls into the sector fans which receive it and from which it is removed by self-priming pump. When cake-drying is sought, air is drawn through cake; next in order, the air-blowback shoe serves as the valve to shut off vacuum and admit blowback air pressure to discharge the cake on filter surface. This pressure is said to be from 5 to 40 in. of water and suited to the lightweight, smooth synthetic fiber cloths that are preferred for this operation. Cake discharge is effected by this blowback without the use of scraper or strings. No figures have been submitted for the moisture content of thin cakes so discharged.

Recognition must be given to the air leakage between the somewhat loosely fitting shoe and the inner surface of the drum. This entails extra vacuum-pump capacity above that used in conventional filtering and dewatering operations. Figures have been supplied that show this excess pump capacity, in terms of free air, is in the order of 5 to 10 cu. ft./min./sq. ft. of filter area. For finer solids the excess pump capacity is said to be greater than for coarser crystal products.

Applications. Increased filter capacity is claimed for this filter by operation at higher drum speeds, made possible by discharge of thin cakes, thus keeping the filter medium clean. Its operation has permitted the discharge of thin cakes in the filtering of various slow-filtering suspensions.

Synthetic-fiber filter covers with smooth, tight weaves are well suited to this operation.

The uses of this filter are limited by the rather small filter area, (140 sq. ft. maximum), and wherever wood construction may be required. Also, the need for steady recirculation of feed pulp is harmful to flocculated slurries. Pulp level also is fixed by the point at which the discharge shoe is set, thus preventing easy variation in the submergence level of the pulp in filter. Among the uses that have been found for this filter are in the production of penicillin, clay slurry, dyestuffs and pigments, suspensions of plastics, and natural-fiber pulps.

Oliver Pipe-line Strainers. It is often necessary to provide a constant supply of dependably clarified solution or water in the system of plant piping. If the clarifying or filtering device need not be large, while the necessary recurrent methods must not interfere with constant service nor be complicated in operation, Oliver pipe-line strainers are suitable. These convenient small units are arranged for installation in parallel, one being switched from service for cleaning of the strainer surface, while the companion unit is returned to use.

Figures 84*D* and 84*E* illustrate operative methods by the cross-section in the assembly drawing, as well as the simple layout for the sequence cleaning, straining, and

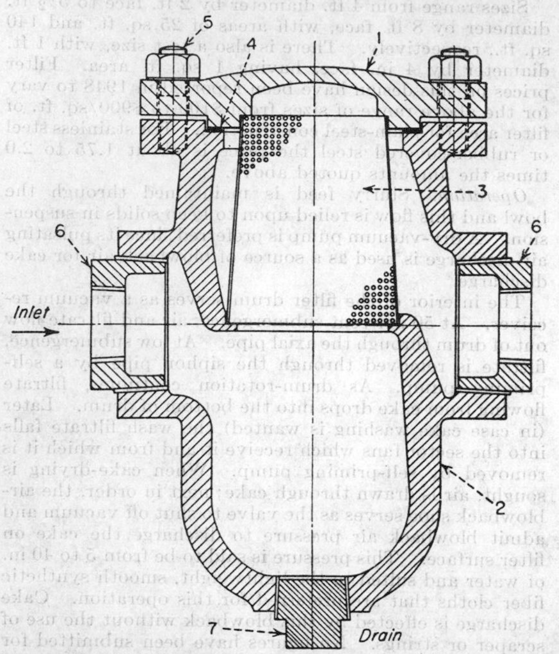

FIG. 84*D*. Oliver standard spray-water strainer. Cast-iron body with brass screen, 925 perforations/in.

1. Hood
2. Base
3. Screen
4. Gasket
5. Studs
6. Bushing
7. Plug

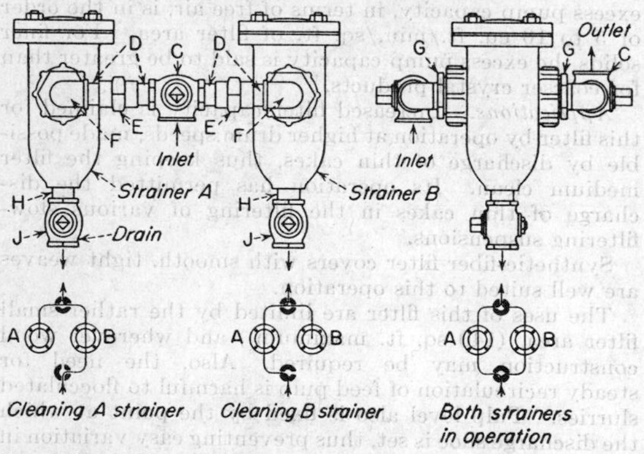

FIG. 84*E*. Oliver double spray-water strainer. Double-strainer arrangement.

A, *B*. Strainers
C. Three-way valve
D. Nipple
E. Union
F. Street ell
G. Bushing
H. Nipple
J. Plug-cock

switching of a unit requiring cleansing. Materials of construction are iron, steel, and alloys, if needed, while the strainer surface may be either woven fabrics or punched sheets. All these units are of small size, since the volumes handled are in all cases rather small.

Streamline Filter. This filter (Fig. 85), also known as the **edge filter**, differs in operating principle as well as mechanical design from other types of filters.

Closely compressed disks of specially prepared paper are used as the filter medium, the filtrate passing by edgewise filtration between, not through, these disks, while the solids collect upon the outer edges of the disks which form a hollow column or "pack" when assembled in a unit. Applications of vacuum to the inside of the pack draws the oil through it, all traces of sludge, carbon, and other solids being retained on the outer edges of the paper disks in the form of a cake. Discharge of this cake is made at the end of the day's operation by reversal of flow, using compressed air.

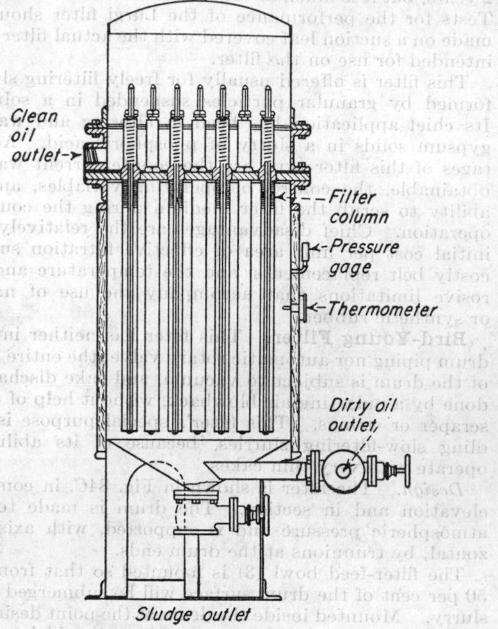

FIG. 85. Streamline filter (section).

Applications of this filter have been found in purification and dehydration of insulating oils and the like. Dehydration occurs through the joint effect of heating the oil followed by subjecting it to vacuum.

Mention should be made of the various small individual filter units mounted on internal-combustion engines for clarifying continuously the crankcase oil. These are usually provided with a cotton-flannel tabular labyrinth into which the dirty oil is pumped. The impurities are retained within the cotton tube, while clarified oil passing through the fabric walls of the tube is returned to the crankcase.

Corrosion-resisting Construction*

Filters are employed under a great variety of conditions, and in chemical work it is necessary to construct them of corrosion-resisting materials capable of withstanding attack by caustic solutions, alkalies, or acids. Each substance or solution to be handled presents a problem calling for special materials of construction.

For caustic solutions and alkalies, filters are built of cast iron, using alloys such as stainless steel and monel for bolts, nuts, fittings, and the filter medium.

For acid conditions, plate-and-frame presses and vacuum filters are made of wood. Fittings may be

* See Sec. 21.

lead-covered or made of special alloys chosen for their resistance to corrosion under the given conditions.

Lead is used as an internal lining for pressure filters of the closed-shell type and for tanks of vacuum filters, the drums of which may be cast lead. Rubber is suitable under severe acid conditions, but it is limited in its applications by the required temperature. Synthetic plastics may also be used.

For filter mediums, wool and woven-wire monel cloth are extensively used, and in special cases asbestos fiber, fiber glass, nitrated cotton, or rubber cloths are used.

Advice from filter manufacturers should always be obtained when considering the filtration of corrosive materials, since they constantly conduct research to prolong the life of their equipment.

Pulp Agitation

To obtain maximum filter output, the feed must be maintained as a homogeneous mixture. If the material to be filtered contains particles varying in size and density, segregation takes place unless there is sufficient agitation.

In shell-type pressure filters, segregation is prevented by maintaining a circulation of excess feed through the filter. When the proportion of solids in the feed is small, this is essential for the formation of even cakes.

The rotating-disk vacuum filter (American) utilizes feed circulation to provide agitation by forcing the feed through inlets in the bottom of the tank directly under each disk, the overflow returning to the feed-supply tank. This gives excellent results with materials of high specific gravity and of uneven size.

In rotary-drum vacuum filters (Oliver), agitators in the form of longitudinal rakes oscillating under the drum in the tank are most satisfactory.

For materials that dewater quickly causing thickening of the feed in the filter tank the oscillating agitator may have hollow rakes through which the returned filtrate is forced, thereby preventing segregation. The amount of filtrate used for this purpose is adjusted to maintain a balance between solids and liquids to give the required density of filter feed.

Vacuum leaf filters of the Moore and the Butters types use air lifts at the sides of the tank between the leaves as a means of circulating the feed and preventing settlement; or a centrifugal pump may be used.

In the Dorrco filter, agitators are never used and segregation is encouraged as an aid to filtration. This is a result of applying the filter medium on the interior of a drum acting as its own pulp container.

Range in Operating Vacuum

The vacuum registered when a filter is operating satisfactorily is an indication of the resistance offered by the filter cake and the filter medium to the passage of air. For instance, a crystalline solid like grainer salt, when filtered from the mother liquor of brine, requires the passage of a large volume of air through the cake, yet the gage will indicate only 2 to 4 in. mercury. In this case, the thick cake and the filter medium offer very low resistance to the passage of as much as 40 cu. ft. free air/min./sq. ft. filter area.

Precipitated calcium sulfate is of much finer grain size and therefore offers a correspondingly increased resistance. It does not require large volumes, usually 1 to 3 cu. ft./min./sq. ft. filter area but the vacuum gage will indicate 10 to 14 in. mercury.

Cane-sugar mud is so finely divided that the passage of air is restricted to very small volumes, about ⅓ cu. ft. free air/min./sq. ft. filter area. The vacuum readings are from 25 to 27 in. mercury.

Vacuum readings serve to indicate the friction encountered in drawing a given amount of filtrate and air through the filter cake and filter medium.

The range of vacuum employed depends upon barometric conditions at the location of the equipment.

Temperature.
Character of material being handled.
Size of vacuum pump in relation to filter equipment.

Industrial Applications

Cyanide Pulp. The recovery of gold and silver from ores is usually effected by grinding the crude ore to 90 per cent −200 mesh and dissolving its precious-metal content with sodium cyanide solution. Still finer grinding is now done in some cases up to 90 per cent −325 mesh. The thickened pulp resulting from this operation is usually filtered and washed on continuous vacuum filters for a thorough recovery of gold and silver.

Although the ore is usually quartz, there is often enough clay and other semiplastic material present to reduce the porosity of the mass when filtered. Therefore, the capacity of such filters varies directly with the proportion of clean quartz in the pulp and, in some cases, capacities of 2000 lb. and upward dry weight per square foot of filter area have been handled on continuous filters in cyanide-slime plants when treating clean quartz ores.

The oxidized and clayey ores will lower such capacities to 450 to 600 lb./sq. ft. filter area/24 hr.

Cyanide-slime filtration requires both filtration and thorough washing of the filter cake before discharge to remove the dissolved gold and silver, and for the latter either, or both, sprays or drip-wash equipment are provided. Wash equipment must be effective and subject to observation, as the character of the washing is revealed by the appearance of the cake surface during the operation.

The pulp fed to cyanide filters is usually 40 to 50 per cent solids, while discharged-cake moistures vary with the nature of the crude ore, ranging in most cases from 15 to 25 per cent.

As the cyanide solutions handled are alkaline with lime, the filter cloths become gradually clogged with calcuim carbonate, and this requires at intervals the use of hydrochloric acid solution wash to remove the coating on the cloth and to maintain its porosity. Cotton fiber covers are universally used for this work, and the filters are built of steel, cast iron, and wood. No brass is used as it is attacked by cyanide solution.

Metallurgical Flotation Concentrates. These products are obtained by the flotation of ore ground to a fineness varing from 65 per cent passing 200 mesh to as much as 90 per cent passing 350 mesh. Usually over 50 per cent of the particles composing the feed are metallic sulfides such as galena (PbS), zinc blende or sphalerite (ZnS), chalcopyrite ($CuFeS_2$), of high specific gravity 7.5, 4.1 and 4.2, respectively. The capacity of the filter will be determined by the fineness of the particles, their specific gravity, the density of the feed, and the amount of gangue of non-metallic material present in the form of slime.

Capacities of 300 lb./sq. ft./day are obtained upon material containing considerable gangue and slime; 1400 lb. or more is obtained on the same material, even with finer grinding, provided a clean concentrate is made.

The object of filtration in this case is simply to dewater the concentrate preparatory to smelting, and cake moistures run from 7.5 to 16 per cent. Pressure filters are unsuitable because the proportion of solids in the feed is high, necessitating frequent opening and discharging and thus making operation and maintenance charges exorbitant.

Rotary vacuum filters, with their continuous discharge of cake, handle such material with minimum of attention and low operating and maintenance charges.

Metallurgical Gravity Concentrates and Sands. These products are obtained by water classification of ores in which a part of the mineral particles is liberated without fine grinding. Screen analyses of the feed selected at random show 10 to 15 per cent held on 20 mesh and 34 to 72 per cent held on 65 mesh.

With such coarse feed it is difficult for a rotary vacuum filter to pick up and hold a cake against the action of gravity when the position of the filter medium becomes such that the undrained cake is on the underside and liable to fall off by reason of its weight. Suitable types of vacuum filters give high capacity ranging from 5 to 35 tons/sq. ft. filter area/day when handling such feed. The Oliver horizontal rotary filter is best suited for this use.

Cane-sugar Sirups. For the clarification of liquor and sirups in sugar refineries, pressure filters are especially suitable and, conversely, vacuum filters are unsuitable. This is because the proportion of solids in suspension is so low as to require a filter aid to enable clarification to be effected. The proportion of filter aid added is about three times the weight of solid to be removed and this makes a reasonable thickness of cake. Plate-and-frame presses and shell-type pressure filters are used.

Paper Pulp. This material, because of its fibrous nature, is free-filtering and is readily handled on open-mesh wire screen. It is bulky, being of low specific gravity, and the quantity to be handled by any unit is large. At the same time, large volumes of water or solution must be removed, and therefore the only filters suitable are vacuum filters with continuous discharge. There are many stages in the manufacture of pulp and paper at which filtration is performed, such as in washing the stock after cooking in the digesters, deckering after screening, thickening prior to bleaching, washing after bleaching, and to recover fiber from paper-machine waste white water.

The output obtained varies with the nature of the work and is from 200 to 1200 lb./sq. ft. filter area/day, and the water, or solution filtered, will vary from 1½ to 20 gal./sq. ft./min.

Sewage Sludge. All types of sewage sludges produced at municipal sewage-treatment plants such as raw, activated, digested, and various mixtures are now being dewatered successfully by means of vacuum filters. Dewatering in this manner is impractical without sludge conditioning, and some sludges are difficult to dewater even after treatment. The suspended particles are finely divided and are compressible and distorted by pressure to such an extent that ⅛ to ½ in. is the usual limit of cake thickness.

Vacuum filters are particularly suitable for handling this material since they accomplish dewatering at a relatively low pressure: 21 in. mercury. Under this pressure (10½ lb./sq. in.) there is least distortion of the compressible feed particles. A fair filtrate is obtained, and the cakes are automatically discharged.

Filter feed ranging from 1 to 10 per cent cake-forming solids gives a capacity of 25 to 250 lb. dry weight cake/sq. ft. filter area/24-hr. day. Cake moistures are 65 to 85 per cent, depending on the type and character of the sludge handled. Raw sludge cake is usually incinerated; activated is incinerated or dried for fertilizer; and digested is incinerated or spread on land for soil conditioning.

Cement Slurry. In Portland-cement manufacture, the raw material is ground either with or without addition of water, and the wet-ground material, known as slurry, is the type of feed used in most cement plants today. Whether a wet or a dry mixture is used, it must then be calcined and burned in a rotary kiln, the water content in the wet mix requiring the use of fuel to evaporate it during the process. To avoid this use of fuel, filtration of the wet slurry before burning has been widely adopted.

The ground material in wet cement slurry usually is chiefly limestone with some shale or clay for the alumina constituent, but there are also plants operating on marl, which is a natural mixture, and on blast-furnace slag and limestone. In all these cases, filter plants have been successfully used to reduce the water in the feed to the cement kilns. The wet-ground material usually enters the filters with 30 to 50 per cent water content, and the filter cake discharged from the filter will range from 17 to 25 per cent moisture. The moisture content in the feed to the filters is usually kept as low as is compatible with the easy handling of the pulp in the pumps.

The filtration of such pulp on vacuum filters shows wide ranges in capacity, as do individual plants. This is caused by variations in raw material from the quarry, so that cement-filter installations should be calculated with prudent factors. Usual filtering rates are 400 to 1000 lb. dry weight/sq. ft. filter area/24 hr.; under exceptional circumstances, rates are considerably higher.

Besides the fuel economy noted, the use of filters enables a given size of cement kiln to burn more clinker than before and thus to increase the capacity without an increase of kiln installation. This provides a reduction in cost per unit of product and is a valuable factor in the economy of cement-plant operation.

In many ways the filtration of cement slurry resembles cyanide-pulp filtration, since both are finely ground mineral pulps in aqueous suspension, and the filter cakes produced are similar. The density of cement slurries prevents segregation of solids in the suspension, which may occur in more dilute mixtures.

Salt and Crystals. This class of material is being separated from accompanying solution and dried with success on continuous vacuum filters.

Formerly, this work was done by centrifuges of various types, but the continuous vacuum filter has now been widely adopted for this service.

The materials handled under this heading may be any practically slime-free aggregate of non-deformable solids whose particle size is usually in a narrow range which assures uniform cake formation. There are cases, however, where heterogeneous masses with wide range of sizes are successfully filtered on salt-type filters.

This plan has been adapted in practice for sandy pulps and for many crystalline precipitates in various solutions.

The ratio of solids to liquids in the filter feed does not affect the filter performance greatly, as is the case with plastic pulps; and the moisture in the discharged cake is always low.

Capacities are usually high: in the case of sodium chloride, 4 to 6 tons/sq. ft. filter area/24 hr. being capacities often reached. In special cases much higher capacities have been reached in regular work.

The liquid in the cake is removed by the use of vacuum and heated air drawn through in the cake while on the filter. The salt is therefore readily discharged at 2 to 2.5 per cent water content and, by special equipment, can be discharged from the filter below 1 per cent water; "bone-dry" salt is now being produced on Oliver-Robison top-feed filters.

Filters of this type usually require cast-iron construction with woven metal-cloth covers and differ from conventional design by small drum dimensions together with large internal conduits in the drum. Vacuum pumps used handle large volumes of air at low-gage readings, 5 in. Hg, or less, being usual.

Beet-sugar Plant Filtration (Saccharate and Carbonation Mud). The recovery of dissolved sugar from beet juice includes the use of filters at two stages: first, in the separation of the carbonated beet juice from the insoluble impurities and lime sludge produced by carbonation. In this stage, the filters remove a clear first and second carbonation juice and wash the resulting lime cake to a low sucrose content in a single filtration.

The second stage consists of the separation and washing of calcium saccharate in the Steffens process. This is a two-stage operation wherein one filtration removes the trisaccharate, and, after further lime treatment a second filtration removes the mono- and disaccharate. The latter step completes the removal of sugar from the beet molasses, since the first precipitation as trisaccharate does not remove the entire sugar content.

Before continuous vacuum filters became the standard for beet-sugar factories, pressure filtration was the rule, but economies gained in substituting continuous vacuum filters in this work have made the use of pressure filters for this duty a rarity.

Summarizing, the results of continuous vacuum filtration in beet-sugar factories are:

Minimum sucrose left in washed lime cake.
Minimum impurities left in washed saccharate cake.
Minimum use of wash water and evaporation cost.
Minimum cost for filter-cloth renewals and acid treatment.
Minimum labor requirements and use of skilled labor unnecessary.
Clarity of filtered carbonation juice is high.

For a typical Steffens A factory installation, using 3.75 per cent CaO addition, a vacuum filter on carbonation mud will handle about 1.35 equivalent tons sliced beets/sq. ft. filter area/24 hr.; and a "milk-of lime" plant, at 2.50 per cent CaO addition, will handle 2.00 equivalent tons; variations in the amount of lime affect the capacities of the filter proportionately.

Calcium saccharate filtration is based directly upon the amount of molasses treated and gives a rate of about 190 lb. molasses/sq. ft. filter area/24 hr. on tricalcium saccharate.

The filter feed on carbonation-mud-filter units is usually about 40° Brix. or about 17.5 per cent cake-forming solids; the pulp is filtered and washed on external drum filters to 1.5 to 2.0 per cent sucrose on CaO. To effect this, as little as 110 per cent wash water is used in terms of wet cake.

The feed to the tricalcium saccharate filters is about 12.5° Brix. with 6.4 per cent sugar, and the wash water from the filter contains about 0.70 to 0.75 per cent sugar.

Petroleum Products. The major problems involving filtration in the refining and processing of petroleum oils include:

1. Contact filtration of contacted petroleum products such as lubricating stocks.
2. Clarification of cracking-still residuum.
3. Filtration of chilled oils to remove wax.
4. Miscellaneous problems, including secondary clarification after contacting, or the "polishing" of products to ensure absolute freedom from solids just prior to shipment, emulsion breaking, etc.

In item 1, finely divided bleaching clay is added (frequently after acid-treating the oil), thoroughly agitated with the oil at the required temperature, and then removed by filtration which produces the required color. The amount of clay employed and the temperatures used vary with the different oils. Usually from 1 to 3 lb. clay/bbl. gasoline will suffice, while in the case of high-viscosity lubricating oils from 4 to 15 per cent of clay (based on the weight of oil) may be necessary. Shell-type filters are generally employed, as the percentage of clay is either very low, as in the case of gasoline, or else the temperatures are near the flash point of the oil (or the naphtha used in cutting back the more viscous oils prior to filtration). Rates per square foot of filter area will vary from as much as 75 gal. on gasoline with little clay, to as little as 2 to 3 gal. on very viscous oils. Continuous removal of contact clay from lubricating oils is now more effectively done by precoat filters. Continuous drum filters, such as the vacuum precoat, are also finding application on problems in item 4.

Not all cracking-still residuums have been successfully filtered. On certain residuums, rates as high as 7 gal./sq. ft. filter area are being obtained. In some cases filter-aid material is being used. Shell-type filters are generally employed as temperatures frequently range as high as 650°F.

Item 3, as noted above, includes the recovery of marketable paraffin wax and also (secondly) the filtration of oils, using a special solvent, for the primary purpose of lowering the cold test. In the first case, special high-pressure plate filters with heavy canvas-filter pads are generally employed, and rates are low and pressures range from 300 to 400 lb./sq. in. In lowering the pour point of oils, either continuous vacuum filters or shell-type pressure filters are employed, depending upon the requirements of the individual problem.

Shell-type filters have generally been employed in division 4, as the percentage of solids is usually very low and filter aids are frequently employed.

Settled Cane-mud (Cachaza) Filtration. In producing sugar from sugar cane in defecation mills making raw sugar, and also in defecation-sulfitation mills making plantation white sugar, the common practice is to settle the hot treated sugar juice; to draw the clarifier juice from the upper parts of the settling tanks; and likewise to draw the settled muds from the lower parts of the settling tanks. These muds are subsequently filtered, often after additional "liming" and sometimes settling. This filtration is decidedly difficult in the large majority of cane mills.

The rapidity with which filter cloths become clogged has resulted in the retention of the plate-and-frame type of filter, although there are some notable installations of leaf-type pressure filters, particularly in Java and in the Philippine Islands. A plate-and-frame filter-press station in a sugar mill is usually the highest loss, highest cost, most unsightly, and most objectionable station in the mill.

In recent years continuous vacuum filters are rapidly replacing plate- and frame-press installations; seldom are plate-and-frame presses installed now in a new factory, preference being given to vacuum filters. The method now generally adopted is to send the settled muds either from open defecators or continuous clarifiers after hot liming treatment to specially designed vacuum drum filters using perforated metal plates as the filter medium, the filtrate being divided into two parts, the first or cloudy filtrate being returned to process and the second or main filtrate going direct to the evaporators.

Filter capacities vary between wide limits in different localities and in different mills in the same locality, being determined to a large extent by factors outside the control of the operating staff, such as the nature of soil and the climatic conditions, and also by other factors under control, such as chemical control and mill conditions. The two primary requisites for vacuum-filter operation are relatively thick muds and sufficient fine cane fiber in the muds for formation of a filter cake. Capacities will ordinarily be from a minimum of 5 to a maximum of 10 tons (2000 lb./ton) cane/sq. ft. filter area/24 hr.

The cake on vacuum filters is washed in the usual

manner for the removal of sugar solution, and no difficulty is experienced in reducing the sucrose content of the cake to a maximum of 1.25 per cent. With care, the sucrose content can in all cases be reduced to the point where no further advantages are gained by further lowering the sucrose content.

The advantages of vacuum filters are:

1. Continuous and automatic filter operation.
2. Absolute elimination of leaks and unknown losses at filter station.
3. Complete change from the most unsightly and objectionable station to a clean and entirely presentable station.
4. Large reduction in filter area.
5. Large reduction in space required for filter station.
6. Entire elimination of filter-cloth troubles and frequent washing and renewals; when a punched-plate filter medium is used, no treatment is required from beginning to end of grinding season, and the estimated life of the medium is 4 years.
7. Minimum possible loss of sucrose in filter cake.
8. Large reduction in labor requirements.
9. Minimum dilution of juice owing to consistent low wash-water consumption. This results in decreased fuel requirements and in increased capacity of evaporators.
10. Steadily consistent results because the human element is largely obviated.
11. Rapid, efficient filtration with practical elimination of inversion losses and maintenance of best possible purities.
12. Practical elimination of overtaxing the filters by increasing the milling capacity without providing additional filters. Vacuum filters can be speeded up to a certain point to provide for increased capacity with small loss in efficiency, but the time required for washing the filter cake cannot be eliminated or reduced as with plate-and-frame filters, nor is it advantageous to the native labor employed on tropical sugar plantations to forget the washing, or otherwise mistreat the filter station, and convenience in running sugar juice to waste is entirely prevented.

Food Products. In the field of food production nearly every important product undergoes filtration at some stage. Important instances are starch and corn sirup derived from conversion of corn. Another field is in the manufacture of beer and wine, while the fruit juice, softdrink, and vinegar industries supply many other applications. Starch is handled on rotary disk filters, and corn sirup finds the vacuum precoat filter useful. The liquids in the varied beverage industries are usually handled to best advantage by some type of pressure-filter unit with or without use of filter aid. Specially prepared asbestos is sometimes used, in pads or in bulk, as the filter medium.

The **Sparkler** filter is a closed cylindrical metal container within which are fixed separate horizontal annular leaves, 8 to 33 in. in diameter, around a hollow vertical shaft. The upper surfaces of the leaves are precoated before filtering begins. The filtrate passes through the leaves and out through the hollow shaft. After filtering, the set of leaves is lifted out, cleaned, and replaced in the filter. The Sparkler finds use in pharmaceutical and beverage plants.

The **Niagara** filter is a closed cylindrical metal container in which is placed a set of vertical metal leaves with wire-cloth covers. The leaves, which are precoated before filtering, may have separate drainage outlets or may discharge into a common header. They are made with tubular rims and are separately removable from the shell.

The **Republic** filter finds especial use in the manufacture of sterile products. It uses a filter medium of specially formed asbestos sheets placed in plates and frames, or precoated wire cloth in the form of pressure cylinders.

The **Oliver pressure** filter (Fig. 86) has a cylindrical shell enclosing vertical leaves, each having a ground joint outlet for filtrate. The leaves may be covered with wire cloth or other fabrics and may or may not be precoated. Sizes from 30 to 110 sq. ft. filter area are available.

Caustic Soda. After thickening, the calcium carbonate, precipitated in the manufacture of caustic soda by the lime-soda ash process, is generally dewatered and washed on vacuum filters.

This precipitate filters quite readily from a feed moisture of 50 to 70 per cent moisture to a cake containing from 30 to 40 per cent moisture. Capacity ranges from 500 to 1200 lb./sq. ft./24 hr., and the vacuum requirements correspond to about 6 cu. ft. free air/min./sq. ft. filtration area.

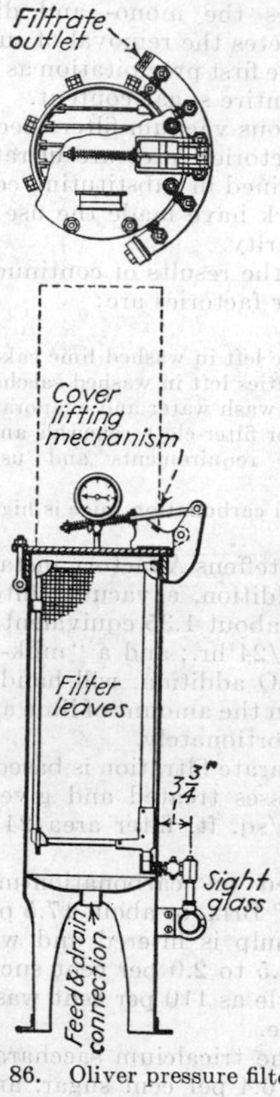

Fig. 86. Oliver pressure filter.

Metallic filter mediums are used. Iron and steel must be used for the filter proper.

Phosphoric Acid. A precipitate of anhydrous calcium sulfate is formed in the manufacture of phosphoric acid by digesting phosphate rock with sulfuric acid. The filtration and washing of this precipitate on vacuum filters at 150° to 165°F. require the use of special corrosion-resisting materials for the filtration mediums and for the filter proper when handling 20° to 35°Be. acid. Operation of a recent phosphoric acid filter plant shows that a synthetic plastic textile gives best results on a lead-protected Oliver horizontal rotary vacuum filter.

The precipitate filters readily from a moisture content of 65 per cent in the filter feed to a cake moisture of 30 to 35 per cent. Capacity ranges from 2000 to 4000 lb. dry $CaSO_4$/sq. ft.

Pigments. Lithopone is a fine white pigment of about 70 per cent barium sulfate and 30 per cent zinc sulfide, simultaneously precipitated by mixing solutions of barium sulfide and zinc sulfate. Vacuum filters are used for dewatering the thickened raw lithopone and the finished lithopone after calcination and grinding.

Lithopone is representative of the group of chemically precipitated mineral pigments. It filters slowly, is extremely finely divided, and must be filtered in the form of a very thin cake.

The thickened feed generally enters the vacuum filter at 50 to 70 per cent moisture and is discharged at 35 to 50 per cent moisture; capacity ranges from 200 to 400 lb./sq. ft./24 hr.

Corrosion-resisting construction must be used, particularly in the filtration of the raw lithopone. Approved materials of construction are wood, monel metal, aluminum-zinc alloys, and rubber-covered iron and steel.

Precipitated barium sulfate, white lead and titanium-base pigments display filtration characteristics similar to those of lithopone.

Power Used on Filters. Vacuum-filter operation requires a vacuum pump which is the chief consumer of power. The pump size varies widely with the porosity of the cake formed on the filter and will range from about ½ cu. ft. pump displacement/sq. ft. filter area for dense cakes of slowly filterable pulps up to as high as 40 to 60 cu. ft./sq. ft. for crystalline materials. Vacuum-gage readings are nearly always found at inverse ratio with the volume of the pump displacement, which avoids excessive power costs for high displacement.

"Dry" vacuum pumps (piston type) are preferred, although rotary-vacuum pumps as well as steam jets are used in some special cases.

No concise table for such power consumptions can be compiled, the experience of the filter manufacturer being the safest criterion for the choice of pump.

Rotation of the filter itself consumes very little power, the larger size rotary vacuum-drum filters not requiring more than about 0.005 hp./sq. ft. filter area, while the filtrate pumps are chosen for volume of filtrate handled and the net pumping head.

Pressure filters may be operated by gravity head on the filter-feed line, or by centrifugal pumps whose size depends on the volume to be handled in a given time; and power depends upon volume handled, size of filter, and operating range of pressure required.

Unit Costs. The continuous-rotary-vacuum-filter costs vary widely with the kind and the amount of feed handled. Although consistent figures over long periods with the rotary-vacuum filter have, in a given case, shown very low costs ($0.025/dry ton of filter cake), it is likely that most continuous vacuum filters will have cost figures ranging from $0.05 to $0.15/ton, dependent on the tonnage handled and the material itself.

In general, a dense feed to the filter, an absence of pasty or colloidal solids in the feed, and an increase of pulp temperatures serve to increase the filtering rates and to lower the unit filtering costs, while the converse increases them.

Filtering costs should include only those operations directly pertaining to this step; some plants include other operations therein, which explains some surprisingly high "filtering costs" in otherwise well-supervised plants.

If the "fixed-charges" burden is included in the cost item, the amount due to that should be separately indicated.

The cost of pressure filtration varies widely with the materials handled; and, since such installations are usually small, compared with continuous vacuum filters, the costs are consequently higher, but the difficult kind of work performed justifies the higher operating costs in most cases.

Floor Space Available. The floor space available affects the choice of type of filter in some cases. For example, if a large filter area is required to handle the product, and the floor space available is restricted, there is at once an advantage in the disk type of vacuum filter, other factors being approximately equal. The saving in building costs and heating expense is often a considerable factor.

Plate-and-frame presses occupy less floor space than shell-type pressure filters for the same filter area. Shell-type filters require more room for opening and closing; but, if large units are employed, space may be economized by installing the twin-type units.

Cost of Equipment. Prices on filters vary so widely, depending upon size, type of filter, and material of construction, that it is impractical to set a unit price common to all such varieties.

In the case of continuous rotary vacuum filters, a typical price per square foot of filter area (not including wash apparatus or accessories) would, in the case of a

Table 5. Range of Filter Sizes

	Area, Sq. Ft.
Oliver	
1-ft. diam. × 1-ft. face	3
3-ft. diam. × 6-in. face to 4-ft. face	4– 36
4-ft. diam. × 2-ft. face to 6-ft. face	25– 75
5-ft. various in. diam. × 4-ft. face to 12-ft. face.	65–200
6-ft. diam. × 1-ft. face to 12-ft. face	18–226
8-ft. diam. × 6-ft. face to 14-ft. face	150–350
11½-ft. diam. × 10-ft. face to 16-ft. face	360–570
14-ft. diam. × 14-ft. face to 18-ft. face	610–790
American	
4-ft. diam. × 1 disk to 4 disk	22– 86
6-ft. diam. × 1 disk to 6 disk	50– 300
8 ft. 6-in. diam. × 2 disk to 10 disk	185 925
12-ft. 6-in. diam. × 5 disk to 12 disk	1000–2400
Sweetland	
No. 1	8½
No. 2	47
No. 5	191
No. 7	261
No. 10	538
No. 12	1020
Kelly	
No. 30	33
No. 50	50
No. 250	270
No. 450	459
No. 650	652
No. 900	918
No. 1300	1304
Dorrco	
4-ft. diam. × 1-ft. and 2-ft. face width	12– 24
6-ft. diam. × 2-ft. and 3-ft. face width	37– 55
8-ft. diam. × 3-ft. to 14-ft. face width	74–345
10-ft. diam. × 4-ft. to 16-ft. face width	124–494
12-ft. diam. × 5-ft. to 16-ft. face width	190–596
14-ft. diam. × 16-ft. to 18-ft. face width	696–783

plain dewatering type, be about $16 to $18/sq. ft. filter area in the smaller sizes, and about $10 to $12/sq. ft. in the larger sizes.

For the special designs required in the wood-construction filters for acid-resisting use, and those built for work with caustic material, the smaller sizes cost about $30 to $40/sq. ft. filter area, and the larger sizes about $20/sq. ft. filter area. These prices are suitable for rough preliminary estimations only. Cost per ton/capacity is more logical than per square foot area.

Cost of accessory vacuum equipment, pumps, etc., for vacuum-filter installations will range around 25 to 33 per cent of the filter cost, additional. All costs mentioned herein are on the basis of quotation f.o.b. point of manufacture, and of data from before the Second World War.

Corrosion. When corrosive material is handled, vacuum filters are more affected than other types, because more parts are exposed and the passage of air through the filter increases oxidation and corrosion. Pressure filters are not exposed to the same extent and because of simplicity lend themselves more readily to corrosion-resisting construction.

The use of special alloys, however, makes it possible to construct filters to suit given conditions of corrosion.

Table 6. Factors Affecting Selection of Type of Filter and Character of Pulps Handled

Typical materials	Character	In. Hg vacuum or lb. pressure	Approx. filter capacity, lb./sq. ft./day	Type of filter suitable		
				Plate and frame	Shell type	Continuous vacuum
Cyanide slime	Finely ground quartz ores	18–25 in.	400– 2,000			X
Flotation concentrates	Minerals, finely ground	18–25 in.	400– 1,800			X
Gravity concentrates and sand	Metallic and non-metallic minerals almost free from slime	2– 6 in.	10,000–70,000	..	..	X
Cement slurry	Finely ground limestone and shale, or clay, etc.	18–25 in.	400– 2,000	..	..	X
Pulp and paper	Free-filtering fibers	6–20 in.	200–1,200 and 1½–20 gal. water/sq. ft./min.	..	..	X
Crystals, salt, etc.	Granular, crystalline	2– 6 in.	3,000–12,000	..	..	X
Cane-sugar-liquor clarification, beverages, etc.	Sirups and solution with small percentage of solids with filter aid	40–50 lb.	36–1,400 gal./sq. ft./day	X	X	
Pigments	Smeary, sticky, finely divided, non-crystalline	20–27 in.	200– 500			X
Sewage sludge	Colloidal and slimy	40–50 lb. 22–24 in.	Batch operation 25 to 250	X	X	X
Varnish	Cloudy viscous liquid, filter aid used for clarification. Filtered hot	15–16 lb.	5 gal./sq. ft./hr.	X	..	
Mineral oils, with or without wax	Removal of bleaching clay from petroleum products. 1 to 20% clay used	50 lb. max. pressure	3–30 gal./sq. ft./hr. (lubricating oils) 25–75 gal./sq. ft./hr. (gasoline)	.. ..	X X	
Cane mud	Vegetable fiber and cane juice			..	..	X

Filter Auxiliaries

Pressure Pumps. For supplying feed to pressure filters, centrifugal and plunger pumps and, sometimes, hydrostatic head pumps are used. Montejus, or pressure tanks, are used in special cases.

Open-impeller centrifugal pumps are recommended where it is desired to start filtration with low pressure and to increase the pressure as the cake thickness increases.

Plunger pumps deliver feed at a regular rate, depending upon the pump displacement. They give high initial pressure which may be maintained throughout the cycle.

Diaphragm pumps can be used. This is a metal case, into which pulp is admitted, and then expelled by movement of a flexible rubber diaphragm. In the old-style diaphragm pump, a rigid mechanical attachment moved the diaphragm. The modern type, known as "O.D.S." pump, uses a diaphragm which is actuated solely by compressed air and has no mechanical attachment for operation.

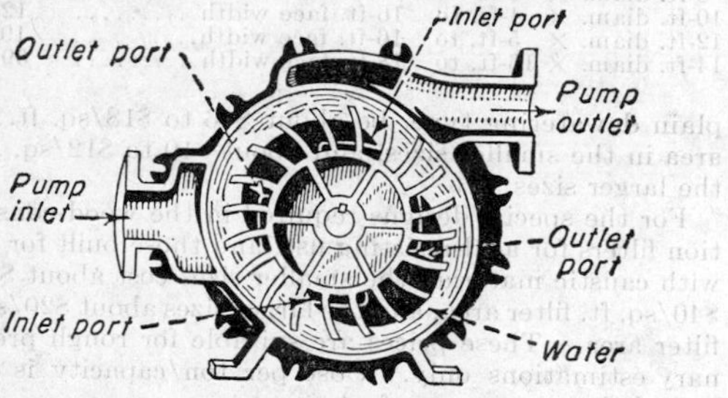

FIG. 87. Nash vacuum pump.

Montejus were largely used in the early days of pressure filtration. A monteju is a pressure tank which may be filled with the filter feed, after which compressed air is admitted at the top and forces the feed through an outlet pipe leading to the pressure filter. Montejus are installed in pairs so that the feed can be supplied steadily, one being filled while the other is being emptied.

Vacuum Pumps. There are two classes of vacuum pumps, wet and dry. The former handle both the filtrate and the entrained air while the latter handle the air only. Wet vacuum pumps are usually of the rotary type of which the Nash Hytor and Connersville cycloidal are examples.

The Nash pump (Figs. 87 and 88) is well adapted for handling large volumes of air with considerable liquid at a moderate vacuum (16 in. Hg) and is much used in the pulp-and-paper industry where this type of service is required.

The Connersville cycloidal type handles large volumes of air at low vacuum (2 to 6 in. Hg) with entrained liquid at lower power consumption. This type of pump is particularly useful when filtering granular or crystalline products such as salt, and the air is either heated or is at normal temperature. Reciprocating dry-vacuum pumps are in common use. If steam is cheap, steam-jet ejectors are used as pumps.

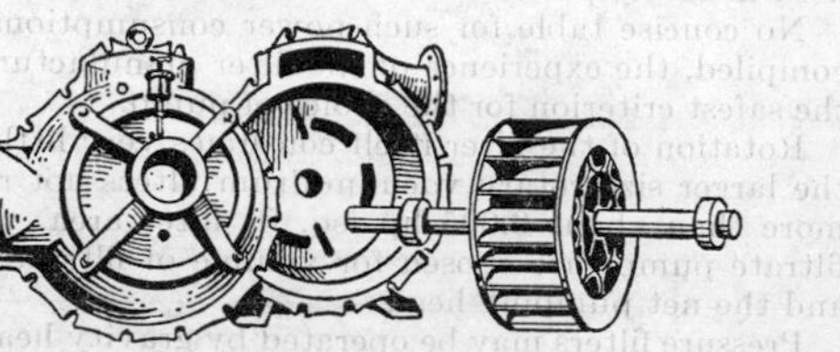

FIG. 88. Nash vacuum pump (dismantled).

Receivers or Separators. When the filtrate and air from a vacuum filter are discharged separately, a receiver or separator is used, consisting of a cylindrical tank usually installed with its long axis vertical (Fig. 89). The incoming filtrate and air enter at the side, air is withdrawn through the vacuum line at the top, and filtrate drains away or is pumped out through a connection at the bottom.

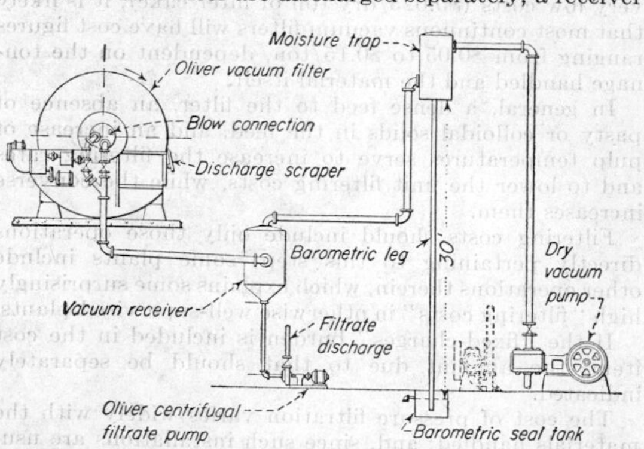

FIG. 89. Vacuum filtration unit. (*Oliver United Filters, Inc.*)

Moisture Traps and Condensers. If a dry vacuum pump is used, it is necessary to install a moisture trap or condenser (Fig. 90), the purpose being to prevent filtrate or condensed moisture from entering the pump. A moisture trap suffices when filtering at normal temperatures, but a condenser is necessary to maintain the vacuum-pump efficiency when operating at temperatures approaching the flash point of the filtrate at the vacuum employed.

A trap is merely a small receiver, and a condenser is similar but is equipped with cold-water showers and baffles. The installation may be made at a height allowing at least 30 ft. vertical between the bottom of the trap or condenser and the seal pit in which the discharge drainpipe terminates (Figs. 89 and 90). If preferred,

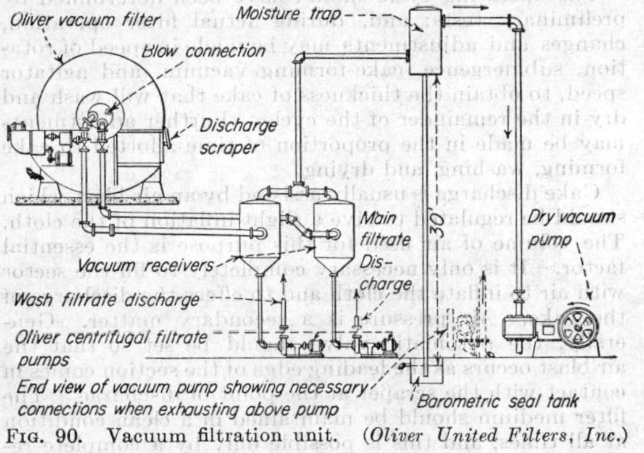

FIG. 90. Vacuum filtration unit. (*Oliver United Filters, Inc.*)

the outlet of the drainpipe may be connected to the suction of the filtrate pump.

Filtrate Pumps. These pumps are usually centrifugals of the closed-impeller type. They pump filtrate from the receiver and discharge it at the desired height. They should be specially designed, since they are called upon to handle both water and air under vacuum and usually discharge against positive hydraulic heads. The manufacturer of the filter equipment should always be consulted as to the design of pump best suited to the equipment involved.

Barometric Legs. Where the product is free-filtering and the amount of liquid to be removed is large, the use of vacuum pumps may be avoided by discharging the filtrate through a barometric leg. This consists of a vertical pipe which discharges into a seal box, the overflow from which goes to waste. The length of vertical discharge should equal the height of the water barometer, and if a vacuum pump is employed to handle the air from the filter there is no danger of water being drawn into the system.

The vacuum created depends upon the velocity of the water in the pipe; therefore the pipe diameter and the height must be adjusted to give the filtrate sufficient velocity to draw out the entrained air, thus creating a vacuum at the filter. The leg should be as nearly vertical as possible.

In pulp and paper mills many filter installations operate with barometric discharge of the filtrate. It is the simplest way to handle filtrate and air collectively and avoid installation and operating expense of vacuum and filtrate pumps; usually 10 to 20 in. vacuum is obtained by this method.

Blowers. Most materials when handled by vacuum filters require an air blast to assist the discharge of the cake. The volume of air used is small, and usually a low pressure is sufficient. Where compressed-air service is not available a small air compressor, or a rotary blower, may be employed. Air compressors, if used, are of the ordinary displacement pump types, but rotary blowers are best suited for pressures under 3 lb./sq. in.

The disk type of filter requires only a few ounces pressure, and an air-injector blower will give this (Fig. 91).

The pulsating air blow used for the Dorrco filter merely requires a rotating valve in the air line to produce air pulsations under the filter medium.

The function of an air blow is quickly to fill the section with air at sufficient pressure to flex the cloth, and to

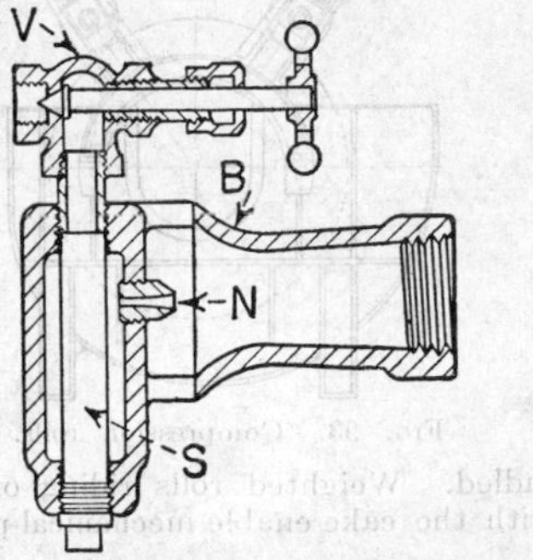

FIG. 91. Air-injector blower. (*New Jersey Meter Co.*)

issue through the pores of cloth, thereby dislodging the cake and freeing the fine particles which might otherwise remain and cause cloth blinding.

Flappers. A flapper is an appliance used with the external-drum type of vacuum filter to reduce the cake moisture (Fig. 92). It is installed above, and parallel to, the drum axis and consists of a shaft to which are attached two pieces of heavy fabric of such length that they strike the cake a regular succession of blows when the shaft is rotated. The blows cause a rearrangement of particles in the cake, closing the cracks and liberating moisture, which is then drawn through into the vacuum system.

To prevent the cake from being dislodged by the blows of the flapper, a piece of heavy fabric is fastened so that

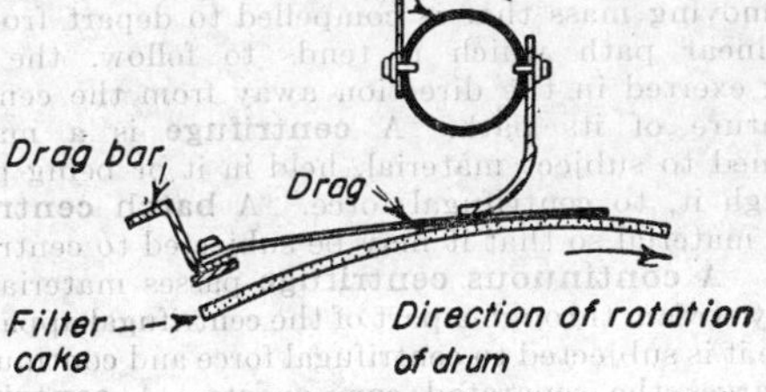

FIG. 92. Flapper. (*Oliver United Filters, Inc.*)

it drags upon the cake as the drum revolves and receives the blows given by the flapper. The flapper reduces the moisture from two to four units, which may be 10 to 20 per cent of that in the unflapped cake.

Mechanical Vibrators. Different devices have been designed to accomplish the cake puddling effect of the flapper by a more compact mechanism with a more controllable principle. This employs, in one case, a slight recurrent lifting and lowering of the cake layer; in another, a movable element "slides" on the cake to give an "ironing" effect, with resultant moisture reduction.

Compression Belts. A cake-compression belt is an endless belt that rides over a system of rollers mounted in a frame above the drum of an external-drum type of vacuum filter. The belt comes in contact with the cake at a point on the ascending side of the drum and leaves at the descending side above the cake discharge. It is driven by contact with the cake, and lateral adjustment is provided by a tracking roller. The belt may be either porous or impervious, depending upon the material being handled. Weighted rolls riding on the belt in contact with the cake enable mechanical pressures to be applied.

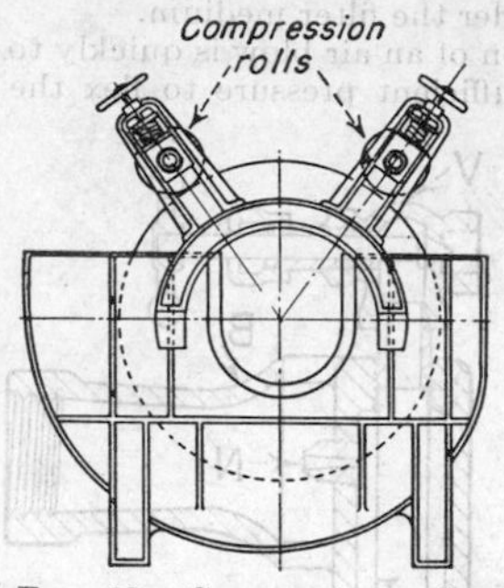

FIG. 93. Compression rolls.

The use of this appliance tends to close cracks in the cake, and in some instances the belt may be applied soon enough to prevent their formation. This reduces the amount of air drawn through the cake and reduces the vacuum-pump displacement. Moisture is reduced, since all the air is drawn through the cake and cannot short-circuit through cracks. Pressure from the rolls expels additional moisture. This arrangement is useful when handling pigments and other finely divided materials.

Compression Rolls. Compression rolls without belts are used for reducing the moisture in crude bicarbonate of soda, paper pulp, etc. (Fig. 93).

Filter Operation

In operating pressure filters, the process is intermittent and the product obtained from each cycle when compared with that previously obtained indicates whether there has been a change of conditions. During the cycle, however, the operator should control the feed pressure and note the clarity and quantity of the filtrate issuing and any other factors that influence the particular problem.

In vacuum filtration there are advantages in that the controlling factors are visible and accessible at all times. The feed is in an open container, and cake thickness can be controlled by regulating the submergence, vacuum, and speed of rotation of the drum or disk.

The operating cycle should have been determined by preliminary tests; and, during actual filter operation, changes and adjustments may be made in speed of rotation, submergence, cake-forming vacuum, and agitator speed, to obtain the thickness of cake that will wash and dry in the remainder of the cycle. Further adjustments may be made in the proportion of time allotted to cake forming, washing, and drying.

Cake discharge is usually assisted by an air blast which should be regulated to give a slight inflation of the cloth. The volume of air used for this purpose is the essential factor. It is only necessary completely to fill the sector with air to inflate the cloth and to effect the discharge of the cake. Air pressure is a secondary matter. Generally, the automatic valve should be set so that the air blast occurs as the leading edge of the section comes in contact with the scraper at the point of discharge. The filter medium should be maintained in a clean condition at all times, and this is possible only by a complete removal of cake at each revolution. If the material being filtered tends to blind the medium, it will be necessary to wash it at regular intervals to maintain capacity. Experience operating a filter upon a specific problem is the best guide.

CENTRIFUGES

BY A. E. FLOWERS

Revised by Selden H. Hull

Definitions. A **centrifugal force** is produced by any moving mass that is compelled to depart from the rectilinear path which it tends to follow, the force being exerted in the direction away from the center of curvature of its path. A **centrifuge** is a machine designed to subject material, held in it or being passed through it, to centrifugal force. A **batch centrifuge** holds material so that it may be subjected to centrifugal force. A **continuous centrifuge** passes material in a steady stream through a part of the centrifugal apparatus, where it is subjected to centrifugal force and continuously discharges the separated components. A **centrifugal clarifier** subjects a mass or stream of liquid to centrifugal force to remove solid or liquid contaminants. A **centrifugal separator** subjects a mass or stream of mixed liquids to centrifugal force, thereby separating them. A **centrifugal purifier** removes centrifugally foreign contaminants (such as water or dirt) from a liquid passed through it. A **test-tube centrifuge** carries cups holding graduated tubes containing liquids to be subjected to centrifugal forces of known amounts for definite times, for measurement of the components and their separability. A **basket centrifuge** holds a mass of material (such as clothes, masses of crystals, etc.) and centrifugally removes the water or other liquid. They usually have large baskets or bowls and are therefore operated at moderate speeds. The term *centrifugal dryer* though often used, is less appropriate, since "drying" is, usually incomplete. The basket may be "perforated" to discharge the liquid through its walls or be the "overflow" or "imperforate" type to collect solids from a slurry fed to it and discharge liquid over an inner lip or "curb." A **centrifugal filter** carries a filtering medium (cloth, paper, or metal screen) to catch and hold solids while allowing liquids to pass through and be discharged.

THEORY

Force

The force on a particle, compelled to move in a circular path, is determined by the "acceleration" toward the center, or rate of change of velocity direction away from the linear path any rotating particle tends to follow.

The centrifugal force may therefore be expressed as

$$F = \frac{W}{g} r\omega^2 \qquad (1)$$

where F is centrifugal force, g.; W is weight of particle, g.; r is radius of curvature of path, cm.; ω is angular velocity, rad./sec.; g is acceleration of gravity, usually taken as 981 cm./sec.^{2}.

Also

$$F = \frac{WV^2}{gr} \tag{2}$$

where V is peripheral velocity, cm./sec., or

$$F = \frac{\pi^2}{900}\frac{W}{g} r\ (\text{r.p.m.})^2 \tag{3}$$

$$F = 1.118 Wr\ (\text{r.p.m.})^2\ 10^{-5}$$

where the angular velocity is expressed in revolutions per minute (r.p.m.).

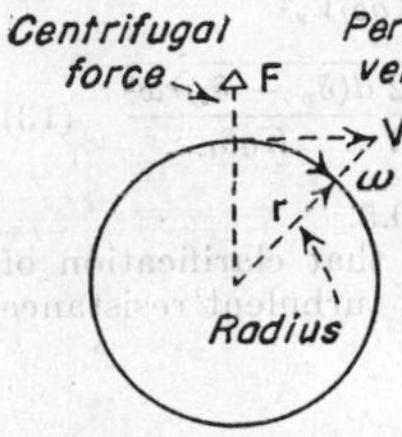

FIG. 94. Derivation of centrifugal force.

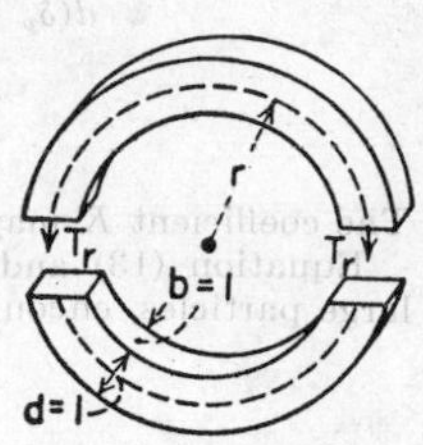

FIG. 95. Derivation of centrifugal force on a ring section.

The force on a ring section, or cylindrical element, of the bowl, or vessel, required to hold the material being processed, due to its own mass, may be expressed approximately as follows:

$$T_r = \frac{\delta_r}{g} r^2\omega^2 \tag{4}$$

or

$$T_r = \frac{\delta_r}{g} V^2 \tag{5}$$

$$T_r = \frac{\pi^2}{900}\frac{\delta_r}{g} r^2\ (\text{r.p.m.})^2 \tag{6}$$

where T_r is the unit stress in the ring section, g./sq. cm.; δ_r is the specific gravity of the wall material, g./cu. cm.; r is the mean radius of the ring, cm.

The force on a liquid at various depths results in a unit liquid pressure which may be expressed as follows:

$$P_l = \frac{\delta_l}{2g}\omega^2(r_2^2 - r_1^2) \tag{7}$$

where r_2 is the radius at any point in the liquid bed for which the pressure is to be calculated; r_1 is the radius of the inner liquid surface.

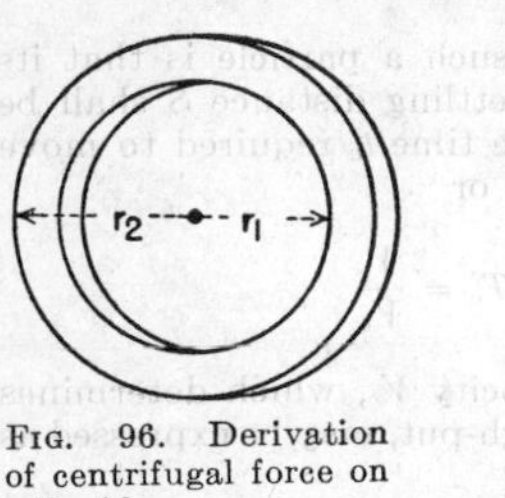

FIG. 96. Derivation of centrifugal force on a liquid.

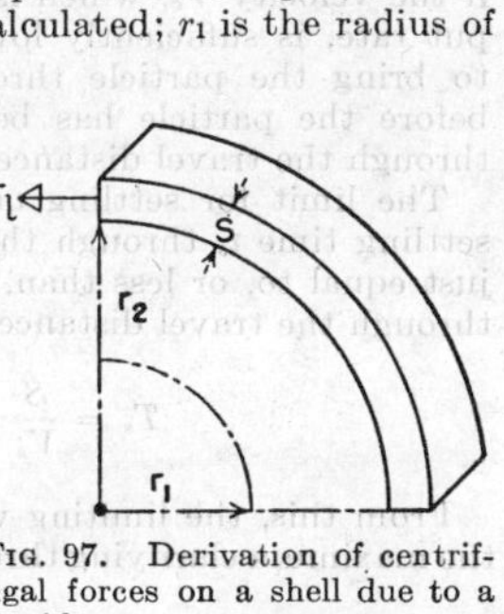

FIG. 97. Derivation of centrifugal forces on a shell due to a liquid.

If the radial thickness of the bowl-shell wall is expressed by the symbol s then the average unit tension produced by the liquid pressure P_l at the inner wall of the bowl shell is

$$T_l = \frac{\delta_l r_2}{2gs}\omega^2(r_2^2 - r_1^2) \tag{8}$$

or

$$T_l = \frac{\delta_l V^2}{2gr_2 s}(r_2^2 - r_1^2) \tag{9}$$

or

$$T_l = \frac{\pi^2\delta_l}{1800gs} r_2(r_2^2 - r_1^2)(\text{r.p.m.})^2 \tag{10}$$

The total stress in the bowl-shell wall is the sum of that due to the wall material and that due to the liquid bed, assuming s small in comparison with r_2, so that r_2 may be substituted for r in Eq. (4).

$$T_b = T_r + T_l = \frac{r^2\omega^2}{g}\left[r_2\delta_r + \frac{\delta_l}{2s}(r_2^2 - r_1^2)\right] \tag{11}$$

The forces on opposing columns of liquids of different specific gravities being centrifugally separated determines the radial position of the cylindrical surface of division (usually called the neutral zone) between the liquids. The heavier liquid, discharging over a weir at a greater radius than that for the lighter liquid, fills the outer part of the bowl above a partition wall (or top disk) and inward below it to the neutral zone. The lighter liquid fills the space from the neutral zone inward to its discharge weir.

The top disk is usually made as large as the size of the bowl shell will permit in order to allow a wide range of positions for the neutral zone.

The position of this neutral zone is an important factor in the performance of a centrifuge. When the neutral zone is near the center, the lighter component is exposed to only a small amount of centrifugal force while the heavier component is exposed to a much greater amount. A cream separator is an example of this type in which the object is to leave the smallest possible quantity of fat in the skim milk.

When the neutral zone is near the larger diameter of the bowl, the greater effect is on the lighter component, as in an oil purifier in which the object is to remove the last possible traces of heavier impurities.

The feed to the separating space of a centrifugal bowl should be as near the neutral zone as possible.

The position of the neutral zone can be controlled by adjusting the radius of the discharge for *either* of the components. It may also be varied by allowing one or both components to rotate faster or more slowly than the bowl during their radial travel. If a component moving outward slips radially, it will rotate more slowly than the bowl and so be acted on by a lower centrifugal force. If one moving inward slips, it will rotate faster and be affected by a greater centrifugal force which will oppose its inward flow. Either of these slippages will displace the neutral zone from its "balanced column" position.

The expression for the relation of the position of the neutral zone or balance circle for radial flow (no slippage) and negligible fluid flow friction may be developed as follows:

The static pressure on the bed of light liquid L between the inner discharge lip and the neutral zone must equal that of the bed of heavy liquid H, where the two lips allow a free discharge into receiving covers or vessels under equal pressures.

From Eq. (7), the liquid pressures may therefore be stated as

$$P_L\ (\text{at the neutral zone } r_n) = P_H$$

Therefore

$$\frac{\delta_{li}\omega^2}{2g}(r_n^2 - r_i^2) = \frac{\delta_{lo}\omega^2}{2g}(r_n^2 - r_o^2)$$

$$\delta_{li}(r_n^2 - r_i^2) = \delta_{lo}(r_n^2 - r_o^2)$$

$$r_n = \sqrt{\frac{\delta_{lo}r_o^2 - \delta_{li}r_i^2}{\delta_{lo} - \delta_{li}}} = \sqrt{\frac{r_o^2 - r_i^2(\delta_{li}/\delta_{lo})}{1 - \delta_{li}/\delta_{lo}}} \tag{12}$$

The difference in the densities of the two liquids exerts the major effect on the position and stability of the balance circle. If this difference is too small, separation becomes difficult through sensitiveness to other, though small, disturbing influences; or even impossible, though separations may be effected easily with differences in densities as small as 3 per cent, and a fraction of 1 per cent by special constructions.

The expression in Eq. (12) shows that the position of the neutral zone radius r_n may be adjusted by increasing or decreasing the radius of either the inner lip r_i or the outer lip r_0 to compensate for differences in the specific gravity ratio of the lighter to that of the heavier liquid. Both are made use of in different types of bowls, though more often the adjustment of the radius of the outer lip for the heavier component is the one employed.

The forces producing the through-put or pumping capacity have no necessary connection with the separating or the clarifying capacity (though often confused with them), as it may be seen from Fig. 98 that a

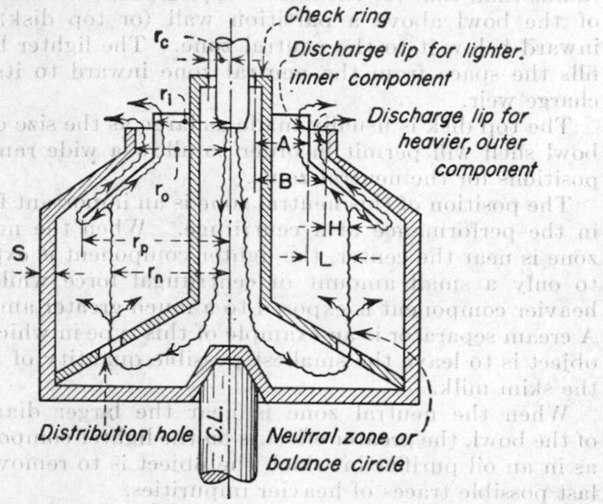

FIG. 98. Diagrammatic representation of the principle of a separating bowl.

bowl must function as a centrifugal pump with a pumping head determined by the radial columns or heads A and B between the check ring, whose radius r_c determines the overflow capacity, in relation to the radii of the inner and outer lips r_i and r_o. Thus the head for the lighter liquid is $A = r_i - r_c$, and the head for the heavier liquid is $B = r_o - r_c$.

The pumping capacity is finally determined by these forces and by the fluid friction-resisting effects, mainly turbulent friction. The volumetric pumping capacity varies little, if at all, with the specific gravity of the liquid, as the pumping force increases along with the turbulent resistance to flow.

It is obvious, however, that the passages into and through the bowl should be ample in cross section and free from unnecessary sharp angles, turns, constrictions, and pockets.

It is undesirable to have the pumping capacity very greatly in excess of the separating or clarifying capacity, since this may result in only partly filling the bowl or the introduction of air drawn in with the feed.

Liquid-solid Separation (Clarification)

Settling Rate for Large Particles—Turbulent Motion. The centrifugal force acting on a large particle moving through a liquid of different net specific gravity brings the particle to a terminal velocity at which the resisting force, because of the hurling of the liquid out of the path of the particle, equals the centrifugal force. The conditions assume that the size and velocity of settling of the particle are sufficiently great to render the viscous resistance of the liquid negligible compared with the turbulent resistance. Thus we have

$$F = R_t$$

$$\frac{(W_p - W_l)}{g} r\omega^2 = K\frac{\pi}{4}\,\delta_l d^2 V_s^2$$

$$\frac{\left(\delta_p \frac{\pi}{6} d^3 - \delta_l \frac{\pi}{6} d^3\right)}{g} r\omega^2 = K\frac{\pi}{4}\,\delta_l d^2 V_s^2$$

$$d(\delta_p - \delta_l) r\omega^2 = \frac{3}{2} K g \delta_l V_s^2$$

$$V_s = \sqrt{\frac{2}{3}\,\frac{d(\delta_p - \delta_l) r\omega^2}{K g \delta_l}} \tag{13}$$

The coefficient K may be taken as 0.5.

Equation (13) and Fig. 99 show that clarification of large particles, encountering mainly turbulent resistance

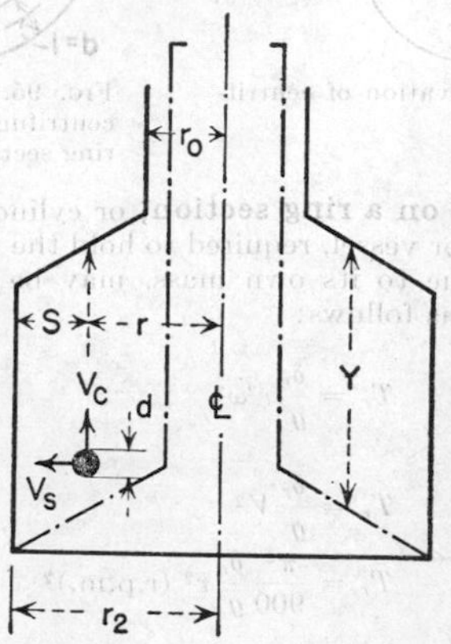

FIG. 99. Diagram for expressions for the through-put of continuous centrifuges.

to settling, is easier for large particles of high density in a light liquid. Equation (13) may be condensed by the use of the symbol D to

$$V_s = D r^{1/2} \omega \tag{14}$$

A particle will be removed from the stream of liquid if the velocity V_c, which is determined by the through-put rate, is sufficiently low for the settling velocity V_s to bring the particle through the settling distance S before the particle has been swept on out in passing through the travel distance Y.

The limit for settling out such a particle is that its settling time t_s through the settling distance S shall be just equal to, or less than, the time t_c required to move through the travel distance Y, or

$$T_s = \frac{S}{V_s} \lesseqgtr T_c = \frac{Y}{V_c}$$

From this, the limiting velocity V_c, which determines the maximum clarifying through-put, may be expressed as

$$V_c = \frac{Y}{S} V_s$$

This shows that the clarifying through-put capacity is directly proportional to the travel distance and inversely as the settling distance. The value of the settling velocity V_s is determined by the centrifugal forces, the excess of the particle density above that of the liquid and the **turbulent** resistance of the liquid to motion through it. The value for the average settling velocity may be obtained by integrating Eq. (14) and substituted above for V_s. This gives

$$V_c = \frac{Y}{S} \times \frac{2}{3}\,\frac{(r_2^{3/2} - r_0^{3/2})}{(r_2 - r_0)} \times \sqrt{\frac{2}{3}\,\frac{d\omega^2(\delta_p - \delta_l)}{kg\delta_l}} \qquad (15)$$

The capacity or volume of liquid passed through a bowl is the product of the through-put velocity and the section area for flow. In this simplified case

$$C = V_c\pi(r_2^2 - r_0^2)$$

Substituting from Eq. (15) and noting that $S = (r_2 - r_0)$ in this simplified case,

$$C = \frac{2}{3}\,\frac{Y\pi(r_2^2 - r_0^2)(r_2^{3/2} - r_0^{3/2})}{(r_2 - r_0)^2}\sqrt{\frac{2}{3}\,\frac{d\omega^2(\delta_p - \delta_l)}{kg\delta_l}} \qquad (16)$$

If $r_0/r_2 \doteq 0$ may be considered negligible compared with r_2, then the capacity for complete separation of particles of diameter equal to or greater than d is

$$C = \frac{2}{3}\pi Y r_2^{3/2}\omega\sqrt{\frac{2}{3}\,\frac{d(\delta_p - \delta_l)}{kg\delta_l}} \qquad (17)$$

Settling Rate for Small Particles—Viscous Resistance. The settling rate for small particles is determined by the viscous resistance of the liquid medium to the motion of particles settling through it. Stokes has shown that this resistance may be expressed in grams as follows:

$$f_z = \frac{3\pi z d V_s}{g}$$

where z is the absolute viscosity of the liquid medium in poises, and the other symbols have their previously given definitions.

The velocity of settling of such a small particle increases until the forces of viscous resistance equal the centrifugal settling force.

$$F = f_z$$

$$\frac{\pi}{6g}d^3(\delta_p - \delta_l)r\omega^2 = \frac{(3\pi z d V_s)}{g}$$

and from this

$$V_s = \frac{d^2}{18z}(\delta_p - \delta_l)r\omega^2 \qquad (18)$$

The clarifying through-put for these very small particles is here also determined by the limiting velocity V_c which would just permit settling out through the settling distance S before passing through the travel distance Y. Therefore,

$$V_c \overline{\gtrless} \frac{Y}{S}V_s$$

Integration to get the average velocity of settling V_s, caused by the centrifugal force at different radial positions for the settling particle, gives

$$V_c = \frac{Y}{36sz}d^2(\delta_p - \delta_l)\left[\frac{(r_2^2 - r_0^2)}{(r_2 - r_0)}\right]\omega^2 \qquad (19)$$

The capacity of the bowl, or volume passed through is as follows:

$$C = V_c \times (\text{section area for flow}) = V_c\pi(r_2^2 - r_0^2)$$

and since in this simple case $S = (r_2 - r_0)$

$$C = \frac{\pi Y d^2\omega^2(\delta_p - \delta_l)(r_2^2 - r_0^2)^2}{36z(r_2 - r_0)^2} \qquad (20)$$

If $\frac{r_0}{r_2} \doteq 0$

$$C = \frac{\pi Y d^2\omega^2(\delta_p - \delta_l)r_2^2}{36z} \qquad (21)$$

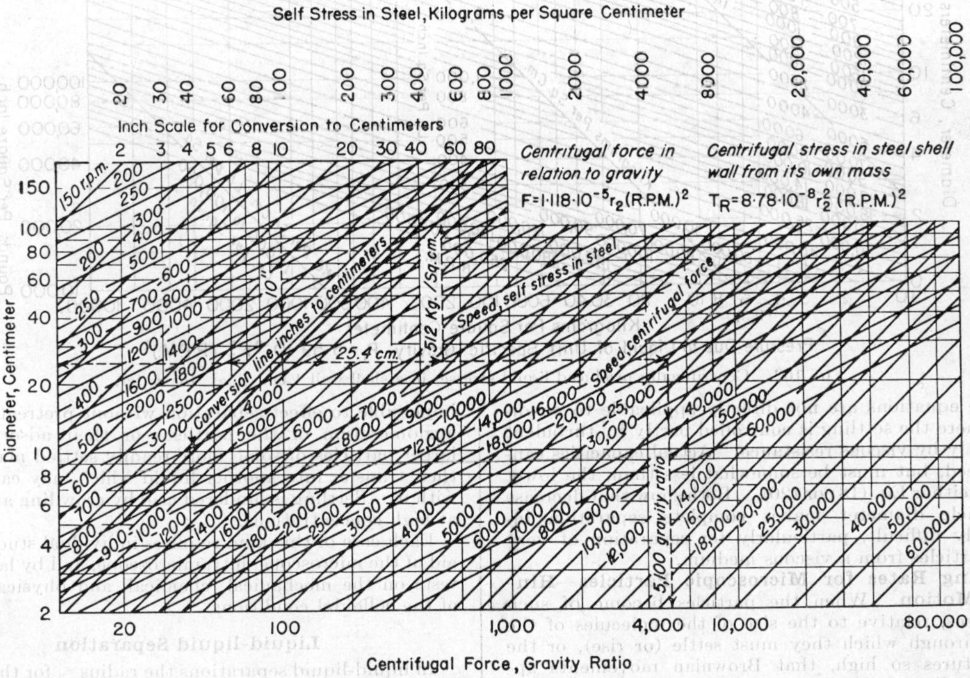

FIG. 100. Centrifugal force–gravity ratio and self stresses in rotating rings.

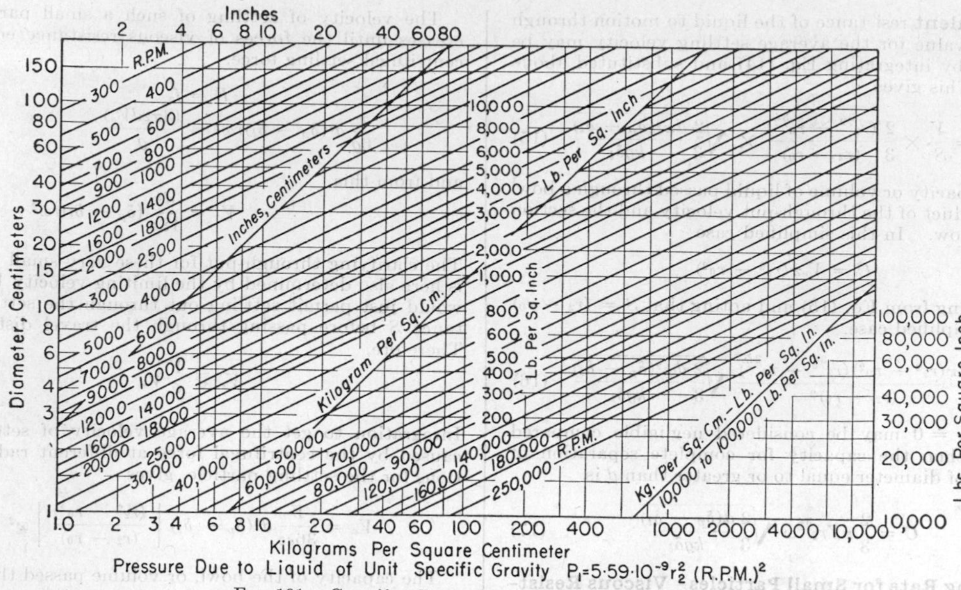

Fig. 101. Centrifugally produced liquid pressure for water.

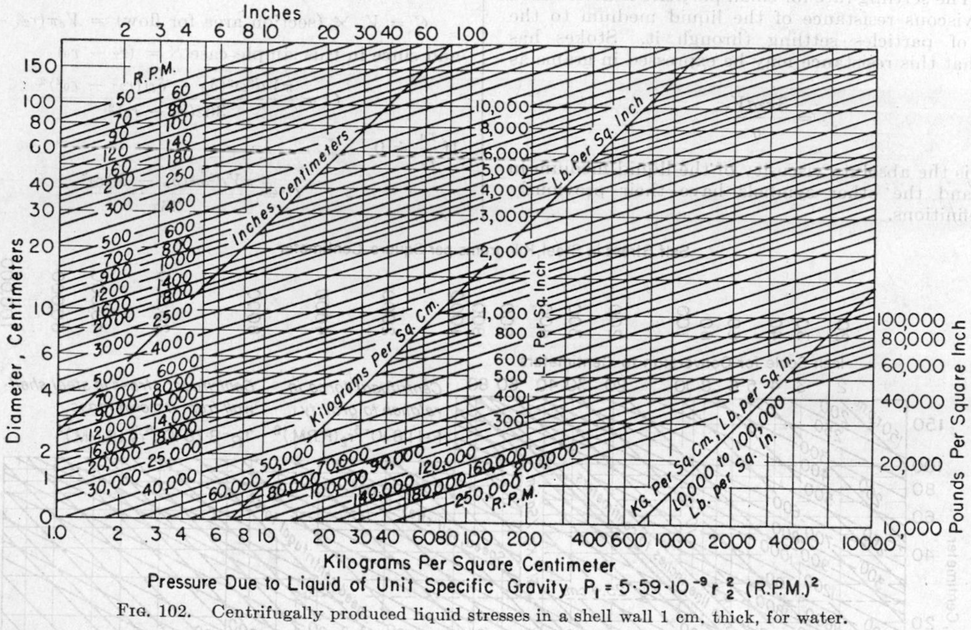

Fig. 102. Centrifugally produced liquid stresses in a shell wall 1 cm. thick, for water.

These equations are not directly applicable to conditions where the settling is controlled partly by turbulent and partly by viscous resistance. Actual capacities cannot exceed, but must be somewhat less than, the lower value of either Eq. (18) or (20). Both show that decrease of particle diameter makes complete separation increasingly difficult, particularly for separation of very small particles from a viscous medium.

Settling Rates for Microscopic Particles—Hindered Motion. When the particles become of such small sizes, relative to the size of the molecules of the liquid through which they must settle (or rise), or the temperatures so high, that Brownian movements appreciably disturb the settling, it may become practically impossible to effect separation without pretreatment for agglomeration of the particles or their adsorption on larger particles of some more readily settled material, or their transfer into another liquid which may carry them with it. Settling is hindered also by crowding and ceases at high concentrations.

Each case of this kind requires individual study by the aid of the microscope, ultramicroscope, and by laboratory tests on the mechanical, chemical, and physical factors of the colloidal conditions.

Liquid-liquid Separation

In liquid-liquid separations the radius r_n for the neutral zone, given in Eq. (12), also shows that the heavier liquid is subjected to the separating effect of the outer part of the bowl beyond r_n and conversely the lighter liquid by the inner part. The neutral zone should be controlled by the radius of the discharge lips to a *small* value where the higher degree of separation is desired for the heavier liquid, and to a *large* value where this is desired for the lighter liquid.

The feed should always be at, or close to, the neutral zone radius to avoid recontaminating one or the other of the separated components.

It is desirable that the position of the neutral zone should shift as little as possible with change of proportion of one component to the other and of through-put.

To secure stability of the neutral zone, the flow should not be needlessly restricted either in or out of the bowl and should be guided in radial directions so that slippage is minimized.

Liquid-liquid-solid Separation

Where two liquids are to be separated from each other and at the same time the separation of solids is to be effected, the bowl must be provided not only with two discharge lips placed at different radial distances from the center line of rotation suitable for the difference in specific gravity of the liquids but also with a sludge space to hold the solid sediment. If a sludge space is provided, both kinds of separations may take place simultaneously, provided only that the solid sediment is of higher specific gravity than the heavier liquid. If it should happen that the solid is intermediate in specific gravity, then it will not rise through, or be discharged with, the lighter component but will "float" on the heavier component and unless unusual constructions are provided will clog at the neutral zone. The equations already given apply, however, to their respective cases of liquid-solid and liquid-liquid separation.

General Design

Hollow bowls were naturally the first type developed for centrifuges. They have the advantage of simplicity and, when built in long tubular form of relatively small diameter, allow the attainment of high centrifugal forces and reasonably high ratios of travel to settling distance.

The inherent limitations are (1) that the settling distance must be nearly equal to the radius, (2) that the liquid near the center gets exposed to only a small centrifugal force, and (3) that sediment collected reduces the maximum centrifugal force on the liquid and, by reducing the flow area, increases the liquid velocity and so reduces the time of exposure to centrifugal force. A hollow bowl without radial vanes is subject to shift of neutral zone with feed rate or proportion of liquids. Bowls of these types are made with diameters up to 12 cm. and for speeds of 15,000 r.p.m.

Bowls with shallow settling spaces have been devised using a number of constructions such as concentric cylinders, spiral leaves, and cone-shaped disks. For settling large heavy particles out of much lighter liquids, bowls with generally one (though sometimes more) concentric cylindrical inner shell still find a considerable field of usefulness because of their simplicity of construction and, therefore, cleaning. The vast majority of separators use cone-shaped disks with thin spacer calks to get very shallow settling spaces, long travel paths, and a moderate-velocity, guided flow for the liquid. The calk or spacer thicknesses are selected so as to be two or more times as great as the diameter of the solid particles or globules of liquid to be separated. Larger values for the spacing calks are better for those cases where large proportions of solids must be removed, and thin calks should be used when only very small proportions of one component, *i.e.*, fractions of 1 per cent, are to be removed from the liquid and where the viscosities are low.

The cones are usually made appreciably smaller in diameter than the inner wall of the bowl shell so as to provide a sediment-holding space for such very heavy sediments as cannot be swept on out along with the heavier component liquid. Sediment may completely fill the space beyond the disk edges without altering the clarifying or separating capacity.

It is relatively easy to guide the flow radially out and back toward the center to the discharge lips by spacer strips or ribs between the disks in radial planes and so maintain a position for the balance circle dividing the two liquids that shifts little with ratio of components, through-put, etc.

The half angle of the cone-shaped disk may be 30 to 50 deg., chosen so as to get a short radial settling distance, without unduly increasing the height of the bowl or getting the angle so small that sediment will not slide along the cone.

The disadvantages of bowls with shallow settling spaces are (1) the number of internal parts to be handled and cleaned, (2) the weight of the parts, and (3) the cost.

The *discharge lips* in any design should be close to the center in order to reduce the power, the frothing of the discharged liquids, and the air entrainment. A certain proportion of power, up to about 40 per cent of the kinetic energy of the discharge, may be conserved by directing the discharges backward (as in a reaction turbine). This also reduces the froth and air entrainment and, by reducing the air pressures induced by the jet action of the discharges, reduces entrainment by windage, of spray from one receiving cover into the other.

APPARATUS

Liquid-solid Separators

Dryers, so called, of the centrifugal type have found almost universal use in laundries and in dry-cleaning plants for the removal of the major part (down to about 20 per cent) of the water or dry-cleaning fluid from cloth. The remainder cannot be removed centrifugally and must be evaporated. Such dryers are made in a large variety of sizes from about 18-in.-diameter "baskets" up to 60 in. The cloths are packed against the perforated walls and require about 5 to 10 min. centrifuging from start to stop. It is important that the cloths be laid smoothly and packed evenly from the inside wall to the width of the curb. If carelessly packed in lumpy wads the liquid removal is obstructed, and serious unbalance and shaking of the centrifuge, or even of the foundations, may occur. The holding volume per foot of vertical depth of basket is given by one of the curves of Fig. 111.

Cloth and similar materials normally carry 6 to 14 per cent moisture so that the removal of water is nearly complete, only a small proportion being left for removal by evaporation. Where dry-cleaning solvents such as petroleum spirits are to be removed, the centrifugal separation should be made as complete as possible, because the residues left by heating or by hot air blowing over the clothes may leave undesirable odors.

Basket centrifuges with overflow over the inner lip of the curb, or through the perforated walls of the basket, or through screens, find a considerable use in chemical plants for recovery of crystals precipitated from their mother liquor. The screening medium may be laid over a grid work or corrugated backing to allow a free flow-discharge path for the liquid after passing through the screen. Monel or other corrosion-resisting metal wires may also be woven in close mesh for screening out fine crystals. Such wire cloth may be "rolled" to flatten the wires and to reduce the mesh opening. The centrifugal drier is desirable where a very complete removal of mother liquor is desirable and where washing of the crystals with water or other liquids is to be resorted to for further purification. For very coarse nonporous particles, 5 to 10 mm. minimum dimension in any direction, only a fraction of 1 per cent by weight of a free-flowing liquid may be left after centrifuging. This is representative of what may be done in removing

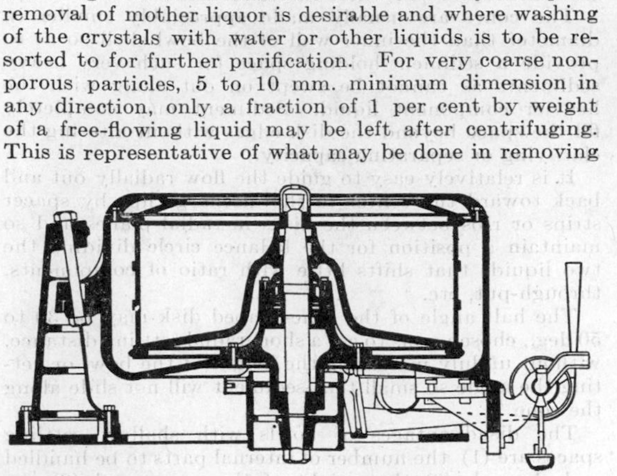

FIG. 103. Center-slung centrifugal.

cutting oil from machine-tool chips even with flat or curled long-strip chips. Finer particles necessarily carry more liquid with them in proportion to the surface-volume relation which increases linearly with decrease of dimension of particles as shown below for the calculated values for spheres, whether assumed stacked with centers in line with each other, or nested, with the centers of the spheres of alternate layers displaced a distance equal to the radius of a sphere.

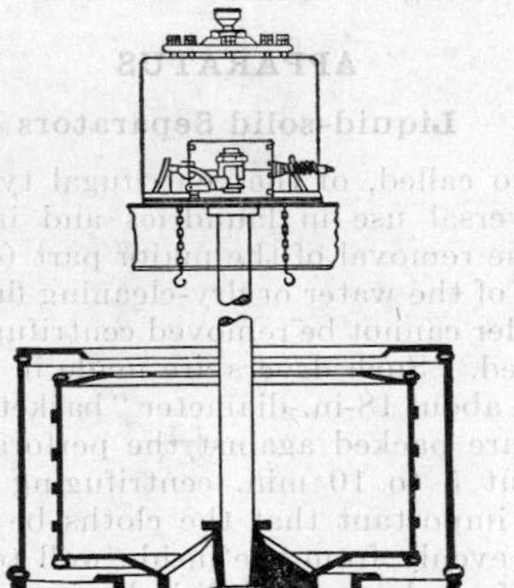

FIG. 104. Suspended centrifugal.

High-viscosity liquids may give thicker films and retain more liquid, unless the time or centrifugal force is increased. Fine slimy precipitates may retain 70 per cent or more liquid, even when centrifuged in the smaller diameter, high-speed baskets which give the greatest centrifugal effect for a given stress on the wall material.

Baskets up to 60 in. in diameter, at speeds up to 500 r.p.m., are used to get the advantage of large load capacity and reduce the time and cost of loading and labor for operating. Sprays for washing the material in the basket are frequently employed with the spray nozzles set in positions to discharge against the inner wall of the "cake."

It is usually unsafe to feed a slurry to the bowl while running, unless a bowl is supported from below or a very sturdy spindle is employed and a spring mounting for the stationary frame provided, such as will allow a considerable unbalance of the bowl, without undue strains on the foundation. Loading and unloading are usually done with the bowl standing still. Bottom plates are often provided which can be lifted or tilted out of the way so that the load may drop or be pushed down and out of the bowl into chutes or boxes on wheels.

Mechanical unloaders, having knives or scoops which are swung into place and pick up the solids while the bowl is still *slowly* moving, can be used with some materials. If health or other risks exist, these offer advantages, but some materials cannot be so handled. The scrapers may leave too much of the solids on the walls, and, for hard or tightly packed solids, it is often difficult to get sufficient rigidity in the scraper or in the bowl support to allow the use of scrapers without "digging in" or jamming, particularly if the bowl has not been slowed down sufficiently or the operator is careless. This is also a difficulty with sticky or hard slimy deposits.

Horizontally mounted basket centrifuges have recently been developed with two bearing supports of great rigidity, which allow automatic unloading while at full speed by scraper knives. These may have fully automatic controls that regulate the feed, rinse, drying time with feed shutoff, and unloading.

Centrifuges for continuous discharge of dried solids as well as for the discharge of the separated liquid, while running at full speed, have been built and will operate for a few easily handled materials, such as the dewatering of washed coal reclaimed from river beds by hydraulic pump-suction dredges, and of some mine slurries.

The solid particles must be fairly large and must be rather hard and of the type that roll or slide easily, without sticking or packing, such as hard coal, stone, sand, etc. Furthermore, the liquid must be quite fluid, with little tendency to adhere to the surfaces of the solids.

A vast amount of time and money has been spent in an attempt to build successful centrifuges with continuous discharge at full speed of sludges, slimes, sticky materials such as sugar crystals, sewage residues, and slurry from mining operations. Attempts to scrape off such materials from the walls or surfaces upon which they have been deposited or to employ some kind of a conveyor mechanism have met with limited success because of the great frictional forces that must be overcome to slide the materials along after they have once been deposited, thus allowing the full centrifugal force—many times that of gravity—to press them against the receiving surfaces. Also, most of these materials tend to gum, pack, or jam ahead of the scrapers. Finally, there is the difficulty of balancing the slime and in balancing and in driving the conveyor parts whose several elements are also subjected to centrifugal forces, holding them together with great force so that the frictional

Table 1. Surface-volume Relations

Diameter of sphere, cm.	Number of spheres/cc. centers in line	Number of spheres/cc. nested fully	Volume of spheres/cc. centers in line	Volume of spheres/cc. nested fully	Surface area/sphere	Surface area/cc. of spheres with centers in line	Surface area/cc. of spheres nested fully
1.0	1.	1.41	0.5236	0.740	3.14	3.14	4.45
0.1	10^3	1.41×10^3	.5236	.740	3.14×10^{-2}	3.14×10	4.45×10
.01	10^6	1.41×10^6	.5236	.740	3.14×10^{-4}	3.1×10^2	4.45×10^2
.001	10^9	1.41×10^9	.5236	.740	3.14×10^{-6}	3.14×10^3	4.45×10^3
.000,1	10^{12}	1.41×10^{12}	.5236	.740	3.14×10^{-8}	3.14×10^4	4.45×10^4
.000,01	10^{15}	1.41×10^{15}	.5236	.740	3.14×10^{-10}	3.14×10^5	4.45×10^5
.000,001	10^{18}	1.41×10^{18}	.5236	.740	3.14×10^{-12}	3.14×10^6	4.45×10^6

resistances to movement are tremendously greater than when the bowl is standing still.

The type of construction using two concentric cones run at slightly different rotational speeds by means of a differential positive gear drive has had a certain amount of success. This type feeds a granular material that does not stick together or stick to the perforated walls of the outer cone and dumps the solids at the lower end (large-diameter part) of the cones, while the liquid is discharged radially through the perforations of the outer cone. The materials must be fed in at the smaller cone end and discharged at the larger end, which is usually at the bottom, employing a vertical shaft machine. Spiral scrapers are rigidly mounted on the inner cone and nearly touch the inner-surface walls of the outer cone from which the granular solids are to be scraped, and the direction of the spiral and the differential of speed must be so chosen as to cause the solids to move to the larger, or discharge, end of the cones. If the axial cone angle is large, centrifugal forces will help roll the solids along, and the scraper spiral may serve to hold the particles back, until dewatered, rather than to push them ahead.

Attempts to dewater sewage slimes, remove molasses from sugar, reclaim yeast from wort, and similar separations by mechanically operated scrapers for solid removal have had a limited success. The principal recent improvement in these types of machines has been the development of differential gear drives that would maintain the differential speed of the scrapers by reaction of gear trains between the bowl and scraper running at the bowl speed. When the torque between the two full-speed elements, required to maintain the small differential speed, is supplied to each part through a train of gears, each train is subjected to the scraping forces *multiplied* by the speed-reduction ratio.

Hollow-bowl clarifiers, having a low centrifugal force with no interior parts except the three to four wings which force the liquid to rotate with the bowl and the feed passages to bring the incoming materials to the outer parts of the bowl, serve sufficiently well for easily removed solids. A useful and simple improvement consists of employing an inner cylindrical shell, with a diameter about 70 per cent of that of the outer shell, so as to give equal holding volumes and through-put velocities in the two shells. The incoming material is fed first to the inner shell, preferably through a feed tube with a side or radially directed outlet so that each side or quarter of the bowl is fed, in turn, equal amounts, so as to ensure equal distribution, maximum purification, and uniform loading of sediment. The inner shell picks up the coarser and more readily separated sediment and then allows the liquid to pass to the outer shell for the removal of the finer sediment. Such bowls serve to remove coarse, unground pigment particles in paints, varnishes, and lacquers. The degree of clarification of pigments, etc., may be controlled either by change of through-put or by adjusting the bowl to a lower speed, so as to avoid throwing out too much of the pigment or requiring too frequent stoppage for bowl cleaning.

Hollow high-speed tubular bowls with high centrifugal forces over 10,000 g. may be used for difficult separations such as clear varnish and other cases of semicolloidal particles.

Disk-bowl clarifiers have usually cone-shaped separating disks of somewhat smaller diameter than the bowl shell so that sediment may collect without packing between, and thus impairing, the clarifying effectiveness of the disks. The feed channels must bring the incoming materials near to the outer edges of the disks and evenly distribute them to and around them.

Disk bowls may be used for difficult clarification problems such as clear varnish and the removal of semicolloidal particles. They are also used for clarifying whole milk to remove particles of dirt picked up during milking or in handling and transhipping the milk to the milk plant. Clarification also removes such undesirable materials as scarf-skin particles from the udder of the cow and occasional pus or blood cells. All these materials are collected as slime in the sediment space, out of the main path of the milk, and so successive contamination of further quantities of milk, which must necessarily occur with filters, is avoided.

Concentrator bowls with nozzle discharge of the solids afford a means of recovering the solids along with a much reduced proportion of the liquid. The nozzle passages lead from the outer part of the bowl, where the solids tend to collect but are swept out along with the portion of the liquid which discharges through the nozzles. The inner surfaces of the bowl shell are usually shaped somewhat like a beehive in order to provide a slope to the nozzle-passage entrances for the purpose of facilitating the sliding along of the solids. The nozzles sometimes discharge directly at the bowl diameter. However, since this may require very small diameter nozzles, which easily clog, and consume much power, the passages often lead to nozzles placed near the center of rotation, and the nozzles may be placed to discharge nearly tangentially, and backward, to recover energy, reduce frothing, and reduce spatter.

Bowls for yeast are built in sizes up to 440 mm. diameter and are usually provided with cone-shaped separating disks. They are generally run at lower speeds than similar separators.

The degree of concentration, or ratio of the (constant) volume discharged by the nozzles to the total volume fed to the bowl, may be regulated by decreasing the nozzle opening, provided, however, that the quantity of solids in the feed liquor and the ratio of solids in the concentrate are not so high as to decrease unduly the fluidity of the concentrate.

The solids in the concentrate may be raised to 50 or 65 per cent without too great flow resistance. The percentage concentration is therefore further limited by the ratio of solids in the feed, since all the solids appear in the concentrate. For instance, if 50 per cent solids is taken as a reasonable limit for free flow, a feed containing 5 per cent solids may be concentrated into 10 per cent of the feed, but if the feed contains 10 per cent solids it would not be possible to concentrate to less than 20 per cent of the feed volume.

Such types of concentrators are almost universally used to concentrate yeast cells into a small portion of the wort in which they were grown, recovering the yeast.

Valve bowls are provided with nozzles or orifices closed by valves mounted on the inside wall. These valves may be opened automatically by means of springs, actuated by a lowered speed, or by liquid-filled floats, which empty directly through the nozzle when liquid feed to the inner end is decreased by accumulation of sludge. Instead of separate valves and nozzles, the whole bowl may be opened by mechanical devices, such as levers, or by feeding a control liquid to a pocket mounted on the bowl and letting the centrifugal force on the liquid in the pocket open the bowl at its periphery. The separate valves may also be opened by a rod concentric with the shaft, connected by appropriate rods and levers to the valve element. If the valve or the whole bowl is opened while running, a considerable amount of energy is dissipated in the discharge, and such machines require many times as much power as the usual types. Provision must also be made for replaceable wearing surfaces in the covers to take the impact of the jets and for steep sliding slopes or mechanical conveyors for removal of the sludges.

Development of various kinds of centrifugal apparatus

for the continuous discharge of solids has been particularly active in recent years.

Liquid-liquid Separators

Concentrators, or ratio separators, are built with two liquid discharges at different radii suitable for the usually small difference in effective specific gravities, and with non-radially guided flow of either or both components in their respective paths toward the center and toward the outer wall of the bowl, so that a certain but controllable proportion of the feed will pass out by each of the two discharge paths. Practically all such concentrators are now built with cone-shaped separating disks and find their greatest field of application as "cream separators," in which field centrifugal separators have superseded all other methods throughout the world, a million or more centrifugal cream-separator machines being produced annually, which is perhaps one hundred or more times the number produced for all other purposes combined. The reasons are not only that the centrifugal separator can produce a cream with a butterfat content of any desired value from about 15 to 60 per cent (or even 80 per cent if adapted for the purpose) with a negligible loss in the skim milk (whereas allowing cream to rise by gravity on standing will seldom produce more than about 20 per cent butterfat cream and incur losses of ¼ per cent or more), but will also do so instantly, without the delay and consequent souring that may occur on "standing" cream.

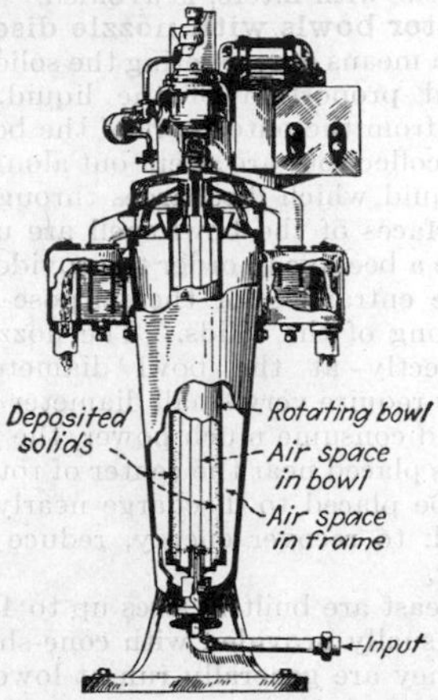

FIG. 105. Tubular-bowl high-speed centrifugal.

Cream is an emulsion of butter oil in milk, and the "cream" is merely a higher concentration of this emulsion than that in fresh whole milk. Whole milk may contain from 3 to 6 per cent of butterfat; "ordinary cream," as obtained for table use by pouring off the cream that rises on standing may have 18 per cent (or less) butterfat. "Heavy cream" is produced by centrifugal separation with 40 per cent fat, which is cut back with milk for "light cream." The name **separator** is firmly established by usage although the term "concentrator" much more truly describes the function performed, of concentrating **all** the emulsion, *i.e.*, the cream in a part of the milk. The cream separator or concentrator is so built that, if fed a homogeneous liquid, such as water, a portion will be discharged through **each** of the two spouts. The proportion of the total passing through the cream discharge may be regulated by making the cream discharge through the hole in a hollow screw, set in a radial direction, so that by screwing it toward the center the inner end picks up the cream at a point nearer the center where there is a lessened centrifuge pumping head, and therefore a smaller quantity discharged, resulting in a "richer" cream, *i.e.*, carrying less milk. This method of control is quite widely used on small separators. The control may also be put on the outer discharge, either by narrowing the opening through which the skim milk passes, or by arranging one or more screws so that they can be set either to allow or to stop slippage. Slippage of the heavier component in its progress toward the center to reach its discharge opening allows the liquid to retain some of the peripheral velocity it had at the larger diameter and so "run ahead" of the inner parts of the bowl thereby building up a centrifugal back pressure which tends to limit the flow in this path. Thus adjustable or permanent radial baffles or ribs in the flow to the discharge lips, by reducing slippage, increase rather than decrease flow.

The distribution holes in the disks are placed near the center so as to apply the major separating effect to the skim milk which constitutes the heavier outer component.

Concentrator-type bowls are quite suitable for constant through-put concentration of a liquid with a practically constant ratio of a lighter component or of an emulsion into a portion of the heavier liquid. They are not suited to separations where the proportion of one component, or the feed rate, varies during a run.

True separator bowls are provided with two discharge diameters, an inner and outer, with radial guide ribs throughout, between the cone-shaped separating disks or in the body of the hollow bowl and also in the return path of the heavier component past the top-disk skirt back to its discharge lip. It should be possible to

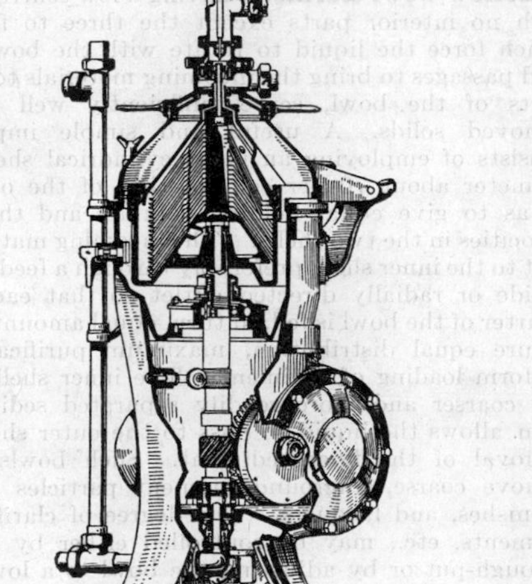
FIG. 106. Closed feed and discharge disk bowl centrifuge.

vary the proportion of one component from 0 to 100 per cent of the total at any through-put rate without breakover of either component into the discharge passage intended for the other. The adjustment of the balance circle and the setting for different ratios of specific gravities are usually accomplished by changing the lip for the outer discharge. A number of rings or disks with center holes of different diameters are provided which may be readily installed by unscrewing a nut which grips the outer rim of the discharge ring firmly against a rubber ring or other gasket, to prevent leakage. Both small-diameter tubular bowls at high rotational speeds and cone-shaped disk bowls of larger diameter at lower speeds are used for separators; each has certain inherent characteristics, merits, and limitations, which have been discussed under the heading General Design (p. 997).

Liquid-liquid-solid Separators

Concentrating separators with nozzles discharge the solids along with a portion of the heavier of two liquids being simultaneously separated from a lighter liquid. The nozzles are fed by passages from the outer part of the bowl. Two liquid-discharge lips and means to adjust the discharge diameter of one or the other to suit the specific gravity difference of the two liquids are provided. The heavier liquid component must be present in the feed, in proportions greater than that required for the nozzles so that an excess remains to fill up to the outer lip. This may be provided either by recirculating a portion of the heavier liquid or by its presence in excess in the feed. Commercial applications of this type of separation appear to be increasing.

Separator bowls with sludge pockets or sludge-holding spaces require a little modification of the liquid-liquid separators, namely, that of making the cone-shaped separating disks appreciably smaller in diameter than the bowl shell, or in allowing a portion of the hollow bowl to be filled up, with the consequent reduction of separating effect due to the progressive reduction of the effective diameter as filling occurs and the increased through-put velocity V_c as the path area for liquid flow is constricted. The bowl must be stopped to clean out the sludge; consequently this method is peculiarly adapted to those cases where only relatively small amounts of solids must be removed.

Traveling-cushion bowls have fed to them a liquid heavier than the liquid to be clarified but near enough to that of the solids or emulsions which are to be swept out of the bowl to cause their movement through the bowl rather than pocketing them in the sludge spaces next to the bowl-shell wall.

This is an attempt to substitute a sweeping-out liquid for the mechanically driven continuous slime-discharge bowls. The difficulties of proportioning the densities and of keeping the slimes from either floating upon or sinking through the traveling cushion of liquid are very great, and commercial applications have been rare. An exception is the method that was widely used in the manufacture of bright stock, where a carrier liquid, usually hot water, immiscible in oil or wax, was injected into the wax space just ahead of the discharge ring, serving the double purpose of a traveling cushion, as far as it reached into the bowl, and of melting the chilled wax. Melting of wax was further assisted by spraying hot water on the bowl top to be thrown into the wax cover along with the wax.

Special Types of Centrifuges

Test Apparatus. Test-tube centrifuges with heads carrying two, four, or more cups, so shaped as to hold graduated test tubes containing two or more liquids, or liquids and sediments, are employed to effect a separation or sedimentation and to give easily read volumetric proportions of the different components.

A wide variety of forms and shapes of test tubes are in use (with corresponding shapes for the holding cups), such as cylindrical glass bottles, *i.e.*, the 4-oz. oil sample bottle, small and large laboratory test tubes, pear-shaped vessels with a small cylindrical extension for reading small proportions, and cylindrical vessels with a cone-shaped lower end to give readable values for very small volumes of sediment or separated liquids.

The standard speed prescribed by the latest revision of A.S.T.M. standard method of test, Designation D 96 for water and sediment, is 1500 r.p.m. with 100-cc. test tubes supported in cups so that the tips swing in a 15- to 17-in. circle. This gives a centrifugal force at the extreme tip of the tube of 500 to 550 times gravity, and less for the other parts of the test tube. These forces are not sufficient to effect some separations or to break some emulsions; therefore, materials that tend to break emulsions and diluents, solvents, or precipitants are often added (see the latest revision of A.S.T.M. Method for Precipitation Number of Lubricating Oils, Designation D 91).

Specially designed and sturdily proportioned machines can be built for speeds up to 6000 r.p.m., using the standard A.S.T.M. 100-cc. cone-shaped test tube.

The test-tube centrifuge gives a means of observing the relative ease of centrifugal separation and a means of measuring the performance of commercial separators, for separation of water from insulating oils, etc., and for clarification of sediments.

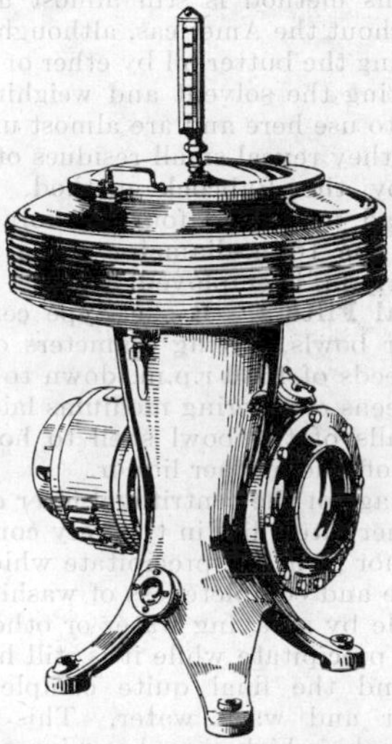

FIG. 107. Test-tube centrifuge.

It must be observed that, as it is the *volume* which is measured, it becomes necessary to consider the extent to which sediments are "compacted" by long-continued centrifuging. It is quite possible to observe, with repeated centrifuging of a single given sample, a quick accumulation of sediment, then no further increment for several repeated centrifuging periods, followed by the accumulation of quite a large volume of sediment (presumably much more difficult to throw down than the first sediment observed) and, finally, the compacting of this last sediment into a smaller volume.

The final volume is, however, larger than the value for the previously observed quantity of quickly settled, coarse or heavy, easily precipitated sediment. When testing an unknown material, the centrifuging should be prolonged beyond the time for the first observed, "apparently" constant volume, unless complete clarification of the liquid, or other observations, confirm that all sediment has been removed from the sample and the sediment compacted down to its final volume. The final compacted volume is generally some function of the speed and time of centrifuging. Sediments differ markedly in compacting, and speed and time may alter the volume by 2 to 1 or more.

In the precipitation test, applied to lubricating oils, the settling of the precipitate is quickly accomplished in the centrifuge test tube and the **volume** of the precipitate read as prescribed by A.S.T.M. Method D 91. In addition the solvent and oil may be decanted and the precipitate collected, dissolved, and **weighed,** which obviates difficulties and inaccuracies in the calibration of the smaller volumes of the test-tube cone-shaped tip and of the compacting of the precipitate, as well as eliminating any possible effect from solids, etc., accidentally already present in the oil sample and therefore not actually precipitated by the precipitation naphtha.

The Babcock butterfat-milk tester is a special form of test-tube centrifuge in which the cylindrical cups are constructed to hold a special form of test bottle with a small-diameter neck attached to the upper part of the test bottle and a filling tube passing through the shoulder alongside the neck and leading to the bottom of the test bottle.

This small neck catches the butter oil rising to the surface after sulfuric acid has destroyed the cream-globule film and allows reading values from 0.01 to 0.25 per cent. This method is still almost universally employed throughout the Americas, although various methods of extracting the butter oil by ether or other solvents, then evaporating the solvent and weighing the residue, are coming into use here and are almost universally used abroad, since they reveal small residues of butterfat, not measurable by the Babcock method. The Babcock centrifuge head may hold four or more up to 32 test bottles. For the 16-in.-diameter swing circle, a speed of about 850 r.p.m. is employed.

Centrifugal Filters. Basket-type centrifuges, with large-diameter bowls, having diameters of 18 in. up to 60 in. and speeds of 1500 r.p.m. down to 500 r.p.m. are used with screens or filtering mediums laid on the inside perforated walls of the bowl shell to hold precipitates and to throw off the mother liquor.

The advantage of the centrifugal filter over plate-and-frame and other filters lies in the very complete removal of mother liquor from the precipitate which centrifuging gives, the ease and completeness of washing the precipitate obtainable by spraying water or other wash liquids on the bed of precipitate while it is still held in the running bowl, and the final quite complete removal of mother liquor and wash water. This gives a quite clean "dry" cake which may have important chemical advantages over plate-and-frame filters, and rotary, pressure, and vacuum filters for certain processes.

Closed-cover, Feed, and Discharge Machines. Gastight, or nearly gastight, covers are provided where spray or fog from the discharges needs to be kept from the room, or where vapor losses must be prevented on account of fire, health, or other hazards. The joints in the covers may be snugly fitted or even made quite tight by gaskets. Flexible metal tubes are provided for the feed and discharge pipes. Inert gases may be fed to the inside of the covers; but, as all liquids tend to dissolve more or less gas, as well as carry away considerable volumes of gas entrapped as froth, large quantities of the inert gas may be needed. Special designs make it possible to hold a vacuum in the bowl chamber and in the covers which receive the discharge.

Liquid-sealed Discharges. It is possible to discharge one or both the liquids from a centrifuge bowl by the equivalent of a small centrifugal pump in which the housing rotates with the bowl and what would be the impeller is held against rotation, so that the liquid under the centrifugal pressure of the "pump" housing receives the liquid at the housing diameter and flows *toward* the center and out through the hollow center of the "shaft" or through a concentric tube surrounding it. Clarification of a single liquid is readily carried out with a single closed discharge. Separation of two liquids with both discharges closed may be effected wherever the two components have a constant proportion to each other, as in cream separators.

There is, however, the drawback that enclosure is not entirely complete, owing to running clearances, and some air froth is still produced. In skim milk this may be several per cent by volume, but it is much less than the 30 to 40 per cent produced by open-discharge separators. A considerable amount of power is consumed in liquid outlets of this type.

Bearing-type sealed passages may be provided by mounting the discharge chamber on the bowl neck as if it were a bearing, but flexibly supporting the chamber on a corrugated metal bellows or on a flexible diaphragm to allow it to adjust itself easily to the position of the high-speed rotating bowl. This construction is feasible for the clarification of liquids that will provide a lubricating effect, suitable for the light side pressure on the bearing-type seal and its high speed and large diameter.

Clarifiers with both inlet and outlet passages sealed by bearing-type connections are usually made with the bowl mounted on a hollow shaft with a bearing at each end so as to allow the feed to come in at one end of the hollow-shaft end and pass out through the other. The bearings may be combined with the feed or discharge passages or may be separate therefrom. If an oil or other material that will lubricate the bearings is to be clarified, it is much simpler, less costly, and more economical of power to combine the bearings with the inlet and outlet passages. For certain conditions of service, where the clarifier is subjected to severe mechanical shocks and vibrations, the double-bearing support offers material advantages over the more common type in which the bowl is mounted on the end of the spindle beyond its bearing support. The bearing is slipped, cap fashion, over the end of the hollow shaft, and the liquid connection is made through a flexible hose screwed on the end of the bearing. The bearing is supported on a spring cushion to allow the bowl to rotate freely about its center of gravity, without undue side pressure on the bearing. The leakage of liquid past the inner end of the bearing along the shaft is controlled by reverse-helical pumping grooves. At the speeds prevailing in centrifugal clarifiers of 6000 to 9000 r.p.m., such helical pumping grooves will prevent leakage with internal pressures on the liquid of about 2 to 3 atm. above atmospheric pressure.

Clarifiers of this type are suitable for use on locomotives and in marine service on Diesel engines. Little or no water is likely to be present in the lubricating oil in Diesel engines, or, if present, would be held with the sludge. The clarification removes worn-metal particles, dust from the air, and carbon and sludge produced by oxidation, polymerization, or cracking of the oil.

Hermetic Separators and Clarifiers. Recently, completely closed-feed and closed-discharge cream separators and milk or other clarifiers have been developed having a hollow spindle feed to a bowl bolted tightly to the spindle and with closed-feed and discharge gaskets made of elastic compounds, shaped like the cup seals of hydraulic presses. See Fig. 106.

Accessory Apparatus

The pipes (see pp. 396–412), used for the feed and discharge connections, must be ample in size, not only because of the power lost in pumping through them against undue friction, but also because turbulence may cause emulsification and thus greatly hamper the separations.

Pumps (see pp. 1414–1439) for handling liquids to be centrifuged should preferably be of the positive-displacement rotary type and run at moderate speeds. Centrifugal pumps are usually undesirable, not only because of the variation of delivery with back pressure, but also because of the violent agitation and tendency to produce emulsification, which they cause. Where, however, a mixing of two liquids is needed to promote a rapid chemical reaction, or a washing process prior to centrifugal separation, a centrifugal-type pump may advantageously be used to accomplish or assist in accomplishing, the mixing. Piston pumps have the disadvantage of intermittent or varying rates of flow with each stroke (unless built with several cylinders), as a pulsating flow decreases the separating ability of the centrifugal bowl.

Regulation of pump delivery of constant-displacement pumps is best accomplished by gate-type valves, or still better by cocks, on the **suction side** of the pumps, since gate valves and cocks give a more definite and reproducible area of opening for a given setting of the handle than can be obtained with globe valves. The control of liquid fed to the pump by restricting the inlet valve opening, thereby producing pressure drop at the valve and vacuum on the pump, is usually better than choking the outlet valve and recirculating through a by-pass relief valve. The latter method builds up back pressure on the pump and increases the power needed to dirve it. Also a recirculation of the liquid is not always desirable as it may promote emulsification.

The pump displacement for removing the centrifuged liquids from free discharge types of bowls should be from 25 to 50 per cent in excess of the liquid volume to allow for froth, air, etc., carried along with the centrifuge discharge.

Presettling tanks may perform an important function in settling out large amounts of coarse or easily separated materials prior to centrifuging, leaving only the smaller quantities of the more difficult part of the seperation to be effected by the centrifuge. If water in large quantities is to be separated from an oil, a decanting tank may be used with a partition wall reaching to a point near the bottom, but leaving a passage under the baffle for water to pass into a water compartment from which it may flow out over a weir, slightly lower in height than the weir for the oil overflow on the other side of the partition wall or baffle. The residues of moisture left by a centrifugal separator in a purified oil are smaller when a smaller ratio of water to oil is present in the feed. In all cases the centrifugal separator should be relieved of coarse work and employed for the finer or more difficult separations.

Heaters or coolers are frequently necessary to give a suitable fluidity or more suitable surface tension to the liquids to be processed, or to gain a differential in specific gravities. It is important that the heater mass and temperatures be kept low to avoid overrun in controlling (or overheating on shutting off) liquid flow, and that ample heat-transfer surfaces be provided in order to lower the maximum surface temperatures. **Radiant electric heaters** are undesirable because the heater units must be brought to a high temperature (*i.e.*, a red heat or higher) to get adequate radiating effects and because, upon shutting off the liquid flow, even with simultaneous disconnection of the electric supply, the temperature equilibrium reached involves excessive temperatures in the liquid.

Strip resistance heaters are more suitable, either immersed in the liquid or attached to the walls of the vessels or pipes. Steam or hot water taken directly from a boiler or from an auxiliary, electrically heated vessel and passed through heating coils, double-pipe heaters, or jackets offers a degree of safety from excessive temperatures not obtainable even with electric resistance heaters.

The rating of the heater required depends directly upon the liquid through-put, the temperature rise desired, and the specific heat of the liquid, so that for a through-put of 1000 gal./hr. each degree Fahrenheit rise requires a net heat input of 2440 watts for water and about 1220 watts for oil. For 1000 l./hr., 1160 watts is needed for each degree rise centigrade for water, and about 580 watts for oil.

The unit heat-transfer rate from pipe walls may be 5 to 6 watts/sq. cm. for heating water, but should be only 1.5 to 2 watts/sq. cm. for oil, to avoid overheating the oil film next to the wall. In any case, the average flow velocities through the pipes should be about 200 ft./min. (1 m./sec.).

Rates of heat transfer from the surface of the pipe wall (in contact with oil) to oil may be 15 to 20 B.t.u./(sq. ft.)-(hr.)(°F.) difference. For hot water to oil with coil or jacket heating, the coefficient of heat transfer may be 30 to 35, postulated, however, not on wall temperatures but on the temperatures of water and oil, and steam will transmit to oil in pipe coils about 60 to 70 B.t.u./(sq. ft.)-(hr.)(°F.), while steam in pipe coils heating oil in tanks will transmit only about 50.

Air-separating traps are often desirable in the centrifugal discharge connections to relieve the discharge pipes and pumps of air and so increase their carrying capacity. No fixed rules can be laid down, as the section area for air relief depends not only on the quantity of discharge and the ratio of air froth to be eliminated but upon the viscosity and, therefore, the temperature of the liquid.

As illustrations it may be said that, for a flow of 20 gal./min., a dry cleaner's naphtha may need only an 8-in.-diameter vessel, while electric insulating oil may need a 30-in.-diameter vessel.

Feed regulators (see pp. 1370–1376) are generally of the float type and operate either a balanced valve of the butterfly or of the piston type to reduce or even to stop the flow from the feed-pipe line as the float reaches a predetermined level in a supply cup or tank.

One point of advantage of such a method is that the rate of feed automatically decreases with increase of viscosity, which may therefore compensate to some extent for the increased difficulty of separation in the bowl on account of the greater viscosity.

When the supply is taken from a low head tank, the float may be made to control the rate of feed quite adequately by the simple expedient of allowing the top of the float to rise against the open end of the faucet or the end of the supply pipe.

A very satisfactory method of feed regulation is the employment of positive-displacement pumps whose speed can be set at any desired value to give a corresponding through-put to the centrifugal. This avoids all probability of emulsification and allows the scheduling of runs as well as standardizing the performance of the centrifuge.

Processing and Preparation

Settling (and Settling Combined with Decanting). This is a simple and often useful process for the preparation of materials to be centrifuged. If a heavier liquid, such as water, is present in large quantities with oil, a settling process may be very advantageous in reducing the load on the centrifuge and also in improving its performance. A float carrying an intake port leading into a flexible hose or into a pipe with flexible couplings, so that only that part is drawn off which is at the top surface and therefore most completely clarified, is better than multiple draw-off cocks set at different levels.

A sight glass in the pipe line should be provided so that the character and amount of liquid being drawn off may be observed.

Where settling of hot liquids is to be accomplished, convection currents should be avoided. Heating, or even maintenance of temperature, by coils, promotes convection currents and is therefore less suitable than jacket heating. All surfaces should be well heat-insulated, as the ideal condition for settling would be the maintenance of an absolutely unchanged temperature throughout the settling tank.

Where sediments or slimes are to be settled, the settling tanks should either have flat or plane-slanted bottoms to allow the use of mechanical scrapers, or should be cone-bottom tanks with cone slopes **greater** than the angle of repose of the sediment, so that opening the draw-off valve will give complete sediment drainage. Draw-offs should be closed by cocks set close to the cone of the tank or by gate valves, so that clots may be readily cleared by push rods.

Coagulating may be a very useful or even an essential step prior to centrifuging, to enhance the possible through-put rate and to improve the separation. If the particles also coalesce, the gain in differential densities usually more than counterbalances the slightly smaller dimensions of the new size of the coalesced particle. Whether coalesced or merely agglomerated, the settling rate and centrifugal separation rate are increased. In many cases, heating alone will coagulate. In others, washing or cooking, or digesting under pressure, with water or other liquids will bring about coagulation. The addition of some kinds of chemicals that do not injuriously affect the materials may be employed to reduce or to neutralize the electrical charge on the dispersed colloidal particles, thus allowing them to unite. Coagulation may also be carried out so as to form a kind of mat which, in settling, entangles and carries down with it all other particles. Disturbances such as convection currents must be avoided.

Partial centrifuging makes possible the separation of three liquid mixtures by carrying out first, a partial separation which removes one liquid from the mixture and, second, by recentrifuging the mixture to separate the two remaining constituents. Generally it is somewhat easier to carry out the first step so as to discharge the mixture at the inner lip and one, only, at the outer lip. The proportion and gravity differentials of each of the three constituents determine the methods that may be used, for if two are closely alike in gravity they may be made to discharge together more readily at one lip, and if one of the three is present in quite large proportions it may be made to sweep along with it small portions of a second constituent that would otherwise lag and clog.

Partial centrifuging may also be applied by centrifuging, first, at a very high through-put rate for a partial separation or clarification, followed by a rerun at a low rate for a high degree of centrifugal purification. Another method is to pass the material through a large-diameter basket-type centrifugal for separation of large proportions of the more readily removed materials, followed by processing in a small high-speed centrifugal for a final purification.

Adsorbent Treatments (see pp. 885 to 916). Adsorbent treatments with carbon or other adsorbents may be used to pick up colloidal particles which may then be removed by settling, filtering, or centrifuging. If percolating beds are used, relatively large mesh particles must be used (usually not finer than 30 to 60 mesh) to get reasonably high liquid flow rates through them, thereby entailing a lower surface/volume relation and lower effectiveness of the adsorbent but somewhat smaller losses of liquid. If stirred in, "fines" may be used, which is more effective, but which usually entails some loss of the liquid, which either is thrown away with the spent absorbent or, if extracted, is of low grade. Adsorbents may remove antioxidants and corrosion inhibitors from oils, and this effect must be borne in mind if they are used in oil treatment. A wide variety of natural, processed, and trade-named materials are used for such purposes, such as diatomaceous and siliceous earths, fuller's earth, betonite, silica gel, adsorbent carbons, Sil-o-Cel, Filtrol, Darco, etc. (see Sec. 14, p. 885*ff*.).

Washing and Chemical Treatments. These are often important and sometimes essential processing aids prior to centrifuging. Washing may be partly a purely mechanical action for removal of impurities but often has chemical or physical effects. For instance, water may extract water-soluble acidic deterioration products from oils, particularly when freshly produced, by preferential, or partial, solubility in water. The subsequent centrifugal separation of the wash water thereupon removes these deterioration products which had been in solution in the oil and so are not otherwise removable by centrifuging, since centrifuging does not separate materials in solution from their solvents. Washing may also carry forward a chemical reaction to a point where the intermediate is converted into an end product, water combination, emulsion, or other condition, where centrifuging can become effective. Deterioration products in oils may thus be quickly carried on over into the sludge or the sludge-and-emulsion stage and then thus removed by centrifuging. Fuller's earth and some other such materials which have an affinity for (or adsorbent action upon) water or other wash liquid may, upon the diffusing of such materials through the liquid to be treated, have their mass or their gravity so increased as to promote greatly their ease of separation.

Temperature Effects. When mixtures of liquids are heated or cooled, the differences in rates of volumetric expansion may increase the gravity differential between the components and thus greatly facilitate their separation. In some cases it may reverse the relative gravities. Heating is frequently used to reduce the viscosity to a value that will permit easy separation. This, however, increases the Brownian movements and may increase the difficulty of separating out very small particles.

Solubility of one component in another also increases with rise of temperature; and oil, centrifuged at high temperatures, may retain small quantities of moisture. Hence insulating oils should be centrifuged at the lowest temperature that will give suitable viscosity, usually not above 30° to 40°C. More viscous oils may be heated till the viscosity is reduced to 300 sec. Saybolt Universal (about 65 centistokes).

Emulsion breaking can often be facilitated by heating to a temperature that gives a suitable low viscosity. Carefully heating to the point where the surface film is broken by partial evaporation of the inner phase but without excessive evaporation or frothing may also be employed. This effect may be obtained by evaporation caused by a vacuum, but is hard to control to prevent frothing. Breaking of emulsions may also be caused by freezing, long standing, and chemical methods such as addition of salts or acids.

Machine-operating Features

Starting devices and driving mechanisms must take into account the large moment of inertia of the bowl and the high speed to which any centrifugal bowl must be brought and be sturdy in design with ample capacity to start the bowl smoothly and to bring it quickly up to speed without jerk or shock.

Light bowls, such as the tubular type, are usually driven by a flexible fabric belt, and, as their inertia is small, they can easily be brought up to full speed by letting the belt slip during this short acceleration period.

Belt-driven machines with bowls of great mass are often started by allowing belt slip to occur at first to a greater, and later to a less, extent while the machine is being brought up to speed. This is the simplest method possible, requiring no special mechanism and giving very good results for infrequent starts, but it usually needs the personal attention of an operator for the several minutes required to attain full speed and may entail undue wear on the belt.

Centrifugal friction clutches with the weights driven positively give an extremely smooth start, and the wearing pieces may be of brake lining or other readily renewable materials. Springs may be used to hold back the friction weights until the motor or the pulley gets up to a predetermined fraction of full speed, which is a considerable advantage for induction motors, particularly single-phase motors with split-phase starting coils.

Automatic electric controllers, which regulate the motor speed so as to bring the motor and centrifuge smoothly and gradually up to full speed together in a predeterminable time cycle with a regulated power demand on the supply lines, are quite advantageous for basket-type centrifuges on large production schedules requiring definitely laid-out schedules for time for loading, time to bring up to speed, time at full speed, time for slowing down, and time at standstill for unloading. Usually in such a cycle the time for accelerating to full speed is greater than any of the others and so is the most important to control. This method is the most economical of power, as none need be wasted in friction or in attendant's time and so, for some uses, well justifies its high first cost.

Driving mechanisms may be direct belt from a motor or line shaft to the bowl spindle pulley or by worm wheel and worm, or, more rarely, by bevel gears. Basket centrifuges use the belt drive almost exclusively, but the smaller high-speed centrifuges may use either, both giving quite adequate service results. For either type of drive the principal requirement is that the size for belts or gear teeth must be amply proportioned in order to give adequate wear resistance and to offset the handicap of the high speeds involved.

Lubrication of vertical spindles carrying bowls mounted on the upper end requires special provisions for the feed and maintenance of a lubricating-oil film. A bearing whose lower end can be immersed in oil may have helical oil-pumping grooves on either the journal or the bearing to raise the oil to the top end. Such grooves may give high pumping pressures; *e.g.*, a 7 m./sec. surface velocity can be made to develop a pressure of 2 atm. The edges of oil grooves must be chamfered to spread the oil peripherally. Such grooves serve also to wash grit out without scoring the surfaces. The top bearing of such a machine is subjected to severe operating conditions.

If, as is usually the case, full speed is above the critical speed, the bowl which rotates about its form center at low speed changes over as it passes through the critical speed to rotate about its center of gravity. This changeover is the usual cause for noise and rough running while coming up to speed. At full speeds, above the critical speed, the spindle will weave slightly and a flexible bearing support must be used if excessive bearing stresses are to be avoided.

The flexible support should be well damped to reduce the amplitude and continuance of vibrations.

The attachment of tightly closed lubricating-oil tubes to provide feed of lubricant under pressure presents certain difficulties because of the vibration of the bearing as it follows the spindle positions. On this account, oil is usually fed to a recess in the upper end of the bearing by an oil-feed tube which is not fastened to the bearing mechanically but is freely supported from the frame in a position to drop the oil into the recess. Either vertical or helical grooves may be used to conduct and to distribute oil to all parts of the bearing. A cap or cover should be fastened to the upper end of the bearing with a hole just large enough to clear the spindle surface to prevent windage from blowing the oil fed off the top end of the bearing. The surface speeds are so high that oils of low viscosity would be satisfactory for continued operation at full speed, but, owing to the need for a lubricating film at the lower speeds during starting and stopping and the need for maintaining a supporting film even after some wear has taken place, the "medium" or "heavy" oils are more suitable considering the all-round conditions of service, particularly considering that vertical centrifuge bearings seldom have a complete oil film.

Balance and vibration have a major effect on wear and performance. Bowls are dynamically balanced when built so that not only the bowl runs smoothly but also the spindle, in order that the frame may not suffer severe vibrating forces. The frame is usually supported upon rubber cushions or held by heavy springs to relieve stress upon the foundation bolts and foundation. Machines with bowls running at speeds of 6000 r.p.m. or more are balanced so that frames and other parts do not have amplitudes of vibration greater than fractions of 0.001 in. Basket centrifuges with speeds of 500 to 1500 r.p.m. may have higher vibration amplitudes. In such machines, which are often loaded and unloaded every few minutes and where the load masses cannot readily be disposed so as to eliminate unbalance, the suspension of the frame on springs is of considerable advantage.

Hollow tubular bowls, having a small mass relative to the frame and suspended on the drive shaft, may be run in frames solidly bolted to the foundation. Guide bushings and semiflexible means for damping side swing are usually provided.

Wear and corrosion present difficulties, if they occur to an appreciable extent in the bowl, since this is the most expensive and carefully made part of a centrifuge. Such changes may not only alter the performance but change the balance or mechanically weaken the parts. Corrosion-resisting materials, such as monel, chrome-iron, chromium-nickel-iron, or protective coatings, such as tin, plated by the hot dipping process, and rubber coatings, are used to meet special conditions.

Energy and Power Requirements

The stored energy in the moving parts may be expressed as

$$E = \tfrac{1}{2} I \omega^2 \qquad (22)$$

where E is the energy, kg.-m.; I is the polar moment of inertia, kg.m.; ω is the angular velocity, rad./sec.

This stored energy is usually quite considerable and may take the full input from the motor or line shaft for periods of 1 to 10 min.

Equation (22) may be restated in a form to give the approximate values for the energy stored in rings of 1-cm. square section and different radii for the mean section, at various speeds, assuming the wall section to have a specific gravity of 7.83, the same as steel. This form is

$$E = 2.75 \times 10^{-9} r^3 \text{ (r.p.m.)}^2 \qquad (23)$$

where E is the stored energy, kg.-m.; r is the mean radius of the section, cm.; r.p.m. is the speed in revolutions per minute.

The accompanying curves (Fig. 108) shows the range of values for stored energy.

The power lost in friction of the bearings and of the driving mechanism of underdriven machines is mainly in the top bearing on the high-speed spindle, next to the bowl, which may represent as much as half of the total machine friction because of its high speed and the large diameter needed to get a sturdy construction for the support of the bowl. Next in order of magnitude is usually the lower bearing, which is generally of much smaller diameter than the top bearing. A worm and worm wheel have generally a relatively small friction loss. A belt drive may likewise have a high efficiency if the belt is not too thick and stiff, and if of adequate width for the load to be carried. The friction in other bearings, if any, is usually a small factor in the friction losses. All these considerations point to the need for attention and supervision on the part of the operator to see that the high-speed spindle bearings (particularly the top bearing) are kept well lubricated and are inspected regularly so that they can be repaired or replaced immediately, if required.

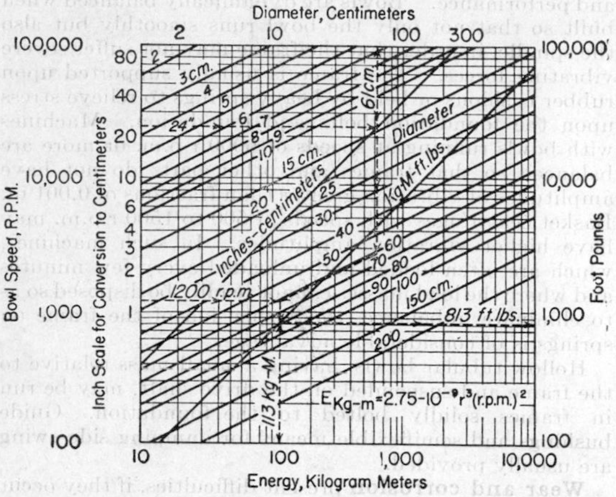

FIG. 108. Energy in rotating steel rings per square centimeter cross section.

The power lost in accelerating the through-put and in the interior passages is one of the inherent factors incident to centrifuging. In a batch centrifuge of the basket or centrifugal-dryer type, where the bowl is loaded, brought up to speed for the centrifugal treatment, drying, etc., and then stopped (usually by brakes employed to save time), the energy stored is irretrievably lost. In such cases the acceleration of the load along with the bowl is calculated by the methods already described for that stored in the bowl. In continuous centrifuges the liquid fed into the bowl must be continuously accelerated up to the linear velocity of the larger diameter part of the bowl through which it passes (if the liquid is to be subjected to the full centrifugal effect). The power required for load acceleration may be expressed as

$$L = \frac{1}{2}\frac{G_2}{g}V^2 = \frac{1}{2}\frac{G_2}{g}r^2\omega^2 \qquad (24)$$

where L = power, kg.-m./sec.
G_2 = through-put kg./sec.
r = radius, m., of the largest part of the bowl with which the liquid rotates at the full speed of this part.
ω = angular velocity, rad./sec.
g = acceleration of gravity, m./sec./sec. = 9.81.

The power may also be expressed in horsepower as

$$\text{Hp.} = 5.167 \times 10^{-9} G_3 R^2 \text{ (r.p.m.)}^2 \qquad (25)$$

where G_3 is through-put, lb./min.; and R is the radius, ft.

If slippage does not occur, the energy imparted to the liquid at the maximum radius it reaches in the bowl is recoverable on its return toward the center line before its exit over the discharge lip. The energy in the "discharged" liquid or liquids at the point of discharge is that corresponding to the radii of the discharge "lips" (rather than the radii of the largest part of the bowl they may pass through), plus that small quantity of energy represented by the liquid velocity radially outward for emergence. This latter quantity is relatively insignificant and is in any case supplied by the pumping pressure due to the centrifugal pumping "head" between the check ring and discharge-lip radius, or radii, and so may best be included with the energy required to overcome liquid friction through the interior passages.

If the discharge lips are provided with shapes for discharge passages which direct the stream "backward," the reaction tends to drive the bowl forward and so a considerable fraction (about half) of the energy in the liquid discharge may be recovered depending upon the tangential angle and the discharge-passage friction.

The liquid friction in the interior passages absorbs an amount of energy which may be as large as that in the liquid discharge.

The friction head is of course supplied by the centrifugal pumping head, produced by the difference in radii of the check ring near the center and the discharge lip. No part of this energy can be "recovered," although the value may be kept small by designing passages of ample size and with smooth curves, so that a small "head" will supply it and so allow the use of discharge lips of small radii.

If a discharge lip (or weir) leads to nozzles discharging in a radial or partly radial direction, a separate determination may be made for the energy required by calculating the centrifugal head between the outer and inner radial positions of entrance and exit of the nozzle.

Relation of Size, Driving Power, and Costs

Cream Separators. The accompanying set of curves (Fig. 109) gives representative data for the through-put of liquid-liquid, continuous, centrifugal separators and customers' prices as cream separators, of the standard type using a simple belt drive and the corresponding data for the net horsepower required for operation at full liquid load. (For motor costs see pp. 1767, 1770.)

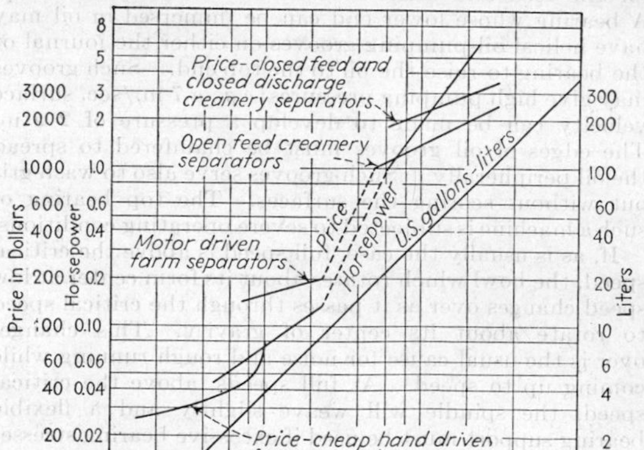

FIG. 109. Milk separators—relation of rating, price, and power requirements.

Oil Purifiers. The set of curves (Fig. 110) gives values for the maximum through-put of continuous liquid-liquid centrifuges and customers' prices, as standard oil purifiers, and the corresponding data for the driving power of the standard types. Special features and accessories are not included in these price figures as the addition of motors, double pumps for inlet and return.

Table 2. Cost Estimates for Centrifuging

Type of service	Hand-driven farm-size cream separator	Power-driven dairy-plant cream separator	Lubricating-oil purifier electric power station	Centrifugal laundry dryer extractor, 48 in.	Sugar centrifugal sugar refinery, 48 in.
Capacity, gal./min	1.6	22.0	15	3–200 lb. loads/hr.	10 tons/hr.
Power requirements, hp	0.09	4.0	4	7½	50
Daily operating schedule, hr	0.67	9.0	24	7	24
Daily attention, man-hr	1.0	1.0	1	2	8 (1 man, 3 machines)
Annual operation, hr	243	3,285	8,780	1,941	7,200
Annual labor time, man-hr	365	365	365	533	2,400
Annual through-put	23,350 gal.	4,350,000 lb.	7,884,000 lb.	1,074,000 lb.	72,000 tons
First cost	$122.50	$2800.00	$2750.00	$2900.00	$8,000.00
Annual fixed costs:					
Obsolescence and depreciation	1/15 $ 8.16	1/10 $ 280.00	1/10 $ 275.00	1/10 $ 290.00	1/10 $ 800.00
Interest	6% 7.35	6% 168.00	8% 165.00	6% 174.00	6% 480.00
Insurance and taxes	2% 2.45	2% 56.00	2% 55.00	2% 58.00	2% 160.00
Subtotal	$ 17.96	$ 504.00	$ 495.00	$ 522.00	$1440.00
Unit fixed costs	$ 0.77	$ 0.116	$ 0.0628	$ 0.486	$ 0.020
Annual variable costs:					
Cost per man-hr	$ 0.30	$ 0.70	$ 0.70	$ 0.60	$ 0.65
Cost per hp.-hr		0.025	0.01	0.02	0.005
Labor cost	$109.50	$ 255.50	$ 255.50	$ 319.80	$1560.00
Power cost		328.50	350.20	218.00	5400.00
Repairs and supplies cost	8.20	100.00	75.00	75.00	150.00
Subtotal	$117.70	$ 684.00	$ 680.70	$ 612.80	$7110.00
Unit variable costs	$ 5.05	$ 0.157	$ 0.0864	$ 0.572	$ 0.0988
Total annual costs	$135.66	$1188.20	$1175.70	$1134.80	$8550.00
Total unit costs	$ 5.82	$ 0.273	$ 0.1492	$ 1.058	$ 0.119
Unit for costs	1000 gal.	1000 gal.	1000 gal.	1000 lb.	2000 lb. ton

heaters, and supplementary blotter presses, such as are used for insulating oil, all mounted upon a single movable platform or truck would add costs about equal to that of the centrifuge alone.

Basket Centrifuges. Basket centrifuges cannot readily be rated in terms of through-put but are usually rated by basket diameters. The curves in Fig. 111 give

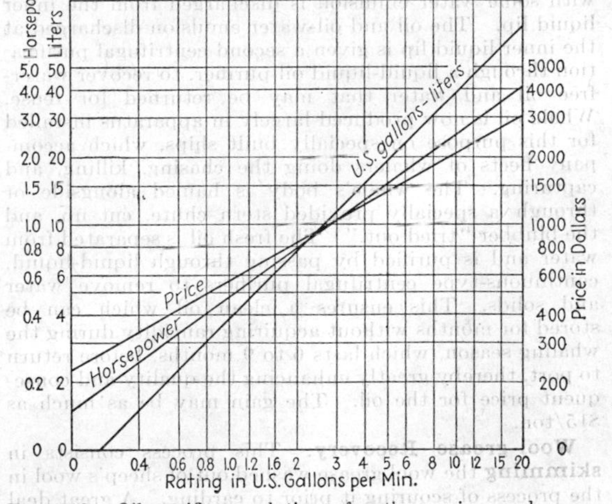

FIG. 110. Oil purifiers—relation of rating, price, and power requirements.

representative figures, showing the relation of holding capacity, price, and the horsepower required for such standard work, as laundry "drying," dry cleaner's solvent drying, and the handling of simple chemicals. The through-puts must be estimated from the drying characteristics of the materials handled, the "dryness" required and the operating schedule that may be maintained under the labor conditions, and other work that the operator may need to attend to.

Costs of centrifuging may be estimated for typical cases, based on assumptions as to operating schedules and other factors, representative of the usual conditions of service. Five such cases are illustrated in Table 2, namely, a medium-sized hand-driven cream separator as used on a farm, a large creamery size, power-driven cream separator, a large oil purifier, as used for the

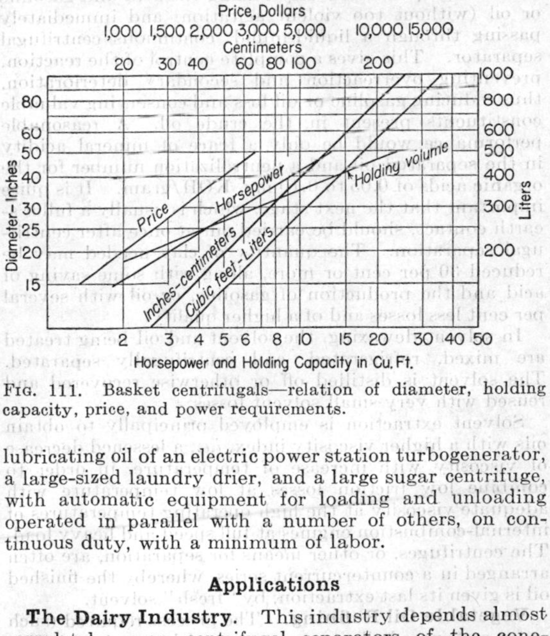

FIG. 111. Basket centrifugals—relation of diameter, holding capacity, price, and power requirements.

lubricating oil of an electric power station turbogenerator, a large-sized laundry drier, and a large sugar centrifuge, with automatic equipment for loading and unloading operated in parallel with a number of others, on continuous duty, with a minimum of labor.

Applications

The Dairy Industry. This industry depends almost completely upon centrifugal separators of the cone-shaped, separating disk type in all its operations for obtaining cream for table use and for making butter. The principal classes of uses are:

1. Farm milk separators to obtain cream from whole milk for shipment or for butter making.
2. Factory milk separators to obtain cream for table use, for butter making, and for ice-cream manufacture.
3. Factory whey separators to recover butterfat from whey after making cheese.
4. Factory milk clarifiers to remove sediment from milk before bottling for distribution to customers.
5. Factory, gathered-cream separators to skim old or gathered cream.

The **farm milk separators** are usually small in capacity, *i.e.*, less than 3 gal./min., most frequently run by hand crank, although motor or gasoline-engine belt drives are available. A reasonable standard of performance is a butterfat loss in the skim milk not over 0.025 per cent by Babcock test, or about 0.085 per cent by Röse-Gottlieb or other extraction methods.

Factory separators are all power-driven by belt from a line shaft, electric motor, or steam-turbine wheel. A reasonable standard of performance is a butterfat loss of 0.01 per cent by Babcock test or 0.05 per cent Röse-Gottlieb.

Petroleum Refineries. The principal uses for centrifuges in petroleum refining have been for bright-stock dewaxing and the separation of sludge from acid treating. Solvent dewaxing and solvent extraction processes have also used various types of special centrifuges for the separation.

In the bright-stock process the slowly chilled oil, previously diluted with a solvent naphtha, is passed through the centrifuge, where the chilled amorphous wax discharges at the outer bowl lip and is swept continuously away by hot water. Removal of the naphtha by distillation yields an oil with a cold test only about 30° to 50°F. higher than the temperature to which the mixture had been chilled.

In the acid treating of gasoline or lubricating oil, the acid and acid sludge will separate by gravity in one or two days but may be separated in a few minutes after contacting by continuously mixing the acid and gasoline or oil (without too violent agitation) and immediately passing through a liquid-liquid, continuous centrifugal separator. This gives a complete control of the reaction, preventing overreaction and secondary deterioration, thus reducing gasoline or oil loss and conserving valuable constituents present in the crude oil. A reasonable performance would be only a trace of mineral acidity in the separated oil and a neutralization number for the organic acids of 0.05 to 0.10 mg. KOH/gram. It is quite important that the next stage, which is usually a fuller's-earth contact, should be carried out at once after centrifugal separation. The quantity of clay needed may be reduced 30 per cent or more, along with some saving of acid and the production of gasoline, or oil with several per cent less losses and of a higher quality.

In solvent dewaxing, the solvent and oil being treated are mixed, refrigerated, and centrifugally separated. The solvent is distilled off or otherwise recovered and reused with very small solvent losses.

Solvent extraction is employed principally to obtain oils with a higher viscosity index, *i.e.*, a lessened decrease of viscosity with increase of temperature, in order to combine low friction losses at low temperature with adequate viscosity at the high operating temperatures of internal-combustion engines at full speed and heavy load. The centrifuges, or other means for separation, are often arranged in a countercurrent series. whereby the finished oil is given its last extraction, by "fresh" solvent.

Vegetable Oil Refining. The process required (such as for cottonseed oil) is a treatment with caustic soda in a water solution to saponify organic acids, remove gums, etc., which may be accomplished by mixing in a moderate-size, continuous reaction vessel and passing the mixture at once through a liquid-liquid, continuous separating-type centrifugal separator. The soap stock is discharged at the outer lip and pure oil at the inner lip. The oil loss in refining depends on the free fatty-acid content, which varies with age and other conditions in the cottonseed-oil mills, and on crop conditions. The loss ranges between values of about 6 to 10 per cent of the raw oil with kettle refining and gravity settling, but with centrifugal separation the loss may be 2 per cent less, *i.e.*, 4 to 8 per cent of the raw oil. A better color is obtained in the refined oil and lower values for moisture and free fatty-acid content.

In **palm oil** and **olive oil**, the problem consists largely of getting a quick separation from water or saps to avoid fermentation and rancidity in the oil, resulting from standing while settling by gravity. The recovery of oil is increased 3 to 5 per cent, and all may be equal to "first pressing" or "first grade," instead of about half "second grade" and up to 5 per cent of "third grade." Cast iron and tinned steel are suitable for bowls and covers, but all copper and brass must be avoided.

In **orange** and **lemon oil,** the problem consists of skimming the very small quantities of oil released from the rinds by crushing and expressing, from the water and juices. Monel or chromium-iron alloy bowls and covers are desirable to avoid corrosion.

Soybean oil requires not only a refining process but also degumming, which may be accomplished either by solvent extraction or by a hydrolyzing process. In either case a centrifuge provides a quick and complete recovery of high-grade oil.

In both **fish oil** and **whale oil,** the problem consists of recovering the oil after cooking or "trying out" free of water and free of shreds of flesh, bone, scales, etc. A nozzle type of liquid-liquid discharge bowl is used for the first-stage separation, so that the solids are discharged through the nozzles along with enough oil-free water to sweep them along and out of the receiving cover. Oil-free water is discharged over the outer liquid lip, and oil with some water emulsion is discharged from the inner liquid lip. The oil and oil-water emulsion discharged at the inner liquid lip is given a second centrifugal purification through a liquid-liquid oil purifier, to recover water-free oil and water that may be returned for reuse. Whale oil is now produced largely in apparatus installed for this purpose on specially built ships, which accompany fleets of whalers doing the chasing, killing, and capturing. The whale's body is hauled alongside or through a specially provided stern chute, cut up, and the blubber "tried out." The fresh oil is separated from water and is purified by passing through liquid-liquid, continuous-type centrifugal purifiers to remove water and solids. This ensures a clean oil which can be stored for months without acquiring rancidity during the whaling season, which lasts 6 to 9 months, before return to port, thereby greatly enhancing the quality and consequent price for the oil. The gain may be as much as $15/ton.

Wool-grease Recovery. This process consists in **skimming** the wool grease washed out of sheep's wool in the process of scouring it prior to carding. A great deal of dirt, sand, etc., is also washed out, requiring settling or decanting vats or the use of basket-type batch centrifuges, prior to skimming the wool grease out by means of continuous, liquid-liquid centrifugal separators. The wool grease, particularly if quite clean and light in color, has some value, but it is even more important to remove the grease from the wash water before it is run off into sewers or water courses, to avoid stream pollution. For this reason, skimming types of bowls have been used in order to give the maximum purification to the water by removing the last traces of grease.

Valve-type bowls of the liquid-liquid solid type are now being used in wool-scouring processes, in which the solids are continuously discharged through the valves, the cleaned and degreased wash water is discharged at the outer discharge lip (to be returned for reuse in washing), and a wet grease is discharged at the inner lip for recovery of grease. This wet grease is usually given a second centrifugal separation, to remove the water and further to purify it preparatory to refining to pure lanolin or the production of fatty acids.

Lubricating-oil Purification. One characteristic requirement common to all cases of lubricating-oil purification is the need for a centrifugal purifier in constant operation upon a continuously recirculated portion of the lubricating oil, whereby an amount equal to the whole amount of oil in the system is separated from any water that may be present and purified of sediment and solids of all kinds every few hours. A liquid-liquid, continuous-separator type of bowl is employed in all cases except for small Diesel engines where water is practically always absent, or in locomotive or marine Diesels where water is not available for water washing the oil. This requirement of constant centrifugal purification is not so much for the purpose of saving oil, important as that is for marine installations, as to reduce wear on the turbines and engines and to avoid stoppages, unexpected outages or interruptions of service, and repair costs.

It is important in all continuous, by-pass purification systems to keep in mind that the criterion of the purification should be the condition of the oil in the whole system and not the purification effect of a single passage through the purifier. In general, and with few exceptions, the centrifugal purifier should be operated at a high through-put rate, even though at a somewhat lessened contrast between the feed and the purified oil discharged from the purifier.

The larger number of passes more than makes up for the somewhat smaller purification coefficient (provided that the sediment is not too small in size to be removable). It may be shown that the optimum through-put rate is

$$C = \sqrt{\frac{s}{u}} \qquad (26)$$

where s is the rate of sediment production, and u is the ratio of sediment in the outgoing oil to that in the feed at unit through-put.

Steam turbine oil is subjected to heat in the bearings, air contact, water contact, and cooling in the water-cooled cooling coils. The heating and air contact promote oxidation, with the production of organic acidity; and in the later reaction stages, sludge. These reactions, in turn, decrease the ease of separation of oil from such water as may come from steam-gland leaks, cooling-water leaks from water-cooled bearings or in water-cooled oil coolers, or condensation on surfaces or by absorption, solution, and reprecipitation from the atmosphere, or even, as a result of the chemical reactions in the oil itself, from dissolved oxygen.

The emulsion sludge as well as the soluble sludges (metallic soaps, etc.) tend to precipitate on cooling and thus to coat the surfaces of water-cooling coils, to interfere with heat transfer, and to involve difficult cleaning operations for their removal. On the other hand, the *freshly* produced organic acidity is many times as soluble in water as in oil and so may readily be washed out by mixing a small stream of boiling water with the oil and immediately separating this water in a continuous liquid-liquid centrifugal separator. The continuous centrifugal separator further protects the turbine from danger of water leaks into the oil and removes worn bearing particles and dirt from the oil. A reasonable maximum performance for a centrifugal purifier on wet turbine oil is that at which it leaves a moisture content of 0.05 per cent and correspondingly less at lower capacities. The neutralization number and steam emulsion number may become stabilized, but at values that vary markedly with the oil, the operating temperature, and other service conditions. Neutralization numbers of 0.5 to 1.5 mg. KOH/g. oil and steam emulsion numbers of 300 to 600 sec. are frequently found in steam-turbine oil in service, when centrifugally purified by a continuous by-pass system. In such a system there should be present only traces of sediment or sludge.

Diesel-engine lubricating oil seems to be slightly subjected to oxidation but is markedly affected by cracking or burning, owing to the high temperatures, and accumulates worn metallic particles owing to the high bearing pressures and the reciprocating motion, as well as dirt from the combustion-air and breather connections. Some residues come from the fuel oil, and sometimes fuel dilution occurs through failure to fire, owing to lack of compression, bad mixture, moisture in the fuel oil, etc. The sediment in the oil consists largely of finely divided carbon, semicolloidal in size, along with small proportions of worn metal and metallic oxides, sand, dirt, etc. The latter types of sediment are readily removable by centrifugal clarifiers for liquid-solid clarification, but the carbonaceous material requires a high degree of clarification for satisfactory removal. By suitably large centrifugal clarifiers on a continuous by-pass purification, the total sediment in the oil in the system may become stabilized at values of 1.0 to 1.5 per cent. It appears undesirable to allow the sediment to reach values much over 2.0 per cent.

If hot water is available, washing the oil with water may be resorted to, using water proportions up to 40 per cent, and then immediately passing the well-mixed water and oil through a liquid-liquid, continuous centrifugal separator, thereby flushing most of the finely divided carbon out of the oil and the bowl along with the water and pocketing only the dense sediment in the bowl. This sediment may be cleaned out of the bowl by hand upon stoppage of the centrifuge. Hot-water washing markedly improves the centrifugal purification of the oil.

Rolling-mill oil usually has large amounts of water splashed in from the cooling water on the rolls and large amounts of mill scale and dirt. Purification requires a liquid-liquid, continuous-separating type of centrifugal bowl, with sludge pockets of considerable size and with gravity presettling tanks for catching the major part of the water and the coarser sediment. A reasonable performance for the maximum through-put appears to be not more than 0.050 to 0.200 per cent of moisture and 0.05 per cent sediment left in the purified oil, depending upon oil temperature and viscosity, and correspondingly lower values at less than maximum through-put.

Gasoline-engine crankcase oil purification is carried out by centrifugal purifiers mounted alongside the engine and operating continuously on a part or all of the lubricating oil being pumped to the engine bearings in only a few types of automobiles at the present time, because of the moderate size of such engines and difficulty of getting adequate attention to cleaning or making repairs by drivers or garage repair men.

Batch reclamation on oil dumped from the crankcases of fleets of cars or trucks is quite common. If the quantities and conditions are such that the oil is to be used only as fuel, then a simple purification to remove free water and coarse sediment is adequate, which can be done by standing and gravity settling on a small scale where ample time is available, or instantly in liquid-

liquid, continuous-separating types of centrifugals provided with sludge pockets of adequate size.

If the oil is to be reused in gasoline engines, water washing and treatment with neutralizing chemicals, such as sodium silicate and trisodium phosphate, may be carried out first, followed by settling or centrifugal separation, with subsequent removal of dilution in some cases and of fuller's-earth treatment for restoration of color in others. These last two steps are quite often considered unnecessary, since some dilution is sure to occur in service, and color is of itself no indication of service value. Furthermore it should be borne in mind that dilution, unless quite excessive, is not altogether a disadvantage since it performs an automatic self-regulating function. Since dilution is generally greatest in cold weather, it renders starting easier and delivery of oil through the lubricating tubes more certain, while in hot weather the lessened dilution causes higher actual viscosities and therefore better adaptation to high temperatures in the bearings and on the cylinder walls.

Block-test lubricating oil needs purification, primarily for the removal of worn metallic particles incident to the running in of newly made parts and of chips and filings and shop dirt left on the parts. Their removal avoids cutting and scoring of parts and hastens the running-in operation. Gasoline engines are quite usually driven by gas and not by gasoline during the block-test run, so that dilution is not a factor. A continuous, by-pass purification, using a liquid-liquid, continuous-separating type of bowl, with sludge pockets of ample size to hold the solid sediment is usually employed. Water, spilled or leaking from the jacket water-cooling system, is often present and a stream of clean water may be passed through the centrifugal along with the oil to assist in its purification by water washing the oil.

Diesel-fuel purification requires the removal of water and solid sediment. Even small proportions of water, particularly if unevenly distributed, will cause misfiring and even stoppage of the engine. In marine service, where sea-water leakage to the oil storage tanks is almost sure to occur, since these tanks are usually placed next to or against the outer hull, or where sea water is pumped in to provide ballast in emptied oil tanks, water separation is of great importance. Solid-sediment removal is desirable in order to reduce piston and cylinder or cylinder-liner wear and clogging fuel-injection nozzles. Water may be separated partly by baffled, gravity settling tanks, but sediment is not removable to any appreciable extent in this way. The liquid-liquid, continuous-separating type of centrifugal, provided with ample sludge-holding pockets, will remove both water and sediment and may be built to operate even during fairly heavy seas.

A reasonable standard of performance for a centrifugal purifier is somewhat difficult to state, because of the wide range of viscosities and specific gravities of Diesel fuel.

For crudes and residues of 300 sec. Saybolt Furol at 50°C. (about 640 centistokes) the moisture residue may be kept to between 0.5 and 1.0 per cent, provided that the oil is heated close to the boiling point of water. In any case, fairly effective centrifugal separation is obtainable only when the viscosity at the temperature of separation is not much greater than 300 sec. Saybolt Universal (*i.e.*, about 65 centistokes).

The sediment under similar conditions as to temperature and viscosity at the temperature of centrifuging may be kept within values of 0.05 per cent. It is well to bear in mind, in regard to ash content, that some ash in oils is due to mineral compounds held in solution and so is not separable by any means.

Transformer-oil purification requires principally the removal of the last minute traces of moisture. Even new oil that has been thoroughly dried at the refinery picks up some moisture in shipment, either by solution from atmospheric humidity or condensation on the upper walls of the containers, or it is produced chemically by the reaction of the oxygen dissolved in the oil from the air and combining with constituents of the oil. Some small amounts of pipe scale, rust, etc., may be picked up also, but this is easily separated in the course of the removal of moisture in a liquid-liquid type of continuous centrifuge. The advantage of the centrifuge over the blotter press, which functions not as a filter but as an absorber of moisture, lies in the fact that the centrifuge continues to operate indefinitely at constant and full effectiveness, while the blotter press works with a decreasing effectiveness and at some unpredictable moment must have the blotter papers exchanged for new or dried ones. The blotter paper may break and allow oil to go through unpurified for some time, since such a break is not visible to the operator. A reasonable standard of performance at maximum through-put of the centrifuge is the leaving of not more than 0.001 per cent of moisture in new oil at temperatures not much above that of the room. Owing to the solubility of water in insulating oils, which may amount to 10 to 30 p.p.m. by weight at room temperatures and ten or more times this in oil at 80°C., it is futile to centrifuge insulating oils at elevated temperatures, except as a preliminary step to break any emulsion that might be present in a used oil, to be followed, after cooling and standing long enough to precipitate the dissolved water, by a second centrifuging cold. An alternative is to centrifuge, once only, at the ambient temperature, but at such a reduced through-put as will give the desired purification. Even if centrifuged at elevated temperatures resulting in a brilliantly clear oil with less than 0.001 per cent of *free* water and having a dielectric breakdown of 25,000 to 40,000 volts between 1-in. disks separated by a gap of 0.100 in., *just after being centrifuged*, such oils will, *upon cooling*, precipitate out some of the dissolved moisture, and the oil will become cloudy and may show only a few thousand volts breakdown value. After several years of service, transformer oils develop organic acidity and finally sludges from oxidation and other reactions. The sludge tends to deposit upon, and so to cause overheating of, the coils by interfering with their dissipation of heat. The organic acidity cannot be **removed** by centrifuging but may be **reduced** by water washing and then centrifuging, or by treating with mild alkalies, such as sodium silicate or trisodium phosphate solutions and then centrifuging. The sludge and water present in old used transformer oil is directly removable by centrifuging, and this simple mechanical separation is at present the general practice in electric power systems. The treatment with neutralizing chemicals, followed by centrifugal separation which restores the oil to its original neutralization number, may be followed by contacting with 1 to 2 per cent by weight of fuller's earth. The diffusion of a small amount of hot water into the oil facilitates the settling of the fuller's earth and then, after settling, the upper layers may be decanted off by means of a floating draw-off pipe to a centrifugal separator for final purification. This results in the production of a reclaimed oil that is as good as, or better than, new oil for reuse.

Electric switch oil is subjected to the heat of the electric arc between the opening contacts of the electric switch, producing gases (principally hydrogen) and finely divided carbon. Water and metallic particles may also be present which are readily removed by a liquid-liquid type of continuous centrifuge, leaving, however, some semicolloidal carbon in the oil which may then be readily removed by subsequent passage through a paper-blotter press dressed with two or more thicknesses of dry paper.

Cable oil requires purification for the removal of the last traces of moisture and sediment, in order to obtain a homogeneous insulating material combining high dielectric breakdown strength, high electric resistivity, low energy loss under alternating electric stresses, and freedom from tendency to change or to deteriorate in service.

A liquid-liquid type of continuous centrifugal separator may be used for the first-stage separation to remove moisture and sediments, waxes, etc., followed by passage through a blotter press dressed with two or more thicknesses of blotter filter paper. The centrifugal purification should remove all the free water, particularly if carried out at about room temperature so that change of the blotter papers which absorb moisture may be rendered very infrequent. However, blotter papers, particularly when newly installed, allow some fibers to be swept off the back side and into the oil, so that a considerable portion of the oil first passed through the press should be returned to the system and rerun.

Varnish and Lacquer Clarification. Clear varnish needs clarification for removal of dirt and fine sediment which otherwise spots or breaks the smooth shiny surface of the dried varnish. A high clarification coefficient is necessary, resulting in a need for very effective types of liquid-solid, clarifying types of bowls and run at relatively low through-puts to obtain the maximum possible clarifying effect. It is not feasible to state in words a standard for a reasonable performance, since this is judged by flowing a small quantity of the clarified varnish over a small piece of flawless plate glass and observing the surface at an angle which reflects the light obliquely. Under such a test the clarified varnish should show no specks or at most only two specks on a plat 10 by 6 in., while if more than four specks show, the varnish should be rejected.

Pigment lacquer clarification is desirable to remove not only the dirt and sediment that may have been picked up by the lacquer and, more particularly, the coarser ground pigment particles which may be reclaimed for regrinding but also partly dried surface films, etc. Only a moderate clarification coefficient is needed, because of the high specific gravity of the pigments and their relatively large size. A simple type of hollow bowl is generally used, though sometimes multiple concentric cylindrical shells are provided. The machine is frequently equipped with an adjustable drive for speeds between one-third and full speed. By adjusting the speed and through-put, any desired degree of clarification may be obtained. The glass-plate or metal-panel method is used for judging the setting and degree of clarification to be employed for each grade or kind of pigmented varnish.

Dry cleaners' solvent clarifiers have been used for many years to remove continuously the dirt and sediment picked up from the clothes in the washer, by-passing the solvent continuously while the washing is in progress through a liquid-solid type of centrifugal clarifier. Simple forms of bowls have generally been used, of either the double-concentric, cylindrical-shell, medium-speed type or the single-shell, tubular, high-speed bowl. The moderate-speed double-shell bowl has been most widely used. The practice abroad has been to use bowls with several concentric cylindrical shells, or bowls with conical disks much smaller in diameter than the shell, to give adequate dirt-holding space.

A typical plant would have one or two washers, 36 in. in diameter and 54 in. long, each of which would have with it a centrifugal clarifier with a bowl having a dirt-holding volume of 330 cu. in. (5400 cc.).

Dry cleaners use "dryers" of the centrifugal-basket type to remove the solvent held in the washed clothes when removed from the washer. The clothes retain a weight of solvent equal to 1¼ to 1¾ times the normal dry weight of the goods, which should be removed mechanically rather than by being evaporated off, not only to conserve solvent but also to avoid leaving solvent and other odors on the clothes. The basket centrifuge should leave not more than 15 to 30 per cent of solvent by weight for removal in the heated tumbling dryer or the cabinet air-blower type of steam-heated dryer to be evaporated off. (This "dry cleaning" does not remove the normal water content present in cloth, although it may be reduced somewhat, *i.e.*, be 8 per cent instead of 10 per cent.)

A typical basket dryer would be about 26 in. in diameter, with a basket depth about half the diameter and run at speeds of 700 to 900 r.p.m. Such a basket will hold up to 90 lb. of clothes (normal dry weight) and require about 5 to 6 min. for centrifuging and about 3 min. for loading and 2 min. for unloading, thus making a cycle in 10 to 11 min. and allowing about four to five loads per hour, when allowances are made for the operator to attend to other duties when not loading or unloading.

Laundry drying must be carried out in two stages, first the mechanical removal, second drying by heat. When removed from the washer, clothes hold two or three times their "dry" weight of water, which may be reduced to 20 to 30 per cent in a centrifugal extractor but is usually reduced only to 50 per cent since ironing or pressing requires some moisture. Dry clothes normally contain 6 to 14 per cent moisture and, if dried below this, become harsh and brittle. A piece of cloth does not seem wet to the hand until it carries about one-third of its dry weight of moisture. Wet-wash laundries operate on a larger scale than dry-cleaning plants and use the largest sizes of extractors, up to 72 in. diameter. Power is required mainly for acceleration, which may take four times that to maintain full speed.

Sugar refineries employ centrifugal driers of the basket type for the removal of sirup from the crystallized sugar after concentration in evaporators.

The requirements of large-scale, low-cost, competitive production have led to the use of the largest possible sizes, up to 60 in. in diameter, with holding capacity of about 15 cu. ft. In addition, the operation must be as nearly automatic as possible and involve a minimum of labor, attention, and cost. For these reasons, mechanical aids of all kinds find a great field of usefulness, such as time switches for starting and accelerating, running and stopping the driving motors, means for filling the basket, mechanical unloaders, dumping mechanisms, etc.

Breweries grow yeast cells in mash tubs or copper vats of large size, such as 12,000 gal. (50,000 l.) with a yeast-cell content which may, with certain grain yeasts, be 3 per cent by volume but with other mashes may reach 9 to 10 per cent. Whether grown for alcohol or beer, or for the yeast itself, it is desirable to terminate the growth definitely and quickly and to separate completely beer and yeast instead of using the relatively slow, gravity settling methods formerly employed. This is now accomplished by centrifugal concentrator bowls, employing nozzles connected to passages leading from the bowl-shell wall so as to pick up and sweep out along with a small proportion of the liquid, all the yeast cells, which, being heavier than the wort, are thrown to the outermost parts of the bowl, while the wort travels toward the center between cone-shaped separating disks and is discharged through a top connection near the center.

Such machines are built in sizes up to about 15 in. diameter (38 cm.), operated at speeds of 5000 r.p.m., with through-puts of 85 gal. (320.1)/min. on 3 per cent yeast and 60 gal. (230 l.)/min. on 9 per cent yeast, and requiring about 5 hp. for drive. The concentration possible depends inversely upon the yeast richness, because a concentrate of about 60 per cent by volume of yeast cells will not flow readily enough to be easily discharged through the nozzles. On this basis, a practical limit would be a concentration to 5 per cent of the original volume for a 3 per cent yeast and to about 15 per cent of a 9 per cent yeast.

After the wort in which the yeast grew has been separated, it is customary to dilute the yeast concentrate to the original volume with cold clean water and then to reconcentrate by running through a centrifugal separator a second time, before passing the concentrate to filters for final recovery.

Rubber latex as taken from the trees consists of about 30 to 40 per cent by weight of rubber particles dispersed in a milklike sap. The rubber particles have a lower specific gravity than the serum, and although they will coagulate on standing and can be made to coagulate very readily by acidification, their concentration is far less readily effected than is the case with cream from cow's milk, in spite of the similarity of dimension of the dispersed particles (about 3μ), and their closely similar specific-gravity differentials.

The upper limit of concentration of about 73 per cent at which the concentrate of rubber latex practically ceases to flow is appreciably lower than that for cream, which is about 80 per cent.

Concentration to about 60 per cent is desirable for the two reasons of saving in shipping weight and the better quality of product when made from centrifugally concentrated latex.

In spite of the presumable ease of concentration, practical results from centrifugal concentration have only recently been obtained. This has been done by specially designing liquid-liquid concentrator bowls with quite high separating coefficients through the use of extra high speeds and with appropriate internal-passage constructions.

Latex concentrators are expected to produce a concentrate with at least 60 per cent dry rubber-content solids, equivalent to 61 to 62 per cent total solids, and not over 12 to 14 per cent solids in the serum.

It is worth noting that centrifugal concentration also removes considerable proportions of water-soluble proteins. This may be accentuated by remixing with water and recentrifuging, one or more times. Rubber prepared by centrifuging has improved mechanical and electrical properties; recentrifuging further improves the electrical properties markedly.

Chemical plants make use of basket centrifuges for the recovery of precipitates or crystals from mother or wash liquors in a great many processes such as caustic slimes after evaporation, ammonium sulfate, dinitrobenzol, anthraquinone, nitrotoluene, etc. Wherever a relatively liquid-free cake is desirable, or a spray wash is needed, the basket centrifuge offers advantages over the filter press or rotary filter, often resulting in half the liquid-residue percentage otherwise attainable. It is important to keep in mind that the smaller diameter baskets permit higher centrifugal separating effects, which may offset their smaller load-holding space and consequently greater labor and operating cost per unit of cake.

A typical operating schedule employing a 40-in.-diameter perforated basket, holding about 7 cu. ft. or 500 to 600 lb. and taking 10 hp., would be as follows:

Loading	2–3 min.
Accelerating to full speed, 900 r.p.m.	3–4 min.
Running at full speed	1 min.
Slowing down (with brakes)	2 min.
Unloading (bottom dumping)	2–4 min.

The number of loads run through must allow for some "lost time," division of attendance with other duties, etc., so that from three to five loads may be run per hour.

Machine-shop chip-handling and cutting-oil reclamation requires the use of both a basket centrifuge for the removal of the cutting oil from the chips and one or two liquid-liquid, continuous-separating bowls with ample sediment-holding capacity for purification of the oil.

The chips may need to be broken into small pieces first, to facilitate handling and to reduce their volume, for which one of the several types of swing-hammer or ring-roll crushers is suitable. In this way, wiry, spiral chips may be broken so as to get 50 to 60 lb./cu. ft., instead of about 10.

When loaded into a typical 40-in. basket centrifuge taking 350 lb. per load, run by a 10-hp. motor, the centrifuging schedule may be as follows:

Loading	5.5 min.
Accelerating	1.5
Running at full speed	3.5
Slowing down (with brake)	1.0
Unloading	5.0
	16.5 min.

With such a schedule from 2½ to 3½ loads may be run per hour, and with a spare basket this may be increased to 4 to 6 loads per hour.

Metal chips hold from 20 to 25 per cent of their weight of oil and after centrifuging only 1 to 2 per cent, thus effecting a 90 to 96 per cent recovery of oil (60 gal. oil/ton chips).

This oil is dirty, septic, and full of fine metallic particles. In order to sterilize as well as to wash out dirt and chips, water may be added and the oil and water heated to boiling and then allowed to settle overnight to separate the water and sediment, which may be drawn off from the bottom of a cone-bottomed settling tank. The oil may be drawn off from the top and after heating may be passed through a liquid-liquid, continuous-separating type of centrifugal bowl provided with a large sludge-holding space to retain the fine solids. A plant handling 5 tons of chips per day and recovering 300 gal. oil would have a storage and heating tank for the dirty oil with a volume of about 450 gal. for the oil and water and a small accessory tank with heating coils and thermostatic control to maintain a uniform temperature of 70°C. (158°F.) on the oil as it passes into a centrifugal separator, requiring 2 hp. for drive, which will separate the water and sediment at a through-put rate of 3 to 4 gal./min., leaving not more than 0.05 per cent water and 0.5 per cent sediment in the purified and sterilized oil.

Machine-shop grinding requires the use of a coolant and lubricant on the grinding wheel and the piece being ground. It is not possible to obtain the best finish without a very clean coolant. The coolants used may be special oils, or special oils or soap compounds dispersed in water. Centrifugal clarifiers have recently been applied to these coolants, resulting in much smoother surfaces, in absence of scratch lines that are difficult to remove in later polishing operations, and in material reductions in grinder-wheel wear.

Activated sewage sludge may be concentrated by valve-bowl clarifiers which discharge a concentrated sludge containing 6 to 10 per cent of solids by dry weight and a clarified water containing not more than 0.05 to 0.10 per cent, suitable for return to the secondary tanks. The concentrated sludge is passed into the digester tank. The increased concentration would allow the use of smaller and less expensive digesters and seems also to yield a greater amount of gas.

Miscellaneous applications include a large number of varied kinds of cases, requiring one or another of the types of separations already described, of which a few may be listed as examples, such as the recovery of seed oysters in oyster hatcheries, the recovery of blood serum free from red corpuscles in slaughterhouses, the clarification of viscose in the rayon industry, the clarification of medicinal preparations, the purification of essential oils, the recovery of caffeine from coffee extract, the recovery of oil and the purification of wash water in washing oily rags for reuse in machine shops, the recovery and separation of oil from steam-engine condensate, the clarification of fruit juices, etc. Wherever a liquid-liquid separation or a liquid-solid clarification, or some combination of these, needs to be carried out to a high degree of perfection, centrifuging should be given consideration.

DUST AND MIST COLLECTION

BY C. E. LAPPLE

REFERENCES: *General.* Anderson, *Trans. Am. Inst. Chem. Engrs.*, **16**, Part I, 69 (1924). DallaValle, "Micromeritics," Pitman, New York, 1948. Drinker and Hatch, "Industrial Dust," McGraw-Hill, New York, 1936. Gibbs, *J. Soc. Chem. Ind.*, **41**, 189T (1922); "Clouds and Smokes," Blakiston, Philadelphia, 1924. Knowles, *Trans. Am. Soc. Heating Ventilating Engrs.*, **24**, 165 (1918). Lapple, *Heating, Piping Air Conditioning*, **16**, 410, 464, 578, 635 (1944); **17**, 611 (1945); **18**, 108 (1946). Larcombe, *Mining Mag.*, **66**, 143, 206, 256 (1942); *Ind. Chemist*, **18**, 433, 477 (1942); **19**, 25 (1943). Meldau, "Der Industriestaub," Ver. deut. Ing. Verlag, Berlin, 1926. Miller, *Chem. & Met.*, **45**, 132 (1938). Powers, *Rock Products*, **45**, 58 (February, 1942); **45**, 46 (April, 1942); **45**, 48, 49, 51 (September, 1942); **46**, 66 (March, 1943); **46**, 70, 72 (June, 1943); **47**, 74, 94 (May, 1944); **47**, 50 (July, 1944); **48**, 92, 94 (June, 1945); **48**, 84, 94 (July, 1945). Roberts, *Power*, **83**, 345, 392 (1939). Anon., *Sheet Metal Worker*, **31**, 51 (January, 1940); 22 (February, 1940); 29 (May, 1940); 24, 25, 38 (June, 1940); 30 (August, 1940).

Properties of Particle Dispersoids. Brown, Hartmann, and Nagy, *U.S. Bur. Mines*, R.I. 3722, October, 1943; R.I. 3751, May, 1944; I.C. 7183, September, 1941; I.C. 7309. January, 1945. National Fire Protection Association, "National Fire Codes for the Prevention of Dust Explosions," Boston, 1944. Trostel and Frevert, *Chem. & Met.*, **30**, 141 (1924). Utah State Department of Health, Division of Industrial Hygiene, "Useful Criteria in the Identification of Certain Occupational Health Hazards," 1945.

Particle Measurements. Anderson, *Trans. Am. Inst. Chem. Engrs.*, **34**, 589 (1938). British Standards Institution, "The Method of Testing Dust Extraction Plants and the Emission of Solids from Chimneys of Electric Power Stations," London, April, 1940. Hardie, *Trans. Am. Soc. Mech. Engrs.*, **59**, 355 (1937). Heywood, *Proc. Inst. Mech. Engrs. (London)*, **140**, 257 (1938); *Engineering*, **146**, 492 (1938). Schweyer and Work, Am. Soc. Testing Materials, "Symposium on New Methods for Particle Size Determination in the Subsieve Range," pp. 1–23, Mar. 4, 1941.

Particle Dynamics. Fletcher, *Phys. Rev.*, **33** (1), 81 (1911). Lapple and Shepherd, *Ind. Eng. Chem.*, **32**, 605 (1940). Prandtl and Tietjens (translated by Den Hartog), "Applied Hydro- and Aeromechanics," Chaps. II, V, McGraw-Hill, New York, 1934. Rouse, "Fluid Mechanics for Hydraulic Engineers," Chaps. I, X, XI, McGraw-Hill, New York, 1938. Schiller, "Handbuch der Experimentalphysik," vol. 4, Part 2, p. 337, Akademische Verlagsgesellschaft, Leipzig, 1932. Schiller and Naumann, *Z. Ver. deut. Ing.*, **77**, 318 (1933). Steinour, *Ind. Eng. Chem.*, **36**, 618, 840, 901 (1944). Wadell, *Physics*, **5**, 281 (1934). Wasser, *Physik. Z.*, **34**, 257 (1933). Anon. (Bosanquet, Carey, and Stairmand), *Engineering (London)*, **150**, 441 (1940).

Impingement Separators. Albrecht, *Physik. Z.*, **32**, 48 (1931). Fairs, *Chem. Trade J.*, **115**, 571 (1944); *Ind. Chemist*, **20**, 637 (1944); *Trans. Inst. Chem. Engrs. (London)*, **22**, 110 (1944). Houghton and Radford, *Trans. Am. Inst. Chem. Engrs.*, **35**, 427 (1939). Langmuir and Blodgett, *U.S. Army Air Forces Technical Report*, No. 5418, Feb. 19, 1946 (U.S. Dept. of Commerce, Office of Technical Services PB 27565). Sell, *Forsch. Gebiete Ingenieurw.*, **2**, Forschungsheft 347, August, 1931. Witzmann, *Kolloid Z.*, **95**, 102 (1941); *Z. Electrochem.*, **46**, 313 (1940).

Cyclones. Alden, "Design of Industrial Exhaust Systems," Chap. VI, Industrial Press, New York, 1940. Anderson, *Chem. & Met.*, **40**, 525 (1933). Drijvor, *Wärme*, **60**, 333 (1937). Feifel, *Forsch. Gebiete Ingenieurw.*, **9**, 68, 306 (1938); **10**, 212 (1939); *Arch. Wärmewirt.* **20**, 15 (1939). Miller and Lissman, "Calculation of Cyclone Pressure Drop," paper presented at December, 1940, meeting of Am. Soc. Mech. Engrs., New York (not published). Parent, *Trans. Am. Inst. Chem. Engrs.*, **42**, 989 (1946). Pollak and Work, *Trans. Am. Soc. Mech. Engrs.*, **64**, 31 (1942). Rosin, Rammler, and Intelmann, *Z. Ver. deut. Ing.*, **76**, 433 (1932). Shepherd and Lapple, *Ind. Eng. Chem.*, **31**, 972 (1939); **32**, 1246 (1940). Van Tongeran, *Mech. Eng.*, **57**, 753 (1935). Wellmann, *Feuerungstech.*, **26**, 137 (1938). Whiton, *Trans. Am. Soc. Mech. Engrs.*, **63**, 213 (1941); *Mech. Eng.*, **65**, 885 (1943); *Power Plant Eng.*, **47**, 92 (December, 1943); *Power*, **88**, 20 (1944); **75**, 344 (1932); *Chem. & Met.*, **39**, 150 (1932).

Packed-bed Collectors. Wells and Fogg, *U.S. Bur. Mines Bull.* 184, p. 159, 1920. Anon. (Lynch), *Fuel Economist*, **12**, 47 (October, 1936).

Cloth Collectors. Capwell, *Gas*, **15**, 31 (August, 1939). Carman, *Trans. Inst. Chem. Engrs. (London)*, **15**, 150 (1937). Carman and Arnell, *Can. J. Research*, **26**(A), 128 (1948). Hallows and O'Hara, Metallurgy of Lead and Zinc, *Trans. Am. Inst. Mining Met. Engrs.*, **121**, 299 (1936). Kling, *Blast Furnace and Steel Plant*, **34**, 1257 (1946). Manegold, *Kolloid Z.*, **81**, 269 (1937). Mumford, Markson, and Ravese, *Trans. Am. Soc. Mech. Engrs.*, **62**, 271 (1940). Weischaus, *Chem. Eng.*, **54**, 113 (August, 1947). Williams, Hatch, and Greenburg, *Heating, Piping Air Conditioning*, **12**, 259 (1940).

Scrubbers. Bransky and Diwoky, *Refiner Natural Gasoline Mfr.*, **19**, 191 (1940). Collins, Seaborne, and Anthony, *Paper Trade J.*, **124**, 45 (June 5, 1947), **125**, 45 (Jan. 15, 1948). Harmon, *J. Inst. Fuel*, **12**, 514 (1938). Junge, *Iron Age*, **78**, 542, 602 (1906). Kinney, *Blast Furnace Steel Plant*, **31**, 113 (January, 1943). Kleinschmidt, *Chem. & Met. Eng.*, **46**, 487 (1939). Kleinschmidt and Anthony, *Trans. Am. Soc. Mech. Engrs.*, **63**, 349 (1941). Pearson, Nonhebel, and Ulander, *J. Inst. Fuel*, **8**, 119 (1935); *Combustion*, **6**, 10 (April, 1935). Piazza, *Anales soc. cient. argentina*, **137**, 49 (February, 1944). Anon., *Engineering*, **77**, 831 (1929); **82**, 12 (1931).

Electrical Precipitators. Anderson, *Physics*, **3**, 23 (July, 1932). Beaver, *Trans. Am. Inst. Chem. Engrs.*, **42**, 251 (1946). Bradley, *Trans. Am. Inst. Elec. Engrs.*, **34**, Part I, 421 (1915). Cree, *Am. Gas. J.*, **162**, 27 (March, 1945). Deutsch, *Ann. Physik*, **68** (4), 335 (1922); **9** (5), 249 (1931); **10** (5), 847 (1931). Drinker, Thomson, and Fitchet, *J. Ind. Hyg.*, **5**, 162 (September, 1923). Ladenburg, *Ann. Physik*, **4** (5), 863 (1930). Ladenburg and Tietze, *Ann. Physik*, **6** (5), 581 (1930). Landolt, *Chem. & Met.*, **25**, 428 (1921); **29**, 588 (1923). Loeb, "International Critical Tables," vol. 6, p. 107, McGraw-Hill, New York, 1929. Loeb, "Fundamental Processes of Electrical Discharge in Gases," Wiley, New York, 1939. Mierdel, *Z. tech. Physik*, **13**, 564 (1932). Mierdel and Seeliger, *Trans. Faraday Soc.*, **32**, 1284 (1936). Nesbit, *Trans. Am. Inst. Elec. Engrs.*, **34**, Part I, 405 (1915). Peek, "Dielectric Phenomena in High-voltage Engineering," McGraw-Hill, New York, 1929. Penney, *Elec. Engineering*, **56**, 159 (1937). Rueder, *Z. Ver. deut. Ing.*, Verfahrungstechnik No. 3, 83 (1942). Schmidt, *Trans. Am. Inst. Chem. Engrs.*, **21**, 11 (1928). Schmidt and Anderson, *Elec. Engineering*, **57**, 332 (1938). Thornton, *Phil. Mag.*, **28** (7), 666 (1939). Welch, Metallurgy of Lead and Zinc, *Trans. Am. Inst. Mining Met. Engrs.*, **121**, 304 (1936). Whitehead, "Dielectric Phenomena—Electrical Discharge in Gases," Van Nostrand, New York, 1927.

Air Filters. American Railroads Association, Mechanical Division, Report on Relative Performance of Air Filters, Jan. 15, 1938, Chicago. Am. Soc. Heating Ventilating Engrs., "Heating, Ventilating, Air Conditioning Guide," Chap. 33, p. 595, 1947. Carrier, Cherne, and Grant, "Modern Air Conditioning, Heating, and Ventilating," Chap. XI, Pitman, New York, 1940. Chase, *Paper Trade J.*, **111** (21), 29 (Nov. 21, 1940). DallaValle, *Heating & Ventilating*, **41**, 54 (April, 1944); **41**, 58 (May, 1944). Frank, *Heating, Piping Air Conditioning*, **1**, 380 (1929); **3**, 378 (1931). Munkelt, *Refrig. Eng.*, **47**, 21, 32, 40 (January, 1944). Rowley and Jordan, *Heating, Piping Air Conditioning*, **7**, 293 (1935); **10**, 539 (1938); **11**, 388, 633 (1939); **12**, 61, 699 (1940); **13**, 99, 246, 304, 524 (1941); **14**, 19, 438 (1942); **15**, 487 (1943).

Nomenclature

Except where special units are specifically indicated in the text, any consistent system of units may be employed. The c.g.s. and English units are given below merely by way of example because they represent the most common of such systems used. Where special units are called for in the text, only those units specified below, in the column "Special units," may be employed. In equations employing electrical quantities, only the c.g.s. system may be used, since there are no comparable electrical units in the English system.

Symbol	Definition	System of consistent units: Metric (c.g.s.)	System of consistent units: English	Special units
A_c	Cyclone inlet area = B_cH_c for cyclone with rectangular inlet	sq. cm.	sq. ft.	
A_e	Area of collecting electrode (side on which particles collect only)	sq. cm.	sq. ft.	
A_p	Area of particle projected on plane normal to direction of flow or motion = $(\pi D_p^2/4)$ for spherical particles	sq. cm.	sq. ft.	
B_c	Width of rectangular cyclone inlet duct	cm.	ft.	
B_e	Spacing between wire and plate or rod and curtain, or between parallel plates in electrical precipitators	cm.	ft.	
B_s	Width of gravity settling chamber	cm.	ft.	
c_d	Dust concentration in inlet or approach gas stream			grains/cu. ft
C	Over-all drag coefficient = $F_d/(\rho u^2/2)(A_p)$ = $4g_LD_p(\rho_s - \rho)/3\rho u^2$ for spherical particles	dimensionless	dimensionless	
D_b	Representative dimension or diameter of body impinged upon	cm.	ft.	
D_c	Cyclone diameter	cm.	ft.	
D_d	Outside diameter of wire or discharge electrode of concentric-cylinder type electrical precipitator	cm.	ft.	
D_e	Cyclone gas exit duct diameter	cm.	ft.	
D_L	Diameter of drops	cm.	ft.	
D_p	Diameter of particle	cm.	ft.	
$D_{p\mu}$	Diameter of particle			microns
D_{pc}	Cut size, diameter of particles of which 50% of those present are collected	cm.	ft.	
$D_{pc\mu}$	Cut size, diameter of particles of which 50% of those present are collected			microns
$D_{p,\min}$	Minimum diameter of particle which is completely collected	cm.		
D_{ps}	Equivalent Stokes's law diameter of particle having a terminal settling velocity of u_t	cm.	ft.	
$D_{ps\mu}$	Equivalent Stokes's law diameter of particle having a terminal settling velocity of u_t			microns
D_t	Inside diameter of collecting tube of concentric-cylinder type electrical precipitator	cm.	ft.	
e	Natural or Naperian logarithmic base	2.718 . . .	2.718 . . .	
E	Electrostatic potential difference	statvolts		
E_c	Electrostatic potential difference required for corona discharge to commence	statvolts		
E_s	Electrostatic potential difference required for sparking to commence	statvolts		
F_{cv}	Cyclone friction loss, expressed as number of cyclone inlet velocity heads, based on area A_c	dimensionless	dimensionless	dimensionless
F_d	Drag or resistance to motion of a body in a fluid	dynes	poundals	
g_c	Conversion factor	$980.6\left(\frac{\text{g. mass}}{\text{g. force}}\right)\left(\frac{\text{cm.}}{\text{sec.}^2}\right)$	$32.17\left(\frac{\text{lb. mass}}{\text{lb. force}}\right)\left(\frac{\text{ft.}}{\text{sec.}^2}\right)$	
g_L	Local acceleration due to gravity	(cm./sec.)/sec.	(ft./sec.)/sec.	
h_{vi}	Cyclone inlet velocity head			in. water
H_c	Height of rectangular cyclone inlet duct	cm.	ft.	
H_s	Height of gravity settling chamber	cm.	ft.	
I	Electrical current per unit of electrode length	statamp./cm.		
k_ρ	Gas density relative to its density at 0°C., 760 mm.	dimensionless	dimensionless	dimensionless
k_{sg}	True particle specific gravity referred to water at 4°C.			dimensionless
K	Empirical proportionality constant, for cyclone pressure drop or friction loss	dimensionless	dimensionless	dimensionless
K_c	Proportionality constant, for cloth pressure drop			$\left[\frac{\text{(in. water)}}{\text{(centipoise)(ft./min.)}}\right]$
K_d	Proportionality constant, for pressure drop through collected dust			$\left[\frac{\text{(in. water)}}{\text{(centipoise)(grain/sq. ft.)(ft./min.)}}\right]$
K_e	Electrical precipitator constant	(sec./cm.)	(sec./ft.)	
K_o	"Energy-distance" constant for electrical discharge in gases	cm.		
K_m	Stokes-Cunningham correction factor	dimensionless	dimensionless	dimensionless
K_{me}	Proportionality factor in Stokes-Cunningham correction factor	dimensionless	dimensionless	dimensionless
K_s	Proportionality factor in "slip flow" correction factor	dimensionless	dimensionless	dimensionless
L_e	Length of collecting electrode in direction of gas flow	cm.	ft.	
L_s	Length of gravity settling chamber in direction of gas flow	cm.	ft.	
ln	Logarithm to the base e; natural logarithm	dimensionless	dimensionless	dimensionless
m_c	Dust capacity of air filter cell			lb. mass
m_p	Mass of particle	g.	lb. mass	
m_s	Mass of solids per unit of cloth area			grains/sq. ft.
M	Molecular weight	g./mole	lb./mole	
n	Exponent	dimensionless	dimensionless	dimensionless
N	Number of gas molecules in a mole	6.06×10^{23} molecules/g. mole	2.76×10^{26} molecules/lb. mole	
N_e	"Effective" number of turns made by gas stream in a cyclone separator	dimensionless	dimensionless	dimensionless

Nomenclature—(*Concluded*)

Symbol	Definition	System of consistent units: Metric (c.g.s.)	System of consistent units: English	Special units
N_e	Number of elementary electrical charges acquired by a particle	dimensionless		
N_{Re}	Reynolds number = $(D_p\rho u/\mu)$ or $(D_p\rho u_t/\mu)$	dimensionless	dimensionless	
N_s	Number of shelves parallel to gas flow in gravity settling chamber	dimensionless	dimensionless	dimensionless
N_t	Number of turns made by gas stream in a cyclone separator	dimensionless	dimensionless	dimensionless
Δp_i	Pressure drop			in. water
Δp_{cv}	Pressure drop, expressed as number of cyclone inlet velocity heads, based on area A_c	dimensionless	dimensionless	dimensionless
q	Gas-flow rate	cc./sec.	cu. ft./sec.	
q_m	Gas-flow rate per air filter cell			cu. ft./min.
r	Radius; distance from center line of cyclone separator; distance from center line of concentric cylinder electrical precipitator	cm.	ft.	
R	Gas constant	$84{,}800\left[\frac{(\text{cm.-g. force})}{(\text{g.-mole})(°\text{C.})}\right]$	$1546\left[\frac{(\text{ft.-lb. force})}{(\text{lb.-mole})(°\text{F.})}\right]$	
t	Time	sec.	sec.	
t_d	Length of operating cycle			days
t_m	Time			min.
T	Absolute gas temperature	°K. or °C. abs.	°R. or °F. abs.	
u	Relative velocity between particle and main body of fluid	cm./sec.	ft./sec.	
u_e	Velocity of migration of particle toward collecting electrode	cm./sec.	ft./sec.	
u_t	Terminal settling velocity of particle under action of gravity	cm./sec.	ft./sec.	ft./sec.
u_{ts}	Terminal settling velocity of particle as calculated from Stokes's law	cm./sec.	ft./sec.	
V_c	Cyclone inlet velocity, average, based on area A_c	cm./sec.	ft./sec.	ft./sec.
V_e	Average velocity of gas flowing through electrical precipitator	cm./sec.	ft./sec.	
V_f	Superficial velocity of gas through cloth			ft./min.
V_o	Average velocity of dust-laden gas	cm./sec.	ft./sec.	
V_s	Average gas velocity in gravity settling chamber	cm./sec.	ft./sec.	
V_{ct}	Tangential gas-velocity component in a cyclone	cm./sec.	ft./sec.	
x	Ratio of volumetric liquid circulation rate to volumetric gas-flow rate	(cc./sec.)/(cc./sec.)	(cu. ft./sec.)/(cu. ft./sec.)	
y	Distance, normal to gas stream, through which liquid is sprayed	cm.	ft.	
δ	Dielectric constant	dimensionless		
δ_o	Dielectric constant at 0°C., 760 mm.	dimensionless		
ϵ	Elementary electrical charge	4.80×10^{-10} electrostatic units		
ϵ_v	Fraction voids in bed of solids	dimensionless	dimensionless	dimensionless
ζ	$=\left[1+2\frac{(\delta-1)}{(\delta+2)}\right]$; ranges from a value of 1 for materials with a dielectric constant of 1 to 3 for conductors	dimensionless		
η	Collection efficiency, weight fraction of entering dispersoid collected	dimensionless	dimensionless	dimensionless
η_t	Target efficiency, fraction of dispersoid in swept volume collected on target	dimensionless	dimensionless	dimensionless
λ_i	Ionic mobility	(cm./sec.)/(statvolt/cm.)		
λ_m	Mean free path of gas molecules	cm.	ft.	
λ_p	Particle mobility = $u_e/\mathcal{E}$	(cm./sec.)/(statvolt/cm.)		
μ	Fluid viscosity	poise	(lb. mass)/(ft.)(sec.)	
μ_c	Fluid viscosity			centipoise
ρ	Fluid density	g./cc.	lb. mass/cu. ft.	lb. mass/cu. ft.
ρ_s	True (*not* bulk) density of solids; also density of liquid drops	g./cc.	lb. mass/cu. ft.	lb. mass/cu. ft.
σ	Ion density	number/cc.		
σ_{avg}	Average ion density	number/cc.		
ϕ	Cumulative weight fraction larger than size	dimensionless	dimensionless	dimensionless
ϕ_s	Particle-shape factor (number less than 1.0)	dimensionless	dimensionless	dimensionless
$\mathcal{E}$	Electrostatic potential gradient	statvolts/cm.		
$\mathcal{E}_c$	Electrostatic potential gradient required for corona discharge to commence	statvolts/cm.		
$\mathcal{E}_i$	Average electrostatic potential gradient in ionization stage	statvolts/cm.		
$\mathcal{E}_o$	Electrical breakdown constant for gas	statvolts/cm.		
$\mathcal{E}_p$	Average electrostatic potential gradient in collection stage	statvolts/cm.		
$\mathcal{E}_s$	Electrostatic potential gradient required for sparking to commence	statvolts/cm.		

Conversion factors:

Multiply statvolts (or e.s.u.) by 300 to obtain volts.
Multiply statamperes (or e.s.u.) by 3.34×10^{-10} to obtain amperes.
Multiply centimeters by 10,000 to obtain microns.
Multiply inches by 25,400 to obtain microns.
Multiply pounds by 7000 to obtain grains.
Multiply centiposes by 0.000672 to obtain lb./(ft.)(sec.).

Silverman, *Heating & Ventilating*, **42**, 85 (December, 1945); **43**, 74 (January, 1946); **43**, 96 (February, 1946). Stacey, *J. Am. Soc. Heating Ventilating Engrs.*, **27**, 355 (1921). Wechsberg, *Heating, Piping Air Conditioning*, **5**, 217 (1933). Ziel and Sleik, *Heating, Piping Air Conditioning*, **15**, 367 (1943).

Miscellaneous Collectors. Blacktin, *J. Soc. Chem. Ind.*, **58**, 334 (1939). Gottschalk and St. Clair, *Mining and Met.*, **18**, 244 (1937). Howard, Fume Arrester, U.S. Patent 896,111 (1908). Jones, *J. Acoust. Soc. Am.*, **18**, 371 (1946). Porter, *Chem. Eng.*, **55**, 100, 101, 115 (March, 1948). St. Clair, U.S. Bur. Mines, R. I. 3400 (May, 1938). St. Clair, Spendlove, and Potter, U.S. Bur. Mines, R. I. 4218 (March, 1948). Watson, *Trans. Faraday Soc.*, **32**, 1073 (1936). Anon, *Power Plant Eng.*, **48**, 90 (December, 1944).

PURPOSE OF DUST AND MIST COLLECTION

Dust and mist collection is concerned with the removal or collection of solid or liquid dispersoids in gases for purposes of:

1. Nuisance elimination—as in cleaning of ventilation air or fly ash removal from power-plant combustion gases.
2. Equipment-maintenance reduction—as in filtration of engine-intake air or pyrites furnace-gas treatment prior to its entry to a chamber sulfuric acid system.
3. Safety- or health-hazard elimination—as in collection of siliceous and metallic dusts around grinding and drilling equipment and in some metallurgical operations and flour dusts from milling or bagging operations.
4. Product-quality improvement—as in air cleaning in the production of pharmaceutical products and photographic film.
5. Recovery of a valuable product—as in collection of dusts from dryers and smelters.
6. Powdered-product collection—as in pneumatic conveying; the spray drying of milk, eggs, and soap; and the manufacture of high-purity zinc oxide and carbon black.

PROPERTIES OF PARTICLE DISPERSOIDS

As a prerequisite to the design of industrial control equipment, an understanding of the fundamental properties and characteristics of gas dispersoids is necessary.

Particle Size. The primary distinguishing feature of gas dispersoids is particle size. The most widely used unit of particle size is the micron, defined as $\frac{1}{1000}$ mm. (1/25,400 in.), which is often designated by the symbol μ. This symbol should not be confused with the same symbol μ used for viscosity or the symbol mμ (millimicron) often used to designate $\frac{1}{1000}$ micron. Particle size of a gas dispersoid is usually taken as the average or equivalent diameter of the particle in the United States, though some writers (particularly German) specify particle size by the radius. Another common method is to designate the screen mesh (see p. 963) that has an aperture corresponding to the particle diameter. This method may lead to confusion unless the screen scale involved is specified.

Particle Classification. Considerable confusion exists regarding the terminology of gas dispersoids. Gas dispersoids are generally classified according to their method of formation and particle size and may be placed in two general categories: "mechanical dispersoids" and "condensed dispersoids" [Anderson, *Trans. Am. Inst. Chem. Engrs.*, **16**, Part I, 69 (1924)].

Mechanical dispersoids are formed by comminution, decrepitation, or disintegration of larger masses of material, or by grinding of solids or spraying of liquids; and they usually involve a wide particle-size distribution. They may be classified further as "dust" or "spray," referring to solids and liquids, respectively. Condensed dispersoids are formed by condensation of the vapor phase or as the product of a vapor-phase reaction and are relatively closely sized. Solid and liquid particles formed in this manner are termed "fume" and "mist," respectively. Each of these methods results in particles of a certain size range, as indicated in Table 1. The demarcation, however, is not so sharp on a particle-size basis as is indicated in Table 1, which gives a representative interpretation of the most probable case. Some dusts may be as fine as 0.1 μ and some mists as large as 20 μ. Condensed dispersoids generally tend to flocculate or agglomerate to form particles of larger size, which may actually exceed the size of some mechanical dispersoids, as in the formation of rain from fog. "Smoke" is a term that usually applies to a fume or mist formed by combustion.

Table 1. Classification and Particle-size Range of Gas Dispersoids

Classification	Approximate Particle Diameter, μ
Mechanical dispersoid:	
Dust	Greater than 1
Spray	Greater than 10
Condensed dispersoid:	
Fume	Less than 1
Mist	Less than 10

For purposes of conciseness and simplicity, the terms "dust" and "mist" will arbitrarily be used to indicate any solid or liquid dispersoid, respectively, except where the further distinctions discussed above are important.

In Table 2 are listed reported particle sizes or size ranges of some common dusts. Natural and process dusts are generally produced in a wide particle-size range and may vary widely with the specific conditions of their formation. Hence these values should be regarded only as average "order of magnitude" quantities, and wide variations from these values are possible. Table 3 relates particle size to some common properties.

Table 2. Approximate Average Diameter of Some Common Dusts and Mists

Dust or mist	Approximate particle diameter, μ	Dust or mist	Approximate particle diameter, μ
Alkali fume	1 – 5	Pigments	0.2– 2
Ammonium chloride fume	0.1– 1	Silica dust	1 – 10
Atmospheric dust	0.5	Smelter dust	0.1–100
Atmospheric fog	2 –15	Sprayed zinc dust	15
Cement dust	40	Sulfuric acid mist	0.5– 15
Coal dust	5 –10	Talc	10
Condensed zinc dust	2	Tobacco smoke	0.2
Flour-mill dust	15	Zinc oxide fume	0.05

Table 3. Approximate Dimensions of Some Common Items

Item	Dimension, μ
Limit of visibility with naked eye:	
Absolute limit	10
Probable limit	40
Diameter of large molecules	0.005
Wave length of visible spectrum:	
Violet light	0.4
Red light	0.7
Wave length of X rays	0.01–0.001
Diameter of human hair	50–200

Explosion Hazards. Most mists and dusts that are capable of uniting with oxygen or reacting in any other way to form a gaseous product are a potential explosion hazard. Such explosions usually occur in two stages: a light primary explosion, resulting from ignition of dust in suspension; followed by a possible devastating secondary explosion, resulting from ignition of large quantities of dust raised by the primary explosion from duct-work sediment or nearby equipment or storage bins. Explosions may be set off by open flames, friction sparks, static electricity, faulty electrical equipment, heat, or spontaneous combustion. As with vapor, upper and lower explosive dust concentrations exist. Concentrations below about 5 gr./cu. ft. are usually considered safe [DallaValle, "Micromeritics," Pitman, New York, 1948. Trostel and Frevert, *Chem. & Met.*, **30**, 141 (1924)]. There are insufficient data to define any upper explosive dust concentration limits reliably. Dusts larger than 35 Tyler mesh are usually not considered explosive unless they are chemically unstable, such as metallic acetylides. Oxygen concentrations below 8 per cent will usually not support combustion. Further discussions of the explosive characteristics of dusts and explosion prevention are presented by Lapple [*Heating, Piping and Air Conditioning*, **16**, 410 (1944)]; Brown, Hartmann, and Nagy (*U.S. Bur. Mines*, R.I. 3722, October, 1943. R.I. 3751, May, 1944; I.C. 7183, September, 1941; I.C. 7309, January, 1945); National Fire Protection Association ("National Fire Codes for the Prevention of Dust Explosions," Boston, 1944); and Trostel and Fervert (*op. cit.*).

Health Hazards. Disability from exposure to dust can be classified into two categories: (1) the pneumoconoises, such as silicosis and asbestosis, which are caused only by dust inhalation; and (2) toxic effects, which are caused by breathing, swallowing, or absorption through the skin or tissues, such as general heavy-metal poisoning, metal-fume fever, and allergic and irritant reactions. Further information will be found in Drinker and Hatch ("Industrial Dust," McGraw-Hill, New York, 1936); Lapple (*op. cit.*); and Utah State Department of Health, Division of Industrial Hygiene ("Useful Criteria in the Identification of Certain Occupational Health Hazards," 1945).

Miscellaneous Properties. For many processes in which dusts are handled, the bulk density, porosity, and pouring properties are of importance, as in packaging powders, unloading storage bins, transporting dusts, and in catalyst or filter beds. Gas dispersoids in the subsieve range also exhibit specific flocculating tendencies, may develop high electrical charges, and have special optical properties. These are discussed by DallaValle (*op. cit.*).

The settling velocity of dusts and mists constitutes one of the most important properties from the viewpoint of design of dust- and mist-collection equipment and will be considered in detail in a subsequent paragraph.

PARTICLE MEASUREMENTS

In order to determine the degree of or justification for dust- and mist-control measures, it is often necessary to evaluate the concentration and characteristics of the dust or mist dispersoid.

Atmospheric-pollution Measurements. The dustfall measurement is one of the common methods for obtaining a relative long-period evaluation of atmospheric pollution. Stack-smoke densities are often graded visually by means of the Ringelmann chart. Equipment for local atmospheric-dust-concentration measurements fall into four general types: (1) the impinger, (2) the hot-wire or thermal precipitator (3) the a.c. precipitator, and (4) the thimble or filter. Further descriptions and information on these measurements will be found in Drinker and Hatch (*op. cit.*) and Lapple [*Heating, Piping and Air Conditioning*, **16**, 464 (1944)].

Process-gas Sampling. In sampling process gases either to determine dust concentration or to obtain a representative dust sample, it is necessary to take special precautions to avoid inertial segregation of the particles. To prevent such classification a traverse of the duct must be made, and at each point the sampling nozzle must face directly into the gas stream with the velocity in the mouth of the nozzle equal to the local gas velocity at that point. If the sampling velocity is too high, the dust sample will contain a lower concentration of dust than the main stream with a greater percentage of fine particles; if the sampling velocity is too low, the dust sample will contain a higher concentration of dust with a greater percentage of coarse particles [Lapple, *Heating, Piping and Air Conditioning*, **16**, 578 (1944). Anderson, *Trans. Am. Inst. Chem. Engrs.*, **34**, 589 (1938). Hardie, *Trans. Am. Soc. Mech. Engrs.*, **59**, 355 (1937)]. Sampling nozzles have been constructed in which velocity balancing is automatically achieved (Lapple, *op. cit.* British Standards Institution, "The Method of Testing Dust Extraction Plants and the Emission of Solids from Chimneys of Electric Power Stations," London, April, 1940. Hardie, *op. cit.*).

Particle-size Analysis. Dust control is generally concerned with the particle-size range of 0.1 to 100 μ. Since most dusts cover a wide particle-size range, it is necessary to specify size in the form of a size distribution. The available methods of particle-size analysis are discussed on pp. 1111–1113 and in Lapple, *Heating, Piping and Air Conditioning*, **16**, 580, 635 (1944); **17**, 611 (1945); Heywood [*Proc. Inst. Mech. Engrs.* (*London*), **140**, 257 (1938); *Engineering*, **146**, 492 (1938)]; and Schweyer and Work (Am. Soc. Testing Materials, "Symposium on New Methods for Particle Size Determination in the Subsieve Range," Mar. 4, 1941). They may be classified as sieving, microscopic, elutriation, sedimentation, and miscellaneous. The miscellaneous methods, including turbidimetric, permeability, gas adsorption and X-ray diffraction measurements, generally measure particle surface. With the exception of cloth-collector design, which is concerned with the air permeability of dust, most dust-control work is concerned with a weight-size distribution defined in terms of the particle-settling-velocity characteristics. For this reason elutriation and sedimentation methods are preferred in this field. In general, it is recommended that, for dust-control work, a qualitative microscopic examination or a screen analysis be made first. Frequently these offer a sufficient basis for selecting proper collection equipment. Where a detailed distribution is required, it is recommended that the pipette or hydrometer-sedimentation method be used when over 30 per cent of the dust is finer than 10 μ; for a coarser dust, an air-elutriation (*e.g.*, a Roller analysis) method is preferred.

Particle-size Representation. Particle-size distribution may be presented on either a frequency or a cumulative basis. The various available methods are discussed and compared in detail in Lapple, *Heating, Piping and Air Conditioning*, **18**, 108 (1946). For dust-control work, particle-size distributions are most conveniently presented as a logarithmic probability cumulative plot. This involves plotting particle size against the cumulative weight per cent of material larger or smaller than that size, as determined by a particle-size analysis, on logarithmic-probability graph paper (Codex Book Co., Catalogue Item 3228).

PARTICLE DYNAMICS

The fundamentals of particle motion are generally expressed in a drag coefficient C vs. Reynolds number N_{Re} plot. Such curves have been determined fairly completely for spheres, disks, cylinders, and miscellaneous shapes [Lapple and Shepherd, *Ind. Eng. Chem.*, **32**, 605 (1940). Prandtl and Tietjens (translated by Den Hartog), "Applied Hydro- and Aeromechanics," Chaps. II, V, McGraw-Hill, New York, 1934. Rouse, "Fluid Mechanics for Hydraulic Engineers," Chaps. I, X, XI, McGraw-Hill, New York, 1938. Schiller, "Handbuch der Experimentalphysik, vol. 4, Part 2, pp. 337, Akademische Verlagsgesellschaft, Leipzig, 1932. Wadell, *Physics*, **5**, 281 (1934)]. Figure 112 presents average curves for spheres, disks, and cylinders. In defining the drag coefficient and Reynolds number, the velocity term u is the relative velocity between the particle and the bulk of the fluid. Except for extraneous effects, such as turbulence, it makes no difference whether the fluid

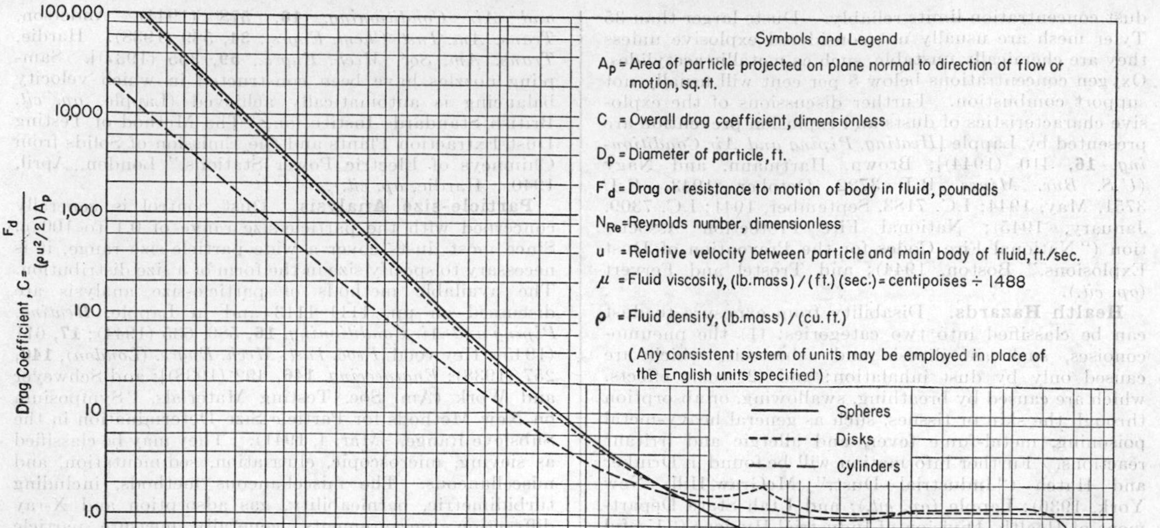

Fig. 112. Drag coefficient for spheres, disks, and cylinders.

Table 4. Drag Coefficient and Related Functions for Spherical Particles*

N_{Re}	C	CN_{Re}	CN^2_{Re}	C/N_{Re}	(u_t/u_{ts}) or $(24/CN_{Re})$	(D_{ps}/D_p) or $(\sqrt{24/CN_{Re}})$	$(D_p u_{ts}\rho/\mu)$ or $(CN^2_{Re}/24)$	$(D_{ps}u_{t}\rho/\mu)$ or $(\sqrt{24N_{Re}/C})$	Special units† $(D_{p_\mu} u_{ts}\rho/\mu_c)$	$(D_{ps_\mu} u_{t}\rho/\mu_c)$
0.1	240	24.0	2.4	2,400	1.000	1.000	0.10	0.10	20.5	20.5
.2	120	24.0	4.8	600	1.000	1.000	.20	.20	41.0	41.0
.3	80	24.0	7.2	267	1.000	1.000	.30	.30	61.5	61.5
.5	49.5	24.8	12.4	99.0	0.968	0.984	.52	.49	106.0	101.0
.7	36.5	25.6	17.9	52.1	.937	.968	.75	.68	153.0	139.0
1.0	26.5	26.5	26.5	26.5	.905	.951	1.10	.95	226	195.0
2	14.6	29.2	58.4	7.3	.822	.906	2.43	1.81	499	371
3	10.4	31.2	93.7	3.47	.769	.876	3.90	2.63	800	540
5	6.9	34.5	173	1.38	.695	.834	7.20	4.17	1.48×10^3	855
7	5.3	37.1	260	0.757	.647	.804	10.83	5.63	2.22	1,153
10	4.1	41.0	410	.410	.585	.765	17.08	7.65	3.50	1,569
20	2.55	51.0	1.02×10^3	.1275	.471	.686	42.5	13.72	8.71	2,820
30	2.00	60.0	1.80	.0667	.400	.633	75.0	19.00	15.38	3,900
50	1.50	75.0	3.75	.0300	.320	.566	156.2	28.3	32.0	5,800
70	1.27	89.0	6.23	.0181	.270	.520	260	36.4	53.2	7,460
100	1.07	107	10.7	.0107	.224	.473	446	47.3	91.5	9,700
200	0.77	154	30.8	3.85×10^{-3}	.156	.395	1.28×10^3	79.0	263	16,200
300	.65	195	58.5	2.17	.123	.351	2.44	105.3	500	21,600
500	.55	275	138	1.10	.0872	.295	5.75	147.6	1.18×10^6	30,200
700	.50	350	245	0.714	.0686	.262	10.20	183.5	2.09	37,600
1,000	.46	460	460	.460	.0522	.2283	19.2	228.3	3.93	46,900
2,000	.42	840	1.68×10^6	.210	.0286	.1692	70.0	338	14.35	69,300
3,000	.40	1,200	3.60	.1333	.0200	.1414	150.0	424	30.8	87,000
5,000	.385	1,920	9.60	.0770	.0125	.1118	400	559	82.0	114,600
7,000	.390	2,730	19.1	.0557	.00879	.0937	796	656	163.0	134,600
10,000	.405	4,050	40.5	.0405	.00592	.0770	1.69×10^6	770	346	160,000
20,000	.45	9,000	180	.0225	.00267	.0516	7.50	1,032	1.54×10^9	212,000
30,000	.47	14,200	426	.0157	.00169	.0411	17.7	1,233	3.64	253,000
50,000	.49	24,500	1.23×10^9	9.80×10^{-6}	.000980	.0313	51.2	1,565	10.50	321,000
70,000	.50	35,000	2.45	7.14	.000685	.0262	102.0	1,835	20.9	376,000
100,000	.48	48,000	4.8	4.80	.000500	.0224	200.0	2,240	41.0	459,000
200,000	.42	84,000	16.8	2.10	.000286	.0169	700	3,380	143.5	693,000
300,000	.20	60,000	18.0	0.667	.000400	.0200	750	6,000	153.8	1,230,000
400,000	.084	33,600	13.4	.210	.000715	.0267	562	10,680	114.4	2,190,000
600,000	.10	60,000	36.0	.1667	.000400	.0200	1.50×10^9	12,000	308	2,460,000
1,000,000	.13	130,000	130	.1300	.000185	.0136	5.41	13,600	1.11×10^{12}	2,790,000
3,000,000	.20	600,000	1.8×10^{12}	.0667	.000040	.0063	7.50	18,900	15.38	3,880,000

* These are average values based on the combined data of Allen, Arnold, Bacon and Reid, Dorr, and Roberts, Flachsbart, Jacobs, Knodel, Krey, Liebster, Lunnon, McKeehan, Millikan, Schmidt, Schmiedel, Shakespear, Silvey, and Wieselsberger. For values at N_{Re} less than 0.1, $C = 24/N_{Re}$.

† Units are D_{p_μ}, D_{ps_μ} = particle diameter, μ; u_t, u_{ts} = settling velocity, ft./sec.; ρ = fluid density, lb./cu. ft.; μ_c = fluid viscosity, centipoises. These values are obtainable by multiplying the respective units given in the previous two columns by 205.0.

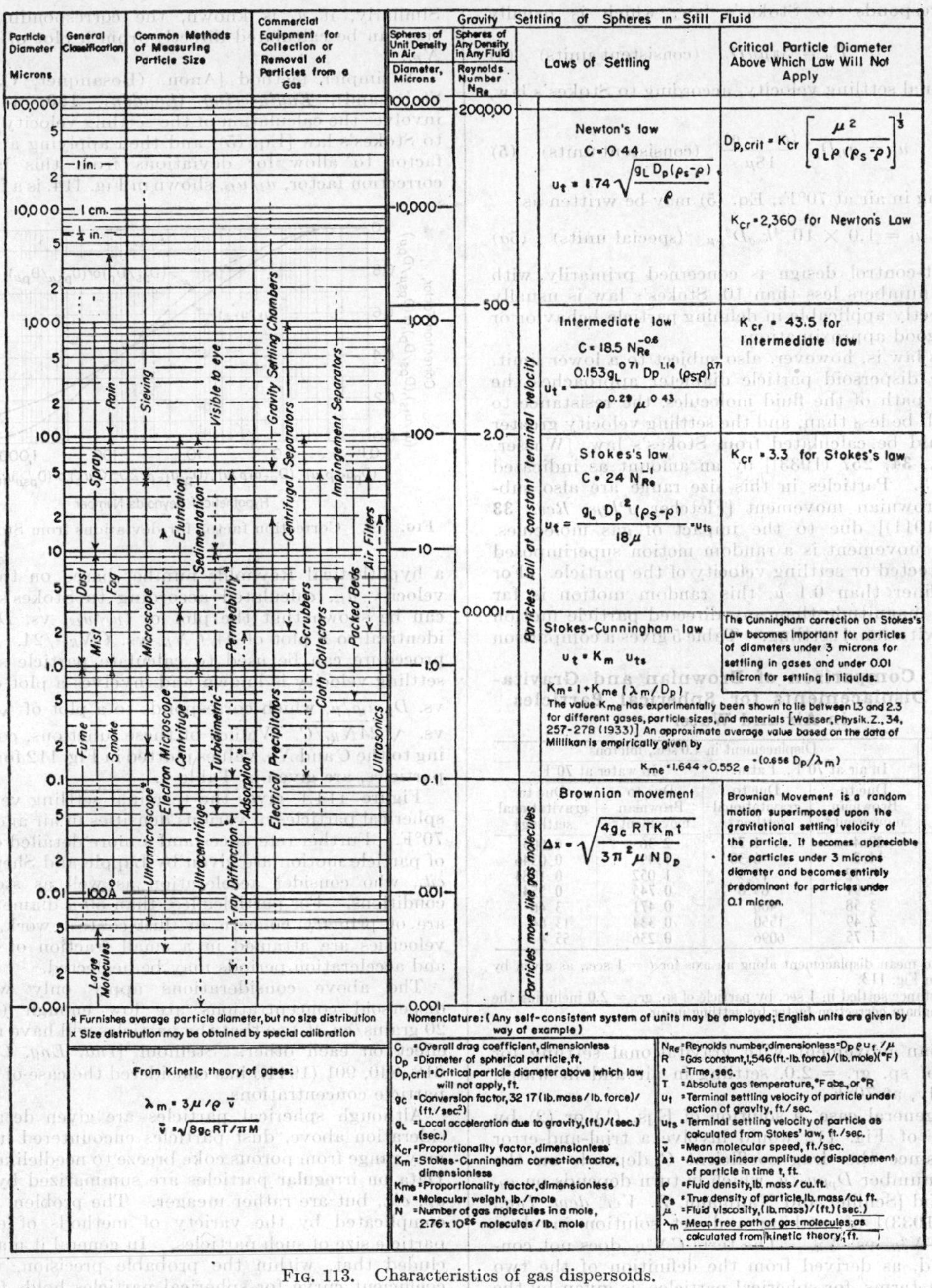

FIG. 113. Characteristics of gas dispersoids.

moves past the particle or whether the particle moves through the fluid.

A particle falling under the action of gravity will accelerate until the frictional drag of the fluid just balances the gravitational acceleration, after which it will continue to fall at a constant velocity known as the terminal or free-settling velocity u_t as given by the following equation:

$$u_t = \sqrt{\frac{2g_L m_p(\rho_s - \rho)}{\rho\rho_s A_p C}} \qquad \text{(consistent units)} \qquad (1)$$

For spherical particles, Eq. (1) becomes

$$u_t = \sqrt{\frac{4g_L D_p(\rho_s - \rho)}{3\rho C}} \qquad \text{(consistent units)} \qquad (2)$$

The various portions of the general drag-coefficient curve for spherical particles may be represented by three analytical relationships, as indicated in Fig. 113. For Reynolds numbers less than 2,

$$CN_{Re} = 24 \qquad \text{(consistent units)} \qquad (3)$$

This corresponds to Stokes's law, which is usually written

$$F_d = 3\pi\mu u D_p \quad \text{(consistent units)} \quad (4)$$

The terminal settling velocity, according to Stokes's law, becomes

$$u_t = g_L D_p^2 \frac{(\rho_s - \rho)}{18\mu} \quad \text{(consistent units)} \quad (5)$$

For settling in air at 70°F., Eq. (5) may be written as

$$u_t = 1.0 \times 10^{-4} k_{sg} D^2_{p\mu} \quad \text{(special units)} \quad (5a)$$

Since dust-control design is concerned primarily with Reynolds numbers less than 10, Stokes's law is usually either directly applicable in defining particle behavior or affords a good approximation.

Stokes's law is, however, also subject to a lower limit. When the dispersoid particle diameter approaches the mean free path of the fluid molecules, the resistance to motion will be less than, and the settling velocity greater than, would be calculated from Stokes's law [Wasser, *Physik. Z.*, **34**, 257 (1933)] by an amount as indicated in Fig. 113. Particles in this size range are also subject to Brownian movement [Fletcher, *Phys. Rev.*, **33** (1), 81 (1911)] due to the impact of gas molecules. Brownian movement is a random motion superimposed on any directed or settling velocity of the particle. For particles finer than 0.1 μ, this random motion is far greater in magnitude than any directed particle motion due to gravitational settling. Table 5 gives a comparison of Brownian movement with gravitational settling for particles of sp. gr. = 2.0, settling in air and in water, respectively, at 70°F.

Table 5. Comparison of Brownian and Gravitational Displacements for Spherical Particles in Air and Water

Particle diameter, microns	Displacement in 1.0 sec., microns: In air at 70°F., 1 atm.: Due to Brownian movement*	In air at 70°F., 1 atm.: Due to gravitational settling†	In water at 70°F.: Due to Brownian movement*	In water at 70°F.: Due to gravitational settling†
0.1	29.4	1.73	2.36	0.005
0.25	14.2	6.30	1.49	0.0346
0.5	8.92	19.9	1.052	0.1384
1.0	5.91	69.6	0.745	0.554
2.5	3.58	400	0.471	3.46
5.0	2.49	1550	0.334	13.84
10.0	1.75	6096	0.236	55.4

* This is the mean displacement along an axis for $t = 1$ sec., as given by the equation in Fig. 113.

† This is distance settled in 1 sec. by particle of sp. gr. = 2.0 including the Stokes-Cunningham correction factor for settling in air.

In the general case, a solution of Eqs. (1) or (2) by direct use of Fig. 112 would involve a trial-and-error solution, since the drag coefficient C depends on the Reynolds number $D_p\rho u_t/\mu$, which in turn depends on u_t. One method [Schiller and Naumann, *Z. Ver. deut. Ing.*, **77**, 318 (1933)] of making a direct solution involves a plot of CN^2_{Re} vs. N_{Re}. The term CN^2_{Re} does not contain u_t and, as derived from the definition of the two component terms, for spherical particles, is given by the following expression:

$$CN_{Re}^2 = 4g_L D_p^3 \rho \frac{(\rho_s - \rho)}{3\pi^2} \quad \text{(consistent units)} \quad (6)$$

From the value of CN_{Re}^2, the corresponding value of N_{Re} is determined from the plot. The terminal velocity u_t can then be calculated directly from N_{Re}

$$u_t = \frac{\mu N_{Re}}{\rho D_p} \quad \text{(consistent units)} \quad (7)$$

Similarly, if u_t is known, the corresponding particle size can be calculated directly from a plot of C/N_{Re} vs. N_{Re}.

A simpler method [Anon. (Bosanquet, Carey, and Stairmand), *Engineering (London)*, **150**, 441 (1940)] involves the calculation of the settling velocity according to Stokes's law [Eq. (5)] and then applying a correction factor to allow for deviations from this law. This correction factor, u_t/u_{ts}, shown in Fig. 114, is a function of

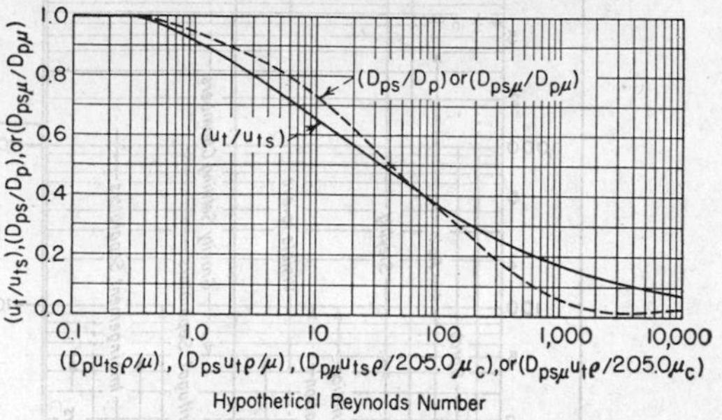

FIG. 114. Correction factor for deviations from Stokes's law.

a hypothetical Reynolds number based on the settling velocity u_{ts}, calculated according to Stokes's law. It can be shown that the plot of (u_t/u_{ts}) vs. $D_p u_{ts}\rho/\mu$ is identical to a plot of $24/CN_{Re}$ vs. $CN_{Re}^2/24$. A similar procedure can be used to calculate particle size if the settling velocity is known and involves a plot of D_{ps}/D_p vs. $D_{ps}u_t\rho/\mu$, which corresponds to a plot of $\sqrt{24/CN_{Re}}$ vs. $\sqrt{24N_{Re}/C}$. Values of these functions, corresponding to the C and N_{Re} values plotted in Fig. 112 for spherical particles, are given in Table 4.

Figure 114*A* gives the terminal settling velocities of spherical particles of various densities in air and water at 70°F. Further references and a more detailed discussion of particle motion are given by Lapple and Shepherd, *op. cit.*, who consider acceleration as well as steady-state conditions. For particles less than 50 μ diameter, which are of primary concern in dust-control work, terminal velocities are attained in a small fraction of a second, and acceleration periods may be neglected.

The above considerations apply only where the dispersoid concentrations are low enough (less than 20 grains/cu. ft.) so that the particles will have no mutual effect on each other. Steinour [*Ind. Eng. Chem.*, **36**, 618, 840, 901 (1944)] has considered the case of very high particle concentrations.

Although spherical particles are given detailed consideration above, dust particles encountered in practice may range from porous coke breeze to needlelike crystals. Data on irregular particles are summarized by Schiller, *op. cit.*, but are rather meager. The problem is further complicated by the variety of methods of expressing particle size of such particles. In general it may be concluded that, within the probable precision, the drag-coefficient curve for spherical particles holds fairly well for irregular particles of not too extreme shape for values of N_{Re} less than 50 where the diameter as measured by screen, elutriation, microscope, or otherwise is taken as the diameter of an equivalent sphere. For Reynolds numbers greater than 50, the drag coefficient for irregular particles levels off rapidly to a constant value averaging about 1.2, with variations of approximately twofold from the average. For very irregular particles large discrepancies may be encountered, and experimental determinations of settling velocity should be resorted to.

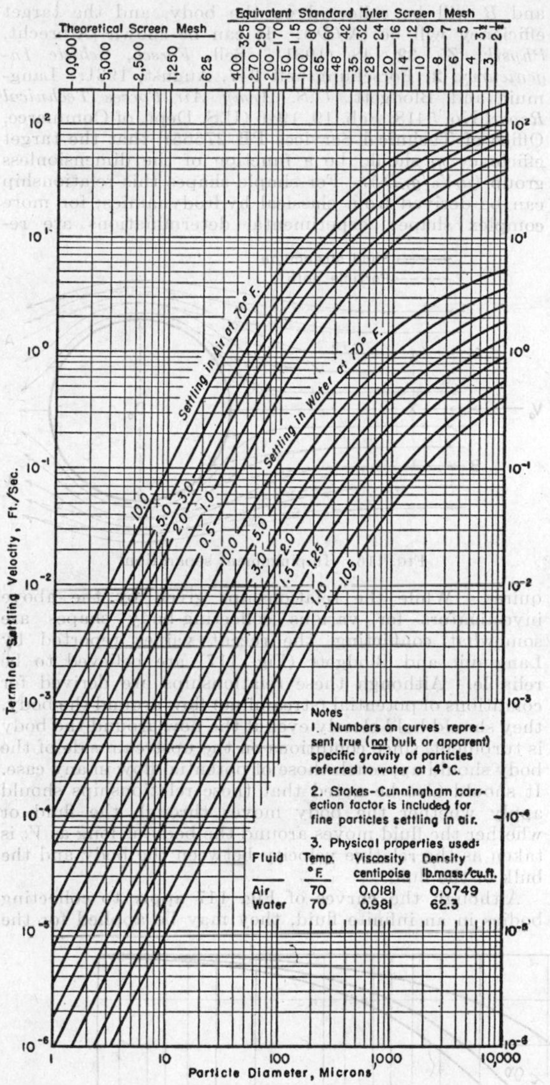

FIG. 114A. Terminal velocities of spherical particles of different density settling in air and water at 70°F. under the action of gravity.

COLLECTION EQUIPMENT

The forces or mechanisms utilized for dust collection may be classified as (1) gravitational, (2) inertial, (3) filtration, (4) electrostatic, (5) physiochemical, (6) thermal, and (7) sonic. Since some forms of dust-collection equipment utilize more than one of these forces, it is more convenient to classify dust-collection equipment according to type rather than according to the underlying operating principle.

The performance of a dust collector is termed collection efficiency and is generally expressed as a weight ratio of dust (or mist) collected to dust entering the apparatus. Some prefer to evaluate collection efficiency entirely by the dust concentration in the exit gases regardless of inlet concentrations. In either case collector efficiency is in itself not a specific characteristic of a given collector, depending on operating conditions as well as on the physical properties of the particular dust (or mist) handled.

Except where specifically noted otherwise, the cost data presented in the following paragraphs are based on information obtained in the years 1945 to 1946.

Gravity Settling Chambers. The gravity settling chamber is probably the simplest and earliest type of dust-collection equipment, consisting of a chamber in which the gas velocity is reduced to enable dust to settle out by the action of gravity. Its simplicity lends it to almost any type of construction. Practically, however, its industrial utility is limited to removing particles larger than 325 mesh (43 μ diameter). For removing smaller particles, the required chamber size is generally excessive.

Gravity collectors are generally built in the form of long, empty, horizontal, rectangular chambers with an inlet at one end and an outlet at the side or top of the other end. Assuming a low degree of turbulence relative to the settling velocity of the dust particle in question, the performance of a gravity settling chamber is given by

$$\eta = \frac{u_t L_s}{H_s V_s} = \frac{u_t B_s L_s}{q} \text{ (for } \eta \leqq 1.0) \quad \text{(consistent units)} \quad (8)$$

Expressing u_t in terms of particle size (equivalent spherical diameter), the smallest particle that can be completely separated out corresponds to $\eta = 1.0$ and, assuming Stokes's law, is given by

$$D_{p,\min} = \sqrt{\frac{18\mu H_s V_s}{g_L L_s(\rho_s - \rho)}} = \sqrt{\frac{18\mu q}{g_L B_s L_s(\rho_s - \rho)}} \quad \text{(consistent units)} \quad (9)$$

For a given volumetric air-flow rate, the collection efficiency depends on the total plan cross section of the chamber and is independent of the height. The height need be made only large enough so that the gas velocity V_s in the chamber is not so high as to cause reentrainment of separated dust. Generally V_s should not exceed about 10 ft./sec.

Curtains, rods, or screens are sometimes suspended in the chamber to minimize eddy currents and reduce turbulence. They also serve to increase collection by impingement and may be equipped with rappers. It is essential that the gas be well distributed laterally on entering the settling chamber; the vertical distribution is not so important. This can be achieved by gradual inlet transitions, guide vanes, or distributing screens or perforated plates. The gas outlet can be abrupt without appreciable effect on performance.

Although alternately arranged vertical baffle plates have been used in settling chambers on the theory of subjecting particles to changes of direction, their use in the simple gravity chamber is detrimental since inertial forces are generally small at the velocities encountered in such chambers. Horizontal plates arranged as shelves within the chamber, however, will give a marked improvement in collection, as indicated in the following equation:

$$\eta = \frac{N_s u_t B_s L_s}{q} \text{ (for } \eta \leqq 1.0) \quad \text{(consistent units)} \quad (10)$$

$$D_{p,\min} = \sqrt{\frac{18\mu N_s q}{g_L B_s L_s(\rho_s - \rho)}} \quad \text{(consistent units)} \quad (11)$$

This arrangement is known as the Howard dust chamber [Fume Arrester, U.S. Patent 896,111 (1908)] and is shown in Fig. 115. It can be used to remove particles as small as 10 μ in diameter. The disadvantage of the unit lies in the difficulty of cleaning due to the close shelf spacing and warpage at elevated temperatures. This is particularly serious at dust loadings over 1 grain/cu. ft.

The pressure drop through a settling chamber is usually small, consisting primarily of entrance and exit losses. Because of its structural simplicity and low pressure drop, a gravity settling chamber finds its most frequent application on natural-draft exhausts from kilns and furnaces. Since it is not subject to abrasion

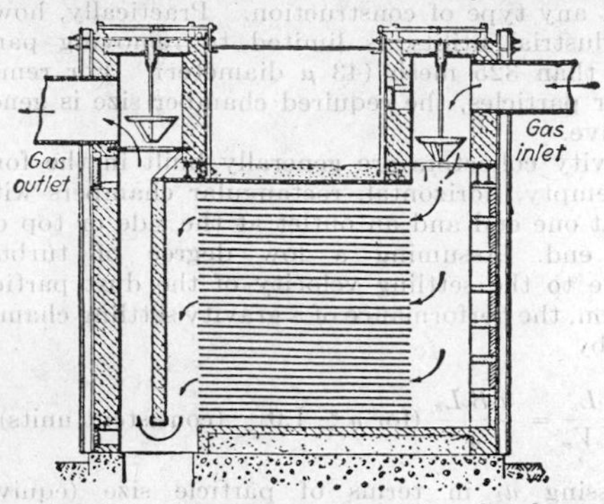

FIG. 115. Howard gravity settling chamber.

due to the velocities employed, a gravity settling chamber may also be utilized as a precleaner to remove very coarse particles in order to minimize abrasion on subsequent equipment.

Impingement Separators. When a dust-laden fluid impinges on a body, the fluid will be deflected around the body, whereas the dust particles, by virtue of their greater inertia, will tend to be collected on the surface of the body. The basic principles of impingement separators can be presented in terms of so-called "target" efficiencies. Target efficiency represents the fraction of particles in the fluid volume swept by the body which will impinge on the body. Thus, for flow around a cylinder, as shown in Fig. 116, all particles that are initially carried in the fluid between streamlines A and B will be collected on the body, and the target efficiency will be (X/D_b). It can be shown [Albrecht, *Physik. Z.*, **32**, 48 (1931). Sell, *Forsch. Gebiete Ingenieurw.*, **2**, Forschungsheft 347, August, 1931. Langmuir and Blodgett, *U.S. Army Air Forces Technical Report* No. 5418, Feb. 19, 1946 (U.S. Dept. of Commerce, Office of Technical Services PB 27565)] that the target efficiency η_t should be a function of the dimensionless group $(u_t V_0/g_L D_b)$. For simple shapes this relationship can be derived from classical hydrodynamics; for more complex shapes, experimental determinations are re-

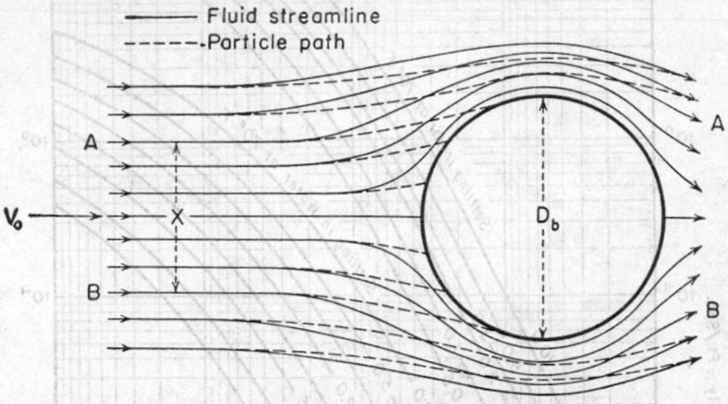

FIG. 116. Impingement separation.

quired. While the relationships given by the above investigators for various collecting-body shapes are somewhat conflicting, the recent values reported by Langmuir and Blodgett (Fig. 117) are believed to be reliable. Although these relationships are derived for conditions of potential (streamline) flow around the body, they should hold closely even if the flow around the body is turbulent, since conditions on the upstream side of the body should approach those of potential flow in any case. It should also be noted that these relationships should apply whether the body moves through the fluid or whether the fluid moves around the body as long as V_o is taken as the relative velocity between the body and the bulk of the fluid.

Although the curves of Fig. 117 apply to collecting bodies in an infinite fluid, they may be applied for the

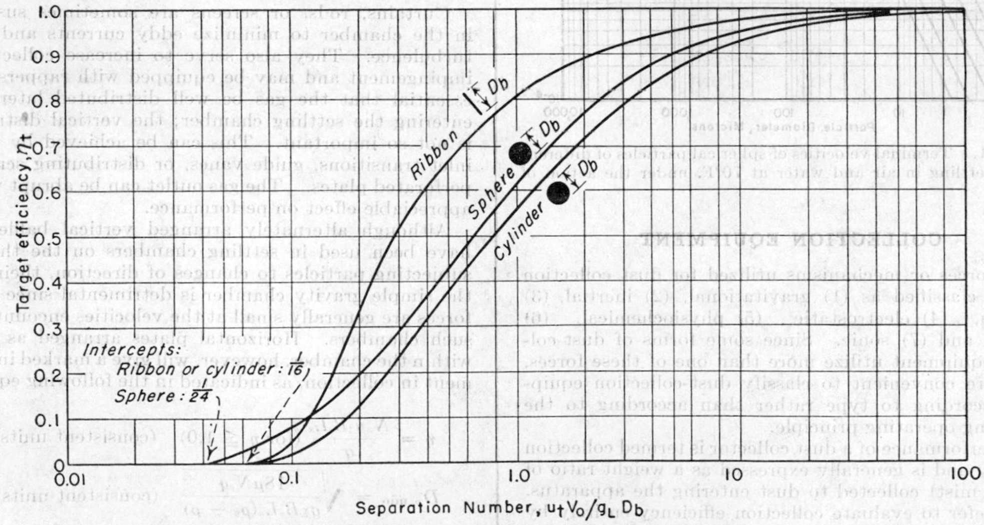

FIG. 117. Target efficiency of spheres, cylinders, and ribbons. The curves apply for conditions where Stokes's law holds for the motion of the particle. Langmuir and Blodgett have also presented similar relationships for cases where Stokes's law does not apply. [*Langmuir and Blodgett, U.S. Army Air Forces Technical Report* 5418, *Feb.* 19, 1946 (*U.S. Dept. of Commerce, Office of Technical Services PB* 27565).]

direct calculation of collection efficiency of units employing such bodies in parallel and series, provided that adjacent collecting members are not so close as to cause an appreciable distortion of the flow pattern. This is substantially the case in many types of air filters and liquid scrubbers. Where collecting members are relatively close, these curves would give a conservative approximation of collection efficiency. For collecting members having shapes differing widely from those shown in Fig. 117, additional experimental determinations are desirable, although order-of-magnitude estimates can be made by judicious interpolation of these curves.

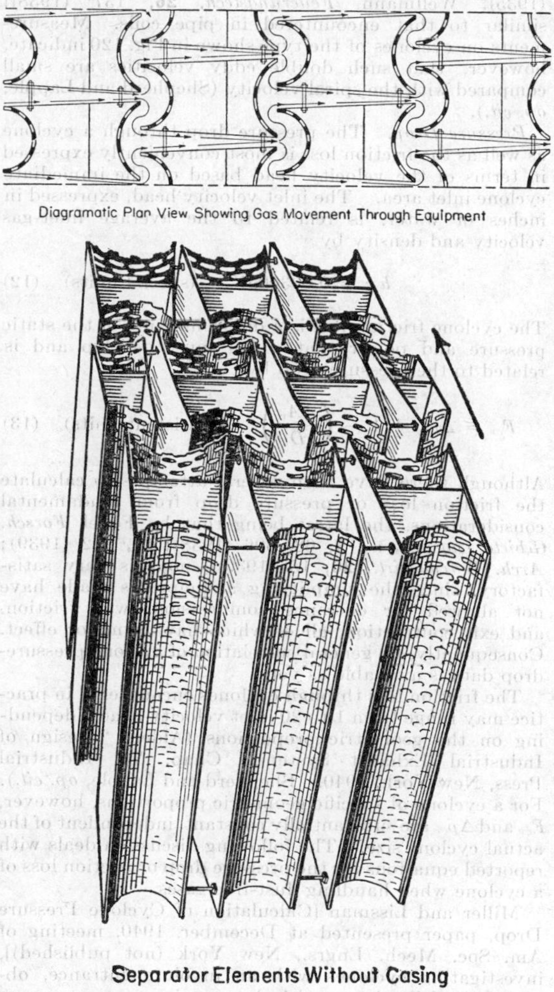

Diagramatic Plan View Showing Gas Movement Through Equipment
Separator Elements Without Casing

FIG. 118. Reverse nozzle impingement separator. *(By-Products Recoveries, Inc.)*

In Fig. 118 is shown a typical commercial impingement collector. In general, impingement collectors are designed for a pressure drop in the range of 0.1 to 1.5 in water, depending on the type and application, and are limited to removing dusts that are predominantly larger than 10 to 20 μ diameter. Rappers are sometimes provided to shake the collected dust off the collecting bodies at definite intervals. The chief advantage of such units lies in their greater adaptability to existing flues or ducts than other types of collectors. They may be used at elevated temperatures but not if the dust becomes tacky, although for the latter case some types are available in which circulating-water films are used to keep the elements clear. These are discussed in the section on scrubbers. Purchase costs of impingement units are generally in the range of $0.10 to $0.20/cu. ft./min. of gas handled for installations of over 10,000 cu. ft./min. capacity in steel (prices in 1945–1946). Further description will be found in Powers [*Rock Products*, **46**, 70, 72 (June, 1943)] and Roberts [*Power*, **83**, 345, 392 (1939)].

The Calder-Fox scrubber [Fairs, *Chem. Trade J.*, **115**, 571 (1944); *Ind. Chemist*, **20**, 637 (1944); *Trans. Inst. Chem. Engrs.* (*London*), **22**, 110 (1944)] shown in Fig. 119, is a British development frequently used for

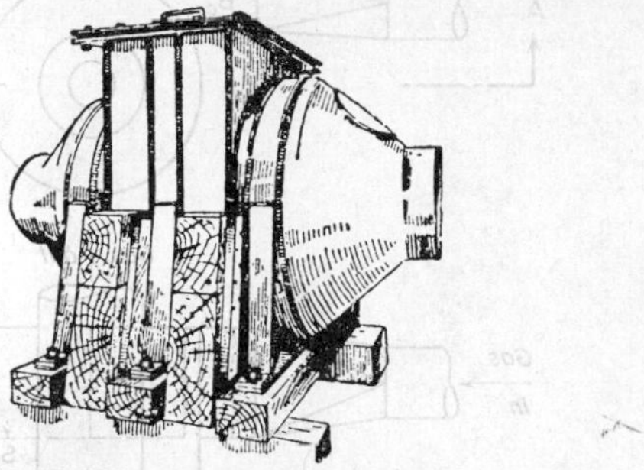

FIG. 119. Calder-Fox scrubber. *(Chance & Hunt, Ltd.)*

removing sulfuric acid mist. It is essentially an impingement separator employing two groups of perforated plates in series. The first "agglomerator" group consists of an "orifice" plate followed by an "impact" plate with staggered orifices and separated by a spacer. This group is followed by a group of three "collector" plates with aligned orifices. In the designs developed for sulfuric acid concentrator exit gases, the plates are made of 8 lb. lead (approximately $\frac{1}{8}$ in. thick). The orifice plate contains $\frac{1}{8}$-in. holes on $\frac{3}{8}$- or $\frac{1}{2}$-in. centers, square pitch, and the impact plate contains $\frac{1}{4}$-in. holes on the same centers and is set $\frac{1}{8}$ in. away from the orifice plate. The collector plates have $\frac{1}{12}$-in. holes and are spaced $\frac{1}{8}$ in. apart. A glass unit has also been developed in which the perforated plates are replaced with strips of glass. In this unit the orifice-plate slots are $\frac{1}{16}$ in. wide and the impact slots $\frac{1}{8}$ in. wide, spaced on $\frac{5}{16}$-in. center lines. These plates are set $\frac{1}{16}$ in. apart, and no collector plates are used. A third type has been developed for use with wax mists in which the impact plate is replaced with a cooled imperforate rotating drum.

The Calder-Fox unit is only effective for mist particles larger than 2 μ diameter. Orifice velocities are generally on the order of 50 to 100 ft./sec. but should not exceed about 150 ft./sec. at atmospheric pressure in order to avoid excessive reentrainment. The pressure drop is approximately 2.0 to 2.5 orifice velocity heads.

Cyclone Separators. The most widely used type of dust-collection equipment is the cyclone, in which dust-laden gas enters a cylindrical or conical chamber tangentially at one or more points and leaves through a central opening (Fig. 120). The dust particles, by virtue of their inertia, will tend to move toward the outside separator wall from which they are led into a receiver. A cyclone is essentially a settling chamber in which gravitational acceleration is replaced by centrifugal acceleration. At operating conditions commonly employed, the centrifugal separating force or acceleration may range from five times gravity in very large diameter, low resistance cyclones, to 2500 times gravity in very small, high-resistance units. The immediate entrance to a cyclone is usually rectangular.

Fields of Application. Cyclone collectors offer one of the least expensive means of dust or mist collection from both an operating and an investment viewpoint. Cyclones have been employed to remove solids and liquids from gases and solids from liquids and have been operated at temperatures as high as 1000°C. and pressures as high as 500 atm. Cyclones for removing solids or liquids from gases are generally applicable when particles of over 5 μ (0.0002 in.) diameter are involved. Unless very small cyclones are used, the efficiency will be low if much of the suspended material is finer than 5 μ. In collecting

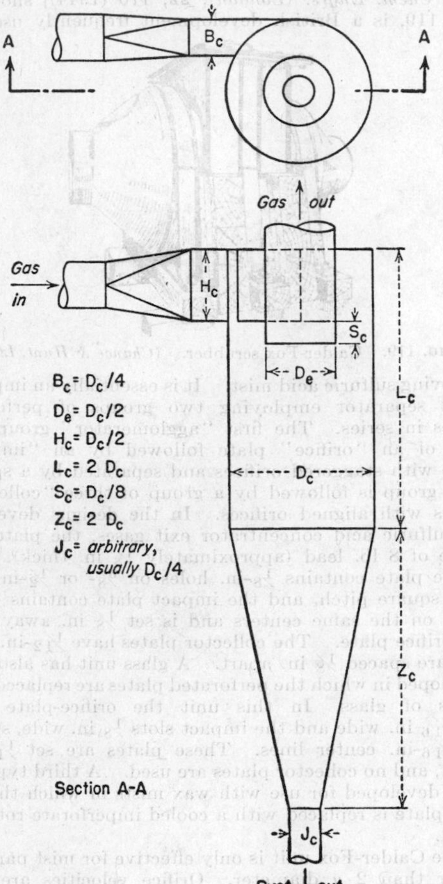

FIG. 120. Cyclone separator proportions.

particles of over 200 μ diameter, cyclones may be used, but gravity settling chambers are usually satisfactory and less subject to abrasion. In special cases where the dust shows a high degree of agglomeration, or where high dust concentrations (over 100 grains/cu. ft.) are involved, cyclones will remove dusts having a much smaller particle size. In certain cases efficiencies as high as 98 per cent have been realized on dusts having an ultimate particle size of 0.1 to 2.0 μ because of the predominant effect of agglomeration.

Flow Pattern. In a cyclone the gas path involves a double vortex with the gas spiraling downward at the outside and upward at the inside. When the gas enters the cyclone, its velocity undergoes a redistribution so that the tangential component of velocity increases with decreasing radius as expressed by $V_{ct} \sim r^{-n}$. The spiral velocity in a cyclone may reach a value several times the average inlet-gas velocity. Theoretical considerations indicate that n should be equal to 1.0 in the absence of wall friction. Actual measurements [Shepherd and Lapple, *Ind. Eng. Chem.*, **31**, 972 (1939); **32**, 1246 (1940)], however, indicate that n may range from 0.5 to 0.7 over a large portion of the cyclone radius. At the wall the gas velocity approaches zero, whereas it reaches a maximum at a certain radius, decreasing rapidly at smaller radii.

Superimposed on the "double spiral," there may be a "double eddy" [Van Tongeran, *Mech. Eng.*, **57**, 753 (1935). Wellmann, *Feuerungstech*, **26**, 137 (1938)] similar to that encountered in pipe coils. Measurements on cyclones of the type shown in Fig. 120 indicate, however, that such double eddy velocities are small compared with the spiral velocity (Shepherd and Lapple, *op. cit.*).

Pressure Drop. The pressure drop through a cyclone as well as the friction loss is most conveniently expressed in terms of the velocity head based on the immediate cyclone inlet area. The inlet velocity head, expressed in inches of water, is related to the average inlet-gas velocity and density by

$$h_{vi} = 0.00300\rho V_c^2 \quad \text{(special units)} \quad (12)$$

The cyclone friction loss is a direct measure of the static pressure and power that a fan must develop and is related to the pressure drop by

$$F_{cv} = \Delta p_{cv} + 1 - \left(\frac{4A_c}{\pi D_e^2}\right)^2 \quad \text{(consistent units)} \quad (13)$$

Although there have been several attempts to calculate the friction loss or pressure drop from fundamental considerations, the latest being that by Feifel [*Forsch. Gebiete Ingenieurw.*, **9**, 68, 306 (1938); **10**, 212 (1939); *Arch. Wärmewirt.*, **20**, 15 (1939)], none is very satisfactory, since the simplifying assumptions made have not allowed for entrance compression, wall friction, and exit contraction, all of which have a major effect. Consequently, no general correlation of cyclone pressure-drop data is available as yet.

The friction loss through cyclones encountered in practice may range from 1 to 20 inlet velocity heads, depending on the geometric proportions (Alden, "Design of Industrial Exhaust System," Chap. VI, Industrial Press, New York, 1940. Shepherd and Lapple, *op. cit.*). For a cyclone of specific geometric proportions, however, F_{cv} and Δp_{cv} are substantially constant, independent of the actual cyclone size. The following discussion deals with reported equations for the pressure drop or friction loss of a cyclone when handling dust-free gases.

Miller and Lissman [Calculation of Cyclone Pressure Drop, paper presented at December, 1940, meeting of Am. Soc. Mech. Engrs., New York (not published)], investigating cyclones with an involute entrance, obtained the following empirical expression:

$$\Delta p_{cv} = K\left(\frac{D_c}{D_e}\right)^2 \quad \text{(consistent units)} \quad (14)$$

The value of K was found to be substantially constant with a value of 3.2 over the following range in proportions: $(B_c/D_c) = \frac{1}{8}$ to $\frac{3}{8}$; $(H_c/D_c) \cong 1.0$; $(D_e/D_c) = \frac{1}{4}$ to $\frac{3}{4}$. For smaller values of (D_e/D_c), the value of K increased, while, for smaller values of (B_c/D_c), it decreased. In these tests D_c, D_e, and B_c were varied but not H_c.

Shepherd and Lapple (*op. cit.*), investigating cyclones

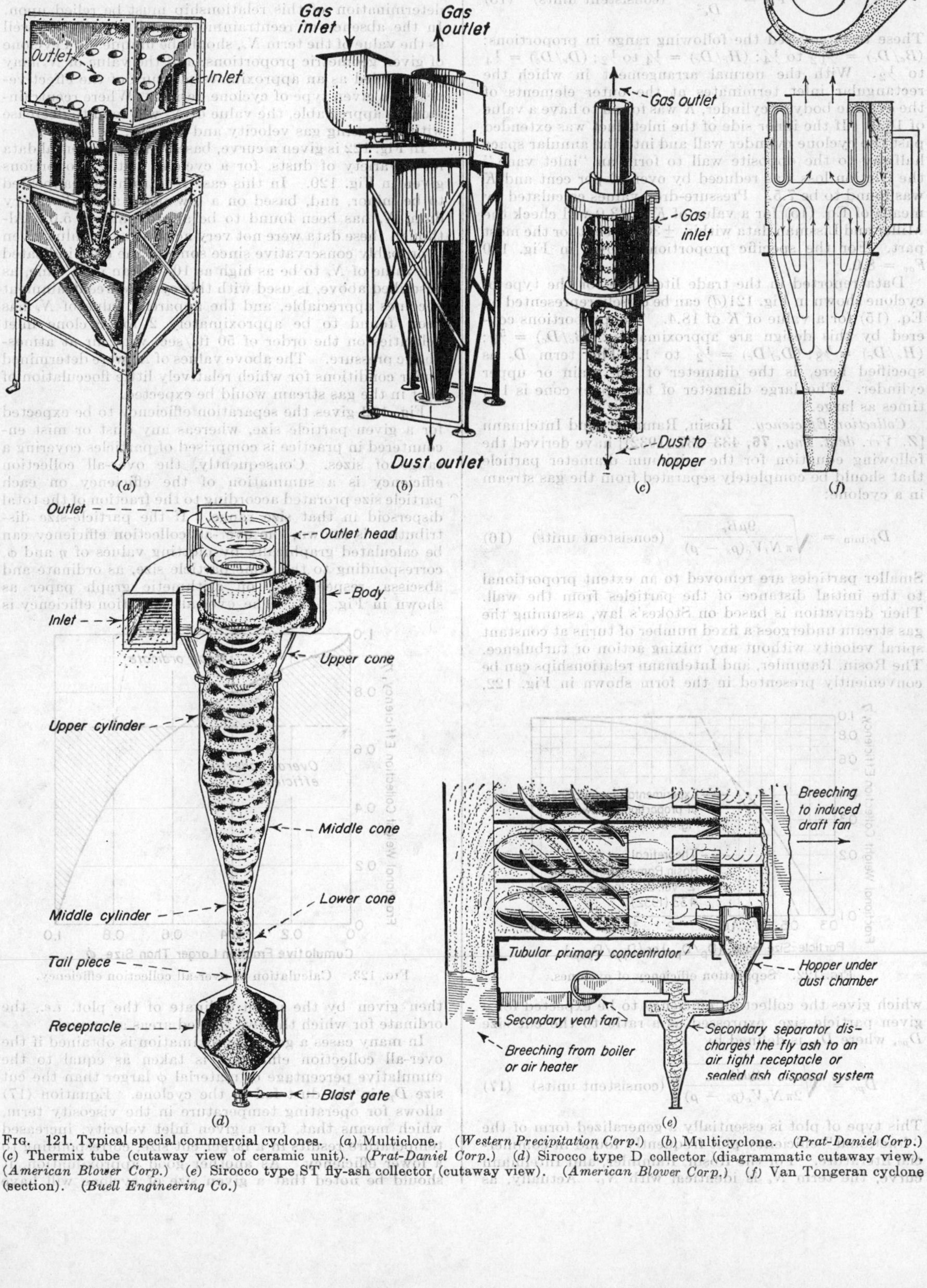

FIG. 121. Typical special commercial cyclones. (a) Multiclone. (*Western Precipitation Corp.*) (b) Multicyclone. (*Prat-Daniel Corp.*) (c) Thermix tube (cutaway view of ceramic unit). (*Prat-Daniel Corp.*) (d) Sirocco type D collector (diagrammatic cutaway view). (*American Blower Corp.*) (e) Sirocco type ST fly-ash collector (cutaway view). (*American Blower Corp.*) (f) Van Tongeran cyclone (section). (*Buell Engineering Co.*)

of the general type shown in Fig. 120, obtained the following empirical expression:

$$F_{cv} = \frac{KB_cH_c}{D_e^2} \quad \text{(consistent units)} \quad (15)$$

These tests covered the following range in proportions: $(B_c/D_c) = 1/12$ to $1/4$; $(H_c/D_c) = 1/4$ to $1/2$; $(D_e/D_c) = 1/4$ to $1/2$. With the normal arrangement in which the rectangular inlet terminates at the outer elements of the cyclone body or cylinder, K was found to have a value of 16.0. If the inner side of the inlet duct was extended past the cyclone cylinder wall and into the annular space halfway to the opposite wall to form an "inlet vane," the friction loss was reduced by over 50 per cent and K was found to be 7.5. Pressure-drop values calculated by means of Eq. (15) for a value of K of 13.0 will check the Miller and Lissman data within ± 30 per cent for the most part. For the specific proportions shown in Fig. 120 $F_{cv} = 8.0$.

Data reported in the trade literature for the type of cyclone shown in Fig. 121(*d*) can be closely represented by Eq. (15) for a value of K of 18.4. The proportions covered by this design are approximately: $(B_c/D_c) = 5/8$; $(H_c/D_c) = 5/8$; $(D_e/D_c) = 1/2$ to 1. The term D_c as specified here, is the diameter of the main or upper cylinder. The large diameter of the upper cone is $1\frac{5}{8}$ times as large.

Collection Efficiency. Rosin, Rammler, and Intelmann [*Z. Ver. deut. Ing.*, **76,** 433–437 (1932)] have derived the following equation for the minimum diameter particle that should be completely separated from the gas stream in a cyclone:

$$D_{p,\min} = \sqrt{\frac{9\mu B_c}{\pi N_t V_c(\rho_s - \rho)}} \quad \text{(consistent units)} \quad (16)$$

Smaller particles are removed to an extent proportional to the initial distance of the particles from the wall. Their derivation is based on Stokes's law, assuming the gas stream undergoes a fixed number of turns at constant spiral velocity without any mixing action or turbulence. The Rosin, Rammler, and Intelmann relationships can be conveniently presented in the form shown in Fig. 122,

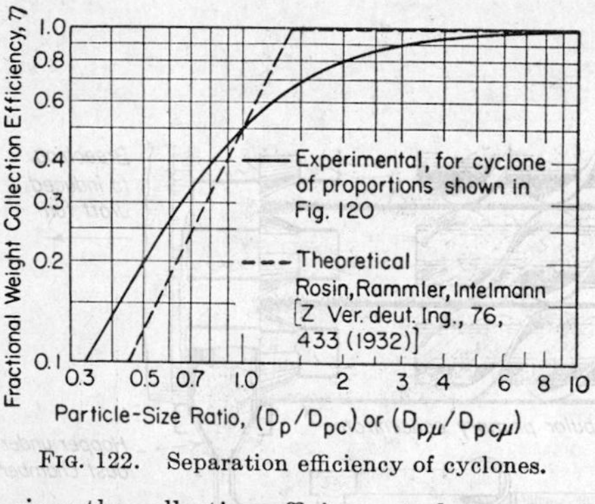

FIG. 122. Separation efficiency of cyclones.

which gives the collection efficiency to be expected for a given particle size, expressed as a ratio to the cut size D_{pc}, where D_{pc} is defined by

$$D_{pc} = \sqrt{\frac{9\mu B_c}{2\pi N_e V_c(\rho_s - \rho)}} \quad \text{(consistent units)} \quad (17)$$

This type of plot is essentially a generalized form of the "fractional" efficiency plot frequently found in commercial literature. For the Rosin, Rammler, and Intelmann curve, the term N_e is identical with N_t. Actually, as previously described, however, the flow pattern is considerably more complex, and the separation or classification efficiency curve is not nearly so sharp as would be predicted by this curve. Consequently, an experimental determination of this relationship must be relied upon. In the absence of reentrainment, such a curve, as well as the value of the term N_e, should be unique for a cyclone of given geometric proportions, and the value of N_e may be regarded as an approximate measure of the effectiveness of a given type of cyclone design. Where reentrainment is appreciable, the value of N_e will tend to decrease with increasing gas velocity and density.

In Fig. 122 is given a curve, based on experimental data for a variety of dusts, for a cyclone of the proportions given in Fig. 120. In this case reentrainment appeared to be minor, and, based on a few plant and laboratory data, N_e has been found to be approximately 5.0. Although these data were not very accurate, the value given is probably conservative since some of the data indicated the value of N_e to be as high as 10. If an inlet vane, as described above, is used with this cyclone, reentrainment becomes appreciable, and the apparent value of N_e has been found to be approximately 2 for cyclone inlet velocities on the order of 50 ft./sec. with air at atmospheric pressure. The above values of N_e were determined under conditions for which relatively little flocculation of dust in the gas stream would be expected.

Fig. 122 gives the separation efficiency to be expected for a given particle size, whereas any dust or mist encountered in practice is comprised of particles covering a range of sizes. Consequently, the over-all collection efficiency is a summation of the efficiency on each particle size prorated according to the fraction of the total dispersoid in that size range. If the particle-size distribution is known, the over-all collection efficiency can be calculated graphically by plotting values of η and ϕ, corresponding to the same particle size, as ordinate and abscissa, respectively, on arithmetic graph paper as shown in Fig. 123. The over-all collection efficiency is

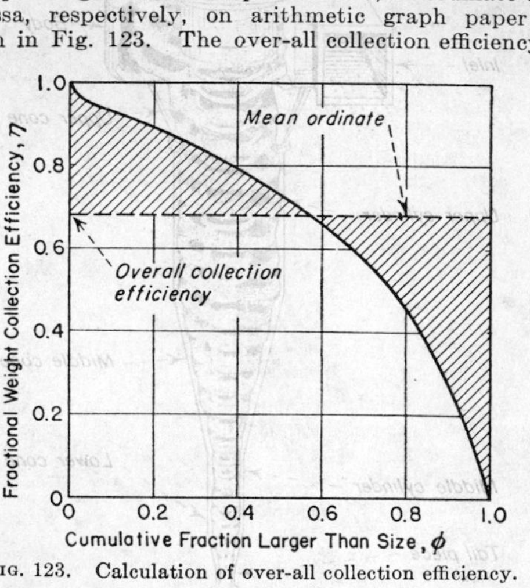

FIG. 123. Calculation of over-all collection efficiency.

then given by the mean ordinate of the plot, *i.e.*, the ordinate for which the two shaded areas are equal.

In many cases a good approximation is obtained if the over-all collection efficiency is taken as equal to the cumulative percentage of material ϕ larger than the cut size D_{pc} in the dust fed to the cyclone. Equation (17) allows for operating temperature in the viscosity term, which means that, for a given inlet velocity, increased temperature results in a larger cut size, corresponding to a lower efficiency. As another good approximation, it should be noted that a given size of cyclone will have substantially the same collection efficiency at any temperature, provided that the pressure drop is the same, because of counterbalancing effects of gas density and viscosity.

Table 6 gives experimental collection-efficiency data reported by Anderson for geometrically similar cyclones of the type shown in Fig. 121(*a*). These will serve to illustrate the order of magnitude of the collection efficiency to be expected for various particle sizes.

Cyclone Design Factors. Cyclones are generally designed to meet specified pressure-drop limitations. For ordinary installations, operating at approximately atmospheric pressure, fan limitations generally dictate a maximum allowable pressure drop corresponding to a cyclone inlet velocity in the range of 20 to 70 ft./sec. Consequently, cyclones are usually designed for an inlet velocity of 50 ft./sec., though this need not be strictly adhered to.

Table 6. Experimental Cyclone Collection Efficiencies*

Cyclone inlet velocity: 44 ft./sec.
Cyclone pressure drop: 4 in. water
Inlet dust concentration: 2–5 grain/cu. ft.
Specific gravity of dust: 3.0 g./cc.
Cyclone proportions: $B_c \cong D_c/6$
Gas: atmospheric air

Inlet Dust Particle-size Analysis

Particle Diam., μ	Cumulative % Larger Than Size
5	74
10	64
20	43

Cyclone diam., in.	Dust collected, %		
6	Total 90	$-5\ \mu$ 66	$+5\ \mu$ 98
9	Total 83	$-10\ \mu$ 60	$+10\ \mu$ 99
24	Total 70	$-20\ \mu$ 47	$+20\ \mu$ 98

* Data reported by Anderson in Perry, "Chemical Engineers' Handbook," 2d ed., p. 1860, McGraw-Hill, New York, 1941.

In the removal of dusts, the collection efficiency can be changed to only a relatively small amount by a variation in the operating conditions. The primary design factor that can be utilized to control collection efficiency is the cyclone diameter; a smaller diameter unit operating at a fixed pressure drop having the higher efficiency [Anderson, *Chem. & Met.*, **40**, 525 (1933). Drijver, *Wärme*, **60**, 333 (1937). Whiton, *Power*, **75**, 344 (1932); *Chem. & Met.*, **39**, 150 (1932)]. Small-diameter cyclones, however, will require a multiple of units in parallel for a specified capacity. In such cases the individual cyclones can discharge the dust into a common receiving hopper [Whiton, *Trans. Am. Soc. Mech. Engrs.*, **63**, 213 (1941)]. The final design involves a compromise between collection efficiency and complexity of equipment. It is customary to design a single cyclone for a given capacity, resorting to multiple parallel units only if the predicted collection efficiency is inadequate for a single unit. Cyclones in series are generally not justified. Exceptions to this are cases where the dust is fine and has a relatively uniform particle-size distribution, and where the dust is present in a highly flocculated state (see p. 1043). In the latter case efficiencies predicted on the basis of ultimate particle-size distribution will be highly conservative. Also, although efficiency is normally increased by increasing the gas through-put (Drijver, *op. cit.*), in such cases the reverse may be true because of the deflocculating effect of higher velocities. Similarly, design proportion variations that result in increased collection efficiency with dispersed dusts may be detrimental in the case of flocculated dusts. Insufficient data are available to permit any generalization for this case, however. The flocculation factor is probably also the chief cause for inconsistency of data reported in the literature.

Reducing the gas-outlet duct diameter will increase both collection efficiency and pressure drop. Increasing the length of a cyclone is generally conceded to increase collection efficiency, though there are no reliable supporting data. There is also no reliable information on the effect of inlet proportions or cone angle on collection efficiency. It is essential that the inlet transition be relatively gradual in order to avoid excessive pressure drop due to gas jetting into the cyclone chamber.

A cyclone will operate equally well on the suction or pressure side of a fan if the dust receiver is airtight. Probably the greatest single cause for poor cyclone performance, however, is the leakage of air into the dust outlet of the cyclone. A slight air leak at this point can result in a tremendous drop in collection efficiency, particularly with fine dusts. For a cyclone under pressure, air leakage at this point is objectionable primarily from the local dust nuisance created. For batch operation, an airtight hopper or receiver may be used. For continuous withdrawal of collected dust, a rotary star valve, a double-lock valve, or a screw conveyor may be used, the latter only with fine dusts; a liquid-seal leg is generally used for mist or spray collectors. Special pneumatic unloading devices can also be used with dusts. In any case it is essential that sufficient unloading and receiving capacity be provided to prevent collected material from accumulating in the cyclone cone.

Generally cone and disk baffles, helical guide vanes, straightening vanes, etc., placed inside a cyclone, will have a detrimental effect on performance. A few of these devices do have some merit, however, under special circumstances. Although an inlet vane (see p. 1026) will reduce pressure drop, it causes a correspondingly greater reduction in collection efficiency. Its use is recommended only where collection efficiency is normally so high as to be a secondary consideration and where it is desired to decrease the resistance of an existing cyclone system for purposes of increased air-handling capacity or where floor-space or headroom requirements are controlling factors. If an inlet vane is used, it is advantageous to increase the gas-exit-duct length inside the cyclone chamber. A disk or cone baffle located beneath the gas-outlet duct may be beneficial if air in-leakage at the dust outlet cannot be avoided. A heavy chain suspended from the gas-outlet duct has been found beneficial to minimize dust build-up on the cyclone walls. Such a chain should be suspended from a swivel so that it is free to rotate without twisting. At present there are no known devices that will recover the gas spiral-velocity energy in the gas-outlet duct. Substantially all devices that have been reported to reduce pressure drop do so by reducing spiral velocities in the cyclone chamber and consequently result in reduced collection efficiency.

At low dust loadings the pressure in the dust receiver of a single cyclone will generally be lower than in the gas-outlet duct. Increased dust loadings will increase the pressure in the dust receiver. Such devices as cones, disks, inlet vane, etc., will generally cause the pressure in the dust receiver to exceed that in the gas-outlet duct. A cyclone will operate as well in a horizontal position as in a vertical position. However, departure from the normal vertical position results in an increasing tendency to plug the dust outlet. If the dust outlet becomes plugged, collection efficiency will, of course, be low. If the cyclone exit duct must be reduced to tie in with proposed duct sizes, the transition should be made at least five diameters downstream from the cyclone and preferably after a bend. In the event that the transition must be made closer to the cyclone, a Greek cross should be installed in the transition piece in order to avoid excessive pressure drop.

Increased dust loadings will result in both decreased pressure drop and increased collection efficiency (Drijver, *op. cit.* Shepherd and Lapple, *op cit.*). At dust loadings of over 200 grain/cu. ft., the pressure drop may be as low as half that calculated in the absence of dust.

Commercial Equipment. Simple cyclones are available in a wide variety of shapes ranging from long slender units similar to that shown in Fig. 120 to short large-diameter units. The body may be conical or cylindrical, and entrances may be involute or tangential and round or rectangular.

In Fig. 121 are shown some of the special types of commercial cyclones. The Sirocco Type D cyclone has an exit-duct collar that can be changed to increase or decrease collection efficiency with a corresponding increase or decrease in pressure drop. The Multicyclone is furnished in multiple units of 2 or 3 ft. diameter cyclones, each containing an adjustable inlet damper to compensate for variations in gas through-put. In the Multiclone a spiral motion is imparted to the gas by annular vanes, and it is furnished in multiple units of 6 and 9 in. diameter. The Van Tongeran cyclone claims to utilize the "double eddy" for increased collection efficiency by providing a by-pass from the top to the conical portion of the cyclone. The Thermix unit comprises 6-in.-diameter units in parallel and is available in ceramic as well as metallic construction. The Siroco fly ash collector consists of horizontal tubes, approximately 15 in. diameter and 6 ft. long, in parallel. The gas is given a spiral motion by vanes located in the inlet at the left, and the dust is concentrated in a small portion of gas which is recirculated through a smaller secondary cyclone separator for final collection of the dust. A recent development is the Aerotec tube [Parent, *Trans. Am. Inst. Chem. Engrs.*, **42,** 989 (1946)] consisting of multiple units of 2 or 3 in. diameter similar in shape to the Thermix tube. Smaller diameter units are also made for special applications. Recently tests have been reported [Anon., *Power Plant Eng.*, **48,** 90 (December, 1944)] on a modification of the type of unit shown in Fig. 121(*c*), in which the gas outlet was provided with an annular skimmer arrangement. Dust concentrated near the wall of the gas-outlet duct was carried off into the skimmer, together with a small amount of gas, and led to a two-stage electrical precipitator (Fig. 140) for final collection.

Cyclone Costs. The purchase cost of simple single-unit cyclones of 16 to 20 gage steel construction ranges from \$0.04 to \$0.08/(cu. ft./min.) of gas handled or \$0.25 to \$0.40/lb. steel. Special types of commercial cyclones generally involve a purchase cost in the range of \$0.15 to \$0.30/(cu. ft./min.) for steel construction. A large part of the higher cost in these cases is due to heavier construction supplied. Very small diameter multiple-cyclone units in steel construction cost on the order of \$0.25 to \$0.50/(cu. ft./min.) of gas handled (prices of 1945–1946).

Entrainment Separation. Cyclones may also be used to separate liquid droplets from gases. In this connection they find their most frequent application in the removal of mechanical entrainment from such equipment as evaporators and distillation and absorption columns. Although they can be used for the separation of mists formed by vapor-phase reaction or condensation, the collection efficiency will generally be low, except with very small diameter units, because of the small particle size involved. In the case of mechanical entrainment, the particles are predominantly larger than 100 μ diameter, and complete separation is a simple problem. Any inefficiency on the part of a cyclone in such service is due not to a failure to collect all the spray on the wall but to a reentrainment or wall creep of the liquid film after the spray has been deposited on the cyclone wall [Pollak and Work, *Trans. Am. Soc. Mech. Engrs.*, **64,** 31 (1942)].

In a cyclone of the type shown in Fig. 120, the liquid film is carried across the cyclone roof to the gas-outlet duct by the action of the "double eddy." The film will then run down the outside of this duct and be carried out of the cyclone. In the case of dusts such action will not occur, since the dust cannot adhere to the exit dust wall as a mobile film. With liquids this action can be minimized by providing either a concentric cylindrical shield around the gas-outlet duct or a conical skirt fastened to the outside of the gas-outlet duct some distance above its mouth. These arrangements avoid carry-over by providing a drip point before the liquid is exposed to the outlet-gas stream. In either case the cyclone diameter should be enlarged over that of a corresponding dust cyclone to avoid direct impingement of the inlet-gas stream on the shield or skirt. A cyclone of the type shown in Fig. 120 equipped in this fashion should remove over 98 per cent of mechanical entrainment, provided that the inlet-gas velocity is maintained less than 150 ft./sec. for operation with air at atmospheric pressure. For high gas densities, lower limiting velocities will apply. Also, if the liquid loadings are very high, as in flash evaporators, lower inlet velocities should be used, since the high liquid loading will upset the spiral path of the gas in the cyclone and accentuate reentrainment tendencies. If such special devices are omitted, the carry-over from the cyclone may increase five to tenfold.

For the separation of mechanical entrainment, cyclones arranged in parallel will serve no useful purpose unless headroom is a prime consideration. With cyclones in series, however, very high degrees of spray elimination can be achieved.

Mechanical Centrifugal Separators. A number of collectors are commercially available in which the centrifugal field is supplied by a rotating member. Typical units are shown in Figs. 124(*a*) and (*b*). In the unit shown in Fig. 124(*a*), the exhauster or fan and dust collector are combined as a single unit. The blades are especially shaped to direct the separated dust into an annular slot leading to the collection hopper while the cleaned gas continues to the scroll. The unit shown in Fig. 124(*b*) is usually used on the inlet side of a fan with the rotor connected to the fan shaft. The dust-laden gas enters on the periphery of the scroll, passing radially inward through the rotor and out the center, which point is normally coincident with the fan inlet port. Dust thrown to the scroll wall is concentrated in a small stream of gas which is by-passed through a cyclone collector, where the dust is finally collected.

Although no comparative data are available, the collection efficiency of units of this type is probably comparable with that of the single-unit, high-pressure-drop cyclone installation. The clearances are smaller and the centrifugal fields higher than in a cyclone, but these are probably compensated for by the shorter gas path and greater degree of turbulence with its inherent reentrainment tendency. The chief advantage of these units lies in their compactness, which may be a prime consideration for large installations or plants requiring a large number of individual collectors. Caution should be exercised when attempting to apply this type of unit to a dust that shows a marked tendency to build up on solid surfaces, because of the high maintenance costs that may be encountered from plugging and rotor unbalancing. The purchase cost of steel units of this type, exclusive of motor, drive, and optional auxiliaries, is in the range of \$0.10 to \$0.30/(cu. ft./min.) of gas handled, depending on type and size (prices of 1945–1946).

Miscellaneous Inertial Separators. Several types of collectors of the inertial type are available that do not specifically fall into any of the previous classes. These range from a simple high-velocity gas-reversal chamber to a rotary unit containing closely spaced parallel sinusoidal plates. Some of these are shown in Fig. 125(*a*), (*b*), and (*c*).

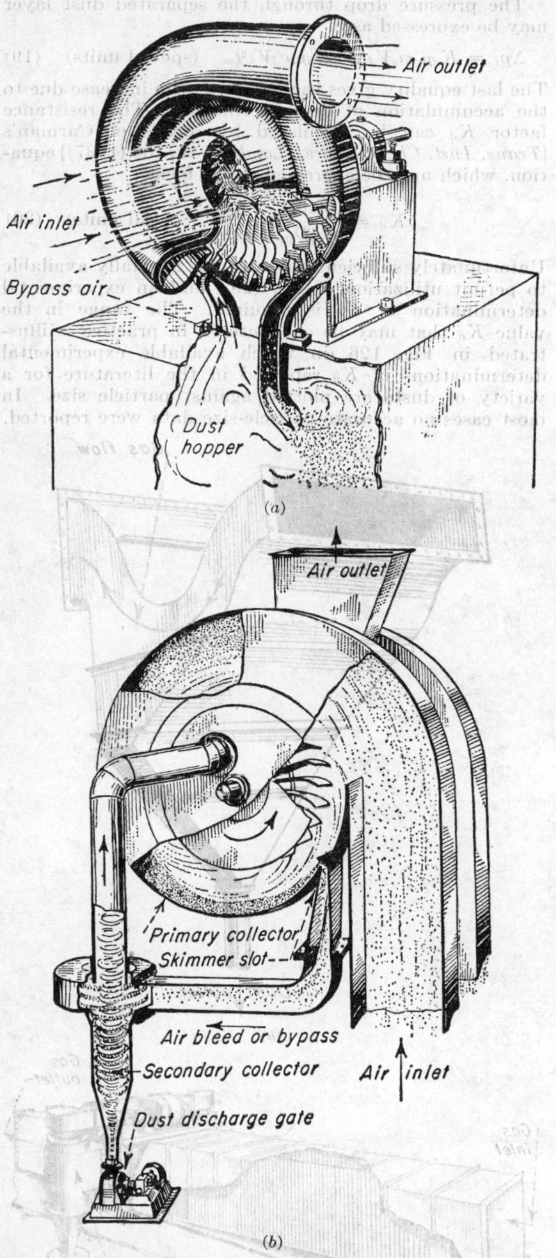

FIG. 124. Typical mechanical centrifugal separators. (*a*) Type D Rotoclone (cutaway view). (*American Air Filter Co.*) (*b*) Sirocco cinder fan (cutaway view). (*American Blower Corp.*)

Most steam or compressed-air entrainment separators are of the inertial type and are intended for the removal of droplets larger than 20 μ diameter. They are available in a wide variety of forms, though most of them employ impingement elements or zigzag parallel plates. Some are cyclonic in action, and a few contain rotating members driven by the gas velocity.

Packed-bed Separators. Although no dust collectors employing beds of packed granular solids are available as standard commercial units, such equipment has been designed for specific applications. The most common type in the chemical industry is the coke box (Wells and Fogg, *U.S. Bur. Mines Bull.* 184, pp. 159–164, 1920), frequently used in the manufacture of sulfuric acid. These collectors generally consist of a tile-, brick-, or lead-lined chamber packed with closely sized coke, with horizontal, upward, or downward gas flow. Coke sizes may range from 1/40 to 1/2 in. diameter, the finer sizes being used in large boxes operating at low superficial velocity to remove very fine mist.

In one common type, the boxes are circular with either upflow or downflow of gas through a 2- to 6-ft. depth of graded coke supported on a tile or brick grid. The effective portion of the bed consists of 1/40- to 1/20-in. coke with thin layers of coarser grades above and below the fine coke for support. Superficial gas velocities may range from 2 to 10 ft./min. with pressure drops in the range of 1 to 10 in. water. For a 30-ton/day sulfuric acid plant, such a box will be on the order of 30 ft. diameter. The cost will be on the same order as that of a Cottrell in this service, but maintenance costs will be less. However, it is subject to gradual plugging, depending on the amount of dust in the gas stream. The sulfuric acid mist handled in such a box has a particle size in the range of 0.5 to 3 μ, and collection efficiencies as high as 99.9 per cent have been obtained with the finer coke sizes operated at low superficial velocity. Increased coke depths and decreased gas velocity or coke particle size will result in increased collection efficiency.

Coarse coke operated at higher superficial gas velocities has been employed for applications involving a coarser mist. Granite, quartz, and sand have also been used as packing materials in some cases.

A relatively recent development is the Lynch granular filter [Anon. (Lynch), *Fuel Economist*, **12**, 47 (October, 1936)], which utilizes a bed of gravel. The gravel is withdrawn from the bottom of the bed continuously and passed over a screen to remove collected dust before the gravel is returned to the top of the bed. Superficial gas velocities employed are on the order of 3 ft./sec. with bed depths of 1 to 4 ft. The gravel ranges from 1/2 to 1 in. diameter, and the pressure drop is on the order of 1 in. water. Steel units have been used up to 850°F., and temperatures up to 1500°F. can be handled with high-chrome steel, and up to 2000°F. with brick.

The separation of dust by packed solids is due to (1) gravity settling, (2) Brownian movement, (3) impingement, (4) interception, and (5) electrostatic separation. Although no conclusive evidence is available, it is doubtful whether electrostatic precipitation, resulting from electrification of the bed or particles due to the relative motion, is an appreciable factor in most cases, since there are no reports that indicate a marked effect of gas humidity. With fine packing, operating at low gas velocities, gravity settling and Brownian movement will be entirely predominant, and collection efficiency would be expected to decrease as the gas velocity increased. With coarse packing, operating at high gas velocities, separation by impingement and interception is controlling, and increased gas velocities would be expected to increase collection efficiency, provided that the gas velocity is not so high as to reentrain collected material. This is in accord with general observation, though almost no data are available that will permit any quantitative evaluations to be made.

Cloth Collectors. In cloth collectors (also known as bag filters), the dust-laden gases are passed through a woven fabric which "filters" out the dust, allowing the gases to pass on. Actually the separation is not a simple filtration, since the pores in the cloth are usually many times the size of the particles separated. When the dust-laden gases first pass through the cloth, the efficiency of separation will be low until enough particles have been removed to build up what corresponds to a "precoat" in the fabric pores. This initial deposition of dust takes place because of interception and impingement on the cloth fibers and by gravity settling and Brownian movement in the pores. With dusts normally encountered in industrial processes, this precoat layer will form in a few minutes, usually only a matter of seconds. Once this layer has formed, the efficiency of separation will usually be well over 99 per cent.

Although direct filtration plays a more important role at this stage, it is not the sole means of separation, since the pores in the collected dust may be considerably larger than some of the dust particles separated. In special cases, *e.g.*, fresh tobacco smoke in a room, the particle size and concentration are so small that an excessive time would be required to form a precoat layer. Normally, however, failure to achieve a very high collection efficiency is due entirely to improper equipment maintenance, from such sources as torn bags, poorly installed bags, or stretched bags.

For conditions encountered in practice, the flow through both the cloth and the collected dust will be streamline in character. The pressure drop through the cloth can be expressed by

$$\Delta p_i = K_c \mu_c V_f \quad \text{(special units)} \qquad (18)$$

where K_c is a constant dependent on the nature of the cloth. Values of K_c for specific clean cloths are given in Table 7. When the pores of the cloth become filled with dust, the value of K_c may be over ten times that for the clean cloth. For dry dusts, however, if the dust layer on the cloth is deeper than about 1⁄16 in. (on the order of 0.1 lb. dust/sq. ft. cloth), the over-all pressure drop across the dust-laden cloth will usually be that across the dust layer, the drop across the cloth itself, including the dust in the pores, being comparatively negligible.

The pressure drop through the separated dust layer may be expressed as

$$\Delta p_i = K_d \mu_c m_s V_f = K_d \mu_c c_d V_f^2 t_m \quad \text{(special units)} \qquad (19)$$

The last equality gives the pressure-drop increase due to the accumulation of dust in time t_m. The resistance factor K_d can be calculated by means of Carman's [*Trans. Inst. Chem. Engrs. (London)*, **15**, 150 (1937)] equation, which may be expressed in the form

$$K_d = \frac{160.0(1 - \epsilon_v)}{\phi_s^2 D_{p\mu}^2 \rho_s \epsilon_v^3} \quad \text{(special units)} \qquad (20)$$

Unfortunately sufficient data are not normally available to permit utilization of Eq. (20), and an experimental determination of K_d is required. The range in the value K_d that may be encountered in practice is illustrated in Fig. 126, in which available experimental determinations of K_d reported in the literature for a variety of dusts are plotted against particle size. In most cases no accurate particle-size data were reported,

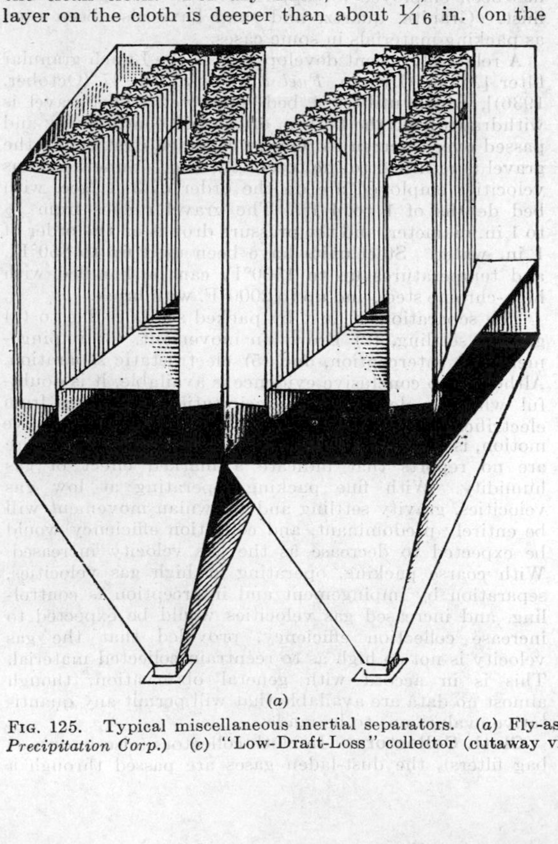

(a)

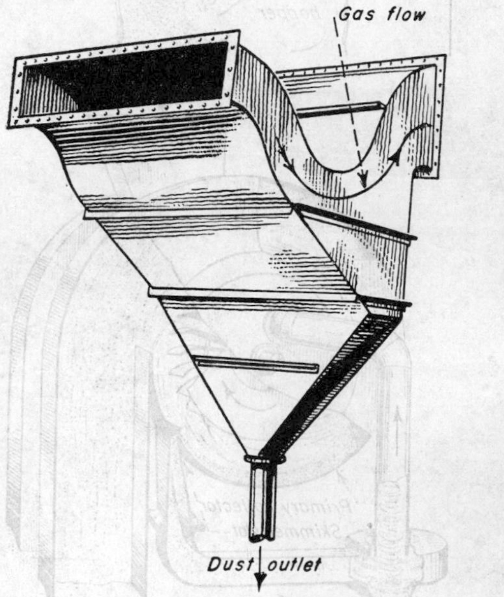

(b)

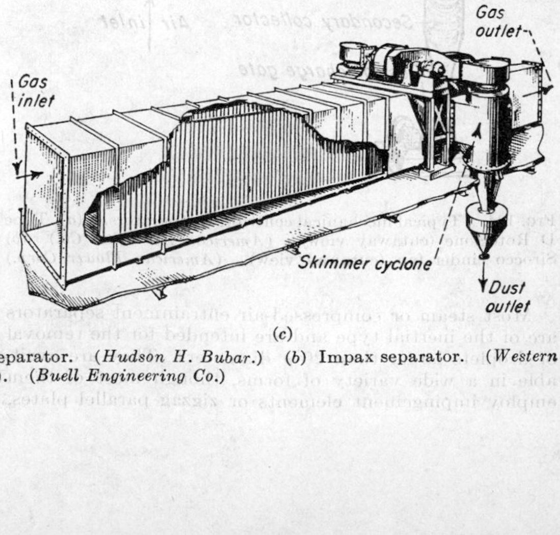

(c)

FIG. 125. Typical miscellaneous inertial separators. (*a*) Fly-ash separator. (*Hudson H. Bubar.*) (*b*) Impax separator. (*Western Precipitation Corp.*) (*c*) "Low-Draft-Loss" collector (cutaway view). (*Buell Engineering Co.*)

Table 7. Resistance Factors for Typical Fabrics

Fabric type	Material	Description: Threads/in.	Nominal weight	Weight, oz./sq. yd.	Thread† diam., in.	Pore size,† in.	K_c‡
Osnaburg (clean)	Cotton	32 × 28			0.02	0.01	0.51
Osnaburg (soiled)*	Cotton	32 × 28					4.80
Drill	Cotton	68 × 40	1.85 yd./lb.	5.28	.01	.01	0.093
Sateen (unnapped)	Cotton	96 × 56	1.55 yd./lb.	6.88	.009	.007	.27
Sateen (unnapped)	Cotton	96 × 64	1.32 yd./lb.	8.23	.01	.005	.88
Sateen (unnapped)	Cotton	96 × 60	1.12 yd./lb.		.012		1.63
Sateen (unnapped)	Cotton	96 × 56	1.05 yd./lb.	10.2	.011	.004	1.12
Wool	Wool		.....	.....	.....	.....	0.25
Wool	Wool	40 × 50	16 oz./yd.	11.5	.014	.....	.33
Glass	Glass	32 × 28	.....	.....	.03	.....	1.60
Tackle twill	Nylon	72 × 196	.....	.....	.010	.....	0.66
Sailcloth	Nylon	130 × 130	.....	.....	.007	.....	1.66
Smoothtex screen	Nickel	(300 mesh)	.....	.....	.....	.....	0.16

* Cloth, similar to previous one, that had been in service and contained dust in pores although free of surface accumulation.

† Estimates based on microscopic examination.

‡ Measured with atmospheric air. This value will be a constant only for streamline flow which, in these measurements, was the case for values of $\rho V_f/\mu_c$ of less than approximately 100. For larger values of $\rho V_f/\mu_c$, K_c tends to increase, since turbulence sets in and kinetic-energy losses become predominant.

$$K_c = \frac{\Delta p_i}{\mu_c V_f}$$

where Δp_i = pressure drop, in. water; μ_c = gas viscosity, centipoises; V_f = superficial gas velocity through cloth, ft./min.; ρ = gas density, lb. mass/cu. ft.

and the curves represent the estimated range of particle size involved. The Williams, Hatch, and Greenburg data comprise a wide variety of dusts, and only the approximate limits enclosing these data are shown. Also included are curves predicted from Eq. (20) for specific values of ϕ_s, ρ_s, and ϵ_v. It is apparent from these

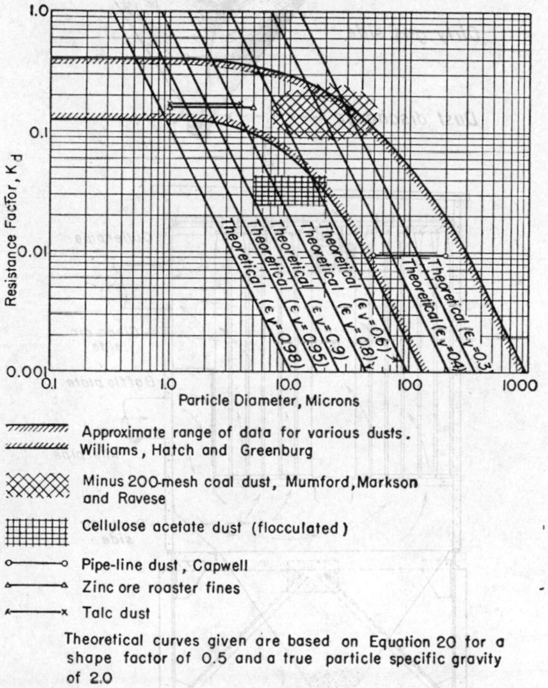

FIG. 126. Resistance factors for dust layers. [*Williams, Hatch, and Greenburg, Heating, Piping and Air Conditioning*, **12,** 259 (1940); *Mumford, Markson, and Ravese, Trans. Am. Soc. Mech. Engrs.*, **62,** 271 (1940); *Capwell, Gas*, **15,** 31 (*August*, 1939).]

curves that smaller particles tend toward higher values of ϵ_v, which is also borne out by the observation that fine dusts have a lower bulk density than coarser fractions of the same material, apparently because of the greater surface forces involved. For sizes under 10 μ, the value of K_d appears to become constant, increased voids compensating for the reduction in size. For coarse dusts, K_d varies approximately inversely as the square of the particle diameter, which implies that the voidage (or bulk density) does not change with particle size.

Equation (20) applies only where the mean free path of the gas molecules is small compared to the particle size of the dust particles. Where the particle size approaches the mean free path of the gas molecules, a correction

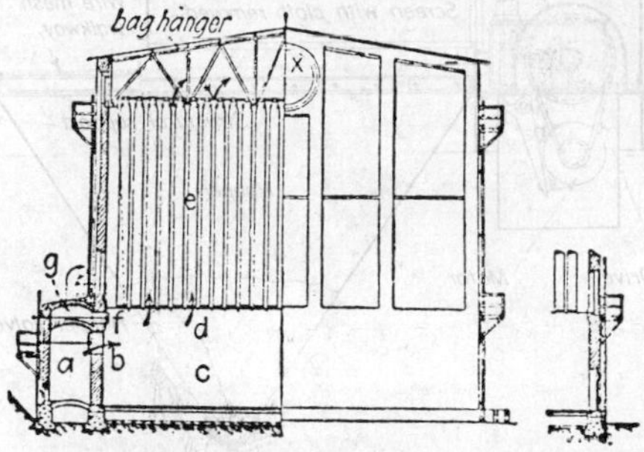

FIG. 127. Typical bag house.

factor must be applied to allow for the so-called "slip flow" and the resistance factor K_d will be less than given by Eq. (20). For atmospheric pressure work, this correction factor becomes appreciable when the particle size of the collected dust is less than 5μ diameter. To correct for slip flow, the value K_d calculated from Eq. (20) must be divided by the factor

$$\left[1 + K_s\left(\frac{1 - \epsilon_v}{\epsilon_v}\right)\left(\frac{\lambda_m}{\phi_s D_p}\right)\right] \text{ (consistent units)}$$

The term K_s is essentially a constant having the approximate value 15. [Carman and Arnell, *Can. J. Research*, **26(A)**, 128 (1948)] when λ_m is calculated from kinetic theory (Fig. 113/XV). The relative constancy of experimental values of K_d at particle sizes under 5μ (Fig. 126) is due to the "slip-flow" correction factor in addition to the increased voidages encountered with smaller particles.

The older baghouses were usually placed on the pressure side of the fan and contained relatively large bags, a typical size being 18 in. diameter by 30 ft. long [Hallows and O'Harra, "Metallurgy of Lead and Zinc," *Trans. Am. Inst. Mining Met. Engrs.*, **121,** 299 (1936)]. These bags are mounted vertically over a dust-receiving hopper (see Fig. 127) and air, entering either at the top or bottom, is discharged through the cloth directly to the atmosphere. Ratings of these units are generally in the range of 0.2 to 1.0 (cu. ft./min.)/sq. ft. cloth area (*i.e.*, superficial cloth velocities are 0.2 to 1.0 ft./min.). Bag life is usually greater than 2 years, although a much shorter life may be obtained depending on the service. The bags are shaken, usually manually, every 2 to 12 hr. Large installations of this type cost on the order of $0.50/sq. ft. cloth area and are usually homemade.

The more recent trend has been toward mechanical filters, which are available as standard commercial units. These comprise two general types. One utilizes cloth envelopes supported by screens (see Fig. 128(*a*) and (*b*)); the other uses either oval or round vertically mounted bags (see Fig. 128(*c*) and (*d*)), usually 5 to 8 in. in diameter and 8 to 17 ft. long. Access platforms for bag maintenance are usually provided on the clean-air side. These units may be shaken manually, although, except for the smaller sizes, motor shaking is generally provided.

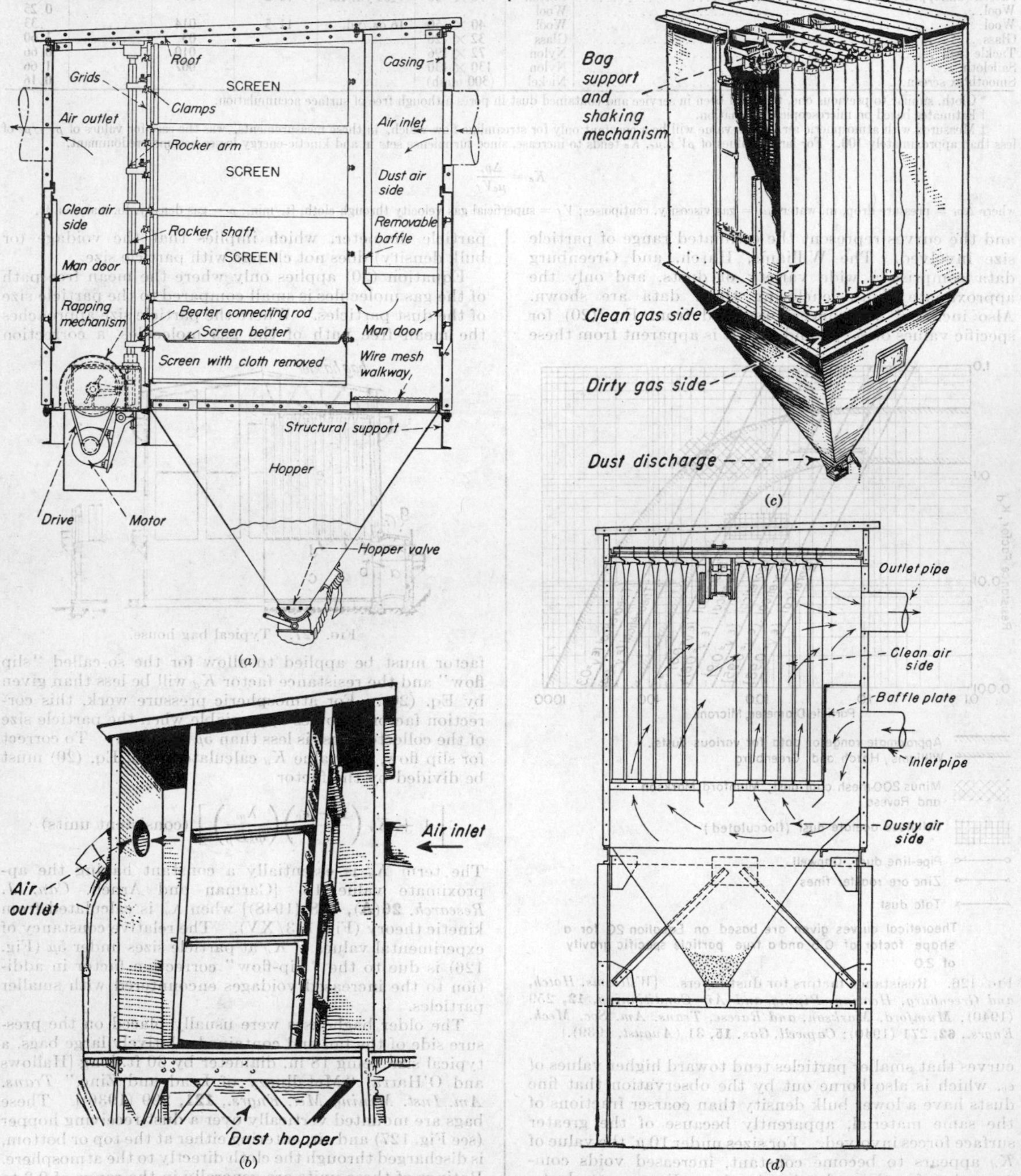

FIG. 128. Typical cloth filters. (*a*) Screen or envelope type (sectional view). (*Pangborn Corp.*) (*b*) Screen or envelope type (cutaway view). (*W. W. Sly Mfg. Co.*) (*c*) Bag type (cutaway view). (*Northern Blower Co.*) (*d*) Bag type (sectional view). (*American

Complete automatic operation is also possible by providing a timer, shaker motor, and air- or motor-operated gas-discharge valves. The small sizes (under 1000 sq. ft. of cloth area) are available as "unit" filters and are shipped completely assembled. The purchase cost of these unit filters is on the order of $0.50 to $1.50/sq. ft. cloth area. Portable unit filters are also available. Large units are built up of standardized rectangular sections in parallel. Each section contains on the order of 1000 to 2000 sq. ft. of cloth, and the sections are assembled in the field to form a single filter housing. In this manner, the filter can be partitioned so that one or more sections at a time can be cut out of service for shaking or general maintenance. Additional capacity can be provided at a later date by adding more sections. The purchase cost of the filter housing and hopper (14 to 16 gage steel) with bags is in the range of $0.30 to $0.70/sq. ft. of cloth area, depending on size and type. Other cost items are: steel supports and outside platforms, 7 to 15 per cent additional; shaker motor and switch, 7 to 10 per cent additional. Other major auxiliary equipment that may be required includes exhauster, exhauster motor, rotary hopper air lock or screw conveyor with drive and motor. The purchase cost of fully automatic filters, complete with shakers, air-reversing valves, timer, manifolds, and supports, but without main fan and motor, is in the range of $1.00 to $2.00/sq. ft. of cloth area, depending on type and size. Normally these filters are furnished with rectangular housings, since these are more economical. However, circular units are available where greater strength is required as in pressure or vacuum service [see Fig. 128(*e*)] (prices of 1945–1946).

A special type of filter is also available which may be classed as automatic and is widely used in grain milling. Cloth bags are arranged radially on a hollow shaft. The air may be passed in from the outside or outward from the shaft, as in the type shown in Fig. 128(*f*). The arrangement is rotated slowly; and, when a given series of bags is over a stagnant-gas area, the bags are rapped with a hammer, which dislodges the dust into a conveyor. A reverse air flow may be provided to assist cleaning. These units are available in 300 to 1500 sq. ft. capacity at a cost of $1.60 to $0.60/sq. ft. of cloth area, respectively. Another type of automatic unit [Fig. 144(*a*)] utilizes a replaceable porous-paper filter medium and is limited in application to relatively low dust concentrations unless a very coarse or fluffy dust is involved (also see p. 1047).

Mechanical filters generally have a pressure drop of 2 to 6 in. water and are rated at 1 to 8 (cu. ft./min.)/sq. ft. cloth area. For very fine dusts or high dust loadings the rating should not exceed 3 (cu. ft./min.)/sq. ft. cloth, and it may be desirable to reduce the rating to $\frac{1}{2}$. For very fine, tacky dusts, bag-type filters are better than screen or envelope types, because of more effective shaking provisions. When dust loadings are high, cyclone precleaners may be employed to reduce the load on the filter.

Ordinary mechanical filters may be shaken every $\frac{1}{4}$ to 8 hr., depending on the service. A manometer connected across the filter is useful in determining when the filter should be shaken. Fully automatic filters may be shaken every 2 min. It is essential that the gas flow through the filter be stopped when shaking in order to permit the dust to fall off. With very fine dust, it may even be necessary to equalize the pressure across the cloth [Mumford, Markson, and Ravese, *Trans. Am. Soc. Mech. Engrs.*, **62**, 271 (1940)]. In practice this can be accomplished without interrupting the operation by cutting one section out of service at a time. In automatic filters this operation involves closing the dampers, shaking the filter units, either pneumatically or mechanically, sometimes accompanied by a reverse flow of cleaned gas through the filter, and, lastly, reopening the dampers. For compressed-air-operated automatic filters, this entire operation may take only 2 to 10 sec. For the ordinary mechanical filters equipped for automatic control, the operation may take as long as 3 min.

The Hersey filter (Fig. 129) is a recent development, in which cleaning is accomplished continuously or periodically, without shutting off the gas flow, by a ring that travels up and down the outer surface of the bag. Compressed air is blown from a $\frac{1}{32}$-in. slot on the inner edge of the ring through the filter fabric in a direction opposite to the flow of the dust-laden gas. This dislodges the dust and allows it to drop into a collecting hopper. The filter bag is made of $\frac{1}{16}$- to $\frac{1}{8}$-in. wool felt in diameters ranging from 8 to 36 in. and lengths up to 40 ft. Gas-handling capacities are high, in the range of 5 to 30 cu. ft./min./sq. ft. of cloth area at a pressure drop in the range of 2 to 6 in. water. Compressed-air requirements, usually furnished by a separate blower at a pressure of $1\frac{1}{2}$ lb./sq. in. gage are on the order of 0.01 to 0.1 cu. ft./cu. ft. of dust-laden gas. Dust loadings as high as 2 lb. dust/lb. gas have been successfully handled. It is reported that collection efficiency is very high (over 99.9 per cent), even during cleaning and that damp dusts can be collected. The purchase cost for a given capacity is comparable to that for the fully automatic mechanical filter described

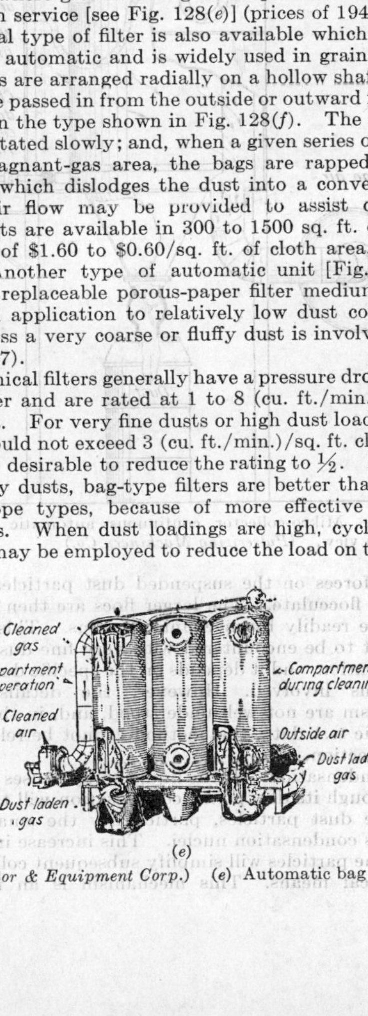

(e)

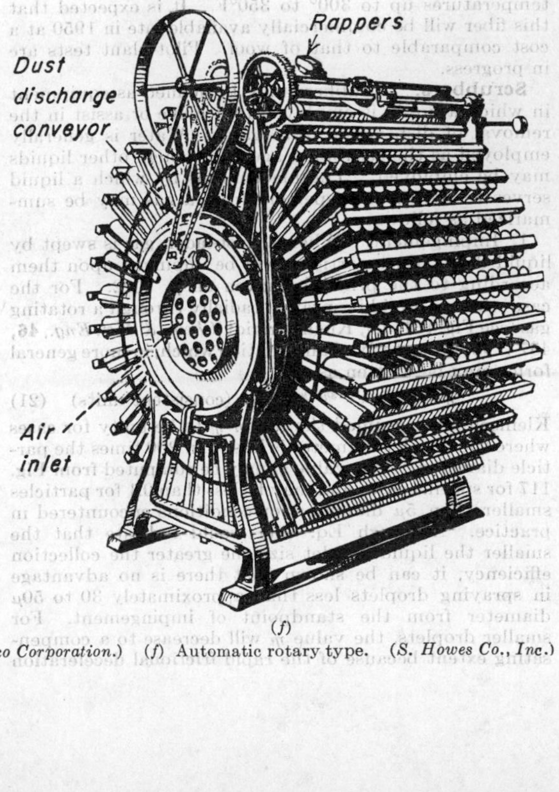

(f)

Wheelabrator & Equipment Corp.) (e) Automatic bag type. (Dracco Corporation.) (f) Automatic rotary type. (S. Howes Co., Inc.)

above. Both suction- and pressure-type units are available.

Filter fabrics are generally cotton sateens of 1.08 to 1.32 yd./lb. weight. For other services, wool, glass, asbestos, or metal may also be used. The maximum recommended operating temperatures are 190°, 235°, 650°, and 650°F. for cotton, wool, glass, and asbestos, respectively, and it is generally advisable to stay well below these limits for good bag life. Bag life varies widely, depending on the frequency of shaking, the nature of the dust, and the gas characteristics. For mechanical filters it is generally 6 months to 2 years, although wider variations are found in specific instances. Wool is generally used for mildly acidic atmospheres, whereas cotton is used for neutral or slightly alkaline conditions. Whereas both glass and asbestos have been used for higher temperature service, results have not been very satisfactory because of excessive bag failure. A recent report on blast-furnace gas cleaning claims that a means of successfully employing a combination glass-asbestos fabric has been developed [Kling, *Blast Furnace and Steel Plant*, **34**, 1257 (1946)]. Cotton sateen costs approximately $0.06/sq. ft., and the cost of wool and glass is approximately four and eight times as much, respectively. When subject to spark ignition, cotton or wool fabrics may be fireproofed by proper impregnation (Mumford, Markson, and Ravese, *op. cit.*).

It is generally economical to be conservative in specifying cloth area. Since pressure drop for a given service varies as the square of the velocity through the cloth, greater cloth area results in considerable reduction in shaking frequency and in increased bag life; incremental cloth costs are also relatively small, particularly in small installations. In operation it is essential that the gas be kept above its dew point to avoid plugging of the bag pores. Cloth filters, however, have been successfully used in steam atmospheres, such as those encountered in vacuum driers. In such cases the housing is generally steam-chased.

Recently a synthetic acrylic fiber, "Orlon," has been developed which is resistant to acidic atmospheres and at temperatures up to 300° to 350°F. It is expected that this fiber will be commercially available late in 1950 at a cost comparable to that of wool. Pilot-plant tests are in progress.

Scrubbers. Scrubbers may be defined as equipment in which a liquid is employed to achieve or assist in the removal of dispersoids from gases. Water is generally employed as the scrubbing agent, although other liquids may be employed. The mechanisms by which a liquid serves to remove a dispersoid from a gas may be summarized as follows:

1. Impingement. When a dust-laden gas is swept by liquid drops, dust particles will be impinged upon them according to the principles given on p. 1022. For the case where a liquid is sprayed radially through a rotating gas (see Fig. 131(*a*)), Kleinschmidt [*Chem. & Met. Eng.*, **46**, 487 (1939)] has derived an equation which, in more general form, may be written as

$$\eta = 1 - e^{-(3\eta_t xy/2D_L)} \quad \text{(consistent units)} \qquad (21)$$

Kleinschmidt assumed the factor η_t to be unity for cases where the droplet diameter is less than 200 times the particle diameter. The value η_t may be computed from Fig. 117 for specific cases and may be less than 0.1 for particles smaller than 5μ diameter for conditions encountered in practice. Although Eq. (21) would indicate that the smaller the liquid-droplet size the greater the collection efficiency, it can be shown that there is no advantage in spraying droplets less than approximately 30 to 50μ diameter from the standpoint of impingement. For smaller droplets, the value η_t will decrease to a compensating extent because of the rapid frictional deceleration of such droplets. Actually finer droplets will be advantageous as the result of another separating mechanism, diffusion, discussed below. Equation (21) may be applied to gravity spray chambers as well as centrifugal scrubbers. For gravity chambers y is generally the height of the chamber; for simple centrifugal chambers y is the radius of the chamber.

2. Diffusion. Dust particles interdispersed among liquid droplets in a gas stream will be deposited on the liquid drops by Brownian diffusion or motion. This mechanism will be predominant in the collection of submicron particles and may be appreciable for dust particles in the range up to 5μ diameter. Diffusion as the result of fluid turbulence may also be a significant factor in dust deposition. Diffusion is claimed to be the controlling action in the recently developed venturi dust and mist scrubber [Johnstone and Roberts, *Ind. Eng. Chem.*, **41**, 2417 (1949)].

3. Humidification. The humidification of a gas by the introduction of a liquid spray may alter the electrostatic

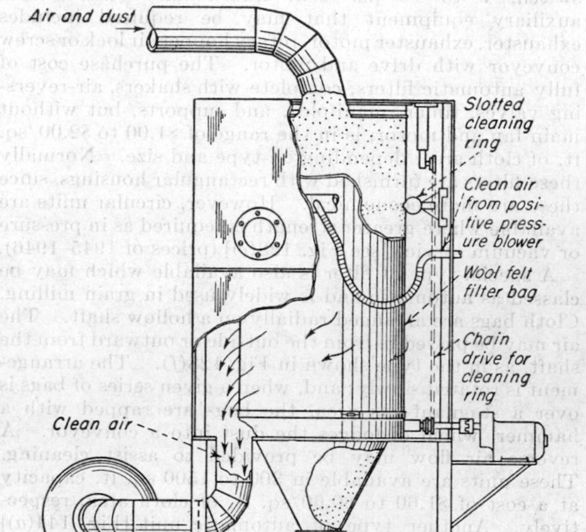

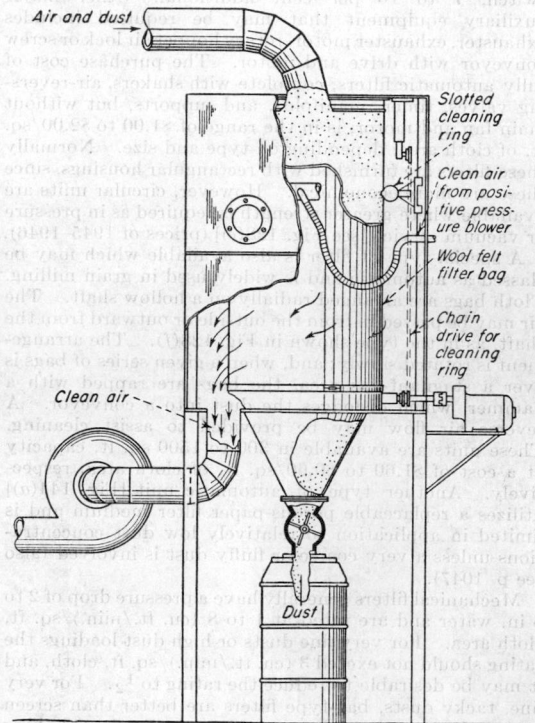

FIG. 129. Mikro-collector continuous automatic bag filter, schematic view. (*Pulverizing Machinery Co.*)

surface forces on the suspended dust particles, causing them to flocculate. The larger flocs are then separated out more readily by mechanical means. This action is most apt to be encountered with very fine dusts at high concentration, and it depends on the specific dust and gas conditions involved. However, the details of this mechanism are not well understood, and, in the absence of specific test data, this method cannot be relied on for dust collection.

4. Condensation. If the liquid spray causes the gas to pass through its dew point, condensation will take place, with the dust particles, particularly the smaller ones, acting as condensation nuclei. This increase in effective size of the particles will simplify subsequent collection by mechanical means. This mechanism is an important

factor only with initially hot gases containing relatively small concentrations (< 1.0 grain/cu. ft.) of dust. To secure an appreciable increase in effective particle size with higher dust concentrations would involve an unlikely amount of condensation.

5. Wetting. Contrary to prevalent beliefs, the wetting characteristics of the liquid do not necessarily play a major role in the actual process of dust collection. Wetting agents, however, may serve to avoid reentrainment of dust particles once they have been impinged on liquid droplets. Reports on the effect of wetting agents are conflicting; but, in general, improvements in collection efficiency resulting from the use of wetting agents have been comparatively minor in magnitude. Further study is required, however, before any definite conclusions can be drawn.

6. Gas Partition. If a gas is passed through a liquid or foam, the gas is segregated into small elements in which the clearances between suspended particles and the surrounding liquid film are relatively small. In such cases separation may be achieved by particle displacements due to gravitational settling and Brownian movement within the bubble, the liquid acting merely as a receiving surface. The mere bubbling of a gas through a liquid is not usually very effective, even with high columns unless the gas is maintained highly dispersed in the liquid.

7. Dust Disposal. In some types of scrubbers the liquid is not dispersed in the gas but flows as a film over collecting surfaces. Aside from humidification or condensation effects, the action of the liquid in such cases is purely one of sweeping the collecting surface free of and avoiding reentrainment of collected dust, the actual collection taking place as the result of the specific mechanical action involved.

8. Electrostatic Precipitation. It is possible that the water droplets may become electrically charged because of the rupture of the water streams and impact on the scrubber walls. This factor may play a role in dust precipitation; but the underlying mechanism is not well understood, and it is questionable whether the effect is very important except in special cases.

Commercial equipment is available that combines scrubbing with almost every other method of dust and mist collection. Unfortunately, however, few comparative performance data are available. In general the simpler types of scrubbers are not very effective in collecting particles finer than about 5 μ diameter. The more effective scrubbers, however, will permit collection of particles as small as about 1 to 2 μ, but collection efficiency drops off very rapidly for finer particles unless flocculation or condensation is a controlling factor. Only a few of the common currently available commercial units are described below. Others are described by Knowles [*Trans. Am. Soc. Heating Ventilating Engrs.*, **24,** 165 (1918)], Meldau ("Der Industriestaub," Ver. deut. Ing. Verlag, Berlin, 1926), Miller [*Chem. & Met.*, **45,** 132 (1938)], Powers [*Rock Products*, **47,** 50 (July, 1944); **48,** 92, 94 (June, 1945)], and Roberts [*Power*, **83,** 345 (1939)].

Chamber Scrubbers. In the conventional air washer the gas is passed horizontally through banks of sprays which may be directed downward, upward, into the gas stream, or with the gas stream. A set of zigzag eliminator plates is provided at the outlet, and frequently such plates are placed between banks of sprays. The pressure drop is generally between 0.1 and 0.5 in. water, the water consumption is 0.5 to 2 (gal./min.)/(1000 cu. ft./min.) gas handled; and superficial gas velocities are on the order of 5 ft./sec. Process gas coolers, in which gases are passed either up or down through a cylindrical tank with sprays located near the top, are also in this category. Such towers may also contain partitions, alternately spaced on opposite sides, with sprays between them. The water jet scrubber shown in Fig. 130 may be considered a modified form of this class. Units of this type are capable of developing a draft up to 8 in. water and have a water consumption on the order of 50 to 100 (gal./min.)/1000 cu. ft. of gas handled for a draft of 1 in. water. In addition to the jet, a water-gas separating device, which may be a simple gas-reversal chamber, must be provided on the outlet side. The jet units are available as standard items in cast iron, steel, lead, lead-lined steel, rubber-lined cast iron, stoneware, and Haveg. Units in other metals or alloys are also available as special orders. Purchase costs for steel eductors are generally in the range of \$0.15 to \$1.50/(cu. ft./min.) of gas handled, depending on the required draft and size of the installation. Sizes ranging from 3 to 72 in. vapor-inlet diameter are available with capacities up to 50,000 cu. ft./min. There is also available a scrubber unit in which the gas is passed through banks of nozzles, consisting of venturi throat pieces and spray nozzles. The water consumption is on the order of 15 (gal./min.)/(1000 cu. ft./min.), but no appreciable draft is developed.

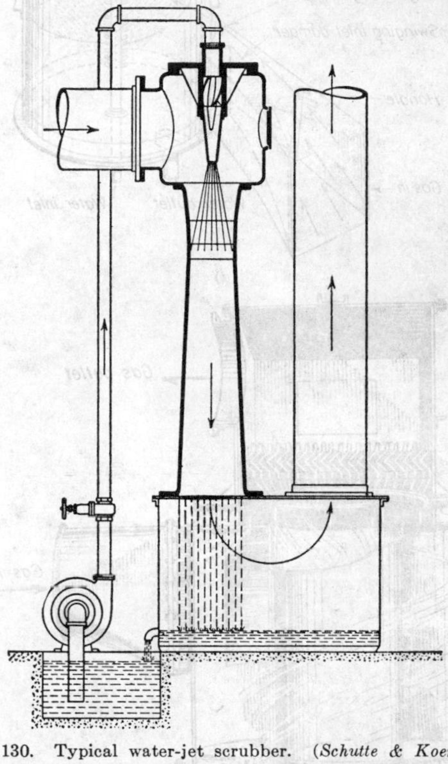

FIG. 130. Typical water-jet scrubber. (*Schutte & Koerting.*)

A recent development that may be placed in this category is the venturi scrubber. This unit involves the introduction of low-pressure water (about 5 lb./sq. in. gage) at the throat of a venturi, at which the gas velocities are in the range of 200 to 300 ft./sec. The water and scrubbed dust or mist are then collected in a cyclone spray separator following the venturi. The pressure drop across the venturi and cyclone is on the order of 15 in. water; water rates are on the order of 3 (gal./min.)/(1000 cu. ft./min.) of gas handled. Very high collection efficiencies on very fine dusts have been reported [Collins, Seaborne, and Anthony, *Paper Trade J.*, **124,** 45 (June 5, 1947); **125,** 45 (Jan. 15, 1948)].

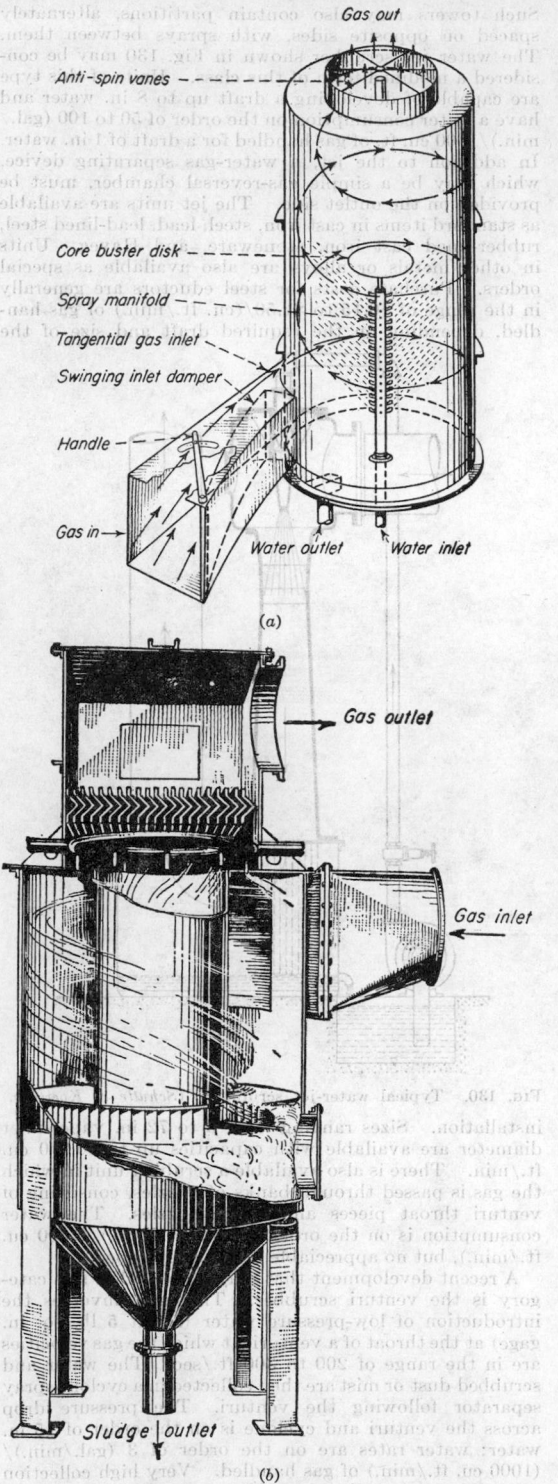

Cyclone Scrubbers. In Fig. 131(*a*), (*b*), and (*c*) are shown typical units in which scrubbing is combined with cyclonic action. In the type shown in Fig. 131(*a*), the central vertical manifold contains nozzles that spray radially across the gas stream and are designed to give as fine a droplet as practicable [Kleinschmidt, *op. cit.* Kleinschmidt and Anthony, *Trans. Am. Soc. Mech. Engrs.*, **63**, 349 (1941)]. In the other two types the liquid dispersion is obtained during passage through the so-called "disintegrator" plates. Several "disintegrator" sections in series may be supplied. In cyclone scrubber units, superficial tower velocities are generally in the range of 4 to 8 ft./sec., and pressure drops usually range from 2 to 8 in. water, while the water circulation is in the range of 3 to 10 (gal./min.)/(1000 cu. ft./min.) of gas handled. The purchase cost of steel units complete

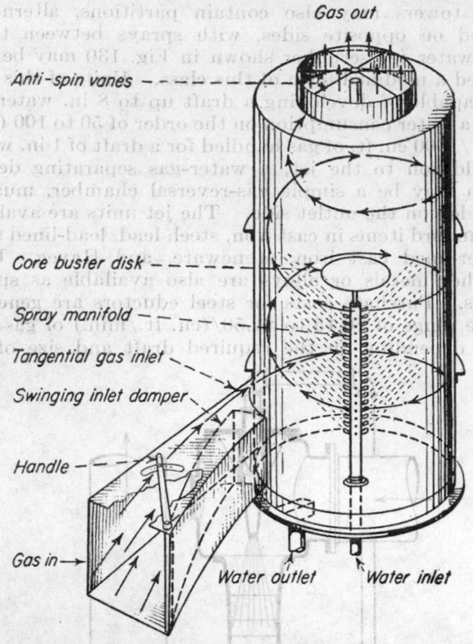

(*a*)

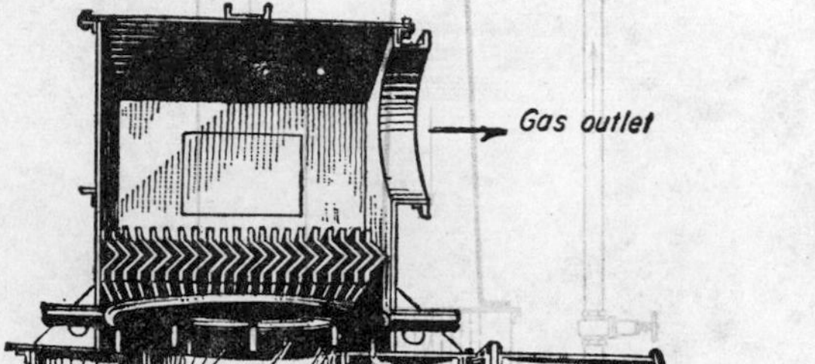

(*b*)

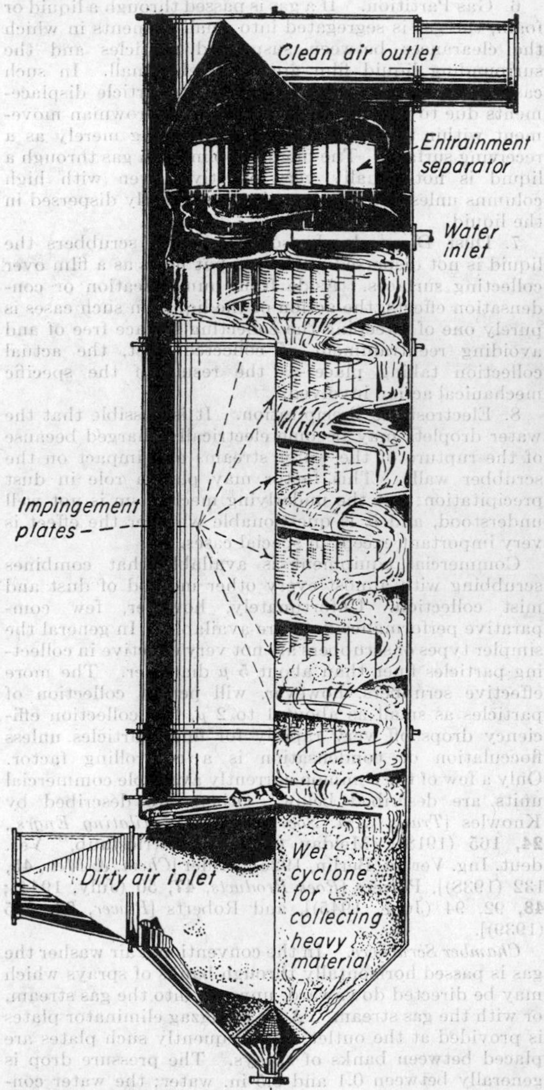

(*c*)

FIG. 131. Typical cyclone-type scrubbers. (*a*) Pease-Anthony (diagrammatic view). [*Kleinschmidt and Anthony, Trans. Am. Soc. Mech. Engrs.*, **63**, 349 (1941); *Pease-Anthony Equipment Co.*] (*b*) Hydro-clone (cutaway view). (*Whiting Corp.*) (*c*) Multiwash (cutaway view). (*Claude B. Schneible Co.*)

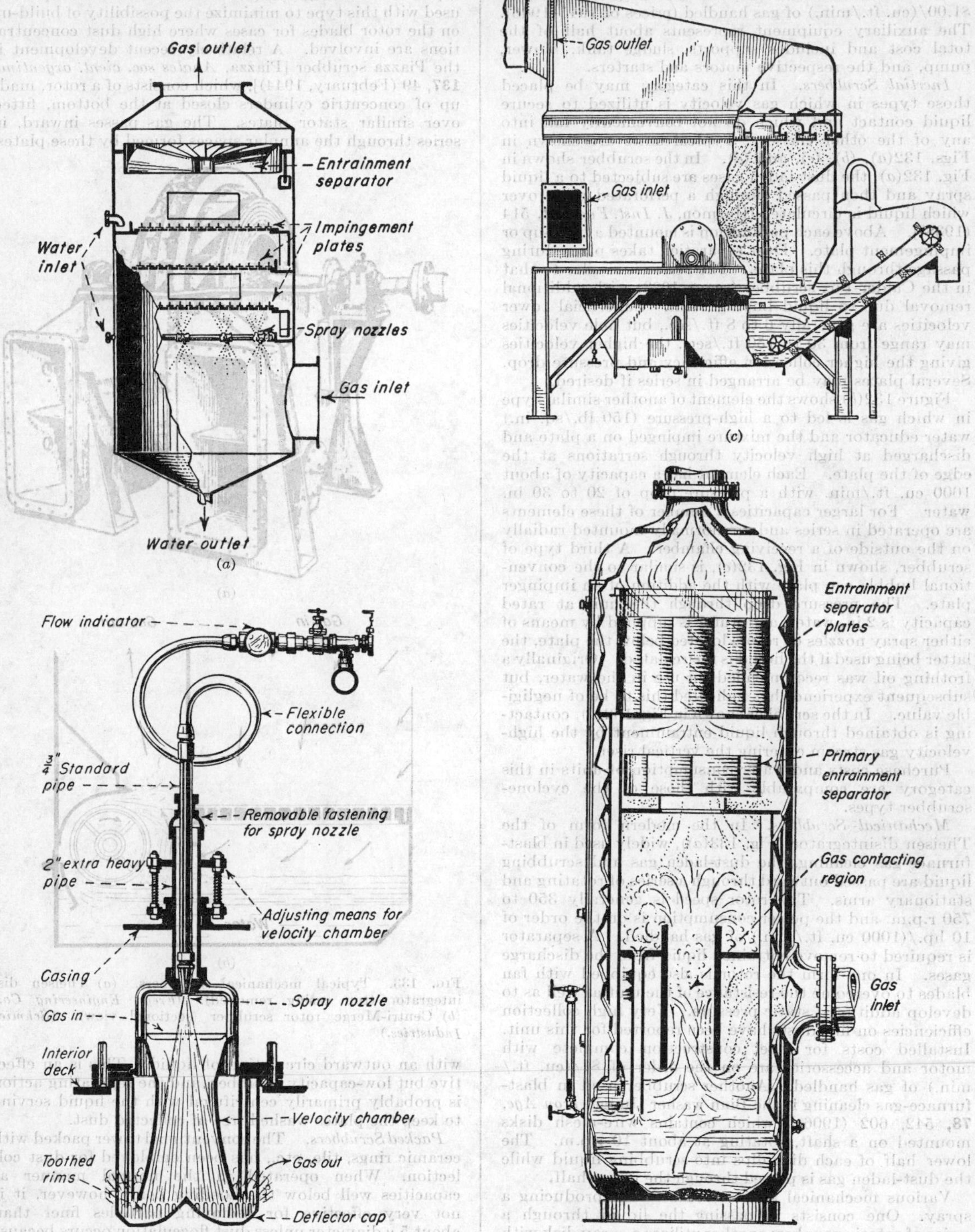

FIG. 132. Typical inertial-type scrubbers. (*a*) Peabody scrubber. (*Peabody Engineering Corp.*). (*b*) Brassert disintegrator nozzle. [*Kinney, Blast Furnace Steel Plant*, **31,** 113 (*January*, 1943); *S. P. Kinney Engineers, Inc.*] (*c*) Bubble-cap scrubber with spray rotor (cutaway view). (*The C. O. Bartlett & Snow Co.*) (*d*) Liquid vortex contactor. (*Blaw-Knox Co.*)

with auxiliaries is generally in the range of \$0.30 to \$1.00/(cu. ft./min.) of gas handled (prices of 1945–1946). The auxiliary equipment represents about half of the total cost and includes supports, sludge tank, blower, pump, and the respective motors and starters.

Inertial Scrubbers. In this category may be placed those types in which gas velocity is utilized to secure liquid contact and which do not conveniently fall into any of the other classes. Typical units are shown in Figs. 132(*a*), (*b*), (*c*), and (*d*). In the scrubber shown in Fig. 132(*a*), the dust-laden gases are subjected to a liquid spray and then passed through a perforated plate over which liquid is circulated [Harmon, *J. Inst. Fuel*, **12**, 514 (1938)]. Above each perforation is mounted a back-up or impingement plate. Dust separation takes place during passage through this plate. The action is similar to that in the Calder-Fox scrubber (see p. 1023) with additional removal due to liquid impingement. Superficial tower velocities are generally 6 to 8 ft./sec., but hole velocities may range from 30 to 150 ft./sec., the higher velocities giving the higher collection efficiency and pressure drop. Several plates may be arranged in series if desired.

Figure 132(*b*) shows the element of another similar type in which gas is led to a high-pressure (150 lb./sq. in.) water educator and the mixture impinged on a plate and discharged at high velocity through serrations at the edge of the plate. Each element has a capacity of about 1000 cu. ft./min. with a pressure drop of 20 to 30 in. water. For larger capacities a number of these elements are operated in series and are normally mounted radially on the outside of a receiving chamber. A third type of scrubber, shown in Fig. 132(*c*), is similar to the conventional bubble-cap plate with the addition of an impinger plate. The pressure drop through the unit at rated capacity is 2 in. water, and water is supplied by means of either spray nozzles or rotors located below the plate, the latter being used if the liquid is recirculated. Originally a frothing oil was recommended for use in the water, but subsequent experience has indicated this to be of negligible value. In the scrubber shown in Fig. 132(*d*), contacting is obtained through liquid entrainment by the high-velocity gas stream entering the vertical riser.

Purchase costs and water consumption of units in this category are comparable with those of the cyclone-scrubber types.

Mechanical Scrubbers. In the modern form of the Theisen disintegrator (Fig. 133(*a*)), widely used in blast-furnace-gas cleaning, the dust-laden gas and scrubbing liquid are passed outward through a series of rotating and stationary arms. The rotor speed is generally 350 to 750 r.p.m. and the power consumption is on the order of 10 hp./(1000 cu. ft./min.) of gas handled. A separator is required to remove entrained liquid from the discharge gases. In one form the rotor is also equipped with fan blades to overcome the resistance of the unit as well as to develop additional static pressure. Very high collection efficiencies on fine dusts have been reported for this unit. Installed costs for steel construction complete with motor and accessories are on the order of \$1/(cu. ft./min.) of gas handled. Another scrubber used in blast-furnace-gas cleaning is the Bian washer [Junge, *Iron Age*, **78,** 542, 602 (1906)], which contains wire-mesh disks mounted on a shaft, rotating at about 10 r.p.m. The lower half of each disk dips into scrubbing liquid while the dust-laden gas is passed through the upper half.

Various mechanical devices are used for producing a spray. One consists in passing the liquid through a series of rotating spokes; another utilizes a spray disk with radial ridges. In Fig. 133(*b*) is shown a type that consists of a rotating shaft that dips below the liquid surface.

A number of units are available in which liquid is sprayed into a mechanical centrifugal separator of the type shown in Fig. 124(*a*). Precleaners are frequently used with this type to minimize the possibility of build-up on the rotor blades for cases where high dust concentrations are involved. A relatively recent development is the Piazza scrubber [Piazza, *Anales soc. cient. argentina*, **137,** 49 (February, 1944)], which consists of a rotor, made up of concentric cylinders closed at the bottom, fitted over similar stator plates. The gas passes inward, in series through the annular spaces formed by these plates, with an outward circulation of liquid. This is an effective but low-capacity scrubber, and the separating action is probably primarily centrifugal with the liquid serving to keep the plates washed free of collected dust.

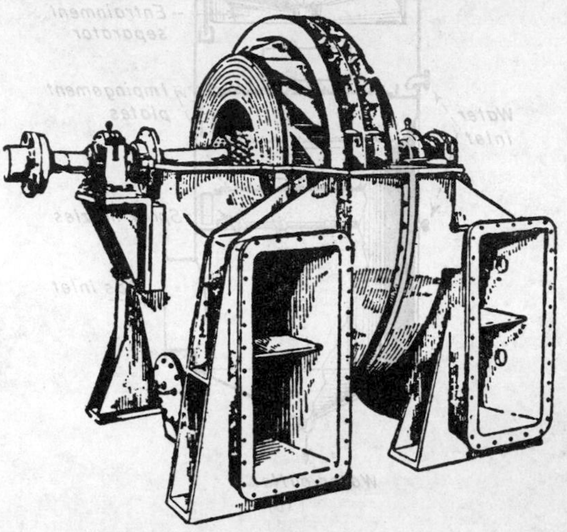

(*a*)

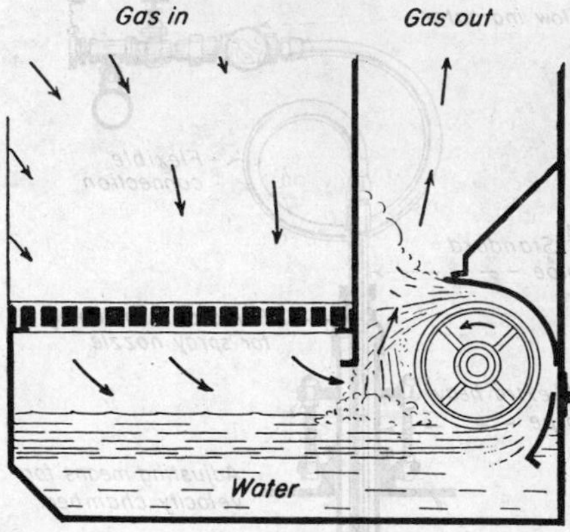

(*b*)

FIG. 133. Typical mechanical scrubbers. (*a*) Theisen disintegrator (top casing removed). (*Freyn Engineering Co.*) (*b*) Centri-Merge rotor scrubber (sectional view). (*Schmieg Industries.*)

Packed Scrubbers. The conventional tower packed with ceramic rings, tile, etc., has been employed for dust collection. When operated in the normal manner at capacities well below the flooding point, however, it is not very effective for removing particles finer than about 5 μ diameter unless dust flocculation occurs because of humidification or condensation. The action would be expected to be primarily one of impingement on the packing with the liquid serving to keep the dust washed off and to avoid reentrainment of dust. Another serious limitation is the possibility of blockage due to dust build-up. A grid packed tower is not particularly subject to the latter limitation, and results of successful operation have been reported [Pearson, Nonhebel, and Ulander, *J. Inst. Fuel,* **8,** 119 (1935); *Combustion,* **6,** 10 (April, 1935)].

Film Scrubbers. This group is intended to include those units in which the sole contact between the gas and liquid is through a liquid film on a collecting surface. Except for collection due to humidification and condensation, the liquid serves merely to keep the collecting surface free of solids and prevent reentrainment of collected dust. In one type the dust-laden gases are passed horizontally across inclined threaded tubes down which water is circulated. In another [Anon., *Engineering,* **77,** 831 (1929); **82,** 12 (1931)] water is allowed to flow down vertical rectangular impingement elements with concave sides.

Miscellaneous Scrubbers. Simple bubbling of gas through a bed of liquid is generally ineffective for the removal of fine dust, because liquid contact is not very good and because gas dispersion is poor, and retention time is relatively short. Good gas dispersion can be obtained by mechanical means or by use of a proper gas sparger. Bransky and Diwoky [*Refiner Natural Gasoline Mfr.,* **19,** 191 (1940)] have reported successful collection of sulfuric acid mist by passing the mist-laden gases up through a deep bed of frothing liquid.

A wide variety of spray-booth scrubbers are commercially available. These consist of a hooded entrance followed by a liquid contacting section, comprising baffles or rotor sprays. The leading face of the rear wall is generally covered with a water film to avoid build-up of sprayed material.

The number and variety of available scrubbers is so great and performance data are so meager that it is extremely difficult to give any specific design recommendations at the present time. Scrubbers are specifically limited to applications where the gases may be cooled and where it is permissible to collect the dust as a slurry, the latter being a particularly important point in the chemical industry. Where these are possible, the proper selection and design of a wet scrubber can give very high collection efficiencies on very fine dust. Wet collection alleviates possibilities of local atmospheric nuisance in disposing of the collected dust but generally complicates the final disposition problem unless the sludge can be returned to the process or sent to a ditch. Wet collection may also intensify corrosion problems.

Electrical Precipitators. When particles suspended in a gas are exposed to gas ions in an electrostatic field, they will become charged and migrate under the action of the field. The functional mechanisms of electrical precipitation may be listed as follows:

1. Gas ionization.
2. Particle collection.
   a. Production of electrostatic field to cause charging and migration of dust particles.
   b. Gas retention to permit particle migration to a collection surface.
   c. Prevention of reentrainment of collected particles.
   d. Removal of collected particles from the equipment.

There are two general classes of electrical precipitators: (1) single-stage, in which ionization and collection are combined; (2) two-stage, in which ionization is achieved in one portion of the equipment, followed by collection in another. Various types in each class differ essentially in the details by which each function is accomplished.

The underlying theory presented in the following paragraphs assumes that the dust concentration is small, since only very incomplete evaluations for conditions of high dust concentration have been made.

Field Strength. Whereas the applied potential or voltage is the quantity commonly known, it is the field strength that determines behavior in an electrostatic field. When the current flow is low (*i.e.,* before the onset of spark or corona discharge), these are related by the following equations for two common forms of electrodes:

Parallel plates:

$$\mathcal{E} = \frac{E}{B_e} \quad \text{(metric units)} \tag{22}$$

Concentric cylinders (wire-in-cylinder):

$$\mathcal{E} = \frac{E}{r \ln (D_t/D_d)} \quad \text{(metric units)} \tag{23}$$

The field strength is uniform between parallel plates, whereas it varies in the space between concentric cylinders, being the highest at the surface of the central cylinder. After corona sets in, the current flow will become appreciable. The field strength near the center electrode will be less than given by Eq. (23), and that in the major portion of the clearance space will be greater and more uniform [see Eqs. (28) and (30)].

Potential and Ionization. In order to obtain gas ionization, it is necessary to exceed, at least locally, the electrical breakdown strength of the gas. Corona is the name applied to such a local discharge that fails to propagate itself. Sparking is essentially an advanced stage of corona in which complete breakdown of the gas occurs along a given path. Since corona represents a local breakdown, it can occur only in a non-uniform electrical field (Whitehead, "Dielectric Phenomena—Electrical Discharge in Gases," p. 40, Van Nostrand, New York, 1927). Consequently, for parallel plates, only sparking occurs at a field strength or potential difference given by the empirical expressions

$$\mathcal{E}_s = \mathcal{E}_o k_\rho \left[1 + \left(\frac{K_o}{k_\rho B_e}\right)\right] \quad \text{(metric units)} \tag{24}$$

$$E_s = \mathcal{E}_o k_\rho B_e + K_o \mathcal{E}_o \quad \text{(metric units)} \tag{25}$$

For air in the range of $k_\rho B_e$ from 0.1 to 2, $\mathcal{E}_o = 111.2$ and $K_o = 0.048$. Thornton [*Phil. Mag.,* **28** (7), 666 (1939)] gives values for other gases. For concentric cylinders (Loeb, "Fundamental Processes of Electrical Discharge in Gases," Wiley, New York, 1939. Peek, "Dielectric Phenomena in High-voltage Engineering," McGraw-Hill, New York, 1929. Whitehead, *op. cit.*) corona sets in at the central wire when

$$\mathcal{E}_c = \mathcal{E}_o k_\rho \left(1 + \sqrt{\frac{K_o}{k_\rho D_d}}\right) \quad \text{(metric units)} \tag{26}$$

$$E_c = \left(\frac{\mathcal{E}_o k_\rho D_d}{2}\right)\left(1 + \sqrt{\frac{K_o}{k_\rho D_d}}\right) \ln \left(\frac{D_t}{D_d}\right) \quad \text{(metric units)} \tag{27}$$

For air approximate values are $\mathcal{E}_o = 110$, $K_o = 0.18$. Corona, however, will only set in if $(D_t/D_d) > 2.718$. If this ratio is less than 2.718, no corona occurs, and only sparking will result, following the laws given by Eqs. (26) and (27) (Peek, *op. cit.*).

In practice, precipitators are usually operated at the highest voltage practicable without sparking, since this increases both the particle charge and the electrical precipitating field. The sparking potential is generally higher with a negative charge on the discharge electrode and is less erratic in behavior than a positive corona discharge. It is the consensus, however, that ozone formation with a positive discharge is considerably less than with a negative discharge, although there are some reports to the contrary. For these reasons negative discharge is generally used in industrial precipitators, and a positive discharge is utilized in air-conditioning applications. In Table 8 are given some typical values for the sparking potential for the case of small wires in pipes of various sizes. The sparking potential varies approximately directly as the density of the gas but is very sensitive to the character of any material collected on the electrodes. Even small amounts of poorly conducting material on the electrodes may markedly lower the sparking voltage. For positive polarity of the discharge electrode, the sparking voltage will be very much lower.

Current Flow. Corona discharge is accompanied by a relatively small flow of electrical current, whereas sparking involves a large flow or power surge and cannot be tolerated for application to electrical precipitation. Besides disruptive effects on the electrical equipment and electrodes, sparking will result in low collection efficiency because of reduction in applied voltage, redispersion of collected dust, and current channeling. Although an exact calculation can be made for the current flow for a d.c. potential applied between concentric cylinders, the following simpler expression, based on the assumption of a constant space charge or ion density, gives a good approximation of corona current [Ladenburg, *Ann. Physik*, **4** (5), 863 (1930)]:

$$I = \frac{8\lambda_i E(E - E_c)}{D_t^2 \ln\left(\frac{D_t}{D_d}\right)} \quad \text{(metric units)} \quad (28)$$

and the average space charge is given by (Whitehead, *op. cit.*)

$$\sigma_{avg} = \frac{4(E - E_c)}{\pi D_t^2 \epsilon} \quad \text{(metric units)} \quad (29)$$

Table 8. Sparking Potentials* (Small Wire Concentric in Pipe)

Pipe diameter, in.	Sparking potential,† volts	
	Peak	Root mean square
4	59,000	45,000
6	76,000	58,000
9	90,000	69,000
12	100,000	77,000

* Data reported by Anderson in Perry, "Chemical Engineers' Handbook," 2d ed., p. 1873, McGraw-Hill, New York, 1941.

† For gases at atmospheric pressure, 100°F., containing water vapor, air, CO_2, and mist, and negative-discharge–electrode polarity.

In the space outside the immediate vicinity of corona discharge, the field strength is sensibly constant, and an average value is given by

$$\mathcal{E} = \sqrt{\frac{2I}{\lambda_i}} \quad \text{(metric units)} \quad (30)$$

which applies if the potential difference is above the critical potential required for corona discharge so that an appreciable current flows.

Ionic mobilities are given by Loeb ("International Critical Tables," vol. 6, p. 107, McGraw-Hill, New York, 1929). For air at 0°C., 760 mm. Hg, $\lambda_i = 624$ (cm./sec.)/(statvolt/cm.) for negative ions. Positive ions usually have a slightly lower mobility. Loeb ("Fundamental Processes of Electrical Discharge in Gases," p. 62, Wiley, New York, 1939) gives a theoretical expression for ionic mobility of gases which is probably good to within ±50 per cent.

$$\lambda_i = \frac{100.0}{k\rho\sqrt{(\delta_o - 1)M}} \quad \text{(metric units)} \quad (31)$$

In general, ionic mobilities are inversely proportional to gas density. Ionic velocities in the usual electrostatic precipitator are on the order of 100 ft./sec.

Electric Wind. By virtue of the momentum transfer from gas ions moving in the electrical field to the surrounding gas molecules, a gas circulation is set up between the electrodes, known as the "electric" or "ionic" wind. For conditions encountered in electrical precipitators, the velocity of this circulation is on the order of 2 ft./sec. Also, as a result of this momentum transfer, the pressure at the collecting electrode is slightly higher than at the discharge electrode (Whitehead, *op. cit.*, p. 167).

Charging of Particles. [Deutsch, *Ann. Physik*, **68** (4), 335 (1922); **9** (5), 249 (1931); **10** (5), 847 (1931). Ladenburg, *op. cit*, Mierdel, *Z. tech. Physik*, **13**, 564 (1932).] Three forces act on a gas ion in the vicinity of a particle: attractive forces due to the field strength and the ionic image; and repulsive forces due to the Coulomb effect. For spherical particles larger than 1 μ diam., the ionic image effect is negligible, and charging will continue until the other two forces balance according to the equation

$$N_o = \left(\frac{\zeta \mathcal{E} D_p^2}{4\epsilon}\right)\left(\frac{\pi\sigma\epsilon\lambda_i t}{1 + \pi\sigma\epsilon\lambda_i t}\right) \quad \text{(metric units)} \quad (32)$$

The ultimate charge acquired by the particle is given by

$$N_o = \frac{\zeta \mathcal{E} D_p^2}{4\epsilon} \quad \text{(metric units)} \quad (33)$$

and is very nearly attained in a fraction of a second. For particles smaller than 1 μ diameter, the initial charging will occur according to Eq. (32). However, owing to the ionic-image effect, the ultimate charge will be considerably greater because of penetration resulting from the kinetic energy of the gas ions. For charging times of the order encountered in electrical precipitation, the ultimate charge acquired by spherical particles smaller than about 1 μ diameter may be approximated (±30 per cent) by the empirical expression

$$N_o = 3.4 \times 10^3 D_p T \quad \text{(metric units)} \quad (34)$$

Values of N_o for various sized particles are listed in Table 9 for 70°F., $\zeta = 2$, and $\mathcal{E} = 10$ statvolts/cm.

Particle Mobility. By equating the electrical force acting on a particle to the resistance due to air friction, as expressed by Stokes's law, the particle velocity or mobility may be expressed by

(*a*) For particles larger than 1 μ diameter:

$$\lambda_p = \left(\frac{u_e}{\mathcal{E}_p}\right) = \frac{\zeta D_p \mathcal{E}_i K_m}{12\pi\mu} \quad \text{(metric units)} \quad (35)$$

(*b*) For particles smaller than 1 μ diameter:

$$\lambda_p = \left(\frac{u_e}{\mathcal{E}_p}\right) = \frac{360 K_m \epsilon T}{\mu} \quad \text{(metric units)} \quad (36)$$

For single-stage precipitators, $\mathcal{E}_i$ and $\mathcal{E}_p$ may be considered as essentially equal. It is apparent from Eq. (36) that the mobility in an electrical field will be almost the same for all particles smaller than about 1 μ diameter, and hence, in the absence of reentrainment, collection efficiency should be almost independent of particle size in this range. Very small particles will actually have a greater mobility because of the Stokes-Cunningham correction factor. Values of u_e are listed in Table 9 for 70°F., $\zeta = 2$, and $\mathcal{E} = \mathcal{E}_i = \mathcal{E}_p = 10$ statvolts/cm.

Table 9. Charge and Motion of Spherical Particles in an Electrical Field

For $\zeta = 2$, and $\mathcal{E} = \mathcal{E}_i = \mathcal{E}_p = 10$ statvolts/cm.

Particle diam., μ	Number of elementary electrical charges, N_o	Particle migration velocity,* u_e, ft./sec.
0.1	10	0.27
.25	25	.15
.5	50	.12
1.0	105	.11
2.5	655	.26
5.0	2,620	.50
10.0	10,470	.98
25.0	65,500	2.40

* Includes Stokes-Cunningham correction factor.

Collection Efficiency. Although the actual particle mobilities may be considerably greater than would be calculated on the above basis because of the action of the electric wind in single-stage precipitators, the latter acts in a compensating fashion, and the over-all effect of the electric wind is probably to provide an equalization of particle concentration between the electrodes similar to the action of normal turbulence (Mierdel, *op. cit.*). On this basis Deutsch (*op. cit.*) has derived the following eqs. for collection efficiency, the form of which had previously been suggested by Anderson on the basis of experimental data:

$$\eta = 1 - e^{-(u_e A_e/q)} = 1 - e^{-K_e u_e} \quad \text{(consistent units)} \quad (37)$$

For the concentric-cylinder (or wire-in-cylinder) type of precipitator, $K_e = 4L_e/D_t V_e$; for rod-curtain or wire-plate types, $K_e = L_e/B_e V_e$. Strictly speaking, Eq. (37) applies only for a given particle size, and the over-all efficiency must be obtained by an integration process for a specific dust distribution, as described on p. 1026. However, over limited ranges of performance conditions, Eq. (37) has been found to give a good approximation of the over-all collection efficiency, with the term for particle migration velocity representing an empirical average value. Such values, calculated from over-all collection-efficiency measurements, are given in Table 10 for specific installations.

For two-stage precipitators with close collecting-plate spacings (Figs. 139, 140), the gas flow is substantially streamline, and no electric wind exists. Consequently, neglecting reentrainment, collection efficiency may be expressed as [Penney, *Elec. Engineering*, **56**, 159 (1937)]

$$\eta = \frac{u_e L_e}{V_e B_e} \quad \text{(consistent units)} \quad (38)$$

Table 10. Performance Data on Typical Single-stage Electrical Precipitator Installations*

Type of precipitator	Type of dust	Gas volume, cu. ft./min.	Average gas velocity, ft./sec.	Collecting electrode area, sq. ft.	Over-all collection efficiency, %	Average particle migration velocity, ft./sec.
Rod curtain.....	Smelter fume	180,000	6	44,400	85	0.13
Tulip type.....	Gypsum from kiln	25,000	3.5	3,800	99.7	.64
Perforated plate.	Fly ash	108,000	6	10,900	91	.40
Rod curtain....	Cement	204,000	9.5	26,000	91	.31

* Courtesy Research Corp.

which holds for values of $\eta \leqq 1.0$. In practice, however, extraneous factors may cause the actual efficiency to approach a relationship of the type given by Eq. (37).

In general, increased pressure increases precipitation efficiency, although a somewhat higher potential is required, because it reduces ion mobility and hence increases the potential required for corona and sparking. Increased temperature reduces collection efficiency because ion mobility is increased, lowering critical potentials, and because gas viscosity is increased, reducing migration velocities.

Application. The theoretical considerations expounded above should be used only for order-of-magnitude estimates, since a number of extraneous factors may enter into the actual performance. Corona discharge actually takes place from local but mobile points rather than in an even distribution along the discharge electrode. In actual installations rectified alternating current is employed. Hence the electrical field is not fixed but varies continuously, depending on the wave form of the rectifier, although Schmidt and Anderson [*Elec. Engineering,* **57,** 332 (1938)] report that the wave form is not a critical factor. Allowances for high dust concentrations have not been fully studied, although Deutsch (*op. cit.*) has presented a theoretical approach. In addition, irregularities on the discharge electrode will result in local discharges. Such irregularities can readily result from dust incrustation on the discharge electrodes due to charging of particles with opposite polarity within the thin but appreciable glow or ionization layer surrounding this electrode. Very high dust loadings increase the potential difference required for corona and reduce the current due to the space charge of the particles. This tends to reduce the average particle charge and reduces collection efficiency. This can be compensated for by increasing the potential difference when high dust loadings are involved.

If the collected dust is not a good conductor, a high potential may develop across this dust layer. This not only reduces the potential across the gas stream but may result in spark discharge with resultant back ionization and reentrainment of dust. Schmidt and Anderson (*op. cit.*) and Anderson [*Physics,* **3,** 23 (July, 1932)] claim that this may be a controlling factor in the collection of all but conducting dust or mists. They state that an increase in relative humidity of 5 per cent may double the precipitation rate because of its effect on the conductivity of the collected dust layer. Mierdel and Seeliger [*Trans. Faraday Soc.,* **32,** 1284 (1936)] further discuss the problem and suggest the following remedies:

1. Avoid thick dust accumulation by using perforated or slotted electrodes.
2. Add moisture or conductive salts to increase the conductivity of the collected dust layer.
3. Change the wave form, using, for example, an alternating current superimposed on the direct current.

Beaver [*Trans. Am. Inst. Chem. Engrs.,* **42,** 251 (1946)] based upon work done by White, indicates that the critical resistivity above which such difficulties may be anticipated is on the order of 10^{10} ohm-cm. Steam, sodium chloride, and ammonia have all been successfully used as conditioning agents to increase conductivity in specific cases. Conditioning as used in the metallurgical field is discussed by Welch [Metallurgy of Lead and Zinc, *Trans. Am. Inst. Mining Met. Engrs.,* **121,** 304 (1936)]. In the great majority of applications, however, the resistivity of the dust and water-vapor content of the gases is such that good precipitation is obtained without the addition of any such agents (Beaver, *op. cit.*).

In order to achieve a maximum collection efficiency, electrical precipitators are operated as close to the sparking voltage as practicable without actually sparking. The following gives the order of magnitude of current, field strength, and ion density usually encountered in practice:

$I = 3 \times 10^3$ to 3×10^4 statamp./cm. (0.001 to 0.01 milliamp./cm.)
$\mathcal{E} = 5$ to 20 statvolts/cm. [(1500 to 6000 v./cm.)]
$\sigma = 10^8$ to 10^9/cc.

Single-stage Precipitators. The single-stage type of unit, commonly known as a Cottrell precipitator, is most generally used for dust or mist collection from industrial process gases. The corona discharge is maintained throughout the precipitator and, besides providing initial ionization, also serves to prevent redispersion of precipitated dust and recharges neutralized or discharged particle ions. Cottrell precipitators may be divided into two main classes, the so-called plate type (Fig. 134), in

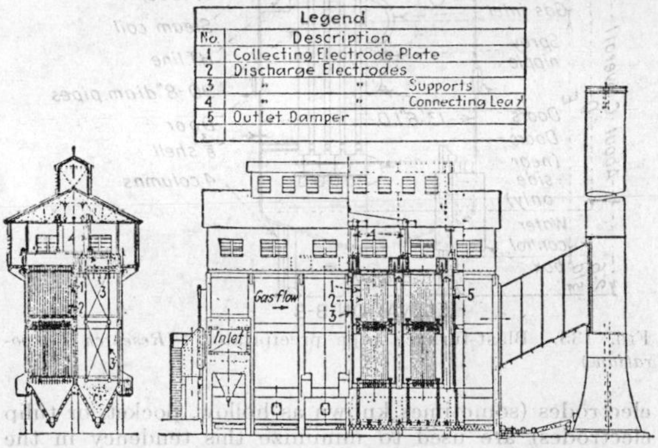

FIG. 134. Horizontal-flow plate precipitator used in cement plant. (*Western Precipitation Corp.*)

which the collecting electrodes consist of parallel plates, screens, or rows of rods, chains, or wires; and the pipe type (Fig. 135), in which the collecting electrodes consist of a nest of parallel pipes which may be square, round, or any other shape. The discharge or precipitating electrodes in each case are wires or rods, either round or edged, which are placed midway between the collecting electrodes or in the center of the pipes and may be either parallel or perpendicular to the gas flow in the case of plate precipitators. Where the collecting electrodes are screens, or rows of rods or wires, the gases are usually passed parallel to the plane of each but may also be passed through it. In pipe precipitators, the gas flow is generally vertically up through the pipe, although downflow is not unusual. The pipe-type precipitator is usually used for the removal of liquid particles and volatilized fumes [Beaver, *op. cit.* Cree, *Am. Gas J.,* **162,** 27 (March, 1945)], and the plate type is used mainly on dusts. In the pipe type, the discharge electrodes are usually suspended from an insulated support and kept taut by a weight at the bottom. Cree (*op. cit.*) discusses the application of electrical precipitators to tar removal in the gas industry.

Except where liquid dispersoids are being collected, or, in the case of film precipitators, where a liquid is circulated over the collecting electrode surface (Fig. 136), thus continuously removing the precipitated material, the collected dust is dislodged from the electrodes either periodically or continuously by mechanical rapping or scraping, which may be performed automatically or manually. Perforated-plate or rod-curtain precipitators are frequently rapped without shutting off the gas flow and with the electrodes energized. This procedure, however, results in a tendency for reentrainment of collected dust. Sectional or composite-plate collecting

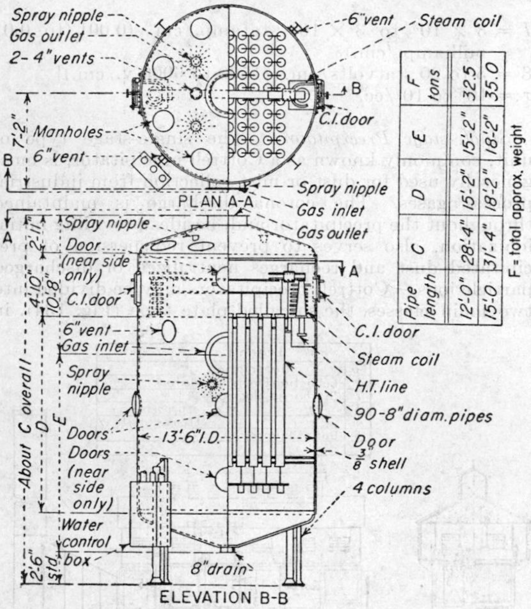

Pipe length	C	D	E	F tons
12'-0"	28'-4"	15'-2"	15'-2"	32.5
15'-0	31'-4"	18'-2"	18'-2"	35.0

FIG. 135. Blast-furnace pipe precipitator. (*Research Corporation.*)

electrodes (sometimes known as hollow, pocket, or tulip electrodes) are used to minimize this tendency in the continuous removal of the precipitated material, provided that it is free-flowing. These are generally designed for vertical gas flow and comprise a collecting electrode containing a dead air space and provided with horizontal protruding slots that guide the dust into this space (see Fig. 137) although some types use horizontal flow.

Semiconductors, such as concrete reinforced with conducting rods, are sometimes used as collecting electrodes for gases in which there is a tendency to disruptive discharge at a potential difference below that required for efficient precipitation. The resistance of the electrode tends to suppress the discharge and thereby stabilizes the electric field. In this case dust may be removed by dragging scrapper chains across the concrete slab, usually with the gas flow shut off. This type is sometimes known as a "graded resistance" precipitator because of the spacing of the reinforcing rods relative to the discharge electrodes in order to provide a maximum electrode resistance across the largest air gap [Schmidt, *Trans. Am. Inst. Chem. Engrs.*, **21**, 11 (1928)]. This type generally permits greater capacity and greater dust accumulation than other types, the dust in some cases being allowed to build up until it drops off by its own weight. However, it is not effective in its intended

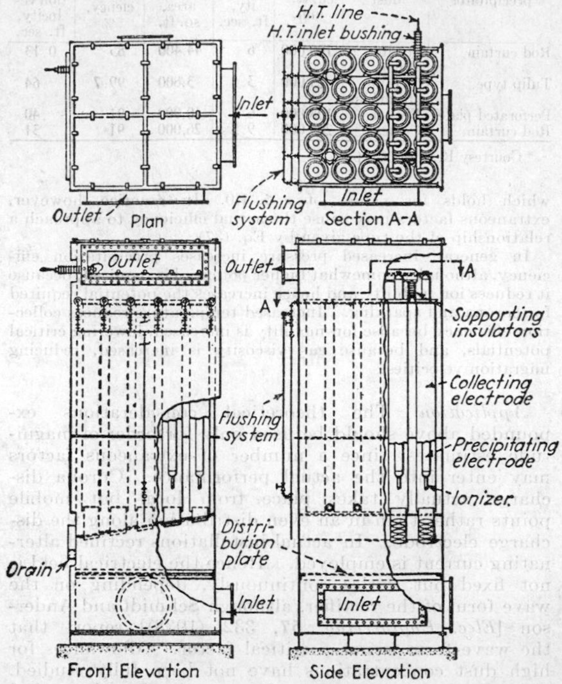

FIG. 136. Two-stage water-film pipe precipitator. (*Western Precipitation Corp.*)

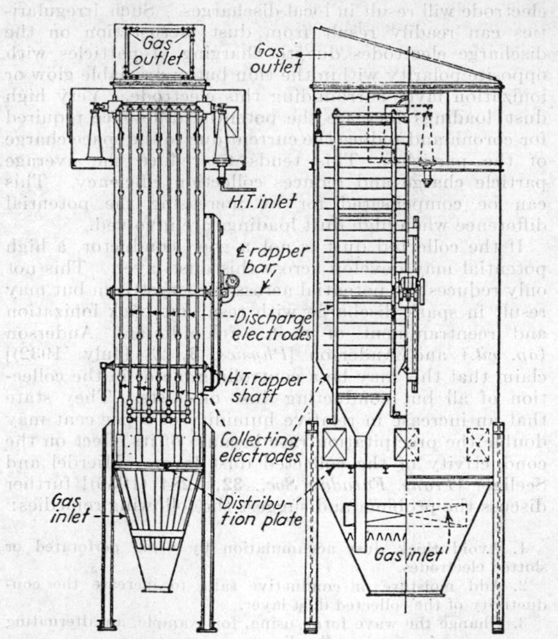

FIG. 137. Vertical-flow heavy-duty plate precipitator. (*Western Precipitation Corp.*)

capacity for very conductive gases or collected materials, since surface creepage tends to destroy the graded resistance effect. Although collecting electrodes are usually metallic, carbon has been used for special corrosive service. Where the precipitated material is a liquid that forms a conducting film, insulators such as glass, terra cotta, or wood have also been used as collecting electrodes.

The choice of size, shape, and type of electrode is based on economic considerations and is usually determined by the characteristics of the gas and suspended matter and by mechanical considerations, such as flue arrangement, the available space, and previous experience with the electrodes on similar problems. The spacing between collecting electrodes in plate-type precipitators and the pipe diameter in pipe-type precipitators usually range from 6 to 15 in. The smaller the spacing the lower the necessary voltage and over-all equipment size, but the greater the difficulties involved in maintaining proper alignment and resulting from disturbances due to collected material. Large spacings are usually associated with high dust concentration in order to minimize spark-over due to dust build-up. For very high dust concentrations, such as those encountered in fluid catalyst plants, it is advantageous to use greater spacings in the first half of the precipitator than in the second half. Precipitators, especially of the plate type, are frequently built with groups of collecting electrodes in series in a common housing. Collecting electrodes are generally on the order of 3 to 6 ft. wide and 10 to 18 ft. high in plate-type precipitators and 6 to 15 ft. high in pipe types. It is essential for good collection efficiency that the gas be evenly distributed across the various electrode elements. Although this can be achieved by proper gas-inlet transitions and guide vanes, perforated plates or screens located on the upstream side of the electrodes are generally used for distribution. Perforated plates of screens located on the downstream side may be used in special cases.

Electrical precipitators are generally designed for collection efficiency in the range of 90 to 99.9 per cent. It is essential, however, that the units be properly maintained in order to achieve the required collection efficiency. Electrical power consumption is generally 0.2 to 0.6 kw. per 1000 cu. ft./min. of gas handled, and the pressure drop across the precipitator unit is usually less than 0.5 in. water, ranging from 1/4 to 1 in. and representing primarily distributor and entrance-exit losses. Applied potentials range from 30,000 to 100,000 volts. Gas velocities are generally in the range of 3 to 10 ft./sec., and rentention times are in the range of 1 to 6 sec. However, velocities must be kept below a certain limit, depending on the collected material, in order to avoid mechanical reentrainment of collected material.

Electrical precipitators are generally energized by rectified alternating current of commercial frequency. The voltage is stepped up to the required value by means of a high-voltage transformer and then rectified. Synchronous motor-driven mechanical rectifiers, together with the necessary switchboard and regulating instruments, are generally used with single-stage precipitators, except for the smaller sizes, since these are ordinarily the least expensive to build and operate for the power loads involved. Electron, mercury-arc, and contact rectifiers, however, are also used. A comparatively recent development is half-wave rectification, with resultant reduction in electrical-equipment requirements and power consumption. Wave-smoothing filters are seldom used for single-stage precipitators since pulsating rectified current appears to give better performance. The electrical equipment is usually housed in a separate substation, which can be located either adjacent to the precipitator or at some remote distance. In certain cases the electrical equipment may be included in a separate compartment of a common shell of the precipitator unit (Fig. 138). The transformer and rectifying equipment may also be housed in a cabinet located near or on the precipitator with the controls located at some convenient operating point.

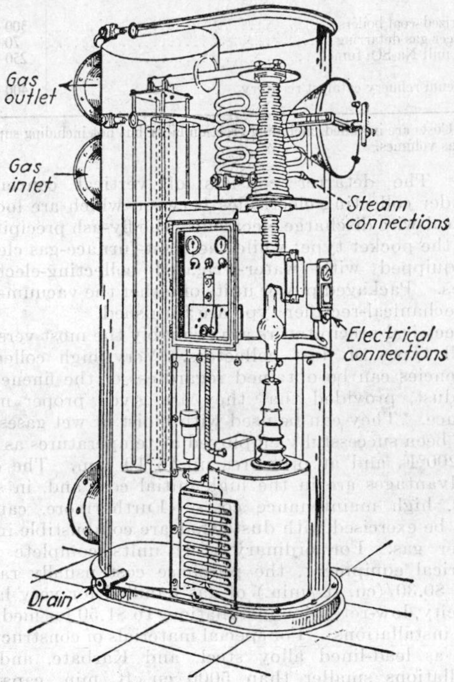

FIG. 138. Pipe precipitator with built-in electron-tube power unit. (*Research Corporation.*)

Electrode insulators must also be designed for a particular service. The properties of the dust or mist and gas determine their design as well as the physical details of the installation. Conducting mists require special allowances such as oil seals, energized shielding cups, or air bleeds. With saturated gas, steam coils are frequently used to prevent condensation on the electrodes.

Typical applications in the chemical field (Beaver, *op. cit.*) include detarring of manufactured gas, removal of acid mist and impurities in contact sulfuric acid plants, recovery of phosphoric acid mists, removal of dusts in gases from roasters, sintering machines, calciners, cement and lime kilns, blast furnaces, carbon-black furnaces, regenerators on fluid catalyst units, chemical recovery furnaces in soda and sulfate pulp mills, and gypsum kettles. Figure 137 shows a vertical-flow steel-plate type precipitator similar to a type used for catalyst-dust collection in certain fluid-catalyst plants. The application in the carbon-black industry is unique in that the electrical precipitator acts to flocculate the carbon particles, which are subsequently collected in low-velocity cyclones. By not attempting to collect the particles in the precipitator, higher velocities may be used in the electrical precipitator with a correspondingly lower investment cost. This is also a typical case where low-velocity cyclones and cyclones in series are advantageous.

Recently the "Elex" precipitator, Swiss counterpart of the Cottrell precipitator, has been introduced on the American market (Koppers Co.). Three types are avail-

Table 11. Approximate Range of Investment Costs for Specific Cottrell Precipitator Applications (1945)

Application	Gas temp., °F.	Type of precipitator	Type of rectifier	Structural material	Collection efficiency, %	Investment cost* $/(cu. ft./min.)
Metallurgical fume	250–300	Plate or rod curtain	Mechanical	Steel	90 –98	0.80–$2.50
Sulfuric acid mist	100–200	Pipe	Mechanical or Kenetron	Lead	95 –99	2.00–$3.50
Blast-furnace gas	100	Pipe (water-flushed)	Mechanical	Steel	95 –98	0.90–$1.50
Pulverized-coal boilers	300–600	Plate	Mechanical	Steel	95 –98	0.30–$0.60
Producer gas detarring	70–120	Pipe	Mechanical	Steel	95 –98	1.00–$1.50
Paper mill Na_2SO_4 fume	250–300	Plate or rod curtain	Mechanical	Steel, concrete, or ceramic shell	90 –95	0.65–$1.25
Petroleum refinery catalyst recovery	400–600	Plate	Mechanical or Kenetron	Steel	99.0–99.8	1.50–$2.50

* Costs are installed, including erection labor but not including supports, flues, etc., the higher values applying for low gas capacities and the lower values for high gas volumes.

able. The detarrer consists of vertical concentric-cylinder collecting electrodes between which are located vertical-wire discharge electrodes; the fly-ash precipitator is of the pocket type; while the blast-furnace-gas cleaner is equipped with water-irrigated collecting-electrode plates. Packaged power units of either the vacuum-tube or mechanical-rectifier type are furnished.

Electrical precipitators are probably the most versatile of all types of dust collectors. Very high collection efficiencies can be obtained regardless of the fineness of the dust, provided that they are given proper maintenance. They can be used with moist or wet gases and have been successfully employed at temperatures as high as 1200°F. and at pressures up to 10 atm. The chief disadvantages are in the high initial cost and, in some cases, high maintenance costs. Furthermore, caution must be exercised with dusts that are combustible in the carrier gas. For ordinary steel units complete with electrical equipment, the purchase cost usually ranges from $0.30/(cu. ft./min.) of gas handled for very large-capacity, low-retention installations to $1.50 for medium-sized installations. For special materials of construction, such as lead-lined alloy steel, and Karbate, and for installations smaller than 5000 cu. ft./min. capacity, costs may run as high as $10/(cu. ft./min.). Schmidt (*op. cit.*) gives some comparative investment costs and operating labor requirements as shown in Table 12. Landolt [*Chem. & Met.*, **29**, 588 (1923)] reports investment costs ranging for the most part from $0.50 to $3.75/(cu. ft./min.) and annual operating costs of $0.15 to $1.00/(cu. ft./min.) of gas handled, averaging about 30 per cent of the investment cost.

In Table 11 are given some approximate current investment cost ranges for some specific Cottrell precipitator applications.

Table 12. Comparative Cost Figures for Cottrell Precipitators on Cement-kiln Gases*

Basis: Treating 100,000 cu. ft./min. at 90 % collection efficiency

Type of precipitator	Space required, cu. ft.	Cost	Operating power, kw.	Labor, man-hr./24-hr. day
Iron pipe electrodes	57,000	$100,000	30	48
Iron plate electrodes	50,000	80,000	20	24
Graded resistance electrodes	35,000	55,000	10	12

* Schmidt, *Trans. Am. Inst. Chem. Engrs.*, **21**, 11 (1928).

Two-stage Precipitators. In two-stage precipitators corona discharge takes place in the first stage between two electrodes having a non-uniform field (see Fig. 139). This is generally obtained by a fine-wire discharge electrode and a large-diameter receiving electrode. In this stage the potential difference must be above that required for corona discharge. The second stage involves a relatively uniform electrostatic field in which charged particles are caused to migrate to a collecting surface. This stage usually consists of either alternately charged parallel plates or concentric cylinders with relatively close clearance compared with their diameters. The only voltage requirement in this stage is that no sparking occur, though higher voltages will result in increased collection efficiency. Since collection occurs in the absence of corona discharge, there is no way of recharging reentrained and discharged particles. Consequently, some means must be provided for avoiding reentrainment of particles from the collecting surface. It is also essential that there be sufficient time and mixing between the first and second stages to secure distribution of gas ions across the gas stream and proper charging of the dust particles.

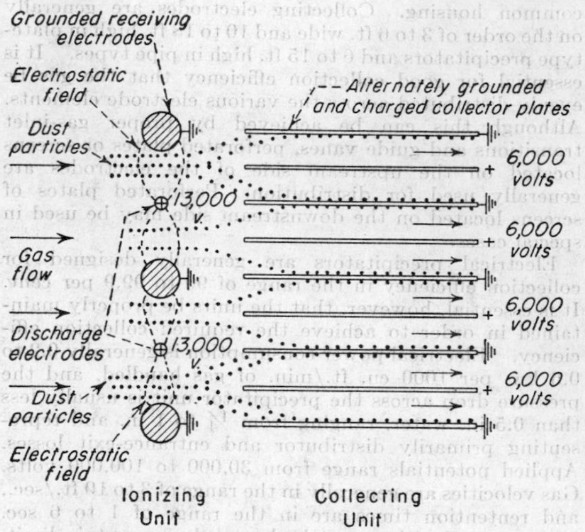

FIG. 139. Two-stage electrical precipitation principle.

In Fig. 136 is shown one of the earlier types of two-stage precipitators used for cleaning process gases. In this unit the ionizing electrode in the first stage is simply a small-diameter extension of the precipitating electrode of the second stage. Reentrainment is avoided and continuous cleaning of the collecting electrode is achieved by circulating a water film down the inside of the collecting electrode.

The large-scale application of two-stage precipitators is, however, a comparatively recent development that has taken place in the air-conditioning field. Several units of the same general type are on the market, a typical one being shown in Fig. 140, whose primary application has been in the cleaning of atmospheric air. In these units the ionizing and collecting stages are built in separate sections of standardized size, and multiple units of each stage are assembled in parallel to meet a specific capacity. The ionizer unit is generally built up of vertical grounded tubes, circular or streamline in cross section and 1¼ in.

nominal diameter, spaced about 3½ in. apart. Between these tubes are stretched parallel discharge or ionizer electrodes, consisting of approximately 0.008-in. diameter tungsten wire. The collecting unit, located on the downstream side of the ionizer unit, consists of 20 B.W.G. plates arranged parallel to the gas flow (see Fig. 139), usually vertical, with alternate plates grounded. These plates are spaced on approximately a 5/16-in. pitch and are about 12 to 18 in. deep.

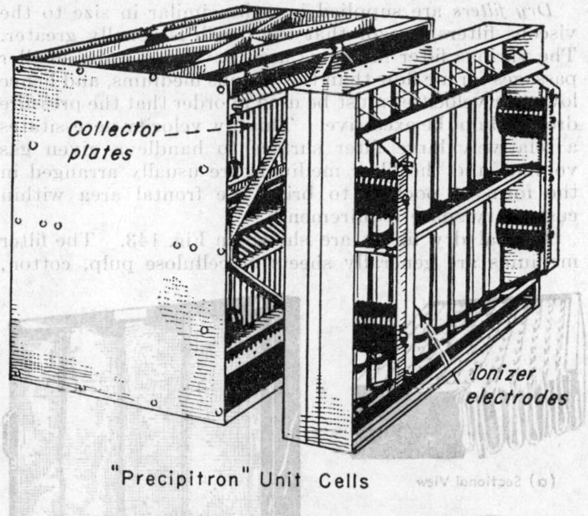

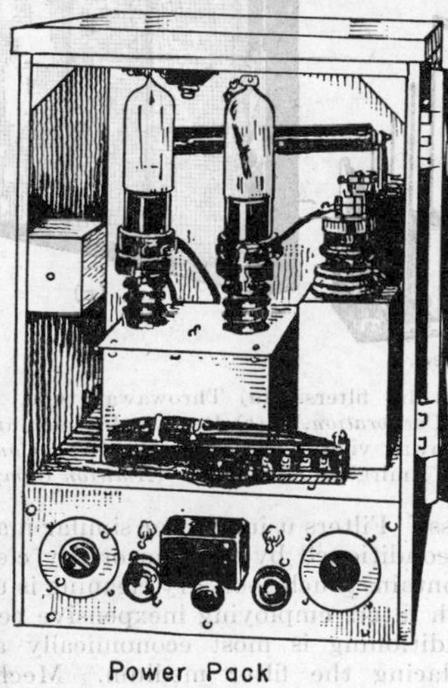

FIG. 140. Typical two-stage electrical precipitator used in air conditioning. (*Westinghouse Electric Corp.*)

Both ionizer and collector sections are generally made in 24- and 36-in.-wide units, the heights being variable depending on the specific type or make. The plates are generally either zinc-coated steel or aluminum. A d.c. potential of 13,000 v. is applied between the ionizer wire and tubes, the wires being positive. A positive d.c. potential of 6000 volts is applied between adjacent collector plates. The necessary voltages for the ionizer and collector plates are obtained through a compact, self-contained vacuum-tube rectifier unit operated directly off of a 110-volt a.c. supply line.

The plates are coated with a viscous oil to avoid re-entrainment of collected dust. When the dust build-up exceeds a depth of approximately 1/16 in., the plate sections must be taken out, washed, and reoiled. Automatic means for cleaning and reoiling in place are also available. Depending on dust concentrations, cleaning may be required every 2 weeks to 3 months. Installations are usually provided with guarded doors that automatically cut off the power when the unit is entered. Where poor approach conditions are involved, perforated-plate air distributors may be employed. The units are rated at 85 to 90 per cent efficiency (U.S. Bureau of Standards Discoloration Test Method) at superficial velocities in the range of 300 to 500 ft./min. Electrical power consumption is approximately 0.02 kw./(1000 cu. ft./min.), and pressure drops range from 0.1 to 0.2 in. water. Purchase costs, complete with power packs and frames, but excluding housings, generally range from $0.10 to $0.25/(cu. ft./min.) of air handled, depending on the type and size of installation (prices of 1945–1946).

A unit is available in which electrostatic precipitation is combined with a dry-air filter of the type shown in Fig. 143(*b*). In another unit an electrostatic field is superimposed on an automatic filter of the type shown in Fig. 144(*b*). In this case the ionizer wires are located on the leading face of the unit, and the collecting electrodes consist of alternate stationary and rotating parallel plates. Cleaning in this case is automatic and continuous. Both these units cost in the range of $0.10 to $0.30/(cu. ft./min.), or approximately three to four times as much as the comparable simple units without the electrostatic provisions.

Although intended primarily for air-conditioning applications, these units have been successfully applied to the collection of relatively non-conducting mists such as oil. However, other process applications have been limited largely to experimental installations. The large cost advantage of these units over the Cottrell precipitator lies in the smaller equipment size made possible by the close plate spacing, in the lower power consumption due to the two-stage operation, and primarily in the mass production of standardized units. In process applications, the close plate spacing is objectionable because of the relatively high dust concentrations involved. Special material or weight requirements for the structural members may eliminate the mass-production advantage except for individual wide applications. Consequently, application to process gases would appear to be limited. A possible outstanding field of application may be that of sulfuric acid mist collection although there are, to date, no commercial installations. The chief difficulty encountered in this case is the short-circuiting of the insulation.

Alternating-current Precipitators. High-voltage alternating current may be employed for electrical precipitation. Corona discharge will result in a net rectification, provided that no spark gaps are used in series with the precipitator. The equipment capacity for a given efficiency is considerably lower than for direct current, however. In addition, difficulties due to induced high-frequency currents may be encountered. The simplicity of an a.c. system, on the other hand, has permitted very satisfactory adaptation for laboratory and sampling purposes [Drinker, Thomson, and Fitchet, *J. Ind. Hyg.*, **5**, 162 (September, 1923)].

Air Filters. The equipment previously described is intended primarily for the treatment of process dusts. Air filters are employed in the elimination of atmospheric dust. The difference in application is not so much one of quality of dust as it is of quantity. Process dust concentrations may run as high as several hundred grains per cubic foot, although usually not exceeding 20 grain. Atmospheric-dust concentrations are generally below 5 grain/1000 cu. ft. Table 13 gives average atmospheric-dust concentrations that may be expected in various districts.

In the elimination of atmospheric dust, no attempt to recover the dust is usually made. Air washers may also be employed for cleaning air, but these are installed primarily for humidifying or cooling the air, and dust removal is of only secondary importance.

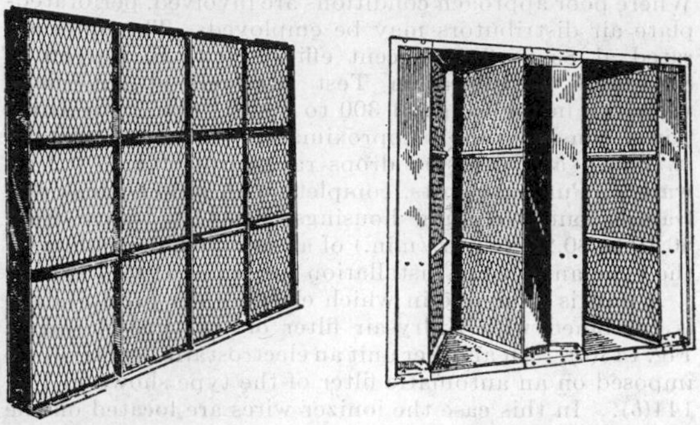

(a) (b)

FIG. 141. Typical types of filter-bank installation. (a) Flat or L-type installation. (b) V-type installation.

Air filters may be classified in three groups on the basis of the type of filter medium employed: viscous, dry, or automatic.

Viscous filters are so called because the filter medium is coated with a viscous material to retain the dust. The filters are supplied in units of convenient size (generally of the order of 20- by 20-in. face area) to facilitate installation, maintenance, and cleaning. Each unit consists of an interchangeable cell or replaceable filter pad and a substantial frame that may be bolted to the frames of other similar units to form an airtight partition between the source of dusty air and its destination (Fig. 141).

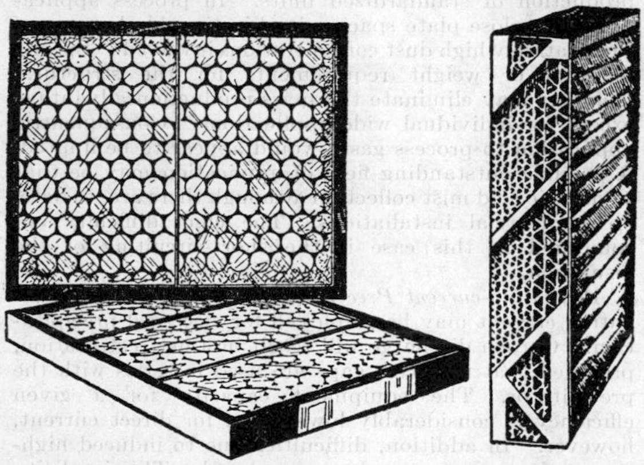

(a) (b)

FIG. 142. Typical viscous unit filters. (a) Throwaway type, Dustop. (*Owens Corning Fiberglas Corporation.*) (b) Cleanable type, Air-Maze Type B, cutaway open-end view. (*Air-Maze Corporation.*)

Felt liners are sometimes used to make the assembly of individual cells airtight. Some types of cell may not have a separate cell frame but are clamped directly to the superstructure.

Typical commercial viscous-filter units are shown in Fig. 142. The filter pad may consist of one of a wide variety of materials, including glass fibers, animal hairs, wood shavings, corrugated fiberboard, split wire, or metal screening. These are coated with a dust-collecting liquid, such as mineral oil, and chemicals of high viscosity and flash point to act as a dust holder.

In the matter of servicing or reconditioning, the viscous-type filters fall into two classes. With most units employing a metallic medium, the unit cells are taken out and washed or steamed, reoiled, and then placed back in service. With the other type, the cell or cell pad is discarded, once the maximum allowable dust load has accumulated, and is replaced by a new one.

Dry filters are supplied in units similar in size to the viscous filters, except that the depth is usually greater. The various filter mediums used have, as a rule, smaller passages for air flow than the viscous mediums, and hence lower air velocities must be used in order that the pressure drop will not be excessive. This low velocity necessitates a relatively large filter surface to handle a given gas volume, and the filter mediums are usually arranged in the form of pockets to bring the frontal area within customary space requirements.

Typical dry filters are shown in Fig. 143. The filter mediums are generally sheets of cellulose pulp, cotton,

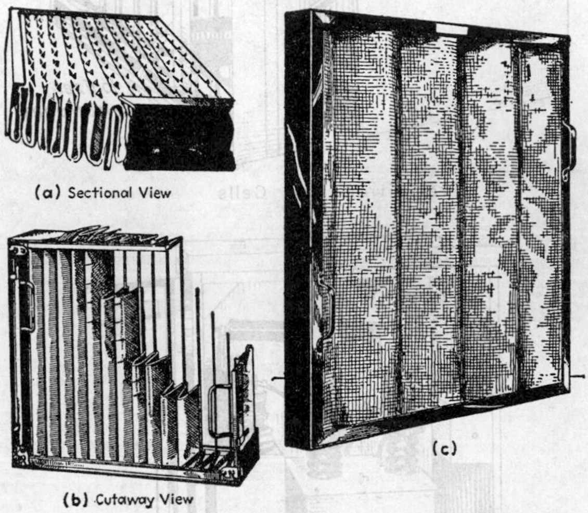

FIG. 143. Typical dry filters. (a) Throwaway type, Airplex. (*Davies Air Filter Corporation.*) (b) Replaceable medium type, Airmat PL-24, cutaway view. (*American Air Filter Company.*) (c) Cleanable type, Amirglass Sawtooth. (*Amirton Company.*)

felt, or spun glass. Filters using felt or similar materials are generally reconditioned by vacuum or dry cleaning. Where the air contains much soot, dry cleaning is usually necessary. With filters employing inexpensive cellulose mediums, reconditioning is most economically accomplished by replacing the filter medium. Mechanical loading devices are often supplied to replace the filter sheets in the frames. In some cases the complete unit cell is discarded when the maximum dust load is reached.

Automatic filters might readily be classed under one or another of the previous groups, since they employ either a viscous-coated or dry-filter medium. They form a distinctive group in the air-filter field, however, in that the cleaning operation is essentially continuous and automatic. Most commercial automatic filters are of the viscous type and consist of perforated, crimped, or woven metallic screens in series [see Fig. 144(*b*) and (*c*)]. The apertures are graded so that the air first meets the larger openings and is subjected to the finest filtering action just before it leaves. The screen curtains are drawn around in a vertical direction, either continuously or intermittently. The oil bath serves to rinse out the dust and coat the screen with a fresh film of oil. The dust is then allowed to settle out as a sludge in the bottom of the hopper. Such filters may be furnished with a hand crank or motor drive as desired.

The Airmat dust arrestor [Fig. 144(*a*)] is a dry automatic filter. It can, however, be considered automatic only when it is applied to dusts that are relatively non-sticky and easily shaken off. It is also used as a dust collector rather than an air filter, since it can handle relatively high dust loads. The air flow must be stopped or diverted, however, when the filter is vibrated.

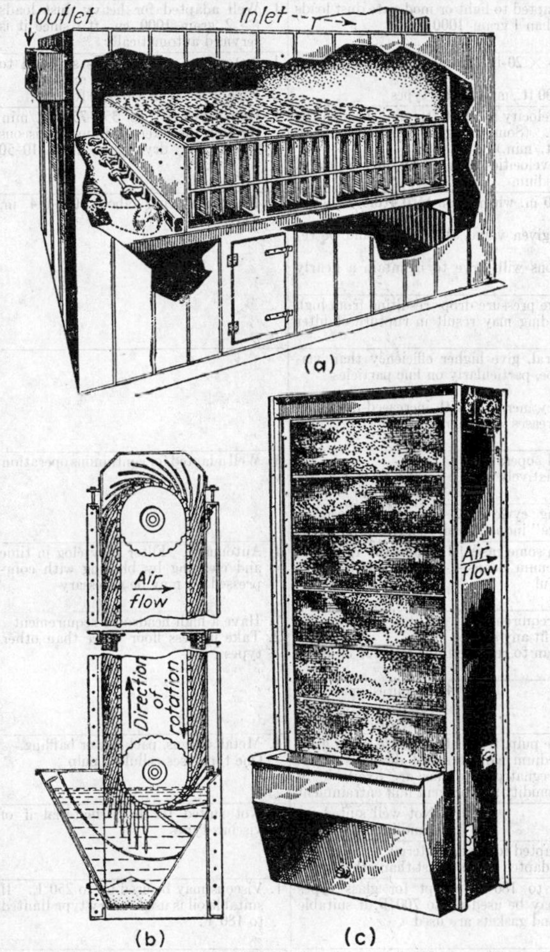

FIG. 144. Typical automatic air filters. (*a*) Dry type, Airmat Dust Arrestor. (*American Air Filter Company.*) (*b*) Multi-Panel. (*American Air Filter Company.*) (*c*) Stay-new Model A. (*Dollinger Corp.*)

General design and performance of air filters. The characteristics of the various types of air filters are compared in Table 14. The pressure drop through a filter increases as dust accumulates. The filter should be replaced when the pressure drop starts to increase rapidly, or else the air capacity will decrease. The maximum allowable pressure drop ranges from about 0.20 to 0.50 in. water, depending on the type of filter medium. The cleanliness of the filtered air may also suffer if cleaning or replacement of dirty air filters is neglected. The dust loading of the air handled generally determines the life of unit filters. For this reason automatic filters become increasingly attractive as the dust concentration of the air to be cleaned increases, since dust capacity is not usually an important item with such filters. The operating cycle or life of a unit filter may be estimated from the following formula:

$$t_d = \frac{500{,}000 m_c}{\eta c_d q_m} \quad \text{(special units)} \qquad (39)$$

This equation is based on a material balance of dust entering and leaving the filter cell. Average dust concentrations encountered in various types of localities are given in Table 13. Dust capacities for unit filters gener

Table 13. Average Atmospheric-dust Concentrations

1 grain/1000 cu. ft. = 2.3 mg./cu. m. = 0.065 mg./cu. ft.

District	Dust Concentration, Grain/1000 Cu. Ft.
Rural or suburban	0.2– 0.4
Metropolitan	0.4– 0.8
Industrial	0.8– 1.5
Dusty factories or mines	4.0–80.0

ally range from 0.5 to 4.0 lb. for a standard 20- by 20-in. unit. In general, the life of viscous-type filters handling average city air may range from 2 to 5 months, while for dry-type filters it will range from 1 to 3 months. For the average chemical plant the life will be from one-half to one-third that of the same filter handling "average" city air and may at times be considerably less because of the higher dust loadings involved.

The matter of a proper schedule for servicing filters cannot be too strongly stressed if satisfactory operation is to be obtained. The over-all time cycle for reconditioning washable viscous filters is generally about 24 hr. Unless they are allowed to drain sufficiently, an entrainment of oil by the filtered air may result.

In viscous air filters, dust collection is achieved by the impingement of particles on the filter surface, according to the principles given on p. 1022, with the viscous coating serving to prevent reentrainment of separated dust. Collection efficiency generally increases as the gas velocity increases unless the velocity becomes so high as to reentrain the dust together with the viscous coating. Efficiency also tends to decrease with increasing dust accumulation because of a saturation of the viscous coating with dust.

In dry-air filters, gravity settling, Brownian movement, and impingement probably all play a role in the collection process, with direct filtration entering only toward the end of the cycle. Increased dust accumulation increases collection efficiency unless the pressure drop becomes so high as to stretch or rupture the fabric. Increased air velocity generally decreases efficiency because of reentrainment.

Ordinary air filters will have a negligibly low efficiency in separating particles below 1 μ diameter. For particle sizes over 10 μ, the filter efficiency for most makes is generally over 85 per cent. An oil impregnation of dry-filter mediums has also proved useful in eliminating any possibility of lint carry-over from the medium itself. Many of the dry mediums have also been fireproofed by suitable treatment.

Power costs may be computed by the following formula with reasonable accuracy (based on an over-all fan-motor efficiency of 63 per cent):

$$\text{Hp.} = \frac{\text{cu. ft. per min.} \times \text{resistance (in. water)}}{4000} \qquad (40)$$

The additional power required for the operation of automatic filters is proportionately very small and may be neglected.

The approximate purchase-cost ranges for commercial steel filters are given in Table 14. For special service, filters may be obtained of stainless steel, galvanized iron, cadmium-plated steel, brass, or aluminum. The American Railroads Association, Mechanical Division (Report on Relative Performance of Air Filters, Jan. 15, 1938, Chicago) and Rowley and Jordan [*Heating, Piping Air Conditioning*, **7**, 293–299 (1935); **10**, 539 (1938); **11**, 388, 633 (1939); **12**, 61, 699 (1940); **13**, 99, 246, 304, 524 (1941); **14**, 19, 438 (1942); **15**, 487 (1943)] report especially complete quantitative and comparative performance data, and Carrier, Cherne, and Grant ("Modern Air

Table 14. Comparative Air-filter Characteristics

	Unit filters				
	Viscous type		Dry type		Automatic filters
	Cleanable	Throwaway	Throwaway	Cleanable	
Dust capacity	1. Well adapted for heavy dust loads (up to 2 grain/1000 cu. ft.) due to high dust capacity		1. Well adapted to light or moderate dust loads of less than 1 grain/1000 cu. ft.		1. Well adapted for heavy dust loads (> 2 grain/1000 cu. ft.) since it is serviced automatically
Filter size		1. Common size of unit filter is 20- × 20-in. face area handling 800 cu. ft./min. at rated capacity 2. Face velocity is generally 300–400 ft./min. for all types			1. Automatic viscous units supplied to handle 1000 cu. ft./min. and over 2. Face velocity is 350–750 ft./min.
Air velocity	1. Rated velocity is 300–400 ft./min. through the filter medium 2. Entrainment of oil may occur at very high velocities		1. Rated velocity is 10–50 ft./min. through the medium. (Some dry glass types run as high as 300 ft./min.) 2. Higher velocities may result in rupture of filter medium		1. Rated velocity is 350–750 ft./min through the filter medium for viscous types. For dry types, it is 10–50 ft./min.
Resistance		1. Resistance ranges from 0.05–0.30 in. when clean to 0.4–0.5 in. when dirty 2. When the resistance exceeds a given value, the cells should be replaced or reconditioned 3. Cycling cells in large installations will serve to maintain a nearly constant resistance			1. Resistance runs about 0.3–0.4 in. water
	4. High resistance due to excessive dust loading results in channeling and poor efficiency		4. Excessive pressure drops resulting from high dust loading may result in rupture of filter medium		
Efficiency	1. Commercial makes are found in a variety of efficiencies, these depending roughly on filter resistance for similar types of medium 2. Efficiency decreases with increased dust load and increases with increased velocity up to certain limits		1. In general, give higher efficiency than viscous type, particularly on fine particles 2. Efficiency increases with increased dust load and decreases with increased velocity		
Operating cycle		1. Well adapted for short-period operations (less than 10 hr./day) due to relatively low investment cost			1. Well adapted for continuous operation
	2. Operating cycle is 1–2 months for general "average" industrial air conditioning		2. Operating cycle is 2–4 weeks for general "average" industrial air conditioning		
Method of cleaning	1. Washed with steam, hot water, or solvents and given fresh oil coating	1. Filter cell replaced. Life may in some cases be lengthened by shaking or vacuum cleaning, but this is not often successful		1. Vacuum cleaned, blown with compressed air, or dry cleaned	1. Automatic. Filter may clog in time and cleaning by blowing with compressed air may be necessary
Space requirement		1. Well adapted for low headroom requirements 2. Form of banks can be chosen to fit any shaped space 3. Space should be allowed for a man to remove filter cells for cleaning or replacement			1. Have a high headroom requirement 2. Take up less floor space than other types
	4. Requires space for washing, reoiling, and draining tanks			4. Requires space for mechanical loader in some cases	
Type of filter medium	1. Crimped, split, or woven metal, glass fibers, wood shavings, hair—all oil coated		1. Cellulose pulp, felt, cotton gauze, spun glass 2. Dry medium cannot stand direct wetting. Oil-impregnated mediums are available to resist humidity and prevent fluff entrainment		1. Metal screens, packing, or baffling. One type uses cellulose pulp
Character of dust	1. Not well suited for linty materials			1. Not well suited for handling oily dusts	1. Not suited for linty material if of viscous types
	2. Well adapted for make-up air and granular materials		2. Well adapted for linty material 3. Better adapted for fine dust than other types		
Temperature limitations	1. All metal types may be used up as high as 250°F. if suitable oil or grease is used. Those utilizing cellulosic materials are limited to 180°F.		1. Limited to 180°F. except for glass types which may be used up to 700°F. if suitable frames and gaskets are used		1. Viscous may be used up to 250°F. If suitable oil is used. Dry type limited to 180°F.
Initial cost	1. Higher first cost than throwaway	1. First cost is relatively low. Frames are generally permanent but cells are replaced		1. Higher first cost than throwaway	1. Highest first cost of all
Operating cost		1. Power costs are comparable for all unit-type filters			1. Power costs are somewhat higher for automatic types
	2. Labor cost to remove, clean, and replace the cells is comparable with the cost of replacement of throwaway-type cells for medium-sized installations of 10,000–50,000 cu. ft./min.	2. Replacement costs are comparable with labor costs for cleanable types for medium-sized installations (10,000–50,000 cu. ft./min.)		2. Same as for Viscous Cleanable	2. Little labor required to inspect, replace oil, or hand crank filter
	3. Maintenance exclusive of filter cells and depreciation are roughly comparable and vary more between different commercial makes in these groups than between different types				3. Depreciation and maintenance costs are the highest for automatic filters
Purchase cost (including frames), cost per 1000 cu. ft./min.	$10 to $25	$5 to $10	$5 to $10 for complete throwaway unit $8 to $30 for units with replaceable medium	$7 to $40	$50 to $100

Conditioning, Heating, and Ventilating," Chap. XI, Pitman, New York, 1940) give more detailed installation and operating-cost information.

Miscellaneous Air-filter Equipment. In Fig. 145 is shown a typical unit in which water is sprayed over a glass-fiber filter cell to combine the functions of dust removal and humidification. The wetted cells are followed by entrainment eliminators which may consist of 1-in. thick glass-fiber mats or zigzag metal plates. Water is circulated at a rate of approximately 3 (gal./min.)/(1000 cu. ft./min.) of air handled and at a pressure of about 6 lb./sq. in. gage, while superficial cell gas velocities are on the order of 300 ft./min. The purchase cost of such units is on the order of $50 to $100/(1000 cu. ft./min.) of air handled, not including piping, duct work, or auxiliary equipment.

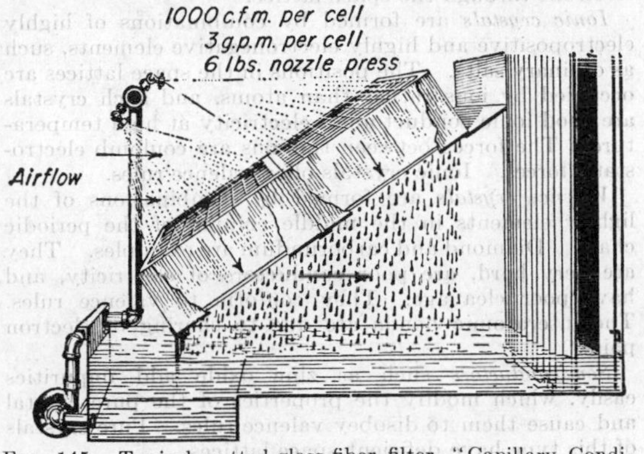

FIG. 145. Typical wetted-glass-fiber filter, "Capillary Conditioner." (*Air & Refrigeration Corp.*)

Complete evaporative cooling installations of this type will cost about $150 to $300/(1000 cu. ft./min.) of air handled (prices of 1945–1946).

A number of units have been developed in which air cleaning is achieved by means of an electrostatic field either by itself or in conjunction with either a dry or viscous coated filtering medium. These units are considerably more effective than dry or viscous air filters in removing particles smaller than 5 μ diameter, and are discussed in detail on pp. 1044 to 1045.

Air filters are available in special forms for use on engine and compressor air intakes as well as for removing dust, scale, and oil in compressed-air or process pipe lines (so-called "pipe-line" filters). A variety of domestic filter units are available for use in window ventilators, room conditioners, or central air-circulation systems. A domestic electrostatic air-cleaning unit, similar to the industrial types discussed on p. 1045, has currently been announced.

A relatively recent development in the air-conditioning field is the use of activated carbon for odor removal [Munkelt, *Refrig. Eng.*, **47**, 21, 32, 40 (January, 1944). Ziel and Sleik, *Heating, Piping and Air Conditioning*, **15**, 367 (1943)], in order to economize on heating costs by reducing the required quantity of outside make-up air. The air is passed through canisters in parallel, as shown in Fig. 146. Each canister is approximately 4 in. diameter by 11 in. high and contains a cylindrical layer of 6- to 14-mesh activated carbon. Carbon bed layers range from 3/8 to 3/4 in. in thickness. The canisters are rated at 25 cu. ft./min. air at a pressure drop of approximately 1/8 in. water. The activated-carbon assembly is generally preceded by a conventional air filter to minimize frequency of reconditioning due to the accumulation of solids. The individual canisters cost from $3 to $5 each and are available as assemblies, including manifold plates but not including casing or erection, at a cost of approximately $100 to $200/(1000 cu. ft./min.) of air handled. Reactivation involves removal and treatment of the canisters at high temperature at a cost of one-fourth to one-fifth of the original investment.

Miscellaneous Collectors. High intensity acoustic vibrations will cause collision and thereby tend to flocculate fumes and mists, whereupon they can be readily collected in conventional apparatus. There is an optimum frequency, generally in the range of 1000 to 10,000 cycles/sec., below and above which effective flocculation will not occur, depending on the size and density of the particle and the viscosity and density of the medium. The U.S. Bureau of Mines has sponsored extensive investigations in the field of ultrasonics [St.

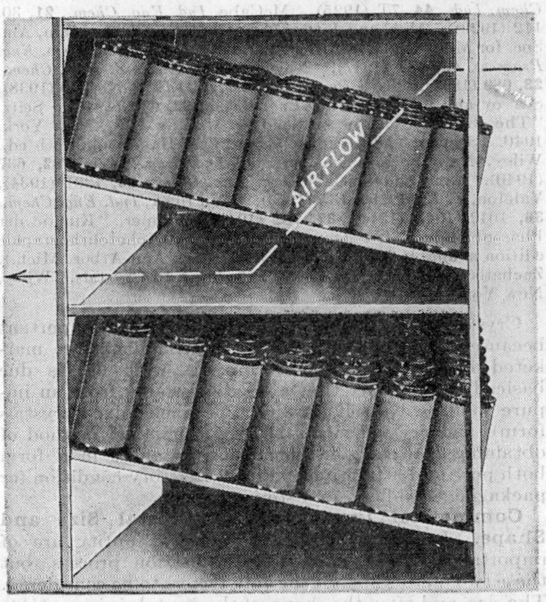

FIG. 146. Activated-carbon-canister filter for odor control. (*W. B. Connor Engineering Corp.*)

Clair *et al.*, U.S. Bur. Mines, R.I. 3400, May, 1938; R.I. 4218, March, 1948; *Mining and Met.*, **18**, 244 (1937)]. It was not until an economical source of acoustic energy was developed recently, however, that commercial application of this principle became feasible. The gas-siren generator (Ultrasonic Corp., Cambridge, Mass.) has permitted the conversion of compressed-gas energy into acoustic energy at 50 to 70 per cent efficiency in units developing up to 100 kw. acoustic output. With this type of generator a compressed-gas pressure of 6 lb./sq. in. gage is normally used for atmospheric-pressure applications and frequency can be varied continuously. A total range in frequency from 1000 to 200,000 cycles/sec. can be covered in several steps. The power consumption for aerosol agglomeration is normally in the range of 2 to 5 kw./(1000 cu. ft./min.) of gas handled and the cost of a complete installation in steel is on the order of $1.00/(cu. ft./min.) for installations in the range of 50,000 cu. ft./min. capacity. Sonic precipitation is limited to cases where the particle concentration is greater than 1 grain/cu. ft. For cases involving a lower particle concentration, sonic precipitation is only feasible if the particle concentration is raised by the introduction of additional particles as with the injection of steam. Pilot-plant or full-scale installations have been made or are under construction for the collection of carbon black, black liquor fumes, sulfuric acid mist, pulp-mill salt cake and soda cement dust, smelter fumes, and fly ash, and development work is under way on the precipitation of fog from airport atmospheres. Further details are given by Jones [*J. Acoust. Soc. Am.*, **48**, 371 (1946)] and Porter [*Chem. Eng.*, **55**, 100, 101, 115 (March, 1948)].

Molecular impacts tend to repel dispersoids away from a heated body. In thermal precipitation this principle is utilized to clean gas of suspended particles by passing the gas through or over a heated grid at low velocity [Blacktin, *J. Soc. Chem. Ind.*, **58**, 334 (1939). Watson, *Trans. Faraday Soc.*, **32**, 1073 (1936)]. Although this method has not yet been applied to industrial-dust collection, it has been very successful for atmospheric-dust sampling.

CRYSTALLIZATION

BY WARREN L. McCABE

References: Badger and Seavoy, "Heat Transfer and Crystallization," Swenson Evaporator Co., Harvey, Ill. (1946). Berthaud, *J. chim. phys.*, **10**, 625 (1912). Griffiths, *J. Soc. Chem. Ind.*, **44**, 7T (1925). McCabe, *Ind. Eng. Chem.*, **21**, 30, 112 (1929). Mehl and Jetters, Age Hardening Symposium, Am. Soc. for Metals, 1939. Miller, Phillips, and Saeman, *Chem. Eng. Progress*, **43**, 667 (1947). Noyes and Whitney, *Z. phys. Chem.*, **23**, 689 (1897). Ross, *Pacific Chem. Met. Inds.*, **2**, No. 3, 9 (1938). Seavoy and Caldwell, *Ind. Eng. Chem.*, **32**, 633 (1940). Seitz, "The Modern Theory of Solids," McGraw-Hill, New York, 1940. Spencer and Meade, "Cane Sugar Handbook," 8th ed., Wiley, New York, 1944. Svanoe, *Ind. Eng. Chem.*, **32**, 637 (1940). Ting and McCabe, *Ind. Eng. Chem.*, **26**, 1201 (1934). Valeton, *Z. Kryst.*, **59**, 135 (1923). Van Hook, *Ind. Eng. Chem.*, **36**, 1042, 1048 (1944); **37**, 782 (1945). Volmer, "Kinetic der Phasenbildung," Dresden and Leipzig, 1939 (photolithographic edition available from Edwards Bros., Inc., Ann Arbor, Mich.). Zachariasen, "Theory of X-ray Diffraction in Crystals," Wiley, New York, 1944.

Crystallization, as an industrial process, is important because of the great variety of materials that are marketed in the crystalline form. Its wide use is due basically to the fact that a crystal forming from an impure solution is itself pure (except when mixed crystals form), and crystallization affords a practical method of obtaining concentrated chemical substances in a form both pure and attractive and in satisfactory condition for packaging, handling, and storing.

Commercial Importance of Crystal Size and Shape. It is obvious that yield and purity are of importance in operating a crystallization process, but these two factors are not the only factors to be considered. The size, and often the shape, of the crystals is important; it is especially necessary that the crystals be of uniform size. Uniformity of size is important for satisfactory appearance and tends to prevent caking, allows easy washing, and results in uniform behavior in use. Large crystals are often demanded, although such demands are not usually justified by any real advantage of large crystals as compared with medium-sized ones. In some cases a definite shape is also required (*i.e.*, needles rather than plates or cubes). Strong single (non-aggregated) crystals not easily broken are required for some purposes.

THEORY OF CRYSTALLIZATION

A crystal is the most highly organized type of nonliving matter. It is characterized by the fact that its constituent parts (atoms or ions) are arranged in orderly array in so-called **space lattices**. The interatomic distances in a crystal of any definite material are constant and characteristic of the material.

Types of Crystals. (See Seitz, "The Modern Theory of Solids," McGraw-Hill, New York, 1940.) Crystalline solids can be classified into five main types. The types vary in the type and strength of the bond between the constituent atoms or ions, and in electrical, magnetic, and mechanical properties.

Metals are formed from the atoms of the electropositive elements. In alloys the atoms of each constituent may either occupy definite positions in the lattice, or each position in the lattice may be occupied by various kinds of atoms in turn. The excellent electrical and thermal conductivities of metals are due to the motions of free electrons through the space lattice.

Ionic crystals are formed by combinations of highly electropositive and highly electronegative elements, such as ordinary salts. The positions in the space lattices are occupied by ions rather than atoms, and such crystals are good ionic conductors of electricity at high temperature. The forces between the ions are coulomb electrostatic forces. Ionic crystals obey valence rules.

Valence crystals are formed by combinations of the lighter elements in the middle column of the periodic chart. Diamond and carborundum are examples. They are very hard, are poor conductors of electricity, and have poor cleavage. They conform to valence rules. The interatomic bonds are due to sharing of electron pairs.

Semiconductors such as zinc oxide add impurities easily, which modify the properties of the pure crystal and cause them to disobey valence rules. Pure crystals of this type have deficient space lattices.

Molecular crystals are characterized by weak bonds that consist essentially of weak residual forces of the van der Waals type. They have low melting points. Organic solids and elements with completed electron shells form such crystals. They are relatively soft and weak. Many can be evaporated molecularly.

Many crystals have characteristics and properties that place them in intermediate positions with respect to the above classification.

Crystal Forms. Law of Haüy. As a result of the space-lattice arrangement of the atoms composing them, crystals, if allowed to form without hindrance from outside bodies, appear in definite polyhedral shapes and exhibit varying degrees of symmetry. It has been found that, although the relative development of the different faces of two crystals of the same material may be widely different, the interfacial angles of corresponding faces of the two crystals are all equal and characteristic of that substance. This is the law of Haüy.

Isomorphism. A generalization that for some time seemed at variance with Haüy's law is the law of isomorphism, which states that in certain series of chemically similar substances the crystals are of the same crystalline form. Until refined methods were available for the measurement of crystal angles it was thought that isomorphic materials gave crystals with the same angles. It has been established, however, that there are small but regular differences in the corresponding angles of isomorphous substances. These differences are of the same sort that exist among other properties of elements in the same periodic group.

Crystallographic Systems. Since the crystals of a definite substance all show the same interfacial angles in spite of wide differences in the extent of development of the faces, crystal forms are classified on the basis of the angles. For example, consider a definite crystal. Take a point and draw lines through this point normal to the faces of the crystal, or to the faces produced. The resulting sheaf of lines is a function of the crystal angles, and the form of the crystal is reflected in the orientation of the lines in this sheaf. Choose three important faces as axial planes. These faces are always taken parallel to planes of symmetry, if there be such. The three intersections of the axial planes determine three non-parallel lines, and the three lines parallel to these intersections drawn through an arbitrary point of intersection are called the axes of the crystal. These axes may be all mutually perpendicular; two of them may be perpendicular to the third, but not perpendicular to each other; they may be equally inclined to each other, with the angle of inclination different from 90 or 60 deg.; or they may be mutually inclined with three different angles, all differing from 90 or 60 deg.

In addition to the three axial faces, a fourth fundamental face, intersecting the three axes, is chosen. The lengths of the segments so cut off from the three axes are expressed as ratios, the length of one of them being taken as unity. The lengths of the axes so determined may be equal for all axes; equal for two axes, but unequal for the third; or unequal for all three.

One class of crystals, showing a hexagonal cross section with 60-deg. angles between normals to the hexagonal sides, is most conveniently referred to four axes instead of the usual three. Three of the axes are at 60 deg. to each other and in the same plane, and the fourth is perpendicular to the plane of the other three.

The combinations of angles and lengths of the axes give rise to seven classes of crystals. These classes are:

1. **The Triclinic System.** Three mutually inclined and unequal axes, all three angles unequal, and other than 90, 60, or 30 deg.
2. **The Monoclinic System.** Three unequal axes, two of which are inclined, but the third is perpendicular to the other two.
3. **The Orthorhombic System.** Three unequal rectangular axes.
4. **The Tetragonal System.** Three rectangular axes, two of which are equal and different in length from the third.
5. **The Trigonal System.** Three equal and equally inclined axes.
6. **The Hexagonal System.** Three equal coplanar axes, inclined to 60 deg. to each other, and a fourth axis different in length from the other three and perpendicular to them.
7. **The Cubic System.** Three equal rectangular axes.

Liquid Crystals. While most substances are obtained in one (or more) crystalline forms, as liquid, and as vapor, some organic substances form another phase intermediate between crystal and liquid. This new phase is called the **liquid-crystal phase.** [See Friedel, *Ann. phys.*, **18** (9), 272 (1922) and also Alexander, "Colloid Chemistry," Chap. 3, Reinhold, New York, 1926.] Materials forming this phase melt at a definite temperature from a solid state to a cloudy viscous liquid, which, when further heated to some higher definite temperature, is transformed to a clear limpid liquid, which is regarded as the true liquid phase. The liquid-crystal phase is birefringent, *i.e.*, is anisotropic. The following substances are among the more than 250 substances which form liquid crystals: ethyl para-azoxybenzoate, ammonium oleate, para-azoxyphenetol, cholesteryl acetate.

Yield of Crystallization Process. In many cases the process of crystallization is slow and the final mother liquor is in contact with a sufficiently large crystal surface so that the concentration of the mother liquor is substantially that of a saturated solution at the final temperature of the process. In such a case the yield of the process is calculated from the composition of the initial solution and the solubility of the material at the final temperature. If appreciable evaporation has taken place during the process, this must, of course, be known or estimated.

Solubility data are ordinarily given as parts by weight of anhydrous material per 100 parts by weight of total solvent, whether the crop contains water of crystallization or not.

When the rate of crystal growth is slow, a considerable time may be required to reach equilibrium. This is especially true where the solution is very viscous, or where the crystals collect in the bottom of the vessel so there is little crystal surface exposed to the supersaturated solution. In such cases the final mother liquor from the process may retain appreciable supersaturation, and the actual yield will be less than that calculated from the solubility curve unless considerable time is allowed for equilibrium to be reached. At any rate, the assumption that the mother liquor is a saturated solution gives the maximum yield of crystals that can be expected. The actual crop, after removal from the crystallizer, will in general retain some adhering mother liquor, which will give an increased weight.

In case the solid product is in the anhydrous form, the calculation of the yield is simple since the solid phase contains no water.

When the crop is hydrated, account must be taken of the water of crystallization in the crystals, since this water is withdrawn from the mother liquor and is not available for retaining the solute in solution.

The following formula can be used to calculate the theoretical yield of a crystallization process. It is valid for either hydrated or anhydrous crystals and assumes only that the mother liquor is saturated with solute at the final temperature, though this last restriction is removed if S (see below) is taken as the actual concentration of solute in the mother liquor at the end of the process. Equation (1) gives the weight of crystals as they exist in the final magma.

$$C = R\,\frac{100w_0 - S(H_0 - E)}{100 - S(R - 1)} \tag{1}$$

where C = weight of crystals in final magma.

$R = \dfrac{\text{molecular weight of hydrated solute}}{\text{molecular weight of anhydrous solute}}$.

S = solubility (parts by weight anhydrous solute per 100 parts by weight total solvent) of material at final temperature.

w_0 = weight of anhydrous solute in original batch.

H_0 = total weight of solvent in batch at the beginning of the process.

E = evaporation during the process.

Example 1. A 30 per cent solution of Na_2CO_3 weighing 10,000 lb. is cooled slowly to 20°C. The crystals formed are sal-soda ($Na_2CO_3.10H_2O$). The solubility of Na_2CO_3 at 20°C. is 21.5 parts of anhydrous salt per 100 parts of water. During cooling 3 per cent of the weight of the original solution is lost by evaporation. What is the weight of $Na_2CO_3.10H_2O$ formed?

Solution. Since the molecular weight of $Na_2CO_3.10H_2O$ is 286.2, and that of Na_2CO_3 is 106, $R = 286.2/106.0 = 2.70$. Also, $w_0 = (0.30)(10{,}000) = 3000$ lb.; the evaporation is $(0.03)(10{,}000) = 300$ lb.; and, therefore, $H_0 - E = 10{,}000 - 3000 - 300 = 6700$ lb. The weight of the crop is, by Eq. (1)

$$C = 2.70\left[\frac{100 \times 3000 - 21.5 \times 6700}{100 - 21.5(2.70 - 1.0)}\right]$$
$$= 6636 \text{ lb.}$$

The solubility data available in the literature are often old and inexact, even for pure substances. Solubilities may be influenced by impurities and by variations in the pH of the solution. Slight errors in the solubility of heavily hydrated solutes are sometimes magnified into larger errors in the calculated yields of such salts. If the yield of a crystallization process must be known with considerable precision, it is desirable to determine the solubility experimentally for the actual solute and solvent involved.

Purity of the Product. Although a crystal itself is necessarily pure, it retains mother liquor when removed from the final magma, and the adhering mother liquor will carry its share of the impurities present in the mother liquor. If the retained mother liquor is dried on the crystal, contamination will result.

In practice, crystals usually are centrifuged or filtered. Centrifuging leaves mother liquor amounting to 2 to 5 per cent of the weight of the crystals. Large uniform crystals from low-viscosity mother liquors will retain a minimum proportion of mother liquor, while non-uniform small crystals from viscous solutions will retain a considerably larger proportion. Comparable statements apply to the filtration of crystals. It is common practice to wash the crystals on the centrifuge or filter with fresh solvent; in principle such washing can reduce impurities to below almost any arbitrary figure. Purity can also be improved by recrystallization, but this method is not usually so satisfactory as that of properly washing the crystals.

Heat Effects in a Crystallization Process. The heat effect of a crystallization process is calculated by means of a heat balance. Such a balance can be computed by two methods: the individual heat effects, such as sensible heats, latent heats, and heats of crystallization, can be computed and combined into a balance equation, or an enthalpy balance can be taken in which the total enthalpy of all leaving streams minus the total enthalpy of all entering streams is equal to the heat absorbed from external sources by the process.

In the first method, the heat removed from the crystallizing solution by external means is equal to the sum of the sensible heat lost by the cooling solution and the heat evolved in the formation of the crystalline crop (heat of crystallization) minus the radiation losses and minus the heat of vaporization of solvent evaporated during the process.

Heat of Crystallization. In heat-balance calculations on crystallization processes the heat of crystallization is usually important. The heat of crystallization is the latent heat accompanying the precipitation of crystals from a saturated solution. Ordinarily the heat of crystallization is exothermic; it varies with both concentration and temperature. Rigorously the heat of crystallization is related to the heat of dilution of the solution and the heat of solution of the crystal. The heat of solution is the heat evolved when a unit mass of solid is dissolved in a very large amount of water, and such data are quite plentiful. A table of heats of solution is given on p. 246. Heats of dilution, on the other hand, are scarce, especially for concentrated solutions, and it is usual to use the negative value of the heat of solution for the heat of crystallization. This is equivalent to neglecting heats of dilution. Ordinarily, the heat of dilution is small in comparison with that of solution, and the approximation is justified. Furthermore, the neglect of the heat of dilution leads to a conservative result because the heat of dilution is usually a heat evolution by the solution.

Example 2. The heat absorbed when 1 g.-mole of $MgSO_4.7H_2O$ is dissolved isothermally at 18°C. in a large amount of water is 3180 cal. (p. 247). What is the heat of crystallization of 1 lb. of $MgSO_4.7H_2O$ if heat of solution effects are negligible?

Solution. The molecular weight of $MgSO_4.7H_2O$ is 246.5. Since 1 cal./g.-mole = 1.8 B.t.u./lb.-mole, the heat of crystallization of $MgSO_4.7H_2O$ is $\frac{(3180)(1.8)}{246.5} = 23$ B.t.u./lb.

Enthalpy-concentration Chart. The enthalpy method of calculating heat balances over crystallization processes is facilitated by the use of the enthalpy-concentration chart so constructed as to show the solid phases [Bošnjaković, *Z. ges. Kälte-Ind.*, **39**, 182 (1932)]. This method rigorously accounts for the heats of dilution and is very simple arithmetically, once the chart has been constructed. The disadvantages of the enthalpy-concentration chart are as follows: (1) considerable data are required for its construction, and these data are often not available; (2) the initial construction of the chart is time-consuming and not justified for a single calculation. For substances commonly crystallized and for which adequate data are available, the enthalpy-concentration chart has considerable utility.

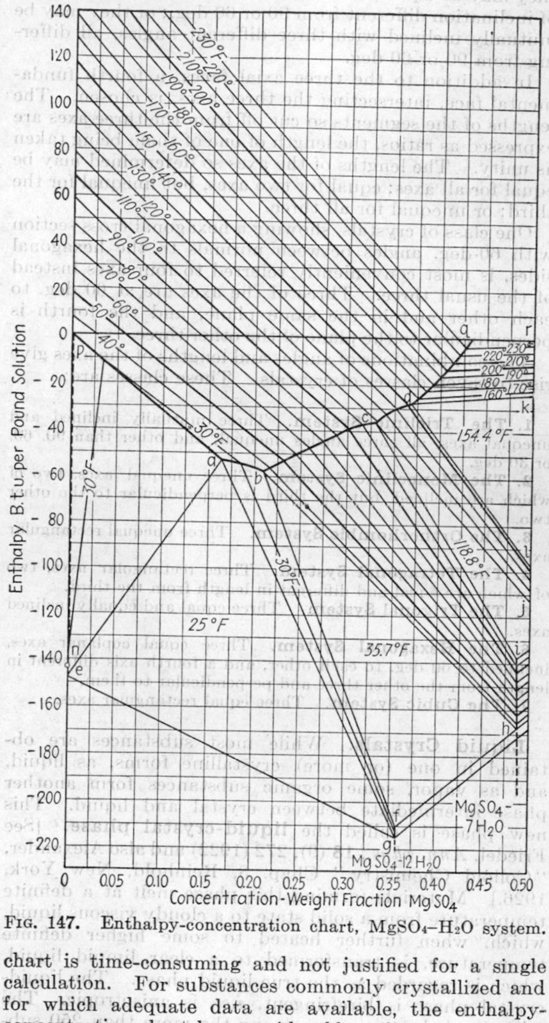

FIG. 147. Enthalpy-concentration chart, $MgSO_4$–H_2O system.

An enthalpy-concentration chart for the system $MgSO_4.H_2O$ is shown in Fig. 147. The use of the chart in heat-balance calculations involving solutions has been described [McCabe, *Trans. Am. Inst. Chem. Engrs.*, **31**, 129 (1935)]. In Fig. 148 the enthalpies of the solid phases from zero to 50 per cent $MgSO_4$ are shown, and the diagram can be correlated with the ordinary phase diagram shown in Fig. 148. The line *pa* represents the freezing points of ice from solutions of $MgSO_4$. Point *a* is the eutectic, and line *abcdq* is the solubility curve of various hydrates. Line *ab* is the solubility curve for $MgSO_4.12H_2O$, *bc* is the solubility curve for $MgSO_4.7H_2O$, *cd* is the solubility curve for $MgSO_4.6H_2O$, and *dq* is a portion of the curve for $MgSO_4.H_2O$. The area *aep* of Fig. 147 represents the enthalpies of all equilibrium mixtures of ice and $MgSO_4$ solution. The isothermal (25°F.) triangle *age* gives the enthalpies of all combinations of ice and partly solidified eutectic and of $MgSO_4.12H_2O$ and partly solidified eutectic. Area *abfg* contains the enthalpy-concentration coordinates of all magmas consisting of $MgSO_4.12H_2O$ crystals and its mother liquor. The isothermal (35.7°F.) area *bhf* represents the isothermal transformation of $MgSO_4.7H_2O$ to $MgSO_4.12H_2O$, and this area represents mixtures consisting of a saturated solution of concentration 21 per cent, solid $MgSO_4.7H_2O$, and solid $MgSO_4.12H_2O$. The area *cihb* represents all magmas of $MgSO_4.7H_2O$ (Epsom salt) and its mother liquor. The isothermal (118.8°F.) area *cji* represents mixtures consisting of a saturated solution containing 35

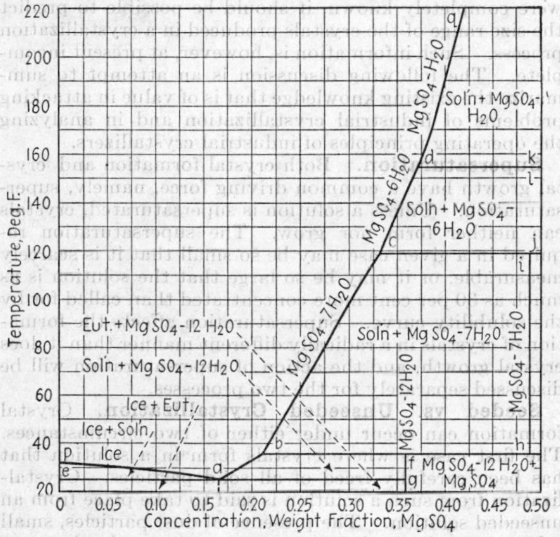

FIG. 148. Phase diagram, $MgSO_4$–H_2O system.

per cent $MgSO_4$, solid $MgSO_4.6H_2O$, and solid $MgSO_4$.-$7H_2O$. Area *dljc* represents magmas of $MgSO_4.6H_2O$ and its mother liquor. The isothermal (154.4°F.) area *dkl* represents mixtures consisting of a saturated solution containing 37 per cent $MgSO_4$, solid $MgSO_4.H_2O$, and solid $MgSO_4.6H_2O$. Area *qrkd* is a part of the field representing saturated solutions in equilibrium with $MgSO_4.H_2O$. Except for the isotherms in the liquid solution field and the solubility and freezing-point curves, all lines on the enthalpy-concentration chart are straight.

A useful basic construction applicable to all enthalpy-concentration charts is shown in Fig. 148. If the materials represented by points *A* and *B* in the chart are combined to form the material represented by *C*, the heat absorbed per unit weight of *C* is represented by the vertical line segment *CD*, measured above the straight line *AB* [Merkel, *Z. Ver. deut. Ing.*, **72**, 109 (1928)].

Example 3. 10,000 lb. of a 32.5 per cent $MgSO_4$ solution at 120°F. is cooled without appreciable evaporation to 70°F. in a crystallizer. How much heat must be removed from the solution, and what weight of $MgSO_4.7H_2O$ crystals will form?

Solution. The crystals and mother liquor are represented by terminals of the straight isothermal line for 70°F. in the field *cihb* of Fig. 147. The initial solution is represented by the point in the undersaturated solution field on the 120°F. isotherm at a concentration of 0.325. The magma must have an average concentration of 0.325 and a temperature of 70°F. From Fig. 147, the coordinates of the four points are as follows:

	Temp., °F.	Conc.	Enthalpy
Original solution	120	0.325	−33.0
Crystals	70	.488	−51.0
Mother liquor	70	.259	−47.3
Magma	70	.325	−78.4

The heat removed from the solution is

$$10{,}000(-33.0 + 78.4) = 454{,}000 \text{ B.t.u.}$$

The same result is obtained if the "basic construction" is used. The original solution can be considered the result of a combination of mother liquor and crystals, and the vertical distance to the point of the original solutions above the line connecting the points representing the mother liquor and crystals is the heat absorbed due to such a combination, or the heat evolved from the separation of solution into crystals and mother liquor.

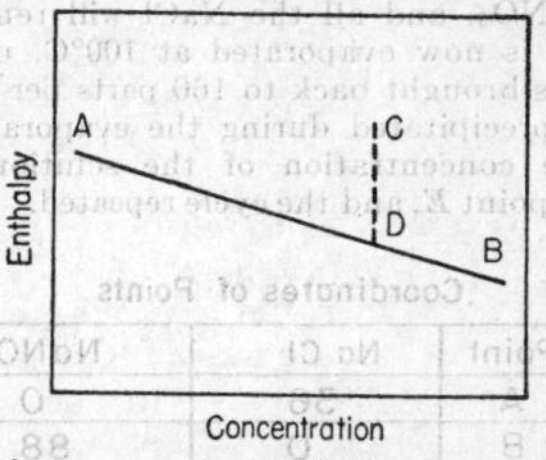

FIG. 149. Basic construction–enthalpy-concentration chart.

The yield of crystals is easily obtained by applying the "lever-arm principle" commonly used in calculations involving equilibrium diagrams.

$$\text{Weight of crystals} = 10{,}000\,\frac{0.325 - 0.259}{0.488 - 0.259} = 2880 \text{ lb.}$$

Fractional Crystallization. When two or more solutes are present in a solution, it is often possible to crystallize one of the solutes and leave the others in solution. Usually, such fractional crystallization methods are based on differences in the solubilities of the solutes.

It is a highly important fact that the solubility of a material in a solution of another solute is, in general, quite widely different from its solubility in the pure solvent. Thus, the solubility of sodium chloride at 20°C. is 36 parts per 100 parts water, and that of sodium nitrate is 88 parts per 100 parts water, but a solution saturated at 20°C. with respect to both of these salts will contain only 25 parts sodium chloride and 59 parts sodium nitrate per 100 parts water.

The mutual solubilities of the above two salts can be shown diagrammatically, as in Fig. 150. The solubilities are plotted for two different temperatures: line *DEF* is for 100°C., at which temperature the solubility of NaCl is 40 parts per 100 parts water, and that of $NaNO_3$ is 176 parts per 100 parts water. A solution saturated at 100°C. with both salts contains 17 parts NaCl and 160 parts $NaNO_3$ per 100 parts water. Points *D*, *E*, and *F* are plotted from these data, and, in the absence of more detailed data, the lines *DE* and *EF* are considered straight. The line *ACB* is the corresponding solubility curve for 20°C., and the line *EC* shows the variation with temperature of the composition of a solution saturated with both components. (Badger and Baker "Inorganic Chemical Technology," 1st ed., p. 82, McGraw-Hill, New York, 1928.)

If a solution at 100°C. has a composition represented by a point on the line *DE*, the solution is saturated with respect to NaCl but not with respect to $NaNO_3$; while if the composition of the solution is represented by a point on line *EF*, the solution is saturated with $NaNO_3$ but not with NaCl. (For a detailed treatment of such solubility relationships, see Blasdale "Equilibria in Saturated Salt Solutions," Reinhold, New York, 1927.)

Example 4. As an illustration of fractional crystallization, consider the separation of $NaNO_3$ and NaCl from a solution saturated at 100°C. with both salts and therefore represented by point *E*. If a basis of 100 lb. water is taken, the solution contains 17 lb. NaCl and 160 lb. $NaNO_3$. Suppose the solution is cooled to 20°C. The solution becomes supersaturated with respect to $NaNO_3$, and crystallization of the latter should take place. The composition of the solution moves along the path *EG*. At 20°C., if equilibrium is reached, the composition of the solution is that represented by point *G*. If line *CB* is considered straight, the abscissa of *G* can be calculated from similar triangles as follows:

$$\text{Parts } NaNO_3 = 59 + \frac{(88 - 59)(25 - 17)}{25} = 68.3$$

On cooling along the line *EG*, there will separate 160 − 68.3 = 91.7 lb. $NaNO_3$, and all the NaCl will remain in solution. If the solution is now evaporated at 100°C. until the $NaNO_3$ concentration is brought back to 160 parts per 100 parts water, NaCl will be precipitated during the evaporation and can be removed. The concentration of the solution will again be represented by point *E*, and the cycle repeated. On each cooling,

Coordinates of Points

Point	Na Cl	$NaNO_3$
A	36	0
B	0	88
C	25	59
D	40	0
E	17	160
F	0	176
G	17	68

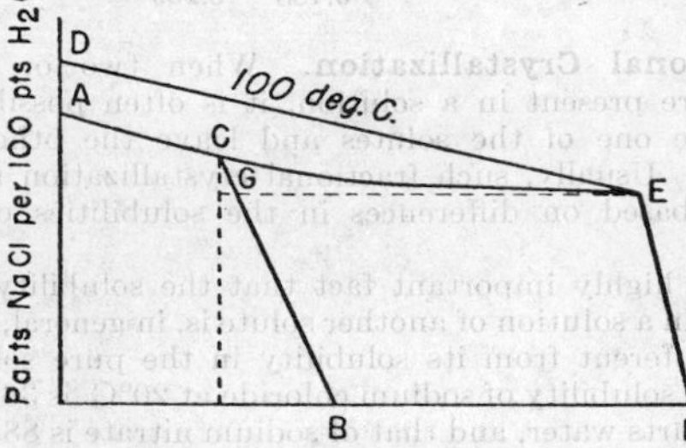

FIG. 150. Fractional crystallization of NaCl and $NaNO_3$.

(91.7/160)(100) = 57.3 per cent of the nitrate in solution will crystallize, and, on each evaporation, the same percentage of the chloride will be precipitated. Various modifications of the method can be used. For example, the amount of water in the batch can be kept constant and the solution resaturated at 100°C. with fresh $NaNO_3$ after each cooling. The hot solvent will act as a selective solvent for the nitrate, since it is impoverished in $NaNO_3$, but saturated with respect to NaCl. Nitrate can therefore be dissolved and chloride left behind. The dissolved nitrate is recovered in the cooling part of the cycle.

Another method of fractional crystallization is sometimes used, where advantage is taken of different crystallization rates. Thus, a solution saturated with borax and potassium chloride will, in the absence of borax seed crystals, precipitate only potassium chloride on rapid cooling. The borax remains behind as a supersaturated solution, and the potassium chloride crystals can be removed before the slower borax crystallization starts.

CRYSTAL FORMATION

There are obviously two steps involved in the preparation of crystalline matter from a solution. The crystals must first form and then grow. The theory can therefore be conveniently considered under three heads: (1) the formation of crystalline nuclei, (2) their resulting growth, and (3) the interrelation between formation and growth.

Certain qualitative facts are apparent. If the concentrations of the initial solution and of the final mother liquor are fixed the total weight of the crystalline crop is also fixed. The distribution of this weight, however, will depend on the relationship of the two processes of formation and of growth. If new crystals form continuously and rapidly during the process, the crop will consist of many small crystals, while if but a few nuclei form at the start, and if the resulting precipitation occurs uniformly on these nuclei without secondary nucleus formation, a crop of large uniform crystals must result. Intermediate cases of simultaneous formation and growth will of course, result in intermediate average size and also a non-uniform grain, since the older crystals will be larger than the younger ones.

If the laws and data for crystal formation and growth were completely known, it should be possible to predict the size range of the crystals produced in a crystallization process. Such information is, however, at present incomplete. The following discussion is an attempt to summarize the existing knowledge that is of value in attacking problems of industrial crystallization and in analyzing the operating principles of industrial crystallizers.

Supersaturation. Both crystal formation and crystal growth have a common driving force, namely, supersaturation. Unless a solution is supersaturated, crystals can neither form nor grow. The supersaturation required in a given case may be so small that it is scarcely measurable, or it may be so large that the solution is as much as 30 per cent more concentrated than called for by the solubility curve. Supersaturation affects the formation of crystals in a radically different manner than it does crystal growth, and the action of supersaturation will be discussed separately for the two processes.

Seeded vs. Unseeded Crystallization. Crystal formation can occur under either of two circumstances. The first case is where crystals form in a solution that has been carefully freed of all solid particles. Crystallization from such a solution is said to take place from an unseeded solution. The presence of dust particles, small solute crystals, or in some cases crystals of other materials, may lead to a second type of crystal formation, namely, that from seeded solutions. In practice, the seeded case is the more important. Except for closed batch crystallizers in which the solutions are heated well above the saturation temperatures before being sent to the crystallizers, completely unseeded solutions are not usually encountered. Batch vacuum crystallizers and special crystallizers, such as sugar-boiling apparatus, fall into the category of equipment that operates on unseeded solutions. On the other hand, crystallizers in which the solution has access to the atmosphere of the plant will in all probability be seeded by the plant dust which invariably will carry tiny crystals and dust particles and which will inoculate the solution. Even a solution that is unseeded at the start of the process becomes seeded the moment crystals form, and at all stages beyond the initial nucleation the process is of the seeded type.

In the case of an unseeded solution, it is possible in viscous solutions of relatively high molecular weight, such as those of sugars, to maintain a highly supersaturated solution indefinitely without the formation of nuclei. To obtain this result the solution must be carefully prepared, must be entirely free from dust particles, and must be "sterilized" or heated well above its saturation temperature for a time in a completely airtight container. Materials of low molecular weight which form solutions of moderate viscosities cannot usually support supersaturations of any great magnitude for an indefinite period. There are undoubtedly borderline solutes to which a definite answer cannot be given concerning their ability to stay in supersaturated solutions indefinitely.

Miers' Theory. Miers and his collaborators, following an earlier suggestion by Ostwald, have elaborated the theory that there is a definite relationship between the concentration and the temperature at which crystals will spontaneously form in an initially unseeded solution. This relationship takes the form of a so-called supersolubility curve which is roughly parallel to the usual solubility curve and is located in the supersaturated field. The Miers theory, which has considerable experimental support, states that under normal conditions appreciable spontaneous formation of crystals will not occur in the area between the solubility curve and the supersolubility curve but that, whenever, the concentration is brought into the labile field in which concentrations are higher than those corresponding to the supersolubility curve, sudden and copious nucleation will occur.

These relationships are shown diagrammatically in Fig. 151 wherein concentration is plotted against temperature.

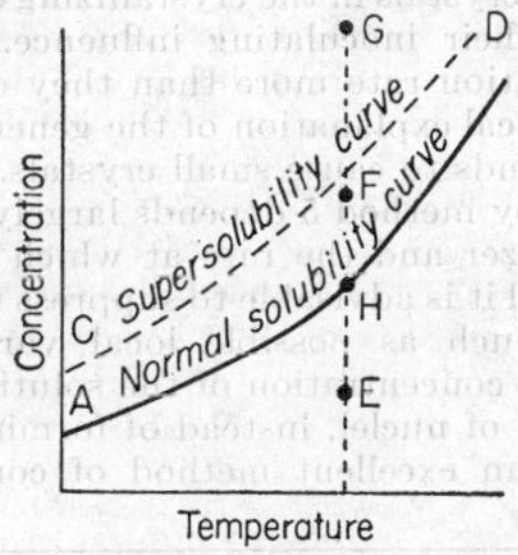

FIG. 151. Diagrammatic representation of Miers' theory.

The normal solubility curve is shown as line AB, and the supersolubility curve is the dotted line CD. A solution at such a temperature and concentration that it is represented by a point below AB (point E, for example) is undersaturated. A solution represented by point F is metastable and will drop to concentration H if seed crystals are added to it, but it will remain at F unless seeds are present. A solution represented by point G will spontaneously crystallize, as its concentration will fall to that of point H.

In spite of considerable controversy concerning such questions as to the effects of mechanical stimulation and the effect of probability, the Miers type of supersolubility curve has been a useful concept in the analysis of crystal formation problems from unseeded solutions.

Most of Miers' work was carried out on solutions that were initially unseeded. Since most industrial cases concern seeded solutions, it is important to determine whether or not seeded solutions can be maintained for appreciable lengths of time under such conditions that nuclei other than the seeds themselves do not form. Work carried out on the supersolubility relationships of seeded solutions has shown that, if a solution initially without seeds is cooled at a definite rate and if definite quantities of a definite size of seeds are added when the concentration of the solution reaches saturation, there is a definite degree of undercooling corresponding to the formation of new nuclei. Other reproducible supersaturation curves are found at which the rate of formation of crystals becomes a maximum. Results of such experiments on $MgSO_4.7H_2O$ are shown in Fig. 152 [Ting and McCabe, *Ind. Eng. Chem.*, **26**, 1201 (1934)]. The t_1 curves show where new nuclei first formed, and the t_2 curves show where the rate of formation reaches a maximum. Supersaturated solutions of KCl show the same type of behavior. Curves of the type shown in the figure are quite readily reproducible, whereas the supersaturation curves of the Miers type are difficult to reproduce except under carefully controlled conditions, as it is difficult to prevent the effects of fortuitous seeding. The supersaturations obtained in seeded solutions are definitely considerably less than those found in crystallizing unseeded solutions.

These results indicate that the ability of crystals to inoculate a solution and to cause the formation of new crystals is an important fundamental factor in crystallization.

Curves such as those of Fig. 152 suggest the question whether or not it is possible to maintain a supersaturated solution in contact with seed crystals for an indefinite period without the formation of new nuclei. Experimental results indicate that it is not possible, even at low supersaturations, to suppress crystal formation indefinitely in a seeded solution. The lower the supersaturation, the slower will be the formation of new nuclei, and the longer will be the time that will elapse before new nuclei are formed, but, if time enough is allowed, present knowledge indicates that sooner or later new nuclei will form. At very low supersaturations, where the material available for crystal formation is small, only a few nuclei can be formed before the solution reaches equilibrium.

If the rate of nucleus formation is very low and if the rate of growth is reasonably high in comparison with the formation rate, it is possible to maintain a seeded supersaturated solution without nucleation long enough so that the original seeds will grow to the desired size before the new nuclei appear. The crystallization of sugar is an example of such a case. In sugar crystallization the nuclei are formed almost instantly by a sudden increase in supersaturation. Once the initial batch of nuclei is formed, supersaturation is maintained at a point low enough that the formation of additional nuclei does not take place before the original crystals have been allowed to grow to the desired size. Apparently, this method is not possible in the case where nucleation tends to be too rapid to allow time enough for the original seeds to grow to the desired size.

Methods of Forming Crystals in Solutions. In any crystallization process the nuclei formation should be under control. In a batch process, if uniform crystals are desired, it is advisable to form as large a proportion as possible of the crystals at the same time, even if additional nucleation cannot be altogether prevented. Otherwise, a non-uniform crop will be obtained. In a continuous process the number of nuclei formed per unit time will be continuous and uniform and must equal the number of crystals that are withdrawn per unit time from the crystallizer. In a continuous crystallizer either most of the nuclei should form within a narrow zone in the unit so that all nuclei can receive the same time of growth (the method applicable to the Swenson-Walker crystallizer, p. 1063, for example) or there must be a classifying action in the crystallizer which will retain the small crystals under treatment until they have grown to the proper size before they are removed from the unit (as is done in the continuous vacuum crystallizer, p. 1065, for example). Nucleation can be accomplished either by adding the desired number of nuclei, usually in the form of crushed crystals, to the crystallizer when the solution is either saturated or supersaturated, or by forming nuclei *in situ*.

New nuclei may originate *in situ* in one or more of the following ways:

1. By spontaneous nucleation from unseeded solutions. In this case an unseeded solution must be cooled into the labile region as shown by the Miers curves.

2. By attrition of existing crystals. If crystals are agitated vigorously, small corners and fragments may

be broken from existing crystals; such fragments and mutilated crystals quickly repair themselves, and the fragments become new nuclei.

3. Mechanical impact. Mechanical impact in a supersaturated solution has been shown to cause nucleation [Young, *J. Am. Chem. Soc.*, **33**, 148, 162 (1911)]. Vigorous stirring, the collision of crystals in the solution, with each other or with the walls of the crystallizers, may cause the formation of some new nuclei. This formation is over and above that resulting from the mechanical fracture of existing crystals. Its importance in industrial crystallization is questionable.

4. New crystals are formed because of the inoculating influences of crystals already present. This method of crystal formation is probably the most important single method and is the method that is subject to the most accurate control.

5. Local variations in the concentration of the solution may cause nucleation in restricting zones. For example, the withdrawal of heat through the containing wall will cause temperature gradients near the wall which can increase the supersaturation enough to accelerate nucleation. Evaporation from the surface may result in abnormally high concentrations in the solution at the surface and lead to nucleation. Even surfaces at solution temperature sometimes appear to catalyze nucleation near them.

In general, the above causes of nucleation are inextricably interwoven, and it is usually not possible to separate them completely in any given case. It is possible, however, to emphasize or to suppress individual nucleation effects and thereby to facilitate control.

Thus, method 1 is essentially an uncontrolled formation method. In general, it is preferable to use method 4 if possible, rather than method 1. Figure 163 [Seavoy and Caldwell, *Ind. Eng. Chem.*, **32**, 633 (1940)] shows the temperature-concentration history of a batch crystallizer operated two ways: first, with an unseeded solution and second, with a seeded solution. In the second case a high supersaturation, which causes uncontrolled nucleation, is prevented.

The formation of crystals by attrition (method 2) should be suppressed as much as possible. Such formation occurs at the expense of existing perfect crystals and is not subject to adequate control. For practical control purposes, methods 3 and 4 can be considered as equivalent. The mechanical impact of stirrers or of crystals on each other accounts for only a small proportion of the total nucleation and depends on stirring rate and the number and size of crystals existing in the equipment at any given time. The much more important inoculating effect is determined by these same variables. Increased stirring rate, for example, brings about more uniform distribution of crystals in the crystallizing solution, which will increase their inoculating influence. Such effects increase nucleation rate more than they do growth rate and are the logical explanation of the general observation that stirring tends to cause small crystals.

Nucleation by method 5 depends largely on the design of the crystallizer and the rate at which it is operated. For best control it is advisable to suppress this method by reducing as much as possible local variations in the temperature or concentration of the solution.

The addition of nuclei, instead of forming them in the crystallizer is an excellent method of controlling sugar

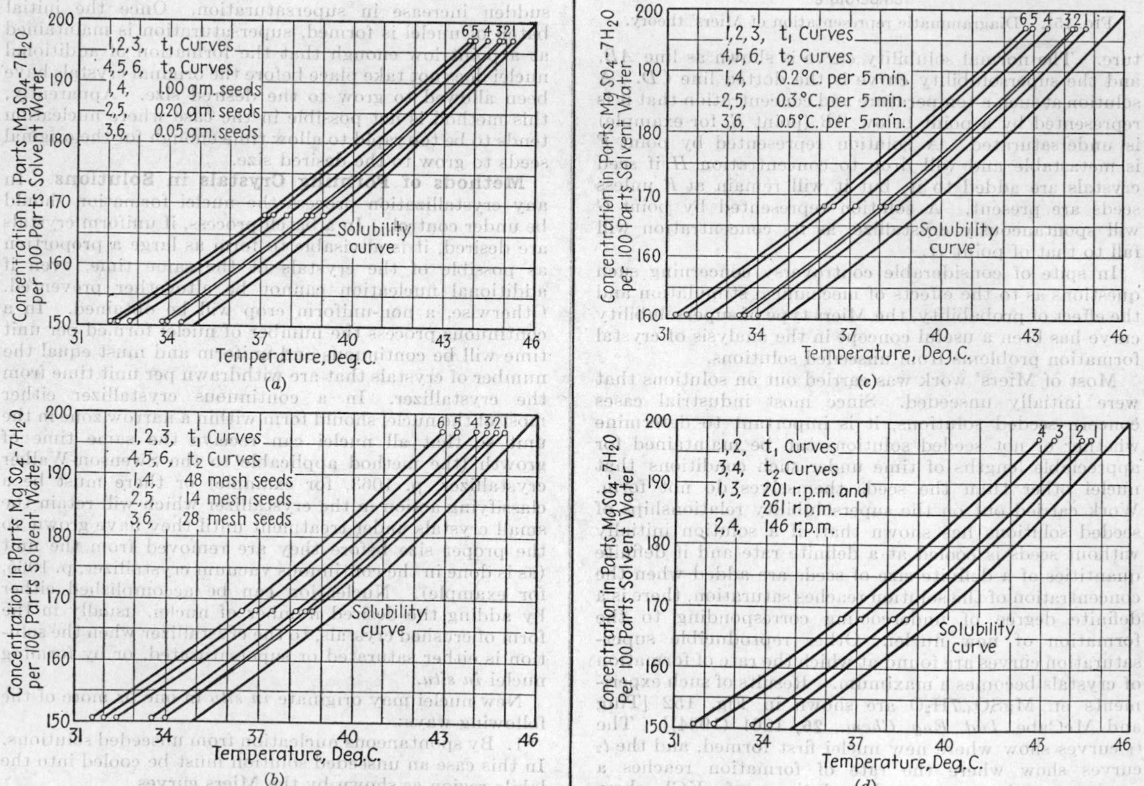

FIG 152. Supersaturation curves, seeded solutions of $MgSO_4.7H_2O$. (*a*) Effect of weight of seed crystals. Grams seeds/1000 g. solution. (*b*) Effect of size of seed crystals. (*c*) Effect of cooling rate. (*d*) Effect of stirring speed.

boiling (Spencer and Meade, "Cane Sugar Handbook," 8th ed., pp. 175–230, Wiley, New York, 1944).

Effects of Impurities. Impurities in the solution may inhibit the formation of new nuclei. The effect of a given impurity cannot be predicted and must be found by experiment. In general, high-molecular-weight materials seem to be the most effective inhibitors.

Kinetics of Nucleation. Recent theoretical work by Volmer, Stanski, Becker, and others has provided a semiquantitative basis for an understanding of the factors that control the rates of nucleation and growth. An outline of these ideas, adapted from Mehl and Jetters (Am. Soc. for Metals, 1939) follows:

Neither nucleation nor growth can occur unless the precipitated substance has a lower thermodynamic potential after precipitation than before. In a solution this means that nucleation and growth occur only in a supersaturated solution. The over-all driving force is the negative value of the free-energy difference

$$\Delta F_1 = F - \mu \tag{2}$$

where F = molal free energy of crystallized material based on zero specific surface.
μ = molal chemical potential of the same component in the solution.

The Gibbs-Thomson law shows that F increases with decreasing particle size and is very large for very small particles. It is convenient to divide the ΔF between a small particle and its mother phase into two parts:

$$\Delta F = \Delta F_1 + \Delta F_2 \tag{3}$$

where ΔF_1 = over-all free-energy change, not including surface effects.
ΔF_2 = difference between free energy of a very large particle and that of the small particle under consideration.

If the solution is supersaturated, ΔF_1 is negative. It decreases with supersaturation, and is proportional to the bulk number of atoms in the crystal. It is therefore proportional to the cube of the linear size. On the other hand, ΔF_2 is positive and is proportional to the number of molecules in the surface of the crystal. It is therefore proportional to the square of the linear size. As the crystal size increases, the free-energy change ΔF increases to a maximum and then decreases. The maximum value of ΔF corresponds to a definite crystal size, called the "critical size." A crystal smaller than the critical size will tend to dissolve, and one larger than the critical size can grow. The value of ΔF corresponding to the critical size represents an energy barrier that must be surmounted by a nucleus before it is stable. It is equal to the work of nucleation. The energy required can come only from momentary and local fluctuations of both concentration and energy. The energy fluctuations are statistical in nature and are of the usual kinetic type that gives rise to homogeneous reactions. The concentration fluctuation requires transport by molecular diffusion of the requisite number of molecules close enough to one another to form a nucleus large enough to equal or exceed the critical size. Becker proposes the following equation for nucleation rate:

$$\frac{dN}{dt} = ce^{(-Q/kT)}e^{-A(T)/kT} \tag{4}$$

where dN/dt = nucleation rate, number/unit volume/unit time.
Q = activation energy for diffusion.
$A(T)$ = work required to form surface of nucleus.
T = absolute temperature.
c = a constant.
k = Boltzman constant.

Several important qualitative deductions can be made from Eq. (4). The work of nucleation $A(T)$ increases markedly with decrease in supersaturation and is infinite at the saturation curve. The term $e^{-Q/kT}$ decreases with increase in supersaturation. The N vs. T curve has a pronounced maximum that corresponds to a definite supersaturation. Also, the shape of the N vs. T curve is such that the value of N is very low for an appreciable supersaturation but increases rapidly when a definite supersaturation is reached. The well-known metastable region and the Miers supersaturation curve are explained by the N vs. T relation. Actually, the supersaturation curve represents a zone where N increases rapidly with T, rather than a sharp boundary. Also, if time enough is allowed, nucleation will eventually occur at any supersaturation.

The state of a crystal at the boundary is abnormal because of the unbalanced forces acting on the surface atoms. The surface molecules can possess a lower activation energy for diffusion. Also, the presence of the solid interface can affect the molecules in the mother liquor in such a manner as to affect the terms of Eq. (4) and increase the nucleation rate N. Such interface behavior may account for the inoculating effect obtained by seed crystals, or in some cases by foreign particles, in supersaturated solutions.

Figure 153 adapted from Mehl and Jetter (*loc. cit.*) shows diagrammatically and qualitatively how the free-

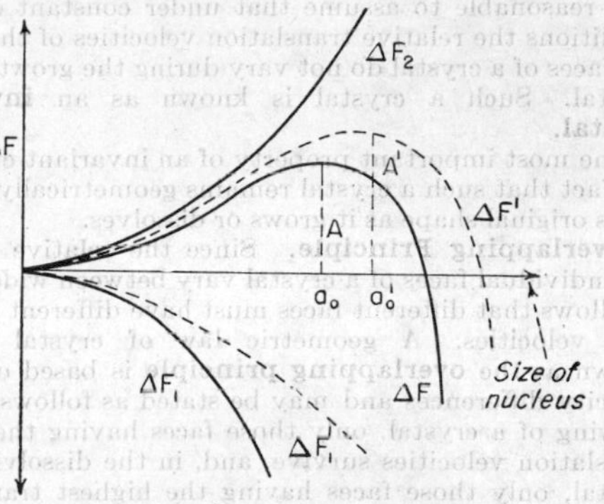

FIG. 153. Free-energy effects of nucleation—qualitative relationships of free energy vs. particle size for nucleation.
ΔF_1 = free-energy excess, between large particles and solute in solution.
ΔF_2 = free-energy excess, between surface of crystal and bulk of crystal.
ΔF = over-all free-energy excess, between nucleus and solute in solution.
A = work of nucleation.
a_o = critical nucleus size.
Dotted curves and quantities A' and a_o' refer to lower supersaturation than that corresponding to full curves and quantities A and a_o. (*After Mehl and Jetter.*)

energy differences of Eq. (3) vary with particle size. The dotted curves correspond to a lower supersaturation than that pertaining to the full curves. The higher the supersaturation, the smaller is the size of the critical nucleus.

GEOMETRY OF CRYSTAL GROWTH

As a preliminary to the questions involving the rate of growth of crystals, certain facts regarding the geometry of crystal growth are important.

Parallel Displacement of Faces. Translation Velocities

Geometrically, a crystal is a solid bounded by planes. The shape and size of such a solid are a function of the interfacial angles and of the linear dimensions of the faces. As a result of the constancy of its interfacial angles, each face of a growing or dissolving crystal, as it moves away from or toward the center of the crystal, is always parallel to its original position. This fact is known as the *principle of the parallel displacement of faces.* The rate at which a face moves in a direction perpendicular to its original position is called the **translation velocity** of that face.

Crystal Habit. From an industrial point of view, the term "crystal habit" refers to the relative sizes of the faces of a crystal. No general law controlling crystal habit has been discovered. This property is easily affected by conditions of crystal formation and growth. It is very difficult to prepare perfect crystals with all faces of the same form equally developed. Small amounts of foreign substances will often completely change the crystal habit of a material. For example, sodium chloride crystallizes as cubes from a pure solution but forms octahedra if precipitated from a solution containing urea. The selective adsorption of dyes by the different faces of a crystal can greatly modify the habit of the crystal (see France, "Colloid Symposium Annual," vol. 7, pp. 59–87, Wiley, New York, 1930). Phenomena of this kind are so general that the prediction of crystal habit is difficult.

Invariant Crystals. Although it is impossible at present to predict the crystal habit of a definite material, it is reasonable to assume that under constant external conditions the relative translation velocities of the different faces of a crystal do not vary during the growth of the crystal. Such a crystal is known as an **invariant crystal.**

The most important property of an invariant crystal is the fact that such a crystal remains geometrically similar to its original shape as it grows or dissolves.

Overlapping Principle. Since the relative sizes of the individual faces of a crystal vary between wide limits, it follows that different faces must have different translation velocities. A geometric law of crystal growth known as the **overlapping principle** is based on these velocity differences and may be stated as follows: In the growing of a crystal, only those faces having the lowest translation velocities survive, and, in the dissolving of a crystal, only those faces having the highest translation velocities survive. For example, consider cross sections of a growing crystal, as in Fig. 154. The polygons shown

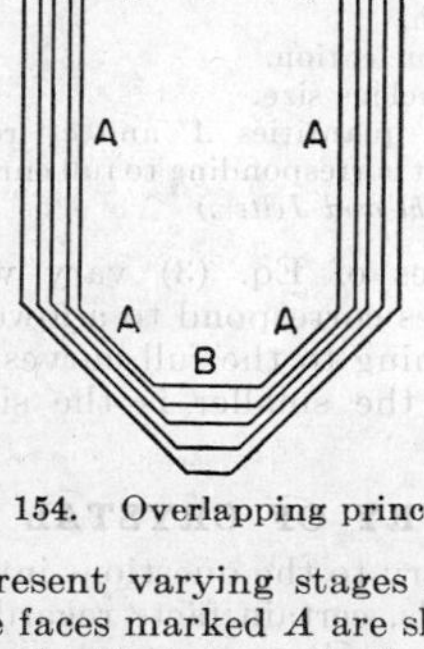

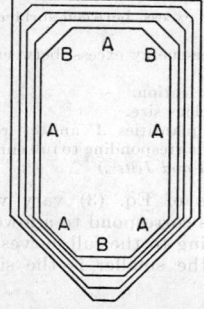

FIG. 154. Overlapping principle.

in the figure represent varying stages in the growth of the crystal. The faces marked A are slow-growing faces (low translation velocities), and the faces marked B are fast growing (high translation velocities). It is apparent from Fig. 154 that the faster B faces tend to disappear, as they are "overlapped" by the slower A faces.

It has been shown [Valeton, *Z. Kryst.*, **59,** 135–169 (1923)] that the overlapping principle, if combined with the principle of the parallel displacement of faces, makes it possible to predict the final shape of a crystal when the initial shape and the relative translation velocities of the faces are known.

Variation of Translation Velocities. The translation velocities of the faces of an invariant crystal are not all equal, unless the crystal is a regular geometric solid. The smaller faces have greater rates of growth, measured in weight per unit area per unit time, than have the larger faces. This difference in rate has been found experimentally.

CRYSTAL GROWTH

The earliest theory of crystal shape was the thermodynamic theory of Gibbs and Curie. This theory stated that a crystal, as it grows, chooses that form compatible with the symmetry of the crystal that gives the minimum surface energy. For commercial sizes, however, the surface-energy differences between a thermodynamically stable crystal and one that is unstable because of its shape are too small to be of importance.

Rate of Solution. The first theories of the rate of crystal growth consisted of attempts to consider crystallization as the reverse of solution. Earlier work by Noyes and Whitney [*Z. physik. Chem.*, **23,** 689 (1897)] and others led to the establishment of a theory of the rate of solution that has been applied to a number of other heterogeneous reaction-velocity problems. The theory assumes that such reactions are controlled by the rate of diffusion of the reactants and products to and from the solid-liquid interface. Any reactions taking place at the interface are very rapid in comparison with the diffusion rates. The diffusional resistance is confined to a comparatively thin film surrounding the solid, since convection currents equalize the concentrations in the bulk of the liquid. Low viscosity and bigorous stirring attenuate the film, decrease the diffusional resistance, and increase the reaction rate; while high viscosity and poor agitation decrease the reaction rate because the films are then relatively thick and the diffusion process is slow.

The Berthoud-Valeton Theory [*J. chim. phys.*, **10,** 625 (1912), *Z. Kryst.*, **59,** 335 (1923); **60,** 1 (1924)]. This theory assumes that the diffusional process is followed in series with a first-order interfacial reaction and the net rate of crystallization depends on both reactions. If the concentration of a saturated solution is C_0, that of the bulk of the solution is C, and that of the solution in contact with the crystal surface is C', the rate of material deposition is

$$\frac{dW}{d\theta} = k'S(C' - C_0) = kS(C - C') \tag{5}$$

where W is the weight, θ is the time, k is the mass transfer from the bulk of the solution to the face of the crystal, k' is the rate of reaction constant of the interfacial reaction, and S is the surface area,

If C' is eliminated from Eq. (5), the result may be written

$$\frac{dW}{d\theta} = \frac{S(C - C_0)}{\dfrac{1}{k'} + \dfrac{1}{k}} \tag{6}$$

There are three corollaries of Eq. (6):

1. The over-all reaction is of the first order, since for any particular face at constant temperature $1/(1/k' + 1/k)$ is a constant. (This will not be true, however, if the interface reaction is not first order.)

2. Although the reaction is one of first order, the value of k' varies from face to face, and so the rates of growth of the individual faces of the same crystal can vary, in

spite of the fact that k (the diffusion film coefficient) is constant for all the faces.

3. The relative magnitudes of k' and k may vary in different cases. If k is large in comparison with k', diffusion is an unimportant factor and the surface reaction controls the process of crystallization. If k' is large in comparison with k, the surface reaction has but little influence on the rate, and the diffusion then is the controlling factor. If k' is very large in comparison with k, the theory becomes identical with the Noyes-Whitney theory.

Effect of Impurities on Crystal Growth. Small amounts of impurities may have an important effect on the rate of crystal growth, just as they do on the rate of crystal formation. Apparently, the inhibition of growth is due to the adsorption of the impurity on the crystal face. No general rule governing these phenomena has been discovered. The amount of impurity adsorbed depends not only on the material and the impurity but varies from face to face of the same crystal. The face that adsorbs the greatest amount of impurity will have the lowest translation velocities and hence will increase in size relative to the other faces, in accordance with the overlapping principle (France, "Colloid Symposium Annual," vol, 7, pp. 59–87, Wiley, New York, 1930).

The ΔL Law. It has been shown [McCabe, *Ind. Eng. Chem.*, **21**, 30, 112 (1929)] that all geometrically similar crystals of the same material suspended in the same solution grow at the same rate, if the growth is measured as the increase in length of geometrically corresponding distances on all the crystals. If ΔL is the increase in linear dimension of one crystal, it is at the same time equal to the increase in the corresponding dimension of each of the other crystals and is independent of the initial size of any of the original crystals, provided that all crystals in the suspension are treated exactly alike.

Calculation of Screen Analysis of Product from That of Seeds. The ΔL law gives a solution to the following problem:

Problem. Given a saturated solution in which is suspended a known weight of seed crystals of known screen analysis, and assuming this solution to be cooled under known conditions, what will be the weight and screen analysis of the crystals at the end of the process, if there is negligible formation of new nuclei?

The weight of material precipitated is calculated in the same manner as is the yield of a crystallization process (see p. 1051). The distribution of the precipitating material takes place in accordance with the ΔL law. If D is the size of the opening of a sieve that will just pass a given crystal, then D and L can be considered proportional, or

$$\alpha D = L \qquad (7)$$

and

$$\alpha\, \Delta D = \Delta L \qquad (8)$$

where α is a proportionality constant that is identical for all crystals of the batch. If ΔL is the same for all crystals, ΔD is also the same for all crystals. It has been shown that

$$W_P = \int_0^{W_s} \left(1 + \frac{\Delta D}{D_s}\right)^3 dW_s \qquad (9)$$

where W_P is the weight of product obtained from W_s g. seed crystals, D_s and W_s are the coordinates of the cumulative screen-analysis curve (where W_s is total weight retained on the screen of opening size D_s), and ΔD is defined in Eq. (8).

The steps in the calculation of the yield and screen analysis of the product of a crystal-growth process are:

1. Calculate the theoretical yield from the ratio of seeds to solution and the solubility change of the material during the process.

2. Assume a value of ΔD, and calculate the weight of the product that corresponds to this assumed value by integrating Eq. (9) over the range $W_s = 0$ to $W_s = 100$.

3. If the value of W_P calculated in step 2 does not check that calculated in step 1, adjust ΔD by trial and error, until fair agreement is reached between the weights of product calculated by the two methods.

4. Using the correct value of ΔD as determined in step 3, plot the integral curve of Eq. (9).

5. Plot W_P against D_P. This plot can be constructed from the data at hand at this point, since the integral curve, plotted in step 4, gives the relationship between W_P and W_s; the screen-analysis curve of the seeds exhibits D_s as a function of W_s; and $D_P = D_s + \Delta D$.

6. From the curve of W_P against D_P that was plotted in step 5, read off the values of W_P that correspond to the sizes of the various screen openings, and convert the values to percentages of the entire product. The result is the cumulative screen analysis of the product. The differential analysis is easily derived from the cumulative analysis by subtraction.

Example 5. A Swenson-Walker (see p. 1063) crystallizer is cooling a potassium chloride solution. It discharges a saturated solution at 60°C. The discharge consists of 1750 lb. saturated solution and 165 lb. crystals per hour. The screen analysis of the crystals is as follows:

Table 1. Screen Analysis of Seeds

Meshes/in.	Size of screen opening, cm.	Screen analysis, %	
		Differential	Cumulative
On 12	0.1397	0.0	0.0
14	.1168	.1	.1
16	.0991	2.9	3.0
20	.0833	12.7	15.7
24	.0701	13.0	28.7
28	.0589	25.8	54.5
32	.0495	19.6	74.1
35	.0417	13.3	87.4
42	.0351	6.3	93.7
48	.0295	3.6	97.3
60	.0246	1.0	98.3
65	.0208	1.2	99.5
Through 100	(0.0175)*	0.5	100.0

* Estimated.

What would be (*a*) the weight and (*b*) the screen analysis of the product if additional sections were added to the crystallizer to cool the above product from 60° to 30°C.?

The solubility of potassium chloride at 60°C. is 45.0 parts per 100 parts water, and the solubility at 30°C. is 37.35 parts per 100 parts water.

Solution. The answer to question *a*, namely, the new crop weight, is computed from the solubility data. Since the potassium chloride that precipitates is in the anhydrous form, and since the solubilities given above are based on weight of water, the increased yield due to the cooling from 60° to 30°C. is 45.0 − 37.35 = 7.65 lb./100 lb. water. Since the crystallizer handles (100/145)(1750) = 1206 lb. water, the increase in yield is (12.06)(7.65) = 92.3 lb. The product from the new section will then be 92.3 + 165 = 257.3 lb. crystals per hour. This is the answer to question *a*.

In order to determine the screen analysis of the new product (question *b*), the above is applied. The screen analysis of the seed crystals is given in the data of the problem. The second and fourth columns of Table 2 are D_s and W_s, respectively, based on 100 lb. seed crystals. The product is (257.3/165)(100) = 156 lb. on this same basis. Step 1 of the procedure has been completed.

The second step is the trial-and-error determination of the correct value of ΔD that corresponds to the required weight increase of 56 lb./100 lb. seeds. This is done by assigning a value to ΔD, evaluating the integral of Eq. (9) between the limits $W_s = 100$ and $W_s = 0$, and thereby calculating a corresponding value of W_P. The required value of ΔD is that value that gives a value of W_P of 156.

Table 2. Coordinates of W_s vs. $(1 + 0.009/D_s)^3$ Curve

W_s	D_s	$\left(1+\frac{0.009}{D_s}\right)^3$	W_s	D_s	$\left(1+\frac{0.009}{D_s}\right)^3$
0.0	0.1397	1.21	50	0.0611	1.51
.1	.1168	1.25	60	.0563	1.56
3.0	.0991	1.30	70	.0514	1.62
5.0	.0960	1.31	80	.0466	1.70
10.0	.0901	1.33	90	.0391	1.86
20.0	.0789	1.38	95	.0330	2.06
30.0	.0695	1.44	98	.0270	2.37
40.0	.0653	1.47	100	.0175	3.47

Without reproducing the preliminary trials, it is found that if ΔD is 0.009, the yield requirement is closely met. This value will be shown to be correct in the next step.

The third step is to substitute $\Delta D = 0.009$ in Eq. (9) and to calculate the integral curve of the equation. This is done by plotting values of $(1 + 0.009/D_s)_3$ as ordinates against values of W_s as abscissas. The relationship between D_s and W_s is obtained by plotting the second column of the screen-analysis table (Table 1) against the fourth column. This curve is curve *A* of Fig. 156. The coordinates of the $(1 + 0.009/D_s)_3$ vs. W_s curve are as shown above, and the curve is shown in Fig. 156.

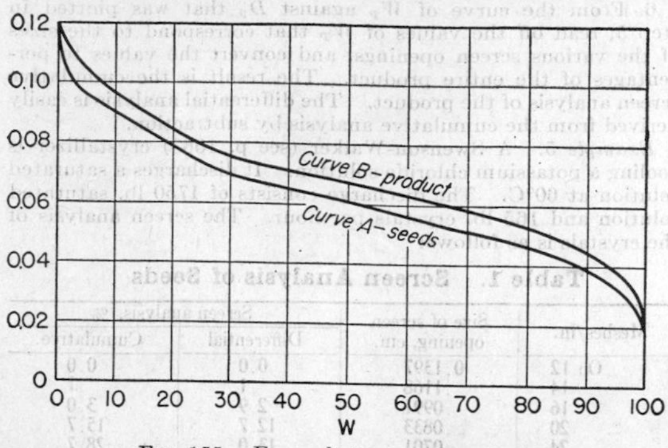

FIG. 155. Example 5: screen analyses.

The integral curve of Eq. (9) is obtained by determining the area bounded by the curve of Fig. 156, the *X*-axis, the ordinate $X = 0$, and the ordinate $X = W_s$, where W_s varies from 0 to 100. For example, the point on the integral curve with an abscissa of $W_s = 50$ is equal to the area *abcda* of Fig. 156.

The coordinates of points on the integral curve, which are corresponding values of W_p and W_s, are shown in Table 3.

It will be seen that when $\Delta D = 0.009$, $W_p = 157.4$, which is a satisfactory check with the theoretical yield of 156. The plot of W_p vs. W_s is shown in Fig. 157.

The next step, the determination of W_p vs. D_p (the size distribution of the product), can now be carried out. It is convenient to assign values of D_p equal to the actual sizes of

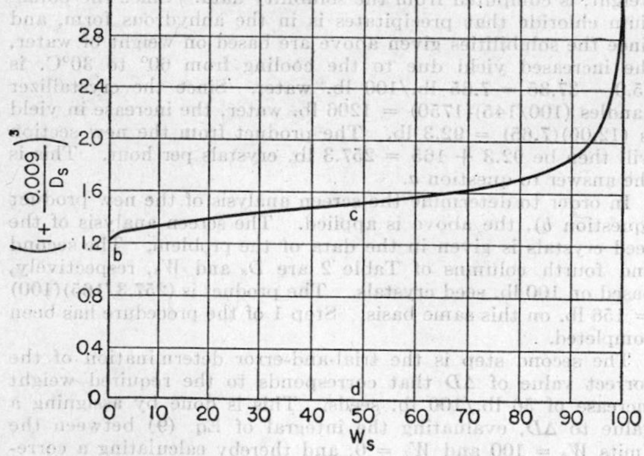

FIG. 156. Example 5: graphical integration.

openings of the screens, obtain the corresponding values of D_s by subtracting 0.009 from these D_p values, read the corresponding values of W_s from curve *A* of Fig. 155, and then read the values

Table 3. Relation between W_p and W_s

W_s	W_p	W_s	W_p
5	6.4	70	101.5
10	13.0	80	118.0
20	26.6	85	126.7
30	40.8	90	135.8
40	55.4	95	145.5
50	70.3	97.5	151.0
60	85.7	100	157.4

of W_p from Fig. 157. The cumulative and differential analyses are then readily calculated. The procedure is shown in Table 5.

Limitations of ΔL Law. It must be understood that the above analysis fails completely in any case where crystals are given preferential treatment based on size. If larger crystals have a higher velocity relative to the solution than do smaller ones, the larger crystals will grow more and the smaller crystals less, than is called for by the ΔL law. If there is a classifying action in the crystallizer by which small crystals are retained longer than large crystals, the ΔL law does not apply. It is applicable only if all crystals, regardless of size, are treated exactly alike, as to both conditions of growth and time of growth. It also applies only to the growth of existing crystals and is not concerned with nucleation.

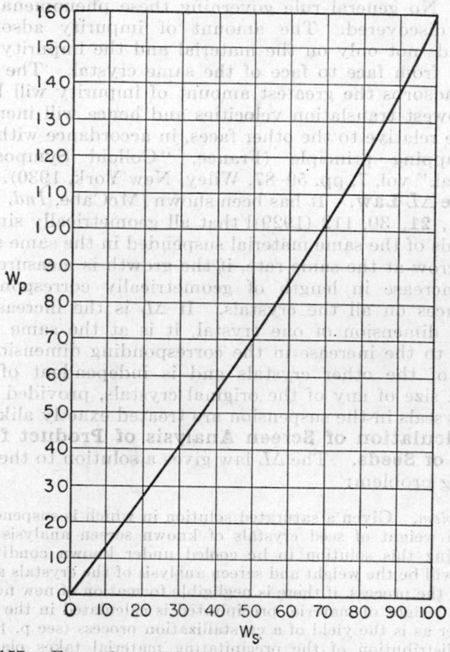

FIG. 157. Example 5: Weight of product (W_p) vs. weight of seeds (W_s).

The effect of environment on the rate of growth is confined to the mass-transfer coefficient *k* of Eq. (6). There is no effect of size on the interfacial reaction constant *k'*, although this factor will vary from face to face of a single crystal. Recent unpublished work has shown that coefficient *k* is important in the ordinary velocity range characteristic of agitated crystallizers. The larger crystals, which have a larger relative velocity with respect

Table 4. Screen Analysis of Product

Meshes/in.	D_p	$D_s = D_p - 0.009$	W_s	W_p	Screen analysis of product: Cumulative	Differential
On 12	0.1397	0.1307	0.0	0.0	0.0	0.0
14	.1168	.1078	.6	1.0	.6	.6
16	.0991	.0901	10.0	13.0	8.3	7.7
20	.0833	.0743	23.7	32.4	20.8	12.5
24	.0701	.0611	49.9	70.3	44.6	23.8
28	.0589	.0499	73.2	107.3	68.2	23.6
32	.0495	.0405	88.5	134.0	85.1	16.9
35	.0417	.0327	96.1	145.7	92.5	7.4
42	.0351	.0261	96.2	152.5	96.8	4.3
48	.0295	.0205	99.5	156.2	99.2	2.4
60	.0246	.0156	100.0	157.4	100.0	0.8

The cumulative analysis of the product is shown as curve *B*, Fig. 155.

to the solution, grow considerably faster than do the smaller crystals in contact with the same solution. The ΔL of the larger crystals, under such conditions, is distinctly greater than that of the smaller ones. Practically, the difference in growth rates between large and small crystals is advantageous, since the difference favors the production of large crystals.

Simultaneous Growth and Formation of Crystals. In most industrial crystallizations growth and nucleation occur simultaneously. The result of the combined process on the size distribution of the crystals depends on the relative rates of formation and growth. Actual experimental data for such rates are practically nonexistent, and at present only qualitative reasoning can be applied. Such reasoning may, however, be useful in analyzing the effects of the crystallization factors in industrial processes.

For aid in such an analysis, hypothetical curves such as those shown in Fig. 158 are useful. Figure 158 shows two

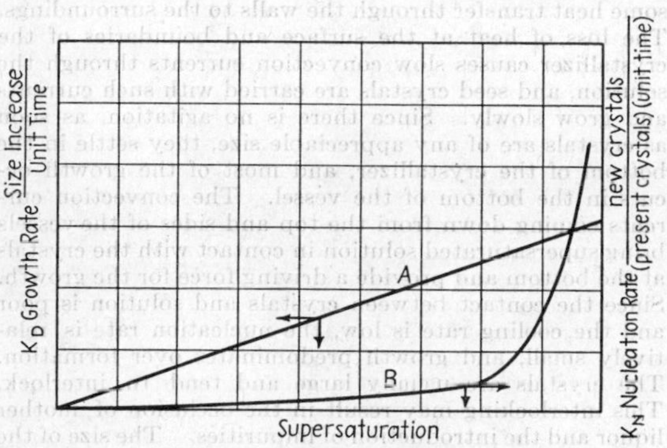

FIG. 158. Hypothetical rate of growth and nucleation curves.

rate curves, both plotted against supersaturation. Curve A shows the growth rate k_Q expressed as a coefficient giving the rate of increase of linear dimension per unit time. Curve B shows the nucleation rate expressed as a coefficient k_N defined as the number of new crystals produced per unit time per existing crystal. On the assumption that growth is approximately a first-order reaction, curve A is straight; on the other hand, the work of Miers and Ting and others indicates that the nucleation rate curve B will start low and remain low in the metastable region but will rise sharply at supersaturations corresponding roughly to the point where the supersaturation curves found by Ting are located. As mentioned above, however, the nucleation rate is not necessarily zero at low supersaturations.

Figure 158 shows that at low supersaturations growth will tend to predominate over nucleation, although both rates will be low. At high supersaturations, especially those in which the nucleation rate has passed the break point in the curve, nucleation will predominate strongly over growth.

The positions of curves such as those shown in Fig. 158 will undoubtedly vary considerably with changes in such operating variables as stirring rate, temperature, concentration, and size and number of crystals present in the solution. For example, the position of curve A can be expected to depend primarily on viscosity and stirring conditions and, perhaps to some extent, on concentration. In agitated crystallizers, curve A will usually be steeper for the larger crystals of a batch than for the smaller. The position of curve B will undoubtedly be influenced by the number of nuclei present in a unit volume of solution, probably by the area of the crystals present, perhaps by the relative motion of the crystals of the various sizes through the solution, and possibly by the rate of stirring. If nucleation at vapor-liquid interfaces or in other localities in a crystallizer occurs, curve B will also be affected by such nucleation. Both curves can be strongly influenced by the presence of impurities.

In a batch crystallization wherein the supersaturation varies with time, the final result of the crystallization process would depend upon an integrated effect along curves such as those shown in Fig. 158. Present knowledge of crystallization is not sufficient to provide a quantitative basis for such an integration. In a continuous crystallizer where conditions are constant at any given point in the apparatus, the rates at this point will be constant and correspond to a definite abscissa on Fig. 158. The higher the supersaturation at a given point, the more will crystal formation tend to predominate over crystal growth, whereas at low supersaturations growth will predominate over formation, but, in general, it may not be possible to eliminate nucleation entirely.

In general, except for such factors as local concentration gradients, the rate curves of Fig. 158 are independent of the type of apparatus used. If large crystals are desired, low supersaturations must be used or too many nuclei will form, regardless of the type of crystallizer used. Low supersaturations necessarily mean low rates of deposition, and large crystals can be made only at the expense of low volumetric crystallizer capacity. Certain types of equipment tend to produce low supersaturations and retain crystals under growth conditions for extended times and therefore give relatively large crystals at low capacity.

Usually, the objective of the crystallizer operator is to achieve a maximum growth rate consistent with a low nucleation rate. Obviously, such an operation should be conducted at a supersaturation just short of the zone of rapid rise of the nucleation rate, approximately as shown by the arrows on curve B in Fig. 158.

It is not always satisfactory to operate under such conditions, however, since factors other than capacity may enter. In some cases agglomeration tends to occur at supersaturations just short of that corresponding to a rapid increase in nucleation. Also, Miller, Phillip, and Saeman (*Chem. Eng. Progress,* **43,** 667 (1947) report that, in a pilot-plant study of ammonium nitrate crystallization in a Krystal type of equipment, the crystals formed at maximum capacity were much weaker than those formed at a capacity corresponding to a lower supersaturation.

The interpretation of the performance of actual crystallization equipment in terms of hypothetical rate curves will be given later.

CRYSTALLIZATION APPARATUS

To force crystallization, it is necessary to maintain the crystallizing solution in a supersaturated condition. The means chosen for producing and maintaining supersaturation usually depend on the solubility-temperature relation of the substance to be crystallized [Svanoe, *Ind. Eng. Chem.*, **32**, 637 (1940)]. Some solutes, such as sodium chloride, have a very small positive temperature coefficient of solubility. Others, such as anhydrous sodium sulfate and sodium carbonate monohydrate, have negative coefficients and become less soluble as the temperature is increased. In such cases supersaturation must be developed by evaporation. Other solutes, *e.g.*, Glauber's salt, Epsom salt, copperas, and hypo, have large positive temperature coefficients of solubility, and cooling without evaporation can produce the required supersaturation. In intermediate cases some evaporation in addition to cooling may be advisable to build up supersaturation.

The application of cooling as compared with evaporation is also important with respect to the yield of the process. If the solubility curves is such that the concentration of the final mother liquor is low in comparison with that of the initial solution, cooling without substantial evaporation will give a satisfactory yield per pass, and the amount of mother liquor recycled will be small. If, however, there is but little solubility change with temperature, the yield per pass through the crystallizer will be small if there is no evaporation and a large quantity of mother liquor must be recycled per unit weight of product. In such a case evaporation should be used, either as such or in combination with cooling. If the water balance of the entire process requires an evaporation step somewhere in the cycle, any evaporation accomplished in the crystallizer also reduces the evaporator load.

A classification of crystallization equipment based on the means used to develop supersaturation and to control yield per pass is

1. Supersaturation produced by cooling without substantial evaporation:

a. Atmospheric cooling by natural convection. Examples: tank crystalizers, Wulff-Bock crystallizer.

b. Cooling by liquid cooling medium, absorbing heat through metal surface. Examples: agitated batch crystallizer, Howard crystallizer, double-pipe crystallizer, Swenson-Walker crystallizer, Krystal cooling crystallizer.

2. Supersaturation produced by evaporation without substantial cooling, where the heat for the evaporation is transferred to the solution through metal surfaces. Examples: crystallizing evaporators, Krystal evaporator crystallizer.

3. Supersaturation produced by adiabatic evaporation and cooling. Example: vacuum crystallizers.

Crystallizers may also be classified in the following manners: batch vs. continuous, agitated or non-agitated, classifying or non-classifying. Classifying crystallizers function in such a manner that crystals are retained in the crystallizer until they have reached a minimum size before discharge.

Tank Crystallization. Common practice in producing crystals has been to prepare hot, nearly saturated solutions and to cool them, by natural convection, in open rectangular tanks. Little or no attempt is made to seed these tanks, to provide agitation, or to control the crystallization during the process. Sometimes rods or strings are hung in the tanks to give the crystals additional surface on which to grow and to keep at least a part of the product out of the sediment that might collect in the bottom.

When the tanks have cooled sufficiently, which is usually a matter of several days, any remaining mother liquor is drained off and the crystals are removed by hand. This involves much labor and often results in the inclusion, with the crystals, of any impurities that have settled to the bottom of the tank. The floor space and labor required and the amount of material tied up in the process are large. For moderate capacities, ordinary bathtubs of cheap quality are often very satisfactory for batch crystallizers.

The following data from a typical batch crystallization are available (G. M. Darby, Dorr Co., New York, private communication). The material crystallized was copperas ($FeSO_4.7H_2O$). The concentrated liquor was placed in rectangular crystallization tanks but 21 ft. 6 in. long, 10 ft. 8 in. wide, and 1 ft. deep. The solution was allowed to cool for from 48 to 96 hr., and the crystals removed. About 6 hr. were required to drain a tank; remove the crystals, and refill. The following data pertain to this process:

Table 5. Tank Crystallization of Copperas

Total time of cycle, hr.*	Solution received, gal.	Sp. gr. of solution received	$FeSO_4$-$7H_2O$ in solution, lb.	Crop, lb.	Mother liquor, gal.	Sp. gr. of mother liquor
48	1,630	1.400 at 62°C.	11,171	7,275	835	1.240 at 15°C.
60	1,630	1.395 at 60°C.	11,085	7,365	945	1.247 at 19°C.
72	1,630	1.935 at 53°C.	10,994	8,550	815	1.210 at 10°C.
96	1,660	1.400 at 51°C.	11,254	8,787	950	1.201 at 9°C.

* Time of cooling = total time—6 hr. (approx).

The mechanism of crystallization in a tank crystallizer can be visualized as follows: Ordinarily such crystallizers are open to the atmosphere. They lose a considerable proportion of their heat by evaporation and by convection at the surface. In all probability such crystallizers are seeded to some extent by the atmosphere which will contain seeds and dust from previous crystallizations of the same material. The evaporation at the surface develops local supersaturation at the surface, and nucleation tends to occur at that point. Nucleation may also occur near the wall of the crystallizer because there will be some heat transfer through the walls to the surroundings. The loss of heat at the surface and boundaries of the crystallizer causes slow convection currents through the solution, and seed crystals are carried with such currents and grow slowly. Since there is no agitation, as soon as crystals are of any appreciable size, they settle in the bottom of the crystallizer, and most of the growth occurs in the bottom of the vessel. The convection currents coming down from the top and sides of the vessels bring supersaturated solution in contact with the crystals at the bottom and provide a driving force for the growth. Since the contact between crystals and solution is poor and the cooling rate is low, the nucleation rate is relatively small, and growth predominates over formation. The crystals are usually large and tend to interlock. This interlocking may result in the occlusion of mother liquor and the introduction of impurities. The size of the individual crystals is variable because of the lack of control of the convection currents and the lack of agitation. The capacity of such a crystallizer is low because of the low rate of heat transfer obtainable by atmospheric cooling. The evaporation from the surface is sensitive to the relative humidity of the air, and at times in summer weather, when the relative humidity and air temperature are both high, production from such equipment may be very low.

In the operation of naturally cooled tank crystallizers, there is no way to control either nucleation or growth except by using suitable lagging, or by varying the ratio of tank surface to tank volume. The size of the unit can be so chosen that the rate of heat loss corresponds roughly to the cooling time necessary to give the desired crystal size.

The Wulff-Bock Crystallizer. The Wulff-Bock [Griffiths, *J. Soc. Chem. Ind.*, **44**, 7*T* (1925)] type of crystallizer has been widely used in Germany and in England but has not been used extensively in the United States. It consists of a shallow trough set at a slight inclination and mounted on rollers so that it can be rocked from side to side. At frequent intervals along its length are partitions extending part way across, so that the liquid, instead of flowing directly from one end to the other, flows in a zigzag path. Cooling is by natural convection, and the crystallizer is continuously operated.

The slow rate of cooling inherent in the Wulff-Bock crystallizer results in a relatively low capacity, but this crystallizer gives uniform crystals of unusually large size.

Agitated Batch Crystallizers. Figure 159 shows an agitated batch crystallizer. Water is circulated through the cooling coils, and the solution is agitated by the propellers on the central shaft. This agitation performs two functions: (1) It increases the rate of heat transfer and keeps the temperature of the solution more nearly uniform, and (2) by keeping the fine crystals in suspension it gives them an opportunity to grow uniformly instead of forming large crystals or aggregates. Further, the agitation, combined with the more rapid cooling, results in the formation of a large number of nuclei as compared with the tank methods, and therefore the product of this operation is not only more uniform but also very much finer than that from the older tanks. The difficulties with this apparatus are: first, that it is a batch or discontinuous method; and, second, that the solubility is least in the stagnant film on the surface of the cooling coils. Consequently crystal growth is most rapid at this point, and the coils rapidly build up with a mass of crystals which decreases the rate of heat transfer.

A certain amount of control can be exercised on the crystallization process occurring in a batch, artificially cooled crystallizer by varying the rate of cooling. As

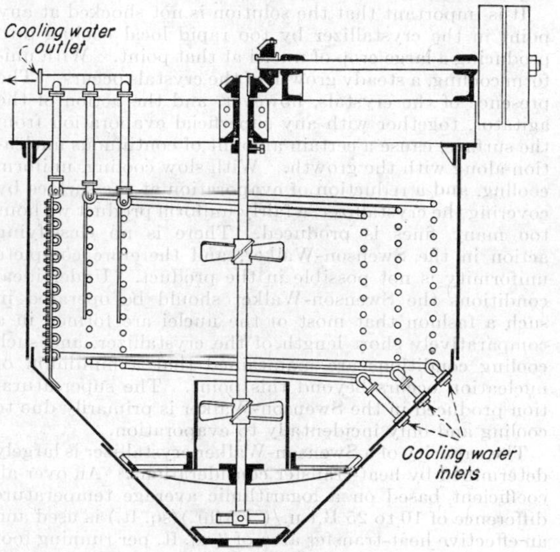

FIG. 159. Agitated batch crystallizer.

long as the solution is undersaturated, it may be cooled as rapidly as the temperature and rate of the cooling water and the area of the cooling surface permit. When the solution becomes supersaturated, the rate of temperature decrease should be retarded in order that the labile region is not entered. It is advisable to seed the solution at this point to prevent the uncontrolled initial nucleation characteristic of unseeded solutions. Once there is available considerable crystal surface for growth, the rate of heat removal can be increased, provided that the coils are not so badly salted that effective heat transmission is prevented. The rate of temperature decrease may be very small during the period of maximum crystallization rate, even if the rate of heat transfer is good, because of the necessity of withdrawing from the solution the latent heat of crystallization.

The Howard Crystallizer. This crystallizer (Fig. 160) consists essentially of a vertical conical device through which solution flows in an upward direction. The upper end of the crystallizer is the wide part of the cone. A concentric outer conical chamber serves as a cooling water channel. Crystals that are suspended in the upward flowing stream of solution must grow to such a size that they will settle through the fastest part of the stream of solution at the apex of the cone (the bottom of the crystallizer) before they can escape. By regulating the velocity of flow at the bottom of the crystallizer, the size of the product is controlled. On the other hand, the cross section of the top of the crystallizer is large, the velocity of the solution is low, and the smaller crystals are not carried over the top. The apparatus functions both as a crystallizer and as a hydraulic classification device. The concentration of the solution is maintained by the inflow of strong solution from a storage tank, and the product is withdrawn continuously in a steam of mother liquor.

Since the Howard crystallizer is a continuous one, conditions in it can reach a steady state. As the solution flows up through the crystallizing cone, its supersaturation increases because of the cooling brought about by the inner cooling cone C. Nucleation should start at a fairly definite point in the crystallizer. Nuclei will tend to drop through the incoming solution because of the action of gravity and are thereby continually contacted with

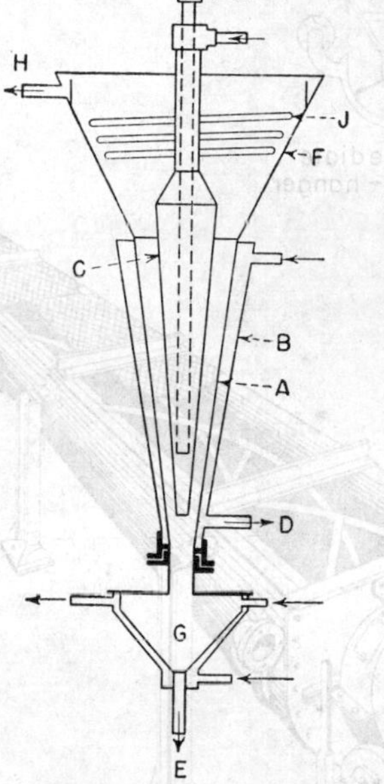

FIG. 160. Howard crystallizer.

fresh supersaturated solution. Growth therefore occurs, and, when the crystal is large enough, it will settle into vessel G. Nucleation due to the inoculating effect of existing crystals can accompany the growth in the same zone in which the growth is taking place.

The Double-pipe Crystallizer. [Seavoy and Caldwell, *Ind. Eng. Chem.*, **32**, 628 (1940).] This crystallizer consists essentially of a double-pipe heat exchanger fitted with internal helical ribbons. The cooling liquid passes through the annular space between the two pipes, and the crystallizing solution is pumped through the inner tube countercurrent to the cooling liquid. The helical ribbons act as scrapers to keep crystals from building up on the cooling surface. The scrapers make contact with the inner wall of the inside pipe. This crystallizer is ordinarily used in a continuous batch manner. It is placed in series with a large tank containing the solution to be cooled, and the solution is pumped through the double-pipe unit at a rate sufficient to ensure an adequate heat-transfer rate.

The Swenson-Walker Crystallizer. In the United States the most successful continuous crystallizer using a liquid cooling medium is the Swenson-Walker crystallizer. This crystallizer is shown in Fig. 161 (Caldwell and Seavoy, *loc. cit.*). It consists of an open trough 24 in. wide, with a semi-cylindrical bottom, a water jacket welded to the outside of the trough, and a slow-speed, long-pitch, spiral agitator set close to the bottom of the trough, but not so close as to make contact with the trough. This apparatus is ordinarily built in units 10 ft. long, and a number of units may be joined together to give increased capacity. Forty feet is the maximum length usually driven from one shaft; and, if lengths greater than this are desired, it is usual to arrange several such crystallizers, one above the other, and allow the solution to cascade from one bank to the other.

The hot concentrated solution to be crystallized is fed at one end of the trough, and cooling water usually flows through the jackets, countercurrent to the solution. In order to control crystal size, it is sometimes desirable to introduce an extra amount of water into certain sections. When conditions are properly adjusted, nuclei begin to

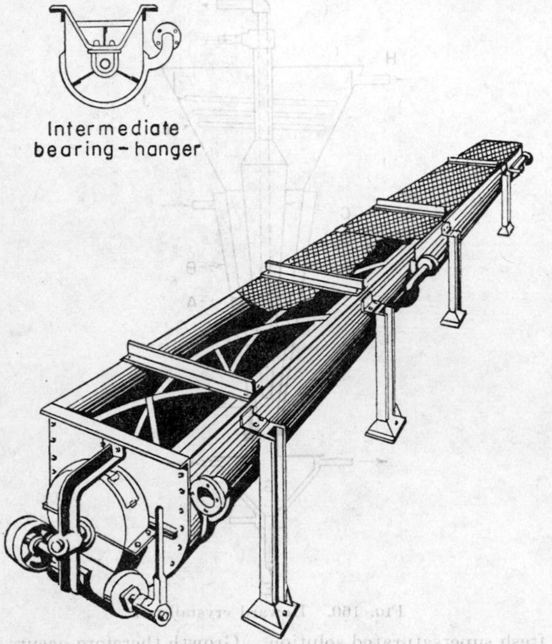

FIG. 161. Swenson-Walker crystallizer. [*Seavoy and Caldwell, Ind. Eng. Chem.*, **32,** 632 (1940).]

form a short distance from the point where feed is introduced; these nuclei grow regularly as the solution passes down the length of the crystallizer. The function of the spiral stirrer is not especially that of either agitation or conveying the crystals. Its purposes are (1) to prevent an accumulation of crystals on the cooling surface and (2) to lift the crystals which have already been formed and shower them down through the solution. In this manner the crystals grow while they are freely suspended in the liquid and therefore are usually fairly perfect individuals, reasonably uniform in size, and free from inclusions or aggregations.

At the end of the crystallizer there may be an overflow gate where crystals and mother liquor together overflow to a draining table or drain box, from which the mother liquor is returned to the process and the wet crystals are raked to a centrifuge. In other cases, a short section of an inclined-screw conveyor lifts the crystals out of the solution and delivers them to the centrifuge, while the mother liquor overflows at a convenient point. The advantages of this type over tank crystallization are high capacity, large saving in floor space and in material in process, and especially a saving in labor. To reach low final temperatures, the cooling can be obtained by the use of refrigerated brine instead of cooling water.

The mechanism of crystallization in the Swenson-Walker crystallizer is as follows: Preferably the incoming solution should be slightly superheated above its supersaturated temperature. It should be cooled uniformly, especially through the range of temperature where nucleation first occurs. It is possible to seed the solution artificially, but ordinarily seeding is restricted to the fortuitous seeding from the atmosphere. If the solution is not seeded and is also protected from atmospheric seeding, it will tend to supercool well into the supersaturated region and on reaching the Miers curve will tend to form in an uncontrolled fashion too many nuclei. Once the nucleation begins, either through fortuitous seeding or by entering the Miers labile region, crystals and solution flow together through the remaining crystallizer length.

It is important that the solution is not shocked at any point in the crystallizer by too rapid local cooling into producing a large crop of nuclei at that point. With uniform cooling, a steady growth of the crystals occurs. The presence of the crystals, however, and the action of the agitator, together with any superficial evaporation from the surface, cause a certain amount of continuous nucleation along with the growth. With slow cooling, uniform cooling, and a reduction of evaporation at the surface by covering the crystallizer, a fairly uniform product without too many fines is produced. There is no classifying action in the Swenson-Walker, and therefore complete uniformity is not possible in the product. Under ideal conditions the Swenson-Walker should be operated in such a fashion that most of the nuclei are formed in a comparatively short length of the crystallizer, and such cooling conditions are maintained that a minimum of nucleation occurs beyond this point. The supersaturation produced in the Swenson-Walker is primarily due to cooling and only incidentally to evaporation.

The capacity of a Swenson-Walker crystallizer is largely determined by heat-transfer considerations. An over-all coefficient based on a logarithmic average temperature difference of 10 to 25 B.t.u./(°F.)(hr.)(sq. ft.) is used and an effective heat-transfer area of 3 sq. ft. per running foot of crystallizer assumed. The number of units to be used in parallel depends on the total capacity desired. For most inorganic salts such as trisodium phosphate, Glauber's salt, etc., a production of from 5 to 15 tons per day per unit can be obtained.

Example 6. A Swenson-Walker crystallizer is to cool a 23 per cent solution of Na_3PO_4 from a temperature of 104° to 77°F. During the cooling, $Na_3PO_4.12H_2O$ is crystallized. It is desired to produce 500 lb. product/hr. The solubility of Na_3PO_4 at 77°F. is 15.5 parts anhydrous salt/100 parts total water. The specific heat of the solution can be taken as 0.77, and the heat of crystallization of 1 lb. product is 63 B.t.u./lb. Cooling water is to enter the crystallizer jacket at 60°F. and is to leave at 68°F. The over-all heat-transfer coefficient is 25 B.t.u./(sq. ft.)(hr.)(°F.). What length of crystallizer should be used?

Solution. The weight of solution per hour that will give 500 lb. of crystals is calculated with the aid of Eq. (1). The molecular weight of $Na_3PO_4.12H_2O$ is 380.2, and that of Na_3PO_4 is 164.0; $R = 380.2/164.0 = 2.32$. If a basis of 100 lb. original solution is chosen, $w_0 = 23.0$; $S = 15.5$; and $H = 100 - 23.0 = 77.0$ lb. The product obtained from 100 lb. solution is, by Eq. (1),

$$C = 2.32 \frac{(100)(23.0) - (15.5)(77.0)}{100 - 15.5(2.32 - 1.0)} = 32.2 \text{ lb.}$$

In order that 500 lb./hr. of crop be obtained, a feed of (100/32.2)(500) = 1550 lb. is necessary. The heat to be removed from the crystallizing solution is:

To cool solution (1550)(0.77)(104 − 77)	= 32,200 B.t.u./hr.
To crystallize: (500)(63)	= 31,500 B.t.u./hr.
Total = q	= 63,700 B.t.u./hr.

The logarithmic mean temperature drop is

$$(\Delta t)_m = \frac{(104 - 68) - (77 - 60)}{2.303 \log \dfrac{104 - 68}{77 - 60}} = 25°\text{F.}$$

The length of crystallizer is

$$\frac{q}{3U(\Delta t)_m} = \frac{63{,}700}{(3)(25)(25)} = 34 \text{ ft.}$$

Four 10-ft. sections should be used.

Crystallizing Evaporators. The development of supersaturation by means of evaporation without substantial cooling is often carried out in equipment that has the physical characteristics of an evaporator and, in fact, is designed essentially as an evaporator largely because the essential engineering problem is one of heat transfer. Usually the equipment employed is so nearly like that used in ordinary evaporation that it is considered to be an evaporator, although the crystallization may be the more difficult of the two parts of the problem. In the evaporation of a salting liquor (*e.g.*, the precipitation of NaCl from brine in the common-salt industry) the crystallization is usually incidental to the evaporation, and no particular control of size is exercised. On the other hand, the crystallizing of sugar is carried out in a vacuum evaporator, but the control is based entirely on building a correct crystal. In this case the operator brings the sirup to a definite density, shocks out the desired number of nuclei, and grows them to the correct size without forming new crystallization centers. The control is exercised entirely by varying the vacuum and steam supply.

A special case of evaporative crystallization is the salt grainer (Badger and Baker, "Inorganic Chemical Technology," 2d ed. pp. 15–18), McGraw-Hill, New York, 1941. In the salt grainer the solution is kept hot and supersaturation is developed by evaporation rather than by cooling. Nucleation occurs at the surface of the brine, the nuclei tend to be retained at the surface by surface-tension effects and to form hopper-shaped crystal agglomerates, which, when large enough, break away, drop to the bottom of the grainer, and are raked out by slow-moving rakes.

Vacuum Crystallizer. Assume that a warm, saturated solution is fed to a lagged closed vessel that is maintained under a vacuum and that the solution is fed in such a way that it reaches the surface of the liquid in the crystallizer. The solution will have a definite boiling temperature under the vacuum existing in the vessel. If this temperature is *less* than that of the feed solution, the solution will spontaneously and adiabatically cool to the boiling temperature corresponding to the vacuum existing in the vessel and reach equilibrium with respect to the vapor in the crystallizer. If the solubility of the solute decreases as the temperature decreases, the cooling will result in crystallization, not only because of the cooling of the solution, but also because of the evaporation of some of the solvent, as the heat evolved in the cooling and crystallizing of the solution must appear as latent heat of evaporation. Such a crystallizer is a vacuum crystallizer.

In a vacuum crystallizer it is necessary to ensure that the incoming solution reaches the surface and can thereby flash to equilibrium with the vapor in the crystallizer. Artificial circulation is necessary to accomplish this; otherwise the feed will tend to short-circuit to the discharge, especially if the discharge is at the bottom of the vessel. Under the low absolute pressures commonly used in this equipment, a foot of hydrostatic head causes a decided increase in boiling point of the solution; if short-circuiting occurs, little or no supersaturation will be produced in the short-circuited solution, and it will not deposit solid material.

The vacuum crystallizer has become very successful during the past few years and continues to replace many of the older types of the mechanical crystallizer. Four forms of the vacuum crystallizer are shown in Figs. 162*A*, *B*, *C*, and *D* [Seavoy and Caldwell, *Ind. Eng. Chem.*, **32**, 627 (1940)]. Figure 162*A* represents a simple type of batch vacuum crystallizer. Propeller agitators are used to develop a swirling action in the crystallizer. In case of temperature in the crystallizer is high enough so that the vapor from it is condensable by the cooling water available, the vapors from the crystallizer pass directly to a condenser as shown in Fig. 162*A*. In most cases, however, the cooling water is too warm to condense the vapor leaving a vacuum crystallizer at the desired temperature of operation in the crystallizer. In this case a

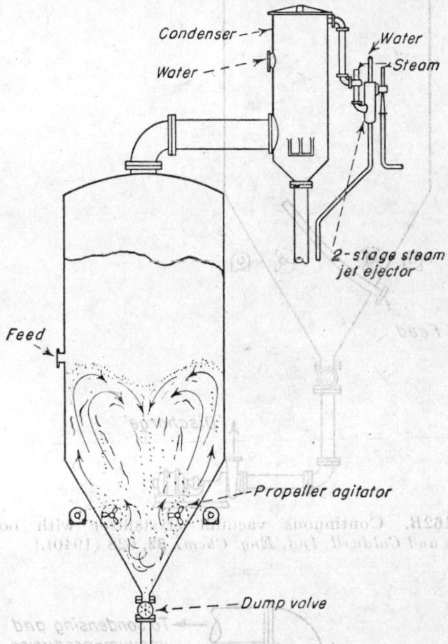

Fig. 162*A*. Batch vacuum crystallizer without booster. [*Seavoy and Caldwell, Ind. Eng. Chem.*, **32,** 628 (1940).]

steam-jet booster is used to compress the vapors to a point where they can be condensed by the cooling water available. The air and non-condensable gases from the condenser are commonly ejected to the atmosphere by further steam-jet equipment. A second type of vacuum crystallizer, operated by such a booster, is shown in Fig. 162*B*. This type is operated continuously. The feed tube is so positioned that the feed solution is forced to the surface and flashes to equilibrium with the vapor. Propeller agitators aid in keeping the crystals in suspension and in preventing short circuiting.

The forms of crystallizer shown in Figs. 162*C* and 162*D* can be operated either as batch or as continuous units. The combination of propeller and draft tube shown in Fig. 162*C* is effective in preventing short circuiting of the feed. The form shown in Fig. 162*D* has an external circulating pump which takes suction from the side of the crystallizer and discharges tangentially into the cone. The agitators can be omitted from this unit. The feed is introduced into the circulation stream, and the circulation stream must be large enough that the mixed stream is not so supersaturated that it is in the labile condition.

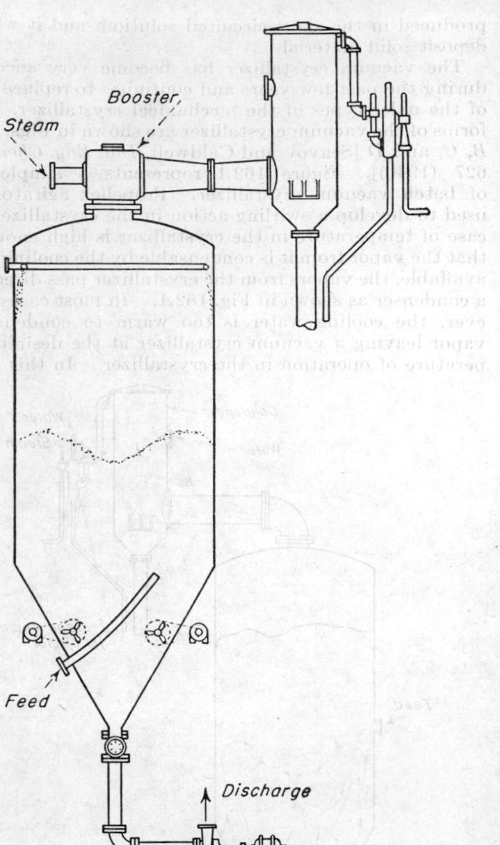

FIG. 162*B*. Continuous vacuum crystallizer with booster. [*Seavoy and Caldwell, Ind. Eng. Chem.*, **32**, 628 (1940).]

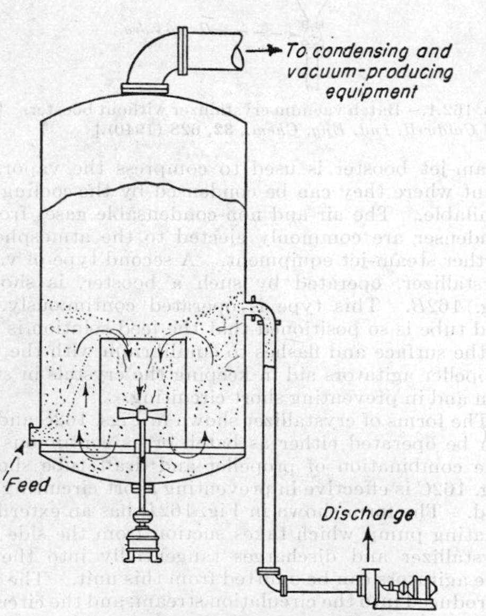

FIG. 162*C*. Continuous vacuum crystallizer with vertical agitator. [*Seavoy and Caldwell, Ind. Eng. Chem.*, **32**, 629 (1940).]

Batch operation has the advantage of low steam consumption if a steam-jet booster is used. The steam required by a steam-jet booster to remove 1 lb. of low-pressure vapor from the crystallizer increases rapidly as the pressure in the crystallizer is reduced. In batch cooling much of the vapor is removed at a relatively high pressure because the solution charged to the crystallizer is essentially hot and only at the end of the batch is the full pressure differential over the booster required and only at the end of the process is the maximum steam consumption called for. The average steam consumption is therefore considerably lower than is the case for a continuous crystallizer where all vapor must be removed under conditions where the steam consumption is a maximum. On the other hand, a continuous crystallizer has the advantages of lower first cost per unit of capacity, ease of control, and constant mass of crystals in the unit. It is easier in the continuous unit to maintain supersaturations outside of the labile field. The concentration vs.

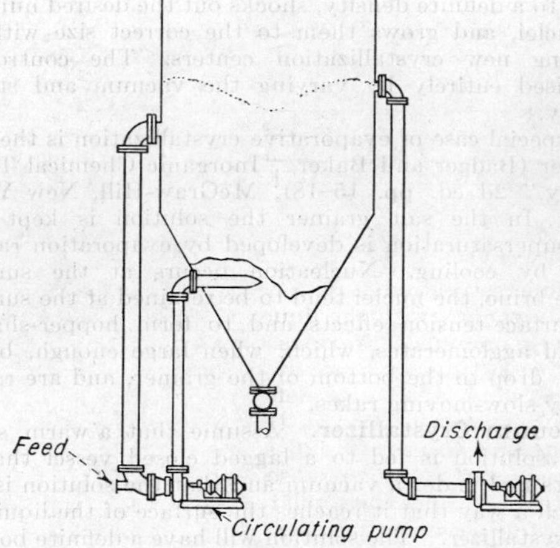

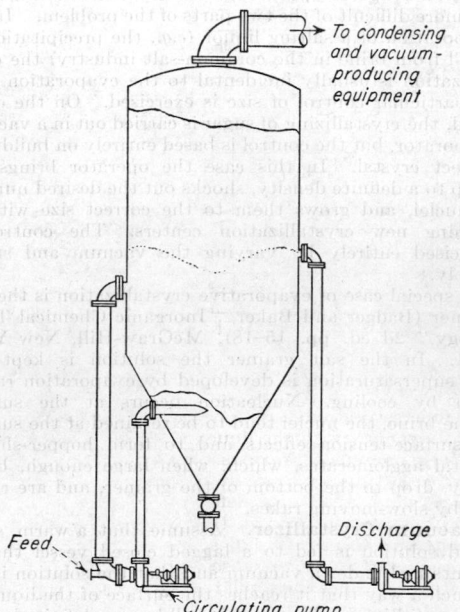

FIG. 162*D*. Continuous vacuum crystallizer with pump circulation. [*Seavoy and Caldwell, Ind. Eng. Chem.*, **32**, 629 (1940).]

time relationship for a typical batch case is shown in Fig. 163.

Continuous vacuum crystallizers can be operated to give a steam economy approaching that of batch operation by passing the solution through a series of units, each provided with its own booster. In the first stages the boosters are operating under conditions where the steam consumption is low and only the last unit calls for a maximum steam consumption per pound of vapor removed. In this way the steam consumption of a continuous unit may be reduced, but only at the expense of higher investment costs.

Analysis of Vacuum Crystallizer Operation. The action of a continuous vacuum crystallizer in forming and growing crystals may be visualized as follows:

Assume that the agitation and circulation in such a crystallizer are adequate to ensure homogeneous conditions throughout the crystallizing liquid. Assume also that the crystallizer is fed constantly with solution of a given concentration and temperature and the that pressure in the crystallizer is constant.

Under steady-state conditions there must be a definite and constant number of crystals in the crystallizer at all times. These crystals will have a definite size distribution, and a steady state is reached if the rate of formation and growth does not change with time. Assume also that the operation of the crystallizer is such that a crystal will not be able to leave the crystallizer until it has reached a definite size. Under steady-state conditions the entering solution immediately flashes to equilibrium temperature. The flash results in a certain amount of evaporation and considerable cooling, which develops in the solution of definite supersolubility. The solution is mixed quickly with the bulk of the solution in the crystallizer and a definite supersaturation can be assumed to exist in the bulk of the liquid. It is the relatively high supersaturation contributed by the flashed solution that maintains the supersaturation at a definite average value in competition with the efforts of the crystals to destroy

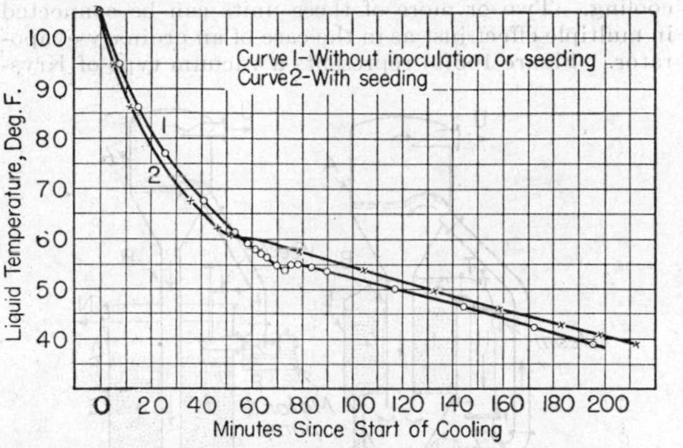

FIG. 163. Typical batch vacuum-crystallizer cooling curves. [*Seavoy and Caldwell, Ind. Eng. Chem.*, **32**, 633 (1940).]

the supersaturation. If the flashed solution is blended rapidly enough with the liquor in the crystallizer, it will not form new nuclei previous to its incorporation with the bulk of the liquid. Since at all times there is a definite number of crystals of a definite size range, and also a definite supersaturation in the solution, there will be a definite rate of growth and a definite rate of formation corresponding to a given abscissa on a curve such as that of Fig. 158. Since there will be formed a constant number of crystals per unit time, the rate of formation of crystals must be equal to the rate at which crystals are withdrawn from the crystallizer. The rate of nucleation must adjust itself in such a manner that the number of crystals formed per unit time must equal the number of crystals withdrawn per unit time. Also, the rate of growth must adjust itself in such a manner that the crystals formed will grow to the desired size in the time available for their growth. If the rate curves are of the form shown in Fig. 158, there is only one supersaturation that will give both a correct rate of formation and a correct rate of growth. For larger crystals, growth rate rather than nucleation rate must be emphasized, and a low supersaturation will exist in the crystallizer. For small crystals, nucleation will increase relative to growth, and larger supersaturation will be necessary. Rate of withdrawal (assuming that adequate flashing and vapor liberation can be maintained) controls all these factors automatically. If the rate of withdrawal from the crystallizer is slow, the crystals withdrawn will be large. The capacity of the crystallizer will decrease, of course, as the size of the crystals is increased.

Yield from a Vacuum Crystallizer. The calculation of the yield obtainable in a vacuum crystallizer depends upon the method used in the heat-balance calculations. A vacuum crystallizer operates essentially adiabatically. The heat liberated by the solution on cooling to the equilibrium temperature and the heat of crystallization are available for vaporizing water from the solution, and these thermal effects must balance. If the enthalpy-concentration chart is used, the total enthalpy of the vapors and magma leaving the crystallizer must equal the total enthalpy of the feed solution entering the unit.

In case the heat items are computed individually, which is the method to be used when the enthalpy-concentration chart is not available, the evaporation can be calculated by means of Eq. (10)

$$E = \frac{(w_0 + H_0)(c)(\Delta t)[100 - S(R - 1)] + q_c R(100 w_0 - S H_0)}{L_w[100 - S(R - 1)]q_c R S} \tag{10}$$

where w_0 is the weight of anhydrous solute; H_0 is the total weight of solvent in the feed solution; c is the specific heat of the feed solution; Δt is the temperature range through which the solution is cooled (temperature of feed to temperature of discharge); q_c is the heat of crystallization per unit weight of crystal; L_w is the latent heat of evaporation from the solution; S is the anhydrous solubility in parts solute per 100 parts total solvent; and R is the ratio of the molecular weight of the crystals to that of the anhydrous salt. When the value of E is known, the yield is calculated by means of Eq. (1). Calculation for batch vacuum crystallizers can be based on Eq. (10) by deviding the cooling range into steps and applying the equation to each step.

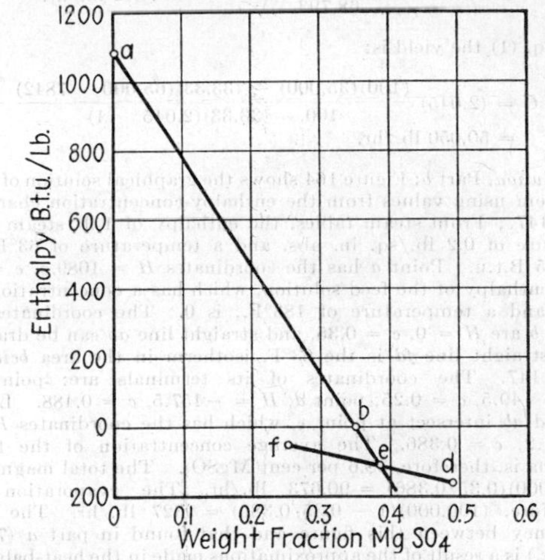

FIG. 164. Solution to Example 7b.

If the enthalpy-concentration chart is used, the simplest method of vacuum crystallizer calculation is one based on the "basic construction" applicable to the chart. Since the process is adiabatic, a single straight line on the chart must pass through the three points on the chart that represent feed solution, magma, and vapor, respectively. Furthermore, the point representing the magma must lie on the straight-line isotherm lying in the magma field and corresponding to the temperature of the vapor and magma. The point representing the magma is found by locating the intersection of these two lines as shown in Fig. 164. The ratio of crystals to mother liquor in the magma is found from the intersection by applying the lever-arm principle to the two line segments on the straight-line isotherm.

Example 7. A continuous vacuum crystallizer is fed with 100,000 lb./hr. of a 35 per cent solution of $MgSO_4$ at a temperature of 183°F. An absolute pressure of 0.2 lb./sq. in. is maintained in the crystallizer by the booster, and the solution has a 10°F. elevation in boiling point.

Calculate the yield of $MgSO_4.7H_2O$, and the evaporation for this crystallizer:

a. By means of Eqs. (1) and (10).

b. By means of Fig. 147.

Solution. Part *a*: For this take

$$c = 0.77 \text{ B.t.u./(°F.)(lb.)}$$
$$L_W = 1080 \text{ B.t.u./lb.}$$
$$q_c = 23 \text{ B.t.u./lb. } MgSO_4.7H_2O$$

The temperature of boiling water at 0.2 lb./sq. in. abs. is 53°F., and the equilibrium temperature of the solution in the crystallizer is 53 + 10 = 63°F.

The remaining numerical values to be substituted in Eqs. (1) and (10) are:

w_0 = (0.35)(100,000) = 35,000 lb./hr.

$$R = \frac{\text{mole wt. } MgSO_4.7H_2O}{\text{mole wt. } MgSO_4} = 2.045$$

S = solubility of anhydrous $MgSO_4$ at 63°F. = 33.33 parts $MgSO_4$/100 parts total water (Fig. 148)

Δt = 183 − 63 = 120°F.

H_0 = 100,000 − 35,000 = 65,000 lb./hr.

then $(w_0 + H_0)(c)(\Delta t)[100 - S(R - 1)]$
$= (100,000)(0.77)(120)[100 - 33.33(2.045 - 1)]$
$= 6.022 \times 10^8$

$(q_c)(R)(100w_0 - SH_0) = (23)(2.045)[(100)(35,000) - (33.33)(65,000)] = 0.62 \times 10^8$

$L_W[(100 - S(R - 1)] + q_cRS = 1080[100 - 33.33(2.045 - 1)] - (23)(2.045)(33.33) = 68,793$

By Eq. (10) the evaporation is:

$$E = \frac{(6.022)(10^8) - 0.627 \times 10^8}{68,793} = 7842 \text{ lb./hr.}$$

By Eq. (1) the yield is:

$$C = (2.045)\frac{(100)(35,000) - (33.33)(65,000 - 7842)}{100 - (33.33)(2.045 - 1)}$$
$$= 50,050 \text{ lb./hr.}$$

Solution, Part *b*: Figure 164 shows the graphical solution of this problem using values from the enthalpy-concentration chart of Fig. 147. From steam tables, the enthalpy of 1 lb. steam at a pressure of 0.2 lb./sq. in. abs. and a temperature of 63°F. is 1089.5 B.t.u. Point *a* has the coordinates H = 1089.5, c = 0. The enthalpy of the feed solution, which has a concentration of 0.35 and a temperature of 183°F., is 0. The coordinates of point *b* are H = 0, c = 0.35, and straight line *ab* can be drawn. The straight line *fd* is the 63°F. isotherm in the area *bcih* of Fig. 147. The coordinates of its terminals are: point *f*, H = −49.5, c = 0.25; point *d*, H = −157.5, c = 0.488. Lines *fd* and *ab* intersect at point *e*, which has the coordinates H = −111.2, c = 0.386. The average concentration of the final magma is, therefore, 38.6 per cent $MgSO_4$. The total magma is (100,000)(0.35/0.386) = 90,673 lb./hr. The evaporation is, therefore, (100,000)(1 − 0.35/0.386) = 9327 lb./hr. The discrepancy between this figure and that found in part *a* (7842 lb./hr.) is a result of the approximations made in the heat-balance items used in part *a*.

The fraction of the magma that is crystalline is

$$\frac{0.386 - 0.25}{0.488 - 0.25} = 0.57$$

and the yield is (0.571)(90,673) = 51,770 lb./hr.

The Krystal Classifying Crystallizer. This equipment, also known as the Jeremiassen or Oslo crystallizer, is characterized by the fact that the supersaturation is produced in a circulating stream, and the supersaturation is developed in one part of the unit and released in another. In the crystallizing element itself, the supersaturated solution flows up through a bed of forming and growing crystals and provides a classifying action. Three types of Krystal equipment are shown in Figs. 165 and 166 [Svanoe, *Ind. Eng. Chem.*, **32**, 636 (1940)]. These three types differ primarily in the means used for developing supersaturation. In Fig. 165(*a*) the supersaturation is obtained by heating the circulating stream while under a static head great enough to prevent its vaporization and flashing of the heated solution in vessel *A*. The vapor released by the flash is removed through pipe *U*. The solution, supersaturated with respect to the temperature existing in crystallizing vessel *E*, leaves the flash vessel *A*, passes up through a screen in the bottom of vessel *E*, contacts crystals above the screen and loses its supersaturation while in contact with them. The overflow stream, leaving vessel *A* and passing to heater *H*, should be practically saturated. Feed solution is mixed with this solution at *T*. Crystals are drawn off continuously or periodically through discharge *M*. Pump *F*, driven by the motor forces the circulating stream through its circuit. Heater *H* is steam-heated. This type of crystallizer is used in cases where the supersaturation must be developed entirely by evaporation and not by cooling. Two or more of these units can be connected in multiple effect just as in the case of an ordinary evaporator. Figure 165(*b*) represents a vacuum type of Krystal unit.

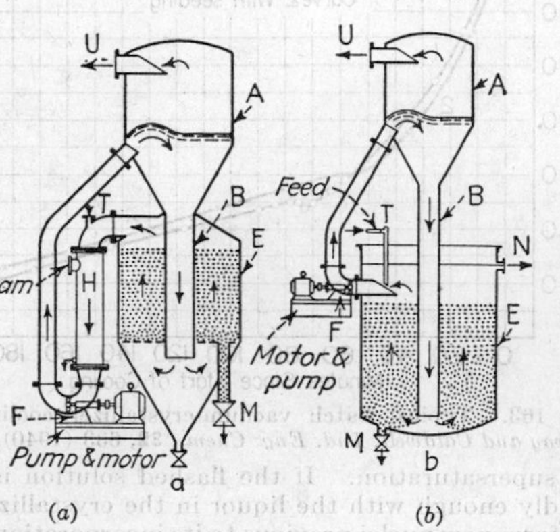

FIG. 165. Krystal crystallizer. (*a*) Evaporator crystallizer. (*b*) Vacuum crystallizer.

This unit is a true vacuum crystallizer, in that the supersaturation is developed by adiabatic pressure reduction on the hot, concentrated, feed solution. The feed enters at *T*, is incorporated in a circulating cycle stream, and the combined stream flashed in flash vessel *A*. Supersaturated solution passing through pipe *B* contacts growing crystals in crystallizer *E*, and the flow of liquid in vessel *E* performs a classifying action. Mother liquor is drawn off at *N* and the magma drawn off at *M*. Nucleation can occur in vessel *E* by the inoculation of the solution by the existing crystals and by any impact of the crystals on each other and on the wall of the vessel. In continuous operation the rate of nucleation must equal the number of finished crystals withdrawn as product.

The modification shown in Fig. 166 develops supersaturation entirely by liquid cooling. The circulating stream passes through the tubes of cooler *H*, and supersaturated solution flows through pipe *B* to the bottom of crystallizer *E*. The feed, which should be warm and concentrated, enters at *T*. It is incorporated immediately in the circulating stream, the combined stream is cooled in *H*, and supersaturation is thereby obtained. The diluting of the incoming feed with a comparatively large circulating stream of mother liquor allows the solution to be cooled in cooler *H* without entering the labile region and thereby allows this cooling to be accomplished without nucleation until the solution comes in contact with the crystals in *E*. Vessel *G* can be used to remove very small nuclei that reach the upper layers of the crystallizer E. If small nuclei are continually removed, the average size of the crystal crop is increased. The action of vessel G has substantially the same effect as a reduction in the rate of nucleation curve shown in Fig. 158. Mother liquor leaves the crystallizer at M.

In the travel of the supersaturated solution up through the crystal mass, the supersaturation, and therefore the rate of crystal growth, decreases from bottom to top. The average supersaturation and the average rate of growth at the top are both considerably lower than the supersaturation and growth rate at the bottom of the mass at the point of entrance of the supersaturated solution.

Miller, Phillip, and Saeman (*loc. cit.*) obtained significantly greater capacity in the crystallization of ammonium nitrate in a Krystal unit [Fig. 165(*b*)] by conducting, through an internal pipe, supersaturated solution from the bottom of pipe B to the top of the crystal bed.

In all forms of the Krystal evaporator the theory of the metastable region of a supersaturated solution is applied; nucleation is not obtained in the solution while developing

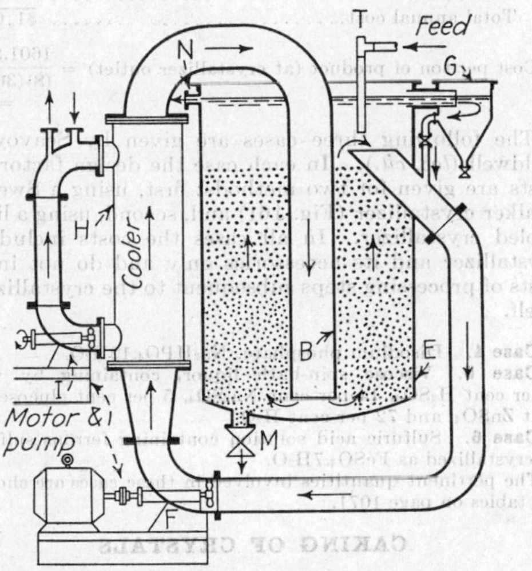

FIG. 166. Cooling crystallizer. [*Svanoe, Ind. Eng. Chem.*, **32**, 638 (1940).]

its supersaturation in flash vessel A or cooler H because the supersaturation developed in these parts is kept low enough so that the labile region of the unseeded solution is not entered. This can be done because of the large amount of recirculating solution. A small supersaturation developed in a large recirculating stream makes available to the crystallizer enough potential precipitating material to force crystallization in vessel E. In vessel E, however, the presence of the nuclei results, not only in the growth, but also in nucleation due to inoculation, impact, surface, and attrition to supply the nuclei required.

Comparison of Vacuum and Liquid-cooled Crystallizers. In cases where the necessary supersaturation can be developed by cooling, the choice of a crystallizer usually lies between a mechanical unit, cooled by transmitting heat through a metal wall, and a vacuum unit has no heat-transfer surface in the crystallizer itself. In such a comparison the Swenson-Walker may be taken as representative of the mechanical type, and the vacuum crystallizers shown in Fig. 162, as representative of the vacuum type [Seavoy and Caldwell, *Ind. Eng. Chem.*, **32**, 631 (1940)].

The main difference between the vacuum and mechanical crystallizers is that the heat is removed in the vacuum crystallizer without passing it through a heating surface. This gives the vacuum crystallizer several important advantages. The absence of large heat-transfer surfaces results in a lower first cost for the vacuum crystallizer and also allows the crystallizer to be built of corrosion-resisting materials. The absence of the cooling surface also eliminates the growth of crystals on a metal surface from which they must be removed mechanically during operation. The limitations of the vacuum crystallizer are: (1) it usually requires a steam-jet booster to obtain the low temperature desired, and the steam consumption may be large; (2) magma densities may be too heavy to circulate freely in the crystallizer; (3) it may not be possible to attain the desired final temperature. The first limitation of high steam consumption has been discussed. Magma density may limit the cooling region available because of the difficulty of discharge and the difficulties of developing adequate circulation in the crystallizer. The critical magma density is approximately 50 to 55 per cent by weight when the density of the crystals is not greatly different from the density of the mother liquor and 35 to 40 per cent by weight when the differences in densities are relatively large.

The difficulty of attaining the desired final temperature may be due to the inability of the booster to exhaust at a low enough pressure. The commercial limit of suction pressure obtainable with such equipment is about 0.11 in. Hg absolute. This corresponds to a water boiling point of 20°F. If the solution possesses an appreciable boiling-point elevation, the minimum temperature obtainable is increased by the amount of boiling-point elevation. For very high boiling-point elevations, such as those encountered in crystallizing caustic soda, vacuum crystallization is out of the question.

Two further disadvantages of the vacuum crystallizer are: (1) it requires more headroom than does a mechanical-type crystallizer and (2) rubber linings, which are commonly used for corrosive solutions, are unsatisfactory for hot acid solutions.

The mechanical crystallizer has the advantage in not requiring vacuum-producing equipment in being independent of the vapor pressure of the solution, and in requiring no steam. It can also handle stiffer magmas than can the vacuum crystallizer because free circulation is not required.

The mechanical crystallizer is frequently limited by the materials of construction. The materials used must be such that they allow the transfer of heat and must be structurally strong enough to resist the erosion by the crystals and the action of the agitators. The first cost of mechanical crystallizers for large capacities is relatively high even when they can be constructed of steel. They are subject also to the difficulties arising from the fouling of the cooling surface on the water side if the water is hard. The water side of a mechanical crystallizer cannot usually be cleaned readily.

In most cases where either type can be used, costs favor the vacuum crystallizer.

Prilling. A combination spray-drying and crystallizing technique has recently been developed for the production of ammonium nitrate for fertilizer. Hot concentrated ammonium nitrate solution is sprayed into a tower, into the bottom of which is blown atmospheric air. The ammonium nitrate crystallizes into agglomerates, or "prills," which are conditioned with diatomaceous earth to reduce caking and are bagged for use.

Satisfactory ammonium nitrate crystals have been produced in Krystal equipment. The product so made is in the form of single crystals rather than aggregates (Miller, Phillip, and Laeman, *loc. cit.*).

Cost of Crystallization Equipment. The cost of a crystallizing tank or an agitated batch crystallizer will approach that of an ordinary tank with or without cooling coils and stirrers. The cost of the crystallizing equipment in such cases is a small item in the total cost of the operation, because of the heavy labor charges necessary to handle the product of such crystallizers.

Prices of Swenson-Walker crystallizers are as follows (G. E. Seavoy, personal communication):

Total length, ft.	Number of decks	Length per deck, ft.	Cost/lin. ft.
20	1	20	$90
40	1	40	68
60	2	30	70
90	1	90	67
120	3	40	60
240– 360	4– 6	40	58
480– 800	12–20	40	55
1000–1200	25–30	40	54

The cost of a vacuum crystallizer is largely affected by the cost of the vacuum-producing equipment, since the balance of the construction is ordinary tank and piping practice. If, however, special metals or rubber-lined equipment is necessary, the fabrication costs of the material used must, of course be used. A detailed discussion of other factors that influence the costs of crystallizing in a vacuum crystallizer is given by Seavoy and Caldwell (*loc. cit.*).

Crystallization Costs. It is hardly possible to give general cost curves showing the costs of crystallizing the materials that are prepared commercially in the crystalline state. The materials, the conditions of crystallization, and the methods of operation are too varied to give general costs. Each individual case should be investigated as it arises. All comparative costs should be made on the basis of a complete crystalline product, properly screened, dried, and ready for packaging. The following specific examples may be of interest:

Case 1. Crystallization of trisodium phosphate, in naturally cooled vats. Production, 31 tons/24 hr. The equipment consists of 80 tanks, each 6 by 12 by 2 ft. The time of cooling is 5 days. The first cost of each tank was $150, and accessory costs were $2000. The total investment was (80)(150) + 2000 = $14,000. Operation, 300 days/year.

Annual Costs:

Depreciation at 5 % per year = (14,000)(0.05)	$ 700
Average interest at 5 %, allowing for interest earned by depreciation reserve = (21/20)(14,000)(0.05/2)	368
Repairs, maintenance, and lubrication	200
Floor rent, 10,080 sq. ft. at $0.20/sq. ft. = (10,800)(0.20)	2,016
Power, 20 kw. at $0.01/kw.-hr. = (20)(24)(300)(0.01)	1,440
Interest on material in process, at 5 % interest and $0.025/lb. = (5)(2000)(31)(0.05)(0.025)	387
Labor, 188 man-hr./day at $1.00/hr. = (188)(300)(0.50)	56,400
Total annual costs	$61,511

Cost per ton of product (not including centrifuging, drying, screening, etc.) $= \frac{61{,}511}{(300)(31)} = \$6.63.$

Case 2. Trisodium phosphate, in Swenson-Walker crystallizer. Production, 31 tons/24 hr. The equipment consists of two units, each 40 ft. long. The investment was $19,000. Operation is 300 days/year.

Annual Costs:

Depreciation at 12½ % per year = (0.125)(19,000)	$ 2,375
Average interest at 5 %, allowing for interest earned by depreciation reserve = (9/8)(19,000)(0.05/2)	535
Repairs, maintenance, and lubrication	500
Floor rent, 1500 sq. ft. at $0.20/sq. ft. = (1500)(0.20)	300
Interest on material in process, at $0.025/lb. and 5 % interest = (62,000)(0.025)(0.06)	77
Power, 15 kw. at $0.01/kw.-hr. = (15)(24)(300)(0.01)	1,080
Water	1,000
Labor, 24 man-hr./day at $1.20 and 14 man-hr./day at $1.00/hr. = (24 × 1.20 + 14 × 1.00)(300)	12,840
Total annual costs	$18,707

Cost per ton of product (not including process costs subsequent to crystallizer) $= \frac{18{,}707}{(31)(300)} = \$2.01.$

It should be noted that the crystals produced in Case 2 are small (but more uniform and regular) than those obtained in Case 1. Costs of centrifuging, drying, screening, reworking of off-size crystals, etc., will be less for Case 2 than for Case 1. If large crystals are demanded, the more expensive vat system would be necessary.

Case 3. Glauber's salt, in a Swenson-Walker crystallizer. Production, 8 tons in 24 hr. The equipment consists of one 20-ft. and one 30-ft. crystallizer. The investment was $2500. Operation is 300 days/year.

Annual Costs:

Depreciation at 10 % per year = (0.10)(2500)	$ 250.00
Average interest, allowing for interest earned on depreciation reserve = (11/10)(2500)(0.05/2)	68.80
Repairs and maintenance	120.00
Power, 2.25 kw. at $0.0125/kw.-hr. = (2.25)(24)(300)(0.0125)	203.00
Water, including fixed and operating costs on pumps	600.00
Labor (casual inspection only) 1 hr./day at $1.20/hr. = (1.20)(300)	360.00
Total annual cost	$1,601.80

Cost per ton of product (at crystallizer outlet) $= \frac{1601.80}{(8)(300)} = \$0.67$

The following three cases are given by Seavoy and Caldwell (*loc. cit.*). In each case the design factors and costs are given for two methods: first, using a Swenson-Walker crystallizer (Fig. 161) and, second, using a liquid-cooled crystallizer. In all cases the costs include the crystallizer and its accessories only and do not include costs of processing steps subsequent to the crystallization itself.

Case 4. Disodium phosphate, $Na_2HPO_4.12H_2O$.

Case 5. Viscose spin-bath liquor, containing by weight 8 per cent H_2SO_4, 14 per cent Na_2SO_4, 5 per cent glucose, 1 per cent $ZnSO_4$, and 72 per cent H_2O.

Case 6. Sulfuric acid solution containing ferrous sulfate to be crystallized as $FeSO_4.7H_2O$.

The pertinent quantities involved in these cases are shown in the tables on page 1071.

CAKING OF CRYSTALS

A problem that is often met in handling crystalline products is their tendency to cake or bind together. This is often troublesome in bulk storage or in barrelled products but is most serious in those cases where crystals are sold in small packages. The difficulty may exist in degrees, varying from loose aggregates that fall apart between the fingers, to solid lumps that can be crushed only by considerable force. The demand of the average consumer that the material shall flow freely from the package makes the prevention of caking a serious problem for the manufacturer.

Critical Humidity. Just as the vapor pressure of water is fixed by its temperature, so the vapor pressure of any solution is fixed by its temperature at an amount somewhat lower than the vapor pressure of water at that temperature. If a saturated solution is brought into contact with air in which the partial pressure of water is less than the vapor pressure of the solution, the solution will evaporate. On the other hand, if the air contains more moisture than this limiting amount, the solution will absorb water until it is so dilute that its vapor pressure is equal to the partial pressure of the moisture of the air with which it is in contact. If a crystal of a soluble salt is in contact with air that contains less water than would be in equilibrium with the saturated solution, the crystal must stay dry, because if it were surrounded

Engineering and Cost Data

	4		5		6	
	Liquid cooled	Vacuum	Liquid cooled	Vacuum	Liquid cooled	Vacuum
Type of crystallizer	Swenson-Walker	Three stage	Swenson-Walker	Single stage	Double pipe	Single stage
Type of operation	Cont.	Cont.	Cont.	Batch	Cont.	Batch
Cooling water, gal./min	360	300	None*	2,900	180†	675
Power for agitation, driver, etc., including pumping, condenser water, hp	70	52	50	80	15	30
Power for refrigeration and brine pumping, hp			630		160	
Steam consumption, lb./hr		650		8,000		2,000
Refrigeration brine, gal./min			535		100	
Brine refrigeration, tons/day			335		85	
Water evaporated, lb./day		84,000		94,000		35,000
Materials of construction	Steel and cast iron	Steel	Nickel and nickel-clad steel	Rubber-lined steel	Steel and lead	Rubber-lined steel
Approx. space (length, width, height)	100′ × 50′ × 10′	30′ × 10′ × 40′	100′ × 50′ × 10′	45′ × 20′ × 55′	30′ × 10′ × 10′	20′ × 15′ × 55′
Installation cost‡	$52,000	$28,000	$140,000	$65,000	$30,000	$18,000
Operating costs, cost per year. Interest and depreciation 15%	$7,800	$4,200	$21,000	$9,750	$4,500	$2,700
Cooling water at 1¢/1000 gal	1,868	1,556		15,040	933	3,500
Power at 1¢/kw.-hr	4,510	3,350	4,840	5,150	1,450	2,160
Refrigeration at 1.4 kw.-hr. per ton refrigeration			30,250		8,225	
Steam at 30¢/1000 lb		1,685		20,750		5,184
Labor at $1.20/man-hr.§	9,130	4,570	18,260	9,130	4,570	4,570
Maintenance	3,120	700	8,400	1,950	3,000	540
Gross operating cost	26,428	16,061	82,750	61,770	22,678	18,654
Credit for water evaporated		3,629¶		12,700‖		4,730‖
Net operating cost	$26,428	$12,432	$82,750	$49,070	$22,678	$13,924

* Not including cooling water for ammonia condenser.
† Cool part way with water. Not including cooling for ammonia condenser.
‡ Exclusive of building and foundations.
§ In original article, 70¢/man-hr.
¶ Triple effect economy: 0.4 lb. steam/lb. evaporation.
‖ Single effect economy: 1.25 lb. steam/lb. evaporation.

with a film of solution, that solution would necessarily evaporate. On the other hand, if the crystal is brought into contact with air containing more moisture than would be in equilibrium with its saturated solution, then the crystal will become damp and in time will absorb water until it is completely dissolved, and the solution is so dilute that it is in equilibrium with the air.

Operating Conditions and Thermal Data

Item	Case 4	Case 5	Case 6
Feed to crystallizer, gal./24 hr	80,000	150,000	42,000
Crystals, lb./24 hr	300,000	225,000	125,000
Feed density, °Be	32	26	50
Feed temperature, °F	190	105	140
Final temperature, °F	100	41	50
Initial crystallization temperature, °F	142	58	90
Boiling-point elevation of feed solution, °F	10	8	11
Boiling-point elevation of mother liquor, °F	8	9	15
Available cooling water temp. (max.), °F	80	85	80
Available steam pressure, lb./sq. in	125	125	125
Specific heat of solution and mother liquor (assumed)	0.85	0.85	0.85
Specific heat of crystals	.40	.35	.30
Heat of crystallization, B.t.u./lb. crystal	70	105	29
Available refrigerated brine, °F		10	20

In the range of temperatures around ordinary room temperature, the vapor pressure of a given solution varies with temperature in such a way that it is nearly a constant percentage of the vapor pressure of water at the same temperature. Saturated sodium chloride, for instance, has a vapor pressure approximately 80 per cent of that of water at the same temperature. If sodium chloride, therefore, is brought into contact with air of more than 80 per cent relative humidity it will absorb moisture, while if it is brought into contact with air of less than 80 per cent relative humidity it will stay dry. From this follows the conception of **critical humidity** of a solid salt. This is the humidity above which it will always become damp and below which it will always stay dry. If the crystal should be coated with impurities derived from the mother liquor from which it was separated (in the case of sodium chloride such impurities would be calcium and magnesium chlorides), this may result in a critical humidity higher or lower than that of the pure salt, according to whether the impurities give solutions having greater or less vapor pressures than that of the salt in question. Consequently, the critical humidity of a commercial grade of a crystalline material may differ appreciably from the critical humidity of the pure substance.

Prevention of Caking. Suppose a sample of sodium chloride is exposed for a short time to an atmosphere more moist than its critical humidity and then that it is removed to an atmosphere less moist than its critical humidity. During the first period it will absorb more moisture, and during the second period it will lose this moisture. If the crystals are large, so that there are relatively few points of contact and there is a large free volume between the crystals, there will probably be no appreciable bonding of the crystals due to this solution and reevaporation, if the time of exposure is not too great. If, on the other hand, the crystals are fine, or have a small percentage of voids, or are in contact with a moist atmosphere for a long time, sufficient moisture may be absorbed to fill the voids entirely with saturated solution; and when this has been reevaporated the crystals will lock into a solid mass. Consequently, to prevent the caking of such salts, the following conditions are desirable: (1) the highest possible critical humidity; (2) a product containing uniform grains with the maximum percentage of voids and the fewest possible points of contact; (3) a coating of powdery inert material that can absorb reasonable amounts of moisture.

The first condition (maximum critical humidity) is often met by removing impurities, such as calcium chloride in the case of common salt, free acid where a salt is formed in acid solutions, etc. It often happens that the impurities have a lower critical humidity than the product desired, although this is entirely accidental. To increase the percentage of voids, it is not necessary to produce larger crystals but to produce a more uniform mixture. For a given crystal form, and for absolutely uniform crystals as to size, the percentage of voids is the same no matter what the size of the crystals. A variation in particle size, however, rapidly decreases the percentage of voids. On the other hand, a fine product has more points of contact per unit volume than a coarse one and, hence, a greater tendency to cake. The third remedy is not always applicable. Illustrations of its use are dusting of table salt with magnesia or tricalcium phosphate and the dusting of flake calcium chloride (25 per cent H_2O) with anhydrous calcium chloride.

Some hydrated salts have a melting point so near room temperature that they may sometimes be stored under conditions where fusion begins. Here again the same considerations hold, for, if the percentage of voids is large or the points of contact between adjacent crystals few, the amount of fused material may not be sufficient to lock crystals together on resolidification. If, because of extremely fine crystal size or a mixture of sizes, the percentage of voids is too far reduced or the number of points of contact too greatly increased, the crystals may be firmly locked on resolidification. In this case, also, caking may be partly prevented by dusting the crystals with powdered material. In the case of hydrated salts, this powdery material may be produced from the salt itself by drying under such conditions that a very thin surface layer is dehydrated.

MISCELLANEOUS METHODS OF MECHANICAL SEPARATION AND CONCENTRATION

BY C. E. BERRY

Miscellaneous Methods of Mechanical Separation. The operation of expression has the same purpose as filtration but is distinguished from it because the pressure is applied by moving the retaining walls instead of pumping material into a fixed space. The pressure employed for expression generally is higher than that used for filtration. The principles of expression are not so fully developed.

Expression has found wide application in the production of vegetable and fish oils, fruit and vegetable juices, and in the recovery of juice from sugar cane. The operation might find more application as an adjunct to filtration or to replace it for certain materials for which a more thorough removal of liquid from the cake is desirable than can be obtained by simple vacuum or pressure filtration alone.

Sweating is a specialized purification operation analogous to the inverse of crystallization and takes advantage of the phenomenon of the depression of the freezing point of a compound because of the presence of a soluble impurity. The separation of such compounds as ortho- and paranitrotoluene is done effectively by the sweating operation.

Concentration of Materials. Jigging, tabling, sink-and-float, and elutriation are methods of mechanical separation in which concentration usually is accomplished by employing principles that take advantage of differences in specific gravity and size of the particles or lumps being separated. A valuable material is concentrated and separated from gangue or less valuable material. Froth flotation is a common method of flotation in which mechanical separation is effected by causing grains of one or more materials selectively to float to the surface of a suspension because of bubbles of air attached to them. The grains are held in a froth which can be removed from the surface of the suspension. The grains of dissimilar materials, to which bubbles do not adhere, remain in suspension and can be recovered by sedimentation or other means.

Electrical methods for separating materials usually are classed as magnetic or electrostatic. In the former method, differences in the magnetic permeability of materials allows the separation or concentration to be made. Materials capable of receiving surface charges to different degrees can be separated in an electrostatic field.

Concentration has wider application in the mineral dressing industry than in the chemical industry and properly is not given extensive treatment in a chemical engineers' handbook. Comprehensive treatment of the subject can be found in the following: Taggart, "Handbook of Mineral Dressing," Wiley, New York, 1945; Richards and Locke, "Textbook of Ore Dressing," 3d ed., McGraw-Hill, New York, 1940; Gaudin, "Principles of Mineral Dressing," McGraw-Hill, New York, 1939.

EXPRESSION

By C. Fred Gurnham

REFERENCES: Schönfeld, "Chemie und Technologie der Fette und Fettprodukte," vol. 1, Chemie und Gewinnung der Fette, Verlag Julius Springer, Vienna, 1936. Koo, Expression of Vegetable Oils, *Ind. Eng. Chem.*, **34**, 342 (March, 1942). Bailey, "Industrial Oil and Fat Products," pp. 469–479, Interscience Publishers, New York, 1945. Dickey and Bryden, "Theory and Practice of Filtration," Chap. 9, Reinhold, New York, 1946.

Definition. Expression is the separation of liquid from a two-phase solid-liquid system by compression of the system under conditions that permit the liquid to escape while the solid is retained between the compressing surfaces. Expression is distinguished from filtration in that the pressure is applied by movement of the retaining walls instead of by pumping the material into a fixed space.

Purpose. Expression has the same purpose as filtration, namely, to separate liquid and solid phases from a mechanical mixture of the two. In filtration operations, the original mixture is sufficiently fluid to be pumpable; in expression this is not usually the case, and the material may appear entirely solid. Expression is therefore employed to separate systems that are not readily pumpable. It is also used instead of filtration when a more thorough removal of liquid from the cake is desired.

Expression Equipment

Hydraulic presses of the batch type are used for the expression of materials to the lowest possible liquid content. The principal types of batch press are the box, platen, pot, curb, and cage presses. For continuous operation, the screw press or expeller is used, and for some materials roller mills are satisfactory.

Batch Presses. *Box Press.* The box press (Fig. 167) consists of a series of steel boxes fitting between the fixed and movable heads of a vertical hydraulic press. The material to be expressed is wrapped in canvas bags or cloths and placed in the boxes, each bag lying on a perforated mat over a grid of drainage channels and being covered and enclosed by the next higher box. The series of loaded boxes is compressed as a unit under hydraulic pressure. The horizontal surfaces of the boxes are corrugated to prevent creeping and tearing of the cloths.

Most of the common vegetable oils are produced by expression processes, in either hydraulic presses or continuous expellers, although in some cases solvent extraction is preferred. As a rule, expression is used for edible oils, and industrial oils may be produced by expression or by extraction. The production of cottonseed oil is typical of hydraulic-press operation. A 15-box press will handle 8 tons of conditioned cottonseed meats in a 24-hr. period, reducing the oil content from 30 to about 6 per cent. The operating cycle is from 20 to 30 min. per batch. The press is first closed rapidly under low pressure until flow of oil starts at about 200 lb./sq. in. pressure on the cake; then high-pressure fluid is used to close the press slowly to the maximum of 1600 lb./sq. in. pressure (4000 lb./sq. in. hydraulic-fluid pressure). The maximum pressure is continued for a few minutes to permit drainage.

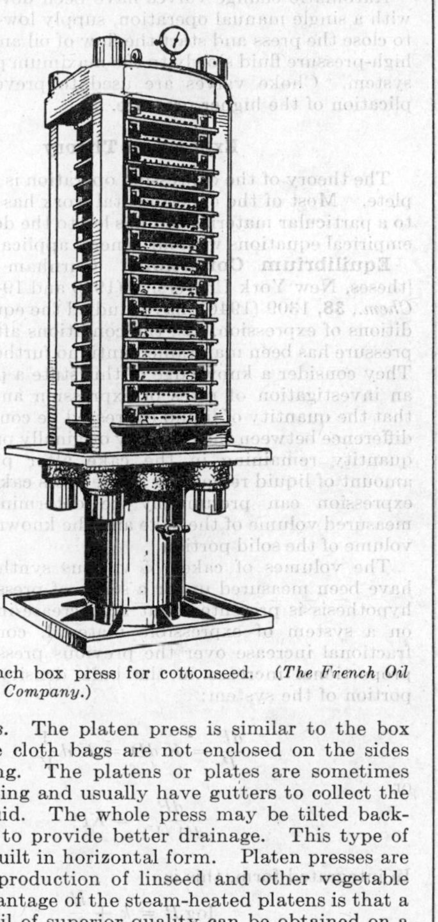

Fig. 167. French box press for cottonseed. (*The French Oil Mill Machinery Company.*)

Platen Press. The platen press is similar to the box press, but the cloth bags are not enclosed on the sides during pressing. The platens or plates are sometimes cored for heating and usually have gutters to collect the expressed liquid. The whole press may be tilted backward slightly to provide better drainage. This type of press is also built in horizontal form. Platen presses are used for the production of linseed and other vegetable oils. An advantage of the steam-heated platens is that a cold-pressed oil of superior quality can be obtained on a first pressing, followed by a further yield of poorer quality hot-pressed oil.

Pot Press. In the pot press, the material to be pressed is enclosed in a cylindrical pot, with filter pads or screens beneath and on top, and is compressed by a ram entering from above. The filter cloths are laid flat and cover only the top and bottom of the material; hence they are not subjected to stretching or tearing as in the box and platen presses. Also, because the material is entirely enclosed, it may be of very oily or even liquid consistency. In practice a series of pots is used in one press, the bottom of each pot serving as the ram for the pot below.

The largest use of pot presses is in the chocolate industry; but they find application for pressing olives, palm and similar nuts and for separating liquid from slushy materials such as chemical products. The Carver 12-pot horizontal press is commonly used in the manufacture of cocoa. Chocolate liquor, produced by grinding cocoa nibs, is pumped to the press chambers, which are then closed under a pressure of 6000 lb./sq. in. The cocoa butter is expressed, leaving a cake of cocoa powder. From 500 to 600 lb. of chocolate liquor are charged per batch, about 50 per cent of which is cocoa butter. Stainless-steel screens are used as filter mediums.

A development of the pot press is the Carver combined filter and hydraulic press, which is used as a conventional filter press until the chambers are full and is then closed under hydraulic pressure to obtain a further yield of liquor and a drier cake. This press is used in the manufacture

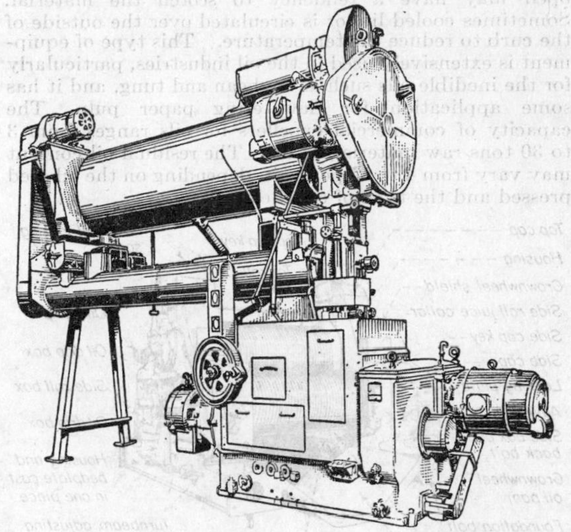

Fig. 168. Anderson expeller with oil-seed tempering unit. (*The V. D. Anderson Company.*)

of cocoa and cocoa butter, in the recovery of crystals from mother liquors, and for separating chemical precipitates. The original material must be pumpable, but there is the possibility of recycling a portion of the filtrate if necessary to obtain this condition.

Curb Press. In the curb press, the material to be expressed is enclosed in a cylinder constructed of wooden slats or beveled steel bars, or even of perforated steel plate. Compression of the material by a solid ram causes the liquid to escape through the walls of the cylinder and flow to collecting channels at the base. Since no filter cloths are used, this type of press is best suited to the expression of fibrous non-oily materials, and the expressed liquid may contain some solids. Curb presses are used in the production of cider and other fruit and vegetable juices, frequently with a screw mechanism instead of hydraulic pressure. They have been used for expressing olive oil, fish oils, and other oils that do not require high pressures, and for dewatering and recovering grease from garbage prior to incineration.

Cage Press. The cage press is similar to the curb press except that the inside of the cylinder has fine grooves extending lengthwise and leading through the cylinder walls to larger drainage channels. It is suitable for oilier and less fibrous materials than the curb press. Intermediate drain plates and cloths are sometimes used within the cake. Cage presses are not extensively used in America but are employed in Europe for expressing castor beans and copra.

Continuous Presses. *Screw Press.* The continuous screw press, typified by the Anderson expeller (Fig. 168) consists of a rotating screw fitting closely inside a horizontal slotted curb. The curb and screw may be tapered toward the discharge end in order to increase the pressure on the material, or this may be achieved by varying the pitch on the screw in a uniform cylinder. The discharge end of the curb is partly closed by a cone or other device that can be adjusted to change the size of the opening and thus to vary the pressure on the material. Rotation of the screw moves the material forward; and, as the pressure increases, liquid is expelled and escapes through the lengthwise slots of the curb. The operation is continuous, and the labor and other operating costs are lower than for hydraulic pressing; but the separation is generally not so thorough, and the frictional heat developed may have a tendency to scorch the material. Sometimes cooled liquor is circulated over the outside of the curb to reduce the temperature. This type of equipment is extensively used in the oil industries, particularly for the inedible oils such as soybean and tung, and it has some application for dewatering paper pulp. The capacity of commercial expellers for oils ranges from 3 to 30 tons raw material/24 hr. The residual oil content may vary from 4 to 15 per cent, depending on the oil seed pressed and the type of expeller.

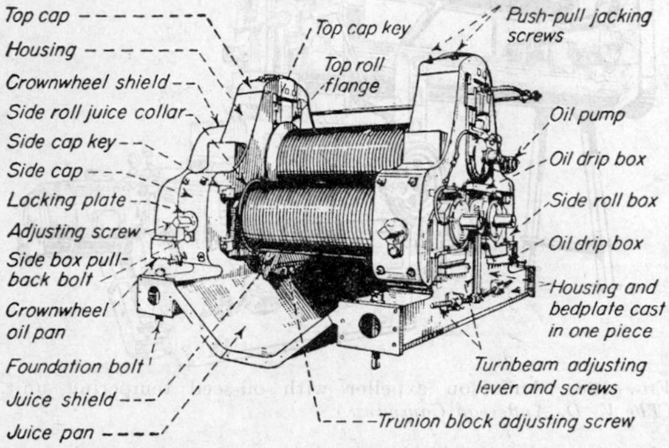

FIG. 169. Three-roll sugar mill. (*The Farrel-Birmingham Co.*)

Roller Mills. Continuous roller mills (Fig. 169), as used in the cane-sugar industry, combine a mechanical breaking and crushing action with the application of pressure to express juice. Three-roll mills are common, with the top roll located above and between the other two, and pressing against them by means of hydraulic rams at each end. The material is squeezed between the top roll and the first roll and is then directed by a turnplate into the nip of the top and second rolls for a second pressing. The rolls are made of a special mixture of cast iron of open grain and rough texture and are corrugated or grooved in various patterns. A feed roll is sometimes used to force-feed the first pair of rolls, thus permitting the use of a smaller mill opening or a higher rate of feed.

In the cane-sugar industry, trains of four to seven of these mills are used, with the blanket of crushed cane carried between them by apron conveyors. The cane is first crushed dry; but, at selected points in the later mills, water or weak liquor is added as a spray or bath, to improve the sugar recovery. This process is known as maceration and is equivalent to a solvent extraction or leaching of the cane, combined with expression in the mills.

Auxiliary Equipment. Hydraulic pumps of various capacities and pressures are used, frequently with high- and low-pressure units on the same drive. Large-capacity low-pressure pumps are required for closing the presses and for miscellaneous services such as cake forming and opening the presses. Low-capacity high-pressure fluid is needed for the final pressing. Sometimes the hydraulic fluid is the same as the liquid being expressed, to avoid possible contamination.

Accumulators are used to reduce the load on the pumps, by building up a reservoir of pressure fluid when the pumps are not serving the presses directly. Weighted accumulators contain a cylinder and piston, which carry a load of iron, stone, or concrete. Air-ballasted accumulators have the advantages of lessened line shock and lighter weight and hence simpler foundations. Separate accumulators are needed for the high- and low-pressure systems.

Automatic change valves have been developed which, with a single manual operation, supply low-pressure fluid to close the press and start the flow of oil and then supply high-pressure fluid slowly to the maximum pressure of the system. Choke valves are used to prevent rapid application of the higher pressure.

Expression Theory

The theory of the expression operation is far from complete. Most of the experimental work has been devoted to a particular material and has led to the development of empirical equations without general application.

Equilibrium Conditions. Gurnham and Masson [theses, New York University (1940 and 1942); *Ind. Eng. Chem.*, **38**, 1309 (1946)] have studied the equilibrium conditions of expression, *i.e.*, the conditions after a constant pressure has been maintained until no further flow occurs. They consider a knowledge of this state a prerequisite to an investigation of rates of expression and recommend that the quantity of liquid expressed be considered as the difference between the quantity originally present and the quantity remaining in the cake after pressing. The amount of liquid remaining in the press cake after partial expression can presumably be determined from the measured volume of the cake and the known or estimated volume of the solid portion.

The volumes of cakes of various synthetic mixtures have been measured under a series of pressures, and the hypothesis is presented that an increase in the pressure on a system of expressible material, considered as a fractional increase over the previous pressure, causes a proportional increase in the bulk density of the solid portion of the system:

$$\frac{dP}{P} = K\,dD = K\,d\frac{1}{V}$$

or

$$\frac{dP}{d(1/V)} = KP$$

In integrated form, this is

$$\log P = k + \frac{k'}{V}$$

where P is the pressure on the system, D is the bulk density of the solid portion of the system, V is the specific volume of the system based on the solid content, and K, k and k' are constants depending on the nature of the material and the conditions of expression. These investigations were carried out in the Carver test cylinder, which is equivalent to a pot press of 1 sq. in. area, and in the cage press of 10 sq. in. area. The materials tested included cotton fiber, woolen yarn, wool felts, asbestos fiber, paper pulp, wood sawdust, and other fibrous substances, pressed in dry condition and wetted with water, oils, or other liquids. The pressures used ranged from 250 to 20,000 lb./sq. in. Most of Deerr's data on sugar cane and bagasse, discussed below, also confirm this theory.

Empirical Equations. Deerr investigated the expression of juice from sugar cane and bagasse, using apparatus equivalent to small pot presses [*Hawaiian*

Sugar Planters' Assoc., Expt. Sta., Agr. & Chem. Series, Bulls. **22** (1908); **30** (1910) and **38** (1912)]. He suggested the formula

$$V = \frac{C}{P^n}$$

in which V is the volume of cake (fiber plus unexpressed juice) under pressure P, C is a constant for a given experiment, and n is a constant or a function of P depending on pressing conditions. This equation is entirely empirical, but it fits the observed data with fair accuracy. Using this equation, Deerr has made theoretical calculations of the pressures on the rolls and of the work done in expression. It has been observed that the equation of Gurnham and Masson fits Deerr's original data with some accuracy, with a few explainable exceptions.

Koo and coworkers have studied the expression of seven different oils, over a range of pressures, temperatures, pressing times, and moisture contents, using a laboratory cage press of 70 cu. in. capacity [Koo and Chen, *Ind. Research (China)*, **6**, 9 (1937). Koo, *J. Chem. Eng. China*, **4**, 15, 207 (1937); **5**, 47, 69 (1938); **7**, 1, 23 (1940); **8**, 1, 5 (1941); *Ind. Eng. Chem.*, **34**, 342 (1942)]. They have developed a formula of the type

$$W = CW_o \frac{\sqrt{P}\,\sqrt[6]{\theta}}{\nu^a}$$

in which W is the weight of oil expressed, W_o the weight of oil in the original material, C a "press constant" depending on the type of material, P the pressure, θ the time of pressing, ν the kinematic viscosity of the oil at the press temperature, and a the viscosity exponent, which depends on the type of oil seed. Pressures from 1000 to 4500 lb./sq. in. were employed, with temperatures from 15° to 125°C., giving a fifteenfold variation in kinematic viscosity, and pressing times from ½ to 9 hr. The moisture content of the seed was shown to be important, the optimum being between 5 and 13 per cent depending on the type of material and the temperature. The constants shown in Table 1 were given, for P in lb./sq. in., θ in hr., ν in stokes (c.g.s. units) measured at press temperature.

Table 1. Constants for Expression Equation

Oil seed	Initial oil content W_o, %	a	C
Castor bean, decorticated	64.19	1/12	0.00921
Cottonseed	34.7	¼	.00535
Peanut, decorticated	51.88	1/6	.00751
Rapeseed	42.18	1/6	.00583
Sesame seed	52.95	1/12	.00835
Soybean	19.5	¼	.00450
Tung nut, decorticated	64.53	1/6	.00907

SWEATING

By W. S. Calcott

In sweating, advantage is taken of the well-known phenomenon of the depression of the freezing point of a compound by soluble impurities. It can be regarded as analogous to the inverse of crystallization. In the purifying of a compound by crystallization, it is brought into the liquid phase by solution or fusion and then is cooled. The desired compound is obtained in the solid phase; the impurities are concentrated in the liquid phase. In the process of sweating, the impure material is converted into the solid phase. If the temperature is raised slowly, a liquid phase forms, comprising the impurities and a certain percentage of the higher melting material being operated upon, which is allowed to drain off as formed. The temperature is gradually raised to substantially the melting point of the pure material being sought. Any solid phase remaining at this temperature obviously must consist of a substantially pure compound, since the presence of even small quantities of impurities would depress the freezing point measurably. Sweating can therefore be used for the production of very high quality material.

Equipment. From the principle involved, it is obviously necessary that close predetermined temperature control be maintained throughout the entire operation. Dealing with the solid phase only, no agitation is possible. The equipment used therefore consists usually of a jacketed vessel in which are immersed batteries of plates through which the temperature-control medium can be passed as well as through the jacket surrounding the entire vessel. The plates are placed vertically, 1 to 2 in. apart, with a "drainage space" of the order of 6 in. deep being left below the bottom of the plates. In order to maintain thermal equilibrium at all times, the rate of increase of temperature must be quite slow, and the time cycle therefore long, ranging from 2 days to 1 week. The sweating pans must therefore be of considerable capacity in order to obtain a reasonable rate of production, 50,000 to 75,000 lb. being a normal size.

Operation. In operation, the material to be separated is run into the sweat pan molten, the cooling liquid circulation is started, and the entire mass is frozen. Control is then ordinarily turned over to a cam-operated thermoregulator which increases the temperature at a predetermined rate. The bottom draw-off is opened, and, as soon as the freezing point of the eutectic is reached, the material below the freezing plates liquefies and is drawn off, freeing the drainage space to receive the sweatings from the crystals above. The operation then proceeds under control of the cam-operated thermoregulator until the temperature comes as close to the melting point of the higher melting constituent as is necessary for the degree of purity required. The temperature is then raised above the melting point of this constituent. It is liquefied and drawn off to a separate container.

Application. It is obvious from the basic principle of this procedure that, by working very close to the melting point of the pure compound, a very high degree of purity can be obtained in the resulting end product. It can obviously be applied only to compounds that are stable at their melting points. It is particularly useful in certain cases where fractional distillation yields a certain percentage of one of the two constituents of the mixture and an azeotrope, except in the rare case where the composition of the azeotrope is identical with that of the eutectic. Such mixtures as ortho- and paranitrotoluene, ortho- and nitrochlorobenzene, and dinitrotoluenes are commonly treated by this sweating procedure.

JIGGING

By Fred D. DeVaney

References: Taggart, The Mechanism of Jigging, *Mining Tech.*, March, 1943. Vedensky, Use of Jigs in Placer Mining, *Miner*, May, 1938. British Patents 438,888 of 1936 to S. Dawson Ware; and 452,664 of 1937 to W. Ruoss. Oke, A Simple Hydraulic Jig, *Mining Mag.*, **54**, 207 (1936). Teddy, Jigs on Tin Dredges, *Bull. Dredging Assoc. Southern Malaya*, rev. in *Mining Mag.*, April, 1935, p. 244. Hardy-Smith, Jigs, *Proc. Australasian Inst. Mining & Met.*, No. 105, p. 1, 1937.

A jig is a mechanical device used for separating materials of different specific gravities by the pulsation of a stream of liquid flowing through a bed of the materials. The liquid pulsates or "jigs" up and down, causing the heavy material to work down to the bottom of the bed, and the lighter material to rise to the top. Each product is then drawn off separately.

Jigging is one of the oldest processes used for concentrating heavy mineral from the lighter gangue and for separating coal from its heavier contaminants. Jigs are simple in operation and can be constructed locally with a low first cost. Power and water consumption are high, and tailing losses on metallic ores are usually high, with the result that the use of jigging is now somewhat limited in this field. Jigging is widely used in the concentration of coal. In 1942, over 50,000,000 tons of coal was concentrated by jigs in the United States. It is used to a more limited extent in treating the lead-zinc ores of the Mid-Continent field, iron ores, and some heavy non-metallic ores like barite. A relatively new type of

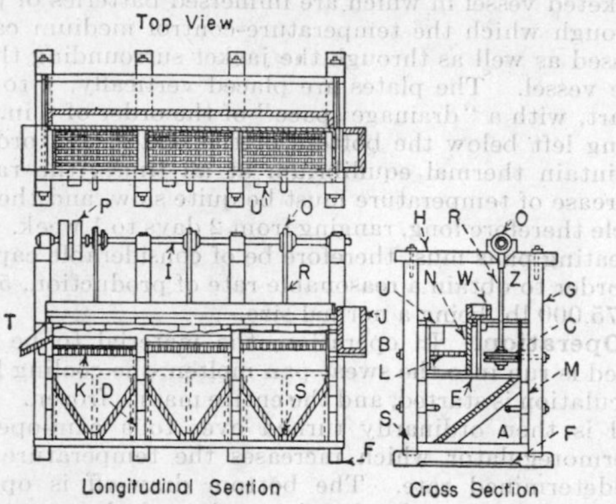

FIG. 170. Harz jig.

high-speed jig is extensively used in the recovery of fine values from placer gold and tin and tungsten deposits, and for recovering a portion of the metallic values liberated in ball-mill grinding circuits. Jigging has been superseded in many milling operations by the adoption of the sink-float process or by fine grinding followed by flotation.

Types of Jigs. There are two principal types of jigs. In the first type, the sieve is stationary, and water is forced up through the screen. One of the simpler forms of this type of jig is called the *Harz* jig and is shown in Fig. 170 (Simons, "Ore Dressing: Principles and Practice," McGraw-Hill, New York, 1924).

A is a tank the upper portion of which is divided in two sections, plunger and sieve, by the partition marked *E*. The lower portion of the tank is called the "hutch." *L* is the screen that supports the mineral. *K* is a wooden frame with crossbars. *C* is the plunger, which is given a reciprocating motion by the eccentric *O*. *S* is the hutch spigot. *U* is the cup or submerged weir under which the concentrate is allowed to pass for the separation of heavy from light mineral, which passes out of the jig at *T*.

A modern form of the fixed-sieve-type jig is the *Jeffrey* air-operated (*Baum*) jig shown in Fig. 171, which is used extensively in coal washing. In this jig the pulsations are caused by alternately applying and exhausting air pressure at about 2½ lb./sq. in. from the pulsion chamber. The amount of refuse rejected is controlled automatically by a "flash float," and this refuse is ejected positively from the screen compartment by a ratchet-operated star gate. Such jigs customarily are built with a number of compartments. Each compartment or cell rejects waste material together with some coal. These middlings can be crushed, recirculated, and some of the coal recovered. Another type of air jig, known as the *Conset*, is used in concentrating iron ores. In this jig, pulsation and suction are obtained by the alternate inflation and deflation of a rubber tube in each hutch compartment. In Mesabi practice, as much as 70 per cent of the weight of the unsized ¾-in. feed is removed at some mills.

In the second type of jig, the sieve moves up and down in a tank of water, the one shown in Fig. 172 being a *Hancock* jig. Item 2 is the cup for concentrate or middlings, 5 is the tray to carry the screens; 6 and 7 are supports for the tray; 8 is the camshaft; 9, a three-point cam; 10, the drive pulley; 11, the flywheel; 12, the rocker arm; 13, the rocker-arm shaft; and 14, the links connecting the rocker arms. The screen tray is given a combined horizontal and vertical reciprocating motion, which causes the ore to pass rapidly over the screens.

Capacity. The *Cooley* jig, which is a variation of the *Harz* jig is used in the Mid-Continent zinc district, has six compartments, each 42 in. wide and 48 in. long. The capacity of each such unit is 25 to 30 tons/hr. when treating a feed approximately ½ in. and finer in size. The capacity of the *Jeffrey-Baum* jig is approximately 3 tons/hr./sq. ft. active screen area when it is treating coal crushed to 4 in. For finer sizes of coal the capacity is smaller. In Mesabi practice the *Conset* jig treats iron ore at the rate of 1 long ton/(hr.)(sq. ft.) screen surface of a smaller than ¾-in. feed. The capacity of the *Hancock* jig varies with feed from 100 to 500 tons/24 hr. For the *Harz* jig, Wiard ("The Theory and Practice of Ore Dressing," McGraw-Hill, New York, 1915) gave the capacity *C*, in tons/(24 hr.)(sq. in.) screen area as $C = \sqrt{d}/100$,

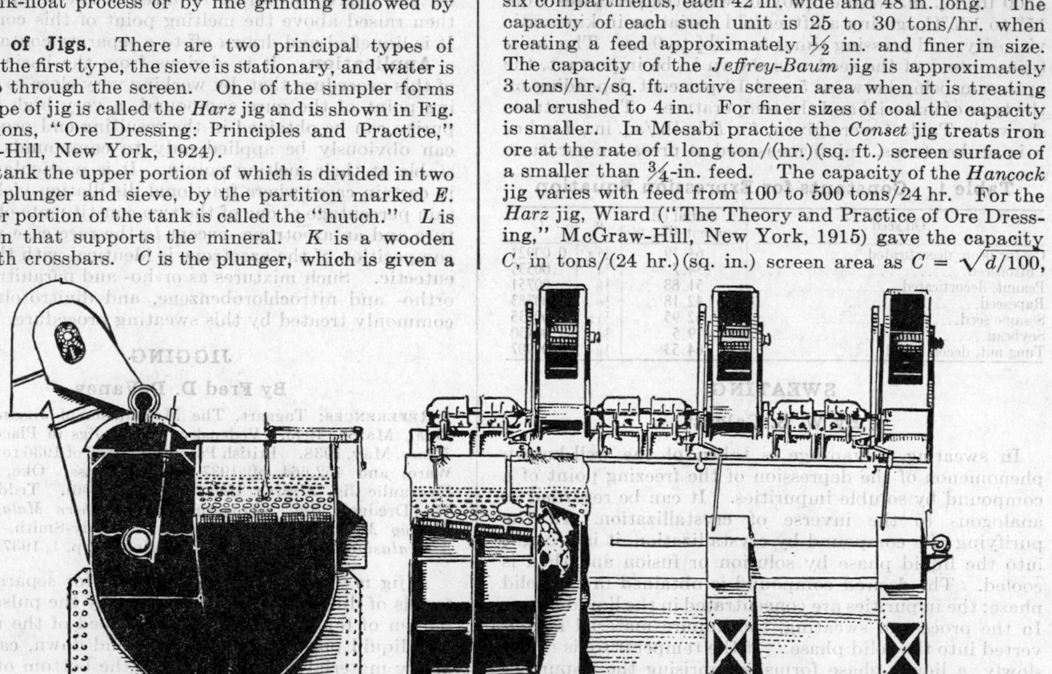

FIG. 171. Jeffrey (Baum type) coal jig.

where d equals average diameter, mm., of the feed to the jig.

Power Requirements. The power required in jigging depends on the screen area, the size of material treated, the percentage of opening in the jig screen, the depth of the bed, the length of stroke, and the number of strokes per minute. The power required for plunger-type jigs treating ½-in. material is about 0.1 hp./sq. ft. jig screen surface.

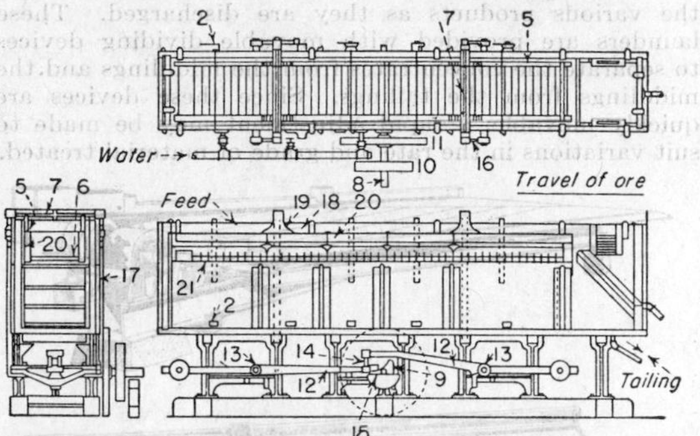

FIG. 172. Hancock jig.

Water Consumption. Jigs require much water. In most installations, the Harz-type jig uses 1500 to 2500 gal. water/ton material treated; and, although the sieve jigs, such as the Hancock, use only about one-half this amount, nevertheless the quantity is still considerable.

Jig Feed. In coal washing, jigging is practiced at some collieries on unsized material, crushed through a 6-in. screen. In metal-milling practice, jigging is now seldom employed on material coarser than ¾ in. Shaking tables usually are considered more efficient than jigs in treating material finer than 2 mm. (10 mesh). In some mills, jigs are used to secure flow-sheet simplicity. Jigs, except when extremely heavy minerals are treated, such as gold, galena, cassiterite, or tungsten minerals, recover only a small percentage of the sizes finer than 65 mesh (¼ mm.).

Jigs for Placer Gold. The commercial utilization of jigs in the recovery of placer gold started about 1914 and resulted in the design of some new jigs that may have some place also in coarse roughing of other heavy minerals. The concentration of placer gold by jigging is successful because of the extreme difference in specific gravity between the gold and the gangue materials. The high capacity of these new jigs recommends them for treating the low-grade material that must be handled in tremendous quantity in most placer operations. The use of jigs in this field was described by Malozemoff [Jigging Applied to Gold Dredging, *Eng. Mining J.*, **138**, 34 (1937)]. Such jigs are made by the Pan-American Engineering Corp., Berkeley, Calif.; The Southwestern Engineering Co., Los Angeles, Calif.; and the Denver Equipment Co., Denver, Colo.

The Denver Mineral Jig. This jig has a number of special features that have been adapted to conditions found particularly in the grinding circuit, *i.e.*, dilution control and the handling of unsized feed. These units are easy to operate and are built in four sizes to handle practically any tonnage. The jig (Fig. 173) consists of two compartments and is of all-steel welded construction, to minimize the space required. In the front of the unit are the two screen compartments over which the entire mill discharge passes; in the rear are the plungers actuated by a walking beam and adjustable eccentric. The plungers are sealed with special rubber diaphragms to give positive displacement of water or solution. A rotating valve is synchronized with the movement of the walking beam so that solution is added only on one part of the stroke of the plungers to counteract the suction normally created. This gives a water action, closely approximating the perfect theoretical jigging action, with pulsations upward and periods of free settling between, with minimum suction. The speed of different jig sizes will vary according to the material being treated; average speed is 300 to 350 ft./min. and a stroke of approximately ¼ in.

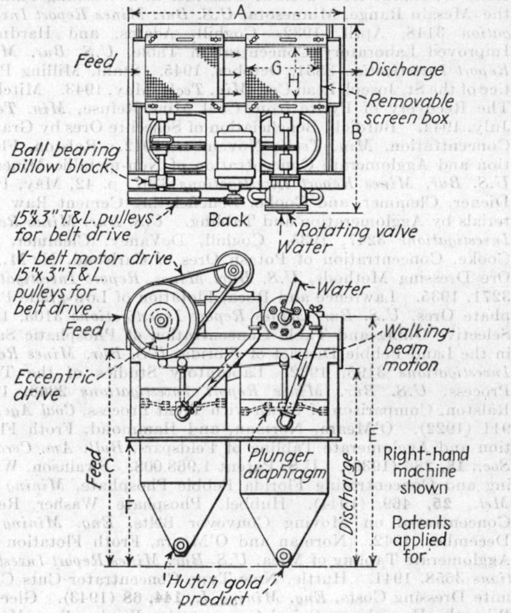

FIG. 173. Denver mineral jig.

The screen compartments are removable, because they must be cleaned up at 10- to 60-day periods, as in any jig operation. A spare compartment is furnished so that the change can be made without shutting down the grinding circuit. A special non-blinding screen is used that practically eliminates clogging.

The metallic gold values and the high-grade concentrates are finer than the screen openings and pass into the hutch below, which has steeply sloping sides, thus allowing clean discharge of this valuable high-grade material. The discharge plug of each hutch can be locked securely to prevent theft. The tonnage shown in Table 2 is that of the grinding-mill initial feed, the jig also handling the circulating load.

Table 2. Denver Mineral Jig Dimensions and Data*

Size, in.	Capacity, tons	A	B	C	D	E	F	G	H	Motor hp.	Shipping weight, lb.	
											Belt	Motor
8 × 12	15– 45	2′11″	2′11″	2′10″	2′7¼″	3′3⅝″	1′6¼″	12″	8″	¾	900	975
12 × 18	50– 200	4′ 0″	3′ 6″	3′ 8⅝″	3′5⅛″	4′3⅛″	2′1¼″	18″	12″	1	1500	1625
16 × 24	200–500	5′ 0″	4′ 3″	3′10⅛″	3′4⅝″	4′3⅜″	2′3¼″	24″	16″	1½	1950	2050
24 × 36	500–1200	7′ 1″	6′ 0″	5′ 0″	4′5″	5′5″	3′6″	36″	24″	2	3000	3150

* Courtesy of the Denver Equipment Co.

Cost. The operating cost of jigging varies in a range that runs from 6 to 25¢/ton feed, depending on the number of jigs in operation, size of feed, and nature of the material.

TABLING

By Fred D. DeVaney

References: Coghill and O'Meara, Milling Methods and Costs at a Flat River (Mo.) Mill, *U.S. Bur. Mines Circ.* 6658, October, 1932. DeVaney and Coghill, Concentration of the Rake Discharge from a Bowl Classifier in a Washing Plant of the Mesabi Range, Minnesota, *U.S. Bur. Mines Report Investigation* 3148, April, 1932. Coghill, Adams, and Hardman, Improved Laboratory Concentration Table, *U.S. Bur. Mines Report Investigation* 3831, October, 1945. Stahl, Milling Practice of the St. Joseph Lead Co., *Min. Tech.*, May, 1943. Mitchell, The Recovery of Pyrite from Coal Mine Refuse, *Min. Tech.*, July, 1944. Burdick, Beneficiation of Scheelite Ores by Gravity Concentration, *Min. Tech.*, November, 1942. Ralston, Flotation and Agglomerate Concentration of Non-metallic Minerals, *U.S. Bur. Mines Report Investigations* 3397, p. 42, May, 1938. Diener, Clemmer, and Cooke, Beneficiating Cement Raw Materials by Agglomeration and Tabling. *U.S. Bur. Mines Report Investigations* 3247, 1935. Coghill, DeVaney, Clemmer, and Cooke, Concentration of Potash Ores of Carlsbad, N. M., by Ore Dressing Methods, *U.S. Bur. Mines, Report Investigations* 3271, 1935. Lawrence and Roca, Flotation of Low-grade Phosphate Ores, *U.S. Bur. Mines Report Investigations* 3105, 1931. Selective Oiling and Table Concentration of Phosphatic Sands in the Land Pebble District of Florida, *U.S. Bur. Mines Report Investigations* 3195, 1932. Laboratory Studies of the Trent Process, *U.S. Bur. Mines Report Investigations* 2263, 1921. Ralston, Comparison of Froth with Trent Process, *Coal Age*, **22,** 911 (1922). O'Meara, Norman, and Hammond, Froth Flotation and Agglomerate Tabling of Feldspars, *Bull. Am. Ceramic Soc.*, **18,** 286 (1939). U.S. Patent 1,968,008. Swainson, Washing and Concentrating Florida Pebble Phosphate, *Mining and Met.*, **25,** 469 (1944). Hubbel, Phosphate Washer Reject Concentrated on Moving Conveyor Belts, *Eng. Mining J.*, December, 1942. Norman and O'Meara, Froth Flotation and Agglomerate Tabling of Mica, *U.S. Bur. Mines Report Investigations* 3558, 1941. Huttle, New Type Concentrator Cuts Chromite Dressing Costs, *Eng. Mining J.*, **144,** 68 (1943). Gleeson, Why the Humphreys Spiral Concentrator Works, *Eng. Mining J.*, **146,** 85 (1945). Hubbard and Humphreys, Where Spirals Replaced Tables and Flotation Cells, *Eng. Mining J.*, **146,** 82 (1945). Brown and Erck, Humphrey Spiral Concentration on Mesabi Range Ore, *Mining Engineering*, **1,** 187 (1949).

Wet Tables. Tabling is a concentration process whereby a separation between two or more minerals is effected by flowing a pulp across a riffled plane surface inclined slightly from the horizontal, differentially shaken in the direction of the long axis and washed with an even flow of water at right angles to the direction of motion. A separation between two or more minerals depends mainly on the difference in specific gravity between the minerals and to a lesser degree on the shape and size of the particles. The process is best suited for the concentration of ore and coal where there is a considerable difference between the effective specific gravity (sp. gr. mineral minus sp. gr. water) of the valuable and the waste material. Tables treat metallic ores effectively in the size range from 6 to 150 mesh but can be used to treat lighter materials such as coal of a considerably larger size.

Shaking tables were developed first about 1896 by Wilfley for concentrating metallic ores, and the peak of their development coincided with the installation of the first oil flotation plants about 1914. Since that time the field of these two more or less competitive processes has been fairly well defined. Tabling is best suited for the treatment of material containing only one valuable mineral that is free at a granular size and where a considerable difference exists between the effective specific gravities of the mineral constituent. Flotation has been found to be best in treating complex ores containing several valuable minerals, those requiring fine grinding for liberation, and those having small gravity differentials.

The heaviest particles in a table feed are the least affected by the current of water washing down over the tables, and they collect in the riffles along which they move to the end of the table. The lighter materials ride above the heavy minerals and tend to be washed over the riffles to the low side of the table. Suitable launders are placed at the end at the low side of the table to catch the various products as they are discharged. These launders are provided with movable dividing devices to separate the concentrates from the middlings and the middlings from the tailings. Since these devices are quickly movable, a rapid adjustment may be made to suit variations in the rate and grade of material treated.

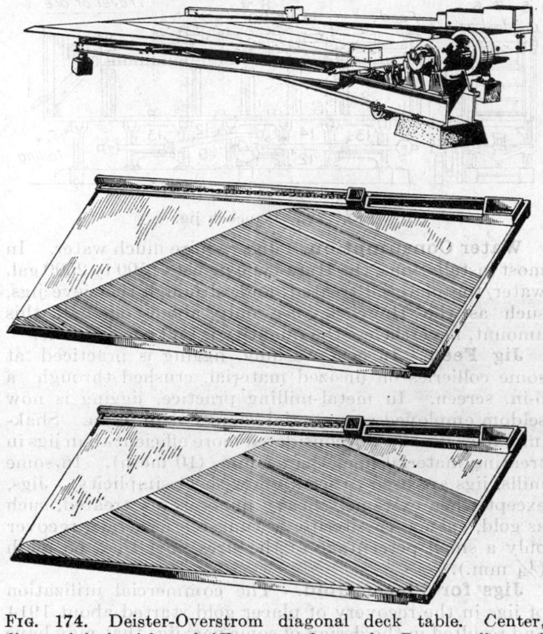

Fig. 174. Deister-Overstrom diagonal deck table. Center, diagonal deck with pool riffle system for sand. Bottom, diagonal deck with pool riffle system for fine sand and slime.

It seldom is possible in tabling to make a sharp separation of the feed into a high-grade concentrate and a low-grade tailing with one pass. Some material of intermediate grade is almost invariably present as a band between these products, and it is customary to return such middlings either with or without additional grinding to the head of the circuit for retreatment. The amount of middling recirculated may amount to 25 per cent of weight of the feed to the table.

Tables usually are surfaced either with heavy battleship linoleum or with rubber. The riffles may be a clear grade of sugar pine or may be rubber strips. Such riffles are usually 3/8 in. wide and taper from the feed end of the table to the discharge end. If the table is used for concentrating coarse material (−8 mesh), the riffles may be as high as 1 in. at the feed end. For fine material the riffles are not over 1/4 in. at the feed end of the table. Almost every mill operator employs a different style of riffling a table, which he believes best for his particular separation. The usual method of riffling is shown in Fig. 174.

If the object of tabling is to produce as clean a concentrate as possible, a diagonal area in the upper discharge side corner is left unriffled. This area is known as the cleaning deck. If the table is to be used in making only a rough concentrate and a finished tailing, the riffling is extended by many operators. Tables are provided with adjustable tilting devices so that the transverse slope may be varied. The head motion is such that the deck reverses its direction with a maximum velocity at one end and a minimum velocity at the other end of the stroke. It is the quickness of the return that causes the material to migrate toward the discharge end. The length of stroke may be adjusted. A longer stroke is required for coarse material than for fine material. This will vary from 1¼ in. for coarse material to ½ in. for fines. Modern tables operate at considerably higher speed than formerly, the range being from 270 for coarse to 350 strokes/min. for fines.

General information for standard-size tables operating on various sized feed is shown in Table 3. The No. 6 table of the Deister Concentrator Co. (Fort Wayne, Ind.) has a deck approximately 6 ft. wide and 14 ft. long. The No. 7 table used primarily for coal work is approximately 8 ft. wide and 16 ft. long. In modern practice, each table is driven by a separate motor through a V-belt drive. Because of the inertia in starting a table, a larger motor must be installed than is needed for running conditions. The installed horsepower is from 1.5 to 3 hp., and power consumed in operation is 0.5 to 1.25 hp.

An essential factor for good table operation is that the rate of feed must be uniform. No one factor will cause more trouble to the table operator than to have a surging feed. The feed to tables may be unsized, or it may be either screened or hydraulically classified.

It is now recognized generally that a feed sized by screening, or hydraulically, will give superior metallurgical results, as compared with an unsized feed, and will permit simultaneously a higher rate of feed. Since close screen sizing of fines is expensive, present practice at most large plants is to classify the feed hydraulically into a number of increments and table each on a separate table [see Coghill and O'Meara, Milling Methods and Costs at a Flat River (Mo.) Mill, *U.S. Bur. Mines Circ.* 6658, October, 1932].

Tabling is a relatively cheap operation and can be done in large installations for 5 to 10¢/ton feed. If the feed is uniform, one operator can take care of 50 to 70 tables. Labor is the principal item of cost. Power requirements and maintenance are both low. The installed cost of a table including supports and launders is from $1500 to $3000. One of the disadvantages of a tabling installation is the relatively large floor space required for the tonnage treated. Its main advantage is that, in the size range for which it is suited, it is a cheap and effective method of concentrating simple ores and coal.

Spiral Concentrators. A new device to separate particles by size or specific gravity is the Humphreys spiral concentrator. It accomplishes the same results as do wet tables. Advantages claimed for spirals are that the floor space required is much less than that needed for wet tables of equal capacity, that there are no moving parts, and that the separation performed is much less sensitive to variations in quantity or grade of feed. The spiral concentrator is a spiral channel, having three to five turns and a curved cross section. The spiral bowls have an outside diameter of 30 in. and a pitch of 13 in./turn.

Water and solids flow down the channel, and a combination of forces produces a double eddy effect, which is common in curved conduits. Grains of small size and highest specific gravity "hug the pole," whereas grains of progressively larger size or lesser specific gravity swing out farther and farther away from the pole. Slime suspended in the wash water follows the outside peripheral lip of the spiral.

Ports set at intervals in the bottom of the channel draw off the flow of heavy fine grains, which becomes the concentrate. Concentrate drawn off from ports in the final turn becomes a middling.

The slurry to be fed to the spirals is maintained in an inverted cone-shaped mixing tank by means of a water jet directed down to the apex of the cone, and it is pumped through a central pipe that is split and resplit until the slurry is divided equally among the spirals, of which scores may be used in an individual installation. The maintenance of the slurry and the equal distribution of it are essential factors in the satisfactory operation of the spirals.

The spiral concentrator was used first to concentrate chromite found in marine sands in Oregon. The heavy minerals found in sands near Jacksonville, Fla., are being concentrated by spirals for recovery of ilmenite and rutile; and after these minerals are collected by dry methods from the heavy mineral concentrate, the residue is put back over other spirals for concentrating zircon.

Spirals were installed in 1948 at one Mesabi mill for recovering −6-mesh iron.

Each spiral has a capacity of 1000 to 8000 lb. new feed/hr., depending upon grain size and percentage of heavy mineral present. Labor requirements are very low. Maintenance is limited to pumps and pipe lines.

Dry Tables. A relatively recent development is the Sutton, Steele, and Steele dry table (U.S. Patents 1,574,637; 1,632,520; and 2,137,678), which has a shaking motion somewhat similar to that of a wet table, except that the direction of motion is inclined upward from the horizontal and, instead of water acting as the medium of distribution, a blast of air is driven through a perforated deck. The table has application in cases where it is desirable to treat material dry, either because of water shortage or because it is undesirable to wet the materials. The table supplements other dry methods of concentration, such as electrostatic and electromagnetic methods. An advantage of the table is the ability to handle material coarser than that treated on most wet tables. Ores as coarse as ¼ in. and coal as coarse as 3 in. can be treated.

Close sizing is necessary to give good results, and until recently this has militated against adoption of the table for fine sizes, owing to the difficulties of screening most ores dry below about 40 mesh. The development of dry methods of sizing with the wind-tunnel, or Schramm, system may promote adoption of the table to the treatment of many ores and materials now handled by other equipment.

Table 3. Generalized Operating Data
Superduty Diagonal-deck Concentrating Table

Table No.	Feed	Feed size	Feed capacity, tons/hr.	Speed, r.p.m.	Stroke, in.	Water with feed, gal./min.	Dressing water, gal./min.	Size of deck
6	Ore	¼ in.–35 mesh	2.0 –10.0	275	1.25	30–150	10–100	6′5″ × 14′1″
6	Ore	35–150 mesh	1.0 – 2.5	285	0.75	16– 40	5– 20	6′5″ × 14′1″
6	Ore	Minus 150 mesh	0.25– 1.0	300	.50	3– 12	3– 10	6′5″ × 14′1″
7	Coal	1½ in.	15.0–25.0	270	1.25	125–210	55– 90	8′¼″ × 16′9¼″
7	Coal	¾ in.	10.0–15.0	280	1.00	60– 85	20– 35	8′¼″ × 16′9¼″
7	Coal	½ in.	7.5–12.0	285	1.00	42– 65	18– 31	8′¼″ × 16′9¼″
7	Coal	⅛ in.	5.0– 7.5	290	0.75	28– 42	12– 18	8′¼″ × 16′9¼″
7	Coal	1⁄16 in.	3.0– 5.0	290	.75	15– 28	9– 12	8′¼″ × 16′9¼″

Dry tables are used commercially in the separation of many types of minerals and in the cleaning of industrial materials such as seeds, cork, bagasse, fiber, nuts, wood chips, and coffee. One interesting use is in the sorting of silicon carbide by grain shapes. Flat and splintery grains are removed from others of more nearly equal dimensions. Approximately 12,000,000 tons of coal was cleaned in the United States in 1942 by air tables.

Agglomeration Tabling. Agglomeration tabling is a process whereby selective flocculation or agglomeration of grains of one mineral in an aggregate is caused by the addition of an agglomerating agent in a conditioning cell or in the ball-mill circuit, the slurry containing the agglomerated grains then being fed across gravity tables. The larger size, the oil-filmed surface, and the feathery texture of the floccules cause them to be washed over the side of the table by the current of cross water while the unflocculated discrete particles remain on the table and are carried off the end in the position followed normally by the concentrate in the usual table feed. An oiled particle will tend to ride on the surface of the water and thus is more readily carried across the side of the table than an unoiled particle. Agglomeration tabling has had more application in the concentration of phosphate minerals than in any other field, although successful tests have been run on limestone, potash, mica, and other ores.

The process is limited to granular material in the size range from 10 to 100 mesh. In this respect it differs from flotation which functions best on material 48 mesh and finer. For best results the material should be well deslimed and should be conditioned with the agglomerating reagents at a high percentage of solids, 65 per cent or greater. A collector is used that will selectively film the mineral to be agglomerated. In phosphate and limestone practice, this collector is usually a cheap fatty acid such as talloel. In potash separation the use of an alkyl sulfate (Emulsol's x-1) has been used to film sylvite (KCl). A bulk oil is always used in addition to the collector to give body to the film and to assist in forming agglomerules. In Florida practice, it is customary to use 0.3 to 0.5 lb./ton talloel and 4 to 5 lb./ton of a 22°Bé. fuel oil. Operating data for the agglomerate tabling of phosphate and potash ore is shown on Table 4.

Table 4. Operating Data
Agglomerate Tabling of Phosphate and Potash Ore

Type of table	Feed size	Feed capacity, tons/hr.	Table speed, r.p.m.	Table stroke, in.	Water with feed, gal./min.	Dressing water, gal./min.	Size of deck
No. 6 Super-duty Diagonal deck.	10–48 mesh	2.5–3.5	295	1.0	20–40	8–15	6′5″ × 14′1″

Apparatus other than concentrating tables may be used to give a similar separation on material conditioned in the same manner. At the plant operated by the Coronet Phosphate Co. (*Eng. Mining J.*, p. 51, December, 1942) such a separation is made on a wide moving belt, and the agglomerated particles are caused to overflow the sides of the belt by impinging a fine sharp spray of water on the mass of material at the center of the belt. At the plant of Swift & Co. (*Rock Products*, vol. 43, p. 27), a separation is affected by feeding the conditioned feed to an inclined submerged screen.

Agglomerate tabling works best on simple ores consisting of two free minerals. It has several advantages over the usual tabling method in that it can be used to separate two minerals the difference in specific gravity of which is so small that an effective separation cannot be made by gravity separation alone. Tables treating an agglomerated feed have a considerable larger capacity than tables using untreated feeds, since the capacity of a table treating an agglomerated feed is limited only by the carrying capacity of the riffles. Disadvantages of the method that must be considered are the cost of the reagents used and the fact that, if the mineral fraction filmed is the one to be sold, the oily film may be objectionable and must be burned off.

CONCENTRATION OF ORES BY SINK-AND-FLOAT METHODS

By Robert Ammon

References: Taggart, "Handbook of Mineral Dressing," Wiley, New York, 1945. Gaudin, "Principles of Mineral Dressing," McGraw-Hill, New York, 1939. Richards and Locke, "Textbook of Ore Dressing," McGraw-Hill, New York, 1940. Davis, Heavy Density Processes for Coal Beneficiation, *Mining Congr. J.*, **32**, 33 (June, 1946). Allan and Trostler, Recent Progress in Sink-and-float, *Trans. Can. Inst. Mining Met.*, **43**, 248 (1940). DeVaney and Shelton, Properties of Suspension Mediums for Sink-and-float Concentration, *U.S. Bur. Mines Report Investigations* 3469R, 1940. Holmes, Sink-and-float Processes, *Mine and Quarry Eng.*, **5**, 249 (1940). Holt, Sink-and-float Separation Applied Successfully on the Mesabi, *Eng. Mining J.*, **141**, 33 (September, 1940). Swainson, Falconer, and Walker, Some Recent Applications of Heavy-media Separation Processes, *Am. Inst. Mining Met. Engrs.*, *Tech. Pub.* 1600, 1943. Just, New Heavy Media Mill at Leadville, *Eng. Mining J.*, **144**, 68, 121 (1943). Knuckey, The Huntington-Heberlein Sink-and-float Process, *Am. Inst. Mining Met. Engrs.*, *Tech. Pub.* 1609, 1943. Vogel, Heavy-media Separation Plant of the Barton Mines Corp., *Am. Inst. Mining Met. Engrs.*, *Tech. Pub.*, 1578, 1943. Rockwood, Sink-and-float Process Applied to Fluorspar, *Rock Products*, **47**, 56 (November, 1944). Driessen, Cleaning Coal by Heavy Liquids, *J. Inst. Fuel*, **12**, 327 (1939); also *Coll. Guard.*, **158**, 783, 832 (1939); The Use of Centrifugal Force for Cleaning Fine Coal and Heavy Liquids and Suspension with Special Reference to the Cyclone Washer, *J. Inst. Fuel*, **19**, 33 (1945). Geer and Yancey, Preliminary American Tests of a Cyclone Coal Washer Developed in the Netherlands, A.I.M.M.E. T.P. 2136, *Coal Technol.*, Feb., 1947. Beall, Recent Developments in Heavy-Density Separation, *Min. & Met.*, **29**, 488 (1948). Hyer, Heavy-Density Separation—a Review of Its Literature, *Quart. Colo. School Mines*, **43**, No. 1 (January, 1948).

The sink-and-float process, known as the heavy-media separation process, is used in the beneficiation of coals; metallic mineral ores, both ferrous and non-ferrous; and non-metallic mineral ores. In heavy-media separation processes, the separation of heavy minerals is made through the use of ferrous media of high specific gravity, such as ferrosilicon and magnetite, or non-ferrous media such as galena. When ground to suitable fineness and mixed with water in correct proportion, these high-specific-gravity solids furnish a medium that closely duplicates a true heavy liquid as regards fluidity and stability in a range of gravities from as low as 1.25 to as high as 3.4. As a consequence, no rising currents of water are necessary to assist in the separation of sink from float, nor is it necessary to supply strong mechanical agitation to maintain the medium in suspension. With such a medium, it is therefore possible to secure separations matching those obtained in a heavy liquid.

Scope of Heavy-media Separation Processes

General Field of Application. In general it may be stated that heavy-media separation processes will treat coarse fractions of any ore in which the valuable constituents have an appreciable difference in specific gravity from the worthless gangue. The practical limiting grain size is about 10 mesh, although in cases of extreme gravity differences grains as fine as 48 mesh can be treated, as will be discussed later. The bulk of the grains must be fragments of individual crystals. The difference in specific gravity can be less than that required for efficient separations by jigging or tabling.

The heavy-media separation processes are capable of performing one of three main functions, depending on the character of the material being treated: (1) The rejection of a waste product to leave an enriched product for further concentration by other methods, in most cases after further reduction in size; (2) the production of finished concentrate and a rejectable waste in one operation; and (3) the production of a finished concentrate and a low-grade reject for additional treatment.

Range of Sizes Amenable to Heavy-media Separation. Although it is feasible in some instances to handle run-of-mine ore without presizing, the efficiency of separation drops off on sizes below 48 mesh. For the treatment of sizes up to 3¼ in., a separatory cone may be used. Other apparatus is available for treatment of sizes larger than 3¼ in.

Range of Separating Gravities Obtainable. Highly efficient separations have been made on coal at specific gravities ranging down to 1.25, using a magnetite medium, and up to 2.20. For the range 2.20 to 2.90, mixtures of magnetite and ferrosilicon are used. Above 2.85, ferrosilicon alone is used, and separations up to as high as 3.40 specific gravity have been made commercially.

Characteristics of Magnetic Media. Magnetite (Fe_3O_4) is a naturally occurring mineral found in abundance in many countries and is mined as an iron ore. Ferrosilicon (Fe 85 per cent, Si 15 per cent) is a furnace product obtainable from various suppliers in several grades of suitable fineness.

Preparation of Feed. Primary slimes, and fines that are not amenable to sink-and-float separation, are objectionable because they dilute the medium and increase its viscosity. Such slimes and fines should be removed so far as practical prior to heavy-media separation, unless special means are provided in the medium-cleaning circuit to handle these fines, as discussed later.

Heavy-media Separations. Referring to Fig. 175, it will be noted that the finer sizes are removed from the feed by the primary washing screen 1 before it enters the separatory cone 2. In the separatory cone, which is filled with ferrous medium of preselected gravity, the low-gravity minerals float on the surface of the medium and are removed by overflowing a weir located opposite the point of entry of feed. The heavier gravity minerals sinking through the medium are continually removed from the cone by means of "air lift" 3. The latter, as will be noted, is, roughly, a J-shaped pipe of suitable size connected to the bottom of the cone. Jets of compressed air introduced at the bottom of the vertical section of this air lift serve to elevate the sink particles in an effective and economical manner.

Removal of Medium from the Separation Products. The float-and-sink products discharged from the cone go to drainage and washing screens 4 and 5. In the particular flow sheet shown in Fig. 175, the float-and-sink products are treated on single drainage and washing screens divided longitudinally by a partition to keep the products separated. In large-sized plants, separate drainage and washing screens are provided to handle the float-and-sink products. The purpose of the drainage screen 4, is, as the name implies, to drain off and remove the medium from the float-and-sink products. Normally, more than 90 per cent of the medium discharged from the cone with the float-and-sink products drains off on the

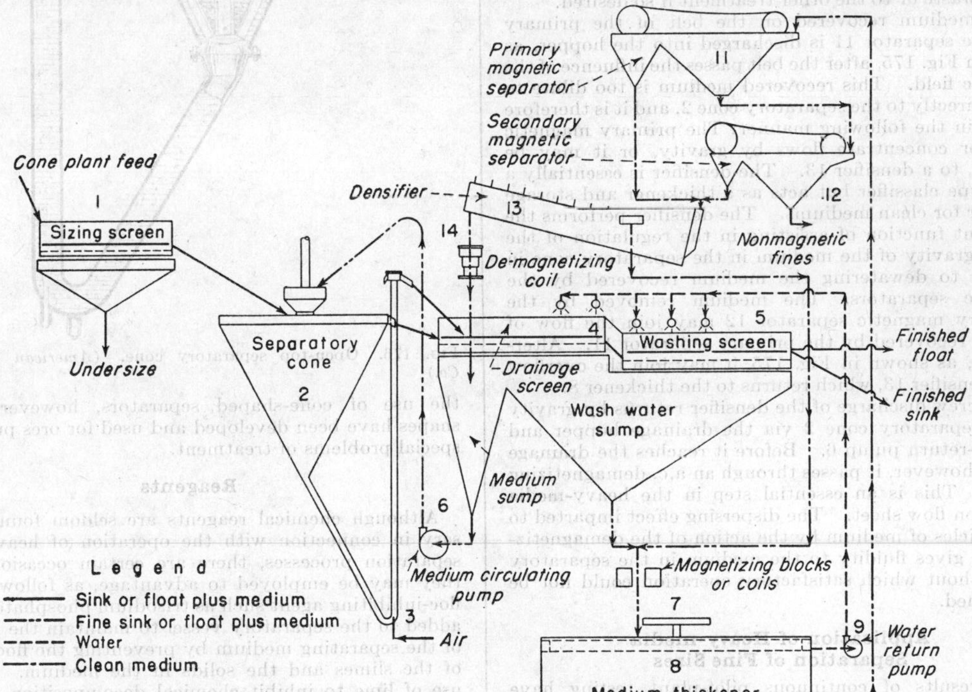

FIG. 175. Heavy-media separation, standard flow sheet. (*American Cyanamid Co.*)

screen or screens mentioned and is returned directly back to the cone without further treatment by means of one or more centrifugal pumps 6. The float-and-sink products pass along the drainage screen or screens 4 to the washing screen or screens 5, where substantially complete removal of the adhering medium is accomplished by means of water sprays, as indicated in Fig. 175. The screen oversize products are discharged as concentrate or tailing.

Reclamation and Cleaning of the Medium for Reuse. The undersize from the washing screen 5 first flows by gravity or is pumped to a medium-reclamation thickener 8. Just before it enters the thickener the diluted medium passes between a set of magnetizing blocks 7. The latter serves to change the charge on the discrete ferrosilicon or magnetite particles whereby they become mutually attracted and flocculate. The net result of this is faster settling of the particles in the thickener, with the consequent advantage of requiring less thickener area and depth than would otherwise be necessary.

The magnetic cleaning circuit, as shown in Fig. 175, comprises a primary (11) and a secondary (12) Crockett type magnetic separator. These separators are of the belt type, the magnet assembly being mounted just above the lower section of the belt. Feed is added under the belt as shown and the fines, not immediately picked up by the action of the magnets and carried by the rubber belt, are dropped and eliminated through the hopper, as shown. This tailing, or reject, goes by gravity to the secondary magnetic separator 12 for additional treatment in order to recover any traces of medium not picked up in the primary separator. The tailing from the secondary separator 12 goes to waste or to the other treatment if so desired.

The medium recovered on the belt of the primary magnetic separator 11 is discharged into the hopper, as shown in Fig. 175, after the belt passes the influence of the magnetic field. This recovered medium is too dilute to return directly to the separatory cone 2, and it is therefore treated in the following manner: The primary magnetic separator concentrate flows by gravity, or it may be pumped, to a densifier 13. The densifier is essentially a screw-type classifier but acts as a thickener and storage reservoir for clean mediums. The densifier performs the important function of assisting in the regulation of the specific gravity of the medium in the separatory cone in addition to dewatering the medium recovered by the magnetic separators. The medium removed by the secondary magnetic separator 12 may join the flow of medium recovered by the primary separator 11. Alternatively, as shown in Fig. 175, it may join the overflow of the densifier 13, which returns to the thickener 8.

The screw discharge of the densifier returns by gravity to the separatory cone 2 via the drainage hopper and medium-return pump 6. Before it reaches the drainage hopper, however, it passes through an a.c. demagnetizing coil 14. This is an essential step in the heavy-media separation flow sheet. The dispersing effect imparted to the particles of medium by the action of the demagnetizing coils gives fluidity to the medium in the separatory cone without which satisfactory operation could not be maintained.

Application of Heavy-media Separation of Fine Sizes

The results of continuous pilot-plant testing have shown the possibility of extending the application of heavy-mediums separation to the treatment of material as fine as 28 to 48 mesh. For the treatment of sizes below 10 mesh, a modified flow scheme has been developed. Essentially, this flow scheme involves the use of dual magnetic cleaning circuits wherein the undersize float and the undersize sink products from the screens are separately subjected to magnetic separation. The magnetic separators recover the medium for reuse, and the tailings from these separations are either a finished float or a finished sink product, as the case may be.

Separatory Cones

The heavy-media separatory cone shown in Fig. 175 is illustrated in more detail in Fig. 176. This open-top cone has been found most suitable for separations where a large amount of float has to be removed. In other cases, particularly for the treatment of iron ores where a finished sink product is required and where the amount of float to be removed is relatively smaller in proportion to the sink, a closed-top cone, as shown in Fig. 177, is used. A third type of cone employs an inside air lift. The scope of the heavy-media separation processes is not limited to the use of cone-shaped separators, however. Other shapes have been developed and used for ores presenting special problems of treatment.

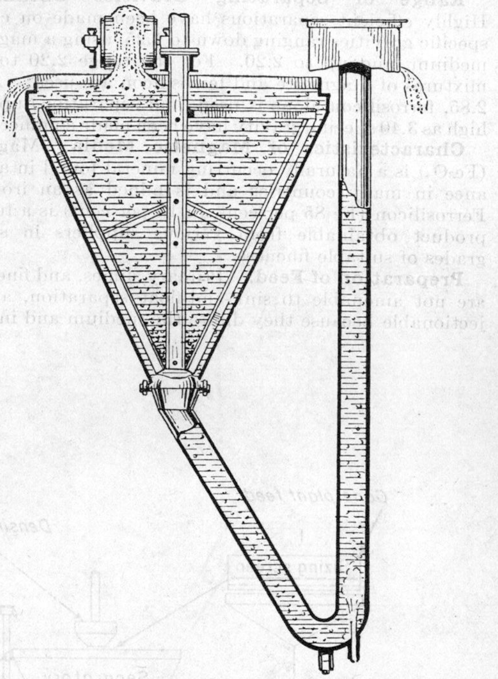

FIG. 176. Open-top separatory cone. (*American Cyanamid Co.*)

Reagents

Although chemical reagents are seldom found necessary in connection with the operation of heavy-media separation processes, there are certain occasions when they may be employed to advantage, as follows: (1) A floc-inhibiting agent such as trisodium phosphate may be added to the separatory vessel to maintain the liquidity of the separating medium by preventing the flocculation of the slimes and the solids in the medium. (2) The use of lime to inhibit chemical decomposition of ferrosilicon in water, especially sea water. Lime is particularly advantageous when ferrosilicon is used in the heavy-mediums separation of carbonate ores. The addition of lime prevents formation of bicarbonates, which, in the presence of ferrosilicon, have a tendency to produce a jelly that increases the viscosity of the medium. (3) The use of chrome salts, such as sodium dichromate.

perform the useful function of preventing caking of certain magnetic ferrous mediums, such as metallic iron, during storage in a thickened or largely dewatered condition. They also act as inhibitors or antioxidants in the use of this class of mediums.

The Dutch State Mines Cyclone Separator Processes

Recently a new process was developed for separating fine mineral or coal particles in the size range of minus ¼ in. to as fine as 65 mesh. In this process, which was originally developed by Dutch State Mines, in Limburg, Holland, a Cyclone Separator of novel design is used to

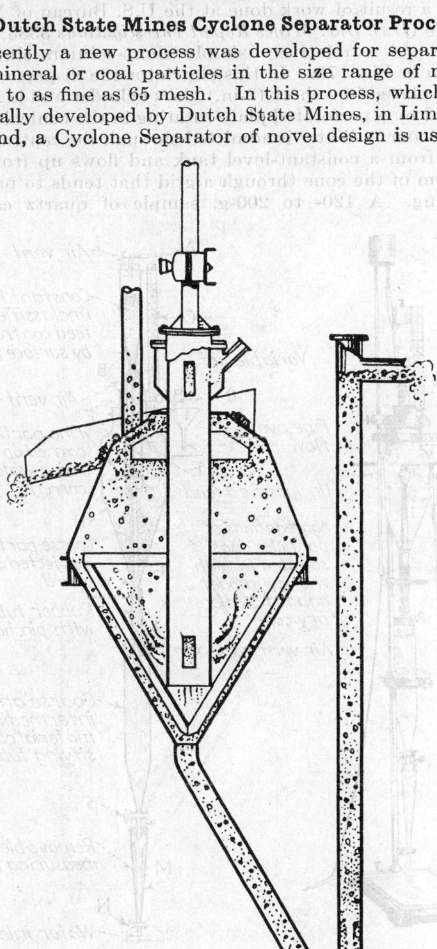

FIG. 177. Closed-top separatory cone. (*American Cyanamid Co.*)

assist the separation of the fine particles which are fed together with a heavy medium of suitable specific gravity under pressure to the cyclone.

Construction and Operation of Cyclone Separator. The material to be treated is suspended in a very fine medium and this pulp is fed tangentially through the feed inlet via (1) of Fig. 177*A* to the short cylindrical section (2). The media may be of various types as described below. The short cylindrical section (2) also carries the central "vortex finder" (3), which prevents short-circuiting within the Cyclone. Separation is made in the cone-shaped part of the Cyclone (4) by the action of centrifugal and centripital forces. The heavier portion of the material treated leaves the Cyclone at the apex opening (5) and the lighter portion leaves at the overflow top orifice (6).

The pressure at which the pulp is introduced into the Cyclone is the principal means of controlling the forces within the Cyclone and therefore will vary with the size range of the feed. In general, separation in the finer size ranges will require higher pressure than that required for coarse material. Operating pressures will vary from 20 to 40 lb./sq. in. Such pressures are readily reached and maintained by the use of a centrifugal pump.

Medium Used in the Cyclone Separator. The medium may be exogenous, *i.e.*, material not present in the coal or other ore to be treated, or it may be autogenous and therefore obtained from the raw coal or ore.

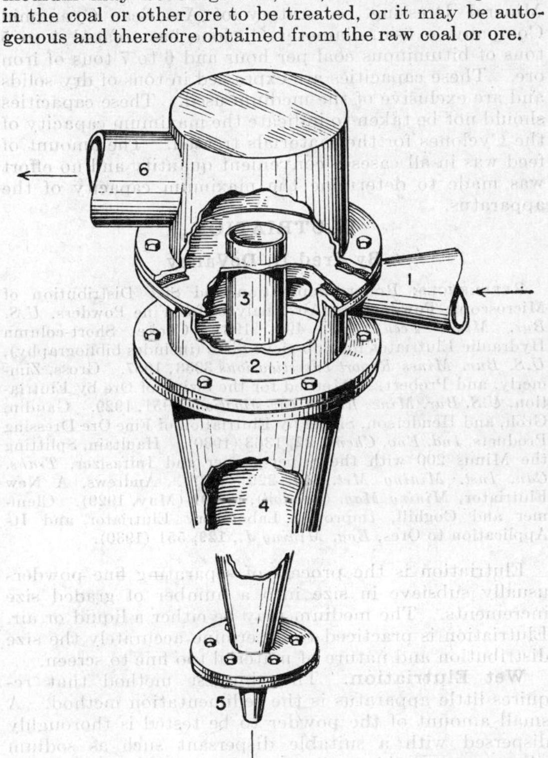

FIG. 177*A*. The Dutch State Mines cyclone separator.

In treating most materials an exogenous medium is preferred and the most efficient appear to be of the magnetic type. Magnetite (a natural iron ore of the composition Fe_3O_4) is used when separation is made in the lower specific-gravity ranges and ferrosilicon (a common iron alloy) is used for separation in the higher ranges.

Excellent results have been obtained by using autogenous media in treating such materials as Mesabi iron ores. In these applications the autogenous media was prepared from the heavy constituents of the ore itself.

Methods of Cleaning Medium. It will be evident that the choice of a medium, aside from the important considerations of its availability, specific gravity, and cost, will depend on whether it can be cleaned by a process proven by commercial use to be satisfactory and economical. Such processes must also be capable of returning the medium to the Cyclone at the proper specific gravity and fineness and without excessive losses. The heavy-media separation processes for cleaning and reusing ferrous media fortunately meet all the requirements.

Capacity of the Cyclone Separator. The capacity of the Cyclone in terms of dry solids depends on the character and the size range of the materials to be treated, *i.e.*, the specific gravity of the material, the specific gravity at which the desired separation is accomplished, and the percentage of near gravity material in the feed. When coarse material is treated, the feed rate will be higher than when finer material is treated. It may be useful to state that a Cyclone of 350 mm. (13.75 in.) diameter has operated in Holland treating a semianthracite coal at a capacity of 20 tons of dry solids per hour and a 6-in.-diameter cyclone on the same type of feed at a capacity of 5 tons of dry solids per hour. In testing work in the Mineral Dressing Laboratory of American Cyanamid Company, a 6-in.-diameter Cyclone has treated 4 to 5 tons of bituminous coal per hour and 6 to 7 tons of iron ore. These capacities are expressed in tons of dry solids and are exclusive of the medium used. These capacities should not be taken to indicate the maximum capacity of the Cyclones for the materials treated. The amount of feed was in all cases a convenient quantity and no effort was made to determine the maximum capacity of the apparatus.

ELUTRIATION

By Fred D. DeVaney

REFERENCES: Roller, Separation and Size Distribution of Microscopic Powders, an Air Analyzer for Fine Powders, *U.S. Bur. Mines Tech. Pub.* 490, 1931. Cooke, Short-column Hydraulic Elutriator for Sub-sieve Sizes (includes bibliography), *U.S. Bur. Mines Report Investigations* 3333, 1937. Gross, Zimmerly, and Probert, A Method for the Sizing of Ore by Elutriation, *U.S. Bur. Mines Report Investigations* 2951, 1929. Gaudin, Groh, and Henderson, Sizing by Elutriation of Fine Ore Dressing Products, *Ind. Eng. Chem.*, **22,** 1363 (1930). Haultain, Splitting the Minus 200 with the Super-panner and Infrasizer, *Trans. Can. Inst. Mining Met.*, **40,** 229 (1937). Andrews, A New Elutriator, *Mining Mag. (London)*, p. 301 (May, 1929). Clemmer and Coghill, Improved Laboratory Elutriator and Its Application to Ores, *Eng. Mining J.*, **129,** 551 (1930).

Elutriation is the process of separating fine powders usually subsieve in size into a number of graded size increments. The medium may be either a liquid or air. Elutriation is practiced to determine accurately the size distribution and nature of material too fine to screen.

Wet Elutriation. The simplest method that requires little apparatus is the sedimentation method. A small amount of the powder to be tested is thoroughly dispersed with a suitable dispersant such as sodium silicate, gum arabic, or lignin sulfonate in a glass cylinder or beaker. From Stokes's law, the time required for a material of given specific gravity and a given size to settle a given distance can be calculated. The desired fine fraction can then be siphoned off. Repeated fractionations are required to ensure an almost complete removal of the fine fraction from the settled material. Any number of size increments can be secured by adjusting the settling time. The size of the particles in any increment should be checked with a microscope. The method is rather tedious and is not very precise but can be used if more elaborate equipment is not available. In any elutriation process the sizing action is a function of the specific gravity of the material. If two minerals of different specific gravity are elutriated, any specific increment will contain finer particles of the heavy mineral than of the lighter mineral.

The Andrews kinetic elutriator (Fig. 178) is rather complicated but is said to give very accurate results. Elutriation is continued for a number of hours until no particles can be seen in the overflow tubes from one vessel to the next. The vessels are then emptied and the contents filtered, dried, weighed, and then examined microscopically for size, using a calibrated grid ocular or an ocular with an accurate scale reading to hundredths of a millimeter. It may be unnecessary to describe the grain sizes, referring simply to the fractions collected with a given hydraulic velocity. The weight of the several fractions collected shows the size distribution in the sample tested. Any fraction that is lost in the final overflow is recognized by the difference in total weight of the fractions as compared with the original weight.

As a result of work done at the U.S. Bureau of Mines, Cooke (*U.S. Bur. Mines Report Investigations* 3333, 1937) developed a relatively simple short-column elutriator shown in Fig. 179. In this elutriator the dispersed sample is placed in a short (2-in.) glass cylinder and is slowly stirred by a perforated plate rotating in the conical metal lower cone. Water is admitted through the central glass tube from a constant-level tank and flows up from the bottom of the cone through a grid that tends to prevent swirling. A 120- to 200-g. sample of quartz can be subdivided into six fractions between 52 and 9.3 μ in 8 hr. and 10 fractions between 52 and 2.3 μ in less than 60 hr.

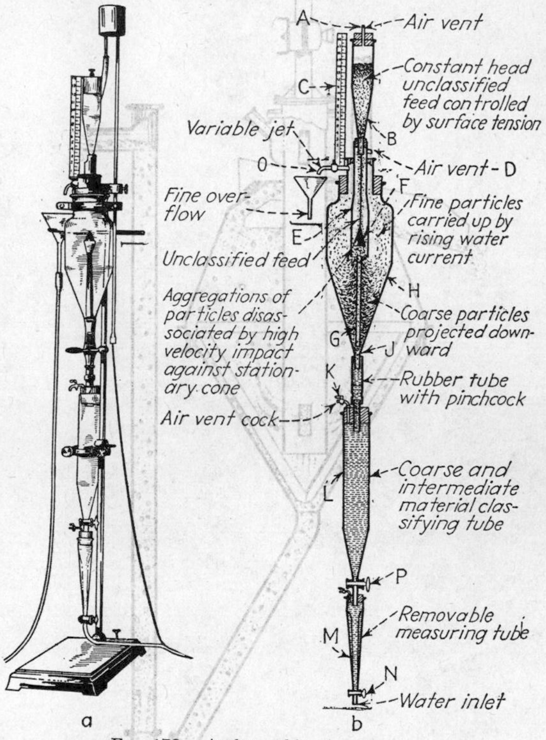

FIG. 178. Andrews kinetic elutriator.

Wet elutriation frequently is employed for size classification in closed-circuit wet-grinding operations (see p. 930).

See also Sec. 16 for other wet methods for particle-size analysis.

Dry Elutriation. Equipment for and the theory of dry elutriation are described by Traxler and Baum [Determination of Particle Size Distribution in Mineral Powders by Air Elutriation, *Rock Products*, **37,** 44 (1934)]. It is necessary in the size analysis of materials subject to hydration—such as cement or lime or to solution as in the case of water-soluble materials. For dry elutriation, complete dispersion of the particles must be accomplished, which in the Traxler and Baum apparatus is done by means of an air jet. Temperature and humidity conditions are controlled and maintained constant throughout each test.

See also Sec. XVI for other particle-size measurement methods.

Haultain Infrasizer. This instrument (Fig. 180) developed at the University of Toronto and marketed by

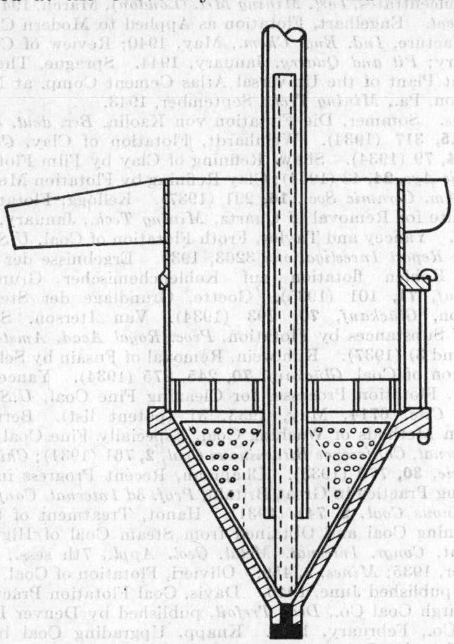

FIG. 179. Cooke short-column elutriator.

Infrasizer Limited, Toronto, is used in many laboratories. It consists of a series of conical stainless-steel tubes of varying size. Diffusion of the particles is effected by impinging the air carrying the particles against a ball around which the particles pass in a high-velocity stream.

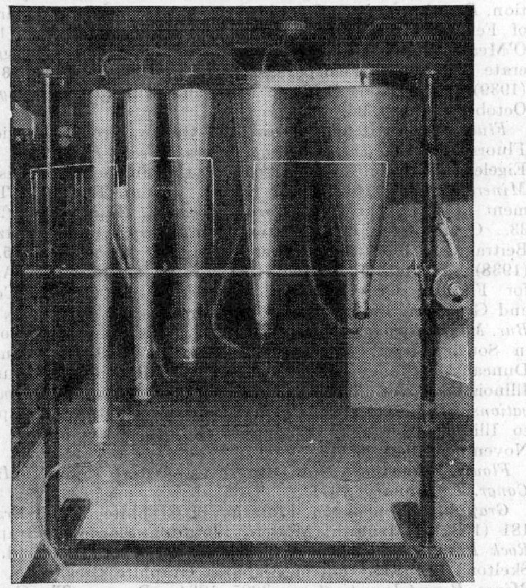

FIG. 180. Haultain Infrasizer for elutriation of fine dry powders. (*Infrasizer Ltd., Toronto.*)

No attempt is made to secure a uniform upward flow of air across the cross section of the tubes, but rather a condition of uniform turbulence is sought.

The tubes are connected in series, and their size is so regulated that the particles from each tube bear a definite geometric size ratio one to another. The cones are mechanically tapped to prevent particles from adhering to their walls. The design of the machine in most common use produces seven products, the finest of which is essentially smaller than 10 μ in size. It treats 100 g./hr. with a maximum charge of 400 g. for relatively coarse materials and 200 g. for very fine material. Through an instrument panel the volume of air and the length of the run are automatically controlled. The machine requires 5 cu. ft. air/min. at not less than 30 lb./sq. in.

FROTH FLOTATION

By Fred D. DeVaney

REFERENCES: Taggart, Taylor, and Ince, Experiments with Flotation Reagents, *Trans. Am. Inst. Mining Eng.*, vol. 87, 1930; *Am. Inst. Mining Engrs., Tech. Pub.* 204. Sproule, Some Fundamentals of Flotation, *Can. Mining J.*, **57,** 582 (1936). Whiting, Modern Flotation Reagents, Their Classes and Uses, *Mining Met.*, **19,** 185 (1938). Shorey, Patek, and Roland, Desliming Ore Pulps with Sodium Silicate as a Deflocculator, *Am. Inst. Mining Engrs., Tech. Pub.* 559, Sept. 15, 1934. Gillies, The Story of the Bubble, *Can. Inst. Mining Met.*, 1935, p. 349. Wark, Principles of Flotation, *Australasian Inst. Mining Met.*, 103, 346 (1938). Gaudin, "Flotation," p. 552, McGraw-Hill, New York, 1932. Giudice, Depressors and Protective Agents in Conditioning for Flotation, *Eng. Mining J.*, **135,** 350 (1934). Giudice, Fundamentals of Flotation in the Light of Recent Research, *Eng. Mining J.*, **135,** 152 (1934). Giudice, Collection in Flotation, *Eng. Mining J.*, **135,** 213 (1934). Taggart, Taylor, and Knoll, Chemical Reactions in Flotation, *Am. Inst. Mining Engrs., Tech. Pub.* 312, February, 1930. Wark and Cox, An Experimental Study of the Influence of Cyanide, Alkalis and Copper Sulfate on the Effect of Sulfur-bearing Collectors at Mineral Surfaces, *Am. Inst. Mining Engrs., Tech. Pub.* 574, October, 1934. Conception of Adsorption Applied to Flotation Reagents, *Am. Inst. Mining Engrs., Tech. Pub.* 732, February, 1937. Wark and Sutherland, Influence of the Anion on Air-mineral Contact in Presence of Collectors of Xanthate Type, *Mining Tech.*, November, 1939. Shepard, Experiments on the Cause of Bubble Attachment in Flotation, *Mining Met.*, **17,** 339 (1936). Gaudin, Groh, and Henderson, Effect of Particle Size on Flotation, *Am. Inst. Mining Engrs., Tech. Pub.* 414, May, 1931. Brighton, Burgener, and Gross, Depression by Cyanide in Flotation Circuits, *Eng. Mining J.*, **133,** 276 (1932). Wark and Cox, An Experimental Study of the Effect of Xanthates on Contact Angles at Mineral Surfaces, *Am. Inst. Mining Engrs., Tech. Pub.* 461, February, 1932. Ralston and Hunter, Activation of Sphalerite for Flotation, *Am. Inst. Mining Engrs., Tech. Pub.* 248, October, 1929. Ralston, King, and Tartaron, Copper Sulfate as Flotation Activator for Sphalerite, *Am. Inst. Mining Engrs., Tech. Pub.* 247, October, 1929. Weinig and Carpenter, The Trend of Flotation, *Colo. School Mines Quart.*, vol. 32, 1937. Kraeber and Boppel, Über die Wirkung von Metallsalzen beider Schwimmaufbereitung oxydischer Mineralien, *Metall Erz*, **31,** 417 (1934). Norman and Ralston, Conditioning Surfaces for Froth Flotation, *Am. Inst. Mining Engrs., Tech. Pub.* 1074, published in *Mining Tech.*, May, 1939. Dean and Hersberger, New Flotation Reagents, *Am. Inst. Mining Met. Engrs.*, **134,** 81 (1939). Gaudin and Vincent, Observations on the Magnitude of Contact Angles and Their Significance in Flotation Phenomena, *Mining Tech.*, November, 1940. Trotter, Controlling Flotation Reagents Automatically, *Eng. Mining J.*, November, 1941, p. 45. Clemmer and Clemmons, An Improved Flotation Test Cell, *Eng. Mining J.*, March, 1943, p. 72. Fitt, Thomas, and Taggart, The Nature of Dispersed Mineral in Flotation Pulps, *Mining Tech.*, March, 1943. Hassialis, Organic Sulphides as Oily Collectors, *Mining Tech.*, May, 1943. Bourke, Automatic pH Control Applied to Flotation, *Eng. Mining J.*, August, 1944, p. 76. Fahrenwald and Prater, The Size, Number and Mineral Carrying Efficacy of Bubbles in Flotation, *Pub. Idaho Bur. Mines Geol.*, May, 1944. Rose, The Controversial Art of Flotation, *Mining Tech.*, March, 1944. Myers and Lewis, Flotation Machines at the Tennessee Copper Co., *Mining Tech.*, March, 1944. Rose, Flotation as a Power Process, *Mining Tech.*, March, 1944. Flotation Index, A Bibliography of Articles on Flotation (published by Great Western Div., The Dow Chemical Co., San Francisco), 1946. Kellogg and

Vasques-Rosos, Amine Flotation of Sphalerite-galena Ores, *Mining Tech.*, November, 1945. Arbitor, Kellog, and Taggart, The Mechanism of Collection of Metals and Metallic Sulphides by Amines and Amine Salts, *Mining Tech.*, May, 1943. DeVaney, Flotation, *Ind. Eng. Chem.*, vol. 38, January, 1946.

Non-sulfide flotation. Clemmer and O'Meara, Flotation and Depression of Non-sulphides—Calcite, Silica and Silicates, Fluorspar, Barite, Apatite and Tungsten Minerals, *U.S. Bur. Mines, Report Investigations* 3239, p. 9, June, 1934. Halbich, Über Neuartige Schwimmittel, *Metall Erz*, **30,** (21), 1 (1933) (use of alcohol sulfates). Growing Use of Flotation for Non-metallic Minerals, *Mining Met.*, March, 1935, p. 129. Beitrag zur Flotation Nicht sulfidischer Mineralien, *Metall Erz*, **27,** 527 (1930). Coghill and Clemmer, Soap Flotation of the Non-sulfides, *Am. Inst. Mining Met. Engrs.*, *Tech. Pub.* 445, 1932; *Eng. Mining J.*, **133,** 136 (1932). Cullen and Lavers, Flotation as Applied to the Chemical Industry, *Trans. Inst. Chem. Engrs.* (*London*), Jan. 15, 1936, reviewed in *Mining J.* (*London*), **192,** 65, 85, 104 (1936). Dean and Hersberger, New Flotation Reagents, *Am. Inst. Mining Engrs.*, *Tech. Pub.* 605, 1935. Patek, Relative Flotability of the Silicate Minerals, *Trans. Am. Inst. Mining Engrs.*, **112,** 486 (1934); *Am. Inst. Mining Met. Engrs.*, *Tech. Pub.* 564. Patek, Colloidal Depressors in Soap Flotation, *Eng. Mining J.*, **137,** 558 (1936). Dean, Clemmer, and Cooke, Use of Wetting Agents in Flotation, *U.S. Bur. Mines, Report Investigations* 3333, p. 3, February, 1937. Taggart, Flotation Application to Non-metallics, *Eng. Mining J.*, **137,** 90 (1936). Ralston, Flotation and Agglomerate Concentration of Nonmetallic Minerals, *U.S. Bur. Mines, Report Investigations* 3397, May, 1938. Taggart and Arbiter, Collector Coatings in Soap Flotation, *Mining Tech.*. July, 1943. Perruche, Flotation of Nonmetallic Minerals and Various Substances, *LaNature*, 1941, p. 411; *Chem. Abst.* 37:2964. Falconer and Crawford, Froth Flotation of Some Non-sulphide Minerals of Strategic Importance, *Am. Inst. Mining Met. Engrs. Tech. Pub.* 1754, 1944. Taggart and Arbiter, The Chemistry of Collection of Non-metallic Minerals by Amine-type Collectors, *Mining Tech.*, May, 1944. Barr, Machines for Non-metallic Flotation, *Mining Tech.*, September, 1945.

Apatite. Luyken and Bierbrauer, Flotative Recovery of Apatite, *Metall Erz.* **26,** 197, 202 (1929); *Mitt. Kaiser-Wilhelm Inst. Eisenforsch. Düsseldorf*, **10,** 317 (1928).

Arsenic. Yasyukevich and Kahn, Experiments on Scorodite Flotation, *Tzvetnuie Metall*, 1934, No. 7, p. 26.

Barite. Timm *et al.*, Barite-bearing Mill Tailing from Kamloops Homestake Mine, Ltd., Jamieson Creek, B.C., *Can. Dept. Mines, Geol. Branch, Rept.* 774, p. 80, 1937. The Flotative Separation of a Mixture of Barite and Dolomite—the Differential flotation of Earth Alkali Minerals, *Berg- u. Hüttenmänn. Jahrb.*, **81,** 139 (1933). Hammann, Flotation of Heavy Spar—a Preview, *Metall Erz*, **30,** 455 (1933). O'Meara and Coe, Froth Flotation of Southern Barite Ores, *Am. Inst. Mining Engrs.*, *Tech. Pub.* 678, 1936. Glembotskii, Flotation of Barite at the Salair Dressing Plant, *Gorno-Obogatitel. Zhur.*, **1937,** No. 5, p. 13. Rankin, Laurence. Davis, Houston, and McMurray, Concentration Tests on Tennessee Valley Barite, *Am. Inst. Mining Engrs.*, *Tech. Pub.* 880, 1938. Norman and Lindsey, Flotation of Barite from Magnet, Core Ark., *Mining Tech.*, May, 1941.

Barite and dolomite. Ringe and Bierbrauger, The Flotative Separation of a Mixture of Barite and Dolomite—the Differential Flotation of Earth Alkali Minerals, *Berg- u. Hüttenmänn. Jahrb.*, **81,** 139 (1933).

Bauxite. Lottermoser and Rumpelt, Studies on the Flotation of Bauxite from Near Bodayk (Hungary), *Kolloid Beihefte*, **35,** 372 (1932). Gandrud and DeVaney, Bauxite: Sink-and-float Fractionations and Flotation Experiments, *U.S. Bur. Mines Bull.* 312, 1929. Preliminary Examination of Low-grade Bauxite, with Particular Reference to Flotation, *Bur. Mines, Report Investigations* 2906, 1928. Clemmer, Clemmons, and Stacy, Preliminary Report on the Flotation of Bauxite, *U.S. Bur. Mines Report Investigations* 3586, 1941. Runke and O'Meara, Beneficiation of Arkansas Bauxite, *Trans. Am. Inst. Mining Met. Engrs.*, **159,** 218 (1944).

Beryl. Gisler, Selective Flotation of Beryl, Master's Thesis, University of Utah, May 15, 1936.

Borax. Flotation of Boric Acid and Borax from the Products of Treatment of the Indera Boracite, *J. Applied Chem.* (*U.S.S.R.*), **10,** 845 (1937). A Flotation Process for the Separation of Borax and Boric Acid from a Mixture of Salts, *J. Chem. Ind.* (*U.S.S.R.*), **12,** 277 (1935).

Calcium carbonate. See Limestone and Cement.

Cassiterite. Flotation of Cassiterite, *Arch. Erzbergbau Erzauf bereit. Metallhüttenw.*, **2,** 1 (1932). Pol'kin and Boldyrev, The Flotation of Cassiterite, *Gorno- Obogatitel. Zhur.*, No. 6, p. 38, 1936. Michel, The Flotation of Pyrite and Arsenopyrite in Tin Ore Concentrates, *Inst. Mining Met.* (*London*), March, 1941.

Cement. Engelhart, Flotation as Applied to Modern Cement Manufacture, *Ind. Eng. Chem.*, May, 1940; Review of Cement Industry; *Pit and Quarry*, January, 1944. Sprague, The New Cement Plant of the Universal Atlas Cement Comp. at Northhampton, Pa., *Mining Tech.*, September, 1943.

Clays. Sommer, Die Flotation von Kaolin, *Ber. deut. keram. Ges.*, **15,** 317 (1934). Weinhardt, Flotation of Clay, *Ceramic Age*, **24,** 79 (1934). Shaw, Refining of Clay by Film Flotation, *Ceramic Age*, **24,** 43 (1934); Clay Refining by Flotation Methods, *Bull. Am. Ceramic Soc.*, **16,** 291 (1937). Kellogg, Flotation of Kaolinite for Removal of Quartz, *Mining Tech.*, January, 1945.

Coal. Yancey and Taylor, Froth Flotation of Coal, *U.S. Bur. Mines, Report Investigations* 3263, 1934. Ergebnisse der selektiven Kohlen flotation auf Kohlenchemischer Grundlage, *Glückauf*, **71,** 101 (1935). Goette, Grundlage der Steinkohl flotation, *Glückauf*, **70,** 293 (1934). Van Iterson, Separation of Substances by Flotation, *Proc. Royal Acad. Amsterdam*, **40** (2 and 3) (1937). Kuhlwein, Removal of Fusain by Selective Flotation of Coal, *Glückauf*, **70,** 245, 275 (1934). Yancey and Taylor, Flotation Processes for Cleaning Fine Coal, *U.S. Bur. Mines Circ.* 6714, May, 1933, 31 (patent list). Berthelot, Modern Methods of Washing Coal, Especially Fine Coal, *Prof. 3d Internat. Conference Bituminous Coal*, **2,** 761 (1931); *Chimie & industrie*, **30,** 770 (1933). Chapman, Recent Progress in Coal Cleaning Practice in Great Britain, *Prof. 3d Internat. Conference Bituminous Coal*, **2,** 741 (1931). Hanot, Treatment of Slimes Containing Coal and Obtained from Steam Coal of High Ash Content, *Congr. Internat. Metal. Geol. Appl.*, 7th sess., Paris, October, 1935; *Mines*, **2,** 459. Olivieri, Flotation of Coal, Univ. of Ill., published June, 1940. Davis, Coal Flotation Practice of Pittsburgh Coal Co., *Deco Trefoil*, published by Denver Equipment Co., February, 1941. Knapp, Upgrading Coal by the Froth-flotation Process, *Colliery Eng.*, **20,** 49, 64 (1943). Goette Froth-flotation of Coal, *Glückauf*, **77,** 707; *Chem. Abst.* 37-3246. Parton, Flotation of Anthracite Silt, *Mining Congr. J.*, June, 1944.

Diatomite. Norman and Ralston, Purification of Diatomite by Froth Flotation, *Mining Tech.*, May, 1940.

Feldspar. Gerty, Preparation of German Feldspar Ore by Flotation, *Ber. deut. keram. Ges.*, **17,** 526 (1936). Ziergiebel and Gerth, Feldspar Separated from German Rocks by Flotation, *Ber. deut. keram. Ges.*, **15,** 517 (1934). Iverson, Separation of Feldspar from Quartz, *Eng. Mining J.*, **133,** 227 (1932). O'Meara, Norman, and Hammond, Froth Flotation and Agglomerate Tabling of Feldspars, *Bull. Am. Ceramic Soc.*, **18,** 286 (1939). New Kona Feldspar Mine and Mill, *Mining Congr. J.*, October, 1945, p. 94.

Fluorspar. Mitchell, Gross, and Oehler, Froth Flotation of Fluorspar, *Am. Inst. Mining Engrs.*, *Tech. Pub.* 999, 1939. Eigeles, Flotation of Fluorspar of the Solonetchoie District, *Mineral Suire'e*, **10** (10), 42 (1935). Anon., Trends in Treatment of Fluorspar Ores, *Gorno-Obogatitel. Zhur.*, 1936 (2), p. 33. Gerth, Über die Flotation von Flusspat und Quartz, ein Beitrag zur Flotation polarer Nichterze, *Metall Erz*, **35,** 314 (1938). Oxidation Products of Paraffin as Flotation Agents for Fluorite Ores. *Mineral Suire'e*, **9,** 44 (1934). Coghill and Greeman, Flotation of Fluorspar Ores for Acid Spar, *U.S. Bur. Mines, Report Investigations*, 2877, 1928. Sinclair, Fluorspar in South Africa, *Mining Mag.*, **55,** 265 (1936). Clemmer, Duncan, DeVaney, and Guggenheim, Flotation of Southern Illinois Lead-zinc Fluorspar Ore, *U.S. Bur. Mines, Report Investigations*, 3437, 1939. Snook, Record Activity Brings Prosperity to Illinois—Kentucky Fluorspar District, *Mining and Met.*, November, 1940, p. 513.

Flour. Flotation Now Improves Wheat Flour, *Mining Congr. J.*, January, 1941.

Graphite. VonSchoen, Flotation of Graphite, *Metall Erz*, **29,** 181 (1932). Menardi, Modern Flotation Plant for Graphite, *Rock Products*, **31,** 74 (1928). Gandrud, Coe, Benefield, and Skelton, The Flotation of Alabama Graphite Ores, *U.S. Bur. Mines, Report Investigations*, 3225, 1934. Parsons, The Concentration of Flake Graphite Ores, *Can. Dept. Mines, Mines Branch, Mem. Ser.* 26, 1926. Clemmer, Smith, Clemmons, and Stacy, Flotation of Weathered Graphitic Schists for Crucible Flake, *Geol. Survey of Alabama*, *Bull.* 49, 1941.

Gypsum. Keck and Jasberg, A Study of the Flotative Properties of Gypsum, *Am. Inst. Mining Engrs.*, *Tech. Pub.* 762, 1927.

Clemmer and DeVaney, Cationic Reagents in the Flotation of Silica from Gypsum Ores, *U.S. Bur. Mines, Report Investigations* 3553, 1941.

Halite. See also Potash Ores. Kuzin, Separation of Halite from Sylvinite by Flotation, *J. Applied Chem.* (*U.S.S.R.*), **10,** 457 (1937). Guyer and Perren, Separation of Water Soluble Salts by Flotation, *Helv. Chim. Acta.*, **25** (1942); *Chem. Abst.* 37:6097.

Hematite. See also Iron Ores. Keck, Eggleston, and Lowry, A Study of the Flotation Properties of Hematite, *Am. Inst. Mining Engrs., Tech. Pub.* 763, 1937.

Ilmenite. McMurray, Froth Flotation of a North Carolina Ilmenite Ore, *Mining Tech.*, January, 1944.

Iron ores. Keck and Jasberg, A Study of the Flotative Properties of Magnetite, *Am. Inst. Mining Engrs., Tech. Pub.* 801, 1937. Keck, Eggleston, and Lowry, A Study of the Flotative Properties of Hematite, *Am. Inst. Mining Engrs., Tech. Pub.* 763, 1937. Clemmer *et al.*, Beneficiation of Iron Ore by Flotation, *U.S. Bur. Mines, Report Investigations*, 3799, 1945. Falconer, New Reagents and Methods Mark Ore Dressing Advance, *Eng. Mining J.*, February, 1946, p. 104. Brown and Tartatron, Concentration of Oxidized Iron Ores, U.S. Patents 2,364,777 and 2,364,778.

Kyanite. O'Meara and Gandrud, Concentration of Georgia Kyanite, *Am. Inst. Mining Eng., Contrib.* 98, 1936 (see also *Tech. Pub.* 605). Concentration of Kyanite from Death Rapids, B.C., *Can. Dept. Mines, Mines Branch, Bull.* 736, *Rept.* 471, p. 238, 1932.

Limestone. See also Cement. Lee, Flotation of Limestone from Siliceous Gangue, *U.S. Bur. Mines, Report Investigations*, 2744, 1926. Miller and Breerwood, Flotation Processing of Limestone, *Am. Inst. Mining Engrs., Tech. Pub.* 606, 1935. Flotation of Limestone, *Mine and Quarry Eng.*, p. 294, October, 1940.

Lithium. Searles Lake Major Lithium Source, *Eng. Mining J.*, March, 1945, p. 93.

Magnesite. Doerner and Harris, Concentration of Low Grade Magnesite Ores by Flotation, *Wash. State Coll. Mining Exp. Sta., Bull.* P-1, June, 1938. Sinkinson and Michaelson, Flotation of California Magnesites, *Am. Inst. Mining Engrs., Tech. Pub.* 733, 1936. Gerth and Baumgarten, Pretreatment of Magnesite Deposits at Zobten, *Chem.-Ztg.*, **60,** 177 (1936); see also *Tech. Pub.* 723. Clemmer, Doerner, and DeVaney, Experimental Flotation of Washington Magnesite Ores, *Mining Tech.*, March, 1940. Weinig, Magnesite Ore Treatment, U.S. Patents 2,363,029 to 2,363,031, 2,363,104.

Magnetite. A Study of the Flotative Properties of Magnetite, *Am. Inst. Mining Engrs., Tech. Pub.* 801, 1937. Scott *et al.*, Amine Flotation of Gangue from Magnetite Concentrates, *Mining Tech.*, November, 1945.

Manganese. DeVaney and Clemmer, Floating of Carbonate and Oxide Manganese Ores, *Eng. Mining J.*, **128,** 506 (1929). Gaudin and Behrens, A Manganese Recovery Problem and Its Solution, *Eng. Mining J.*, **138,** 40 (1937). Basmanov, Flotation of Manganese Slimes at Chiaturi, A Preliminary Communication, *Gorno-Obogatitel. Zhur.*, 1936 (3), p. 11. Norcross, Cuban Manganese Milling Methods, *Mining J.*, Feb. 29, 1940. Concentration and Nodulizing "Pink" Manganese at Anaconda, *Mining World*, January, 1943.

Mica. Norman and O'Meara, Froth Flotation and Agglomerate Tabling of Mica, *U.S. Bur. Mines, Report Investigations*, 3558, March, 1941.

Nepheline. Nephelite and Vanadium Concentrates from Khibin Apatite-nephelite Ore, *Novosti Tekniki Ser. Gornorudnaya Prom.*, **193** (18), 9. Smirnov and Sheblovnskii, Concentration Apatite-nephelite Ore of the Khiba District, *Inst. Mekhanischeskoi Obrabotko Poleznuikl Meklanubr*, 1930, Part 1, p. 184.

Non-metallics. See Non-sulphides.

Oxide ores. See Non-sulphides.

Pigments. Merz, Pigment Purification, U.S. Patent 2,144,115, Jan. 17, 1939.

Paper. Booth, Clarification of White Water by Froth Flotation, U.S. Patent 2,347,147, 1944. Light, De-inking Waste Paper—Flotation Methods for Removal of Printing Ink, *Chem. Age*, Mar. 9, 1940, p. 111.

Phosphate. Barr, Development and Application of Phosphate Flotation, *Ind. Eng. Chem.*, **26,** 811 (1934). O'Meara and Pamplin, Selective Oiling and Table Concentration of Phosphate Sands in the Land-pebble District of Florida, *U.S. Bur. Mines, Report Investigations*, 3195, 1932. Lawrence and Roca, Flotation of Low-grade Phosphate Ores, *U.S. Bur. Mines, Report Investigations*, 3105, 1931. Lawrence and DeVaney, Flotation of Low-grade Phosphate Ores, *U.S. Bur. Mines, Report Investigations*, 2860, 1928. Belash, Selective Flotation of Phosphate Minerals, *Mineral Suire'e*, **11** (6), 49 (1936); **9** (6), 36 (1934). Pamplin, Ore Dressing Practice with Florida Pebble Phosphate, *Am. Inst. Mining Engrs., Tech. Pub.* 881, January, 1938. Swainson, Washing and Concentrating Florida Pebble Phosphate, *Eng. Mining J.*, October, 1944, p. 469.

Phosphorites. The Flotation of Phosphorites from the Soch Deposits in Baschkir Republic, *Goruni Zuhr.*, **110** (5), 58 (1934).

Potash ores. Coghill, DeVaney, Clemmer, and Cooke, Concentration of the Potash Ores of Carlsbad, N.M., by Ore Dressing Methods, *U.S. Bur. Mines, Report Investigations*, February, 1935. Flotation of Langbeinite from the Potash Field of New Mexico and Texas, *U.S. Bur. Mines, Report Investigations*, 3300, 1936. Kunzin, Flotation of Solykamsk Sylvinite Ores, *Kali* (*U.S.S.R.*), 1937, p. 17. Russian Patent 41,508, Jan. 7, 1934. Preobrajenski, Outline for the Utilization of Residues from the Working Over of Crude Sylvite, *J. Inst. Mech. Working Useful Minerals; Mining Concentrating J.* (*U.S.S.R.*), 1932 (No. 1), p. 29. See Sodium and potassium chlorides. Cole, Concentration of Sylvinite Ores, U.S. Patents 2,355,365 and 2,365,805 (1944). Weinig, Concentration of Sylvinite Ores, U.S. Patent 2,349,393 (1944).

Quartz. Gerth and Baumgarten, Flotation of Quartz, *Metall Erz*, **35,** 314 (1938). Dasher and Ralston, New Methods of Cleaning Glass Sands, *Am. Ceramic Soc. Bull.*, June, 1941.

Resin—rubber. Grace, Klassen, and Watson, Extraction of Resin Rubber Gum from Milkweed in Canada, *Deco Trefoil*, August, 1944.

Seeds. Earle, Process for the Separation of (Plant) Seeds by Froth Flotation, U.S. Patent 2,155,219 (1939).

Selenium. Belash, Flotation of Selenium from Sulphuric Acid Plant Slimes, Redkie *Metal.*, **5** (5), 15 (1936).

Sewage. Hansen and Gotoas, Sewage Treatment by Flotation, *Sewage Works J.*, p. 242, 1943.

Silica. See also Non-sulphides. Patek, Relative Floatability of the Silica Minerals, *Am. Inst. Mining Engrs., Tech. Pub.* 564, 1934; *Trans. Am. Inst. Mining Engrs.*, **112,** 486, 508 (1934).

Silver. Leaver and Wolf, Some Factors Affecting the Flotation of Silver Minerals, *U.S. Bur. Mines, Report Investigations*, 3436, 1937.

Soap flotation. Coghill, Soap Flotation, *Eng. Mining J.*, **135,** 125 (1934).

Sodium fluoride. Eigeles, Separation of Sodium Fluoride from the Melt by Flotation, *Mineral Suire'e*, **9** (2), 39 (1934).

Sodium and potassium chlorides. Kuzin, Separation of Sodium and Potassium Chlorides, Sulphates, and Nitrates by Flotation, *J. Applied Chem.* (*U.S.S.R.*), **9,** 818 (1936) (abs. in French on p. 833). See also Potash ores.

Spodumene. Norman and Gieseke, Beneficiation of Spodumene Rock by Froth Flotation, *Mining Tech.*, March, 1940.

Sulfur. Hazen, Recovering Sulphur, *Eng. Mining J.*, **127,** 830 (1929); Recovery of Sulphur from Surface Deposit, *Mining J.* (*Phoenix, Ariz.*), **13,** 7 (1930).

Strontium. Fine and O'Meara, Bureau Seeks Strontium from Domestic Ore, *Eng. Mining J.*, December, 1944, pp. 93–95.

Talc. Norman, O'Meara, and Baumert, Froth Flotation of Talc Ores from Gouverneur, N.Y., *Bull. Am. Ceramic Soc.*, **18** (8), 202–297 (1939). Carnochan and Rogers, Experimental Tests on Madoc Talc for the Separation of Dolomite, *Can. Dept. Mines, Mines Branch. Bull.* 736, *Rept.* 469, pp. 231–234, 1932. Trauffer, Froth Flotation Economically Recovers Valuable Material from Talc Waste, *Pit and Quarry*, **31,** 28 (1939). Talc Flotation, *Ceramic Age*, **27,** 41 (1936). Franz, Flotation of Talc, *Bull. Amer. Ceramic Soc.*, December, 1944.

Talc-magnesite. Clemmer and Cooke, Flotation of Vermont Talc-magnesite Ores, *U.S. Bur. Mines, Report Investigations*, 3314, 1936.

Tinstone. See Cassiterite.

Tourmaline. Bayula, Concentration of Tourmaline from the Tailings of Kluchevskii Ores, Novosti Tekhniki. *Ser. Gornorudnaya Prom.*, **3** (22), 11 (1935).

Tungsten. Gorodetskii, The Flotation of Scheelite, *Gorno-Obogatitel, Zhur.*, 1937 (4), p. 37. Winkler, Flotation of Tinstone and Wolframite, *Metall Erz*, **32,** 181 (1935). Nevada Massachusetts Co., *Eng. Mining J.*, July, 1945, p. 81. Cumings, Beneficiation of Some British Columbia Tungsten Ores, *Can. Mining Met. Bull.*, February, 1943, p. 47.

Zircon ores. Corbett, Concentrating Zircon Ores, British Patent 416,018, Feb. 12, 1934; British Patent 406,043, Feb. 12, 1934; U.S. Patent 2,082,383, June 1, 1937.

General Principles. Flotation is a process whereby the grains of one or more minerals, or chemical compounds in a pulp or slurry, are selectively caused to rise to the surface in a cell or tank by the action of bubbles of air. The grains are caught in a froth formed on the surface of the tank and are removed with the froth, while the grains that do not rise remain in the slurry and are drawn off the bottom of the cell or tank.

Flotation first came into commercial use in about 1914 for separating non-ferrous sulfide ores. Its use was restricted for many years to the flotation of metal sulfides and such elements as sulfur and graphite. Its use has now spread to the concentration of many non-metallic ores, to the separation of various salts from brines, and to the purification of various industrial waste products. There is a good possibility that any two minerals or chemicals may be separated one from the other by this method, provided that they differ in having either an unlike metal or acid radical. Suitable flotation methods have been worked out for many mineral separations, but relatively little has been done in other branches. In 1942, more than 150,000,000 tons of ores were concentrated by the flotation process in the United States.

Floatability is a property of solids, and some solids are more easily floatable than others. Sulfur, graphite, and sulfides of the metals are easily floatable, whereas oxides, silica, and silicates are not so readily floatable. Floatability is a surface phenomenon. The nature of the film on the outside of the particle is the controlling factor. The selective filming of grains of one mineral in an aggregate by a specific reagent promotes floatability of these grains in preference to the others.

Reagents that will film certain minerals are known as collectors; those which induce a froth are frothers; and those which assist in the selective separation of one solid from another by depressing one, or inhibiting its flotation, are depressers. Reagents that disperse slime coatings on grains, thus favoring filming, are deflocculating agents. Acids and alkalies are added to control the pH. Various inorganic reagents are used for special purposes, particularly for activating or assisting other reagents.

Collectors. The collectors used most commonly in the flotation of the native metals, such as gold or copper, the sulfides, arsenides, tellurides, and sulfosalts, are the xanthates, dio-thio-phosphates, alpha naphthylamine, thiocarbamates, etc. A curious fact is that many rubber accelerators are collectors of sulfide minerals in froth flotation.

Collectors used for non-sulfide minerals are oleic and other fatty acids and soaps, fatty alcohol sulfates, and mineral and coal oils (usually added in an emulsified form); and for the oxides and silicates the collectors are the so-called "cationic" reagents. Kerosene and coal oils are used as collectors in coal flotation.

Cationic Reagents. Since the development of cationic reagents is new, a description is pertinent.

Lenher, in U.S. Patent 2,132,902 of Oct. 11, 1938, recognized that reagents that have the surface-active constituent in the positive ion will flocculate and collect minerals that are not flocculated by the reagents such as oleic acid or soaps, in which the surface active ingredient is the negative ion. In the patent a large number of such chemicals were disclosed, principally quaternary ammonium compounds.

This permitted a classification of the minerals into two groups, positive and negative. Those which are positive presumably carry a positive electric charge on the mineral surfaces, and those which are negative carry a negative charge. The character of the charge is probably relative to the charge of the collecting reagent and is affected by acidity of the pulp and other factors. One of the most important minerals that is negative is the mineral quartz, which in all previous work had been floated, if at all, only with the greatest difficulty. Minerals that are negatively charged may occur together, but one that has a stronger charge can be collected away from the other.

The use of cationic collectors have since been investigated by many agencies. Commercial installations are now in use for floating quartz from barite, quartz from phosphate rock, and sylvite, KCl, from halite, NaCl, and others. Many separations are possible. Dean and Hersberger list many cationic reagents. The two types of cationic collectors that have shown greatest promise are the halogen-substituted quaternary ammonium compounds and the aliphatic amines. Of these the aliphatic amines are in most favor, because they are somewhat cheaper than the quaternary compounds. For silicate minerals containing 5 per cent or more of water such as talc, pyrophyllite, sericite clays, and weathered mica, Norman found that short-chain amines such as di-n-butyl were effective. In floating quartz, an amine corresponding to the lauryl amine is usually found to be the most effective. When amines are used as collectors, they are added as either the acetate or the hydrochloride, since the longer chain amines have a very low solubility. The collective power of some of the cationic collectors is remarkable. It has been demonstrated that, with the proper conditioning agents, 1.0 lb. of lauryl amine hydrochloride will film and float 12,000 lb. of minus 65 mesh quartz. As more is learned regarding the use of cationic collectors it becomes apparent that it is dangerous to generalize too much on reagent behaviour. Each ore usually requires much testing to determine the proper grind, length of carbon chain of the collector, pH of the circuit, conditioning reagents, pulp density, and degree of desliming, if any, required. PH control is very important, and the change from an alkaline to an acid circuit may result in a complete reversal of the separation.

Frothers. Agents used to produce a froth are pine oil, cresylic acid, and the branched-chain alcohols sold by duPont under the designation "B" series, or a mixture of these alcohols with fuel or pine oil sold by American Cyanamid as their "AC series." Pine oil gives a brittle active froth and cresylic acid a somewhat tougher froth. The higher alcohols are preferred for some flotation separations because they have very slight collecting power and thus allow better control.

Depressers. Sodium cyanide is used to prevent the flotation of sphalerite, ZnS, and pyrite during the flotation of galena, PbS, in mixed lead and zinc ores. Lime is added also to depress the pyrite; in copper-lead separation, cyanide and lime depress the copper sulfides. Chromate salts are used to depress galena and to permit copper minerals to be floated away from lead.

Quebracho and tannin are used to prevent the flotation of calcite in the fatty acid flotation of fluorspar. Sodium hexametaphosphate is a strong depressant for hematite in the soap flotation of quartz in a strongly alkaline circuit.

Deflocculating Agents. Caustic soda, soda ash, sodium silicate, sodium hexametaphosphate, and calcium lignin sulfonate serve to disperse slime and to clean mineral surfaces.

Activation and Miscellaneous Reagents. Lime is the most common reagent used in sulfide mineral flotation to raise the pH. Sulfuric acid is used when an acid circuit is desired. The consumption of lime for this purpose exceeds 50,000,000 lb./year. Soda ash is used in the flotation of galena unless pyrite is present, when lime is used. Copper sulfate is added in zinc sulfide flotation as a promoter.

With lead-zinc ores the lead is first floated, the zinc is then activated by the addition of copper sulfate, more collector is added, and the zinc is floated. In non-sulfide flotation a large number of reagents have been used including starches, dextrines, gums, various phosphates, sodium fluoride, and others. Solvents for some water-insoluble reagents are employed as are emulsifying reagents for oils, fatty acids, etc. The alcohol sulfates are suitable for this purpose.

Flotation Machines. A large number of machines of many types have been designed. Some are pneumatic and others entirely mechanical. Still others are mechanical with compressed air introduced through a porous diaphragm in the bottom of the cell or at the base of a rapidly revolving impeller. Two types of machine in common use are illustrated in Figs. 181 and 182. Figure

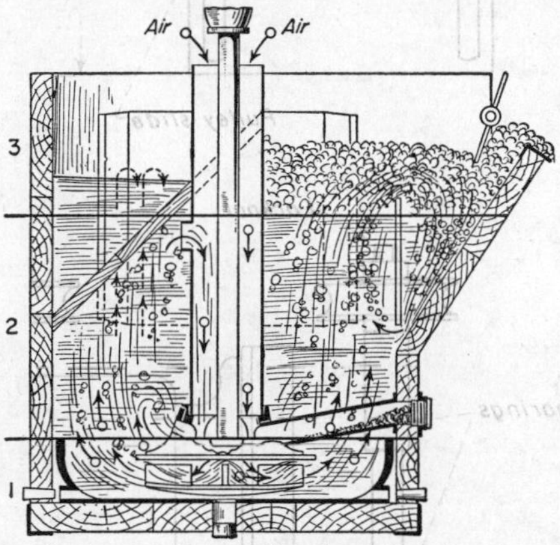

FIG. 181. Section of a Denver Sub-A cell (showing three zones). (*Denver Equipment Co.*)

183 shows a 500-g. laboratory test machine designed by Clemmer of the U.S. Bureau of Mines.

General Operations. Ores must be ground to a point of complete or nearly complete liberation; and, even though this can be accomplished by coarse crushing, grinding to finer than 20 mesh is necessary in all cases and to finer than 48 mesh in most. Grinding is done in closed circuit in ball or rod mills in series with classifiers. Pulp density is important and requires close control. In a few cases warm water is used. The character of the water is important. Usually better results are obtained with purer water. Hard water causes slime flocculation and in flotation with soaps and oleic acids increases greatly the reagent consumption and decreases selectivity. In the flotation of potash salts and other water-soluble chemicals requiring the use of brines, salt-water soaps, and brine-resistant reagents, such as Avirol (sodium octyl sulfate), are used as are also certain of the cationic reagents such as lauryl amine hydrochloride.

In most instances, superior flotation results are obtained by conditioning the reagents with the ore before the flotation step. With heavy collectors these are sometimes added to the grinding circuits to ensure dispersion. For proper selectivity, a definite time contact is sometimes required between reagent and ore, and this is usually secured by agitating the reagent and ore pulp in a "conditioner" consisting of a tank with a vertical impeller.

Flotation machines are built in multiple units, and the flow of the pulp through various units is adjusted for the best results. Common practice is to feed the pulp to several cells known as "roughers," which produce a barren tailing and a low-grade concentrate. The concentrate is treated, sometimes after regrinding, in "cleaner" cells and "recleaner" cells for final concentration. The tailings from the cleaner and recleaner cells are recirculated back through the system or concentrated separately in additional cells. Regrinding of these middlings is necessary in many ores. The concentrate is usually, but not always, collected in the froth from the cells.

In cationic flotation the froth may be the waste material and the underflow the concentrate. A diagrammatic flow sheet of the treatment of a simple ore by flotation is shown in Fig. 184.

Flotation costs vary enormously with the type of feed. In most cases of sulfide mineral flotation the grinding is the principal item of expense. In large mills, treating thousands of tons per day, the total mill cost varies between 40¢ to $1.25/ton. In small mills, treating more limited tonnages, costs are higher.

The actual cost of the flotation operation itself is determined mainly by the type and amount of reagent required. For simple ores the reagent cost may be only 4 or 5¢/ton. On complex ores, particularly on non-sulfide ores where a large amount of material must be floated, the reagent cost may be as much as 50¢/ton. Power required is in the order of from 1 to 6 kw.-hr./ton ore, depending on the rapidity of the float and the number of cleanings necessary.

The cost of a concentrating plant using flotation of a capacity of 500 tons/day is roughly $1200/ton daily capacity. Smaller mills cost more in proportion and larger mills less. The grinding equipment costs half or more of the total in most mills.

Steffensen Flotation Machine. Cells are of unit construction and consist of an inverted pyramidal tank with a rectangular top (Fig. 182). Air is blown from header pipe 1 into cells through the pipe and diffused through openings in annular cylinder 2. Air follows outside pipe 3 and is caught and further dispersed through the opening in open-bottomed cylinder 4. Small-bubble froth is obtained by further diffusion

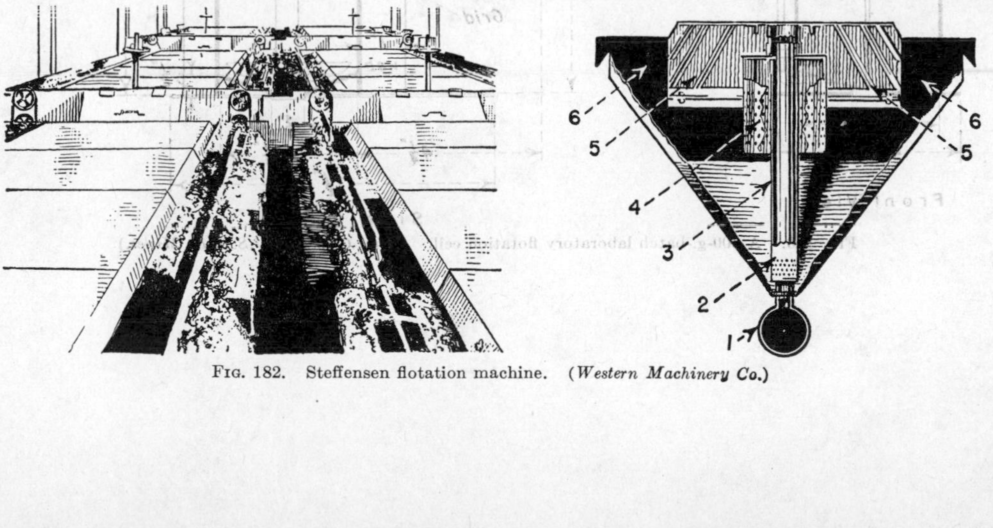

FIG. 182. Steffensen flotation machine. (*Western Machinery Co.*)

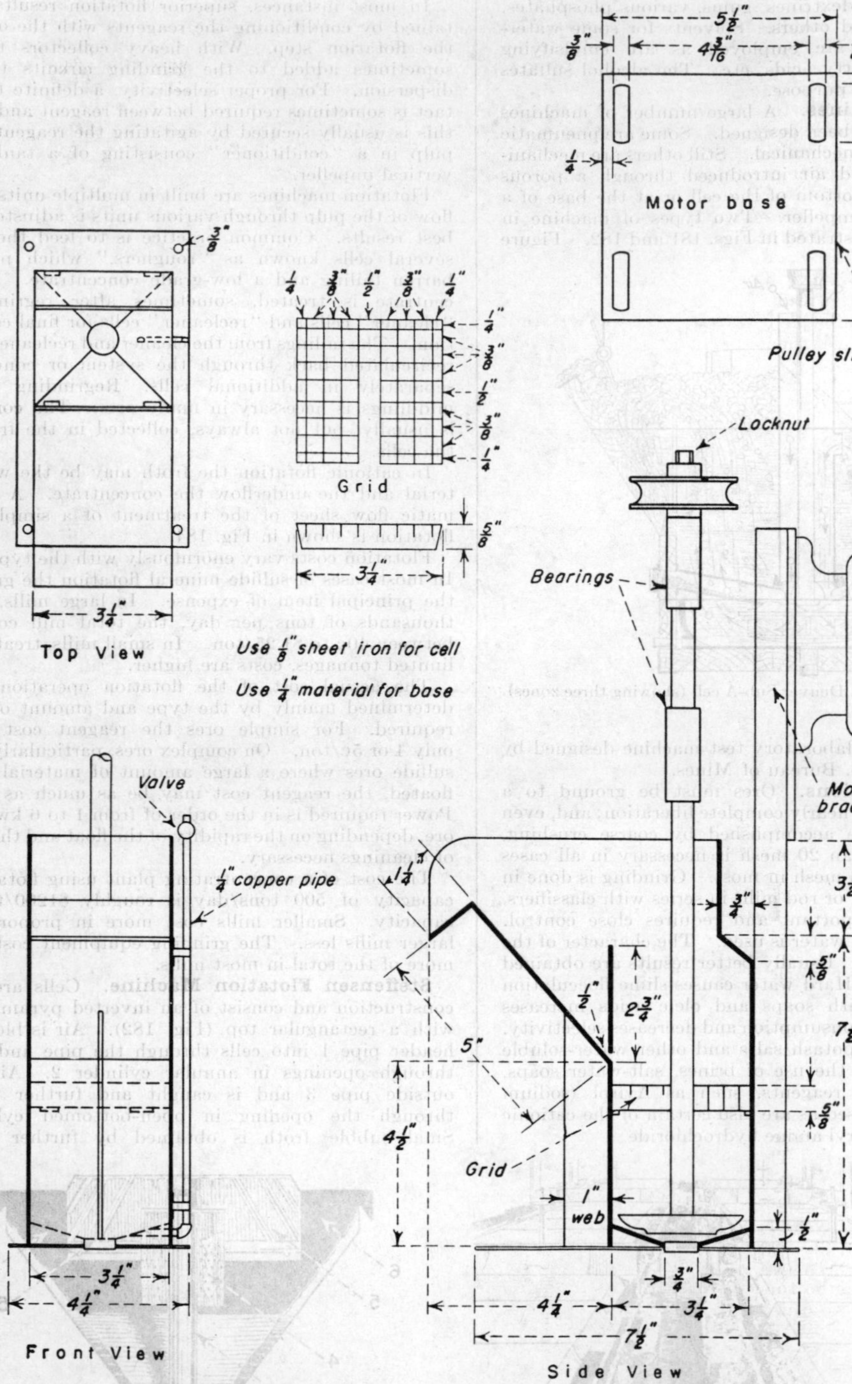

FIG. 183. A 500-g.-batch laboratory flotation cell. (*After Clemmer, U.S. Bur. Mines.*)

through holes in baffles 5. Froth collects at 6 and is removed by rotating paddles. The flow of material between cells is controlled by weirs. The cell tanks may be made of steel, wood, or concrete. Because the admission of the air at the bottom of an inverted cone, there is little tendency for such cells to sand up.

Because it is impossible to generalize as to the type of reagents and operating conditions that will give optimum results for all ores, a rather complete set of references is given at the beginning of this subsection. The references discuss the flotation technique for specific substances.

Denver Sub-A (Fahrenwald) Flotation Machine—Operation of the Denver Sub-A Cell (Fig. 181). 1. *Mixing and Aeration Zone.* The pulp flows into the cell by gravity through the feed pipe, dropping directly on top of the rotating impeller below the stationary hood. As the pulp cascades over the impeller blades, it is thrown

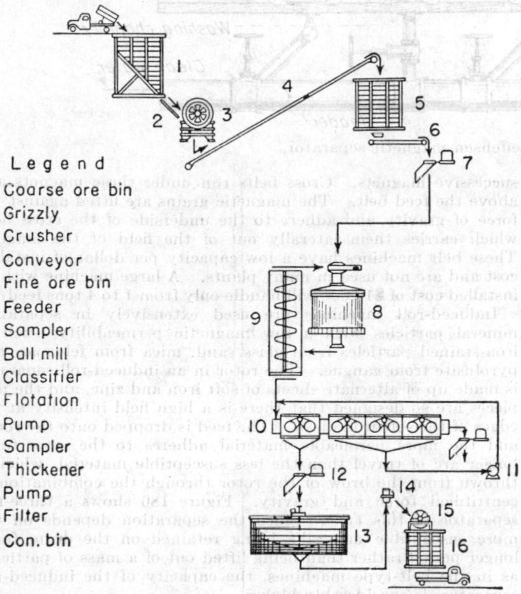

Fig. 184. Flowsheet of a simple flotation mill. (*Galigher Co., Salt Lake City.*)

outward and upward by the centrifugal force of the impeller. The space between the rotating blades of the impeller and the stationary hood permits part of the pulp to cascade over the impeller blades. This creates a positive suction through the ejector principle, drawing large and controlled quantities of air down the standpipe into the heart of the cell. This action thoroughly mixes the pulp and air, producing a live pulp fully aerated with very small air bubbles. These exceedingly small, intimately diffused air bubbles support the largest number of mineral particles.

2. *Separation Zone.* In locating the impeller below the stationary hood at the bottom of the cell, agitating and mixing are confined to this zone. In the central or separation zone the action is quiet and cross currents are eliminated, thus preventing the dropping or knocking of the mineral load from the supporting air bubble, which is very important. In this zone the mineral-laden air bubbles separate from the worthless gangue, and the middling product finds its way back into the agitation zone through the recirculation holes in the top of the stationary hood.

3. *Concentrate Zone.* In the concentrate or top zone, the material being enriched is partly separated by a baffle from the spitz or concentrate discharge side of the machine. The cell action at this point is very quiet, and the mineral-laden concentrate moves forward and is quickly removed by the paddle shaft (note direct path of mineral).

MAGNETIC SEPARATION

By Fred D. DeVaney

References: Dean and Davis, Magnetic Separation of Ores, *U.S. Bur. Mines Bull.* 425, 1941. Cooke, Microscopic Structure and Concentratability of the Important Iron Ores of the United States, *U.S. Bur. Mines Bull.* 391, 1936. Davis, Magnetic Roasting of Iron Ore, *Univ. Minn. Bull.* 13, Vol. 42, 1937. Magnetic Concentration of Ores, Dean and Davis, *Trans. Am. Inst. Mining Met. Engrs.*, vol. 112, 1934. Frantz and Jarman, Magnetic Beneficiation of Non-metallics, *Trans. Am. Inst. Mining Met. Engrs.*, vol. 102, 1932. Hatfield, The Action of Alternating and Moving Magnetic Fields upon Particles of Magnetic Substances, *Proc. Phys. Soc. (London)*, vol. 46, 1934. Schilling and Johnson, Separation of Hematite by Hysteretic Repulsion, *Am. Inst. Mining Met. Engrs. Tech. Pub.* 654, 1936. Taggart, "Handbook of Ore Dressing," Wiley, New York, 1945. Gaudin and Spedden, Magnetic Separation of Sulfide Minerals, *Mining Tech.*, January, 1943.

All mineral and metallic substances are permeable to some extent when placed in a magnetic field. Those substances which are attracted by a magnetic field are called paramagnetic, and the relatively few that are repulsed are classified as diamagnetic. The paramagnetic minerals usually are classified for reasons of practical consideration into three classes, namely, strongly magnetic or ferromagnetic (magnetite, ilmenite, etc.), weakly magnetic (hematite, pyrolusite, etc.), and non-magnetic (quartz, calcite, etc.).

The art of separating one substance from another substance by means of a magnetic field is called magnetic separation. The art and the machines developed for making such separations are highly specialized. The principal uses to which magnetic separation are being put and some of the types of machine being used are shown in outline form below:

I. For removal of ferromagnetic or other strongly magnetic materials.

A. When the amount of strongly magnetic material forms only a small part of the feed.

1. *Dry Methods.* Magnetic pulleys, or suspended poles over moving belts or inclined chutes are used to remove "tramp iron," magnetite, or other magnetic materials present in the product or purposely or unavoidably introduced. Permanent or electromagnets are placed below chutes in some installations, as in flour mills to cause stray bits of iron fragments to remain on the chutes, which are cleaned periodically. Some agricultural products are cleaned by mixing them with iron filings. Seeds with rough surfaces to which the filings adhere are removed from others with smooth surfaces.

2. *Wet Methods.* The removal of small amounts of iron-bearing impurities from slurries of ceramic products, pigments, dyes, etc., is one of the most important applications of magnetic methods to the chemical industry. Several machines are on the market. One made by the S. G. Frantz Co. of New York is called a *Ferro-Filter.* The Dings Magnetic Separator Co. of Milwaukee makes one known as the *De-ironer.* These machines have an electromagnet. Screens through which the slurry flows are magnetized, and the iron particles adhere to the screens. To avoid interrupting the flow of the slurry when the screens are cleaned, the installations are made in duplicate or triplicate. The slurry feeds for a given interval through one unit and is then switched automatically to the second unit. The current is interrupted from the first, and the screens are cleaned by flushing water through them which is run to waste. The Magnetic Products Co. of Trenton, N.J., makes a unit called a *Separmag.* The "screen" is a box of tacks, or stainless-steel attractors, which are magnetized by an Alnico permanent magnet.

B. When the amount of strongly magnetic material is a relatively large proportion of the feed.

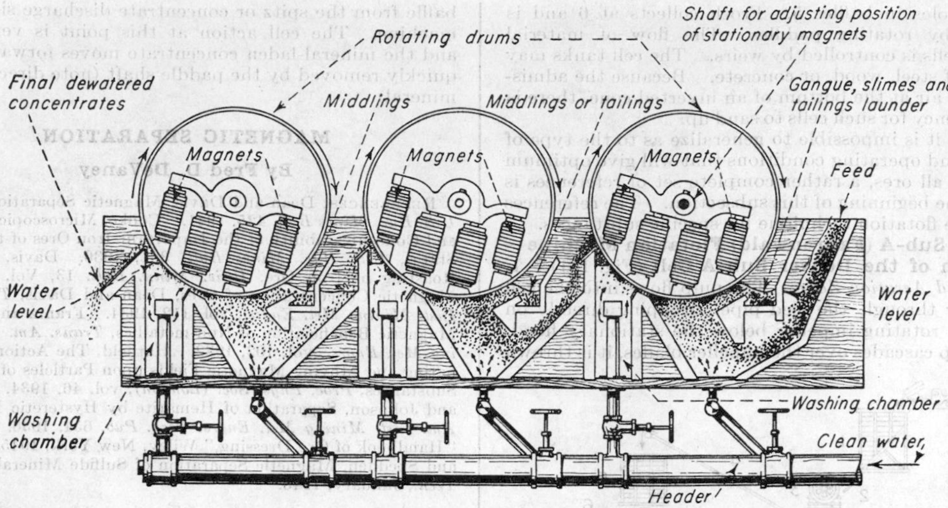

FIG. 185. Three-drum Jeffrey-Steffensen magnetic separator.

1. *Dry Methods.* Belt machines and rotary separators were formerly widely used for the dry separation of magnetite, Fe_3O_4, from gangue; but, because of their dustiness and the tendency for non-magnetic gangue to become physically entrained between strongly polarized particles, their use has been superseded in many instances by wet separators.

2. *Wet Methods.* Suspended-belt machines such as the *Crockett* and *Linney* are widely used at mills concentrating magnetite iron ores. Such separators are also used in cleaning the ferrosilicon or magnetite suspension medium used in sink-and-float separating plants. In the *Crockett* machine (made by the Dings Magnetic Separator Co. of Milwaukee), the material is fed as a pulp into one end of the machine and comes into contact with the lower surface of a belt suspended between two pulleys. This belt is partly submerged in water and passes under a number of strong enclosed electromagnets. The magnetic portion of the treated material is attracted to the belt surface because of the magnetic field exerted by the overlying magnets. The non-magnetic material is not attracted and drops out below the surface of the water and is withdrawn from the first compartment of the tank. The magnetic particles adhere to the surface of the belt and are further concentrated by passage through the water and under magnets of different polarity, after which they are carried above the water level and are discharged at a point just out of the magnetic field. The capacity of such separators on magnetic iron ore is high. In roughing service for treating material in the size range of 8 mesh, the capacity of the Crockett separator is approximately 15 tons feed/hr./ft. magnet width. For treating material 100 mesh and finer, the capacity is about 5 tons feed/hr./ft. magnet width.

Another type of modern separator used in concentrating strongly magnetic materials is the *Jeffrey-Steffensen.* This machine, which is shown in Fig. 185, is of the rotary type and is particularly well adapted to the treatment of fine materials. The mineral pulp is fed into the feed trough which carries it to the bottom on the separator where a rising current of water suspends and carries the solids upward toward the rotating drum, which is partly submerged. The concentrates are cleaned by the winnowing action caused by the alternate polarity of the pole pieces and by the countercurrent washing of water admitted through openings. The machine shown in Fig. 185 is the type used for making a finished concentrate from fine material. For roughing purposes, single- or double-drum units are employed.

II. For removal of moderately magnetic or weakly magnetic materials.

A. Dry Methods. Two types of machine are available. A belt-type machine, such as the *Wetherill,* is adapted only to the separation of minerals that are moderately magnetic such as ilmenite, franklinite, and pyrrhotite. Some operators prefer the belt machines for these separations, particularly in cases where the desired mineral is associated with another mineral that is itself magnetic. They believe the selectivity is greater.

The material is fed into the machine on a belt and under successive magnets. Cross belts run under these magnets and above the feed belt. The magnetic grains are lifted against the force of gravity and adhere to the underside of the cross belt, which carries them laterally out of the field of the magnet. These belt machines have a low capacity per dollar of installed cost and are not used in many plants. A large machine with an installed cost of $10,000 will handle only from 1 to 4 tons feed/hr.

Induced-roll machines are used extensively in separating mineral particles with a low magnetic permeability, such as iron-stained particles from glass sand, mica from feldspar, and pyrolusite from gangue. The rotor in an induced-roll separator is made up of alternate sheets of soft iron and zinc, and the pole pieces are so designed that there is a high field intensity at the edges of the soft iron disks. The feed is dropped onto the rotor, and the more permeable material adheres to the rotor for a longer arc of travel than the less susceptible material, which is thrown from the brow of the rotor through the combination of centrifugal force and gravity. Figure 186 shows a three-roll separator of this type. Since the separation depends on the more susceptible mineral's being retained on the drum for a longer period rather than being lifted out of a mass of particles, as in the belt-type machines, the capacity of the induced-roll separators is considerably higher.

Induced-roll separators are made ordinarily with a number of rolls, and the fields for the individual rolls are designed to give a field that will separate the desired minerals. Several separations between minerals of different magnetic susceptibilities may be made on the same machine. The capacity of these machines depends on the length of the rotor, the size of the feed, and the degree of selectivity required. The energizing current for a machine making four separations at the rate of 2 tons/hr. is 100 watt-hr./ton. The requirement for a machine making five separations with a capacity of 3 tons/hr. may require more than 300 watt-hr. Most machines are made with a double tier of rotors, the magnets being activated by the same coils. Such machines are made by the Dings Magnetic Separator Co. and by the Stearns Co., both of Milwaukee, and by the Exolon Co. of Tonawanda, N.Y.

All dry high-intensity magnetic separators have the disadvantage that the feed must be thoroughly dry and should preferably be sized for best efficiency. They will not treat material coarser than 8 mesh and finer than 200 mesh.

B. Wet Methods. No continuous wet high-intensity magnetic separators are in commercial use in the United States. Surface-tension effects are so much greater than the magnetic susceptibility of the weakly magnetic minerals as to make an effective design difficult. Several types, such as the *Graniggi* and the magnetic ring-and-drum machine, have been used in Germany and Sweden (see Dean, Magnetic Separation of Ores, *U.S. Bur. Mines, Bull.* 425, 1941).

III. For extremely weak paramagnetic or for "diamagnetic materials." Only one machine has been built for separating exceedingly weakly magnetic materials from grains that are diamagnetic. This is the *Franz*, isodynamic magnetic separator, made by the S. G. Frantz Co. of New York; the present models of this separator have capacities of only 50 to perhaps 200 lb./hr. Nevertheless, the separations of supposedly "non-magnetic" materials that can be made on this machine are rather remarkable. The machine is essentially a laboratory unit and has not yet been brought to commercial development.

IV. Special magnetic machines and processes. Hundreds of special machines have been built, among which those operating with an alternating current are perhaps the most interesting. For a description of this type of separator, see Dean (*op. cit.*).

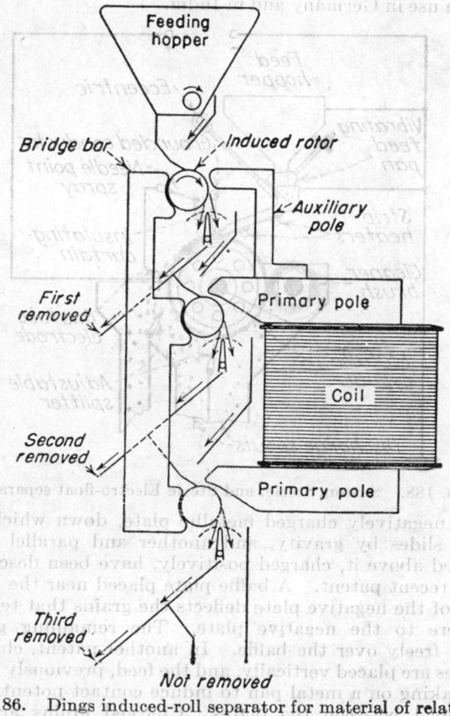

FIG. 186. Dings induced-roll separator for material of relatively low magnetic susceptibility.

In general, it may be said that wet methods are applicable to material of finer grain size than are dry methods, but then only for the separation of strongly magnetic particles. Dry feeds must be granular, free-flowing, and dust-free.

Magnetic Roasting. Many minerals of low magnetic susceptibility may be roasted and made strongly magnetic. Hematite or limonite, when roasted in a reducing atmosphere for about 30 min. at 500° to 600°C., is converted to the strongly magnetic mineral magnetite. Siderite, $FeCO_3$, may likewise be converted to magnetite by roasting in an inert or mildly oxidizing atmosphere at 700° to 800°C. Pyrite, FeS_2, may be converted into the strongly magnetic mineral pyrrhotite, $Fe_{11}S_{12}$, by a controlled oxidizing roast.

ELECTROSTATIC METHODS OF CONCENTRATION

By J. L. Gillson

REFERENCES: Johnson, Electrostatic Separation, *Am. Inst. Mining Engrs., Tech. Paper* 877, February, 1938; *Eng. Mining J.*, **138**, September, 1938, pp. 37, 51; October, 1938, pp. 42, 43, 52; December, 1938, p. 41. U.S. Patents 1,020,063, Mar. 12, 1912, Process of Electrostatic Separation; 1,017,701, Feb. 20, 1912, Electrostatic Separator, both issued to Sutton, Steele, and Steele. Fraas and Ralston, Discussion of Electrostatic Separation at Meeting of Am. Inst. Mining Eng., *Trans. Am. Inst. Mining Eng.*, **134**, 419 (1939). Johnson, Recovery of Materials by Electrostatic Separation, *Chem. & Met. Eng.*, **50**, 130 (November, 1943). Fraas and Rolston, Electrostatic Separator for Fine Powders, *U.S. Bur. Mines Report Investigations*, 3677, December, 1942. Fraas, The Conductance Electrostatic Separator, *Mining Tech.*, *Tech. Pub.* 1511, September, 1942.

Principle.* The principle of electrostatic separation is based on the fact that, if one or more of the materials in a granular mixture can receive a surface charge on or just before entering an electrostatic field, the grains of that material will be repelled from the active electrode or attracted toward it, depending upon the sign of the charge. By causing such grains to fall into separate chutes from other grains not so affected, a separation or concentration results.

Actually, the word electrostatic may be a misnomer. As an example, a highly charged electrode is surrounded by a corona that glows in the dark with a faint blue light. This corona marks the zone of rapid leakage in the electrostatic field; hence the charge is not entirely static.

The ability of the grains to receive and hold a charge varies with the material. Some materials can take a charge from friction, as in the case of the ebony or hard-rubber rod used in elementary courses in physics to demonstrate static electricity. Also familiar is the spark one obtains when touching a metal object, such as a doorknob, after having slid one's feet over a carpet. The chain trailing on the ground behind an oil tank truck is placed there to discharge static electricity produced by friction. The charge resulting from friction is called "contact potential."

In addition to the charge resulting from friction, non-conductors can be given a surface electrical charge if they are passed through a strong electrostatic field. Even grains carrying a surface frictional charge may take on a charge of opposite sign if passed through an electrostatic field of sufficient intensity.

If electrical conductors have received a charge, it is immediately dissipated when they come in contact with the electrode of opposite sign.

The separation of materials of widely differing conductivity is fairly simple, whereas the separation of conductors from other conductors or of non-conductors from other non-conductors is more difficult.

Types of Equipment. *Rotor Type.* The most popular form of machine used to separate materials by electrostatic methods consists of a feeding rotor, which is itself an electrode and is called the separating electrode, usually grounded. This rotor discharges the feed as a curtain in front of one or more charged electrodes. The electrostatic field lies mainly between the two electrodes.

The design of the rotor type of machine is varied dependent upon the type of separation to be performed. The separation of conductors from non-conductors and of non-conductors from each other takes advantage in each case of a separate electrostatic effect. Figure 187 illustrates the differences between the two effects.

In Fig. 187*A*, material is fed onto a rotor, which is a grounded electrode. With the rotation of the rotor, the grains enter a field of a corona discharge caused by an active electrode, which is either a comb (needlepoint) electrode, or a fine wire. Either of these types cannot hold the charge given them, and electricity "leaks" to the grounded electrode as movement of charged ions of gas.

The bombardment of these ions "pins" the grain to the grounded electrode; but, with the rotation of the rotor, as soon as the grains leave the electrostatic field, the "good"

* The author obtained much information and many suggestions from Dr. Oliver C. Ralston, of the U.S. Bureau of Mines, and J. Hall Carpenter, of Jacksonville, Fla.

conductors lose their charges and fall away from the rotor along the path of normal trajectory induced by the rotation of the rotor. Poor conductors and non-conductors adhere to the electrode until they are brushed off by a brush in the "8 o'clock" position on the rotor. Thus conductors are separated readily from poor conductors and non-conductors.

In Fig. 187*B*, material that has been passed over a vibrating feeder to induce a frictional charge is spread on

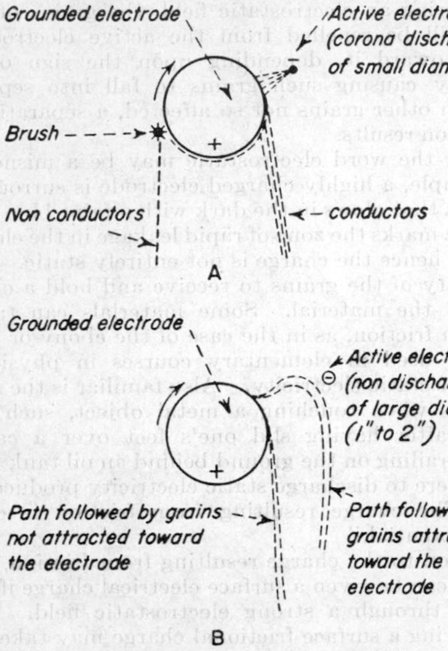

FIG. 187. Principle of rotor-type electrostatic separator.

the rotor, which is a grounded electrode. With rotation of the rotor, the material enters an electrostatic field surrounding an active but non-discharging electrode. This is a large-diameter rod, or a glass tube that can carry the voltage impressed on it without leakage.

Since there is no bombardment of ions, there is no pinning of any of the grains to the rotor; rather there is an attraction of those grains which have a contact potential of opposite electric sign toward the active electrode. They are therefore pulled away from the path of normal trajectory and fall into a separate chute.

In one type of machine (Fig. 188), an attempt to take advantage of both effects has been made by putting two types of active electrodes on it, both the comb type (discharging) and a glass tube (non-discharging).

Obviously, reversing the polarity may reverse the position that a grain takes in the electrostatic field. Actually, most non-conductors are reversible to a greater or less extent.

Grains must not adhere to the charged electrode, since this would distort and eventually insulate the field. Thus the active non-discharging electrodes, if made of metal, rotate with a brush behind the electrode to clean them or, as in one machine, consist of a glass tube (which is an insulator) in which is a gas conducting the current, as in a neon-lighting tube.

The forces acting against the attraction or repulsion are the force of gravity and the tangential velocity given to the grains by the rotation of the rotor. It is obvious that the relative mass of the individual grains must be within fairly close limits; moderately close sizing of the feeds to electrostatic machines is therefore necessary. Furthermore, material so fine as to float in air or to be blown about by air currents or material that becomes compacted cannot be treated, *i.e.*, the feed must be free-flowing. The size and speed of rotation of the rotor is important, since these factors control the length of time the grains are in the electrostatic field and the trajectory of the grains from the rotor.

Toboggan Type. A toboggan form of conductance separator was favored in the early days. It is merely a different mechanical way of getting particles into and out of electrostatic fields. It is built of sheet metal in plane sheets and since there are no moving parts, bearing trouble is avoided. Highly developed forms of toboggans are in use in Germany and in India.

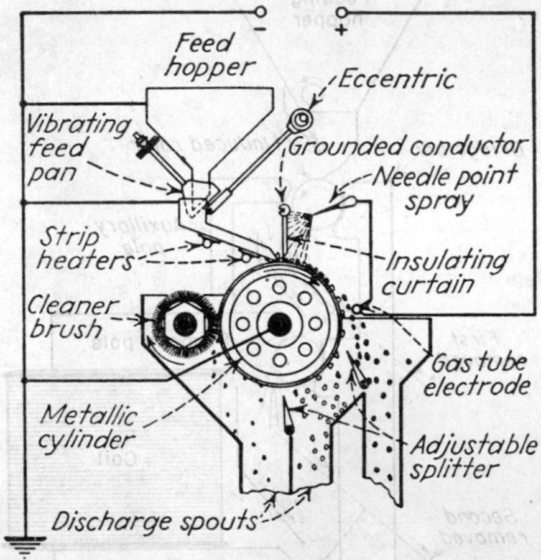

FIG. 188. Sutton, Steele, and Steele Electro-float separator.

A negatively charged metallic plate, down which the feed slides by gravity, and another and parallel plate placed above it, charged positively, have been described in a recent patent. A baffle plate placed near the lower end of the negative plate deflects the grains that tend to adhere to the negative plate. The remaining grains slide freely over the baffle. In another patent, charged plates are placed vertically, and the feed, previously given a shaking on a metal pan to induce contact potential, is dropped between the plates. Charged grains are deflected from their free-falling path.

Table Type. An ingenious new design has been developed by the Foote Mineral Company of Philadelphia. A shaking table identical with those used in wet gravity concentration is covered with a metal deck, which is grounded. This platform provides a means for advancing a thinly spread layer of feed in a uniform way over approximately 75 sq. ft. of surface area. Two to 4 in. above the table is a charged grid, consisting of a closely spaced series of metal troughs. These troughs shake at right angles to the shaking direction of the table below them. The grains in the feed which have received a surface frictional charge are repelled from the grounded plate while in the electrostatic field induced by the charged grid, and they bounce freely into the air. Many are caught in a trough as they fall, and the others are caught in another trough, on the second, third, or fiftieth bounce. The lateral shake of the multiple troughs empties them off the side of the table.

The energy which causes hundreds of pounds of mineral per hour to bounce and rebounce, six or more inches into the air, is supplied by the work done in shaking the grains on the metal surface, thus inducing the surface frictional charge.

Commercial Machines. Four commercial rotor-type machines now on the United States market differ in detail. The machine of the Ritter Products Co. of

Rochester, N.Y., resembles the original Huff machine devised in 1902 and consists of a series of rotors superimposed in a tier, each with a rotating active electrode of non-discharging type. Material not rejected by the first electrode passes to the second rotor and is again exposed to a field with the same or reversed polarity. The Sutton, Steele, and Steele machine has two electrodes (Fig. 188), one a needlepoint corona discharge electrode that is designed to convect a strong charge onto the mineral grains and a second gas-filled glass-tube non-discharging electrode having the same polarity as the needlepoint electrode that forms a pure electrostatic field. The third machine, marketed by the Feldspathic Research Co., preconditions the feed by passing it quickly through a chamber in which it is exposed to a warm atmosphere of HF of about 1 per cent strength. The electrostatic separator itself is similar to that of the old Huff machine. Presumably the HF cleans the mineral surfaces of inhibiting films, and it may give the grain surfaces a positive or negative charge. Tests have proved that the subsequent attraction or repulsion of the grains so treated is more marked. The fourth machine, recently on the market, is made by the Carpco Engineering Company of Jacksonville, Florida. This machine is made in units of four rotors lying parallel in a horizontal plane, and the rotors work simultaneously on a common feed. Each rotor is 60 in. long and 6 in. in diameter, and the four handle about 10 tons per hour of feed. The rotor was a "beam"-type electrode developed by the inventor, J. Hall Carpenter.

Effect of Humidity. To control humidity conditions, which are important in the separation of non-conductors, the machines are operated in an air-conditioned atmosphere. Tests have proved that a relative humidity below 35 per cent is necessary for the effective separation of some non-conductors. In most cases, the air conditioning is obtained by heating the feed to a temperatures.

Power Generation. High voltages are developed by a.c. generator sets with mechanical rectification with a cross-arm rectifier or by vacuum tubes. A simple generator is the McCutcheon (made by the Communications Measurement Laboratory of New York) in which a transformer steps up a primary plant circuit and rectifies it by means of four R.C.A. No. 878 tubes. Rectox Bridge circuits are in use also. The Carpco Engineering Company builds its own rectifying unit employing electronic rectification.

Voltages used vary from 10,000 to 100,000, but the amperage is very low, from $\frac{1}{20}$ to 2.5 milliamp. High voltages present physical hazards, and an electrostatic machine should be limited to low current value so that there is no danger to the operators. The wave form of the rectified current is important. A sharply peaked wave is desired but half wave can be used. In vacuum-tube rectification, the primary circuit should be 50- or 60-cycle.

Application of Electrostatic Methods. The principal application of electrostatic methods so far is on agricultural products, but a number of ores are being treated by the method. The separation of zircon, rutile, and ilmenite is difficult by other means because of similar physical properties of the minerals, and two operations are being conducted abroad, two in Florida, one in New Jersey, one in New York, and one in Pennsylvania, using electrostatic equipment for concentrating these minerals. The separation of feldspar and quartz for the ceramic industry is being done electrostatically by one of the domestic feldspar companies; the advantage of the process over froth flotation is that it avoids subsequent drying. A large plant in Florida has cleaned rock-phosphate concentrates electrostatically, and it has been used with success on flake graphite. In Germany, coal has been cleaned by electrostatic methods, and the Ritter machine has been used to size coal. In general, it can be said that electrostatic methods are applicable as an adjunct to other dry methods of concentration where it is desirable or necessary to keep feeds dry throughout the concentration process and in a few special cases where other methods are not successful.

Electrostatic machines are effective on ores or other granular aggregates in the range from about 10 to 200 mesh (but feeds must be dust-free); and for clean separation the grains must be free, *i.e.*, individual particles of a single mineral or substance. The inability to handle material finer than about 200 mesh restricts the application of the method. Most ores must be ground to 20 to 60 mesh to cause nearly complete liberation of the particles. By that grinding, from 20 to 40 per cent of the feed is unnecessarily, but inevitably, ground finer than 200 mesh. Fine particles fly about in the air. Some tests have been made on equipment to conduct the separation of fines in reduced atmospheres, but this increases the tendency for arcing across from electrode to ground, which would inhibit separation.

Capacity of Electrostatic Machines. Electrostatic machines have large capacities varying from 100 to 500 lb./hr./ft. rotor length. The horsepower required is low, running from 1 to 2 hp./machine, depending upon the number of rotors required. The floor space is small, but the working height of an 8- to 12-rotor machine will run from 12 to 20 ft. With a multiple-rotor machine, the upper rotors are used for roughing, the intermediate rotors for scavenging, and the lower rotors for cleaning. The passage of material from rotor to rotor must therefore be adaptable, permitting the discharge from either the front or the back of the machine to reach the next or a succeeding rotor. A single operator can handle 3 to 30 machines, depending upon whether feeds are continuous or whether units are installed to handle successively batches of different products accumulating in surge bins.

Patent Art. A useful source of information on electrostatic separation is the patent art. Recent U.S. patents are as follows:

2,392,044	2,388,731	2,382,122	2,361,946
2,357,658	2,314,940	2,314,939	2,306,105
2,305,872	2,258,767	2,246,253	2,245,200
2,235,304	2,225,096	2,216,254	2,213,510
2,198,972	2,197,865	2,197,864	2,187,637
2,180,804	2,168,681	2,154,682	2,135,716
2,129,161	2,127,664	2,123,301	2,090,418
2,090,418	2,071,460		

SAMPLING

BY J. W. STILLMAN

REFERENCES: Jones, Sampling of Materials: I. General Principles Involved, *Metal Ind.* (*London*), **32,** 585 (1928); II. Gases and Liquids, *ibid.*, 609; III. Sampling of Water, Oil, Slimes, and Powdered Substances, *ibid.*, **33,** 3 (1928); IV. Sampling of Coal, *ibid.*, 27; V. Sampling of Ores and Concentrates, *ibid.*, 125; VI. Sampling of Ores, Metals and Alloys, *ibid.*, 199. Baule and Benedetti-Pichler, Sampling of Granular Materials, *Z. anal. chem.*, **74,** 442 (1928). Grumell, A Decade of Sampling,

Am. Inst. Mining Met. Engrs., Tech. Pub. 1044, 1939. Munch and Bidwell, What Constitutes an Adequate Sample? *J.A.O.A.C.*, **11**, 220 (1928). Standard Methods for the Sampling and Analyzing of Aluminum and Certain Aluminum Alloys, Aluminum Research Inst., 1939. Calkins *et al.*, Testing of Coal and Coke, *Intern. Cong. Testing Materials* 1927, II, 641. Grumell, Report on the Sampling of Coal with Special Reference to the Size-weight-ratio Theory. Crawford and Reed, Notes on Sampling and Analysis for Ash Content, *Brit. Standards Inst., London*, 1938. Standard Methods of Laboratory Sampling and Analysis of Coal and Coke, A.S.T.M., D 271–48. Bushell, The Sampling of Coal, *J. Chem. Met. Mining Soc. S. Africa*, **37**, 361, 494, 499, 566 (1937). "Methods of the Chemists of the U.S. Steel Corp. for the Sampling and Analysis of Coal, Coke and By-products," 3d ed., 1930. Fieldner and Selvig, Review of Standard Methods Used in Various Countries for Sampling and Analysis of Solid Fuels, *Fuel*, **17**, 266 (1938); Notes on the Sampling and Analysis of Coal, *U.S. Bur. Mines, Tech. Paper* 586, 1938. Sampling of Coal Tar and Its Products, *Brit. Standards Inst.*, 1938. Matthews *et al.*, Porous Solid Filters for Sampling Industrial Dusts, *Bull. Inst. Mining Met.*, No. 386, 1936. Briscoe *et al.*, The Sampling of Industrial Dusts by Means of the "Labyrinth," *Bull. Inst. Mining Met.*, No. 393, 1937. Littlefield and Shrenk, Dust Sampling with the Bureau of Mines Midget Impinger, Using a New Hand-operated Pump, *U.S. Bur. Mines, Report Investigations* 3387, 1938. Richardson *et al.*, Standard Methods for the Sampling and Analysis of Commercial Fats and Oils, *Ind. Eng. Chem.*, **18**, 1346 (1926). Thatcher, Sampling of Boiler Flue Gases, *Power*, **64**, 774 (1926). "Methods of the Chemists of the U.S. Steel Corp. for the Sampling and Analysis of Gases," 3d ed., 1927. "Sampling and Analysis of Carbon and Alloy Steels—Chemists Committee of the Subsidiaries of the U.S. Steel Corp.," Reinhold, 1939. "Recommended Methods for Sampling and Analysis of Cast Ferrous Metals and Alloys," 2d ed., Brit. Cast Iron Research Assoc., 1933. "Methods of the Chemists of Subsidiary Companies of the U.S. Steel Corp. for the Sampling and Analysis of Pig Iron," 3d ed., 1934. Zwicker, Note on the Significance of Sampling for the Chemical Analysis of Metal Alloys, *Chem.-Ztg.*, **53**, 546 (1929). Wittka, Sampling and Measuring the Amount of Oil from Tank Ships, *Allgem. Oel-Fett-Ztg.*, **33**, 470 (1936). Smither *et al.*, Standard Methods for the Sampling and Analysis of Commercial Soap and Soap Products, *Ind. Eng. Chem.*, anal. ed., **9**, 2 (1937). Fuller *et al.*, Sampling and Testing Lacquer Solvents and Diluents, *Proc. Am. Soc. Testing Materials*, **27**, Part 1, 870 (1927).

Sampling is the process of obtaining a small amount of material which shall be as nearly representative as possible of the whole mass of material that is being considered. This process usually is made up of several separate steps: (1) the collection of a comparatively large amount of material which has been selected in a systematic manner from different parts of the mass; (2) if the material is composed of solid particles, the crushing and grinding of the portion collected as described under (1) in order to reduce the size of aggregates and to provide a certain amount of mixing; and (3) the separation of this comparatively large portion into two parts by subdivision, such as by quartering, so that one part provides sufficient material for the required analysis or test and at the same time has the same average composition as the large portion before subdivision.

Since the final sample is in most cases to be used for test purposes the results of which will determine the use to which the entire mass may be put, it is obvious that all precautions which aid in making this sample representative of the original material are thoroughly justified. An analysis or test, however efficiently it is carried out, will be rendered valueless if the sample has been improperly taken or prepared. Methods of sampling have been devised with due consideration for the laws of probability and of averages. They should be applied by someone who understands the scientific aspects of sampling and who comprehends the objectives of the analysis or test to be made on the sample. Any directions that are given here are intended to supplement the experience of the sampler and guide him in selecting methods that are applicable.

In taking the gross sample, careful consideration should be given to the present condition of the material. Such questions as the following should be answered and will decide the number and location of portions to be taken to provide the gross sample:

Is the surface layer the same as the mass underneath, or has it been changed by exposure to the weather or other external conditions? Has there been segregation of coarse and fine particles or of materials of different specific gravities? If the material has been transported has there been segregation? When the material is a mixture of liquids, or of liquids and solids, there is a tendency to segregation.

These questions call attention to some of the difficulties that have to be overcome in sampling heterogeneous materials. Only homogeneous materials, of which very few are met in practice, can be sampled at random and a representative portion obtained.

Once the gross sample has been collected, the amount of material actually required for the analysis determines the extent to which grinding, mixing, and subdivision shall be carried. Just as much care is required in these operations as for assembling the gross sample.

Where materials are bought and sold on specification, both buyer and seller are interested in obtaining an estimate of the material with respect to its specified properties. It is the customary practice to have the sampling done under conditions set by agreement. At one time, samples are taken for the buyer and seller and one or two extra portions are retained for referee samples in case of disagreement between the buyer and seller. It simplifies matters in such cases to take one gross sample and then prepare the necessary small portions at the same time and under identical conditions.

As has already been mentioned, it is easy to sample only in the case of homogeneous materials such as gases, true solutions, and finely divided solids. The difficult problems of sampling are encountered with solid materials which are practically always heterogeneous in nature, and hence this general discussion is concerned mainly with solids. Liquids and gases will be taken up in special paragraphs below. If a material can be rendered homogeneous by thorough mixing, then any part of it can be taken as a representative sample. Usually, however, with solid particles the attainment of homogeneity is difficult if not impossible, and, for that reason, in practice the different methods of sampling described below are necessary.

During the transportation of material in railroad cars, trucks, and the like, the finer materials tend to settle to the bottom leaving the larger particles on top. Certain materials are subject to oxidation when exposed to the air. The resulting oxide will, of course, be greater in amount on the surface of the mass of material. At the same time, if this oxide is a fine powder which is easily removed, it will penetrate to a greater or less extent into the pile of material as the result of erosion. These examples are typical of many heterogeneous materials which are encountered in practice, and the methods of sampling must be so designed that as far as possible the relative proportions of coarse and fine, of metal and oxide, etc., will be the same in the gross sample and in the mass of material. Once the gross sample is collected, it can be crushed, ground, shredded, etc., to provide a more homogeneous mass for subdivision to the final sample. The greater the difference in size or other characteristics between the components of the material, the larger should be the gross sample taken.

When it is necessary to combine samples to make a representative composite sample, weights of the portions entering into the composite must bear the same ratio to each other as do the weights of the initial materials sampled.

Since ideal conditions are never realized in actual practice, a great deal of time has been spent on a study of the theory of sampling, and statistical methods have been used to assist in formulating rules for the taking of samples with due consideration for the characteristics of the material being sampled and for the requirements of the tests to be applied to the sample. For a complete discussion of this approach, reference should be made to Taggart ("Handbook of Mineral Dressing," Wiley, New York, 1945) and Dodge and Romig ("Sampling Inspection Tables," Wiley, New York, 1944).

Methods of Sampling

Sampling is carried out according to two general methods: hand sampling and mechanical or automatic sampling. The former, as its name implies, involves the taking of the sample by an operator using a simple tool for the purpose. On large lots this method is slow and expensive and in all cases puts great responsibility on the individual operator. In mechanical sampling a certain predetermined portion of the material is taken continuously or at regular intervals.

Grab Sampling. Grab sampling is the simplest method of sampling and is subject to the greatest inaccuracy. It consists in taking small equal portions, either at random or at regular intervals, by hand or with a scoop or shovel. The advantages of grab sampling are its economy and the speed with which it may be carried out. The disadvantage is that it is difficult to have all components truly represented when small portions are taken, especially if the material is lumpy or the sizes of particles are not uniform. The smaller the particles of the material to be sampled by this method the more accurate will be the sample. In general, grab sampling should be applied only when the material is practically homogeneous and only when approximate accuracy is required. The following typical examples of grab sampling may be indicative of the usefulness of this method: (1) In unloading a tank car of acid, a side-arm outlet from the main unloading pipe is opened, say every 2 min., and a small portion of 50 to 100 cc. is allowed to run out into a container. This method will give for an ordinary tank car a gross sample of approximately 5 gal. which may be thoroughly mixed and reduced to the proper amount for analysis. (2) As a mixed fertilizer, ready for bagging, passes down a chute to a bin, a portion may be taken periodically with a scoop and the accumulated sample for the day mixed and worked up for analysis. (3) A large pile of material may be sampled by taking a shovel or scoop full at different parts of the entire exposed surface of the pile. The points where portions are taken should be spaced at regular intervals. One method of laying out a pile for sampling is to use a rope with knots tied in it at regular intervals corresponding to the distance between locations for sampling. The rope is thrown across the pile, and a portion is taken from directly below each knot. If the character of the material in the pile varies with the depth, a better plan is to take samples at regular intervals over the new face of the pile which is formed as the material is being moved away. The resulting gross sample is subdivided in the usual way.

The following methods of **grab sampling** are abstracted from the *Journal of Industrial and Engineering Chemistry*, **1**, 107 (1909) and describe methods employed by the U.S. Steel Corporation.

Car Sampling. Samples must be taken uniformly over the surface of cars by selecting a minimum of 15 places at regular intervals. These may be selected according to the parallel or zigzag system. If the former is adopted, a convenient method is to use a net with the intersecting ropes knotted at the desired locations for sampling. A sample is taken below each knot. In the zigzag system, samples are taken at regular intervals along a line drawn from one corner of the car across to a point on the opposite side about one-third of the car length from the starting point, then back to the other side at a point about two-thirds of the car length from the starting point, and then across the car to the corner diagonally opposite the first corner.

When lumps are encountered at the designated points where samples are to be taken, small portions of each lump must be chipped off. When rock occurs it must also be sampled as ore, *i.e.*, a proportionate amount of it must be taken, representing the area for which the sample is taken, together with an amount of adjacent material to bring the whole to the amount taken at each sampling point. A gross sample is taken for each 10 cars or less.

Boat Sampling. If the material is located in cone-shaped piles, portions are taken at regular intervals (about one shovel length apart) starting at a point about two shovel lengths from the side of the boat and proceeding along a line up across the apex of the cone and down the opposite side.

When a certain amount of material has been unloaded leaving the remainder with a newly exposed face, samples can be taken at regular intervals on this face. Starting at a distance of two shovel lengths from the side of the boat, the sampler proceeds up the face of the pile taking a sample at every shovel length up to the top. The next vertical line is measured four shovel lengths from the first, and in a similar manner the whole face is covered.

Round Sampling. In round sampling, one-third of the gross sample is taken, after 5 or 6 ft. of the face of the material has been exposed, in the manner described in the above paragraph. When the grabs have removed all the material that can be reached, a second round of portions comprising two-thirds of the gross sample is taken along the face or faces that remain.

Coning and Quartering. Coning and quartering is one of the most familiar forms of hand sampling and is used generally in subdividing a gross sample to give the retained portion. It may be used on lots of material amounting to not more than 50 tons in which the particles do not exceed 2 in. in diameter. The method may be described as follows: The material to be sampled is piled into a conical heap and then spread out into a circular cake. The cake is divided into quarters, and two of the diagonally opposite quarters are taken as the sample while the two remaining quarters are rejected. The two quarters taken as a sample are collected together, and the procedure of coning and quartering is repeated until a lot of the material of the desired size is obtained. A complete discussion of the application of coning and quartering together with illustrations of the different steps is given in Fig. 189 and p. 1101.

During these operations care should be taken that the material is not contaminated by anything on the floor or that part of the sample is not lost through cracks or openings in the floor. Preferably the floor should be swept clean and the operation of coning and quartering carried out with the floor covered with paper or some other suitable material. The advantages of coning and quartering are that few tools are required to carry out the method and the procedure is applicable to all kinds of solid materials. Among the disadvantages of this method are that it is expensive because frequent handling of the material is required and that it does not give an accurately representative sample. The larger sizes of material roll down the sides of the cone and collect around the base, and pieces of intermediate size arrange themselves on the slope of the pile according to their size with the large

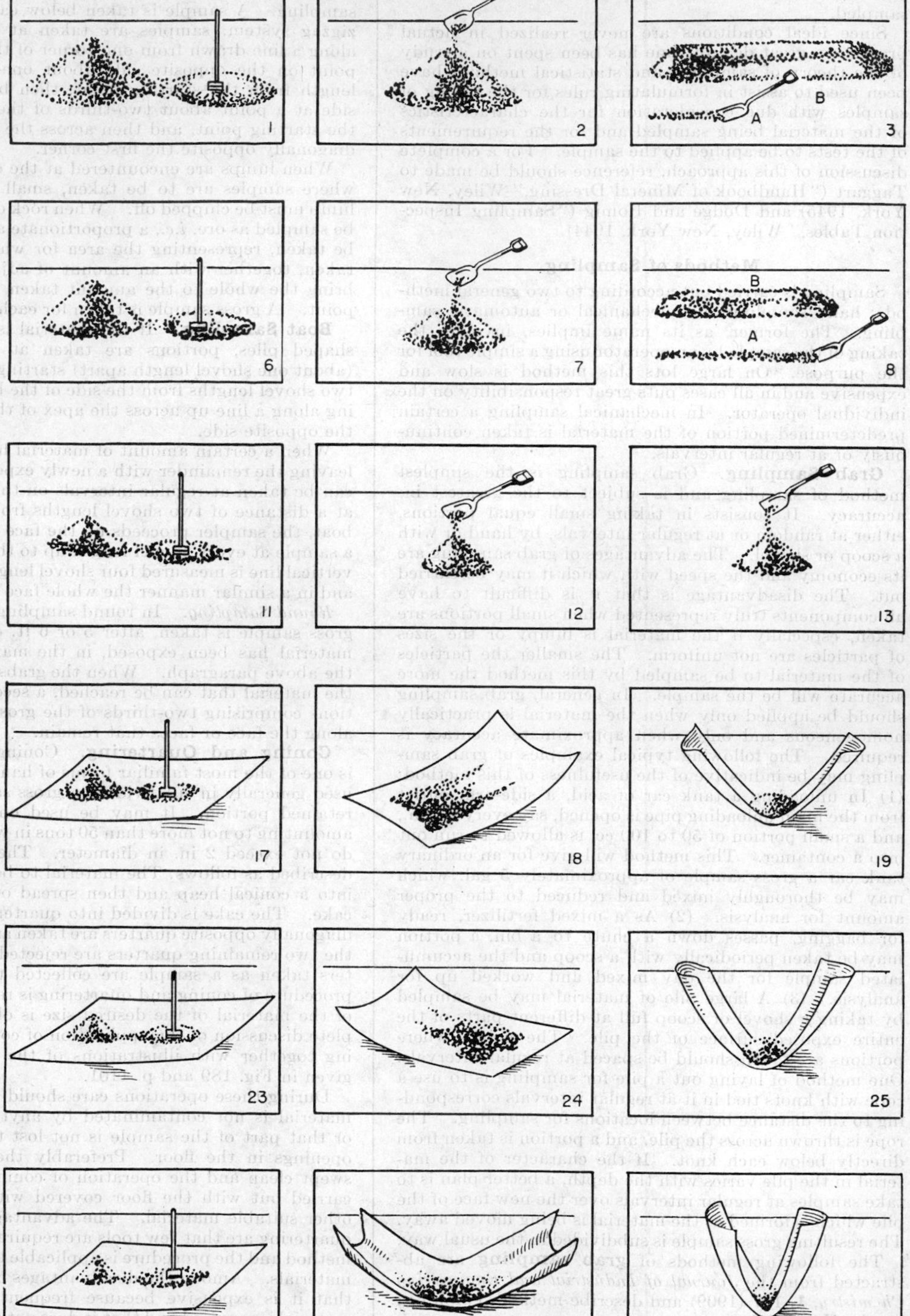

Fig. 189. Sampling method

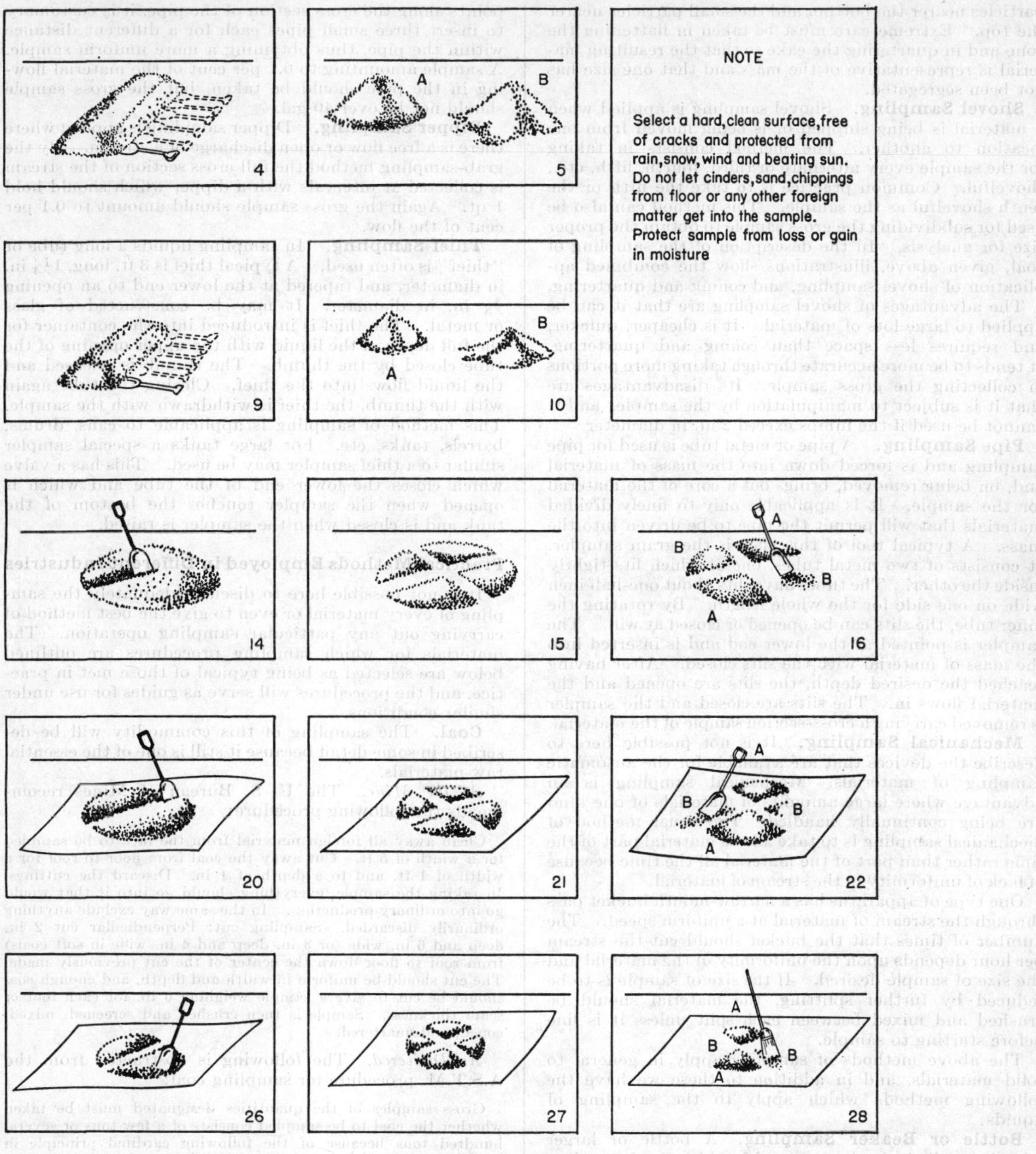
4
A
B
5
NOTE
Select a hard, clean surface, free of cracks and protected from rain, snow, wind and beating sun. Do not let cinders, sand, chippings from floor or any other foreign matter get into the sample. Protect sample from loss or gain in moisture
9
B
A
10
14
15
A
A
16
20
21
A
B
B
B
A
22
26
27
A
B
B
A
28
32
33
A
B
B
A
34

by coning and quartering.

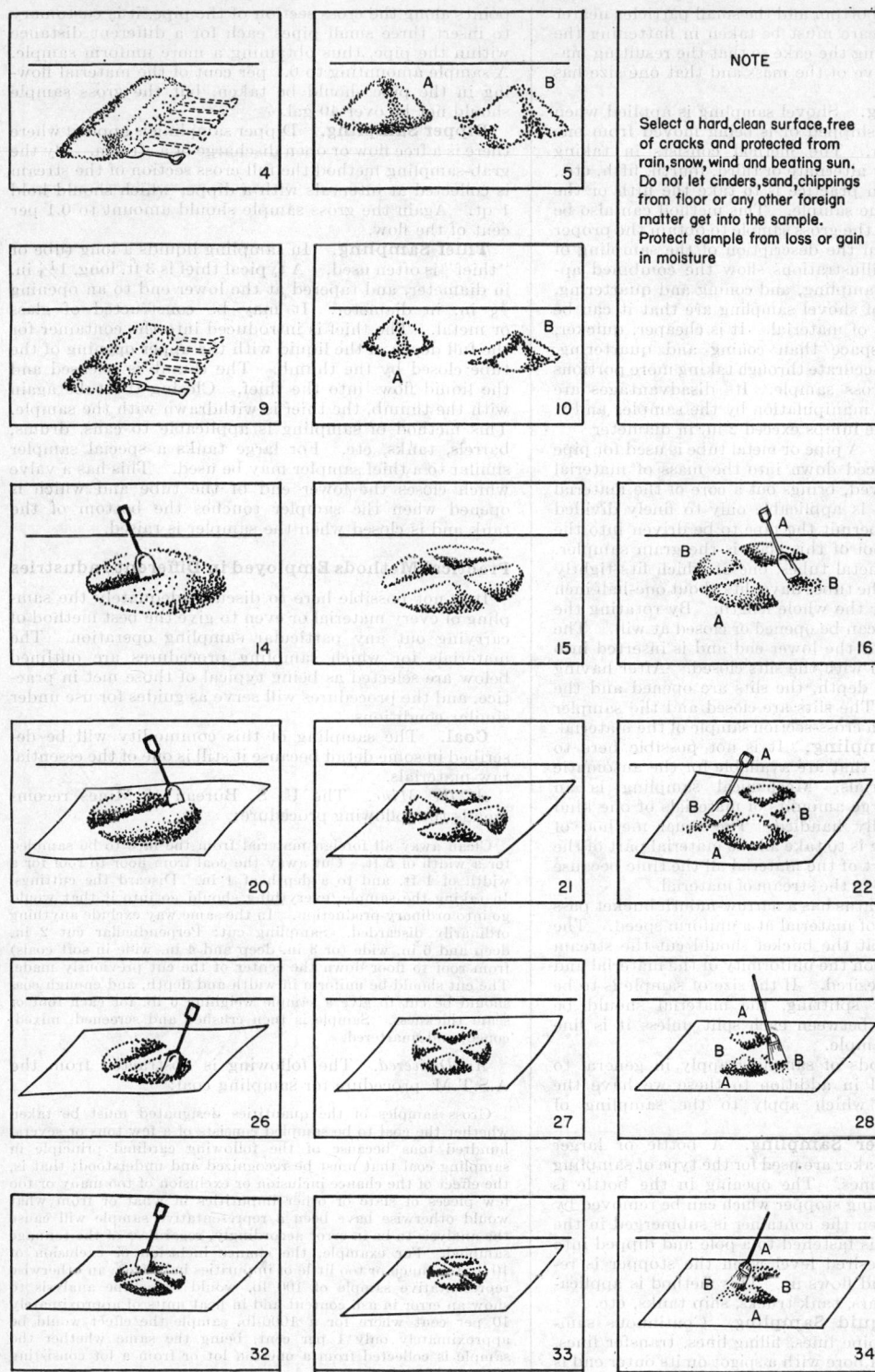

particles nearer the bottom and the small particles nearer the top. Extreme care must be taken in flattening the cone and in quartering the cake so that the resulting material is representative of the mass and that one size has not been segregated.

Shovel Sampling. Shovel sampling is applied when a material is being shipped or is being moved from one location to another. The method consists in taking for the sample every alternate or third, fourth, fifth, etc., shovelful. Common practice is to take the fifth or the tenth shovelful as the sample. This method can also be used for subdividing the gross sample to obtain the proper size for analysis. In the description of the sampling of coal, given above, illustrations show the combined application of shovel sampling, and coning and quartering.

The advantages of shovel sampling are that it can be applied to large lots of material. It is cheaper, quicker, and requires less space than coning and quartering. It tends to be more accurate through taking more portions in collecting the gross sample. Its disadvantages are that it is subject to manipulation by the sampler and it cannot be used if the lumps exceed 2 in. in diameter.

Pipe Sampling. A pipe or metal tube is used for pipe sampling and is forced down into the mass of material and, on being removed, brings out a core of the material for the sample. It is applicable only to finely divided materials that will permit the pipe to be driven into the mass. A typical tool of this class is the grain sampler. It consists of two metal tubes, one of which fits tightly inside the other. The tubes have slits about one-half inch wide on one side for the whole length. By rotating the inner tube, the slits can be opened or closed at will. The sampler is pointed at the lower end and is inserted into the mass of material with the slits closed. After having reached the desired depth, the slits are opened and the material flows in. The slits are closed and the sampler is removed carrying a cross-section sample of the material.

Mechanical Sampling. It is not possible here to describe the devices that are available for the automatic sampling of materials. Mechanical sampling is an advantage where large amounts of materials of one kind are being continually handled. The usual method of mechanical sampling is to take all the material part of the time rather than part of the material all the time because of lack of uniformity in the stream of material.

One type of apparatus has a narrow-mouth bucket pass through the stream of material at a uniform speed. The number of times that the bucket should cut the stream per hour depends upon the uniformity of the material and the size of sample desired. If the size of sample is to be reduced by further splitting, the material should be crushed and mixed between each split unless it is fine before starting to sample.

The above methods of sampling apply in general to solid materials, and in addition to these we have the following methods which apply to the sampling of liquids:

Bottle or Beaker Sampling. A bottle or larger container called a beaker are used for the type of sampling known by their names. The opening in the bottle is closed by a tight-fitting stopper which can be removed by means of a wire when the container is submerged in the liquid. The bottle is fastened to a pole and dipped into the liquid to the desired level when the stopper is removed and the liquid flows in. This method is applicable to tanks, tank cars, tank trucks, ship tanks, etc.

Continuous Liquid Sampling. Continuous sampling is applied to pipe lines, filling lines, transfer lines, etc. A pipe of small bore with a spigot on its outer end is inserted through the wall of the pipe line, and in this way a small part of the liquid stream is diverted for a sample. Since the flow of a stream is not uniform at different points along the cross section of the pipe, it is customary to insert three small pipes each for a different distance within the pipe, thus obtaining a more uniform sample. A sample amounting to 0.1 per cent of the material flowing in the pipe should be taken, but the gross sample should not be over 40 gal.

Dipper Sampling. Dipper sampling is applied where there is a free flow or open discharge of a stream. By the grab-sampling method the full cross section of the stream is collected at intervals with a dipper which should hold 1 qt. Again the gross sample should amount to 0.1 per cent of the flow.

Thief Sampling. In sampling liquids a long tube or "thief" is often used. A typical thief is 3 ft. long, $1\frac{1}{4}$ in. in diameter, and tapered at the lower end to an opening $\frac{3}{8}$ in. in diameter. It may be constructed of glass or metal. The thief is introduced into the container for the full depth of the liquid with the upper opening of the tube closed by the thumb. The thumb is released and the liquid flows into the thief. Closing the tube again with the thumb, the thief is withdrawn with the sample. This method of sampling is applicable to cans, drums, barrels, tanks, etc. For large tanks a special sampler similar to a thief sampler may be used. This has a valve which closes the lower end of the tube and which is opened when the sampler touches the bottom of the tank and is closed when the sampler is raised.

Practical Methods Employed in Different Industries

It is not possible here to discuss adequately the sampling of every material or even to give the best method of carrying out any particular sampling operation. The materials for which sampling procedures are outlined below are selected as being typical of those met in practice, and the procedures will serve as guides for use under similar conditions.

Coal. The sampling of this commodity will be described in some detail because it still is one of the essential raw materials.

At the Mine. The U. S. Bureau of Mines recommends the following procedure:

Clean away all foreign material from the face to be sampled for a width of 5 ft. Cut away the coal from floor to roof for a width of 1 ft. and to a depth of 1 in. Discard the cuttings. In taking the sample, everything should go into it that would go into ordinary production. In the same way exclude anything ordinarily discarded. Sampling cut: Perpendicular cut 2 in. deep and 6 in. wide (or 3 in. deep and 4 in. wide in soft coals) from roof to floor down the center of the cut previously made. The cut should be uniform in width and depth, and enough coal should be cut to give a sample weighing 6 lb. for each foot of seam thickness. Sample is then crushed and screened, mixed, coned, and quartered.

As Delivered. The following is abstracted from the A.S.T.M. procedure for sampling coal:

Gross samples of the quantities designated must be taken whether the coal to be sampled consists of a few tons or several hundred tons because of the following cardinal principle in sampling coal that must be recognized and understood; that is, the effect of the chance inclusion or exclusion of too many or too few pieces of slate or other impurities in what or from what would otherwise have been a representative sample will cause the analysis to be in error accordingly, regardless of the tonnage sampled. For example, the chance inclusion or exclusion of 10 lb. too much or too little of impurities in or from an otherwise representative sample of 100 lb. would cause the analysis to show an error in ash content and in heat units of approximately 10 per cent where for a 1000-lb. sample the effect would be approximately only 1 per cent, being the same whether the sample is collected from a one-ton lot or from a lot consisting of several hundred tons.

Coal should be sampled as it is being loaded or unloaded from railroad cars or other conveyances, or when discharged from bins, etc. Samples from the surface of piles or bins are generally unreliable. To collect samples, a shovel or similar tool shall be used to take equal portions. For small sizes of coal, increments of 5 to 10 lb. may be taken, but, with lump or run-of-the-mine coal, increments should be at least 10 to 30 lb. The increments shall be regularly and systematically collected so that the entire quantity of coal sampled will be represented proportionately in the gross sample, and with such frequency that a gross sample of the required amount shall be collected. The standard gross sample shall be not less than 1000 lb. except that, for slack coal and small sizes of anthracite, when impurities are not present in abnormal quantities, a gross sample of approximately 500 lb. shall be considered sufficient. If the coal contains an unusual amount of impurities, a gross sample of 1500 lb. or more shall be collected. The gross sample should contain the same proportion of lump coal, fine coal and impurities as contained in the mass. A gross sample should be taken for every 500 tons or less unless special agreement is made otherwise.

After the gross sample has been collected, it shall be systematically crushed, mixed, and reduced in quantity to convenient size for the transmission to the laboratory. The sample should be protected from loss or contamination during these operations. The progressive reduction in the weight of the sample to the quantities given in Table 1 shall be carried out according to the steps which are illustrated in Fig. 189 (Plate V, *U. S. Bur. Mines, Bull.* 116).

Table 1

Weight of Sample to Be Divided, Lb.	Largest Size of Coal and Impurities Allowable in Sample before Division, In.
1000, or over	1
500	3/4
250	1/2
125	3/8
60	1/4
30	3/16, or to pass a 4760-μ (No. 4) sieve.

If it is necessary to sample a carload of coal before it is unloaded, cut a trench 2 ft. wide by 2 ft. deep along the two center axes of the car. The sample is then taken at intervals along the newly exposed faces and bottoms of the trenches.

Sampling of Coal Which Is Classified According to Ash Content. Statistical studies of the sampling of coal have shown that it is not always necessary to take as large gross samples as are specified above in order to obtain a reasonable accuracy in the test to be made. These studies have also shown that the ash content is a satisfactory criterion for the classification of coals. The Brit. Standards Inst. and the A.S.T.M. (D 492-48) have set up tables showing the size of gross sample to be taken for different coals classified according to ash content and according to size. The A.S.T.M. method provides two procedures, one for special accuracy and one which in 95 cases out of 100 will give an accuracy of ±10 per cent in the ash content. As this new method of selecting the size of sample becomes familiar, it will no doubt displace the older method described above.

Iron Ore. *Piles.* Divide surface of pile into equal areas and select equal volume of material from each area. From all fines, select portion from center of area. From all lumps break off small pieces. From lumps and fines take proper proportion of both. Pile should be sampled at intervals of 2 to 10 ft. As suggested above, a rope with knots tied in it is convenient for locating sampling spots.

Cars. A minimum of 24 samples should be taken. Composites may be made to include up to 10 cars.

Boats. One-half pound samples may be taken at several points below and at equal distance from the apex of conical piles; or at intervals along newly exposed faces as ore is unloaded.

Pig Iron. *In Furnace.* Sampled while molten with spoon or ladle and poured into mold or on an iron plate. A sample is taken from the middle of each ladle of iron in the cast. Equal quantities from each portion are combined into one sample. Test pieces may be drilled or crushed.

From Car or Storage. Not less than three pigs shall be taken to represent any lot or shipment, and for lots of more than 30 tons one pig shall be taken for each 10 tons of iron. The pigs shall be selected by some means such as the knotted-rope system that eliminates the element of personal choice. Three to seven pigs taken in the order selected shall constitute a unit sample.

The sample for analysis shall be collected, after the proper discard, by drilling each pig or portion of pig with a properly sharpened and hardened, flat-bead, 5/8 or 3/4 in. high-speed tool-steel drill in a direction at right angles to the long axis of the pig and at such a feed as to form a minimum amount of fine drillings. The first drillings shall be discarded, and only the drillings collected after the drill has penetrated 1/4 in. shall be reserved for the sample. The drilling shall extend to within 1/4 in. of the opposite surface of the pig. Suitable precautions shall be taken to collect all the drillings, fine as well as coarse particles, and to avoid contamination of the sample in any way. The sample for analysis shall be composed of equal portions by weight of the drillings from the pigs forming the unit sample, and the weight of the combined sample shall be not less than 75 g. The drillings shall be thoroughly mixed, as by gentle grinding in a suitable mortar until all pass the No. 20 (840 μ) sieve. If desired, the sample may be carefully divided by coning and quartering to provide portions for buyer and seller.

Steel. One sample per heat is taken at a time when one-half has been poured by holding a spoon under the stream from the ladle while it is slackened. The sample is poured into a mold and drilled with 3/4-in. drill, using no lubricant. All drillings are discarded until the outer edge of the drill is buried.

Metals and Alloys. In general, metals and alloys are sampled by drilling with a 1/2-in. drill, using no lubricant. Any surface contamination should be removed and discarded before the sample is collected. It is often convenient to use a template in order that the piece can be drilled in a systematic manner.

Cement. According to the A.S.T.M. recommendations:

For Individual Samples. If sampled in cars, one test sample shall be taken from each 50 bbl. or less. If sampled in bins, one sample shall represent each 200 bbl.

For Composite Samples. If sampled in cars, one sample shall be taken from one sack in each 40 sacks (or 1 bbl. in each 10 bbl.) and combined to form one test sample. If sampled in bins or warehouses, one test sample shall represent not more than 200 bbl.

Soft Solids. Three sets of borings 3/4-in. in diameter are taken through the depth of the material.

Lumpy Solids. Sample taken by the grab method taking approximately 0.1 per cent of the lot but not less than 50 lb. nor more than 1000 lb.

Petroleum Products.* *For Tanks and Tank Cars.* Four types of liquid samples are recognized: all-level sample, upper sample, middle sample, and lower sample. An all-level sample is obtained when a bottle unstoppered is lowered through the liquid to the bottom and back to the surface at such a rate that the bottle is just filled

* Abstracted from recommendations of the A.S.T.M. D 270-33.

when it reaches the surface. Samples at the different levels are taken by lowering a stoppered bottle to the desired level, releasing the stopper and allowing the liquid at that level to fill the bottle. A sample taken at a point 10 per cent below the top is an upper sample, and a sample taken at a point 10 per cent above the bottom is a lower sample. The middle sample is taken at a depth of 50 per cent.

In making a composite sample for a tank, samples from the different levels should be combined as shown in Table 2.

Table 2

Sample	Vessels of uniform cross section	Horizontal cylindrical tanks (full)
Upper	1 part	1 part
Middle	3 parts	8 parts
Lower	1 part	1 part

Individual samples taken with a beaker, bottle, or dipper should amount to 1 qt. Composite samples should be 5 qt. for vessels of uniform cross section and 10 qt. for horizontal cylindrical tanks. Gross liquid mixed-cargo samples are taken from the various ship's tanks in multiples of 5 or 10 qt. depending upon the shape of the tank. Liquid samples taken by the continuous or dipper method should be 0.1 per cent of the amount shipped but not less than 5 gal. or more than 40 gal.

For Pipe Lines. The method of continuous sampling described on p. 1100 is used most frequently in the petroleum industry for the sampling of materials flowing in pipe lines. If the material to be sampled is semiliquid, the lines and receiver are warmed to keep the material just above the liquefying temperature. The gross continuous sample, as collected and amounting to not over 40 gal. as specified above, is thoroughly mixed and sampled with a thief to get a 1 qt. sample.

Lacquer Solvents. *From a tank car,* a ½-gal. gross sample in small portions of not over 1 qt. each is taken from near the top and near the bottom using the bottle method.

From drums, at least 5 per cent of the packages are sampled using a thief at center of the drum and taking not less than ½ pt. from each drum. Gross sample should be not less than 1 qt.

Water. *Reservoir.* A stoppered bottle is submerged to the required depth and the stopper released. The bottle should be rinsed several times before the sample is taken.

Stream. Immerse the bottle in the stream, making an effort not to disturb silt or other solid material on bottom or sides.

For bacteriological tests the sample should be collected in a bottle that has been thoroughly sterilized by steaming, and special precautions should be taken that the sample is not contaminated.

Molten Metal. Dip the sample with a ladle, breaking through the slag and holding the ladle until it reaches the temperature of the mass.

Gases. The sampling of gases is comparatively easy, since in many cases they are homogeneous. Two kinds of samples are recognized: accumulative, which is taken continuously over a period of time (½ to 24 hr.); and control, which is taken for less than ½ hr.

Where a gas is known to be homogeneous in the cross section, the sample may be withdrawn through a pet cock inserted in the wall of the container. If the gas is under pressure, the sample may be released through the pet cock. Otherwise it may be withdrawn by means of an aspirator bottle, filled with water, mercury, or other suitable liquid.

Flue Gas. If one constituent of flue gas, *e.g.*, CO_2, is to be determined continuously and automatically, a permanent fixture is connected in the flue to conduct the gas to the CO_2 recorder.

If a complete analysis of the gas is to be made, consideration must be given to the fact that the gases move with different velocities at different points in the cross section of the flue. For sampling the gas under such conditions, a perforated pipe is used which enters the flue with an airtight connection and extends entirely across the flue. Perforations in the pipe should be equal distances apart; and, to obtain an equal flow through all openings, their combined area should be less than the cross-sectional area of the pipe—3:4 being considered a safe ratio. The gas is then withdrawn through the pipe by means of suction until the desired sample has been collected.

Air or Gas in Confined Space. Fill a 2-oz. bottle with mercury, let stand for 2 min., and then pour the mercury back into the stock bottle, thus filling the bottle with air at that spot.

In special cases where an instantaneous undiluted sample is required, an evacuated tube is used. The tube is pumped out and the stopcock is opened where the sample is to be taken.

Dust- or Fume-laden Air. The air is filtered through a specially prepared filter such as a Gooch crucible containing filter paper, and the residue obtained from a known volume of air is weighed. Or the air may be projected on a glass slide with some means of causing the dust to adhere with suitable provision for observing the increase in the deposit.

General Rules for Sampling

1. Sample to be taken by or under direct supervision of a person qualified by experience to recognize that sample is satisfactory for the subsequent test to be made.
2. Select the most appropriate method of sampling with due regard to the kind of material and the conditions of storage or handling.
3. Watch for special conditions that make for nonhomogeneity in the material such as weathering of the outside of a pile of material, segregation of sizes, or more than one layer in liquids.
4. Take all necessary precautions to avoid contamination of sample after it has been taken.
5. Label the sample clearly with all necessary information to designate the source from which it was taken.

SECTION 16

SIZE REDUCTION AND SIZE ENLARGEMENT

BY

C. E. Berry, B. S., Chemical Engineer, E. I. duPont de Nemours & Co., Inc., Wilmington, Del.; Member, American Chemical Society, American Institute of Chemical Engineers, American Institute of Mining and Metallurgical Engineers. (Crushing and Grinding. Editor: Miscellaneous Methods of Size Reduction and of Size Enlargement)

Thos. B. Dorris, Ph. D., Chief Chemical Engineer, Sprout, Waldron & Co., Muncy, Pa.; Member, American Chemical Society, National Society of Professional Engineers, Technical Association of the Pulp and Paper Industry; Fellow, American Institute of Chemists. (Cutting)

J. H. Foote, Assistant Research Director, Pulverizing Machinery Co., Summit, N.J. (Machining)

W. M. Sheldon, B. S., Director of Research, Pulverizing Machinery Co., Summit, N.J. (Machining)

D. J. Van Marle, Ch. E., Chemical Engineer, Buflovak Equipment Division of Blaw-Knox Co., Buffalo, N.Y.; Member, American Chemical Society. (Flaking)

J. I. Yellott, B. S., M. M. E., Director of Research, Locomotive Development Committee, Bituminous Coal Research, Inc., Baltimore, Md.; Member, American Society of Mechanical Engineers. (Explosive Disintegration)

C. K. Sloan, A. B., M. S., Ph. D., Research Chemist, E. I. duPont de Nemours & Co., Inc., Wilmington, Del.; Member, American Chemical Society. (Emulsification, Flocculation)

R. V. Kleinschmidt, A. B., A. M., S. B., S. D., Professor of the Practice of Mechanical Engineering, Harvard Graduate School of Engineering; Consulting Engineer, Stoneham, Mass.; Member, American Society of Mechanical Engineers, American Physical Society, American Association for the Advancement of Science. (Theory of Dispersion of Liquid Droplets)

H. G. Houghton, B. S., S. M., D. Sc., Professor of Meteorology and Head of Department, Massachusetts Institute of Technology, Cambridge, Mass.; Member, American Physical Society, American Meteorological Society, American Association for the Advancement of Science, American Geophysical Union. (Spray Nozzles)

S. A. Miller, B. S., Ph. D., Associate Professor of Chemical Engineering, Department of Chemical Engineering, University of Kansas, Lawrence, Kans.; Member, American Chemical Society, American Institute of Chemical Engineers, American Society for Engineering Education, American Association for the Advancement of Science. (Gas Dispersion)

Charles M. Fields, A. B., Technical Project Engineer, Shellmar Products Corp., Zanesville, Ohio; Member, American Chemical Society. (Extrusion)

P. W. Crane, Ch. E., M. S., Asst. Dir. of Sales, Polychemicals Department, E. I. du Pont de Nemours & Co., Inc., Wilmington, Del.; Member, American Chemical Society. (Extrusion)

J. E. Teagarden, Ph. D., Polychemicals Department, E. I. duPont de Nemours & Co., Inc., Wilmington, Del. (Molding of Plastics)

A. F. Randolph, B. S., Ch. E., Asst. to Dir. of Sales, Polychemicals Dept., E. I. duPont de Nemours & Co., Inc., Wilmington, Del.; Member, American Chemical Society, Society of Chemical Industry, Faraday Society. (Molding of Plastics)

W. E. Rahm, Technical Field Service, Polychemicals Dept., E. I. duPont de Nemours & Co., Inc., Arlington, N.J. (Molding of Plastics)

Lawrence H. Bailey, B. S., M. S., Chief Engineer, F. J. Stokes Machine Co., Philadelphia, Pa. Awarded Stevens Institute Medal for Achievement in Powder Metallurgy. (Compacting)

Gilbert E. Seil,* M. S., Ph. D., Late Technical Consultant, Day & Zimmermann, Inc., Philadelphia, Pa.; Member, American Chemical Society, American Institute of Chemical Engineers, American Institute of Mining and Metallurgical Engineers, American Society for Testing Materials, National Research Council, British Ceramic Society; Fellow, Mellon Institute, American Institute of Chemists, American Ceramic Society. (Size Enlargement by Fusion)

* Deceased.

CONTENTS

CRUSHING AND GRINDING

	Page
References	1107
Principles of Crushing and Grinding	1110
Introduction	1110
Crushing and Grinding	1110
Properties of Solids	1110
Objectives	1111
Product Specifications	1111
Particle-size Measurement	1111
Methods	1111
Sieving	1111
Microscopic	1112
Elutriation	1112
Gravity Sedimentation	1112
Comparison of Methods	1112
Centrifugal Sedimentation	1112

PAGE
Permeability........ 1112
Modifications........ 1112
Turbidimeter........ 1113
Adsorption........ 1113
Shape Factor in Size Measurement........ 1113
Particle-size Representation........ 1113
Analytical Relationships........ 1113
Average Particle Size........ 1113
Reduction Ratio........ 1113
Specific Surface........ 1113
Grindability........ 1114
Hardness........ 1114
Soft Materials........ 1114
Hard Materials........ 1114
Grindability Methods........ 1114
Ball Mill........ 1114
Hardgrove Machine........ 1114
Indexes Compared........ 1114
Other Grindability Tests........ 1114
Work Required for Size Reduction........ 1114
Theory........ 1114
Kick's Law........ 1114
Rittinger's Law........ 1114
Generalized Relation........ 1114
Application........ 1114
Grinding Efficiency........ 1115
Theoretical........ 1115
Measured........ 1115
Efficiency Coefficients........ 1115
Dispersing Agents and Grinding Aids........ 1115
Wet Grinding........ 1115
Dry Grinding........ 1115
Classification and Selection of Equipment........ 1116
Classification........ 1116
Selection........ 1116
Size Reduction Combined with Size Classification........ 1117
Continuous Open-circuit Operation........ 1117
Continuous Closed-circuit Operation........ 1118
Dry vs. Wet Grinding........ 1118
Types of Size Classifiers........ 1118
Use of Size Classifiers........ 1118
Independent........ 1118
External........ 1118
Internal........ 1118
Size Reduction Combined with Other Operations........ 1119
Mixing and Blending........ 1119
Heat Transfer........ 1119
Heating and Cooling........ 1119
Dehydrating........ 1119
Drying........ 1119
Processing........ 1119
Cleaning and Concentrating........ 1119
Safety........ 1119
Metal Powders........ 1119
Non-metal Powders........ 1120
Description of Crushers........ 1120
Jaw Crushers........ 1120
Design and Operation........ 1120
Performance........ 1125
Type H........ 1125
Fine Reduction........ 1125
Universal........ 1125
Crusher Product Size Distribution........ 1125
Gyratory Crushers........ 1125
Design........ 1125
Operation........ 1126
Performance........ 1126
Superior McCully........ 1126
Type R........ 1126
Type T Bulldog........ 1126
Gearless Gyratory........ 1126
Telsmith Breaker........ 1126
Cone Crushers........ 1126
Design and Operation........ 1126
Performance........ 1126
Symons........ 1126
Telsmith........ 1126
Pan Crushers........ 1126
Design and Operation........ 1126
Performance........ 1127
Bonnot........ 1127
Smooth-roll Crushers........ 1127

PAGE
Design and Operation........ 1127
Performance........ 1127
Corrugated- and Toothed-roll Crushers........ 1127
Design and Operation........ 1127
Performance........ 1128
Fairmount........ 1128
Jeffrey........ 1128
Sawtooth........ 1128
Rotary Crushers........ 1129
Design and Operation........ 1129
Performance........ 1129
Bartlett and Snow........ 1129
Horizontal........ 1129
Hammer Crusher........ 1129
Design and Operation........ 1129
Double Impeller........ 1129
Performance........ 1129
Super-Jumbo........ 1129
Jeffrey........ 1129
Impactor........ 1129
American Ring........ 1129
Stedman Type B........ 1129
Description of Grinders or Grinding Mills........ 1130
Ball, Pebble, Rod, Tube, and Compartment Mills........ 1130
Design of Tumbling Mills........ 1130
Batch........ 1130
Continuous........ 1130
Batch and Continuous........ 1130
Conical Mill........ 1130
Vibrating Mill........ 1130
Operation of Tumbling Mills........ 1130
Ball Action........ 1130
Grinding Medium........ 1130
Mill Speed........ 1130
Material and Ball Charges........ 1130
General Considerations........ 1131
Performance of Tumbling Mills........ 1132
Selection of Mill........ 1132
Capacity and Power Consumption........ 1132
Performance of Proprietary Equipment........ 1132
Allis-Chalmers Mfg. Co........ 1132
Ball Mill........ 1132
Compeb........ 1132
Preliminator........ 1132
Ball Peb........ 1132
Rod Mill........ 1133
F. L. Smidth & Co........ 1133
Kominuter........ 1133
Unikom........ 1133
Unidan........ 1133
Pyrator........ 1133
The Mine and Smelter Supply Co........ 1133
Marcy Ball Mill........ 1133
Open End Rod Mill........ 1133
Hardinge Co........ 1133
Conical Mill........ 1133
Ball Mill........ 1134
Foster Wheeler Corp........ 1134
Ball mill........ 1134
Aerofall Mills Ltd........ 1134
Aerofall Mill........ 1134
Particle-size Classifiers Used with Grinding Mills........ 1135
External Classifiers........ 1135
Whizzer........ 1135
Spinner........ 1135
Gayco........ 1135
Whirlwind........ 1135
Superfine and Loop........ 1136
Internal Classifiers........ 1136
Whizzer........ 1136
Vacuum Multi-vane........ 1136
Ring-roller Mills........ 1137
Roller Mills........ 1137
Ring-roller Mills without Internal Classification........ 1137
Sturtevant........ 1137
Kent Maxecon........ 1137
Ring-roller Mills with Internal Screen Classification........ 1137
Bradley-Hercules........ 1137
Junior Hercules........ 1137
Griffin........ 1137
Ring-roller Mills with Internal Air Classification........ 1138
Babcock & Wilcox........ 1138

PAGE
Raymond........ 1138
Williams........ 1138
Bowl........ 1138
Hammer Mills........ 1139
Hammer Mills without Internal Air Classifiers........ 1139
Helix-Seal........ 1139
Stedman Type A........ 1140
Stedman Disintegrator........ 1140
Jeffrey Type A........ 1140
Mikro-Pulverizer........ 1140
S. P. Mikro-Pulverizer........ 1140
Raymond Screen........ 1140
Jay Bee........ 1141
Blue Streak Dual Screen........ 1141
Riley Atrita........ 1141
Aero........ 1141
Whirlwind........ 1141
Rietz Disintegrator........ 1141
Hammer Mills with Internal Air Classifiers........ 1141
Imp........ 1141
Automatic........ 1142
Vertical........ 1142
Mikro-Atomizer........ 1143
Limited........ 1143
Disk Attrition Mills........ 1143
Sprout-Waldron........ 1143
Bauer........ 1144
Frigidisc........ 1144
Buhrstone Mills........ 1144
Colloid Mills........ 1145
Fluid Energy or Jet Mills........ 1145
Micronizer........ 1145
Reductionizer........ 1145
Blaw-Knox Jet Pulverizer........ 1146
Fluid Energy Reduction Mill........ 1146
Flash Pulverizer........ 1146
Nozzle Pulverizer........ 1147
Anger Mill........ 1147
Industrial Applications of Grinding Mills........ 1147
Milling of Cereals and Other Vegetable Products........ 1147
Flour and Feed Meal........ 1147
Soybeans, Soybean Cake, and Other Pressed Cakes........ 1148
Cocoa Powder........ 1148
Starch and Other Flours........ 1148
Hay and Other Herbage........ 1148
Dried Fruit and Vegetables........ 1148
Metalliferous Ores........ 1148
Comparison of Roll, Rod Mill, and Ball Mill Products 1149
Non-metallic Minerals........ 1149
Silica and Feldspar........ 1149
Talc and Soapstone........ 1150
Clays and Kaolins........ 1150
Non-metallic Carbonates and Sulfates........ 1150
Fluorspar........ 1151
Asbestos and Mica........ 1151
The Fertilizer Industry........ 1151
Oyster Shells and Lime Rock........ 1151
Phosphates........ 1151
Basic Slag........ 1152
The Cement, Lime, and Gypsum Industries........ 1152
Portland Cement........ 1152
Closed-circuit Grinding of Cement Clinker........ 1153
Lime........ 1154
Gypsum........ 1154
Coal, Coke, and Other Carbon Products........ 1154
Bituminous Coal........ 1154
Anthracite Coal........ 1156
Coke........ 1156
Pitch........ 1156
Natural Graphite........ 1156
Mineral Black........ 1156
Bone Black........ 1156
Decolorizing Carbons........ 1156
Charcoal........ 1156
Gilsonite........ 1156
Carbon Mixtures........ 1156
Lampblack........ 1156
Pigments, Chemicals, and Insecticides........ 1157
Dry Colors and Dyestuffs........ 1157
White Pigments........ 1157
Mineral Pigments........ 1157
Lead Oxides........ 1157

PAGE
Chemicals........ 1157
Calcium Arsenate........ 1158
Rock Salt and Alum........ 1158
Monocalcium Phosphate........ 1158
Sodium Bicarbonate........ 1158
Derris and Pyrethrum........ 1158
D.D.T........ 1158
Sulfur........ 1158
Drugs, Pharmaceuticals, and Spices........ 1158
Resins, Gums, Waxes, and Molding Powders........ 1159
Resins........ 1159
Hard Rubber........ 1159
Molding Powders........ 1159
Grinding of Soaps........ 1160

MISCELLANEOUS METHODS OF SIZE REDUCTION

Introduction........ 1160
Phase Relationship of the Size-reduction Operations........ 1160
Cutting........ 1160
Rotary Knife Cutters........ 1160
Abbé Engineering........ 1161
Ball & Jewell........ 1161
Mercer-Robinson........ 1161
Paul O. Abbé........ 1161
Sprout, Waldron........ 1161
Wiley Laboratory Mill........ 1161
Wolf........ 1161
Precision Knife Cutters........ 1161
Stokes........ 1162
Taylor-Stiles........ 1162
Slitting Cutters........ 1162
Camachine........ 1162
Machining........ 1162
Filing Machines........ 1163
Saws........ 1163
Shaving Machines........ 1163
Milling Machines........ 1163
Turning........ 1163
Mikro-Chipper........ 1163
Flaking........ 1164
Drum Flakers........ 1164
Applications of Drum Flakers........ 1164
Operating Factors........ 1165
Drum Cooling........ 1165
Drum Capacity and Flake Thickness........ 1165
Heat-transfer Rates........ 1165
Costs........ 1166
Auxiliary Equipment........ 1166
Explosive Disintegration........ 1166
Masonite Process........ 1166
Puffing of Materials........ 1166
Explosive Shattering........ 1166
Winterschall-Schmalfeldt Process........ 1167
Emulsification........ 1167
Characteristics of Emulsions........ 1167
Emulsification Equipment........ 1167
Emulsion Stability........ 1167
Homogenizers........ 1167
Viscolizer........ 1167
Gaulin........ 1168
Flow Master........ 1168
Kom-bi-nator........ 1168
Versator........ 1168
Sonic Oscillator........ 1169
Colloid Mills........ 1169
Charlotte........ 1169
Premier........ 1169
Noblewood........ 1169
Gaulin........ 1169
Applications of Colloid Mills and Homogenizers........ 1169
Spraying........ 1169
Theory of Dispersion of Liquid Droplets........ 1169
Mechanism........ 1170
Energy Relations........ 1170
Dispersion........ 1170
Spray Nozzles........ 1170
Pressure Nozzles........ 1171
Hollow-cone Nozzles........ 1171
Solid-cone Nozzles........ 1171
Fan Nozzles........ 1171

PAGE
Impact........ 1172
"Fog" Nozzles for Firefighting........ 1172
Materials of Construction........ 1172
Discharge Rates of Typical Pressure Nozzles........ 1172
Rotating Nozzles........ 1172
Gas-atomizing Nozzles........ 1173
Applications........ 1173
Paint Spraying........ 1173
Humidification........ 1173
Oil Burners........ 1174
Drop Size........ 1174
Drop-size Distributions........ 1174
Common Applications........ 1175
Gas Dispersion........ 1175
Objectives........ 1175
Gas-liquid Contacting........ 1175
Agitation........ 1176
Foam Production........ 1176
Methods........ 1176
Spargers........ 1176
Simple Bubblers........ 1176
Porous Septa........ 1177
Precipitation and Generation Methods........ 1178
Precipitation........ 1178
Generation........ 1178
Fluid Attrition Systems........ 1178
Nozzles and Pipe-line Contactors........ 1178
Cascade Systems........ 1178
Mechanical Agitators........ 1178
Mixco........ 1179
Turbo-gas-absorber........ 1179
Gas Entraining Capacity........ 1179

SIZE ENLARGEMENT

Introduction........ 1179
Extrusion........ 1180
Methods........ 1180
Extrusion of Metals........ 1180
Equipment........ 1180
Operation........ 1181
Extrusion of Plastics........ 1181
Equipment........ 1181
Operation........ 1182
Wire Covering........ 1183
Molding of Plastics........ 1183
Definitions........ 1184

PAGE
Casting........ 1184
Swaging........ 1184
Shaping........ 1184
Die Pressing........ 1184
Laminating........ 1184
Extruding........ 1184
Molding........ 1184
Compression Molding........ 1184
Transfer Molding........ 1185
Injection Molding........ 1185
Jet Molding........ 1186
Comparison of Methods........ 1186
Compacting........ 1186
Granulation........ 1186
With a Binding Agent........ 1186
Granulating Stage........ 1187
Fertilizers........ 1187
Granulation of Briquettes........ 1187
Granulation by Fusion........ 1187
Granulation by Spray Drying........ 1188
Cabot Spheronizing Process........ 1188
Briquetting........ 1188
Pelleting........ 1189
Tableting........ 1189
Compressing (Powder Metallurgy)........ 1189
Flocculation........ 1190
Flocculation and Deflocculation........ 1190
Enlargement of Effective Size by Flocculation........ 1190
Equipment for Flocculation and Dewatering........ 1190
Dorrco Flocculator........ 1190
Permutit Precipitator........ 1190
Dorrco Tray Thickener........ 1190
Hardinge Spiral Clarifier........ 1190
Size Enlargement by Fusion........ 1191
Definitions........ 1191
Nodulizing........ 1191
Pelletizing........ 1191
Agglomerating........ 1191
Sintering........ 1191
Briquetting........ 1191
Product Characteristics........ 1191
Rotary Methods........ 1191
Equipment Capacity........ 1191
Operating Temperature........ 1191
Feed Conditions........ 1191
Static Methods........ 1191
Comparison of Nodulizing and Sintering Processes........ 1192

CRUSHING AND GRINDING*

BY C. E. BERRY

REFERENCES:† *Crushing and Grinding. General.* Work, Crushing and Grinding, *Ind. Eng. Chem.*, **42,** 26 (1950). Dalla-Valle, "Micromeritics," 2d ed., Pitman, New York, 1948. Epstein, Statistical Aspects of Fracture Problems, *J. Applied Phys.*, **19,** 140 (1948). Fisher and Hollomon, A Statistical Theory of Fracture, *Metals Tech.*, **14,** *Tech. Pub.* 2218 (Aug., 1947). Podszus, The Fundamentals and Laws of Comminution with Special Reference to Metals, *Arch. Metallkunde*, **1,** 318 (1947). Various Authors, Milling and Concentration, *Trans. Am. Inst. Mining Met. Engrs.*, **169,** 1946. Poncelet, Fracture and Comminution of Brittle Solids, *Trans. Am. Inst. Mining Met. Engrs.*, **169,** 37 (1946). Sack, Extension of Griffith's Theory of Rupture to Three Dimensions, *Proc. Phys. Soc. (London)*, **58,** 729 (1946). Taylor, Mechanism of Fracture of Glass and Similar Brittle Solids, *J. Applied Phys.*, **18,** 943 (1947). Taggart, "Handbook of Mineral Dressing," Wiley, New York, 1945. Dickenson, Development of the Stone Breaker, *Cement, Lime, and Gravel*, **20,** 78 (1945). McLaren, Fundamentals of Grinding, *Can. Mining J.*, **64,** 705 (1943); **65,** 153 (1944); The Ball Mill, *ibid.*, **65,** 21 (1944); Discussion of the Practice of Tube Mill Grinding, *ibid.*, **65,** 153 (1944). Counselman, Ore Concentration and Milling, *Mining and Met.*, **24,** 59 (1943). Bell, The Law of Crushing, *Am. Inst. Mining Met. Engrs.*, *Tech. Pub.* 1415, 1942. Benedict, Ore Concentration and Milling, *Mining and Met.*, **23,** 71 (1942). Bowen, Advances in Milling Methods; Reduction Crushing; and Hatch, Current Fine Grinding Practice and Recent Developments, *Mining Congr. J.*, **28,** 65 (1942). Waeser, Progress in Grinding, *Chem. Fabrik*, **14,** 39 (1941). Farrant, A Review of Certain Unit Processes in the Reduction of Materials, *Trans. Inst. Chem. Engrs. (London)*, **18,** 56 (1940). Fahrenwald, Progress in Milling Practice and Equipment, *Mining Congr. J.*, **26,** 17 (1940). Richards and Locke, "Textbook of Ore Dressing," 3d ed., McGraw-Hill, New York, 1940. Gaudin, "Principles of Mineral Dressing," McGraw-Hill, New York, 1939. Sheppard, Primary Crushing: Summary of Field Tests, *U.S. Bur. Mines, Report Investigations*, 3132, 1939. Anon., Pulverizing 200 Process Materials, *Chem. & Met. Eng.*, **45,** 241 (May, 1938). Bond, The Reduction Ratio Curves for Crushing and Grinding, *Eng. Mining J.*, **139,** 48 (1938). Coghill and DeVaney, Conclusions from Experiments on Grinding, *Bull. Missouri School of Mines and Metallurgy*, **13,** 1938. Gross, Crushing and Grinding, *U.S. Bur. Mines Bull.* 402, 1938. Work, Developments in Crushing and Pulverizing Equipment, *Trans. Am. Inst. Mining Met. Engrs.*, **34,** 101 (1938); Factors Influencing Particle Size and Shape in Grinding, *Bull. Am. Ceramic Soc.*, **17,** 1 (1938). A Group of Articles on Size Reduction by Several Authors, *Chem. & Met. Eng.*, **45,** 226 (May, 1938). Blanc, The Laws of Crushing and Grinding, *Rev. ind. minérale*, **386,** 35 (1937); Theoretical Study of Crushing of Hard Materials, *ibid.*, **435,** 106 (1939). Hönig, Fundamentals of Grinding, *Forschungsheft*, **378** (1936); Supplement to Part B of *Forsch. Gebiete Ingenieurw.* (Ver. deut. Ing., Berlin). Fahrenwald, Some Fine Grinding Fundamentals, *Trans. Am. Inst. Mining Met. Engrs.*, **112,** 88 (1935). Miller, "Crushers for Stone and Ore," Van Nostrand, New York, 1935. Various Authors, Milling Methods, *Trans. Am. Inst. Mining Met. Engrs.*, 1935. Carey, Crushing and Grinding, *Trans. Inst. Chem. Engrs. (London)*, **12,** 179 (1934). Carey and Bosanquet, Study of Crushing Brittle Solids, *J. Soc. Glass Tech.*, **17,** 384T (1933). Work, Simplify Your Process by Combining Drying and Grinding Operations, *Chem. & Met. Eng.*, **40,** 306 (1933); An Analysis of Crushing and Pulverizing Processes, *Am. Soc. Mech. Engrs.*, RP-55 (1932). Seymour, Secondary Crushing Machinery, *Crushing and Grinding*, **1,** 35 (1931). Fahrenwald, Grinding and Classification: I. Batch Grinding, *U.S. Bur. Mines, Rept. Investigations*, 2989, 1930. Fahrenwald, Grinding and Classification: II. Batch Closed-circuit Grinding, *U.S. Bur. Mines, Rept. Investigations*, 2990, 1930. Gaudin, An Investigation of Crushing Phenomena, *Trans. Am. Inst. Mining Met. Engrs.*, **73,** 253 (1926). Griffith, The Phenomena of Rupture and Flow in Solids, *Phil. Trans. Royal Soc. (London)*, **221A,** 163 (1921).

Grindability. Anon., D409-37T, Tentative Method of Test for Grindability of Coal by the Hardgrove-Machine Method, *Am. Soc. Testing Materials, Standards*, **1944** III, 1174. Bond, Standard Grindability Tests Tabulated, *Mining Tech.*, **11,** *Tech. Pub.* 2180 (July, 1947). Bond and Maxson, Standard Grindability Tests and Calculations, *Mining Tech.*, **7,** *Tech. Pub.* 1579 (March, 1943). Coe and Coghill, Contrasts in Grinding Characteristics of Mineral Products, *U.S. Bur. Mines, Rept. Investigations*, 3704 (1943). Engelhardt, Grindability and Surface Energy of Solids, *Naturwissenshaften*, **33,** 195 (1946). Henglein, Work of Subdivision of Substances, *Chem. Ztg.*, **68,** 23 (1944). Romer, A Study of the Grindability of Coal and the Fineness of Pulverized Coal When Using the Lea-Nurse Air Permeability Method for Evaluating the Subsieve Fractions, *Proc. Am. Soc. Testing Materials*, **41,** 1152 (1941). Brunjes, Grindability Index Determination by Ball Mill Method, *Combustion*, **11,** 31 (1940). Hardgrove, The Effect of Grindability on Particle Size Distribution, *Trans. Am. Inst. Chem. Engrs.*, **34,** 131 (1938). Yancey and Geer, Further Investigation of Methods for Estimating the Grindability of Coal, *Am. Inst. Mining Met. Engrs.*, *Contrib.* 94, 1936. Sloman and Barnhart, Relative Grindability of Coal, *Trans. Am. Soc. Mech. Engrs.*, **56,** 773 (1934). Hardgrove, Grindability of Coal, *Trans. Am. Soc. Mech. Engrs.*, *Fuels Steam Power*, **54,** 37 (1932).

Work and efficiency. Bond, Crushing Tests, *Mining Tech.*, **10,** *Tech. Pub.* 1895 (1946). Prentice, Crushing and Grinding Efficiencies, *Bull. Inst. Mining Met. (London)* 477, 1946. Taggart, New Units of Crusher Capacity and Crusher Efficiency, *Mining Tech.*, **5,** *Tech. Pub.* 1297 (1941). Bond and Maxson, Grindability and Grinding Characteristics of Ores, *Trans. Am. Inst. Mining Met. Engrs.*, **134,** 296 (1939). DeVaney and Coghill, Use of the Coercimeter in Grinding Tests, *Am. Inst. Mining Met. Engrs.*. *Tech. Pub.* 862, 1938. Wilson, A Method for Estimating the Efficiency of Pulverizers, *Trans. Am. Inst. Mining Met. Engrs.*, **129,** 170 (1938). Fahrenwald *et al.*, Ball Mill Studies. II. Thermal Determinations of Ball Mill Efficiency, *Am. Inst. Mining Met. Engrs.*, *Tech. Pub.* 416 (February, 1931). Martin, Laws of Fine Grinding, *Crushing and Grinding*, **1,** 5 (1931). Gross and Zimmerley, Crushing and Grinding. III. Relation of Work Input to Surface Produced in Crushing Quartz, *Trans. Am. Inst. Mining Met. Eng.*, **87,** 35 (1930). Gaudin, Gross, and Zimmerley, The So-called Kick Law Applied to Fine Grinding, *Mining and Met.*, **10,** 447 (1929). Shaw, A Survey of Research on Rock Crushing, *Rock Products*, **32,** 70 (July, 1929). Coghill, Evaluating Grinding Efficiency by Graphical Methods, *Eng. Mining J.*, **126,** 934 (1928). Martin, Recent Research in the Science of Fine Grinding, *J. Soc. Chem. Ind. (London)*, **45,** 160 (1926). Martin *et al.*, Researches on the Theory of Fine Grinding, *Trans. Ceramic Soc. (London)*, **23,** 61 (1924); *ibid.*, **25,** 51, 63, 226, 240 (1925–6); *ibid.*, **26,** 21, 45 (1927); *ibid.*, **27,** 247, 259, 284 (1928). Herman, The Laws of Crushing, *Eng. Mining J.-Press*, **115,** 498 (1923). Gates, Kick vs. Rittinger, An Experimental Investigation in Rock Crushing, *Trans. Am. Inst. Mining Met. Engrs.*, **52,** 875 (1915). Gates, The Crushing-Surface Diagram, *Eng. Mining J.*, **95,** 1039 (1913).

Grinding aids. Anon., Chemical Method Aids in Size Reduction, *Chem. Industries*, **54,** 700 (1944). Dorris, How Solid Carbon Dioxide Assists in Grinding Low-Melting Waxy or Plastic Solids, *Chem. & Met. Eng.*, **51,** 114 (July, 1944). Price, Surface Active Agents, *Am. Ink Maker*, **22,** 21 (1944). Dawley, Grinding Aids for Portland Cement, *Pit & Quarry*, **36,** 57 (July, 1943). Fisher and Jerome, Pigment-dispersion with Surface Active

* The crushing and grinding section of the first and second editions was prepared by S. B. Kanowitz.

† These references supplement those given in the text and serve as a guide to the literature for those seeking additional sources of information. The references are divided by subjects. The newest reference is listed first for each subject. See also the bibliography in Gross, *U.S. Bur. Mines, Bull.* 402, 1938.

Special thanks are due H. J. Kamack for preparing the references.

Agents, *Ind. Eng. Chem.* **35,** 336 (1943). Anon., Evaluation of Surface Active Agents in Pigment Grinding, *Paint Oil & Chem. Rev.*, **102,** 70 (1940). Kennedy and Mardulier, Effect of Dispersion on Cement Raw Materials, *Rock Products*, **44,** 76 (August, 1941). Sweitzer and Craig, Colloidal Carbon as a Grinding Aid in Portland Cement, *Ind. Eng. Chem.*, **32,** 751 (1940). Rockwood, Aids to Clinker Grinding by Use of Dispersing Agent, *Rock Products*, **42,** 38 (May, 1939). Kanowitz, Low Temperature Grinding, *Chem. & Met. Eng.*, **45,** 236 (May, 1938). Manson, Dry Grinding of Enamel and Its Influence on Its Working Conditions, *J. Am. Ceramic Soc.*, **21,** 316 (1938). Kennedy, Portland Cement: Effects of Catalysis and Dispersion, *Ind. Eng. Chem.*, **28,** 963 (1936).

Selection of equipment. Gray and Elliot, Screening and Classification in Grinding Circuits, *Trans. Can. Inst. Mining Met.*, **49,** 466 (1946). Gorman, On Selecting Crusher, *Mining Congr. J.*, **26,** 68 (1940). Reid, Considerations Affecting Selection of Pulverizers, *Combustion*, **10,** 26 (1938). Kennedy, Selection of Ore-crushing and Grinding Equipment, *Mining and Met.*, **17,** 139 (1936). Farrell, Balancing Equipment in Crushed Stone Quarries, *Rock Products*, **37,** 34 (May, 1934). Holman, Crushing and Grinding Appliances—The Connection between Type and Purpose, *Trans. Inst. Chem. Engrs. (London)*, **12,** 186 (1934). Farrant, Types of Modern Grinding Mills, *Crushing and Grinding*, **1,** 55 (1931).

Size classification. Koren, Pulverizers with Air Separation and Air Drying, *Ind. Eng. Chem.*, **30,** 909 (1938). Hardinge, Air Classification in Pulverizing, *Ind. Eng. Chem.*, **26,** 1139 (1934). Rammler and Prockat, Air Separator Tube Mills, *Zement*, **23,** 557 (1934). Newton, A Study of Classification Calculations, *Rock Products*, **35,** 26 (August, 1932). Wright, Recent Developments in the Application of Dry Fine Grinding with Air Classification, *Rock Products*, **33,** 57 (July, 1930).

Safety. Hartmann, Recent Research on the Explosibility of Dust Dispersions, *Ind. Eng. Chem.*, **40,** 752 (1948). Hartmann and Greenwald, The Explosibility of Metal-Powder Dust Clouds, *Mining and Met.*, **26,** 331 (1945). Hartmann and Nagy, Inflammability and Explosibility of Powders Used in the Plastics Industry, *U.S. Bur. Mines, Rept. Investigations*, 3751 (1944). Twiss and McGowan, Dust Explosions, *India-Rubber J.*, **107,** 292 (1944). Hartmann, Nagy, and Brown, Inflammability and Explosibility of Metal Powders, *U.S. Bur. Mines, Rept. Investigations*, 3722 (1943). Keicher, Dust Explosions during the Grinding of Synthetic Resins, *Arbeitschutz*, **1943,** 234 (1943). Anon., Laboratory Studies of the Inflammability of Coal Dusts, *U.S. Bur. Mines, Bull.* 389, 1935. Brown and Hanson, Venting Dust Explosions, *Chem. & Met. Eng.*, **40,** 116 (March, 1933). Brown, The Value of Inert Gas as a Preventive of Dust Explosions in Grinding Equipment, *U.S. Dept. Agr., Tech. Bull.* **74** (1928).

Crushers. Edwards, Jaw Type Crusher with Impact Hammer Action, *Rock Products*, **47,** 80 (November, 1944). Miller and Badger, Roller Crushers, *British Coal Utilization Research Assoc. Bull.* **3,** 619 (1939). Cadena, Grinding with a Hammer Mill at Norris Dam, *Trans. Am. Inst. Mining Met. Engrs.*, **129,** 185 (1938). Bernhard, Bell Shaped Heads and Concaves for Gyratory Crushers, *Mining and Met.*, **13,** 107 (1932).

Ball and rod mills. Norman and Loeb, Wear Tests on Grinding Balls, *Mining Tech.*, **12,** *Tech. Pub.* 2319 (May, 1948). Myers, Michaelson, and Bond, Rod Milling: Plant and Laboratory Data, *Mining Tech.*, **11,** *Tech. Pub.* 2175 (July, 1947). Myers and Tower, Symposium on Milling Devices and Practices, *Mining Tech.*, **11,** *Tech Pub.* 2162 (May, 1947). Timmermans, Factors in Relative Wear of Grinding Ball Sizes, *Eng. Mining J.*, **148,** 78 (May, 1947). Berry, Wear Resistance of Domestic Materials for Pebble Mill Linings, *Mining Tech.*, **10,** *Tech. Pub.* 1948 (March, 1946). Baines, Ball and Pebble Mills, *Chem. Age*, **53,** 390 (1945). Banks, Increasing Efficiency of Fine Grinding, *Mining Tech.*, **9,** *Tech. Pub.* 1890 (1945). Davis, Pulp Densities within Operating Ball Mills, *Mining Tech.*, **9,** *Tech. Pub.* 1843 (1945). Metcalf, Review of Modern Ball Milling, *Mine and Quarry Eng.*, **10,** 3 (1945). Metz, Recent Developments for Grinding Ceramic Materials in Pebble, Ball, and Tube Mills, *Bull. Am. Ceramic Soc.*, **24,** 357 (1945). Michaelson, Determination of Ball-Mill Size from Grindability Data, *Mining Tech.*, **9,** *Tech. Pub.* 1844 (May, 1945). Ramsay, How to Operate a Grinding Circuit, *Eng. Mining J.*, **146,** 96 (1945). Stahl, Short Rod Grinding in Ball Mills, *Mining Tech.*, **9,** *Tech. Pub.* 1821 (1945). Prentice and Davis, Ball Wear in Cylindrical Mills, *Mining Tech.*, **8,** *Tech. Pub.* 1736 (1944). Bond, Wear and Size Distribution of Grinding Balls, *Trans. Am. Inst. Mining Met. Engrs.*, **153,** 373 (1943). Howes, Ball-Mill Liners, *Mining Tech.*, **7,** *Tech. Pub.* 1577 (March, 1943). Bachmann, The Development of the Vibration Mill, *Chem. Tech.*, **15,** 195 (1942). Duggan, Climax Milling Practice, *Mining Tech.*, **6,** *Tech. Pub.* 1456 (March, 1942). Creyke and Webb, Ratio of Water to Solids in Cylinder Grinding, *Trans. Ceramic Soc. (London)*, **40,** 55 (1941). Bond and Agthe, Ball Coating in Grinding, *Mining Tech.*, **4,** *Tech. Pub.* 1160 (March, 1940). Carey, Robey, and Heywood, Centrifugal Ball Mill, *Engineering*, **149,** 378 (1940). Redd, Getting the Best Production from a Ball or Pebble Mill, *Bull. Am. Ceramic Soc.*, **19,** 253 (1940). Wardell, Ball, Rod, and Tube Mills, *Mine and Quarry Eng.*, **5,** 255 (1940). Fahrenwald, Older Ore Dressing Practices Restudied, *Eng. Mining J.*, **140,** 73 (1939). Andreasen, Berg, and Kjaer, Colloid Grinding with a Ball Mill, *Kolloid Z.*, **82,** 37 (1938). Bond, Measuring the Circulating Load. Useful Formulas for Wet and Dry Grinding, *Rock Products*, **41,** 64 (January, 1938). Underwood, Multiple Use of Pebble and Ball Mills, *Ind. Eng. Chem.*, **30,** 905 (1938). Withington, Ball, Rod, and Tube Mills, *Ind. Eng. Chem.*, **30,** 897 (1938). Coghill and DeVaney, Ball Mill Grinding, *U.S. Bur. Mines, Tech. Pub.* 581, 1937. Metz, Grinding Ceramic Materials in Ball, Pebble, Rod, and Tube Mills, *Bull. Am. Ceramic Soc.*, **16,** 461 (1937). Chandler, Relationship of Mill Charge to Surface Area of Cement, *Rock Products*, **38,** 38 (September, 1935). Coghill, DeVaney, and O'Heara, Advantages of Ball Mills of Large Diameters and Advantages of Improved Bearings, *Trans. Am. Inst. Mech. Engrs.*, **112,** 79 (1934). Bennett, Rubber Linings for Rotary Grinding Mills, *Mining and Met.*, **14,** 399 (1933). Fahrenwald and Lee, Ball Mill Studies, *Am. Inst. Mining Met. Engrs., Tech. Pub.* 375, 1931. Gilbert, Grinding Plant Research, *Rock Products*, **34,** 39 (November, 1931); **35,** 27 (February, 1932); **35,** 32 (March, 1932); **35,** 40 (April, 1932); **35,** 24 (June, 1932); **35,** 35 (July, 1932); **35,** 23 (August, 1932); **35,** 23 (October, 1932); **37,** 31 (June, 1934). Gow, Campbell, and Coghill, A Laboratory Investigation of Ball Milling, *Trans. Am. Inst. Mining Met. Engrs.*, **87,** 51 (1930).

Fluid energy grinding. Anon., Developments in Pulverizer Design, *Power Plant Eng.*, **51,** 101 (1947). Murphy, Ross, and Sharpe, Non-Mechanical Methods of Size Reduction, *Brit. Coal Utilization and Research Assoc. Bull.*, **11,** 221 (1947). Anon., Some Novel Methods of Coal Pulverization, *Eng. Boiler House Rev.*, **61,** 147 (1946). Berry, Modern Machines for Dry Size Reduction in Fine Size Range, *Ind. Eng. Chem.*, **38,** 672 (1946).

Paint grinding. Peingault, Grinding Processes in the Paint Industry, *Peintures, pigments, vernis*, **23,** 270 (1947). Mills, Pigment Dispersion by Ball and Pebble Mills, *Natl. Paint Bull.*, **10,** 5 (1946). Bonney, Roller Mill Grinding, A Review, *Official Digest Federation Paint & Varnish Production Clubs*, 237, 345 (1944). Olson, Pebble and Ball Mill Grinding, A Review, *ibid.*, 340. Fischer, The Evolution of Mills for Grinding, *Interchem. Rev.*, **3,** 91 (1944). Draper, Wet Grinding Mills, *Paint Manuf.*, **12,** 196 (1942); Roll Mill Grinding, *ibid.*, **13,** 20 (1943); Ball Mills and Ball Mill Grinding, *ibid.*, **14,** 7 (1944). Fischer, Dispersion of Pigments by Ball and Pebble Mills, *Ind. Eng. Chem.*, **33,** 1465 (1941). Kunze, Grinding of Dye Pigments, *Paint Varnish Production Mgr.*, **21,** 124 (1941). Anon., Lacquer Paste Grinding on Three Roller Mill, *Am. Paint J. Convention Daily*, **25,** 25 (1940). Kendall, Modern Ball and Pebble Mill Technique, *J. Oil Colour Chemists' Assoc.*, **15,** 66 (1932).

Industrial applications. Myers and Lewis, Fine Crushing with a Rod Mill at the Tennessee Copper Company, *Mining Tech.*, **10,** *Tech. Pub.* 2041 (1946). Slegten, Belgian Experiments in Clinker Grinding, *Rock Products*, **49,** 60 (September, 1946). Anon., Mills for Clinker Grinding, *Cement and Lime Manuf.*, **18,** 1 (1945). Biddulph, Milling of Enamels, *Foundry Trade J.*, **75,** 343 (1945). Hales, Sherritt Gordon Crushing Plant, *Western Miner*, **18,** 31 (1945). Freeman *et al.*, Effect of Moisture on Grinding of Tung Kernels and Solvent Extraction of Meal, *Oil and Soap J.*, **21,** 328 (1944). McLaren, The Hollinger Crushing and Grinding Plant, *Can. Mining J.*, **65,** 363 (1944). Anon., Feldspar Mining Plant in North Carolina, *Eng. Mining J.*, **144,** 73 (1943). Anon., Reducing Grinding Costs, *Rock Products*, **45,** 64 (January, 1942). Jones and Holland, Groundwood at 5400 Surface Feet per Minute, *Pulp Paper Mag. Can.* **43,** 141 (1942). Barker and Lewis, Developments in Ball-Mill Grinding Practices at New Cornelia, *Am. Inst. Mining Met. Engrs., Tech. Pub.* 1361, 1941. Dean, Seed Crushing, *Oil Colour Trades J.*, **100,** 449 (1941). Kivari, Milling at Permanente Cement Plant, *Am. Inst. Mining Met. Engrs., Tech. Pub.* 1359, 1941. The Staff, Fine Grinding Investigations at Lake Shore Mines, *Trans. Can. Inst. Mining Met.*, **43,** 299 (1940). Carnochan, Some Modern Methods in Milling of Industrial Materials, *Trans. Can. Inst. Mining Met.*, **42,** 29 (1939). Fellows and McLaughlin, Studies of Milling and Its Effect on Properties of Porcelain Enamel Slips, *J. Am. Ceramic Soc.*, **22**, 260 (1939). Anderson, Coal Crushing Equipment, *Iron and Coal Trades Rev.*, **136**, 405 (1938). Rockwood, A Brief Resume of Trends on Grinding in the Cement Industry, *Rock Products*, **41**, 60 (January, 1938). Tuck, Rod Mill Practice at Ray Mines Division, Kennecott Copper Corp., *Am. Inst. Mining Met. Engrs., Tech. Pub.* 994, 1938. Tyler, Technology and Economics of Ground Mica, *Am. Inst. Mining Met. Engrs., Tech. Pub.* 889, 1938. Holland *et al.*, Groundwood Studies: II. Effect of Process Variables in Grinding Mechanical Pulp, *Can. Pulp Paper Assoc.*, **1935**, 96 (1935).

Particle-size measurement. General. DallaValle, "Micromeritics," Chap. 4, Methods of Particle Size Measurement, 2d ed., Pitman, New York, 1948. Several Authors, Symposium on Particle Size Analysis, *Inst. Chem. Engrs. (London)*, 1947. Heywood, Application of Sizing Analysis to Mill Practice, *Bull. Inst. Mining Met. (London)* 477 (March, 1946). Lapple, Mist and Dust Collection in Industry and Buildings, *Heating, Piping and Air Conditioning*, **16**, 635 (1944); **17**, 611 (1945); **18**, 108 (1946). Taggart, "Handbook of Mineral Dressing," Sec. 19, p. 99, Wiley, New York, 1945. Hawksley, Particle Size Measurement, *British Coal Utilization Research Assoc. Bull.*, **8**, 245 (1944). Piriani, Fine Powders and Particles, *ibid.*, 361. Harvey, Particle Size Analysis, *Interchemical Review*, **3**, 59 (1944). Kendall, Determination of Pigment Particle Size, *Am. Ink Maker*, **21**, 21 (1943); *ibid.*, **22**, 27 (1944). Anon., Apparatus for Particle Size Analysis, *Engineering*, **154**, 141 (1942). Dasher, Directions for Laboratory Mineral Sizing, *U.S. Bur. Mines, Circ.* 7224, 1942. Schweyer, Particle Size Studies, *Ind. Eng. Chem.*, anal. ed., **14**, 622 (1942). Schweyer, Particle Size Studies. Properties of Finely Divided Materials, *Chem. Rev.*, **31**, 295 (1942). Am. Soc. Testing Materials, Symposium on New Methods for Particle Size Determination in the Subsieve Size Range, 1941. Berg, Particle Size Distribution, *Kolloid-Beihefte*, **53**, 149 (1941). Heywood, Measurement of the Fineness of Powdered Materials, *Proc. Am. Inst. Mech. Engrs.*, **140**, 257 (1938). Roller, A Classification of Methods of Mechanical Analysis of Particulate Materials, *Proc. Am. Soc. Testing Materials*, **37**, Part II, 675 (1937). Andreasen, The Grinding of Materials, *Kolloidchem. Beihefte*, **27**, 349 (1928).

Sieving. DallaValle, "Micromeritics," Chap. 5, Theory of Sieving and Grading of Materials, 2d ed., Pitman, New York, 1948. Heywood, A Study of Sizing Analysis by Sieving, *Bull. Inst. Mining Met. (London)* 477, March, 1946. Anon., E11-39, Standard Specifications for Sieves for Testing Purposes, *Am. Soc. Testing Materials, Standards*, **1946**, Part IIIA, 730. Brewer, Comparison of Test Sieves of Different Countries, *U.S. Bur. Mines, Rept. Investigations*, 3766, 1944. Gaudin, "Principles of Mineral Dressing," Chap. 3, Laboratory Sizing, McGraw-Hill, New York, 1939. Weber and Moran, A Precise Method for Sieving Analysis, *Ind. Eng. Chem.*, anal. ed., **10**, 180 (1938). McCalman, The Accuracy of Sieve Testing, *Ind. Chemist*, **13**, 464 (1937); The Manufacture of Test Sieves, *ibid.*, **14**, 64 (1938); An Investigation of British Standard Test Sieves, *ibid.*, **14**, 231 (1938); **15**, 161 (1939). Anon., Screen Testing of Ores, *Am. Standards Assoc. M*-5-1932. Porter, An Experimental Study of Sieving, *Nat. Research Council Can., Rept.* 22, 1928.

Microscopy. Watson, Particle Size Determinations with Electron Microscopes, *Anal. Chem.*, **20**, 576 (1948). Heywood, A Comparison of Methods of Measuring Microscopical Particles, *Bull. Inst. Mining Met. (London)*, 477, 1946. Anon., E20-33T, Tentative Method for Particle Size Distribution of Subsieve Size Particulate Substances, *Am. Soc. Testing Materials, Standards*, **1946**, Part IIIA, 1219. Martin, Use and Limitations of Methods of Particle Size Measurement, *Paint Manuf.*, **13**, 73 (1943). Fairs, The Use of the Microscope in Particle Size Analysis, *Chemistry & Industry (London)*, **62**, 374 (1943). Amberg, Size Determination with the Haemycytometer, *J. Am. Ceramic Soc.*, **19**, 207 (1936). Dunn, Microscopic Measurements for the Determination of Particle Size of Pigments and Powders, *Ind. Eng. Chem.*, anal. ed., **2**, 59 (1930). Perrott and Kinney, The Meaning and Microscopic Measurement of Average Particle Size, *J. Am. Ceramic Soc.*, **6**, 417 (1923). Green, A Photomicrographic Method for the Determination of Particle Size of Paint and Rubber Pigments, *J. Franklin Inst.*, **192**, 637 (1921). McCartney, Determination of the Size Distribution of Fine Coal Particles by the Electron Microscope, *U.S. Bur. Mines, Rept. Investigations*, 3827, 1945. Green and Fullam, Some Applications of the High Resolving Power of the Electron Microscope, *J. Applied Phys.*, **14**, 332 (1943). Anon., E20-48T, Tentative Recommended Practice for Analysis by Microscopical Methods for Particle Size Distribution of Particulate Substances of Subsieve Sizes, *Am. Soc. Testing Materials, Standards*, **1948**, Part IIIB, 283.

Elutriation. Katan, An Air Elutriator for Use with Coherent Dusts, *Chemistry & Industry (London)* **1948**, 119 (1948). Haultain, Splitting the Minus 200 with the Superpanner and Infrasizer, *Trans. Can. Inst. Mining Met.*, **40**, 229 (1937). Traxler and Baum, Determination of Particle Size Distribution in Mineral Powders by Air Elutriation, *Rock Products*, **37**, 44 (June, 1934). Roller, Measurement of Particle Size with an Accurate Air Analyzer, *Proc. Am. Soc. Testing Materials*, **32**, Part II, 607 (1932). Roller, Separation and Size Distribution of Microscopic Particles, *U.S. Bur. Mines, Tech. Pub.* 490 (1931). Gaudin, Groh, and Henderson, Sizing by Elutriation of Fine Ore-dressing Products, *Ind. Eng. Chem.*, **22**, 1363 (1930). Andrews, A New Elutriator, *Mining Mag. (London)*, **28**, 232 (1936). Gonell, An Air Separator for Determining Size Analysis of Dusty Materials, *Z. Ver. deut. Ing.*, **72**, 945 (1928). Schöne, New Apparatus for Elutriation Analyses, *Z. anal. Chem.*, **7**, 29 (1868).

Gravity sedimentation. Barrett, A Method of Particle Size Determination of Solids, Cement, etc., by Means of a Chainomatic Specific Gravity Balance, *Proc. Am. Soc. Testing Materials*, **46**, 1355 (1946). Anon., D422–39, Standard Method of Mechanical Analysis of Soils, *Am. Soc. Testing Materials, Standards*, **1946**, Part II, 652. Berg, The Submerged Weight Method of Determining Size Distribution, *Ber. deut. keram. Ges.*, **23**, 271 (1942). Klein, Improved Hydrometer Method for Use in Fineness Determinations, *Proc. Am. Soc. Testing Materials*, **41**, 953 (1941). Loomis, Grain Size of Whiteware Clays as Determined by the Andreasen Pipette, *J. Am. Ceramic Soc.*, **21**, 393 (1938). Andreasen and Berg, Examples of the Use of the Pipette Method of Particle Size Analysis, *Z. angew. Chem.*, **48**, 283 (1935). Puri, A New Type of Hydrometer for the Mechanical Analysis of Soils, *Soil Science*, **33**, 241 (1932). Bouyoucos, The Hydrometer as a New Method for the Mechanical Analysis of Soils, *Soil Science*, **23**, 343 (1927). Bishop, A Sedimentation Method for the Determination of the Particle Size of Finely Divided Materials Such as Hydrated Lime, *Bur. Standards J. Research*, **12**, 173 (1934). Andreasen, An Apparatus for Particle Size Analysis and an Investigation with It, *Kolloid Z.*, **49**, 253 (1929). Svedberg and Rinde, The Determination of the Distribution of Size of Particles in Disperse Systems, *J. Am. Chem. Soc.*, **45**, 943 (1923). Odén, A New Method for Determination of Particle Size Distribution, *Kolloid Z.*, **18**, 33 (1916). Goodhue and Smith, The Particle Size of Insecticidal Dusts—A New Differential Manometer Type Sedimentation Apparatus, *Ind. Eng. Chem.*, anal. ed., **8**, 469 (1936). Knapp, New Apparatus for Determination of Size Distribution of Particles in Fine Powders, *Ind. Eng. Chem.*, anal. ed., **6**, 66 (1934). Duncombe and Withrow, The Kelly Tube and the Sedimentation of Portland Cement, *J. Phys. Chem.*, **36**, 31 (1932). Lukirsky and Kosman, A Method of Measuring the Size of Particles, *J. Soc. Chem. Ind. (London)*, **46**, 21 (1927). Kraemer and Stamm, A New Method for the Determination of the Distribution of Size of Particles in Emulsions, *J. Am. Chem. Soc.*, **46**, 2709 (1924). Kelly, Determination of Distribution of Particle Size, *Ind. Eng. Chem.*, **16**, 928 (1924). Carey and Stairmand, Size Analysis by Photographic Sedimentation, *Trans. Inst. Chem. Engrs. (London)*, **16**, 57 (1938). Travis, Measurement of Average Particle Size by Sedimentation and Other Physical Means, *Am. Soc. Testing Materials Bull.* 102, 29 (1940).

Centrifugation. Robison and Martin, Beaker-Type Centrifugal Sedimentation of Subsieve Solid-Liquid Dispersions, *J. Phys. Chem.*, **52**, 854 (1948). Jacobsen and Sullivan, Centrifugal Sedimentation Method for Particle Size Distribution, *Ind. Eng. Chem.*, anal. ed., **18**, 360 (1946). Brown, Particle Size Distribution by Centrifugal Sedimentation, *J. Phys. Chem.*, **48**, 246 (1944). Dana, A Pipette Method of Size Analysis for the Centrifuge, *J. Sediment. Petrol.*, **13**, 21 (1943). Martin, Particle Size Distributions of Pigment Suspensions—Determination with a Beaker-Type Centrifuge, *Ind. Eng. Chem.*, anal. ed., **11**, 471 (1939). Norton and Speil, Measurement of Particle Size in Clays, *J. Am. Ceramic Soc.*, **21**, 89 (1938). Hauser and Reed, Studies in Thixotropy: I. Development of a New Method for Measuring Particle Size Distribution in Colloidal Systems, *J. Phys. Chem.*, **40**, 1169 (1936). Marshall, Studies in the Degree of Dispersion of Clays: I. Notes on the Technique and Accuracy of Mechanical Analysis Using the Centrifuge, *J. Soc. Chem. Ind. (London)*, **50**, 444 (1931). Svedberg and Nichols, Determination of Size and Distribution of Size of Particle by Centrifugal Methods, *J. Am. Chem. Soc.*, **45**, 2910 (1923).

Permeability. Pechukas and Gage, Rapid Method for Determining Specific Surface of Fine Powders, *Ind. Eng. Chem.*, anal.

ed., **18,** 370 (1946). Rigden, The Specific Surface of Powders: A Modification of the Air Permeability Method for Rapid Routine Testing, *J. Soc. Chem. Ind.* (*London*), **62,** A1 (1943). Blaine, Air Permeability Fineness Apparatus, *Am. Soc. Testing Materials Bull.* 123, 51 (1943). Blaine, Studies of the Measurement of Specific Surface by Air Permeability, *Am. Soc. Testing Materials Bull.* 108, 17 (1941). Gooden and Smith, Measuring Average Particle Diameter of Powders, *Ind. Eng. Chem.*, anal ed., **12,** 479 (1940). Lea and Nurse, The Specific Surface of Fine Powders, *J. Soc. Chem. Ind.* (*London*), **58,** 277 (1939). Carman, The Determination of the Specific Surface of Powders. *J. Soc. Chem. Ind.* (*London*), **47,** 225 (1938). Arnell, Surface Area Measurements, *Can. J. Research*, **24A,** 103 (1946).

Turbidimetry. Anon., C115-42, Standard Method of Test for Fineness of Portland Cement by the Turbidimeter, *Am. Soc. Testing Materials, Standards,* **1946,** Part II, 62. Klein, A Suspension Turbidimeter for Determination of Specific Surface of Granular Materials, *Proc. Am. Soc. Testing Materials*, **34,** Part II, 303 (1934). Stutz and Pfund, Relative Method for Determining Particle Size of Pigments, *Ind. Eng. Chem.*, **19,** 51 (1927).

Adsorption. Anderson and Emmett, Measurement of Carbon Black Particles by the Electron Microscope and Low Temperature Nitrogen Adsorption Isotherms, *J. Applied Phys.*, **19,** 367 (1948). Jura and Harkins, Surfaces of Solids. XI. Determination of the Decrease of Free Surface Energy of the Solid by an Adsorbed Film, *J. Am. Chem. Soc.*, **66,** 1356 (1944). Kraemer (editor), "Advances in Colloid Science," Chap. 1, Emmett, The Measurement of the Surface Areas of Finely Divided or Porous Solids by Low Temperature Adsorption Isotherms, Interscience Publishers, New York, 1942. Emmett and DeWitt, Determination of Surface Areas, *Ind. Eng. Chem.*, anal. ed., **13,** 28 (1941). Smith and Green, Measuring the Specific Surfaces of Particulate Substances, *Ind. Eng. Chem.*, anal. ed., **14,** 382 (1942). Emmett and Brunauer, The Use of Low Temperature van der Waals Adsorption Isotherms in Determining the Surface Area of Iron Synthetic Ammonia Catalysts, *J. Am. Chem. Soc.*, **59,** 1553 (1937). Harkins and Gans, An Adsorption Method for the Determination of the Area of a Powder, *J. Am. Chem. Soc.*, **53,** 2804 (1931).

Miscellaneous. Bailey, Particle Size by Spectral Transmission, *Ind. Eng. Chem.*, anal. ed., **18,** 365 (1946). Guinier, Determination of the Size of Submicroscopic Particles by X-rays, *U.S. Bur. Mines, Inf. Circ.* 7391, 1946. Zimens, Surface Determination and Diffusion Measurements by Radioactive Noble Gases, *Z. physik. Chem.*, **A191,** 1 (1942); **A192,** 1 (1943). Harkins and Jura, An Absolute Method for the Determination of the Area of a Fine Crystalline Powder, *J. Chem. Phys.*, **11,** 430 (1943). Jones, Particle Size Measurements by the X-ray Method, *J. Sci. Instruments*, **18,** 157 (1941). Gottschalk, Coercive Force of, Magnetic Powders, *U.S. Bur. Mines, Rept. Investigations*, 3268, 83, 1935.

Representation of size distribution. Gaudin and Hukki, Principles of Comminution: Size and Surface Distribution, *Trans. Am. Inst. Mining Met. Engrs.*, **169,** 67 (1946). Lapple, Mist and Dust Collection in Industry and Buildings, *Heating, Piping and Air Conditioning*, **18,** 108 (1946). Griffith, Size Distribution of Particles, *Can. J. Research*, **A21,** 57 (1943). Austin, Methods of Representing Distribution of Particle Size, *Ind. Eng. Chem.*, anal. ed., **11,** 334 (1939). Otto, A Modified Probability Graph for the Interpretation of Mechanical Analyses of Sediments, *J. Sediment, Petrol.*, **9,** 62 (1939). Hazen, Storage To Be Provided in Impounding Reservoirs for Municipal Water Supply, *Trans. Am. Soc. Civil Engrs.*, **77,** 1539 (1914). Roller, Law of Size Distribution and Statistical Description of Particulate Materials, *J. Franklin Inst.*, **223,** 609 (1937). Bennett, Broken Coal, *J. Inst. Fuel* (*London*), **10,** 22 (1936). Hatch, Determination of Average Particle Size from the Screen Analysis of Non-Uniform Particulate Substances, *J. Franklin Inst.*, **215,** 27 (1933). Rosin and Rammler, The Laws Governing the Fineness of Powdered Coal, *J. Inst. Fuel* (*London*), **7,** 29 (1933). Schuhmann, Principles of Comminution: I. Size Distribution and Surface Calculations, *Mining Tech.*, **4,** *Tech. Pub.* 1189 (1940). Hatch and Choate, Statistical Description of the Size Properties of Non-Uniform Particulate Substances, *J. Franklin Inst.*, **207,** 369 (1929). Gaudin, An Investigation of Crushing Phenomena, *Trans. Am. Inst. Mining Met. Engrs.*, **73,** 253 (1926).

PRINCIPLES OF CRUSHING AND GRINDING

Introduction

Crushing and Grinding. The broad field of size reduction can be divided into crushing, grinding, cutting, machining, flaking, emulsification, spraying, and gas dispersion:

Size reduction					
Crushing	Grinding		Miscellaneous		
Primary	Pulverizing	Disintegration	Cutting	Emulsification	Gas dispersion
Secondary	Coarse	Coarse	Machining	Spraying	
	Fine	Fine	Flaking		

The term "grinding" has become generic in common usage. It refers to both pulverizing and disintegration. The operations differ because of the nature of the feed material, its size, and the reduction ratio that can be attained. These characteristics, among others, fix the design of the equipment to be employed. The size limitations of the various operations are shown in Table 2 on p. 1117. The distinction between pulverizing and disintegration lies in the physical homogeneity of the material being handled. Disintegration applies to the reduction in size of aggregates of soft particles that are feebly bonded, and it is implied that there is no change in the size of the ultimate particles of the mass [Work, Am. Soc. Mech. Engrs., RP-55-6, 75 (1932)].

Properties of Solids. A single particle or lump has linear size, surface, hardness, and structure. The linear size may be the diameter in case of a sphere, an edge length for a cube, or some fictitious average linear dimension in case of an irregular-shaped lump. The surface is the exterior of most particles, although some have interstitial surface. The surface is readily calculated for cubes and spheres but must be estimated for irregular shapes in most cases. Hardness is measured by the conventional scratch criterion. Structure may be homogeneous or heterogeneous. A mixture of particles such as those in a powder has particle-size distribution, surface, specific surface, limiting particle size, and grindability. Particle-size distribution is the functional relation of the distribution with respect to size of individual particles in the powder. The surface is a summation of the individual grain surfaces, and specific surface is the surface of a unit of weight or volume. Limiting particle size is the size of the largest or smallest particles in the powder. Grindability is a measure of the grinding characteristics of a material. A number of grindability methods subject a sample of sized material to a miniature grinding operation with controlled energy input and a measured size reduction, in which one of these is fixed and the other is used as an index of grindability.

The ability of a material to withstand reduction generally depends upon its hardness, but its structure and the manner in which it is fractured frequently are important factors [Gaudin, *Trans. Am. Inst. Mining Met. Engrs.*, **73,** 253 (1926)]. A foliated talc or a flake graphite is reduced easily to a certain size beyond which further reduction is accomplished only with difficulty. Other factors affecting the grinding characteristics of a material are water of combination, hygroscopicity, tendency to flocculate and agglomerate, combustibility, and sensitiveness to changes in temperature. Glauber's salt, for instance, gives off water of crystallization at a comparatively low temperature, causing clogging of equipment; calcium chloride is so hygroscopic that it may actually dissolve in the moisture absorbed; synthetic resins and gums become soft and plastic beyond a critical temperature; other materials may burn or char; certain chemicals and dyestuffs are unstable and may ignite or explode if the temperature is excessive; and many mineral pigments such as ochers and siennas tend to change in color at elevated temperature.

Moisture in material being pulverized is a commonly encountered factor having a marked effect on the performance of equipment. The curves in Fig. 1 show decreasing production rate with increasing moisture content [Work, *Chem. & Met. Eng.*, **40**, 306 (1933)]. All three materials were being ground to 99.9 per cent through a No. 200 sieve. The fineness to which a material is ground has a marked effect on its production rate. Figure 2 shows the relation among capacity, power, and cost of pulverizing coal to different finenesses.

Objectives. The two immediate objectives of crushing and grinding are to obtain products meeting maximum or minimum limiting size specifications, or both, and to produce materials to meet specific-surface requirements. Sometimes certain lump-shape requirements have to be met in crushing operations. The ultimate objectives are numerous. The production of surface to obtain completeness and rapidity of chemical reaction often is a requirement. The production of the surface within maximum and minimum size limits is important in many instances.

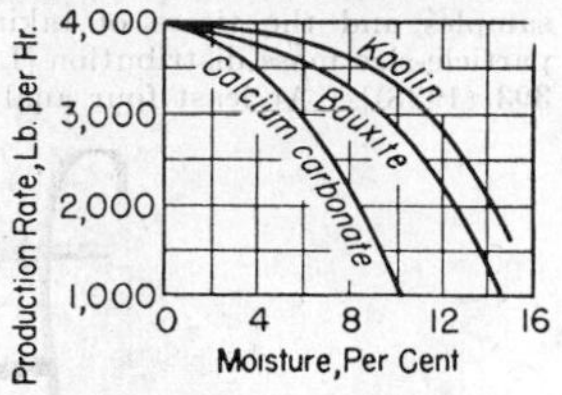

FIG. 1. Effect of moisture on production rate of a pulverizer. [*Work, Chem. & Met. Eng.*, **40**, 306 (1933).]

Product Specifications. The most complete description of a powder is given by its particle-size distribution, which can be plotted in the form of a frequency- or a cumulative-distribution graph, as shown in Fig. 3. The frequency-distribution curves form the smoothed outline of the histograms obtained by graphical differentiation of the cumulative-distribution curves. The latter are the integrated forms of the former. The powder *A* has a narrower size range for the bulk of its mass than does *B*. Both materials have the same size at the point marked with the arrow, and it would be possible to have any

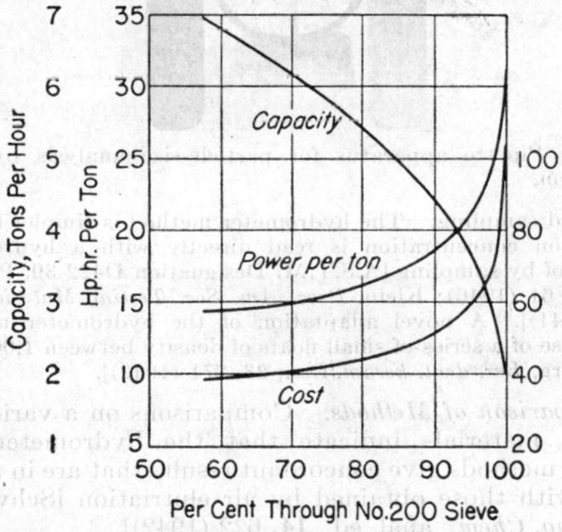

FIG. 2. The variation in capacity, power, and cost of grinding relative to the fineness of product.

number of distribution curves passing through this point. The statement of a single mesh size through which the bulk of a powder will pass does not describe it sufficiently in most cases. The statement that "85 per cent by weight will pass through a No. 60 sieve," for example, is a limiting size specification. Two points are a minimum number for determining size distribution. Such a specification might be "85 per cent through a No. 60 and 5 per cent through a No. 325 sieve." Complete particle-size analysis allows the complete particle-size distribution to be obtained. From this, the specific surface can be calculated. The specific surface varies inversely as the particle size.

It is common in some industries, such as Portland-cement manufacturing, to specify the specific surface of the dry pulverized product. Limiting size specifications are not required but remain within close tolerances for a given surface specification because of the operating consistency of the equipment and uniformity of clinker in grinding.

Equipment manufacturer's performance tabulations generally are based on approximately 85 per cent by weight of product being smaller than the specified size if a statement of the percentage is not given.

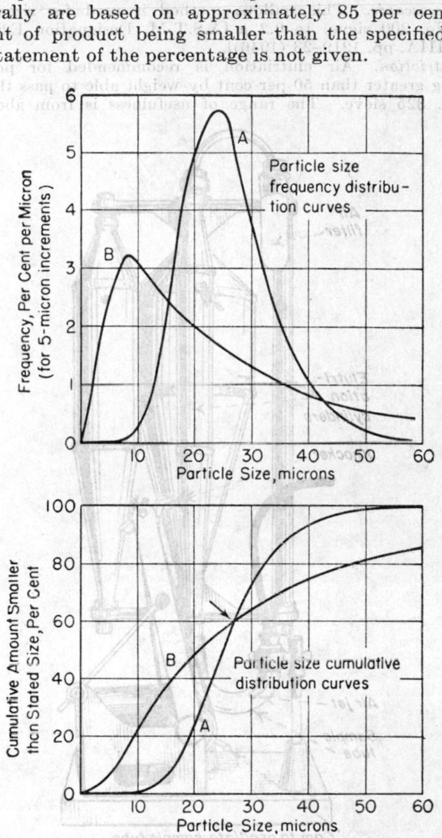

FIG. 3. Particle-size distribution curves for simple powders.

Particle-size Measurement

Size distribution is a fundamental attribute of particulate materials, and measurement methods are essential for characterizing the results of size-reduction operations.

Methods. These include sieving, microscopic, elutriation, gravity sedimentation, centrifugal sedimentation, permeability, turbidimetric, and adsorption methods (Schweyer and Work, Am. Soc. Testing Materials Symposium on New Methods for Particle Size Determination in the Subsieve Range, pp. 1–22, 1941).

Sieving. Coarse sieves cover the size range from 4¼ to ¼ in. (No. 3 sieve) and fine sieves from 0.223 (No. 3½ sieve) to 0.0015 in. (37 μ or No. 400 sieve) size; testing sieve specifications and calibration methods are available [A.S.T.M. Designation E11-39, Part IIIA, pp. 730–36 (1946)]. In size-analysis work, the apertures are understood to be square in a sieve and round in a screen [A.S.T.M. Designation E13-42, Part IIIA, p. 748 (1946)]. The use of testing sieves is described in the W. S. Tyler Co. catalogue 53. Dry sieving is used unless the material is so fine that flocculation interferes. In such cases wet sieving is employed. A combination of the two procedures is often advantageous; wet sieving to remove slimes or material able to pass through a No. 325 sieve, for example, and dry sieving on the dried, retained portion (Methods for Screen Testing of Ores, Am. Standards Assoc., M5-32). Material smaller than No. 325 is determined by difference.

Standard procedures are given for sieve analyses of specific materials such as A.S.T.M. Designations B214-46T, C92-46, C136-46, D197-30, D293-29, D392-38, D502-39 for metal powders, refractories, concrete aggregate, powdered coal, coke, molding powders, and soaps, respectively.

Microscopic. This tedious method is used for sizes from 74 (No. 200 sieve) to 0.2 μ [A.S.T.M. Designation E20-33T, Part IIIA, pp. 1219–23 (1946)].

Elutriation. Air elutriation is recommended for powders having greater than 50 per cent by weight able to pass through a No. 325 sieve. The range of usefulness is from about 74

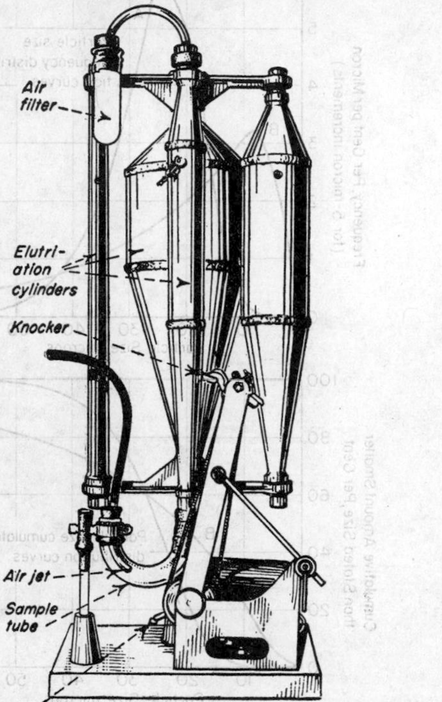

FIG. 4. Air elutriator for particle-size analysis. (*Roller Analyzer.*)

(No. 200 sieve) to 5 μ, depending on the density and dispersibility of the material being fractionated. This method may be exemplified by the Roller analyzer (Roller, *U.S. Bur. Mines, Tech. Pub.*, 490, 1931). Materials are separated into closely sized fractions (< 5, 5–10, 10–20, 20–40, 40–80, > 80, for example) by carrying upward in a controlled stream of air those particles too small to settle against the upward velocity and removing them from the air stream for weighing in a paper filter, as shown in Fig. 4. A sample having a bulk volume of about 25 cc. is required. The method is based on Stokes's law:

$$D = \sqrt{\frac{18\mu u \times 10^8}{(\rho_s - \rho_f)g}}$$

where μ is viscosity, poises; u is air velocity, cm./sec.; ρ_s is density of solid particle, g./cc.; ρ_f is density of fluid; D is diameter of a sphere, microns; and g is acceleration due to gravity, (cm./sec.)/sec. For non-spherical particles, D is the equivalent diameter of a sphere rising at the same rate as the particle. Corrections for deviation from Stokes's law may be necessary at the large end of the size range (see p. 1019). (See p. 1085 for a description of the Haultain Infrasizer and p. 1084 for an account of the Cooke short-column hydraulic elutriator.)

Gravity Sedimentation. Sedimentation methods are useful in the size range from 20 to 2.0 μ in a gravitational field with water as the suspending fluid. Materials that are so finely divided that they are not readily dispersed in the air elutriator usually can be dispersed in water or other liquid if an appropriate dispersing agent is employed. The pipette and hydrometer methods are the most acceptable applications of the gravity-sedimentation method. In the Andreasen modification of the pipette method [Andreasen, *Kolloid Z.*, **49,** 253 (1929)], the thoroughly dispersed material is allowed to settle and definite volumes of suspension are withdrawn from a fixed level at predetermined time intervals corresponding to the desired size limits (2.5, 5.0, 10.0, and 20 μ, for example) as calculated from Stokes's law (Fig. 5). The weight of the dried residues in the samples and the times of taking them are used to calculate particle-size mass distribution [Loomis, *J. Am. Ceramic Soc.*, **21,** 393 (1938)]. At least four analyses can be made at a time by

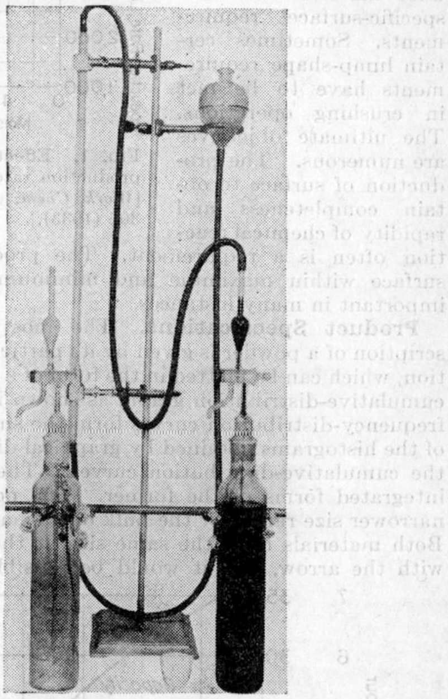

FIG. 5. Pipette apparatus for particle-size analysis by sedimentation.

staggered sampling. The hydrometer method is simpler in that suspension concentration is read directly with a hydrometer instead of by sampling [A.S.T.M. Designation D422-39, Part II, pp. 652–61 (1946); Klein, *Proc. Am. Soc. Testing Materials*, **41,** 953 (1941)]. A novel adaptation of the hydrometer method makes use of a series of small floats of density between 1.001 and 1.01 [Berg, *Ber. deut. keram. Ges.*, **23,** 271 (1942)].

Comparison of Methods. Comparisons on a variety of ground materials indicate that the hydrometer and pipette methods give concordant results that are in agreement with those obtained by air elutriation [Schweyer, *Ind. Eng. Chem.*, anal. ed., **14,** 622 (1942)].

Centrifugal Sedimentation. The particle-size range can be extended to 0.01 μ by use of centrifugal fields and modifications of the hydrometer, pipette, and Odén methods [Norton and Speil, *J. Am. Ceramic Soc.*, **21,** 89 (1938); Dana, *J. Sediment. Petrol.*. **13,** 21 (1943); Jacobsen and Sullivan, *Ind. Eng. Chem.*, anal. ed. **18,** 360 (1946)].

Permeability. The air-permeability method gives only average particle size, not particle-size-distribution information, and is useful in the range from 50 to 0.1 μ. The method is based on the relation between specific surface of packed particles and their permeability. [Carman, *J. Soc. Chem. Ind.* (*London*), **57,** 225 (1938)]. The "Sub Sieve Sizer" apparatus is available commercially for taking advantage of the air permeability method [Gooden and Smith, *Ind. Eng. Chem.*, anal. ed., **12,** 479 (1940)].

Modifications. The above-described methods are believed to be those most useful in connection with crushing and grinding work, but there are many modifications and variations; some of them are described in the references at the beginning of the section.

Turbidimeter. Turbidimetric methods are based on the property of suspensions of fine particles to affect the transmission of light through them [Stutz and Pfund, *Ind. Eng. Chem.*, **19**, 51 (1927)]. Use of the turbidimeter has been standardized for Portland cement [A.S.T.M. Designation C115-42, Part II, pp. 62–9 (1946)].

Adsorption. The gas-adsorption method has application for measurement of the surface of finely divided solids having specific surfaces of the order of 1×10^4 to 100×10^4 sq. cm./g. The amount of a gas such as nitrogen adsorbed in a unimolecular layer on the surface of the solid is measured [Emmett and Brunauer, *J. Am. Chem. Soc.*, **59**, 1553 (1937)]. A method that does not require the assumption of a molecular area for the gas molecules has been described [Jura and Harkins, *J. Am. Chem. Soc.*, **66**, 1356 (1944)]. There is close agreement between these methods.

Shape Factor in Size Measurement. Particle shape should be considered because it affects the relation between sizes measured by different means and the evaluation of specific surface from particle-size-distribution information. The surface shape factor k_a for spheres is 3.14 ($3.14 = \pi = A/D^2$); the volume shape factor k_v is 0.52 ($0.52 = \pi/6 = V/D^3$); and the specific-surface shape factor k_s is 6 ($6 = AD/V = k_a/k_v$). Values for nonspherical particles such as crushed quartz are $k_a = 2.5$, $k_v = 0.27$, and $k_s = 9.3$ for a linear dimension which is the straight line dividing a particle lying on a microscope slide into two apparently equal parts [Martin, *Trans. Brit. Ceramic Soc.*, **27**, 285 (1928)].

Although volume shape factors for flaky materials such as mica are as small as 0.003, factors for many crushed materials lie between 0.2 and 0.4 based on mean projected diameters of the particles [Heywood, *Proc. Inst. Mech. Engrs. (London)*, **140**, 257 (1938)]; and from these it can be shown from Heywood's data that the ratio of sedimentation size to sieve-aperture size should be 0.86 to 0.95, respectively. However, comparisons of No. 325 sieve analyses with sedimentation analyses show 53 μ to be the sedimentation size corresponding to the No. 325 sieve (44-μ aperture), a ratio of 1.2 [Schweyer, *Chem. Rev.*, **31**, 295 (1942)].

Particle-size Representation

Analytical Relationships. A number of equations have been proposed to correlate the quantity of a particulate material with its particle size to obtain a distribution relationship [Austin, *Ind. Eng. Chem.*, anal. ed., **11**, 334 (1939)], but it is unlikely that many practical powders comply with them throughout their complete size range. The relations do have value, however, within the limits of their application. The logarithmic probability relation has general usefulness, and plotting paper designed by Hazen is available (Codex Book Co., Norwood, Mass., Logarithmic Probability Paper No. 3228). Many a cumulative particle-size distribution can be represented by a line that is approximately straight or by two straight lines on logarithmic probability paper. The mathematical treatment of this relationship to obtain the properties of particulate materials has been published [Hatch and Choate, *J. Franklin Inst.*, **207**, 369 (1929). Hatch, *J. Franklin Inst.*, **215**, 27 (1933)]. A comparison of a number of analytical relationships is shown in Fig. 6 for a hypothetical powder [see Lapple, *Heating, Piping Air Conditioning*, **18**, 108 (1946)].

Average Particle Size. The average particle size of a powder may be obtained from its distribution relationship (Hatch, *loc. cit.*), and it can have a number of values depending on the property to be accented: weight or volume, surface, and specific surface (Hatch and Choate, *loc. cit.*; A.S.T.M. Designation E 20-33T, Part IIIA, pp. 1219–23 (1946). Volume and specific-surface average sizes are expressed, respectively, as $\sqrt[3]{\Sigma\,\Delta W/\Sigma\,\Delta W\,D_m^{-3}}$ and $\Sigma\,\Delta W/\Sigma\,\Delta W\,D_m^{-1}$ on a weight basis; and as $\sqrt[3]{\Sigma\,\Delta n\,D_m^3/\Sigma\,\Delta n}$ and $\Sigma\,\Delta n\,D_m^3/\Sigma\,\Delta n\,D_m^2$ on a number basis; where ΔW and Δn are incremental weight and number of particles, respectively, and D_m is the mean size of the increment.

When determining size distribution, a powder should be divided into size increments bearing definite relation to each other. The accepted practice for testing sieves is that the ratio of the length of aperture of a sieve to that of the next finer sieve in the series is $\sqrt{2}$. This geometrical progression can be extended indefinitely to molecular size. Ordinal numbers can be assigned to the size increments beginning with 1 for a mean aperture of $1 \times 10^{-4.07}$ μ and extending through 41 and 48, for example, for mean apertures of 89 and 1000 μ, respectively. The

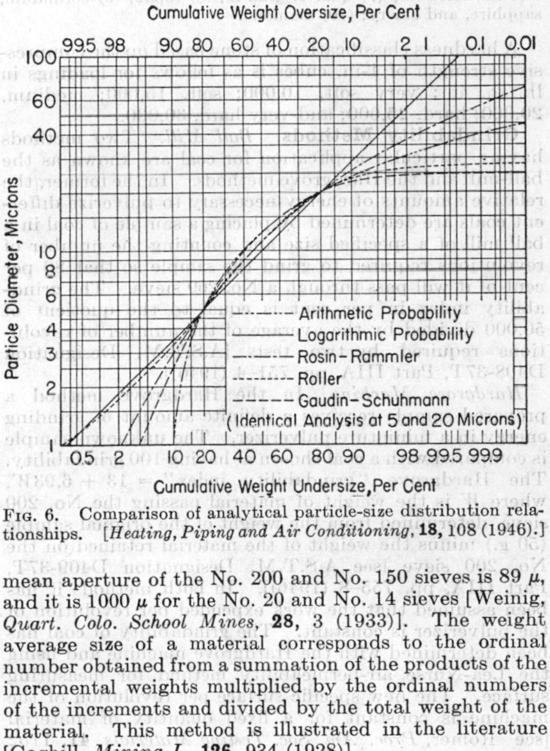

FIG. 6. Comparison of analytical particle-size distribution relationships. [*Heating, Piping and Air Conditioning*, **18**, 108 (1946).]

mean aperture of the No. 200 and No. 150 sieves is 89 μ, and it is 1000 μ for the No. 20 and No. 14 sieves [Weinig, *Quart. Colo. School Mines*, **28**, 3 (1933)]. The weight average size of a material corresponds to the ordinal number obtained from a summation of the products of the incremental weights multiplied by the ordinal numbers of the increments and divided by the total weight of the material. This method is illustrated in the literature [Coghill, *Mining J.*, **126**, 934 (1928)].

Reduction Ratio. The reduction ratio obtained for size-reduction operations is more properly represented as the ratio of average size of feed to average size of product rather than as the ratio of maximum sizes or of arbitrary sizes for which the same percentage is smaller in the feed and product.

Specific Surface. This can be expressed as surface per unit weight or volume. It can be obtained from the distribution relationship of a particulate material (Hatch, *loc. cit.*) or with the aid of ordinal numbers (Coghill, *loc. cit.*). In general, it can be expressed as $s = (k_s/\rho)(\Sigma\,\Delta W\,D_m^{-1}/\Sigma\,\Delta W)$ where s is specific surface in sq. cm./g., k_s is specific-surface shape factor and is equal to 6 for spheres, ρ is density of the material in g./cc., ΔW is incremental weight in g., and D_m is the mean size of the increment in cm.

Grindability

Hardness. The hardness of a mineral as measured by the Moh scale is a criterion of its resistance to crushing [Fahrenwald, *Trans. Am. Inst. Mining. Met. Engrs.*, **112**, 88 (1934)]. It is a fairly good indication of the abrasive character of the mineral, a factor that determines the wear on the grinding mediums. Arranged in increasing order of hardness, the scale is as follows: 1, talc; 2, gypsum; 3, calcite; 4, fluorite; 5, apatite; 6, feldspar; 7, quartz; 8, topaz; 9, corundum; 10, diamond.

For the purpose being considered, the tabulation may be expanded; and the materials of hardness 1 to 4, inclusive, may be classed as "soft" and the others as "hard":

Soft Materials. (1) Talc, dried filter-press cakes, waxes, soapstone, and aggregated salt crystals; (2) gypsum, rock salt, crystalline salts in general, graphite, and soft coal; (3) calcite, marble, soft limestone, barytes, chalk, and brimstone; (4) fluorite, soft phosphate, magnesite, and limestone.

Hard Materials. (5) Apatite, hard phosphate, hard limestone, chromite, and bauxite; (6) feldspar, ilmenite, orthoclase, and hornblendes; (7) quartz, granite; (8) topaz; (9) corundum, sapphire, and emery; (10) diamond.

A hardness classification of stone based on the compressive strength of 1-in. cubes is as follows for loadings in lb./sq. in.: very soft, 10,000; soft, 15,000; medium, 20,000; hard, 25,000; and very hard, 30,000.

Grindability Methods. *Ball Mill.* Two methods having particular application for coal are known as the ball-mill and the Hardgrove methods. In the former, the relative amounts of energy necessary to pulverize different coals are determined by placing a sample of coal in a ball mill of a specified size and counting the number of revolutions required to grind the sample so that 80 per cent of it will pass through a No. 200 sieve. The grindability index in per cent is equal to the quotient of 50,000 divided by the average of the number of revolutions required by two tests [A.S.T.M. Designation D408-37T, Part IIIA, pp. 751–4 (1946)].

Hardgrove Machine. In the Hardgrove method a prepared sample receives a definite amount of grinding energy in a miniature pulverizer. The unknown sample is compared with a coal chosen as having 100 grindability. The Hardgrove "grindability index" = $13 + 6.93W$, where W is the weight of material passing the No. 200 sieve, determined from the weight of the original sample (50 g.) minus the weight of the material retained on the No. 200 sieve [see A.S.T.M. Designation D409-37T, Part IIIA, pp. 755–8 (1946)]. In both methods it has been assumed that the work expended per revolution of the pulverizer is constant. The grindability of coal has been determined with the Hardgrove machine and using the Lea-Nurse air-permeability method for measuring surface. The new specific surface per revolution of the machine is constant for a fixed quantity of material [see Romer, *Proc. Am. Soc. Testing Materials*, **41**, 1152 (1941)].

Indexes Compared. A comparison of the indexes is:

Hardgrove	20	30	40	50	60	70	80	90	100	110
Ball mill	14	21	28	36	44	52	60	70	80	90

The use of these methods has not come into general use for materials other than coal because of the diversity of grinding equipment that might be used and the dissimilar physical and chemical characteristics of materials to be pulverized. However, grindability values have been reported for many materials [Hardgrove, *Trans. Am. Inst. Chem. Engrs.*, **34**, 131 (1938)].

Other Grindability Tests. Standard ball-mill and rod-mill grindability tests have been made on many ores and minerals (Bond and Maxson, *Mining Tech., Tech. Pub.*, 1579, 1943). The chief purpose of the standard grindability tests is to estimate the size of mill needed to produce a specified tonnage, and the power requirement for grinding. Ball-mill tests are made at a simulated 250 per cent circulating load. Tests are conducted at the mesh size to which the material is to be ground in practice. The grindability of the sample is the number of net grams of screen undersize produced per revolution of the mill. In the case of grinding to No. 48 sieve fineness, values of 0.304 and 9.26 are reported as the grindabilities for petroleum coke and graphite, respectively. The standard rod-mill tests are made at a simulated 100 per cent circulating load, and the results are reported in a similar manner. Values of 9.26 and 59.5 are given for quartzite and barite, respectively, when grinding to No. 20 sieve fineness.

The net work in hp.-hr. required to grind 1 ton of a material to a maximum limiting size in a channel-roller pulverizer may be taken as its grindability. Many industrially important materials have been tested in this manner (Coe and Coghill, *U.S. Bur. Mines, Report Investigations*, 3704, 1943); no correlation is shown among specific gravity, mineralogical hardness, and grinding resistance.

Equipment manufacturers maintain laboratories in which grindability tests are made to determine the suitability of their machines.

Work Required for Size Reduction

Theory. *Kick's Law.* On the basis of stress-analysis theory for plastic deformation within the elastic limit, the work required for crushing a given quantity of material is constant for the same reduction ratio, irrespective of the original size [Kick, "Das Gesetz der proportionalen Widerstande und seine Anwendung," Leipzig, 1885. See also Stadler, *Trans. Inst. Mining Met. (London)*, **19**, 471, 509 (1910); **20**, 420 (1911)]. The law can be written

$$E = C \log \frac{D_1}{D_2} \tag{1}$$

where (D_1/D_2) is the size-reduction ratio and E is the work done.

Rittinger's Law. Another theoretical analysis states that the work consumed for reduction of particle size is directly proportional to the new surface produced (Rittinger, "Lehrbuch der Aufbereitungskunde," Ernst and Korn, Berlin, 1867).

$$E = C'(s_2 - s_1) \tag{2}$$

where s_2 and s_1 are the specific surface values after and before size reduction, respectively; but $s = (k_s/\rho)(D^2/D^3)$; so that

$$E = C''\left(\frac{1}{D_2} - \frac{1}{D_1}\right) \tag{3}$$

Generalized Relation. A differential equation for both cases (Walker, Lewis, McAdams, and Gilliland, "Principles of Chemical Engineering," 3d ed., McGraw-Hill, New York, 1937) is

$$dE = \frac{C\,dD}{D^n} \tag{4}$$

Solutions of the equation for n equal to 1 and 2 result in the Kick and Rittinger laws, respectively. For $n > 1$, the solution is

$$E = \left(\frac{C}{n-1}\right)\left(\frac{1}{D_2^{\,n-1}} - \frac{1}{D_1^{\,n-1}}\right) \tag{5}$$

Application. The greater applicability of the Rittinger relation has been demonstrated by various investigators [Gross and Zimmerley, *Trans. Am. Inst.*

Mining Met. Engrs., **87**, 35 (1930). See also Bond and Maxson, *Trans. Am. Inst. Mining Met. Engrs.*, **134**, 296 (1939)]. Equations (1) and (3) contain constants that are reciprocal efficiency coefficients. The efficiency coefficient $(n - 1)/C$ is variable in Eq. (5) and may have greater usefulness over wide ranges of size.

In the use of the equations, the material will possess a distribution of sizes before and after grinding; and, for Eq. (3), $1/D = \Sigma(\Delta W/D_m)/\Sigma(\Delta W)$, where ΔW is the mass of a closely sized increment of the material and D_m is the average of the size limits of the increment. A sample of material can be divided into orderly sized increments by a number of particle-size measurement methods. In the case of testing sieves, a geometric series is used such that the edge length of the square openings of the sieve through which the increment of material passes is $\sqrt{2}$ or 1.4 times the edge length of the openings in the retaining sieve (see also p. 1111). The chief difficulty encountered is the assignment of a value to D_m for the portion of material too fine to be subdivided and in which a major portion of the surface may be present.

If the equation of part of the size distribution is known and it is assumed to extend indefinitely into the small sizes, the specific surface can be calculated. This can be done for a number of representations of particle-size distribution [Gaudin, *Trans. Am. Inst. Mining Met. Engrs.*, **73**, 253 (1926). Hatch, *J. Franklin Inst.*, **215**, 27 (1933). Rosin and Rammler, *J. Inst. Fuel*, **7**, 29 (1933)].

The specific surface (sq. cm./g.) and the size (cm.) of a particle having the same specific surface as the whole powder can be determined directly by use of the air-permeability method [Lea and Nurse, *J. Soc. Chem. Ind.*, **58**, 277 (1939). Gooden and Smith, *Ind. Eng. Chem.*, anal. ed., **12**, 479 (1940)]. The Lea-Nurse apparatus has been used to determine the surface of coal pulverized in the Hardgrove machine (see p. 1114), which is shown to operate in accordance with Rittinger's law [Romer, *Proc. Am. Soc. Testing Materials*, **41**, 1152 (1941)]. Gas-adsorption methods may also be used for measuring the surface of pulverized materials (Gaudin and Hukki, *Mining Tech.*, **8**, *Tech. Pub.*, 1779, 1944).

Grinding Efficiency

Theoretical. The energy efficiency of grinding operations is 0.06 to 1 per cent based on theoretical values of the surface energy of quartz [Martin, *Trans. Inst. Chem. Engrs. (London)*, **4**, 42 (1926); Gaudin, *Trans. Am. Inst. Mining Met. Engrs.*, **73**, 253 (1926)].

Measured. Grinding efficiency ranges from 6 to 25 per cent, based on thermal measurements (Fahrenwald et. al., *Mining Tech.*, *Tech. Pub.*, 416, 1931). Drop-weight devices have been used to obtain ratios of surface produced to work input under idealized conditions and the performance of commercial equipment compared with them; efficiencies of 25 to 60 per cent have been shown [Wilson, *Mining Tech.*, *Tech. Pub.*, 810, 1937; Bond and Maxson, *Trans. Am. Inst. Mining Met. Engrs.*, **134**, 296 (1939)].

Efficiency Coefficients. A practical requirement is that least power be consumed for the amount of crushing or grinding desired. For this purpose the input can be stated in ft.-lb./min. (or in horsepower) and the output in new surface as sq. ft./min., to obtain an efficiency coefficient of sq. ft./ft.-lb. which should be constant for a given material, for various reduction ratios, to satisfy Rittinger's law. Values in the range of 0.02 to 0.05 sq. ft./ft.-lb. have been reported for materials such as gold and copper ores, Portland-cement clinker, petroleum coke, and pyrite when using a twin-ball pendulum impact device for crushing. The values might be halved for commercial installations (Bond, *Mining Tech.*, **10**, *Tech. Pub.*, 1895, 1946). A graphical method, based on the Rittinger law, compares crushing efficiencies [Gates, *Eng. & Mining J.*, **95**, 1039 (1913)]. If grinding to a limiting size is the criterion, the efficiency coefficient may be stated as the ratio of tons per hour output able to pass through a specified sieve to the input in horsepower. The ratio is then tons/hp.-hr. The value of this coefficient is between about 0.02 and 0.1 for wet ball-mill pulverizing hard to medium-hard minerals to No. 200 sieve size.

Dispersing Agents and Grinding Aids

Wet Grinding. The limiting product fineness in dry grinding operations is that being obtained when particles begin to cake on the grinding mediums and walls of the equipment. Wet grinding can be employed to disperse the particles so that further reduction in size can take place, but it is restricted in many instances because of flocculation of particles smaller than about 10 μ (see p. 1118). In the production of finely divided solids, commonly used dispersing agents are silicates, phosphates, and alkyl aryl sulfonic acids ("Daxad" types). Less than 1 per cent of the agent is used on a solids basis. The function of dispersing agents in water is explained by the "double-layer" theory (Ward, "Colloids," p. 27, Interscience Publishers, Inc., New York, 1946).

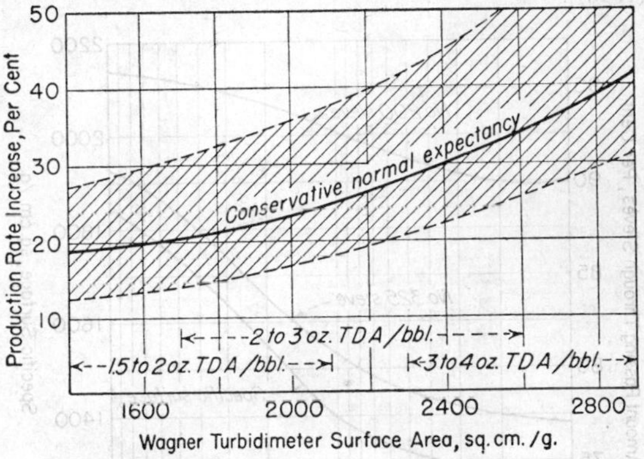

FIG. 7. Use of TDA grinding aid on cement clinker. (*Dewey and Almy Chemical Co.*)

The use of surface-active agents in "paint grinding" is common practice [Anon., *Paint, Oil & Chem. Rev.*, **102**, 70 (1940). Fischer and Jerome, *Ind. Eng. Chem.*, **35**, 336 (1943). Price, *Am. Ink Maker*, **22**, 21, 45 (1944)].

Dry Grinding. The practice of adding dispersing agents to wet-grinding circuits has its counterpart, which is known as the use of grinding aids, for dry pulverizing and disintegration. Such materials as RDA (aryl alkyl sulfonic acid) and TDA (a mixture of triethanolamine salts and soluble calcium salts of modified lignin sulfonic acids) are used for raw materials and finished cement pulverizing, respectively [Kennedy and Mardulier, *Rock Products*, **44**, 78 (1941), Kennedy, *Ind. Eng. Chem.*, **28**, 963 (1936)]. For equal fineness (specific surface) the production rate is 20 to 40 per cent greater when using 0.03 to 0.07 per cent of TDA (solids basis) in the form of a 10 to 15 per cent water solution added continuously to ball or tube mills (Fig. 7). Maximum-production-rate increases are obtained in mills that normally experience ball coating; lesser increases are obtained where no ball coating is experienced. Air-classifier efficiency is also improved with TDA. Application for RDA as a grinding aid has been found in the dry pulverizing of silica and graphite.

Grinding aids also give a finer product for a fixed production rate as well as an increased production rate for a fixed fineness. Colloidal carbon is valuable for dry pulverizing [Sweitzer and Craig, *Ind. Eng. Chem.*, **32**, 751 (1940)]. The effect of increasing additions of carbon on fineness of product is shown in Fig. 8. Without the use of carbon, the specific surface is 1270 sq. cm./g., and the No. 200 and No. 325 sieve finenesses are 90.0 and 77.3 per cent, respectively. Coal may be used as a grinding aid (Bond and Agthe, *Mining Tech., Tech. Pub.*, 1160, 1940). Carbon and coal, however, are not attractive for cement grinding because of their deleterious effect on durability. Small amounts of water added to ball mills grinding dry-process enamels serve to prevent packing of the powder on the mill walls and balls [Manson, *J. Am. Ceramic Soc.*, **21**, 316 (1938)].

Advantages have been shown for Vinsol resin, cod oil, beef tallow, and aluminum stearate as grinding aids for cement clinker [Dawley, *Pit and Quarry*, **36**, 57 July, (1943)].

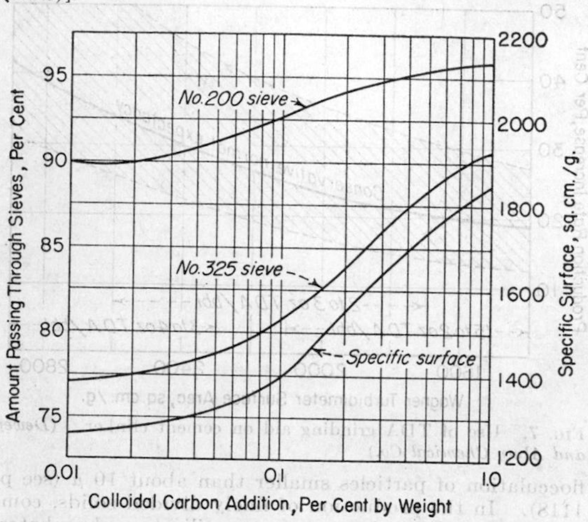

FIG. 8. Effect on cement fineness of colloidal carbon used as a grinding aid. [*Ind. Eng. Chem.*, **32**, 751 (1940).]

The use of ammonium salts or urea as grinding aids for graphite is disclosed in British Patent 564,418. The usefulness of glycerol and wood-resin grinding aids for cement is described in U.S. Patents 2,203,809, 2,225,146, and 2,225,149. Oleic acid is described as a grinding aid for zinc blende in U.S. Patent 1,985,076. The use of polar, non-polar agents to increase the production rate and improve the fineness of dry white pigments is disclosed in U.S. Patent 2,274,521.

Classification and Selection of Equipment

A wide variety of equipment is available differing in details of design or having characteristic points of advantage. The chief reasons for lack of standardization are the variety of products to be ground and product qualities demanded as well as the limited amount of useful grinding theory. In addition to the properties of the materials (see p. 1110), the economic balance between investment cost and operating cost is important in selecting equipment.

Classification. Units of equipment can be classified as (1) "crushers" and (2) "grinders" or "grinding mills," according to the commonly used nomenclature. Among grinders, pulverizers and disintegrators overlap so widely that it is impractical to distinguish between them. Hence "grinders" is the generic term for both. The classification does not imply that a grinding mill cannot be used for crushing, however. A classification of crushing and grinding equipment is given in Table 1.

Table 1. Classification of Crushing and Grinding Equipment*

	Comments
Crushers:	
A. Jaw	Blake and Dodge are common types. Blake for high production and nonclogging; Dodge for low production, intermittent service and higher reduction ratio. See Fig. 17 for Blake type.
B. Gyratory	Replacing jaws as primary crushers. Three general types are suspended. supported, and fixed-spindle. See Fig. 18.
C. Cone	This is a type of gyratory crusher. Successor to disk crusher. See Fig. 15.
D. Pan	Edge runner and chaser mills are variations. Can be operated wet or dry. See Fig. 20.
E. Roll 1. Smooth 2. Toothed	Smooth rolls for hard and soft materials; toothed rolls for soft materials. Toothed rolls available in single, double, and multiple arrangement. See Fig. 19.
F. Rotary	Suitable for coarse crushing of soft materials. See Fig. 21.
G. Hammer	Heavy construction with breaker plates, sturdy hammers, bar grate discharge. Operates at low speeds. See Fig. 22.
Grinders, mills, or grinding mills:	
H. Ball 1. Ball 2. Pebble 3. Rod 4. Tube 5. Compartment	Length equal to diameter with large balls for crushing. See Fig. 23. Length equal to diameter for batch grinding. Pebble mills used to avoid iron contamination. Length 2–3 times diameter for open- or closed-circuit operation. Rod mill produces minimum "fines"; can operate with moist material; rods instead of balls; length is twice diameter. Ball, pebble, and rod mills operate wet or dry. Tube and compartment mills have length greater than twice the diameter. Compartment mill is tube mill divided into zones with decreasing size of balls from feed end. Frequently operated in closed circuit with size classifier.
I. Ring-roller 1. Vertical ring 2. Horizontal ring *a.* Bowl *b.* Ball and ring	Generally incorporate air or screen classification. Adapted to drying while grinding. Either the ring may rotate, or the rollers revolve. Spring pressure or centrifugal force may hold rollers against ring. See Figs. 16 and 28 for ring-roller and ball-and-ring types.
J. Hammer 1. Rigid hammer 2. Swing hammer 3. Ring hammer 4. Multiple-cage 5. Disk (toothed plates)	Also called impact mills. See Fig. 27. Smaller clearance between grate bars than for crushing and higher speeds. Often have perforated plate or screen discharge. Called beater mills with rigid hammers or beaters. May have horizontal or vertical shaft. See Fig. 25. Multiple-cage disintegrator and toothed plate disk mill are in reality hammer or impact mills. See Figs. 30 and 24.
K. Disk (essentially smooth plates) 1. Single runner 2. Double runner 3. Buhrstone	Double runner gives higher relative speeds. See Fig. 24. Also known as attrition mills or plate mills. Buhrstone mill is in this category. See Fig. 29. Can be operated wet or dry. Colloid mills may be included in this class.
L. Fluid energy (jet mills) 1. Gas *a.* Air *b.* Steam	No moving parts in mill. Heat-sensitive materials can be pulverized with cooled air. Steam is superheated to avoid condensation on product. See Fig. 26.

* See pp. 1121–24 for typical illustrations.

Selection. A guide to the selection of equipment may be based on feed size and hardness (p. 1114), as shown in Table 2. It should be emphasized that Table 2 is merely a guide and that exceptions can be found in practice. An arbitrary distinction is made between crushing and grinding, in that feed materials from a lump size of

60 to ¼ in. require crushing and feed materials of smaller size require pulverizing or disintegration. Primary and secondary crushing each have two stages, coarse and fine.

Table 2. Guide to Selection of Crushing and Grinding Equipment

Size-reduction operation	Hardness of material	Size*: Range of feeds, in.† Max.	Range of feeds, in.† Min.	Range of products, in.† Max.	Range of products, in.† Min.	Reduction ratio‡	Type of equipment§
Crushing:							
Primary...	Hard	60	12	20	4	3:1	*A, B, C*
		20	4	5	1	4:1	*A, B, C*
Secondary..	Hard	5	1	1	0.2	5:1	*A, B, C, D, E*₁
		1.3	0.25	0.185 (4)	0.033 (20)	7:1	*D, E*₁*, H, I*
	Soft	20	4	2	0.4	10:1	*C, D, E*₂*, F, G*
Grinding:							
Pulverizing:							
Coarse...	Hard	0.185 (4)	0.033 (20)	0.023 (28)	0.003 (200)	10:1	*H, I*
Fine.....	Hard	0.046 (14)	0.0058 (100)	0.003 (200)	0.00039 (1250)¶	15:1	*H, I, L*
Disintegration:‖							
Coarse...	Soft	0.5	0.065 (10)	0.023 (28)	0.003 (200)	20:1	*H, I, J, K*
Fine.....	Soft	0.156 (5)	0.0195 (32)	0.003 (200)	0.00039 (1250)¶	50:1	*H, I, J, K, L*

* 85% by weight smaller than the size given.
† Sieve number in parentheses.
‡ Higher reduction ratios for closed-circuit operations.
§ See Table 1.
¶ 10 μ.
‖ And pulverizing soft materials.

Pulverizing and disintegration differ chiefly in the homogeneity of the materials handled (see p. 1110). There are two stages in each case. The reduction ratio increases as the size decreases for each operation, and a low reduction ratio is characteristic of crushing. The ratio is high for disintegration and pulverizing of soft materials [Anon., Pulverizing 200 Process Materials, *Chem. & Met. Eng.*, **45**, 241 (1938)].

Size Reduction Combined with Size Classification

Grinding systems are batch or continuous in operation. Most operations are continuous; the outstanding excep-

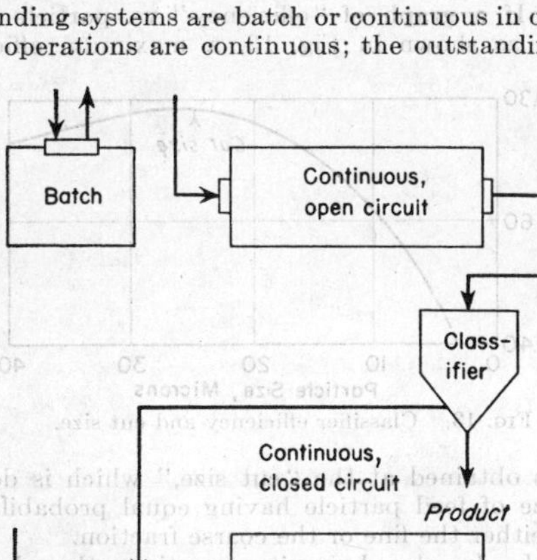

FIG. 9. Grinding systems.

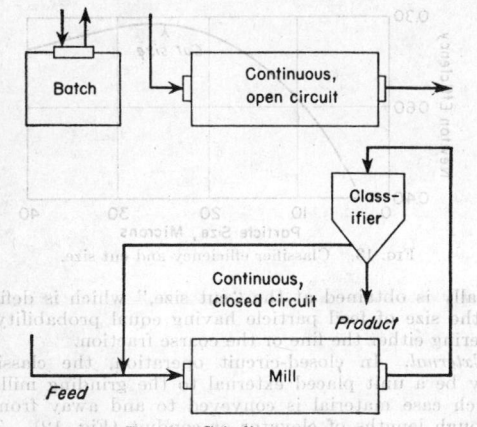

tions are those using batch ball or pebble mills. Continuous operation is accomplished in open or closed circuit, as illustrated in Fig. 9.

Continuous Open-circuit Operation. As the raw material moves through a tube mill 6 ft. in diameter by 26 ft. long, the greater portion is reduced to the required size in the first few feet, as illustrated in Fig. 10. A 6-ft.-long ball mill might have been used in closed circuit with an external classifier to obtain a fine product meeting the required limiting size specifications, the coarse product being returned to the feed end of the mill. The product from the closed-circuit system would be expected to have lower specific surface than that from the open-circuit operation. If high specific surface is required, the full length of the tube mill would be indicated to be required.

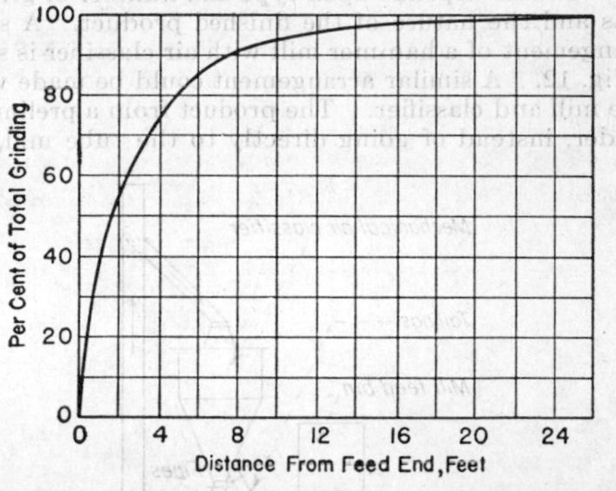

FIG. 10. Grinding accomplished at various distances from the feed end of a tube mill.

Continuous Closed-circuit Operation. Most crushing and grinding equipment can be operated in closed circuit with size classifiers. Material from a size-reduction machine is conveyed to a classifier in which large particles are removed and returned to the machine and particles of the desired size, and smaller, are discharged as product. By this procedure a product is obtained with more uniform size distribution than would be obtained by batch or continuous open-circuit operation to the same maximum limiting size. This is shown graphically in Fig. 11 [see also Work, *Bull. Am. Ceramic Soc.*, **17**, 1 (1938)]. Internal size classification plays an

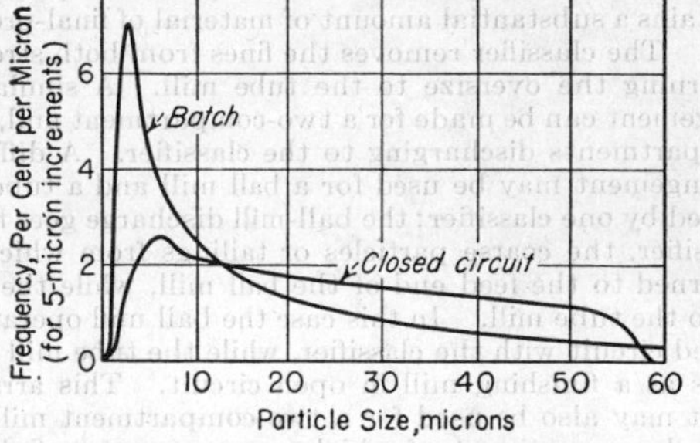

FIG. 11. Comparison of size distributions of products from batch and from continuous closed-circuit grinding systems.

essential role in the functioning of machines for dry grinding in the fine size range; particles must be kept in the grinding zone until they are as small as required in the finished product; then they should be allowed to discharge.

Operating economy is the object of closed-circuit grinding to meet a limiting size specification. The production rate at a given limiting size is greater and the product specific surface is less for closed-circuit than for batch or open-circuit operation. Coarse material returned to a mill by a classifier is the circulating load, and its rate may be 100 to 1000 per cent of the production rate. Up to the limits of ability of a mill to grind material without choking, the production rate increases with increase in circulating load, but at a diminishing rate until a constant value is reached. The per cent circulating load can be calculated from size analyses of the feed, fine product, and coarse product of the classifier in a closed-circuit grinding system [Bond, *Rock Products*, **41**, 64 (January, 1938)].

There are many possible arrangements for connecting classifiers in closed circuit with grinding equipment; the most suitable depends upon type and number of grinding units and the nature of the finished product. A simple arrangement of a hammer mill with air classifier is shown in Fig. 12. A similar arrangement could be made with a tube mill and classifier. The product from a preliminary grinder, instead of going directly to the tube mill, may

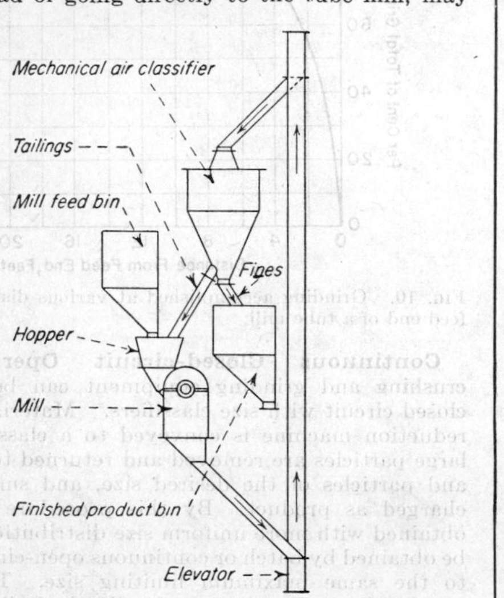

FIG. 12. A hammer mill in closed circuit with an air classifier.

first be passed to a classifier together with the tube-mill discharge if the product of the preliminary grinder contains a substantial amount of material of final-product size. The classifier removes the fines from both streams, returning the oversize to the tube mill. A similar arrangement can be made for a two-compartment mill, both compartments discharging to the classifier. A different arrangement may be used for a ball mill and a tube mill served by one classifier: the ball-mill discharge goes to the classifier, the coarse particles or tailings from which are returned to the feed end of the ball mill, while the fines go to the tube mill. In this case the ball mill operates in closed circuit with the classifier, while the tube mill functions as a finishing mill in open circuit. This arrangement may also be used for a two-compartment mill, the second compartment of which may serve as a finishing mill.

In an arrangement used for a three-compartment mill with single classifier, tailings and fines from the classifier may be returned to any compartment, depending upon the amount of fines required in the finished product. For a three-compartment mill with two classifiers, the following arrangement may be used: The first classifier receives the discharge from the first compartment and the second classifier the discharge from the second compartment and the fines from the first classifier; tailings from the two classifiers are the feed for the second compartment, while the fines from the second compartment go to the third compartment, which functions as finishing mill. Many other flow sheets may be used for compartment mills as well as for separate units.

Dry vs. Wet Grinding. Ball mills have a large field of application for wet grinding in closed circuit with size classifiers. The reader is referred to pp. 930 to 937 for detailed information. If the presence of liquid with the finished product is not objectionable or the feed is moist or wet, wet grinding generally is preferable to dry grinding. In the fine dry pulverizing or disintegration ranges, surface forces come into action to cause flocculation and "cushioning," with a resulting inefficient use of energy. This condition limits the application of dry grinding and makes wet-grinding operations necessary in many instances. Other factors that influence the preference are availability of water, relative investment required, cost of drying, and operating costs as influenced by local conditions.

Types of Size Classifiers. These can be divided into gravity and centrifugal types. Screens, sieves, and grizzlies are used for size classification with crushers. Sieves can be used in wet grinding, but the more commonly encountered unit is of a hydraulic type such that particles are classified according to their size and density (see pp. 956 to 958 on screening equipment and pp. 923 to 927 on wet classifiers, as well as Chap. VII on Industrial Screening and Chap. IX on Classification in Gaudin, "Principles of Mineral Dressing," McGraw-Hill, New York, 1939). Centrifugal hydraulic-size classifiers are employed to a limited extent; their use in the wet-grinding systems for cement is an example.

Centrifugal air classifiers are used at fineness levels for which dry screens are impractical. Dry screening becomes impractical in the No. 35 to No. 100 sieve range, depending on the nature of the material and the production rate that is required.

Use of Size Classifiers. *Independent.* Classifiers may be used independently of grinding equipment to classify a particulate material into two or more fractions more closely sized than the feed material (Fig. 39). The performance of a size classifier can be calculated for any size, between the size limits of the feed distribution, from particle-size analyses of the feed and fractions [Newton, *Rock Products*, **35**, 26 (August, 1932). See also p. 960]. If a graph of "efficiency" vs. particle size is prepared, as shown in Fig. 13, a maximum efficiency

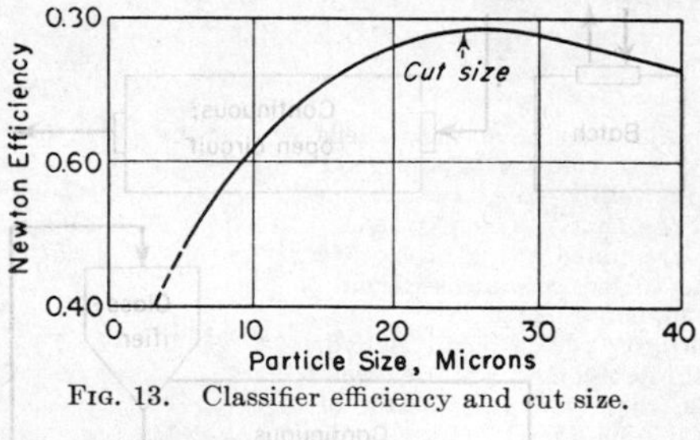

FIG. 13. Classifier efficiency and cut size.

usually is obtained at the "cut size," which is defined as the size of feed particle having equal probability of entering either the fine or the coarse fraction.

External. In closed-circuit operation, the classifier may be a unit placed external to the grinding mill, in which case material is conveyed to and away from it through lengths of elevator or conduit (Fig. 12). The grinding mill may be air-swept, and the discharge from it may be conveyed pneumatically to a cyclone collector and to the air-classifier unit, or directly to the latter (Fig. 40).

Internal. Many closed-circuit operations incorporate grinding and classification functions in a single housing or so closely integrated that the classifier is termed of an internal type (Fig. 41). Internal classification has particular application for dry grinding; material remains in suspension and is conveyed pneumatically throughout its residence in the mill.

A distinction that can be made between internal and external classification is that the circulating load cannot be isolated or measured readily in the former case.

Size Reduction Combined with Other Operations

Mixing and Blending. (Also see Sec. 17.) Batch ball mills with low ball charges can be used in the dry mixing or standardizing of dyes, pigments, colors, and insecticides to incorporate wetting agents and inert extenders. The rotation of the mill is favorable for many types of mixing, and the presence of balls tends to break up centrifuging of the charge and to promote lateral motion. The mixing of liquids is favored by the use of a charge of small balls. Disk mills and other high-speed disintegration equipment are useful for final intensive blending of insectide compositions, earth colors, and a variety of finely divided materials that tend to agglomerate in ribbon and conical blenders. The intensive action of a disk mill will break open these agglomerates of the components of a mixture and produce an intimate blend of ultimate particles.

Mills with air-classification units may be equipped so that the circulating air can be conditioned by mixing with hot or cold air or gases introduced into the mill or by dehumidification to prepare the air for the grinding of hygroscopic materials. Liquid sprays, or gases, may be injected into the mill or air steam, for mixing with the material being pulverized, to effect chemical reaction or surface treatment.

Heat Transfer. *Heating and Cooling.* Some materials are easier to disintegrate at elevated temperatures, even when thoroughly dry, because they disperse and flow more freely than at lower temperatures. Heat-sensitive materials with low softening temperatures are amenable to pulverizing if proper temperature control is exercised. Compositions containing fats and waxes are pulverized and blended readily, if refrigerated air is introduced into their grinding systems [Kanowitz, *Chem. & Met. Eng.*, **45**, 236 (1938)]. Patents have been granted for the use of dry ice to be mixed with materials before and during pulverizing to prevent melting and sticking together of the particles (U.S. Patents 1,739,761 and 2,098,798. See also p. 1159).

Ball mills may have their shells and heads jacketed for the flow of hot or cold fluids, and good heat transfer to the material within the mill is obtained, provided that the material does not cake on the heat-transfer surfaces. In some cases mills without jackets are placed over direct fire or erected inside a furnace with the cooled bearings external to it [Underwood, *Ind. Eng. Chem.*, **30**, 905 (1938)].

Dehydrating. The term "dehydration" is used to denote the removal of combined water as well as free moisture, such as, for example, the removal of water of crystallization from copper sulfate, $CuSO_4.5H_2O$. This is done by introducing hot gas into a closed-circuit grinding system, as illustrated in Fig. 42 on p. 1138.

Drying. Many materials can be ground to better advantage when dry, and the introduction of heated air into a system employing air conveying, or classification, serves to increase the productivity of the equipment (Fig. 1).

The drying of materials while they are being pulverized or disintegrated is known variously as "flash" or "dispersion" drying. A generic term is "pneumatic conveying" drying (Sec. 13). A flash-drying system can be used for raw materials of moderate moisture content and also for precipitated products in the form of wet sludges or cakes coming from filters or centrifuges. The method of conditioning the air is the same whether the mill is of the ball, ring-roller, or hammer-mill type used for heating, cooling, dehydrating, or drying. Flash drying is described on pp. 834*ff*. Data for the grinding and drying of bauxite in a ring-roller mill are given in Table 3.

Table 3. Operating Data for Grinding and Drying of Bauxite in a Ring-roller Mill

Initial moisture, %	9.75
Final moisture, %	0.75
Feed, lb./hr	12,560
Product, lb./hr	11,420
Moisture evaporated, lb	1,140
Temperature of gases entering mill, °F	700
Temperature of gases leaving mill, °F	170
Temperature of feed, °F	70
Temperature of material leaving mill, °F	150
Oil consumed, gal	15.3
Heating value of oil, B.t.u./gal	142,000
Thermal efficiency, %	68.5
Total power for drying and pulverizing, hp	105
Power for drying, hp	10
Final product, % through No. 100 sieve	90

Processing. Ball and pebble mills, batch or continuous, offer considerable opportunity for combining a number of processing steps that include grinding (Underwood, *loc. cit.*).

Vacuum drying and pulverizing of heat-sensitive materials is possible; chemical and physical reactions can be carried out; agitation, evaporation, chemical reaction, drying, and pulverizing can be accomplished in successive stages; and fibrous materials, for example, can be loosened by dry grinding, impregnated with hot liquid resin, vacuum-dried, chilled to increase friability, and pulverized and discharged as a fine dry molding powder.

Cleaning and Concentrating. Pulverizers equipped with air-classification apparatus for closed-circuit operation, or followed by air classifiers, can serve to separate components of mixtures because of differences in specific gravity and particle size. The removal of impurities by this means is known as cleaning or concentrating. Screens are used to separate coarse particles, not easily pulverized, from fine particles of the component that is pulverized readily. Throwout boxes are built into some mills for collection of grossly oversize foreign material. Magnetic separators frequently are employed to remove tramp magnetic solids from the feed to high-speed hammer and disk mills.

Sand may be removed from clay, and refractory impurities separated from hydrated lime. Phosphate rock may be freed from such impurities as clays and silt. The various oxides of lead, copper, and other metals often contain a certain amount of unoxidized material. When such materials are fed to a pulverizer and air-classification arrangement, the relatively soft oxide is reduced to fine powder and separated from the metal. To separate lead from dross, skimmings, and scrap battery plates, the material is first coarsely crushed in a high-speed hammer mill discharging to a screen. Oversize, almost all metallic lead, goes to the melting pot; undersize, a mixture of metal and oxide, goes to the pulverizer.

In the cleaning and concentrating of clays, chalks, and marls, water flotation is frequently used. The product is then dried and disintegrated in hammer mills. If further cleaning is desired, the washed product may be processed in a series of air classifiers, the tailings from one classifier being fed into the next. Tailings from the last classifier are either discarded or fed to the water flotation system. Some clays give a sufficiently fine product by dry pulverization and air classification.

Lime, after hydration, usually contains all the impurities of the original lime, such as sand, gravel, core, and clinker. The mixture from the hydrator is processed in a hammer-mill pulverizer with a throwout, or in a series of classifiers, the tailings from the last classifier being used for agricultural purposes. The results obtained in processing hydrated lime are given in Table 4.

Certain phosphate sands are cleaned and concentrated to remove sand and silt. Table 5 gives the results obtained in processing dry phosphatic sand in a Raymond automatic pulverizer with throwout.

Safety. *Metal Powders.* These present a hazard because of their flammability. Their combustion is favored during grinding operations in which ball, hammer, or ring-roller mills are employed and during which a high grinding temperature may be reached.

Table 4. Cleaning and Concentrating Hydrated Lime

Size of classifier, diam., ft	12
Power for classifier, hp	25
Power for pulverizer, hp	7.5
Power for pulverizer fan, hp	40
Feed to classifying system, tons/hr	8
Production of hydrated lime, tons/hr	7.2
Tailings, tons/hr	0.8

Chemical Analysis

	Feed	Finished product	Tailings
Calcium and magnesium hydrate, %	88.95	98.80	5.12
Silica, %	3.16	0.15	28.61
Iron and aluminum oxides, %	2.26	.12	20.42
Calcium and magnesium carbonates, %	2.63	.26	23.15
Calcium and magnesium oxides, %	3.00	.67	22.70
Fineness, % through No. 200 sieve	90	99.5	4.5

Isolation of the mills, the use of non-sparking materials of construction, and magnetic separators to remove foreign magnetic material from the feed are useful precautions (Hartmann, Nagy, and Brown, *U.S. Bur. Mines, Report Investigations*, 3722, 1943).

Many finely divided metal powders in suspension in air are potential explosion hazards, and causes for ignition of such dust clouds are numerous [Hartmann and Greenwald, *Mining & Met.*, **26**, 331 (1945)]. The concentration of the dust in air and its particle size are important factors that determine its explosibility. Below the lower limiting concentration, no explosion can result because the heat of combustion is insufficient to propagate it. Above a maximum limiting concentration, an explosion cannot be produced because insufficient oxygen is available. The finer the particles, the more easily is ignition accomplished and the more rapid is the rate of combustion. This is illustrated in Fig. 14.

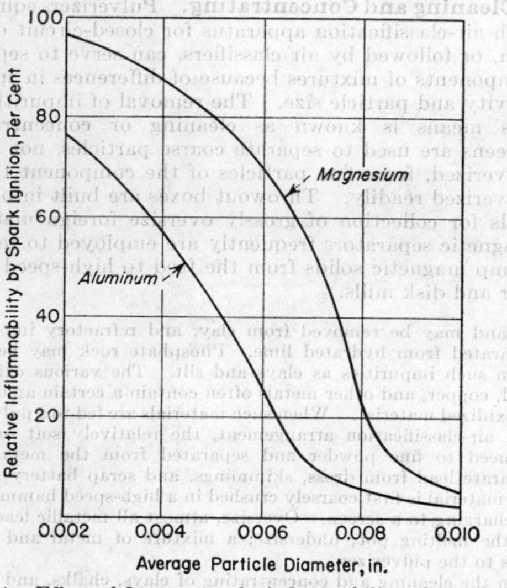

FIG. 14. Effect of fineness on the inflammability of metal powders. (*U.S. Bureau of Mines, Report Investigations.*, 3722, 1943.)

Non-metal Powders. Such materials as sulfur, starch, wood flour, cereal dust, dextrin, coal, pitch, hard rubber, and plastics are potential hazards when in finely divided form. Explosions and fires may be initiated by discharges of static electricity, sparks from flames, hot surfaces, and spontaneous combustion. Reduction of the oxygen content of air present in grinding systems is a means for preventing dust explosions in the equipment (Brown, *U.S. Dept. Agr., Tech. Bull.* 74, 1928). Maintenance of oxygen content below 12 per cent should be safe for most materials, but 8 per cent is recommended for sulfur grinding.

Table 5. Cleaning and Concentrating Phosphate Sand

Size of automatic pulverizer	No. 3
Feed, tons/hr	8.5
Capacity (of throwout), tons/hr	5.25
Fines, tons/hr	3.25
Power required by pulverizer, hp	15
Power required by fan, hp	40

Chemical Analysis

	Feed	Throwout product*	Fines
$Ca_3(PO_4)_2$ content, %	52	68	26
$Fe_2O_3 + Al_2O_3$ content, %	12.5	6	23

* The throwout product contains most of the valuable material and is the product required.

The use of inert gas has particular adaptation to pulverizers equipped with air classification; flue gas can be used for this purpose, and it is mixed with the air normally present in a system (see p. 1158 for sulfur grinding). Despite the protection afforded by use of inert gas, equipment should be provided with explosion vents, and structures should be designed with venting in mind [Brown and Hanson, *Chem. & Met. Eng.*, **40,** 116 (1933)].

Hard rubber presents a fire hazard when reduced on steam-heated rolls (see p. 1159). Its dust is explosive [Twiss and McGowan, *India-Rubber J.*, **107,** 292 (1944)]. Inert gas can be introduced at 110° to 130°F. into systems operating in closed circuit with sieves or air classifiers.

Many synthetic resins and plastics are hazardous in the finely divided state (Hartmann and Nagy, *U.S. Bur. Mines, Report Investigations,* 3751, 1944).

An annual publication, "National Fire Codes for the Prevention of Dust Explosions," is available from the National Fire Protection Association, Boston, Mass., and should be of interest to those handling hazardous powders (see also Sec. 30).

DESCRIPTION OF CRUSHERS

Jaw Crushers. (Fig. 17.) *Design and Operation.* These may be divided into three main groups (Fig. 31): the *Blake*, with movable jaw pivoted at the top, giving greatest movement to the smallest lumps; the *Dodge*, with the movable jaw pivoted at the bottom, giving greatest movement to the largest lumps; and modifications of the two, giving nearly equal movement to all sizes. The Blake has a removable crushing plate, usually corrugated, fixed in a vertical position at the front end of a hollow rectangular frame. A similar plate, at a suitable angle, is attached to a swinging lever (movable jaw) suspended from a shaft resting in the sides of the frame. Movement is accomplished through a knuckle action by the rising and falling of a second lever (pitman) carried by an eccentric shaft. The vertical movement is communicated horizontally to the jaw by two plates (toggles).

The Dodge type has the advantage of a larger feed opening for the same cost as a Blake, and it is useful for low-production intermittent service to produce a uniform product in sizes having smaller than 11- by 15-in. feed opening.

The setting of a jaw crusher may be stated as being the close or the wide opening between the moving jaws at the outlet end. The reciprocating motion of the jaws causes the opening to vary between close and wide. Specifications usually are based on close settings. The setting is adjustable.

The *Fine Reduction* (*Allis-Chalmers*) and *Kue-Ken Balanced Jaw* (*Straub Manufacturing Co.*) principles (Fig. 31) fall into the third category. The crusher having the former principle is a single toggle machine having its swing jaw mounted directly on the eccentric shaft so that it receives a downward as well as a forward motion. The lower end of the swing jaw is held in position against the toggle by a tension rod. The *Kue-Ken* balanced-jaw crusher has two opposed jaws, both swinging freely like two pendulums. Material is crushed without rubbing as the jaws move toward each other.

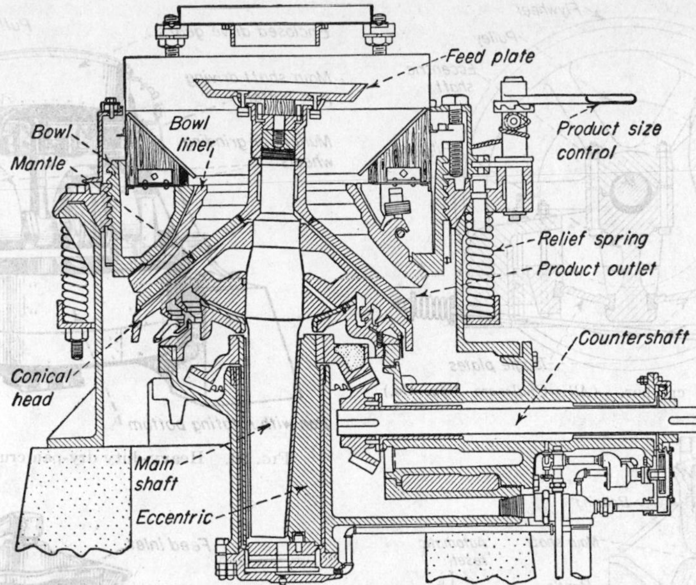

Fig. 15. Symons standard cone crusher. (*Nordberg Mfg. Co.*)

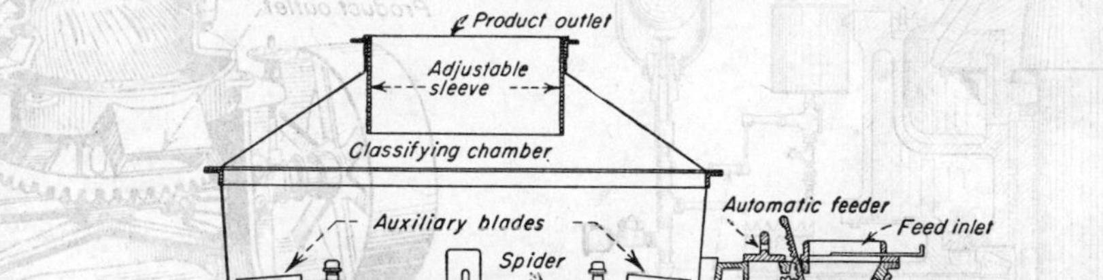

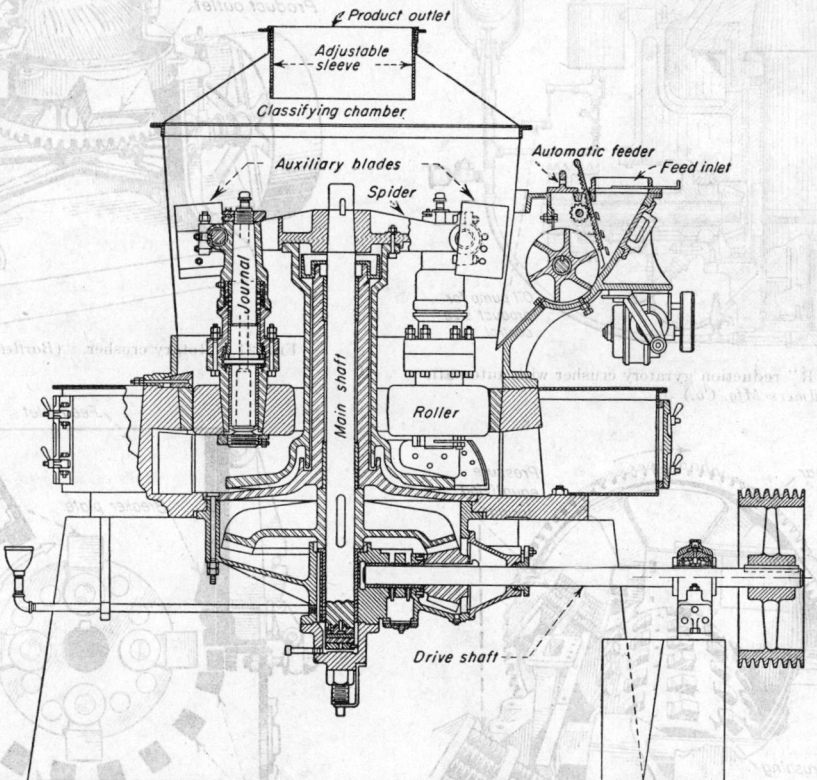

Fig. 16. Low-side ring-roller mill. (*Raymond Pulverizer Division, Combustion Engineering Co.*)

Classification of Crushing and Grinding Equipment see (Table 1).

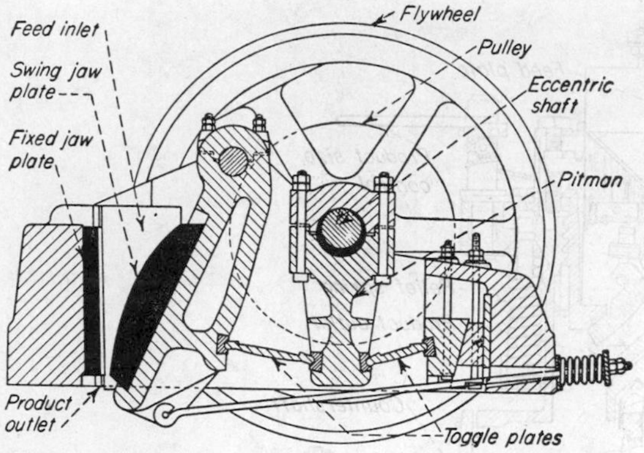

FIG. 17. Blake-type jaw crusher. (*Allis-Chalmers Mfg. Co.*)

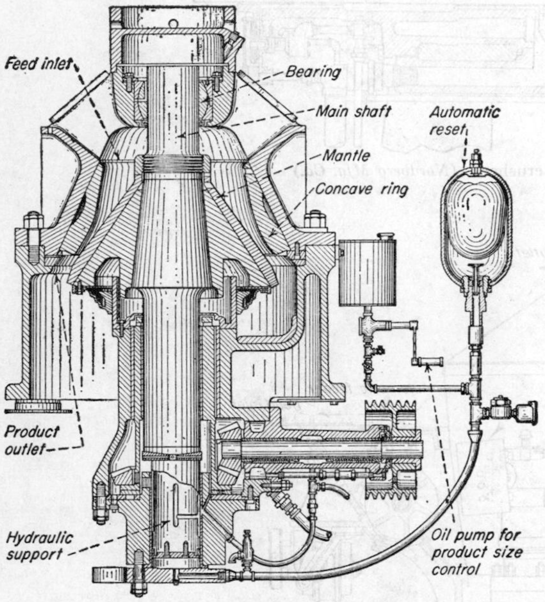

FIG. 18. Type "R" reduction gyratory crusher with automatic reset. (*Allis-Chalmers Mfg. Co.*)

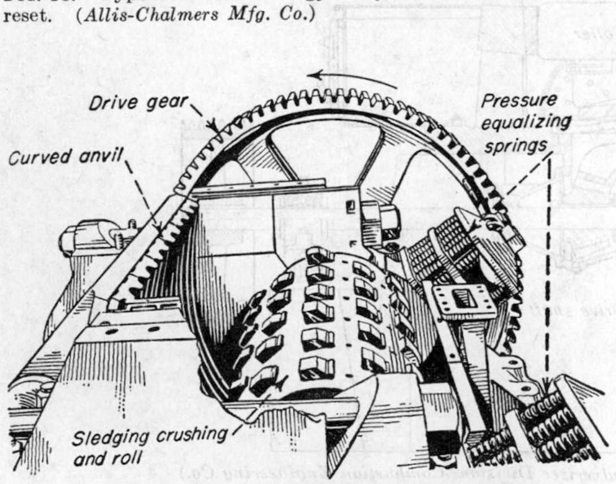

FIG. 19. Fairmount single-roll-type crusher. (*Allis-Chalmers Mfg. Co.*)

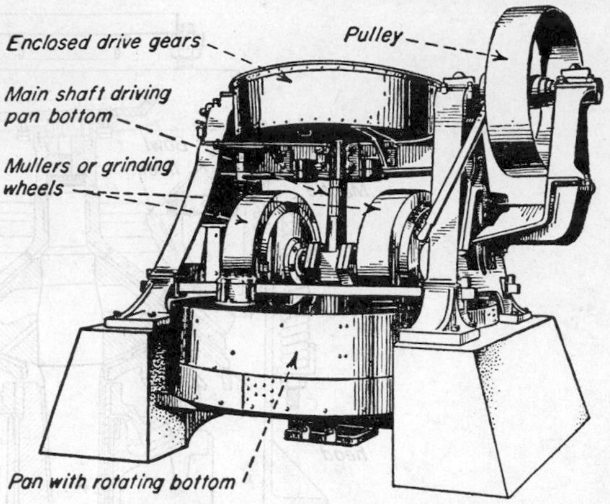

FIG. 20. Heavy-duty dry-pan crusher. (*Bonnot Co.*)

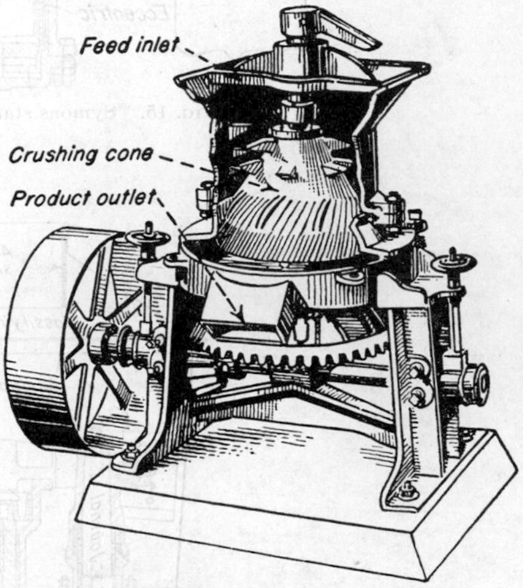

FIG. 21. Rotary crusher. (*Bartlett & Snow.*)

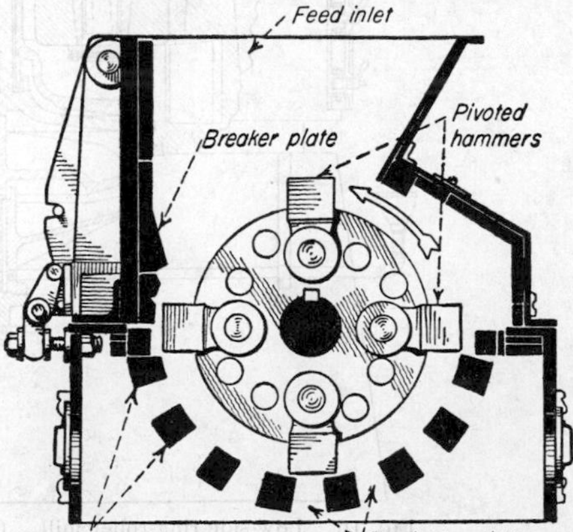

FIG. 22. Hammer crusher. (*The Jeffrey Mfg. Co.*)

Classification of Crushing and Grinding Equipment see (Table 1).

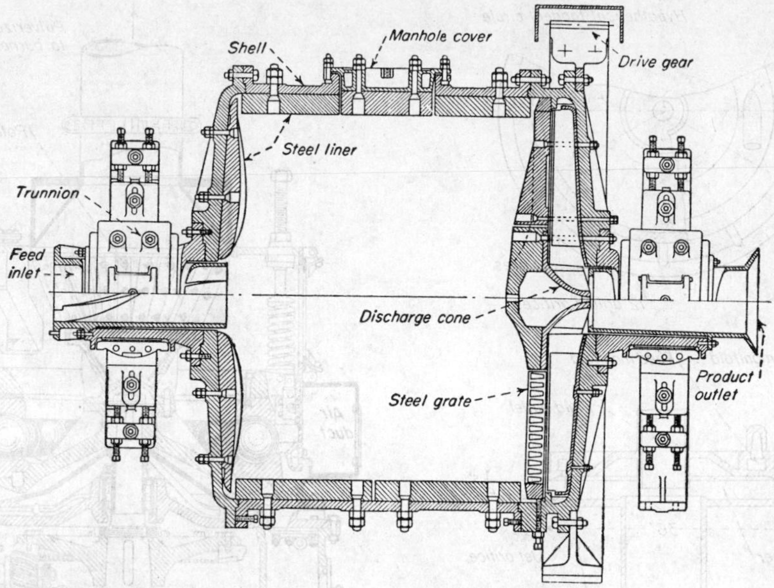

FIG. 23. Marcy grate-type continuous ball mill. (*Mine & Smelter Supply Co.*)

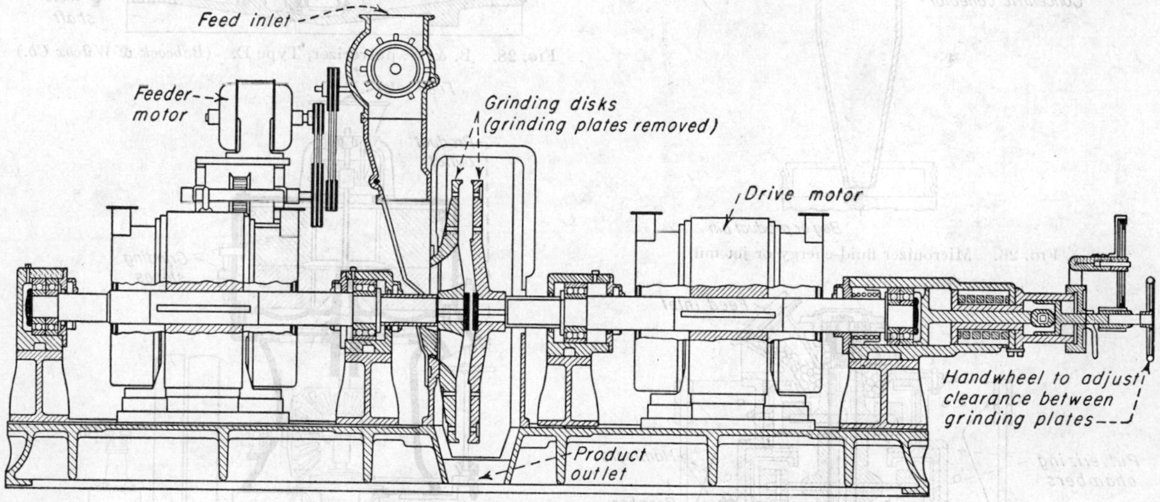

FIG. 24. Double-runner attrition mill. (*Sprout, Waldron & Co.*)

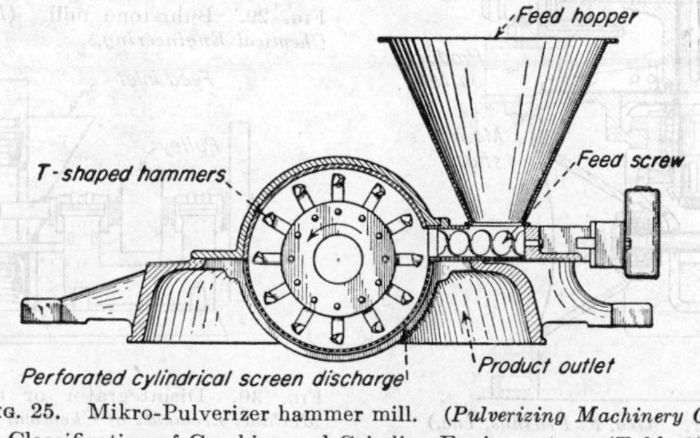

FIG. 25. Mikro-Pulverizer hammer mill. (*Pulverizing Machinery Co.*)
Classification of Crushing and Grinding Equipment see (Table 1).

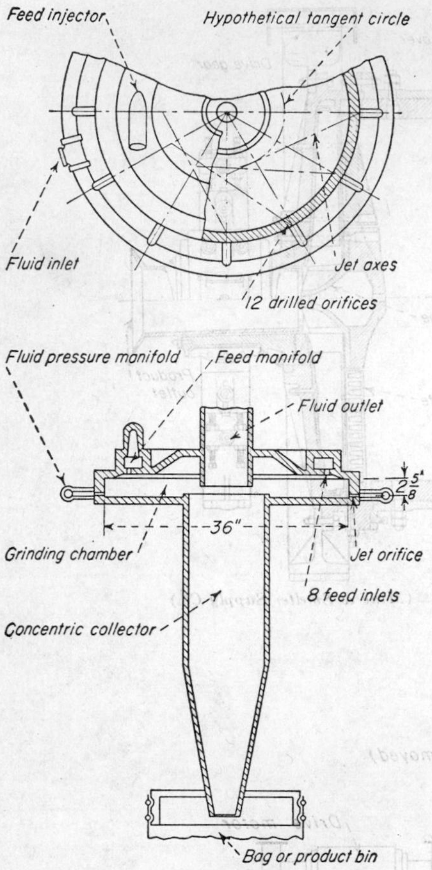

FIG. 26. Micronizer fluid-energy or jet mill.

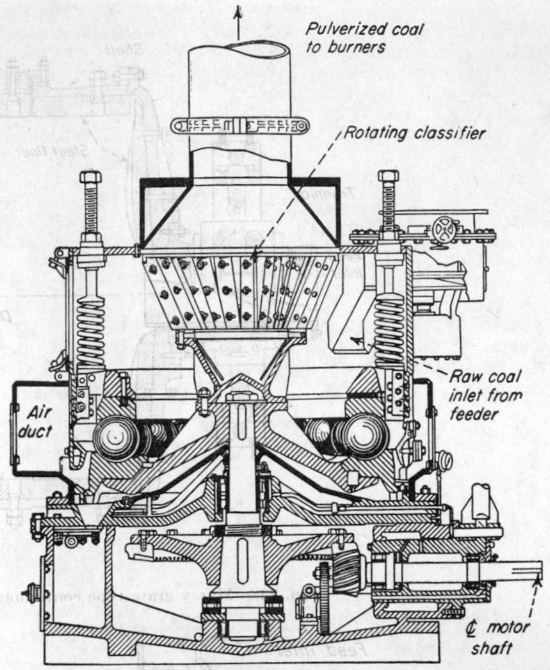

FIG. 28. B. & W. pulverizer, Type E. (*Babcock & Wilcox Co.*)

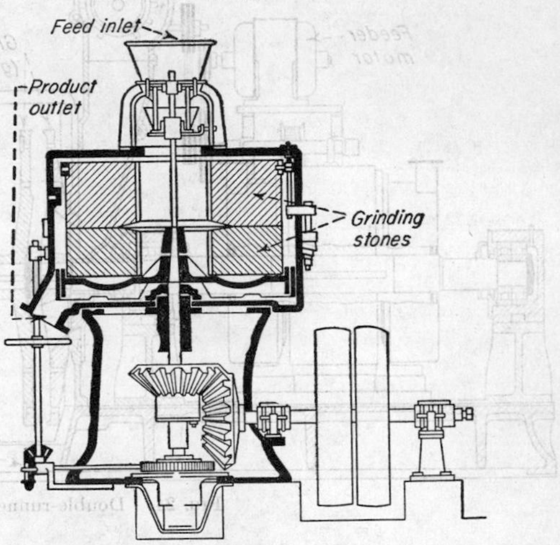

FIG. 29. Buhrstone mill. (*Badger and McCabe, Elements of Chemical Engineering.*)

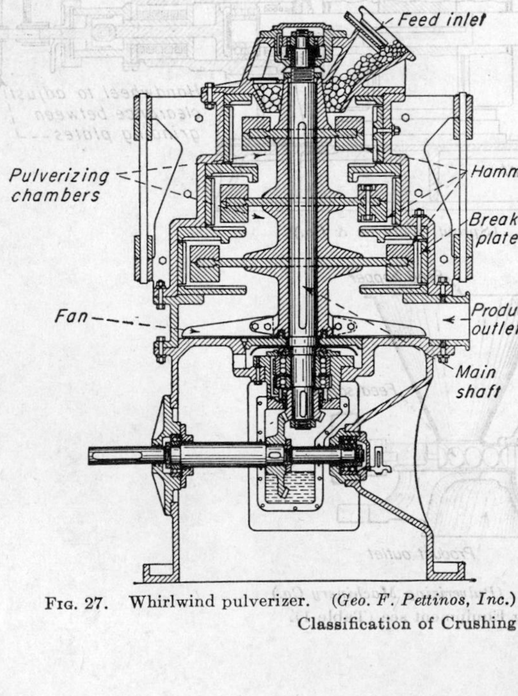

FIG. 27. Whirlwind pulverizer. (*Geo. F. Pettinos, Inc.*)

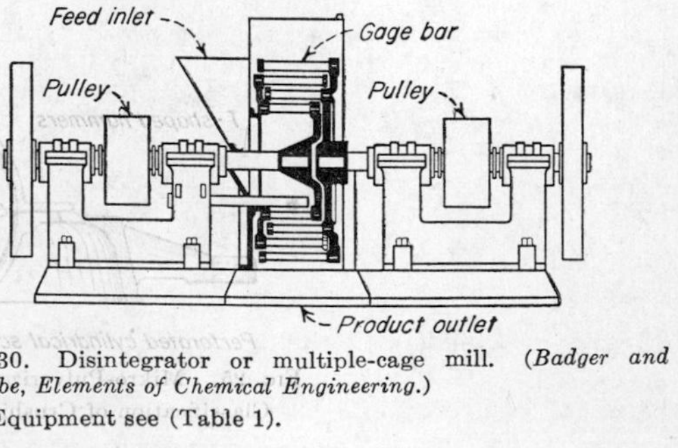

FIG. 30. Disintegrator or multiple-cage mill. (*Badger and McCabe, Elements of Chemical Engineering.*)

Classification of Crushing and Grinding Equipment see (Table 1).

Performance. Jaw crushers are applied to primary crushing of hard materials, and are usually followed by other types of crushers. In smaller sizes they are used as single-stage machines. The Blake type is available with receiving opening as large as 66 by 86 in.

The *Type H* crusher (*Traylor*) is distinguished by a welded-plate construction, two-piece pitman, and curved jaw plates. It is available in 10 sizes from 8- by 12-in. to 30- by 42-in. openings with capacities varying from 4 to 275 tons/hr. when producing minus ⅞ to minus 5 in. stone. The maximum power requirements of the sizes are from 10 to 100 hp. The *Type HB* crusher is equipped with a rod-type *Bulldog Pitman* with safety device. It is available in five sizes from 36- by 42-in. to 56- by 72-in. openings with capacities varying from 120 to 640 tons/hr. when producing minus 2- to minus 9-in. stone. The maximum power requirements of the sizes are from 115 to 250 hp. The *Type S* crusher is designed to withstand the most severe crushing duty. It is available in seven sizes

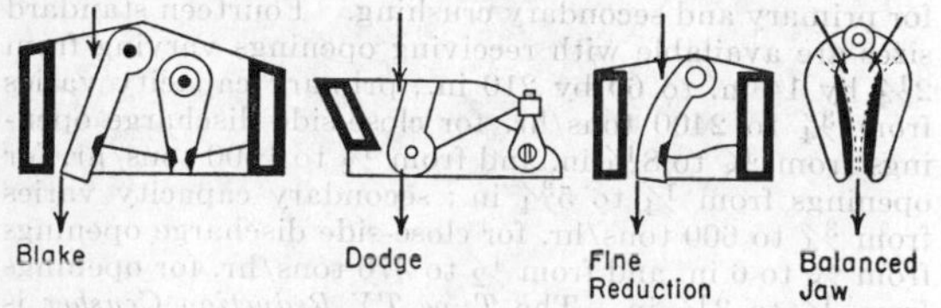

FIG. 31. Jaw-crusher designs.

from 36- by 42-in. to 66- by 86-in. openings with capacities varying from 120 to 1100 tons/hr. when producing minus 2- to minus 11-in. stone. The maximum power requirements of these sizes are 115 to 300 hp.

The capacities and power consumption of a *Dodge* (*Allis-Chalmers*) type jaw crusher operating on a tough ore that offers considerable resistance to reduction are shown in Table 6.

Table 6. Performance Data of Dodge (Allis-Chalmers) Jaw Crusher

Size of opening, in.	Capacity, tons/hr. Size of product				Approx. hp. required
	½ in.	¾ in.	1 in.	1½ in.	
4 × 6	¼	½	1	..	3
7 × 9	..	1	2	3	6
8 × 12	..	1½	3	4	10
11 × 15	..	2	4	6	15

The *Fine Reduction* jaw crusher is suitable for producing finished ball-mill feed in one pass. The character of material being crushed determines to a large extent the screen analysis of the crusher product. With a screened feed, approximately 90 per cent of the product will pass a round hole in a flat testing screen corresponding to the open setting and approximately 50 per cent corresponding to the closed setting. With a feed opening of 24 by 10 in., a 25-hp. drive, and ½-in. stroke; a capacity of 6 tons/hr. (1 ton = 20 cu. ft.) can be obtained with a close setting of ¼ in. Capacity increases directly with setting up to 1½ in.

The *Universal* jaw crusher (*Universal Engineering Corp.*) is a combination of the Dodge and Blake types. A high eccentric above the feed hopper and a radial toggle action at the bottom produce both horizontal and vertical actions of the movable jaw. The jaw moves forward and downward, tending to force the feed and force the discharge. The pitman that carries the moving jaw plate cannot drop directly away from the stationary jaw, and therefore large pieces of stone will not pass through until they are crushed. In this overhead eccentric force-feed crusher, there are two crushing blows to each revolution of the shaft, a primary blow at the top where most needed and a secondary or finishing stroke at the bottom. Table 7 gives performance data for several types and sizes of this crusher. Crushers 1 to 4 are used largely as prebreakers in rock quarries and mining industries where a large feed opening and a high ratio of reduction are required. Crushers 5 to 7 are well adapted for crushing to ¾ in. and finer and will produce a fine uniform product for concrete construction and work of a similar nature. Crushers 8 to 11 are for producing small sizes of stone or ore in moderate capacities. Crushers 1 to 7 may be used very advantageously in the production of a suitable feed size for a pulverizer.

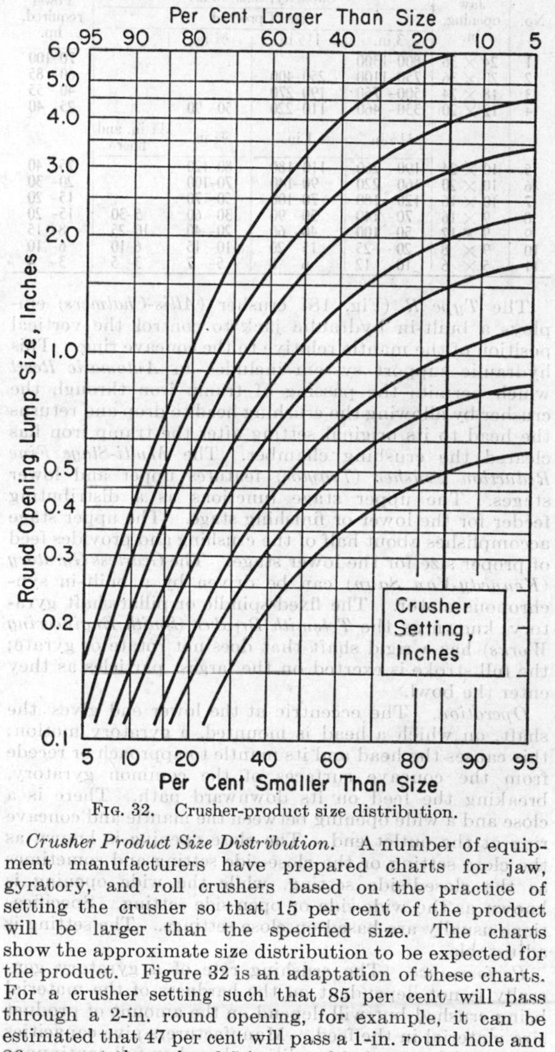

FIG. 32. Crusher-product size distribution.

Crusher Product Size Distribution. A number of equipment manufacturers have prepared charts for jaw, gyratory, and roll crushers based on the practice of setting the crusher so that 15 per cent of the product will be larger than the specified size. The charts show the approximate size distribution to be expected for the product. Figure 32 is an adaptation of these charts. For a crusher setting such that 85 per cent will pass through a 2-in. round opening, for example, it can be estimated that 47 per cent will pass a 1-in. round hole and 26 per cent through a 0.5-in. round hole.

Gyratory Crushers. (Fig. 18.) *Design.* The housing of a primary gyratory crusher has the shape of the frustums of two cones placed together with the narrow sections in the center. Crushing is done in the upper half, while the lower half houses the driving mechanisms, eccentric, etc. Both the mantle and the concave ring of secondary gyratory crushers generally are curved to minimize wear and eliminate packing between them [Bernhard, *Mining & Met.*, **13**, 107 (1932)].

A distinction can be made between primary types and secondary or reduction types of gyratory crushers in that the upper cone of the latter, if any, is merely a receiving opening with both crushing and driving mechanisms in the lower housing. The three general types of gyratory crusher are the suspended-spindle, the supported-spindle, and the fixed-spindle types. Primary gyratories are designated by the size of feed opening and secondary or reduction crushers by the diameter of the head in feet and inches.

Table 7. Performance Data for Universal Jaw Crushers

No.	Jaw opening, in.	Capacity, tons/10 hr. Size of product				Power required, hp.
		3 in.	1½ in.	¾ in.		
1	24 × 36	800–1300	......	......	.....	70–100
2	20 × 36	750–1100	250–400	......	.....	60–85
3	18 × 24	500–750	190–270	......	.....	40–55
4	12 × 20	330–460	110–220	50–90	.....	25–40
		1½ in.	1 in.	¾ in.	¼ in. and finer	
5	10 × 24	190–260	110–180	80–120	.....	25–40
6	10 × 20	160–220	90–140	70–100	.....	20–30
7	10 × 16	120–180	70–100	50–70	.....	15–20
8	9 × 16	70–140	50–90	30–60	5–30	15–20
9	9 × 12	50–100	40–60	20–40	10–25	8–15
10	9 × 8	20–25	15–20	10–15	6–10	6–10
11	5 × 6	10–12	6–8	5–7	3–5	3–4

The *Type R* (Fig. 18) crusher (*Allis-Chalmers*) employs a built-in hydraulic jack to control the vertical position of the mantle relative to the concave ring. This hydraulic support system includes an *Automatic Reset* which permits the passing of tramp iron through the crusher by allowing the crushing head to drop and returns the head to its original setting after the tramp iron has cleared the crushing chamber. The *Multi-Stage Fine Reduction Crusher* (*Traylor*) features upper and lower stages. The upper stage functions as a distributing feeder for the lower or finishing stage. The upper stage accomplishes about half of the crushing and provides feed of proper size for the lower stage. The *Gearless Gyratory* (*Kennedy-Van Saun*) can be driven by a built-in synchronous motor. The fixed-spindle or pillar-shaft gyratory, known as the *Telsmith Breaker* (*Smith Engineering Works*) has a rigid shaft that does not rotate or gyrate; the full stroke is exerted on the largest particles as they enter the bowl.

Operation. The eccentric at the lower end gives the shaft, on which a head is mounted, a gyratory motion; this causes the head and its mantle to approach or recede from the concave surfaces of the common gyratory, breaking the feed on its downward path. There is a close and a wide opening between the mantle and concave ring at the outlet end. The close opening is known as the close setting or the close-side setting and sometimes as the closed-side setting, while the wide opening is known as the wide-side or open-side setting. Specifications usually are based on close settings. The setting is adjustable.

Performance. The crushing rate of a gyratory generally is not dependent on the hardness of the material being crushed but will depend on the amount of product size material in the feed. Manufacturers give capacities estimated for average conditions based on full continuous feed of quarry- or mine-run material weighing 100 lb./cu. ft. after crushing. Gyratory crushers are used for high-capacity primary crushing, or to follow primary crushers of either the jaw or gyratory type.

The *Superior McCully* (*Allis Chalmers*) crusher is made in 11 sizes; each size has two feed openings ranging from 8¾ by 35 in. to 59¾ by 196 in. Horsepower ranges from 15 to 500. Minimum settings range from ⅞ to 7 in. with capacity from 30 to 1070 tons/hr. Maximum settings range from 2⅜ to 10½ in. with capacity from 51 to 2120 tons/hr. The 18-in. *Fine Reduction Superior McCully* crusher's feed opening is 18 by 68 in. Settings range from 1½ to 4 in. with corresponding capacity of 245 to 734 tons/hr., and horsepower ranges from 150 to 200.

Type R crushers are made in four sizes. Each size can be equipped with either standard or fine reduction concaves. With standard concaves, feed openings range from 3 to 8 in. with minimum close-side settings 3/16 to ⅜ in. and maximum close-side settings of 1¼ to 3¼ in. Capacity varies from 10 to 340 tons/hr. With the fine reduction concaves, feed openings range from 2 to 5 in. with minimum close-side settings ⅛ to 3/16 in. and maximum close-side settings of ⅜ to ⅞ in. Capacity varies from 7 to 207 tons/hr. Horsepower ranges from 25 to 150.

Approximately 60 to 65 per cent of a *Type R* product will pass through a square opening of a testing sieve equal to the "close-side" opening of the crusher and 95 to 100 per cent of the "wide-side" opening.

The *Type T Bulldog* (*Traylor*) gyratory crusher is used for primary and secondary crushing. Fourteen standard sizes are available with receiving openings varying from 2¼ by 14 in. to 60 by 210 in.; primary capacity varies from ¾ to 2400 tons/hr. for close-side discharge openings from ⅜ to 8¼ in. and from ½ to 1600 tons/hr. for openings from ¼ to 5¾ in.; secondary capacity varies from ¾ to 600 tons/hr. for close-side discharge openings from ⅜ to 6 in. and from ½ to 370 tons/hr. for openings from ¼ to 3½ in. The *Type TY Reduction Crusher* is made in six sizes with receiving openings varying from 3 by 15 in. to 22 by 66 in. Capacities range from 4 tons/hr. with the close side set to ⅛ in. in the smallest crusher up to 590 tons/hr. with the discharge opening set to 3½ in. in the largest crusher.

Gearless Gyratory crushers are available with feed-opening sizes from 3 by 8 in. to 66 by 235 in., crush to from ½ to 10 in. at ½ to 3600 tons/hr., and require from 1 to 250 hp. The fine crushers are built with feed openings from 1¾ to 14 in. wide.

Seven standard sizes of *Telsmith Breaker* are made, from 15 to 20 hp. to 100 to 125 hp., and receiving opening from 6¾ by 35 in. to 25 by 106 in. Capacities range from 17 to 18 tons/hr. for the smallest size to 300 to 350 tons/hr. for the largest, discharge opening from 1 to 4 in. The Telsmith reduction or secondary crusher has a large bowl, much wider at the bottom than at the top, allowing free escape of the material. Four standard sizes are made, 25 to 65 hp., 5- to 8-in. feed opening, ⅞- to 1⅝-in. discharge opening, with capacities ranging from 18 to 21 to 85 to 100 tons/hr.

Cone Crushers. (Fig. 15.) *Design and Operation.* The cone or conical head, gyrated by means of an eccentric driven through gears and a countershaft, is supported from the base. Heavy springs hold the upper frame fixed; when choked by overfeeding or tramp iron, the springs allow the upper frame to rise at the point of stress so that the material can be discharged. The conical head gyrates in much the same manner as for the gyratory crushers, but the cone travels a greater distance and gyrates faster. Material receives a series of rapid blows as it passes through the crushing cavity.

Performance. The cone crusher is a secondary or reduction crusher. The two common types are the *Symons* (*Nordberg Mfg. Co.*) and the *Telsmith* (*Smith Engineering Works*). Performance characteristics are given in Table 8. In addition to the *Standard* crusher (Fig. 15), *Symons Short Head Cone Crushers* are available for still greater reduction in size.

Pan Crushers. (Fig. 20.) *Design and Operation.* The pan crusher consists of one or more grinding wheels or mullers revolving in a pan; the pan may remain stationary and the mullers be driven, or the pan may be driven while the mullers revolve by friction. In some types the mullers are made of stone; in others they are of stone or of iron equipped with steel tires. Iron scrapers or plows at a proper angle feed the material under the mullers.

In the *Bonnot Dry Pan* (Fig. 20), the clearance between the mullers and bottom of the pan can be regulated. The pan bottom rotates and has a central, solid crushing ring as well as an outer ring of screen plates with openings from 1/16 to 1/2 in., as required.

Table 8. Operating Characteristics for Cone Crushers (Symons Standard)

Size of crusher, ft.	Width of feed opening, in.	Capacity, tons/hr. Discharge setting, in.					Hp.
		3/8	1/2	3/4	1	1 1/2	
2	2 1/4	20	25	35	...	...	25–30
3	3 7/8	35	40	70	...	...	50–60
4	5	60	80	120	150	...	75–100
5 1/2	7 1/8	..	..	200	275	340	150–200
7	10	..	..	330	450	600	250–300

Performance. The dry pan is useful for crushing medium hard and soft materials such as clays, shales, cinders, and soft minerals such as barytes. Materials fed should normally be 3 in. or smaller, and a product can be delivered able to pass No. 4 to No. 16 sieves, depending on the hardness of the material. Finer products can be obtained by operating a pan in closed circuit with a vibrating screen. High reduction ratio with low power and maintenance are features of pan crushers.

The Bonnot dry pan is available from 5 to 10 ft. pan diameter with mullers ranging from 28 to 62 in. in diameter with 5- to 18-in. face and 1800 to 30,000 lb. weight per pair. Power ranges from 15 to 75 hp. or from 1 to 5 hp./ton of product. Production rate varies from 1 to 50 tons/hr. according to pan size and hardness of material as well as fineness of feed and product.

Smooth-roll Crushers. *Design and Operation.* Two rolls of the same diameter are revolved toward each other at the same speed. One of the shafts moves in fixed bearings, the other in movable bearings. The distance between the rolls is adjustable, and a nest of powerful springs holds the movable roll to the clearance that has been set.

The tension springs exert pressures on the rolls up to 6000 lb./lin. in. of roll face for light duty to as high as 40,000 lb./lin. in. for heavy duty. This is equivalent to crushing strengths of 18,000 to 120,000 lb./sq. in. based on effective face length equal to one-third of actual length. Automatic lateral adjusting mechanisms can be provided to move the fixed roll from side to side to minimize annular corrugation and flanging.

The angle of nip, the angle formed by the tangents to the roll faces at the point of contact with a particle to be crushed, is determined by $\cos (N/2) = (r + a)/(r + b)$, where r = radius of rolls, a = one-half distance between rolls, b = radius of particle, and N = angle of nip. The angle of nip varies for different operations, but seldom exceeds 30°. The required roll diameter is determined by the maximum size of feed that can be nipped without slippage: $b_{max} = 0.04r + a$; all dimensions usually are in inches.

The peripheral speed at which rolls normally operate is from 200 to 1200 ft./min., occasionally as high as 1500. The economical range of reduction usually is limited to a No. 12 to No. 16 sieve product. For crushing coarse material, the roll speed should be less than for fine material. For soft and brittle materials, higher speeds can be used. A reduction ratio of 4 should not be exceeded for hard materials. For large pieces of hard materials, 3 or 2.5 gives better results. For small feed material, however, about one-third or one-quarter of the size the rolls will nip, a reduction ratio of 8 may be made if a quantity of fines is not objectionable.

The capacity increases with the length and the diameter of the roll. When the rolls are kept full, the crushing is done not only by the action of the rolls but by the attrition between the particles themselves. This is called choke crushing. In free crushing the rolls are fed at such a rate that each particle is crushed and ejected before the next is nipped. Free crushing produces a larger proportion of coarser sizes and is generally more advantageous, whereas choke crushing is resorted to for the production of a fine product if other types of crushers are not found more suitable.

The following procedure may be used to determine operating characteristics of rolls, bearing in mind that capacity is influenced by the character of the feed, fineness of reduction, and the required manner of operation. Capacity is in direct ratio to width and peripheral speed and may be calculated by the following formula: $C = TWS/1728$, where C = capacity, cu. ft./min.; T = distance between rolls, in.; W = width of rolls, in.; and S = peripheral speed, in./min. This gives the theoretical capacity and is based on the rolls discharging a continuous, solid, uniform ribbon of material. Because of irregularity in the feed, the actual capacity may vary between 25 and 35 per cent of theoretical.

Performance. The chart shown in Fig. 33 can be used to find smooth-roll sizes, capacities, speeds, and power requirements for average conditions under which rolls operate. Although the chart has been prepared carefully and is based on many years of practical experience, the results obtained for a particular application must be considered approximate.

An example of the use of the chart can be based on the following conditions: hard-rock feed of minus 1 in. slot size, open circuit operation, 1/4-in. roll spacing, 20 short tons/hr. average and 24 short tons/hr. maximum production rate, with rate controlled by feeder from bin.

1. Follow down vertical line from 1-in. feed size. First roll size diagonal intersect is 36-in. roll which means a 36-in. roll is the smallest which will nip 1-in. material. Use 42-in. roll for better nipping and longer roll-shell life.
2. Follow horizontal line from intersection of vertical line and 42-in. roll diagonal. A peripheral speed of 700 ft./min. or 64 r.p.m. is indicated.
3. Follow parabolic line from intersection of vertical line and 42-in. roll diagonal. A roll with a 16-in. face should crush 25.6 short tons/hr. with 33 1/3 per cent of the theoretical ribbon. Always select standard-face rolls.
4. Horsepower can be determined by interpolation. In this case, a 25-hp. motor or two 15-hp. motors should be recommended.

Standard diameters for smooth-shell crushing rolls are 12, 18, 24, 30, 36, 42, 48, 54, 60, 64, 72, and 78 in. The length of the roll face is approximately one-half to one-fourth of the diameter. Standard roll lengths are 10, 12, 14, 16, 18, 20, 24, 30, and 36 in., but not all these lengths are available for all diameters.

Corrugated- and Toothed-roll Crushers. *Design and Operation.* These terms are used to describe a variety of machines consisting of one or more cylinders rotating in a horizontal plane; the rolls may be toothed or corrugated, the same or different sizes; and they may run at the same or differential speeds. Crushing rolls can be fitted with hardened shells of various designs, corrugated, grooved, or smooth. The size of feed to a two-roll smooth-shell crusher is smaller than to a corrugated- or toothed-roll crusher of the same roll diameter. The simplest type of toothed-roll crusher consists of a single roll operating against a breaker; generally it is used for coarse crushing.

The Fairmount crusher (Allis-Chalmers) contains a fixed element or curved anvil held in position in relation to the frame by heavy steel tie rods coacting against a

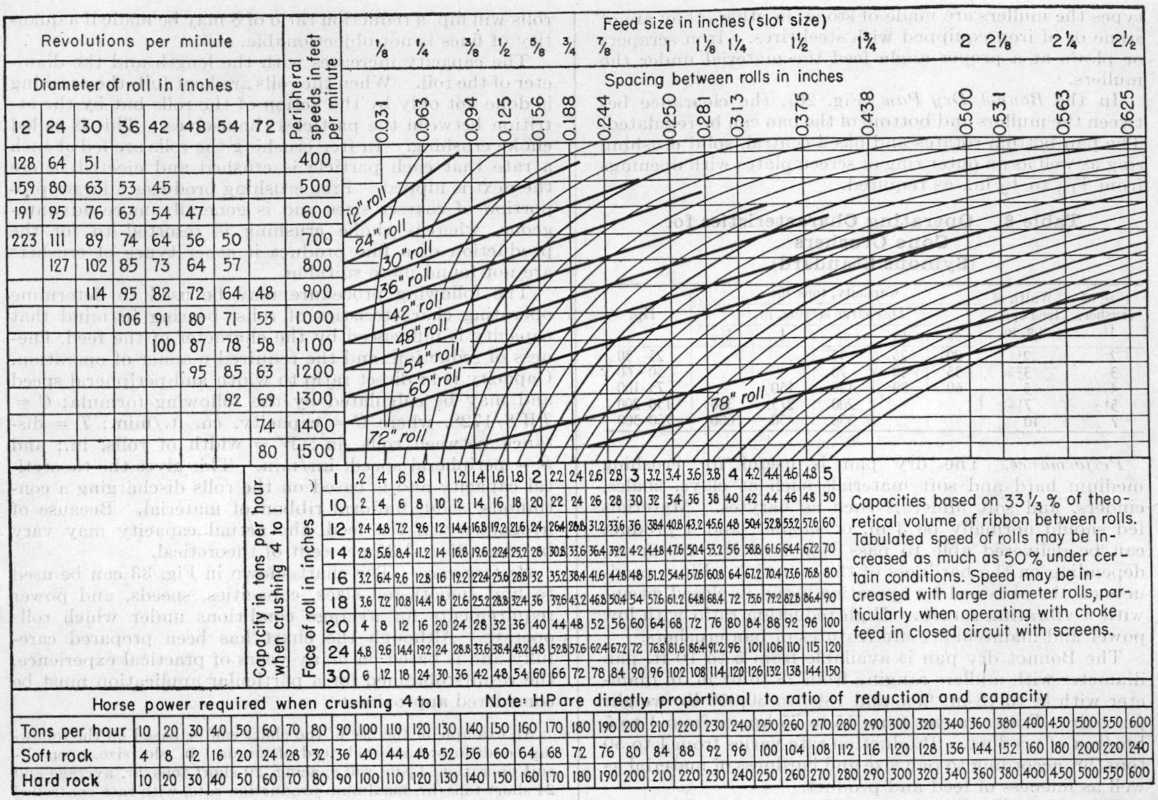

FIG. 33. Size, speed, capacity, and power required for smooth crushing rolls under average conditions. (*Allis-Chalmers Mfg. Co.*)

powerful nest of springs, which are provided for taking up or equalizing excessive pressures (Fig. 19). The single roll is so located with relation to the anvil that preliminary fracturing of large pieces of rock is effected by the sledging action of the teeth on the roll. Secondary crushing is accomplished by direct nipping of the fragments between the roll projections and the anvil.

For many purposes the single-roll crusher is as satisfactory as the larger multiple-roll units. When crushing coal the *Bartlett and Snow single-roll unit* should be confined to conditions where a product 1¼ in. or larger is desired. If the capacity must be larger or the size of the product must be small, the two- or four-roll crushers give more uniform size of product.

The *Sawtooth Crusher* (*Sprout, Waldron & Co.*) has two shafts geared together at differential speeds normally in the ratio of 2¼ to 1. Each shaft carries sawtooth and spacer assemblies. The size of product can be controlled by the spacing of the saws.

Performance. The *Fairmount single-roll crusher* is adapted to laminated stone deposits and to sedimentary formations in which the bedding planes produce shabby stone rather than cubical pieces when quarried. Soft materials such as limestones, dolomites, phosphate rock, cement rock of the Lehigh valley district, shale, and similar deposits provide suitable feed for this toothed-roll crusher. The compressive strength of the rock should not exceed 15,000 lb./sq. in. Operating data are given in Table 9.

The *Jeffrey single-roll crusher* (*Jeffrey Mfg. Co.*) is available in sizes having openings from 18 by 18 in. to 36 by 54 in. with capacities from 25 to 700 tons/hr. on hard

Table 9. Operating Data for Standard Fairmount Crusher

Crusher size, in.	Approx. max. feed thickness, in.	Approx. cap., tons/hr. for discharge opening, in.					Hp.	Roll speed, r.p.m.
		2	3	4	5	6		
24 × 48	14	90	135	180	. . .	. . .	75–100	58
24 × 60	14	115	170	230	. . .	. . .	100–125	58
36 × 60	24	. . .	170	230	290	345	180–220	39

bituminous coal. The power requirements vary from 10 to 125 hp., depending on the size of the crusher.

Performance characteristics for toothed-roll crushers on bituminous coal are given in Table 10.

Table 10. Performance Characteristics for Bartlett and Snow Two-roll Crushers

Roll size, in.		Capacity, tons/hr. for various size products, in.				Hp. at given capacities for various sizes				Max. feed lump size, in.
Diam.	Width	1	1¼	1½	2	1	1¼	1½	2	Average hardness
26	24	50	65	75	100	20	22	24	27	14
26	36	75	95	110	150	30	33	35	40	14
30	36	85	115	140	180	35	40	45	50	16
37	36	100	140	160	210	40	48	50	60	18
37	48	130	190	210	280	50	60	65	75	18

The sawtooth crusher is available in a number of sizes. The action is largely tearing rather than compressing, leading to a minimum of heating, far fewer fines, and a somewhat lower power requirement than characterizes other types of preliminary breaking. Models are available for sheets up to 3 in. thickness and 60 in. width. Length is not a consideration and can be continuous.

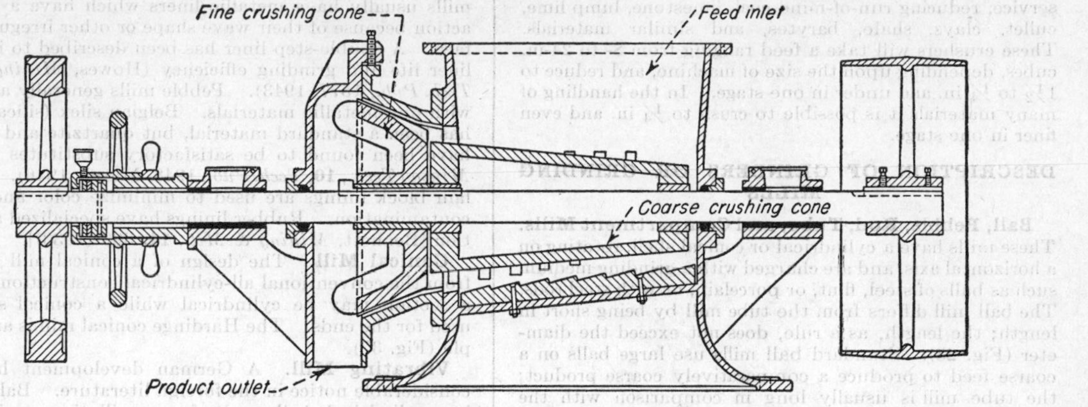

FIG. 34. Horizontal crusher. (*Sprout, Waldron & Co.*)

Sawtooth crushers are also used for friable lump stocks up to 6-in. ring size. Applications include the processing of press cakes, phenolic plastics, alkali cellulose sheets, sheet glue, naphthalene, resins, sulfur, bark, lump pitch, calcium chloride, and asphalt floor tiling.

Rotary Crushers. (Fig. 21.) *Design and Operation.* A shaft, usually vertical, carries a cone with large teeth at the top for initial crushing and small teeth or furrows at the bottom for finer crushing. The shell enclosing the cone has corresponding teeth. The clearance between cone and shell is adjustable by raising or lowering the cone. The *Horizontal Crusher* (*Sprout, Waldron & Co.*) has its cone supported on a horizontal shaft for preliminary crushing; final crushing takes place between close-fitting sections at the base of the cone (Fig. 34). Clearance is adjusted with a handwheel. The lower head-room required by the horizontal crusher is an advantage.

Performance. Rotary crushers are successfully applied to such materials as burned lime, gypsum, phosphate rock, clay, and filter-press cake. A *No. 4 Bartlett and Snow* crusher (Fig. 21) operating on burned lime or gypsum has a production rate of 3 to 5 tons/hr. between ½ and ¼ in.; it requires 8 to 10 hp. The horizontal crusher is used for friable material such as pitch, rosin, mica, coconut shells, and compacted inorganic salts. Its output ranges from 1 to 10 tons/hr. to a product that may all be able to pass through a No. 5 sieve and contain material finer than a No. 100 sieve depending on the feed quality. Power is less than 15 hp.

Hammer Crusher. (Fig. 22.) *Design and Operation.* Pivoted hammers are mounted on a horizontal shaft, and crushing takes place by impact between the hammers and breaker plates. A cylindrical grating may be positioned beneath the rotor. It retains material until reduced to a size small enough to pass between the bars of the grating. A number of hammer crushers are symmetrically designed so that the direction of rotation can be reversed to distribute wear evenly on the hammers and breaker plates. The size of the product can be regulated by changing the spacing of the grate bars and also by lengthening or shortening the hammers. Speed varies from 500 to 1800 r.p.m., depending on size of the machine. The *Double Impeller Crusher* (*New Holland Machine Co.*) is a hammer mill with two rotors, each mounting three impeller bars. Crushing takes place by impact with the bars and breaker plates. High reduction ratio is claimed.

Performance. The characteristics of the *Super-Jumbo Crusher* (*Williams Patent Crusher & Pulverizer Co.*) for handling limestone are given in Table 11.

Table 11. Performance of Williams Hammer Crusher

Size No.	Throat opening, in.	Capacity, tons/hr. 2 in.	1¼ in.	¾ in.	Hp.
4	24 × 16	125	100	60	100
6	30 × 20	210	165	100	175
8	30 × 26	310	250	150	250
10	30 × 30	400	330	200	350

The *Jeffrey hammer crusher* (*Jeffrey Mfg. Co.*) is available in *Type A* for general purposes and *Type B* for fine-reduction severe duty. Performance is given in Table 12 for crushing limestone.

Table 12. Performance of Jeffrey Type B Hammer Crusher

Size of machine, in.	Production rate, tons/hr., for bar opening, in. ⅛	1	Sieve	% smaller than size* for bar opening, in. ⅛	1
20 × 12	2	5	½ in.	..	99
36 × 24	12	30	¼ in.		91
36 × 60	35	80	No. 10	96	54
42 × 66	55	150	No. 100	49	17

* 24-in. machine at 1600 r.p.m.

Feed should not exceed 6-in. lump size for 36-in. and smaller machines or 12-in. lump size for larger than 36-in. machines.

The *Impactor* (*Pennsylvania Crusher Co.*) is a reversible hammer crusher without a discharge grating or cage. It is used for size reduction of cement rock, limestone, and shale in large tonnages for cement mills and in limestone and gypsum operations. The *American Ring Crusher* (*American Pulverizer Co.*) features a rotor assembly with loose crushing rings, held outwardly by centrifugal force, which crush by impact. Performance on run-of-mine bituminous coal is given in Table 13. The rotor speed is 600 r.p.m. for 1½- and 1¼-in. product; 720 r.p.m. for 1-, ¾-, and ½-in. product. The maximum feed lump size is 10 to 28 in., depending on the size of machine.

Table 13. Performance of American Ring Crusher

Number of mill	Capacity, tons/hr. 1½ in.	1¼ in.	1 in.	¾ in.	½ in.	Hp.
15 S	80	50– 65	45– 50	35– 40	25– 30	20– 25
24 S	150	130–140	100–120	80–100	60– 70	30– 40
38 S	300	225–250	200–235	175–200	125–150	60– 75
48 S	400	350–375	330–350	300–325	225–250	100–125
60 S	500	425–450	400–425	350–375	275–300	150–175

The *Stedman Type B Heavy-Duty Crushers* (*Stedman's Foundry & Machine Works*) are designed for heavy-duty service, reducing run-of-mine coal, limestone, lump lime, cullet, clays, shale, barytes, and similar materials. These crushers will take a feed ranging from 8- to 24-in. cubes, depending upon the size of machine, and reduce to 1½ to ¼ in. and under in one stage. In the handling of many materials it is possible to crush to ¼ in. and even finer in one stage.

DESCRIPTION OF GRINDERS OR GRINDING MILLS

Ball, Pebble, Rod, Tube, and Compartment Mills. These mills have a cylindrical or conical shell, rotating on a horizontal axis, and are charged with a grinding medium such as balls of steel, flint, or porcelain, or with steel rods. The ball mill differs from the tube mill by being short in length; the length, as a rule, does not exceed the diameter (Fig. 23). Standard ball mills use large balls on a coarse feed to produce a comparatively coarse product; the tube mill is usually long in comparison with the diameter, uses smaller balls, and produces a finer product. The compartment mill, a combination of the above types, consists of a cylinder divided into two or more sections by perforated partitions; preliminary grinding takes place at one end and finish grinding at the discharge end. Rod mills are similar to tube mills but use rods as the grinding medium, depending on line contact instead of point contact for grinding. Rod mills deliver a more uniform and more granular product than other revolving mills, thus minimizing the percentage of fines, which are detrimental in some industries. The pebble mill is a tube mill with flint pebbles as the grinding medium and is lined with silex or other non-metallic liners.

Tumbling mills is a generic name sometimes used in referring to ball, pebble, rod, tube, and compartment mills, because of the action of the grinding medium.

The ball mill and the pebble mill are simple to operate and versatile in use. A steel- or stone-lined cylindrical steel shell, containing a charge of steel balls or stone pebbles, is rotated horizontally about its axis so that size reduction or pulverization is effected by the tumbling of the balls or pebbles on the material between them. The mills may be operated wet or dry, in either batch or open-circuit use or in closed circuit with size classifiers (see p. 1118 and pp. 930–937).

Design of Tumbling Mills. *Batch.* The conventional type of batch mill consists of a cylindrical steel shell with flat steel-flanged heads. Openings are provided through which the grinding medium and the ground material can be loaded or discharged. Mill length is equal to or less than the diameter [Coghill, DeVaney, and O'Heara, *Trans. Am. Inst. Mech. Engrs.*, **112**, 79 (1934)]. The discharge opening is at a point opposite the loading manhole and for wet grinding usually is fitted with a valve. One or more vents are provided to release any pressure developed in the mill, to introduce inert gas, or to supply pressure to assist discharge of the mill. In dry grinding, the material is discharged into a hood through a grate over the manhole, while the mill rotates. Jackets can be provided for heating or cooling.

Continuous. Continuous mills are more sturdily built than the batch mill. Material is fed and discharged through hollow trunnions at opposite ends of the mill (Fig. 23). A grate or diaphragm just inside the discharge end may be employed to regulate the slurry level in wet grinding and thus control the retention time. In the case of air-swept mills, provision is made for blowing air in at one end and removing the ground material in air suspension at the same or other end. The rod mill and compartment mill are variations of the tube mill.

Batch and Continuous. The ball mill usually is equipped with horizontal baffles, if the lining is smooth, to key the charge to the wall and prevent slippage. Ball mills usually have metallic liners which have a baffling action because of their wave shape or other irregular surface. A double-step liner has been described to increase liner life and grinding efficiency (Howes, *Mining Tech., Tech. Pub.*, 1577, 1943). Pebble mills generally are lined with non-metallic materials. Belgian silex (silica) block has been a standard material, but quartzite and granite have been found to be satisfactory substitutes (Berry, *Mining Tech.*, **10**, *Tech. Pub.*, 1948, March, 1946). Porcelain block linings are used to minimize color and metal contamination. Rubber linings have specialized application [Bennett, *Mining & Met.*, **14**, 399 (1933)].

Conical Mill. The design of a conical mill departs from the conventional all-cylindrical construction in that a section may be cylindrical while a conical shape is used for the ends. The Hardinge conical mill is an example (Fig. 37).

Vibrating Mill. A German development has had considerable notice in the foreign literature. Ball action in a cylindrical shell results from oscillating or vibrating the shell [Bachmann, *Chem. Tech.*, **15**, 195 (1942)].

Operation of Tumbling Mills. *Ball Action.* Cascading and cataracting are the terms applied to the motion of the grinding medium. The former applies to the rolling of balls or pebbles from the top to the bottom of the heap, and the latter refers to the throwing of the balls through the air to the toe of the heap. Ball action has been studied and given mathematical consideration [Gow, Campbell, and Coghill, *Trans. Am. Inst. Mining Met. Engrs.*, **87**, 51 (1930)].

Grinding Medium. The chief factors determining the size of grinding balls are the fineness of the material being ground and the maintenance cost for the ball charge. A coarse feed requires a larger ball than a fine feed; a relation has been proposed: $D_b^2 = KD_p$, where D_b is the ball diameter; D_p is the size of the coarser feed particles, both in inches; and K is a grindability constant varying from 55 for hard to 35 for soft materials (Coghill and Devaney, *U.S. Bur. Mines, Tech Pub.*, 581, 1937).

The need for a calculated ball-size feed distribution is open to question; however, methods have been proposed for calculating a rationed ball charge [Bond, *Trans. Am. Inst. Mining Met. Engrs.*, **153**, 373 (1943)].

The lives and efficiencies of non-metallic grinding mediums have been compared [Metz, *Bull. Am. Ceramic Soc.*, **24**, 357 (1945)].

A graded charge of rods results from wear in a rod mill. Rod diameter may range from 4 to 1 in., for example. A new rod load usually is patterned after a used one found to give good results.

Mill Speed. The criterion by which the ball action in mills of various sizes may be compared is the concept of "critical speed." It is the theoretical speed at which the centrifugal force on a ball in contact with the mill shell at the height of its path equals the force on it due to gravity: $N_c = 76.6/\sqrt{D}$, where N_c is the critical speed in r.p.m., and D is the diameter of the mill in feet for a ball diameter that is small with respect to the mill diameter.

Actual mill speeds vary from 65 to 80 per cent of critical. It might be generalized that 65 to 70 per cent is required for fine wet grinding in viscous suspension, 70 to 75 per cent for fine wet grinding in low-viscosity suspensions and for fine dry grinding, and 75 to 80 per cent for wet and dry grinding of large particles up to ½-in. size. The speeds might be increased 5 per cent of critical for unbaffled mills, in which slippage of the ball charge might be experienced.

The chart shown in Fig. 35 can be used to find the percentage of critical speed for mills of various diameters if the actual speed is known.

Material and Ball Charges. The load ot a grinding medium can be expressed in terms of the percentage of the

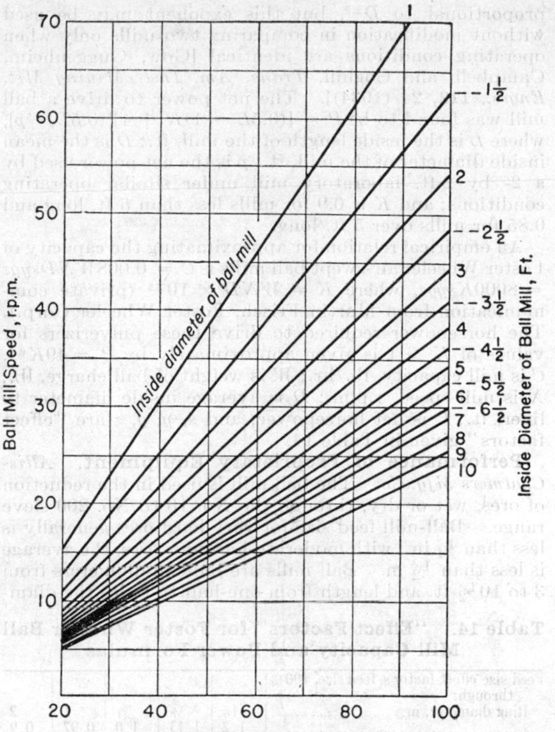

FIG. 35. Ball-mill speed.

volume of the mill that it occupies, *i.e.*, a bulk volume of balls half filling a mill is a 50 per cent ball charge. The void space in a bulk volume of balls is approximately 38 per cent; it is 40 per cent for pebbles. Steel balls have a bulk density of approximately 300 lb./cu. ft.; stone pebbles of 100 lb./cu. ft.

As an aid in finding ball weights and cylindrical-ball-mill volumes, the chart shown in Fig. 36 can be used. A mill with a volume of 98.5 cu. ft., for example, should contain 11,800 lb. of steel balls for a 40 per cent charge, and the void volume would be 15 cu. ft.

The amount of material in a mill can be expressed conveniently as the ratio of its volume to that of the voids in the ball charge. This is known as the material-to-void ratio. If the solid material and its suspending medium (water, air, etc.) just fills the ball voids, the M/V ratio is 1, for example. Grinding-medium charges vary from 20 to 50 per cent in practice and M/V ratios from 1 to 5.

The solids concentration in a pebble-mill slurry may be critical with respect to best grinding efficiency [Creyke and Webb, *Trans. Brit. Ceramic Soc.*, **40**, 55 (1941)].

Control of pulp level to obtain high circulating load is accomplished by use of grate discharge mills. In one case an 18 per cent increase in capacity resulting from conversion of an overflow mill to a grate discharge mill despite a loss of 10 per cent of the mill volume due to the change. The grates allowed passage of sufficient pulp to maintain the circulating load at 400 per cent (Duggan, *Mining Tech.*, *Tech. Pub.*, 1456, March, 1942).

General Considerations. The controlling factors conceded to govern the ore-grinding efficiency of cylindrical mills are as follows:

1. Speed of mill affects capacity, also liner and ball wear, in direct proportion up to 85 per cent of critical speed.
2. Ball charge equal to 50 per cent of the mill volume gives the maximum capacity.

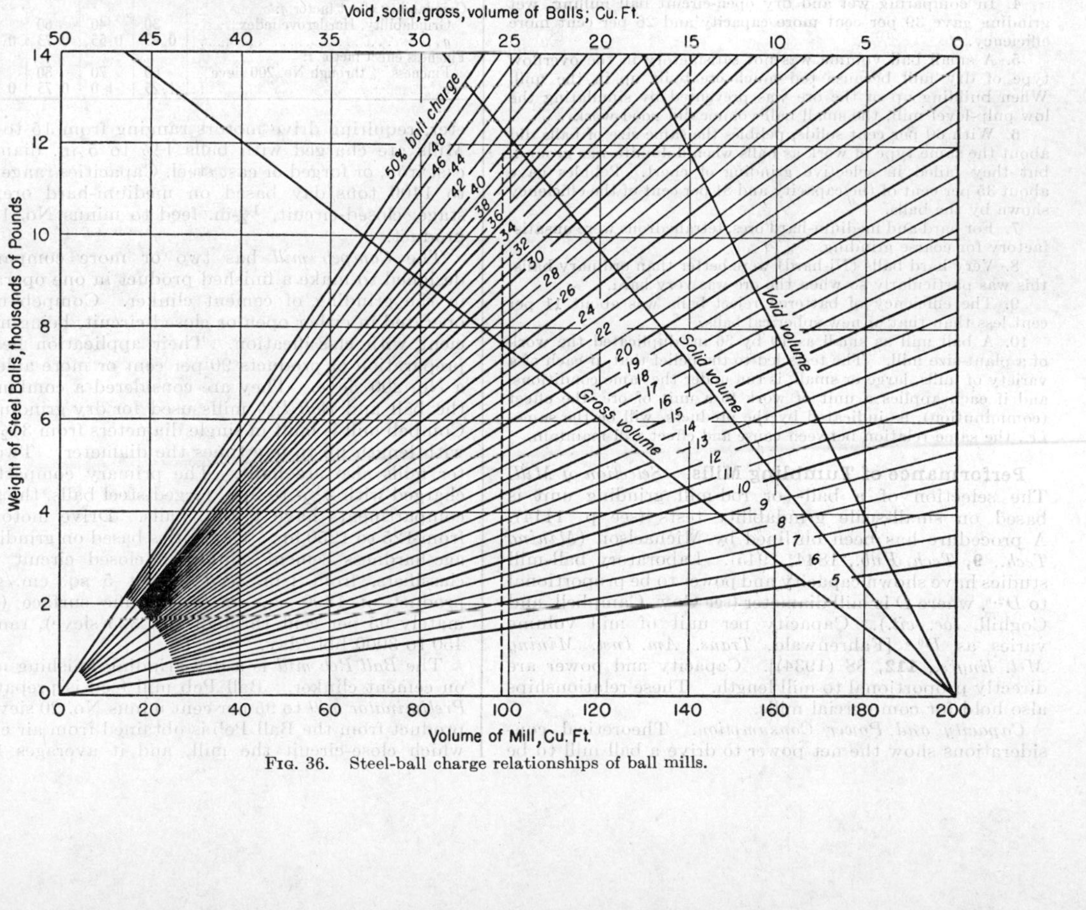

FIG. 36. Steel-ball charge relationships of ball mills.

3. Minimum-size balls capable of grinding the feed give maximum efficiency.
4. Grooved liners of the wave type have found much favor among operators.
5. Classifier efficiency becomes more important in multiple-stage grinding.
6. Higher circulating loads tend to increase production and decrease the amount of unwanted fine material.
7. Low-level or grate discharge has increased grinding capacity over the center or overflow discharge, but liner, grate, and charge wear is higher.
8. Ratio of solids to liquids in the mill must be considered on the basis of ore gravity and volumetric relation.

Experimental evidence presented in a paper by Coghill and DeVaney (Ball Mill Grinding, *U.S. Bur. Mines, Tech. Pub.*, 581, 1937) causes the authors to draw the following conclusions:

1. In wet-batch ball milling with ore charges from 200 to 350 lb. (about 75 lb. of ore was required to fill the interstices of the balls at rest) and speeds from 30 to 80 per cent critical, the slow speed gave the same type of grinding as high speed. Heavy ore charges yielded a little more selective grinding of the coarse particles than light charges. Best capacities were obtained with light charges, and slightly better efficiencies were obtained with heavy ore charges. To split hairs about efficiencies at various speeds the reader will have to study the table and be his own judge.
2. Some of the characteristics of dry-batch ball milling were unlike those of wet grinding. In the dry work, efficiency as well as capacity was best with the light ore charge. Power decreased with decrease in the amount of ore in the mill; in wet grinding it increased with a decrease in the amount of ore in the mill. In dry grinding high speed was more efficient than low speed.
3. In comparing wet and dry grinding the tests were paired so that all the set variables were the same, except pulp consistency (wet or dry). With an intermediate weight of ore charge, selective grinding was of the same degree; with a heavy ore charge, wet grinding was more selective, and with the light ore charge, dry grinding was more selective.
4. In comparing wet and dry open-circuit ball milling, wet grinding gave 39 per cent more capacity and 26 per cent more efficiency.
5. A small ball volume was not satisfactory in the overflow type of dry mill because too much ore built up in the mill. When building up of the ore was prevented by simulating the low pulp-level mill, the small ball volume did good work.
6. With 60 per cent solids, pebbles the same size of balls did about the same type of work as balls when dolomite was ground, but they failed in selective grinding of chert. Pebbles gave about 35 per cent of the capacity and 81 per cent of the efficiency shown by the balls.
7. For hard and medium-hard ores, tetrahedrons were unsatisfactory for coarse grinding.
8. Very hard balls (Ni-hard) were better than ordinary balls; this was particularly so when the ore was very hard.
9. The efficiency of battered reject balls was about 11 per cent less than that of new spherical balls.
10. A ball mill as small as 19 by 36 in. duplicated the work of a plant-size mill. The tests led to the belief that, if each of a variety of mills, large or small, is run under the same conditions, and if each applies a unit of work to a unit of ore, the effect (comminution), as indicated by the products, will be the same; *i.e.*, the same relation between cause and effect will maintain.

Performance of Tumbling Mills. *Selection of Mill.* The selection of a ball- or rod-mill grinding unit is based on small-scale grindability tests (see p. 1114). A procedure has been outlined by Michaelson (*Mining Tech.*, **9**, *Tech. Pub.*, 1844, 1945). Laboratory ball-mill studies have shown capacity and power to be proportional to $D^{2.6}$, where D is mill diameter (see Gow, Campbell, and Coghill, *loc. cit.*). Capacity per unit of mill volume varies as $D^{0.6}$ [Fahrenwald, *Trans. Am. Inst. Mining Met. Engrs.*, **112**, 88 (1934)]. Capacity and power are directly proportional to mill length. These relationships also hold for commercial mills.

Capacity and Power Consumption. Theoretical considerations show the net power to drive a ball mill to be proportional to $D^{2.5}$, but this exponent may be used without modification in comparing two mills only when operating conditions are identical [Gow, Guggenheim, Campbell, and Coghill, *Trans. Am. Inst. Mining Met. Engrs.*, **112**, 24 (1934)]. The net power to drive a ball mill was found to be $P = [(0.5L - 1)K + 1][(0.5D)^{2.5}p]$, where L is the inside length of the mill, ft.; D is the mean inside diameter of the mill, ft.; p is the net power used by a 2- by 2-ft. laboratory mill under similar operating conditions; and K is 0.9 for mills less than 5 ft. long and 0.85 for mills over 5 ft. long.

An empirical relation for approximating the capacity of Foster Wheeler air-swept ball mills is $C = 0.008WNDsygz = 8000Ksygz$, where $K = WND \times 10^{-6}$ (private communication from Martin Frisch, Foster Wheeler Corp.). The horsepower required to drive these pulverizers for values of $K > 1$ is given approximately by $P = 49K^{0.9}$. C is mill capacity, lb./hr.; W is weight of ball charge, lb.; N is mill speed, r.p.m.; D is average inside diameter to liner, ft.; P is net horsepower; and s, y, g, z are "effect factors" given in Table 14.

Performance of Proprietary Equipment. *Allis-Chalmers Mfg. Co.* The ball mill is used in the reduction of ores, wet or dry, through the No. 10 to No. 200 sieve range. Ball-mill feed size for very hard ores generally is less than ¼ in.; with moderate-hardness ores, the average is less than ½ in. Ball mills are built in diameters from 3 to 10½ ft. and length from one-half to twice the diameter, requiring drive motors ranging from 15 to 800 hp. Mills are charged with balls 1½ to 5 in. diameter, of cast iron, or forged or cast steel. Capacities range from 16 to 1400 tons/day based on medium-hard ore, single-stage, closed-circuit, ½-in. feed to minus No. 100 sieve product.

Table 14. "Effect Factors" for Foster Wheeler Ball Mill Capacity and Power Formulas

Feed size effect factor s, feed size, 100% through:					
Ring diameter, in.	1/16	1/8	3/4	1	2
s	1.2	1.13	1.0	0.97	0.9
Moisture effect factor y:					
Surface moisture, %	0	3	6	9	15
y	1.0	1.0	0.92	0.88	0.65
Grindability effect factor g:					
Grindability, Hardgrove index	30	40	60	80	100
g	0.43	0.55	0.73	0.83	1.0
Fineness effect factor z:					
Fineness % through No. 200 sieve	60	70	80	90	99
z	1.25	1.0	0.75	0.49	0.22

The *Compeb mill* has two or more compartments, designed to make a finished product in one operation, as in the grinding of cement clinker. Compeb mills are operated in either open or closed circuit, being more efficient with classification. Their application lies in the preparation of products 90 per cent or more able to pass a No. 200 sieve. They are considered a combination of the ball and *Ball Peb* mills used for dry grinding. The Compeb mill is built in single diameters from 3½ to 8 ft., with length up to five times the diameter. Two-diameter mills also are built. The primary compartment is charged with 2½- to 5-in. forged-steel balls, the finishing compartments have 1–2-in. balls. Drive motors range from 125 to 1250 hp. Capacities based on grinding average-hardness cement clinker in closed circuit with air classifiers, from 1-in. feed (about 5 sq. cm./g.) to a product of 1800 sq. cm./g. specific surface (approximately 95 per cent through No. 325 sieve), range from 450 to 3000 bbl./day.

The *Ball Peb mill* is a dry-grinding finishing mill used on cement clinker. Ball Peb mill feed is prepared by a *Preliminator mill* to 95 per cent minus No. 20 sieve. The product from the Ball Peb is obtained from air classifiers which close-circuit the mill, and it averages 1800 sq.

cm./g. specific surface (approximately 95 per cent minus No. 325 sieve), or finer. Capacities range from 850 to 4200 bbl./day. Ball Peb mills are built 3 to 8 ft. in diameter and in lengths four times the diameter. Longer mills are multicompartment, the charge in each compartment having a different size in order to grind the feed to that compartment more efficiently. The charge consists of forged-steel balls ¾ to 1½ in. in diameter. Drive motors range from 40 to 900 hp.

The *Allis-Chalmers rod mill* is a wet or dry grinder operating most efficiently in the size range from No. 10 to No. 35 sieve size. Less efficient operation results in the No. 35 to No. 65 sieve range, where operation overlaps the ball mill. Rod mills are best applied for primary grinding where the production of extremely small particles or sliming is undesirable. Large-diameter rod mills are used as intermediate stage of crushing and in instances have effectively displaced crushing rolls for preparing ball-mill feed. Such applications are warranted only for medium-hard and soft ores. It is a selective grinder, particularly on heterogeneous ores. Rod mills will receive ⅞-in. and larger feed of moderate-hardness ores; but extreme wear on feed-end mill liners and differential wear on rods result, requiring larger diameter rods, which are less efficient. Rod mills are built from 3 to 9½ ft. diameter, length equal to or greater than the diameter, with a maximum of about 16 ft. Overflow and peripheral discharge mills are used. The grinding medium consists of steel rods 2 to 4½ in. diameter, a few inches shorter than the mill. Drive motors from 15 to 600 hp. are used. Capacities on medium-hard ores through No. 35 sieve range from 50 to 1600 tons/day in open circuit.

The F. L. Smidth & Co. Tumbling mills built by this company find their principal use in cement plants; the best known types are the *Kominuter*, *Unikom*, *Unidan*, and *Pyrator*. For dry operation, the mills usually are fed by a table or cradle feeder; for wet operation, feeding is from a slurry trough by means of an orifice or scoop. The mills may have silex (stone) or *dragpeb* (steel) linings. Flint pebbles are used with silex linings and *cylpebs* (a cylindrical grinding medium) with dragpeb lining.

The *Kominuter* is a screen-type mill operated in closed circuit, usually the first in a two-stage unit where reduction is carried only to the size of a rather coarse mesh, which is fed to tube mills for final grinding. In wet grinding, water is added at the feed hopper, passing the slurry produced through the mill as in dry grinding.

The *Unikom* is a four-compartment mill, a section with large diameter forming the first, or granulating, compartment, with a section of small diameter divided into three compartments. The enlarged section is fitted with liner plates and a special screening arrangement which by-passes fines to the first compartment in the second section and returns oversize to the granulating compartment for further reduction. The first chamber of the second section is equipped with grinding balls, the following compartments with cylpebs, graded downward in size.

The *Unidan* mill is a compartment mill, with three or more compartments, equally well suited for wet and dry grinding. Balls are used in the first compartment, which is also equipped with liners. The compartments for fine grinding are equipped with special rings and steel-alloy lining and with cylpebs graded downward in size toward the discharge end. An added feature of the mill is a special screen arrangement mounted within the mill body; the material does not leave the mill body until finally discharged at the outlet end.

The *Pyrator* mill is used for granulating, pulverizing, and drying damp material in a single unit; it consists of a two-compartment tube mill comprising a ball chamber with liners; a combined screening and ball-separating partition, and a fine-grinding chamber with special ring and alloy lining and cylpebs as the grinding medium. The steel balls are heated and have the double capacity of acting as grinding medium and supplying the heat required for drying. Hot air circulates around the mill body, which is provided with a jacket to retain the hot air in closed circuit, and the mill is provided with means for removing the balls, which are elevated, heated in a hot-air furnace, and returned to the feed end. When the material does not contain an excessive amount of moisture, this unit is very efficient, compact, and economical.

The Mine and Smelter Supply Co. The *Marcy* ball mill (Fig. 23), used extensively for wet and dry grinding of ores, will take feed as coarse as 2 in. and grind to No. 200 sieve size in closed circuit with a classifier. Discharge grates are used to give a rapid change of mill content with a high circulating load. Performance is given in Table 15. The *Open End rod mill* is designed for a heavy, revolving rod mass and a discharge pulp level below the rod mass, giving rapid passage through the mill. A specially designed discharge housing mounted independently of the mill and with a hinged door permits easy access for inspection, relining, and charging of rods. It takes a 1-in. feed, reducing it in one pass to No. 8 to No. 20 sieve size. The uniform discharge product results from the fact that the low-pulp-line mill does not make a displacement product, since the difference in elevation between feed and discharge ensures rapid removal of the finished product. For a finer product of No. 60 to 80 sieve size, the mill is closed-circuited with screens or classifiers. Performance on 1-in. medium-hard material is shown in Table 16.

Hardinge Company. The Hardinge Conical Mill, shown in Fig. 37, is used extensively for both wet and dry grinding in open and closed circuits. The conical mill is similar to the cylindrical mill in that it consists of a drum rotating about its horizontal axis and operating in much the same way, but unlike the cylindrical mill, it has conical ends instead of straight ends. Ball segregation takes place and roughly proportions the energy to the work per-

Table 15. Performance of Marcy Ball Mills

Size, ft.	Ball charge, tons	Hp. to run	Mill speed, r.p.m.	Capacity, tons/24 hr. (based on medium-hard ore)								
				No. 8 sieve* 20% −200	No. 20 sieve 35% −200	No. 35 sieve 50% −200	No. 48 sieve 60% −200	No. 65 sieve 70% −200	No. 80 sieve 80% −200	No. 100 sieve 85% −200	No. 150 sieve 93% −200	No. 200 sieve 97% −200
3 × 2	0.85	5– 7	35	19	15	12	10	8	6½	5	4	3
4 × 3	2.73	20– 24	30	80	64	53	45	36	28	22	18	14
5 × 4	5.25	44– 50	27	180	145	120	102	82	63	51	41	32
6 × 4½	8.90	85– 95	24	375	300	250	210	170	135	105	85	66
7 × 5	13 10	135–150	22½	640	510	425	360	290	225	180	145	113
8 × 6	20.2	220–245	21	1100	885	735	625	500	390	310	250	195
9 × 7	30.0	345–380	20	1800	1450	1200	1020	815	635	505	410	315
10 × 10	56.50	700–750	18	3680	2960	2450	2100	1700	1325	1050	850	655
12 × 12	90.5	1260–1345	16.4	7125	5725	4750	4070	3290	2570	2035	1650	1275

* Sieve through which substantially all the material can pass.

formed, the large balls assembling in the cylinder at the feed end of the mill where the diameter is largest, while the smaller balls arrange themselves in decreasing sizes toward the discharge end of the mill.

Hardinge Ball Mills are lined with metallic liners of the wedge-bar type, or of the wave or ribbed type. The wearing bar of the wedge-bar type of lining serves the purpose of lifting the mass of balls and material as well as holding the liner plates in place. *Hardinge Pebble Mills* may be lined with adamant silica, silex, porcelain, or any other non-metallic lining required for the operation. Hardinge wet grinding mills are supplied with discharge arrangements for high, medium, or low pulp levels, the use of which depends on the particular problem under consideration. A suitable grate is used which will permit carrying a maximum ball charge and pulp load in a given-size mill and it also keeps the balls from spilling out of the mill and prevents an accumulation of tramp oversize at the discharge end of the mill. For dry grinding a vertical grate with low-pulp-level discharge vanes is used.

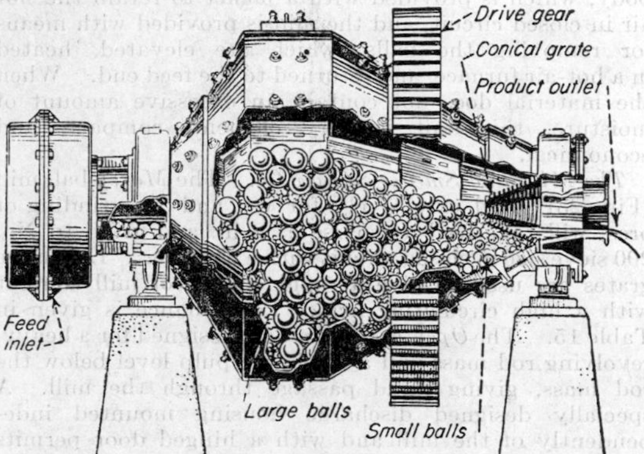

FIG. 37. Hardinge conical mill. (*Hardinge Co.*)

Mill feeders attached to the feed trunnion of the conical mill and used to pass the feed into the mill without backspill are of several types. A feed chute is generally used for dry grinding, this consisting of an inclined chute sealed at the outer edge of the trunnion, and down which the material slides to pass through the trunnion and into the mill. A screw feeder may also be used when dry grinding, consisting of a short section of screw conveyor which extends part way into the opening in the feed trunnion and conveys the material into the mill. For wet grinding, several different types of feeders are available; the scoop feeder attached to and rotating with the mill trunnion and which dips into a stationary box to pick up the material and pass it into the mill; a drum feeder attached to and rotating with the feed trunnion, having a central opening into which the material is fed, and an internal deflector or lifter to pass the material through the trunnion into the mill; or a combination drum and scoop feeder, where the new feed to the mill is fed through the central opening of the drum while the scoop picks up the oversize being returned from a classifier to a scoop box well below the center line of the mill. The mill feeder must be able to handle any quantity of material which the mill may be capable of grinding, and in addition, a circulating load which may be as high as 1000 per cent of the new feed rate. The dry-grinding performance of Hardinge mills on materials of average hardness is given in Table 17.

Regulating feeders are built in two basic designs, *i.e.*, constant-volume and constant-weight feeders. Typical of the constant-volume type is the *Hardinge Disc Feeder*, which consists of a circular hopper which is generally fastened to the bottom of the feed bin, the feed sliding from the hopper on to the center of a revolving disk, and then being scraped from the disk by one or two adjustable scrapers into the mill feeder. The quantity of material may be regulated by varying the speed of the feeder or by adjusting the scraper to vary the quantity of material being scraped from the disk. The disk feeder is suitable for damp materials or for a feed having large lumps. *The Hardinge Constant Weight Feeder*, or *Feedometer*, is a feeder of the constant-weight type which will aid the operator in securing maximum over-all mill efficiency and maintaining a record of the material fed to the mill. It controls the feed to the mill by a constant weight rather than a constant volume, regardless of bin segregation or changes in the physical characteristics of the material being fed. The Hardinge Constant Weight Feeder consists of an endless traveling belt, mounted on a structural-steel frame, the whole of which is suspended on pivots below a feeder hopper. A gate controlling the material at the front of the feeder hopper is also pivoted and connected to the frame through a linkage, and after the feeder is once set for the correct weight of the material on the feeder belt, any change in position of the feeder frame moves the gate to maintain a constant weight of the material on the belt. The quantity of material fed by this feeder is changed by means of a variable-speed drive. An improved design of this feeder is the Hardinge Feedometer, consisting of the same basic features but including instruments to indicate and record the quantity of material fed.

Table 16. Performance of Marcy Rod Mills

Size, ft.	Rod charge, tons	Hp. to run	Mill speed, r.p.m.	Capacity, tons/24 hr.				
				No. 8 sieve	No. 20 sieve	No. 35 sieve	No. 48 sieve	No. 65 sieve
2 × 4	0.9	4– 6	38	20	15	12	10	7
3 × 6	3.6	18– 22	30	105	80	65	50	40
4 × 8	7.6	44– 48	25	240	180	145	120	90
5 × 10	14.5	85– 95	21	525	390	315	260	195
6 × 12	24.1	135–150	17½	855	640	510	425	320
7 × 15	42.1	225–250	15	1600	1200	965	800	600
8 × 12	43.4	230–250	13.2	1675	1250	1000	830	625
9 × 12	54.7	310–340	12.5	2240	1680	1350	1115	835

Table 17. Performance of Hardinge Ball and Pebble Mills

Size of mill*	Approx. weight, lb.			Speed, r.p.m.	Hp. to run	Capacity, tons per 24 hr.		
	Mill	Lining	Balls			1½ in. to 90%, through No. 100 sieve	Closed circuit with air classifier: ¾ in. to 90%, through No. 200 sieve	Closed circuit with air classifier: ½ in. to 98%, through No. 325 sieve
2′ × 8″	900	375	400	40	1	4	3	1½
4½′ × 24″	8,100	5,400	4,500	28	25	48	36	18
6′ × 48″	17,000	12,000	15,000	25	70	144	108	54
8′ × 48″	27,000	23,000	31,000	21	160	360	252	126
10′ × 66″	51,000	35,000	65,000	18	350	840	600	300
			Pebbles			Open circuit, ½-in. to No. 10 sieve size	Closed circuit, ½ in. to No. 48 sieve size	
2′ × 8″	900	400	175	42	½	2	1½	
4½′ × 24″	8,000	2,300	2,400	30	12	15	9	
6′ × 48″	12,000	5,000	8,500	27	30	54	36	
8′ × 48″	17,000	14,000	14,000	24	62	120	84	
10′ × 66″	32,000	20,000	28,000	18	160	336	216	

* Diameter by length of cylindrical section.

Foster Wheeler Corp. This ball mill is built for dry pulverizing fuels and other materials. The mill may be fitted with a single classifier at one end or with two classifiers, one at each end. Preheated air for drying the material while it is undergoing pulverization enters the mill through the trunnion at one end and, as it passes through the mill, picks up pulverized material and carries it through the trunnion at the other end into a spiral-flow classifier. The oversize particles rejected in the classifier mix with the feed that is introduced into the classifier, and the mixture is conveyed through the trunnion into the grinding zone by means of a ribbon conveyor direct-driven from the mill. The ratio of recirculating oversize to the feed may be as high as 6 and as low as 1.5. The hot oversize, which has lost most of its moisture, tends to coat the wet incoming feed particles and blot off the free surface moisture. The effect is to reduce the sensitivity of the pulverizing process to moisture. The feed is, and the product discharge and air temperature may be, controlled automatically, each factor independently of the others. The rate of feed is controlled by an automatic device actuated by the level of material within the mill. The variation of this level is held within close limits. The rate of product discharge is controlled by varying the air flow through the mill. Mill characteristics are given in Table 18.

Table 18. Characteristics of Foster Wheeler Ball Mills*

Size	1	3	5	7
Nominal inside diam., ft. (approx. = D)	5	7	8.5	10
Max. ball charge, lb., W	13,000	28,000	48,000	80,000
Max. $WND \times 10^{-6}$ (max. K)	1.9	4.4	8.5	15.0
Max. speed, r.p.m. (max. N)	28.5	23.4	21.2	19.3

* See capacity and power formulas on p. 1132.

Aerofall Mills Limited (*Toronto, Canada*). The *Aerofall Mill* is a dry, combined crushing and grinding unit, capable of taking run-of-mine or quarry-sized material (less than 12-in. size in the smaller units and less than 18-in. size in the larger units) and reducing it in a single operation to as fine as 99.5 per cent able to pass a No. 325 sieve for some materials. Air classification is employed and the product may be collected by wet or dry cyclones or by cloth bag-type filters. Depending on the nature of the material to be treated, the unit may be operated using the material itself as the crushing and grinding medium, or it may be operated with a charge of steel or tungsten carbide balls. Only a small charge of either kind of ball is used, 2.5 per cent of the mill volume, as their function is that of impact reduction only.

The Aerofall mill has been employed successfully on both hard and soft ores. Barytes ore has been treated to obtain a product having 99.5 per cent able to pass a No. 325 sieve with a run-of-mine feed of 12-in. lump size and smaller, using the material itself as its own crushing and grinding medium. A unit of 9-ft. diameter and 4-ft. length produces 50 tons/24 hr. and requires 100 hp. exclusive of the classifier. At a rate of 125 tons/24 hr. the product fineness is 75 per cent able to pass through a No. 325 sieve.

Units are available in sizes from 5 to 16 ft. diameter and it is expected that units as large as 25 ft. diameter will be constructed.

Particle-size Classifiers Used with Grinding Mills. Ball mills or tube mills can be operated in closed circuit with external air classifiers with or without air sweeping being employed, as shown in Fig. 9. If air sweeping is employed, a cyclone separator may be placed between mill and classifier. (The principles of size reduction combined with size classification are discussed on pp. 1117 to 1119.) Likewise other types of grinding mill can be operated in closed circuit with external size classifiers (Fig. 12), as will be described at appropriate places on succeeding pages. However, many types of grinders are air-swept and are so closely coupled with their classifiers that the latter are termed internal classifiers.

Some equipment manufacturers refer to their air classifiers as "separators," but this is generic usage. For the sake of consistency and to conform with the generally accepted terminology for hydraulic and mechanical devices for the same purpose the term "air classifier," or simply "classifier," is used for a pneumatic device to separate a material into two or more fractions, each more closely sized than the feed material.

External Classifiers. The *Whizzer* (*Raymond Pulverizer Division, Combustion Engineering Co.*), *Spinner* (*Williams Patent Crusher & Pulverizer Co.*), *Gayco* (*Universal Road Machinery Co.*), and *Whirlwind* (*Sturtevant Mill Co.*) classifiers are examples. They may be used independently as well as externally, *i.e.*, in closed circuit with grinding mills. An external-classifier arrangement in which conveying is accomplished with an elevator is shown in Fig. 38. The *Reversed Current air classifiers*

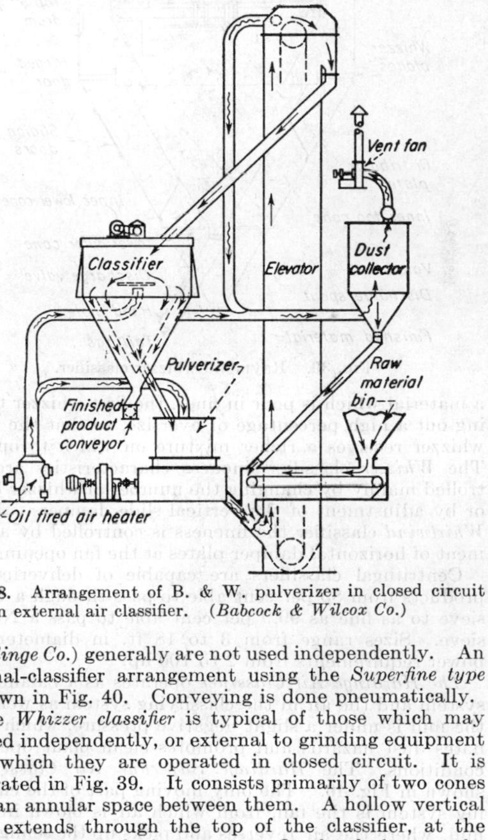

FIG. 38. Arrangement of B. & W. pulverizer in closed circuit with an external air classifier. (*Babcock & Wilcox Co.*)

(*Hardinge Co.*) generally are not used independently. An external-classifier arrangement using the *Superfine type* is shown in Fig. 40. Conveying is done pneumatically.

The *Whizzer classifier* is typical of those which may be used independently, or external to grinding equipment with which they are operated in closed circuit. It is illustrated in Fig. 39. It consists primarily of two cones with an annular space between them. A hollow vertical shaft extends through the top of the classifier, at the center, down into the inner cone. On the shaft, near the top, is mounted the main fan; below the fan is the whizzer, and below the whizzer the distributing plate. As this plate rotates at a high velocity, the centrifugal effect on the material throws it radially, in a uniform stream, in the space between the edge of the plate and the inner surface of the inner cone, into the path of an upblast of air. The dust-laden air passes up through the whizzer where the coarser particles are eliminated, and discharges through the fan into the annular space between the cones. The rotating fan blades throw the material to the inner surface of the outer cone, and it is discharged through a valve in the bottom of the cone, while the air returns to the inside of the inner cone through the portholes formed by the deflector blades or vanes, placed radially around the inner cone. Material discarded by the whizzer drops to the bottom of the inner cone and is spouted from the classifier as tailings.

The whizzer type of air classifier often is built with two whizzers one above the other. The purpose of the two whizzers is to obtain a finer product than is possible with the single-whizzer type. It is in fact a two-stage classifier. This is of particular advantage when classifying

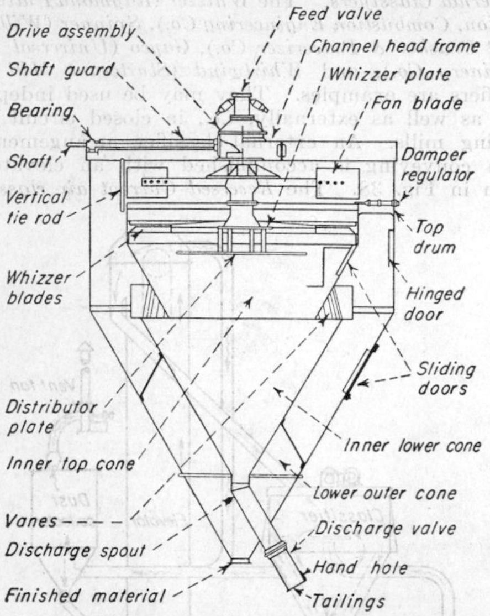

FIG. 39. Raymond whizzer classifier.

a material which is poor in fines, the first whizzer throwing out a high percentage of oversize, so that the upper whizzer receives a richer mixture on which to operate. The *Whizzer* classifier fineness characteristics are controlled mainly by changing the number of whizzer blades or by adjustment of the vertical slide dampers. In the *Whirlwind* classifier the fineness is controlled by adjustment of horizontal damper plates at the fan opening.

Centrifugal classifiers are capable of delivering fine products from 85 per cent able to pass through a No. 60 sieve to as fine as 99.9 per cent able to pass a No. 400 sieve. Sizes range from 3 to 18 ft. in diameter with power requirements from 2 to 100 hp.

The Hardinge Air Classifier operates as a balanced air system and the air in the classifying system as well as in the mill is under a slight negative pressure, which eliminates dust hazards and promotes clean plant operating conditions. The *Hardinge Superfine Air Classifier* is shown in Fig. 40. The only moving part of the classifying system is the fan, from which air is blown into the mill, where the air reverses and picks up the semiground material and conveys it through the uptake pipe to the Superfine Classifier. A partial classification of the coarse and fine material is made between the outer and inner cones of the classifier by means of reduced velocity of the air. The remaining material and the air then pass through the ports into the inner cone of the classifier. The ports give a centrifugal action to the air and material and further classifying is made in the inner cone, the coarse material being thrown to the outside of the cone and sliding down and out through the bottom of the inner cone, where it joins the coarse material previously dropped between the two cones and all is then returned to the feed end of the mill for further grinding. The fine material or final product is carried by the air out the top of the Superfine Classifier and into the product collector, where product and air are separated by centrifugal force. The product is discharged from the product collector at the bottom airlock, while the air from the top of the product collector is returned to the inlet side of the fan, thus completing the cycle. The discharge of the product collector may be placed any reasonable distance up to 100 ft. above the mill. In order to maintain a negative pressure in the air-classifying system and to prevent dusting, it is necessary to vent a sufficient amount of air to overcome air leakage into the system. This vent air may be discharged directly to the atmosphere or, if desired for economic, hazard, or nuisance reasons, it may be discharged into a bag-type dust collector and the dust in the vent air recovered. Hardinge Air Classifiers are of two types, the *Superfine Classifier* for products from No. 60 to No. 400 sieve size, and the *Loop Classifier* for products from No. 10 to No. 100 sieve size.

Internal Classifiers. A typical application of the *Whizzer classifier* to internal use is with the ring-roller mill. Its design and method of installation are shown in Fig. 41. The whizzer disks rotate in a horizontal plane. They may consist of one or two disks in number, each one fitted with multiwhizzer blades. The whizzer is driven through a variable drive, and the fineness of the finished product is regulated by the speed of the whizzer. The faster the whizzer rotates the more particles are thrown out and the finer becomes the finished product. The raw material passes up from the grinding surfaces vertically and, before passing out of the machine at the top, is compelled to pass between the rotating-whizzer blades. The oversize is returned to the periphery and dropped back for further grinding while the desired fines pass up through the whizzer blades into the duct leading to the cyclone collector.

Vacuum Multi-vane air classifiers of the inverted double-cone type are employed with *Raymond ring-roller*

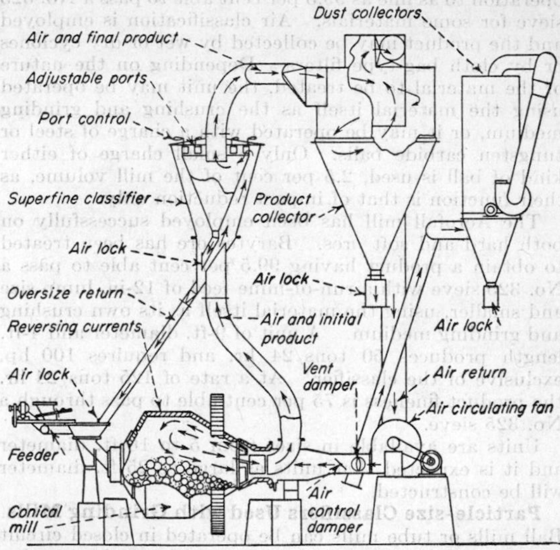

FIG. 40. Hardinge conical mill with Reversed Current air classifier of the Superfine type.

mills where products of moderate fineness are being made. They occupy a position the same as shown for the *Whizzer classifier* in Fig. 41. The air containing the pulverized product travels upward between the two cones and enters the inner cone, guided by vertical and adjustable deflector vanes. The coarsest product is obtained with the vanes in a radial position and the finest with the vanes at the most extreme angle with the radial setting.

The bottom of the inner cone is fitted with a flap valve for returning oversize material to the mill. Fine product is discharged through a suction sleeve centrally located in the top of the classifier.

Additional applications of internal size classification to various types of grinding mill will appear in the text.

Ring-roller Mills. (Fig. 16.) These are equipped with rollers that operate in conjunction with grinding rings. Grinding takes place between the surfaces of the grinding elements, *i.e.*, the ring and rollers. Pressure may be applied with heavy springs or by centrifugal force of the rollers against the ring. Either the ring or

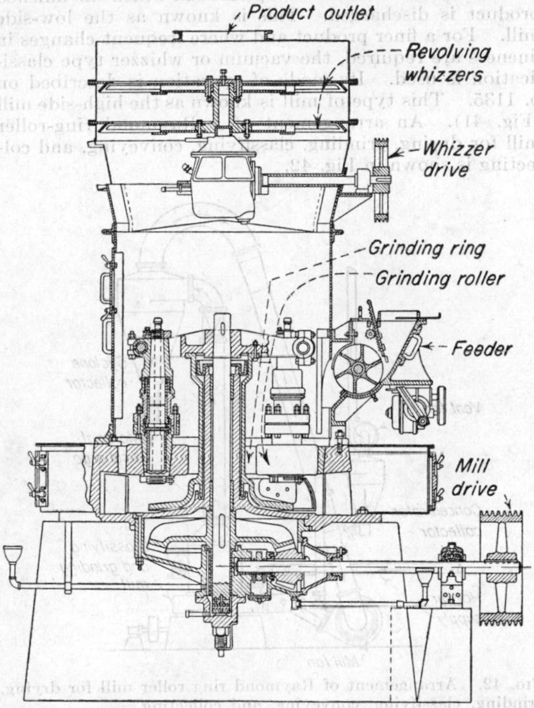

FIG. 41. Raymond high-side mill with internal whizzer classifier.

rollers may be stationary. The grinding ring may be in a vertical or horizontal position. Ring-roller mills also are referred to as ring-roll mills or roller mills. The ball and ring (Fig. 28), and bowl mills (Fig. 43) are types of ring-roller mill.

Ring-roller mills should be distinguished from roller mills. The latter are used for such operations as paint grinding and flour milling. Paint-grinding roller mills consist of two to five smooth rollers (sometimes called rolls) operating at differential speeds. A paste is fed between the first two or low-speed rollers and is discharged from the final or high-speed roller by a scraping blade. The paste passes from the surface of one roller to that of the next because of the differential speed, which also applies shear stress to the film of material passing between the rollers. A three-roller mill is illustrated in Fig. 50. Roller mills for producing flour from grain consist of one or more pairs of rollers. The rollers (or rolls) in each pair run toward each other at different speeds. Material is fed between the pairs of rollers in series, and reduction takes place on each pair. Grooved rollers are used for crushing the grain and smooth rollers for final milling of flour.

Ring-roller Mills without Internal Classification. The *Sturtevant mill* has a concave vertical grinding ring and is used for non-metallics, especially phosphate rock. A No. 1 mill with external air classifier grinds 2 to 4 tons/hr. of limestone or phosphate rock to 90 per cent through a No. 80 sieve. The *Kent Maxecon mill* is used for bauxite, coke, limestone, magnesite, and phosphate rock. It requires a 25- to 50-hp. drive and grinds at 1000 lb./sq. in. roller pressure. Capacity in closed circuit with external screen or air classifier is 4 tons/hr. of phosphate rock for acidulation or 10 tons/hr. of limestone for agricultural use.

Ring-roller Mills with Internal Screen Classification. The *Bradley Hercules Three-roller Mill* (*Bradley Pulverizer Co.*) is used for semifine grinding of materials such as cement rock, cement clinker, limestone, phosphate rock, phosphate rock clinker, etc. The grinding action of the Bradley Hercules Mill is that of the three rollers being revolved around and against a steel die ring at a speed which, through centrifugal force, creates a pressure between the rollers and the die, where they come in contact with the material to be pulverized. In this manner the material is reduced within the mill to the desired fineness, after which it is discharged through an internal screen surrounding the grinding chamber, thence through ports in the base of the mill to a screw conveyor installed in the foundation of the mill to an elevator or direct to storage as may be desired.

The Bradley Hercules Mill is a large-capacity pulverizer, capable of producing as much as 25 to 50 tons/hr. when grinding average-hardness dry limestone, or 135 to 150 barrels/hr. of cement clinker. Finished product will average 90, 60, and 50 per cent able to pass through Nos. 20, 100, and 200 sieves, respectively. It is designed to take material 2 in. and under for feed, reducing same to the desired fineness in a single operation, without auxiliary machinery, all the material passing through the mill being of a uniform fineness. The fineness of the finished product discharging from the mill is determined by the mesh of the screen installed on the mill, which is so designed that it can be quickly changed to a screen of another mesh size, so that the operator may secure whatever fineness of finished product desired, keeping in mind the limitations of the mill. In this manner, users are enabled to produce products of various finenesses. This mill has proved to be very useful when grinding materials to fineness of No. 20 sieve, which is the fineness best suited to the requirements of the Portland cement manufacturers, who use tube mills for the final grinding of their product. Due to the great pressure exerted by the grinding rollers revolving against the die ring, this type of machine, when grinding to 100 per cent passing a No. 20 sieve, produces a large proportion of Nos. 60, 100, and 200 sieve and finer material. Motors of 300 or 350 hp. are used, depending on the kind of material being ground.

The *Type "B" Junior Hercules Mill* operates in the same manner and on the same materials as the large Hercules Mill but with capacities of 5 to 12 tons/hr.

The *Griffin Mill* (*Bradley Pulverizer Co.*) has a single roller revolving against and around a horizontal grinding ring. The ground material is discharged through an internal screen by the aid of an air current resulting from the speedily revolving roller. Griffin Mills are used for the reduction of materials similar to those mentioned above with capacities of 1 to 6 tons/hr. Design characteristics

Table 19. Design Data for Ring-roller Mills with Internal Screen Classification

Design Characteristics	Bradley-Hercules	Type "B" Junior Hercules Mill	Griffin Giant	Griffin Junior Giant
Weight of mill, lb..........	60,000	19,650	26,000	13,000
Speed of mill, r.p.m.........	130	165	175	210
Diameter of roller, in........	22	16½	24	18
Diameter of ring, in.........	66	42	40	30
Weight of roller head, lb....	750	400		
Motor size, hp..............	350	100	100	50

for Bradley mills are given in Table 19. Applications are described on p. 1152.

Ring-roller Mills with Internal Air Classification. The *Babcock & Wilcox pulverizers, Type B, 100-Series* consist of a single row of balls operating between a stationary bottom ring and a rotating top ring. The *Type B, 200- and 300-Series,* are designed with multiple rows of balls to produce maximum capacity in the space occupied. The *200-Series pulverizer* consists of two rows of balls, one above the other. The top and bottom rings are stationary with the intermediate ring rotating. Externally adjustable springs load the grinding elements to the pressure required. The *300-Series pulverizers* are the same as the 200-Series except that a third row of balls has been added inside of the top row of the 200-Series to increase the capacity still further.

In operation, wet raw feed is admitted to the center of the pulverizer and is fed through the upper balls by centrifugal force, then through the lower row by gravity. Preheated air carries the partly pulverized material up to the rotating internal classifier. The finished product passes through the classifier with the carrying air, and on out of the pulverizer. Oversize material is returned by gravity to the grinding elements. For grinding non-combustible materials, the pulverizer is arranged to discharge the material by gravity after it passes through the lower row of balls, and it operates in closed circuit with an external classifier. Preheated air can be used in the closed-circuit system to dry material as it is being pulverized. A typical arrangement of a closed-circuit unit is shown in Fig. 38. The *200- and 300-Series pulverizers* are used as air-swept units for grinding large quantities of coal for either direct firing or storage. As closed-circuit units, they are used in drying and grinding cement raw materials, agricultural limestone, chrome ore, phosphate rock, and other materials. They are built in nine sizes with capacities up to 45 tons/hr. Pulverizers of the B. & W. type may be used in either pressure or suction systems.

The *B. & W. pulverizer, Type E,* consists of a single row of balls operating between a rotating bottom ring and a stationary top ring (Fig. 28). Externally adjusted springs apply pressure to the top ring to give the required loading for proper pulverization. In operation wet raw coal is admitted inside the ball row and is fed through the grinding elements by centrifugal force. The partly pulverized coal is picked up outside the ball row by preheated air and carried to the rotating centrifugal-type classifier in the upper part of the pulverizer. Coal that is pulverized passes through the classifier with the air, and out of the pulverizer, while the oversize coal is returned by gravity to the grinding elements. The classifier is designed with adjustable blades so that the fineness of pulverization can be varied by adjusting the classifier. The *Type E pulverizer* is particularly suited to the direct firing of rotary kilns and industrial furnaces where close temperature control is required and long periods of continuous operation are essential. It is built in seventeen sizes with capacities up to 14 tons/hr.

B. & W. pulverizers operate successfully in circulating systems in which a single pulverizer fires two or more kilns or furnaces by distribution through a circulating loop, as shown in Fig. 58.

The *Raymond* ring-roller mill (Figs. 16 and 41) is of the internal air-classification type. The base of the mill carries the grinding ring, rigidly fixed in the base and lying in the horizontal plane. Underneath the grinding ring are tangential air ports through which the air enters the grinding chamber. A vertical shaft with a bevel gear near the bottom and resting on a thrust bearing is driven by a horizontal shaft through a pinion. Keyed rigidly to the shaft near the top is a spider which carries the roller journals. These journals have rollers on the bottom rotating on their own bearings while traveling around the ring. Two or more journals are pivotally suspended by trunnions fastened at the top of the journal housing and supported in the arms of the spider. They hang almost vertically, so that when the mill is at rest the rolls press only lightly against the grinding ring.

The method of classification used with Raymond mills depends on the fineness desired. If a medium-fine product is required, up to 85 or 90 per cent through a No. 100 sieve, a single-cone air classifier is used (see Fig. 16). This consists of a housing surrounding the grinding elements with an outlet on top through which the finished product is discharged. This is known as the low-side mill. For a finer product and where frequent changes in fineness are required, the vacuum or whizzer type classification is used. Its mode of operation is described on p. 1135. This type of mill is known as the high-side mill (Fig. 41). An arrangement of a Raymond ring-roller mill for drying, grinding, classifying, conveying, and collecting is shown in Fig. 42.

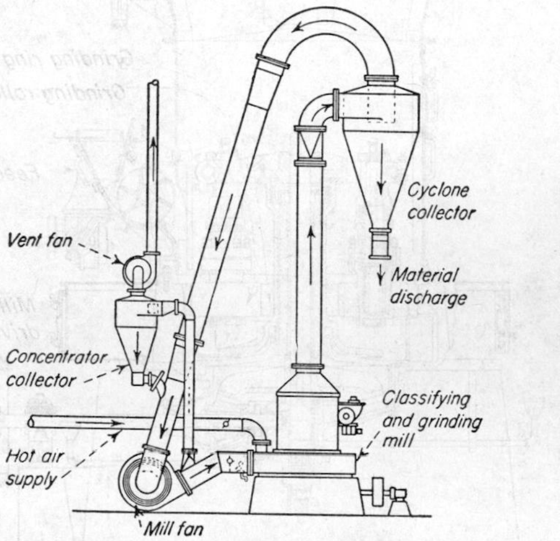

FIG. 42. Arrangement of Raymond ring roller mill for drying, grinding, classifying, conveying, and collecting.

The Raymond mill with air classification is used extensively for the grinding of coal, coke, limestone, barytes, gypsum, phosphate rock, sulfur, bauxite, bentonite, and practically all the non-metallics except the siliceous materials of an abrasive character. It is also used to a large extent in the grinding of the chemical and mineral pigments. Raymond mills are available in seven sizes with from three to five rollers and having grinding-ring diameters of 30, 42, 50, 54, 60, 66, and 73 in.

The *Williams* ring-roller mill (*Williams Patent Crusher & Pulverizer Co.*) can be supplied with an internal classifier of the rotating-blade type known as the *Spinner air classifier* or with a double-cone classifier.

The *Raymond Bowl Mill* is a departure from the design of the standard ring-roller mill. In this style of mill the journals that carry the grinding rollers are stationary while the grinding ring rotates. The grinding pressure is produced by means of springs, which may be adjusted to give the required pressure, and the distance between the rollers and the ring may be set to any predetermined clearance. The rollers do not touch the ring, there being no metal-to-metal contact between the grinding surfaces. Figure 43 shows the construction of the bowl mill. The grinding ring is carried on the lip of a rotary bowl. The raw material from the feeder drops on the bowl where,

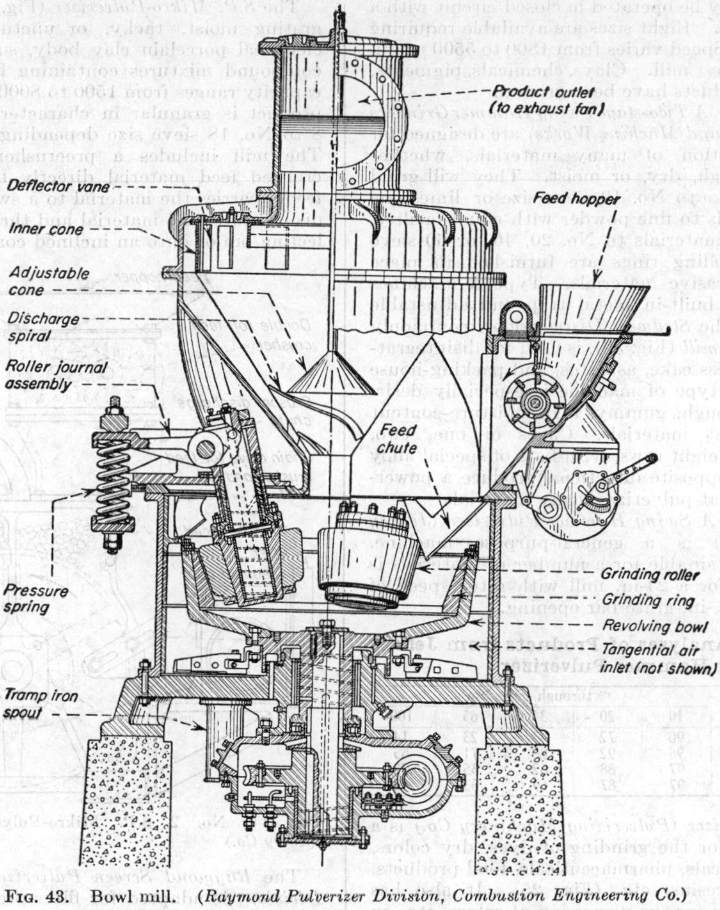

FIG. 43. Bowl mill. (*Raymond Pulverizer Division, Combustion Engineering Co.*)

owing to the centrifugal force of rotation, it is forced to the periphery and, owing to the angle of the ring, it is forced upward between the ring and the rollers, where it is pulverized. The action of the tapered rollers on the tapered ring causes the pulverized material to work upward and out of the grinding chamber into an upblast of air. The air with the pulverized material passes up into a classifier of the double-cone vacuum type (see p. 1136). Here the required fines are removed and the oversize dropped back to the bowl, where it is mixed with the raw feed. This type of mill is primarily used for pulverizing coal and blowing it directly into industrial furnaces or rotary kilns. Tramp iron and other extraneous hard materials are thrown out of the mill automatically through a spout.

Hammer Mills. (Figs. 25 and 27.) Hammer mills for pulverizing and disintegration are operated at high speeds. The rotor shaft may be vertical or horizontal, generally the latter. The shaft carries hammers, sometimes called beaters. The hammers may be T-shaped elements, bars, or rings fixed or pivoted to the shaft or to disks fixed to the shaft. The rotor runs in a housing containing grinding plates or liners. The clearance maintained between the liners and rotor is important with respect to the fineness of product. A cylindrical screen or grating usually encloses all or part of the rotor. The fineness of product can be regulated by changing rotor speed, feed rate, or clearance between hammers and grinding plates, as well as by changing the number and type of hammers used and the size of discharge openings.

The screen or grating discharge for a hammer mill serves as an internal classifier, but its limited area does not permit effective usage when small apertures are required. To meet critical maximum size specifications in the intermediate size range, the hammer mill may be operated in closed circuit with external screens of larger area than could be employed in the mill itself. The mill discharge screen then has large apertures to retain grossly oversize material in the grinding zone.

The grinding action results from impact and attrition between lumps or particles of the material being ground, and the grinding elements. The hammer mill is made in a great many types and sizes and can be used on a greater variety of soft materials than any other type of machine. It is capable of taking ½-in. feed material and reducing it to a product substantially all able to pass a No. 200 sieve. For producing materials in the fine-size range, it may be operated in conjunction with external air classifiers. Such an arrangement is shown in Fig. 12. A number of machines have internal air classifiers.

Hammer Mills without Internal Air Classifiers. The *Williams Helix-Seal mill* (*Williams Patent Crusher & Pulverizer Co.*) is suitable for fine pulverizing, disintegration, and shredding. Twist hammers and chisel hammers may be used, the latter for tearing and shredding. A preliminary breaker can be provided for mounting on the feed hopper. It may be operated in closed circuit with a *Spinner air classifier*. Eight sizes are available requiring from 5 to 100 hp. Speed varies from 1800 to 5500 r.p.m. from largest to smallest mill. Clays, chemicals, pigments, drugs, and food products have been ground.

The *Stedman Type A Two-stage Swing Hammer Grinders* (*Stedman's Foundry and Machine Works*) are designed for the uniform reduction of many materials, whether friable, fibrous, tough, dry, or moist. They will grind greasy crackling cake to No. 10 sieve size or limestone and similar materials to fine powder with equal facility. They grind friable materials to No. 20, 40, or 60 sieve size and finer. Rolling rings are furnished in place of hammers for abrasive materials. Type A machines are equipped with built-in metal trap and adjustable grinding plates. The *Stedman Disintegrator*, commonly referred to as a *cage mill* (Fig. 30), is used for disintegrating clays, colors, press cake, asbestos, and packing-house by-products. This type of machine is especially desirable for handling tough, gummy, high-moisture-content or low-melting-point materials. Cages of one, two, three, four, six, and eight rows, with bars of special alloy steel, revolving in opposite directions, produce a powerful impact action that pulverizes many materials.

The *Jeffrey Type A Swing Hammer Pulverizer* (*Jeffrey Manufacturing Co.*) is a general-purpose machine. Product fineness attainable for a number of materials is given in Table 20 for a 24-in. mill with rotor speed of 1600 r.p.m. and a ⅛-in. grate bar opening.

Table 20. Sieve Analyses of Products from Jeffrey Swing Hammer Pulverizer

Material	% through sieve No.				
	10	20	35	65	100
Alum cake	90	72	43	23	14
Burned lime	98	92	80	71	65
Rock salt	97	88	62	35	22
Gypsum	97	87	69	53	42

The *Mikro-Pulverizer* (*Pulverizing Machinery Co.*) is a hammer mill used for the grinding of dyes, dry colors, carbon black, chemicals, pharmaceuticals, food products, molding powders, resins, etc. (Fig. 25). It also has extensive application for the dispersion of colors, etc., in mixes such as cosmetics, cold-water paints, and insecticides. In addition to dry grinding, it is used for the reslurrying of filter-press cake and the grinding of slurries. Special hammers, feeding devices, and housings make possible a wide variation in the fineness and character of its products. These range from 99.9 per cent able to pass a No. 325 sieve in the case of soft press cakes to molding powders that pass No. 14 but contain a minimum of fines passing a No. 80 sieve. Feed material should preferably be smaller than 1½ in. If larger, crushers, which may be either an integral part of the feeder or separately mounted, are supplied to reduce the feed material to a desirable size. Mikro-Pulverizers are made in five sizes, as shown in Table 21. The *Bantam* size, while extensively used in small production work, is designed expressly for laboratory and pilot-plant work. Grinding-performance data obtained with it may be translated directly into the performance to be expected from the larger units.

Table 21. Mikro-Pulverizer Performance

Pulverizer no.	Rotor diam., in.	Max. speed, r.p.m.	Hp.	Average Capacities, lb./hr.		
				6X sugar	Clay and graphite water slurry	Dry colors
Bantam	5	16,000	¾–1	60	75	70
1	8	9,600	3–5	400	550	500
2	12	6,900	7½–10	1200	1600	1500
3	18	4,600	20–40	3600	4800	4500
4	24	3,600	30–60	4800	6400	6000

The *S.P. Mikro-Pulverizer* (Fig. 44) is used for disintegrating moist, tacky, or unctuous materials such as electrical porcelain clay body, soaps, D.D.T., molding-compound mixtures containing fibrous fillers, etc. Its capacity ranges from 1500 to 8000 lb./hr. at 10 hp. The product is granular in character and ranges from No. 8 to No. 18 sieve size depending upon the rotor speed. The mill includes a precrusher that discharges the crushed feed material directly to a roll feeder. This feeder carries the material to a swing hammer rotor that disintegrates the material and throws it either into a collecting bin or onto an inclined conveyor belt.

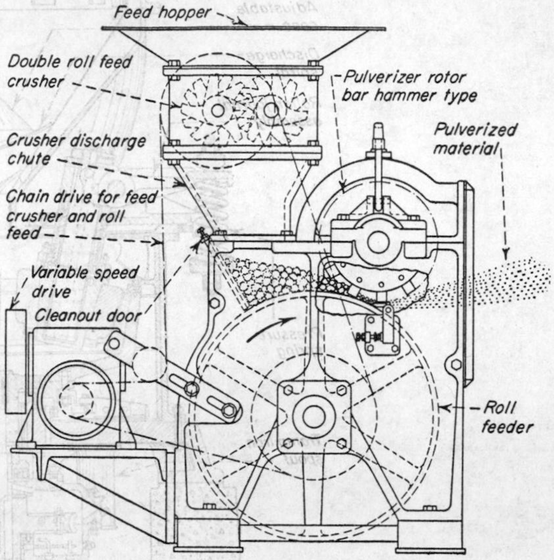

FIG. 44. No. 2 S.P. Mikro-Pulverizer. (*Pulverizing Machinery Co.*)

The *Raymond Screen Pulverizer* (*Raymond Pulverizer Division*) is adapted for filter-press products, chemicals, colors, and dyes. It is easily cleaned when changing from one color or chemical to another. A screw conveyor in the bottom of the hopper forces the feed into the grinding chamber uniformly against the tips of the hammers. This screw is designed to pack the material, making it act as a seal, thus ensuring dustless operation. Performance characteristics are given in Table 22. Larger sizes, in combination with air classifiers, are used for asbestos, mica, and similar fibrous and flaky materials. They cannot be cleaned so easily as the smaller sizes and are therefore used where changes of material are less frequent.

Although the screen mill is inexpensive, easy to operate, and efficient in its grinding range, it cannot be classed as a fine-grinding mill by present-day standards.

Table 22. Performance of Raymond Screen Pulverizer

Material	Capacity, lb./hr.	Fineness	Hp.
Para red	800	97% through No. 200 sieve	5
Tartaric acid	600	99% through No. 60 sieve	5
Soap powder	2000	No. 30 sieve	7.5
Ultramarine blue	1500	98% through No. 200 sieve	7
Boric acid	1000	92% through No. 100 sieve	6.5
Malted milk	1500	No. 20 sieve	5
Bismarck brown	400	98% through No. 200 sieve	5
Sugar	650	90% through No. 200 sieve	5

The finest round perforations in screen commercially available are 0.023 in. diameter. With any moderately tough material, a few hundredths per cent will be found in the finished product close to this screen size, which is approximately that of the No. 30 sieve. Even with readily pulverized materials, oversize will begin to show on the No. 40 or 50 sieves. These fine screens wear rapidly and plug up readily. A heat-sensitive material may readily be damaged attempting to use them. However, a substantial amount of material able to pass a No. 200 sieve is obtained when the feed is composed of soft aggregates of minute particles or of readily friable material, even with discharge screens having perforations larger than 0.023 in. diameter.

The *Jay Bee Pulverizer* (*J. B. Sedberry Co.*) is of the screen hammer-mill type. It has a horizontal shaft, and its peripheral speed is high. Disks are mounted on the rotor, between which are pivoted double-hinged hammers. Usually it is operated with a pneumatic conveyor discharge. Some of its applications are listed in Table 23.

Table 23. Performance of Jay Bee Pulverizer

Material	Type and size	Screen size, in.	Capacity, lb./hr.	Hp.
Asbestos	No. 2 Standard	3/32	900	25
Bones, dry	No. 3 Junior	1/16	3400	15
Chicle	No. 2 Junior	3/4	300	10
Chili pods	No. 2 Junior	3/32	300	10
Cocoa-bean shells	No. 3 direct-connected	1/20	2700	50
Copra	No. 2 Standard	3/16	2000	25
Glue	No. 1 Standard	1/32	1500	20
Malt	Midget mill	1/64	250	3
Leather scrap	No. 2 direct-connected	7/16	200	35
Mace	Midget mill			
	First grinding	1/4		
	Second grinding	1/16	175	3
Nutmeg	No. 3 Junior			
	First grinding	3/4		
	Second grinding	1/4	170	10

The *Blue Streak Dual Screen Pulverizer* (*Prater Pulverizer Co.*) is used for the grinding of resins, chemical salts, plastic scrap, food products, and similar materials to a granular uniform powder of No. 30 or No. 40 sieve fineness. Feed enters opposite ends of the rotor and undergoes three stages of size reduction by hammers of decreasing size. Two perforated screens cover more than 70 per cent of the area of the final sizing drum through which the product passes.

The *Riley Atrita* unit (*Riley Stoker Corp.*) pulverizer for coal is available in three single types and two duplex types. Capacities vary from 2500 lb./hr. for the smallest unit to 15,000 lb./hr. for the largest duplex unit. This type of pulverizer utilizes a series of swing hammers pivoted to the rotor hub, around which is a stationary grid, cut away at one section so that foreign material is thrown out. After passing through this first effect, the coal is carried in a current of air into the second effect, which contains alternate rows of moving and stationary pegs, where most of the pulverizing is done. Leaving the second effect, the coal is passed through a rejector, a number of scooplike blades on the main shaft, where the heaviest particles are thrown back into the pulverizing compartment, permitting the passage of the finer particles only, which enter the fan inlet and are carried into the furnace. Hot air can be introduced into the machine for drying the coal. Air at 300°F. dries coal with 8 per cent moisture down to about 1 per cent.

The *Aero* (*Foster Wheeler Corp.*) unit pulverizer is used for coal, pitch, and coke, blowing the ground material directly into the furnace. The housing is divided into two or three short cylindrical pulverizing chambers. Primary air is admitted at the feed end and between the last chamber and fan. The horizontal shaft carries disks to which hammers are fixed, a set for each chamber. Coal is pulverized by impact and attrition. Annular baffles between the chambers of increasing diameter cause particles to be retained until properly reduced in size for discharge from the final chamber in suspension in the air stream. Hot gases can be introduced to dry the fuel being pulverized. Refractory material such as tramp iron is removed in the first pulverizing chamber and eliminated through a tramp-iron pocket.

The *Whirlwind Pulverizer* (*George F. Pettinos, Inc.*) has its hammers mounted on a vertical shaft and consists of three grinding stages: top, intermediate, and bottom (Fig. 27). Beneath the bottom or third stage is the discharge chamber from which the finished product is removed by means of a two-blade fan mounted on the vertical shaft. When grinding clays or lime this machine has a capacity of 1.5 to 3 tons/hr. and requires 25 to 50 hp.

The *Rietz Disintegrator* (*Process Machinery Co.*) was designed for size reduction of dry or wet materials and for sticky or gummy services. While having the appearance of a vertical hammer mill, its action most nearly parallels that of plate attrition mills. The hammers, which are fixed and do no swing, are wide-faced with tips in close opposition to the surface of the screen which completely surrounds the rotor. Two half-cylindrical screens, clamped or bolted at their flanges, form a cutting and sizing cylinder of practically 360 deg. open area, which aids in producing any given grind with a minimum temperature rise. Selective grinding is possible because the bottom of the disintegration chamber may be left open to allow a secondary discharge or it may be fitted with an adjustable mechanism to regulate the discharge of disintegration-resistant material, thus producing true differential disintegration. Applications include grinding and pulverization of dry, oily, or gummy solids in cereal, food, and food by-products industries, trituration or pulping, impact mixing and homogenization in food, industrial- and chemical-process industries, such as processing of corn and potato starch, vegetable-oil seed, viscose service for cellophane and rayon, and wood pulp for paper.

Hammer Mills with Internal Air Classifiers. A detailed description of two modern machines of this type, the *Mikro-Atomizer* and the *Raymond Vertical Mill*, has been published [see Berry, *Ind. Eng. Chem.* **38**, 672 (1946)].

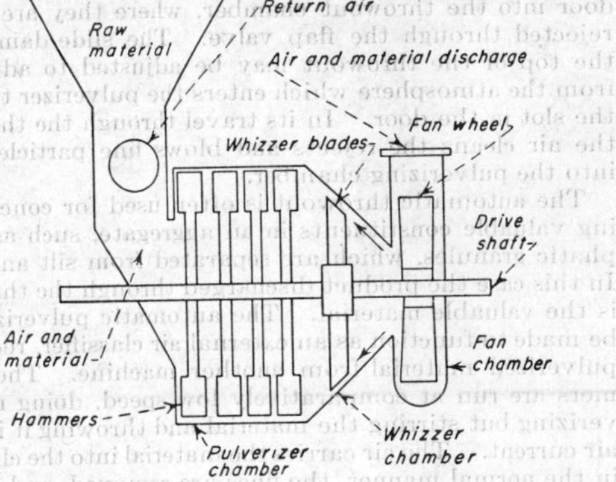

FIG. 45. Whizzer air classification applied to Raymond Imp mill.

The *Imp Pulverizer* (*Raymond Pulverizer Division, Combustion Engineering Co.*) is an air-swept hammer mill, as illustrated in Fig. 45. This machine is made in many sizes from the smallest, having one row of hammers using 10 hp., to the largest size, with six rows of hammers and requiring 100 hp. to drive it. The machines are equipped with a hopper below which is the star feeder, actuated by a pawl-and-ratchet mechanism. A vent from the top of the return-air pipe passes through a tubular dust collector, which makes the system dustless.

A fan is placed on one end of the hammer shaft; between the fan and the hammers is the whizzer consisting of two or more thin blades with tips tapered to conform to the housing. Distance between blades and housing is regulated by moving the whizzer along the shaft. As the whizzer is moved toward the hammers, a coarser product results. The action of the whizzer is that of a fan wheel opposing the action of the main fan. With minimum clearing between blades and housing, a maximum countercurrent is set up at the periphery in the direction indicated by the arrows. An air current accompanying the feed dropping into the pulverizing chamber carries the pulverized material through the clearing toward the fan intake. As the centrifugal force is greater on the coarser particles, their radial velocity will exceed the lateral; hence they are thrown to the periphery and deflected to the hammers by the countercurrent, while the finer particles discharge through the fan intake. The classified product passes through the fan and is blown to a cyclone collector, where it is discharged into bins or containers. The air goes back to the pulverizer, completing the cycle.

The *Automatic Pulverizer* (*Raymond Pulverizer Division*) is a hammer-type machine equipped with air classifier of the vacuum multivane type (see p. 1136) or the double whizzer type. It has a horizontal shaft on which may be mounted one or more disks fitted with hammers. On the door of the pulverizing chamber is mounted an automatic throwout, the function of which is to remove refractory materials contained in the feed, such as sand and gravel from clay. A star feeder with pawl-and-ratchet mechanism receives the raw material from a stock bin and drops it into the pulverizing chamber, on top of which is mounted the air classifier. The air enters the pulverizing chamber at the rear and removes the pulverized material. Particles of proper fineness are blown to the cyclone, which discharges to bins or containers, while oversize is returned to the pulverizer through the bottom valve of the inner cone. Impurities accumulate in the grinding chamber until they are picked up by the rapidly revolving hammers and thrown through the slot on the door into the throwout chamber, where they are finally rejected through the flap valve. The slide damper on the top of the throwout may be adjusted to admit air from the atmosphere which enters the pulverizer through the slot in the door. In its travel through the throwout the air cleans the rejects and blows fine particles back into the pulverizing chamber.

The automatic throwout is often used for concentrating valuable constituents in an aggregate, such as phosphatic granules, which are separated from silt and clay. In this case the product discharged through the throwout is the valuable material. The automatic pulverizer can be made to function as an external air classifier, receiving pulverized material from another machine. The hammers are run at comparatively low speed, doing no pulverizing but stirring the material and throwing it into an air current. The air carries the material into the classifier in the normal manner, the fines are removed, and the rejects eliminated through the throwout.

The *Raymond Vertical Mill* rotating components are carried on its vertical shaft. They are the grinding element, double whizzer classifier, and fan, as shown in Fig. 46. The grinding element at the bottom of the shaft carries bar-shaped hammers free to swing from the fixed end. Grinding of material takes place when it falls on the rotating grinding element and is accelerated to high speed in an upward spiral path as a result of the air entering at the bottom of the shaft. As the material rises to the double whizzer classifier, its rotational velocity is increased, and coarse particles are concentrated along the wall of the chamber because of the centrifugal force acting on them. The coarse particles are continually returned to the grinding element while the fine particles pass between the radial blades of the whizzer classifier and are carried in the air stream through the fan and discharge port. The fine particles are separated from the air stream by a cyclone collector into a suitable container. The air discharged by the cyclone can be returned to the machine in any desired proportion or be vented to a cloth-bag collector.

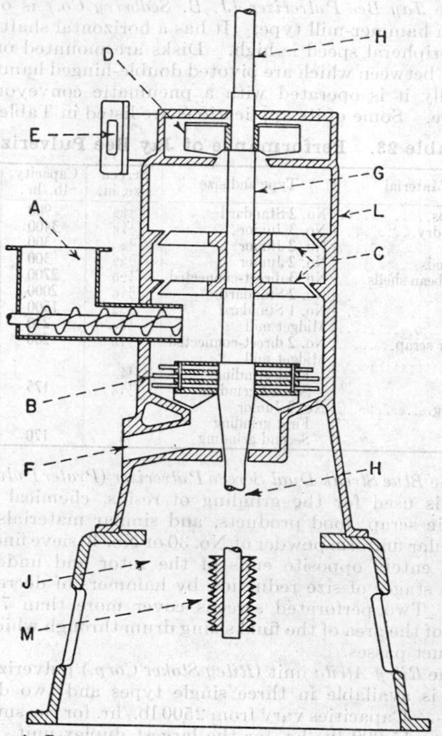

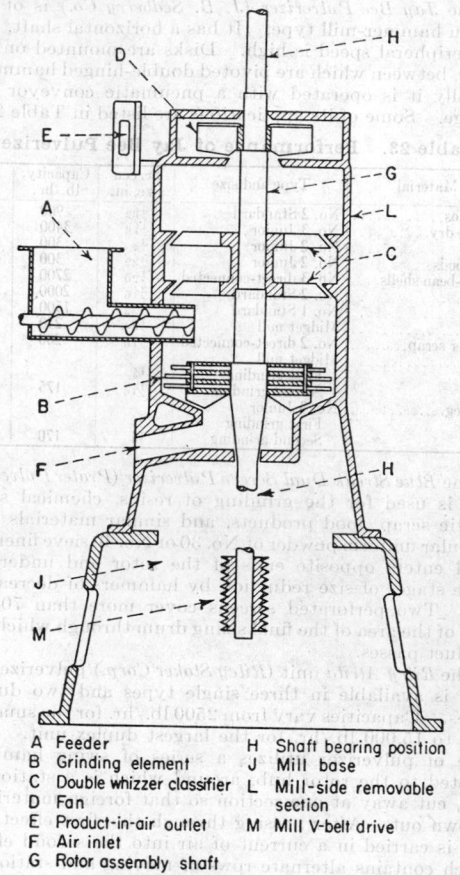

A Feeder
B Grinding element
C Double whizzer classifier
D Fan
E Product-in-air outlet
F Air inlet
G Rotor assembly shaft
H Shaft bearing position
J Mill base
L Mill-side removable section
M Mill V-belt drive

FIG. 46. Raymond vertical mill. [*Ind. Eng. Chem.*, **38**, 672 (1946).]

Machines are available with rotor diameters of 18 and 35 in., driven by 20- and 100-hp. motors, respectively. The larger mill is directly connected to a vertical motor. The nominal rotor speed for the 18-in. Raymond vertical mill is 6500 r.p.m., and 3600 r.p.m. for the 35-in. machine. In general, the relative production rates for the two machines are directly proportional to the power applied. Product fineness is controlled by the number of blades used in each bank of the whizzer classifier and by the size of the fan wheel. Materials of the smallest particle size generally are produced when the maximum number of whizzer blades and the smallest fan are used.

The field of application of the Raymond vertical mill is for producing materials that range in size from those having 99 per cent passing a No. 325 sieve to those having 99 per cent smaller than 5 to 10 μ, depending on the state of aggregation of the feed. A production rate of 500 lb./hr. is achieved with a chemical in an 18-in. machine consuming 18 hp. when the product is substantially smaller than 15 μ. In a talc operation on a 35-in. machine requiring 100 hp., a production rate of 400 lb./hr. is obtained when the product is 95 per cent smaller than 10 μ. At a production rate of 3000 lb./hr., a sample of the product leaves only a trace of talc on a No. 325 testing sieve.

The rotating elements carried on the shaft of the *Mikro-Atomizer* (*Pulverizing Machinery Co.*) are the hammers, classifier wheels, and fans (Fig. 47). Material,

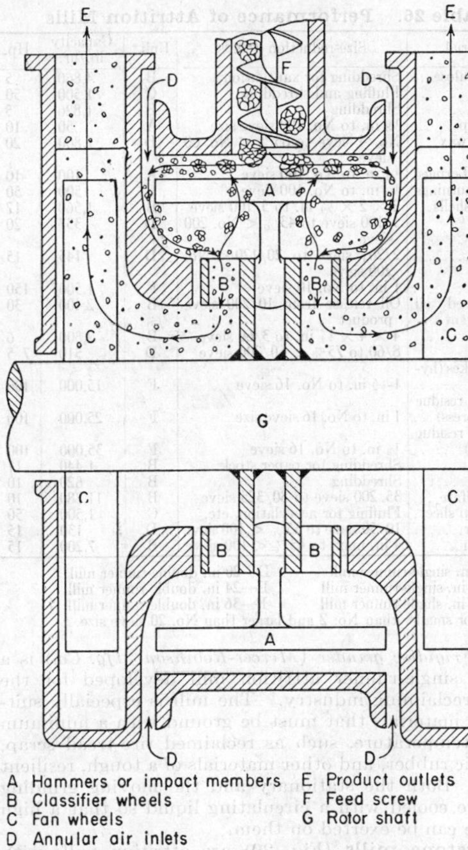

FIG. 47. Mikro-Atomizer operating principle. [*Ind. Eng. Chem.*, **38**, 672 (1946).]

coming in contact with the T-shaped hammers, is divided into two streams in opposite directions and flows in spiral paths away from the middle and toward the ends of the rotor while grinding takes place. Air entering the annular inlets creates a dispersion of the particles leaving the ends of the rotor. The dispersion of particles travels to the classifier wheels, which contain closely spaced radial vanes at their peripheries. Centrifugal force and aerodynamic drag are opposing forces on the particles between the vanes of the classifier wheels. When the centrifugal force on a particle is the larger force, it is returned to the grinding zone. When the force due to aerodynamic drag on a particle is the larger, the particle passes between the vanes of the classifier wheel, around the rotor shaft, into the fans, and out one of the product outlets, which generally converge into a single conduit. The fine particles are separated from the air stream by a cyclone collector, cloth-bag filter, or a combination of the two. High rotor or classifier-wheel speed and low air rate are the conditions for producing materials of smallest particle size. The particle size of the product also can be controlled by variation in the width of the vanes in the classifier wheels. With other factors remaining fixed, a finer product is obtained with wide vanes than with narrow vanes.

Machines are available in three sizes with characteristics given in Table 24.

Table 24. Mikro-Atomizer Operating Characteristics

Machine No.	Rotor diam., in.	Max. rotor speed, r.p.m.	Hp.	Relative capacity
5	8	9600	5	1
6	12	7000	20	5
8	24	3600	75	20

Feed size is limited to smaller than ¾ in. Typical figures for fineness and production rate for a number of materials are given in Table 25.

Table 25. Performance of No. 6 Mikro-Atomizer

Material	Particle size, μ		Production rate, lb./hr.
	Avg.	Max.	
Sugar	19	40	500
Calcium carbonate	5	25	600
Nickel carbonate	5	20	650
Nickel carbonate	2.5	10	300
Lead oxide	2	5	1250
Dry colors	4	15	500

The *Limited Mill* (*Schutz-O'Neill Co.*) consists of two or more beater plates on which are mounted fixed hammers which pulverize the material against a shell, which usually is corrugated. Sometimes a perforated shell, similar to a grater, is used. This is of advantage in pulverizing tough fibrous material. Fineness of the finished product is regulated by a cone with attached blades or whizzers. The pulverized material travels longitudinally toward the cone and whizzer. When the cone plate is moved into the mill, a coarser product is obtained. With the plate in the extreme outer position, flush with the end of the mill, only the fines pass through, while the coarser particles are thrown into a tailings groove on top of the grinding chamber, at the end of which a rotary valve returns it to the feed end of the mill. For very fine grinding and for extremely hard material, a perforated plate mounted on the shaft in front of the beater is employed. A row of holes in the plate, varying from ½ to ⅛ in. in diameter, provides an exit for the finely ground material and relieves the air pressure, thus permitting cooler operation. The mill is made in five sizes from 16 to 28 in. diameter, requiring 10 to 50 hp. for speeds from 3500 to 2600 r.p.m. Applications of the Limited mill are given in Table 56 on p. 1159.

Disk Attrition Mills. (Fig. 24.) The disk or attrition mill is a modern counterpart of the early buhrstone mill. Stones are replaced by steel disks mounting interchangeable metal or abrasive grinding plates rotating at higher speeds, thus permitting a much broader range of application. Grinding takes place between the plates, which may operate in a vertical or horizontal plane. One or both disks may be rotated; if both, then in opposite directions. The assembly, comprising a shaft, disk, and grinding plate, is called a runner. Feed material enters at *F* (Fig. 48), passes between the grinding plates *C* as indicated by the arrow *D*, and is discharged at *E*. The grinding plates are bolted to the disks *A* and *B*; the distance between them is adjustable.

The *Sprout-Waldron Attrition Mill* is available in single- and double-runner models with production sizes based on 16- to 36-in.-diameter disks and with power ranging up to 400 hp. By use of a variety of plates and shell constructions these units are represented in such installations as coarse granulating, pulverizing, and shredding. A single-runner model, such as that shown in Fig. 49, having plates with concentric circular rows of projecting spikes on the rotating plate meshing with those on the stationary plate, acts much like a hammer mill, the spikes being the fixed hammers, and can serve for such applications.

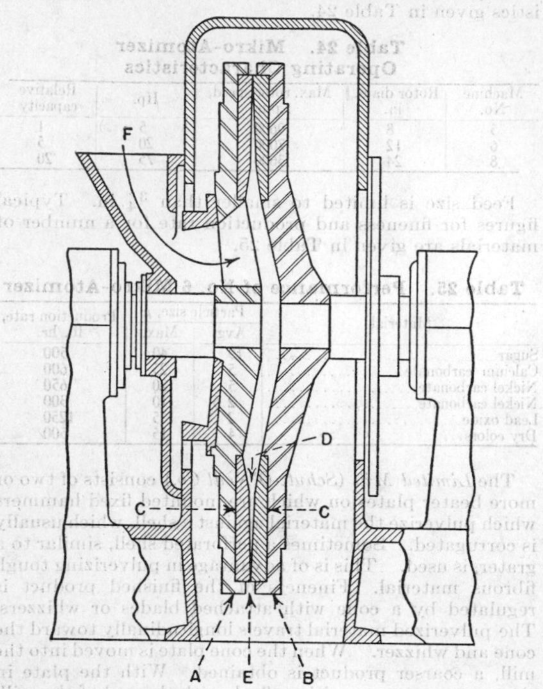

FIG. 48. Disk mill principle. (*The Bauer Bros. Co.*)

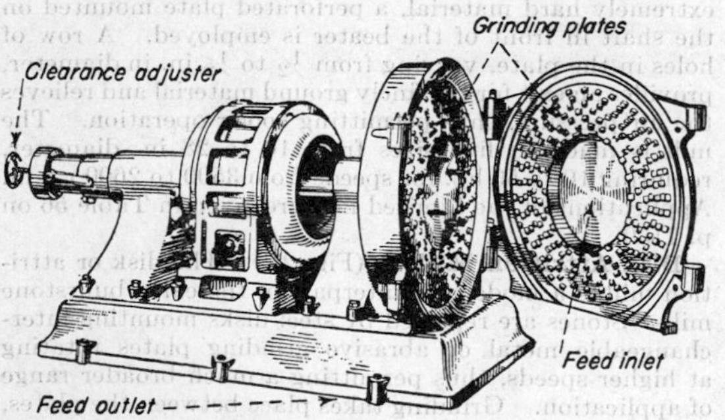

FIG. 49. Single-runner attrition mill. (*Sprout, Waldron & Co.*)

Bauer Double Disk Mills (*The Bauer Bros. Co.*) are used for grinding of fibrous and non-fibrous substances, fluffing of fibrous materials, intensive mixing of fine powders, and hydration of cellular materials (Fig. 48). Six sizes are made with disk diameters from 8 to 36 in. and power from 5 to 400 hp.

In general, single-runner mills are used for the same purposes as double-runner mills, excepting that they will accept a coarser feed stock, their range of reduction for a given material is more limited, and they offer correspondingly higher outputs at lower power. In addition, there are a number of applications unique to this unit such as fluffing of sheet pulp from continuous rolls, to which the inlet provisions of a double-runner mill are not suited. The same range of plate types can be used on both single- and double-runner mills. While spike-tooth plates can be used in certain applications to simulate hammer-mill action, they are more generally applied to specialized tasks involving tearing, shredding, or controlled shattering, as in dehulling. The performance data presented in Table 26 typify the applications of the attrition mill.

Table 26. Performance of Attrition Mills

Material	Size-reduction details	Unit*	Capacity lb./hr.	Hp.
Alkali cellulose...	Shredding for xanthation	B	4,860	5
Asbestos.........	Fluffing and shredding	C	1,500	50
Bagasse..........	Shredding	B	1,826	5
Bronze chips......	1/8 in. to No. 100 sieve size	A	50	10
Carnauba wax....	No. 4 sieve to 65% < No. 60 sieve	D	1,800	20
Cast-iron borings.	1/4 in. to No. 100 sieve	A	100	10
Cast-iron turnings	1/4 in. to No. 100 sieve	E	500	50
Cocoanut shells...	2 × 2 × 1/4 in. to 5/100 sieve	B	1,560	17
	5/100 sieve to 43% < No. 200 sieve	D	337	20
Cork.............	2/20† sieve to 20/120 < No. 200 sieve	D	145	15
Corn cobs........	1 in. to No. 10 sieve	F	1,500	150
Cotton seed oil and solvent	Oil release from 10/200 sieve product	B	2,400	30
Mica.............	4 × 4 × 1/4 in. to 3/60 sieve	B	2,800	6
	8/60 to 75% < 60/200 sieve	D	510	7.5
Oil-seed cakes (hydraulic)........	1-1/2 in. to No. 16 sieve	F	15,000	100
Oil-seed residue (screw press)...	1 in. to No. 16 sieve size	F	25,000	100
Oil-seed residue (solvent).......	1/4 in. to No. 16 sieve	F	35,000	100
Rags.............	Shredding for paper stock	B	1,440	11
Ramie............	Shredding	B	820	10
Sodium sulfate....	35/200 sieve to 80/325 sieve	B	11,880	10
Sulfite pulp sheet.	Fluffing for acetylation, etc.	C	1,500	50
Wood flour.......	10/50 sieve to 35% < 100 sieve	D	130	15
Wood rosin.......	4 in. max. to 45% < 100 sieve	B	7,200	15

* A—8 in. single-runner mill; B—24 in. single-runner mill; C—36 in. single-runner mill; D—20 in. double-runner mill; E—24 in. double-runner mill; F—36 in. double-runner mill

† 2/20, or smaller than No. 2 and larger than No. 20 sieve size.

The *Frigidisc grinder* (*Mercer-Robinson Mfg. Co.*) is a rugged, single-runner attrition mill developed for the rubber-reclaiming industry. The mill is especially suitable for materials that must be ground with a minimum rise in temperature, such as reclaimed tire-tread scrap, synthetic rubber, and other materials of a tough, resilient nature. Both the stationary and the moving grinding disks are cooled with a circulating liquid so that a high pressure can be exerted on them.

Buhrstone mills (Fig. 29) are attrition mills with hard circular stones serving as grinding mediums, generally French, American, or Esopus buhrstones; rock emery or combination of French buhr and esopus or of pebble grit and emery rock are also used. Buhrstone mills are used extensively in the milling industry for grinding cereals and grains. Other uses include grinding of minerals, colors, spices, and mineral pigments; reduction of cork and sawdust; wet grinding of enamels, mica, starch, drop blocks, and polishing rouge. Feed enters the mill through a center hole in one of the stones. It is distributed between the stone faces and ground while working its way to the periphery. The buhrstone or stone mill for "paint grinding" is being replaced by the roller mill (Fig. 50).

The *Sturtevant Rock-Emery Mill* (*Sturtevant Mill Co.*) is made in two types, a horizontal and a vertical; the former is made in one size and has 42-in. stones; it requires 15 to 20 hp. Four standard sizes are made of the vertical mill, with 24-, 30-, 36-, and 42-in. stones, requiring 12 to 15, 18 to 20, 30 to 35, and 45 to 80 hp., respectively. This mill is used for grinding coal and coke for foundry facing, shale, talc, siennas, ochers, and umbers. For very fine grinding it is operated in closed circuit with air classifiers.

Colloid Mills. These might be classified as a type of disk or attrition mill. Modern colloid mills fall into three main groups: the hammer, the smooth-surface, and the rough-surface types. They find use for the size reduction of soft solids and the dispersion of finely divided solid particles in liquid mediums. A colloid-mill preparation of emulsions is described on pp. 1167 to 1169.

Fluid Energy or Jet Mills. A detailed description of mills of this type, which includes the Micronizer, the Reductionizer, and the Eagle Mill, has been published [Berry, *Ind. Eng. Chem.*, **38**, 672 (1946)].

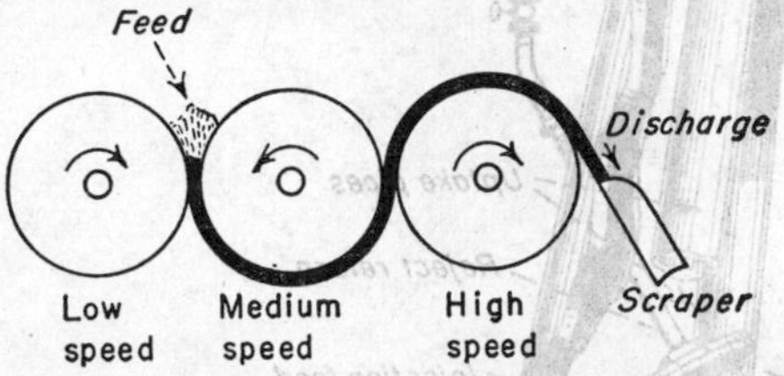

FIG. 50. Roller mill for paint grinding.

The *Micronizer* (*The Micronizer Co.*) consists of a shallow circular grinding chamber wherein the material to be pulverized is acted upon by a number of gaseous fluid jets issuing through orifices spaced around the periphery of the chamber, as shown in Fig. 26. The gaseous jets generally are superheated steam supplied at pressures from 100 to 500 lb./sq. in. and at a temperature up to 800°F. superheat. Compressed air at 100 lb./sq. in. pressure, or higher, may be utilized instead of steam. The pulverizing chamber varies from 2 to 48 in. in diameter and from 1 to 2½ in. in axial height.

Jet orifices are drilled through the peripheral wall of the grinding chamber and vary in number from 3 to 16, equally spaced around the chamber; the orifices vary in diameter up to ¼ in. They are generally placed in a tangential position so that entering steam or air will promote the rotation of the material to be pulverized in one direction. When a fluid of high energy content is introduced into the small pulverizing chamber, the fluid pressure is transferred into velocity head by expansion to substantially atmospheric pressure. This causes a high-speed rotation of the contents of the grinding chamber. The centrifugal force of rotation causes the material to concentrate at the periphery where the jets are introduced. Since most of the energy of the fluid jets is dissipated near the point of entry, intense local velocity gradients and intense interactions are set up within the circulating material. A great deal of reduction of the material is thus caused by the impinging of the particles upon each other, which tends to reduce the wear on the pulverizer housing.

The gaseous fluid supplying the grinding energy is withdrawn at an inward point, tending to cause the dust-laden gas to travel spirally. The smaller particles are carried out with the gas, and the coarser particles thrown to the periphery are subjected to further reduction. Thus the grinding chamber also serves as an internal classifier.

The outlet from the grinding chamber leads directly into a concentric centrifugal collector. This collector receives the material as it is traveling in a high-velocity rotary motion which is conducive to the separation of the material from the fluid so that about 85 to 95 per cent of the product is collected in the concentric collector.

The feed size should be smaller than ¼ in. Production rate, fluid consumption, and fineness figures are shown in Table 27.

Table 27. Micronizer Performance*

Material	Product avg. size, μ	Feed		Fluid consumption, lb. fluid/lb. solid	
		Size, sieve No.	Rate, lb./hr.	Air	Steam
Ceylon graphite	2	3	200	...	8.5
Cryolite	3	60	900	...	4.0
Limestone	3.5	80	1000	...	4.0
Hard talc	3.5	20	1000	...	4.0
Silica gel	5.5	8	500	...	3.5
Soft talc	6.5	20	1800	...	2.5
Barytes	3.5	40	1800	...	2.2
Bituminous coal	2	10	1300	...	1.2
Copal resin	5	2	600	7.5	
Wolframite ore	5.5	10	800	5.6	
Sulfur	3.5	3	1300	3.5	

* *Ind. Eng. Chem.*, **38**, 672 (1946).

The *Reductionizer* (*Reduction Engineering Corp.*) reduces solid materials to small size by the impact and attrition between particles in a tube having at least one curved portion into which fluid energy is introduced in the form of superheated steam or compressed air. Coarse material is transported by a venturi injector into the grinding tube *CBA* (Fig. 51). Fluid under pressure enters the tube through a series of nozzles while both fluid and finished product are discharged through a cyclone or other collector. The top tube turn acts as an internal classifier. Although saturated steam is employed in some operations, usually it is superheated to 500° to 650°F. Pressure of air or steam at the nozzles generally is 100 lb./sq. in. gage, or higher.

Reductionizers having 1- to 8-in. tube diameters are in use. Compressed air is employed in amounts from 30 to 2,000 cu. ft./min., and steam at rates from 100 to 5000 lb./hr. A feed larger than No. 100 sieve size generally is not recommended. Production rates range from 5 to 4600 lb./hr. Materials that can be ground are pigments, drugs, waxes, insecticides, talc, graphite, phosphates, calcium sulfate, silica, clay, minerals, cosmetic powders, organic colors, etc.

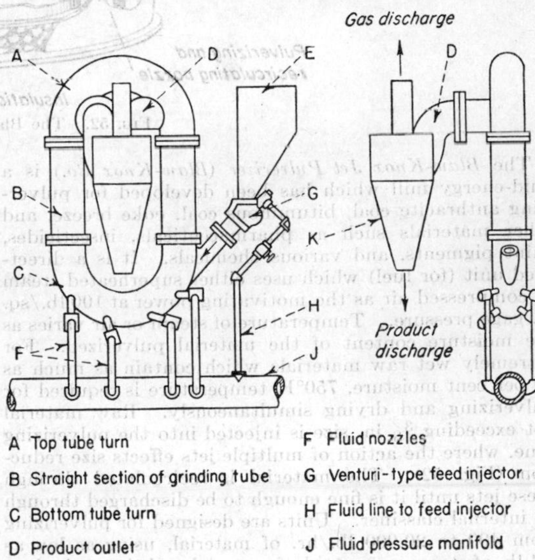

FIG. 51. Reductionizer fluid-energy or jet mill. [*Ind. Eng. Chem.*, **38**, 672 (1946).]

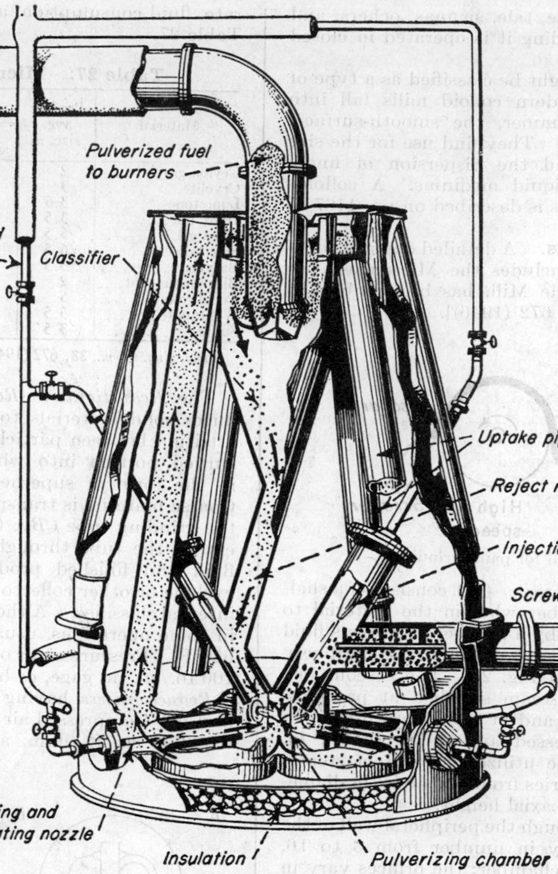

FIG. 52. The Blaw-Knox jet pulverizer.

The *Blaw-Knox Jet Pulverizer* (*Blaw-Knox Co.*) is a fluid-energy mill which has been developed for pulverizing anthracite coal, bituminous coal, coke breeze, and other materials such as pharmaceuticals, insecticides, paint pigments, and various chemicals. It is a direct-fired unit (for fuel) which uses either superheated steam or compressed air as the motivating power at 100 lb./sq. in. gage pressure. Temperature of steam or air varies as the moisture content of the material pulverized. For extremely wet raw materials which contain as much as 20 per cent moisture, 750°F. temperature is required for pulverizing and drying simultaneously. Raw material not exceeding 3/8 in. size is injected into the pulverizing zone, where the action of multiple jets effects size reduction (Fig. 52). The material is recirculated through these jets until it is fine enough to be discharged through an internal classifier. Units are designed for pulverizing from 400 to 20,000 lb./hr. of material, using as low as 0.4 lb. of steam per pound of material and the equivalent of 15 kw.-hr./ton for compressed-air operation. Fuel is pulverized to 90 per cent through a No. 200 sieve for combustion purposes.

The *Fluid Energy Reduction Mill* (*C. H. Wheeler Manufacturing Co.*) has a reduction chamber into which fluid such as air or steam may be introduced through suitable nozzles [see *Power*, **92**, 753 (1948)]. The nozzles are so directed that they set up a system velocity in the mill circuit. The usual range of nozzle inlet pressure is from 75 to 200 lb./sq. in. Fluid temperatures from subzero to 1000°F. are selected to meet the physical and chemical requirements of the material being processed. As the material, usually 4-mesh or finer, enters the reduction chamber, it is entrained by the stream of circulating fluid. The violent jet action in the reduction chamber reduces the particle size by impact and attrition. Particles ascend an up stack into the classification zone where part of the circulating fluid is subjected to a severe change of direction, and flows out of the mill with entrained fine particles of the required size, while the remaining fluid and coarser particles are recirculated through a down stack to the reduction chamber for further size reduction and classification. The average particle size of the finished product can be as low as a fraction of a micron. While over-all costs of grinding vary with different materials and particle-size requirements, they fall generally well within permissible economic limits.

A process called *Flash Pulverization* for drying and comminuting minerals was introduced in 1945 by the research staff of the Institute of Gas Technology [Yellott and Singh, *Power Plant Eng.*, **49**, 82 (Dec. 1945)]. It was found that a wide variety of friable materials, ranging from coal and coke breeze to oil shale, could be pulverized by introducing them into a stream of gas at moderate pressure and causing the streaming entrainment to pass through a nozzle or orifice into a zone of lower pressure. Originally conceived as a continuous explosion process, it was later demonstrated that the size reduction was actually accomplished by the impact of the particles upon the nozzle walls, and by attrition among the particles as they passed at high velocity and in very turbulent flow through the nozzle. When superheated steam was used as the motive fluid, the product was dried to less than 1 per cent moisture content for an initial content of 7 per cent.

The operating principle of the flash pulverizer is shown in Fig. 53, which resembles closely the first experimental apparatus. Crushed coal, charged into a hopper, is fed into a steam line at a controlled rate. The temperature and pressure of the steam are also controlled, and the rate of steam flow is dependent upon the size of the nozzle, the pressure and temperature of the steam, and the amount of coal which is fed per unit weight of steam. Because of its simplicity and its adaptability to operation with a high discharge pressure, the *Nozzle Pulverizer*, using the flash pulverization principle, has been studied intensively for gas-turbine applications by the Locomotive Development Committee. Tests at Johns Hopkins University showed that the relatively coarse product coming from the nozzle could be greatly improved by causing the high-velocity steam to strike against a target. It was also found that the pulverization was improved by causing the pressure drop to take place in several stages. Because of its resemblance to certain types of liquid disintegrators, the combination of the nozzle and target has been designated as the "coal atomizer" (see Yellott and Singh, *loc. cit.*). It has been found also that the fineness of the product from this combination can be considerably improved by the addition of a simple attrition section, such as, for example, a series of pipe bends. The target is subject to considerable abrasion, but boron carbide has been found to make a suitable material. The fineness with this combination depends primarily upon the air-coal ratio, and to a lesser extent upon the pressure ratio across the nozzle. With coal of low grindability, a fineness of 5 per cent retained on a No. 100 sieve can be attained with a pressure ratio of 3 to 1, and an air-coal ratio of 2.5 to 1.

A number of nozzle pulverizers have been used abroad. Most important among these is the *Anger Mill*, which has been used widely in German power plants. A description of this and other types of nozzle pulverizer is given by Frisch ("Pulverized Fuel for the Gas Turbine," a paper presented on the panel Improved Application of Coal Burning Equipment, at the Annual Meeting, A.S.M.E., New York, Dec. 3, 1946).

INDUSTRIAL APPLICATIONS OF GRINDING MILLS

Milling of Cereals and Other Vegetable Products. *Flour and Feed Meal.* The roller mill is the one most widely used for grinding wheat and rye into high-grade flour. A typical mill used for this purpose is fitted with two pairs of rolls, capable of making two separate reductions. After each reduction the product is taken to a bolting machine to separate the fine flour, the coarse product being returned for further reduction. The rolls run toward each other at different speeds in order to produce a rubbing action. Grooved rolls are used for crushing the grain, cleaning up the bran, and for grinding corn. For grinding to a finished product, smooth rolls are employed. Feed is supplied at the top where a vibratory shaker spreads it out in a thin stream across the full width of the rolls. For best results the feed should be regular, continuous, and even from one end of the roll to the other.

Rolls are made with various types of corrugation, special corrugations being used where certain results are desired. Two standard types are most generally used; the dull and the sharp, the former mainly on wheat and rye, and the latter for corn and feed. Under ordinary conditions, a sharp roll is used against a sharp roll for very tough wheat; a sharp fast roll against a dull slow roll for moderately tough wheat; a dull fast roll against a sharp slow roll for slightly brittle wheat; and a dull roll against a dull roll for very brittle wheat. The speed ratio usually is 2½ to 1 for corrugated rolls and 1¼ to 1 for smooth rolls.

Milling of wheat is not only a question of grinding and sifting, but it also involves proper preparation of the wheat prior to grinding. As the grain arrives in the mill, it contains sticks, straw, string, sand, and other materials that must be removed. It is first passed through a receiving separator with three superimposed screens, where cleaning is aided by suction. A cleaner separation is obtained with the milling separator which is used in larger mills. When the grain is very dirty, a washer and drier are used. A wheat steamer is often used advantageously. Uniform heating mellows the wheat berry and so conditions it that moisture can be added in whatever quantities desired in the milling. The grinding rolls become effective, and further granulation is secured; bolting is also made easier. A product can thus be maintained at a uniform standard, irrespective of season or the condition of the original wheat.

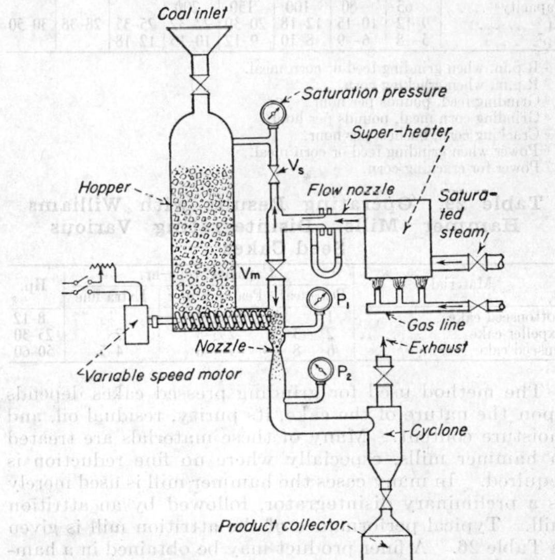

FIG. 53. Schematic diagram of the Flash Pulverizer.

The cereal milling industry produces a great variety of products besides wheat flour: such as meal, corn flour, graham flour, barley, oats, rye, buckwheat, hominy grits, and whole-wheat flour. In addition to roller mills, buhrstone and attrition mills are used extensively, particularly when the whole grain is to be ground. Besides the double-roller mill used for flour, single-roller mills are used, principally for cracking corn and rolling oats. Two-pair high and three-pair high roller mills are used to a great extent in the feed industries, the former principally for coarse feed, such as screenings; the latter for all cereal grains for table use and also for various grains for feed. This mill can be used in combination, the top pair for cracking corn, the middle pair for finishing into coarse feed or corn meal, and the bottom pair exclusively for rolling oats. The various feedstuffs also are ground on swing hammer mills. Performance of a single-runner attrition mill is given in Table 28.

Soybeans, Soybean Cake, and Other Pressed Cakes. Soybeans are ground in about the same manner as the various grains, depending on the nature of the product desired. Roller mills, driers, and bolting reels may be used. After granulation on rolls the granules are generally treated in presses to remove the oil. The product from the presses goes to attrition mills or flour rolls and then to bolters, depending upon whether the finished product is to be a feed meal or a flour. If the whole cake is to be pulverized without removal of fibrous particles, it may be ground in a hammer mill, with or without air classification. A 20-hp. hammer mill with air classifier, grinding pressed cake, had a capacity of 300 lb./hr., 90 per cent through No. 200 sieve; a 20-hp. screen-hammer mill grinding to 1/16-in. screen produced 1000 lb./hr.

Table 28. Operating Characteristics of a Single-runner Robinson Attrition Mill, Grinding Grain

	Size of mill							
	16 in.	18 in.	20 in.	24 in.	26 in.	30 in.	32 in.	36 in.
Speed, r.p.m.[1]	2500	2250	2200	1800	1600	1400	1300	1200
Speed, r.p.m.[2]	1000	950	900	800	750			
Capacity[3]	1200	1600	2000	3300	4000	5000	5300	6300
Capacity[4]	1200	1300	1500	1900	1900	2200		
Capacity[5]	65	80	100	150	200			
Hp[6]	9–12	10–15	12–18	20–30	22–32	25–35	28–38	30–50
Hp[7]	5– 8	6– 9	8–10	9–12	10–15	12–18		

[1] R.p.m. when grinding feed or corn meal.
[2] R.p.m. when cracking corn.
[3] Grinding feed, pounds per hour.
[4] Grinding corn meal, pounds per hour.
[5] Cracking corn, bushels per hour.
[6] Power when grinding feed or corn meal.
[7] Power for cracking corn.

Table 29. Operating Results with Williams Hammer Mills, Disintegrating Various Seed Cakes

Material	Capacity, tons/hr.			Hp.
	Pea meal	Pea and finer	Extra fine	
Cottonseed cake	1	3/4–1	1/2	8–12
Expeller cake	2 1/2–3	2 1/2–3	2	25–30
Linseed cake	6 –8	5 –6	4–5	50–60

The method used for grinding pressed cakes depends upon the nature of the cake, its purity, residual oil, and moisture content. Many of these materials are treated in hammer mills, especially where no fine reduction is required. In many cases the hammer mill is used merely as a preliminary disintegrator, followed by an attrition mill. Typical performance of the attrition mill is given in Table 26. A finer product may be obtained in a hammer mill in closed circuit with an external screen or classifier. Table 29 gives the results obtained with hammer mills disintegrating linseed cake, cottonseed cake, and an expeller cake.

Cocoa Powder. Pulverization of cocoa powder involves processes for heating and cooling the powder in its travel through the equipment. Besides producing a powder of uniform fineness the proper color must be obtained, generally a dark tan or brown, approaching a chocolate color. This color is obtained by heating the cocoa moderately during the grinding, whereby it darkens to the desired color; if the powder is then chilled suddenly, the dark color remains permanently. The cake from the presses usually contains 15 to 25 per cent butterfat. Hammer mills are used in closed circuit with external air classifiers. The mills are swept with chilled air, and refrigerated air is introduced at the entrance to the cyclone product collector. The air is partly recirculated.

Starch and Other Flours. Grinding of starch is not particularly difficult, but precautions must be taken against explosions; starches must not come in contact with hot surfaces, sparks, or flame when suspended in air. Where a product of medium fineness is required, a hammer mill of the screen type is employed. For finer products the air-classifying pulverizer of the hammer type, or a screen-hammer mill in closed circuit with an external air classifier, is used. Potato flour, tapioca, banana, and similar flours are handled in this manner. Table 30 gives the capacity of different types of pulverizers.

Table 30. Pulverizer Capacity for Grinding Starch and Flour

	Cornstarch	Potato starch	Tapioca flour
	Screen mill	Screen mill	Hammer mill (with air classification)
Capacity, lb./hr	3000	2000	2000
% through No. 200 sieve	75	92	98
Hp	15	7.5	40
	Hammer mill (with air classifier)	Hammer mill (with air classifier)	Hammer mill (with air classifier)
Capacity, lb./hr	4000	1000	1500
% through No. 200 sieve	95	98	90
Hp	45	7.5	25
	Screen mill (external air classifier)		
Capacity, lb./hr	3500		
% through No. 200 sieve	90		
Hp	30		

Hay and Other Herbage. Loose hay, velvet beans, alfalfa, corn fodder, bean straw, and other herbage are ground in hammer mills equipped with perforated screens. The Williams hay cutter is equipped with special feeder, and a belt conveyor carries it under a draper which compresses it and forces it into the cutting mechanism. The material is cut between high-speed chisel-shaped hammers and a stationary cutting knife. The fineness is regulated by the size of the openings in the screen or cage. A mill equipped with a 0.5-in. cage when grinding hay at the rate of 6 to 8 tons/hr. requires 125 to 150 hp. When coarse alfalfa is further reduced to a fine meal, a hammer mill producing 2 to 2.5 tons/hr. requires 100 to 125 hp.

Dried Fruits and Vegetables. Pulverization of dried fruits and vegetables is greatly affected by the residual moisture and by the drying process used. Certain drying processes embrittle the material, giving a product that is easily ground. Swing hammer mills, with slotted or perforated screens or cage bars, are most generally used. A recent development is the use of an air-classification pulverizer for dehydrating and pulverizing wet waste fruit pulp. Hot air is introduced into the machine, and the moisture is removed while the material is being pulverized and conveyed to the bagging machine.

Metalliferous Ores. Grinding is one of the major problems in milling practice and one of the main items of expense. Mill manufacturers, operators, and engineers find it necessary to compare grinding practice in one plant with that of another, attempting to evaluate circuits and practices [The Staff, *Trans. Can. Inst. Mining Met.*, **43**, 299 (1940)]. Modern milling practice involves the closed-circuit principle, and this factor must be taken into account in such standardized test procedures as relate to the grindability of different ores (see p. 1114).

Crushing rolls and rod and ball mills are used extensively. Most ores are heterogeneous; and, when they are ground in closed circuit in ball mills, the softer fraction tends to be reduced faster than the harder portion, causing the quantity of the latter in the circulating load to increase. The rod mill tends to grind a heterogeneous material differentially such that the soft portion is ground faster than the hard fraction. The rod mill on a homogeneous material gives a product like that from a crushing roll, lacking the fines ordinarily produced in ball mills. This characteristic reduces losses of mineral due to sliming and permits a higher recovery, particularly in the No. 10- to No. 48-sieve range.

A comparison of roll, rod-mill, and ball-mill products of a homogeneous ore is shown in Fig. 54. The curves show the narrower frequency distribution for the rod mill and roll crusher. The ore is ground to 11 per cent passing a No. 200 sieve in each case, but 10.2 per cent is retained on a No. 14 sieve in case of the ball mill in comparison with 1.3 and 0.2 per cent for the rod mill and rolls. Reduction of the ball-mill product to pass a No. 14 sieve would cause the amount smaller than No. 200 to be increased.

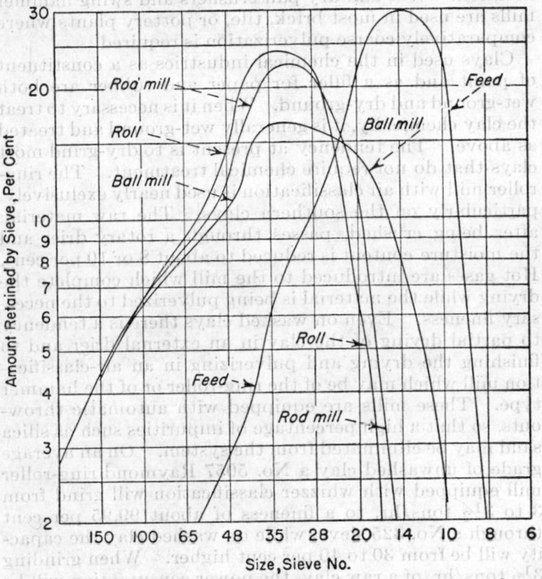

FIG. 54. Comparison of roll-, rod-, and ball-mill products. (*Allis-Chalmers Mfg. Co.*)

Non-metallic Minerals. Dry grinding generally is used; wet grinding is resorted to where certain impurities have to be removed, such as iron oxide or fine grit, and where washing imparts certain desirable properties to the finished product. The system of wet grinding and classification chosen depends upon the nature of the material. Very hard materials are ground in ball mills with hydraulic classifiers, the finished product going to a filter and drier. After drying, the cake generally has to be broken up in some type of disintegrator.

It is often possible, particularly with a soft material like clay, to employ dry grinding with a series of air classifiers, at the same time eliminating impurities such as grit and sand. Nearly every type of pulverizer is used in the grinding of non-metallics in open or closed circuit. The objective may be to obtain many grades of the same material by tying air classifiers and screens into the grinding system to remove various-sized products. Choice of equipment generally depends on (1) hardness and (2) contaminations. Most of the refractory silicates are ground in ball mills and tube mills in continuous or batch operation. Silex or flint lining and balls of flint, porcelain, and similar materials are used where iron contamination must be avoided. Capacity of any system decreases rapidly with increasing fineness of the material; this applies particularly to non-metallics, where extreme fineness is usually required.

Silica and *feldspar* are ground in silex-lined mills with flint balls. Feldspar for ceramic and chemical industries is ground finer than for the glass industry. The following description of a feldspar mill is abstracted from *U.S. Bur. Mines, Circ. No.* 6488. The fine-grinding department consists of three silex-lined Hardinge mill units with flint pebbles, 27 to 28 r.p.m., each fed from a 50-ton surge bin by James automatic belt feeders. Unit 1 has an 8- by 4-ft. Hardinge mill, discharging by an elevator to a Gayco air classifier or a James vibrating screen, according to the kind of product desired, a fine (No. 120 to 250 sieve) or a coarse (No. 20 sieve). An intermediate product (No. 40 to 100 sieve) is made by removing fines in the Gayco and screening the coarse. A tube mill is used in connection with unit 1 for fine grinding; it receives about half the oversize from the Gayco, the other half returning to the Hardinge mill. Plant capacity is increased about 8 per cent by the use of the tube mill. Finished products from screen and classifier are elevated to the top of the plant, sampled automatically, and delivered through chutes to five 50-ton bins. Fine-grinding units 2 and 3 are similar to unit 1, with the exception that the Hardinge mills are smaller, 8 by 3 ft., and a double Hum-mer screen is used in conjunction with a Sturtevant air classifier. All chutes and feeders are lined with silex or rubber.

As a spar free from fines is desired by the glass industry, crushing is done with rolls. Discharge from the Reliance jaw crushers, about ½-in. size, or from the drier, is transferred by conveyor with magnetic-head pulley to a bucket elevator discharging into a 125-ton surge bin feeding Sturtevant rolls set ⅜ in. apart. The discharge is elevated to a double-deck Hum-mer screen with ¼-in. and 20-mesh screens. Oversize on the coarser screen is returned to the rolls; intermediate goes to a second Sturtevant set $^3/_{16}$ in. apart, and undersize through 20 mesh, which is now finished as to maximum size and is carried by belt conveyor and elevator to a Gayco air classifier for removal of fines. Discharge from the second rolls is elevated to a second double-deck Hum-mer with 8- and 20-mesh screens; oversize from the coarser is returned to the rolls, intermediate to a third set of Sturtevant rolls set close, and the undersize through 20 mesh to the Gayco.

Discharge from the third set of rolls is elevated to a third double-deck Hum-mer with 10- and 20-mesh decks. Oversize on 10 mesh, principally mica, is removed as a by-product, the intermediate goes back to the rolls, and undersize through 20-mesh joins the similar products from the preceding screens on their way to the Gayco. A small amount of smaller than No. 140 sieve material in the undersize through 20 mesh is finally removed in the Gayco. The oversize from the Gayco is elevated to a Hum-mer and separated into two sizes, -20 $+40$ mesh, and -40 $+140$ mesh, which are held in 70-ton feed bins and run over two Johnson induction separators, one machine for each size.

Figures 55*A* and 55*B* give the comparative sieve analysis of the two units as installed. The Hardinge mill, just discussed, had not been designed for granulating. It was a very fine grinding mill arranged only for trunnion overflow. It was therefore not possible to pass the spar through the mill fast enough to prevent overgrinding. As a result the Hardinge product shows a much higher percentage of the undesirable material passing No. 200 sieve. Figure 55*C* gives the results of a Hardinge mill equipped with a vertical grate and discharge flights for granular grinding, which shows that the products from the two systems (*B* and *C*) are more nearly the same.

With slight modifications the systems may be used to produce a fine grade or granules from the following materials: quartz, slate, marble, corundum, carborundum, tripoli, pumice, and volcanic ash. Practically all abrasive silicates are handled in ball and tube mills followed by air classifiers. Table 31 gives the results obtained

Table 31. Grinding Refractory Siliceous Materials in Pebble Mills

	Feldspar	Silica sand	Enamel frit	Grog
Size of mill	8′ × 60″	8′ × 48″	4½′ × 16″	5′ × 22″
Feed size	2″	20 mesh	⅛″	1½″
Size of product	99% through No. 200 sieve	98% through No. 325 sieve	97% through No. 100 sieve	95% through No. 10 sieve
Capacity, tons/hr.	1.75	1.25	0.225	5
Power for mill, hp.	68	58	8.5	28
Power for auxiliaries, hp.	21	20		
Pebble load, lb.	10,000	12,000	2000	2800
Speed of mill, r.p.m.	22	18	30	30
Moisture, %	1	1	0	1
Type of classifier	Hardinge	Air	Trommel screen on mill	
Lining and grinding mediums	Flint blocks and flint pebbles			Steel

with Hardinge pebble mills, grinding several siliceous refractory materials.

Talc and soapstone are generally easily pulverized, although certain fibrous and foliated talcs may offer great resistance to reduction to impalpable powder. Arranged according to the resistance offered to fine pulverization, talcs from various sources may be listed as

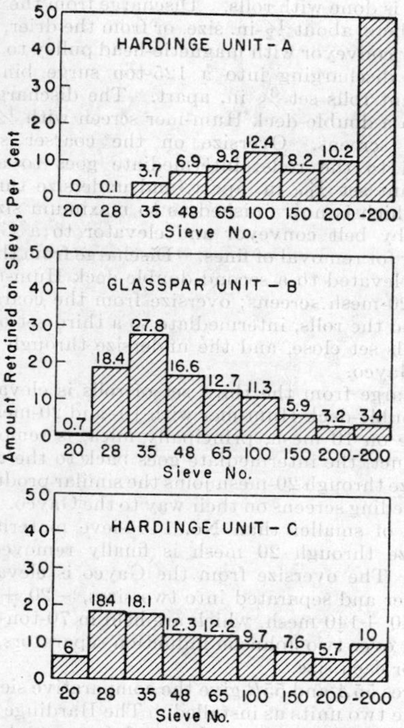

FIG. 55. Comparative sieve analyses of granular products from different types of installations.

follows: Italy, Manchuria, Russia, France, California, Vermont, Georgia, South Africa, India, Quebec, North Carolina, Ontario, Virginia, North Carolina (pyrophyllite), Massachusetts (foliate), New York (Natural Bridge), New York (Gouverneur).

For a ring-roller mill receiving 1-in. feed, production rates range from 6000 to 3000 lb./hr. for 80 hp. grinding to 99 to 99.5 per cent able to pass a No. 200 sieve.

U.S. Bureau of Mines, Bull. 213 gives an excellent description of mining and processing talc and soapstone, which may be applied to many of the other non-metallics. Tube mills, 150 hp. each, lined with silex or porcelain, are set in two lines of four each in tandem. Each line has a capacity of 2 tons/hr. Closed-circuit grinding with air classifiers frequently is employed. Batch grinding is also used. Batch pebble mills are operated for a certain length of time up to 8 hr. for the finer grades, before dumping. In Talcville, N.Y., 6- to 8-ft. pebble mills are used, 30 to 35 hp., charged with 1 ton talc and 3 tons flint pebbles and rotated for 4 to 7 hr. at 22½ to 23 r.p.m.

Clays and Kaolins. A large percentage of clays and kaolins are washed and water-floated, after which they are filter-pressed, dried, and disintegrated in hammer mills, followed by a series of air classifiers. This is of special advantage when the clay contains impurities such as fine mica flakes which are not removed by water flotation. Wet and dry pan crushers and swing hammer mills are used in most brick, tile, or pottery plants where comparatively coarse pulverization is required.

Clays used in the chemical industries as a constituent of paint and as a filler for paper and rubber are both wet-ground and dry-ground. When it is necessary to treat the clay chemically, it is generally wet-ground and treated as above. The tendency at present is to dry-grind most clays that do not require chemical treatment. The ring-roller mill with air classification is used nearly exclusively, particularly on the southern clays. The raw material, after being crushed, passes through a rotary drier and the moisture content is reduced to about 8 or 10 per cent. Hot gases are introduced to the mill which complete the drying while the material is being pulverized to the necessary fineness. Even on washed clays there is a tendency to partial drying of the clay in an external drier and to finishing the drying and pulverizing in an air-classification mill which may be of the ring-roller or of the hammer type. These mills are equipped with automatic throw-outs, so that a high percentage of impurities such as silica sand may be eliminated from the system. On an average grade of unwashed clay a No. 5057 Raymond ring-roller mill equipped with whizzer classification will grind from 3 to 3½ tons/hr. to a fineness of about 99.95 per cent through a No. 325 sieve, while on washed clay the capacity will be from 30 to 40 per cent higher. When grinding 3½ tons/hr. of a raw clay, the power consumption will be about 100 hp., and it takes about 750 cu. ft. of natural gas containing 1000 B.t.u./cu. ft. to dry the clay from 10 per cent moisture down to about 1 per cent.

Non-metallic Carbonates and Sulfates. Non-metallic carbonates include limestone, calcite, marble, marls, chalk, dolomite, and magnesite; the most important sulfates are barite, celestite, anhydrite, and gypsum; these are used as fillers in paint, paper, and rubber. Gypsum and anhydrite are treated under the heading The Cement, Lime, and Gypsum Industries (see p. 1152).

Table 32 gives the capacity and power requirement for a Raymond No. 4227 ring-roller mill grinding typical limestone to various degrees of fineness. The capacity increases in the ratio of 1.6, 2.0, and 2.7 for the No. 4237, 5047, and 5057 mills, while the power requirement increases in the ratio of 1.2, 1.6, and 1.8, respectively. These results may be applied to practically the entire group. When a material is very soft, such as high-grade barite, the capacities may be about 25 to 35 per cent higher. Of the carbonates, magnesite is generally the

Table 32. Performance of Ring-roller Mill Grinding Limestone

Fineness	Capacity, lb./hr.	Hp.
75% through No. 200 sieve	4200–4500	55–65
85% through No. 200 sieve	3700–4000	53–55
95% through No. 200 sieve	2600–3000	50–53
99% through No. 200 sieve	2000–2300	47–50
99% through No. 300 sieve	1200–1300	45–47
99% through No. 325 sieve	1000–1100	42–45
99.5% through No. 325 sieve	900–1000	40–42

hardest to pulverize. This material is often calcined and pulverized in the same manner as lime. Dead-burned magnesite is treated as cement clinker.

Fluorspar. A No. 4237 Raymond ring-roller mill grinding fluorspar had a capacity of 3700 lb./hr., 95 per cent through No. 200 sieve, with a power consumption of 32 hp. on the mill and 23 hp. on the fan, a total of 22.3 hp.-hr./ton. Magnesite generally the most difficult of the carbonates to pulverize, grinds similarly to fluorspar. When calcined it is ground in the same manner as quicklime. Dead-burned magnesite is handled like cement clinker. Table 33 gives the results obtained in a Hardinge mill, wet-grinding barytes and limestone to be used as a paint filler.

Table 33. Wet Grinding of Barytes and Limestone in Hardinge Mill

	Limestone	Barytes
Size of mill	8′ × 48″	7′ × 36″
Size of feed	1½″	1½″
Fineness of product, sieve No.	325	325
Capacity, tons/hr.	¾	2
Mill speed, r.p.m.	18	22
Power for mill, hp.	40	25
Classifier system	Cone	Drag
Moisture in mill, %	30	28
Type of mill lining	Flint	Flint
Grinding mediums	Coarse limestone	Lump barytes

Asbestos and Mica. The choice of crusher for asbestos depends on whether a long or a short fiber is desired. Crushing is done in slow stages to preserve as much as possible of the fiber length. Primary crushers employed are usually of the jaw type with secondary crushers of the smaller jaw type. Small gyratories and corrugated rolls are also used. With some grades a third reduction may be required. After drying and crushing to about 2 in., the asbestos rock goes to the so-called fiberizing machines, which reduce the rock, liberate the fiber, and split it into fine and coarse fiber. There are different types of fiberizers, the swing hammer mill, the Jumbo, and the Laurie and Pharo cyclones. The *Jumbo* consists of a cylindrical shell surrounding a shaft with six pairs of arms placed at 6-in. intervals and disposed crosswise to each other. The arms are of heavy steel bars with chilled iron beaters, the faces of which are constructed on an angle. The *Laurie* cyclone consists of two beaters of the screw-propeller type, driven in opposite directions at 1700 to 2000 r.p.m., in a cast-iron chamber. The *Pharo* cyclone was designed to overcome the tearing effect on the fiber, one of the objections to the Laurie. It is of the same general type, but the hood above the discharging end is cut off immediately above the latter, and the crushing blades, or beaters, of which the paddles are one right and the other left, rotate in the same direction.

As the material reaching the mill contains a large amount of freed asbestos, classification of the fiber begins immediately. It is first put through a screening trommel, the fines are discharged on a shaking screen, and the overflow—all above 1½ in.—falls into one of the fiberizing devices which discharges on the same screen. The latter is slightly inclined and is made from wire or perforated plates. It has an oscillating movement which, apart from the sizing of the rock and eliminating the sand, causes the fiberized asbestos to rise to the top. The liberated fiber is taken up by a fan, while the overflow falls into a second fiberizing machine, which discharges, like the first, on a screen, where the asbestos is again lifted by a fan, and so on until the rock is practically entirely pulverized. Tailings free from asbestos go to the dump.

Asbestos is often pulverized. This is the case when it is used for molded products. The pulverizing is usually accomplished by passing the material through a series of buhrstones or by using a high-speed screen mill with air-transport system. A mill with a 1/64-in. screen pulverized 400 lb./hr. with 13-hp. power consumption. Certain impurities, such as sand, gravel, and hard fiber, may be removed by using an air-classification pulverizer with automatic throwout (see p. 1120).

The micas, as a class, are difficult to grind to a fine powder; one exception is disintegrated schist, in which the mica occurs in minute flakes. The material pulverized is generally the waste from production of sheets and scrap from punching and trimming. Arranged in order of increasing resistance to grinding to a fine powder, micas from various sources may be classified as follows: Madagascar, Ontario, Quebec, Manchuria, India, New Hampshire, North Carolina, South Africa, Russia, Brazil.

Mica is pulverized wet or dry; the wet-ground product is the more desirable, as it retains its luster to a high degree. When ground wet, it is first passed through revolving screens with a constant stream of water; it is then ground in wooden chaser mills at a slow rate and graded after drying, by passing through a series of bolting reels, the finest reel being about 200 mesh. A modification of this process is used in certain European countries, where the mica is ground in chaser mills and buhrstone mills. The water with the ground mica is passed over screens and thus graded. After pressing out the water, the solids are dried and disintegrated in a double-cage mill. Table 34 gives data obtained in wet-grinding a Manchurian mica.

For dry grinding, hammer mills equipped with an air transport system are generally used. The material, after dropping through a perforated screen into the intake of an exhauster, is collected in a cyclone followed by bolting reels. Table 35 gives the operating characteristics grinding North Carolina mica in a high-speed hammer mill.

Table 34. Wet Grinding Manchurian Mica

Amount passed through mill, lb./hr.	400
Total power consumption, including pumps, screens, mill elevators, and conveyor, kw.	60
Power consumption, kw.-hr./ton product	300
Screen analysis, feed to bolting reels:	
% on No. 20 sieve	5
% on No. 50 sieve	18
% on No. 80 sieve	17
% on No. 100 sieve	12
% on No. 200 sieve	19
Through No. 200 sieve	29

Table 35. Grinding Mica in High-speed Hammer Mill

Size of motor (direct-connected), hp.	60
Size of feed	Scrap
Production, lb./hr.	950

Screen analysis of discharge, % on sieve:

No. 20	No. 40	No. 60	No. 80	No. 100	No. 150	through No. 150
1 %	15 %	22 %	16 %	10 %	11 %	25 %

The Fertilizer Industry. Many of the materials used in the fertilizer industry are pulverized, such as those serving as sources for calcium, phosphorus, potassium, and nitrogen. The most commonly used for their lime content are limestone, oyster shells, marls, lime, and, to a small extent, gypsum. Limestone is generally ground in hammer mills, ring-roller mills, and ball mills. Fineness required varies greatly from No. 10 sieve to 75 per cent through No. 100 sieve.

Oyster Shells and Lime Rock. Operating characteristics for hammer mills grinding oyster shells and burned lime for agricultural purposes are given in Table 36.

Phosphates. Phosphate rock is generally pulverized for one of two major purposes: for direct application to the soil, or for acidulation with sulfuric acid in the manufacture of acid or superphosphate, phosphoric acid, and

Table 36. Operating Data Grinding Oyster Shells and Burned Lime in Hammer Mills

Type of mill	Material	Size, in.	Capacity, tons/hr.	Hp.
Jeffrey	Oyster shells	15 × 8	0.5– 0.75	8
		20 × 12	1 – 1.5	12
		24 × 18	2 – 3	20
		30 × 24	4 – 5	30
		36 × 24	8 –10	40
Stedman	Burned lime	12 × 9	1.5	8
		20 × 12	4	20
		24 × 20	8	40
		30 × 30	12	60
		36 × 36	20	100

the various phosphates. Table 37 gives the data for pulverizing phosphate materials in a Raymond ring-roller mill equipped with internal air classification, and also in ring-roller mills equipped with external air classification units.

Table 37. Grinding Phosphate Materials

	Production rate, tons/hr.			
	(a) 90–95% through No. 60 sieve; 50–55% through No. 200 sieve	(b) 90–95% through No. 200 sieve	Hp.	
			a	b
Algiers	7.5	...	13	
Arkansas block rock	6.5	4	15	19
Belgium	8.5	...	12	
Bohemia (apatite)	6.5	...	17	
Canada (apatite)	6	3	18	26
Egypt	6.5	...	17	
Florida (pebble)	6.5	3.5	17	24
Florida (hard rock)	7	4	17	24
Florida (soft rock)	...	5.5	..	15
Idaho	7	4	14	19
Kentucky	7.5	5	12	17
Morocco	8	...	12	
Pacific and Indian oceans:				
Angaur Island	8.5	...	12	
Christmas Island	8	...	12	
Marshall Islands	8.5	...	12	
Makatea Island	7	...	13	
Nauru Island	7	...	13	
Ocean Island	7.5	...	13	
Russia (Podolian)	6.5	...	13	
Tennessee (blue rock)	7	4.5	14	18
Tennessee (brown rock)	7.5	5	12	17
Tennessee (gray rock)	8	5	12	17
Tennessee (phosphatic limestone)	7	4.5	14	18
South Carolina	6.5	3.5	17	24
Tunis	8	...	12	

The coarse material (*a*) is used for acidulation, and the fine material (*b*) for direct application to the soil.

Some average results obtained in grinding various organic and inorganic raw materials for fertilizers are given in Table 38.

Inorganic salts seldom require fine pulverization, but they frequently become lumpy. In such a case they are passed through a double-cage mill or a hammer mill with

Table 38. Results Obtained in Grinding Raw Materials for Fertilizers

Material	Hammer mill	Type or size	Bar opening, in.	Capacity, lb./hr.	Hp.
Acid phosphate	Stedman	36 in.		12,000	40
Steamed bone	Jeffrey	A-30 × 24 in.	1/8	10,000	40
Dry kelp	Williams	Shredder 2	1/4	12,000	35
Guano (Peruvian)	Jeffrey	A-24 × 18 in.	1/8	7,000	40

the screen or cage bars removed. This is done with ammonium sulfate from by-product ovens and sodium nitrate. When used as an ingredient fertilizer the latter is generally mixed with other raw materials, and the mixture is later disintegrated. The various potassium salts used in fertilizers are generally shipped ready for use, but if they have become caked in transit they are broken in a disintegrator.

Basic slag is often used as a source of phosphorus. Its grinding resistance depends largely upon the way it has been cooled, slowly cooled slag generally being more easily pulverized. The most common method for grinding basic slag is in a ball mill followed by a tube mill or a compartment mill. Both systems are in closed circuit with the air classifier. A 7- by 5-ft. mill, requiring 125 hp., operating with a 14-ft., 30 hp. classifier, gave a capacity of 5 tons/hr. from the classifier, 95 per cent through a No. 200 sieve. Mill product was 68 per cent through a No. 200 sieve, and circulating load 100 per cent.

The Cement, Lime, and Gypsum Industries. *Portland cement.* This is manufactured by wet and dry processes. In the latter the material is dried in rotary driers prior to grinding. It is then burned and the clinker pulverized. In the wet process the materials are mixed and ground, a certain amount of water is added, giving a mill discharge with up to 40 per cent water. The slurry is calcined in the usual manner and the clinker ground as in the dry process.

Crushing in the dry process is done in gyratories, jaw crushers, hammer mills, or ring-roller mills. Pulverizing the raw mix is done in mills of the ring-roller type with or without tube mills, in ball mills followed by tube mills, and in compartment mills. Air classification is standard practice. Production for a 250-hp. Bradley-Hercules, operating on a 2-in. feed, was about 40 tons/hr., 51.8 per cent through No. 200 sieve, 60.2 per cent through No. 100 sieve, 69.9 through No. 50, 74.6 through No. 40, 81.4 through No. 30 and 90.1 per cent through No. 20 sieve. A 5- by 22-ft. Allis-Chalmers two-compartment compeb mill, 200 hp., 26 r.p.m., operating on 1-in. feed, had an hourly capacity of 9 tons, 78 per cent through No. 200 sieve. An 8-ft. by 60-in., 345-hp. Hardinge conical mill, 18 r.p.m., operating on 3/8-in. feed with 2 per cent moisture had an hourly capacity of 26 tons, 91 per cent through No. 200 sieve; type of air classifier, 12 ft. superfine, with 145-hp. fan; lining and balls, steel. Performance of Griffin mills on cement and allied materials is given in Table 39.

Table 39. Performance of Griffin Mills on Cement and Allied Materials

Material	Output/hr.	Fineness of product	Screen on mill, mesh
Giant:			
Limestone	4.5–5 tons	83.2% through No. 200 sieve	30 and 35
Cement clinker preliminary to tube mill	26 bbl.	62%–64% through No. 200 sieve	6
White cement clinker.	35 bbl.	72% through No. 100 sieve; 54% through No. 200 sieve	6 and 8
Limestone	3–3.5 tons	96.5% through No. 100 sieve; 82.5 through No. 200 sieve	35
Cement clinker	12–14 bbl.	96.6% through No. 100 sieve; 82.8 through No. 200 sieve	35
Junior Giant:			
Coal	1.5–2 tons	88%–90% through No. 100 sieve	30 and 35
Clinker	5.5–6 bbl.	80%–82% through No. 200 sieve	35
Coke	0.5–0.75 tons	90% through No. 100 sieve	35
Limestone	2 tons	82%–85% through No. 100 sieve	16

A flow sheet for a 9½- by 8- by 40-ft. compeb mill and two classifiers is shown in Fig. 56. Fifty tons of raw material enter the mill. At the first compartment a peripheral screen removes the large oversize material and returns it to the front end of the mill. The final product from this compartment (38 per cent through No. 200 sieve) passes to an air classifier which divides the initial 50 tons into fines and tailings. Fines are considered to be about 13 tons testing 92 per cent through No. 200 sieve, tailings about 37 tons testing 18 per cent through No. 200 sieve. The tailings enter the second compartment of the mill; the product from this compartment enters the second classifier which removes the fines, the tailings entering the third compartment, the discharge from which is likewise fed to the second classifier.

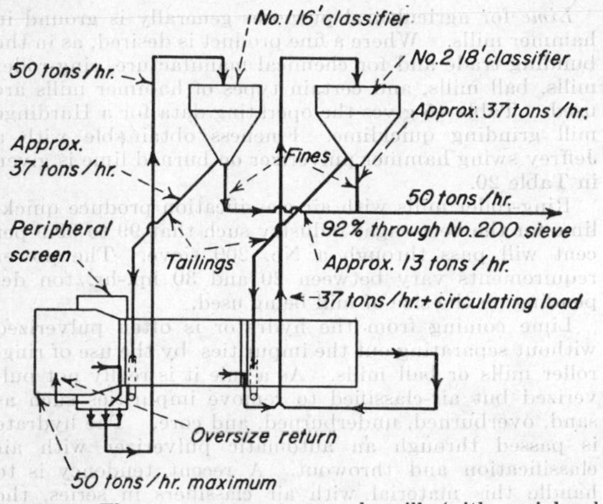

FIG. 56. Three-compartment compeb mill (with peripheral screen on first compartment) in closed circuit with two air classifiers grinding and classifying the raw mix in a cement plant.

Wet grinding of clinker in an 8- by 4-ft., 150-hp. Hardinge mill, 17 r.p.m., with trommel screen, metal lining, and steel balls, grinding an 8-mesh feed, showed an hourly capacity of 32 tons, 100 per cent through No. 10 sieve; ball consumption 0.24 lb./ton. A 7- by 26-ft. Allis-Chalmers two-compartment compeb mill, 500 hp., 20 r.p.m., with Dorr rake in closed circuit with the primary compartment and Dorr bowl in closed circuit with the secondary compartment, gave the operating characteristics shown in Table 40.

Table 40. Operating Data for Mill with Hydraulic Classifier

	Open circuit	Closed circuit
Capacity, bbl./hr.	70.1	167
Power consumption, kw.-hr./bbl.	5.5	3.01
Loss of grinding medium, lb./bbl.		
In primary compartment	0.086	0.057
In secondary compartment	0.321	0.112
Fineness of product, % through No. 200 sieve	8	97
% through No. 100 sieve	95	
Charge in primary compartment	13 tons balls, 2¼–4 in.	
Charge in secondary compartment	39 tons balls, ¾–⅞ in.	

Clinker is ground by the same equipment in both processes; ball mills, tube mills, and compartment mills are used. Mills of the ring-roller type, such as the Bradley-Hercules, are sometimes used as preliminary to a tube mill. In recent years closed-circuit systems have been favored. A 7- by 24-ft. Allis-Chalmers mill with 185-kw. power consumption in open circuit gave a capacity of 46 bbl./hr. cement, 97.9 per cent through No. 200 sieve; clinker feed 115°F., discharge 174°F. An 8- by 7- by 40-ft. three-compartment mill, in open circuit, operating on 2½-in. clinker, produced 2000 bbl/24 hr. Percentages through No. 200 sieve at the end of first, second, and third compartment were 34, 60, and 88 to 90, respectively; the first compartment using 2- to 4-in. balls, the second, 1¼- to 2-in. balls or cylpebs, and the third ⅝-in. cylpebs.

The results obtained in grinding clinker in a Bradley-Hercules as preliminary mill are given in Table 41. Discharge from the Bradley-Hercules goes to intermediate mills, and the discharge of the three mills goes to a fourth mill for final grinding.

Table 41. Operating Data for Hercules Mill, Grinding Cement Clinker

Preliminary mill	Bradley-Hercules with 9-mesh screen
Fineness of product	97% through No. 20 sieve, 65.5% through No. 100 sieve, 51% through No. 200 sieve
Production, bbl./hr.	75.5
Horsepower/bbl. cement	4.64
Intermediate mills	6 × 22 ft. with 15% load of grinding mediums
Fineness of output	95.6% through No. 200 sieve
Temperature of clinker, °F.	240
Size of fourth mill for finishing	6'-6" × 22'
Grinding mediums, %	15
Fineness of finished product	97.6% through No. 200 sieve 96% through No. 300 sieve
Temperature of mill, °F.	257
Horsepower/bbl. cement	3.97

Closed-circuit Grinding of Cement Clinker. Figure 57 shows a compeb mill with two classifiers. Undersize from the peripheral screen goes to the first classifier, which makes fines going to the fourth compartment and

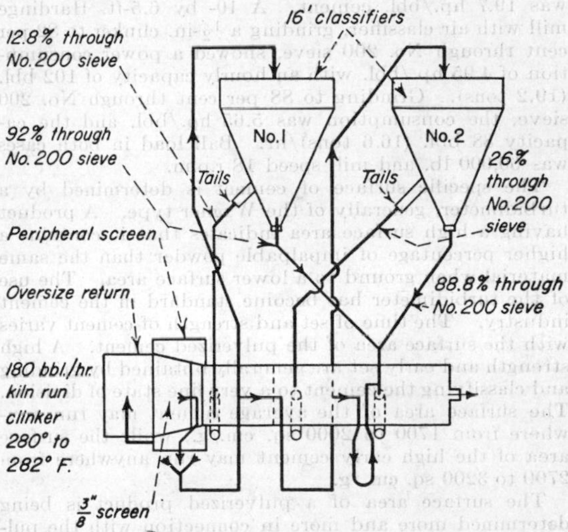

FIG. 57. Four-compartment compeb mill (with peripheral screen on first compartment) in closed circuit with two air classifiers grinding and classifying cement clinker.

tailings going to the second compartment, the latter discharging to classifier No. 2. Here two products are made, fines for the fourth compartment, tailings for the third. Table 42 gives the results obtained when operating this system.

Table 42. Operating Data, Grinding Cement Clinker in Closed Circuit

Size of mill, ft.	9.5 × 8 × 40
Size of mill motor, hp.	1100
Capacity, kiln-run clinker fed to mill, bbl.*/hr.	180
Temperature of kiln-run clinker, °F.	280

Fineness of feed to the two classifiers:

	Classifier 1, %	Classifier 2, %
No. 10 sieve	95.2	
No. 20 sieve	82.0	98.4
No. 28 sieve	73.2	97.6
No. 48 sieve	56.8	91.2
No. 100 sieve	42.4	68.4
No. 200 sieve	32.8	46.4

Fineness of finished product, through No. 200 sieve, %	90.2
Temperature of finished cement, °F.	250

* The capacity at times was 250 bbl./hr., with a fineness of 88% through No. 200 sieve.

It has been found that the time of setting and the strength of many cements vary with the fineness to which the cement has been ground. A fineness between 94 and 98 per cent through No. 325 sieve is often required to obtain a cement having the desired properties. An air classifier properly connected with the mills will often decrease the power required. Table 43 gives the results obtained when grinding cement to a fine state of division in order to produce a superior product.

Table 43. Operating Data, Grinding Cement to a Very Fine Product

Type and size of mill, ft	Three-compartment, 7.5 × 43
Diam. classifier, ft	18
Hourly capacity, bbl	46
Feed to classifier, % through No. 200 sieve	98.7
Tailings from classifier, % through No. 200 sieve	83.2
Finished product, % through No. 200 sieve	99.8
through No. 325 sieve	97.9
−30μ	92.1
−25μ	83.8
−20μ	72.9
Total power required, hp	746
Horsepower/bbl. of product	12.1

When this cement was ground in open circuit to produce a comparable material, the power required was 19.7 hp./bbl. cement. A 10- by 6.5-ft. Hardinge mill with air classifier, grinding a ½-in. clinker to 82 per cent through No. 200 sieve, showed a power consumption of 4.95 hp./bbl. with an hourly capacity of 102 bbl. (19.2 tons). Grinding to 88 per cent through No. 200 sieve, the consumption was 5.65 hp./bbl. and the capacity 88 bbl. (16.6 tons)/hr. Ball load in both cases was 60,000 lb. and mill speed 18 r.p.m.

The specific surface of cement is determined by a turbidimeter, generally of the Wagner type. A product having a high surface area indicates that it contains a higher percentage of impalpable powder than the same material when ground to a lower surface area. The use of the turbidimeter has become standard in the cement industry. The time of set and strength of cement varies with the surface area of the pulverized cement. A high strength and early set are generally obtained by grinding and classifying the cement to a very fine state of division. The surface area of the average cement may run anywhere from 1700 to 2000 sq. cm./g., while the surface area of the high early cement may run anywhere from 2700 to 3200 sq. cm./g.

The surface area of a pulverized product is being determined more and more in connection with the pulverization of many other minerals. A product should, of course, not be ground to a surface much higher than is absolutely necessary since the power for pulverization and classification increases at a very rapid rate as the surface area increases. This is indicated in Table 44, which shows the power for grinding slate used as a filler for rubber or as an ingredient in asphalt. For power at 1 ¢/kw.-hr., the total cost would be two to three times the power cost.

Table 44. Pulverizing Slate to Different Surface Areas

Surface area, sq. cm./g.	Production rate, tons/hr.	Power, kw.-hr./ton	Power ratio
1700	4.30	21.0	1.0
1850	3.40	25.0	1.2
2140	2.38	33.0	1.6
2275	2.00	38.0	1.8
2500	1.54	49.0	2.3
2800	1.10	68.0	3.2
2910	0.90	84.0	4.0
3100	.70	107.0	5.1
3300	.55	135.0	6.4
3500	.39	192.0	9.2
3660	.29	258.0	12.3
3910	.22	341.0	16.2
4060	.18	417.0	19.9

Table 45. Operating Data, Grinding Quicklime in Ball Mill

Size of mill, ft	7 × 3
Type of classifier	Hardinge air classifier
Size of feed, % through No. 100 sieve	90
Capacity, tons/hr	12.5
Power required for mill, hp	100
Power required for auxiliaries, hp	35
Ball load, lb	20,000
Speed of mill, r.p.m	20

Lime for agricultural purposes generally is ground in hammer mills. Where a fine product is desired, as in the building trade and for chemical manufacture, ring-roller mills, ball mills, and certain types of hammer mills are used. Table 45 gives the operating data for a Hardinge mill grinding quicklime. Fineness obtainable with a Jeffrey swing hammer pulverizer on burned lime is given in Table 20.

Ring-roller mills with air classification produce quicklime for the beet-sugar industry such that 99 to 99.9 per cent will pass through a No. 200 sieve. The power requirements vary between 20 and 30 hp.-hr./ton depending on the size of mill being used.

Lime coming from the hydrator is often pulverized without separating out the impurities by the use of ring-roller mills or ball mills. As a rule it is really not pulverized but air-classified to remove impurities such as sand, overburned, underburned, and core. The hydrate is passed through an automatic pulverizer with air classification and throwout. A recent tendency is to handle this material with air classifiers in series, the tailings from the final classifier being discarded or sold for agricultural use. A modification of this system is to feed the tailings to an automatic pulverizer with throwout for final cleaning. Very clean tailings are thus obtained (see p. 1119).

Gypsum is usually calcined in kettles or rotary calciners, after reduction to a fineness varying from 75 to 95 to 98 per cent through No. 200 sieve. Mills of the ring-roller type, equipped with air classification, and, to a lesser extent, ball mills are used, although buhrstones are still in use in many of the older plants.

In the following list gypsum from various sources has been arranged in order of increasing resistance to grinding: Iowa, New York, Nova Scotia, Kansas, Michigan, Ohio, Wyoming, Virginia, Texas, Nevada, and Montana. Gypsum calcined in a rotary kiln is usually pulverized after calcination. Ring-roller mills and tube mills with air classification are sometimes employed, but more often intermittently operating tube mills are used.

Coal, Coke, and Other Carbon Products. *Bituminous Coal.* The grinding characteristics of bituminous coal are affected by impurities contained, such as inherent ash, slate, gravel, sand, and sulfur balls. The grindability of a coal is determined by grinding it in a standard laboratory mill and comparing the results with the results obtained under identical conditions on a coal selected as a standard. This standard coal is a low-volatile coal from Jerome Mines, Upper Kittaning bed, Somerset Co., Pa., and is assumed to have a grindability of 100. Thus a coal with a grindability of 125 could be pulverized more easily than the standard, while a coal with a grindability of 70 would be more difficult to grind. Grindability and grindability methods are discussed on p. 1114. The capacity obtained with a mill will be a function of grindability (Hardgrove, Advance Paper presented at the Semiannual Meeting of Am. Soc. Mech. Engrs., Chicago, June, 1933).

There are two general methods of burning pulverized coal, the unit system, in which the coal is blown directly into the furnace as it is pulverized; and the central grinding system, in which the coal is stored in a bin before it is used.

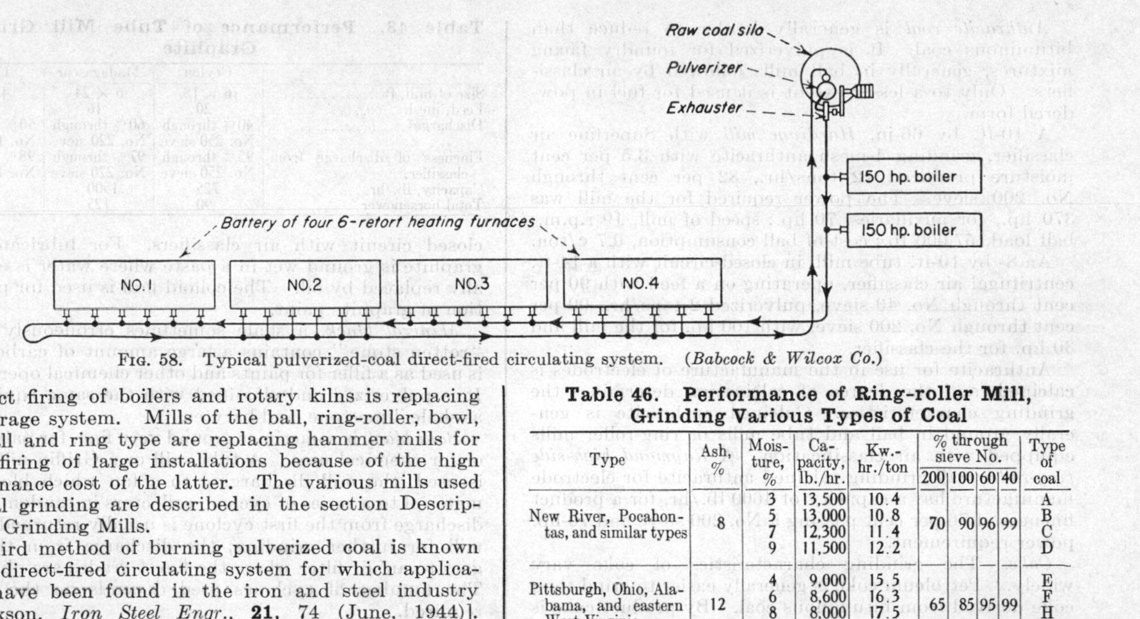

FIG. 58. Typical pulverized-coal direct-fired circulating system. (*Babcock & Wilcox Co.*)

Direct firing of boilers and rotary kilns is replacing the storage system. Mills of the ball, ring-roller, bowl, and ball and ring type are replacing hammer mills for direct firing of large installations because of the high maintenance cost of the latter. The various mills used for coal grinding are described in the section Description of Grinding Mills.

A third method of burning pulverized coal is known as the direct-fired circulating system for which applications have been found in the iron and steel industry [Wilcoxson, *Iron Steel Engr.*, **21**, 74 (June, 1944)]. This new method should be of interest to the chemical industry. This system is designed so that a pipe carries

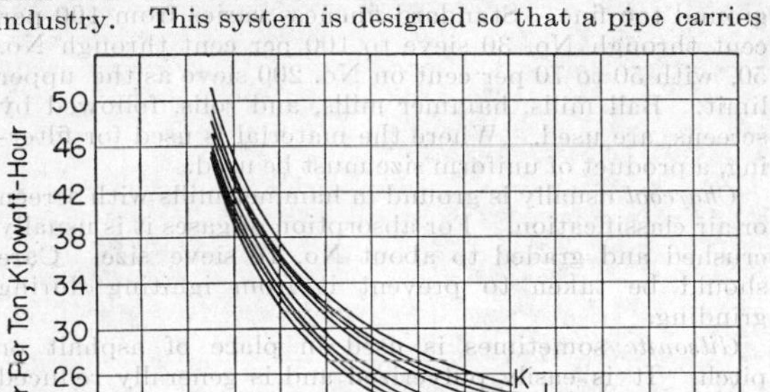

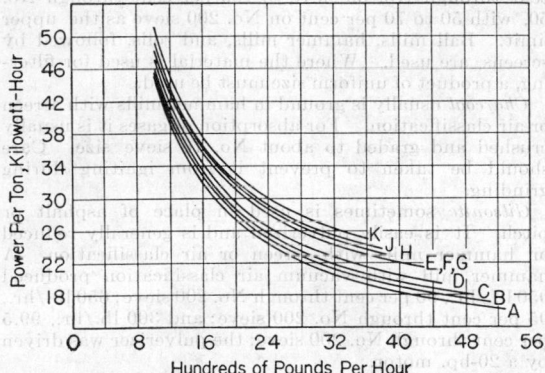

FIG. 59. The variation in power relative to capacity of a Raymond Imp hammer mill when grinding various grades of bituminous coal.

the pulverized coal and its primary air in a continuous loop arranged so that take-offs can be made to a number of furnaces or boilers. The advantages of a direct-fired system are obtained by heating a series of operations, no one of which could justify direct coal firing. A typical arrangement is shown in Fig. 58.

The *Kennedy air-swept tube* mill is of relatively short length. No screens are used either inside or out; the fine coal is mixed thoroughly with air and air-floated to the burners. The mill rotates slowly at a speed of 19 to 50 r.p.m., depending on the size. Hot air is used if drying is desired.

The curves shown in Fig. 59 give the performance of a *Raymond Imp hammer mill* when used for direct firing and grinding various grades of bituminous coal. Letters *A*, *B*, etc., refer to the type of coal pulverized, as given in Table 46. The same letter is used for different coals where they have the same grinding characteristics. Table 46 gives the performance of a *Raymond ring-roller mill* when used as a unit pulverizer.

Table 46. Performance of Ring-roller Mill, Grinding Various Types of Coal

Type	Ash, %	Moisture, %	Capacity, lb./hr.	Kw.-hr./ton	% through sieve No. 200	100	60	40	Type of coal
New River, Pocahontas, and similar types	8	3	13,500	10.4	70	90	96	99	A
		5	13,000	10.8					B
		7	12,300	11.4					C
		9	11,500	12.2					D
Pittsburgh, Ohio, Alabama, and eastern West Virginia	12	4	9,000	15.5	65	85	95	99	E
		6	8,600	16.3					F
		8	8,000	17.5					H
		10	7,200	19.5					K
Illinois and Indiana	15	10	9,000	15.5	65	85	95	99	G
		12	8,600	16.3					H
		14	8,000	17.5					J
		16	7,200	19.5					K
Texas and Dakota lignites	14	30	9,000	15.5	65	85	95	99	G
		32	8,600	16.3					H
		34	8,000	17.5					J
		36	7,200	19.5					K
Eastern Kentucky and eastern West Virginia	9	3	8,500	17.6	65	85	95	99	G
		5	8,300	18.0					H
		7	7,800	19.2					J
		9	7,000	21.4					K

The *Raymond bowl mill* is used primarily for pulverizing coal and blowing it directly into industrial furnaces or rotary kilns. Hot gases may be used for drying the coal while it is being pulverized. Table 47 gives pertinent data on the operation of a bowl mill when grinding a typical grade of coal having a grindability of 55 to 60. The data are based on grinding coal to a fineness of approximately 80 per cent to pass a No. 200 sieve. Thermostatic control through a tempering device may be used to maintain a constant temperature in the mill irrespective of the moisture in the coal.

Table 47. Performance of Bowl Mills Grinding Bituminous Coal

Mill size	Capacity, lb./hr.	Air, lb./min.	Power, kw.-hr./ton
271	2,520	130	18.4
312	2,980	150	18.0
352	3,740	170	17.5
372	5,200	200	17.0
412	6,720	250	15.9
452	8,125	310	15.9
492	11,150	400	15.4
493	14,100	450	14.9
532	17,100	550	14.2
533	20,000	600	13.3
593	23,700	700	12.7
613	27,500	820	12.7
673	34,900	960	12.7

The *B. & W. Type E pulverizer* is used for direct-firing industrial furnaces and has application in direct-fired circulating systems, as shown in Fig. 58 and described on p. 1138.

Anthracite coal is generally harder to reduce than bituminous coal. It is pulverized for foundry facing mixtures, generally in ball mills followed by air classifiers. Only to a lesser extent is it used for fuel in powdered form.

A 10-ft. by 66-in. *Hardinge mill* with Superfine air classifier, grinding 4-mesh anthracite with 3.5 per cent moisture produced 12 tons/hr., 82 per cent through No. 200 sieve. The power required for the mill was 370 hp., for auxiliaries, 70 hp.; speed of mill, 19 r.p.m.; ball load, 57,000 lb.; cost of ball consumption, 0.7 ¢/ton.

An 8- by 10-ft. tube mill, in closed circuit with a 14-ft. centrifugal air classifier, operating on a feed with 90 per cent through No. 40 sieve, pulverized 2 tons/hr., 99 per cent through No. 200 sieve, with 100 hp. for the mill and 30 hp. for the classifier.

Anthracite for use in the manufacture of electrodes is calcined, and the degree of calcination determines the grinding characteristics. Calcined anthracite is generally ground in ball and tube mills or ring-roller mills equipped with air classification. A *Raymond high-side ring-roller mill* grinding calcined anthracite for electrode manufacture has a capacity of 4600 lb./hr. for a product fineness of 76 per cent passing a No. 200 sieve and 70-hp. power requirement.

Coke. The grinding characteristics of coke vary widely. Petroleum coke is generally easier to grind than coke derived from bituminous coal. By-product coke is hard and abrasive, while certain foundry and retort coke is extremely hard to grind. For certain purposes it may be necessary to produce a uniform granule with minimum fines. This is best accomplished in rod or ball mills in closed circuit with screens. Hourly capacity of a 4- by 10-ft. rod mill with screens, operating on by-product-coke breeze was 9 tons, 100 per cent through No. 10 sieve, and 73 per cent on No. 200 sieve; power requirement 40 hp.

Petroleum coke is generally pulverized for manufacture of electrodes; ring-roller mills with air classification and tube mills are generally used. A No. 5057 *Raymond ring-roller mill* gave an hourly output of 3.8 tons, 78.5 per cent through No. 200 sieve, with 90 hp.

Pitch may be pulverized as a fuel or for other commercial purposes; in the former case the unit system of burning is generally employed, and the same equipment is used as described for coal. The grinding characteristics vary with the melting point, which may be anywhere from 50° to 175°C.

Natural graphite may be divided into three grades in respect to grinding characteristics: flake, crystalline, and amorphous. Flake is generally most difficult to reduce to fine powder, and the crystalline variety is the most abrasive. The following is an arrangement of graphites according to origin, in order of increasing resistance to grinding: Mexico (Sonora), Michigan, Rhode Island, Korea, Ontario, East Siberia, New York, Quebec, Alabama, Madagascar, and Ceylon. Graphite is generally ground in ball mills, tube mills, and in ring-roller mills, with or without air classification. For large capacities, ball and tube mills are used, particularly on the flake and crystalline varieties. Performance of a tube mill with vacuum air classifier in closed circuit is given in Table 48.

A *Raymond ring-roller mill*, grinding Mexican graphite, gave a capacity of 2500 lb./hr., 99.1 per cent through No. 200 sieve, with a power consumption of 75 hp.

Micronizer performance on natural graphite is given in Table 27. Graphite for pencils has 47, 83, 91, and 94 per cent by weight smaller than 4, 9, 18, and 31 μ, respectively, when ground in the *Eagle fluid energy mill* [*Ind. Eng. Chem.*, **38**, 672–678 (1946)].

Artificial graphite has been ground in ball mills in closed circuit with air classifiers. For lubricants the graphite is ground wet in a paste where water is eventually replaced by oil. The colloid mill is used for production of graphite paint.

Table 48. Performance of Tube Mill Grinding Graphite

	Ceylon	Madagascar	Korea
Size of mill, ft.	6 × 18	6 × 24	4 × 8
Feed, mesh	20	16	25
Discharge	40% through No. 250 sieve	60% through No. 220 sieve	50% through No. 100 sieve
Fineness of discharge from classifier.	95% through No. 250 sieve	97% through No. 220 sieve	98% through No. 100 sieve
Capacity, lb./hr.	725	1500	800
Total horsepower	90	125	70

Mineral black, a shale sometimes erroneously called "rotten stone," contains a large amount of carbon and is used as a filler for paints and other chemical operations. It is pulverized and classified with the same equipment as shale, limestone, and barytes.

Bone black is sometimes ground very fine, for paint, ink, or for chemical uses. A tube mill or a Griffin mill often is used, the mill discharging to a fan which blows the material to a series of cyclone collectors in tandem. The discharge from the first cyclone is usually returned to the mill for further grinding; the discharge from the last goes to an air filter where the finest grades are obtained. The number of cyclones used depends on the grades required.

Decolorizing carbons of vegetable origin should not be ground too fine. Standard fineness varies from 100 per cent through No. 30 sieve to 100 per cent through No. 50, with 50 to 70 per cent on No. 200 sieve as the upper limit. Ball mills, hammer mills, and rolls, followed by screens, are used. Where the material is used for filtering, a product of uniform size must be used.

Charcoal usually is ground in hammer mills with screen or air classification. For absorption of gases it is usually crushed and graded to about No. 16 sieve size. Care should be taken to prevent it from igniting during grinding.

Gilsonite sometimes is used in place of asphalt or pitch. It is easily pulverized and is generally reduced on hammer mills with screen or air classification. A hammer mill with vacuum air classification produced 950 lb./hr., 90 per cent through No. 200 sieve; 650 lb./hr., 95 per cent through No. 200 sieve; and 300 lb./hr., 99.5 per cent through No. 200 sieve; the pulverizer was driven by a 20-hp. motor.

Carbon mixtures (green mix) are generally made from flour of petroleum coke, graphite, and lampblack, mixed with a binder such as pitch; solvents such as benzol are incorporated in the mix. After cooling, the mixture is caked and therefore generally reground. Table 49 gives the results of several grinding mills used in the carbon brush industry for grinding such mixtures.

Table 49. Grinding Carbon Mixtures

	Hammer mill	Ring-roller mill
Material in mix	Graphite	Lampblack, pitch, coke
Capacity, lb./hr.	700	1350
Fineness, % through No. 200 sieve	65	62.5
Power required, hp	20	30
Type of classification	Vacuum, air	Vacuum, air

Lampblack when manufactured is usually very fine. The gas is passed through baffled chambers or ducts in which the various grades are precipitated; the coarser grades are often pulverized for the carbon brush industry. Grinding may be done in ball mills, hammer mills, or ring-roller mills, with or without air classification. Where an extremely fine product is required, the same system as described for bone black may be used. A hammer mill equipped with air classification ground about 200 lb./hr. to a fineness of 95 per cent through No. 200 sieve, with a power consumption of 20 hp.

Pigments, Chemicals, and Insecticides. Most chemicals, dry colors, and dyes offer little resistance to disintegration, but other difficulties incidental to grinding arise. Chief among these is the agglomeration, or balling, of the pulverized material. This occurs particularly with certain precipitates or filter-press products.

Dry colors and dyestuffs generally are pulverized in hammer mills (see Tables 21 and 25). For very small productions the jar mill is probably the most practicable, as it is easily cleaned when changing from one color to another. A train of any number of such jar mills can be operated from one drive, each producing a different material. These mills are particularly adapted for wet grinding, although they are also used frequently for dry grinding. However, in the latter case, there is a tendency for the material to coat the balls and to adhere to the lining, thus decreasing the efficiency of the grinding operation.

Most colors are not ground very finely in a dry state, inasmuch as at a later stage they are ground wet in pebble mills or on roller mills with the proper vehicle. Some dyes, however, are ground very fine; if very hard and crystalline, they are reduced in an air-classification pulverizer or in a screen hammer mill in closed circuit with an air classifier.

White Pigments. A few of the basic chemical pigments used in the manufacture of paints are, however, pulverized in large quantities, and, as no necessity exists for cleaning the pulverizing equipment, they are often ground in large ring-roller mills equipped with air classification. Some operating data for grinding several of the basic pigments are given in Table 50. Micronizer performance for a number of extenders is given in Table 27.

Table 50. Grinding Pigments in Ring-roller Mills

Pigment	Capacity, lb./hr.	Power, hp.-hr./ton	Fineness of product, % on No. 325 sieve
Barium sulfate	4500	28	1
Zinc sulfide	5000	32	1
Barium carbonate	4600	29	0.9
Lithopone	5000	25	.1
Titanium with barium base	2000	63	.1
Titanium with calcium base	2200	57	.08
Titanium dioxide	1800	70	.05
Titanium with aluminum silicate	2800	45	.08
White lead	5500	23	.5

Figure 60 shows the grinding characteristics of a mill equipped with whizzer classification when grinding a titanium pigment to different finenesses.

Mineral pigments, such as ochers, umbers, siennas, and red oxides of iron, were for many years ground on stone mills and bolted. When the materials were wet-ground and classified, they were filtered-pressed, dried, and disintegrated in a cage mill or a hammer mill with screen classification. This is still the process when the material is water-floated. When the oxides are ground dry, ring-roller and hammer mills with air classification have replaced a great many stone mills. Even when buhrstones are used, bolting has generally been replaced by air classification. Some of the red pigments are still ground on buhrstones as the idea prevails in some quarters that the stones make a smoother and richer colored material. A 36-in. buhrstone in closed circuit with an 8-ft. air classifier produced 300 lb./hr. of ground hematite, 99.2 per cent through No. 325 sieve, with 15 hp. on the stone. Table 51 gives operating characteristics of a Raymond No. 5057 high-side mill grinding various oxides of iron.

Power consumption includes power required for grinding, classifying, and conveying product to bins above

Table 51. Grinding Iron Oxides in a Ring-roller Mill

Material	Fineness	Capacity, lb./hr.	Total hp.-hr./ton
Raw sienna	99% through No. 200 sieve	5950	23.5
Burned sienna	99.5% through No. 200 sieve	5800	22.1
Raw umber	99% through No. 200 sieve	5200	26.9
Burned umber	99.5% through No. 200 sieve	5400	25.9
Natural ocher	99.9% through No. 200 sieve	4500	31.0
Iron oxide (ore)	99% through No. 325 sieve	3100	45.0
Iron oxide (precipitated)	99.9% through No. 325 sieve	1800	72.5

the baggers. A 4.5-ft. by 16-in. Hardinge conical mill in closed circuit with classifier, grinding 50-mesh iron oxide with 33 per cent moisture for the paint trade showed a capacity of 25 tons/24 hr., 100 per cent through No. 200 sieve. Power consumption was 20 hp., mill speed 30 r.p.m., ball load 4000 lb.

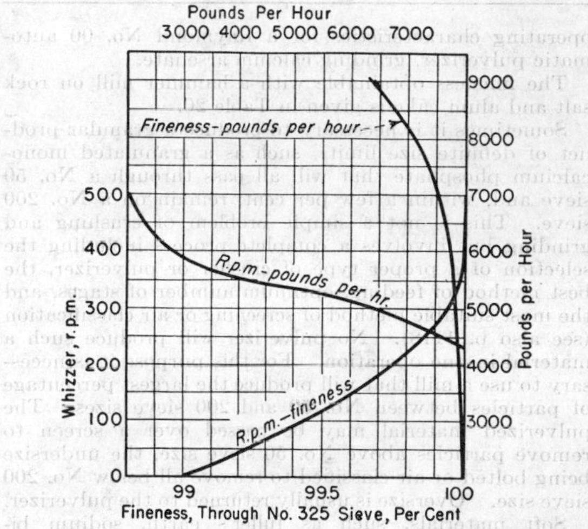

FIG. 60. Grinding characteristics of a ring-roller mill with whizzer classifier when grinding a titanium pigment.

Lead oxides generally are first ground in high-speed automatic pulverizers with air classification and automatic throwout. This method is particularly adapted to grinding of incompletely oxidized materials and those containing an appreciable amount of metallic lead, which is eliminated through the throwout. The objective in the production of certain lead oxides is to obtain a product of lowest possible apparent density. Different types of mills produce oxides of varying densities, even though the sieve analysis may be quite similar. How the type of mill used and the grinding method applied affect the apparent density of the product may be seen from the following test: The lead oxide was first passed through a pulverizer with throwout, for removal of the metallic lead. It next went to a ring-roller mill equipped with air classification and was subjected to extremely fine grinding. Finally, the product from the ring-roller mill went to an Imp pulverizer equipped with whizzer classification. This pulverizer both changed the particle shape and reduced the apparent density of the material. The results are given in Table 52.

Chemicals. A high-speed pulverizer, with air classification and air conveying, gives a calcium arsenate product possessing proper physical properties. Table 53 gives the

Table 52. Data Showing the Influence of the Type of Mill upon Apparent Density of Lead Oxide

Automatic Pulverizer	
Power (total) required by pulverizer system, hp.	75
Production, lb./hr.	4000
Apparent density, g./cu. in.	38
Ring-roller Mill	
Power required by mill, hp.	75
Capacity, lb./hr.	4000
Apparent density, g./cu. in.	27
Imp Pulverizer	
Power required by pulverizer, hp.	75
Production, lb./hr.	1900
Apparent density, g./cu. in.	18

Table 53. Performance of Automatic Pulverizer Grinding Calcium Arsenate

Fineness, % through No. 200 sieve	Apparent density, cu. in./lb.	Production, lb./hr.	Hp.
90	81.5	1300	28
94	86.0	1100	28
96	90.6	950	27
98	94.2	750	27
99	98.7	625	25
99.5	104.8	550	24
99.9	110.0	475	22
99.99	124.8	400	20

operating characteristics for a Raymond No. 00 automatic pulverizer, grinding calcium arsenate.

The fineness obtainable with a hammer mill on rock salt and alum cake is given in Table 20.

Sometimes it is necessary to produce a granular product of definite size limits, such as a granulated monocalcium phosphate that will all pass through a No. 50 sieve and, within a few per cent, remain on a No. 200 sieve. This is not a simple problem of crushing and grinding but involves a complete process including the selection of a proper type of crusher or pulverizer, the best method of feeding, optimum number of stages, and the most suitable method of screening or air classification (see also p. 1118). No pulverizer will produce such a material in one operation. For this purpose it is necessary to use a mill that will produce the largest percentage of particles between No. 50 and 200 sieve sizes. The pulverized material may be passed over a screen to remove particles above No. 50 sieve size, the undersize being bolted or air classified to remove all below No. 200 sieve size. Oversize is usually returned to the pulverizer.

Soft materials, such as fuller's earth, sodium bicarbonate, and monocalcium phosphate, are generally ground in flour mills to obtain a finely divided product. A ratio of reduction of 2.5 to 1 or even 1.5 to 1 is generally used; the material passes through a series of rolls with bolting reels. Sometimes air classifiers are used in the circuit for removal of fines. Table 54 gives the results obtained in granulating soft materials.

Table 54. Granulation of Soft Materials

	Fuller's earth	Sodium bicarbonate
Size of product required, sieve No.	−80 +150	−120 +200
Type of mill used	Two roller	Two roller
Number of mills in series	12	5
Size of rolls, in.	7 × 16	7 × 20
Capacity, lb./hr.	3500	2500
Recovery from original feed, %	80	55
Horsepower required, total	80	30

Derris and Pyrethrum. A great many insecticides are made by extracting the toxic agent from derris root and pyrethrum flowers. The use of these materials has become extensive. The term "derris root" is used rather indiscriminately for different species of roots containing rotenone as the toxic agent. Amongst these roots are derris root, cube root, timbo root, and barbasco. A 50-in. ring-roller mill will produce about 700 lb./hr. of cube root, about 600 lb./hr. of derris root, and about the same amount of the other roots. The fineness to which this material is reduced is about 95 per cent through No. 200 sieve. The power consumption is from 125 to 150 kw.-hr./ton of finished product. The same type of equipment is used for pulverizing pyrethrum flowers. The capacity on pyrethrum flowers when grinding to a fineness of 95 per cent through No. 200 sieve is about 25 per cent less than that obtained on derris root. The power is 160 to 175 kw.-hr./ton.

D.D.T. Lump D.D.T. can be reduced in size in hammer mills. A granular product ranging from No. 8 to No. 18 sieve size is obtained with the *S.P. Mikro-Pulverizer* (see p. 1140). Finer products require screen-discharge hammer mills, and cooling of the material is necessary to prevent fusion during grinding. A product having 60 to 70 per cent able to pass a No. 100 sieve is obtained with an *Imp pulverizer* at a rate of 20 lb./hp.-hr.

In many instances D.D.T. is ground with an inert extender to produce compositions having 10 to 50 per cent of the active ingredient. Grinding characteristics vary with the nature and quality of the D.D.T. as well as the nature and fineness of the extender. A *Raymond ring-roller mill* will handle most of these compositions. As an example, a fineness of 99 per cent smaller than 20 μ with a surface mean diameter of 4.8 μ is obtained at a rate of 30 lb./hp.-hr. Hammer mills with internal classification such as the *Raymond vertical mill* and *Mikro-Atomizer* have application for fine grinding of D.D.T. compositions, as do also the fluid-energy jet mills.

Sulfur. The ring-roller mill arranged as shown in Fig. 42 can be used for the fine grinding of sulfur. Inert gases are supplied instead of hot air (see p. 1120 for use of inert gas). Performance of a *Raymond No. 5057 ring-roller mill* is given in Table 55. If power cost is 2 ¢/kw.-hr., the total cost might be three to four times the power cost and include labor, inert gas, maintenance, and fixed charges.

Table 55. Grinding Sulfur

Fineness, % through	sieve No.	Capacity, tons/hr.	Power, kw.-hr./ton
60	200	13.50	6.6
70	200	11.50	7.7
80	200	9.00	9.8
90	200	6.50	13.0
90	325	4.50	17.0
95	325	3.75	19.2
97	325	3.50	20.6
99	325	3.00	24.0
99.5	325	2.25	32.5
99.7	325	1.75	39.0
99.9	325	1.50	45.5
99.95	325	1.25	55.0
99.99	325	1.00	67.0

Drugs, Pharmaceuticals, and Spices. Nearly all drugs and pharmaceuticals are ground in small quantities, sometimes so small that mortar and pestle or hand-operated jar mills are used. Materials of fibrous character are first passed through a rotary cutter or high-speed hammer mill; when a very fine product is required, stone mills or attrition mills are used. Many materials, such as tonka, vanilla beans, and rose leaves, are ground in batches in pebble mills or in small ball mills consisting of a narrow cylinder with one large ball, the diameter of which equals the width of the cylinder. Oily seeds, such as olibanum, mustard, and cochineal, when pulverized very fine, are usually ground in a pounder mill, similar to a stamp mill. Mustard, in pulp-paste or semipaste form, is generally ground in horizontal stone mills.

A mill of different type, fitted with grinding plates, is used extensively for grinding drugs and spices. One mill of this type is the *Quaker City* grinding mill. Many spices are first reduced on breakers and crackers, fitted with corrugated steel cones, plates, or buhrstones. The fine product from these breakers is usually a good feed for a finishing mill.

Performance of a 20-in. Schutz O'Neill Limited mill at 3200 r.p.m. is given in Table 56. Jay Bee pulverizer applications are given in Table 23.

When drugs, chemicals, and spices are to be granulated, they are usually ground by roller mills in closed circuit with bolting reel and sifter. The mill most frequently used is a four-roller mill. The upper roll pair may be set rather open, the product passes to the lower pair and then to a sifter, where fines are removed. Tailings, after passing through a coarse screen to remove shreds, threads of fiber, and similar objectionable material, are returned to the upper rolls. A 9- by 18-in., four-roller mill, grinding pepper to No. 30 sieve size, showed an hourly capacity of 600 lb., with pulley speed 450 r.p.m. and 15 hp., including sifter and elevators

Resins, Gums, Waxes, and Molding Powders. The grinding characteristics of the various resins, gums, waxes, hard rubbers, and molding powders depend greatly upon their softening temperatures. When a finely divided product is required, it is often necessary to use a water-jacketed mill or a pulverizer with air classifier in which cooled air is introduced into the system. Not all waxes can be ground, inasmuch as some of them are soft at the temperatures obtainable. However, a great many of them can be powdered if precautions are taken to prevent overheating. Hammer and cage mills are used for this purpose. Some low-softening-temperature resins can be ground by mixing with 30 to 100 per cent by weight of Dry Ice before grinding. Refrigerated air sometimes is introduced into the hammer mill to prevent softening and agglomeration [Dorris, How Solid Carbon Dioxide Assists in Grinding Low-melting Waxy or Plastic Solids, *Chem. & Met. Eng.*, **51**, 114 (July, 1944). See also p. 1119].

Table 56. Performance of Schutz O'Neill Limited Mill

Material	Production, lb./hr.	Hp.
Pyrethrum flowers	35– 45	15–20
Quassia	25– 30	15–20
Senna leaves	100–125	20–25
Buchu	100–110	20–25
Stramonium	125–150	20–25
Bay leaves	125–150	20–25
Celery seed	125–150	15–20
Tragacanth gum	50– 60	15–20
China twigs	75– 80	20–25
Cassia	90–100	20–25
Ginger	90–100	20–25
Cloves	125–150	20–25
Pepper, white	200–225	20–25
Pepper, Singapore	225–250	20–25
Gelatin	60– 70	15–20
Cayenne pepper	100–125	20–25

Most gums and resins, natural or artificial, when used in the paint, varnish, or plastic industries, are not ground very fine, and any hammer or cage mill will produce a suitable product. Typical performance of the attrition mill is given in Table 26. Roll crushers will often give a sufficiently fine product. Certain resins used in the phenolic resin industries must be pulverized very fine; pebble mills, cooled with water or brine, in closed circuit with an air classifier, are used. In general, one may say that the grinding of the thermoplastic resins, or of the thermosetting resins after they have set, is difficult except in the very coarse range. In the case, however, of the thermosetting resins, before setting, a very high fineness may be obtained readily, although these materials, particularly when ground on a hammer mill, may cake up very rapidly after leaving the mill. Phenol-formaldehyde resins have been ground in the *Raymond Imp Mill* to about 80 per cent through No. 325 sieve at a rate of 30 lb./hp.-hr. This is about 99 per cent through No. 100 sieve.

Hard rubber is one of the few combustible materials which is generally ground on heavy steam-heated rollers. The raw material passes to a series of rolls in closed circuit with screens and air classifiers. Farrel-Birmingham rolls are used extensively for this work. There is a differential in the roll diameters, and the particular size best suited for the average hard rubber is one having rolls with 13 and 17 in. diameters and 20-in. face. The motor should be separated from the grinder by a fire wall. It is also desirable to run these machines at rather low speed and low differential between the rolls because it is very easy to overheat hard rubber in grinding, making it smolder, which necessitates the shutting down of the grinder until it cools off before clearing out the charred material. The performance of a series of rolls grinding hard rubber, producing a finished product through an air

Table 57. Grinding Hard Rubber

Number of roller mills	3
Size of each mill, in.	13 and 17 × 20
Motor on each mill, hp.	50
Size of vacuum air classifier, ft.	4.5
Size of motor on classifier fan, hp.	20
Fineness of feed to classifier, % through No. 100 sieve	32
Fineness of product from classifier, % through No. 100 sieve	95
Production, lb./hr.	250

classifier is given in Table 57. A larger production could probably be attained, but operation at the lower rate is advisable to prevent generation of an excessive amount of heat.

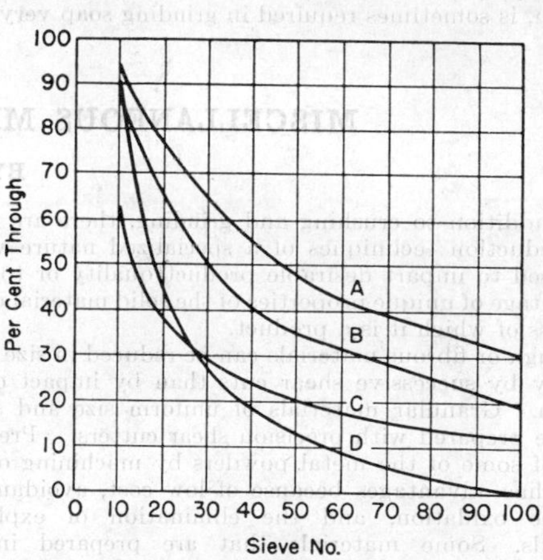

FIG. 61. Sieve analysis of molding powders produced by various installations.

Molding Powders. Specifications for molding powders vary widely, from a No. 8 to a No. 60 sieve product; generally the coarser products are No. 12, 14, or 20 sieve material. Specifications usually prescribe a minimum of fines (below No. 100 and No. 200 sieve). For most purposes the ideal molding powder would consist of particles testing smaller than No. 20 and larger than No. 100 sieve size. Molding powders are produced with hammer mills, either of the screen type or equipped with air classifiers.

Curves *A* and *B* in Fig. 61 give the screen analysis of a molding powder produced with screen pulverizers fitted with an 8-mesh screen. Curve *C* gives the data obtained with an air-classification pulverizer unit operated to give a minimum of No. 100 sieve material, which amounted to only 12 per cent. This material was passed over an 8-mesh screen to remove oversize, and the resulting product passed through an air classifier to remove the No. 100 sieve size particles. Curve *D* gives the screen analysis of the final granular product.

The following material may be ground at ordinary temperatures if only the regular commercial fineness is required: amber, arabac, tragacanth, rosin, olibanum, gum benzoin, myrrh, guaiacum, and montan wax. If a finer product is required, hammer mills or attrition mills in closed circuit, with screens or air classifiers, are used. Attrition-mill performance is given in Table 26.

Grinding of Soaps. Soaps in a finely divided form may be classified as soap powder, powdered soap, and chips or flakes. The term soap powder is applied to a granular product, No. 12 to No. 16 sieve size with a certain amount of fines, which is produced in hammer mills with perforated or slotted screens. A *No.* 2 *Mikro-Pulverizer* gives a production of 2750 lb./hr. analyzing a slight trace on No. 16 sieve and 19 per cent through a No. 80 sieve using 4 hp. for a moisture content of 19 per cent. The capacity drops to about 1000 lb./hr. at 25 per cent moisture content for the same fineness. Soaps of the type used for tooth powder show a production rate of 400 lb./hr., require 10 hp. and analyze about 98.5 per cent passing a No. 100 sieve. Lump soaps, which are mixtures of soap and caustic containing 20 per cent moisture, give a capacity of 1500 lb./hr. at 9 hp. for 98 per cent through a No. 100 sieve.

Powdered soap is a finely ground powder, with 99 per cent or more through No. 200 sieve. Grinding to this fineness, a No. 00 Raymond automatic pulverizer will handle 300 to 350 lb./hr. with 15 to 17 hp. Cooling, generally by introduction of cold air in the air-classifying system, is sometimes required in grinding soap very fine. Grinding in closed circuit, with screens or air classification often is advantageous, giving a granular product and preventing overheating.

Pulverizing of the metallic soaps, stearates, palmitates, resinates, laurates, and erucates is not difficult using modern equipment with provision for keeping the material cool and in rapid motion. Batch grinding is not practicable, as the material tends to cake, particularly if a fine product is needed. Oleates are usually most troublesome, as they tend to become plastic and creamy. Listed in order of their resistance to pulverization some of the metallic soaps are: lead, silver, zinc, copper, and nickel stearate; zinc, lead and copper palmitate; lead copper, and zinc laurate; silver, mercury, and lead erucate; and silver and lead oleate.

The oleates and erucates are best pulverized by multicage mills; laurates and palmitates in cage mills and also in hammer mills if particularly fine division is not required; stearates may generally be pulverized in multicage mills, screen mills, and air-classification hammer mills. Table 58 gives the operating characteristics of hammer mills when grinding zinc stearate and aluminum stearate to a finely divided powder.

Table 58. Performance of Screen-type Hammer Mill Grinding Zinc and Aluminum Stearates

	Zinc stearate	Aluminum stearate
Capacity, lb./hr.	500	300
Fineness, % through No. 325 sieve	60	70
In closed circuit with air classifier:		
Capacity, lb./hr.	100	75
Fineness, % through No. 325 sieve	99.5	99.7
Horsepower required	25	20

MISCELLANEOUS METHODS OF SIZE REDUCTION

BY C. E. BERRY

In addition to crushing and grinding, there are other size-reduction techniques of a specialized nature which are used to impart desirable product quality or to take advantage of unique properties of the solid material or the process of which it is a product.

Tough or fibrous materials can be reduced in size more readily by successive shear cuts than by impact or attrition. Granular materials of uniform size and shape can be prepared with precision shear cutters. Preparation of some of the metal powders by machining operations has advantages because of low cost, avoidance of surface oxidation, and the elimination of explosion hazards. Some materials that are prepared in the molten state are converted advantageously to flake form by cooling a thin layer continuously on the surface of a rotating drum. Thus massive cooling and subsequent pulverizing are avoided. Certain materials, particularly those of a fibrous nature, may be disintegrated by first saturating them with a gas or a vapor at high pressure and then releasing the pressure as rapidly as possible.

The size reduction of liquids in liquid mediums is known as emulsification or homogenization. Such operations are as truly ones of size reduction as is the pulverization of solids. Likewise, spraying and atomizing are processes of size reduction of a liquid in a gaseous phase, and aeration or gas dispersion are operations of size reduction of a gas in a liquid phase. The latter is particularly true in such manipulations in which the identity of the small bubbles of gas is retained because of the high viscosity of the liquid or its low surface tension. Since all gases are miscible, examples of the size reduction of a gas in a gaseous phase cannot be cited. The phase relationships for the size-reduction operations are shown in Table 59.

Table 59. Phase Relationship of the Size-reduction Operations

Continuous phase / Discontinuous phase	Liquid	Gas
Solid	Grinding	Grinding, cutting, machining, flaking, explosive disintegration
Liquid	Emulsification	Spraying
Gas	Gas dispersion	

CUTTING

By Thos. B. Dorris

Rotary Knife Cutters. Certain materials, notably those which are tough or fibrous, are reduced best by successive shear cuts rather than by attrition or impact. In such cases a rotary cutter is an ideal choice, provided that the width of the feed stock does not exceed the cutter knife length and the thickness is limited to less than 1 in. The usual arrangement includes a central rotor with knives placed uniformly at the periphery so as to cut against stationary knives, which with screens and hopper inlet constitute a cylinder encasing the rotor. Product size is determined by the screen openings; fines are controlled in part by rotor speed and the number of knife cuts per revolution. With screen openings 20 mesh and finer, a pneumatic product-collecting system is generally required. Certain materials can be cut to pass through openings as fine as 80 mesh by such a unit.

The data shown in Table 60 are for a unit having a 10-in. rotor and a knife length of 18 in. and consisting of five fly and five bed knives, operating at 920 r.p.m. The data are typical of the application of this method of size reduction.

Table 60. Performance of Rotary Knife Cutter

Material	Screen opening	Feed rate, lb./hr.	Hp.	Air	Remarks on product
Amosite asbestos pencils.	1½″	1000	11	Yes	Finer fiber bundles average length 2″
Cellophane bags..	11⁄32″	200	10	Yes	Finer than 5⁄16″
Cork............	3⁄16″	525	16	Yes	90% 4/24* sieve
Chemical cotton..	60 mesh	120	15	Yes	Flock; 35% under No. 100 sieve
Leather scrap....	¾″	600	20	Yes	Precutting before shredding
Fiberglas	3⁄16″	300	18	Yes	1″ (approx.) lengths
Waste paper.....	5⁄16″	338	13	Yes	Through No. 4 sieve and finer
Sheet pulp.......	40 mesh	150	15	Yes	Flock; 85%, 40/100 sieve
Tenite scrap.....	5⁄16″	340	12	No	Granulated for reuse
Vinylite scrap....	7⁄32″	300	15	Yes	35%, 6/10 sieve; granular
⅛″ Geon sheet...	5⁄16″	540	11	No	99%, 4/20 sieve; for molding granules
Cotton rags......	¾″	500	11	Yes	No linting
Buna scrap......	10 mesh	264	12	Yes	Granular
Neoprene scrap..	30 mesh	90	14	Yes	20°F. temperature rise
Soft-wood chips..	⅛″	960	12	Yes	90%, 10/50 sieve
Hard-wood chips.	1⁄16″	290	11	Yes	83%, 20/100 sieve

* 90 per cent 4/24 sieve, *i.e.*, 90 per cent is through No. 4 and on No. 24 sieve.

The cutters described below are equipped with shaft seals and outboard antifriction bearings. In general they can be obtained in steel or in stainless steel or other corrosion-resistant metal. Shear-cut rotor knives leading to a marked lowering of shock load are obtainable in all models.

Abbé (*Engineering Co.*) rotary cutters are made in six standard sizes and with capacities ranging from 25 to 3000 lb./hr. requiring 1 to 50 hp. Abbé models are well adapted to fibrous material, such as asbestos, coconut shells, paper, and leather, and are used in many places to precut stock for attrition mills and buhrstone mills.

Ball & Jewell Improved Patented rotary cutters. The latest models of this manufacturer are designed with smooth crevice-free interiors and are readily dismantled for thorough cleaning, as is required for applications with colored plastics. They have a top center hopper with protecting offset for feeding; three to five shear-cut rotor knives; four to six stationary knives equally spaced in the two upper quadrants; readily removable single-sheet semicircular screens in bottom half of cylinder. These models feature compact direct or V-belt drives and self-contained discharge bins. Sizes range from a 300-lb./hr. laboratory unit using 2 hp. to the No. 2 unit requiring 25 to 40 hp.

Mercer-Robinson Unique heavy-duty cutters. The frame of this model is an all-welded steel cage; the top feed hopper is offset from center and with four or five stationary knives completes the upper half of the cylinder. Readily removable filler plates between stationary knives provide immediate access to knives and interior for cleaning and adjustment. The lower half of the cylinder is one-piece semicircular screen. Split rotor knives (herringbone arrangement) provide shear, without conveying contents against ends. Two series of units are available: 12-in.-diameter rotor with 12- and 18-in. knife lengths; 16-in.-diameter rotor with 18-, 24-, and 30-in. knife lengths. Unit weights range from 700 to 1900 lb. and motor requirements from 10 to 30 hp.

Paul O. Abbé Master rotary cutters. Open rotor assembly with five shear-cut knives mounted on three knife heads along cutter shaft; six stationary knives evenly divided in two upper quadrants with top center hopper between; semicircular bottom one-piece screen. Fly knives are in a fixed position on the rotor with adjustment by stationary knives only from the exterior of the cutter. The unit is readily accessible for complete cleaning. Model specifications are given in Table 61.

Table 61. Rotary Cutter Specifications

Machine No.	Floor space required, in.	Shipping weight, lb.	Speed, r.p.m.	Hp.	Screen size, in.
0	37 × 17	500	900–1200	2–5	10 × 17
1	54 × 34	1,500	600–900	5–15	20 × 24
2	68 × 42	4,000	600–900	15–40	20 × 28
2½	96 × 39	6,000	500–800	20–45	35 × 36
3	102 × 43	12,000	500–750	30–60	51 × 30

Sprout, Waldron Heavy-Duty rotary knife cutters. This manufacturer offers two series of units as follows: (1) 10-in.-diameter, 920 r.p.m. rotor units with 18-, 24-, and 30-in. knife lengths mounted in a cast frame (iron, steel, stainless); and (2) 20-in.-diameter, 750 r.p.m. rotor units of 10- and 30-in. knife lengths with all-welded steel frame. A conventional arrangement of the largest of these units (Model F-11) is illustrated by Fig. 62.

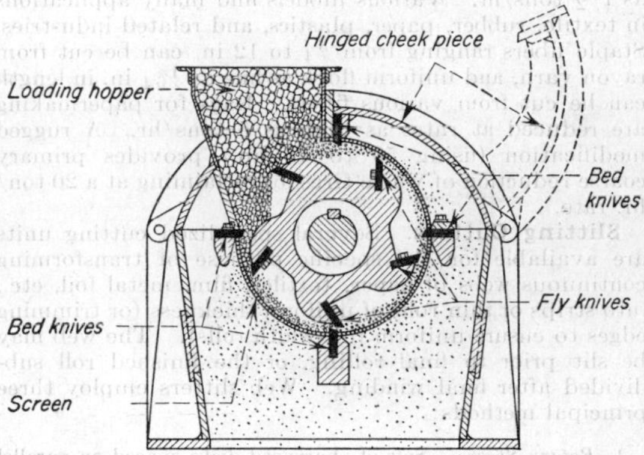

FIG. 62. Fabricated-steel rotary-knife cutter. (*Sprout, Waldron & Co.*)

These basic units are varied considerably according to application. Entry of feed is by hopper, slot, or compression-feed rolls. Generally five rotor knives are specified, and these are set at a slight angle with the shaft to provide shear cuts with direction reversed on alternate knives to avoid conveying the charge against the end of the cutter. Two to seven stationary knives may be specified alternating with screen sections around the cage to provide maximum discharge area and to keep fines at a minimum. Variations in construction permit such widely different applications as sheet-plastic granulation, flocking, tobacco-leaf threshing, etc. Models are powered by 10- to 60-hp. motors through V-belts and employ shear-pin safety hubs.

Wiley Laboratory Mill (Arthur H. Thomas Co.) is essentially a small-scale rotary knife cutter. The mill is available in several models, some large enough to serve for limited-scale production operations.

Wolf rotary cutter. This unit is designed for cutting or cubing material of about 1-in. ring size as a maximum to give a 3- to 6-mesh product. It is unique in that the rotor knives are composed essentially of four longitudinal sections with successive sections arranged stepwise from one end of the rotor to the other. For fine granulation, a 60-tooth corrugated rotor is substituted. These units are built in six standard sizes with capacities ranging from ¼ to 5 tons/hr.

Precision Knife Cutters. These differ from the conventional units described above in that a feeder synchronized with the rotor knives carries the stock over a single bed knife where it is cut to the exact size required. No screens are needed. Sheet stock is transformed to strips of uniform width; continuous strands into fibers (or fiber bundles) of very uniform length.

Stokes Universal fiber cutter. A small precision unit for producing staple fibers (wet or dry) to uniform lengths is available. It has four interchangeable rotors synchronized with variable-speed rolls to assure any product from 8 in. to flock. For cutting flock a special hood can be supplied to which a suction collector unit may be attached. Model 72-E, 16 by 18 by 49 in. (height) at ½ hp. handles four to eight strands simultaneously and cuts rayon to 1½-in. lengths at a 750 to 1000 lb./hr. rate.

Taylor-Stiles Precision Cutters. In addition to obtaining uniformity by synchronizing feeder and rotor speeds, this company offers a *Tandem* arrangement which is essentially two cutters generally set at right angles to each other. Uniform strips from plastic sheets can be diced by these knives to cubes of extreme uniformity and regularity, virtually no fines are produced. Cloth can be cut to squares, rectangles, or diamond shapes, at rates as high as 1½ tons/hr. Various models find many applications in textile, rubber, paper, plastics, and related industries. Staple fibers ranging from ¾ to 12 in. can be cut from rayon yarn, and uniform flock as fine as 1/64 in. in length can be cut from various fibers. Rags for papermaking are reduced at rates as high as 5 tons/hr. A rugged modification (using 50 to 75 hp.) provides primary coarse reduction of whole tires for reclaiming at a 20 ton/hr. rate.

Slitting Cutters. Several specialized cutting units are available for the specific purpose of transforming continuous webs of paper, textiles, film, metal foil, etc., into strips or thin rolls of uniform thickness (or trimming edges to ensure uniform full-width rolls). The web may be slit prior to final rolling, or the finished roll subdivided after final winding. Web slitters employ three principal methods:

1. *Rotary Shears.* Sets of sharpened disks spaced on parallel power-driven shafts above and below moving sheet and overlapping slightly at the point of cutting contact (Fig. 63).
2. *Top Slitters.* Thin steel disks with serrated periphery, spaced on powered shaft operating against winding roll (Fig. 64).
3. *Score Cutters.* V-edged slitting wheels idling under pressure against driven platen roll with hardened steel surface over which the sheet travels (Fig. 65).

Rotary shear pairs must be kept in perfect alignment with disks sharp and ground to uniform diameter to avoid fraying and uneven cutting. The top slitter is not applicable to heavy sheets, and its saw action is a source of dust. The score-cut method, depending on pressure rather than shear, avoids fraying, does not require highly sharpened disks, and is applicable to both thin and heavy

Table 62. Applications of Score-cut Method

Model	Working width, in.	Min. slit width, in.	Speed, ft./min., max.	Min. hp.	Application
26-6PL	25 – 55	¼	250	5	Pressure-type adhesive tape
6-2C	42½– 62½	½	150–450	5	Lined rubber tape or belting, plain or frictioned
26-4D	32 – 50	¼	250	5	Friction tape for electrical insulation
10-11	42 – 82	⅜	1500	10	Plain, waxed, gummed, and coated paper and board
18	73 –142	2	1200	20–40	Board stock from paper machine
18	73 –162	2	2000	20–40	Winding paper directly from high-speed paper machine
24-7	26 – 50	⅜	250	5	Cellophane, Kodapak, Pliofilm, etc.

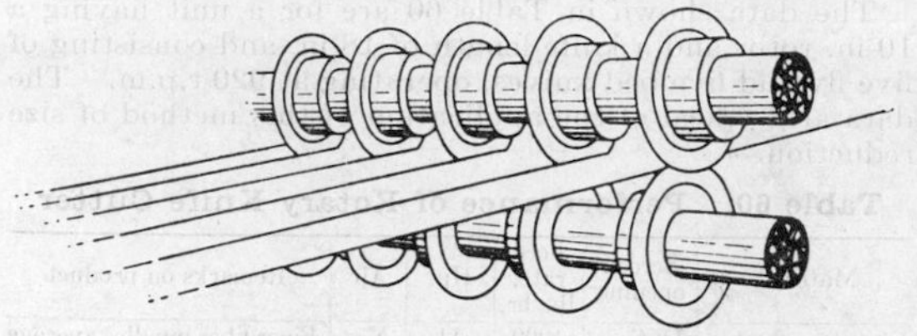

FIG. 63. "Rotary shears" mounted on opposite power-driven shafts.

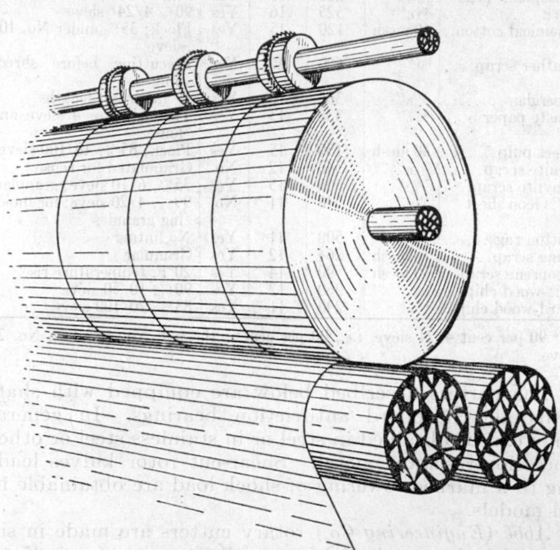

FIG. 64. "Top Slitter"—material slit after it is on the roll.

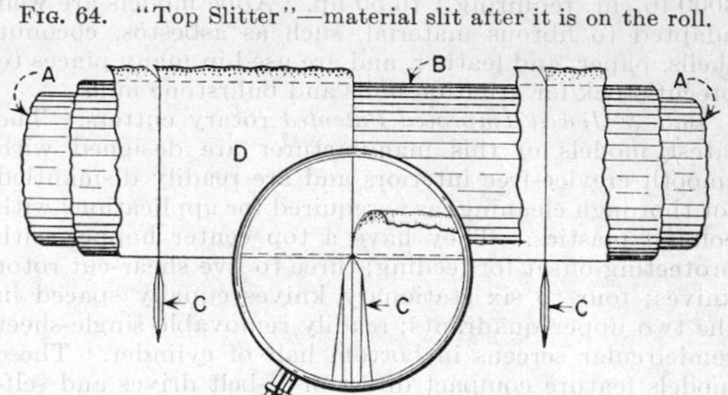

FIG. 65. Camachine score-cut method.

sheets whether heavily coated or filled, surfaced with adhesive, etc.

Camachine (*Cameron Machine Co.*) offers a series of web slitting units integrated with winding equipment to produce rolls of uniform density with straight walls free from frayed edges. Most of these units are of the score-cut type. Their applications are characterized by the examples given in Table 62.

The *Camachine* score-cut method is illustrated in Fig. 65. The core shaft for mounting sleeve sections of the slitter roller is shown at *A*. *B* is the flint-hard polished steel surface of the slitter roller, and *C* is the ball-bearing-mounted slitter wheel. *D* is the material being slit.

MACHINING

By J. H. Foote and W. M. Sheldon

REFERENCES: Jones, "Principles of Powder Metallurgy," Longmans, New York, 1937. Wulff, "Powder Metallurgy,"

The American Society for Metals, 1942. Allen, *Steel*, **104**, 43 (1939). Comstock, *Iron Age*, **143**, 40 (1939); *Metal Progress*, **35**, 343, 347, 465, 576 (1939). Delmonte, *Modern Plastics*, **16**, 49 (1939). Hardy, *Eng. Mining J.*, **140**, 85 (1939); *Metal Progress*, **35**, 171 (1939); **36**, 57 (1939). Schlecht and Trageser, *Chem. Fabrik*, **12**, 243 (1939). Also editorial report in *Metal Progress*, **33**, 263 (1938).

Preparation of powders directly by machining of cast metal or plastic billets has been found to be advantageous from the standpoint of production costs in the case of the softer metals such as tin, lead, zinc, magnesium, aluminum, solder, and antimony and of various plastics. In addition, it eliminates the explosion hazard that is encountered in the production of certain powders by the atomizing of molten metal. A further advantage is the elimination of surface oxidation that may occur, as in lead alloys.

It is possible in most cases to produce powders down to 8 mesh directly from billets by a machining operation. When powders finer than 8 mesh are required, it is necessary to use suitable grinding equipment in conjunction with screens.

Chips and turnings as produced by the normal machine-shop operations can be used in some chemical operations. These turnings may be further reduced by grinding to sizes finer than about 4 mesh.

In general it will be found that the thinner the chip produced by this billet machining operation, the more easily it can be reduced to a fine powder in the pulverizing equipment. Of even greater importance is the development of fracture lines in the chips. This is accomplished by the cutting tools, which must be so designed as to give the chips the greatest possible distortion as they are removed from the billet by the cutting tool. It has been found that most soft metals cannot be ground with any degree of success if these fracture lines are absent. Lack of sufficient fracture lines is evidenced by the tendency of the chips to form themselves into ungrindable spheres when subjected to pulverization. Normally it is not practical to use lubricants or coolants in the chipping operation because of the difficulty and expense of removing them from the chips. Heavy flat chips tend to produce high-density powders, and low-density powders are usually obtained from thin curled chips.

The more commonly used methods of producing chips are as follows:

Filing Machines. This type of machine consists of a rotating built-up cylinder of circular files. The material to be chipped is generally in the form of extruded rods that are held by suitable means against the surface of the files. This machine produces a fine chip of needle shape having a large surface area that lends itself to further size reduction. The production by this machine is relatively low, and the temperature rise in the material produced is moderate. Care should be exercised in selecting a file cut so that the teeth will not blind with the type of material being chipped.

Saws. Saws are practical for producing chips that are to be used as they come from the saw. They are not readily grindable. Trouble may be experienced in the blinding of the saw teeth, especially when saws having fine teeth are used or when the softer metals are cut. The size and character of the chips may be altered by varying the thickness of the saw blades and the number and shape of teeth and their "set."

Shaving Machines. Shaving is an operation that can be used on any type of material. In this operation special cutters may be used to obtain products of varying characteristics. The biggest drawback of shaving processes is the fact that considerable time is lost when the cutter returns and the stock is advanced into position for the next cutting stroke. With this method it is possible to obtain long curly, stringy chips like wood excelsior. These machines are capable of developing internal fracture lines for efficient grindability.

Milling Machines. Milling operations employing either standard milling cutters or cutters of special design have proved fairly satisfactory for producing chips. The billet as presented to the milling cutters may be held stationery or be revolved. If straight-tooth milling cutters are used, a large and very tightly curled chip is obtained, which, when ground, produces a particle having a shape like a sliver that is difficult to screen. However, chips made with fine-tooth cutters are friable and lend themselves to subsequent size reduction. The only disadvantage is that these fine cutters blind easily and generate considerable heat. Cutting speeds as high as 3000 ft./min. have been used with success. The maintenance of sharp cutting edges is of the greatest importance. Very hard cutting tools of tungsten carbide or similar alloys have resulted in many cases in an increase of 1000 per cent in the life measured in operating hours, over cutters made of tool steel.

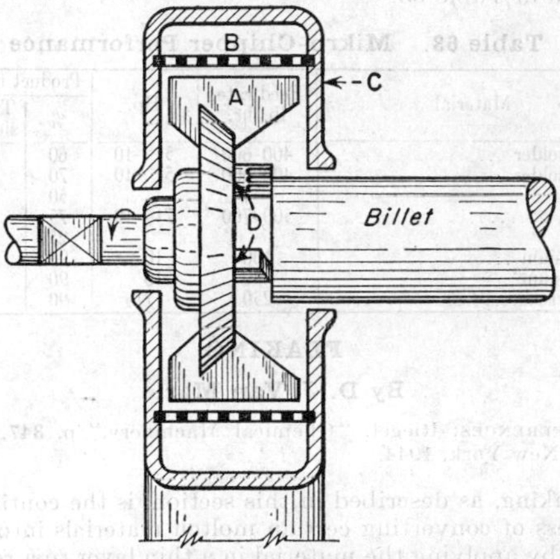

FIG. 66. Mikro-Chipper. (*Pulverizing Machinery Co.*)

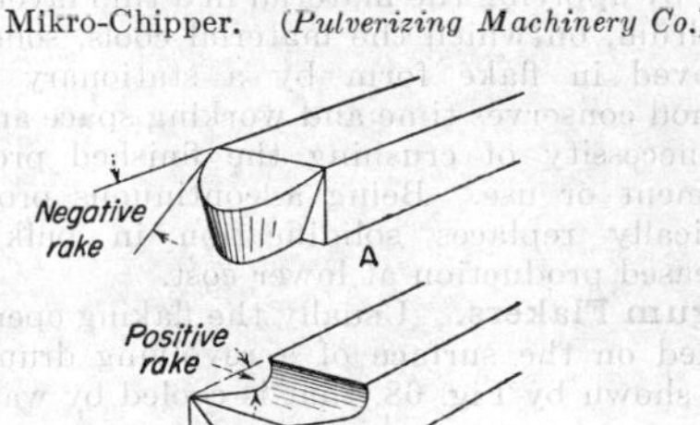

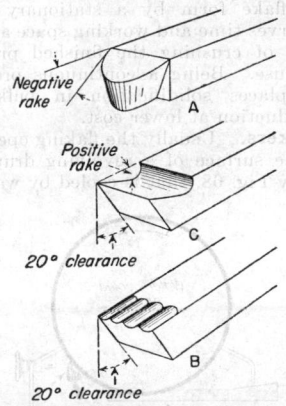

FIG. 67. Cutters for Mikro-Chipper.

Turning. This is an effective method of producing chips or turnings suitable to further size reduction. It has had widespread use in recent years. A typical machine employing this method is the *Mikro-Chipper* shown in Fig. 66, which consists of a high-speed spindle mounting a cutter face plate. The material, in billet form, to be chipped by the cutters *D* is mounted on a carriage which advances at a predetermined speed. The chipped material is broken by blades *A* on the periphery of the face plate and discharged through a screen *B* held in a volute casing *C*. Since the chips are thus broken, they are given a reduction equivalent to a primary grinding stage, thus producing a size distribution of which as high as 50 per cent is material able to pass through a No. 200 sieve.

It is desirable to use cutters of tungsten carbide or similar alloys that permit cutting speeds as high as 7500 ft./min. without resorting to the use of coolants or lubricants. Figure 67 illustrates three types of cutters that have been used successfully in commercial operations. Cutter *A* has been particularly successful in chipping soft solder and materials of similar nature. Cutter *B* has given excellent results when chipping less ductile materials such as tin, zinc, antimony, magnesium, etc. Cutter *C* gives best results when chipping plastics. Care should be exercised that the clearance angle is sufficient to ensure that the cutter does not rub on the billet surface being cut.

Performance characteristics for the Mikro-Chipper are shown in Table 63.

Table 63. Mikro-Chipper Performance

Material	Feed rate, lb./hr.	Hp.	Product fineness %	Product fineness Through sieve No.
50:50 solder	400–600	5 –10	60	50
20:80 solder	400–600	5 –10	70	50
			50	100
Tin	500–700	7½–10	75	100
			25	200
Magnesium	100	10	90	20
Magnesium	30	10	90	50
Polystyrene	250	7½	90	12

FLAKING

By D. J. Van Marle

REFERENCES: Riegel, "Chemical Machinery," p. 347, Reinhold, New York, 1944.

Flaking, as described in this section, is the continuous process of converting certain molten materials into flake form by applying the material in a thin layer to a revolving drum, on which the material cools, solidifies, and is removed in flake form by a stationary knife. This method conserves time and working space and eliminates the necessity of crushing the finished product before shipment or use. Being a continuous process, it economically replaces solidification in bulk and offers increased production at lower cost.

Drum Flakers. Usually the flaking operation is performed on the surface of a revolving drum, similar to that shown by Fig. 68, that is cooled by water, brine, or

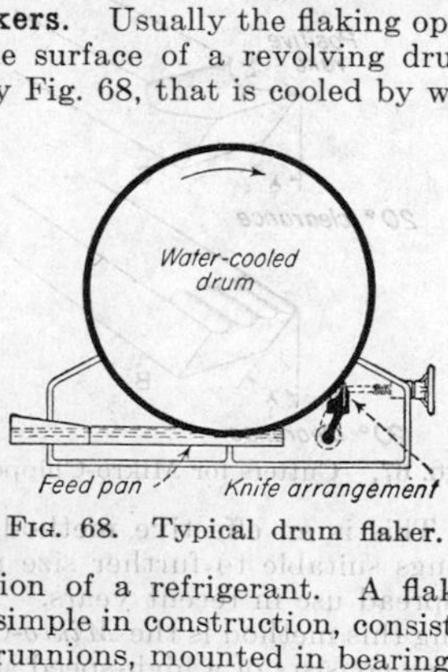

FIG. 68. Typical drum flaker.

direct expansion of a refrigerant. A flaker or cooling drum is very simple in construction, consisting of a drum with hollow trunnions, mounted in bearings. The drum dips in a shallow, heated pan filled with the liquid. A knife or doctor held firmly against the drum removes the product from the surface in solid form. In most cases the layer of solid material at this point is sufficiently brittle to break into flakes or grains. Since the reduction to flake size is obtained by the chipping action of the knife, size of the flakes is not very regular, varying with the properties of the product and the method of operation.

For products which do not adhere readily to a cold drum or which drop off too quickly, a double-drum flaker is preferred to a single-drum machine. A double-drum flaker, shown by Fig. 69, consists of two drums

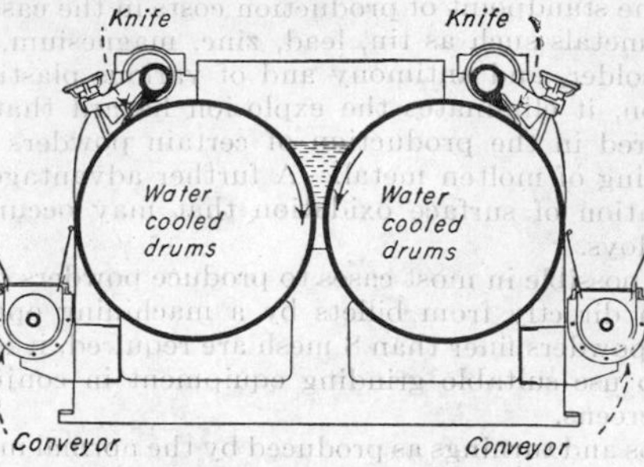

FIG. 69. Double-drum flaker.

placed close together and revolved in opposite directions. The liquid is fed between the drums, being prevented from flowing out at the end of the drums by dam plates or endboards held against the heads. With the drums turning toward each other at the center, a layer is formed on the drum surfaces, the thickness of which is determined by the clearance between the drums. This thickness is limited by the maximum clearance beyond which the liquid begins to run through. There is danger of incrustation of the endboards or of solidification of the liquid in the valley between the drums. If the drums revolve away from each other at the center a heavier layer can be formed.

Drums are usually made of cast iron. Bronze, nickel, stainless steel, or chromium plate are also used. The drums are provided with an internal baffle which leaves an annular space at the periphery of the drum through which the cooling medium flows, entering through one trunnion and leaving through the other; or they are provided with internal spray nozzles. The pan can be made of any suitable metal and can be steam-jacketed or provided with a heating coil for products whose melting points are not too high. Otherwise it is heated directly, generally by gas, to prevent any solidification of the product in the pan. Sometimes the pan is made adjustable or able to be dumped quickly to prevent freezing of the liquid in case of shutdowns. An overflow pipe also can be provided to permit circulation of the liquid through the pan and to maintain a constant level of the liquid. This level can be regulated by providing the overflow pipe with an adjustable nipple or by means of a liquid-level control of the capacitance or other suitable type.

When used on organic materials, drum flakers often require a dust- and fumetight ventilated enclosure, with doors and observation glasses to provide accessibility to the pan and knife, as well as observation of the liquid level.

Revolving tables and traveling metallic belts have been used for flaking. Water is sprayed against the bottom to carry off the heat. These designs lend themselves better to positive control of the flake thickness and to the production of very thick flakes.

Applications of Drum Flakers. Operation and construction of the flaking machine are so simple that it can be applied to practically all chemical products that have a definite but not too low a melting point, both inorganic and organic, including salts that melt and dissolve completely in their water of crystallization. Such products include caustic soda with a melting point of 318°C., caustic potash, sulfur, 70 to 75 per cent calcium chloride, magnesium chloride, 60 per cent sodium sulfide, sodium acetate (U.S. Patent 1,911,479), trisodium phosphate, sulfur, beta naphthol (U.S. Patent 1,594,390), benzidine (U.S. Patent 1,591,688), naphthylamines, phthalic anhydride (U.S. Patent 1,817,304), and paradichlorbenzene with a melting point of only 53°C. Flaked ice is made by the same operation on a specially designed flaking machine, described on p. 1699. Calcium chloride flakes are liable to cake to such an extent that it has been found desirable to subject them to surface dehydration (U.S. Patent 1,527,121). The flaking operation can also be applied to products of more indefinite composition such as asphalt, pitch, paraffin, various waxes, and stearic acid. Although not strictly a flaking operation, the cooling of lard on a drum surface cooled by artificial refrigeration may be included [*Chem. & Met. Eng.*, **31**, 699 (1924)].

Operating Factors. Depending on the individual properties of the material and conditions under which the flaking operation is carried out, the cooled product will be in the form of flakes, generally of irregular shape, or more or less granular. Some products adhere more strongly to the drum surface than others and require considerable knife tension for their removal. The degree of adherence also varies with the metal of which the drum is made, as well as with the polished condition of the drum surface, affecting thereby the size of the flakes to some extent as well as the capacity. For one product this was increased from 45 to 55 lb./(hr.)(sq. ft.) by changing from a cast-iron to a bronze drum. Wetting of the drum surface may assist in removing the solid product as has been proposed for niter cake (U.S. Patent 1,312,430).

Flakes become thicker if depth of the liquid is increased, or if drum speed, cooling-water temperature, or liquid temperature are reduced. Best results are obtained when the temperature of the liquid in the pan is close to the solidifying point in order to make the material solidify quickly when it comes in contact with the drum. Knife maintenance is important, especially if the drum is made of nickel, stainless steel, or other ductile metals which are scored readily. As the knife wears, a heel forms at the edge which should be removed by daily honing or frequent grinding. Otherwise more and more tension may have to be applied to remove the flakes, which increases the wear of the knife and may result in serious damage to the drum, even of a surface as hard as that of a chromium-plated drum.

Drum Cooling. Generally, it is not necessary to maintain a uniform temperature of the cooling water because a large enough temperature difference exists at the point where the flakes are removed from the drum surface. The water, in most cases, flows parallel to the axis of the drum, while the temperature of the material varies along the circumference. Therefore only a small temperature increase of the water can be allowed and a liberal flow of water must be maintained. With spray nozzles, this temperature gradient of the water parallel to the axis is eliminated. The water is applied to the internal drum surface more positively and with greater velocity, resulting in a higher rate of heat transfer and increased capacity. Since spray nozzles require a considerable supply of water, recirculation is advisable if water economy is important. If water is recirculated through an overflow tank, water temperature can be kept at a maximum consistent with satisfactory operation and can be kept uniform throughout the year by regulating the amount of makeup water. Occasionally the water temperature must be more closely controlled, better results being obtained with water that is not too cold. In that case the water can be circulated through an overflow tank and only enough cold water admitted to maintain the desired temperature. Such a procedure is recommended if the material, such as DDT, is subject to supercooling and does not solidify rapidly enough, when cooled too quickly well below the solidifying point. If cold water of uniform temperature is required, or water economy becomes important, the necessary flow may be provided by recirculation of the water through a tubular or vacuum cooling system. Low temperature is required only for products of very low melting point. Ordinary seasonal fluctuation in cooling-water temperature, in most cases, is not objectionable. Such fluctuation does have its effect on capacity, sometimes causing a difference of about 20 per cent between summer and winter production. If necessary, drum speed can be adjusted to compensate for the variation in water temperature. Insufficient flow of water, leaving the drum at too high a temperature, affects the size of the flakes. In case of a hydrated inorganic salt, with a melting point of 77°C., flakes were much larger when the cooling water left the drum at 20°C. than when it left at 40°C. Flakes were slightly thinner at higher water temperature.

Increased level of the liquid increases the length of travel of the drum through the liquid, thereby giving an opportunity for a larger amount of the liquid to adhere to the drum, increasing the flake thickness and capacity.

Drum Capacity and Flake Thickness. Unit capacity of flakers varies a great deal. In view of the moderate values of latent heat of fusion compared with latent heat of evaporation and of specific heat in the solid state, the amount of heat to be transferred through the drum surface is not very great. Consequently, capacity is high, varying from as low as 10 lb./(hr.)(sq. ft.) of drum surface to as high as 150. Drum speed has a greater effect on flake thickness than on capacity for the simple reason that greater speed reduces the time of contact between a point on the drum surface and liquid. This results in a decrease in the amount of product adhering to the surface at any point, making a thinner flake. At the same time hourly capacity increases. Speed of the drum may therefore be regulated to produce a flake of desired thickness or to obtain maximum capacity. Depending on the product, drum speed varies from 1 to 20 r.p.m. With one product a capacity of 10 lb./(hr.)(sq. ft.) was obtained at 0.2 r.p.m. producing flakes 1/8 in. thick. At 1 r.p.m. the thickness was reduced to about 1/64 in., but capacity was increased to 14 lb./(hr.)(sq. ft.). Another product at 1 r.p.m. formed a thick sheet at a capacity of 7.5 lb./(hr.)(sq. ft.), and at 4 r.p.m. flakes were produced while capacity was increased to 12.5. lb./(hr.)(sq. ft.).

Table 64 illustrates the influence of these factors in flaking an organic chemical.

Table 64. Factors Influencing Flaker Performance

Drum speed, r.p.m.	Product in, °F.	Product out, °F.	Water in, °F.	Water out, °F.	Water flow, gal./min.	Capacity, lb./hr.	Flake thickness, in.
10½	291	95	40	52	12	568	0.015
10½	293	113	40	70	4.1	550	.014
10½	293	126	78	95	7.8	503	.013
5½	298	104	43	52	9.6	440	.017
5½	302	113	60	69	9.3	341	.015
5½	302	140	80	90	8.2	342	Rough

Temperature of the flakes often is fairly high, 40° to 80°C. It varies with the drum speed, increasing in one case from 55°C. at 7 r.p.m. to 70°C. at 9 r.p.m.

Heat-transfer Rates. In regard to heat transfer, it is practically impossible to establish a definite heat-transfer coefficient because of the irregular variation of temperature around the circumference of the drums and the increase of temperature of the cooling water parallel to the drum axis in drums with internal baffles. Even if an average water temperature is taken, no average temperature difference can be calculated. Three definite stages occur in the conversion of the liquid into the solid state. First the liquid is cooled to the solidification point. In the next stage solidification takes place at constant temperature with absorption of the latent heat of fusion by the cooling water. Finally the solid product is cooled. Under actual conditions these stages possibly overlap, the layer next to the drum surface being a stage ahead of the outside layer. Heat transfer, for this reason, can be expressed best in B.t.u. per hour per square foot of drum surface. It is easily determined by measuring the flow and the temperature increase of the water. On this basis heat transfer was found to be about 2500 B.t.u./(hr.)(sq. ft.) for a wax with a melting point of 80°C., fed at 90°C., and cooled to 45°C. For a hydrated salt melting at 77°C., fed at 95°C., and cooled to 55°C., heat transfer was 5400 B.t.u./(hr.)(sq. ft.), and for an organic chemical with a melting point of 130°C., fed at 170°C., cooled to 65°C., heat transfer was about 6500 B.t.u./(hr.)(sq. ft.). Heat-transfer rates of 30,000 to 40,000 B.t.u./(hr.)(sq. ft.) may be obtained with products of high melting point, such as caustic soda.

Costs. Owing to the high capacity of drum flakers, flaking cost per pound of product is low, consisting of a small amount of heat to keep the product in the pan in a liquid condition; power to revolve the drum amounting to 0.1 to 0.15 hp./sq. ft. drum surface; power to pump the liquid to the pan if necessary and to pump the water through the drum; and water to carry off the heat liberated. Operation is practically automatic as far as the flaker itself is concerned. Labor is required only to package the flakes. Cost of the equipment is very reasonable in view of the high production obtained. In cast-iron construction this cost varies from $60 to $70/sq. ft. drum surface in accordance with the size of the machine.

Auxiliary Equipment. Flaking apparatus is practically self-contained. Auxiliary equipment is of standard design. If gravity flow of the liquid to the feed pan is not possible, a pump must be provided for the transfer of the liquid. Submerged centrifugal pumps give satisfactory service for the purpose. In case a constant level is maintained in the feed pan by means of an overflow pipe, a circulating pump is provided. Preferably the flaker is placed at a higher level than the liquid storage tank, liquid being pumped into the feed pan and the overflow returned to the storage tank by gravity.

On the discharge side of the flaker, automatic packing and weighing machinery can be employed to advantage. A breaker attachment can be made part of the flaker if it is necessary to reduce the size of the flakes as they are removed from the drum surface. Where artificial refrigeration is needed, standard refrigerating practice can be applied.

Many flakers are fitted with screw conveyors for transporting flakes to the delivery point. These conveyors may help to break up overly large flakes, such as 3 by 4 in., if they occur, but also may cause undesirable fines.

EXPLOSIVE DISINTEGRATION

By J. I. Yellott

REFERENCES: Dean and Gross, Explosive Shattering of Minerals, *U.S. Bur. Mines Report Investigations*, 3118, 1932; 3201, 1933; Progress Reports, Metallurgical Division, *U.S. Bur. Mines Report Investigations*, 3223, 1934; 3268, 1935; 3306, 1936; 3331, 1937. Gross, Crushing and Grinding, *U.S. Bur. Mines Bull.* 402. Poulter and Wilson, The Permeability of Glass and Fused Quartz to Ether, Alcohol, and Water at High Pressure, *Phys. Rev.*, **40**, 872 (June, 1932). Godwin, Coal Pulverization by Internal Explosion, *Armour Research Foundation Report*, August, 1939. Meigs, Explosion, Unit Operation of the Process Industries, *Chem. & Met. Eng.*, **48**, 122 (1941). American Gas Association, Low Investment Cost Gas Production Processes, New York, 1946.

It has long been known that certain materials, particularly those of a fibrous nature, can be disintegrated by first saturating them with a gas or vapor at high pressure, and then releasing that pressure as rapidly as possible. Many references to such processes, particularly in the pulp and paper field, are to be found in the patent literature. More recently an entire industry has been developed by the Masonite Co. in the field of wood products.

Masonite Process. Wood can be disintegrated by an intermittent explosion process. Mason, in U.S. Patent 1,578,609, disclosed the general principle of subjecting wood chips to steam at pressures ranging up to 1500 lb./sq. in. within an enclosed vessel of appropriate strength. After a proper amount of time is allowed for the steam to penetrate and thoroughly heat the wood chips, the batch is discharged by opening a valve at the bottom of the vessel. This allows the steam pressure to force the wood chips into a region where atmospheric pressure is maintained. By this process the wood can be quite thoroughly disintegrated into a mass of separate fibers. The fibers can be recombined into the desired shape through the application of heat and high pressure.

Puffing of Materials. Going to the other extreme in pressure release, the food-products industry is also well acquainted with the process of "puffing" cereals, using batch-type processes with steam at moderate pressure. A leading exponent of this method of preparation uses the slogan, "Food shot from guns." Meigs describes the modification of materials such as grains, scrap rubber, and meat by the explosion technique as well as the separation of hulls and germs from kernels of corn, and the shelling of walnuts.

Explosive Shattering. The U.S. Bureau of Mines, through the work of Gross and Dean, has carried out intensive investigations on the size reduction of ores by various intermittent explosion processes. The early reports of these workers indicated a belief that the water used in the process played an important part, but their later investigations indicate their opinion that it is the rapid, almost explosive, release of the pressure surrounding the ore particles which brings about their disintegration. Pulverization was accomplished in these investigations by filling a cylinder with crushed ore and then subjecting this ore to high-pressure steam. A drawing of the apparatus, as reproduced from *U.S. Bur. Mines Bull.* 402, is shown in Fig. 70. Material enters the explosion chamber through the plug valve and leaves through the quick-opening discharge valve. The cycle is such that the cam closes the discharge valve and stops. Material enters through the plug valve, which is properly synchronized with the discharge valve. Steam enters the chamber, after which the cam operates and the explosion occurs.

Experiments were made at pressures up to 3000 lb./sq. in. The steam apparently entered the pores and interstices of the material. When the pressure in the chamber was released as suddenly as possible by the opening of the valve, the steam that was trapped in the pores, being unable to emerge with sufficient rapidity, expanded within the ore, causing it to shatter along its planes of cleavage.

Carrying this process to an extreme, Poulter has shown that glass rods, when subjected to extremely high pressure in an atmosphere of ether or alcohol, can be shattered by a sudden release of that pressure. In the course of a research in explosive shattering, Godwin of the Armour Research Foundation learned in 1938 that coal can be partly pulverized by saturating it with superheated steam under high pressure (1500 lb./sq. in.) and then releasing that pressure as rapidly as possible by the opening of a valve or the rupture of a diaphragm.

Winterschall-Schmalfeldt Process. A number of somewhat similar processes have been reported in the literature. When a material with a very high internal moisture content must be pulverized, it has been shown in the German literature that a considerable degree of shattering can be accomplished by suddenly introducing this material into a stream of hot gas. By

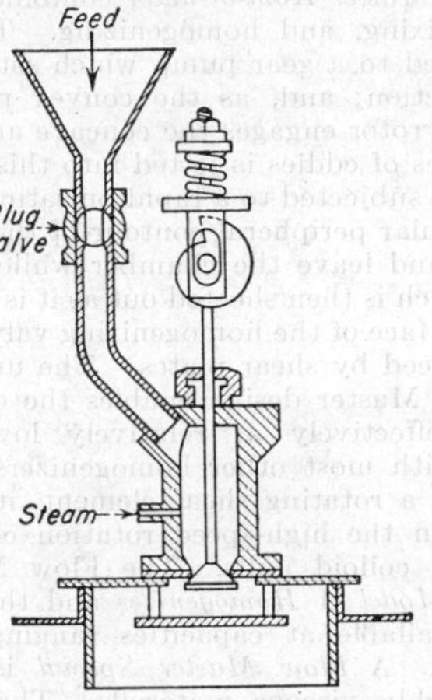

FIG. 70. Sectional diagram of explosive-shattering machine. [*U.S. Bur. Mines*, **402** (1938).]

generating steam within the pores of the material, pulverization is accomplished. The Winterschall-Schmalfeldt process (German Patent 686, 761) is a commercial application of this method of comminution.

EMULSIFICATION

By C. K. Sloan

REFERENCES: Berkman and Egloff, "Emulsions and Foams," Reinhold, New York, 1941. Clayton, "The Theory of Emulsions and Their Technical Treatment," Blakiston, Philadelphia, 1943. Alexander, "Colloid Chemistry," vol. VI, Reinhold, New York, 1946.

Characteristics of Emulsions. Emulsification is that process of size reduction in which two or more immiscible liquids are intimately mixed, one as the dispersed or discontinuous phase, the other being the dispersing or continuous medium. In most emulsions of interest, one of the immiscible phases is usually aqueous in nature. Milk is an example of an oil-in-water type of emulsion, while mayonnaise and unworked butter are examples of water-in-oil emulsions.

The particle size of the dispersed phase of most common emulsions is of the order of 1 to 10 μ diameter. Such particles are not visible to the eye even in the coarser emulsions where the size is in the upper part of this size range. However, even the particles of the finer emulsions in the 1-μ range can be readily resolved under a microscope at moderate magnification (at about 500×).

Emulsification Equipment. The design of specialized equipment for the production of fine emulsions emphasizes the principle of subjecting the liquid mixture to a vigorous shearing action rather than to the impacting action often stressed in many processes for the size reduction of solids. Such equipment for manufacture of emulsions includes colloid mills and homogenizers. The colloid mill subjects the liquid mixture to the shearing action by passing the fluid between two surfaces that move at high velocity and at close clearance with respect to each other. In the homogenizer, this shearing action usually is attained by forcing the liquid under pressure through small orifices or between closely clearing but relatively fixed surfaces. Recently, homogenizers have been developed for applying this shearing action by other means. A more detailed description of the several types of emulsification equipment is given below.

Emulsion Stability. Comminution of particle size is more readily accomplished in an emulsion than it is in a suspension because the dispersed phase being worked on in the emulsion is liquid rather than solid. Conversely, maintenance of this finely divided condition after cessation of the application of the mechanical force to the emulsion is relatively more difficult because the liquid globules of the emulsion are subject to complete coalescence on contact, whereas the individual particles of a suspension maintain their identity even after close contact with each other.

The ease of size reduction in emulsification and the maintenance of this condition (stability) are influenced in large measure by a number of factors, including:

1. Interfacial tension liquid-liquid.
2. Electrical charge on the individual globules.
3. Viscosity of film at liquid-liquid interface.

Surface-active agents are widely used in emulsion preparation, their function being to decrease the interfacial tension between the liquid phases and/or to increase the electrical charge on the individual particles. The polar character of the surface-active agent may even determine the type of emulsion, *i.e.*, whether it is oil-in-water or water-in-oil. The so-called protective colloid or emulsoid type of agent, such as gelatin, agar, or polyvinyl alcohol, plays an important role in the preparation and maintenance of the finely divided liquid droplets. In addition to their polar action, these hydrophilic materials impart desirable viscosity to the system, particularly to the viscous film at the liquid-liquid boundary. In restraining particle-to-particle contact, the agents tend to promote a condition of deflocculation. General directions as to the use of these emulsification agents are given in the literature on emulsions, but application to specific systems of interest usually must depend on experimentation with the materials and equipment available.

Homogenizers. The term "homogenization" describes the process of putting incompatible or immiscible components into a stabilized suspension in a liquid medium. Many types of equipment have been developed to perform this function for accomplishing the dispersion of either liquids or relatively soft solids.

Most conventional types of homogenizers function by passing the product under pressure between closely clearing but relatively fixed surfaces. The high velocity, hydraulic shear, pressure release, and impact rend the dispersed phase into a very fine state of subdivision of the order of 1 μ in diameter (about 1/25,000 in.). The machines are rugged in construction and are capable of continuous high-volume production. The machines were designed originally for the homogenization of milk and dairy products, but application has developed in many other fields.

Most of the machines function by pumping the fluid mixture under high pressure to a narrow opening between a valve plug and its seat, the size of the opening being controllable. In the Cherry-Burrell *Viscolizer* and in the *Gaulin Homogenizer*, the fluid is actuated by a reciprocating pump comprising a series of multiple plungers operated by an eccentric driving shaft to force the liquid through the valve openings. The product develops a very high velocity as it passes through the extremely small opening. As a result of this velocity and of the frictional drag of material in actual contact with the surfaces of the valve plug and seat, marked internal shear develops within the product, tending to break down the particle size of the dispersed phase. An explosive effect also occurs as the particles are released from the zone of high pressure to that of atmospheric pressure. The dispersion is then completed by impact of the outlet streams. The valves are designed to impart desirable homogenization characteristics. The *Split Flo* valve of the *Viscolizer* shown in Fig. 71 is said to permit the use of somewhat

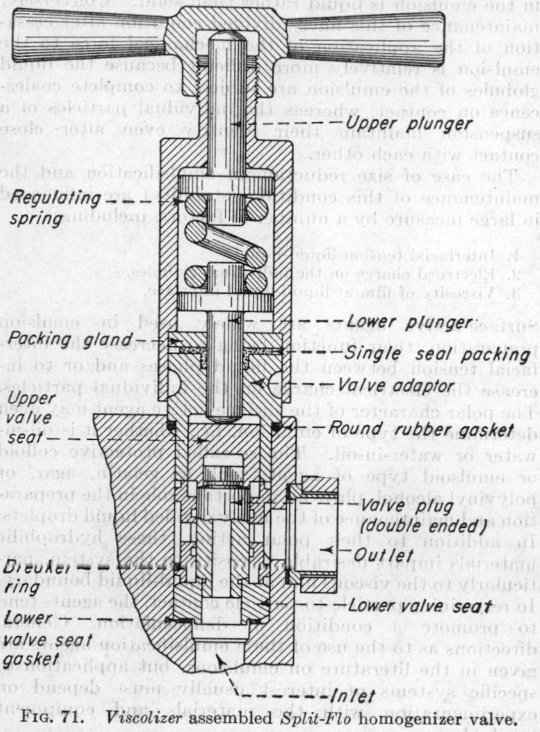

FIG. 71. *Viscolizer* assembled *Split-Flo* homogenizer valve.

lower pressures, *i.e.*, of the order of 1000 to 3000 lb./sq. in. The "two-stage" valve of the Gaulin homogenizer is said to increase the effectiveness in preventing clumping and coalescence in the product. Consideration is given to sanitary requirements in the design of the machines. Material coming in contact with the product is made of either stainless steel or alloy.

The *Viscolizer Model* 41 is available in sizes having capacities ranging from 300 to 1500 gal/hr. *Junior Viscolizer* units are available at 75, 125, and 200 gal./hr. capacities. These smaller units are equipped with automatic viscosity control and are said to be conveniently cleanable. The Gaulin homogenizer is available at capacities ranging from 200 to 2500 gal./hr. A *Manton-Gaulin Laboratory Homogenizer* is available with a capacity of 25 gal./hr. and is said to be able to handle small quantities of material of the order of 1 pt. of product. This laboratory mill is compact and rugged and is constructed of stainless material. It can be actuated by a 1.5- to 2-hp. motor operating a small plunger at 95 r.p.m. to give pressures of the order of 4000 to 5000 lb./sq. in. The pressure is controllable by a handwheel.

The *Marco Flow Master Homogenizer* utilizes a gear pump for applying pressure to the product. The machine has application in the manufacture of emulsions of two or more liquid phases as well as for the size reduction and dispersion of many soft solid materials in a liquid phase. In addition to the action of forcing the material through the homogenizing valve, the meshing of the gears serves to macerate the solid material. Although this device is not suitable for dry grinding, it is said to be able to grind to micron size any product that contains elements that will act as the vehicle to carry the solids through the multitude of shearing actions that take place in the grinding action of the head. This action is particularly suitable in preparation of products in pulp form.

The *Flow Master Kom-bi-nator* combines the action of grinding, mixing, and homogenizing. The material is first subjected to a gear pump which sets up a swirling turbulent action; and, as the convex portion of each tooth in one rotor engages the concave area of the other rotor, a series of eddies is jetted into this swirling body. The liquid is subjected to a rapid pulsating action caused by the irregular peripheral contour of the rotor teeth as they enter and leave the chamber while delivering the product, which is then sheeted out as it is forced between the seat and face of the homogenizing valve. This valve can be replaced by shear plates. The unique character of the Flow Master design enables the operation to be conducted effectively at relatively low pressures as compared with most other homogenizers. Although it may employ a rotating shear element, its action is not dependent on the high-speed rotation characteristic of conventional colloid mills. The Flow Master line includes the *Model A Homogenizer* and the *Kom-bi-nator* that are available at capacities ranging from 200 to 1000 gal/hr. A *Flow Master Special* is available for handling highly viscous materials. The units are designed for sanitary requirements and are constructed of stainless materials.

In contrast to the conventional design of homogenizers comprising a device for forcing the product through a small opening, other machines have recently been announced that employ other devices for obtaining the turbulent action desired in promoting homogenization. One of these is the *Cornell Versator*, which comprises a vacuum or pressure chamber in which a 26-in.-diameter open-bowl disk rotates at high speed (900 to 1800 r.p.m.). The feed material enters through a 2½-in. line and is introduced at the center of the rapidly spinning disk by a spring-loaded film-forming ring which can be set to apply the film to the disk at an initial thickness of a few thousandths of an inch. The centrifugal forces tend to throw the applied material out radially and up the disk. Inasmuch as the material wets the disk, it also tends to rotate with the disk. The applied material therefore travels around the disk in ever-widening spirals, reaching maximum velocity and minimum thickness at the outer edge of the spinning disk. In its path uphill, the film thickness is rapidly and progressively attenuated, tapering from a few thousandths of an inch near the center to only a fraction of a thousandth of an inch at the periphery. With the high spinning velocities, the frictional drag is pronounced, and large shearing forces are created without the application of mechanical pressure.

Flow within the film is extremely turbulent, even though the film is, for all practical purposes, two-dimensional or "all surface." This turbulent action in the film of the Versator enables the machine to be used for several operations, including mixing, emulsifying, and homogenizing. The machine may also be used as a means of aerating or deaerating by adjustment of the internal atmosphere. The *Versator* requires a minimum feed of about 7 gal./min. but has been operated at as high as 100 gal./min. For batch operation, as little as 10 to 15 gal. material may be used. The power requirement is largely determined by the viscosity of the material being treated. For material of the viscosity of heavy lubricating oil, a 7½-hp. motor is sufficient; whereas a 25-hp. motor may be required for processing semiplastic material.

The *Submarine Signal Sonic Oscillator* (*Raytheon Mfg. Co.*) employs a different principle and has been developed for the homogenization of milk. Research on the effect of powerful vibrators for production of underwater signals has been applied to the sonic treatment of milk for improving digestibility. The Sonic Oscillator consists of a stainless-steel diaphragm approximately 2 ft. in diameter, which is vibrated electromagnetically at a frequency of 360 vibrations/sec. A stainless-steel cover plate is mounted over and close to the diaphragm, forming a chamber therewith. The milk to be treated is introduced into this chamber at its outer edge, from which it flows radially across the diaphragm to the outlet opening at the center. In so doing, it is subjected to the intense vibration of the diaphragm at the rate of 250 gal./hr. The motor generator requires about 5 kw. for operation.

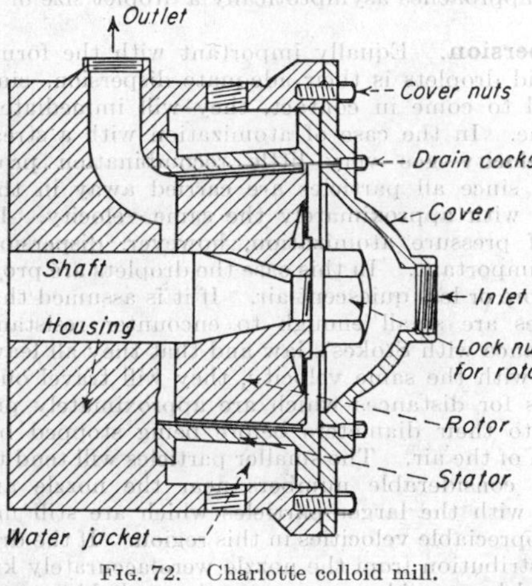

FIG. 72. Charlotte colloid mill.

Colloid Mills. In colloid mills, size reduction is accomplished by gravity feed of the material to a narrow opening between two surfaces that move at high speed with respect to each other. In some mills the opening is adjustable, and in others it is fixed. The smallest openings are of the order of 0.001 in. Most mills have a rotor and stator. The mills may differ in the character of the surfaces. Some are grooved and others are smooth.

The *Charlotte* mill has a grooved conical rotor that moves within a grooved stator at a clearance regulated by a calibrated external adjustment. A sectional drawing is shown in Fig. 72. The whirling currents set up within the grooves subject the product to both hydraulic shear and impact. All models operate at 3600 r.p.m. The following sizes are available:

Hp.	Capacity, Gal./Hr.
3	20– 50
7	50– 100
20	100– 400
50	400–1000
75	1000–5000

A laboratory model W-10 operates at 1 hp. with a capacity of 1 to 50 gal./hr. The mills are available in several materials, including stainless steel, nickel, monel, bronze, and cast iron. A special model ND is designed for mayonnaise and salad oils. Sanitary models are available for processing foodstuffs.

The rotor of the *Premier* mill is shaped like the frustum of a cone. Surfaces are smooth, and adjustment of the clearance can be made from 0.001 in. upward. The mill is jacketed for temperature control. Direct-connected liquid-type mills are available with 15- and 21-in. rotors. These mills operate at 3600 r.p.m. at capacities up to 3000 gal./hr. They are powered up to 100 hp. The direct-connected paste-type mills have 6- and 12-in. rotors operating at speeds of 1800 to 3600 r.p.m. The power requirement and capacity vary considerably, depending on the material processed. The range of power is 1½ to 50 hp. and of capacity is 5 to 1000 gal./hr. The geared-head high-speed paste-type mills have 6- and 10-in. rotors and operate at speeds up to 9000 r.p.m. with power of 3 to 30 hp. For pilot-plant operation, the *Premier Three-Purpose Laboratory Mill* is available with 3- and 4-in. rotors. These mills are belt-driven and operate at 7200 to 17,000 r.p.m. with capacities of 5 to 150 gal./hr. *Laboratory Mill No.* 200 has a 2-in. rotor and operates at 13,500 r.p.m. at a power of 1½ hp. and capacity of 2 to 25 gal./hr.

The *Noblewood Mill* employs on a large scale the principle of the colloid mill, *i.e.*, high-speed rubbing or fluid shear. The mill has a grooved rotor 26 in. in diameter and 12 in. wide which operates at a shaft speed of 1800 r.p.m. and a surface speed of 12,000 ft./min. The opening between rotor and stator is 0.0007 to 0.0013 in. The mill utilizes an 800-hp. motor and is used to convert daily 10 to 100 tons of ground wood waste into fibrous pulp for newsprint paper. The output is 30 to 600 gal./min.

The *Gaulin Colloid Mill* has a smooth rotor, shaped like a discus. Entering material is first thrown outward along the disk and then around the edge and inward, thus giving a *two-stage* action. The gap setting between the rotor and housing is adjustable down to 0.001 in. The rotor is made of stainless steel and operates at 3600 r.p.m. The mill is jacketed for temperature control. Operating characteristics are given in Table 65.

Table 65. Colloid Mill Operating Characteristics

Rotor size, in.	Hp.	Capacity, gal./hr., at 0.001-in. setting	
		Water	Mineral oil
2	1½	27	8
4	5	60	18
8	10	120	36

Applications of Colloid Mills and Homogenizers. Although, as has been mentioned, homogenizers were originally used commercially for processing dairy products, use has spread to a number of industries. Many branches of the food industry now use such equipment to improve their products. Purées, food pastes, food pulps, sirups, sauces, and the like, are prepared at many plants by homogenization—with improvement in appearance, storage stability, and the palatability that accompanies the increase in surface of the dispersed phase. Many pharmaceutical materials, including ointments, oil emulsions, creams, shaving soaps, lotions, and suspensions such as milk of magnesia are improved by such treatment. The components of preparations containing essential oils and perfumes are more effective if the system is homogenized. Use in the chemical industry has shown the advantage that often appears with increase in interfacial area between immiscible phases. Wax emulsions, asphalt emulsions, color pastes, and the blending of oils and greases are examples of such applications.

SPRAYING

Theory of Dispersion of Liquid Droplets

By R. V. Kleinschmidt

REFERENCES: A very complete résumé of this subject as applied particularly to fuel atomization in internal-combustion engines is given by Castleman, *Bur. Standards J. Research*, **6**, 369

(1931). Other references: Plateau, "Statique expérimentale et théoretique, etc.," Paris, 1873; Scheubel, *Wiss. Ges. Luftfahrt Jahrsbuch*, 1927, p. 140.

Mechanism. The basic mechanism of droplet formation consists in drawing out the liquid into a slender stream or filament. Lord Rayleigh [*Proc. London Math. Soc.*, **10**, 4 (1879) and "Theory of Sound," Chap. XX] has shown that a liquid cylinder is unstable and that any slight displacement will cause it to neck down in places and bulge out in others, eventually collapsing into droplets. Just at the moment of collapse a second set of very much smaller droplets forms from the last filaments of liquid connecting the primary droplets. The phenomenon may be readily observed in a thin stream of water from a faucet and is sufficiently regular to be observed by stroboscopic light. The existence of the secondary droplets may also be demonstrated by passing such a stream through a gentle breeze which will deflect the secondary droplets so that they may be caught and studied.

The direct formation of streams of liquid thin enough to produce fine sprays is not usually practical; hence secondary actions are resorted to. Two of these methods are as follows:

1. Filaments of liquid are dragged out by the impingement of high-velocity turbulent air or steam jets on the liquid surface. In this case a primary droplet is dragged away from the surface, carrying behind it a filament which may be stretched out by the velocity and turbulence of the gas to a very fine thread before it collapses to droplets. Such action produces a wide range of particle sizes, many of them exceedingly fine.

2. The other common method of forming filaments is to spread the liquid out into a thin sheet. This sheet then draws up into a filament on its free edge, and this filament, in turn, breaks down into droplets. Such sheets or films of liquid, when projected into air at a high velocity, often exhibit another and extremely interesting phenomenon. The rapid relative motion of the film of liquid and the surrounding air sets up turbulence in the air which, reacting on the liquid film, causes it to wave or flap, just like a flag flying in the breeze. At its free edge, this flapping often becomes so violent that the film actually rolls up and joins itself into a tube which breaks away from the sheet and, being unstable, as a solid filament is, necks down and breaks into droplets. In this case, however, the droplets are hollow, having enclosed a considerable amount of air. Often these particles also enclose other smaller particles previously formed. These hollow particles are usually very thin-walled and present a large surface, which is frequently desirable.

Energy Relations. The energy required to form a liquid into droplets is composed mainly of three parts: (1) energy required to form surface against surface tension, which is simply the surface tension times the additional surface formed. In the case of water at room temperature, the net energy per unit volume required to form droplets 1 μ in diameter, corresponds to a pressure on the liquid of only 0.005 lb./sq. in. (2) Since the time during which droplet formation usually takes place is very short, often a few microseconds, the rate of deformation of the liquid is very high, and viscous forces become enormous. The energy required to produce this deformation is therefore appreciable although not readily computed. (3) There is energy lost due to inefficient application of energy to the fluid. When the energy is supplied directly to the liquid by a pump and released in a well-designed nozzle, this last efficiency is probably high, but, in the case of air or steam jets impinging on the liquid, the transfer of energy is relatively low. Offsetting this is the fact that very large amounts of energy can be applied in a compressible fluid such as air or steam, as compared with that stored in a liquid under pressure. The latter is simply 144 PV ft.-lb./lb. liquid (where P = pressure, lb./sq. in., and V = specific volume of liquid, cu. ft./lb.) or $144P/62.3$, or $2.3P$ for water at room temperature. The available energy of 1 lb. saturated steam at 60 lb./sq. in. gage expanding to atmospheric pressure is 36,300 ft.-lb. In order to concentrate this energy in 1 lb. liquid water, a pressure of over 15,000 lb./sq. in. would be required. Although, in general, increasing amounts of energy applied to the atomization of a liquid tend to produce finer and finer particles, there is probably a theoretical and certainly a practical limit to which this can be carried. As the filaments of liquid become finer and finer, their rate of collapse increases rapidly owing to the high surface energy per unit volume of liquid. The rate and amount of deformation required also increase. Sauter (*Forsch. Gebiete Ingenieurw.*, 1928, No. 312) found that at high air speeds atomization in a certain type of nozzle approaches asymptotically a droplet size of about 6 μ.

Dispersion. Equally important with the formation of liquid droplets is their adequate dispersion, since, if allowed to come in contact, they will immediately recombine. In the case of atomization with a stream of high-velocity gas very little recombination probably occurs, since all particles are carried away in the gas stream with approximately the same velocity. In the case of pressure atomization, however, dispersion becomes important. In this case the droplets are projected into more or less quiescent air. If it is assumed that the particles are small enough to encounter resistance in accordance with Stokes's law and that they all leave the nozzle with the same velocity, they will travel out into the gas for distances which are approximately proportional to their diameters before being stopped by the friction of the air. The smaller particles will tend to collect in considerable numbers near the nozzle and to collide with the larger particles which are still moving with appreciable velocities in this region. If the droplet-size distribution from the nozzle were accurately known, it would be possible to compute the probable amount of recombination due to collisions by the methods outlined by Kleinschmidt [*Chem. & Met. Eng.*, **46**, 487 (1939)]. This is seldom possible, and the only practical value of such considerations is to indicate the importance of maintaining an adequate flow of gas past the nozzle to remove the droplets as formed.

Spray Nozzles

By H. G. Houghton

REFERENCES: Anon., Design of Spray Nozzles, *Engineering (London)*, **159**, 21, 61, 103 (1945). Castleman, The Mechanism of the Atomization of Liquids, *Bur. Standards J. Research*, **6**, 369 (1931). Doble, Design of Spray Nozzles for Outputs up to 1800 Gallons Per Hour, *Proc. Inst. Mech. Engrs. (London)*, **157**, 103 (1947). Doble and Halton, Application of Cyclone Theories to Centrifugal Spray Nozzles, *Proc. Inst. Mech. Engrs. (London)*, **157**, 111 (1947). Fogler and Kleinschmidt, Spray Drying, *Ind. Eng. Chem.*, **30**, 1372 (1938). Lewis *et al.*, Atomization of Liquids in High Velocity Gas Streams, *Ind. Eng. Chem.*, **40**, 67 (1948). Merrington and Richardson, The Break-up of Liquid Jets, *Proc. Roy. Soc.*, **A59**, 1 (1947).

A spray nozzle is a device for breaking up a liquid into drops. The applications of spray nozzles are numerous and varied, and consequently a large number of different forms are in use. All spray nozzles may be classified under one of the following types:

1. *Pressure nozzles* in which the fluid is under pressure and is broken up by its inherent instability and its impact

on the atmosphere or by its impact on another jet or a fixed plate.

2. *Rotating nozzles* in which the fluid is fed at low pressure to the center of a rapidly rotating disk or cup. Centrifugal force causes the fluid to be broken up into drops.

3. *Gas-atomizing nozzles* in which the fluid is subjected to the disrupting effect of a high-velocity jet of gas. There are several forms of each of these types in common use, and these will now be described in turn.

Pressure Nozzles

Hollow-cone Nozzles. The pressure nozzles find the widest field of application and are available in a variety of forms and sizes. The most common of these is the so-called hollow-cone nozzle. In this nozzle the fluid is fed into a whirl chamber through tangential passages or through a fixed spiral so that it acquires a rapid rotation. The orifice is placed on the axis of the whirl chamber, and the fluid exits in the form of a hollow, conical sheet which then breaks up into drops. Such nozzles are illustrated in Figs. 73, 74, and 75. Hollow-cone nozzles

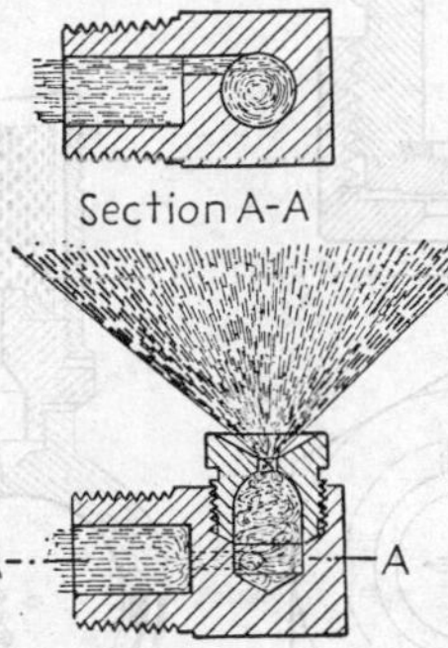

FIG. 73. A small hollow-cone nozzle of the tangential type. (*Spray Engineering Co.*)

are made with orifices from 0.02 to 2 in. in diameter with corresponding discharge rates of from less than 0.01 to more than 200 gal./min. The larger sizes are used for cooling ponds, for washing gravel and sand, aerating water, etc., and are usually operated at relatively low pressures. Smaller nozzles may be used for spray drying, air washers and humidifiers, oil burners, gas absorption, etc., and are usually operated at somewhat higher pressures. In common with all pressure nozzles the capacity of a given nozzle is nearly proportional to the square root of the pressure except at extremely high pressures where friction limits the discharge. Operating pressures do not usually exceed 300 lb./sq. in. except in special cases such as milk-powdering sprays where pressures of from 1000 to 7000 lb./sq. in. are used. For a given design of nozzle the discharge at constant pressure is approximately proportional to the area of the orifice, although the orifice does not run full. The discharge does not vary much with the viscosity of the fluid, until the viscosity is more than ten times that of water, although the drop size is somewhat altered, as will be pointed out below. The included angle of the spray cone usually increases slowly with pressure to a maximum and then decreases, but it is largely determined by the proportions of the nozzle. A spiral with a short pitch produces a wide-angle spray, and conversely a large pitch spiral gives a small included angle. The angle may be from 15 to 135 deg., but it is not always possible to obtain stock nozzles of a desired angle when the pressure and discharge rate are also fixed. Nozzles with a small included angle tend to produce a solid-cone rather than a hollow-cone spray.

Solid-cone Nozzles. The solid-cone nozzle is a modification of the hollow-cone nozzle which is used when complete coverage of a fixed area is desired. Such nozzles are used for certain washing applications, for cooling and aerating water, and for other purposes where the more uniform spatial distribution of the drops is advantageous. The construction and operation of a typical solid-cone nozzle are illustrated in Fig. 76. The nozzle is essentially a hollow-cone nozzle with the addition of an axial jet which strikes the rotating fluid just within the orifice. The break-up is largely due to this impact and the resulting turbulence. The fluid appears to leave the orifice in drop form, whereas in a hollow-cone nozzle a short conical fluid sheet which breaks up outside the orifice is usually observed. To obtain a uniform spatial distribution, it is necessary to design the nozzle so that the proper relation exists between the amount of liquid fed to the center jet, the amount which is rotated, and the orifice size. Normally, more of the fluid is given a rotary motion than is passed through the axial jet. A separate feed line may be connected to the axial jet

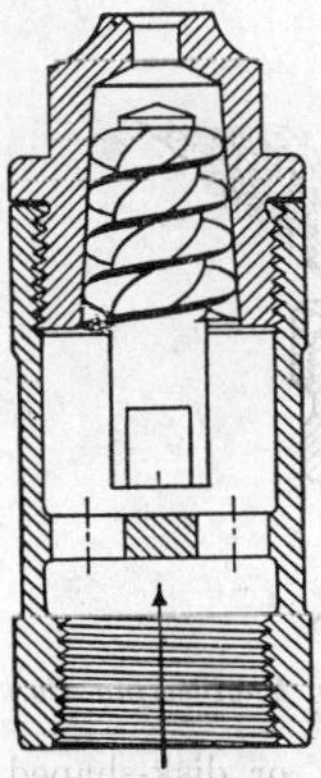

FIG. 74. A small hollow-cone nozzle of the fixed-spiral type. (*Schutte and Koerting Co.*)

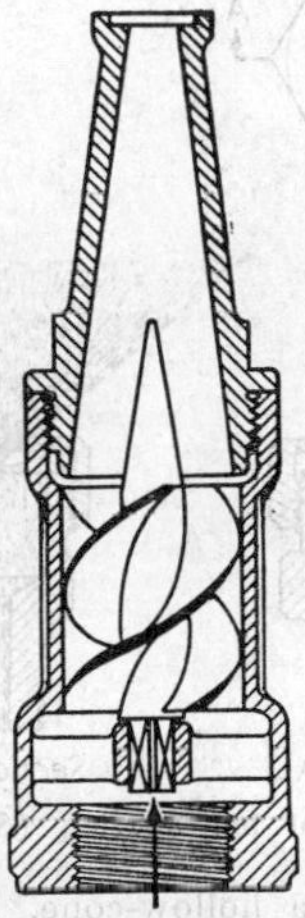

FIG. 75. A large hollow-cone nozzle. (*Schutte and Koerting Co.*)

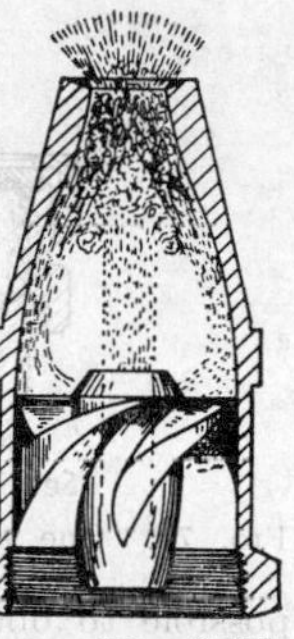

FIG. 76. Solid-cone nozzle. (*Spray Engineering Co.*)

so that two liquids or a liquid and a gas can be intimately mixed. This is often useful for chemical applications.

The included angle of the solid-cone spray is a function of the design of the nozzle and is nearly independent of pressure. Various commercial solid-cone nozzles produce cones with included angles of from 30 to 100 deg. As indicated above, hollow-cone nozzles with small included angles (less than about 30 deg.) give a solid-cone spray without the addition of a center jet. By special design, it is possible to produce a solid-cone spray without a center jet with included angles as large as 100 deg. Solid-cone nozzles are not usually available in such small sizes as are hollow-cone nozzles, but stock sizes have discharge rates from less than 1 gal./min. to several hundred gallons per minute.

Fan Nozzles. A third form of pressure nozzle is the so-called fan nozzle. By means of milled cuts or channels on the rear face of the orifice plate, and sometimes an elongated orifice, or by means of two inclined jets, the fluid is caused to exit in the form of a flat fan-shaped fluid sheet which then breaks up into drops. Typical fan nozzles are shown in Figs. 77 and 78. Owing to surface tension, the edges of the sheet are usually bounded by solid streams or "horns," particularly in the smaller sizes, which may comprise from one-fourth to one-half of the total amount of liquid sprayed. These streams break up into larger drops than the central sheet. The horns are usually not so pronounced in the larger sizes, and for included angles of spray that are less than about 50 deg. Fan nozzles are useful when it is desired to distribute the spray along a line such as in washing, cleaning, coating, or cooling material in a continuous process. The included angle of the fan is from 10 to 130 deg. in standard nozzles, and capacities range from 0.1 to 20 gal./min.

Impact Nozzles. Another type of nozzle which is used for certain special purposes is the impact nozzle. A solid stream of fluid under pressure is caused to strike a fixed surface or another similar stream. By a proper orientation and shape of the plate or by varying the size and direction of the two fluid streams, it is

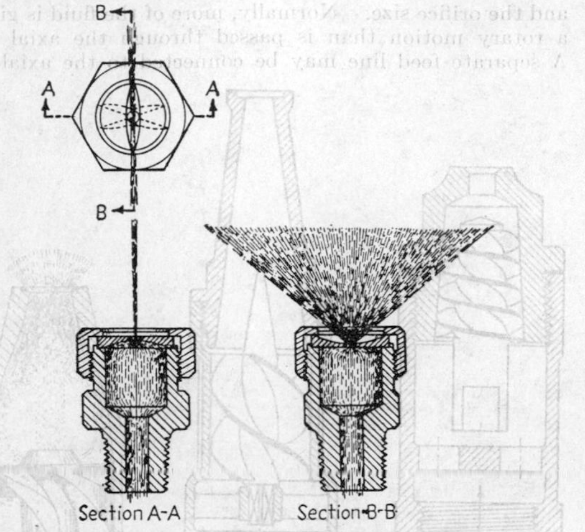

FIG. 77. One type of fan nozzle. (*Spray Engineering Co.*)

possible to obtain a hollow-cone, fan-, or disk-shaped fluid sheet. It has been found possible to produce drops of more uniform size with an impact nozzle than with other types of pressure nozzles if laminar flow is maintained. It is extremely difficult to produce laminar flow in other types of pressure nozzles because of their essential interior parts. The orifices of impact nozzles, on the other hand, may be designed to produce laminar flow if proper precautions are taken. Such laminar-flow impact nozzles are applicable to continuous operations such as gas-washing and chemical reactions between a liquid and a gas, in which the more uniform drop size results in an over-all saving in spite of greater nozzle cost.

Small impact nozzles of the type shown in Fig. 79 are often used in air-moistening equipment.

"Fog" Nozzles for Firefighting. A number of special spray nozzles for fighting fires, particularly oil fires, are now on the market. These are usually pressure-type nozzles designed to produce a dense blanket or "fog" of relatively small water drops. The fire-extinguishing effect is primarily due to the cooling of the burning gases partly by contact with the water drops but principally by the evaporation of the drops. A relatively small amount of water is used, in comparison with an ordinary fire hose, thus reducing flooding and the consequent spreading of flaming liquids. It is common to use a multiple spray head comprising a number of individual nozzles of any of the usual types. This serves to produce small drops and also to form a continuous blanket of spray in a relatively large volume. Such nozzles are operated at pressures of from 50 to 200 lb./sq. in. and discharge up to 200 gal./min.

Pressure nozzles as a class are relatively simple, small, and inexpensive, and they usually consume less power than other types. They may be used with all fluids that have a viscosity less than about 300 to 500 sec. Saybolt and that do not contain solid particles larger than the passages in the nozzle.

Materials of Construction. Pressure nozzles are commonly furnished in cast iron and cast brass or bronze in the larger sizes and in steel, brass, and bronze in the smaller sizes. When corrosion or erosion is important, the nozzles may be formed from any material that can be either machined, cast, or molded. Some of the more common special materials are stainless steel, monel metal, hard lead, ceramics, hard rubber, and glass. When erosion is an important consideration, tips of

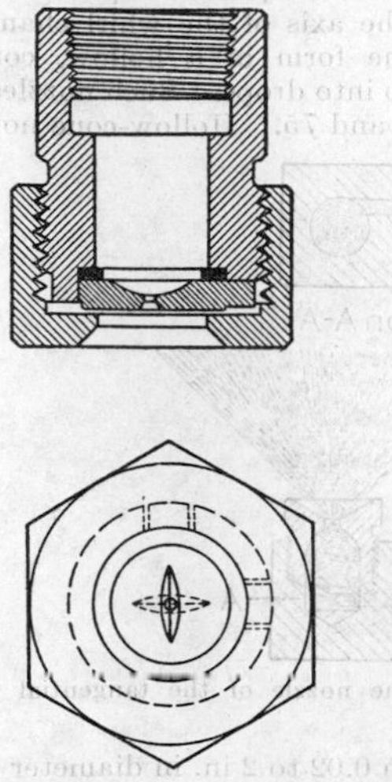

FIG. 78. Another design of fan nozzle. (*Schutte and Koerting Co.*)

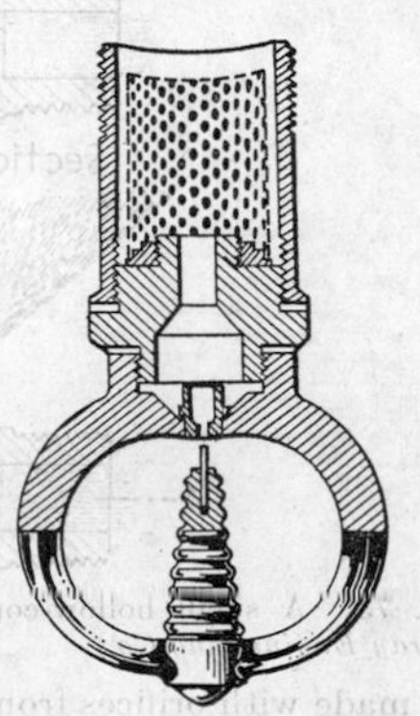

FIG. 79. A small impact nozzle as used in a direct humidifier. (*Parks-Cramer Co.*)

stellite or other hard alloys may be used. Monel-metal nozzles are particularly useful for high-temperature applications. Typical pressure-capacity data for standard pressure nozzles are contained in Table 66.

Table 66. Discharge Rates and Included Angle of Spray of Typical Pressure Nozzles*

Nozzle type	Orifice diameter, in.	Discharge, gal./min., and included angle of spray							
		10 lb./sq. in.		25 lb./sq. in.		50 lb./sq. in.		100 lb./sq. in.	
		Discharge	Angle, deg.	Discharge	Angle, deg.	Discharge	Angle, deg.	Discharge	Angle, deg.
Hollow cone	0.046	......	..	0.10	65	0.135	68	0.183	75
	.140	0.535	82	0.81	88	1.10	90	1.50	93
	.218	1.25	83	1.88	86	2.55	89	3.45	92
	.375	7.2	62	11.8	70	16.5	70		
Solid cone	.047	......	..	0.167	65	0.235	70	0.34	70
	.188	1.60	55	2.46	58	3.42	60	4.78	60
	.250	3.35	65	5.40	70	7.50	70	10.4	75
	.500	17.5	86	27.5	84	38.7	73		
Fan	.031	0.085	40	0.132	90	0.182	110	0.252	110
	.093	0.70	70	1.12	76	1.57	80	2.25	80
	.187	2.25	50	3.70	59	5.35	65	7.70	65
	.375	9.50	66	15.40	74	22.10	75	30.75	75

* Data furnished through the courtesy of the Spray Engineering Co.

Rotating Nozzles

The essential part of a rotating nozzle is a disk or cup which is usually directly connected to an electric motor. The fluid to be sprayed is fed under low pressure to the center of the rotating disk. Various forms of disks are used in an attempt to improve the spraying characteristics. Vanes are often attached to the periphery of the disk or mounted separately a short distance from the periphery to assist the break-up or to remove some of the larger drops. The rotary nozzle is particularly useful for spraying viscous liquids, slurries, and liquids containing solid particles which would clog other nozzles. They are also used in some air washers, in small air-moistening units, and in domestic oil burners. The spray is distributed in all directions in the plane of the disk, and this is often a disadvantage. Disk speeds depend on the application and size of the nozzle and vary from a few hundred to several thousand r.p.m. The quantity of fluid sprayed may be readily controlled over wide limits. Small units may spray only a fraction of a gallon per hour for air moistening, while some large units operate at a discharge rate of 100 gal./min. The size of the drops produced may be varied by changing the speed of rotation and the discharge rate, high speeds and low discharge rates giving smaller drops. Rotating nozzles commonly require somewhat more power to operate than a pressure nozzle for a given application. This is probably due to friction losses between the fluid and the disk and between the fluid and the air. In addition, a pump must often be used to deliver the fluid to the disk. The rotating nozzle is relatively large and expensive and is not commonly used for purposes to which pressure nozzles are equally applicable.

Gas-atomizing Nozzles

In a gas-atomizing nozzle the liquid is broken up by impingement with a high-velocity stream of gas, usually air or steam. The fluid may be fed under pressure, low gravity head, or sucked up by the injector action of the gas stream. The contact between the fluid and the gas may take place entirely outside the nozzle or within a chamber from which the spray exits through an orifice. The shape of the cloud of spray may be controlled by the shape of the orifice in the internal mixing types and by additional gas jets in the external mixing type.

Applications. Gas-atomizing nozzles are used when very small drops are desired. They are also capable of spraying more viscous fluids than pressure nozzles. They are commonly used for spray painting, for air and material moistening, for the application of insecticides, and in oil burners. Except for the oil-burner application the discharge rate of gas-atomizing nozzles is small, seldom exceeding 10 gal./hr. Considerably more power is required to spray at a given rate with a gas-atomizing nozzle than with a pressure nozzle because the fluid is much more finely divided.

Paint Spraying. The application of paint by spraying has become common because of the speed of application and of drying and the ease with which the film thickness may be controlled. Practically any type of paint or lacquer can be sprayed if the proper equipment and thinner are used. Atomizers for paint spraying are available in a wide variety of types to meet the various needs. External mixing is more common, but internal mixing is also used. It is often desirable to have a fan-shaped spray for painting. In external mixing nozzles this is accomplished by two external air jets which impinge on the spray from opposite sides and flatten it out. The degree of flattening may be controlled by varying the pressure of the forming jets. The amount of paint sprayed is commonly controlled by adjusting the travel of the control valve. A typical paint spray gun is shown in Fig. 80. This is an external-mixing type with forming jets to produce a fan spray. Paint spray guns are operated at air pressures of from 20 to 80 lb./sq. in., but 40 to 60 lb./sq. in. is the usual range. The air pressure required depends on the type of gun, the rate of spray, and the viscosity of the paint. Production-type spray guns usually require from 4 to 8 cu. ft. free air/min. at a pressure of 40 lb./sq. in. The paint is fed to the gun by suction, gravity, or pressure. When small amounts of paint are to be sprayed, the paint is usually contained

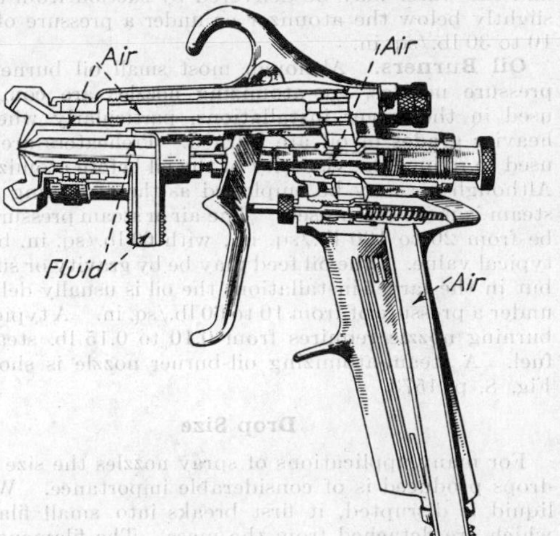

FIG. 80. Typical paint spray gun. (*De Vilbiss Co.*)

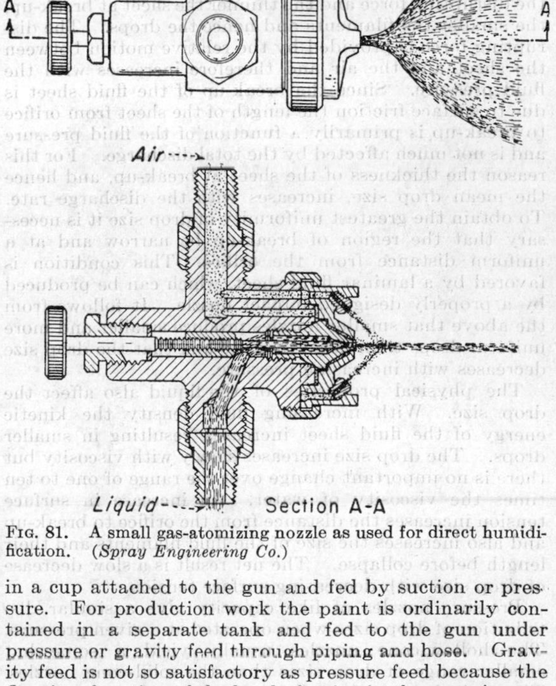

FIG. 81. A small gas-atomizing nozzle as used for direct humidification. (*Spray Engineering Co.*)

in a cup attached to the gun and fed by suction or pressure. For production work the paint is ordinarily contained in a separate tank and fed to the gun under pressure or gravity feed through piping and hose. Gravity feed is not so satisfactory as pressure feed because the flow is a function of the level of paint in the container.

Humidification. Gas-atomizing nozzles are often used for direct humidification in plants where controlled humidity is required, such as textile and paper mills. A typical unit is shown in Fig. 81. A number of atomizers are mounted on the supply pipes, which are attached to the ceiling. The nozzles are arranged to spray horizontally and are adjusted so that the spray will be completely evaporated before reaching the floor or machinery beneath them. Humidifying nozzles spray from 1 to 10 gal./hr. and require from 40 to 100 cu. ft. free air/gal. water. The air pressure is usually about 30 lb./sq. in., and the water may be delivered by suction from a level slightly below the atomizer or under a pressure of from 10 to 30 lb./sq. in.

Oil Burners. Although most small oil burners use pressure nozzles, gas-atomizing nozzles are commonly used in the larger installations, particularly when the heavier grades of oil are burned. Preheaters are often used to reduce the viscosity of the oil before atomization. Although air may be employed as the atomizing agent, steam is more often used. The air or steam pressure may be from 20 to 100 lb./sq. in., with 60 lb./sq. in. being a typical value. The oil feed may be by gravity or suction, but in the larger installations the oil is usually delivered under a pressure of from 10 to 60 lb./sq. in. A typical oil-burning nozzle requires from 0.10 to 0.15 lb. steam/lb. fuel. A steam-atomizing oil-burner nozzle is shown in Fig. 8, p. 1573.

Drop Size

For many applications of spray nozzles the size of the drops produced is of considerable importance. When a liquid is disrupted, it first breaks into small filaments which are detached from the mass. The filaments contract because of surface tension, and an oscillation is set up which results in the formation of one or more large drops and several smaller drops from each filament. Since the filaments are not all the same size, a considerable range of drop sizes invariably results. The greater the disrupting force and the thinner the sheet at break-up, the smaller the filaments and hence the drops. The disrupting force is provided by the relative motion between the sheet and the air and therefore increases with the fluid pressure. Since the break-up of the fluid sheet is due to surface friction the length of the sheet from orifice to break-up is primarily a function of the fluid pressure and is not much affected by the total discharge. For this reason the thickness of the sheet at break-up, and hence the mean drop size, increases with the discharge rate. To obtain the greatest uniformity of drop size it is necessary that the region of break-up be narrow and at a uniform distance from the orifice. This condition is favored by a laminar fluid sheet, which can be produced by a properly designed impact nozzle. It follows from the above that smaller nozzles produce smaller and more uniform drops than larger nozzles and that the drop size decreases with increasing pressure.

The physical properties of the liquid also affect the drop size. With increasing fluid density the kinetic energy of the fluid sheet increases, resulting in smaller drops. The drop size increases slowly with viscosity but there is no important change over the range of one to ten times the viscosity of water. An increase in surface tension increases the distance from the orifice to break-up and also increases the size of the fluid filaments and their length before collapse. The net result is a slow decrease of drop size with decreasing surface tension.

Pressure nozzles of like capacity give a similar distribution of drop sizes when operated at a given pressure. The hollow-cone nozzles usually yield a somewhat smaller range of drop sizes than the solid-cone nozzles. The central sheet of small fan nozzles is also particularly good in this respect, but the "horns" at the edge of the sheet break up into much larger drops so that the advantage is largely lost. As already noted, the smallest range of drop size at a given pressure and capacity will be formed by an impact nozzle designed to produce a laminar fluid sheet. For applications which require the smallest possible range of drop sizes a large number of small nozzles should be used in preference to a few nozzles of large capacity. If a maximum number of small drops is required, nozzles of the smallest size practicable should be used and operated at the highest possible pressure.

Table 67. Drop-size Distributions Produced by Three Hollow-cone Nozzles of the Same Design

Nominal drop diam., μ	Number of drops in each size group					
	0.063-in. orifice diam.			0.086-in. orifice diam.		0.128-in. orifice diam.
	50 lb./sq. in.	100 lb./sq. in.	200 lb./sq. in.	100 lb./sq. in.	200 lb./sq. in.	200 lb./sq. in.
10	375	800	1700	100	300	100
25	200	280	580	60	150	50
50	160	180	260	41	100	45
100	50	60	70	26	34	27
150	27	31	35	14	18	15
200	19	23	27	9	12	11
300	8	9	11	5	8	6
400	2	4	4	4	7	3
500	1	1	....	2	1	2
600	1	...	....	1		1

NOTE: 1 $\mu = 10^{-4}$ cm. = 0.0000394 in. The nominal diameter is the mid-diameter of a drop group which includes a finite range of sizes. The "25" group includes drops from 17.5 to 37.5 μ, the "50" group contains drops from 37.5 to 75 μ, etc. The number of drops has been adjusted in each case so that the total amount of fluid sprayed is the same for each size distribution.

Because of the wide range of drop sizes formed by a spray nozzle, it is difficult to define an average drop size which will be significant for all purposes. It is usually better to have detailed information on the frequency distribution of the drop sizes. It is not practicable to present complete data of this sort here because of the large number of variables involved. However, a few typical drop-size distributions which illustrate the effect of pressure variations and of the nozzle size are given in Table 67. These figures may be taken as a fair example of the performance of pressure nozzles. If liquids of different physical properties are sprayed, the same size distributions will be obtained but at different pressures.

For the graphical representation of drop-size distribution it is convenient to make use of the Rosin equation:

$$F_d = e^{-bd^n}$$

where F_d is the fraction of the mass of the sample contained in drops of diameter greater than d, and b and n are constants. Most size distributions appear to follow this equation, which yields a straight line if log d is plotted against $\log \log \frac{1}{F_d}$. The slope of the line is a measure of the breadth of the distribution.

For a number of applications the total surface area of the spray drops per unit volume of fluid sprayed is a significant factor. For a given nozzle design the exposed surface area in square feet per gallon sprayed is a function of the fluid pressure and the discharge rate. This relationship is illustrated in Fig. 82. These curves are based on a relatively small amount of data and for one type of hollow-cone nozzle. Although other standard nozzle types would probably give somewhat different results, the data of Fig. 82 may be taken as reasonably typical of the performance of well-designed nozzles.

From the standpoint of drop size, the only difference between a rotating nozzle and a pressure nozzle is that in the former the liquid is formed into a thin sheet of suitable velocity by centrifugal force instead of by direct fluid pressure. As a result, the drop-size distribution of a rotating nozzle is quite similar to that of a pressure nozzle. Because there are no small passages in a rotating nozzle, the flow is maintained at much higher viscosities than in a pressure nozzle. The effect of variations in viscosity, density, and surface tension on the drop size has not been investigated, and it is probable that somewhat different laws apply. It is undoubtedly true, however, that high rotational speed, low viscosity, and a small discharge will tend to produce small drops and a minimum range of drop size.

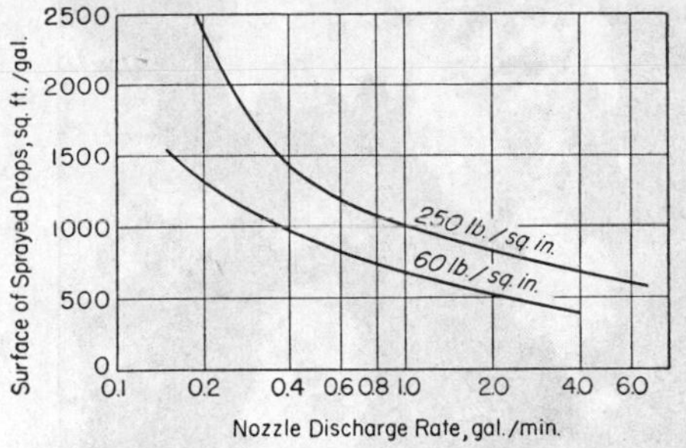

FIG. 82. Typical performance of well-designed nozzles.

As ordinarily operated, gas-atomizing nozzles produce much smaller drops than pressure or rotating nozzles. Although the smaller atomizing nozzles tend to give a somewhat narrower range of drop sizes than the larger nozzles, the size factor is not so important as in the case of the pressure nozzles. The determining factor is the relation between the quantities of gas and of fluid. When insufficient gas is used, large drops are formed which are readily visible since they are projected well beyond the cloud of small drops. The quantities of liquid and of gas should always be adjusted so that no such large drops are in evidence. Assuming that such proper operation is maintained, the drop size can be controlled by varying the gas pressure, higher pressures yielding smaller drops. The fluid pressure has little effect on the drop size and for the most part only determines the quantity of fluid delivered, which is also a function of the gas pressure. An empirical equation for the mean drop size produced by a gas-atomizing nozzle has been given by Nukiyama and Tanasawa [*Trans. Soc. Mech. Engrs. (Japan)*, **5**, 18, 63 (1939)]. This equation, which is given below, is generally applicable when the liquid density is between 0.7 and 1.2 g./cc., the surface tension is from 19 to 73 dynes/cm., the viscosity is between 0.003 and 0.5 poises, and the gas velocity is subsonic.

$$D_o = \frac{1920}{v}\sqrt{\frac{\sigma}{\rho}} + 597\left(\frac{\mu}{\sqrt{\sigma\rho}}\right)^{0.45}\left(\frac{1000Q_1}{Q_2}\right)^{1.5}$$

where D_o is the mean drop diameter in microns (defined as a drop with the same ratio of volume to surface as the total sum of all drops formed); v is the relative velocity of the gas with respect to the fluid in ft/sec.; σ is the surface tension in dynes/cm.; ρ is the fluid density in g./cc.; μ is the fluid viscosity in poises; and Q_1/Q_2 is the ratio of liquid volume to gas volume. Since Q_1/Q_2 is usually small, D_o is determined primarily by the first term. This means that D_o is nearly independent of viscosity over the range of applicability of the equation.

For purposes of comparison with Table 67 a drop-size distribution of a small air-atomizing nozzle is given in Table 68.

Table 68. Drop-size Distribution of a Small Atomizing Nozzle

Drop diam., μ	Number of drops	Drop diam., μ	Number of drops
2	390,000	35	1,730
5	340,000	40	1,080
10	165,000	45	650
15	40,200	50	430
20	11,680	60	350
25	4,970	70	220
30	2,160		

NOTE: The fluid pressure and the gas pressure were each 15 lb./sq. in. The total quantity of fluid represented by this size distribution is the same as that in Table 67, so that the numbers of drops are directly comparable.

Table 69. Common Applications for Spray Nozzles

First number:* type most used.
Second number: type frequently used.
Third number: type sometimes used.

Types

1. Solid-cone wide-angle spray.
2. Hollow-cone wide-angle spray.
3. Narrow-angle spray.
4. Pressure atomizing spray.
5. Tangential spray.
6. Flat spray.
7. Deflector or impact spray.
8. Air- or gas-atomizing spray.
9. Rotating-disk spray.

Pressure Nozzles

Cooling circulating water for condenser (5, 1, 6)
Spray-type condensers
Aerating and purifying water supplies (5, 1, 6)
Scrubbing and washing gases (1, 3, 9)
Humidification and dehumidification (4, 8, 3)
Spray refrigeration (5, 1)
Gas absorption and adsorption (1, 3, 5)
Spray drying (4, 8)
Chemical processes where a large free surface is required (1, 4, 8)
Distributing oil over the fuel bed in gas machines (1)
Enriching gas with a liquid distillate (1, 4)
Oil burners (4, 8, 9)
Desuperheaters (4)
Washing or coating materials in process (4, 2, 8)
Washing liquids (1, 4)
Washing automobiles, railway coaches, etc. (6, 3)
Washing coal, sand, gravel, etc. (2, 6)
Beating down foam (1, 3, 6)
Cooling mill rolls (1, 4, 6)
Descaling hot billets (3, 6)
Quenching coke and pig iron (5, 1)
Settling dust (4, 1)
Applying insecticides, weed killers, etc. (1, 7, 8)
Applying asphalt to highways (1, 7, 6)
Fire protection (7)
Ornamental sprays

Rotating Nozzles

Spraying viscous liquids and slurries (7, 9, 8)
Oil burners (4, 8, 9)
Small air moisteners (8)
Spray drying (4, 8)
Air washing (4, 1, 9)

Gas-atomizing Nozzles

Spray painting (8)
Oil burners (4, 8, 9)
Spray drying (4, 8)
Air moistening (8, 4)
Moistening materials with water or other fluids (8, 4)
Spraying small quantities of insecticides, etc. (8, 4)
Metal coating (8)
Applying cements, refractories, etc. (8, 7)

* Classification kindly supplied by S. G. Ketterer.

GAS DISPERSION

By S. A. Miller

REFERENCES: For discussion and extensive bibliography on foam formation, see Berkman and Egloff, "Emulsions and Foams," pp. 112–152, Reinhold, New York, 1941. For a comprehensive treatment of froth flotation, see Taggart, "Handbook of Mineral Dressing," Sec. 12, pp. 52–84, Wiley, New York, 1945. For a review of methods of producing cellular elastomers, see Gould in "Symposium on Application of Synthetic Rubbers," pp. 90–103, American Society for Testing Materials, 1944.

Objectives of Gas Dispersion

The dispersion of gas as bubbles in a liquid or in a plastic mass is effected for one of the following purposes: (1) gas-liquid contacting (to promote absorption or stripping), (2) agitation of the liquid phase, and (3) foam production.

Gas-liquid Contacting. Usually this is accomplished with conventional columns or with spray absorbers (Sec. 10). For systems containing solids or tar

(a)

(b)

FIG. 83. Comparison of bubbles produced by a porous septum and by a perforated-pipe sparger in water at 70°F. (a) Grade 30 *Carbocell* diffuser operating under a pressure differential of 13.7 in. of water. (b) *Karbate* pipe perforated with 1⁄16-in. holes on 1-in. centers. (*National Carbon Co.*)

likely to plug columns, for absorptions accompanied by strongly exothermic reactions, or for treatments involving a readily soluble gas or a condensable vapor, however, gas dispersers may be used to advantage, particularly when the major resistance to interphase transfer is in the liquid film.

Agitation. Agitation by a stream of gas bubbles (usually air) rising through a liquid is employed in tanks of such large volume or of such unsymmetrical shape as to make mechanical agitation ineffective or expensive. Gas spargers may replace mechanical agitators also for simple blending operations involving a liquid of low volatility or for applications in which it is difficult to seal around an agitator shaft.

Foam Production. This is important in froth flotation separations, in the manufacture of cellular elastomers and plastics, and in certain special applications (food products, fire extinguishers). Berkman and Egloff ("Emulsions and Foams," pp. 112–152, Reinhold, New York, 1941) have pointed out that foam is produced only in systems possessing the proper combination of surface tension, viscosity, and volatility. From the standpoint of gas comminution, foam production requires the creation of small bubbles in a liquid capable of sustaining foam.

Methods of Gas Dispersion

In general, the problem of dispersing gas in a liquid may be attacked in two ways: (1) the gas is introduced into the liquid initially in the form of bubbles of the desired size; (2) a massive bubble or stream of gas is disintegrated within the liquid. The former is accomplished by sparging, precipitation, or generation; the latter by fluid attrition or mechanical agitation, the gas supply being either pressure-fed or self-induced. Gas dispersion may be assisted by addition agents that reduce the surface tension of the liquid or prevent coalescence of the bubbles.

Spargers. *Simple Bubblers.* The simplest method of dispersing gas in a liquid contained in a tank is to introduce the gas through an open-end standpipe, a horizontal perforated pipe, or a perforated plate at the bottom of the tank. Although the size of the bubbles will be a function of the diameter of the orifice through which the gas is introduced at low rates, at ordinary gassing rates relatively large bubbles will be produced regardless of the size of the orifice. Eversole, Wagner, and Stackhouse [*Ind. Eng. Chem.*, **33**, 1459 (1941)] reported that at a bubbling frequency of 40 to 50 per sec., bubbles 0.23 cm. in diameter were produced by capillaries with diameters ranging from 0.0138 to 0.0340 cm.

Perforated-pipe or -plate spargers usually have orifices 1⁄8 to 1⁄2 in. in diameter. A perforated-pipe sparger should be so designed that the pressure drop across the individual orifices is large compared with the pressure drop down the length of the pipe; otherwise, the orifices most remote from the gas supply may not function.

Simple spargers are used as agitators for large tanks,

principally in the cement and oil industries. No complete data on their operation are available. Kauffman [*Chem. & Met. Eng.*, **37**, 178–180 (1930)] reported the following air rates for various degrees of agitation in a tank containing 9 ft. of liquid:

Degree of Agitation	Air Rate, Cu. Ft./(Sq. Ft. Tank Cross Section)(Min.)
Moderate	0.65
Complete	1.3
Violent	3.1

For a liquid depth of 3 ft. he recommended that the above rates be doubled.

According to Glinkov [*Compt. rend. acad. sci. U.R.S.S.*, **51**, (2), 99 (1946)], the turbulence induced by bubbles liberated beneath a liquid may be characterized by a

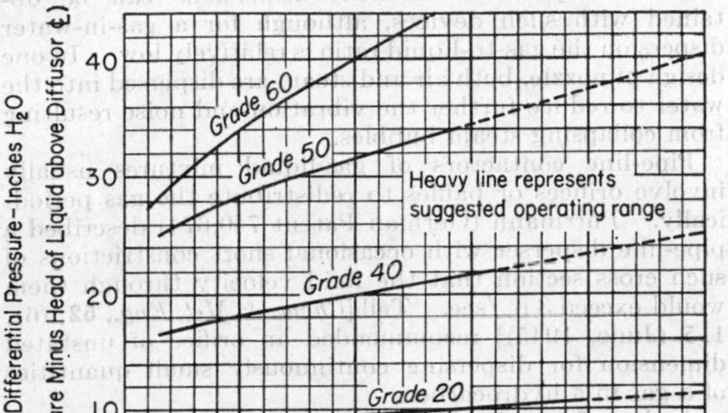

FIG. 84. Pressure drop across *Carbocell* diffusers submerged in water at 70°F. (*National Carbon Co.*)

dimensionless index G, which is the product of the Reynolds and Froude numbers (based on bubble diameter and velocity of bubble ascent) and the group $\rho_b q/\rho S u_b$, where q is the volumetric rate of gas liberation, cu. m./sec.; S is cross section of the liquid perpendicular to the direction of bubble rise, sq. m.; u_b is the velocity of bubble rise, m./sec.; and ρ and ρ_b are the densities of the gas-free liquid and of the gas, respectively, kg./cu. m. Glinkov correlated the data from "boiling" slag baths of commercial size as a plot of the inverse Prandtl number against G.

Open-end pipes sometimes are used to introduce gas to chlorinators with no mechanical agitation. In cases of low chemical reaction rate, they produce an adequate dispersion; for most absorption applications, however, they are inadequate. Cooper, Fernstrom, and Miller [*Ind. Eng. Chem.*, **36**, 504 (1944)] found that less than 1 per cent of the oxygen was absorbed from air bubbled at 0.54 cu. ft./min. from a ¼-in. tube through 9.5 in. of aqueous sodium sulfite, whereas 42 per cent was absorbed when the solution was agitated vigorously.

Porous Septa. Porous plates, tubes, disks, or other shapes are made by bonding together carefully sized particles of carbon, ceramics, or metal. The resulting septa may be used as spargers to produce much smaller bubbles than will result from a simple bubbler. Figure 83 shows a comparison of the bubbles emitted by a perforated-pipe sparger and by a *Carbocell* septum. The size of the bubbles formed is proportional not only to the pore diameter but also to the pressure drop across the septum. At high gas rates coalescence occurs on the surface of the septum, and poor gas dispersion results.

Table 70 lists typical grades of porous carbon, silica, and stainless steel commercially available. The air

Table 70. Characteristics of Porous Septa

Grade	Avg. % porosity	Avg. pore diam., μ	Air-permeability data: Diaphragm thickness, in.	Pressure differential, in. water	Air flow, cu. ft./(sq. ft.)(min.)
Carbocell porous carbon*					
60	48	33	1	2	....
50	48	48	1	2	....
40	48	69	1	2	4.0
30	48	99	1	2	8.5
20	48	140	1	2	17.0
10	48	190	1	2	33.0
Filtros porous silica†					
H	26.0	55	1.5	2	1.5
E	28.8	110	1.5	2	6
S	31.1	130	1.5	2	10
R	33.7	150	1.5	2	15
C	30.2	200	1.5	2	24
B	33.8	250	1.5	2	40
A	36.5	300	1.5	2	60
Micro Metallic stainless steel‡					
H	....	5	0.125	1	1.3
G	....	10	.125	1	2.2
F	....	20	.125	1	3.6
E	....	35	.125	1	13
D	....	65	.125	1	43

* Data by courtesy of National Carbon Co.
† Data by courtesy of Filtros Inc.
‡ Data by courtesy of Micro Metallic Co. Similar septa made from other metals are available.

permeabilities indicate the relative resistances of the various grades, but they may not be used in designing a disperser for submerged operation; the resistance of the septum increases when it is wet. National Carbon Co. [Catalog Section M-8900, 1946] reported air permeabilities for water-submerged porous carbon of some of the grades listed in the table. These data, determined with septa ⅝ in. thick in water at 70°F., are shown in Fig. 84. The best dispersions are obtained within the range of the solid lines. The variation of capacity with septum thickness is shown in Fig. 85.

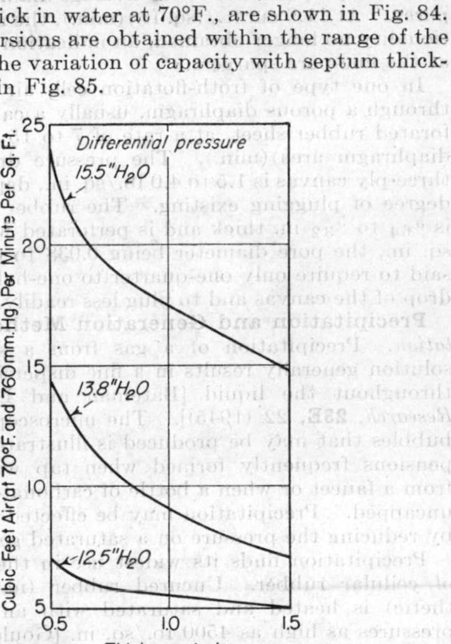

FIG. 85. Relation between septum thickness and air permeability (wet). Grade 30 *Carbocell* submerged in water at 70°F. (*National Carbon Co.*)

Slabs of porous material are installed by grouting or welding together to form a diaphragm, usually horizontal. Tubes are prone to produce coalesced gas at rates high enough to cause bubbling from their lower faces, but they have the advantage of being demountable for cleaning or replacement (U.S. Patent 2,328,655). Roe [*Sewage Works J.*, **18**, 878 (1945)] claimed that silicon carbide tubes are superior to horizontal plates, principally because of the wiping action of the liquid circulating past the tube. He reported respective maximum capacities of 5 and 3 cu. ft. gas/(sq. ft.)(min.) for a horizontal tube and a horizontal plate of the same material (unspecified grade). Mounting a flat-plate porous sparger vertically instead of horizontally seriously reduces the effectiveness of the sparger for three reasons: (1) the gas is distributed over a reduced cross section; (2) at normal rates, the lower portion of the sparger may not operate because of difference in hydrostatic head; (3) there is a marked tendency for bubbles to coalesce along the sparger surface. Bone (M.S. thesis in chemical engineering, University of Kansas, 1948) found that the oxygen-sulfite solution coefficient identified with a 1⅛- by 4-in. rectangular porous carbon sparger was 26 to 41 per cent lower for vertical than for horizontal operation of the sparger, the greatest reduction occurring when the long dimension was vertical.

Porous dispersers are used chiefly to promote gas absorption, particularly in sewage aeration tanks. At low gas rates they are believed to be about as effective as mechanical agitators, and they are superior to simple spargers. Malony (private communication, May, 1942) found that in the absorption of carbon dioxide from a 9 per cent mixture with air into a 12-in. depth of 3*N* aqueous diethanolamine, a No. 20 *Carbocell* tube delivering 8.3 cu. ft./(sq. ft.)(min.) gave a coefficient of 3.3 lb.-moles/(hr.)(cu. ft.)(atm.), whereas a ⅜-in. open-end sparger gave a coefficient of 0.7 lb.-mole/(hr.)(cu. ft.)(atm.). Bone (*loc. cit.*) reported that *Carbocell* tubes and plates delivering 2.4 cu. ft./(sq. ft.)(min.) produced oxygen sulfite solution coefficients that were inversely proportional to the average diameter of septum pores. At 11.9 cu. ft./(sq. ft.)(min.), the inverse proportionality no longer obtained, an indication of coalescence at the fine-pore spargers.

In one type of froth-flotation cell, air is distributed through a porous diaphragm, usually a canvas or a perforated rubber sheet, at a rate of 7 to 15 cu. ft./(sq. ft. diaphragm area)(min.). The pressure drop through a three-ply canvas is 1.5 to 4.0 lb./sq. in., depending on the degree of plugging existing. The rubber sheeting used is 5⁄64 to 5⁄32 in. thick and is perforated with 200 holes/sq. in., the pore diameter being 0.038 to 0.045 in.; it is said to require only one-quarter to one-half the pressure drop of the canvas and to plug less readily.

Precipitation and Generation Methods. *Precipitation.* Precipitation of a gas from a supersaturated solution generally results in a fine dispersion of bubbles throughout the liquid [Bateman and Lang, *Can. J. Research*, **23E**, 22 (1945)]. The microscopic size of the bubbles that may be produced is illustrated by the suspensions frequently formed when tap water is drawn from a faucet or when a bottle of carbonated beverage is uncapped. Precipitation may be effected by heating or by reducing the pressure on a saturated gas solution.

Precipitation finds its widest use in the manufacture of cellular rubber. Uncured rubber (natural or synthetic) is heated and saturated with an inert gas at pressures as high as 4500 lb./sq. in. [Gould, *Rubber Age (N.Y.)*, **54**, 526; **55**, 65 (1944)]. The pressure is released before vulcanization, permitting liberation of the dissolved gas within the rubber and expansion of the elastomer. Taylor (U.S. Patent 2,372,695) applied this method to the manufacture of expanded thermoplastic materials by the use of a volatile solvent at 210°C. and 3000 lb./sq. in. pressure.

Generation. Fine, well-dispersed bubbles are produced if a dissolved or finely divided suspended material is decomposed to yield a gas. An example is the foam produced by a soda-acid fire extinguisher.

Generation methods are employed to prepare cellular elastomers or thermoplastics to which the resulting products of decomposition are not harmful. A number of patented "blowing agents" are used, the most common being ammonium bicarbonate, calcium carbonate, ammonium nitrite, and diazoamino derivatives. Colin-Russ [*Chem. Trade J.*, **115**, 631, 634 (1944)] suggested gas-saturated leather charcoal as a blowing agent. In every case, the blowing agent is compounded into the elastomer prior to its curing, the particles being uniformly dispersed throughout the plastic mass before gas generation occurs.

Fluid Attrition Systems. *Nozzles and Pipe-line Contactors.* The turbulence developed during the rapid flow of fluid through a nozzle or a pipe sometimes is utilized to disperse a gas in a liquid. Steam-water mixers of the venturi-nozzle type are manufactured by several companies. Excellent dispersions can be obtained with such devices, although for a gas-in-water dispersion the gas-to-liquid ratio is relatively low. In one design of nozzle, both air and steam are dispersed into the water to reduce further the vibration and noise resulting from collapsing steam bubbles.

Pipe-line contactors of gas-liquid mixtures usually involve orifices or baffles to redistribute the gas periodically. Pfirrmann (German Patent 740,674) described a pipe-line disperser with occasional short constrictions of such cross section that the fluid velocity through them would exceed 3 ft./sec. Tell [*Chem. & Met. Eng.*, **52**, (6), 115 (June, 1945)] recommended an orifice of unstated dimension for dispersing continuously small quantities of a gas in a hydrocarbon.

A downflow pipe-line disperser for air and sewage was developed by Nordell (U.S. Patent 2,374,772). Initial dispersion was accomplished by a vortex above the downpipe into which the air was entrained as small bubbles.

Nozzles and flow mixers may be used for gas-liquid contacting only where cocurrent flow is permissible.

Cascade Systems. A stream of liquid falling through a gas into a pool will entrain, under the proper conditions, approximately its own volume of gas and will disperse the entrained gas into the pool. This principle was first employed in cascade-type froth flotation machines.

Cooper (U.S. Patent 2,398,345) described a gravity cascade system designed for scrubbing a gas with the cascading liquid. Mertes (U.S. Patent 2,128,311) reported that a solution containing a ferrous compound, when discharged vertically downward with a velocity of 40 ft./sec. through air from a nozzle into a pool less than 3 in. below it, entrained sufficient air for rapid oxidation of the ferrous salt.

No design data are available for cascade systems.

Mechanical Agitators. Mechanical agitators, usually rotating impellers, constitute the most flexible and the most generally effective gas dispersers known. Turbines, paddles, or propellers may be used. In general, a turbine with a solid web is preferable, although not essential. For the best dispersions and the highest capacities, baffles or stator elements must be employed to minimize liquid swirl and increase the possible power input and shear rate at the agitator.

Mechanical agitators have long been used as foam producers for froth flotation and for the preparation of such foods as meringue and whipped cream. The froth-flotation cell most widely used today, the so-called subaeration machine, applies vigorous agitation to a mixture of pulp and a controlled quantity of air which is introduced directly at the impeller. The air may be fed through a hollow agitator shaft or through a sparger pipe, and the supply may be pressure-fed or self-induced. The cross section of a typical subaeration machine is shown in Fig. 181 of Sec. 15. Subaeration cells are operated at impeller peripheral speeds between 1450 and 2150 ft./min. and at power inputs equivalent to 2500 to 7500 ft.-lb./(min.)(cu. ft. cell volume). Froth flotation is discussed in Sec. 15, and operating and design details of flotation machines are given by Taggart ("Handbook of Mineral Dressing," Sec. 12, pp. 52–108, Wiley, New York, 1945).

Cooper, Fernstrom, and Miller (*loc. cit.*) investigated the absorption of oxygen from air into aqueous sodium sulfite contained in baffled tanks agitated with flat paddles or vaned-disk dispersers. They showed that it is possible to represent the performance of geometrically

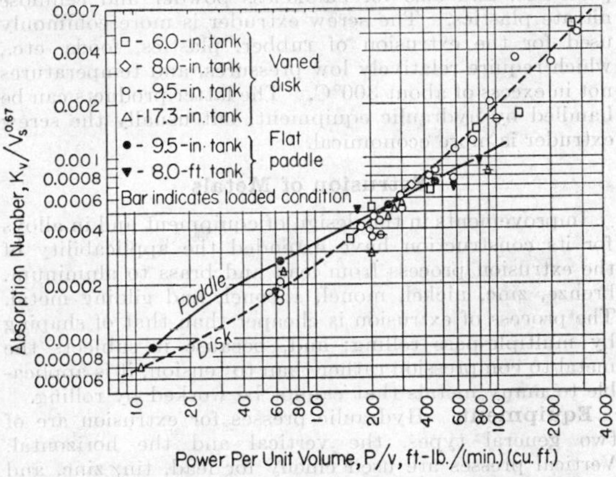

Fig. 86. Performance of mechanical agitator-dispersers. Absorption of oxygen from air into an aqueous solution originally 1.2N with respect to sodium sulfite (Na_2SO_3). Average temperature, 30°C. [*From Ind. Eng. Chem.*, **36,** 504 (1944), *by permission.*]

similar equipment, regardless of size, by the correlation of Fig. 86, where K_v is an over-all volumetric absorption coefficient, lb.-moles/(hr.)(cu. ft.)(atm.); and V_s is the superficial gas velocity based on inlet gas volume and cross section of the tank, ft./hr. The driving force is the logarithmic mean of the entering and leaving partial pressures of the solute gas. The power involved is that delivered to the agitator while gas is being dispersed, usually less by 30 to 70 per cent than the no-gas power at the same impeller speed. No quantitative means of predicting the reduction of power with gas rate is known, and final agitator speed must be adjusted in the field to provide the power desired.

Foust, Mack, and Rushton [*Ind. Eng. Chem.*, **36,** 517 (1944)] investigated the dispersion of air in water by a Mixing Equipment Company *Mixco* turbine. They also reported a reduction in agitator power at a given speed with increasing air rate, and found this reduction to be related systematically to the gas-liquid contact time by the equation

$$\theta = c\left(\frac{P}{vF}\right)^{0.47}$$

where F is the superficial gas velocity based on the tank cross section, ft./sec.; P is the power input to the agitator, hp.; v is the volume of air-free water, cu. ft.; and θ is the average contact time per unit depth of liquid, sec./ft. The value of the constant c lies between 1.26 and 1.65, depending on the degree of baffling, agitator placement, and liquid depth. If the hold-up volume were predictable, this correlation would permit the estimation of the power requirement of a geometrically similar impeller at any gas rate.

In general, there is for each power level of a mechanical agitator-disperser a limiting gas-feed rate which may not be exceeded without danger of causing the tank to foam over. This limiting velocity has not been defined; at reasonable power levels, however, it is not likely to be higher than 25 cu. ft./(sq. ft. tank cross section)(min.), and at power intensities above 2000 ft.-lb./(min.)(cu. ft.) it may be 25 per cent of this maximum or less.

The *Turbo-Gas-Absorber*, manufactured by the Turbo-Mixer Corp. and described in Sec. 17, is a mechanical disperser designed to induce its own gas supply. Estimates of its gas-entraining capacity and of maximum effective gas rates for its operation as a pressure-fed disperser are given in Table 71. For gas rates greater

Table 71. Turbo-Gas-Absorber Gas-entraining Capacity*

Impeller diam., in.	Maximum gas-entraining capacity, cu. ft./min.	Maximum gas rate for pressure-fed operation, cu. ft./min.
4	2	3
6	10	15
9	15	20
12	25	40
18	60	100
22	100	180
27	150	300
42		800
60		3000

* Data by courtesy of the Turbo-Mixer Corp., a division of General American Transportation Corp.

than 150 cu. ft./min., a pressure-fed installation must be used. The two largest impellers are designed to operate only with a positive-pressure gas feed.

A propeller operating in a draft tube or any impeller operating near the surface of liquid in a baffled tank may act as a self-entrainer. Greenup and Johnston (U.S. Patent 2,324,988) used the latter type of installation to prepare latex foam in the production of sponge rubber. There are no known design data for such equipment.

SIZE ENLARGEMENT

By C. E. Berry

Size enlargement, as contrasted with size reduction, deals with the production of portions of matter having volumes greater than that of the component particles. Size enlargement may also be defined to include operations in which volumes do not change, but a linear dimension is increased such as happens in the extrusion of a metal billet or the melt-spinning of a filament.

In the extrusion and molding of plastics, solid granules of the material are brought to a softened state so that they fuse into a continuous mass and take the shape of a die or mold. The finished article is a homogeneous composite of many grains of the original material. Many chemical compounds and elements can be converted from a finely divided state to larger size pieces by processes of compacting. The use of binding agents may be required as well as the application of pressure, although the agitation of a dry powder sometimes is sufficient for the formation of agglomerates. Sturdier pieces result from the application of pressure.

The success of filtration and sedimentation operations in which fine particles are involved depends on the formation of loosely held clusters or flocculates that act like large particles. In the metallurgical industries it is advantageous frequently to change finely divided dusts and concentrates to larger pieces by partial fusion of one of the components so that the particles are bound together after cooling.

EXTRUSION

By Charles M. Fields and P. W. Crane

REFERENCES: Colombel, Extrusion of Metals, *Rolling Mill J.*, **5**, 355 (May, 1931); **5**, 419 (June, 1931); **5**, 479 (July, 1931); **5**, 539 (August, 1931); **5**, 599 (September, 1931); **5**, 667 (October, 1931); **5**, 719 (November, 1931). Cotter and Clark, *Metals Tech. Tech. Pub.*, 1850, September, 1945. Delmonte, "Plastics in Engineering," Penton, Cleveland, 1949. DuBois, "Plastics," American Technical Society, Chicago, 1945. Genders, Extrusion of Metals, *Metal Ind. (London)*, **40**, 345 (March, 1932). Pearson, "The Extrusion of Metals," Wiley, New York, 1944. Plastes, "Plastics in Industry," Chemical Publishing Company, Brooklyn, 1941. Barron, "Modern Plastics," Chapman & Hall, London, 1946. Simonds, Weith, and Ellis, "Handbook of Plastics," Van Nostrand, New York, 1949. Peters, Extrusions Push Ahead, *Scientific American*, **174**, 53 (February, 1946). Intemann, Extrusion of Vinyl Compounds in the Wire and Cable Industry, *Wire and Wire Products*, **18**, 618, 622, 645 (1943). Linzer, Extruding Thermal Plastics, *Modern Plastics Encyclopedia*, 555 (1946). Anon., Extruding Thermosets, *Modern Plastics Encyclopedia*, 560 (1946). Extruding, and Extruding Machines, *Modern Plastics Encyclopedia*, 486*ff*., 1011 (1950). E. I. du Pont de Nemours & Co., Polychemicals Dept., Extrusion of Nylon, *Technical Service Bull.*, 23, 1950.

Extrusion is the process of forcing material in plastic condition through a suitable orifice in order to produce continuously a body of uniform desired cross section.

The process of extrusion has been regarded inevitably as an art rather than as a science, because the complexity of the conditions involved has made it impossible to analyze by any accurate mathematical treatment. In particular, the relation between the shape of the orifice and the cross section of the product is indefinite, being influenced to such an extent by characteristics of the material being extruded and by the conditions of the extrusion. Accordingly, the design of an orifice is a matter of trial and error.

In order to be extruded, the material must be put into a plastic condition, ordinarily by heat, but in some cases with the aid of a solvent or by adding a plasticizer. Then the material, in a closed vessel, is subjected to pressure in order to force it through the orifice.

The process of extrusion has been applied to soft metals (*e.g.*, lead, tin, copper, brass, aluminum, magnesium, and various alloys), rubber, plastics, smokeless powder, foods (*e.g.*, macaroni, spaghetti, candy), soaps, cosmetics, ceramics, ore-flotation concentrates, and pastes of carbon.

Methods of Extrusion

There are two general methods of extrusion. The first, an intermittent process, involves loading a charge of material into a cylinder and extruding it by application of the pressure of a hydraulic ram. The second and more recently developed method utilizes an Archimedean screw to provide the pressure, and permits of continuous loading. The hydraulic press is used principally for metals, which require high temperatures and high pressures, and also for smokeless powder and cellulose nitrate plastics. The screw extruder is more commonly used for the extrusion of rubber, plastics, foods, etc., which require relatively low pressures, and temperatures not in excess of about 300°C. The latter products can be handled by hydraulic equipment, but usually the screw extruder is more economical.

Extrusion of Metals

Improvements in the design of equipment and in alloys for its construction have extended the applicability of the extrusion process from lead and brass to aluminum, bronze, zinc, nickel, monel, Inconel, and gilding metal. The process of extrusion is cheaper than that of shaping by multiple-pass rolling; and, because it subjects the metal to compression rather than to tension, it is applicable to many metals that cannot be worked by rolling.

Equipment. Hydraulic presses for extrusion are of two general types, the vertical and the horizontal. Vertical presses are used chiefly for lead, tin, zinc, and other metals that are extruded at temperatures close to their melting points and require only moderately low pressure. Vertical presses have the advantage of occupying less floor space than is needed for horizontal presses. Few vertical presses are made to apply pressure greater than 12,000 tons total.

Horizontal presses are more generally used and will be covered here in some detail. The essential parts of a horizontal press (Fig. 87), which are mounted on a suitable frame *A*, include a container *B*, which holds the billet *C* of metal to be extruded, a die *D*, which forms the front end of the container and provides the orifice for extrusion, a die holder *E*, which is ordinarily locked in place on the frame by a wedge device, and a hydraulic

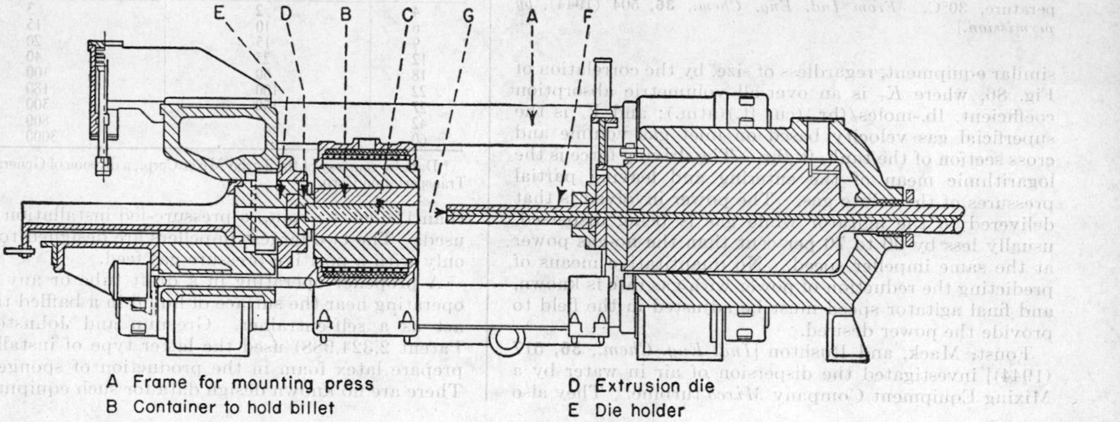

FIG. 87. Horizontal hydraulic extrusion press.

ram *F*, which enters the rear end of the container. A runout table placed in front of the die supports the extruded metal as it comes out of the die.

Modern hydraulic machines are equipped with automatic controls of temperature and pressure and with mechanical means of disposing of the stump or butt which is left in the container after the billet has been extruded.

Horizontal presses range in capacity from 500 to 55,000 tons. The hydraulic cylinders, which operate at 3000 to 5000 lb./sq. in. may be as large as 48 in. in diameter and exert pressures as high as 100,000 lb./sq. in. on the billet. Pressures as high as this may be necessary in order to initiate the extrusion; but, once the operation has been started, less pressure is required because of the heat developed by friction within the billet and against the surfaces of the die. The size of the press depends on the size of the cross section to be extruded, and of course the diameter of the billet must be greater than the largest diameter of the extruded shape. The rate of extrusion ranges from a few inches per minute to several feet per minute, depending upon the metal being extruded and upon the relation in size between the press and the extruded cross section. The body and the liner of the container are made of special alloy steels which withstand the high temperatures of the operation. The billet is preheated, and the container and the die are preheated and then maintained at the proper temperature by gas flames.

For the extrusion of tubes, a second or piercing ram *G* is required. This operates concentrically within the extrusion ram, as presently to be described. Ordinarily the extrusion of tubing requires a larger unit pressure then does the extrusion of solid shapes, because of the greater area of friction with the die and piercing ram and also because of the cooling of the billet by the piercing ram.

Operation. A billet of metal of a diameter slightly less than that of the container is heated, generally in a reducing or inert atmosphere, and placed in the preheated container. The ram is then advanced to create pressure on the billet. Between the ram and the billet, however, is often interposed a dummy of diameter slightly less than that of the container, so that, as the ram progresses in extruding the metal, it will leave undisturbed the oxidized surface of the billet, which will remain as a shell against the walls of the container.

The pressure of the ram upon the hot and plastic metal forces it out continuously through the die to form a continuous body of the desired cross section. The operation is stopped when about 90 per cent of the metal has thus been extruded, since the remainder would give defective material, because of inclusions of oxide from the ends and other surfaces of the billet. The ram is withdrawn, the die block is removed, and the stump or butt of metal remaining attached to the die is severed by sawing or shearing. The thin shell of the billet remaining in the container is then pushed out by the pressure of the ram against an interposed dummy of nearly the full diameter of the container, and the container is cleaned for the next loading.

For the production of tubing, the heated billet in the container is compressed by the ram until it fills the entire cross section of the container. Then the main ram is withdrawn slightly and the piercing ram, which operates through the center of the main ram, is brought forward so as to push through the billet a mandrel which finally emerges at the front end to form the center mandrel of the orifice of the die, and remains in that position throughout the extrusion, which is effected, as before, by the pressure of the main ram.

The extruded metal can subsequently be further processed, *e.g.*, straightened, die-drawn, annealed, tempered, or heat-treated, as required. For example, tubing of small diameter may be made by extruding tubing of larger diameter and then drawing and annealing it.

For different metals, different temperatures and pressures will be required. The temperature of the billet must be sufficiently high to put the material into a plastic condition without melting it, and closer control of temperature is required with some metals than with others (Table 72).

Table 72. Temperature and Pressure Required for Extrusion of Metals

Metal	Billet temp., °C.	Extruding pressure, lb./sq. in.
Lead	350– 400	35,000– 50,000
Copper	1000–1050	150,000
Brass	1600–1700	40,000– 60,000
Aluminum:		
Soft	400– 450	50,000– 60,000
Hard	400– 450	80,000–125,000

Hydraulic pressure can be furnished by individual pumps; but, since ordinarily not more than 40 per cent of the operating cycle is devoted to actual extrusion, it is usually more economical to operate several machines from a common accumulator system. This has the further advantage of facilitating adjustment of the pressure applied to the ram. An auxiliary low-pressure hydraulic system is desirable for moving the rams for other purposes than actual extrusion.

Extrusion of Plastics

The present discussion is concerned primarily with the extrusion of plastics in the condition of relatively stiff doughs, which may or may not contain volatile solvent. If a plastic can be sufficiently softened by heat, with the assistance of any non-volatile plasticizer that it may contain, then there is no point in using a volatile solvent and having to provide for the removal of it afterward from the extruded product. In some cases, however, a volatile solvent must be used because without it the plastic cannot be sufficiently softened by temperatures below those which will cause decomposition.

Plastics with which solvent is ordinarily used include those of cellulose nitrate, casein, and polyvinyl alcohol, and some compositions of cellulose acetate. Plastics with which a volatile solvent is ordinarily not required include those of ethyl cellulose, cellulose acetate butyrate, acrylic resins, vinyl copolymers, polystyrene, vinylidene chloride copolymers, polyethylene and, usually, cellulose acetate.

Extrusion of two other types should be mentioned at this point, but will not be described in detail. One of these is the production of wide thin sheeting of certain materials, notably plasticized polyvinyl butyral for safety-glass interlayer, through extrusion of a dough containing a rather large percentage of solvent, by pumping it with a gear pump into a chamber provided with a long slit orifice, from which the sheet emerges into equipment that removes the volatile solvent. The other is the melt-spinning of nylon into fine and coarse filaments in equipment that comprises in sequence a melting pot, a pump, and screw extruder of special type, a spinneret, and a quenching bath. For the technique of extruding nylon either unsupported or as a sheathing on wire, see the bulletin cited in the references on p. 1180.

Equipment for Extrusion of Doughs of Plastic. Cellulose nitrate plastic and cellulose acetate plastic, containing solvent to assist in softening them, are in some cases extruded by the use of hydraulic presses similar in principle and general construction to those described above for metals. For cellulose nitrate, in particular, hydraulic equipment, employing a compacted billet of the plastic, is often preferred to the alternative screw equipment, for the special reason that cellulose nitrate is subject to decomposition by heat and that in the hydraulic equipment there is less risk that the trapping and compression of pockets of air may cause a dangerous local rise in temperature. In general, however, the plastics are handled in screw equipment.

A screw extruder for plastics (Fig. 88) comprises a cylinder *A*, a screw *B* rotating within the cylinder, and a die *C* attached to the forward end of the cylinder to provide the orifice for extrusion. Also necessary is means *D* for supporting the extruded product as it issues from the orifice in soft and deformable condition, and for cooling it and carrying it away.

The cylinder consists of an accurately bored steel tube which is jacketed, usually in several sections *E*, for control of temperatures by the circulation of hot water, steam, hot oil, or Dowtherm. Recently some success has been had in heating with electric resistance units instead of by hot fluid.

In a modern machine the cylinder is equipped with a hardened corrosion-resistant liner. At the feed end of the cylinder a slot or hopper *F* provides for continuously loading plastic into the cylinder. The discharge end is closed by a removable head *G* to which the die is attached.

The design of the screw is a very important factor. The screw must carry the plastic from the feed end through to the discharge end of the cylinder. During this passage the material, if loaded cold, must become heated, or, if loaded hot, must be maintained at an optimum temperature. Aided by heat, the screw must serve to compress and compact the plastic, so that the fragments or granules put in at the feed end become converted into a continuous mass, free from air, before reaching the discharge end. Furthermore the screw is ordinarily counted upon to provide some mixing action, so as to ensure the homogeneity of the mass with respect not only to composition and color, but also to temperature, since inequality of temperature within the mass will result in an irregularity in its consistency which will be carried through as irregularity or lumpiness in the extruded product.

All these functions of the screw are influenced by its design, and the screw that is found best for one plastic is likely to be unsatisfactory with another one. Hence a variety of screws will be needed for handling a variety of materials, and the design of each one is based more upon experience than upon any theoretical consideration.

The compressional effect of the screw is promoted by either progressively increasing the root diameter of the thread or progressively reducing its pitch. The screw must be massive enough to withstand the high torque encountered under certain operating conditions. The screw is driven through a reduction drive and thrust bearing assembly. Usually a variable-speed drive is provided. Sometimes a strainer plate *H* and screens *J* are placed between the end of the screw and the orifice of the die in order to increase the working pressure in the screw as well as to hold back any foreign matter accidentally present.

The head, which closes the front end of the cylinder, is attached to it by means of heavy bolts. The die is usually inserted in the head by means of a threaded die carrier, which makes possible the use of various die blocks, all of which must be interchanged in the same holder. The head section must be built to withstand pressures as high as 20,000 lb./sq. in. without being deformed.

For simple symmetrical shapes such as rods, ribbons, and tubing, the orifice of the die is generally almost the same as the cross section of the desired product, but orifices for producing extruded goods of irregular cross section cannot be accurately designed in advance. Such orifices must usually be worked out by actual trial of dies made of soft metal, which can be altered by filing until the desired result is obtained.

The extruded material coming from the die must ordinarily be supported until it has been hardened by cooling.

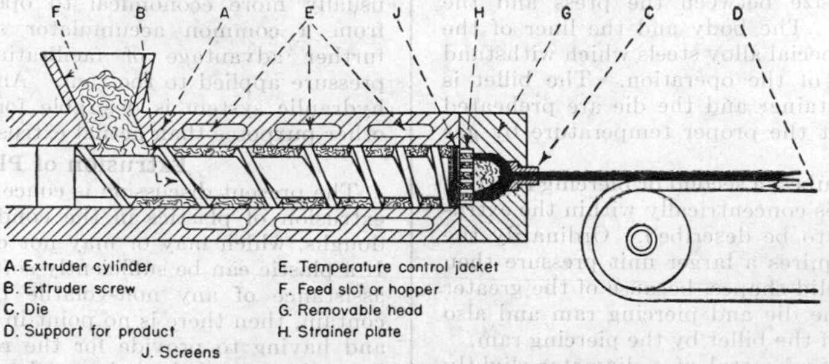

FIG. 88. Screw extruder for plastics.

In some cases this support may take the form of an inclined chute, down which the material slides as it is formed. In other cases, take-off rolls or conveyor belts may be used.

Operation. The plastic to be extruded is usually in granular form. It is loaded into a feed hopper, whence it falls into the cylinder to provide a continuous feed. The material moves forward in the thread of the screw as a result of the relative drag of the cylinder and the advancing action of the thread. As it advances, it becomes heated by the heated walls of the cylinder and also by the friction developed within it as a result of the combination of the rotation of the screw and the compression due to the packing action of the screw. Thus, as the material advances through the cylinder toward the die, it becomes transformed from a mass of granules into a continuous homogeneous plastic dough which is readily forced through the orifice and accepts the desired shape.

It is very important that the conditions be maintained uniform in all respects throughout a run of a given material, in order that the product may be uniform. To facilitate this control it is desirable that the temperature of the equipment be controllable independently in different parts, and it is for this reason that the cylinder is ordinarily provided with several independent heating units. The die also should have its own independent control of temperature, and it is frequently desirable that the screw itself be cored for circulation of fluid for control of its temperature.

As the material emerges from the orifice of the die, it passes unsupported across a gap before gaining the support of the take-off mechanism. In this gap it is subject to deformation, which may be minimized by so controlling its temperature that it is not softer than is consistent with an economical rate of production. In an extreme case it may be necessary to minimize deformation by arranging the orifice to extrude the material vertically downward and by providing a quenching bath of liquid to cool the material almost immediately upon its emergence from the orifice.

Some plastics are sensitive to moisture, in that when not almost completely dry they cannot be extruded without the occurrence of bubbles. Such materials must usually be dried just before being fed into the extruder, and for them a heated feed hopper is desirable also.

General information on the extrusion of certain plastics, without solvent, in a screw machine is given in Table 73. Note that the nylons therein are those which are extruded by the conventional plastics technique, and not those which are spun from a liquid melt.

Average productive capacities of screw extrusion machines for a typical plastic are shown in Table 74.

Table 73. General Information on the Extrusion of Plastics

Plastic	Extrusion temp., °F.	Stability at extrusion temp.	Consistency at extrusion temp.	Extruded cross sections made
Cellulose acetate	340–500	Good	Semimelt	Sheets, rods, tubes, shapes
Cellulose acetate butyrate	340–500	Good	Semimelt	Sheets, rods, tubes, shapes
Ethyl cellulose	340–500	Good	Semimelt	Sheets, rods, tubes, shapes
Polyvinyl chloride	300–350	Fair	Dough	Sheets, rods, tubes, shapes
Polystyrene	375–500	Good	Rubbery	Sheets, rods, tubes, shapes
Acrylic resins	250–350	Good	Rubbery	Sheets, rods, tubes, shapes
Vinylidene chloride resins	300–350	Fair	Semimelt	Sheets, small cross sections, filaments
Polythene (polyethylene)	300–500	Good	Dough	All shapes and filaments
Nylons	400–600	Fair	Semimelt	Sheets, small cross sections, filaments

Table 74. Average Productive Capacities of Screw Extrusion Machines for 0.25-in. Rod of Polyvinyl Chloride (Sp. Gr. 1.35)

Diameter of Screw, In.	Productive Capacity, Lb./Hr.
2	30– 50
2.5	40– 75
3.25	75–125
3.5	95–150
4.5	150–200
6	250–300

Wire Covering

Closely allied to the extrusion of continuous shapes is the procedure of covering wire with insulating material. Equipment for this purpose is similar to that just described except that the head of the machine is so designed that a wire is introduced into the mass of softened plastic and withdrawn through the die in such a way that the extruded plastic encloses the wire. Equipment must be provided for unreeling the bare wire, and in some cases for preheating it before it enters the die. The coated wire is cooled in a trough of water or by a spray of water. The wire is pulled through the die and ultimately wound upon spools by the action of a capstan. The equipment must be so aligned that the bare wire is fed exactly through the center of the orifice, so that the thickness of the coating of plastic applied to the wire will be uniform. The thickness of the coating is governed by the relation between the diameter of the orifice, the rate of feed of plastic by the screw, and the linear rate of travel of the wire. It is customary to drive wire-covering machines with d.c. motors in order that the speeds of the screw and the capstan can be accurately controlled over wide ranges. Coatings in thicknesses from a few thousandths of an inch to as great as 1 in. can be applied to wire by this method.

MOLDING OF PLASTICS

By J. E. Teagarden, A. F. Randolph, and W. E. Rahm

REFERENCES: *General.* Anon., Extrusion Molding, *Modern Plastics*, **23**, 103 (1945). Anon., Molding Acrylics, *Plastics (Chicago)*, **4**, 52 (1946). Clarke, "Molding and Casting," Standard Arts Press, Baltimore, 1946. Crawford and Nathanson, Impression Molding, *Modern Plastics*, **23**, 161 (1946). Dearle, "Plastic Molding and Plant Management," Chemical Publishing, Brooklyn, 1944. DeBell, Developments in Moulding Equipment, *Brit. Plastics*, **16**, 485 (1944). Delmonte, "Plastics in Engineering," Penton Publishing, Cleveland, 1949. Donohue, Processing Phenolic Molding Materials, *Modern Plastics*, **23**, 140 (1945). DuBois and Pribble, "Plastics Mold Engineering," American Technical Society, Chicago, 1945. Freund, Heat Problems in Molding, *Plastics (Chicago)*, **3**, 72 (1945). Halliday, Quality Control in Plastic Molding, part 1, *Plastics (London)*, **11**, 667 (1947); subsequent parts monthly through year 1948. Hantz, Molding Equipment and Accessories, *Ind. Plastics*, **1**, 14 (1946). Hantz, Cold Molded Plastics, *Plastics (Chicago)*, **4**, 46 (1946). Anon., Modern Plastics Encyclopedia, 215*ff.*, New York, 1950. Peterson, Estimating Molding Costs, *Plastics (Chicago)*, **7**, 60 (1947). Rahm, L. F., "Plastic Molding," McGraw-Hill, New York, 1933. Rahm, L. F., Developments in Plastics Molding Equipment, *Mech. Eng.*, **60**, 117 (1938). Sachs and Snyder, "Plastics Mold Design," Murray Hill Books, New York, 1937. Sampson, Molding for Quantity and Quality, *Modern Plastics*, **23**, 133 (1945). Sayre, What's Wrong with Molding Equipment?, *Plastics (Chicago)*, **1**, 64 (1944). Simonds, Weith, and Ellis, "Handbook of Plastics," Van Nostrand, New York, 1949. The Society of the Plastics Industry, "SPI Handbook," SPI New York, 1947. Stannett, Molding Cellulose Acetate Plastics, *Plastics (London)*, **12**, 174 (1948); **12**, 231 (1948). Thayer, "Plastics Mold Designing," American Industrial Publishers, Cleveland, 1946.

Injection Molding. Anon., Continuous Injection Molding, *Plastics (Chicago)*, **1**, 74 (1944). Anon., Injection Molding, Modern Plastics Encyclopedia, 191*ff.*, New York, 1950. E. I. du Pont de Nemours & Co., Inc., Polychemicals Dept., Injection Molding of Nylon FM10001, Technical Service Bull. 8B, March, 1950; "Lucite" Acrylic Injection Molding Powders, Technical Service Bull. 5, April 15, 1949. Hall, "Tri-dyne" (Two-Stage Injection) Process, *Modern Plastics*, **25**, 114 (1948). Macht, Rahm, W. E., and Paine, Injection Molding, *Ind. Eng. Chem.*, **33**, 563 (1941). Robb, Injection Mold Design, *Plastics (Chicago)*, **4**, 68 (1946); **4**, 44 (1946); **5**, 52 (1946). Smith, Injection Molding of Thermoplastics, *Trans. Inst. Plastics Ind. (London)*, 54 (1947). Spaulding, Development of Lower Cost Injection Molds, Society of the Plastics Industry paper, May 20, 1948. Thomas, "Injection Molding of Plastics," Reinhold, New York, 1947. Whitehead, Injection Molding Large Pieces, Society of the Plastics Industry paper, May 20, 1948.

Compression Molding. Anon., Compression Molding, Modern Plastics Encyclopedia, 171*ff.*, New York, 1950. Anon., Magazine Molding and How It Is Done, *Modern Plastics*, **23**, 144 (1946). Anon., Compression-Molded Thermoplastic Laminates and Sheets, *Modern Plastics*, **25**, 118 (1948). Clark, Plastics by Compression Molding, *Ind. Plastics*, **1**, 20 (1946). Ferguson, A New Method of Moulding Thermoplastics, *Brit. Plastics*, **16**, 430 (1944). Massopust and Potts, Control of Compression Equipment, *Modern Plastics*, **22**, 122 (1944). Prince, Compression Mold Design, *Ind. Plastics*, **1**, 16 (1946).

Transfer Molding. Anon., Transfer Molding, Modern Plastics Encyclopedia, 179*ff.*, New York, 1950. Moxness, Transfer Molding Equipment, *Ind. Plastics*, **1**, 22 (1946). Norris, Transfer Molding, Practical Application of the Closed Mold Method, *Plastics (Chicago)*, **8**, 16 (1948). Pribble, Transfer Molds—Why and How, *Ind. Plastics*, **1**, 12 (1946). Robb, How Transfer Molding Affects Plastics Parts Design, *Product Eng.*, 129 (Nov. 1946). Smith, High Speed Molding of Thermosetting Resins, *Plastics (Chicago)*, **2**, 70 (1945). Wilcox, Pot Design for Transfer Molds, *Plastics (Chicago)*, **4**, 92 (1946).

Jet Molding. Buchanan, Jet Moulding, *Brit. Plastics*, **16**, 66 (1944).

Blow Molding. Anon., Blow Molding, Modern Plastics Encyclopedia, 259*ff.*, New York, 1950.

Pulp Molding. Anon., Pulp Molding, Modern Plastics Encyclopedia, 271*ff.*, New York, 1950.

Solvent Molding. Anon., Solvent Molding, Modern Plastics Encyclopedia, 247–248, New York, 1950.

Cold Molding. Anon., Cold Molding Compounds, Modern Plastics Encyclopedia, 673*ff.*, New York, 1950.

Definitions. In technology and the arts, the word "molding" is used to designate many different processes. In connection with plastics the word has come to have a specialized meaning, to designate only one of the several techniques by which these materials are shaped while in a plastic or a liquid condition. Among these techniques the most important are:

Casting. Pouring material in liquid condition into a mold and causing it to harden therein.

Swaging. Stretching or drawing a softened sheet of a thermoplastic to a bowl shape.

Shaping. Stretching or drawing as applied to large sheets. (The common designation of this operation as "forming" leads to ambiguity; "drawing" is a correct term also.)

Die-pressing. Pressing a softened blank of a thermoplastic to a desired shape in a confining die.

Laminating. Simultaneously shaping and welding together a plurality of sheets, *e.g.*, of cloth impregnated with plastic; or glass-plastic-glass for safety glass.

Extruding. Forcing softened material through an orifice for production of rods, tubes, sheets, etc.

Molding. The conversion of a mass of discrete solid particles into integral articles of desired shape by application of heat to soften the material and of pressure to weld it together and to shape the resulting plastic or liquid mass in a mold. The present article is limited to molding as thus defined.

Plastics fall into two general classes in accordance with their behavior when heated. The thermoplastics, represented by plastics of cellulose nitrate, cellulose acetate, and methyl methacrylate, can be softened by heat, repeatedly if desired, remain soft as long as they are hot, and are hardened by being cooled again to normal temperatures. The heat-setting, or thermosetting, plastics, represented by phenol-formaldehyde and urea-formaldehyde, are those which can be softened by heat but are hardened by continued exposure to heat and cannot thereafter be usefully softened by temperatures below those which will decompose them.

This difference in behavior necessitates differences in techniques for molding, as will be seen below in connection with the description of the principal methods of molding of plastics, namely, compression, transfer, injection, and jet.

Compression Molding. Compression molding is the oldest type of molding. Typical equipment, as illustrated diagrammatically in Fig. 89, comprises a hydraulic press, and, attached to the platens of the press, the matching top (male) and bottom (female) portions of a mold which when closed provides an accurately machined and polished cavity corresponding to the article to be molded. The body of the mold is cored for circulation of fluid.

At the start of the operation the mold is heated by circulation of steam. Into the lower part of the mold is placed an accurately weighed or measured amount of granular or powdered molding compound, sufficient to form the molded article, plus any necessary small allowance for overflow. As the mold is closed, the heat softens the molding compound, and its individual particles become welded together into a continuous plastic mass. The pressure upon this mass causes it to flow and take the shape of the cavity of the mold.

It is necessary next to harden this mass while it is held in shape by the pressure. If the molding material is thermoplastic, it must be hardened by being chilled. To this end, the circulation of steam is cut off, and cold water is circulated until the molded article is made rigid enough to be removed. It is released by opening the mold. If the plastic is of the heat-setting type, the hardening is done by continuing the heating of it in the mold, to effect the chemical change, or "cure," which converts it from a plastic to a rigid condition. The mold is not chilled, but is merely opened when the cure has been completed, and the molded article is removed still hot.

The molded article duplicates the finish of the mold cavity, and therefore if the mold is in good polished condition the molded article requires no further treatment except the removal of a thin fin or flash by tumbling or buffing.

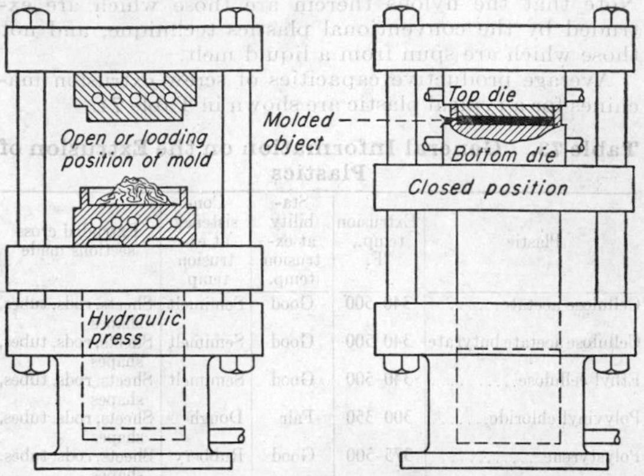

FIG. 89. Typical compression-molding equipment.

By compression molding, articles may be made ranging in weight from a fraction of a gram up to several pounds. For all but very large articles it is customary to use multiple-cavity molds so as to reduce the molding cost per article. The number of cavities in such a mold will be that which will give the smallest sum of (1) the cost, per article, of amortization of the mold, which, for a given total number of articles produced, is increased by increasing the number of cavities; and (2) the labor cost and other operating cost, per article, which is decreased by increasing the number of cavities.

Compression molding is done almost entirely in hydraulic presses, which range in capacity (total pressure upon the mold) from a few tons to several thousand tons. The pressure required per square inch of projected area of the cavity or cavities is usually around 3000 lb.

Accurate charges of molding powder may be obtained by weighing, but with many materials it is more economical to run the powder through a tableting machine which automatically measures out each charge and compresses it into a firm tablet or preform. Multiple-cavity molds can be quickly loaded with preforms by the use of a dropping board.

A dropping board is a device consisting of two sheets of metal or of plastic, or wooden boards, arranged to slide one over the other and provided with openings slightly larger than the preforms and spaced to correspond with the cavities of the mold. With the two members out of register, preforms are laid in the openings of the upper member and rest upon the lower. This loading of the device is done during the curing of the preceding charge in the mold. When the mold has been opened and emptied, the dropping board is put into position over the cavities of the mold, the lower member is pushed into register with the upper, and the preforms fall into the cavities.

Inserts of metal, to serve as electrical contacts or as means of attachment or assembly, can be placed on suitable supports in the mold so that they will become enclosed and anchored firmly in the molded article.

In some instances the productive capacity of a mold can be increased by preheating the preforms, by means of steam tables, infrared lamps, or electrical resistance heaters, or by high-frequency heating devices. Of these the last are particularly effective in thoroughly heating a thick preform. Preheating shortens the time required in the mold, particularly by heat-setting plastics, while of course it cannot reduce the time required for the chilling of thermoplastics.

Transfer Molding. In the compression molding of articles requiring delicate inserts or having deep holes formed by slender pins in the mold, the inserts and pins are sometimes bent by the lateral thrust of the molding compound as it flows under the vertical thrust of the press. The process of transfer molding eliminates this trouble.

In transfer molding (Fig. 90), the plastic, usually a heat-setting one, is heated to soft condition in a chamber

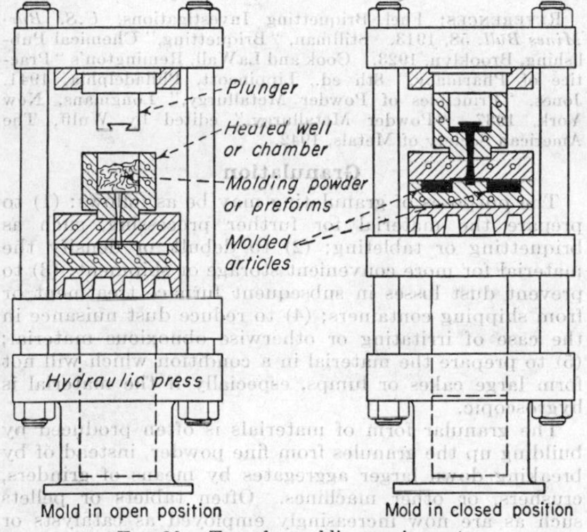

FIG. 90. Transfer-molding equipment.

or well, located directly above the heated mold, and is forced, by the pressure of a plunger, through an orifice into the mold, which is already closed. Distortion of pins and inserts is avoided by the soft condition of the material as it flows around them and by the fact that they are not subjected to heavy pressure until the mold is full.

In the mold, the article is hardened by continued application of heat, if of heat-setting type, or by chilling, if a thermoplastic.

The amount of a heat-setting molding material put into the well must be sufficient only to fill the mold and to provide a slight excess to ensure maintenance of pressure on the molded article during its cure. Any amount above this is wasted, since it hardens in the well and cannot be used as part of the next loading. Provision is made for ejecting what material remains in the well, so that the well is empty for the next loading.

The weighed charge in loose granular form, or the measured charge as a preform, can be preheated before being put into the well.

The plunger or offset type of transfer molding (Fig. 91) differs from the usual type in that the mold is clamped by the main ram of the press and the transfer is effected by a separate plunger.

Injection Molding. The equipment and procedure for injection molding may be described with reference to Fig. 92.

A thermoplastic molding material in granular form is fed into one end of a heated cylinder, through which it is forced by a plunger and from which it is ejected, in softened condition, through an orifice into a closed mold. The amount of molding material in the heating cylinder

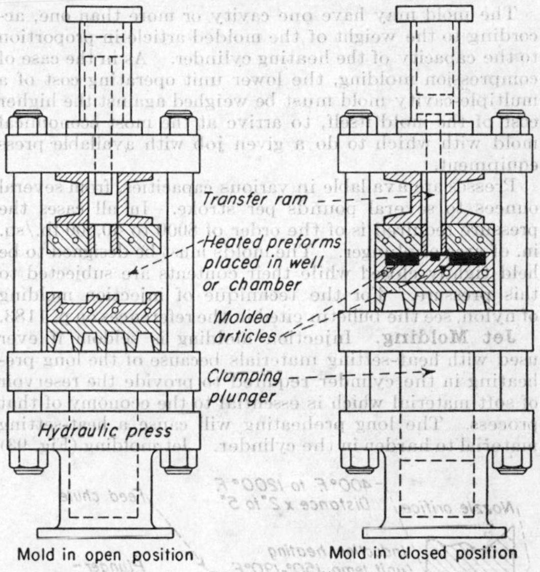

FIG. 91. Plunger-molding press.

is a multiple of that required to fill the mold and is kept constant by the fact that on each stroke of the plunger the amount pushed out into the mold is balanced by the amount fed in at the other end by a metering device. The material progresses from the feed end to the discharge end, becoming gradually heated and thereby softened, and the capacity of the cylinder is a sufficiently large multiple of the capacity of the mold to provide time

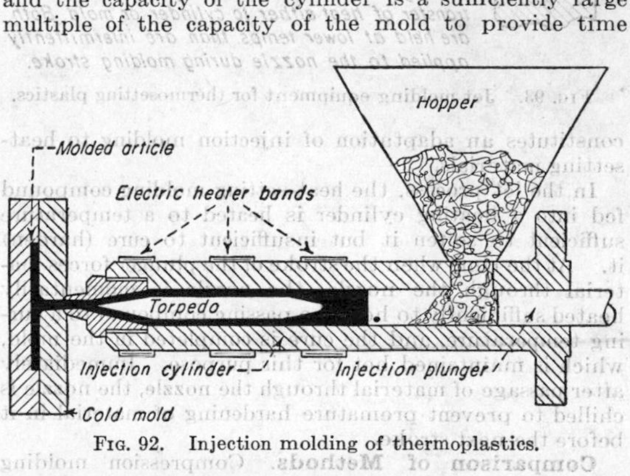

FIG. 92. Injection molding of thermoplastics.

enough for this heating to be accomplished. Thus each time the mold is emptied and reclosed there is soft compound ready to be pushed into it, and the cycle of the operation is the time required successively to fill the mold, to cool its contents under pressure of the plunger, to open it for ejection of the molded article, and to close it again. Of this cycle the principal part is that of cooling the material in the mold. This is made as short as possible by keeping the mold always at the lowest temperature consistent with filling it completely with a thoroughly welded mass free from objectionable strains.

The step of opening the mold is accompanied by a movement of the mold away from the cylinder, which causes the connecting stem of material to be parted at its point of least diameter. The molded article as ejected carries this stem or sprue, which must then be trimmed off. These trimmings, being thermoplastic, can be reused.

The mold may have one cavity or more than one, according to the weight of the molded article in proportion to the capacity of the heating cylinder. As in the case of compression molding, the lower unit operating cost of a multiple-cavity mold must be weighed against the higher cost of the mold itself, to arrive at the most economical mold with which to do a given job with available press equipment.

Presses are available in various capacities, from several ounces to several pounds per stroke. In all cases the pressure required is of the order of 8000 to 40,000 lb./sq. in. of area of plunger. The molds must be designed to be held tightly closed while their contents are subjected to this pressure. For the technique of injection molding of nylon, see the bulletin cited in the references on p. 1183.

Jet Molding. Injection molding is seldom if ever used with heat-setting materials because of the long preheating in the cylinder required to provide the reservoir of soft material which is essential to the economy of that process. The long preheating will cause a heat-setting material to harden in the cylinder. Jet molding (Fig. 93)

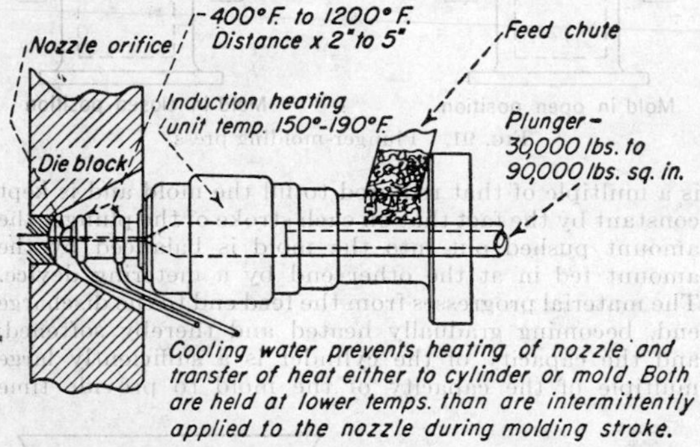

FIG. 93. Jet molding equipment for thermosetting plastics.

constitutes an adaptation of injection molding to heat-setting materials.

In the jet process, the heat-setting molding compound fed into a heating cylinder is heated to a temperature sufficient to soften it but insufficient to cure (harden) it. At the time when the stroke of the plunger forces material through the nozzle, the nozzle is momentarily heated sufficiently to heat the passing compound to a curing temperature, and the cure is completed in the mold, which is maintained hot for this purpose. Immediately after passage of material through the nozzle, the nozzle is chilled to prevent premature hardening of material in it before the next stroke.

Comparison of Methods. Compression molding offers the advantage of relatively inexpensive press equipment and is adaptable to the production of a large variety of molded articles. It is used with heat-setting materials, but seldom with thermoplastics because they require a relatively long operating cycle of alternately heating and cooling the mold, and also because thermoplastics in general are not adapted to the operation of preforming, which in the case of heat-setting compounds effects considerable economy in measuring and loading. Compression molding is particularly advantageous in the molding of large runs from heat-setting material, where the number of articles to be made is large enough to carry the cost of a mold having a large number of cavities, with which the labor cost per article becomes very small.

The less economical procedure of transfer molding is used, chiefly with heat-setting materials, for making articles having deep holes or delicate inserts, or configurations not amenable to compression molding.

Injection molding is generally applicable to thermoplastics and in particular provides a means of molding articles of complicated shape and thin wall. The relatively short molding cycle and the availability of presses of capacities great enough to fill multiple-cavity molds help to compensate for the carrying cost of the relatively expensive press equipment and for the higher cost of thermoplastics.

Jet molding adapts injection molding to heat-setting materials.

COMPACTING

By Lawrence H. Bailey

REFERENCES: Fuel Briquetting Investigations, *U.S. Bur. Mines Bull.* 58, 1913. Stillman, "Briquetting," Chemical Publishing, Brooklyn, 1923. Cook and LaWall, Remington's "Practice of Pharmacy," 8th ed., Lippincott, Philadelphia, 1941. Jones, "Principles of Powder Metallurgy," Longmans, New York, 1937. "Powder Metallurgy," edited by Wulff, The American Society of Metals, 1942.

Granulation

The purposes of granulating may be as follows: (1) to prepare the material for further processing, such as briquetting or tableting; (2) to debulk or densify the material for more convenient storage or shipment; (3) to prevent dust losses in subsequent furnace treatment or from shipping containers; (4) to reduce dust nuisance in the case of irritating or otherwise obnoxious material; (5) to prepare the material in a condition which will not form large cakes or lumps, especially if the material is hygroscopic.

The granular form of materials is often produced by building up the granules from fine powder, instead of by breaking down larger aggregates by means of grinders, crushers, or other machines. Often tablets or pellets such as are now increasingly employed as catalysts or catalyst carriers can be formed much more readily from granules than from powders; indeed, sometimes they cannot be formed from the powdered form at all.

There are several general methods of producing granules from powder: (1) moistening with water, solvents, or binding solutions with subsequent screening, grinding, or rotary-drying operations, either in the damp condition or in the dry condition, or both; (2) briquetting of the fine powder, followed by grinding or screening; (3) fusing to produce the necessary adhesion, giving granules directly in some cases, or in other cases, producing a mass, sheet, or film which can be broken down to the granular condition; (4) preparing the powder in the form of a suspension in liquid, where it is sometimes possible and desirable to reduce it in one operation to the granule stage by spray drying; (5) treating by the "Spheronizing" process, which is purely mechanical and usually requiring no addition agents.

Granulation with a Binding Agent. In the first of the above methods, the required adhesion is obtained from, or brought into effect by, the liquid used for granulating. With water-soluble materials or mixtures containing such material, the addition of water alone may be sufficient, as in the case of sugars, salts, extracts, etc. If water will not develop the required adhesiveness, a solvent for the material, or one or more ingredients of the mixture, may be used. If none of the ingredients develop adhesiveness by the addition of water or solvent, a binder is introduced by dissolving it in the water or solvent used for wetting up the powder. In some cases, a dried binder in powdered form is mixed with the powder being treated, and the mixture is then moistened to activate the binder. In general, the dry binder subsequently moistened is not so effective as a solution of the same binder, but in some cases this method gives the desired results and may effect some economies in processing.

In some cases the material being granulated may develop too much adhesiveness. It may not mix smoothly with the granulation solution, or it may become excessively sticky, producing unmanageable masses. In such cases the wetting power of the solvent may be reduced by mixing it with other solvents; *e.g.*, water may be mixed with alcohol when handling sticky water-soluble materials. In using alcohol-soluble materials, the alcohol may be diluted with water or other solvents to reduce its solvent power.

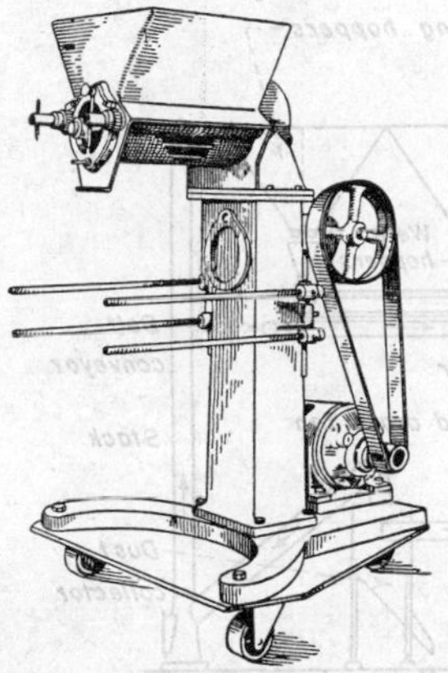

FIG. 94. Oscillating granulator, assembly. (*F. J. Stokes Machine Co.*)

Some of the materials that have been used in aqueous binding solution for granulation are sugars, glue, gelatin, dextrins, gums, starch, flour, molasses, and sulfite waste liquor; and, with other solvents, shellac, waxes, lacquers, etc. Among binders introduced dry and later activated are pulverized sugar, spray-dried glucose, malt extract, or almost any of the above-mentioned water-soluble binders. As stated before, however, a larger percentage of the binder is required than when it is introduced in solution.

Granulating Stage. After the damp mixture is made, it must be processed to produce the required granular condition. The mixture is usually dried down to a proper moisture content and then crushed or ground in equipment that will give the desired screen analysis, using for this purpose a suitable mill, grinder, or granulator. One usual requirement in the finished granulation is a minimum of fine dust. Also the machine must not develop enough heat to make the material gum up and necessitate a stoppage of operation. A granulator consisting of a set of bars arranged in cylindrical form and oscillating over a screen of suitable mesh usually produces the desired results in the case of fairly coarse granules, say 20 mesh or over. Such an oscillating granulation is shown by Figs. 94 and 95. With smaller mesh size this type of machine is rather slow, and other types of grinders are more effective, such as plate or cone mills or crushing rolls. With larger mesh screens, the capacity increases greatly and may run as high as 1000 to 2000 lb./hr. using 6- and 4-mesh screens.

One very effective method of handling the damp material is to granulate through a coarse screen, as above, and effect further reduction after drying. This produces a minimum of fines.

Granulation of Fertilizers. Many fertilizer products are granulated to eliminate excessive caking and lumping of the material before it is used and to provide the product in a convenient form. A typical large-scale process of granulation of fertilizers is the Davidson process described by Mackall and Shoeld [*Chem. & Met. Eng.*, **47**, 102 (1940)]. As shown by Fig. 96, the material to be granulated is first conditioned by moistening and then fed directly to a rotary drier. By carefully controlling the moisture content of the conditioned material, which requires constant visual attention, the rotary drier will form granules of the desired form. Hardesty and Ross [*Chem. & Ind.*, **58**, 885 (1939)] found that the plasticity and fineness of particle size of a material determine the ease with which it can be granulated by the rotary-drying method.

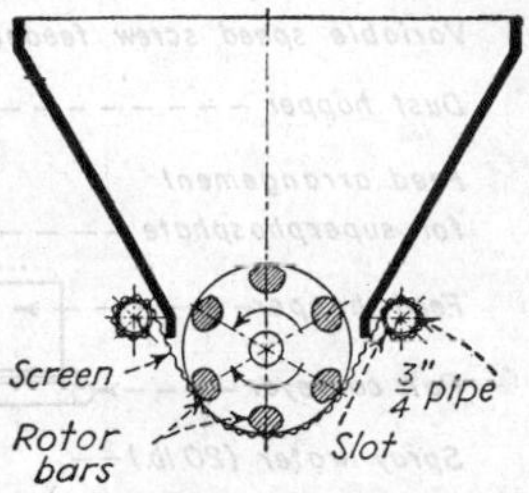

FIG. 95. Oscillating granulator, schematic section. (*F. J. Stokes Machine Co.*)

Ammonium nitrate fertilizer often is produced by the spray granulating process or with graining kettles. Conditioning treatments can be employed with advantage. Miller *et al.* [*Ind. Eng. Chem.*, **38**, 709 (1946)] report the results of a plant-scale study of the conditioning of granular ammonium nitrate produced in batch graining kettles.

Granulation of Briquettes. The second method of granulating is to briquette the fine powder, using machines especially designed for this work, and then to break the briquettes down to the required sizes, sifting out the fines if they are undesirable and rebriquetting them in a subsequent operation. This method was originally developed in the pharmaceutical industry when working with material such as aspirin or other mixtures which react when wet and which could, therefore, not be granulated by wet methods. It is now used in producing granular plastic molding powders and various granular materials. It is also used for reducing dust nuisance with materials such as sodium or calcium hypochlorite.

The briquettes or "slugs" are made in small sizes, usually not over 1¼ in. diameter, suitable for feeding to the granulator previously mentioned, and are produced at high speed by single or double types of rotary compressing machines obtaining capacities up to 700 pieces/min. The method of grinding or breaking down the briquettes may be quite critical. Grinders in most cases produce too much fine material, especially if a hammer or high-speed type is used. In general, the oscillating granulator, described and illustrated previously, produces the most satisfactory result. In one carefully investigated case, 50 per cent more usable granules were produced by the granulator than by another type of grinder.

Granulation by Fusion. The third, or fusion, method of granulation is used in a few highly specialized industries. Mixtures containing pitch or wax binders can be agglomerated by means of heating and then can be reduced to granules after cooling. A familiar example is found in the manufacture of plastic molding powders. A mixture of the resin with wood flour or other filler is passed through friction rolls to develop adhesiveness and produce a sheet material which, after cooling, is reduced to the familiar granular form. See also Size Enlargement by Fusion, p. 1191.

The fusion method is used for producing granular effervescent salts. When a mixture of bicarbonate and citric acid is heated, the citric acid melts in its crystal water and causes adherence of the particles. Rapid cooling arrests reaction or effervescence. The resulting material is then put through a screen and dried to remove the remainder of the water of crystallization, producing a stable material that will effervesce readily and smoothly when placed in water. Several other materials containing fusible crystal salts can be handled in a similar manner.

Waxes and fusible organic or inorganic material of various kinds can be granulated, flaked or chipped by chilling the melted material in the form of a film on the surface of a cooled revolving drum and removing by means of a knife. The character of the flake or chip can be varied by changing the temperature of the liquid material or of the cooling drum or by the method of sharpening the knife. In some cases the material can be taken off in sheet form and then reduced to flake form. A few of the materials handled by these methods are as follows: naphthalene, beta naphthol, caustic soda, potash, cyanides, carnauba wax, sulfonated detergents, lead acetate, and trisodium phosphate (see Flaking, p. 1164).

Metal powders for use in powder metallurgy, which in their original state are too fine for satisfactory feeding to the die by gravity, can be put in a granular form by running a thin layer of powder through a light sintering operation to produce a slight adhesion of the particles and subsequently running the material through a mill or granulator to produce the required particle size. This method is effective on the so-called carbonyl powders of iron and of noble metals such as iridium and osmium.

Granulation by Spray Drying. The fourth, or spray-drying, method is used to a certain extent in the ceramic industry where the raw materials are wet-ground as an aqueous suspension. It has the advantage of permitting complete dewatering and granulation in one operation but is probably applicable only to industries where grinding or mixing as an aqueous suspension is desirable. For further information on spray drying, see Sec. 13.

Cabot Spheronizing Process. The fifth, or "Spheronizing," method is a more recent development and was originally used for handling carbon black, but it is also applicable to other finely divided materials such as pigments and dyes. Usually no addition agents are required, and the Spheronized material breaks down to the original particle size when mixed with other ingredients. In this process, the material is subjected to turbulent mechanical action in either horizontal or vertical types of mixers with specially designed agitators and it gradually agglomerates into spherical form with a large reduction in bulk density.

The process can usually be speeded up by using small spherical "starters" sifted from a previous batch. The batch should consist of approximately 50 per cent starters and 50 per cent fine powder. Non-agglomerative powders may be made susceptible to this treatment by reducing the particle size by ball milling previous to the Spheronizing operation. Proper conditions for operation must be determined experimentally. If a charge is too light it may fail to agglomerate and if the charge is too heavy the shearing action of the agitator may tend to break agglomerations previously formed. The time required is usually 1 to 2 hr., and the process must be watched carefully, since further agitation after the granules are completely formed may break them down again.

The process is under intensive investigation to increase its applicability and to put its operation on a more scientific basis. The process and equipment are covered by U.S. Patents Reissue 19,750, 2,120,540, 2,316,043, 2,120,541 and others. The James Russell Engineering Works, Inc., of Boston, Mass., are licensees under these patents for the building of equipment except for the carbon-black industry.

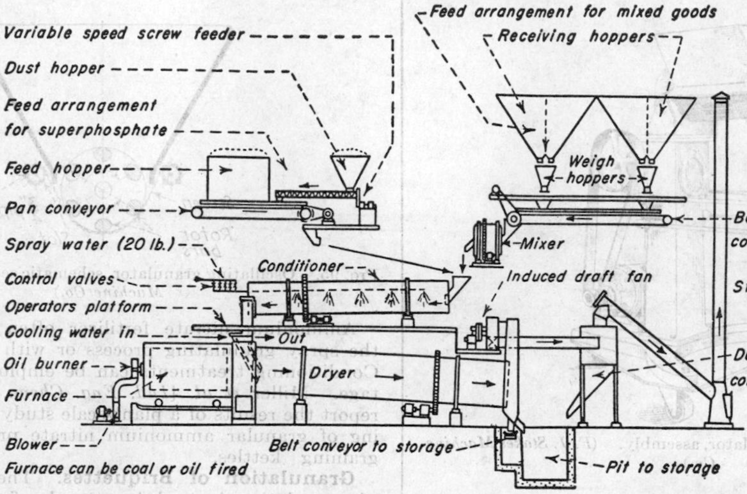

FIG. 96. Apparatus for granulation of fertilizers. [*Chem. & Met. Eng.*, **47**, 103 (1940).]

Briquetting

When the granular material produced is to be briquetted it is sometimes important to have a considerable percentage of smaller granules and some fine material, as material of the properly diversified screen sizes will make a stronger briquette than one made from more uniform granules. In briquetting it may be necessary to add a lubricant so that the compressed material will not stick to the punch faces or to the die but rather will be assisted in ejection from the die. Commonly used lubricants are oils and oil mixtures, powdered waxy materials, powdered soap, talcum powder, metallic stearates, and boric acid. In general, only a very small percentage of the lubricant is required, especially if it is an extremely fine powder. In nearly all cases, the lubricant is added to the otherwise finished granulation.

Large high-production briquettes such as fuel, ore, and scrap-metal briquettes are made on specially designed roll-type presses requiring hot pitch binders or large presses with single- or multiple-die cavities. Binders are usually required except in the case of scrap metal. Subsequent waterproofing may be required if the briquettes made with soluble binders are to be exposed to the weather.

Pelleting

Pellet mills are designed to agglomerate permeable free-flowing materials into pellet form. The material is pressed through a perforated die. The die is ring-shaped and rollers are caused to operate on the inside of the ring in such a manner as to cause materials to be pressed through the die. A knife on the exterior of the ring die cuts the extruded cylindrical material into pellets. A machine of this type is manufactured by the Sprout, Waldron Co. Although originally designed for pelleting animal feeds, it is now being adapted to the handling of many other products.

Pelleting of dusts, fumes, flotation concentrates, and the like can be accomplished in a rotating-drum device known as the *Dwight-Lloyd Segregating Pelletizer.* A production rate of 2 tons/hr. can be expected on zinc fume when using a drum 3 ft. in diameter and 4½-ft. long. The pelletizer produces a sized product within the limits of most requirements.

Tableting

Many materials are more conveniently used in various industries when produced in the form of compressed tablets. Early tablet machines were developed for producing small tablets for the pharmaceutical trade, single-punch machines being used for moderate production and rotary types for large-scale production. These machines and larger machines developed on similar principles were soon used on industrial products, including compressed candy mints, mothballs and other insect repellents, washing compounds, laundry blue, starch cubes, food products, chemical catalyzers, explosives, fireworks, water-color paints, dyes, abrasives, plastic preforms, cold-molded parts, ceramic products, and electrical products such as resistors, contacts, iron cores, spacer disks, getters, glass beads, battery components, and generator brushes.

Materials to be compressed into tablets usually require some simple treatment to ensure proper operation on the compressing machine. Fine powders do not feed readily and tend to entrap air which may cause cracking or lamination of the pieces. Most materials have to be granulated by one of the methods outlined above. To prevent the material from sticking to the inside of the die or the punch faces, lubricants such as oil mixtures or fine dry lubricants such as stearic acid, talc, graphite, or special waxes are used.

Tablets are produced at rates of 20 to 100 pieces/min. on single-stroke machines and up to 500 or 1000 pieces/min. on rotary types and double rotary types. Special rotary machines making small tablets in large quantities, such as are required for chemical catalyzers, are operated at speeds up to 4000/min.

Compressing (Powder Metallurgy)

In powder-metallurgy processes, metal powders including small percentages of lubricant are compressed into the desired shape and are then sintered in a furnace with a non-oxidizing atmosphere to develop metallic properties. Since the temperature usually is far below the melting point, there are no changes in shape and size other than a slight shrinkage or, in some cases, a slight expansion. Subsequent sizing or coining of the sintered pieces may be required.

FIG. 97. Powder metallurgy press used chiefly for porous bearings. (*F. J. Stokes Machine Co.*)

Modern powder-metallurgy methods were first developed for producing metals such as tungsten. The method was soon applied to the manufacture of other products:

1. Porous metals for self-lubricating bearings.
2. Unusual alloys such as silver and tungsten for electric contacts, and copper and carbon for brushes.
3. Combining metals and non-metals, as in radio cores and friction materials.
4. Hard materials in metal matrices, such as the various carbides in cobalt in the manufacture of carbide tools or diamonds in copper for diamond grinding wheels.
5. Alloys with poor casting or shaping qualities, such as Alnico used for permanent magnets.
6. Manufacture of small machine parts, where in many cases the process may compete with screw machines, precision casting, die casting, or other mechanical methods.

The metal powders used are produced by several methods, including electrolysis, spraying, reduction of oxides, and other chemical methods. Hammered powders are generally not well suited to powder-metallurgy operation.

Machines previously used for making tablets or briquettes have found wide application in the production of parts by powder metallurgy. Several new types of presses, both mechanically and hydraulically operated, have been developed specially for this work, so that more complicated pieces can be produced and better density distribution in the compressed piece can be obtained. Nearly all the presses used are of the double-acting cam-operated type, and extra punches or specially constructed dies are used when more complicated pieces are to be made. A powder-metallurgy press for producing porous bearings is illustrated in Fig. 97. Small parts requiring high production are made in rotary-type presses, which also give compression both from above and below.

FLOCCULATION

By C. K. Sloan

REFERENCES: Alexander, "Colloid Chemistry," vol. VI, p. 782, Reinhold, New York, 1946. Kite and Fischer, Sedimentation and Hydraulic Classification, *Ind. Eng. Chem.*, **38,** 16 (1946). Yancey *et al.*, Flocculation as an Aid in the Clarification of Coal Washery Water, *U.S. Bur. Mines, Report Investigations*, 3494, 1940. Hay, New Concept of Coagulation, *Water and Sewage Works*, **93,** 225 (1946). Donald, Sedimentation and Flocculation, *Trans. Inst. Chem. Eng. (London)*, **18,** 24 (1940). Samuels, Theoretical and Commercial Aspects of Flocculation, *Trans. Inst. Chem. Eng. (London)*, **16,** 47 (1938).

Two common methods employed in the separation of finely divided particles from a fluid medium involve (1) filtration and (2) gravity sedimentation. The size of the individual particles of many materials is so small that such methods are not feasible unless the individual particles exist in the size-enlarged form of clusters. In many systems such as pigment slurries, clay suspensions, coal slurries, oil-in-water emulsions, mineral slimes, turbid water, and the like, the individual particles are too small to be restrained by ordinary filtration mediums and are too small to settle by gravity within a reasonable period of time. Such particles must be conditioned so that they form the larger units that will be restrained by filter mediums or will respond to gravity sedimentation.

Flocculation and Deflocculation. The effect of degree of dispersion of the individual particles with respect to each other is illustrated by the accompanying diagrams indicating the behavior of flocculated and deflocculated systems on sedimentation rate and volume. In the deflocculated systems *D*, of Fig. 98, the individual particles are free to move independently of each other, and they therefore settle very slowly, if at all. In the flocculated systems *F* of Fig. 98, the individual particles are attracted to each other by mild but persistent forces, and the resulting size enlargement causes settling to proceed at a faster rate and makes a more voluminous sedimentation layer.

If the individual particles of the dispersed phase of a deflocculated system are small (*i.e.*, of the order of 1 μ or less), sedimentation is usually too slow to be practical. Furthermore, such particles are difficult to separate from the dispersion medium by filtration because they pass through the pores of the filter. Conversely, separation of such particles from a flocculated system is readily accomplished by employing either filtration or gravity separation.

Enlargement of Effective Size by Flocculation. Problems of separating particles from a deflocculated system usually involve a treatment or a conditioning of the individual particles so that they become flocculated. This is most generally accomplished by the use of agents that modify either the interfacial tension between the phases or modify the electrical charges on the individual particles. Surface-active agents and electrolytes, particularly those electrolytes which furnish polyvalent ions or change the pH of the system, are often successfully used in bringing about the desired change in degree of dispersion. Examples of commercial use of such agents to reduce the stabilizing deflocculating forces, thereby permitting flocculation and separation, include (1) use of trivalent aluminum and iron ions to flocculate the particles responsible for the turbidity of many water supplies, (2) use of starch to flocculate highly deflocculated coal slurries, (3) use of glue to clarify strong alum solutions, (4) use of hydrochloric acid to flocculate clay slurries, (5) flocculation of magnesium hydroxide in brine solution, (6) flocculation of metallurgical slimes by adjustment of pH to the isoelectric point, (7) clarification of coal washery water, (8) addition of surface-active agents to break water-in-oil petroleum emulsions, and (9) coagulation of rubber latex by electrolyte treatment.

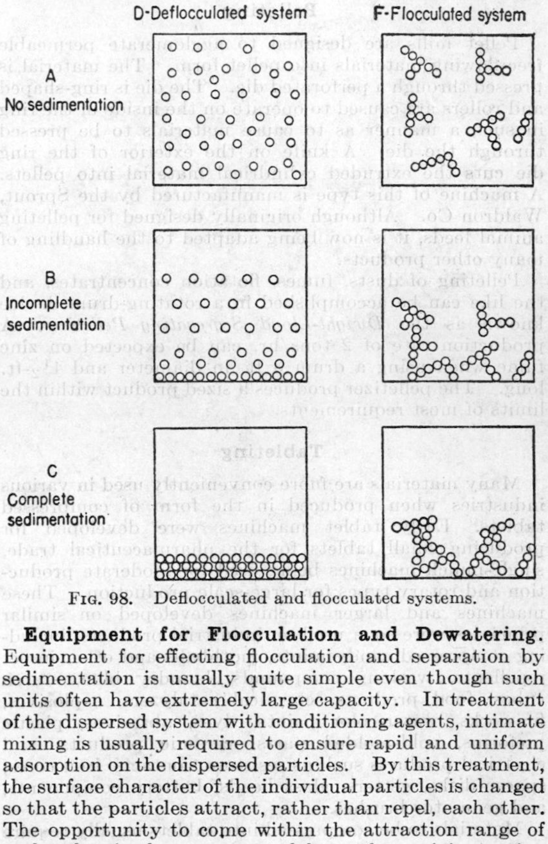

FIG. 98. Deflocculated and flocculated systems.

Equipment for Flocculation and Dewatering. Equipment for effecting flocculation and separation by sedimentation is usually quite simple even though such units often have extremely large capacity. In treatment of the dispersed system with conditioning agents, intimate mixing is usually required to ensure rapid and uniform adsorption on the dispersed particles. By this treatment, the surface character of the individual particles is changed so that the particles attract, rather than repel, each other. The opportunity to come within the attraction range of each other is then augmented by a slow mixing action which permits more rapid flocculation. The flocculated system is then subjected to sedimentation in a container under as nearly stagnant conditions as practical. In water-purification plants, these three operations are spoken of as (1) mixing, (2) flocculation, and (3) sedimentation. Equipment is designed to give the desired critical degree of turbulence for each of the three steps in the process. Equipment for handling such operations includes the *Dorrco Flocculator* and the *Permutit (Spaulding) Precipitator*.

For the further dewatering of sedimented material, a special type of equipment is available for thickening the settled flocculated material. Apparatus such as the *Dorr Tray Thickener* and the *Hardinge Spiral Clarifier* (see also Sedimentation in Sec. 15) employs the principle of slowly raking the sedimented material at a critical rate, which breaks up the fragile flocculated structure of the sediment to permit closer packing without resuspending the dispersed phase. The action is that of converting a system in a condition as shown in Diagram *FC* of Fig. 98 to a condition shown in *DC*.

SIZE ENLARGEMENT BY FUSION

By Gilbert E. Seil

REFERENCES: Greenawalt, The Sintering Process and Some Recent Developments, *Am. Inst. Mining Met. Engrs., Tech. Pub.*, 963, 1938. Shallock, Sintering of Iron Ore, *Iron Steel Engr.*, **18**, 59 (1941). Agnew, Sinter and Blast-furnace Thermal Principles, *Steel*, **113**, 110, 262, 282 (1943). Filer, Sintering of Ore and Blast-furnace Flue Dust, Pittsburgh Carnegie Library, 1944. Seil, Nodulizing Iron Ore, *Iron Age*, **153**, 40 (1944).

Size enlargement by fusion is employed in the preparation of finely divided materials so that they can be used in equipment where high-velocity gases tend to remove small particles from the zone of reaction or where the rate of chemical reaction requires retardation. Many methods have been advocated for converting small particles into larger pieces by fusion. The methods may be divided into two types. One may be called the "static" method in which there is no relative movement between the charge and the equipment in which the partial fusion takes place. In the second method there is a real and continued relative movement between the material treated and the equipment in which the size enlargement takes place. Blast-furnace flue dust, ore concentrates, manganese dioxide ore, magnesia from sea water and brines, magnesite, and dolomite are treated by fusion processes.

Definitions. The following definitions are generally accepted in industry:

Nodulizing. To form in a rotary kiln a mass of large, nearly spherical bodies from fine particles by raising the temperature rapidly to create a liquid phase, allowing the mixture of solids moistened with high-temperature liquid to roll into balls, and cooling until the liquid solidifies to form nodules.

Pelletizing. To make spherical bodies from fine particles by adding a liquid phase at atmospheric temperature, rolling the liquid-coated solids into balls, and drying the outer surface.

Agglomerating. To make any shape of bodies in a rotary device by creating a liquid phase at high temperature but during a relatively long period, coating the solids with the liquid phase and cooling until the liquid phase is frozen.

Sintering. To bed materials with sufficient fuel to cause the formation of liquid phase while the fuel burns, cooling and then crushing the mass. The bed requires a large percentage of coarse material for porosity. This allows the air for burning the fuel to pass through the bed. Very fine particles and high moisture content interfere with the efficiency of the process.

Briquetting. To press in a mold a fine material mixed with a bond or to extrude a mixture of fine material and bond through a die. The formed body may be dried or dried and heated to form a ceramic bond.

Product Characteristics. These products vary considerably in characteristics such as density, chemical reactivity, ability to withstand rough handling, ability to withstand thermal shock, deformation at varying temperatures, and in porosity. It is therefore essential to determine what is required of the finished product before the method for size enlargement is selected. Porosity is an important characteristic of the enlarged particles. Percentage of porosity means very little, since per cent porosity may cover a variety of pore sizes, pore shapes, and wall thicknesses. Control of the size of the finished product can be accomplished by liberating the liquid phase at a predetermined space in the kiln and allowing sufficient time with the proper temperature gradient for the fine materials to roll up into the desired size.

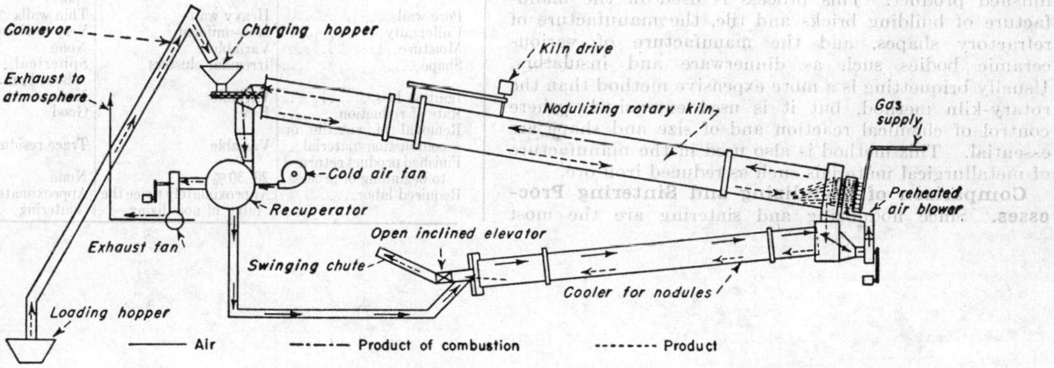

FIG. 99. Diagrammatic arrangement of nodules plant.

Rotary Methods. Nodulizing, pelletizing, and agglomerating are accomplished in rotary devices, such as kilns or driers. The product from cement kilns is known as "clinker." A diagrammatic arrangement of a nodules plant is shown in Fig. 99. The relative movement, the thickness of the bed, and the temperature gradient determine the physical characteristics such as size, porosity, and shape of the enlarged bodies.

Equipment Capacity. The rotary equipment can be designed for a wide range of capacity, since there is equipment in operation as small as 1 ft. in diameter by 18 ft. long and as large as 12 ft. in diameter by 500 ft. long. The capacities of these kilns vary from 200 lb./day to 5000 tons/day.

Operating Temperature. The temperatures at which these operations take place vary from 70° to 3800°F. Except for the cement industry, rotary kilns are not widely used because of the difficulty experienced with mud rings at the charging end of the kiln and fused rings and scale at the hot zones. By the application of sound combustion principles and by the use of directional flames so placed that the heat liberation is directly on the charge instead of on the refractories in the kiln, it has been possible to operate at exceedingly high temperatures for periods of years.

Feed Conditions. The feed to the rotary device used for pelletizing, nodulizing, and agglomerating can vary as to size and moisture in all degrees.

Static Methods. Static methods of size enlargement are divided into two subgroups: (1) sintering, in which there is no accurate sizing or dimensional control of the finished product, and (2) briquetting, in which there is definite sizing and dimensional control of the finished product. In sintering, the materials to be enlarged in size and fuel are mixed and placed on a grate. They are ignited on the top surface while air is being drawn through the bed. The fuel is burned out of the mixture, and the heat fuses a portion of the charge, causing it to freeze together on cooling. The product of this operation is non-uniform in size and porosity and varies in hardness and all other physical properties from one portion of the bed to another. The capacity of sintering equipment usually is exceedingly large, the maintenance is high, and the process can be used only with relatively coarse material having a controlled moisture content.

In briquetting, the fine material is mixed with a bonding compound and is formed into a predetermined shape and size by application of pressure. The briquettes are dried or heated, depending upon the type of bond used. If a ceramic bond is desired, the briquettes are heated either in a periodic or in a tunnel kiln to yield a suitable finished product. This process is used in the manufacture of building bricks and tile, the manufacture of refractory shapes, and the manufacture of various ceramic bodies such as dinnerware and insulators. Usually briquetting is a more expensive method than the rotary-kiln method, but it is used extensively where control of chemical reaction and of size and shape are essential. This method is also used in the manufacture of metallurgical materials such as reduced iron ore.

Comparison of Nodulizing and Sintering Processes. Since nodulizing and sintering are the most important methods of size enlargement by fusion used at the present time, a comparison of the two processes is given in Table 75.

Table 75. Comparison of Nodulizing and Sintering for Plants of the Same Capacity

Item	Sintering plant	Nodulizing plant
Capital costs	Higher than nodulizing	Less than sintering
Operating costs	Higher than nodulizing	Less than sintering
Maintenance	High	Very low
Engineering supervision	Constant	Intermittent
Operation	Semicontinuous	Continuous
Feed	Minimum variation—coarse with minimum of fines with controlled moisture and with solid fuel	All variations—any size, wet or dry, coarse or fine, with or without carbon
Fuel:		
Type	Solid fuel with gas ignition	Any fuel depending upon conditions
Cost	Lower than nodulizing	Usually higher than sintering
Finished product:		
Weight/cu. ft. (iron ore)	100 lb. or less/cu. ft.	140–150 lb./cu. ft.
Porosity	Large open pores	Uniform fine connected pores
Pore walls	Heavy walls	Thin walls
Uniformity	Non-uniform	Very uniform
Moisture	Variable	None
Shape	Irregular clusters	Spherical—averaging about ½ in. in diameter
Ignition loss	Variable	None
Rate of reduction	Fair	Good
Removal of volatile or combustion material	Variable	Trace residual
Finished product returned to beginning of process	20–30%	None
Required labor	Approximately twice the labor of nodulizing	Approximately half of sintering

SECTION 17

MIXING OF MATERIAL

INCLUDING GASES, LIQUIDS, PASTES, PLASTICS, AND SOLIDS*

BY

Kenneth S. Valentine, A. B., Ch. E., New York Manager, The Patterson Foundry and Machine Co.; Member, American Institute of Chemical Engineers.

Gordon MacLean, Ch. E., Technical Director, Allied Foods Co.; Member, American Chemical Society, American Institute of Chemical Engineers.

CONTENTS

FUNDAMENTALS OF MIXING

PAGE

General Objectives of Mixing 1195
Practical Objectives of Mixing 1196
1. Simple Physical Mixture 1196
2. Physical Change 1196
3. Dispersion 1196
4. Promotion of a Reaction 1196
Physical Factors in Mixing 1197
1. Consistency, or Apparent Viscosity, at Mixing Velocities 1197
Viscosity 1197
Plasticity 1197
Mobility 1197
Pseudoplasticity or Thixotropy 1197
Apparent Viscosity or Consistency 1199
Examples of Degrees of Consistency 1202
2. Specific Gravity and Relative Gravity 1202
3. Other Physical Properties of the Materials before or during Mixing 1202
Ease of Wetting 1202
Surface Tension 1202
Particle Size 1202
Temperature Effect of the Addition 1202
Variation of Consistency or Viscosity during Mixing 1202
4. Relative Proportion of the Ingredients and the Order of Their Addition to the Mix 1202

TYPES OF MIXERS

A. Flow Mixers 1203
1. Jet Mixers 1203
2. Injectors 1203
3. Orifice-columns or Turbulence Mixers 1203
4. Circulating Mixing Systems 1203
5. Centrifugal Pumps 1204
6. Spray and Packed Towers 1204
B. Paddle or Arm Mixers 1204
7. Straight Arm or Blade Paddle Mixer 1205
7*A*. Rake Mixer 1205
8. Gate Type 1205
9. Paddles with Intermeshing Stationary Fingers 1205
10. Horseshoe Type 1205
11. Traveling Paddle 1205
12. Rotating Pans with Offset Blades 1206
13. Double-motion Paddle 1206
14. Planetary-motion Paddle 1206
15. Whipper or Emulsifier 1206
16. Air-lift Agitator 1207
17. Kneader 1207
C. Propeller Mixers, Including a Few of the Helical Type. 1208
18. Propeller as a Gas Mixer 1208
19. Propeller with Vertical Shaft 1208
20. Angular, Off-center Propeller—Top-entering 1208
21. Propeller in Side of Tank 1209
22. Propeller in Draft Tube 1209
23. Pug Mill 1209

* The authors are indebted to Professor Shelby A. Miller for his constructive suggestions in the present revision and for contributing certain additions to this Section.

PAGE

24. Soap Crutcher 1210
25. Ribbon (Double-helical) Mixer 1210
D. Turbines or Centrifugal-impeller Mixers 1210
26. Turbine Blower or Centrifugal Fan 1210
27. Simple Turbine Mixer 1210
28. Turbine Mixer with Stationary Deflecting Blades ... 1210
29. Turbodisperser 1211
30. The Continuous Turbomixer 1211
31. Turbo-gas Absorber or Oxidizer 1211
E. Tumbling Mixers 1212
32. Tumbling Barrel 1212
33. Double-cone Mixer 1212
34. Mushroom Mixer 1212
F. Miscellaneous Types 1212
35. Colloid Mill 1212
36. Homogenizer 1213
37. Votator 1213
38. Revolving-cone Mixer 1213
38*A*. The Feld Gas Scrubber 1213
39. Mixing Rolls 1214
40. Pan Mixer 1214

FITTING THE MIXER TO THE OPERATION

Recommended Equipment

1. Mixing Gases with Gases 1215
Desirable Types 1215
2. Mixing Liquids with Liquids 1215
Notes on Tables 1 and 2 1215
Comparison of Mixers for Liquids with Liquids 1215
3. Mixing Liquids and Gases 1216
Notes on Table 3 1216
Comparison of Types 1217
4. Mixing Liquids and Solids 1217
Special Cases 1217
Notes on Tables 4 and 5 1217
Comparison of Types 1217
Continuous Mixing of Liquids and Solids 1220
Continuous Flow, Series, and Countercurrent Tanks . 1220
Proportioning 1220
5. Mixing of Pastes, Plastics, and Doughy Masses 1220
Analysis of Mechanical Actions Involved 1220
6. Mixing of Solids with Solids (also Solids with Gases) ... 1221

HEAT TRANSFER IN MIXERS

Coefficients in Agitated Vessels 1221
Heating Means 1223
Water or Steam 1223
Circulating Systems 1223
Direct Fire 1224
Indirect Fire 1224
Hot Oil 1224
Dowtherm 1224
Mercury Vapor 1224
Electricity 1224
Cooling Means 1224
Evaporation of a Volatile Liquid 1224
Vacuum Cooling 1224

POWER REQUIREMENTS OF MIXERS FOR LIQUIDS

	Page
General Method of Correlation	1224
Paddles	1225
Turbines	1225
Propellers	1225
General	1226

THE TRANSMISSION OF POWER TO MIXERS

Prime Movers	1226
Power Transmission to Mixer	1227

FROM LABORATORY TO PLANT

	Page
Translation	1228
Relative Importance of Mixing Factors	1228
Scale-up of Model Agitator Designs	1228
Heat Transfer	1229
Dissolving Solids	1229
Contacting Immiscible Liquids	1229
Contacting Liquids and Gases	1229
Other Functions of Liquid Mixing	1229
Plastic Masses and Dry Solids	1229

CONTINUOUS VS. BATCH MIXING

Considerations for Deciding Whether Batch or Continuous Mixing	1230
Method for Determining Degree of Completion of Reaction in Continuous Mixing Systems	1230

MIXING OF MATERIAL

References: See other references in text of Sec. 14. Also Bingham, "Fluidity and Plasticity," McGraw-Hill, New York, Hatchek, "The Viscosity of Liquids," J. P. Bell, London, 1926. Carpenter, "Mechanical Mixing Machinery," Benn, London 1925. Seymour, "Agitating, Stirring and Kneading Machinery," Benn, London, 1925. Hixson and coworkers, *Ind. Eng. Chem.*, **23**, 923, 1002, 1160 (1931); **25**, 1196 (1933); **29**, 927 (1937); **33**, 478 (1941); **34**, 120, 194 (1942); **36**, 528 (1944). *Trans. Am. Inst. Chem. Engrs.*, **31**, 113 (1935). Brothman and coworkers, *Chem. & Met. Eng.*, **46**, 633 (1939); **50** (3), 108; **50** (5), 111; **50** (8), 113 (1943); **52** (5), 126 (1945). Chilton, Drew, and Jebens, Heat Transfer Coefficients in Agitated Vessels, *Ind. Eng. Chem.*, **36**, 510 (1944).

The subject of *mixing* has been one of the most difficult of the unit operations of chemical engineering to submit to scientific analysis. To date there has been developed no formula or equation that can be used to calculate the degree or speed of mixing under a given set of conditions. Steps have been taken to improve this situation, however, notably by Hixson, White, Büche, Brothman, Chilton, and their respective coworkers.

It is sometimes said that the power plant input to a mixer alone gives a true measure of the thoroughness of mixing because a definite amount of work is required to mix the particles of material within the container. However, this can *never* be true in practice, because of the immeasurable interferences—cross currents, eddies, etc. which are set up (even in the mixing of plastics and solids) within the container. Consequently, a tremendous amount of power might be consumed in producing a very vigorous local action with good mixing around the mixing element but with no action at all outside of this zone because the energy has been dissipated in producing local interferences.

Because of the fact that the mixing art is so empirical, and because of the almost infinite variety of substances to be mixed, the number of mixes which have been developed is enormous. Some are good and some are poor, but there are few standards. Each industry has developed mixers peculiarly for its own use. Such diversification is not only unnecessary, but it is the greatest obstacle to a sound coordination of knowledge of the subject. When mixing is viewed from the standpoint of the physical characteristics of the materials to be mixed, analysis will indicate that about 40 distinct types will satisfactorily cover every mixing operation in every industry.

The main purpose of this section, therefore, is to classify all mixing problems according to the materials to be mixed and to recommend a type or types of mixer to be used for each of these problems.

FUNDAMENTALS OF MIXING

General Objectives of Mixing

Whether we are dealing with liquids, solids, or gases, or any combination of these phases, the fundamental object to be accompanied by theoretically perfect mixing is always the same. It can be stated as follows: *In all cases, two or more materials existing either separately or in an unevenly mixed condition are, by mixing, to be put into such a condition that each particle of any one material lies as nearly adjacent as possible to a particle of each of the other materials* (see Fig. *A*). In practice these perfect results are never obtained (except in the blending of miscible fluids of low viscosity), and it may be stated that in some few cases such theoretically perfect mixing would be undesirable. The result of mixing may be a blend, a dispersion (suspension, or emulsion), a solution, or a chemical reaction, and in industrial work the desired quality of the finished product in almost every case is the largest factor in determining the required thoroughness of mixing. For instance, in preparing cod-liver oil emulsion (with water), the most intimate mixture is the most desirable because this product must never show separation after being sold, whereas in treating (temporarily emulsifying) certain gasolines with caustic soda solution a too intimate mixture is to be avoided because subsequent rapid separation is necessary.

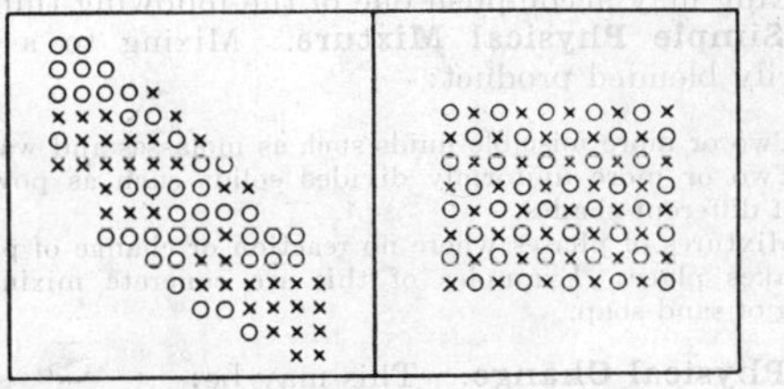

Unmixed — Two Materials Showing Uneven Particle Distribution; Mixed — Two Materials Showing Even Particle Distribution

Fig. *A*. The effect of mixing.

Two general rules must be fulfilled in solving every mixing problem.

Rule 1. A degree of mixing sufficient to yield the desired results must be produced. This rule refers only to microscopic and ultramicroscopic characteristics of the mixture.

Rule 2. A satisfactory rate and direction of motion of the **entire** body of material (however remote from the mixing element some portions are) must be established and maintained, so that **all** the material within the container may be mixed to the desired degree within the optimum time. This rule refers to characteristics of the mixture that may be observed with the eye.

Considering Rule 1, intimacy or degree of mixing depends on the differential rate of flow of the various constituents of the mixture. This differing rate is produced either by direct physical contact between the ingredients of the mixture and the mixer itself (of which the container must always be considered a part), or by the state of motion imparted to the materials by the mixing element, or by both of these. Shear, either directly or through the intermediate steps of momentum and impact, is the most important factor in translating the forces generated by the mixer into differential rate of flow. The greatest intimacy is generally produced in the vicinity of the mixing element, because the most vigorous motion occurs in that region.

A substantial part of the total power input is expended in bringing about this intimacy of mixing, while the remainder is utilized in maintaining the necessary general flow.

When Rule 2 is fulfilled, the entire contents of the container will be evenly distributed and in proper condition to be repeatedly subjected to the more intense mixing action required by Rule 1. Stratification and settling will render the performance of an otherwise excellent mixer quite worthless. Unless both horizontal and vertical flows are sufficient, and unless all the material in the container is moved frequently into the zone of intensified action, whatever mixing, dispersion, solution or reaction takes place there will be **completely** nullified because of insufficient mixing in remote parts of the container. This point cannot be too strongly stressed, for it is often neglected, with disastrous results to yields, time, and power consumed. In cases of liquid mixing, too much turbulence interferes with the proper operation of Rule 2, because so much of the momentum imparted by the mixing element to the liquid is lost in local eddies and interferences that the residual momentum is insufficient to carry with any vigor to the extremities of the container.

Under Types of Mixers, the fundamental principles of design as applied to the solution of typical problems will be discussed.

Practical Objectives of Mixing

Mixing may accomplish one of the following things:

1. Simple Physical Mixture. Mixing to a satisfactorily blended product:

a. Two or more miscible fluids such as molasses and water.

b. Two or more uniformly divided solids such as powdered dyes of different shades.

c. Mixtures of phases where no reaction or change of particle size takes place. Examples of this are concrete mixing and mixing of sand soap.

2. Physical Change. This may be:

a. Dissolution. The passing of gas, liquid, or solid into another phase by solution, such as the dissolving of chlorine, butanol, or salt in water; the deodorization of vegetable oil by blowing superheated steam through it; leaching; etc. The work of Hixson *et al.* at Columbia University forms a valuable and practical addition to the subject of dissolving solids in liquids [Performance of Agitators in Liquid-solid Chemical Systems, *Ind. Eng. Chem.*, **25,** 1196 (1933); **33,** 478 (1941)]. These references constitute important verifications of the "Cube Root Law" of Hixson and Crowell [*Ind. Eng. Chem.*, **23,** 923, 1002, 1160 (1931)].

b. Precipitation. The formation of crystalline or amorphous solids from a supersaturated solution.

c. Evaporation. Removal of all or part of the constituents of a mixture by conversion to the gaseous phase.

d. Extraction. Considered as the passage of material across a liquid-liquid interface, as in the purification of lubricating oils by the "solvent" processes.

e. Adsorption. The selective removal of minor constituents by surface phenomena, as in the bleaching of oils by fuller's earth, the decolorization of sirup, etc., by means of vegetable carbon.

f. Flocculation. The collection of a precipitate into flocs for the purpose of settling or filtration.

3. Dispersion. Mixing to a quasi-homogeneous product.

a. Two or more immiscible fluids.

b. One or more fluids with finely divided solids.

Common Examples. a. Mayonnaise (liquid in liquid); marshmallow whip (gas in liquid).

b. Paints (solid in liquid); ore flotation (solids, immiscible liquids, and gas in liquid).

The continuous phase is the external phase, while the discontinuous (or disperse) phase is the internal phase. If the internal phase is liquid, the dispersion is an emulsion; if it is solid, the dispersion is a suspension; and if it is gas, it is a foam.

For practical purposes the degree of dispersion of the internal phase desired in a finished product varies considerably.

Figure *B* shows a comparison of the ingredients of ice cream before and after homogenization. Both *a* and *b* appear about the same to the eye, but, under the microscope, *b* is seen to consist of much finer and more uniformly sized droplets than *a*. Therefore *b* will produce a more uniform finished product because the butterfat particles have been so completely dispersed that all danger of subsequent agglomeration has been removed.

The particle size of the dispersed phase is measured in microns. (1 micron = 1μ = 0.001 mm.) So-called suspensions or emulsions have a particle diameter greater than 0.1μ. Colloidal solutions have particle diameters between 0.1μ and 0.001μ.

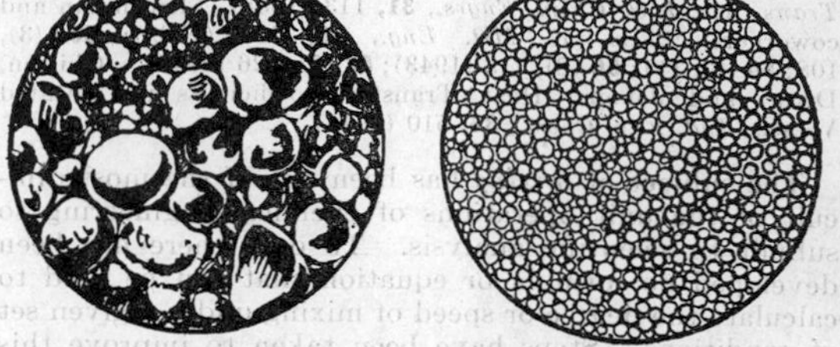

(a) (b)

Fig. *B*. Ice-cream mix before and after homogenizing. (*a*) Rough dispersion. (*b*) Fine dispersion.

Emulsions may be classed as permanent and temporary. In Fig. *B*, *b* would probably be permanent, while *a* could be temporary.

To form dispersions or emulsions that will not *segregate*, one or more of these conditions must be fulfilled:

a. The specific gravity of the phases must be made equal or gravity separation prevented by overcrowding in dispersed phase.

b. The particles must be in Brownian movement because of their small size.

c. If neither of the former conditions can be fulfilled the viscosity of the outer phase must be high enough to make settling negligibly slow or

d. The outer phase must have a slight set (curve *ABC* on Fig. *G*.

To prevent *separation* of dispersions by aggregation of the internal phase, one or more of the following conditions must be fulfilled:

e. The particles of the internal phase must be protected from contacting each other by a film of a third material, either a finely divided solid or a protective colloid, adsorbed by the interface.

f. Same as (*b*) above.

g. The outer phase must have a strong set (butter-holding brine).

h. The specific gravities of the two phases should not be widely different.

Taking examples from successful commercial emulsions or dispersions, we find various combinations of these conditions. The oil in mayonnaise is kept from segregating by (*a*) and from separating out by (*e*) and (*h*). A high-grade non-settling paint gets this property from (*b*), (*c*), (*d*), and (*e*). Scott's Emulsion of cod-liver oil relies on (*c*) and (*e*). It will be noted that the importance of (*e*) is being recognized in the paint industry, and modern formulation takes account of this action.

4. Promotion of a Reaction. Mixing with its consequent blending or dispersion is often necessary to cause or hasten a reaction. With intensive mixing, it is possible in some cases to lower the temperature or pressure at which a given reaction will occur. Examples are: (*a*) neutralizing, (*b*) halogenating, (*c*) hydrogenating oils, (*d*) hydrolyzing starches to sugars, etc. Furthermore, by thorough mixing, undesirable side reactions are prevented by eliminating the danger of local excess of reagents or local overheating.

Physical Factors in Mixing

The physical factors that play a part in all mixing operations are:

1. The consistency, or apparent viscosity, of the mixture at mixing velocities. This is the most important physical factor (see also Sec. 5, pp. 369–374).
2. The specific gravity of the mixture and the relative gravities of each phase.
3. Other physical properties of the materials before, or during, mixing.
4. Relative proportion of the materials and their order of addition to the mixture.

1. Consistency, or Apparent Viscosity, at Mixing Velocities. These important terms, for the purposes of this section, will be considered synonymous. Therefore they require discussion, and the curves in Figs. *C*, *G*, *H*, *I* and *J*, constitute a very necessary part of this discussion. The terms viscosity, plasticity, pseudoplasticity or thixotropy, and inverted plasticity must also be explained.

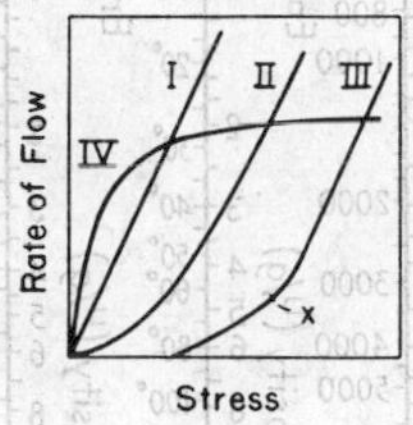

Fig. *C*. Typical consistency curves. I. True viscosity. II. Pseudoplasticity. III. Plasticity. IV. Inverted plasticity.

The curves in Fig. *H* were obtained on the Gardner mobilometer, illustrated in Fig. *F*. The other curves were obtained with the modified Stormer viscometer illustrated in Fig. *E*.

Viscosity may be defined roughly as that property of a fluid which causes it to resist flow. **Fluidity** is the inverse of viscosity. Viscosity is defined by the relationship

$$F = \mu A \frac{dv}{ds}$$

where F is the force exerted tangentially on a plane of area A to produce a velocity gradient dv/ds perpendicular to the plane at its point of application. The coefficient μ, a property of the fluid to which the force is applied, is called the viscosity. The c.g.s. unit of viscosity is the poise; μ has the value of 1 poise if, when $A = 1$ sq. cm. and $dv/ds = 1$ (cm./sec.)/cm., $F = 1$ dyne. The centipoise (0.01 poise) is in more common use than the poise. The English unit of viscosity is lb./(ft.)(sec.) = 0.0672 poise.

Figure *D* gives a means of transposing some of the ordinarily used units to the poise scale. Figure *D*1 offers a method of converting falling-ball-viscosity determinations to poises. With the aid of this chart, viscosities can be determined with a variety of diameters of ball and cylinder. Follow the number progression with a straightedge.

Referring to Fig. *C*, curve I, it is seen that, for a truly viscous substance or mixture, the rate of flow, or rate of shear (expressed, for instance, in centimeters per second), starts from zero and increases at a constant rate as the stress increases. The rate of flow-stress curve, therefore, is a straight line, and the cotangent of the angle which any of these lines makes with the base, as *AOD*, Fig. *G*, indicates the viscosity. Many substances and mixtures have properties approaching such true viscosity; for example, water, glucose solutions, gasoline, or glycerol (see also Figs. *H* and *I*). These are mostly pure liquids, true solutions, or dilute suspensions. It must be assumed that the molecules, colloidal particles, or larger particles here are loosely related to each other. There is no interlocking or overcrowding.

Plasticity differs from viscosity only in the fact that, when a constantly increasing force is imposed on a material or a mixture, a definite yield point must be reached before flow is established. **Mobility** expresses the opposite of plasticity.

The flow of a material which approaches a plastic substance or mixture, is represented by curve III (Fig. *C*). The origin is not zero, for a definite minimum stress, known as the "yield point," is required to start this material flowing, but from that point on it appears to flow as does a viscous material, with constantly increased rate of flow as the stress is increased up to the point where there is a rather sharp upward bend in the curve at x. A theoretically true plastic would not bend at x but would continue in a straight line. The 56 per cent asbestine (magnesium silicate)-in-linseed-oil curve on Fig. *G* is an example of true plasticity.

It may be said that there are few, if any, true plastics, though many materials, such as modeling clay, synthetic resins during molding, cheese, etc., approach this state. It must here be assumed that there is an interlocking arrangement of the molecules, colloidal particles, or larger particles which refuses to change until the yield point is reached, when the interlocking structure breaks down completely.

Pseudoplasticity or Thixotropy. *Non-Newtonian Fluids.* Many materials or mixtures are characterized by flow curves which are combinations of the above two types (curves I and III). Curve II on Fig. *C* illustrates so-called pseudoplasticity [Williamson, *Ind. Eng. Chem.*, **21**, 1108 (1929); Williamson and Heckert, *Ind. Eng. Chem.*, **23**, 667 (1931)]. Such materials are called non-Newtonian fluids. A **pseudoplastic** fluid is one whose apparent viscosity or consistency decreases instantaneously with increasing rate of shear. Though starting from the origin, the curve for a pseudoplastic shows a very slow increase in rate of flow (rate of shear) for moderately low shearing stresses. In other words, the lower portion of curve II stays down near the abscissa and in so doing indicates plastic characteristics, though no definite yield point is evident; but as the shearing stress is further increased we begin to get a sharply climbing curve, which finally becomes a straight line when sufficiently high shearing stress is imposed. In this upper range, viscous flow is evident. In Fig. *G*, the curves for (*a*) 60 per cent lithopone in linseed oil, (*b*) Scott's Emulsion, (*c*) 40 per cent aluminum powder in linseed oil, and (*d*) Hellman's mayonnaise illustrate this property in varying degrees. Figure *H* is the result of a study made on the same materials in a Gardner mobilometer, employing higher percentages of solids to give pastes of high consistencies. The types of curves in both Fig. *H* and Fig. *G* are much the same, indicating similar properties of mixtures of like materials, regardless of the proportion of each in the mixture, *i.e.*, regardless of whether the mixture is a liquid or a paste. The relatively high "apparent viscosity" indicated by the lower portion of the curves is known as "false body."

Total false body may be expressed as $\frac{(\eta_0 - \eta_\infty)}{\eta_\infty}$ where η_0 = apparent viscosity at zero rate of shear, and η_∞ = apparent viscosity at infinite rate of shear (Wil-

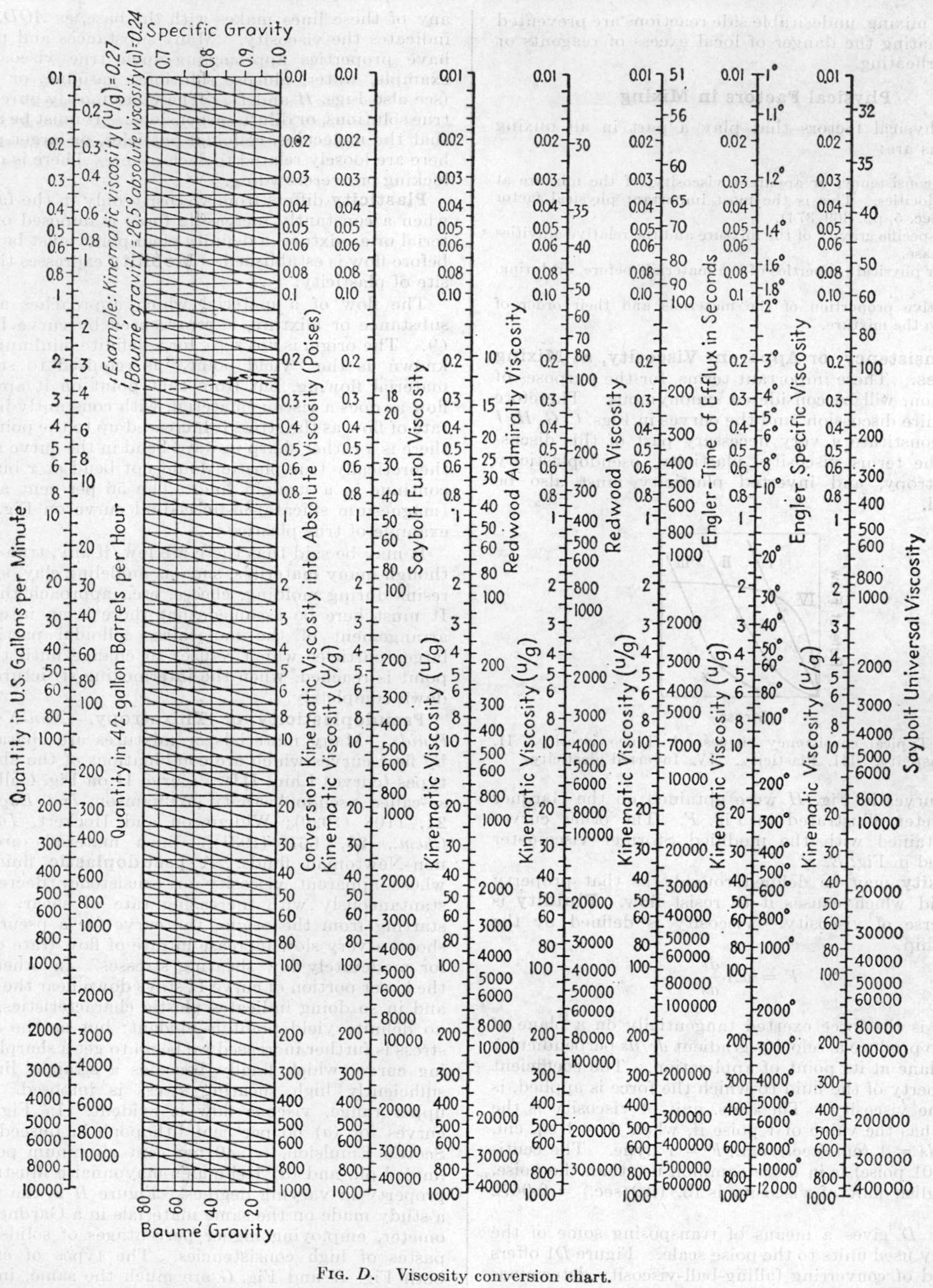

FIG. *D*. Viscosity conversion chart.

liamson, *loc. cit.*). This gradually accelerated rate of flow (shear) as the stress increases is probably due to a progressive orientation of particles to parallelism with the lines of liquid flow and the consequent release of interlocked liquid. Pseudoplasticity is particularly evident in some colloidal dispersions and in some mixtures of solids with liquids, such as certain pyroxylin lacquers, paints, paper-pulp suspensions, heavy gypsum slurries, etc.

A **thixotropic** fluid is one whose apparent viscosity decreases with time to some minimum value at any constant rate of shear. Conversely, when the shearing stress is removed, the apparent viscosity of a thixotrope increases with time to the original value at zero rate of shear. The consistency of such a material is thus a function of its past history.

A **dilatant** fluid, or inverted plastic, is one whose apparent viscosity increases instantaneously with increas-

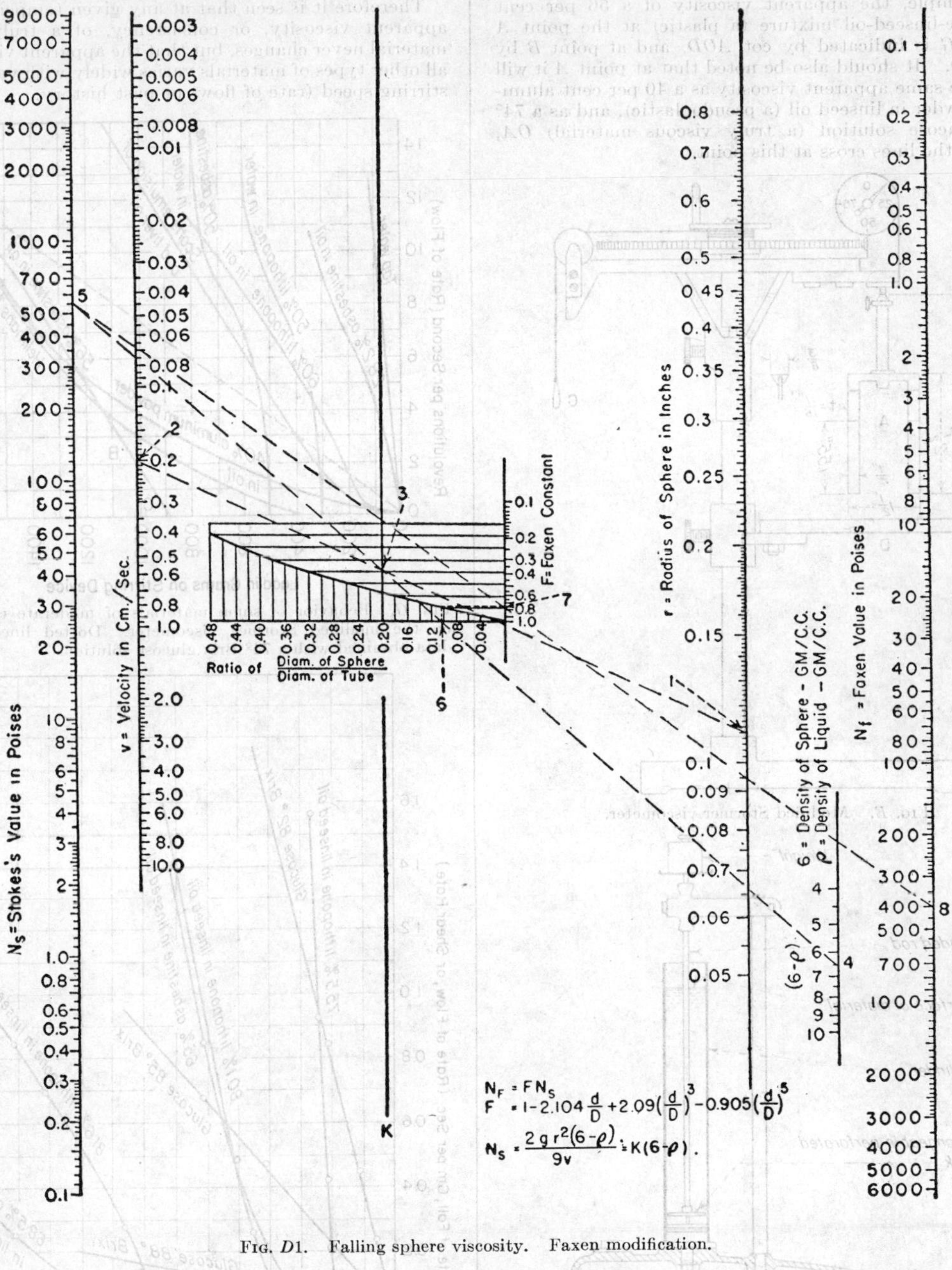

FIG. D1. Falling sphere viscosity. Faxen modification.

ing rate of shear. Heavy starch suspensions and quicksand possess this property (see Fig. J). Inverted plasticity is illustrated by Curve IV.

A **rheopectic** fluid is one whose apparent viscosity increases with time to some maximum value at any constant rate of shear.

It is necessary to describe the above properties of materials before a good picture of **consistency** can be arrived at. It must be fully realized that the consistency of practically all materials and mixtures varies with conditions such as rate of flow, temperature, and pressure. As far as mixing is concerned, it is necessary to define the term "consistency" in some useful way, and, to be practical, a standard means for measuring consistency must be described. These points are strongly stressed because consistency is the most important single factor in any mixing problem.

Apparent viscosity or consistency, for the purpose of this discussion, is considered to be the ratio of shearing stress to rate of shear at the point on the stress-flow curve which represents the agitation condition of the material. It is an index of the resistance of the material to flow. It may be used to characterize any fluid, whether truly viscous, plastic, or non-Newtonian.

For example, the apparent viscosity of a 56 per cent asbestine-linseed-oil mixture (a plastic) at the point *A* in Fig. *G* is indicated by cot *AOD*, and at point *B* by cot *BOD*. It should also be noted that at point *A* it will have the same apparent viscosity as a 40 per cent aluminum powder in linseed oil (a pseudoplastic), and as a 74° Brix glucose solution (a truly viscous material) *OA*, because the lines cross at this point.

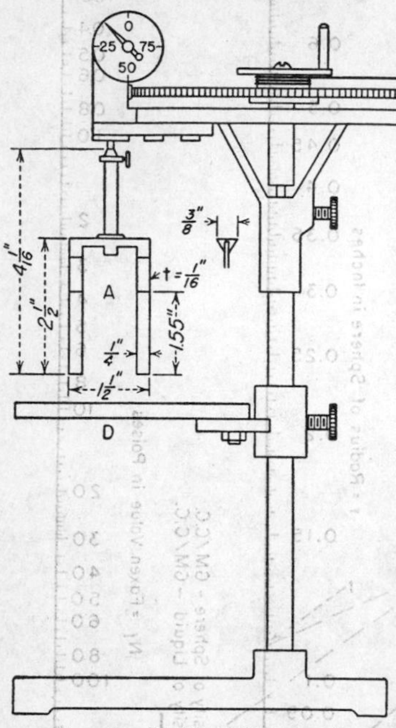

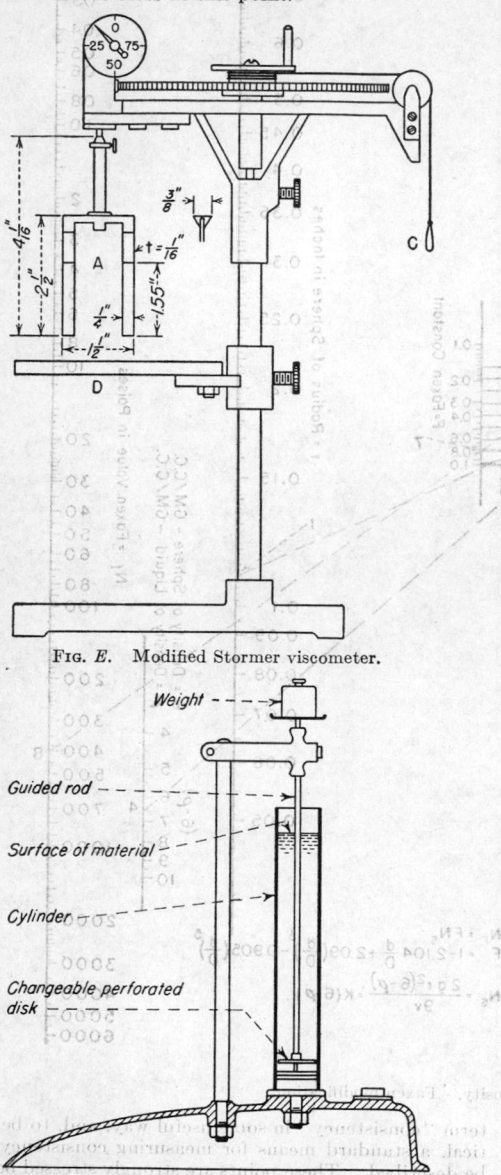

FIG. *E*. Modified Stormer viscometer.

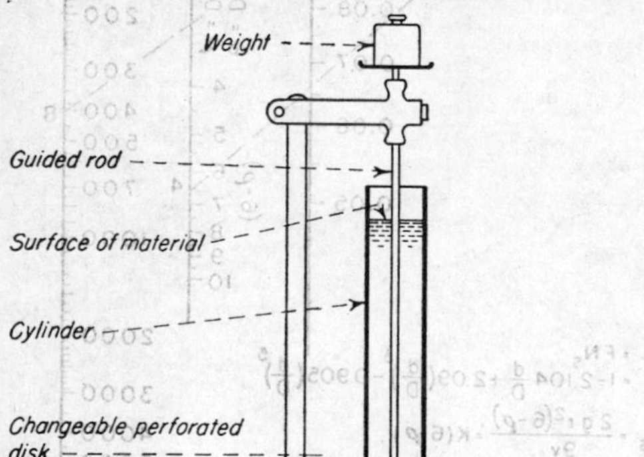

FIG. *F*. Gardner mobilometer.

However, at point *B*, which represents $2\frac{1}{2}$ r.p.s. of the stirrer, the apparent viscosity of the 56 per cent asbestine-linseed-oil mixture is now far greater than that of the 74° Brix glucose solution *OA*, when stirred at the same number of r.p.s., since cot *BOD* is far greater than cot *AOD*. Also, the apparent viscosity of the 40 per cent aluminum-powder linseed-oil mixture now lies between the other two.

Therefore it is seen that at any given temperature the apparent viscosity, or consistency, of a truly viscous material never changes, but that the apparent viscosity of all other types of materials varies widely with variation of stirring speed (rate of flow) or past history.

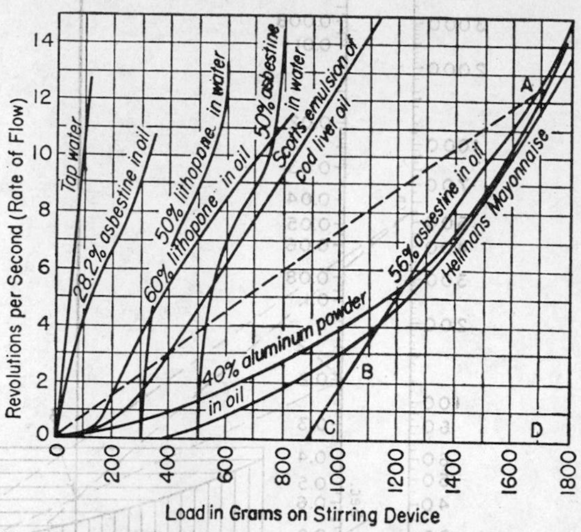

FIG. *G*. Behavior of some materials of moderate consistency on the modified Stormer viscometer. Dotted line refers to data obtained with a 74° Brix glucose solution.

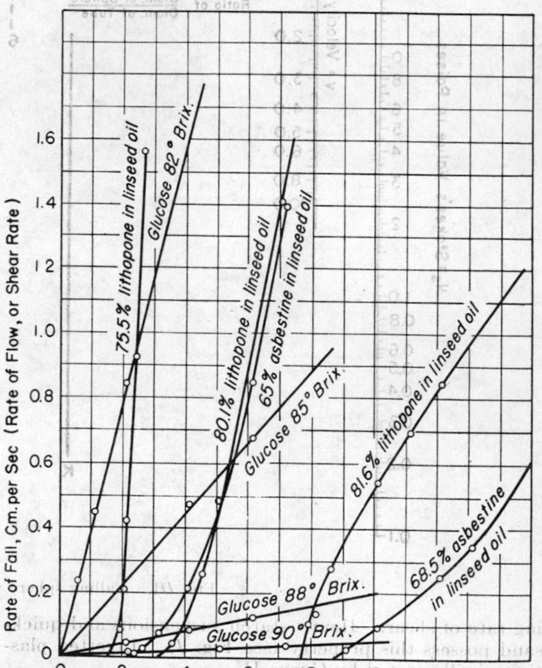

FIG. *H*. Behavior of some of the same materials in a Fisher-Gardner mobilometer with a four-hole disk.

Figure *I* relates variable data of this type to the absolute system so that the apparent viscosity of any material under any given set of mixing conditions may be expressed in centipoises. For example, for all practical purposes, a 13.8 per cent solution of 21 second nitrocotton is found to exhibit true viscosity at all loads and all rates of flow, this viscosity being about 10,500 centipoises. As another example, the point *A* on Fig. *I* is seen to be about 2300 centipoises, and point *B* in 9400 centipoises. Therefore this shows a great increase of consistency of 56 per cent asbestine as the stirring speed drops from 7.1 to 2 r.p.s.

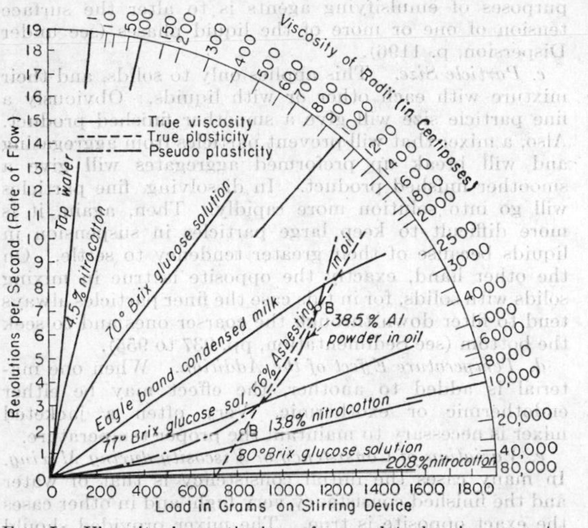

FIG. *I*. Viscosity in centipoises of several truly viscous materials, one true plastic (55 per cent asbestine in oil) and one pseudoplastic (38.5 per cent aluminum powder in oil) at varying rates of stirring (modified Stormer viscometer).

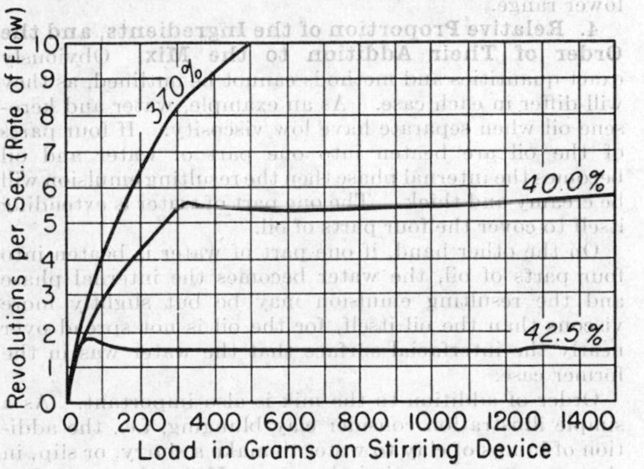

FIG. *J*. Inverted plasticity illustrated by starch suspensions (modified Stormer viscometer).

The modified Stormer, Brookfield, and MacMichael viscometers have been successfully and widely used in the measurement of fluid consistencies. Price-Jones [*J. Oil & Colour Chemists' Assoc.*, **26,** 3 (1943)] has obtained interesting practical results with the modified Couette viscometer. For high consistencies the Gardner mobilometer (Fig. *F*) and the Brabender plastograph are useful. A correct interpretation of the data obtained with such instruments assists in deciding:

1. The type of mixer necessary to establish and maintain the flow required.
2. The size of mixing elements necessary.
3. The optimum speed at which to turn the mixer.
4. The power required at that speed.

For example, a propeller- or turbine-type mixer would be useless on a 42 per cent cornstarch suspension (Fig. *J*) because it must operate by propelling the liquid at a fairly high rate of flow which would be unobtainable in this material. In this case a slow-moving paddle, arm, or helical ribbon type would be more effective. On the other hand, a propeller- or turbine-type mixer of ample size in proportion to the total volume of mix would do well on materials like the 56 per cent asbestine-in-oil in Fig. *G* because the comparatively high rates of flow under which they operate would reduce the apparent viscosity and thus produce a freer flowing mixture. For practical purposes, the point *C* on the 56 per cent asbestine curve has this significance: at any part of the container where a force (per unit area) corresponding to this point, or any lesser force, is exerted, there will be *no motion* of the material; hence mixing will cease.

Other comments on Figs. *G*, *H*, *I*, and *J*:

It is significant to note the strong plastic tendency of lithopone in water as compared with lithopone in oil.

It is also significant to note the much higher yield point of 56 per cent asbestine as compared with 60 per cent lithopone in the same grade of linseed oil.

All curves shown are drawn in accordance with the definition of viscosity, but for "eye" comparisons of this kind the abscissa should really be the "grams weight."

Using different coordinates, higher consistency materials can be pictured more accurately than Fig. *I* permits. The grade of materials used in the preparation of these curves may be identified as follows:

Nitrocellulose solutions from 21-second cotton	Hercules Powder Co.
Cornstarch	Corn Products Refining Co.
Glucose	Corn Products Refining Co.
Lithopone	Krebs Pigment & Color Corp.
Asbestine	Commercial
Linseed oil	Spencer Kellogg Co.

Figure *K* illustrates a practical application of the modified Stormer viscometer in the testing of paint consistency by referring to standard curves prepared by

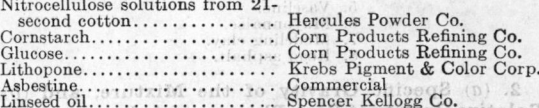

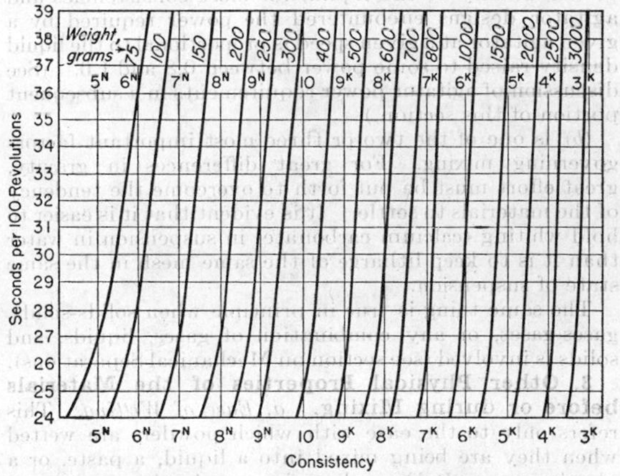

FIG. *K*. Standard curves for use in classifying paints (modified Stormer viscometer).

testing known mixtures. The paint to be tested is preferably contained in a pint can and is thoroughly stirred to make sure that all pigment is in suspension. It is placed on the stand of the instrument and the stand raised until the surface of the paint in the container rises to the marks on the spatula-like blades. Weights are applied and readings taken until a reading falls between 24 and 36 sec. per 100 revolutions of the spatulas. This reading is then spotted on the chart and translated into consistency by dropping vertically to the horizontal scale. The consistency is merely a superimposed arbitrary scale in which 10 consistency represents a medium-bodied paint, 9K to 3K progressively thicker paints, and 9N to 5N progressively thinner paints. The usual types of ready-mixed paints seldom run thicker than about 6K or thinner than about 6N. The instrument does not function with pastes thicker than about 3K.

The authors are greatly indebted to W. W. Heckert of E. I. duPont de Nemours & Co. for the data from which the curves in Figs. *G*, *I*, and *J* were prepared, to E. S. Steinbring of the same company for the data pertaining to Fig. *K* and to O. F. Redd, Technical Director of The Patterson Foundry and Machine Co. for the Gardner mobilometer data shown in Fig. *H*.

Examples of Degrees of Consistency. The following tables express ranges of consistency in terms of materials with which everyone is familiar. (Consider them taken at low rates of shear in all cases.)

Liquids at 68°F.	Approximate Viscosity, Centipoises
1. Ether	0.1
2. Water	1.0
3. Kerosene	10
4. Medium motor oil, S.A.E. 10	100
5. Glycerin or castor oil	1,000
6. Blue Label Karo corn sirup	10,000
7. Molasses	100,000
8. Confectioners' glucose	1,000,000 (or more)

Pastes and Plastics
(In order of increasing consistency)

1. Thickened gravy.
2. Cream sauce.
3. Tomato catsup.
4. Butter at 72°F.
5. Vaseline.
6. Mayonnaise.
7. Modeling clay.
8. Road asphalt.

2. (*a*) Specific Gravity of the Mixture, and (*b*) Relative Gravity of Each Phase. (*a*) is directly related to the power required by the agitator in most mixing applications. Although the power drawn by a mixer operating in a liquid of very high viscosity is independent of the density of the liquid, for most consistencies and agitator designs encountered the power required by a given agitator at a given speed is proportional to the liquid density raised to some power between 0.8 and 1.0. (See discussion of agitator power requirements in a subsequent portion of this section.)

(*b*) is one of the two or three most important factors governing mixing. For great differences in gravity, great effort must be put forth to overcome the tendency of the materials to settle. It is evident that it is easier to hold whiting (calcium carbonate) in suspension in water than it is to keep litharge of the same mesh in the same state of suspension.

The same thing is true in principle when solids-solids, gases-gases, or any combination of gases, liquids, and solids is involved (see section on Mechanical Separations).

3. Other Physical Properties of the Materials before or during Mixing. *a. Ease of Wetting.* This refers only to the ease with which powders are wetted when they are being mixed into a liquid, a paste, or a plastic mass. It is much easier to mix clay into water than to mix zinc stearate into water. It has been shown that solids are more easily wetted by the liquids that are closest to them in chemical structure. For example, zinc stearate would be much more easily wetted by alcohol than by water, because the stearate and the alcohol are both organic compounds. In zinc stearate the long aliphatic chain far overbalances the inorganic zinc with respect to facility of wetting.

One of the most difficult substances to wet thoroughly with water or even with oil is carbon black. This material resists wetting because of the tremendous surface of its particles, which consequently occlude much air. This air must be forced out before the particles can be thoroughly wetted.

Adhesiveness, or stickiness of solid particles, also adds to the difficulty of wetting these particles. Often a very dry powder which is difficult to wet may be incorporated by lowering the surface tension of the liquid.

b. Surface Tension. This greatly influences the particle size and permanence of emulsions, as well as the bubble size of gas dispersions in liquids. One of the chief purposes of emulsifying agents is to alter the surface tension of one or more of the liquid phases (see under Dispersion, p. 1196).

c. Particle Size. This applies only to solids, and their mixture with each other or with liquids. Obviously a fine particle size will give a smoother finished product. Also, a mixer that will prevent particles from aggregating and will break up preformed aggregates will give a smoother finished product. In dissolving, fine particles will go into solution more rapidly. Then, again, it is more difficult to keep large particles in suspension in liquids because of their greater tendency to settle. On the other hand, exactly the opposite is true in mixing solids with solids, for in this case the finer particles always tend to filter down through the coarser ones and to seek the bottom (see Sedimentation, pp. 937 to 955).

d. Temperature Effect of the Addition. When one material is added to another, the effect may be either endothermic or exothermic. Very often a jacketed mixer is necessary to maintain the proper temperature.

e. Variation of Consistency or Viscosity during Mixing. In many cases the initial consistency is that of water and the finished consistency very high, and in other cases the exact opposite is true. The mixer provided should be efficient over the whole range, and it does not necessarily follow that every mixer which is satisfactory for high consistency or viscosity is also satisfactory for the lower range.

4. Relative Proportion of the Ingredients, and the Order of Their Addition to the Mix. Obviously exact quantities and methods cannot be outlined, as they will differ in each case. As an example, water and kerosene oil when separate have low viscosity. If four parts of the oil are beaten into one part of water and oil becomes the internal phase then the resulting emulsion will be creamy and thick. The one part of water is extending itself to cover the four parts of oil.

On the other hand, if one part of water is beaten into four parts of oil, the water becomes the internal phase and the resulting emulsion may be but slightly more viscous than the oil itself, for the oil is not spread over nearly the interfacial surface that the water was in the former case.

Order of addition to the mix is also important. As a simple illustration consider clay blunging, *i.e.*, the addition of lumps of clay to water to make a slurry, or slip, in the cement or ceramic industries. If the clay were first put in the container, and the water then added, mixing, if not impossible, would at least require excessive power through the thick stages. Therefore, it obviously is best to start with the water and add the clay while stirring. In other cases the procedure is equally important but not so obvious.

TYPES OF MIXERS

In the proper design of mixers, not only the mixing element but also the shape of the container must be considered. A very fine mixing element in the wrong vessel may be utterly useless. Furthermore, the exact result to be attained should be kept in mind so that ample mixing may be provided to obtain that result with a large factor of safety. Usually the additional cost required by this extra provision is trifling compared with the cost of all the equipment involved in a process.

Since mixing occupies a place at the very heart of the

process, it is important to do it well. A properly designed mixer may avert a bottleneck in the plant.

The variety of devices used for mixing is extremely large, and many of them have no claims to distinction. Before mixing technology can advance very far, it will be necessary to recognize certain fundamental forms around which our studies and our knowledge may be built. This, of course, does not preclude the future development of new and better forms, but it does give a basis for a certain amount of standardization which is now vitally needed.

Mixers may be grouped under five primary classifications: (*A*) flow mixers; (*B*) paddle or arm mixers; (*C*) propeller or helical mixers; (*D*) turbine or centrifugal-impeller mixers; (*E*) a few miscellaneous types. These may, in turn, be divided so that about 40 truly useful and practical types will cover the entire field of mixing.

A. Flow Mixers

The materials are practically always pumped through this type of mixer, and the mixing effect is produced by interference with the flow. They are used only in continuous or circulating systems for the thorough mixing of miscible fluids. They are rarely used for the mixing of two phases where extreme intimacy is desired. The word "turbulence" does not necessarily imply satisfactory mixing.

1. Jet mixers, such as oxyhydrogen torches, rely on the impingement of a jet against another jet, usually with both jets fed under pressure. This mixer is sometimes used for liquid mixing but finds its greatest application in the mixing of combustible gases just before ignition. See Fig. 1.

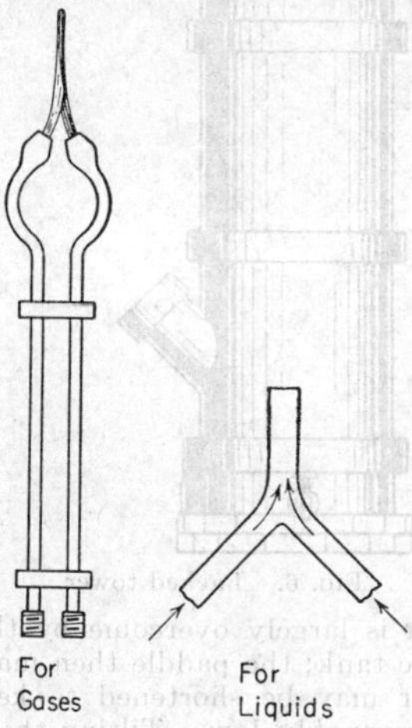

FIG. 1. Jet mixers.

2. Injectors consist essentially of a main pipe and an auxiliary pipe, jet, nozzle, tube, or orifice, through which a second ingredient is injected into the main stream. This simple and inexpensive type of mixer is widely used for the mixing in any proportions of gases with gases, gases with liquids, and liquids with liquids. Bunsen burners, oil burners, spray guns, cement guns, carburetors, atomizers, and nozzle mixers (see Fig. 2) for the mixing of immiscible liquids, are all examples of this type. Either gas or liquid may be the main ingredient. In some cases the velocity of flow in the main pipe induces a flow of material in the auxiliary pipe. In other cases, material is fed through the auxiliary pipe under sufficiently high pressure and velocity to cause the flow through the main pipe. This may be material recirculated from the tank itself by means of an outside pump. A requisite of rapid, thorough mixing in this type is that the mass velocity in the auxiliary stream be considerably higher than in the main stream. Chilton and Genereaux [*Trans. Am. Inst. Chem. Engrs.*, **25,** 102 (1930)] found that, when mixing two gases with this type of mixer, good mixing can be obtained by making the mass velocity of the added stream two or three times that in the main stream.

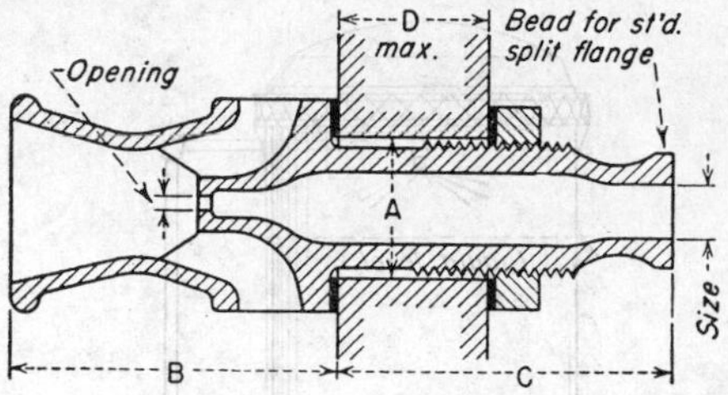

FIG. 2. The injector mixer.

3. Orifice columns or **turbulence mixers,** largely used for the continuous treating of petroleum distillates, may be like *A* in Fig. 3, which shows a simple orifice column, or like *B*, which shows a Duriron nozzle carefully designed to give maximum turbulence. They rely for their action on the translation of pressure into turbulent velocity, finding many applications where viscosity is low enough to allow reactions to be completed with the very short holding time available. Both types are very simple to install.

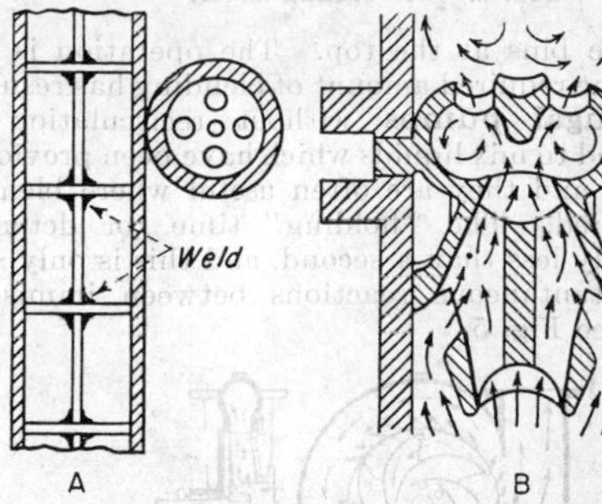

FIG. 3. Baffle-plate and orifice-column mixers.

4. Circulating mixing systems, such as the air lift, "vomit" tubes, long draft tubes, and outside circulating pumps, are usually used to produce a slow turnover of the contents of large tanks by means of comparatively small mixing units. In practically all these circulating types, a very small proportion of the material is being agitated at one time, making them unsuitable where continued intimacy of mix is desired. They are never useful where rapid, thorough mixing is required. Other materials, such as gases, liquids, or slurries, may be introduced in the riser or pump to ensure preliminary absorption or mixing before being discharged into the main tank. See Fig. 4.

Gas Spargers are of this general type. Compressed gas, usually air, may be bubbled into a liquid to produce mixing. The sparger may be a simple open-end pipe, a perforated pipe or plate, or a porous septum. The stream of gas emerging from the orifices of the sparger and rising through the liquid produces circulation of the material that is capable of blending miscible liquids or maintaining suspensions of solids of low settling velocity in a liquid. Spargers are useful in tanks in which, because of their large size or for other reasons, mechanical agitators cannot be used. Their agitating action is mild, and they must be avoided if volatility losses are important. Air must not be used, furthermore, if there is likelihood of undesirable oxidation occurring.

Circulating systems are also used for blending large amounts of solids, usually in excess of 1000 cu. ft. The simplest form consists of two or more bins feeding through automatically regulated feed valves to a belt or conveyor. The material is then elevated and redistributed

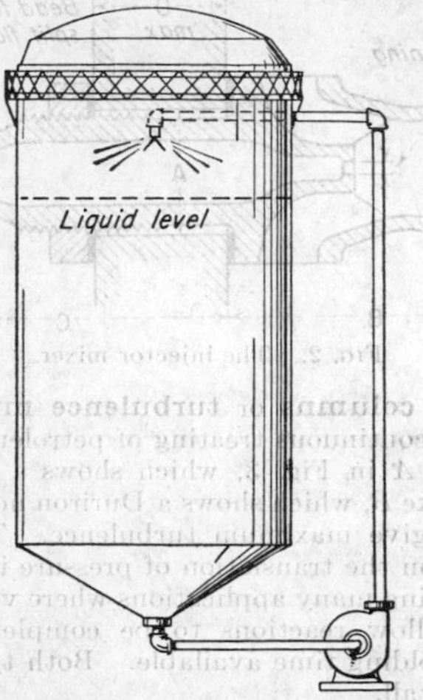

Fig. 4. Circulating mixer.

equally to the bins at the top. The operation is continued until the required amount of blending has resulted.

5. Centrifugal pumps without recirculation are sometimes used to mix liquids which have been previously proportioned, and they are often useful where blending alone is desired. The "holding" time (or detention time) is usually less than a second, and this is only sufficient for instantaneous reactions between immiscible materials. See Fig. 5.

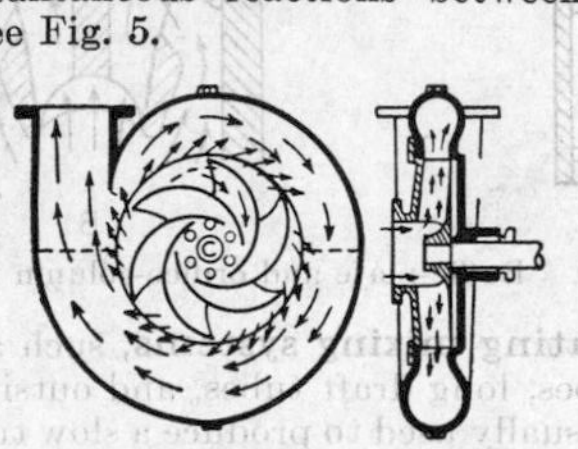

Fig. 5. Centrifugal pump.

6. Spray and packed towers, while used most commonly for the absorption of a pure gas in a liquid or for the removal of some part of mixed gases, are also finding increased application in removing a constituent of a liquid mixture by means of an immiscible liquid of higher or lower specific gravity. Countercurrent operation is the rule for this type of equipment, which has much to do with its success in many applications.

Packed towers are not desirable where there is any tendency to form a precipitate, for the cleaning problem is usually serious. See Fig. 6.

B. Paddle or Arm Mixers

This is probably the oldest type of mixer and consists essentially of one or more blades, horizontal, vertical, or diagonal, fastened to a horizontal, vertical, or diagonal shaft (axis) and rotated axially (though not always centrally) within the container. Thus the material to be mixed is actually pushed, or carried, around in a circular path. In thin liquids in unbaffled containers, paddles always impart a swirling motion to the entire contents of the container. In all cases, that material directly in the path of the blades is always pushed faster than that lying between the blades. This factor has the greatest influence in changing the relationship of successive laminae (or strata) parallel to the blades with respect to each other. Having accomplished this important step, however, paddles lack effective means of producing, perpendicular to the blades, forces that would cut through these strata and cause them to mix with each other. This is their greatest shortcoming.

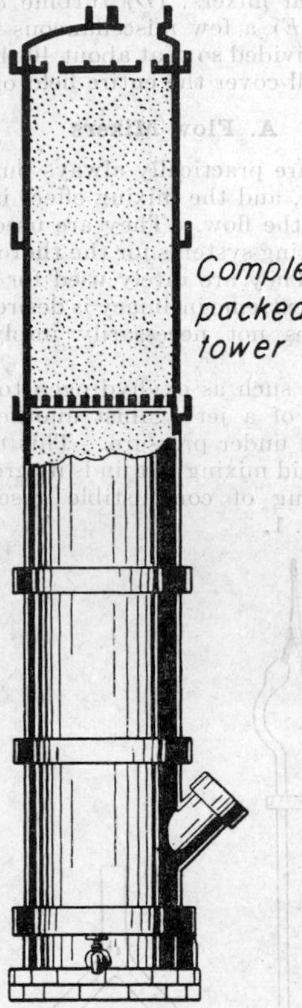

Fig. 6. Packed tower.

Stratification is largely overcome by the installation of baffles in the tank; the paddle then may be operated more slowly or may be shortened to keep the power requirement reasonably low. Tilting the paddle blade (Fig. 7) increases axial flow in a baffled tank, but has practically no effect in a low-viscosity liquid in an unbaffled tank.

Paddle mixers, or arm mixers, are more widely used than any other type, because (1) they are oldest, best known, and first to be thought of; (2) often they can be homemade; (3) the first cost is usually quite low; (4) above all, on many kinds of work they are entirely satisfactory. For instance, for the mixing or kneading of heavy pastes or plastics (or doughs) the arm type (Fig. 17) is indispensable. However, where stratification may easily occur, as in the suspending of fairly heavy solids in light liquids or the mixing of light pastes or liquids of considerable viscosity, a paddle mixer, no matter how carefully it is designed, is comparatively inefficient both as to power consumed and as to quality of results.

7. Straight Arm or Blade Paddle Mixer. This is a common form of mixer and may be either horizontal or vertical. The blades may be either flat or tilted to produce an up or down thrust on the liquid. It is worth noting that in the latter case the result is more nearly that of a propeller than a paddle. Badger, Wood, and Whittemore [*Chem. & Met. Eng.*, **27,** 1176 (1922)] experimented with blades tilted at 45 deg. and found that under the best conditions this mixer took three times as long as and 25 per cent more power than a propeller required to produce the same degree and speed of mixing in the same container. Nevertheless, it was justly concluded

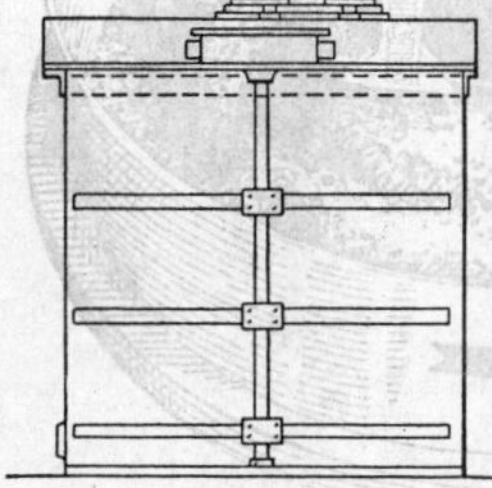

FIG. 7. Simple paddle mixer.

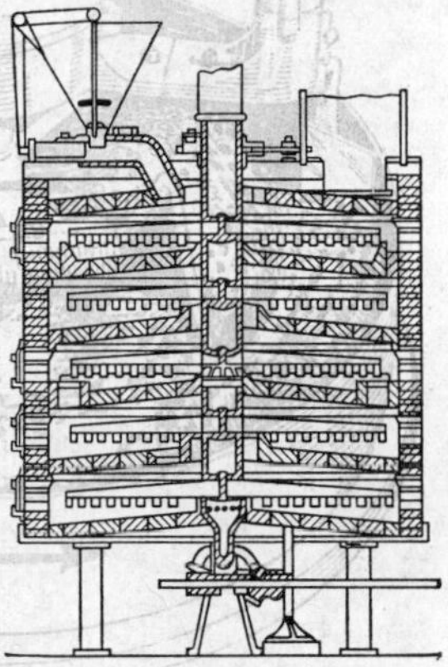

FIG. 7*A*. Rake mixer (Herreshoff furnace).

that this paddle type was entirely satisfactory in many cases where the requirements were not too severe.

7A. Rake Mixer. The rake mixer is a modified straight paddle type. Figure 7*A* shows the Herreshoff furnace, used for roasting ores, the object being for the slowly rotating rakes to bring fresh material constantly to the surface and, incidentally, to break up lumps. Many rake mixtures are built with one set of rakes only, rather than the several superimposed sets shown. A type with much longer rakes is used for turning over grain in malting operations.

8. Gate Type. This type covers many designs of which Fig. 8 is an example. It is questionable whether

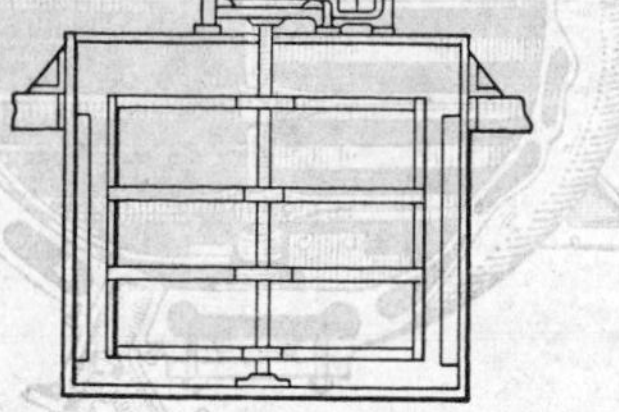

FIG. 8. Gate mixer.

or not a combination of horizontal, vertical, and sometimes diagonal blades improves the mixing, but it is often used where structural strength is desired.

9. Paddles with Intermeshing Stationary Fingers. This type may be horizontal or vertical. In thin liquids the stationary fingers tend to prevent swirl of the entire mass and also to direct currents more or less at right angles to the fingers, thus aiding mixing. This type is also used in the mixing of heavier liquids, pastes, and doughs such as paints, starch paste, and sizes, and in this case the stationary blades aid in the stretching, shearing, folding over, and consequent mixing of these materials. See Figs. 9*a* and 9*b*.

10. Horseshoe Type. This is used in kettles, usually for heavy duty such as grease mixing, caustic fusions, cake dough, etc. A distinctive feature is that the mixing element always conforms to the walls of the container, sweeping or actually scraping them free from pasty or

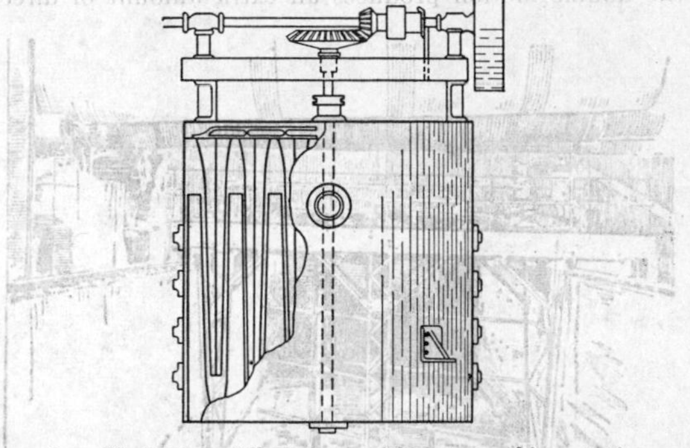

FIG. 9*a*. Vertical paddle mixer with intermeshing stationary fingers.

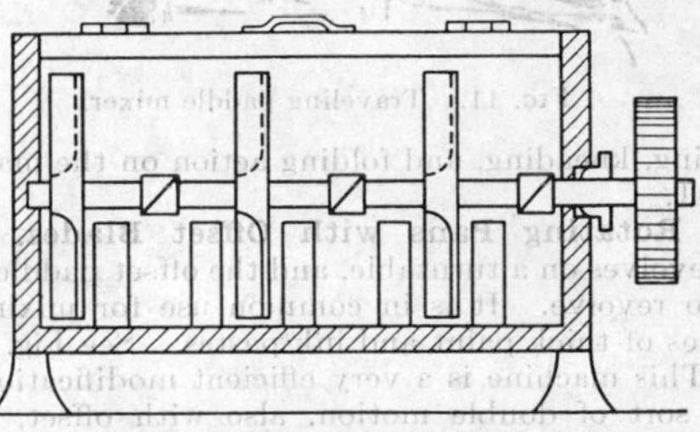

FIG. 9*b*. Intermeshing fingers in horizontal tank.

solid material that might otherwise cake upon them. It is particularly important to prevent this caking in mixtures which burn when locally overheated, or in cases where the walls must be kept clean to permit good heat transfer. Therefore this type (as well as other types mentioned later) is widely used in jacketed kettles or on

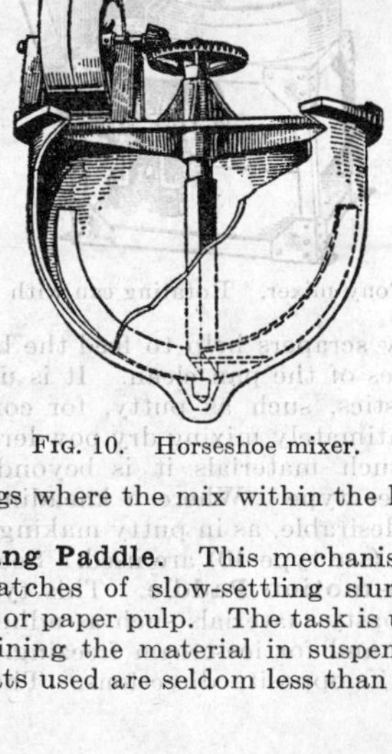

FIG. 10. Horseshoe mixer.

furnace settings where the mix within the kettle is thick. See Fig. 10.

11. Traveling Paddle. This mechanism is used for very large batches of slow-settling slurries, such as cement slurry or paper pulp. The task is usually a matter of maintaining the material in suspension, and the sizes of the vats used are seldom less than 40 ft. long by

25 ft. wide by 20 ft. high and run up to more than 150 ft. long by 40 ft. wide by 30 ft. high. This type is also adapted to circular tanks.

A traveling mechanism, either paddle propeller or turbine, is the only one which will operate successfully on large quantities of slurries and pulps.

Types 12, 13, 14, and 17 which follow represent double-motion paddle or arm mixers and are used for pastes and plastics. In these cases there is little flow produced, and the double motion produces an extra amount of direct

FIG. 11. Traveling paddle mixer.

shearing, kneading, and folding action on the mass. See Fig. 11.

12. Rotating Pans with Offset Blades. *a.* The can revolves on a turntable, and the offset paddles within it also revolve. It is in common use for mixing small batches of thick paint and ink pastes. See Fig. 12*a*.

b. This machine is a very efficient modification of the same sort of double motion, also with offset, rotating blades, and rotating pan. The plowlike blades, mounted with springs, fit the bottom of the pan very closely,

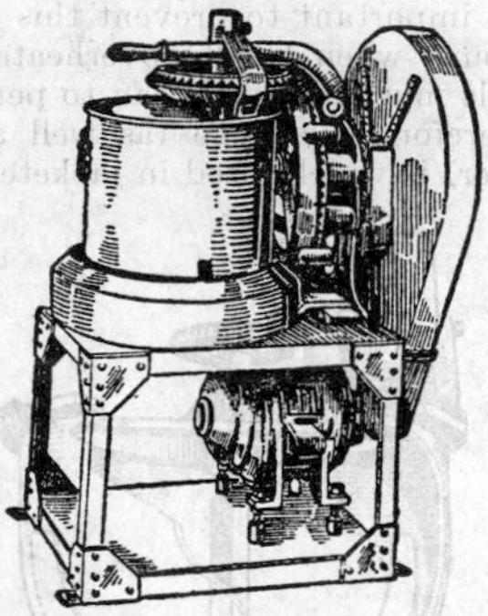

FIG. 12*a*. Pony mixer. Rotating can with offset blades.

and stationary scrapers help to feed the blades and also scrape the sides of the pan clean. It is used for mixing pastes or plastics, such as putty, for concrete mixing, and also for intimately mixing dry powdered or granular solids. For such materials it is beyond doubt faster than any other type. Where a kneading or a smearing action is desirable, as in putty making, mullers as in the pan mixer (see type 40) are used. See Fig. 12*b*.

13. Double-motion Paddle. This type is used extensively for pasty materials such as adhesives, greases, and cosmetics and for ice-cream freezing. Two sets of blades rotate in opposite directions. The outer sweep is often provided with scraper blades which keep the container wall clean. This results in improved heat transfer, making it possible to heat or to cool batches in as little as one-quarter of the time necessary in vessels equipped with non-scraping agitators. This mixing action is probably not to be excelled for the type of work mentioned above. See Fig. 13.

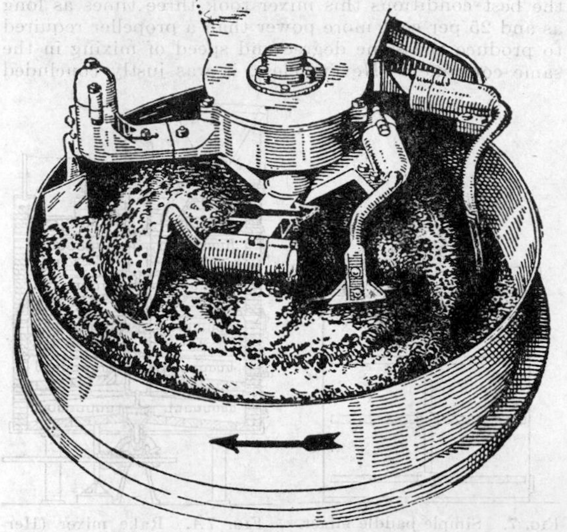

FIG. 12*b*. Rotating pan mixer.

14. Planetary-motion Paddle. This is not unrelated to the traveling paddle of type 11. A paddle rotates on a shaft which is located off center in the kettle or container, and at the same time the shaft revolves around the center or axis of the kettle. This planetary motion causes the action to visit every portion of the kettle in turn, giving thorough local mixing and carrying

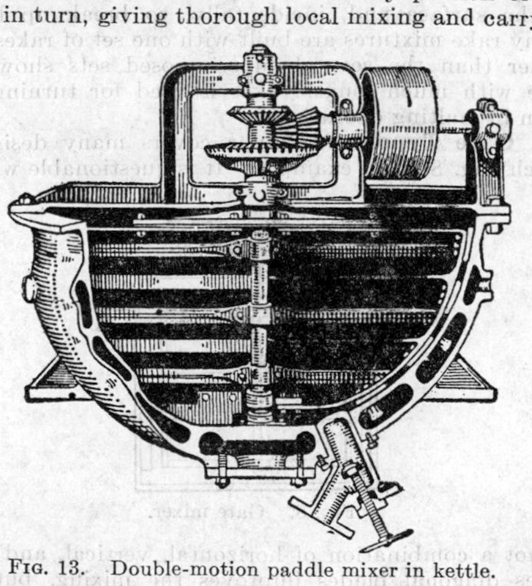

FIG. 13. Double-motion paddle mixer in kettle.

particles forward in overlapping cycloidal paths, thereby producing intermixing. This type is widely used for pastes and doughs, particularly in the food industries, in the manufacture of cake batters, mayonnaise, etc. See Fig. 14.

15. Whipper or Emulsifier. A familiar example of this type is an egg beater. The device, whatever its form, is always run at high speed, and because of the actual **beating** together of the two fluids, a fine state of division, or emulsification, is produced. Often it has two intermeshing grids rotating in opposite directions. It is used extensively for the preparation of whipped cream (liquid and gas), mayonnaise (immiscible liquids), etc. See Fig. 15.

16. Air-lift Agitator. Air forces the slurry up the central tube to the overhead rotating distributor pipe. The slurry flows out and is distributed over the surface.

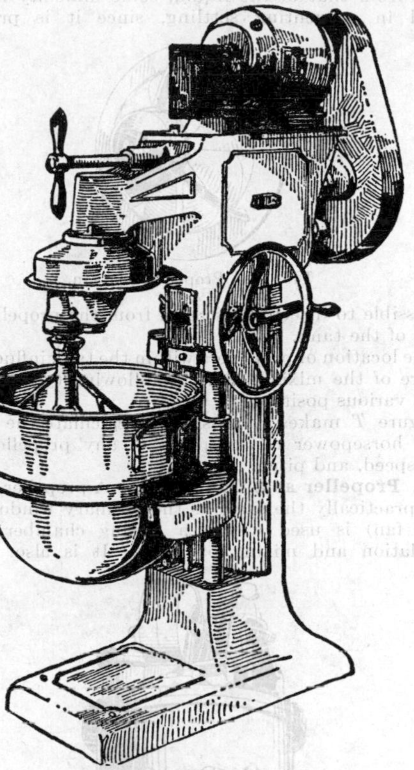

FIG. 14. Double-motion paddle mixer (baker's type).

FIG. 15. Whipper or emulsifier.

The paddle at the bottom is also equipped with an air pipe along its entire length for freeing it when stuck in settled slurry.

This type is useful for maintaining large masses of slurry in suspension. The mixing action, if any is desired, is very slow. Sizes range as a rule from 20 ft. diameter by 12 ft. high, up to twenty times that volume. See Fig. 16.

17. The kneader, with two arms or blades rotating in opposite directions in a container with a divided trough or saddle, is used for mixing thick, plastic, gummy, doughy masses. Of all 40 types of mixers, type 17 is most nearly indispensable to its particular field of materials. Heavy "sigma" blades, slightly helical, rotating oppositely across the trough division, simultaneously effect transportation, kneading, tearing, stretching, folding. (See pp. 1220 to 1221 for fuller description of the action.) These blades or arms are sometimes toothed or serrated to intensify the tearing action, as in pulp shredders. In certain cases the blades may be more like

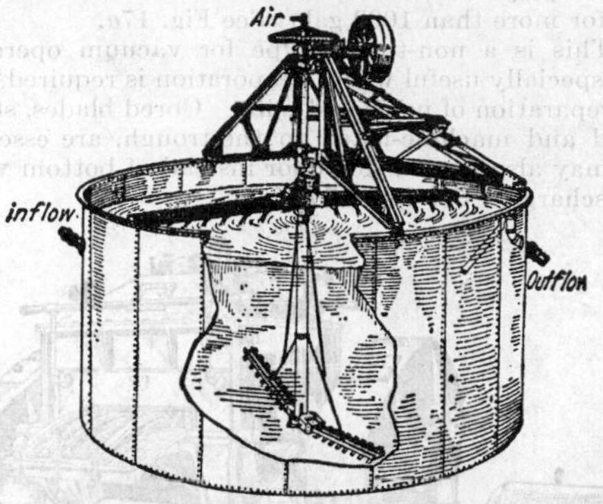

FIG. 16. Air-lift agitator with scrapers.

figure 8's than sigmas in order to double the action per revolution. The two blades are often made so that they overlap to produce a better interchange of material from one blade and trough to the other. On the other hand, non-overlapping blades running at differential speeds may be employed, and in this case they are usually adjusted with slight clearance so that they clean each other in passing and also produce more positive shear, like scissors. For fine work, especially that which involves heat transfer, the blades are machine-fitted to the

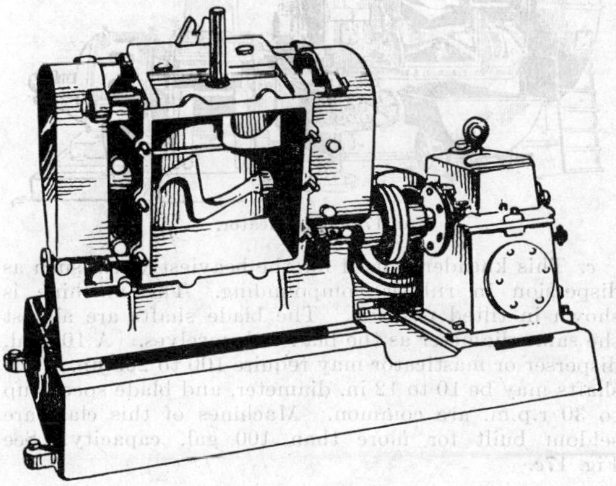

FIG. 17a. Kneader.

trough with as little as 0.001-in. clearance at working temperatures. For greater heat transfer, hollow blades are used.

Because of the difficulty of discharging plastic materials, the majority of kneaders are made to empty by tilting, the tilt being either hand- or power-operated. Other machines empty through a large door at the side of one trough. In cases where the mixture is sufficiently fluid or granular at the end of the process, valves in the bottom of each trough are best adapted for easy and rapid discharge.

a. This is a kneader for general purposes, bread dough representing the mean consistency. A 100-gal. machine requires from 5 to 80 hp. A typical trough is 38 in. long by 32 in. wide by 28 in. high. The sigma blade shafts and the blades themselves vary from 3 to 7 in. diameter as power increases. Bread dough requires about 15 hp. and a heavy asphalt-asbestos mastic about 60 hp. Blade speeds of 20 to 40 r.p.m. are common. Where a differential speed is used, a ratio of 3 to 2 or even 6 to 7 is often employed. Machines of this class are not often built for more than 1000 gal. See Fig. 17*a*.

b. This is a non-tilting type for vacuum operation. It is especially useful where evaporation is required, as in the preparation of powdered milk. Cored blades, steam-heated and machine-fitted to the trough, are essential. This may also have a side door instead of bottom valves for discharge. See Fig. 17*b*.

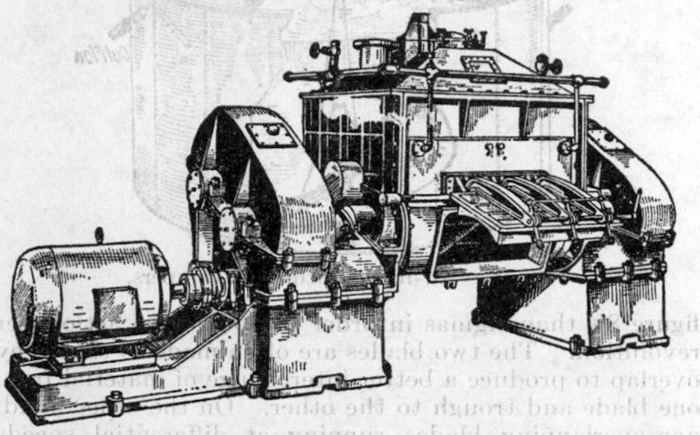

FIG. 17*b*. Non-tilting vacuum kneader.

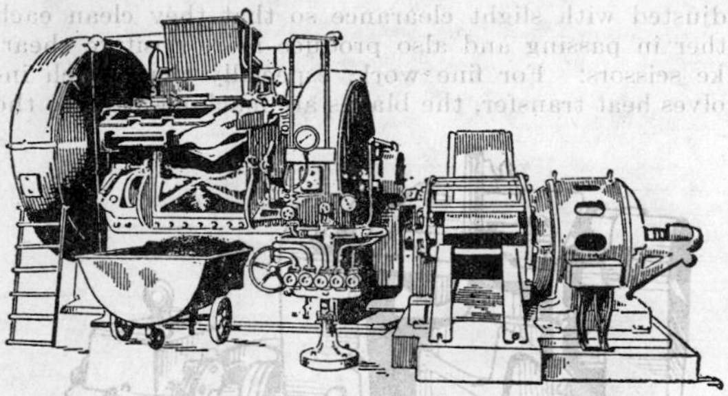

FIG. 17*c*. Masticator.

c. This kneader is used for the heaviest work, such as dispersion in rubber compounding. The machine is shown in tilted position. The blade shafts are almost the same diameter as the blades themselves. A 100-gal. disperser or masticator may require 100 to 200 hp. The shafts may be 10 to 12 in. diameter, and blade speeds up to 30 r.p.m. are common. Machines of this class are seldom built for more than 100 gal. capacity. See Fig. 17*c*.

C. Propeller Mixers, Including a Few of the Helical Type

Propeller mixers furnish an inexpensive, simple, and compact means for mixing in a wide variety of cases. Their mixing action follows from the fact that the revolving helical blades constantly push forward what is to all intents and purposes a continuous cylinder of material, although "slip" induces currents which modify considerably this cylindrical form. Since the propeller causes a cylinder of material to move in a straight line, the shape of container itself will govern the subsequent disposition of this stream. For this reason the shape of the container is particularly important in this case, and yet this factor is often neglected. Figure 22 illustrates how a container may be shaped and a draft tube used to improve propeller mixing.

Propellers are most effective for liquids not over 2000 centipoises apparent viscosity, with or without the presence of light solids, though useful up to 4000 centipoises. Where the specific gravity of the solids differs substantially from that of the liquid, some difficulty is experienced in preventing settling, since it is practically impossible to direct the stream from the propeller to all parts of the tank.

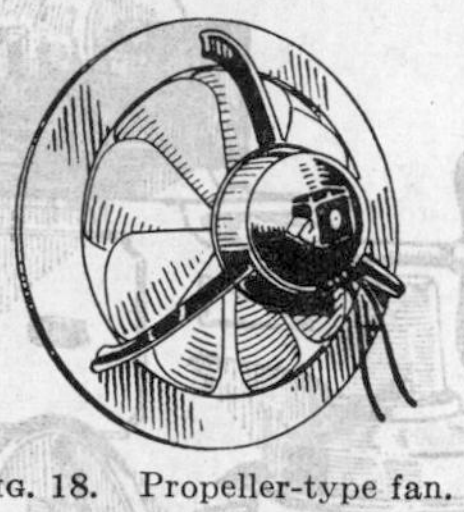

FIG. 18. Propeller-type fan.

The location of propellers within the tank influences the nature of the mixing, and the following types illustrate these various positions.

Figure *T* makes it possible to calculate the approximate horsepower consumption of any propeller, given size, speed, and pitch.

18. Propeller as a Gas Mixer. A propeller, disk, or fan (practically the same as the ordinary window ventilator fan) is used within a mixing chamber to give circulation and mixing of gases. It is also used for mixing gases under continuous-flow conditions. See Fig. 18.

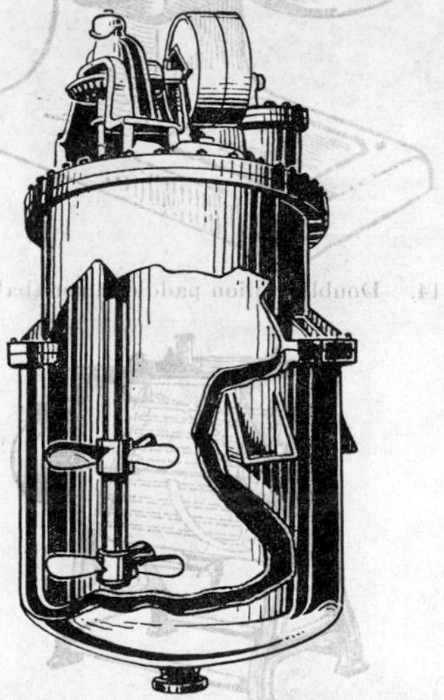

FIG. 19. Propeller mixer, push-pull type.

19. Propeller with Vertical Shaft. These are used in combinations of one, two, or more propellers on the shaft. The propellers may all thrust upward, all thrust downward, or operate on a push-pull basis; the push-pull combination is usually the most desirable for small tanks. See Fig. 19.

20. Angular, Off-center Propeller—Top-entering. This type is mounted on or near the side of the tank with its shaft inclined from the vertical. Usually the shaft is not in the plane of a tank diameter. For viscosities up to 300 centipoises, a direct-connected shaft operating at full motor speed may be used; for higher viscosities, however, a geared machine should be employed. The smaller mixers of this type (⅛-1 hp.) are obtainable as portable units to be clamped to the side of a tank. They are compact and convenient. See Fig. 20.

21. Propeller in Side of Tank. This type is usually located non-radially. The motion produced is a swirl which gradually brings the entire tank contents into the influence of the propellers. This motion is most useful for large batches of light liquid, such as

FIG. 20. Portable-type propeller mixer.

gasoline or aqueous solutions, where great speed of mixing is not necessary. In such cases this type gives satisfactory blending up to 200,000 gal. capacity and is one of the best means of mixing very large tanks of light liquids. In these large tanks it is usually desirable to use two or more units at intervals around the periphery. See Fig. 21.

In the paper industry, propellers on horizontal shafts are used for circulation with incidental mixing of stock in large stock chests, built with one or more mid-feathers. The propeller may be freely installed in one of the aisles,

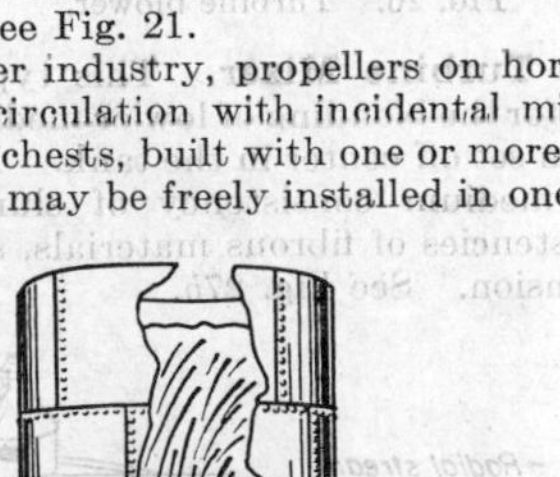

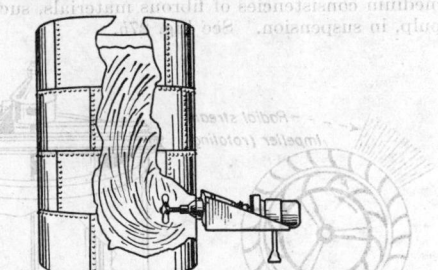

FIG. 21. Propeller mixer. Side-entering installation.

or it may be located in a large hole in a partition placed across the aisle; somewhat like the fan, Fig. 18.

Side propeller agitators vary in size from ½ to 50 hp., with propellers from 4 to 84 in. A good stuffingbox and strong bearings must be provided.

For blending liquids of considerable difference of specific gravity, a high-velocity vertical stream is desirable, and for this purpose the propeller may be replaced by a centrifugal impeller.

22. Propeller in Draft Tube. One or more propellers are surrounded by a tube, which usually has small clearance from the propeller tips. The tube serves to guide fluid through the propeller, appreciably overcoming side slippage of currents. Provided that the tank is well shaped, a thorough circulation at a rapid rate with consequent uniformity of mixing action takes place. This type is probably the most effective axial circulator among propeller mixers and is related to type 24. It is used mainly on light or moderately viscous liquids where intimate mixing is desired. See Fig. 22.

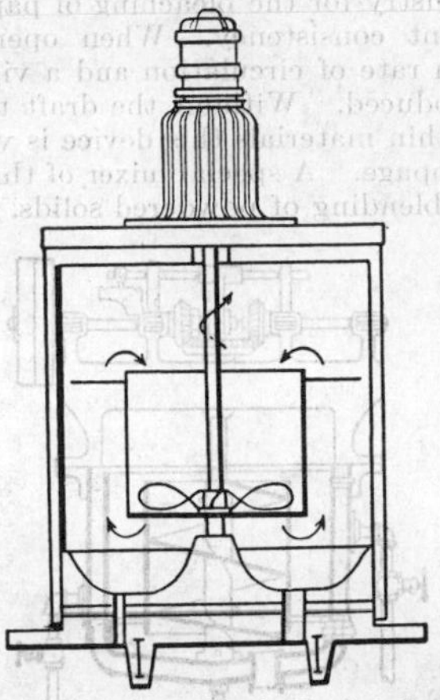

FIG. 22. Propeller in draft tube.

23. Pug Mill. This machine is indispensable to the ceramic and related industries for securing a thorough mixing of very heavy clay masses, usually on a continuous basis. The unmixed or partly mixed ingredients are fed at one end of a trough or cylinder, usually enclosed to withstand heavy pressure, within which a series of very short and stout paddles are revolving. These paddles,

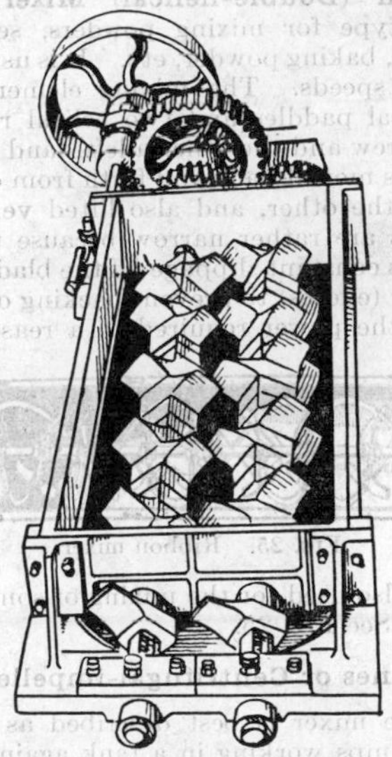

FIG. 23. Open-top pug mill.

tilted to approximate the form of propellers, transport the mixture gradually to the other end of the trough, cutting and kneading it constantly in transit. It is often discharged by extrusion through one or more holes. The entire operation is known as "pugging." Pug mills may be either vertical or horizontal. Because very heavy slips are handled, the power consumption is high. See Fig. 23.

24. Soap Crutcher. This consists of a continuous helix in a draft tube which fits closely around the screw. The course of the material is usually upward through the helix and downward on the outside of the tube. Its conveying action is well adapted to pastes of the consistency of soap, and it is universally used in the soap industry as a mixer. It is also useful for other pasty or fibrous materials. Huge mixers of this type are used in the paper industry for the bleaching of paper pulp at 16 to 18 per cent consistency. When operated at high speeds, a high rate of circulation and a vigorous mixing action are produced. Without the draft tube it is ineffective. On thin materials this device is very inefficient because of slippage. A special mixer of this type is used for the rapid blending of powdered solids. It is usually

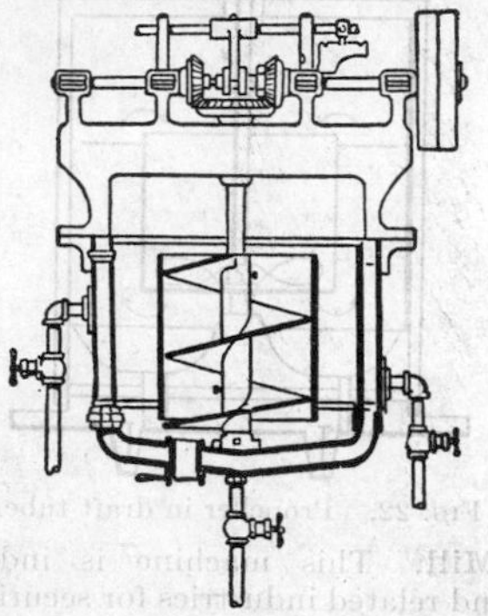

FIG. 24. Soap crutcher.

fitted with a stationary horizontal disk located just above the helix for the reduction of aggregates by a rubbing action. See Fig. 24.

25. Ribbon (Double-helical) Mixer. This is a satisfactory type for mixing powders, self-rising flour mixtures, talc, baking powder, etc. It is usually operated at moderate speeds. The mixing element consists of several vertical paddles and two helical ribbons, one a right-hand screw and the other a left-hand screw, so that the material is moved back and forth from one end of the container to the other, and also lifted vertically. The ribbon blades are rather narrow because the mixing of solids requires constant slippage off the blades to promote local turnover (eddies) to prevent packing of the material and to hold the power required to a reasonable figure.

FIG. 25. Ribbon mixer.

This type is also used for the mixing of some moderately thin pastes. See Fig. 25.

D. Turbines or Centrifugal-impeller Mixers

The turbine mixer is best described as one or more centrifugal pumps working in a tank against practically no back pressure. As is evident from Figs. 27*a* and 28, material enters the impeller axially through the central opening. The material is accelerated by the vanes and is discharged more or less tangentially from the impeller and at fairly high velocity. A curved stationary deflecting-blade ring, which deflects these tangential currents to a radial direction, may be used. The entire direction change from vertical to horizontal and radial is thus accomplished smoothly with the smallest possible loss of kinetic energy, and, as a result, the radial currents are still traveling at high velocity when they reach the remote parts of the container. Thus, Rule 2 (p. 1195) is especially well fulfilled by the turbine type. Rule 1 is also fulfilled, since the discharge from the impeller is scattered along the radii in literally an infinite number of directions and the process is repeated many times a minute. The entire contents of the tank are kept in vigorous and well-directed motion.

When two or more impellers are used, the vertical currents are as shown in Fig. 28.

The power required by a turbine mixer is approximately one-thirtieth that required by an outside circulating centrifugal pump delivering the same volume of liquid, and the mixer impeller revolves at a moderate speed. For instance, a 36-in. rotor turns at about 75 r.p.m.

Turbine mixers are especially useful for mixing viscous liquids or heavy slurries, for suspending heavy solids, for rapid dissolving, for good dispersions, and for mixing in irregularly shaped containers.

26. Turbine Blower or Centrifugal Fan. This type will mix gases very intimately when the gases are supplied in the desired proportions on continuous work. It is also used for batch mixing of gases, being located inside or outside the mixing chamber. It will handle large volumes with low power consumption. See Fig. 26.

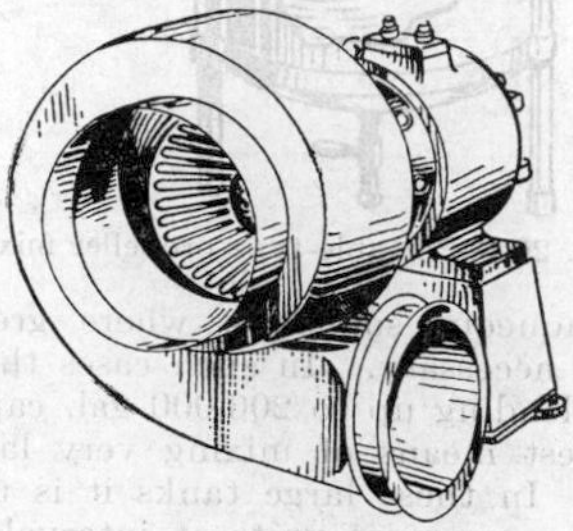

FIG. 26. Turbine blower.

27. Simple Turbine Mixer. This type is particularly desirable for the blending of low or medium viscosity, especially when set off center in the tank. It is also good for low and medium consistency of slurries and for medium consistencies of fibrous materials, such as paper pulp, in suspension. See Fig. 27*b*.

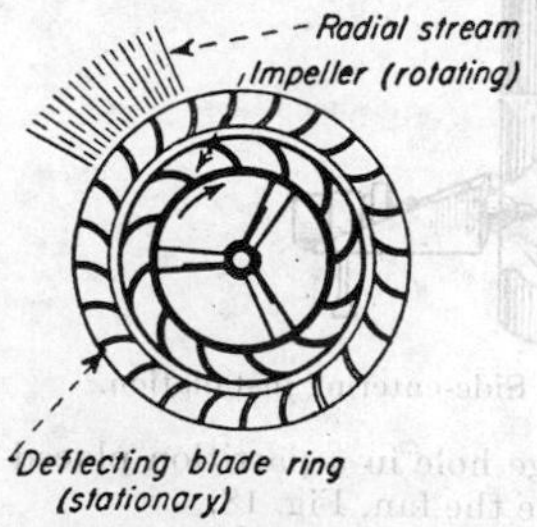

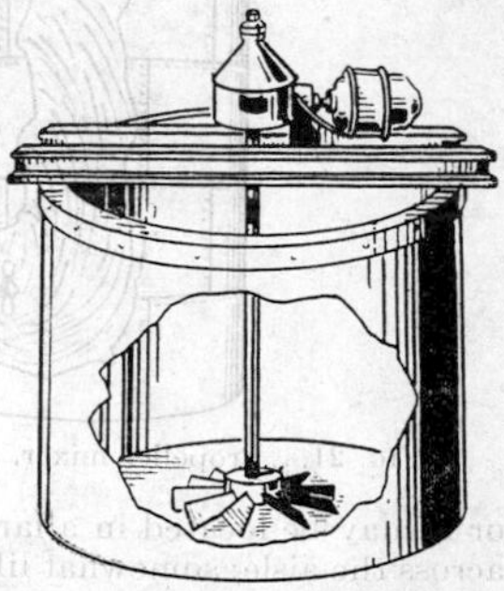

FIG. 27*a*. Typical turbine mixer. FIG. 27*b*. Simple turbine mixer.

28. Turbine Mixer with Stationary Deflecting Blades. With this type one or more impellers may be used. It is characterized by intense shearing action at the impeller, pronounced radial-tangential flow out through the stator element, and good circulation at points remote from the impeller. It is thus best suited for low- or medium-viscosity materials, since in a high-viscosity liquid the circulation currents are seriously damped by the deflecting ring. Where an impeller is used at the bottom of a container, a dished bottom is desirable to help direct the flow upward from the mixing element. See Fig. 28.

Special designs comprising only a few vertical rectangular blades set at a small angle to the tangent of their working circle produce a vigorous cutting, dispersing action when seen at high peripheral speeds.

Type 28 is better than any other mixer for use in irregularly shaped containers, *i.e.*, rectangular tanks, horizontal cylinder tanks, etc., because the radial flow penetrates to the extreme corners.

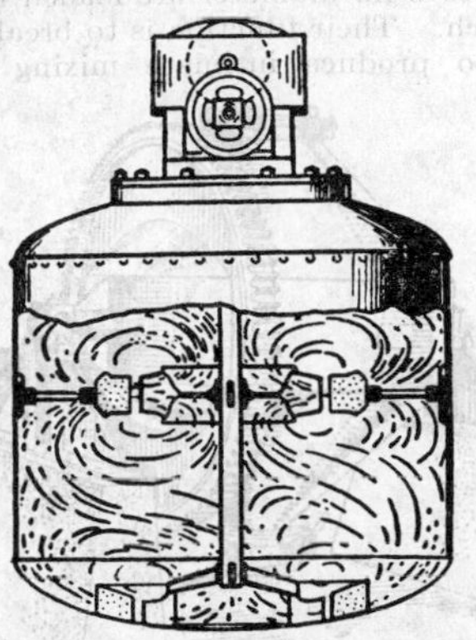

FIG. 28. Turbine mixer with stationary deflecting blades.

29. Turbodisperser. This type consists of a centrifugal turbine impeller which rotates with a screen or perforated plate interposed between impeller and stationary deflecting blades. The turbine blades come close to the screen. The high degree of shear, the extruding action, and the high flow, all contribute toward the accomplishment of dispersions and the dissolving of types of material which are difficult in simpler types of mixer. See Fig. 29.

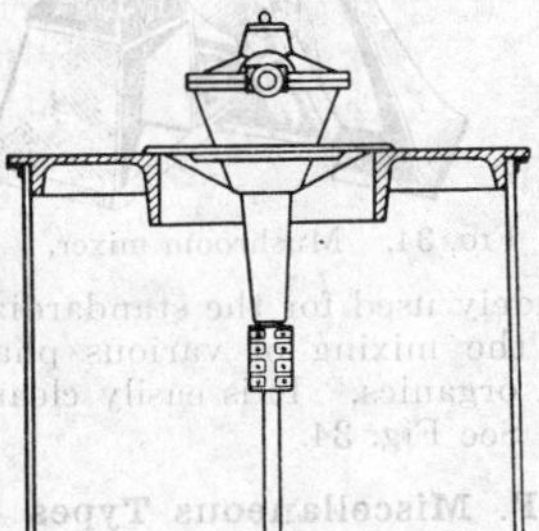

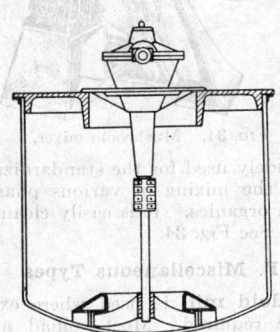

FIG. 29. Turbine disperser.

30. The continuous turbomixer has all the advantages of type 28 but, owing to the continuous flow and the division of the volume into three compartments in series, smaller equipment is allowable for a given production. The main stream has proportioned into it the solid, liquid, or gas with which it is to be treated and passes progressively from one compartment to the next, thus eliminating short circuiting. The tremendous circulation through the centrifugal impellers assures the greatest contact between the materials, so that the resultant rates of reacting, dissolving, or contacting are high. The type of impeller and stator used in each stage is determined by the type of work to be done. A 200-gal. size will dissolve 100 tons of salt per day practically to saturation, while a 100-gal. machine is capable of mixing 15,000 gal. asphalt and naphtha per hour. See Fig. 30.

A modification of this type permits high rate of countercurrent flow to be maintained without other aid than the feed pumps to the system.

31. Turbo-gas absorber is used for promoting contact between gases and liquids. Hydrogenations,

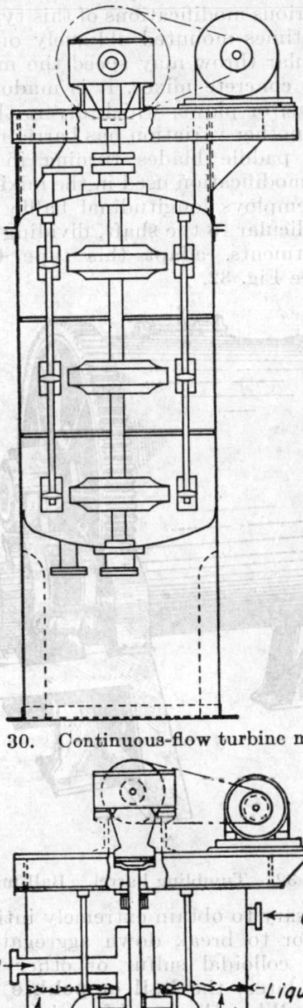

FIG. 30. Continuous-flow turbine mixer.

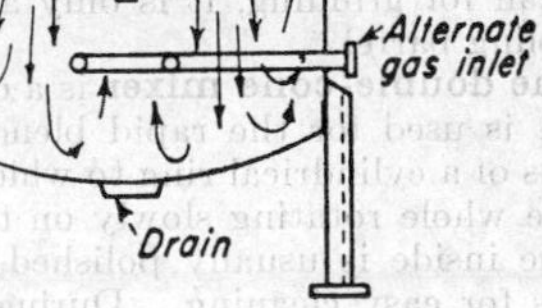

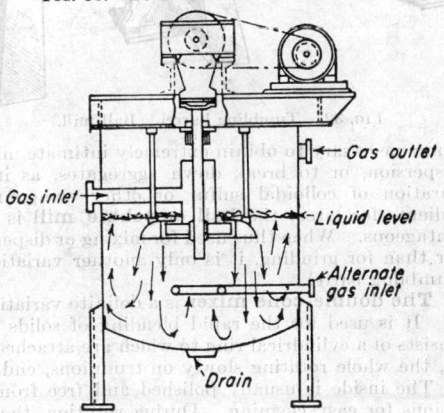

FIG. 31. Turbine gas absorber.

oxidations, chlorinations, purifications, etc., are greatly stimulated by the long gas path through the liquid under violent agitation. The constant distortion of the bubbles of the gas together with the continual exchange of liquid at the interface account for the high efficiency of this type. In some applications the gas is self-induced at the surface, while in others pressure gas is fed to the lower of a series of the absorbers on a single shaft and recirculated from above the surface by the self-induction of the top impeller. Improvements in this type have made it relatively insensitive to moderate changes in the liquid level. See Fig. 31.

E. Tumbling Mixers

32. The tumbling barrel is simple but useful. It consists of a barrel, mounted on a horizontal shaft and rotating with the shaft. Tumbling the barrel over and over mixes the contents. It is extensively used for mixing powders and for all concrete mixing. For types of work involving two or three phases with such widely differing materials as stones, powders, and water, it has no equal. Various modifications of this type exist. The barrel is sometimes mounted obliquely on the shaft, so that the irregular throw may speed the mixing. Sometimes, as in a concrete mixer, it is made with internal baffles, scrapers, or plows, which divert the contents to the outlet. Another variation has barrel rotating in one direction and paddle blades turning in the opposite direction. A modification used in the mixing of hair felt, for example, employs longitudinal baffle blades. Disk baffles perpendicular to the shaft, dividing the body into series compartments, adapt this type to continuous operation. See Fig. 32.

FIG. 32. Tumbling barrel. Ball mill.

Where necessary to obtain extremely intimate mixing, or dispersion, or to break down aggregates, as in the preparation of colloidal sulfur or other compounding ingredients for latex, the **ball** or **pebble mill** is most advantageous. When thus used for mixing or dispersing, rather than for grinding, it is only another variation of the tumbling barrel.

33. The double-cone mixer is a definite variation in form. It is used for the rapid blending of solids only. It consists of a cylindrical ring to which are attached two cones, the whole rotating slowly on trunnions, end over end. The inside is usually polished and free from obstructions for easy cleaning. During rotation, the bottom cone is tilted to a point where the angle of repose of the contents is exceeded. The surface layers then roll down toward the opposite cone, followed quickly by the entire mass, which slips rapidly into the other cone, now near the bottom position. Striking against the conical walls, much material is deflected toward the center and thence upward through the remainder of the mass. Because no two particles take parallel paths and, further, because there is a great difference in the velocities of various particles, homogeneity quickly results. Ten minutes is usually sufficient time for thorough blending of any materials. The mixer is quickly loaded or discharged, a positive-seating, quick-acting, dust-tight discharge valve being provided. A magnetic brake on the drive stops the machine in any position, and an electrical inching mechanism permits it to be brought slowly around to the correct point for charging and discharging. This type is widely used for the mixing of solid granules or powders where speed or great cleanliness is required, *e.g.*, for dry color standardization or for the blending of colored resin products. Power consumption is not over 1.5 hp./1000 lb. of contents. See Fig. 33.

34. The mushroom mixer is the third distinct tumbling type. A flat, covered bowl is mounted on an inclined shaft and rotated. Three to eight heavy metal balls from 3 to 6 in. diameter are loaded into the mixer with the batch. Their function is to break down aggregates and to produce intimate mixing by shearing.

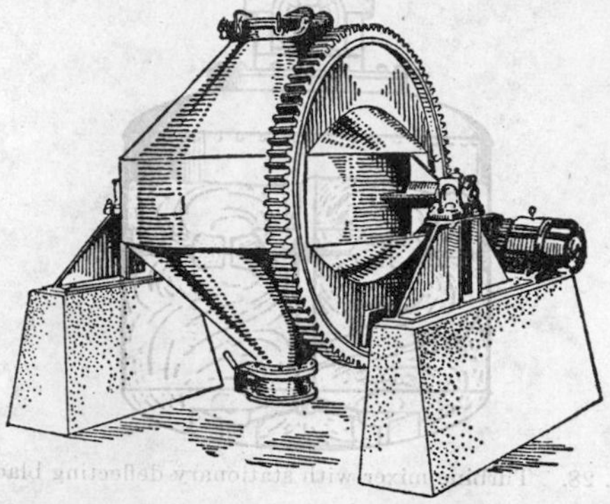

FIG. 33. Double-cone mixer.

FIG. 34. Mushroom mixer.

This type is widely used for the standardization of dyestuffs and for the mixing of various pharmaceuticals, botanicals, and organics. It is easily cleaned and must be dust-tight. See Fig. 34.

F. Miscellaneous Types

35. The colloid mill is used where extremely fine dispersions are required. Most colloid mills are the same in principle, though they may differ in details of construction. As in Fig. 35, the materials to be dispersed are fed between a very rapidly revolving solid rotor and its casing, which it clears by 0.001 in. or less. The rotor may or may not be grooved, and it may or may not be conical. The material is subjected to intense shear and intense centrifugal force, and the combination acts to make excellent dispersions. The material is usually premixed in an ordinary mixer, and this coarser dispersion is then reduced by passing it through the mill. Owing to the electric charge imparted to the particles, and the small size to which they are reduced, emulsions can usually be made with very little stabilizer. Pigments can be dispersed in oils to the original ground particle size, but it is doubtful if actual grinding takes place. Colloid mills have the advantage of giving continuous flow but the disadvantage of a high first cost, high power requirements, and a heating effect on the material. On some types of work where the maximum degree of dispersion is required, nothing else has replaced them.

36. The homogenizer may be described as a positive high-pressure pump in which the pressure is released radially past a disk or valve which is tightly pressed against the end of the discharge pipe by means of a spring. Homogenizing is often done at pressures of 1000 lb./sq. in.

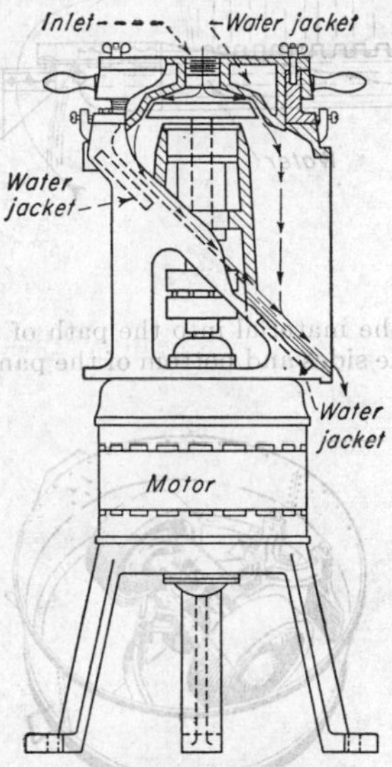

Fig. 35. Colloid mill.

and above. On some products, a finer break-up is obtained by passing the material through a second valve in series with the first valve. The valves are commonly constructed of agate, but today there is evidenced a preference for the use of very hard non-corrosive metals such as Hastelloy and the chrome-nickel steels. The homogenizer is used for breaking up the butterfat in ice-cream mixes, evaporated milk, and other food products, and for the manufacture of emulsions. It cannot be used with materials having any abrasive action. Its disadvantages are about the same as those of type 35, but

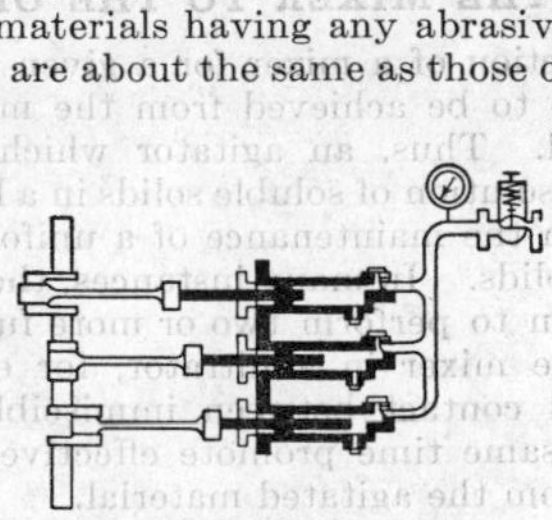

Fig. 36. Homogenizer.

here again it is doing some types of work which no other machine has been able to do. See Fig. 36.

37. Votator. This type is widely used today where fast heat transfer is needed together with a smooth finished product, usually of high consistency, such as paraffin wax, lard, ice cream, etc. As shown in Fig. 37, this precision-built machine consists of a jacketed tube within which a shaft with scrapers rotates at high speed. The diameter of the shaft is about three-quarters of the tube diameter, leaving only a narrow annular space, through which the material in process passes rapidly. The jacket space is narrow so that high velocity of heating or cooling medium is obtained. Because of both extra-good removal of film and high velocities along both inside and outside of the tube, the Votator produces the highest known heat-transfer coefficients in the processing of high-consistency material (see pp. 1221 to 1223).

38. Revolving-cone Mixer. This type usually consists of one or more hollow, truncated cones rotating axially, as shown in Fig. 38. Narrow vertical internal

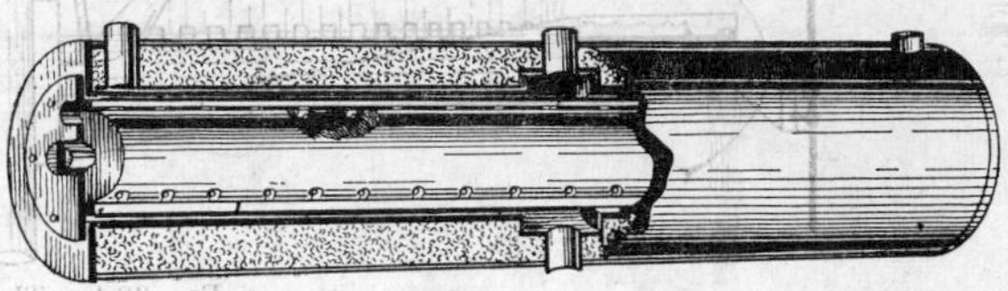

Fig. 37. Votator.

vanes are employed, fastened just inside the cone and extending throughout its height. In some cases where more vigorous shear or radial flow is required, the vanes project beyond the large end of the cone. Cones are used in either upright or inverted position.

The type is most useful in agitating materials of high apparent viscosity or consistency, especially those exhibiting pseudoplasticity or thixotropy, because the material

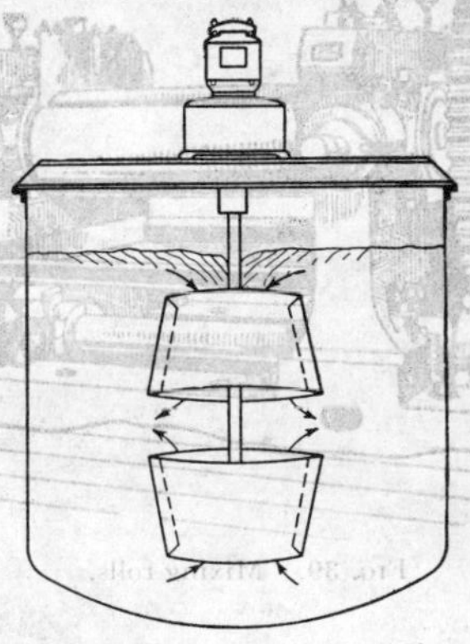

Fig. 38. Revolving-cone mixer.

is actually transported over an appreciable distance and kept under direct shear for an extended time interval as it travels from top to bottom of the cone.

38A. The Feld Gas Scrubber. The Feld scrubber, illustrated in Fig. 38*A*, is a special form of revolving-cone mixer. It operates by throwing screens of liquid across the path of an incoming gas for the purpose of removing certain materials from the gas, either to purify the gas or to dissolve it in the liquid. It is built up of several superimposed elements. Each of these consists essentially of a tray of liquid around which the gas passes, and a rapidly revolving conical frustum. Liquid travels from the tray up the cone by centrifugal force and is discharged as a spray. Passing through this spray, the upflowing gas is scrubbed.

39. Mixing Rolls. This type consists of two rolls, usually turning at different speeds, between which the materials to be mixed are passed. A kneading, tearing, stretching, folding, and shearing action is produced. They are used in certain cases where an exceptionally intimate mixture of a solid with a liquid is desired, *e.g.*, in

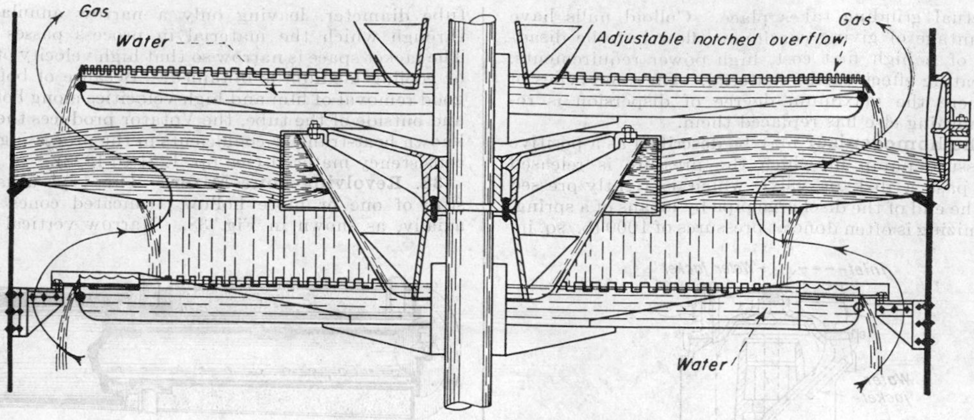

FIG. 38A. The Feld gas scrubber.

printing-ink manufacture. They are also used for the heaviest types of work in which mixing is possible, *e.g.*, mixing fillers into rubber and blending rubber stocks. The rolls are often corrugated to afford a better grip on the material. See Fig. 39.

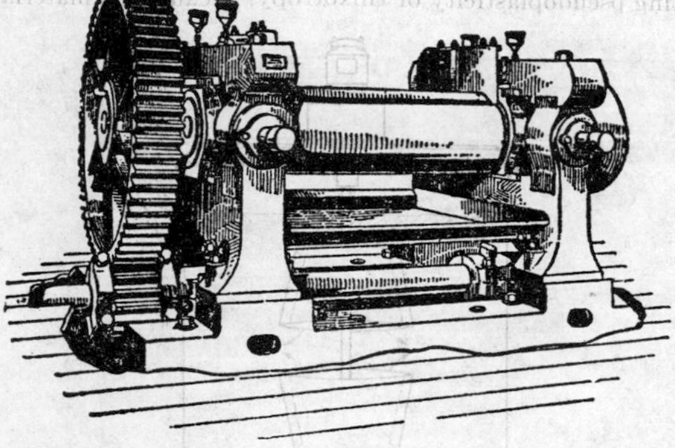

FIG. 39. Mixing rolls.

40. Pan mixer, sometimes known as putty chaser, edge runner, or muller mixer; this type is widely used for the manufacture of putty, *i.e.*, whiting mixed with linseed oil, for clay mixtures, and for other similar operations on plastic and on semidry materials, such as foundry sand. It is also used for the intimate mixing of dry materials, where the breakdown of aggregates or the coating of one solid particle with other solid particles is desired. An enclosed, jacketed form is sometimes used for blending various ingredients of a mixture, with subsequent concentration by evaporation, often under vacuum, as in powder metallurgy. Built on an ancient grinding principle, having one or more large wheels (or mullers) rolling around in a pan, together with scraper blades or plows, it combines a kneading, grinding, and mixing action, giving thereby very intimate mixtures. As the speed of rotation is usually slow, the power required is not excessive. The rollers are usually steel, though sometimes stone; they are more often heavy than light. Since the mullers have a wide face, there is constant twisting or shear on the line of contact between the muller face and the material next to the bottom of the pan. The scrapers, or knives, rotating with the rollers around the central axis, deflect the material into the path of the rollers and also scrape the sides and bottom of the pan. See Fig. 40.

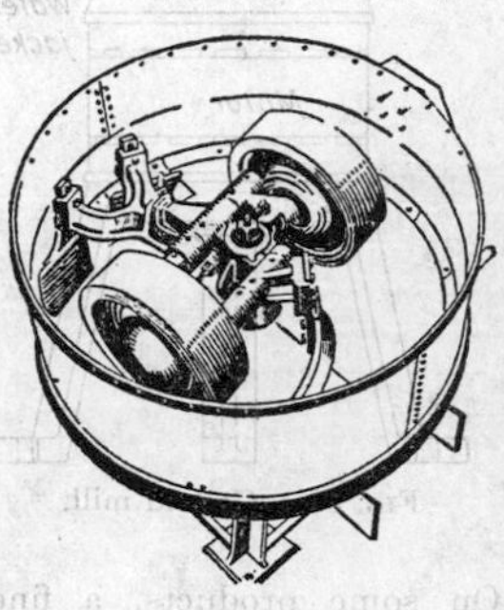

FIG. 40. Pan mixer.

In 5-ft.-diameter pans the mullers may be 30 in. diameter and may weigh from 100 to 400 lb. each. There are some putty mixers with 10-ft. pans and 6-ft.-diameter mullers.

FITTING THE MIXER TO THE OPERATION

In the selection of a mixer for a given operation, the specific result to be achieved from the mixing must be kept in mind. Thus, an agitator which satisfactorily promotes the solution of soluble solids in a liquid may not be effective in the maintenance of a uniform suspension of insoluble solids. In many instances, the agitator may be called upon to perform two or more functions simultaneously: the mixer in a nitrator, for example, must produce good contact between immiscible liquids and must at the same time promote effective heat transfer within and from the agitated material.

The selection of the appropriate mixer will depend also on the medium in which it is to operate: whether a gas, a liquid, or a mass of particulate solids; whether homogeneous or heterogeneous with respect to phases; whether of low or high consistency.

The complete design of an agitator requires the specification of the type of device, the size and proportions of the mixing element, the location of the element within the mixing container, and the operating conditions for the agitator. In general, the optimum mixer, like the optimum equipment for other unit operations, is that which will accomplish the desired result with the most economical combination of fixed and operating charges. In actual practice, one usually selects the mixer which assures effective performance and which is most attractive from the standpoints of installed cost and maintenance, since the energy expended in mixing is frequently only a small part of that required by the process as a whole. The careful designer will scrutinize the energy requirements, nevertheless, particularly for operations (such as the mixing of dough) demanding high-power mixers.

The type or types, the dimensions, and the position of the agitator for the operation are determined first from broad, empirical principles. The operating conditions then are specified; these are less easily generalized, and are derived from the experience of the engineer or the equipment vendor, or from model tests. The scale-up of model tests is discussed elsewhere in this section.

In Tables 1 to 5, mixing problems have been classified according to the materials to be mixed, both by phases (gaseous, liquid, and solid and any combination thereof) and by consistency. A subclassification according to degree of mixing has been employed. The types of mixers best suited to a given case are indicated by numbers which refer to the descriptive items of pp. 1202 to 1214.

When several alternate types are given, that type considered least satisfactory will be given first. Many types of mixers now in use for certain operations have been intentionally omitted because it is believed that the types listed are usually preferable.

Limits of size of batch, particle size, degree of dispersion, consistency, etc., are also given, **although it must be realized that these limits are in no way fixed,** and there are no rigid rules covering any of this work. The limits are empirical but appear to be entirely reasonable in the light of experience.

The general type only can be indicated. For questions in individual cases regarding power, exact size, price, etc., of mixer, the manufacturers should be consulted. Representative manufacturers of the most usual types of mixers have been listed by Smith, *Chem. Ind.*, **64,** 399, March (1949).

1. Mixing Gases with Gases

Mixers of jet (1), injector (2), baffle-column (3), propeller (18), or turbine (26) types are universally used for gas mixing.

The mixing of gases is not ordinarily considered a difficult operation. Gases are readily moved and can be made to flow together without elaborate equipment. Nearly all mixing of gases is done as a continuous process by any of the above-mentioned types, but more especially by types 1 and 2, which are the simplest forms.

Where batch mixing is desired, especially in cases where great differences in specific gravity exist, a mechanical type such as a propeller (fan) within the container is advisable (type 18).

2. Mixing Liquids with Liquids

Notes on Tables 1 and 2.

1. As the viscosity increases, the size of the so-called "small" batch becomes smaller, so that at 200,000 centipoises we should not expect to handle over 5000 gal. per batch. For intermediate points, construct a centipoise-gallons graph using 100 centipoises, 20,000 gal., and 200,000 centipoises, 5000 gal. Connect these two points by a straight line and interpolate.

2. The same considerations apply to all other limits of quantity and degree here given. There are no hard and fast rules to be applied. They will vary for different cases. The limits are given here as reasonable suggestions.

3. Emulsifications or dispersions require sufficient time of exposure of the liquids involved to the action of the agitator. This is fixed by cycle length in a batch process and by through-put rate, degree of recirculation, or by number of agitators in series in a continuous process.

4. Continuous dispersions must always be done rapidly.

Comparison of Mixers for Liquids with Liquids.

Compressed air or other gas may be sparged into a liquid to produce mild but satisfactory agitation for the blending of miscible liquids. It is frequently used for agitating laboratory baths or for other small-size installations, such as photographic developer tanks [Ives and Kunz, *J. Soc. Motion Picture Engrs.*, **34,** 364 (1940)]. It may also be employed in plant-scale equipment in which it is undesirable or impractical to install a mechanical mixer. Kauffman [*Chem. & Met. Eng.*, **37,** 178 (1930)] reported that 0.65–3.1 cu. ft. of air/(sq. ft. of tank cross section)(min.) are required to produce moderate to violent agitation in 9 ft. of liquid. The obvious disadvantages of gas sparging are loss of vapors and danger of oxidation when air is used. The agitation produced is usually too mild to move immiscible liquids of appreciable density difference into good contact.

Outside circulation offers a method of mixing liquids in tanks without putting a mechanism in the tank. Heavy liquids can be pumped from the bottom and distributed over the top. Power costs are higher than for internal mixing, but the pumps are usually already present for other purposes and hence are available without extra cost. This kind of mixing is always slow.

Paddles, if they are long and slow-moving in unbaffled tanks, produce relatively mild agitation, but are satisfactory blenders of miscible liquids. They are less effective than turbines for high-viscosity blends, as they permit stratification and consequently neglect some portion of the container, and are not recommended for mixing immiscible liquids. Wood, Whittemore, and Badger [*Chem. & Met. Eng.*, **27,** 1176 (1922)] showed complete mixing of brine and water in a 600-gal. tank in less than 1 min. at low power input to a long paddle.

Short paddles with straight blades (blade diameter < one-half the tank diameter) and **paddles in baffled tanks** produce pronounced axial flow and vigorous agitation in low-viscosity liquids. Miller and Mann [*Trans. Am. Inst. Chem. Engrs.*, **40,** 709 (1944)] produced excellent dispersions of kerosene and water by means of short two- and four-blade paddles in unbaffled tanks (paddle-to-tank diameter ratio of 0.33). The power required was about 1 hp./1000 gal. of mixture.

NOTE. For continuous processes the above types should be avoided.

Propellers are particularly valuable for large-quantity blending of light liquids and for smaller quantities of liquids of medium viscosity. Because their mixing action is rapid, they may be used for both batch and continuous work. Their first cost is moderate. If they are mounted off-center and angularly they are excellent contactors of low-viscosity immiscible liquids where not too fine a break-up is desired. They should be avoided for high-viscosity liquids.

Turbines, if properly designed, will mix all the classes listed in Tables 1 and 2 and probably excel all types for mixing viscous liquids. A turbine operating within a stator will excel all other types except a colloid mill and a homogenizer in producing fine dispersions. This is particularly true of the turbodisperser (30). Furthermore they are the fastest of all types, and their power consumption is moderate. Because of the ability of a

Table 1. Mixing Liquids with Liquids.* Preferable Types of Mixers for Miscible Liquids

Basis of operation	Speed of operation	Consistency: Thin, to 100 cp. Example: blending gasoline	Consistency: Medium, to 2500 cp. Example: blending lube fuel, or Diesel oils	Consistency: High viscosity, to 200,000 cp. Example: blending pyroxylin bases
Small batch High viscosity, to 5000 gal. Thin and medium, to 20,000 gal.	Slow Thin and medium, 15 min. High viscosity, 1 hr.	Air Outside circulation (4) Paddles (9) Propellers (20,21) Turbines (28)	As for thin	Paddles (9, 10, 12) Turbines (27)
	Fast Thin and medium, 30 sec. High viscosity, 10 min.	As above	Paddles (7) Propellers (20, 21) Turbines (27, 28)	Turbines (27)
Large batch High viscosity, to 20,000 gal. Thin and medium to 200,000 gal.	Slow Thin and medium, 15 min. to 3 hr. High viscosity	As above	As for thin	
	Fast Thin and medium, 2 to 30 min.	Propellers (21) Turbines (27, 28)	As for thin	
Continuous Any desired gallonage through-put	Slow Up to 6 hr. as in sewage treatment	Air (special design) Paddles (7, 9) Turbines (28)		
	Fast 1 to 60 sec.	Injectors (2) Orifice column (3) Pumps (5) Propellers (22) Turbines (30)	Pumps (5) Turbines (30)	Turbines (30)

* The least desirable mixer is listed first. cp. = centipoises.

properly designed turbine-type mixer to approximate instantaneous mixing, it is the most satisfactory type for continuous work. As a partial offset to these advantages the structure of a turbine is more complicated (making it more difficult to clean) and its first cost is usually somewhat higher.

Jets, orifices, turbulent-flow tubes, and pumps are excellent for the continuous blending of low-viscosity liquids. Where space is available for its installation, a turbulent-flow section of pipe 50 diameters in length is perhaps the simplest of all mixers. The designer should be assured, however, that the flow is turbulent. Jets and orifice columns will provide fairly coarse dispersions of immiscible liquids with economy, but fine dispersions may involve unattractively high pressure drop and power. These types operate best at one capacity, and the results obtained are variable when this is changed. The centrifugal pump is a good contactor, but it is difficult to control.

Colloid mills and homogenizers will produce the maximum degree of dispersion and, although not strictly mixers, are included here because they represent the apparatus which will produce the smallest particle size on most materials. They are not very satisfactory on high viscosities. Because of the high rate of shear developed, they use a large amount of power, ranging from 20 to 50 hp./100 gal./hr.

3. Mixing Liquids and Gases (Table 3)

Notes on Table 3.

1. Reactions between gases and liquids always require mixing except where gas under pressure, fed to the bottom of the tank itself, supplies sufficient action.
2. The liquids are assumed of low viscosity (say, under 100 centipoises). Cases where viscosities are higher are too rare and need not be treated here.
3. If chemical reaction rate between a liquid and dissolved gas controls the rate of absorption, sufficient contact time must be provided for the desired degree of reaction. In batch equipment this means longer cycles; in continuous equipment, it means a reduced rate of through-put, longer path of travel, or multistage contactors. In either case, the contacting need be sufficient only to keep the liquid saturated with the gas.
4. In continuous systems the best results are generally obtained by countercurrent flow, though sometimes parallel flow is used.
5. In many cases where high efficiency is desired, two or more towers, turbo-absorbers, etc., are used in series.
6. Equipment can be obtained to handle any desired quantities of gas and liquid, except in the case of very high pressure work. Here the size is limited by the mechanical strength of the containers.

Comparison of Types. The conditions favoring the use of mechanical gas-liquid contactors over conven-

Table 2. Mixing Liquids with Liquids.* Preferable Types of Mixers for Immiscible Liquids

Basis of operation	Thin, to 100 cp.: Coarse (visible droplets) Example: caustic wash of gasoline	Thin, to 100 cp.: Fine (invisible droplets) Example: acid treatment of gasoline	Thin, to 100 cp.: Finest (Brownian movement) Example: ice-cream mix, Commercial emulsions	Viscous, to 200,000 cp.: Coarse (visible droplets) Unusual case	Viscous, to 200,000 cp.: Fine (invisible droplets) Example: H_2SO_4 treatment of heavy lube oil	Viscous, to 200,000 cp.: Finest (Brownian movement) Example: malt extract with cod-liver oil
Small Up to 1000 gal.	Paddles (9, 10) Outside circulation (4) Air Propellers (19, 20) Turbines (28)	Whippers (15) Propellers (20, 21, 22) Turbines (28)	Whippers (15) Propellers (20, 21, 22) Turbines (28)		Paddles (13, 14) Turbines (28)	Turbines (28) Paddles (13, 14) Turbodisperser (29)
Large Up to 20,000 gal.	Outside circulation (4) Air Propellers (19, 21) Turbines (28)	Propellers (20, 21) Turbines (28)	Propellers (20, 21) Turbines (28)		Turbines (28) Paddles (13, 14)	
Continuous Any desired gallonage throughout	Nozzle (2) Orifice (3) Propellers (22) Turbines (30)	Orifice (3) Nozzle (2) Pumps (5) Propellers (22) Turbines (30)	Colloid mill (35) Homogenizer (36)		Pumps (5) Turbines (30)	Turbines (30) Turbodisperser (29) Colloid mill (35)

* The least desirable mixer is listed first. cp. = centipoises.

Table 3. Mixing of Liquids and Gases.* Preferable Types of Mixers for Intimate Mixing

One or more liquid phases and one gas, with or without solids present	Batch (both liquid and gas)	High pressure, up to 2000 lb./sq. in. Example: hydrogenation	Turbo-absorber (31) Rolling bomb
		Low pressure or open, up to 50 lb./sq. in. Example: chlorination	Bubbling gas through tank Recirculation through towers (6) Recirculation through nozzles (2) Turbo-absorber (31)
	Continuous (both liquid and gas)	High pressure (as above) Example: washing sulfur out of natural gas	Injector (2) Towers (6) Bubble columns Continuous turbomixer (30)
		Low pressure or open (as above) Example: sulfite liquor preparation	Cascades Injector (2) Continuous turbomixer (30) Towers (6)
One or more liquid phases and two or more gases, with or without solids present	Liquid-batch (gas continuous)	Pressure or open Example: removing CO_2 from air	Bubbling gas through tank Turbo-absorber (31) Recirculation through towers (6) Recirculation through nozzles (2) Recirculation over cascades Scrubbers (38*A*)
	Both continuous	Pressure or open Example: washing illuminating gas with water	Cascades Injector (2) Towers (6) Bubble columns Scrubbers (38*A*)

* The least desirable mixer is listed first.

tional gas absorption equipment are outlined in Section 10. This section should be consulted for a more detailed discussion of spargers and agitated contactors, and Secs. 9, 11, and 15 should be consulted for information regarding bubble-plate and packed columns and spray towers.

For autoclave work or for reaction-controlling processes such as hydrogenation and amidation, the **turbo-absorber** is excellent because it not only draws bottom-entering gas into the impeller but also recirculates large volumes of freeboard gas through the impeller and distributes it downwardly throughout the liquid in a finely divided state. Caution must be observed, however, that the liquid level does not change sufficiently to reduce the entraining capacity of the device. Repair and maintenance costs are relatively low.

2. The **continuous turbomixer and absorber** has also a very high absorption rate on continuous work, because of the fine dispersion and long path of travel of the gases. However, it has the disadvantage that this path of travel is not so long as it is, say, in a 40-ft. tower. though the tower is more expensive to construct. To overcome this, when necessary, mixers may be used in series.

A **vaned disk,** which is essentially an impeller, operating in a baffled tank or within a stator element is an effective gas-liquid mixer and may be used for applications which require an agitated contactor and in which recirculation of freeboard gas is impracticable or undesirable. The vanes may be either on the surface of the disk or at its periphery, and may be radial, tangential, or V-shaped.

Porous spargers compare favorably in contacting performance with agitated contactors for some applications involving no plugging solids and permitting mild liquid agitation. **Simple bubblers,** however, may be used only for very rapid absorptions (*e.g.*, the absorption of chlorine in caustic soda).

For the treatment of large volumes of gas with small volumes of liquid (scrubbing operations, humidification), **scrubbers, packed or bubble-plate columns,** or **spray towers** are useful. The latter are effective only when liquid-film resistance is small. Towers and nozzles must be avoided in cases where plugging by solids may occur.

4. Mixing Liquids and Solids (Tables 4 and 5)

Special Cases. The tables are based on conditions involving one liquid and one solid.

In cases where one liquid phase and two or more solids are to be mixed, the tables will also serve if we bear in mind that we must choose the mixer that will best mix the solid of highest gravity. An example of this is found in the manufacture of abrasives.

In cases where two or more liquid phases and one or more solids are to be mixed, the degree of dispersion desired in the liquid phases is usually the governing factor, *e.g.*, metal-polishing preparations. But this is not always true, especially if a very heavy solid is present. For example, in the preparation of aniline from nitrobenzene, iron borings, and HCl solution, we have two liquid phases and a very dense solid phase. Here the turbine type is clearly indicated for most efficient action.

Notes on Tables 4 and 5. The mixing of liquids and solids is by far the greatest and most complex category commonly encountered. The number of combinations that may result is infinite. Therefore in this case, least of all, can anyone impose definite limits on size of batch, consistency of mixture, degree of break-up, time of reaction, etc. Certain points can be indicated, however, and certain examples can be given that may help others to get a better picture of their own particular situation.

1. **Small scale** in this division is as follows: 25,000 gal. at thin consistency, and less than 10 per cent of that amount at very high consistency (as 100,000 centipoises apparent viscosity); with **large scale,** anything over the above. For **intermediate points,** construct a centipoise-gallon graph, connecting the points 1 centipoise, 25,000 gal., and 100,000 centipoises, 2500 gal., and interpolate.

2. The specific gravity of solids plays an important part in determining the type of mixer. The specific gravity of the liquid also must be considered, though it is not so important because it usually varies much less than that of the solid. For instance, common liquid extremes are gasoline, etc., at 0.7 sp. gr. and sulfuric acid at 1.84, whereas solids often vary from 0.9 to 7.0 Note that the limits given in the table refer to the **sp. gr. difference** between liquid and solid.

Comparison of Types.

Paddles, when correctly designed, are satisfactory for liquids of thin and medium consistency (1–10,000 cp.) and for low-gravity solids, except for very rapid or intimate mixing such as in rapid solution or in the preparation of fine dispersions. Where applicable, paddles are economical because (*a*) of their low first cost and, (*b*) when run at low speeds, their power requirements are moderate. They will produce intimate mixtures of thick materials of all gravity differences, if sufficient time is allowed. For this, the power consumption is very high. They are also satisfactory for slow dissolving of fibrous, crystalline, and amorphous solids and for maintaining fibrous materials in fairly uniform suspension. They are not at all useful for maintaining high-gravity solids in suspension in liquids of thin and medium consistency as they have not enough sustained lifting power, and they

Table 4. Mixing of Liquids and Solids.* Preferable Types of Mixers for Small-scale Batch Mixing

Consistency	Character of solids	Examples S.S. = simple suspension I.D. = intimate dispersion D. = dissolving	Simple suspension or Rough dispersion or Slow precipitation or Slow leaching	Intimate mix or dispersion or fast precipitate or fast leach: Ordinary	Intimate mix or dispersion or fast precipitate or fast leach: Maximum	Dissolving: Slow	Dissolving: Fast
Thin Having an apparent viscosity up to 100 cp. when placed under high rate of shear	Low-gravity difference (not over 1)	S.S. $CaCO_3$ in water I.D. Wax in water (for polish) D. Salts in water	Paddles (7, 8, 9, 10) Outside circulation (4) Air Propellers (19, 20, 21)	Paddles (9, 10, 11) Propellers (19, 20, 21, 22) Turbines (27, 28)	Propellers (22) Turbines (28) Turbo-disperser (29)		Paddles (9) Turbines (28) Propellers (21, 22)
	High-gravity difference (over 1)	S.S. Sand in water I.D. Pigment washing D. Metals in acid	Propellers (19, 21) Turbines (27, 28)	Propellers (20, 21, 22) Turbines (27, 28)	Propellers (22) Turbines (28) Turbo-disperser (29)		Paddles (9) Turbines (28) Propellers (21, 22)
	Fibrous (as cellulose)	S.S. 2% paper pulp in water I.D. Mixing same with dye D. Nitrocellulose in solvents	Air Outside circulation (4) Paddles (7) Propellers (19, 21) Turbines (27, 28) Revolving cone (38)	Paddles (7, 9) Propellers (19, 21) Turbines (27, 28)	Propellers (19, 21, 22) Turbines (27, 28) Beaters	Outside circulation (4) Paddles (9, 10, 13, 14)	Paddles (9) Turbines (28) Propellers (21, 22)
Medium Having an apparent viscosity up to 2500 cp. when placed under high rate of shear	Low-gravity difference (not over 1)	S.S. $Mg(OH)_2$ in water I.D. Clay in lube oils D. Gum dissolving	As for thin	As for thin	Tumbling barrel (32) Propellers (19, 22) Turbines (28) Turbodisperser (29)	Tumbling barrel (32) Propellers (19, 20, 21, 22)	Paddles (9) Turbines (28)
	High-gravity difference (over 1)	S.S. HgI_2 in alcohol I.D. Lead pigment in linseed D. Litharge in NaOH	As for thin	As for thin	Ball mill (32) Propellers (19, 22) Turbines (28) Turbodisperser (29)		Paddles (9) Turbines (28)
	Fibrous	S.S. 5% paper pulp in water I.D. Mixing same with dye D. Nitrocellulose in solvents	As for thin	As for thin	Propellers (22) Beaters Turbines (27) Revolving cone (38)		Paddles (9) Turbines (28)
Thick Apparent viscosity over 2500 cp. and up to 200,000 cp. when placed under high rate of shear	Low-gravity difference (not over 1)	S.S. Bentonite in water I.D. Slate powder in asphalt D. Rubber in gasoline	Paddles (8, 9, 10) Propellers (19, 20, 21, 22) Turbines (27, 28)	Paddles (9, 12, 13, 14) Turbines (27, 28)	Ball mill (32) Paddles (9, 12, 13, 14 Turbines (28) Turbodisperser (29)		Turbines (28)
	High-gravity difference (over 1)	S.S. Flat wall paint I.D. Lead pigment in lacquer D. Plasticizer in leather dopes	Propellers (19, 20, 21, 22) Paddles (8, 9, 10) Turbines (27, 28)	Paddles (9, 12, 13, 14) Turbines (27, 28)	Ball mill (32) Paddles, (9, 12, 13, 14) Turbines (28) Turbodisperser (29)		Turbines (28)
	Fibrous	S.S. 7% paper pulp in water I.D. Mixing same with dye D. Nitrocellulose with solvents	Propellers (20, 21) Revolving cone (38)	Propellers (21)	Paddles (9)		Turbines (28)

* The least desirable mixer is listed first.

are also impractical with fibrous material of high consistency, as the torque becomes too great.

In suspending granular solids in a liquid, White and Sumerford found [*Chem. & Met. Eng.*, **43**, 370 (1936)] that at a given paddle speed the best suspension is obtained with a paddle length slightly less than half the tank diameter, without baffles. The clearance of the paddle from the bottom should equal the paddle width. For a given power input, the optimum size of paddle would be slightly smaller than this. The superiority of this size of paddle over others was found to be independent of tank size, paddle speed, and size and amount of sand.

Propellers are very useful over a wide range of liquid and solid mixtures. They are satisfactory for suspension and for intimate, though not the most intimate, mixing of materials of thin and medium consistency of all kinds in batches of all sizes. They are not satisfactory on large batches of heavy materials because of size limitation, nor are they good for intimate mixtures of high consistency except those of fibrous solids. They are faster than paddles for most dissolving operations. The first cost of the various types of propellers is comparatively low, but their power requirement is moderate.

Turbine mixers surpass all other types in speed of mixing and dissolving and in intimacy of mixing. Their power consumption is not excessive, being as a rule considerably lower than that for propellers. They are entirely satisfactory on materials of all consistencies or apparent viscosities here considered and on batches of all sizes. Because of their speed and thoroughness they are the best type for continuous mixing or dissolving opera-

Table 5. Mixing of Liquids and Solids.* Preferable Types of Mixers for Large-scale Batch Mixing

Consistency	Character of solids	Examples S.S. = simple suspension I.D. = intimate dispersion D. = dissolving	Simple suspension or Rough dispersion or Slow precipitation or Slow leaching	Intimate mix or dispersion or fast precipitation or fast leach: Ordinary	Intimate mix or dispersion or fast precipitation or fast leach: Maximum	Dissolving: Slow	Dissolving: Fast
Thin (see Table 4)	Low-gravity difference (not over 1)	S.S. Sugar clarification I.D. $Al(OH)_3$ precipitation D. Ice in water	Air Outside circulation (4) Paddles (7, 8, 9) Propellers (19, 21, 22) Turbines (27, 28)	Paddles (9) Propellers (22)	Turbines (28) Propellers (22)		Turbines (27, 28) Propellers (21, 22)
	High-gravity difference (over 1)	S.S. Leaching phosphate rock I.D. *Blanc-fixe* precipitation D. Sodium sulfate	Propellers (21, 22) Turbines (27, 28)	Propellers (22) Turbines (28)	Turbines (28)		Turbines (27, 28)
	Fibrous (as cellulose)	S.S. 2% paper pulp in water I.D. Mixing same with alum D. Nitrocellulose in solvents	Air Paddles (7, 8, 9) Propellers (21, 22) Turbines (27, 28)	Propellers (22)	Turbines (28)	Outside circulation (4) Paddles (7, 9) Propellers (21, 22) Turbines (27, 28) Traveling paddles (11)	Turbines (27, 28)
Medium (see Table 4)	Low-gravity difference (not over 1)	S.S. Starch washing I.D. Clay blunging D. Leaching diatomaceous earth	Paddles (7, 8, 9) Propellers (19, 21, 22) Traveling paddles (11) Air lift (16)	As for thin	As for thin		Turbines (27, 28) Propellers (21, 22)
	High-gravity difference (over 1)	S.S. Metallurgical slurries I.D. Cyanide process (gold) D. Barytes bleaching	Turbines (27, 28) Air lift (16)	As for thin	As for thin		Turbines (28)
	Fibrous (as cellulose)	S.S. 5% paper pulp in water I.D. Mixing same with bleach D. Cellulose acetate in acetone	Paddles (7, 9) Propellers (21, 22) Turbines (27) Revolving cones (38) Traveling paddles (11)	As for thin	As for thin	Outside circulation (4) Paddles (7, 9) Propellers (21, 22) Turbines (27, 28) Traveling paddles (11)	Turbines (28)
Thick (see Table 4)	Low-gravity difference (not over 1)	S.S. Separate digestion sewage sludge I.D. Casting slip D. Grain mash	Paddles (7, 8, 9) Turbines (27, 28) Air lift (16) Traveling paddles (11)	Paddles (9) Turbines (27) Rake mixer (7A) Traveling paddle (11)			
	High-gravity difference (over 1)	S.S. Cement slurries I.D. Asphalt filling D. Leaching copper oxide	Paddles (9) Turbines (27, 28) Air lift (16) Traveling paddles (11)	Paddles (9) Turbines (27, 28) Traveling paddles (11)		Paddles (7, 9) Turbines (27, 28) Air lift (16)	
	Fibrous	S.S. 7% paper pulp in water I.D. Asbestos in asphalt D. Cellulose acetate in acetone	Propellers (19, 21, 22) Revolving cones (38) Traveling paddles (11)	Paddles (7, 9) Propellers (19, 21, 22) Traveling paddles (11)			

* The least desirable mixer is listed first.

tions. As a partial offset to these advantages, their first cost is somewhat higher than that of paddles or propellers, although this is not true in all cases. Being based on a centrifugal principle, and hence a true disperser, or scatterer, the turbine type (especially in the turbo-disperser style) is very efficient for intimate dispersions of all sizes in batch or continuous operations. This is also true of dissolving processes.

Revolving cones are superior to other types for maintaining uniform suspensions of fibrous solids, particularly where the consistency of the slurry is high. Underwood [*Paper Ind. and Paper World*, **18**, 639 (1938)] reported the ability of the cone to uniformize completely 4,000 gal. of 6.5 per cent paper stock in 1 to 3 minutes.

Traveling mixers are useful for huge batches, where nothing else could be used (*e.g.*, in tanks 100 ft. long by 40 ft. wide by 30 ft. high), for they accomplish sufficient turnover of the contents with rather low power consumption. They are particularly useful on cement slurries, paper pulp, and other similar materials.

Outside circulation has a limited usefulness and has been discussed before under Liquids and Liquids.

Air or gas sparging is useful in tanks unsuited to mechanical agitators because of their large size or peculiar shape. It is satisfactory for maintaining solids of low settling velocity in suspension; for example, it is used extensively to mix cement slurries. Stanclift (M.S. thesis in chemical engineering, University of Kansas, 1948) reported that the position and design of the sparger is critical if the best suspensions are to be obtained; in small cylindrical tanks, the sparger should be arc-shaped concentric with the tank, and should be located about two-thirds of the distance from the center to the wall of the tank. It should occupy no more than two quadrants to cause the whole contents of the tank to circulate vertically. Stanclift was able to effect quite complete suspensions of particles with settling velocities up to 0.25 ft./sec. by using no more than 3 cu. ft. of air/(sq. ft. of tank cross section)(min.). Gas sparging may not be used where volatility losses are important, and air must be avoided if oxidation is dangerous. It should be used only in liquids of relatively low viscosity.

The air-lift agitator (Dorr) is good for large batches of material, especially high-gravity slurries, where it accomplishes a slow, thorough turnover of this material.

The ball mill, or tumbling mixer with balls, is mentioned because of its ability to produce very intimate dispersions, especially in the paint industry. This is not necessarily a grinding action but is simply a separation of the flocs of pigment particles by impact. Hence it performs an intimate mixing operation. The ball mill is also occasionally useful for slow solution of nitrocellulose and gums in solvents.

Continuous Mixing of Liquids and Solids.

Tables 1 to 5 indicate the mixers to be used for large- and small-batch mixing. What mixers should be used for continuous work?

For the case of the continuous mixing of simple suspensions where approximately uniform distribution, etc., is satisfactory, use the same equipment as for batch, for both large and small scale.

For all the other cases, fast action—as nearly instantaneous as is possible—is implied, and usually the smallest possible container with the greatest possible through-put is the most desirable. Turbines are usually the most satisfactory for all cases of continuous mixing where intimate mixing or dispersion, fast dissolving, or precipitation is required, because they produce the longest path of travel and the greatest amount of recirculation within a container of given size and within a given holding time. Type 30 should be used where the flow and required holding time allow, but the usual practice for high flow rates is the use of two or more separate tanks in series. **Countercurrent flow** offers many advantages in some applications and can often be accomplished in turbine installations without the use of intermediate pumps. Many continuous-flow operations require the use of almost exact quantities of the ingredients, making it necessary to find some means of accurate proportioning.

Proportioning. Many continuous-flow operations require the use of almost exact quantities of ingredients, making some means of accurate proportioning necessary. For dry feed both constant-volume and constant-weight feeders are available while for liquids reliance is often placed on meters and pumps. Specialized equipment from which exact proportioning may be obtained has been developed. It is essential that no entrained air or gas reach a volumetric proportioning device, for it cannot discriminate between gas and liquid in its measurement of uniform volumes.

5. Mixing of Pastes, Plastics, and Doughy Masses

Strictly speaking, all materials or mixtures which possess the properties of plasticity (pseudoplasticity, inverted plasticity) as discussed on pp. 1197–1199, fall into this class. However, from a practical viewpoint, some of these materials are so thin that they must be considered in the liquid-liquid, or the solid-liquid class, and have already been discussed under 3 and 4. Therefore, it may be said that pastes, plastics, and doughy masses are those materials or mixtures, the consistency, or apparent viscosity, of which ranges from 200,000 centipoises to several million centipoises. They include Nos. 4 to 8 in the list of pastes and plastics given on p. 1202, and also such mixtures as greases, bentonite solutions, dough, putty, and countless others. In evaluating their consistency, a Gardner Mobilometer may be used.

The most difficult task in the whole field of mixing is presented by these thick, doughy, sticky materials. Their yield points are usually high, and this means that substantial force must always be applied before any shear, or motion, takes place. For the same reason, the flow of these materials is limited, and the mixing is achieved through a *stretching*, a *folding*, a *kneading*, or a *tearing* action, or the most desirable combination of these actions. The object of these actions is stated exactly in Rules 1 and 2 on p. 1195. Since there is little flow, the particles of one constituent must be *forced* between the particles of other constituents until the whole mass is in a thoroughly mixed condition and all surfaces of solid particles are thoroughly wetted. Furthermore, every particle within the container must be brought to the place where it may best be subjected to this force.

As the consistency (apparent viscosity) of these pastes and plastics becomes greater, ranging from 200,000 centipoises to several million centipoises, increasingly heavier mechanisms must be used. In general, however, except for types 39 and 40, they will all be found to be multibladed paddle, or arm, types. The following types are arranged in the order of the consistencies they are capable of handling 7, 9, 25, 12*a*, 13, 14, 12*b*, 40, 24, 17*a*, 17*c*, 39.

In comparison with mixers for more fluid mixtures, the sizes of these heavy-duty machines are usually small. *i.e.*, seldom over 1000 gal. and usually smaller. Nevertheless, a large amount of time and power is consumed for the operation. For example, a machine of type 17*a*, designed to mix 300 gal. of a plastic, modeling clay, requires 75 hp. and an hour to do the work. The dispersion type (17*c*) often requires 200 hp./100 gal., for example, on rubber compounding. The main thing is that this machine is thoroughly successful at its task. Types 12*b* and 40 would also satisfactorily mix a batch of modeling clay.

Analysis of Mechanical Actions Involved. In order to analyze the mechanical actions involved in mixing materials of this class, consider the preparation of bread dough in a mixer of type 17*a*. The raw materials are a limited amount of water or milk, other minor ingredients, and flour which is full of air. The water must displace the air and wet the entire surface of each particle of flour. When completely mixed, both materials must be uniformly distributed with respect to each other in that form which is known as dough. The mixing elements and the container itself must be able to perform various functions. First, the elements must *transport* material from one end of the container to the other, and back again. The design makes this possible, because the mixer arms, or blades, usually of the sigma form, are not parallel to the axis upon which they rotate, but are pitched so that material is pushed back and forth. Second, the elements must *knead* the material by pressing it against the walls of the container and against contiguous material. This pressure tends to force water between the flour particles, an action that displaces the air that has been present. It also causes shear, which brings a new relationship to different portions of the partly mixed material. Third, as the mixer arms rotate, they *tear* loose portions of the mass, carrying these portions to other parts of the container, thus redistributing the contents. Fourth, as the mass becomes coherent, *stretching* takes place. That is, the mixer arm grips a portion of the material, stretching it as a rubber band is stretched. The tension to which this portion of material is subjected, with the concomitant compression, which occurs at right angles to the other force and is analogous to a kneading action, is one of the chief factors in working the water into the flour. Fifth, the stretched material is then *folded* over on itself or on fresh material, and thus a realignment of material is brought about. These actions occur repeatedly until the whole is mixed.

Each of these actions depends for its effectiveness on the amount of shear or transportation it produces. A combination of all these actions is the usual thing in mixing pastes, plastics, and doughy masses, and all the types listed here are capable of a similar performance on suitable consistencies.

6. Mixing of Solids with Solids (Also Solids with Gases)

The problems encountered depend on relative size, shape, and gravity of particles. When two kinds of solid particles to be mixed are the same size but of different gravity, naturally the heavier seek the bottom. When of the same gravity but of different size, the smaller particles seek the bottom. So also do round, smooth particles, while the jagged or polyhedral ones seek the top of the mass. In mixing solids, these natural separating tendencies must be overcome, and this is invariably done by some means which lifts material from the bottom to the top of the mass, the resulting voids being filled from above by gravity. Simultaneously the means must also produce horizontal transportation in at least two opposite directions.

Types 4, circulating system; 12*b*, rotating pans with offset blades; 25, the helical ribbon mixer; 32, the tumbling barrel; 33, the double-cone mixer; 34, the mushroom mixer; 7*A*, the rake mixer; and 40, the putty chaser, or pan mixer, are used. The rake mixer and the tumbling barrel as in the xanthation of cellulose are also used for mixing solids with gases.

For mixing of batches larger than 1000 cu. ft. (as in the large-scale preparation of molding powders), circulating systems or large tumbling barrels are preferable, though no ideal method has yet been discovered. Circulating systems represent a mechanical quartering method and therefore are slow. In such sizes tumbling barrels, because they do not produce good end-to-end flow of material, even when provided with internal baffles, give incomplete mixing unless excessive time is consumed.

For batches of ordinary size, or up to 1000 cu. ft., the double-cone mixer is the most successful type yet devised. Large or small batches are mixed in 15 min. or less. Power is low, wear is negligible, and discharge is rapid and leaves a smooth inside surface which is easily brushed clean. Thus it is useful where extreme cleanliness is required or where the mixture is colored, as in the mixing of dry colors. The ribbon mixer is the most common type because it is older and somewhat lower in first cost but is slower than the cone mixer for the same power consumption and is not quite so thorough nor is it so easily cleaned. The tumbling barrel is less effective than either of the two just mentioned.

Whenever necessary to obtain extremely intimate mixing, involving the breaking down of aggregates or the coating of one solid material with another, as in the standardizing of dyestuffs with salt or the coating of resin granules with color, a mixer producing a smearing or shearing action must be used. The types with mullers, as 12*b* and 40, are good. A tumbling barrel containing a few balls or pebbles is often satisfactory. The mushroom mixer with balls is a modified type of tumbling barrel finding special use in the dye and pharmaceutical industries. The usual blending requires 1.0 to 1.5 hp./1000 lb. solids.

It should always be remembered that mixtures of solids of different size, shape, or density tend to reseparate easily. This means that solid mixtures should be further processed or packaged as soon as possible and with the minimum of handling after they leave the mixer to avoid segregation.

Lacey [*Trans. Inst. Chem. Engrs.* (*London*), **21,** 53 (1943); *Chem. Age* (*London*), **52,** 119, 145 (1945)] and Beaudry [*Chem. Eng.*, **55** (7), 112 (1948)] have presented statistical methods for characterizing the uniformity of mixtures of particulate solids.

HEAT TRANSFER IN MIXERS

Efficient heat transfer in kettles and tanks is very closely related to the amount of mixing provided for these same containers. Where plant operations are being scientifically handled, it is axiomatic that good agitation is not only desirable, but absolutely necessary, to increase heat transfer and to prevent local overheating or cooling. In many instances the heat transfer requirement may be the factor which determines the design of the agitator used.

Heating or cooling of tanks or kettles is employed

1. To control the speed and extent of a chemical reaction, such as in the sulfonation of an oil, nitration, catalytic reaction, etc.
2. To accomplish a physical change, such as evaporation, pasting, crystallization, dissolution or melting, to change consistency or to promote blending.
3. To aid in dispersing one material in another.
4. To control the temperature of the tank contents.

For heating, the means encountered in the chemical industry today are: (1) direct or indirect firing; (2) water or steam; (3) hot oil; (4) Dowtherm (diphenyl or diphenyl oxide or a mixture of these), liquid, or vapor; (5) mercury vapor; (6) electric heaters, contact, immersion, or radiant.

For cooling, the means employed are: (1) air; (2) an evaporant, such as liquid ammonia; (3) water or brine; (4) oil; (5) Dowtherm; (6) vacuum; (7) direct ice addition.

Coefficients in Agitated Vessels

The resistance of the container wall should be considered in the design, particularly when a relatively poor thermal conductor like glass enamel is involved. The finish on the metal surface influences the resistance; for this reason, wherever practical, both surfaces of the wall should be smooth and free from pits and ridges. Almost always, however, the major problem is the resistance of the fluid films on one or both sides of the container or coil walls. For good heat transfer these films must be kept thin; usually this is accomplished by high fluid velocities past the heat-transfer surfaces. Only as the local film resistances are kept low can a satisfactorily high over-all coefficient of heat transfer [U, B.t.u./(sq.ft.)(hr.)(°F.)] be obtained. For good heat transfer, these films must be thin or even negligible. The conductivity of most of the mediums mentioned above, as well as on most of the materials being mixed, is only a very small fraction of the conductivity of the metals of the container; therefore a much greater difference of potential (in terms of heat) is required to force the heat through $\frac{1}{100}$ or $\frac{1}{50}$ in. of film than to force it through $\frac{1}{2}$ in. of metal (see Sec. 6 on Heat Transfer for detailed discussion). It cannot be too strongly stressed that, in designing a good unit, every feasible means for reducing both films should be considered so that the highest desired over-all rate of heat transfer, herein called U, can be obtained. U = over-all B.t.u./(sq. ft.)(°F.)(hr.).

A higher coefficient, U, results, in practice, in,

1. An improved product because of prevention of local overheating.

2. Longer life of equipment for same reason.
3. A faster process (see Figs. *L* and *M*).
4. Less square feet of heat-transfer surface.
5. Saving of fuel due to reduction of heat losses.
6. Less medium to circulate, with consequent saving of power.

Satisfactory rates, *U*, are obtained by maintaining a substantial velocity of liquid or vapor along both heat-transfer surfaces. For liquids these speeds may economically range from 200 to 400 ft./min., and for vapor, much higher. In coils these velocities are fairly easily attained—a reason why coils are at times preferred to jackets. Where the heat-transfer medium is a liquid, jackets with baffles and preferably with spiral baffles should be employed if a high rate is important. For further improvement the space between container shell and jacket shell should be narrow, *i.e.*, 1 to 3 in.; otherwise the rate of heat transfer may be surprisingly poor. For example, in a case where a viscous milk product was being cooled in a jacketed tank, a plain, unbaffled jacket gave a *U* of 35, whereas the same unit, with jacket spirally baffled to maintain a cold-water velocity of 200 ft./min. gave a *U* of 55, using the same amount of cooling water at the same inlet temperature. Figure 13 shows a kettle with baffled jacket and scrapers. The fabrication of jacket baffles requires care to avoid leaks past the baffles; a relatively small leak can result in serious short-circuiting and reduction in velocity.

Instead of jacket baffles, tangential nozzles may be used to inject the jacket fluid and impart to it a sufficiently high velocity. Such nozzles are particularly useful in the jackets of glass-lined tanks, where the welding of spiral baffles to the shell is not permissible. Greene [*Glass Lining*, **17** (4), 9 (1948)] has described the use of jacket nozzles, and reported that they will improve a water-to-water *U* from 75 to 150.

In fired kettles, best results are obtained by arranging the setting so that hot gases sweep around the bottom and sides of a kettle in a gradually ascending spiral, for this causes the cooler gas film against the kettle to be greatly attenuated.

Inside the tank, unless the contents are a boiling, low-viscosity liquid, the wall film must be thinned by fluid attrition (produced by an agitator) or by actual scraping (for high-consistency material).

For low- and medium-viscosity liquids **paddles, propellers,** and **turbines** are effective heat-transfer promoters for the kettle wall and for properly placed helical coils. Baffles and stator elements should be avoided, and coil supports should be kept small to minimize their baffling action. Paddles are effective because of the high-velocity swirl and gentle axial flow which they produce. Propellers are effective because of their ability to produce strong axial flow which can be directed past the heat-transfer surfaces. Turbines combine the swirl of paddles and the axial flow of propellers and produce stronger radial components than either of the others. The displacement capacities of propellers and turbines have been measured by Rushton, Mack, and Everett [*Trans. Am. Inst. Chem. Engrs.*, **42,** 441 (1946)], who reported that there is no reliable method of estimating the flow rate actually produced by such agitators from the calculated figure based on theoretical displacement and speed.

Chilton, Drew, and Jebens [*Ind. Eng. Chem.*, **36,** 510 (1944)] found that coil and jacket film coefficients in agitated unbaffled vessels could be correlated by the use of Prandtl, Nusselt, and Reynolds numbers based on agitator diameter and speed, tank or coil diameter, and agitated-fluid properties. They reported film coefficients on the order of 1200 B.t.u./(hr.)(sq.ft.)(°F.) obtainable with heating water agitated by a simple flat paddle. Rushton, Lichtmann, and Mahoney [*Ind. Eng. Chem.*, **40,** 1082 (1948)] reported agitated-side film coefficients of the same order for vertical heating pipes acting as baffles in water agitated by a flat-bladed turbine; if good heat transfer must be achieved in a baffled vessel, this appears to be an effective method of accomplishing it.

High-viscosity liquids may be handled by **gate** or **horseshoe agitators** which conform closely to the contours of the container. If the viscosity is very high

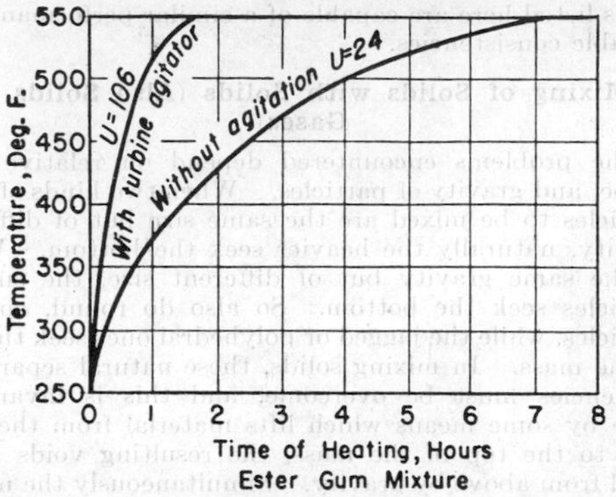

FIG. *L.*

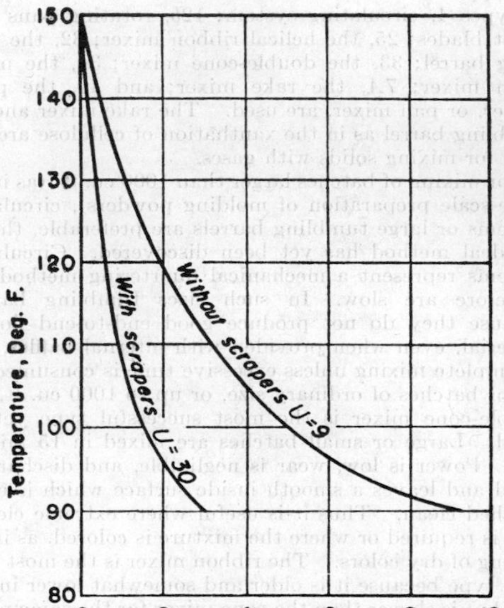

FIG. *M.*

(as in the cooling of a viscous liquid) or if material tends to cake on the heat-transfer surfaces, coils should be avoided and an agitator which scrapes the vessel wall should be used. Hinged or spring scraper blades affixed to agitator arms effectively remove the adhering material and constantly expose fresh heat-transfer surface (see Fig. 13). In cases involving very sluggish materials, a double-motion mixing element, as in Fig. 21, is desirable to transport material from the center of the vessel to the walls.

Figures *L* and *M* show the effect of good agitation on film removal with consequent improvement of heat trans-

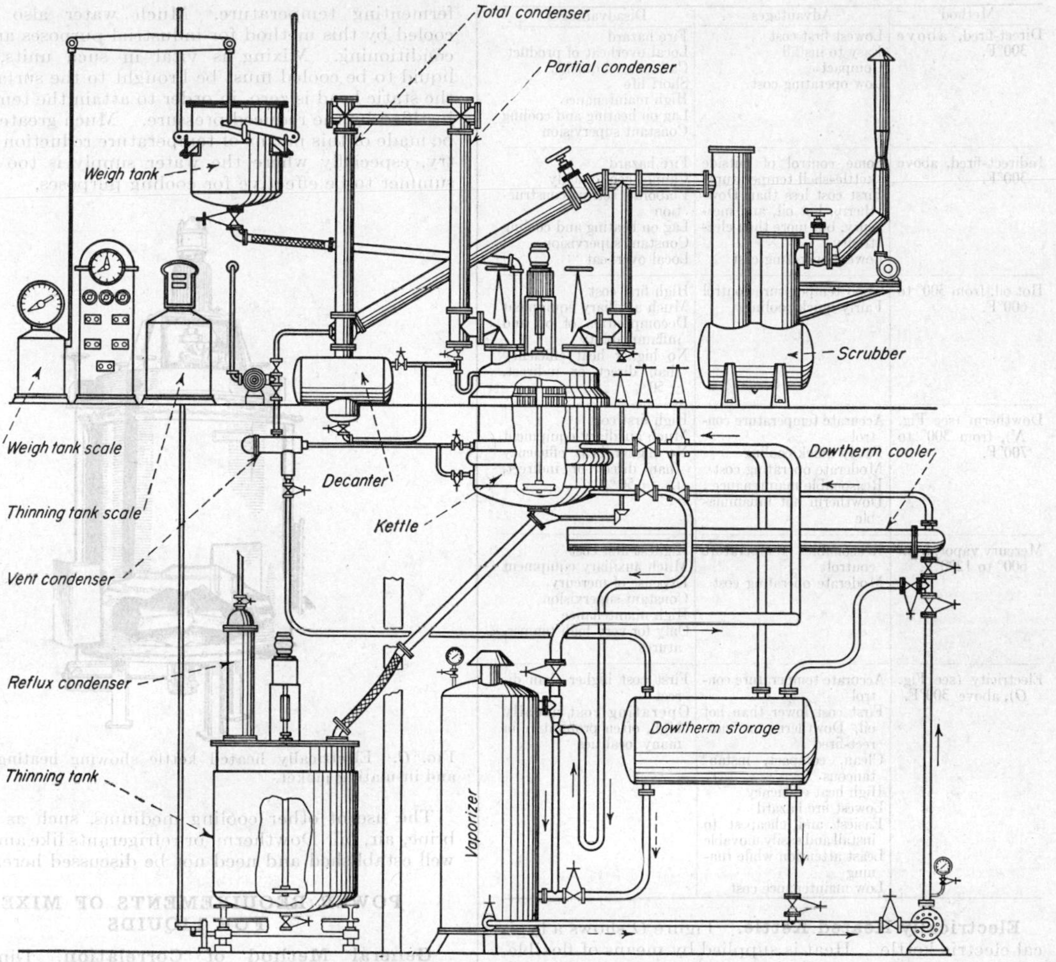

FIG. *N*. Alkyd resin plant with Dowtherm heating system.

fer. For instance, Fig. *L* shows the time of heating an ester gum in a 1000-gal. kettle from 250°F. to the top temperature of 550°F. In this case the temperature on the outside of the kettle was around 700° to 750°F. Without agitation a coefficient of heat transfer of approximately 24 is obtained. Satisfactory agitation raises the U to 106, approximately a fourfold improvement. Figure *M* illustrates the cooling of chocolate mixture in 1000-gal. batches from 150° to 90°F. with scraping and non-scraping stirrers and with jacket cooling water initially at 70° to 75°F. The over-all coefficient is tripled by the use of scrapers.

Additional performance data for scraping agitators in high-consistency materials have been reported by Huggins [*Ind. Eng. Chem.*, **23**, 749 (1931)] and by Laughlin [*Trans. Am. Inst. Chem. Engrs.*, **36**, 345 (1940)]. The latter described the use of a kettle equipped with a scraping agitator to convert in one continuous operation a 40 per cent aqueous solution to a dry, crystalline solid ready for packaging.

The Votator (Fig. 37) has produced the highest known rates of heat transfer in the processing of high-consistency materials. For example, in making starch paste for salad dressing continuously, producing 1800 lb./hr., a U of 583 has been obtained when heating the mixture from 76° to 196°F. On cooling the thickened mass back to 80°F., a U of 316 was produced. This is more than five times the rate that has previously been possible in any other device. Houlton [*Ind. Eng. Chem.*, **36**, 522 (1944)] has given heat-transfer data for a Votator as a function of scraper speed.

Heating Means

Water or Steam. Almost invariably, water or steam in jackets or coils is used for heating in any operation where the desired temperature may be so attained. The advantages are evident.

Only in higher temperature work does the question of proper selection of heating method arise. No method is universally applicable. The location of the plant often becomes the deciding factor.

Circulating Systems. Figure *N* illustrates a typical circulating system employing Dowtherm vapor for heating the kettle and Dowtherm liquid for the subsequent cooling. In this case, condensate returns to the boiler by gravity. Where this is not feasible, a return pump is employed. Cooling Dowtherm is handled from the receiver by a separate pump to the kettle and thence back through the water-cooled heat exchanger.

Method	Advantages	Disadvantages
Direct-fired, above 300°F.	Lowest first cost Easy to install Compact Low operating cost	Fire hazard Local overheat of product Cannot cool quickly Short life High maintenance Lag on heating and cooling Constant supervision
Indirect-fired, above 300°F.	Some control of outside kettle-shell temperature First cost less than Dowtherm, hot oil, and mercury, but more than electricity Lowest operating cost	Fire hazard Cannot cool quickly Elaborate setting construction Lag on heating and cooling Constant supervision Local overheat
Hot oil, from 300° to 600°F.	Good temperature control Fairly quick cooling	High first cost Much auxiliary equipment Decomposition of oil and inflammability No higher heat efficiency than direct or indirect, ± 50%
Dowtherm (see Fig. *N*), from 300° to 700°F.	Accurate temperature control Fairly quick cooling Moderate operating cost Reasonable maintenance Dowtherm not inflammable	High first cost Much auxiliary equipment No higher heat efficiency than direct or indirect, *i.e.*, ± 50%
Mercury vapor, from 600° to 1200°F.	Reasonable temperature control Moderate operating cost	Highest first cost Much auxiliary equipment Expense of mercury Constant supervision High maintenance Only for very high temperatures
Electricity (see Fig. *O*), above 300°F.	Accurate temperature control First cost lower than hot oil, Dowtherm, or indirect-fired Clean, compact, instantaneous High heat efficiency Lowest fire hazard Easiest and cheapest to install and easily movable Least attention while running Low maintenance cost	First cost higher than direct Operating cost usually high, often prohibitive in many localities

Electrically Heated Kettle. Figure *O* shows a typical electric kettle. Heat is supplied by means of flexible, spirally formed, metal ribbons, centrally located inside of hoods (or pipes) solidly welded to the kettle shell, the ribbons being prevented from touching the pipes by means of porcelain spacers. The elements themselves may be heated to any temperature from 300° to 1300°F., producing temperatures approximately 50° to 100° lower within the kettle itself. Radiation is prevented by means of a double steel jacket carrying 6 in. of insulation. Between the insulating jacket and the tank is a substantial air space with a blower for air cooling. Electric elements are usually arranged in two or three banks for zone heating, and these may be run separately or together.

Cooling Means

A surprising amount of cooling is still done in some chemical industries by the direct addition of crushed ice, but in many others this method is outlawed by the attendant dilution. A satisfactory substitute may be seen in the chilling of lubricating oil through the addition of liquid propane and its subsequent evaporation. This agent boils off at room temperature, but low temperatures may be attained by the application of vacuum, thus lowering the boiling point. By the application of suitable vacuum, water itself becomes the evaporant, and increasing use is being made of steam jet ejectors for this purpose. Most of the large distilleries in this country are using this method for cooling corn mashes part way to the fermenting temperature. Much water also is being cooled by this method for industrial purposes and for air conditioning. Mixing is vital in such units, for the liquid to be cooled must be brought to the surface where the static head is zero, in order to attain the temperature justified by the reduced pressure. Much greater use will be made of this means of temperature reduction in industry, especially where the water supply is too warm in summer to be effective for cooling purposes.

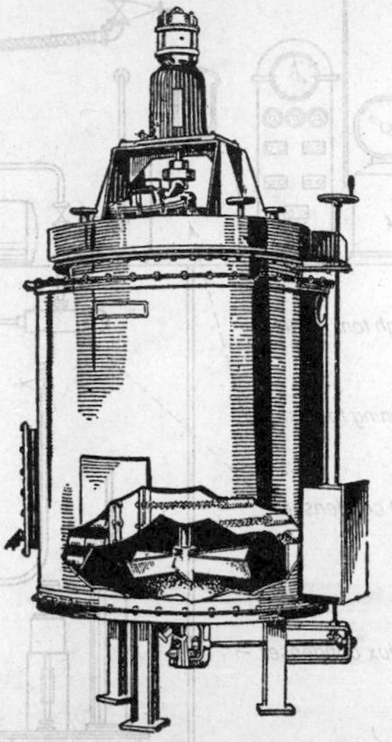

FIG. *O*. Electrically heated kettle showing heating elements and insulating jacket.

The use of other cooling mediums, such as water or brine, air, oil, Dowtherm, or refrigerants like ammonia, is well established and need not be discussed here.

POWER REQUIREMENTS OF MIXERS FOR LIQUIDS

General Method of Correlation. Dimensional analysis predicts the correlation of the power requirements of an agitator in a homogeneous fluid by means of a number of dimensionless groups: a power function, $\phi = g_cP/L^5\rho n^3$; a modified Reynolds number, $N_{Re} = L^2n\rho/\mu$; and a series of shape factors, (W/L), (D/L), (H/L),

where D = tank diameter, ft.
g_c = gravitational conversion factor, (lb. mass) (ft.)/(lb. force)(sec.)(sec.).
H = liquid depth, ft.
L = agitator diameter, ft.
n = agitator speed, r.p.s.
P = agitator power, ft.-lb./sec.
W = agitator height, ft.
μ = liquid viscosity, lb./(ft.)(sec.)
ρ = liquid density, lb./cu. ft.
ϕ = power function, dimensionless, value varies with agitator type and proportions.

Any other set of consistent units may be employed. If geometrically similar models are used, the shape factors may be bulked into a single constant with the power function. This type of correlation has been confirmed by a number of investigators for a variety of impellers.

If the material being agitated is a heterogeneous fluid, the same correlation is valid provided that the effective density and viscosity of the mixture are employed. The correct *density* is the average property of the mixture if it is well mixed, a value which usually can be computed from the densities of the components. The correct *viscosity* varies with the case, as follows:

1. *Low-consistency solid-liquid mixtures.* The viscosity of the liquid in which the solids are suspended may be used.

2. *High-consistency solid-liquid mixtures.* These are non-Newtonian fluids and should be treated as such (items 5 and 6, below).

3. *Two-phase mixtures of immiscible liquids.* A geometric mean viscosity,

$$\mu_G = \mu_X{}^x \cdot \mu_Y{}^y,$$

should be used, where μ_X and μ_Y are the absolute viscosities of phases X and Y, respectively, and x and y are the respective volume fractions of these phases in the mixture [Miller and Mann, *loc. cit.;* Olney and Carlson, *Chem. Eng. Progress*, **43**, 473 (1947)].

4. *Suspensions of gases in liquids.* The viscosity of the pure liquid should be used. The power correlation may not be valid if gas is being introduced directly at the agitator. See p. 1211.

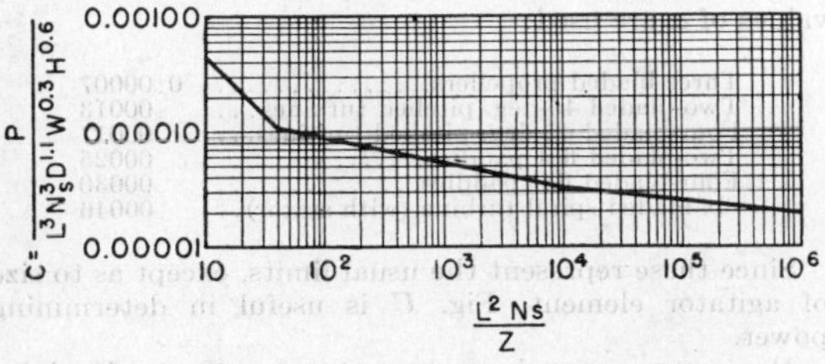

FIG. *P*. Graph for determining power coefficient.

5. *Non-Newtonian mixtures, pseudoplastic and thixotropic.* The viscosity should be computed from the measured power input to a laboratory model of the plant mixer to be used operating under satisfactory conditions of agitation. The result will be conservative, in that the power required by the large-scale agitator may be less than predicted.

6. *Non-Newtonian mixtures, dilatant and rheopectic.* No method is known for predicting the effective viscosity under conditions of agitation; it must be measured during mixing in the equipment of the size and design to be employed.

Paddles.* In the case of cylindrical tanks with unimpeded swirl, a prediction of the power input to flat-paddle agitators can be made by the use of the data of White and Sumerford [*Chem. & Met. Eng.*, **43**, 370 (1936)]. In the higher ranges of paddle size and/or speed, White found the power input to be related to the paddle dimensions, speed, and properties of the liquid by the following equation:

$$P = cL^3 sN^3 D^{1.1} W^{0.3} H^{0.6} \qquad (1)$$

where P = power input, hp.
c = power coefficient, non-dimensional.
D = vessel diameter, ft.
H = liquid depth, ft.
L = paddle length, ft.
N = paddle speed, r.p.s.
s = liquid density, lb./cu. ft.
W = paddle width, ft.
z = absolute viscosity, lb./(ft.)(sec.)

The power coefficient c they found to be related to a modified Reynolds number, just as the friction factor in pipe-line flow of fluids is related to the true Reynolds number. Figure P is a plot of c vs. the modified Reynolds number L^2Ns/z. This correlation they found to hold very well for tanks without baffles, having paddles with a length one-third or more of the tank diameter and with a width less than one-sixth the length. The shaft must be on the center line of the tank and the depth of the liquid not more than 50 per cent greater than the tank diameter.

* H. W. Bellas contributed to this section.

To find the power input to a given paddle agitator at a given speed, the modified Reynolds number must first be determined. By reference to Fig. P, the power coefficient c for the system is determined, and from it and Eq. (1) the power input P is calculated.

In designing a paddle agitator for a given purpose, a procedure is outlined by Büche [*Z. Ver. deut. Ing.*, **81**, 1065 (1937)]. The shape and speed of the agitator are first determined by small-scale tests, and the modified Reynolds number $(L_L)^2N_{Ls/z}$ for the small tank is calculated. The plant-size vessel and agitator are then made geometrically similar to the laboratory model, and a speed N_P is chosen dependent on the laboratory Reynolds number and the linear scale factor, as follows:

$(L_L)^2N_{Ls/z}$	N_P/N_L
< 1.0	1.0
1.0–2000	$(L_L/L_P)^{0.56}$
> 2000	$(L_L/L_P)^{0.67}$

The Reynolds number in the plant equipment can then be calculated and the power input predicted by the use of Fig. P and Eq. (1). The basis for this method is Büche's finding that the degree of agitation in a similar system is dependent upon power input per unit volume of liquid. Brothman and Kaplan [*Chem. & Met. Eng.*, **46**, 633, 639 (1939)] have also found that under like conditions the degree of mixing is related to the power input per unit volume and have plotted it vs. a "shear index," which is a measure of the work output of the paddle per unit input.

The White-Sumerford correlation is now generally known to predict power requirements which are too high, but it is presented as the only general correlation available. The power computed from it may be reduced safely by 30 to 60 per cent as Reynolds number increases from 1000 to 100,000 (Zarker, M.S. thesis in chemical engineering, University of Kansas, 1948).

No general data are available for the power requirements of paddles in baffled tanks, but it is known that the power may be at least four times that for the unbaffled paddle. Data for specific designs are given by Hixson and Baum [*Ind. Eng. Chem.*, **34**, 194 (1942)] and by Mack and Kroll [*Chem. Eng. Progress*, **44**, 189 (1948)].

The best method of predicting the power requirement of a pitched paddle from data for a simple flat paddle is that of Stern (M.S. thesis in chemical engineering, M.I.T., 1941). He recommends the application of the factor $e^{(\sin x - 1)}$ to the power required for a flat paddle of the same blade width (measured in the plane of the paddle) and operating at the same speed as the pitched paddle, where x is the angle of paddle pitch measured from the plane of rotation.

Turbines. Figures Q, R, and S show how power consumption of turbine mixers varies with changes in size of turbines, and with peripheral speed. Additional specific data have been given by Olney and Carlson (*loc. cit.*). Bissell and coworkers [*Chem. Eng. Progress*, **43**, 649 (1947)] have reported power-increase factors for the addition of baffles to a tank agitated by a turbine or by a propeller.

Propellers. Figure T indicates power requirements of propellers over a full range of sizes. In using Fig. T to determine the proper size of motor for an installation, the indicated horsepower can be multiplied by the factor 0.6 to get the approximate actual power required. This represents the pumping efficiency of the propeller. It will be smaller with the two-blade and larger with

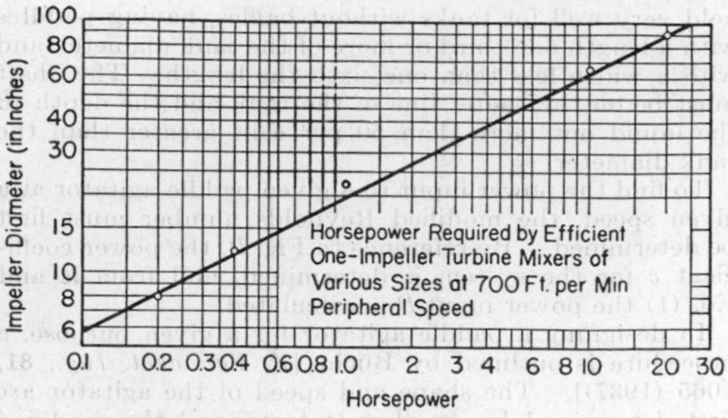

FIG. *Q*. Power requirement for turbine mixer.

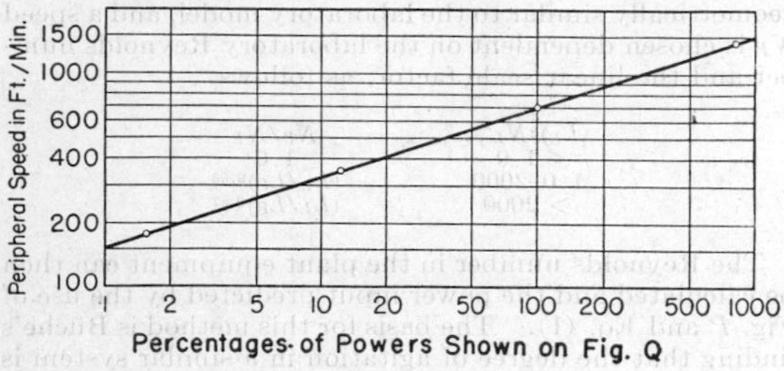

FIG. *R*.

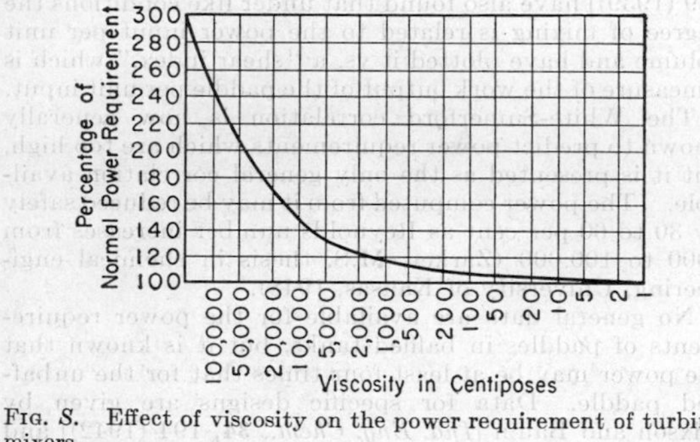

FIG. *S*. Effect of viscosity on the power requirement of turbine mixers.

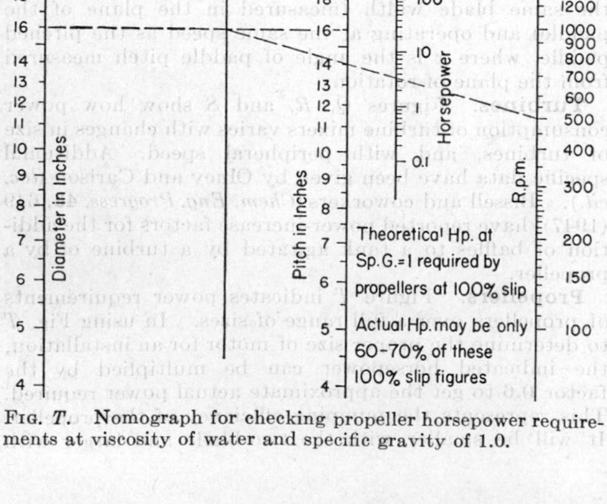

FIG. *T*. Nomograph for checking propeller horsepower requirements at viscosity of water and specific gravity of 1.0.

the four-blade type. Miller and Rushton [*Ind. Eng. Chem.*, **36**, 500 (1944)] have given data for marine propellers of two different sizes.

General. A number of investigators have reported power data for a variety of specific agitator designs. Many of these have been summarized as power-function plots by Martin [*Trans. Am. Inst. Chem. Engrs.*, **42**, 777 (1946)] and in nomographic form by Olney and Carlson [*loc. cit.*]. An attempt to generalize many of these specific results has been made by Hixson and Baum [*loc. cit.*], who plotted power-function correction factors for various deviations from an arbitrary standard shape against Reynolds number. Olney and Carlson [*loc. cit.*] have generalized power data in the form of the equation

$$\text{Hp.} = \alpha L^{4.70} n^{2.85} \rho^{0.85} \mu^{0.15}$$

where Hp. is the horsepower required, α is a parameter depending on impeller design with the same dimensions as the power function, and the remaining symbols are as defined above under General Method of Correlation. This solution to this equation, which is valid only for $N_{Re} > 75$, is shown in Fig. *U*, wherein the following values of α are used:

Three-bladed propellers	0.00007
Two-bladed 45-deg. pitched turbines	.00013
Four-bladed 45-deg. pitched turbines	.00017
Two-bladed flat paddles	.00025
Four-bladed flat paddles	.00030
Six-bladed spiral turbine (with stator)	.00046

Since these represent the usual limits, except as to size of agitator element, Fig. *U* is useful in determining power.

However, *power is never a true measure of mixing efficiency*, when different mixing devices are being compared. For example, Olney and Carlson (*loc. cit.*) found that, for equal mixing in a two-phase system, the spiral-blade turbine was many times more efficient than an arrowhead impeller. First, it was impossible for the arrowhead impeller to produce as good an emulsion as the spiral-blade impeller, even when the former was operated with exorbitant power consumption. This means that the spiral-blade turbine with stator is capable of a more intensive shearing action. Second, to produce equivalent emulsions (though not the best), the arrowhead impeller required ten to twenty times the power absorbed by the spiral turbine.

Figure *V* is a nomograph useful for computing the theoretical horsepower required to bring a flow of water to a given velocity.

THE TRANSMISSION OF POWER TO MIXERS*

Prime Movers. In recent years the tendency has been for individual drives, wherever possible, to replace the line-shaft method of power transmission. Neat installations, flexible in operation and easily maintained at low cost, may be made. Individual drives, therefore, are heartily recommended wherever working conditions permit. The use of totally enclosed and explosion-proof motors makes this possible today in places where formerly it was not to be considered because of dirt, water, or flammable materials.

In many mixing operations the exact degree of break-up or dispersion, as well as other desired results, is controlled largely by the speed of the mixer. To obtain the proper adjustment, a variable-speed motor is often used. Steam turbines have also been employed with excellent effect for this purpose. Air motors and hydraulic motors have been introduced for variable-speed operation.

* See section on Mechanical Power Transmission, pp. 1660*ff*.

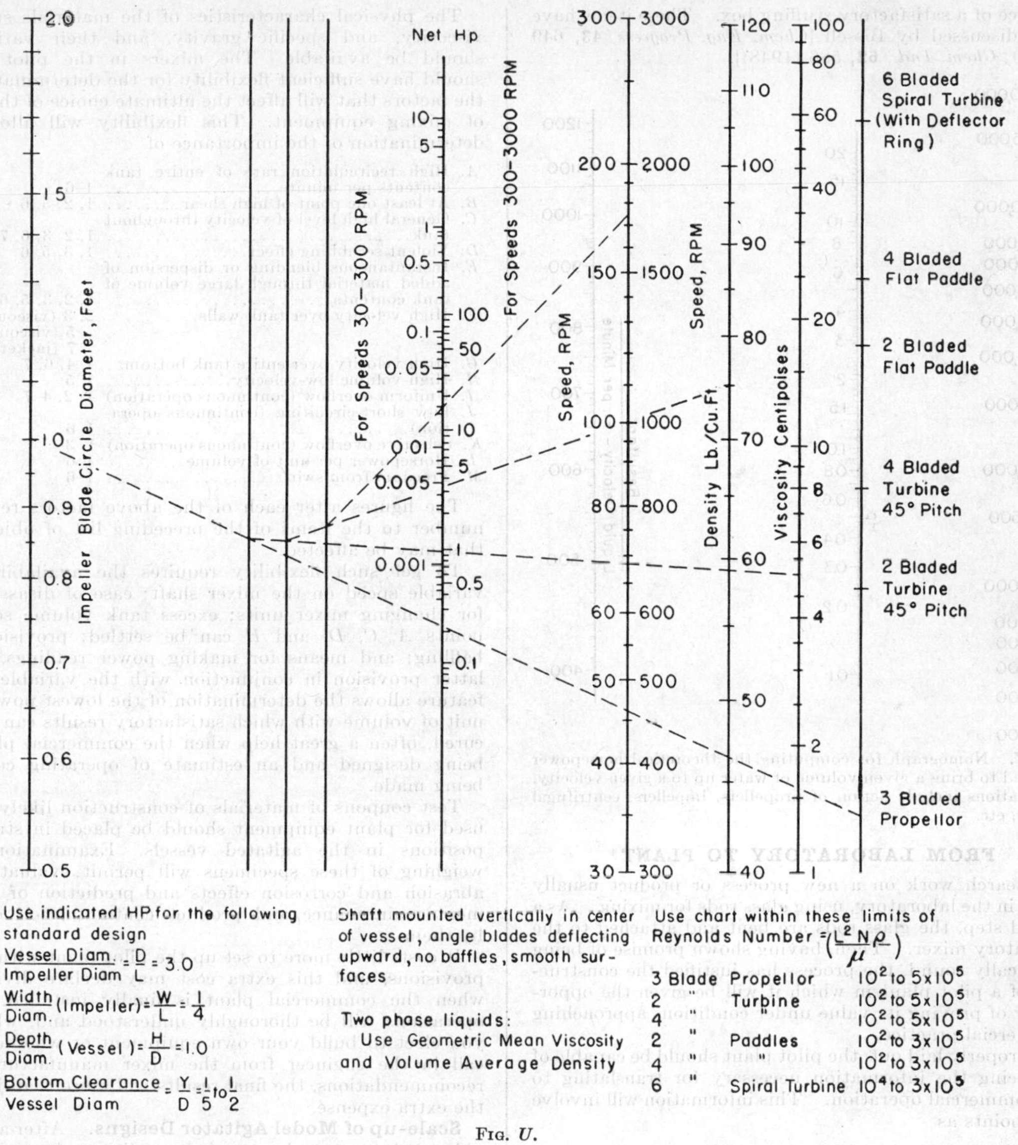

FIG. *U.*

Power Transmission to Mixer. It is then necessary to transmit the power to the mixer shaft so that it will run at the required speed. (Individual drives only will be considered.)

The following methods are usually used (numbers refer to illustrations in section, Types of Mixers):

1. Direct-connected (20). This gives full motor speed.
2. Gear-reduction units with built-in or separate motors (7, 21, 22).
3. Connected through multi-V-belt (14, 30).
4. Connected through chain (17c).
5. Connected through flat belt with pulley (8, 9).

Of all these methods the gear-reduction unit has gained the widest popularity, and justly so. With reliable prime movers, it has no disadvantage. The first cost is moderate; it is compact and easy to maintain and to operate. However, in cases where the horsepower is above 60 or where severe starting shock may be experienced, a V-belt or chain drive is preferable.

In the transmission of power to a mixer, variable speed is often obtained by the interposition of a Reeves drive, a Link-belt PIV drive, or other similar device between motor and mixer.

The above considerations apply to cases where the mixing elements are mounted on either horizontal or vertical shafts. However, where vertical shafts are used, the vertical motor-reducer unit is preferred to the right-angle drive unit in present practice where headroom permits.

Adequate bearing design and shaft support are essential to the dependable, low-maintenance operation of mixers. Side-entering agitators and agitators in sealed tanks, furthermore, require care in the design and maintenance of a satisfactory stuffing box. These items have been discussed by Bissell [*Chem. Eng. Progress*, **43**, 649 (1947); *Chem. Ind.*, **62**, 586 (1948)].

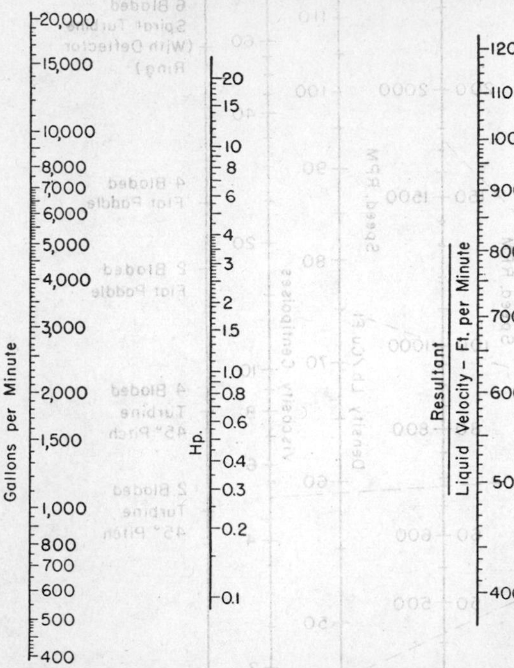

FIG. V. Nomograph for computing the theoretical horsepower required to bring a given volume of water up to a given velocity. Applications include action of propellers, impellers, centrifugal pumps, etc.

FROM LABORATORY TO PLANT*

Research work on a new process or product usually starts in the laboratory, using glass rods for mixing. As a second step, the glass rods are bent and attached to the laboratory mixer. Then, having shown promise of being chemically sound, the process has justified the construction of a pilot plant in which it will be given the opportunity of proving its value under conditions approaching commercial practice.

If properly laid out, the pilot plant should be capable of producing the information necessary for translating to full commercial operation. This information will involve such points as

a. The allowable materials of construction.
b. The unit operations involved.
c. The limitations on batch size; or
d. The possibility of continuous-flow operation.
e. The requirements to be met by the commercial mixing equipment.
f. The necessity for and the extent of heating or cooling.

At the very start of pilot-plant design it is necessary to take into account the objectives of the mixing **operations**, such as

1. Chemical reaction.
2. Blending.
3. Dissolution or washing.
4. Physical change.
5. Dispersion.
6. Adsorption.
7. Heat transfer.

* Taken in part from an article published in *Ind. Eng. Chem.*, **30**, 489 (1938) through courtesy of the publishers and coauthor E. J. Lyons of the Turbomixer Corp.

The physical characteristics of the materials such as viscosity, and specific gravity, and their variations should be available. The mixers in the pilot plant should have sufficient flexibility for the determination of the factors that will affect the ultimate choice of the type of mixing equipment. This flexibility will allow the determination of the importance of

A. High recirculation rate of entire tank contents per minute.................. 1–6
B. At least one point of high shear........ 1, 2, 5, 6
C. General high level of velocity throughout tank.................................. 1, 2, 3, 6, 7 (coils)
D. Violent scrubbing effect............... 1, 3, 5, 6
E. Instantaneous blending or dispersion of added material through large volume of tank contents.......................... 1, 2, 3, 5, 6
F. High velocity over tank walls.......... 2, 3 (viscous), 5 (viscous), 7 (jacketed)
G. High velocity over entire tank bottom.. 1–4, 6, 7
H. High-volume low-velocity.............. 1–5
I. Uniform overflow (continuous operation) 1, 2, 4–7
J. Low short-circuiting (continuous operation)................................ 1–6
K. Selective overflow (continuous operation) 1, 3
L. Horsepower per unit of volume......... 1–6
M. Freedom from swirl.................. 1–6

The figures after each of the above factors refer by number to the items of the preceding list of objectives that may be affected.

To get such flexibility requires the availability of variable speed on the mixer shaft; ease of disassembly for changing mixer units; excess tank volume so that points *A*, *C*, *D*, and *H* can be settled; provision for baffling; and means for making power readings. The latter provision in conjunction with the variable-speed feature allows the determination of the lowest power per unit of volume with which satisfactory results can be secured, often a great help when the commercial plant is being designed and an estimate of operating costs is being made.

Test coupons of materials of construction likely to be used for plant equipment should be placed in strategic positions in the agitated vessels. Examination and weighing of these specimens will permit evaluation of abrasion and corrosion effects and prediction of equipment maintenance and product contamination caused by them.

It costs little more to set up the pilot plant with these provisions, and this extra cost may declare dividends when the commercial plant is finally installed. The operations will be thoroughly understood and, whether you elect to build your own equipment or whether you call in the engineer from the mixer manufacturer for recommendations, the final results will more than justify the extra expense.

Scale-up of Model Agitator Designs. After acceptable equipment and operating conditions for agitation have been determined in the laboratory or pilot plant, the process design engineer must design or specify plant mixers which will perform the same as the small units. The prediction of the power required by a plant agitator of any size operating at any speed in any fluid whose properties are known is not difficult if power data are available for a model of any size, since the general correlations discussed in the preceding paragraphs can be applied. To specify the plant design which will give the same functional performance as the laboratory mixer, however, is less simple.

Certain empirical methods of scaling up agitator designs for constant performance have been developed for some agitation functions and, when these are available, they should be used. All of these methods postulate geometrical similarity between equipment pieces of different sizes, and the laboratory mixers should be of such design, therefore, as to be easily copied on a larger

		Mixing unit operation	Yield or thorough	Physical properties	Speed or completion	Cost of operation	Cost of material	Typical % use
Catalytic processes. Hydrogenation and other gas-liquid reactions. Neutralization, Precipitation, Esterification, Sulfonation, Nitration, etc.	1	Chemical reactions	1	2	3	5	4	33
Autoclave feeds. Blending successive batches for uniformity of product. Uniform heating during continuous flow. Flash mixing, Large tank blending, Dilutions, etc.	2	Blending	1	2	4	3		24
Of salts, Nitrocotton, Cellulose acetate, Pigment pastes, Sugar, etc., Washing of acids, Alkalies, Salts or organic materials out of solids, Leaching	3	Dissolution washing	1	4	2	3		23
Flocculation, Breaking up of sinter, Breaking down or development of plasticity, Change of viscosity, Cellulose acetate precipitation, Crystallization, Repulping	4	Physical change	2	1	4	3		7
Emulsification Treating of oils, Tinting of lacquers and paints, Asphalt filling, Air fluffing of mayonnaise, Soap, etc.	5	Dispersion	1	2	3	4	5	6
Flotation, Decolorizing carbon or clay treatments, Removal of colloids with immiscible liquids or gases	6	Adsorption	1	1	4	3	2	5
With coils, jacketed kettles, or by means of vacuum, etc.	7	Heat transfer			1	3	2	2

Yield or thoroughness. Usually the main aim of the operation; Consequently the No. 1 point, Secondary to none.

Physical properties. Usually refers to product, So it is highly important, It covers size of crystals, Type of precipitate, Viscosity, etc.

Speed or completion Small unit, Quick cycle Vs. large unit, long cycle, Continuous flow with small tank and short holding time Vs. large tank and long holding time

Cost of operation. Efficient Vs inefficient mixtures, Covering hp.-hours, Time, Direct labor, Maintenance, Floor space, Auxiliary equipment, Amortization, etc.

Cost of material. Secondary to thoroughness and physical properties, but of first importance when sales prices are low and competition keen.

100% Taken from industrial flow sheets

FIG. W. Evaluation chart. Starting with a series of typical industrial flow sheets, the mixing operations are placed in seven classes of unit operations. Five mixing factors are then arbitrarily evaluated with respect to each unit operation.

scale. The scale-up methods are summarized below.

Heat Transfer. The correlation of Chilton, Drew, and Jebens [*loc. cit.*] requires that the Reynolds numbers in large and small equipment be proportional to a linear dimension of the equipment for the same film coefficient to obtain in both. Usually a larger coefficient is required in the larger equipment because of the less favorable ratio of heat-transfer area to tank volume, and the correlation permits the selection of a mixer speed to provide this larger coefficient.

Dissolving Solids. Hixson and Baum [*Ind. Eng. Chem.*, **34**, 478 (1941)] showed that the same mass-transfer coefficient is obtained in large and small equipment if the peripheral speed of the agitators is the same. Büche [*Z. Ver. deut. Ing.*, **81,** 1065 (1937)], on the other hand, found equal coefficients with application of equal power per unit volume of agitated liquid. Since the latter is the more conservative of the two methods, it is recommended.

Contacting Immiscible Liquids. According to Miller and Mann [*loc. cit.*], the same degree of dispersion is achieved in large and small units with application of equal power per unit volume.

Contacting Liquids and Gases. As discussed in Secs. 10 and 15, equal absorption coefficients result from application of equal power per unit volume.

Other Functions of Liquid Mixing. For functions for which no scale-up plan has been reported, such as suspending solids and blending liquids, the basis of equal power per unit volume is recommended. This is supported by the general developments of Brothman and Kaplan [*Chem. & Met. Eng.*, **46**, 633, 639 (1939)]. Basing his deductions on the usual shape of an agitation power-function plot, Büche [*loc. cit.*] proposed a method of computing the speed of a plant mixer geometrically similar to a laboratory mixer of known speed if the two are to apply equal power per unit volume. The Reynolds number for the laboratory mixer, $(L_L)^2N\rho/\mu$, is calculated, and the ratio of plant-mixer speed to laboratory-mixer speed, N_P/N_L, is calculated according to the following table:

$(L_L)^2N_P/\mu$	N_P/N_L
< 50	1.0
$50 - 100{,}000$	$(L_L/L_P)^{0.56}$
$> 100{,}000$	$(L_L/L_P)^{0.67}$

It should be noted that equal power per unit volume is not a satisfactory general index to performance if two agitators of different types or designs are being compared.

Plastic Masses and Dry Solids. No methods of scaling up mixers handling these materials have been published.

CONTINUOUS VS. BATCH MIXING

Often the laboratory or pilot plant is called on to collect data to aid in making the decision as to whether a process is to operate by batch or by continuous flow. This section will give a general method for obtaining and interpreting this information.

Continuous flow is of doubtful benefit

1. If a slow reaction must go to completion.
2. If volumes are small or time cycles are long or both.
3. If operation is intermittent, as in plants running only 8 hr./day.
4. If complex temperature or pressure cycles must be followed.
5. If undesirable bacterial action can take place under the operating conditions.

Continuous flow is worth consideration

1. If the reaction does not need to go to completion.
2. If volumes are large or time cycles are short.
3. If the materials are of low intrinsic value.
4. If countercurrent operation is feasible.
5. If a considerable excess of one reactant is allowable.
6. If precipitations, solutions, or extractions are involved.
7. If 24-hr. operation is planned.
8. If much laboratory work is called for to determine end points of each batch.
9. If, for some reason, a considerable fraction of the batch time cycle must be taken up in filling and emptying the tanks.

Although the term "reaction" has been used to express the operation going on, this actually can be any of the Practical Objectives of Mixing as given on pp. 1196–1197.

If continuous flow appears feasible, it may be well to read MacMullin and Weber [*Trans. Am. Inst. Chem. Engrs.* **31**, 409 (1934–5); or *Chem. & Met. Eng.*, **42**, 254 (1935); *ibid.*, **52** (5), 101 (1945)]; also Olsen and Lyons [*Chem. & Met. Eng.*, **52** (5), 118 (1945)] for a better understanding of the problem.

Tests in commercial-size equipment are more reliable than on laboratory or pilot-plant scale. The tendency in small equipment is to use excessive power for agitation, which usually gives better results than can possibly be expected from the full-scale equipment. A comparison of the horsepower per gallon (excluding that required for the drives, belts, etc.) will show whether the test figure is reasonable.

It is always vital to know the degree of completion of the "reaction" under a given set of conditions. The method given below is based on Case 5 of MacMullin and Weber (*Trans. Am. Inst. Chem. Engrs.*, **31**, 409 (1934–5) and is used with their permission. At any time t in the operation, the degree of completion of the "reaction" D must be known.

From the laboratory or pilot-plant test data a curve is drawn similar to Fig. *X*, in which the abscissa is D, the degree of completion of the "reaction," expressed in fractions, and the ordinate t is the time in seconds, minutes, or hours. The curve shown in Fig. *X* represents a true first-order reaction, but in this method the order of the reaction is immaterial since predictions will be based on actual performance rather than on mathematical considerations.

For the purpose of illustration, the final plant will operate with these fixed conditions:

F = flow, gal./min. = 100 gal./min.
t = time, = 7 min. for complete reaction (see Fig. X)
n = number of tanks in series = 3
v = working volume in each tank = 333⅓ gal.

For satisfactory results, nv must always be greater than Ft.

It is obvious that the stream from the last tank in series will be composed of material that has been held any time from years down to the minimum possible transit time through the tanks, possibly less than 1 sec. The next step is to determine just how long each portion of the stream has been held, with respect to D, the degree of completion of the "reaction." That which has been held for 7 min. or longer will have been completely "reacted."

What portion of the total through-put has been completely "reacted"? What various degrees of partial

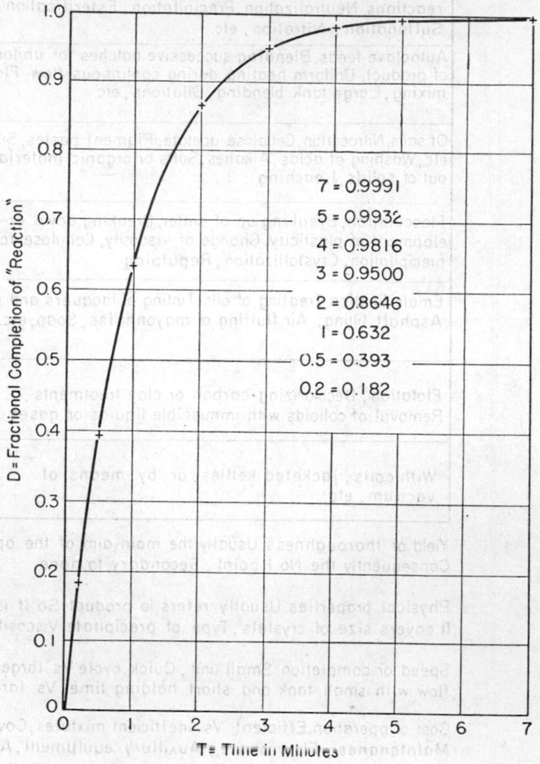

FIG. *X*.

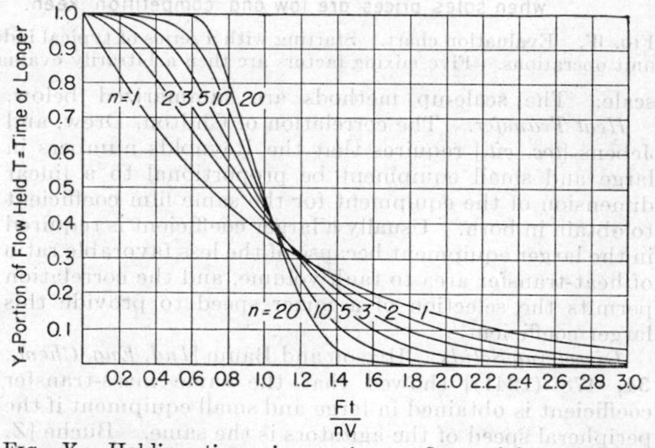

FIG. *Y*. Holding times in continuous-flow mixing tanks in series. [*MacMullin and Weber, Trans. Am. Inst. Chem. Engrs.*, **31**, 409 (1934–5), *reproduced by permission of authors.*]

completion of "reaction" have been attained by portions of the remainder of the through-put?

In Fig. *Y*, y, the part of the total flow held for any given time t or longer is plotted against Ft/nv for a number of different values of n, the number of tanks in series. In this example $\frac{Ft}{nv} = \frac{100 \times 7}{3 \times 333\frac{1}{3}} = 0.7$. Then Ft/nv =

0.7 is followed up to the curve $n = 3$, and y is read as 0.65. This means that 65 per cent of the flow from the third of a series of three tanks has been retained in the

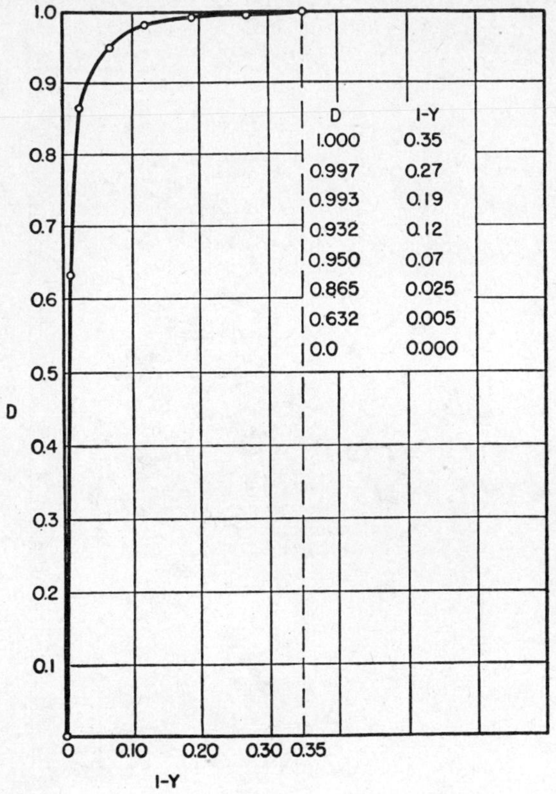

FIG. Z.

tank for 7 min. or longer and thus has been completely reacted.

To determine what has happened to the remaining 35 per cent, which is incompletely "reacted," because it has been retained in the system for an insufficient time, Table 6 is first prepared, using values of D from Fig. X and y from Fig. Y.

Table 6

t, min.	$\frac{Ft}{nv}$	y (from Fig. Y)	$1 - y$ calculated	D (from Fig. X)
7	0.7	0.65	0.35	1.000
6	.6	.73	.27	0.997
5	.5	.81	.19	.993
4	.4	.88	.12	.982
3	.3	.93	.07	.950
2	.2	.975	.025	.865
1	.1	.995	.005	.632
0	0	1.00	0	0

Figure Z shows D plotted against $1 - y$. The area under the curve represents the degree of completion of the incompletely reacted 35 per cent. This area can be calculated, can be read with a planimeter, or, most easily, can be read by weight of paper if a chemical balance is available.

$$\frac{\text{Weight of paper below the curve}}{\text{Weight of entire graph}} = D = 0.9674 \text{ in the example}$$

Hence 35 per cent of the total through-put is 96.74 per cent completely "reacted," and 65 per cent is 100 per cent completely "reacted." Therefore the over-all D = $(0.9674 \times 0.35) + (1.000 \times 0.65) = 0.9886$.

It is also worth while to note that, unless instantaneous separation of the "reactants" takes place at the outlet from the tanks, the "reaction" will continue in the settling tanks or whatever equipment follows the mixers. Therefore, it is possible that this figure of 98.86 per cent completion will undoubtedly be improved by the time separation takes place.

If D = 98.86 per cent is satisfactory, then the problem is solved; but, if it is too low or unnecessarily high, then the problem should be refigured with a somewhat higher or lower volume V assumed, or with more or fewer stages n, until a satisfactory solution is reached.

Rapid Reactions. In case the "reaction" time is so fast that no satisfactory curve can be drawn for Fig. Z, the problem is well solved by designing the commercial equipment so that the minimum transit time of any drop through the system is equal to the complete reaction time. If the agitation in the system is equal to that in the experimental equipment, the problem is then solved

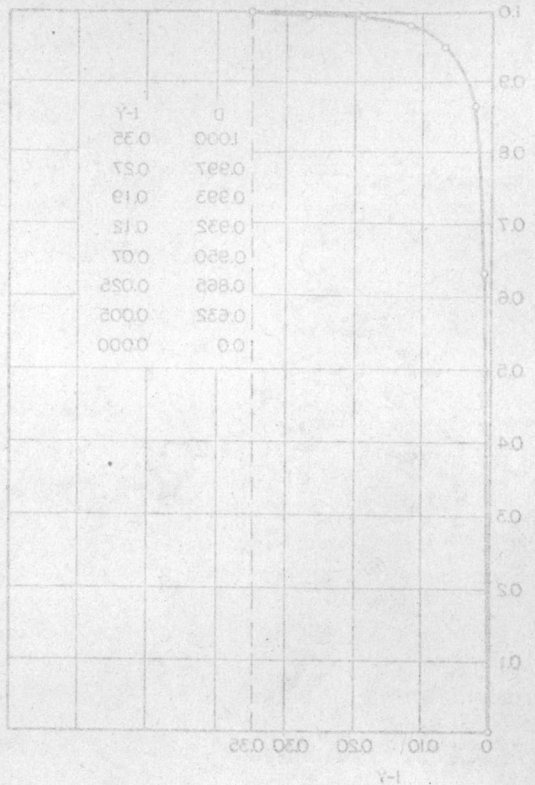

SECTION 18

HIGH-PRESSURE TECHNIQUE

BY

Barnett F. Dodge, D. Sc., Professor and Chairman of Department of Chemical Engineering, Yale University; Member, American Institute of Chemical Engineers, American Chemical Society, American Association for the Advancement of Science, American Society for Engineering Education.

CONTENTS

	Page
Applications	1234
Selection of Materials	1234
Design and Construction of Pressure Vessels	1237
General	1237
Thin-walled cylinders	1237
Stress Distribution in Wall of a Thick Cylinder	1238
Theories of Elastic Failure	1239
Design for Superpressures	1240
Effect of Temperature in Design	1240
Methods for Improving Stress Distribution in Thick-walled Cylinders	1241
Types of Construction	1242
Joints and Closures	1243
Various Types of Joints	1243
Vessel Closures	1248
Piping, Valves, and Fittings	1250
Sight Glasses and Windows	1252
Electric Lead Seals	1253
Flowmeters	1254
Pressure Measurement	1254
Reaction Vessels	1256
Autoclaves	1256
Continuous Reactors	1257
Pumps	1258
Compressors	1258
Safety	1261

HIGH-PRESSURE TECHNIQUE

REFERENCES: Bridgman, "The Physics of High Pressure," Macmillan, New York, 1931. Eucken and Jacob, Editors, 'Der Chemie-Ingenieur," vol. 3, Part 4, Newitt, Schneider, Natta, and Roberti, "High Pressure Operations," Akademische Verlagsgesellschaft M.B.H., Leipzig, 1939. Newitt, "The Design of High Pressure Plant and the Properties of Fluids at High Pressures," Oxford, New York, 1940. Dodge, High Pressure Processes, Chap. IV of "Roger's Manual of Industrial Chemistry," Furnas, Editor, 6th ed., Van Nostrand, New York, 1942. Newitt, The Design of Vessels to Withstand High Internal Pressures, *Trans. Inst. Chem. Engrs.* (*London*), **14**, 85 (1936). A.P.I.-A.S.M.E. Code for the Design, Construction, Inspection and Repair of Unfired Pressure Vessels for Petroleum Liquids and Gases, 1943 ed. with supplements. Rules for Construction of Unfired Pressure Vessels, A.S.M.E. Boiler Construction Code, Sec. VIII (with addenda), Am. Soc. Mech. Engrs., New York, 1943. Curtis, High Pressure Equipment and Technique, Chap. X of "Fixed Nitrogen," Chemical Catalog Co., New York, 1932. Tongue, "The Design and Construction of High Pressure Chemical Plant," Chapman & Hall, London, 1934. Hesse and Rushton, "Process Equipment Design," Van Nostrand, New York, 1945. Symposium on the Effect of Temperature on the Properties of Metals, Am. Soc. Testing Materials and Am. Soc. Mech. Engrs., 1932. Mechanical Properties of Metals and Alloys, *Bur. Standards Circ.*, C447, 1943. Bridgman, Recent Work in the Field of High Pressures, *Rev. Modern Phys.*, **18**, 1 (1946) (A comprehensive review of work in the whole field between 1930 and 1945 with 674 references). Bone, Newitt, and Townend, "Gaseous Combustion at High Pressures," Longmans, London, 1929. Tongue, High Pressure Chemical Engineering Equipment of the Chemical Research Laboratory, Teddington, *Trans. Inst. Chem. Engrs.* (*London*), **8**, 81 (1930). Keyes, Methods and Procedures Used in the Mass. Institute of Technology Program of Investigation of the Pressures and Volumes of Water to 460°C. *Proc. Am. Acad. Arts Sci.*, **68**, 505 (1933). Meigs, Closures for High Pressure Vessels, *Trans. Am. Inst. Chem. Engrs.*, **39**, 769 (1943). "Compilation of Available High-temperature Creep Characteristics of Metals and Alloys," Am. Soc. Testing Materials and Am. Soc. Mech. Engrs., 1938. Ernst, Equipment for High Pressure Reactions, *Ind. Eng. Chem.*, **18**, 664 (1926). Ernst, Edwards, and Reed, A Direct Synthetic Ammonia Plant, *Ind. Eng. Chem.*, **17**, 775 (1925). Special high-pressure, High temperature issue, *Chem. & Met. Eng.*, September, 1930. Seibel, Factors in High-pressure Design, *Ind. Eng. Chem.*, **29**, 414 (1937). Clark *et al.*, Bench-scale Equipment and Techniques. High Pressure Reactions, *Ind. Eng. Chem.*, **39**, 1555 (1947). Reynolds, A Survey of High Pressure Designs in Germany, FIAT Final Report 1067. Krase, Design and Construction of High Pressure Compressors and Reaction Equipment, FIAT Final Report 611. High Pressure, *Chem. Eng.*, p. 107, Aug. (1949).

APPLICATIONS

The term "high pressure" will be taken to refer to any pressure above a lower limit of about 50 atm. (approx. 750 lb./sq. in.). Application of high pressures in chemical industry is relatively new and may be said to date from about 1913 when the industrial synthesis of ammonia from the elements and the Burton process for oil cracking were first put into practice. The latter process was operated at pressures well below 750 lb./sq. in. but nevertheless was the beginning of the many applications of high pressure in the petroleum industry. Pressures up to 1500 atm. are now in large-scale use in the chemical industry and further research in the laboratories of universities, research foundations, and industrial concerns may be expected to lead to further applications in this range and probably to some important applications at still higher pressures. The technique for carrying out reactions under pressures up to 10,000 atm. is fairly well developed, on a small scale at least, but no important industrial use for such pressures has yet been found. Pressures as high as 3,000,000 lb./sq. in. have been produced in research laboratories and used to a limited extent in the study of physical properties. Pressures of the order of 50,000 to 75,000 lb./sq. in. are commonly developed in heavy artillery; pressures up to 150,000 lb./sq. in. are used in the autofrettage process of strengthening gun barrels; and even pressures as high as 500,000 lb./sq. in. may be said to have some industrial importance, since pressures of this order are estimated to be developed in lubricants of some gears and bearings.

The uses of high pressure in chemical industry may be classified under the following heads:

1. The production or maintenance of a liquid phase. Examples: liquefaction of oxygen and similar gases, separation of hydrogen and helium from gas mixtures, liquid-phase cracking of petroleum hydrocarbons, hydrolysis of chlorbenzene to produce phenol, hydrogenation of hydrocarbons, and use of retrograde condensation in petroleum production.
2. Shifting chemical equilibrium. Examples: synthesis of ammonia from the elements, methanol and higher alcohols from CO and H_2, aliphatic acids from CO and the alcohol, and higher alcohols from fatty acids.
3. Storage of gases. Example: shipping of O_2, N_2, CO_2, H_2, He, and other common gases in high-pressure cylinders.
4. Increase of chemical reaction rate. Examples: This is intimately tied up with class 2, and most of the same examples will illustrate this use of pressure. In certain cases, such as petroleum reforming, alkylation, and oxidation of hydrocarbons under pressure, relative reaction rates are varied by pressure so that certain desired reactions are favored and the end products are due both to shifting equilibrium and the effect of pressure on rates.
5. Increase of gas solubility. Examples: scrubbing gas mixtures to remove CO and CO_2, and recovery of gasoline from natural gas.
6. Separation of liquids from solids. Examples: pressing vegetable oils from the seed or other natural product; dehydration of pulps.
7. Compacting of powders, extrusion, and related pressing operation on solids. Examples: molding of plastics, powder metallurgy, making briquettes.

In the discussion to follow, techniques used in both the experimental laboratory and in large-scale plants will be presented without any attempt at making a clear distinction between them; this would in fact be next to impossible.

SELECTION OF MATERIALS

Five important factors should be kept continually in mind when selecting a metal for high-pressure service, namely, (1) the working pressure, (2) the working temperature, (3) the size of the vessel, (4) the nature of the process with particular reference to the corrosive action of materials, and (5) the stress conditions to be encountered, particularly whether static or dynamic. Steel is practically the only material available at reasonable cost with the necessary physical properties to withstand the mechanical stresses imposed by high pressures, though copper and bronzes are sometimes used on a small scale for pressure vessels and for tubing to convey fluids. It is well to emphasize that a steel to be used for high-pressure work should be free from cracks and undue segregation. This should be checked by examining suitable etched samples from the bar or forging. Where corrosion resistance is required or to protect from deleterious catalytic action, the steel vessel or pipe is fitted with a liner of copper, nickel, monel, lead, aluminum, or silver. For example, carbon monoxide reacts with iron at elevated temperature and pressure to form the volatile iron carbonyls. Usually the attack is not sufficient to cause serious weakening of the vessel wall, but the iron carbonyl may poison catalysts and cause other harmful effects. The best protection against this action is the use of non-ferrous metal liners. The action of hydrogen in causing embrittlement as a result of attack along grain boundaries is sometimes a factor, and this may be minimized by the use of certain alloy steels or by keeping the steel walls of the vessel at as low a temperature as possible. In cases where the vessel is subjected to an unusually high temperature, *i.e.*, about 500°C. along with high pressure of hydrogen, *e.g.*, in the Claude converters for ammonia and alcohol synthesis, a chromium-iron-nickel alloy known as BTG alloy and containing 60 per cent Ni, 12 per cent Cr, 2.5 per cent W, and the rest iron, is used. Another alloy that is sometimes used for simultaneous resistance to high temperature and high pressure is nichrome, 80 to 85 per cent Ni, 20 to 15 per cent Cr. An alloy steel extensively used in the coal hydrogenation process in Germany in which exposure to hydrogen at high pressure and 500°C. was encountered had the following composition by per cent: 0.20 C, 0.30–0.40 Mn, 0.20–0.35 Si, 2.50–3.50 Cr, 0.35–0.45 W, 0.35–0.45 Mo and 0.70–0.85 V. It was reported to have an ultimate tensile of 100,000 to 110,000 lb./sq. in. and a yield strength of 70,000 to 75,000 lb./sq. in. at room temperature. In its pilot plant for this same process the Bureau of Mines has found 18-8 stainless steel type 347 to be very satisfactory.

Plain carbon steels may be used for some high-pressure applications particularly when temperatures do not exceed 400°C., but usually alloy steels are preferred because of the better physical properties that may be developed by heat-treatment and the greater uniformity of properties obtainable in large forgings. Furthermore, it is possible to develop higher strengths in them without so great an increase in brittleness. A low alloy steel has approximately twice the yield strength of a plain carbon steel of the same carbon content and usually a higher ratio of yield to ultimate strength. For this and other reasons, alloy steels are frequently preferred and may be more economical even in cases where plain carbon steels would give satisfactory service. Such alloy steels contain nickel and chromium along with small amounts of vanadium, molybdenum, or tungsten.

In this country, a chrome-vanadium steel of 0.3 to 0.4 per cent C, 1.0 to 2.0 per cent Cr, and 0.2 per cent V was originally favored for large forged vessels to be used for ammonia and alcohol syntheses. It is heat-treated to approximately the following physical properties; ultimate tensile 100,000 lb./sq. in.; elastic limit* 75,000 lb./sq. in.; 20 per cent elongation in 2 in.; and 40 per cent reduction in area. In England, a nickel-chromium-molybdenum steel (Vibrac) is extensively used. A typical analysis and set of properties are C 0.25 to 0.30, Si 0.15, Mn 0.60, Cr 0.6 to 0.7, Ni 2.50, Mo 0.6; ultimate tensile 125,000; elastic limit 100,000; 24 per cent elongation in 2 in.; 65 per cent reduction in area; Izod impact strength 55 ft.-lb.† These properties are for small forgings and will be somewhat reduced for large ones. As a rough guide, the following tabulation lists a range of values of tensile properties for a considerable number of large forged vessels in both plain carbon and alloy steels that were built in this country.

	Yield point, lb./sq. in.	Ultimate tensile strength, lb./sq. in.	% elongation in 2 in.	% reduction in area
Plain carbon........	30,000–40,000	60,000– 80,000	18–27	30–42
Alloy...............	45,000–70,000	70,000–100,000	15–25	30–45

For laboratory experimentation at very high pressures, a nickel-chromium steel (S.A.E. 3250) has been considered satisfactory [Poulter, *Phys. Rev.*, **40**, 877 (1932)]. Of course, many other steels are satisfactory and even preferred for special cases. One experimenter (British Intelligence Objectives Services Final Report No. 609) prefers plain carbon steels to alloy steels for pressures up to 15,000 atm. It is practically impossible to give general statements of any value about materials to use, because every case is special to some extent. The best procedure is to discuss the problem in as much detail as possible with the steel manufacturer.

Another point to be considered when tensile properties are specified is that there may be a considerable difference in the properties shown by tests cut from a forging, depending on the orientation of the test bar. In forgings of this type, the ductility values obtained with a test bar located in the transverse direction (*i.e.*, at right angles to the ingot axis) are usually somewhat lower than those taken in a longitudinal direction. Consequently, when specifying requirements, the user should indicate clearly the location and direction of the test. The majority of specifications for this type of forging call for transverse specimens. A logical basis for choosing the orientation of the test bar would be to take it, where practical, in the direction in which the maximum stress occurs in service. In the cases considered in this discussion, that would be in the transverse or tangential direction. Also, the user and the manufacturer should agree on the method of testing, and there should be a thorough understanding as to how the elastic property of the material is to be recorded, *i.e.*, as yield strength, yield point, elastic limit, or proportional limit.

Hydrogen causes decarburization of steels, the rate increasing rapidly with the temperature and the partial pressure of the hydrogen. The result may be a serious weakening of the material because of fissuring, which tends to follow grain boundaries. Nitrogen also attacks steel because of the formation of nitrides of iron and of most of the alloying elements used in steels, but the reaction becomes appreciable only at elevated temperatures and is probably not of much practical importance. Plain carbon steels are particularly susceptible to such attack; and, where conditions are such as to lead to attack by either or both of these gases, alloy steels should be used. At pressures of the order of 3000 atm. and above, hydrogen attack may cause failures even at room temperature. There is very little reliable information on this subject and until more is available one must proceed with caution in using hydrogen-containing gases at these pressures.

Low alloy steels, generally with 1.5 to 6 per cent chromium and other alloying elements such as vanadium, nickel, molybdenum, and tungsten are more resistant to hydrogen attack. Of course, the higher chromium alloys and the austenitic 18-8 Cr-Ni steels are even more resistant, but they are less desirable from strength and economic standpoints. For the mixture of gases used in ammonia synthesis, a chrome-vanadium steel whose composition was given above has been widely used. Laboratory tests indicated that a chrome-tungsten steel (0.58 C, 0.55 Cr, 1.6 W) was also one of the best. The recent trend, however, is toward a design that keeps the

* For practical design purposes, no clear distinction has been made between elastic limit, proportional limit, and yield strength, the latter quantity being the one usually obtained from the stress-strain diagram. In fact it is doubtful if most structural materials have a true proportional limit or elastic limit, and the terms are generally used loosely for a particular value of yield strength.

† Newitt, The Design of Vessels to Withstand High Internal Pressures, *Trans. Inst. Chem. Engrs.* (*London*), **14**, 85 (1936).

shells of large converters for such reactions as the ammonia synthesis at a low enough temperature so that no attack by the gases occurs, with the result that only the physical properties and ease of forging need to be considered. Chromium also gives increased resistance to air oxidation, and even the low Cr alloys are much better than plain carbon steels. High chromium irons and the 18-8 type of alloy have excellent resistance to oxidation. For those cases where creep resistance is important, high chromium irons or stainless steels of the 18 Cr, 8 Ni type are desirable. Molybdenum and tungsten in small percentages are particularly valuable alloying elements for increasing creep resistance.

Higher tensile strengths than those cited above can be developed by suitable heat-treatment, especially in the high carbon and the alloy steels, but this is always accompanied by a loss in ductility and an increase in brittleness, the effect being more pronounced in the plain carbon steels than in the alloy steels. Ultimate strengths as high as 300,000 lb./sq. in. are easily obtainable in small bars by suitable heat-treatment, but experience with large vessels would indicate that 125,000 lb./sq. in. is about the present practical limit. For experimental work on a small scale, however, the higher tensiles are frequently used. Although the belief is firmly held by most designers of pressure equipment that the high tensile and less ductile materials are to be avoided, there is actually very little factual basis for it. There is no doubt, however, that ductility plays a very important role in the readjustments that prevent stress concentrations and is a property that should be given considerable weight, especially when dealing with thick-walled vessels.

The effect of temperature on the physical and also the chemical properties may be important in many cases, and a brief discussion of this is in order. The properties generally stated are for room temperature. As temperature is increased, the tensile strength and elastic limit decrease. No definite point can be given at which the effect becomes noticeable because it depends on the metal, its heat-treatment, and other factors; but, generally speaking, it lies in the range 600° to 900°F. Furthermore, the phenomenon of creep—the slow continuous distortion under constant load below that producing prompt fracture—becomes a factor that must be considered in design.

Some typical figures showing the effect of temperature on the tensile properties of three typical steels are given in Table 1.

At low temperatures most steels become less ductile and more brittle, and their resistance to impact becomes very poor. The effect depends on the type of steel and its treatment but becomes noticeable with plain carbon and low alloy steels at temperatures of the order of −50°C. and even higher in some cases. For the very low temperatures encountered in the liquefaction of the so-called permanent gases, nickel steels containing about 3 per cent nickel or the 18-8 stainless steels are recommended. The 18-8 austenitic steels are not adversely affected in any of their common tensile properties at the lowest test temperatures so far used (about −300°F.). Copper and copper alloys, monel, and nickel are also suitable materials of construction at these extreme temperatures. The ordinary lead-tin solders can also be used with safety at these low temperatures. In fact the tensile properties of such soft metals are greatly improved at low temperatures.

The behavior of steels when overstrained, *i.e.*, stressed beyond the elastic limit, is of importance, particularly in connection with the process of "autofrettage," and a brief discussion of deformation will be given with the aid of Fig. 1, which shows the two typical types of stress-strain curves encountered in steels.

Table 1. Effect of Temperature on the Tensile Properties of Steel

a. Nickel-chromium-tungsten Steel*
0.46% C, 26.5% Ni, 14.0% Cr, 3.6% W

Temp., °F	Ultimate strength, lb./sq. in.	Elongation in 2 in. %	Reduction in area, %	Limiting creep stress, lb./sq. in.†
65	102,000	33.5	42.5	
930	92,000	27.0	33.0	
1110	78,000	22.0	29.0	24,600
1290	56,000	21.0	28.0	13,400
1470	39,000	39.0	43.0	4,500

b. Plain Carbon Steel, 0.15% C

Temp., °F	Ultimate strength, lb./sq. in.	Yield strength, lb./sq. in.	Elongation in 2 in., %	Reduction in area, %	Creep stress, 0.01% in 1000 hr.
80	62,400	42,000	36	67	
750	58,000	24,600	34	67	
800	......	......	..	..	18,500
900	45,500	23,500	38	71	12,800
1000	36,500	20,100	42	77	2,700
1200	20,000	10,200	54	89	290
1400	9,000	3,800	70	77	

c. Chrome-molybdenum Steel, 1.2% Cr, 0.5% Mo

Temp., °F.	Ultimate strength, lb./sq. in.	Yield strength, lb./sq. in.	Elongation in 2 in., %	Reduction in area, %	Creep stress, 0.01% in 1000 hr.
85	146,000	108,000	17	42	
750	130,000	97,000	20	59	
800	....	....	..	..	47,000
1000	100,000	80,200	22	74	7,800
1200	62,800	31,800	29	88	220

* Newitt, "The Design of High Pressure Plant and the Properties of Fluids at High Pressures," Oxford, New York, 1940.

† Creep rate not stated but believed to be 0.0001 per cent in 24 hr.

Figure 1(*a*) illustrates the curve of a steel exhibiting the so-called "drop of the beam," and Fig. 1(*b*) illustrates the curve of a steel having a continuously curved plastic zone. Upon initial application of load, each steel deforms

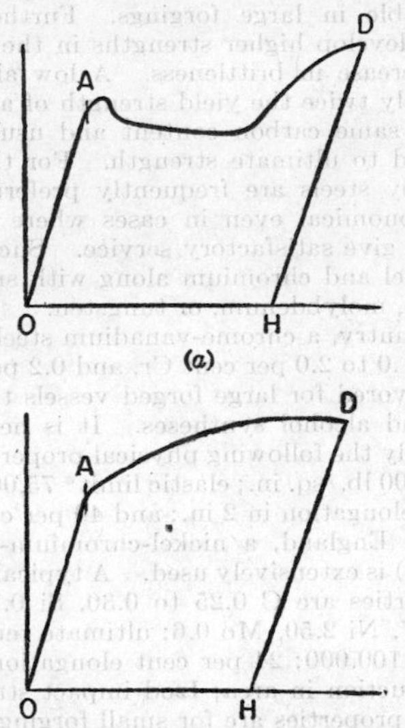

Fig. 1. Behavior of a high-tensile steel when overstrained. Stress is plotted along the ordinate and strain along the abscissa.

elastically up to point *A* on the curves. Upon release of the load within this range, the specimen will return to its original dimensions. Beyond *A*, plastic deformation sets in. The shape of the curve beyond *A* is a measure of the manner and rate at which the steel work hardens in relation to the rate at which the specimen cross section decreases because of lateral contraction. The "drop of the beam" is really fictitious, and its shape depends on the rate at which the specimen is pulled and on the elastic constant of the machine on which the specimen is tested. It nonetheless indicates a steel with plastic formability different from that of a steel of type 1*b*. Upon release of the load at any point beyond *A*, such as *D*, the slope of the curve to no load essentially follows the slope of the initial elastic load curve. *OH* is a measure of the total plastic strain. The net effect of overstrain is to reduce the elastic limit greatly, increase the yield point, and reduce the elongation in proportion to the total amount of overstrain. If the steel is subjected to a low-temperature heat-treatment in the range 200° to 400°C., the elastic properties are recovered, the elastic limit being increased to at least the value of the overtensioning stress and generally higher.

For further details on overstrain, refer to Macrae ("Overstrain of Metals and Its Application to the Auto-Frettage Process of Cylinder and Gun Construction." H.M. Stationery Office, London, 1930). A condensed but detailed treatment of the theory and practice of autofrettage is given by Wintsch [*Schweizer Arch. angew. Wiss. Tech.*, **9**, 81 (1943)]. For further details on the properties of steels, especially for high-pressure high-temperature service, see Newitt ("The Design of High Pressure Plant and the Properties of Fluids at High Pressures," Oxford, New York, 1940), Tongue ("The Design and Construction of High Pressure Chemical Plant," Chapman & Hall, London, 1934), the Symposium on the Effect of Temperature on The Properties of Metals (Am. Soc. Testing Materials and Am. Soc. Mech. Engrs., 1932) and J. L. Cox, *Trans. Am. Inst. Chem. Engrs.*, **29**, 43 (1933). Also, the following references will be found useful: *Trans. Chem. Eng. Congr. World Power Conference*, **1**, 1, 66 (1937); and *Ind. Eng. Chem.*, **28**, 1366 (1928).

DESIGN AND CONSTRUCTION OF PRESSURE VESSELS

General. High-pressure design is concerned not only with pressure vessels for the storage of fluids and the carrying out of chemical reactions or physical operations but also with the equipment for generating pressures. To narrow the field somewhat, the treatment will be restricted to vessels where static conditions prevail; and consequently the question of fatigue, which is important in the design of equipment subjected to dynamic stresses, such as pumps and compressors, can be omitted.

The design of a pressure vessel depends on a great many factors and conditions, and in this discussion it will be possible to mention only a few of the most important ones and to enter into details on only one or two.

The following list shows some of the most important factors that need to be considered when the design of a pressure vessel is undertaken:

1. Dimensions, *i.e.*, diameters, length, etc., and limitations on them.
2. Operating pressure.
3. Operating temperature.
4. Temperature differential in walls.
5. Method of heating or cooling.
6. Corrosive nature of reacting materials and products.
7. Type of operation, *i.e.*, batch or continuous.
8. Number and size of openings in vessel and heads.
9. Method of agitation.
10. Vertical or horizontal installation.
11. Available materials, their tensile properties and cost.
12. Type of construction, *i.e.*, forged, welded, riveted, etc.

In this section only the general principles governing the relation between working pressure and the wall thickness of a cylindrical vessel will be considered under the head of design. Actually, design of a pressure vessel involves a great many other practical details, such as design of heads, openings and closures, safety valves, and welding efficiencies, which cannot be treated. For such details in connection with pressures up to a maximum of 2500 to 3000 lb./sq. in., the two generally accepted pressure-vessel codes [A.P.I.-A.S.M.E. Code for the Design, Construction, Inspection and Repair of Unfired Pressure Vessels for Petroleum Liquids and Gases, 1943 ed. with supplements. Rules for Construction of Unfired Pressure Vessels, A.S.M.E. Boiler Construction Code, Sec. VIII (with addenda), Am. Soc. Mech. Engrs., New York, 1943] should be consulted. For higher pressures, there are as yet no accepted codes, and the engineer must rely wholly on general principles guided by whatever experience is available to him.

In the treatment of the pressure-stress relations in cylinders to be developed in the next paragraphs, the following assumptions are involved: (1) there is no temperature gradient in the wall of the vessel; (2) the applied pressure is a steady, non-fluctuating one; (3) there is no effect of scale, *i.e.*, the stress depends on the diameter ratio and is independent of the diameter itself; (4) the material is isotropic; (5) there are no initial stresses in the material before application of a pressure; and (6) temperatures are low enough so that creep can be neglected.

The complex system of stresses existing in the walls of a closed cylindrical vessel under pressure (internal or external or a combination) is generally reduced to a system of three principal stresses acting at right angles to one another. These are (see Fig. 2): (1) a tangential

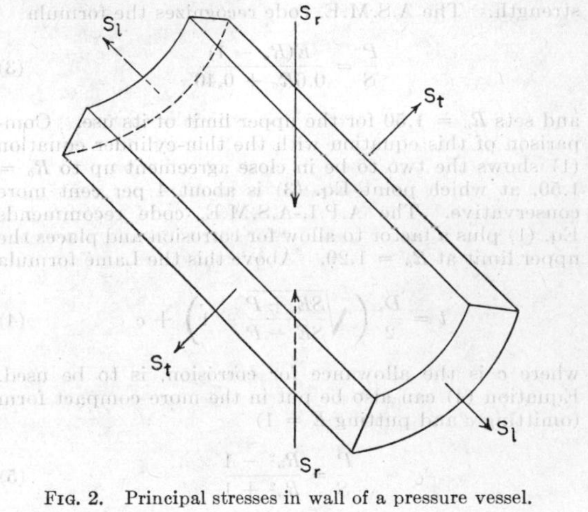

FIG. 2. Principal stresses in wall of a pressure vessel.

or hoop stress S_t, (2) a radial stress S_r, and (3) a longitudinal stress S_l. In the case of a cylinder subjected to internal pressure, S_t and S_l are tensile stresses and S_r is a compressive stress. S_l is generally so small in relation to S_t that it does not need to be considered in connection with the failure of a vessel. Of course, in the case of wire-wound vessels, for example, this would not be true and in fact S_l becomes the limiting factor.

Thin-walled Cylinders. In the case of a thin-walled cylinder, *i.e.*, one where R_o (ratio of outside diameter to inside diameter) is < 1.20, we can focus attention on the hoop stress and assume that this stress is uniformly distributed throughout the wall cross section. In such a cylinder it can readily be shown that the longitudinal stress is only one-half of the hoop stress and so is never controlling. The symbol S without a subscript will always refer to a hoop stress. This assumption leads to the simple relation (external pressure assumed to be zero)

$$t = \frac{PD_m}{2SE} \tag{1}$$

where P = internal pressure, lb./sq. in.
D_m = mean diameter, in.
t = wall thickness, in.
E = efficiency of welded or other joints.
or, in another form, when $E = 1$,

$$\frac{P}{S} = \frac{2(R_o - 1)}{R_o + 1} \tag{2}$$

This relation has been verified many times by testing vessels to destruction, and for all practical purposes it not only holds to the elastic limit but will correctly predict the bursting strength as well. When S equals the elastic limit as determined by the usual tensile test, elastic failure will begin, and at higher pressures the wall will yield and show a permanent set. When S equals the ultimate strength as determined by the usual tensile test, failure by fracture of the wall will occur. In determining the necessary wall thickness for a given vessel and internal pressure, an allowable stress is chosen for S. This is some fraction of either the elastic limit or the ultimate strength, and the reciprocal of this fraction is called a factor of safety. The choice of such a factor is quite arbitrary and depends on the particular conditions under which the vessel is to be used. Common values are 2 to 2.5 based on the elastic limit and 4 to 5 based on the ultimate strength.* The two pressure-vessel codes recognize only a factor of safety based on the ultimate strength. The A.S.M.E. code recognizes the formula

$$\frac{P}{S} = \frac{E(R_o - 1)}{0.6R_o + 0.40} \tag{3}$$

and sets $R_o = 1.50$ for the upper limit of its use. Comparison of this equation with the thin-cylinder equation (1) shows the two to be in close agreement up to R_o = 1.50, at which point Eq. (3) is about 4 per cent more conservative. The A.P.I.-A.S.M.E. code recommends Eq. (1) plus a factor to allow for corrosion and places the upper limit at $R_o = 1.20$. Above this the Lamé formula

$$t = \frac{D_i}{2}\left(\sqrt{\frac{SE + P}{SE - P}} - 1\right) + c \tag{4}$$

where c is the allowance for corrosion, is to be used. Equation (4) can also be put in the more compact form (omitting c and putting $E = 1$)

$$\frac{P}{S} = \frac{R_o^2 - 1}{R_o^2 + 1} \tag{5}$$

Stress Distribution in Wall of a Thick Cylinder. The various thin-cylinder formulas assume a uniform hoop stress across the wall of the cylinder. This is actually far from true in a thick-walled cylinder, as the following analysis will show:

Consider a cylinder of diameters D_o and D_i (Fig. 3). The condition for static equilibrium in a thin section dr at radius r is

$$S_t = -S_r - r\frac{dS_r}{dr} \tag{6}$$

The assumption that the longitudinal stress is constant across the wall leads to the relation

$$S_t - S_r = \text{const.} = 2a \tag{7}$$

Combining (6) and (7) and integrating,

$$S_r = \frac{c}{r^2} - a \tag{8}$$

$$S_t = \frac{c}{r^2} + a \tag{9}$$

Since $S_r = P_1$ (inside pressure) when $r = r_i$ and $S_r = P_2$ (outside pressure) when $r = r_o$, both c and a can be

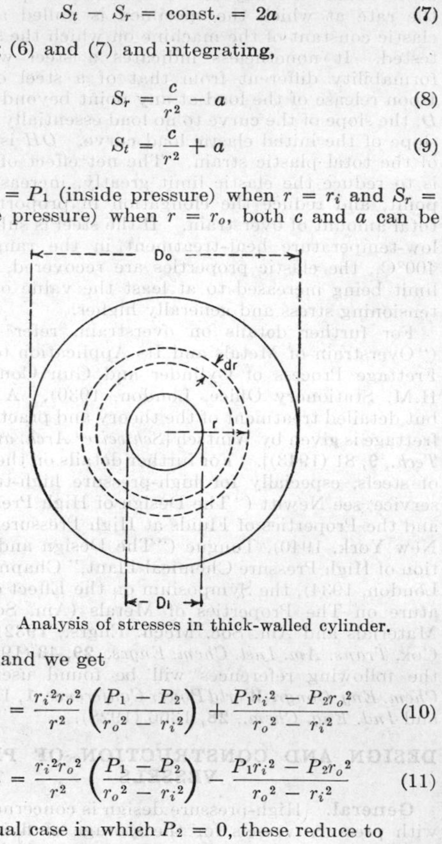

FIG. 3. Analysis of stresses in thick-walled cylinder.

evaluated, and we get

$$S_t = \frac{r_i^2 r_o^2}{r^2}\left(\frac{P_1 - P_2}{r_o^2 - r_i^2}\right) + \frac{P_1 r_i^2 - P_2 r_o^2}{r_o^2 - r_i^2} \tag{10}$$

and

$$S_r = \frac{r_i^2 r_o^2}{r^2}\left(\frac{P_1 - P_2}{r_o^2 - r_i^2}\right) - \frac{P_1 r_i^2 - P_2 r_o^2}{r_o^2 - r_i^2} \tag{11}$$

For the usual case in which $P_2 = 0$, these reduce to

$$S_t = P_1 \frac{r_i^2}{r^2}\left(\frac{r_o^2 + r^2}{r_o^2 - r_i^2}\right) \tag{12}$$

$$S_r = P_1 \frac{r_i^2}{r^2}\left(\frac{r_o^2 - r^2}{r_o^2 - r_i^2}\right) \tag{13}$$

or, letting $R_o = r_o/r_i$ and $R = r/r_i$, we have

$$S_t = \frac{P_1}{R^2}\left[\frac{R_o^2 + R^2}{R_o^2 - 1}\right] \tag{14}$$

and

$$S_r = \frac{P_1}{R^2}\left[\frac{R_o^2 - R^2}{R_o^2 - 1}\right] \tag{15}$$

When $R = 1$, S_t and S_r have maximum values given by

$$S_{t\max} = P_1\left[\frac{R_o^2 + 1}{R_o^2 - 1}\right] \tag{16}$$

$$S_{r\max} = P_1 \tag{17}$$

In Fig. 4, S/P_1 is plotted as a function of wall thickness expressed as R. S_t/P_1, being a tension, is shown as positive, and S_r/P_1 is made negative since it is a compression. It will be noted that the hoop stress drops off sharply from a maximum value at the inner wall to a value at the outer wall of only about one-fifth that at the inner wall. With a larger ratio of diameters the drop from the maximum becomes sharper, and the ratio of the two extreme values is greater. Figure 4 shows that, if a maximum allowable stress is fixed, the allowable internal pressure rises rapidly at first as the wall thickness is increased

* These figures are based primarily on use of plain carbon steels and are not very logical for alloy steels where the ratio of elastic limit to ultimate strength is quite different. The author believes that the elastic limit or preferably the yield strength should always be used as the design criterion of failure, but this is contrary to present code requirements.

but beyond a diameter ratio of 4 to 5 there is very little gained by increasing the wall thickness.

Equations (14), (15), and related ones have generally been attributed to Lamé, and any one of them as well as Eqs. (4) and (5) may be referred to as the Lamé equation.

Although the longitudinal stress in a closed-end cylinder is seldom controlling, there are occasions when

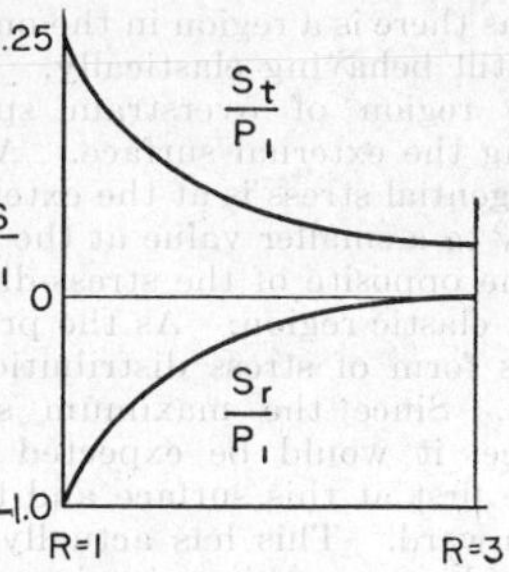

Fig. 4. Variation of tangential and radial stresses in a cylinder wall.

one wishes to calculate it. Based on the assumption of uniform distribution, it is given by the equation

$$S_l = \frac{P_1}{R_o^2 - 1} \tag{18}$$

Theories of Elastic Failure. In order to design a safe vessel it has generally been considered necessary to keep the stresses well below a value that could cause elastic failure at the inner wall of the vessel. It is therefore necessary to have some criterion of elastic failure; and, when this is combined with an equation such as (14) relating the tangential stress to the geometry of the vessel, a design formula is obtained.

There are many such theories, but five of the commonest are designated as follows:

1. Maximum-principal-stress theory.
2. Maximum-shear-stress theory.
3. Maximum-strain theory.
4. Maximum-strain-energy theory.
5. Constant-energy-of-distortion theory (Von Mises-Hencky; also known as shear-strain-energy theory).

According to (1), failure occurs when $S_{t\max} = F$ where F = tensile stress at the elastic limit of the material, as determined in the usual tensile test (actually yield strength for a given offset is determined). Combining this criterion with (16), one obtains

$$\frac{P_1}{F} = \frac{R_o^2 - 1}{R_o^2 + 1} \tag{19}$$

For a safe design it is necessary to choose an allowable stress S less than F, and the design equation becomes

$$\frac{P_1}{S} = \frac{R_o^2 - 1}{R_o^2 + 1} \tag{20}$$

The other theories lead to the following relations:

Maximum shear:

$$\frac{P_1}{S} = \frac{R_o^2 - 1}{2R_o^2} \tag{21}$$

Maximum strain:

$$\frac{P_1}{S} = \frac{R_o^2 - 1}{(1+m)R_o^2 + (1-m)} \tag{22}$$

where m is the Poisson ratio. For $m = 0.25$, (22) becomes

$$\frac{P_1}{S} = \frac{R_o^2 - 1}{1.25R_o^2 + 0.75} \tag{23}$$

Formulas based on the maximum-strain theory are also known by the name of Clavarino.

Maximum-strain-energy:

$$\frac{P_1}{S} = \frac{2(R_o^2 - 1)}{\sqrt{10R_o^4 + 6}} \tag{24}$$

(also based on $m = 0.25$).

Constant-energy-of-distortion:

$$\frac{P_1}{S} = \frac{R_o^2 - 1}{\sqrt{3}\,R_o^2} \tag{25}$$

The question as to which of these relations is the best to use can be decided only by experiment, and data in this field are extremely limited. It is probable that the relation to use will depend on the material; in fact there is some experimental support of the assertion that the maximum-principal-stress theory best describes the behavior of brittle materials and the maximum-shear-stress relation or the constant-energy-of-distortion relation is best for high-tensile alloy steel. It should be noted that, of the four relations applying to thick-walled cylinders, the maximum-shear relation is the most conservative and the maximum-principal-stress one is the least conservative.

It should be emphasized again that relations (19) through (25) are all based on elastic breakdown at the inner wall of a cylindrical vessel. When stressed beyond this point, plastic flow occurs in the inner layers and a cylindrical boundary surface between a plastic zone and an elastic zone is considered to exist which moves toward the outer surface as internal pressure increases. The following simple relation is sometimes assumed to apply to such a state of partial plasticity:

$$p = F\left(\frac{R_o^2 - n^2}{2R_o^2} + \ln n\right) \tag{26}$$

where n = radius of plastic-elastic boundary ÷ bore radius.

F = a yield strength whose value depends on a particular theory. Based on the Von Mises-Hencky theory, $F = \frac{2}{\sqrt{3}} f$ where f is the yield stress in pure tension.

When $n = R_o$, that is, 100 per cent plasticity, this equation reduces to

$$p = F \ln R_o \tag{27}$$

This equation with F as the ultimate tensile strength is sometimes used to obtain a rough value for the pressure that will cause bursting of a thick-walled cylinder.

It may be well to caution the reader about putting too much faith in any of the above equations. They all involve assumptions which, for the most part, either have not been tested or were introduced to simplify the analysis and consequently they ignore some factors which may be of importance in a specific case.

There is a need for many more experimental data in this field, though it is admittedly difficult and expensive to perform the necessary experiments, since it involves testing vessels to destruction. Furthermore, an adequate coverage of the various important variables such as diameter ratio, type of steel, heat-treatment, and type of construction (*i.e.*, whether monobloc, compound, autofrettaged, etc.) would require the destruction of a considerable number of vessels.

Further details on the stress relations and theories of elastic failure in thick-walled vessels may be found in Newitt ["The Design of High Pressure Plant and the Properties of Fluids at High Pressures," Oxford, New York, 1940; The Design of Vessels to Withstand High Internal Pressures, *Trans. Inst. Chem. Engrs. (London)*, **14,** 85 (1936)], in Tongue "The Design and Construction of High Pressure Chemical Plant," Chapman & Hall, London, 1934, Nadai, "Plasticity," McGraw-Hill, New York, 1931, and in a paper by Comstock [*Trans. Am. Inst. Chem. Engrs.*, **39,** 299 (1943)].

Design for Superpressures. The maximum pressure that a cylinder, even with walls of infinite thickness, will stand before elastic failure occurs is equal to the elastic limit by Eq. (19), and to one-half the elastic limit by Eq. (21). The other two relations give intermediate values. This means that the maximum pressure for a monobloc vessel even with the best steels is somewhere in the range of 2000 to 4000 atm. It may also be noted that, based on Eq. (20), there would be little gain in making $R_o > 4$. Actually, in some experimental work, pressures much higher than this have been used and with apparent safety. In view of this fact and the fact that higher pressures may soon become important in industry, it is desirable to consider briefly what happens to metals when strained beyond their elastic limit and what kind of a relation might be used for design at such elevated pressures.

When a metal specimen is tested in a simple tensile test, once it starts to yield it would appear at first thought that it should break without any further increase in stress. Actually it requires a considerably higher stress to cause ultimate failure, and this is due to the fact that as the metal flows its strength increases, the phenomenon commonly known as "strain hardening." It is this phenomenon which makes it possible for a thick-walled vessel to withstand pressures in excess of those that cause elastic breakdown. In a simple tensile test, the ultimate strength is seldom more than twice the elastic limit, but in a cylinder the ratio of the pressure that will cause bursting to the elastic breakdown pressure is believed to increase steadily with increase in diameter ratio. This is illustrated in Fig. 5, which is based on the calcu-

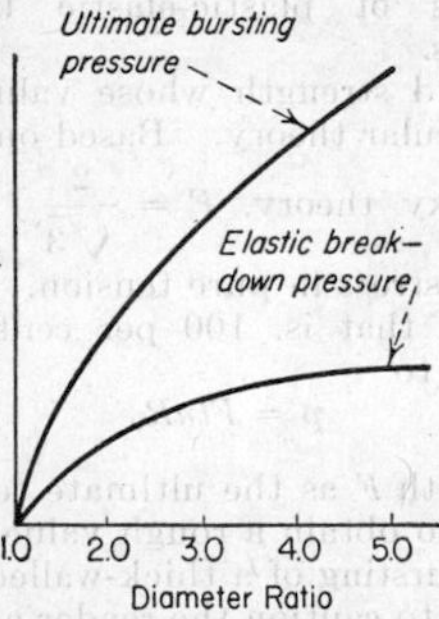

FIG. 5. Relation of bursting pressure to elastic-breakdown pressure.

lations of Manning [*Engineering*, **159,** 101, 182 (1945)]. For the basis of the calculation of the bursting pressure, reference must be made to the original paper. It will suffice to note here that it is based on the theory that the relation between maximum shear stress and maximum shear strain in a thick-walled cylinder is the same as that for a torsion test on a circular specimen. The only experimental data required are a stress-strain relation for the material of the vessel obtained in a torsion test.

Manning's theory throws an interesting light on the manner in which failure occurs in a thick-walled cylinder. As the pressure is increased, the elastic limit is first reached at the inner surface of the cylinder. As overstrain occurs, Manning's calculations show that the tangential stress falls in the inner layers and then rises to a maximum and finally decreases again as the external layers are approached. This form of a stress distribution exists as long as there is a region in the outer layers where the metal is still behaving elastically. As the pressure increases, the region of overstrain spreads outward, finally reaching the external surface. At this point the maximum tangential stress is at the exterior surface, and it falls steadily to a smaller value at the internal surface. This is just the opposite of the stress-distribution curve for the wholly elastic region. As the pressure is further increased, this form of stress distribution persists until failure occurs. Since the maximum stress is at the outside surface, it would be expected that the vessel would rupture first at this surface and the break would then spread inward. This has actually been observed, though the evidence on this point is quite meager. It may be well to note at this point that Manning's calculations apply only for the case of static loads very slowly applied; in other words, equilibrium is assumed.

In concluding this brief discussion of the case where the pressures are high enough so that the elastic limit is exceeded, one may summarize the situation as far as design is concerned by saying that an apparently sound basis for the design of vessels to withstand pressures of 3000 atm. and higher is available, that it requires data from torsion tests on the material to be used, but that it still needs the confirmation of actual tests on vessels under a variety of conditions to put it on a firm, unassailable basis.

Effect of Temperature in Design. Two quite distinct temperature effects will be considered. The first is the slow yield—commonly called creep—that occurs with all ductile materials under continued load but becomes a serious factor in design only above certain temperatures. The other effect is that due to the stresses set up on the walls of a vessel by virtue of a temperature gradient. Only a very brief treatment can be accorded each of these.

The creep of metals is a function of stress, temperature, and time. The usual test consists in determining the rate of yield at a constant temperature and stress, and data are commonly correlated by means of elongation vs. time curves with stress and temperature as parameters. The creep rate is obtained from a portion of the curve which is nearly linear and which can be established in tests lasting for at least several hundred hours.

Compilations of creep-test data such as those given in the Sympsosium on the Effect of Temperature on the Properties of Metals (Am. Soc. Testing Materials and Am. Soc. Mech. Engrs., 1932) and in Mechanical Properties of Metals and Alloys (*Bur. Standards Circ.*, C447, 1943) present values of the stress (designated creep stress or creep limit) which produces a designated creep rate for 1000 hr. This creep rate is expressed as a percentage elongation of the specimen, and common values are 0.01, 0.10, and 1.0 per cent. Actual rupture would not occur until elongations of considerably greater magnitude had been attained, but there are, of course, various safety and structural considerations that would set a much lower limit to the allowable creep.

There are no universally accepted methods of using creep data in design nor have any methods been covered in the two common codes for unfired pressure vessels. It is a matter of combining good judgment with the available data. Some figure for an allowable creep in a given period of time must be arrived at from a knowledge of the particular installations; and, from the available figures on creep limit, one can estimate whether a particular material will stay within the limit fixed. This involves the assumption that the creep rate found from the relatively short time tests will remain constant for the whole period of use of the vessel. For example, a figure of 1 per cent creep in 100,000 hr. of continuous load has sometimes been taken as a suitable limit. Suppose that the stress under load is 15,000 lb./sq. in., then obviously a material for which the creep stress is 10,000 lb./sq. in. at an elongation of 0.1 per cent per 1000 hr. would not be suitable. Since creep rates depend on a large number of factors the effects of many of which are not well known, the available data must be used with due caution and generous factors of safety.

For good creep resistance, one generally turns to chromium steels or other chromium alloys. For example, the well-known 18-8 chromium-nickel alloy has excellent creep properties.

Vessels that are being heated or cooled externally may have comparatively large temperature gradients in the walls. If heat is being transferred in from the outside, the effect of the temperature gradient alone is to decrease the hoop stress in the outer layers of the wall and increase it in the inner layers. Since the temperature stress is added to that caused by the internal pressure, the result is a higher stress at the inside diameter than if no temperature gradient existed, and consequently a heavier wall must be used. When heat is being transferred to the outside through the walls, the effect is just the reverse, the stress due to temperature tending to bring about a more uniform distribution, and hence a thinner wall can be used.

The calculation of the stresses due to a temperature gradient is based on the assumption of a steady flow of heat through the walls, and this leads to an expression for the temperature as a function of the radius. Combining this equation with the usual stress-strain relation and the equations of equilibrium leads to expressions for the three principal stresses due to temperature. The resulting expressions for the tangential stress at the inside and outside surfaces, respectively, are

$$S_{ti} = \frac{\alpha E\,\Delta t}{2(1-m)}\left(\frac{2R_0^2}{R_0^2-1} - \frac{1}{\ln R_0}\right) \qquad (28)$$

$$S_{to} = \frac{\alpha E\,\Delta t}{2(1-m)}\left(\frac{2}{R_0^2-1} - \frac{1}{\ln R_0}\right) \qquad (29)$$

where Δt is the temperature difference through the wall and α is the linear coefficient of expansion. The other symbols are the same as previously used. Similar expressions are readily obtained for the other two principal stresses.

The magnitude of these temperature stresses is much greater than generally realized. For example, a steel cylinder with an internal diameter of $1\frac{1}{4}$ in. and an external diameter of 6 in. and a Δt of 425°F. across the wall would develop a tensile stress of almost 90,000 lb./sq. in. at the inner wall, and a compression of about 35,000 lb./sq. in. at the outer wall when heated externally. In internal heating these stresses would be the same in magnitude but reversed in sign. To these values must be added algebraically the stresses due to pressure alone. For further details, see Newitt [*Trans. Inst. Chem. Engrs. (London)*, **14**, 85 (1930)] and Luster [*Trans. Am. Soc. Mech. Engrs., Fuel Steam Power*, **53**, 12, 161 (1931)].

Methods for Improving Stress Distribution in Thick-walled Cylinders. Since there is a considerable variation in the magnitude of the stress across the wall of a thick-walled cylinder when at a uniform temperature, resulting in a poor utilization of the outer layers of metal, it is clear that any means of making the stress more uniform or at least lowering the hoop stress at the inner wall and raising it at the outer wall would be reflected in greater strength for a given wall thickness or a thinner wall for a given pressure. Four principal methods have been developed to accomplish this purpose. They may be listed as follows:

1. Compound vessels by shrinking.
2. Wire-wound vessels.
3. Autofrettage or "self-hooping."
4. Built-up or laminated walls.

The effect of a temperature gradient in producing stresses in the walls of a cylinder suggests the use of a controlled gradient to superimpose stresses upon those due to the internal pressure and thereby obtain a more uniform distribution. This has not been used to the author's knowledge. 1 and 2 are old techniques that have been used especially in the construction of gun barrels but have had very limited application to vessels for industrial use.

When a cylinder having an inside diameter slightly less than the outside diameter of a second cylinder is heated to expand it (or the inner cylinder cooled to contract it) so that it can be shrunk over the latter, a set of initial hoop stresses is set up which are compressive in the inner cylinder and tensional in the outer one, due to the fact that the shrinkage of the outer cylinder acts like an external pressure on the inner cylinder and an internal pressure on the outer cylinder. When a pressure is applied to the compound cylinder, the resultant stresses are the algebraic sum of the shrinkage stresses and the stresses due to the pressure alone. This situation is illustrated diagrammatically in Fig. 6. It will be noted

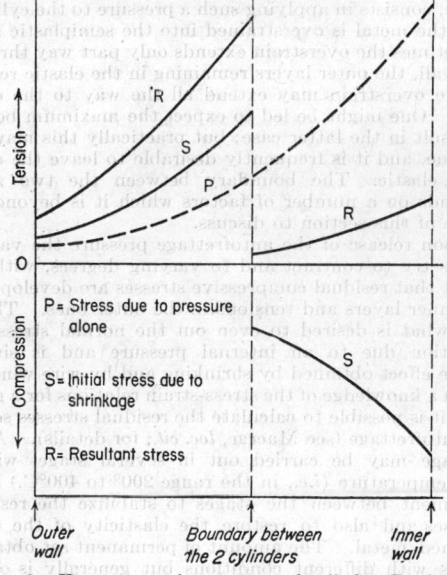

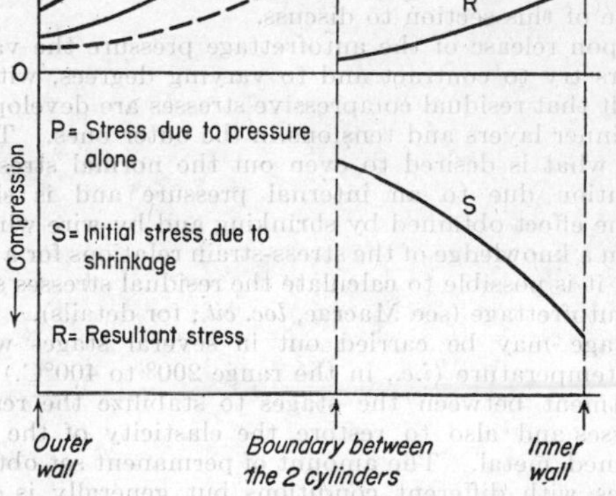

FIG. 6. Hoop stresses in a compound cylinder (Duplex).

that the effect of the shrinkage is to decrease the tangential stress in the inner layers and increase it in the outer layers over what it would have been in a single cylinder of the same dimensions (curve *P*).

The calculation of the resultant stresses in a compound cylinder after application of a pressure can be made by first calculating the initial stresses due to shrinkage from a knowledge of the amount of shrinkage and the dimensions of the two cylinders. Then the stresses due to pressure alone are calculated for the compound cylinder just as if it were a simple cylinder. The resultant stresses are simply the algebraic summation of the stresses due to these two independent effects. (For further details, consult Newitt, "The Design of High Pressure Plant and the Properties of Fluids at High Pressures," Oxford, New York, 1940.)

This method has been used on a laboratory scale with success, but the very accurate machining required has militated against its application to large vessels. However it has been used successfully for some large, 1000-atm. vessels. It has also been successfully applied in the construction of pump blocks and intensifier cylinders where fatigue is an important factor. Other advantages of this method of construction are that two different steels may be used and that the development of tensile properties by heat-treatment can be better controlled in a thinner wall.

By winding a cylinder with wire or ribbon at high tension it is possible to accomplish the same end, namely, to set up initial compressive tangential stresses in the original cylinder which will thus be covered by layers of wire in which there is an initial tension. When the hoop stresses introduced when pressure is applied are added algebraically to the initial stresses due to the winding, the resultant stresses are somewhat similar to those obtained by shrinkage, and the net result is a more favorable stress distribution throughout the total wall than in the equivalent simple cylinder. (For a quantitative treatment of this method, see Newitt, *op. cit.*) It should be noted that wire winding does not add to the longitudinal strength of a vessel, and this may become the limiting factor in such a vessel.

The process of autofrettage, which is actually little more than a cold-working scheme with a high-sounding name, consists in applying such a pressure to the cylinder that the metal is overstrained into the semiplastic zone. Sometimes the overstrain extends only part way through the wall, the outer layers remaining in the elastic region, or the overstrain may extend all the way to the outer wall. One might be led to expect the maximum benefit to result in the latter case; but practically this may not be true, and it is frequently desirable to leave the outer zone elastic. The boundary between the two zones depends on a number of factors which it is beyond the scope of this section to discuss.

Upon release of the autofrettage pressure the various layers try to contract and to varying degrees, with the result that residual compressive stresses are developed in the inner layers and tensions in the outer ones. This is just what is desired to even out the normal stress distribution due to an internal pressure and is similar to the effect obtained by shrinking and by wire winding. From a knowledge of the stress-strain relations for a given steel it is possible to calculate the residual stresses set up by autofrettage (see Macrae, *loc. cit.*, for details). Autofrettage may be carried out in several stages with a low-temperature (*i.e.*, in the range 200° to 400°C.) heat-treatment between the stages to stabilize the residual stresses and also to restore the elasticity of the overstrained metal. The amount of permanent set obtained varies with different conditions but generally is of the order of 2.5 to 6.0 per cent at the inside diameter.

The autofrettage method seems to be more practical for large vessels than the other two and was used considerably during the warf or gun-barrel construction. If pressures in the chemical industry are pushed above the present limits, it is probable that autofrettage will find important application.

In his work at pressures of 12,000 atm. and higher, Bridgman ("The Physics of High Pressure," Macmillan, New York, 1931) has prepared his vessels by an autofrettage or work-hardening treatment using lead to transmit the pressure. The bore is machined undersize, then stretched by pressure, and finally reamed to the desired size. (For a recent treatment of autofrettage with illustrative examples, see Wintsch, *op. cit.*)

A relatively recent development is the multilayer vessel, pioneered in this country by the A. O. Smith Corp. of Milwaukee. Such a vessel is constructed by starting with a relatively thin-walled cylinder and wrapping successive layers of sheet steel (about 1/4 in. or less) around it. Each layer is wrapped and tightened, and the longitudinal seam is welded before the next layer is put on. A complete pressure vessel is made by welding together as many of these multilayer cylindrical sections as are needed to obtain the required length and then welding heads on the ends. Vessels have been made with a maximum wall thickness of 8 1/2 in., and the majority were for a maximum working pressure of 5000 lb./sq. in. or less, but it is believed that a few vessels have been built for pressures as high as 10,000 lb./sq. in. The present technique of construction is probably good to considerably higher pressures, though, since thicker walls are used, more uncertainty is introduced in the welding of heads and circumferential joints. The sheet used has a tensile strength of about 75,000 lb./sq. in., but there seems to be no reason why a material of much higher tensile strength (at least 125,000) cannot be used.

Within these limits, the technique has been thoroughly demonstrated by the successful construction of several thousand vessels. By controlling the tension of the wrapping, it is possible to produce a pattern of initial stress in the walls such that the superposition of stresses due to internal pressure will produce a uniform stress across the walls at the working pressure, and under these conditions the thin-cylinder formula can be used to calculate the wall stress. This has apparently been verified by a number of tests, some carried to destruction of the vessel. This is one of the advantages claimed for this method of construction. Others are that the inner tube can be of various metals chosen for corrosion resistance and the wrapped layers of the proper steel for the best strength characteristics; that large vessels can be built up in the field from the wrapped sections; that one is not limited in the size of the final vessel by the size of the largest available ingot, as in the case of forged vessels; and that, if failure does occur, there is no shattering or fragmentation.

In closing this discussion, it may be well to point out that, whereas the various methods discussed for improving the stress distribution in thick-walled cylinders will materially increase the pressure at which elastic failure occurs, there is no assurance that the ultimate bursting strength will be increased at all. This is a point on which there is little direct evidence, and many tests to destruction are needed to shed light on it.

Types of Construction. Pressure vessels are almost invariably cylindrical in shape (spherical vessels are also used in a few cases notably for large storage vessels at relatively low pressures) and are made in various ways. One of the oldest is to roll a plate into a cylinder and rivet the seam. The ends are then closed by heads preformed from plate and riveted to the cylinder. Although some riveted vessels are still made, the more usual type of construction is to weld the longitudinal seams and the girth seams where the heads are joined, particularly for pressures above 300 lb./sq. in. Welded vessels are the commonest type of large vessel for pressures up to about 1500 lb./sq. in. For small vessels, usually referred to as "cylinders" and commonly used for the storage and transportation of compressed gases at pressures up to 3000 lb./sq. in., but also for storage cylinders or other purposes at somewhat higher working pressures, a seamless construction is preferred. Such vessels are commonly made by a cupping and drawing process. A heated ingot of metal is pressed to a cup shape and then drawn over a mandrel through a series of dies until the desired dimensions are obtained.

For large vessels to operate at pressures in excess of about 1500 to 2000 lb./sq. in. one should use either a forged vessel or a multilayer vessel, as described in the previous discussion. The forging process starts with a cast ingot of alloy steel of the necessary size to yield a finished vessel of the desired dimensions. The size of the finished vessel is limited to about 100 tons by the size of the largest ingot of steel that can be made. The ingot is heated and pierced through by a punch to make a hole for insertion of a mandrel. The pierced billet is then heated again, placed on a water-cooled mandrel, and pressed on a large hydraulic press to increase the length and the inside diameter. This operation is repeated a number of times until the billet has been brought to approximately the final dimensions desired. It is then finished by machining. For some further details on the construction of large forged vessels, Tongue (*op. cit.*) may be consulted and also a paper by Brown [*Chem. & Met. Eng.*, **46,** 353 (1939)]. This paper also gives some useful details on calculations of stresses in high-pressure vessels.

Laminated vessels made by the A. O. Smith Corp. method have been discussed. For further details on them, the following references are suggested: [Jasper and Scudder; Multi-layer Construction of Thick-wall Pressure Vessels, *Trans. Am. Inst. Chem. Engrs.*, **37,** 885 (1941)]; and Scudder [Modern Pressure Vessel Design and Construction Materials, *Petroleum Engr.* (May 1944)].

The use of wire winding to build a pressure vessel has already been discussed; and, since the method has not been used to any extent for large vessels, no further details will be given. It is of interest to note, however, that small vessels for experimental work on gas explosions at pressures up to 15,000 atm. have been made by this method (Bone, Newitt, and Townend, "Gaseous Combustion at High Pressure," Longmans, London, 1929). Mention should be made, however, of a method of construction developed in Germany just prior to the war and rather widely used during the war. It is a variant of wire winding but uses, instead of wire, a metal strip (sometimes known as "Wickel" band) about 3 to 4 in. wide of the shape shown in Fig. 7. The first layer is wound

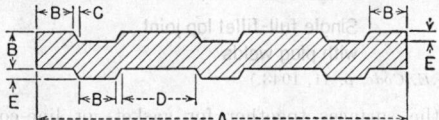

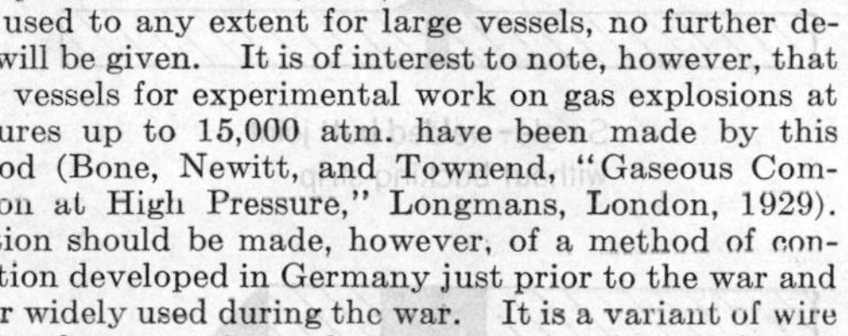

FIG. 7. Section of strip used in Schierenbeck method of construction of pressure vessels. Approximate dimensions: *A*, 3 1/16 in.; *B*, 5/16 in.; *C*, 0.04 in.; *D*, 21/32 in.; *E*, 5/64 in.

into machined grooves on the face of a hollow cylindrical core, and the subsequent layers fit together as tongue-and-groove joints. The strip is heated electrically just before winding and then is cooled by air jets as it is wound. This is used to raise the elastic limit of the material and control the initial stresses in the wall of the built-up vessel. By this method it is claimed that a vessel can be built up which, unlike the wire-wound vessels, has longitudinal strength. Flanges can be built up by the winding process, and it is said that the wound section can be drilled into for bolt holes and otherwise treated as if it were solid metal. One can also use solid-metal flanges grooved to thread onto the winding. An assembled vessel is shown in cross section in Fig. 8.

The process was developed by Schierenbeck at Oppau, and a considerable number of vessels have been built and used successfully, some as large as 60 ft. long and 4 ft. in diameter. They were used for pressures up to 10,000 lb./sq. in., and small vessels for experimental work to 60,000 lb./sq. in. were constructed. The method is said to have all the advantages of wire winding, at the same time that it eliminates the chief disadvantage; and over the laminated-sheet type of vessel it has the advantage of not requiring the welded girth seams necessary for a long vessel. It also eliminates the troublesome head welds. For further details see Krase, FIAT Final Report 611, Design and Construction of High Pressure Compressors and Reaction Equipment, Dec. 12, 1945. Also see Clark, FIAT Final Report 577.

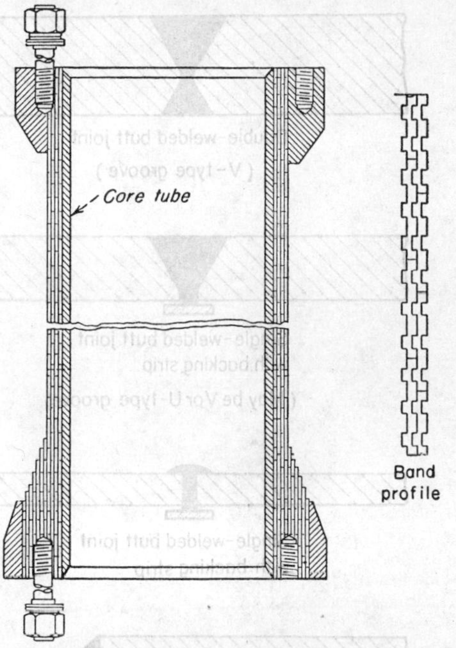

FIG. 8. Assembly of banded pressure vessel.

JOINTS AND CLOSURES

Various Types of Joints. There are six general methods of making a pressuretight joint between two metal surfaces, namely, (1) riveting, (2) welding, (3) brazing or soldering, (4) screw threads, (5) packing and gaskets, and (6) line contact. See Secs. 5 and 21 on gaskets and packing, in general.

Method 1 may be dismissed without further discussion, since it is almost never used for pressures in the range covered in this section. If further details should be desired, consult the A.P.I.-A.S.M.E. Code for the Design, Construction, Inspection and Repair of Unfired Pressure Vessels for Petroleum Liquids and Gases (*op. cit.*), the A.S.M.E. Boiler Construction Code (*op. cit.*), and Hesse and Rushton ("Process Equipment Design," Van Nostrand, New York, 1945).

Autogenous or fusion welding, either by the oxy-acetylene torch or the electric arc, is a highly developed art that is widely used for seams in plates, fastening heads to vessels, connecting pipes and fittings to a vessel, joining pipe and tubing, and for various other purposes. Figure 9 illustrates various kinds of welds that are permitted by the A.P.I.-A.S.M.E. code. Welding has been used for pressures up to 30,000 lb./sq. in., but this is not to be taken as an upper limit. So many factors enter into the choice of a method of joining metals that no hard-and-fast rules can be given. For large vessels, especially at pressures above 5000 lb./sq. in., a seamless construction would generally be preferred, but for other purposes welded joints may be used to higher pressures with entire satisfaction. When done by an experienced welder with proper selection of welding-rod material and proper subsequent heat-treatment, if necessary, to relieve local stresses, a welded joint is practically as strong as the adjacent metal. The A.P.I.-A.S.M.E. code is conservative on this point and allows weld efficiencies [to be used for E in Eq. (1), for example] from 55 to 80 per cent depending on the type of weld. Both the codes cover the rules for good welding practice in considerable detail and should be consulted for further information on specific

amounts of silver, of which there are several grades, melting in the range 1200° to 1600°F. Recently low-temperature brazing alloys of copper and phosphorus have been introduced that melt in the silver-solder range. One objection to all hard-soldering or brazing is that the high temperatures required may cause some weakening of the materials being joined.

Screw threads with a taper, such as the standard pipe thread, can be used to make a tight seal when put together with a lubricant but are recommended only for moderate pressures. The chief function of screw threads in making pressuretight joints is to provide the force to

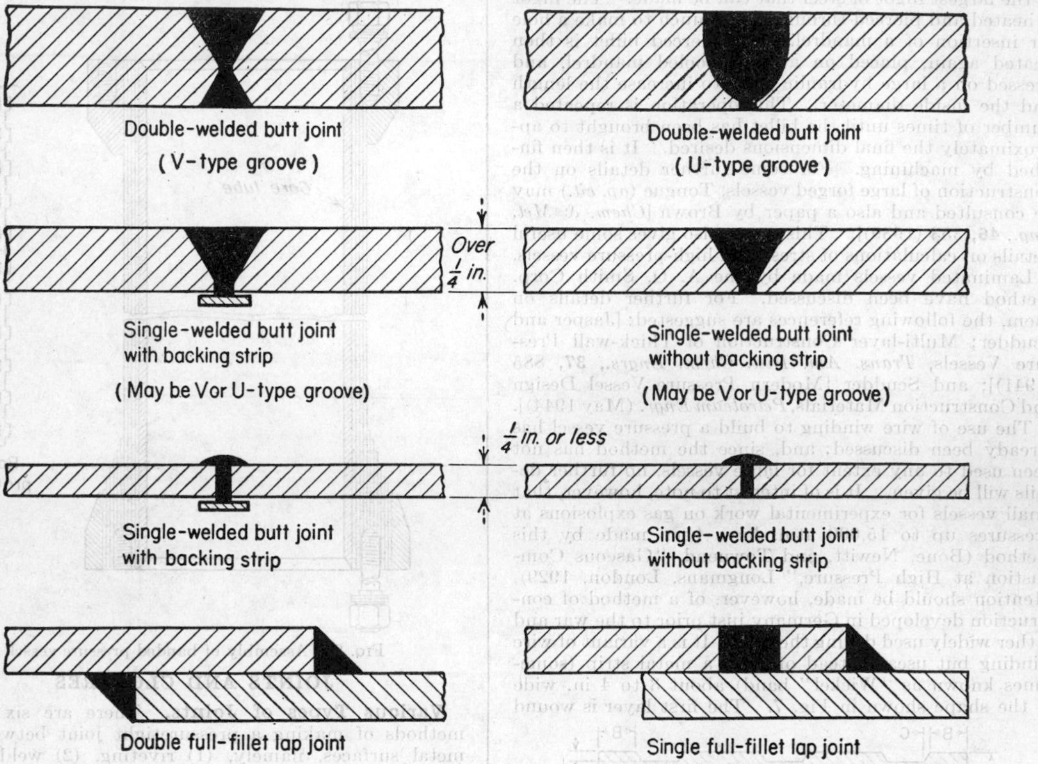

FIG. 9. Types of welds. (*A.P.I.-A.S.M.E. Code*, p. 41, 1943.)

points [see also Spraragen, *Ind. Eng. Chem.*, **29**, 366 (1937)].

Small joints with soft solder (lead-tin) are satisfactory when not to be subjected to any elevated temperature and not depended on for much mechanical strength. The parts to be joined should fit with a small clearance and generally should be "tinned," *i.e.*, precoated with solder and fitted together at a temperature above the melting point of the solder. The best example is a threaded joint in which the small clearance between the male and female threads is filled with solder. In such a joint the thread takes the load due to pressure and the solder is merely depended on for a fluidtight seal. A joint of this kind is serviceable to 1000 to 2000 atm., but above that the solder may be squeezed out. For greater mechanical strength in cases where the solder metal itself must carry the load, brazing or hard soldering is satisfactory for small parts. Brazing has the advantage over welding that it can be done at a much lower temperature and the hard or silver solders can be used at still lower temperatures. These are copper alloys containing varying

hold the surfaces together for gasket- or line-contact joints. The thread itself is not expected to prevent leakage of fluid.

In designing threads for bolts or other fastening elements, it is commonly assumed that failure would occur by shear of the metal at the base of the thread, and the necessary length of thread is calculated on this basis. A good rule of thumb is to make the threaded length of the bolt or nut equal to its diameter. This provides ample margin of safety for all ordinary applications. In figuring the stresses in threads or the shanks of bolts, it is well to make an allowance for the stresses due to initial tightening, which may be as much as 50 per cent of the working stress due to the pressure and is superimposed on the latter.

Buttress threads are desirable whenever the thread is subjected to a heavy thrust as in the head closures of vessels, the thrust being taken against the thread surface that is horizontal. Threads subjected to a combination of high pressure and elevated temperature are very apt to seize. If it is desired to unscrew such a threaded joint, it should be lubricated when made up. A graphite grease such as Gredag is suitable, or a mixture of tallow and graphite is sometimes used. A method of avoiding the use of threads by a breechblock type of mechanism will be mentioned later.

The principle governing methods 5 and 6 is a very simple one, namely, that the pressure between the faces of the joint must always be greater than the internal pressure to be resisted. If this condition is maintained, the joint will not leak. The practical application of this simple principle takes many forms, some of the most important of which will be discussed in the following paragraphs. The simplest adaptation of this principle is to place a deformable ring of soft material, or a gasket,

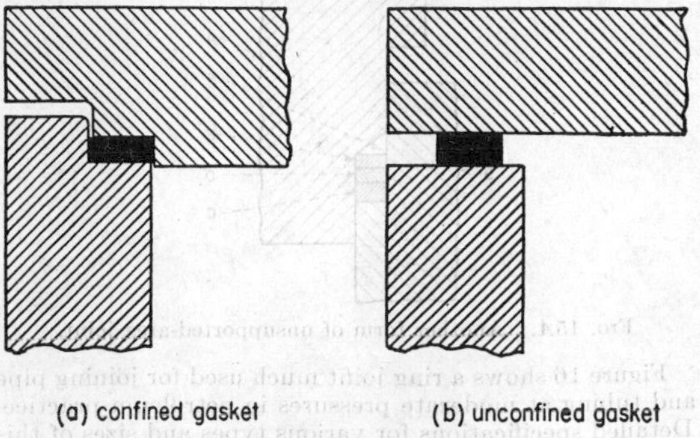

FIG. 10. Types of gaskets.

between the two surfaces. The gasket may be unconfined as in Fig. 10(*b*), *i.e.*, free to move laterally in both directions, or confined as in Fig. 10(*a*) and prevented from flowing except into the small clearances. The confined gasket is easier to make and keep tight and is less liable to be blown out, but it requires more careful machining of the parts to be joined and is not so easy to disassemble. An unconfined gasket, on the other hand, is apt to leak if the joint is subjected to temperature changes and is to be avoided for such applications. The "tongue-and-groove" joint shown in Fig. 11 is one form of the confined-gasket joint.

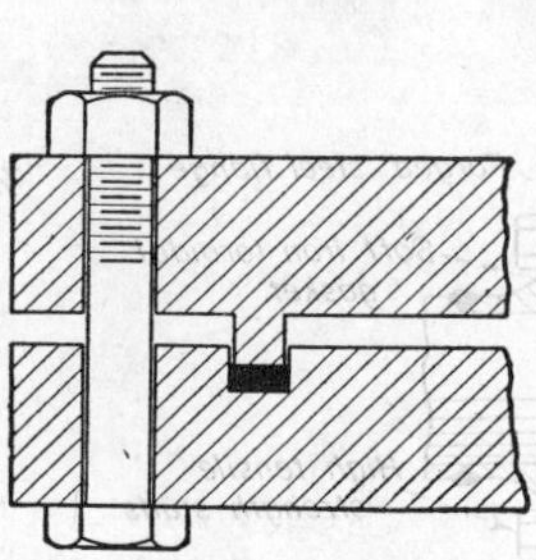
FIG. 11. Tongue-and-groove joint.

Gaskets may be made from a wide variety of materials including most of the metals, rubber, leather, hard fiber, silicones and various composite materials. For high pressure and especially when accompanied by elevated temperatures, copper is probably the most generally serviceable material though some experimenters have found monel to be superior. The gasket must be soft enough to flow and fill the irregularities on the surface but not too soft so that the requisite sealing pressure cannot be developed. A small V-groove in both bearing surfaces helps materially in maintaining a tight joint. Copper gaskets should be annealed by heating to redness and quenching before use. For some helpful rules in designing copper gaskets see a paper by Edwards [*Chem. & Met. Eng.*, **44**, 134 (1937)]. For other details on gaskets, see papers by Perry [*Chem. & Met. Eng.*, **41**, 194 (1934)] and Sandstrom (*ibid.*, 130).

Recently various fluorocarbons, of which Teflon is one example, have become available as packing and gasket materials. They are especially valuable where chemical inertness and resistance to elevated temperatures are desired.

The gasket joints just discussed are simple to make but suffer from two serious drawbacks: (1) they are not satisfactory when the joint has to be frequently opened and closed; and (2) the force necessary to tighten is considerable, and the initial tension (or compression) in the members that take up the load must be added to that caused by the internal pressure. The latter disadvantage leads to difficult problems in the design of heads of large vessels and to trouble in trying to seal the joint so that it will remain tight at the full working pressure. To lessen these difficulties, the self-sealing principle, to be discussed presently, is now used almost universally for large joints.

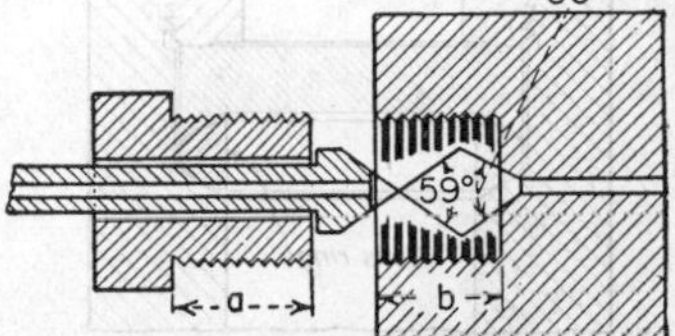

FIG. 12. Cone joint.

When a curved surface and a plane surface meet or two plane surfaces meet at an angle, "line contact" results. Actually, of course, the metal is deformed slightly and a narrow band results, but by such a contact a high pressure can be produced at the joint by the application of a relatively small force. The cone joint shown in Fig. 12 is based on this principle. The cone on the nipple has a 58- to 59-deg. angle, and it seats in a 60-deg. cone. Dimension *a* should be somewhat greater than *b* and the latter long enough to allow at least four threads to be engaged. This type of joint is very satisfactory for small tubing unions, perhaps up to ½-in. size, and is good at least to 2000 atm. The nipple must be joined to a tube or pipe by soldering or welding. This can be avoided by the joint shown in Fig. 13 in which the male cone is made directly on the tubing and the nut works against a left-hand threaded sleeve to pull the surfaces together. The disadvantage of this joint is that the tube is weakened by the thread cut in the wall. (For detailed

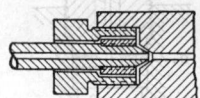

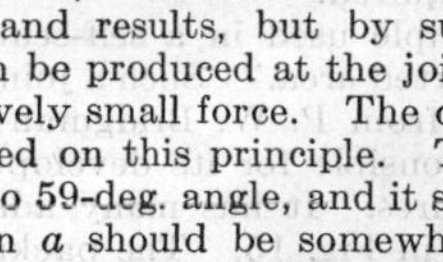
FIG. 13. Cone turned on tube end.

drawings of these joints and various other small high-pressure fittings, see Curtis High Pressure Equipment and Technique, Chap. X of "Fixed Nitrogen" Chemical Catalog Co., New York, 1932.)

Another adaptation of this principle is the lens-ring joint shown in Fig. 14. The spherical surfaces of the hardened steel ring make line contact with conical surfaces on the ends of tubing to be joined or, as in the figure, between a tube and a closure. This joint is very useful in small-scale apparatus. It is easy to assemble and dismantle, and this can be done many times. The joint is good to at least 5000 atm. and has been used on large-scale equipment. It is also partly self-sealing, since, as the pressure increases, the ring expands and wedges more tightly between the conical surfaces. This effect can be augmented by hollowing out the ring so as to make it more flexible. The joint can be made between two parallel surfaces, in which case it has no self-sealing ability. This type of joint is not very satisfactory for alternating pressures where failure due to fatigue will occur.

A further development of the lens ring is the wave-ring joint, used in England but not to any extent in this country. The ring has two spherical surfaces, each of which makes line contact with one of the surfaces to be joined. The internal pressure acts wholly in a radial direction, and hence the joint is completely self-sealing. In order to have some pressure for an initial seal, the ring is made a few thousandths of an inch larger than the socket into which it fits, and it is sprung into place. By making it of softer metal (brass or copper) than the surfaces to be joined, the deformation is confined to the ring, which is easily replacable. The wave-ring has been

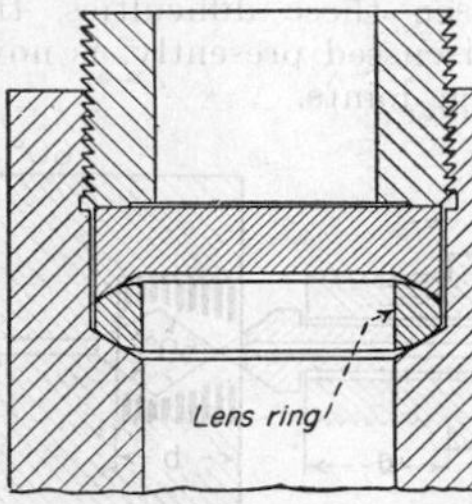

FIG. 14. Lens-ring joint. (*F. H. Bramwell, U.S. Patent* 1,722,623.)

used for both small and large joints and to pressures of 180,000 lb./sq. in. A disadvantage is the very careful workmanship required.

Another principle used in a self-sealing joint is that of the "unsupported area." Such a joint is often called a Bridgman joint from P. W. Bridgman of Harvard who was chiefly responsible for its development and use at very high pressures. It has many adaptations, one of which is shown in Fig. 15. The packing ring *a* is held between the nipple *c* and nut *b*. The nut need only be tightened lightly to compress the packing initially, after which the thrust due to internal pressure compresses the packing. Since the total thrust acts over the whole area of the opening in the wall and the area of the packing ring is less by the area of the inlet tube (the unsupported area), the pressure developed in the packing is

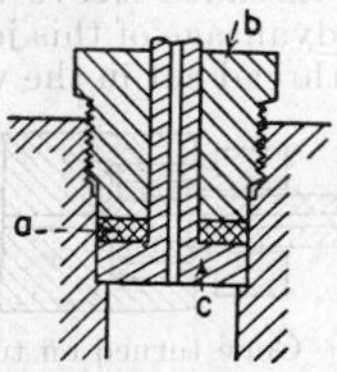

FIG. 15. Bridgman "unsupported-area" joint.

always higher than the internal pressure, and the joint will not leak at any pressure until the pressure gets so high that it fails due to "pinching off" of the tube, which occurs at a pressure very roughly of the order of the ultimate tensile strength of the material of the tube (Bridgman, *op. cit.*). The best packing material is soft rubber, but for elevated temperatures rings of copper, aluminum, or silver may be used. In this case, however, the packed surfaces must be carefully polished to prevent "freezing" of the joint, and a somewhat greater unsupported area should be provided to obtain a greater pressure to flow the metals. In any case the clearances must be very small to prevent the gasket material from flowing out of the joint.

An interesting and useful adaptation of the unsupported-area principle is shown in Fig. 15*A*. *A* and *B* are the two parts whose surfaces are to be sealed; *C*, *D*, and *E* are rings of hard steel, lead, and soft steel, respectively. The two parts are brought together initially with only enough force to flow the lead ring and form an initial seal. As the internal pressure builds up the top ring is pushed against the surface lying at an angle to the horizontal and is deformed. As long as the triangular space *F* above this top ring remains unfilled there is an unsupported area and the pressure acting to flow the soft steel is greater than the internal pressure in the vessel, and sealing is effected. This type of joint is simple to construct and is usable to very high pressures.

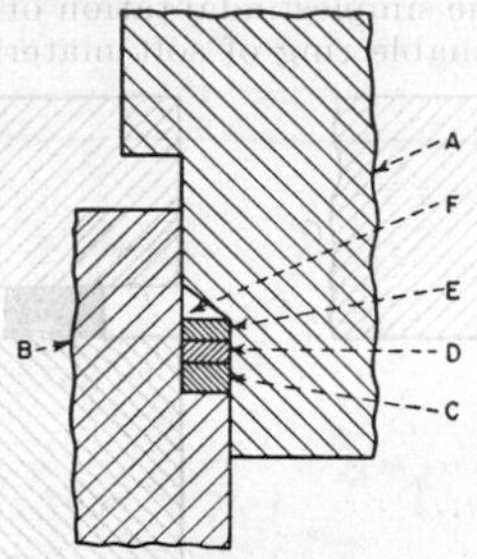

FIG. 15*A*. Another form of unsupported-area joint.

Figure 16 shows a ring joint much used for joining pipe and tubing at moderate pressures in petroleum practice. Detailed specifications for various types and sizes of this joint are covered by A.P.I. standards.

The connection shown in Fig. 17 is useful for introducing tubing into a vessel either for heat-transfer purposes or for a thermocouple well. The seal is formed by the loose cone sleeve being forced on the outside against an edge of the seat in the wall of the vessel, and on the inside the thin edge is squeezed tightly around the tube.

The double-cone joint used in compression unions for small copper tubing also finds application in high-pressure unions. One form is shown in Fig. 18. This type of joint has been used with very high pressures, at least 150,000 lb./sq. in.

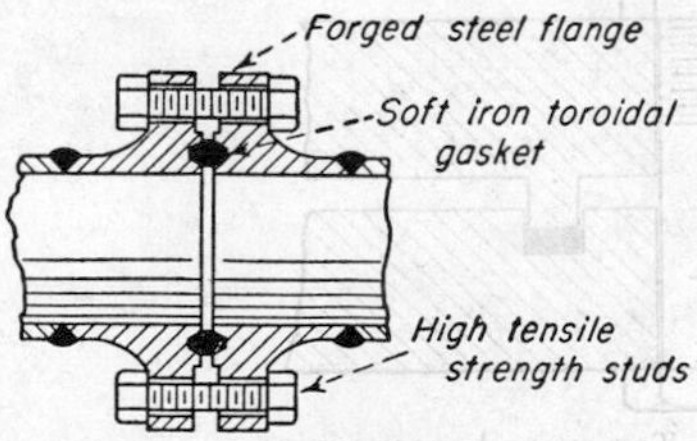

FIG. 16. Ring joint.

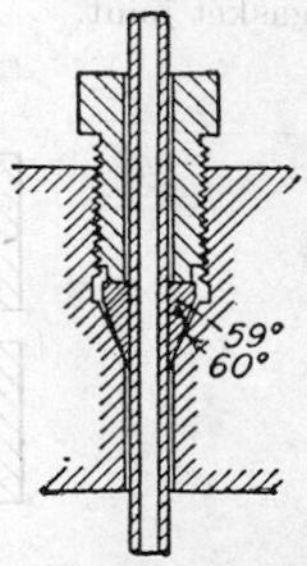

FIG. 17. Connection for introducing tubing into a vessel.

One of the difficulties with any joint subjected to changes in temperature is that it is apt to become loosened due to unequal expansion and it may leak. To avoid this a spring-loaded connection has been developed by the Bureau of Mines and found very satisfactory [Clark, *et al., Ind. Eng. Chem.*, **39**, 1555 (1947)]. Figure 19 is taken from this paper.

The joints so far described can all be used for a variety of purposes, such as joining two tubes, joining tubes to valves or other fittings, connecting tubes to heads or walls of pressure vessels, or closing the vessels themselves. In all these cases, both of the surfaces of the joint are stationary. Frequently one has the problem of sealing

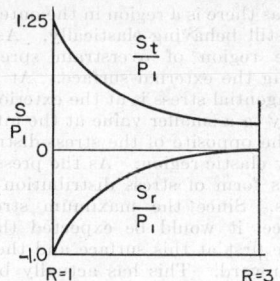

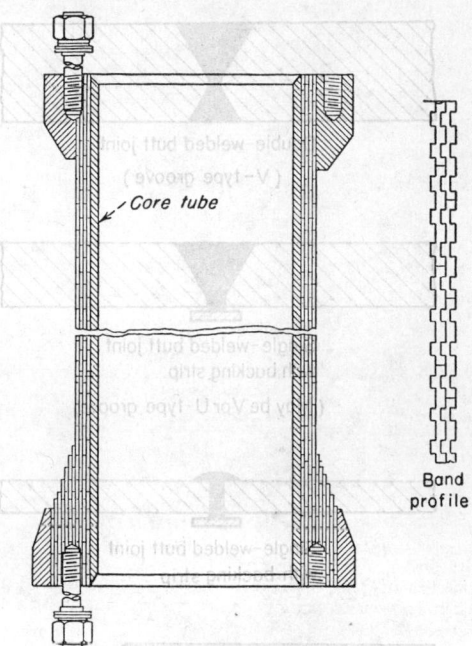

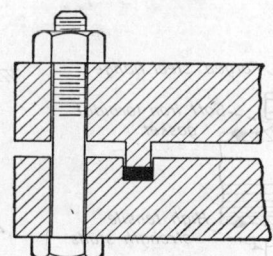

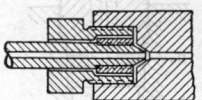

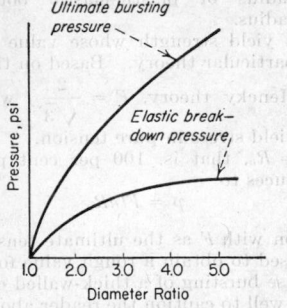

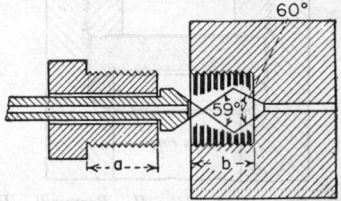

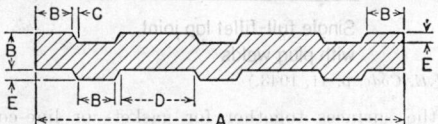

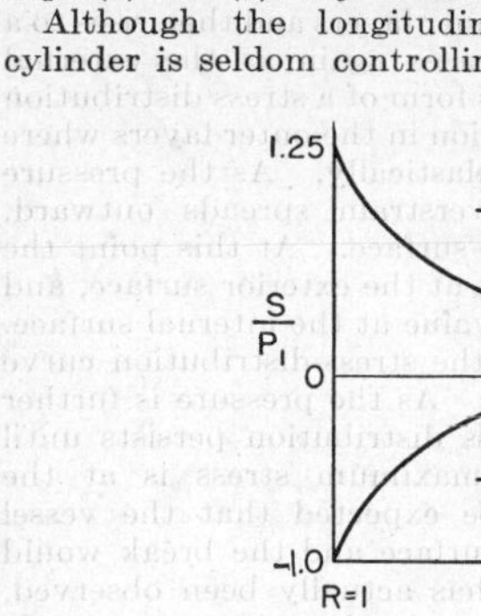

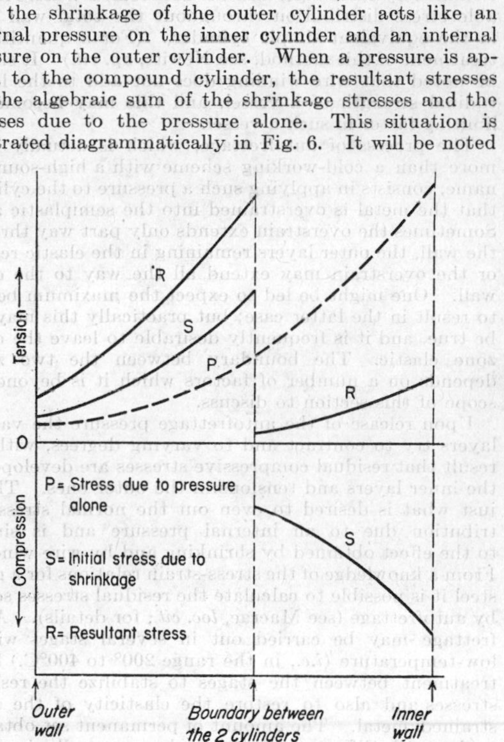

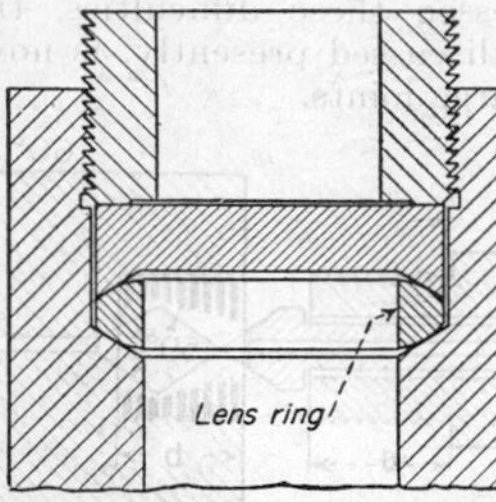

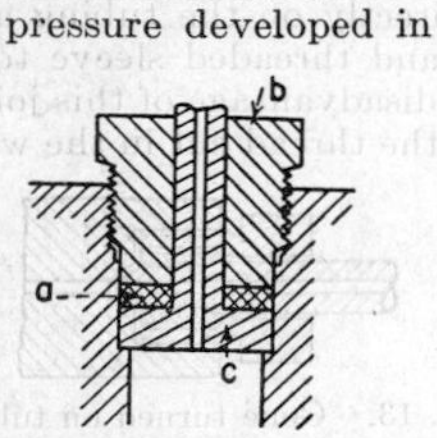

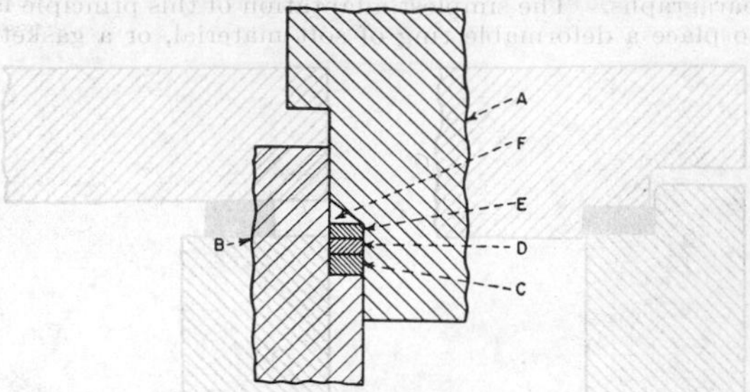

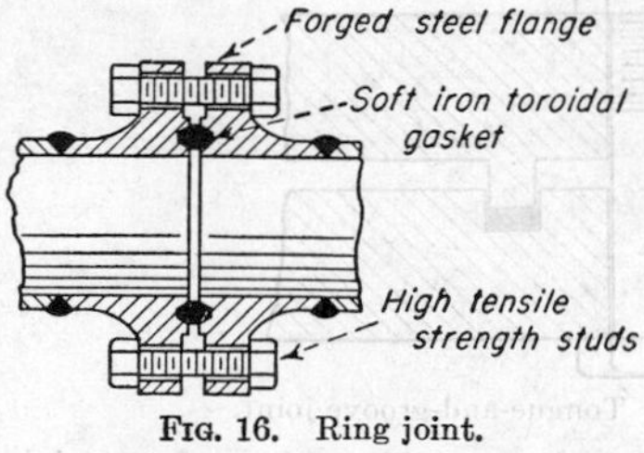

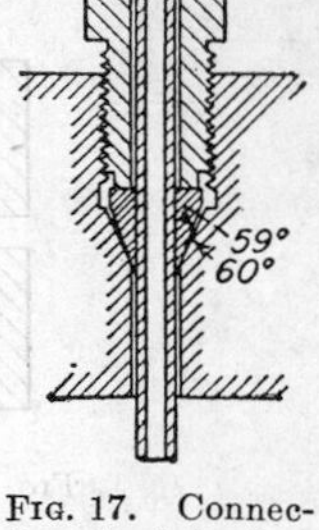

between one fixed and one moving surface as in a valve stem, a piston rod, or a plunger. This is usually handled by the conventional "stuffingbox," and many types of packing are available, depending on the particular conditions. We shall confine the discussion here to a few special methods of packing that have proved useful for high pressures.

In hydraulic work, specially formed leather rings shown in Fig. 20 find wide use. In Fig. 21 is shown a pump plunger packed with a U-leather. Packings of this type are self-sealing and have been used to pressures of the order of 150,000 lb./sq. in. For pressures of this order and higher, the piston packing shown in Fig. 22 based on

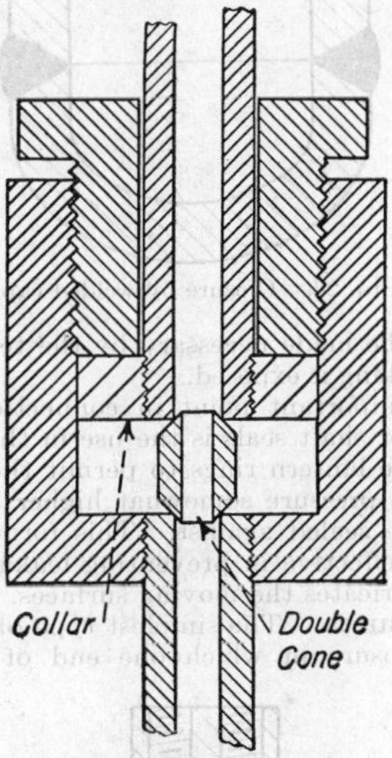

Fig. 18. Double-cone joint.

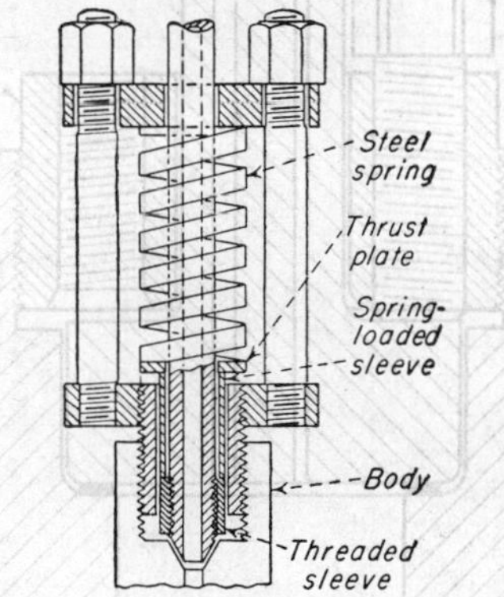

Fig. 19. Spring-loaded connection.

the unsupported-area principle has been used in experimental work. *D* is the packing, confined between two steel rings *C*, and *A* is the piece with an unsupported area. The very simple plunger-packing method shown in Fig. 23, due to Poulter, is quite successful. Both these packing methods have been used where very high pressures are to be developed by a single stroke of the plunger. They have not been adapted to continuously operating pumps.

It is of interest to note in passing that the plungers for use at very high pressures are usually made of tool steel, heat-treated and quenched to almost glass hardness. This is necessary to obtain the high compressive strengths necessary to prevent deformation of the plunger, causing it to stick in the cylinder. Because of this extreme hardness, the metal is very brittle, and care must be taken to avoid any bending forces.

Piston-rod or stirrer-shaft packings are commonly made in three types: (1) fibrous, (2) semimetallic, and (3)

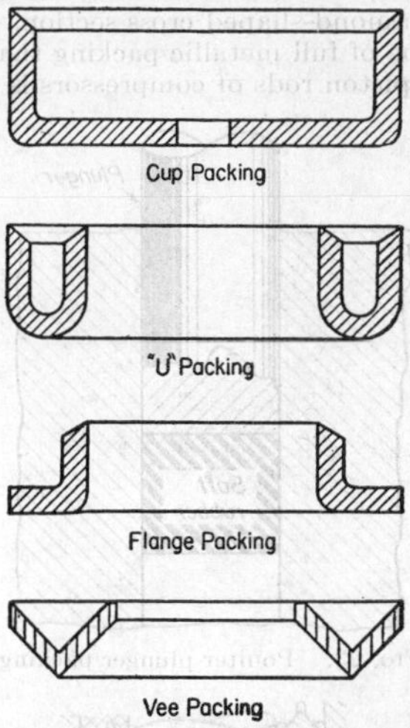

Fig. 20. Leather packing rings.

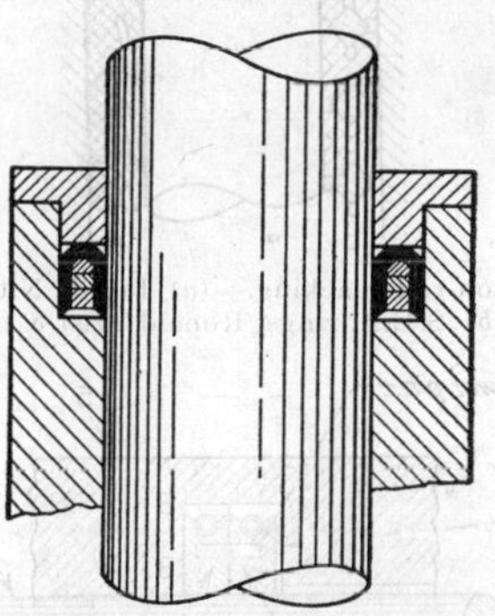

Fig. 21. U-leather packing on a pump plunger. (*Graton & Knight Co.*)

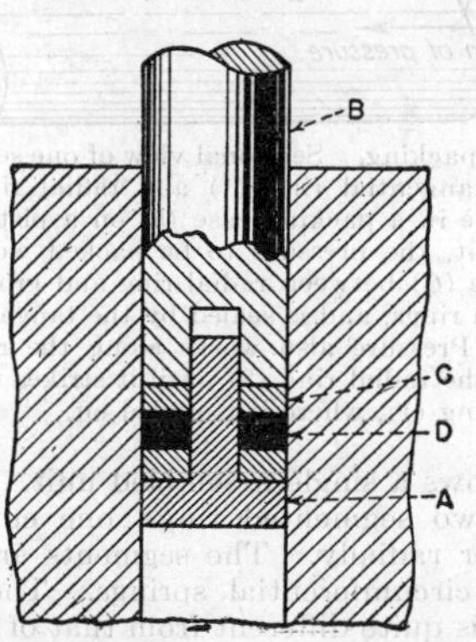

Fig. 22. Plunger packing based on unsupported-area principle.

metallic. Type 1 is made in the form of rings or coils from such materials as cotton, flax, asbestos, rubber, and the like; whereas type 2 generally consists of shredded soft metals compounded with a lubricant. Both depend for their action on being deformed and squeezed against the surfaces to be packed by pressure of a gland. Type 3 may consist of soft metal rings alternating with steel rings, and again pressure from a gland deforms the soft rings to make the seal. A packing of this type is shown at *a* in Fig. 24. It consists of alternate rings of a bearing metal and steel, the former having an X-shaped and the latter a diamond-shaped cross section.

Another type of full metallic packing that is especially useful for the piston rods of compressors is shown in Fig. 25. This shows a single assembled unit. This packing consists of two segmented rings, one cut tangentially and the other radially. The segments are lightly held together by circumferential springs. The principle of this packing is quite different from that of the types just previously discussed, because there is no deformation of the segments of the rings. Since the rings are flexible, however, the seal is made by the pressure forcing the segments into a close fit on the rod and against the end of the packing box. Wear is automatically compensated for up to a certain limit.

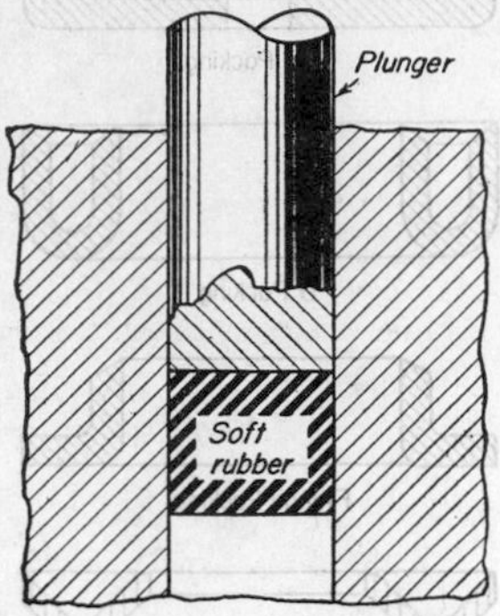

Fig. 23. Poulter plunger packing.

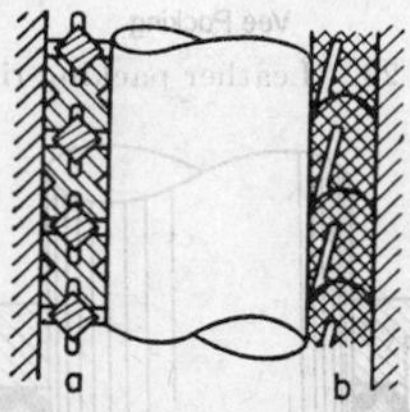

Fig. 24. Piston-rod packing. (*a*) Fixed Nitrogen Research Laboratory. (*b*) S.E.A. rings, Ronald Trist & Co., Ltd.

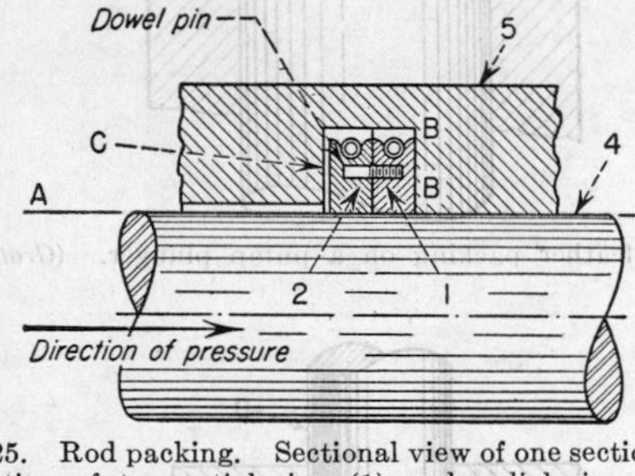

Fig. 25. Rod packing. Sectional view of one section of packing consisting of tangential ring (1) and radial ring (2) installed within a groove in a packing case (5) on a piston rod (4). As indicated in cut, the pressure to be packed works up through clearance space (*C*) between radial ring and groove, around the outside of both rings, and is sealed by the tangential ring (1) on surface *B-B*. Pressure also works along the rod through the radial cuts in the radial ring (2) until it strikes the leading edge of tangential ring (1), where it is sealed off.

The packing shown at *b* in Fig. 24 consists of the S.E.A. rings made by Trist and Co. of London and is used for the piston rods of gas compressors as well as of hydraulic pumps. They are of rubber composition, and their sealing action is similar to that of U-leathers. Such packings are sometimes referred to as "chevron" packings.

It is to be noted that, for gas-compressor piston rods and for autoclave-stirrer shafts, the metallic type of packing may be made necessary by the temperatures to which the packing is exposed.

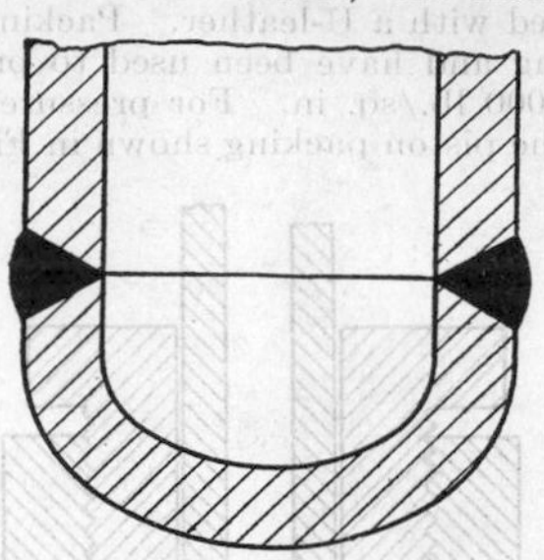

Fig. 26. Closure by welded cap.

One very important point in connection with high-pressure rod or shaft seals is the use in the packing box of one or more lantern rings to permit the introduction of oil under a pressure somewhat higher than the fluid pressure to be sealed against. This forms a hydraulic seal which is effective in preventing leakage and at the same time lubricates the moving surfaces.

Vessel Closures. The simplest type of closure is the permanent closure in which one end of the vessel is permanently closed by a solid end or a welded cap or plug. For example, when a vessel is formed by drilling and boring a bar, or forged from an ingot, one end may be left solid. Another method of closure is to weld a disk or a screwed plug in the end of a tube or to butt-weld a cap to the tube as in Fig. 26. When the head is to be removed, many types of closures have been used, and it is possible to illustrate only a few of them. The type shown in Fig. 27 is suitable for small vessels constructed from a

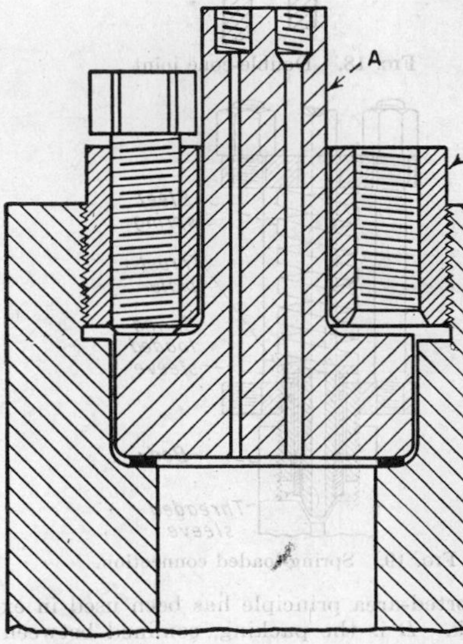

Fig. 27. Removable head closure for small vessel.

tube or a drilled bar. The head A is pressed against a gasket by a circle of bolts threaded through ring B, which in turn is threaded into the wall of the vessel. To facilitate its removal, the retaining ring is usually provided with interrupted threads so that, by turning the ring only a part of one revolution, it may be lifted out. A

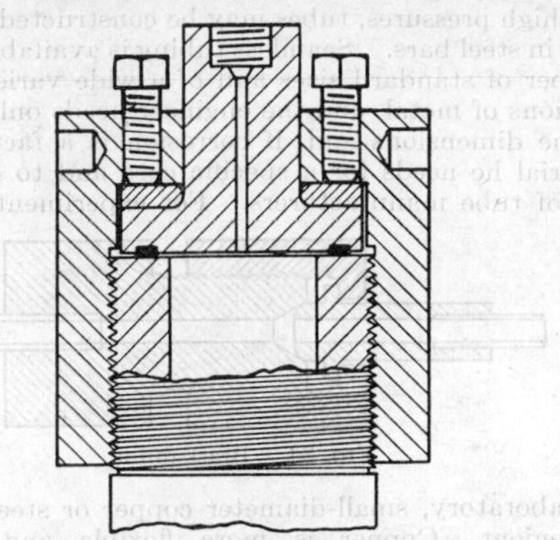

FIG. 28. Removable head closure for small vessel.

variation of this closure is to dispense with the bolts and change the ring to a threaded nut which bears directly on the flange of the closure piece. Another variation is shown in Fig. 28, in which a cap is screwed over the outside of the tube and a ring of screws threaded through the cap presses on a hardened ring which in turn bears

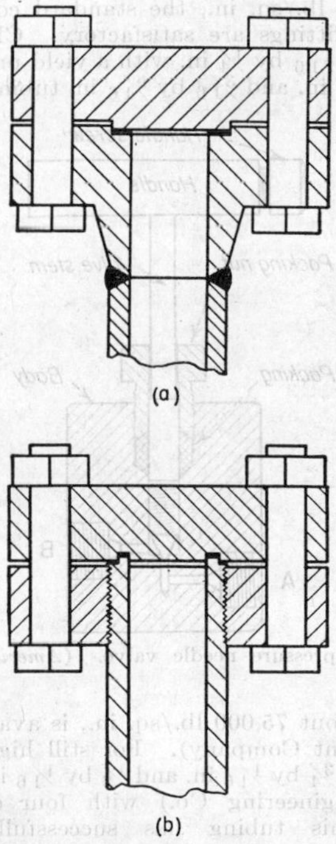

FIG. 29. Two types of bolted-flange closures

on the flange of the head. Figure 29 shows two common types using bolted flanges; in (a), the lower flange is welded to the tube forming the main part of the vessel, and in (b) it is screwed to the outside of the tube.

In large forged vessels the walls of the cylinder are generally made heavier at the two ends, as shown in Fig. 30, in order to accommodate the bolts without serious loss of strength. This shows a self-sealing type of closure which is very desirable in a large vessel. A different type of self-sealing head is shown in Fig. 31. Here the head is hollowed out as shown at A, and the internal pressure acts to bend the wall B down so that it presses

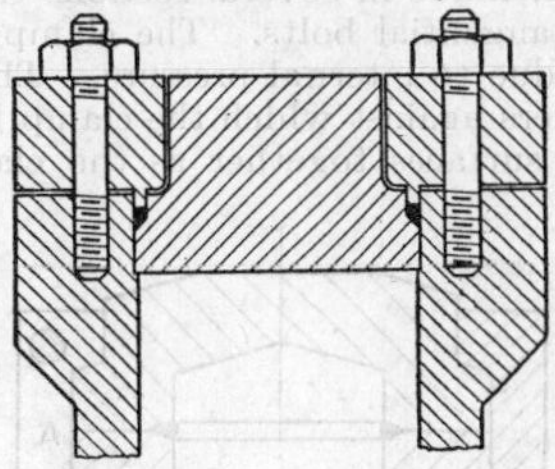

FIG. 30. Self-sealing closure for large pressure vessel.

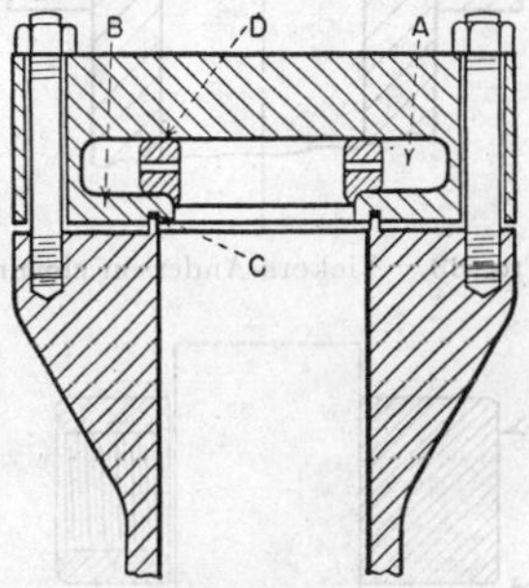

FIG. 31. Special form of self-sealing gasket closure. (*Midvale Steel Co.*)

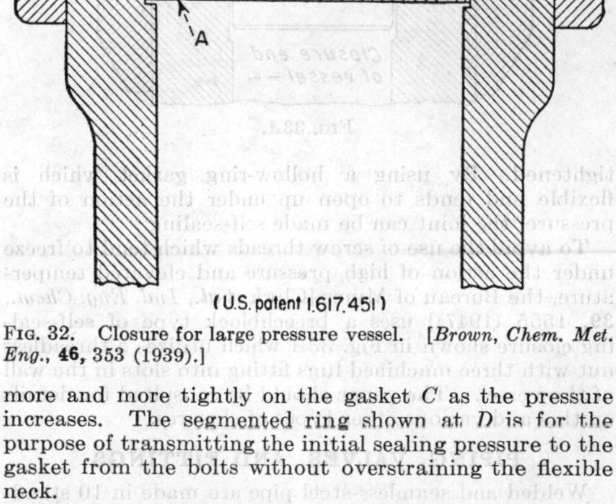

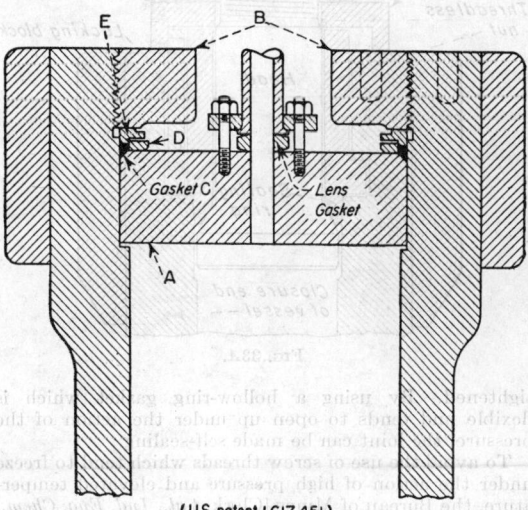

(U.S. patent 1,617,451)

FIG. 32. Closure for large pressure vessel. [*Brown, Chem. Met. Eng.*, **46**, 353 (1939).]

more and more tightly on the gasket C as the pressure increases. The segmented ring shown at D is for the purpose of transmitting the initial sealing pressure to the gasket from the bolts without overstraining the flexible neck.

Another self-sealing closure is illustrated in Fig. 32. The thrust due to internal pressure acting on the floating head A is taken by the large retaining ring B threaded into the walls of the forged vessel. The gasket C is confined by means of two removable rings D and E which are easily replaceable if they become deformed.

Figure 33 shows the Vickers-Anderson joint, which has been used in England on large converters. It is intended to reduce the bulkiness of large flange joints by eliminating the bolts. The two parts to be joined are surrounded by a clamp A, made in several sections that are bolted together by tangential bolts. The clamp takes up the entire thrust due to internal pressure. The slight angle on the shoulders against which the clamp bears tends to pull the two surfaces together as the clamp bolts are tightened. By using a hollow-ring gasket which is flexible and tends to open up under the action of the pressure, the joint can be made self-sealing.

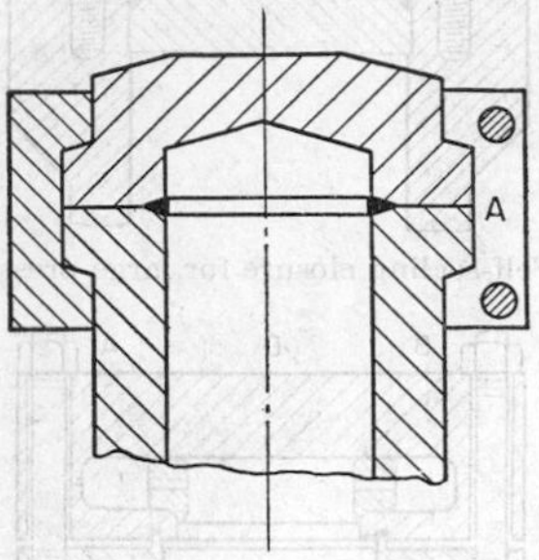

FIG. 33. Vickers-Anderson closure.

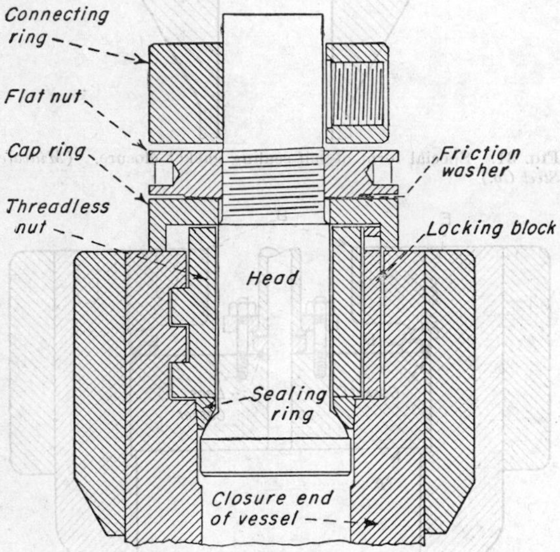

FIG. 33A.

To avoid the use of screw threads which tend to freeze under the action of high pressure and elevated temperature, the Bureau of Mines [Clark *et al.*, *Ind. Eng. Chem.*, **39,** 1555 (1947)] uses a breechblock type of self-sealing closure shown in Fig. 33A which utilizes a threadless nut with three machined lugs fitting into slots in the wall of the vessel. The paper should be consulted for details on this and various other types of closures.

PIPING, VALVES, AND FITTINGS

Welded and seamless-steel pipe are made in 10 standard weights and thicknesses denoted by numbers from 10 to 160, the latter having the heaviest wall. The old designations of "extra strong" and "double extra strong" are no longer used for pipe standards. Since the schedule number is approximately equal to 1000 P/S, where P is internal pressure and S the allowable fiber stress, it is evident that the heaviest pipe can be used with safety only to pressures of the order of 2500 to 5000 lb./sq. in., depending on various factors. Above these pressures, the best piping material is seamless drawn tubing; or, for very high pressures, tubes may be constructed by drilling holes in steel bars. Seamless tubing is available in a large number of standard sizes and of a wide variety of compositions of metal, and the engineer needs only to decide on the dimensions and, if corrosion is a factor, on the material he needs for a specific case and to consult the lists of tube manufacturers. For experimental work in the laboratory, small-diameter copper or steel tubes are convenient. Copper is more flexible and easier to handle and, with proper wall thickness (necessary thickness may be roughly calculated by the methods previously given, assuming 50,000 lb./sq. in. for the ultimate strength), can be used to 15,000 lb./sq. in. Joints can be made by soft or hard solder, and for a union the tube can be soldered into a brass nipple with spherical surface on the end, as shown in Fig. 34, which in turn is forced into a conical seat by the hexagon nut. For pressures up to about 5000 lb./sq. in., the standard compression or flared-tubing fittings are satisfactory. Chrome-molybdenum tubing $\frac{1}{16}$ by $\frac{1}{4}$ in. with a yield pressure of over 100,000 lb./sq. in. and $\frac{3}{16}$ by $\frac{9}{16}$ in. tubing with a yield pressure of about 75,000 lb./sq. in., is available (American Instrument Company). For still higher pressures triplex tubing $\frac{3}{4}$ by $\frac{1}{16}$ in. and $\frac{1}{2}$ by $\frac{1}{16}$ in. is available (Harwood Engineering Co.) with four different core materials. This tubing has successfully withstood 200,000 lb./sq. in. Fittings and valves for use with all the above tubes are also available.

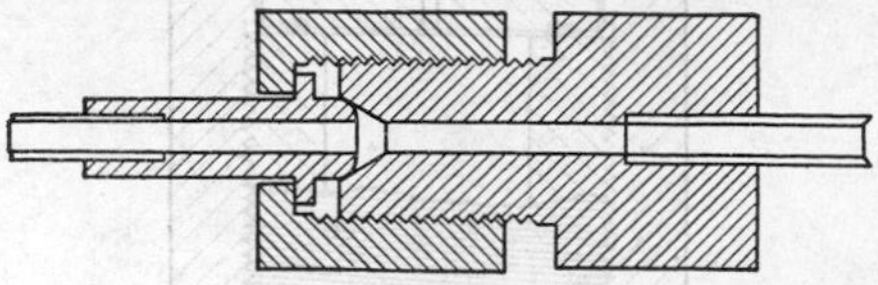

FIG. 34. Brass union.

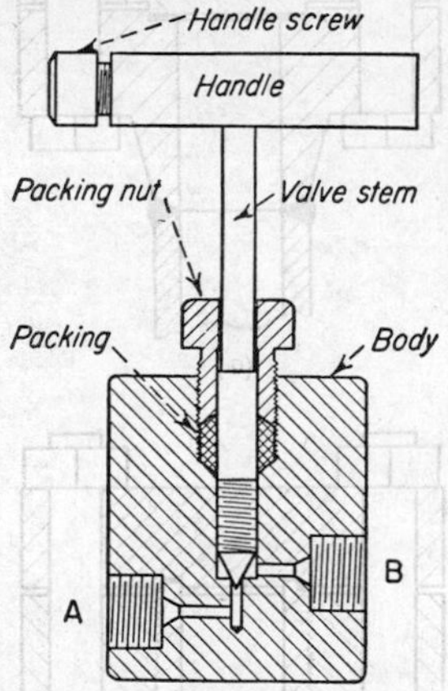

FIG. 35. High-pressure needle valve. (*American Instrument Co.*)

Various types and sizes of standard forged tees, unions, crosses, elbows, valves, and other fittings are available for pressures up to about 6000 lb./sq. in.; and in some cases they are available as special items for somewhat higher pressures, perhaps up to 15,000 lb./sq. in. One should consult the catalogues of such companies as the Crane

Co. or the Walworth Co. and others for further details. Standard specifications for piping systems to be used under pressures above 15 lb./sq. in. gage are given in "American Standard Code for Pressure Piping," American Standards Association, 1942, with two supplements; published by A.S.M.E., New York.

For higher pressures fittings are not standard and are usually machined from solid bar stock or from forgings. Valves and fittings useful for laboratory work are shown in Figs. 35 and 36. The valve-stem packing may be an

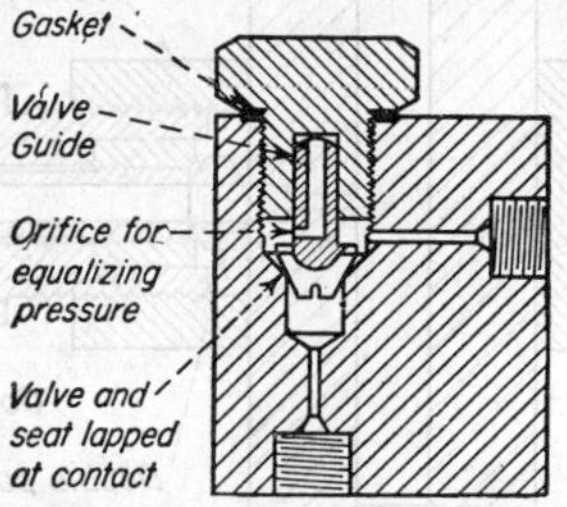

FIG. 36. High-pressure check valve. (*American Instrument Co.*)

asbestos cord impregnated with graphite or a shredded soft metal mixed with graphite or leather rings soaked in tallow or waxes or Teflon. The valve stem ends in a hardened and ground cone which seats on a shoulder of softer metal. The lower part of the valve stem is of slightly larger diameter than the upper part and is large enough so that it will not pass through the stuffingbox nut. This is an important safety feature, because the valve stem is most likely to break near the point or the thread and the high pressure would eject the portion above the break with high velocity if it were not retained in this way.

A valve that can be repacked under pressure is shown in Fig. 37. This is made possible by the double cone on the stem, the upper cone engaging a seat when the stem is screwed up, thus making a pressuretight seal with the valve open.

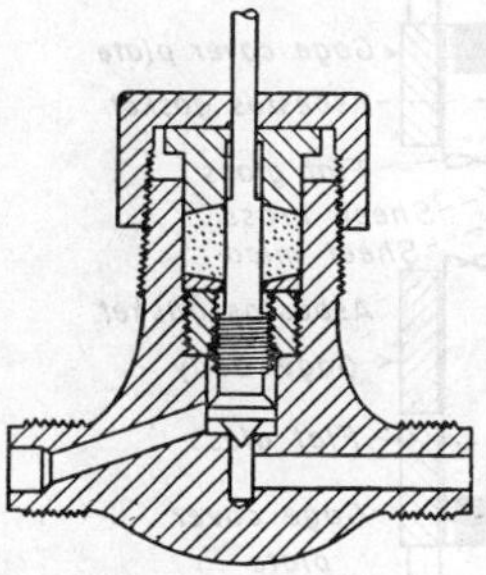

FIG. 37. Forged valve for repacking under pressure.

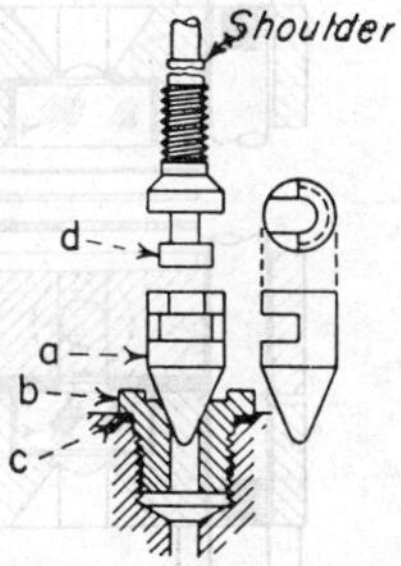

FIG. 38. Two-piece valve stem and removable seat.

For service where the valve seat is subjected to considerable wear by corrosion or abrasion, it is desirable to make the seat removable. One design of removable seat is shown in Fig. 38. The seat threads into the body of the valve and is made tight by a gasket at *c*. Another special feature is that the stem is made in two pieces, the upper one bearing a collar *d* which slides into a groove in the lower piece *a*. Because of this construction, the needle point does not turn as it is seated, and less scoring of the seat should result. Clark *et al.* (*loc. cit.*) describe, with a drawing, a valve that has given good service under severe erosive conditions. It's main novel features are a removable seat and stem insert made of Kenna metal.

The valves described above are not well suited to close regulation of flow rate. For this purpose it is better to use a valve with long tapered cones on the stem and the seat so that the area of the opening changes more slowly as the stem turns. Another design for flow control is shown in Fig. 39, in which the stem is a lapped fit in the valve cylinder. The resistance to flow is a function of the distance the stem is inserted into the cylinder, being proportional to the length of the annulus. A conical seat at the upper end of the cylinder provides for complete closure.

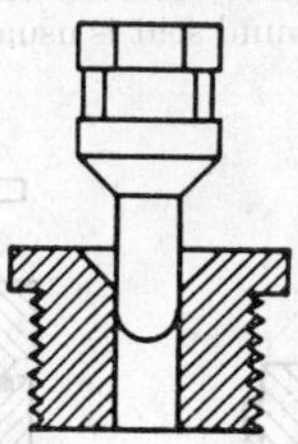

FIG. 39. Valve for close control of flow. (*Mathieson Alkali Works.*)

At very high pressures, the friction in screw threads may become so great that valves with a stem operated by a screw are hard to use and some other means of moving the stem is desirable. Valves operated by oil or water pressure through pistons or diaphragms have been developed for this purpose. However, screw-operated valves of small size have been used at least to 125,000 lb./sq. in. The effect of the friction can be minimized by a thrust bearing or by extending the stem on through the valve which requires the use of another stuffing box.

Valves for use at elevated temperatures are sometimes made with an extension on the upper part of the body to

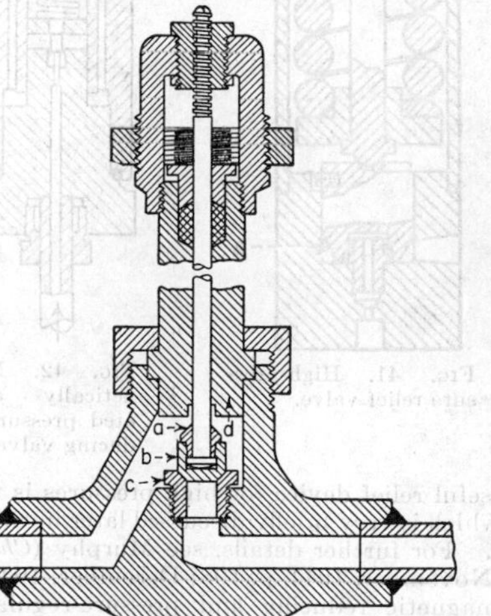

FIG. 40. Valve for use at elevated temperatures.

allow removal of the thread and the packing to a lower temperature region. One design is shown in Fig. 40. This valve also has a removable seat *c* and can be repacked in the open position as the gasket *a* seats against a shoulder in *d*. Valves for this service are sometimes provided with jackets for water cooling the upper end or fins for air cooling.

To avoid trouble from the rusting of valve stems and seats, these parts are often made of special alloys such as monel or stainless steels.

Figure 41 shows a high-pressure relief valve, the special feature of which is that it can relieve the pressure many times and still reseat and be gastight, a feature not found in the ordinary valves with ground seats. The gastight seal is made by a very narrow steel ring machined on the end of the piece *A* and ground to a smooth surface. This thin annulus presses into a softer disk of copper and makes its own seat. [For further details, see Ernst and Reed, *Mech. Eng.*, **48,** 595 (1926).] Actually, however, the usual relief valve is not expected to open very often, and for the great majority of cases the more common and cheaper type with ground seat is usually satisfactory.

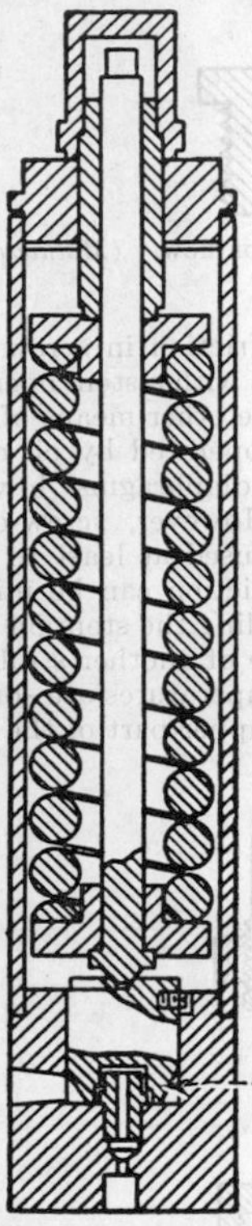

FIG. 41. High-pressure relief valve.

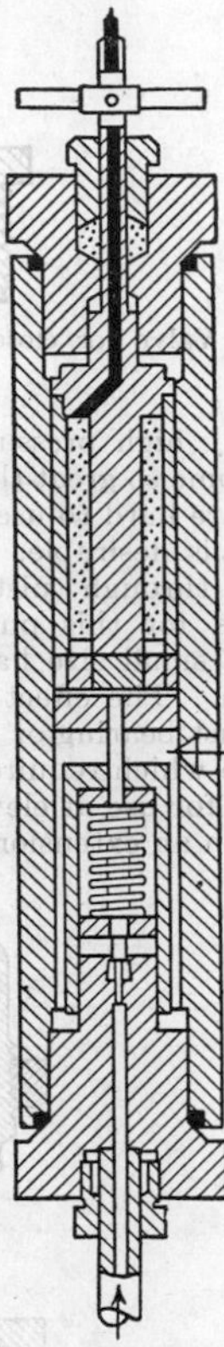

FIG. 42. Magnetically operated pressure-reducing valve.

A useful relief device for high pressures is the rupture disk, which is very briefly discussed later in the section on safety. For further details, see Murphy (*Chem. & Met. Eng.*, November, 1944, p. 108; December, 1944, p. 99).

A magnetic reducing and pressure-regulating valve useful in experimental work is shown in Fig. 42. The valve is normally closed, the stem being held on its seat (a steel cone on the needle stem presses into a softer metal shoulder) by the spring. When the pressure reaches a predetermined point, an adjustable electric contact on a Bourdon gage closes and the solenoid in the valve is energized and the armature raised, lifting the needle from the seat. When contact is broken by a fall in pressure, the current ceases to flow, and the spring reseats the valve. Such a device will maintain the pressure constant on the low side to a few per cent [Larson and Karrer, *Ind. Eng. Chem.*, **14,** 1012 (1922)].

Connections of tubing to fittings and valves and to heads or walls of vessels can be made in any one of the various ways already discussed under the subject of joints. One special type of connection useful in laboratory work is also worth mentioning. This is the joint-ring connection shown in Fig. 43, which enables one to make connection to the wall of a pressure vessel without

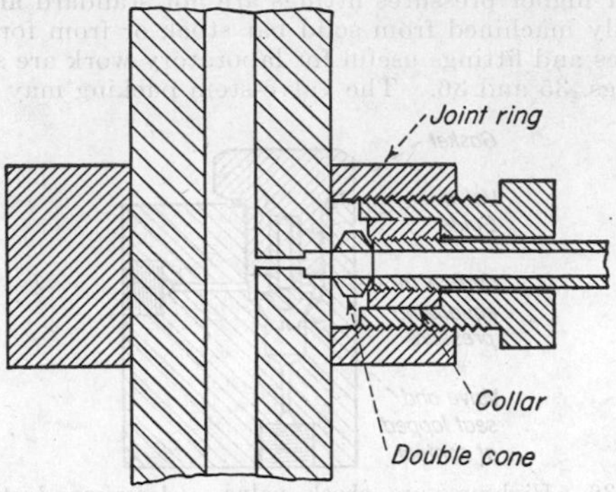

FIG. 43. Joint-ring connector.

appreciably weakening the wall. This is very useful for small vessels at very high pressures.

SIGHT GLASSES AND WINDOWS

On some occasions, it is desirable to see into high-pressure equipment, especially in experimental work. In plant operation, the chief need arises in connection with the location of liquid levels. The ordinary gage glass using cylindrical glass tubing is unreliable. A sight glass much used in high-pressure steam practice is shown in Fig. 44. The hot water is prevented from coming in contact with the flat glass plates by thin sheets of mica acting as diaphragms to transmit the pressure, and hence the glass itself does not have to form the seal. By this construction the gage can be tightened by the bolts

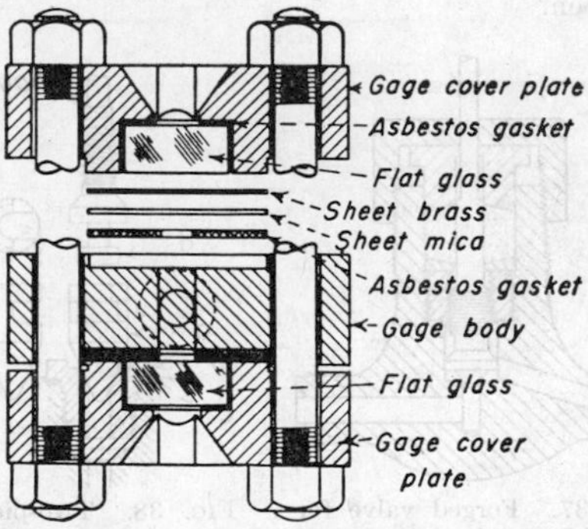

FIG. 44. High-pressure sight glass. (*Diamond Power Specialty Co.*)

without putting any strain on the glass, all the stress being transmitted from one part of the body to the other through the thin sheets of brass and mica and the gasket. When under pressure the glass acts as a support for the thin diaphragms of brass and mica and is relatively free to expand and contract under temperature changes.

Sight glasses of similar construction with celluloid sheets in place of mica have been used to about 5000 lb./sq. in. at ordinary temperatures. For higher pressures the quartz window shown in Fig. 45 has been used. The quartz cone *a* is ground into the steel shell, held in place by the threaded collar *b* and made tight by the gasket *c*. A very simple window that has been used to pressures as

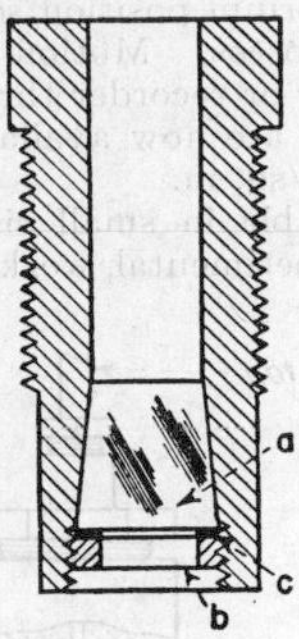

Fig. 45. High-pressure quartz window. (*Bone, Newitt, and Townend, "Gaseous Combustion at High Pressures," by permission of Longmans, Green & Co., London,* 1929.)

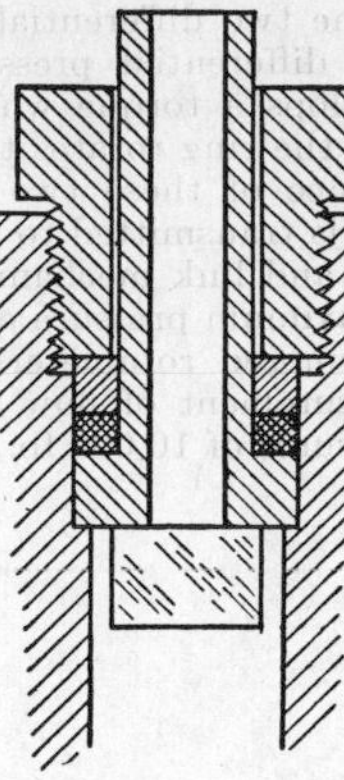

Fig. 46. High-pressure cylindrical window.

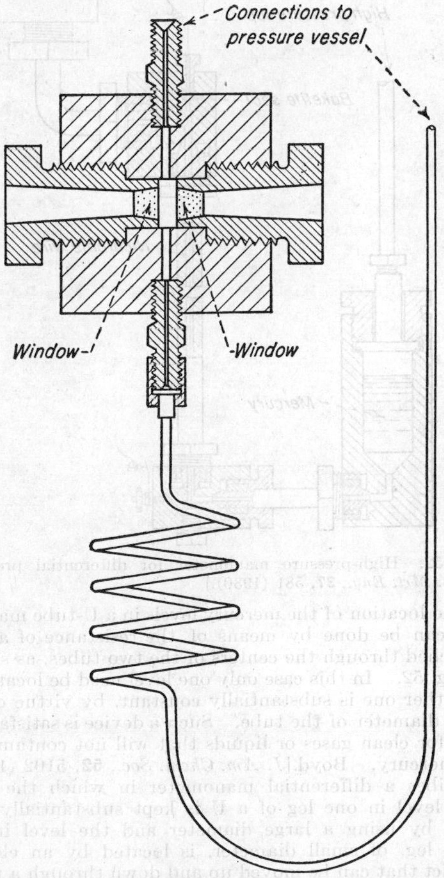

Fig. 47. Liquid-level indicator. (*Newitt, "The Design of High Pressure Plant and the Properties of Fluids at High Pressure," Oxford, New York,* 1940.)

high as 30,000 atm. [Poulter, *Phys. Rev.*, **35**, 297 (1930)] is illustrated in Fig. 46. This particular form is due to Bridgman (*op. cit.*). The cylindrical disk of glass or quartz is merely fastened to the mouth of the steel nipple by an adhesive such as Canada balsam. The glass and the steel should both be ground optically flat. The steel nipple is made tight in the wall of the vessel by a joint of the unsupported-area type.

Application of a window such as that of Fig. 45 to a liquid-level indicator is illustrated in Fig. 47. The indicator is connected to the pressure vessel by a flexible coil of tubing so that the sight glass may be raised or lowered to bring the liquid level within the window aperture.

ELECTRIC LEAD SEALS

For bringing out electric leads from heaters or electrically operated instruments inside of pressure vessels, various types of insulated electrode connectors have been devised, some of which are shown in Figs. 48, 49,

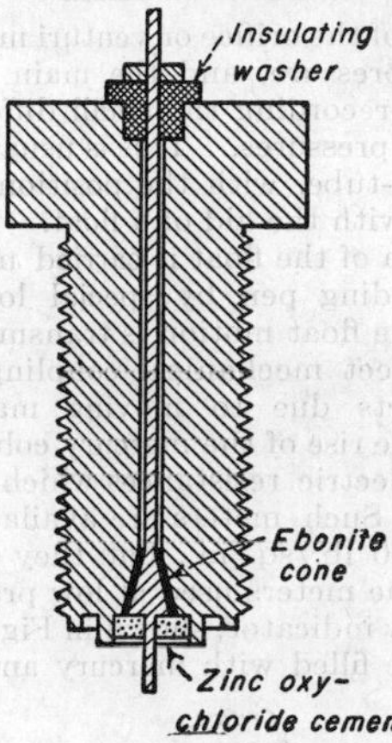

Fig. 48. Electric-lead seal.

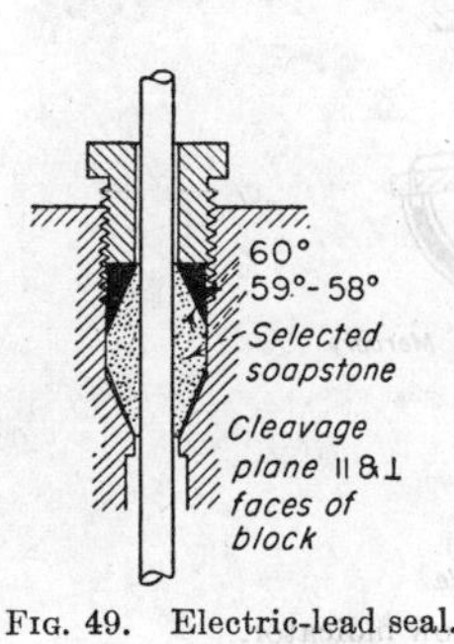

Fig. 49. Electric-lead seal.

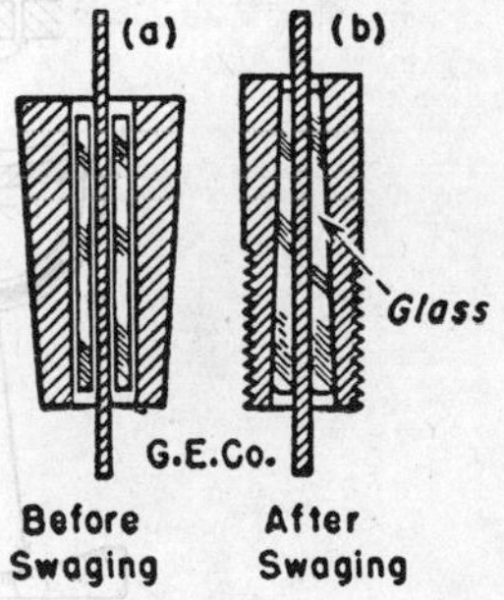

Fig. 50. Electric-lead seal.

and 50. In Fig. 48 the ebonite cone (probably various plastic materials could also be used) makes the gastight seal between the electrode and the steel plug, and the upper end of the electrode is insulated by the washer as shown. In assembly the cone is softened by heat and pulled up tightly between the two metal surfaces by the nut on top of the washer. To complete the seal, the recess below the cone is filled with zinc oxychloride cement. This type of lead seal is good only for relatively low temperatures.

The type shown in Fig. 49 uses a soapstone double cone (some plastics such as polystyrene may also be used for lower temperature service) to form the seal between the electrode and the metal plug and is good for elevated temperatures. The pressure limit of both these seals is not known, but they have been used to 1000 atm. Figure 50 shows a seal developed by the General Electric Co. for vacuum use. It has been used in high-pressure experiments at least to 5000 lb./sq. in. and would probably go much higher. Such seals are very convenient, because they can be purchased ready for use at a small price. In manufacture the pieces are assembled as at *a* and then heated and swaged until the glass is conical in shape and firmly welded to the metal wall to form a gastight seal. The metal case may then be threaded for screwing into the apparatus. They can be used up to temperatures below the softening point of the glass. A somewhat similar seal is described by Welbergen [*J. Sci. Inst.*, **10**, 247 (1933)].

For pressures much above 1000 atm, the types of insulated leads just described are probably not suitable, and recourse may be had to one developed by Bridgman (*op. cit.*) and used by him to about 20,000 atm.

FLOWMETERS

The principle of the orifice or venturi meter may be used to almost any pressure, and the main difficulty comes in indicating or recording the small differential pressure at high absolute pressures. This is usually accomplished by a mercury U-tube, with the position of the mercury being indicated with the aid of a float. In some types of meter the motion of the float is carried mechanically to a pointer or recording pen by special low-friction seals. In other cases the float motion is transmitted to the outside without direct mechanical coupling by the use of inductance effects due to moving magnets. In still another meter the rise of the mercury column is measured by a series of electric resistances which are successively short-circuited. Such meters are available for pressures up to about 5000 lb./sq. in., but they do not differ in principle from the meters used at low pressures.

The Bosch flow indicator, shown in Fig. 51, consists of a semicircular tube filled with mercury and balanced on a

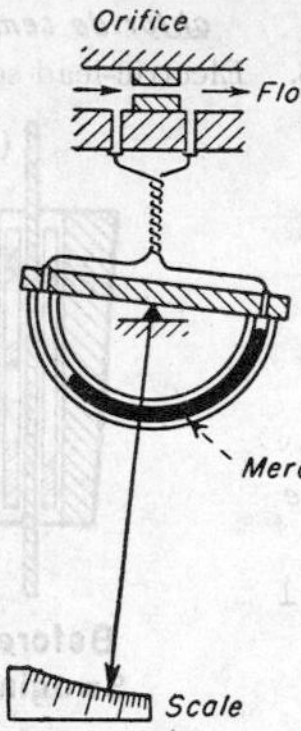

FIG. 51. Bosch flow indicator.

knife-edge. Flexible capillary tubing connects the two differential pressure taps to the two ends of the mercury column. The displacement of the mercury causes the tube to tilt and indicate the flow on a scale. The pressure leads are sufficiently flexible so that they interfere very little with the free motion of the semicircular tube. Such a meter is not made as a standard unit in this country as far as the author knows, but it can readily be constructed either in this form or simply as a U-tube which is mounted so that it is free to turn on an axis.

A somewhat similar instrument, known as a Ring Balance Meter, is marketed by the Hagan Corporation. The differential pressure developed in an orifice or flow nozzle is transmitted to a hollow ring containing a sealing fluid (usually mercury) and mounted on a knife-edge to permit rotation about the axis of the ring. The ring is divided into two pressure compartments by a partition at the top and these two compartments are connected to the two differential-pressure leads by flexible tubing. The differential pressure acting on the ring partition develops a torque which is resisted by a leaf spring so that the ring rotates to an equilibrium position set by a balance of these two opposing forces. Motion of the ring is transmitted to an indicator or recorder through a cam and link mechanism. Rings are now available for a maximum pressure of 10,000 lb./sq. in.

Armored rotameters are available in small sizes for measurement of flow rates in experimental work up to pressures of 10,000 lb./sq. in.

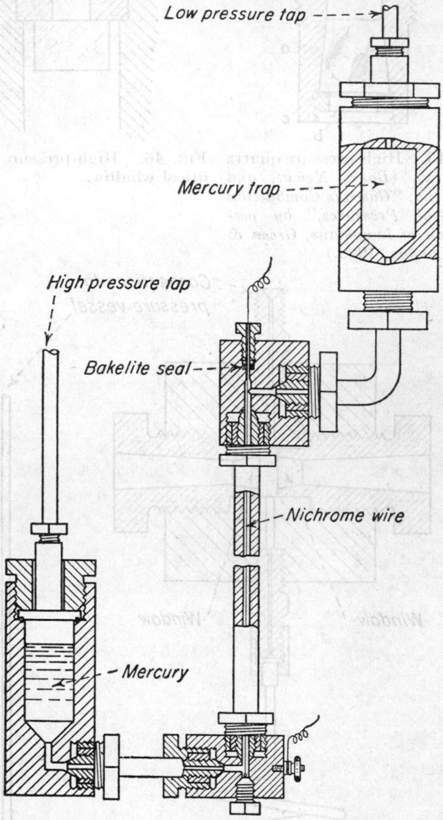

FIG. 52. High-pressure manometer for differential pressure. [*Chem. Met. Eng.*, **37**, 581 (1930).]

The location of the mercury levels in a U-tube manometer can be done by means of the resistance of a wire stretched through the centers of the two tubes, as shown in Fig. 52. In this case only one level need be located as the other one is substantially constant, by virtue of the large diameter of the tube. Such a device is satisfactory only for clean gases or liquids that will not contaminate the mercury. Boyd [*J. Am. Chem. Soc.*, **52**, 5102 (1930)] describes a differential manometer in which the mercury level in one leg of a U is kept substantially constant by using a large diameter and the level in the other leg, of small diameter, is located by an electric contact that can be moved up and down through a packing gland.

PRESSURE MEASUREMENT

The most usual method of measuring pressure is by means of the Bourdon-tube pressure gage, which consists of a flattened bronze or steel tube bent into an arc. As pressure is applied the tube tends to straighten, and this movement is transmitted to a dial through a suitable magnification train. Bourdon tubes for high pressures are made of steel. Since so much depends upon them, only tubes made according to the most exacting standards and carefully aged by the manufacturer should be used. It is customary to use the gages at one-half the maximum pressure provided on the scale on fluctuating pressures and two-thirds of that pressure when the pressure is steady. If a Bourdon tube is overranged and subjected to a pressure higher than that at which it was stressed in the aging process, permanent distortion may occur, making recalibration necessary.

Gages that are in continuous use, and especially those subjected to constant rapid fluctuations of pressure, should be checked at frequent intervals. A convenient way of doing this is to have an accurate master gage that can be attached for comparison somewhere in the line to the regular gage. At regular intervals the master gage should be checked against a dead-weight gage. The

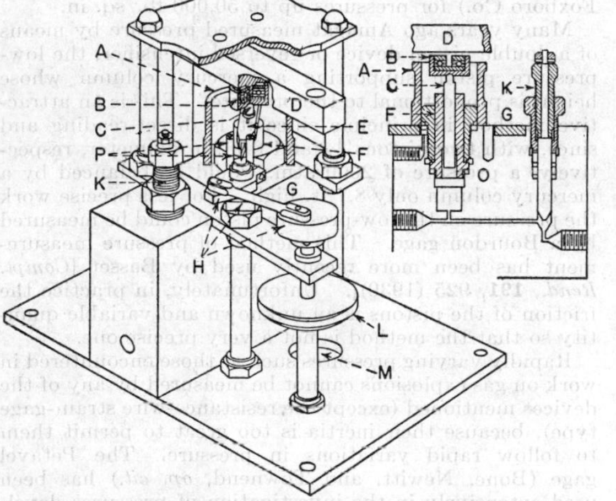

FIG. 53. Dead-weight piston gage. *B*, ball bearing; *C*, piston; *D*, crossbar for scale pan; *F*, cylinder; *K*, shut-off valve; *L*, pulley for driving the oscillating mechanism. [*Keyes, Proc. Am. Acad. Arts, Sci.*, **68**, 530 (1933).]

Bourdon gage is quite satisfactory for pressures up to about 2000 atm., provided that an accuracy of 2 to 3 per cent is sufficient. Such gages are available with maximum scale readings of about 100,000 lb./sq. in.

For more precise measurement of pressure as needed in research or for checking other types of gages, the dead-weight piston gage is commonly used. This is very simple in principle, consisting merely of a cylinder with a very accurately fitted piston which is loaded above by weights. The load is balanced by the pressure of oil that is injected into the cylinder beneath the piston by a suitable pump. The oil pressure is in turn balanced by the pressure to be measured, usually through a mercury U-tube, the mercury level being used to indicate the balance through an electric contact device. The clearance between piston and cylinder is so small that the oil leak is slight even at high pressures and is compensated by pumping in more oil intermittently. To reduce the tendency of the piston to stick in the cylinder, the gage is usually equipped with a device for oscillating or rotating the piston.

One form of gage and its oscillating mechanism are described in a paper by Keyes [*Proc. Am. Acad. Arts Sci.*, **68**, 530 (1932–1933)], and some data on its performance are included. Figure 53 is reproduced from this paper. Four piston-and-cylinder combinations of different diameters are used to cover the range from several to 1200 atm. The piston gage is absolute, since the pressure is calculated from a knowledge of the weight pressing on the piston and the average area of the piston and cylinder, which can be measured very accurately. The method is precise and can readily be used to about 1500 atm. For the precise comparison of such a gage with a mercury column, see a paper by Keyes and Dewey [*J. Optical Soc. Am.* and *Rev. Sci. Instruments*, **14**, 491 (1927)]. Pressures higher than this introduce difficulties; because, even with a piston as small as ⅛ in. diameter, the weight required at 2000 atm. is 350 lb. and above 3000 atm. pressure, leakage past the piston becomes a serious factor.

In spite of the difficulties, Bridgman has used a dead-weight gage to about 12,000 atm. It had a piston only ¹⁄₁₆ in. in diameter and was so designed that the high pressure was exerted on both sides of the cylinder at its lower end to prevent expansion of the inside diameter (see Fig. 54). It is of interest to note that most organic

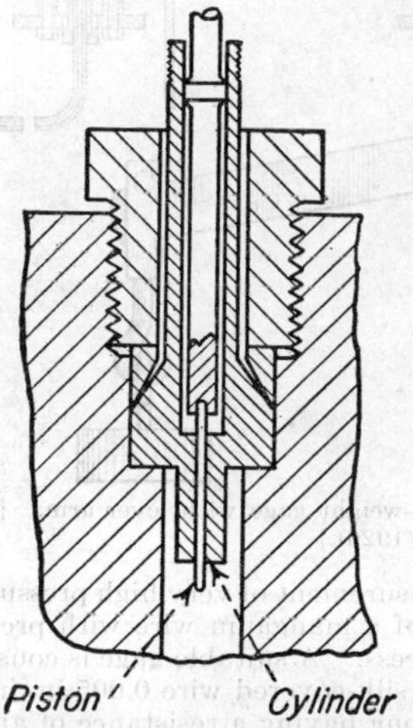

FIG. 54. Bridgman design of dead-weight gage for very high pressures. (*Bridgman, "The Physics of High Pressure," by permission of The Macmillan Company, New York*, 1931.)

liquids solidify at room temperature under such a pressure and that the fluid used for transmitting the pressure is a mixture of glycerin and water. The American Instrument Co. supplies a piston gage for a maximum pressure of 100,000 lb./sq. in.

Various other schemes have been used to adapt the piston gage to higher pressures or to permit the use of less weight at lower pressures. One scheme shown diagrammatically in Fig. 55 involves the use of a lever to multiply the force due to the weights. This is satisfactory but is less precise. Another method, illustrated in Fig. 56, uses a differential piston, the net result of which is that the pressure acts only on an area equal to the difference in the areas of the two ends of the differential piston. This also involves a sacrifice in accuracy, since the effective area is a difference of two larger ones that are measured.

Early measurements of high pressures were made with mercury columns, and these are still used for calibrating piston and other types of gages. Amagat, in his work on the compressibility of gases, used a column about 900 ft. long erected in a mine shaft. To reduce the height of the column, short columns may be connected in series. Such a column at a Leiden laboratory consisted of 15 columns each 3.14 m. long and could measure pressures up to 60 atm.

The constants of a piston gage may be checked by a standard reference pressure. A convenient one is the vapor pressure of carbon dioxide at 0°C., which is 34.401 atm. For very high pressures, a convenient reference point for checking gages is the freezing point of mercury, which is 7400 atm. at 0°C.

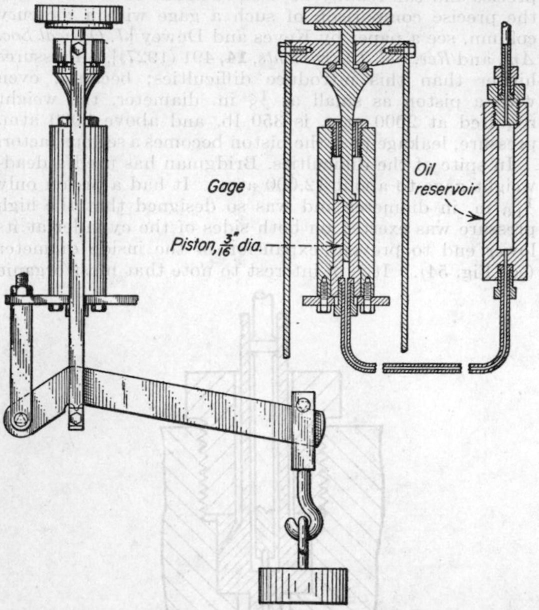

FIG. 55. Dead-weight gage with lever arm. [*Ernst, Ind. Eng. Chem.*, **18**, 666 (1926).]

For the measurement of very high pressures, the change of resistance of a manganin wire with pressure has been used with success. A suitable gage is constructed from a coil of double silk-covered wire 0.005 in. in diameter and about 20 ft. long having a resistance of about 120 ohms. The wire is wound non-inductively on a cylindrical core about 3/4 in. in diameter. Since the temperature coefficient of resistance of manganin is very small, no special precautions for maintaining constant temperature are necessary. The relationship between pressure and resistance has been found to be linear up to 12,000 atm., and by extrapolation of the straight line the gage has been used to 20,000 atm.

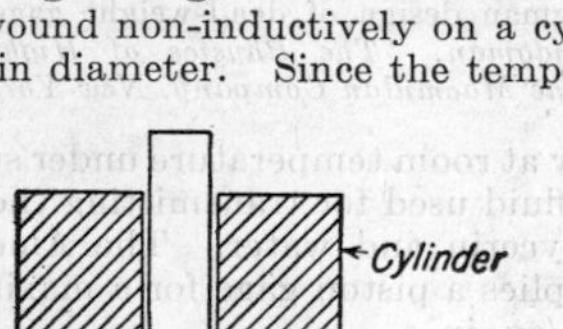

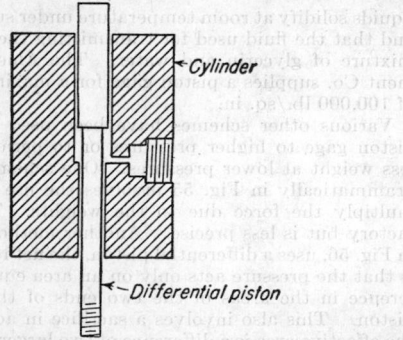

FIG. 56. Differential-piston type of dead-weight gage. (*Bridgman, "The Physics of High Pressures," by permission of The Macmillan Company, New York,* 1931.)

Since a pressure change of 1000 atm. will change the resistance only 0.23 per cent, it is clear that very precise resistance measurements must be made. This has been one objection to the use of the resistance gage for anything but laboratory measurements. However, at least one instrument company is experimenting with manganin wire-pressure indicators and recorders, and it is likely that such instruments for industrial use will be on the market in the near future.

Measurement of the strain in a tube subjected to pressure has also been used to determine pressures up to about 100,000 lb./sq. in. The strain can be measured by one of the sensitive dial strain gages; or one of the recently developed wire strain gages may be used, in which the strain changes the resistance of the wire and this is measured by a Wheatstone-bridge circuit. This, is, of course, not an absolute method but must be calibrated by another instrument such as the piston gage. Present indications are that the pressure-strain relation within the elastic limit is linear as called for by Hooke's law and quite reproducible. Pressure indicators and recorders using this principle are now available (The Foxboro Co.) for pressures up to 50,000 lb./sq. in.

Many years ago Amagat measured pressure by means of a double piston device or reversed intensifier, the low-pressure piston supporting a mercury column whose height is proportional to the pressure. This is an attractive method in principle since it is direct-reading and since, with two pistons 1/4 and 6 in. in diameter, respectively, a pressure of 2000 atm. would be balanced by a mercury column only 8.7 ft. high. For less precise work the pressure on the low-pressure piston could be measured by a Bourdon gage. This method of pressure measurement has been more recently used by Basset [*Compt. Rend.*, **191**, 925 (1930)]. Unfortunately, in practice the friction of the pistons is an unknown and variable quantity so that the method is not a very precise one.

Rapidly varying pressures such as those encountered in work on gas explosions cannot be measured by any of the devices mentioned (except the resistance-wire strain-gage type), because their inertia is too great to permit them to follow rapid variations in pressure. The Petavel gage (Bone, Newitt, and Townend, *op. cit.*) has been used extensively in the investigation of pressures developed in explosions. In principle, it consists of a piston whose motion is resisted by a metal tubular spring, the slight movements of which are magnified by an optical lever. More recently, the use of piezoelectric crystals has been developed into a very satisfactory device for measuring rapidly varying pressures. Indications or records of the pressure vs. time are obtained from a cathode-ray oscillograph. Such a device, good to pressures up to 5000 lb./sq. in., is on the market. Also pressure gages based on the use of SR-4 electric strain gages can be used to follow rapidly fluctuating pressures.

REACTION VESSELS

High-pressure reaction vessels may be conveniently classified into two general groups: (1) batch reactors or autoclaves, and (2) continuous reactors or converters.

Autoclaves. These are generally vertical cylindrical vessels equipped with an agitator and either an external jacket or internal coils for heating or cooling. Figure 57 shows an autoclave for 900 lb./sq. in. provided with internal coils for cooling or heating, electric strip heaters on the outside, and turbine agitators. The autoclave vessel is generally constructed of welded plate with hemispherical, A.S.M.E., or semiellipsoidal heads and a flange-and-gasket type of closure, as shown in the figure. The construction of welded autoclaves is covered by the two codes previously mentioned. There are practically no standard autoclaves that one can purchase "off the shelf" except possibly some of the small laboratory ones. Each one is a special design to suit the particular needs, and the designer needs as complete information as possible on the reaction to be carried out, its kinetics, and the action of the various reactants and products on materials of construction. Laboratory autoclaves of the general

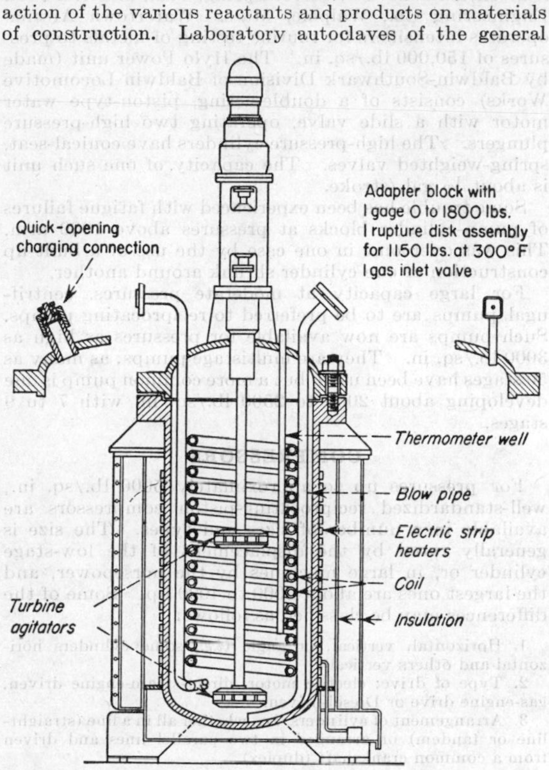

FIG. 57. Autoclave for 900 lb./sq. in. pressure. [*Gooch, Ind. Eng. Chem.*, **35**, 935 (1943).]

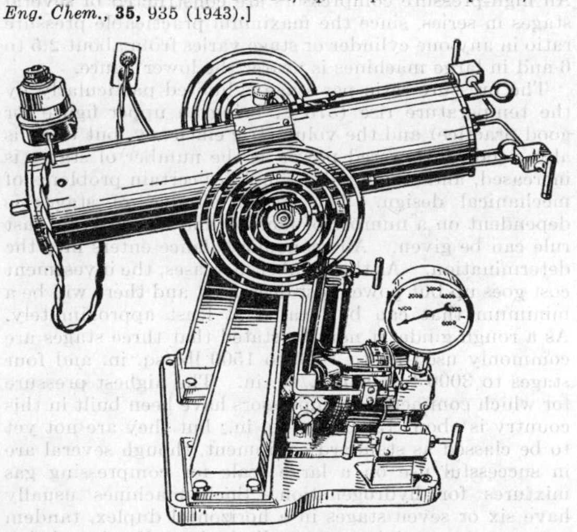

FIG. 58. Rocking type of autoclave. (*American Instrument Co.*)

type of that shown in Fig. 57 but constructed from a forging or bored from a solid bar of metal are available for pressures up to 25,000 lb./sq. in.

Laboratory autoclaves are frequently agitated by shaking or rocking mechanisms. One commonly used type is pictured in Fig. 58. This is oscillated about 60 times/min. and is electrically heated by an external resistance heater. The pressure connection to the bomb for measuring the pressure or bleeding off gases is made through a flat spiral of tubing to obtain the necessary flexibility. McMillan and Krase [*Ind. Eng. Chem.*, **24**, 1001 (1932)] describe a laboratory autoclave with a high-speed stirrer which is useful for certain types of reaction. The distinctive feature of this autoclave is the enclosure of the electric motor for the agitator within the pressure vessel, thus avoiding the difficult problem of packing the shaft at high speeds of rotation. Tongue (*op. cit.*) describes a laboratory autoclave used for studying reactions in liquids at pressures up to 12,000 atm.

Autoclaves are generally used for reactions that have to be carried out at some elevated temperature, and the method of heating is an important feature of the design. Depending on the temperature desired and various other factors, the following heating methods are used:

1. Hot water for systems sensitive to temperature and which must not be heated above 212°F. or some lower temperature.
2. Steam for the range 212° to 350°F.
3. Direct firing with various fuels.
4. Indirect firing, *i.e.*, the fuel is burned in a separate furnace and the combustion products circulated around the autoclave.
5. Hot oil.
6. Dowtherm.
7. Electric heating, either by strip heaters or by induction.
8. Circulating molten salt.
9. Mercury vapor.

The advantages and disadvantages of these various heating methods and some details on design, on materials, and on various types of agitators and their drives will be found in a paper by Gooch [*Ind. Eng. Chem.*, **35**, 927–946 (1943)]. Other references for further details on autoclaves are Eucken and Jacob (Editors, Vol. 3, Part 4 of Der Chemie-Ingenieur, "High Pressure Operations, Leipzig, 1939); and Tongue (*op. cit.*).

Continuous Reactors. These are generally long cylindrical tubes which are either forged for the highest pressures (about 1500 atm. is the highest pressure used industrially at the present time) or constructed of seamless tubing or, for somewhat lower pressures, are constructed of welded plate. Usually the reaction is carried out over a solid catalyst, as in the case of ammonia or alcohol synthesis; but, in the case of urea synthesis from CO_2 and NH_3, the reaction is carried out non-catalytically in the liquid phase. In the case of exothermic gas-phase reactions such as ammonia synthesis, the entering gases are heated to the reaction temperature by heat exchange inside the converter with the exit gases. There must be an auxiliary heater for starting, and this may be external or an internal electric heater.

Some of the factors governing the design of the vessel and types of head closures for such vessels have already been discussed above. One further point is worth mentioning here. Although many of the high-pressure reactions are carried on at temperatures in the range 300° to 500°C., the walls of the pressure vessel can be maintained at much lower temperatures by the simple expedient of allowing the incoming cold gas to flow along the walls before entering the exchanger to be heated by the hot off gases. The catalyst is placed in a separate thin-walled container, and the annulus between it and the wall of the pressure-resisting shell forms the passage for the incoming gases that cool the shell. The entire assembly of internal parts, including the catalyst chamber, the heat exchanger, starting heater, thermocouple wells, etc., is generally suspended from the head so that it is removable as a unit when the head is removed.

The design of the internal assembly of a catalyst reactor is highly specialized, depending on a number of factors, and cannot be considered in this section. Some further details may be obtained from Curtis (*op. cit.*); Tongue (*op. cit.*); Ernst [Equipment for High Pressure Reactions, *Ind. Eng. Chem.*, **18**, 664 (1926)]; and Ernst, Edwards, and Reed [A Direct Synthetic Ammonia Plant, *Ind. Eng. Chem.*, **17**, 775 (1925)].

PUMPS

Pumps for high-pressure service are usually reciprocating-piston or plunger types, driven by an electric motor through a chain or belt or by steam or gas engines. There are usually two or more cylinders in parallel to give a more even discharge and a better balancing of the forces. Cylinders may be either vertical or horizontal. The general arrangement of cylinder, valves, and plunger in a simple, plunger type of pump is shown in Fig. 59.

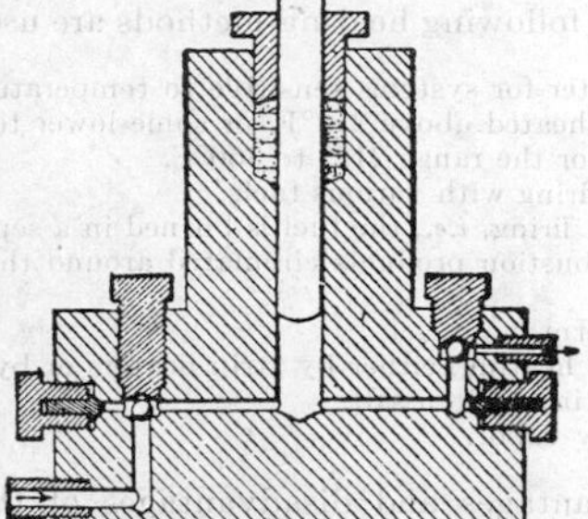

FIG. 59. Hydraulic plunger pump.

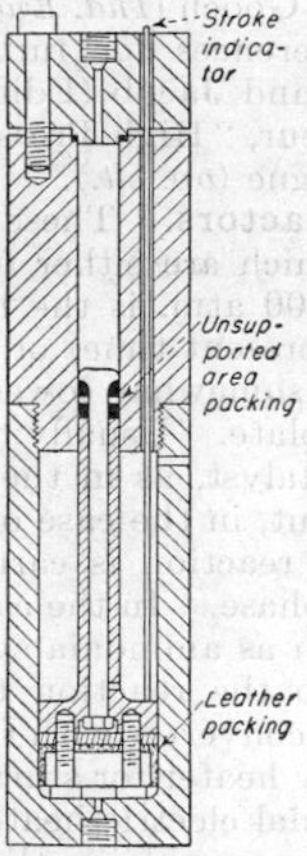

FIG. 60. Pressure intensifier. (*American Instrument Co.*)

This particular pump has ball valves, but some prefer the conical type of valve checks. Large pumps for pressures up to 35,000 lb./sq. in. and a maximum horsepower of 500 are available as standard equipment. For higher pressures, small hand-operated pumps made for laboratory or testing use will give somewhat higher pressures—to about 60,000 lb./sq. in. For still higher pressures, some form of intensifier is frequently used. This is merely a double-piston device, one form of which is shown in Fig. 60. The lower and larger piston is driven by pressure from a pump, and the smaller piston will then produce a higher pressure, the ratio of the two being approximately as the ratio of the areas of the two pistons. The high-pressure piston is packed with a ring of rubber or neoprene.

Small intensifiers are commercially available for pressures up to about 100,000 to 125,000 lb./sq. in. Like the one shown in Fig. 60, they are single-stroke pumps but can be made to repeat the strokes and give a continuous pumping action. Small intensifiers are made by American Instrument Co., Silver Spring, Md., and Harwood Engineering Co., Walpole, Mass. Watertown Arsenal operates intensifiers for routine testing of metals to pressures of 150,000 lb./sq. in. The Hylo Power unit (made by Baldwin-Southwark Division of Baldwin Locomotive Works) consists of a double-acting, piston-type water motor with a slide valve, operating two high-pressure plungers. The high-pressure cylinders have conical-seat, spring-weighted valves. The capacity of one such unit is about ½ gal./stroke.

Some trouble has been experienced with fatigue failures of pump-cylinder blocks at pressures above 1000 atm. This was overcome in one case by the use of a built-up construction of one cylinder shrunk around another.

For large capacity at moderate pressures, centrifugal pumps are to be preferred to reciprocating pumps. Such pumps are now available for pressures as high as 3000 lb./sq. in. They are multistage pumps; as many as 50 stages have been used, but a more common pump is one developing about 2000 to 2500 lb./sq. in. with 7 to 9 stages.

COMPRESSORS

For pressures up to approximately 5000 lb./sq. in., well-standardized reciprocating-piston compressors are available in a number of sizes and types. The size is generally rated by the displacement of the low-stage cylinder or, in large machines by the horsepower, and the largest ones are about 3000 to 4000 hp. Some of the differences may be classified as follows:

1. Horizontal, vertical, or angle (*i.e.*, some cylinders horizontal and others vertical).
2. Type of drive: electric motor, direct-steam-engine driven, gas-engine drive or Diesel-driven.
3. Arrangement of cylinders, *i.e.*, whether all in a line (straight-line or tandem) or arranged in two parallel lines and driven from a common crankshaft (duplex).

All high-pressure compressors are constructed of several stages in series, since the maximum practicable pressure ratio in any one cylinder or stage varies from about 2.5 to 6 and in large machines is nearer the lower figure.

The pressure ratio per stage is limited particularly by the temperature rise (375°F. being an upper figure for good practice) and the volumetric efficiency, but there is also a gain in lowered power as the number of stages is increased, and staging also simplifies certain problems of mechanical design. The actual number of stages is dependent on a number of factors, and no hard-and-fast rule can be given. An economic balance enters into the determination. As the number increases, the investment cost goes up but power cost decreases, and there will be a minimum that can be located at least approximately. As a rough guide, it may be stated that three stages are commonly used up to 1000 to 1500 lb./sq. in. and four stages to 3000 to 5000 lb./sq. in. The highest pressure for which commercial compressors have been built in this country is about 15,000 lb./sq. in.; but they are not yet to be classed as standard equipment, though several are in successful use on a large scale for compressing gas mixtures for hydrogenation. Such machines usually have six or seven stages in a horizontal duplex, tandem arrangement and have very limited application, being used almost exclusively in one process for synthesis of ammonia and alcohols. The Sulzer 1000-atm. compressor, made in Switzerland, is seven-stage and is distinguished by the fact that the last two stages have vertical cylinders and that the gas pistons are driven hydraulically by oil pistons.

At least two companies in this country have built small five-stage laboratory compressors for a top pressure of

15,000 lb./sq. in., one a straight-line machine and the other a duplex with the first, second, and fourth stages on one side and the third and fifth on the other. A photograph of the latter compressor is shown in Fig. 61.

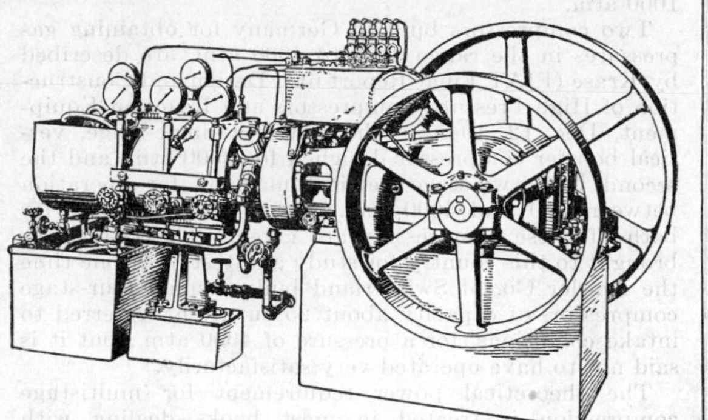

FIG. 61. Norwalk five-stage, 1000-atm. compressor. (*The Norwalk Co.*)

Before the war a popular 1000-atm. compressor for experimental work was the five-stage straight-line machine of Andreas Hofer of Mülheim-Ruhr, Germany. A cross section of it is shown in Fig. 62. It is fairly typical of the small compressors for this pressure. The usual cylinder arrangement, starting from the crank end, is second, first, third, fourth, and fifth stages. The second stage is put first so that the only externally packed joint will tend to leak gas out rather than air in, an important safety feature. The first stage is double-acting, and all the others single-acting.

For experimental work at pressures above 5000 lb./sq. in., several laboratories have made use of the hydraulic-compression system shown in Fig. 63, which is relatively cheap and simple and permits the attainment of gas pressures equal to the pressure obtainable from a hydraulic pump. The system is operated as follows to obtain a fairly continuous flow of gas at 15,000 lb./sq. in. from an initial supply at 4500 lb./sq. in.: One of the cylinders, *A*, is filled with water which is then displaced by gas at the available pressure, after which the waste-line valve is closed. Water is then pumped in until the pressure reaches the desired higher pressure. Meanwhile cylinder *B*, previously pressured up in this way, is supplying the gas at, say 1000 atm., the pressure being kept constant by pumping water in. When *B* is filled with water (a float and electric contact indicator may be used) the gas-line valves are closed and the water drained out; meanwhile cylinder *A* supplies the flow of gas at the desired pressure. This cycle is then repeated. For an intermittent supply of high-pressure gas, *B* may be used purely as a storage cylinder and all the compressing done in *A*.

In some cases it is desirable to avoid contact of the gas with water or oil. A mercury-piston compressor designed by Michel for 2500 atm. is shown in Fig. 64. Oil is pumped into the lower of the two cylinders, originally filled with mercury, displacing it into the upper cylinder originally filled with gas at some high pressure. The rising mercury column compresses the gas; and, when the desired pressure is reached, a valve in the line to the high-

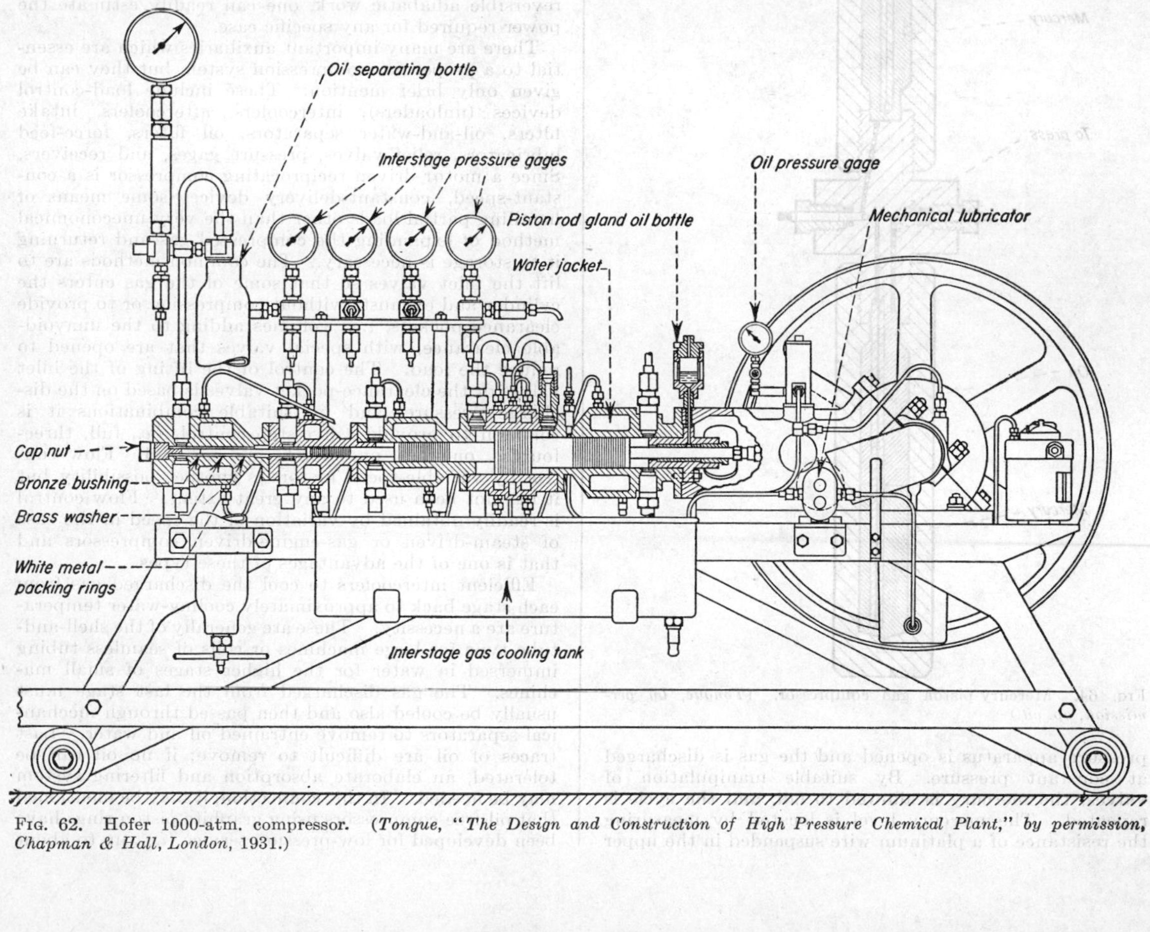

FIG. 62. Hofer 1000-atm. compressor. (*Tongue, "The Design and Construction of High Pressure Chemical Plant," by permission, Chapman & Hall, London,* 1931.)

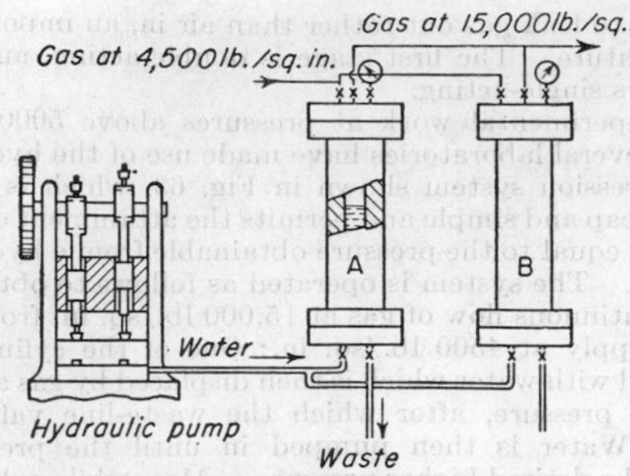

FIG. 63. Hydraulic compression system.

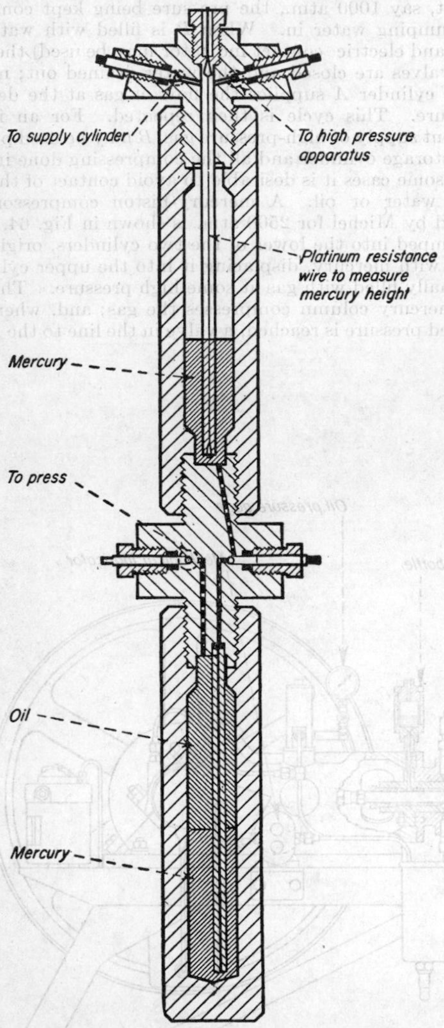

FIG. 64. Mercury-piston gas compressor. (*Tongue, by permission, op. cit.*)

pressure apparatus is opened and the gas is discharged at constant pressure. By suitable manipulation of valves, the mercury is allowed to fall, and the cycle is repeated. The mercury level is located by measuring the resistance of a platinum wire suspended in the upper cylinder. Compressors of this type have been used commercially at 1500 atm.

As far as the author is aware no compressors have been constructed in this country for pressures higher than 1000 atm.

Two compressors built in Germany for obtaining gas pressures in the range 3000 to 4000 atm. are described by Krase (FIAT Final Report 611, Design and Construction of High Pressure Compressors and Reaction Equipment, Dec. 12, 1945). The first is a single-stage, vertical booster compressor designed for 3000 atm. and the second is a two-stage vertical machine for operation between 300 and 4000 atm. Some detailed prints for both of these compressors are given. The latter was brought to this country for study and test. At one time the Amsler Co. of Switzerland built a small four-stage compressor, of capacity about 25 cu. ft./hr. referred to intake conditions, for a pressure of 4000 atm., but it is said not to have operated very satisfactorily.

The theoretical power requirement for multistage compression is treated in most books dealing with engineering thermodynamics (*e.g.*, Dodge, "Chemical Engineering Thermodynamics," Chap. VII, McGraw-Hill, New York, 1944; and also on pp. 303 and 1439 in this handbook) and will not be considered here. Mechanical efficiencies will run about 88 to 90 per cent for large steam-driven compressors and somewhat higher for power-driven ones. Compression efficiency (theoretical adiabatic power ÷ indicated horsepower) will average about 90 per cent for well-designed large machines. A good round figure for the over-all efficiency is 80 per cent; and, combining this figure with the theoretical calculated reversible adiabatic work, one can readily estimate the power required for any specific case.

There are many important auxiliaries which are essential to a proper gas-compression system but they can be given only brief mention. These include load-control devices (unloaders), intercoolers, aftercoolers, intake filters, oil-and-water separators, oil filters, force-feed lubricators, relief valves, pressure gages, and receivers. Since a motor-driven reciprocating compressor is a constant-speed, constant-delivery device, some means of handling partial loads other than the very uneconomical method of expanding the compressed gas and returning it to storage is necessary. The common methods are to lift the inlet valves so that some of the gas enters the cylinder and exhausts without compression or to provide clearance pockets, *i.e.*, volumes adding to the unavoidable clearance, with special valves that are opened to reduce the load. The control of the lifting of the inlet valves or the clearance-pocket valves is based on the discharge pressure, and by suitable combinations it is common to provide a five-step control, *i.e.*, full, three-fourths, one-half, one-fourth, and no load. Flow control by variable-speed motors is another possibility, but it has not been used to any great extent. Flow control is readily obtained by variation of the speed in the case of steam-driven or gas-engine-driven compressors and that is one of the advantages of these types.

Efficient intercoolers to cool the discharged gas from each stage back to approximately cooling-water temperature are a necessity. These are generally of the shell-and-tube type for large machines or coils of seamless tubing immersed in water for the higher stages of small machines. The gas discharged from the last stage must usually be cooled also and then passed through mechanical separators to remove entrained oil and water. Last traces of oil are difficult to remove; if no oil can be tolerated, an elaborate absorption and filtering system may be needed. In this connection, it may be mentioned that oil-free compressors using graphite piston rings have been developed for low-pressure service (*i.e.*, up to about

300 lb./sq. in.). Attempts have been made to adapt the principle to machines for higher pressures, but they have met with only partial success. Such a compressor would be useful for compressing oxygen into storage cylinders, since no oil whatever can be tolerated when compressing oxygen. The present practice is to lubricate the cylinders with water or a soap solution, but this is not too satisfactory.

Relief valves and pressure gages on every stage are essential for operating control and safety. A high-pressure receiver or storage vessel is usually desirable to smooth out the pulsations and give a steady pressure.

Compressor cylinders, if provided with pistons (*i.e.*, rather than plungers), must be lubricated, and the same is true of stuffingboxes. This is done by means of small hydraulic pumps driven from the compressor shaft or piston rod. The lubrication of the stuffingboxes through lantern glands is also important in maintaining gastight seals.

Many designs of valves are used, but all have one common feature, especially in the lower stages, namely, that they are of the flat-plate type giving large area of opening for low lift. They generally consist of a series of rings covering several annular ports or rectangular plates covering ports of the same shape.

In a high-pressure catalytic process such as ammonia or alcohol synthesis, the gas is recirculated to the catalyst converter after having been cooled and otherwise treated to recover the product. The circulating pumps for this service are essentially single-stage reciprocating booster compressors operating at a small relative pressure difference, usually from 5 to 20 per cent of the absolute pressure.

In designing a high-pressure compresssor, one should take account of the deviation of the gas from the ideal gas law, which is considerable for a gas such as CO_2, even at pressures of only 1000 lb./sq. in., and is appreciable for the more permanent gases like nitrogen and hydrogen at pressures above 300 atm. Thus, for N_2 at 1000 atm. and room temperature, the actual volume is about twice the ideal, and calculations based on the assumption of an ideal gas would be seriously in error. For some specific examples, see Newitt ["The Design of High Pressure Plant and the Properties of Fluids at High Pressures," Oxford, New York, 1940; The Design of Vessels to Withstand High Internal Pressures, *Trans. Inst. Chem. Engrs. (London)*, **14**, 85 (1936)] and Kleinschmidt [*Trans. Am. Inst. Chem. Engrs.*, **29**, 88 (1933)].

SAFETY

When reasonable precautions are taken, high-pressure plants can be made as safe as any. It seems worth while to enumerate a few hazards peculiar to work at high pressures, and attention should be called to some rules.

1. At least one fatality and several accidents are known to have resulted from split Bourdon tubes. The fatality is believed to have been the result of a rush of compressed inflammable gas into a Bourdon tube containing air. The minor explosion that occurred split the tube, and a particle flying from the gage case was responsible for the death of a man. To permit the escape of gases, the cases of Bourdon gages should be provided with large vent openings covered with paper or foil for dust protection. Whenever possible the gage glass should be removed and the face left open; or, for dust protection, the glass should be replaced with thick transparent sheeting or with safety glass so that the danger from flying particles will be avoided. Bourdon gages should be placed above eye level. In some laboratories, it is required that Bourdon gages be placed behind a barrier and read only by their reflection in mirrors so that the observer is at all times out of the direct line of danger. The usual procedures should be followed, such as the use of oil seals to prevent corrosion, throttling down to prevent fatigue, and the use of built-in checks to prevent the pressure from rising or falling too suddenly.

2. With flammable gases at high pressures, the Armstrong effect is a possible source of trouble: when a gas containing finely divided liquid or solid particles passes at high velocity over an insulated metallic object, that object becomes electrically charged. The charge may become sufficiently great to cause a spark, which in turn ignites the gas.

3. When metallic containers rupture, frictional effects may cause very high local temperatures and the ignition of flammable gases.

4. It is rarely possible to cool a direct-fired vessel and its setting in a hurry. Such apparatus should be followed by a by-pass leading to the atmosphere so that, if a failure or a plug occurs farther along in the process, the by-pass can be opened and the material kept running through the direct-fired apparatus while it cools.

5. Where it is not possible to use a self-seating safety valve, it is sometimes possible to have two safety valves, one set slightly above the operating pressure with a shutoff valve between it and the body to be protected, and the other set somewhat higher and without the shutoff valve. When the lower safety valve releases, it can be reseated with the shutoff valve closed, while close observation and the higher safety valve provide the necessary security.

6. A variety of pressure relief that has met with favor because of its positive action and because it cannot readily be prevented from blowing off when the pressure exceeds a certain value is the rupture disk, which consists of a plate held in place over an opening by means of flanges. When the pressure rises too high, the disk ruptures. The disadvantages of the method are that the entire contents of the pressure system are discharged and that corrosion may so weaken the disk that, unless frequently inspected and renewed, it may burst at too low a pressure; on the other hand, since the working pressure produces stresses close to the yielding stress of the material of the disks, they tend to become strain-hardened with continued use and may then fail to relieve the pressure at the desired point.

It should be noted that, because of the usual relation between yield strength and ultimate bursting strength, a rupture disk is generally designed to relieve at about 1.5 times the working pressure when the latter is steady or at 1.75 to 2.0 times for fluctuating or pulsating pressure. However these ratios may vary appreciably depending on conditions. In one case at 1000 atm. the ratios were 1.30 and 1.40, respectively. For details on the calculation, characteristics, and uses of rupture disks, reference is made to the paper by Murphy (*Chem. & Met. Eng.*, November, 1944, p. 108; December, 1944, p. 99).

7. Carbon monoxide rapidly attacks pure nickel within a limited temperature range forming nickel carbonyl, and even attacks iron, though much more slowly. When carbon monoxide is to be used, the proper selection of alloys is important. It is also necessary to warn against the use of mercury in any apparatus in which it can come in contact with copper, brass, or other metals with which it can amalgamate. The general belief that mercury does not wet steel or form alloys with it is incorrect; Bridgman found that mercury can wet surfaces made by breaking steel in mercury. This is of great importance where there is any possibility of a crack in a mercury container opening slightly under pressure, allowing the mercury to wet the steel and start to undermine the walls. Above about 5000 atm., mercury will penetrate steel and eventually will cause rupture even without any initial cracks.

8. Some catalysts, notably very active nickel or iron

powders, are highly pyrophoric. Where there might be a danger of their leaking through valves or stuffingboxes, drying out, and glowing, it is desirable to provide a constant stream of water to wash them away so that any inflammable vapors or gases that accompany them may not catch fire.

9. The larger and hotter pieces of high-pressure equipment may be placed to advantage behind heavy barricades; for, in case of failure, heavy walls and light roofs give protection to the surroundings. At the same time it should be urged that anything permitting the formation of gas pockets is to be avoided; the buildings in which high-pressure apparatus is housed should permit the free passage of air through all parts, especially up under the roof. Where poisonous or flammable gases are used, additional protection for the operators should be provided by the installation of forced ventilation.

Some prefer work screens of heavy rope suspended from supports as protection against flying pieces of equipment. These have the advantage that the equipment behind is easier to get at and service. Tubing should be anchored securely at frequent intervals to prevent whipping in case of a break.

10. Since leaks are a serious fire or health hazard, they should be repaired as promptly as possible, especially since erosion quickly makes the leak worse and may cut into the metal so severely as to require replacement of a portion of the apparatus.

11. Welded apparatus for low-temperature service should be carefully constructed and properly heat-treated. The selection of appropriate construction materials for the temperature range involved is of primary importance, and the proper heat-treatment to develop weld ductility should be proved by notch impact tests on samples of the same material welded in the identical manner to be employed in the high-pressure equipment. Several split welds in high-pressure equipment cooled with liquid air have shown how serious a cause of accident this might be.

12. Oxygen cannot be compressed with safety in the presence of oil [Hersey, A Study of the Oxygen-oil Explosion Hazard, *J. Am. Soc. Naval Engrs.*, **36**, 231 (1924)], so that water must be used as a lubricant. It has been found that, when oxygen is admitted rapidly at high pressures into a space containing a bit of oily material, explosions may ensue. The explosion in the Bourdon tube mentioned above may have been of this nature. Every reason indicates that pressure should always be built up slowly.

13. A word of caution is needed because of a common factory usage according to which *oxygen* is called *air*. Since compressed air may be used with oil at pressures much higher than are safe with oxygen, the substitution of oxygen when compressed air was desired has been the cause of accidents.

14. The periodic inspection of high-pressure equipment is a very real necessity and is a matter of routine in all high-pressure plants. Apparatus should be constructed to permit inspection of the inside surfaces and measurement of the outer dimensions. Dimensions should be recorded in such a way that creep or deformation will be easily detected. Sections of piping should be removed and inspected for changes in dimensions and for the growth of longitudinal cracks formed in drawing. High-pressure equipment should be initially and then periodically given a hydrostatic test at 1.5 to 2 times the working pressure.

15. Too rapid rise in either pressure or temperature should be avoided. As shown in a previous section, large temperature gradients in the wall of a vessel lead to high stress.

16. Pressure joints except the glands of stuffingboxes should not be tightened under pressure.

17. Safety valves and rupture disks should be piped so that the discharge is removed to a place where it can do no harm. The discharge line should be of sufficient size so that no serious interference with free discharge occurs.

18. It is always desirable to have more than one person present in a given work space when poisonous gases are being handled. It may be noted that canary birds are useful detectors of poisonous gases, being more susceptible to low concentrations of them than human beings.

19. Gas-storage systems must be carefully designed and supervised to avoid the production of unknown and potentially dangerous mixtures. Analyses should be checked frequently; and one must always be sure of the composition of a gas before using it.

SECTION 19
PROCESS CONTROL*

BY

Richard W. Porter, B.S., Editorial Director, *The Paper Industry.* Member, American Chemical Society, Technical Association of the Pulp and Paper Industry (Chairman, Mill Instrument Control Committee), Instrument Society of America.

Douglas M. Considine, B.S., Manager Technical Section, Sales Department, Brown Instruments Division, Minneapolis-Honeywell Regulator Company. Member, American Institute of Chemical Engineers, American Chemical Society.

CONTENTS

PAGE

Process Instrumentation 1265
Instruments and Their Functions 1265
Terminology 1266
Process Variables and Their Measurement 1266
Elements of Measurement 1266
Indicating and Recording 1267
Indicating Devices 1267
Recording Mechanisms 1267
Types of Records 1267
Signaling 1268
Remote Indicating and Recording 1268
Temperature Measurement 1269
Temperature Scales 1269
Temperature Measurement Based on Physical Phenomena 1270
Bimetallic Thermometers 1270
Liquid-in-glass Thermometers 1270
Pressure-filled Expansion Thermometers 1270
Seger Cones 1272
Temperature Measurement Based on Electrical Phenomena 1272
Thermocouples 1272
Thermocouple Selection 1273
Protecting Tubes 1273
Resistance Thermometers 1274
Radiation Pyrometry 1274
Optical Pyrometers 1275
Radiation Pyrometers 1275
Electric Measuring Instruments 1276
Millivoltmeters 1276
Galvanometers 1276
Potentiometers 1276
Null-balance Potentiometers 1277
Deflection Potentiometers 1277
Self-balancing Potentiometers 1277
Pressure and Vacuum Measurement 1279
Pressure Gages 1279
Manometers 1279
Liquid-sealed Bell 1279
Bourdon Tube 1279
Spiral Element 1279
Helix 1279
Bellows 1280
Volumetric Pressure Gage 1280
Pulsation Dampeners 1280
High Vacuum 1280
McLeod Gage 1280

* First Edition, this section prepared by Henry L. Young, B.S., Aridye Corp., and Theodore R. Olive, A.B., Associate Editor, *Chemical Engineering;* Member, American Society Mechanical Engineers.

Second Edition, this section revised by Henry S. Winnicki, B.S., Chemical Engineer, Westvaco Chemical Division, Food Machinery Co.; Junior Member, American Institute of Chemical Engineers.

Special acknowledgment is made to Theodore R. Olive, Associate Editor, *Chemical Engineering,* for his help and advice in planning and preparing this section. Acknowledgment is made to Victor F. Hansen, E. I. duPont de Nemours & Co., and Ralph H. Munch, Monsanto Chemical Co., for reviewing and commenting on the manuscript.

PAGE

Thermal Gages 1281
Ionization Gage 1281
Flow Measurement 1282
Positive-displacement Meters 1282
Rate-of-flow Meters 1283
Orifice Plates 1283
Location of Pressure Taps 1284
Flow Nozzles 1285
Venturi Tubes 1285
Installation of Orifices 1285
Differential Flowmeters 1285
Mercury-float Manometer 1285
Floating-bell Manometer 1286
Ring-balance Manometer 1286
Bellows-type Meter 1286
Electric Meters 1286
Flow Integrators 1287
Installation of Differential Meters 1287
Seals and Purges 1287
Pitot Tubes 1287
Weir Meters 1288
Area Meters 1288
Rotameters 1288
Piston and Gate Meters 1288
Current Meters 1288
Liquid-level Measurement 1289
Buoyancy 1289
Hydrostatic Pressure 1289
Differential Pressure 1289
Differential Temperature 1290
Electric Methods 1290
Level of Solids 1290
Weighing and Weight Control 1291
Weighing and Feeding Devices 1291
Scales for Basic Weighing 1291
Automatic Weights 1292
Pivoted Belt Feeders 1292
Loss-in-weight Feeders 1293
Weigh Batching Systems 1293
Conveyor Batching Scales 1294
Continuous Weigh Checking 1294
Totalizing Conveyor Scales 1294
Miscellaneous Quantity and Rate Variables 1295
Thickness 1295
Speed 1295
Measurement of Physical and Chemical Characteristics 1296
Density, Specific Gravity 1296
Liquid Density 1297
Hydrometers 1297
Density of Gases 1297
Gas-density Balance 1297
Humidity 1298
Relative Humidity 1298
Hygrometers 1298
Psychrometers 1298
Absolute Humidity 1299
Dew-point Recorder 1299
Moisture Content of Solids 1300
Viscosity and Consistency 1300

	Page
Viscosity of Fluids	1301
Consistency Control	1301
Calorific Value and Combustion	1301
Color	1302
Color Comparator	1302
Smoke and Fume Detector	1302
Electrical Conductivity	1302
Thermal Conductivity	1303
Chemical Absorption	1303
Orsat Analyzer	1304
Automatic Orsat	1304
X-ray Diffraction	1304
Principles	1304
Geiger-Muller Counter Tube	1304
Ultraviolet and Infrared Spectrometry	1305
Principles	1305
Ultraviolet Spectrometers	1305
Infrared Spectrometer	1306
Emission Spectrometry	1306
Mass Spectrometry	1306
Mass Spectrometer	1306
Leak Detector	1307
Polarography	1307
Dropping-mercury Electrodes	1307
Hydrogen-ion Concentration	1307
pH Scale	1308
pH vs. Normality	1308
Buffer Action	1308
Measuring Electrodes	1309
Reference Electrodes	1309
Recording and Controlling pH	1309
Oxidation-reduction Potential	1309
Fundamentals of Automatic Control	1309
Elements of Process Cotrol	1310
Elements of Automatic Controllers	1310
Operation of Automatic Controllers	1310
Automatic Control Characteristics	1311
Process-control Characteristics	1312
Process Lags	1312
Capacity Lag	1313
Transfer Lag	1313
Dead Time	1314
Controller Lags	1314
Self Regulation	1314
Types of Automatic Control Action	1315
Two-position	1315
Proportional Position	1316
Floating	1317
Proportional Plus Reset	1317
Proportional Plus Rate	1319
Proportional Plus Reset Plus Rate	1319
Automatic Control Mechanisms	1320
Pneumatically Operated Controllers	1320
Electrical Controllers	1323
Telemetering and Remote Transmission	1324
Pneumatic	1324
Hydraulic	1324
Electric	1325
Ratio Controllers	1325
Cycle Controllers	1326
Final Control Elements	1326
Control Elements	1326
Control Valves	1326
Valve Characteristics	1327
Valve Selection and Installation	1327
Rotary Valves	1328
Butterfly Valves and Dampers	1328
Louvers	1328
Gate Valves and Slide Dampers	1328
Power Units	1328
Pneumatic Diaphragm Motor	1328
Power Cylinder	1328
Valve Positioner	1329
Solenoids	1329
Electric Motors	1329
Electric Proportioning Motors	1329
Automatic-control-system Applications	1329
Maintenance and Supervision of Instruments	1336
Instrument Service Departments	1336
Classified List of A.S.M.E. Automatic-control Terms	1337
Glossary of Automatic-control Terms	1337

PROCESS CONTROL

REFERENCES: Rhodes, "Industrial Instruments for Measurement and Control," McGraw-Hill, New York, 1941. Olive, Measurement and Control of Process Variables, *Chem. & Met. Eng.*, May, 1943, pp. 97–144. Eckman, "Principles of Industrial Process Control," Wiley, New York, 1945. Smith, "Automatic Control Engineering," McGraw-Hill, New York, 1944. American Institute of Physics, Symposium, "Temperature, Its Measurement and Control in Science and Industry," Reinhold, New York, 1941. Gess and Irwin, "Flow Meter Engineering Handbook," Brown Instrument Co., 1946. "Flow Meter Engineering," Foxboro Co., 1945. Considine, "Industrial Weighing," Reinhold, New York, 1948. Werey, "Instrumentation and Automatic Control in the Oil Refining Industry," Brown Instrument Co., 1941 (out of print). Wood and Cork, "Pyrometry," 2d ed., McGraw-Hill, New York, 1941. Miller, "Lectures on Instrumentation," Department of Chemistry and Chemical Engineering, Case School of Applied Science, Cleveland, 1941. The New York State Vocational and Practical Arts Association, "Instruments and Process Control," Delmar Publishers, Inc., Albany, N.Y., 1945.

Special reference is made to the review by Theodore R. Olive entitled Instruments for Measuring and Controlling Process Variables, *Chem. & Met. Eng.*, May, 1943, pp. 108. Both text material and drawings from this article have been used freely in the preparation of this section of the handbook.

An extensive bibliography on this subject appears in Smith, "Automatic Control Engineering," pp. 348–359, McGraw-Hill, New York, 1944. It includes early papers as far back as 1868 on the subject and extends partly through 1944. Two engineering societies are extremely active in this field. First is the A.S.M.E. Industrial Instruments and Regulators Division, papers of which are usually published in the society's transactions. More recently formed (1945) is the Instrument Society of America. Official organ for I.S.A. is the monthly periodical, *Instruments*, Instruments Publishing Co., Pittsburgh, Pa. Other periodicals devoted solely to this subject are the *Review of Scientific Instruments*, American Institute of Physics, New York, and *Instrumentation*, Brown Instruments Divn., Philadelphia, Pa., and *Taylor Technology*, Taylor Instrument Co., Rochester, N. Y.

PROCESS INSTRUMENTATION

All chemical-engineering operations depend on the measurement and control of process variables. Instrumentation has come to be an integral part of industrial processes, and instruments have ceased to be regarded as auxiliary equipment. Automatic control is one of the cornerstones of continuous processes, and the development of such processes and of such control have largely gone hand in hand. In fact, process control is often classed as a unit operation, and a most important one; for without reliable control methods, manual or automatic, process industries could not operate. Although batch processes can sometimes be operated with a minimum of instruments for the purpose of guiding operators, it must be emphasized that operation of many modern continuous processes would be practically impossible without adequate instrumentation. Continuous processes such as petroleum refining require that each step of the process be carried out under closely controlled conditions at all times. Instrumentation in the process industries, therefore, can no longer be regarded as a convenience; rather, it must be considered an absolute necessity.

Although a number of present chemical processes cannot be operated without instruments, every properly designed application will result in improved operation. Foremost is the improvement of product quality and uniformity. Increased interest in direct measurement of product quality has led to large-scale industrial application of methods previously utilized only in the laboratory. Direct cost reduction from the use of automatic-control instruments results from savings in manual labor. Increased equipment capacity results from uniformity of flow and uniform control of process variables. Elimination of surges enables the equipment to be operated nearer to its ultimate capacity and for a given throughput permits smaller process equipment to be used.

In the final analysis, then, the control of a chemical process or unit operation is a problem of engineering economics. Each part of the investment in process equipment, including instruments, must be justified in terms of its contribution to producing a salable product at a profit. Therefore, a reasonable investment-cost balance must be found between complete instrumentation, which improves both product quality and operating efficiency, and a neglect of automatic control, which leads to constant loss from operating troubles and degraded products. This balance is best attained where development and design engineers understand the fundamentals of instrumentation and process controllability. Thus, from research and development through process design to plant operation, it is the chemical engineer's responsibility to know intimately the basic tenets of measurement and control.

Instruments and Their Functions. Industrial instruments fall into two general classes, namely, (1) those which measure and (2) those which measure and control. A measuring instrument may be used by itself, or it may be combined with a control device to form an automatic controller. Although a majority of industrial instruments are used for control purposes, they are not all automatic, since many measuring instruments are employed primarily as a guide for manual control. Recording instruments are utilized as a check on operations and, in the case of flowmeters and weighing devices, are often used for determining energy and material balances.

Different measuring instruments must be employed for different process variables as well as for different conditions and ranges of values of a given variable. Thus, for measuring temperature, over a range from −330° to 5000°F., there are a number of radically different measuring instruments such as radiation pyrometers, resistance thermometers, and pressure-type thermometers. Included among the variables that require individual types of measuring means are temperature, pressure, fluid flow, pH, specific gravity, viscosity, absorption, and a number of others.

On the other hand, the same controller mechanisms may be applied to control almost any variable automatically. Since a controller is motivated only by change or deviation of a process variable, as expressed by a measuring element, it follows that the exact nature of neither the variable nor the measuring instrument makes any significant difference to the controller mechanism. However, a wide variety of controller mechanisms are necessary to meet the requirements of various process characteristics. Different types of control action can thus be applied to the same variable or to different variables, depending on the exact nature of the process and the equipment in which the process is carried out.

Not only do industrial instruments measure and control process variables, but they also have a number of other functions. In addition to producing a measurement that may not be visible in the case of some controllers, they may indicate, record, and sometimes totalize the measurement. Sometimes they control at a particular value of a variable, in other cases they alter the control point according to a definite time schedule. They may be used to start various parts of the process at definite times and to control the duration of various operations. If desired, instruments can issue a warning or shut down the process in case a dangerous or otherwise undesirable condition is arising. They may be used to control one variable in a definite relation to another. Finally, instruments are available for transmitting indications and control impulses over considerable distances without undue lag or loss of accuracy.

Terminology. In general, measurement of variables rests on well-known physical and chemical laws and can be readily expressed in mathematical terms. Measuring devices have been used for a long time, and the principles involved are fairly well understood. On the contrary, the mathematical relationships involved in process control are complex and in some cases not fully developed. Automatic process control has been largely developed in the past twenty-five years; and, despite the fact that it involves the use of a large number of intricate and highly developed mechanisms, its application to industrial processes has often rested on an empirical basis. One of the detriments to ready understanding of automatic-control principles has been the extensive terminology that grew larger and more complex as more complicated control mechanisms were developed.

Perhaps the most significant advance in simplifying and standardizing automatic-control terminology was the publication of a list of classified instrument terms and definitions prepared and adopted by the A.S.M.E. Industrial Instruments and Regulators Division Committee on Terminology. This list includes most of the terms used to describe automatic process control and defines the various elements and characteristics involved in process instrumentation. Since a uniform terminology is important, the classified list of A.S.M.E. standard terms is given at the end of this section, p. 1337, together with a glossary that defines both A.S.M.E. terms and many not included in this list. The glossary should be especially helpful when reference is made to the relatively large volume of published literature in which a wide variety of terms has been used. In general, an attempt has been made to use the standard terminology, wherever possible, in this section.

PROCESS VARIABLES AND THEIR MEASUREMENT

A *process variable* is any condition or state of the process material or of its environment that is subject to change. In determining how a process shall be controlled, it is important to isolate all the process variables that will be encountered and to determine which are the independent variables and which will influence the process results enough to require control. Certain variables inherently will remain within suitable limits and so need not be controlled. Some will be found to be dependent on others fluctuating in definite relation to other variables or groups of variables and will not require independent control, provided that the variables on which they are dependent are themselves controlled. The measured variable of the process is usually not an end in itself but is an indication of the state of balance of the process and sometimes is just an indication of a reaction rate within the process. Measurement of the variable is the basis for control action. Consequently it is necessary to determine whether or not the measured variable actually represents the condition of balance of the process. Since the purpose of the measuring element is to detect any change or deviation in the controlled variable, the measured value of the variable must have a definite relation to the state of the process.

Process variables may be classified in a number of different ways. One of the most convenient is on the basis of whether they are affected by (1) the *energy state* of the material, (2) the *quantity or flow-rate* relations of the several materials in the process, or (3) the *physical and chemical characteristics* of the material. These include the following:

Energy variables.
- Temperature.
- Pressure and vacuum.
- Electricity.
- Sound.
- Radiation.

Quantity and rate variables.
- Fluid flow.
- Liquid level.
- Weight.
- Thickness.
- Speed.

Physical and chemical characteristics.
- Density and specific gravity.
- Humidity.
- Moisture content of solids.
- Viscosity and consistency.
- Calorific value and combustion.
- Color.
- Electrical conductivity.
- Thermal conductivity.
- Chemical absorption.
- Refractive index.
- X-ray diffraction.
- Ultraviolet and infrared absorption.
- Emission spectrum.
- Mass spectrum.
- Polarity.
- Hydrogen-ion concentration.
- Oxidation-reduction potential.

Elements of Measurement

A measuring instrument has three functional elements, namely, (1) a *primary element* such as a pH electrode, thermometer bulb, or orifice plate to detect changes in the magnitude of the controlled variable; (2) a *transmitting means* such as a capillary tube, lead wire, or piping for connecting the detecting element to (3) the *measuring element* such as a bourdon tube, manometer, or potentiometer. A measuring instrument may be self-operated or power-operated. A pressure thermometer is self-operated and utilizes the energy developed by the thermal system to indicate, record, or motivate a control mechanism. A potentiometer, however, is power-operated in that it utilizes an auxiliary source of power to amplify the output of a thermocouple or other primary element in order to indicate, record, or operate a controller.

One of the most important requisites for good control, regardless of the form of measurement, is *sensitivity* and responsiveness of the primary element. If a thermocouple or bulb of a temperature instrument is slow in its pick-up of thermal changes, because of either poor design, poor construction, excessive mass, excessive lagging, or improper location, good control will be difficult to obtain. Sluggishness of response within the instrument proper can also cause control trouble. The pen positioning and recording mechanism should be as fast as the detecting element. Regardless of the type of primary indication, whether it be temperature, pressure, flow, or any of the other functions, the same precautions relating to responsiveness of the primary element are applicable. Measurement of the variable is the basis for control action, since the response of the controller depends on the accurate, rapid detection of changes. No controller can be better than its measuring system.

Indicating and Recording

Before discussing the various mechanisms for measuring and controlling process variables, it would be well to cover briefly certain functions common to all instruments. Except for self-operated and other non-indicating controllers, most instruments indicate or record the value of the particular variable measured. In fact, the purpose of industrial instruments is to indicate, record, or control many of the variables encountered in manufacturing processes. Specific applications have individual requirements; *e.g.*, a simple indicator may serve efficiently the needs of one application, while another application may simultaneously require continuous indication, a permanent record, and automatic control.

Indicating Devices. The simplest indicating devices are the *sight-gage glass*, the *mercury-in-glass thermometer*, and the *glass manometer*. Inclined-tube manometers are used to measure and indicate small pressures, and glass rotameters are widely used to indicate flow. Essential to any indicator is a *calibrated scale* over which a *pointer* may travel to indicate the instantaneous value of the measured process variable. The pointer is usually attached by linkages or by rack and pinion to the measuring element. Scales may be straight in the case of a strip-chart instrument, or they may be concentric or eccentric depending on the particular instrument design. Where exceptional accuracy is required on potentiometer-type instruments a *vernier*-type scale may be used. Here the moving scale is linked to the measuring element.

Recording Mechanisms. As with indicators, it is possible to position a *pen* or other recording device on a moving paper *chart* to plot a continuous record of the fluctuations of the variable against time. Recording devices are comprised of a pen or marking device, a chart, and a unit to drive the chart. In most round-chart instruments, the pen is positioned by a mechanical linkage, as in Fig. 1. The chart drive may be a spring-

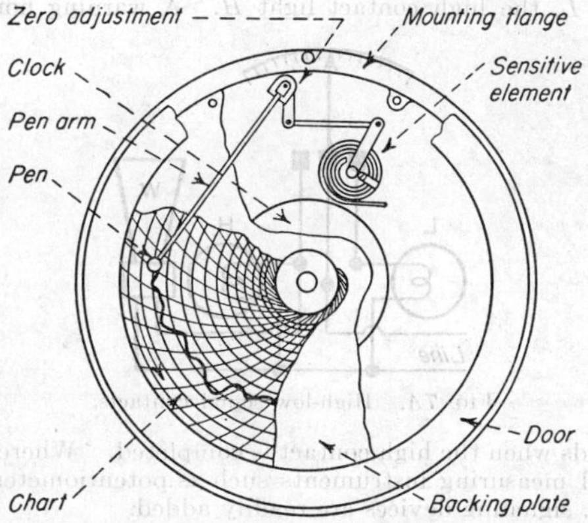

FIG. 1. Round chart recorder.

wound clock or an electric clock. *Round-chart recorders* may be either single or multi-record; and, depending on the instrument, the charts usually vary in size from 8 to 12 in. Usually not more than three records can be put on an individual chart, because standard-size instrument cases will not usually accommodate more than three measuring systems. *Strip-chart* instruments, however, may record from one to over a hundred different temperatures. Chart speeds as high as 120 in./hr. may be used on instruments that record a large number of values. Fast switching action too is necessary in this sort of application, since the instrument is switched successively to each of the primary elements.

Relatively few mechanisms are widely used compared with the many that have been developed. Earliest instrument records were made by tracing a needle on a smoked plate. Wax engravings, pin pricks, spark records, sensitized papers, and various other means of recording all have their place but are seldom if ever used in industrial plant instruments. Where records are

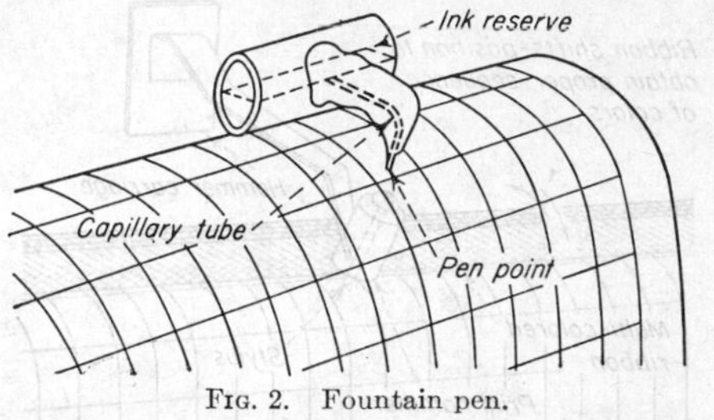

FIG. 2. Fountain pen.

not kept on file, round-chart instruments may use cleanable plastic charts instead of paper.

Types of Recorders. Instrument recordings can be classified as (1) fluid-ink records as produced by a pen constantly in contact with a chart and (2) a dotted or printed record formed by the periodic impression of an inked ribbon against the chart surface. Three common types of pen used to obtain fluid-ink records are the fountain pen, V-pen, and bucket pen. The fountain pen is used on strip-chart instruments, and the V-pen and bucket pen are used on circular-chart instruments. With the *fountain pen*, illustrated in Fig. 2, a 30-day

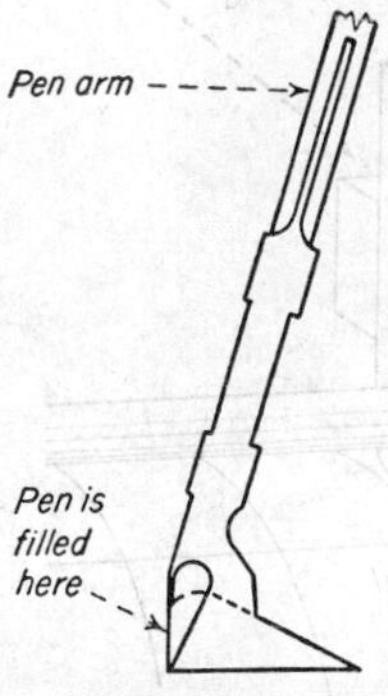

FIG. 3. V-pen.

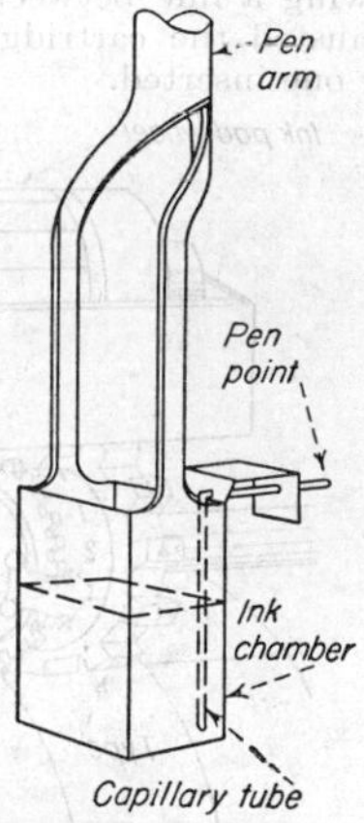

FIG. 4. Bucket pen.

supply of ink may be held in a glass reservoir. A thin capillary glass tube runs from the reservoir to a metal point, constantly feeding the point with a fresh ink supply. The pen, mounted in a carriage, is moved across the chart by means of a drive shaft. The *V-pen* (see Fig. 3) is similar to a common steel writing pen, except that a chamber is provided for a moderate supply of ink. The *bucket pen*, shown in Fig. 4, contains a larger reservoir for ink and operates by capillary action, similar to the fountain pen previously described.

Where cycling of the pen exists, especially in flow and pressure installations, there is a tendency for the ink to produce a blotty record. In some cases, this effect can be eliminated by the use of a faster chart speed. It is also desirable to eliminate the pulsating effect by the use of a damping means in the instrument proper. There are three main causes for pen replacement, namely, (1) wear, (2) corrosion, and (3) damage. The wear on a pen is usually noticed by the production of a broader record. The pen edges should be inspected frequently for wear. Corrosion is usually evidenced by a rough pen edge and by undue clogging of the pen. Under normal use and with reasonable care, pens should require replacement only at infrequent intervals.

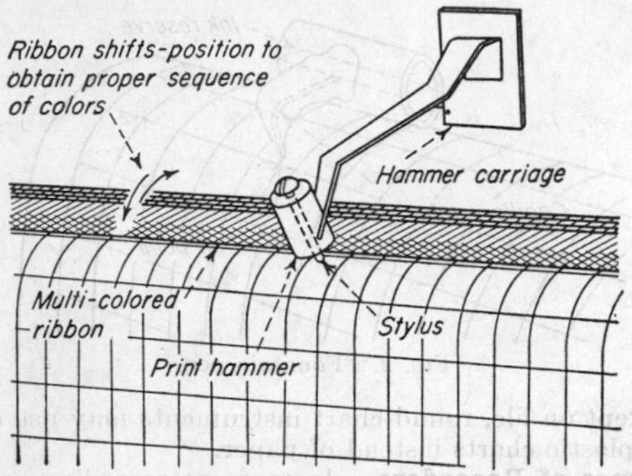

FIG. 5. Print hammer.

More recently the ball-point pen has come into use and is often recommended for recording rapidly fluctuating variables, especially when the lines drawn are closely spaced. The ball-point pen draws a fine line that dries immediately and that does not smear even when successive lines overlap. In contrast to the standard glass fountain pen, the ball-point pen never floods nor cuts through the chart. The ball-point pen is capable of drawing a line between 3000 and 5000 ft. long. When exhausted the cartridge can be readily removed and a new one inserted.

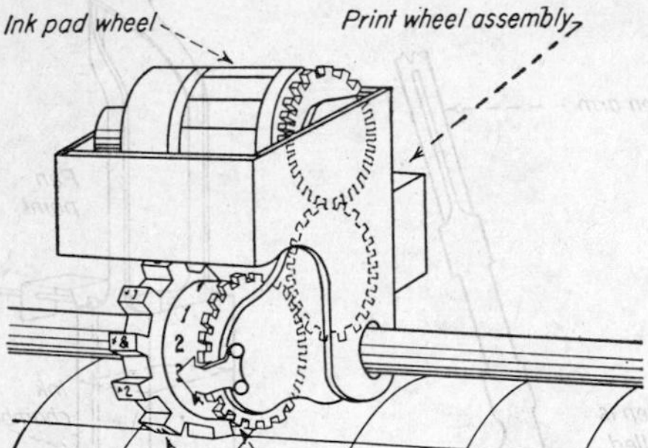

FIG. 6. Print-wheel mechanism.

Devices for effecting printed records take three basic forms, namely, the print hammer, print wheel, and depressor-bar printer. All are restricted to use on strip-chart instruments. The *print hammer* (Fig. 5) consists essentially of a hammer, with stylus on one end, mounted on an arm, which in turn is attached to the hammer carriage. This carriage is moved up and down scale by means of the drive shaft. A multicolored ribbon is employed so that records of different colors can be obtained. Records are produced by the impact of the print hammer on the ribbon proper. The *print-wheel* mechanism (Fig. 6) consists essentially of two parts, namely, the print wheel proper, and the ink-pad wheel. the two being connected by means of a gear so that the proper synchronism between type and color will be obtained. The print-wheel assembly rides up and down scale on a drive shaft, similar to the print hammer. Recordings are made periodically; and, each time a record is made, the proper type on the wheel comes in contact with the ink pads. Print wheels may operate on a fixed cycle or, with high-speed instruments, may be synchronized with the balancing mechanism. The *depressor-bar printer* (Fig. 7) is peculiar to recording millivoltmeters. This mechanism comprises a moving galvanometer pointer;

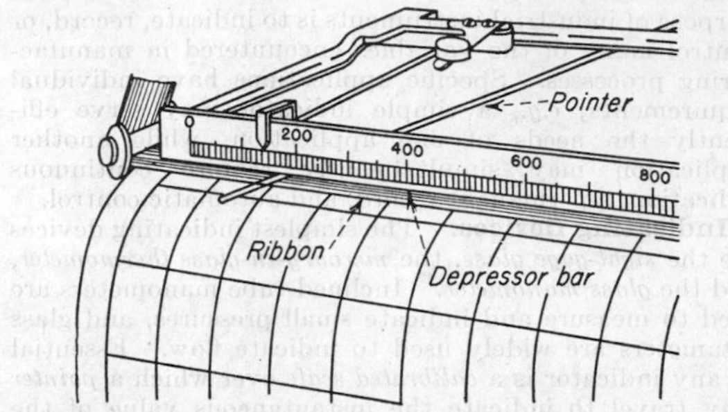

FIG. 7. Depressor-bar printer.

an inked ribbon, which rides between the pointer and the chart; and a depressor bar, which periodically clamps the pointer and presses it against the inked ribbon to produce a mark on the chart.

Signaling. In many processes where automatic control of a given variable is not desired, it is still necessary to warn the operator if the variable threatens to go beyond acceptable limits or to assume dangerous proportions. In such cases it is a simple matter to provide electrical contacts on an indicating or recording instrument to illuminate lights that show high, intermediate, or low values of the variable, to sound a warning signal, or to shut down the process in case of danger. Figure 7*A* shows the addition of signaling and alarm contacts to an indicating pointer. Here the low contact illuminates light *L*, the high contact light *H*. A warning horn *W*

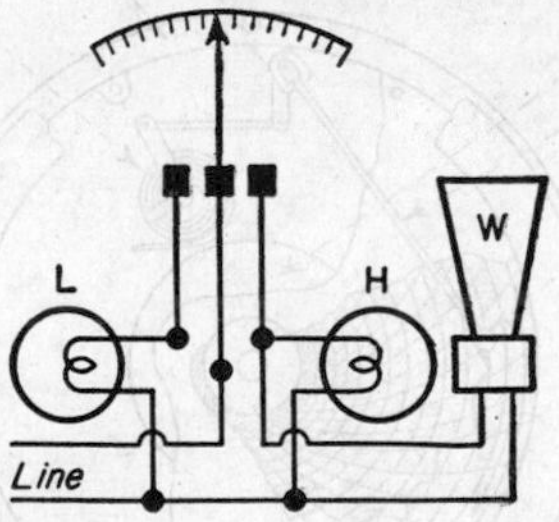

FIG. 7*A*. High-low signal contacts.

sounds when the high contact is completed. Where electrical measuring instruments such as potentiometers are used, signaling devices are readily added.

Remote Indicating and Recording. It is often desirable to produce or reproduce the indication or record of a measuring unit at a distance. A definite trend in modern instrumentation practice is to concentrate on a central panel board all the instruments and controls involved in a chemical process. In many cases this is a practical necessity, because the instrument itself may be located in an inaccessible position such as at the top of a fractionating tower or in a hazardous area. Such an arrangement makes it possible for an operator to check his whole process in a fraction of the time required to go from one instrument to another when they are scattered throughout the plant. Most types of industrial measuring instruments may be located a moderate distance from the primary detecting element, but this is often not sufficient. Therefore, a number of systems have been developed for this purpose; but, since they are quite closely related to control mechanisms, they are subsequently described under Telemetering and Remote Transmission (p. 1324).

TEMPERATURE MEASUREMENT

Temperature is an important factor in processing, because many properties of the substances that enter processes are vitally affected by temperature. Most substances, for example, change their physical state, *i.e.*, from solids to liquids, or from liquids to vapors, and vice versa, at precise temperatures. Advantage of these physical phenomena is taken in processing wherein freezing or boiling points form the basis for separating materials. The rate of chemical reaction between substances is also very dependent upon temperature, the rate of reaction increasing with an increasing temperature. As a rule, in many homogeneous systems, the rate increases two or threefold per each 10°C. rise in temperature. Solubility of solids in liquids and of gases in liquids is also dependent upon temperature. In general, the solubility of solids increases with an increasing temperature, and the solubility of gases in liquids increases with a decreasing temperature.

Temperature is a thermopotential, comparable with a pressure head or an electrical voltage. It is the condition of matter that determines the flow of heat between bodies. In conjunction with specific heat, it is a measure of the amount of heat energy contained in a material. Temperature cannot be measured directly but must be inferred from the property of the material or from those of another material in equilibrium with it. It may be inferred from the expansion of solids, liquids, or gases; from the vapor tension of a liquid; from the electrical resistance of materials, usually solids; from the intensity of the total radiation or of a particular band of wave length of radiation given off by the hot body; from the value of an e.m.f. created at the junction of two dissimilar metals; and from the changes of state of solids, liquids, or gases.

Temperature Scales. Temperature cannot be measured by absolute standards. It must be determined in relation to the temperature of standard bodies under conditions of temperature known to be constant and reproducible. Thus a body that will, when placed in the same environment, neither lose heat to nor gain heat from melting ice at atmospheric pressure is arbitrarily assigned a temperature of 0°C., 32°F., 0° Réaumur, 273° Kelvin (abs. °C.), or 492° Rankine (abs. °F. = °F. + 460°). Similarly, a body that will neither gain nor lose heat in contact with saturated steam at atmospheric pressure (760 mm. Hg) is assigned a temperature of 100°C., 212°F., 80°. Réaumur, 373° Kelvin, or 672° Rankine. The range of temperature between the lower and upper standard temperature in the various cases is 100°C., 180°F., 80° Réaumur, 100° Kelvin, and 180° Rankine. The degrees in each case are equal divisions in the entire range. Temperatures above or below the standard range are located by extrapolation.

Any given temperature on one scale is related to the corresponding temperature on any other scale by the fact that the range of units between the freezing and boiling temperatures (of water), although expressed differently on the various scales, refers to exactly the same temperature difference. On any scale the number of degrees from the freezing point to a given temperature, divided by the number of degrees from the freezing point to the boiling point, is equal to the corresponding ratio for the corresponding temperature range on any other scale. The relations C./100 = (F. − 32)/180 = Réaumur/80 = (Kelvin − 273)/100 = (Rankine − 492)/180 therefore constitute the fundamental and easiest method for converting temperatures on one scale to those of another.

Thermodynamic Temperature Scale. From a consideration of the part played by temperature in the theoretical Carnot cycle, Kelvin proposed a fundamental temperature scale that is independent of the properties of actual matter. On this scale the absolute temperature of the freezing and boiling points of pure water at a pressure of 1 atm. are proportional to the heat rejected and absorbed by a reversible thermodynamic engine operating between these temperatures, which are then in the ratio of 273 to 373 on the centigrade scale. No actual thermometer gives results that correspond exactly to this scale, although the *constant-volume gas thermometer* follows it very closely when used with the "permanent" gases such as helium, hydrogen, and nitrogen. By correcting for the known deviations of these gases from the laws of perfect gases, it is found that only small corrections are required and that these are generally insignificant for engineering purposes. The constant-volume gas ther-

Table 1. Fixed Points for Thermometer and Pyrometer Standardization

Substance	Phase change	Temperature	
		°C.	°F.
Helium	Melts	<−271	<−456
Hydrogen	Boils	−253	−423
Oxygen	Melts	−227	−377
Nitrogen	Boils	−196	−321
Oxygen	Boils	−183	−297
Isopentane	Melts	−160	−256
Methyl cyclohexane	Melts	−126	−195
Carbon bisulfide	Melts	−112	−170
Toluene	Melts	−95.0	−139.0
Carbon dioxide	Sublimes	−78.5	−109.3
Chloroform	Melts	−63.5	−82.3
Mercury	Melts	−38.9	−38.0
Carbon tetrachloride	Melts	−22.9	−9.2
Water	Melts	00.0	+32.0
Glauber's salt	Melts	+32.4	90.3
Acetylene dichloride	Boils	55.0	131.0
Ethyl alcohol	Boils	78.3	172.9
Water	Boils	100.0	212.0
Toluene	Boils	110.0	230.0
Chlorobenzene	Boils	132.0	269.6
Brombenzene	Boils	156.6	313.9
Aniline	Boils	184.5	364.1
Nitrobenzene	Boils	209.0	408.2
Tin	Melts	231.9	449.4
Diphenyl	Boils	254.6	490.3
Naphthol (α)	Boils	278.0	532.4
Diphenylamine	Boils	302.0	575.6
Lead	Melts	327.4	621.3
Mercury	Boils	357.3	675.1
Potassium dichromate	Melts	397.5	747.5
Zinc	Melts	419.4	786.9
Sulfur	Boils	444.6	832.3
Lead chloride	Melts	501.0	933.8
Calcium nitrate	Melts	561.0	1041.8
Antimony	Melts	630.0	1166.0
Aluminum	Melts	658.7	1217.7
Manganous sulfate	Melts	700.0	1292.0
Potassium chloride	Melts	770.3	1418.5
Sodium chloride	Melts	800.4	1472.7
Sodium carbonate	Melts	852.0	1565.6
Sodium sulfate	Melts	884.7	1624.5
Silver	Melts	960.5	1760.9
Gold	Melts	1063	1945
Potassium sulfate	Melts	1069	1956
Copper	Melts	1083	1981
Stannic oxide	Melts	1127	2061
Lithium silicate	Melts	1201	2194
Barium fluoride	Melts	1280	2336
Nickel	Melts	1452	2646
Cobalt	Melts	1480	2696
Iron	Melts	1530	2786
Palladium	Melts	1549	2820
Platinum	Melts	1755	3191
Alumina	Melts	2000	3632
Tungsten	Melts	3400	6152

mometer is used only as a fundamental or experimental instrument because of its inconvenience. Other more practical instruments have been devised and calibrated in terms of fixed points established by its use. A series of such fixed points appears in Table 1, p. 1269.

Two general classes of devices are employed to measure temperature, namely, (1) those depending on physical changes of a solid, liquid, or gas; and (2) those depending on electrical phenomena. In the first group is included the bimetallic or solid-expansion thermometer; and the liquid-in-glass and pressure-filled thermometers, which depend on fluid expansion. The latter is most commonly used in automatic control systems. Electrical devices include thermocouples, resistance thermometers, radiation pyrometers, and optical pyrometers. Temperature ranges normally covered by common measuring devices are as follows:

Measuring Instrument	Temperature Range, °F.
Bimetallic thermometers	0 to 1000
Mercury-in-glass thermometers	35 to + 750
Mercury-pressure thermometers	40 to +1000
Gas-pressure thermometers	− 125 to + 800
Vapor-pressure thermometers	− 40 to + 600
Resistance thermometers	− 330 to +1800
Base-metal thermocouples	0 to +2100
Noble-metal thermocouples	0 to +2900
Total-radiation pyrometers	0 to +3200
Optical pyrometers	+1200 to +5000

Temperature Measurement Based on Physical Phenomena

Bimetallic thermometers depend on the differential expansion of dissimilar metals. Thermometer bimetal consists of sheets of metal of relatively high and low coefficients of expansion such as brass, monel, or steel, and iron nickel-iron, or invar laminated by welding, brazing, soldering, or even riveting. It may be used in strip form or coiled into spirals or helixes. As the temperature of the bimetal changes, the system deflects to one side or the other. The motion may operate only an indicating device, or it may be used to actuate a control device. This type of unit is most commonly used in thermostats. Figure 8 represents a modern type of bimetallic thermom-

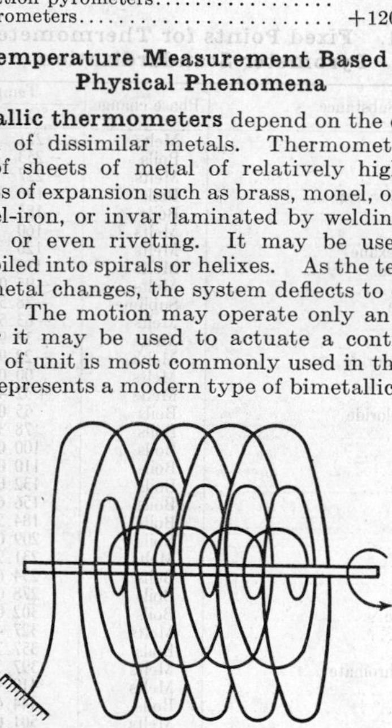

FIG. 8. Bimetallic thermometer.

eter (Weston) consisting of a multiple-layer helix in which a large deflection is secured with a light and sensitive bimetal because of its extreme length. One end of the coil is anchored, and the other end is attached to the pointer shaft. Ranges up to 1000°F. and accuracy of 1½ per cent are possible. Usually bimetals are shorter and heavier, resulting in greater lag and lower accuracy.

Liquid-in-glass thermometers responding to the temperature expansion of the liquid, usually mercury, have wide use. Mercury-filled glass thermometers are available for the range −35° to 750°F. Using special types of borosilicate glass, mercury thermometers, filled with nitrogen under pressure, are satisfactory up to temperatures as high as 1200°F. For temperatures below the freezing point of mercury, other liquids may be used, such as toluol, alcohol, or pentane. Glass thermometers are calibrated either for total immersion or for partial immersion. Total-immersion thermometers, when insufficiently immersed, will read low for temperatures above ambient and high for those below ambient. Stem correction in degrees is $K_n(T - t)$, where n = number of degree divisions of emergent mercury column, T = bulb temperature, and t = mean temperature of the emergent column. K = 0.00009 for Fahrenheit and 0.00016 for centigrade scales. Figure 9 is a cross section

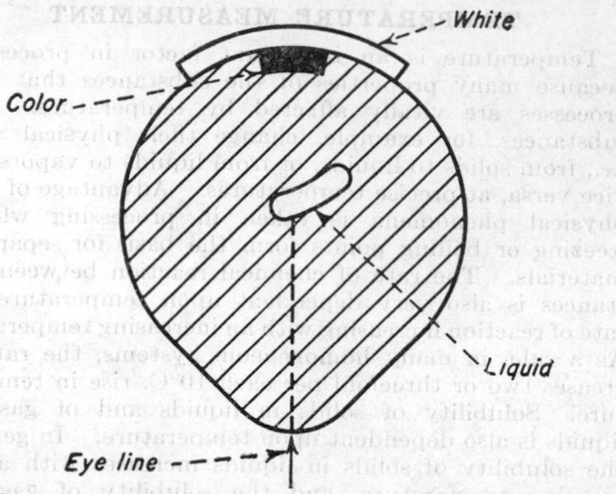

FIG. 9. Liquid-in-glass thermometer (cross section).

of one type of improved thermometer tubing in which the liquid column is reflected in the colored backing, whereas reflection in the tubing itself makes the backing invisible above the top of the liquid column.

Figures 10 and 11 show two forms of *industrial thermometers*. These are partial-immersion thermometers, usually employing mercury, protected with a glass-covered metal scale and a metal tube over the bulb. Such thermometers are available in various lengths with scale

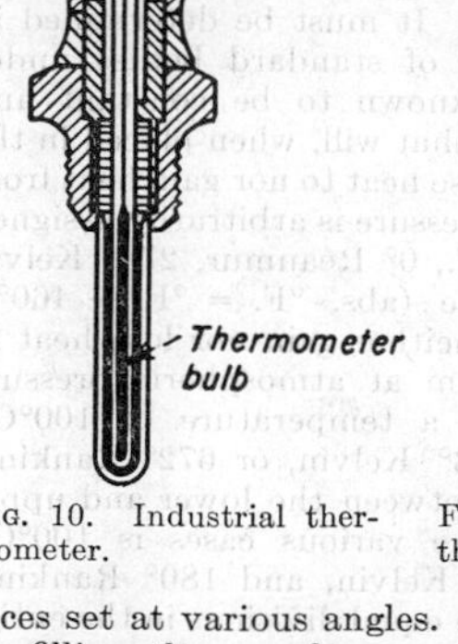

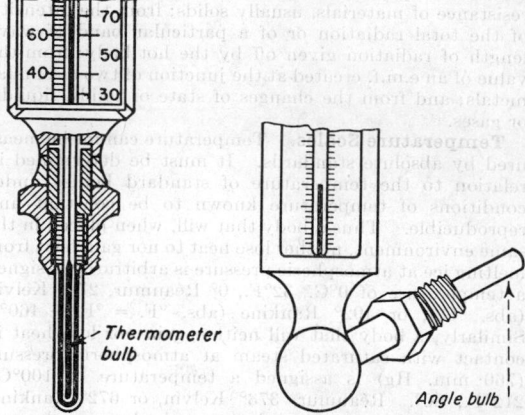

FIG. 10. Industrial thermometer.

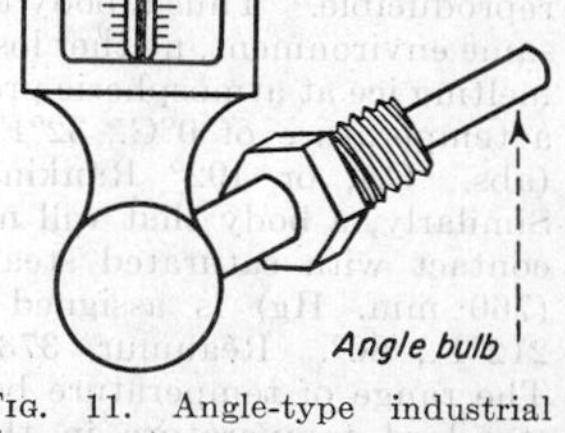

FIG. 11. Angle-type industrial thermometer.

faces set at various angles. Temperature lag is reduced by filling the annular space between the bulb and the protecting tube with a heat-transfer medium such as oil, mercury, copper powder, graphite, or low-melting metal. For measuring the temperature of gases, the protecting tube is perforated to decrease temperature lag.

Pressure-filled Expansion Thermometers. An industrial thermometer of this type is illustrated in Fig. 12). Thermometers of the pressure-element type utilize the thermal expansion of fluid with increase in temperature to provide indication of the temperature. When the fluid is confined within a small element, the expansion of the fluid with temperature increases the pressure inside the element. This pressure, which is proportional to the temperature, is measured to give the temperature reading. Figure 13 illustrates the general construction of pressure thermometers. The system consists of a *bulb*, often inclosed in a *protecting socket*, connected by *capillary*

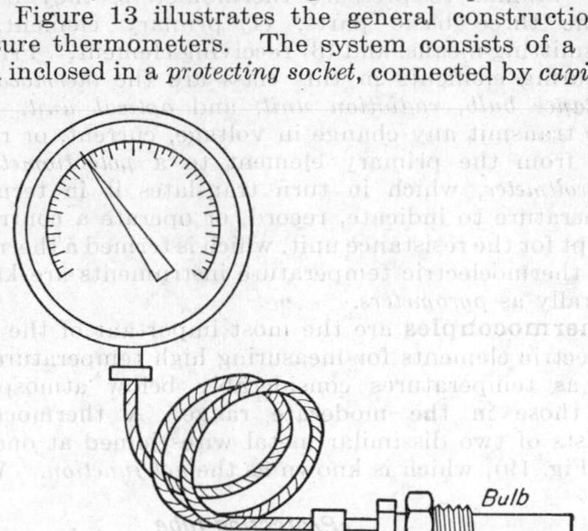

FIG. 12. Pressure-filled expansion thermometer.

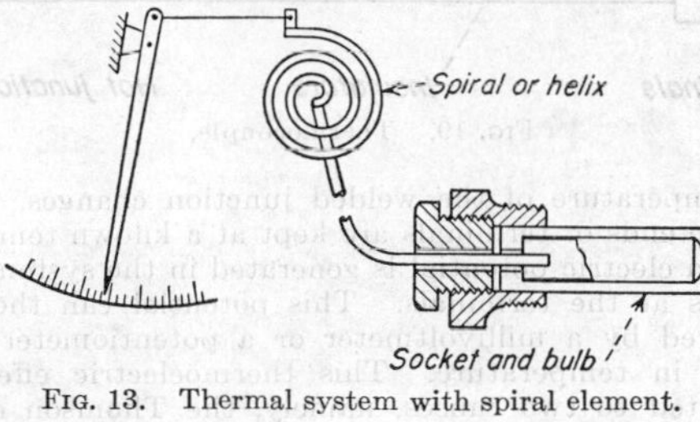

FIG. 13. Thermal system with spiral element.

tubing to an element capable of expanding or otherwise altering its dimensions under increasing pressure. The *receiving element* usually consists of a metallic tube that has been flattened and bent to form a spiral (Fig. 13), bourdon tube (Fig. 14), or helix (Fig. 15). Since one end is sealed and the other end is fixed in position, the element tends to straighten or uncoil with increasing pressure. This causes the sealed end to move in an arc and by means of linkages to position an indicat-

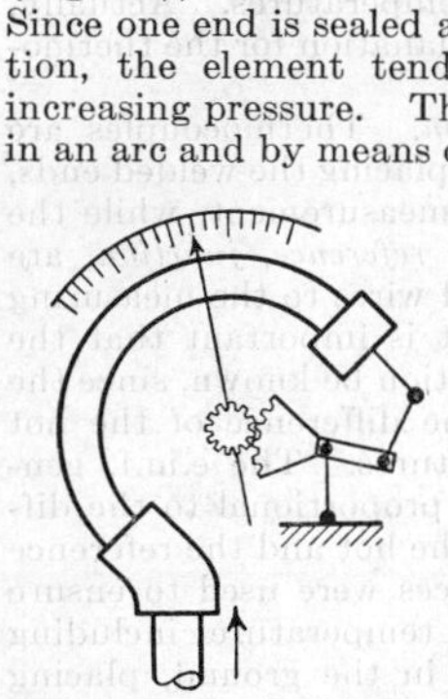

FIG. 14. Bourdon tube.

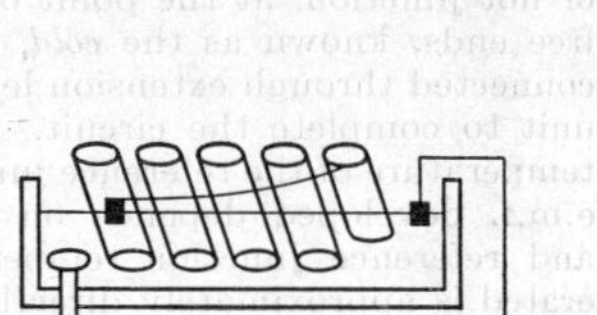

FIG. 15. Helix element.

ing arm or recording pen or to operate a control mechanism. The receiving element may also consist of a bellows (Fig. 16) or diaphragm. Three common types of pressure thermometers are used, namely, the liquid-pressure, the gas-pressure, and the vapor-pressure thermometer.

Liquid-pressure thermometers are usually filled with mercury, although organic fluids such as hydrocarbons may also be used. Liquid expansion in a pressure thermometer results in an approximately linear relationship between temperature and the movement of the receiving element following the law of volumetric expansion expressed by $V_2 = V_1(1 + Bt)$, where V_2 = final volume, V_1 = initial volume, t = temperature change in degrees absolute, and B = coefficient of expansion. This permits a temperature scale with equal graduations. Mercury-filled thermometers are satisfactory over the range of $-40°$ to $+1000°$F.; and organic fillings may be available in a range of $-170°$ to $+500°$F.

A difficulty encountered with pressure-type thermometers is the false indication resulting from volume changes

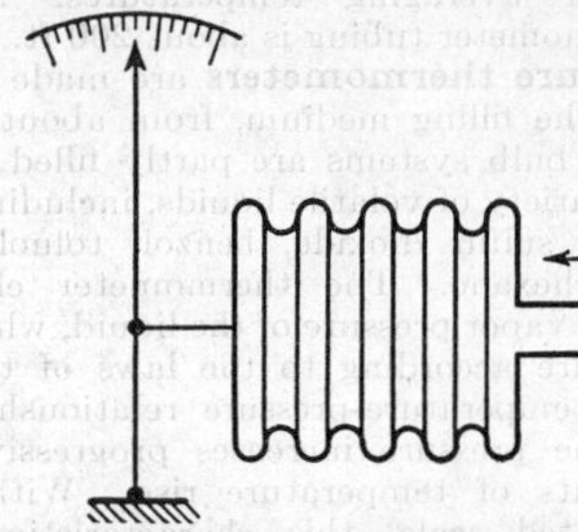

FIG. 16. Bellows.

of the fluid in the capillary in case of ambient temperature variations along the tube. This is particularly true of mercury-filled thermometers. Uncompensated mercury-pressure thermometers are usually limited to a maximum capillary length of 25 ft. Compensation for ambient temperature changes is generally necessary if the capillary tubing is more than a few feet long, particularly with the organic fillings. For case compensation, a bimetallic element connected to the measuring spring offsets the spring motion occurring as a result of spring temperature changes, or spring and tubing temperature changes (assuming tubing always at spring temperature) (see Fig. 17*a*). The tubing may be compensated separately and more exactly by the invar-wire method (Fig. 17*b*). By proper choice of tube and wire dimensions the differ-

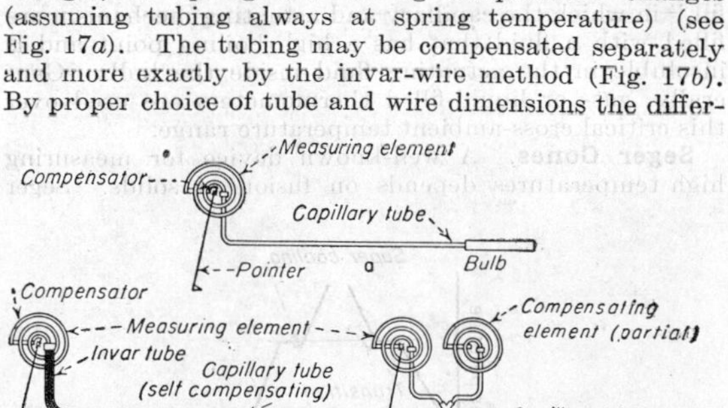

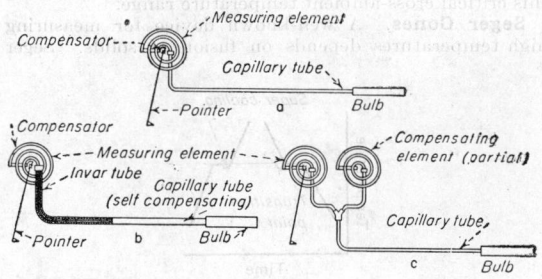

FIG. 17. Types of ambient-temperature compensation.

ential expansion between the tube and wire may be made just to offset the expansion of the filling medium in that tubing. Complete compensation is also secured by the parallel-system method in which the case and tubing errors of the measuring system are exactly offset in a second filled system, identical with the first except for omission of the bulb (see Fig. 17*c*). Compensated mercury-filled thermometers may have capillary-tube lengths up to about 200 ft.

Gas-pressure thermometers are generally satisfactory over ranges between $-125°$ and $+800°$F. Although the most common filling medium is nitrogen, other inert gases may be used. Gas-filled thermometers operate in accordance with Charles's law, $P_1/P_2 = T_1/T_2$, where P_1 = initial pressure absolute, P_2 = final pressure absolute, T_1 = initial temperature absolute, and

T_2 = final temperature absolute. Thus the temperature scale is uniformly graduated, although, where extreme accuracy is required, deviation from the ideal gas law must be compensated for. Gas-filled thermometers are affected by ambient conditions along the capillary tube and can be compensated in the same manner as described for mercury thermometers. However, gas-filled thermometers are often compensated by providing a large bulb volume in comparison with the volume of the capillary tubing and the actuating element. Large bulbs are usually satisfactory, since this type of thermometer is often used for averaging temperatures. Maximum length for thermometer tubing is about 200 ft.

Vapor-pressure thermometers are made in ranges depending on the filling medium, from about −40° to +600°F. The bulb systems are partly filled with any one of a wide variety of volatile liquids, including methyl chloride, ether, sulfur dioxide, benzol, toluol, butane, propane, and hexane. The thermometer element is actuated by the vapor pressure of the liquid, which varies with temperature according to the laws of thermodynamics. The temperature-pressure relationship is not linear, since the pressure increases progressively with equal increments of temperature rise. With an unequally graduated scale, this characteristic is often utilized to increase readability in the operating range. Moderately small bulbs and high responsiveness are characteristic of vapor-pressure thermometers. The maximum tubing length is usually 250 ft. Although vapor-pressure thermometers do not require compensation, they have certain limitations with regard to the cross-ambient effect. When the temperature to be measured is near the ambient temperature, the speed of response is slowed down by the necessity for vaporizing, or liquefying the fluid in the capillary when the measured temperature crosses the atmospheric or the ambient temperature. This is sometimes overcome by a "double fill," in which the capillary and actuating mechanism are filled with a fluid that has a high boiling point and is insoluble in the actuating fluid inside the bulb. Generally, gas- or liquid-filled thermometers are used over this critical cross-ambient temperature range.

Seger Cones. A well-known device for measuring high temperatures depends on fusion of solids. Seger

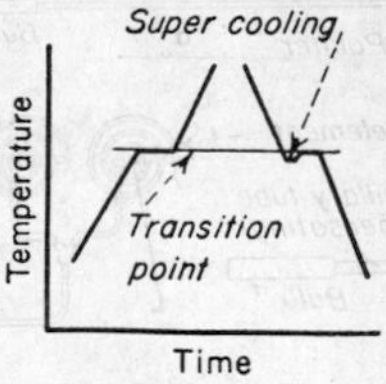

FIG. 18. Transition-point location.

cones, which are small pyramids made up of a mixture of oxides, are useful for the approximate measurement of temperatures by steps averaging about 36° from 1100° to 3700°F. Their principal use is in ceramics, where they are considered indispensable, not because of high-temperature accuracy, but because their temperature-time performance is similar to that of the materials being fired. Also, it is easy to distribute them throughout a kiln or furnace charged to study temperature gradients.

Melting points of metals are often used for standardizing pyrometers. In using this method, the readings of pyrometer should be plotted against time, with the temperature of the metal batch both ascending and descending to the transition point, as in Fig. 18. This assures its correct location.

Temperature Measurement Based on Electrical Phenomena

Thermoelectric instruments depend on a change of electrical characteristics to indicate changes in temperature. Similar to pressure thermometers, they consist of the three basic parts, (1) primary element, (2) transmitting means, and (3) receiving element. Primary measuring elements in this class are the *thermocouple, resistance bulb, radiation unit,* and *optical unit. Lead wires* transmit any change in voltage, current, or resistance from the primary element to a *potentiometer* or *millivoltmeter,* which in turn translates it in terms of temperature to indicate, record, or operate a controller. Except for the resistance unit, which is termed a thermometer, thermoelectric temperature instruments are known generally as *pyrometers.*

Thermocouples are the most important of the thermoelectric elements for measuring high temperatures, as well as temperatures considerably below atmospheric and those in the moderate range. A thermocouple consists of two dissimilar metal wires joined at one end (see Fig. 19), which is known as the *hot junction.* When

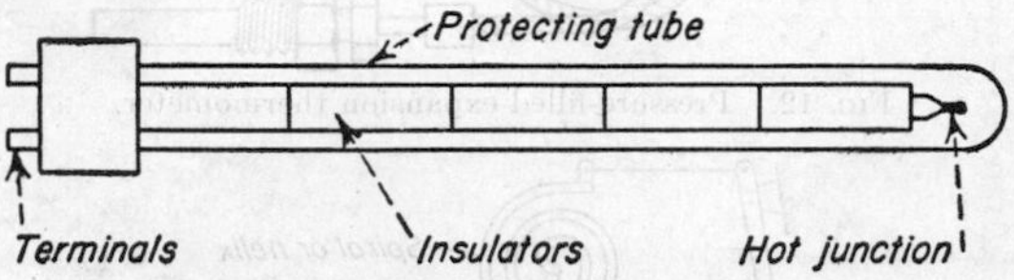

FIG. 19. Thermocouple.

the temperature of the welded junction changes, while the free ends or terminals are kept at a known temperature, an electric potential is generated in the system and appears at the terminals. This potential can then be measured by a millivoltmeter or a potentiometer calibrated in temperature. This thermoelectric effect is attributed to two causes, namely, the Thomson e.m.f. and the Peltier e.m.f. The Peltier e.m.f. is that portion of the total e.m.f. caused by potential difference at the junction of the two dissimilar wires. The Thomson e.m.f. is that portion of the total caused by the temperature gradient over a single section of homogeneous wire when the ends are at different temperatures. Actually, there is no perfect theoretical explanation for the thermoelectric effect.

Reference-junction Compensation. Thermocouples are used to measure temperature by placing the welded ends, or hot junction, at the point of measurement, while the free ends, known as the *cold, or reference, junction,* are connected through extension lead wires to the measuring unit to complete the circuit. It is important that the temperature of the reference junction be known, since the e.m.f. developed depends on the difference of the hot and reference junction temperatures. The e.m.f. generated is approximately directly proportional to the difference in temperature between the hot and the reference junction. Formerly, many devices were used to ensure a constant reference junction temperature including burial of the reference junction in the ground, placing it in a thermostatically controlled compartment, or keeping it at the temperature of melting ice. At present, most installations extend the reference junction to the measuring instrument by use of *compensating lead wires,* which in the case of base-metal couples are usually flexible multistrand wires of the same material as the corresponding thermocouple elements. In the case of noble-metal couples, *extension leads* are used that have substantially the same thermoelectric characteristics as the noble metals used. By this means, it is possible to make any necessary reference-junction corrections in the instrument case by use of bimetal to shift the instrument zero; by use of nickel or copper resistors in one arm of a potentiometer bridge circuit; or by some equivalent method.

Several different combinations of metals are used for thermocouples, depending on temperature limits. Since the desirable e.m.f.-temperature difference is a straight-line relation, it is important to select a combination of metals that will closely approach this. Other requirements are reproducibility within limits, constancy of calibration, resistance to oxidation and corrosion, and high thermoelectric power. The following table shows the common thermocouples, their composition and common operating-temperature range:

Thermocouple symbol	Positive	Negative	Temperature range, °F.
Q.R.	Platinum	Platinum-rhodium (90% Pt, 10% Rh)	1300 to 2900
M.A.	Chromel (90% Ni, 10% Cr)	Alumel (95% Ni, 2% Al, 3% Mn, 1% Si)	0 to 2100
I.C.	Iron	Constantan (60% Cu, 40% Ni)	0 to 1600
C.C.	Copper	Constantan (60% Cu, 40% Ni)	−300 to +600

Figure 20 shows the temperature-e.m.f. relation between platinum and various other metals. At any temperature, the vertical distance between any two curves is the e.m.f. for a thermocouple using those two

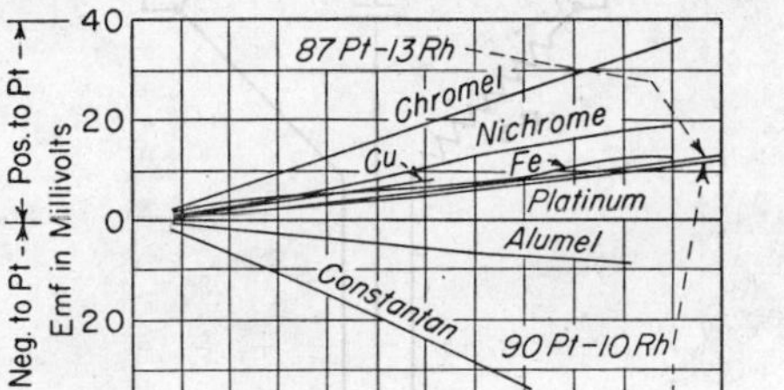

FIG. 20. Temperature-e.m.f. relation between metals (based on platinum).

wires. Noble-metal couples are used for laboratory measurement requiring highest precision in the range of 500° to 3000°F., for calibrating base-metal couples, and in industrial applications for temperatures between 1300° and 2900°F. Cost, however, is prohibitive for most industrial use.

Thermocouple Selection. Three major factors must be considered when selecting a thermocouple, namely, (1) required *speed of response* to change in temperature; (2) *depth of immersion*, or distance that the thermocouple must extend into the process equipment; and (3) *life*, or time the thermocouple will be expected to retain acceptable accuracy of calibration.

Response of a thermocouple is the speed or rate at which it will follow changes in temperature—or simply the time required by the thermocouple to react to a given temperature change. The rate of response increases as the diameter or mass of the thermocouple decreases. Therefore, small-diameter thermocouple wire and tubes should be used on applications where it is essential to detect small temperature changes rapidly, provided that such light wire may be expected to give reasonable life.

Depth of immersion of a thermocouple into a process determines the relationship between the true temperature of the process and the temperature reached by the thermocouple. If the depth of immersion is insufficient, the thermocouple will be lower in temperature than the process because of conduction of heat along the wires as well as the walls of the protecting tube. Therefore, for accuracy of measurement, the thermocouple should extend sufficiently into the process or furnace so that the conduction error is negligible. Although no given rule holds for all sets of conditions, it is generally recommended that the thermocouple be immersed for a minimum distance from inside the vessel wall equivalent to approximately four times the outside diameter of the protecting tube. Where the thermocouple is installed either through a water-cooled furnace wall or through the walls of a muffle furnace, the immersion should be approximately ten times the tube diameter. The temperature of the couple head should never exceed 250°F.

The life of a thermocouple may be defined as the duration of time over which the thermocouple retains acceptable accuracy of calibration. In general, the life of a thermocouple increases with the diameter of the wire and decreases with increasing temperatures. At elevated temperatures, heavy gage wires more ably resist deterioration and contamination, factors that eventually cause a permanent shift in thermocouple calibration. Thermocouples are sometimes used without protecting tubes. The practice should be followed only where temperature conditions are mild and the atmosphere is free from contaminating gases. The type of atmosphere to which a thermocouple is exposed greatly affects its useful life. An industrial-process atmosphere is either oxidizing or reducing. An *oxidizing atmosphere* is one that contains free oxygen, which tends to combine with some materials, causing scaling, peeling, etc. In a *reducing atmosphere*, on the other hand, there is a deficiency of oxygen, a condition that tends to change the composition of certain materials that are exposed to it. Thermocouples must withstand the action of the process atmosphere or be thoroughly protected from it.

In general, *iron-constantan* thermocouples perform best in reducing atmospheres. *Chromel-alumel* thermocouples, conversely, perform best in oxidizing atmospheres. *Copper-constantan* thermocouples are generally used in low-temperature work where atmospheric conditions are not critical, and they are preferable where moisture is high. *Platinum–platinum rhodium* thermocouples, when suitably protected, can be used in either oxidizing or reducing atmospheres.

Practically all thermocouples used for industrial temperature measurement require **protecting tubes** or wells (Fig. 19). If they are not protected, the corrosive action of most atmospheres and liquids at high temperatures or mechanical damage may cause early failure of the thermocouple. Protection is of three sorts: (1) insulation for the wires, (2) primary protection tube to avoid contamination, and (3) secondary protection tube to protect the primary tube. All three are desirable for severe conditions. All thermocouples require insulation. This is provided by means of a single- or double-hole porcelain or lava tubes through which the wires are threaded. Over this assembly the primary tube is slipped and joined to a terminal head by some form of gas-tight connection. Secondary tubes may be either joined to the terminal head or supported by the apparatus. Materials used in this service depending on specific conditions of temperature and corrosive medium, include glass, porcelain, fused silica, silicon carbide, graphite, clay, nickel, steel, iron, and other materials and alloys.

Thermocouples are usually purchased completely assembled and ready to use. However, thermocouple wire may be purchased and the thermocouples made up as needed. The hot junction of the couple is usually made by wrapping one wire about the other for a short distance and then joining the wires together with an electric arc or by oxyacetylene gas welding. The wires are threaded through insulating bushings which consist of either single-hole or double-hole porcelain or lava tubes. *Calibrating and checking* thermocouples should take place at frequent intervals to ensure accurate measurement of temperature. If the couple can be removed from service, it is best to check against a standard thermocouple at one or more temperatures in a muffle furnace or in a molten bath of metal or salts. Molten tin covered with graphite to prevent oxidation is often used. Use of muffle furnaces is considered best practice. If thermocouples must be checked in place, the hot junction of the standard thermocouple must be as close as possible to that of the couple being checked. Contaminated couples can only be checked in place because varying the thermal gradient will vary the couple output.

When a single thermocouple is used, it is connected directly to the measuring instrument, but there are times when it is desirable to connect more than one thermocouple to the instrument. *Multiple thermocouples* are employed for (1) amplifying very small e.m.f. generated at low temperatures, (2) obtaining average temperatures, and (3) obtaining several temperatures in a system. Thermocouples are connected in series to amplify thermoelectric output. Cold or reference junctions must be held at constant temperature. The magnitude of the e.m.f. produced will increase directly with the number of thermocouples. *Average temperatures* are usually obtained by connecting the couples in parallel. Where several temperatures in a system are to be measured intermittently, the thermocouples are connected to the instrument through a *rotary switch* which permits rapid reading of all temperatures. To provide *cold-junction compensation*, a cold-junction box fitted with a booster thermocouple having extension leads is used. Any change in temperature in the cold-junction box will vary the e.m.f. of the booster couple and exactly compensate for this variable at the instrument.

Installation. Best results are obtained when thermocouples are properly installed. Certain rules are worthy of mention: Do not locate the couple in the direct path of a flame. Do locate it where the average temperature will be measured. Do locate the couple where the hot end may be seen from a door of the furnace. Do extend the couple in the furnace so that the hot junction is entirely within the temperature zone to be measured. Do not insert a porcelain protecting tube suddenly into a hot furnace, or it may crack. Do support horizontally mounted thermocouples where the temperature is above the softening point of the protecting tube to prevent sagging.

Resistance thermometers depend on the principle that the electrical resistance of a metallic conductor changes with temperature and that this resistance-temperature relationship is sufficiently constant to permit accurate measurement of temperature. The temperature-sensitive element is a carefully constructed coil of wire with high temperature coefficient of resistance, usually nickel or platinum, installed as one arm of a wheatstone-bridge circuit, as shown in Fig. 21. Resistances R_1, R_2, R_3 are fixed and of a metal with low temperature coefficient of resistance, such as manganin. These three resistances and the bulb usually have the same resistance at room temperature. In operation, any unbalance of galvanometer g, due to variation in resistance r, with temperature change, can be compensated by moving slide wire S, which is calibrated in temperature. The temperature can also be read directly by the deflection of a millivoltmeter substituted for g, without the use of a slide wire. The first, or null, method is most commonly employed and uses a self-balancing instrument for adjusting the slide wire until the galvanometer deflection is zero.

Nickel is the most common material for resistance bulbs and may be used over a range from −150° to +300°F. Platinum wire is employed for greater accuracy and for temperatures up to 1800°F. *Low-temperature* bulbs are often wound on bakelite spools, while *high-temperature* bulbs may be wound on the core of mica or glass, after which the coil assembly is enclosed in a metal tube or sheet. Accuracy of the resistance thermometer is inherently better than that of other temperature-measuring devices. Since resistance bulbs are extremely sensitive to contamination and corrosion, which change the resistance, they are installed in protecting tubes or wells. It is essential, too, that all resistance changes in the surface occur in the bulb or that *compensation* be made for the lead wires. One type of compensation is to use a three-wire lead-line arrangement so that equal lengths of lead wire are thrown into each arm of the bridge. Thus any variations in resistance will balance out, causing no change in calibration. Lead

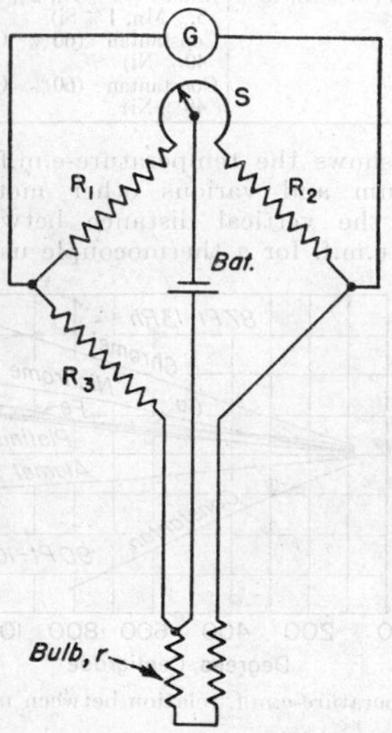

FIG. 21. Resistance thermometer circuit.

lines up to 1000 ft. long are usually made of 16 gage B. & S. copper wire, while 1000- to 1600-ft. lead wires should be of 14 gage copper wire. Speed of response is generally lower than for a thermocouple because of the greater mass of the resistance bulb.

Radiation Pyrometry. Temperatures above the upper limit of thermocouples are determined by measuring the radiation of hot objects. Actually, it has been common practice to use radiation instruments for certain types of applications ranging in temperature from 800°F. and up to 5000°F. However, radiation pyrometers are now available for measuring as low as 100°F. Two methods are based on measurement of radiation, the *optical pyrometer*, which measures the intensity of a particular wave length given off by a hot object (monochromatic); and the *radiation pyrometer*, which measures the intensity of total radiation. The relationship of temperature to rate of radiation is expressed by the Stefan-Boltzmann law, which shows that the total radiant energy transmitted between two bodies of different temperature (under black-body conditions) is proportional to the fourth power of their absolute temperature. A law formulated by Wien and reformulated more exactly by Planck shows that the intensity of radiation of any given wave length is a complex relation between the wave length and the absolute temperature. Theoretically, radiation measurement depends for accuracy on the assumption of *black-body conditions.* A black body is one that absorbs all incident radiation and reflects and transmits none, radiating only in proportion to its temperature. Most materials are not thorough black bodies. However, many industrial furnaces closely approach black-body conditions, and objects within such furnaces will radiate approximately as if they were true black bodies. Where black-body conditions do not exist, the instrument may be sighted on the interior of a tube of refractory material such as fused quartz or porcelain which projects into the zone where temperature is to be measured. This arrangement will closely approximate black-body conditions. Presence of oxides, such as scale on iron billets, helps to reduce errors in measuring the temperature of a non-black body. Where large errors result, it is possible to correct by means of curves that show the relation between true and apparent temperatures. The presence of smoke, flame, and dust may introduce large errors.

Optical pyrometers depend upon a comparison of the black-body temperature of two radiating bodies made by comparing the intensity of light of a given wave length given off by each. A common method of accomplishing this is by use of the *disappearing-filament pyrometer*, which requires *visual comparison*, one type of which is shown in Fig. 22. This instrument, known as

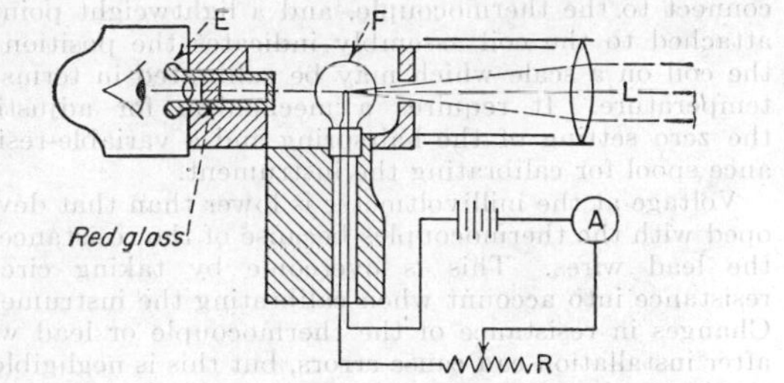

FIG. 22. Optical pyrometer—disappearing-filament type.

the Morse type (Leeds & Northrup), requires an incandescent filament for comparison. Light from the radiating body enters through lens L and is viewed through the eyepiece E. Temperature of the filament F is altered by means of adjustable resistance R until the filament disappears against the background of the hot object, at which point the filament current measured by the milliammeter A is a measure of the black-body temperature. A red-glass filter interposed between the eyepiece and the filament facilitates comparison of intensity without interference caused by color differences.

Several other optical pyrometers also require visual comparisons. In the wedge type (Fischer Scientific Co.), the filament current remains constant, and the apparent intensity of the radiator body is varied to match the filament by rotating a wedge of absorbing glass. The thickness of the wedge at the point where the filament just disappears is a measure of the temperature. The polarizing type of optical pyrometer (Wanner) uses a direct-vision spectroscope and polarizing prisms to isolate monochromatic radiation. The operator, in looking in the eyepiece, sees two half disks, one monochromatic light from the standardized electric light, and the other from the radiating body. Rotation of the analyzer for the beam from the standardized lamp decreases its intensity until a "match" is obtained between the two half disks. Rotation is taken as a measure of the intensity of radiation from the source in comparison with the standard, and hence as a measure of the temperature. The lamp is standardized periodically by setting on a standard amyl acetate lamp.

Automatic optical pyrometers not requiring visual comparison have been made possible by use of photoelectric cells. The Optimatic (Brown) uses two photoelectric cells, one to view the temperature to be measured, and the other to maintain automatically a relative value of conductivity proportional to the first cells illuminated by a standard lamp. Current in the standard lamp is a measure of temperature. This is shown in Fig. 23. Photocells P_1 and P_2, respectively, measure light from the object and from a standard lamp used for comparison. The cells are part of an a.c. bridge circuit that uses an amplifier tube instead of a galvanometer. In operation, light from the heated bodies strikes the exposed photocell P_1, varying its resistance and unbalancing the bridge. A

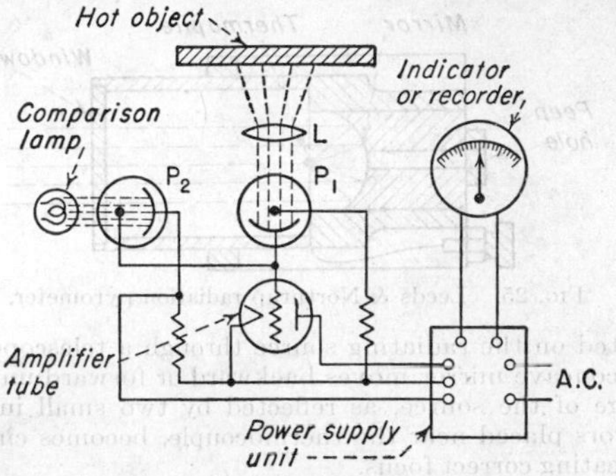

FIG. 23. Automatic optical pyrometer.

change in the output of the amplifying tube changes the flow of current to the comparison lamp so as to rebalance the bridge. A meter then indicates or records the lamp-circuit current in terms of the measured temperature. Since the reaction of photocells is instantaneous, this instrument is capable of measuring and recording temperatures of moving objects such as billets and slabs in steel mills.

Radiation Pyrometers, of which types are shown in Figs. 24 to 26, have an advantage over optical pyrometers (except the photoelectric pyrometer) in not requiring visual comparison. Instruments of this type are made in two general forms, the fixed-focus and the adjustable-focus type. Most radiation pyrometers make use of thermocouples to measure radiant energy emanating from a hot body. The radiation from a particular area of the hot body is concentrated on the hot junction

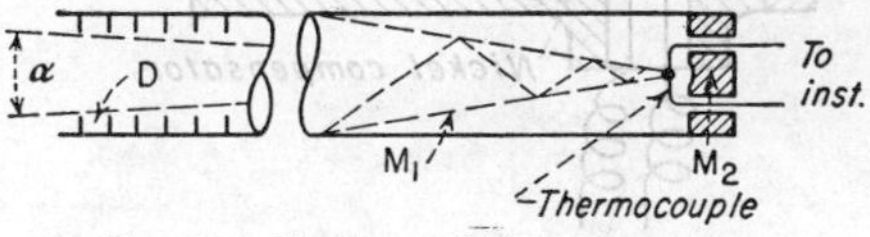

FIG. 24. Thwing radiation pyrometer.

of the thermocouple or *thermopile* (group of thermocouples in series), and the thermocouple potential indicates the intensity of all radiation reaching it.

In Fig. 24, a fixed-focusing type (Thwing) is shown. This consists of a tube containing non-reflecting surfaces D at the entrance, a conical mirror M_1, and a concave mirror M_2, which concentrate the radiation on the thermocouple. The quantity of radiation reaching the thermocouple is independent of the distance of the hot body, provided that the area of the latter is large enough to fill the angle defined by the diameter of the entrance opening, and the length of the tube to the thermocouple.

Figure 25 shows the mirror-type Rayotube (Leeds & Northrup), in which a mirror focuses rays from the hot object through a window onto a radial thermopile. A peephole is provided for aiming on the object. A somewhat similar instrument is shown in Fig. 26. This is an adjustable-focus instrument, the Radiamatic (Brown), which uses a lens instead of a mirror to concentrate rays on the center of a radial thermopile. This instrument has a nickel compensating coil near the cold junctions of the thermopile, which is shunted across the leads to compensate for variations of the cold-junction temperature. It also employs a peephole for aiming. Temperatures as low as 100°F. may be measured with the Radiamatic, in which a special infrared transmitting lens is used. Another adjustable-focus instrument (Frery) uses a concave mirror to focus radiation on a small thermocouple, which is protected from direct radiation. The instrument is

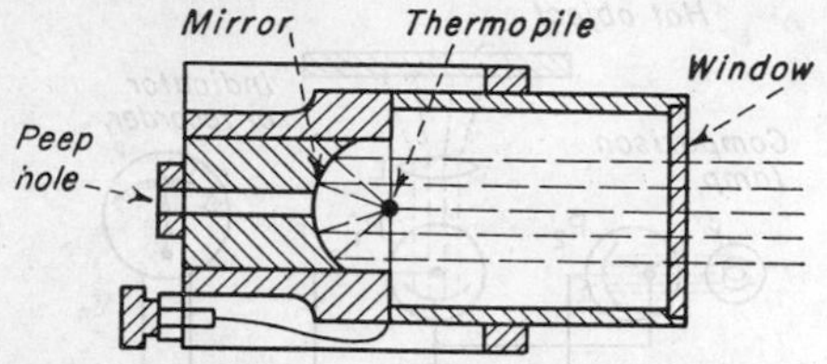

FIG. 25. Leeds & Northrup radiation pyrometer.

sighted on the radiating source through a telescope, and the concave mirror moves backward or forward until the image of the source, as reflected by two small inclined mirrors placed near the thermocouple, becomes circular, indicating correct focus.

As previously mentioned, non-black-body conditions can be readily compensated for, and it is common practice to use a closed tube of refractory material in which the instrument is sighted. Although this corrects for non-black-body errors, it slows the response of the instrument. Since thermocouples are used, it is necessary to provide cold-junction compensation. When the instrument is located in an area of high ambient temperature, it is often cooled by use of a water jacket, or a cold air stream may

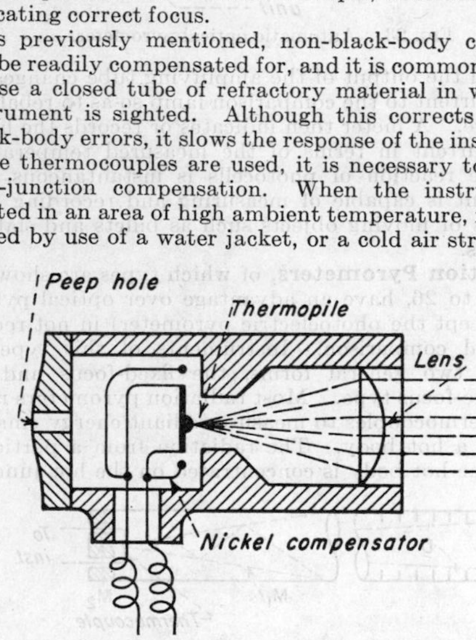

FIG. 26. Brown radiation pyrometer.

be used. Air jets may also be used to keep the sighting tube free of smoke or flame. While the radiation-type unit has the advantage of having no physical contact with the object whose temperature is being measured, it has the disadvantage of large errors when radiation from the object must pass through smoke or dust of appreciable concentration. The range of radiation pyrometers may extend from 100°F. to the highest industrial temperature. Ordinary glass-lens units may not be used below 800°F. Calibration of both optical and radiation pyrometers is accomplished by sighting on a furnace of known temperature under black-body conditions.

Electric Measuring Instruments. Two types of instruments, millivoltmeters and potentiometers, are commonly used to measure the small voltages or changes in potential generated by thermoelectric primary elements, as well as other electric variables. **Millivoltmeters** are simpler and less expensive and are extensively used, although the potentiometer in its various forms is more versatile and overcomes the basic limitations and disadvantages of the millivoltmeter. Most millivoltmeters for this service operate on the principle of the *d'Arsonval galvanometer* (Fig. 27), in which the movement of a coil in a constant magnetic field is proportional to the electrical current passing through the coil, which in turn is proportional to the e.m.f. set up by a thermocouple. It consists essentially of a movable coil restrained by hair springs mounted on pivots between the

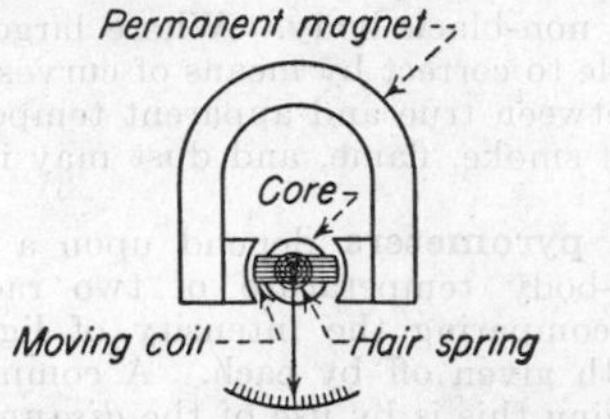

FIG. 27. D'Arsonval galvanometer.

poles of a permanent magnet. The ends of the coil connect to the thermocouple, and a lightweight pointer attached to the coil assembly indicates the position of the coil on a scale which may be calibrated in terms of temperature. It requires a mechanism for adjusting the zero setting of the hairspring and a variable-resistance spool for calibrating the instrument.

Voltage at the millivoltmeter is lower than that developed with the thermocouples because of the resistance of the lead wires. This is overcome by taking circuit resistance into account when calibrating the instrument. Changes in resistance of the thermocouple or lead wire after installation can cause errors, but this is negligible if the internal resistance of the instrument is high (about 100 to 1) compared with that of the lead wires and thermocouple. As previously pointed out, a thermocouple must have a reference junction at substantially constant temperature. Since a stable temperature is most likely to be found at the instrument, the reference junction is usually located within the instrument case. Instruments not otherwise compensated are provided with a zero adjustment for setting the pointer to take care of a known reference-junction temperature. However, where changes in ambient temperature occur, variations in reference-junction temperature may be compensated for by use of a *bimetallic spiral* to readjust the zero point and correct the error.

The millivoltmeter requires no auxiliary power and operates from the e.m.f. generated by the thermocouple or other primary element. Its chief disadvantages include (1) inaccuracy due to variations in resistance of the leads and other parts of the circuit, (2) inaccuracies due to sensitivity of the mechanism to vibration. Friction must be kept to a minimum because of the small amount of power available for operation. Pivots and bearings are a source of trouble because of misalignment or deterioration caused by vibration. However, when properly applied and installed, the millivoltmeter gives excellent service and is therefore widely used for measurement and control. The millivoltmeter may be of either the indicating or the recording type, but it is essentially an indicating instrument.

Potentiometers. Inherently a more versatile and accurate instrument than the millivoltmeter, the potentiometer in its various forms is widely used in industry to measure temperature, pH, conductivity, and other variables that can be measured by electrical means. The principal advantage of the potentiometer circuit over the millivoltmeter is that its measurements are unaffected by variations in the external resistance of the circuit, as in lead wires, connections, etc. Potentiometer circuits are either *fully balanced* (null) or *partly balanced* (deflection); but, compared with the null-type, applications of the deflection instrument are negligible. In the *null-type potentiometer* circuit, the unknown e.m.f. is balanced against an equal and opposite known e.m.f. developed by a battery. Figure 28 shows an elementary potentiometer circuit. In the upper circuit, the battery produces a potential drop in the *slide wire* which may be adjusted by resistance *R*. In the lower circuit, an opposite potential is produced by a standard cell of definitely known voltage, and the drop in the slide wire can be made equal and opposite to that produced by the battery by adjustment of the slider, this point being indicated by a zero reading of the galvanometer. A thermocouple can then be substituted for the standard cell, and the galvanometer again brought to zero by adjusting the slider. The relation between the first and second slider readings is a measure of the distance between the thermocouple potential and that of the standard cell, and the slide-wire position may be calibrated in terms of the temperature of the hot junction corresponding to the various potentials.

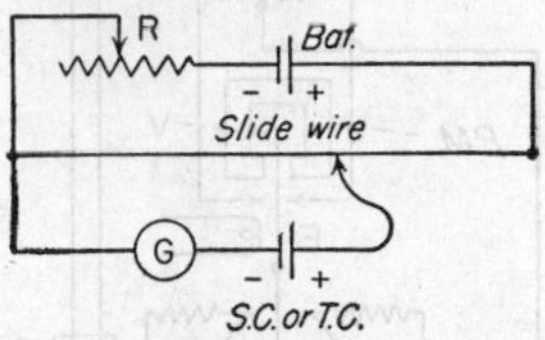

Fig. 28. Elementary potentiometer circuit.

In practice, potential drop across the slide wire, supplied by a dry-cell battery, must be kept constant by overcoming the constant current drain. This is done by means of a *standard cell* and an *adjustable resistor* in the battery circuit. Voltage of the standard cell is compared periodically with the battery circuit, which is then readjusted to the necessary voltage. Standardization may be either manual or automatic. Since the cold junction is at the instrument, it is necessary to correct for variations in ambient temperature. A number of methods are used, both manual and automatic. The most common is an automatically compensated circuit utilizing a nickel resistance coil which has a high thermal coefficient of resistance. Thus changes in temperature vary the resistance of the nickel coil to compensate the circuit.

Although it is infrequently used in industry, a brief description of the *deflection potentiometer* is in order. The basic circuit of the deflection potentiometer is shown in Fig. 29. Only a part of the thermocouple potential is balanced against the battery potential, with the remainder being indicated by the deflection of the galvanometer. This circuit is sometimes employed in indicating instruments where rather small temperature fluctuations are expected. The slide-wire contacts represent resistance values in steps of some magnitude. The galvanometer therefore cannot be balanced by adjustment of the slide wire, and the residual e.m.f. deflects the galvanometer. Each contact point on the slide wire might, for example, represent 10° temperature difference, while the deflection of the galvanometer would show temperature variations from zero to 10°.

Self-balancing Potentiometers. Although some potentiometers, especially portable models, require manual balancing of the circuit, most industrial instruments are self-balancing. Figure 30 shows a self-balancing potentiometer circuit with *manual standardization* and *automatic reference-junction compensation.* By means of the switch shown, the standard cell is thrown into the circuit to permit standardizing the dry battery with resistance *R*. Then the thermocouple is thrown into the circuit for temperature measurement. Resistances 1, 2, and 3 are of manganin, and resistance 4 is of nickel for reference-junction compensation. An automatic balancing mechanism detects unbalance in the galvanometer *G* and adjusts the slide wire by means of the motor drive *M*. Most potentiometers are equipped with *fail-safe* protection, in which a resistor is hooked up between one side of the potentiometer and the slide wire. If the thermocouple fails, the pen of the instrument will go to one end of the scale, and current to the controller will be cut off.

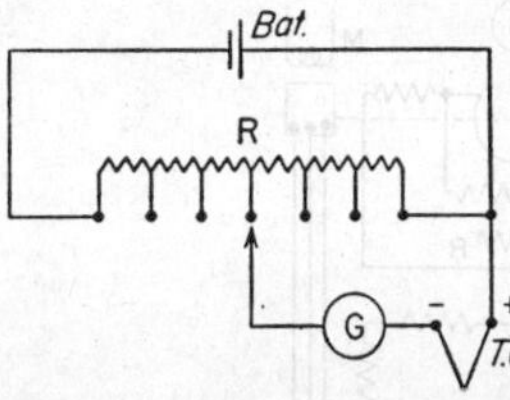

Fig. 29. Deflection potentiometer circuit.

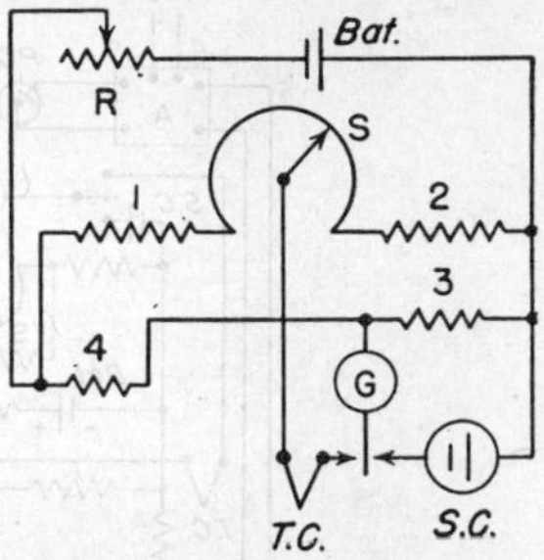

Fig. 30. Self-balancing potentiometer circuit.

Numerous mechanisms are used for detecting and rebalancing potentiometer circuits. Most common in the past and still widely used is the *periodic-balance potentiometer*, which uses a galvanometer to detect any unbalance of voltages. When the temperature changes, the electrical circuit is unbalanced and the galvanometer pointer deflects from its zero position, indicating the amount and direction of unbalance. A lever system or step-table arrangement detects the position of the galvanometer pointer while it is momentarily clamped in its deflected position. A drum on which a slide wire is wound is then turned in a direction to overcome the unbalance. The pointer is freed, and if unbalance still exists the cycle is repeated until the circuit is balanced. The mechanism is operated by a continuously running motor which drives the slide-wire contactor by means of a clutch arrangement.

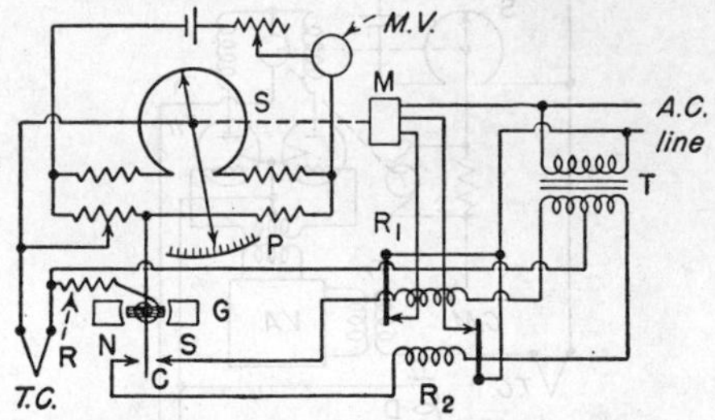

Fig. 31. Bristol pyrometer circuit.

A number of potentiometer balancing arrangements have been developed to eliminate the time delays in detecting and correcting unbalance inherent in the periodic balance instrument. These *semicontinuous and continuous balance potentiometers* are being widely used to meet the requirements of applications where fast action is a prime factor. Several of the instruments are described here. Figure 31 is a simplified version of the Pyromaster circuit (Bristol) which employs a contacting galvanometer to balance the circuit. *MV* represents a millivoltmeter for standardizing the circuit, *S* the slide wire, *P* the chart pointer, *G* the galvanometer, and *C* the galvanometer contact. *R* is a resistance which as soon as contact takes place automatically increases the galvanometer current to increase the strength of contact in the proper direction. When contact is made, indicating unbalance in one direction or the other, the appropriate relay R_1 or R_2 is energized, energizing either the forward or reversing windings on the motor *M*. The motor then runs to reposition the slider on the slide wire until contact is no longer made at *C*.

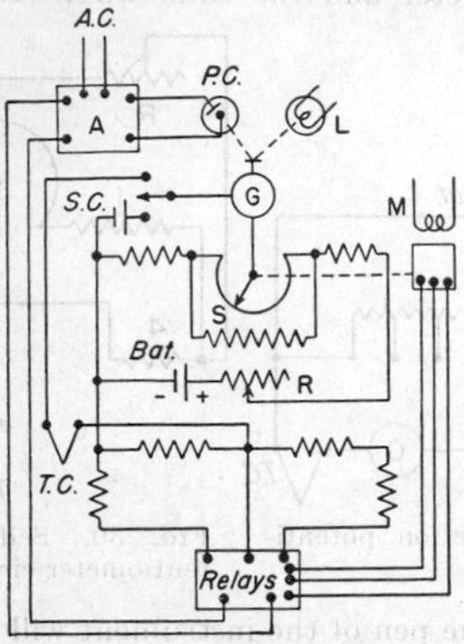

FIG. 32. Tagliabue Celectray.

Another system of potentiometer balancing is found in the Celectray (Tagliabue) system. This instrument (Fig. 32) employs a mirror galvanometer *G* which reflects a beam of light on or off a photocell *PC* to control the balancing motor. In this diagram *A* is an electronic amplifier for the photocell current, *L* is an incandescent lamp, *M* is the balancing motor, *S* is the slide wire, *R* is the potentiometer standardizing resistance, *SC* is the standard cell which may be substituted in the circuit for the thermocouple for standardizing, and "relays" represent a pair of relays energized by the amplifier output, one

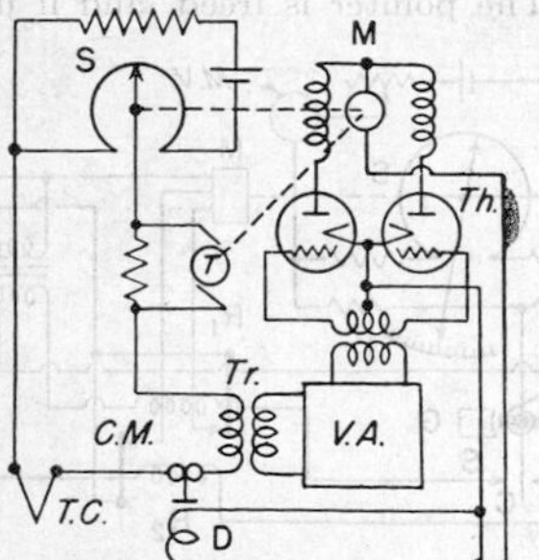

FIG. 33. Leeds & Northrup Speedomax.

controlling the forward and the other the reversing winding of the motor. When both relays are closed, the motor moves the slide-wire contact upscale; and, when both are open, it moves it downscale. One relay incorporates a time delay to avoid energizing the motor with rapid swings of the galvanometer.

No galvanometer is employed in the Speedomax (Leeds & Northrup) illustrated in Fig. 33. *S* is the slide wire of the potentiometer, the contact of which is positioned by motor *M*. Potential produced by the thermocouple is put through a carbon microphone *CM*, which is vibrated by a microphone drive *D*. This produces an interrupted current in transformer *Tr*, the a.c. component of which is put through a voltage amplifier *VA*. The output of this amplifier in turn is fed to two thyratron tubes *Th*. The voltage of these tubes comes from the same supply line that feeds the microphone vibrator. When the thermocouple current changes, unbalancing the potentiometer, current may flow in either direction of current flow, and this phase in turn determines which thyratron supplies current to the motor and, hence, which direction the motor will turn. The slide-wire contact is then moved in the proper direction to achieve balance. To avoid overshooting, the drive motor is coupled to a tachometer magneto *T*, the potential of which is applied to the thermocouple circuit in the direction to oppose the thermocouple potential. This circuit is rapid, permitting the pen to traverse a 10-in. chart in 2 sec. without overshooting. A modification of this instrument is the Speedomax G, which differs from the standard Speedomax in that it employs a converter and amplifier for operating the balancing motor instead of thyratron tubes.

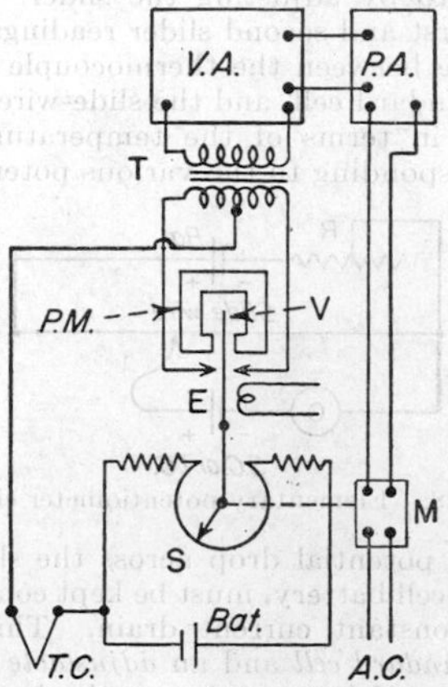

FIG. 34. Brown continuous-balance potentiometer circuit.

A simplified version of the Continuous-Balance potentiometer (Brown) is shown in Fig. 34. Here slide-wire contact *S* is positioned by motor *M* through the action of the vibrator *V*. A current flow in the thermocouple circuit, due to unbalance of the potentiometer, may be in either direction. The vibrator *V* is vibrated at 60 cycles between the poles of a permanent magnet *PM* by means of an energizing coil *E*. By means of the two contacts shown, this direct current is alternated through the primary of transformer *T*, and the alternating and secondary current is then applied to an electronic voltage amplifier *VA*, and power amplifier *PA*, the output of which is applied to the motor *M* to achieve balance. As in the system illustrated in Fig. 33, the direction of unbalance determines the phase relation between the generated alternating voltage and the a.c. supply voltage. Rising temperature produces an in-phase relation, and falling temperature a 180-deg. out-of-phase relation. The motor used is a two-phase reversible motor, and the direction of its rotation depends on whether one phase is lagging or leading the other. Hence the motor selects the proper direction of rotation to return the slide-wire contact toward the balance point. Pen speeds permit full travel across an 11-in. strip chart in as little as 2½ sec.

The Dynalog (Foxboro) is still another electronic potentiometer. Continuous balancing is provided by a simple rotating variable air capacitor driven by a double-solenoid type drive, which moves only when the instrument is rebalancing. Full-scale pen travel in as little as

1 sec. on a standard 12-in. circular chart provides rapid recording.

PRESSURE AND VACUUM MEASUREMENT

Pressure-measuring devices are extremely important in any study of instruments, not only because of the innumerable applications in the process industries but also because the principles involved form the basis for other instruments, such as pressure thermometers, liquid-level gages, and differential flowmeters. Pressure is a *force per unit area.* In most cases it is measured directly by balancing against a known force rather than by inferential methods, such as are required for temperature measuring. The known force may be that of a liquid column, a spring, or a weight-loaded piston, or of a spring-loaded diaphragm or other element capable of quantitative distortion under the application of pressure. Inferential methods are sometimes used, such as that of measuring the thermal conductivity of the material, or measuring the charge produced by piezoelectric crystal under pressure. Pressure measurements are based on a normal atmospheric pressure of 14.70 lb./sq. in. as the datum line, with forces greater than this termed gage pressure and those below called vacuum. Pressure measured above zero pressure as the datum line is termed absolute pressure. Use of the strain gage is becoming increasingly important.

Pressure Gages. The simplest form of pressure-measuring device is the liquid column, such as mercury barometers and U-tubes. Figure 35(*a*) shows the basic

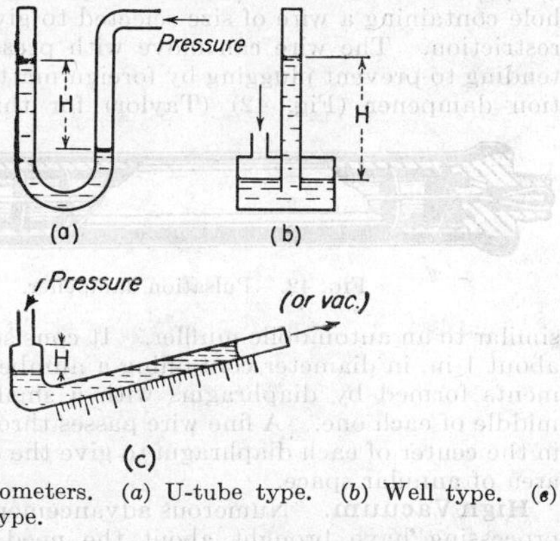

FIG. 35. Manometers. (*a*) U-tube type. (*b*) Well type. (*c*) Inclined-tube type.

U-tube manometer in which, with the density of the manometer fluid (mercury, water, glycerin, light oil, etc.) known, the head H is a measure of the pressure in comparison with that of the atmosphere. A simple *well-type manometer* with H, again a measure of pressure, is seen in Fig. 35(*b*). For the measurement of small pressures or vacuum, the *inclined-tube manometer* in Fig. 35(*c*) is often used, since the head can be multiplied several times by the sloping liquid column. One leg of the U-tube makes an angle of 5 to 10 deg. with the horizontal, so that a 7- to 10-in. movement is obtained for pressure changes of 1 in. head of water. The inclined tube, however, is not readily adapted to recording instruments, and thus for small pressure a *liquid-sealed bell* is often employed. In this case (Fig. 36), the weight of the bell is shown supported by a spring, the pressure raising the bell and counteracting a part of the spring tension. A spring-loaded *slack-diaphragm* element, as in Fig. 37, is often used for low pressure and draft measurements. These units are well adapted for multiple mounting with up to 12 indicating units available in a single case. Diaphragms are usually made of flexible leather.

Perhaps the most important single pressure-measuring element in common use is the bourdon tube and its variations. The *bourdon tube* (Fig. 14), which consists of a thin flat metal tube (elliptical in cross section) sealed at one end, is in itself a spring that tends to straighten itself by an amount proportional to the increase in pressure. Conversely, vacuum causes it to curl up. The movement is transmitted through a linkage which, in the common

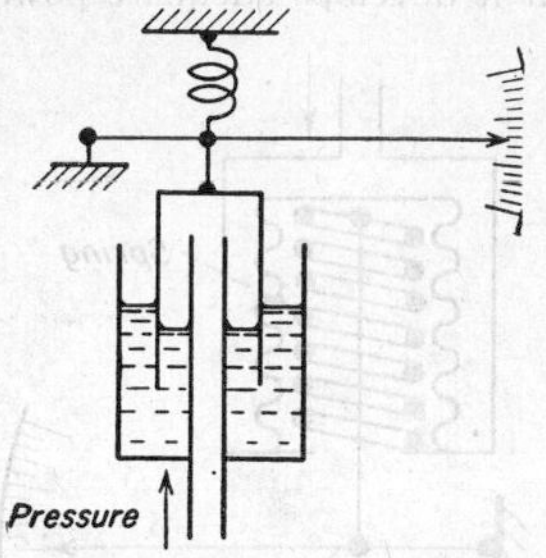

FIG. 36. Liquid-sealed bell.

pressure gage, is amplified by a rack and pinion to actuate a pointer. The shape, material, and wall thickness depend on the pressure it is designed to measure. For ranges up to about 1000 lb./sq. in., seamless drawn bronze tubing is generally used. Higher pressures up to 10,000 lb./sq. in., and where pressures of corrosive fluids are measured, require use of alloy steel. Protection again pulsating pressures is desirable in some applica-

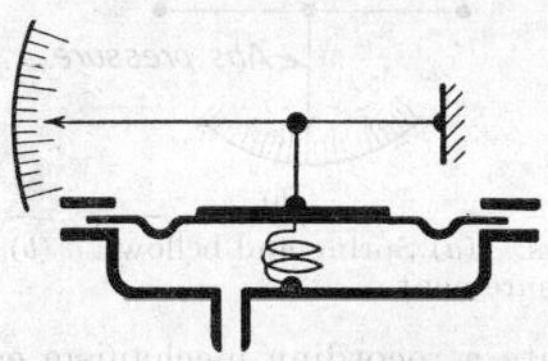

FIG. 37. Slack diaphragm.

tions. This may be provided by using an element constructed of two bourdon tubes opposing each other, arranged to move a single pointer through a double linkage. Commonly used are pulsation dampeners or pressure snubbers of various types, described in a subsequent paragraph. Protection against corrosive fluids may be supplied by a liquid seal or a diaphragm seal mounted below the element. A single coil trap that traps condensate in the gage may be used to protect the

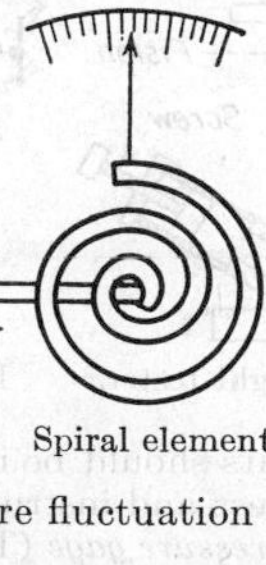

FIG. 38. Spiral element.

element from temperature fluctuation in measuring pressures of live steam.

The most important variations of the bourdon tube are the *spiral* element shown in Fig. 38 and the *helical* type of element in Fig. 15. Used extensively in recorders for pressures up to 1000 lb./sq. in. and for pressure thermometers, the spiral and helix provide greater deflection for a given pressure change than the bourdon tube to eliminate the need for a multiplying mechanism. The center shaft transmits the deflection of the helix to the pointer or a pen arm, while the spiral may be linked direct. These elements can withstand fairly high overloads without damage.

Widely employed for comparatively low pressures (up to 100 lb./sq. in.) and for absolute pressures is the *bellows* type of element utilizing a metallic bellows with or without a spring. The important advantage of this type of element is that it develops adequate power at low pressures to activate a recording mechanism easily. Figure 18 shows a bellows element, but greater accuracy of calibration may be obtained by using a *spring and bellows*, as in Fig. 39(*a*). For measurement of vacuum or absolute pressure, two bellows are utilized, as shown in Fig. 39(*b*). One bellows is evacuated to substantially zero pressure, and the other is hooked up to the vacuum to be measured. The indication by balance between the two bellows is thus corrected for atmospheric pressure variations and can be calibrated in terms of absolute pressure.

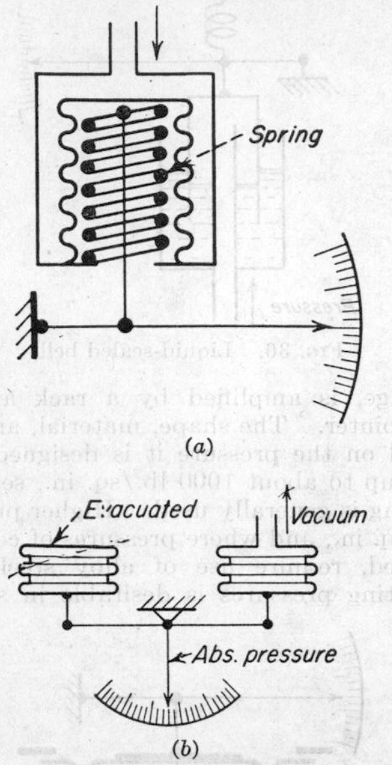

FIG. 39. Bellows. (*a*) Spring and bellows. (*b*) Double bellows for vacuum measurement.

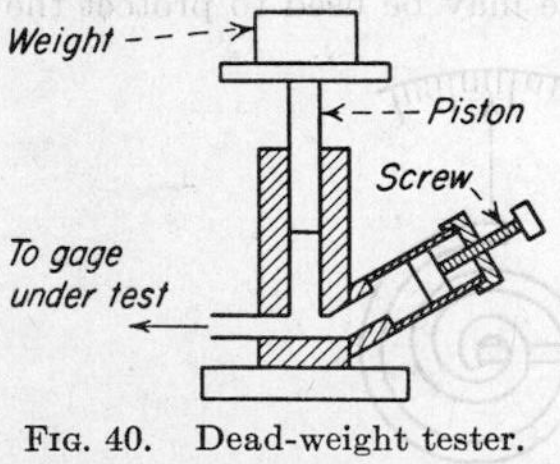

FIG. 40. Dead-weight tester.

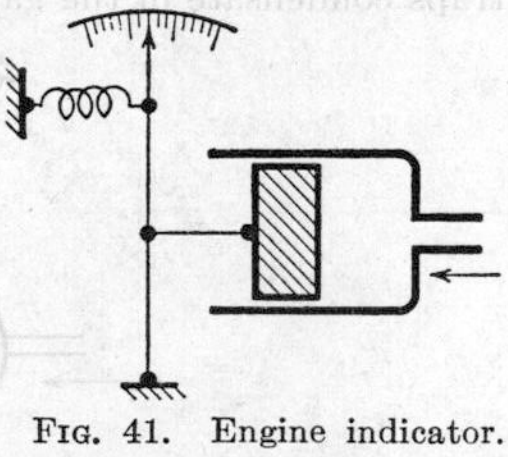

FIG. 41. Engine indicator.

Vacuum instruments should be installed with large short lines between process and instrument.

The *volumetric pressure gage* (Taylor) uses a compressible bulb containing mercury, which is connected to a bourdon tube by means of temperature-compensated tubing. It is used in applications such as homogenizers where it is impractical to allow the fluid to enter the measuring element. Diaphragm seals may also be used in such application. The sketch of Fig. 40 shows a *dead-weight tester*, which is the fundamental method of producing pressures for testing of pressure gages. Here a piston of known area weighted with a known weight produces a known pressure in the cylinder which can be communicated to a pressure gage. In use, the screw-operated piston is used to force liquid into the cylinder so that the pressure piston is raised. To avoid friction effects, measurements are made with the weight tables turning slowly. An occasionally used method of pressure measurements is an *engine indicator*, as shown in Fig. 41, which is a piston-and-spring combination.

High pressures above 1000 atm. may be measured inferentially with a manganin gage. A bridge circuit made of 5-mil-diameter manganin wire is inclosed in a block so that all arms will be at about the same temperature. Each arm of the bridge is adjusted to have a resistance of 600 ohms, and the resistors are annealed for about 12 hr. at 125°C. Two opposite arms of the bridge are submitted to the pressure to be measured, which is transmitted hydraulically by means of oil from the process to the instrument. Resistance of the manganin wire is proportional to the pressure, and the resistance can therefore be considered an indication of the pressure.

Pulsation Dampeners. Rapidly fluctuating pressures are ordinarily objectionable not only because they make chart readings difficult, but because they cause undue wear on the instrument. Pulsation dampeners may be used to smooth out the fluctuations. A deadener of fine tubing in the form of a coil inserted in the line is sometimes effective. In one type of pressure snubber (Ray), the restriction is the annular passage formed by a hole containing a wire of size selected to give the desired restriction. The wire can move with pressure changes, tending to prevent plugging by foreign matter. A pulsation dampener (Fig. 42) (Taylor) for this purpose is similar to an automobile muffler. It consists of a tube of about 1 in. in diameter containing a number of compartments formed by diaphragms with a small hole in the middle of each one. A fine wire passes through the holes in the center of each diaphragm to give the correct size of area of annular space.

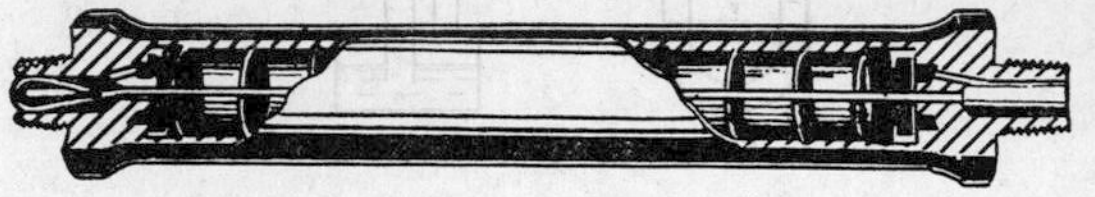

FIG. 42. Pulsation dampener.

High Vacuum. Numerous advancements in vacuum processing have brought about the need for accurate gages for measuring and recording high vacuums from 1 mm., equivalent to 1000 μ, down to 1 μ mercury pressure absolute. Although devices for measuring such low pressures have long been available for laboratory use, only in recent years have they been designed for industrial applications.

The ultimate standard for vacuum measurement is the *McLeod gage* from which all other types of vacuum gages are commonly calibrated. This device involves direct measurement of vacuum in terms of absolute pressure of mercury. One type of McLeod gage (Stokes) (Fig. 43) operates on a swivel from charging position *a* to measuring position *b* and cuts off a definite volume of rarefied gas at the unknown pressure of the vacuum system. The trapped gas is compressed to a smaller volume at a higher pressure, the value of which is equal to the difference in level between the mercury in the center tube *C* and the compensating capillary. From Boyle's law a suitable scale can be calibrated to represent the pressure value in the vacuum system. This unit can be used for recording by means of a resistance-type electrical measuring system which indicates the height of mercury in the measuring

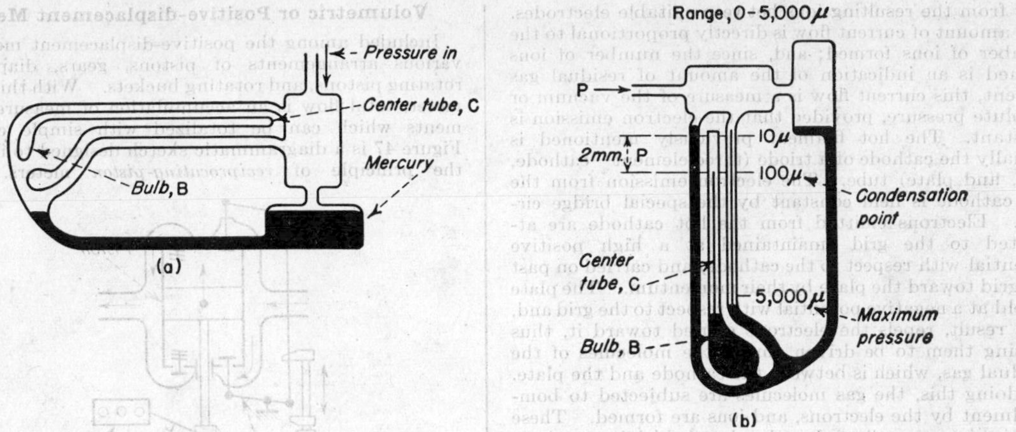

FIG. 43. McLeod gage (Stokes).

tube by changing the resistance in accordance with changes in the mercury level. The resistance value is, in effect, a measurement of the pressure in the vacuum system. Obviously this type of gage can measure only at intervals. It has a wide range from a fraction of a micron to 5000 μ. Since Boyle's law applies only to non-condensible gases, where water vapor or other condensible gases are present, a false reading will result to indicate a higher vacuum than actually exists. This source of error may be eliminated by a chemical trap to absorb the condensible vapor or a freezing trap utilizing dry ice or liquid air.

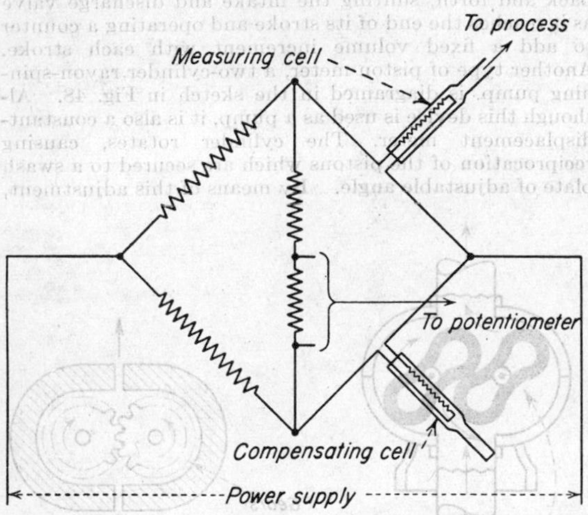

FIG. 44. Pirani circuit.

Thermal Gages. The *Pirani gage* operates on the principle that the heat loss from a filament of resistance wire through which there is a constant flow of electric current is a function of the pressure surrounding the filament. The heat loss, from conduction and convection, is dependent on the number of gaseous molecules surrounding the filament. At low pressures, from 1 μ to 1 mm. mercury, there is considerable change in temperature. Figure 44 represents a Pirani circuit. The open-ended tube is exposed to the vacuum system where pressure is to be measured. The closed-off tube is sealed off at a pressure lower than 1 μ and acts to compensate for temperature variations.

Another type of thermal gage is shown in Fig. 45. Four filaments are each continuously and uniformly heated by currents from separate windings of the power transformer and are maintained at constant values by resistances in series with each filament circuit. Two filaments are in a sealed reference chamber at a pressure of 1 μ mercury. The other two are placed in the vacuum system to be measured. A small sensitive thermocouple

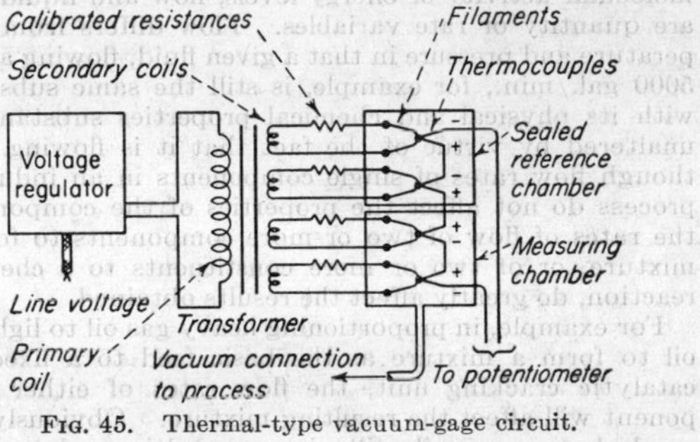

FIG. 45. Thermal-type vacuum-gage circuit.

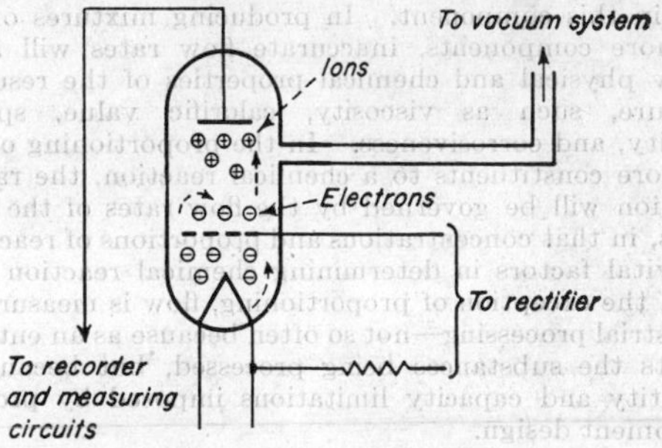

FIG. 46. Ionization gage.

is attached to each filament, with all four thermocouples hooked up so that the e.m.f. from one set of couples in the reference chamber tends to cancel the e.m.f. from the other set of couples, leaving only a residual e.m.f. to be measured. This, then, is a measure of the difference in pressure between the vacuum system and the reference cell.

Operation of the *ionization gage* (see Fig. 46) is based on the ability of electrons emitted from a hot filament to bombard the molecules of the residual gas in an evacuated system and thereby form an electric-current flow from the resulting ions between suitable electrodes. The amount of current flow is directly proportional to the number of ions formed; and, since the number of ions formed is an indication of the amount of residual gas present, this current flow is a measure of the vacuum or absolute pressure, provided that the electron emission is constant. The hot filament previously mentioned is actually the cathode of a triode (three-element—cathode, grid, and plate) tube. The electron emission from the hot cathode is held constant by the special bridge circuit. Electrons emitted from the hot cathode are attracted to the grid (maintained at a high positive potential with respect to the cathode) and carried on past the grid toward the plate by their momentum. The plate is held at a negative potential with respect to the grid and, as a result, repels the electrons carried toward it, thus causing them to be driven among the molecules of the residual gas, which is between the cathode and the plate. By doing this, the gas molecules are subjected to bombardment by the electrons, and ions are formed. These positive ions are collected on the plate (which is negative). As a result, a current flow to the plate occurs which is proportional to the number of ions formed. Thus the intensity of the current flow is a measure of the absolute pressure in the system.

FLOW MEASUREMENT

Although temperature and pressure may be referred to as energy variables in that they are manifestations of molecular activity or energy levels, flow and liquid level are quantity or rate variables. Flow differs from temperature and pressure in that a given fluid, flowing at 5 or 5000 gal./min., for example, is still the same substance with its physical and chemical properties substantially unaltered by virtue of the fact that it is flowing. Although flow rates of single components in an industrial process do not affect the properties of the components, the rates of flow of two or more components to form a mixture, or of two or more constituents to a chemical reaction, do greatly affect the results obtained.

For example, in proportioning heavy gas oil to light gas oil to form a mixture as the basic feed to a fixed-bed catalytic cracking unit, the flow rates of either component will affect the resulting mixture. Obviously, too much heavy gas oil will give a resulting mixture too rich in this component. In producing mixtures of two or more components, inaccurate flow rates will affect many physical and chemical properties of the resulting mixture, such as viscosity, calorific value, specific gravity, and corrosiveness. In the proportioning of two or more constituents to a chemical reaction, the rate of reaction will be governed by the flow rates of the reactants, in that concentrations and proportions of reactants are vital factors in determining chemical reaction rate. With the exception of proportioning, flow is measured in industrial processing—not so often because as an entity it affects the substances being processed, but because of quantity and capacity limitations imposed by process-equipment design.

Fluid flow is one of the most important process variables and may be measured by direct or inferential means. Direct measurement, seldom used where automatic control is desired, is accomplished with volumetric devices, such as positive-displacement meters. Here the flowing fluid is divided into definite unit volumes by the primary element, and the units are then counted or totalized automatically. Inferential methods, however, measure the instantaneous value of rate of flow. Instruments in this class actually determine the flow rate by measurement of related variables, such as differential pressure or head, orifice areas, velocity, or addition of energy; and they are most widely used in the process industries.

Volumetric or Positive-displacement Meters

Included among the positive-displacement meters are various arrangements of pistons, gears, diaphragms, rotating pistons, and rotating buckets. With this type of meter, total flow is an accumulation of measured increments which can be totalized with simple counters. Figure 47 is a diagrammatic sketch designed to illustrate the principle of *reciprocating-piston* meters. Under

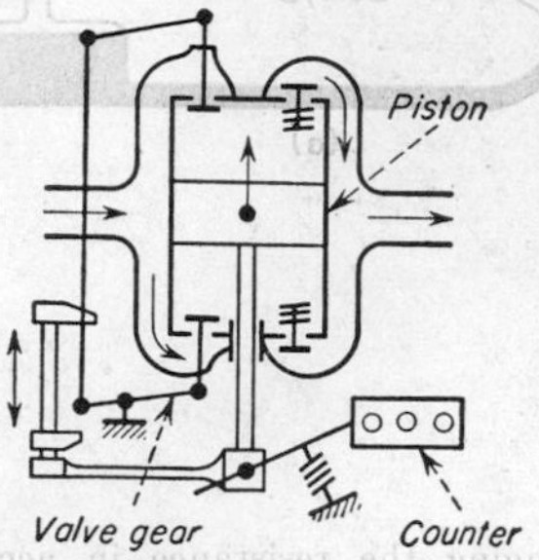

FIG. 47. Reciprocating-piston meter.

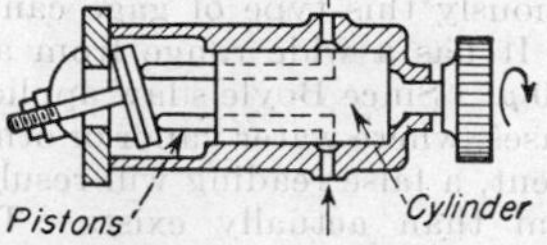

FIG. 48. Constant-displacement pump.

pressure of the fluid being metered, the piston shuttles back and forth, shifting the intake and discharge valve as it reaches the end of its stroke and operating a counter to add a fixed volume increment with each stroke. Another type of piston meter, a two-cylinder rayon-spinning pump, is diagramed in the sketch in Fig. 48. Although this device is used as a pump, it is also a constant-displacement meter. The cylinder rotates, causing reciprocation of the pistons which are secured to a swash plate of adjustable angle. By means of this adjustment,

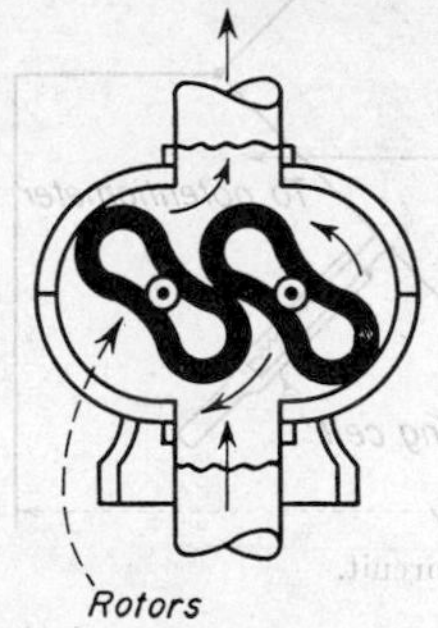

FIG. 49. Rotating-impeller-type meter.

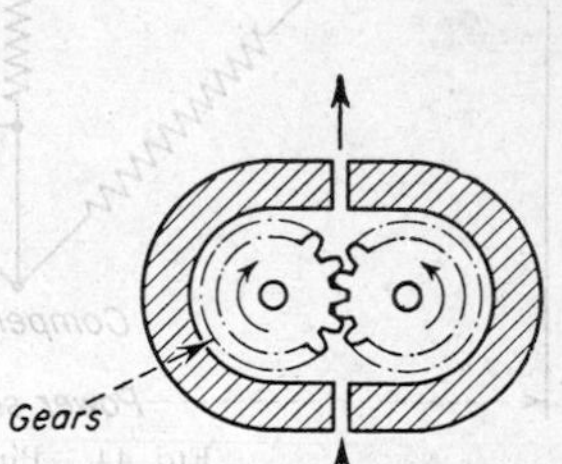

FIG. 50. Gear-pump-type meter.

delivery of the pump per revolution can be adjusted precisely. A type of meter employing *rotating impellers* for positive displacement is illustrated in Fig. 49. This type (Roots) may be considered the limit of a *gear pump* shown in Fig. 50, which can also be used as a positive-displacement meter. The common *nutating-disk* meter used for domestic water metering and shown in Fig. 51, uses a form of rotating piston. The tilted disk is cut by a stationary vertical partition near the discharge, which causes the space above and below the disk to form separate chambers. The pressure of water passing through then causes the disk to describe a continuous nutating action, causing rotation of a counter.

A number of elementary types of constant-volume meters in addition to those mentioned have been also used. The primitive *tilting-box* type shown in Fig. 52 is probably not used today but has been important in the past. The *condensate meter* shown in Fig. 53 is, however, a modern development of this idea. The vanes shown here in cross section are closed at the ends to form buckets which rotate at a rate directly proportional to the

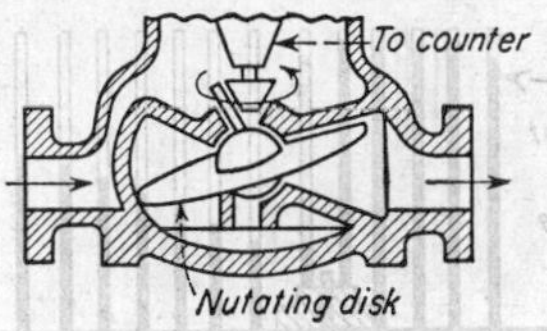

FIG. 51. Nutating-disk meter.

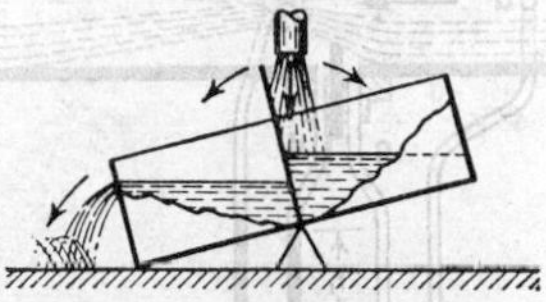

FIG. 52. Tilting-box meter.

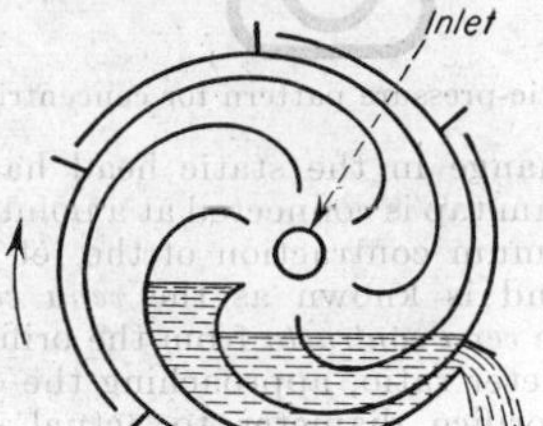

FIG. 53. Condensate meter.

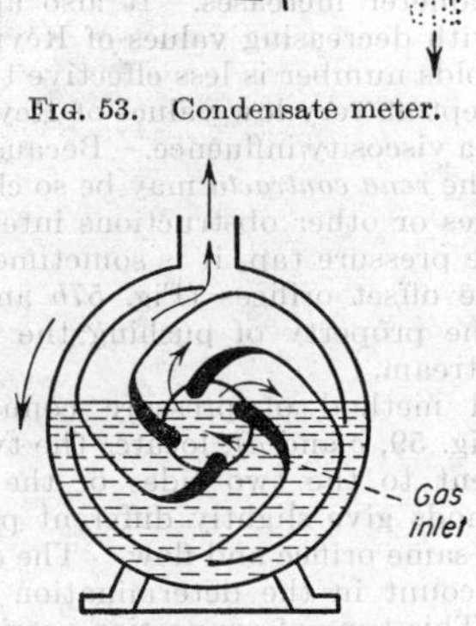

FIG. 54. Water-sealed gas meter.

flow. A similar principle is employed in the *water-sealed gas meter*, frequently used in gas plants. This meter, shown in Fig. 54, uses water contained in the lower half of a casing as a sealing and valving agent to control the passage of gas into the various compartments which are caused to rotate by the pressure of the gas. A different, more complex principle is used in the ordinary domestic *diaphragm gas meter*. Figure 55 attempts to illustrate the principle. This type of meter is difficult to portray in simple sketch, and the diagram is merely suggestive. Two partitions and two diaphragms form four chambers numbered 1, 2, 3, and 4. Reciprocation of the diaphragms operates cranks which in turn rotate a single crankshaft driving the counter and the slide valve. which alternately connect the chambers with the endless end of the box and with the discharge. Figure 56 illustrates the *meter prover*, which is merely a small water-sealed gas holder in which a known volume of gas at adjustable pressure can be stored. Since the amount of gas it discharges can be determined exactly, this device is often used for meter testing and calibration.

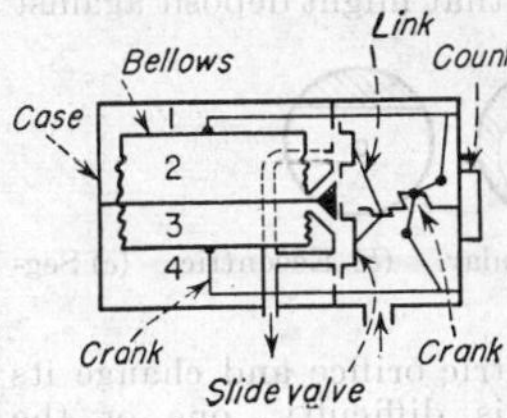

FIG. 55. Diaphragm gas meter.

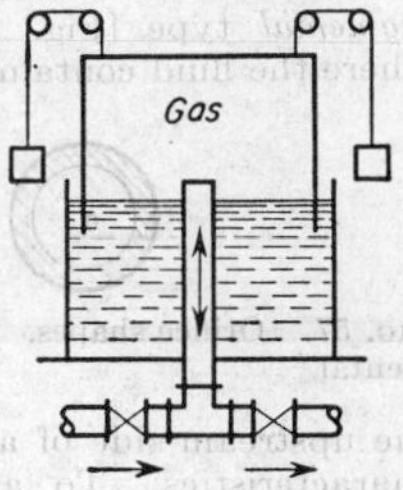

FIG. 56. Water-sealed gas holder.

Rate-of-flow Meters

Inferential methods used to measure rate of flow are more important to the process industries than are the volumetric methods. Most flow control, for example, is accomplished with differential flowmeters. Based on the general law for the flow of fluids, rate of flow in closed conduits or pipes can be determined by measuring the differential pressure or head across a restriction of constant area, by measuring the velocity head directly, and by measuring the area with a constant differential pressure across a restriction in flow. Generally, flowmeters consist of a primary device such as an *orifice* to produce *differential pressure*, and a unit such as a manometer to measure and to translate the differential pressure in terms of flow rate.

Bernoulli's theorem (p. 375) is a corollary of the law of conservation of energy, which states that the total energy of unit weight of a continuous mass of fluid passing through an isolated system remains constant from point to point along its line of flow, without regard to what may be done with the fluid within the system. An isolated system precludes the addition of energy from the outside or the removal of energy to the outside. Hence, except for changes due to friction or volume variation which change the heat content of the system, the total mechanical energy of the fluid will remain constant from point to point. Since variation in the heat content can be compensated or made practically negligible, these facts form the basis of three methods of flow measurement.

In the first, the velocity of the fluid is changed momentarily as it passes through a constriction, and the resulting change in static head is taken as a measure of the velocity and hence of the volume rate of flow. The devices for producing this change in static head are as follows: *Orifices, flow nozzles,* and *venturi tubes.* In the second method, the differential pressure remains constant, and the constriction area varies with the flow rate in what is called an *area meter*. Third, but less important, is the *pitot tube*, which measures velocity head directly without appreciably affecting the flow pattern of the fluid being measured. The pitot tube measures the difference between the impact pressure head and the static pressure head. Also depending on head pressure is the *weir meter*, which determines the flow rate of liquid flowing in an open channel by measuring the height of the crest above a dam over which the liquid is flowing or above the bottom of a notch cut in the dam through which the flow passes.

Orifice Plates. These are the most common form of restriction used to produce differential pressures, because they are easy to reproduce, low in cost, simple to install, and easily changed. Pressure taps are located on either side of the orifice plate so that the differential pressure between the upstream and downstream sides can be measured. The orifice is most often a circular hole in a metal diaphragm $\frac{1}{32}$ to $\frac{1}{16}$ in. (preferably the lower) thick [Fig. 57(*a*)], placed concentrically between flanges in the flow pipe. Occasionally it is preferable to use an offset orifice of either the *eccentric* [Fig. 57(*b*)] or the *segmental* type [Fig. 57(*c*)]. These orifices are useful where the fluid contains solids that might deposit against

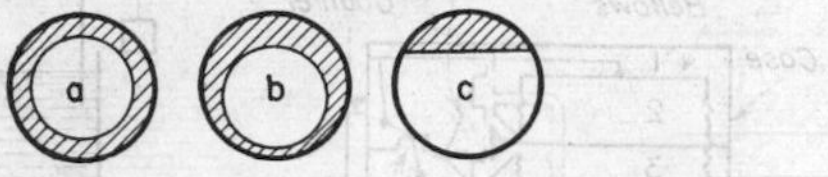

FIG. 57. Orifice shapes. (*a*) Circular. (*b*) Eccentric. (*c*) Segmental.

the upstream side of a concentric orifice and change its characteristics. To avoid this difficulty, one or the other may be used with the unobstructed side down so as to prevent solids from accumulating. Condensate in steam or vapor lines is often disposed of by a small hole in the orifice plate on the bottom side. For conditions of widely varying flow, adjustable orifices of the segmental type can be used. One type resembles a gate valve in construction, except that a micrometer screw is used to adjust the position of the plate, and pressure taps are installed, as in Fig. 58.

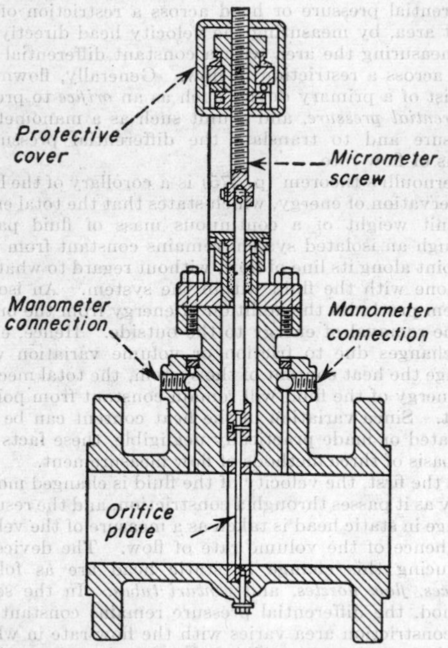

FIG. 58. Adjustable orifice.

Action of a concentric orifice may be observed in Fig. 59. Sudden constriction of the flow causes the jet to contract below the orifice, reducing the flow to considerably less than the theoretical. In accordance with Bernoulli's theorem, any change in the velocity through the orifice must change the static head at points on either side of the orifice. For any pair of such points, the static-pressure difference will vary as the difference of the squares of the velocities at and before constriction, and it is this fact which is used in determining the flow rate. But there is also a static-pressure variation from point to point on either side of the orifice. An indication of the character of this variation is given by the vertical lines in Fig. 59. It will be noted that the differential pressure h across the orifice will vary slightly depending on the points chosen for the pressure connections. The pressure connections must therefore be made for any orifice at the same points used in calibrating the orifice.

Location of Pressure Taps. There are three principal methods of locating the pressure connections. In the first the upstream tap is connected at a point a (Fig. 59)

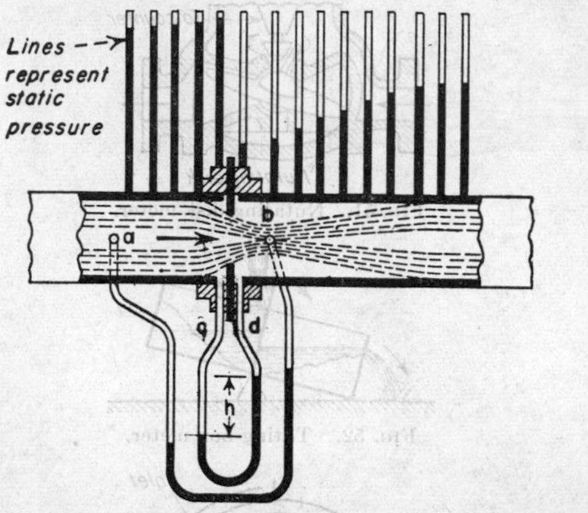

FIG. 59. Static-pressure pattern for concentric orifice plate.

before any change in the static head has taken place. The downstream tap is connected at a point b, which is the point of maximum contraction of the jet and minimum static head and is known as the *vena contracta*. The distance of the *vena contracta* from the orifice plate varies with the diameter ratio, approaching the orifice plate as the ratio of orifice diameter to actual (not nominal) inside-pipe diameter increases. It also approaches the orifice plate with decreasing values of Reynolds number, but the Reynolds number is less effective than the diameter ratio, except at very low values of Reynolds number where there is a viscosity influence. Because with certain orifice ratios the *vena contracta* may be so close to the orifice that flanges or other obstructions interfere with the location of the pressure tap, it is sometimes desirable to use one of the offset orifices (Fig. 57*b* and *c*), both of which have the property of pushing the *vena contracta* farther downstream.

The second method of pressure connection, called *flange taps* (Fig. 59, *c* and *d*), locates the two taps in the flanges adjacent to the two sides of the orifice plate. The two methods give slightly different pressure differentials for the same orifice and flow. The differences are taken into account in the determination of the orifice coefficients. This type of connection sacrifices a certain amount of pressure differential, but for most installations it is just as satisfactory as the *vena contracta*.

A third method uses *pipe-line taps* in which the upstream tap is located $2\frac{1}{2}$ pipe diameters above the orifice, and the downstream tap 8 diameters below the orifice to measure only the permanent pressure drop. Although this method permits considerable tolerance in tap location, the differential in this case is considerably smaller than that obtained by either of the two foregoing methods, and pipe surface has a greater effect on discharge coefficients, creating inaccuracies. This tap arrangement is not generally used for large pipe installations but is often used with small pipe of less than 3 in. in diameter where it is difficult to place taps of adequate size in the space available on a small pipe. It may also be used where a large flow must be measured with a meter of limited range, since there is a lower differential than with the other two tap arrangements.

It was noted above in connection with Fig. 59 that the jet below an orifice contracts and reduces the flow to less than the theoretical. At the *vena contracta*, the area of the jet is about 62 per cent of the orifice area. Also friction through the orifice reduces the flow 1 or 2 per cent below the theoretical. The net result is that a sharp-edged orifice will pass only about 61 per cent of the theoretical quantity of fluid. This value, 0.61 is called the discharge coefficient. Its exact value varies slightly, depending on the ratio or orifice diameter to internal pipe diameter and the Reynolds number, and it must therefore be determined experimentally. In practice, the discharge coefficient may represent more than simply the percentage of theoretical flow that an orifice will

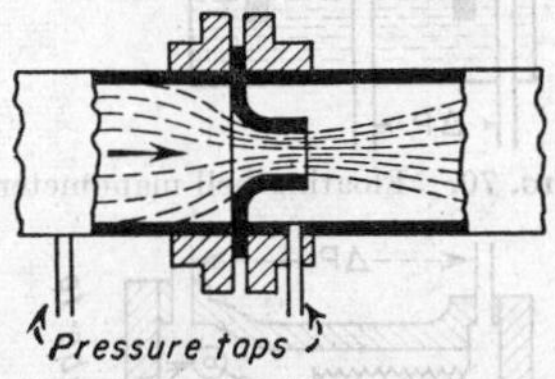

FIG. 60. Flow nozzle.

pass. It is also used to compensate for variation in the location of the pressure taps and frequently includes a quantity known as the "velocity-of-approach factor." These additional functions account in part for the variations in coefficient curves that are published by different investigators.

Flow Nozzles. For very high flow velocity, an ordinary orifice will not be sufficiently accurate. An orifice ratio d/D of more than 0.6 to 0.7 is usually not recommended, because inequalities in the pipe may give rise to inaccuracies. Where the flow is high, this difficulty may be surmounted by using a well-rounded orifice or flow nozzle which, for the same differential, is capable of passing considerably more fluid than a plate orifice of the same diameter. A flow nozzle is a rounded, funnel-like aperture similar to Fig. 60. Because the stream is constricted gradually, there is little or no contraction below the nozzle. Hence the discharge coefficient will

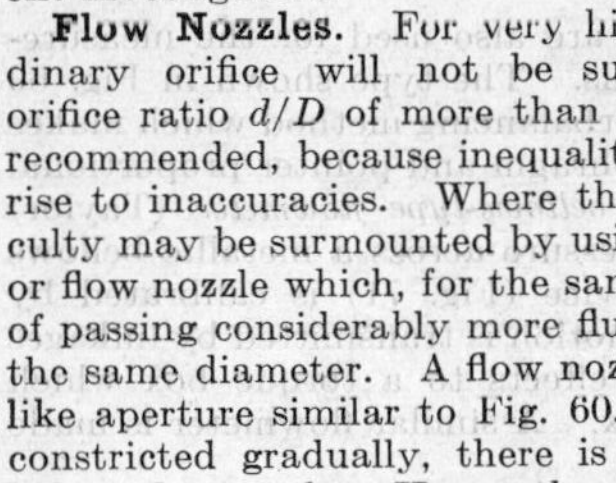

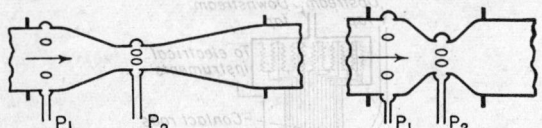

FIG. 61. Venturi tube.

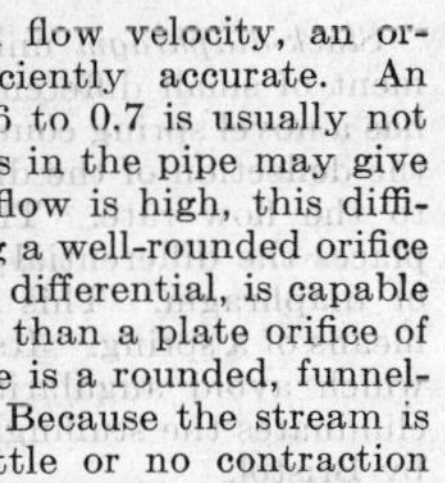

FIG. 62. Venturi nozzle.

range from 0.97 to 0.99 depending on smoothness and the accuracy of the design. For maximum accuracy, each different design of flow nozzle must be individually calibrated. Poor designs may have coefficients as low as 0.90.

Venturi Tubes. Figure 61 illustrates this type of construction (see pp. 400, 406), which is the most accurate, has a discharge coefficient close to unity, and is capable of nearly complete downstream pressure recovery. The venturi nozzle (Fig. 62) is similar to the venturi tube but approximates the flow nozzle in operating characteristics. These are used mainly where large flows must be measured and when maximum pressure recovery is desired.

Installation of Orifices. Careless installation can easily spoil the accuracy of a good orifice meter. For best results the orifice should be placed in a long, straight run of pipe with no change of direction or obstruction, such as a valve or fitting, closer than 10 to 50 pipe diameters on the upstream side; or 1 or 2 pipe diameters on the downstream side. It may, however, be placed somewhat closer to a large-radius bend. Particularly if curves preceding the orifice are not all in one plane, straightening vanes (Fig. 63) should be placed in front of the orifice not less than 6 pipe diameters upstream. Such vanes cut the pipe for a short distance into a number of smaller straight passages and tend to eliminate turbulence and swirling. Pulsating flow particularly must be avoided. Orifices must be kept clean and sharp. If solids build up around the upstream periphery, the velocity will be increased and the accuracy badly impaired. Orifice fittings are available with which it is not

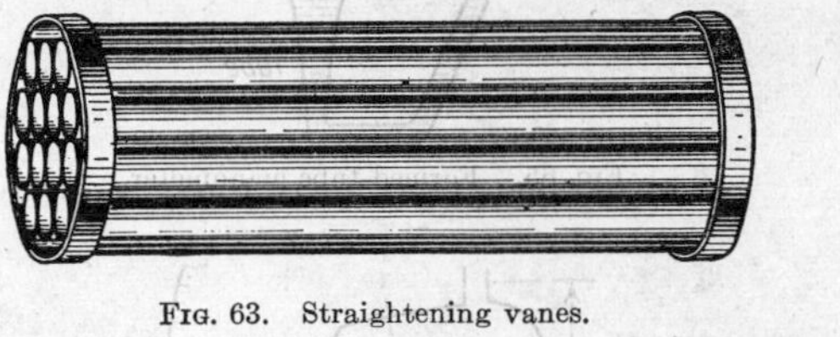
FIG. 63. Straightening vanes.

necessary to shut down the line when orifice plates are removed to be cleaned or changed.

Differential Flowmeters. These are essentially pressure-measuring devices for translating differential pressures caused by orifices or other restrictions into flow rates. Since flow varies as the square root of the differential, flow-graduated meters must use non-uniformly divided square-root charts or must be designed to extract the square root of the pressure differential automatically. Where total flow is to be ascertained, the meter must sum up the succession of instantaneous flow increments by solving the integral of *qdt*.

Manometers are most widely used. Glass-tube manometers are sometimes used as indicating meters, especially in laboratory or pilot-plant work. Various types are employed industrially. Manometer flowmeters usually operate from a metal float in an enlarged leg of a

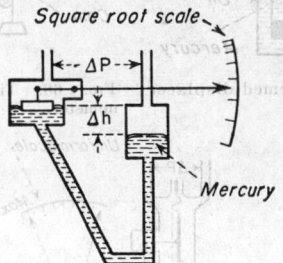

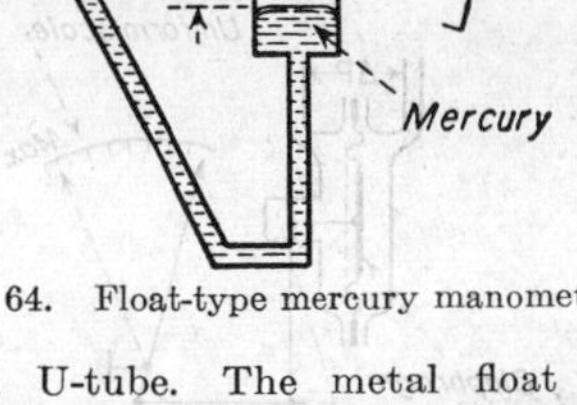

FIG. 64. Float-type mercury manometer.

mercury-filled U-tube. The metal float follows the changes in level of the mercury in the chamber as the differential pressure changes with rate of flow. Motion of the float is transmitted through a pressuretight shaft to the indicator or recording pen in mechanical meters. Electrical meters transmit the motion of the float by induction coils or other electrical means. Figure 64 shows the common *float-type mercury manometer* which gives a direct measure of the pressure differential p, which is proportional to the square of the flow rate. The scale calibrated in flow is unevenly divided. Many methods are used for extracting the square root of the differential. One common method is to interpose a shaped cam between the float and the pointer. Another useful method, shown in Fig. 65, is to employ a specially formed tube for one leg of the manometer. Other methods are shown in Figs. 66 to 69.

For lower pressure ranges, *floating-bell manometers* are often used, as in Fig. 70. In order to extract the square root of the differential with a bell-type instrument, a formed bell may be used as in Fig. 65 (Bailey). A uni-

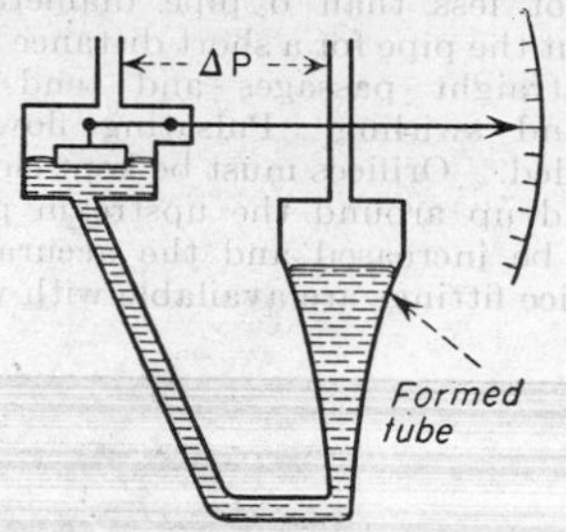

FIG. 65. Formed-tube manometer.

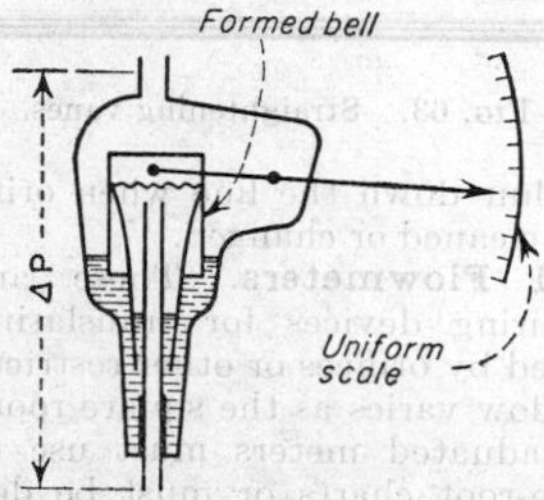

FIG. 66. Formed-bell manometer.

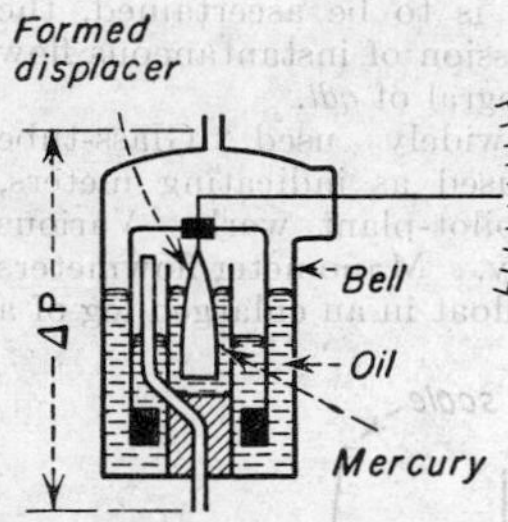

FIG. 67. Formed-displacer meter.

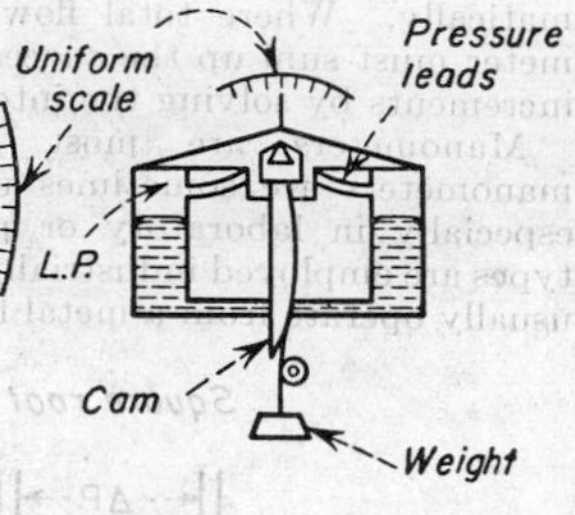

FIG. 68. Tilting-type manometer.

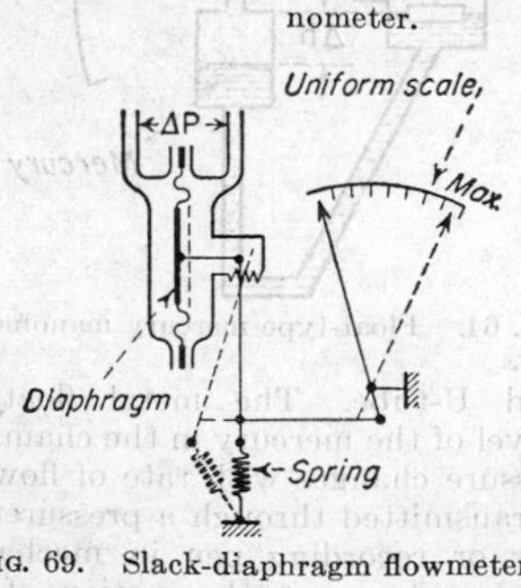

FIG. 69. Slack-diaphragm flowmeter.

form scale is thus achieved. Another method of extracting the square root with a bell-type instrument is shown in Fig. 67. This, for lower differential pressures, uses a formed displacer in a mercury well to give a variable counterbalancing to the bell. Still another manometer, shown in Fig. 68, is of the *tilting type* (Cochrane) in which the manometer is suspended on a pivot, tipping in proportion to the differential. This type can be made to tilt in proportion to the flow rate rather than the differential by the addition of a shaped cam bearing against a weighted tape. Similar in principle is the *ring-balance* flowmeter where the manometer is balanced on a rotating shaft or knife-edge bearings. The differential pressure displaces the mercury in the manometer, and the manometer rotation is transmitted directly to the indicator. It has a wide range and is readily adjusted by means of weights or by adjusting the spring of a spring-loaded ring balance.

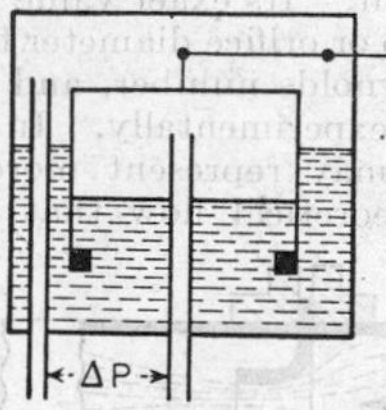

FIG. 70. Floating-bell manometer.

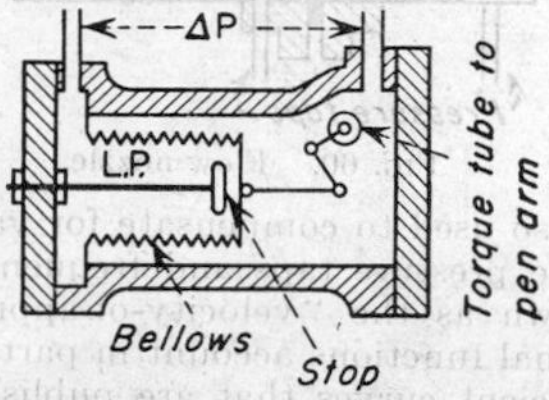

FIG. 71. Bellows-type flowmeter.

Slack-diaphragm units are also used for the measurement of small differentials. The type shown in Fig. 69 has a novel spring counterbalancing method which makes the deflection of the diaphragm and pointer proportional to the flow rate. The *bellows-type flowmeter* (Taylor) places the differential pressure across a metallic bellows or diaphragm. This device (Fig. 71) is calibrated by means of a spring. Its motion is transmitted by linkages which avoid angularity effects to a torque box which eliminates the stuffingbox. A similar flowmeter is made by Bristol.

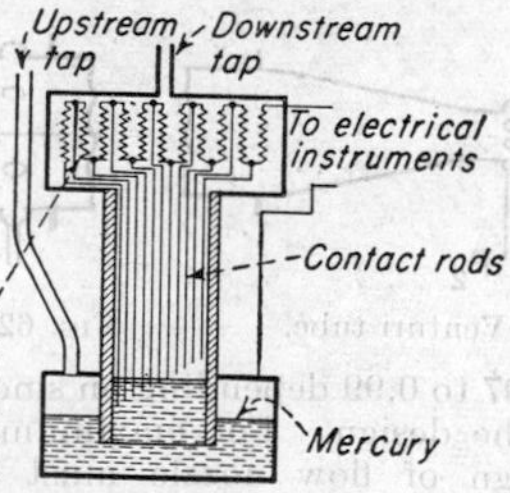

FIG. 72. Resistance (electrical) flowmeter.

Electric meters transmit manometer response to differential pressure changes out of the float chamber to the indicating mechanism. The induction-bridge meter has an armature inside the float chamber which is positioned by the float. The inductance coil is on the outside of the chamber and forms an inductance bridge with a receiving coil at the indicating or recording instrument. The bridge may be balanced by its own electrical force (self-operated) or by a potentiometer system (power-operated). A *resistance electrical* flowmeter (Republic) (Fig. 72) makes use of an ascending spiral of metal rods with which the rising mercury in the low-pressure leg of the manometer successively makes contact. The rods are connected to resistance coils so that the total resistance is proportional to the height of mercury in the manometer. By making the spiral of varying pitch, correction is made for the square-root relation between flow and differential pressure, and the chart divisions are uniform. With this instrument, a modified watt-hour meter may be used as an integrator or totalizer of flow.

Flow Integrators. A number of methods are used for totalizing flow. Figure 73 shows a simple integrator in which a square-root extracting cam positioned by the manometer float limits the motion of a reciprocating link which drives a counter C through a ratchet. Figure 74 illustrates diagrammatically a common type of integrator having a movable guard over a ratchet wheel which is

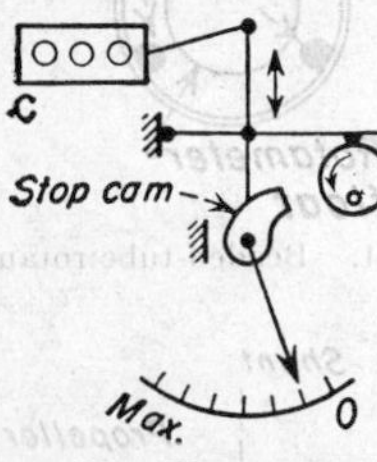

FIG. 73. Flowmeter integrator.

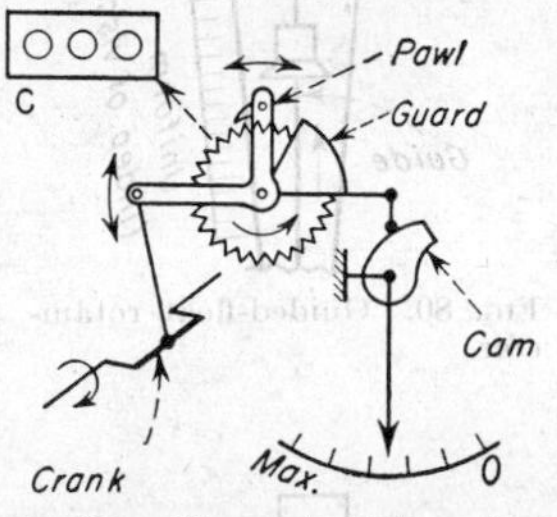

FIG. 74. Flowmeter integrator.

positioned by a square-root cam to limit the motion of an oscillating pawl which drives the ratchet wheel. The number of teeth by which the ratchet is advanced each oscillation thus becomes proportional to the flow rate. The wheel drives a counter C.

Another principle of square-root extraction which makes possible a simple method of integration has recently appeared in two modifications. That shown in Fig. 75 employs a flyball governor (Leeds & Northrup), and another type employs a gyroscope. Gyroscopes and flyball governors give a force proportional to the square of their speed and so can be used to balance a differential pressure, which itself is proportional to the square of the flow rate. The manometer in Fig. 75, designated as M, is a tilting type, maintained level by the force of flyball

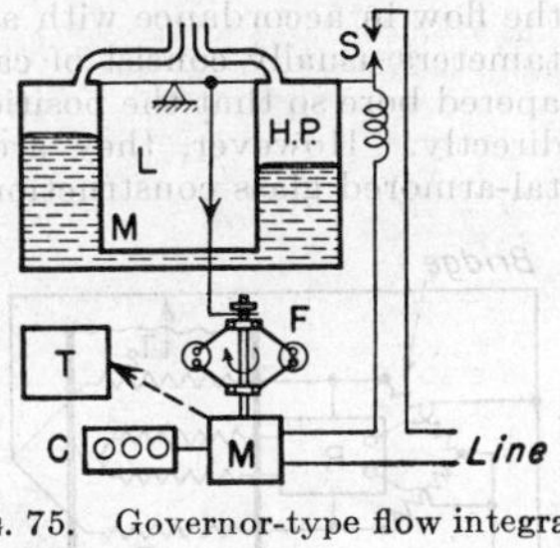

FIG. 75. Governor-type flow integrator.

governor F, which is driven by a motor. When the manometer tilts down to the left, contact S energizes the motor, which increases the flyball speed until link L has restored the balance and opened the contact. Thus the motor alternately speeds up and slows down, maintaining a speed closely proportional to the flow rate. A counter C totalizes the motor revolutions in terms of flow, and a tachometer T operates a flow-rate indicator or recorder.

Installation of Differential-type Flowmeters. Piping between orifice taps and the flowmeter is important, since improper piping arrangement accounts for many faulty flowmeter installations. Piping must be free of leaks and must be arranged to prevent trapping of air or gas at points that will disturb the pressure differential. This is especially important where operating pressures are of high magnitude. The distance between the orifice and the flowmeter is dependent on the slope required in the lines and the inertia of the system. Connecting piping should have a slope of at least 2 in./ft. For satisfactory operation, the distance between the flowmeter and the orifice should not exceed 100 ft. for steam and liquids or 50 ft. for gas measurement. Inertia and friction of the system together with the difficulties of removing air from the lines are the limiting conditions.

Seals and Purges for Orifice Meters. Where a condensible vapor such as steam is being metered, it is necessary to provide condenser or seal pots above the meter on each line between the orifice and the manometer, as shown in Fig. 76(*b*). These condenser pots are usually constructed of steel pipe, are about 4 in. in diameter and 10 in. long, and are installed to maintain substantially equal head of the condensed fluid on each leg of the manometer. Where a fluid is metered that might corrode the manometer or might foul the manometer liquid, it is necessary to provide a liquid seal, as shown in Fig. 76(*a*). Purge systems may be used for protection of the meter. This is done by passing a small flow of inert gas or liquid, too small to upset the flow measurement, into the system through each manometer leg. This prevents the fluid being measured from reaching the manometer. Air, steam, water, and oil are common purge fluids.

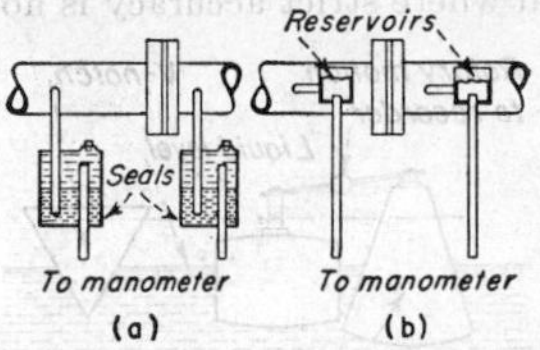

FIG. 76. Seals for orifice meters. (*a*) Liquid seals. (*b*) Condenser pots.

Pitot Tubes. Pitot tubes (see p. 397) are much used for portable and exploratory work but, because of the low differentials produced and the difficulties of calibration and clogging, are seldom used in permanent industrial installations. This method of flow measurement depends on changing the velocity head as in orifices and venturi tubes, and upon the measurement of the difference between the impact and static heads of the fluid. In measuring this difference, the two pressures are connected to the two sides of a manometer as in Fig. 77. The reading is the difference, or velocity head, $V^2/2g$ multiplied by a calibration factor, when the velocity can be calculated.

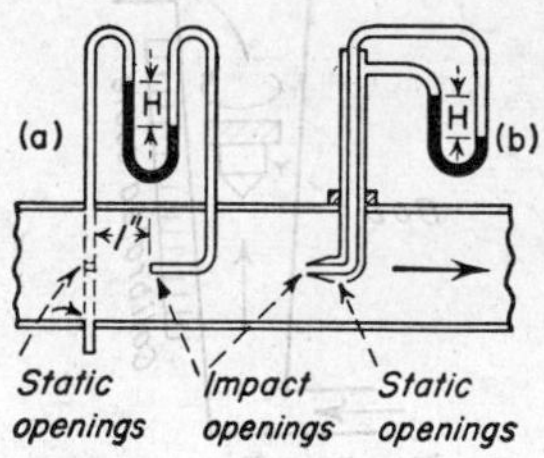

FIG. 77. Pitot tubes.

In using a pitot tube, account must be taken of the fact that velocity within the pipe is not uniform throughout the cross section. Velocity near the wall of a pipe, when there are no disturbances (valves, fittings, etc.) within 50 pipe diameters of the pitot tube, is about one-half the center velocity. The average velocity is at about one-fourth the radius from the pipe wall. The center velocity is about 20 per cent greater than the mean velocity on which the flow rate must be based. It cannot, however, be assumed that this is strictly the case when the pipe interior is rough or contains disturbing elements. In accurate work, particularly for permanent installation, it is necessary to explore the pipe at 10 to 20 points to determine the location of mean velocity or to determine the percentage decrease from center to mean velocity (velocity factor).

Weir meters are sometimes used to measure flow rates of fluid in open channels. One such meter shown in Fig. 78 employs shaped displacers over a V-notch weir which are positioned by buoyancy rather than by floating to position a pointer on a uniform scale. Weir meters are frequently used where strict accuracy is not required and

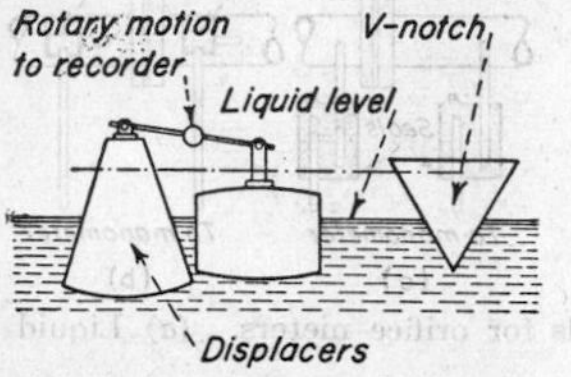

FIG. 78. Weir meter.

in many cases consist of a wooden weir box and a liquid-level gage or recorder. The indicator scale or chart can be calibrated in terms of a rate of flow and the flow totalized from the chart with a planimeter. Weirs are extensively used to measure slurries and fluids containing solids in suspension.

Area Meters.* The area flowmeter operates on the principle that there is a different orifice area for each rate of flow and that consequently the differential pressure is constant. The most important area meter is the rotameter (Fig. 79), in which a float is supported in the flowing stream by the differential pressure. Flow area is the annular opening between the float and the flow chamber. As the flow increases, the float moves upward, thereby increasing the area of the opening until the differential pressure just balances the weight of the float.

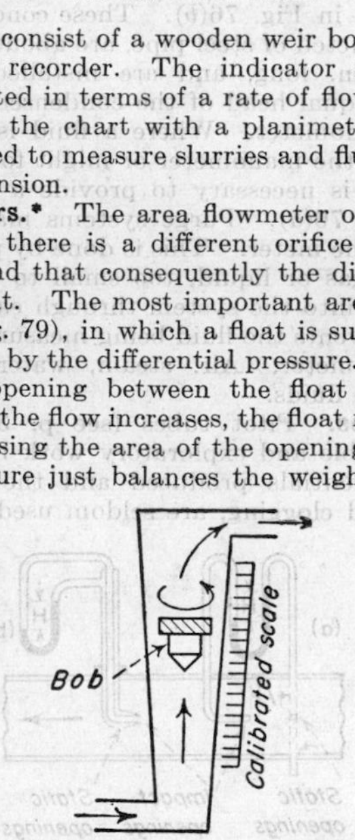

FIG. 79. Rotameter.

The float then assumes a position exactly proportional to the flow rate. The type shown in the sketch uses a rotating float or bob to produce the variable orifice but rides at a level that varies with viscosity and must be calibrated for the particular fluid handled. The type shown in Fig. 80 (Fischer & Porter) eliminates the effect of viscosity by use of a guided bob that has a sharp edge at the bottom. Predictable calibrations for this type are obtained from coefficient charts. An important variation is the beaded-tube *rotameter*. Here the tube contains three glass guides for guiding the float (Fig. 81). This eliminates any sidewise movement which might occur in the meter of Fig. 80. Most important, it permits float visibility where turbid fluids are measured. The rotameter is particularly well suited for measuring small flows; and, since the usual construction is of glass, it is useful to measure most corrosive liquids and liquids that must be free of any metallic contamination. It is used to control flow rate and for automatic ratio control.

Several other types of area meters are also used, such as the one in Fig. 82, which has a tapered plug rising in an orifice. *Piston and gate types* of variable orifices are also used. Scale calibrations of area flowmeters are linear only if the increase in area is linear with various float

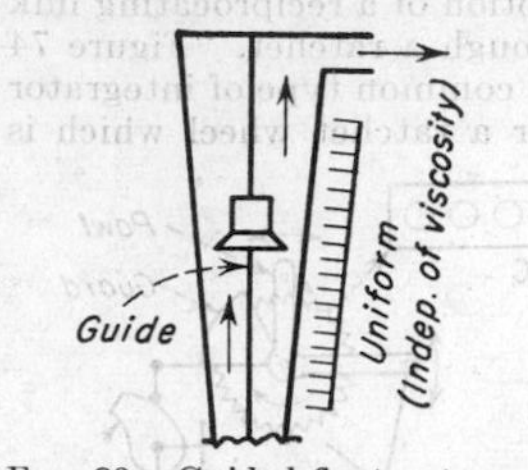

FIG. 80. Guided-float rotameter.

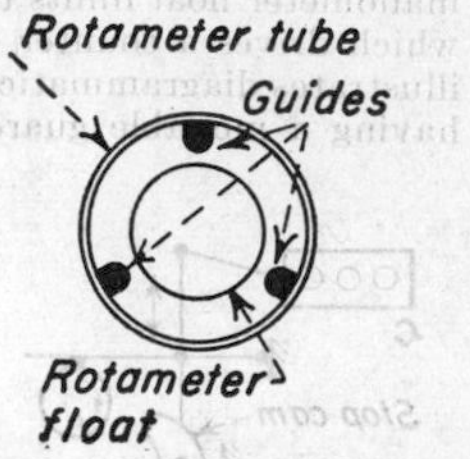

FIG. 81. Beaded-tube rotameter.

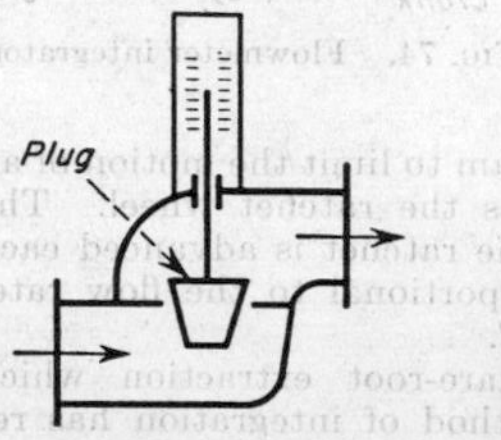

FIG. 82. Piston-type area meter.

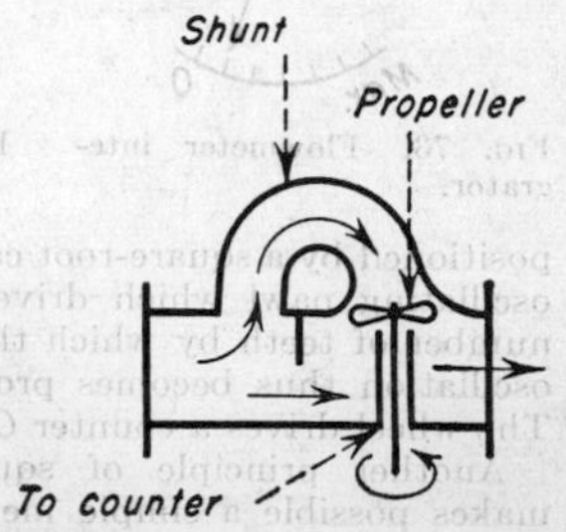

FIG. 83. Current meter on shunt meter.

positions. Area flowmeters of both rotameter and piston types often employ inductance bridge or other electrical or magnetic means such as a magnetic coupling (Fischer & Porter) to transmit the float position to an indicating, recording, or control instrument. However, rotameter flow control may be effected by use of photoelectric cells that regulate the flow in accordance with a selected float position. Rotameters usually consist of calibrated glass tubes with a tapered bore so that the position of the float can be read directly. However, they are available in metal and metal-armored glass construction as well.

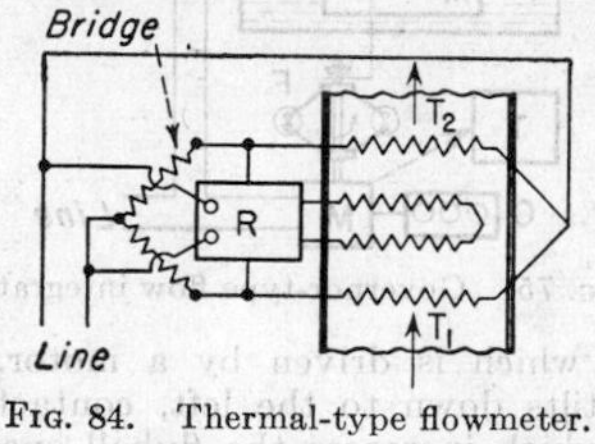

FIG. 84. Thermal-type flowmeter.

Medium to large flows are often handled with current meters in which the energy of the flowing stream is used to rotate some type of propeller or turbine at a rate directly proportional to the flow rate. The simplest type is the cup anemometer used in meteorology. One type of **current meter** (Builders Iron Foundry, Fig. 83), which is often used for steam and gases, is the shunt meter, in which a proportional part of the flow is forced through a shunt around an orifice, being measured in terms of the whole flow by means of a small propeller that drives a counter through a magnetic transmission and hydraulic drag.

* See p. 408.

The so-called method of mixtures can also be used for flow measurement; *i.e.*, energy or some detectable substance can be added in known amount to a flow of material, and its concentration analyzed downstream. Practically the only device applying this principle industrially is the energy-adding meter (Cutler-Hammer), which is used in the gas industry. As shown in Fig. 84, this meter adds electric heat to the flowing gas to give a fixed temperature rise between resistance thermometers T_1 and T_2, recording the energy added in terms of the amount of flow.

LIQUID-LEVEL MEASUREMENT

Measurement of liquid level is one of the fundamental measurements most frequently encountered in the process industries. Measurement of liquid level in a vessel may be made simply as a check on the supply of material on hand, as a means of determining the amount of liquid to be delivered to a process, or it may be the primary measurement in a control system designed to maintain level in a vessel that is a part of a continuous process. An important factor is the shape of the vessel in which liquid level is to be measured. The degree of accuracy depends on the vessel shape, since a tall vessel of small diameter can be more accurately measured than a flat vessel of large diameter. Conversely, where liquid level is to be controlled, it may be desirable to have a vessel with large horizontal cross section, since this will provide capacity to the controlled system. Obviously, the shape of the vessel will affect the sensitivity of the liquid-level instrument in volume measurement but will also indicate the type of instrument most desirable for the location.

Liquid level is measured both by *direct* and by *inferential means.* The first class includes the *gage stick* and *gage glass*, *floats*, and *buoyancy-type displacers*, and the method of weighing a small floating tank containing liquid at the same level as the main tank. Among the inferential methods is the common one of measuring the *hydrostatic head* of the liquid above a suitable reference point, and an inferential method used in boilers which depends on the difference in temperature between the steam and the water in the boiler drum. Float measurement is often the simplest type for small or medium ranges of level fluctuation, or for control. For wide fluctuations, the hydrostatic pressure head of liquid at a suitable reference point is usually measured, either directly by any suitable type of pressure-responsive mechanism, or indirectly, by measuring the pressure of an air column of equivalent pressure head.

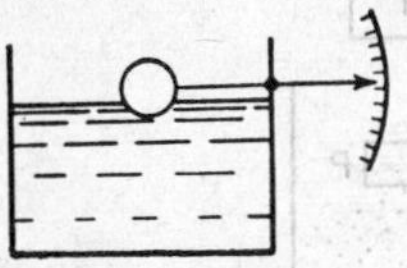

FIG. 85. Float indicator.

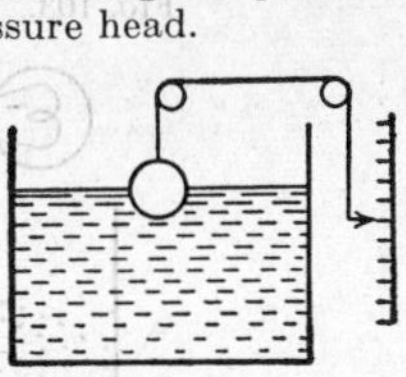

FIG. 86. Float, rope, and pulley indicator.

Figures 85 to 88 illustrate *float-type* methods. Figure 85 shows a simple float and pointer; Fig. 86 shows a simple float, rope, and pulley indicator; and Fig. 87 shows simple float-type non-indicating controller. Figure 88 illustrates interface level indication with a float that floats on the heavier of two immiscible liquids and sinks in the lighter. Interface level can also be indicated by the *buoyancy-type* level indicator illustrated in Fig. 89. In this last sketch, a displacer counterbalanced by a spring rises not with the level but by an amount proportional to the level changes. Depending on spring and displacer characteristics, any desired range of level variations can be handled with a small change in displacer position. Figure 90 illustrates a method frequently used with vessels under pressure. A small spring-counterbalanced weigh tank connected flexibly to the main tank indicates by its level a quantity proportional to the main tank level.

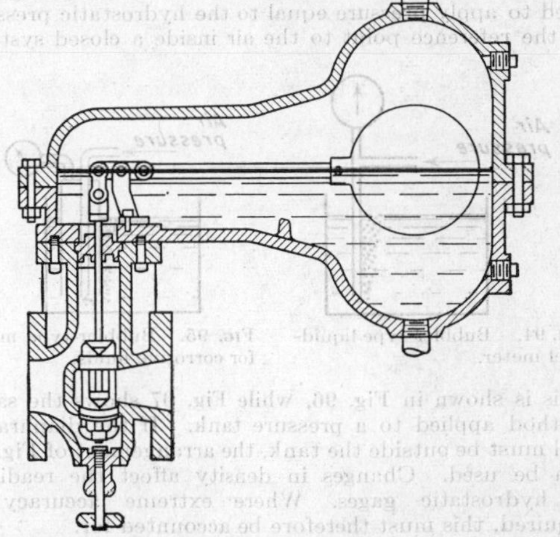

FIG. 87. Float-type non-indicating liquid-level controller.

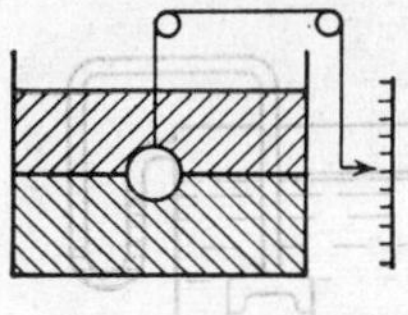

FIG. 88. Interface level indicator.

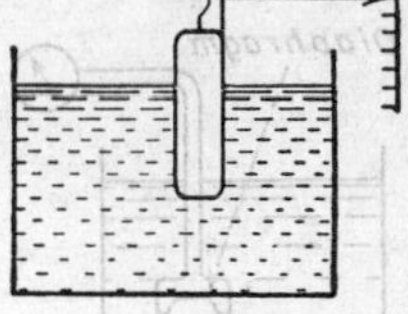

FIG. 89. Buoyancy-type level indicator.

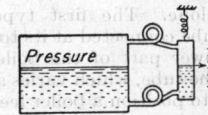

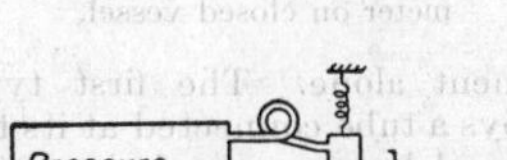

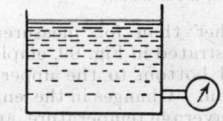

FIG. 90. Liquid level in tank under pressure.

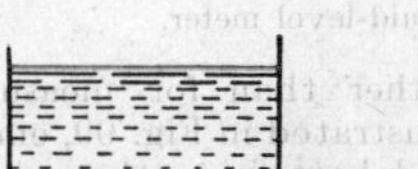

FIG. 91. Hydrostatic head meter.

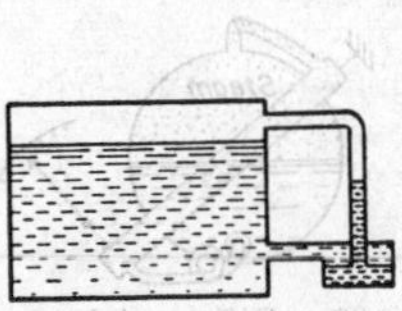

FIG. 92. Differential-pressure manometer-type level meter.

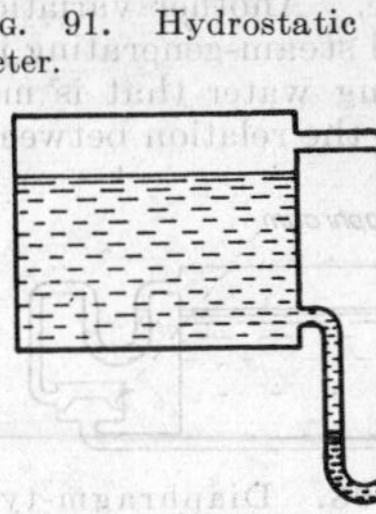

FIG. 93. Manometer-type liquid-level meter.

Hydrostatic-pressure methods are illustrated in Figs. 91 to 98. In Fig. 91 the hydrostatic head is measured directly in terms of level in an open tank by means of a pressure gage at the reference point. In a closed system under pressure, a *differential-pressure* manometer can be used to give a reading proportional to the level. As shown in Fig. 92, the manometer is placed at the reference point. However, the manometer can be placed below the tank as in Fig. 93, but the extra head thus introduced must be allowed for. Another way of getting at the hydrostatic head, as shown in Fig. 94, is to measure the pressure necessary to force air down a pipe until it bubbles from the bottom at the desired reference point. A variation of this idea often used for corrosive liquids is shown in Fig. 95. Frequently, instead of bubbling air into the tank, a *closed bell* with a flexible diaphragm is used to apply pressure equal to the hydrostatic pressure at the reference point to the air inside a closed system.

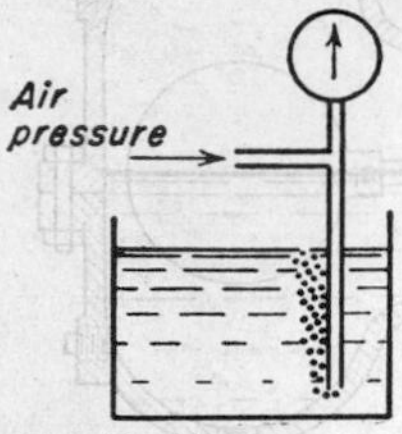

FIG. 94. Bubbler-type liquid-level meter.

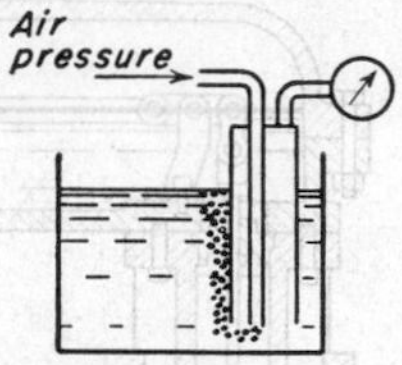

FIG. 95. Bubbler-type meter for corrosive fluids.

This is shown in Fig. 96, while Fig. 97 shows the same method applied to a pressure tank. If the diaphragm bell must be outside the tank, the arrangement of Fig. 98 can be used. Changes in density affect the readings of hydrostatic gages. Where extreme accuracy is required, this must therefore be accounted for.

It was mentioned that a *differential-temperature* method of level measurement can be used in boiler drums. Three types are employed, principally for level control

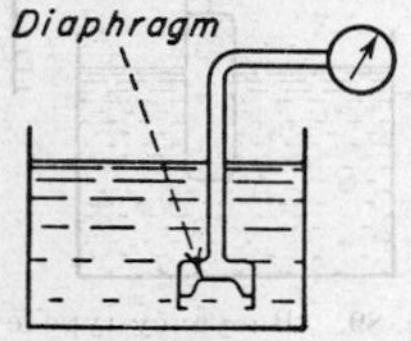

FIG. 96. Diaphragm-type liquid-level meter.

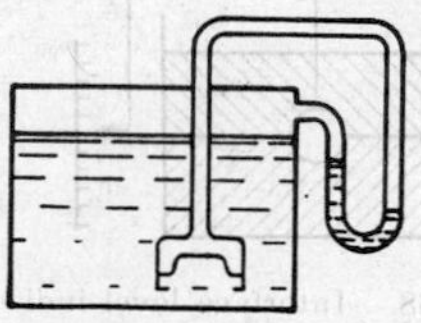

FIG. 97. Diaphragm-type meter on closed vessel.

rather than for measurement alone. The first type, illustrated in Fig. 99, employs a tube connected at its top and bottom to the upper and lower part of the boiler drum. Changes in the length of the tube, which varies as its average temperature, are used to position a boiler feed valve. Another variation of this scheme is to employ a small steam-generating jacket around the first tube, containing water that is more or less vaporized depending upon the relation between steam space and water space in

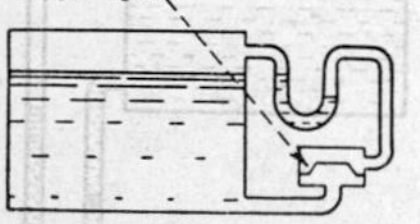

FIG. 98. Diaphragm-type meter arranged outside of tank.

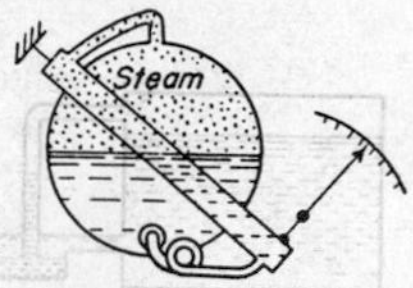

FIG. 99. Differential-temperature method of boiler water level.

the inner tube. Pressure in the small generator can be used to operate a pressure gage or to position a boiler feed valve. This arrangement is shown in Fig. 100. A third method measures the average tube temperature by a series of thermocouples.

Several *electrical methods* of level measurement are also available but are used chiefly for level control at one point or between two points without intermediate measurement. The first, shown in Fig. 101, employs *electrodes* to detect the level in a conducting liquid. With an electronic amplifier the level of liquids of negligible conduction can be detected. Figure 102(*a*) shows an electronic method using *condenser plates* connected into an electronic oscillator circuit and arranged on either side of a gage glass. When water rises between the plates, their capacitance is changed. This causes the oscillator to operate a relay to start or stop a pump or to light indicat-

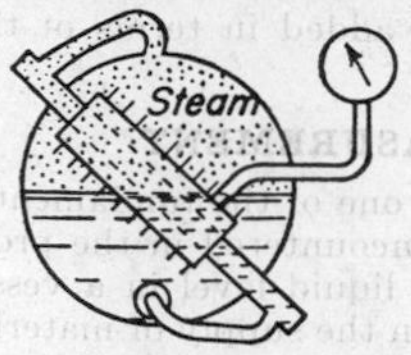

FIG. 100. Boiler water-level meter.

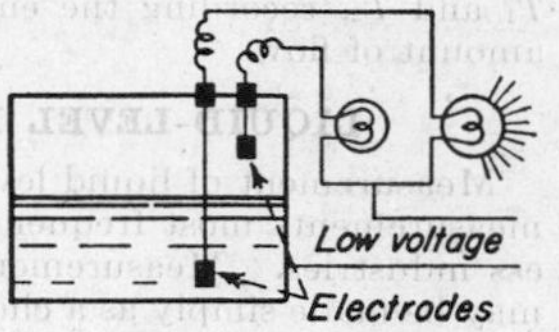

FIG. 101. Electrode-type liquid-level meter.

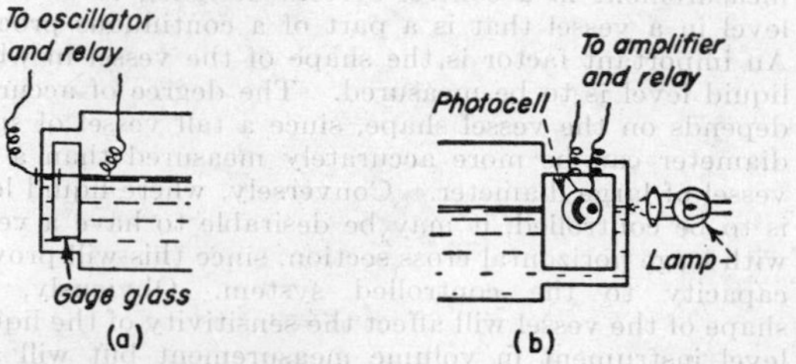

FIG. 102. Electronic liquid-level meters. (*a*) Capacitance type. (*b*) Photoelectric device.

ing lights. A second method accomplishing the same result is to use a photocell illuminated by a light source placed on the far side of the gage glass, shown in Fig. 102(*b*).

Level of Solids. Several methods have been developed for measuring the level of solids stored in bins. One method (Bindicator) is shown in Fig. 103. A flexible

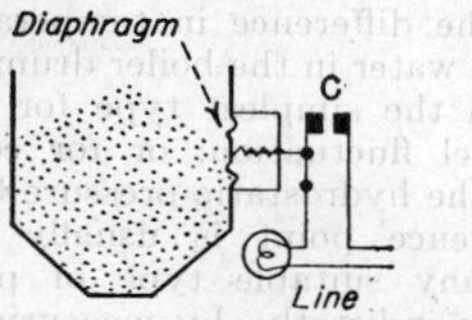

FIG. 103. Bin level indicator.

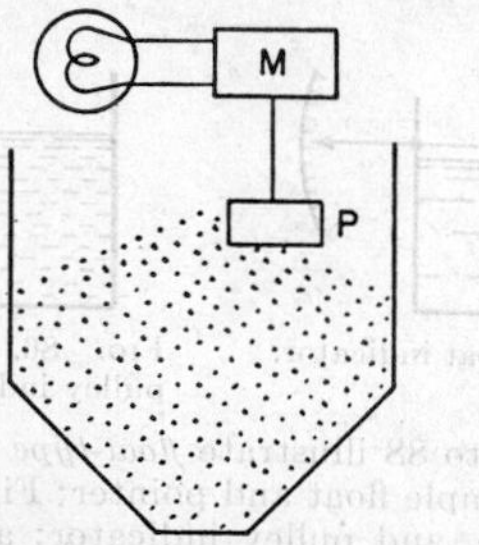

FIG. 104. Rotating-paddle-type solids-level meter.

diaphragm mounted in the side of the bin wall operates contacts C when material rests against the diaphragm. Another solids-level detector (Fuller), shown in Fig. 104, uses a rotating paddle P driven by a motor. When the solids rise high enough to stop the paddle, the motor M rotates about its shaft and closes a contact to light a signal light. These methods require an indicator at every level at which the height of the material is to be detected. A third method (Dravo), shown in Fig. 105, gives a semicontinuous indication of level at any point in the bin. By a system of cables and pulleys, a weight is moved upward and downward by a winch *W*. When the weight strikes bottom, a slack detector *D* opens contact *C*, stopping the winch, which, after a time delay, pulls the weight to the top and again lowers it. The points at which the pointer reverses on scale *S* represent the succession of levels of material.

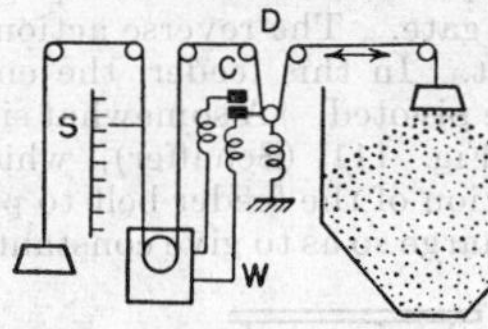

FIG. 105. Periodic indication of bin level.

WEIGHING AND WEIGHT CONTROL

Weighing in the process industries is of principal importance to (1) process control and (2) cost control. In process control, the major applications of scales and weigh feeders are in (*a*) continuous feeding, (*b*) continuous proportioning, (*c*) batching, and (*d*) product testing. In cost control, scales and fluid flowmeters are the accountant's most practical means for collecting data regarding the flow of materials through his plant. In this connection, weighing is important to (1) receiving raw materials, (2) controlling inventories, (3) making interdepartmental transfers, (4) scheduling production, (5) paying wages, (6) packaging, and (7) shipping the final products and by-products.

Weighing and feeding devices take numerous forms, but nearly all are designed around the lever-type scale. A scale is an instrument for measuring and balancing forces (usually weight) and incorporates the following essential components: (1) a means by which the load can be detected and supported during weighing, generally in the form of a platform, tank, section of conveyor, platter, hook, or other convenient method for supporting or containing the load; (2) a means for transmitting the force of the load to the counterbalancing force—in all but the equal-arm balance (Fig. 106) and simple spring scale (Fig. 107), this transmitting means takes the form of a lever system that reduces the load force to a value where it can be counterbalanced conveniently; and (3) a means for producing a counterbalancing force sufficient to balance the load and to indicate that balance. This counterbalancing force may constitute: (*a*) the resistance of a spring to added tension or compression, (*b*) the effect of a manually movable poise placed on a balance beam, (*c*) the effect of an automatically balancing pendulum, and (*d*) the effect of an automatically balancing plummet suspended in a vessel containing water, mercury, or other suitable liquid

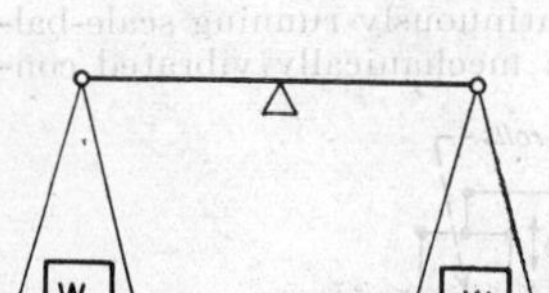

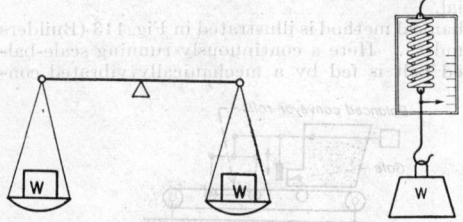

FIG. 106. Equal arm balance.

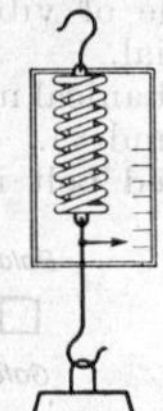

FIG. 107. Spring scale.

(hydrostatic principle). Few industrial scales employ spring counterbalancing. The automatically balancing pendulum is rapidly displacing the manually balanced beam because of the numerous advantages of automatic balancing, including: (1) a savings of the operator's time, (2) reduction in weighing errors by almost entirely eliminating the human element, (3) greater reading accuracy from the high visibility of widely spaced dial graduations, (4) automatic computation of prices, percentages, etc., and (5) adaptability to automatic weight-control systems. The latter advantage is of extreme importance where continuous operations are preferred over batch operations.

Scales for basic weighing, with their average upper weighing limits, are listed in Table 1. Basic weighing can be defined as simple, direct, elementary weighing in which the load is manually placed on and removed from the scale. A scale that automatically reaches a balance and indicates the weight without any manual manipulations, such as adjusting poises along a beam, is known as an automatic scale. Many scales of this type are used for basic weighing.

The modern automatic scale takes the general form of the *portable scale* illustrated in Fig. 108, with a large

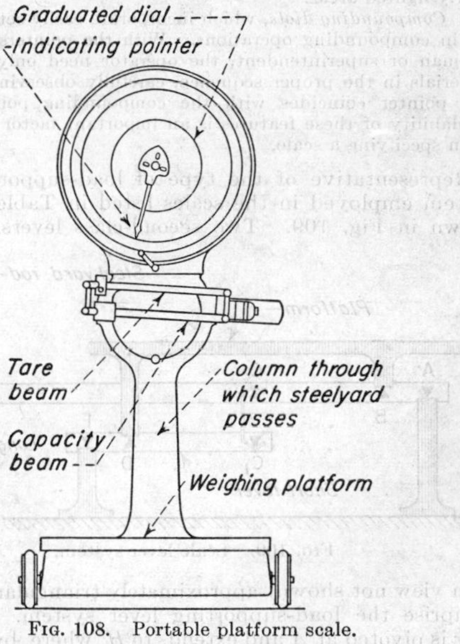

FIG. 108. Portable platform scale.

circular indicating dial. The automatic pendulum counterbalancing mechanism is contained in the scale head in the rear of the dial. Most of the scales listed in Table 1 are available with numerous accessories that enhance their value and efficiency. These accessories include:

1. *Automatic printing mechanisms*, which produce a permanent record of each weighing on a ticket or tape.
2. *Counting attachments*, which render possible use of the scale as an instrument for counting large numbers of uniform parts (pellets, pills, etc.).
3. *Double indication*, in which the scale dial is graduated on both sides for convenient reading.
4. *One-spot indication*, in which the conventional indicating pointer is replaced by a light metal disk around the periphery of which are stamped the weight graduations. Only that part of the disk showing the weight of the load on the scale is visible at one time. Since all graduations are horizontal, this feature avoids the necessity of reading figures at an angle; and, with the reading lens placed at the most convenient and natural position for reading, parallax errors are eliminated.

5. *Remote indication or telemetering* of weight readings, which is especially desirable where the weighing must be carried out in a section of the plant unsuitable for on-the-spot record keeping. Remote indication usually is accomplished by means of two Selsyn motors; see p. 1325.

Table 1. Average Weighing Limits of Typical Industrial Scales

Type	Upper Weighing Limit, Lb.
Equal-arm scale	5
Fan scale	50
Cylinder scale	70
Hanging scale	150
Bench dial scale	200
Portable-platform scale	1,600
Floor scale	6,500
Overhead-track scale	12,500
Built-in scale	21,500
Suspension-hopper scale	26,000
Truck scale	60,000
Truck-and-trailer scale	100,000
Railroad-track scale	335,000

6. *Graphic recording* made possible by the attachment of a circular-chart or strip-chart recording instrument to the head of the scale. Connection to the scale lever system is made by a lightweight link.

7. *Dial and chart illumination*, which is especially desirable where the scale must be read from a considerable distance and in poorly lighted areas.

8. *Compounding dials*, which incorporate easily set pointers to aid in compounding operations. With the pointers set by the foreman or superintendent, the operator need only weigh out materials in the proper sequence, carefully observing when the dial pointer coincides with the compounding pointers. The availability of these features is an important factor to consider when specifying a scale.

Representative of the type of load-supporting lever system employed in the scales listed in Table 1 is that shown in Fig. 109. Two second-class levers, both (in

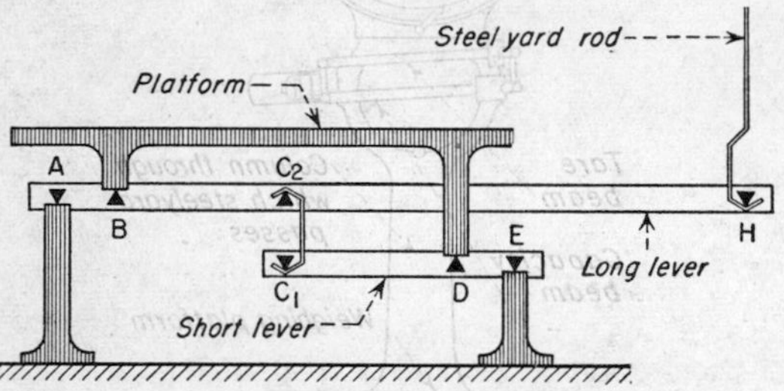

FIG. 109. Scale lever system.

plan view not shown) approximately triangular in shape, comprise the load-supporting lever system. The *long lever* is pivoted at *A* and extends to *H*, where, by means of a *steelyard rod*, it is connected to the counterbalancing end of the scale. The load from the platform is applied to the long lever at point *B*. The *short lever* is pivoted at point *E* and supports the platform at point *D*. The two levers are connected by means of link C_1C_2. So that the scale will weigh accurately, regardless of the position the load assumes on the platform, the pivot distances of the levers must bear the following relationship: $AB{:}AC_2 = ED{:}EC_1$.

Automatic weighers, in addition to coming to balance automatically, further eliminate manual operations by providing a means to move material to and from the weighing mechanism. The principal fields of application for automatic weighers include: (1) continuous weigh feeding, (2) continuous proportioning, (3) weigh batching, (4) continuous weigh checking (5) packaging, and (6) totalizing loads carried by conveyors.

Continuous weigh feeders can be classified into two major groups: (1) the pivoted-belt and (2) the loss-in-weight hopper.

Pivoted-belt feeders comprise a feed hopper and an endless traveling belt mounted on a pivoted frame, an adjustable weight that counterbalances the load on the belt, and a means for continuously and automatically adjusting the feed of material to the belt. In the weigh feeder illustrated in Fig. 110 (Hardinge), the rate of flow of material to the belt is controlled by changing the position of a gate over the outlet of the feed hopper. The feed gate is linked mechanically to the pivoted-belt frame so that a downward tilt of the belt, resulting from the load running too heavy, will automatically cause partial closing of the feed gate. The reverse action occurs when the load runs light. In this feeder, the entire belt and counterweights are pivoted. A somewhat similar method is illustrated in Fig. 111 (Schaffer), which utilizes a scale-balanced section of the feeder belt to position a gate at the hopper discharge so as to give constant belt loading.

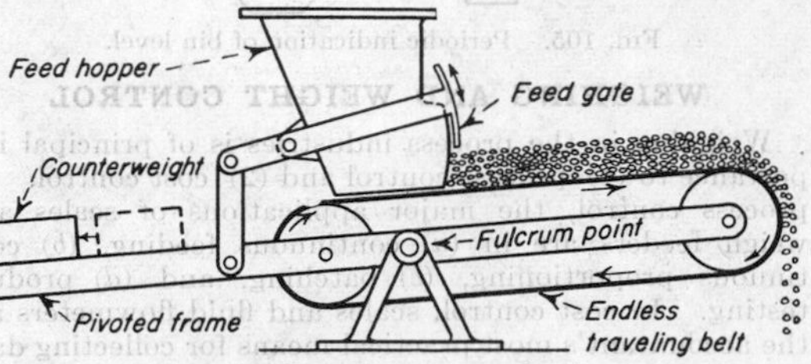

FIG. 110. Hardinge weigh feeder.

Another weigh feeder (Jeffrey), illustrated in Fig. 112, employs an electrically controlled vibrating deck to charge material onto the pivoted belt. When properly adjusted to the desired feeding rate, the scale beam is level, and neither the over- nor the undercontact is made. When the belt tilts downward, indicating that the flow of material is excessive, the overcontact at the end of the scale beam makes, which actuates a rheostat, decreasing the amplitude of vibrations in the deck and hence throttling the flow of material to the unit. When the belt tilts upward, indicating a deficiency in the flow of material, the undercontact makes, which increases the amplitude of vibrations, resulting in an increased flow of material.

A mechanical method is illustrated in Fig. 113 (Builders Iron Foundry). Here a continuously running scale-balanced feed belt is fed by a mechanically vibrated con-

FIG. 111. Schaffer feeder belt system.

veyor, the amplitude of which depends on the feed rate desired. This amplitude is adjusted continuously by means of a resilient wedge hung from the weigh beam, which is interposed between two jaw plates, one attached to a rapid reciprocating mechanism, and the other to the vibrating feed trough.

The pivoted-belt feeders just described are available with two types of adjustments for changing the rate of feed, namely, (1) a constant-speed belt with adjustable counterpoise or (2) a variable-speed belt with constant counterpoise. The first arrangement is most commonly used, since it is the simplest and least expensive. The adjustment can easily be made by the operator. The second arrangement is employed where adjustments must be made from a remote point and usually is used where the proportioning of several ingredients is required.

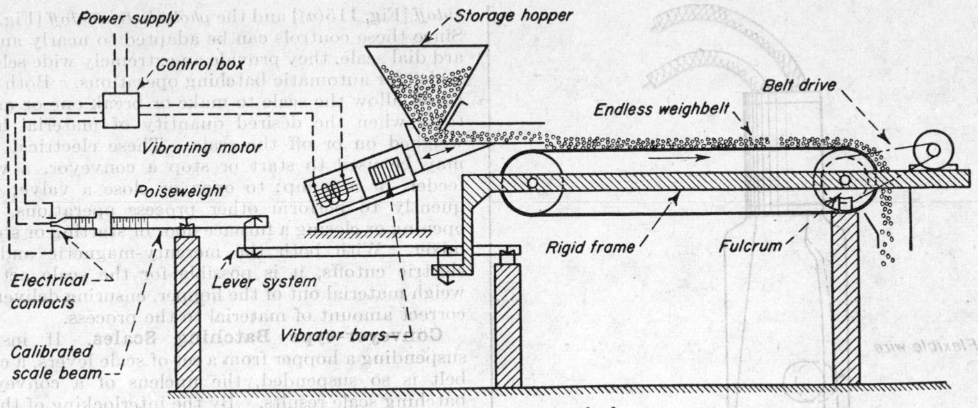

FIG. 112. Jeffrey vibrating feeder.

Pivoted-belt feeders are available with numerous accessory devices and refinements, including: (1) counters that totalize the flow; (2) recorders that provide permanent records of the flow rate; (3) no-load cutoffs that automatically stop the feeder and sound an alarm if the material supply becomes exhausted; (4) flush control (Syntron), which automatically corrects for sudden flushing of materials from the feed hopper; and (5) attachments for proportioning liquids with solid materials.

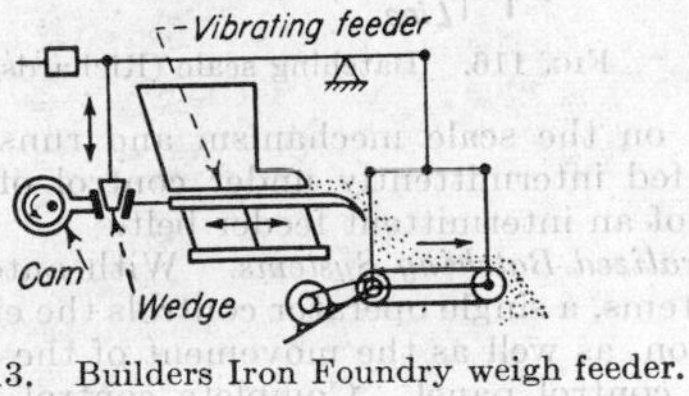

FIG. 113. Builders Iron Foundry weigh feeder.

Loss-in-weight feeders are based upon the rate of weight loss of a hopper or tank rather than upon the instantaneous weight of a moving conveyor belt. In essence, the feeder comprises a hopper or tank suspended from scale levers, a means for throttling the flow of material from the hopper or tank, and a scale beam with electrically driven counterpoise which fixes the rate of flow from the unit. The hopper must be recharged periodically. As the counterpoise slowly travels along the beam, it tends to unbalance the beam at the rate at which feeding is desired. The scale and beam can be brought back to balance only by causing material to flow from the hopper or tank, resulting in a loss of weight. Various means are used to translate the scale beam unbalance into movement of the hopper gate or tank valve. In one system, this unbalance is detected by the flapper mechanism or free vane of a pneumatic control instrument. The capacity of this type of feeder is limited only by the scale lever system that supports the hopper or tank.

Weigh Feeders Are Not Scales. It is important to note that the devices just described are termed *weigh feeders* and not *scales*. There is an important distinction between these two terms in that weigh feeders are not recognized by weights and measures officials as scales and therefore cannot be sealed. Feeders utilize the weight of a flowing material to make adjustments in feeding so that a relatively constant flow—on a weight basis—is obtained. Note, however, that these feeders do not weigh first and then feed, but feed first and then check and adjust that feed through weighing. Although many of these feeders cause a substantially constant flow over a given period, with the over- and underweight increment of feed averaging out, it cannot be stated that the flowing material has been truly weighed, and, hence, feeders cannot be classified as scales proper. For feeding critical processes, weigh feeders may be checked periodically by collecting a stream of material over a timed period and weighing the collected material on an accurate scale.

Continuous proportioning of solids can be accomplished by interlocking the actions of two or more weigh feeders. A common application of this arrangement is the proportioning of gypsum to clinker and of limestone to shale in the manufacture of cement

Proportioning of liquids by weight also can be accomplished through the interlocking of two loss-in-weight feeders in which tanks containing the liquids are suspended from the feeder lever systems.

Weigh batching systems generally comprise two or more automatic hopper scales. The essential components of an automatic hopper scale include: (1) a storage space, usually in the form of a hopper or bunker, for supplying material to the scale, (2) a means for feeding material to the scale, (3) an automatic device to start and stop materials running into the scale, with provision for dribbling the last amounts of feed into the weigh hopper, (4) a weigh hopper, suspended from or supported by the scale lever system, (5) a device for continuously counterbalancing and indicating the load in the weigh hopper to the process. Figure 114 shows an *automatic hopper scale*

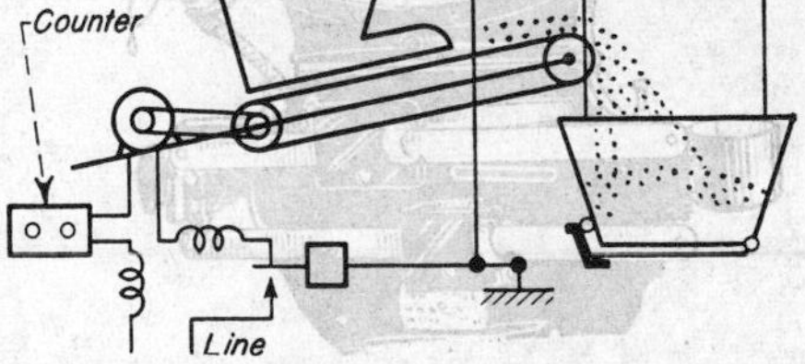

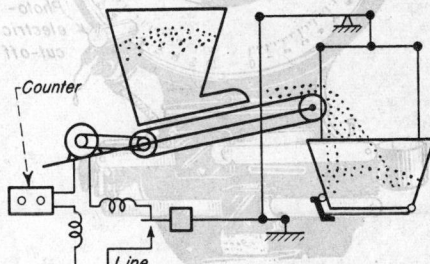

FIG. 114. Automatic hopper scale.

for lump or granular solids. The scale-balanced hopper moves a scale beam which makes or breaks contact, controlling a feeder conveyor. When the set weight is reached in the hopper, the feeder stops, the hopper bottom opens, discharges, and closes, and the cycle repeats recording the number of cycles on a counter. The principal methods for feeding materials to the weigh hopper include the gate feed, screw feed, conveyor-belt feed, and vibrating feed. The latter type is best suited to close

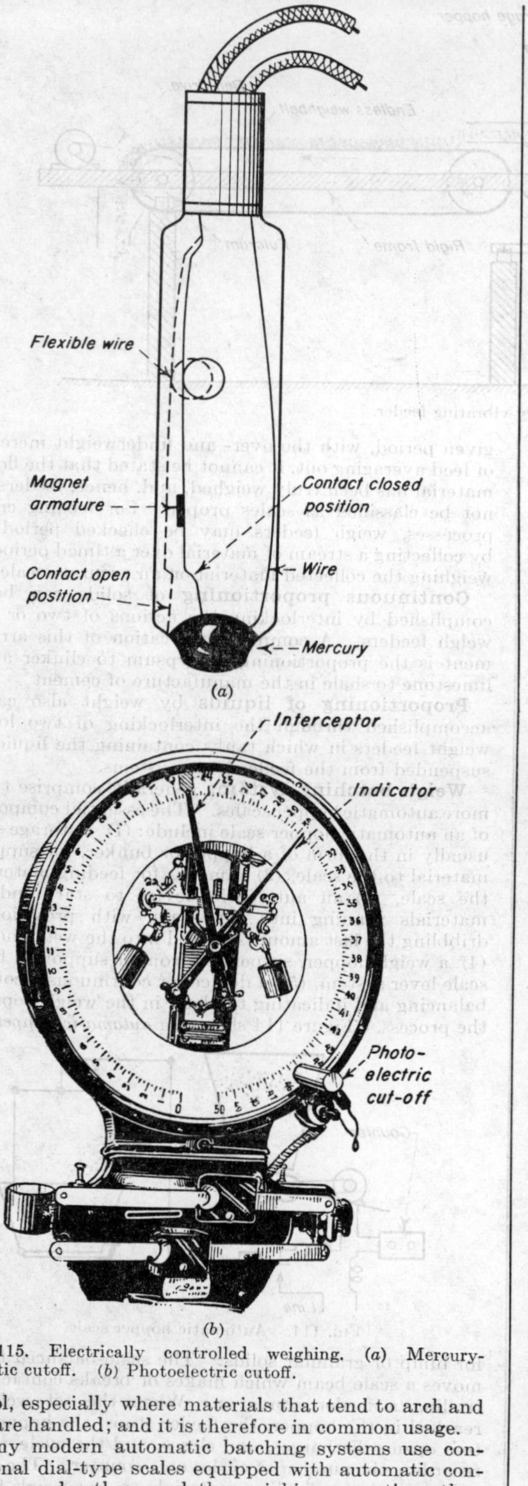

FIG. 115. Electrically controlled weighing. (*a*) Mercury-magnetic cutoff. (*b*) Photoelectric cutoff.

control, especially where materials that tend to arch and flush are handled; and it is therefore in common usage.

Many modern automatic batching systems use conventional dial-type scales equipped with automatic controls to render them and the weighing operations they perform almost totally automatic. The two general types of electrical control used are the *mercury-magnetic cutoff* [Fig. 115(*a*)] and the *photoelectric cutoff* [Fig. 115(*b*)]. Since these controls can be adapted to nearly any standard dial scale, they provide an extremely wide selection of scales for automatic batching operations. Both types of cutoff allow the scale to make or break one or more contacts when the desired quantity of material has been weighed on or off the scale. These electrical contacts may be used to start or stop a conveyor, a vibrating feeder, or a pump; to open or close a valve; and frequently to perform other process operations, such as opening or closing a furnace door or starting or stopping a mixer. With both the mercury-magnetic and photoelectric cutoffs, it is possible for the scale actually to weigh material out of the hopper, ensuring delivery of the correct amount of material to the process.

Conveyor-type Batching Scales. If instead of suspending a hopper from a set of scale levers, a conveyor belt is so suspended, the nucleus of a conveyor-type batching scale results. By the interlocking of the action of a weigh conveyor with a feed conveyor, the weighing action of an automatic hopper scale is substantially duplicated. In Fig. 116 (Richardson), the weigh belt is

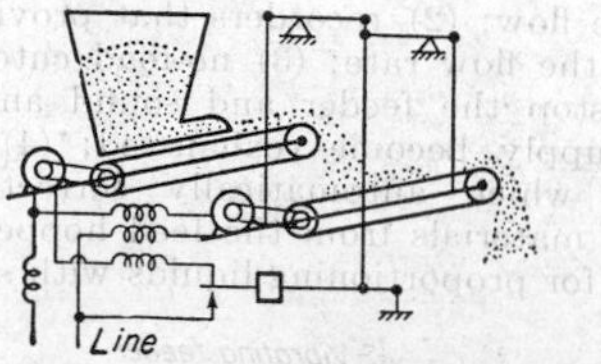

FIG. 116. Batching scale (Richardson).

carried on the scale mechanism and runs continuously but is fed intermittently under control of the scale by means of an intermittent feeder belt.

Centralized Batching Systems. With automatic batching systems, a single operator controls the entire batching operation, as well as the movement of the batch, from a central control panel. Complete control is maintained over the movement of materials from the time they leave the supply bins to their delivery to the process. *Packaging by weight* often is important to meet legal requirements, insofar as the weight of material in the package must equal or exceed that marked on the package, and to prevent shipment of materials in excess of that required, since it is easy to package profits unless close control over this operation is exercised. Packaging scales vary from small even-arm types with a capacity of a few pounds to large hopper scales with a capacity of several hundred pounds.

Continuous Weigh Checking. Automatic scales also may weigh a continuous moving sheet of material and thus provide a means of checking material density, thickness, gage, etc. If the weight is not within desired tolerances, electric signal contacts are made to warn the operator of an unsatisfactory condition. Scales of this type are used in the sheeting and calendering of rubber and plastics and in the coating of fabrics. The scale illustrated in Fig. 117 consists of: (1) three rollers, supported by a rigid frame and lying in the same horizontal plane, the middle roller being the live or weigh roller, (2) scale levers supporting the weigh roller, (3) a scale beam with movable poise for adjusting the scale, (4) an over-under indicating head, and (5) electrical contacts which operate warning signal lights or alarms.

Totalizing Conveyor Scales. Weighing materials while they continuously move on a conveyor to or from a process, permits close materials accounting without affecting the continuity or speed of the process. A section of conveyor is supported by a weigh roller, similarly to the checking scale just described. Counterbalance of the

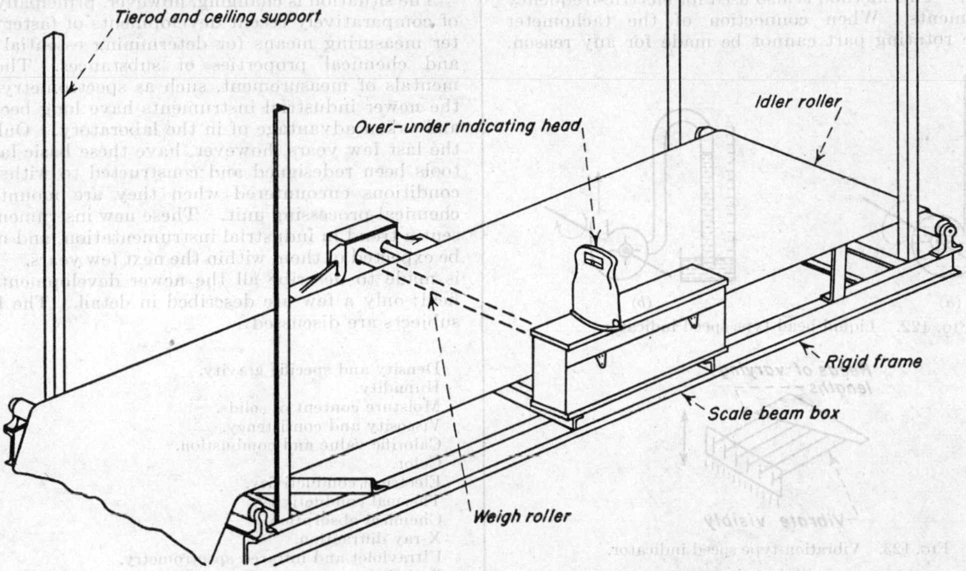

FIG. 117. Scales for weighing a continuously moving sheet.

load is accomplished through an automatically balancing pendulum or by means of a cylindrical steel float in a pot of mercury (hydraulic principle). Integration of the load passing over the scale can be accomplished either electrically or mechanically. In either case, it is important that the integrating mechanism be compensated for changes in belt speed.

MISCELLANEOUS QUANTITY AND RATE VARIABLES

Thickness. Occasionally the thickness of material in motion is a process variable that requires measurement or control. Various methods are used, depending on the properties of the material. Direct *calipering* by a fixed

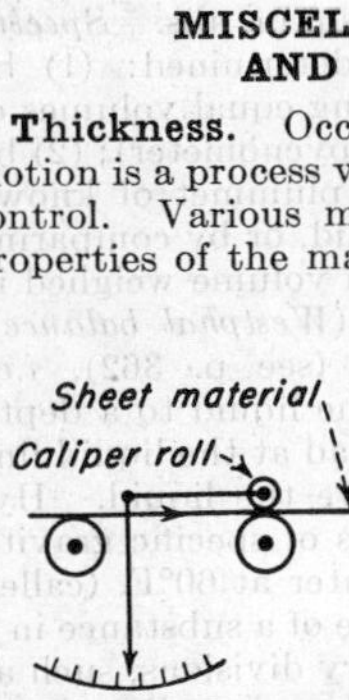

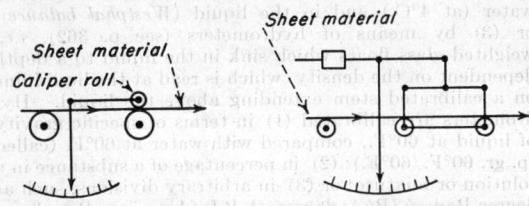

FIG. 118. Sheet-calipering device.

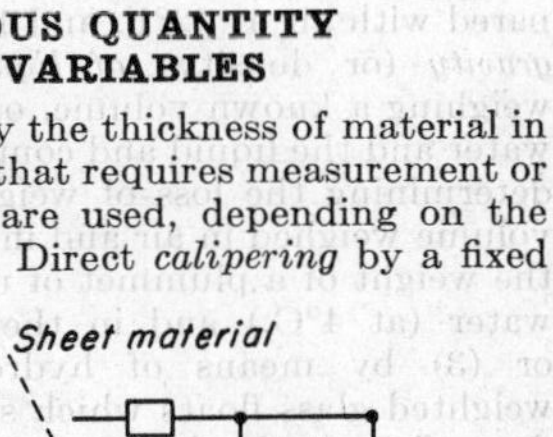

FIG. 119. Caliper measurement by weight.

and movable measuring element, as in Fig. 118, is one method of measuring the thickness of moving sheet material. Another is to pass the material between condenser plates where it acts as the dielectric, its thickness being a function of the indicated capacity of the condenser. Materials can also be weighed in transit for thickness determination, as in the case of the moving sheet of the material in Fig. 119 or Fig. 117, provided that the width is fairly constant. A number of laboratory methods for thickness measurement are also available, including interference methods, which can be used for extremely thin transparent films and coatings; and the magnetic thickness gage for determining the thickness of non-magnetic coatings on magnetic-base materials, which gives a response proportional to the distance by which the instrument and the magnetic base of the coating are separated.

Speed. Rotational speeds of various pieces of process equipment often require measurement and usually require control. Tachometers for speed measurement operate on both direct and inferential principles. The direct type of instrument, which counts the number of revolutions in a given time, cannot be used for continuous speed measurement, although an automatic type that would give such indications periodically could be designed

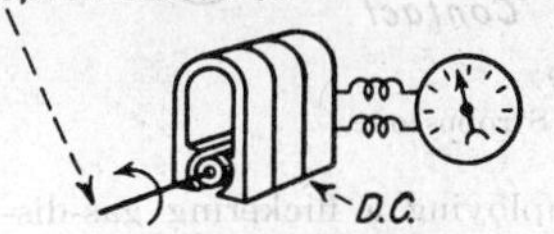

FIG. 120. Magneto tachometer.

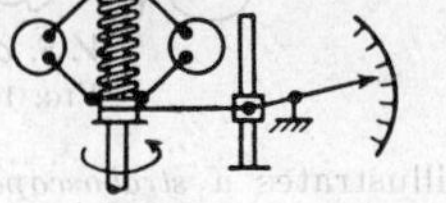

FIG. 121. Flyball governor.

if desired. Continuous methods of speed measurement are all inferential.

The most common type is the *magneto tachometer* illustrated in Fig. 120, in which a small d.c. magneto operates a voltmeter calibrated in terms of speed. Other common methods include the *magnetic-drag type* used in automobile speedometers and the *flyball-governor* type illustrated in Fig. 121. The force is proportional to the square of the speed, and this type therefore gives a non-uniform scale. Less used methods depend on the pressure produced by some sort of pump or blower. For example, in Fig. 122(*a*) a small *centrifugal pump* produces a liquid column head which can be calibrated in terms of rotative speed of the pump. A similar scheme is to use a *blower*, as in Fig. 122(*b*), to draw a liquid column up to a height bearing a calibratable relation to the speed. Another method (Fig. 123) uses a group of *metal reeds* of different vibration periods. This type operates simply by contact with the running machine, one or more reeds vibrating visibly at a rate equal to the vibration period of the machine. This method is also used for electric-frequency measurements. When connection of the tachometer with the rotating part cannot be made for any reason,

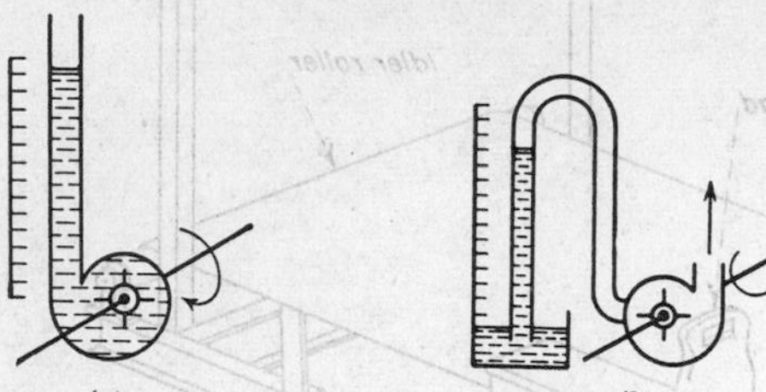

(a) (b)

FIG. 122. Liquid-head-type speed indicator.

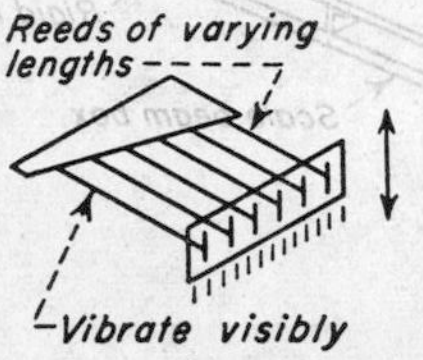

FIG. 123. Vibration-type speed indicator.

stroboscopes are often used. These operate by use of either a rotating shutter or a flashing light with which to view the rotating part. When synchronism or a multiple of synchronism is obtained, the rotating part appears to be stationary, at which point the speed of the shutter or the frequency of the light is measured. Figure 124

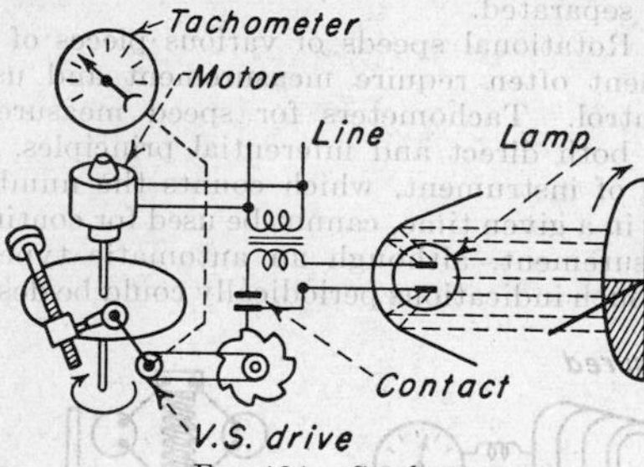

FIG. 124. Stroboscope.

illustrates a *stroboscope* employing a flickering gas-discharge lamp synchronized with the rotating object.

MEASUREMENT OF PHYSICAL AND CHEMICAL CHARACTERISTICS

A class of variables intimately concerned with the physical and chemical characteristics of a substance is increasing rapidly in importance relative to the others. Physical properties, such as density and viscosity, and chemical characteristics, such as actual composition or analysis and explosibility and flammability, often are the ultimate objective of measurement and automatic control. In paint and resin manufacture, for example, it is the viscosity of the final product that is regarded as of utmost importance. Often it is more practical, however, to control the resin kettle with temperature as a basis rather than continuously measuring and basing kettle control on product viscosity. Control of a fractionating column usually is based upon some temperature or pressure simply because there are practical difficulties in continuously measuring the actual chemical composition of a product stream.

The situation is changing, however, principally because of comparatively recent developments of faster and better measuring means for determining essential physical and chemical properties of substances. The fundamentals of measurement, such as spectrometry, used in the newer industrial instruments have long been known and taken advantage of in the laboratory. Only within the last few years, however, have these basic laboratory tools been redesigned and constructed to withstand the conditions encountered when they are mounted on a chemical-processing unit. These new instruments represent a trend in industrial instrumentation, and much can be expected of them within the next few years. No effort is made to describe all the newer developments in this field; only a few are described in detail. The following subjects are discussed:

Density and specific gravity.
Humidity.
Moisture content of solids.
Viscosity and consistency.
Calorific value and combustion.
Color.
Electrical conductivity.
Thermal conductivity.
Chemical absorption.
X-ray diffraction.
Ultraviolet and infrared spectrometry.
Emission spectrometry.
Mass spectrometry.
Polarography.
Hydrogen-ion concentration.
Oxidation-reduction potential.

Density, Specific Gravity

Density is defined as weight per unit volume. Specific gravity of liquids and solids is the density compared with that of water at 4°C.; and, of gases, it is the density compared with air at 32°F. and 14.7 lb./sq. in. abs. *Specific gravity* (or density) *of liquids* is determined: (1) by weighing a known volume, or weighing equal volumes of water and the liquid and comparing (pycnometer); (2) by determining the loss of weight of a plummet of known volume weighed in air and in the liquid, or by comparing the weight of a plummet of unknown volume weighed in water (at 4°C.) and in the liquid (*Westphal balance*); or (3) by means of hydrometers (see p. 362), *i.e.*, weighted glass floats which sink in the liquid to a depth dependent on the density, which is read at the liquid line on a calibrated stem extending above the liquid. Hydrometers are calibrated (1) in terms of specific gravity of liquid at 60°F., compared with water at 60°F. (called sp. gr. 60°F./60°F.); (2) in percentage of a substance in a solution or mixture; or (3) in arbitrary divisions, such as degree Baumé (Bé.); degrees A.P.I. (American Petroleum Institute); degrees Twaddell (Tw.), used in England; or degrees Brix (also called Fisher).

For liquids lighter than water,

$$°\text{Bé.} = \frac{140}{\text{sp. gr. } 60°\text{F.}/60°\text{F.}} - 130$$

$$°\text{A.P.I.}^{*} = \frac{141.5}{\text{sp. gr. } 70°\text{F.}/60°\text{F.}} - 131.5$$

$$°\text{Brix} = \frac{400}{\text{sp. gr. } 60°\text{F.}/60°\text{F.}} - 400$$

* Used in the United States principally for petroleum products.

For liquids heavier than water,

$$°\text{Bé.} = 145 - \frac{145}{\text{sp. gr. } 60°\text{F.}/60°\text{F.}}$$

$$°\text{Tw.} = 200(\text{sp. gr.} - 1)$$

$$°\text{Brix}^* = \text{arbitrary graduation}$$

To correct hydrometer readings for departures of the liquid from the calibration temperature, the coefficient of expansion of the liquid near the working temperature must be known.

Density of Liquids. Important methods for measuring the density of liquids are illustrated in Figs. 125 to 129. In Fig. 125, the weight of a fixed volume of liquid

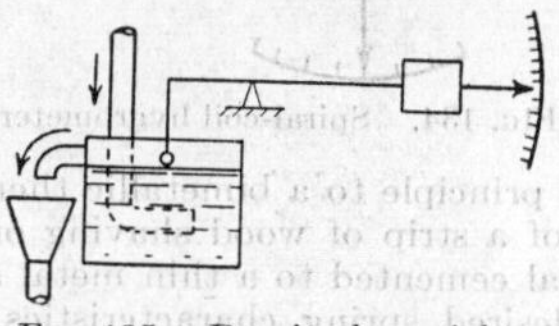

FIG. 125. Density by weight.

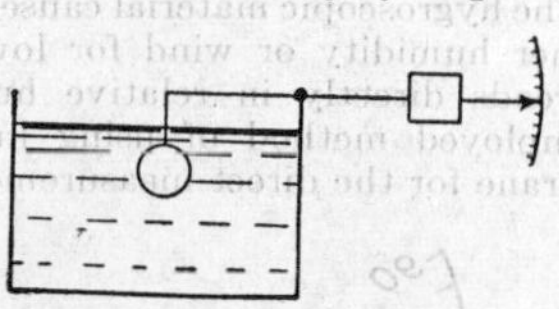

FIG. 126. Density by displacement.

is determined directly. Figure 126 illustrates the method of determining liquid density by determining the weight of the liquid displaced by an object suspended on a scale beam. The loss of weight when the object is submerged is the weight of a volume of liquid equal to the volume of the displacer. In Fig. 127 is illustrated the *hydrometer*, which measures directly by displacement that volume of liquid which is equal to the weight of the hydrometer. However, calibration is in terms of density or specific gravity, rather than weight. Figure 128 illustrates an

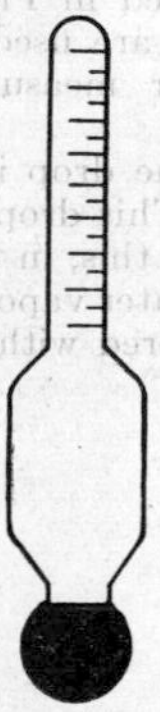

FIG. 127. Hydrometer.

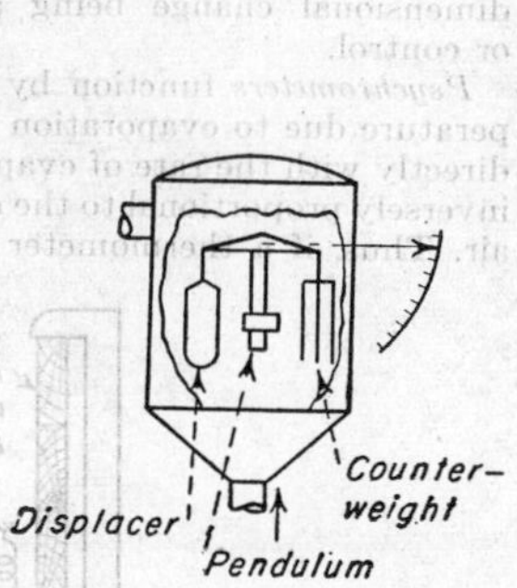

FIG. 128. Automatic hydrometer.

automatic density meter or hydrometer (Bailey), which employs a balanced beam carrying a large-volume displacer at one end, balanced against a small-volume counterweight at the other. A pendulum regulates the movement for a given change in density. To correct for temperature variations, the large-volume displacers contain a sample of the test liquid. Another method of measuring liquid density is to measure the differential pressure produced by two air-bubble pipes of different lengths in the liquid. This method (Hays) is illustrated in Fig. 129. The differential pressure H is proportional to the liquid density.

Continuous indication of specific gravity of fluids is possible with a specific-gravity indicator made by Schutte & Koerting. This consists of a special modification of a standard rotameter body. The liquid enters the bottom, passing up through the glass tube to a level which is kept constant by an overflow pipe. A standard hydrom-

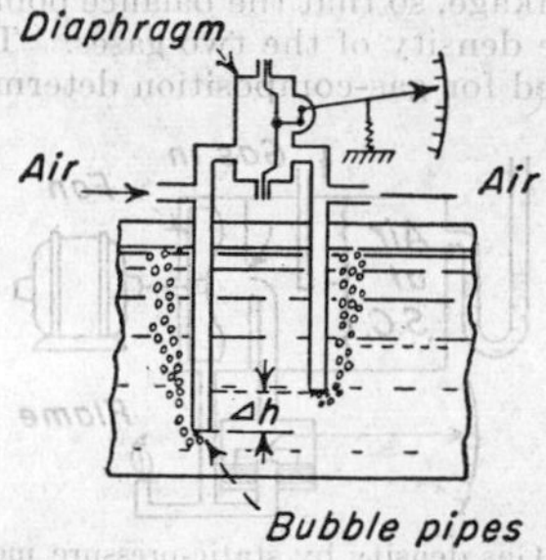

FIG. 129. Differential-pressure density meter.

eter of any given range can be placed inside the tube, where it indicates the specific gravity of the flowing liquid.

Density of Gases. A variety of methods are available for measurement of gas density. The density of gases may be inferred from the viscous drag produced by a moving or rotating column of the gas, or from the increase in kinetic pressure of the gas between the suction and blade tips of a rotating radial impeller. More commonly, however, a device capable of weighing a column of the gas is used. Gas-density measuring devices are illustrated in Figs. 130 to 133. Figure 130 shows the *gas-density balance* (Edwards) in which a scale-balanced displacer

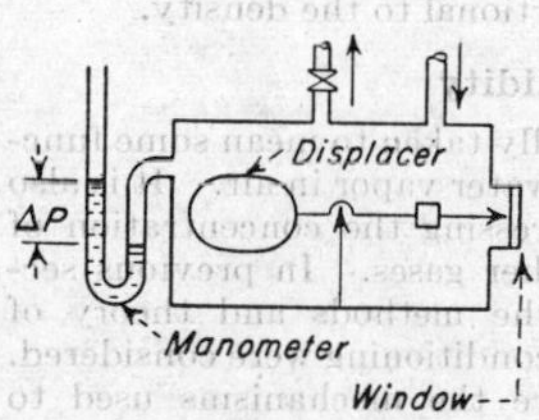

FIG. 130. Gas-density balance.

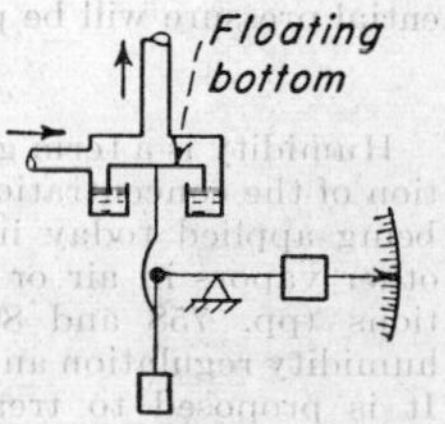

FIG. 131. Gas-density balance.

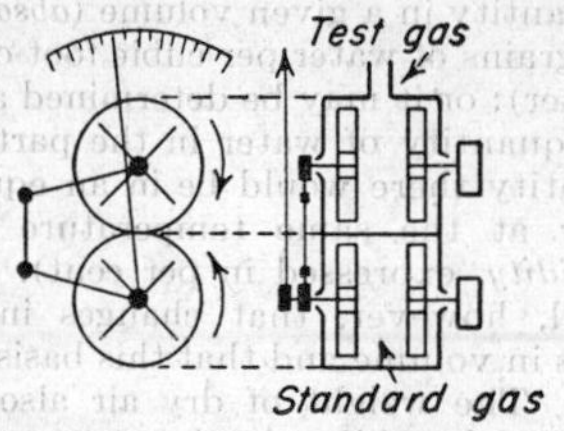

FIG. 132. Viscous-drag-type gas-density meter.

enclosed within a box is brought to a balance point as viewed through a window, using a standard gas. The test gas then is passed into the box and its pressure adjusted to give the same buoyancy as indicated by the pointer. The difference in pressure between the standard and test gases as shown by a manometer then is a measure of the difference between their densities. Another method of weighing the gas which operates directly is shown in Fig. 131. The gas balance (Alpha-Lux) weighs a tall column of gas at atmospheric pressure by means of a

* So graduated that 1° Brix = 1 per cent sugar in solution; used as a saccharimeter.

floating liquid-sealed bottom to the test chamber, which is balanced by a scale beam. The *viscous-drag method* of density measurement is found in the Ranarex instrument (Permutit) shown in Fig. 132. The instrument incorporates two chambers, one containing the standard gas, with the test gas flowing through the other. In each chamber are two paddles; the rear paddles are driven in opposite directions at the same speed by a motor. The whirling column of gas in each chamber tends to rotate the front paddles, which are tied together by a linkage, so that the balance point is a measure of the relative density of the two gases. This method is frequently used for gas-composition determination.

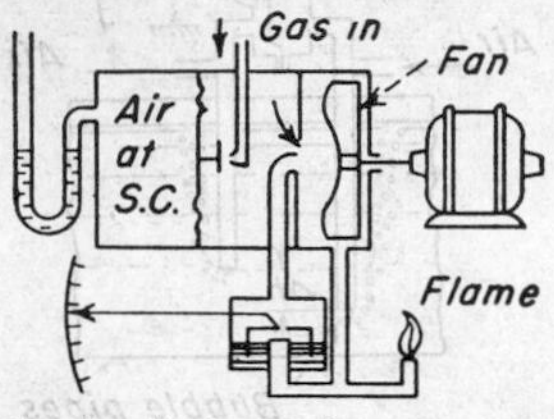

FIG. 133. Gas density by static-pressure measurement.

Figure 133 shows the Metric Gravitometer, which measures gas density in terms of the increase in static pressure as the gas passes from the center to the blade tips of a fan. Air is sealed under standard conditions in the left-hand chamber. The flexible diaphragm controls a reducing valve to admit test gas at the same pressure as the standard gas at the existing temperature. The test gas flows continuously through the instrument, the issuing jet being ignited. Test gas is drawn into a fan, the pressures at the fan inlet and blade tip being measured by a liquid-sealed bell manometer. Still another method is to pass the gas at a constant flow rate through an orifice. If the flow is constant, the differential pressure will be proportional to the density.

Humidity

Humidity is a term generally taken to mean some function of the concentration of water vapor in air. It is also being applied today in expressing the concentration of other vapors in air or in other gases. In previous sections (pp. 758 and 800), the methods and theory of humidity regulation and air conditioning were considered. It is proposed to treat here the mechanisms used to measure humidity.

The amount of water in the air may be determined as the actual quantity in a given volume (*absolute humidity*, expressed in grains of water per cubic foot of air or grams per cubic meter); or it may be determined as the relation between the quantity of water in the particular sample and the quantity there would be in an equal volume of saturated air at the same temperature and pressure (*relative humidity*, expressed in per cent). It is well to bear in mind, however, that changes in temperature cause changes in volume and that this basis is therefore a shifting one. The weight of dry air also is used as a basis. Absolute humidity is then expressed as grains or pounds per pound of dry air and *percentage absolute humidity* as the relation between the weight of water per pound of dry air in the sample and the weight of water per pound of dry air if it were saturated at the same temperature and pressure. The percentage relative humidity and the percentage absolute humidity will not agree and will differ more widely at higher temperatures.

Relative humidity is the commonly accepted measure in industrial work. The numerous instruments developed to determine it may be divided into two groups: hygrometers and psychrometers.

Hygrometers (see pp. 775*ff.*) depend for their operation on the expansion and contraction of certain hygroscopic substances with variations in the moisture content. This moisture content changes as the humidity of the surrounding air varies. The materials commonly used are wood, paper, silk, animal membranes, and hair. Figure 134 shows a spiral-coil hygrometer similar in appearance and principle to a bimetallic thermometer. It is composed of a strip of wood shaving or other hygroscopic material cemented to a thin metal strip or screen having the desired spring characteristics. Changes in dimension of the hygroscopic material cause the coil to unwind for higher humidity or wind for lower humidity. The device reads directly in relative humidity. The commonly employed method of using human hair or animal membrane for the direct measurement of relative humidity is illustrated in Fig. 135. Another method using a hygroscopic material is illustrated in Fig. 136, where wood blocks cut across the grain are used, their dimensional change being amplified for measurement or control.

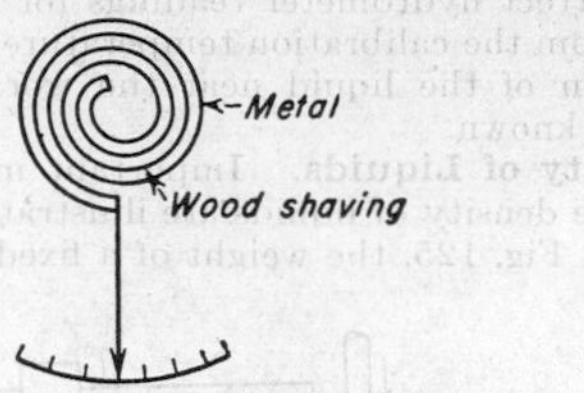

FIG. 134. Spiral-coil hygrometer.

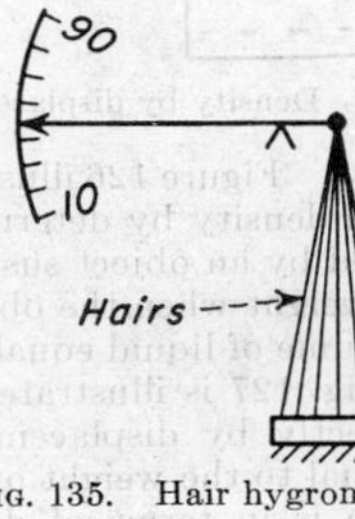

FIG. 135. Hair hygrometer.

Psychrometers function by reason of the drop in temperature due to evaporation of water. This drop varies directly with the rate of evaporation, and this, in turn, is inversely proportional to the amount of water vapor in the air. Thus, if a thermometer bulb is covered with a thin film of water and air is passed rapidly over it, the temperature will fall a definite amount dependent on the initial temperature and the amount of moisture in the air, provided that the air velocity is sufficiently great. The minimum usually is set at 15 ft./sec. From the initial, or dry-bulb, temperature and the difference or depression of the wet bulb, the relative humidity can be calculated (see p. 775). Intermittent or check readings are usually

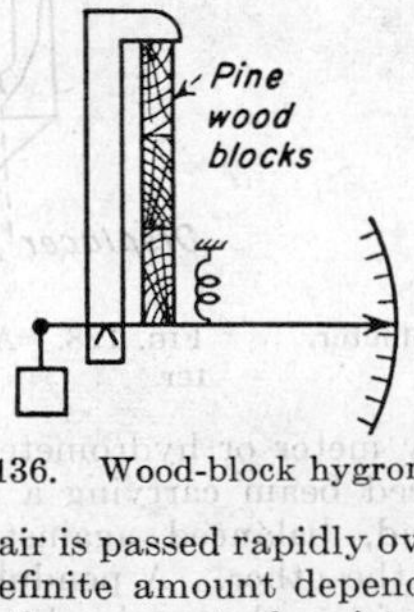

FIG. 136. Wood-block hygrometer.

made with a *sling psychrometer* consisting of two glass-stem thermometers mounted in some mechanism for whirling rapidly, one bulb being covered with a moistened cotton wick, as shown in Fig. 137.

Records of humidity can be obtained by the use of two-pen thermometers (see Fig. 138) of the types described on p. 1267, with one of the bulbs moist and the other dry. For this reason, these psychrometers are frequently termed *wet-and-dry-bulb thermometers.* If the air circulation at the bulbs is not sufficiently great, a small motor fan may be used to suck air over the bulbs. Means for keeping the bulb moist have been the subject of much research. The most common is the cotton wick, a piece of cotton that has been boiled to remove any sizing or grease. The thickness does not seem to be so important as the porosity and ability to absorb water. Since this is dependent on the cleanliness, wicks must be changed frequently, the period varying with the amount of dirt and lint in the air and the amount of dissolved solids left behind by the water. The wick is made sufficiently large to cover the bulb and to dip into a water reservoir mounted beneath it. The water is maintained

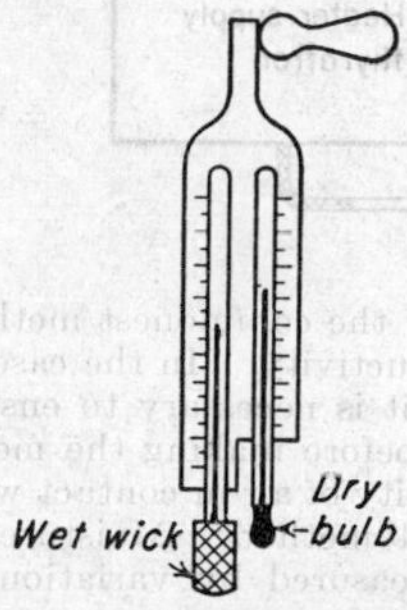

FIG. 137. Sling psychrometer.

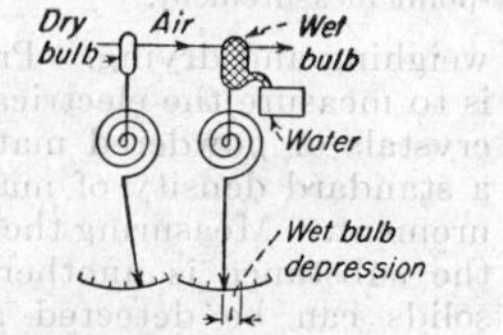

FIG. 138. Wet-and-dry-bulb thermometers.

at a constant level by a float valve, an overflow, or an inverted bottle. With an overflow, a needle valve should be connected in the water line to provide close adjustment. A method of eliminating the changing of wicks is an alundum sleeve that fits over the bulb. Water is admitted to the inside and seeps through the pores. When this is properly installed and operated, the solids from the water as well as dirt from the process fall off or are easily wiped off.

The humidity controller (Parks-Cramer) illustrated in Fig. 139 employs no wick. Air is sucked over the dry bulb by the spray, which saturates it before it passes over the wet bulb. The spray water must be at the approximate temperature of the wet bulb. One direct-reading relative-humidity recorder operates from wet-and-dry-bulb resistance-thermometer elements (Leeds and Northrup). A complex self-balancing Wheatstone-bridge circuit interprets the two separate resistance values in relative-humidity percentage.

Wet- and dry-bulb thermometers are limited to temperatures between the freezing and boiling points of water. In addition, certain errors are encountered in both the lower and the upper portions of this range. These are not of great consequence where only comparative readings are desired in order to duplicate conditions.

Automatic control of humidity (see p. 786) can be accomplished by the use of two-element temperature units, Fig. 138, of the types described on pp. 1320 to 1326. The dry-bulb instrument or element then controls the temperature by a valve in the steam line to the heating coil. The wet bulb controls a valve in the water or steam line to the spray. Or, in circulating-air driers, a damper may be shifted to govern the amount of saturated air taken from the circuit, and fresh air may be admitted. Another method is to circulate the air through a chamber where it is cooled and saturated at the proper dew point and then heated to the desired temperature.

A comparatively new development in hygroscopic measurement is illustrated in Fig. 140. Here a double coil *H* of palladium wire is wound on an insulating core, the wires being coated with a film of hygroscopic lithium chloride suspended in a water-soluble coating. The electrical conductivity of the coating varies with the relative humidity and is measured by a microammeter *M*, provided with a.c. energy through a transformer *T* and a full-wave copper oxide rectifier *R*. To secure a wide range, several such sensitive elements must be used in series, each different from the others in over-all resistance. One instrument of this type (American Instrument Co.) is available with electronic recorders and electric or pneumatic control.

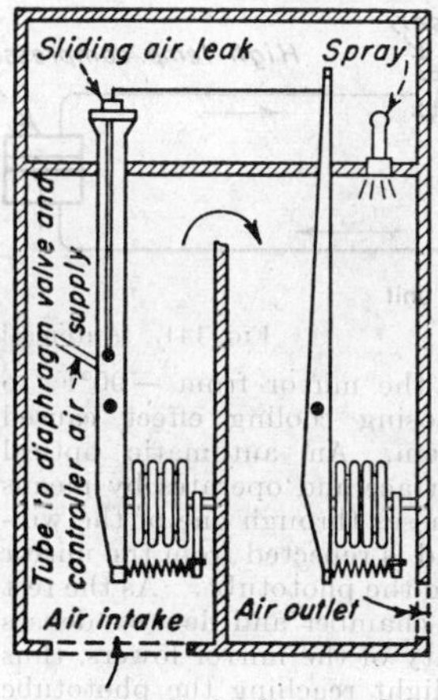

FIG. 139. Parks-Cramer humidity controller.

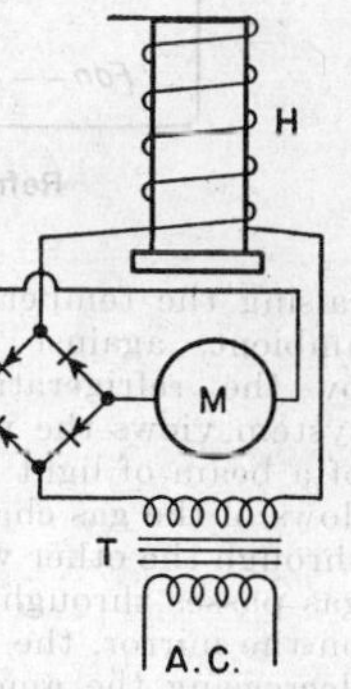

FIG. 140. Conductivity-type hygrometer.

Absolute Humidity (see p. 759). The actual moisture content of the air may be determined by absorbing the water vapor from a known volume of air or by observing the temperature at which dew forms on a polished metallic or mirror surface. Instruments of the latter type are known as dew-point hygrometers.

Dew-point Recorder. Referring to Fig. 141, which is a schematic diagram of one type (General Electric) of automatic dew-point recorder, note that the gas under test enters the gas chamber where it comes in contact with a thin metallic mirror to which a thermocouple is attached. The mirror is refrigerated to −90°F. Consequently, moisture from the gas under test forms on the surface, thus fogging the mirror. Coincident with the formation of fog or dew, the mirror is heated to the temperature at which the dew vanishes. This temperature is the dew point of the gas under test. The electronically controlled heater maintains the mirror at this temperature until the dew point of the gas under test changes, at which time the mirror temperature is changed to coincide with the formation of dew and is held at this new temperature. By a recording of the output voltage from the thermocouple attached to the mirror, by means of a potentiometer, a continuous record of the dew point of the gas is obtained.

In operation, the mirror is constantly refrigerated by a gas flow against the back of the mirror. The refrigerating gas flow is cooled to −90°F. by the action of a two-stage compressor and a dual refrigerating system. A heater mounted directly behind the mirror is capable of

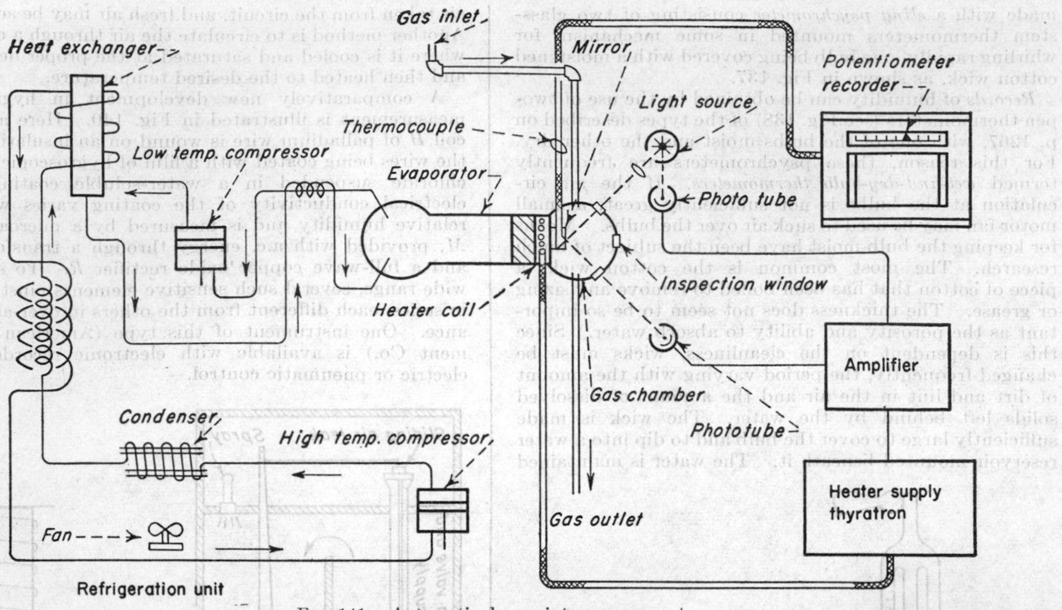

FIG. 141. Automatic dew-point measurement.

raising the temperature of the mirror from −90°F. to ambient, against the opposing cooling effect caused by the refrigerating system. An automatic optical system views the mirror surface and operates by means of a beam of light which passes through one of the windows of the gas chamber and is reflected from the mirror through the other window to the phototube. As the test gas passes through the gas chamber and dew condenses on the mirror, the reflectivity of the mirror lowers, thus decreasing the amount of light reaching the phototube and initiating a change in the magnitude of the heater current.

To reduce the effect of voltage variations and aging on the phototube characteristics, the lamp also shines on a second phototube that is balanced in the electric circuit against the first tube. The amplified difference between the two phototube currents controls the heater current.

Another instrument for humidity measurement and control is the Dewcel (Foxboro). This operates by bringing a saturated solution of lithium chloride to such a temperature that its partial pressure equals that of the water in the atmosphere surrounding the detecting unit. A thin-walled stainless-steel tube coated with phenol-formaldehyde varnish to insulate it electrically is wound with a layer of Fiberglas woven tape that serves as a wick to retain the lithium chloride solution. This is covered with a double winding of silver wire. Inside the tube is a resistance thermometer to measure equilibrium temperature. The silver wires are connected to a source of 25-volt alternating current through a ballast resistor.

After the unit is started, an equilibrium condition is reached at which point the amount of power converted to heat is just sufficient to maintain the unit at a temperature where the partial pressure of the water vapor over the saturated lithium chloride solution is equal to the partial pressure of water in the atmosphere. Quantity of moisture gained or lost by the wick is small. Dew points between −16° and +160°F. may be measured with this device.

Moisture Content of Solids

The moisture content of solids can be measured in various ways in addition to the fundamental method of weighing and drying. Probably the commonest method is to measure the electrical conductivity. In the case of crystals or powdered materials it is necessary to ensure a standard density of material before making the measurement. Measuring the humidity of air in contact with the substance is another useful method. Moisture in solids can be detected and measured by variation of dielectric constants.

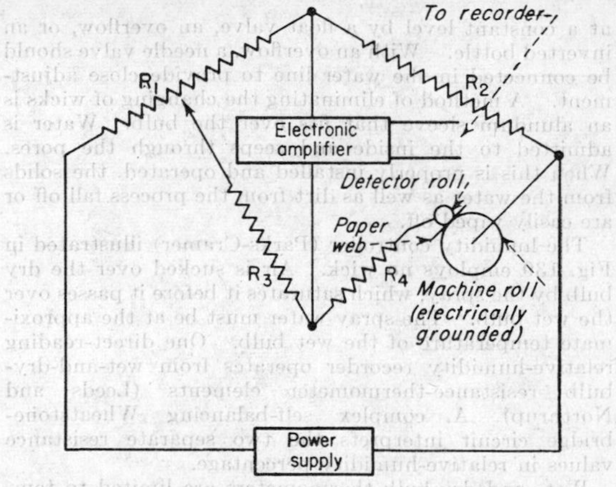

FIG. 142. Moisture measurement of moving sheet.

In one method, moisture is detected by measuring the electrical resistance from a detector roll in direct contact with the continuously moving sheet of paper on a paper machine or warp on a textile slasher. In Fig. 142 is illustrated the basic circuit of this device. The bridge is supplied with power from a standard radio-type power supply consisting of a transformer, rectifier tube, and filter.

Viscosity and Consistency

Viscosity is a measure of the force required to shear a fluid at unit rate. Consistency is a property related to viscosity or plasticity which is encountered in the case of suspensions, such as paper pulp. Suspensions do not exhibit true viscosity, but the consistency can be measured by methods similar to those used for viscosity. Viscosity or consistency can be measured by determining the torque required to rotate a paddle or cylinder in the material or by measuring the rotative speed of a paddle or cylinder driven by a known torque. Other methods include timing the flow of a definite quantity through a short tube or a capillary; measuring the pressure drop through a capillary; or, in the case of liquids, timing the

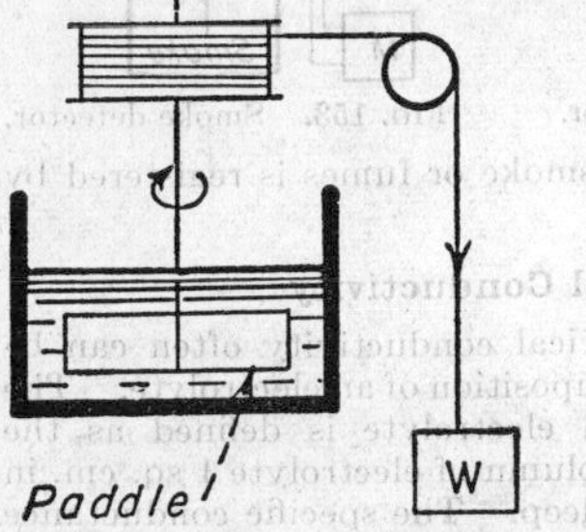

FIG. 143. Constant-speed paddle wheel indicates viscosity.

FIG. 144. Common orifice-type viscometer.

rise of an air bubble through the liquid or timing the fall of a ball or other object.

Examples of viscosity- and consistency-measuring devices are illustrated in Figs. 143 to 148. In Fig. 143, a paddle rotated in a liquid at a known speed by a known force measures viscosity. However, viscosity often is determined by measuring the time required for a definite quantity of flow through a short tube, as shown in Fig. 144. This is the method employed in most viscometers, such as the Saybolt and Engler. For continuous measurement of viscosity, the method shown in Fig. 145 is commonly used. This involves measurement of the pressure drop in a material flowing at a known rate

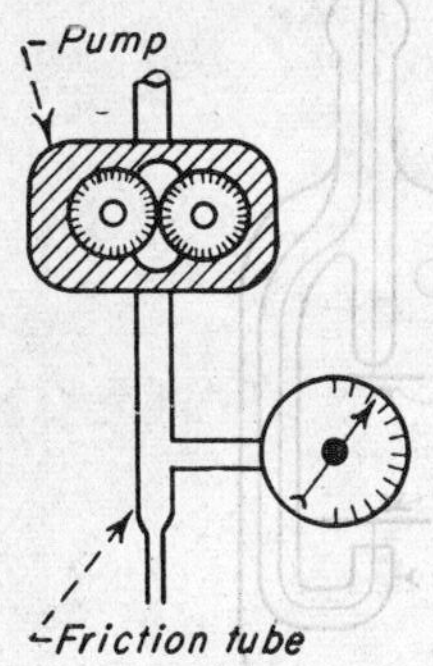

FIG. 145. Continuous viscometer.

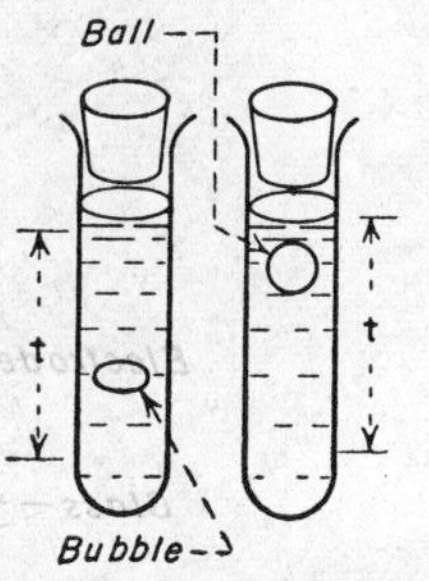

FIG. 146. Rising-bubble and falling-ball viscometers.

through a friction tube. Figure 146 illustrates the rising-bubble and falling-ball methods, which are particularly adaptable to extremely viscous materials. A continuous viscometer (Fischer & Porter) consists of a modification of the rotameter in which two metering floats operate in the same tube. The lower float is viscosity immune and is used for manual adjustment to a specified predetermined flow rate. The second float, which is viscosity-sensitive, rides above the first and indicates viscosity. A built-in thermometer allows temperature corrections to be made. Automatic constant flow can be provided by special attachments.

Consistency-measuring methods are most important in the paper industry, where they generally are employed for automatic control. One such method, shown in Fig. 147, measures the torque required to drive a paddle in a paper-stock suspension in terms of the tension in the tight side of a belt or chain drive. The tension is used to adjust the position of a water-dilution valve. Another method

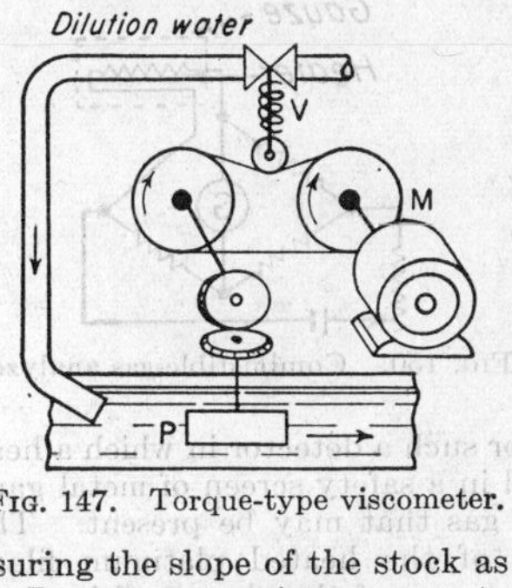

FIG. 147. Torque-type viscometer.

involves measuring the slope of the stock as it flows in a trough. The Brammer consistency controller (Paper and Industrial Appliances) does this by measuring the differential pressure in two air-bubble pipes, as shown in Fig. 148. The ratio of ΔP to L is a measure of the consistency. The differential pressure can readily be applied as a means to control.

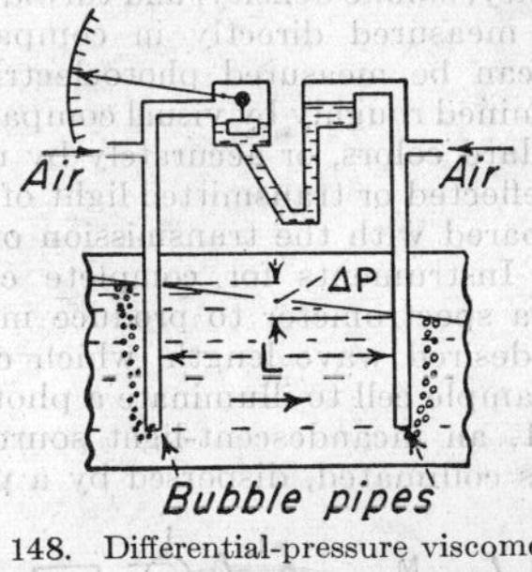

FIG. 148. Differential-pressure viscometer.

Calorific Value and Combustion

Calorific value of combustible materials is determined by burning the material and transferring the heat to a material of known specific heat, such as water or air, after the temperature rise of the heat-absorbing material is measured.

A typical *gas calorimeter* (Cutler-Hammer) is illustrated in Fig. 149. Here the test gas is metered and

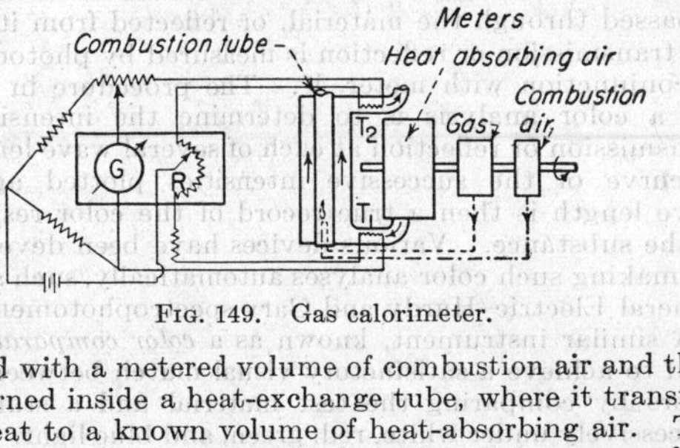

FIG. 149. Gas calorimeter.

mixed with a metered volume of combustion air and then is burned inside a heat-exchange tube, where it transfers its heat to a known volume of heat-absorbing air. The temperature rise of the heat-absorbing air is measured by two resistance thermometers T_1 and T_2 connected into a Wheatstone bridge. A self-balancing galvanometer is used to bring the bridge circuit to balance, the slide-wire position being a measure of the temperature difference between the thermometers and hence a measure of the calorific value.

Closely related to the automatic calorimeter are the instruments for the detection of hazardous concentrations of combustible gases. For example, Fig. 150 shows

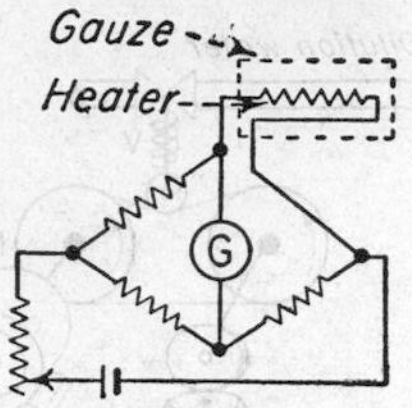

Fig. 150. Combustible-gas analyzer.

the circuit for such a detector in which a heated platinum wire inclosed in a safety screen of metal gauze burns any combustible gas that may be present. The increase in temperature of the heated platinum filament is then measured in terms of the change of its resistance, often calibrated in per cent L.E.L. (lower explosive level). In such instruments, a Wheatstone-bridge circuit of the deflection (unbalanced) type is employed.

Color

Color, opacity, smoke density, and turbidity are factors that can be measured directly in comparison with a standard or can be measured photoelectrically. Color may be determined roughly by visual comparison with one or more standard colors, or accurately by measuring the intensity of reflected or transmitted light of various wave lengths, compared with the transmission or reflection of a standard. Instruments for complete color analyses make use of a spectrometer to produce monochromatic light of any desired wave length, which can be passed through the sample cell to illuminate a photoelectric cell.

In Fig. 151, an incandescent-light source I produces light, which is collimated, dispersed by a prism, and re-

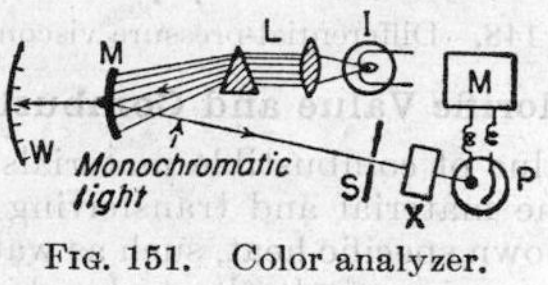

Fig. 151. Color analyzer.

flected by a movable mirror M onto a slit S. The wave length of monochromatic light passing through the slit is determined by the angle of the mirror, which is indicated on wave-length scale W. The monochromatic light is passed through the material, or reflected from it; and the transmission or reflection is measured by photocell P, in conjunction with meter M. The procedure in making a color analysis is to determine the intensity of transmission or reflection at each of several wave lengths. A curve of the successive intensities plotted against wave length is then a true record of the color response of the substance. Various devices have been developed for making such color analyses automatically, such as the General Electric–Hardy and Cary spectrophotometers.

A similar instrument, known as a *color comparator*, is used to achieve a satisfactory visual match between two colors by comparing the test material and a standard successively under white, red, green, and blue light. One type (Westinghouse) is illustrated in Fig. 152. The intensities of reflected light from the sample and from the standard S are compared by photocell P under the different colors, using color filter F. The mirror M reflects the light on the sample (or standard), which is then reflected diffusely from a white diffusing surface inside the enclosure onto the photocell P.

Smoke and fume detectors, as illustrated in Fig. 153, employ a beam of light traversing a duct or chimney from incandescent filament I, through a lens system and windows, to a photocell P, where the decrease in light

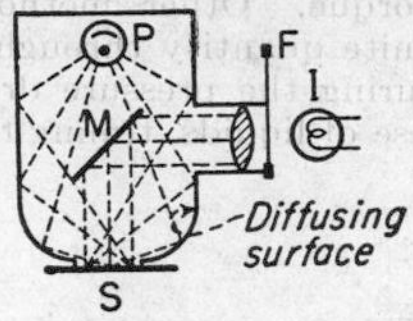

Fig. 152. Color comparator.

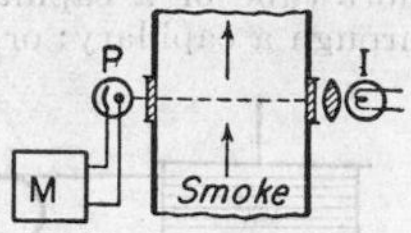

Fig. 153. Smoke detector.

transmission due to the smoke or fumes is registered by meter M.

Electrical Conductivity

The property of electrical conductivity often can be used to determine the composition of an electrolyte. The specific resistance of an electrolyte is defined as the resistance in ohms of a column of electrolyte 1 sq. cm. in cross section and 1 cm. deep. The specific conductance is the reciprocal of specific resistance and is expressed in reciprocal ohms. As distinguished from the determination of hydrogen-ion concentration (pH), the conductivity of a solution is a function of all the ions present in the solution. When the total number of ions increases, the conductivity increases, and vice versa. Hence the concentration in a solution of a single electrolyte, or group of electrolytes, can be readily determined.

A typical cell for determining electrical conductivity of a solution is illustrated in Fig. 154. The two electrodes

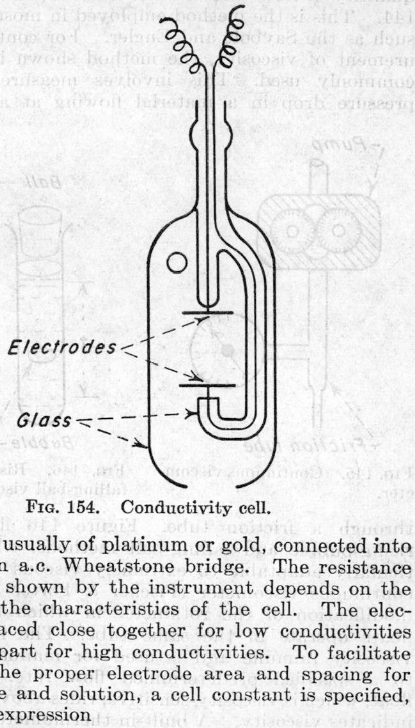

Fig. 154. Conductivity cell.

of the cell are usually of platinum or gold, connected into one arm of an a.c. Wheatstone bridge. The resistance of the cell as shown by the instrument depends on the solution and the characteristics of the cell. The electrodes are placed close together for low conductivities and farther apart for high conductivities. To facilitate selection of the proper electrode area and spacing for a given bridge and solution, a cell constant is specified, given by the expression

$$C = RK$$

in which R is the resistance measured in ohms, or maximum resistance measurable by the bridge; K is the expected minimum conductivity of the electrolyte to be measured; and C is the cell constant. Alternating current is used to reduce polarization. Theoretically, a high frequency is desirable, but the error introduced by using 60-cycle current is no more than ½ per cent and is usually unimportant for industrial purposes. In industrial instruments, an a.c. galvanometer is used either as the indicator or as the null detector of an a.c. bridge circuit, often a self-balancing type. Simple conductivity bridges frequently employ a cathode-ray null indicator for bridge balancing instead of a galvanometer. Conductivity changes to the extent of about 2 per cent per °C. temperature change; it is therefore necessary to use a constant-temperature bath or a reference cell.

A typical electrical-conductivity arrangement is illustrated in Fig. 155. A *conductivity cell* consisting of two

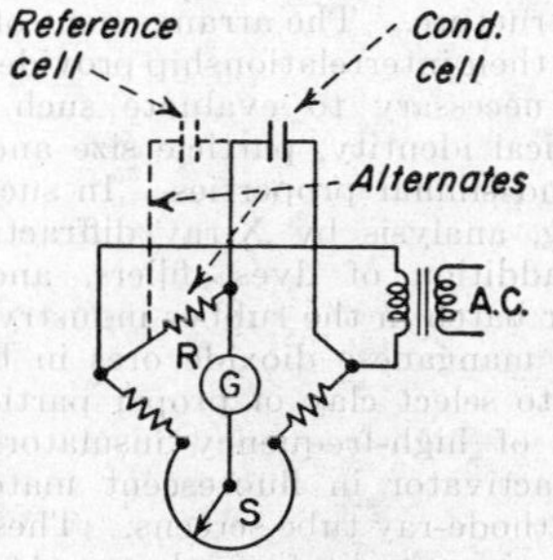

FIG. 155. Conductivity cell.

inert electrodes is placed in the solution under test. This cell is connected in a Wheatstone-bridge circuit operated on alternating current to avoid polarization of the electrodes. Instead of resistance R, a reference cell containing a sample of the liquid of desired concentration can be suspended in the test liquid, but not allowed to mix with it, in order that changes in resistance of the liquid due to temperature may be compensated.

Thermal Conductivity

One of the most used methods of analyzing gas mixtures is to measure the thermal conductivity. The method measures the change in resistance of a heated wire due to the heat lost from the wire through the surrounding atmosphere of test gas. Table 2 shows the ratio of thermal conductivities of certain gases to that of air at 32°F. and at 212°F. (the latter in italic type).

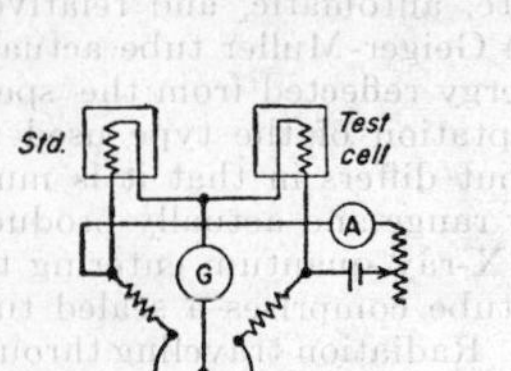

FIG. 156. Thermal-conductivity unit.

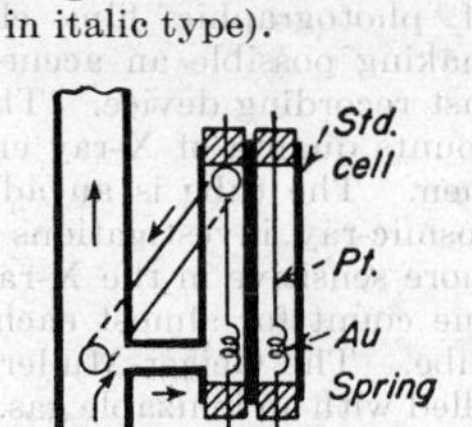

FIG. 157. Thermal-conductivity unit.

A typical thermal-conductivity hook-up is illustrated in Fig. 156. This method employs a pair of *thermal-conductivity cells*, one containing a standard gas, the other the test gas. The comparative rate of heat loss from the heated platinum wires of the two cells is a measure of the relative thermal conductivity of these two gases. The method requires calibration for the type of gas mixture tested and is generally suitable only where the various

Table 2. Thermal Conductivities of Gases Referred to Air*

Figures in roman type, gases and air at 32°F.
Figures in italic type, gases and air at 212°F.

Gas	K_{gas}/K_{air} (32°F)	K_{gas}/K_{air} (212°F, *italic*)
Air	1.000	*1.000*
Ammonia	1.15	*0.98*
Carbon dioxide	0.585	*0.690*
Carbon disulfide	0.312	.....
Carbon monoxide	0.958	.....
Chlorine	0.370	.....
Ethane	0.876	.....
Ethyl alcohol	0.708	.....
Ethylene	0.698	*0.885*
Helium	6.08	*5.53*
Hydrogen	7.35	*6.94*
Hydrogen chloride	0.635	.....
Hydrogen sulfide	0.648	.....
Methane	1.127	.....
Methanol	1.314	.....
Nitrogen	1.015	*1.012*
Oxygen	1.007	*1.034*
Sulfur dioxide	0.415	.....

components of the mixture differ considerably in thermal conductivity. However, it is sometimes possible to separate such components chemically or alter them so as to permit analysis. For example, carbon monoxide has nearly the same thermal conductivity as air, but carbon dioxide differs considerably from air. To analyze a mixture of carbon dioxide, carbon monoxide, and air (Cambridge), the mixture is first passed through one of the two conductivity cells where the concentration of carbon dioxide is measured in terms of resistance change, after which the mixture is passed over a heated carbon rod to burn the carbon monoxide to carbon dioxide. The mixture then is passed through the second cell, and the analysis is determined in terms of the comparison of conductivities in the two cells. Figure 157 illustrates a typical arrangement of standard and test cells for a flowing sample.

An oxygen meter (Hays) based on the magnetic property of oxygen utilizes two thermal-conductivity cells, one of which is located between the poles of a permanent magnet, the other serving as a reference cell. Oxygen in the gas sample is magnetized, increasing the rate of circulation over the heated filament to produce a cooling effect that is proportional to the amount of oxygen in the sample. Actually, this is a thermal-flow meter.

Chemical Absorption

A mixture of gases may be analyzed by measuring the volume decrease of the sample when several components are absorbed, one by one, in various solutions. The principle of the familiar *Orsat analyzer* is the basis of most manual and automatic gas analyzers. The Orsat instrument consists of a measuring burette, a leveling bottle for moving gas and adjusting pressure, and absorption pipettes for removing the several gases. A simplified Orsat analyzer with one pipette is shown in Fig. 158. The leveling bottle is used to draw gas into the measuring burette where its volume is adjusted to atmospheric pressure, after which the displacement water (or other confining fluid), which must be well saturated with the gases, is used to force the mixture successively into the several pipettes. During absorption of any component, when there is no further decrease in volume, this is taken as complete absorption. Total gas volume is measured after each absorption to show the percentage of that gas in the original sample.

In one unit (Hays), gas is drawn into a measuring chamber by tilting the apparatus. The measuring chamber is open to the atmosphere through the outlet line. Tilting in the opposite direction causes the gas to pass into the absorption chamber, at the same time seal-

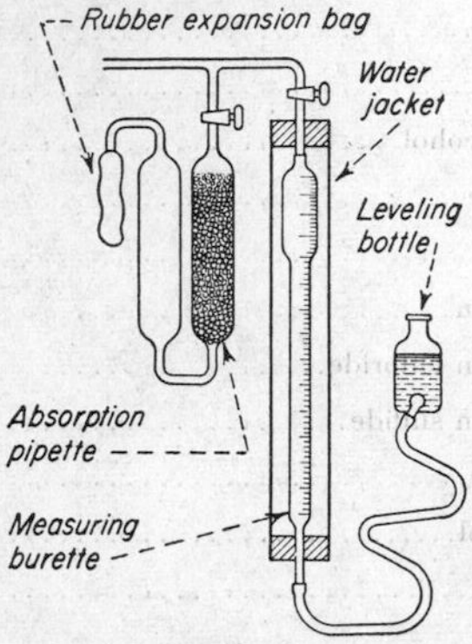

FIG. 158. Orsat analyzer.

ing the outlet line. Absorption of the absorbable gas causes a decrease in pressure and is indicated by the bellows, which is used as a measure of the gas composition. For flue gas, the first absorption pipette contains strong KOH solution (1 part KOH, 2 parts H_2O) for absorbing CO_2; the second pipette contains alkaline pyrogallol (1 part pyrogallol and 3 parts water plus an equal volume of the above KOH solution) for absorbing O_2; and the third pipette contains Cu_2Cl_2 for absorbing CO. This is made by mixing together, in a large bottle containing scrap copper, 150 g. Cu_2Cl_2 and 1 liter 1.12 sp. gr. commercial HCl. The KOH solution absorbs about forty times its volume, the pyrogallate about twice its volume, and the Cu_2Cl_2 solution only about its own volume.

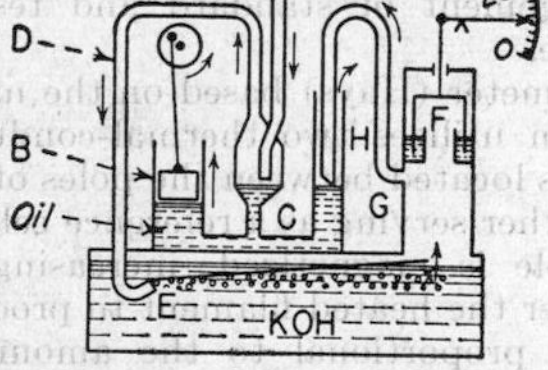

FIG. 159. Automatic orsat.

One type of *automatic Orsat* for carbon dioxide measurement (Republic) is illustrated in Fig. 159. This device employs a power-driven piston to achieve the measuring cycle, whereas others generally use water pressure. Flue gas enters at *A*, after being filtered and dried (or saturated), part of the flow passing to the apparatus. Piston *B* draws in the sample and forces it through the absorption system by rise or fall of oil moved by the piston. In the position shown, a sample has been drawn into measuring chamber *C*. The piston moves down, raising the oil level in *C* and *H*, trapping the sample in *C* at a measured volume and at atmospheric pressure, then pushing it through pipe *D* into the caustic potash tank *E*, under a baffle. The carbon dioxide is absorbed, and the remaining gas rises into liquid-sealed bell *F*, which rises to a height depending on the volume *C*, less the carbon dioxide absorbed. The piston then rises, drawing the spent gas through pipe *G* and into *H* where the next cycle expels it. At the same time a new sample is drawn into *C*, and the cycle repeats.

X-ray Diffraction

The use of X-ray diffraction for analytical purposes has become an extremely important aid to several of the process industries. The time and technique required to conduct an analysis by this method through photographic means have formerly limited its usefulness. However, development of the Geiger-counter X-ray spectrometer has greatly simplified the procedure and reduced the time required to make an analysis.

Basically, the diffraction of X-rays in a solid serves to analyze its structure. The arrangement of atoms within the solid and their interrelationship provide the analytical information necessary to evaluate such properties as simple chemical identity, particle size and distribution, grain size, and similar properties. In such processes as the following, analysis by X-ray diffraction serves (1) to control addition of dyes, fillers, and accelerators to the rubber batch in the rubber industry; (2) to select the required manganese dioxide ores in battery manufacture; (3) to select clay of proper particle size in the manufacture of high-frequency insulators; and (4) to control the activator in fluorescent materials used in lamps and cathode-ray tube screens. These applications are representative of a host of others. At present a tool of the product-development and process-control laboratory, future research may adapt it to continuous process control.

Principle of X-ray Diffraction. When a beam of X rays is directed into a sample undergoing analysis, a series of secondary beams is reflected from the specimen in the form of a diffraction cone. Analysis of this cone with properly sensitized photographic film shows a series of diffraction lines in the form of concentric circles. These lines occur as a result of the cancellation and reinforcement of the reflections from the various atomic planes of the specimen. No two different materials have been found that show identical patterns, and identification by this means is therefore positive. However, identification by the examination of photographic film is somewhat inadequate.

The use of the **Geiger-Muller counter tube** in place of photographic film eliminates these inadequacies, making possible an accurate, automatic, and relatively fast recording device. The Geiger-Muller tube actually counts quanta of X-ray energy reflected from the specimen. The tube is an adaptation of the type used for cosmic-ray investigations but differs in that it is much more sensitive in the X-ray range and actually produces one count for almost each X-ray quantum entering the tube. The Geiger-Muller tube comprises a sealed tube filled with an ionizable gas. Radiation traveling through the gas within the tube results in partial ionization of the gas, which in turn allows an electric current to flow through the tube. This current is amplified and put through an electronic integrating circuit so that it can be evaluated in terms of the intensity of X-radiation.

In order to obtain a diffraction pattern with the counter tube, it is necessary to scan a cross section of the diffraction cone. This is accomplished by rotating the counter tube and the sample in such a manner that the former moves at twice the angular speed of the latter so that the tube will be progressively exposed to the various beams that are reflected from the sample. As illustrated in

Fig. 159*A*, the counter tube is rotated through 90 deg. of arc by means of a motor-driven, calibrated scanning arm (goniometer) whose speed of rotation is precisely coordinated with the chart speed of a recording potentiometer. In this way, an exact space relationship between the representative peaks and dips of reflected radiation is reproduced on the recorder chart.

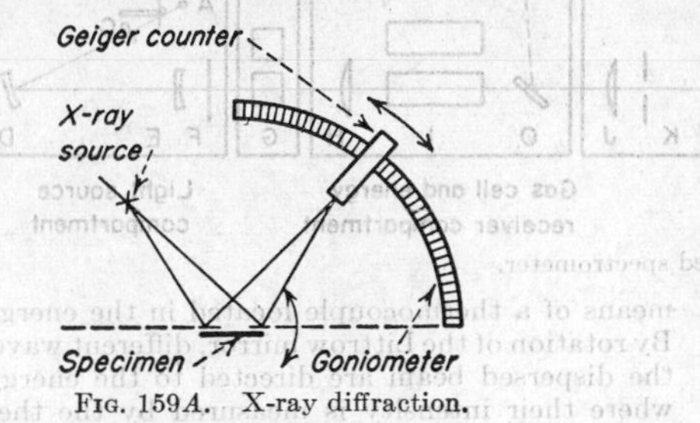

FIG. 159*A*. X-ray diffraction.

Ultraviolet and Infrared Spectrometry

Principles. Spectrometry is based upon the fact that chemical substances, in either gas, liquid, or solid form, absorb light at certain wave lengths. When light of a single color or wave length is passed through a layer of any substance, the light may be transmitted completely through the substance, or it may be partly or completely absorbed, depending upon the chemical identity of the substance and the amount of the substance in the path of the light.

If a plot is made of wave length vs. percentage of light absorbed under various conditions, an irregular curve of hills and valleys will result. The hills correspond to wave lengths or colors for which the substance is relatively opaque; the valleys correspond to wave lengths for which the substance is relatively transparent. In some cases, the hills show sharp absorption peaks; in other cases the absorption maxima are broad.

The absorption curve for any substance is unique. No two chemical substances absorb light in exactly the same fashion. The absorption spectrum of a substance serves as a fingerprint to identify the substance when in pure form. When the substance is mixed with others, usually it will be found that at least some of the absorption peaks of the substance will not coincide with the absorption peaks of other substances.

Substances that appear colored to the eye obviously absorb light in the visible region of the spectrum. Many chemical substances are colorless, absorbing no wave lengths in the visible region. Such substances will nevertheless absorb radiation in the invisible spectral regions, depending upon the types of chemical bonds in the molecules. Thus, molecules that include the benzene ring in their structure, such as benzene, toluene, xylene, and the like, show distinctive absorption patterns in the ultraviolet region. Conjugated diolefins also absorb in the ultraviolet region. Ordinary saturated hydrocarbons, such as propane, and simple unsaturated compounds, such as butylene, do not absorb extensively in the ultraviolet. For the analysis of the latter compounds, the infrared region of the spectrum is utilized. Typical absorption curves for mixtures of butane and isobutane are shown in Fig. 160.

Ultraviolet Spectrometer. Continuous automatic analysis of a flowing gas or liquid now is possible with this instrument and is accomplished by recording the percentage transmission of light through the sample at the frequency selected. Major components of the instrument include a suitable light source, a monochromator, and a measuring system. Automatic analyses of more than one sample flow can be effected by means of electric time-cycle control and associated solenoid valves and fittings in the sample lines.

Instruments for the ultraviolet region are generally referred to as quartz spectrometers, since their transmitting portions in the optical system are of either fused or crystal quartz. Figure 161 shows the Beckman spectrometer. With the proper light source, the instrument can be used between 1000 millimicrons in the near infrared, through the visible range and 200 millimicrons in the ultraviolet. For ultraviolet absorption, a hydrogen-discharge lamp provided with electronic voltage control supplies the radiation. The radiation from *A* in Fig. 161 is projected by a condensing mirror *B* and a diagonal mirror *C* through an entrance slit *D* onto the collimating mirror *E*, which renders the light parallel and reflects it onto the crystal quartz prism *F*. Since the back of this prism is aluminized, the light is reflected back through the prism, undergoing further dispersion. The spectrum so produced is reflected by the collimating mirror and focused on the exit slit, which is above the entrance slit at *D*. If the prism is rotated, light of any wave length within the range of the instrument can be projected through the exit slit, where it passes through the absorption cell *G*, which contains the sample, to a photocell *H*. The cell output is amplified electronically and recorded. For continuous automatic operation, the instrument re-

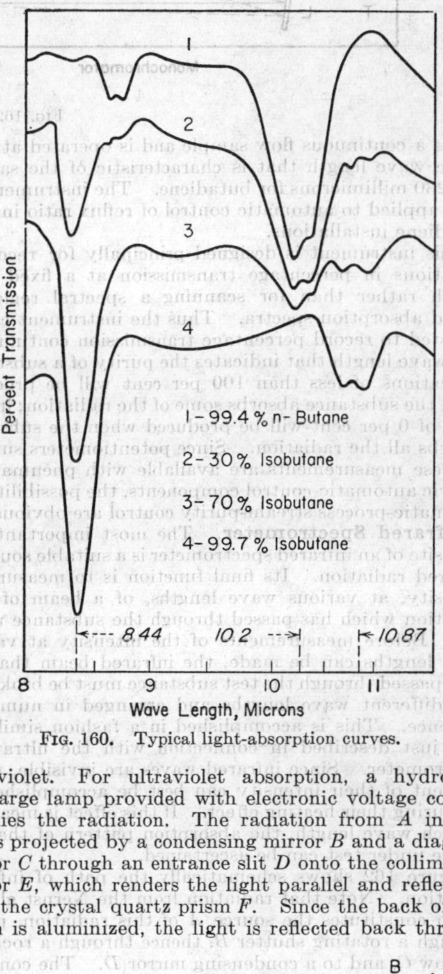

FIG. 160. Typical light-absorption curves.

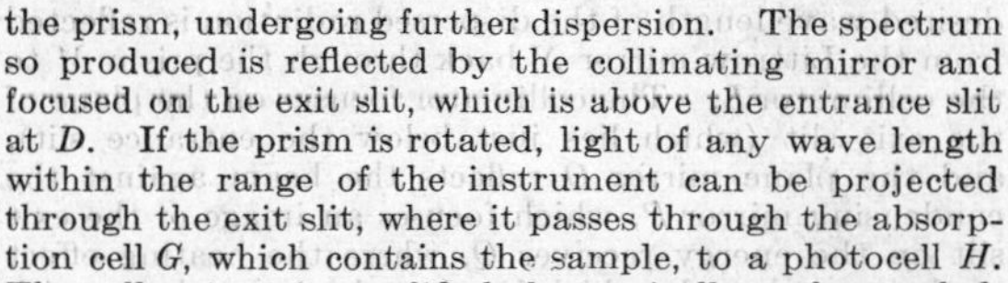

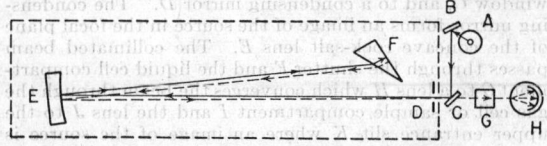

FIG. 161. Ultraviolet spectrometer.

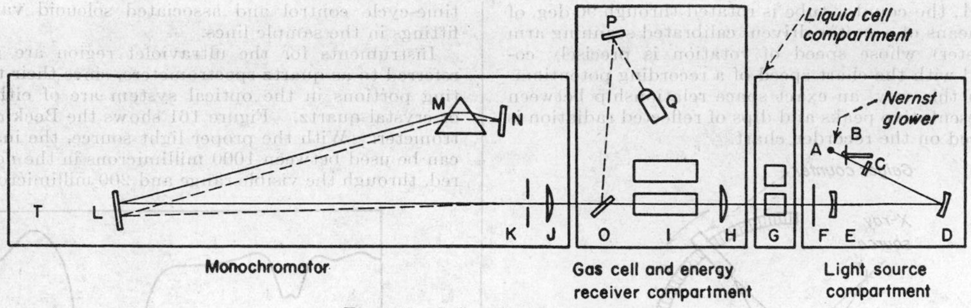

FIG. 162. Infrared spectrometer.

ceives a continuous flow sample and is operated at some single wave length that is characteristic of the sample, *e.g.*, 230 millimicrons for butadiene. The instrument has been applied to automatic control of reflux ratio in some butadiene installations.

This instrument is designed principally for recording variations in percentage transmission at a fixed wave length rather than for scanning a spectral region to record absorption spectra. Thus the instrument can be adjusted to record percentage transmission continuously at a wave length that indicates the purity of a substance. Indications of less than 100 per cent will be produced when the substance absorbs some of the radiation; indications of 0 per cent will be produced when the substance absorbs all the radiation. Since potentiometers suitable to these measurements are available with pneumatic or electric automatic-control components, the possibilities of automatic-process stream-purity control are obvious.

Infrared Spectrometer. The most important prerequisite of an infrared spectrometer is a suitable source of infrared radiation. Its final function is to measure the intensity, at various wave lengths, of a beam of such radiation which has passed through the substance under test. Before measurements of the intensity at various wave lengths can be made, the infrared beam that has been passed through the test substance must be broken up into different wave lengths and arranged in numerical sequence. This is accomplished in a fashion similar to that just described in connection with the ultraviolet spectrometer. Since infrared waves are invisible, measurement of their intensity can best be accomplished by measuring their heating effect. If this effect is measured at each wave length, the absorption pattern of the substance under test can be ascertained.

Figure 162 shows schematically the path of infrared radiation. Note that radiation from the Nernst glower, which constitutes the source *A* of this radiation, passes through a rotating shutter *B*, thence through a rock-salt window *C*, and to a condensing mirror *D*. The condensing mirror forms an image of the source in the focal plane of the concave rock-salt lens *E*. The collimated beam passes through the shutter *F* and the liquid cell compartment *G* to a lens *H* which converges the beam through the gas cell or sample compartment *I* and the lens *J* to the upper entrance slit *K* where an image of the source is formed. Radiation from the slit is collimated by a spherical mirror *L* and dispersed by the prism *M*. The desired wave length of the dispersed radiation is reflected from the Littrow mirror *N* back through the prism *M* to the collimator *L*. The collimator focuses on the plane of the exit slit (which lies just below the entrance slit), and the plane mirror *O* reflects the beam against the condensing mirror *P*, which focuses an image of the exit slit on the energy receiver *Q*, where the heating effect of the wave length, which is selected in accordance with the position of the Littrow mirror *N*, is measured by means of a thermocouple located in the energy receiver. By rotation of the Littrow mirror, different wave lengths of the dispersed beam are directed to the energy receiver, where their intensity is measured by the thermocouple and recorded by a potentiometer. Thus a record of the absorption pattern of the substance under test is obtained. Self-filtering infrared analyzers such as those made by Baird Associates and Leeds & Northrup are of value.

Emission Spectrometry

Used mainly for metallurgical identification and analysis, emission spectrometry employs either a spark, an arc, or a flame for exciting the material to be examined and causing it to emit its characteristic spectrum. The spectrum produced is independent of the excitation (or temperature) except that higher excitation produces a greater number of characteristic lines. The arc spectrum, most widely used, is produced by placing a quantity of the sample in a depression in the lower carbon electrode (anode) and drawing a 2000-volt a.c. or half-wave d.c. arc. Specially purified carbon is employed for this purpose. In some cases, metallic specimens may be cast or machined into anodes and thus used directly. In spark spectrometry, a high-frequency spark is employed, rather than an arc, giving greater excitation and consequently requiring a shorter exposure.

Spectrometers employed in emission spectrometry are entirely similar to those described in earlier paragraphs, employing either a prism or a grating for dispersion. Since spectra of many elements lie in both the ultraviolet and visible ranges of frequency, the optical system is usually of quartz. Recording is by photographic plate or film, and the usual arrangement is to photograph a considerable number of spectrum bands on a single plate. Emission spectrometry is comparatively rapid.

Flame emission is of little interest except in certain specialized applications such as that used for the rapid routine determination of sodium and potassium salts in solution in the presence of other elements. Difficult by chemical methods, this analysis can be accomplished in 2 min. for 3 per cent accuracy, or in 20 min. for 1 per cent accuracy. The sample is aspirated from a beaker and sprayed into a gas flame. A strong characteristic frequency for each of sodium and potassium is isolated by means of two appropriate color filters and applied to two photocells, one for sodium, the other for potassium. A switch enables either cell to indicate on the meter.

Mass Spectrometry

A **mass spectrometer** is an instrument for separating and determining the relative abundance of ions of different mass to charge ratio. Its use in gas analysis depends on the fact that positive ions can be produced by bombarding a gas with electrons. If the gas is a compound, not only will ions of mass equal to that of the compound be formed, but so will fragments of the original molecule. The relative abundance of the ions thus formed is characteristic of the compound from which they were formed. A plot of relative abundance of the ions as a function of mass to charge ratio is termed a **mass spectrum.** Mass spectra can be used for qualitative identification and for quantitative analysis of mixtures.

The deflection of ionized components of elements, compounds, or mixtures as caused by an applied electrostatic or electromagnetic force is the principle of the mass spectrometer. The degree of deflection depends on the relative masses of the components. In operation (see Fig. 163), the gaseous or volatile substance is leaked

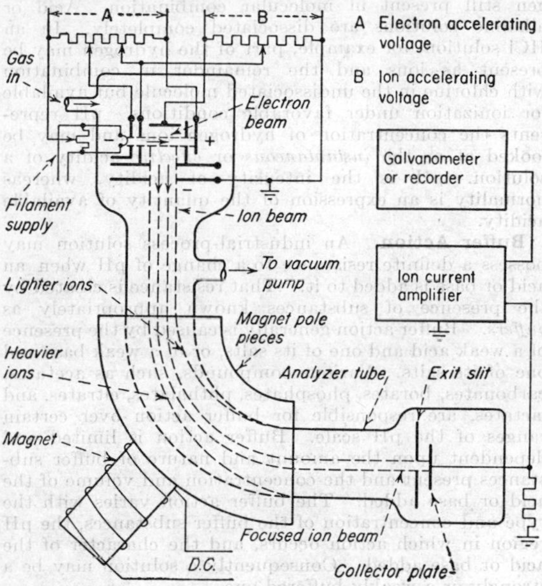

FIG. 163. Mass spectrometer.

across the positive ray behind the cathode. The ions are accelerated by high-voltage field, velocity-filtered between electrically charged plates, and then subjected to a constant magnetic field. Deflected ions follow individual parabolic paths according to masses of respective components they represent. A record is made by means of photographic film. This instrument provides a means for rapid analysis, particularly where distribution of certain components or their presence in relation to others must be quickly ascertained. It is used widely in the study of isotopes.

A special adaptation of the mass spectrometer is the so-called **leak detector,** which played an important part during the war. It is a portable unit capable of detecting concentrations of helium as low as 1 part in 200,000 parts of air, and has a quick response, requiring about 1 sec. to operate. Since it is adjusted to detect helium gas, it is used with a probe of helium to explore the outside of the system or, if the system is surrounded by helium, any leak into the system that will admit helium. The detector consists of a mass spectrometer tube and the necessary electronic circuit, together with a diffusion pump, a rough vacuum pump of the rotary-oil type, a thermocouple vacuum gage, an ionization gage for high-vacuum measurement, a control panel, and the necessary batteries and battery charger.

Polarography

The polarograph is an analytical instrument, both qualitative and quantitative, for determining the chemical constituents of either aqueous or non-aqueous solutions. Although extremely small quantities of material can be detected, it is not limited to microanalysis. In both micro- and macroanalytical work, it replaces tedious and slow wet chemical methods for determining metallic and organic constituents.

The dropping-mercury electrode, illustrated schematically in Fig. 164, comprises a very fine-bore capillary tube

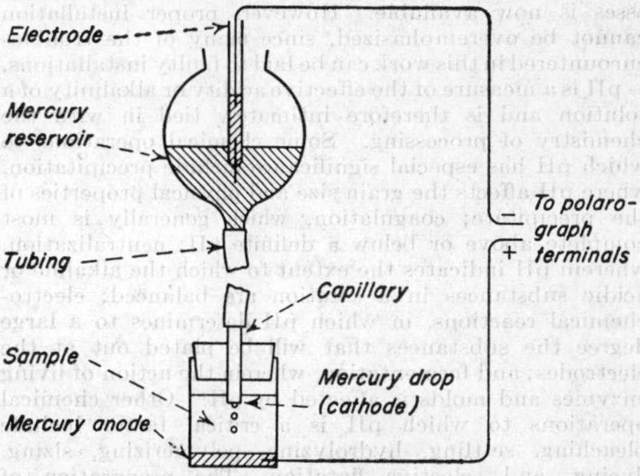

FIG. 164. Dropping-mercury electrode.

connected by a flexible tube to a mercury reservoir and placed in the solution so that very fine drops of mercury are formed beneath the surface. The rate of mercury flow can be adjusted by varying the height of the reservoir and normally amounts to a drop every 1 to 3 sec. The mercury reservoir is connected to one terminal of the polarograph; the other cell terminal is a pool of mercury in the bottom of the solution vessel.

In operation, a few drops of mercury added to the sample act as one electrode of the cell. The dropping mercury electrode next is inserted. An inert gas, usually nitrogen, then is bubbled through the solution to remove atmospheric oxygen, which is reducible and gives a polarographic wave not usually of interest to the analyst. As the voltage impressed on the cell is raised slowly, the current remains nearly constant until a voltage is reached at which a component of the solution is reducible. The current then rises sharply to a new level, at which it remains constant until the voltage is reached at which a second component is reducible. Another sharp rise in current to a new constant value then occurs. These sharp rises in current are referred to as polarographic steps or waves. Since the polarographic steps for various components occur at different voltages, several substances can be determined in one solution on a single polarogram.

The output from the polarographic cell is fed to the recorder so that the displacement of the pen from zero is determined by the amount of current flowing through the cell. The chart is driven by a constant-speed synchronous motor. The polarographic bridge which supplies voltage to the cell is also driven by a synchronous motor so that the chart displacement is a linear function of the voltage applied to the cell, *e.g.*, the time ordinate of the chart becomes, in essence, the voltage ordinate. The record obtained, therefore, is current vs. voltage.

Hydrogen-ion Concentration, or pH

The success of many chemical and bacteriological processes depends upon the accurate measurement and close control of pH. Automatic and continuous pH instrumentation steadily is replacing manual measurement and control. The basic arguments of increased economy, better quality, and greater yields—which generally apply to automatic control—apply equally well to pH. The increasing interest shown by process engineers in pH measurement is due principally to the comparatively recent development of more practical and reliable means of measurement. pH-measuring equipment of rugged design and requiring a minimum of maintenance for installation directly in chemical processes is now available. However, proper installation cannot be overemphasized, since many of the troubles encountered in this work can be laid to faulty installations.

pH is a measure of the effective acidity or alkalinity of a solution and is therefore intimately tied in with the chemistry of processing. Some chemical operations to which pH has especial significance include precipitation, where pH affects the grain size and physical properties of the precipitate; coagulation, which generally is most complete above or below a definite pH; neutralization, wherein pH indicates the extent to which the alkaline or acidic substances in a solution are balanced; electrochemical reactions, in which pH determines to a large degree the substances that will be plated out at the electrodes; and fermentation, wherein the action of living enzymes and molds is affected by pH. Other chemical operations to which pH is a critical factor include bleaching, settling, hydrolyzing, polymerizing, sizing, dyeing, and selective flotation. The preparation of penicillin, tanning of leather, treatment of sewage and wastes, electroplating, and anodizing are a few examples of commercial processes requiring pH measurement and control.

Theory. When pure water dissociates, equal numbers of hydrogen and hydroxyl ions are formed, and it is termed neutral. The concentration of H^+ and OH^- ions in pure water and neutral solutions are each equal to 10^{-7}, 0.0000001, or 1/10,000,000 g. equivalents/l. at 22°C. The use of a negative exponent, decimal, or fractional expression of such concentrations is cumbersome and inconvenient for hydrogen-ion calculations. The pH scale was devised and utilizes the negative exponent as a means of expressing the hydrogen-ion concentration. Thus a neutral solution is termed as having a pH of 7 instead of a hydrogen-ion concentration of 10^{-7} g. equivalents/l. Mathematically, pH is defined as the logarithm of the reciprocal of the hydrogen-ion concentration, shown by $pH = -\log [H^+]$; or $[H^+] = 10^{-pH}$.

When an acid such as HCl dissociates, H^+ and Cl^- ions are formed, and the pH of the solution is dependent upon the degree of dissociation. From Fig. 165, it may be noted that the pH will lie between 0 and 7 for an acid solution. Similarly, a base such as NaOH forms Na^+ and OH^- ions upon dissociation. It also shows that the pH of an alkaline solution will lie between 7 and 14.

The pH scale not only is of advantage in hydrogen-ion calculations, but one can learn to use it intelligently without knowing the derivation of the term, just as one may learn to use the Fahrenheit scale of temperature without knowing the derivation of degrees F. This enables the operator or foreman of a process to make intelligent use of instruments for determining pH and to apply a practical interpretation of the data thus obtained. It must be stressed that the hydrogen-ion concentration of a solution increases by a power of 10 while the pH term decreases by 1 unit. Practically speaking, a solution of pH 5 is ten times as acidic as one of pH 6.

pH vs. Normality. The normality of a solution should not be confused with the pH. Normality represents the *total acidity* of a solution, namely, the hydrogen ions present as ions, plus the ionizable hydrogen still present in molecular combination. Acid or alkaline solutions are dissociated completely. In an HCl solution, for example, part of the hydrogen may be present as ions and the remainder in combination with chlorine in the undissociated molecule but available for ionization under favorable conditions. pH represents the concentration of hydrogen ions and may be looked at as the *instantaneous* or *effective* acidity of a solution. pH is the intensity of acidity, whereas normality is an expression of the quantity of available acidity.

Buffer Action. An industrial-process solution may possess a definite resistance to a change of pH when an acid or base is added to it. That resistance is created by the presence of substances known appropriately as *buffers.* Buffer action generally is caused by the presence of a weak acid and one of its salts, or of a weak base and one of its salts. Various compounds, such as acetates, carbonates, borates, phosphates, phthalates, citrates, and lactates, are responsible for buffer action over certain ranges of the pH scale. Buffer action is limited and dependent upon the amount and nature of buffer substances present and the concentration and volume of the acid or base added. The buffer action varies with the type and concentration of the buffer substances, the pH region in which action occurs, and the character of the acid or base added. Consequently a solution may be a strongly or a weakly buffered one.

A solution containing acetic acid and sodium acetate will present an example of buffer action. If a strong acid, such as HCl is added to such a mixture, the HCl will react with some of the sodium acetate to form slightly ionizable acetic acid, with a corresponding decrease in the amount of free hydrogen ions remaining in solution. Thus the added HCl will cause little or no change in the pH. Similarly, when a base such as NaOH is added to the solution, the NaOH will combine with the acetic acid to produce sodium acetate and water, and no additional hydroxyl ions will be produced to change the pH of the solution.

Methods of Measurement. There are two general methods of pH measurement: (1) the fundamental electrometric method and (2) the derived colorimetric method based on the use of indicators and permanent color standards. Since the second method depends on visual com-

$[H^+]$	10^0	10^{-1}	10^{-2}	10^{-3}	10^{-4}	10^{-5}	10^{-6}	10^{-7}	10^{-8}	10^{-9}	10^{-10}	10^{-11}	10^{-12}	10^{-13}	10^{-14}
pH	0	1	2	3	4	5	6	7	8	9	10	11	12	13	14
pOH	14	13	12	11	10	9	8	7	6	5	4	3	2	1	0
$[OH^-]$	10^{-14}	10^{-13}	10^{-12}	10^{-11}	10^{-10}	10^{-9}	10^{-8}	10^{-7}	10^{-6}	10^{-5}	10^{-4}	10^{-3}	10^{-2}	10^{-1}	10^0

Strong acids — Neutral solutions — Strong bases

FIG. 165. Hydrogen- and hydroxyl-ion-concentration chart.

parison of color, it cannot record or control automatically, but nevertheless it is used to some extent, especially where cost is a major concern.

Electrometric pH Measurement. In this method, two electrodes are immersed in the unknown solution. One electrode—the *measuring* electrode—is sensitive to the presence of hydrogen ions; the other electrode—the reference *electrode*—is required to complete an electric circuit through the solution and to supply a constant potential. The reference electrode is not affected by pH. Since to date it is impossible to measure the potential of a single electrode, the reference electrode must be used. Each electrode is sometimes referred to as a *half cell.* The potential measured is the algebraic sum of the two electrodes and is conveniently measurable with a potentiometer.

Measuring Electrodes. Four outstanding electrodes have been developed that are sensitive to pH, namely, (1) the *hydrogen electrode,* (2) the *quinhydrone electrode,* (3) the *antimony electrode,* and (4) the *glass electrode.* The glass electrode is the modern and most practical means of pH measurement and has practically replaced the other electrodes for industrial applications. When a thin membrane of special glass separates two solutions of different hydrogen-ion concentrations, a potential is created that is a function of the two concentrations. If the pH of one solution is maintained constant, that of the other solution can be determined from the potential produced. A typical glass electrode consists of a glass tube about ½ in. in diameter with a bulbous glass tip. The solution of constant pH is contained inside the tip. Several theories regarding the operation of glass electrodes have been advanced, but their explanation must be left to complete texts on the subject.

Reference Electrodes. A good reference electrode must possess a reasonably constant potential regardless

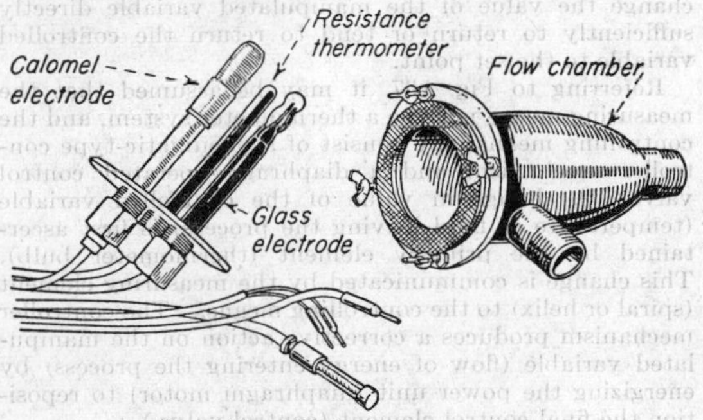

FIG. 166. pH electrode assembly.

of temperature changes, must maintain this potential over long periods of time, must not easily be poisoned by solutions, must be simple and easy to maintain, and must not in itself be affected by changes in pH. Years of experience have proved the calomel electrode, saturated or unsaturated, to be the best answer to the above requirements. The saturated calomel electrode, most widely used, comprises mercury and mercurous chloride (calomel) with saturated potassium chloride as an electrolyte. In a typical electrode, a liquid junction between the electrode and the process solution is established by means of a small hole covered by a ground-glass sleeve. Diffusion of electrolyte from the electrode to the solution is slow, but sufficient to provide excellent electrical conductivity.

Electrode Assemblies. Sturdy construction of electrode assemblies has been responsible for a wider industrial use of pH measurement during the past few years. The flow assembly illustrated in Fig. 166 can be installed directly in process pipe lines and is capable of operating under moderate pressures. Numerous materials of construction, including porcelain-covered iron, rubber, stainless steel, nickel alloys, and glass, are available. An assembly available for immersion in tanks, vats, and other process vessels also is available.

pH Recording and Controlling Equipment. Null-balance potentiometers rather than deflectional instruments are required for pH measurement, since the measuring instrument must not draw any appreciable current from the electrodes. Any current flow would cause polarization at the electrodes, immediately upsetting the accuracy of measurement. Potentiometers are available in four basic ranges: 0 to 7; 3 to 10, 6 to 13, and 2 to 12 pH. Pneumatic and electrical control components available with thermocouple potentiometers are also available with pH potentiometers. Equipment for measuring the e.m.f. of glass electrode cells must be designed to measure the e.m.f. in circuits having a resistance of 10^9 ohms, within a few millivolts.

Oxidation-reduction Potential

The measurement of oxidation-reduction potential is a reasonably well known technique of physical chemistry. However, its application to industrial processes is just emerging from the experimental stage. Basically, oxidation-reduction potential is to oxidation and reduction chemistry what pH is to double decomposition chemistry or neutralization of acids and bases. Kolthoff and Furman* state that, "It is the potential an unattackable metal electrode assumes in a solution relative to that of the normal hydrogen electrode." The measurement of oxidation-reduction potential in practice involves the use of a standard reference electrode, such as a calomel electrode, and a platinum, gold, or other non-attackable metallic electrode. The potential from the electrodes is measured by an instrument comparable with a potentiometer designed for pH measurement but with the scale graduated to read in millivolts rather than in pH units. The measured potential may be expressed as millivolts or as E_h or R_h.

Because oxidation-reduction-potential measurement has been so infrequently used, it is a fertile field for investigation. In addition to concentration control of reducing and oxidizing solutions, it has been used to measure the chlorine content of water to less than 1 p.p.m., using a calomel and a silver–silver chloride electrode. Chemical engineers will visualize many other applications that can be corroborated only by test data.

FUNDAMENTALS OF AUTOMATIC CONTROL†

Essentially, process control consists of maintaining within limits, or altering in a predetermined manner, the energy (and sometimes the material balance) of matter undergoing treatment in a process. The process is controlled automatically by measuring the state of a selected process variable, either continuously or at frequent intervals, and then correcting the input of energy or material to maintain the value of the variable within acceptable limits.

The automatic controller and the process form a control circuit or closed loop that must be considered as a dynamic unit, since a change in one part of the circuit carries through to affect the whole system. Since automatic-control operation is based on deviation of a process variable, then, before the controller can act,

* "Potentiometric Titrations," Wiley, New York, 1931.

† Peters and Olive, Fundamental Principles of Automatic Control, *Chem. & Met. Eng.*, May, 1943, pp. 98–107. Horn, "Graphical Representation and Analysis of Automatic Control Terminology" presented Sept. 18, 1946, at the Pittsburgh, Pa., meeting of Am. Soc. Mech. Engrs. Eckman, "Principles of Industrial Process Control," Wiley, New York, 1945.

a deviation of the measured process variable must occur. Therefore, to produce the desired effect on the process, the process variable selected for control must have a definite relationship to the state of the process. Frequently, industrial processes contain a number of variables and conditions that make it difficult to locate the particular functions whose deviations are significant to the process. Selecting the right process condition or variable for measurement is one of the main problems in applying automatic control (see Process Variables and Their Measurement, p. 1266).

Although a wide variety of mechanisms and devices are used to accomplish automatic process control, most of them fit into a readily defined pattern. There are a definite number of elements to an automatically controlled process. Measuring devices and control mechanisms fall readily into definite classifications, and all process-control systems have certain characteristics that can be overcome by applying particular kinds of control action of which there are a definite number of different types and combinations.

Elements of Process Control

A certain minimum number of separate elements is required to control a process automatically. These elements may be defined as follows: (1) An **automatic controller** is a mechanism that measures the value of a variable quantity or condition and operates to correct or limit **deviation** of this measured value from a selected reference; it includes both the measuring means and the controlling means. (2) The **controlled variable** is that quantity or condition which is measured and controlled. (3) The **controlled medium** is that process energy or material in which a variable is controlled. (4) The **manipulated variable** is that quantity or condition which is varied by the automatic controller to affect the value of the controlled variable. (5) The **control agent** is that process energy or material of which the manipulated variable is a condition or characteristic.

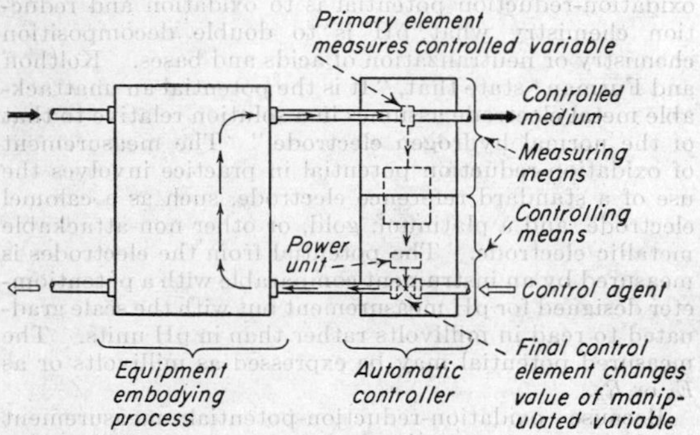

FIG. 167. Elements of automatic control.

The automatic controller comprises a measuring means and a controlling means, as shown in Fig. 167, illustrating a fluid heat-exchange process with automatic temperature control. The measuring means may consist of a thermometer or potentiometer system measuring the temperature of the controlled medium, a fluid leaving the process. The controlling means consists of a mechanism embodying predetermined controller action affecting the condition or characteristic of the control agent or fluid energy entering the process. This action upon the control agent produces an effect that corrects or limits the temperature of the controlled medium.

Process equipment and the automatic controller are installed in such a manner that the initial action within the process and the counteraction of the automatic controller form a closed loop or circuit of action. The process is an operation or series of operations in which the value of a quantity or condition is controlled. It includes all functions that directly or indirectly affect the value of the controlled variable. In Fig. 167, the controlled variable (temperature) of the controlled medium (fluid leaving the process) is thus measured by the automatic controller, which acts in a predetermined manner to vary the value of the manipulated variable (flow) of the control agent (fluid energy entering the process). Action of the automatic controller restores, or tends to restore, the value of the controlled variable to the desired value. The controlled variable is a condition or characteristic of the controlled medium, and the manipulated variable is a condition or characteristic of the control agent.

Elements of Automatic Controllers

Any automatic controller necessarily consists of two essential parts, a measuring means and a controlling means. In many controllers these may be further broken down and five fundamental elements recognized: (1) a primary element that either utilizes or transforms energy from the controlled medium to produce an effect or response governed by change of deviation in the value of the controlled variable; (2) a measuring element that converts the response of the primary element into an indication of the value, error, or deviation of the controlled variable and communicates this value to the control means (sometimes, as in the pressure gage, both 1 and 2 comprise a single functional unit); (3) the controller proper, which detects any deviation of the instantaneous position of the measuring element from the set point and initiates proper corrective action; (4) the power unit, which converts corrective action of the control mechanism into position of the (5) final control element to change the value of the manipulated variable directly sufficiently to return or tend to return the controlled variable to the set point.

Referring to Fig. 167, it may be assumed that the measuring means may be a thermometer system, and the controlling means may consist of a pneumatic-type controller mechanism and a diaphragm-operated control valve. A change in value of the controlled variable (temperature of fluid leaving the process) is first ascertained by the primary element (thermometer bulb). This change is communicated by the measuring element (spiral or helix) to the controlling means. The controller mechanism produces a corrective action on the manipulated variable (flow of energy entering the process) by energizing the power unit (diaphragm motor) to reposition the final control element (control valve.)

Operation of Automatic Controllers

In some controllers, all five elements are not separate and distinct. A *self-operated* controller is illustrated in Fig. 168, showing a diaphragm-type pressure-reducing

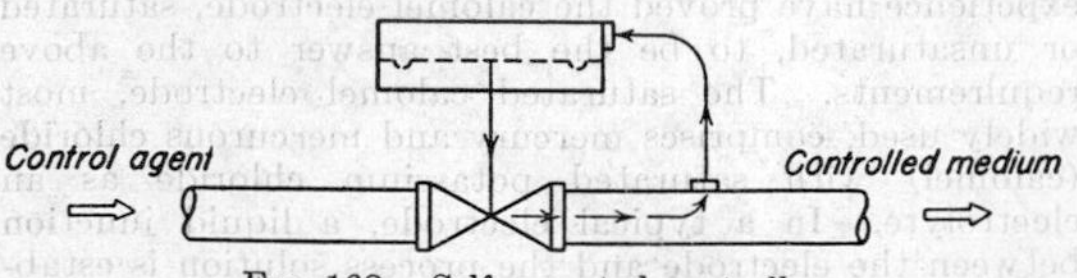

FIG. 168. Self-operated controller.

valve, operated by energy obtained only from the controlled medium through the primary element. A self-operated controller must have both self-operated measuring means and self-operated controlling means. It must derive all its energy through the primary element and may not derive energy separately from the controlled medium to obtain a relay-operated control. The self-operated measuring means of the pressure-reducing valve is the diaphragm, which is also the primary element. The diaphragm measures the pressure of the controlled variable and transforms this pressure to a force. Self-operated controlling means consists of the diaphragm of the measuring means, the valve body, and plug. Diaphragm force directly operates the valve plug, and the relative position of the valve body and plug affects the value of the controlled variable. Note that, in this example, the controlled medium and the control agent are the same.

Relay-operated controllers are illustrated in Figs. 169 and 170. A relay-operated controller must have at least one source of energy other than that derived from the

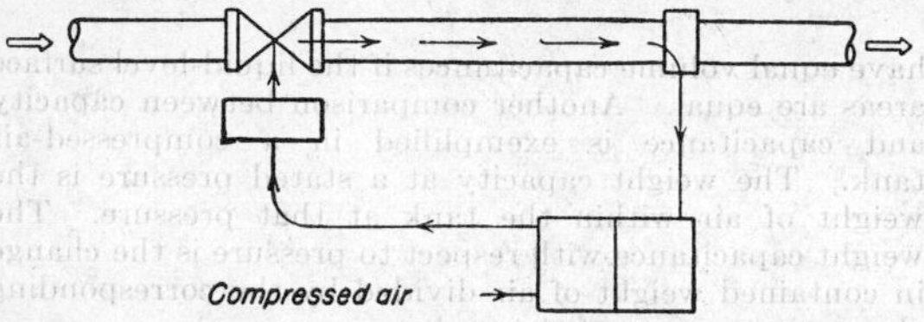

FIG. 169. Self-operating measurement and relay-operated control.

controlled medium through the primary element, to operate the final control element. A relay-operated controller may have either a self-operated measuring means and a relay-operated controlling means or a relay-operated measuring means and a relay-operated controlling means. Figure 169 shows a self-operated measuring means and a relay-operated controlling means. Figure 170 shows a relay-operated measuring means and a relay-operated controlling means. The self-operated measuring means of Fig. 169 may be a millivoltmeter pyrometer or a pressure-actuated thermometer, in which case the energy derived from the controlled medium through the primary element is transformed before the controlling means is actuated. However, this energy is neither supplemented nor amplified by energy from another source. In Fig. 169, if the system were reversed to embody a relay-operated measuring means and a self-

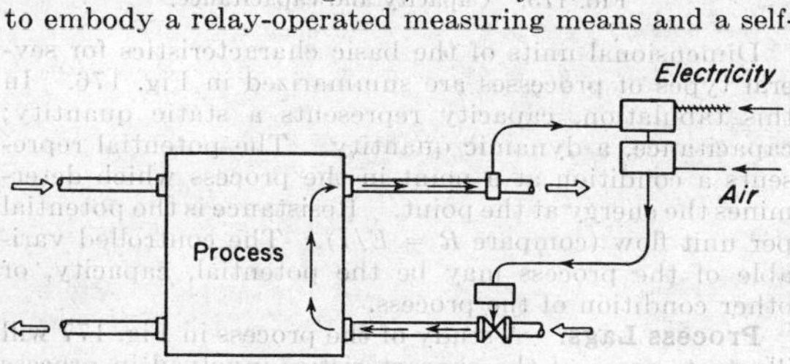

FIG. 170. Relay-operated controller.

operated controlling means, the resultant combination also would be a relay-operated controller.

Automatic-control Characteristics. Closely associated with the characteristics of automatic controllers are the characteristics of automatic control itself. *Cycling,* or hunting, of the controlled variable is illustrated in Fig. 171. It is a continuous change of the controlled variable, alternately increasing and decreasing in value. The time interval between two successive peaks of the cycle remains substantially constant and is termed the period of the cycle. The amplitude of cycling is the maximum departure of the controlled variable from the average value of the cycle. Three types of cycling of the controlled variable are shown in Fig. 171. Cycling in which the amplitude gradually decreases may be obtained, for example, with proportional-speed floating controller action. Cycling in which the amplitude is constant may be obtained with two-position controller action. Cycling in which the amplitude gradually increases is not suitable for automatic control.

The *desired value* of the controlled variable is determined by the operating requirements of the process. Usually the set point of the automatic controller is adjusted so that the control point and the desired value will coincide. The *set point* of an automatic controller is the position of the control-point-setting mechanism translated into units of the controlled variable. Generally, a set point scale is provided. The set point may be varied manually or by automatic means, as in time-cycle or ratio control. The *control point* is the value of

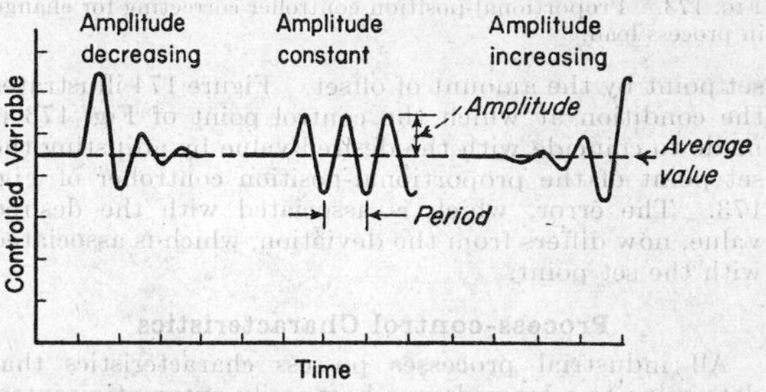

FIG. 171. Cycling of the controlled variable.

the controlled variable maintained by the controller. In some types of automatic controllers, *e.g.*, those with two-position differential-gap or floating-with-neutral controller action, the control point becomes a *control range* of the controlled variable, and the set point is generally selected as the center of this range of values.

Figure 172 illustrates the condition at which the desired value, set point, and control point coincide. This type of controlled-variable record may be obtained by a proportional-plus-reset controller action correcting for a sustained change of process load. The set point was initially adjusted to coincide with the desired value; and the set point and control point coincide because of the floating-type controller action employed. In floating-with-neutral-zone controller action, the control point and

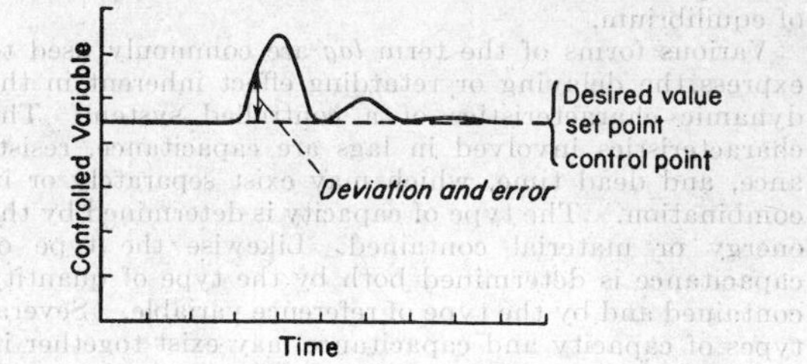

FIG. 172. Record of automatic-reset controller correcting for change of process load.

the set point always coincide. The error and deviation coincide because of their respective associations with the desired value and set point.

Figure 173 illustrates the condition at which the desired value and set point coincide, and the control point differs from the set point by the amount of *offset.* This type of controlled-variable record may be obtained by a proportional-position controller action correcting for a sustained change of process load. The deviation and error coincide because the set point was initially adjusted to coincide with the desired value. In positioning-type controller action, the control point may lie anywhere within a predetermined range of values of the controlled variable. The control point may then differ from the

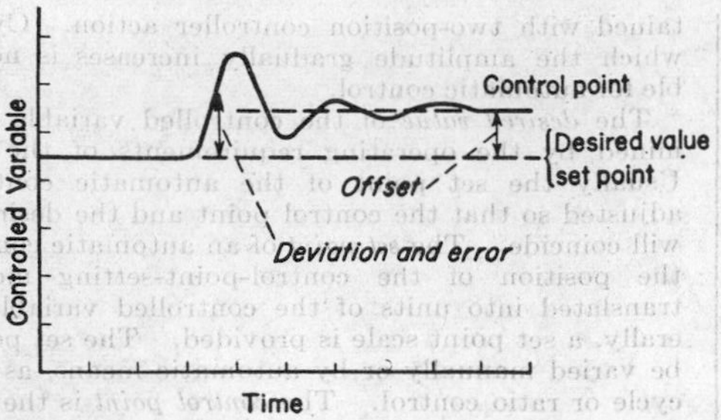

FIG. 173. Proportional-position controller correcting for change in process load.

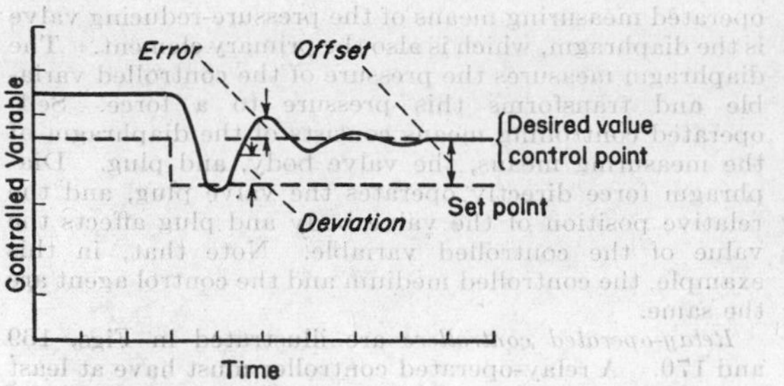

FIG. 174. Offset eliminated by changing set point.

set point by the amount of offset. Figure 174 illustrates the condition at which the control point of Fig. 173 is made to coincide with the desired value by adjusting the set point of the proportional-position controller of Fig. 173. The error, which is associated with the desired value, now differs from the deviation, which is associated with the set point.

Process-control Characteristics

All industrial processes possess characteristics that determine to a large degree how easily automatic control can be applied and the results that can be obtained from automatic control. If all processes were alike, automatic controllers of one design or one type would suffice. However, processes vary widely in the number of characteristics they possess, and the extent to which these characteristics are manifest has led to the development of many mechanisms. Some of the basic characteristics of a process are: (1) *Capacity*—a measure of the maximum quantity of energy or material that can be stored. It is measured in units of quantity. (2) *Capacitance*—the change in quantity per unit change in some reference variable. (3) *Resistance*—opposition to flow measured in units of potential change required to produce unit change in flow. (4) *Dead time*—any definite delay period between two related actions; it is measured in units of time. (5) *Self-regulation* is a sustained reaction inherent in the process which assists or opposes the establishment of equilibrium.

Various forms of the term *lag* are commonly used to express the delaying or retarding effect inherent in the dynamic characteristics of a controlled system. The characteristics involved in lags are capacitance, resistance, and dead time, which may exist separately or in combination. The type of capacity is determined by the energy or material contained. Likewise the type of capacitance is determined both by the type of quantity contained and by the type of reference variable. Several types of capacity and capacitance may exist together in one process. Combinations of resistances and capacitances are frequently found in many industrial processes.

A comparison between capacity and capacitance is illustrated in Fig. 175. In this example, the capacities are equal, the capacitances are unequal, and the liquid-level surface areas do not vary with change in liquid head. The capacity, or "static quantity," is the cubic volume of liquid contained in each vessel. The volume capacitance, or "dynamic quantity," is equivalent to the liquid-level surface area of each vessel. In this example, the capacitance is equal to the change of volume capacity divided by the change of reference variable, the liquid head. If the shape of the vessel causes the liquid-level surface area to vary with change in liquid head, the capacitance will likewise vary with head. Vessels of different shape have equal capacities if the quantities of material contained are equal. Vessels of different shape have equal volume capacitances if the liquid-level surface areas are equal. Another comparison between capacity and capacitance is exemplified in a compressed-air tank. The weight capacity at a stated pressure is the weight of air within the tank at that pressure. The weight capacitance with respect to pressure is the change in contained weight of air divided by the corresponding change in pressure of the tank.

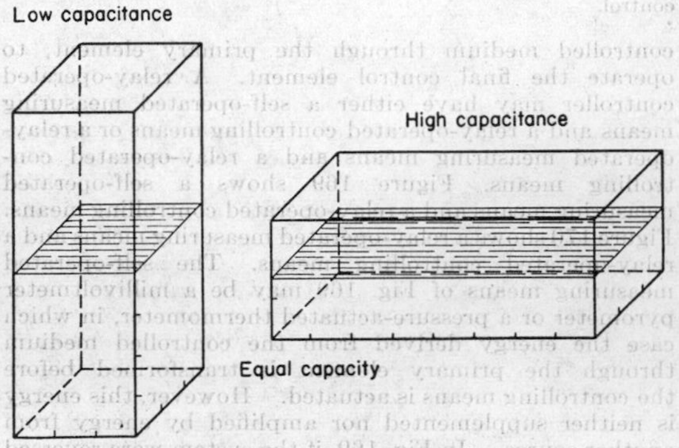

FIG. 175. Capacity and capacitance.

Dimensional units of the basic characteristics for several types of processes are summarized in Fig. 176. In this tabulation, capacity represents a static quantity; capacitance, a dynamic quantity. The potential represents a condition at a point in the process which determines the energy at the point. Resistance is the potential per unit flow (compare $R = E/I$). The controlled variable of the process may be the potential, capacity, or other condition of the process.

Process Lags. A study of the process in Fig. 177 will illustrate some of the characteristics involved in process control. Heat requirements of material being processed constitute the demand, and the heat available in the heating agent represents the supply. Numerous factors can upset the desired balance of energy in the process. On the supply side, for example, the pressure of the heating steam may change. On the demand side, the flow of process material may increase or decrease; its entering temperature may change; radiation to the surroundings may change because of a change in ambient temperature; or a reaction that is taking place in the process may become more or less exothermic or endothermic.

Furthermore, still other factors can affect the rate of heat transfer without a change in demand. Except where the heating is the result of direct contact with the hot substance, or with radiation from it, heat must pass through a barrier separating the demand and supply sides. The thermal resistance of this heat-transfer barrier will determine the temperature potential necessary to force

	Thermal	Pressure	Liquid Level		Electrical
Capacity	Btu.	Cubic Foot	Cubic Foot	Pounds	Coulomb
Potential	Degree	Pounds / Sq. in.	Foot	Foot	Volt
Capacitance	$\frac{\text{Btu.}}{\text{Deg.}}$	$\frac{\text{Cu.Ft.}}{\text{Psi.}}$	$\frac{\text{Cu.Ft.}}{\text{Ft.}}$	$\frac{\text{Lb.}}{\text{Ft.}}$	$\frac{\text{Coulomb}}{\text{Volt}}$ = Farad
Resistance	Deg. / Btu. / Sec.	$\frac{\text{Psi.}}{\text{Cu.Ft./Sec.}}$	$\frac{\text{Feet}}{\text{Cu.Ft./Sec.}}$	$\frac{\text{Feet}}{\text{Lb./Sec.}}$	$\frac{\text{Volt}}{\text{Coulombs / Sec.}}$ = Ohms

FIG. 176. Dimensional units for characteristics of different variables.

a given quantity of heat through a unit area of the barrier in a given time. Hence, changes in the character of the surface, as by sealing or corrosion, can affect the thermal head necessary; while, in the case of a vertical barrier, changes in level on either the supply or the demand sides will affect the area available for heat transfer.

Thus it is clear that it must be possible to detect any unbalance between heat demand and supply, as evidenced by a change in the temperature of the effluent stream, and then to alter the thermal potential in such a way that the new required rate of heat transfer will be obtained promptly. In the case of the ideal process, this is a simple matter. Unfortunately, inherent in most processes are unfavorable lags, which on the one hand delay the discovery of a disturbance and retard the recognition of its magnitude, and on the other hand retard the establishment of a new thermal potential. Further, controllers themselves require more or less time to detect changes and make the necessary corrections, giving rise to the controller lags.

Capacity Lag. The first process lag to be considered is not ordinarily disadvantageous to control. In fact, it is usually an advantage. Referring to Fig. 177, this is the

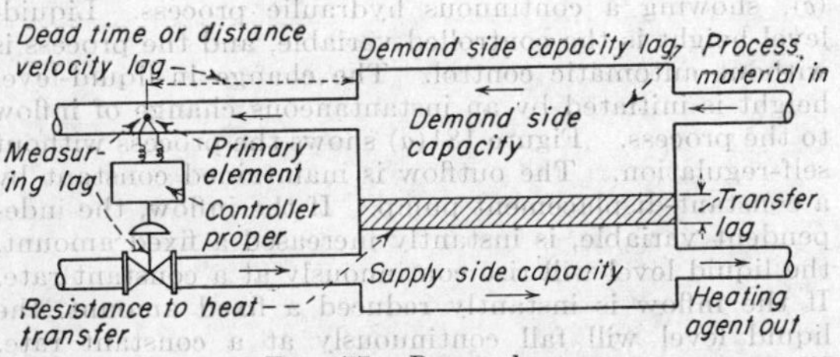

FIG. 177. Process lags.

demand-side capacity lag, which results from the heat storage and consequent "thermal inertia" of the demand side. A high demand-side capacity tends to stabilize the process temperature and prevent rapid departures from the control point. It is disadvantageous only when prompt response to a change in control point is desired. Figure 178 portrays the situation in an uncontrolled process, with demand-side capacity lag only, when a sudden supply change takes place. Curve *a* shows the change in supply, while curves *b* and *c* show the resulting change in temperature. A process with low demand-side capacity becomes stabilized quickly at the new temperature, as in *b*, while one with high demand-side capacity responds slowly, as in *c*. At a given rate of outflow, a receiver containing a large volume of process material thus tends to be more stable than one containing a small volume of material. A jacketed vessel, for example, is more stable than a shell-and-tube heat exchanger, which in turn is more stable than a concentric pipe heater.

Transfer Lag. Where the thermal inertia of the demand side is ordinarily favorable, the reverse is true of the supply-side capacity and its thermal inertia. The supply-side capacity can be considered the sum of all conditions on the heat-supply side which tend to stabilize

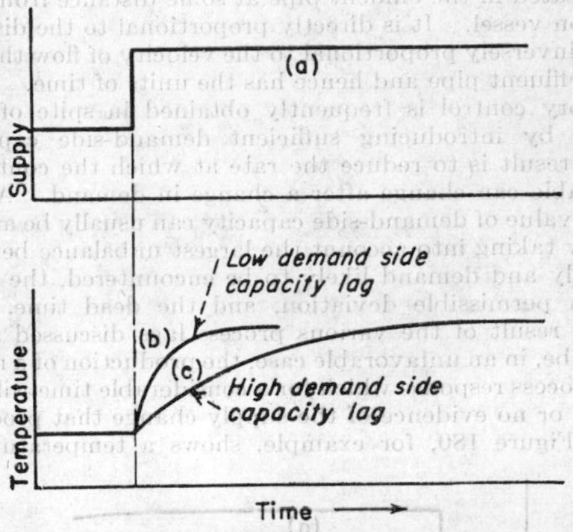

FIG. 178. Capacity lag.

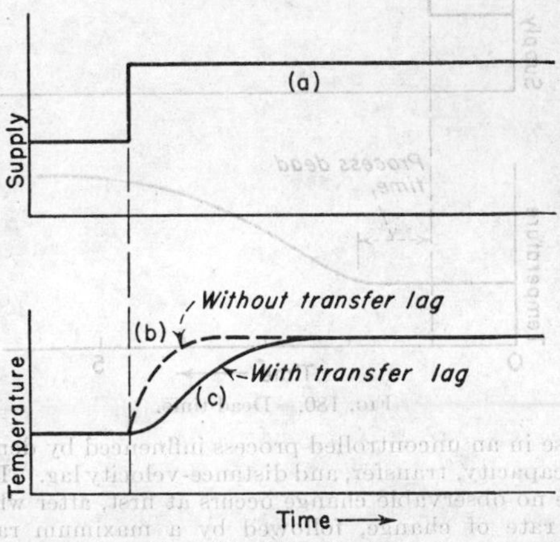

FIG. 179. Transfer lag.

the available rate of heat transfer and make a change to a higher or lower heat-transfer rate either difficult or time-consuming. For example, time is required to change the temperature of the heating medium. Furthermore, any barrier between supply and demand sides adds thermal capacity as well as thermal resistance and, if its capacity is high, acts as a thermal flywheel which must itself gain or lose considerable energy before a new desired rate of heat transfer can be achieved. Thus supply-side capacity and the thermal resistance of a heat-transfer barrier result in another lag, known as transfer lag, which is the retardation in establishing a new heat-transfer rate following a change in supply potential. The curves of Fig. 179 portray the situation in an uncontrolled process having both demand-side capacity lag and transfer lag. Curve *a* shows the change in supply, *b* the response to a supply change with demand-side capacity lag only, and *c* the response when both lags are present.

Transfer lag, as defined above, is always unfavorable to control, since it limits the rate at which a change in supply rate can be made effective on the demand side. The result, of course, is a tendency to cause overshooting in a controlled process.

Dead Time. The third type of process lag is dead time or distance-velocity lag, because it is a delay (not a retardation) representing the time required to transport a material over a given distance at a given velocity. Its relation to the other lags is indicated in Fig. 177. This lag occurs, for example, where the sensitive element is located in the effluent pipe at some distance from a reaction vessel. It is directly proportional to the distance and inversely proportional to the velocity of flow through the effluent pipe and hence has the units of time. Satisfactory control is frequently obtained in spite of dead time by introducing sufficient demand-side capacity. The result is to reduce the rate at which the controlled variable can change after a change in demand. A suitable value of demand-side capacity can usually be arrived at by taking into account the largest unbalance between supply and demand likely to be encountered, the maximum permissible deviation, and the dead time. The total result of the various process lags discussed above may be, in an unfavorable case, the production of a record of process response which for a considerable time will give little or no evidence of the supply change that produced it. Figure 180, for example, shows a temperature response in an uncontrolled process influenced by demand-side capacity, transfer, and distance-velocity lag. In this curve no observable change occurs at first, after which a slow rate of change, followed by a maximum rate of change, takes place. The period in which a change cannot be ascertained is usually called the process dead time. Obviously, the deviation cannot possibly be checked for a period equal to the dead time, even though a correction is made the instant the temperature begins to deviate from normal.

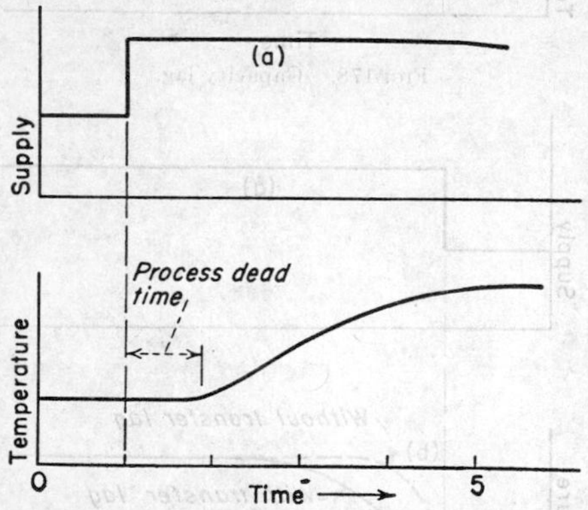

FIG. 180. Dead time.

Controller Lags. One more type of lag, which pertains to the controller rather than to the process, must also be recognized. This lag is actually the resultant of a group of lags, some of which are retardations and some actual delays. They may be referred to as controller lag and measuring lag (see Fig. 177). *Measuring lag* is a retardation in response of the measuring means to a change in the measured variable. An instantaneous change of temperature, for example, at a thermometer bulb does not cause instantaneous, complete response of the controller, but a short time is required for the controller pen or pointer to indicate the full change. Consequently, the indicated or recorded temperature or pressure lags behind its actual magnitude. *Controller lag* is a retardation in response of a control valve or other element to a change in pen or pointer position at the controller. For example, a pneumatic control valve requires a small time to change its position in response to pressure changes because of the volume that must be filled to a higher pressure. Also, an electric-motor-operated control valve requires a specified time to traverse full stroke. The actual magnitudes of measuring and controller lags are of no significance except when considered in relation to response rates called for by changes in the process. These lags may or may not have appreciable effect with a particular mechanism, depending on the application. In some cases the lags of the controller predominate, usually because of the nature of the required measuring mechanism, and a relatively refined form of controller must be used to overcome the effects of its own lags.

Generally capacity is favorable because of its leveling-out function and ability to absorb and slow down large changes made in valve position. Transfer lag is a disadvantage because it causes a slow initial response and requires that the controller makes slow corrective action. Dead time is detrimental to good control, for it prevents all response to controller action and delays any correction of changes in controlled variable until the dead-time period has elapsed.

Self-regulation. An important characteristic of processes which bears on the ease with which they may be controlled is their degree of tendency toward self-regulation. Self-regulation is illustrated in Fig. 181(*a*), (*b*), and (*c*), showing a continuous hydraulic process. Liquid-level height is the controlled variable, and the process is without automatic control. The change in liquid-level height is initiated by an instantaneous change of inflow to the process. Figure 181(*a*) shows the process without self-regulation. The outflow is maintained constant by a constant-displacement pump. If the inflow, the independent variable, is instantly increased a fixed amount, the liquid level will rise continuously at a constant rate. If the inflow is instantly reduced a fixed amount, the liquid level will fall continuously at a constant rate. Figure 181(*b*) illustrates the process with positive self-regulation. The outflow passes through a fixed restriction. If the inflow is instantly increased a fixed amount, the liquid level will rise at a decreasing rate as the result of increased outflow due to the constantly increasing liquid head. If the inflow is instantly decreased a fixed amount, the liquid level will fall at a decreasing rate as a result of reduced outflow due to the constantly decreasing liquid head on the outflow restriction. Figure 181(*c*) shows the process with positive self-regulation as the inflow is instantly increased a fixed amount, and negative self-regulation as the inflow is instantly decreased a fixed amount. A constant-displacement pump regulates the outflow. In this example, the nature of the self-regulation is governed by the non-linear capacitance of the process due to the shape of the vessel.

Advantage of the basic process characteristics can be taken in designing equipment so that those characteristics desirable and conducive to good automatic control will be incorporated and those characteristics which hinder automatic control will be reduced or eliminated. Equipment design with these factors in mind can result in substantial savings in the investment in automatic-control equipment and at the same time result in the production of a higher quality product because of the accuracy in control obtainable.

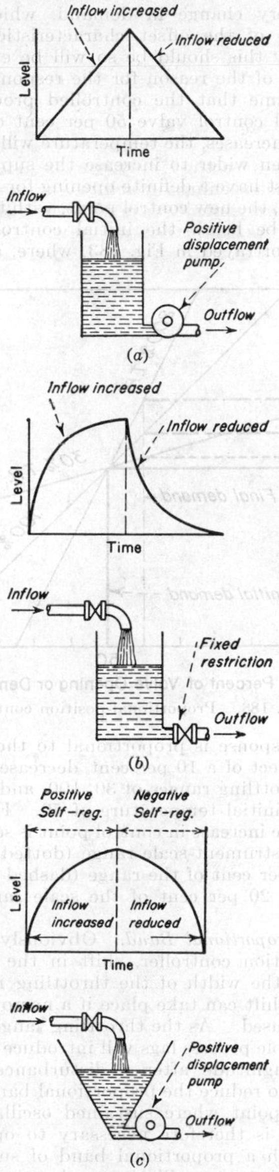

FIG. 181. Self-regulation. (*a*) No self-regulation. (*b*) Positive self-regulation. (*c*) Positive and negative self-regulation.

Types of Automatic Control Action

It is the problem of a controller to measure changes in the controlled condition and to make corrections based on the measurements that will return the condition of the variable as quickly as possible toward the control point. In the simpler processes, which have little lag aside from demand-side capacity lag, the control problem is easy and a simple controller will do an excellent job. When transfer lag and dead time are involved to an appreciable extent, a more complicated control mechanism is necessary. The more refined types of controller initiate corrective actions in accordance with the sense and amount of the deviation, the rate of change of the deviation, and sometimes also the rate of change of the rate of change of the deviation. In the analyses that follow, the process is assumed to be initially in equilibrium between demand and supply, after which a sudden demand change takes place.

Methods of control most generally used today may roughly be classified into (1) those types which provide one or more definite rates of flow of the control agent; and (2) those types which provide a continuous range of flow rates of the supply medium, a suitable rate being selected by the controller in some predetermined relation to the deviation or rate of change of the deviation from the control point. The main types are:

Two-position.
Proportional-position.
Floating.
Proportional-plus-reset.
Proportional-plus-rate.
Proportional-plus-reset-plus-rate.

Two-position Control. The simplest type of controller, the two-position controller, is one in which the valve is adjusted to take either of two positions, a high value greater than the maximum demand, or a low value less than the minimum demand. It includes as a special case on-and-off controllers in which the control valve is either wide open or tight shut. Another special case is the multiposition type of controller in which the controller may select, say, a "low" high or a "high" high rate of supply flow if the deviation is slightly low, or considerably low; or a "high" low or a "low" low if the reverse conditions maintain.

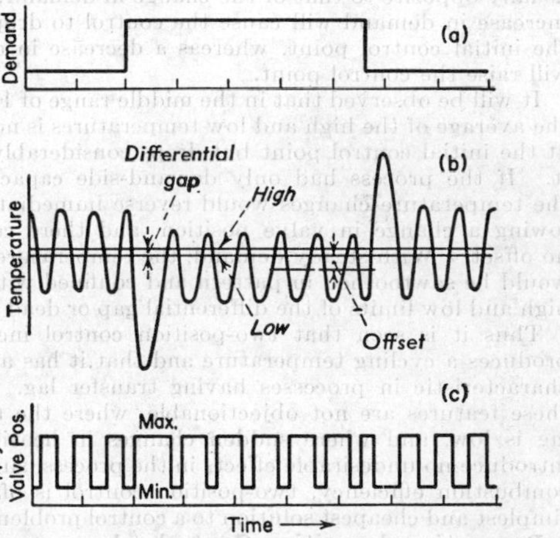

FIG. 182. Responses of two-position controller before and after a load change.

The two-position controller is simple and cheap and gives excellent results in many processes having a fairly large demand-side capacity lag but no other lags of importance. A common type is the contact-making thermometer or pyrometer which opens a solenoid fuel valve to a suitable upper limit whenever the temperature is below the control point, de-energizing the valve, and allowing it to close to a suitable lower limit whenever the temperature rises above the control point. To avoid too frequent valve changes in a response process, such controllers employ either high and low contacts to give a differential gap or zone of suitable width.

In Fig. 182 are charted the characteristic responses of a

two-position controller before and after a change in load. The controller is confronted by considerable transfer lag. A demand change is considered to be the ultimate change in heat requirement of the process corresponding to a sudden change in the rate of flow of water through a tank. The supply is considered as the output of the heater.

Referring now to Fig. 182, curve *a* shows the demand, *b* the temperature, and *c* the supply or valve position. Because of the transfer lag, the controlled temperature continues to fall after the heater is energized on a downswing and continues to rise after it is de-energized on an upswing. If the transfer lag were less, the swings would also be less. With demand-side capacity lag only, the direction of the temperature trend would instantly respond to change in rate of heating, and there would be no overshooting.

In the earlier part of the record, the temperature swings equally above the high and below the low limits of the differential gap. This is because the heat requirement is such that it is satisfied by the application of heat at the high rate 50 per cent of the time and at the low rate the remaining 50 per cent of the time. In the middle portion of the record range, demand increases. This means that the higher supply rate will now be closer to the new demand, while the lower supply rate will be further removed. After a transient overswing, therefore, the controller settles down to the new demand with its swing only slightly above the high control range limit but considerably below the lower limit.

This introduces an important effect in automatic control called offset or droop. Types of controllers that exhibit this characteristic operate to maintain either the average or the instantaneous temperature within a band rather than at a single value. The magnitude of the offset is a function of the demand, and its direction is usually opposite to that of the change in demand; *i.e.*, an increase in demand will cause the control to drop below the initial control point, whereas a decrease in demand will raise the control point.

It will be observed that in the middle range of Fig. 182, the average of the high and low temperatures is no longer at the initial control point but drops considerably below it. If the process had only demand-side capacity lag, the temperature changes would reverse immediately following a change in valve position, and there could be no offset. With steady demand, the temperature record would be sawtoothed in pattern and confined within the high and low limits of the differential gap or dead zone.

Thus it is seen that two-position control inevitably produces a cycling temperature and that it has an offset characteristic in processes having transfer lag. Where these features are not objectionable, where the transfer lag is low, and where sudden changes in heating rate introduce no undesirable effects in the process, such as in combustion efficiency, two-position control is often the simplest and cheapest solution to a control problem.

Proportional-position Control. In many processes continuous cycling of the valve is undesirable, and a type of control in which the flow rate is continuously adjustable is preferred. One of the simplest of these is the so-called proportional-position or proportional control in which the controller selects a definite and different position for the final control valve for every temperature within the working range. This method of control introduces the concept of proportional band or throttling range, which is the range of values of the measuring element necessary to cause the final control valve to move from full shut to full open. The proportional band is generally expressed as a percentage of the full range of the control instrument and may be anything from a fraction of a per cent to several hundred per cent of the full range. Since the width of the proportional band has an important effect on the results obtained with the same controller applied to different processes, its width is generally adjustable in modern proportional instruments.

The fact that each position of the control valve is tied to a definite temperature with controllers of this type means that the instrument will control at a different point following every change in demand, which is another manifestation of the offset characteristic already discussed. Why this should be so will be evident from a consideration of the reason for the response of the controller. Assume that the controlled process is steady with the final control valve 50 per cent open. If now the demand increases, the temperature will drop, and the valve will open wider to increase the supply. Because the valve must have a definite opening for each temperature, however, the new control point, with the wider open valve, must be below the initial control point. The situation is portrayed in Fig. 183, where, assuming that

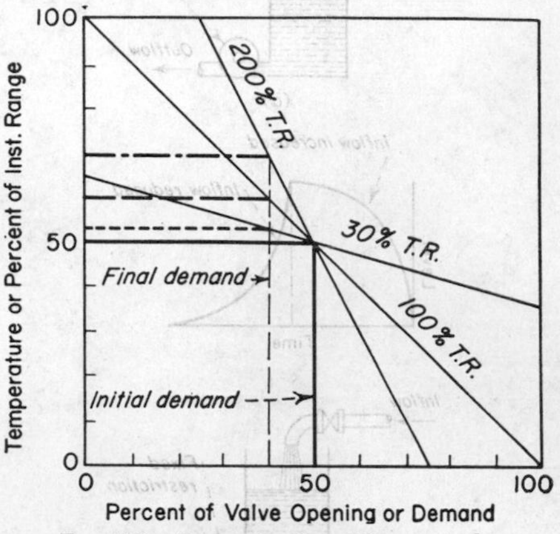

FIG. 183. Proportional-position control.

the supply response is proportional to the valve movements, the effect of a 10 per cent decrease in demand is shown for throttling ranges of 30, 100, and 200 per cent, assuming an initial temperature of 50. For the 30 per cent range, the increase in control point is seen to be 3 per cent of the instrument-scale range (dotted line); for 100 per cent, 10 per cent of the range (dashed line); and for 200 per cent, 20 per cent of the scale range (dot-dash line).

Effect of Proportional Band. Obviously, with a proportional-position controller, shift in the control point is limited to the width of the throttling range, so that only a small shift can take place if a narrower throttling range can be used. As the throttling range is narrowed, any unfavorable process lags will introduce oscillations of increasing magnitude after a disturbance. Successive adjustments to reduce the proportional band still further will reach a point where sustained oscillations will be obtained. It is therefore necessary to operate such a controller with a proportional band of sufficient width to avoid continuous oscillation under the worst conditions that are likely to occur in the process. This increases the possible offset with demand changes, but deviation from the desired control point can be limited by manually changing the relationship between the measured temperature and the valve position.

The curves of Fig. 184 show the effect of the width of the throttling range on both the amount of the offset and the stability. Curve *a* shows a sudden change in demand, while curves *b* to *d* show the resulting response of temperature (or valve position, which is proportional to it) for progressively narrower throttling ranges. In curve *d*, the offset is slight, but it will be observed that the temperature has developed a slight sustained oscillation at the end of the record. The wide range adjustment of *b* would be satisfactory if only slight changes in demand were to be encountered.

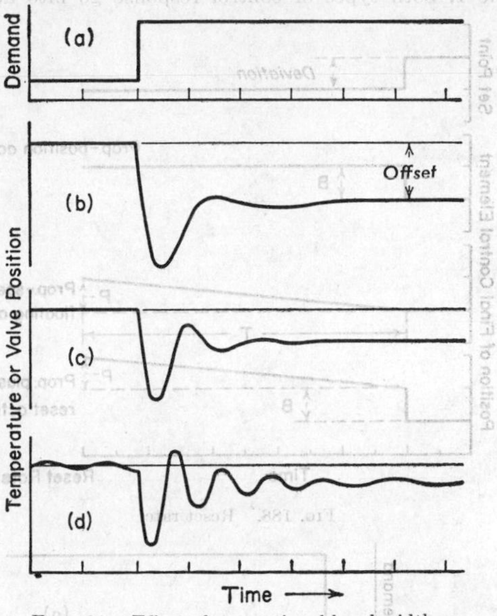

FIG. 184. Effect of proportional band width.

Floating Control. In another class of controller generally known as floating, the position of the final control valve bears no fixed relation to the temperature but is changed continuously in the proper direction whenever the temperature deviates from the control point. Ordinarily, the rate of valve movement is made slow enough to avoid reaching one of the limits of its travel before it has been arrested by the return of the temperature to the control point. Since the valve remains in whatever position its movement is arrested, having no relation to the temperature, this instrument has no offset. On the other hand, to avoid constant cycling of the temperature, floating controllers of the constant-valve-speed type generally are provided with a neutral zone. No control actions can take place, regardless of temperature changes, as long as the temperature remains within this neutral zone.

Single-speed, multispeed, and *proportional-speed floating* control actions are all employed although for temperature control they are generally used in combination with proportional-position action, as explained later. A single-speed floating controller moves the valve continuously at constant speed as long as the temperature remains outside the dead zone. It is represented by controllers using a motor-operated final control valve and high and low control contacts in fixed position. The valve opens slowly but continuously, as long as the "high" contact is made; then, upon reversal of the temperature trend, it remains in the position where the contact was broken until temperature has passed through the neutral zone and completed the low contact. The two-speed floating controller employs two valve speeds, a low speed if the deviation is slight, and a higher speed if the deviation is large. This is easily accomplished by the use of two high and two low contacts, the low-speed contacts either inserting resistance in the motor circuit or reducing the average motor speed by other suitable means. The proportional-speed floating controller differs from the foregoing types in two particulars: first, the valve speed is made proportional to the amount of temperature deviation; and second, the controller has little or no neutral zone in most cases.

The *floating rate* of a proportional-speed floating controller action is illustrated in Fig. 185. It is the rate of change of position of the final control element, divided by the controlled-variable deviation initiating the corrective action. Floating rate is commonly expressed in per cent of full range motion of the final control element per minute per per cent deviation.

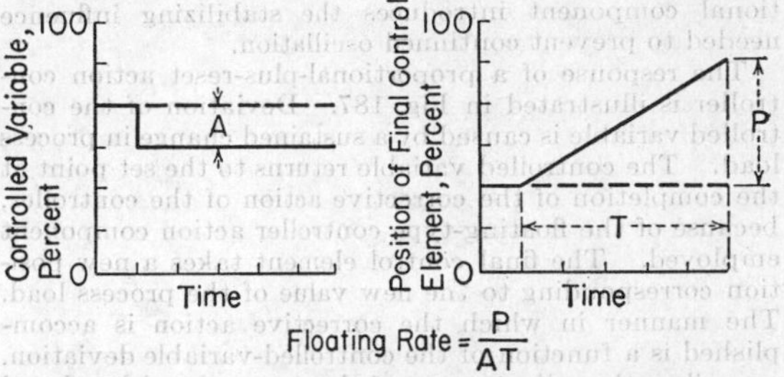

FIG. 185. Proportional-speed floating controller action.

The way valve movement varies with deviation from the control point is shown for the three floating types in Fig. 186. Curve *a* for single-speed and curve *b* for two-speed floating controllers have a neutral zone of temperature in which valve movement cannot take place, whereas the proportional-speed type shown by curve *c* has no neutral zone. Controllers of this type are not ordinarily satisfactory except in the virtual absence of process lags, which explains why they are seldom used in temperature control. Even demand-side capacity lag is unfavorable in this case, and successful application depends on the use of an extremely low valve speed.

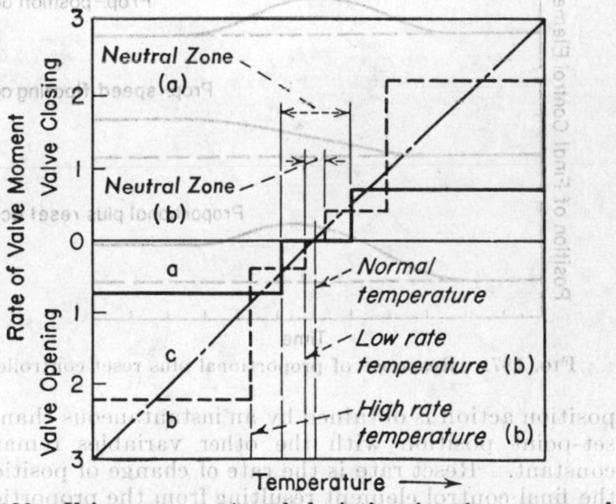

FIG. 186. Valve movement and deviation for three types of floating control.

However, this type of control has become extremely important as an adjunct to the proportional-position controller where it has the function of resetting the control point automatically after demand changes so as to eliminate the offset ordinarily present in the proportional type. In such instruments the proportional-speed floating form is usually employed.

Proportional-plus-reset Control. In proportional control, the final control-value position is determined by the temperature, which means that the rate of valve movement is proportional to the rate of change of the deviation of the temperature from the set point.

In a proportional-speed floating controller, however, the ultimate valve position bears no relation to the temperature, the valve being moved at a rate proportional to the magnitude (not the rate of change) of the deviation. Therefore, if the two types of valve action can be superimposed, so that the valve is moved at a rate that is the sum of a rate proportional to the deviation from the control point and the rate of change of this deviation, a method of control can be achieved that has the stability of proportional controller and the invariable control point of the floating controller. In this type, the floating component requires no neutral zone, since the proportional component introduces the stabilizing influence needed to prevent continued oscillation.

The response of a proportional-plus-reset action controller is illustrated in Fig. 187. Deviation of the controlled variable is caused by a sustained change in process load. The controlled variable returns to the set point at the completion of the corrective action of the controller, because of the floating-type controller action component employed. The final control element takes a new position corresponding to the new value of the process load. The manner in which the corrective action is accomplished is a function of the controlled-variable deviation, as well as the adjustments of the proportional band and the reset rate. The proportional-band adjustment simultaneously affects the proportional-speed floating action in such a manner that the reset rate remains constant at its set value.

The *reset rate* of a proportional-plus-reset controller is illustrated in Fig. 188. In this illustration, proportional-

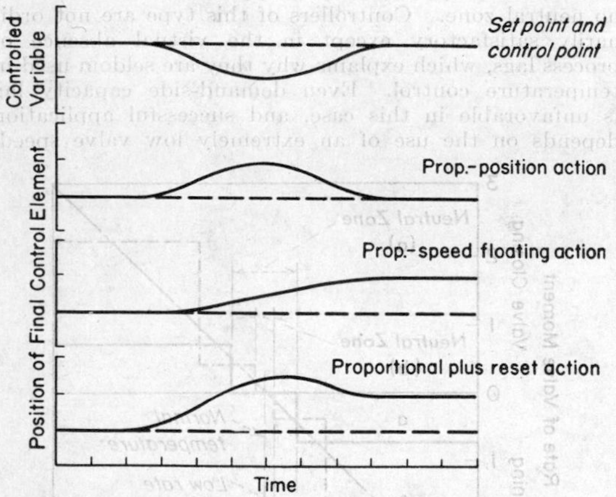

FIG. 187. Response of proportional plus reset controller.

position action is obtained by an instantaneous change in set-point position with the other variables remaining constant. Reset rate is the rate of change of position of the final control element resulting from the proportional-speed floating-action component, divided by the change in position of the final control element resulting from the proportional-position action component, with the same deviation in both cases. The two controller actions are initiated simultaneously. The reset rate is fixed by a manually set adjustment; it is independent of controlled-variable deviation and the proportional-band adjustment of the controller.

The action that takes place after a sudden load change with a proportional-plus-floating controller is analyzed in the curves of Fig. 189. Assume that a sudden change in demand takes place at time 1 as in curve (*a*). With a well-adjusted proportional-plus-floating controller, the temperature might be brought back to the control point, as in curve (*b*). Curve (*c*) shows the component of the valve movement due to proportional control alone, while curve (*d*) shows that due to floating control alone. Curve (*e*), their sum, shows how the control stabilizes at the new demand rate. When the demand change takes place at time 1, both types of control response go into action.

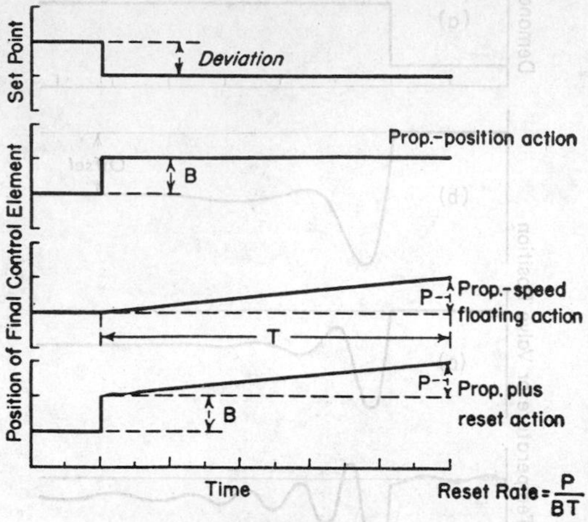

FIG. 188. Reset rate.

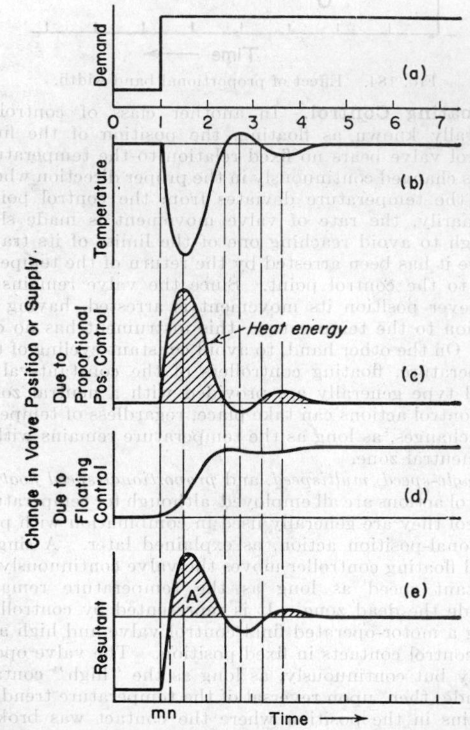

FIG. 189. Load change with proportional-plus-floating controller.

The proportional control action, curve (*c*), serves to increase the valve opening while the temperature is decreasing and to decrease it while the temperature is increasing but has no permanent effect in determining the final valve position, its final effect on the valve position being zero after the temperature has returned to the control point. The floating control action, curve (*d*) operating the valve at a rate that is always proportional to the deviation, opens the valve as long as the temperature is low and closes it while it is high, eventually leaving the valve opening at that required for the new demand rate.

Since curves (*c*) to (*e*) are rate-of-supply curves (on the assumption that changes in rate of supply are always proportional to changes in valve opening), the areas under the curves can be taken as measures of quantities of heat supplied. In Fig. 189, area *A* represents a quantity of heat, supplied in excess of the new steady demand require-

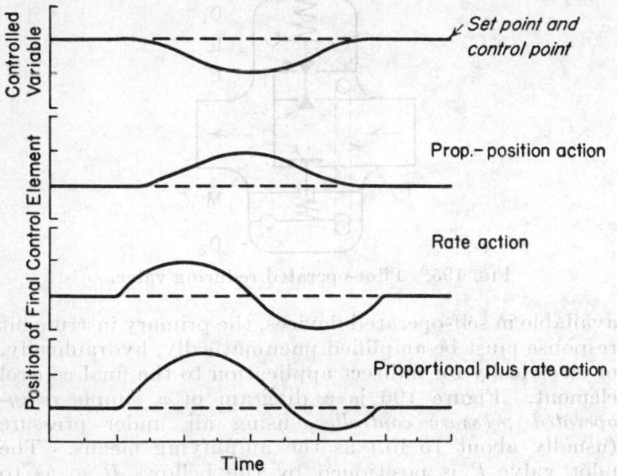

FIG. 190. Response of proportional-plus-rate action controller.

ment, which serves to replace stored energy lost during period *m-n* when the demand exceeded the supply and also to increase the energy stored in the heater to that required for the new thermal head. It will be noted from curve (*b*), representing the temperature, that the controller has permitted no deviation of the control point.

Proportional-plus-rate action is that in which proportional-position action and rate or second-derivative action are combined. The rate-action component employs a derivative action that maintains a linear relation between the first derivative or rate of change of the controlled variable and the position of the final control element. The response of a proportional-plus-rate action

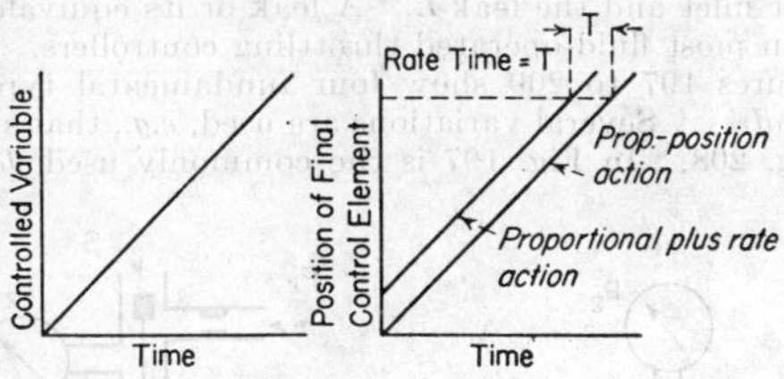

FIG. 191. Rate time.

controller is illustrated in Fig. 190. The deviation of the controlled variable is caused by a temporary process-load change of short duration. The controlled variable returns to the set point, and the final control element returns to its original position. If the process-load change were sustained, the controlled variable would line out at a new control point, which would differ from the set point by the amount of offset. The manner in which the corrective action is accomplished is a function of the controlled-variable deviation as well as of the adjustments of the proportional band and rate time.

Rate time of a proportional-plus-rate action controller is illustrated in Fig. 191. It is the advance in position of the final control element resulting from the rate-action component. The advance is measured in units of time, generally minutes, and is equal to the time required for a selected motion of the final control element due to the effect of proportional-position action alone, minus the time required for the same motion due to the effect of proportional-position plus rate actions, with the same rate of change of the controlled variable in both cases. The rate action and the proportional-position action are initiated at the same instant by the deviation of the controlled variable.

Proportional-plus-reset-plus-rate Control. The response of a proportional-plus-reset-plus-rate action controller is illustrated in Fig. 192. The deviation of the

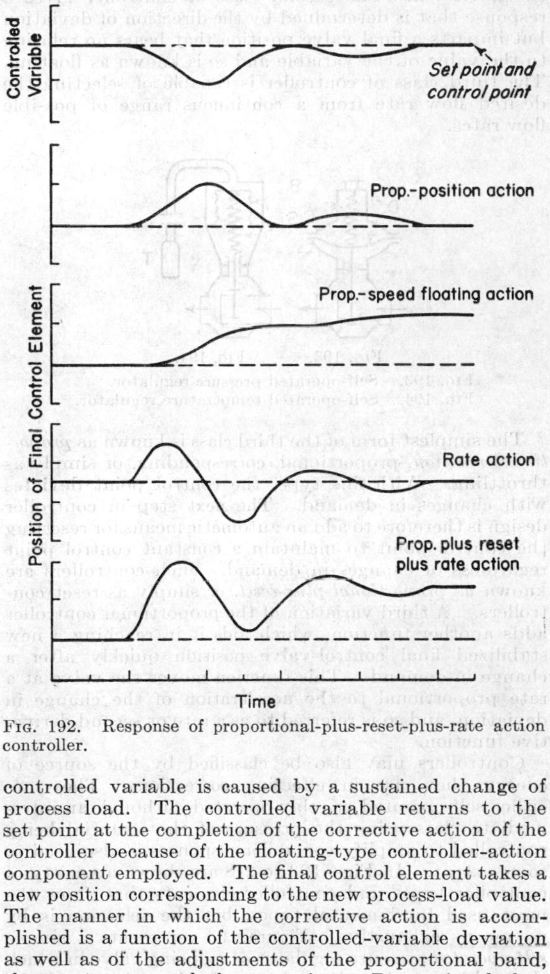

FIG. 192. Response of proportional-plus-reset-plus-rate action controller.

controlled variable is caused by a sustained change of process load. The controlled variable returns to the set point at the completion of the corrective action of the controller because of the floating-type controller-action component employed. The final control element takes a new position corresponding to the new process-load value. The manner in which the corrective action is accomplished is a function of the controlled-variable deviation as well as of the adjustments of the proportional band, the reset rate, and the rate time. Proportional-plus-reset-plus-rate controller action is that in which there is a predetermined relation between the position of the final control element and (1) the deviation of the controlled variable, (2) the time integral of the deviation, and (3) the rate of change of the deviation. Proportional-position, reset, and rate controller-action components occur simultaneously and are initiated by the controlled-variable deviation. The proportional-band value of the proportional-position action component directly affects the magnitude of the floating and rate controller actions. In automatic controllers having proportional-plus-reset-plus-rate action, a reset-rate adjustment and a rate-time adjustment, the proportional band may be adjusted without affecting the set value of the reset rate or the rate time.

AUTOMATIC CONTROL MECHANISMS

In classifying controllers according to the type of response, three main variations, as previously discussed, are noted. In the first are fixed-position controllers, which respond to the need for a correction by producing a certain fixed rate of flow. In this group are two-position (on-and-off) controllers, which give a maximum or minimum flow rate, and multiposition controllers, which can select the best of several possible flow rates, depending on the deviation. The second class of controller gives a response that is determined by the direction of deviation but imparts a final valve position that bears no relation to the value of the variable and so is known as floating. The third class of controller is capable of selecting the desired flow rate from a continuous range of possible flow rates.

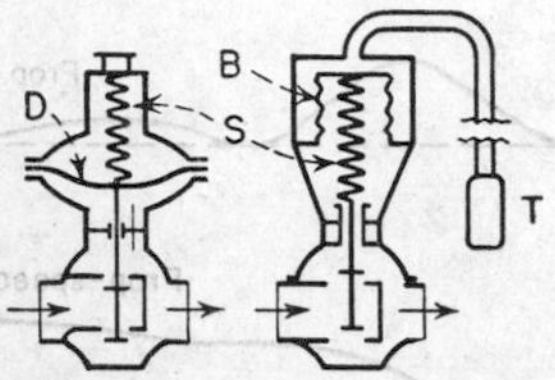

FIG. 193. FIG. 194.

FIG. 193. Self-operated pressure regulator.
FIG. 194. Self-operated temperature regulator.

The simplest form of the third class is known as *proportional-position*, proportional, corresponding, or simply as throttling. With this type, the control point deviates with changes in demand. The next step in controller design is therefore to add an automatic means for resetting the control point to maintain a constant control point regardless of changes in demand. Such controllers are known as *proportional-plus-reset*, or simply as reset controllers. A third variation of the proportional controller adds another function, which aids it in reaching a new stabilized final control-valve position quickly after a change in demand. This function moves the valve at a rate proportional to the acceleration of the change in deviation, and so is referred to as a rate or second derivative function.

Controllers may also be classified by the source of energy they use in effecting corrections. They are either self-operated or relay-operated, although an intermediate class using the pressure of the material being controlled to amplify controller responses is also used to some extent. In Figs. 193 to 218 are shown a variety of controller types and controller elements that show the most used fundamental types, but the selection is by no means exhaustive. Figure 193 represents a simple *self-operated* pressure regulator or reducing valve in which the valve opening depends on the interaction of the downstream pressure on diaphragm D, and the force of the opposing loading spring S. Adjustment of S adjusts the control point. A similar principle is met in self-acting temperature regulators such as Fig. 194. The thermometer system, consisting of bulb T and bellows B, is filled with a volatile liquid or a gas. Working against the adjusting spring, it positions the valve to control the flow of fuel or steam and so maintain the temperature.

The principle of the intermediate type of controller of the pilot- or relay-controlled, self-operated type, is illustrated by Fig. 195, which diagrams a pilot-operated reducing valve. Pilot valve P allows some of the high-pressure material to leak through L to the low-pressure side, applying the intermediate pressure between the two pilot needles through tube T to the under side of diaphragm D_2, which controls the main valve M. The pilot position in turn is controlled by the interaction of the downstream pressure acting on diaphragm D_1, and the force of the diaphram-opposing spring.

Pneumatically Operated Controller. For most control problems where more power is needed than is

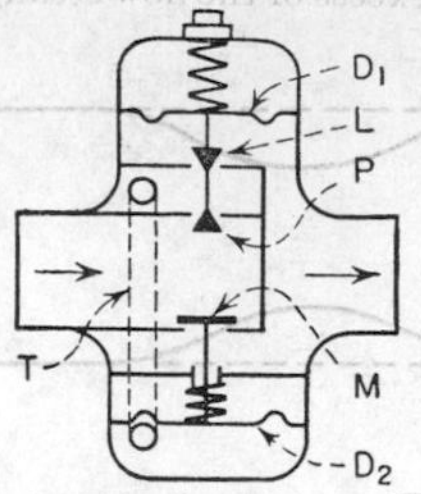

FIG. 195. Pilot-operated reducing valve.

available in self-operated devices, the primary instrument response must be amplified pneumatically, hydraulically, or electrically for indirect application to the final control element. Figure 196 is a diagram of a simple *relay-operated pressure controller*, using air under pressure (usually about 15 lb.) as the amplifying means. The pilot valve P is positioned by the bellows B so as to throttle the air pressure applied to the final control valve diaphragm and so give the desired opening of the control

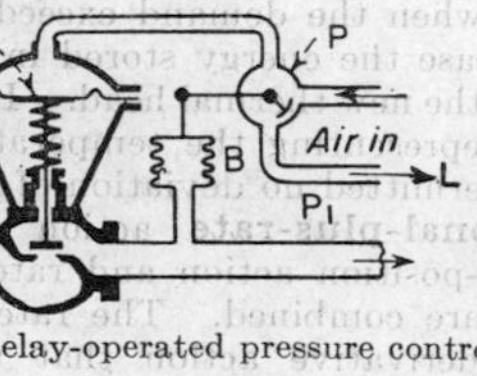

FIG. 196. Relay-operated pressure controller.

valve. For clarity a rotary-type three-way pilot valve is shown. It is obvious how pressure P_2 can be controlled by choosing the proper relation between the opening of the air inlet and the leak L. A leak or its equivalent is used in most fluid-operated throttling controllers.

Figures 197 to 200 show four fundamental types of *pilot valves*. Several variations are used, *e.g.*, that shown at Fig. 208. In Fig. 197 is the commonly used *flapper*

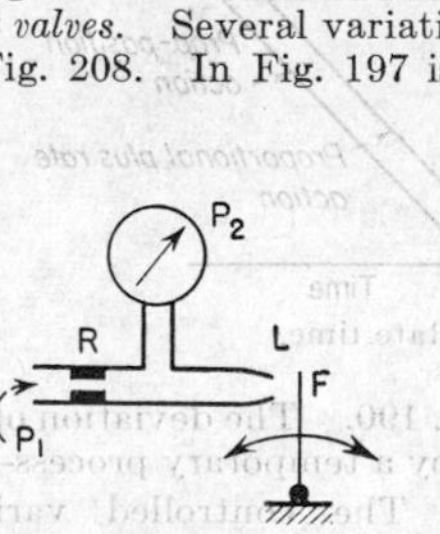

FIG. 197. Flapper-type pilot valve.

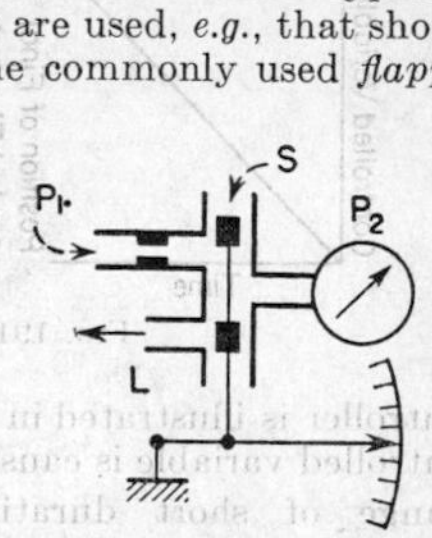

FIG. 198. Slide-valve pilot valve.

type in which air inlet pressure P_1 is throttled first through a restriction R and then through a nozzle leak L, under control of flapper F, which is positioned by the instrument pointer. The resultant pressure P_2 is proportional to the nozzle opening and hence is related to the position of the pointer. Figure 198 shows a *slide valve* S, the position of which determines the relation of inlet and leak and hence controls P_2. A restriction is necessary. The common *ball-valve* type is diagramed in Fig. 199. The position of the ball determines the amount of leak at L and also introduces the necessary restriction, thus determining P_2. This is similar to the slide-valve type. In Fig. 200 is the *jet-type* used by Askania. Depending on the position of the jet, more or less of the velocity pressure of a liquid issuing from the pipe under pump pressure is converted to static pressure in the receiving nozzle. The jet is positioned by the measuring instrument.

The pressure P_2 produced by any of these pilot devices may be applied to diaphragms, pistons, or bellows to position the final control device. Most common is the so-called diaphragm motor diagramed in Fig. 196, which is used to operate valves and dampers. Since such a diaphragm is opposed by a spring, a definite opening (within limits of friction) will be produced by a definite pressure P_2. Pistons and bellows may also be opposed by a spring. An alternative method is shown in Fig. 201, which shows a slide valve S controlling a piston. Without the lever F, which is shown dotted, there would be no relation between piston position and the measured quantity that positioned the slide valve, for the piston would move to one extreme or the other unless arrested by neutralization or opposite movement of the slide valve. Addition of F, which is called a follow-up, causes the piston movement to reposition the slide valve to stop the movement. Thus a definite instrument change will produce a definite piston movement. Such an arrangement, either with or without the follow-up, is often used for positioning large valves. In program control of valve opening and shutting, the follow-up is not required. It is needed only where throttling valve action is used. The jet-pipe method of positioning a piston is shown in Fig. 202. Instead of a single receiving nozzle as in Fig. 200, two are used.

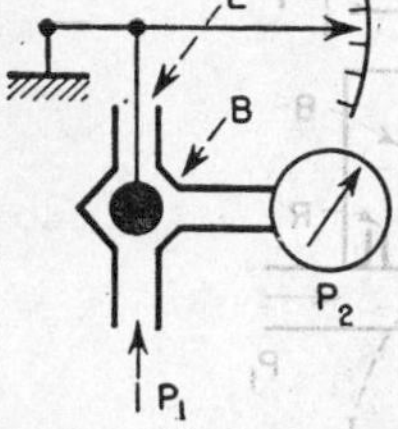

FIG. 199. Ball-type pilot valve.

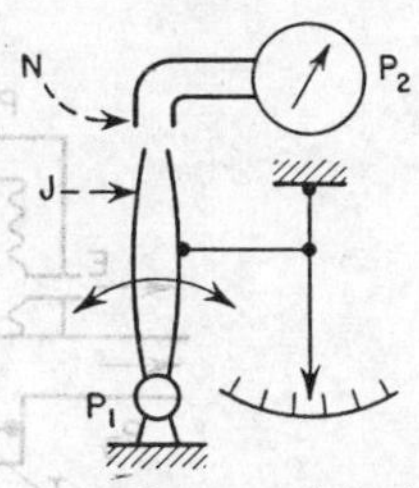

FIG. 200. Jet-type pilot valve.

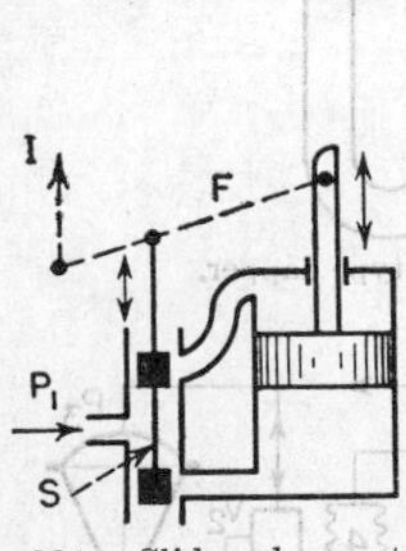

FIG. 201. Slide valve controlling a piston.

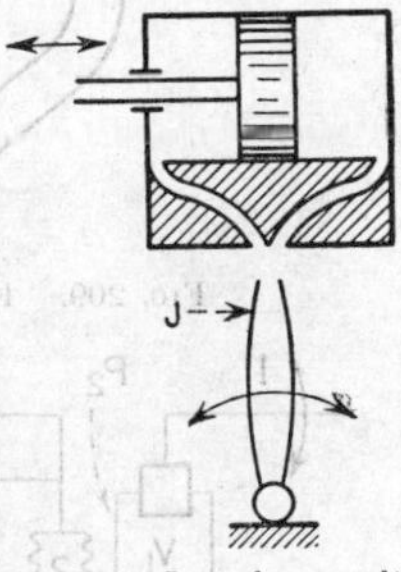

FIG. 202. Jet pipe positions a piston.

The development of a pneumatic pilot system to encompass the various types of control mentioned in the opening portion of this section, as given in Fig. 203, shows the same arrangement as in Fig. 199, namely, a ball pilot valve P.V. which receives air at pressure P_1 and throttles it to P_2 for use in controlling the opening of a final control valve. In this form, the pilot valve is capable of proportioning, but it has a slight reaction on the measuring system and so is not of highest accuracy. The reaction on the measuring system can be much reduced and the amplification increased by interposing the relay system shown in the Fig. 204. The instrument pointer here positions a flapper F before a nozzle N, causing bellows B to control the position of the ball valve. A restriction R must be used in the nozzle supply line. Air-supply pressure is reduced from P_1 to P_2 by the ball valve. In this type, such a small movement of the flapper is needed to change P_2 from maximum to minimum that such an arrangement is difficult to use for throttling types of control. Instead, it is used for two-position control.

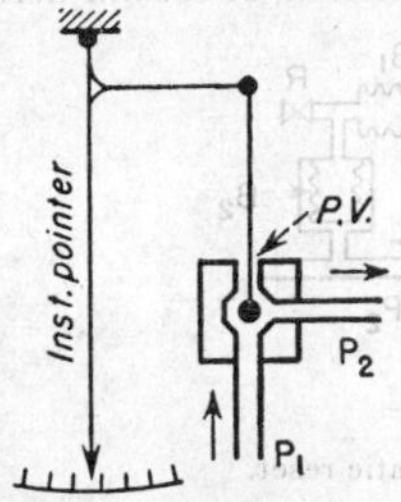

FIG. 203. Pilot valve.

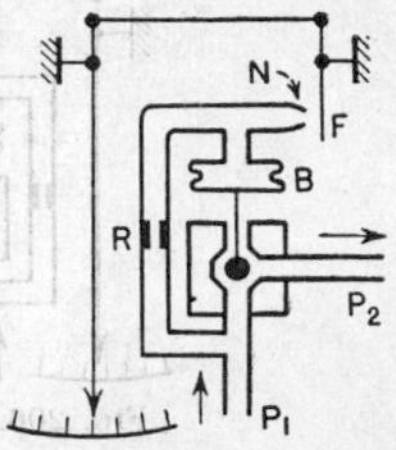

FIG. 204. Pilot valve and relay.

Figure 205 shows how Fig. 204 can be modified for proportional-position or throttling control by the addition of a follow-up, similar in principle to its mechanical counterpart in Fig. 201. This is accomplished by adding a throttling bellows B, which is positioned by the controlled pressure P_2. As this bellows carries a moving pivot for the flapper, it opposes the initial change in flapper position and after a pressure change produces a definite new flapper position, which is proportional to the new instrument-pointer position. Thus proportionality between the measured quantity and pressure P_2 is obtained, but the flapper will assume a slightly different position for every value of the instrument position, so that every value of the controlled load will give a different control point; *i.e.*, the instrument will have offset or droop. How much offset it has will depend on the amount of pointer movement needed to full-stroke the final control valve, a characteristic that is adjustable in most throttling types of controller.

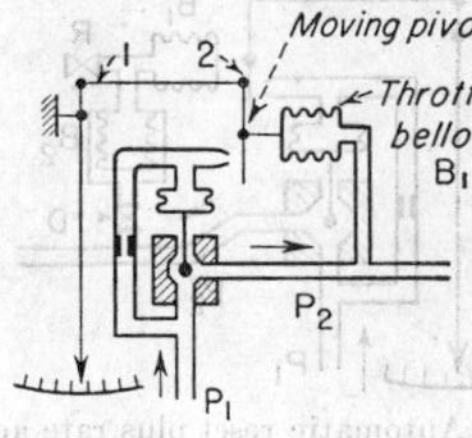

FIG. 205. Proportional-position control.

Many controllers get away from the difficulty of offset or droop by the use of automatic reset or proportional-speed floating action, the next step shown in Fig. 206. In Fig. 201, rapid movement of the piston (without the follow-up) would give two-position or on-off control; while very slow movement, which would allow the controlled condition to be affected quickly enough to arrest the piston before it reached the end of the stroke, would give floating control, which, as previously noted, has no relation between the measured value of the variable and the position of the final control element. To compensate a proportional-position controller for its offset or droop characteristic, it is only necessary to add a floating component, as in Fig. 206. The addition consists of a second bellows B_2 and a reset valve R to control the rate of reset. The initial effect of a movement of the flapper is exactly the same as with Fig. 205, *i.e.*, B_1 moves the moving pivot of the flapper so as to oppose the initial pressure change. At that point, however, the slow equalization of the pressure in the two bellows with that of the atmosphere begins to take effect, and a second movement of B_1 takes place, opposite to the first, thus resetting the flapper to a slightly greater opening or closing than the initial movement.

Addition of rate or second-derivative action is shown in Fig. 207. This consists of a restriction, or second-derivative, valve D, which is inserted between B_2 and the controlled pressure system for the purpose of slowing down the follow-up action of B_1. Thus an initial change in the flapper produces an immediate large change in the controlled pressure, but this is brought back gradually to the desired final pressure as the controlled pressure leaks through D. The two valves, R for setting the reset rate and D for setting the time required for taking off the second-derivative action, are used to tune the controller to the characteristics of the particular process with which it is to be used. Sometimes, especially for use in batch processes, second-derivative action is used without reset on a proportional controller, which means elimination of valve R.

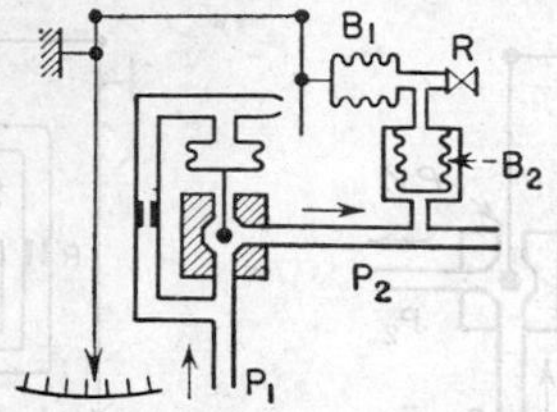

FIG. 206. Automatic reset.

Figures 203 to 207 show the principles used in most actual pneumatic controllers, although many different equivalent devices are adopted for obtaining the desired results. Features present in actual controllers, but not shown in the sketches, are the methods for altering the control point and changing the proportional band or throttling range. The former would be adjusted by some means of changing the length of link 1 in Fig. 205, and the latter by moving pivot 2 closer to or farther from pivot 3.

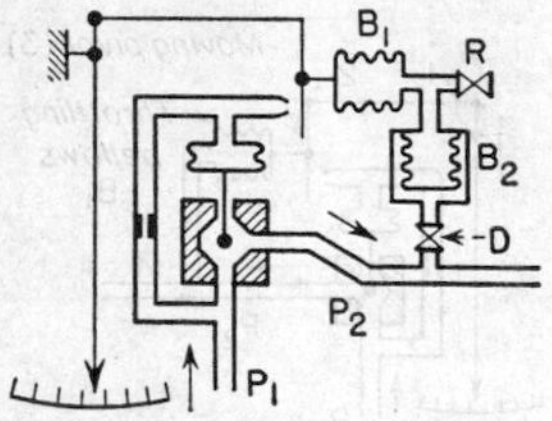

FIG. 207. Automatic reset plus rate action.

Some of the special variations of the pneumatic methods discussed are shown in Fig. 208 to 210. Figure 208 is a non-bleed type of relay air valve (Brown) which is claimed to have high sensitivity but to require a minimum of air, the only leak being that at the flapper. Air at pressure P_1 passes through a restriction and into the space outside a double bellows B, flowing to the nozzle N where the pressure in the nozzle system is controlled by a flapper F_1. If the bellows is in the balanced position, the exhaust tube T is closed by flapper F_2, which also closes the inlet port I. Pressure is then at P_3 in the output system. Assume a change in position of flapper F_1, which raises the nozzle pressure P_2. This forces tube T and flapper F_2 downward, opening I and raising P_3 until the force against the small bellows balances that against the larger bellows, when I is closed at the new value of P_3. On the other hand, a reduction in nozzle pressure lifts tube T, exposing the exhaust port and reducing P_3 to its new

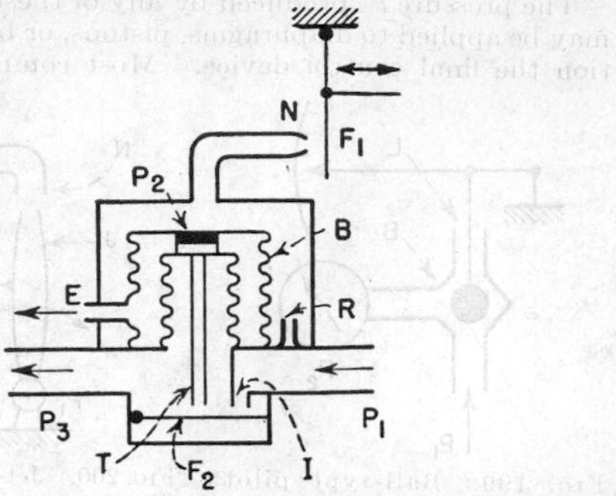

FIG. 208. Non-bleed air-valve relay.

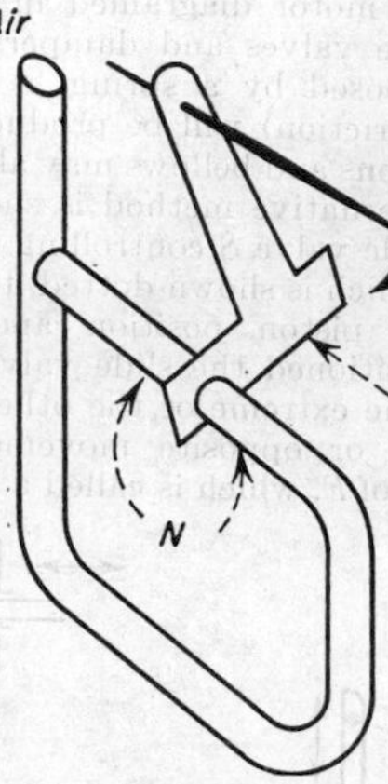

FIG. 209. Free-vane-type flapper.

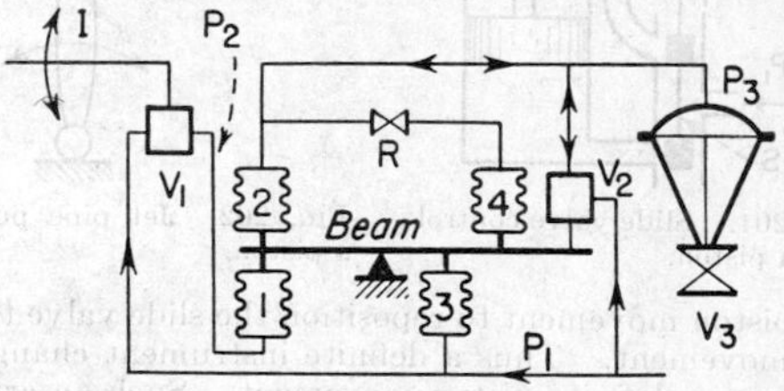

FIG. 210. Pneumatic balance system.

pressure corresponding to the position of flapper F_1. A follow-up not shown but equivalent to that of Fig. 206, is used to give a definite position to flapper F_1.

The free-vane type of flapper (Bristol) shown in Fig. 209 has been developed to reduce the reaction on the instrument pointer encountered with the type of flapper shown in earlier sketches. As is self-evident from the drawing, a vane moves between two opposing air nozzles without contact, to control the nozzle pressure.

In certain air-operated controllers having an electrical measuring system (Leeds & Northrup), the pneumatic balance system shown in Fig. 210 is used. The instrument I has a cam that positions primary air valve V_1, to control the pressure on bellows 1, which bears against a balance beam. Bellows 3 achieves an initial balance of the beam which positions secondary air valve V_2. The output pressure P_3 is applied to the final control valve V_3 and also to bellows 2, which acts initially as a follow-up; but, as the pressure between 2 and 4 is equalized through reset valve R, this follow-up action is taken off, and the final position of V_2 is therefore reset to the requirements for the new demand.

Electrical Controllers. Electricity is often used as the medium for translating control impulses into final responses. Its action is akin to other amplifying methods in that only a light contact in the primary instrument is needed to control a circuit; although, if the controlled circuit carries more than a minimum of current, some

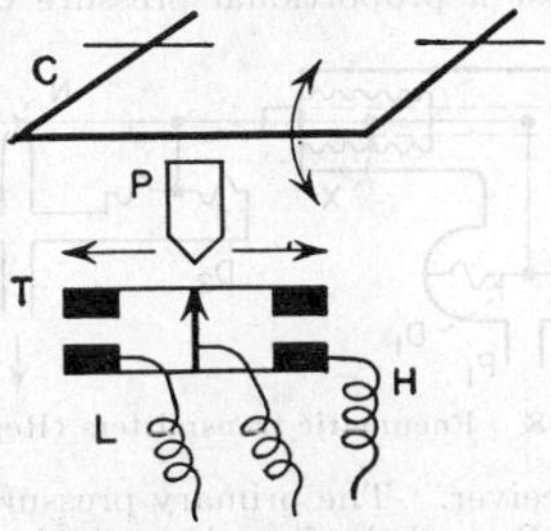

FIG. 211. Mechanically operated relay.

form of relay positioned either mechanically or by the primary current is needed. One method of closing controller contacts mechanically to overcome the low power available from the measuring instrument is shown in Fig. 211. Here the instrument pointer P moves over a tilting contact table T and is periodically depressed by a chopper bar C. The control point is at the table pivot so that deviation of the controlled variable, evidenced by the position of the pointer either side of the center, will tip the table one way or the other and close either the high contact H or the low contact L. Many other mechanically operated devices are used, generally to close mercury switches or snap-acting open switches.

Another method for offsetting the weak primary action of the measuring instrument, which is one of several ways to use electronics for this purpose, is shown in Fig. 212. Here the instrument pointer carries a light metal vane

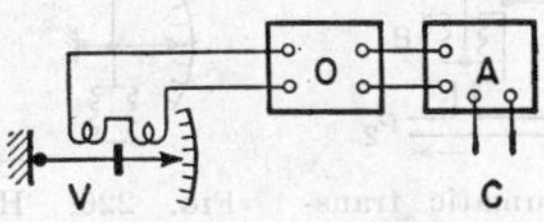

FIG. 212. Electronic relay.

V, which moves into or out of an oscillator O. Presence of the vane between the coils changes the oscillator tuning to operate the amplifier relay A and close the control circuit C.

Where two-position control is used, contacting primary instruments can open and close valves or other final control units. If throttling valve action is needed, some method of obtaining valve action proportional to the measured variable is needed. The most usual method is to employ a primary instrument in which the indication of the variable can be expressed as position of a slider on a rheostat or slide-wire. Then it is only necessary to use a Wheatstone-bridge circuit with a second slide-wire or rheostat positioned by the valve motor to attain a proportional position for the valve. Figures 213 to 215 show different developments of this idea, comparable with the development of the pneumatic controller shown in Figs. 205 and 206. Figure 213 shows a simple proportional hook-up, which has an offset characteristic similar to Fig. 205. The instrument positions slide-wire S_1; and, through action of the relay R and valve motor M, the final control valve is moved and with it slide-wire S_2, until the bridge is again in balance. To add proportional-band or throttling range adjustment, some arrangement equivalent to that shown in Fig. 214 is used (Brown), with two more resistances and contacts C_1 and C_2. These can be varied to change the resistance of the motor slide-wire and so to permit the bridge to balance with any desired proportionality between the movements of S_1 and S_2. Another refinement is shown in Fig. 215, for the pur-

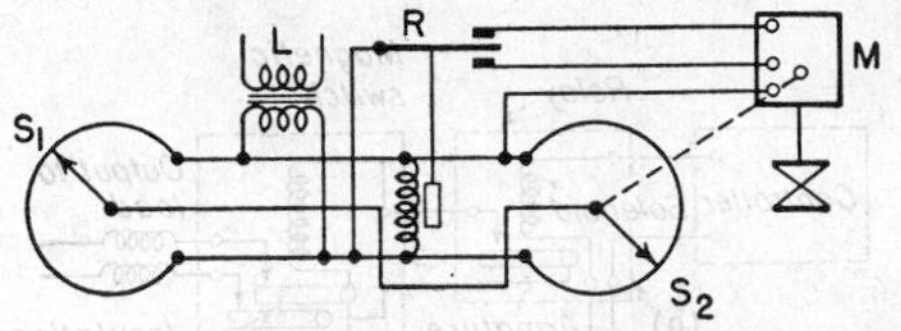

FIG. 213. Simple proportional controller.

pose of adding reset to the circuit. Since resetting a controller is equivalent to moving the proportional band up- or downscale, this can be accomplished by the addition of another pair of resistances and contacts, C_3 and C_4. When the contacts are moved simultaneously, they change the resistance at one end of the motor slide-wire and at the other end of the instrument slide-wire, thus achieving a relative shift in balancing points for the two slide-wires. This shift can be handled manually but is also performed automatically by a resetting motor in the instruments of several manufacturers.

Electrical relays are required when the current to operate the control valve or other device is greater than

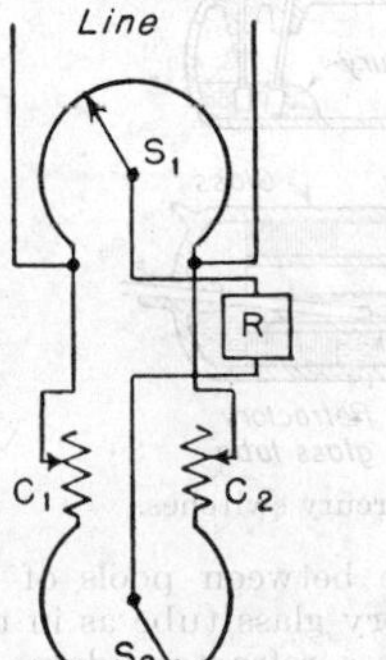

FIG. 214. Proportional-position controller.

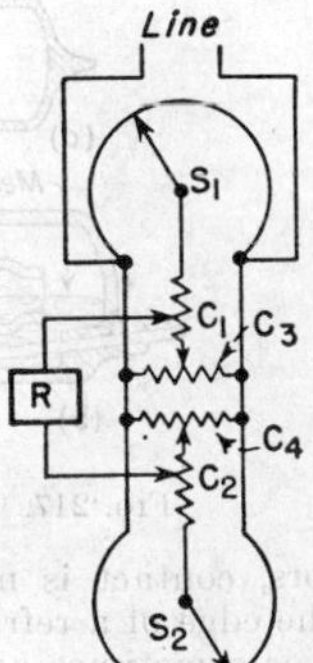

FIG. 215. Automatic reset controller.

can be carried by the instrument contacts. Figure 216 shows two common types of relay. That in *a* is a single-contact type used with single-contact controllers. The control circuit is completed only so long as the instrument circuit is completed. The type of relay shown in *b* is used with double-contact instruments. This differs from *a* in including a holding contact so that the control circuit remains completed until the instrument contact is made in the opposite direction. This type of relay is used to prevent chattering even when the current demand of the control valve or other device is below the permissible limit for the instrument contacts. Instead of operating open contacts, some types of relay tilt mercury contact switches by solenoid action. In others the solenoid displaces a mercury pool which makes contact.

For very low contact currents, microswitches or electronic relays are available. Electronic relays are used extensively in high-speed electronic instruments and especially where currents less than a few milliamperes are used.

Mercury Switches. On instruments controlling electric heating, direct operation of mercury contactors is common. These contactors are glass tubes evacuated, or filled with inert gas, containing a small amount of mercury and two or more contact points. Tilting makes or breaks the contact. Figure 217 shows two types of mercury-contactor switch, the first, *a*, a type for carrying 2 to 10 amp., and the second, *b*, for carrying from 10 to 50 or 60 amp. In the second and similar heavy-duty

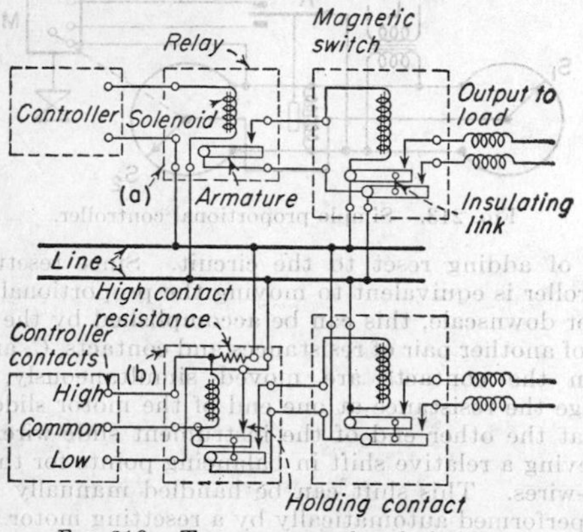

FIG. 216. Two common electrical relays.

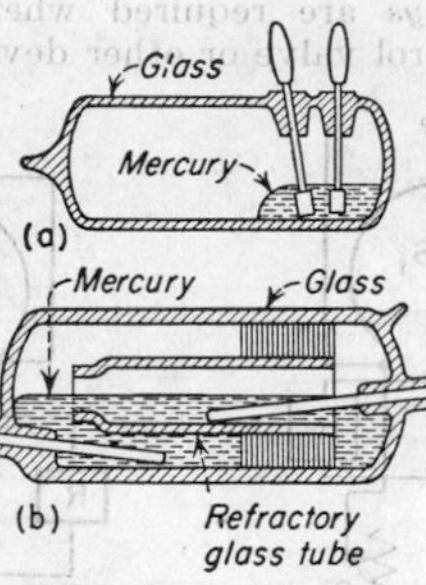

FIG. 217. Mercury switches.

contactors, contact is made between pools of mercury across the edge of a refractory glass tube as in the illustration, or sometimes across a refractory dam. Metal-enclosed types are made in sizes to 200-amp. rating.

Telemetering and Remote Transmission

Often it is necessary to locate the primary measuring instrument at a considerable distance from the controller or central instrument panel. With most types of industrial instruments, except those which measure an electrical quantity proportional to the variable, there is a definite limit to the acceptable distance between the instrument's sensitive element and the indicator, recorder, or controller. On the other hand, there is a tendency toward the centralization of indicators and recorders as well as of manual operating controls, which requires that many instruments be farther from their sensitive elements than good practice allows. Without special transmitting means, too much time lag would be introduced, or the primary elements would have insufficient power to transmit the indication with the requisite accuracy.

Numerous telemetering methods have therefore been developed, employing pneumatic, hydraulic, and electrical methods. *Pneumatic telemeters* convert the quantity that is to be transmitted into an accurately proportional pressure which is transmitted to the receiving instrument through a tube of diameter sufficient to avoid undue lag. *Hydraulic transmissions*, employing an incompressible fluid, displace a certain amount of liquid at the sending end and transmit an equal quantity to the receiver, where the quantity transmitted positions the instrument. *Electrical telemeters* convert the sending impulse into some kind of electrical quantity such as potential, current, or frequency.

Pneumatic Telemeters. One type of pneumatic telemeter (Republic) is shown in Fig. 218. This method balances the force produced by the primary indication directly against a proportional pressure of air which is

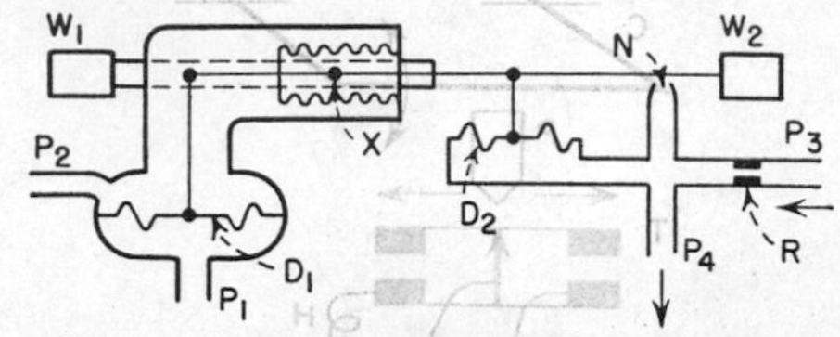

FIG. 218. Pneumatic transmitters (Republic).

sent to the receiver. The primary pressure P_1 is applied to diaphragm D_1, and the force is applied to a scale beam counterweighted by W_1 and W_2 and pivoted at X. If the primary impulse is a differential pressure, then a bellows is used to seal the shaft, as shown. The beam controls a leak from nozzle N of air at pressure P_3, which flows to the nozzle through a restriction R. The reduced pressure P_4 is then applied to diaphragm D_2, where it balances the scale, and to the receiving instrument. This pressure bears a definite pressure relation to P_1, depending on the characteristics of the beam, counterweights, and diaphragms.

Another type of pneumatic transmitter (Brown) is shown in Fig. 219. This device usually is employed in conjunction with a relay valve of the sort shown in Fig. 208. Flapper F is controlled by the instrument pointer and serves to increase or decrease the flow of air from a

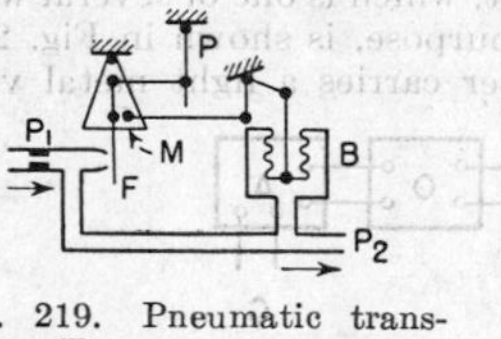

FIG. 219. Pneumatic transmitter (Brown).

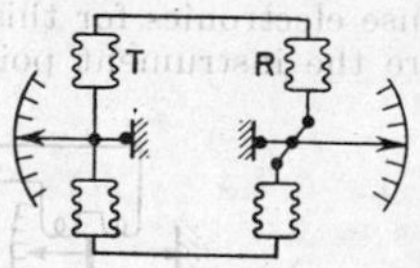

FIG. 220. Hydraulic transmission.

nozzle. Air at pressure P_1 flows through a restriction and thence to the nozzle. The reduced pressure P_2 is then applied to the receiving instrument as a pressure proportional to the primary indication. The method of achieving proportionality is to apply pressure P to a bellows B to which the moving pivot M of the flapper is attached. If the flapper moves toward the nozzle, raising the pressure of P_3, this in turn compresses the bellows, moves M to the right, and tends to pull the flapper away from the nozzle, so that a definite flapper position is attained for each pointer position.

Hydraulic Transmission. In Fig. 220 is shown one type of hydraulic transmission (Liquidometer) used in transmitting level indications. The double bellows system is used for temperature compensation. Changes in the volume of liquid in the capillaries offset each other through the linkage shown.

Electric Telemeters. Half a dozen or more different principles are used in electric telemetering devices. Figure 221 is a diagram of the circuit of a pair of self-synchronous (Selsyn) motors, which have three-phase stators S_1 and S_2 connected together and the two single-phase rotors R_1 and R_2 connected to the same line. If a primary impulse is applied to rotor R_1, the receiving rotor

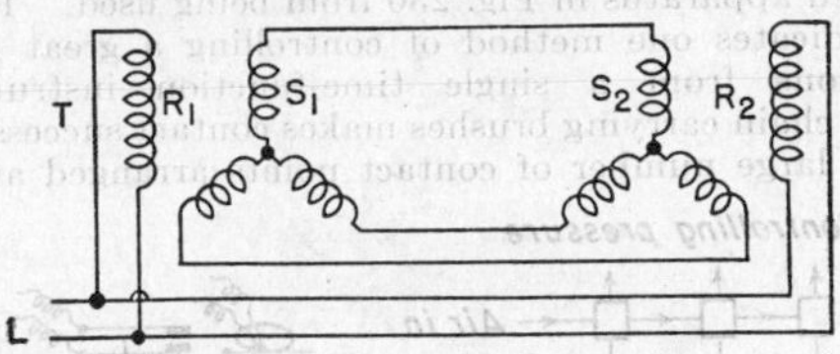

FIG. 221. Selsyn motor telemeter.

R_2 will turn by an equal amount. The rotors remain stationary unless turned by an external force.

The simple rheostat and ammeter, as used in automobile gasoline gages, has some industrial application, but more used are methods in which the rheostat is part of a Wheatstone-bridge circuit. Figure 222 shows such a setup. Here the transmitting pointer positions a contact on a slide-wire, and a null-type galvanometer bal-

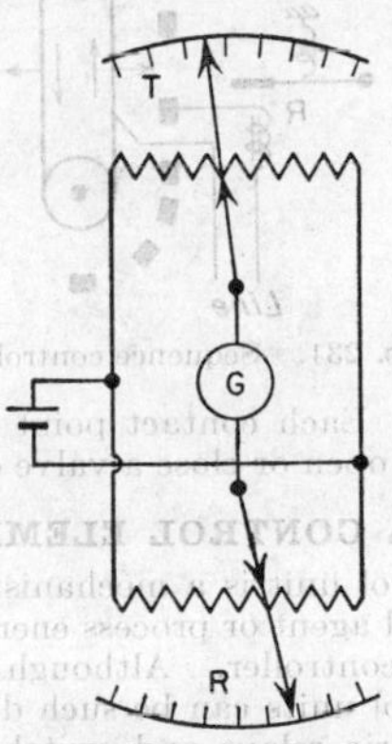

FIG. 222. Bridge-circuit telemeter.

ances the bridge by causing the receiving pointer to move to the proper point on the receiving slide-wire. A modification of this method is the induction balance, shown in Fig. 223, which is a self-balancing a.c. bridge. A pair of coils in series at the transmitting end, C_1 and C_2, is connected with a similar pair, C_3 and C_4, at the receiving end. A balanced iron core at each end seeks the same vertical position so that, if one is positioned by the

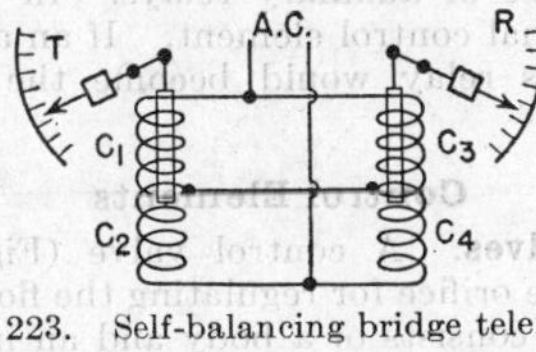

FIG. 223. Self-balancing bridge telemeter.

transmitter pointer, the other positions the receiving pointer.

Another method is for the transmitter end to send electrical impulses of duration proportioned to the magnitude of the quantity to be telemetered. One variation of this idea is shown in Fig. 224. This system (Bailey) uses two synchronized cams driven by clock motors and synchronized by a method not shown. These close contacts at each end, C_1 and C_2, for a period proportional to the position of the pointers, P_1 in the transmitter T, and P_2 in the receiver R. If the duration of contact at each end is not the same, the reversing motor in the receiver drives the receiver pointer to the proper point. When both shading coils A and B in the motor are either shorted or open, the motor cannot run; but, if one is shorted and the other open, because of unequal contact, the motor runs until both contact times are equal.

A new method of telemetering is suitable for remote indication and control in a wide variety of applications. This device (Allis-Chalmers) uses direct current of two-phase sinusoidal character and consists of a rotary transmitter, either manually or automatically actuated, and any desired number of receivers up to the wattage output rating of the transmitter. As the transmitter is rotated, all receivers follow its rotation with an accuracy

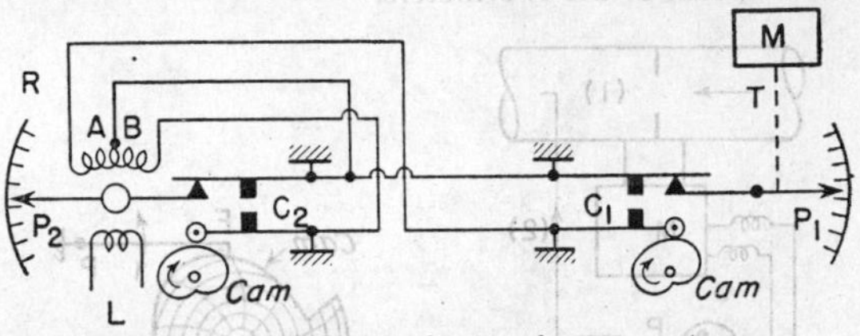

FIG. 224. Synchronized-cam telemeter system.

of plus or minus 1 deg. Telemetering methods are being used also for remote operation to centralize manual control. Various ways are possible, *e.g.*, by the use of self-synchronous motors. One can be geared to a valve handwheel, while the other is used to drive a distant valve. Pneumatic telemeters can be used to operate remote diaphragm valves. Or an electric-motor-operated valve can have its push button at the point of use, while some form of position-indicating telemeter shows the valve position.

Ratio Controllers

It is often desirable, especially in continuous processes, to measure one variable and to control another variable in a desired relation to the first by resetting the control point for the second variable. This is most commonly used to control the flow of two fluids. Pneumatic ratio control systems are perhaps the most common where the primary instrument measures a variable flow and transmits an air pressure to reset the control point of the secondary controller. The secondary controller then controls the magnitude of the secondary flow to the valve called for by its control point. Figure 225 shows an

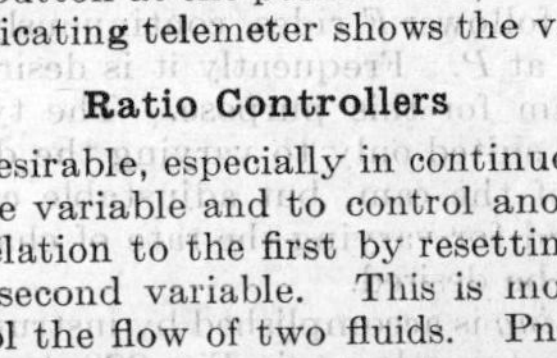

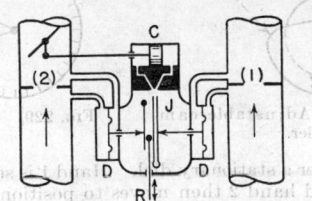

FIG. 225. Ratio controller.

Askania ratio controller. Flow of 1 is measured and adjusts the flow of 2. Swinging jet pipe J squirts oil under pressure against adjacent openings communicating with opposite ends of cylinder C. Differential diaphragms D position the jet pipe. When the flows are at the desired ratio, the jet pipe is centered, and the piston does not move. If they are not in proper relation, the jet pipe moves, moving the piston and damper. The ratio is adjusted by vertical movement of the pivot point on the upper end of ratio rod R.

Similar methods can be employed with other types of flow controllers. Most types provide a continuously adjustable ratio ranging from zero to as high as 200 or 300 per cent. In addition, several other methods of providing one flow in fixed relation to another have been developed for the feeding of comparatively small quantities of chemicals into a larger flow. For example, Fig. 226 indicates how a proportioning pump P can be paced by a meter that is measuring the primary flow 1. With this arrangement, the meter generally causes one stroke of P for every 1, 10, or 100 units of flow 1, thus injecting a definite volume of flow 2. Ratio control can be accomplished with volumetric meters in both the primary and secondary flow lines. A differential gear mechanism connects the two meters, and a control valve in the secondary line is positioned by the differences of rotational speeds of the two meters.

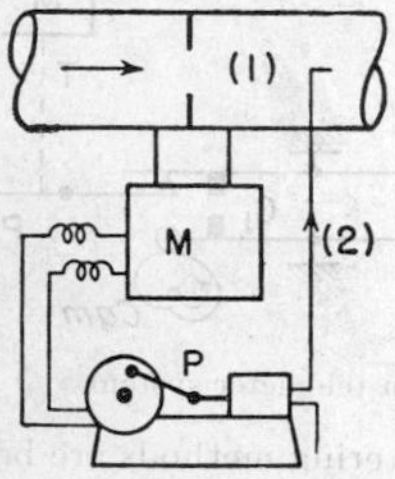

Fig. 226. Proportioning pump for ratio control of flow.

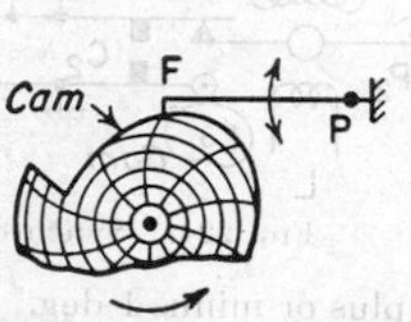

Fig. 227. Cam-type cycle controller.

Cycle Controllers

Cycle control is the automatic changing of the control point of a variable as a function of time. It is usually accomplished as in Fig. 227 by means of a cut cam on which a cam follower F rides, continuously resetting the control point at P. Frequently it is desirable to use an adjustable cam for this purpose. The type illustrated in Fig. 228 is suited only to varying the duration of the low portion of the cam, but adjustable cams have also been developed for varying the rate of change of control point as may be desired.

Interval timing is accomplished by instruments of many types. In the type shown in Fig. 229, two hands, 1 and

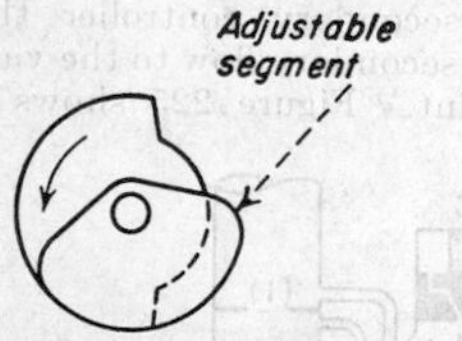

Fig. 228. Adjustable-cam cycle controller.

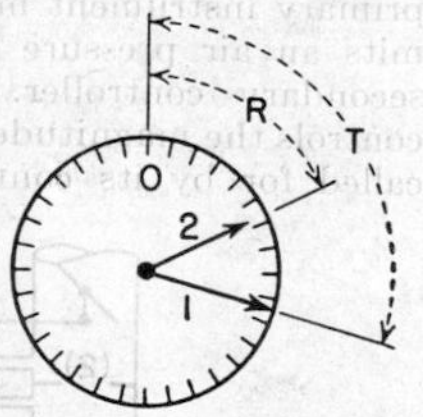

Fig. 229. Interval timer.

2, move over a stationary dial. Hand 1 is set to the total time T, and hand 2 then moves to position 1 and starts the process. It travels counterclockwise at uniform speed, stopping the process at zero. At any intervening time, a position such as 2 shows the remaining time R. After reaching zero, hand 2 may be reset either automatically or manually.

In automatic operation, or *programing control*, a clock-driven shaft carries a cam for each operation that is to be started and stopped, the cams being cut to give the desired sequence and duration of various operations, as in Fig. 230. Here cams 1, 2, and 3 are operating pilot air valves, cam 4 a mercury switch, and cam 5 an open-contact switch. The switches and pilot valves can control various circuits and process control valves. By combination of automatic control of the variables of the process and automatic starting and stopping of the various operations, it is often possible to eliminate manual operation entirely.

It is sometimes necessary to control large numbers of valves by automatic operation, which prevents the cam-operated apparatus in Fig. 230 from being used. Figure 231 indicates one method of controlling a great many operations from a single time-function instrument. Here a chain carrying brushes makes contact successively with a large number of contact points arranged at any

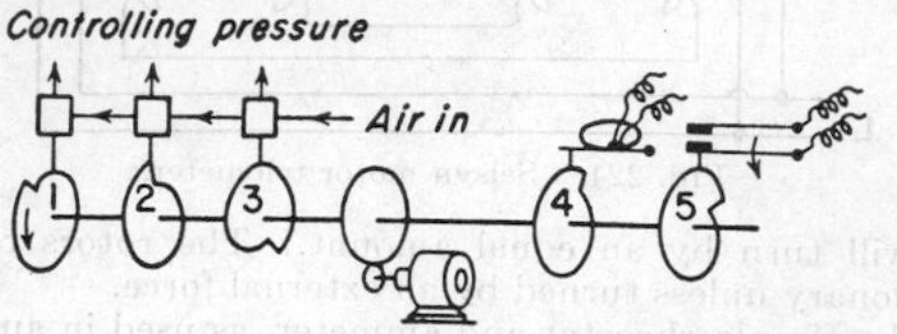

Fig. 230. Sequence controller.

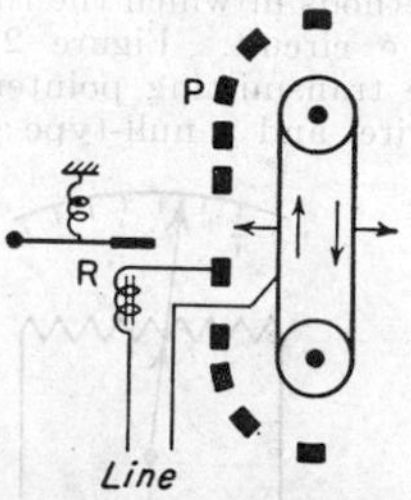

Fig. 231. Sequence controller.

desired interval. Each contact point is connected to a relay arranged to open or close a valve or damper.

FINAL CONTROL ELEMENTS*

The final control unit is a mechanism that varies the flow of the control agent or process energy in response to the action of the controller. Although usually a type of valve, final control units can be such devices as louvers, dampers, or electric relays and switches, depending on how the control is accomplished. The control agent may be air, gas, or liquid, or it may be electricity. In a large measure the control agent determines the type of final element required for a particular application. Control units usually consist of the control element proper and a power unit operated by the controller. An interesting example of a final control element is the control switch of a pyrometer controller for electric furnaces, where all current to the furnace is carried by the control switch without the use of auxiliary relays. In this case, the switch is the final control element. If an auxiliary relay were used, this relay would become the final control element.

Control Elements

Control Valves. A control valve (Fig. 232) is in effect a variable orifice for regulating the flow of the control agent. It consists of a body and an inner valve or plug which moves over a seat or port to change the area of opening through the valve. Valves may be designed for *rotary stem motion* or for *sliding stem motion*, the latter being far more widely used to control industrial processes. Inner valves of the sliding-stem type may be *single-*

* Ross, Significance of Design in Motorized Control Valves of the Sliding Stem Type, *Instrumentation*, January–February, 1946, pp. 16–22. Olsen, Selection Factors and Operating Characteristics of Control Valves, *Chem. & Met. Eng.*, May, 1943, pp. 132–136.

seated with only one port and one seat, as in Fig. 233, or may be *double-seated* with both ports opened and closed simultaneously, as shown in Fig. 232. The *double-seated* or *balanced valve* is pressure-balanced so that the forces of pressure differential on the valve stem are opposed. Resultant thrust on the inner valve is thereby reduced to a minimum regardless of the pressure differential on the plug. However, the double-seated valve cannot be tightly shut off because of clearances provided for temperature expansion of material of the valve body and plug. On the other hand, where tight shutoff is required, single-seated valves must be used.

Valve Characteristics. A variety of shapes of valve plugs have been developed to achieve different flow-lift characteristics for different process requirements. Most commonly used in on-off or two-position service is the simple *bevel plug* or disk valve (Fig. 233), which consists

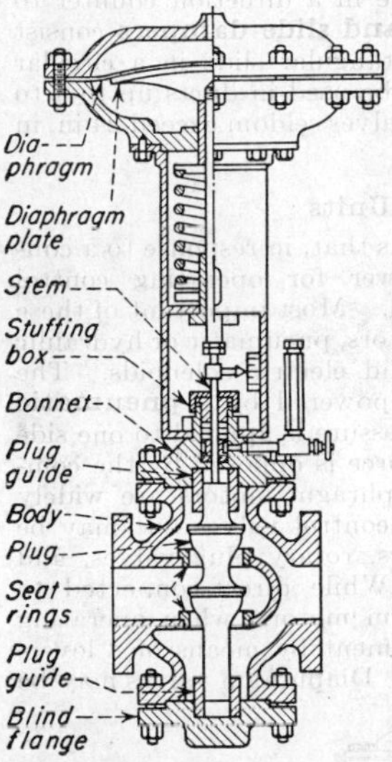

FIG. 232. Double-seated parabolic plug valve.

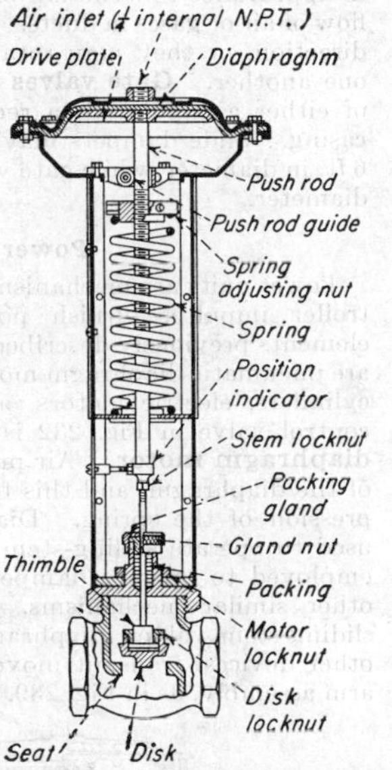

FIG. 233. Single-seated bevel-disk valve.

of a flat disk seated against a circular opening. Two-position control usually requires single-seated valves for low and moderate pressures and double-seated valves for larger sizes and higher pressures. This type of valve can be used for processes that can be controlled with a high sensitivity (narrow proportional band) controller, since here it is not necessary to control with fine increments of flow.

Although a substantial part of all control-valve requirements can be satisfied with the bevel-plug type of valve, the so-called *characterized valves* provide more uniform increments of flow for equal increments of valve lift. The most common of these is the *V-port* valve shown in Figs. 234 and 235. V-port plugs may have all ports made in the shape of straight-sided triangles, or they can be shaped to obtain different flow-lift characteristics. *Ratio or parabolic plug valves* (Fig. 232) and the specially shaped V-port valves come close to having equal or percentage flow-lift characteristics. Characterized valves are widely used on applications where flow of the control agent is throttled in relatively small increments and are most useful on applications involving a combination of long dead time and large load change (low sensitivity, wide proportional band). Although considerable emphasis has been placed on the advantages of characterized valves, there is a growing tendency to scrutinize application requirements more closely to avoid using these more expensive mechanisms where they are not dictated by the process. Generally, characterized valves are advantageous for use with flow controllers in continuous processes, since the process can be easily upset by fluctuating flow rates due to faulty control.

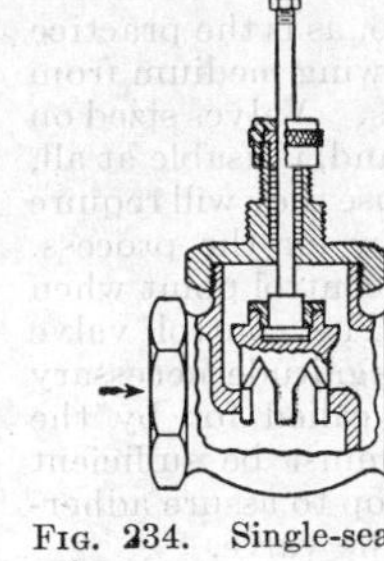

FIG. 234. Single-seat V-port valve.

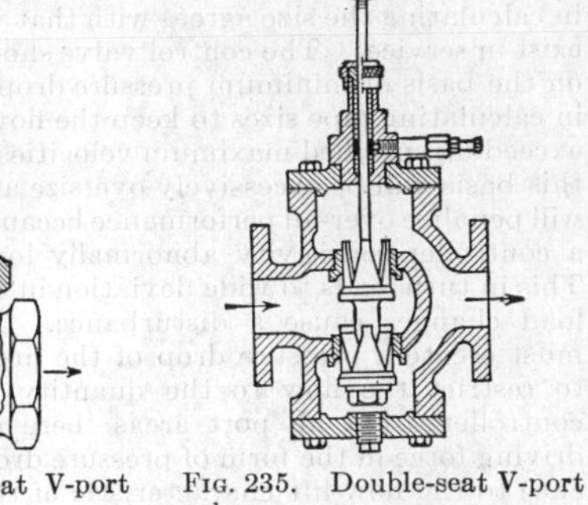

FIG. 235. Double-seat V-port valve.

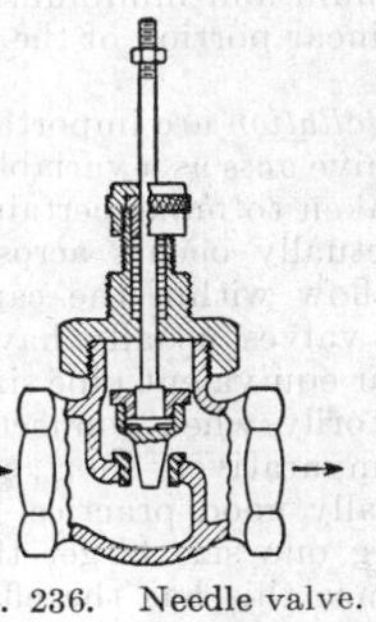

FIG. 236. Needle valve.

Valves for controlling small flows are often required. Most common is the needle valve in Fig. 236. Another type is extremely useful (Fig. 237) (Hammel-Dahl). It consists of a close-fitting piston-like plug in a straight-walled valve seat. The surface of the plug contains one or more V-shaped notches that may become larger as the valve is opened. This type of valve can handle flows small enough to require capillary action without wire drawing. Slurries are often handled by the Saunders-

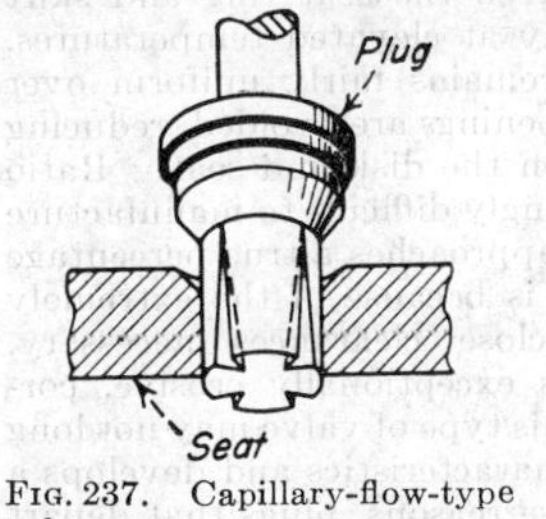

FIG. 237. Capillary-flow-type valve.

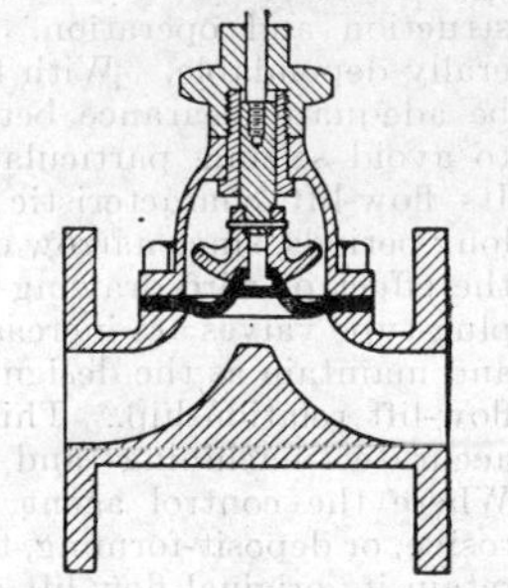

FIG. 238. Saunders valve.

type valve shown in Fig. 238. Even this type of valve is subject to becoming fouled. Fouling and plugging can be prevented by hooking the valve up to a cycle controller that will open it full for flushing out before assuming its controlled position. The cycle controller should be independent of the controller that regulates the flow.

Valve Selection and Installation. Incorrect sizing of valves perhaps more than any other single factor can contribute to poor over-all performance of an automatic control system. Too often, oversize valves are selected, because of either an improper analysis of the conditions under which the valve is to operate or to a lack of knowledge of the true capacity and flow-lift characteristic of the valve being selected. Two factors are of principal importance when sizing a valve.

First, it should be certain that the pressure drop used in calculating the size agrees with that which will actually exist in service. The control valve should not be selected on the basis of minimum pressure drop, as is the practice in calculating pipe sizes to keep the flowing medium from exceeding specified maximum velocities. Valves sized on this basis will be excessively oversize and, if usable at all, will penalize over-all performance because they will require a controller sensitivity abnormally low for the process. This in turn leads to wide deviation in control point when load changes cause a disturbance. The control valve must create a pressure drop of the magnitude necessary to restrict the flow to the quantity called for by the controller. For all port areas there must be sufficient driving force in the form of pressure drop to assure adherence to the flow-lift characteristic of the valve.

Second, maximum and minimum flow rates should be well within the linear portion of the flow-lift curve of the valve.

Piping and installation are important, since in any control circuit the valve acts as a variable orifice, and precaution should be taken to make certain that the maximum pressure drop actually occurs across the disk and seat for all rates of flow within the capacity of the valve. Percentage-type valves usually have maximum capacities less than their equivalent pipe size and therefore may perform satisfactorily when connecting piping is of the same size and comparatively short, such as 50 pipe diameters. It is usually good practice, however, to employ connecting piping one size larger than the valve body ports call for, since this has the effect of confining the pressure drop at the valve plug, with the result that the flow characteristics of the valve are preserved, enabling the flow to change exactly in accordance with the dictates of the controller itself. It is important, too, in the selection of valves to make sure that the controller can utilize a characterized valve. If a controller with nonlinear characteristics is hooked up to a valve with linear characteristics, the system as a whole will be non-linear.

From the maintenance standpoint, the bevel-plug type of valve is most reliable because of its simple construction and operation. V-port valves are also generally dependable. With this type of valve, there must be adequate clearance between the seat ring and skirt to avoid seizing, particularly at elevated temperatures. Its flow-lift characteristic remains fairly uniform over long periods, since narrow openings are avoided, reducing the effect of wire drawing on the disk and seat. Ratio plug-type valves are increasingly difficult to manufacture and maintain as the design approaches a true percentage flow-lift relationship. This is because of the extremely accurate machining and close clearances necessary. Where the control agent is exceptionally erosive, corrosive, or deposit-forming, this type of valve may not long retain its original flow-lift characteristics and develops a tendency to stick. For these reasons, plugs that depart further from the percentage flow-lift relationship and therefore do not of necessity have to be held to such close tolerances in manufacture are used extensively.

Materials of construction are of prime importance. It is possible to obtain valves made of various corrosion-resistant alloys. By far the majority of control valves are supplied with cast-iron or steel bodies. Valve plugs, stems, and seat rings may be bronze, stainless steel, or other alloys, depending on the service.

The **rotary valve** consists of a cylindrical body containing a plug that may have a rectangular or V-port with characteristics similar to those of corresponding slide-stem valves. Rotary motion uncovers the plug port to permit flow through the valve. Characterization may be obtained by the linkage arrangement between the power unit and the valve. **Butterfly valves and dampers** usually consist of a vane rotating about an axis inside a circular or rectangular casing. It is operated by a shaft projecting through the casing. Butterfly valves can best control the flow of liquids or gases with low pressure differentials and are usually unsuitable for high pressure differentials. Butterfly dampers are mostly used to control flow of gases in ducts. One type of butterfly valve has its body arranged so that the vane is rotated past a V-slot in the body. **Louvers** consist of a number of adjacent rectangular vanes somewhat similar in appearance to Venetian blinds and are used to control flow of air or gases in ducts. They can rotate in the same direction, or they may rotate in a direction counter to one another. **Gate valves and slide dampers** consist of either a circular or a rectangular slide in a circular casing. Slide dampers may be used in ducts up to 4 to 6 ft. in diameter, while gate valves seldom exceed 24 in. in diameter.

Power Units

Power units are mechanisms that, in response to a controller impulse, furnish power for operating control elements previously described. Most important of these are pneumatic diaphragm motors, pneumatic or hydraulic cylinders, electric motors, and electric solenoids. The control valve in Fig. 232 is powered by a **pneumatic diaphragm motor.** Air pressure is applied to one side of the diaphragm, and this force is opposed by the compression of the spring. Diaphragm motors are widely used to operate sliding-stem control valves but may be employed to operate dampers, rotary plug valves, and other similar mechanisms. While direct-connected to sliding-stem valves, diaphragm motors, when operating other devices, transmit movement by means of a lever-arm assembly, as in Fig. 239. Diaphragm valves may be

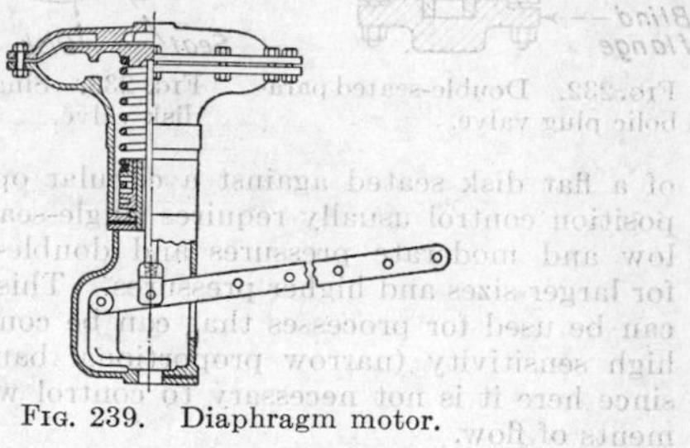

FIG. 239. Diaphragm motor.

either *direct-acting,* closing on an increase in pressure, or *reverse-acting* to open on an increase in pressure. If air pressure fails, the spring will cause the valve to open or close, depending on whether the valve is direct- or reverse-acting. A pneumatic or hydraulic **power cylinder** is a cylinder containing a double-acting piston for operating control elements that require more power and longer stroke than is available with diaphragm motors. However, when hooked up to large dampers, the weight of the element itself should be counterbalanced so that the power unit will have to overcome only the friction of the element.

Neither the diaphragm motor nor the power cylinder, when subject to minute changes in pressure from the controller, will accurately position the control element against the appreciable resistance forces in the control element. Where accurate control is necessary, a position controller or valve positioner is used. A **valve positioner,** employed on a diaphragm motor, applies a large air pressure to the diaphragm when it is moving in one direction and almost entirely relieves the air pressure when moving in the opposite direction. When the diaphragm has moved to its proper position, the pressure is reduced to the value necessary to hold that position. This device, then, counteracts the effects of stem friction, diaphragm and spring characteristics, and value thrust to reduce hysteresis and controller dead time. Air pressure from the control instrument positions a bellows and through a differential lever system attached to the valve stem positions a pilot valve on the air-supply line to the diaphragm to maintain the valve in the exact position called for by the controller. In another form, the bellows positions a baffle against which is operated a nozzle attached to the valve stem. A pilot valve operated by the nozzle air pressure controls the supply air to the valve diaphragm. Thus, while a change in controlled air pressure from the instrument might be of the order of 2 or 3 lb./sq. in., the air pressure to the diaphragm motor or power cylinder could be the full supply pressure of 20 lb./sq. in. or higher, depending on the limitations of the unit.

Solenoids are usually applied to sliding-stem valves for on-off or two-position control. Although most solenoid valves are of the single-seated, bevel plug type, solenoids may also be used to actuate balanced valves and needle valves. The valve plug is held either in the open or shut position, and the solenoid must overcome the force of the spring to move the plug to the opposite extreme. The solenoid, either integral with the current or attached outside and linked to the stem, operates when energized by the control current. In some types the solenoid operates a pilot valve which permits the pressure acting upon the top of the valve to be exerted to the underside when in the open position. If the pilot valve opening is made larger than the opening admitting downstream pressure when in the closed position, a pressure differential will exist upon the two sides of the valve, which is then caused to open, when the pilot valve is opened. Most makers supply these solenoid valves in sizes not over 3 in. for pressures up to 150 lb./sq. in., although specially constructed valves, which make use of the pressure of the fluid handled to operate the valves, have been built in sizes to 12 in. and 400 lb./sq. in. pressure. Solenoid valves are made for both alternating and direct current. Solenoid valves are also used as pilots for fluid-control devices. Where valves are very large or exceptionally fast action is required from remote valves, this is a convenient method. Although such valves ordinarily draw current that can be handled by the controller contacts, it is customary, to avoid chatter due to frequent secondary contact, to use a double-contact controller and a relay.

Electric motors are frequently used to operate control elements and are built for two-position, floating, and proportioning control. The two position motor may drive the valve open and allow a spring to close the valve, or it may drive the valve both open and closed. Two-position valves may be operated full open and full shut, or they may operate between two adjustable settings. They include butterfly, gate, and balanced valves, operated singly or in pairs. For fuel control, a single motor often operates a fuel valve and an air valve simultaneously. In one type of motor valve, motor functioning moves the valve disk to its other extreme with each opposite control impulse made by the controller. In such valves a motor is connected through a train of gears to a crank or cam which moves the valve stem through its entire travel in one-half revolution. As the half revolution is completed, a limit switch is opened, at the same time closing contacts connected to the second pair of controller contacts, so that a full revolution of the valve camshaft is completed on the next opposite control impulse. Open-and-shut motor valves are customarily by-passed to take care of minimum flow requirements while the valve itself is closed. Such valves are often used for remote manual control. Motor units are available for attaching to standard valves.

Electric proportioning motors are of a number of types. Usually, a reversing motor operates the valve, running in one direction for a "high" contact, remaining stationary as long as the controller is not making contact, and running in the opposite direction as soon as a "low" contact is made. For greater lag, a compensating, step-by-step, or delayed-action device must be inserted in the control circuit to avoid overshooting the control point. Without some such device, even a small lag will result in "hunting." In step-by-step valves, a "high" contact will increase the valve opening slightly, whereupon the mechanism disconnects the motor from the valve for a brief adjustable interval and waits to determine whether the valve adjustment is sufficient. If it is not sufficient, adjustments continue at intervals until the control point is reached. More complicated devices are designed to make valve changes of decreasing magnitude as the control point is approached or even to make a negative correction shortly before reaching the control point in order to avoid overshooting. Proportional-position control is secured in motor valves (most often those used with potentiometer instruments) by using corresponding resistors and slide-wire contacts in both motor mechanism and instrument. These resistors are connected in a bridge circuit with a sensitive polarized magnetic or electronic relay to detect unbalance and to operate the valve motor in a direction to rebalance the bridge. Thus the valve is maintained in a position corresponding to the value of the measured variable, a proportional-position control results.

Electric motor valves have an advantage generally not obtainable in pneumatic or hydraulic systems in that thrust caused by the unbalanced forces of a plug-type valve is not a problem because of the large gear reduction between the valve and motor. If the power fails, the valve will remain in its last position. Similarly, the valve-stem friction caused by the packing gland around the stem generally does not affect the operation of the motor valve. The coast of the motor can be compensated by damping the action of the motor as it nears the balance point.

In some electric valves the actual valve-stem operation is hydraulic, the motor driving an oil pump which in turn furnishes oil to the valve-operating cylinder. In the larger sizes, electrohydraulic operation may be cheaper and far more durable.

AUTOMATIC-CONTROL-SYSTEM APPLICATIONS

Previous parts of this section on instrumentation point out the many variations that enter into process control. Instrumentation engineering involves an evaluation of all these variations, including process characteristics and automatic controller characteristics. Once these variations are evaluated, the engineer must select the instruments with characteristics that most appropriately meet the requirements of the process, not forgetting the cost of the installation. Some idea of the numerous ways in which instruments and controls can be applied to a process can be obtained from the examples that follow.

Figure 240 illustrates a simple *open-kettle temperature-control system.* The measuring means comprises a pressure-type thermometer bulb located through the side of the vessel. Internal pressure in the thermal system is transmitted to a measuring spiral within the instrument case by means of armored capillary tubing. Automatic control is accomplished pneumatically by positioning a diaphragm motor valve on the steam-supply line to the heating coils.

In order to increase production, it may be desirable to heat the contents of the kettle as quickly as possible to a given temperature and then to hold the kettle at that temperature. The heating time is a controllable factor

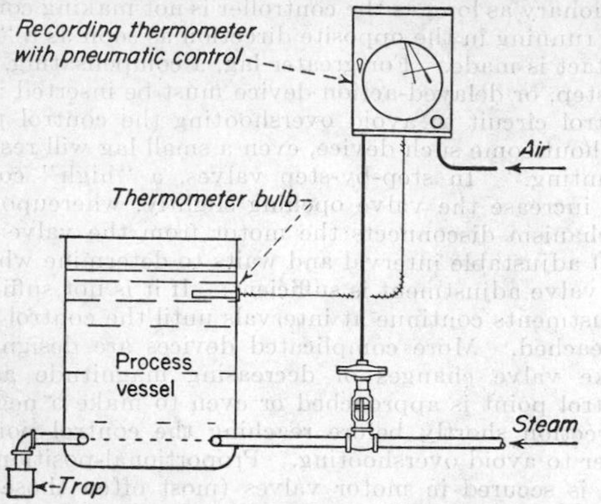

FIG. 240. Simple batch-temperature control.

that should be reduced to a minimum because it is unproductive time. This means that, during the heating time, a large heat input is required and that, when the control temperature is reached, a much smaller heat input, sometimes only sufficient to supply radiation losses, is required. To meet the demands of such operations, two control valves, as illustrated in Fig. 241, may be used in parallel, a large valve and a smaller valve. If the controlled-output air pressure from the instrument varies from 0 to 15 lb./sq. in. with 7.5 lb./sq. in. at the

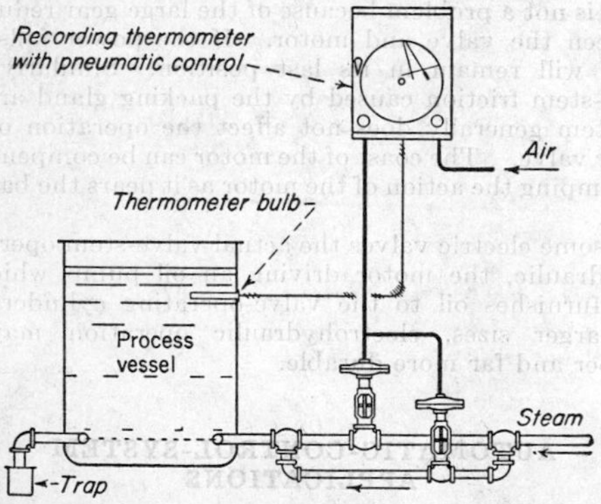

FIG. 241. Two control valves operate over different ranges of instrument air pressure.

control point, then the large valve will be designed to throttle from an open to a closed position when the diaphragm air pressure is varied from 0 to 7.5 lb./sq. in., and the smaller valve will be designed to proportion over the entire range from 0 to 15 lb./sq. in. Thus both valves will be open in the early stages of heating. The large valve will move toward a closed position and will be closed when the set point is reached, and the smaller valve will be throttling in mid-position at the set point.

In some operations, it is desirable to employ a *jacketed kettle* and circulate the heating or cooling medium through the jacket. Figure 242 is a schematic diagram of a temperature controller applied to such an installation. In this case a three-way valve is used to direct any part or all of the flow of the heating medium through the jacket. That which does not go through the jacket is by-passed around it. Three-way valves in such applications permit a large quantity to pass through the jacket for quick heating and a small quantity for holding the temperature at the set point, similar to the two-valve application of Fig. 241, and they do not materially restrict the total flow of heating medium being controlled.

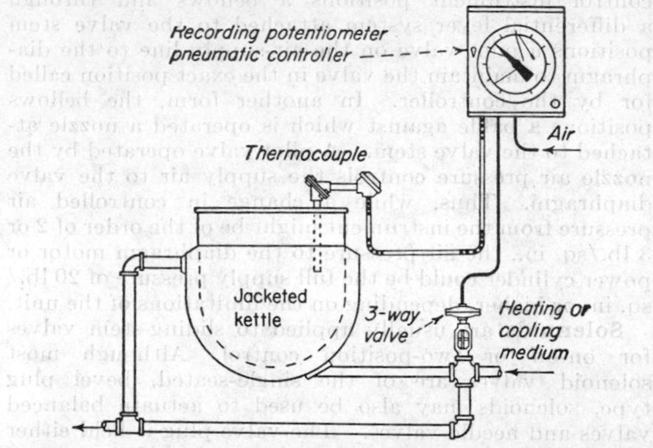

FIG. 242. Three-way valve permits by-passing control medium.

Often it is desirable to bring an open kettle up to a given temperature, hold it at that temperature for a given length of time, and then shut off the heating medium and discharge the contents of the vessel. A process-control system to perform these operations is shown in Fig. 243. A pressure switch in the air line to the steam valve closes when the temperature reaches the set point. The pressure switch starts a timer; and, at the end of the period for which the timer is set, the three-way solenoid valves in the air lines to the steam valve and to the discharge valve are operated. Both the steam valve and the dis-

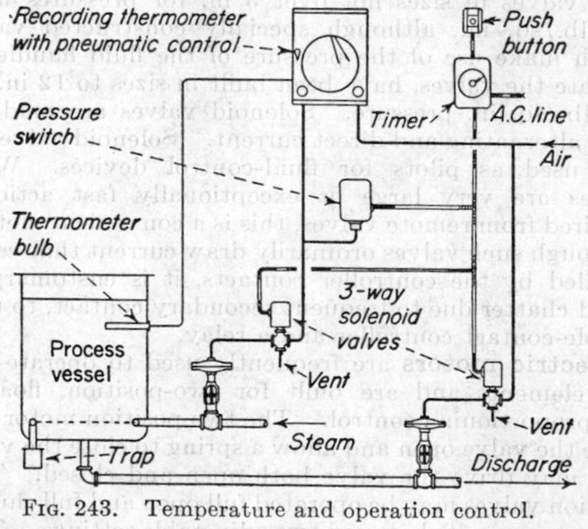

FIG. 243. Temperature and operation control.

charge valve are normally closed so that they will move to a closed position upon air failure. Therefore, at the end of the timing period, the three-way solenoid valve in the air line to the steam valve closes and bleeds the air from the diaphragm, and the steam valve closes; also the three-way valve in the air line to the discharge valve closes its exhaust port and admits air to the diaphragm from the supply line to open the discharge valve. The steam valve remains closed, and the discharge valve remains open. When the vessel is recharged, the operator pushes the start button, which momentarily opens the electric current supply. This automatically resets the timer to its starting position and opens the steam valve and closes the discharge valve. In the meantime, while the vessel was being emptied, the temperature dropped below the set point, and the pressure switch opened. The vessel now can be recharged to repeat the heating and holding cycle. In plants where a number of vessels are under the supervision of one operator, it may be

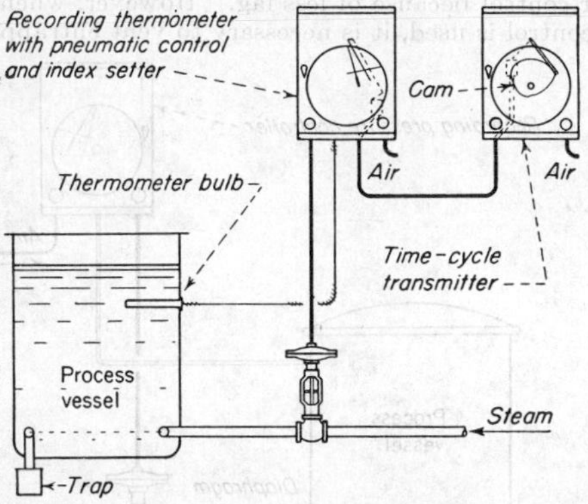

FIG. 244. Cycle control.

desirable to incorporate signal lights with the automatic equipment to indicate when the vessels are being emptied.

In some operations, it is desirable that the *temperature follow a predetermined program.* Figure 244 is a schematic diagram of such an installation in which the control index of the recorder is moved by varying air pressure from the transmitter. The output air pressure from the transmitter varies in direct proportion to the radius of the cam mounted on it. Thus the control instrument automatically controls to the time-temperature pattern of the cam. The cams are usually constructed of aluminum

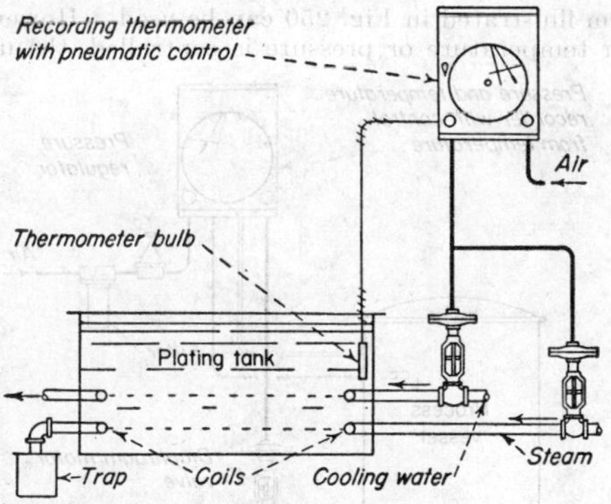

FIG. 245. Plating-tank control.

sheets. Graduations are printed on the aluminum disk similar to the graduations printed on a paper chart. This makes it possible to keep transmitter disks on hand and to cut new cams from them as required. The index of the control instrument can be set manually for operations when an automatic time-temperature program is not required.

Where there is energy dissipation within a vessel, as with *exothermic reactions,* it often is necessary to remove heat from the process to prevent the temperature from becoming excessive. Chromium plating is an example of a process in which energy is dissipated in the bath because of the passage of electric current. Figure 245 illustrates the application of temperature control to a plating tank. The bath contains both heating (steam) and cooling (water) coils. The valves in these lines are connected to the output air line from the control instrument. One valve opens with an increase in air pressure, and the other valve closes with an increases in air pressure. The valve springs are so adjusted that one valve is open at 0 and closed at 8 lb./sq. in., while the other valve is open at 15 and closed at 9 lb./sq. in. Thus, when the air pressure on the diaphragms is between 8 and 9 lb./sq. in., both valves will be closed. The instrument-control mechanism is adjusted to a narrow proportioning band with an

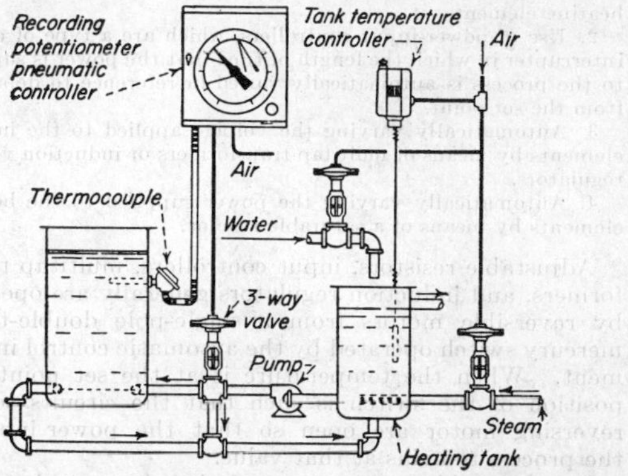

FIG. 246. Control medium circulator from reservoir to process.

output air pressure of 8.5 lb./sq. in. at the set point. When the temperature rises above the set point, the steam valve is closed, and the water valve opens. When the temperature drops below the set point, the water valve is closed and the steam valve opens.

Another method of obtaining both *heating and cooling* in a control system is shown in Fig. 246. In this system, a tank of hot water is automatically held at the set point by a separate controller that regulates the flow of steam and water to the tank. The tank is provided with an overflow. A pump circulates hot water from the tank

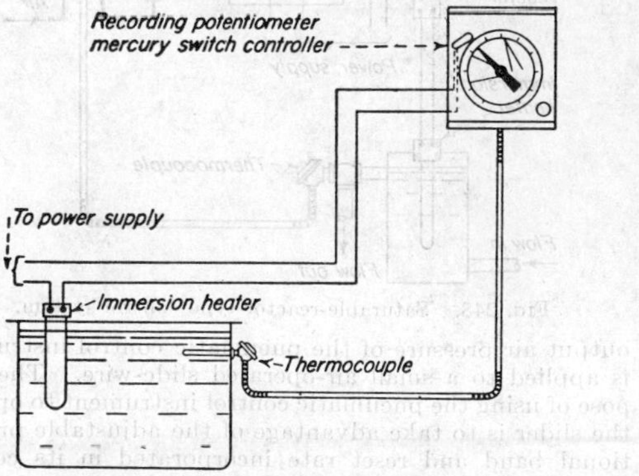

FIG. 247. Control of immersion heater.

through a closed-coil heater in the process vessel or through a jacket surrounding the vessel. The process-temperature controller operates a three-way valve to regulate the flow of hot water to the process or by-pass it back to the auxiliary tank. With a high rate of circulation, very close control can be obtained with this method, and it possesses the advantage of limiting the temperature of heating medium where high temperatures would cause decomposition. The steam and water valve arrangement on the heating tank is similar to Fig. 245.

Figure 247 illustrates an open vessel in which the liquid

contents are *heated by means of an electric immersion element*. The power supply to the heating element is turned on and off automatically by means of a mercury switch in the control instrument. Mercury switches such as this are designed to break at about 30 amp. at 115 volts. For greater power requirements, an auxiliary contactor is necessary.

A number of other systems have been devised to regulate the power input to electrically heated processes, including:

1. Automatic adjustment of resistances in series with the heating elements.
2. Use of power-input controllers, which are a type of circuit interrupter in which the length of time that the power is supplied to the process is automatically varied in reference to departure from the set point.
3 Automatically varying the voltage applied to the heating elements by means of multitap transformers or induction voltage regulators.
4. Automatically varying the power supplied to the heating elements by means of a saturable reactor.

Adjustable resistors, input controllers, multitap transformers, and induction regulators generally are operated by reversible motors from a single-pole double-throw mercury switch operated by the automatic control instrument. When the temperature is at the set point, the position of the switch is such that the circuits to the reversing motor are open so that the power input to the process remains at that value.

The saturable-reactor type of electric control system receives its impulse from the change in position of the sliding contact on a resistor or from the change in direct current from the electron tubes controlled by the process.

In Fig. 248, a thermocouple is located in the discharge of an electrically heated continuous fluid heater. The

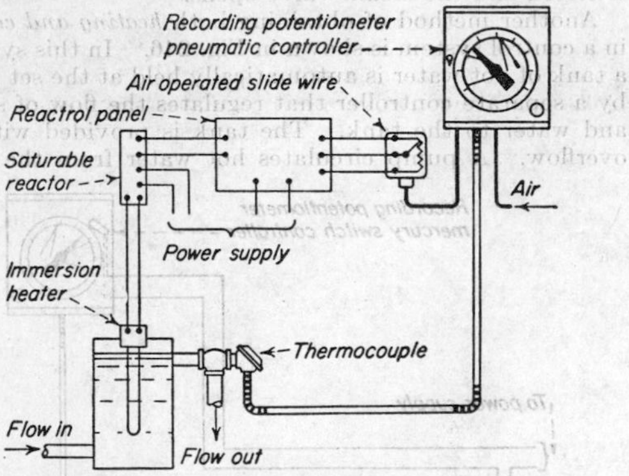

Fig. 248. Saturable-reactor-type control system.

output air pressure of the pneumatic control instrument is applied to a small air-operated slide-wire. The purpose of using the pneumatic control instrument to operate the slider is to take advantage of the adjustable proportional band and reset rate incorporated in its control system. These control characteristics are necessary for good control on continuous processes in which the through-put varies.

The potential unbalance caused by the change in position of the sliding contact is amplified electronically to regulate the amount of direct current in the saturating winding of a saturable-core reactor. The a.c. winding is connected in series with the heating element and the power supply. The d.c. saturating winding is connected to a d.c. supply with suitable regulating means such as a rheostat or electronic tubes. Large variation in the alternating current can be obtained by variation of small direct currents in the saturating winding.

In most heating operations in *closed vessels*, such as autoclaves, the upper space in the vessel is filled with steam. In such cases, a pressure controller may be used for regulating the flow of the heating medium. The steam pressure is an indirect measure of the temperature. Thus pressure controllers may be used in place of temperature controllers and have the advantages of lower cost, simplicity of operation and installation, and often better control because of less lag. However, when pressure control is used, it is necessary to vent entrapped air

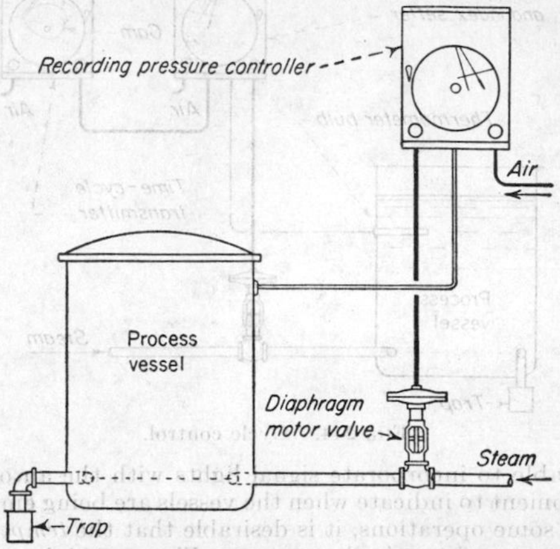

Fig. 249. Simple pressure-control system.

or other gases released during the heating-up period. Pressure control in lieu of temperature control may not be satisfactory for processes requiring the introduction of inert or other gases.

Figure 249 shows a pressure controller applied to a closed vessel. If direct temperature control is required, because of the presence of inert gases or other reasons, the system illustrated in Fig. 250 can be used. However, if either temperature or pressure is controlled, the uncon-

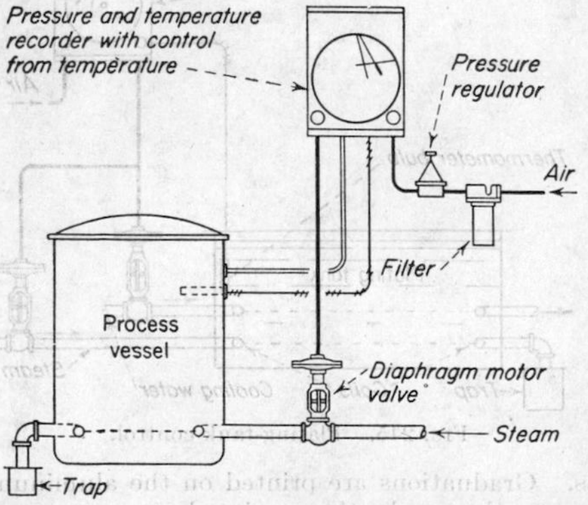

Fig. 250. Pressure- or temperature-control system.

trolled temperature or pressure can be recorded on the same chart where desired.

Suppose that the upper space in a closed vessel is filled with steam and that a pressure recorder connected to it reads 100 lb./sq. in. If a thermometer bulb partly filled with water and sealed is inserted in the same space and is connected to read on the same chart, it will also read 100 lb./sq. in. This combination has been used to detect the presence of gases in closed vessels, such as paper digesters, by noting whether the pressure pen records the same value as the temperature pen. When other vapors are present, as in distillation columns, the thermometer bulb can be partly filled with the condensed vapors, and the difference between the temperature and pressure pens will then be a measure of the composition of the vapor in the column compared with the vapor in the thermometer bulb.

The pressure in closed vessels can be reduced *below atmospheric pressure* by means of a steam-jet ejector. As a rule, steam-jet ejectors will not operate satisfactorily

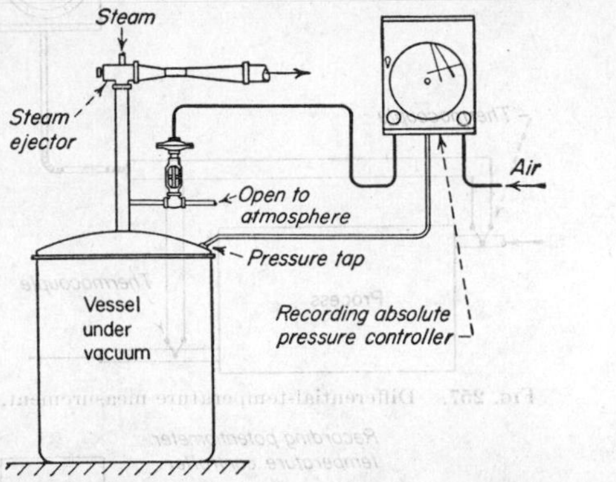

FIG. 251. Control of steam-jet vacuum system.

over a wide range of steam flow. If the steam flow to the ejector is set at its optimum value, good control can be obtained by regulating a valve opening to atmosphere. Such an installation is illustrated in Fig. 251. In low-pressure operations, it is often necessary to use absolute pressure controllers to obtain consistent results. Absolute pressure controllers compensate for variations in atmospheric pressure.

Pneumatic transmission of flow, temperature, pressure, and liquid-level measurements from process locations to a central control board was originally developed to avoid the use of electricity for this purpose in plant areas where explosive atmospheres are possible. A further advantage

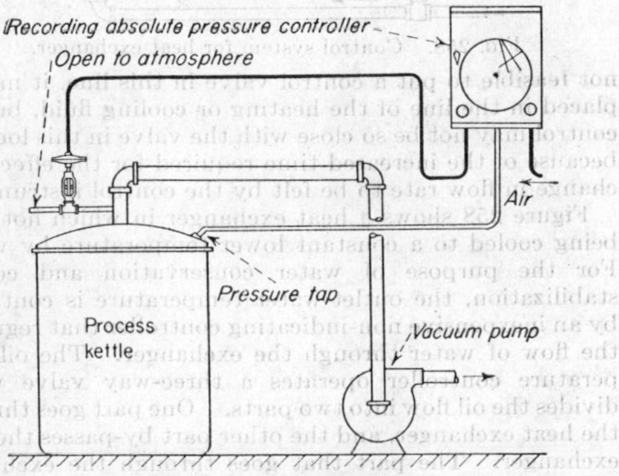

FIG. 252. Vacuum control.

is the complete isolation from the control board of all process piping containing high pressures or hazardous fluids. A pressure-control system using pneumatic transmission is shown in Fig. 253. The indicating pressure transmitter is located near the high-pressure line on the processing unit and transmits the pressure reading to the recording control instrument, which operates the control valve in the pipe line. To reduce lag, a transmitter equipped with control is often used.

Automatic control of flow and liquid level represents the largest field of application for *differential pressure controllers* in the process industries. Figure 254 is a schematic diagram of a typical flow-control installation in which the control valve is located in the flow line. A differential mercury manometer connected across the orifice transmits the movement of its mercury electrically to a recording control instrument, which in turn operates the valve in the flow line to maintain a constant differential pressure across the orifice and consequently a uniform rate of flow in the pipe.

Often it may be desirable to control automatically *one rate of flow in direct proportion to another rate of flow.*

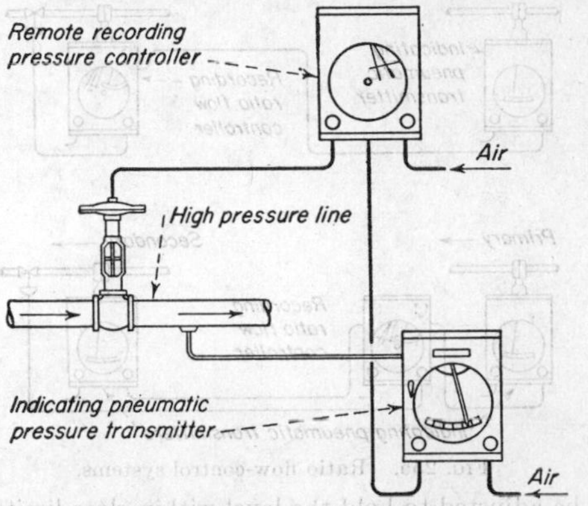

FIG. 253. Pneumatic transmission for remote control.

A schematic diagram showing how this can be accomplished is given in Fig. 255. In the upper diagram, a meter body or manometer connected across the orifice in the primary flow line operates a pneumatic mechanism that develops an air pressure in direct proportion to the differential pressure across the orifice. This air pressure is transmitted to the control instrument connected across the orifice in the secondary flow line, where, by means of a bellows-operated mechanism, it moves the control index. Thus the set point of the controller regulating the secondary flow is moved in proportion to the primary flow. The

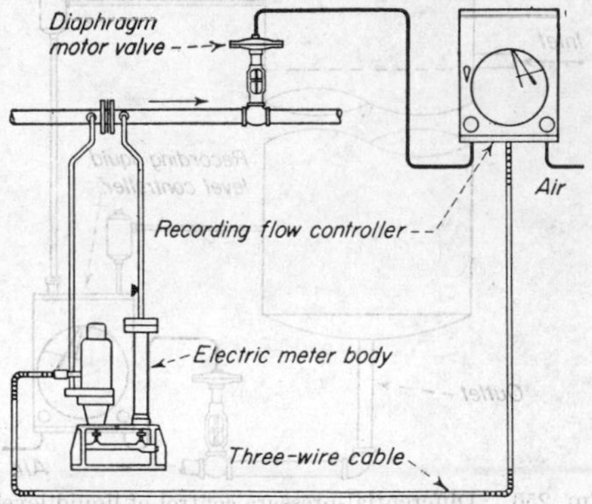

FIG. 254. Flow-control system.

mechanism in the secondary-flow controller is adjustable so that the ratio of the two flows can be varied. Maximum and minimum limits can also be set for the secondary flow rate.

The lower diagram (Fig. 255) shows how pneumatic transmission can be used in a *ratio flow-control system* to permit locating the recording controller remotely with respect to the primary and secondary flow lines. It is possible by using equipment similar to that shown in Fig. 255 to reset the control index of a flow controller from variations in liquid level or to reset the control index of a pressure controller from variations in temperature or other combinations that may be desirable.

Differential-pressure recorders also are used in connection with continuous filters, screens, and heat exchangers to reveal undesirable operating conditions such as the accumulation of scale or foreign material that affect the efficiency of the equipment.

Figure 256 illustrates the application of a differential-pressure control to maintain the *level in a closed vessel under pressure* automatically. Instruments of this type

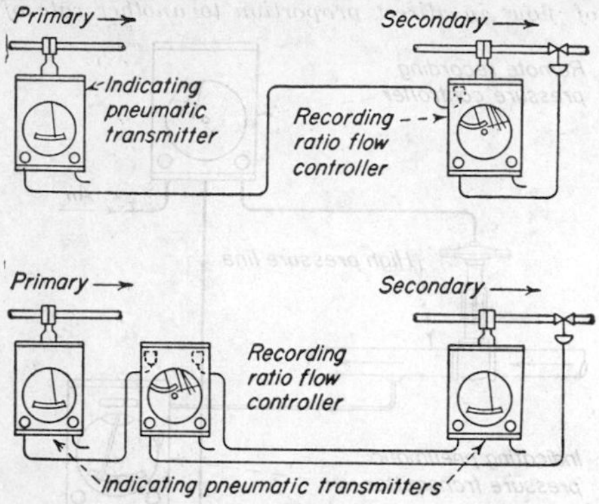

FIG. 255. Ratio flow-control systems.

can be adjusted to hold the level within close limits, but this is not always desirable. For example, the level in an accumulator tank is not important until the tank is quite full or almost empty. For such applications, the controller can be adjusted to permit the level to change between wide limits; and, where these limits are reached, the discharge valve will move to make large changes in the rate of discharge to prevent the tank from filling or becoming empty.

On some occasions it is desirable to *record or control the difference in temperature between two points* in a process or

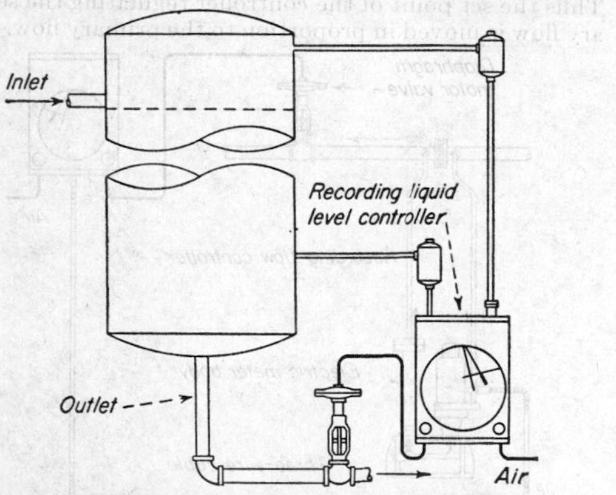

FIG. 256. Differential-pressure control of liquid level.

to control one temperature at a fixed difference with respect to another. This can be accomplished by using two thermocouples connected differentially to a potentiometer or by using two resistance thermometers in a bridge circuit. These are subject to error due to non-linearity of the primary element. The temperature difference across heat exchangers, compressors, centrifugal pumps, or between the heating jacket and the column on adiabatic laboratory stills are a few applications for this arrangement. A diagram of such an application is shown in Fig. 257.

Heat exchangers are used to heat or cool, with heat conservation as a possible dividend. The temperature of the fluid being heated or cooled can be controlled automatically by regulation of the rate of flow of the fluid through the exchanger by means of a temperature controller responsive to the outlet temperature. If it is

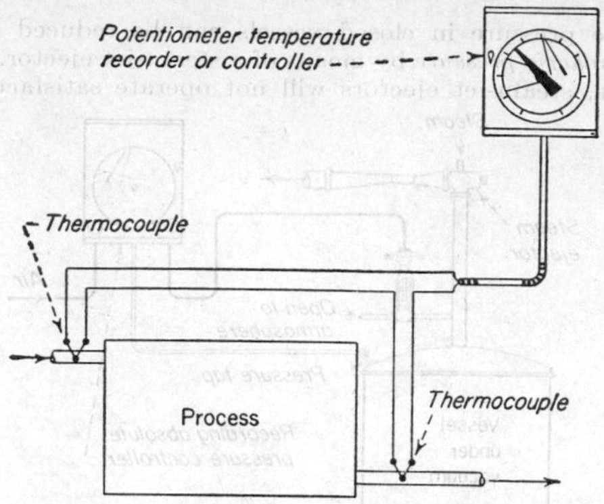

FIG. 257. Differential-temperature measurement.

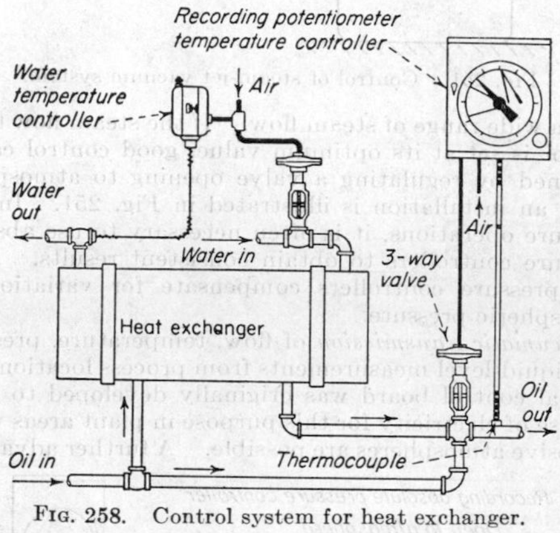

FIG. 258. Control system for heat exchanger.

not feasible to put a control valve in this line, it may be placed in the line of the heating or cooling fluid, but the control may not be so close with the valve in this location because of the increased time required for the effect of a change in flow rate to be felt by the control instrument.

Figure 258 shows a heat exchanger in which hot oil is being cooled to a constant lower temperature by water. For the purpose of water conservation and control stabilization, the outlet-water temperature is controlled by an inexpensive non-indicating controller that regulates the flow of water through the exchanger. The oil-temperature controller operates a three-way valve which divides the oil flow into two parts. One part goes through the heat exchanger, and the other part by-passes the heat exchanger. The part that goes through the exchanger is cooled below the desired final temperature. Upon leaving the exchanger, the cooled oil is mixed with the hot oil by-passed around the exchanger. Thus the temperature controller blends hot oil and cold oil to obtain oil at the desired temperature. In this system, a change in valve position is felt very quickly by the temperature controller, and only a controller that will respond quickly should be used.

In some cases, a liquid is heated by steam in a heat exchanger. If the exchanger has excess capacity, a pressure controller can be used to regulate the flow of steam, and the final temperature of the liquid also can be recorded on the same chart.

Dowtherm vapor is used as a high-temperature heating medium in many of the process industries. Figure 259 is a diagram of an automatic control system applied to an oil-fired Dowtherm vaporizer. A pressure controller responsive to the outlet pressure of the vaporizer regulates the flow of oil and air to the burner. A photoelectric flame detector and pressure switches in the air and oil lines are safety devices to open the circuit of the solenoid valve to shut off the oil supply and sound an alarm when the air pressure is too low, when the oil pressure is too low, or if the flame goes out. On electrically heated Dowtherm vaporizers, the pressure con-

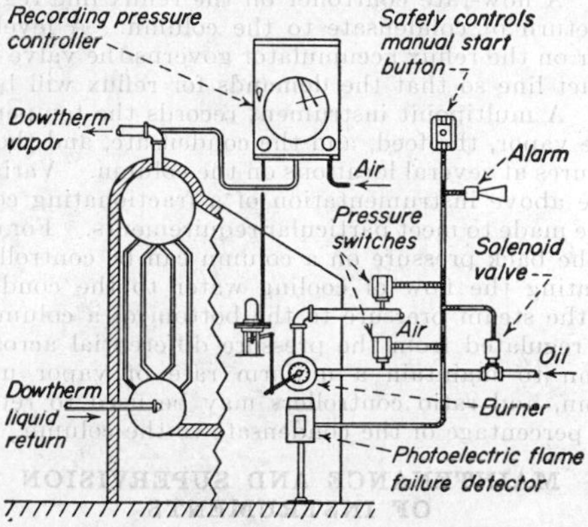

FIG. 259. Dowtherm vaporizer control.

troller can be used to operate a switch to open and close the circuit to the heating elements or to operate the slider on a potential divider, as shown previously for the control of electric heating elements by means of saturable reactors.

Kilns used for drying may be automatically controlled in response to the temperature of the exit material. The material in kilns is usually in the form of small particles so that a stationary thermocouple can be placed in the stream of material leaving the kiln. Care should be taken so that the thermocouple will respond to changes in the temperature of the material and will not be affected by air currents.

Figure 260 is a *kiln-control system* in which the oil burner is regulated from an instrument connected to a

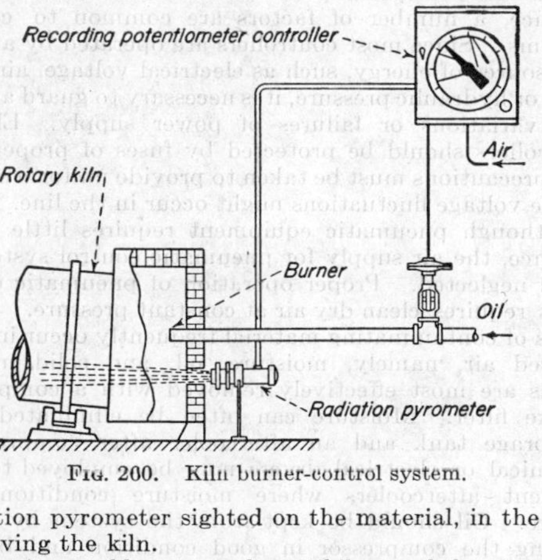

FIG. 200. Kiln burner-control system.

radiation pyrometer sighted on the material, in the kiln, or leaving the kiln.

Dust is a problem around most kilns, and the air-cooled fitting not only cools the radiation head but prevents dust from depositing on the lens. The location of the radiation pyrometer should be selected so that the line of sight does not include any part of the flame.

When the *rate of drying* is to be controlled, wet- and dry-bulb measurements of the air prior to its contact with the material are desirable. If the wet and dry bulbs are located in the air stream leaving the materials, a measure of the moisture from the material is obtained. The rate of drying may be accelerated by introduction of fresh air and retarded by introduction of water vapor. It should be borne in mind that, when control equipment is applied to driers, moisture removed is carried away in the exit air and that the distribution of air in the drier is important

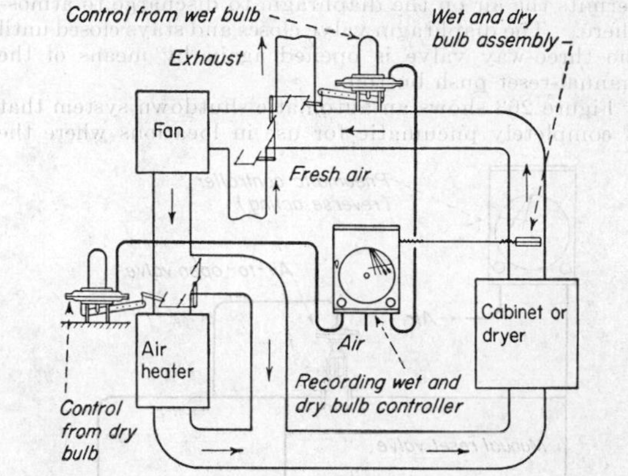

FIG. 261. Recirculating-dryer control system.

from the standpoint of uniform drying. The air velocity across the wet bulb should be at least 15 ft./sec. to ensure sufficient evaporation.

Direct measurement of relative humidity by means of hair elements is restricted to temperatures below 40°C., and they should be used only in clean atmospheres, because grease and dust close the pores of the hair, and it will not respond rapidly to moisture changes.

Thermocouples, particularly pencil-type couples, can be used satisfactorily for wet-bulb measurements. The

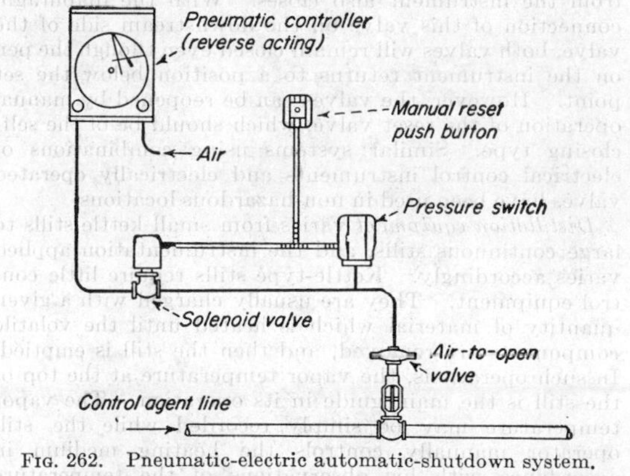

FIG. 262. Pneumatic-electric automatic-shutdown system.

wet bulb should be sufficiently long to preclude erroneous readings due to heat conduction to the hot junction of the thermocouple. When relative humidities above 90 per cent are involved, careful consideration should be given to the sensitivity and accuracy of the instruments, because the wet- and dry-bulb temperatures approach the same value at 100 per cent.

Figure 261 shows a wet- and dry-bulb control installation on a *recirculating dryer*. The sensitive elements are located in the air leaving the dryer. The dry bulb controls dampers either to direct air through the heater or to by-pass it around the heater. The wet bulb controls admission of fresh air and release of moisture-laden air by means of a three-way damper.

At times it is desirable to have a valve close and stay closed after a process has reached a predetermined condition. (*automatic shutdown systems*) In Fig. 262, a pneumatic controller is shown with its control index set at a predetermined value. When the value is reached, the output air pressure of the controller drops below 15 lb./sq. in., and the pressure switch opens the circuit to a three-way valve that shuts off the air to the valve and permits the air on the diaphragm to discharge to atmosphere. The diaphragm valve closes and stays closed until the three-way valve is opened again by means of the manual-reset push button.

Figure 263 shows an automatic shutdown system that is completely pneumatic for use in locations where the use of electricity is considered hazardous. In this system, when the pen reaches the set point, the output air pressure from the instrument decreases. The controlled valve closes, and a small diaphragm valve in the air line from the instrument also closes. With the diaphragm connection of this valve on the downstream side of the valve, both valves will remain closed even though the pen on the instrument returns to a position below the set point. However, the valves can be reopened by manual operation of the reset valve, which should be of the self-closing type. Similar systems using combinations of electrical control instruments and electrically operated valves have been used in non-hazardous locations.

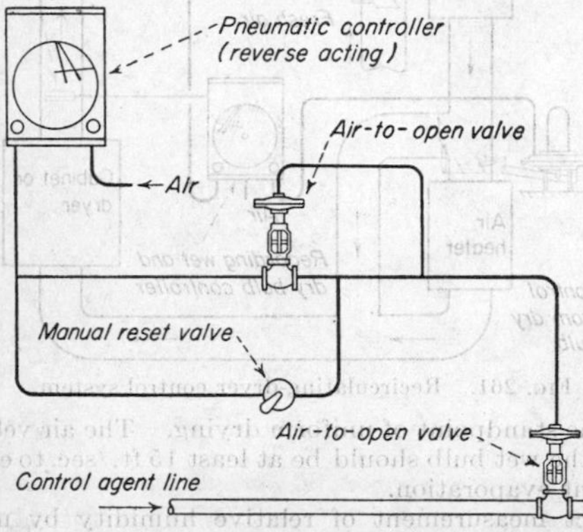

Fig. 263. Pneumatic shutdown system.

Distillation equipment varies from small kettle stills to large continuous stills, and the instrumentation applied varies accordingly. Kettle-type stills require little control equipment. They are usually charged with a given quantity of material which is heated until the volatile components are removed, and then the still is emptied. In such operations, the vapor temperature at the top of the still is the main guide in its operation. The vapor temperature may be simply recorded while the still operator manually controls the heating medium in accordance with his observations of the temperature record, or the temperature recorder may incorporate a control system to control the heating medium automatically or to shut off the heating medium and open the discharge valve when a predetermined temperature is reached. In some continuous still operations, the feed is sufficiently uniform that efficient operation can be obtained with a flow-rate controller on the feed and a flow-rate controller on the steam supply, a temperature recorder on the vapor line, and a liquid-level controller regulating a valve in the discharge from the bottom.

Instrumentation of a fractionating column with a separate reboiler is outlined as follows. A flow-rate controller admits the feed to the column at a uniform rate. The feed could also be brought through a controlled heat exchanger so that it would enter the column at a constant temperature. A pressure controller regulates the release of vapor from the column to maintain a uniform back pressure. A temperature controller resets the control index of a flow controller on the steam line to the reboiler to maintain a uniform temperature of the vapor leaving the reboiler. This method of controlling the steam eliminates variations due to fluctuations in steam pressure. A flow-rate controller on the reflux line regulates the return of condensate to the column. A level controller on the reflux accumulator governs the valve in the product line so that the demands for reflux will be met first. A multipoint instrument records the temperature of the vapor, the feed, and the condensate, and the temperatures at several locations on the column. Variations of the above instrumentation of a fractionating column can be made to meet particular requirements. For example, the back pressure on a column can be controlled by regulating the flow of cooling water to the condenser. Also the steam pressure to the bottom of a column has been regulated from the pressure differential across the column to maintain a uniform rate of vapor up the column, and ratio controllers may be used to return a fixed percentage of the condensate to the column.

MAINTENANCE AND SUPERVISION OF INSTRUMENTS

Since instruments are such an integral part of modern processes, it is essential that they be kept in as perfect a condition as possible. Periodic inspection and careful intelligent maintenance are a prerequisite to satisfactory operation of any control system. The first steps leading to maintenance, however, go back to the design of process and equipment, selection of instruments, and installation. Even with the highest quality of engineering, though, it usually is not possible or feasible for all the factors and conditions surrounding process instrumentation to be favorable. In fact, many installations are necessarily made under conditions far from ideal. Although dirt, wear, and corrosion are detrimental to good control, it still is not always possible to locate instruments in clean corrosion-free atmospheres.

Although each type of measuring and controlling instrument has special problems of installation and maintenance, a number of factors are common to control systems. Since most controllers are operated by an outside source of energy, such as electrical voltage, air pressure, or hydraulic pressure, it is necessary to guard against any variations or failures of power supply. Electric controllers should be protected by fuses of proper size, and precautions must be taken to provide uniform supply where voltage fluctuations might occur in the line.

Although pneumatic equipment requires little maintenance, the air supply for pneumatic control systems is often neglected. Proper operation of pneumatic equipment requires clean dry air at constant pressure. Three types of contaminating material frequently occur in compressed air, namely, moisture, oil, and solid matter. Solids are most effectively removed with a compressor intake filter. Moisture can often be eliminated with a storage tank and an aftercooler after compression. Chemical or electrical dryers may be employed to supplement aftercoolers where moisture conditions are severe. Oil should be kept from the system by maintaining the compressor in good condition and by not overloading. Individual filters should be located at each controller, and an individual regulator should provide uniform air pressures to the instrument.

Instrument Service Departments

Regardless of whether a plant is large or small, responsibility for maintenance of automatic control systems should be centralized at one point in the organization. Only by so doing is it possible to provide regular maintenance and to coordinate all instrument information so that it may be fully utilized.

The main function of an instrument service department is to keep all instruments in good working order. This includes greasing and servicing of control valves, cleaning and oiling instrument mechanisms, inspection of auxiliary equipment, and checking calibrations. Most important are the checks on instrument accuracy, especially on flowmeters measuring critical material balances. New controllers must be adjusted to fit process characteristics. All adjustments of instrument control action should be made by an instrument man, and accurate records should be kept for each change made.

From the standpoint of the process, one of the important functions of an instrument man is to assist in diagnosing process difficulties. Since instruments are part of the process, any process unbalance shows up on the instruments. Often the cooperation of the instrument man and the operator results in rapid determination of such process troubles.

Usually it is considered good practice for the instrument personnel to change charts and ink pens daily. Prompt intelligent inspection of chart records leads to rapid detection of faulty instrument operation. Full records should be kept on all instrument equipment. Often it is necessary to know the date of an installation, the results of a calibration check, or whether any renewal of parts has been made. Instrument difficulties can sometimes be traced to omission of cleaning a filter, inspection of a valve packing, or some similar detail. Usually all information concerning an individual instrument is entered on a form for the purpose of providing a complete history.

Two functions of an instrument department that are often overlooked but are extremely worth while are the improvement of control operation and the development of new instruments. Plant control methods can always be improved by a wide-awake, alert instrument man. Usually, for a given process and a given set of control equipment, there is one best control arrangement for a particular unit. Depending on conditions, various methods are better on different services. It is the instrument engineer's responsibility to improve process control whenever possible.

Although a wide variety of instruments are already available, special process requirements continually arise that lead to the development of new instruments or improvements to old ones. Not in any way supplanting a research or development department function, the instrument service department is a valuable adjunct to research by virtue of its daily contact with everyday instrument problems. Whether or not an instrument department is made up of few or many men, its success will be measured partly, at least, by its success or failure to accomplish these various functions. Much of the progress made during the past 10 years is the result of close cooperation of the user and the manufacturer of instruments.

CLASSIFIED LIST OF AUTOMATIC-CONTROL TERMS*

100 Automatic controllers and control systems
- 101 Automatic controller
- 102 Self-operated controller
- 103 Relay-operated controller
- 104 Desired value
- 105 Set point
- 106 Control point

200 Basic characteristics
- 201 Capacity
- 202 Capacitance
- 203 Resistance
- 204 Dead time

300 Processes, their elements and characteristics
- 301 Process
- 302 Self-regulation
- 303 Controlled variable
- 304 Controlled medium
- 305 Manipulated variable
- 306 Control agent

400 Characteristics of automatic control
- 401 Error
- 402 Deviation
- 403 Offset
- 404 Corrective action
- 405 Cycling

500 Types of automatic-controller action
- 501 Positioning action
- 501*a* Two-position action
- 501*aa* Two-position differential-gap action
- 501*ab* Two-position single-point action
- 501*b* Multiposition action
- 501*c* Proportional-position action
- 501*d* Average-position action
- 502 Integral action
- 502*a* Floating action
- 502*aa* Single-speed floating action
- 502*ab* Multispeed floating action
- 502*ac* Proportional-speed floating action
- 502*ad* Floating average-position action
- 503 Derivative action
- 503*a* Rate action
- 504 Multiple action
- 504*a* Proportional-plus-floating action
- 504*aa* Proportional-plus-reset action
- 504*b* Proportional-plus-derivative action
- 504*c* Proportional-plus-floating-plus-derivative action
- 504*ca* Proportional-plus-reset-plus-rate action

600 Adjustments of automatic-controller action
- 601 Neutral zone
- 602 Differential gap
- 603 Proportional band
- 604 Floating speed
- 605 Floating rate
- 606 Reset rate
- 607 Rate time

700 Elements and characteristics of automatic controllers
- 701 Measuring means
- 701*a* Primary element
- 701*b* Self-operated measuring means
- 701*c* Relay-operated measuring means
- 702 Controlling means
- 702*a* Power unit
- 702*b* Final control element
- 702*c* Self-operated controlling means
- 702*d* Relay-operated controlling means

* Defined by the Terminology Committee of the Industrial Instruments and Regulators Division, Am. Soc. Mech. Engrs. Definitions appear in alphabetical order in the following glossary.

GLOSSARY OF AUTOMATIC-CONTROL TERMS*

A.S.M.E. standard terms are followed by their classification number.

Anticipatory control. See Rate action.

Automatic control. Automatic maintenance of balanced conditions within a process.

Automatic controller. (101) A mechanism that measures the value of a variable quantity or condition and operates to correct or limit deviation of this measured value from a selected reference. It includes both the measuring means and the controlling means. (Automatic regulator is a synonymous term.)

Automatic regulator. See Automatic controller.

* Automatic Control Terms, prepared by Am. Soc. Mech. Engrs. Industrial Instruments and Regulators Division, Committee on Terminology, *Mech. Eng.*, February, 1946. Automatic Control Terminology, *Chem. & Met. Eng.*, May, 1943, p. 125. Eckman, "Principles of Industrial Process Control," Wiley, New York, 1945. Smith, "Automatic Control Engineering," McGraw-Hill, New York, 1944.

Automatic reset. See Proportional-speed floating action.
Average-position action. (501*d*) Action in which there is a predetermined relation between value of the controlled variable and the time-average position of a final control element which is moved periodically from one of two fixed positions to the other. This controller action is similar to two-position action in which the percentage "time on" of the final control element is dependent upon the value of the controlled variable. The percentage "time on" may have either a fixed or an infinite number of values to correspond to any one of the other positioning-controller actions defined previously.
Booster response. See Rate action.
Capacitance. (202) The change in quantity contained per unit of change in some reference variable. It is measured in units of quantity, divided by the reference variable. The energy or material being contained and the reference variable determine the type of capacitance. Process capacitance may involve different quantities and reference variables, and several types may exist together in one process. The volume capacitance of an open tank with respect to head is the change of volume of stored liquid per unit change of head, which is equivalent in value to the area of the liquid surface. It should be noted that, if the shape of the tank causes the liquid surface area to vary with change of head, the capacitance will likewise vary with head. The weight capacitance of a gas-filled tank with respect to pressure is the change of weight of stored gas per unit change of pressure.
Capacity. (201) A measure of the maximum quantity of energy or material that can be stored. It is measured in units of quantity. The volume capacity of an open tank, for example, is the maximum volume of liquid it will hold without overflowing. The weight capacity of a compressed-air tank is the maximum weight of air it will hold without exceeding safe pressure.
Capacity lag. A process lag (*q.v.*) which results in a retardation of the detection of a change in value of a variable due to energy or material capacity.
Constant-speed floating action. See Single-speed floating action.
Control. See Automatic controller.
Control agent. (306) That process, energy, or material of which the manipulated variable is a condition or characteristic. See example for Manipulated variable.
Control band. See Proportional band.
Control circuit. See Automatic control.
Control effect. See Corrective action.
Control element. See Final control element.
Control-index setting. See Set point.
Control instrument. See Automatic controller.
Control point. (106) The value of the controlled variable which, at any instant, the automatic controller operates to maintain.
Control-point setting. See Set point.
Control response. See Corrective action.
Control setting. See Set point.
Control system. See Automatic control system.
Controlled medium. (304) That process, energy, or material in which a variable is controlled. See example for Controlled variable.
Controlled variable. (303) That quantity or condition which is measured and controlled. The controlled variable is a condition or characteristic of the controlled medium. For example, where temperature of water in a tank is automatically controlled, the controlled variable is temperature, and the controlled medium is water.
Controller adjustment. Manually adjustable characteristic of an automatic controller for varying relationship between controlled variable and controller response.
Controller function. See Corrective action.
Controller lag. Retardation or delay in response of final control element to change in controlled variable at the controller.
Controller response. See Corrective action.
Controlling means. (702) Consists of those elements of an automatic controller which are involved in producing a corrective action.
Conversion response. See Proportional-position action.
Corrective action. (404) Predetermined variation of the manipulated variable initiated by a deviation.
Corresponding control. See Positioning action.
Cycling. (405) A periodic change of the controlled variable from one value to another. ("Oscillation" is a synonymous term.) There are three types of cycling, *i.e.*, cycling in which the amplitude gradually decreases, cycling in which the amplitude is constant, and cycling in which the amplitude gradually increases.
Damping. Effect due to whatever cause, tending to hinder or prevent oscillation.
Damping control. See Rate action.
Dead neutral. See Neutral zone.
Dead-period lag. See Dead time.
Dead spot. See Differential gap or Neutral zone.
Dead time. (204) Any definite delay period between two related actions. It is measured in units of time.
Dead zone. See Differential gap or Neutral zone.
Deflection. See Deviation.
Delay. See Dead time.
Demand change. See Load change.
Departure. See Deviation.
Derivative action. (503) Action in which there is a predetermined relation between a derivative function of the controlled variable and position of a final control element.
Desired condition. See Desired value.
Desired value. (104) The value of the controlled variable which it is desired to maintain.
Deviation. (402) The difference between the instantaneous value of the controlled variable and the value of the controlled variable corresponding with the set point.
Differential gap. (602) Applying to two-position controller action, this is the smallest range of values through which the controlled variable must pass in order to move the final control element in succession to both its fixed positions. Differential gap is commonly expressed in percentage of controller-scale range.
Differentiating control. See Rate action.
Direct-operated controller. See Self-operated controller.
Displacement. See Deviation and offset.
Distance-velocity lag. See Transportation lag and Dead time.
Drift. See Offset.
Drift compensation. See Proportional-speed floating action.
Droop. See Offset.
Droop correction. See Proportional-speed floating action.
Dynamic error. Difference between true value of a quantity or condition changing with time and the value indicated by a measuring means.
Elastic follow-up. See floating action.
Error. (401) The difference between the instantaneous value and the desired value of the controlled variable.
Final control element. (702*b*) That portion of the controlling means which directly changes the value of the manipulated variable.
Finite time lag. See Dead time.
Floating action. (502*a*) Action in which there is a predetermined relation between value of the controlled variable and the rate of motion of a final control element. A neutral zone, in which no motion of the final control element occurs, is often employed in floating-controller action.
Floating average-position action. (502*ad*) Action in which there is a predetermined relation between value of the controlled variable and rate of change of the time-average position of a final control element which is moved periodically from one of two fixed position to the other. This controller action is similar to two-position action in which the percentage "time on" of the final control element is gradually changed at a rate dependent upon the value of the controlled variable. The rate of change of the percentage "time on" may have either a fixed or an infinite number of values to correspond to any one of the other floating controller actions defined previously.
Floating component. See Floating action.
Floating rate. (605) Applying to proportional-speed floating controller action, this is the rate of motion of the final control element corresponding to a specified deviation. Floating rate is commonly expressed in percentage of full-range motion per minute per per cent deviation.
Floating response. See Floating action.
Floating sensitivity. See Floating rate.
Floating speed. (604) Applying to single or multispeed floating controller action, this is the rate of motion of the final control element. Floating speed is commonly expressed in percentage of full-range motion per minute.
Floating time. See Reset rate.
Flow lag. See Dead time.
Follow-up. Device that is used with relay elements in controllers to establish a definite control response for a given change in variable by setting up a counterresponse.

High-low control. See Two-position control.
Hunting. Cycling.
Inactive neutral. See Differential gap.
Index setting. See Set point.
Instrument. Device for measuring or measuring and controlling the values of a process variable.
Integral action. (502) Action in which there is a predetermined relation between an integral function of the controlled variable and position of a final control element.
Integral of deviation. See Integral action.
Integral response. See Proportional-speed floating action.
Integrating control. See Integral action.
Interval. See Dead time.
Inventory. See Capacity.
Inverse minutes. See Reset rate.
Kicker. See Rate action.
Lag. Retardation or delay of one physical condition with respect to some other condition to which it is closely related.
Lapse. See Dead time.
Load change. Change in process conditions which requires a change in the average value of manipulated variable to maintain the controlled variable at the desired value.
Load error. See Offset.
Loss of control point. See Offset.
Manipulated variable. (305) That quantity or condition which is varied by the automatic controller so as to affect the value of the controlled variable. The manipulated variable is a condition or characteristic of the control agent. For example, where a final control element changes the rate of fuel-gas flow to a burner, the manipulated variable is rate of flow, and the control agent is fuel gas.
Master controller. In a metered control system, that automatic controller which adjusts the control point of another automatic controller.
Measured variable. Quantity or condition the value of which is automatically ascertained by an instrument or an automatic controller.
Measurement. Act of ascertaining the value of a quantity or condition.
Measuring element. See Measuring means and Primary element.
Measuring lag. Retardation or delay in response of measuring means of an instrument or an automatic controller to changes in measured variable.
Measuring means. (701) Consists of those elements of an automatic controller which are involved in ascertaining and communicating to the controlling means either the value of the controlled variable, the error, or the deviation.
Measuring system. See Measuring means.
Metered control system. An automatic control system in which the automatic controller operates a second automatic controller for adjusting the value of the manipulated variable.
Mode. Refers to type of control action such as proportional-position mode.
Modulating control. See Proportional-position action.
Multiple action. (504) Action in which two or more controller actions are combined.
Multiposition action. (501*b*) Action in which a final control element is moved to one of three or more predetermined positions, each corresponding to a definite range of values of the controlled variable.
Multispeed floating action. (502*ab*) Action in which a final control element is moved at two or more rates, each corresponding to a definite range of values of the controlled variable.
Neutral zone. (601) A predetermined range of values of the controlled variable in which no corrective action occurs. Neutral zone is commonly expressed in percentage of controller-scale range. A neutral zone is employed in some types of floating controller action.
Non-corresponding. See Floating action.
Normal. See Desired value.
Offset. (403) A sustained deviation due to an inherent characteristic of positioning-controller action. The difference existing at any time between the control point and the value of the controlled variable corresponding with the set point.
On-off action. (501*a*) Action in which a final control element is moved from one of two fixed positions to the other. ("Open and shut action" and "on-off action" are synonymous terms.)
Open-and-shut action. See Two-position action.
Oscillating. See Cycling.
Overshooting. See Cycling.
Per-rate response. See Rate action.
Per-time response. See Floating action.
Pilot-operated controller. See Relay-operated controller.
Pilot valve. Device for controlling the flow of an auxiliary fluid used to amplify the power of a controller measuring system in effecting control.
Positioning action. (501) Action in which there is a predetermined relation between value of the controlled variable and position of a final control element.
Power unit. (702*a*) A portion of the controlling means which applies power for operating the final control element.
Primary element. (701*a*) That portion of the measuring means which first either utilizes or transforms energy from the controlled medium to produce an effect in response to change in the value of the controlled variable. The effect produced by the primary element may be a change of pressure, force, position, electrical potential, or resistance.
Process. (301) A process comprises the collective functions performed in and by the equipment in which a variable is to be controlled. "Equipment," as embodied in this definition, should be understood not to include any automatic-control equipment.
Process lag. Retardation or delay in response of controlled variable at point of measurement to a change in value of manipulated variable.
Process load. Sum, taken at any instant, of the energy or material requirements of the process resulting in a specific value of manipulated variable.
Process reaction rate. Maximum rate of change of controlled variable caused by a specified, sudden change in value of manipulated variable.
Process variable. Quantity or condition associated with a process the value of which is subject to change with time.
Program controller system. See Time-cycle controller system.
Proportional band. (603) Applying to proportional-position controller action, this is the range of values of the controlled variable which corresponds to the full operating range of the final control element. Proportional band is commonly expressed in percentage of controller-scale range or, particularly in the absence of a controller scale, in units of the controlled variable.
Proportional control. See Proportional-position action.
Proportional-plus-derivative action. (504*b*) Action in which proportional-position action and derivative action are combined.
Proportional-plus-floating action. (504*a*) Action in which proportional-position action and floating action are combined.
Proportional-plus-floating-plus-derivative action. (504*c*) Action in which proportional-position action, proportional-speed floating action, and derivative action are combined.
Proportional-plus-reset action. (504*aa*) Action in which proportional-position action and proportional-speed floating action are combined.
Proportional-plus-reset-plus-rate action. (504*ca*) Action in which proportional-position action, proportional-speed floating action, and rate action are combined.
Proportional-position action. (501*c*) Action in which there is a continuous linear relation between value of the controlled variable and position of a final control element.
Proportional-speed floating action. (502*ac*) Action in which there is a continuous linear relation between value of the controlled variable and rate of motion of a final control element.
Rate action. (503*a*) Action in which there is a continuous linear relation between rate of change of the controlled variable and position of a final control element. This controller action maintains a linear relation between first derivative or rate of change of the controlled variable and position of a final control element. This identical controller action may also be considered as maintaining a linear relation between second derivative or rate of the rate of change of the controlled variable and rate of motion of the final control element.
Rate-component control. See Rate action.
Rate-of-change method. See Rate action.
Rate-of-departure component. See Rate action.
Rate of droop correction. See Reset rate.
Rate response. See Rate action.
Rate time. (607) Applying to proportional-plus-rate controller action and proportional-plus-reset-plus-controller action,

this is the time interval by which the rate action advances the effect of the proportional-position action upon the final control element. Rate time is commonly expressed in minutes. It is determined by subtracting (1) the time required for a selected motion of the final control element, due to the combined effect of proportional-position-plus-rate actions, from (2) the time required for the same motion due to the effect of proportional-position action alone, with the same rate of change of the controlled variable in both cases. In automatic controllers having proportional-plus-rate action and a rate-time adjustment, the proportional-band adjustment simultaneously affects the rate action in such a manner that the rate time remains substantially constant at its set value. Similarly, in automatic controllers having proportional-plus-reset-plus-rate action and a rate-time adjustment, the proportional band may be adjusted without affecting the set value of the rate time.

Ratio control system. An automatic control system in which value of the variable is controlled in a predetermined relation to value of another measured variable.

Recovery. Change with time of a controlled variable resulting from a sustained or temporary change in process load.

Regulation. See Offset.

Regulator controller. See Automatic controller.

Relay-operated controller. (103) A controller in which the energy transmitted through the primary element is either supplemented or amplified for operating the final control element by employing energy from another source. This type of automatic controller may have either a self-operated measuring means and a relay-operated controlling means, or a relay-operated measuring means and a self-operated controlling means, or a relay-operated measuring means and a relay-operated controlling means.

Relay-operated controlling means. (702*d*) A controlling means in which the energy transmitted from the measuring means is either supplemented or amplified for operating the final control element by employing additional energy.

Relay-operated measuring means. (701*c*) A measuring means in which the energy transmitted through the primary element is either supplemented or amplified for actuating the controlling means of an automatic controller by employing additional energy.

Reset constant. See Reset rate.

Reset control. See Proportional-speed floating action.

Reset rate. (606) Applying to proportional-plus-reset controller action and proportional-plus-reset-plus-rate controller action, this is the number of times per minute that the effect of the proportional-position action upon the final control element is repeated by the proportional-speed floating action. Reset rate is commonly expressed as a number of "repeats" per minute. It is determined by dividing (1) the travel of the final control element in 1 min. due to the effect of proportional-speed floating action by (2) the travel due to the effect of proportional-position action, with the same deviation in both cases. In automatic controllers having proportional-plus-reset action and a reset-rate adjustment, the proportional-band adjustment simultaneously affects the proportional-speed floating action in such a manner that the reset rate remains substantially constant at its set value. Similarly, in automatic controllers having proportional-plus-reset plus rate action and a reset-rate adjustment, the proportional band may be adjusted without affecting the set value of the reset rate.

Reset response. See Proportional-speed floating action.

Reset sensitivity. See Reset rate.

Reset speed. See Reset rate.

Reset time. See Reset rate.

Resistance. (203) Opposition to flow. It is measured in units of potential change required to produce unit change in flow.

Response characteristic. See Corrective action.

Response delay. See Dead time.

Second derivative. See Rate action.

Secondary controller. An automatic controller in which the control point is automatically and continuously adjusted from an external source.

Self-acting controller. See Self-operated controller.

Self-actuated controller. See Self-operated controller.

Self-operated controller. (102) A controller in which all the energy necessary to operate the final control element is derived from the controlled medium through the primary element. This type of automatic controller must have both self-operating measuring means and self-operated controlling means.

Self-operated controlling means. (702*c*) A controlling means in which all the energy necessary to operate the final control element is derived from the measuring means.

Self-operated measuring means. (701*b*) A measuring means in which all the energy necessary to actuate the controlling means of an automatic controller is derived from the controlled medium through the primary element.

Self-regulation. (302) A sustained reaction inherent in the process which assists or opposes the establishment of equilibrium.

Sensing element. See Primary element.

Sensitive element. See Primary element and Measuring means.

Sensitivity. See Proportional band.

Servo. See Power unit.

Servo-operated controller. See Relay-operated controller.

Set point. (105) The position to which the control-point-setting mechanism is set. Where the automatic controller possesses a set-point scale, the set point is the scale reading translated into units of the controlled variable. Where a setting scale is not provided, the set point is the position of the control-point-setting mechanism translated into units of the controlled variable. In some types of automatic controllers, *e.g.*, those with two-position differential gap, floating with neutral or proportional-position action, the set point is related to the position of a range of values of the controlled variable. The set point is then generally selected as the center of this range of values. The set point may be varied manually or by automatic means, such as in time-schedule or ratio control.

Set value. See Set point.

Setting. See Set point.

Single-speed floating action. (502*aa*) Action in which a final control element is moved at a single rate.

Speed of reset. See Reset rate.

Stability. State of controlled variable in which the variable does not cycle, or cycles with decreasing amplitude.

Static error. Difference between true value of a quantity or condition not changing with time, and the value indicated by a measuring means.

Supply change. See Load change.

Swinging. See Cycling.

Three-position action. See Multiposition action.

Throttling action. See Proportional-position action.

Throttling band. See Proportional band.

Throttling range. See Proportional band.

Time response. See Rate action.

Time-cycle control system. An automatic control system in which value of the controlled variable is controlled in a predetermined relation to time.

Transfer lag. Retardation, not delay, in response of controlled variable caused by existence of distributed capacity or two or more separated capacities in a controlled system.

Transmission lag. Retardation or delay caused in transmitting a measurement of variable from primary element to controller. Transportation lag or distance-velocity lag is a direct delay or postponement of the beginning of a change in a given process variable, following an instantaneous change in a related variable that affects it at some other point in the process. See Dead time.

Two-position action. (501*a*) Action in which a final control element is moved from one of two fixed positions to the other. ("Open and shut action" and "on-off action" are synonymous terms.)

Two-position differential-gap action. (501*aa*) Action in which a final control element is moved from one of two fixed positions to the other when the controlled variable reaches a predetermined value from one direction and subsequently is moved to the other position only after the variable has passed in the opposite direction through a range of values to a second predetermined value.

Two-position single-point action. (501*ab*) Action in which a final control element is moved from one of two fixed positions to the other at a single value of controlled variable. The differential gap of this type of two-position action is zero. Such controller action may also be considered as proportional-position action in which the proportional band is zero, or as floating action with zero neutral zone and with infinite floating speed.

Upset. See Load change.

Variable. See Process variable.

Variable-speed reset. See Proportional-speed floating action.

Velocity-distance lag. See Transportation lag and Dead time.

SECTION 20

MOVEMENT AND STORAGE OF MATERIALS

BY

Wilbur G. Hudson, M.E., Consulting Engineer; Member, American Society of Mechanical Engineers; formerly Chief Engineer, Link Belt Co. (Material-handling Equipment)

Fred A. Miller, Materials Handling Consultant, E. I. duPont de Nemours & Co. (Industrial Trucks, Tractors, and Trailers)

F. L. Lucker, M.E., Professional Engineer of Pennsylvania, formerly, Special Representative to Chemical Industry, Ingersoll-Rand Co. (Movement of Liquids and Gases, Compression of Gas)

Reed W. Hyde, Metallurgical Engineer; President, Sintering Machinery Corp. Professional Engineer of New Jersey; Member, American Institute Mining and Metallurgical Engineers, Mining and Metallurgical Society of America. (Feeders and Feeding Mechanisms)

R. W. Lahey, B.S., American Cyanamid Co.; Chairman of Metal Packages Committee of Manufacturing Chemists' Association; Member of Committees D-6 and D-10, American Society for Testing Materials; Member of the Label and Precautionary Information Committee, Manufacturing Chemists' Association; formerly, Consultant on Packaging to the War Production Board; formerly, Consultant on Packaging to the Tropical Deterioration Committee, Office of Scientific Research and Development. (Packaging Equipment)

CONTENTS

MATERIAL-HANDLING EQUIPMENT

PAGE

General Factors Affecting the Selection of Conveying and Elevating Equipment 1343
Conveyor Drives 1343
Motor Types 1343
  Electrical Characteristics 1343
  Motor Protection 1344
Screw Conveyors 1344
  Lumpy Material 1345
  Power Requirements 1345
  Applications 1345
Flight Conveyors 1347
  Power Determination 1347
  Applications 1348
Apron Conveyors 1348
  Power Requirements 1348
Bucket Elevators 1348
  Applications 1349
Skip Hoists 1349
  Speeds 1350
  Power 1350
  Applications 1350
Continuous-flow Conveyors 1351
Belt Conveyors 1354
Carriers 1357
Pneumatic Conveyors 1360
Hydraulic Conveyors 1360
Vibrating Feeders, Screens, Conveyors 1361
Chutes 1361
Industrial Trucks, Tractors, and Trailers 1362
  Stevedore Trucks 1363
  Factory Trucks 1363
  Stackers 1363
  Trailers and Tractors 1363
  Skids 1363
  Jack-lift Trucks 1364
  Powered Low-lift Platform Trucks 1366
  Powered High-lift Platform Trucks 1366
  Fork Truck 1366
  Truck Capacity 1366
  Operating and Safety Features 1367
  Pallets 1367
  Pallet and Load Dimensions 1368
  Operating Aisles 1369
  Hand Pallet Trucks 1369
  Building Requirements 1369

PAGE

  Loading-platform Heights 1369
  Special Applications 1369
  Job Analysis 1369

FEEDERS AND FEEDING MECHANISMS

Feeders 1370
Undercut-gate Feeder 1370
Lifting-gate Feeder 1370
Screw-conveyor Feeder 1370
Roll Feeder 1370
Rotating Vane or Pocket Feeder 1371
Revolving Plate or Table Feeder 1371
Apron-conveyor Feeder 1371
Plunger Feeder 1371
Reciprocating-plate Feeder 1372
Shaker Feeder 1372
Vibrating Feeder 1372
Ross Feeder 1372
Flight-conveyor Feeder 1372
Constant-weight Feeders 1373
Reagent Feeders 1374
Feeding Gases 1375
Proportioning 1375

STORAGE OF MATERIALS

Solids 1376
Liquids 1378
Gases 1380

PACKAGING AND FILLING EQUIPMENT

Definitions 1381
Purpose 1381
Storage in Bulk or in Containers 1382
Dry Material 1382
Storage-bin Construction for Dry Materials 1382
Weighing 1382
Weighing and Bagging Equipment 1382
  Auger Packers 1384
  Hand Weighing 1385
  Valve Bag Packers 1386
Bag-closing Equipment 1388
  Bag Holders 1389
  Liner Insertion 1390
Filling and Weighing Equipment for Packing Dry Materials into Drums, Barrels, and Kegs 1392

	Page
Liner Insertion	1394
Agitators	1394
Filling and Weighing Equipment for Packing Liquids into Drums or Tight Barrels	1395
Packaging Equipment for Shelf Packages	1397
Bottle-cleaning Equipment	1397
Air Cleaner for Bottles	1397
Filling Equipment for Liquids	1397
Bottle and Jar Closing	1399
Filling Equipment for Dry Products	1400
Folding Carton, Forming, Sealing, Lining, and Cartoning Equipment	1404
Package Wrapping	1406
Small-bag Filling and Closing	1410
Collapsible-tube Filling and Closing	1411
Labeling Machines	1412

PUMPING OF LIQUIDS

Pumping Equipment	1414
Capacity	1415
Velocity	1415
Viscosity	1415
Friction Head	1415
Velocity Head	1415
Total Static Head	1415
Total Suction Lift	1415
Total Suction Head	1415
Total Discharge Head	1415
Total Head	1415
Net Positive Suction Head	1415
Suction Limitations on a Pump	1415
Work Performed in Pumping	1416
Centrifugal Pumps	1417
Action of a Centrifugal Pump	1417
Centrifugal-pump Characteristics	1418
Single-stage Centrifugal Pumps	1418
Chemical Pumps	1418
Close-coupled Pumps	1419
Special-application Pumps	1419
Multistage Centrifugal Pumps	1419
Propeller Pumps	1419
Selecting a Pump	1420
Packing and Shaft Seals	1420
Smothering Glands	1423
Mechanical Seals	1423
Pump Materials	1424
Materials for Pumping Various Liquids	1424
Friction Losses of Liquid Flowing through Pipe	1426
Reciprocating Pumps	1426
Volumetric Efficiency	1431
Indicated Efficiency	1431
Mechanical Efficiency	1431
Total Efficiency	1431
Suction Lift	1431
Types of Reciprocating Pumps	1436
Simplex, Double-acting	1456
Duplex, Double-acting	1436
Triplex, Single-acting	1456
Triplex, Double-acting	1436
Choice of a Reciprocating Pump	1436
Metering and Proportioning Pumps	1437
Rotary Pumps	1437
Handling Liquids with Fluid Pressure	1438
The Air Lift	1438
Displacement Pumps	1439
Jet Pumps	1439

COMPRESSION OF GAS

Compressors	1439
Reciprocating Compressors	1443
Single-acting	1443
Multistage	1443
Control Devices	1444
High-pressure Compressors	1446
Piston-rod Packing	1446
Fans	1447
Centrifugal Fans	1448
Straight-blade Fans	1448
Forward-curved-blade Fans	1448
Backward-curved-blade Fans	1448
Axial-flow Fans	1448
Disk Type	1448
Propeller Type	1448
Fan Performance	1449
Selection of Fans	1450
Turboblowers and Turbocompressors	1450
Rotary Blowers and Compressors	1452
Steam-jet Ejectors	1453
Condensation Processes	1455
Batch Distillation	1455
Vacuum Shelf Drying	1455
Evaporators and Vacuum Pans	1456
Cooling and Freezing	1456
Moistening and Impregnating	1456

MATERIAL-HANDLING EQUIPMENT

BY WILBUR G. HUDSON

General Factors Affecting the Selection of Conveying and Elevating Equipment. Pulverized and granular materials that are not actively corrosive, abrasive, or sticky are handled easily by any of the standard bucket elevators, flight conveyors, apron conveyors, continuous-flow conveyor-elevators, skip hoists, belt conveyors, and screw conveyors. In this class, among others, are:

Arsenic salts	Lead salts
Bentonite	Pebble lime
Bone meal	Limestone dust
Bran	Malt
Pulverized chalk	Pulverized mica
Clays	Petroleum coke
Coal	Salt
Coffee beans	Salt cake
Ground cork	Sawdust
Cornmeal	Soap flakes
Copra	Soda ash
Flaxseed	Sodium aluminum sulfate
Flour	Starch
Clean fly ash	Sugar
Fuller's earth	Talc
Granular glue	Wheat and other grains
Flake graphite	Wood chips
Calcined gypsum	Zinc oxide

More difficult are such materials as acid phosphate because it is sticky; ammonium nitrate because it is very hygroscopic and must not come in contact with copper, zinc, or brass; flaked magnesium chloride because it is hygroscopic and fragile; and aluminum oxide because it is extremely abrasive. Among other abrasives are borax, crushed feldspar, flue dirt, ground quartz, and sand. Such materials require specially constructed handling equipment to withstand their destructive action.

A very few materials, of which wet sewage sludge is outstanding, are extremely difficult to handle by any type of machine with the possible exception of the skip hoist, because they are abrasive, corrosive, and very sticky.

Some materials are extremely friable and must be handled gently. Thus a pneumatic conveyor, a high-speed centrifugal discharge elevator, or a scraper-flight conveyor, would not be specified for pelletized carbon black, granular fuller's earth, or soap flakes if degradation is objectionable.

The metallic dusts are difficult to convey because they work into and stiffen the chain joints until they cannot articulate. Thus a belt-and-bucket elevator would be preferable to a chain-and-bucket elevator or a continuous-flow machine which has the chain embedded in the material. Table 1 lists a variety of materials and indicates the types of equipment usually employed for each. Of course, the rate of handling must be taken into consideration. If we are to handle salt at 10 tons/hr., a screw conveyor would probably be specified rather than a belt conveyor costing several times as much; but, if the capacity is to be 250 tons/hr., the screw conveyor is not to be recommended, and a belt conveyor should be used.

Conveyor Drives. Mechanical conveyors and elevators are almost invariably driven by an electric motor connected with the head shaft through a suitable speed-reduction mechanism. The motor and gearing must be protected against injury should the machine be jammed or overloaded. Elevators and steeply inclined conveyors must be protected against reversal should the overload release function.

Small high-speed motors cost least and may have the speed-reduction gearing mounted integral with the frame. The motorized reducer or the motor may be connected by V-belt or silent chain to a herringbone reducer, which in turn is connected to the head shaft with a further speed reduction of 4 to 1 or 5 to 1. This drive usually is through a short-pitch bushed-steel roller chain, but the chain speed should not exceed 1000 ft./min.

If a variable-speed drive is required, it may be secured by a Reeves or P.I.V. transmission, preferably introduced where the r.p.m. will be high so the torque will be a minimum—with due regard to the maximum allowable speed of the speed changer. It is important that this unit be selected with a rating equal to the rating of the motor rather than according to the power requirement of the conveyor.

Motor Types. D.c. motors are rather unusual. A.c. motors, three phase, 60 cycle, and 220 or 440 volts, are almost invariably specified. The motor should be suited to the conditions under which it will operate. If the surroundings are clean and reasonably free from dust, the open, general-purpose motor will serve. If the motor will operate in a wet atmosphere or in vapors, a splash-proof motor should be specified. For extremely bad conditions, the totally enclosed motor is advisable. Finally, there is the explosionproof motor used where there are explosive gases or dusts. This type has metal-to-metal joints and shaft seals; and, though some gas or dust may enter the motor, it is very strongly made, and the flame of an internal explosion cannot penetrate to the outside.

The *electrical characteristics* must be suited to the power requirements of the conveyor. Some machines require extremely heavy starting torque. Some do not. Charts *A* to *D* of Fig. 1 show the general characteristics of a.c. motors.

Squirrel-cage motors (chart *A*) are essentially constant-speed motors, with better than 150 per cent starting torque on full voltage and better than 250 per cent momentary pull-out torque. There is 5 per cent slip between no-load and full-load speeds. The starting current is high so that the motor is applied with moderate margin of starting torque. It is low in cost and frequently applied to elevators and conveyors of moderate power requirements and minimum starting inertia.

Double-squirrel-cage motors (chart *B*) have 240 per cent starting torque with 400 per cent starting current. Maximum running torque is 200 per cent with 5 per cent slip. They are used with machines having heavy starting inertia, such as belt conveyors of extreme length, crushers, large-capacity elevators, and inclined heavy-duty flight conveyors. Controls are simpler than for the single squirrel-cage motor.

The high-resistance-rotor motor (chart *C*) has a maximum torque at about 10 per cent speed and high power factor at starting. The high rotor resistance gives high slip at full load and a wide range of speeds with changing loads.

Table 1. Recommended Material-handling Equipment

Material	Weight/ cu. ft.	Type of machine: Flight conveyor	Apron conveyor	Steel screw	Cast-iron screw	Continuous flow	Belt conveyor	Pivoted bucket conveyor	Skip hoist	Pneumatic conveyor	Slow-speed bucket elevator	High-speed bucket elevator	Vibrating conveyor
Ashes:													
Wet	45–55	*a*	*d*	....	Yes	*d*	*d*	Yes	Yes	....	....	Yes	
Dry	35–40	*a*	*d*	Yes	Yes	*d*	*d*	Yes	Yes	Yes	....	Yes	
Cement	85–100	....	....	Yes		Yes	Yes	....	*b*	Yes	Yes	Yes	*h*
Chemicals:													
Abrasive	......	*a*	Yes	....	Yes	....	Yes	Yes	Yes	....	Yes	Yes	Yes
Corrosive					Yes		Yes	Yes	Yes		Yes	Yes	Yes
Sticky		....	....	....	*c*	....			Yes				
Clays	35–55	Yes	Yes	Yes	Yes	Yes *h*	Yes	*d*	*d*	Yes	Yes	Yes	
Coal:													
Anthracite, *k*	45–55	Yes	Yes	Yes	Yes	Yes	*j*	Yes	Yes	Yes	Yes	Yes	Yes
Bituminous, *k*	50	Yes	Yes	Yes	Yes	Yes	*j*	Yes	Yes	Yes	Yes	Yes	Yes
Coke, *k*	40	....	Yes	....	Yes	....	*e*	Yes	Yes	....	Yes	*f*	Yes
Coke breeze	30	....	Yes	....	Yes	....	Yes	Yes	Yes	Yes	Yes	Yes	Yes
Copra, *k*	35–40	*d*	*d*	Yes	*d*	Yes	Yes	*d*	*d*	Yes	Yes	Yes	
Cottonseed	25–30			Yes	...	....	Yes	*d*		Yes		Yes	
Flaxseed	45	Yes	*d*	Yes	*d*	Yes	Yes	*d*	....	Yes	Yes	Yes	
Fly ash (clean)	30–45	*d*		Yes	Yes	Yes	*d*	Yes	*d*	Yes	....	Yes	
Grains, *k*	25–60	*d*		Yes	...	Yes	Yes	....	....	*f*	Yes		
Heavy ores (lumpy)		Yes	Yes			....	Yes	Yes	Yes	....	Yes	....	Yes
Light ores	......	Yes	Yes	Yes	Yes	*g*	Yes		Yes	Yes	Yes	Yes	Yes
Metallic dusts, *k*		*d*	*d*	Yes	...	....	Yes	Yes	*b*	*d*	Yes	Yes	
Rock			Yes		....	....	Yes		Yes		Yes		
Salt	45–80	Yes	Yes	Yes	....	Yes	Yes	*d*	*d*	Yes	Yes	Yes	*d*
Salt cake	75–95	Yes	*d*	*d*	Yes	Yes	Yes	Yes	*d*	Yes	Yes	Yes	
Sand and gravel			*d*	*d*	Yes	....	Yes	*d*	Yes	....	....	Yes	*h*
Sawdust	15–20	*d*	*d*	Yes	...	Yes	Yes	....	*d*	Yes	....	Yes	
Sewage sludge, wet, *l*		....	....	....	*c*	....	....	....	Yes				
Soda ash	25–65	Yes	*d*	Yes	Yes	Yes	Yes	....	*d*	Yes	Yes	Yes	
Soybeans, *k*	45			Yes	...	Yes	Yes	....	....	Yes	Yes	Yes	
Soybean meal, *k*	40			Yes	...	Yes		....		Yes, *k*	Yes	Yes	
Starch, *k*	25–40		....	Yes		Yes	Yes	....	....	*f*	Yes	Yes	
Sugar, *k*	55	....	*d*	Yes		Yes	Yes	....	....	*f*	Yes	Yes	
Sulfur, *k*	55–80		Yes	Yes	....	Yes	Yes	....	Yes	*i*	Yes	Yes	
Wood chips	20			Yes		Yes	Yes	....	....	....	Yes	Yes	
Extremely hot materials			Yes	....	Yes	....		Yes	Yes				

Packages, bags, crates, boxes, etc., are usually handled by roller conveyors, tray elevators, flat sliding belts, tractors, and trams.

NOTES:
a. Drag-chain type only.
b. With provision for suppressing dust.
c. Ribbon type best.
d. Possible but unusual.
e. If not red-hot.
f. Not if grain suffers from rough handling.
g. If not too abrasive.
h. May give trouble.
i. Static electricity may give trouble.
j. Special synthetic belt necessary if oil-sprayed.
k. Dust is explosive.
l. Difficult with any conveyor.

The slip-ring motor has a wound rotor with the conductors brought out to slip rings to permit the insertion of resistance to increase the starting torque and decrease the starting current. It is suited to crushers, skip hoists, heavy-duty screens, etc.

Synchronous motors (chart *D*) are infrequently specified in connection with material-handling equipment. They are intended for such constant loads as fans and centrifugal pumps. As now made, their starting torque is up to 250 per cent.

Motor Protection. Motors are protected against sustained overloads by cutouts having a thermal relay whose timed action prevents interruptions due to momentary peaks. However, since the inertia of the spinning motor would cause heavy strain and probably breakage of the moving parts of an elevator or conveyor, it is possible to provide shear-pin protection between driving and driven members. Shear-pin protection has no value in a belt conveyor, since damage to a belt may occur without overload.

If two or more conveying units are in tandem and driven by independent motors, the shear-pin protection must be so arranged that, when it lets go, all motors will stop. Otherwise, since the motor released by the shearing of the pin continues on the circuit, the interlocked motor or motors of the tandem units continue to discharge their loads into the stalled machine.

To prevent reversal of a stalled and released elevator or steeply inclined conveyor, a solenoid brake is provided that is normally held off by the motor current, a differential brake, or a pivoted pawl which blocks the chain should it reverse a pitch or two.

Screw Conveyors

The standard screw conveyor consists of a steel helix or helicoid mounted on a spindle and suspended in a sheet-steel U-trough. The helix is rotated at moderate speed by a motorized reducer. The conveyor may be either right-hand or left-hand. Sometimes one section is right-hand and another section left-hand. Figure 2 shows the method of designation. At the discharge end, a right-hand conveyor rotates clockwise.

Standard sizes range from 4 in. diameter of helix up to 24 in. Handling capacities are shown in Fig. 3. It will be noted that the percentage of loading decreases as the material is heavier, more abrasive, or more destructive in its action on bearings or trough. If the material is actively abrasive or corrosive, as illustrated by wet ashes, the helix and trough may be of cast iron.

Usually a screw conveyor is horizontal, but it may be inclined. The capacity is reduced sharply as the inclination increases; thus a standard-pitch screw inclined at 25 deg. has 45 per cent of its capacity when horizontal.

Lumpy Material. The diameter of the helix determines the size of lumps the conveyor can handle safely. If the material is all lumps:

Helix diameter, in.	4	6	10	14	18	24
Maximum lumps, in.	¼	½	¾	1¼	2	2½

If the lumps constitute only 25 per cent of the total, these lump sizes may be doubled, *e.g.*, a 14-in. conveyor can then handle material containing lumps up to 2½ in. If the material is sticky or stringy, *e.g.*, hot tar or asphalt, a ribbon helix is specified.

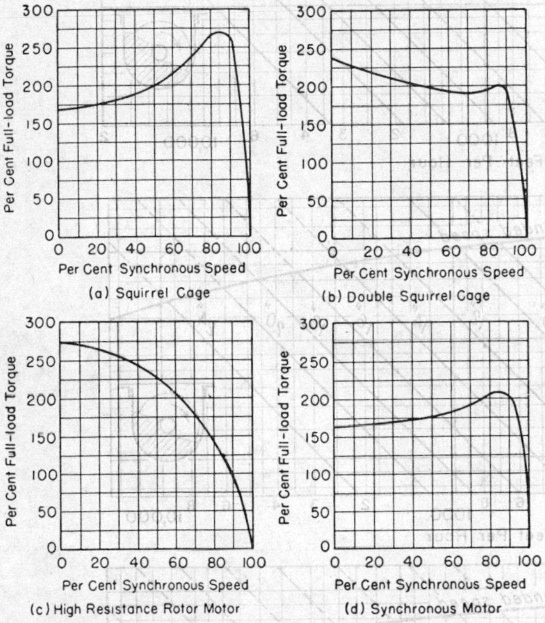

FIG. 1. A.c. motor characteristics. (*a*) Squirrel cage. (*b*) Double squirrel cage. (*c*) High-resistance rotor motor. (*d*) Synchronous motor.

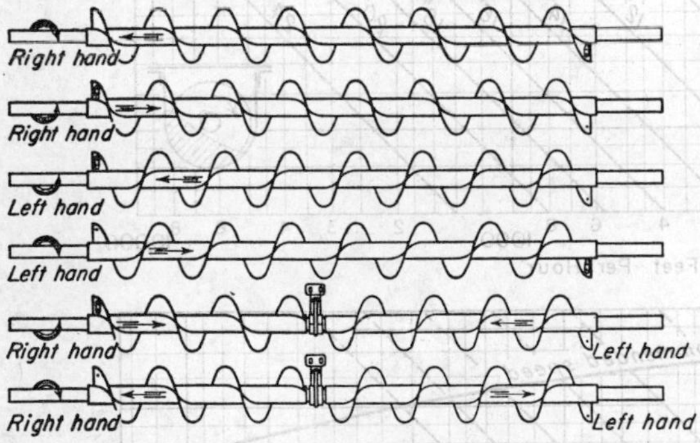

FIG. 2. Method of designating right-hand and left-hand screw conveyors.

Frequently the trough is open, as formed by two side plates with the helix between. Then the material drags across that already in a bin and discharges over the brow of the pile thus formed. Wheat farmers in fact, use an "auger," a helix without any trough, letting the wheat form its own trough.

If a screw conveyor is fed by gravity flow from a bin or tank, the section beneath the opening is given a shorter pitch or smaller diameter to reduce the rate of feed to what the conveyor beyond that point can handle without overloading.

Power Requirements. Power depends on the frictional resistance or drag of the material, the length of the conveyor, and the handling rate. The method of determination and the factors for drag are given in Table 2.

Applications. If the capacity is within the range of the standard sizes and the distance or length is not too great as regards torsion on the shaft, the screw conveyor provides the simplest and least costly machine for handling flowable granular or pulverized materials over a horizontal or slightly inclined path. It can easily compete with the smaller flight conveyors. It is not an

Table 2. Screw Conveyors—Capacity Charts for Horizontal Conveyors

F = material factor

Class *a* materials: light fine non-abrasive material, free-flowing, 30 to 40 lb./cu. ft.

Class *b* materials: medium-weight non-abrasive materials, granular or small lumps mixed with fines, weight up to 50 lb./cu. ft.

Class *c* materials: non-abrasive or semiabrasive materials, granular or small lumps mixed with fines, weights 40 to 75 lb./cu. ft.

Class *d* materials: semiabrasive or abrasive materials, fines, granular or small lumps mixed with fines, weights 50 to 100 lb./cu. ft.

Class *e* includes highly abrasive, lumpy, or stringy materials that must not come in contact with the bearings. Some other type of conveyor may be better

Class *a* ($F = 0.4$)	Class *b* ($F = 0.6$)	Class *c* ($F = 1$)	Class *d* (F as noted)	Class *e*
Barley† Dry brewer's grains Coal, pulverized Cornmeal† Cottonseed meal Flaxseed Flour† Lime, pulverized Malt† Rice*,† Wheat†	Alum, fine Beans, soy† Coal, fines and slack Cocoa beans† Coffee beans*,† Corn, shelled† Corn grits Gelatin, granular*,† Graphite flakes Lime, hydrated	Alum, lumpy† Borax Wet brewer's grains Charcoal Coal, sized Coal, lignite Cocoa† Cork, ground Fly ash, clean Lime, unslacked Milk, dried† Paper pulp Paper stock Salt, coarse or fine† Sludge, sewage Soap, pulverized Soda ash Starch† Sugar, refined	Bauxite, 1.8 Bone meal, 1.8 Carbon black, 1.6 Cement, 1.4 Chalk, 1.4 Clay, 2 Fluorspar, 2 Gypsum, crushed, 1.6 Lead oxides, 1 Lime, pebble, 1.3 Limestone dust, 1.6 Acid phosphate, damp, 7% moisture, 1.4 Sand, dry, 2 Shale, crushed, 1.8 Slate, crushed, 1.6 Sugar, raw, 1.8 Sulfur, 1.6 Zinc oxide, 1.6	Ashes, 4 Flue dirt, 3.5 Quartz, pulverized, 2.5 Sand, silica, 2.0

* To reduce degradation by keeping material low in trough, it is sometimes advisable to use *c* or *d* lines with corresponding reduction in capacity rating.

† Oil must be kept from contact with material by use of oilless bushings.

Horsepower at drive shaft:

$$\text{Horsepower} = \frac{CLWF}{33{,}000}$$

where C = capacity of conveyor, cu. ft./min.; L = length, ft.; W = weight of material, lb./cu. ft.; and F = material factor (see above).

If length exceeds 100 ft., add 10 to 15 per cent.

Motor size:

If horsepower is less than 2, multiply above result by 2.

If horsepower is less than 4, multiply above result by 1.5.

If conveyor is gravity-loaded from bin or hopper, allow ½ to 1 hp.

Screw-conveyor Charts. The charts in Fig. 3 (Link Belt Co.) show the advisable maximum speed for each class of material, *a*, *b*, *c*, *d*, and the capacity in cu. ft./hr. at the r.p.m. indicated. Class *e* use 50 per cent of chart *d* capacities and keep maximun speed below 40 r.p.m.

Illustration of the method of using power formula and charts:

Problem: A screw conveyor will be required to convey slack coal at 25 tons/hr., horizontally a distance of 80 ft. Determine the motor horsepower and size and speed of conveyor.

Slack coal is in Class *b*; therefore $F = 0.6$. 25 tons/hr. = 1000 cu. ft./hr. or 17 cu. ft./min., approximately. By the formula,

$$\text{Motor hp.} = \frac{17 \times 80 \times 50 \times 0.6}{33{,}000} = 1.24$$

Allow 90 per cent for efficiency of the drive = 1.38. The horsepower is less than 2; therefore multiply by 2 = 2.76. Use 3-hp. motor.

Referring to the chart for Class *b*, we can use for 1000 cu. ft./hr., a 10-in. screw at 64 r.p.m. If the slack contains many lumps, we should use a 14-in. conveyor at 37 r.p.m.

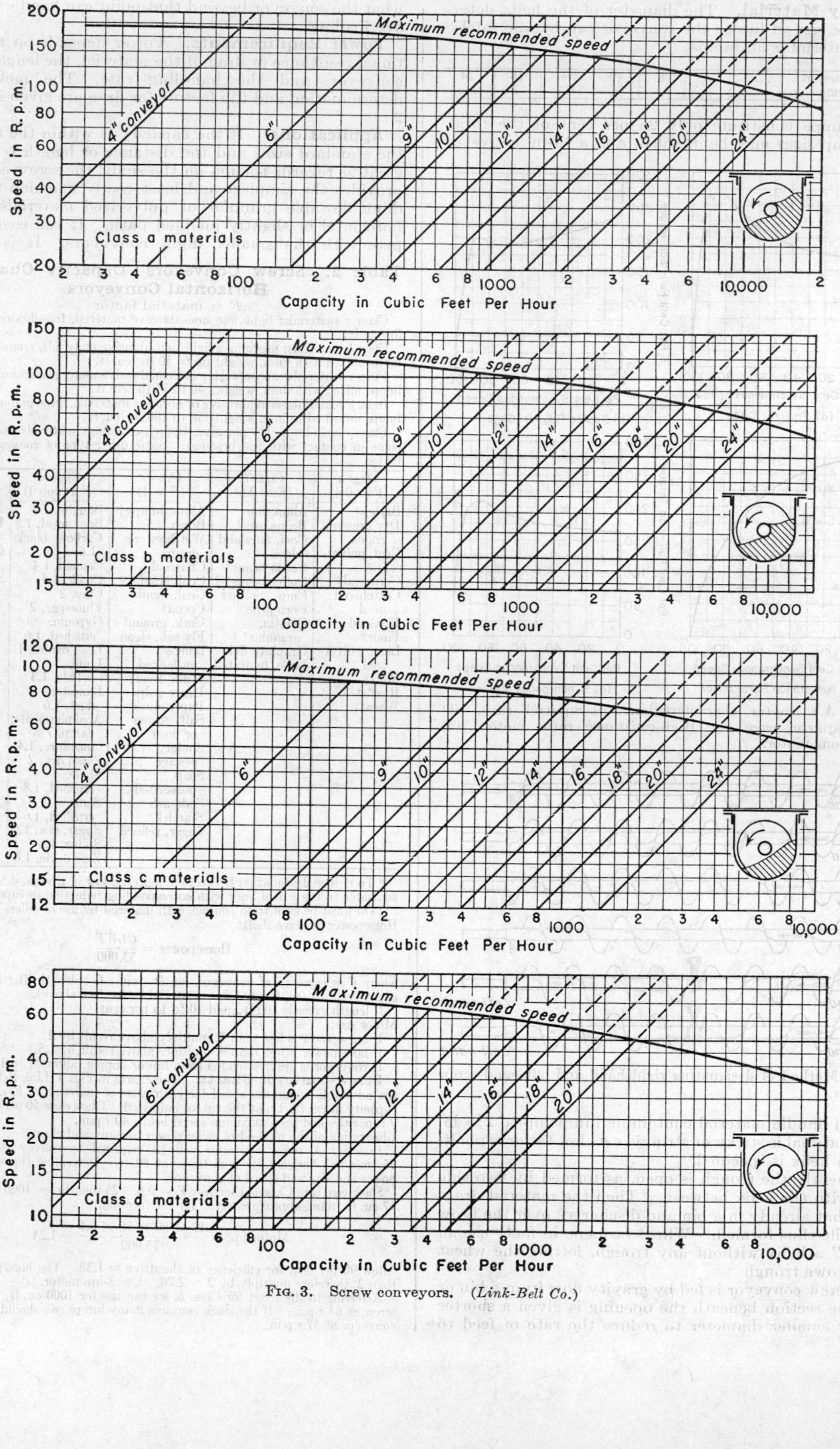

FIG. 3. Screw conveyors. (*Link-Belt Co.*)

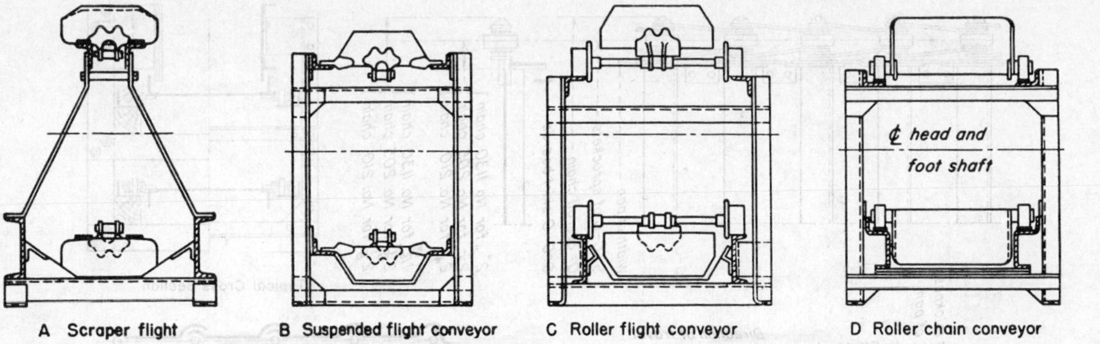

FIG. 4. Standard types of flight conveyors. *A*. Scraper flight conveyor for small capacities and moderate lengths between centers. *B*. Suspended flight conveyor for larger capacities, moderate lengths between centers, and for materials not containing large lumps. *C* and *D*. Roller flight and roller chain conveyors for heavy duty. If the material contains large lumps, *D* is preferable as the chains do not restrict the feed to the conveyor. The power loss in the machine is ⅓ in *C* or *D* as compared with *A* or *B*.

alternative for the heavy-duty flight or belt conveyor of large capacity.

Flight Conveyors

The flight conveyor consists of an endless chain or twin chains to which are attached spaced pushers or flights, a trough, and the usual head and foot assemblies. The standard types are shown in Fig. 4. If the capacity and length are comparatively small, the flights are pulled along the trough (the scraper flight conveyor). For longer lengths and greater capacities, the flights are suspended clear of the trough by sliding shoes (the suspended flight conveyor). For still larger installations where power consumption warrants consideration, the flights are carried on flanged rollers (the roller flight conveyor), or chain and flights are carried clear of the trough by roller chains (the roller-chain suspended flight conveyor).

The conveyor may be inclined up to a point where the load begins to flow backward over the tops of the flights. The limit is about 30 deg. The capacity becomes smaller as the slope increases.

The power required is the sum of that required to pull the empty conveyor plus that required to drag the material along the trough. The latter depends on the coefficient of friction of the material against the trough plate—usually of steel—and the conveyor is therefore not well suited to a material such as wet sand. The friction coefficients of several materials are given in Table 3. The last five are materials that can be handled to better advantage by some other type of conveyor.

Table 3. Friction Coefficients against Steel Plate

Material	Coefficient	Material	Coefficient
Clay	0.60–0.70	Sawdust, wet	0.60
Coal, anthracite	0.33	Soda ash	.65
Coal, bituminous	.59	Starch	.60
Cork, granular	.65	Sugar, granular	.67
Grains	0.30–0.40	Wood chips	.35
Heavy chemicals	0.40–0.70	Ashes, damp	.70
Sliding conveyor parts	0.33	Cement	.93
Hog fuel, dry	.60	Coke, lumpy	.36
Lime, hydrated	.65	Magnesium chloride	.70
Limestone, pulverized	.58	Sand, wet	1.00

Power Determination. The chain pull is first determined. In a horizontal conveyor, this is given by the formula

$$C_p = (2 \times W \times L \times F) + (W_1 \times L \times F_1)$$

where C_p = chain pull, lb.
W = weight/lin. ft. of the conveying element, lb.
L = length, ft., from head to foot shafts.
F = coefficient of friction for sliding or rolling parts of conveyor.
W_1 = weight of material/lin. ft. of conveyor, lb.
F_1 = coefficient of friction between material and trough (Table 3).

In the roller flight or roller-chain conveyor, the factor F is 0.1. For the scraper flight conveyor, F is 0.33. If the conveyor is steeply inclined, the downward pull of the return run will reduce the power requirement but not the chain pull. For a conveyor inclined at θ deg. the chain pull is

$$C_p = (W \times L)[(\cos\theta \times F) + \sin\theta] + (W_1 \times L)[(\cos\theta \times F_1) + \sin\theta] + (W \times L)[(\cos\theta \times F) - \sin\theta]$$

It is the pull required for the equivalent horizontal path plus that required to lift the load. The last item becomes a negative amount for steep slopes. In a double-strand conveyor, the above gives the total pull on the two chains.

The *motor horsepower* is the net turning effort at the head sprocket plus 10 per cent for friction in the bearings plus 10 per cent for loss in the reduction gearing plus 10 per cent if there are momentary surges. Calling this the "gross turning effort,"

$$\text{Motor hp.} = \frac{\text{gross turning effort} \times \text{speed, ft./min.}}{33{,}000}$$

Since the flight conveyor cannot control the rate of flow to it, a suitable feeder must be provided and interlocked with the conveyor either mechanically or electrically, so that the feeder stops when the conveyor stops.

Table 4. Capacities of Horizontal Flight Conveyors Handling Coal and Size of Lumps Handled

Flight width and depth, in.	Amount of material/ft. of conveyor, cu. ft.	Approx. capacity for 50 lb. material at 100 ft./min.,* tons/hr.	Lump size, in. (lumps not to exceed 10% of total volume)	
			Single strand	Double strand
12 × 6	0.40	60	3½	4
15 × 6	.49	73	4½	5
18 × 6	.56	84	5	6
24 × 8	1.16	174	...	10
30 × 10	1.60	240	...	14
36 × 12	2.40	360	...	16

* If the conveyor is inclined, multiply the horizontal capacity by 0.90 for 20 to 25 deg., by 0.80 for 25 to 30 deg., by 0.70 for 30 to 36 deg.

Retarding Conveyors. When it is required to convey material "downhill," and where the slope does not exceed 35 deg. or so, the flight conveyor serves well. The motor

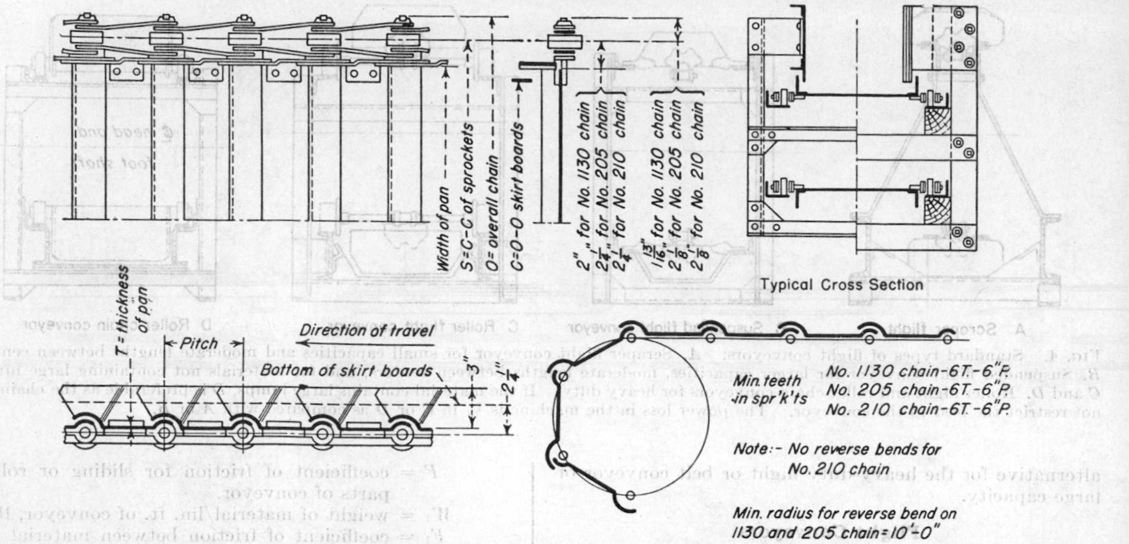

FIG. 5. "S-A" steel apron conveyor-style "C." Double beaded, shallow type, chains at end. (*Stephens-Adamson Mfg. Co.*)

then is located at the upper end and may serve as the speed control by regenerative braking, pumping current back into the line. Usually it is necessary to provide a special type of automatic brake that will hold the conveyor stationary when the current is shut off but will not function should overhauling the motor reduce the current to a negligible amount.

Applications. The flight conveyor is a simple rugged machine capable of maintaining heavy peak loads without damage, withstands tough treatment and neglect without serious results, has reasonably low maintenance costs, and is a reliable unit. It costs about twice as much as a screw conveyor but may have much greater length and handling capacity. It may or may not be better than a belt conveyor. It can function on a steeper slope than a belt and can handle hot materials that would destroy rubber.

Apron Conveyors

The apron conveyor has overlapping pans supported between two strands of roller chain or sometimes mounted upon the chains. Usually high side plates or guards restrict the width of the load and prevent spillage over the edges of the apron; moreover they permit loading to a considerable depth.

Apron conveyors are commonly used as feeders, as from beneath a track hopper to an elevator or conveyor. The speed then is usually between 15 and 30 ft./min. Table 5 gives the capacities for horizontal apron conveyors and feeders and the maximum lumps that may be handled, assuming that the percentage of lumps is large.

Table 5. Capacities of Apron Conveyors and Feeders
Speed 20 ft./min.
Material weighing 50 lb./cu. ft.

Width of apron between skirts, in.	Depth of material on apron, in.	Capacity, tons/hr.	Max. lump size, in.
24	12	45	4–6
30	12	56	6–8
36	12	68	8–12
42	12	79	8–14
48	12	90	8–18
60	12	113	12–18
60	24	225	15–28

Since the load is carried upon an apron, only rolling friction must be overcome, and the power required is therefore much less than in a conveyor that scrapes the load along a trough (Fig. 5). There is, however, some added resistance because of side rub of the material against the guards. A fair average is given in Table 6.

Table 6. Drag of Material against Skirt Boards per Foot of Conveyor (Pounds)

Coal	6
Coke	3
Hydrated lime	8
Pulverized limestone	8
Wood chips	3

Power Requirements. The horsepower is

$$\text{Chain pull} \times \frac{\text{speed, ft./min.}}{33{,}000}$$

The chain pull in pounds is

$$C_p = RF \times (2W_e + W_m \times L) + D_s$$

where W_e = weight of one run of apron.
W_m = weight material/ft. apron.
L = distance between centers.
D_s = skirt drag (Table 6) multiplied by length between centers.
RF = rolling friction and usually may be taken as 0.1.

If the apron is inclined upward, allowance must be made for the power required to lift the load. The capacity in tons per hr. may be approximated by taking 75 per cent of the cross section of the load in sq. ft., multiplying by the weight/cu. ft./lb. and the speed in ft./hr., and dividing by 2000.

Bucket Elevators

Bucket elevators consist of head and foot assemblies and an endless chain, twin chains, or a belt to which are attached buckets for elevating pulverized, granular, or lumpy materials along a vertical or steeply inclined path. Usually the elevator is enclosed in a dustproof casing. The standard types are as follows:

High-speed centrifugal-discharge elevator (Fig. 6A) with a single strand of chain or a belt and with the buckets so spaced as to throw out the material by centrifugal action as they round the head wheel. The buckets scoop up their loads when rounding the foot wheel within a suitably curved boot. This type is lowest in first cost and is specified for small and medium capacities. However, the belt-and-bucket centrifugal-discharge elevator may be applied to the largest capacities if the material is pulverized or granular.

Slow-speed perfect-discharge elevator (Fig. 6*B*), a double-strand elevator with the chains snubbed back after rounding the head sprockets to give almost complete upturn. It is therefore nicely adapted to materials that are sluggish and do not discharge readily. Like the

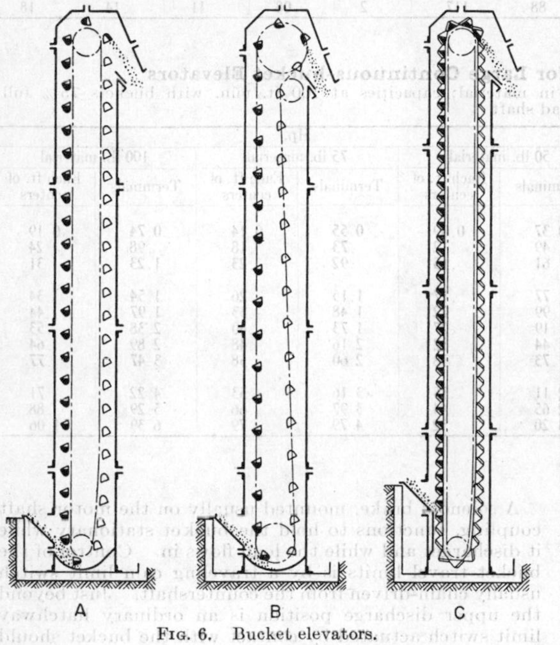

FIG. 6. Bucket elevators.

centrifugal-discharge elevator, it loads by scooping up the material.

Slow-speed continuous-bucket elevator (Fig. 6*C*), with either single or double strands of chain, or a belt. The elevator is coupled to a feeder, and the buckets are loaded after they line up for the ascent. This type is specified for capacities ranging from moderate to very large. Because of its slow speed, it is well suited to fragile materials.

Continuous-bucket chain-and-bucket elevators are preferred by many to high-speed centrifugal-discharge elevators from a maintenance viewpoint; but the reverse is true for belt-and-bucket elevators, because there is less chance that the material will wedge in between buckets and belt.

The motor horsepower of any of the above elevators may be approximated as twice the tons per hour multiplied by the lift in feet and divided by 1000.

The gravity-discharge elevator (Fig. 7) is a moderate-speed double-strand machine, usually having a vertical run followed by a horizontal run. In the horizontal run the buckets travel through a trough and so function as do the flights in a flight conveyor. Since there is no transfer at the top of the lift, the material suffers less breakage than when handled by an elevator and a tandem conveyor. However, if the horizontal run has considerable length, the machine costs much more than an elevator and conveyor.

The power requirement of this type of elevator may be approximated by adding the power for the corresponding vertical elevator to that required for a roller-chain flight conveyor with a length equal to the upper horizontal run.

Applications. Bucket elevators will handle materials that are not too sticky to discharge or that have no lumps too large for the buckets. If the lift is extremely high, a skip hoist may be a better application, as involving less weight and costing much less. If the capacity is large and the material is abrasive, an inclined belt may be preferable, although it will cost more if the supports are

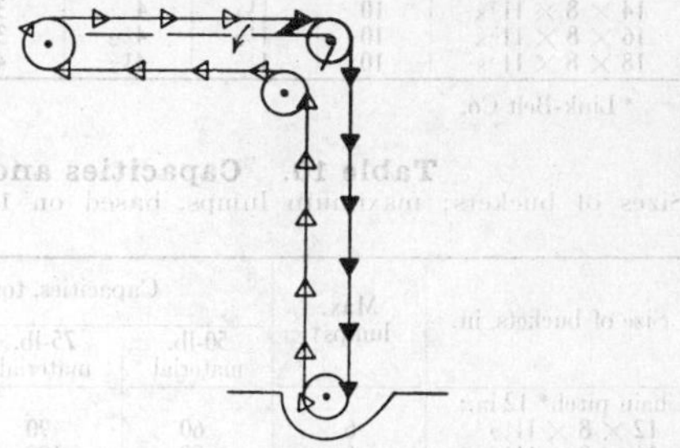

FIG. 7. Gravity-discharge elevator.

included. The maintenance cost for the inclined belt will usually be less, especially if the material is abrasive or corrosive. If the material is hotter than 300°F., the elevator or skip hoist is best.

Elevators must be protected against reversal of motion should the current fail or the drive shear pin let go.

A vertical path is better than an inclined path, since the supporting guides are eliminated.

Table 7. Capacity of Vertical Centrifugal-discharge Elevators and Maximum Size of Lumps

Sizes of buckets; maximum lumps; capacities, buckets 75% filled

Bucket spacing, in.	Size, length by width, in.	Speed, ft./min.	Max. lumps		Capacity, tons/hr.		
			All lumps	10% lumps	35-lb. material	50-lb. material	100-lb. material
13	6 × 4	225	½	2½	5	7	14
16	8 × 5	230	¾	3	9	13	27
16	10 × 6	230	1	3½	16	23	47
18	12 × 7	268	1¼	4	27	38	77
18	14 × 7	268	1¼	4	32	46	92
19	16 × 8	262	1½	4½	44	63	127

Horsepower: Motor horsepower for centrifugal-discharge elevators closely approximates

$$\frac{\text{Tons/hr.} \times 2 \times \text{lift, ft.}}{1000}$$

Table 8. Capacities and Horsepower of Gravity-discharge Elevators, Handling Coal*

(Used primarily for bituminous coal)

Capacities in tons/hr. at 100 ft./min. Buckets loaded to 80% of level full

Size of bucket, in. $L \times W$	Capacity, tons/hr. at 100 ft./min. Bucket spacing, in.			Hp.† with material at 50 lb./cu. ft. Spacing of buckets, in.					
				Per 10-ft. vertical lift			Per 100-ft. horizontal run		
	18	24	36	18	24	36	18	24	36
16 × 15	46	35	23	0.59	0.44	0.30	5.32	4.24	3.04
20 × 15	58	44	29	.74	.56	.37	6.32	4.97	3.54
24 × 15	70	52	35	.90	.67	.45	7.34	5.74	4.04
20 × 20	104	..	52	1.30	....	.66	9.20	....	4.85
24 × 20	125	..	63	1.60	....	.80	10.92	....	5.74
30 × 20	159	..	79	2.00	....	1.00	13.70	....	7.08
36 × 20	191	..	95	2.42	....	1.21	16.30	....	8.40

* Link-Belt Co.

† Add 5% to horsepower for each bend in path of loaded run.

Skip Hoists

The skip hoist consists of a bucket raised, dumped, and lowered by a cable attached to a winding machine. If the capacity is large, twin or balanced buckets—one

Table 9. Capacities and Horsepower for Smaller Sizes of Continuous-bucket Elevator*

Bucket sizes; maximum lumps; capacities; hp. Capacities with buckets 75 % full; speed, 125 ft./min.

Buckets		Max. lumps		Capacity, tons/hr.				Hp. required for each ft. centers				
Size	Gage of steel	All lumps	10% lumps	35-lb. material	50-lb. material	75-lb. material	100-lb. material	For terminals	35-lb. material	50-lb. material	75-lb. material	100-lb. material
8 × 5 × 7¾	12	¾	2½	12	17	25	34	0.1	0.02	0.03	0.04	0.05
10 × 5 × 7¾	12	¾	2½	15	21	32	42	.1	.03	.04	.05	.07
10 × 7 × 11⅝	12	1	3	19	27	41	54	.1	.03	.04	.07	.09
12 × 7 × 11⅝	10	1	3	23	32	49	65	.1	.04	.05	.08	.10
14 × 7 × 11⅝	10	1	3	26	38	57	76	.1	.04	.06	.09	.12
12 × 8 × 11⅝	10	1¼	4	27	39	58	78	.2	.05	.07	.11	.13
14 × 8 × 11⅝	10	1¼	4	32	45	68	91	.2	.05	.08	.12	.14
16 × 8 × 11⅝	10	1½	4½	36	52	78	104	.2	.07	.09	.13	.16
18 × 8 × 11⅝	10	1½	4½	41	58	88	117	.2	.08	.11	.14	.18

* Link-Belt Co.

Table 10. Capacities and Horsepower for Large Continuous-bucket Elevators

Sizes of buckets; maximum lumps, based on 10–20% of lumps in material; capacities at 100 ft./min. with buckets 75% full; hp. at head shaft

Size of buckets, in.	Max. lumps†	Capacities, tons/hr.			Hp.					
					50 lb. material		75 lb. material		100 lb. material	
		50-lb. material	75-lb. material	100-lb. material	Terminals	Each ft. of centers	Terminals	Each ft. of centers	Terminals	Each ft. of centers
Chain pitch* 12 in.:										
12 × 8 × 11½	6	60	90	120	0.37	0.09	0.55	0.14	0.74	0.19
16 × 8 × 11½	6	80	120	160	.49	.12	.73	.18	.98	.24
20 × 8 × 11½	6	100	150	200	.61	.15	.92	.23	1.23	.31
Cha n pitch 18 in.:										
16 × 12 × 17½	8	115	175	230	.77	.17	1.15	.26	1.54	.34
20 × 12 × 17½	8	145	220	290	.99	.22	1.48	.33	1.97	.44
24 × 12 × 17½	8	175	260	345	1.19	.27	1.73	.40	2.38	.53
30 × 12 × 17½	8	215	320	425	1.44	.32	2.16	.48	2.89	.64
36 × 12 × 17½	8	255	385	510	1.73	.38	2.60	.58	3.47	.77
Chain pitch 24 in.:										
24 × 17 × 23½	10	230	345	460	2.11	.35	3.16	.53	4.22	.71
30 × 17 × 23½	10	285	430	570	2.65	.44	3.97	.66	5.29	.88
36 × 17 × 23½	10	345	520	690	3.20	.52	4.79	.79	6.39	1.06

* These are the long-pitch engineering chains.
† Not over 20% large lumps.

rising as the other descends—are operated by a single winding machine. The winding machine (Fig. 8) is a carefully designed unit, usually with antifriction bearings and herringbone gears running in oil. The types (classified according to method of control) are *semi-automatic*, in which, when the starting button is pressed, the bucket makes one round trip and stops in loading position; and *full-automatic*, in which the bucket continues to function until it is stopped by the attendant.

Table 11. Capacities and Maximum Lump Sizes for Centrifugal-discharge Belt and Bucket Elevators

Bucket spacing, in.	Size, length by width	Speed, ft./min.	Max. lumps		Capacity, tons/hr.		
			All lumps	10% lumps	35 lb. material	50 lb. material	100 lb. material
13	6 × 4	225	½	2½	5	7	14
16	8 × 5	258	⅔	3	11	15	30
16	10 × 6	258	1	3½	18	26	52
18	12 × 7	298	1¼	4	30	42	85
18	14 × 7	298	1¼	4	36	52	103
18	16 × 8	298	1½	4½	53	114	152

Link-Belt recommends 32-oz. belt up to and including 10-in. buckets.
Horsepower: Motor horsepower for centrifugal-discharge elevators closely approximates

$$\frac{\text{Tons/hr.} \times 2 \times \text{lift, ft.}}{1000}$$

The semiautomatic skip is used where the loading is irregular, as when material is delivered by wheelbarrows. When the bucket is full, the operator starts it upward.

Full-automatic control is used for continuous operation, as when material is to be removed from a track hopper. Loading then is automatic (Fig. 9), by means of a pivoted chute thrust upward by the bucket as it starts to ascend and brought down as the bucket nears the end of its return trip. The material flows in until it reaches its angle of repose; then, after an interval determined by a thermal relay, the loaded bucket again starts upward.

A solenoid brake, mounted usually on the motor shaft coupling, functions to hold the bucket stationary while it discharges and while the load flows in. Control of the bucket travel limits is by a traveling cam limit switch usually chain-driven from the countershaft. Just beyond the upper discharge position is an ordinary hatchway limit switch actuated by contact with the bucket should it overtravel a foot or two. If the bucket descends too far, it bottoms, and a slack cable switch (Fig. 10) functions. Normally the bucket does not bottom but is suspended on the cable while loading.

Speeds. A further classification of industrial skip hoists is *slow speed*, up to about 80 ft./min.; *medium speed*, up to about 140 ft./min.; and *high speed*, up to about 260 ft./min.

Semiautomatic skips are slow- or medium-speed as a rule. High-speed skips may have a two-speed motor to slow the bucket down automatically as it approaches the automatic loader or the tilting point, thus reducing shock.

Power. The horsepower of the hoist motor depends on the capacity and type of skip. If the bucket is counterweighted, *i.e.*, with a counterweight moving in the opposite direction by a separate cable from the drum, the power requirement is substantially reduced. The counterweight equals the weight of the empty bucket plus one-half the load, and the maximum torque in either direction therefore corresponds to one-half the load. Table 12 (Link-Belt Co.) gives the capacities, power requirements, speeds, etc., for skips of the various types. RS½, RSI, etc., refer to the size of their standard winding machines (Fig. 8).

Applications. The skip hoist has advantages if the lift is high or if the material is corrosive or abrasive, contains large lumps, or is very hot. It has few moving parts, and maintenance cost is low.

The skip hoist with supporting structure, where the lift and capacity are moderate to large, costs less than an inclined-belt or flight conveyor with supports, or a bucket elevator. If the lift and capacity are comparatively small, either an inclined conveyor or an elevator may be the better selection.

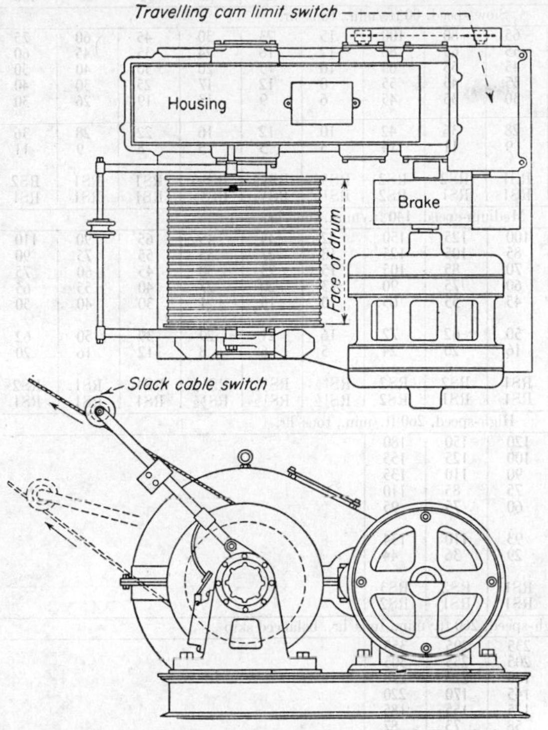

FIG. 8. Skip-hoist winding machine. (*Link-Belt Co.*)

Size	Drum, in.		Weight (without motor and brake), lb.	Max. unbalanced cable pull, lb. ($T_1 - T_2$)			Max. total pull allowed ($T_1 + T_2$) lb.
	Pitch diam.	Face		80 ft./min.	140 ft./min.	260 ft./min.	
RS-½	24	18	3,930	5,700	5,500	5,000	8,000
RS-1	30	27	5,860	7,850	7,500	7,000	16,000
RS-2	36	27	8,000	12,700	11,300	8,700	22,000
RS-3	48	30	15,000	21,500	20,000	17,000	35,000

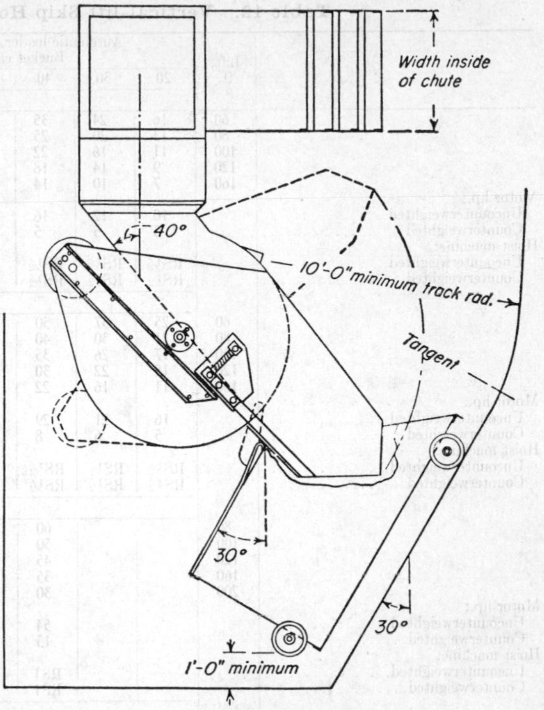

FIG. 9. The automatic loader.

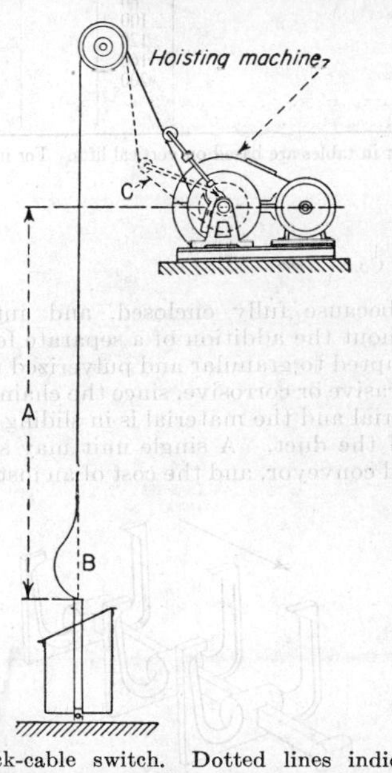

FIG. 10. Slack-cable switch. Dotted lines indicate line of cable under action of slack-cable device. Sufficient accumulation of slack cable at *C* trips switch and prevents operation of skip hoist until slack is removed and switch is reset.

Continuous-flow Conveyors

Continuous-flow conveyors are of two types: (1) with skeleton or framelike flights as exemplified alone by the Redler (Fig. 11, Stephens Adamson Manufacturing Co.), in which the material moves because resistance to motion of the mass is less than the effort required to pull the flights through the mass and (2) with full flights, in which the material moves quite as in a flight conveyor, but as a continuing core completely filling the duct. The principal machines of type 2 are (Fig. 12):

The *Master Flow* (Gifford Wood Co.), in which the lateral thrust of the unit loads is reduced by the position and form of the flights and discharge of residual material is facilitated by the cantilever extension of the links to which the flights are attached.

The *Bulk Flo* (Link-Belt Co.), characterized by peaked flights which facilitate discharge of residual material by their front and rear sloping surfaces.

The *Uni Flo* (Chain Belt Co.), in which the flights are pivoted at the outer edge and tilted upward by a rotating cam at the point of outflow or discharge.

The *Mass Flow* (Jeffrey Manufacturing Co.), in which suspended traylike flights are swung upward by a fixed guide for residual discharge.

Continuous-flow conveyor-elevators are characterized by slow speed, preferably not more than 60 ft./min., large capacity compared with space required, dustless

Table 12. Vertical-lift Skip Hoists Handling Coal (50 Lb./Cu. Ft.)*

	Lift, ft.	Automatic loader, full-automatic control Bucket capacity, cu. ft.							Automatic loader, semiautomatic control Bucket capacity, cu. ft.					
		20	30	40	60	80	100	120	20	30	40	60	80	100
		Slow-speed, 80 ft./min., tons/hr.												
	60	16	24	35	50	65	80	100	15	23	30	45	60	75
	80	13	20	25	40	55	65	80	12	18	24	35	45	60
	100	11	16	22	30	45	55	65	10	15	20	30	40	50
	120	9	14	18	25	35	45	55	8	12	17	25	30	40
	160	7	10	14	21	30	35	45	6	9	13	19	26	30
Motor hp.:														
Uncounterweighted		10	12	16	22	28	36	42	10	12	16	22	28	36
Counterweighted		5	5	5	8	9	11	14	5	5	5	8	9	11
Hoist machine:														
Uncounterweighted		RS½	RS½	RS½	RS1	RS1	RS2	RS2	RS½	RS½	RS½	RS1	RS1	RS2
Counterweighted		RS½	RS½	RS½	RS1	RS1	RS1	RS2	RS½	RS½	RS½	RS1	RS1	RS1
		Medium-speed, 140 ft./min., tons/hr.												
	60	25	37	50	75	100	125	150	22	35	45	65	90	110
	80	21	30	40	65	85	105	125	18	27	35	55	75	90
	100	17	26	35	50	70	85	105	15	23	30	45	60	75
	120	15	22	30	45	60	75	90	13	20	27	40	55	65
	160	11	16	22	35	45	55	65	10	15	21	30	40	50
Motor hp.:														
Uncounterweighted		16	21	29	39	50	62	72	16	21	29	39	50	62
Counterweighted		5	6	8	12	16	20	24	5	6	8	12	16	20
Hoist machine:														
Uncounterweighted		RS½	RS½	RS½	RS1	RS1	RS2	RS2	RS½	RS½	RS½	RS1	RS1	RS2
Counterweighted		RS½	RS½	RS½	RS1	RS1	RS1	RS2	RS½	RS½	RS½	RS1	RS1	RS1
		High-speed, 260 ft./min., tons/hr.												
	80			60	90	120	150	180						
	100			50	75	100	125	155						
	120			45	65	90	110	135						
	160			35	55	75	85	110						
	200			30	45	60	75	95						
Motor hp.:														
Uncounterweighted				54	73	93	110	134						
Counterweighted				15	22	29	36	44						
Hoist machine:														
Uncounterweighted				RS1	RS1	RS1	RS2	RS3						
Counterweighted				RS1	RS1	RS1	RS1	RS2						
		High-speed, 260 ft./min., tons/hr., balanced skip												
	80			115	175	235	295	355						
	100			100	155	205	255	305						
	120			90	135	180	225	270						
	160			75	110	145	170	220						
	200			60	90	125	155	185						
Motor hp.				29	44	58	73	87						
Hoist machine				RS1	RS1	RS1	RS1	RS2						

Horsepower in tables are based on vertical lifts. For inclined runs, multiply horsepowers by the following:

80 deg.	0.985
70 deg.	.940
60 deg.	.866
50 deg.	.766

* Link-Belt Co.

operation because fully enclosed, and automatic feed control without the addition of a separate feeder. They are best adapted to granular and pulverized materials not actively abrasive or corrosive, since the chain is embedded in the material and the material is in sliding contact with the walls of the duct. A single unit may serve as both elevator and conveyor, and the cost of an installation may

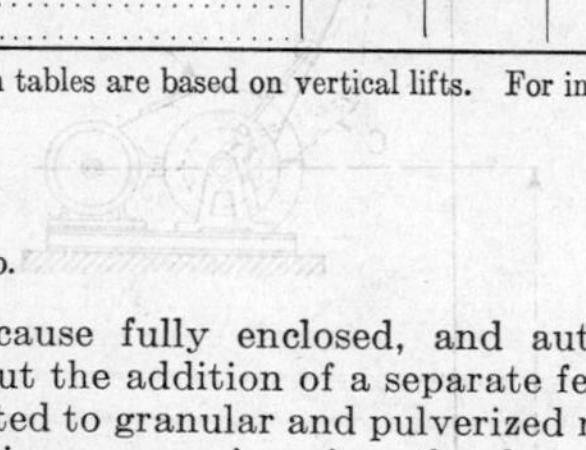

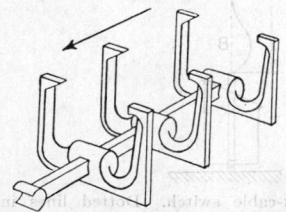

FIG. 11. Redler skeleton flights.

therefore be substantially lower than for the combination of a feeder, bucket elevator, and conveyor. If the path is a straight horizontal or inclined run, the continuous-flow conveyor will cost as much or more than a flight conveyor and at least twice as much as a screw conveyor.

Continuous-flow installations usually are of moderate capacity—40 tons/hr. or less—though some have been installed for capacities upward of 200 tons/hr. The capacity is a function of the cross section of the duct, the speed of the element, and the weight per cubic foot of the material. Roughly the capacity will correspond to 80 per cent of the volume swept through. Thus, if the cross section of the duct is 1 sq. ft., the speed 40 ft./min., and the weight 50 lb./cu. ft., the capacity will approximate $0.80 \times 1 \times 40 \times 60 \times 50/2000 = 48$ tons/hr.

The power requirement depends on the distance or lift, the handling rate, and the coefficient of friction of the material against the walls of the duct. The continuous-flow elevator requires more power than a bucket elevator. As a conveyor, if loaded to capacity of the duct, it requires more power than a flight conveyor. If partly loaded, it requires the same power as the corresponding flight conveyor—which it then resembles in its functioning.

The standard duct cross sections correspond quite closely in the machines mentioned. Figure 13 shows the approximate capacities in tons per hour for the standard duct sections for different weights of materials and speeds. Figure 14 gives the motor horsepower formulas for the Bulk Flo. The factor C is given in Table 13. The formulas will apply quite closely to the other machines listed.

The continuous-flow conveyor or elevator operating at moderate speed, properly applied, and handling material not characterized as destructive or with unbreak-

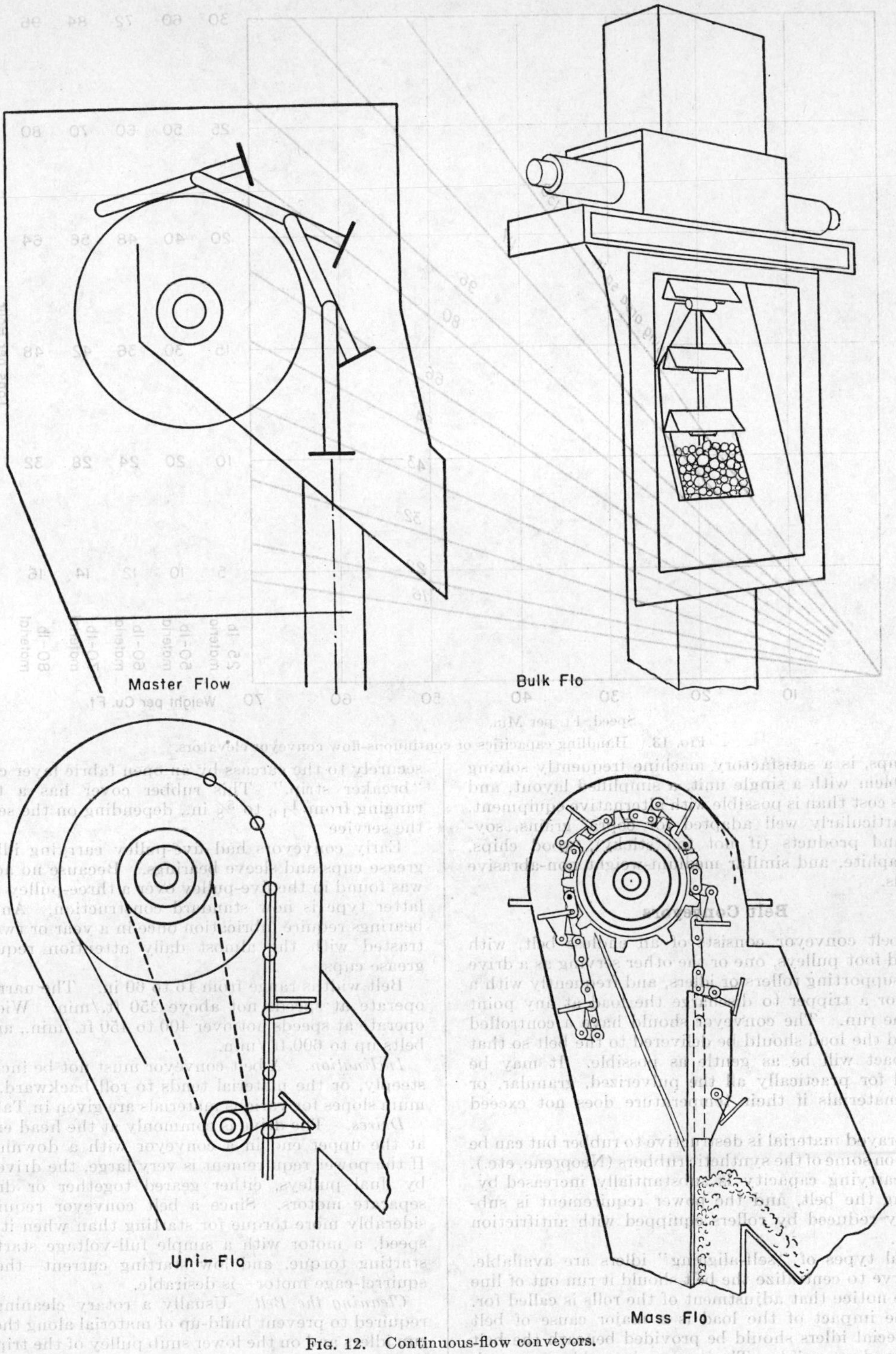

FIG. 12. Continuous-flow conveyors.

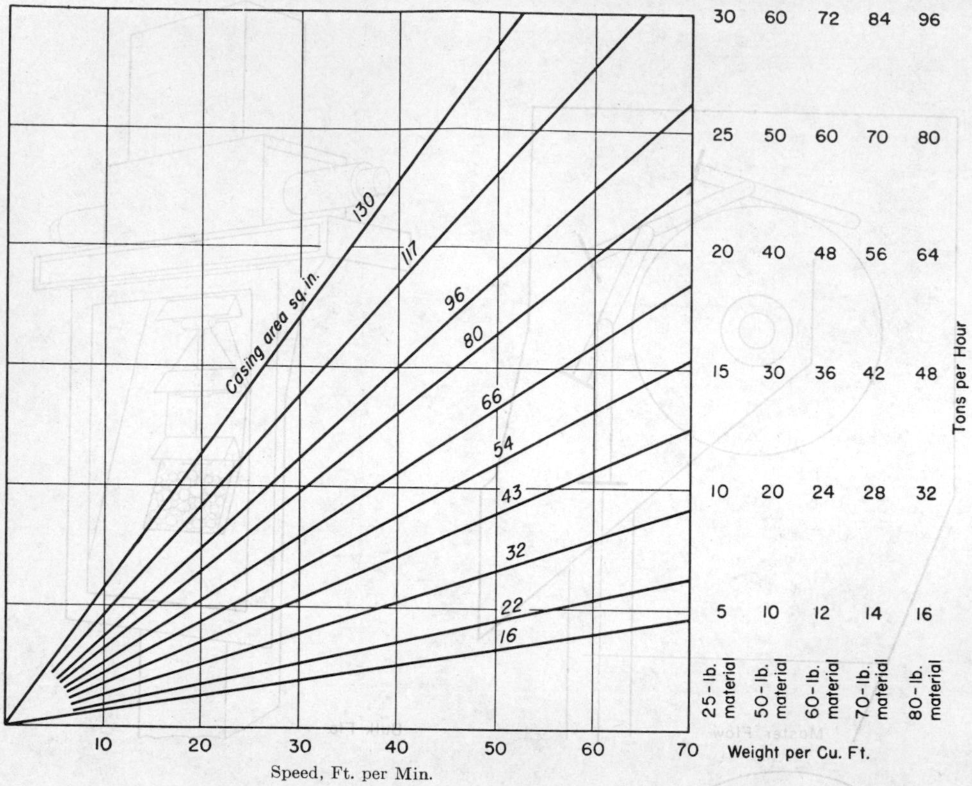

FIG. 13. Handling capacities of continuous-flow conveyor elevators.

able lumps, is a satisfactory machine frequently solving the problem with a single unit, a simplified layout, and with less cost than is possible with alternative equipment. It is particularly well adapted for copra, grains, soybeans and products (if not too sticky), wood chips, flake graphite, and similar medium-weight non-abrasive materials.

Belt Conveyors

The belt conveyor consists of an endless belt, with head and foot pulleys, one or the other serving as a drive pulley, supporting rollers or idlers, and frequently with a scraper or a tripper to discharge the load at any point along the run. The conveyor should have a controlled feed, and the load should be delivered to the belt so that the impact will be as gentle as possible. It may be specified for practically all the pulverized, granular, or lumpy materials if their temperature does not exceed 250°F.

Oil-sprayed material is destructive to rubber but can be handled on some of the synthetic rubbers (Neoprene, etc.).

The carrying capacity is substantially increased by troughing the belt, and the power requirement is substantially reduced by rollers equipped with antifriction bearings.

Several types of "self-aligning" idlers are available. They serve to centralize the belt should it run out of line and give notice that adjustment of the rolls is called for. Since the impact of the load is a major cause of belt wear, special idlers should be provided beneath the belt at the loading point. These may be rubber-covered, cushion-tired, or pneumatic-tired.

Conveyor belts usually have three or more plys of cotton duck, referred to as the "carcass," embedded in rubber. The cover on the carrying side is cemented securely to the carcass by an open fabric layer called the "breaker strip." This rubber cover has a thickness ranging from $\frac{1}{16}$ to $\frac{5}{8}$ in., depending on the severity of the service.

Early conveyors had five-pulley carrying idlers with grease cups and sleeve bearings. Because no advantage was found in the five-pulley over a three-pulley idler, the latter type is now standard construction. Antifriction bearings require lubrication once in a year or two as contrasted with the almost daily attention required for grease cups.

Belt widths range from 16 to 60 in. The narrow belts operate at speeds not above 250 ft./min. Wider belts operate at speeds not over 400 to 450 ft./min., and 60-in. belts up to 600 ft./min.

Inclination. A belt conveyor must not be inclined too steeply, or the material tends to roll backward. Maximum slopes for various materials are given in Table 14.

Drives. The drive is commonly at the head end but is at the upper end in a conveyor with a downhill slope. If the power requirement is very large, the drive may be by dual pulleys, either geared together or driven by separate motors. Since a belt conveyor requires considerably more torque for starting than when it is up to speed, a motor with a simple full-voltage starter, high starting torque, and low starting current—the double squirrel-cage motor—is desirable.

Cleaning the Belt. Usually a rotary cleaning belt is required to prevent build-up of material along the return-run idlers and on the lower snub pulley of the tripper. If the rotary cleaner is one that requires considerable pressure against the belt to function effectively, it requires from 5 to 10 hp. Rotary brushes also are liable to build up solidly with material, and if sharp particles become embedded in the mass the belt may be cut. Some ma-

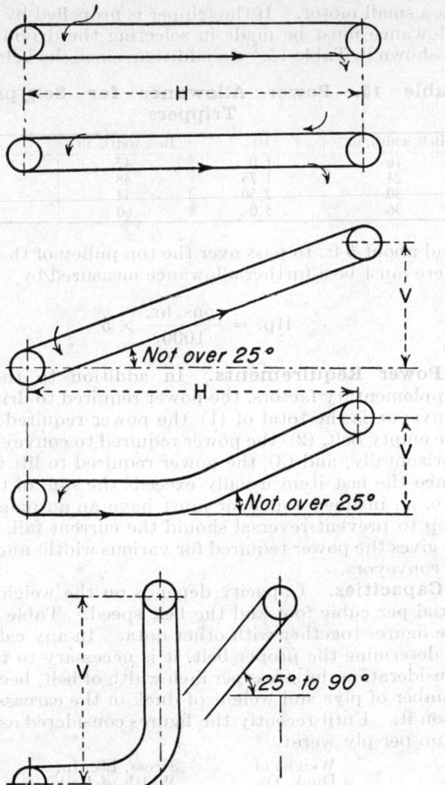

Motor Hp. = TPH x H x 0.002 x C

Motor Hp. = TPH x H x 0.004 x C

Motor Hp = TPH x (H x 0.002 + V x 0.001) x C

Motor Hp. = TPH x (H x 0.0032 + V x 0.003) x C

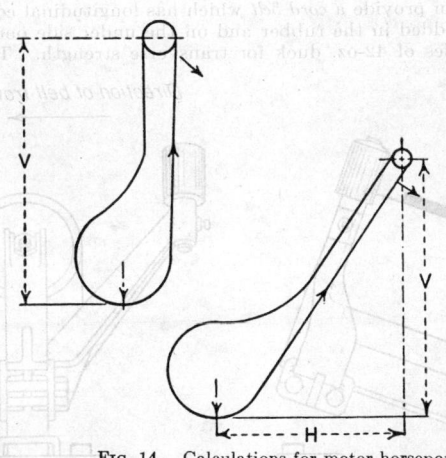

Motor Hp. = TPH x (V + 3) x 0.003 x C

Motor Hp. = TPH x (H x 0.002 + V x 0.003) x C

FIG. 14. Calculations for motor horsepower for Bulk Flow. See Table 13. (*Link-Belt Co.*)

Table 13. Material Factors for Formulas in Fig. 14

$C = 1$	Bran
	Coffee, ground
	Cocoa beans
	Flaxseed
	Graphite flakes
	Nut kernels
	Soap flakes
	Soybeans
	Coconut, shredded
	Copra, ground
$C = 1.2$	Beans
	Coal ¼ × 0, ¾ slack
	Coffee beans
	Charcoal
	Copra cake
	Flour 1.2–1.5
	Sawdust 1.2–1.5
	Soybean meal 1.2–1.5
	Wheat
	Wood chips, dry
$C = 1.5$	Talc
	Starch, pulverized
	Salt, pulverized 1.5–2
	Salt, rock
	Wood chips, wet
$C = 2$	Clay, 2–2.5
	Fly ash
	Lime, pebble
	Starch, granular
	Sugar, granular
	Soda ash, light
	Sugar, pulverized 2–2.5
	Zinc oxide
$C = 2.2–2.5$	Alum
	Borax
	Cork, ground
	Lime, packs
	Soda ash, heavy
	Tobacco scraps
	Limestone, pulverized

Table 14. Maximum Slopes, Deg., Recommended for Various Materials

Bulk chemicals, 18	Damp sand, 20
Coal, bituminous slack, 20	Dry sand, 16
Coal mine run, 18	Crushed ore, 20
Grains, 18	Wood chips, 28
Bank gravel, 18	Lime, 20
Washed gravel, 12	

terials cannot be cleaned off by a rotary brush, and more effective means must be employed, such as the multiple scraper shown in Fig. 16. For extremely sticky material such as wet clay, diagonal scrapers of thin stainless steel may be best.

Trippers. Trippers may be hand-propelled by a crank, or self-propelled either by power derived from the belt or by a small motor. If the tripper is propelled by the belt, allowance must be made in selecting the driving motor, as shown in Table 15. In addition, since the belt lifts the

Table 15. Power Allowance for Self-propelled Trippers

Belt width, in.	Hp.	Belt width, in.	Hp.
16	1.0	42	4
24	1.75	48	5
30	2.50	54	6
36	3.0	60	7

load about 5 ft. to pass over the top pulley of the tripper, there must be a further allowance measured by

$$\text{Hp.} = \frac{\text{tons/hr.}}{1000} \times 5$$

Power Requirements. In addition to the above supplementary factors, the power required to drive a belt conveyor is the total of (1) the power required to drive the empty belt, (2) the power required to convey the load horizontally, and (3) the power required to lift the load. Since the last item usually exceeds the sum of the other two, an inclined conveyor must have an automatic back stop to prevent reversal should the current fail. Figure 17 gives the power required for various widths and lengths of conveyors.

Capacities. Capacity depends on the weight of material per cubic foot and the belt speed. Table 16 gives the figures together with other data. In any calculation to determine the proper belt, it is necessary to take into consideration the stress per inch width of belt, because the number of plys and weight of duck in the carcass depend upon it. Until recently the figures considered as a maximum per ply were:

Weight of Duck, Oz.	Stress, Lb./In. Width of Belt
28	25
32	30
36	35
42	45

Extensive research since the war has produced special belt constructions which permit far greater stresses than given in the table above. The B. F. Goodrich Company can provide a *cord belt* which has longitudinal cords embedded in the rubber and on the under side one or two plies of 42-oz. duck for transverse strength. The cord

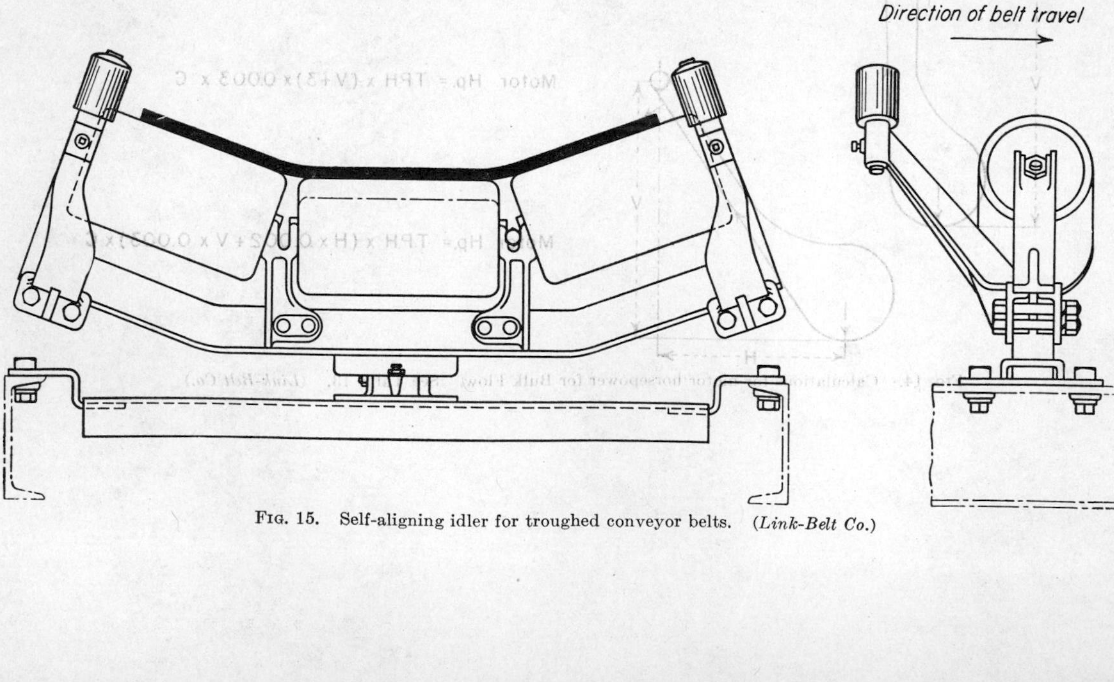

FIG. 15. Self-aligning idler for troughed conveyor belts. (*Link-Belt Co.*)

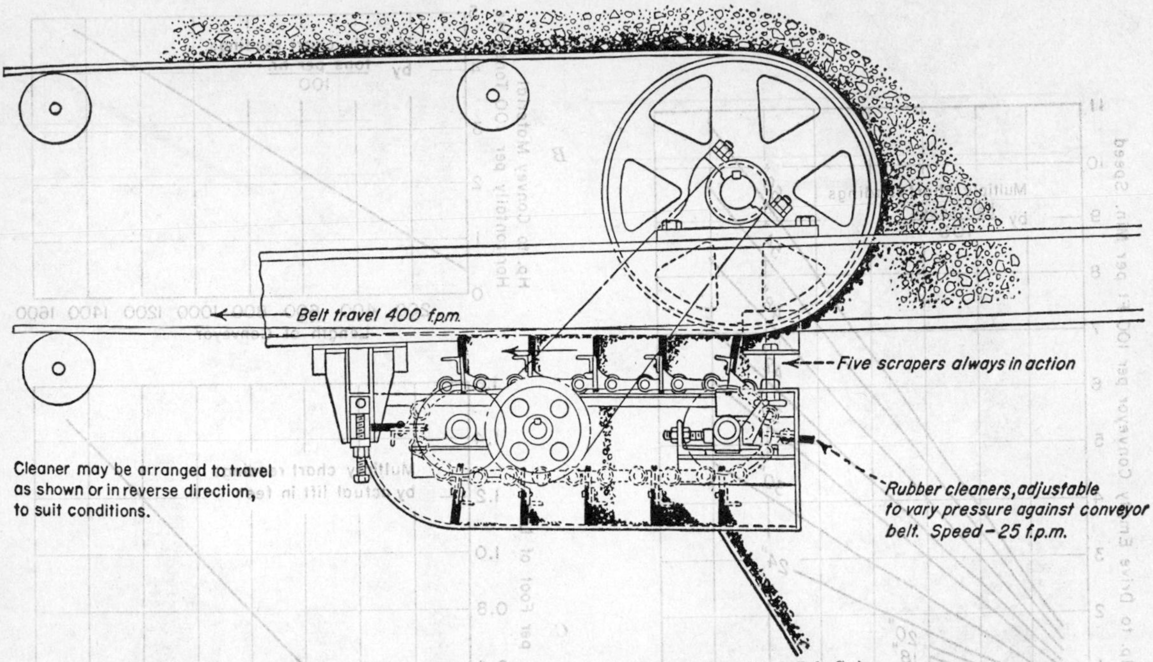
Belt travel 400 f.p.m.
Five scrapers always in action
Cleaner may be arranged to travel as shown or in reverse direction, to suit conditions.
Rubber cleaners, adjustable to vary pressure against conveyor belt. Speed – 25 f.p.m.

FIG. 16. Hudson belt cleaner for sticky materials. (*Link-Belt Co.*)

belts are standardized for 50, 70, and 100 lb./in. of width per ply (cord layers). The United States Rubber Company can provide a belt with carcass made of a combination of nylon and specially treated duck fabric rated for 85, 110, and 150 lb./in. of width per ply. The Goodyear Tire and Rubber Company can provide a wire-cable belt rated for 1000 to 3000 lb./in. of belt width, and a belt with single-layer large-diameter cords, to be specified where extremely high tensions are not required. If a conveyor of great length and belt tension is under consideration, the above manufacturers should be consulted to determine whether these belts should be specified notwithstanding their higher cost.

Applications. The belt conveyor with supporting structure costs more than a flight conveyor, bucket elevator, or skip hoist, but it has reached such a high state of development that it is the outstanding machine for very large capacities, especially when the distance between centers is great. It cannot withstand misuse or neglect; but, when properly designed and serviced, it assures long life with very low maintenance costs.

Sliding Belts. Instead of troughed belts carried on rollers, a belt sliding on a deck may be advisable. These are seen frequently in the shipping room of department stores for handling miscellaneous parcels, in post offices for handling mail bags, in chemical plants for handling miscellaneous light waste, etc.

The decking is preferably of maple strips. If of steel, the deck should be perforated at intervals to relieve the vacuum effect between the bottom of the belt and the deck. Stitched duck belts or Balata belts are best. The speed should be as low as possible. The return run as a rule is carried on 4-in. straight face rollers. Since the belt is in sliding contact with the deck instead of supported on rollers with antifriction bearings, the power requirement is much higher.

Carriers

Carriers of the chain-and-bucket type are of two kinds, those intended for handling materials around a path in a vertical plane, and those for handling material in a horizontal runaround path.

For a runaround path in a vertical plane, the pivoted-bucket carrier, consisting of pivoted buckets suspended between two strands of roller chain, is used. The buckets overlap but do not interfere at the bends if attached to

Table 16. Belt Conveyors; Capacities, Speeds, and Sizes of Pieces*

Width of belt, in.	Belt ply		Cross section of load, sq. ft.	Cu. ft./hr. at 100 ft./min.	Capacities, tons/hr. at 100 ft./min., at weight/cu. ft. of material of				Max. advisable belt speed, ft./min.	Medium size of material, in.	
	Min.	Max.			50 lb.	75 lb.	100 lb.	150 lb.		80% under	20% not over
12	3	4	0.084	504	12.6	18.9	25.2	37.8	300	1½	2
14	3	5	.114	686	17.1	25.6	34.3	51.3	300	2	3
16	3	5	.149	896	22.4	33.6	44.8	67.2	300	2½	4
18	4	6	.189	1,134	28.3	42.5	56.7	84.9	350	3	5
20	4	6	.233	1,400	35.0	52.5	70.0	105.0	350	3½	6
24	4	6	.336	2,016	50.4	75.6	100.8	151.2	400	4½	8
30	5	7	.525	3,150	78.7	118.1	157.5	236.2	500	7	12
36	6	8	.751	4,536	113.4	170.1	226.8	340.2	600	8	16
42	6	9	1.029	6,174	154.3	231.5	308.7	463.0	600	10	20
48	6	10	1.333	8,064	201.6	302.4	402.2	604.8	600	12	24
54	8	11	1.701	10,206	255.1	382.6	510.2	765.3	600	13	26
60	8	12	2.100	12,600	315.0	472.5	630.0	945.0	600	14	28

* Link-Belt Co.

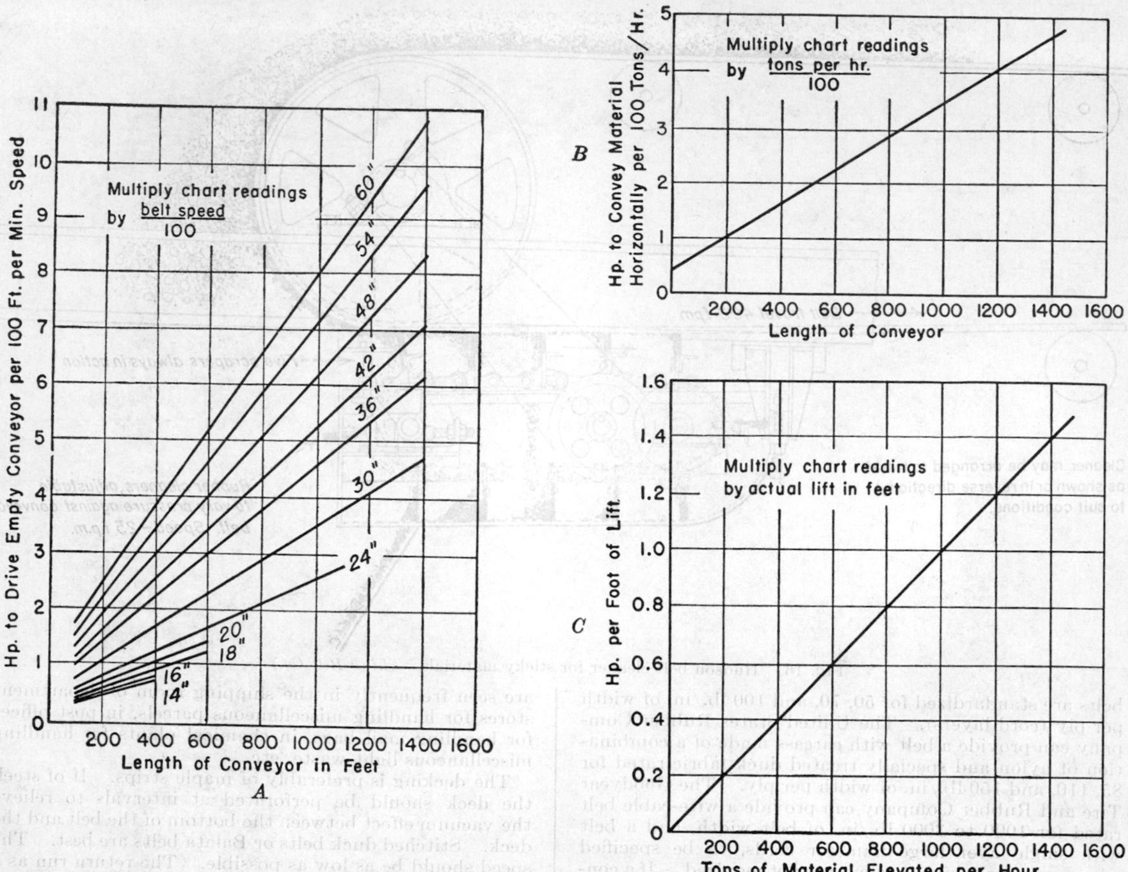

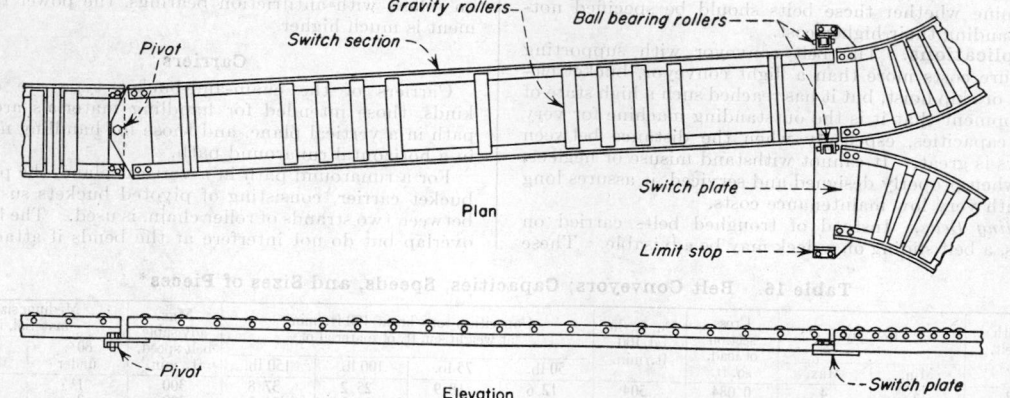

FIG. 17. Belt conveyors. Horsepower to drive a belt conveyor is the sum of power to drive the empty belt (*A*) + power to convey the material horizontally (*B*) + power to lift the load, if the conveyor is inclined (*C*).

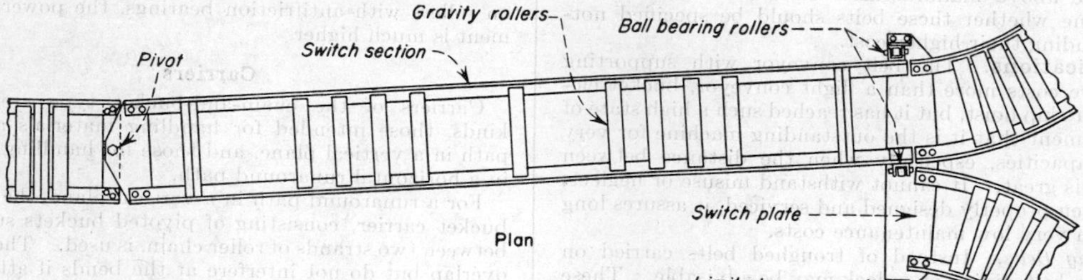

FIG. 18. Gravity roller conveyor, showing switch section, indicating flexibility of such equipment.

rearward cantilever extensions of the chain links. The buckets are loaded by a feeder at any point along the lower run and discharged by a stationary or movable tripper along the upper run which tilts the buckets through an arc of about 130 deg. The carrier is essentially a slow-speed machine operated at about 50 ft./min. Standard capacities of the Peck carrier (Link-Belt Co.) are given by Table 17. For a given capacity, a carrier with long pitch and narrow buckets will cost less than a carrier with shorter pitch and wider buckets, since a substantial part of the cost is in the twin roller chains.

Power Requirements. The manufacturers will furnish on request charts giving the power requirements of their machines. As illustrating the very low power requirement of this type of conveyor, Fig. 19 charts the power for the Peck carrier with 18 in. pitch and bucket widths of

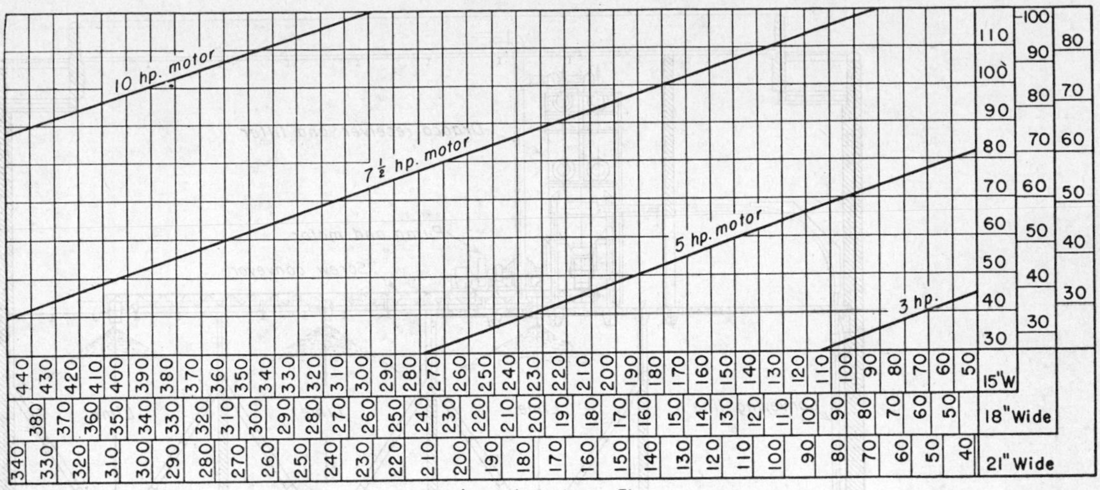

FIG. 19. Motor determinations for pivoted bucket carriers of 18 in. pitch, bucket widths 15, 18, and 21 in. (*Link-Belt Co.*)

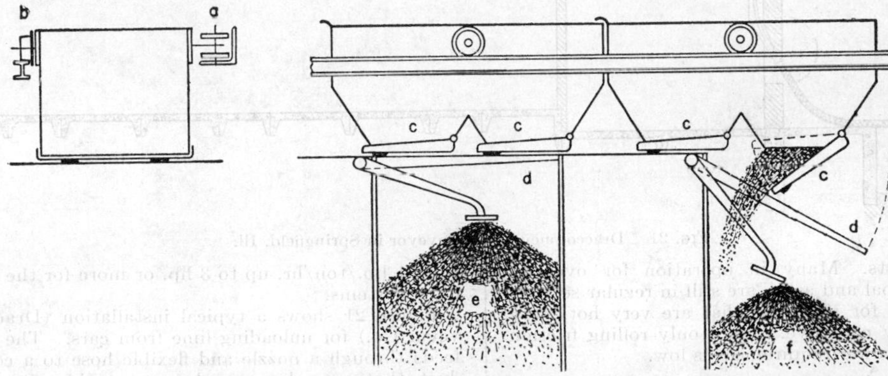

FIG. 20. The Davis Horizontal Runaround Carrier. Specifically for fragile material. *a*. Chain. *b*. Supporting rollers. *c*. Outlet trays. *d*. Gap bridging lever actuated by lug on every twelfth bucket. *e*. Chute filled and buckets crossing without discharge. *f*. Partially filled chute receiving discharge. Lever connections to lug not shown.

15, 18, and 21 in. for horizontal centers of 50 to 400 ft. and vertical centers of 30 to 100 ft. when operating at full capacity with material weighing 50 lb./cu. ft.

Table 17. Capacities of Peck Carriers with Material Weighing 50 lb./Cu. Ft. at the Speeds Noted

For materials at other weights the capacities are proportional

Bucket pitch × width, in.	Capacity, tons/hr. 50 lb. material	Speed, ft./min.
18 × 15	15– 20	30–40
18 × 18	20– 25	30–40
18 × 21	25– 30	30–40
24 × 18	35– 45	40–50
24 × 24	50– 60	40–50
24 × 30	60– 75	40–50
24 × 36	70– 90	40–50
30 × 24	80–105	45–60
30 × 30	95–130	45–60
30 × 36	115–155	45–60
36 × 36	160–255	50–80

Horizontal Runaround Carriers. Until recently there has been no suitable horizontal runaround carrier for fragile materials. If the material is not fragile, the runaround Redler conveyor fills the requirements, but degradation results from dragging material along a trough. The current carrier is the invention of N. L. Davis (Fig. 20). It was developed for the continuous distribution of magnesium chloride flakes in the manufacture of metallic magnesium.

Rectangular buckets are attached to a single strand of roller chain suitably guided around terminal sprockets. The buckets have hinged flaps, normally held horizontal by the trough plate. As the buckets cross the outlet chutes, the flaps tilt downward to discharge. To prevent degradation, which would occur if the flaps dragged across the material in a filled chute, a pivoted $<$ piece is arranged as shown. The upper leg of the $<$ forms a continuation of the trough when in the raised position. The lower leg terminates in a plate. The $<$ is lifted and dropped by a lug on every twelfth bucket. When the level of the load in the chute nears the top, the plate resting on it causes the upper leg to remain horizontal so that the bucket flaps do not discharge to that chute until the load therein has lowered. Degradation of this extremely fragile material was thus almost completely eliminated.

The Link-Belt Co. will furnish on request a booklet showing applications of this carrier.

Applications. The pivoted-bucket carrier is a costly but remarkably efficient conveyor with extremely low

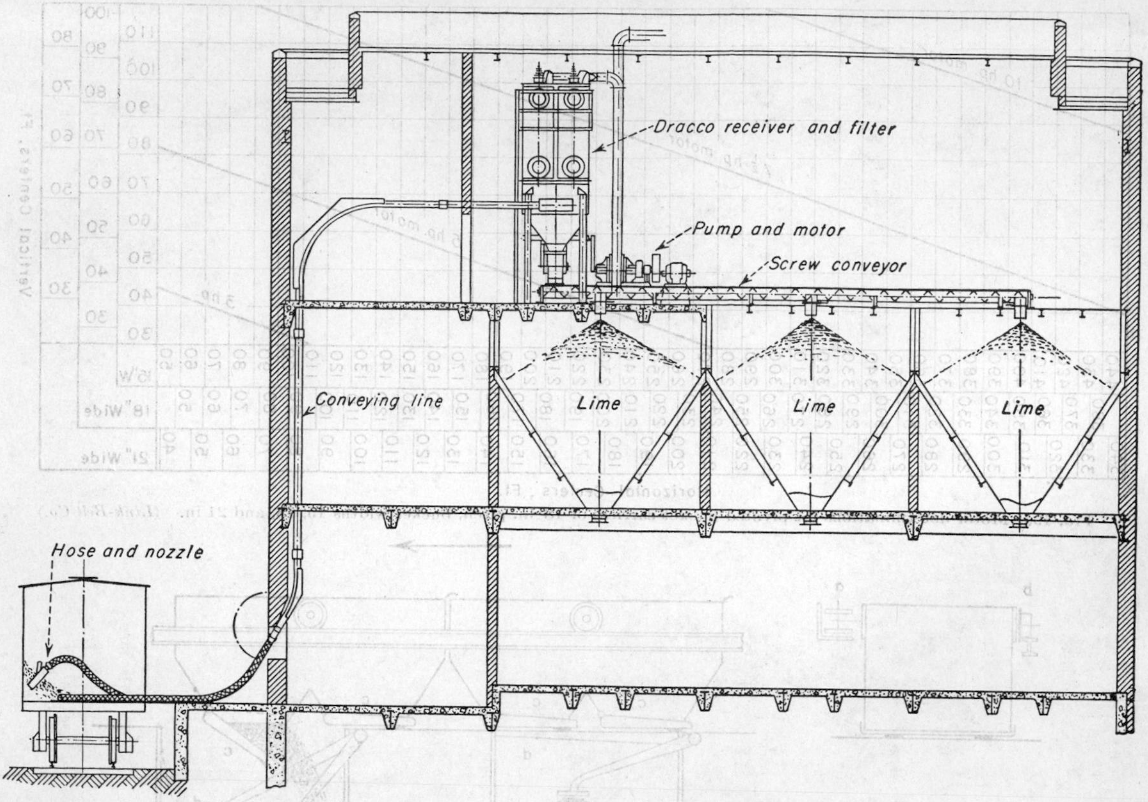

FIG. 21. Dracco pneumatic conveyor in Springfield, Ill.

maintenance costs. Many in operation for over 40 years handling coal and ashes are still in regular service. It is well suited for materials that are very hot, very abrasive, or very corrosive. Since only rolling friction is involved, the power requirement is low.

Pneumatic Conveyors

The pneumatic conveyor consists of a duct or pipe in which an air stream is maintained by the propulsive action of a steam jet, a blower, or a fan. There are two types, usually referred to as the pressure system and the suction system.

In the pressure system a blower forces air through the duct, and the material is fed to the duct through a rotary air-lock gate. In the suction system the exhauster draws the air through the duct and through a separator and dust collector. The material feeds out from the separator through an air-lock gate.

The air stream may be maintained by a steam jet located at the upturn of the duct. The steam expands as it issues from the nozzle, causing a partial vacuum behind it and a positive push beyond the jet. The modern steam pneumatic system locates an annular jet between a centrifugal receiver and a centrifugal dust collector. The receiver has a flap gate actuated by a relay that functions to shut off the steam periodically and open the gate so that the batch of ashes that has entered during the preceding minute or two may discharge into a bin beneath. The steam pressure should be at least 75 lb./sq. in.

The pneumatic conveyor requires considerably more power per ton per hour moved than any type of mechanical conveyor. For industrial installations, the ratio is from 1 hp./ton/hr. up to 3 hp. or more for the ash-handling systems.

Figure 21 shows a typical installation (Dracco Conveyor Co.) for unloading lime from cars. The material feeds through a nozzle and flexible hose to a conveying duct that extends upward to a combination receiver and dust collector. The material feeds through an air-lock gate to a screw conveyor for distribution to the storage bins. The handling rate for such a system with single nozzle will range from 15 to 30 tons/hr.

Applications. The pneumatic conveyor is suitable for granular and pulverized materials that are free-flowing and will not pack. Materials such as wet ashes, acid phosphate, and magnesium chloride will build up at the elbows and choke the duct. Although the power requirement is very high, the pneumatic conveyor can solve problems in material handling that a mechanical conveyor cannot attempt, such as when the path is tortuous, when the material is at a very high temperature, or when it is required to handle material from cars with absolute elimination of dust. One great advantage is that no moving parts are in the vicinity of the operator and there is practically no risk of personal injury even with an unskilled man. As a rule the handling capacities are moderate, up to 40 tons/hr. or so; but at European ports, where the pneumatic conveyor is used for unloading cargoes of grain, capacities upward of 200 tons/hr. are not unusual.

Hydraulic Conveyors

Industrial applications of hydraulic conveyors are chiefly in connection with the disposal of ashes from steam plants.

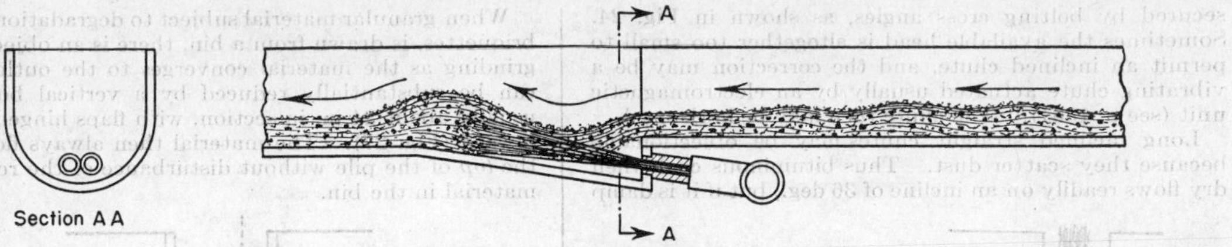

FIG. 22. Booster-jets hydraulic sluice.

A high-velocity jet flushes the ashes from the pits to the conveying duct, which is either a cast-iron pipe or an open trough. Booster jets at intervals of about 50 ft. flush the ashes to a sump from which the water may be drained out and the ashes removed by grab bucket. An alternative is to add make-up water and pump the thinned mixture to a disposal point. If the conditions permit, the ashes can be flushed directly for a distance of 400 ft. or so.

Although requiring considerable power per ton of ash removed, the hydraulic method is often preferred, since all danger to the operators is eliminated, there are no moving parts, operation is dustless, and the handling rate may be very large if the necessary volume of water is available. Usually 5 lb. water/lb. ash is required. With 1000 gal./min. at 100 lb. pressure, 1 ton/min. may be disposed of. This includes the make-up water for pumping. Thus the conveyor need operate but a few minutes to take care of the output from a medium plant.

Coal handling by hydraulic conveyor is possible but has the objectional feature that, after the water drains off, the coal will not flow from a bunker or bin. For stocking out a ground-storage reserve, if the location is such that the runoff will not cause complications, the hydraulic method sometimes offers a convenient and low-cost installation. A 10- or 12-in. light steel pipe is located to receive the discharge from the coal elevator and is extended out over the storage area with a downward slope of about 10 deg. A water jet introduced at the receiving end will convey coal in about the proportion of 10 per cent of the volume of water, if the pipe is straight.

Assuming a pipe 200 ft. long and a contemplated handling rate of 40 tons/hr., the coal volume is 1600 cu. ft./hr., and the volume of water is therefore 17,600 cu. ft./hr., or 293 cu. ft./min. The head loss is 8.8 ft. in 200 ft., and the weight of mixture is about 64 lb./cu. ft. The theoretical horsepower is $(293 \times 64 \times 8.8) \div 33{,}000 =$ 5 water hp., *i.e.*, a 10-hp. motor.

Vibrating Feeders, Screens, Conveyors

If a horizontal plate is subjected to vertical vibrations, pulverized granular or lumpy materials upon it will be thrown upward repeatedly more or less in synchronism with the pulsations. If the plate is inclined downward, there is a progressive forward movement, and the result is the very effective *vibrating feeder*. The rate of feed may be changed by varying the intensity of the vibrations or by changing the slope of the plate.

If the plate is perforated, it is the *vibrating screen*. The particles bounce upward, and eventually small particles drop through. If the vibrations are rectilinear but inclined at an angle with the plate, the material is thrown forward as it rebounds, and the plate, then in the form of a trough, forms the *vibrating conveyor*. It need not be inclined and in fact may have an upward slope of 5 deg. or less.

Vibrating feeders and screens are actuated by (1) a high-speed unbalanced pulley, (2) two *unbalanced pulleys* geared together so that the weights give a resultant rectilinear thrust, (3) electromagnetic vibrations from one or more solenoids, (4) electromagnetic vibrations by a reciprocating armature mounted between two pole pieces, or (5) vibrations by the slower pulsations secured by a crankshaft or eccentric shaft connected to the plate then mounted on rearward-inclined spring supports. This last forms the "grasshopper conveyor." The trough may have considerable length and provides an excellent conveyor for extremely hot materials, small castings, corrosives, etc.

The vibrating feeder actuated by solenoid is now considered preferable to the well-known reciprocating feeder because of adjustment of the rate of feed, elimination of lubrication, and extremely small power requirement.

The vibrations induced by an unbalanced pulley are not too effective for a feeder because of the circular cycle induced in the plate, as contrasted with the propulsive effect of straight-line pulsation. However, as a vibrating screen it does function very well.

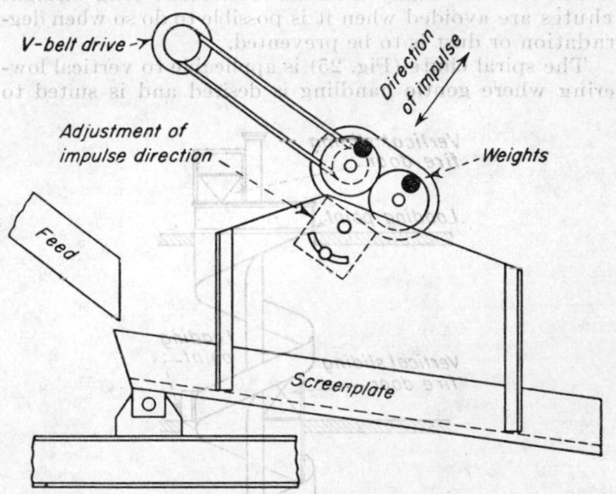

FIG. 23. Vibrating screen actuated by geared unbalanced pulleys.

One limitation in the application of vibrating plates (feeders and conveyors) is that the material must be dry or nearly dry. When it is wet, the water is squeezed out against the plate and the pulverized material tends to adhere.

The inclined vibrating screen is an excellent dewaterer. The vibrations effectually clear surface water from the particles, and it passes through the perforations with the undersize, while the dewatered lumpy or granular material passes out over the end. The device is widely used for dewatering coal in washery plants.

Chutes

The simplest chute—the straight trough—may by trial be found to be on too flat a slope so that the material will back up, or on too steep a slope so that the material will accelerate. If the slope is too flat, the difficulty often may be corrected by flooring the chute with a thin sheet of stainless steel or aluminum or with plate glass. If it is too steep, the equivalent of a reduced slope may be secured by bolting cross angles, as shown in Fig. 24. Sometimes the available head is altogether too small to permit an inclined chute, and the correction may be a vibrating chute actuated usually by an electromagnetic unit (see p. 1361). The chute then may be horizontal.

Long inclined straight chutes may be objectionable because they scatter dust. Thus bituminous coal when dry flows readily on an incline of 36 deg., but if it is damp

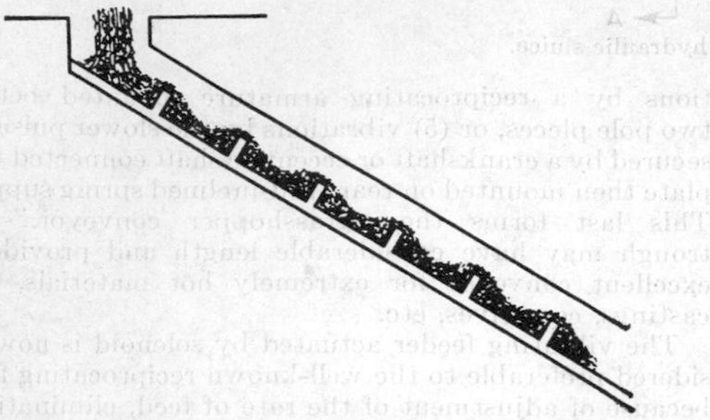

FIG. 24. Baffle plates reduce rate of flow on steep chute.

or if the chute is rusted it will not flow. So it is usual to specify a 43- to 45-deg. slope. Coal will flow on this slope even when quite damp but will build up speed and shatter or scatter dust. Because this same difficulty holds true for many granular chemicals, long straight chutes are avoided when it is possible to do so when degradation or dust is to be prevented.

The spiral chute (Fig. 25) is applicable to vertical lowering where gentle handling is desired and is suited to

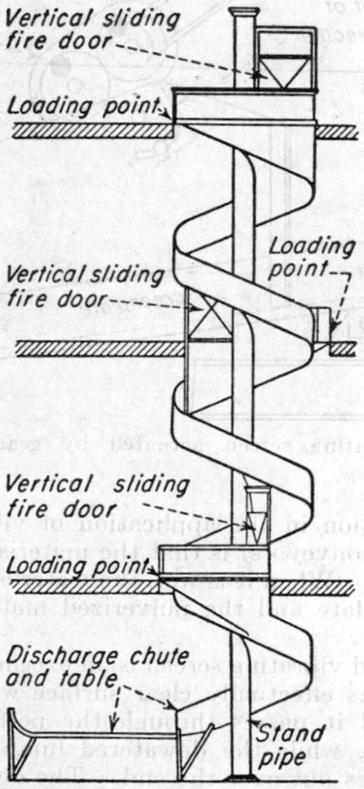

FIG. 25. Open-type single-blade spiral chute. Double-blade and enclosed types are available.

either bulk or packaged material. Bulk material skims around the outer circumferential plate until it reaches the mass already in the bin, then discharges radially inward. An alternative, frequently used in anthracite-coal pockets, is a vertical box chute with steps projecting alternately from two opposite sides, as in Fig. 26. The material cascades with little degradation.

When granular material subject to degradation, such as briquettes, is drawn from a bin, there is an objectionable grinding as the material converges to the outlet. This can be substantially reduced by a vertical box chute, usually about 24 in. in section, with flaps hinged to open inward (Fig. 27). The material then always flows from the *top* of the pile without disturbance of the rest of the material in the bin.

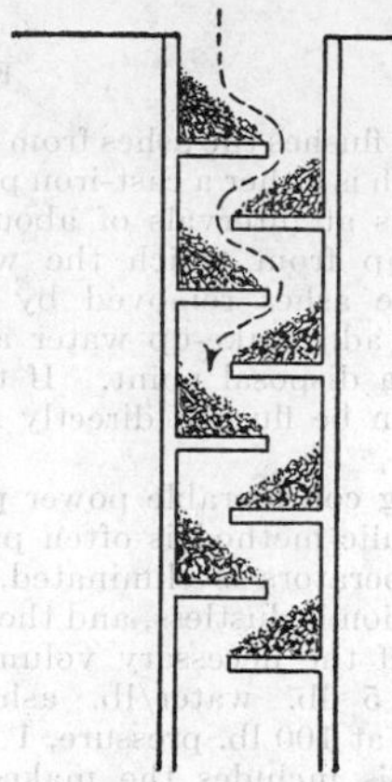

FIG. 26. Trickler chute.

Recent experience indicates that aluminum is a better material than sheet steel for chutes handling a wide variety of somewhat abrasive granular and pulverized materials; and, since its cost will probably be much lower than heretofore, it will no doubt come into general use. Also now available is stainless-clad sheet steel, a sheet of open-hearth steel with a facing of stainless steel. If chute wear is excessive, it may be reduced by lining the bottom plate with a grid or grating. The material fills up the interspaces and thereafter flows over the surface

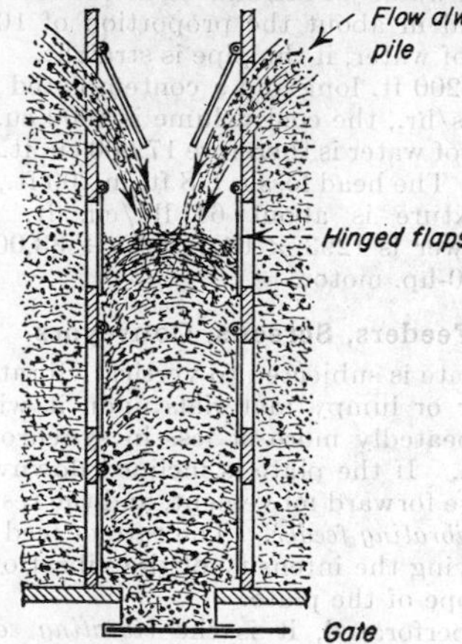

FIG. 27. Outflow chute to reduce degradation.

thus formed. However, it may then be necessary to increase the slope.

INDUSTRIAL TRUCKS, TRACTORS, AND TRAILERS

By Fred A. Miller

Container and Miscellaneous Parts Handling

In the engineering of industrial operations many installations have been designed for thorough mechanization of all production steps up to and beyond the packaging of the product. Increasing costs for both labor and space emphasize the need for efficient low-cost handling. The following paragraphs describe the handling devices commonly used to reduce materials handling costs.

Stevedore Trucks. The two-wheel stevedore truck (Fig. 28) is frequently selected for operations requiring the handling of a few packages per day. This inexpensive

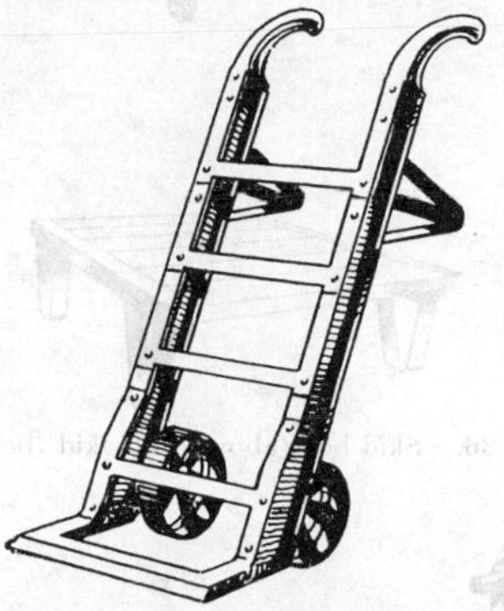

FIG. 28. Stevedore truck.

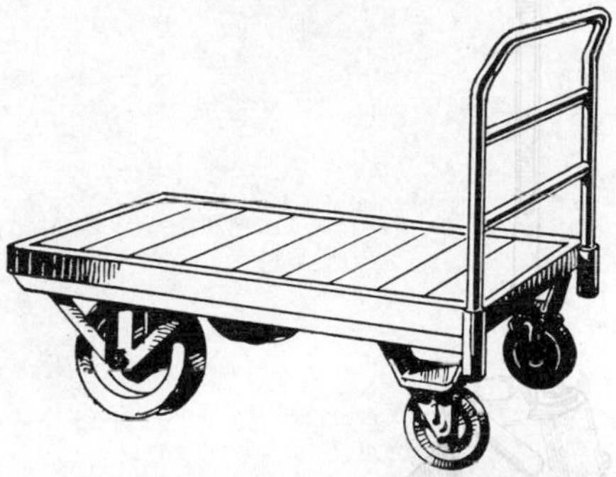

FIG. 29. Four-wheel factory floor truck.

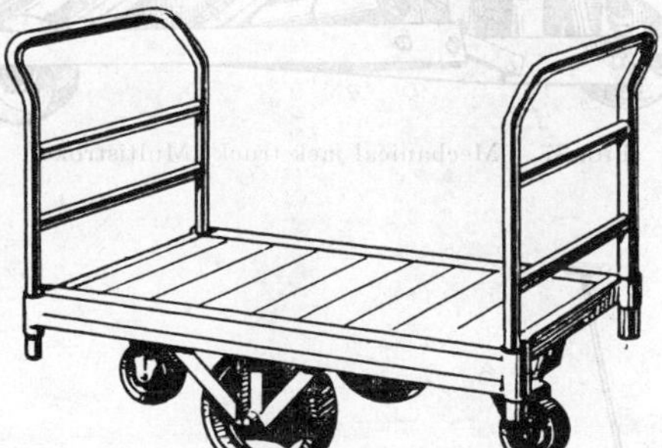

FIG. 30. Four-wheel factory truck.

device increases the handling capacity of the individual laborer. If the volume is considerable, it will be found that the use of this equipment makes for high labor costs, and more efficient equipment may be easily justified.

Factory Trucks. The four-wheel factory truck (Figs. 29 and 30) provides a carrying capacity of four to ten times that of the two-wheel truck, but each package must be lifted into place and removed. Hence the gain in carrying capacity is partly offset by an increased unit loading time. This truck may be profitably employed for infrequent movements of bulky or unwieldy items, and it is also used in conjunction with tote boxes (or other containers) for some retail stock-picking operatings.

Stackers. In manual handling systems, it is frequently necessary to stack the material to a considerable height in order to conserve floor space. Hand stacking is costly. The portable stacking machine (Fig. 31) was developed to reduce the cost of stacking and permit

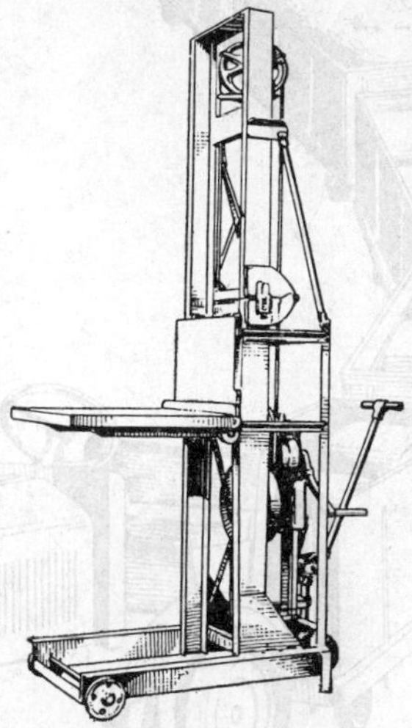

FIG. 31. Portable stacking machine.

higher piling. However, most engineers consider these units hazardous, because the usual operation requires the manual transfer of containers from the elevated platform to the pile. No suitable protection is provided for the operator, who must balance himself some distance off the floor on the stacker platform and on the pile of containers while he is moving the load into place.

Trailers and Tractors. The next step in speeding the movement of materials, particularly with long hauls and ramps, is the application of coupling devices to the factory trucks and the use of power-driven tractors to move trailer trains (Fig. 32). The adoption of the tractor and trailer system enormously reduces the cost of material movement but does not alter the high cost of loading the material on the floor truck or trailer and unloading it. Trailers are of three principal types, all having four wheels and differing mainly in the method of wheel mounting. They are (1) caster, (2) fifth-wheel, and (3) four-wheel steer (Figs. 33, 34, and 35). The caster trailer has two swiveling wheels in front and the rear wheels are mounted on a fixed axle. It is the type of trailer best adapted to either power haulage in trains or local manual handling indoors. The fifth-wheel trailer has one fixed axle and one swiveling axle with a tongue attached for coupling to a tractor or tandem trailer. The four-wheel-steer trailer has all four wheels mounted on steering knuckles connected by link bars, so that all four wheels steer.

Skids. The first step in reducing the cost of loading and unloading material was the adoption of the skid lift system using live, semilive, or dead skids or skid boxes. In practice the material is piled on the skid (Fig. 36) at the receiving location or the last processing or manufacturing point, transported, stored, withdrawn from storage, and moved to the point of use or delivery without rehandling of the individual packages. This is the "unit load" system.

Jack-lift Trucks. The mechanical (and later hydraulic) jack-lift truck (Fig. 37) was developed to move unit loads. Where the semilive type of skid (Fig. 38) is employed, it could be easily moved with the aid of a

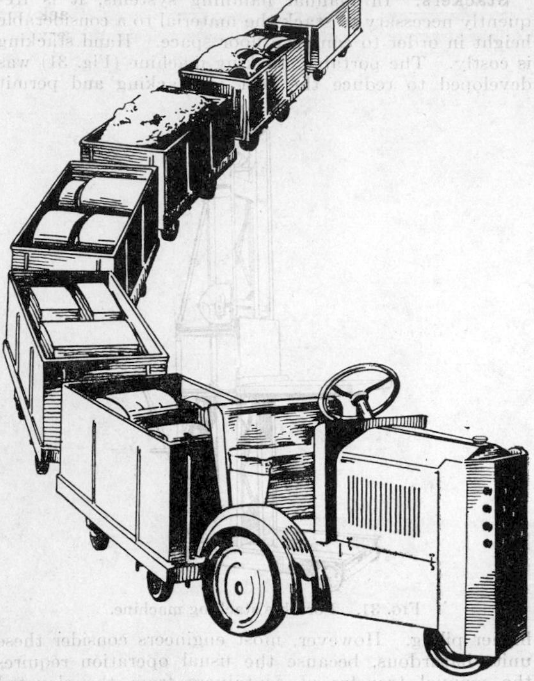

FIG. 32. Tractor-trailer train.

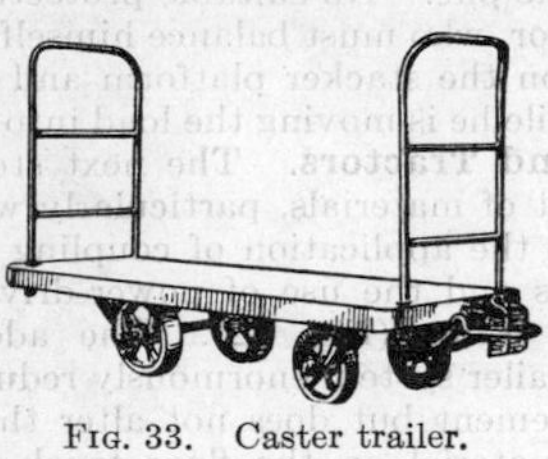

FIG. 33. Caster trailer.

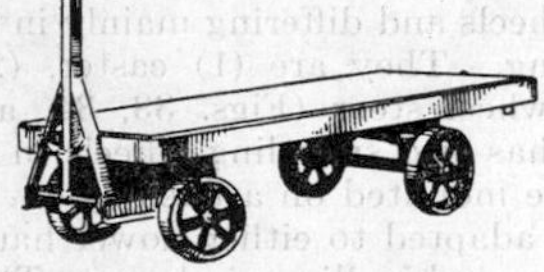

FIG. 34. Fifth-wheel trailer.

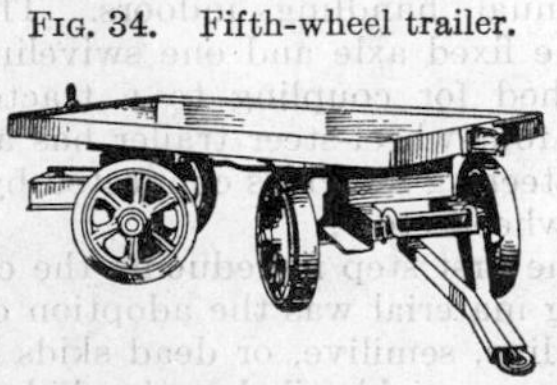

FIG. 35. Four-wheel-steer trailer.

simple lever on wheels. Where heavy loads or many skids are handled, the dead skid is usually employed, and movement of individual skid loads is accomplished by the jack-lift truck, which is a four-wheel unit with a hori-

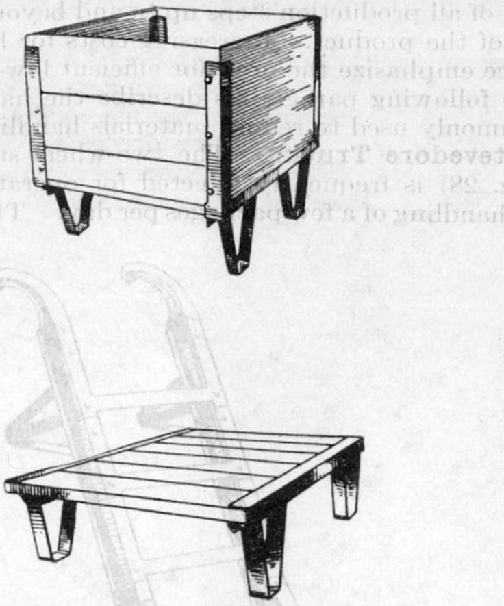

FIG. 36. Skid box (above) and skid (below).

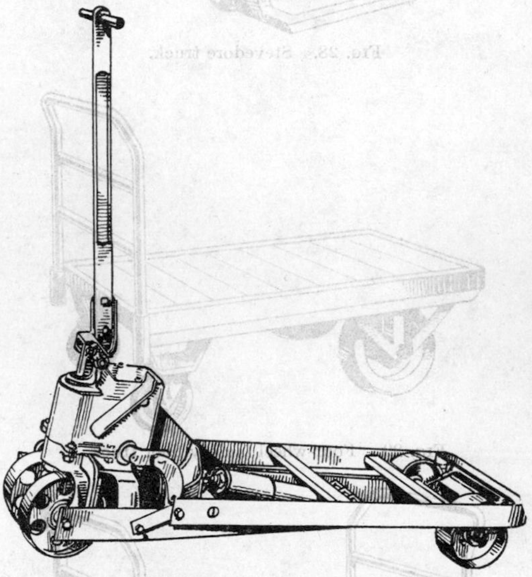

FIG. 37. Mechanical jack truck (Multistroke).

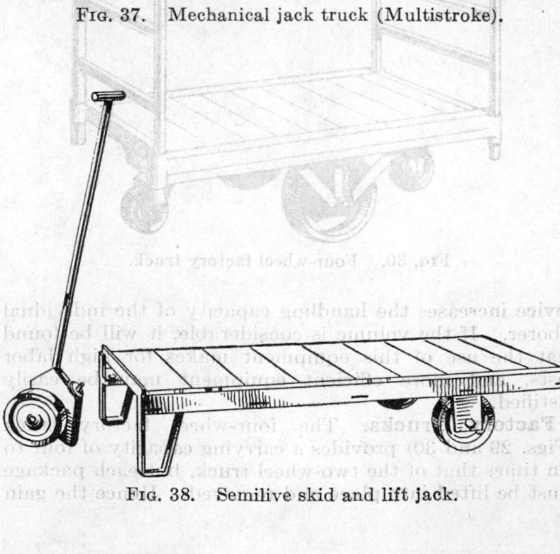

FIG. 38. Semilive skid and lift jack.

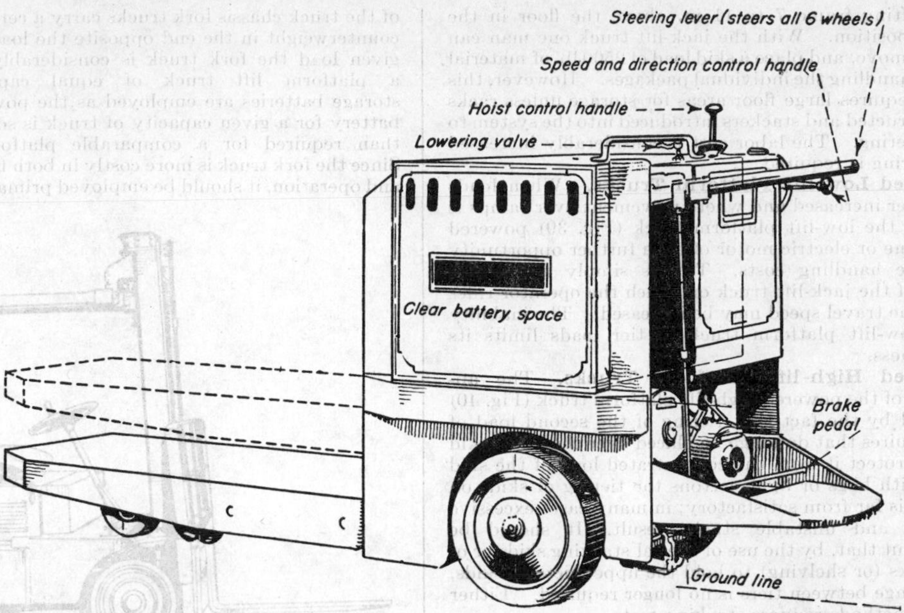

FIG. 39. Powered low-lift platform truck.

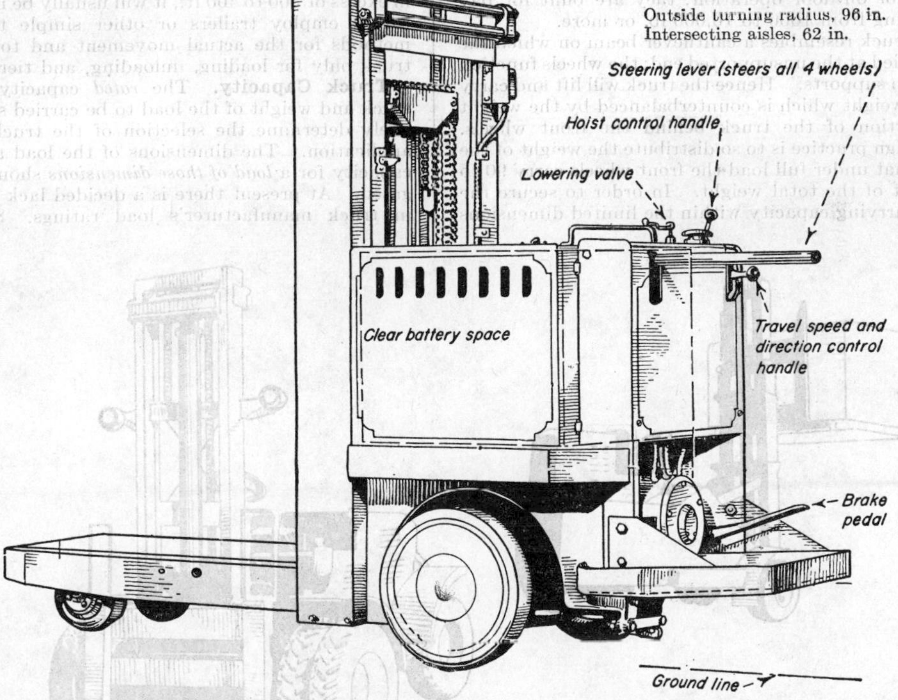

FIG. 40. Powered high-lift platform truck.

zontal lifting frame 7 to 11 in. above the floor in the lowered position. With the jack-lift truck one man can pick up, move, and place a skid load of 2500 lb. of material without handling the individual packages. However, this system requires large floor areas for storage unless racks are constructed and stackers introduced into the system to permit tiering. The labor cost is considerably increased when tiering is required.

Powered Low-lift Platform Trucks. When loads are further increased and where movement over ramps is required, the low-lift platform truck (Fig. 39) powered by gasoline or electric motor offers a further opportunity to reduce handling costs. This is simply a powered version of the jack-lift truck on which the operator rides so that the travel speed may be increased. The inability of the low-lift platform truck to tier loads limits its effectiveness.

Powered High-lift Platform Trucks. The application of the powered high-lift platform truck (Fig. 40) is limited by the fact that tiering of the second load of skids requires that dunnage be placed over the lower skid load to protect it from the concentrated load of the skid legs. With bags or light cartons the tiering of skids on dunnage is far from satisfactory; in many cases excessive pressures and unstable stacks result. It should be pointed out that, by the use of special stacking skids or of fixed racks (or shelving) to hold the upper tiers of skids, the dunnage between tiers is no longer required. Either of these methods reduces stacking costs.

Fork Truck. The fork truck with its extended forks picks up a unit load placed on an inexpensive pallet, transports, and stacks or tiers its own load some 12 to 16 ft. high—thus more fully utilizing the cubic content of any given storage space. There are solid-tired units (Fig. 41) for indoor service, and pneumatic-tired units (Fig. 42) for outdoor operation; they are built for pay loads ranging from 1000 to 10,000 lb. or more.

A fork truck resembles a cantilever beam on which the load is carried at the unsupported end; the wheels function as the beam supports. Hence the truck will lift and carry only that weight which is counterbalanced by the weight of the portion of the truck behind the front wheels. Actual design practice is to so distribute the weight of the machine that under full load the front wheels carry 90 to 95 per cent of the total weight. In order to secure the required carrying capacity within the limited dimensions of the truck chassis fork trucks carry a certain amount of counterweight in the end opposite the load. Thus for a given load the fork truck is considerably heavier than a platform lift truck of equal capacity. Where storage batteries are employed as the power source, the battery for a given capacity of truck is somewhat larger than required for a comparable platform lift truck. Since the fork truck is more costly in both first investment and operation, it should be employed primarily as a means

FIG. 41. Fork truck.

for tiering unit loads and not as a means for merely transporting loads from place to place. With frequent hauls in excess of 300 to 400 ft., it will usually be more economical to employ trailers or other simple transportation methods for the actual movement and to use the fork truck only for loading, unloading, and tiering (Fig. 43).

Truck Capacity. The *rated* capacity of the fork truck and weight of the load to be carried should not entirely determine the selection of the truck for a given application. The dimensions of the load and the truck capacity for a *load of those dimensions* should be investigated. At present there is a decided lack of uniformity in truck manufacturer's load ratings. Some makers

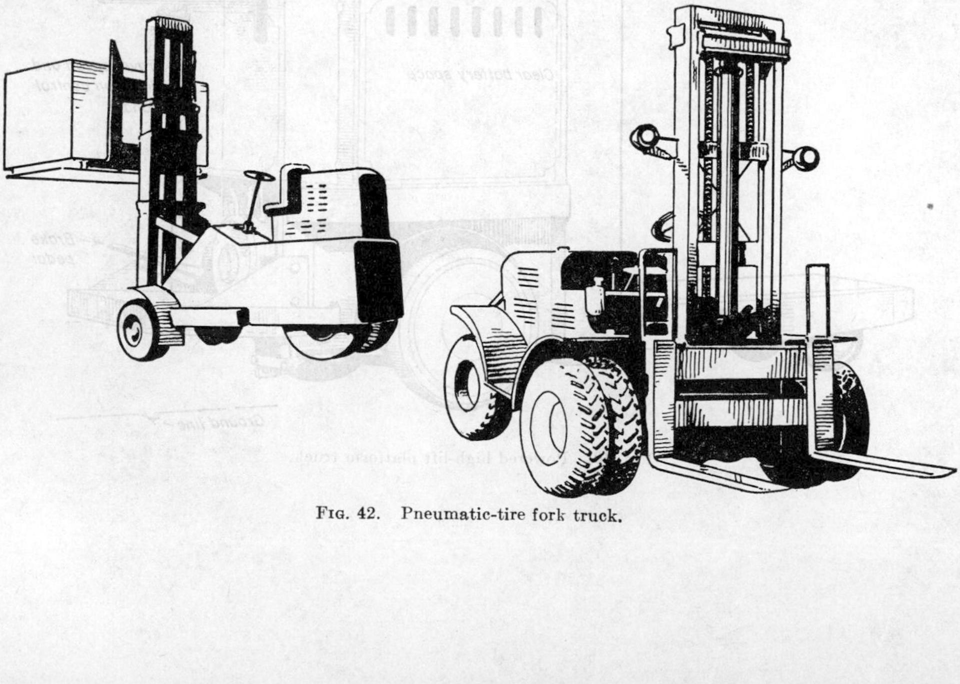

FIG. 42. Pneumatic-tire fork truck.

quote ratings based on 30-in. load length, others on 48-in. and even 60-in. load length on trucks of 4000 lb. or lower nominal ratings, so that the actual load-carrying capacity may vary greatly. In each case, the inch-pound rating based on *the distance from the load center to the front axle* should determine the carrying capacity for the actual load length. The capacity may be determined from the following, where X = capacity under the actual load.

$$X = \frac{d + L/2}{d + L'/2} \times \text{manufacturer's load rating in pounds}$$

where d = distance from front axle to load side of the vertical part of forks, L = load length for manufacturer's load rating, and L' = actual load length.

Note that some manufacturers quote the distance to the center of the load (which is $L/2$) and others quote the load length in establishing their ratings. None of the makers quotes the distance from the axle to the vertical face of the fork, but the dimension may be obtained from outline drawings of the trucks.

FIG. 43. Trailer-train pallet transportation.

Operating and Safety Features. Practical experience has demonstrated that the most desirable location for the operator is in a sitting position in the center of the truck facing the forks. Fatigue is reduced and greater protection is assured because he is protected on all sides by the truck structure. Good visibility in all directions is another advantage of this design; placing the operator close to the forks helps him to spot and pick up loads accurately.

A frequently employed safety device is a canopy over the operator's position. This is a sturdy structure designed to provide protection from falling packages without restricting visibility. To prevent containers from sliding off the pallet when the forks are tilted toward the operator, a fork back rest should be provided that is proportioned to conform to the loads.

Pallets. Pallets used in conjunction with the fork truck may be of wood, metal, paper, or other materials. At the present time, the nailed wood pallet is commonly used. Pallets may be two-way entry, four-way entry, or even eight-way entry, single face or double face. The usual design is the two-way entry type, which simply means that the forks may be inserted in two ends of the pallet and the other two sides of the pallet are blanked off by the stringers. The four-way pallet has been developed for exceptional cases where it is necessary to operate from one aisle in placing the pallet and from another aisle, at 90 deg. to the first, in removing it. This rather rare type of pallet has fork openings on all four sides. The eight-way pallet is a further development of the four-way pallet which permits inserting the forks at an angle of 45 deg. to the edge of the deck and represents a development of rather doubtful value from an economic standpoint.

Single-face pallets have a top deck supported by stringers which rest directly on the floor or, in the case of the upper tiers, on the load beneath. Although the cost is lower, the single-face pallet is limited to applications where the material is not subject to damage from concentrated load of the stringers. In effect the single-face pallet is a skid where each pair of legs has been replaced by a stringer.

The more commonly used double-face pallet is constructed with a top and bottom deck separated by the stringers. Recent designs have employed two (for very small pallets), three, and five stringers per pallet. The three stringer-type is by far the most commonly used design. Stringers may be either flush with the outer edge of the decks or inset from 3½ to 5 in. to permit use of sling bars for hoisting. Unless hoisting operations are definitely required, the flush stringer design is desirable because breakage of deck boards is reduced by this design. Wood pallets used in the Eastern United States are generally of hardwood such as oak, beech, birch, hickory, or maple for the decks or planking, with stringers of the same material or of longleaf yellow pine. For the pallet sizes commonly used, 1-in. hardwood lumber is sufficient for the deck boards with 2 by 4 stock for the stringers. On the West coast the softwood pallet of douglas fir is widely used. The softwood pallet has usually been constructed of 1½- or 2-in. top deck boards, 1-in. bottom boards, and 3 by 4 or 4 by 4 stringers. In sizes usually employed, this pallet is entirely too heavy to be moved by one man (which is not the case with the hardwood pallet); and, since the cost is about the same, it offers no particular advantage. In either case the most economical method of assembling the pallet is with cement-coated helical-thread wedge-point nails driven through holes drilled in the deck boards. For the hardwood pallet, the No. 6 nail 2½ in. long is recommended. Naturally a longer nail is required for the softwood design. The cost of the nailed wood pallet should be one-fourth to one-tenth of the cost of a wood and steel skid platform of approximately the same dimensions.

It should be noted that, where hand lift trucks are employed for moving pallets, it is necessary to provide certain openings in the bottom deck of the pallet in order that the rear wheels of the hand lift truck may come in contact with the floor after insertion in the pallet. This design is shown in Fig. 44.

Although the above concerns the nailed-wood pallet, many other varieties are available. Bolted wood and combination wood and metal pallets have been manufactured as well as pallets of sheet steel, steel wire, and aluminum alloys. A comparatively recent development is the laminated-paper pallet developed primarily for one-time use in shipping. Further developments in low-cost pallets for one-time use may be expected.

The widespread adoption of the fork-truck pallet system is due to the fact that it permits mechanized tiering and transporting of the pallet load without the use of dunnage, special racks, or other devices to protect the lower loads, since the solid lower decks of the double-face pallet may be placed readily on top of any previous load if the same size and type of package is used or if the lower pallet load has a level top. With an even distribution of the load, no damage is caused to the individual container. A much safer stacking condition is obtained because the pallet serves as a tying-in medium. In the

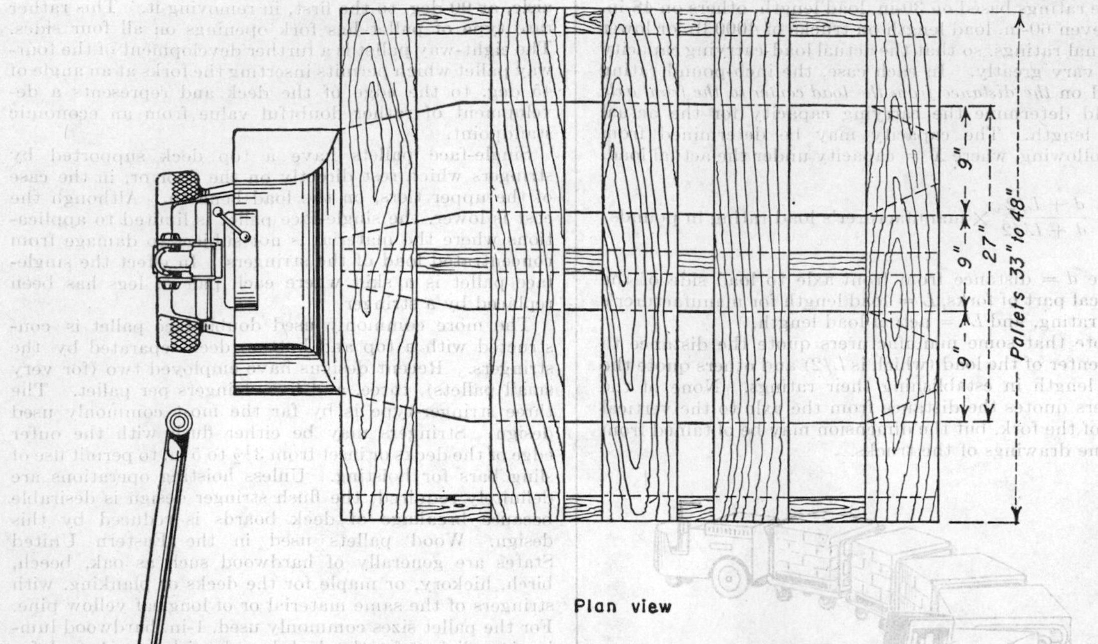

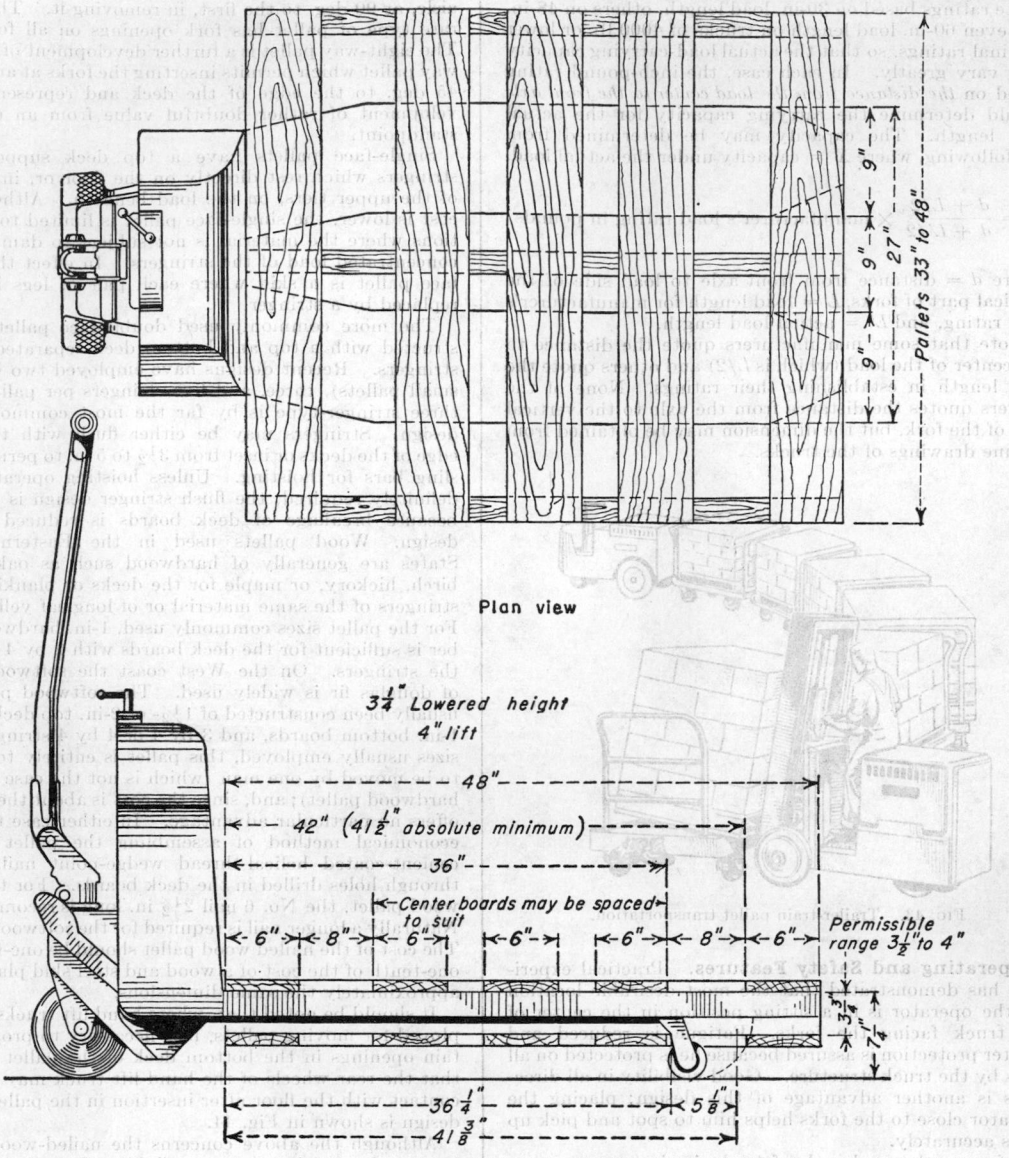

FIG. 44. Pallet and hydraulic pallet hand truck.

case of some materials in freshly packed bags, it may be found necessary to allow the palletized bags to settle for several hours or to apply positive pressure by mechanical means to remove entrained air and compact the load before tiering. In general, no difficulty is experienced in palletizing materials packed in substantial cartons, wood boxes, barrels, or drums.

Pallet and Load Dimensions. There are an infinite number of possible dimensions for pallets. Generally speaking, the larger the pallet, the more stable the stack. On the other hand, the large pallet requires a high-capacity fork truck and consequently a wider operating aisle. For general service the 48-in.-square pallet represents a size that is a compromise between the extremes of dimensions and is a practical size for stowing in railway boxcars. Although the railroads at present will not accept pallets as dunnage, all industries are vitally interested in effecting the economies of this method and eventually hope to arrange for shipments in this manner without paying freight on the pallets for either shipping or returning empties. This means that heavier carloadings are economically obtainable and that individual packages would not be handled many times before arriving at the customer's plant but would be placed on the pallet once and remain there until ready to be used. Aside from the economics effected, a substantial reduction in damage claims would result from fewer re-handlings of the individual containers.

By adherence to 48-in.-square pallets, an interchange with other companies' units might eventually be arranged to eliminate the need for returning all empties to their original source. Another possible solution would be for the railroads to furnish boxcars with 40 or 50 pallets where shipments can be palletized. These units could then be loaded or replaced with previously loaded units taken from storage.

Where only one commodity or group of commodities is to be stored and the dimensions do not coincide with the 48-in.-square pallet it is frequently possible to design a pallet and to lay out pallet patterns where the width of pallet is maintained at 48 in. and the length is altered slightly. Naturally enough, in many applications a pallet size entirely different from the 48-in.-square design is unquestionably justified from an economic viewpoint.

Operating Aisles. The generally accepted practice in industry is to arrange the unit loads in the form of a cube approximately 4 ft. in size. A load of these dimensions does not restrict the truck operator's visibility and may be safely transported. Paper cartons and bags of all descriptions are usually stacked three pallets high; whereas wood boxes, other firm containers, drums, and barrels can be stacked four high with a reasonable safety factor for stability in storage. The width of operating aisles in a storage area using fork trucks and 4-ft.-square pallets is dependent on the pay-load capacity, as shown:

Up to 2000-lb. load	10 ft. 6 in. aisle
Up to 4000-lb. load	12 ft. 6 in. aisle
Up to 6000-lb. load and over	14 ft. 0 in. aisle

Hand Pallet Trucks. In fork-truck installations, interruptions to materials movement will often occur at filling points, segregation centers, and loading docks when fork trucks are busy at some other operation. To alleviate this condition it may be found desirable to provide a hand-operated pallet truck (Fig. 45) for

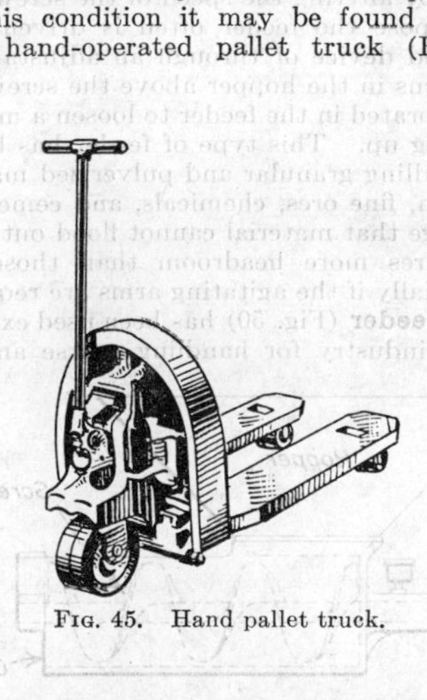

Fig. 45. Hand pallet truck.

temporary transfer of unit loads of 2000 lb. or under, and a power-propelled hand-operated pallet truck (Fig. 46) for heavier loads. The hand pallet truck and power-propelled hand pallet truck are designed so that the rear wheels and truck frame are inserted between the upper and lower decks of the pallet, as shown in Fig. 44. A series of booster rollers on the truck frame near the rear wheels facilitates the placing of the truck in the pallet. In ordering these trucks it is advisable to specify a fork or frame length several inches longer than the actual pallet dimension to allow operating clearance with over-hanging pallet loads.

Building Requirements. In planning the installation of a fork-truck–pallet system, several other physical limitations of this equipment require investigation. Since the combined weight of the fork truck and load is considerably in excess of that of a comparable platform lift truck with load and most of this weight is concentrated on two wheels, it is important that the floor construction be checked to determine whether it will safely carry the concentrated loads. Also, in the case of the hydraulic-lift fork truck, it will be found that the inner mast starts to rise above the rest of the structure after the forks have been elevated a comparatively short distance. In cases where headroom is restricted in parts of the operating area, the matter of over-all height when stacking requires careful investigation.

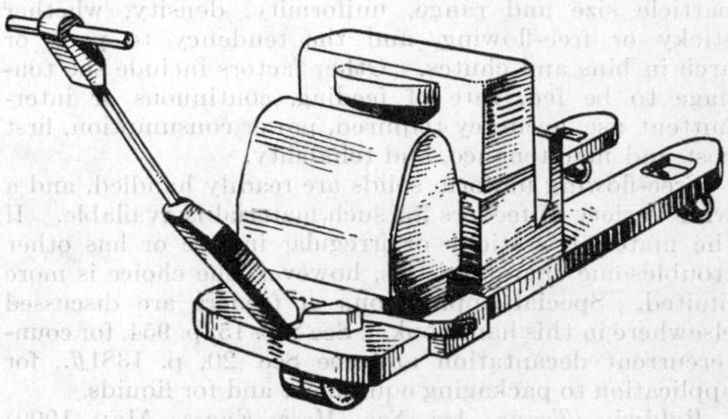

Fig. 46. Hand electric pallet truck.

Loading-platform Heights. To complete an engineering survey that resolves itself to the use of a pallet and fork truck system, it would be necessary to provide the correct height for loading platforms for both trucks and boxcars. Since trailer-truck floors are approximately 52 in. when empty and 46 in. when loaded, a 48-in.-high loading platform has been found to be most advantageous. For boxcar platforms the height from the top of the rail to the floor should be 3 ft. 7 in. This average-height figure has been arrived at by actual measurement over a period of years and has recently been adopted by the A.R.A. as a standard for car manufacturers.

Special Applications. Many modifications have been applied to industrial trucks and tractors. Among these are chisel forks for handling bales without pallets, scoop buckets for bulk materials, booms for miscellaneous heavy loads, and power rotating devices for handling roll paper.

Job Analysis. Selection and recommendation of materials-handling facilities discussed in this chapter should be based on providing equipment that is reasonably safe in operation and designed to minimize fatigue to the handler. A momentary effort of from 70 to 100 lb. and a sustained effort of not more than 50 lb. is considered to be a good working standard to maintain satisfactory morale. When volume and distance to be hauled are such that more than one operator is required, power equipment should be considered. Finally, an intelligent analysis of storage and handling costs must necessarily take into account the over-all investment in building space and equipment as well as direct labor expenditures and will effectively demonstrate the need for efficient loading, unloading, and tiering devices which utilize more of the cubic content of the building without increasing labor costs.

FEEDERS AND FEEDING MECHANISMS

BY REED W. HYDE

Feeders. The development of continuous processes and automatic machines created a need for automatic feeding devices which has resulted in a large variety of feeders for solids, liquids, and gases. Liquid and gaseous materials are generally uniform in character and may be handled readily in pumps, fans, pressure vessels, and the like. Feeder design in these fields has therefore been largely a matter of coordinating fluid-moving equipment with measuring and indicating devices such as meters, thermometers, and photoelectric cells.

Solids, however, vary so greatly in character that many different types of feeders and auxiliary equipment have been developed for handling them. In selecting a feeder for a particular service the physical state of the material to be fed must be considered carefully—particle size and range, uniformity, density, whether sticky or free-flowing, and the tendency to pack or arch in bins and chutes. Other factors include the tonnage to be fed, rate of feeding, continuous or intermittent use, accuracy required, power consumption, first cost and maintenance, and reliability.

Free-flowing uniform solids are readily handled, and a wide variety of feeders for such material is available. If the material is sticky or irregular in size or has other troublesome characteristics, however, the choice is more limited. Special applications of feeders are discussed elsewhere in this handbook. See Sec. 15, p. 954, for countercurrent decantation and see Sec. 20, p. 1381*ff*., for application to packaging equipment and for liquids.

Baldwin (*Trans. Am. Soc. Mech. Engrs.*, May, 1909) made an analysis of the principles used in the design of automatic bulk feeders for solids and found 11 general types. To these should be added vibrating, chain, and flight feeders. Figures 47 to 55, inclusive, and much of the accompanying discussion have been reproduced from Baldwin's presentation, which has been rearranged to include swinging-plate and plunger feeders

The undercut-gate feeder (Fig. 47) is an arc-shaped gate pivoted to swing beneath the hopper outlet and

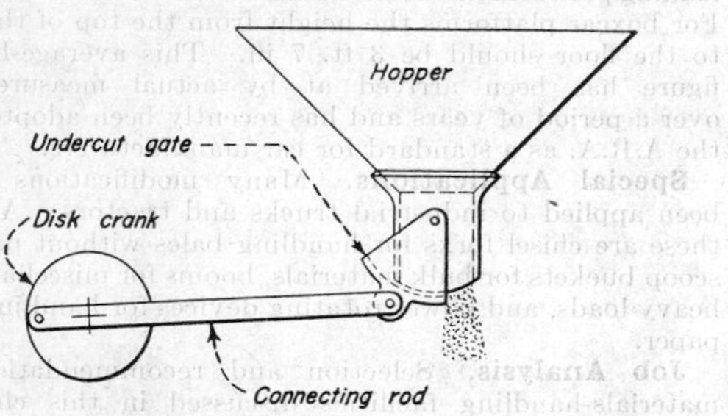

FIG. 47. The undercut-gate feeder.

moved back and forth by an eccentric or crank to open and close the feed opening. A simple relatively inexpensive device requiring little headroom, this type of feeder works best on free-flowing granular or moderately sized material. In an inclined, open-top chute, it is often used to control the flow of mine-run ores. Since the discharge is intermittent, it is well adapted to feeding chain or bucket conveyors, the strokes being timed with the buckets. The capacity may be changed only by varying the length or number of strokes. For the drive an adjustable crank is preferable, and no decided advantage accrues from the quick return obtained with the more costly eccentric drive. The so-called **stirrup feeder** is a variant using a transverse baffle above the swinging gate.

The lifting-gate feeder (Fig. 48), a slightly more elaborate device, also gives an intermittent feed suitable for chain or bucket conveyor or elevator. The chute is hinged beneath one side of the hopper opening so that

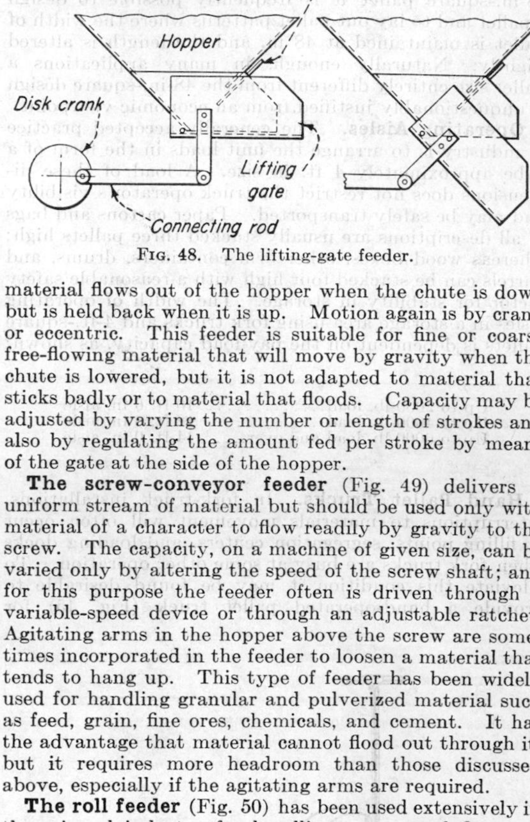

FIG. 48. The lifting-gate feeder.

material flows out of the hopper when the chute is down but is held back when it is up. Motion again is by crank or eccentric. This feeder is suitable for fine or coarse free-flowing material that will move by gravity when the chute is lowered, but it is not adapted to material that sticks badly or to material that floods. Capacity may be adjusted by varying the number or length of strokes and also by regulating the amount fed per stroke by means of the gate at the side of the hopper.

The screw-conveyor feeder (Fig. 49) delivers a uniform stream of material but should be used only with material of a character to flow readily by gravity to the screw. The capacity, on a machine of given size, can be varied only by altering the speed of the screw shaft; and for this purpose the feeder often is driven through a variable-speed device or through an adjustable ratchet. Agitating arms in the hopper above the screw are sometimes incorporated in the feeder to loosen a material that tends to hang up. This type of feeder has been widely used for handling granular and pulverized material such as feed, grain, fine ores, chemicals, and cement. It has the advantage that material cannot flood out through it, but it requires more headroom than those discussed above, especially if the agitating arms are required.

The roll feeder (Fig. 50) has been used extensively in the mineral industry for handling coarse and fine ma-

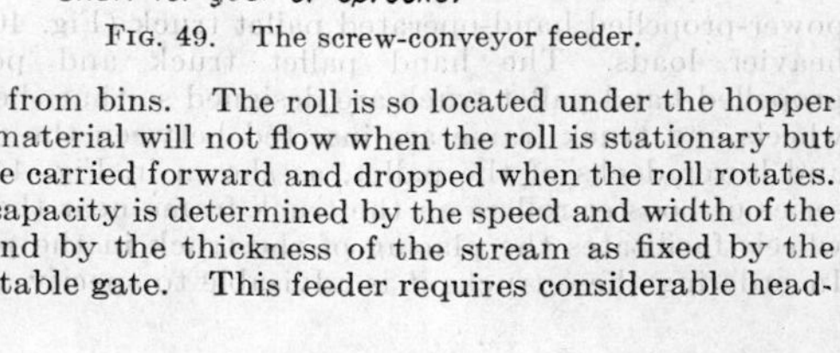

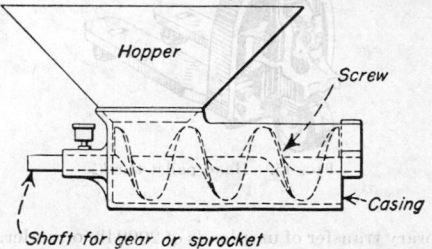

FIG. 49. The screw-conveyor feeder.

terial from bins. The roll is so located under the hopper that material will not flow when the roll is stationary but will be carried forward and dropped when the roll rotates. The capacity is determined by the speed and width of the roll and by the thickness of the stream as fixed by the adjustable gate. This feeder requires considerable headroom, a roll 6 to 8 ft. in diameter sometimes being required for coarse run-of-mine material. The fact that it provides a moving bottom under the bin, which can be as wide as desired, makes it useful particularly on materials that tend to stick or hang up in the bin, the movement of the roll helping to drag the material forward and out.

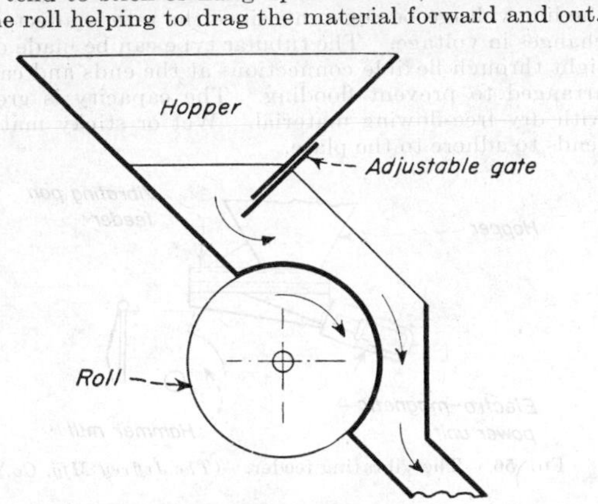

Fig. 50. The roll feeder.

In **the rotating-vane** or **pocket feeder** (Fig. 51) a number of pockets or segments are formed by the vanes or paddles mounted on the shaft. As the feeder rotates, the pocket is filled when in the upper quadrant, carries the material around, and discharges it at the lower quadrant. It may act both as a feeder and as a measuring device. The capacity is changed by varying the speed of rotation. It requires more headroom than the gate feeders but gives a more even feed, which can be made practically uniform by the use of a vane set on a slight spiral along the shaft. It is used for free-flowing material. Wet sticky material tends to pack and stick in the pockets. The *pocket feeder* or *"revolving-door"*

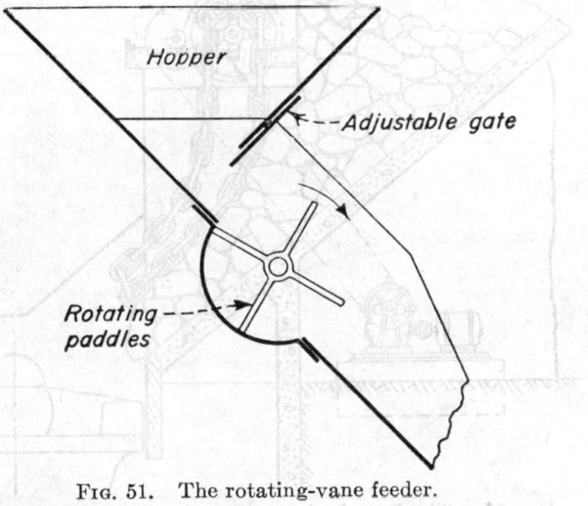

Fig. 51. The rotating-vane feeder.

feeder is a variant in which the feeder is tightly housed to permit delivery against pressure or vacuum. The feeder can readily be made tight to serve as an air-lock gate.

The revolving-plate or **table feeder** (Fig. 52) consists of a disk or table, generally horizontal, beneath the hopper outlet and driven by gears from above (as shown) or from below. As the table revolves the material is drawn from the hopper to be scraped off by the skirt board. The feed can be regulated by the height of the adjusting gate, the position of the skirt board or cutoff blade, or (less readily) by changing the speed of the table. This feeder may be used with material fine or coarse, wet or dry. Sticky material is handled quite well, as the skirt board pushes it off the table into the chute. Material that floods is difficult to control. Little headroom is required, since material is delivered from the feeder practically at the level of the hopper bottom.

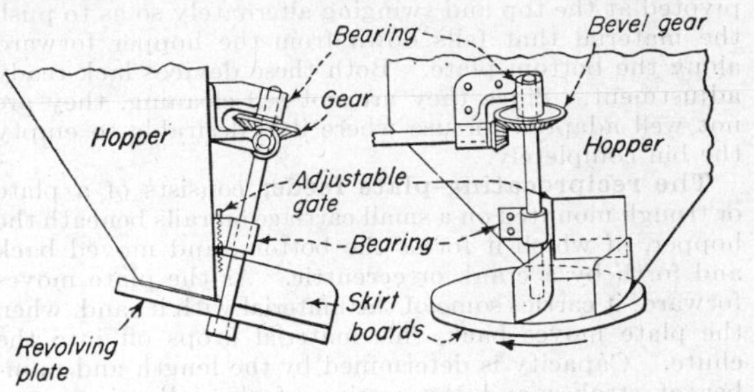

Fig. 52. The revolving-table feeder.

The apron (conveyor) feeder (Fig. 53) may be any of the various types of apron flights or conveyor belt supported on idlers. Capacity is fixed by the width and speed of the belt or apron, and the position of the regulating gate, the latter being the usual means for regulating the feed. A wide variety of apron, pan, and belt feeders has been designed. They are in frequent use, since they

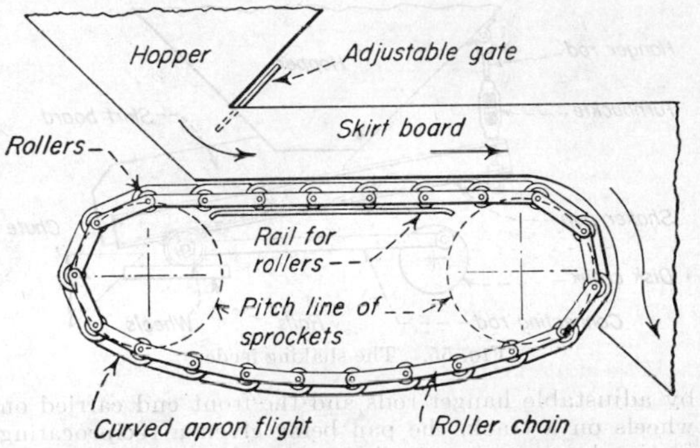

Fig. 53. The apron-conveyor feeder.

permit ready adjustment of feed to a degree sufficient for most purposes, give a uniform feed, will handle materials varying greatly in size and feeding characteristics, can be made as wide or as long as desired, and can deliver the material at a considerable distance from the feeding point. For wet sticky material the steel-plate apron with its corrugated surface seems to be more effective than the smooth-surfaced belt conveyor.

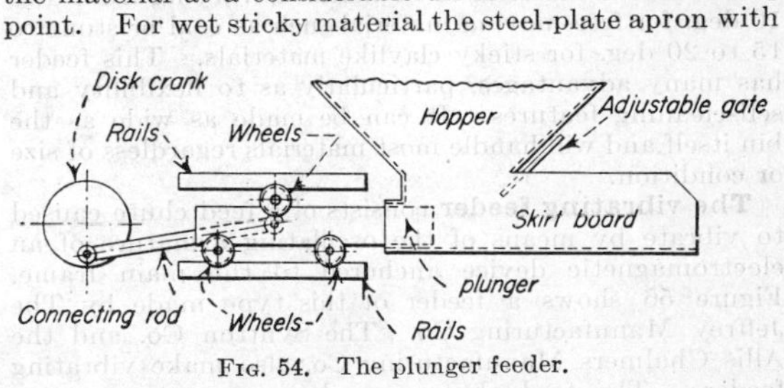

Fig. 54. The plunger feeder.

The plunger feeder (Fig. 54) consists of a plunger or piston which is moved back and forth in a trough at the bottom of the hopper. As the plunger is drawn back by the crank, the material drops down into the trough to be pushed out on the forward stroke. The feed is intermittent, but two or more plungers, moving alternately, may be used to approximate a continuous stream, Capacity is fixed by the number and length of the strokes. In operation it is somewhat similar to the **swinging-plate feeder,** which has been used for moving coal and similar materials of all sizes. This consists of two plates, pivoted at the top and swinging alternately so as to push the material that falls down from the hopper forward along the bottom plate. Both these devices lack ready adjustment. Since they are not self-cleaning, they are not well adapted for use where it is desirable to empty the bin completely.

The reciprocating-plate feeder consists of a plate or trough mounted on a small carriage on rails beneath the hopper, of which it forms the bottom, and moved back and forth by a crank or eccentric. As the plate moves forward, it carries some of the material with it; and, when the plate moves back, this material drops off into the chute. Capacity is determined by the length and number of strokes and the setting of the adjusting gate. Adjustment (except of the gate) is difficult, and this feeder, like the plunger feeder, will not clear itself. It is suitable for non-abrasive, lumpy material, such as coal, but not for sticky material or material that floods. It requires little headroom and is non-jamming because the plate will continue to slide under a blocked lump.

The shaker feeder (Fig. 55) consists of a shaker pan mounted under the hopper with the back end suspended

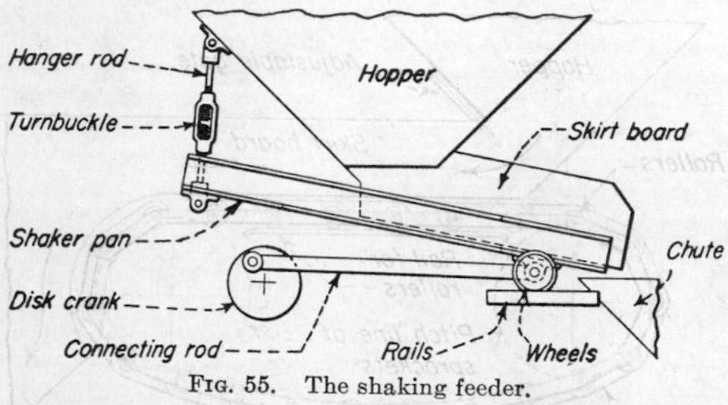

Fig. 55. The shaking feeder.

by adjustable hanger rods and the front end carried on wheels on a track, the pan being given a reciprocating motion by a crank. As the pan is shaken the material is moved forward and finally drops off the end into the chute. The number of strokes is not readily changed and usually is left at about 75 per minute. Length of stroke may be varied from 4 to 12 in. or more. The angle of inclination depends on the capacity desired and on the nature of the material to be handled, varying from 8 to 10 deg. for free-moving material such as coal or stone to 15 to 20 deg. for sticky claylike materials. This feeder has many advantages, particularly as to flexibility and self-cleaning features. It can be made as wide as the bin itself and will handle most materials regardless of size or condition.

The vibrating feeder consists of a feed chute caused to vibrate by means of the oscillating armature of an electromagnetic device anchored to the main frame. Figure 56 shows a feeder of this type made by The Jeffrey Manufacturing Co. The Syntron Co. and the Allis Chalmers Manufacturing Co. also make vibrating feeders. The feed chute may be an open pan or a closed tube; it may be jacketed for heating, drying, or cooling; or a screen deck may be incorporated. Materials varying from dust to large chunks and from clean granular material to mixtures of earth, sand, and rock have been handled. Capacities range from ounces to 1250 tons/hr. Power consumption is said to be low, 40 watts being reported for sand fed at the rate of 1500 lb./hr. Capacity is regulated by adjusting the current input, thereby controlling the pull of the electromagnet and thus the length of stroke. It is possible, therefore. to regulate the flow from a distant point or through automatic devices. For uniform feeding it is necessary that the line voltage be constant, since the rate is affected by changes in voltage. The tubular type can be made dust-tight through flexible connections at the ends and can be arranged to prevent flooding. The capacity is greater with dry free-flowing material. Wet or sticky material tends to adhere to the plate.

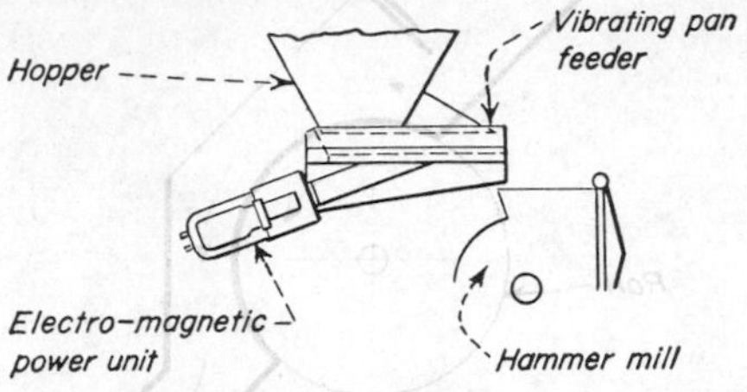

Fig. 56. The vibrating feeder. (*The Jeffrey Mfg. Co.*)

The Ross feeder (Fig. 57), a patented device made by the Ross Screen and Feeder Co., is used extensively in the mineral industries for handling ore, rock, and other materials. It consists of a curtain of heavy endless chain driven by an overhead tumbler. The chains are so suspended as to lie on the material and travel with it as it flows down the chute, the chain regulating the rate of flow irrespective of the size of the material. These chains are made in widths of 6 in. to 20 ft., capacities from 1/4 to 2000 tons/hr., and in styles fitted with ship-anchor chain ranging from small steamer to battleship size. Crushed material can be fed from storage with an

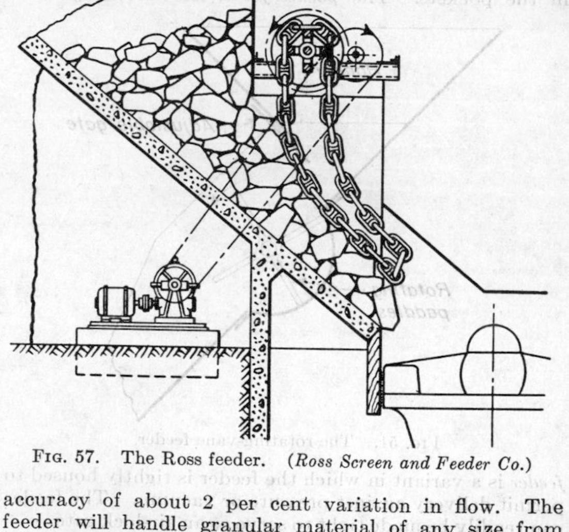

Fig. 57. The Ross feeder. (*Ross Screen and Feeder Co.*)

accuracy of about 2 per cent variation in flow. The feeder will handle granular material of any size from primary steam shovel to coal slack.

The flight-conveyor feeder consists of a series of flights, vertical scrapers, plows, or chains (forming in effect a drag conveyor) moved in a stationary trough. The trough may be annular, in which case the flights travel in a circular path; or more usually it is straight line with the flights driven by sprockets at the ends and returning underneath the trough, following a path similar to that of the apron conveyor (Fig. 53). The flights drag the material forward from beneath the hopper and drop it into the chute in an even stream. Capacity is varied by changing the speed of travel of the flights, the drive often being through a variable-speed device. This type of feeder can readily be made the full width of the bin and can be made dust-tight, advantages in handling material that sticks or floods. By using flights of suitable shape to suit the commodity a great variety of materials can be handled.

Constant-weight Feeders. The feeders described above all feed a more or less accurately regulated volume of material, but without regard to variations in density. The **feeding of a constant or controlled weight** of material has also been developed by a combination of volume feeders with weighing mechanisms. The Hardinge Constant Weight feeder (see p. 1292) pivots a belt feeder against a counterweight in such a manner as to open or close a gate on the feed hopper to maintain the flow at constant weight. The Schaeffer Poidometer uses only the loaded part of the belt to operate a gate on the feed hopper. The Richardson Conveyoweigh is a combined weighing feeder which uses two belts and a scale element. In the Jeffrey Waytrol feeder, an electrically actuated vibrating deck charges an endless belt mounted on scale beams the counterpoise of which acts through electrical contacts and rheostats to vary the amplitude of vibrations of the deck and thus maintain a given load on the belt. Omega Machine Co. makes a somewhat similar gravimetric feeder in which the feeding tray or deck is vibrated mechanically to a degree regulated by the movement of the scale beam from which the endless belt is suspended, thereby maintaining a uniform feed on the belt.

The continuous coupling between the weighing device and the bin-gate operating device of most constant-weight feeders results in a "hunting" effect. Each time the load crossing the scale varies, as it frequently does because of irregularities in flow from the bin, the gate continues to open (or close) until an overload is built up, reaches the scale, and causes the gate to start closing. This continual fluctuation results in an average rather than a constant weight of feed, which can vary appreciably from the desired rate at any given time. This is avoided in the Transportofeeder of the Sintering Machinery Corp., in which a change in feed reaching the scale results in the gate being moved to a limited degree only and allows time for the resulting change in feed to reach the scale before another movement of the gate takes place. Instead of the rather wide alternations of over and under feed rate, a very constant rate is maintained. The Transportofeeder is an integrating scale which shows the true weight of material passing over it as well as maintains a constant weight of feed.

Other types work on the "loss in weight" principle, as by utilizing a hopper or tank on a scale with the scale beam equipped with electrically driven counterpoise and actuating a throttling means by which the rate of flow is regulated. Accessories are available for use with feeders, such as counters, recorders, no-load cutoffs, and proportioning devices. The Hardinge "Electric Ear" is a device for controlling feed to a grinding mill by the sound of the mill. When the microphone near the mill picks up a characteristic overload sound, the device slows down the feed and conversely opens it when the underload sound is heard. Used in conjunction with a wet-grinding mill the relay can be made to operate a solenoid valve in the water-supply line in addition to controlling the dry feed to the mill and thus maintain a constant pulp density in mill and classifier. The electric eye has been used extensively in connection with automatic batch weighing and for other purposes. The possibilities for controlling feeding devices through such instruments seem almost unlimited (see also Sec. 19, Process Control).

Getting material to flow from a bin is a frequent problem, particularly with wet sticky materials or those which tend to arch and pack or flood. Solids confined by the walls of a bin support part of the mass against the walls and so tend to form an arch that holds back or "hangs up" the material above it. When the arch breaks, the material, especially if very fine, is likely to flood out in a strong stream. The degree to which this occurs depends largely on the material itself, on the relative proportions of the bin, the size and arrangement of hopper outlets, and other factors. The character of the material itself usually must be accepted as it comes, but a choice as to bin and feeder arrangement and style is generally possible. Corners in a bin form starting points for arching; and for this reason, as well as for structural advantages, the round bin or conical hopper is preferable to a square one. Projecting ledges, tie rods, or other construction that interferes with free movement of material should be avoided. When the material does hang up it usually can be started moving by a jarring action, such as by mechanical, magnetic, or pneumatic vibrators attached to or acting through the walls of the bin. Double-walled bins, the inner wall flexed by compressed air, have been used. In small bins or hoppers a suspended sheet of 24-gage stainless steel often will prevent arching of materials that otherwise hang up. Air jets, intermittently operated, near the bottom of the bin often are effective. For small hoppers or light materials, mechanical agitators may be placed inside the bin to stir the material. The hoppered portion of the bin should be kept at an angle well above the angle of repose of the material and, wherever possible, the steeper the better. Louvers or poke holes in the lower sides of the bin are frequently used with sticky ores to allow barring down of material that will not otherwise move. Figure 58 shows two types of louvered bins in successful use.

The bottom opening of the bin should be as large as possible, at least three times the size of the largest lump. Figure 59 shows an arrangement for discharging difficult

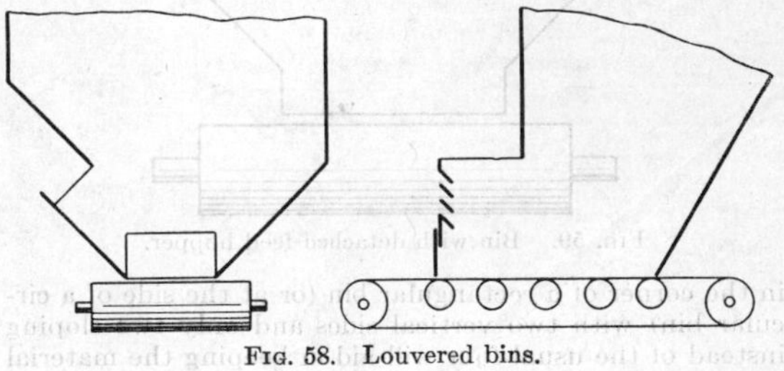

Fig. 58. Louvered bins.

material from a bin in which the bin itself has a large opening for free flow while beneath it is a small hopper with a more restricted opening to a feeder; space between the bin opening and the hopper permits barring when necessary.

The apron, roll, revolving table, scraper, and chain feeders, which in effect are moving bottoms beneath the bin outlet, have a tendency to pull the material out as they travel and thus to promote flow from the bin. Heavy apron feeders, the full width of the bin, are often used in feeding such material as wet flotation concentrate; on shallow bins, or with less difficultly flowing material, the belt feeder is useful. An upward tilt to the feeder helps to agitate and move the ore and thereby aids the flow. In extreme cases it may be necessary to make a "live bottom" under the bin, its full width, and to flare out the sides of the bin to prevent the material from hanging up. The bottom may be a belt or apron conveyor, flight conveyor, etc., of a width equal to the full width of the bin, or it may be a series of smaller conveyors, such as screw conveyors, which together cover the entire bottom. The effect is to leave no support on the floor of the bin on which the material can hang. Flaring the walls likewise reduces the possibility of arching against the sides, since any slight motion of the mass breaks it away from the supporting walls.

Use of a number of bottom-discharge openings is helpful in preventing bridging, but it introduces a larger number of feeders to be regulated. If the bin outlet is a sloping chute, the gate can be set 3 to 4 ft. from the edge of the bin, to give room for barring. Placing the outlet

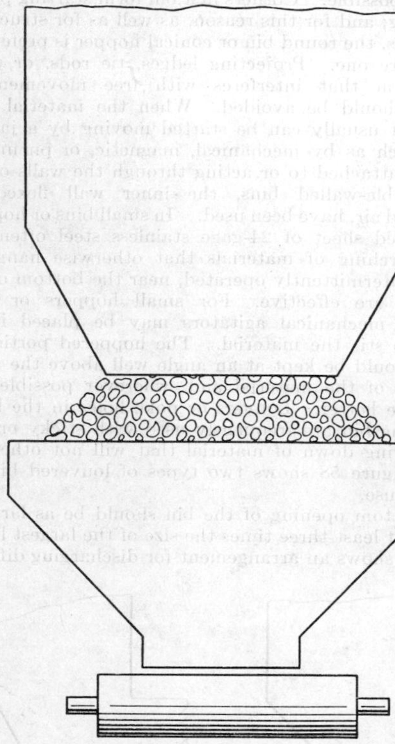

FIG. 59. Bin with detached feed hopper.

in the corner of a rectangular bin (or at the side of a circular bin) with two vertical sides and only two sloping instead of the usual four will aid in keeping the material moving, since there is less tendency to hang against the vertical sides.

When fine dry material that has hung up in a bin finally breaks loose and starts to flow, it is likely to come out in a fluid stream that floods out over the ordinary feeder and is difficult to stop except by tight-sealing mechanisms. For this sort of material the screw feeder and rotary pocket feeder are particularly useful because they readily can be made tight. Special forms of vibrating feeders and flight feeders are also adapted to this service. For material that is likely to flood, The Jeffrey Manufacturing Co. makes a rotary bin check valve (patented) consisting of a conical hopper attached below the bin, having a revolving agitator blade within it which moves the material and drops it into a lower compartment where an eight-bladed rotor sweeps it horizontally 180 deg. to the discharge chute.

Reagent feeders, using various of the principles discussed above, are designed to feed accurately into a processing operation small proportions of solid, liquid, or gaseous reagents. Usually they are interconnected with some device for measuring the flow of the principal material, so that a constant proportion of reagent is fed regardless of fluctuation in the rate of flow of the principal material. The screw, vibrator, belt, revolving-table, and rotary-pocket types are well adapted for feeding solid reagents. To prevent irregularities in feed, some device for ensuring flow of the chemical from the hopper down into the feeder is generally used, such as agitating blades or electrical or mechanical vibrators. Wallace and Tierman Co., Inc., make a feeder having an automatically flexed hopper and oscillating feed spout for this purpose, and Omega Machine Co. has an oscillating hopper of adjustable stroke feeding over a sliding tray.

For feeding liquid reagents special types of pumps are often used. These may be regulated to feed a given quantity on a fixed time cycle at controlled intervals, or they may be operated in conjunction with metering devices to feed the reagent at a rate proportional to the flow of the principal liquid. The Adjust-O-Feeder of Proportioneers, Inc., Providence, R.I., is an example. Another, a plunger-type adjustable stroke pump, is made by the Milton Roy Company, Philadelphia. This feeds corrosive chemicals, viscous materials, slurries, and other liquids in precisely controlled volume and at pressures up to 20,000 lb./sq. in. Capacities range from 1 pt./hr. to 50 gal./min. A number of pumps, each feeding a different reagent at its own rate, can be driven from a common motor, the speed of the group as a whole being variable over a wide range and under manual or automatic control.

Other liquid feeders operate on the dipper or cup principle. In one such type, made by Denver Equipment Co., a number of small cups, suspended ferris-wheel fashion, pick up reagent from a well or tank and are discharged by an adjustable tripper into a trough which can be set to receive the flow from one or all of the cups. The Geary feeder, made by the Galigher Co., of Salt Lake, has a cup which is raised along a guideway from the reagent tank by a crank, adjustable to tilt the cup by varying degrees to empty just the desired amount of reagent at each stroke.

The Omega Machine Co's. Rotodip meter has a metering wheel carrying eight curved dippers which pick up liquid as they revolve in a constant-level tank and discharge it through side openings at the hub, giving practically continuous discharge. Regulation is through variable-speed drive with a feeding range of 100 to 1. A pulley dipping into a tank of reagent with a scraper blade set to remove a desired part of the liquid carried over on the rim is a simple device used for feeding flotation reagents. By varying the speed of the pulley or the width of the scraper blade in contact with the rim the feed can be adjusted within a considerable range.

Among other devices is the lampwick drip. This is not so satisfactory as the adjustable siphon, however (see Sec. 15, p. 954), or as the swinging spout shown in Fig. 60. Details of this device for handling corrosive liquids at rates from 30 cc./min. upward are given in *Eng. Mining J.*, May, 1935. As the spout descends in step with the level of liquid in the tank, the head H remains constant regardless of the level of liquid in the tank, so that feeding proceeds at a constant rate. Rate may be adjusted by varying the effective head H through a change in setting of the involute V on the pulley shaft supporting float F and counterweight W.

The adjustable siphon is utilized in an ingenious manner to regulate the flow of reagent in the Massco-Adams reagent feeder of the Mine and Smelter Supply Co., Denver, Colo. The siphon is supported on a movable frame which can be tilted by means of a handscrew to raise or lower the discharge end of the siphon and thus regulate the flow of solution. Feed rates are said to range from 1 to 100 cc./min. on oils and from 5 to 500 cc./min. on water-soluble reagents.

A subdivision of flow in launders can be obtained by the use of gates or vanes or by causing the liquid to fall vertically onto the apex of a distributing cone. For an accurate division a revolving distributor is used. This consists of a bowl with weir discharge, revolving above any desired number of annular pockets. Each pocket supplies an equal part of the original stream through a pipe or launder. A simple and fairly accurate distributor consists of a small tank with the desired number of

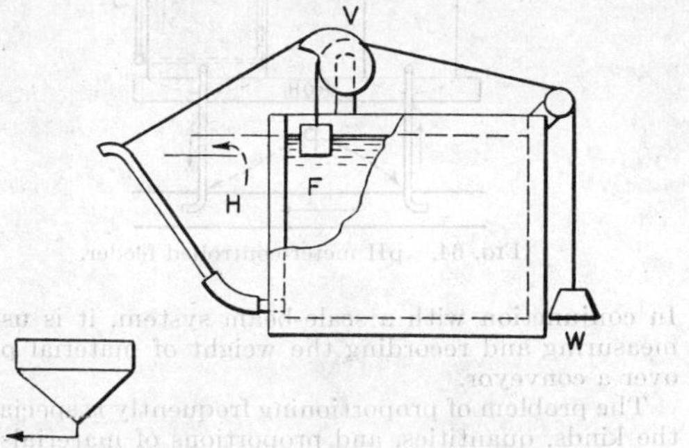

FIG. 60. Reagent feeder for corrosive liquids.

outlet pipes through the sides at the same elevation. The liquid is supplied through a central well from which it overflows evenly to the outlet pipes. Greater accuracy is obtained by maintaining a constant head of liquid in the distributing tank.

Feeding gases, such as chlorine and ammonia, has become an important industrial operation having extensive application in the purification of water, sewage treatment, etc. For this purpose the Wallace and Tiernan Co., Inc., produce various types of equipment, some of which automatically proportion the supply of chlorine to the flow of water to be treated. Figure 62 indicates the sectional detail of the tray and injector of their visible vacuum chlorinator. Chlorine supply is controlled by the vacuum produced by an auxiliary water supply flowing through an injector. This draws the chlorine into the injector where it is dissolved in the water, this solution being conveyed to the water supply being treated. The device has various special features such as an orifice meter for measuring the chlorine and other controls which are in part indicated on the diagram.

Proportioning, or the coordination of feeding devices to deliver two or more moving streams of solids, liquids, or gases in constant ratio, or in desired variation, has become a highly developed art. In its simplest form it consists of means for measuring the flow of a principal material together with means for feeding one or more secondary materials in proportion to that flow. For example, a simple device for proportioning zinc dust to cyanide solution is shown in Fig. 63. The solution falling into the tripper box causes the box to tilt and to move the

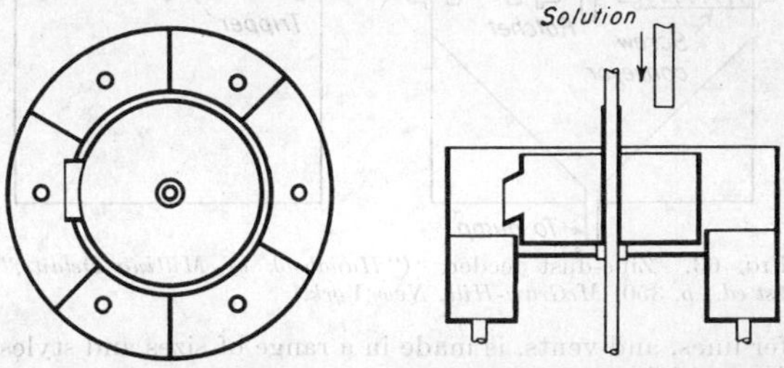

FIG. 61. Rotary distributor. (*"Handbook of Milling Details", 1st ed., p. 265, McGraw-Hill, New York.*)

shaft, through the ratchet and pawl. This turns the screw feeder a certain amount and thus feeds a quantity of zinc dust into the quantity of solution discharged by the tripper box.

More elaborate mechanisms are required for proportioning more difficult materials or where greater accuracy is required. For treating boiler feed water, for example, metering pumps may be regulated to feed conditioning chemicals at a rate corresponding to the water supply. If the flow of water is relatively constant, a time-cycle regulation may be sufficient, the chemical pumps being operated to deliver a fixed quantity per hour, corresponding to the average water flow. If flow is uneven, however, the water supply is generally metered, and the chemical pumps are operated automatically at rates corresponding to the water flow at the time. By similar use of a pH meter, the water treatment can be proportioned both to the rate of water flow and to the pH of the water. The H-O-H feeder, made by D. W. Haering & Co., Inc., of Chicago (Fig. 64) is a combined liquid feeder and proportioner having an accuracy of 0.5 per cent over a wide

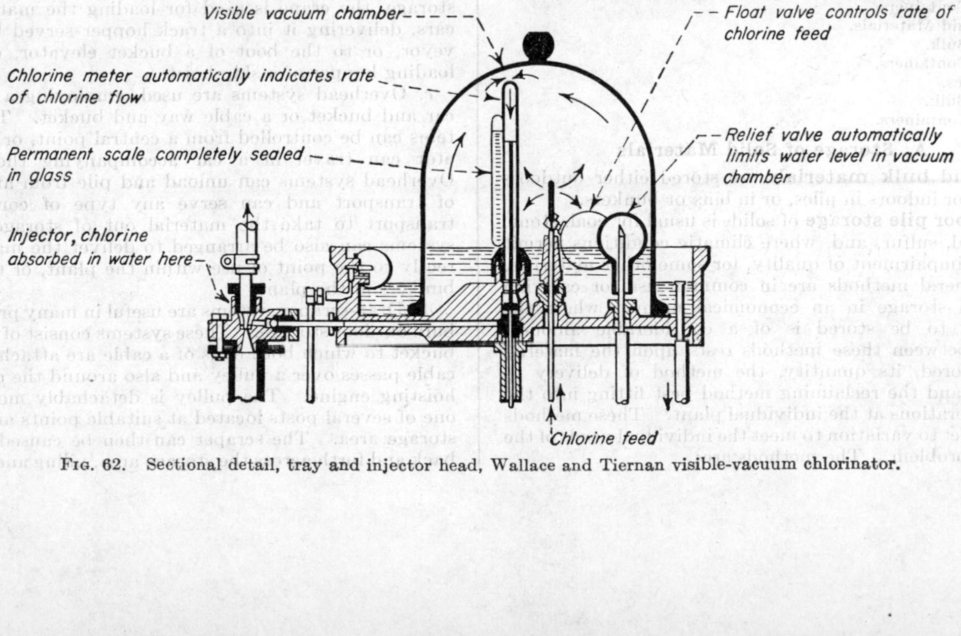

FIG. 62. Sectional detail, tray and injector head, Wallace and Tiernan visible-vacuum chlorinator.

range of flow. Pipe-line liquid enters the pitot tube and displaces some of the colored fluid piston *F* through the needle valve *N* and sight glass *S*. This drives a corresponding quantity of the reagent *R* out through the other pitot tube *P* to mingle with the pipe-line liquid. The device, fitted with the necessary check valves, trans-

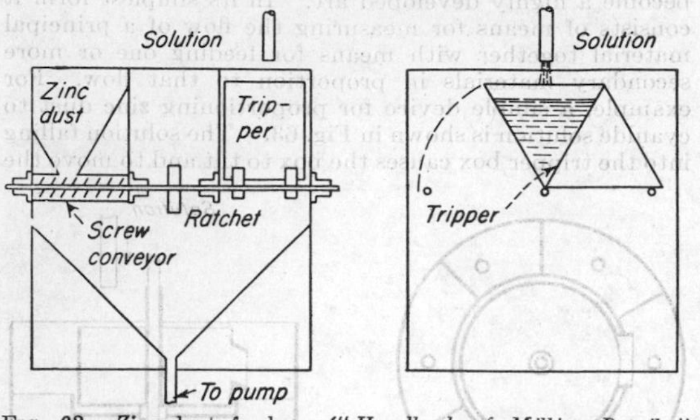

FIG. 63. Zinc-dust feeder. ("*Handbook of Milling Details*," *1st ed., p. 350, McGraw-Hill, New York.*)

fer lines, and vents, is made in a range of sizes and styles for special reagents.

For feeding chemical solutions by means of their metering pumps in proportion to the flow of another liquid, Proportioneers, Inc., of Providence, R. I., make special devices operated by meters, pumps, and similar instruments in the main circuit. The Chronoflo, made by Builders-Providence, Inc., Providence, R. I., is a time-impulse device for measuring flow which has many interesting applications to proportional feeding. It can be used to regulate solid or liquid reagent feeders in ratio to the flow of a liquid in a pipe line or of a solid material on a conveyor, or to regulate a variable-speed drive to maintain a constant feed of bulk material on a conveyor.

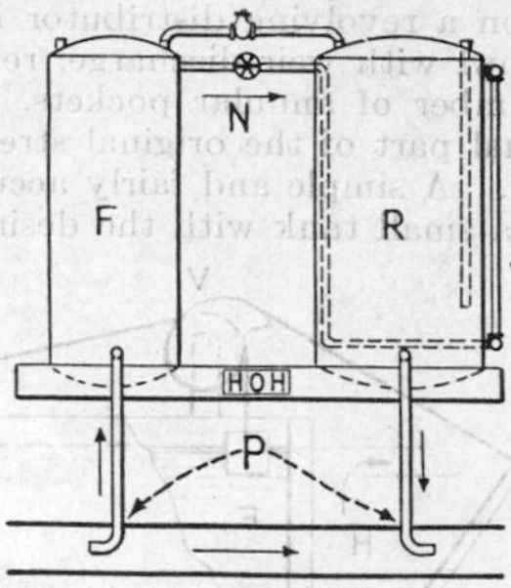

FIG. 64. pH meter-controlled feeder.

In conjunction with a scale-beam system, it is used for measuring and recording the weight of material passing over a conveyor.

The problem of proportioning frequently is special as to the kinds, quantities, and proportions of materials to be handled. Manufacturers catalogues and bulletins contain much information, especially as to sizes, capacities, power requirements, etc., of their equipment, and should be studied in connection with a contemplated installation.

STORAGE OF MATERIALS

All problems relating to the handling and movement of materials also involve in some measure the storage of materials, except for some few cases where the movement is incidental to the process, or where the movement of raw materials or finished products is direct. It is therefore necessary to give some consideration to the storage methods commonly employed in the chemical engineering industries and an indication of how the movement into, and out of, storage can be conveniently made.

The materials commonly stored have the following characteristics:

A. Solid materials.
  1. Bulk.
  2. Containers.

B. Liquid Materials.
  1. Bulk.
  2. Containers.

C. Gases.
  1. Bulk.
  2. Containers.

A. Storage of Solid Materials

1. Solid bulk materials are stored either outdoors in piles; or indoors in piles, or in bins or bunkers.

Outdoor pile storage of solids is usual for coal, stone, ore, wood, sulfur, and, where climatic conditions permit without impairment of quality, for some other materials. Four general methods are in common use for carrying out such storage in an economical manner where the quantity to be stored is of a considerable amount. Choice between these methods rests upon the material to be stored, its quantity, the method of delivery to storage, and the reclaiming method best fitting into the other operations at the individual plant. These methods are subject to variation to meet the individual needs of the storage problem. The methods are:

a. Storage in a pile under a traveling bridge, or gantry, crane with handling into, and out of, storage by means of a traveling bucket operated from the crane. This bucket can unload from cars, barges, or other means of transport and can deliver to some means of transport, directly to a secondary storage in the plant, to a conveyor, or directly to such equipment as furnaces.

b. Storage in piles on either side of a track, served by a locomotive crane. The crane is used to place the material in storage, taking it from a car, truck, or barge, or from a track hopper in which it has been dumped from hopper-bottom cars, or to which it has been delivered by some other means of handling. In reclaiming from storage, the crane is used for loading the material into cars, delivering it into a track hopper served by a conveyor, or to the boot of a bucket elevator, or to the loading hopper of a skip hoist.

c. Overhead systems are used, employing a monorail car and bucket or a cable way and bucket. These systems can be controlled from a central point, or the operator can travel in a car accompanying the bucket. Overhead systems can unload and pile from any means of transport and can serve any type of conveyor or transport to take the material out of storage. These systems can also be arranged to deliver the material directly to the point of use within the plant, or to bins or bunkers in the plant.

d. Drag-scraper systems are useful in many problems of bulk outdoor storage. These systems consist of a scraper bucket to which both ends of a cable are attached. The cable passes over a pulley and also around the drum of a hoisting engine. The pulley is detachably mounted on one of several posts located at suitable points around the storage area. The scraper can then be caused to move back and forth across the storage area, piling and reclaiming material. Drag scrapers usually deliver to a conveyor, a bucket elevator, or skip hoist.

Pile storage for wood, used in paper mills and other industries, differs from the general cases mentioned. Such storage is carried out with conveying devices that have been specially developed for the purpose.

Contents of outdoor storage piles cannot be calculated exactly unless the pile is within walls and is kept at an even thickness, as is sometimes done. Large random piles can be estimated with sufficient accuracy for most purposes by obtaining the contours of sections evenly spaced over the storage area, say 5 to 10 ft. apart, and then figuring the cubic contents of the material between sections. Such sections are easily determined by the use of surveying instruments.

Smaller piles are frequently estimated by means of soundings, but this method leaves much to be desired as to accuracy, whatever its convenience.

Indoor pile storage is usually employed for materials that must be kept dry or protected from the weather for some reason. Among these are ceramic materials, scrap metals, glass sand, agricultural products, minerals, and chemical products. The usual method is to have a large storage shed or room within one of the buildings of the plant for each material; or, when quantities are relatively small, to divide a single storage space into several sections by means of partitions. Frequently such storage spaces are served by an overhead crane or a monorail crane, which is used for both storing and reclaiming the materials. Such a crane can serve the processing equipment direct where that is convenient, or conveyors may be interposed when such an arrangement is better.

Indoor bulk storage may be served entirely by conveyors. For example, two belt conveyors can be used, one at the top of the storage space for taking material from an elevator and distributing it to the storage spaces, and the other at the bottom, beneath the floor, fed through a feeding device or loading hopper, for reclaiming. Another conveyor application useful in such cases is the pivoted-bucket carrier, which will serve for both storing and reclaiming.

Where the quantities stored are small, it may not be economical to handle otherwise than with manual labor.

Intermediate quantities in bulk storage of dry solids are often served by electrical industrial trucks or tractors and trailers. The trucks or the trailers, for this purpose, can be equipped with dump bodies, which serve to eliminate much hand labor that would otherwise be necessary.

Quantities in indoor pile storage are easily estimated by obtaining the contour of the pile by soundings. This method is not accurate but serves most purposes. Where piles can be kept level, *i.e.*, evenly distributed over the whole storage floor, more accurate estimates can be made from the dimensions of the storage space and the height of the pile on the wall. Where this method is used, it is well to have an indicating scale in feet and fractions of feet painted on the wall.

Indoor storage of bulk solids in bins or bunkers is frequently employed. Particularly where such storage can be arranged to feed by gravity to the production equipment or shipping facilities, this system finds an economical application. Hopper bottoms or parabolic bottoms are usual for such bins or bunkers. The material will then feed directly to a chute or spout by gravity, whenever the outlet gate is opened. In the storage, by this method, of some materials that are likely to clog in the outlet, it is customary to install mechanical or air agitators, designed to keep the material loose and free-flowing. The chute or spout can feed direct to the equipment as is usually the case with furnaces; or it can charge a larry car or similar transport device which serves to move the material to the point of use, as is usual in gas or coke practice. Bins or bunkers can also be arranged to feed to a belt or other conveyor, or to load into industrial trucks or cars.

Material is placed in bins or bunkers through the open top, or through gates in a closed top. One method in common use for this purpose is to employ an overhead crane or monorail crane with a bucket. A belt conveyor passing over the top of the storage, with suitable unloading arrangements, and fed by a bucket elevator or skip hoist, is another common method of storing. A third method employs a pivoted-bucket conveyor, which can also be used for reclaiming the material.

Indoor storage of bulk solids in silos is an alternate method that finds much favor in some industries, particularly in the lime, cement, and ceramics fields. Silos are cylindrical structures of concrete, tile, steel, or wood. Their advantages lie in the economy of floor space occupied for a given amount of storage and their relatively inexpensive construction. Compared with bins and bunkers, silos are less adaptable as to possible locations about the plant, and they have a relatively greater height for the amount of storage space provided. Because of this great height, and because the bottoms are seldom raised far above the ground level, silos must usually be loaded by means of an elevator or skip hoist and unloaded by means of a conveyor or by trucks working in a pit.

Where several silos are in use, they can be grouped; and a belt conveyor can be arranged to take material from an elevating device and unload it into any one of the silos desired. Grouped silos frequently are spaced about 2 ft. apart and connected by curtain walls, thus providing a fifth, or interspace, bin for each group of four. Reclaiming from silos can best be done by an arrangement of conveyors suited to the movements of the materials reclaimed, or by means of wheeled transport fed from hopper gates under the silos.

Quantities in storage in bins, bunkers, or silos are figured with sufficient accuracy for most purposes from the known dimensions of the storage and the depth of the material stored. Another method, more accurate than the first, is to weigh the material as it is stored and then to weigh the amounts taken from storage. This method incurs considerable first cost for the purchase of weighing devices but is often justified where a close check on quantities is desired.

2. Solid materials stored in containers present few problems not already considered in this section on the movement of materials. Such containers are usually boxes, bags, barrels, or steel drums. The storage methods employed usually consist of piling or storing these containers outdoors when the material stored and the weather conditions permit, or indoors in other cases.

One of the most useful methods of handling materials into and out of such storage employs the electrical industrial truck, equipped with the high-lift or tiering feature when stacks or piles are to be used, and, if the loads to be handled are heavy, equipped with a crane.

Another method widely used employs an overhead crane or monorail. This is particularly useful for handling bags, several of which can be taken at one time in a rope sling.

For barrels and drums, much special equipment is available, including barrel elevators and barrel stackers.

In storing these containers within buildings, care should be taken not to overload the storage floors.

Quantities of barrels, bags, and drums in storage can be easily estimated by counting the number in some unit of space and then extending the count to cover the whole space occupied, making proper allowance for those areas where the space is not filled to an even level.

No mention has been made of the storage of finished solid products in containers. In general, such materials will be stored in the manner that is described under Liquid Materials Stored in Containers (see below).

B. Storage of Liquid Materials*

1. Bulk storage of liquid materials employs reservoirs where the material can be exposed to the elements, and tanks for all other purposes. The opportunities for using reservoirs are few, because most liquids must be protected from contamination or dilution. Water is the liquid commonly stored in this way in chemical engineering industries. Where the quantities dealt with are large, the reservoir becomes a problem for the civil engineer, and adequate assistance of this nature should be employed. For small reservoirs, concrete-walled excavations or concrete tanks without tops, either sunken in the ground or raised above it, are in common use. Such tanks should be constructed with reinforced walls adequate to hold the contents when the reservoir is completely filled. The concrete should be waterproofed so as to prevent any possibility of leaking. Quantities stored in this way can be easily figured from the known dimensions of the storage and the measured depth of the liquid.

Storage of the liquid materials of chemical engineering industry is usually carried out in tanks. Such tanks are classified as *vertical*, *horizontal*, *rectangular*, or *spherical*. Vertical tanks are cylindrical tanks with the axis in a vertical plane. Horizontal tanks are cylindrical tanks with the axis in a horizontal plane. Rectangular and spherical tanks are shaped as indicated by the names. Vertical tanks are most commonly used for outdoor storage, in tank fields, for such materials as petroleum, tar, and asphalt. These tanks are also sometimes used for indoor storage. Another frequent use for vertical tanks is for elevated water storage. Horizontal and rectangular tanks are the most useful types for all sorts of storage of liquids. Spherical tanks are used for the storage, generally outdoors, of volatile liquids, such as natural-gas gasoline, which are likely to develop a pressure in the tank due to the volatilization of part of the liquid.

Tanks are constructed of metal, generally steel plate, wood staves, or concrete. In cases where the liquid being stored is of a nature to attack the material of the tank, it may be lined with rubber or some other protective material, or coated on the inside with some resistant coating. Wood-stave and concrete tanks are frequently treated to prevent leaking.

Filling and drawing from liquid-storage tanks are usually done with pumps. In many cases, however, tanks can be so located that one or both of these operations can be accomplished by gravity. Obviously, tanks can be readily connected by pipe lines to any point in the plant.

Quantities stored in tanks can be readily determined from the dimensions of the tank and the depth of the liquid in the tank which can be obtained by means of a liquid-level gage or by sounding. The horizontal cross-sectional area of a vertical tank is the area of the circle having the same diameter as the tank, while that of a rectangular tank is the length multiplied by the breadth.

When figuring stored quantities from the dimensions of a tank, the simplest method is to divide the volume in cubic inches by 231, the number of cubic inches in 1 gal.

With horizontal tanks the contents will be influenced by the shape of the tank ends, which may be convex, straight, or concave. For ordinary estimating, the effect of the ends may be neglected. The content of the horizontal tank, when filled, is found by multiplying the area of the end of the tank by the length and dividing by 231. When the tank is only partly filled, a deduction must be made from the total contents of the tank equal to the unfilled portion above the liquid. If the tank is filled to a point above the axis, then subtract from the total contents an amount equal to the contents of a space having the same length as the tank and an end area equal to the unwetted segment of the tank end. If the tank is filled to a point below the axis, then the contents become equal to that obtained by multiplying

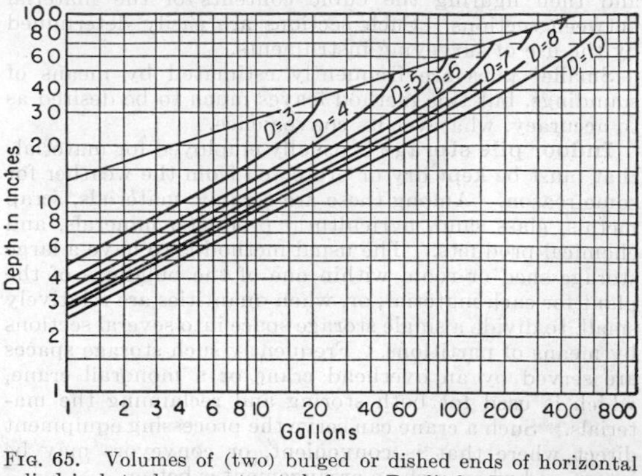

FIG. 65. Volumes of (two) bulged or dished ends of horizontal cylindrical tanks, constructed from Doolittle Formula (*Ind. Eng. Chem.*, March, 1928):

$$V \text{ (both ends)} = 0.0009328\, h^2(3r - h) \text{ gal.}$$
h = depth, in.
r = radius of tank, in.

In curves D = diameter of tank in feet.

the length of the tank by the area of the wetted segment of the end of the tank.

When it is desired to figure the contents of a horizontal tank with curved ends more accurately, it becomes necessary to determine the radius of the curvature or to obtain this from the tank maker. This is often inconvenient, and the alternate method of calibration is recommended. This consists in completely filling the tank with water, then drawing off the contents in steps, reducing the depth of the liquid by even amounts, and accurately measuring the quantity of liquid withdrawn for each of these unit reductions in depth. In this way a table can be compiled giving the tank contents for each unit of depth of liquid in the tank. In practice, with such a table and a liquid-level gage, the quantity in a tank can be readily obtained.

The calibration method recommended for horizontal tanks is also best for use with spherical tanks, as its use avoids the recurrence of tedious calculations.

2. Liquid materials stored in containers are generally packed in barrels, kegs, drums, cans, or in glass containers. The storage of the barrels, kegs, and drums will be the same as that described for solids packed in similar containers. For cans, one of the most convenient methods is to stack the cans several tiers high on palettes or skid platforms and then handle them into and out of storage by means of electric or hand-operated industrial lift trucks. Where small cans are to be stacked to a considerable height, machinery developed for service in food canneries will be found useful for both stacking and reclaiming.

Storage for glass containers, such as carboys or bottles, may be either on skids or palettes, or directly on the floor. Movement into and out of storage is best carried

* See p. 1573.

Table 1. Capacities (U.S. Gallons) of Rectangular Tanks for Each Foot of Liquid

231 cu. in. = 1 U.S. gal.;* 1 cu. ft. = 1728 cu. in. = 7.4805 U.S. gal.

Tank width, in.	Tank length													
	6″ 0.5′	12″ 1′	18″ 1′-6″	24″ 2′	30″ 2′-6″	36″ 3′	42″ 3′-6″	48″ 4′	54″ 4′-6″	60″ 5′	66″ 5′-6″	72″ 6′	78″ 6′-6″	84″ 7′
6	1.87	3.74	5.61	7.48	9.35	11.22	13.09	14.96	16.83	18.70	20.57	22.44	24.31	26.18
12	3.74	7.48	11.22	14.96	18.70	22.44	26.18	29.92	33.66	37.40	41.14	44.88	48.62	52.36
18	5.61	11.22	16.83	22.44	28.05	33.66	39.27	44.88	50.49	56.10	61.71	67.32	72.93	78.55
24	7.48	14.96	22.44	29.92	37.40	44.88	52.36	59.84	67.32	74.81	82.29	89.77	97.25	104.73
30	9.35	18.70	28.05	37.40	46.75	56.10	65.45	74.81	84.16	93.51	102.86	112.21	121.56	130.91
36	11.22	22.44	33.66	44.88	56.10	67.32	78.55	89.77	100.99	112.21	123.43	134.65	145.87	157.09
42	13.09	26.18	39.27	52.36	65.45	78.55	91.64	104.73	117.82	130.91	144.00	157.09	170.18	183.27
48	14.96	29.92	44.88	59.84	74.81	89.77	104.73	119.69	134.65	149.61	164.57	179.53	194.49	209.45
54	16.83	33.66	50.49	67.32	84.16	100.99	117.82	134.65	151.48	168.31	185.14	201.97	218.81	235.64
60	18.70	37.40	56.10	74.81	93.51	112.21	130.91	149.61	168.31	187.01	205.71	224.42	243.12	261.82
66	20.57	41.14	61.71	82.29	102.86	123.43	144.00	164.57	185.14	205.71	226.29	246.86	267.43	288.00
72	22.44	44.88	67.32	89.77	112.21	134.65	157.09	179.53	201.97	224.42	246.86	269.30	291.74	314.18
78	24.31	48.62	72.94	97.25	121.56	145.87	170.18	194.49	218.81	243.12	267.43	291.74	316.05	340.36
84	26.18	52.36	78.55	104.73	130.91	157.09	183.27	209.45	235.64	261.82	288.00	314.18	340.36	366.55
90	28.05	56.10	84.16	112.21	140.26	168.31	196.36	224.42	252.47	280.52	308.58	336.62	364.68	392.73
96	29.92	59.84	89.77	119.69	149.61	179.53	209.45	239.38	269.30	299.22	329.14	359.06	388.99	418.91

Tank width, in.	Tank length													
	90″ 7′-6″	96″ 8′	102″ 8′-6″	108″ 9′	114″ 9′-6″	120″ 10′	126″ 10′-6″	132″ 11′	138″ 11′-6″	144″ 12′	150″ 12′-6″	156″ 13′	162″ 13′-6″	168″ 14′
6	28.05	29.92	31.79	33.66	35.53	37.40	39.27	41.14	43.01	44.88	46.75	48.62	50.49	52.36
12	56.10	59.84	63.58	67.32	71.06	74.81	78.55	82.29	86.03	89.77	93.51	97.25	100.99	104.73
18	84.16	89.77	95.38	100.99	106.60	112.21	117.82	123.43	129.04	134.65	140.26	145.87	151.48	157.09
24	112.21	119.69	127.17	134.65	142.13	149.61	157.09	164.57	172.05	179.53	187.01	194.49	201.97	209.45
30	140.26	149.61	158.96	168.31	177.66	187.01	196.36	205.71	215.06	224.42	233.77	243.12	252.47	261.82
36	168.31	179.53	190.75	201.97	213.19	224.42	235.64	246.86	258.08	269.30	280.52	291.74	302.96	314.18
42	196.36	209.45	222.55	235.64	248.73	261.82	274.91	288.00	301.09	314.18	327.27	340.36	353.45	366.55
48	224.42	239.38	254.34	269.30	284.26	299.22	314.18	329.14	344.10	359.07	374.03	388.99	403.95	418.91
54	252.47	269.30	286.13	302.96	319.79	336.62	353.45	370.29	387.12	403.95	420.78	437.61	454.44	471.27
60	280.52	299.22	317.92	336.62	355.32	374.03	392.73	411.43	430.13	448.83	467.53	486.23	504.94	523.64
66	308.57	329.14	349.71	370.29	390.86	411.43	432.00	452.57	473.14	493.71	514.29	534.86	555.43	576.00
72	336.62	359.06	381.51	403.95	426.39	448.83	471.27	493.71	516.16	538.60	561.04	583.48	605.92	628.36
78	364.68	388.99	413.30	437.61	461.92	486.23	510.55	534.86	559.17	583.48	607.79	632.10	656.42	680.73
84	392.73	418.91	445.09	471.27	497.45	523.64	549.82	576.00	602.18	628.36	654.55	680.73	706.91	733.09
90	420.78	448.83	476.88	504.94	532.99	561.04	589.09	617.14	645.19	673.25	701.30	729.35	757.40	785.45
96	448.83	478.75	508.68	538.60	568.52	598.44	628.36	658.29	688.21	718.13	748.05	777.97	807.90	837.82

* For Imperial gallons, multiply above capacities by 1.2.

Table 2. Capacity of Pipes and Cylindrical Tanks (Square Bottom) of Various Diameters

In gallons per foot of length

Feet	Inches											
	0	1	2	3	4	5	6	7	8	9	10	11
0		0.0408	0.1632	0.3672	0.6528	1.020	1.469	1.999	2.611	3.305	4.080	4.937
1	5.875	6.895	8.00	9.18	10.44	11.79	13.22	14.73	16.32	17.99	19.75	21.58
2	23.50	25.50	27.58	29.74	31.99	34.31	36.72	39.21	41.78	44.43	47.16	49.98
3	52.88	55.86	58.92	62.06	65.28	68.58	71.97	75.44	78.99	82.62	86.33	90.13
4	94.00	97.96	102.0	106.1	110.3	114.6	119.0	123.4	127.9	132.6	137.3	142.0
5	146.9	151.8	156.8	161.9	167.1	172.4	177.7	183.2	188.7	194.2	199.9	205.7
6	211.5	217.4	223.4	229.5	235.7	241.9	248.2	254.6	261.1	267.7	274.3	281.1

Feet	Inches				Feet	Inches				Feet	Inches			
	0	3	6	9		0	3	6	9		0	3	6	9
7	287.9	308.8	330.5	352.9	16	1504	1551	1600	1648	25	3672	3746	3820	3896
8	376.0	399.9	424.5	449.8	17	1698	1748	1799	1851	26	3972	4048	4126	4204
9	475.9	502.7	530.2	558.5	18	1904	1957	2011	2065	27	4283	4363	4443	4524
10	587.5	617.3	647.7	679.0	19	2121	2177	2234	2292	28	4606	4689	4772	4856
11	710.9	743.6	777.0	811.1	20	2350	2409	2469	2530	29	4941	5027	5113	5200
12	846.0	881.6	918.0	955.1	21	2591	2653	2716	2779	30	5288	5376	5465	5555
13	992.9	1031	1071	1111	22	2844	2909	2974	3041	31	5646	5737	5830	5923
14	1152	1193	1235	1278	23	3108	3176	3245	3314	32	6016	6111	6206	6302
15	1322	1366	1412	1457	24	3384	3455	3527	3599	33	6398	6495	6593	6692

One gallon = 0.1337 cu. ft.; 1 cu. ft. = 7.4805 gal.

out by means of hand-operated or electrical industrial trucks. Special bodies are available for these trucks for handling carboys efficiently.

Quantities in storage, of large-sized containers for liquids, are quickly estimated by counting the number on a unit area and then extending this quantity to cover the whole storage. With cans or other small containers, such a method is not so convenient. In this case it is better to use some kind of indicators in the stack. Such indicators can, for instance, be used to mark every gross in the stack, and then quantities are obtained by counting the number of indicators and multiplying by the unit.

Floor loadings must, of course, be considered with any kind of storage in buildings, but they are particularly important with small containers such as cans because of the ease with which such containers can be stacked to a height that overloads the floor. Ordinarily, building floors can be loaded to 150 lb./sq. ft. Mill-type buildings and reinforced-concrete structures are usually constructed for loads up to 250 lb./sq. ft. In many cases, buildings of special construction are good for higher loadings. In any case, the chemical engineer should ascertain the load that the floors of the buildings in his plant can safely support and take pains to see that this loading is not exceeded.

2. *Bulged or Dished Ends.* For horizontal cylindrical tanks, with bulged or dished ends, Table 3 should be used in conjunction with Fig. 65. This chart gives the

volumes of the bulged or dished ends for tanks of diameters of 3, 4, 5, 6, 7, 8, and 10 ft. corresponding to various depths. The volumes read from this chart must be added, for bulged ends, to (or subtracted from, for dished ends) the volumes read from Table 3.

Table 3. Capacities of Horizontal Cylindrical Tanks

Method of Calculation

1. *Flat Ends.* Assume diameter = 48 in., and length = 120 in. Required the volume (in U.S. gallons) of shaded area of figure (*ACE*) of height 10 in.

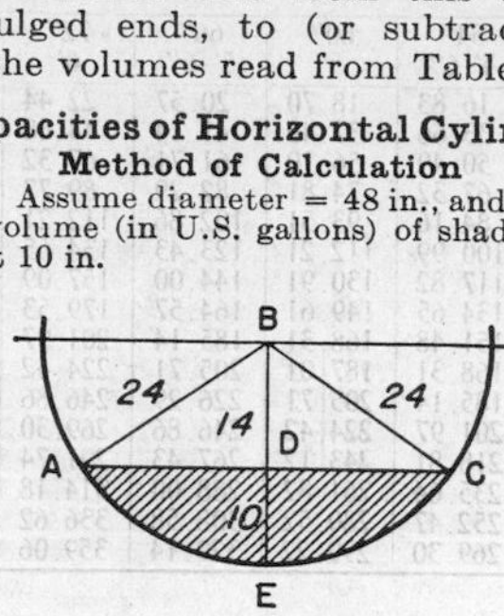

FIG. 65*A*. Dimensions used in calculation of capacities of Horizontal Cylindrical Tanks.

Area ACE = area $ABCE$ − area ABC.

Area $ABCE = \left(\frac{2\ \angle ABD}{360}\right) \times$ area of circle, and $\angle ABD$ is found from its cosine which is $^{14}\!/_{24}$.

$$\therefore \angle ABD = 54.25°$$

$$\text{Area } ABC = 14 \times 24 \times \sin \angle ABD$$

or

$$14 \times 24 \times \sin 54.25 = 14 \times 24 \times .8116 = 272.7$$

$$\text{Area } ABCE = \frac{108.50}{360} \times \pi \times (24)^2 = 545.4.$$

$$\therefore \text{Area } ACE = 545.4 - 272.7 = 272.7$$

$$\therefore \text{Vol. (U.S. gallons) per foot of length} = \frac{272.7 \times 12}{231} = 14.17$$

See the table attached.

To calculate the contents of a tank over one-half full, let h be the depth of the unfilled part of the tank. Find the volume corresponding to this depth from the table and subtract it from the volume of the full tank as given in the last column of each horizontal line of the table.

C. Storage of Gases

1. Gases stored in bulk are usually stored in **holders** or **tanks.** Holders are of two types: telescopic and stationary.

Telescopic gas holders are the familiar type of gasometer, exclusively employed until recently for the storage of city gas supplies. These holders consist of a number of concentrically arranged steel plate bands or rings. The bottom ring has the largest diameter, is fixed in position, and has a gastight bottom. The top ring is the smallest in diameter and is furnished with a tight top. Intermediate rings are evenly graded in diameter between the top and bottom rings. Where these rings fit one within the other, liquid seals are provided to prevent the escape of gas. The whole structure is carried within an open supporting frame, upon the columns of which rest guide wheels or rollers attached to the rings, holding them in a fixed horizontal position with relation to the frame and to each other. These rings are, however, free to move up and down, impelled by the gas pressure within the holder.

Such telescopic holders are constructed in very large capacities, as well as in small sizes, holders of 10,000,000 and 15,000,000 cu. ft. capacity being not uncommon. Since these holders usually involve a considerable investment, the proper procedure for obtaining one is to obtain competitive bids upon specifications from several manufacturers who specialize in this equipment.

Stationary gas holders are a more recent development than the telescopic type. These are sometimes called "waterless" holders, which term is hardly appropriate, as a liquid seal must also be employed in this design. Holders of this type consist of cylindrical or polygonal tanks with vertical axes. The bottom and sides are gastight, while the top, usually rounded or slightly conical, is vented. Within the tank is a horizontal partition, carried in a liquid seal at the point

Table 3. Capacities of Horizontal Cylindrical Tanks—(*Concluded*)

Contents given in U.S. gallons for 1 ft. of length

Flat ends

Diameter of tank, in.	Depth of liquid, in.																
	3	6	9	12	15	18	21	24	27	30	33	36	39	42	45	48	51
12	1.15	2.94	4.73	**5.88**													
18	1.45	3.86	6.61	9.36	11.77	**13.22**											
24	1.70	4.60	8.05	11.75	15.45	18.90	21.80	**23.50**									
30	1.91	5.23	9.27	13.72	18.36	23.00	27.45	31.49	34.81	**36.72**							
36	2.12	5.79	10.34	15.43	20.85	26.44	32.03	37.45	42.54	47.09	50.76	**52.88**					
42	2.28	6.31	11.31	16.97	23.07	29.46	35.99	42.51	48.90	55.00	60.66	65.66	69.69	**71.97**			
48	2.45	6.78	12.20	18.38	25.10	32.20	39.54	47.01	54.47	61.81	68.91	75.63	81.81	87.23	91.56	**94.01**	
54	2.60	7.22	13.04	19.68	26.97	34.72	42.80	51.08	59.49	67.90	76.18	84.26	92.01	99.30	105.94	111.76	116.38
60	2.75	7.64	13.82	20.91	28.72	37.06	45.82	54.88	64.11	73.45	82.78	92.01	101.07	109.83	118.17	125.98	133.07
66	2.89	8.04	14.56	22.07	30.37	39.28	48.65	58.39	68.42	78.59	88.87	99.14	109.31	119.34	129.08	138.45	147.36
72	3.02	8.42	15.26	23.17	31.92	41.36	51.32	61.71	72.45	83.41	94.54	105.76	116.98	128.11	139.07	149.81	160.20
78	3.15	8.78	15.94	24.21	33.41	43.34	53.86	64.87	76.27	87.97	99.90	111.97	124.12	136.27	148.34	160.27	171.97
84	3.26	9.12	16.57	25.24	34.85	45.24	56.29	67.87	79.91	92.30	104.98	117.85	130.87	143.95	157.03	170.05	182.92
90	3.43	9.46	17.20	26.20	36.21	47.05	58.61	70.75	83.39	96.43	109.81	123.46	137.28	151.22	165.25	179.27	193.21
96	3.50	9.79	17.80	27.13	37.52	48.81	60.84	73.52	86.73	100.39	114.44	128.79	143.40	158.17	173.07	188.01	202.95
102	3.61	10.10	18.37	28.01	39.00	50.49	62.99	76.18	89.94	104.20	118.89	133.92	149.25	164.81	180.53	196.36	212.25
108	3.71	10.39	18.94	28.90	40.03	52.14	65.09	78.74	93.04	107.87	123.17	138.87	154.89	171.19	187.71	204.37	221.15

Diameter of tank, in.	Depth of liquid, in.																
	54	57	60	63	66	69	72	75	78	81	84	87	90	93	96	102	108
12																	
18																	
24																	
30																	
36																	
42																	
48																	
54	**118.98**																
60	139.25	144.14	**146.89**														
66	155.66	163.17	169.69	174.84	**177.73**												
72	170.16	179.60	188.35	196.26	203.10	208.50	**211.52**										
78	183.37	194.38	204.90	214.83	224.03	232.30	239.46	245.09	**248.24**								
84	195.60	207.99	220.03	231.61	242.66	253.05	262.66	271.33	278.78	284.64	**287.90**						
90	207.03	220.68	234.06	247.10	259.74	271.88	283.44	294.28	304.29	313.29	321.03	327.06	**330.49**				
96	217.85	232.62	247.23	261.58	275.63	289.29	302.50	315.18	327.21	338.50	348.89	358.22	366.23	372.52	**376.02**		
102	228.12	243.95	259.67	275.23	290.56	305.59	320.28	334.54	348.30	361.49	373.99	385.48	396.47	406.11	414.38	**424.48**	
108	238.05	254.75	271.53	288.19	304.71	321.01	337.03	352.73	368.03	382.86	397.16	410.81	423.76	435.87	447.00	465.51	**475.90**

Table 4. Capacities of Vertical Cylindrical Tanks in U.S. Gallons

Diameter		Gal./ft.	Diameter		Gal./ft.	Diameter		Gal./ft.
Ft.	In.	depth	Ft.	In.	depth	Ft.	In.	depth
0	0		10	0	587.52	20	0	2350.1
	3	0.37		3	617.26		3	2409.2
	6	1.47		6	647.74		6	2469.1
	9	3.31		9	678.95		9	2529.6
1	0	5.88	11	0	710.90	21	0	2591.0
	3	9.18		3	743.58		3	2653.0
	6	13.22		6	776.99		6	2715.8
	9	17.99		9	811.14		9	2779.3
2	0	23.50	12	0	846.03	22	0	2843.6
	3	29.74		3	881.64		3	2908.6
	6	36.72		6	918.00		6	2974.3
	9	44.43		9	955.08		9	3040.8
3	0	52.88	13	0	992.91	23	0	3108.0
	3	62.06		3	1031.5		3	3175.9
	6	71.97		6	1070.8		6	3244.6
	9	82.62		9	1110.8		9	3314.0
4	0	94.00	14	0	1151.5	24	0	3384.1
	3	106.12		3	1193.0		3	3455.0
	6	118.97		6	1235.3		6	3526.6
	9	132.56		9	1278.2		9	3598.9
5	0	146.88	15	0	1321.9	25	0	3672.0
	3	161.93		3	1366.3		3	3745.8
	6	177.72		6	1411.5		6	3820.3
	9	194.25		9	1457.4		9	3895.6
6	0	211.51	16	0	1504.1	26	0	3971.6
	3	229.50		3	1551.4		3	4048.4
	6	248.23		6	1599.5		6	4125.9
	9	267.69		9	1648.4		9	4204.1
7	0	287.88	17	0	1697.9	27	0	4283.0
	3	308.81		3	1748.2		3	4362.7
	6	330.48		6	1799.3		6	4443.1
	9	352.88		9	1851.1		9	4524.3
8	0	376.01	18	0	1903.6	28	0	4606.1
	3	399.88		3	1956.8		3	4688.8
	6	424.48		6	2010.8		6	4772.1
	9	449.82		9	2065.5		9	4856.2
9	0	475.89	19	0	2120.9	29	0	4941.0
	3	502.70		3	2177.1		3	5026.6
	6	530.24		6	2234.0		6	5112.9
	9	558.51		9	2291.7		9	5199.9

where its circumference is in contact with the holder wall and arranged to float up or down on top of the gas. As gas is forced into the tank, the partition is raised by the gas pressure created, and, as gas is taken from the tank, the partition automatically lowers, maintaining even pressure in the gas.

Bulk storage of gas is usually carried out in tanks of the types just described, which permit the pressure to be maintained fairly constant. Where storage is carried out in ordinary tanks, arrangement must be made to prevent the gas from becoming diluted with air, and also to prevent too great a pressure from building up within the tank. Ordinarily, such tanks are used only for compressed air, in which case they are called *air receivers*.

Gas storage in holders or tanks is accomplished by means of compressors or blowers, taking the gas from its source and feeding to the storage through pipe lines. Reclaiming of gas from storage is also done by means of a compressor or blower, or by utilizing the gas pressure in the storage when this is great enough. Such reclaiming is done through piping systems. Frequently the gas pressure in the storage tank is sufficient to serve partly, and it can then be supplemented by additional pressure imparted to the gas by means of a blower or compressor.

Quantities of gas stored in holders must be calculated for some standard conditions, because of the changes that occur in gas volumes with variations in pressure and temperature. The gas placed in storage and that removed from storage may be metered in one of many types of gas meters available, and in this way a running check may be kept on the quantity stored.

2. Gas stored in containers is usually stored in what are called gas "cylinders." These are heavily constructed tanks of drum or bottle shape. Heavy construction is used in order that the gas may be stored under considerable pressure, thus permitting a relatively large amount of gas (when considered from the standpoint of its volume under ordinary pressures) to be stored in a relatively small container. Container storage for gas is usually employed for the storage and shipment of such gases as oxygen, carbon dioxide, acetylene, propane, butane, and hydrogen. It provides a convenient method for gas shipment. When such gas cylinders are stored in the plant, their relatively great weight, due to the heavy construction used, should be kept in mind so that storage floors will not be overloaded.

PACKAGING AND FILLING EQUIPMENT*

BY R. W. LAHEY

Definitions. Packaging equipment includes all machinery that may be used in preparing a product for shipment. It includes equipment for filling, weighing, measuring and counting, closing and sealing, agitating, conveying, labeling, wrapping, lining, etc.

The term *package* refers to shelf or retail packages that are packed in shipping containers. Included in this group are bottles, jars, cans, boxes, bags, etc.

The term *bulk containers*, as used here, excludes all so-called "retail" packages and refers to the "commercial" containers, such as large bags of 50 lb. minimum capacity, drums, barrels, and kegs. In general, such containers serve as both packages and shipping containers.

Purpose. It is the function of packaging equipment to prepare a product for shipment. It is unfortunate that a plant engineer cannot choose equipment by a formula, but most of these problems can be solved if all the following factors receive proper and detailed consideration:

1. Physical characteristics of product.
   a. Variation in specific gravity.
   b. Degree of gravity flow.
   c. Effect of moisture absorption.
   d. Effect of temperature change.
2. Plant layout.
3. Maximum and minimum production rates.
4. Choice of container.
5. Factory storage in bulk (bins or tanks) or in containers.

In most instances these factors vary for each product and for each production installation. This requires special consideration for every problem, and often existing machinery must be altered to fit a special situation. A general statement of the known machinery and processes is probably of greater utility than any other form of presentation.

There are so many types of packaging equipment that it is impossible to cover the whole field. The more important types are listed as a general guide. There is an urgent need for the accumulation of these data by an agency or some central bureau which the engineer can consult with the assurance that he has up-to-date and complete information.

* The data on equipment for packaging small containers have been assembled through the able assistance of W. F. Deveneau, National Folding Box Co., and of The Packaging Institute.

Storage in Bulk or in Containers. After the container for a new product is chosen, the next question that must be settled before packaging machinery can be considered is to determine whether the product is to be stored at the factory in bulk or in containers.

If the quality of the product is not affected, it is more economical to store finished products in bulk and pack them in containers just before shipping. It is obvious that this system requires less handling. Less storage space is required, and operations can be conducted with a smaller inventory of filled and empty containers. The shorter storage period is beneficial in preserving the appearance of the packages and shipping containers. Even under the best of storage conditions, coatings on steel containers become marred, stenciling is often obliterated, and labels become dirty and torn.

Disadvantages of bulk storage include cost of tanks or bins for storage and possible extra cost of additional packing and filling equipment that may be required for peak shipments if the business is seasonal.

In the rock-products and the fertilizer industries where the products are inexpensive and extra handling costs are liable to make the difference between profit and loss, the almost universal practice is either to store in bulk or to manufacture only against shipping orders.

Storage-bin Construction for Dry Materials. The classification "dry materials" includes all degrees of particle size from the finest powder to large lumps varying in degree of flow from free-flowing to, and including, those products which will not flow. Some free-flowing materials lose this desirable characteristic on absorption of moisture in humid weather or on excessive changes in temperature.

Free-flowing Materials. Even though products are stored in containers, there must be an intermediate bulk storage just before the packing operation to absorb the fluctuations in both production and packaging.

The basic requirement for accurate weighing is to provide a uniform stream of material to the packaging equipment. Pulsating or uneven flow will seriously handicap proper functioning of packing equipment, and it is therefore of paramount importance to design storage bins or tanks so that this uniform feed will result.

Usually the best shape for a storage bin for even gravity flow of material is conical. Bins with two straight adjacent sides eliminates any footing for the arching of the product on two sides. For some types of products, this shape gives best results. Under any conditions, avoid square or angled corners where material has an opportunity to pack and then arch. The inside of the bin should be free from any projections, braces, or stay rods that interfere with the free movement of the product.

Since the fundamental function of a storage bin is to take up the slack between manufacturing and packaging, it is obvious that the level of material in the bin will fluctuate, causing differences in the rate of flow of material.

In connecting the outlet of a storage bin to a filling machine of any kind, it is always desirable to avoid the possibility that the variation of the head load in the storage bin will directly affect the uniformity of the flow. This is sometimes accomplished by baffles placed in the discharge spout, but more frequently by various types of offset connections and mechanical means such as are shown in the various sketches. Variations in head load can be controlled by the use of Bin-Dicators, which provide positive operation of signals and alarms and automatically control conveyors, elevators, feeders, weighing devices, and other machinery. They are diaphragm-operated mercury switches for use with bulk dry materials (see Fig. 67).

Some type of agitation in the bin or controlled feeding to the packaging machine is often required. The type of material that arches and then surges can probably be handled by agitation in the bin to prevent arching, while the strictly non-free-flowing type will require a mechanical feeding attachment between the bin and the packer which will regulate and mechanically convey the proper amount of material. Manufacturers of weighing and bagging equipment have devices for regulating flow of materials, and they should be consulted on this problem.

Weighing. In the bulk packaging field, two systems of weighing are in general use:

Net Weighing. As the name implies, this system weighs the material before it is packed into the container. The product flows into a bucket or hopper which is attached to a balanced scale beam. When the bucket is filled, the movement of the scale beam actuates a shutoff to stop the flow of material. There is a gate for discharge at the bottom of the bucket which, when opened, either automatically or manually dumps the material into the container.

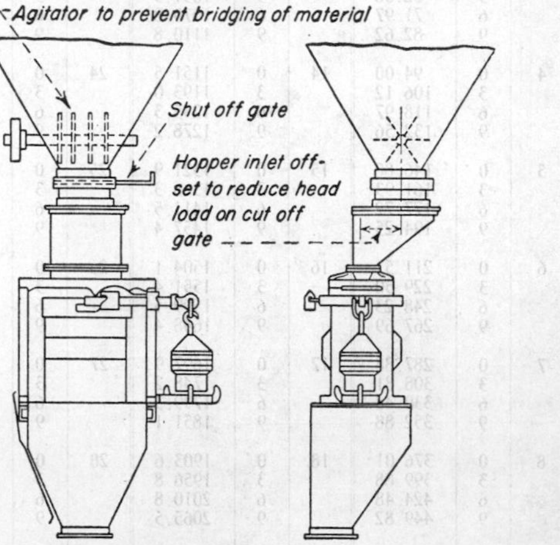

FIG. 66. Automatic scale showing connection to bin outlet, with agitator, to prevent bridging of material.

Gross Weighing. The bag is attached to the weighing mechanism, and a shutoff gate is closed when the desired amount has flowed into the container. This shutoff gate is operated either automatically or manually.

Other measuring systems, such as the volume method and a combination of the volume and weighing method, are largely used for small retail packages.

Weighing and Bagging Equipment. *Automatic Scales.* The choice of scales depends on the rate of production and the flow characteristics of the product. Automatic scales are best adapted for volume production, and they can be arranged in batteries of two or more to take care of large volume. The bagging capacity of an automatic scale varies from 5 to 8 bags of 100-lb. capacity/min. depending on the flow of the product. Automatic scales are built for commodities of special physical characteristics and usually cannot be adapted to commodities of widely different characteristics. Scales can be built to handle most types of dry material.

Automatic scales are often used in conjunction with equipment for packing products into containers such as valve bags where the openings are small and some device must be used to carry the material through the valves into the bags. Another instance of this is the use of auger packers, which are used to pack material that must be compressed.

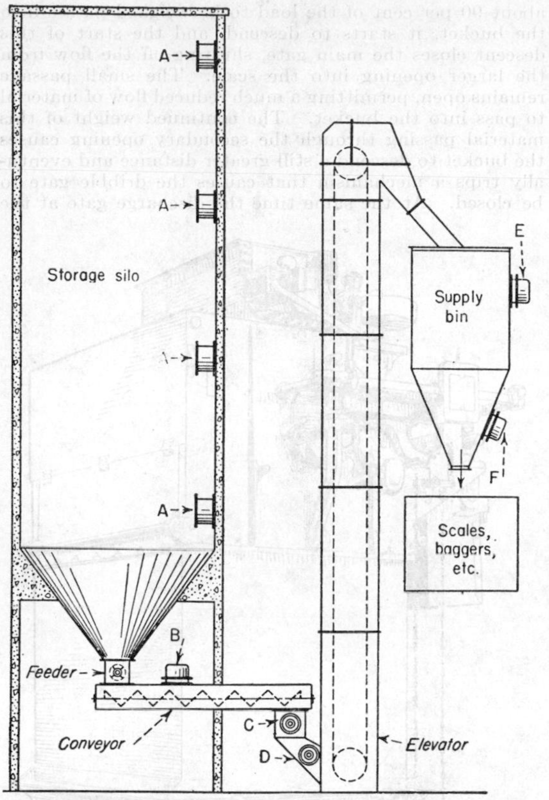

FIG. 67. Several types of Bin-Dicator level indicators. *A*. Bin-Dicators mounted at various elevations in storage silos to indicate level of material by operating visual signals at a central control point. (Bin-Dicators may be mounted either inside or outside of bin.) *B*. Bin-Dicator mounted on cover of screw conveyor to serve one or more of following purposes: 1. Indicate overfeed or choking of conveyor, by lights, horns, or bells located at central control point. 2. Indicate lack of feed in same manner. 3. Electrically control speed of feeder to provide uniform flow of material through system. 4. Automatically control (through solenoid valve) injection of compressed air into material in bin, to provide uniform and full feed of material. *C*. Bin-Dicator mounted at top of elevator boot to serve either or both of following purposes: 1. Indicate overfeed or choking of elevator, by lights, horns, or bells located at central control point. 2. Automatically shut off feeding conveyor in case of overfeed or choking. *D*. Bin-Dicator mounted on lower part of elevator boot, to indicate lack of feed, by lights or other signals located at central points. *E*. Bin-Dicator mounted near top of supply bin to serve one or more of the following purposes: 1. Indicate when bin is almost full, by lights or other signals located at central control point. 2. Automatically shut off conveying and elevating machinery when bin is almost full. 3. In conjunction with "*F*," automatically provide continuous supply of material to supply bin. (See *E* and *F* below.) *F*. Bin-Dicator mounted near bottom of supply bin to serve one or more of following purposes: 1. Indicate when bin is almost empty, by lights or signals located at central control point. 2. Automatically shut off weighing devices or packing machines before supply bin becomes empty—thereby preventing errors in weights due to lack of supply. 3. In conjunction with *E*, automatically provide continuous supply of material to supply bin. (See *E* and *F* below.) *E*. and *F*. Two Bin-Dicators mounted on supply bin to automatically provide continuous supply of material. To accomplish this purpose, the Bin-Dicators are connected to the control circuits of the motors driving the feeding, conveying, and elevating machinery. The conveying system will automatically start when the material level drops to point *F*, and automatically stop when it rises to point *E*.

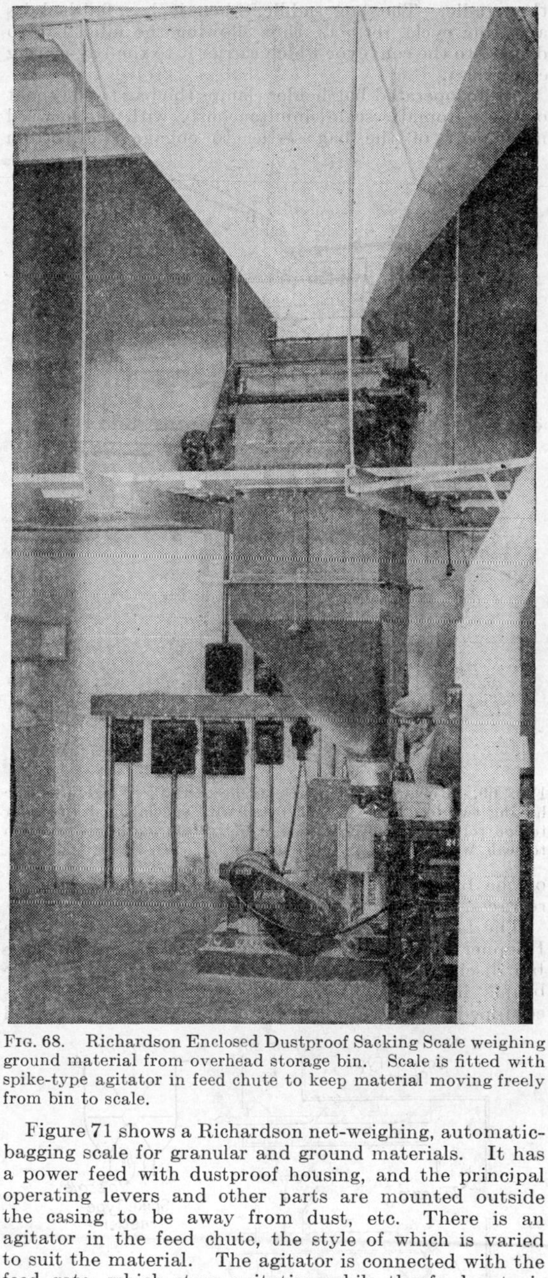

FIG. 68. Richardson Enclosed Dustproof Sacking Scale weighing ground material from overhead storage bin. Scale is fitted with spike-type agitator in feed chute to keep material moving freely from bin to scale.

Figure 71 shows a Richardson net-weighing, automatic-bagging scale for granular and ground materials. It has a power feed with dustproof housing, and the principal operating levers and other parts are mounted outside the casing to be away from dust, etc. There is an agitator in the feed chute, the style of which is varied to suit the material. The agitator is connected with the feed gate, which stops agitation while the feed gate is closed. It has a hand pull and an automatic discharge mechanism. The manufacturer claims that it will weigh batches of 50 to 300 lb. as required and, depending on the free-flowing characteristics of the product, will pack from 1 to more than 12 bags of 100-lb. capacity/min.

The Richardson Scale Co. has an interesting combination of automatic-scale and bag-settling device called an "Oscillating Packer" (see Fig. 72). This packer oscillates the bag saddle by a crank action on a revolving shaft causing the bag to weave. The crankshaft is operated by ⅓-hp. motor driving through a V-belt and speed reducer, mounted on a carriage that also carries the cradle. The bag saddle, when it has finished its agitating cycle, retracts, thus allowing the filled bag to drop onto the conveyor which carries it to the bag-closing equipment.

An air-operated bag holder clamps the bag to the spout of the automatic scale simultaneously with the forward movement of the bag settler to engage the bottom

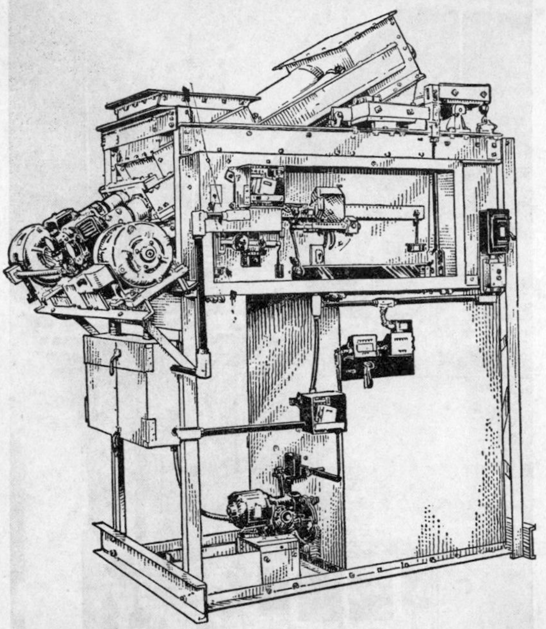

FIG. 69. Richardson Duo-Screw Feed Bulk Weighing Scale having two feed screws, each fitted with special flush governor to control delivery of powdered or pulverized materials from bin to scale weigh hoppers.

of the bag. The bag holder releases as the bag settler retracts.

The hopper shown at the top of the illustration of the Hoepner scale (Fig. 74) is connected to the outlet of the bin discharge. The bucket itself is suspended on scale beams in the usual manner. This intake hopper is equipped with two openings, one a large size intended

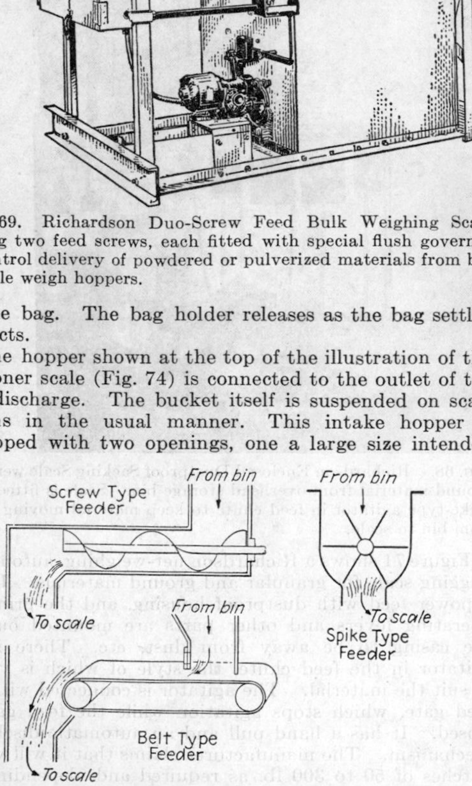

FIG. 70. Installations for proper feeding of automatic scales.

for the flow of the bulk stream and the other a small one intended for the dribble stream. At each of these openings there is a gate operated by the bucket. The bottom of the scale bucket is equipped with a gate that is also controlled by the movement of the bucket itself.

With the bottom gate closed and the top gates open, the flow of material passes into the bucket. When about 90 per cent of the load to be weighed passes into the bucket, it starts to descend, and the start of this descent closes the main gate, shutting off the flow from the larger opening into the scale. The small passage remains open, permitting a much reduced flow of material to pass into the bucket. The continued weight of this material passing through the secondary opening causes the bucket to descend a still greater distance and eventually trips a mechanism that causes the dribble gate to be closed. At the same time the discharge gate at the

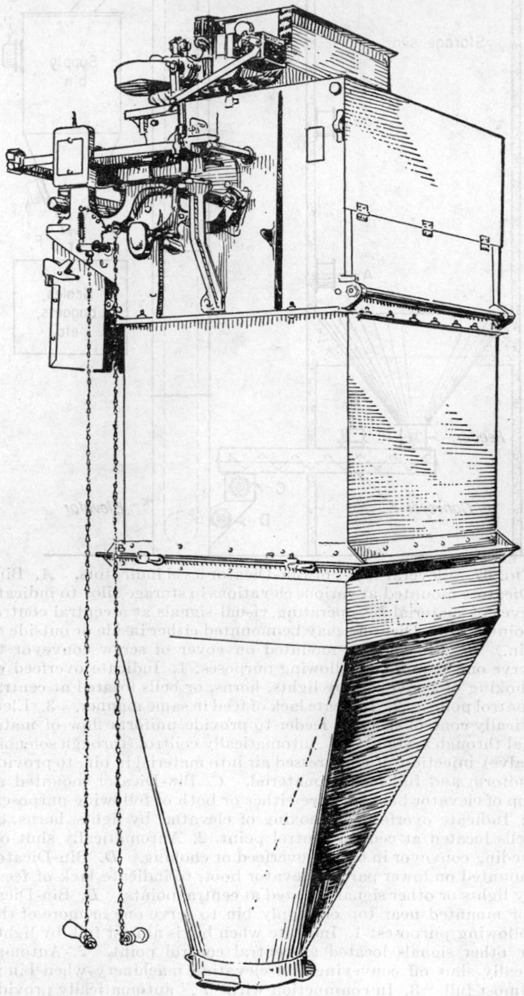

FIG. 71. Richardson Automatic Scale, fully enclosed type.

bottom of the bucket is opened, permitting the material to flow into the container beneath the discharge spout of the scale. It is possible to change the weight of the material being weighed in the scale by the use of different poise weights, much the same as in the ordinary platform scale.

Auger Packers. Certain materials will not settle by vibration or shaking. These products have to be forced into the container by means of an auger or screw which creates sufficient pressure to deaerate the material and compress it. The compacting sometimes takes place before the material is discharged into the container. The compressing type of packer is usually used in con-

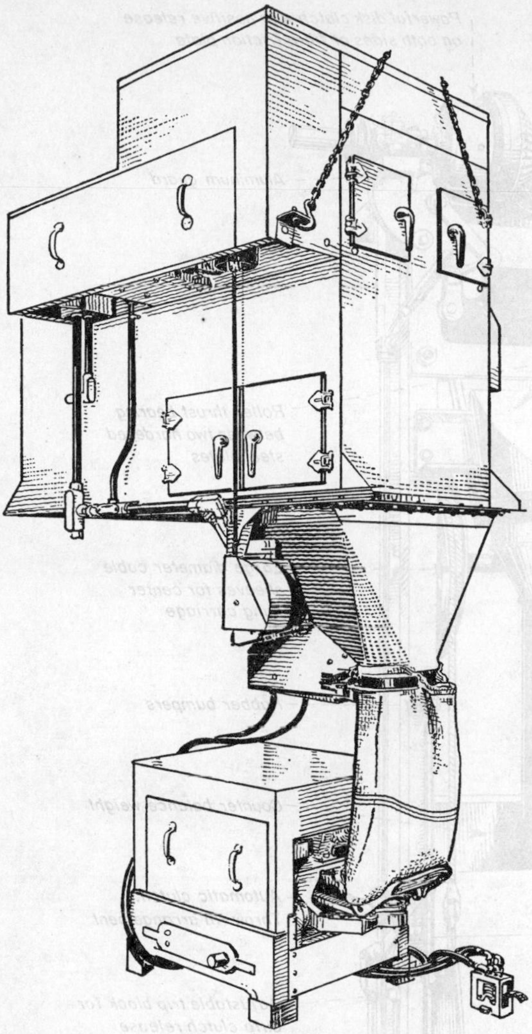

FIG. 72. Richardson Oscillating Packer.

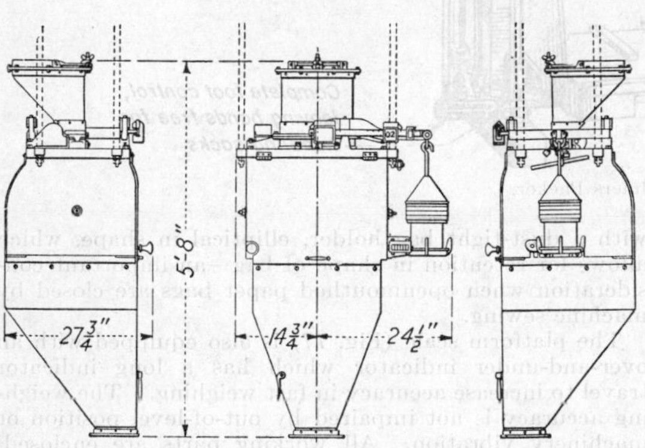

FIG. 73. Hoepner O2 scale, enclosed type.

junction with automatic scales. They slow up the operation of packing and are used where the character of the product requires it. They have been used for packing such materials as flour, gypsum, cement, lampblack, clay products, dried milk, and some chemicals.

The Allis-Chalmers auger packer (Fig. 75) is fed from an overhead bin. After the bag has been slipped over the auger tube, the platform is released and raised automatically. When it reaches the top position, the clutch is automatically engaged, causing the auger to force the material into the container. The platform is forced down against the resistance of a friction brake until the desired weight is in the bag. The clutch is then automatically disengaged, and the auger stops grinding. The

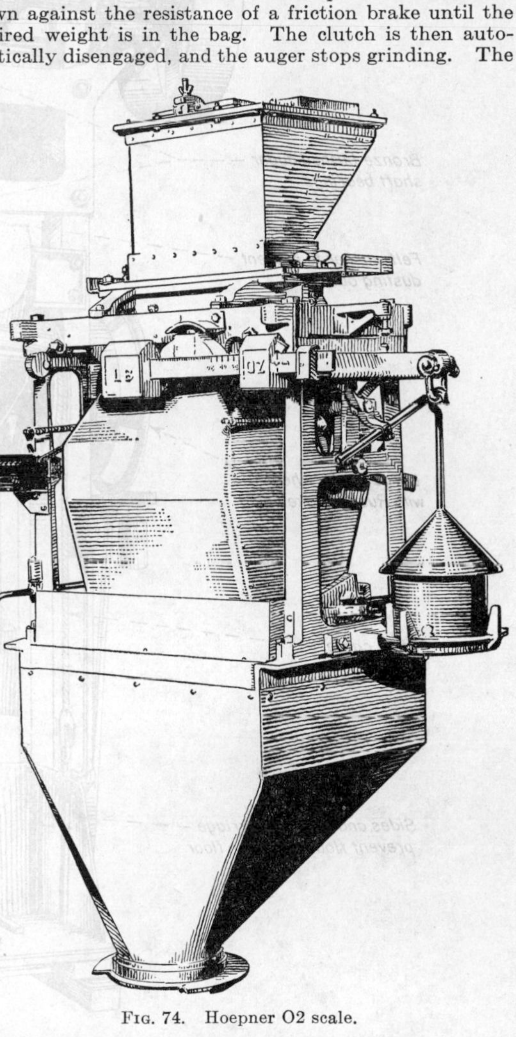

FIG. 74. Hoepner O2 scale.

brake is released by a foot pedal, and the platform drops to allow the removal of the filled bag. The brake is provided with an adjustment by which fine adjustments can be made, thus making it possible to control weights to within a few ounces tolerance.

Hand Weighing. In this operation the container is attached to a scale, and the operator closes the filling gate of the hopper as the container is filled to the desired height or weight. The accuracy of this operation depends upon the skill of the operator but, on the average, is not so accurate as automatic weighing. Often the container is filled to the approximate amount required, and then the container is transferred to a platform scale where the weight is adjusted by the "put-and-take" method. Additional speed is obtained, but two operators are required.

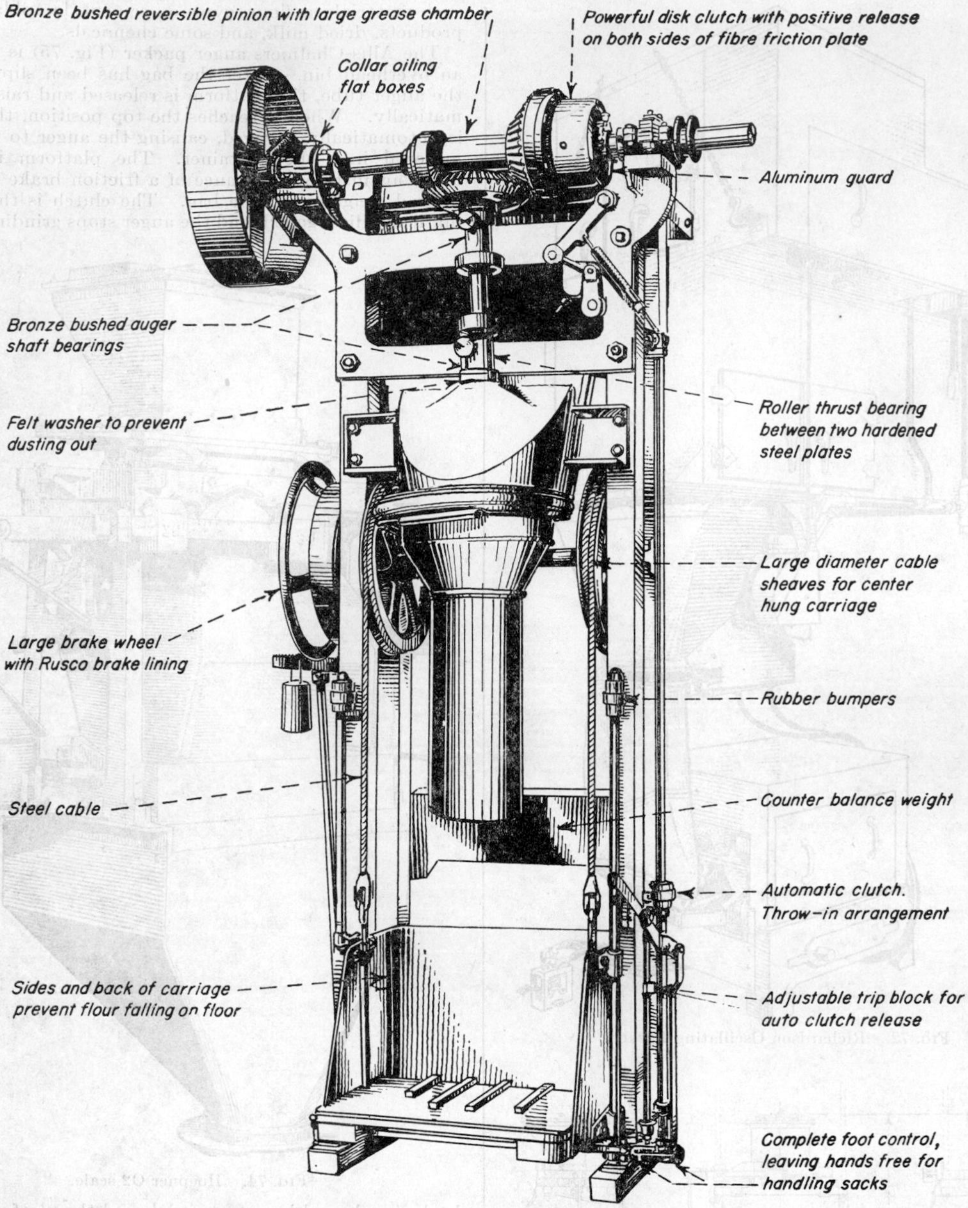

FIG. 75. Allis-Chalmers Packer.

The practice of filling predetermined weights into containers is fast replacing the custom of filling each container as full as possible. The advantage of this is the reduction in human error of filling and reduced cost of record keeping, billing, etc.

The most efficient type of equipment for predetermined weighing is the "over-and-under" type of indicating scales because they increase accuracy and speed of filling.

The sacking scale in Fig. 76 is equipped with the over-and-under indicator. It will handle loads up to 200 lb. at a rate of speed of 3 to 6 bags/min. It is designed to fit directly under a storage hopper outlet and occupies a total space of 48 by 21 by 58 in. It can be floor-mounted, swing-mounted, or suspended. This scale is equipped with a dust-tight bag holder, elliptical in shape, which allows for retention in shape of bag—an important consideration when openmouthed paper bags are closed by machine sewing.

The platform scale (Fig. 77) is also equipped with an over-and-under indicator which has a long indicator travel to increase accuracy in fast weighing. The weighing accuracy is not impaired by out-of-level position or machinery vibration. All working parts are enclosed, and the rail attached to the platform supports the bags.

St. Regis Valve-bag-filling Machines. The introduction of multiwall paper bags has led to the development of several filling and weighing machines of particular interest. There are several types of equipment that

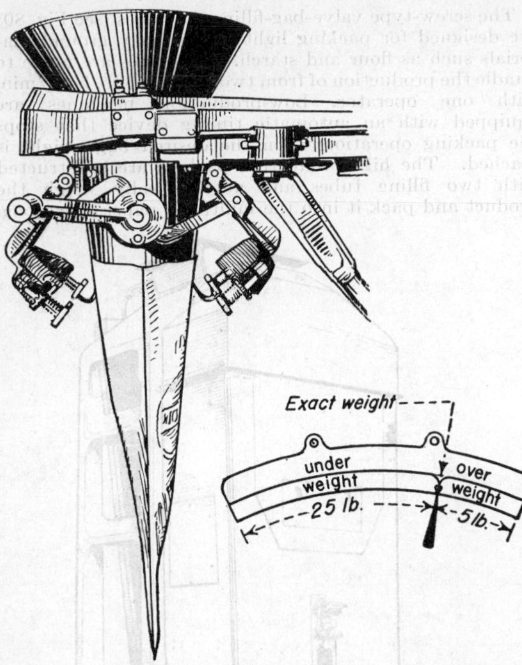

FIG. 76. Dustite Holder. (*The Exact Weight Scale Co.*)

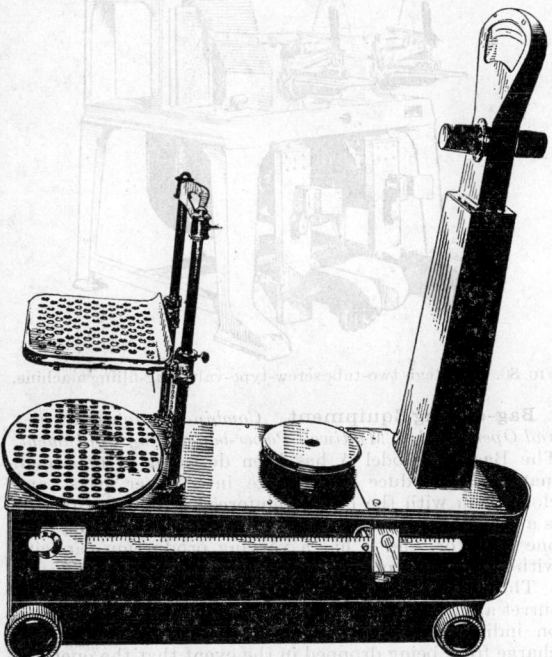

FIG. 77. Check or hand-weighing scale, 150-lb. capacity.

will weigh and pack into valve bags a wide variety of powdered and granular products. Valve bags are closed at both ends save for a small opening or "valve" in one corner through which the material is injected. This valve is so designed as to be self-closing because of the internal pressure of the contents as the bag is withdrawn from the filling tube. For granular products the valves are equipped with paper "sleeves," which, folded manually, prevent sifting.

Valve-bag Packers

Type	Typical products	Type scale	Weight of package, lb.	Approx. production, one operator, tons/hr.
Standard impeller packer	Cement Lime Limestone Plaster	Gross weighed Gross weighed Gross weighed Gross weighed	94 50 50–100 80–100	15–70* * * *
Vertical impeller-shaft packer	Talc Clay Bicarbonate of soda	Gross weighed Gross weighed Gross weighed	100 100 100	10–15*
Belt packer (single tube) (301-FB)	Bicarbonate of soda Trisodium phosphate Flake graphite Granular soda ash Chicken grits	Gross weighed Gross weighed Gross weighed Gross weighed Gross weighed	100 100 100 100 100	10–20*
Belt packer (1-tube) (301-PB)	Dog food	Preweighed	25–100	9–15*
Belt packer (2-tube) (160-FB)	Fertilizer	Gross weighed	25–100	20–40*
Belt packer (2-tube) (152-PB)	Calcium chloride Sugar Salt Charcoal Alum	Preweighed Preweighed Preweighed Preweighed Preweighed	100 25–100 25–100 25–100 25–100	15–25 on 100 lb.* two tube
Screw packer (2-tube) (402-PS)	Flour Calcimine Starch	Preweighed Preweighed Preweighed	100 100 100	12–18*
Screw packer (1-tube)	Flour Starch	Gross and Preweighed	25–100	3–9*

* Production varies, depending on type of installation, size of bags being packed, physical properties of product, etc.

Machines used for the filling of valve bags are of the impeller, belt, and screw types and are equipped with either preweighing or gross-weighing scales. The above tabulation shows the available types of valve-bag-filling machines with their rates of production.

The standard impeller-type machine consists essentially of a continuously running shaft equipped with feeder blades which, mounted in a housing, force the material through the filling tube into the valve bag.

The other impeller-type machines (horizontal and vertical shaft) are similar in principle, the differences being in the arrangement of the impeller blades and that the impeller shaft is started and stopped at the beginning and end of the filling cycle for each bag rather than turning continuously.

The belt-type machines (see Fig. 78) are used primarily for granular products and especially for crystalline products whose physical characteristics may be changed in passing through an impeller-type machine.

With the belt-type machines (see Fig. 79) the packing principle is as follows: The material flows or is fed from a hopper over the packing machine through an inlet leading to a deep groove in a rotating pulley to which it is confined by an endless belt pressing against the outer flanges of the pulley. As the pulley rotates (at a peripheral speed of from 1500 to 2100 ft./min.), the material is pressed against the belt by centrifugal force and is rapidly accelerated. After rotating through an arc of 90 deg., the material is discharged at a tangent to the pulley through the filling tube and into the bag.

On the two-tube gross-weighing machines of the belt type, a settler arrangement is provided to agitate the bag and settle the contents during a part of the filling cycle. On the preweighing machine of the belt type the

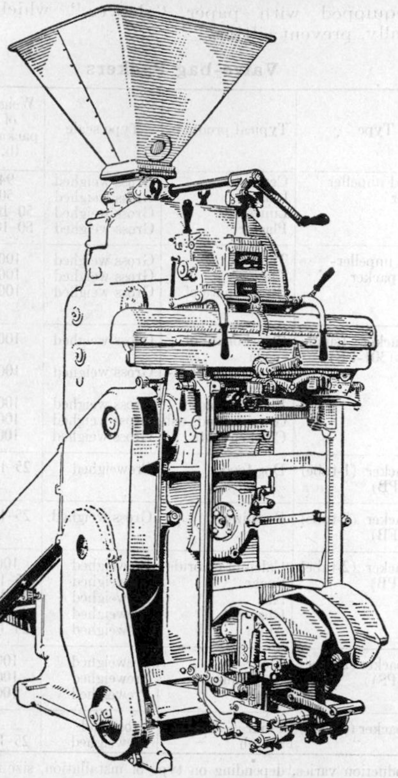

FIG. 78. Shifting-tube belt packer.

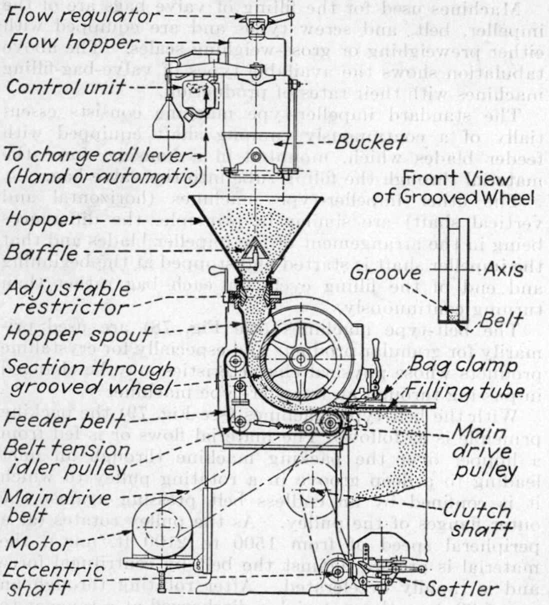

FIG. 79. Belt-type packer.

settler provides such agitation during the complete filling cycle. Belt-type machines are ordinarily equipped to pack 100-lb. bags, but special attachments are furnished for the filling of 25-, 50-, and 80-lb. bags as well.

The screw-type valve-bag-filling machines (see Fig. 80) are designed for packing lightweight finely ground materials such as flour and starch. Machines are made to handle the production of from two to six 100-lb. bags/min. with one operator. Low-production machines are equipped with an automatic timing device that stops the packing operation when the desired bag weight is reached. The high-production packers are constructed with two filling tubes and automatically weigh the product and pack it into the multiwall paper valve bag.

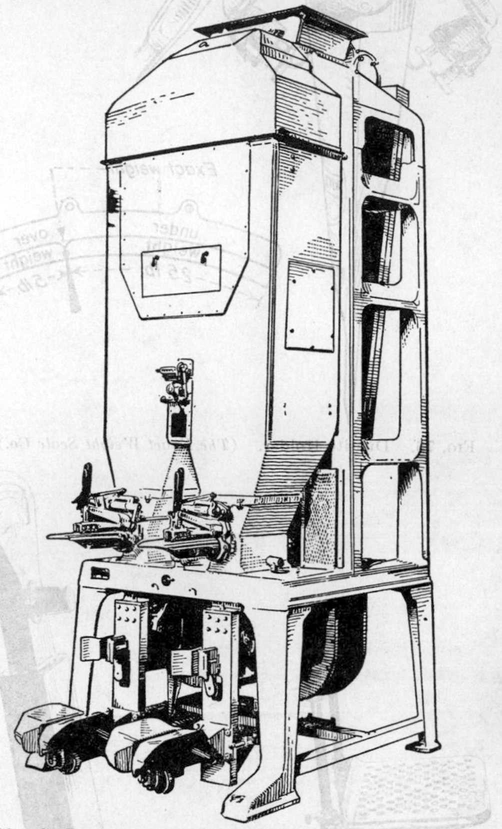

FIG. 80. St. Regis two-tube screw-type-valve bag-filling machine.

Bag-closing Equipment. *Combined Automatic Scales and Openmouthed Multiwall Paper-bag-closing Equipment.* The Bagpak Model A has been designed to net-weigh material, introduce the charge into paper bags, and close them with the heavy reinforced stitch closure. It is a multistation fully automatic machine, requiring only one operator, and it has a varying production capacity with a maximum of over 1000 bags of 100 lb. capacity/hr.

This machine (see Figs. 81 and 82) has a revolving turret around the periphery of which the bags are carried on individual spouts. Automatic devices prevent the charge from being dropped in the event that the operator fails to insert a bag. Further mechanisms jog or settle the material in the bag. Automatic bag-gripping carrier chains divert the filled bag from the turret and guide it through a sewing unit which applies the heavy reinforcing stitch and then through a taping mechanism which folds a strip of heavy kraft paper over the end and glues and presses it into place. The sewing unit and taping mechanism are mounted over a straight-away bag conveyor at the discharge end of the machine. This

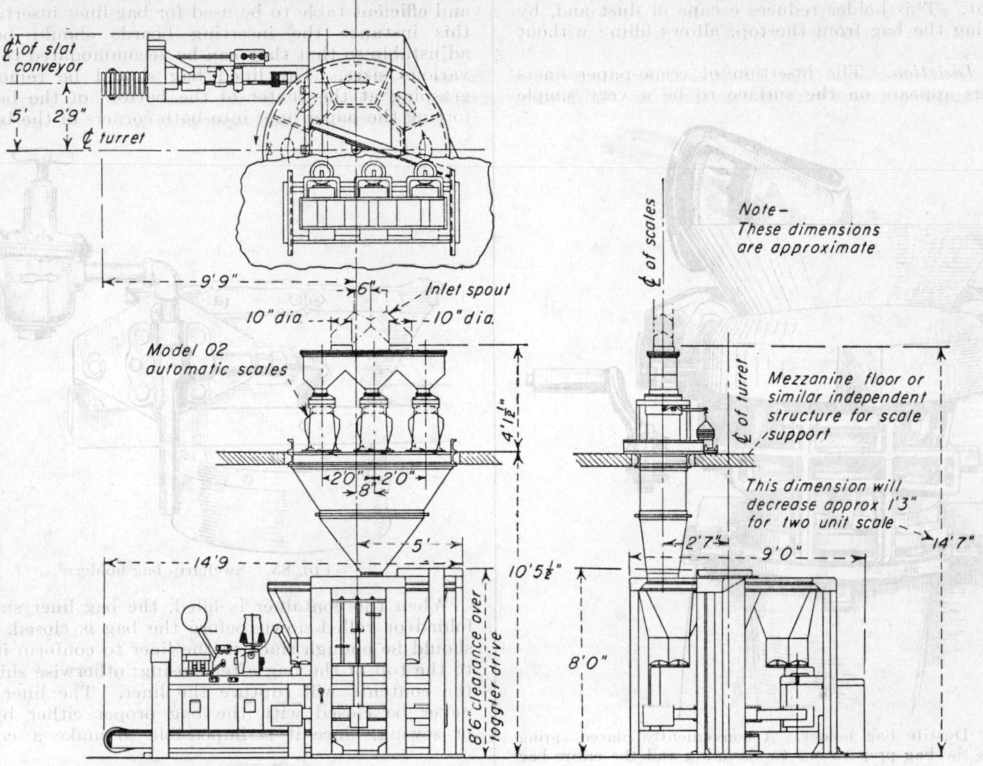

FIG. 81. Bagpak Model A.

conveyor delivers the filled and closed bag to take-away conveyors or hand trucks.

Sewing Unit (Fig. 82). Hoepner No. 150 heavy-duty type, gear-driven from main drive shaft. Sewing head equipped to apply various types of double- or single-stitch seam, including the Cushion Stitch Closure.

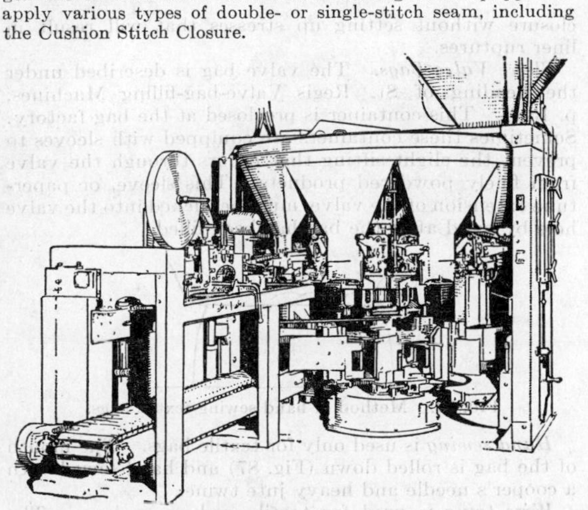

FIG. 82. Bagpak Model A.

Conveyors. Circular polished-steel conveyor underneath turret—slat type on the straightaway, separate-motor-driven.

Controls. Continuous operation motivated by two control switches.

Scales. This machine is equipped with two or three scales, dependent upon the production speed required. These are the Hoepner automatic bucket scales. Individual scale adjustments can be made without shutting down machine. On such materials as are not readily free-flowing, scale feeders are required. They ensure a steady stream of material at all times, thus preventing inaccuracies in weighing.

Portability. This machine is not portable.

Motors. The straightaway conveyor, the entire closing unit, and the revolving turret are driven by a 5-hp. motor. The settling devices are driven by 2- and 1-hp. motors. Scale feeders, when required, are driven by a motor the size of which is dependent upon the material handled but in any case will not be over 5 hp. All motors are of dustproof type.

Bag Holders. The problem of holding bags on automatic and manually operated bag hoppers often requires special equipment. The round hopper with a holding device, usually a ring, is most often used for textile bags. Sometimes the hopper operator merely holds the textile bag over the lip of the hopper with one hand while the bag is filling.

The most desirable type of bag holder for paper bags is oval in shape; and, where the bag must be suspended above the floor, grippers are used, as shown in Figs. 83 and 84. It is important to preserve the oval contour of paper bags if they are to be closed by machine sewing. Otherwise, difficulties will be encountered in that operation.

Another type holds the container by means of inflating a rubber tube that is attached to the periphery of the bagging spout (see Fig. 85). It uses 10 to 15 lb./sq. in. air pressure, and one valve inflates and exhausts the air. It is adjustable for different sizes by adjusting a pressure regulator. This device is available in diameters from 6 to 16 in. in two widths. The smaller width (3¼ in.) handles bags varying not more than 2½ in. in width and up to 100 lb. in weight. The large size (8 in. wide) handles bags that vary 6 in. in width and up to 400 lb. in weight. This holder reduces escape of dust and, by suspending the bag from the top, allows filling without voids.

Liner Insertion. The insertion of crepe-paper liners into bags appears on the surface to be a very simple

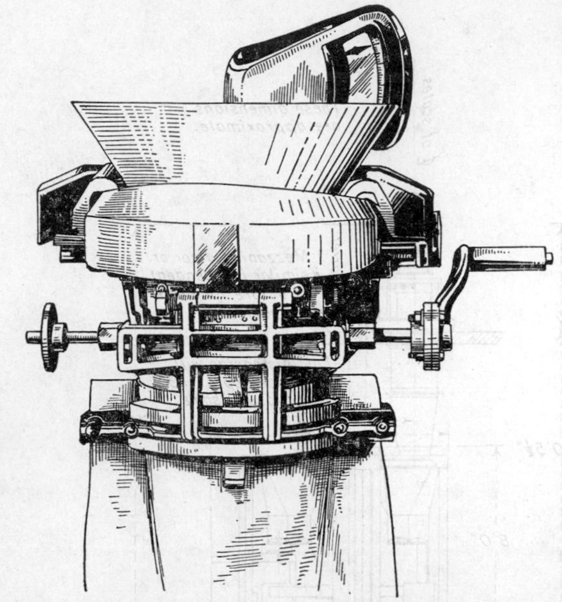

Fig. 83. Dustite bag holder. A conveniently placed spring clip holds the bag preparatory to clamping and the entire bag opening is completely closed and held tight in one operation.

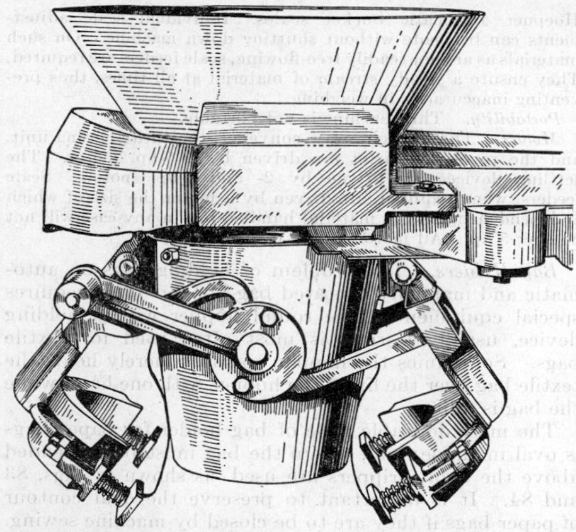

Fig. 84. Side view of Dustite bag holder with bag removed.

operation; but, if the paper liner is to be of any value, it must be properly and carefully inserted. Bearing in mind that the function of the liner is to prevent sifting and to eliminate contamination of the product by the container, it must depend for its strength on the container. This, therefore, requires that the liner be inserted to conform completely to the contour of the outer container, or else liner breakage will result from the filling operation or the abuses of handling and transportation. The illustration (see Fig. 86) shows a simple and efficient table to be used for bag-liner insertion. In this instance the inserting boards should be made adjustable so that they can be accommodated to bags of various sizes. The lined bag should be removed by grasping at the center of the bottom of the bag, thus forcing the paper liner into both corners of the bag.

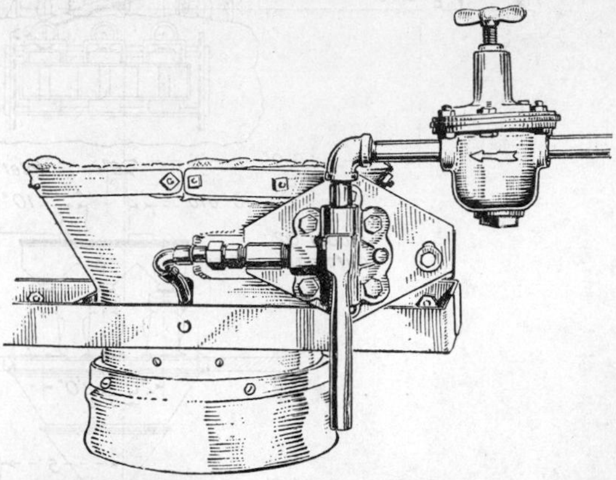

Fig. 85. Swellgrip bag holder.

When the container is filled, the bag liner should be folded or rolled down before the bag is closed. There should be enough slack in the liner to conform in shape to the top of the bag after closing; otherwise shifting of the contents will rupture the liner. The liner should never be closed with the bag proper either by tying or sewing, since it is impossible to make a combined

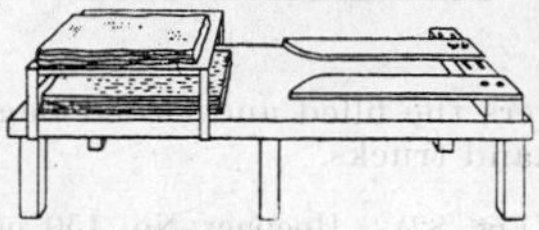

Fig. 86. Table for bag-liner insertion.

closure without setting up stresses that will result in liner ruptures.

Sleeve Valve Bags. The valve bag is described under the heading of St. Regis Valve-bag-filling Machines, p. 1386. This container is preclosed at the bag factory. Sometimes these containers are equipped with sleeves to prevent the slight sifting that occurs through the valve from finely powdered products. This sleeve, or paper-tube extension of the valve, must be tucked into the valve hole by hand after the bag has been filled.

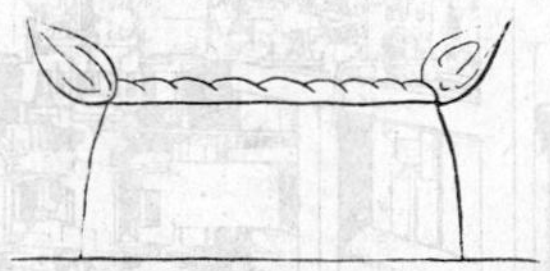

Fig. 87. Method of hand sewing textile bags.

Hand sewing is used only for textile bags. The mouth of the bag is rolled down (Fig. 87) and hand-sewed with a cooper's needle and heavy jute twine.

Wire tying is used for textile and paper bags. The bag mouth is gathered by pleating and the wire tie applied by a twisting tool (see Fig. 88).

Machine Sewing. Both textile and paper bags may be closed by machine sewing. Many ingenious installations of the sewing machine have been devised to meet a great variety of production conditions, varying from a conveyorized production line (taking the output of one or more automatic scales) to the small-volume operations where one operator fills, check-weighs on a platform scale, and closes the bags.

Large textile bags are sewed with either the single-thread chain stitch or the two-thread double-locked stitch. Either stitch can be used for paper bags, but the insertion of a filler cord on the top side of the stitch is recommended to prevent sifting through the needle punctures. This feature does not complicate the operation of the sewing head.

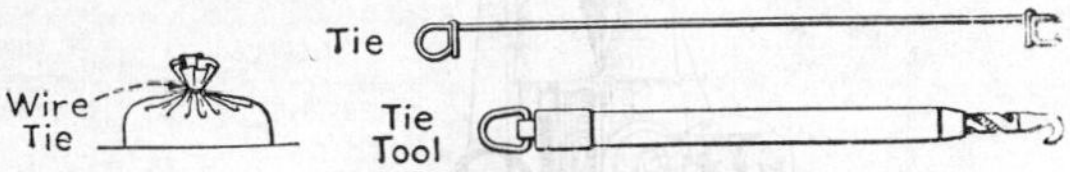

FIG. 88. Method of wire-tying bags.

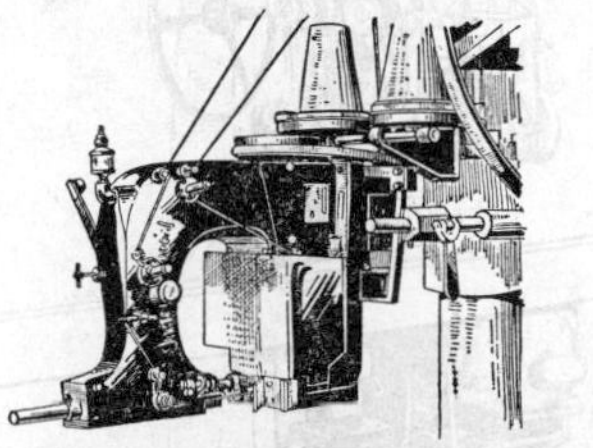
FIG. 89. Heavy-duty sewing head for bags.

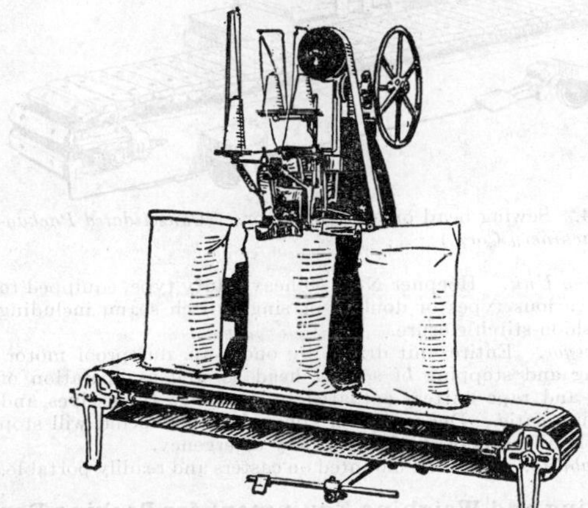
FIG. 90. The Union Special head mounted as described under style 80600 H.

The illustrations (Figs. 89, 90, and 91) show the two different heavy-duty sewing heads that are designed to close heavy textile and multiwall paper bags.

A Union Special sewing head, shown in Fig. 90, which is typical of the heavy-duty class 80600, is used for sewing textile and paper bags. The uses for special styles are as follows:

80600 E—For closing paper bags using a filler cord guided under the presser foot to seal needle punctures. Produces double-locked stitch. Stitch range from 4 to 3 per in.

80600 F—Produces single-thread chain stitch; otherwise same as style 80600 E.

80600 H—For closing multiwall paper bags, using a crepe-paper binding and simultaneously inserting a filler cord. Machine is equipped with an automatic mechanical tape-clipping device; it produces the double-locked stitch.

80600 AC—This machine is recommended for closing bags made from heavy-weight fabric. It produces the double-locked stitch.

80600 AD—Same as style 80600 AC except that it makes the single-thread chain stitch.

The Hoepner sewing head uses the well-known principle of two threads forming a chain stitch. These heads have been particularly designed to meet the rough requirements usually found in plants closing bags as large as 100 lb. or greater. The mechanism of the head is practically entirely enclosed to protect it from the dust and dirt that are invariably found in operations of closing bags of this kind.

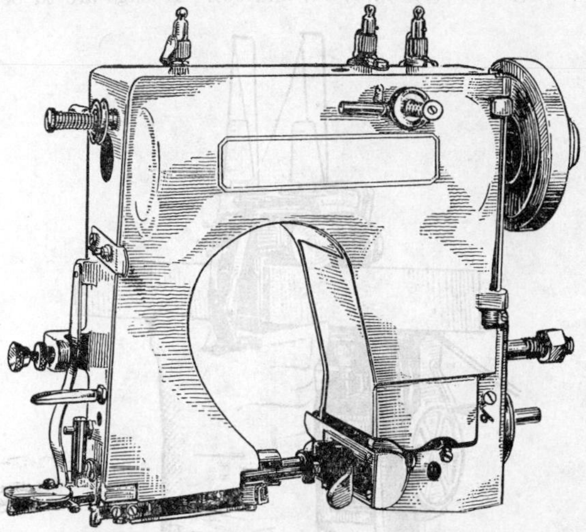
FIG. 91. High-speed Hoepner sewing head, Model 100. (*Consolidated Packaging Machinery Corp.*)

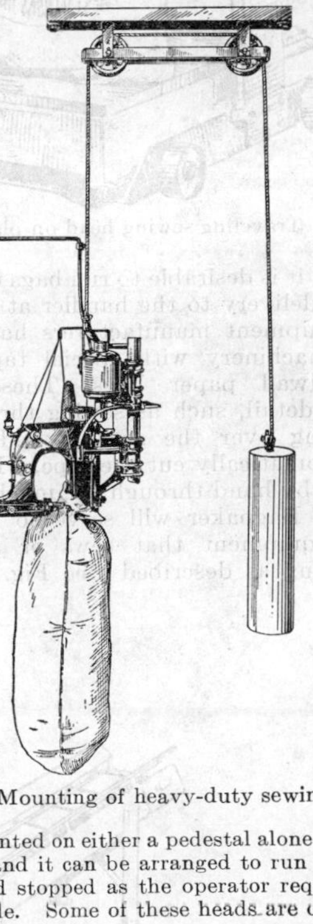
FIG. 92. Mounting of heavy-duty sewing head.

The head is mounted on either a pedestal alone or on a pedestal over a conveyor, and it can be arranged to run continuously or can be started and stopped as the operator requires by merely depressing a treadle. Some of these heads are operated as high as 3200 r.p.m.

These sewing heads can be mounted to fit almost any need. In Fig. 92 is shown a heavy-duty head, mounted on a tandem pulley, cable, and counterweight. The machine is operated by a hand lever which is pulled by the operator across the top of the bag. Figure 93 has a traveling head and suitable base for supporting a platform scale.

For larger production and more efficient sewing, the head is mounted to synchronize with a belt or slat conveyor which carries the bag mouth through the stationary sewing head (Figs. 94, 95, and 96). If bags are to be

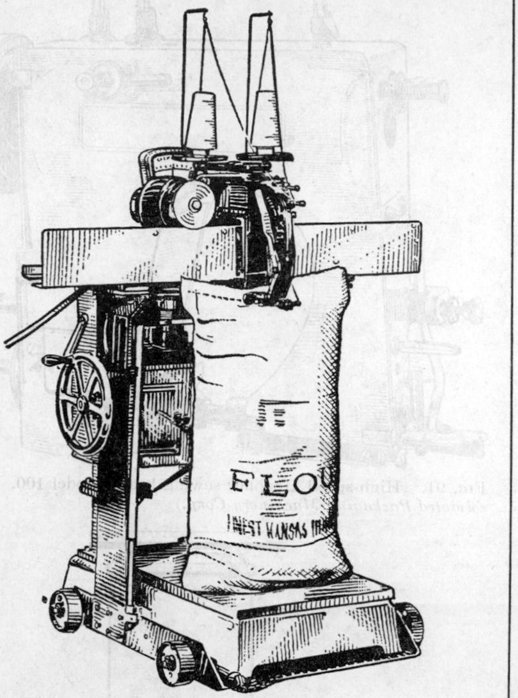

FIG. 93. Traveling sewing head on platform scale.

hand-trucked, it is desirable to run bags up on an inclined conveyor for delivery to the handler at waist height.

Several equipment manufacturers have provided the bag-closing machinery with special taping devices for closing multiwall paper bags. These devices vary somewhat in detail, such as sewing through the creped tape or taping over the sewing, and some of these machines automatically cut the tape. In each case, the bag is guided by hand through the machine.

The D. A. Bagpaker will serve to illustrate paper-bag-closing equipment that sews or sews and tapes over the sewing as described (see Fig. 97). This machine consists of a sewing unit and a tape applicator, mounted in a fixed position over a bag conveyor, which is adjustable vertically for various bag lengths and is designed for use in plants that already have weighing and filling facilities and where only the bag-closing equipment is desired. This machine is designed to handle paper bags of from 25 to 100 lb. capacity.

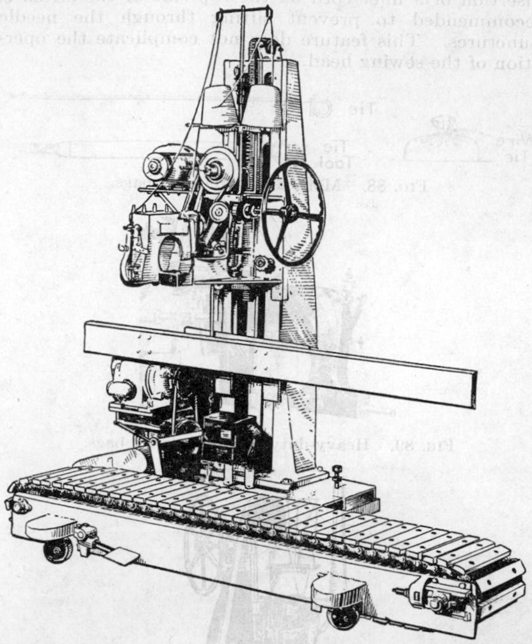

FIG. 94. Sewing head on slat conveyor. (*Consolidated Packaging Machinery Corp.*)

Sewing Unit. Hoepner No. 150 heavy-duty type, equipped to apply various types of double- or single-stitch seam, including the cushion-stitch closure.

Conveyor. Entire unit driven by one 1-hp. dustproof motor Starting and stopping of sewing head as well as operation of thread and tape cutters actuated by the bags themselves and fully automatic. Depression of convenient foot pedal will stop instantly all parts of machine for any emergency.

Portability. Machine mounted on casters and readily portable.

Filling and Weighing Equipment for Packing Dry Materials into Drums, Barrels, and Kegs. Container economy often requires agitation of barrels, large drums, and kegs during filling. This prevents weighing until the approximate weight has been packed, when the

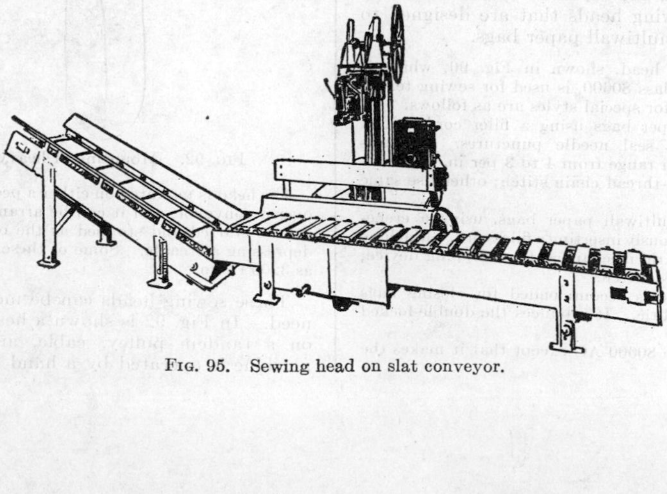

FIG. 95. Sewing head on slat conveyor.

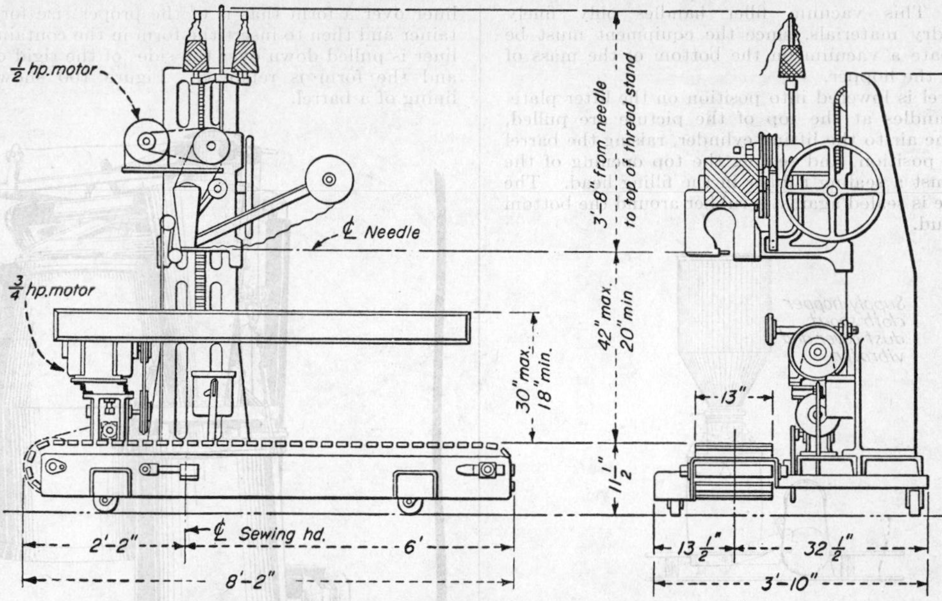

FIG. 96. Sewing head on slat conveyor.

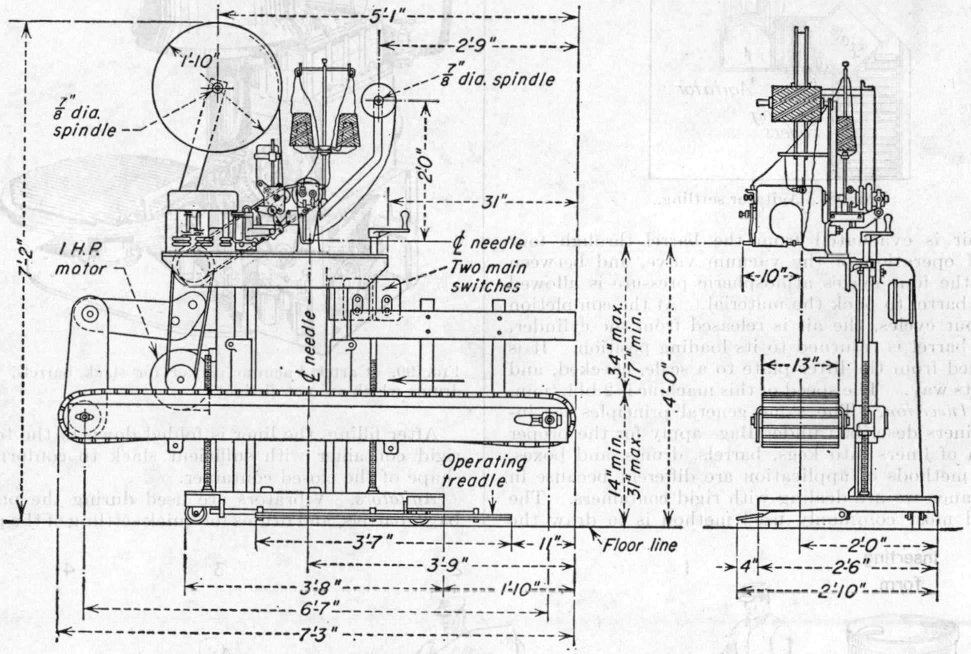

FIG. 97. D. A. Bagpaker sews and tapes closures of multiwall paper bags.

container is removed from below the filling hopper to a platform scale. The correct weight is then adjusted by the put-and-take method.

Automatic scales are occasionally used for weighing and dumping into large rigid containers, but the speed of these devices often makes it difficult to pack, since insufficient time is allowed for settling. The best plan in using automatic scales for barrel filling is to divide the weight to be packed into two or more dumps of the scale and have the container agitated throughout the operation. The layout of Fig. 98 shows an agitator settling the load at the filling spout. After the approximate weight has been packed, the container is moved by roller conveyor to the check-weigh scale and then removed to the warehouse.

A vacuum method of packing can be used for filling drums, barrels, and kegs that are not airtight if they are filled in an airtight shroud. The equipment illustrated (Fig. 99) packs the material solidly in the container without agitation. It packs by volume, always filling to a constant density. The density can be varied by adjusting the amount of vacuum used in the packing operation. This vacuum filler handles only finely powdered dry materials, since the equipment must be able to create a vacuum on the bottom of the mass of material in the hopper.

The barrel is lowered into position on the lifter plate, and the handles at the top of the picture are pulled, releasing the air to the lifting cylinder, raising the barrel into filling position, and sealing the top opening of the barrel against a sealing rubber in the filling head. The lifting plate is sealed against a rubber around the bottom of the shroud.

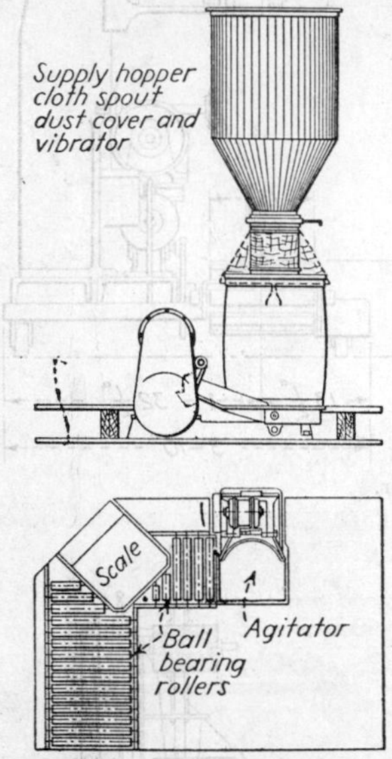

FIG. 98. Agitator settling..

The air is evacuated from the barrel through four cycles of operation of the vacuum valve, and between each of the four cycles atmospheric pressure is allowed into the barrel to pack the material. At the completion of the four cycles, the air is released from the cylinder, and the barrel is returned to its loading position. It is then rolled from the lifter plate to a scale, checked, and sent on its way. The speed of this machine is 2 bbl./min.

Liner Insertion. The same general principles for inserting liners described under Bags apply for the proper insertion of liners into kegs, barrels, drums, and boxes, but the methods of application are different because in this instance we are dealing with rigid containers. The best and most commonly used method is to draw the liner over a form that is of the proper size for the container and then to insert the form in the container. The liner is pulled down over the sides of the rigid container, and the form is removed. Figure 100 shows proper lining of a barrel.

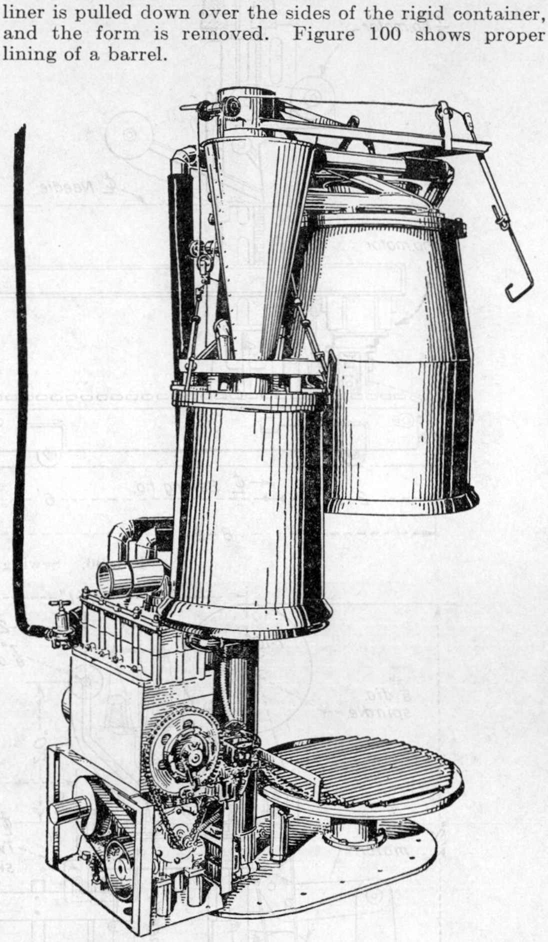

FIG. 99. Carter Vacuum packer for slack barrels, kegs, and drums which are not airtight.

After filling, the liner is folded down in the top of the rigid container with sufficient slack to conform to the shape of the closed container.

Agitators. Vibrators are used during the packing of bags, barrels, and drums for quick settling of the products,

FIG. 100. Method for inserting crepe-paper liners into barrels. 1. Insert the device into the lining as far as it will go. 2. Fold inwardly the ears formed by this operation and insert the lining and device into the barrel. 3. Fold the lining down over the chime of the barrel and remove the device. 4. The bottom of the lining lies smooth in the bottom of the barrel.

thereby effecting economies by reduction in container size. Most vibrators are adaptable for use with barrels, drums, or bags.

The Syntron vibrator, shown in Fig. 101, consists of a vibrating platform with the vibrator mounted under the cushioned top plate. The motion is horizontal in direction at the rate of 3600 vibrations/minute. This is accomplished through the use of a thermionic valve that changes alternating current to pulsating waves with a time interval between.

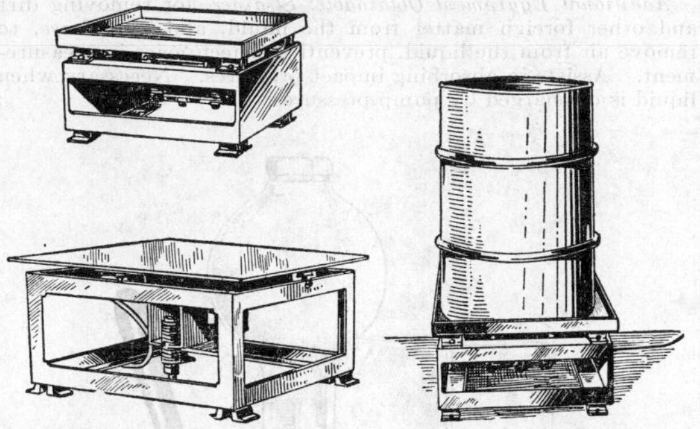

Fig. 101. Syntron vibrator.

Specifications	Models		
	VP-15	VP-55	VP-125
Capacity	50 lb.	300 lb.	750 lb.
Deck dimensions	11 by 11 in.	20 by 20 in.	30 by 30 in.
Over-all height	6 in.	11½ in.	15 in.
Net weight (packing machine)	30 lb.	135 lb.	275 lb.
Shipping weight (machine and controller)	50 lb.	200 lb.	350 lb.

For compacting light fluffy materials that cannot be settled by high-speed horizontal vibration, Syntron offers hydraulic jolters. These vibrators (Fig. 102) operate by water power to lift their deck plates a short distance and drop them with a sharp jolt at a speed of 100 times/min. The jolters require at least 40 lb./sq. in. water pressure to operate to advantage—the higher the pressure above 40 lb./sq. in. the lower the water consumption.

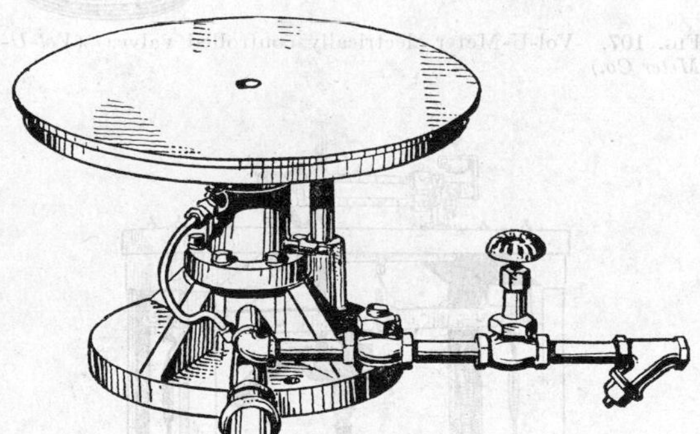

Fig. 102. Syntron hydraulic jolter.

Where no water or only low pressure is available, a small electric-motor-driven closed-circuit pump system can be furnished. Jolters are supplied with pressure valves, operating valves, and short hose connections to take the jolt, or water hammer, off the line.

Model	H.J.-250	H.J.-550
Capacity—lift at 90 lb./sq. in. pressure, lb	275	550
Deck dimensions, in	20 × 20	30 × 30
Over-all height, in	13	15
Base dimensions, in	12 × 12	13 × 13
Water/hr., gal	200	400
Net weight, lb	150	310
Shipping weight, lb	350	450

The Vibrox packer (Figs. 103 and 104) has an oscillating and vibrating motion somewhat similar to the figure 8. It comes equipped with a 1-hp. motor and should be installed at floor level to prevent lifting of containers.

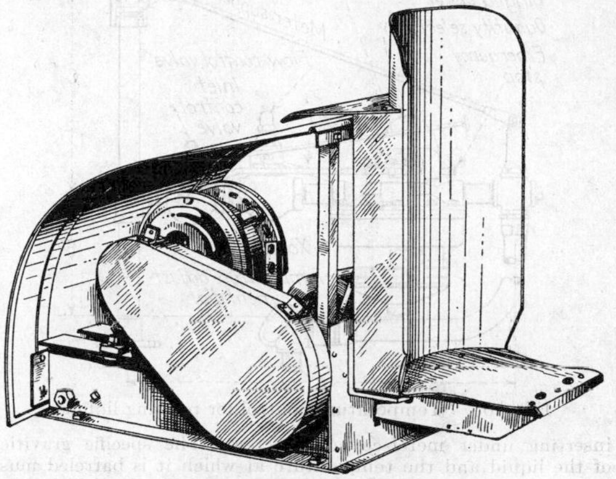

Fig. 103. Vibrox packer.

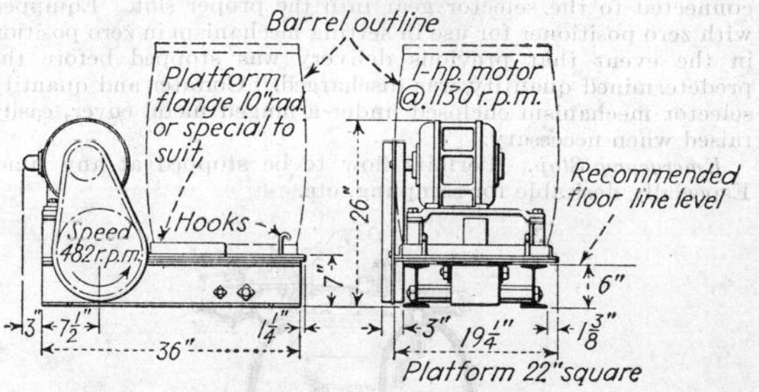

Fig. 104. Vibrox packer.

Filling and Weighing Equipment for Packing Liquids into Drums or Tight Barrels. Liquids are measured either by weight or by volume.

Volume Method. When using the volume method, it is necessary to adjust for expansion or contraction by correcting to a set temperature. The illustrated meter has a temperature-control dial which can be quickly set and corrects volume to a set temperature basis. It automatically stops the flow of liquid when proper volume has been filled into the container. One man can perform all duties: rolling the empty barrel into position, filling, marking, and rolling the filled barrel away.

Sizes. Available with either 1½- or 2-in. meter. Both sizes are equipped with the temperature-control dial for compensating measurement to a 60° basis.

Construction. All exterior parts are built of cast iron; interior parts made of bronze and stainless steel. Cylinders are equipped with bronze liners; pistons are leather-sealed type.

Capacity. 1½-in. size has a maximum capacity of 50 gal./min. 2-in. size has a capacity of 110 gal./min.

Pressure. Both sizes have a maximum capacity of 50 lb./sq. in. In most cases, however, the maximum flow can be secured at much lower pressure.

Range. Either size can be furnished to deliver predetermined quantities in single gallons from 14 to 60. May also be had in the same range for Imperial gallons. However, the meter can be furnished to deliver only all United States gallons or all imperial gallons—not a combination.

Continuous Counter. Records in gallons to 1,000,000, and repeats. Cannot be set back.

Temperature-control Dial. All models are equipped with temperature-control dial for compensating measurement to a 60° basis. Glass-covered metal holder furnished for temperature-adjustment chart. Holder attached and held in place by

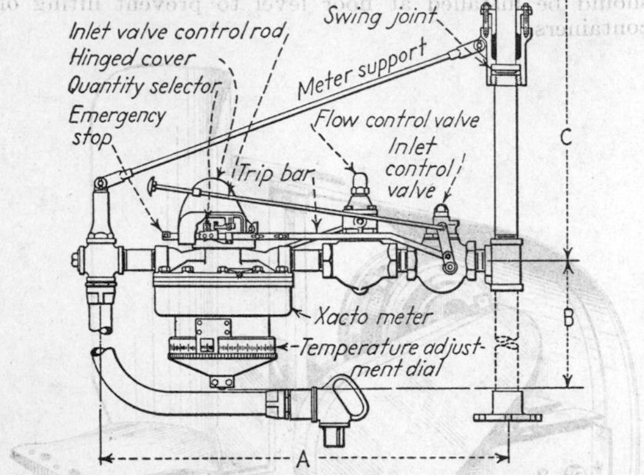

FIG. 105. Temperature control for packing liquids.

inserting under meter-cover capscrew. The specific gravities of the liquid and the temperature at which it is barreled must be specified in order that a proper chart can be furnished.

Quantity Selector. Gears made of steel, set by sliding the bar connected to the selector gear into the proper slot. Equipped with zero positioner for use in setting mechanism in zero position in the event that previous delivery was stopped before the predetermined quantity was discharged. Counter and quantity selector mechanism enclosed under a hinged metal cover, easily raised when necessary.

Emergency Stop. Permits flow to be stopped at any time. Especially desirable for sampling, etc.

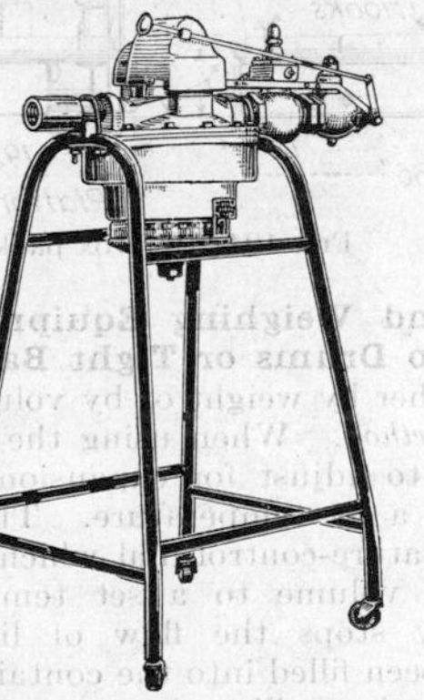

FIG. 106. Control valve for packing liquids.

Control Valve. Made of brass; shuts off flow automatically when the predetermined quantity has been discharged. Manually opened by pushing knob on rod attached to valve. Single-stage type for use where inlet line pressure is 25 lb./sq. in. or less. Double-stage type where pressure is over 25 lb./sq. in. but not in excess of 50 lb./sq. in.

Flow-control Valve. For use on all installations where the inlet pressure is in excess of 25 lb./sq. in.

Hose. Made of oil-resisting, pliable synthetic rubber. Furnished in same size as meter pipe connections.

Nozzles. Made of bronze. Equipped with built-in check valve to prevent dripping. Shuts off instantly when flow stops. Furnished with two tips, 1⅝ and 2⅛ in. for filling barrels with small or large openings.

Swing Joint. Permits unit to be turned in a 360-deg. arc for convenience. Connects to inlet pipe line.

Connections. Inlet, furnished in 1½- or 2-in. pipe sizes.

Discharge. Same size as inlet. An L-type connection permits hose to be attached to end or bottom connection. Pipe plug included.

Additional Equipment Obtainable: Strainer, for removing dirt and other foreign matter from the liquid, and *air release*, to remove air from the liquid, preventing inaccuracies in measurement. Assists in absorbing impact pressures. Necessary when liquid is discharged by pump pressure.

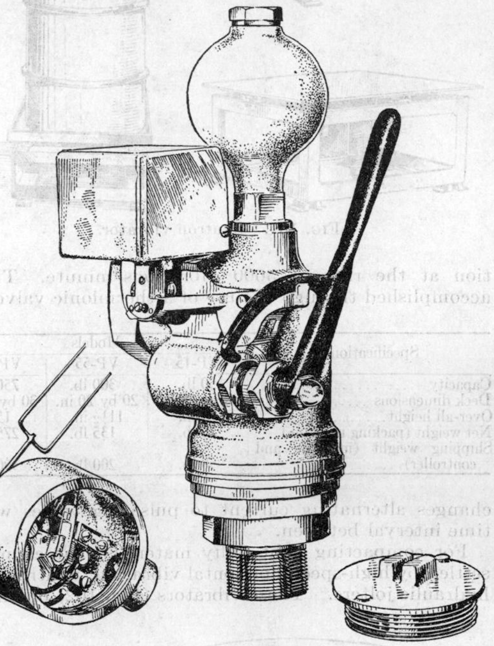

FIG. 107. Vol-U-Meter electrically controlled valve. (*Vol-U-Meter Co.*)

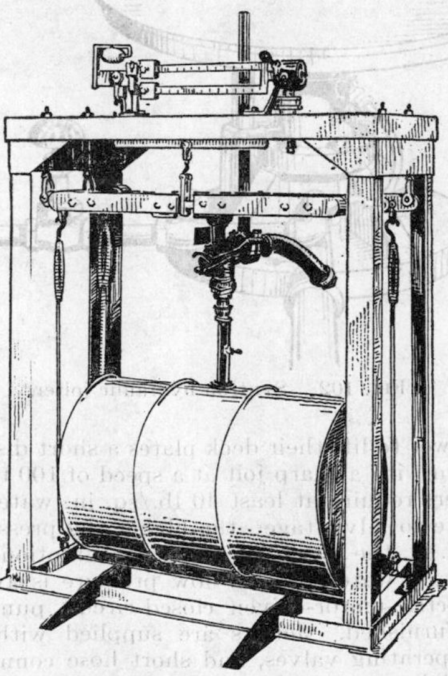

FIG. 108. Vol-U-Meter installation.

Weight Method. Where random weights are used, the container is filled to the desired level and then check-weighed. The container is either filled on a scale or placed on a scale after filling. Where predetermined weights are used, the container must be filled on a scale.

The Vol-U-Meter is an electrically controlled valve (Fig. 107) for filling containers over 5 gal. capacity which can be operated from a light socket. This equipment operates by means of a mercury-tube cutoff switch placed at the extreme end of the scale beam which automatically closes the valve the instant the beam rises.

The complete Vol-U-Meter unit includes:
1 valve.
1 telescopic spout.
1 mercury-tube cutoff switch.
1 yoke for attaching cutoff switch to scale.

Valve specifications:
Intake 2 in., discharge 1½ in.
Spout plain or no-drip.
Valve dimensions 14 in. high, 5 in. wide, 8 in. deep.

Metal:
Bronze, cast-iron, or aluminum valves are standard.
Special alloys can be obtained.

There are several efficient installations of this equipment; those illustrated in Figs. 108 and 109 are the most important.

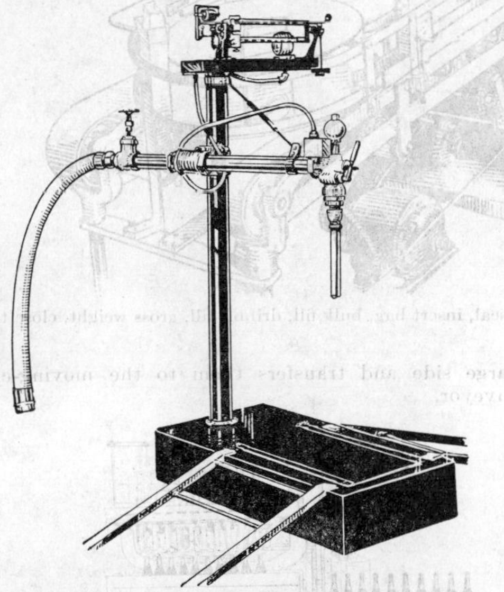

Fig. 109. Vol-U-Meter installation.

PACKAGING EQUIPMENT FOR SHELF PACKAGES

Equipment for forming, filling, sealing, wrapping, conveying, etc., of small or shelf packages is available in a wide variety of types from semiautomatic to fully automatic machines. Packaging lines usually consist of several distinctive operations, each of which may be performed by equipment specially chosen for each such operation. Machines in a packaging line may be supplied by different equipment manufacturers as long as the speeds of such machines are synchronized and proper conveyors are chosen to move the containers between two successive packaging operations (see Fig. 110).

When installation of mechanical equipment is contemplated, it should be understood that the equipment is purchased to perform a mechanical function. The manufacturing organization must therefore be flexible enough to accommodate itself to the machine operation.

Equipment manufacturers should be given every opportunity to recommend specification tolerances for materials and containers to be handled on their equipment. Machines cannot operate efficiently with materials that vary too greatly in size, etc. It is often more economical to pay a little more for materials that are carefully held within acceptable tolerances. Higher production rates can thus be maintained.

Equipment should be chosen that will not be required to operate beyond its rated capacity; otherwise, the quality of its work is liable to fall off materially.

Equipment can be built for almost any type of packaging operation. Often special features must be incorporated into standard equipment for a special installation.

BOTTLE-CLEANING EQUIPMENT

New glass bottles shipped into the bottling plant in fiber shipping containers can generally be satisfactorily cleaned by use of air cleaners. Using from 50 to 80 lb./sq. in. air pressure, machines are available that will clean from 25 bottles per minute for the hand-operated types to more than 120 per minute for the automatic types. Production rates depend upon the size and shapes of the containers and, for the hand-operated types, on the skill of the operators.

Machines are made that rinse bottles by internal spraying. Equipment is also obtainable that will wash and clean bottles with the aid of brushes inside and outside.

Air Cleaner for Bottles (Pneumatic Scale Corp., Fig. 111). This is a fully automatic machine designed to air-clean regular finish glass bottles, jugs, and jars. It will handle containers from 3 to 32 oz. capacity. Containers are fed in an upright position on a drum, then inverted and cleaned with a blast of filtered dry air at 60 lb. pressure. Production varies from 40 to 120 per minute, depending on the size and shape of container. One operator is required to place empty bottles on the intake. This operation may be included as the first operation in a bottle-filling production line.

FILLING EQUIPMENT FOR LIQUIDS

Liquid fillers are usually of the constant-level, weighing, or metering types.

Constant-level fillers operate by pressure, gravity, vacuum, or a combination of vacuum and gravity. When the liquid reaches the proper level in the container, any excess is drawn off through the air-escape tubes, and it is returned to the supply tank. Semiautomatic, fully automatic, and hand-operated machines are on the market.

Metering fillers are usually actuated by piston pumps and are made in hand-operated, semiautomatic, and fully automatic types. They are usually used for handling viscous liquids.

Weighing fillers are usually hand-operated. They consist essentially of a scale beam that actuates a cutoff valve when the scale beam is tipped.

Figure 112 illustrates the automatic Samco vacuum filler of the Pneumatic Scale Corp., used for filling bottles, cans, and jars. It will handle a large variety of liquids and semiliquids. The standard machine can fill containers ranging in size from 1 oz. to 1 qt. The speed of the machine is determined by the size of the container and the number of filling heads built into the unit. The average speed of the standard machine is 60 to 70 containers filled per minute. One operator is required to load empty bottles unless the machine is fed by conveyor from an automatic air cleaner.

The rotary filler illustrated in Fig. 113 is manufactured by the Horix Manufacturing Co. It is a gravity-vacuum type having 32 filling valves with automatic feed and

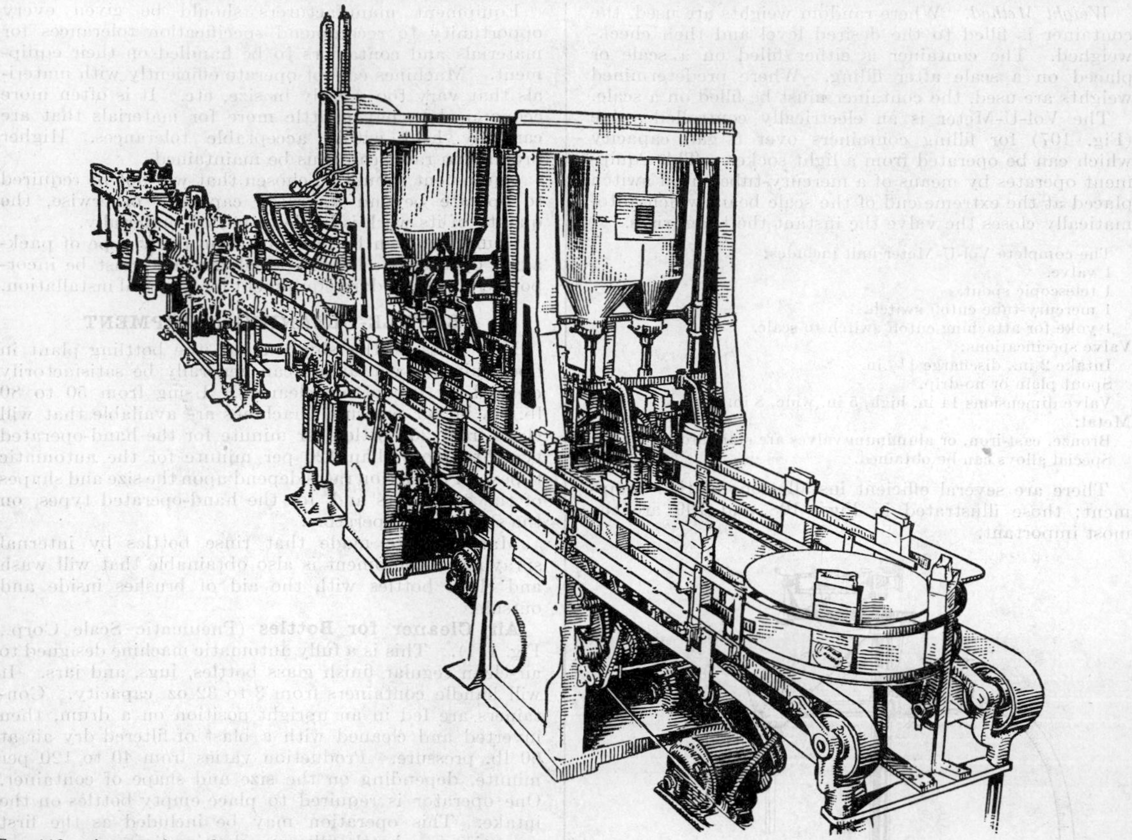

FIG. 110. Automatic packaging line which includes carton feed, bottom seal, insert bag, bulk fill, dribble fill, gross weight, close top of bag. Top sealed-carton production 60/min. (*Stokes & Smith Co.*)

discharge. It has a production capacity of 150 qt./min. Filling valves are of special design, depending on the liquid to be handled, size of container opening, and filling height. Each valve acts automatically and independently, thereby preventing opening unless the container is

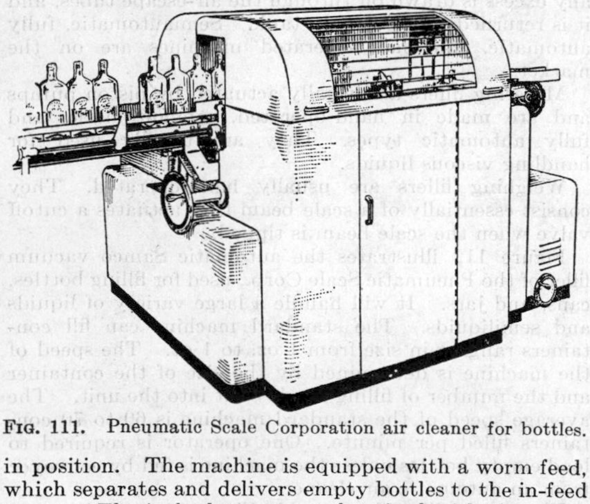

FIG. 111. Pneumatic Scale Corporation air cleaner for bottles.

in position. The machine is equipped with a worm feed, which separates and delivers empty bottles to the in-feed turret. The in-feed turret synchronized with the worm feed positions the empty containers on the lifters. A similar turret removes the filled containers at the discharge side and transfers them to the moving-chain conveyor.

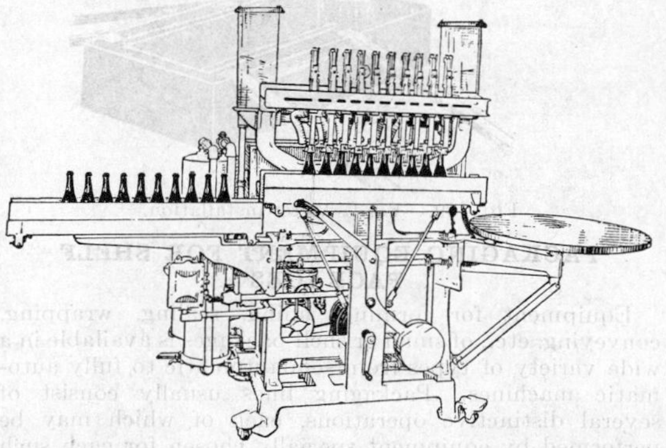

FIG. 112. Pneumatic Scale Corporation "Samco" vacuum filler for liquids.

The Horix straight-line gravity-vacuum filler illustrated in Fig. 114 is equipped with eight valves and has a capacity when used for filling free-flowing liquids of:

4 to 5 gal. containers per minute.
7 to 9 ½-gal. containers per minute.
24 to 28 qt. containers per minute.
40 to 50 pt. containers per minute.

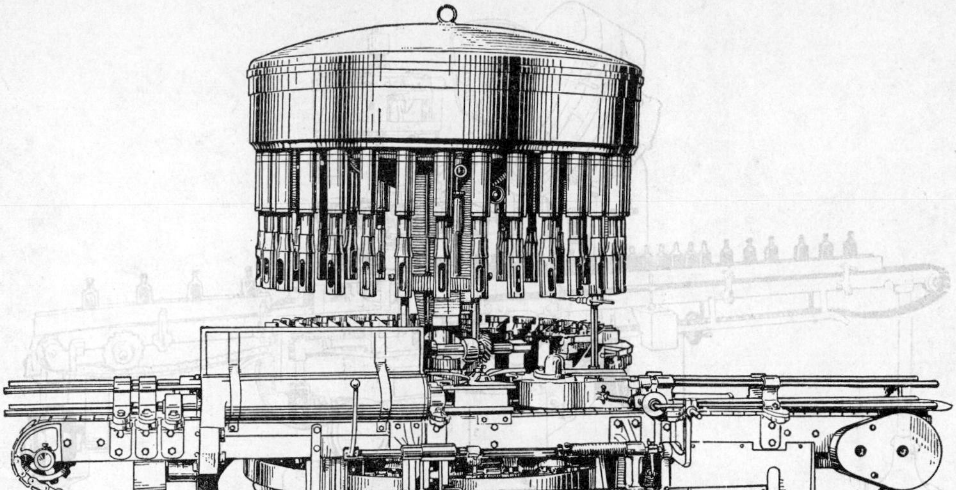

FIG. 113. Horix 32-valve automatic feed and discharge rotary filling machine.

Empty bottles are placed in a row in front of a filling bar. A lever bar actuating the feed bar positions the containers under the filling valves. Bottles are lifted by a counterweighted hand lever until the valves seal the bottle openings and the valves open. The operator positions another row of empties while the bottles are filling. With the lifting of the filling head, the new supply of empties is positioned under the filling valves, and

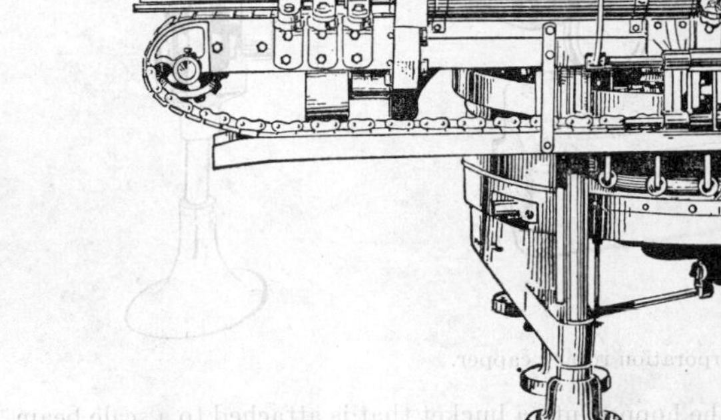

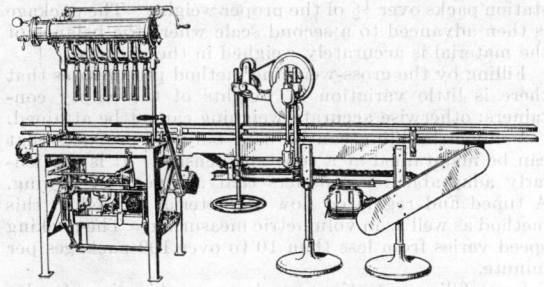

FIG. 114. Horix Model NV-8 straight-line filling machine, with 10-ft. discharge chain conveyor, driven by geared-head motor. Conveyor equipped with semiautomatic corking unit and transfer disk.

the filled containers are simultaneously pushed to the back of the machine.

The Horix portable vacuum filler is hand-operated and can be used for filling light or heavy liquids in any size or style of bottle up to 1 gal. capacity. Figure 115 illustrates a four-filling valve unit. The motor-driven vacuum unit is portable and can be attached to a light socket. A 2-gal. overflow jar, vent, vacuum, and liquid hose 6 ft. long are standard equipment. The rotary vacuum pump is directly connected to the motor.

The filling head is fabricated from aluminum and formed to fit the hand. Filling valves are adjustable to fit various sizes of bottles and are in $\frac{1}{4}$- and $\frac{5}{16}$-in. sizes.

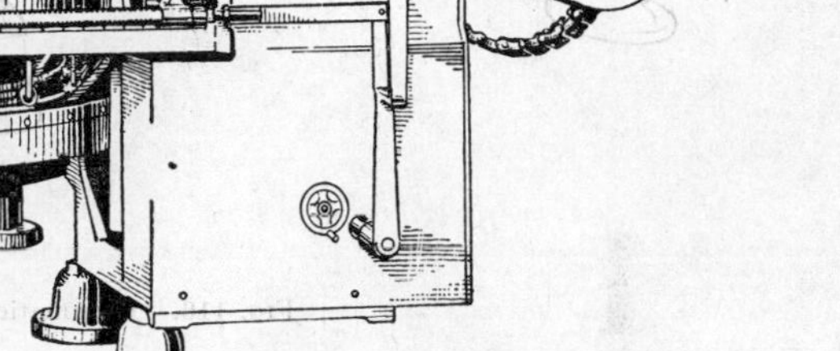

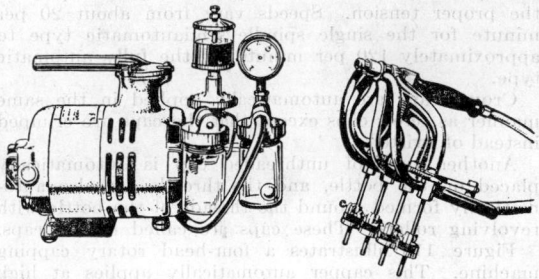

FIG. 115. Horix Model TV portable vacuum filling machine.

Production capacity when used for filling free-flowing liquids is:

6 1-qt. bottles per minute with 2-valve head.
14 1-pt. bottles per minute with 3-valve head.
36 4-oz. bottles per minute with 4-valve head.

BOTTLE AND JAR CLOSING

Equipment of the hand-operated, semiautomatic, and fully automatic types can be used for applying the many types of bottle closures.

Corks, plain or flanged, positioned into the bottleneck can be forced into position mechanically. Fully automatic corking machines have been built for particular applications.

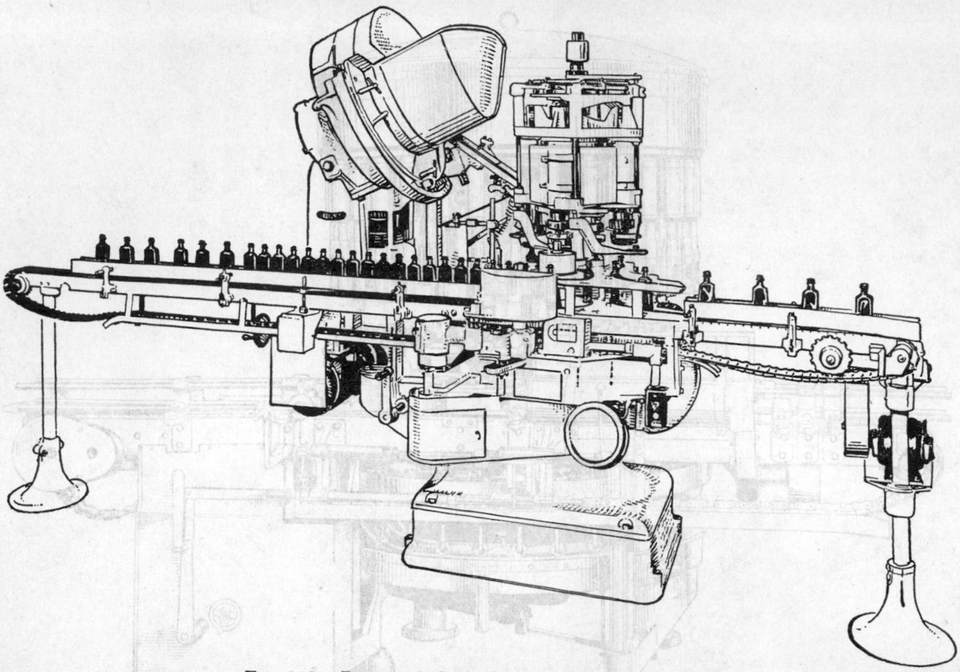

FIG. 116. Pneumatic Scale Corporation rotary capper.

Screw caps must be applied with sufficient force to form a tight seal with the cap liner around the lip of the bottle. It must not be too tight for hand removal. Mechanical tightening of caps eliminates variations in tension. All capping equipment can be adjusted to provide the desired tension.

Caps can be started by hand and tightened automatically. Fully automatic cappers feed the screw caps from a hopper and apply them to the bottles as they are conveyed from the filler, always twisting the caps with the proper tension. Speeds vary from about 20 per minute for the single spindle semiautomatic type to approximately 120 per minute for the fully automatic type.

Crown caps are automatically applied in the same manner as screw caps except that the caps are crimped instead of twisted.

Another type of unthreaded cap is automatically placed on the bottle, and the threads are also automatically formed around the threads of the bottle with revolving rollers. These caps are called roll-on caps.

Figure 116 illustrates a four-head rotary capping machine. This capper automatically applies at high speed turn-on and lug-type closures to a variety of sizes and shapes of bottles, cans, and jars. The machine will handle almost any size of container up to 1 qt. capacity.

The rated production capacity of the machine varies from 50 to 120 per minute, depending on the size of the container. One operator is required to maintain a supply of closures in the cap feed hopper.

FILLING EQUIPMENT FOR DRY PRODUCTS

As in the case of the bulk packaging equipment, machines for filling and measuring dry products can be divided into three types—net weighing, gross weighing, and volumetric measuring.

Net weighing, as the name implies, consists of weighing the product before it is packed into the package. This is accomplished by means of flowing the material from the hopper into a bucket that is attached to a scale beam. When the proper amount of product flows into the bucket, the flow of material from the hopper is shut off, and the bucket dumps its contents into the package.

If the product is not free-flowing, agitation in the hopper or a mechanical feeder may be used. Packaging speeds vary from less than 10 up to about 100 packages per minute.

In gross weighing the product is weighed in its package. When the proper amount of material has been packed into the package, the movement of the scale beam actuates the shutoff gate to the bin. The filling can be speeded up by using two weighing operations. The first station packs over ½ of the proper weight. The package is then advanced to a second scale where the balance of the material is accurately weighed in the package.

Filling by the gross-weighing method presupposes that there is little variation in weights of the empty containers; otherwise accurate weighing cannot be attained.

Volumetric filling can be used only for products that can be maintained at a constant density. It is particularly adaptable for products that are not free-flowing. A timed and regulated flow of material is used in this method as well as in volumetric measuring. The packing speed varies from less than 10 to over 100 packages per minute.

Some filling operations employ a combination of volumetric filling with gross weighing. This type of packing is particularly effective for filling fluffy or aerated products that must be compressed.

Equipment is available for filling such packages as glass jars and bottles, metal and fiber cans, folding cartons and bags, as long as the filling opening is large enough not to restrict too greatly the flow of material.

The Stokes and Smith Co. manufacture a variety of filling machines for measuring and filling dry powdered and granular materials. They are made in single unit, two- and four-station models as well as semiautomatic and fully automatic types covering a production range of from 15 to 120 packages per minute. Various combinations of machines or separate attachments can be supplied such as rotary or volumetric filling devices. Auger-vacuum filling attachments are available for filling fine powders by a combination of the auger and vacuum principles.

Model G-1 is a single-unit Universal Filling machine (Fig. 117). It will fill from ¼ oz. to 5 lb. into cans, jars, boxes, envelopes,

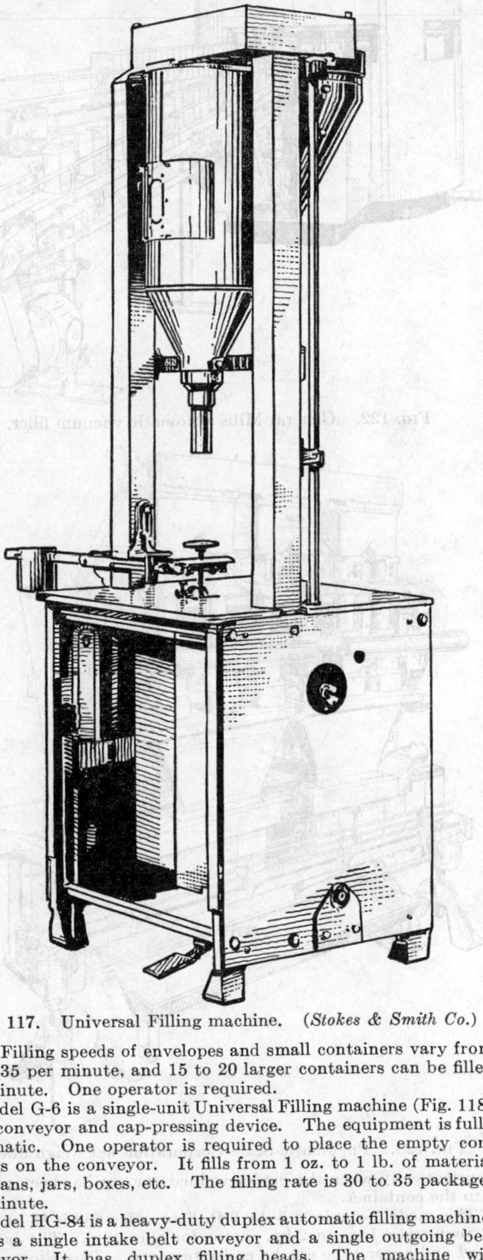

FIG. 117. Universal Filling machine. (*Stokes & Smith Co.*)

etc. Filling speeds of envelopes and small containers vary from 25 to 35 per minute, and 15 to 20 larger containers can be filled per minute. One operator is required.

Model G-6 is a single-unit Universal Filling machine (Fig. 118) with conveyor and cap-pressing device. The equipment is fully automatic. One operator is required to place the empty containers on the conveyor. It fills from 1 oz. to 1 lb. of material into cans, jars, boxes, etc. The filling rate is 30 to 35 packages per minute.

Model HG-84 is a heavy-duty duplex automatic filling machine. It has a single intake belt conveyor and a single outgoing belt conveyor. It has duplex filling heads. The machine will handle from 1 oz. to 1 lb. of material filling into cans, jars, boxes, etc. One operator is required. Production ranges from 60 to 70 packages per minute.

Model HG-86-87 is an automatic four-station heavy-duty filling machine. It has a single intake belt conveyor and a

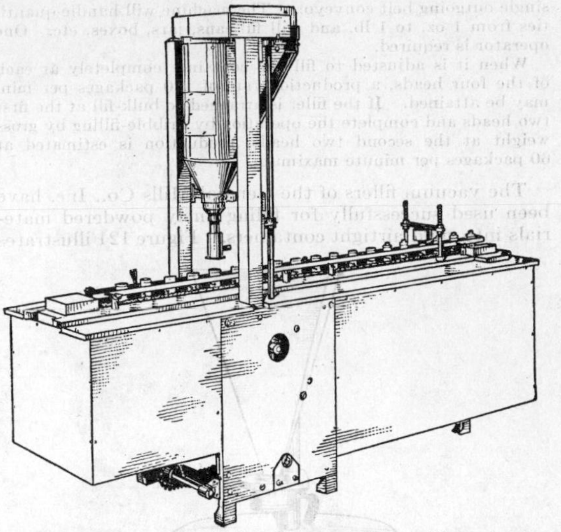

FIG. 118. Stokes & Smith Co. Universal Filler with chain conveyor and cap-pressing device.

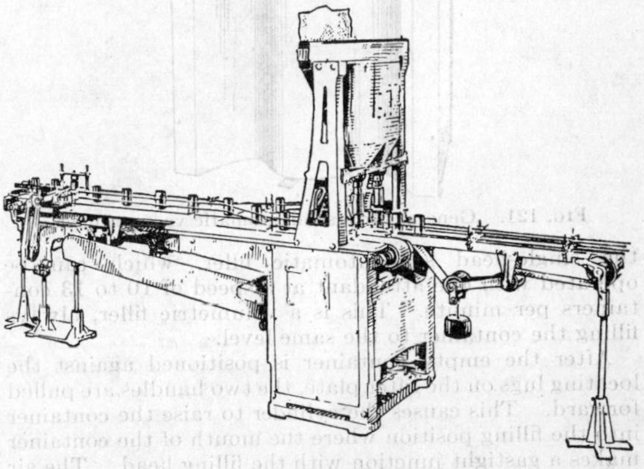

FIG. 119. Stokes & Smith Co. HG-84 automatic duplex filling machine.

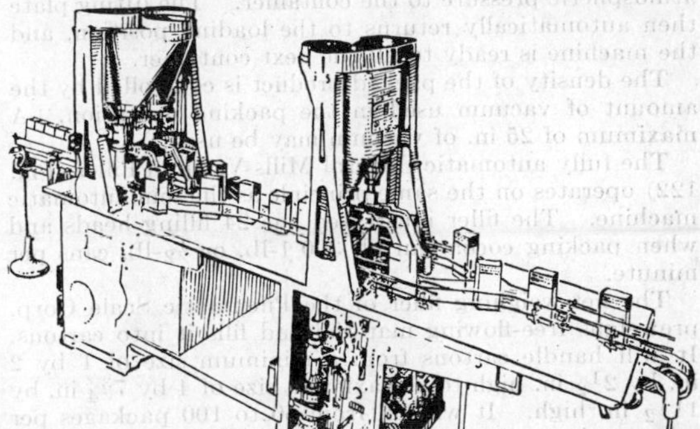

FIG. 120. Stokes & Smith Co. HG-86-87 automatic tandem four-station filling machine. Capacity, 60 to 120/min.

single outgoing belt conveyor. The machine will handle quantities from 1 oz. to 1 lb. and will fill cans, jars, boxes, etc. One operator is required.

When it is adjusted to fill the container completely at each of the four heads, a production rate of 120 packages per min. may be attained. If the filler is arranged to bulk-fill at the first two heads and complete the operation by dribble-filling by gross weight at the second two heads, production is estimated at 60 packages per minute maximum.

The vacuum fillers of the General Mills Co., Inc. have been used successfully for filling finely powdered materials into rigid airtight containers. Figure 121 illustrates

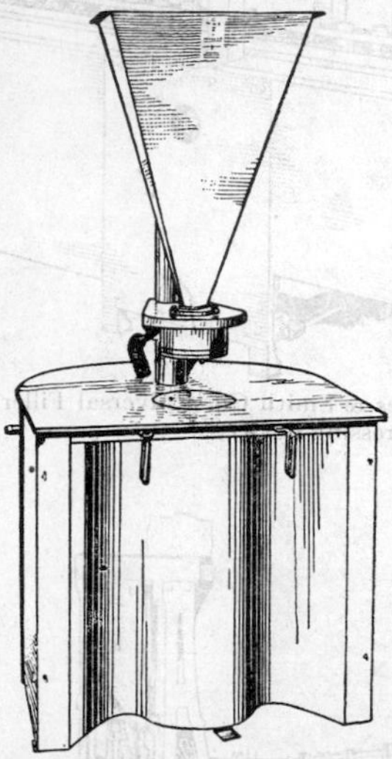

FIG. 121. General Mills semiautomatic vacuum filler.

the single-head semiautomatic filler, which can be operated with one attendant at a speed of 10 to 13 containers per minute. This is a volumetric filler, always filling the container to the same level.

After the empty container is positioned against the locating lugs on the lifter plate, the two handles are pulled forward. This causes the cylinder to raise the container into the filling position where the mouth of the container makes a gastight junction with the filling head. The air is then removed from the container, causing the product to flow into the container. The alternating valve action goes through four cycles of removing air and admitting atmospheric pressure to the container. The lifting plate then automatically returns to the loading position, and the machine is ready to fill the next container.

The density of the packed product is controlled by the amount of vacuum used in the packing operation. A maximum of 25 in. of vacuum may be used.

The fully automatic General Mills Vacuum Filler (Fig. 122) operates on the same principle as the semiautomatic machine. The filler illustrated has 24 filling heads and when packing cocoa can fill 300 1-lb. or ½-lb. cans per minute.

The net-weighing filler of the Pneumatic Scale Corp. preweighs free-flowing material and fills it into cartons. It will handle cartons from a minimum size of 1 by 2 in. by 2½ in. high to a maximum size of 4 by 7¾ in. by 11½ in. high. It will fill from 50 to 100 packages per minute, depending on the package size and the physical characteristics of the product. Figure 123 illustrates the six-scale model.

The Triangle Package Machinery Co. make several types of auger, volumetric, and weigh-scale packers for dry powdered, granular, and lump materials.

Figure 124 illustrates their Model U-1 auger packer used for filling powdered materials into bags, jars, cans, cartons, or bottles. It fills from 1 oz. to 5 lb., and the production is 20 to 30 packages per minute with one operator. The operator places the empty container on the packing platform, which rises to the filling position. The auger starts automatically and shuts off

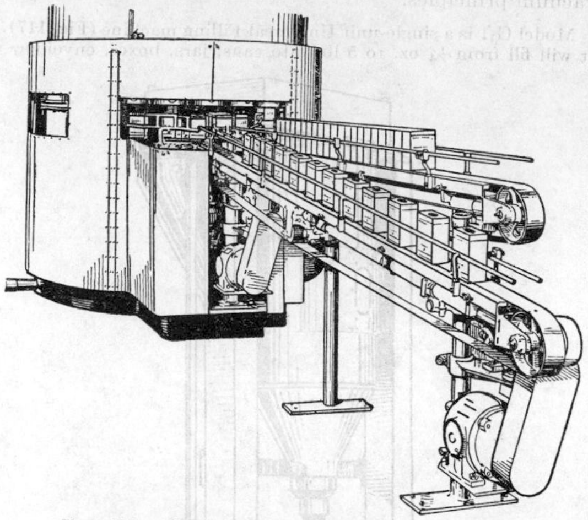

FIG. 122. General Mills automatic vacuum filler.

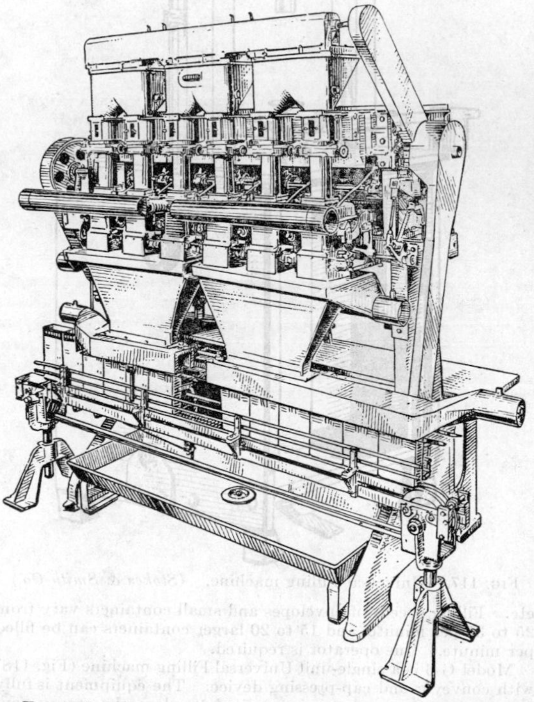

FIG. 123. Pneumatic Scale Corporation net weigh filler.

automatically when the predetermined amount has been packed into the container.

This model occupies a 3- by 3-ft. floor space and is 83 in. high. It requires a 1½- to 2-hp. motor for operation. The packing bowl and auger tube can be easily removed. This allows quick cleaning.

Model SN (Fig. 125) is a heavy-duty auger packer which can be used to pack from 2 to 50 lb. per package. One operator is required, and it will pack from 10 to 20 packages per minute. Its operation closely duplicates that of the Model U-1 machine

The triangle volumetric filler, Model SPA (Fig. 126), packs the volumetrically measured amount with an automatic plunger which assures positive discharge from the measuring chamber. It is available with conveyor for automatic high-speed operation. Production ranges from 30 to 60 packages per minute. Machines can pack from a minimum of ½ oz. to a maximum of 16 oz.,

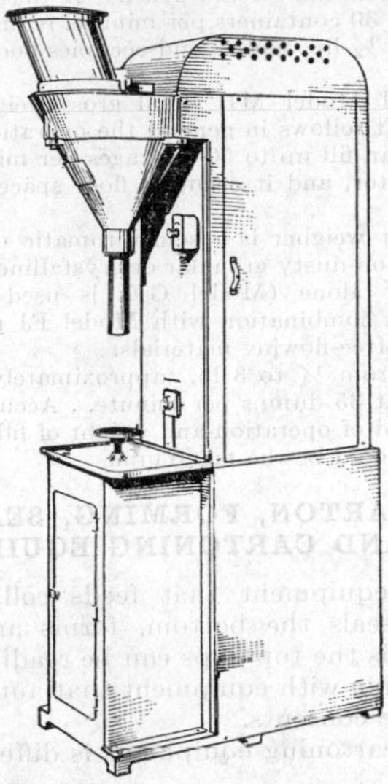

FIG. 124. Triangle Package Machinery Co. auger packer.

depending on the size of the measuring chamber. The normal range of a measuring chamber is a maximum of 1½ times the minimum. Accuracy of 1/32 to ⅛ oz. is claimed by the manufacturers.

The filler requires a ½-hp. motor for operation. It occupies 2½ by 3 ft. of floor space, exclusive of conveyor, and is 6 ft. 6 in. high.

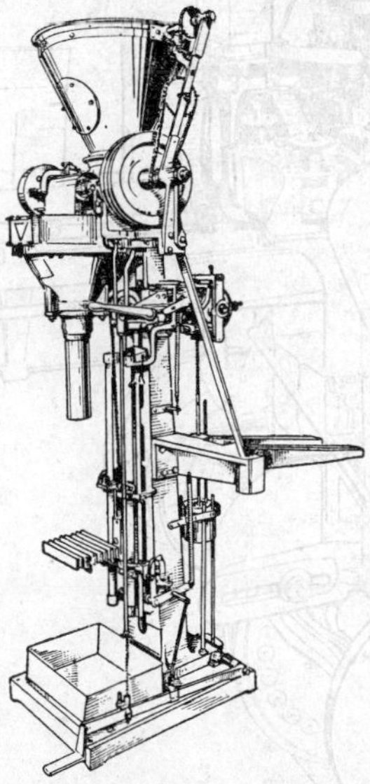

FIG. 125. Triangle Package Machinery Co. heavy-duty auger packer.

The United States Automatic Box Machinery Co. Filling Machines. Model JK (Fig. 127) is an automatic packer having two filling stations. It can operate as a volume filler or as a gross weigher. When the machine is used as a volume packer, empty containers are elevated, causing the actual filling by auger under controlled pressure. When the machine is equipped as a packer-weigher, the main portion of the contents is packed by auger under pressure, and the remainder is packed by weighing at the second station. A vibrating settler can be installed for settling of the contents after filling. The machine will not fill if a container is not at the filling station.

The equipment can fill packages with base dimensions up to 5 by 4 in. and heights ranging from a minimum of ¼ in. to a

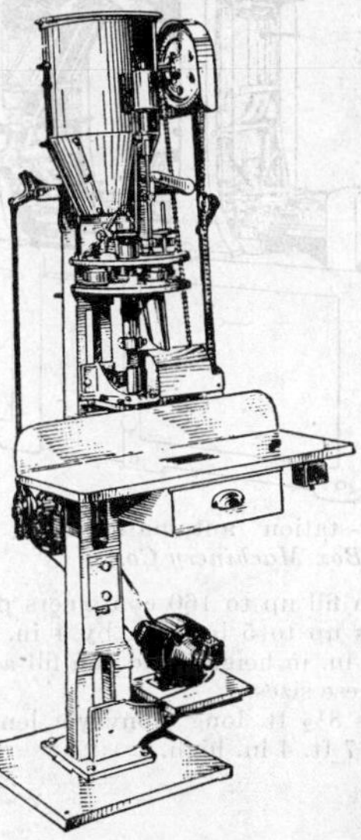

FIG. 126. Triangle Package Machinery Co. volumetric filler.

maximum of 12 in. It will handle any weight of material that can be packed into the package size limits.

The running speed is quoted at 60 packages per minute. The machine is 4 ft. 3 in. long, 3 ft. 4½ in. wide and 7 ft. 4 in. high. The length, including the standard conveyor, is 12 ft.

Model JN, fully automatic packer, filler, and packer-weigher (Fig. 128), is a four-station machine that will pack most types of dry powdered or granular materials into round, square, rectangular, or irregular containers. Entering the machine in

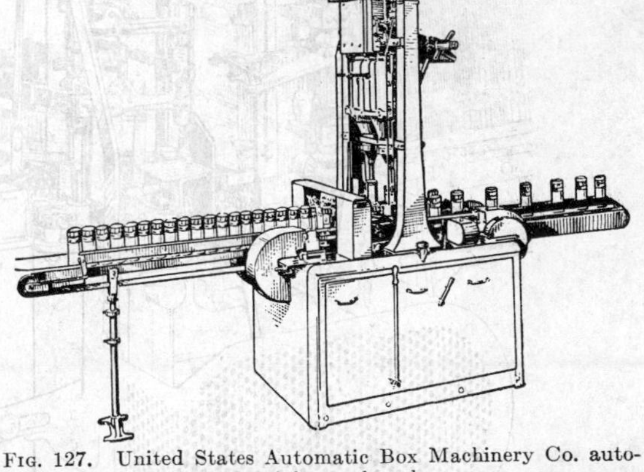

FIG. 127. United States Automatic Box Machinery Co. automatic twin-station gross weigher and packer.

single line, containers are automatically separated into two lines progressively, and, after the filling operation, are regrouped into a single line for hand or automatic closing.

The operation is the same as Model JK. All four stations can be used for gross weighing if reasonable weight tolerances are acceptable. If not, two stations are used for bulk filling and the other two for completing the filling operation by weighing accurately. The machine will not fill if no container is at the filling station.

Use of dust collector hoods and drip-free shutoff gates at filling stations provide practically dustproof operation. Vibrating platforms are available for settling products that require agitation while packing.

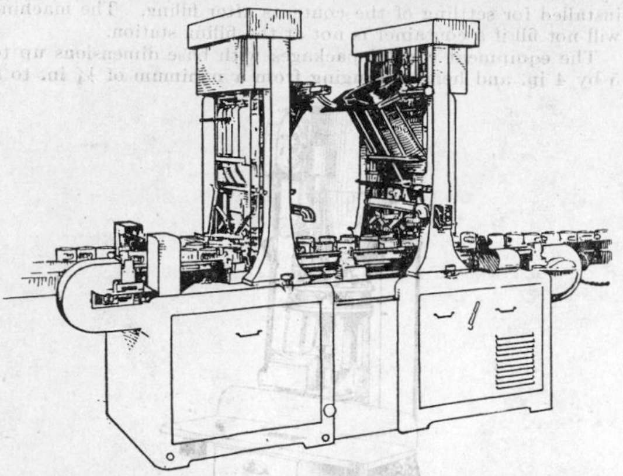

FIG. 128. Four-station automatic filling machine. (*United States Automatic Box Machinery Co.*)

The packer can fill up to 160 containers per minute. It will handle containers up to 5 in. long by 4 in. wide, with heights from ¼ in. to 12 in. in height, and can fill any weight that can be packed into these sizes.

The machine is 8½ ft. long (conveyor length optional), 3 ft. 4½ in. wide, and 7 ft. 4 in. high.

The National Model MG Bond gross weigher is a single-station semiautomatic filler that can be used for packing non-free-flowing powders, and for granular and other dry products. It can be equipped for weighing, packing, or volume filling. It will handle containers ranging in size up to 14 in. high and with base up to 8 in. square. It weighs from a fraction of an ounce up to 7 lb., depending on the density of the product. It will pack from 10 to 30 containers per minute, requires one operator, is operated by a ½-hp. motor, and occupies floor space 3 ft. 4 in. by 6 ft.

The National Model MH Bond gross weigher is a double-station filler. It follows in general the operation of Model MG except that it can fill up to 50 packages per minute, is operated by a ¾-hp. motor, and it occupies floor space 4 ft. 2 in. wide by 6 ft. long.

The Scott net weigher is a semiautomatic unit for weighing and filling dry non-dusty granular or crystalline materials. The gravity weigher alone (Model GE) is used for free-flowing products and in combination with Model FJ power feeder can be used for non-free-flowing materials.

It can pack from ¼ to 3 lb. (approximately) per container. Speed is rated at 35 dumps per minute. Accuracy of weighing varies with speed of operation and weight of fill, but a variation of 1/16 oz./lb. should be the maximum.

FOLDING CARTON, FORMING, SEALING, LINING, AND CARTONING EQUIPMENT

Automatic equipment that feeds collapsed cartons, forms them, seals the bottom, forms and inserts bag liners, and seals the top flaps can be readily combined in a production line with equipment that automatically fills and weighs the contents.

Automatic cartoning equipment is differentiated from

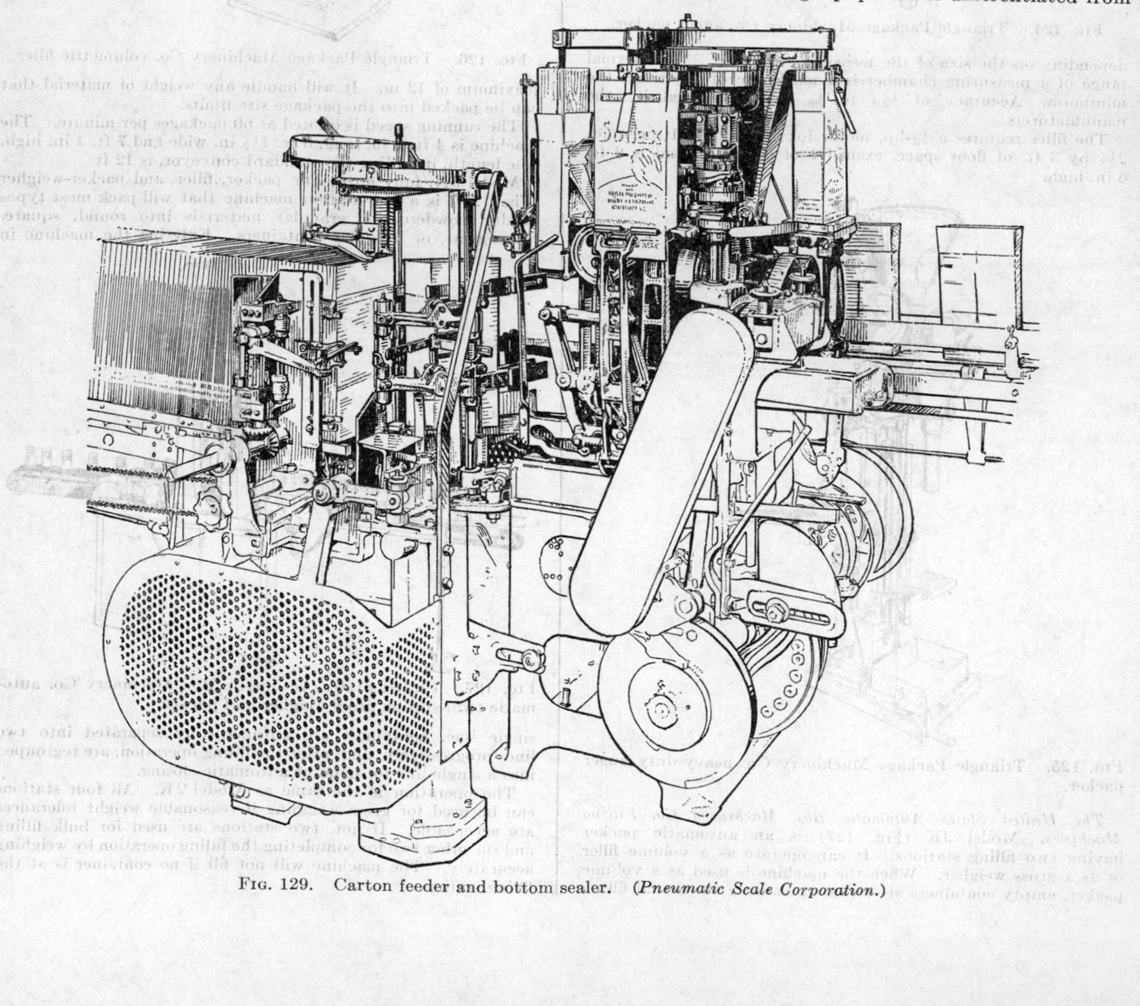

FIG. 129. Carton feeder and bottom sealer. (*Pneumatic Scale Corporation.*)

that described in the preceding paragraph in that this machinery packs products that have been previously packed in bottles, jars, or collapsible tubes, or wrapped, such as razor blades. The equipment feeds the folding carton, sets it up, packs the product in the carton, and closes both ends by either tucking the flaps or sealing them. Some types will also insert instruction leaflets.

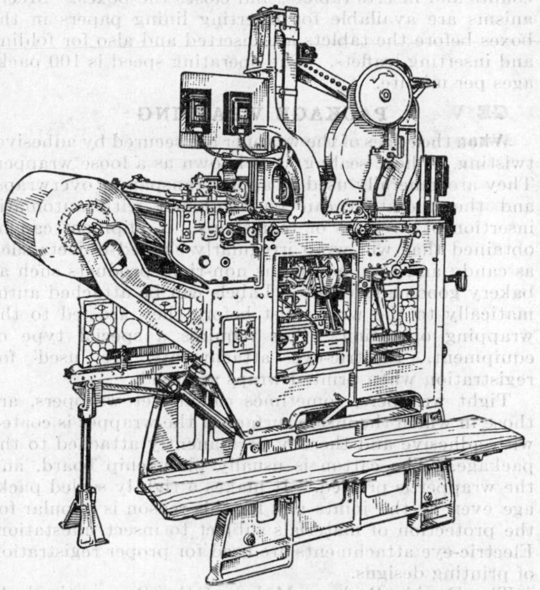

FIG. 130. Pneumatic Scale Corporation bag liner machine for cartons.

The carton feeder and bottom sealer machine (Fig. 129) of the Pneumatic Scale Corp. receives cartons in the flat from a stacked supply magazine. It automatically opens each carton, one at a time, pushing it upward onto a block which holds the carton square while the bottom flaps are glued, folded, and sealed under pressure. After the pressure is released, the block and carton rotate to the last station of the sealing operation where the carton is stripped from the block and carried by conveyor to the next machine.

It can handle cartons from a minimum of 1 by 2 in. by 3 in. high (at the score lines) to a maximum size of 3½ by 7 in. by 12⁷⁄₁₆ in. high (at the score lines plus extended top flap). The production rate is 25 to 40 cartons per minute, depending on the size of the package. One operator is required for occasional refilling of the carton-feeder magazine.

The Pneumatic Scale Corp. has a completely automatic machine for making bag liners and inserting them into empty cartons ready for filling (Fig. 130). The bag paper is fed from a roll cut into sheets, formed around blocks, sealed on the sides, and inserted into the cartons. Glassine and wax-treated paper are most often used for this purpose. The side seams are sealed with glue, heat-sealing, or both.

This equipment will form bags for cartons varying in size from 1½ by 3 in. by 5½ in. high to a maximum of 3½ by 8 in. by 11½ in. high. Measurements are taken between score lines. Production varies from 25 to 30 lined cartons per minute, depending on size. No operator is required other than to replenish the rolls of bag paper.

For sealing the filled carton, Pneumatic Scale Corp. has an automatic top sealer. This machine (Fig. 131) glues and folds the top flaps of the filled carton and then holds the carton under pressure between compression belts and rollers until the glue adhesive has set. Both bag-lined cartons and unlined cartons may be closed on this equipment. When a liner is used, the top sealer interleaves the top of the bag with the carton flaps, closing them simultaneously.

The machine will close cartons ranging in size from a minimum of ¾ by 1¼ in. by 2½ in. high to a maximum of 4⅜ by 7¾ in. by 9⅜ in. high. All measurements are taken between score lines. From 35 to 40 cartons can be sealed per minute, depending on size.

An example of a machine that forms the carton, seals the bottom flaps, forms and inserts a bag liner if required, fills the carton, and seals the top flaps is the Neverstop machine of Stokes and Smith (Fig. 132). The machine is automatic, requiring one attendant for all operations. Filling may be by the gross-weigh, net-weigh, or volume methods. Cartons are in constant motion so that the filling is accomplished by use of a rotating turret.

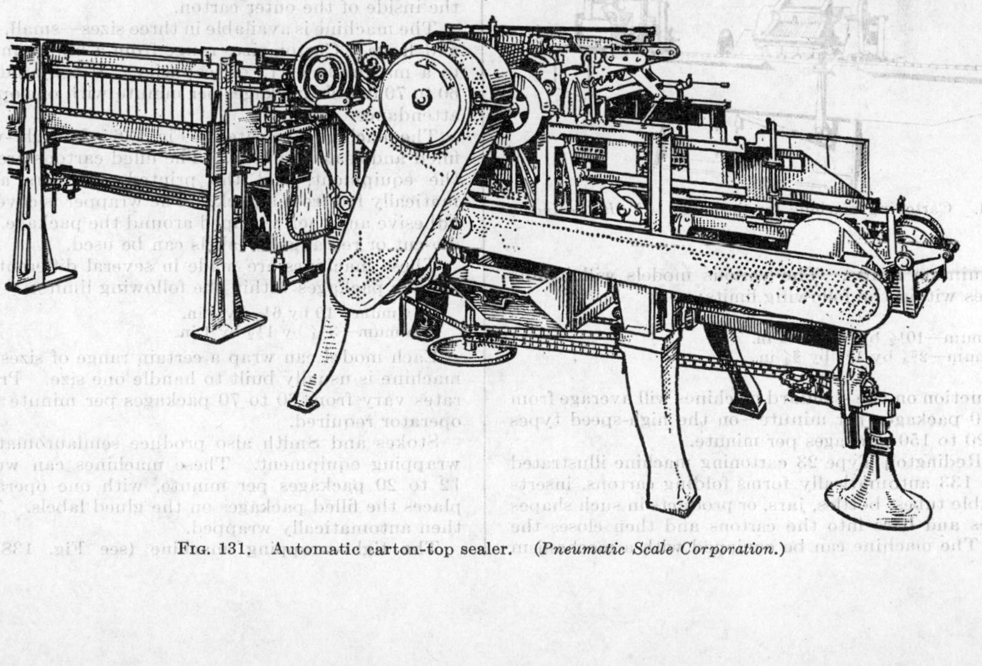

FIG. 131. Automatic carton-top sealer. (*Pneumatic Scale Corporation.*)

Made-up liners are fed and inserted into the carton automatically. Liners for the larger cartons are formed from a roll of paper, the sides sealed and inserted into the carton. The tops of the liner are folded into the top seal of the carton.

The machine is made in several different models, each model taking a range of cartons within certain maximum

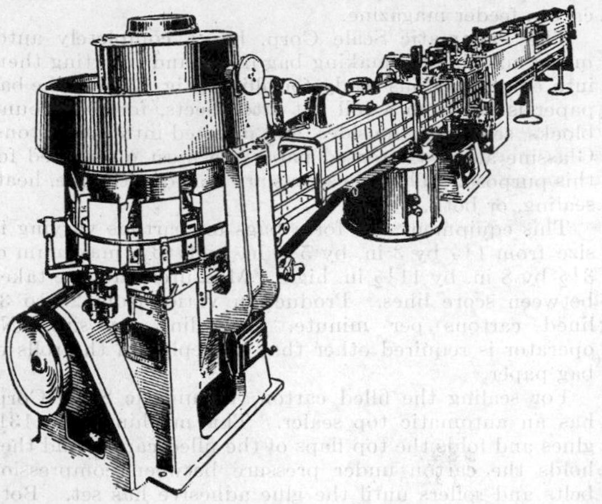

FIG. 132. Stokes & Smith automatic carton-filling and sealing machine.

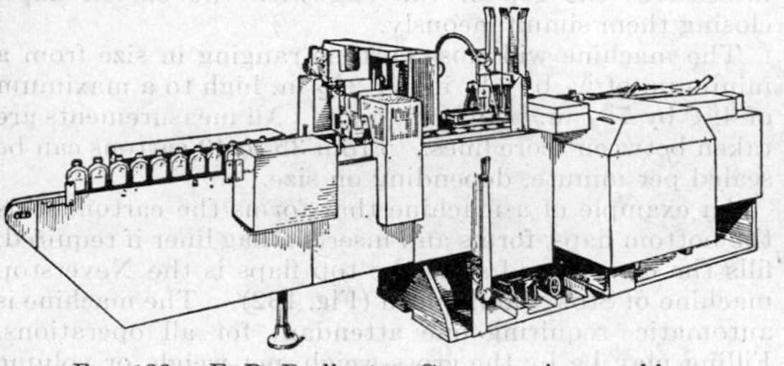

FIG. 133. F. B. Redington Co. cartoning machine.

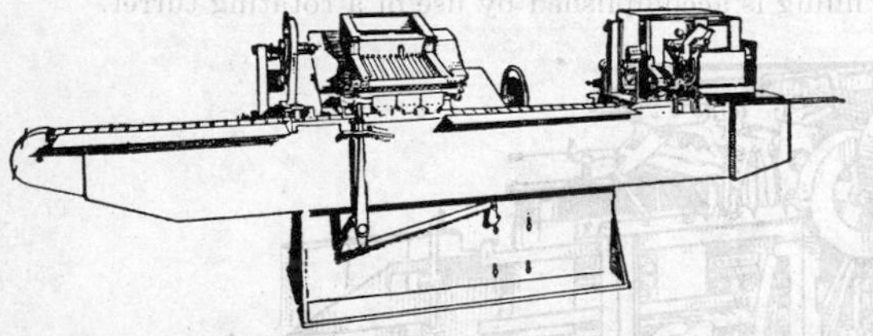

FIG. 134. Cartoning machine for tin boxes. (*F. B. Redington Co.*)

and minimum limits. The various models will handle packages within the following limits:

Maximum—10½ by 6½ by 4 in.
Minimum—2¾ by 1⅝ by ¾ in.

Production on the standard machines will average from 60 to 70 packages per minute—on the high-speed types from 120 to 150 packages per minute.

The Redington Type 23 cartoning machine illustrated in Fig. 133 automatically forms folding cartons, inserts collapsible tubes, bottles, jars, or products in such shapes as cakes and bars into the cartons and then closes the flaps The machine can be equipped with a mechanism for automatically folding and inserting instruction sheets or leaflets as well as a mechanism for placing a corrugated protector around the contents. The speed of operation varies with the product, but the usual production rate is from 100 to 150 packages per minute.

Figure 134 is an ingenious machine, made by F. B. Redington Co., which automatically opens aspirin tins, counts and inserts tablets, and closes the boxes. Mechanisms are available for inserting lining papers in the boxes before the tablets are inserted and also for folding and inserting leaflets. The operating speed is 100 packages per minute.

PACKAGE WRAPPING

When the edges of the wrapper are secured by adhesive, twisting, or heat sealing, it is known as a loose wrapper. They are generally used to apply transparent overwraps, and the operation can be combined with automatic insertion of circulars or tear strips. Equipment can be obtained that will wrap irregularly shaped objects such as candy and soap as well as non-rigid products such as bakery goods and cigars. Labels can be attached automatically to the wraps just before they are fed to the wrapping operation. This requires a special type of equipment. Electric-eye attachments are used for registration when printed wraps are used.

Tight wrappers, sometimes called wet wrappers, are those in which the inside surface of the wrapper is coated with adhesive and thereby permanently attached to the package. The carton is usually plain chip board, and the wrapper is printed. It makes a tightly sealed package even at the joints and for this reason is popular for the protection of materials subject to insect infestation. Electric-eye attachments are used for proper registration of printing designs.

The Double Package Maker of the Pneumatic Scale Corp. (Fig. 135) automatically fabricates a bag within a carton. It forms the bag liner around a block from paper fed from a roll and then glues the side and bottom at succeeding stations. After the inner bag is formed and while it is still on the block, a scored carton blank is fed from a stack and formed about the block. The block continues to rotate to the stations where the operation of gluing the sides and bottom of the carton is completed. The bag liner may be made from wax paper, glassine, or other protective sheets, and if desired it may be glued to the inside of the outer carton.

The machine is available in three sizes—small, medium, and large. Carton sizes range from a minimum of 1 oz. to a maximum of 11 oz. capacity. It can produce from 60 to 70 double packages per minute with one operator in attendance.

The machine illustrated in Fig. 136 is for tight-wrapping filled and sealed cartons. The filled cartons are fed into the equipment, and the printed wrappers are automatically fed from a pack. The wrapper is covered with adhesive and then wrapped around the package. Either die-cut or rectangular labels can be used.

These machines are made in several different models, to take packages within the following limits:

Maximum—10 by 6½ by 4 in.
Minimum—3¾ by 1½ by 1 in.

Each model can wrap a certain range of sizes, but the machine is usually built to handle one size. Production rates vary from 60 to 70 packages per minute with one operator required.

Stokes and Smith also produce semiautomatic tight-wrapping equipment. These machines can wrap from 12 to 20 packages per minute, with one operator who places the filled packages on the glued labels. They are then automatically wrapped.

The tight-wrapping machine (see Fig. 138) of the

FIG. 135. Pneumatic Scale Corporation Double Package Maker.

Pneumatic Scale Corp. feeds the label from either sheet or roll form and applies the wrapper, coated with glue, to one face of the carton. The wrapping operation is then completed automatically. Cartons varying in size from 1½ by 2⅝ in. by 4¼ in. high to boxes 4 by 7½ in. by 10¼ in. high can be wrapped with this equipment. It operates at approximately 60 cartons per minute. One operator is required to refill the wrapper feed.

Model FA of the Package Machinery Co. (Fig. 139) is a popular wrapper that can wrap rigid and flexible

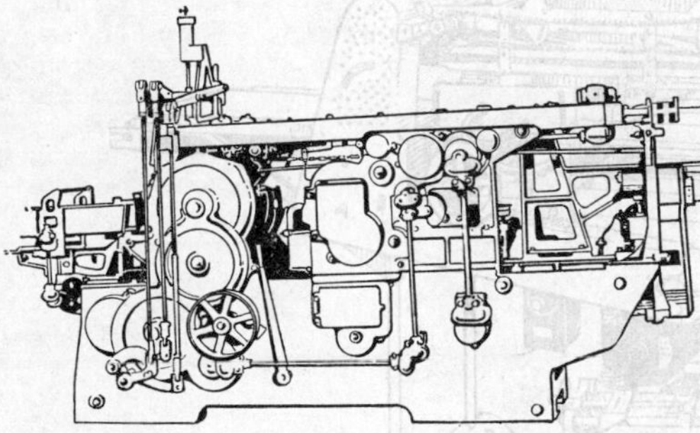

FIG. 136. Stokes & Smith Co. tight wrapping machine.

products, including those packed loose in boats or trays. Packages pass through the machine in a straight line. They are always on an even plane, the wrapping operation being performed without turning the package. This machine can be equipped to wrap with glassine, cellophane, waxed paper, foil, or plain paper as well as various forms of laminated wrapping materials.

The equipment can be provided with an electric eye for accurate registration of printed material fed from a roll. It can also be equipped for sheet feeding.

The machine is readily adjustable to different sizes of packages. Adjustments for all three dimensions are made with handwheels.

Glue pots are so designed that they will handle any type of glue desired as well as Cellosolve for sealing cellophane wrappers.

From 40 to 100 packages can be wrapped per minute, depending on the size of the article to be wrapped. Two

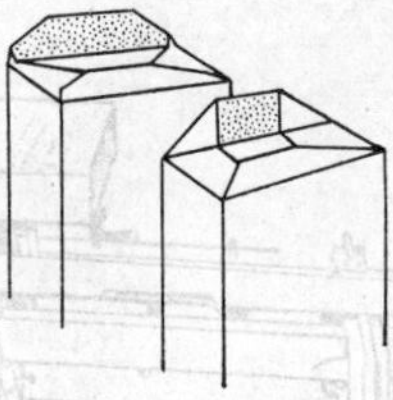

FIG. 137. Two of various types of labels and end folds for which machines may be furnished. Either of those shown may be wrapped on the same machine.

attendants are required—one to feed and one to pack the wrapped packages, unless equipment is provided for automatic feeding and automatic packing into shipping cases.

It will wrap packages with minimum dimensions of 2 in. long by 1⅛ in. wide by ⅜ in. thick to packages of maximum dimensions of 12 in. long by 5⅝ in. wide by 3½ in. thick.

Two other models, FA-2 and FA-4, can wrap packages of somewhat larger sizes, but the wrapping speed is reduced to 30 to 75 per minute.

Model CA-2 produced by Package Machinery Co.

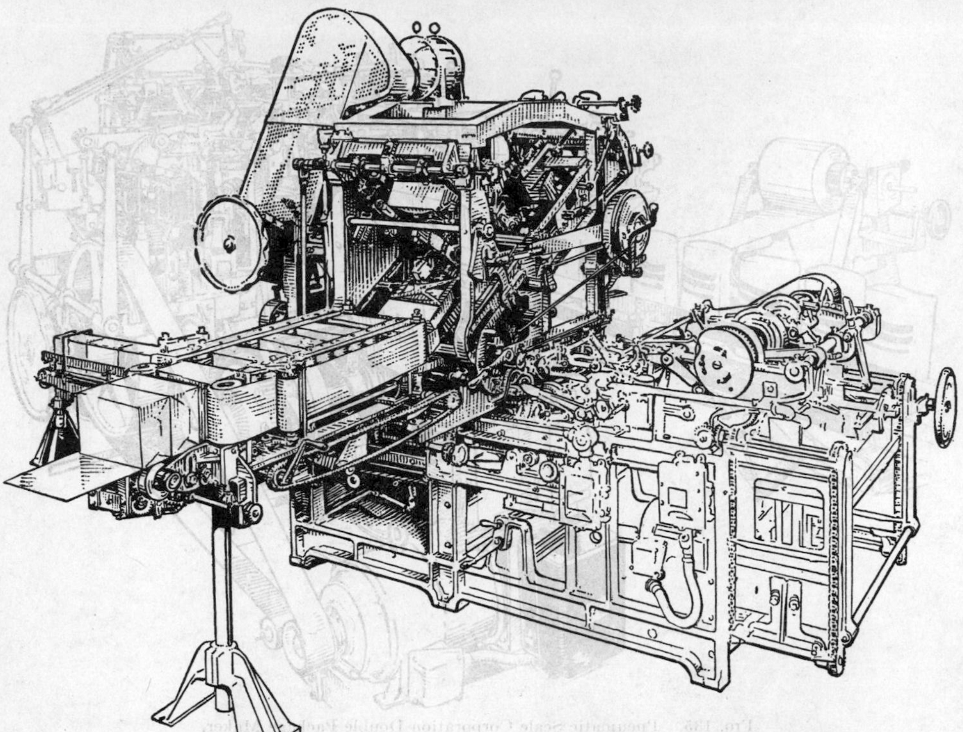

Fig. 138. Tight wrapping machine. (*Pneumatic Scale Corporation.*)

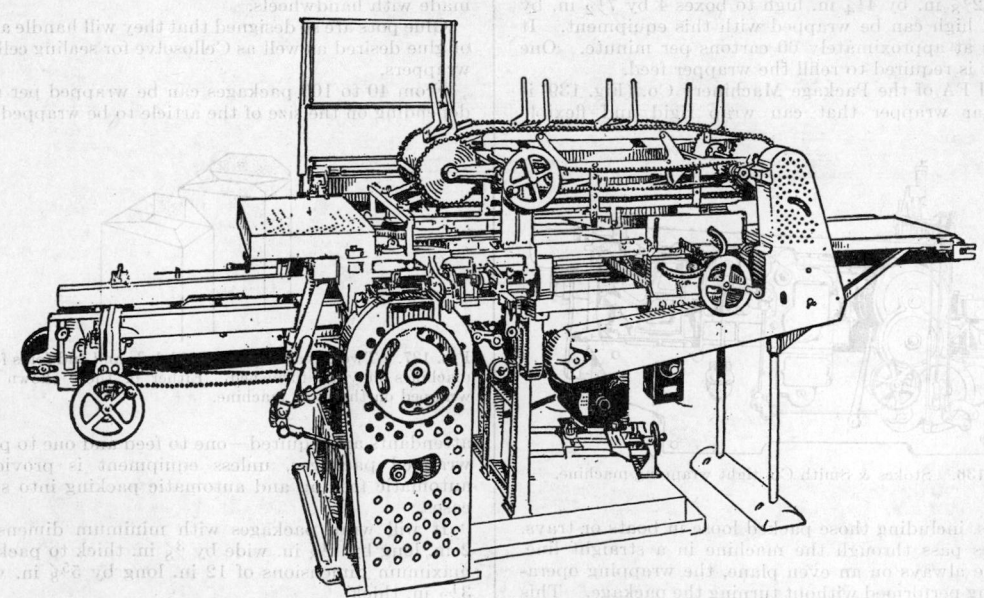

Fig. 139. Package Machinery Company Model FA wrapping machine.

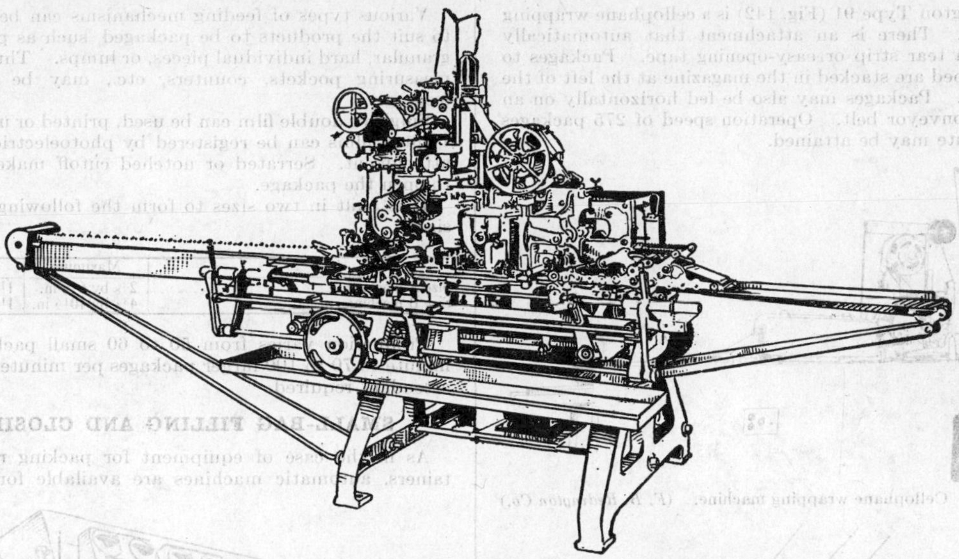

FIG. 140. Package Machinery Company Model CA-2 wrapping machine.

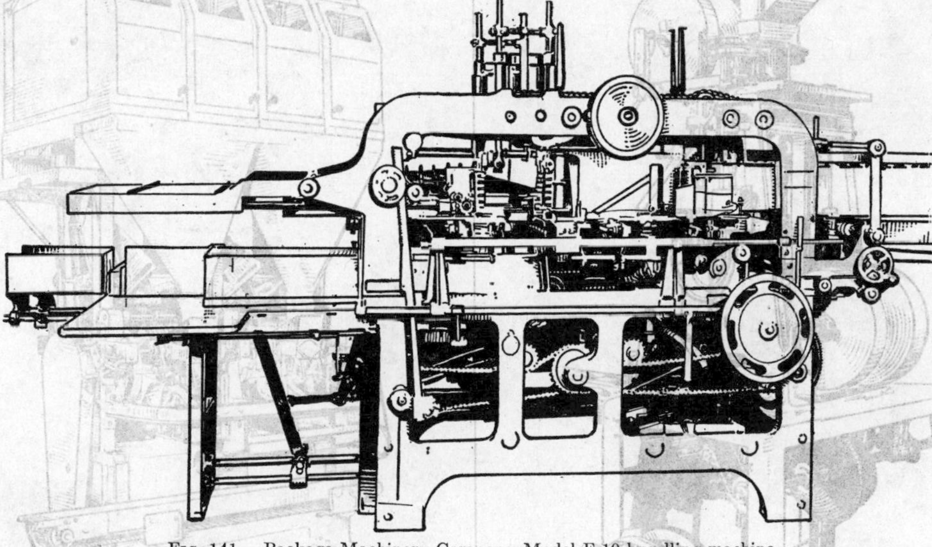

FIG. 141. Package Machinery Company Model F-10 bundling machine.

(Fig. 140) can enclose an article in a single wrap of cellophane, glassine, foil, waxed paper, or similar material. It also can apply an outer wrap in the form of a band or over-all wrap. Printed wrappers can be fed from a roll and registered by electric eye. The machine can also be equipped to insert a card into the wrapping as well as to apply an easy opening tape. It operates at a speed of 80 to 150 wrapped packages per minute, depending on the type of wrapping. In general, the machine is used for wrapping small objects.

Models F-10 and F-6 bundling machines of the Package Machinery Co. wrap a number of boxes with strong paper and apply label end seals which may be printed. A machine with one operator will wrap from 8 to 24 bundles per minute, depending on size. Machines are easily adjustable to accommodate bundles of different sizes.

Model F-10 (Fig. 141) is designed especially for bundling large-sized packages. It will handle various sizes within the following limits: length—7½ to 20 in.; width—6 to 13¼ in.; height—4 to 9 in. The machine can be fed by hand or can be equipped with an assembler. Wrapping paper is fed from a roll, and end seals are automatically applied after the folds have been made. Printed and gummed end seals are fed from a magazine, or they can be fed from a roll.

Model F-6 is used for bundling small packages. It makes bundles within the following size limits: length—6 to 14½ in.; width—2⅛ to 7½ in.; height—1⅝ to 5⅛ in. The machine can be fed from a single belt, where packages will permit it, or it can be equipped with an external assembler, which in turn is fed by a belt. It applies separate cut-to-size end seals.

Redington Type 91 (Fig. 142) is a cellophane wrapping machine. There is an attachment that automatically applies a tear strip or easy-opening tape. Packages to be wrapped are stacked in the magazine at the left of the machine. Packages may also be fed horizontally on an intake conveyor belt. Operation speed of 275 packages per minute may be attained.

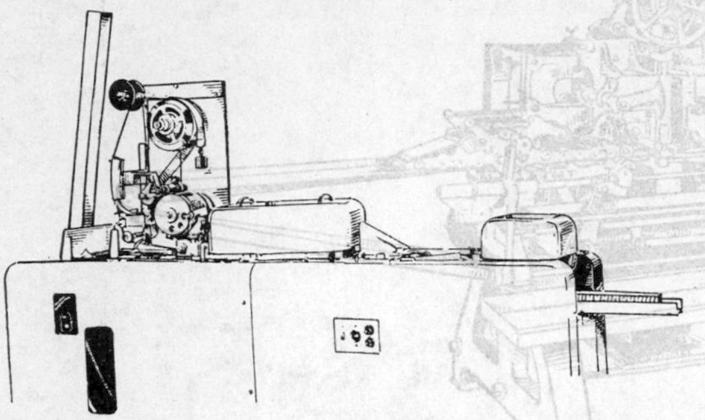

FIG. 142. Cellophane wrapping machine. (*F. B. Redington Co.*)

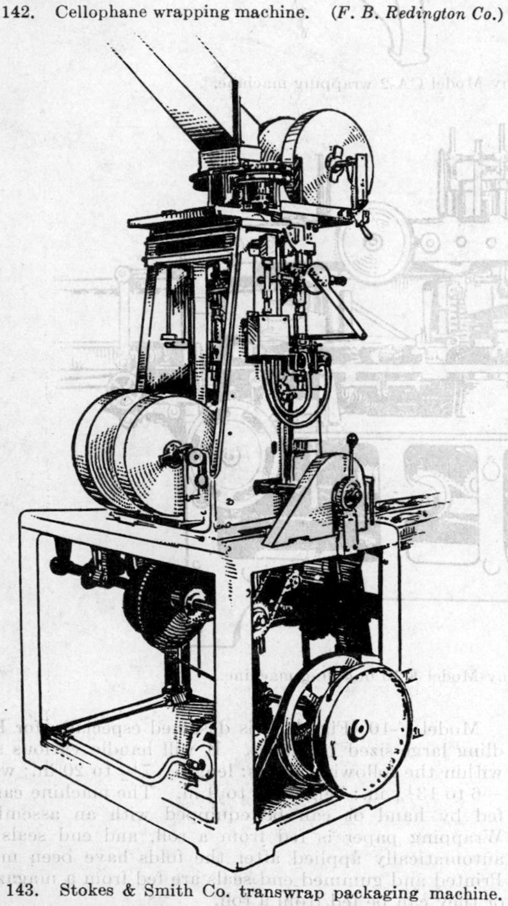

FIG. 143. Stokes & Smith Co. transwrap packaging machine.

Stokes and Smith have a unique machine that forms envelopes from a roll and fills and seals them. It is called the Transwrap Automatic Packaging Machine (Fig. 143). Heat-sealing cellophane, pliofilm, foils, or other materials are fed from a roll. The package is automatically formed by heat-sealing joints. It can form packages of the pillow, gusset, or "Finseal" type.

Various types of feeding mechanisms can be supplied to suit the products to be packaged, such as powdered, granular, hard individual pieces, or lumps. Thus augers, measuring pockets, counters, etc., may be used for feeding.

Single or double film can be used, printed or unprinted. Printed films can be registered by photoelectric registry attachment. Serrated or notched cutoff makes it easy to open the package.

It is built in two sizes to form the following package sizes:

	Maximum	Minimum
Size A machine	$2\frac{3}{8}$ by $6\frac{1}{2}$ in.	$1\frac{1}{4}$ by $1\frac{1}{2}$ in.
Size B machine	$4\frac{3}{4}$ by $10\frac{1}{2}$ in.	$1\frac{13}{16}$ by 3 in.

Production varies from 50 to 60 small packages per minute to 70 to 100 larger packages per minute with one attendant required.

SMALL-BAG FILLING AND CLOSING

As in the case of equipment for packing rigid containers, automatic machines are available for packing

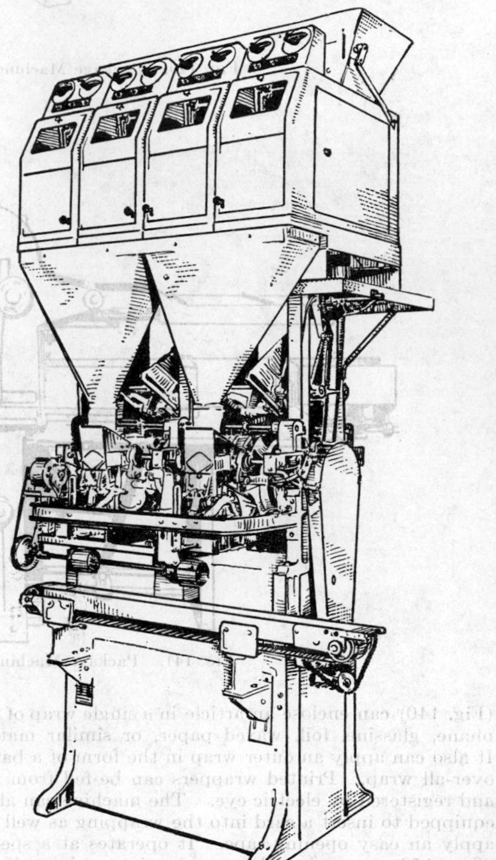

FIG. 144. Bemis Series No. 112 bag feeding and filling machine.

small bags holding up to 25 lb. This equipment will automatically feed bags, form, fill, agitate to settle the contents, close, and pack them in container bags. Multiple-station bag-filling equipment combined with rotary-type pasted closing equipment or fast machine sewing provide high-production packaging lines. Bags can be automatically packed into shipping containers such as cartons or container bags.

Equipment for forming pouch bags of different types are available. Joints are usually formed by heat-sealing. The machines will handle most types of heat-sealing films, including laminated stock and foils.

The Bemis Bro. Bag Co. Deltaseal machines are fully automatic. The feeding and filling machine Series 112 (Fig. 144) feeds bags onto the filling spouts automatically. Charges are preweighed and automatically dumped. The filled bags drop to a belt conveyor and

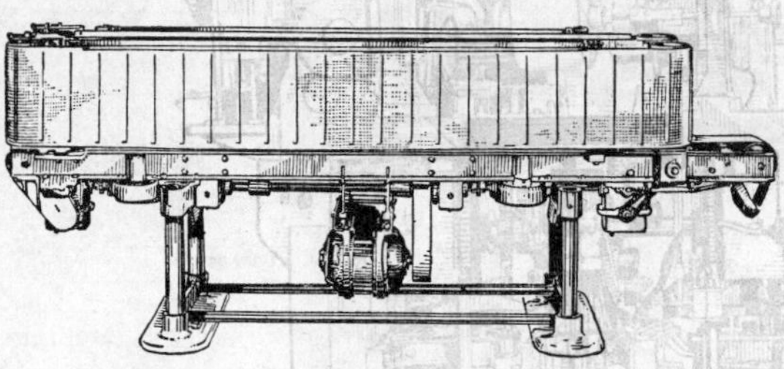

FIG. 145. Bemis Deltaseal shaker for small bags.

then pass through a shaker. This settler has a vibrating belt, and the sides weave back and forth (Fig. 145). Bags then pass through the closing machine (Fig. 146), which folds, pastes, and then seals the tops. Bags then pass to a conveyor, which delivers them to the station where they are packed by hand or automatically into shipping containers. Installations operate at better than 40 bags per minute of 5 lb. capacity. Only one operator is required.

The Union Special Machine Co. conveyor and sewing machine forms the closure for bags holding up to 25 lb. The equipment applies two strips of reinforcing paper tape, sews through the tapes, trims the bag top to provide a neat upper edge, and clips the tape between the bags.

The machine includes a Union Special 60,000 D sewing head combined with a belt conveyor, the speed of which

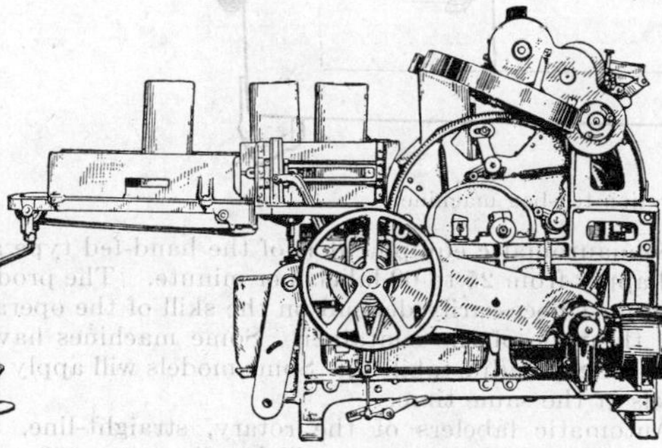

FIG. 146. Bemis Deltaseal bag-closing machine.

is synchronized with the speed of the sewing head. Style 21800N is a portable table-type unit. The conveyor is 8 in. wide and 4 ft. long and is adjustable vertically to accommodate different sizes of bags (Fig. 147). The equipment can be supplied for fully automatic or semiautomatic operation.

Pneumatic Scale Corp. (Fig. 148) has an automatic machine that forms a small bag, fills and seals it, and attaches a stringed tag to the bag. This machine is used for packing tea. Heat-sealing film is fed from a roll, formed into bags, filled with a predetermined quantity of tea, and sealed on three sides by two heat-sealing blocks. After the bags are heat-sealed, they are cut free and dropped into a chute through which they are pushed into the tagging mechanism. As the bags enter the tagging device, string tags are attached and wound around the bags, and the tags are tucked under the string to provide easy packing and prevent tangling.

This equipment will fabricate bags in sizes ranging from a minimum of 2½ by 2 in. to a maximum of 2¾ by 2½ in. It will fabricate, fill, and tag 160 bags per minute. One mechanical operator can supervise two machines, and two packers are required to handle the production of each machine.

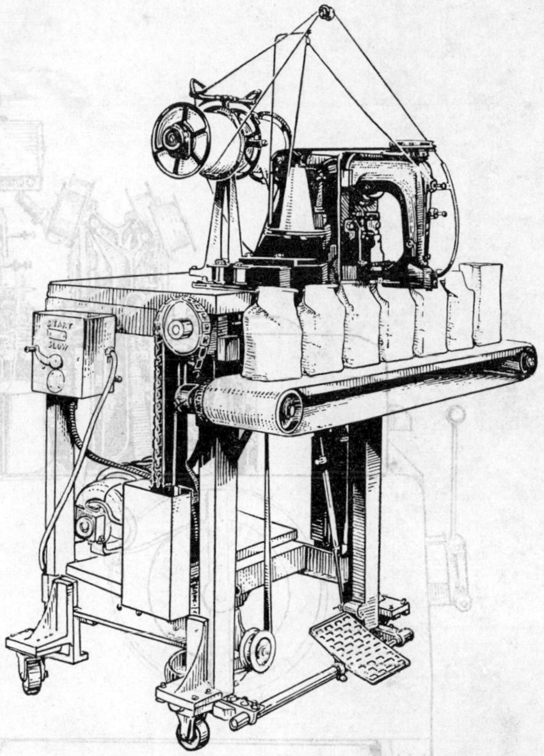

FIG. 147. Union Special Machine Company Style 21800N bag-sewing machine.

COLLAPSIBLE-TUBE FILLING AND CLOSING

Most collapsible-tube-filling machines will not fill tubes larger than 1½ in. diameter and 7 in. long. Production varies from 2 to 3 tubes filled per minute up to 120 tubes per minute for the fully automatic equipment.

Liquids and semiliquids are usually fed with gravity-type fillers. Some pastes can be fed in the same manner if they will flow when heated. They usually require the use of jacketed hoppers, and the filling mechanism is often heated. Usually pastes are filled by the use of a filling pump that starts the filling operation at the bottom of the collapsible tube to prevent air pockets. Agitators of the impeller type are required for some types of pastes in order to force the material down through the hopper to the filling pump.

The ends of the tubes may be sealed by making a double fold to which may be applied a clip where necessary. Clipless closures consist of a quadruple fold that is reinforced by corrugating the fold. There is equipment that hermetically seals the folded joint by the application of heat and pressure.

Attachments to register the printed matter in proper

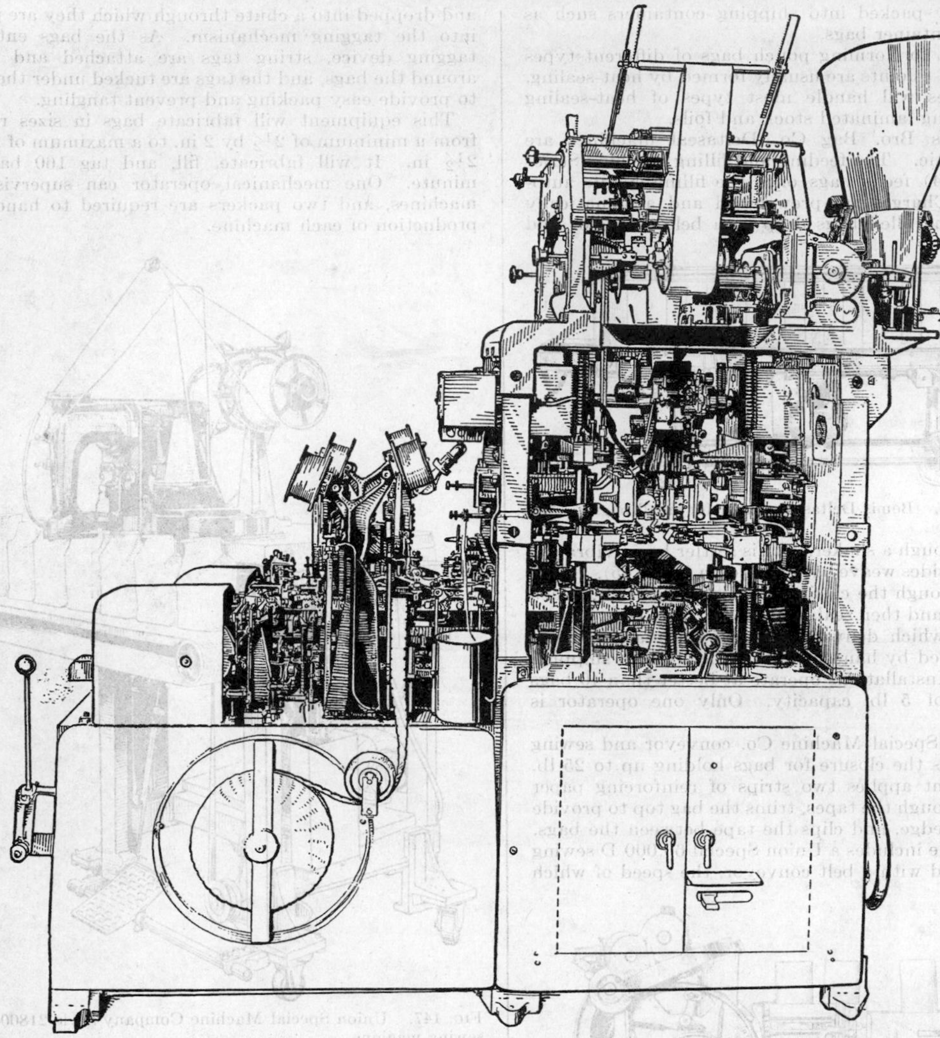

FIG. 148. Pneumatic Scale Corporation tea-bag machine.

relation to the fold are obtainable for use with equipment in which the empty tubes are automatically fed.

The Arenco Machine Co. tube filling and closing machine illustrated in Fig. 149 is hand-fed but filled and closed automatically. The operator places the tubes in the carrying chain. The tubes are then automatically cleaned by air suction, and caps are tightened, filled with measured volumes of paste, closed with any desired type of fold, imprinted with code marking, and finally lifted from the carrying chain.

The equipment will fill any diameter of tube up to 1¾ in. and any length up to 7½ in. Its output is from 30 to 50 tubes per minute with one attendant.

LABELING MACHINES

Labeling machines can be obtained to label bottles, jars, paper and wood boxes, cans, etc. They will apply paper, paper-backed foil, or heavily embossed foil labels.

The methods of application are basically (1) apply adhesive to the labels and then attach them to the container, (2) apply adhesive to container and then attach label to container, and (3) combinations of (1) and (2).

Semiautomatic equipment is of the hand-fed type and will apply from 25 to 60 labels per minute. The production must necessarily depend on the skill of the operator and the type of the container. Some machines have a wide range of adjustability. Some models will apply two labels at the same time.

Automatic labelers of the rotary, straight-line, and straight-line multiple types will handle up to 200 packages per minute. The straight-line type can be used for labeling a wider range of shapes of containers, and they are more accurate in placing the labels on the packages.

Round cans usually have wrap-around labels. Semiautomatic machines usually cover the labels with adhesive before it is placed in contact with the can. Production varies from 5 to 10 packages labeled per minute.

The fully automatic equipment operates at high speed. Cans roll on their sides, picking up paste in spots or lines. The cans, as they roll over a magazine containing labels, pick up their labels. A line of glue is applied to the overlapped end to form a neat sealed joint.

To facilitate application of labels by hand, a number of

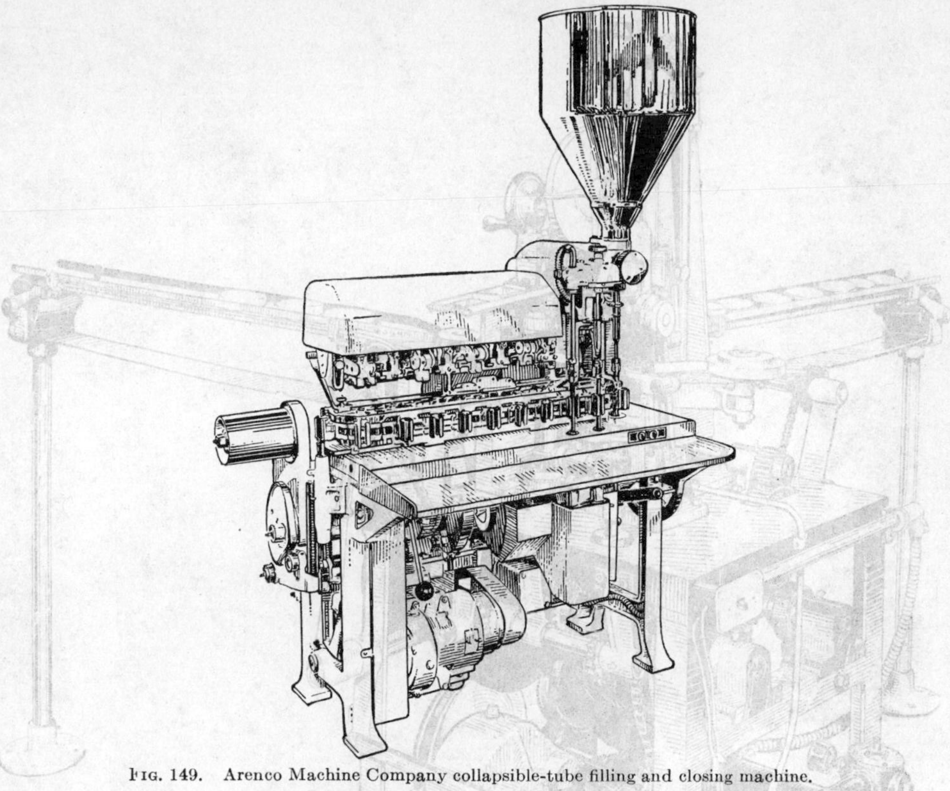

FIG. 149. Arenco Machine Company collapsible-tube filling and closing machine.

FIG. 150. Pneumatic Scale Corporation Duplex front and back labeler.

types of gummers are on the market. These machines apply adhesive to ungummed labels or moisten gummed labels as desired.

Pneumatic Scale Corp. Duplex labeler (Fig. 150) is a good example of an automatic machine. It applies labels to two packages at a time—to the front only or to the front and back simultaneously. The machine will handle containers less than 3 in. wide with labels not running over 3¾ in. wide. It is rated at a capacity of 80 to 120 packages labeled per minute. A part-time attendant is required to replenish labels and adhesive.

A tape sealing machine (Fig. 151) manufactured by Package Machinery Co. has interesting possibilities in the small-packaging field. This equipment automatically applies pressure-sensitive sealing tape to lid-closure fiber or metal cans. It is rated at a speed of 12 to 25 packages per minute.

Round packages varying in diameter between 2½ and

FIG. 151. Package Machinery Company tape sealing machine.

$4\frac{1}{2}$ in. with heights between $\frac{3}{4}$ in. and 6 in. can be sealed with this equipment. It will also handle rectangular packages varying in width between $1\frac{1}{2}$ and 4 in. and height between $\frac{3}{4}$ and 6 in. The width of the tape is limited to $\frac{3}{8}$ and $\frac{1}{2}$ in.

The machine also applies an easy opening tab, projecting at right angles from the outer end of the tape. The operation of the machine is such that each can measures its own tape

PUMPING OF LIQUIDS*

BY F. L. LUCKER

Pumping Equipment. For the purposes of this section, a pump is defined as any device for the transference of liquids. More broadly speaking, pumps also include devices for handling gases, but the subject of the movement of gases is covered in the latter portion of this section.

Liquids are said to be practically incompressible. This is near enough to the truth at low pressures, but at

* Acknowledgment is gratefully made for the assistance of A. W. Loomis, Engineer, Ingersoll-Rand Co. in the preparation of the material on pumps, compressors, fans, etc.

high pressures there is a slight change in density, which in some cases it is necessary to consider.

The pressure existing at any point in a liquid at rest is caused by the pressure exerted on its surface, plus the weight of liquid above the point in question. If the liquid is in an open vessel, the pressure on the surface is atmospheric pressure. Pressure thus exerted on a liquid is equal in all directions and acts perpendicularly to any surfaces in contact with the liquid. All liquid pressures can be thought of as equivalent to a column of liquid of a

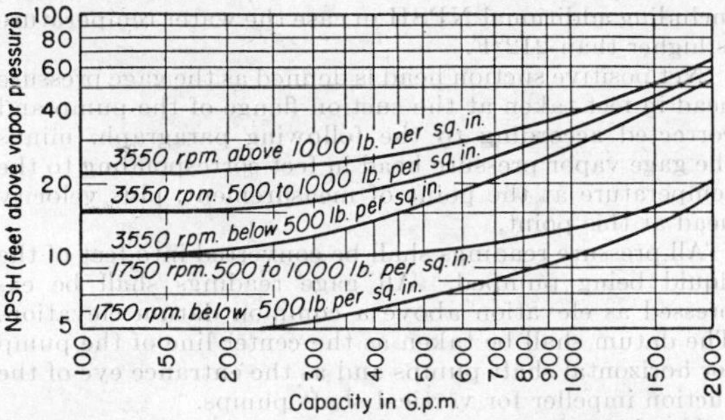

FIG. 152. Net positive suction head, centrifugal hot-water pumps—single suction. Compiled from data by representative companies. The curves apply to water temperatures up to 212°F. For temperatures above 212°F. use temperature correction chart. For speeds within ±25 per cent of those shown correct capacity according to: r.p.m. $\sqrt{\text{gal./min.}}$ = constant. (*By permission of Hydraulic Institute.*)

height sufficient by its weight to produce the pressure in question. In pumping practice, pressures or heads are thus often expressed in feet of liquid.

Capacity is usually expressed as volume per unit of time, as gallons per minute or barrels per hour.

Velocity. Since a liquid is practically incompressible, there is a definite relation between the quantity flowing past a given point in a given time and the velocity of flow. This relation is expressed thus:

$$Q = Av$$

$$v \text{ (for circular conduits)} = \frac{0.4085 \text{ gal./min.}}{d^2} = \frac{0.2859 \text{ bbl./hr.}}{d^2}$$

Viscosity (see Sec. 5 for further information). In flowing liquids the existence of internal friction or the internal resistance to relative motion of the fluid particles must be considered. This resistance is called viscosity. It varies greatly from one liquid to another and decreases with rising temperature. Viscous liquids tend to increase the horsepower required by a pump and to reduce the pump efficiency, head, and capacity, and increase the friction in pipe lines.

Friction head (h_f) is the pressure (feet of liquid) required to overcome the resistance to flow in pipe and fittings.

Velocity head (h_v) is the vertical distance a body would have to fall to acquire the velocity v. It corresponds to the static or pressure head that would cause that velocity.

$$h_v = \frac{v^2}{2g} = 0.0155v^2 \quad \text{approx.}$$

$$h_v \text{ (for circular conduits)} = \frac{0.00259 \text{ gal./min.}^2}{d^4} = \frac{0.00127 \text{ bbl./hr.}^2}{d^4}$$

Velocity head may be converted to equivalent pressure head by suitable means such as a venturi tube. In any flowing liquid, the sum of pressure head and velocity head remains constant.

Total static head (h_{ts}) on a pump is the vertical distance (feet) between the free level of the source of supply and the point of free discharge or the level (or equivalent level) of the free surface of the discharge liquid.

Total suction lift (l_s) is the reading of a mercury column at the suction flange of a pump (corrected to the pump center line* and converted to feet of liquid) minus the velocity head (feet) at the point of column attachment.

$$(l_s)\dagger = (l_{ss}) + (h_f \text{ suction})$$

Total suction head (h_s) is the reading of a gage at the suction flange of a pump (corrected to the pump center line* and converted to feet of liquid) plus the velocity head (feet) at the point of gage attachment.

$$(h_s)\dagger = (h_{ss}) - (h_f \text{ suction})$$

Total discharge head (h_d) is the reading of a pressure gage at the discharge flange of a pump (corrected to the pump center line* and converted to feet of liquid) plus the velocity head (feet) at the point of gage attachment.

$$(h_d)\dagger = (h_{sd}) + (h_f \text{ discharge}) + \text{velocity head loss at discharge of system}$$

Total head (H) on a pump is the total discharge head minus the total suction head or plus the total suction lift.

$$(H)\dagger = (h_d) - (h_s) \quad \text{or} \quad (H)\dagger = (h_d) + (l_s)$$

Net Positive Suction Head (*NPSH*). Net positive suction head is the head available at the entrance or eye of a pump impeller to move and accelerate the water entering the eye. It is the gage pressure at the suction flange of the pump (corrected to the pump center line* and converted to feet of liquid) minus the gage vapor pressure in feet of liquid corresponding to the tem-

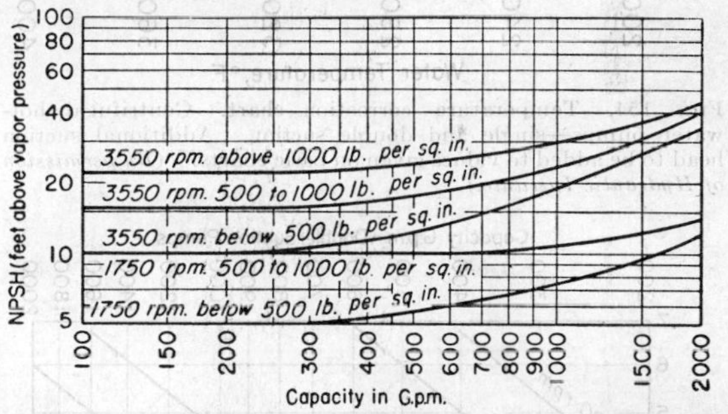

FIG. 153. Net positive suction head, centrifugal hot-water pumps—double suction first stage. Compiled from data by representative companies. The curves apply to water temperatures up to 212°F. For temperatures above 212°F. use the temperature correction chart. For speeds within ±25 per cent of those shown correct capacity according to: r.p.m. $\sqrt{\text{gal./min.}}$ = constant. (*By permission of Hydraulic Institute.*)

perature at the point of measurement) plus the velocity head (feet at the point of measurement).

$$(\text{NPSH})\dagger = (\text{atm.}) + (h_s) - (vp) = (\text{atm.}) + (h_{ss}) - (h_f \text{ suction}) - (vp)$$

or

$$(\text{NPSH})\dagger = (\text{atm.}) - (l_s) - (vp) = (\text{atm.}) - (l_{ss}) - (h_f \text{ suction}) - (vp)$$

When the NPSH required by a pump is known, and it is desired to find the static suction lift that the pump will handle, the following form of the formula is convenient:

$$\text{Max. static suction lift} = (\text{atm.}) - (\text{NPSH}) - (h_f \text{ suction}) - (vp)$$

Suction Limitations on a Pump. The maximum theoretical suction lift would be a distance equivalent in feet of liquid to the atmospheric pressure less the head equivalent to the vapor pressure of the liquid, less the head

* On vertical pumps, the correction should be made to the eye of the suction impeller.

† These formulas apply when estimating before a pump is installed.

equivalent to the pressure of any gas in solution, less the friction head, and less the head equivalent to the entrance loss. Practically, however, the suction lift that can be handled without cavitation and vibration in the pump is somewhat less than the theoretical. The actual suction lift obtainable depends not only on the above factors but

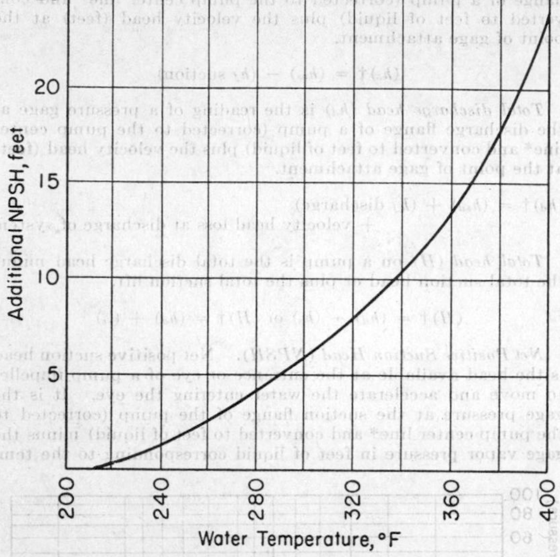

FIG. 154. Temperature correction chart. Centrifugal hot-water pumps—single and double suction. Additional suction head to be added to values given on basic charts. (*By permission of Hydraulic Institute.*)

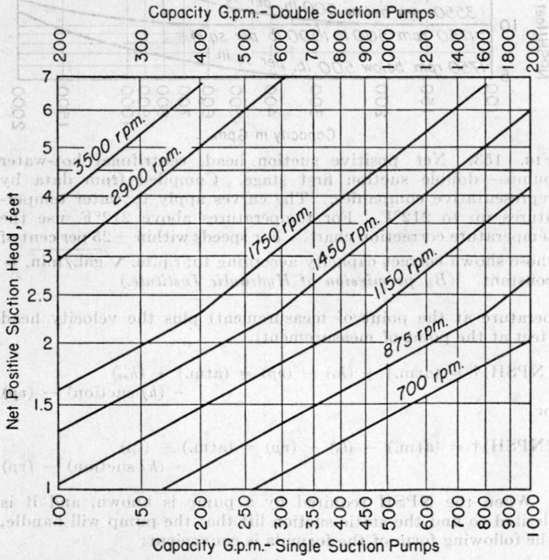

FIG. 155. Capacity and speed limitations for condensate pumps with shaft through eye of impeller. (*By permission of Hydraulic Institute.*)

also on the characteristics of the liquid, the total head, and the pump speed, capacity, and impeller design. Abnormally high suction lifts usually cause serious reductions in the capacity and efficiency of the pump and often lead to serious troubles from vibration and cavitation. Every pump manufacturer knows the suction limitations of each of his pumps. Therefore, when requesting quotations on a pump for a particular job, the suction conditions should be given so that the manufacturer can recommend a proper pump for those conditions.

Net Positive Suction Head Curves for Centrifugal Hot-water Pumps

Hot-water Pumps. These curves (Figs. 152, 153, and 154) give the net positive suction heads in feet (NPSH) above vapor pressure for different capacities and speeds, including additional NPSH in case the water temperature is higher than 212°F.

Net positive suction head is defined as the gage pressure head in feet taken at the suction flange of the pump and corrected according to the following paragraph, minus the gage vapor pressure head in feet corresponding to the temperature at the point of measurement, plus velocity head at this point.

All pressure readings shall be converted into feet of the liquid being pumped. All gage readings shall be expressed as elevation above a common datum elevation. The datum shall be taken as the center line of the pump for horizontal shaft pumps and as the entrance eye of the suction impeller for vertical shaft pumps.

If a hot-water pump takes its suction from a source where the prevailing pressure is equivalent to the vapor pressure corresponding to its temperature, the net positive suction head is defined as the difference in elevation between the liquid level at the source and the pump center line minus the entrance and friction losses in the suction piping.

Example. If the pump handles water of 350°F. temperature with 150 lb./sq. in. gage pressure at the suction nozzle and with 12 ft./sec. velocity, what is the net positive suction head?

$$\text{The vapor pressure for 350°F. water} = 134.6 - 14.7 = 119.9 \text{ lb./sq. in. gage}$$

$$\text{Specific gravity of 350°F. water} = 0.89$$

$$\text{Velocity head, } V^2/2g = 2.22 \text{ ft.}$$

$$\text{Thus NPSH} = \frac{(150 - 119.9)2.31}{0.89} + 2.22 = 80.32 \text{ ft.}$$

The first curve applies to single-suction pumps, the second to double-suction pumps, and the third is the temperature-correction curve.

Work Performed in Pumping. To move a liquid against gravity with a pump, work must be expended. A pump may actually raise the liquid, or it may force it into a pressure vessel or merely give it enough head to overcome pipe friction. No matter what the service required of a pump, all forms of energy imparted to the liquid in performing this service must be accounted for in establishing the work performed. In order that all these forms of energy may be algebraically added, it is customary to express them all in terms of head expressed in feet of liquid.

In order to determine the work required of a pump, it is necessary to know the total dynamic head and the weight of liquid to be pumped in a given time. Usually weight is not given but rather volume in gallons per minute or barrels per hour, along with the density or specific gravity. The weight can be calculated from this information, but this is not necessary, since the horsepower formulas are usually given in terms of gallons per minute or barrels per hour. The theoretical horsepower of a pump is usually called the hydraulic horsepower.

$$\text{Hydraulic hp.} = \frac{\text{gal./min.}(H)(s)8.33}{33{,}000} = \frac{\text{gal./min.}(H)(s)}{3960} = \frac{\text{bbl./hr.}(H)(s)}{5660}$$

$$\text{Hydraulic hp.} = \frac{\text{gal./min.}(H_p)}{1714} = \frac{\text{bbl./hr.}(H_p)}{2450}$$

$$\text{Brake hp.} = \frac{\text{gal./min.}(H)(s)}{(3960 \times \text{pump efficiency})}$$

The actual or brake horsepower of a pump is greater than the theoretical or hydraulic horsepower by the amount of the losses incurred in the pump, through friction leakage, etc. The efficiency of a pump is therefore measured as

$$\text{Pump efficiency} = \frac{\text{hydraulic hp.}}{\text{brake hp.}}$$

Other useful formulas are:

For motor-driven pumps:

$$\text{Kw./1000 gal./hr. liquid} = \frac{0.00315(s)(H)}{(\text{pump efficiency})(\text{motor efficiency})}$$

For direct-acting steam pumps:

$$\text{Duty in terms of ft.-lb./1000 lb. steam} = \frac{\text{theoretical hp. } (1980)1{,}000{,}000}{\text{total steam/hr., lb.}}$$

Sometimes theoretical horsepower for figuring duty is calculated using static head instead of total dynamic head. In this case, duty is in terms of foot-pounds of useful work done per 1000 lb. steam. As given above, the formula is in terms of foot-pounds of actual work done per 1000 lb. steam.

Symbols Used in This Discussion:

A = flow area of conduit, sq. ft.
atm. = atmospheric pressure, ft. liquid, abs.
bbl./hr. = flow in barrels (42 gal.)/hr.
d = inside diameter of circular conduit, in.
g = acceleration due to gravity, ft./sec.^{2} at sea level and 45° latitude; this is 32.174 ft./sec.^{2}, and this value has been used in conversions.
gal./min. = flow in gal./min.
H = total head, ft. liquid.
H_p = total head, lb./sq. in.
h_d = discharge head, ft. liquid.
h_f = friction head, ft. liquid.
h_s = suction head, ft. liquid.
h_{sd} = static discharge head, ft.
h_{ss} = static suction head, ft.
h_{ts} = total static head, ft.
h_v = velocity head, ft.
l_s = suction lift, ft. liquid.
l_{ss} = static suction lift, ft.
Q = quantity of flow, cu. ft./sec.
s = specific gravity (water at 60°F. = 1).
v = velocity of flow, ft./sec.
vp = vapor pressure of liquid, ft. liquid, abs.

Types of pumps that are of importance in the chemical engineering field include:

1. *Centrifugal pumps*, including sump pumps, and propeller or axial- and mixed-flow type pumps, and deep-well pumps.
2. *Reciprocating, or piston pumps*, including steam and power pumps, and diaphragm pumps.
3. *Rotary pumps*.
4. *Air-pressure pumps*, including air lifts; and displacement pumps such as acid eggs, automatic elevators, pneumatic pumps, flammable-liquid pumps, Humphrey gas pumps, and Pulsometers.
5. *Jet pumps*, including injectors and ejectors, operated by steam or liquid.
6. *Hydraulic rams* and Hydrautomats.
7. *Special pumps*, including the screws, the scoop wheel, etc.

CENTRIFUGAL PUMPS

The centrifugal pump is the type most widely used in the chemical industry for transferring liquids of all types—raw materials, materials in manufacture, and finished products—as well as for general services of water supply, boiler feed, condenser circulation, condensate return, etc. They are available through a vast range of sizes; in capacities from 2 or 3 gal./min. up to 100,000 gal./min.; and for discharge heads (pressures) from a few feet up to 3000 lb./sq. in. The size and type best suited to a particular application can be determined only by an engineering study of the problem.

The primary advantages of a centrifugal pump are simplicity, low first cost, uniform (non-pulsating) flow, small floor space, low maintenance expense, quiet operation, and adaptability to use with motor or turbine drive.

A centrifugal pump, in its simplest form, consists of an impeller rotating within a casing. The impeller consists of a number of blades, either open or shrouded, mounted on a shaft that projects outside the casing. Impellers may have their axis of rotation either horizontal or vertical, to suit the work to be done. Closed-type or shrouded impellers are used for most types of work, since they are generally most efficient. Open- or semiopen-type impellers are used for viscous liquids or liquids containing solid materials and on many small pumps for general service. Impellers may be of the single-suction type or double-suction type—single if the liquid enters from one side, double if it enters from both sides.

Casings are of three general types but in any case consist of a chamber in which the impeller rotates, provided with inlet and exit for the liquid being pumped. The simplest form of casing is the circular casing, consisting of an annular chamber around the impeller, no attempt being made to overcome the losses that will arise from eddies and shock when the liquid leaving the impeller at relatively high velocities enters this chamber. Such casings are seldom used.

Volute casings take the form of a volute, increasing in cross-sectional area as the outlet is approached. The volute converts the velocity energy imparted to the liquid by the impeller into pressure energy, with comparatively low losses.

A third type of casing is used in diffuser-type or turbine pumps. In this type, guide vanes or diffusers are interposed between the impeller and the casing chamber. Losses are kept to a minimum in a well-designed pump of this type. This construction is often used in multistage, high-head pumps.

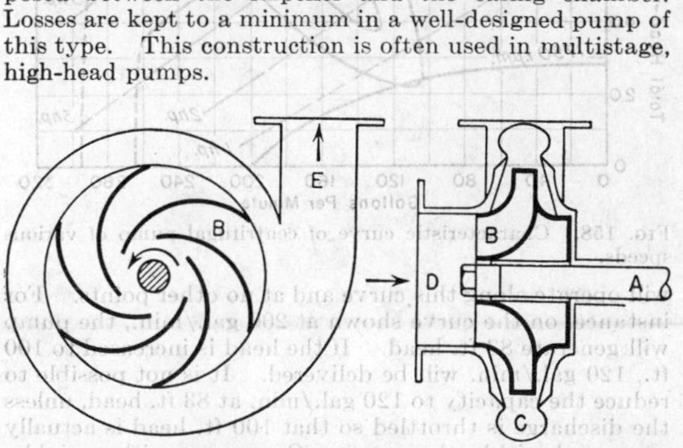

FIG. 156. Diagram of a simple centrifugal pump.

Action of a Centrifugal Pump. Briefly the action of a centrifugal pump may be shown by Fig 156. Power from an outside source is applied to shaft A, rotating the impeller B within the stationary casing C. The blades of the impeller in revolving produce a partial vacuum at the entrance or eye of the impeller. This causes liquid to flow into the impeller from the suction pipe D. This liquid is forced outward along the impeller blades at an increasing velocity. The velocity head it has acquired when it leaves the blade tips is changed to pressure head as the liquid passes into the volute chamber and thence out the discharge *E*.

Centrifugal-pump Characteristics. Figure 157 shows a typical characteristic curve of a centrifugal pump. It is important to note that at any fixed speed the pump

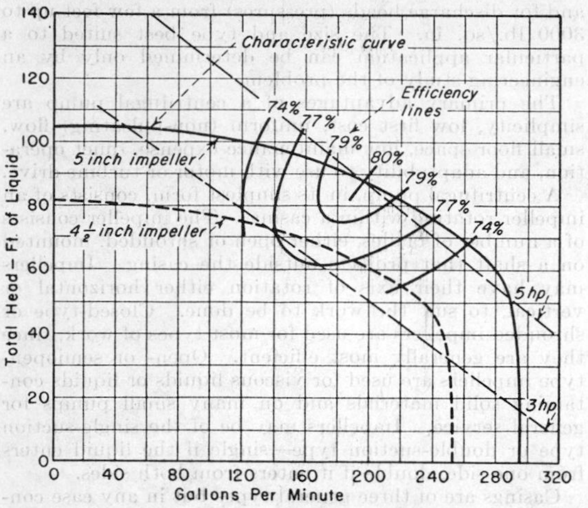

FIG. 157. Characteristic curve of centrifugal pump of constant speed.

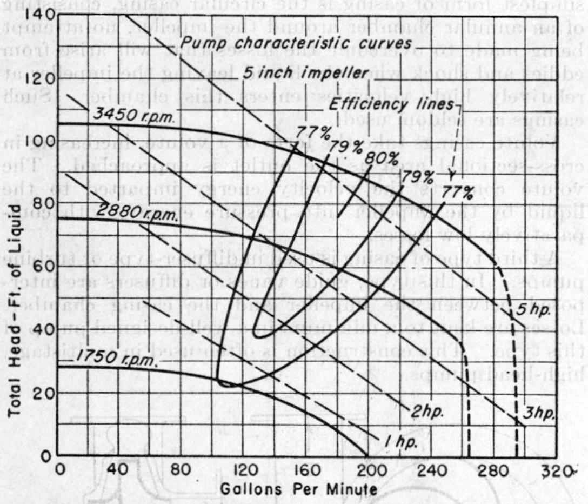

FIG. 158. Characteristic curve of centrifugal pump of various speeds.

will operate along this curve and at no other points. For instance, on the curve shown at 200 gal./min., the pump will generate 83 ft. head. If the head is increased to 100 ft., 120 gal./min. will be delivered. It is not possible to reduce the capacity to 120 gal./min. at 83 ft. head, unless the discharge is throttled so that 100 ft. head is actually generated within the pump. On pumps with variable-speed drivers such as steam turbines, it is possible to change the characteristic curve, as shown by Fig. 158.

Single-stage Centrifugal Pumps

Single-stage centrifugal pumps are available in capacities up to 50,000 gal./min. and over, for heads (pressures) up to 200 ft. or sometimes as high as 350 ft. They are available in a variety of designs for particular services.

Chemical pumps is a term usually applied to single-stage units of simple design for capacities up to about 3000 gal./min. and heads up to about 200 ft. (see Fig. 159). Such units are designed for ease in dismantling, accessibility, with stuffingboxes built especially to handle corrosive liquids, and provisions to prevent any liquid from the stuffingbox being lost or reaching unprotected parts of the pump. They usually have single-suction impellers. Chemical pumps are available in a variety of materials.

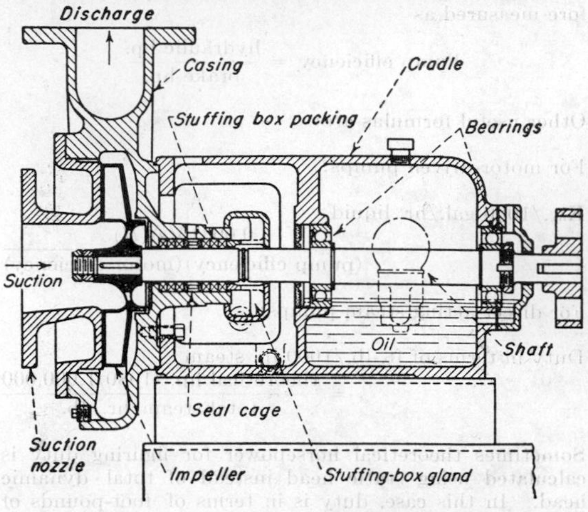

FIG. 159. Ingersoll-Rand chemical pump.

Metal pumps are the most widely used. They may be obtained in iron, bronze, iron with bronze fittings, and a practically unlimited selection of steel alloys (see pp. 1427*ff.* for a list of the most commonly used materials). Pumps are also available in glass, carbon, rubber-clad metal, lead-lined, stoneware, etc., such units usually being used for special purposes. When handling many chemicals, proper pump packing is important. This is treated separately under Packing and Shaft Seals.

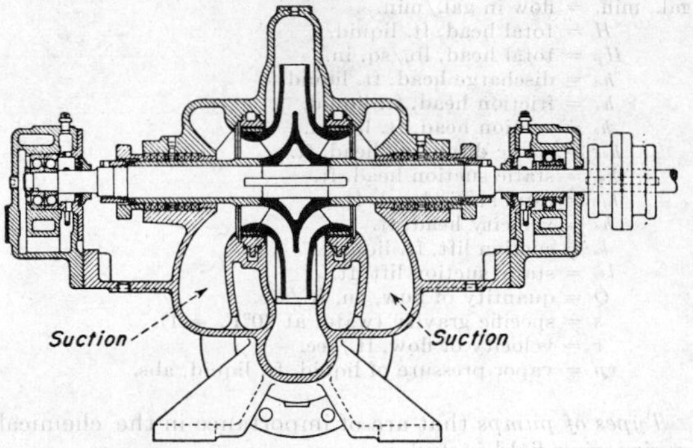

FIG. 160. Single-stage double-suction pump.

Double-suction single-stage pumps (see Fig. 160) are used for general water-supply and circulating service and for chemical service when handling liquids that are non-corrosive to iron or bronze. They are available for capacities from about 25 gal./min. up to as high as 50,000 gal./min. and heads up to 200 ft. or so. Such units are available in iron or bronze or iron with bronze fittings. Other materials increase the cost; and, where such materials are required, a standard-type chemical pump is usually more economical.

Close-coupled Pumps (see Fig. 161). Pumps with built-in electric motor or sometimes steam-turbine-driven (*i.e.*, with pump impeller and driver on the same shaft) are known as close-coupled pumps. Such units are extremely compact and are suitable for a variety of services where standard iron and bronze materials are satisfactory. They are available in capacities up to about 2000 gal./min. for heads up to about 240 ft. Two-stage units in the smaller sizes are available for heads to around 500 ft., and at least one manufacturer builds a vertical four-stage unit.

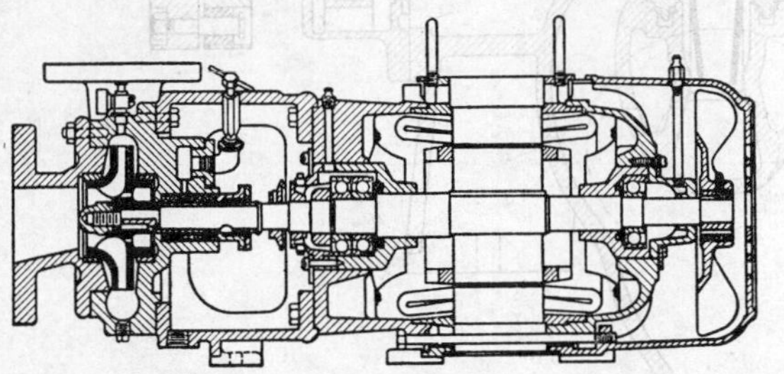

FIG. 161. A close-coupled pump.

Special-application Pumps. A number of pump designs are available for special applications such as handling paper stock, hot liquids, volatile hydrocarbons, etc. Such pumps are basically of the types described but with features designed to meet special needs of the service for which they are to be used. Pump manufacturers can furnish complete information on units for such applications.

Sump pumps are small single-stage pumps used to drain shallow pits or sumps. They usually have single-suction open-type impellers. The pump itself is immersed in the liquid and driven by a shaft from a motor or belt pulley above the pit. A tube or column encases the shaft and forms a connecting support between the pump and the driver and for whatever shaft bearings may be required.

Multistage Centrifugal Pumps

Multistage pumps are in general used for services requiring higher heads (pressures) than can be generated by single-stage pumps. Such services include high-pressure water-supply pumps, fire pumps, boiler-feed pumps, and charge pumps for refinery processes. Such pumps are available for pressures as high as 3000 lb./sq. in. at capacities up to 3000 gal./min. and above.

Multistage pumps may be of the volute type or of the diffuser type. Volute-type pumps (see Fig. 162) usually have single-suction impellers arranged with half of the impeller inlets facing one direction and half in the opposite direction to balance thrust. Some two-, three-, and four-stage (see Fig. 163) units have double-suction impellers; but this construction, except for special-purpose units, makes the casings of impractical size for more stages.

Diffusor-type pumps (see Fig. 164) usually have single-suction impellers arranged with all impeller inlets facing in the same direction and impeller thrust neutralized by a differential-pressure device known as a balancing drum.

Pumps for very high pressures often have an inner volute-type casing or assembly of diffuser units placed within a forged-steel shell or barrel.

Multistage pumps for small or moderate capacity are available in a vertical design which for many services simplifies the piping and installation (see Fig. 165). One manufacturer furnishes such a pump with the motor below the pump. This unit has a mercury seal between the motor and the liquid from the pump.

PROPELLER PUMPS

The term *propeller pump* is applied to units with axial-flow or mixed-flow (*i.e.*, part axial and part centrifugal) impellers. Such units are available in capacities from 100 gal./min. upward for heads up to about 100 ft. per stage. Propeller pumps are usually vertical, although some single-stage horizontal units are built.

A common form of propeller pump has the pump element mounted at the bottom of a column that serves as the discharge pipe (see Fig. 166). Such units are im-

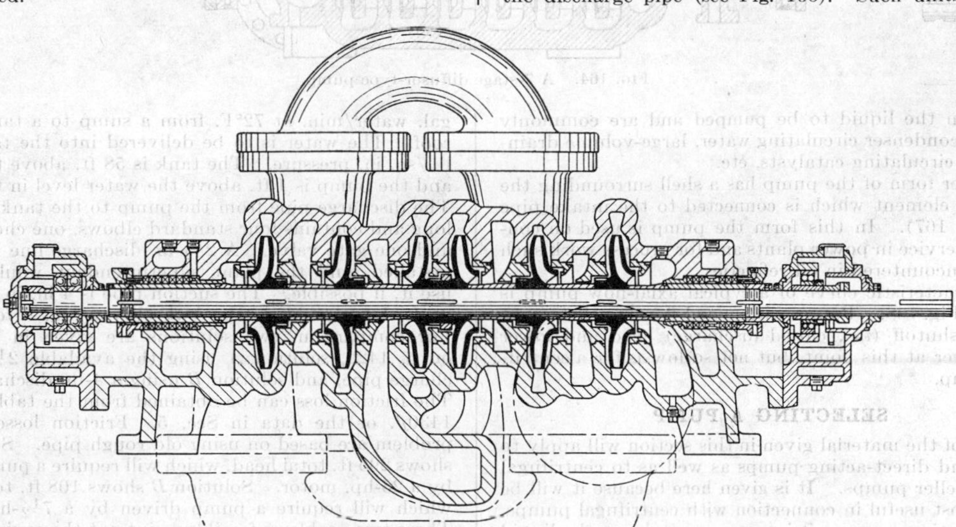

FIG. 162 A 6-stage volute-type pump.

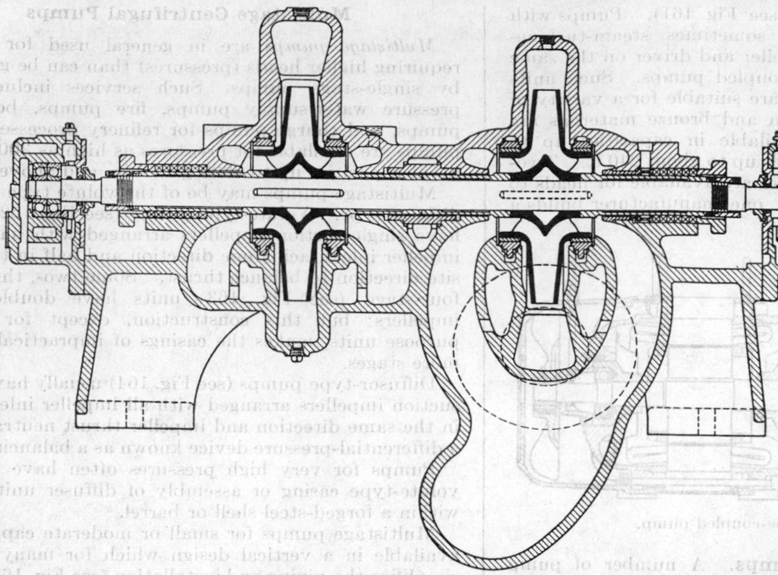

FIG. 163. Two-stage pump with double-suction impellers.

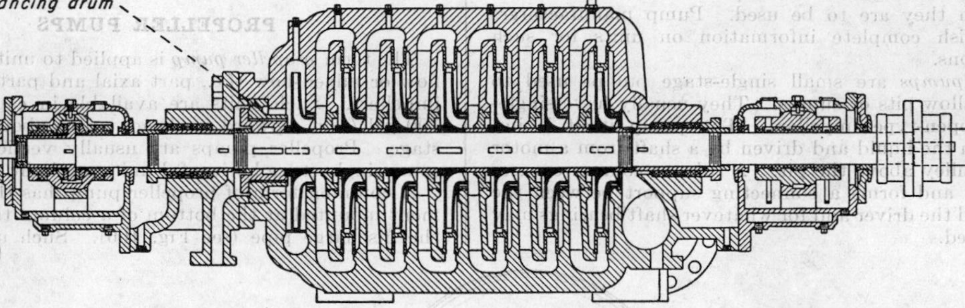

FIG. 164. A 7-stage diffusor-type pump.

mersed in the liquid to be pumped and are commonly used for condenser circulating water, large-volume drainage, and circulating catalysts, etc.

Another form of the pump has a shell surrounding the pumping element which is connected to the intake pipe (see Fig. 167). In this form the pump is used on condensate service in power plants and for process work such as that encountered in oil refineries.

A characteristic curve of a typical axial-flow pump is given in Fig. 168. Pumps with mixed-flow impellers have a lower shutoff (*i.e.*, closed discharge) head and lower horsepower at this point, but not so low as for a centrifugal pump.

SELECTING A PUMP

Some of the material given in this section will apply to rotary and direct-acting pumps as well as to centrifugal and propeller pumps. It is given here because it will be found most useful in connection with centrifugal pumps.

In selecting a pump, it is necessary to know the liquid to be handled; the total discharge head; the capacity; the suction lift or head; and, in most cases, the temperature, viscosity, and vapor pressure of the liquid to be handled.

To illustrate these points, a typical example is given: An industrial plant wishes to install a pump to lift 200 gal. water/min. at 72°F. from a sump to a tank on the roof. The water is to be delivered into the tank at 10 lb./sq. in. pressure. The tank is 58 ft. above the sump, and the pump is 4 ft. above the water level in the sump. The discharge pipe from the pump to the tank is 400 ft. long and contains four standard elbows, one check valve, and one gate valve. A 2½-in. discharge line is already installed, and the plant superintendent would like to use it, if possible. The suction pipe is 4 in. in diameter, 25 ft. long, and contains two elbows and a foot valve. For comparison, two solutions are given in the table on p. 1422, solution *A* using the available 2½-in. discharge pipe, and solution *B* using a 4-in. discharge pipe. The friction loss can be obtained from the tables on pp. 1430*ff.* or the data in Sec. 5. Friction losses in this problem are based on using old rough pipe. Solution *A* shows 295 ft. total head, which will require a pump driven by a 25-hp. motor. Solution *B* shows 108 ft. total head, which will require a pump driven by a 7½-hp. motor. These two problems forcibly point out the savings that a discharge pipe of proper size makes possible.

PACKING AND SHAFT SEALS

Proper packing and shaft seals are items of major importance, particularly for centrifugal pumps used by

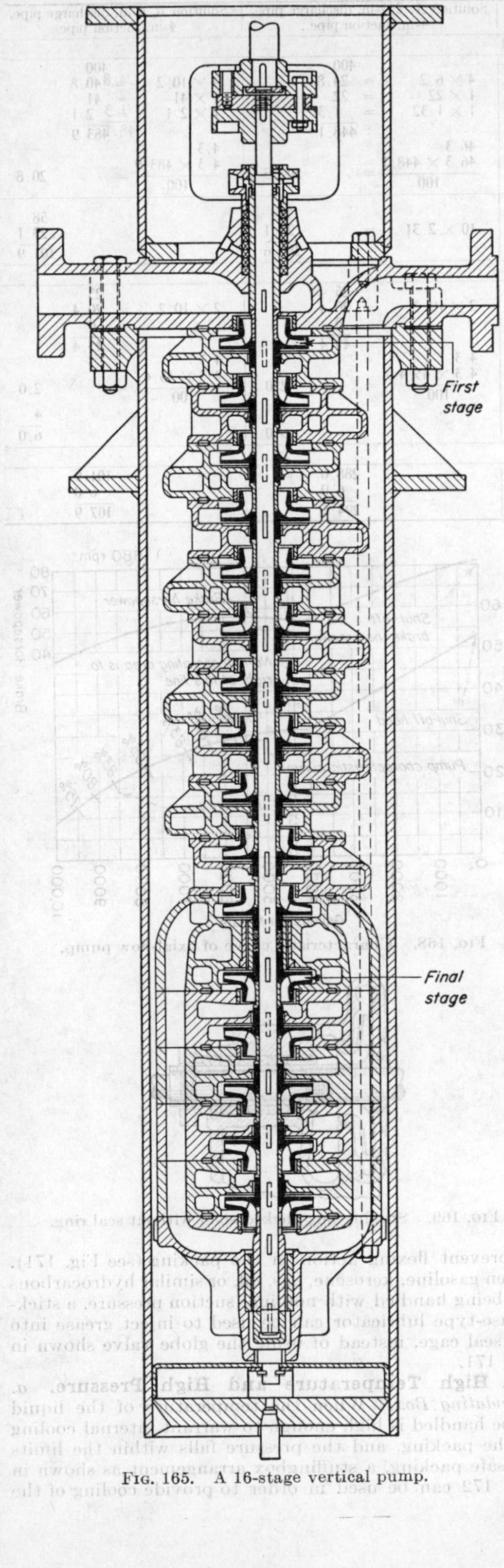

FIG. 165. A 16-stage vertical pump.

the chemical industry. The success of the pumping installation often depends upon the proper selection of packing. Many types and designs are available, some of which are described below.

Stuffingbox arrangements vary with the liquid, the temperature, and the pressure. They can generally be classified as being packed solid or having a sealing cage

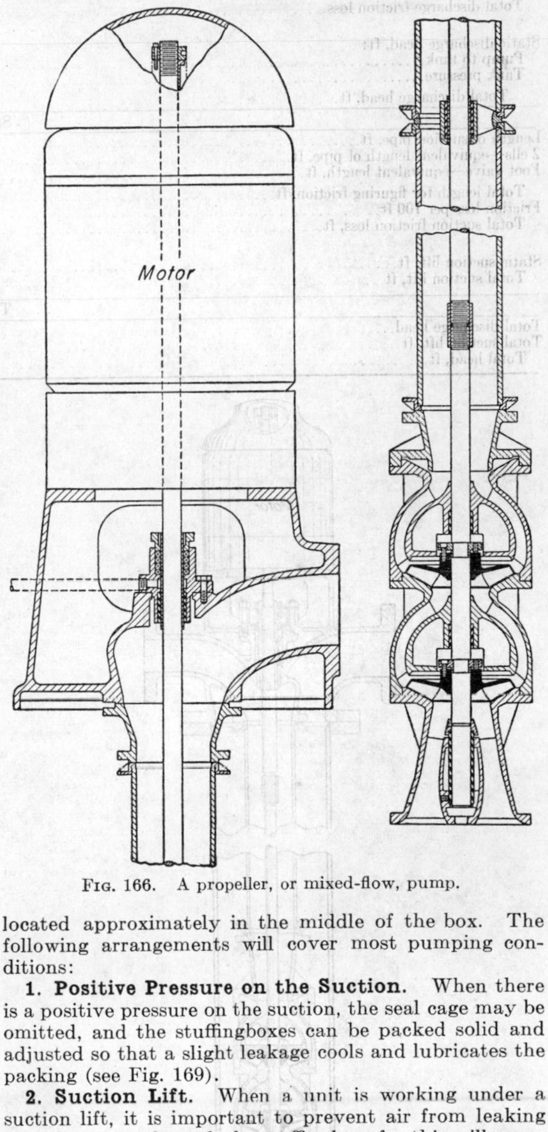

FIG. 166. A propeller, or mixed-flow, pump.

located approximately in the middle of the box. The following arrangements will cover most pumping conditions:

1. Positive Pressure on the Suction. When there is a positive pressure on the suction, the seal cage may be omitted, and the stuffingboxes can be packed solid and adjusted so that a slight leakage cools and lubricates the packing (see Fig. 169).

2. Suction Lift. When a unit is working under a suction lift, it is important to prevent air from leaking into the pump through the stuffingbox, for this will cause a reduction in capacity or loss of prime. To prevent air leakage and to lubricate the packing, sealing liquid is injected at a seal cage usually located at approximately the center of the box. When the sealing medium is the liquid pumped, the arrangement is known as an internal seal (see Fig. 170). When the sealing medium comes from an outside source, it is known as external sealing (see Fig. 171).

3. Flooded Suction. An external or internal seal should be provided in order to cool and lubricate the packing (see Figs. 170 and 171).

	Solution *A*, 2½-in. discharge pipe, 4-in. suction pipe		Solution *B*, 4-in. discharge pipe, 4-in. suction pipe	
Discharge head				
Length of discharge pipe, ft.		400		400
4 ells—equivalent length of pipe, ft.	4 × 6.2 =	24.8	4 × 10.2 =	40.8
1 check valve equivalent length of pipe, ft.	1 × 22 =	22	1 × 41 =	41
1 valve—equivalent length of pipe, ft.	1 × 1.32 =	1.3	1 × 2.1 =	2.1
Total length for figuring friction, ft.		448.1		483.9
Friction loss per 100 ft.	46.3		4.3	
Total discharge friction loss	$\frac{46.3 \times 448.1}{100}$ =	207.5	$\frac{4.3 \times 483.9}{100}$ =	20.8
Static discharge head, ft:				
Pump to tank		58		58
Tank pressure	10 × 2.31 =	23.1		23.1
Total discharge head, ft.		288.6		101.9
Suction lift				
Length of suction pipe, ft.		25		25
2 ells—equivalent length of pipe, ft.	2 × 10.2 =	20.4	2 × 10.2 =	20.4
Foot valve—equivalent length, ft.		0		0
Total length for figuring friction, ft.		45.4		45.4
Friction loss per 100 ft.	4.3		4.3	
Total suction friction loss, ft.	$\frac{4.3 \times 45.4}{100}$ =	2.0	$\frac{4.3 \times 45.4}{100}$ =	2.0
Static suction lift, ft.		4		4
Total suction lift, ft.		6.0		6.0
Total head				
Total discharge head		288.6		101.9
Total suction lift, ft.		6.0		6.0
Total head, ft.		294.6		107.9

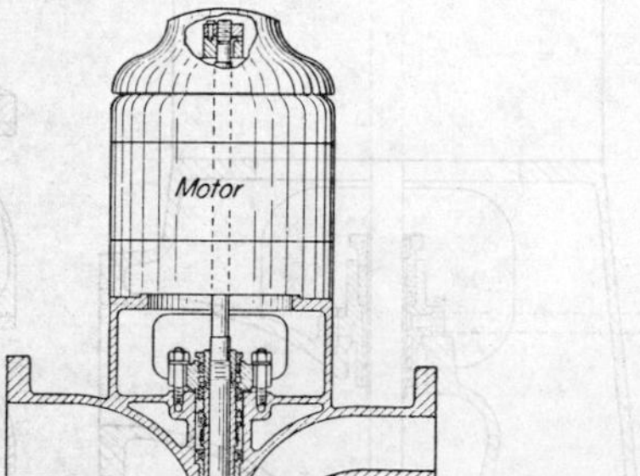

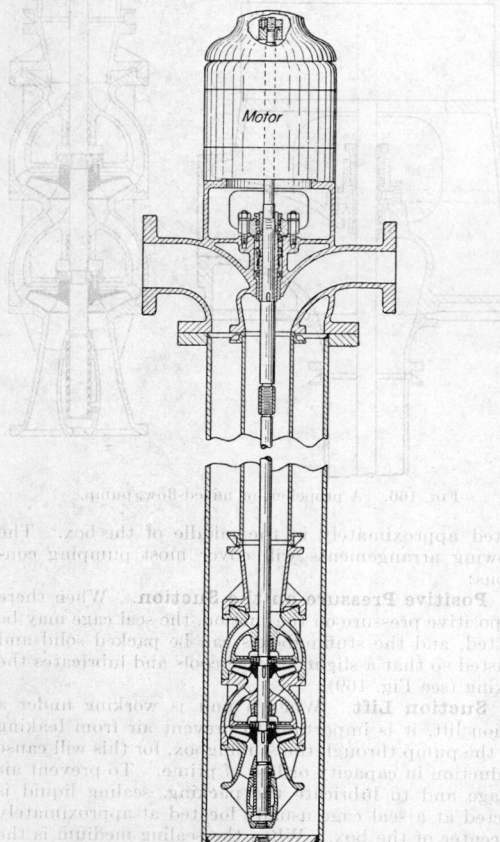

FIG. 167. A 2-stage mixed-flow pump with suction shell.

4. Gritty or Corrosive Liquids. When the liquid handled is gritty or corrosive, the stuffingboxes should be externally sealed with a cool clean liquid at approximately 10 lb./sq. in. above the suction pressure. The sealing liquid should be supplied at constant pressure

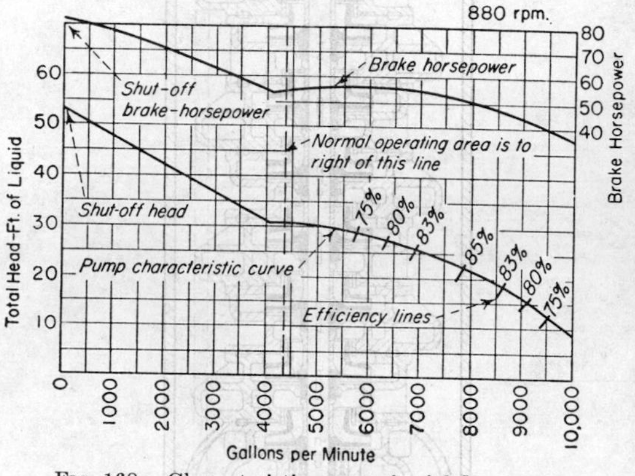

FIG. 168. Characteristic curve of axial-flow pump.

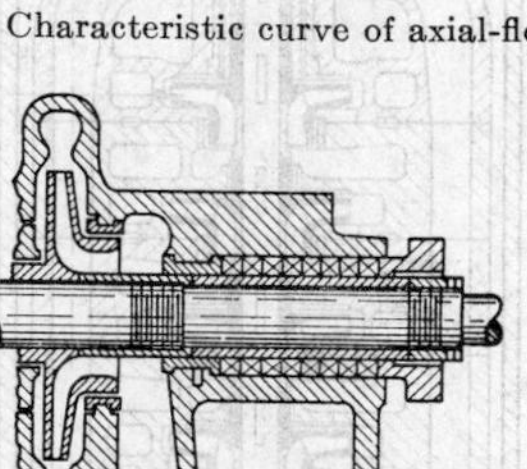

FIG. 169. Stuffing box packed solid without seal ring.

to prevent flexing action in the packing (see Fig. 171). When gasoline, kerosene, gas, oil, or similar hydrocarbons are being handled with nominal suction pressure, a stick-grease-type lubricator can be used to inject grease into the seal cage, instead of using the globe valve shown in Fig. 171.

5. High Temperature and High Pressure. *a. Circulating Box.* When the temperature of the liquid to be handled is high enough to warrant internal cooling of the packing, and the pressure falls within the limits for safe packing, a stuffingbox arrangement as shown in Fig. 172 can be used in order to provide cooling of the packing. Cool clean liquid enters the sealing cage at one connection, circulates through the seal cage, and leaves at another connection.

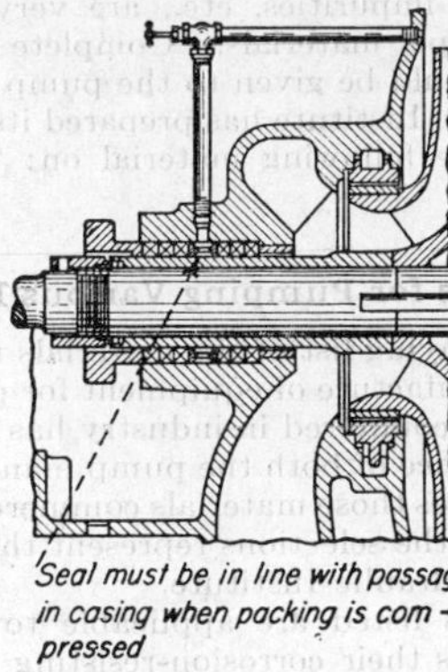

FIG. 170. Stuffing box with internal seal.

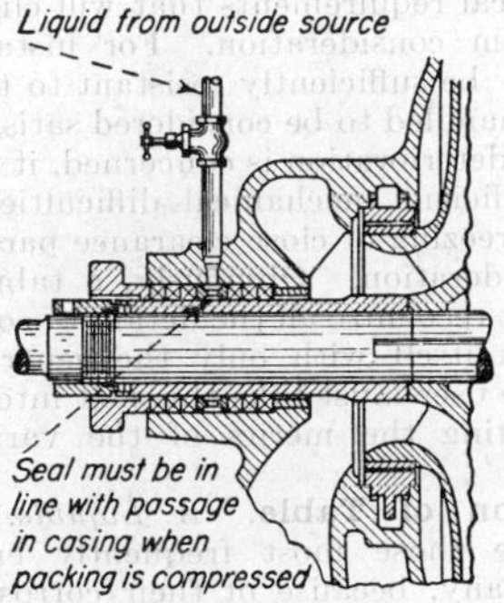

FIG. 171. Stuffing box with external seal system.

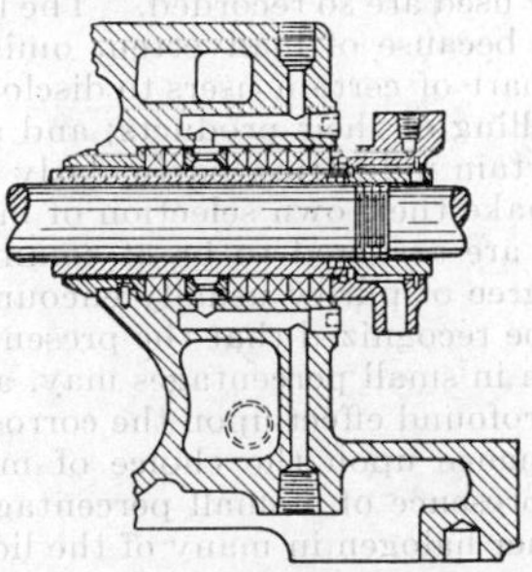

FIG. 172. Circulating-type stuffing box.

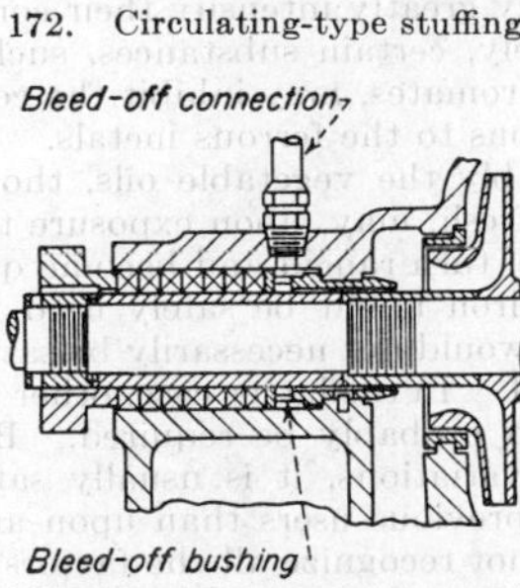

FIG. 173. Stuffing box with bleed-off bushing.

b. Bleed-off Box. It sometimes happens that the suction pressure must be reduced to an amount that can be safely packed in order to ensure satisfactory operation of the stuffingbox. This can be accomplished by locating a bleed-off bushing in the bottom of the stuffingbox, as shown in Fig. 173. The bleed-off bushing can usually be used in conjunction with any of the stuffingbox arrangements previously described. The tapped bleed-off connection must be piped to a low-pressure point in the system. The pressure should never be reduced to such an extent that it will be below the vapor pressure of the liquid at the pumping temperature.

SMOTHERING GLANDS

In the handling of liquids where the lubricating leakage through the packing might result in inflammable gas, noxious fumes, or steam, a smothering-type packing gland may be furnished (see Fig. 174). The water used

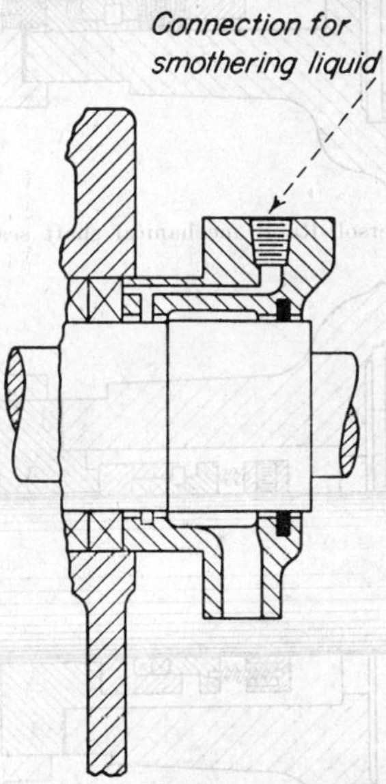

FIG. 174. Smothering-type stuffing-box gland.

for smothering may be taken from a cooling-water line through a flexible hose or dead soft copper tube.

MECHANICAL SEALS

In order to overcome packing difficulties, several types of mechanical sealing devices have been designed and have proved very satisfactory under severe operating conditions. Figures 175, 176, and 177 show mechanical shaft seals for use in chemical pumps.

Mechanical seals consist of a rotating and a stationary element having sealing faces that are brought into sliding contact under a pressure sufficient to prevent escape of the fluid being pumped.

Mechanical seals are generally used where the fluid handled has at least slight lubricating qualities and is free from abrasive and other foreign material that would tend to destroy the sealing faces. Satisfactory operation of a mechanical seal depends on maintaining perfect contact of the sealing faces at all times.

Double Mechanical Seals. Under a few conditions of service it is desirable to provide a double seal to prevent the liquid being handled by the pump from coming in contact with the sealing parts. Pumps handling corrosive and high-temperature liquids may require a double seal. In such cases, non-corrosive or low-temperature liquids are injected between the two seals at a pressure slightly higher than that obtained in the pump chamber next to the inner seal. For double seals, it is necessary to have additional equipment such as sealing liquid pump-tank-pressure regulator and piping to the stuffing boxes.

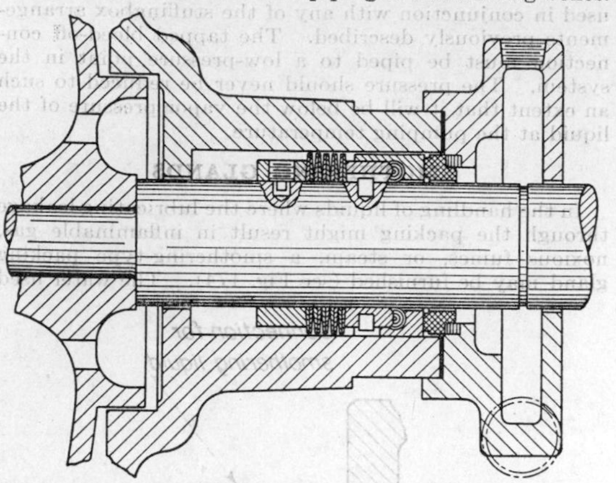

FIG. 175. Ingersoll-Rand mechanical shaft seal for chemical pumps.

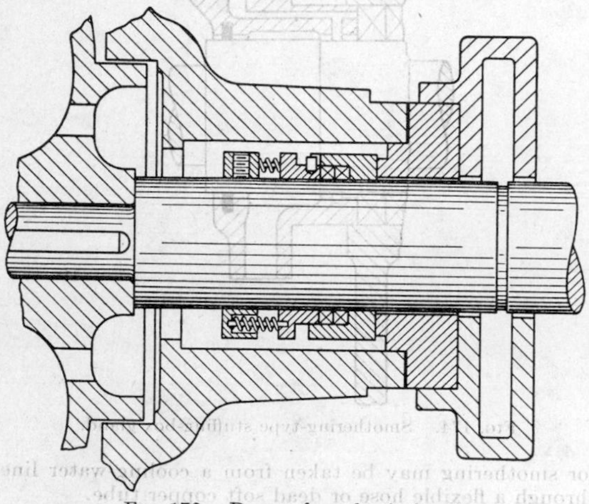

FIG. 176. Duraseal mechanical seal.

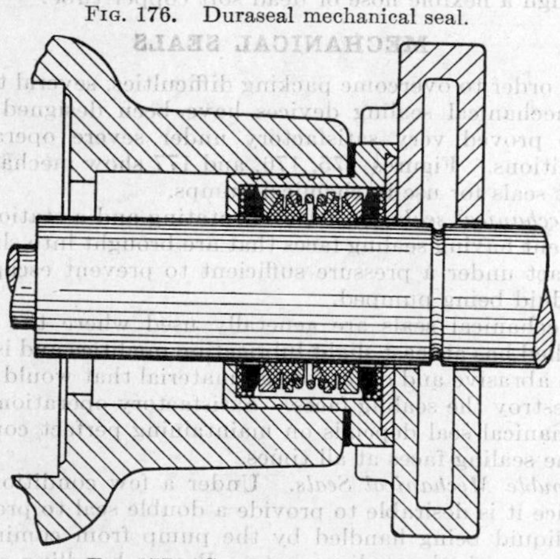

FIG. 177. Syntron mechanical seal.

PUMP MATERIALS

The proper selection of materials for a pump is extremely important, particularly in the chemical industry where a wide variety of corrosive and abrasive liquids are handled. It must be recognized that temperature, presence of abrasive matter, concentration of solution, purity, type of impurities, etc., are very important in determining proper materials. Complete information on these points should be given to the pump manufacturer.

The Hydraulic Institute has prepared its recommendations of and the following material on: "Materials for Pumping.

Materials for Pumping Various Liquids

The accompanying list of the materials most generally used in the manufacture of equipment for pumping many of the liquids encountered in industry has been compiled for the convenience of both the pump manufacturers and users. It includes those materials commercially available up to 1946, and the selections represent the best practice as known to Hydraulic Institute.

The materials listed are applicable to all classes of pumps insofar as their corrosion-resisting characteristics are concerned, but, in some cases, design features may impose physical requirements that will eliminate certain materials from consideration. For instance, although cast iron may be sufficiently resistant to the attack of a liquid being handled to be considered satisfactory insofar as structural deterioration is concerned, it may, however, introduce sufficient mechanical difficulties through the rusting and freezing of close clearance parts to eliminate it from consideration. Obviously, a tabulation of this nature cannot recognize all the details of construction but must concern itself with only the general problem of corrosion; the user must take this fact into consideration when evaluating the merits of the various materials suggested.

Explanation of Table. *A. Liquids.* The liquids tabulated are those most frequently encountered in industry. Many, because of their corrosive characteristics, necessitate the use of special materials, and those most generally used are so recorded. The list is probably not complete because of inadvertent omission; a reluctance on the part of certain users to disclose information as to the handling of their products; and a tendency on the part of certain industries, particularly the petroleum industry, to make their own selection of materials.

The liquids are assumed to be of commercial quality and of the degree of purity usually encountered. However, it must be recognized that the presence of a foreign substance even in small percentages may, and frequently does, have a profound effect upon the corrosiveness of the solution and hence upon the choice of materials. For instance, the presence of a small percentage of a soluble chloride or other halogen in many of the liquids included in the table may greatly intensify their corrosive properties. Conversely, certain substances, such as the chromates and dichromates, may inhibit the corrosive action of many solutions to the ferrous metals. Further, some liquids, noticeably the vegetable oils, though relatively inactive when fresh, may, upon exposure to heat and/or the atmosphere, turn rancid and become quite corrosive. Although cast iron might be safely used with such oils when sweet, it would not necessarily be satisfactory after they had soured. In the latter case, other more resistant materials would probably be required. Because of the effect of such variations, it is usually safer to rely on experiences of previous users than upon any generalization that may not recognize all the factors involved.

Where precise performance data are not available, corrosion tests, using specimens of selected materials suitably exposed to the liquid to be handled, may assist in determining the most satisfactory material or materials to be used. However, the practical limitation of such a test must be recognized. The difficulty of subjecting a single test specimen to the many variables that may exist in a system, such as temperature, velocity, impingement, abrasion, aeration, and galvanic action, any of which may have an important effect on the result, is considerable. Where actual service data are lacking, however, such tests should be encouraged, since much helpful information may be obtained through their use despite their limitations.

B. Condition. In some cases the satisfactory use of a particular material is restricted to a definite temperature and/or concentration range and, where this is known to occur, the limitations are so noted. Since the corrosion rate usually increases with temperature, the latter becomes an important factor in making a material selection. Where the space is left blank, it is assumed the materials listed are suitable over the ranges of concentration and temperature ordinarily encountered.

C. Chemical Symbol. The chemical symbols have been included where available, both as a matter of information and as a means of identification in the event that the name of the liquid is not fully descriptive.

D. Specific Gravity. These data are given where accurate information is available; unless otherwise specified, they apply at room temperature.

E. Materials Used. The materials listed are those most commonly used in the principal parts of the pump, the casing, and impellers. The trim items, such as rings, sleeves, and glands, may differ in composition to best suit some mechanical requirement, but the material so selected should be suitable for the environment encountered.

Because of the many variables that may affect the results, it is not possible in any generalization to say with certainty that any one material will best withstand the corrosive attack of a given liquid; hence, more than one type is usually included. Further, the order of listing does not indicate necessarily relative superiority, because certain factors predominating in one instance may be sufficiently overshadowed in others to reverse the arrangement.

Where standard fitted and/or all-iron pumps are listed, a cast-iron case is assumed. No attempt is made to specify a particular grade of iron, because the actual composition used may vary somewhat to suit each manufacturer's practice. Alloying elements may or may not be used, but unless the total content exceeds 3 per cent, the resulting iron is not considered as having special corrosion-resisting properties in the handling of liquids. Sound castings are, in general, more resistant to attack than spongy ones because of the lesser surface exposed to the liquid. Any alloy addition that promotes soundness is helpful.

Explanatory Remarks. Although the use of trade names has been purposely avoided, an attempt has been made to include those alloys commercially encountered. The listing of the higher priced materials is most evident where the conditions of corrosion and abrasion are severe. They are also included in those cases where even a slight contamination of the liquid, through introduction of metallic salts from the pump, particularly if the latter is not washed out during shutdown periods, may adversely affect the color or characteristics of the product or develop toxicity in the case of foodstuffs.

Some plants prefer to use the lower priced materials, where contamination of the product is not involved, and replace the pumps frequently; while others, with perhaps more stabilized processes, prefer to use the more resistant grades of materials and thus avoid frequent renewals. In some cases, the less resistant materials are used during a process development period while the more resistant materials are installed after the conditions have become standardized and the corrosion problems are recognized.

Many types of bronzes are in use, largely because of the tendency of each manufacturer to adjust the composition to suit his own design and production requirements. Freedom from porosity is usually a more important factor in determining the service life of a casting than the corrosion resistance of the particular composition in which it is made. Since the spread in the corrosion-resisting characteristics of the various bronzes at the velocities normally encountered in centrifugal pumps is not significant, the commercial types may be used more or less interchangeably. The few exceptions to this general statement do not warrant the confusion that would follow an attempt to list several types separately in the tabulation. Hence the bronzes are considered as a single group.

If the liquid being handled is an electrolyte, any combination of dissimilar metals that may tend to promote galvanic action should, where practical, be avoided if deterioration is to be held to a minimum. The rate of corrosion where such couples exist will depend upon such things as the nature of the electrolyte, the current concentration, temperature, velocity, and particularly the relative cathode-anode surface area. Although, as indicated, bronze fittings in a pump handling sea water tend to accelerate the corrosion of iron or steel, nevertheless large pumps so fitted are frequently utilized in such service for economic reasons. In order to reduce the galvanic reaction in a standard fitted pump in salt water and thus improve the life of the iron casing, special-alloy trims are sometimes used.

Cast iron, because of the multiplicity of galvanic couples set up by the graphite flakes and the metallic base, corrodes rapidly when in contact with sea water. Fortunately, the products of corrosion, principally iron oxides, mat with the graphite and, if not mechanically dislodged, remain in place so that the casting retains its original form. The process is known as graphitization. The graphitized layer, as it thickens, acts as a protective coating and progressively decelerates further attack. After a thickness of, say, $\frac{1}{8}$ in. is developed, further attack is slow.

The graphitized coating is relatively soft and, under high or eddying velocities, may be eroded away as rapidly as formed. Under such conditions, the deterioration of the iron is rapid. This explains why iron impellers and wearing rings fail more rapidly than the casing itself and why the former parts are usually made in bronze, as in the standard fitted pump. The bronze resists the attack, and the couple action with the graphitized surface is low.

In large units, where the velocities are low, all-iron pumps may, under favorable conditions, stand up sufficiently well in sea-water applications to justify their use economically. However, when replacement parts become necessary, the new parts, not graphitized, become anodic, and their deterioration may be rapid.

The handling of liquids at high temperatures requires special consideration as to both the types of materials used and the construction detail. The maximum temperature for gray iron authorized by the A.S.M.E. Code for Pressure Vessels is 450°F.; and, although some pumps are handling liquids in excess of this temperature, such a procedure, excepting in the case of internal parts, must be classed as contrary to the generally accepted practice. Carbon steel is normally employed for temperatures up to 750°F. where corrosion is not a serious factor. The use of alloys, incorporating such elements as molybdenum and chromium, becomes desirable for still higher temperatures.

Nomenclature. The standard material classification of pumps follows. Their listing in the material column is abbreviated through the use of the associated initials.

1. Standard fitted.......................... SF
2. All iron.................................. AI
3. All bronze................................ AB
4. Types 4, 5, 6, and 7...................... AA

To simplify recording, the symbol AA is used in those cases where types 4, 5, 6, and 7 would normally be listed. This does not necessarily mean, however, that all are equally effective in all environments. It merely means that each type has been satisfactorily applied in handling that liquid under some, possibly all, conditions.

Other materials, including corrosion-resisting steels, are listed by number in accordance with the following tabulation:

Type No.	Commercial designation	Carbon	Chromium	Nickel	Molybdenum	Remarks
1	A.I.S.I. 410	0.15 max.	11.5–13.5	....	....	Free machining type is 416
	A.C.I. CA14	0.14 max.	11–14	1.0 max.		
2	A.I.S.I. 442	0.35 max.	18–23			
	A.C.I. CB30	0.30 max.	18–22	2.0 max.		
3	A.I.S.I. 446	0.35 max.	23–27			
	A.C.I. CC35	0.35 max.	27–30	3.0 max.		
4	A.I.S.I. 304	0.08 max.	18–20	8–10	....	Free machining type is 303
	A.C.I. CF7	0.07 max.	18–20	8–10	....	Free machining type is CF7Se
5	A.I.S.I. 316	0.10 max.	16–18	10–14	1.75–2.5	
	A.C.I. CF7M	0.07 max.	18–20	8–10	1.5 –3.5	
6	....	0.07 max.	15–28	22–36	1.5 –4.0	Optional elements, Cu, W, Si, Mn, Ti, Cb
7	A series of non-ferrous alloys, of less than 20 per cent iron, containing nickel and chromium or molybdenum, or both, in major amounts, and copper, tungsten, silicon, and manganese in lesser percentages					
8	High silicon iron. 14.25 per cent silicon minimum					
9	Austenitic cast iron, total nickel, chromium, copper contents, 22 per cent minimum					
10	Monel metal					
11	Lead					
12	Non-metallic					
13	Nickel					
14	Steel					

A.I.S.I. = American Iron and Steel Institute.
A.C.I. = Alloy Casting Institute.

Type 6 covers a series of alloys developed by different manufacturers to augment the standard 4 and 5 types and to provide superior corrosion-resisting properties. Type 7, special alloys, covers a wide range of non-ferrous materials of less than 20 per cent iron, containing nickel and chromium or molybdenum, or both, in major amounts, and copper, tungsten, silicon, and manganese in smaller percentages. Since these are special materials for special services, their effective application requires close cooperation among the producer, pump manufacturer, and user, each giving due consideration to the local factors and economic aspects involved. It should be pointed out, however, that a pump of the type 6 material will not necessarily cost more than one in, say, type 4, because a manufacturer, having standardized on the more expensive material, may be able to sell it as a standard product at a lower price than if he constructed a single pump in a less costly material.

FRICTION LOSSES OF LIQUID FLOWING THROUGH PIPE

The loss of head or pressure due to friction, which is experienced when a liquid flows through a pipe, is called the friction head. A vast amount of research has been conducted to determine the amount of friction loss for different conditions. Section 5 contains basic information on friction losses. For convenience in handling pump problems, detailed formulas and friction-loss tables are given in this discussion. The loss of head due to the friction flowing in pipe can be determined for any liquid or gas by means of the Fanning equation derived from Bernoulli's theorem. This equation may be written thus:

$$h = \frac{fLv^2}{2gd}$$

The friction factor f in this equation has been determined experimentally in relation to a dimensionless ratio known as the Reynolds number. Reynolds number is written $R = Dv\rho/\mu$. A curve giving values of the friction factor f is given in Sec. 5. This curve may be used in connection with the formula above or with those below, which are in units widely used by engineers.

$$R = \frac{Dv\rho}{\mu} = \frac{3162q}{dk} = \frac{2214B}{dk}$$

$$h = \frac{fLv^2}{2gD} = \frac{0.03112fLq^2}{d^5} = \frac{0.0153fLB^2}{d^5}$$

Below a Reynolds number of about 2100, the character of flow changes from the usual turbulent type to viscous or streamline flow. Experiments have shown that the surface of the pipe has no influence on the friction factor when flow is of the streamline type. Therefore, for all kinds of pipe when the Reynolds number is less than 2100, a value for f of $64/R$ may be used.

Symbols used in the above formulas are as follows:

B = flow of liquid, bbl. (42 gal.)/hr.
d = inside diameter of circular pipe, in.
D = inside diameter of circular pipe, ft.
f = friction factor, dimensionless.
g = acceleration of gravity, ft./sec.^{2} (taken as 32.174 ft./sec.^{2} in making conversions).
h = head loss due to friction, ft. (equivalent to a column of liquid of a height, ft., giving the pressure necessary to overcome the friction loss).
k = kinematic viscosity, centistokes = z/s.
L = length of pipe line, including equivalent length for loss through fittings, ft. In long lines over about 4000 ft. where loss due to fittings is a small portion of the total, they can usually be neglected.
ρ = density at temperature and pressure at which fluid is flowing, lb./cu. ft.
q = flow of liquid, gal./min.
R = Reynolds number, dimensionless.
s = density, g./c.c. = sp. gr. referred to water at 39.2°F. as 1.
μ = absolute viscosity, lb./ft.-sec. = $z/1488$.
v = velocity of flow, ft./sec.
z = absolute viscosity, centipoises = 1488 μ.

Tables based on these formulas will be found on pp. 1430*ff*. They may be used for any true liquid, but not for suspensions or plastics. For types of pipe other than smooth, clean steel, copper, or brass pipe on which the tables are based, the original formula, together with friction factors from the chart in Sec. 5, should be used.

Materials for Pumping Various Liquids*

Liquid	Condition	Chemical symbol	Specific gravity	Materials commonly used
Acetaldehyde		CH_3CHO	0.78	AI
Acetate solvents				AI, SF, AB, AA
Acetone		CH_3COCH_3	0.79	AI, SF
Acetic anhydride		$(CH_3CO)_2O$	1.08	AA, 8
Acid:				
Acetic	Conc. cold	CH_3COOH	1.055	AA, 8
Acetic	Dil. cold			AB, AA, 8
Acetic	Conc. boiling			5, 6, 7, 8
Acetic	Dil. boiling			5, 6, 7, 8
Arsenic		As_2O_5		AA, 8
Benzoic		C_6H_5COOH		AA
Boric	Aqueous sol.	H_3BO_3		AB, AA, 8
Butyric	Conc.	$CH_3(CH_2)_2CO_2H$	0.96	AA
Carbolic	Conc. (m.p. 106°F.)	C_6H_5OH	1.071	AI, AA
Carbolic	Aqueous sol.			SF, AA
Carbonic	Aqueous sol.	$CO_2 + H_2O$		AB
Chromic	Aqueous sol.	$CrO_3 + H_2O$		AA, 8
Citric	Aqueous sol.	$C_6H_8O_7H_2O$		AB, AA, 8
Fatty (oleic, palmitic, stearic, etc.)				AB, AA
Formic		$HCOOH$	1.2	5, 6, 7
Fruit				AB, AA, 10
Hydrochloric	Com'l. conc.	HCl	1.16 (20°Be.)	7, 8, 12
Hydrochloric	Dil. cold			6, 7, 8, 10, 12, 13
Hydrochloric	Dil. hot			7, 8, 12
Hydrocyanic		HCN	0.70	AI, AA
Hydrofluoric	Anhydrous, with hydrocarbon	$HF + HC$		10, 14
Hydrofluoric	Aqueous sol.	$HF + H_2O$		AB, 10
Hydrofluosilicic		H_2SiF_6		AB, 10
Lactic		$CH_3CHOHCOOH$	1.249	AB, AA, 8
Mine water				AB, AA
Mixed				AI, AA, 8, 14
Muriatic (see Acid, Hydrochloric)				
Naphthenic				AI, AA, 1
Nitric	Conc. boiling	HNO_3	1.41	2, 3, 6, 8
Nitric	Dilute			1, 2, 3, 4, 5, 6, 8
Ortho-phosphoric		H_3PO_4	1.36–1.4	5, 6, 7
Oxalic	Cold	$C_2H_2O_42H_2O$		AA, 8
Oxalic	Hot	$C_2H_2O_42H_2O$		6, 7, 8
Picric		$(NO_2)_3C_6H_2OH$		AA, 8
Pyrogallic		$C_6H_3(OH)_3$		AA
Pyroligneous		$H_2C_2H_3O_2$	1.018–1.03	AB, AA
Sulfuric	$>$ 77% cold	H_2SO_4	1.69–1.835	AI, 6, 7, 8
Sulfuric	65–93% $>$ 175°F.			7, 8
Sulfuric	65–93% $<$ 175°F.			6, 7, 8
Sulfuric	10–65%			6, 7, 8, 11
Sulfuric	$<$ 10%			AB, 6, 7, 8, 10, 11
Sulfuric (oleum)	Fuming	$H_2SO_4 + SO_3$		6, 7, 14
Sulfurous		H_2SO_3		AB, AA, 11
Tannic		$C_{14}H_{10}O_9$		AB, AA, 10
Tartaric	Aqueous sol.	$C_4H_6O_6$		AB, AA, 10
Alcohols				AB, SF
Alum (see Aluminum sulfate and Potash alum)				
Aluminum sulfate	Aqueous sol.	$Al_2(SO_4)_3$		6, 7, 8, 10, 11
Ammonia, aqua		NH_4OH		AI
Ammonium bicarbonate	Aqueous sol.	NH_4HCO_3		AI
Ammonium chloride	Aqueous sol.	NH_4Cl		5, 6, 7, 8, 10
Ammonium nitrate	Aqueous sol.	NH_4NO_3		AI, AA, 10
Ammonium phosphate	Aqueous sol.	$(NH_4)_2HPO_4$		AI, AA, 10
Ammonium sulfate	Aqueous sol.	$(NH_4)_2SO_4$		AI, AA
Ammonium sulfate	With H_2SO_4			AB, 5, 6, 7, 8, 11
Aniline		$C_2H_5NH_2$	1.022	AI, SF
Aniline hydrochloride	Aqueous sol.	$C_6H_5NH_2HCl$		7, 8, 12
Asphalt	Hot		0.98–1.4	AI, 1
Barium chloride	Aqueous sol.	$BaCl_2$		AI, AA
Barium nitrate	Aqueous sol.	$Ba(NO_3)_2$		AI, AA
Beer				AB, 4
Beer wort				AB, 4
Beet juice				AB, 4
Beet pulp				AB, SF, AA
Benzene (see Benzol)				
Benzine (see Petroleum ether)				
Benzol		C_6H_6	0.88	AI, SF
Bichloride of mercury (see Mercuric chloride)				
Black liquor (see Liquors, Pulp mill)				
Bleach solutions (see type)				
Blood				SF, AB
Boiler feed water (see Water, boiler feed)				
Brine:				
Calcium chloride	pH $>$ 8	$CaCl_2$		AI
Calcium chloride	pH $<$ 8			AB, 6, 7, 9, 10
Calcium and mangesium chlorides	Aqueous sol.			AB, 6, 7, 9, 10
Calcium and sodium chloride	Aqueous sol.			AB, 6, 7, 9, 10
Sodium chloride	Under 3% salt, cold	$NaCl$	1.02	AI, AB, 9
Sodium chloride	Over 3% salt, cold		1.02–1.20	AB, AA, 9, 10
Sodium chloride	Over 3% salt, hot			5, 6, 7, 8, 10
Sea water			1.03	AI, SF, AB
Butane		$CH_3(CH_2)_2CH_3$	0.60 at 32°F.	AI, SF, 14
Calcium bisulfite	Paper mill	$Ca(HSO_3)_2$	1.06	5, 6, 7, 11
Calcium chlorate	Aqueous sol.	$Ca(ClO_3)_2 2H_2O$		6, 7, 8, 12
Calcium hypochlorite		$Ca(OCl)_2$		AI, 6, 7, 8
Calcium magnesium chloride (see Brines)				

* Courtesy of Hydraulic Institute, 90 West Street, New York.

Materials for Pumping Various Liquids* —(*Continued*)

Liquid	Condition	Chemical symbol	Specific gravity	Materials commonly used
Cane juice				SF, AB, 9
Carbon bisulfide		CS_2		AI
Carbonate of soda (see Soda ash)				
Carbon tetrachloride	Anhydrous	CCl_4	1.58	AI, SF
Carbon tetrachloride	Plus water			AB, 4
Catsup				AB, AA
Caustic potash (see Potassium hydroxide)				
Caustic soda (see Sodium hydroxide)				
Cellulose acetate				5, 6, 7
Chlorate of lime (see Calcium chlorate)				
Chloride of lime (see Calcium hypochlorite)				
Chlorine water	Depending on conc.			5, 6, 7, 8, 11, 12
Chlorobenzene		C_6H_5Cl	1.1	SF, AB, 4
Chloroform		$CHCl_3$	1.5	AB, AA, 10
Chrome alum	Aqueous sol.	$CrK(SO_4)_2 12H_2O$		6, 7, 8
Condensate (see Water, Distilled)				
Copper ammonium acetate	Aqueous sol.			AI, AA
Copper chloride (cupric)	Aqueous sol.	$CuCl_2$		7, 8, 12
Copper nitrate		$Cu(NO_3)_2$		AA
Copper sulfate, blue vitriol	Aqueous sol.	$CuSO_4$		AA, 8, 11
Copperas, green (see Ferrous sulfate)				
Creosote (see Oil, Creosote)				
Cresol, meta		$CH_3C_6H_4OH$	1.04	AI, 6, 7
Cyanide (see Sodium cyanide and Potassium cyanide)				
Cyanogen	In water	C_2N_2(gas)		AI
Diphenyl		$C_6H_5C_6H_5$		AI, 14
Enamel				AI
Ethanol (see Alcohols)				
Ethylene chloride (dichloride)	Cold	CH_2ClCH_2Cl	1.28	AB, AA, 10
Ferric chloride	Aqueous sol.	$FeCl_3$		7, 8, 12
Ferric sulfate	Aqueous sol.	$Fe_2(SO_4)_3$		AA, 8
Ferrous chloride	Cold, aqueous	$FeCl_2$		7, 8, 12
Ferrous sulfate (green copperas)	Aqueous sol.	$FeSO_4$		5, 6, 7, 8, 10, 11
Formaldehyde		HCHO	1.075–1.081	AB, AA
Fruit juices				AB, AA, 10
Furfural		C_4H_3OCHO	1.16	AI, AB, AA
Gasoline			0.68–0.75	AI, SF
Glauber's salt (see Sodium sulfate)				
Glucose				SF, AB
Glue	Hot			AI, SF
Glue sizing				AB
Glycerol (glycerin)		$C_3H_5(OH)_3$	1.262	AI, SF, AB
Green liquor (see Liquors—pulp mill)				
Heptane		C_7H_{16}	0.69	AI, SF
Hydrogen peroxide	Aqueous sol.	H_2O_2		AA
Hydrogen sulfide	Aqueous sol.	H_2S		AA
Hydrosulfite of soda (see Sodium hydrosulfite)				
Hyposulfite of soda (see Sodium thiosulfate)				
Kaolin slip	Suspension in water			AI, 14
Kaolin slip	Suspension in acid			6, 7, 8
Kerosene (see Oil, Kerosene)				
Lard	Hot			AI, SF
Lead	Molten			AI, 14
Lead acetate (sugar of lead)	Aqueous sol.	$Pb(C_2H_3O_2)_2 3H_2O$		5, 6, 7, 10
Lime water (milk of lime)		$Ca(OH)_2$		AI
Liquors—pulp mill:				
Black				AI, 5, 6, 7, 9, 10, 14
Green				AI, 5, 6, 7, 9, 10, 14
Pink				AI, 5, 6, 7, 9, 10, 14
Sulfite				5, 6, 7, 11
White				AI, 5, 6, 7, 9, 10, 14
Lithium chloride	Aqueous sol.	LiCl		AI
Lye, caustic (see Potassium and Sodium hydroxide)				
Magnesium chloride	Aqueous sol.	$MgCl_2$		6, 7, 8, 12
Magnesium sulfate (Epsom salts)	Aqueous sol.	$MgSO_4$		AI, AA
Manganese chloride	Aqueous sol.	$MnCl_2$		AB, AA, 8
Manganous sulfate	Aqueous sol.	$MnSO_4 4H_2O$		AI, AB, AA
Mash				SF, AB, 4
Mercuric chloride	Very dilute aqueous sol.	$HgCl_2$		5, 6, 7, 8
Mercuric chloride	Com'l. conc. aqueous sol.	$HgCl_2$		7, 8, 12
Mercuric sulfate	In H_2SO_4	$HgSO_4$		6, 7, 8, 12
Mercurous sulfate	In H_2SO_4	Hg_2SO_4		6, 7, 8, 12
Methyl chloride		CH_3Cl	0.92	AI
Methylene chloride		CH_2Cl_2	1.26	AI, 4
Milk			1.028–1.035	4
Milk of lime (see Lime water)				
Mine water (see Acid, Mine water)				
Miscella	(20% Soybean oil and solvent)		0.75	AI
Molasses				SF, AB
Mustard				AB, AA, 8
Naphtha			0.78–0.88	AI, SF
Naphtha, crude			0.92–0.95	AI, SF
Nicotine sulfate		$(C_{10}H_{14}N_2)_2H_2SO_4$		6, 7, 8, 10
Nitre (see Potassium nitrate)				
Nitre cake (see Sodium bisulfate)				
Nitroethane		$CH_3CH_2NO_2$	1.041	AI, SF
Nitromethane		CH_3NO_2	1.139	AI, SF
Oil:				
Coal tar				AI, SF, AA
Coconut			0.905	AI, SF, AB, AA, 10
Creosote			1.04–1.10	AI, SF

* Courtesy of Hydraulic Institute, 90 West Street, New York.

Materials for Pumping Various Liquids*—(*Continued*)

Liquid	Condition	Chemical symbol	Specific gravity	Materials commonly used
Oil: (*Cont.*):				
Crude	Cold			AI, SF
Crude	Hot			14
Essential				AI, SF, AB
Fuel				AI, SF
Kerosene				AI, SF
Linseed			0.94	AI, SF, AB, AA, 10
Lubricating				AI, SF
Mineral				AI, SF
Olive			0.90	AI, SF
Palm			0.895	AI, SF, AB, AA, 10
Quenching			0.912	AI, SF
Rapeseed			0.92	AB, AA, 10
Soybean				AI, SF, AB, AA, 10
Turpentine			0.87	AI, SF
Paraffin	Hot			AI, SF
Perhydrol (see Hydrogen peroxide)				
Peroxide of hydrogen (see Hydrogen peroxide)				
Petroleum ether				AI, SF
Phenol (see Acid, carbolic)				
Photographic developers				AA, 12
Pink liquor (see Liquor—pulp mill)				
Plating solutions	(Varied and complicated; consult pump manufacturers)			
Potash	Plant liquor			AB, AA, 9, 10
Potash alum	Aqueous sol.	$Al_2(SO_4)_3K_2SO_424H_2O$		AB, 5, 6, 7, 8, 9, 10
Potassium bichromate	Aqueous sol.	$K_2Cr_2O_7$		AI
Potassium carbonate	Aqueous sol.	K_2CO_3		AI
Potassium chlorate	Aqueous sol.	$KClO_3$		AA, 8
Potassium chloride	Aqueous sol.	KCl		AB, AA, 10
Potassium cyanide	Aqueous sol.	KCN		AI
Potassium hydroxide	Aqueous sol.	KOH		AI, 1, AA, 9, 10, 13
Potassium nitrate	Aqueous sol.	KNO_3		AI, AA, 1
Potassium sulfate	Aqueous sol.	K_2SO_4		AB, AA
Propane		$CH_3CH_2CH_3$	0.585 at 48°F.	AI, SF, 14
Pyridine		$CH(CHCH)_2N$	0.975	AI
Pyridine sulfate				6, 8, 11
Rhigolene				SF
Rosin (colophony)	Paper mill			AI
Sal ammoniac (see Ammonium chloride)				
Salt cake	Aqueous sol.	Na_2SO_4 + impurities		AB, AA, 8
Salt water (see Brines)				
Sea water (see Brines)				
Sewage				AI, SF, AB
Shellac				AB
Silver nitrate	Aqueous sol.	$AgNO_3$		AA, 8
Sirup (see Sugar)				
Slop, brewery				AI, SF, AB
Slop, distiller's			1.05	AB, AA
Soap liquor				AI
Soda ash	Cold	Na_2CO_3		AI
Soda ash	Hot			AA, 9, 10
Sodium bicarbonate	Aqueous sol.	$NaHCO_3$		AI, AA, 9
Sodium bisulfate	Aqueous sol.	$NaHSO_4$		6, 7, 8, 11
Sodium carbonate (see Soda ash)				
Sodium chlorate	Aqueous sol.	$NaClO_3$		AA, 8
Sodium chloride (see Brines)				
Sodium cyanide	Aqueous sol.	NaCN		AI
Sodium hydroxide	Aqueous sol.	NaOH		AI, AA, 1, 9, 10, 13
Sodium hydrosulfite	Aqueous sol.	$Na_2S_2O_4.2H_2O$		AA, 11
Sodium hypochlorite		NaOCl		6, 7, 8, 12
Sodium hyposulfite (see Sodium thiosulfate)				
Sodium meta silicate				AI
Sodium nitrate	Aqueous sol.	$NaNO_3$		AI, AA, 1
Sodium phosphate:				
Monobasic	Aqueous sol.	NaH_2PO_4		AB, AA
Dibasic	Aqueous sol.	Na_2HPO_4		AI, AB, AA
Tribasic	Aqueous sol.	Na_3PO_4		AI
Meta	Aqueous sol.	$NaPO_3$		AB, AA
Hexameta	Aqueous sol.	$(NaPO_3)_6$		AA
Sodium plumbite	Aqueous sol.			AI
Sodium sulfate	Aqueous sol.	Na_2SO_4		AB, AA
Sodium sulfide	Aqueous sol.	Na_2S		AI, AA, 11
Sodium sulfite	Aqueous sol.	Na_2SO_3		AB, AA, 11
Sodium thiosulfate	Aqueous sol.	$Na_2S_2O_35H_2O$		AA, 12
Stannic chloride	Aqueous sol.	$SnCl_4$		7, 8, 11, 12
Stannous chloride	Aqueous sol.	$SnCl_2$		7, 8, 11, 12
Starch		$(C_6H_{10}O_5)x$		SF, AB
Strontium nitrate	Aqueous sol.	$Sr(NO_3)_2$		AI, 4
Sugar	Aqueous sol.			AB, AA, 9
Sulfite liquors (see Liquors—pulp mill)				
Sulfur	In water	S		AI, AB, 9
Sulfur	Molten	S		AI
Sulfur chloride	Cold	S_2Cl_2		AI, 11
Tallow	Hot		0.895	AI
Tanning liquors				AB, AA, 8, 10
Tar	Hot			AI, 14
Tar and ammonia	In water			AI
Tetrachloride of tin (see Stannic chloride)				
Tetraethyl lead		$Pb(C_2H_5)_4$	1.65	AI, SF
Toluene (toluol)		$CH_3C_6H_5$	0.86	AI, SF

* Courtesy of Hydraulic Institute, 90 West Street, New York.

Materials for Pumping Various Liquids—(*Concluded*)

Liquid	Condition	Chemical symbol	Specific gravity	Materials commonly used
Trichloroethylene		C_2HCl_3	1.47	AI, SF, AB, 4
Urine				AB, AA
Varnish				AI, SF, AB, 4, 10
Vegetable juices				AB, AA, 10
Vinegar				AB, AA, 8
Vitriol, blue (see Copper sulfate)				
Vitriol, green (see Ferrous sulfate)				
Vitriol, oil of (see Acid, sulfuric)				
Vitriol, white (see Zinc sulfate)				
Water, boiler feed High make-up	Not evaporated pH > 8.5 pH < 8.5		1.00	AI SF
Water, boiler feed Low make-up	Evaporated, any pH		1.00	5% Cr, 1, 4, 10
Water, distilled	High purity Condensate		1.00	AB, 4 SF, AB
Water, fresh			1.00	SF
Water, mine (see Acid, Mine water)				
Water, salt and sea (see Brine)				
Whisky				AB, 4
White liquor (see Liquors—pulp mill)				
White water	Paper mill			AI, SF, AB
Wine				AB, 4
Wood pulp (stock)				AI, SF, AB
Wood vinegar (see Acid, Pyroligneous)				
Wort (see Beer wort)				
Xylol (xylene)		$C_6H_4(CH_3)_2$	0.87	AI, SF, AA
Yeast				SF, AB
Zinc chloride	Aqueous sol.	$ZnCl_2$		5, 6, 7, 8
Zinc sulfate	Aqueous sol.	$ZnSO_4$		AB, 5, 6, 7

* Courtesy of Hydraulic Institute, 90 West Street, New York.

RECIPROCATING PUMPS

There are two general classes of piston or reciprocating pumps: steam pumps and power pumps. In general the action of the liquid-transferring parts of these pumps is the same, a cylindrical piston, plunger, or bucket being caused to pass back and forth in a cylinder. The device is equipped with valves for inlet and discharge of the liquid being pumped, and the operation of these valves is related in a definite manner to the motions of the piston.

In considering the operation of a reciprocating pump, several efficiencies must be taken into account. These are volumetric efficiency, hydraulic efficiency, indicated efficiency, and mechanical efficiency.

Volumetric efficiency is the relation of the water actually pumped to that which theoretically should be moved on a basis of the piston displacement. It indicates the percentage loss and, when stated as $1 - \text{effy.}_{\text{vol.}}$, is called the slip. In good practice, slip should not be over 5 per cent. In new pumps, or those kept in good condition, it will be as low as 1 per cent.

Hydraulic efficiency is the ratio of the actual head pumped to the theoretical head and is expressed by the equation

$$\text{Efficiency}_{\text{hydraulic}} = \frac{H}{H + \text{hydraulic losses}}$$

The hydraulic losses are the losses in head in the suction and discharge lines. In the suction line, these consist of (1) velocity head, (2) entrance head, (3) friction in suction pipe, (4) losses in bends, and (5) losses in suction valves. The loss in the discharge line consists of (6) loss in discharge valves, (7) velocity head, and (8) friction in discharge pipe.

Indicated efficiency is the relation of the horsepower required to move the water actually pumped against the total head to the horsepower calculated from the indicator card of the water end.

Mechanical efficiency is the relation of the indicated water horsepower of the pump to the indicated steam horsepower. It shows the losses of power transmission incurred in operating the pump. This efficiency varies

Friction Losses in Pipe; 1 In. (1.049 In. Inside Diam.). Loss, Ft. Liquid/1000 Ft. Pipe*

Figures for clean steel pipe, except those in italics, which are for any pipe

Flow		Kinematic viscosity, centistokes									
U. S. gal./min.	Bbl./hr. (42 gal.)	0.6	1.1	2.1	2.7	4.3	7.4	10.3	13.1	15.7	20.6
		Approx. S.S.U. viscosity									
			31.5	33	35	40	50	60	70	80	100
0.5	0.71	0.07	*0.28*	*0.55*	*0.70*	*1.12*	*1.93*	*2.68*	*3.41*	*4.08*	*5.35*
1	1.4	1.00	1.13	*1.09*	*1.41*	*2.24*	*3.86*	*5.36*	*6.82*	*8.16*	*10.7*
2	2.9	3.60	4.02	4.70	4.7	*4.48*	*7.72*	*10.7*	*13.6*	*16.3*	*21.4*
3	4.3	7.50	8.16	9.48	10.2	10.6	*11.6*	*16.1*	*20.5*	*24.5*	*32.1*
4	5.7	12.5	14.1	15.7	16.5	18.1	*15.4*	*21.5*	*27.3*	*32.6*	*42.8*
5	7.1	18.9	21.4	23.2	24.8	26.5	*19.3*	*26.8*	*34.1*	*40.8*	*53.5*
6	8.6	25.7	29.1	32.6	34.5	37.8	41.3	*32.2*	*40.9*	*49.0*	*64.2*
7	10	34.9	38.5	43.3	45.6	49.5	55.3	*37.5*	*47.7*	*57.2*	*74.9*
8	11.4	44.5	50.2	55.2	58.3	63.5	70.3	74.2	*54.5*	*65.2*	*85.6*
9	12.9	55.6	61.6	68.6	70.7	78.6	87.3	92.0	*61.4*	*73.4*	*96.3*
10	14.3	69.0	74.8	83.6	87.4	95.0	106	111	118	*81.6*	*107*
12	17.1	94.6	103	117	121	133	144	158	166	170	*129*
14	20.0	127	138	154	161	175	194	208	218	225	*150*
16	22.8	163	176	196	205	223	247	264	277	285	302
18	25.7	203	220	242	254	277	305	326	340	350	370
20	28.6	247	265	294	308	335	370	392	412	423	448
25	35.7	374	410	442	463	503	552	597	619	638	676
30	42.9	530	575	618	644	698	775	820	865	886	936
35	50.0	703	755	822	858	925					

Loss in lb./sq. in. = 0.433 (sp. gr.) (figures from table).

* These friction-loss tables are from "Cameron Hydraulic Data," courtesy of Ingersoll-Rand Co.

Friction Losses in Pipe; 1 In. (1.049 In. Inside Diam.). Loss, Ft. Liquid/1000 Ft. Pipe—(*Concluded*)

Flow		Kinematic viscosity, centistokes									
		26.4	32.0	43.2	65.0	108.4	162.3	216.5	325	435	650
U. S. gal./min.	Bbl./hr. (42 gal.)	Approx. S.S.U. viscosity									
		125	150	200	300	500	750	1000	1500	2000	3000
0.1	0.14	*1.37*	*1.66*	*2.25*	*3.38*	*5.65*	*8.45*	*11.3*	*16.9*	*22.6*	*33.8*
.3	.43	*4.12*	*4.98*	*6.75*	*10.2*	*17.0*	*25.3*	*33.8*	*50.7*	*67.8*	*102*
.5	.71	*6.86*	*8.32*	*11.3*	*16.9*	*28.3*	*42.3*	*56.4*	*85*	*113*	*169*
1	1.4	*13.7*	*16.6*	*22.5*	*33.8*	*56.5*	*84.5*	*113*	*169*	*226*	*338*
2	2.9	*27.5*	*33.2*	*45.0*	*67.6*	*113*	*169*	*226*	*338*	*452*	*676*
3	4.3	*41.2*	*49.8*	*67.5*	*102*	*170*	*253*	*338*	*507*	*678*	
4	5.7	*55.0*	*66.5*	*90.0*	*136*	*226*	*338*	*452*	*677*	*904*	
5	7.1	*68.7*	*83.2*	*113*	*169*	*283*	*423*	*564*	*846*		
6	8.6	*82.4*	*99.7*	*135*	*203*	*339*	*507*	*677*			
7	10	*96.2*	*117*	*158*	*237*	*395*	*591*	*790*			
8	11.4	*110*	*133*	*180*	*271*	*452*	*676*	*903*			
9	12.9	*124*	*150*	*203*	*303*	*508*	*760*				
10	14.3	*137*	*167*	*225*	*338*	*565*	*845*				
12	17.1	*165*	*200*	*270*	*406*	*678*					
14	20.0	*192*	*233*	*315*	*474*	*792*					
16	22.8	*220*	*266*	*360*	*541*	*904*					
18	25.7	*248*	*299*	*405*	*609*						
20	28.6	470	*332*	*450*	*677*						

Friction Losses in Pipe; 1½ In. (1.610 In. Inside Diam.). Loss, Ft. Liquid/1000 Ft. Pipe*

Figures for clean steel pipe, except those in italics, which are for any pipe

Flow		Kinematic viscosity, centistokes									
		0.6	1.1	2.1	2.7	4.3	7.4	10.3	13.1	15.7	20.6
U. S. gal./min.	Bbl./hr. (42 gal.)	Approx. S.S.U. viscosity									
			31.5	33	35	40	50	60	70	80	100
1	1.4	0.13	*0.10*	*0.20*	*0.25*	*0.41*	*0.69*	*0.97*	*1.23*	*1.47*	*1.93*
2	2.9	.45	.50	*.39*	*.51*	*.83*	*1.39*	*1.93*	*2.46*	*2.94*	*3.86*
3	4.3	.92	1.04	1.19	1.25	*1.24*	*2.08*	*2.89*	*3.68*	*4.41*	*5.79*
4	5.7	1.55	1.75	1.99	2.10	*1.65*	*2.78*	*3.86*	*4.91*	*5.88*	*7.72*
5	7.1	2.33	2.61	2.97	3.11	*2.06*	*3.47*	*4.82*	*6.14*	*7.35*	*9.65*
6	8.6	3.24	3.54	4.12	4.30	4.63	*4.17*	*5.79*	*7.37*	*8.82*	*11.6*
8	11.4	5.45	6.12	6.92	7.23	7.97	8.85	*7.72*	*9.83*	*11.8*	*15.5*
10	14.3	8.17	9.10	10.4	10.8	12.0	13.5	*9.65*	*12.3*	*14.7*	*19.3*
12	17.1	11.4	12.6	14.4	15.1	16.6	18.6	19.9	*14.7*	*17.6*	*23.2*
15	21.4	17.1	18.9	21.5	22.5	24.8	27.7	29.4	31.1	*21.0*	*29.0*
20	28.6	29.0	32.0	35.8	37.8	41.4	46.1	49.8	52.0	54.0	*38.6*
25	35.7	43.8	48.2	54.0	57.0	62.0	69.0	74.2	77.8	80.7	85.0
30	42.9	62.3	66.3	75.7	78.8	87.0	96.5	103	107	112	117
40	57.1	105	114	127	133	146	162	173	182	188	198
50	71.4	161	173	190	200	216	242	258	270	280	296
60	85.7	224	244	266	278	302	336	358	374	389	410
70	100	300	327	356	372	400	445	473	496	513	541
80	114	386	420	457	472	508	563	604	633	655	692
90	129	484	522	568	588	635	698	745	783	810	847
100	143	593	633	690	715	772	850	898	944	978	

Flow		Kinematic viscosity, centistokes									
		26.4	32.0	43.2	65.0	108.4	162.3	216.5	325	435	650
U. S. gal./min.	Bbl./hr. (42 gal.)	Approx. S.S.U. viscosity									
		125	150	200	300	500	750	1000	1500	2000	3000
1	1.4	*2.47*	*3.00*	*4.14*	*6.09*	*10.2*	*15.2*	*20.3*	*30.4*	*40.8*	*60.9*
2	2.9	*4.95*	*6.00*	*8.28*	*12.2*	*20.3*	*30.4*	*40.6*	*60.8*	*81.5*	*122*
3	4.3	*7.42*	*9.00*	*12.4*	*18.3*	*30.4*	*45.6*	*60.9*	*91.3*	*122*	*183*
4	5.7	*9.90*	*12.0*	*16.6*	*24.4*	*40.6*	*60.8*	*81.2*	*122*	*163*	*244*
5	7.1	*12.4*	*15.0*	*20.7*	*30.4*	*50.7*	*76.0*	*102*	*152*	*204*	*304*
6	8.6	*14.9*	*18.0*	*24.8*	*36.5*	*60.8*	*91.2*	*122*	*183*	*244*	*365*
8	11.4	*19.8*	*24.0*	*33.1*	*48.7*	*81.2*	*122*	*163*	*243*	*326*	*487*
10	14.3	*24.7*	*30.0*	*41.4*	*60.9*	*102*	*152*	*203*	*304*	*408*	*609*
12	17.1	*29.7*	*36.0*	*49.7*	*73.2*	*122*	*182*	*244*	*365*	*490*	*732*
15	21.4	*37.1*	*45.0*	*62.2*	*91.4*	*152*	*228*	*304*	*457*	*612*	*914*
20	28.6	*49.5*	*60.0*	*82.8*	*122*	*203*	*304*	*406*	*608*	*815*	
25	35.7	*61.9*	*75.0*	*103*	*152*	*254*	*380*	*507*	*760*		
30	42.9	124	*90.0*	*124*	*183*	*304*	*456*	*609*	*913*		
40	57.1	208	216	*166*	*244*	*406*	*608*	*812*			
50	71.4	310	324	342	*304*	*507*	*760*				

Loss in lb./sq. in. = 0.433 (sp. gr.) (figures from table).

* These friction-loss tables are from "Cameron Hydraulic Data," courtesy of Ingersoll-Rand Co.

from around 50 per cent for small pumps up to 90 per cent for the larger sizes.

Total efficiency of a steam-driven reciprocating pump is the product of the volumetric, hydraulic, and mechanical efficiencies.

The *suction lift* is theoretically the atmospheric pressure less the vapor pressure of the liquid being pumped. The pump must develop sufficient suction head to overcome this lift.

Reciprocating pumps are usually provided with an air chamber, as shown in Fig. 178. This serves to smooth out irregularities in the discharge of the pump and gives a uniform flow. It is always used with single steam pumps and power pumps. With low-pressure direct-acting duplex steam pumps, it is not needed. With duplex pumps its volume should be four times the displacement of one piston per stroke; with all other types, eight times

Friction Losses in Pipe; 2 In. (2.067 In. Inside Diam.). Loss, Ft. Liquid/1000 Ft. Pipe*

Figures for clean steel pipe, except those in italics, which are for any pipe

Flow		Kinematic viscosity, centistokes									
U. S. gal./min.	Bbl./hr. (42 gal.)	0.6	1.1	2.1	2.7	4.3	7.4	10.3	13.1	15.7	20.6
		Approx. S.S.U. viscosity									
			31.5	33	35	40	50	60	70	80	100
1	1.4	0.04	*0.04*	*0.07*	*0.09*	*0.15*	*0.26*	*0.36*	*0.45*	*0.54*	*0.71*
2	2.9	.13	.15	*.15*	*.19*	*.30*	*.51*	*.71*	*.90*	*1.08*	*1.42*
4	5.7	.45	.52	.60	.63	*.59*	*1.02*	*1.42*	*1.81*	*2.17*	*2.84*
6	8.6	.94	1.07	1.23	1.29	1.43	*1.53*	*2.13*	*2.71*	*3.25*	*4.26*
8	11.4	1.56	1.77	2.05	2.16	2.39	*2.04*	*2.84*	*3.61*	*4.33*	*5.68*
10	14.3	2.34	2.61	3.04	3.24	3.57	*2.56*	*3.56*	*4.52*	*5.42*	*7.11*
12	17.1	3.28	3.70	4.23	4.47	4.90	5.52	*4.27*	*5.43*	*6.51*	*8.53*
14	20.0	4.33	4.86	5.57	5.90	6.47	7.28	*4.98*	*6.33*	*7.59*	*9.96*
16	22.8	5.58	6.18	7.10	7.45	8.20	9.23	9.90	*7.23*	*8.67*	*11.4*
18	25.7	6.84	7.60	8.68	9.20	10.2	11.5	12.2	12.8	*9.76*	*12.8*
20	28.6	8.33	9.25	10.6	11.1	12.3	13.8	14.7	15.5	*10.8*	*14.2*
25	35.7	12.6	14.0	15.7	16.5	18.3	20.6	22.1	23.1	24.2	*17.8*
30	42.9	17.5	19.3	22.0	22.9	25.3	28.4	30.4	32.1	33.3	35.1
35	50.0	23.5	25.8	28.8	30.4	33.2	37.2	40.1	42.1	43.8	46.6
40	57.1	30.1	32.7	37.0	38.6	42.5	47.6	50.9	53.9	55.6	59.0
50	71.4	45.4	49.5	55.4	57.8	63.5	71.0	76.0	80.1	82.5	87.5
60	85.7	64.2	69.3	77.2	81.0	88.0	98.7	106	110	115	121
70	100	85.5	92.3	102	107	118	130	139	146	151	161
80	114	109	118	130	135	147	165	176	184	190	203
90	129	136	147	162	169	183	203	217	227	236	251
100	143	165	178	198	205	223	244	264	276	287	304
110	157	198	213	236	244	264	292	312	328	340	360
120	171	233	252	276	288	309	342	366	383	397	421
130	186	271	292	318	335	357	396	419	441	458	486
140	200	312	337	368	382	408	453	479	505	524	553
150	214	356	384	419	432	464	513	545	572	594	624
160	228	403	432	474	487	525	578	616	643	567	701
170	243	452	482	532	545	587	645	687	715	744	782
180	257	502	535	591	605	650	716	760	792	824	867
190	271	554	597	653	668	715	789	835	870	906	954
200	286	610	661	718	736	785	860	920	958	995	
210	300	670	728	788	807	860	947				
220	314	735	795	860	880	935					
230	328	795	860	930	960						
240	343	865	935								

Flow		Kinematic viscosity, centistokes									
U. S. gal./min.	Bbl./hr. (42 gal.)	26.4	32.0	43.2	65.0	108.4	162.3	216.5	325	435	650
		Approx. S.S.U. viscosity									
		125	150	200	300	500	750	1000	1500	2000	3000
1	1.4	*0.91*	*1.10*	*1.49*	*2.24*	*3.74*	*5.60*	*7.48*	*11.2*	*15.0*	*22.4*
2	2.9	*1.82*	*2.21*	*2.98*	*4.48*	*7.49*	*11.2*	*15.0*	*22.4*	*30.0*	*44.9*
3	4.3	*2.73*	*3.31*	*4.47*	*6.73*	*11.2*	*16.8*	*22.4*	*33.6*	*45.0*	*67.4*
4	5.7	*3.64*	*4.42*	*5.96*	*8.98*	*15.0*	*22.4*	*29.9*	*44.8*	*60.0*	*89.9*
5	7.1	*4.56*	*5.52*	*7.45*	*11.2*	*18.7*	*28.0*	*37.4*	*56.0*	*75.0*	*112*
6	8.6	*5.47*	*6.63*	*8.95*	*13.5*	*22.5*	*33.6*	*44.8*	*67.2*	*90.0*	*135*
7	10.0	*6.38*	*7.73*	*10.4*	*15.7*	*26.2*	*39.2*	*52.3*	*78.4*	*105*	*157*
8	11.4	*7.29*	*8.84*	*11.9*	*18.0*	*30.0*	*44.8*	*59.8*	*89.6*	*120*	*180*
9	12.9	*8.20*	*9.94*	*13.4*	*20.2*	*33.7*	*50.4*	*67.3*	*101*	*135*	*202*
10	14.3	*9.11*	*11.0*	*14.9*	*22.4*	*37.4*	*56.0*	*74.8*	*112*	*150*	*224*
12	17.1	*10.9*	*13.3*	*17.9*	*26.9*	*44.9*	*67.3*	*89.7*	*135*	*180*	*269*
14	20.0	*12.7*	*15.5*	*20.9*	*31.4*	*52.4*	*78.4*	*105*	*157*	*210*	*314*
16	22.8	*14.6*	*17.7*	*23.9*	*35.9*	*59.9*	*89.6*	*120*	*179*	*240*	*359*
18	25.7	*16.4*	*19.9*	*26.8*	*40.3*	*67.4*	*101*	*135*	*202*	*270*	*404*
20	28.6	*18.2*	*22.1*	*29.8*	*44.9*	*74.9*	*112*	*150*	*224*	*300*	*449*
25	35.7	*22.8*	*27.6*	*37.3*	*56.1*	*93.6*	*140*	*187*	*280*	*375*	*562*
30	42.9	*27.3*	*33.1*	*44.7*	*67.3*	*112*	*168*	*224*	*336*	*450*	*674*
35	50.0	*31.9*	*38.7*	*52.2*	*78.5*	*131*	*196*	*262*	*392*	*525*	*786*
40	57.1	50.9	*44.2*	*59.6*	*89.8*	*150*	*224*	*299*	*448*	*600*	*899*
45	64.3	70.7	80.4	*67.1*	*101*	*168*	*252*	*336*	*504*	*675*	
50	71.4	92.5	96.7	*74.5*	*112*	*187*	*280*	*374*	*560*	*750*	
60	85.7	128	133	143	*135*	*225*	*336*	*448*	*672*	*900*	
70	100	168	175	188	*157*	*262*	*392*	*523*	*784*		
80	114	215	224	239	*180*	*300*	*448*	*598*	*896*		
90	129	264	275	294	321	*337*	*504*	*673*			
100	143	320	332	354	390	*374*	*560*	*748*			
110	157	380	395	420	460	*412*	*617*	*823*			
120	171	445	462	490	536	*449*	*673*	*898*			
130	186	512	529	566	615	*487*	*728*				
140	200	582	608	648	701	*524*	*784*				
150	214	661	681	729	795	892	*840*				
160	228	744	769	818	895	995	*896*				
170	243	825	858	915			*952*				
180	257	910	953								
190	271	999									

Loss in lb./sq. in. = 0.433 (sp. gr.) (figures from table).

* These friction-loss tables are from "Cameron Hydraulic Data," courtesy of Ingersoll-Rand Co.

Friction Loss in Pipes; 3 In. (3.068 In. Inside Diam.). Loss, Ft. Liquid/1000 Ft. Pipe*

Figures for clean steel pipe, except those in italics, which are for any pipe

Flow		Kinematic viscosity, centistokes									
U. S. gal./min.	Bbl./hr. (42 gal.)	0.6	1.1	2.1	2.7	4.3	7.4	10.3	13.1	15.7	20.6
		Approx. S.S.U. viscosity									
			31.5	33	35	40	50	60	70	80	100
8	11.4	0.24	0.27	0.31	0.33	*0.24*	*0.42*	*0.59*	*0.74*	*0.89*	*1.18*
10	14.3	.35	.41	.46	.49	.54	*.53*	*.73*	*.93*	*1.11*	*1.47*
15	21.4	.73	.83	.95	1.00	1.11	*.79*	*1.10*	*1.40*	*1.67*	*2.20*
20	28.6	1.25	1.39	1.58	1.67	1.84	2.08	*1.46*	*1.86*	*2.23*	*2.93*
25	35.7	1.86	2.06	2.38	2.50	2.75	3.10	3.33	*2.33*	*2.79*	*3.66*
30	42.9	2.60	2.88	3.30	3.46	3.79	4.26	4.62	4.82	*3.35*	*4.40*
35	50.0	3.43	3.82	4.32	4.55	5.03	5.62	6.05	6.35	6.65	*5.13*
40	57.1	4.40	4.92	5.50	5 80	6.38	7.13	7.71	8.07	8.53	*5.87*
50	71.4	6.65	7.35	8.26	8.60	9.52	10.7	11.5	12.0	12.5	13.6
60	85.7	9.40	10.2	11.5	12.0	13.2	14.7	15.8	16.7	17.3	18.5
70	100	12.4	13.6	15.3	15.8	17.3	19.3	20.9	22.0	22.9	24.3
80	114	16.0	17.3	19.4	20.3	22.0	24.6	26.4	27.9	29.0	30.8
90	129	19.9	21.5	23.9	25.2	27.3	30.5	32.7	34.3	35.8	37.9
100	143	24.0	26.1	29.1	30.3	33.0	37.1	39.4	41.3	43.2	45.8
120	171	34.0	37.0	40.7	42.2	46.7	51.8	54.8	57.7	59.7	63.3
140	200	45.8	49.1	54.0	56.2	61.7	67.3	72.1	75.7	79.0	83.6
160	228	58.2	62.8	69.3	71.5	77.5	85.7	91.5	96.4	99.8	106
180	257	72.9	78.7	86.2	89.1	96.3	106	113	119	123	131
200	286	89.0	95.5	104	108	116	128	136	144	149	158
225	322	111	118	130	135	144	160	170	177	185	195
250	357	135	145	158	164	175	192	204	215	224	235
275	393	163	174	188	196	208	229	243	254	264	281
300	429	192	204	223	230	246	268	285	297	309	326
325	464	223	238	259	266	284	310	330	344	358	378
350	500	256	273	296	306	326	354	377	393	407	433
375	536	290	310	335	348	371	403	429	448	464	485
400	571	330	351	378	392	418	455	483	506	522	547
425	607	370	394	424	439	468	508	539	563	579	609
450	643	413	438	473	488	520	563	595	623	441	678
475	679	456	487	523	541	575	624	662	688	708	746
500	714	502	537	576	596	632	686	722	756	780	820
525	750	550	587	632	652	694	748	790	825	853	897
550	786	604	641	688	709	757	817	860	900	930	973
575	822	655	697	749	773	820	887	933	975		
600	857	710	757	810	835	890	960				

Flow		Kinematic viscosity, centistokes									
U. S. gal./min.	Bbl./hr. (42 gal.)	26.4	32.0	43.2	65.0	108.4	162.3	216.5	325	435	650
		Approx. S.S.U. viscosity									
		125	150	200	300	500	750	1000	1500	2000	3000
4	5.7	*0.75*	*0.91*	*1.23*	*1.85*	*3.08*	*4.62*	*6.16*	*9.25*	*12.4*	*18.5*
6	8.6	*1.13*	*1.37*	*1.84*	*2.77*	*4.62*	*6.92*	*9.24*	*13.9*	*18.5*	*27.7*
8	11.4	*1.50*	*1.82*	*2.45*	*3 70*	*6.16*	*9.23*	*12.3*	*18.5*	*24.7*	*36.0*
10	14.3	*1.88*	*2.28*	*3.06*	*4.62*	*7.70*	*11.5*	*15.4*	*23.1*	*30.9*	*46.2*
12	17.1	*2.25*	*2.73*	*3.68*	*5.55*	*9.24*	*13.8*	*18.5*	*27.7*	*37.1*	*55.5*
14	20.0	*2.63*	*3.18*	*4.29*	*6.47*	*10.8*	*16.2*	*21.5*	*32.3*	*43.3*	*64.7*
16	22.8	*3.00*	*3.64*	*4.90*	*7.39*	*12.3*	*18.5*	*24.6*	*37.0*	*49.5*	*73.9*
18	25.7	*3.38*	*4.09*	*5.52*	*8.31*	*13.9*	*20.8*	*27.7*	*41.6*	*55.6*	*83.2*
20	28.6	*3.76*	*4.55*	*6.13*	*9.24*	*15.4*	*23.1*	*30.8*	*46.2*	*61.8*	*92.4*
25	35.7	*4.69*	*5.69*	*7.67*	*11.5*	*19.3*	*28.8*	*38.5*	*57.7*	*77.3*	*115*
30	42.9	*5.63*	*6.83*	*9.20*	*13.9*	*23.1*	*34.6*	*46.2*	*69.3*	*92.7*	*139*
35	50.0	*6.57*	*7.97*	*10.7*	*16.2*	*27.0*	*40.3*	*53.8*	*80.9*	*108*	*162*
40	57.1	*7.51*	*9.10*	*12.3*	*18.5*	*30.8*	*46.2*	*61.6*	*92.5*	*124*	*185*
50	71.4	*9.39*	*11.4*	*15.3*	*23.1*	*38.5*	*57.7*	*77.0*	*115*	*154*	*231*
60	85.7	19.5	*13.7*	*18.4*	*27.7*	*46.2*	*69.2*	*92.4*	*139*	*185*	*277*
70	100	25.4	26.5	*21.5*	*32.3*	*53.9*	*80.8*	*108*	*162*	*216*	*323*
80	114	32.3	33.8	*24.7*	*37.0*	*61.6*	*92.3*	*123*	*185*	*247*	*369*
90	129	40.0	41.7	44.6	*41.6*	*69.3*	*104*	*139*	*208*	*278*	*416*
100	143	48.2	50.3	53.7	*46.2*	*77.0*	*115*	*154*	*231*	*309*	*462*
120	171	66.7	70.0	74.3	*55.5*	*92.4*	*138*	*185*	*277*	*371*	*555*
140	200	88.0	91.7	98.0	108	*108*	*162*	*215*	*323*	*433*	*647*
160	228	112	116	125	136	*123*	*185*	*246*	*370*	*495*	*739*
180	257	138	144	154	167	*139*	*208*	*277*	*416*	*556*	*832*
200	286	165	174	185	202	*154*	*231*	*308*	*462*	*618*	*924*
225	322	206	214	227	250	279	*260*	*346*	*520*	*696*	
250	357	249	258	275	301	338	*288*	*385*	*577*	*773*	
275	393	295	308	326	358	399	*317*	*423*	*635*	*850*	
300	429	344	359	382	416	462	*346*	*462*	*693*	*927*	
325	464	397	413	440	479	534	*375*	*500*	*751*		
350	500	452	472	504	548	613	669	*538*	*809*		
375	536	515	535	562	619	690	755	*577*	*867*		
400	571	579	602	638	696	770	851	*616*	*925*		
425	607	642	670	712	778	862	936	*654*	*982*		
450	643	715	743	790	865	955					
475	679	787	818	869	949						

Loss in lb./sq. in. = 0.433 (sp. gr.) (figures in table).

* These friction-loss tables are from "Cameron Hydraulic Data," courtesy of Ingersoll-Rand Co.

Friction Losses in Pipe; 4 In. (4.026 In. Inside Diam.). Loss, Ft. Liquid/1000 Ft. Pipe*

Figures for clean steel pipe, except those in italics, which are for any pipe

Flow, U. S. gal./min.	Flow, Bbl./hr. (42 gal.)	Kinematic viscosity, centistokes 0.6	1.1	2.1	2.7	4.3	7.4	10.3	13.1	15.7	20.6
	Approx. S.S.U. viscosity		31.5	33	35	40	50	60	70	80	100
20	28.6	0.32	0.36	0.42	0.44	0.48	0.57	*0.50*	*0.63*	*0.75*	*0.99*
30	42.9	.67	.74	.86	.91	.99	1.16	1.25	*.95*	*1.13*	*1.48*
40	57.1	1.12	1.24	1.43	1.51	1.64	1.92	2.08	2.20	*1.51*	*1.96*
50	71.4	1.69	1.85	2.13	2.23	2.42	2.82	3.06	3.24	3.39	*2.47*
60	85.7	2.37	2.61	2.96	3.12	3.35	3.90	4.24	4.50	4.67	5.00
70	100	3.14	3.46	3.91	4.09	4.39	5.14	5.55	5.89	6.18	6.53
80	114	4.06	4.42	4.95	5.19	5.57	6.49	7.01	7.43	7.77	8.30
90	129	5.06	5.48	6.09	6.43	6.90	8.02	8.70	9.16	9.55	10.2
100	143	6.12	6.67	7.41	7.76	8.35	9.68	10.4	11.1	11.5	12.3
120	171	8.56	9.31	10.3	10.8	11.6	13.4	14.4	15.3	16.0	17.0
140	200	11.4	12.5	13.7	14.3	15.2	17.5	18.9	19.9	20.8	22.2
160	228	14.6	16.0	17.5	18.4	19.4	22.3	24.0	25.3	26.6	28.1
180	257	18.3	19.8	21.7	22.7	24.0	27.5	29.4	31.0	32.4	34.7
200	286	22.2	24.0	26.5	27.6	29.2	33.3	35.6	37.6	39.1	41.5
220	314	26.6	28.6	31.6	32.8	34.7	39.7	42.2	44.5	46.3	49.4
240	343	31.3	33.8	37.2	38.6	40.7	46.2	48.9	51.6	54.2	57.6
260	371	36.4	39.2	42.5	44.5	47.0	53.4	56.7	59.8	62.1	66.1
280	400	41.8	45.1	49.8	51.2	54.0	61.0	64.8	68.3	71.2	75.2
300	429	47.7	51.3	56.4	58.3	61.4	68.5	73.8	77.5	80.6	85.0
350	500	63.7	68.5	75.0	77.8	81.5	91.1	96.9	102	106	112
400	571	81.9	88.0	95.7	99.9	104	117	123	130	135	142
450	643	103	110	119	123	130	145	153	160	167	176
500	714	125	134	146	150	159	176	185	192	200	212
550	786	150	160	174	179	189	209	221	228	238	250
600	857	176	189	204	210	221	244	259	267	276	293
650	929	204	220	236	244	256	283	299	311	319	339
700	1000	236	253	272	281	294	325	342	358	366	387
750	1070	268	287	310	320	334	368	387	404	418	438
800	1140	302	324	350	361	376	415	437	453	468	490
850	1215	340	365	392	404	423	464	489	510	524	547
900	1285	381	405	437	450	470	518	544	563	589	608
950	1360	423	449	483	497	520	575	600	622	633	670
1000	1430	465	494	532	546	575	632	660	685	707	740
1100	1570	555	596	639	655	685	754	785	815	842	884
1200	1715	661	703	750	772	805	885	925	958	985	

Flow, U. S. gal./min.	Flow, Bbl./hr. (42 gal.)	Kinematic viscosity, centistokes 26.4	32.0	43.2	65.0	108.4	162.3	216.5	325	435	650
	Approx. S.S.U. viscosity	125	150	200	300	500	750	1000	1500	2000	3000
15	21.4	*0.95*	*1.15*	*1.55*	*2.34*	*3.91*	*5.85*	*7.80*	*11.7*	*15.7*	*23.4*
20	28.6	*1.27*	*1.54*	*2.07*	*3.12*	*5.21*	*7.80*	*10.4*	*15.6*	*20.9*	*31.2*
30	42.9	*1.90*	*2.30*	*3.11*	*4.68*	*7.82*	*11.7*	*15.0*	*23.4*	*31.3*	*46.8*
40	57.1	*2.54*	*3.08*	*4.15*	*6.25*	*10.4*	*15.6*	*20.8*	*31.2*	*41.8*	*62.5*
50	71.4	*3.17*	*3.84*	*5.18*	*7.81*	*13.0*	*19.5*	*26.0*	*39.0*	*52.2*	*78.1*
60	85.7	*3.80*	*4.61*	*6.22*	*9.37*	*15.6*	*23.4*	*31.2*	*46.8*	*62.7*	*93.7*
70	100	*4.44*	*5.38*	*7.25*	*10.9*	*18.2*	*27.3*	*36.4*	*54.6*	*73.2*	*109*
80	114	8.83	*6.15*	*8.29*	*12.5*	*20.8*	*31.2*	*41.6*	*62.4*	*83.6*	*125*
90	129	10.8	11.4	*9.33*	*14.1*	*23.4*	*35.1*	*46.8*	*70.2*	*94.1*	*141*
100	143	13.0	13.6	*10.4*	*15.6*	*26.0*	*39.0*	*52.0*	*78.0*	*105*	*156*
120	171	18.0	18.8	20.2	*18.8*	*31.2*	*46.8*	*62.4*	*93.7*	*125*	*187*
140	200	23.5	24.7	26.5	*21.9*	*36.4*	*54.6*	*72.8*	*109*	*146*	*218*
160	228	29.9	31.1	33.8	*25.0*	*41.7*	*62.4*	*83.2*	*125*	*167*	*250*
180	257	36.6	38.4	41.2	45.5	*46.9*	*70.2*	*93.6*	*140*	*188*	*281*
200	286	44.3	46.3	49.6	54.7	*52.1*	*78.0*	*104*	*156*	*209*	*312*
220	314	52.5	54.8	58.7	64.5	*57.3*	*85.8*	*114*	*172*	*230*	*343*
240	343	61.1	63.8	68.6	76.1	*62.5*	*93.6*	*125*	*187*	*251*	*375*
260	371	70.2	73.3	78.9	87.0	*67.7*	*101*	*135*	*203*	*272*	*406*
280	400	80.0	83.6	90.2	99.3	*73.0*	*109*	*146*	*218*	*292*	*437*
300	429	90.2	94.6	102	112	126	*117*	*156*	*234*	*313*	*468*
325	464	104	109	117	128	146	*127*	*169*	*254*	*340*	*508*
350	500	118	124	133	147	166	*136*	*182*	*273*	*366*	*547*
375	536	134	140	151	166	187	*146*	*195*	*293*	*392*	*585*
400	571	151	157	169	186	209	*156*	*208*	*312*	*418*	*625*
450	643	186	193	208	229	260	284	*234*	*351*	*470*	*703*
500	714	224	233	250	277	312	342	*260*	*390*	*523*	*781*
550	786	263	275	296	326	367	403	*286*	*429*	*575*	*860*
600	857	309	322	344	382	428	472	505	*468*	*627*	*937*
650	929	358	371	398	438	493	543	583	*507*	*680*	
700	1000	407	426	453	497	564	622	665	*546*	*732*	
750	1070	462	482	511	563	636	700	750	*585*	*784*	
800	1140	519	540	574	631	710	780	838	*624*	*836*	
850	1215	579	601	638	700	793	867	937	*663*	*889*	
900	1285	640	668	708	773	878	965			*941*	
950	1360	703	735	783	852	965				*993*	

Loss in lb./sq. in. = 0.433 (sp. gr.) (figures in table).

* These friction-loss tables are from "Cameron Hydraulic Data," courtesy of Ingersoll-Rand Co.

Friction Losses in Pipe; 6 In. (6.065 In. Inside Diam.). Loss, Ft. Liquid/1000 Ft. Pipe*

Figures for clean steel pipe, except those in italics, which are for any pipe

Flow		Kinematic viscosity, centistokes									
U. S. gal./min.	Bbl./hr. (42 gal.)	0.6	1.1	2.1	2.7	4.3	7.4	10.3	13.1	15.7	20.6
		Approx. S.S.U. viscosity									
			31.5	33	35	40	50	60	70	80	100
75	107	0.47	0.53	0.62	0.65	0.73	0.82	0.89	0.94	0.98	
100	143	.80	.89	1.03	1.09	1.21	1.37	1.49	1.56	1.64	1.75
125	178	1.21	1.33	1.54	1.64	1.80	2.02	2.18	2.32	2.42	2.59
150	214	1.67	1.85	2.16	2.26	2.49	2.80	3.01	3.21	3.35	3.55
175	250	2.23	2.44	2.76	2.92	3.26	3.67	4.00	4.22	4.38	4.70
200	286	2.86	3.13	3.52	3.71	4.19	4.65	5.03	5.33	5.53	5.96
225	322	3.55	3.92	4.34	4.54	5.15	5.76	6.22	6.53	6.84	7.31
250	357	4.32	4.75	5.27	5.50	6.17	6.94	7.45	7.87	8.26	8.77
275	393	5.17	5.63	6.27	6.55	7.29	8.27	8.84	9.29	9.77	10.3
300	429	6.08	6.63	7.32	7.65	8.46	9.60	10.3	10.9	11.4	12.0
350	500	8.19	8.84	9.76	10.2	11.1	12.7	13.6	14.1	14.9	15.9
400	571	10.5	11.3	12.5	13.0	14.1	16.1	17.2	18.2	19.0	20.1
450	643	13.1	14.1	15.5	16.0	17.5	19.7	21.3	22.4	23.3	24.6
500	714	15.9	17.1	18.8	19.5	21.2	23.7	25.8	26.9	28.1	30.0
550	786	19.3	20.5	22.5	23.4	25.2	28.0	30.5	32.1	33.5	35.4
600	857	22.7	24.1	26.3	27.6	29.5	32.8	35.5	37.7	39.1	41.3
650	929	26.3	28.1	30.5	31.7	34.0	37.8	40.7	43.3	45.1	47.5
700	1000	30.1	32.3	35.0	36.5	39.0	43.2	46.5	49.5	51.5	54.2
750	1070	34.3	36.8	39.7	41.3	44.4	48.7	52.2	55.7	58.1	61.5
800	1140	38.8	41.7	44.8	46.6	50.0	54.8	58.3	62.2	65.0	69.0
900	1285	48.7	52.2	55.9	57.8	62.7	67.7	72.5	76.3	80.0	85.5
1000	1430	59.2	63.8	68.3	70.6	75.9	82.0	87.5	91.9	96.5	103
1100	1570	70.8	76.2	81.8	84.8	90.0	98.3	104	108	114	122
1200	1715	84.3	89.7	96.2	99.5	105	115	122	127	134	142
1400	2000	113	120	129	132	140	153	161	167	175	186
1600	2285	146	155	166	171	180	196	206	214	223	233
1800	2570	182	192	206	212	224	244	256	265	275	290
2000	2860	224	238	252	259	273	294	310	320	334	350
2200	3140	268	283	302	309	325	353	371	382	397	415
2400	3430	319	332	356	366	384	412	437	450	465	490
2600	3710	370	388	414	426	447	479	503	523	535	565
2800	4000	428	448	477	492	513	548	577	601	615	643
3000	4285	485	512	542	560	587	627	655	682	701	732
3250	4640	569	593	633	650	681	730	763	789	819	850
3500	5000	657	688	728	750	785	837	875	900	932	967

Flow		Kinematic viscosity, centistokes									
U. S. gal./min.	Bbl./hr. (42 gal.)	26.4	32.0	43.2	65.0	108.4	162.3	216.5	325	435	650
		Approx. S.S.U. viscosity									
		125	150	200	300	500	750	1000	1500	2000	3000
50	71.4	*0.62*	*0.74*	*1.00*	*1.51*	*2.52*	*3.78*	*5.04*	*7.57*	*10.1*	*15.1*
75	107	*.92*	*1.12*	*1.51*	*2.27*	*3.78*	*5.66*	*7.56*	*11.4*	*15.2*	*22.7*
100	143	*1.23*	*1.40*	*2.01*	*3.03*	*5.05*	*7.55*	*10.1*	*15.1*	*20.3*	*30.2*
125	178	2.75	*1.86*	*2.51*	*3.79*	*6.31*	*9.45*	*12.6*	*18.9*	*25.3*	*37.8*
150	214	3.79	4.03	*3.01*	*4.54*	*7.58*	*11.3*	*15.1*	*22.7*	*30.4*	*45.4*
175	250	4.94	5.21	5.58	*5.30*	*8.84*	*13.2*	*17.6*	*26.5*	*35.5*	*53.0*
200	286	6.26	6.56	7.05	*6.06*	*10.1*	*15.1*	*20.2*	*30.3*	*40.5*	*60.6*
225	322	7.76	8.09	8.70	*6.82*	*11.4*	*17.0*	*22.7*	*34.1*	*45.6*	*68.1*
250	357	9.31	9.74	10.4	*7.57*	*12.6*	*18.9*	*25.2*	*37.8*	*50.7*	*75.7*
275	393	11.0	11.5	12.4	13.6	*13.9*	*20.8*	*27.7*	*41.7*	*55.8*	*83.2*
300	429	12.9	13.5	14.5	15.9	*15.1*	*22.6*	*30.2*	*45.4*	*60.9*	*90.8*
350	500	16.9	17.7	19.0	20.8	*17.7*	*26.4*	*35.3*	*53.0*	*71.0*	*106*
400	571	21.4	22.3	23.9	26.3	*20.2*	*30.2*	*40.3*	*60.6*	*81.1*	*121*
450	643	26.3	27.4	29.5	32.4	36.6	*34.0*	*45.3*	*68.2*	*91.3*	*136*
500	714	31.5	33.0	35.6	39.0	44.1	*37.8*	*50.4*	*75.7*	*101*	*151*
550	786	37.2	39.1	41.8	46.2	52.2	*41.6*	*55.4*	*83.3*	*112*	*166*
600	857	43.6	45.7	49.0	53.9	61.0	*45.3*	*60.5*	*90.9*	*122*	*182*
650	929	50.3	52.6	56.4	62.2	69.9	*49.1*	*65.5*	*98.5*	*132*	*197*
700	1000	57.3	59.8	64.0	70.7	80.0	87.5	*70.6*	*106*	*142*	*212*
750	1070	65.0	67.6	72.6	79.9	90.2	99.1	*75.6*	*114*	*152*	*227*
800	1140	72.9	75.8	81.6	89.5	100	111	*80.6*	*121*	*162*	*242*
900	1285	89.8	93.5	99.5	111	124	137	147	*136*	*183*	*272*
1000	1430	109	113	121	134	150	164	176	*151*	*203*	*302*
1100	1570	129	134	143	157	177	195	208	*167*	*223*	*333*
1200	1715	151	157	166	184	206	226	245	*182*	*243*	*363*
1400	2000	199	207	220	241	270	298	320	351	*284*	*424*
1600	2285	251	261	280	305	342	378	404	446	*324*	*484*
1800	2570	309	322	344	375	423	465	497	547	589	*545*
2000	2860	370	385	414	456	511	559	598	658	705	*605*
2200	3140	438	457	493	540	603	662	708	780	833	*666*
2400	3430	512	534	574	630	700	771	827	903	980	*726*
2600	3710	590	615	659	723	812	889	950			*787*
2800	4000	677	703	750	828	923					
3000	4285	766	793	847	935						
3200	4570	856	890	950							

Loss in lb./sq. in. = 0.433 (sp. gr.) (figures in table).

* These friction-loss tables are from "Cameron Hydraulic Data," courtesy of Ingersoll-Rand Co.

Types of Reciprocating Pumps. The ordinary types of reciprocating pumps are four.

1. *Simplex, Double-acting.* These may be direct-acting (*i.e.*, direct-connected to a steam cylinder) or power-driven (through crank and flywheel from the cross-head of a steam engine). Figure 178 is a pump of this type, designed for use at heads up to 200 ft. In this figure, the piston consists of disks A and B, with packing rings C between. A bronze liner for the water cylinder is shown at D. Suction valves are E_1 and E_2. Discharge valves are F_1 and F_2. In the steam end, pilot valve L is operated by a rod, actuated by piston rod M. This pilot operates main valve N to cover or uncover steam ports P.

2. *Duplex, Double-acting.* These pumps differ primarily from those of the simplex type in having two water cylinders whose operation is coordinated. These pumps may be direct-acting, steam-driven, or power-driven with crank and flywheel.

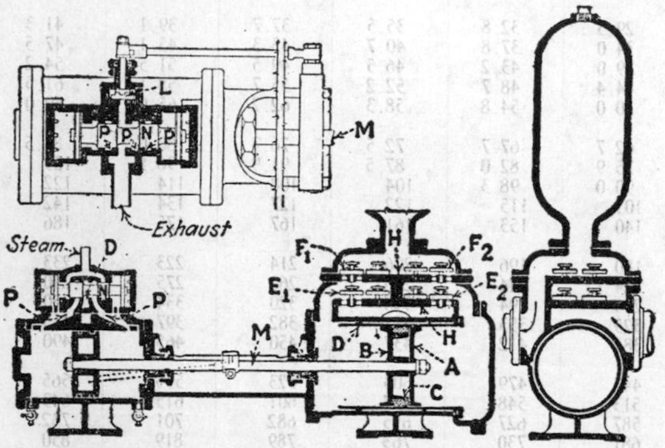

FIG. 178. Water end of a double-acting steam-driven reciprocating pump.

A duplex, outside-end-packed plunger pump with pot valves, of the type used with hydraulic presses and for similar service, is shown in Fig. 179. In this drawing, plunger A is direct-connected to rod B, while plunger C is operated from the rod by means of yokes D, D and tie rods.

3. *Triplex, Single-acting.* These pumps have three single-acting plungers and cylinders and are used to give a uniform flow. They are usually of vertical design. The drive may be from a motor, belt, or steam engine. This is the common type of power pump, an example of which, arranged for belt drive, is shown in Fig. 180, from which the action is readily traced.

4. *Triplex, Double-acting.* This is a double-acting arrangement of the pump discussed under class 3. The design is generally used for horizontal triplex pumps. Other features are similar to class 3.

Choice of a Reciprocating Pump. The advantages of a steam-driven reciprocating pump lie in good efficiency over a wide range of operating conditions and flexibility of capacity, head, and speed. The disadvantages are usually high first cost, large floor space required, noisy operation, and more attention required.

The advantages given for a direct-driven steam pump do not apply to power pumps. A triplex single- or double-acting pump does not have flexibility as to speed or capacity. It does have uniform delivery. Its use is indicated in pumping wells, in handling steady flow of liquids, and in serving machines operated by hydraulic pressure.

Simplex double-acting pumps are most suitable for water service and boiler feed.

Duplex double-acting pumps are universally adaptable. When made of the proper materials, particularly as to cylinder linings and valve material, and when of correct design, these pumps are frequently used for tars, oils, and other viscous liquids.

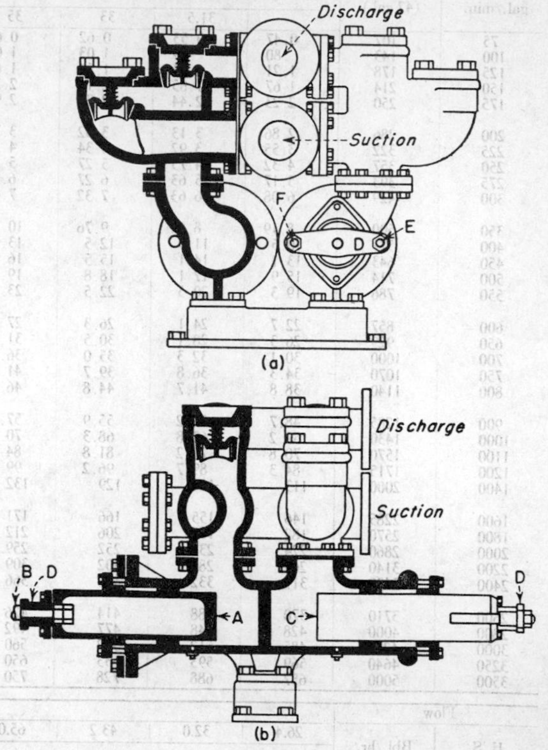

FIG. 179. Duplex double-acting steam-driven plunger pump.

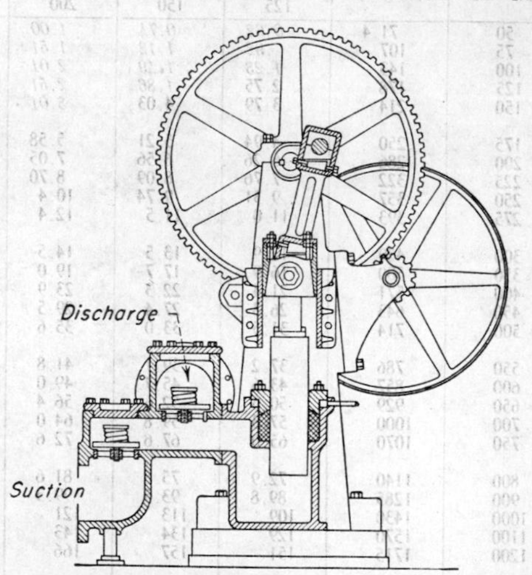

FIG. 180. Triplex single-acting power pump.

Triplex single-acting pumps are suitable for producing uniform flow when conditions are constant. Triplex double-acting pumps are used under the same conditions as triplex single-acting pumps but are generally of horizontal design.

Reciprocating pumps, particularly of the direct-acting

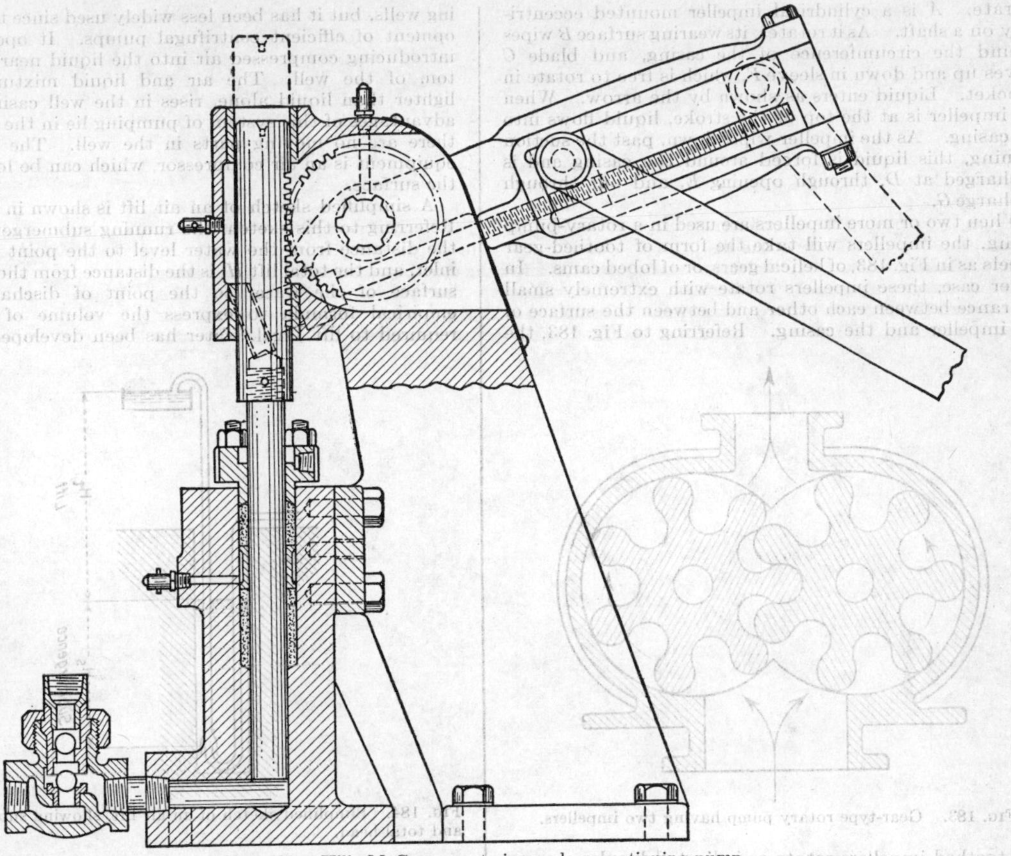

FIG. 181. Hills-McCanna metering and proportioning pump.

steam-driven design, are available in a great variety of designs and sizes, to fill every need for which such a pump is suitable.

Another form of reciprocating pump is known as the diaphragm pump. In these units a flexible diaphragm separates the suction chamber from the discharge chamber. The diaphragm moves up and down through the action of a yoke or rod, and a discharge valve forces the liquid from the suction chamber into the discharge chamber. Diaphragm pumps are used for emptying tanks and sumps and for dewatering excavations, etc.

Metering and Proportioning Pumps. Metering and proportioning pumps are reciprocating pumps with the drive so arranged as to allow the stroke to be varied from zero to the maximum pump stroke. They are designed to measure or control the flow of liquid within an accuracy of approximately 2 per cent with capacities up to approximately 50 gal./min.

When two or more pump units are arranged in multiple, the flow of the different liquids may be varied and proportioned, and the complete unit becomes a *proportioning pump* (see Fig. 181). Capacities may be controlled manually or by automatic control to vary the stroke of the units.

ROTARY PUMPS

Rotary pumps differ from centrifugal and reciprocating pumps in that they will deliver a positive quantity of liquid under conditions of varying head or pressure. These pumps, when built of proper materials, will handle any liquid that contains no grit or abrasive material.

This type of pump consists of a stationary casing in which are located one or more rotating members. When

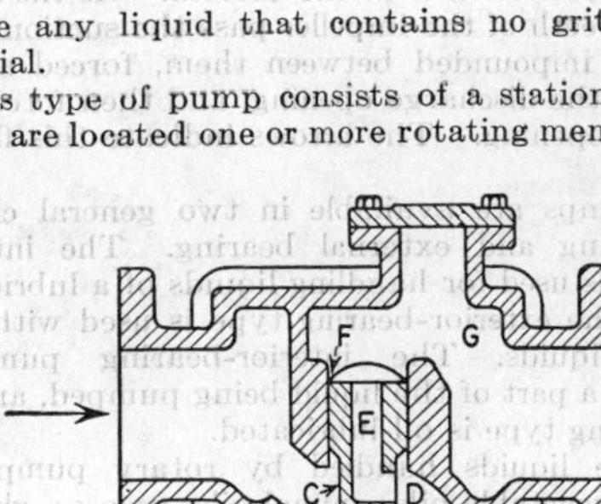

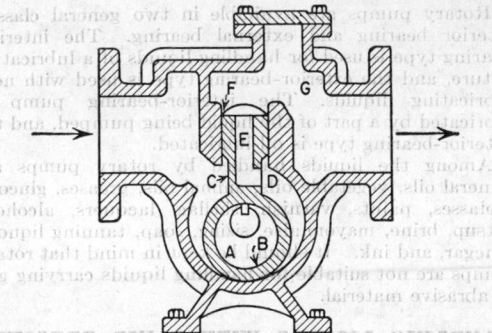

FIG. 182. Single-impeller rotary pump with plunger or reciprocating blade.

one rotating member is used, it is mounted eccentrically on the shaft. The impeller in this type of pump is usually circular in section and is provided with a reciprocating blade or blades or a horizontal abutment. Figure 182 shows the cross section of a pump of this type. Although there are many designs, an explanation of that shown in Fig. 182 will serve to indicate how such pumps operate. *A* is a cylindrical impeller mounted eccentrically on a shaft. As it rotates, its wearing surface *B* wipes around the circumference of the casing, and blade *C* moves up and down in sleeve *F*, which is free to rotate in a socket. Liquid enters as shown by the arrow. When the impeller is at the top of the stroke, liquid flows into the casing. As the impeller wipes down, past the suction opening, this liquid is forced around the casing and is discharged at *D*, through opening *E*, and out through discharge *G*.

When two or more impellers are used in a rotary-pump casing, the impellers will take the form of toothed-gear wheels as in Fig. 183, of helical gears, or of lobed cams. In either case, these impellers rotate with extremely small clearance between each other and between the surface of the impeller and the casing. Referring to Fig. 183, the

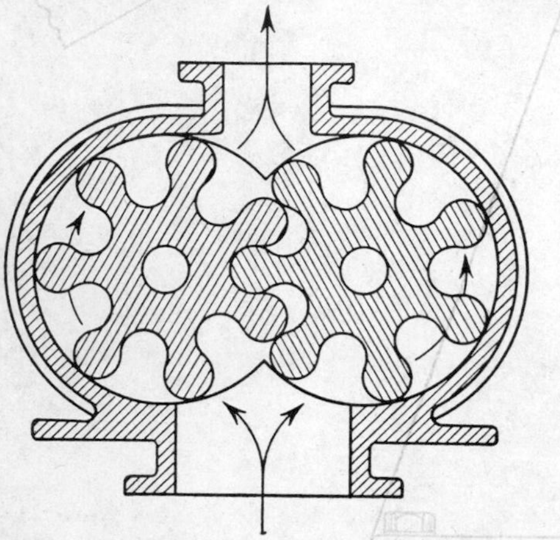

FIG. 183. Gear-type rotary pump having two impellers.

two toothed impellers rotate as indicated by the arrows. The suction connection is at the bottom. As the spaces between the teeth of the impeller pass the suction opening, water is impounded between them, forced around the casing to the discharge opening, and then forced out through this opening. The arrows indicate this flow of water.

Rotary pumps are available in two general classes, interior bearing and external bearing. The interior-bearing type is used for handling liquids of a lubricating nature, and the exterior-bearing type is used with non-lubricating liquids. The interior-bearing pump is lubricated by a part of the liquid being pumped, and the exterior-bearing type is oil-lubricated.

Among the liquids handled by rotary pumps are mineral oils, vegetable oils, animal oils, greases, glucose, molasses, paints, varnish, shellac, lacquers, alcohols, catsup, brine, mayonnaise, sizing, soap, tanning liquors, vinegar, and ink. It should be kept in mind that rotary pumps are not suitable for handling liquids carrying grit or abrasive material.

HANDLING LIQUIDS WITH FLUID PRESSURE

In addition to the liquid-handling devices, such as pumps, that depend on the mechanical action of pistons, plungers, or impellers to move the material, a large group of devices for this purpose employ fluid pressure to move the liquid. This group includes air lifts, acid eggs, jet pumps, pulsometers, and diaphragm pumps among others.

The air lift is a device for raising liquid by means of compressed air. In the past it was widely used for pumping wells, but it has been less widely used since the development of efficient centrifugal pumps. It operates by introducing compressed air into the liquid near the bottom of the well. The air and liquid mixture, being lighter than liquid alone, rises in the well casing. The advantages of this system of pumping lie in the fact that there are no moving parts in the well. The pumping equipment is an air compressor, which can be located on the surface.

A simplified sketch of an air lift is shown in Fig. 184. Referring to this sketch, the running submergence H_s is the distance from the water level to the point of an air inlet, and the total lift H_t is the distance from the working surface of the water to the point of discharge. An empirical formula to express the volume of free air required to lift 1 gal. water has been developed by the

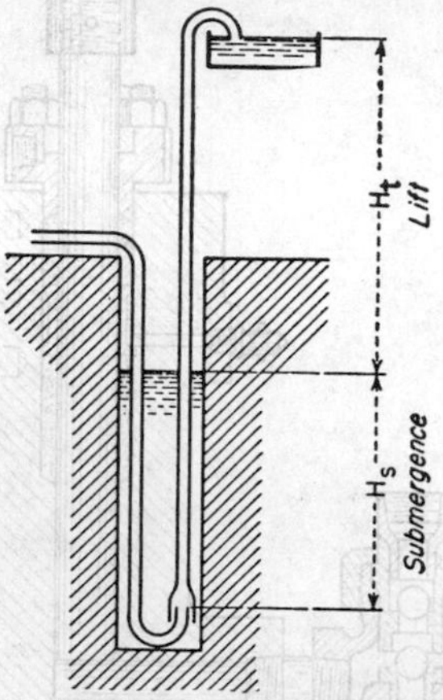

FIG. 184. Simplified sketch of an air lift, showing submergence and total head.

Ingersoll-Rand Co. from practice. According to this formula

$$V_a = 0.8 \frac{H_t}{C \log_e [(H_s + 34)/34]}$$

where V_a is the volume of free air, cu. ft., and *C* is a constant, values for which, varying with the total head, are

H_t, Ft.	*C*
10–60	245
61–200	233
201–500	216
501–650	185
651–750	156

In the design of an air lift, the submergence, as expressed by the ratio $H_s/(H_t + H_s)$ should vary from 0.66 for a lift of 20 ft. to 0.41 for a lift of 500 ft. The results given by this formula approximate those found in practice, but variations occur caused by the design of the foot pieces. The air pressure required to operate an air lift is given by

$$p = B + 0.434s$$

where *p* is in lb./sq. in. abs., *B* is the barometric pressure, lb./sq. in., and *s* is the submergence, ft.

The efficiency of an air lift, *i.e.*, the ratio of the water horsepower to the indicated air horsepower of the compressor, is about 70 per cent, except at starting, when it is considerably lower. Higher air pressure is also required for starting, when the pressure must be equivalent to the height of the water level, at rest, above the end of the air pipe.

DISPLACEMENT PUMPS

Displacement pumps operate by forcing the liquid that is to be moved from one point to another by means of pressure exerted by a gas or vapor. The acid egg is a device of this type frequently encountered by chemical engineers. It consists of an egg-shaped container which can be filled with a charge of the liquid that is to be pumped. This container is fitted with an inlet pipe for the charge, an outlet pipe for the discharge, and a pipe for the admission of compressed air or gas, as illustrated in Fig. 185. Pressure of air or gas on the surface of the

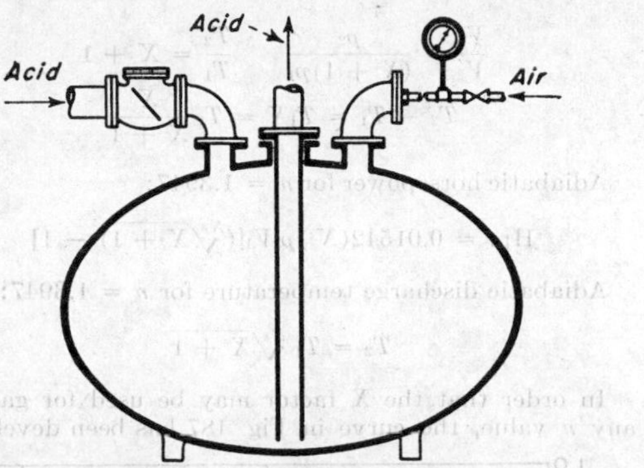

FIG. 185. A semiautomatic form of acid egg, with manually controlled air supply.

liquid forces it out the discharge pipe. Such pumps can be hand-operated or can be arranged for semiautomatic or automatic operation.

Jet pumps are another class of liquid-handling device that makes use of fluid pressure as an operating medium. Ejectors and injectors are the two types of jet pumps of interest to the chemical engineer. The ejector, also called siphon, exhauster, or eductor, is designed for use in operations where the head pumped against is low and is less than the head of the fluid used for pumping. The injector is a special type of jet pump, operated by steam and used for boiler feed and similar services, in which the fluid being pumped is discharged into a space under the same pressure as that of the steam that is used to operate the injector.

Figure 186 shows a simple design of jet pump of the ejector type. The pumping fluid enters through the nozzle at the left and passes through the venturi nozzle at the center and out the discharge opening at the right. As it passes into the venturi nozzle, it develops a suction that causes some of the fluid in the suction chamber to be

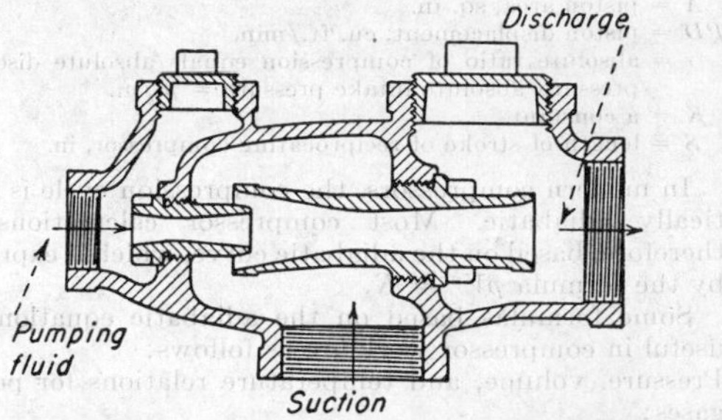

FIG. 186. Simple type of ejector for water service.

taken into the venturi nozzle and, entraining with the stream passing through the discharge, be delivered through this discharge.

The efficiency of an ejector or jet pump is low, being only a few per cent. The head developed by the ejector is also low, except in special types. The device has the disadvantage of diluting the fluid pumped by mixing it with the pumping fluid. In steam injectors for boiler feed and similar services, where the heat of the steam is recovered, the efficiency is close to 100 per cent.

The simple ejector or siphon is widely used, in spite of its low efficiency, for transferring liquids from one tank to another, for lifting acids, alkalies, or solid-containing liquids of an abrasive nature, and for emptying sumps.

COMPRESSION OF GAS

BY F. L. LUCKER

COMPRESSORS

Many applications in the chemical industry require the movement of gases and vapors. When these gases are below atmospheric pressure, the equipment used is known as *vacuum pumps* or *exhausters*. When the gases are above atmospheric pressure, the equipment used is known as *fans, blowers, boosters,* or *compressors.*

This equipment may be divided into the following general classifications:

1. *Fans,* which are widely used for pressures up to approximately ½ lb./sq. in.
2. *Rotary blowers* and *rotary compressors,* which are essentially constant-volume units when operating at constant speed.
3. *Turbo-* or *centrifugal blowers,* which are essentially constant-pressure units and particularly adaptable to large volumes at pressures up to approximately 40 lb./sq. in.
4. *Reciprocating compressors,* which are available in many sizes from a fraction of a horsepower up to 3000 hp. and for pressures from 5 to 25,000 lb./sq. in.
5. *Jets,* which are available for vacuums down to 5 μ.
6. *Boosters* or *circulators,* which are used to compress gas from a given pressure to a higher pressure for use in some particular process. When used to overcome pressure loss in a closed system, these are known as *circulators.*

It should be noted that *boosters, circulators,* and *vacuum pumps* may be of any of the types listed above. Because of variations in the service, they may have valves, seals, etc., of slightly different construction, but they are essentially standard-type compressors.

The fundamental theory for calculating compressor requirements and performance is based upon a *perfect,* or *ideal, gas.* Most gases follow the perfect gas laws closely when compressed to pressures up to 500 lb./sq. in., but many deviate considerably at the higher pressures. Complete information on the gas characteristics must be known before reliable calculations can be made. The following useful equations may be used for all diatomic gases. Further information is given on the theory of gas compression in Sec. 4 and data of the physical characteristics of gases in Sec. 3.

The *fundamental gas law* for perfect gases is $pV = MRT$. Any consistent system of symbols may be used in this formula. However, for convenience in this formula and those that follow, we have used the following symbols:

V = volume of gas, cu. ft.
p = pressure, lb./sq. in. abs.

M = weight of substance under consideration, lb.
R = gas constant (mechanical work done by the expansion of a unit weight of a perfect gas at constant pressure while heat is added to increase its temperature 1°F.).
n = ratio of specific heat at constant pressure to that at constant volume.
N_s = number of stages of compression.
N = number of first-stage pistons of reciprocating compressor.
r.p.m. = revolutions per minute.
T = absolute temperature, °F.

Other symbols to be used in the following formulas include

A = piston area, sq. in.
PD = piston displacement, cu. ft./min.
r = absolute ratio of compression equals absolute discharge pressure/absolute intake pressure = p_2/p_1.
K = a constant.
S = length of stroke of reciprocating compressor, in.

In modern compressors, the compression cycle is practically adiabatic. Most compressor calculations are therefore, based on the adiabatic curve, which is expressed by the formula $pV^n = K$.

Some formulas based on the adiabatic equation and useful in compressor work are as follows:
Pressure, volume, and temperature relations for perfect gases:

$$\frac{p_2}{p_1} = \left(\frac{V_1}{V_2}\right)^n \qquad \frac{T_2}{T_1} = \left(\frac{V_1}{V_2}\right)^{n-1} \qquad \frac{p_2}{p_1} = \left(\frac{T_2}{T_1}\right)^{\frac{n}{n-1}}$$

Theoretical adiabatic horsepower required by a single-stage compressor is calculated by

$$\text{Hp.} = 0.0043636V_1p_1\left(\frac{n}{n-1}\right)\left[\left(\frac{p_2}{p_1}\right)^{\left(\frac{n-1}{n}\right)} - 1\right]$$

Theoretical adiabatic horsepower required by a multistage compressor, assuming equal division of work between the cylinders and intercooling back to intake temperature:

$$\text{Hp.} = N_s \times 0.0043636V_1p_1\left(\frac{n}{n-1}\right)\left[\left(\frac{p_2}{p_1}\right)^{\frac{n-1}{N_s \times n}} - 1\right]$$

Theoretical adiabatic discharge temperature, assuming on multistage machines, equal division of work between the cylinders and intercooling back to intake temperature:

$$T_2 = T_1\left(\frac{p_2}{p_1}\right)^{\frac{n-1}{N_s \times n}}$$

Gas-compressor calculations are sometimes based on the isothermal curve because the calculations are much simpler. This is expressed by the formula $pV = K$. Theoretical isothermal horsepower for any number of stages is expressed as follows:

$$\text{Hp.} = 0.0043636p_1V_1 \log_e \frac{p_2}{p_1}$$

NOTE: Since horsepower is a rate of doing work, V_1 in the above formulas represents the cubic feet of gas per minute at intake conditions.

Piston displacement of reciprocating compressors with single-acting cylinders:

$$PD = \frac{A \times S \times N \times \text{r.p.m.}}{1728}$$

For double-acting cylinders:

$$PD = \frac{A \times S \times N \times \text{r.p.m.}}{874}$$

The above formula for double-acting cylinders makes a reasonable deduction for the volume occupied by the piston rod. Without this deduction, the divisor would be 864.

Air and a number of other gases have a value for n of about 1.39 to 1.41. To simplify calculations for these gases, tables have been made of the expression

$$(r^{\left(\frac{n-1}{n}\right)} - 1)$$

for a value of $n = 1.3947$ or $(r^{0.283} - 1)$. These are known as X factors, and they are given in Table 1.

Using X factors, the adiabatic formulas read as follows:
Adiabatic temperature, pressure, and volume relations for $n = 1.3947$:

$$\frac{V_1}{V_2} = \frac{p_2}{(X+1)p_1} \qquad \frac{T_2}{T_1} = X + 1$$

$$T_2 - T_1 = T_1X = T_2\frac{X}{X+1}$$

Adiabatic horsepower for $n = 1.3947$:

$$\text{Hp.} = 0.01542(N_s)p_1V_1[(\sqrt[N_s]{X+1}) - 1]$$

Adiabatic discharge temperature for $n = 1.3947$:

$$T_2 = T_1\sqrt[N_s]{X+1}$$

In order that the X factor may be used for gases of any n value, the curve in Fig. 187 has been developed.

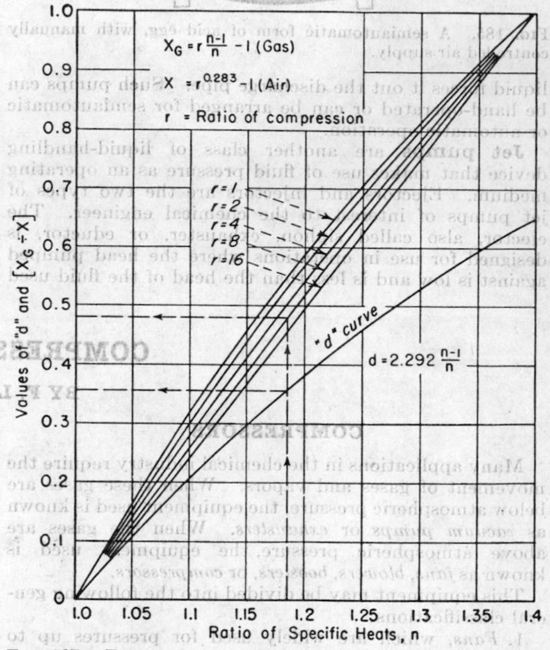

FIG. 187. Factors for use in adiabatic formula. Values of X to be used in finding X_G may be obtained from Table 1. *(By permission of Compressed Air Data.)*

Values of d and X_G/X may be found from this curve. Values of X_G are therefore obtained by multiplying X_G/X from the curve by the values of X found in Table 1.

The following formulas apply:

Adiabatic temperature, pressure, and volume relations:

$$\frac{V_1}{V_2} = \frac{p_2}{(X_G + 1)p_1} \qquad \frac{T_2}{T_1} = X_G + 1$$

$$T_2 - T_1 = T_1X_G = T_2\frac{X_G}{X_G + 1}$$

Table 1. Values of *X* for Normal Air and Perfect Diatomic Gases*

$$X = r^{0.283} - 1$$

r	0	1	2	3	4	5	6	7	8	9
1.00	0.00 000	028	057	085	113	141	169	198	226	254
1.01	.00 282	310	338	366	394	422	450	478	506	534
1.02	.00 562	590	618	646	673	701	729	757	785	812
1.03	.00 840	868	895	923	951	978	006	034	061	089
1.04	.01 116	144	171	199	226	253	281	308	336	363
1.05	.01 390	418	445	472	500	527	554	581	608	636
1.06	.01 663	690	717	744	771	798	825	852	879	906
1.07	.01 933	960	987	014	041	068	095	122	148	175
1.08	.02 202	229	255	282	309	336	362	389	416	442
1.09	.02 469	495	522	549	575	602	628	655	681	708
1.10	.02 734	760	787	813	840	866	892	919	945	971
1.11	.02 997	024	050	076	102	129	155	181	207	233
1.12	.03 259	285	311	337	363	389	415	441	467	493
1.13	.03 519	545	571	597	623	649	675	700	726	752
1.14	.03 778	804	829	855	881	906	932	958	983	009
1.15	.04 035	060	086	111	137	162	188	213	239	264
1.16	.04 290	315	341	366	391	417	442	467	493	518
1.17	.04 543	569	594	619	644	670	695	720	745	770
1.18	.04 796	821	846	871	896	921	946	971	996	021
1.19	.05 046	071	096	121	146	171	196	221	245	270
1.20	.05 295	320	345	370	394	419	444	469	493	518
1.21	.05 543	567	592	617	641	666	691	715	740	764
1.22	.05 789	813	838	862	887	911	936	960	985	009
1.23	.06 034	058	082	107	131	155	180	204	228	253
1.24	.06 277	301	325	350	374	398	422	446	470	495
1.25	.06 519	543	567	591	615	639	663	687	711	735
1.26	.06 759	783	807	831	855	879	903	927	951	974
1.27	.06 998	022	046	070	094	117	141	165	189	212
1.28	.07 236	260	283	307	331	354	378	402	425	449
1.29	.07 472	496	520	543	567	590	614	637	661	684
1.30	.07 708	731	754	778	801	825	848	871	895	918
1.31	.07 941	965	988	011	035	058	081	104	128	151
1.32	.08 174	197	220	243	267	290	313	336	359	382
1.33	.08 405	428	451	474	497	520	543	566	589	612
1.34	.08 635	658	681	704	727	750	773	795	818	841
1.35	.08 864	887	910	932	955	978	001	023	046	069
1.36	.09 092	114	137	160	182	205	228	250	273	295
1.37	.09 318	341	363	386	408	431	453	476	498	521
1.38	.09 543	566	588	611	633	655	678	700	723	745
1.39	.09 767	790	812	834	857	879	901	923	946	968
1.40	.09 990	012	035	057	079	101	123	145	168	190
1.41	.10 212	234	256	278	300	322	344	366	389	411
1.42	.10 433	455	477	499	521	542	564	586	608	630
1.43	.10 652	674	696	718	740	761	783	805	827	849
1.44	.10 871	892	914	936	958	979	001	023	045	066
1.45	.11 088	110	131	153	175	196	218	239	261	283
1.46	.11 304	326	347	369	390	412	433	455	476	498
1.47	.11 520	541	562	584	605	627	648	669	691	712
1.48	.11 734	755	776	798	819	840	862	883	904	925
1.49	.11 947	968	989	010	032	053	074	095	116	138
1.50	.12 159	180	201	222	243	264	286	307	328	349
1.51	.12 370	391	412	433	454	475	496	517	538	559
1.52	.12 580	601	622	643	664	685	706	726	747	768
1.53	.12 789	810	831	852	872	893	914	935	956	977
1.54	.12 997	018	039	060	080	101	122	142	163	184
1.55	.13 205	225	246	266	287	308	328	349	370	390
1.56	.13 411	431	452	472	493	513	534	554	575	595
1.57	.13 616	636	657	677	698	718	739	759	780	800
1.58	.13 820	841	861	881	902	922	942	963	983	003
1.59	.14 024	044	064	085	105	125	145	165	186	206
1.60	.14 226	246	267	287	307	327	347	367	387	408
1.61	.14 428	448	468	488	508	528	548	568	588	608
1.62	.14 628	648	668	688	708	728	748	768	788	808
1.63	.14 828	848	868	888	908	928	948	968	988	007
1.64	.15 027	047	067	087	107	126	146	166	186	206
1.65	.15 225	245	265	284	304	324	344	363	383	403
1.66	.15 423	442	462	481	501	521	540	560	580	599
1.67	.15 619	638	658	678	697	717	736	756	775	795
1.68	.15 814	834	853	873	892	912	931	951	970	990
1.69	.16 009	028	048	067	087	106	125	145	164	184
1.70	.16 203	222	242	261	280	299	319	338	357	377
1.71	.16 396	415	434	454	473	492	511	531	550	569
1.72	.16 588	607	626	646	665	684	703	722	741	760
1.73	.16 780	799	818	837	856	875	894	913	932	951
1.74	.16 970	989	008	027	046	065	084	103	122	141

r	0	1	2	3	4	5	6	7	8	9
1.75	0.17 160	179	198	217	236	255	274	292	311	330
1.76	.17 349	368	387	406	425	443	462	481	500	519
1.77	.17 538	556	575	594	613	631	650	669	688	706
1.78	.17 725	744	762	781	800	818	837	856	874	893
1.79	.17 912	930	949	968	986	005	023	042	061	079
1.80	.18 098	116	135	153	172	191	209	228	246	265
1.81	.18 283	302	320	339	357	376	394	412	431	449
1.82	.18 468	486	505	523	541	560	578	596	615	633
1.83	.18 652	670	688	707	725	743	762	780	798	816
1.84	.18 835	853	871	890	908	926	944	962	981	999
1.85	.19 017	035	054	072	090	108	126	144	163	181
1.86	.19 199	217	235	253	271	289	308	326	344	362
1.87	.19 380	398	416	434	452	470	488	506	524	542
1.88	.19 560	578	596	614	632	650	668	686	704	722
1.89	.19 740	758	776	794	811	829	847	865	883	901
1.90	.19 919	937	954	972	990	008	026	044	061	079
1.91	.20 097	115	133	150	168	186	204	221	239	257
1.92	.20 275	292	310	328	345	363	381	399	416	434
1.93	.20 452	469	487	504	522	540	557	575	593	610
1.94	.20 628	645	663	681	698	716	733	751	768	786
1.95	.20 804	821	839	856	874	891	909	926	944	961
1.96	.20 979	996	013	031	048	066	083	101	118	135
1.97	.21 153	170	188	205	222	240	257	275	292	309
1.98	.21 327	344	361	379	396	413	431	448	465	482
1.99	.21 500	517	534	552	569	586	603	620	638	655
2.00	.21 672	689	707	724	741	758	775	792	810	827
2.01	.21 844	861	878	895	913	930	947	964	981	998
2.02	.22 015	032	049	066	084	101	118	135	152	169
2.03	.22 186	203	220	237	254	271	288	305	322	339
2.04	.22 356	373	390	407	424	441	458	474	491	508
2.05	.22 525	542	559	576	593	610	627	644	660	677
2.06	.22 694	711	728	745	762	778	795	812	829	846
2.07	.22 863	879	896	913	930	946	963	980	997	013
2.08	.23 030	047	064	080	097	114	130	147	164	181
2.09	.23 197	214	231	247	264	281	297	314	331	347
2.10	.23 364	380	397	414	430	447	463	480	497	513
2.11	.23 530	546	563	579	596	613	629	646	662	679
2.12	.23 695	712	728	745	761	778	794	811	827	844
2.13	.23 860	877	893	909	926	942	959	975	992	008
2.14	.24 024	041	057	074	090	106	123	139	155	172
2.15	.24 188	204	221	237	253	270	286	302	319	335
2.16	.24 351	368	384	400	416	433	449	465	481	498
2.17	.24 514	530	546	563	579	595	511	627	644	660
2.18	.24 676	692	708	724	741	757	773	789	805	821
2.19	.24 838	854	870	886	902	918	934	950	966	983
2.20	.24 999	015	031	047	063	079	095	111	127	143
2.21	.25 159	175	191	207	223	239	255	271	287	303
2.22	.25 319	335	351	367	383	399	415	431	447	463
2.23	.25 479	495	511	526	542	558	574	590	606	622
2.24	.25 638	654	669	685	701	717	733	749	765	780
2.25	.25 796	812	828	844	859	875	891	907	923	938
2.26	.25 954	970	986	001	017	033	049	064	080	096
2.27	.26 112	127	143	159	175	190	206	222	237	253
2.28	.26 269	284	300	316	331	347	363	378	394	409
2.29	.26 425	441	456	472	488	503	519	534	550	566
2.30	.26 581	597	612	628	643	659	675	690	706	721
2.31	.26 737	752	768	783	799	814	830	845	861	876
2.32	.26 892	907	923	938	954	969	984	000	015	031
2.33	.27 046	062	077	092	108	123	139	154	169	185
2.34	.27 200	216	231	246	262	277	292	308	323	338
2.35	.27 354	369	384	400	415	430	446	461	476	492
2.36	.27 507	522	538	553	568	583	599	614	629	644
2.37	.27 660	675	690	705	721	736	751	766	781	797
2.38	.27 812	827	842	857	873	888	903	918	933	948
2.39	.27 964	979	994	009	024	039	054	070	085	100
2.40	.28 115	130	145	160	175	190	205	220	236	251
2.41	.28 266	281	296	311	326	341	356	371	386	401
2.42	.28 416	431	446	461	476	491	506	521	536	551
2.43	.28 566	581	596	611	626	641	656	671	686	701
2.44	.28 716	730	745	760	775	790	805	820	835	850
2.45	.28 865	879	894	909	924	939	954	969	984	998
2.46	.29 013	028	043	058	073	087	102	117	132	147
2.47	.29 162	176	191	206	221	235	250	265	280	295
2.48	.29 309	324	339	353	368	383	398	412	427	442
2.49	.29 457	471	486	501	515	530	545	559	574	589

* Printed by permission of Compressed Air Data.

Table 1. Values of *X* for Normal Air and Perfect Diatomic Gases*—(*Continued*)

r	0	1	2	3	4	5	6	7	8	9	r	0	1	2	3	4	5	6	7	8	9
2.50	0.29 604	618	633	647	662	677	691	706	721	735	2.75	0.33 147	161	174	188	202	215	229	243	256	270
2.51	.29 750	765	779	794	808	823	838	852	867	881	2.76	.33 284	297	311	325	338	352	366	379	393	407
2.52	.29 896	911	925	940	954	969	984	998	013	027	2.77	.33 420	434	448	461	475	488	502	516	529	543
2.53	.30 042	056	071	085	100	114	129	144	158	173	2.78	.33 556	570	584	597	611	624	638	651	665	679
2.54	.30 187	202	216	231	245	260	274	289	303	318	2.79	.33 692	706	719	733	746	760	773	787	801	814
2.55	.30 332	346	361	375	390	404	419	433	448	462	2.80	.33 828	841	855	868	882	895	909	922	936	949
2.56	.30 476	491	505	520	534	548	563	577	592	606	2.81	.33 963	976	990	003	017	030	044	057	070	084
2.57	.30 620	635	649	663	678	692	707	721	735	750	2.82	.34 097	111	124	138	151	165	178	191	205	218
2.58	.30 764	778	793	807	821	836	850	864	879	893	2.83	.34 232	245	259	272	285	299	312	326	339	352
2.59	.30 907	921	936	950	964	979	993	007	021	036	2.84	.34 366	379	393	406	419	433	446	459	473	486
2.60	.31 050	064	079	093	107	121	136	150	164	178	2.85	.34 500	513	526	540	553	566	580	593	606	620
2.61	.31 193	207	221	235	249	264	278	292	306	320	2.86	.34 633	646	660	673	686	700	713	726	739	753
2.62	.31 335	349	363	377	391	405	420	434	448	462	2.87	.34 766	779	793	806	819	832	846	859	872	886
2.63	.31 476	490	505	519	533	547	561	575	589	603	2.88	.34 899	912	925	939	952	965	978	991	005	018
2.64	.31 618	632	646	660	674	688	702	716	730	744	2.89	.35 031	044	058	071	084	097	110	124	137	150
2.65	.31 759	773	787	801	815	829	843	857	871	885	2.90	.35 163	176	190	203	216	229	242	255	269	282
2.66	.31 899	913	927	941	955	969	983	997	011	025	2.91	.35 295	308	321	334	347	361	374	387	400	413
2.67	.32 039	053	067	081	095	109	123	137	151	165	2.92	.35 426	439	452	466	479	492	505	518	531	544
2.68	.32 179	193	207	221	235	249	262	276	290	304	2.93	.35 557	570	584	597	610	623	636	649	662	675
2.69	.32 318	332	346	360	374	388	402	416	429	443	2.94	.35 688	701	714	727	740	753	767	780	793	806
2.70	.32 457	471	485	499	513	527	540	554	568	582	2.95	.35 819	832	845	858	871	884	897	910	923	936
2.71	.32 596	610	624	637	651	665	679	693	707	720	2.96	.35 949	962	975	988	001	014	027	040	053	066
2.72	.32 734	748	762	776	789	803	817	831	845	858	2.97	.36 079	092	105	118	131	144	157	169	182	195
2.73	.32 872	886	900	913	927	941	955	968	982	996	2.98	.36 208	221	234	247	260	273	286	299	312	324
2.74	.33 010	023	037	051	065	078	092	106	119	133	2.99	.36 337	350	363	376	389	402	415	428	440	453

r	0	1	2	3	4	5	6	7	8	9
3.0	0.3647	0.3659	0.3672	0.3685	0.3698	0.3711	0.3723	0.3736	0.3749	0.3761
3.1	.3774	.3786	.3799	.3811	.3824	.3836	.3849	.3861	.3874	.3886
3.2	.3898	.3911	.3923	.3935	.3947	.3959	.3971	.3984	.3996	.4008
3.3	.4020	.4032	.4044	.4056	.4068	.4080	.4091	.4103	.4115	.4127
3.4	.4139	.4150	.4162	.4174	4186	.4197	.4209	.4220	.4232	.4244
3.5	.4255	.4267	.4278	.4290	.4301	.4313	.4324	.4335	.4347	.4358
3.6	.4369	.4380	.4392	.4403	.4414	.4425	.4437	.4448	.4459	.4470
3.7	.4481	.4492	.4503	.4514	.4525	.4536	.4547	.4558	.4569	.4580
3.8	.4591	.4602	.4612	.4623	.4634	.4645	.4656	.4666	.4677	.4688
3.9	.4698	.4709	.4720	.4730	.4741	.4752	.4762	.4773	.4783	.4794
4.0	.4804	.4815	.4825	.4835	.4846	.4856	.4867	.4877	.4887	.4898
4.1	.4908	.4918	.4928	.4939	.4949	.4959	.4970	.4980	.4990	.5000
4.2	.5010	.5020	.5030	.5040	.5050	.5060	.5070	.5080	.5090	.5100
4.3	.5110	.5120	.5130	.5140	.5150	.5160	.5170	.5179	.5189	.5199
4.4	.5209	.5219	.5228	.5238	.5248	.5258	.5267	.5277	.5287	.5296
4.5	.5306	.5316	.5325	.5335	.5344	.5354	.5363	.5373	.5382	.5392
4.6	.5401	.5411	.5420	.5430	.5439	.5449	.5458	.5467	.5477	.5486
4.7	.5495	.5505	.5514	.5523	.5533	.5542	.5551	.5560	.5570	.5579
4.8	.5588	.5597	.5606	.5616	.5625	.5634	.5643	.5652	.5661	.5670
4.9	.5679	.5688	.5697	.5706	.5715	.5724	.5733	.5742	.5751	.5760
5.0	.5769	.5778	.5787	.5796	.5805	.5814	.5822	.5831	.5840	.5849
5.1	.5858	.5867	.5875	.5884	.5893	.5902	.5910	.5919	.5928	.5936
5.2	.5945	.5954	.5962	.5971	.5980	.5988	.5997	.6006	.6014	.6023
5.3	.6031	.6040	.6048	.6057	.6065	.6074	.6082	.6091	.6099	.6108
5.4	.6116	.6125	.6133	.6142	.6150	.6159	.6167	.6175	.6184	.6192
5.5	.6200	.6209	.6217	.6225	.6234	.6242	.6250	.6258	.6267	.6275
5.6	.6283	.6291	.6300	.6308	.6316	.6324	.6332	.6340	.6349	.6357
5.7	.6365	.6373	.6381	.6389	.6397	.6405	.6413	.6421	.6430	.6438
5.8	.6446	.6454	.6462	.6470	.6478	.6486	.6494	.6502	.6509	.6517
5.9	.6525	.6533	.6541	.6549	.6557	.6565	.6573	.6581	.6588	.6596
6.0	.6604	.6612	.6620	.6628	.6635	.6643	.6651	.6659	.6666	.6674
6.1	.6682	.6690	.6697	.6705	.6713	.6721	.6729	.6736	.6744	.6752
6.2	.6759	.6767	.6774	.6782	.6789	.6797	.6805	.6812	.6820	.6827
6.3	.6835	.6843	.6850	.6858	.6865	.6873	.6880	.6888	.6895	.6903
6.4	.6910	.6918	.6925	.6933	.6940	.6948	.6955	.6963	.6970	.6978
6.5	.6985	.6992	.7000	.7007	.7014	.7021	.7028	.7036	.7043	.7050
6.6	.7058	.7065	.7073	.7080	.7087	.7095	.7102	.7110	.7117	.7124
6.7	.7131	.7138	.7145	.7153	.7160	.7167	.7174	.7181	.7189	.7196
6.8	.7203	.7210	.7217	.7224	.7232	.7239	.7246	.7253	.7260	.7267
6.9	.7274	.7281	.7288	.7295	.7302	.7309	.7316	.7323	.7330	.7338
7.0	.7345	.7352	.7359	.7366	.7373	.7380	.7386	.7393	.7400	.7407
7.1	.7414	.7421	.7428	.7435	.7442	.7449	.7456	.7463	.7470	.7477
7.2	.7483	.7490	.7497	.7504	.7511	.7518	.7524	.7531	.7538	.7545
7.3	.7552	.7559	.7565	.7572	.7579	.7586	.7592	.7599	.7606	.7613
7.4	.7620	.7626	.7633	.7640	.7646	.7653	.7660	.7666	.7673	.7680
7.5	.7687	.7693	.7700	.7706	.7713	.7720	.7726	.7733	.7740	.7746
7.6	.7753	.7760	.7766	.7773	.7779	.7786	.7792	.7799	.7806	.7812
7.7	.7819	.7825	.7832	.7838	.7845	.7851	.7858	.7864	.7871	.7877
7.8	.7884	.7890	.7897	.7903	.7910	.7916	.7923	.7929	.7936	.7942
7.9	.7949	.7955	.7961	.7968	.7974	.7981	.7987	.7993	.8000	.8006

* Printed by permission of Compressed Air Data.

Table 1. Values of X for Normal Air and Perfect Diatomic Gases*—(*Concluded*)

r	0	1	2	3	4	5	6	7	8	9
8.0	0.8013	0.8019	0.8025	0.8032	0.8038	0.8044	0.8051	0.8057	0.8063	0.8070
8.1	.8076	.8082	.8089	.8095	.8101	.8108	.8114	.8120	.8126	.8133
8.2	.8139	.8145	.8151	.8158	.8164	.8170	.8176	.8183	.8189	.8195
8.3	.8201	.8207	.8214	.8220	.8226	.8232	.8238	.8245	.8251	.8257
8.4	.8263	.8269	.8275	.8281	.8228	.8294	.8300	.8306	.8312	.8318
8.5	.8324	.8330	.8336	.8343	.8349	.8355	.8361	.8367	.8373	.8379
8.6	.8385	.8391	.8397	.8403	.8409	.8415	.8421	.8427	.8433	.8439
8.7	.8445	.8451	.8457	.8463	.8469	.8475	.8481	.8487	.8493	.8499
8.8	.8505	.8511	.8517	.8523	.8529	.8535	.8541	.8547	.8552	.8558
8.9	.8564	.8570	.8576	.8582	.8588	.8594	.8600	.8605	.8611	.8617
9.0	.8623	.8629	.8635	.8641	.8646	.8652	.8658	.8664	.8670	.8676
9.1	.8681	.8687	.8693	.8699	.8705	.8710	.8716	.8722	.8728	.8734
9.2	.8739	.8745	.8751	.8757	.8762	.8768	.8774	.8779	.8785	.8791
9.3	.8797	.8802	.8808	.8814	.8819	.8825	.8831	.8837	.8842	.8848
9.4	.8854	.8859	.8865	.8871	.8876	.8882	.8888	.8893	.8899	.8905
9.5	.8910	.8916	.8921	.8927	.8933	.8938	.8944	.8949	.8955	.8961
9.6	.8966	.8972	.8977	.8983	.8989	.8994	.9000	.9005	.9011	.9016
9.7	.9022	.9028	.9033	.9039	.9044	.9050	.9055	.9061	.9066	.9072
9.8	.9077	.9083	.9088	.9094	.9099	.9105	.9110	.9116	.9121	.9127
9.9	.9132	.9138	.9143	.9149	.9154	.9159	.9165	.9170	.9176	.9181
10.0	.9187	.9192	.9198	.9203	.9208	.9214	.9219	.9225	.9230	.9235
10.1	.9241	.9246	.9252	.9257	.9262	.9268	.9273	.9278	.9284	.9289
10.2	.9295	.9300	.9305	.9311	.9316	.9321	.9327	.9332	.9337	.9343
10.3	.9348	.9353	.9358	.9364	.9369	.9374	.9380	.9385	.9390	.9396
10.4	.9401	.9406	.9411	.9417	.9422	.9427	.9432	.9438	.9443	.9448
10.5	.9453	.9459	.9464	.9469	.9474	.9480	.9485	.9490	.9495	.9500
10.6	.9506	.9511	.9516	.9521	.9526	.9532	.9537	.9542	.9547	.9552
10.7	.9558	.9563	.9568	.9573	.9578	.9583	.9589	.9594	.9599	.9604
10.8	.9609	.9614	.9619	.9625	.9630	.9635	.9640	.9645	.9650	.9655
10.9	.9660	.9665	.9671	.9676	.9681	.9686	.9691	.9696	.9701	.9706
11.0	.9711	.9716	.9721	.9726	.9732	.9737	.9742	.9747	.9752	.9757
11.1	.9762	.9767	.9772	.9777	.9782	.9787	.9792	.9797	.9802	.9807
11.2	.9812	.9817	.9822	.9827	.9832	.9837	.9842	.9847	.9852	.9857
11.3	.9862	.9867	.9872	.9877	.9882	.9887	.9892	.9897	.9902	.9907
11.4	.9912	.9916	.9921	.9926	.9931	.9936	.9941	.9946	.9951	.9956
11.5	.9961	.9966	.9971	.9975	.9980	.9985	.9990	.9995	1.0000	1.0005
11.6	1.0010	1.0015	1.0019	1.0024	1.0029	1.0034	1.0039	1.0044	1.0049	1.0054
11.7	1.0058	1.0063	1.0068	1.0073	1.0078	1.0083	1.0087	1.0092	1.0097	1.0102
11.8	1.0107	1.0112	1.0116	1.0121	1.0126	1.0131	1.0136	1.0140	1.0145	1.0150
11.9	1.0155	1.0160	1.0164	1.0169	1.0174	1.0179	1.0184	1.0188	1.0193	1.0198
12.0	1.0203	1.0207	1.0212	1.0217	1.0222	1.0226	1.0231	1.0236	1.0241	1.0245

* Printed by permission of Compressed Air Data.

NOTE: Taken from Moss and Smith, Engineering Computations for Air and Gases, *Trans. Am. Soc. Mech. Engrs.*, vol. 52, 1930, paper APM-52-8. For nozzles $r = p_1/p_2$. For compressors and exhausters $r = p_2/p_1$.

r	x	r	x	r	x	r	x	r	x	r	x	r	x	r	x	r	x
12.5	1.0428	15.0	1.1520	17.5	1.2479	20.0	1.3345	22.5	1.4136	25.0	1.4867	27.5	1.5546	30.0	1.6183	32.5	1.6783
13.0	1.0666	15.5	1.1720	18.0	1.2659	20.5	1.3509	23.0	1.4287	25.5	1.5006	28.0	1.5678	30.5	1.6306	33.0	1.6899
13.5	1.0887	16.0	1.1916	18.5	1.2835	21.0	1.3669	23.5	1.4435	26.0	1.5144	28.5	1.5794	31.0	1.6434	23.5	1.7014
14.0	1.1103	16.5	1.2108	19.0	1.3008	21.5	1.3828	24.0	1.4581	26.5	1.5280	29.0	1.5933	31.5	1.6647	34.0	1.7127
14.5	1.1314	17.0	1.2295	19.5	1.3189	22.0	1.3983	24.5	1.4725	27.0	1.5414	29.5	1.6059	32.0	1.6666	34.5	1.7240

Values of X from 12.5 to 34.5 calculated by Ingersoll-Rand Co.

Adiabatic horsepower:

$$\text{Hp.} = \frac{0.01 N_s p_1 V_1}{d} [(\sqrt[N_s]{X_G + 1}) - 1]$$

Adiabatic discharge temperature:

$$T_2 = T_1 \sqrt[N_s]{X_G + 1}$$

Reciprocating Compressors

Reciprocating compressors are the type most widely used in the chemical industry. They are furnished for steam-engine, electric-motor, and gas- or diesel-engine drive; and, in a few cases, for turbine drive through reduction gears.

Reciprocating compressors are furnished either *single-stage* or *multistage*. The number of stages is determined by the compression ratio p_2/p_1. The compression ratio per stage is generally limited to 4, although small-sized units are furnished with a compression ratio as high as 8 and even higher.

Single-acting, air-cooled and water-cooled compressors are available in sizes up to about 100 hp. Such units are available in one, two, three, or four stages for pressures as high as 3500 lb./sq. in. These machines are seldom used for gas compression because of the difficulty of preventing gas leakage and contamination of lubricating oil.

The compressors most commonly used for compressing gases have a crosshead to which the connecting rod and piston rod are connected. This provides a straight-line motion for the piston rod and enables adequate packing to be used. Figure 188 illustrates a simple single-stage machine of this type having a double-acting piston. Either single-acting (Fig. 189) or double-acting pistons (Fig. 190) may be used, depending on the size of the machine and the number of stages. In some machines double-acting pistons are used in the first stages and single-acting in the later stages (see Fig. 191).

On **multistage** machines, intercoolers are provided between stages. These are essentially heat exchangers which remove the heat of compression from the gas and reduce its temperature to approximately the temperature existing at the compressor intake. Such cooling reduces the volume of gas going to the high-pressure cylinders, reduces the horsepower required for compression; and,

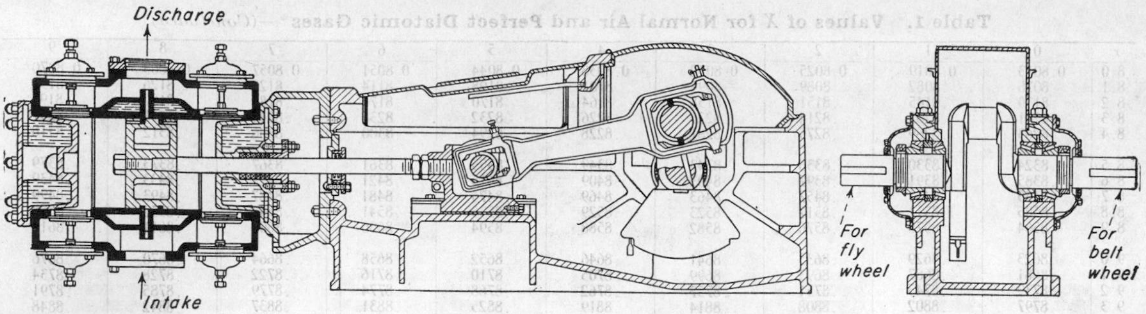

FIG. 188. A typical single-stage water-cooled compressor.

at high pressures, keeps the temperature within safe operating limits.

Figure 192 illustrates a two-stage compressor end such as might be used on the compressor illustrated in Fig. 188 A three-stage compressor end is illustrated in Fig. 191.

Compressors with horizontal cylinders such as illustrated in Figs. 188 to 192 are most commonly used because of their accessibility. However, machines are also built with vertical cylinders and other arrangements such as right angle (one horizontal and one vertical) and V-angle.

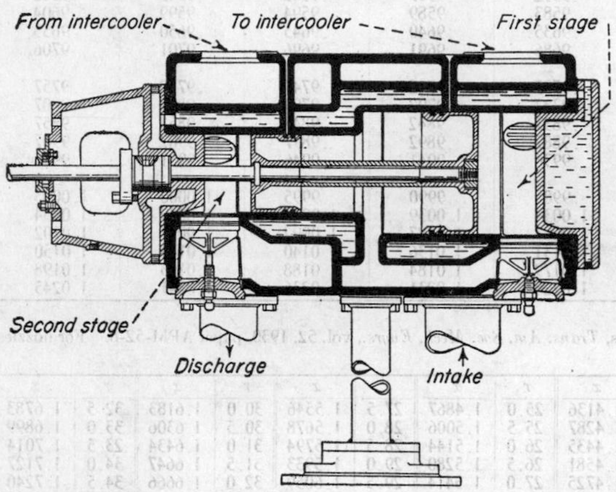

FIG. 189. Two-stage single-acting opposed piston in a single-step-type cylinder.

Compressors up to around 100 hp. usually have a single center-throw crank, as illustrated in Fig. 188. In sizes above this, compressors are commonly of duplex construction with cranks on either end of the shaft (see Fig. 193). Some large synchronous motor-driven units are of four-corner construction, *i.e.*, they are of double-duplex construction with two connecting rods from each of the two crank throws (see Fig. 194). Steam-driven compressors have one or more steam cylinders which are connected directly by piston rod or tie rods to the gas cylinder piston or crosshead (see Fig. 195).

Control Devices. In many installations the usage of gas is irregular, and some means of controlling the output of the compressor is therefore necessary. In other cases constant output is required despite variations in discharge pressure, and the control device must operate to maintain a constant compressor speed. Compressor capacity, speed, or pressure may be varied in accordance with the requirements. The nature of the control device will depend on what function is to be regulated; whether pressure, volume, temperature, or some other factor determines the amount of regulation required; and the type of the compressor driver.

The most common control requirement is regulation of capacity. Many capacity controls, or unloading devices, as they are usually termed, are actuated by the pressure on the discharge side of the compressor. A falling pressure indicates that gas is being used faster than it is being compressed and that more gas is required. A rising pressure indicates that more gas is being com-

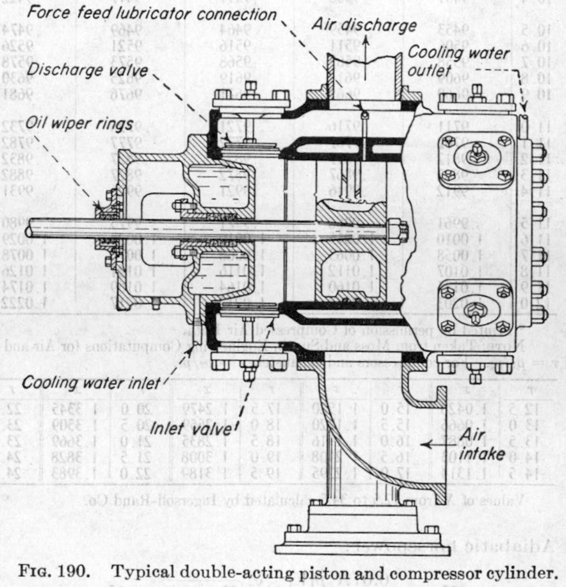

FIG. 190. Typical double-acting piston and compressor cylinder.

pressed than is being used and that less gas is therefore required.

An obvious method of controlling the capacity of a compressor is to vary the speed. This method is applicable to steam-driven compressors and to units driven by internal-combustion engines. In these cases the regulator actuates the steam-admission or fuel-admission valve on the compressor driver and thus controls the speed.

Motor-driven compressors usually operate at constant speed, and other methods of controlling the capacity are necessary. On reciprocating compressors up to about 100 hp., two types of control are usually available. These are automatic-start-and-stop control and constant-speed control.

Automatic-start-and-stop control, as its name implies, stops or starts the compressor by means of a pressure-actuated switch as the gas demand varies. It should be used only when the demand for gas will be intermittent.

Constant-speed control should be used when gas de-

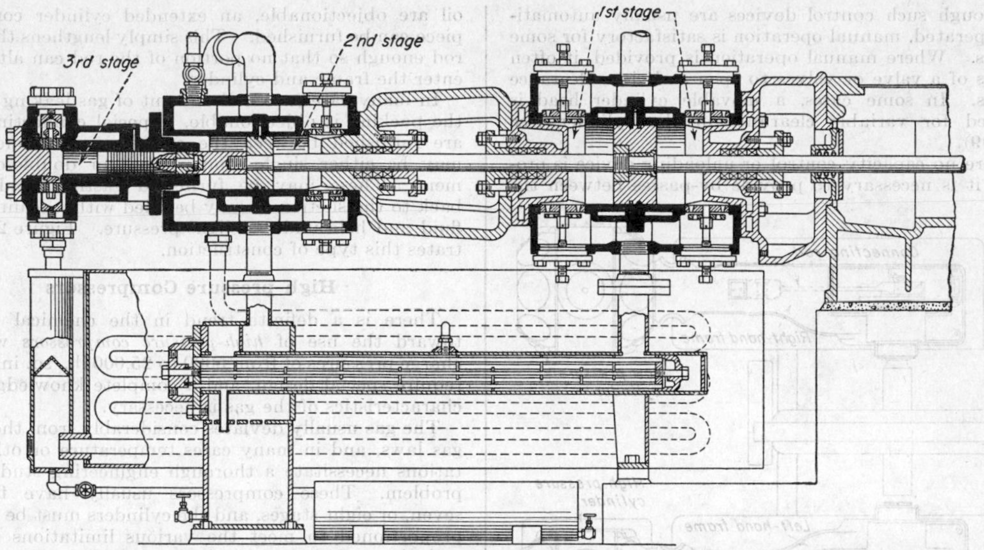

Fig. 191. Three-stage compressor cylinders. (Double-acting first stage, single-acting opposed second and third stages.)

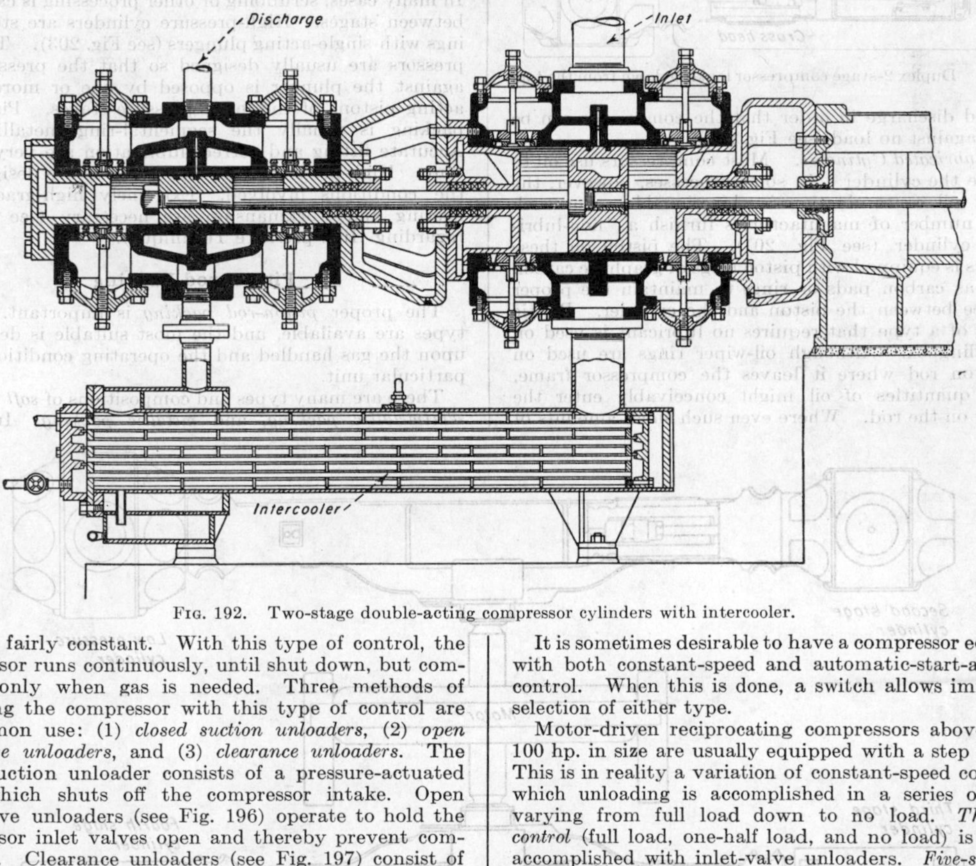

Fig. 192. Two-stage double-acting compressor cylinders with intercooler.

mand is fairly constant. With this type of control, the compressor runs continuously, until shut down, but compresses only when gas is needed. Three methods of unloading the compressor with this type of control are in common use: (1) *closed suction unloaders*, (2) *open inlet-valve unloaders*, and (3) *clearance unloaders*. The closed suction unloader consists of a pressure-actuated valve which shuts off the compressor intake. Open inlet-valve unloaders (see Fig. 196) operate to hold the compressor inlet valves open and thereby prevent compression. Clearance unloaders (see Fig. 197) consist of pockets or small reservoirs which are opened when unloading is desired. The gas is compressed into them on the compression stroke and reexpands into the cylinder on the return stroke, thus preventing the compression of additional gas.

It is sometimes desirable to have a compressor equipped with both constant-speed and automatic-start-and-stop control. When this is done, a switch allows immediate selection of either type.

Motor-driven reciprocating compressors above about 100 hp. in size are usually equipped with a step control. This is in reality a variation of constant-speed control in which unloading is accomplished in a series of steps, varying from full load down to no load. *Three-step control* (full load, one-half load, and no load) is usually accomplished with inlet-valve unloaders. *Five-step control* (full load, three-fourths load, one-half load, one-fourth load, and no load) is accomplished by means of clearance pockets (see Fig. 198). On some makes of machines, inlet-valve and clearance control unloading are used in combination.

Although such control devices are usually automatically operated, manual operation is satisfactory for some services. Where manual operation is provided, it often consists of a valve or valves to open and close clearance pockets. In some cases, a movable cylinder head is provided for variable clearance in the cylinder (see Fig. 199).

Where no capacity control or unloading device is provided, it is necessary to provide by-passes between the

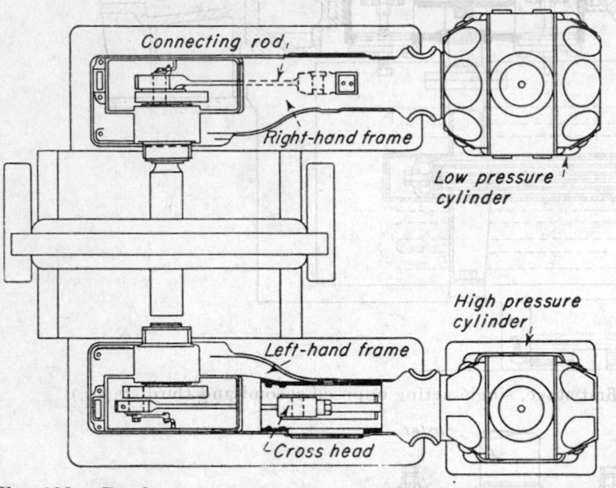

FIG. 193. Duplex 2-stage compressor looking down from the top.

inlet and discharge in order that the compressor can be started against no load (see Fig. 200).

Non-lubricated Cylinders. Most compressors use oil to lubricate the cylinder. In some processes, however, the slightest oil contamination is objectionable. For such cases a number of manufacturers furnish a "non-lubricated" cylinder (see Fig. 201). The piston on these cylinders is equipped with piston rings of graphitic carbon as well as carbon pads or rings to maintain the proper clearance between the piston and the cylinder. Plastic packing of a type that requires no lubricant is used on the stuffingbox. Although oil-wiper rings are used on the piston rod where it leaves the compressor frame, minute quantities of oil might conceivably enter the cylinder on the rod. Where even such small amounts of oil are objectionable, an extended cylinder connecting piece can be furnished. This simply lengthens the piston rod enough so that no portion of the rod can alternately enter the frame and cylinder.

In many cases, a small amount of gas leaking through the packing is objectionable. Special connecting pieces are furnished between the cylinder and frame, which may be either single-compartment or double-compartment. These may be furnished gastight and vented back to the suction or may be filled with a sealing gas or fluid and held under a slight pressure. Figure 202 illustrates this type of construction.

High-pressure Compressors

There is a definite trend in the chemical industry toward the use of *high-pressure compressors* with discharge pressures of from 5000 to 25,000 lb./sq. in. These require special design, and a complete knowledge of the characteristics of the gas is necessary.

The gas usually deviates considerably from the perfect gas laws, and in many cases temperature or other limitations necessitate a thorough engineering study of the problem. These compressors usually have five, six, seven, or eight stages, and the cylinders must be properly proportioned to meet the various limitations involved and also to balance the load among the various stages. In many cases, scrubbing or other processing is carried on between stages. High-pressure cylinders are steel forgings with single-acting plungers (see Fig. 203). The compressors are usually designed so that the pressure load against the plunger is opposed by one or more single-acting pistons of the lower pressure stages. Piston-rod packing is usually the segmental-ring metallic type. Accurate fitting and correct lubrication are very important. High-pressure compressor valves are designed for the conditions involved. Extremely high-grade engineering and workmanship are necessary (see Sec. 18 regarding High-pressure Technique).

Piston-rod Packing

The proper *piston-rod packing* is important. Many types are available, and the most suitable is dependent upon the gas handled and the operating conditions for a particular unit.

There are many types and compositions of *soft packing, semimetallic packing,* and *metallic packing.* In many

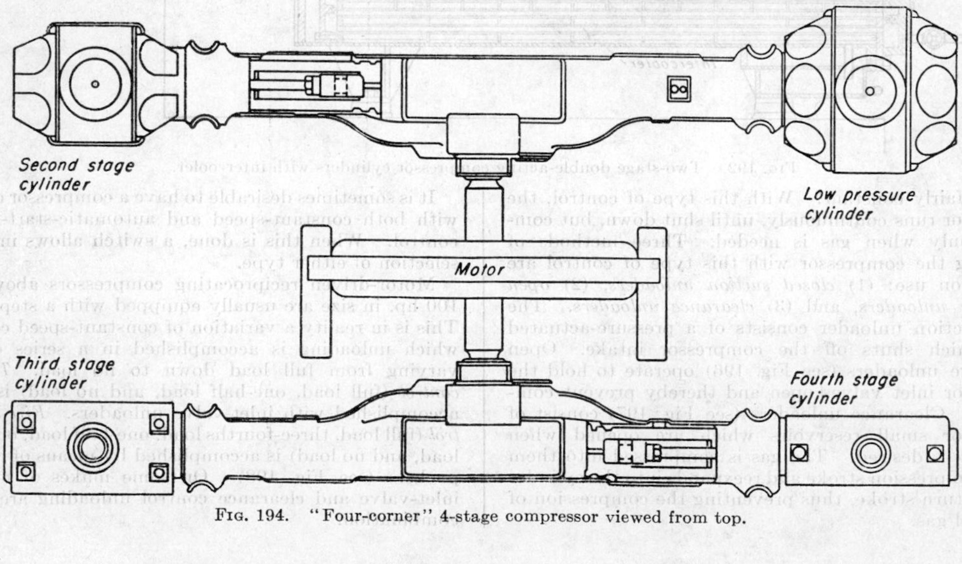

FIG. 194. "Four-corner" 4-stage compressor viewed from top.

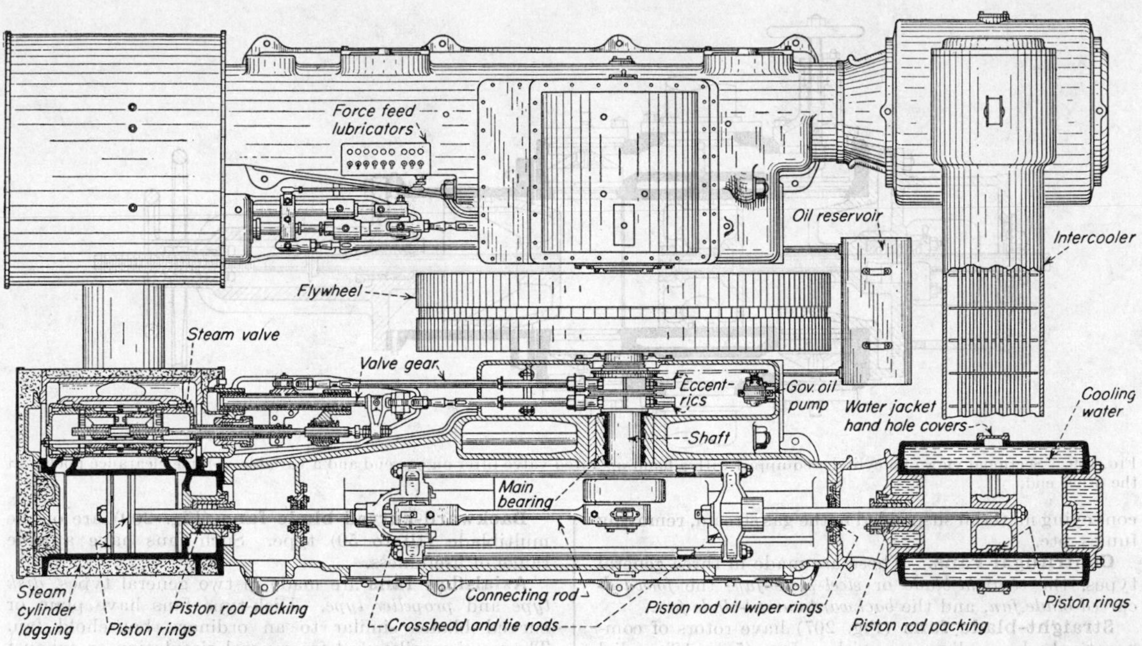

FIG. 195. Steam-driven 2-stage compressor.

cases, metallic packing is to be recommended. A typical low-pressure packing arrangement is shown in Fig. 204. A high-pressure packing arrangement is shown in Fig. 205.

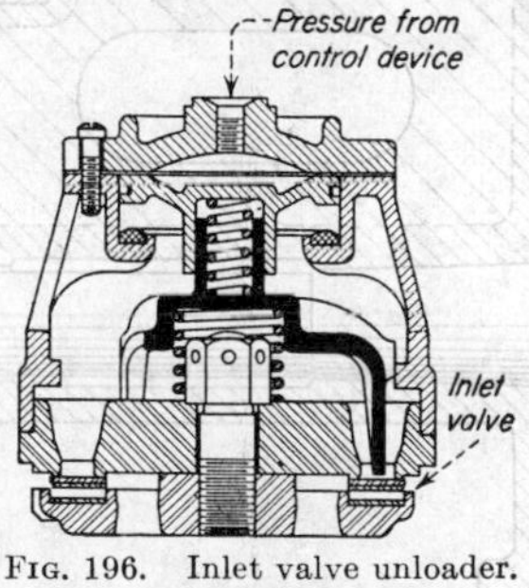

FIG. 196. Inlet valve unloader.

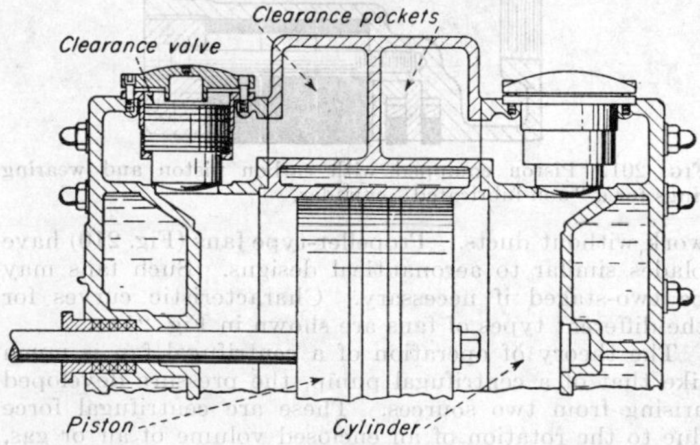

FIG. 197. Ingersoll-Rand "Clearance Control" cylinder.

Where wet, volatile, or dangerous gases are handled or where the service is intermittent, an auxiliary packing gland and soft packing are usually employed (see Fig. 206).

FANS

Fans are used for low pressures, in general, for pressure heads of less than 0.5 lb./sq. in. They are usually

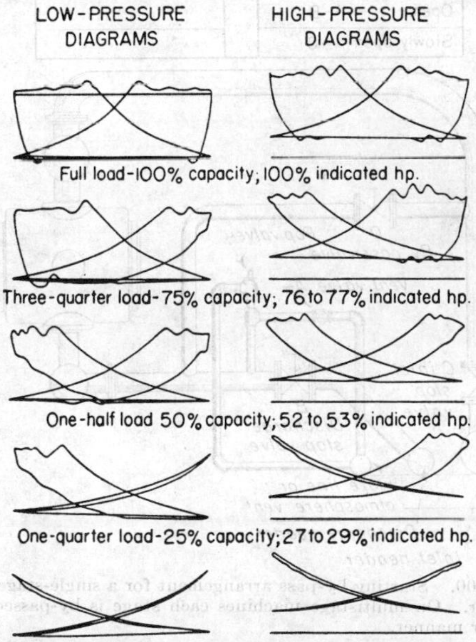

FIG. 198. Actual indicator diagram showing operation of clearance control at five load points of a 2-stage compressor.

classified as of the centrifugal type or of the axial-flow type. Both types are used for ventilating work, supplying draft to boilers and furnaces, moving large volumes of air or gas through ducts, supplying air for drying,

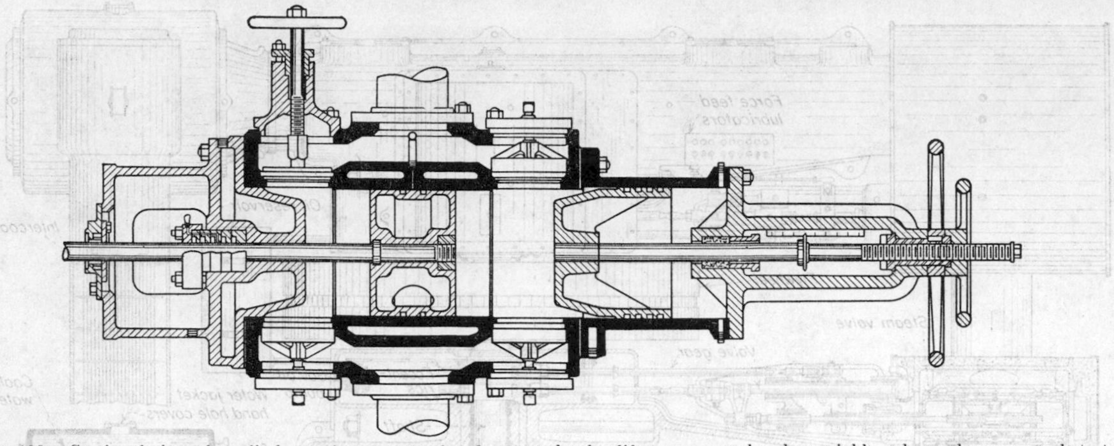

FIG. 199. Sectional view of a cylinder equipped with a hand-operated valve lifter on one end and a variable-volume clearance pocket at the other end.

conveying material suspended in the gas stream, removing fumes, etc.

Centrifugal Fans. These are made in three general types, the *straight-blade* or *steel-plate fan*, the *forward-curved-blade fan*, and the *backward-curved-blade fan*.

Straight-blade fans (Fig. 207) have rotors of comparatively large diameter with a few (5 to 12) radial

Starting compressor	Stopping compressor
Start with A and D open	Close - - - - C
Close - - - - - - D	Close - - - B
Close - - - - - - - A	Open - - - A and D
Open - - - - - - - B	
Slowly open - - - C	

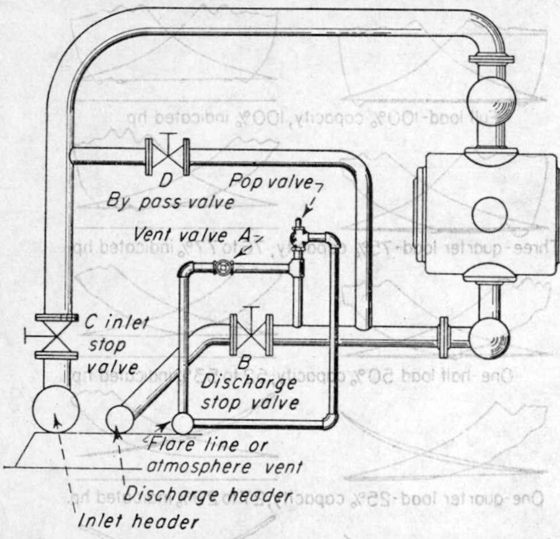

FIG. 200. Starting by-pass arrangement for a single-stage compressor. On multistage machines each stage is by-passed in a similar manner.

blades resembling paddle wheels. These operate at comparatively low speed. They are often used in exhaust work, particularly where wastes are carried in the air stream.

Forward-curved-blade fans (Fig. 208) are usually of the multiblade (20 to 64) "Sirocco" type. The rotors are of smaller diameter, and they operate at higher speeds than straight-blade units.

Backward-curved-blade fans (Fig. 209) are of the multiblade (10 to 50) type. Such fans have a wide range of usefulness.

Axial-flow fans are made in two general types, *disk type* and *propeller type*. Disk-type fans have plain or curved blades similar to an ordinary household fan. They are usually used for general circulation or exhaust

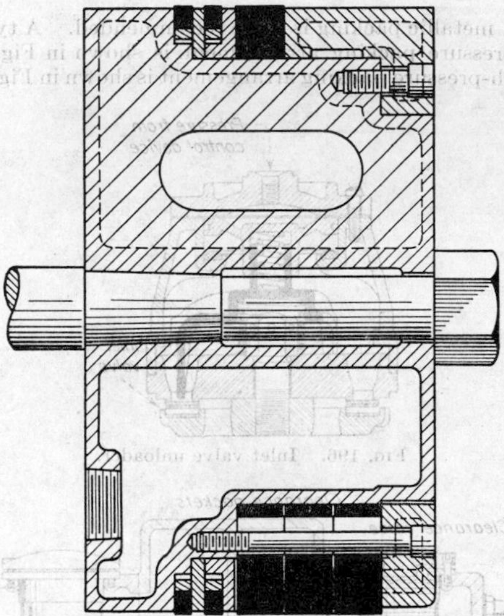

FIG. 201. Piston equipped with carbon piston and wearing rings for a "non-lubricated" cylinder.

work without ducts. Propeller-type fans (Fig. 210) have blades similar to aeronautical designs. Such fans may be two-staged if necessary. Characteristic curves for the different types of fans are shown in Fig. 211.

The theory of operation of a centrifugal fan is much like that of a centrifugal pump, the pressure developed arising from two sources. These are centrifugal force due to the rotation of an enclosed volume of air or gas, and velocity imparted to the air or gas by the blades and partly converted to pressure by the volume or scroll-shaped fan casing.

The centrifugal force developed by the rotor produces a compression of the air or gas which, in fan engineering,

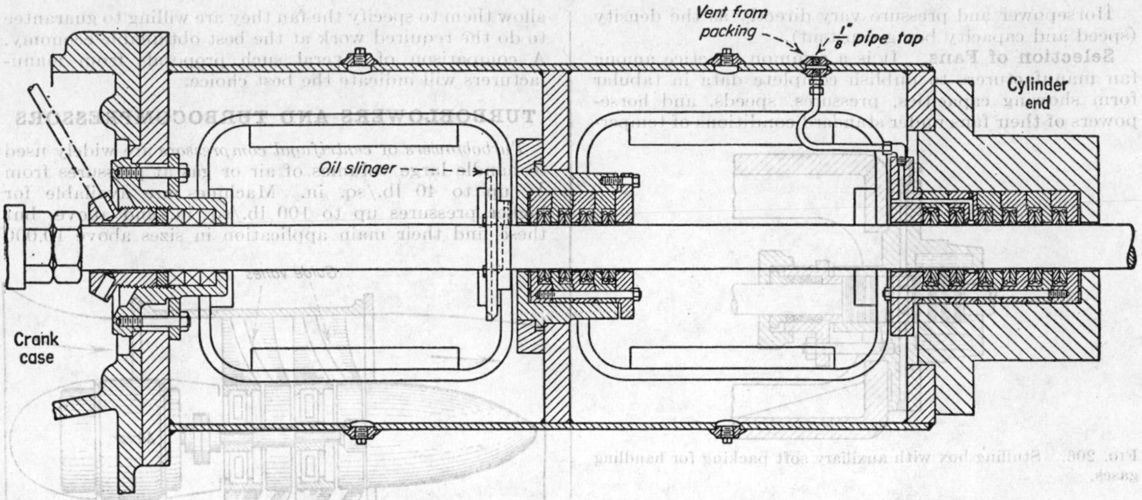

FIG. 202. Double-compartment distance piece between frame and cylinder.

is called the *static pressure.* The amount of this static pressure developed depends upon the ratio of the velocity of the air leaving the tips of the blades to the velocity of the air entering the fan at the heel of the blades. Therefore, the longer the blades, the greater the static pressure developed by the fan.

Operating efficiencies of fans are in the range of 40 to 70 per cent. Operating pressure is the sum of the static

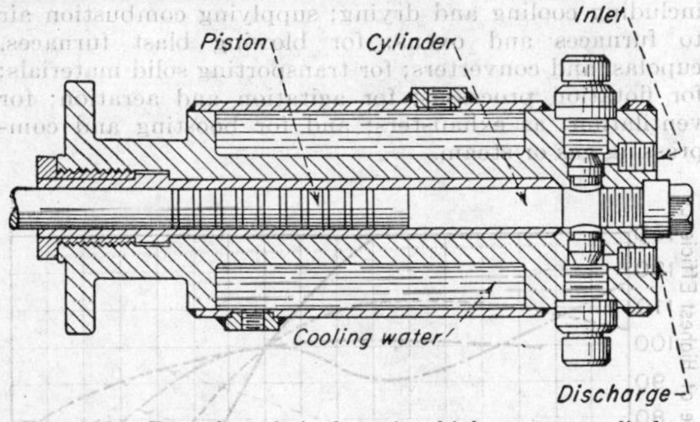

FIG. 203. Forged-steel single-acting high-pressure cylinder.

pressure and the velocity head of the air leaving the fan. It is generally expressed in inches of water gage, or in ounces per square inch.

The air horsepower of a fan is given by

$$\text{Air hp.} = \frac{V(p_2 - p_1)144}{33{,}000} = 0.000157 \times V \times (\text{developed head, in. water})$$

$$\text{Shaft hp.} = \frac{\text{air hp.}}{\text{efficiency}}$$

where V = volume handled, cu. ft./min.

p_1 = inlet pressure, lb./sq. in.

p_2 = discharge pressure, lb./sq. in.

Fan Performance. The performance of a centrifugal fan varies with changes in conditions such as temperature, speed, and density of the gas being handled. It is important to keep this in mind in using the catalogue data of various fan manufacturers, since such data are usually based on assumed standard conditions, such as 70°F. and 29.92 in. barometric pressure, or 68°F. and 50 per cent relative humidity. Corrections must be

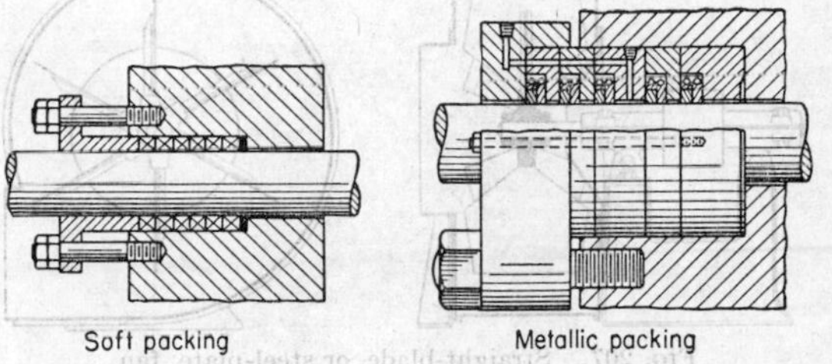

FIG. 204. Typical packing arrangements for low-pressure stuffing boxes.

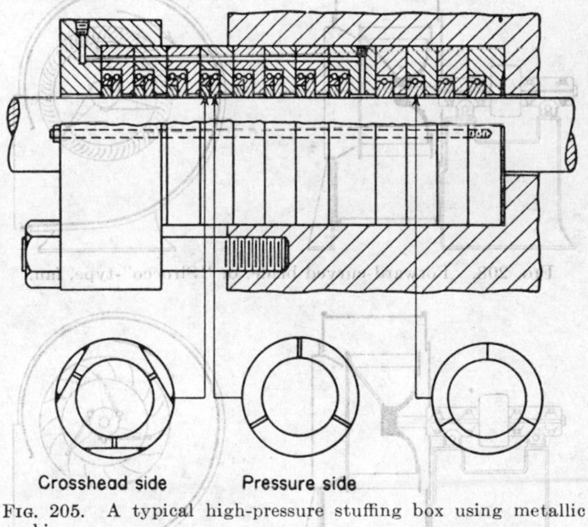

FIG. 205. A typical high-pressure stuffing box using metallic packing.

made for variations from these assumed standards. The usual variations are as follows:

When speed varies:

1. Capacity varies directly as the speed ratio.
2. Pressure varies as the square of the speed ratio.
3. Horsepower varies as the cube of the speed ratio.

When temperature of air or gas varies:

Horsepower and pressure vary inversely as the absolute temperature (speed and capacity being constant).

When density of air or gas varies:

Horsepower and pressure vary directly as the density (speed and capacity being constant).

Selection of Fans. It is a common practice among fan manufacturers to publish complete data in tabular form showing capacities, pressures, speeds, and horsepowers of their fans under standard conditions of temper-

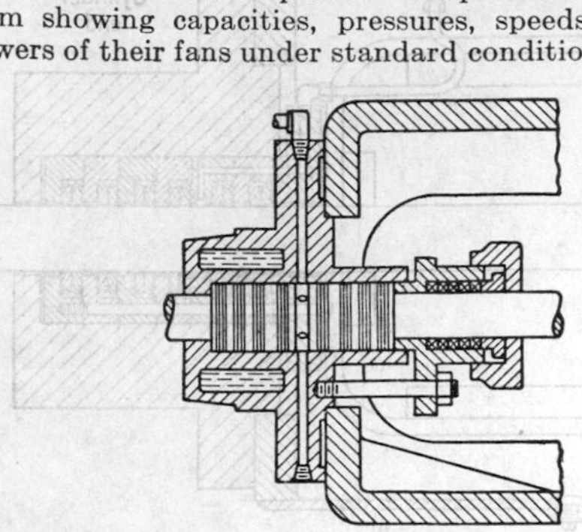

FIG. 206. Stuffing box with auxiliary soft packing for handling gases.

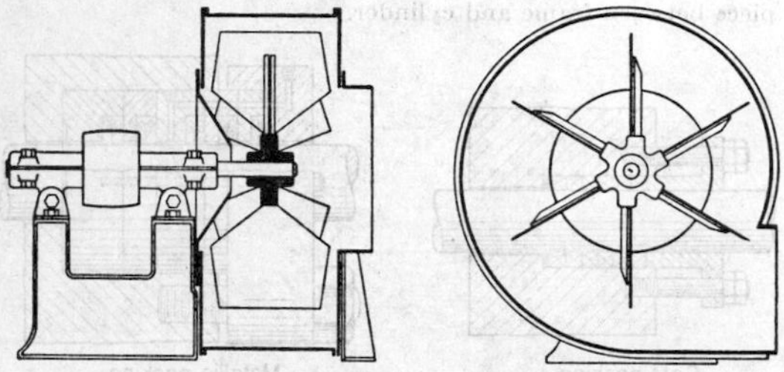

FIG. 207. Straight-blade, or steel-plate, fan.

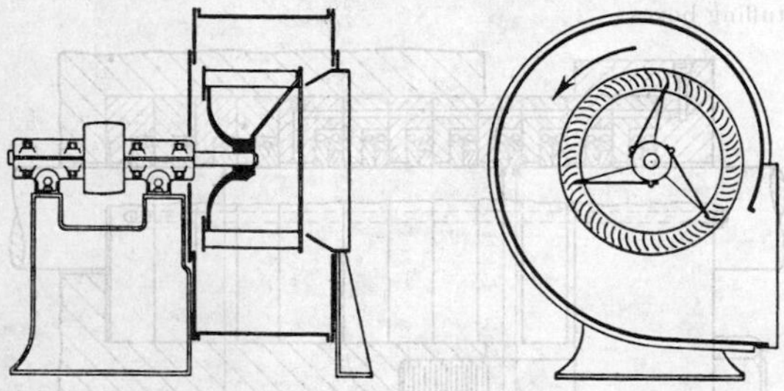

FIG. 208. Forward-curved blade, or "Sirocco"-type, fan.

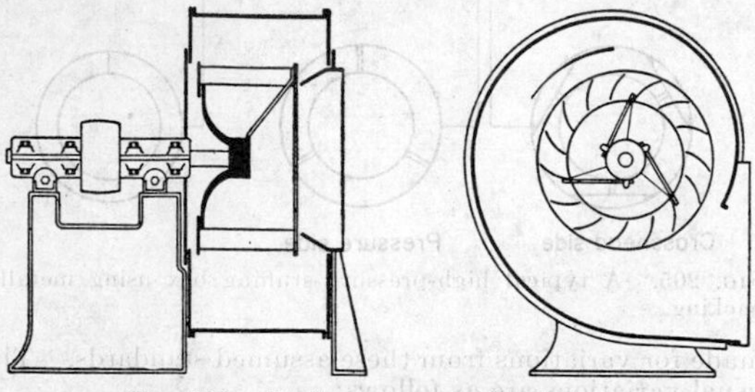

FIG. 209. Backward-curved blade fan.

ature and air density. These tables are of great use to the heating and ventilating engineer and to others who specialize in fan engineering. Those who do not specialize along these lines, including the chemical engineer, should not attempt to select fans from these tables. The proper course to follow is to put full data concerning the job to be done in the hands of fan manufacturers and allow them to specify the fan they are willing to guarantee to do the required work at the best obtainable economy. A comparison of several such proposals from manufacturers will indicate the best choice.

TURBOBLOWERS AND TURBOCOMPRESSORS

Turboblowers or *centrifugal compressors* are widely used to handle large volumes of air or gas at pressures from ½ up to 40 lb./sq. in. Machines are available for higher pressures up to 100 lb./sq. in. and above, but these find their main application in sizes above 10,000

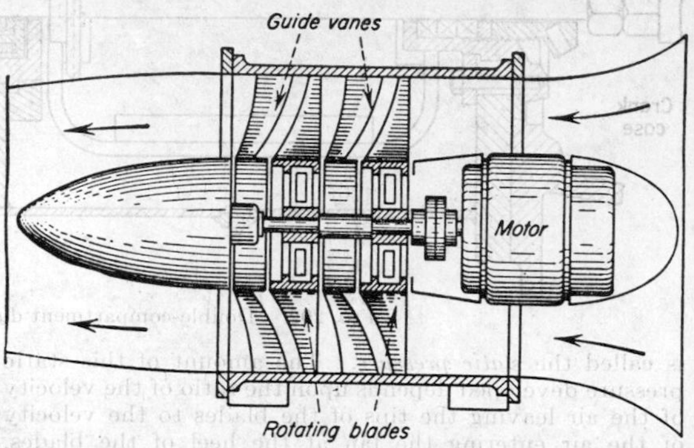

FIG. 210. A two-stage axial-flow fan.

cu. ft./min. For pressures below ½ lb./sq. in., one of the several types of fans is ordinarily selected.

Turboblowers are used for a wide variety of services, including cooling and drying; supplying combustion air to furnaces and ovens; for blowing blast furnaces, cupolas, and converters; for transporting solid materials; for flotation processes; for agitation and aeration; for ventilation; as exhausters; and for boosting and compressing gas or steam.

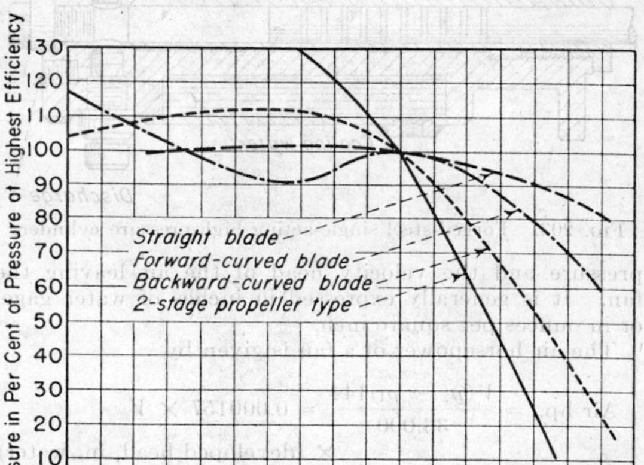

FIG. 211. Approximate characteristics of various types of fans.

The principle of a turboblower is the same as that of a centrifugal pump, which has been described in a previous section, the main difference being that the air or gas handled in a blower is compressible while the liquids handled in a pump are practically incompressible. Inasmuch as the pressure is developed by centrifugal force, it is very important to know the density of the gas to be compressed. Since turboblowers develop a pressure within themselves dependent upon the nature and condition of the gas being handled and virtually independent of the process load, in selecting a proper size of blower, the combination of the most adverse conditions occurring simultaneously must be determined. The conditions to be considered are as follows:

Lowest barometric pressure.
Lowest intake pressure.
Maximum intake temperature.
Highest ratio of specific heats (n value).
Lowest specific gravity.
Maximum *intake* volume.
Maximum discharge pressure.

Most turboblowers operate at speeds of 3500 r.p.m. or higher. They are almost exclusively driven by electric motors or steam turbines. Single-stage blowers, when

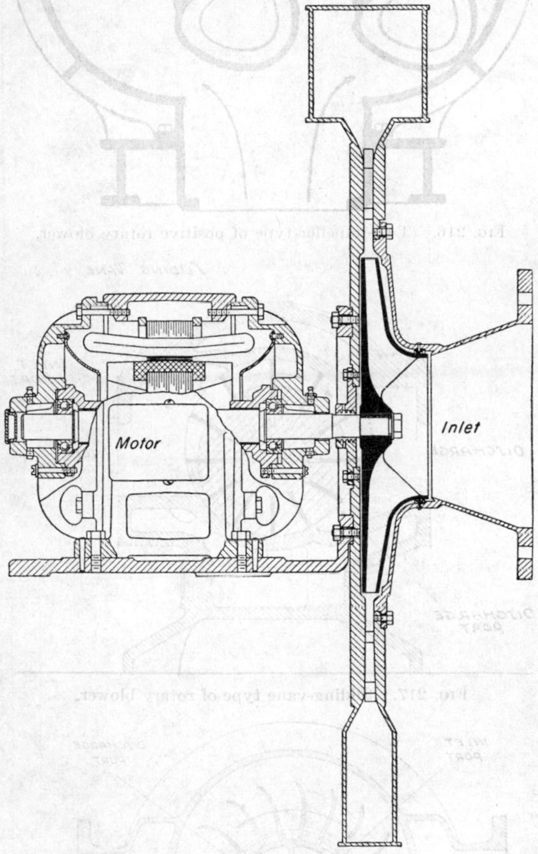

FIG. 212. Single-stage blower.

compressing air or high-density gas, are available to produce a pressure rise of up to 5 or 6 lb./sq. in. (see Fig. 212). From two- to seven-stage units produce pressures from 3 to 40 lb./sq. in. (see Fig. 213). For higher pressures, even more stages are required.

Typical characteristic curves of a constant-speed blower and of a variable-speed blower are shown in Fig. 214. From these curves it will be seen that a turboblower is essentially a constant-pressure machine and that power consumption is almost directly proportional to the volume delivered. For motor-driven blowers, various devices, such as the *Power Wheel*, *hydraulic coupling*, *magnetic coupling*, or wound-rotor motors, may be used to obtain efficient operation at part loads or under adverse operating conditions.

Turboblowers can be equipped with controls to deliver gas at constant discharge pressure, constant suction pressure, constant volume or constant weight.

The usual practice on small- and moderate-capacity units is to insert a *blast gate* in the inlet or discharge line.

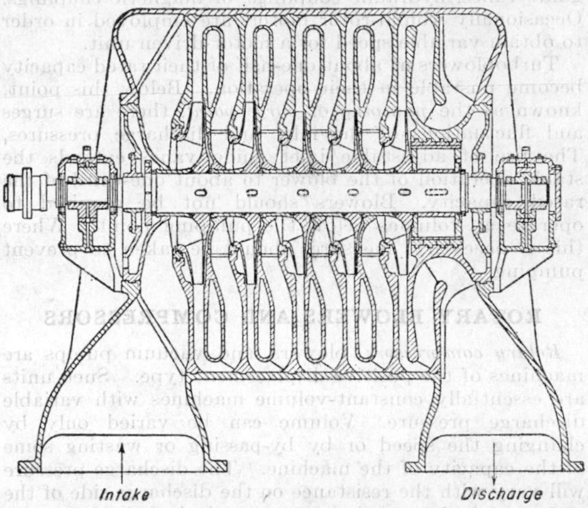

FIG 213. A 5-stage turboblower.

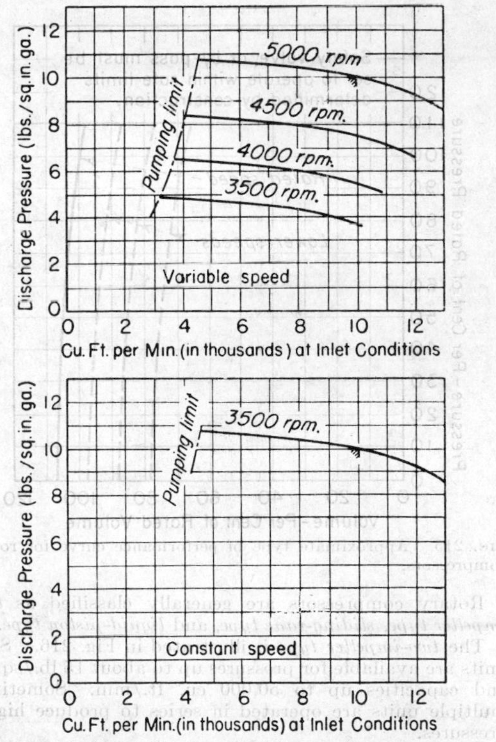

FIG. 214. Characteristic turboblower performance curves.

This can be set manually to control the volume or pressure.

Where required, the blast gates can be automatically controlled through suitable pressure- or volume-measuring devices. The constant-weight control is an automatic control for the blast gate which controls the volume by correcting for both temperature and pressure changes in the gas handled.

On large turbine-driven units, control is usually effected by varying the speed. On large constant-speed units, control is often effected by the use of adjustable inlet guide vanes, hydraulic couplings, or magnetic couplings. Occasionally wound-rotor motors are employed in order to obtain variable speed for a motor-driven unit.

Turboblowers at about one-half of their rated capacity become unstable in their operation. Below this point, known as the *pumping* or *surge point*, there are surges and fluctuations of the inlet and discharge pressures. The use of adjustable inlet guide vanes extends the stable operation of the blower to about one-third of the rated capacity. Blowers should not be required to operate at volumes below the pumping point. Where this is necessary, measures should be taken to prevent pumping.

ROTARY BLOWERS AND COMPRESSORS

Rotary compressors, blowers, and vacuum pumps are machines of the positive-displacement type. Such units are essentially constant-volume machines with variable discharge pressure. Volume can be varied only by changing the speed or by by-passing or wasting some of the capacity of the machine. The discharge pressure will vary with the resistance on the discharge side of the system. A characteristic curve typical of the form produced by these rotary units is shown in Fig. 215.

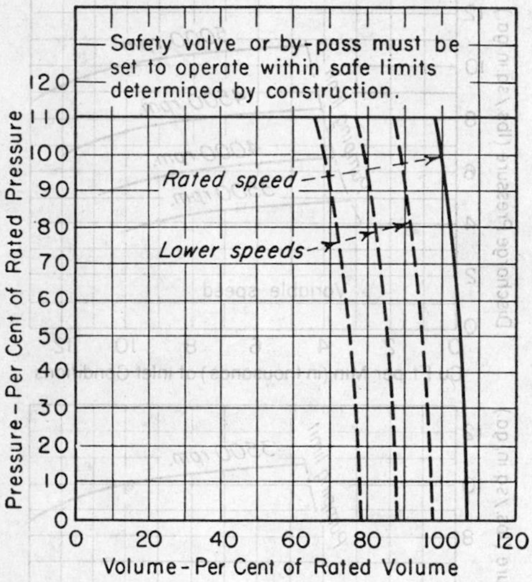

FIG. 215. Approximate type of performance curve for rotary compressors.

Rotary compressors are generally classified as *two-impeller type, sliding-vane type*, and *liquid-piston type.*

The *two-impeller type* is illustrated in Fig. 216. Such units are available for pressures up to about 15 lb./sq. in. and capacities up to 50,000 cu. ft./min. Sometimes multiple units are operated in series to produce higher pressures.

The *sliding-vane type* is illustrated in Fig. 217. These units are offered for pressures up to 125 lb./sq. in. and capacities up to 2000 cu. ft./min. Generally, machines for pressures above 50 lb./sq. in. are built as two-stage units.

The *liquid-piston type* is illustrated in Fig. 218. They are offered as single-stage units for pressures up to about 75 lb./sq. in. and capacities up to about 4000 cu. ft./min. Two-stage units are available for higher pressures. These units have found wide application as vacuum pumps on wet vacuum service.

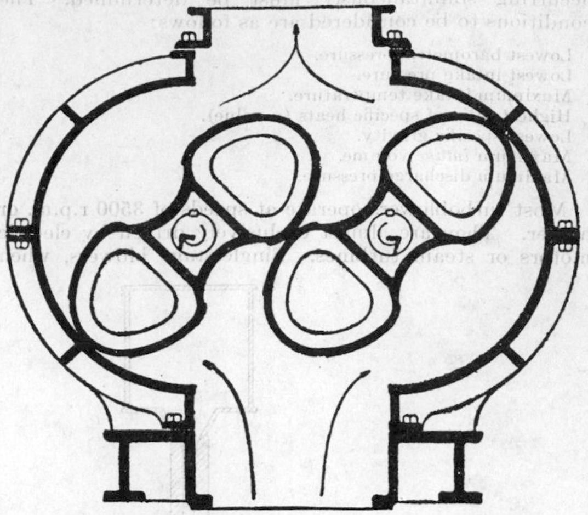

FIG. 216. Two-impeller type of positive rotary blower.

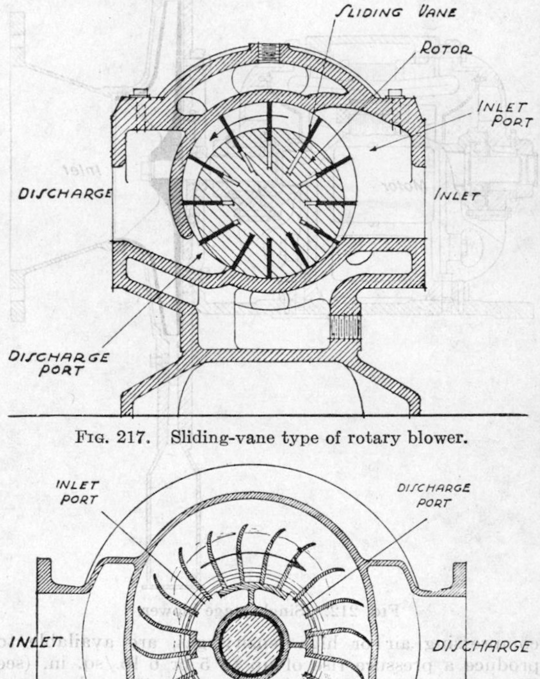

FIG. 217. Sliding-vane type of rotary blower.

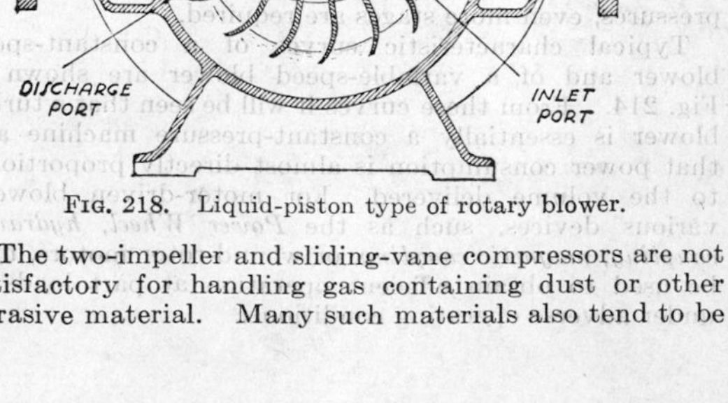

FIG. 218. Liquid-piston type of rotary blower.

The two-impeller and sliding-vane compressors are not satisfactory for handling gas containing dust or other abrasive material. Many such materials also tend to be retained by the liquid in the liquid-piston type. The liquid in the latter type of pump can be selected to prevent contamination of the gas being handled and in some cases can be treated as a part of the process by contact with the liquid in the pump.

STEAM-JET EJECTORS

A *steam-jet ejector* is a device for removing air, gases, or vapors from condensers and vacuum equipment in industrial processes. An ejector is a simplified type of vacuum pump or compressor which has no moving parts such as pistons, valves, or rotor. Figure 219 illustrates a

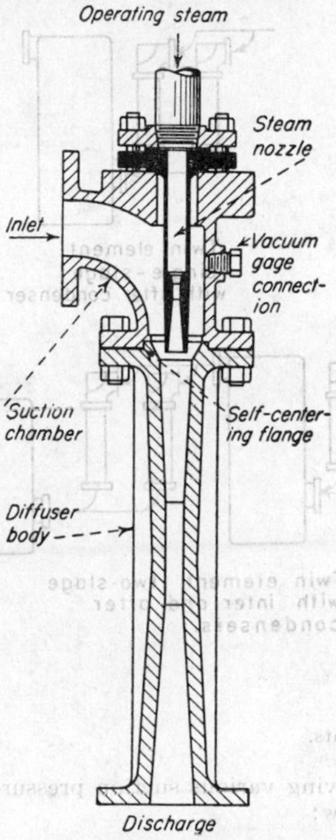

Fig. 219. A typical steam-jet ejector.

steam-jet ejector. It consists essentially of a steam nozzle that discharges a high-velocity jet of steam across a suction chamber that is connected to the vacuum equipment that is to be evacuated. The gas to be evacuated is entrained by the steam and carried into a venturi-shaped diffuser which converts the velocity energy of the steam into pressure energy. Figure 220 shows a large-sized ejector, sometimes called a *booster ejector*, with multiple nozzles.

Two or more steam-jet ejectors may be connected in series or stages. Also, a number of ejectors may be connected in parallel in order to handle larger quantities of gas or vapor. See Fig. 221 for typical ejector arrangements.

Water-cooled condensers of either the direct-contact, *i.e.*, barometric, type or the surface type are usually used between stages of multistage units to condense the operating steam from the preceding stage. This reduces the load on the following ejector stages, thus reducing steam consumption and making possible the use of smaller ejectors. Likewise, a precondenser installed ahead of an ejector reduces the size of the ejector required and the steam consumption if the mixture to be handled contains condensable vapors that can be removed under existing conditions of suction pressure and cooling-water temperature. An aftercondenser is frequently used to condense steam from the final ejector stage, although this does not affect the ejector performance. *Multistage ejectors* are built in commercial sizes to produce absolute pressures as low as 50 μ Hg. Figure 222 illustrates typical ejector-performance curves for single- and multistage ejectors. On multistage units performance will approximate curve A at overloads if the secondary stages have ample capacity. However, if the secondary stages cannot handle all the non-condensed vapors from the first stage, the curve will approximate B or B^1.

Steam-jet ejectors are widely used in the chemical industry for vacuum applications. They are in general more suitable for the lower absolute pressures, *i.e.*, higher vacuum, than reciprocating vacuum pumps, although ejectors can be furnished for any suction and discharge pressure and capacity. In some cases they are designed for compressing steam at pressures above atmospheric pressure. In such cases they are usually called *thermal compressors*.

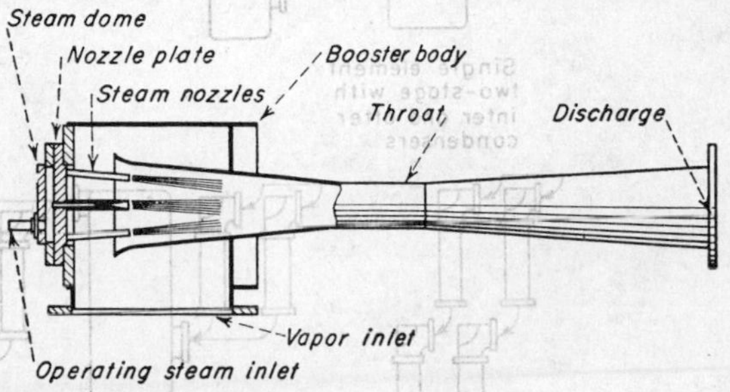

Fig. 220. A multiple-nozzle booster ejector.

The range of popular types of steam-jet ejectors is shown below on the basis of a typical motive steam pressure and cooling-water temperature (100 lb./sq. in. 85°F.), with ejector discharging to a maximum back pressure of 1 lb./sq. in. gage.

Range of Design Suction Pressure, In. or Mm. Hg Abs.	Arrangement of Ejector
30 –3.0 in.	Single-stage
5.0–0.5 in.	Two-stage
50 –1 mm.	Three-stage
5 –0.05 mm.	Four- or five-stage

Note that there is some overlap in range. The choice of the most suitable type for a given application depends upon the following factors:

1. *Steam Pressure.* Ejector selection should be based upon the minimum pressure in the supply line selected to serve the unit.
2. *Water Temperature.* Selection is based on the maximum water temperature.
3. *Suction Pressure and Temperature.* The over-all process requirements should be considered. Selection is usually governed by the minimum suction pressure required (highest vacuum).
4. *Capacity Required.* Again over-all process requirements should be considered, but selection is usually governed by the capacity required at the minimum process pressure.

Ejectors are easy to operate and require very little maintenance. Installation costs are low; and, since they have no moving parts, they have long life, high sustained efficiency, and low maintenance cost. Ejectors are suitable for handling practically any type of gas or vapor. They are also suitable for handling wet or dry mixtures or gases containing sticky or solid matter such as chaff or dust.

Ejectors are available in suitable materials for handling high temperature or corrosive gases. Where the gases

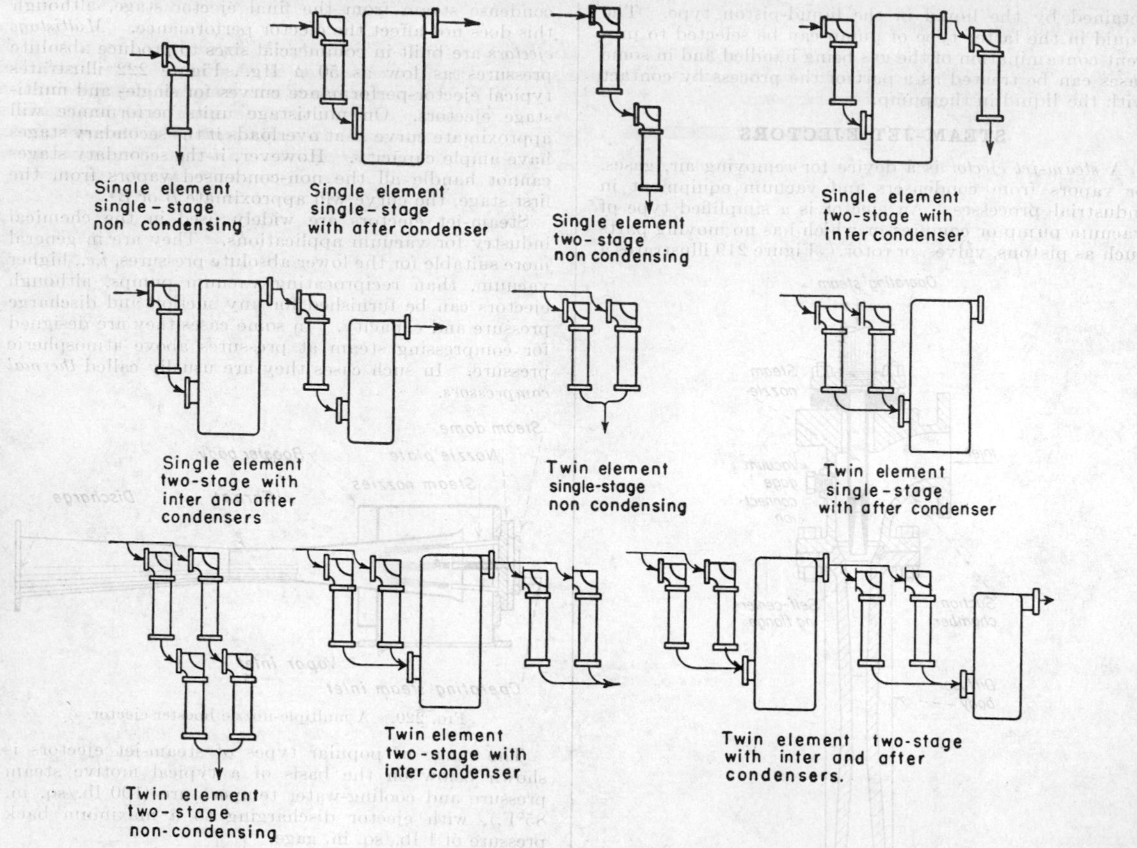

FIG. 221. Common ejector arrangements.

or vapors are not corrosive, the diffuser is usually constructed of cast iron and the steam nozzle of stainless steel. For more corrosive gases and vapors, practically any combination of metals can be used such as bronze,

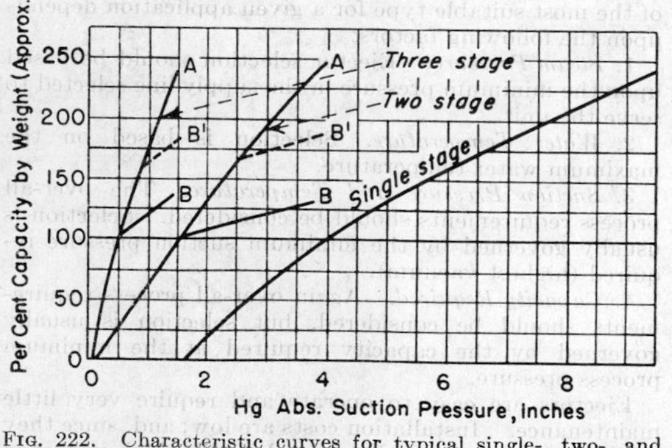

FIG. 222. Characteristic curves for typical single-, two-, and three-stage ejectors.

various stainless-steel alloys, etc. Besides this, ejectors are constructed from solid carbon and from glass.

The capacity of ejectors is usually expressed as pounds per hour. This has proved less subject to confusion and errors than the practice of referring to capacity on the volumetric basis. Typical average steam consumption of ejectors having various suction pressures is shown in the table below:

Average Steam Consumption, Lb./Hr. at 100 Lb. Gage/Sq. In. Pressure

Weight mixture, lb./hr.	% net dry air by weight	Suction pressure, in. Hg abs. 0.5		1.0		1.5	2.0	3.0	4.0		6.0
		3 stage	2 stage	3 stage	2 stage	2 stage	2 stage	2 stage	2 stage	1 stage	1 stage
10	100	73	99	59	70	58	50	42	38	58	36
10	70	59	84	47	60	49	42	35	31	63	39
10	40	45	68	33	47	38	32	26	23	68	41
10	10	24	45	16	28	21	17	14	12	74	42

NOTE: Steam consumption is approximately proportional to capacity.

This table gives the steam consumption in pounds per hour at 100 lb. gage while handling the gas-vapor mixture weighing 10 lb./hr. For larger or smaller capacities, the steam consumption is approximately proportional to the capacity.

Since there are no moving parts, the operating efficiency of an ejector depends entirely on the size and shape of the diffuser, and on steam-nozzle parts and the size of the various inlet and outlet openings. Many factors affect the proper design of these parts, among which are steam pressure, capacity, absolute pressure of gas being evacuated, and presence of foreign materials. The influence of these factors on design has in many cases been determined by test and empirical relationships established. It is therefore important that the ejector manufacturer have as complete information as possible regarding the gas or vapor to be handled and the operating conditions. Among the data required are suction pressure, capacity (usually in pounds per hour), percentage of non-condensables in vapor, presence and amount of solid matter, analysis of gas or vapor, temperature of gas, minimum steam pressure, and maximum cooling-water temperature (for intercondensers, aftercondensers, etc.).

Condensers used between ejector stages, also precondensers and aftercondensers, may be of either the surface type or the direct-contact type. Surface-type units consist of a shell through which the gas or vapor flows. It is partly filled with tubes through which cooling water flows. Such units are used where it is necessary to reclaim the vapor that is condensed. Direct-contact or barometric-type condensers consist of a shell into which the air-vapor mixture is discharged and into which

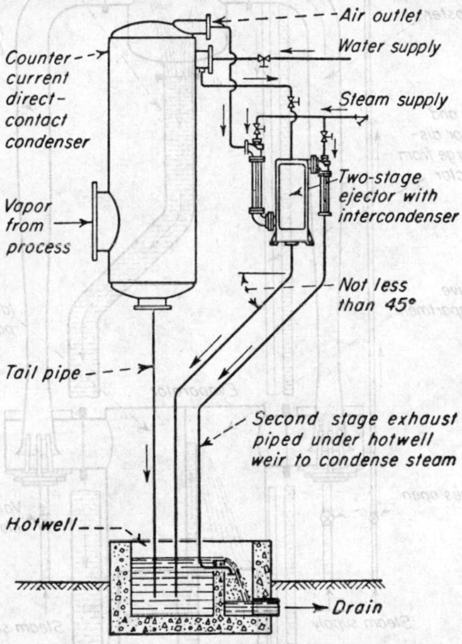

FIG. 223. Barometric condenser and ejector used for condensation of vapor.

the cooling water is directed. The air-vapor mixture and the cooling water are intimately mixed by means of weirs or baffles. The non-condensable vapors are removed from this shell and carried into the following ejector stage. The condensed vapors are carried out with the cooling water. Direct-contact condensers require a barometric leg approximately 34 ft. in height or a suitable removal pump.

In general, steam-jet ejectors may be used wherever it is necessary to produce vacuum or to handle or remove vapors or gases from some operation or process. Specifically, in the chemical, food, and pharmaceutical industries, they are widely applied to such operations as cooling by evaporation, condensing, conveying solid materials, crystallizing, degasifying or deaerating, deodorizing, distillation, drying, evaporating, filtering, freezing, impregnating and moistening, liquid transfer, mixing, rendering, stripping, and subliming. The following drawings illustrate briefly a few of the most common ejector applications.

Condensation Processes (Fig. 223). Barometric condensers are used for condensation where the condensable load is principally water vapor, *i.e.*, steam; where low first cost is desirable; where the condensate is not saved; or where cooling water is dirty and contains trash.

Ejectors remove the non-condensables (such as air) from the condenser and thus maintain the pressure differential that assures flow of steam to the condenser.

Barometric-type condensers are not ordinarily used for vapor mixtures whose main constituents are condensable vapors other than steam, *i.e.*, such products as oil vapors

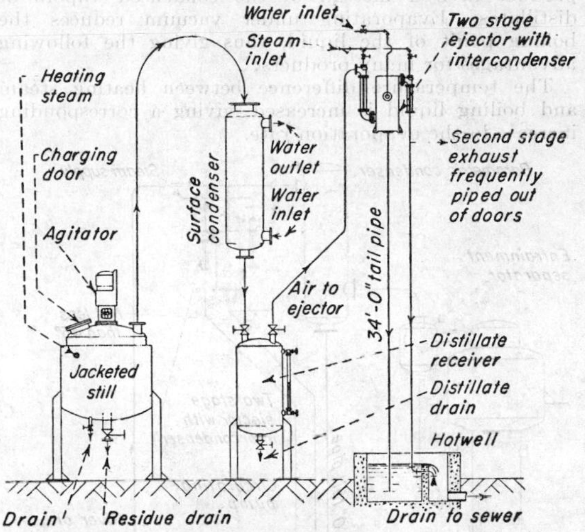

FIG. 224. Ejectors used in connection with batch distillation.

or solvents. For such service, surface-type condensers are normally used. These may or may not operate under vacuum, depending on the requirements of the vapor to be condensed.

Batch Distillation (Fig. 224). In batch distillation processes, operation under vacuum permits lower temperatures and allows separation of products that would be decomposed or altered by distillation at atmos-

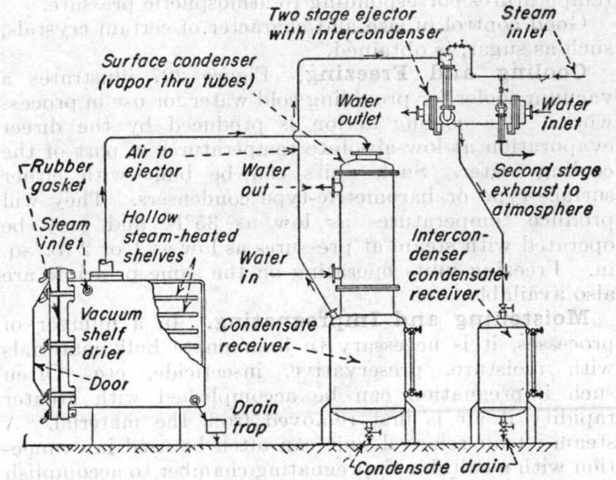

FIG. 225. Ejectors used in connection with vacuum shelf drying.

pheric pressure. It is widely used in the separation of organic chemicals. Ejectors are used to maintain the vacuum.

Vacuum Shelf Drying (Fig. 225). Materials that are temperature-sensitive or easily oxidized are often dried in batches in vacuum shelf driers. The material to be dried is placed on trays on heated shelves, and the moisture is evaporated and passes to a condenser. An air ejector removes the air leakage, thus maintaining the vacuum. Such units have found application, to mention a few, in preparing dyes and dye products, pharmaceutical preparations, dog foods, resins, etc.

Evaporators and Vacuum Pans (Fig. 226). Evaporation is another separation operation similar to distillation, the main difference being that, in evaporation processes, the residue or concentrated solution is the product desired instead of the condensed vapors or distillates. Evaporating under vacuum reduces the boiling point of the liquid, thus giving the following advantages for many products:

The temperature difference between heating steam and boiling liquid is increased, giving a corresponding increase in the evaporation rate.

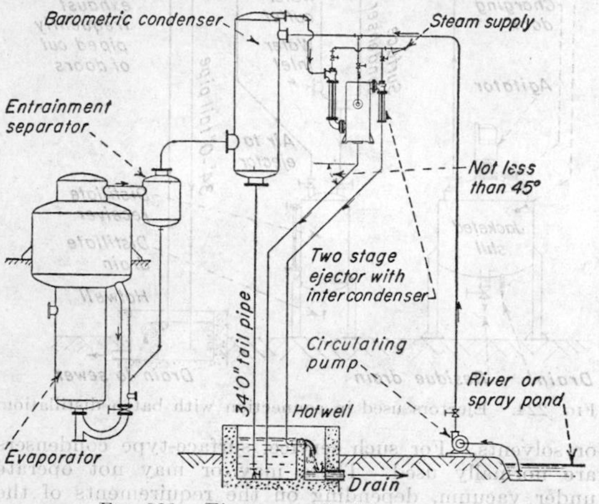

FIG. 226. Ejectors used with evaporator

Low pressure, and hence sometimes cheaper heating steam, may be used.

Many organic materials, such as sugar, milk, glue, and tannin, deteriorate when their solutions are boiled at temperatures corresponding to atmospheric pressure.

Good control of size and character of certain crystals, such as sugar, is obtained.

Cooling and Freezing. Figure 227 illustrates a vacuum cooler for providing cold water for use in process work. The cooling action is produced by the direct evaporation at low absolute temperatures of part of the cooling water. Such units can be built with either surface-type or barometric-type condensers. They will produce temperatures as low as 35°F. and may be operated with steam at pressures as low as 1 or 2 lb./sq. in. Freezing units operating on the same principle are also available.

Moistening and Impregnating. In a number of processes, it is necessary to impregnate bulk materials with moisture, preservative, insecticide, etc. Often such impregnation can be accomplished with greater rapidity if air is first removed from the material. A steam-jet air-removal unit can often be used in connection with a suitable impregnating chamber to accomplish such results. One manufacturer of such devices often uses reciprocating vacuum pumps and steam-jet ejectors in combination to obtain speed and accurate control of the air removal.

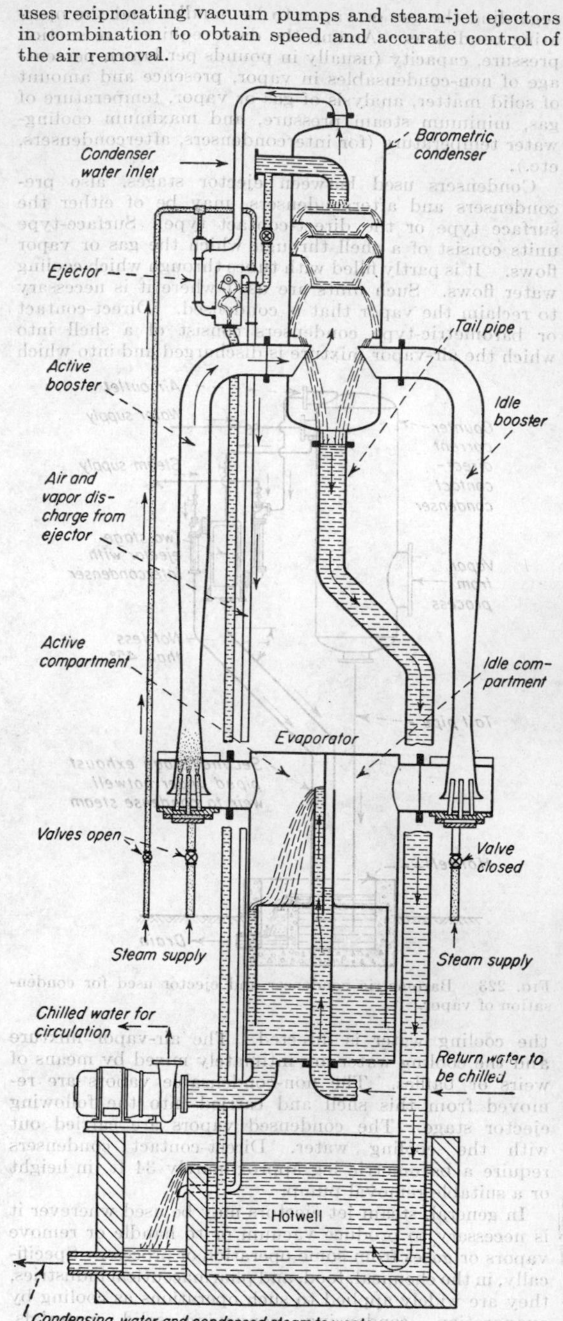

FIG. 227. Ingersoll-Rand steam-jet cooler unit.

SECTION 21

MATERIALS OF CONSTRUCTION*

BY

James A. Lee, B. S., B. A., M. A., Southwestern Representative, *Chemical Engineering;* Member, American Institute of Chemical Engineers, American Chemical Society, American Society for Testing Materials, Technical Association of the Pulp and Paper Industry, Electrochemical Society, Royal Society of Arts.

H. L. Maxwell, Ph. D., Metallurgist, Engineering Department, E. I. duPont de Nemours & Co.

E. C. Fetter, B. S., Managing Editor, *Chemical Engineering.* (Chemical Resistance of Gasket Materials)

H. H. Dunkle, Goetze Division, Johns-Manville Sales Corp. (Chemical Resistance of Gasket Materials)

CONTENTS

	Page
Introduction	1458
The Total Immersion Test	1458
Special Properties of Certain Materials	1460
Chemical Resistance of Constructional Materials	1461
Directory of Materials for Construction of Chemical Equipment	1527
Materials of Construction—Base Metals. Physical Properties and Methods of Fabrication	1539
Steel Numbering System	1543
Precious Metals and Their Alloys	1548
Glass-lined and Enameled Steel	1549
Chemical Stoneware, Porcelain, and Cements	1549
Refractory Materials	1549
Structural Carbon and Graphite	1550
Rubber and Rubberlike Products	1551
Vulcanized Fiber	1551
Woods	1552
Chemical Resistance of Gasket Materials	1554
Concrete for Chemical-tank Construction	1558

* Largely from material published by *Chemical & Metallurgical Engineering.*

MATERIALS OF CONSTRUCTION

Introduction. Materials for chemical-plant construction fall naturally into two general classes, **metals** and **non-metals,** the former class being further divisible into pure metals and solid-solution-type alloys on the one hand and non-homogeneous alloys on the other. The latter distinction is an important one as, for corrosion-resisting purposes, the solid-solution alloys frequently behave as though they were chemical individuals of definite characteristics, while the non-homogeneous type usually has roughly the resistance of the poorest constituent, the resistance being still further diminished by the presence of numerous electrolytic cells. The difference in behavior between properly and improperly heat-treated metals of the stainless-steel class well illustrates this characteristic difference between the solid-solution and the mixed-crystal types.

In the selection of material for the construction of a chemical plant, resistance to the corroding medium is usually the determining factor because the choice would otherwise fall automatically on the cheapest material mechanically suitable, and laboratory corrosion tests are frequently the quickest and most satisfactory means of arriving at a preliminary selection of the most suitable materials to use. Unfortunately, however, it is not possible on the basis of any existing laboratory test to predict with accuracy the behavior of the selected material in the plant. The outstanding difficulty in this connection lies not so much in carrying out the test as in interpreting the results and translating them into terms of plant performance. A laboratory test of the standardized type gives but one factor, the chemical resistance of the proposed material to the corrosive agent, and there are numerous other factors entering into the behavior of the materials in the plant.

The following method of determining this factor, which is the so-called total immersion test, represents an unaccelerated method which has been found to give reasonably concordant results in approximate agreement with results obtained on the large scale when the other variables are taken into account. Various other tests have been proposed and are in use, such as the salt spray, the accelerated electrolytic, the alternate immersion, the aerated total immersion, etc.; but, in view of the numerous other complications entering into the translation of laboratory results into plant results, the simplest test would seem the most desirable for routine preliminary work, reserving the special test methods for special cases. This preliminary test serves quite well to eliminate the materials that obviously cannot be used; further selection among those which apparently can be used can be made on the basis of a knowledge of the properties of the materials concerned and the working conditions, or by constructing larger scale equipment of the proposed material in which the operating conditions can be simulated.

THE TOTAL IMMERSION TEST

The following conditions and factors affecting the results of tests have been investigated:

Test Piece. The shape of the test piece does not affect results within a reasonable range of ratio of length of edge to surface, except for the accelerating effect of the internal segregations exposed on the edge sections. In general the proportion of edge section should be kept low. Fine polishing is unnecessary, although tool marks and oxide coatings should be removed. Many determinations have been made on roughened and smoothed pieces.

Volume of the Corroding Solution. Investigations using volumes of solution ranging from 20 to 200 cc./6.45 sq. cm. of test piece exposed show that the effect of the volume of the corroding solution becomes practically negligible if a relation of volume to square inch of test piece in excess of a certain minimum is used.

Temperature of Corroding Solution. Results of many experiments show that the temperature coefficient is high for all reactions. For theoretical work in determining the law of temperature, it is absolutely necessary to use an accurately regulated thermostat. In applying results to plant practice, similarly careful consideration must be given to temperature control in all specific tests. In comparing different metals, care must be taken that the temperature is the same for all tests. A convenient way to control temperature is to use a constant-boiling liquid.

Time of Exposure. It has been found that results of corrosion tests vary enormously with the time of exposure of the material to the corroding solution. This variation is due to initial electrochemical surface actions such as overvoltage and to the period of time required for the formation of a protective coat. Evidently initial effects of this kind must be neglected if a corrosion factor for use over a long period of time is to be obtained. The logical way is to measure the rate over an interval of time after this initial high corrosion rate has decreased and become constant. For the most careful work this must be done. The experimental data available at present indicate that corrosion should be determined at the end of 48, 96, and 240 hr. Sufficient accuracy for comparative work may usually be obtained by neglecting the corrosion over the first 48 hr. and averaging the rate over the second 48-hr. period.

Cleaning Test Pieces. Many methods have been investigated for removing coatings from the test pieces. They may consist of dissolving the coat or of removing it mechanically. In each case the error due to a particular method has been determined. The first 48-hr. immersion in the corroding solution may be considered a preparation of the test piece for the final test.

Reagents and Apparatus. Corroding Solution. Use 250 cc. of the corroding solution per test piece of given area (4.6 sq. in.) for a fairly rapid corrosion rate (0.01-in. penetration per month). The volume should be increased in proportion for pieces of greater area.

Preparation of the Sample. 1. Metal Test Piece. Size 2 by 1 by 0.1 in. (area 4.6 sq. in.). Dimensions should be accurate to 0.01 in. in order to save the time of measurement and area calculation in the laboratory. Other shapes may be used within a range of ratio of length of edge to total area of test piece between 4 and 8.

2. Preparation. The strip of indicated size may be cut from flat sheet metal or turned from pipe. Tool marks should be removed by successive use of file and emery. Exceedingly fine finishes are unnecessary, but the surface should be clean and reasonably smooth as, for example, finished with 120 grain emery.

Detailed Procedure. Standard Static Corrosion Test. Place the corroding solution in a flask or wide-mouth bottle and bring to the temperature of the test. Maintain this temperature to at least 1°C. in a carefully regulated thermostat. Suspend the test piece upon a glass hook from a stopper of such a size as to fit the flask or the bottle loosely. The stopper must be made of a material that will be unaffected by the corroding solution. Immerse the test piece in the solution when it reaches the proper temperature, closing the bottle or flask with the stopper. If the test is being made at a high temperature, a reflux condenser must be used, taking care to prevent the condensate running directly on to the test piece.

The highest accuracy is desirable. Expose the test piece for 48 hr. to the corroding solution. Remove from the solution, wash thoroughly in a stream of water, and remove any coating. This cleaning may be done by dissolving off the coating (lead sulfate in ammonium acetate solution, lead chloride in hot water, etc.) or by rubbing and scouring with or without a soft powder as a mild abrasive. Do not adopt any method of cleaning until the error due to its use has been determined. Weigh the test piece after thoroughly drying, especially wiping out the hole by which it was suspended on the glass hook. Immerse the test piece in the corroding solution for a second 48-hr. period. Care should be taken that, while drying and making the first weight, no oil or grease is allowed to get on the test piece by handling, since this would give low results. After the second 48-hr. immersion period, clean and weigh again as described above. From the loss during the last 48 hr., calculate the average rate of corrosion ($t = 48$ hr.) as indicated under Calculations.

Run a check test simultaneously. Never place test pieces of different metals in the same container. It is permissible to place two pieces of the same metal in the same container, but they should not be in contact.

In case the metal develops pitting, this factor must be included in the results, since failure in any case will occur when a pit has entirely penetrated the metal. Determine the magnitude of this effect by grinding down on a metallographic grinding set until all the pits have just disappeared and solid metal is reached. The loss in weight during this grinding is determined and the pitting calculated as indicated.

Form for Card File of Corrosion Data

Material
Corrosive agent
Concentration
Temperature
Rate of corrosion, in./mo.
Report of data
Method No.
By
For

Form for Laboratory Data

Metal
Solution
Temperature
Test piece No.
Dimensions
  Length
  Width
  Thickness
Hole diameter
  Area
Weight
  Before test
  After test
  Loss
Time
  Start
  Finish
  Run, hr.
Corrosion rate

Calculations. Standard Static Corrosion Test. If W = loss in weight (in grams) of test piece during the second 48-hr. immersion; A = area of test piece in square inches; S = density of metal in grams per cubic centimeter; and t = time of exposure in hours ($t = 48$):

Then C = rate of chemical corrosion expressed as inches penetrated per month:

$$C = \frac{24 \times 30 \times W}{(2.54)^3 ASt} \text{ or } \frac{43.9W}{ASt}$$

In order to calculate the pitting corrosion, let p = loss in weight (in grams) due to grinding out pits:

Then d = rate of penetration of metal by both normal corrosion over the entire surface and by local action due to pitting:

$$d = \frac{43.9(W + p)}{ASt}$$

Effect of Variables on Corrosion Test. As in other branches of engineering, it is necessary to apply a factor of safety to the results obtained, the factor varying with the degree of confidence in the applicability of the results. Ordinarily, a factor of from 3 to 10 might be considered normal.

Among the more important points which should be considered in attempting to base plant design on laboratory data are the following:

Effect of Oxygen. The test given above does not call for aeration, the reason being that the effect of oxygen is considered a specific one, to be taken into account for each problem. In general, its effect is to increase the rate of corrosion markedly; and, if it is to be present in the plant operation, its effect should be allowed for.

Electrolysis. This is a frequent source of trouble on a large scale. Not only is the use of different metals in the same piece of equipment dangerous, but the effect of cold working may be sufficient to establish potential differences of objectionable magnitude between different parts of the same piece of metal. **Riveting,** even when extreme precautions are taken to have the rivets of identical composition with the sheet, is very likely to establish potential differences. Even such slight working as threading without subsequent annealing has been known to cause rapid failure from electrolysis. The mass of metal in chemical apparatus is ordinarily so great, and the electrical resistance consequently so low, that a very small voltage can cause a very high current. It might also be noted that improper heat-treatment of a solid-solution-type alloy may convert it from the solid-solution type to the non-homogeneous type, with the result that it fails even more rapidly than its poorest constituent. **Welding** also, if not properly done, may leave the weld of a different physical or chemical composition from the body of the sheet and cause the development of stray currents. A simple test for weld homogeneity is to treat a sample of the welded sheet containing a portion of both weld and sheet with a suitable solvent such as nitric acid or aqua regia, till about half the thickness of the metal has been dissolved; if heterogeneity exists, evidence of differential attack should be seen.

Velocity of Corrosive Liquid. Because of increased rate of removal of corrosion products and increased mechanical action, corrosion increases rapidly with increasing velocity of corrosive medium. This is especially true if the flow is turbulent, or if the liquid carries suspended matter. Frequently, slight alterations in design can be made to reduce this factor.

Local Concentration. Both local variations in temperature and crevices that permit the accumulation of corrosion products are capable of allowing the formation of concentration cells, with the result of accelerated local corrosion.

Temperature. In the laboratory, the temperature of the test specimen is that of the liquid in which it is immersed, and the measured temperature is actually that at which the reaction is taking place. In the plant, heat being supplied through the metal to the liquid in many cases, the temperature of the film of (corrosive) liquid on the inside of the vessel may be a number of degrees higher than that registered by the thermometer. As the relation between temperature and corrosion is a logarithmic one, the rate of increase is very rapid. Like other chemical reactions, the speed ordinarily increases two- to threefold for each 10° temperature rise, the actual relation being that of the equation: $\log K = A + \frac{B}{T}$, where K represents the rate of corrosion, and T absolute temperature. This relationship, although expressed mathematically, must be understood to be a qualitative rather than strictly a quantitative one.

Impurities. The effect of impurities in either structural material or corrosive material is so marked (while at the same time it may be either accelerating or decelerating) that for reliable results the actual materials it is proposed to use should be tested and not types of these materials. In other words, it is much more desirable to test the actual plant solution and the actual metal or non-metal than to rely upon a duplication of either. Since as little as 0.01 per cent of certain organic compounds will reduce the rate of solution of steel in sulfuric acid 99.5 per cent, and 0.05 per cent bismuth in lead will increase the rate of corrosion over 1000 per cent under certain conditions, it can be seen how difficult it would be to attempt to duplicate here all the significant constituents.

General. A convenient summary of many of the factors related to corrosion testing may be found in A.S.T.M. A-279-44T, 1946 Book of A.S.T.M. Standards, Part 1-A, p. 1028.

SPECIAL PROPERTIES OF CERTAIN MATERIALS

Metals. Iron withstands anhydrous acids and concentrated solutions of some acids but is attacked by dilute aqueous solutions. It resists alkaline solutions with the exception of very hot, highly concentrated solutions, though its resistance is sufficiently good in the latter case for extensive use.

Copper, as can be seen from its position in the electromotive series of the metals, has little tendency to dissolve in non-oxidizing acids. It is, however, very subject to oxidation and, as the oxide is readily soluble in most acids, is subject to marked attack in the presence of air or other oxidizing agents. Ammonia and amines similarly attack copper under oxidizing conditions. It is less attacked by *strong* caustic solutions than is iron.

Nickel possesses in many respects the characteristics of a seminoble metal. It is acted upon by dilute acids with relative slowness, so that it can frequently be used at a fairly low pH. It resists concentrated caustic solutions very well. It is less affected by ammonia and by the amines than copper and is less rapidly oxidized.

Lead, because of its low tensile strength and elastic limit, is used chiefly in the form of linings. If not carefully anchored in place it is liable to creep slowly, with the resulting formation of buckles, especially if alternately heated and cooled. It is attacked but slowly by most acids and hence is extensively used in chemical-plant construction. Homogeneous lead lining overcomes the tendency of lead to creep and also improves the heat transfer as compared with loose linings.

Aluminum, while belonging to the class of metals which decompose water, is quite resistant to dilute and weak acids because of the presence of a surface film of the relatively inert hydrated oxide. Any material capable of removing the aluminum oxide, however, such as the halogen acids or alkalies, will cause very rapid corrosion of aluminum by exposing the highly reactive metal. Thermometers in aluminum equipment should be carefully protected, as the mercury from a broken bulb, if it comes in intimate contact with the metal, will cause rapid perforation.

Tantalum, though not extensively used in the past, possesses the valuable property of being unattacked by aqueous hydrochloric acid, either hot or cold, and by most other chemical reagents except hydrofluoric acid, fluorides, and sodium hydroxide solutions. Its relatively high cost restricts its use to linings for the most part but makes possible the construction of certain equipment in metal which would otherwise have to be built of a non-metal.

Silver also possesses a fair degree of resistance to the aqueous solutions of the halogen acids. In this case the resistance is due largely to the formation of a protective film of the insoluble halide, so that conditions which favor the recrystallization of the halide, such as the presence of ammonium salts or very high temperatures, materially reduce the life of the equipment. Silver is also widely used for handling and storage of organic acids, particularly at high temperatures.

Magnesium is one of the few materials not attacked by dilute aqueous solutions of hydrofluoric acid.

Antimony, while seldom used because of brittleness, also withstands well the action of aqueous hydrochloric acid.

Alloys. Chromium-iron-nickel alloys (stainless steels) belong to the solid-solution type of alloy when properly heat-treated; when improperly heat-treated, segregation takes place and the resulting electrolysis may reduce the resistance below that of ordinary steel. Roughly, the resistance to acids increases with increasing nickel content.

Bronzes tend to possess the acid resistance of copper with a decreased tendency to oxidation, so that their overall resistance is somewhat greater than that of copper while the mechanical properties are much superior to those of copper.

Brasses, like bronzes, have mechanical properties and workability superior to those of copper. As a class they are subject to dezincification and consequent weakening on exposure to corrosive conditions.

Silicon alloys, such as **Corrosiron, Duriron,** and **Durichlor,** have the high resistance to be expected from the high silicon content and also the brittleness and hardness which are ordinarily associated with the compounds of silicon. The high-silicon alloys must be worked by casting and grinding. These alloys are sensitive to temperature shock. They can be welded, but great care is required in the preheating and cooling.

Non-metals. One important difference between the metals and the non-metals, which greatly affects the design of equipment, is the low thermal conductivity of the latter class. It is frequently difficult or impossible to obtain the necessary heat transfer through the wall of the vessel when working with a non-metal, so that some form of internal heating or cooling must be resorted to, such as coils, electrical heating, or live steam.

Glass possesses the desirable quality of almost perfect resistance to acids of all strengths and at almost all temperatures, hydrofluoric acid being an exception to the rule. The difficulty of obtaining large sizes and the fragility of the material have somewhat limited its application, piping being the chief form of equipment ordinarily used. Glass is also sensitive to sudden changes of temperature and must be carefully protected from them.

Silica ware, while possessing the corrosion resistance of glass in a more marked degree, also possesses all the weaknesses of glass with the exception of its susceptibility to sudden temperature changes.

(Continued on p. 1547.)

Table 1. Chemical Resistance of Constructional Materials*

No.	Material	Ratings	Exposure conditions	Applications
Significance of numbers: 1–3xx Metals 4xx Carbon 5xx Cement 6xx Ceramics 7xx Plastics 8xx Rubber 9xx Wood To identify material further, look up number in Directory of Materials, p. 1527		Ratings are those of materials manufacturers. A, good F, fair V, varies depending on conditions X, unsuitable	° = deg. F. % = concentration Rm. = room temp. Conc. = concentrated Dil. = dilute Sol. = solution -L = lined	B, bodies of pumps and valves I, impellers V, valve trim P, piping T, tanks S, shipping containers C, condensing surfaces H, heating surfaces D, ducts for fumes F, fans and blowers R, tower packing

Acetic Acid

No.	Material	Ratings	Exposure conditions	Applications
3	Admiralty	A		
4	Admiralty	V	....	CH
6	Adnic	A	All %; 70–212°	
10–17	Aluminum	A	All %; rm.	BIPTSDF
		A	Glacial; 212°	BIPTSCHDF
19	Aloyco	A	All to 100%; boil.	BV
		A	Vapors; 30, 100%; hot	BV
22	Ambralloy	F	....	CH
23	Ambralloy	F	....	CH
24	Ambralloy	V	....	CH
29–40	Ampco	A	All %; cold; crude or pure	BIVPTCHDF
		V	Vapors	
51	Berylco	A	Non-aerated	
54	Brass	X		
61	Brass	X		
63	Brass	A-F	....	PCH
66	Bronze	A		
73	Bronze	F	....	VT
74	Bronze	F	....	V
75	Bronze	F	....	VP
76	Bronze	F	....	Instruments
77	Bronze	X		
81	CA-FA20	A	Glacial; boil.	BIVP
82	CA-MM	A	Glacial; boil.	BIVP
86	Cast iron	V	Crude	
86A	Cast iron	V	Crude acid only	
87	Causul	F	150° max.	B
88	Chlorimet	A	All %; any temp.	BIVCHF
89	Chlorimet	A	All %; any temp.	BIVCHF
111	Copper	A		
114	Copper	F	....	PTCHR
117	Copper	V	....	S
118	Copper	A		
119	Corrosiron	A	All %; rm.	BIPHDF
123	Cupro-Ni	F	....	CH
124	Cupro-Ni	A		
139	Durco	A	All %; any temp.	BIVHF
140	Durichlor	A	All %; any temp.	BIVPCHFR
141	Durimet	A	All %; any temp.	BIVHF
142	Durimet	A	All %; any temp.	BIVHF
143	Duriron	A	All %; any temp.	BIVPHFR
148	Everdur	F	....	BIVFR
149	Everdur	F	....	PTCHDR
150	Everdur	F	....	PCH
156	Gold	A	0.5–100%; any temp.	
159	Hastelloy	A	All %; any temp.	BIVPTCH
160	Hastelloy	A	All %; any temp.	BIVPTCHDF
161	Hastelloy	A	All %; any temp.	BIVPTCHDF
162	Hastelloy	A	All %; any temp.	BIVP
163	Stellite	A	All %; any temp.	BIV
165	Stellite	A	All %; any temp.	BIV
184	Inconel	A	5%; aerated; 86°	
		F	5, 50%; unaerated; boil.	C
		F	80; storage	
		A	Glacial; boil.	C
196, 200, 266	Lead	V	Conc. or glacial; unaerated	PT
216	Monel	X	5%; aerated; 86°	
		A	5, 50%; unaerated; boil.	BIVPTCH
		A	80%; storage	
		A	Glacial; boil.	
219	Muntz	X		
224	Nickel	X	5%; aerated; 86°	
		F	5, 50%; unaerated; boil.	BVPTH
		F	80%; storage	
		A	Glacial; boil.	
225	Ni-clad	A	....	BTSC
226	Ni-silver	F	....	Instruments
231	Ni-Resist	X	5%; aerated; 86°	
		X	5%, 50%; unaerated; boil.	
233	NS-5	A	Dil.; impure; 150° max.	V
234	Olympic	A		
235	Olympic	A		
236	Palladium	A	0.5–100%; any temp.	
240	Platinum	A	0.5–100%; any temp.	
242	Ir.-Platinum	A	0.5–100%; any temp.	
244	Rh.-Platinum	A	0.5–100; any temp.	
245	Pyrasteel	V	....	BIVPTSCHF
249	Resistac	A	Glacial, conc. sol. only	BIVCHP
266	Lead	V	In absence of oxygen	PT
268	Silver	A	0.5–100%; any temp.	
270	304-clad	A	Over 15%; boil.	BTSC
		A	All %; 70°	
271	316-clad	A	All %; 70°	BTSC
274	430-clad	A	Over 15%; boil.	BTSC
		A	All conc.; 70°	
275	St. 301	V	All %; rm.	
276	St. 302	V	All %; rm.	
		A	All %; 70°	VTSCDF
		X	Over 15%; boil.	
		X	33%, 100% vapors; hot	
		A	10%, 50%, 100%; boil.	
		F	80%; boil.	
		X	100%; 400°; at 150 lb./sq. in.	
		A	Weak or medium	
		A	Conc.; cold	
		X	Conc.; boil.	
278	St. 303	V	All %; rm.	
279	St. 304	V	All %; rm.	
		A	All %; 70°	VTSCDF
		X	Over 15%; boil.	
		A	20% and 100%; 100°	P
		F	Glacial; boil.	BIVP
280	St. 308	V	All %; rm.	
281	St. 309	V	All %; rm.	
282	St. 310	V	All %; rm.	
		A	5–100%; 70°	P
		A	20% and 100%; 100°	P
283	St. 316	V	All %; boil. (total immersion)	BIVPTCHDF
		A	All %; 70°	
		A	Over 15%; boil.	
		A	33% vapors; hot	
		X	100% vapors; hot	
		A	20% and 100%; 100°	
		A	Glacial; boil.	BIVP
		A	All %; 70° and boil.	
		A	100%; 400°; at 150 lb./sq. in.	
284	St. 317	V	All %; boil. total immersion	BIVPTCHDF
		A	5–100%; 70°	P
		A	20% and 100%; 100°	P
285	St. 321	V	All %; rm.	
286	St. 347	V	All %; rm.	PT
287	St. 403	V	Glacial, 10%; rm.	
288	St. 405	V	Glacial, 10%; rm.	
290	St. 410	V	Glacial, 10%; rm.	
		F	Glacial; boil.	BIVP
		A	10%; 70°	
		A	All %	
291	St. 414	V	Glacial, 10%; rm.	
292	St. 416	V	Glacial, 10%; rm.	
		A	All %	
293	St. 418	V	Glacial, 10%; rm.	
294	St. 420	V	Glacial, 10%; rm.	
		A	All %	
295	St. 430	V	Glacial; boil. (total immersion)	
		V	35%; rm.	
		A	All %; 70°	VTSCHDFP
		X	Over 15%; boil.	
		A	20% and 100%; 100°	P
		A	100%; boil.	
		A	Weak	
		X	Medium or strong	
296	St. 430F	V	Glacial; boil. (total immersion)	
		V	33%; rm.	
297	St. 431	V	Glacial; boil. (total immersion)	
		V	33%; rm.	

* From data supplied by the manufacturers, the editors of *Chemical Engineering* compiled this table showing the resistance of materials to 58 common and troublesome chemicals. Also shown are some of the important applications in which the materials are being used. Now a word of caution. Corrosion involves many more variables than this or any table can possibly recognize. These data should never be construed as anything more than an aid in narrowing the field of materials that are worth investigation. Refer to materials for pumping various liquids on p. 1427.

Table 1. Chemical Resistance of Constructional Materials—(*Continued*)

No.	Material	Ratings	Exposure conditions	Applications
298	St. 440A	V	Glacial; boil. (total immersion)	
		V	33%; rm.	
300	St. 440C	V	Glacial; boil. (total immersion)	
		V	33%; rm.	
301	St. 442	V	Glacial; boil. (total immersion)	
		V	33%; rm.	
303	St. 446	V	Glacial; boil. (total immersion)	
		V	33%; rm.	
309	St. CC-35	A	100%; rm.–boil.	
316	St. CF-20	A	....	BIV
360A	Steel	X		
362	Still Met	A	....	BIVF
364	Stoody	A	10%–conc.; boil.	VF
365	Stoody	A	10%–conc.; boil.	IVF
367	Super Ni	F	....	PTCH
368	Tantalum	A	All %; 392° max.	CH
369	Telnic	A		
380	Tuf-Stuf	F	Cold	BIV
387	Stainless	A	All %	
388	Stainless	A	Weak; cold	
		X	Weak; boil.	
		X	Medium or strong	
389	Stainless	A	All %	
390	Worthite	A	All conditions	BIVPC
392	Wyndaloy	A	All %; 68°	
401	Karbate	A	All %; to boil.	BIVPTCHDR
402	Karbate	A	All %; to boil.	BIVPTCHDR
403	Kempruf	A	All %; any temp.	PTDR
404	Acheson	A	All %; any temp.	PTDR
500	Sul. cement	A	200° max.	D, T, towers, trenches, floors, pipe joints
501	Sul. cement	A	200° max.	Same as above
502	Furan cement	A	360° max.	Same as above
503	Furan cement	A	360° max.	Same as above
504	Phen. cement	A	360° max.	Same as above
505	Phen. cement	A	360° max.	Same as above
506	Silicate cement	F	1600° max.	Same as above
507	Silicate cement	A	1600° max.	Same as above
508	Acichlor	V	Dil. and conc.; 300° max.	
510	Acitite	A	Dil. and conc.; 2500° max.	
513	Asplit	A	All %; 350° max.	
514	Asplit-F	A	All %; 350° max.	
515	Basolit	A	200° max.	T
517	Carboline	A		
518	C-Basolit	A	200° max.	T
521	Causplit	A	All %; 350° max.	
523	Duralon	A	350° max.	BIPTCHDF
524	Durisite	A	350° max.	BIPTCHDF
535	Nukem	A	350° max.	T
536	Nukem	A		T
537	Pecomastic	F	Dil. and conc.; 300° max.	F
538	Penchlor	A	All %; 750° max.	
539	Penchlor	A	All %; 750° max.	
540	Penchlor	A	All %; 500°–2000°	
541	Pennsalt	A	All %; 350° max.	
542	Permanite	A	360° max.	TD
544	Plastite	F	175° max.	PTSD
545	P-Basolit	A	200° max.	T
554	Silastic	A	(A.S.T.M. D-543-43)	
556	Staminite	A	All %; any temp.	TD
559	Thiokol	V	150° max.	
600	Acid brick	A	1200° max.	TDR
603, 612, 615	Glass	A	....	THDR
604	Ceratherm	A	200–400° depending on design	BIPTSCHDFR
606	Stoneware	A	All conditions	BIVPTCHDFR
607	Glass-L	A	600° max.	PTCH
610	Stoneware	A	All %; any temp.	BIVPTCHDR
611	Porcelain	A	All %; any temp.	BIPTDR
614	Glass-L	A	All %; boil.; agitated	BTCH
616	Pyrex	A		
617	Stoneware	A	140–160°	BIPTSCHDFR
618	Vitreo	A	1000° max.	
619	Vitreosil	A		
621	Vycor	A		
700	Ace Saran	A	....	P
703	Compar	X		
704	DC Silicone	A	(A.S.T.M. D-543-43)	HD
713	Haveg	A	All %; rm. and boil.	BIVPTDFR
715	Heresite	A	200° max.	
716	Heresite	A	200° max.	
720	Lamicoid	A		T
723	Nixon	X	5%; 77°	
724	Nixon	A	5%; 77°	S
725	Nixon	A	5%; 77°	S
726	Nukemite	A	5%; rm.	PTSDF
727, 728	Nylon	A	Dilute	
729–732	Nylon	X		
733	Permanite	A	360° max.	BVPTD
736	Pyroflex	V	....	PTCHD
737	Resilon	F	175° max.	PTSD
738	Resistoflex	X		
739	Resistoflex	X		
740	Saran	A	77°	PD
741	Sealon	A	Dil.; 160° max.	BPTDF
		X	Glacial	
742	Teflon	A	Rm.	VP
743	Textolite	A	Dil.-conc.; rm.	T
744, 745	Textolite	F	Dil.; 77–122°; total immersion	BI
		X	Conc.; 77–122°; total immersion	
746	Tygon	V	Low conc.	BIPTSCDF
800	Ace Hd. Rub.	A	....	BIVPTDF
801	Acidseal	A	All %; 150° max.	BIPTDF
805	Butyl	A	30%; rm.	
809	Fairprene	V		
814	G.E. Silicone	A	Dil.; rm.	VIPT
		X	Conc. rm.	
817	Heresite	A	Glacial to 86°	
836	Natural (S)	V		
837	Natural (H)	A		
838	GR-S (S)	V		
839	GR-S (H)	A		
840	GR-A (S)	X		
841	GR-A (H)	V		
842	GR-M (S)	X		
843	GR-P (S)	X		
844	GR-I	V		
848	Silastic	A	(A.S.T.M. D-543-43)	
853	Thiokol	V	150° max.	T
854	Thiokol	X		
856	Vistanex	A	30%; rm.	
913	Redwood	A	40–80%; rm.; 45 years	T
		A	30–40%; rm.; shellacked; food	T
		A	2–20%; 36–165° max.; lithography	
		A	10%; 68–110°; asphalt-L; vinegar	T
		A	10%; 36–80°; asphalt-L; vinegar	T
		A	10%; 75–120°; 23 years; food	T
		A	10%; 70°; 18 years; vinegar	T
		A	6–10%; 60–90°; 13 years; food	T
		A	5–10%; 75°; paraffined; 30 years; food	T
		A	2–5%; 36–80°; painted; photo film	T

Acetic Anhydride

No.	Material	Ratings	Exposure conditions	Applications
3	Admiralty	A		
4	Admiralty	V	....	CH
6	Adnic	A	All %; 70°	
10–17	Aluminum	A	Rm.–boil.	BIPTSCHDF
19	Aloyco	A	90%; 200° aerated or boil.	BV
		A	30, 60%; 200°; quiet	BV
22	Ambraloy	F	....	CH
23	Ambraloy	F	....	CH
24	Ambraloy	V	....	CH
29–40	Ampco	A	....	BIVPTCHDF
54	Brass	X		
61	Brass	X		
63	Brass	AF	....	PCH
66	Bronze	A		
73	Bronze	F	....	VT
74	Bronze	F	....	V
75	Bronze	F	....	VP
76	Bronze	F		
77	Bronze	X		
81	CA-FA20	A	Boil.	BIVP
82	CA-MM	A	Boil.	BIVP
86	Cast iron	V		
86A	Cast iron	V		
88	Chlorimet	A	Any temp.	BIVCHF
89	Chlorimet	A	Any temp.	BIVCHF
111	Copper	A		
114	Copper	F	....	PTCHR

Table 1. Chemical Resistance of Constructional Materials—(*Continued*)

No.	Material	Ratings	Exposure conditions	Applications
118	Copper	A		
123	Cupro-Ni	F	......	CH
124	Cupro-Ni	A		
139	Durco	A	Any temp.	BIVHF
140	Durichlor	A	Any temp.	BIVPCHFR
141	Durimet	A	Any temp.	BIVHF
142	Durimet	A	Any temp.	BIVHF
143	Duriron	A	All % and temp.	BIVPHFR
148	Everdur	F	......	BIVR
149	Everdur	F	......	PTCHDR
150	Everdur	F	......	PCH
156	Gold	A	Any temp.	
159	Hastelloy	A	Any temp.	BIVPTCH
160	Hastelloy	A	Any temp.	BIVPTCHD
161	Hastelloy	A	Any temp.	BIVPTCHD
162	Hastelloy	A	Any temp.	BIV
163	Stellite	A	Any temp.	BIV
165	Stellite	A	Any temp.	BIV
184	Inconel	A	Boil.	
193, 196, 200	Lead	A	In absence of oxygen	PT
216	Monel	A	Boil.	P
219	Muntz	X		
224	Nickel	A	Boil.	
226	Ni-silver	F		
234	Olympic	A		
235	Olympic	A		
236	Palladium	A	Any temp.	
240	Platinum	A	Any temp.	
242	Ir-Platinum	A	Any temp.	
244	Rh-Platinum	A	Any temp.	
245	Pyrasteel	A	......	BIVPTSCHF
268	Silver	A	Any temp.	
275	St. 301	A	......	BIVTSCH
276	St. 302	A	70°–boil.	
		A	Any temp.	VTSCDF
		A	......	BIVTSCHDF
278	St. 303	A	......	IVF
279	St. 304	F	Boil.	BIVP
		A	Any temp.	VTSCDF
		A	......	BIVPTSCHD
280	St. 308	A	......	BIVPTSCHDF
281	St. 309	A	......	BIVTSCHDF
282	St. 310	A	......	BIVPTSCHDF
283	St. 316	A	70°–boil.	
		A	Boil.	BIVP
		A	Any temp.	VTCHDF
		A	......	BIVPTSCHDF
284	St. 317	A	......	BIVPTSCHDF
285	St. 321	A	......	BIVPT
286	St. 347	A	......	BIVPTSCHDF
287	St. 403	F	Rm.	
288	St. 405	F	Rm.	
290	St. 410	X	Boil.	
		F	Rm.	
291	St. 414	F	Rm.	
292	St. 416	F	Rm.	
293	St. 418	F	Rm.	
294	St. 420	F	Rm.	
295	St. 430	X	70°	
		A	Rm.	BIVPTSCDF
296	St. 430F	A	Rm.	IV
297	St. 431	A	Rm.	IV
298	St. 440A	A	Rm.	
300	St. 440C	A	Rm.	IV
301	St. 442	A	Rm.	BIVTSCHDF
303	St. 446	A	Rm.	BIVPTSCHDF
316	St. CF-20	A	......	BIV
318	St. CF-7C	A	90%; rm. to boil.	
319	St. CF-7M	A	90%; rm. to boil.	
360A	Steel	V		
367	Super Ni	F	......	PTCH
368	Tantalum	A	......	Not used commercially
369	Telnic	A		
390	Worthite	A	All conditions	BIV
401	Karbate	A	Up to boil.	
402	Karbate	A	Up to boil.	
403	Kempruf	A	Any temp.	
404	Acheson	A	Any temp.	
500	Sul. cement	A	200° max.	D, T, towers, trenches, floors
501	Sul. cement	A	200° max.	Same as 500
502	Furan cement	A	360° max.	Same as 500
503	Furan cement	A	360° max.	Same as 500
504	Phen. cement	A	360° max.	Same as 500
505	Phen. cement	A	360° max.	Same as 500
506	Silicate cement	F	1600° max.	Same as 500
507	Silicate cement	A	1600° max.	Same as 500
513	Asplit	A	350° max.	
514	Asplit-F	A	350° max.	
515	Basolit	A	200° max.	T
518	C-Basolit	A	200° max.	T
521	Causplit	A	350° max.	
523	Duralon	A	350° max.	BIPTCHF
524	Durisite	A	350° max.	
535	Nukem	A	350° max.	T
536	Nukem	A	......	T
538	Penchlor	A	750° max.	
539	Penchlor	A	750° max.	
540	Penchlor	A	500-2000°	
541	Pennsalt	A	350° max.	
542	Permanite	A	360° max.	TD
544	Plastite	V	175° max.	PTSD
545	P-Basolit	A	200° max.	T
554	Silastic	A	(A.S.T.M. D-543-43)	
559	Thiokol	V	150° max.	
600	Acid brick	A	1200° max.	TDR
603, 612, 615	Glass	A	......	THDR
604	Ceratherm	A	200-400° depending on design	BIPTSCHDFR
606	Stoneware	A	All conditions	BIVPTCHDFR
607	Glass-L	A	600° max.	PTCH
611	Porcelain	A	All %; any temp.	BIPTDR
614	Glass-L	A	Boil	BTCH
616	Pyrex	A		
617	Stoneware	A	140–160°	BIPTSCHDFR
618	Vitreo	A	1000° max.	
619	Vitreosil	A		
621	Vycor	A		
700	Ace Saran	A		P
703	Compar	X		
713	Haveg	A	265° max.	BIVPTDFR
715	Heresite	A	200° max.	
716	Heresite	A	200° max.	
720	Lamicoid	A	......	T
723	Permanite	A	360° max.	BVPDT
737	Resilon	V	175° max.	PTSD
738	Resistoflex	X		
739	Resistoflex	X		
740	Saran	F	77°	PD
742	Teflon	A		
746	Tygon	X		
800	Ace Hd. Rub	A	......	BIVPTDF
801	Acidseal	A	150° max.	BIPTDF
809	Fairprene	F		
817	Heresite	F	86° max.	
836	Natural (S)	V		
837	Natural (H)	A		
838	GR-S (S)	V		
839	GR-S (H)	A		
840	GR-A (S)	X		
841	GR-A (H)	V		
842	GR-M (S)	V		
843	GR-P (S)	X		
844	GR-I	V		
846	Saniprene	A	150° max.	BIPTDF
848	Silastic	A		
849	Superflexite	A	150° max.	BIPTDF
853	Thiokol	V	150° max.	T
854	Thiokol	X		

Alum

No.	Material	Ratings	Exposure conditions	Applications
3	Admiralty	A		
4	Admiralty	F	......	CH
6	Adnic	A	70°	
10–17	Aluminum	AF	Rm.	BIPTSDF
19	Aloyco	A	10%, sat.; boil.	BV
22	Ambraloy	F	......	CH
23	Ambraloy	F	......	CH
24	Ambraloy	F	......	CH
29–40	Ampco	A	......	BIVPTCHDF
51	Berylco	A		
54	Brass	X		
61	Brass	X		
63	Brass	AF	......	PCH
66	Bronze	A		
73	Bronze	F	......	VT
74	Bronze	F	......	V
75	Bronze	F	......	VP
76	Bronze	F		
77	Bronze	X		
81	CA-FA20	A	10%; boil.	BIVP
82	CA-MM	A	10%; boil.	BIVP
86	Cast iron	X		
86A	Cast iron	X		
111	Copper	A		
114	Copper	F	......	PTCHR
118	Copper	A		

Table 1. Chemical Resistance of Constructional Materials—(*Continued*)

No.	Material	Ratings	Exposure conditions	Applications
119	Corrosiron	A	10%; rm.	BIP
123	Cupro-Ni	A	....	CH
124	Cupro-Ni	A		
139	Durco	A	All %; any temp.	BIVHF
140	Durichlor	A	All %; any temp.	BIVPCHFR
141	Durimet	A	All %; any temp.	BIVHF
142	Durimet	A	All %; any temp.	BIVHF
143	Duriron	A	All %; any temp.	BIVPHFR
148	Everdur	F	....	BIVR
149	Everdur	F	....	PTCHDR
150	Everdur	F	....	PCH
156	Gold	A	All %; any temp.	
196, 200, 266	Lead	A	All %	BIPTH
217	Monel-clad	A	....	BTSCH
219	Muntz	X		
225	Ni-clad	A	....	BTSCH
226	Ni-silver	F		
234	Olympic	A		
235	Olympic	A		
236	Palladium	A	All %; any temp.	
240	Platinum	A	All %; any temp.	
242	Ir-Platinum	A	All %; any temp.	
244	Rh-Platinum	A	All %; any temp.	
245	Pyrasteel	A	....	BIVPTSCHDF
268	Silver	A	All %; any temp.	
270	304-clad	A	10% max.	BTSCH
271	316-clad	A	10% max.	BTSCH
275	St. 301	V		
276	St. 302	V		
		A	10%; any temp.	VTSCDF
		F	Sat.; boil.	
278	St. 303	V		
279	St. 304	V		
		A	10%; any temp.	VTSCDF
		A	10%; boil.	BIVP
280	St. 308	V		
281	St. 309	V		
282	St. 310	V		
283	St. 316	A	....	BIVPTSCH
		A	10%; any temp.	VTCHDF
		A	10%; boil.	BIVP
		A	Sat.; boil.	
284	St. 317	A	....	BIVPTSCH
285	St. 321	V		
286	St. 347	V		
287	St. 403	V		
288	St. 405	V		
290	St. 410	V		
		X	10%; boil.	
291	St. 414	V		
292	St. 416	V		
293	St. 418	V		
294	St. 420	V		
295	St. 430	V		
		X	10%; any temp.	
		A	10%; 70°	
		F	10%; boil.	
		X	Sat.; boil.	
296	St. 430F	V		
297	St. 431	V		
298	St. 440A	V		
300	St. 440C	V		
301	St. 442	V		
303	St. 446	V		
316	St. CF-20	A		
318	St. CF-7C	A	10%; rm.–boil.	
319	St. CF-7M	A	10%; rm.–boil.	
360A	Steel	X		
367	Super Ni	A	....	PTCH
368	Tantalum	A	....	Not used commercially
369	Telnic	A		
390	Worthite	A	All conditions	BIV
401	Karbate	A	All %; to boil.	BIPH
402	Karbate	A	All %; to boil.	BIPH
403	Kempruf	A	All %; any temp.	T
404	Acheson	A	All %; any temp.	T
500	Sul. cement	V	200° max.	D, T, towers, trenches, floors, pipe joints
501	Sul. cement	V	200° max.	Same as 500
502	Furan cement	V	360° max.	Same as 500
503	Furan cement	V	360° max.	Same as 500
504	Phen. cement.	V	360° max.	Same as 500
505	Phen. cement	V	360° max.	Same as 500
506	Silicate cement	V	1600° max.	Same as 500
507	Silicate cement	V	1600° max.	Same as 500
513	Asplit	A	All %; 350° max.	
514	Asplit-F	A	All %; 350° max.	
515	Basolit	A	200° max.	T
518	C-Basolit	A	200° max.	T
521	Causplit	A	All %; 350°	
523	Duralon	A	350° max.	BIPTCHDF
524	Durisite	A	350° max.	BIPTCHDF
532	Lumnite	A	10% max.; 90° max.	TD
534	N-series	A	Sat.; 160°	BIVPTCDF
535	Nukem	A	350° max.	T
536	Nukem	F	....	T
538	Penchlor	A	All %; 750° max.	
539	Penchlor	A	All %; 750° max.	
540	Penchlor	A	All %; 500–2000°	
541	Pennsalt	A	All %; 350° max.	
542	Permanite	A	360° max.	TD
544	Plastite	A	175° max.	PTSD
545	P-Basolit	A	200° max.	T
559	Thiokol	A		
600	Acid brick	A	1200° max.	
603, 612, 615	Glass	A	....	THDR
604	Ceratherm	A	200–400° depending on design	BIPTSCHDFR
606	Stoneware	A	All conditions	BIVPTCHDFR
607	Glass-L	A	600° max.	PTCH
611	Porcelain	A	All %; any temp.	BIPTDR
614	Glass-L	A	All conc.; 250° max.; agitated	BTCH
616	Pyrex	A		
617	Stoneware	A	140–160°	BIPTSCHFDR
618	Vitreo	A	1000° max.	
621	Vycor	A		
700	Ace Saran	A	....	P
711	Haveg	A	All %; 300° max.	BIVPTDFR
713	Haveg	A	All %; 300° max.	BIVPTDFR
715	Heresite	A	200° max.	
716	Heresite	A	200° max.	
718	Koroseal	A	All %; 150° max.	TI
720	Lamicoid	A	....	T
726	Nukemite	A	150° max.	PTSDF
733	Permanite	A	360° max.	BVPDT
737	Resilon	A	175° max.	PTSD
742	Teflon	A	Rm.	VP
746	Tygon	A	All %; 180° max.	BIPTSCDF
800	Ace Hd. Rub.	A	....	BIVPTDF
801	Acidseal	A	All %; 150° max.	BIPTSDF
802	Acidseal	A	All %; 150° max.	BIPTSDF
805	Butyl	A	Dry salt; aqueous sol. at rm.	
809	Fairprene	A		
817	Heresite	A	Sat.; 212° max.	
835	Perbunan	A	Dry salt; aqueous sol. at rm.	
836	Natural (S)	A		
837	Natural (H)	A		
838	GR-S (S)	A		
839	GR-S (H)	A		
840	GR-A (S)	A		
841	GR-A (H)	A		
842	GR-M (S)	A		
843	GR-P (S)	X		
844	GR-I	A		
846	Saniprene	A	All %; 150° max.	BIPTSDF
849	Superflexite	A	All %; 150° max.	BIPTSDF
853	Thiokol	A	150° max.	T
854	Thiokol	A	....	PTSCDF
855	Triflex	A	All %; 150° max.	BIPTSDF
856	Vistanex	A	Dry salt; aqueous sol. at rm.	

Aluminum Chloride

No.	Material	Ratings	Exposure conditions	Applications
3	Admiralty	V		
4	Admiralty	F	....	CH
6	Adnic	A	70°	
10–17	Aluminum	V	Sol.; rm.	
22	Ambraloy	F	....	CH
23	Ambraloy	F	....	CH
24	Ambraloy	F	....	CH
29–40	Ampco	A	....	BIVPTCHDF
42	Antaciron	A	All conditions	BIV
54	Brass	X		
61	Brass	X		
63	Brass	FV	....	PCH
66	Bronze	V		
73	Bronze	F	....	VT
74	Bronze	F	....	V
75	Bronze	F	....	VP
76	Bronze	F	....	Instruments
77	Bronze	X		

Table 1. Chemical Resistance of Constructional Materials—(*Continued*)

No.	Material	Ratings	Exposure conditions	Applications
86A	Cast iron	X	Aqueous sol.	
88	Chlorimet	A	All % and temp.	BIVCHF
89	Chlorimet	A	All %; moderate temp.	BIVCHF
111	Copper	V		
114	Copper	F	....	PTCHR
118	Copper	V		
123	Cupro-Ni	F	....	CH
124	Cupro-Ni	V		
139	Durco	A	All % at moderate temp.	BIVHF
140	Durichlor	A	All % and temp.	BIVPCHFR
141	Durimet	A	All % at moderate temp.	BIVHF
142	Durimet	A	All % at rm.	BIVHF
143	Duriron	A	All % and temp.; 140 preferred	BIVPHFR
148	Everdur	F	....	BIVR
149	Everdur	F	....	PTCHDR
150	Everdur	F	....	PCH
156	Gold	A		
159	Hastelloy	A	All %; 160° max.	BIVPT
160	Hastelloy	A	All %; any temp.; also as dry catalyst with organics up to 300°	BIVPTCHDF
161	Hastelloy	A	All %; 125° max.	BIVPT
162	Hastelloy	A	Rm. max.	V
163	Stellite	A	Moist; rm. max.; dry; 300° max.	BIV
165	Stellite	F	Moist; rm. max.; dry; 300° max.	BIV
184	Inconel	V	26%; 65°	
196, 200, 266	Lead	A	....	TC
216	Monel	A	26%; 65°	VT
217	Monel-clad	A	....	BTSCH
219	Muntz	X		
224	Nickel	A	65%; 65°	
226	Ni-silver	F	....	Instruments
231	Ni-Resist	F	26%; 65°	
234	Olympic	V		
235	Olympic	V		
236	Palladium	A		
240	Platinum	A		
242	Ir.-Platinum	A		
244	Rh.-Platinum	A		
245	Pyrasteel	X		
249	Resistac	A	....	BIVCHF
266	Lead	A	....	TC
268	Silver	A		
275	St. 301	X		
276	St. 302	X		
278	St. 303	X		
279	St. 304	X		
280	St. 308	X		
281	St. 309	X		
282	St. 310	X		
283	St. 316	V		
284	St. 317	V		
285	St. 321	X		
286	St. 347	X		
287	St. 403	X		
288	St. 405	X		
290	St. 410	X		
291	St. 414	X		
292	St. 416	X		
293	St. 418	X		
294	St. 420	X		
295	St. 430	X		
296	St. 430F	X		
297	St. 431	X		
298	St. 440A	X		
300	St. 440C	X		
301	St. 442	X		
303	St. 446	X		
316	St. CF-20	A	....	BIV
319	St. CF-7M	A	Sat.; 100°	
360A	Steel	X	Aqueous sol.	
367	Super Ni	F	....	PTCH
368	Tantalum	A	....	Not used commercially
369	Telnic	V		
390	Worthite	F	Cold, weak	BIV
401	Karbate	A	All %; to boil.	BIVPH
402	Karbate	A	All %; to boil.	BIVPH
403	Kempruf	A	All %; any temp.	TR
404	Acheson	A	All %; any temp.	TR
500	Sul. cement	A	Solutions; 200° max.	PT
501	Sul. cement	A	Solutions; 200° max.	PT
502	Furan cement	A	Solutions; 360° max.	PT
503	Furan cement	A	Solutions; 360° max.	PT
504	Phen. cement	A	Solutions; 360° max.	PT
505	Phen. cement	A	Solutions; 360° max.	PT
506	Silicate cement	A	Solutions; 1600° max.	PT
507	Silicate cement	A	Solutions; 1600° max.	PT
513	Asplit	A	All %; 350° max.	
514	Asplit-F	A	All %; 350° max.	
515	Basolit	A	200° max.	T
517	Carboline	A		
518	C-Basolit	A	200° max.	T
521	Causplit	A	All %; 350° max.	
523	Duralon	A	350° max.	BIPTCHDF
524	Durisite	A	350° max.	BIPTCHDF
534	N-series	A	25%; rm.	BIVPTCDF
535	Nukem	A	350° max.	T
536	Nukem	A	....	T
538	Penchlor	A	All %; 750° max.	
539	Penchlor	A	All %; 750° max.	
540	Penchlor	A	All %; 500°–2000°	
541	Pennsalt	A	All %; 350° max.	
542	Permanite	A	....	TD
544	Plastite	A	175° max.	PTSD
545	P-Basolit	A	200° max.	T
554	Silastic	A	(A.S.T.M. D–543–43)	
556	Staminite	A	All %	TD
559	Thiokol	X		
600	Acid brick	A	1200° max.	TDR
603, 612, 615	Glass	A	....	THDR
604	Ceratherm	A	200°–400°, depending on design	BIPTSCHDFR
606	Stoneware	A	All conditions	BIVPTCHDFR
607	Glass-L	A	600° max.	PTCH
610	Stoneware	A	All %; any temp.	BIVPTCHDR
611	Porcelain	A	All %; any temp.	BIPTDR
614	Glass-L	A	All %; 250° max.; agitation	BTCH
616	Pyrex	A		
617	Stoneware	A	140°–160°	BIPTSCHDFR
618	Vitreo	A	1000° max.	
619	Vitreosil	A		
621	Vycor	A		
700	Ace Saran	A	....	P
704	DC Silicone	A	(A.S.T.M. D–543–43)	HD
711	Haveg	A	All %; 300° max.	BIVPTDFR
713	Haveg	A	All %; 300° max.	BIVPTDFR
715	Heresite	A	200° max.	
716	Heresite	A	200° max.	
718	Koroseal	A	All %; 150° max.	TD
720	Lamicoid	A	....	T
726	Nukemite	A	150° max.	PTSDF
733	Permanite	A	360° max.	BVPDT
736	Pyroflex	A	....	PTCHD
737	Resilon	A	175° max.	PTSD
741	Sealon	A	160° max.	BPTDF
742	Teflon	A	Rm.	VP
746	Tygon	A	180° max.	BIPTSCDF
800	Ace Hd. Rub.	A		
801	Acidseal	A	All %; 150° max.	BIPTSDF
802	Acidseal	A	All %; 150° max.	BIPTSDF
805	Butyl	A	Dry salt; aqueous sol.; rm.	
809	Fairprene	A		
817	Heresite	A	Sat.; 212° max.	
835	Perbunan	A	Dry salt; aqueous sol.; rm.	
836	Natural (S)	A		
837	Natural (H)	A		
838	GR-S (S)	A		
839	GR-S (H)	A		
840	GR-A (S)	A		
841	GR-A (H)	A		
842	GR-M (S)	A		
843	GR-P (S)	X		
844	GR-I	A		
846	Saniprene	A	All %; 150° max.	BIPTSDF
848	Silastic	A	(A.S.T.M. D–543–43)	
849	Superflexite	A	All %; 150° max.	BIPTSDF
853	Thiokol	V	150° max.	T
854	Thiokol	V	150° max.	PTSCDF
855	Triflex	A	All %; 150° max.	BIPTSDF
856	Vistanex	A	Dry salt; aqueous sol.; rm.	

Aluminum Sulfate

No.	Material	Ratings	Exposure conditions	Applications
3	Admiralty	A		
4	Admiralty	F	....	CH
6	Adnic	A	70°	
10–17	Aluminum	AF	Rm.	BIPT
		A	Dry powder	
19	Aloyco	A	10%, Sat.; boil.	BV
22	Ambraloy	F	....	CH
23	Ambraloy	F	....	CH
24	Ambraloy	F	....	CH

Table 1. Chemical Resistance of Constructional Materials—(*Continued*)

No.	Material	Ratings	Exposure conditions	Applications
29–40	Ampco	A	....	BIVPTCHDF
54	Brass	X		
61	Brass	X		
63	Brass	FA	....	PCH
66	Bronze	A		
73	Bronze	F	....	VT
74	Bronze	F	....	V
75	Bronze	F	....	VP
76	Bronze	F	....	Instruments
77	Bronze	X		
81	CA-FA20	A	10%; 70°	BIVP
82	CA-MM	A	10%; 70°	BIVP
86A	Cast iron	X		
87	Causul	A	12.5% max.; cold	B
88	Chlorimet	A	All % and temp.	BIVCHF
89	Chlorimet	A	All % and temp.	BIVCHF
111	Copper	A		
114	Copper	F	....	PTCHR
117	Copper	V	....	P
118	Copper	A		
123	Cupro-Ni	A	....	CH
124	Cupro-Ni	A		
139	Durco	A	All % and temp.	BIVHF
140	Durichlor	A	All % and temp.	BIVPCHFR
141	Durimet	A	All % and temp.	BIVHF
142	Durimet	A	All % and temp.	BIVHF
143	Duriron	A	All % and temp.	BIVPHFR
148	Everdur	F	....	BIVR
149	Everdur	F	....	PCHDRT
150	Everdur	F	....	PCH
156	Gold	A		
159	Hastelloy	A	All %; 180° max.	BIVPT
160	Hastelloy	A	All % to boil.	BIVPTCH
161	Hastelloy	A	All % to boil.	BIVPTCH
162	Hastelloy	A	All % to boil.	BIVP
163	Stellite	A	All %; rm.	BIV
165	Stellite	A	All %; rm.	BIV
184	Inconel	A	0.1%; 60°	
		A	Storage; 25%	
196, 200, 266	Lead	A	All %	BIPTH
216	Monel	A	0.1, 5%; 60°	BVPTHD
		A	Storage 25%	
		F	Evaporator; 57%; 240°	
217	Monel-clad	A	....	BTSCH
219	Muntz	X		
224	Nickel	A	0.1%; 60°	T
		A	Storage; 25%	
		X	Evaporator; 57%; 240°	
225	Ni-clad	A	....	BTSCH
226	Ni-silver	F	....	Instruments
231	Ni-Resist	A	0.1%; 60°	B
		F	5%; 60°	
234	Olympic	A		
235	Olympic	A		
236	Palladium	A		
240	Platinum	A		
242	Ir-Platinum	A		
244	Rh-Platinum	A		
245	Pyrasteel	X		
249	Resistac	A	....	BIVCHF
266	Lead	A	All %	
268	Silver	A		
270	304-clad	A	All %; any temp.	BTSCH
271	316-clad	A	All %; any temp.	BTSCH
275	St. 301	A	....	BIVTSCH
276	St. 302	A	....	BIVTSCH
		A	All %; any temp.	VTSCDF
		A	10% neutral; 70°, boil.	
		A	Sat. neutral; 70°	
		F	Sat. neutral; boil.	
278	St. 303	A	....	VP
279	St. 304	A	....	BIVPTSCHD
		A	All %; any temp.	VTSCDF
280	St. 308	A	....	BIVPTSCH
281	St. 309	A	....	BIVTSCH
282	St. 310	A	....	BIVPTSCH
283	St. 316	A	....	BIVPTSCH
		A	All %; any temp.	VTCHDF
		A	10%, sat.; 70°, boil.	
284	St. 317	A	....	BIVPTSCH
285	St. 321	A	....	BIVPT
286	St. 347	A	....	BIVPTSCH
287	St. 403	F	Rm.	
288	St. 405	F	Rm.	
290	St. 410	F	Rm.	
		X	10%; 70°	
291	St. 414	F	Rm.	
292	St. 416	F	Rm.	
293	St. 418	F	Rm.	
294	St. 420	F	Rm.	
295	St. 430	A	....	BIVPTSCH
		X		
		F	10%; sat.; 70°	
		X	10%, sat.; boil.	
296	St. 430F	A	....	IV
297	St. 431	A	....	IV
298	St. 440A	A		
300	St. 440C	A	....	IVF
301	St. 442	A	....	BIVTSCH
303	St. 446	A	....	BIVPTSCH
316	St. CF-20	A	....	BIV
319	St. CF-7M	A	Sat.; rm.–boil.	
360A	Steel	X		
367	Super Ni	A	....	PTCH
369	Telnic	A		
390	Worthite	A	All conditions	BIV
401	Karbate	A	All %; to boil.	BIVPH
402	Karbate	A	All %; to boil.	BIVPH
403	Kempruf	A	All %; any temp.	
404	Acheson	A	All %; any temp.	
500	Sul. cement	A	200° max.	PT
501	Sul. cement	A	200° max.	PT
502	Furan cement	A	360° max.	PT
503	Furan cement	A	360° max.	PT
504	Phen. cement	A	360° max.	PT
505	Phen. cement	A	360° max.	PT
506	Silicate cement	F	1600° max.	T
507	Silicate cement	F	1600° max.	T
513	Asplit	A	All %; 350° max.	
514	Asplit-F	A	All %; 350° max.	
515	Basolit	A	200° max.	T
517	Carboline	A		
518	C-Basolit	A	200° max.	T
521	Causplit	A	All %; 350° max.	
523	Duralon	A	350° max.	BIPTCHDF
524	Durisite	A	350° max.	BIPTCHDF
532	Lumnite	A	150° max.	TD
535	Nukem	A	350° max.	T
536	Nukem	F	....	T
538	Penchlor	A	All %; 750° max.	
539	Penchlor	A	All %; 750° max.	
540	Penchlor	A	All %; 500°–2000°	
541	Pennsalt	A	All %; 350° max.	
542	Permanite	A	....	TD
544	Plastite	A	175° max.	TDPS
545	P-Basolit	A	200° max.	T
554	Silastic	A	A.S.T.M. D-543-43	
556	Staminite	A	All %	TD
559	Thiokol	A	150° max.	
600	Acid brick	A	....	TDR
603, 612, 615	Glass	A	....	THDR
604	Ceratherm	A	200–400°, Depending on design	BIPTSCHDFR
606	Stoneware	A	All conditions	BIVPTCHDFR
607	Glass-L	A	600° max.	PTCH
610	Stoneware	A	All %; any temp.	BIVPTCHDR
611	Porcelain	A	All % and temp.	BIPTDR
614	Glass-L	A	All %; under 302°; agitation	BTCH
616	Pyrex	A		
617	Stoneware	A	140°–160°	BIPTSCHFDR
618	Vitreo	A	1000° max.	
621	Vycor	A		
700	Ace Saran	A	....	P
704	DC Silicone	A	(A.S.T.M. D-543-43)	HD
711	Haveg	A	All %; 300° max.	BIVPTDFR
713	Haveg	A	All %; 300° max.	BIVPTDFR
715	Heresite	A	200° max.	
716	Heresite	A	200° max.	
718	Koroseal	A	All %; hard rubber better for conc. up to 5%	TD
720	Lamicoid	A	....	T
726	Nukemite	A	150° max.	PTSDF
733	Permanite	A	360° max.	BVPDT
736	Pyroflex	A	....	PTCHD
737	Resilon	A	175° max.	PTSD
741	Sealon	A	160° max.	BPTDF
742	Teflon	A	Rm.	VP
746	Tygon	A	180° max.	BIPTSCDF
800	Ace Hd. Rub.	A	....	BIVPTDF
801	Acidseal	A	All %; 150° max.	BIPTSDF
802	Acidseal	A	All %; 150° max.	BIPTSDF
805	Butyl	A	Dry salt; aqueous sol.; rm.	
809	Fairprene	A		
817	Heresite	A	Sat.; 212° max.	
835	Perbunan	A	Dry salt; aqueous sol.; rm.	
836	Natural (S)	A		
837	Natural (H)	A		
838	GR-S (S)	A		
839	GR-S (H)	A		

Table 1. Chemical Resistance of Constructional Materials—*(Continued)*

No.	Material	Ratings	Exposure conditions	Applications
840	GR-A (S)	A		
841	GR-A (H)	A		
842	GR-M (S)	A		
843	GR-P (S)	X		
844	GR-I	A		
846	Saniprene	A	All %; 150° max.	BIPTSDF
848	Silastic	A	(A.S.T.M. D-543-43)	
849	Superflexite	A	All %; 150° max.	BIPTSDF
853	Thiokol	A	150° max.	T
854	Thiokol	A	150° max.	PTSCDF
855	Triflex	A	All %; 150° max.	BIPTSDF
856	Vistanex	A	Dry salt; aqueous sol.; rm.	
913	Redwood	A	33%; rm.; 16 years; paper mill	T
		A	20%; 60-215°; petroleum refinery	T
		A	11%; 60-110°; petroleum refinery	T
		A	2-4%; 50-100°; fiberboard mill	T

Amines

No.	Material	Ratings	Exposure conditions	Applications
3	Admiralty	V		
4	Admiralty	A	....	CH
6	Adnic	V	70°	
10-17	Aluminum	V	All %; rm.-boil.	PCH
22	Ambraloy	F	....	CH
23	Ambraloy	F	....	CH
24	Ambraloy	F	....	CH
54	Brass	V	....	CH
61	Brass	V	....	VCH
63	Brass	FV	....	PCH
66	Bronze	V		
73	Bronze	F	....	VT
74	Bronze	F	....	V
75	Bronze	F	....	PV
76	Bronze	F	....	Instruments
77	Bronze	V	....	PCH
86A	Cast iron	A		
88	Chlorimet	A	All % and temp.	BIVCHF
89	Chlorimet	A	All % and temp.	BIVCHF
111	Copper	V		
114	Copper	F	....	PTCHR
118	Copper	V		
123	Cupro-Ni	F	....	CH
124	Cupro-Ni	V		
139	Durco	A	All % and temp.	BIVHF
140	Durichlor	A	All % and temp.	BIVPCHFR
141	Durimet	A	All % and temp.	BIVHF
142	Durimet	A	All % and temp.	BIVHF
143	Duriron	A	All % and temp.	BIVPHFR
148	Everdur	F	....	BIVR
149	Everdur	F	....	PTCHDR
150	Everdur	F	....	PCH
156	Gold	A	All conc. and temp.	
159	Hastelloy	A	All %; any temp.	BIVPT
160	Hastelloy	A	All %; any temp.	BIVPTCH
161	Hastelloy	A	All %; any temp.	BIVPTCH
162	Hastelloy	A	All %; any temp.	BIVP
163	Stellite	A	All %; any temp.	BIV
165	Stellite	A	All %; any temp.	BIV
219	Muntz	V	....	CH
226	Ni-silver	F	....	Instruments
234	Olympic	V		
235	Olympic	V		
236	Palladium	A		
240	Platinum	A		
242	Ir-Platinum	A		
244	Rh-Platinun	A		
268	Silver	A		
275	St. 301	A	....	BIVTSCH
276	St. 302	A	....	BIVTSCH
278	St. 303	A	....	IV
279	St. 304	A	....	BIVPTSCHD
280	St. 308	A	....	BIVPTSCH
281	St. 309	A	....	BIVTSCH
282	St. 310	A	....	BIVPTSCH
283	St. 316	A	....	BIVPTSCH
284	St. 317	A	....	BIVPTSCH
285	St. 321	A	....	BIVPT
286	St. 347	A	....	BIVPTSCH
287	St. 403	A	....	IV
288	St. 405	A	....	BIVTSCH
290	St. 410	A	....	BIVPTSCH
291	St. 414	A	....	IV
292	St. 416	A	....	IV
293	St. 418	A		
294	St. 420	A		
295	St. 430	A	....	BIVPTSCH
296	St. 430F	A	....	IV
297	St. 431	A	....	IV
298	St. 440A	A		
300	St. 440C	A	....	IV
301	St. 442	A	....	BIVTSCHDF
303	St. 446	A	....	BIVPTSCH
360A	Steel	A		
367	Super Ni	A	....	PTCH
368	Tantalum	V	High pressure and temp. (662° max.)	H
369	Telnic	V		
390	Worthite	A	Girbotol process circulation, typical	BIV
401	Karbate	A	All %; to boil.	BIVP
402	Karbate	A	All %; to boil.	BIVP
403	Kempruf	A	All %; any temp.	R
404	Acheson	A	All %; any temp.	R
500	Sul. cement	V	200° max.	T
501	Sul. cement	V	200° max.	T
502	Furan cement	V	360° max.	T
503	Furan cement	V	360° max.	T
504	Phen. cement	V	360° max.	T
505	Phen. cement	V	360° max.	T
506	Silicate cement	V	1600° max.	T
507	Silicate cement	V	1600° max.	T
523	Duralon	A	350° max.	BIPTCHDF
524	Durisite	A	350° max.	BIPTCHDF
535	Nukem	X		
542	Permanite	A	....	TD
544	Plastite	X		
554	Silastic	A	(A.S.T.M. D-543-43)	
556	Staminite	A	All %	TD
559	Thiokol	X		
600	Acid brick	A	....	TDR
603, 612, 615	Glass	A	....	THDR
604	Ceratherm	A	200-400°, depending on design	BIPTSCHDFR
606	Stoneware	A	Not commonly used	
607	Glass-L	A	600° max.	PTCH
610	Stoneware	A	All %; any temp.	BIVPTCHDR
611	Porcelain	A	All % and temp.	BIPTDR
614	Glass-L	V	....	BTCH
616	Pyrex	A		
617	Stoneware	A	140-160°	BIPTSCHDFR
618	Vitreo	A	1000° max.	
619	Vitreosil	A		
621	Vycor	A		
703	Compar	F	Rm.	TD
704	DC Silicone	A	(A.S.T.M. D-543-43)	HD
713	Haveg	A	All conditions	BIVPTDFR
715	Heresite	A	200° max.	
716	Heresite	A	200° max.	
720	Lamicoid	A	....	T
733	Permanite	A	360° max.	BVPDT
736	Pyroflex	A	....	PTCHD
737	Resilon	X		
738	Resistoflex	F	Conc.; rm.	P
739	Resistoflex Compar	F	Conc.; 140° max.	IVP
741	Sealon	X		
742	Teflon	A	Rm.	VP
746	Tygon	V	....	BIPTSCDF
801	Acidseal	V	All %; 150° max.	BIPTDF
817	Heresite	A	Sat.; 212° max.	
836	Natural (S)	X		
837	Natural (H)	V		
838	GR-S (S)	X		
839	GR-S (H)	V		
840	GR-A (S)	V		
841	GR-A (H)	V		
842	GR-M (S)	X		
843	GR-P (S)	X		
844	GR-I	X		
846	Saniprene	V	All %; 150° max.	BIPTDF
848	Silastic	A	(A.S.T.M. D-543-43)	
849	Superflexite	V	All %; 150° max.	BIPTDF
853	Thiokol	X	150° max.	
854	Thiokol	F	150° max.	PTSCDF

Ammonia

No.	Material	Ratings	Exposure conditions	Applications
3	Admiralty	X		
4	Admiralty	A	Dry	CH
		X	Moist	
6	Adnic	V		
10-17	Aluminum	A	Dry; any temp.	BIPTSCHDF
		AV	All %; rm., boil.	BIPTSCHDF

Table 1. Chemical Resistance of Constructional Materials—(*Continued*)

No.	Material	Ratings	Exposure conditions	Applications
22	Ambraloy	A	Dry	CH
		X	Moist	
23	Ambraloy	A	Dry	CH
		X	Moist	
24	Ambraloy	A	Dry	CH
		X	Moist	
54	Brass	A	Dry	CH
		X	Moist	
61	Brass	A	Dry	VCH
		X	Moist	
63	Brass	A	Dry	PCH
		X	Moist	
66	Bronze	X		
73	Bronze	A	Dry	VT
		X	Moist	
74	Bronze	A	Dry	V
		X	Moist	
75	Bronze	A	Dry	VP
		X	Moist	
76	Bronze	A	Dry	Instruments
		X	Moist	
77	Bronze	A	Dry	PCH
		X	Moist	
81	CA-FA20	A	Conc. sol.; 70°	BIVP
82	CA-MM	V	Conc. sol.; 70°	BIVP
85	Cast iron	A	Aqua	
86	Cast iron	A		
86A	Cast iron	A	Dry or sol.	
88	Chlorimet	A	All % and temp.	BIVCHF
89	Chlorimet	A	All % and temp.	BIVCHF
111	Copper	X		
114	Copper	A	Dry	PTCHR
		X	Moist	
118	Copper	X		
123	Cupro-Ni	A	Dry	CH
		X	Wet	
124	Cupro-Ni	F		
127–133	Dowmetal	A	Liquid or hydroxide	
139	Durco	A	All % and temp.	BIVHF
140	Durichlor	A	All % and temp.	BIVPCHFR
141	Durimet	A	All % and temp.	BIVHF
142	Durimet	A	All % and temp.	BIVHF
143	Duriron	A	All % and temp.	BIVPHFR
148	Everdur	A	Dry	BIVR
		X	Moist	
149	Everdur	A	Dry	PTCHDR
		X	Moist	
150	Everdur	A	Dry	PCH
		X	Moist	
156	Gold	A	All % and temp.	
159	Hastelloy	A	Wet; any temp.	BIVPTCH
160	Hastelloy	A	Wet; any temp.	BIVPTCH
161	Hastelloy	A	Wet; any temp.	BIVPTCH
162	Hastelloy	A	Wet; any temp.	BIVPS
163	Stellite	A	Wet; any temp.	BIV
165	Stellite	A	Wet; any temp.	BIV
184	Inconel	A	All %; 70°	
185	Inc-clad	A	....	BTSCH
216	Monel	A	2.7%; 70°; aerated	(VP, anhyd.) (V, aqua)
		X	10%; 70°; aerated	
217	Monel-clad	A	....	BTSCH
219	Muntz	A	Dry	CH
		X	Moist	
224	Nickel	A	Hydroxide; 1.1%; 70°; aerated	P
		X	Hydroxide; 2.7%; 70°; aerated	
		X	Hydroxide; 10%; 70°; aerated	
225	Ni-clad	A	....	BTSCH
226	Ni-silver	A	Dry	Instruments
		X	Moist	
231	Ni-Resist	A	All %; 70°	(BV, anhyd.) (BIV, aqua)
234	Olympic	X		
235	Olympic	X		
236	Palladium	A	All % and temp.	
240	Platinum	A	All % and temp.	
242	Ir-Platinum	A	All % and temp.	
244	Rh-Platinum	A	All % and temp.	
245	Pyrasteel	A	....	BIVPTSCHF
268	Silver	A	Oxygen-free sol. to 160°	
		X	Over 160°; anhydrous	
270	304-clad	A	All %; 70°	BTSCH
271	316-clad	A	All %; 70°	BTSCH
274	430-clad	A	All %; 70°	BTSCH
275	St. 301	A	....	BIVTSCH

No.	Material	Ratings	Exposure conditions	Applications
276	St. 302	A	....	BIVTSCH
		A	All %; 70°	VTSCDF
		X	Gas; hot	
		A	Sp. gr. 0.91; 70°; boil. gas liquor	
278	St. 303	A	....	TV
279	St. 304	A	....	BIVPTSCHDF
		A	All %; 70°	VTSCDF
		X	Gas; hot	
		A	Conc.; 70°	BIVPT
280	St. 308	A	....	BIVPTSCHDF
281	St. 309	A	....	BIVTSCHDF
282	St. 310	A	....	BIVPTSCHDF
283	St. 316	A	....	BIVPTSCHDF
		A	....	VTCHDF
		A	Conc.; 70°	BIVPT
284	St. 317	A	....	BIVPTSCHDF
285	St. 321	A	....	BIVPT
286	St. 347	A	....	BIVPTSCHDF
287	St. 403	A	....	IV
288	St. 405	A	....	BIVTSCHDF
290	St. 410	A	....	BIVPTSCHDF
		A	Conc.; 70°	BIVP
291	St. 414	A	....	IV
292	St. 416	A	....	IV
293	St. 418	A		
294	St. 420	A		
295	St. 430	A	....	BIVPTSCHDF
		X	Gas; hot	
		A	All %; 70°	VTSCHDF
		A	Sp. gr. 0.91; 70°; boil. gas liquor	
296	St. 430F	A	....	IV
297	St. 431	A	....	IV
298	St. 440A	A		
300	St. 440C	A	....	IV
301	St. 442	A	....	BIVTSCHDF
303	St. 446	A	....	BIVPTSCHDF
313	St. CF-7	A	All %; rm.–boil.	
318	St. CF-7C	A	All %; rm.–boil.	
319	St. CF-7M	A	All %; rm.–boil.	
338	St. CK-25	A	All %; rm.–boil.	
360A	Steel	A	Dry or sol.	
367	Super Ni	A	Dry	PTCH
		V	Moist	PTCH
368	Tantalum	X		
369	Telnic	X		
401	Karbate	A	All %; to boil.	BIVPCHF
402	Karbate	A	All %; to boil.	BIVPCHF
403	Kempruf	A	All %; any temp.	
404	Acheson	A	All %; any temp.	
500	Sul. cement	F	200° max.	T
501	Sul. cement	F	200° max.	T
502	Furan cement	A	360° max.	PT
503	Furan cement	A	360° max.	PT
504	Phen. cement	A	360° max.	PT
505	Phen. cement	A	360° max.	PT
506	Silicate cement	X	1600° max.	
507	Silicate cement	X	1600° max.	
508	Acichlor	A	Dilute and conc.; 300° max.	
510	Acitite	V	Dilute and conc.; 250° max.	
513	Asplit	A	All %; 350° max.	
514	Asplit-F	A	All %; 350° max.	
515	Basolit	X		
518	C-Basolit	X		
521	Causplit	A	All %; 350° max.	
523	Duralon	A	350° max.	BIPTCHDF
524	Durisite	A	350° max.	BIPTCHDF
535	Nukem	A	....	T
536	Nukem	A	....	T
539	Penchlor	A	All %; 750° max.	
541	Pennsalt	A	All %; 350° max.	
542	Permanite	A	....	TD
544	Plastite	V	175° max.	PTSD
545	P-Basolit	X		
554	Silastic	A	(A.S.T.M. D–543–43)	
556	Staminite	X		
559	Thiokol	A	150° max.	
600	Acid brick	A	....	TDR
603, 612, 615	Glass	A	....	THDR
604	Ceratherm	A	200–400°, depending on design	BIPTSCHDFR
606	Stoneware	A	Not commonly used	
607	Glass-L	V	....	PTCH
610	Stoneware	A	All %; any temp.	BIVPTCHDR

Table 1. Chemical Resistance of Constructional Materials—(*Continued*)

No.	Material	Ratings	Exposure conditions	Applications
611	Porcelain	A	All % and temp.	BIPTDR
614	Glass-L	V	....	BTCH
617	Stoneware	A	140–160°	BIPTSCHDFR
618	Vitreo	V	1000° max.	
619	Vitreosil	A		
700	Ace Saran	X		
703	Compar	X		
704	DC Silicone	A	(A.S.T.M. D–543–43)	HD
706	Formica	A	Conc.; rm.	PTF
713	Haveg	A	300° max.	BIVPTDFR
715	Heresite	A	200° max.	
716	Heresite	A	200° max.	
720	Lamicoid	A	....	T
726	Nukemite	A	....	PTSDF
727	Nylon	A		
728	Nylon	A		
729–732	Nylon	A		
733	Permanite	A	360° max.	BVPDT
736	Pyroflex	A	....	PTCHD
737	Resilon	V	175°	PTSD
738	Resistoflex	X		
739	Resistoflex	X		
740	Saran	X	77°	
741	Sealon	A	160° max.	BPTDF
742	Teflon	A	Rm.	VP
746	Tygon	V	180° max.	BIPTSCDF
800	Ace Hd. Rub.	A	....	BIVPTDF
801	Acidseal	A	Up to sat.; 100° max.; in water	BIPTDF
817	Heresite	A	Sat.; 212° max.; aqueous sol.	
836	Natural (S)	A		
837	Natural (H)	A		
838	GR-S (S)	A		
839	GR-S (H)	A		
840	GR-A (S)	A		
841	GR-A (H)	A		
842	GR-M (S)	A		
843	GR-P (S)	X		
844	GR-I	A		
846	Saniprene	A	Up to sat.; 100° max.; in water	BIPTDF
848	Silastic	A	(A.S.T.M. D–543–43)	
849	Superflexite	A	Up to sat.; 100° max.; in water	BIPTDF
853	Thiokol	V	150° max.	T
854	Thiokol	A	150° max.	PTSCDF

Ammonium Chloride

No.	Material	Ratings	Exposure conditions	Applications
3	Admiralty	X		
4	Admiralty	X		
6	Adnic	V		
10–17	Aluminum	V	Sol.; rm.	
		A	Dry powders	
22	Ambraloy	X		
23	Ambraloy	X		
24	Ambraloy	X		
29–40	Ampco	V	If not ammoniacal	BIVPTCHDF
42	Antaciron	A		
54	Brass	X		
61	Brass	X		
63	Brass	X		
66	Bronze	X		
73	Bronze	X		
74	Bronze	X		
75	Bronze	X		
76	Bronze	X		
77	Bronze	X		
85	Cast iron	A		
86	Cast iron	V		
86A	Cast iron	F	Frequently used	
88	Chlorimet	A	All % and temp.	BIVCHF
89	Chlorimet	A	All % and temp	BIVCHF
111	Copper	X		
114	Copper	X		
117	Copper	V	....	P
118	Copper	X		
119	Corrosiron	A	25%; rm.	BIPF
123	Cupro-Ni	X		
124	Cupro-Ni	F		
139	Durco	A	All % and temp.	BIVHF
140	Durichlor	A	All % and temp.	BIVPCHFR
141	Durimet	A	All % and temp.	BIVHF
142	Durimet	A	All % and temp.	BIVHF
143	Duriron	A	All % and temp.	BIVPHFR
148	Everdur	X		
149	Everdur	X		
150	Everdur	X		
156	Gold	A	All % and temp.	
159	Hastelloy	A	All %; 160° max.	BIVPT
160	Hastelloy	A	All % to boil.	BIVPTCH
161	Hastelloy	A	All % to 125°	BIVPT
162	Hastelloy	A	All %; rm. only	BIVP
163	Stellite	A	All %; rm. only	BIV
165	Stellite	F	All %; rm. only	BIV
184	Inconel	A	5%; boil.	
		V	28–40%; boil. in evap., 216°	
216	Monel	A	5%; boil.	BVT
		A	28–40%; boil. in evap., 216°	
217	Monel-clad	A	....	BTSCH
219	Muntz	X		
224	Nickel	A	5%; boil.	H
		A	28–40%; boil. in evap., 216°	
		A	Molten, 660°	
225	Ni-clad	A	....	BTSCH
226	Ni-silver	X		
231	Ni-Resist	A	5%; boil.	BI
		A	28–40%; boil. in evap., 216°	
234	Olympic	X		
235	Olympic	X		
236	Palladium	A	All % and temp.	
240	Platinum	A	All % and temp.	
242	Ir-Platinum	A	All % and temp.	
244	Rh-Platinum	A	All % and temp.	
245	Pyrasteel	V	....	BIVPTSCHF
249	Resistac	A	....	BIVCHF
268	Silver	A	Rm.	
		F	Over rm.	
270	304-clad	A	To 10%; 70°	BTSCH
271	316-clad	A	To 10%; 70°	BTSCH
274	430-clad	A	To 10%; 70°	BTSCH
275	St. 301	V		
276	St. 302	V		
		A	10% max.; 70°	VTSCDF
		A	10%; 20%; 50%; boil.	
278	St. 303	V		
279	St. 304	V		
		A	10% max.; 70°	BTSCDF
		A	Sat. at 70°	P
280	St. 308	V		
281	St. 309	V		
282	St. 310	V		
		A	Sat. at 70°	A
283	St. 316	V		
		A	10% max.; 70°	VTCHDF
		A	10%; 20%; 50%; boil.	
284	St. 317	V		
285	St. 321	V		
286	St. 347	V		
		A	Sat. at 70°	P
287	St. 403	V		
288	St. 405	V		
290	St. 410	V		
291	St. 414	V		
292	St. 416	V		
293	St. 418	V		
294	St. 420	V		
295	St. 430	V		
		A	10% max.; 70°	VTSCHDF
		A	10% boil.	
296	St. 430F	V		
297	St. 431	V		
298	St. 440A	V		
300	St. 440C	V		
301	St. 442	V		
303	St. 446	V		
318	St. CF-7C	A	5%–Conc.; rm.–boil.	
319	St. CF-7M	A	5%–Conc.; rm.–boil.	
360A	Steel	F	....	Frequently used
367	SuperNi	V	....	PTCH
368	Tantalum	A	Boil.; crude and c.p.	H
369	Telnic	X		
390	Worthite	F	Cold; weak	BIV
401	Karbate	A	All %; to boil.	BIVPH
402	Karbate	A	All %; to boil	BIVPH
403	Kempruf	A	All %; any temp.	T
404	Acheson	A	All %; any temp.	T
500	Sul. cement	A	200° max.	PT
501	Sul. cement	A	200° max.	PT
502	Furan cement	A	360° max.	PT
503	Furan cement	A	360° max.	PT
504	Phen. cement	A	360° max.	PT
505	Phen. cement	A	360° max.	PT
506	Silicate cement	F	1600° max.	T
507	Silicate cement	V	1600° max.	T

Table 1. Chemical Resistance of Constructional Materials—(*Continued*)

No.	Material	Ratings	Exposure conditions	Applications
508	Acichlor	A	Dilute and conc.; 300° max.	
510	Acitite	A	Dilute and conc.; 250° max.	
513	Asplit	A	All %; 350° max.	
514	Asplit-F	A	All %; 350° max.	
515	Basolit	A	200° max.	T
517	Carboline	A		
518	C-Basolit	A	200° max.	T
521	Causplit	A	All %; 350° max.	
523	Duralon	A	350° max.	BIPTCHDF
524	Durisite	A	350° max.	BIPTCHDF
535	Nukem	A	....	T
536	Nukem	A	....	T
537	Pecomastic	F	300° max.	
539	Penchlor	A	All %; 750° max.	
541	Pennsalt	A	All %; 350° max.	
542	Permanite	A	....	TD
544	Plastite	A	175° max.	PTSD
545	P-Basolit	A	200° max.	T
554	Silastic	A	(A.S.T.M. D-543-43)	
559	Thiokol	A	150° max.	
600	Acid brick	A	....	TDR
603, 612, 615	Glass	A	....	THDR
604	Ceratherm	A	200-400°, depending on design	BIPTSCHDFR
606	Stoneware	A	All conditions	BIVPTCHDFR
607	Glass-L	A	600° max.	PTCH
610	Stoneware	A	All %; any temp.	BIVPTCHDR
611	Porcelain	A	All % and temp.	BIPTDR
614	Glass-L	A	All %; under 320°; agitated	BTCH
616	Pyrex	A		
617	Stoneware	A	140°-160°	BIPTSCHDFR
618	Vitreo	A	1000° max.	
619	Vitreosil	A		
621	Vycor	A		
700	Ace Saran	A	....	P
704	DC Silicone	A	(A.S.T.M. D-543-43)	HD
707	Formica	A	All %; 212° max.	
711	Haveg	A	All %; 300° max.	BIVPTDFR
713	Haveg	A	All %; 300° max.	BIVPTDFR
715	Heresite	A	200° max.	
716	Heresite	A	200° max.	
720	Lamicoid	A	....	T
726	Nukemite	A	....	PTSDF
733	Permanite	A	360° max.	BVPDT
736	Pyroflex	A	....	PTCHD
737	Resilon	A	175° max.	PTSD
740	Saran	A	77°	
741	Sealon	A	160° max.	BPTDF
742	Teflon	A	Rm.	VP
746	Tygon	A	180° max.	BIPTSCDF
800	Ace Hd. Rub.	A	....	BIVPTDF
801	Acidseal	A	Up to sat.; 150° max.	BIPTDF
805	Butyl	A	Dry salt; aqueous sol.; rm.	
809	Fairprene	A		
817	Heresite	A	Sat.; 212° max.	
835	Perbunan	A	Dry salt; aqueous sol.; rm.	
836	Natural (S)	A		
837	Natural (H)	A		
838	GR-S (S)	A		
839	GR-S (H)	A		
840	GR-A (S)	A		
841	GR-A (H)	A		
842	GR-M (S)	A		
843	GR-P (S)	X		
844	GR-I	A		
846	Saniprene	A	Up to sat.; 150° max.	BIPTDF
848	Silastic	A	(A.S.T.M. D-543-43)	
849	Superflexite	A	Up to sat.; 150° max.	BIPTDF
853	Thiokol	A	150° max.	T
854	Thiokol	A	150° max.	PTSCDF
856	Vistanex	A	Dry salt; aqueous sol.; rm.	

Ammonium Phosphate

No.	Material	Ratings	Exposure conditions	Applications
3	Admiralty	F		
4	Admiralty	V	....	CH
6	Adnic	V	All %; 70°	
10-17	Aluminum	A	3% max.; rm.	
		V	Over 3%; rm.	
22	Ambraloy	V	....	CH
23	Ambraloy	V	....	CH
24	Ambraloy	V	....	CH
29-40	Ampco	F	....	BIVPTSCHDF
54	Brass	X		
61	Brass	X		
63	Brass	VF	....	PCH
66	Bronze	F		
73	Bronze	V	....	VT
74	Bronze	V	....	V
75	Bronze	V	....	VP
76	Bronze	V	....	Instruments
77	Bronze	X		
85	Cast iron	A		
86	Cast iron	V		
86A	Cast iron	A	Metaphosphate	
		F	Monobasic, dibasic	
88	Chlorimet	A	All % and temp.	BIVCHF
89	Chlorimet	A	All % and temp.	BIVCHF
111	Copper	F		
114	Copper	V	....	PTCHR
117	Copper	V	....	P
118	Copper	F		
123	Cupro-Ni	V	....	CH
124	Cupro-Ni	F		
139	Durco	A	All % and temp.	BIVHF
140	Durichlor	A	All % and temp.	BIVPCHFR
141	Durimet	A	All % and temp.	BIVHF
142	Durimet	A	All % and temp.	BIVHF
143	Duriron	A	All % and temp.	BIVPHFR
148	Everdur	V	....	BIVR
149	Everdur	V	....	PTCHDR
150	Everdur	V	....	PCH
159	Hastelloy	A	All % to boil.	BIVPTSCH
160	Hastelloy	A	All % to boil.	BIVPTSCH
161	Hastelloy	A	All % to boil.	BIVPTSCH
162	Hastelloy	A	All % to boil.	BIVP
163	Stellite	A	All % to boil.	BIV
165	Stellite	A	All % to boil.	BIV
184	Inconel	A	To boil.	
196, 200, 266	Lead	A	All %	BIPTH
216	Monel	A	To boil.	
219	Muntz	X		
224	Nickel	A	To boil.	
226	Ni-silver	V	....	Instruments
231	Ni-Resist	A	To boil.	BIVPTSCHF
234	Olympic	F		
235	Olympic	F		
245	Pyrasteel	A	....	B
275	St. 301	A	....	BIVTSCH
276	St. 302	A	....	BIVTSCH
278	St. 303	A	....	IV
279	St. 304	A	....	BIVPTSCHDF
280	St. 308	A	....	BIVPTSCH
281	St. 309	A	....	BIVTSCH
282	St. 310	A	....	BIVPTSCH
283	St. 316	A	....	BIVPTSCH
284	St. 317	A	....	BIVPTSCH
285	St. 321	A	....	BIVPTSCH
286	St. 347	A	....	BIVPTSCH
287	St. 403	A	....	IV
288	St. 405	A	....	BIVTSCH
290	St. 410	A	....	BIVPTSCH
291	St. 414	A	....	IV
292	St. 416	A	....	IV
293	St. 418	A		
294	St. 420	A		
295	St. 430	A	....	BIVPTSCHDF
296	St. 430F	A	....	IV
297	St. 431	A	....	IV
298	St. 440A	A		
300	St. 440C	A	....	IV
301	St. 442	A	....	BIVTSCH
303	St. 446	A	....	BIVPTSCH
360A	Steel	A	Metaphosphate	
		F	Monobasic, dibasic	
367	Super Ni	F	....	PTCH
368	Tantalum	A	....	Not used commercially
369	Telnic	F		
390	Worthite	A	All %; any temp.	BIV
401	Karbate	A	All %; to boil.	
402	Karbate	A	All %; to boil.	
403	Kempruf	A	All %; to boil.	
404	Acheson	A	All %; to boil.	
500	Sul. cement	F	200° max.	PT
501	Sul. cement	F	200° max.	PT
502	Furan cement	A	360° max.	PT
503	Furan cement	A	360° max.	PT
504	Phen. cement	A	360° max.	PT
505	Phen. cement	A	360° max.	PT
506	Silicate cement	X	1600° max.	
507	Silicate cement	F	1600° max.	T
513	Asplit	A	All %; 350° max.	
514	Asplit-F	A	All %; 350° max.	
515	Basolit	A	200° max.	T
518	C-Basolit	A	200° max.	T
521	Causplit	A	All %; 350° max.	

Table 1. Chemical Resistance of Constructional Materials—(*Continued*)

No.	Material	Ratings	Exposure conditions	Applications
523	Duralon	A	350° max.	BIPTCHDF
524	Durisite	A	350° max.	BIPTCHDF
535	Nukem	A	....	T
536	Nukem	A	....	T
539	Penchlor	A	All %; 750° max.	
541	Pennsalt	A	All %; 350° max.	
542	Permanite	A	.	TD
544	Plastite	A	175° max.	PTSD
545	P-Basolit	A	200° max.	T
554	Silastic	A	(A.S.T.M. D-543-43)	
559	Thiokol	A	150° max.	
600	Acid brick	A	....	TDR
603, 612, 615	Glass	A	....	THDR
604	Ceratherm	A	200–400°, depending on design	BIPTSCHDFR
606	Stoneware	A	Not commonly used	
607	Glass-L	V	....	PTCH
610	Stoneware	A	All %; any temp.	BIVPTCHDR
611	Porcelain	V	Dil.; cold	BIPTDR
614	Glass-L	A	Monobasic, meta; under 212°; all %	BTCH
		A	Dibasic; 77°; all %	BTCH
616	Pyrex	A		
617	Stoneware	A	140–160°	BIPTSCHDFR
618	Vitreo	V	1000° max.	
621	Vycor	A		
700	Ace Saran	A	....	P
704	DC Silicone	A	(A.S.T.M. D-543-43)	HD
711	Haveg	A	All %; 300° max.	BIVPTDFR
713	Haveg	A	All %; 300° max.	BIVPTDFR
715	Heresite	A	200° max.	
716	Heresite	A	200° max.	
720	Lamicoid	A	....	T
726	Nukemite	A	....	PTSDF
733	Permanite	A	360° max.	BVPDT
736	Pyroflex	A	....	PTCHD
737	Resilon	A	175° max.	PTSD
741	Sealon	A	160° max.	BPTDF
742	Teflon	A	Rm.	VP
746	Tygon	A	180° max.	BIPTSCDF
800	Ace Hd. Rub.	A	....	BVIPTDF
801	Acidseal	A	Up to sat.; 150° max.	BIPTDF
805	Butyl	A	Dry salt; aqueous sol.; rm.	
809	Fairprene	A		
817	Heresite	A	Sat.; 212° max.	
835	Perbunan	A	Dry salt; aqueous sol.; rm.	
836	Natural (S)	A		
837	Natural (H)	A		
838	GR-S (S)	A		
839	GR-S (H)	A		
840	GR-A (S)	A		
841	GR-A (H)	A		
842	GR-M (S)	A		
843	GR-P (S)	X		
844	GR-I	A		
846	Saniprene	A	Up to sat.; 150° max.	BIPTDF
848	Silastic	A	(A.S.T.M. D-543-43)	
849	Superflexite	A	Up to sat.; 150° max.	BIPTDF
853	Thiokol	A	150° max.	T
854	Thiokol	A	150° max.	PTCSDF
856	Vistanex	A	Dry salt; aqueous sol.; rm.	

Ammonium Sulfate

No.	Material	Ratings	Exposure conditions	Applications
3	Admiralty	F		
4	Admiralty	V	....	CH
6	Adnic	V	All %; 70°	
10–17	Aluminum	A	5% max.; rm.	PT
		A	Dry powders	
19	Aloyco	A	10%, sat.; boil.	BV
		A	Plus free H_2SO_4 at 150°	BV
22	Ambraloy	V	....	CH
23	Ambraloy	V	....	CH
24	Ambraloy	V	....	CH
29–40	Ampco	A	....	BIVPTSCHDF
54	Brass	X		
61	Brass	X		
63	Brass	VF	....	PCH
66	Bronze	F		
73	Bronze	V	....	VT
74	Bronze	V	....	V
75	Bronze	V	....	VP
76	Bronze	V	....	Instruments
77	Bronze	X		
81	CA-FA20	A	10%; boil.	BIVP
82	CA-MM	A	10%; boil.	BIVP
85	Cast iron	A		
86	Cast iron	A		
86A	Cast iron	F		
88	Chlorimet	A	All % and temp.	BIVCHF
89	Chlorimet	A	All % and temp.	BIVCHDF
111	Copper	F		
114	Copper	V	....	PTCHR
117	Copper	V	....	P
118	Copper	F		
119	Corrosiron	A	25%; rm.	BIP
123	Cupro-Ni	V	....	CH
124	Cupro-Ni	F		
139	Durco	A	All % and temp.	BIVHF
140	Durichlor	A	All % and temp.	BIVPCHFR
141	Durimet	A	All % and temp.	BIVHF
142	Durimet	A	All % and temp.	BIVHF
143	Duriron	A	All % and temp.	BIVPHFR
148	Everdur	V	....	BIVP
149	Everdur	V	....	PTCHDR
150	Everdur	V	....	PCH
159	Hastelloy	A	All % to 180°	BIVPT
160	Hastelloy	A	All % to boil.	BIVPTCH
161	Hastelloy	A	All % to boil.	BIVPTCH
162	Hastelloy	A	All % to boil.	BIVP
163	Stellite	A	All %; rm.	BIV
165	Stellite	A	All %; rm.	BIV
184	Inconel	A	Sat. plus 5% sulfuric; 150°	
193, 196, 200, 266	Lead	A	Various %; rm.	PT
216	Monel	A	Sat. plus 5% sulfuric; 150°	BIVPTHDF
217	Monel-clad	A	....	BTSCH
219	Muntz	X		
224	Nickel	F	Sat. plus 5% sulfuric; 150°	PH
225	Ni-clad	A	....	BTSCH
226	Ni silver	V	....	Instruments
231	Ni-Resist	A	Sat. plus 5% sulfuric; 150°	BIV
234	Olympic	F		
235	Olympic	F		
245	Pyrasteel	V	....	BIVPTSCHF
249	Resistac	A	....	BIVCHF
270	304-clad	A	1, 5%; 70°	BTSCH
271	316-clad	A	1, 5%; 70°	BTSCH
274	430-clad	A	1, 5%; 70°	BTSCH
275	St. 301	A	....	BIVTSCH
276	St. 302	A	....	BIVTSCH
		A	1%; 50%; 70°	VTSCDF
		X	10% sat.; boil.	
		A	10%; sat. cold; boil.	
278	St. 303	A	....	IV
279	St. 304	A	....	BIVPTSCHDF
		A	1%; 5%; 70°	VTSCDF
		X	10% sat.; boil.	
		A	10%; boil.	BIVP
280	St. 308	A	....	BIVPTSCH
281	St. 309	A	....	BIVTSCH
282	St. 310	A	....	BIVPTSCH
283	St. 316	A	....	BIVPTSCH
		X	10% sat.; boil.	
		A	10%; sat. cold; boil.	
		A	10%; boil.	BIVP
284	St. 317	A	....	BIVPTSCH
285	St. 321	A	....	BIVPTSCH
286	St. 347	A	....	BIVPTSCH
287	St. 403	F	Rm.	
288	St. 405	F		
290	St. 410	F	Rm.	
		X	10%; boil.	
291	St. 414	F	Rm.	
292	St. 416	F	Rm.	
293	St. 418	F	Rm.	
294	St. 420	F	Rm.	
295	St. 430	A	....	BIVPTSCHDF
		A	1%; 50%; 70°	VTSCHDF
296	St. 430F	A	....	IV
297	St. 431	A	....	IV
298	St. 440A	A		
300	St. 440C	A	....	IV
301	St. 442	A	....	BIVTSCH
303	St. 446	A	....	BIVPTSCH
316	St. CF-20	A	....	BIV
318	St. CF-7C	A	Conc.; rm. boil.	
319	St. CF-7M	A	Conc.; rm. boil.	
360A	Steel	F		
367	Super Ni	F	....	PTCH
368	Tantalum	A	....	Not used commercially
369	Telnic	F		
390	Worthite	A	All %; any temp.	BIV
401	Karbate	A	All %; to boil.	BIVPH
402	Karbate	A	All %; to boil.	BIVPH
403	Kempruf	A	All %; any temp.	
404	Acheson	A	All %; any temp.	
500	Sul. cement	A	200° max.	PT
501	Sul. cement	A	200° max.	PT
502	Furan cement	A	360° max.	PT

Table 1. Chemical Resistance of Constructional Materials—(*Continued*)

No.	Material	Ratings	Exposure conditions	Applications
503	Furan cement	A	360° max.	PT
504	Phen. cement	A	360° max.	PT
505	Phen. cement	A	360° max.	PT
506	Silicate cement	V	1600° max.	T
507	Silicate cement	V	1600° max.	T
513	Asplit	A	All %; 350° max.	
514	Asplit-F	A	All %; 350° max.	
515	Basolit	A	200° max.	T
517	Carboline	A		
518	C-Basolit	A	200° max.	T
521	Causplit	A	All %; 350° max.	
523	Duralon	A	350° max.	BIPTCHDF
524	Durisite	A	350° max.	BIPTCHDF
532	Lumnite	A		TD
535	Nukem	A		T
536	Nukem	A		T
539	Penchlor	A	All %; 750° max.	
541	Pennsalt	A	All %; 350° max.	
542	Permanite	A		TD
544	Plastite	A	175° max.	PTSD
545	P-Basolit	A	200° max.	T
554	Silastic	A	(A.S.T.M. D-543-43)	
559	Thiokol	A	150° max.	
600	Acid brick	A		TDR
603, 612, 615	Glass	A		THDR
604	Ceratherm	A	200–400°, depending on design	BIPTSCHFDR
606	Stoneware	A	All conditions	BIVPTCHFDR
607	Glass-L	A	600° max.	PTCH
610	Stoneware	A	All %; any temp.	BIVPTCHDR
611	Porcelain	A	All % and temp.	BIPTDR
614	Glass-L	A	All %; under 302°; agitation	BTCH
616	Pyrex	A		
617	Stoneware	A	140–160°	BIPTSCHDFR
618	Vitreo	A	1000° max.	
621	Vycor	A		
700	Ace Saran	A		P
704	DC Silicone	A	(A.S.T.M. D-543-43)	HD
711	Haveg	A	All %; 300° max.	BIVPTDFR
713	Haveg	A	All %; 300° max.	BIVPTDFR
715	Heresite	A	200° max.	
716	Heresite	A	200° max.	
718	Koroseal	A	All %.	TD
720	Lamicoid	A		T
726	Nukemite	A		PTSDF
733	Permanite	A	360° max.	BVPDT
736	Pyroflex	A		PTCHD
737	Resilon	A	175° max.	PTSD
740	Saran	A	77°	
741	Sealon	A	160° max.	BPTDF
742	Teflon	A	Rm.	VP
746	Tygon	A	180° max.	BIPTSCDF
800	Ace Hd. Rub.	A		BIVPTDF
801	Acidseal	A	All %; 150° max.	BIPTDF
802	Acidseal	A	All %; 150° max.	BIPTDF
809	Fairprene	A		
817	Heresite	A	Sat.; 212° max.	
836	Natural (S)	A		
837	Natural (H)	A		
838	GR-S (S)	A		
839	GR-S (H)	A		
840	GR-A (S)	A		
841	GR-A (H)	A		
842	GR-M (S)	A		
843	GR-P (S)	X		
844	GR-I	A		
846	Saniprene	A	All %; 150° max.	BIPTDF
848	Silastic	A	(A.S.T.M. D-543-43)	
849	Superflexite	A	All %; 150° max.	BIPTDF
853	Thiokol	A	150° max.	T
854	Thiokol	A	150° max.	PTSCDF
855	Triflex	A	All %; 150° max.	BIPTDF

Benzol

No.	Material	Ratings	Exposure conditions	Applications
3	Admiralty	A		
4	Admiralty	A		CH
6	Adnic	A		
10–17	Aluminum	A	Rm., boil.	BIPTSCHDF
22	Ambraloy	A		CH
23	Ambraloy	A		CH
24	Ambraloy	A		CH
29–40	Ampco	A		BIVPTCHDF
51	Berylco	A		
54	Brass	A		CH
61	Brass	A		VCH
63	Brass	A		PCH
66	Bronze	A		

No.	Material	Ratings	Exposure conditions	Applications
73	Bronze	A		VT
74	Bronze	A		V
75	Bronze	A		VP
76	Bronze	A		Instruments
77	Bronze	A		PCH
85	Cast iron	A		
86	Cast iron	A		
86A	Cast iron	A		
87	Causul	A	All %; any temp.	B
88	Chlorimet	A	All % and temp.	BIVCHF
89	Chlorimet	A	All % and temp.	BIVCHF
111	Copper	A		
114	Copper	A		PTCHR
117	Copper	A		P
118	Copper	A		
123	Cupro-Ni	A		CH
124	Cupro-Ni	A		
127–133	Dowmetal	A		
139	Durco	A	All % and temp.	BIVHF
140	Durichlor	A	All % and temp.	CHFR
141	Durimet	A	All % and temp.	BIVHF
142	Durimet	A	All %; any temp.	BIVHF
143	Duriron	A	All % and temp.	BIVPHFR
148	Everdur	A		BIVR
149	Everdur	A		PTCHDR
150	Everdur	A		PCH
156	Gold	A	Any temp. to cracking temp.	
159	Hastelloy	A	All conc. to boil.	
160	Hastelloy	A	All conc. to boil.	
161	Hastelloy	A	All conc. to boil.	
162	Hastelloy	A	All conc. to boil.	
163	Stellite	A	All conc. to boil.	BIV
165	Stellite	A	All conc. to boil.	BIV
184	Inconel	A	Boil.	
2 6	Monel	A	Boil.	VC
217	Monel-clad	A		BTSCH
2 9	Muntz	A		CH
224	Nickel	A	Boil.	T
225	Ni-clad	A		BTSCH
226	Ni-silver	A		Instruments
231	Ni-Resist	A	Boil.	BI
234	Olympic	A		
235	Olympic	A		
236	Palladium	A	Any temp. to cracking temp.	
240	Platinum	A	Any temp. to cracking temp.	
242	Ir-platinum	A	Any temp. to cracking temp.	
244	Rh-platinum	A	Any temp. to cracking temp.	
245	Pyrasteel	A		BIVPTSCHF
268	Silver	A	Any temp. to cracking temp.	
270	304-clad	A	All %; any temp.	BTSCH
271	316-clad	A	All %; any temp.	BTSCH
274	430-clad	A	All %; any temp.	BTSCH
275	St. 301	A		BIVTSCHDF
276	St. 302	A		BIVTSCHDF
		A	Hot	
		A	Any temp.	VTSCDF
278	St. 303	A		IV
279	St. 304	A		BIVPTSCDF
		A	Any temp.	VTSCDF
280	St. 308	A		BIVPTSCDF
281	St. 309	A		BIVTSCHDF
282	St. 310	A		BIVPTSCHDF
283	St. 316	A		BIVPTSCDF
		A	Hot	
		A	Any temp.	VTCHDF
284	St. 317	A		BIVPTSCDF
285	St. 321	A		BIVPTSC
286	St. 347	A		BIVPTSCH
287	St. 403	A		IV
288	St. 405	A		BIVTSC
290	St. 410	A		BIVPTSCDF
291	St. 414	A		IV
292	St. 416	A		IV
293	St. 418	A		
294	St. 420	A		
295	St. 430	A		BIVPTSCHDF
		A	Any temp.	VTSCHDF
296	St. 430F	A		IV
297	St. 431	A		IV
298	St. 440A	A		
300	St. 440C	A		IV
301	St. 442	A		BIVTSCHDF
303	St. 446	A		BIVPTSCDF
308	St. CB-30	A	High temp.	
360A	Steel	A		
367	Super Ni	A		PTCH

Table 1. Chemical Resistance of Constructional Materials—(*Continued*)

No.	Material	Ratings	Exposure conditions	Applications
368	Tantalum	A	....	Not used commercially
369	Telnic	A		
401	Karbate	A	All %; to boil.	BIVP
402	Karbate	A	All %; to boil.	BIVP
403	Kempruf	A	All %; any temp.	T
404	Acheson	A	All %; any temp.	T
500	Sul. cement	X		
501	Sul. cement	X		
502	Furan cement	A	360° max.	PTD
503	Furan cement	A	360° max.	PTD
504	Phen. cement	A	360° max.	PTD
505	Phen. cement	A	360° max.	PTD
506	Silicate cement	F	1600° max.	TD
507	Silicate cement	F	1600° max.	TD
508	Acichlor	X		
510	Acitite	F	Dilute and conc.; 250° max.	
513	Asplit	A	All %; 350° max.	
514	Asplit-F	A	All %; 350° max.	
515	Basolit	A	....	T
517	Carboline	A		
518	C-Basolit	A	....	T
521	Causplit	A	All %; 350° max.	
523	Duralon	A	350° max.	BIPTCHDF
524	Durisite	A	350° max.	BIPTCHDF
534	N-series	X		
535	Nukem	A	....	T
536	Nukem	A	....	T
537	Pecomastic	X		
538	Penchlor	A	All %; 750° max.	
539	Penchlor	A	All %; 750° max.	
540	Penchlor	A	All %; 500–2000°	
541	Pennsalt	A	All %; 350° max.	
542	Permanite	A	....	TD
544	Plastite	X		
545	P-Basolit	A	....	T
554	Silastic	X	(A.S.T.M. D-543-43)	
556	Staminite	A	All %	T
559	Thiokol	X		
600	Acid brick	A	....	TDR
603, 612, 615	Glass	A	....	THDR
604	Ceratherm	A	200–400°, depending on design	BIPTSCHDFR
606	Stoneware	A	Not commonly used	
607	Glass-L	A	600° max.	PTCH
610	Stoneware	A	All %; any temp.	BIVPTCHDR
611	Porcelain	A	All % and temp.	BIPTDR
614	Glass-L	A	Under 302°; agitation	BTCH
616	Pyrex	A		
617	Stoneware	A	140–160°	BIPTSCHDFR
618	Vitreo	A	1000° max.	
621	Vycor	A		
700	Ace Saran	F	....	P
703	Compar	A	Conc.; −40° to boil.	TD
704	DC Silicone	X	(A.S.T.M. D-543-43)	
713	Haveg	A	All conditions	BIVPTDFR
715	Heresite	A	200° max.	
716	Heresite	A	200° max.	
718	Koroseal	X		
720	Lamicoid	A	....	T
723	Nixon	A	77°	
724	Nixon	A	77°	S
733	Permanite	A	360° max.	BVPDT
736	Pyroflex	V	....	PTCHD
737	Resilon	X		
738	Resistoflex	A	Conc.; to boil.	P
739	Resistoflex	A	Conc.; 275° max.	IVP
740	Saran	V	77°	
741	Sealon	X		
742	Teflon	A	Rm.	VP
744	Textolite	A	77–122°	BI
745	Textolite	A	77–122°	BI
746	Tygon	X		
800	Ace Hd. Rub.	X		
801	Acidseal	X		
802	Acidseal	X		
805	Butyl	X		
817	Heresite	X		
835	Perbunan	V		
836	Natural (S)	X		
837	Natural (H)	X		
838	GR-S (S)	X		
839	GR-S (H)	X		
840	GR-A (S)	X		
841	GR-A (H)	F		
842	GR-M (S)	X		
843	GR-P (S)	V		
844	GR-I	X		
846	Saniprene	X		
848	Silastic	X		
849	Superflexite	X		
853	Thiokol	F	150° max.	T
854	Thiokol	V	150° max.	PTSCDF
855	Triflex	X		
856	Vistanex	X		

Boric Acid

No.	Material	Ratings	Exposure conditions	Applications
3	Admiralty	A		
4	Admiralty	A	....	CH
6	Adnic	A		
10–17	Aluminum	A	All %; rm.	BIPTSDF
		A	Dry powders	
22	Ambraloy	A	....	CH
23	Ambraloy	A	....	CH
24	Ambraloy	A	....	CH
29–40	Ampco	A	....	BIVPTCHDF
51	Berylco	A		
54	Brass	F	....	CH
61	Brass	F	....	VCH
63	Brass	A	....	PCH
66	Bronze	A		
73	Bronze	A	....	VT
74	Bronze	A	....	V
75	Bronze	A	....	VP
76	Bronze	A	....	Instruments
77	Bronze	F	....	PCH
86	Cast iron	X		
86A	Cast iron	X		
88	Chlorimet	A	All % and temp.	BIVCHF
89	Chlorimet	A	All % and temp.	BIVCHF
111	Copper	A		
114	Copper	A	....	PTCHR
117	Copper	V	....	P
118	Copper	A		
123	Cupro-Ni	A	....	CH
124	Cupro-Ni	A		
139	Durco	A	All % and temp.	BIVHF
140	Durichlor	A	All % and temp.	BIVPCHFR
141	Durimet	A	All % and temp.	BIVHF
142	Durimet	A	All % and temp.	BIVHF
143	Duriron	A	All % and temp.	BIVPHFR
148	Everdur	A	....	BIVR
149	Everdur	A	....	PTCHDR
150	Everdur	A	....	PCH
156	Gold	A	Any temp.	
159	Hastelloy	A	All conc. to boil.	
160	Hastelloy	A	All conc. to boil.	
161	Hastelloy	A	All conc. to boil.	
162	Hastelloy	A	All conc. to boil.	
163	Stellite	A	All conc. to boil.	BIV
165	Stellite	A	All conc. to boil.	BIV
196, 200, 266	Lead	V	Various %; rm.; unaerated	BIPTH
216	Monel	A	4%; 104°	BIVPTH
217	Monel-clad	A	....	BTSCH
219	Muntz	F	....	CH
225	Ni-clad	A	....	BTSCH
226	Ni-silver	A	....	Instruments
234	Olympic	A		
235	Olympic	A		
236	Palladium	A	Any temp.	
240	Platinum	A	Any temp.	
242	Ir-Platinum	A	Any temp.	
244	Rh Platinum	A	Any temp.	
245	Pyrasteel	A	....	BIVPTSCHF
268	Silver	A	Any temp.	
270	304-clad	A	....	BTSCH
271	316-clad	A	....	BTSCH
274	430-clad	A	....	BTSCH
275	St. 301	A	....	BIVTSCH
276	St. 302	A	....	BIVTSCH
		A	Conc.; boil.	
278	St. 303	A	....	IV
279	St. 304	A	....	BIVPTSCHDF
280	St. 308	A	....	BIVPTSCH
281	St. 309	A	....	BIVTSCH
282	St. 310	A	....	BIVPTSCH
283	St. 316	A	....	BIVPTSCH
		A	Conc.; boil.	
284	St. 317	A	....	BIVPTSCH
285	St. 321	A	....	BIVPTSCH
286	St. 347	A	....	BIVPTSCH
287	St. 403	A	....	IV
288	St. 405	A	....	BIVTSCH
290	St. 410	A	....	BIVPTSCH
291	St. 414	A	....	IV
292	St. 416	A	....	IV
293	St. 418	A		
294	St. 420	A		

Table 1. Chemical Resistance of Constructional Materials—(*Continued*)

No.	Material	Ratings	Exposure conditions	Applications
295	St. 430	A	....	BIVPTSCHDF
		A	Conc.; boil.	
296	St. 430F	A	....	IV
297	St. 431	A	....	IV
298	St. 440A	A		
300	St. 440C	A	....	IV
301	St. 442	A	....	BIVTSCH
303	St. 446	A	....	BIVPTSCH
316	St. CF-20	A	....	BIV
360A	Steel	X		
367	Super Ni	A	....	PTCH
368	Tantalum	A		Not used commercially
369	Telnic	A		
390	Worthite	A	All %; any temp.; aqueous sol. 210°	BIV
401	Karbate	A	All %; to boil.	P
402	Karbate	A	All %; to boil.	P
403	Kempruf	A	All %; to boil.	
404	Acheson	A	All %; to boil.	
500	Sul. cement	A	200° max.	PT
501	Sul. cement	A	200° max.	PT
502	Furan cement	A	360° max.	PT
503	Furan cement	A	360° max.	PT
504	Phen. cement	A	360° max.	PT
505	Phen. cement	A	360° max.	PT
506	Silicate cement	F	1600° max.	T
507	Silicate cement	A	1600° max.	PT
513	Asplit	A	All %; 350° max.	
514	Asplit-F	A	All %; 350° max.	
515	Basolit	A	200° max.	T
518	C-Basolit	A	200° max.	T
521	Causplit	A	All %; 350° max.	
523	Duralon	A	350° max.	BIPTCHDF
524	Durisite	A	350° max.	BIPTCHDF
535	Nukem	A	350° max.	T
536	Nukem	A	....	T
538	Penchlor	A	All %; 750° max.	
539	Penchlor	A	All %; 750° max.	
540	Penchlor	A	All %; 500–2000°	
541	Pennsalt	A	All %; 350° max.	
542	Permanite	A	....	TD
544	Plastite	A	175° max.	TPSD
545	P-Basolit	A	200° max.	T
554	Silastic	A	(A.S.T.M. D-543-43)	
556	Staminite	A	All %	T
559	Thiokol	A	150° max.	
600	Acid brick	A	....	TDR
603, 612, 615	Glass	A	....	THDR
604	Ceratherm	A	200–400°, depending on design	BIPTSCHDFR
606	Stoneware	A	Not commonly used	
607	Glass-L	A	600° max.	PTCH
610	Stoneware	A	All %; any temp.	BIVPTCHDR
611	Porcelain	A	All % and temp.	BIPTDR
614	Glass-L	A	All %; under 320°; agitation	BTCH
616	Pyrex	A		
617	Stoneware	A	140–160°	BIPTSCHDFR
618	Vitreo	A	1000° max.	
619	Vitreosil	X		
621	Vycor	A		
708	Ace Saran	A	....	P
704	DC Silicone	A	(A.S.T.M. D-543-43)	HD
711	Haveg	A	All conditions	BIVPTDFR
713	Haveg	A	All conditions	BIVPTDFR
715	Heresite	A	200° max.	
716	Heresite	A	200° max.	
718	Koroseal	A	Sat.	TD
726	Lamicoid	A	....	T
724	Nixon	A	Conc.; 77°	S
726	Nukemite	A	150° max.	PTSDF
733	Permanite	A	360° max.	BVPDT
736	Pyroflex	A	....	PTCHD
737	Resilon	A	175° max.	PTSD
740	Saran	A	77°	
741	Sealon	A	160° max.	BPTDF
742	Teflon	A	Rm.	VP
743	Textolite	A	Rm.	T
744	Textolite	F	Dil.; 77–122°	BI
745	Textolite	F	Dil.; 77–122°	BI
746	Tygon	A	....	BIPTSCDF
800	Ace Hd. Rub.	A	....	BIVPTDF
801	Acidseal	A	All %; 150° max.	BIPTSDF
802	Acidseal	A	All %; 150° max.	BIPTSDF
805	Butyl	F	Rm.	
809	Fairprene	A		
814	GE-Silicone	A	Dil.; rm.; 7-day immersion	VPTDF
817	Heresite	A	Sat.; 212° max.	
835	Perbunan	F	Rm.	
836	Natural (S)	A		
837	Natural (H)	A		
838	GR-S (S)	A		
839	GR-S (H)	A		
840	GR-A (S)	A		
841	GR-A (H)	A		
842	GR-M (S)	A		
843	GR-P (S)	V		
844	GR-I	A		
846	Saniprene	A	All %; 150° max.	BIPTSDF
848	Silastic	A	(A.S.T.M. D-543-43)	
849	Superflexite	A	All %; 150° max.	BIPTSDF
853	Thiokol	A	150° max.	T
854	Thiokol	A	150° max.	PTSCDF
855	Triflex	A	All %; 150° max.	BIPTSDF
856	Vistanex	F	Rm.	

Bromine

No.	Material	Ratings	Exposure conditions	Applications
3	Admiralty	A	Dry	
		V	Moist	
4	Admiralty	A	Dry	CH
		V	Moist	
6	Adnic	A	Dry gas	
10–17	Aluminum	X		
22	Ambraloy	A	Dry	CH
		V	Moist	
23	Ambraloy	A	Dry	CH
		V	Moist	
24	Ambraloy	A	Dry	CH
		V	Moist	
29–40	Ampco	V		
54	Brass	A	Dry	CH
		X	Moist	
61	Brass	A	Dry	VCH
		X	Moist	
63	Brass	A	Dry	PCH
		V	Moist	
66	Bronze	A	Dry	
		V	Moist	
73	Bronze	A	Dry	VT
		V	Moist	
74	Bronze	A	Dry	V
		V	Moist	
75	Bronze	A	Dry	VP
		V	Moist	
76	Bronze	A	Dry	Instruments
		V	Moist	
77	Bronze	A	Dry	PCH
		X	Moist	
86A	Cast iron	X		
88	Chlorimet	A	Dry; all temp.	BIVCHF
		X	Wet	
89	Chlorimet	A	Dry; all temp.	BIVCHF
		X	Wet	
111	Copper	A	Dry	
		V	Moist	
114	Copper	A	Dry	PTCHR
		V	Moist	
118	Copper	A	Dry	
		V	Moist	
123	Cupro-Ni	A	Dry	CH
		V	Moist	
124	Cupro-Ni	A	Dry	
		V	Moist	
127–133	Dowmetal	A	Dry	
139	Durco	X		
140	Durichlor	X		
141	Durimet	X		
142	Durimet	X		
143	Duriron	X		
148	Everdur	A	Dry	BIVR
		V	Moist	
149	Everdur	A	Dry	PTCHDR
		V	Moist	
150	Everdur	A	Dry	PCH
		V	Moist	
156	Gold	X		
159	Hastelloy	F	Dry only	
160	Hastelloy	A	HBr; any temp.	BIVPTCH
161	Hastelloy	A	Wet gas; rm. only	BIVPTCH
162	Hastelloy	F	Dry only	
163	Stellite	A	Wet gas; rm. only	BIV
165	Stellite	A	Wet gas; rm. only	BIV
184	Inconel	A	Dry	
		X	Wet	
216	Monel	A	Dry	
		X	Wet	
219	Muntz	A	Dry	CH
		X	Moist	

Table 1. Chemical Resistance of Constructional Materials—*(Continued)*

No.	Material	Ratings	Exposure conditions	Applications
224	Nickel	A	Dry	
		X	Wet	
226	Ni-silver	A	Dry	Instruments
		V	Moist	
231	Ni-Resist	A	Dry	
		X	Wet	
234	Olympic	A	Dry	
		V	Moist	
235	Olympic	A	Dry	
		V	Moist	
240	Platinum	A	Rm.	
		X	Elevated temp.	
242	Ir-Platinum	A	Rm.	
		X	Elevated temp.	
244	Rh-Platinum	A	Rm.	
		X	Elevated temp.	
245	Pyrasteel	A	....	BIVPTSCHF
268	Silver	X		
275	St. 301	X		
276	St. 302	X		
		X	70°	
		X	70°; bromine water	
278	St. 303	X		
279	St. 304	X		
		X	70°; bromine water	
280	St. 308	X		
281	St. 309	X		
282	St. 310	X		
283	St. 316	X		
		X	70°	
		X	70°; bromine water	
284	St. 317	X		
285	St. 321	X		
286	St. 347	X		
287	St. 403	X		
288	St. 405	X		
290	St. 410	X		
291	St. 414	X		
292	St. 416	X		
293	St. 418	X		
294	St. 420	X		
295	St. 430	X		
		X	70°	
		X	70°; bromine water	
296	St. 430F	X		
297	St. 431	X		
298	St. 440A	X		
300	St. 440C	X		
301	St. 442	X		
303	St. 446	X		
360A	Steel	X		
367	Super Ni	A	Dry	PTCH
		V	Moist	
368	Tantalum	A	302° max.	CH
369	Telnic	A	Dry	
		V	Moist	
390	Worthite	X		
401	Karbate	V		
403	Kempruf	V		
500	Sul. cement	F	120° max.	T
501	Sul. cement	F	120° max.	T
502	Furan cement	F	360° max.	T
503	Furan cement	F	360° max.	T
504	Phen. cement	F	360° max.	T
505	Phen. cement	F	360° max.	T
506	Silicate cement	F	1600° max.	T
507	Silicate cement	F	1600° max.	T
513	Asplit	X		
514	Asplit-F	X		
515	Basolit	A	....	T
518	C-Basolit	A	....	T
523	Duralon	X		
524	Durisite	X		
535	Nukem	X		
536	Nukem	A	....	T
538	Penchlor	A	All %; 750° max.	
539	Penchlor	A	All %; 750° max.	
540	Penchlor	A	All %; 500–2000°	
541	Pennsalt	X		
542	Permanite	X		
544	Plastite	V	175° max.	PTSD
545	P-Basolit	A	....	T
554	Silastic	X	(A.S.T.M. D-543-43)	
559	Thiokol	X		
600	Acid brick	A	....	TDR
603, 612, 615	Glass	A	....	THDR
604	Ceratherm	A	200–400°, depending on design	BIPTSCHDFR
606	Stoneware	A	All conditions	BIVPTCHDFR
607	Glass-L	A	600° max.	PTCH
610	Stoneware	A	All %; any temp.	IVPTCHDR

No.	Material	Ratings	Exposure conditions	Applications
611	Porcelain	A	All % and temp.	BIPTDR
614	Glass-L	V	Under 212°; small capacity units	BTCH
616	Pyrex	A		
617	Stoneware	A	140–160°	BIPTSCHDFR
618	Vitreo	V	1000° max.	
619	Vitreosil	A		
621	Vycor	A		
700	Ace Saran	X	Liquid	
703	Compar	X		
704	DC Silicone	X	(A.S.T.M. D-543-43)	
711	Haveg	X		
715	Heresite	A	200° max.	
716	Heresite	A	200° max.	
720	Lamicoid	X		
726	Nukemite	A	....	PTSDF
733	Permanite	X		
736	Pyroflex	V	....	PTCHD
737	Resilon	V	175° max.	PTSD
738	Resistoflex	X		
739	Resistoflex	X		
740	Saran	X	77°	
741	Sealon	A	160° max.	BPTDF
742	Teflon	A	Rm.	VP
746	Tygon	V	180° max.	
805	Butyl	X		
809	Fairprene	X		
817	Heresite	V	Sat.; 212° max.	
835	Perbunan	X		
836	Natural (S)	V		
837	Natural (H)	V		
838	GR-S (S)	V		
839	GR-A (H)	V		
840	GR-A (S)	V		
841	GR-A (II)	V		
842	GR-M (S)	V		
843	GR-P (S)	X		
844	GR-I	V		
848	Silastic	X		
853	Thiokol	X		
854	Thiokol	X		
856	Vistanex	X		

Calcium Chloride

No.	Material	Ratings	Exposure conditions	Applications
3	Admiralty	A		
4	Admiralty	F	Hot	CH
6	Adnic	A		
10–17	Aluminum	AF	Sol.; rm.	PT
22	Ambraloy	F	....	CH
23	Ambraloy	F	....	CH
24	Ambraloy	F	....	CH
29–40	Ampco	F	May be erratic	BIVPTCHDF
51	Berylco	A	Cold	
54	Brass	X		
61	Brass	V	....	VCH
63	Brass	FA	....	PCH
66	Bronze	A		
73	Bronze	F	....	VT
74	Bronze	F	....	V
75	Bronze	F	....	VP
76	Bronze	F	....	Instruments
77	Bronze	V	....	PCH
81	CA-FA20	A	5%; 70°	BIVP
82	CA-MM	A	5%; 70°	BIVP
86	Cast iron	A		
86A	Cast iron	F	Frequently used. Inhibitors useful in reducing corrosion	
87	Causul	A	Cold	B
88	Chlorimet	A	All % and temp.	BIVCHF
89	Chlorimet	A	All % and temp.	BIVCHF
111	Copper	A		
114	Copper	F	....	PTCHR
117	Copper	V	....	P
118	Copper	A		
123	Cupro-Ni	F	....	CH
124	Cupro-Ni	A		
139	Durco	A	All % and temp.	BIVHF
140	Durichlor	A	All % and temp.	BIVPCHFR
141	Durimet	A	All % and temp.	BIVHF
142	Durimet	A	All % and temp.	BIVHF
143	Duriron	A	All % and temp.	BIVPHFR
148	Everdur	F	....	BIVR
149	Everdur	F	....	PTCHDR
150	Everdur	F	....	PCH
159	Hastelloy	A	All % to 160°	BIVPTCH
160	Hastelloy	A	All % to boil.	BIVPTCH
161	Hastelloy	A	All % to 125°	BIVPTCH

Table 1. Chemical Resistance of Constructional Materials—(*Continued*)

No.	Material	Ratings	Exposure conditions	Applications
162	Hastelloy	A	Rm. only	BIVP
163	Stellite	A	Rm. only	BIV
165	Stellite	F	Rm. only	BIV
184	Inconel	A	35% in evaporator; 160–320°	C
185	Inc-clad	A	....	BTSCH
216	Monel	A	35% in evaporator; 160–320°	BIVPTCH
217	Monel-clad	A	....	BTSCH
219	Muntz	X		
224	Nickel	A	35% in evaporator; 160–320°	BIVTCH
225	Ni-clad	A	....	BTSCH
226	Ni-silver	F	....	Instruments
231	Ni-Resist	A	35% in evaporator; 160–320°	BIV
234	Olympic	A		
235	Olympic	A		
245	Pyrasteel	V	....	BIVPTSCHF
270	304-clad	A	All %; 70°	BTSCH
271	316-clad	A	All %; 70°	BTSCH
275	St. 301	V		
276	St. 302	V		
		A	Sat. cold; 212°	
		X	All %; hot	
		A	All %; 70°; alkaline	VTSCDF
278	St. 303	V		
279	St. 304	V		
		X	All %; hot	
		A	All %; 70°; alkaline	VTSCDF
		V	5%; 70°	
280	St. 308	V		
281	St. 309	V		
282	St. 310	V		
283	St. 316	A	....	BIPTSCH
		A	Sat. cold; 212°	
		X	All %; hot	
		A	All %; 70°; alkaline	VTCHDF
		A	5%; 70°	
284	St. 317	A	....	VP
285	St. 321	V		
286	St. 347	V		
287	St. 403	X		
288	St. 405	X		
290	St. 410	X		
		X	5%; 70°	
291	St. 414	X		
292	St. 416	X		
293	St. 418	X		
294	St. 420	X		
295	St. 430	X		
		X	All %; hot	
		X	All %; 70°; alkaline	
296	St. 430F	X		
297	St. 431	X		
298	St. 440A	X		
300	St. 440C	X		
301	St. 442	X		
303	St. 446	X		
313	St. CF-20	A	....	BIV
360A	Steel	F	Frequently used. Inhibitors useful in reducing corrosion	
367	Super Ni	A	....	PTCH
368	Tantalum	A	392° max.	CH
369	Telnic	A		
390	Worthite	A	All %; any temp.	BIV
392	Wyndaloy	A	Very dilute; 68°	
		AV	Moderate dilution; 68°	
		AV	Conc.; 68°	
401	Karbate	A	All %; to boil.	
402	Karbate	A	All %; to boil.	P
403	Kempruf	A	All %; any temp.	
404	Acheson	A	All %; any temp.	
500	Sul. cement	A	200° max.	PT
501	Sul. cement	A	200° max.	PT
502	Furan cement	A	360° max.	PT
503	Furan cement	A	360° max.	PT
504	Phen. cement	A	360° max.	PT
505	Phen. cement	A	360° max.	PT
506	Silicate cement	F	1600° max.	T
507	Silicate cement	F	1600° max.	T
513	Asplit	A	All %; 350° max.	
514	Asplit-F	A	All %; 350° max.	
515	Basolit	A	200° max.	T
517	Carboline	A		
518	C-Basolit	A	200° max.	T
521	Causplit	A	All %; 350° max.	
523	Duralon	F	350° max.	BIPTCHDF
524	Durisite	F	350° max.	BIPTCHDF
534	N-series	A	Sat. at 160°	BIVPTCDFR
535	Nukem	A	....	T
536	Nukem	A	....	T
538	Penchlor	A	All %; 750° max.	
539	Penchlor	A	All %; 750° max.	
540	Penchlor	A	All %; 500–2000°	
541	Pennsalt	A	All %; 350° max.	
542	Permanite	A	....	TD
544	Plastite	A	175° max.	PTSD
545	P-Basolit	A	200° max.	T
554	Silastic	A	(A.S.T.M. D-543-43)	
556	Staminite	A	All %	T
559	Thiokol	A	150° max.	
600	Acid brick	A	....	TDR
603, 612, 615	Glass	A	....	THDR
604	Ceratherm	A	200–400°, depending on design	BIPTSCHDFR
606	Stoneware	A	Not commonly used	
607	Glass-L	A	600° max.	PTCH
610	Stoneware	A	All %; any temp.	BIPTCHDR
611	Porcelain	A	All % and temp.	BIPTDR
614	Glass-L	A	All %; under 302°; agitation	BTCH
616	Pyrex	A		
617	Stoneware	A	140–160°	BIPTSCHDFR
618	Vitreo	A	1000° max.	
621	Vycor	A		
700	Ace Saran	A	....	P
704	DC Silicone	A	(A.S.T.M. D-543-43)	HD
711	Haveg	A	All %; 300° max.	BIVPTDFR
713	Haveg	A	All %; 300° max.	BIVPTDFR
715	Heresite	A	200° max.	
716	Heresite	A	200° max.	
718	Koroseal	A	All %	TD
720	Lamicoid	A		
726	Nukemite	A	170° max.	PTSDF
733	Permanite	A	360° max.	BVPDT
736	Pyroflex	A	....	PTCHD
737	Resilon	A	175° max.	PTSD
740	Saran	A	77°	
741	Sealon	A	160° max.	BPTDF
742	Teflon	A	Rm.	VP
746	Tygon	A	180° max.	BIPTSCDF
800	Ace Hd. Rub.	A	....	BIVPTDF
801	Acidseal	A	All %; 150° max.	BIPTSDF
802	Acidseal	A	All %; 150° max.	BIPTSDF
805	Butyl	A	Dry salt; aqueous sol.; rm.	
809	Fairprene	A		
817	Heresite	A	Sat.; 212° max.	
835	Perbunan	A	Dry salt; aqueous sol.; rm.	
836	Natural (S)	A		
837	Natural (H)	A		
838	GR-S (S)	A		
839	GR-S (H)	A		
840	GR-A (S)	A		
841	GR-A (H)	A		
842	GR-M (S)	A		
843	GR-P (S)	X		
844	GR-I	A		
846	Saniprene	A	All %; 150° max.	BIVPTDF
848	Silastic	A	(A.S.T.M. D-543-43)	
849	Superflexite	A	All %; 150° max.	BIVPTDF
853	Thiokol	A	150° max.	T
854	Thiokol	A	150° max.	PTCSDF
855	Triflex	A	All %; 150° max.	BIVPTDF
856	Vistanex	A	Dry salt; aqueous sol.; rm.	

Calcium Hypochlorite

No.	Material	Ratings	Exposure conditions	Applications
3	Admiralty	V		
4	Admiralty	F	....	CH
6	Adnic	V	Dil. sol.; 70°	
10–17	Aluminum	V	Rm.	
22	Ambraloy	F	....	CH
23	Ambraloy	F	....	CH
24	Ambraloy	F	....	CH
29–40	Ampco	F	May be erratic	BIVPTCHDF
42	Antaciron	A	All conditions	
54	Brass	X		
61	Brass	X		
63	Brass	FV	....	PCH
66	Bronze	V		
73	Bronze	F	....	VT
74	Bronze	F	....	V
75	Bronze	F	....	VP
76	Bronze	F	....	Instruments
77	Bronze	X		
81	CA-FA20	A	2%; 70°	BIVP
82	CA-MM	X	2%; 70°	

Table 1. Chemical Resistance of Constructional Materials—*(Continued)*

No.	Material	Ratings	Exposure conditions	Applications
85	Cast iron	A		
86	Cast iron	V		
86A	Cast iron	V		
88	Chlorimet	X		
89	Chlorimet	A	All %; moderate temp.	BIVCHF
111	Copper	V		
114	Copper	F	...	PTCHR
117	Copper	V	Dil. sol.	P
118	Copper	V		
123	Cupro-Ni	F	...	CH
124	Cupro-Ni	V		
139	Durco	V	All % and temp.	BIVHF
140	Durichlor	A	All % and temp.	BIVPCHFR
141	Durimet	V	All % and temp.	BIVHF
142	Durimet	V	All % and temp.	BIVHF
143	Duriron	A	All % and temp. (140 preferred)	BIVPHFR
148	Everdur	F	...	BIVR
149	Everdur	F	...	PTCHDR
150	Everdur	F	...	PCH
159	Hastelloy	X		
160	Hastelloy	X		
161	Hastelloy	A	Rm. only	BIVPTSC
162	Hastelloy	X		
163	Stellite	A	Rm. only	BIV
165	Stellite	X		
184	Inconel	A	3 g.p.l.* max. free Cl_2; 70°	
		X	Over 3 g.p.l. free Cl_2; 70°	
216	Monel	A	3 g.p.l. max. free Cl_2; 70°	V
		X	Over 3 g.p.l. free Cl_2; 70°	
219	Muntz	X		
224	Nickel	A	3 g.p.l. max. free Cl_2; 70°	
		X	Over 3 g.p.l. free Cl_2; 70°	
226	Ni-silver	F	...	Instruments
231	Ni-Resist	A	0.07%; 60°	BIV
		X	Over 3 g.p.l. free Cl_2; 70°	
233	NS-5	A	Bleach liquor; cold	V
234	Olympic	V		
235	Olympic	V		
240	Platinum	A	All temp.	
242	Ir-Platinum	A	All temp.	
244	Rh-Platinum	A	All temp.	
245	Pyrasteel	V	...	BIVPTSCHF
275	St. 301	V		
276	St. 302	V		
		A	2%; 70°	VTSCDF
278	St. 303	V		
279	St. 304	V		
		A	2%; 70°	VTSCDF
		V	2%; 70°	BIVP
280	St. 308	V		
281	St. 309	V		
282	St. 310	V		
283	St. 316	A	...	BIPTS
		A	2%; 70°	VTCHDF
284	St. 317	A	...	VP
285	St. 321	V		
286	St. 347	V		
287	St. 403	X		
288	St. 405	X		
290	St. 410	X		
		X	2%; 70°	
291	St. 414	X		
292	St. 416	X		
293	St. 418	X		
294	St. 420	X		
295	St. 430	X		
		A	2%; 70°	VTSCHDF
296	St. 430F	X		
297	St. 431	X		
298	St. 440A	X		
300	St. 440C	X		
301	St. 442	X		
303	St. 446	X		
360A	Steel	X		
367	Super Ni	F	...	PTCH
368	Tantalum	A	...	Not used commercially
369	Telnic	V		
390	Worthite	A	All %; any temp.	BIV
401	Karbate	A	70° max.	BIV
402	Karbate	A	70° max.	BIV
403	Kempruf	F	All %; any temp.	
404	Acheson	F	All %; any temp.	
500	Sul. cement	F	Dilute sol.; low temp. only	T
501	Sul. cement	F	Dilute sol.; low temp. only	T
502	Furan cement	F	Dilute sol.; low temp. only	T
503	Furan cement	F	Dilute sol.; low temp. only	T
504	Phen. cement	F	Dilute sol.; low temp. only	T
505	Phen. cement	F	Dilute sol.; low temp. only	T
506	Silicate cement	X		
507	Silicate cement	X		
513	Asplit	A	All %; 350° max.	
514	Asplit-F	A	All %; 350° max.	
515	Basolit	A	...	T
518	C-Basolit	A	...	T
521	Causplit	A	All %; 350° max.	
523	Duralon	V	350° max.	BIPTCHDF
524	Durisite	V	350° max.	BIPTCHDF
535	Nukem	A	350° max.	T
536	Nukem	A	...	T
538	Penchlor	A	All %; 750° max.	
539	Penchlor	A	All %; 750° max.	
540	Penchlor	A	All %; 500–2000°	
541	Pennsalt	A	All %; 350°	
542	Permanite	V	...	TD
544	Plastite	A	175° max.	TPSD
545	P-Basolit	A	...	T
554	Silastic	F	(A.S.T.M. D-543-43)	
559	Thiokol	X		
600	Acid brick	A	...	TDR
603, 612, 615	Glass	A	...	THDR
604	Ceratherm	A	200–400°, depending on design	BIPTCHDFR
606	Stoneware	A	All conditions	BIVPTCHDFR
607	Glass-L	V	...	PTCH
610	Stoneware	A	All %; any temp.	BIVTPCHDR
611	Porcelain	A	All % and temp.	BIPTDR
614	Glass-L	V	Under 77°; agitation	BTCH
616	Pyrex	A		
617	Stoneware	A	140–160°	BIPTSCHDFR
618	Vitreo	A	1000° max.	
621	Vycor	A		
700	Ace Saran	A	...	P
704	DC Silicone	F	(A.S.T.M. D-543-43)	HD
713	Haveg	F	All conditions	BIVPTDFR
715	Heresite	A	5% at 70°	
716	Heresite	A	5% at 70°	
720	Lamicoid	X		
726	Nukemite	A		PTSDF
733	Permanite	V	360° max.	BVPDT
736	Pyroflex	A	...	PTCHD
737	Resilon	A	175° max.	PTSD
740	Saran	F	77°	
741	Sealon	A	160° max.	BPTDF
742	Teflon	A	Rm.	VP
746	Tygon	A	180° max.	BIPTSCDF
800	Ace Hd. Rub.	A	...	BIVPTDF
801	Acidseal	A	All %; 150° max.	BIPTDF
805	Butyl	V		
809	Fairprene	V		
817	Heresite	A	Sat.; 212° max.	
835	Perbunan	V		
836	Natural (S)	V		
837	Natural (H)	V		
838	GR-S (S)	V		
839	GR-S (H)	V		
840	GR-A (S)	V		
841	GR-A (H)	V		
842	GR-M (S)	X		
843	GR-P (S)	X		
844	GR-I	V		
846	Sanipreme	A	All %; 150° max.	BIPTDF
848	Silastic	F	(A.S.T.M. D-543-43)	
849	Superflexite	A	All %; 150° max.	BIPTDF
853	Thiokol	X		
854	Thiokol	F	150° max.	PTSCDF
856	Vistanex	V		

Carbon Tetrachloride

No.	Material	Ratings	Exposure conditions	Applications
3	Admiralty	A	Dry	
		V	Moist	
4	Admiralty	A	Dry	CH
		F	Moist	
6	Adnic	A		
10–17	Aluminum	AF	Rm.	BIPTC
19	Aloyco	A	Boil	BT
22	Ambraloy	A	Dry	CH
		F	Moist	
23	Ambraloy	A	Dry	CH
		F	Moist	
24	Ambraloy	A	Dry	CH
		F	Moist	
29–40	Ampco	A	...	BIVPTCHDF
51	Berylco	A		
54	Brass	A	Dry	CH
		V	Moist	
61	Brass	A	Dry	VCH
		V	Moist	

* g.p.l. = grams per liter.

Table 1. Chemical Resistance of Constructional Materials—(*Continued*)

No.	Material	Ratings	Exposure conditions	Applications
63	Brass	A	Dry	PCH
		FV	Moist	
66	Bronze	A	Dry	
		V	Moist	
73	Bronze	A	Dry	VT
		F	Moist	
74	Bronze	A	Dry	V
		F	Moist	
75	Bronze	A	Dry	VP
		F	Moist	
76	Bronze	A	Dry	Instruments
		F	Moist	
77	Bronze	A	Dry	PCH
		V	Moist	
85	Cast iron	A		
86	Cast iron	V		
86A	Cast iron	A	Dry only	
87	Causul	A	Hot or cold	B
88	Chlorimet	A	All % and temp.	BIVCHF
89	Chlorimet	A	All % and temp.	BIVCHF
111	Copper	A	Dry	
		V	Moist	
114	Copper	A	Dry	PTCHF
		F	Moist	
118	Copper	A	Dry	
		V	Moist	
123	Cupro-Ni	A	Dry	CH
		F	Moist	
124	Cupro-Ni	A	Dry	
		V	Moist	
127–133	Dowmetal	A	Dry	
139	Durco	A	All % and temp.	BIVHF
140	Durichlor	A	All % and temp.	BIVPCHFR
141	Durimet	A	All % and temp.	BIVHF
142	Durimet	A	All % and temp.	BIVHF
143	Duriron	A	All % and temp.	BIVPHFR
148	Everdur	A	Dry	BIVR
		F	Moist	
149	Everdur	A	Dry	PTCHDR
		F	Moist	
150	Everdur	A	Dry	PCH
		F	Moist	
159	Hastelloy	A	Any temp.	
160	Hastelloy	A	Any temp.	
161	Hastelloy	A	Any temp.	
162	Hastelloy	A	Any temp.	V
163	Stellite	A	Any temp.	
165	Stellite	A	Any temp.	
184	Inconel	A	Boil.	
185	Inc-clad	A	Pure	BTSCH
216	Monel	A	Boil.	BIVPTCHDF
217	Monel-clad	A	Pure	BTSCH
219	Muntz	A	Dry	CH
		V	Moist	
224	Nickel	A	Boil.	TCH
225	Ni-clad	A	Pure	BTSCH
226	Ni-silver	A	Dry	Instruments
		X	Moist	
231	Ni-Resist	A	Boil.	BIV
233	NS-5	A	Any temp.	V
234	Olympic	A	Dry	
		V	Moist	
235	Olympic	A	Dry	
		V	Moist	
245	Pyrasteel	V	....	BIVPTSCHF
268	Silver	A	All temp.	
270	304-clad	A	Pure; 70°	BTSCH
271	316-clad	A	Pure; 70°	BTSCH
274	430-clad	A	Pure; 70°	BTSCH
275	St. 301	V		
276	St. 302	V		
		A	Anhydrous; 70°; boil.	
		X	5%; 10%; 70°; aqueous sol.	
		A	Pure; 70°	
278	St. 303	V		
279	St. 304	V		
		X	5%; 10%; 70°; aqueous sol.	
		A	Pure; 70°	VTSCDF
280	St. 308	V		
281	St. 309	V		
282	St. 310	V		
283	St. 316	A	....	BIVPTS
		A	Anhydrous; 70°	
		A	Pure; 70°	VTCHDF
284	St. 317	A	....	VP
285	St. 321	V		
286	St. 347	V		
287	St. 403	X		
288	St. 405	X		
290	St. 410	X		
		A	Anhydrous; 70°	
291	St. 414	X		
292	St. 416	X		
293	St. 418	X		
294	St. 420	X		
295	St. 430	F		
		A	Anhydrous; 70°	
		A	Pure; 70°	VTSCHDF
296	St. 430F	F		
297	St. 431	F		
298	St. 440A	F		
300	St. 440C	F		
301	St. 442	F		
303	St. 446	F		
308	St. CB-30	A	Boil.	
309	St. CC-35	A		
360A	Steel	A	Dry only	
367	Super Ni	A	Dry	PTCH
		F	Moist	
368	Tantalum	A	....	Not used commercially
369	Telnic	A	Dry	
		V	Moist	
390	Worthite	A	50% 184°	BIV
401	Karbate	A	All %; to boil.	BIV
402	Karbate	A	All %; to boil.	BIV
403	Kempruf	A	All %; any temp.	
404	Acheson	A	All %; any temp.	
500	Sul. cement	X		
501	Sul. cement	X		
502	Furan cement	A	360° max.	PT
503	Furan cement	A	360° max.	PT
504	Phen. cement	A	360° max.	PT
505	Phen. cement	A	360° max.	PT
506	Silicate cement	F	1600° max.	T
507	Silicate cement	F	1600° max.	T
508	Acichlor	X		
510	Acitite	F	Dilute and conc.; 250°	
513	Asplit	A	All %; 350° max.	
514	Asplit-F	A	All %; 350° max.	
515	Basolit	X		
518	C-Basolit	X		
521	Causplit	A	All %; 350° max.	
523	Duralon	A	350° max.	BIPTCHDF
524	Durisite	A	350° max.	BIPTCHDF
534	N-series	X		
535	Nukem	A	....	T
536	Nukem	A	....	T
537	Pecomastic	X		
538	Penchlor	A	All %; 750° max.	
539	Penchlor	A	All %; 750° max.	
540	Penchlor	A	All %; 500–2000°	
541	Pennsalt	A	All %; 350° max.	
542	Permanite	A	....	TD
544	Plastite	X		
545	P-Basolit	X		
554	Silastic	X	(A.S.T.M. D-543-43)	
556	Staminite	A	All %	T
559	Thiokol	F	150° max.	
600	Acid brick	A	....	TDR
603, 612, 615	Glass	A	....	THDR
604	Ceratherm	A	200–400°, depending on design	BIPTSCHDFR
606	Stoneware	A	All conditions	BIVPTCHDFR
607	Glass-L	A	600° max.	PTCH
610	Stoneware	A	All %; any temp.	BIVPTCHDFR
611	Porcelain	A	All % and temp.	BIPTDR
614	Glass-L	A	Under 302°; agitation	BTCH
616	Pyrex	A		
617	Stoneware	A	140–160°	BIPTSCHDFR
618	Vitreo	A	1000° max.	
619	Vitreosil	A		
621	Vycor	A		
700	Ace Saran	A	....	P
703	Compar	A	−40° to boil.	TD
704	DC Silicone	X	(A.S.T.M. D-543-43)	
706	Formica	A	Commercial	TD
707	Formica	A	All %; 212° max.	
713	Haveg	A	All conditions	BIVPTDFR
715	Heresite	A	200° max.	
716	Heresite	A	200° max.	
718	Koroseal	X		
720	Lamicoid	A		
723	Nixon	A	C.P.; 77°	S
724	Nixon	A	77°	S
725	Nixon	X		
726	Nukemite	X		

Table 1. Chemical Resistance of Constructional Materials—(*Continued*)

No.	Material	Ratings	Exposure conditions	Applications
733	Permanite	A	360° max.	BVPDT
735	Polythene	F	Rm.; swells slightly	
		X	176°	
736	Pyroflex	V	...	PTCHD
737	Resilon	X		
738	Resistoflex	A	Conc.; to boil	P
739	Resistoflex	A	Conc.; 275° max.	IVP
740	Saran	F	77°	
741	Sealon	V	160° max.	BPTDF
742	Teflon	A	Rm.	VP
746	Tygon	X		
800	Ace Hd. Rub.	X		
801	Acidseal	X		
802	Acidseal	X		
805	Butyl	X		
809	Fairprene	X		
817	Heresite	X		
835	Perbunan	X		
836	Natural (S)	X		
837	Natural (H)	X		
838	GR-S (S)	X		
839	GR-S (H)	X		
840	GR-A (S)	X		
841	GR-A (H)	X		
842	GR-M (S)	X		
843	GR-P (S)	V		
844	GR-I	X		
846	Saniprene	X		
848	Silastic	X		
849	Superflexite	X		
853	Thiokol	A	150° max.	T
854	Thiokol	A	150° max.	PTSCDF
855	Triflex	X		
856	Vistanex	X		

Carbonic Acid

No.	Material	Ratings	Exposure conditions	Applications
3	Admiralty	A		
4	Admiralty	A	...	CH
6	Adnic	A		
10–17	Aluminum	A	All %; any temp.	BIPTCHDFS
22	Ambraloy	A	...	CH
23	Ambraloy	A	...	CH
24	Ambraloy	A	...	CH
29	Ampco	A	...	BIVPTCHDF
54	Brass	A	Dry	CH
61	Brass	A	Dry	VCH
		V	Moist	
63	Brass	A	Dry or moist	PCH
66	Bronze	A		
73	Bronze	A	...	VT
74	Bronze	A	...	V
75	Bronze	A	...	VP
76	Bronze	A	...	Instruments
77	Bronze	A	Dry	PCH
		V	Moist	
86	Cast iron	V	Wet carbon dioxide	
86A	Cast iron	F		
88	Chlorimet	A	All % and temp.	BIVCHF
89	Chlorimet	A	All % and temp.	BIVCHF
111	Copper	A		
114	Copper	A	...	PTCHR
118	Copper	A		
123	Cupro-Ni	A	...	CH
124	Cupro-Ni	A		
139	Durco	A	All % and temp.	BIVHF
140	Durichlor	A	All % and temp.	BIVPCHFR
141	Durimet	A	All % and temp.	BIVHF
142	Durimet	A	All % and temp.	BIVHF
143	Duriron	A	All % and temp.	BIVPHFR
148	Everdur	A	...	BIVR
149	Everdur	A	...	PTCHDR
150	Everdur	A	...	PCH
159	Hastelloy	A	All % and temp.	
160	Hastelloy	A	All % and temp.	
161	Hastelloy	A	All % and temp.	
162	Hastelloy	A	All % and temp.	
163	Stellite	A	All % and temp.	BIV
165	Stellite	A	All % and temp.	BIV
184	Inconel	A	H_2O sat. with CO_2; 450 lb./sq. in.; 60°	PC
216	Monel	A	H_2O sat. with CO_2; 450 lb./sq. in.; 60°	BIVP
219	Muntz	A	Dry	CH
		V	Moist	
224	Nickel	A	H_2O sat. with CO_2; 450 lb./sq. in.; 60°	PH
226	Ni-silver	A	...	Instruments
231	Ni-Resist	A	H_2O sat. with CO_2; 450 lb./sq. in.; 60°	BV
233	NS-5	A	...	V
234	Olympic	A		
235	Olympic	A		
245	Pyrasteel	A	...	BIVPTSCHF
275	St. 301	A	...	BIVTSC
276	St. 302	A	...	BIVTSC
		A	In water	
278	St. 303	A	...	IV
279	St. 304	A	...	BIVPTSCH
		A	Sat. at 70°	P
280	St. 308	A	...	BIVPTSC
281	St. 300	A	...	BIVTSC
282	St. 310	A	...	BIVPTSC
283	St. 316	A	...	BIVPTSC
		A	In water	
		A	Sat. at 70°	P
284	St. 317	A	Sat. at 70°	BIVPTSC
285	St. 321	A	...	BIVPTSC
286	St. 347	A	...	BIVPTSC
287	St. 403	A	...	IV
288	St. 405	A	...	BIVT
290	St. 410	A	...	BIVPTSC
291	St. 414	A	...	IV
292	St. 416	A	...	IV
293	St. 418	A		
294	St. 420	A		
295	St. 430	A	...	BIVPTSC
		A	In water	
		A	Sat. at 70°	P
296	St. 430F	A	...	IV
297	St. 431	A	...	IV
298	St. 440A	A		
300	St. 440C	A	...	IV
301	St. 442	A	...	BIVTSC
303	St. 446	A	...	BIVPTSC
316	St. CF-20	A	...	BIV
360A	Steel	F		
367	SuperNi	A	Dry or moist	PTCH
368	Tantalum	A	...	Not used commercially
369	Telnic	A		
390	Worthite	A	All %; any temp.	BIV
401	Karbate	A	All %; to boil.	
402	Karbate	A	All %; to boil.	
403	Kempruf	A		
404	Acheson	A		
500	Sul. cement	A	200° max.	PT
501	Sul. cement	A	200° max.	PT
502	Furan cement	A	360° max.	PT
503	Furan cement	A	360° max.	PT
504	Phen. cement	A	360° max.	PT
505	Phen. cement	A	360° max.	PT
506	Silicate cement	X		
507	Silicate cement	X		
515	Basolit	A	200° max.	T
518	C-Basolit	A	200° max.	T
523	Duralon	A	350° max.	BIPTCHDF
524	Durisite	A	350° max.	BIPTCHDF
532	Lumnite	A	10% max.; 90° max.	TD
535	Nukem	A	...	T
536	Nukem	A	...	T
542	Permanite	A	...	TD
544	Plastite	A	175° max.	TDPS
545	P-Basolit	A	200° max.	T
554	Silastic	A	(A.S.T.M. D-543-43)	
556	Staminite	A	All %	T
559	Thiokol	A	150° max.	
600	Acid brick	A	...	TDR
603, 612, 615	Glass	A	...	THDR
604	Ceratherm	A	200–400°, depending on design	BIPTSCHDFR
606	Stoneware	A	Not commonly used	
607	Glass-L	A	600° max.	PTCH
610	Stoneware	A	All %; any temp.	BIVPTCHDR
611	Porcelain	A	All % and temp.	BIPTDR
614	Glass-L	V	...	BTCH
616	Pyrex	A		
617	Stoneware	A	140–160°	BIPTSCHDFR
618	Vitreo	A	1000° max.	
621	Vycor	A		
700	Ace Saran	A	...	P
704	DC Silicone	A	(A.S.T.M. D-543-43)	HD
711	Haveg	A	All conditions	BIVPTDFR
713	Haveg	A	All conditions	BIVPTDFR
715	Heresite	A	200° max.	
716	Heresite	A	200° max.	
718	Koroseal	A	...	TD
720	Lamicoid	A		
724	Nixon	A	77°	S
726	Nukemite	A	...	PTSDF
733	Permanite	A	360° max.	BVPDT

Table 1. Chemical Resistance of Constructional Materials—(*Continued*)

No.	Material	Ratings	Exposure conditions	Applications
736	Pyroflex	A	...	PTCHD
737	Resilon	A	175° max.	PTSD
741	Sealon	A	160° max.	BPTDF
742	Teflon	A	Rm.	VP
743	Textolite	A	Rm.	
744	Textolite	F	Dil.; 77–122°	BI
745	Textolite	F	Dil. 77°–122°	BI
746	Tygon	A	...	BIPTSCDF
800	Ace Hd. Rub.	A	...	BIVPTDF
801	Acidseal	A	Sat.; 150° max.; 1 atm. press.	BIPTSDF
802	Acidseal	A	Sat.; 150° max.; 1 atm. press.	BIPTSDF
805	Butyl	A	Rm.	
809	Fairprene	A		
814	G.E. Silicone	A	Weak; rm.	DVPTF
817	Heresite	A	Sat.; 212° max.	
835	Perbunan	A	Rm.	
836	Natural (S)	A		
837	Natural (H)	A		
838	GR-S (S)	A		
839	GR-S (H)	A		
840	GR-A (S)	A		
841	GR-A (H)	A		
842	GR-M (S)	A		
843	GR-P (S)	A		
844	GR-I	A		
846	Saniprene	A	Sat.; 150° max.; 1 atm. press.	BIPTSDF
848	Silastic	A	(A.S.T.M. D-543-43)	
849	Superflexite	A	Sat.; 150° max.; 1 atm. press.	BIPTSDF
853	Thiokol	A	150° max.	T
854	Thiokol	A	150° max.	PTSCDF
855	Triflex	A	Sat.; 150° max.; 1 atm. press.	BIPTSDF
856	Vistanex	A	Rm.	

Chloracetic Acid

No.	Material	Ratings	Exposure conditons	Applications
3	Admiralty	V		
4	Admiralty	V	...	CH
6	Adnic	V	Unaerated	
10–17	Aluminum	V	Rm.	
22	Ambraloy	F	...	CH
23	Ambraloy	F	...	CH
24	Ambraloy	V	...	CH
29–40	Ampco	F	...	BIVPTCHDF
42	Antaciron	A	All conditions	
54	Brass	X		
61	Brass	X		
63	Brass	FV	...	PCH
66	Bronze	V		
73	Bronze	F	...	VT
74	Bronze	F	...	V
75	Bronze	F	...	VP
76	Bronze	F	...	Instruments
77	Bronze	X	...	PCH
86A	Cast iron	X		
111	Copper	V		
114	Copper	F	...	PTCHR
118	Copper	V		
123	Cupro-Ni	F	...	CH
124	Cupro-Ni	V		
139	Durco	X		
140	Durichlor	A	All % and temp.	BIVPCHFR
141	Durimet	X		
142	Durimet	X		
143	Duriron	A	All % and temp.	BIVPHFR
148	Everdur	F	...	BIVR
149	Everdur	F	...	PTCHDR
150	Everdur	F	...	PCH
156	Gold	A		
159	Hastelloy	A	All % to 160°	BIVPT
160	Hastelloy	A	All % to boil.	BIVPTCH
161	Hastelloy	A	All % to 125°	BIVPT
162	Hastelloy	A	Rm. only	BIVP
163	Stellite	A	Rm. only	BIV
165	Stellite	F	Rm. only	BIV
184	Inconel	A	100%; 158°	
216	Monel	A	100%; 158°	V
219	Muntz	X		
224	Nickel	A	100%; 158°	V
226	Ni-silver	F	...	Instruments
234	Olympic	V		
235	Olympic	V		
236	Palladium	A		
240	Platinum	A		
242	Ir-Platinum	A		
244	Rh-Platinum	A		
268	Silver	A		
275	St. 301	X		
276	St. 302	X		
		X	All conc.; any temp.	
278	St. 303	X		
279	St. 304	X		
		X	All conc.; any temp.	
280	St. 308	X		
281	St. 309	X		
282	St. 310	X		
283	St. 316	X	All conc.; any temp.	
285	St. 321	X		
286	St. 347	X		
287	St. 403	X		
288	St. 405	X		
290	St. 410	X		
291	St. 414	X		
292	St. 416	X		
293	St. 418	X		
294	St. 420	X		
295	St. 430	X		
		X	All conc.; any temp.	
296	St. 430F	X		
297	St. 431	X		
298	St. 440A	X		
300	St. 440C	X		
301	St. 442	X		
303	St. 446	X		
360A	Steel	X		
367	Super Ni	F	...	PTCH
368	Tantalum	A	...	Not used commercially
369	Telnic	V		
500	Sul. cement	V	200° max.	T
501	Sul. cement	V	200° max.	T
502	Furan cement	F	360° max.	PT
503	Furan cement	F	360° max.	PT
504	Phen. cement	A	360° max.	PT
505	Phen. cement	A	360° max.	PT
506	Silicate cement	F	1600° max.	PT
507	Silicate cement	F	1600° max.	PT
513	Asplit	A	All %; 350° max.	
514	Asplit-F	A	All %; 350° max.	
515	Basolit	A	200° max.	T
517	Carboline	A		
518	C-Basolit	A	200° max.	T
521	Causplit	A	All %; 350° max.	
523	Duralon	A	350° max.	BIPTCHDF
524	Durisite	A	350° max.	BIPTCHDF
535	Nukem	A	350° max.	T
536	Nukem	F	...	T
538	Penchlor	A	All %; 750° max.	
539	Penchlor	A	All %; 750° max.	
540	Penchlor	A	All %; 500-2000°	
541	Pennsalt	A	All %; 350° max.	
542	Permanite	A	...	TD
544	Plastite	F	175° max.	TDPS
545	P-Basolit	A	200° max.	T
554	Silastic	X	(A.S.T.M. D-543-43)	
559	Thiokol	V	150° max.	
600	Acid brick	A	...	TDR
603, 612, 615	Glass	A	...	THDR
604	Ceratherm	A	200–400°, Depending on design	BIPTSCHDFR
606	Stoneware	A	All conditions	BIVPTCHDFR
610	Stoneware	A	All %; any temp.	BIVPTCHDR
611	Porcelain	A	All % and temp.	BIPTDR
614	Glass-L	A	Under 302°; agitation	BTCH
616	Pyrex	A		
617	Stoneware	A	140–160°	BIPTHSCDFR
618	Vitreo	V	1000° max.	
619	Vitreosil	A		
621	Vycor	A		
703	Compar	X		
704	DC Silicone	X	(A.S.T.M. D-543-43)	
713	Haveg	A	All conditions	BIVPTDFR
715	Heresite	A	200° max.	
716	Heresite	A	200° max.	
733	Permanite	A	360° max.	BVPDT
736	Pyroflex	V	...	PTCHD
737	Resilon	F	175° max.	PTSD
738	Resistoflex	X		
739	Resistoflex	X		
741	Sealon	A	160° max.	BPTDF
742	Teflon	A	Rm.	VP
744	Textolite	X	Conc.; 77–122°	
745	Textolite	X	Conc.; 77–122°	
746	Tygon	X		
801	Acidseal	F	20% max.; 100° max.	BIPTDF

Table 1. Chemical Resistance of Constructional Materials—(*Continued*)

No.	Material	Ratings	Exposure conditions	Applications
809	Fairprene	F		
814	G.E. Silicone	X	Rm.	
817	Heresite	V	Sat.; 212° max.	
836	Natural (S)	X		
837	Natural (H)	V		
838	GR-S (S)	X		
839	GR-S (H)	V		
840	GR-A (S)	X		
841	GR-A (H)	V		
842	GR-M (S)	X		
843	GR-P (S)	X		
844	GR-I	X		
846	Saniprene	F	20% max.; 100° max.	BIPTDF
848	Silastic	X		
849	Superflexite	F	20% max.; 100° max.	BIPTDF
853	Thiokol	V	150° max.	T
854	Thiokol	X		

Chlorine

No.	Material	Ratings	Exposure conditions	Applications
3	Admiralty	A	Dry	
		V	Moist	
4	Admiralty	A	Dry	CH
		V	Moist	
6	Adnic	A	Dry gas	
10–17	Aluminum	X		
22	Ambraloy	A	Dry	CH
		V	Moist	
23	Ambraloy	A	Dry	CH
		V	Moist	
24	Ambraloy	A	Dry	CH
		V	Moist	
29–40	Ampco	F	Dry gas	BIVPTCHDF
54	Brass	A	Dry	CH
		X	Moist	
61	Brass	A	Dry	VCH
		X	Moist	
63	Brass	A	Dry	PCH
		V	Moist	
66	Bronze	A	Dry	
		V	Moist	
73	Bronze	A	Dry	VT
		V	Moist	
74	Bronze	A	Dry	V
		V	Moist	
75	Bronze	A	Dry	VP
		V	Moist	
76	Bronze	A	Dry	Instruments
		V	Moist	
77	Bronze	A	Dry	PCH
		X	Moist	
81	CA-FA 20	A	Dry gas; 70°	BIVP
82	CA-MM	A	Dry gas; 70°	BIVP
86	Cast iron	A	Dry	
86A	Cast iron	A	Dry only	
88	Chlorimet	A	Dry to 1000°	BIVCHF
		X	Wet	
89	Chlorimet	A	Dry; all % to 950°	BIVCHF
		A	Wet; all % at rm.	BIVCF
111	Copper	A	Dry	
		V	Moist	
114	Copper	A	Dry	PTCHR
		V	Moist	
118	Copper	A	Dry	
		V	Moist	
123	Cupro-Ni	A	Dry	CH
		V	Moist	
124	Cupro-Ni	A	Dry	
		V	Moist	
139	Durco	A	Dry to 600°	BIVCHF
140	Durichlor	A	Wet; all %; rm.	BIVPCHFR
141	Durimet	A	Dry to 500°	BIVHF
142	Durimet	A	Dry to 500°	BIVHF
143	Duriron	X		
148	Everdur	A	Dry	BIVR
		V	Moist	
149	Everdur	A	Dry	PTCHDR
		V	Moist	
150	Everdur	A	Dry	PCH
		V	Moist	
184	Inconel	A	Dry; to 1000°	
		X	Wet	
216	Monel	A	Dry; to 800°	BVPTSH
		X	Wet	
217	Monel-clad	A	Dry gas	BTSCH
219	Muntz	A	Dry	CH
		X	Moist	
224	Nickel	A	Dry; to 1000°	VTHD
		X	Wet	
225	Ni-clad	A	Dry gas	BTSCH
226	Ni-silver	A	Dry	Instruments
		V	Moist	
231	Ni-Resist	X	Wet	
234	Olympic	A	Dry	
		V	Moist	
235	Olympic	A	Dry	
		V	Moist	
240	Platinum	A	480° max.	
242	Ir-Platinum	A	480° max.	
244	Rh-Platinum	A	480° max.	
268	Silver	F	Rm.	
271	316-clad	A	Dry gas; 70°	BTSCH
275	St. 301	X		
276	St. 302	X		
		A	Gas, dry; 70°	
		F	Gas, moist; 70°	
		X	Gas, moist; 212°	
		X	Chlorinated water; sat. 70°	
		X	Gas, dry; 70°	
278	St. 303	X		
279	St. 304	X		
		X	Chlorinated water; sat.; 70°	
		X	Gas, dry; 70°	
280	St. 308	X		
281	St. 309	X		
282	St. 310	X		
283	St. 316	X		
		A	Gas, dry; 70°	
		F	Gas, moist; 70°	
		X	Gas, moist; 212°	
		X	Chlorinated water; sat.; 70°	
		V	Gas, dry; 70°	VTCHDF
284	St. 317	X		
285	St. 321	X		
286	St. 347	X		
287	St. 403	X		
288	St. 405	X		
290	St. 410	X		
		X	Dry gas; 70°	
291	St. 414	X		
292	St. 416	X		
293	St. 418	X		
294	St. 420	X		
295	St. 430	X		
		F	Gas, dry; 70°	
		F	Gas, moist; 70°	
		X	Gas, moist; 212°	
		X	Chlorinated water; sat.; 70°	
		X	Gas, dry; 70°	
296	St. 430F	X		
297	St. 431	X		
298	St. 440A	X		
300	St. 440C	X		
301	St. 442	X		
303	St. 446	X		
360A	Steel	A	Dry only	
364	Stoody	A	Sat. vapor	
365	Stoody	F	Sat. vapor	
367	Super Ni	A	Dry	PTCH
		F	Moist	
368	Tantalum	A	302° max.; wet or dry	CH
369	Telnic	A	Dry	
		V	Moist	
390	Worthite	X	Wet	
392	Wyndaloy	V	Aqueous sol.; 68°	
401	Karbate	A	All %; dry; 338° max.	
		V	Moist	P
402	Karbate	A	All %; dry; 338° max.	
		V	Moist	P
403	Kempruf	A	All %; any temp.	
404	Acheson	A	All %; any temp.	
500	Sul. cement	A	200° max.	PT
501	Sul. cement	A	200° max.	PT
502	Furan cement	A	360° max.	PT
503	Furan cement	A	360° max.	PT
504	Phen. cement	A	360° max.	PT
505	Phen. cement	A	360° max.	PT
506	Silicate cement	F	1600° max.	T
507	Silicate cement	A	1600° max.	PT
513	Asplit	X		
514	Asplit-F	X		
515	Basolit	A	....	T
517	Carboline	A		
518	C-Basolit	A	....	T
521	Causplit	X		
523	Duralon	V	350° max.	BIPTCHDF
524	Durisite	V	350° max.	BIPTCHDF
535	Nukem	X		

Table 1. Chemical Resistance of Constructional Materials—(*Continued*)

No.	Material	Ratings	Exposure conditions	Applications
536	Nukem	A	....	T
538	Penchlor	A	All %; 750° max.	
539	Penchlor	A	All %; 750° max.	
540	Penchlor	A	All %; 500°–2000°	
541	Pennsalt	X		
542	Permanite	V	....	TD
544	Plastite	A	175° max.	PTSD
545	P-Basolit	A	....	T
554	Silastic	X	(A.S.T.M. D-543–43)	
559	Thiokol	X		
600	Acid brick	A	....	TDR
603, 612, 615	Glass	A	....	THDR
604	Ceratherm	A	200–400°, depending on design	BIVPTSCHDFR
606	Stoneware		All conditions	BIVPTCHDFR
607	Glass-L	A	600° max.	PTCH
610	Stoneware	A	All %; any temp.	BIVPTCHDR
611	Porcelain	A	All % and temp.	BIPTDR
614	Glass-L	V	....	BTCH
616	Pyrex	A		
617	Stoneware	A	140–160°	BIPTSCHDFR
618	Vitreo	V	1000° max.	
619	Vitreosil	A	....	PH
621	Vycor	A		
700	Ace Saran	X	Liquid	
703	Compar	X		
704	DC Silicone	X		
711	Haveg	A	All conditions	BIVPTDFR
713	Haveg	A	All conditions	BIVPTDFR
715	Heresite	A	70° max.	
716	Heresite	A	70° max.	
720	Lamicoid	X		
726	Nukemite	A	....	PTSDF
733	Permanite	V	360° max.	BVPDT
736	Pyroflex	V	....	PTCHD
737	Resilon	A	175° max.	PTSD
738	Resistoflex	X		
739	Resistoflex	X		
740	Saran	X	77°	
741	Sealon	A	160° max.	BPTDF
742	Teflon	A	Rm.	VP
746	Tygon	F	180° max.	BIPTSCDF
801	Acidseal	A	Sat.; 1 atm. lb./sq. in.; 150° max. (wet only)	BIPTDF
805	Butyl	X		
809	Fairprene	X		
817	Heresite	V	Sat.; 212° max.	
835	Perbunan	X		
836	Natural (S)	V		
837	Natural (H)	V		
838	GR-S (S)	V		
839	GR-S (H)	V		
840	GR-A (S)	V		
841	GR-A (H)	V		
842	GR-M (S)	X		
843	GR-P (S)	X		
844	GR-I	V		
846	Saniprene	A	Sat.; 1 atm. 150° max. (wet only)	BIPTDF
848	Silastic	X		
849	Superflexite	A	Sat.; 1 atm. 150° max. (wet only)	BIPTDF
853	Thiokol	X		
854	Thiokol	F	150° max.	PTSCDF
856	Vistanex	X		

Chromic Acid

No.	Material	Ratings	Exposure conditions	Applications
3	Admiralty	X		
4	Admiralty	X		
6	Adnic	X	Severe attack	
10–17	Aluminum	F	10% max.; rm.	
22	Ambraloy	X		
23	Ambraloy	X		
24	Ambraloy	X		
29–40	Ampco	X		
54	Brass	X		
61	Brass	X		
63	Brass	X		
66	Bronze	X		
73	Bronze	X		
74	Bronze	X		
75	Bronze	X		
76	Bronze	X		
77	Bronze	X		
86	Cast iron	V		
86A	Cast iron	F		
88	Chlorimet	X		
89	Chlorimet	A	All % and temp.	BIVCHF
111	Copper	X		
114	Copper	X		
118	Copper	X		
123	Cupro-Ni	X		
124	Cupro-Ni	X		
127–133	Dowmetal	A	C.p.	
		V	Technical; all %	
139	Durco	A	All % and temp.	BIVHF
140	Durichlor	A	All % and temp.	BIVPCHFR
141	Durimet	A	All % and temp.	BIVHF
142	Durimet	A	All % and temp.	BIVHF
143	Duriron	A	All % and temp.	BIVPHFR
148	Everdur	X		
149	Everdur	X		
150	Everdur	X		
159	Hastelloy	X	All % and temp.	
160	Hastelloy	X	All % and temp.	
161	Hastelloy	A	All % to 160°	BIVPCH
162	Hastelloy	X	All % and temp.	
163	Stellite	A	All % and temp.	BIV
165	Stellite	F		
184	Inconel	A	5%; 86°	
191, 193	Lead	A	Sat.; to boil; aerated; agitated; commercial purity	PTH
216	Monel	A	5%; 86°	VPTD
		X	Sat. aqueous sol.; 250°	
217	Monel-clad	A	....	BTSCH
219	Muntz	X		
224	Nickel	A	5%; 86°	
225	Ni-clad	A	....	BTSCH
226	Ni-silver	X		
231	Ni-Resist	A	5%; 86°	BIV
234	Olympic	X		
235	Olympic	X		
240	Platinum	A		
242	Ir-Platinum	A		
244	Rh-Platinum	A		
245	Pyrasteel	A	....	BIVPTSCHF
270	304-clad	A	5%; 70°	BTSCH
271	316-clad	A	10%; c.p.; boil.	BTSCH
274	430-clad	A	5%	BTSCH
275	St. 301	X		
276	St. 302	X		
		X	50%; not c.p., contains SO_3; boiling	
		A	50%; not c.p., contains SO_3; 70°	
		A	10%; c.p., free of SO_3; 70°; boil.	
		A	50%; c.p., free of SO_3; 70°	
		F	50%; c.p., free of SO_3; boil.	
		A	5%; 70°	VTSCDF
		X	10% c.p.; boil.	
278	St. 303	X		
279	St. 304	X		
		A	5%; 70°	VTSCDF
		X	10%; c.p.; boil.	
280	St. 308	X		
281	St. 309	X		
282	St. 310	X		
283	St. 316	X		
		X	50%; not c.p., contains SO_3; boil.	
		A	50%; not c.p., contains SO_3; 70°	
		A	10%; c.p., free of SO_3; 70°; boil.	
		A	50%; c.p., free of SO_3; 70°	
		X	50%; c.p., free of SO_3; boil.	
		A	5%; 70°	VTCHDF
		A	10%; c.p.; boil.	VTCHDF
284	St. 317	X		
285	St. 321	X		
286	St. 347	X		
287	St. 403	X		
288	St. 405	X		
290	St. 410	X		
291	St. 414	X		
292	St. 416	X		
293	St. 418	X		
294	St. 420	X		
295	St. 430	X		
		X	50%; not c.p., contains SO_3; boil.	
		A	5%; 70°	VTSCHDF
		X	10%; c.p.; boil.	
296	St. 430F	X		
297	St. 431	X		
298	St. 440A	X		
300	St. 440C	X		
301	St. 442	X		

Table 1. Chemical Resistance of Constructional Materials—(*Continued*)

No.	Material	Ratings	Exposure conditions	Applications
303	St. 446	X		
360A	Steel	F		
367	Super Ni	X		
368	Tantalum	A	....	Not used commercially
369	Telnic	X		
390	Worthite	A	Plating sol.; anodizing sol.; 150°	BIVP
401	Karbate	X		
402	Karbate	X		
403	Kempruf	X		
404	Acheson	X		
500	Sul. cement	X		
501	Sul. cement	X		
502	Furan cement	X		
503	Furan cement	X		
504	Phen. cement	X		
505	Phen. cement	X		
506	Silicate cement	A	1600° max.	PT
507	Silicate cement	A	1600° max.	PT
508	Acichlor	F	Dilute and conc.; 300°	
510	Acitite	F	Dilute and conc.; 250°	
513	Asplit	X		
514	Asplit-F	X		
515	Basolit	A	Rm.	T
518	C-Basolit	A	Rm.	T
521	Causplit	X		
523	Duralon	X		
524	Durisite	X		
532	Lumnite	A	10% max.; 90°	
535	Nukem	X		
536	Nukem	A	....	T
537	Pecomastic	A		
538	Penchlor	A	All %; 750° max.	
539	Penchlor	A	All %; 750° max.	
540	Penchlor	A	All %; 500–2000°	
541	Pennsalt	X		
542	Permanite	X		
544	Plastite	F	175° max.	TDSP
545	P-Basolit	A	Rm.	T
556	Staminite	F	Cold	T
559	Thiokol	X		
600	Acid brick	A	....	TDR
603, 612, 615	Glass	A	....	THDR
604	Ceratherm	A	200–400°, depending on design	BIPTSCHDFR
606	Stoneware	A	All conditions	BIVPTCHDFR
607	Glass-L	V	....	PTCH
610	Stoneware	A	All %; any temp.	BIVPCHDR
611	Porcelain	A	All % and temp.	BIPTDR
614	Glass-L	A	All %; under 302°; agitation	BTCH
616	Pyrex	A		
617	Stoneware	A	140–160°	BIPTSCHDFR
618	Vitreo	A	1000° max.	
619	Vitreosil	A		
621	Vycor	A		
703	Compar	X		
704	DC Silicone	A	(A.S.T.M. D-543–43)	HD
711	Haveg	X		
713	Haveg	X		
715	Heresite	A	200° max.	
716	Heresite	A	200° max.	
718	Koroseal	F	35% max.; 130° max.	TD
720	Lamicoid	A		
726	Nukemite	A	Rm.	PTSDF
727	Nylon	X	Conc.	
728	Nylon	X	Conc.	
729–732	Nylon	X	Conc.	
733	Permanite	X		
736	Pyroflex	V	....	PTCHD
737	Resilon	F	175° max.	PTSD
738	Resistoflex	X		
739	Resistoflex	X		
741	Sealon	A	160° max.	BPTDF
742	Teflon	A	Rm.	VP
744	Textolite	X	Conc.; 77–122°	BI
745	Textolite	X	Conc.; 77-122°	BI
746	Tygon	A	180° max.	BIPTSCDF
800	Ace Hd. Rub.	A		
805	Butyl	F	20%; 140° max.	
809	Fairprene	X		
817	Heresite	X		
836	Natural (S)	X		
837	Natural (H)	V		
838	GR-S (S)	X		
839	GR-S (H)	X		
840	GR-A (S)	X		
841	GR-A (H)	V		
842	GR-M (S)	X		
843	GR-P (S)	X		
844	GR-I	A		
848	Silastic	A	(A.S.T.M. D-543–43)	
853	Thiokol	X		
854	Thiokol	F	150° max.	PTSCDF
856	Vistanex	F	20%; 140° max.	

Citric Acid

No.	Material	Ratings	Exposure conditions	Applications
3	Admiralty	A		
4	Admiralty	A	....	CH
6	Adnic	A		
10–17	Aluminum	AF	Rm., 122°	BIPTS
			Dry powders	
19	Aloyco	A	15%, Conc.; boil.	BT
22	Ambraloy	A	....	CH
23	Ambraloy	A	....	CH
24	Ambraloy	A	....	CH
29–40	Ampco	A	....	BIVPTCHDF
51	Berylco	A		
54	Brass	X		
61	Brass	X		
63	Brass	A	....	PCH
66	Bronze	A		
73	Bronze	A	....	VT
74	Bronze	A	....	V
75	Bronze	A	....	VP
76	Bronze	A	....	Instruments
77	Bronze	X		
81	CA-FA20	A	5%; 70°	BIVP
82	CA-MM	A	5%; 70°	BIVP
86	Cast iron	X		
86A	Cast iron	X		
88	Chlorimet	A	All % and temp.	BIVCHF
89	Chlorimet	A	All % and temp.	BIVCHF
111	Copper	A		
114	Copper	A	....	PTCHR
117	Copper	V	....	P
118	Copper	A		
119	Corrosiron	A	25%; rm.; quiet; c.p.	BIP
123	Cupro-Ni	A	....	CH
124	Cupro-Ni	A		
139	Durco	A	All % and temp.	BIVHF
140	Durichlor	A	All % and temp.	BIVPCHFR
141	Durimet	A	All % and temp.	BIVHF
142	Durimet	A	All % and temp.	BIVHF
143	Duriron	A	All % and temp.	BIVPHFR
148	Everdur	A	....	BIVR
149	Everdur	A	....	PTCHDR
150	Everdur	A	....	PCH
156	Gold	A		
159	Hastelloy	A	All % and temp.	BIVP
160	Hastelloy	A	All % and temp.	BIVP
161	Hastelloy	A	All % and temp.	BIVP
162	Hastelloy	A	All % and temp.	BIV
163	Stellite	A	All % and temp.	BIV
165	Stellite	A	All % and temp.	BIV
184	Inconel	A	5%; 70°	
		A	58%; boil.	
185	Inc-clad	A	....	BTSCH
216	Monel	A	5%; 70°	BIVPTCH
		A	58%; boil.	
217	Monel-clad	A	....	BTSCH
219	Muntz	X		
224	Nickel	A	5%; 70°	BIVPTH
		F	58%; boil.	
225	Ni-clad	A		BTSCH
226	Ni-silver	A	....	Instruments
231	Ni-Resist	A	5%; 70°	BIV
234	Olympic	A		
235	Olympic	A		
236	Palladium	A		
240	Platinum	A		
242	Ir-Platinum	A		
244	Rh-Platinum	A		
245	Pyrasteel	A	....	BIVTPSCHF
268	Silver	A		
270	304-clad	A	15% max.; any temp.	BTSCH
271	316-clad	A	Conc.; boil.	BTSCH
274	430-clad	A	15% max.; any temp.	BTSCH
275	St. 301	A	....	BIVTSCH
276	St. 302	A	....	BIVTSCH
		A	10%; 70°; boil.	
		A	25%; 70°	
		F	25%; boil.	
		A	50%; 70°	
		F	50%; boil.	
		A	5% at 45 lb./sq. in. pres.; 275°	
		A	15% max.; any temp.	VTSCDF
		X	Conc.; boil.	

Table 1. Chemical Resistance of Constructional Materials—*(Continued)*

No.	Material	Ratings	Exposure conditions	Applications
278	St. 303	A		IV
279	St. 304	A		BIVPTSCH
		A	15% max.; any temp.	VTSCDF
280	St. 308	A		BIVPTSCH
281	St. 309	A		BIVPTSCH
282	St. 310	A		BIVPTSCH
283	St. 316	A		BIVPTSCH
		A	10%; 70°, boil.	
		A	25%; 70°; boil.	
		A	50%; 70°; boil.	
		A	5% at 45 lb./sq. in. pres.; 275°	
		A	15% max.; any temp.	VTCHDF
		A	Conc.; boil.	VTCHDF
284	St. 317	A		BIVPTSCH
285	St. 321	A		BIVPTSCH
286	St. 347	A		BIVPTSCH
287	St. 403	F	Rm.	
288	St. 405	F	Rm.	
290	St. 410	F	Rm.	
		F	5%; 70°	BIVP
291	St. 414	F	Rm.	
292	St. 416	F	Rm.	
293	St. 418	F	Rm.	
294	St. 420	F	Rm.	
295	St. 420	A		BIVPTSCH
		A	15% max.; any temp.	VTSCHDF
		A	5–15%; 70–100°	P
296	St. 430F	A		IV
297	St. 431	A		IV
298	St. 440A	A		
300	St. 440C	A		IV
301	St. 442	A		BIVTSCH
303	St. 446	A		BIVPTSCH
316	St. CF-20	A		BIV
360A	Steel	X		
368	Super Ni	A		PTCH
369	Tantalum	A		Not used commercially
387	Telnic	A		
390	Worthite	A	All %; any temp.	BIV
401	Karbate	A	All %; to boil.	BIVP
402	Karbate	A	All %; to boil.	BIVP
403	Kempruf	A	All %; to boil.	
404	Acheson	A	All %; to boil.	
500	Sul. cement	A	200° max.	PT
501	Sul. cement	A	200° max.	PT
502	Furan cement	A	360° max.	PT
503	Furan cement	A	360° max.	PT
504	Phen. cement	A	360° max.	PT
505	Phen. cement	A	360° max.	PT
506	Silicate cement	F	1600° max.	T
507	Silicate cement	F	1600° max.	T
513	Asplit	A	All %; 350° max.	
514	Asplit-F	A	All %; 350° max.	
515	Basolit	A	200° max.	T
517	Carboline	A		
518	C-Basolit	A	200° max.	T
521	Causplit	A	All %; 350° max.	
523	Duralon	A	350 max.	BIPTCHDF
524	Durisite	A	350° max.	BIPTCHDF
532	Lumnite	A		TD
535	Nukem	A	350° max.	T
536	Nukem	A		T
538	Penchlor	A	All %; 750° max.	
539	Penchlor	A	All %; 750° max.	
540	Penchlor	A	All %; 500°–2000°	
541	Pennsalt	A	All %; 350° max.	
542	Permanite	A		TD
544	Plastite	A	175° max.	TDPS
545	P-Basolit	A	200° max.	T
554	Silastic	A	A.S.T.M. D-543–43	
559	Thiokol	A	150° max.	
600	Acid brick	A		TDR
603, 612, 615	Glass	A		THDR
604	Ceratherm	A	200–400°, depending on design	BIPTSCHDFR
606	Stoneware	A	All conditions.	BIVPTCHDFR
607	Glass-L	A	600° max.	PTCH
610	Stoneware	A	All %; any temp.	BIVPTCHDR
611	Porcelain	A	All % and temp.	BIPTDR
614	Glass-L	A	All%; under 302°; agitation	BTCH
616	Pyrex	A		
617	Stoneware	A	140–160°	BIPTDFRSCH
618	Vitreo	A	1000° max.	
619	Vitreosil	A		
621	Vycor	A		
700	Ace Saran	A		P
704	DC Silicone	A	(A.S.T.M. D-543–43)	HD
711	Haveg	A	All conditions	BIVPTDFR
713	Haveg	A	All conditions	BIVPTDFR
715	Heresite	A	200° max.	
716	Heresite	A	200° max.	
720	Lamicoid	A		
726	Nukemite	A		PTSDF
727	Nylon	A	Dilute	
728	Nylon	A	Dilute	
729–732	Nylon	X		
733	Permanite	A	360° max.	BVPDT
736	Pyroflex	A		PTCHD
737	Resilon	A	175° max.	PTSD
740	Saran	A	77°	
741	Sealon	A	160° max.	
742	Teflon	A	Rm.	VP
744	Textolite	F	Dil.; 77–122°	BI
745	Textolite	F	Dil.; 77–122°	BI
746	Tygon	A	180° max.	BIPTSCDF
800	Ace Hd. Rub.	A		BIVPTDF
801	Acidseal	A	Sat.; 150° max.	BIPTDF
809	Fairprene	A		
814	G.E. Silicone	A	Dilate; rm.	DVPTF
817	Heresite	A	Sat.; 212° max.	
836	Natural (S)	A		
837	Natural (H)	A		
838	GR-S (S)	A		
839	GR-S (H)	A		
840	GR-A (S)	A		
841	GR-A (H)	A		
842	GR-M (S)	A		
843	GR-P (S)	X		
844	GR-I	A		
846	Saniprene	A	Sat.; 150° max.	BIPTDF
848	Silastic	A	(A.S.T.M. D-543–43)	
849	Superflex	A	Sat.; 150° max.	BIPTDF
853	Thiokol	A	150° max.	T
854	Thiokol	A	150° max.	PTSCDF

Copper Sulfate

No.	Material	Ratings	Exposure conditions	Applications
3	Admiralty	A		
4	Admiralty	F		CH
6	Adnic	A		
10–17	Aluminum	X		
19	Aloyco	A	Plus 10% sulfuric; 150°	BT
22	Ambraloy	F		CH
23	Ambraloy	F		CH
24	Ambraloy	F		CH
29–40	Ampco	F		BIVPTCHDF
54	Brass	X		
61	Brass	X		
63	Brass	AF		PCH
66	Bronze	A		
73	Bronze	F		VT
74	Bronze	F		V
75	Bronze	F		VP
76	Bronze	F		Instruments
77	Bronze	A		
86	Cast iron	X		
86A	Cast iron	X		
88	Chlorimet	X		
89	Chlorimet	A	All % and temp.	BIVCHF
111	Copper	X		
114	Copper	F		PTCHR
117	Copper	V		P
118	Copper	X		
119	Corrosiron	A	25%; rm.; unagitated; c.p.	BIP
123	Cupro-Ni	F		CH
124	Cupro-Ni	A		
139	Durco	A	All % and temp.	BIVHF
140	Durichlor	A	All % and temp.	BIVPCHFR
141	Durimet	A	All % and temp.	BIVHF
142	Durimet	A	All % and temp.	BIVHF
143	Duriron	A	All % and temp.	BIVPHFR
148	Everdur	F		BIVR
149	Everdur	F		PTCHDR
150	Everdur	F		PCH
156	Gold	A		
159	Hastelloy	A	Cuprous; all % to 180°	BIVPT
		X	Cupric; all % and temp.	
160	Hastelloy	A	Cuprous; all % to boil.	BIVPTCH
		X	Cupric; all % and temp.	
161	Hastelloy	A	Cuprous; all % to boil.	BIVPTCH
		A	Cupric; all % to 160°	BIVPTCH
162	Hastelloy	A	Cuprous; all % to boil.	BIVP
		A	Cupric; all % and temp.	
163	Stellite	A	Cuprous; rm. only	BIV
		A	Cupric; all %; rm. only	BIV

Table 1. Chemical Resistance of Constructional Materials—(*Continued*)

No.	Material	Ratings	Exposure conditions	Applications
165	Stellite	A	Cuprous; rm. only	BIV
		A	Cupric; all %; rm. only	BIV
184	Inconel	A	10%; 60°; unaerated	
		A	2–5% plus 2–5% sulfuric; 140°	
185	Inc-clad	A	....	BTSCH
193, 196, 200, 266	Lead	A	Various %; pH below 7; 212° max.	PTH
216	Monel	A	10%; 60°; unaerated	
		F	2–5% plus 2–5% sulfuric; 140°	BIV
217	Monel-clad	A	....	BTSCH
219	Muntz	X		
224	Nickel	A	10%; 60°; unaerated	
		F	2–5% plus 2–5% sulfuric; 140°	
226	Ni-silver	F	....	Instruments
231	Ni-Resist	X	10%; 60°; unaerated	BV
234	Olympic	A		
235	Olympic	A		
236	Palladium	A		
240	Platinum	A		
242	Ir-Platinum	A		
244	Rh-Platinum	A		
245	Pyrasteel	A	....	BIVPTSCHF
268	Silver	A		
270	304-clad	A	Sat.; boil.	
		A	5%; 70°	
271	316-clad	A	Sat.; boil.	
		A	5%; 70°	
274	430-clad	A	5%; 70°	
275	St. 301	A	....	BIVTSCH
276	St. 302	A	....	BIVTSCH
		A	Sat.; boil.	
		A	5% max.; 70°	VTSCDF
		A	Sat.; boil.	VTSCDF
278	St. 303	A	....	IV
279	St. 304	A	....	BIVPTSCH
		A	5% max.; 70°	VTSCDF
		A	Sat.; boil.	VTSCDF
280	St. 308	A	....	BIVTSCH
281	St. 309	A	....	BIVTSCH
282	St. 310	A	....	BIVPTSCH
283	St. 316	A	....	BIVPTSCH
		A	Sat.; boil.	VTCHDF
		A	5% max.; 70°	VTCHDF
284	St. 317	A	....	BIVPTSCH
285	St. 321	A	....	BIVPTSCH
286	St. 347	A	....	BIVPTSCH
287	St. 403	A	....	IV
288	St. 405	A	....	BIVTSCH
290	St. 410	A	....	BIVPTSCH
291	St. 414	A	....	IV
292	St. 416	A	....	IV
293	St. 418	A		
294	St. 420	A		
295	St. 430	A	....	BIVPTSCH
		A	5% max.; 70°	VTSCHDF
		A	Sat.; 70°	P
296	St. 430F	A	....	IV
297	St. 431	A	....	IV
298	St. 440A	A		
300	St. 440C	A	....	IV
301	St. 442	A	....	BIVTSCH
303	St. 446	A	....	BIVPTSCH
316	St. CF-20	A	....	BIV
360A	Steel	X		
367	Super Ni	F	....	PTCH
368	Tantalum	A	....	Not used commercially
369	Telnic	A		
390	Worthite	A	200°; with or without H_2SO_4	BIV
401	Karbate	A	All %; to boil.	P
402	Karbate	A	All %; to boil.	P
403	Kempruf	A	All %; any temp.	
404	Acheson	A	All %; any temp.	
500	Sul. cement	A	200° max.; do not use in electrolytic tanks	PT
501	Sul. cement	A	200° max.; do not use in electrolytic tanks	PT
502	Furan cement	A	360° max.	PT
503	Furan cement	A	360° max.	PT
504	Phen. cement	A	360° max.	PT
505	Phen. cement	A	360° max.	PT
506	Silicate cement	F	1600° max.	PT
507	Silicate cement	F	1600° max.	PT
513	Asplit	A	All %; 350° max.	
514	Asplit-F	A	All %; 350° max.	
515	Basolit	X		
517	Carboline	A		
518	C-Basolit	X		
521	Causplit	A	All %; 350° max.	
523	Duralon	A	350° max.	BIPTCHDF
524	Durisite	A	350° max.	BIPTCHDF
532	Lumnite	A	10%; max. 90° max.	TD
534	N-series	A	20%; rm.	BIVPTCDFR
535	Nukem	A	350° max.	T
536	Nukem	X		
538	Penchlor	A	All %; 750° max.	
539	Penchlor	A	All %; 750° max.	
540	Penchlor	A	All %; 500–2000°	
541	Pennsalt	A	All %; 350° max.	
542	Permanite	A	....	TD
544	Plastite	A	175° max.	TDPS
545	P-Basolit	X		
554	Silastic	A	(A.S.T.M. D–543–43)	
556	Staminite	F	....	T
559	Thiokol	A	150° max.	
600	Acid brick	A	....	TDR
603, 612, 615	Glass	A	....	THDR
604	Ceratherm	A	200–400°, depends on design	BIPTSCHDFR
606	Stoneware	A	All conditions	BIVPTCHDFR
607	Glass-L	A	600° max.	PTCH
610	Stoneware	A	All %; any temp.	BIVPTCHDR
611	Porcelain	A	All % and temp.	BIPTDR
614	Glass-L	A	All %; under 302°; agitation	BTCH
616	Pyrex	A		
617	Stoneware	A	140–160°	BIPTSCHDFR
618	Vitreo	A	1000° max.	
621	Vycor	A		
700	Ace Saran	A	....	P
704	DC Silicone	A	(A.S.T.M. D–543–43)	HD
707	Formica	A	All %; 212° max.	F
711	Haveg	A	All %; 300°	BIVPTDFR
713	Haveg	A	All %; 300°	BIVPTDFR
715	Heresite	A	200° max.	
716	Heresite	A	200° max.	
718	Koroseal	A	All %; 150° max.	TD
720	Lamicoid	A		
726	Nukemite	A	....	PTSDF
733	Permanite	A	360° max.	BVPDT
736	Pyroflex	A	....	PTCHD
737	Resilon	A	175° max.	PTSD
741	Sealon	A	160° max.	BPTDF
742	Teflon	A	Rm.	VP
746	Tygon	A	180° max.	BIPTSCDF
800	Ace Hd. Rub.	A	....	BIVPTDF
801	Acidseal	A	All %; 150° max.	BIPTSDF
802	Acidseal	A	All %; 150° max.	BIPTSDF
805	Butyl	F	Dry salt; aqucous sol.; rm.	
809	Fairprene	A		
817	Heresite	A	Sat.; 212° max.	
835	Perbunan	F	Dry salt; aqueous sol.; rm.	
836	Natural (S)	V		
837	Natural (H)	A		
838	GR-S (S)	V		
839	GR-S (H)	A		
840	GR-A (S)	V		
841	GR-A (H)	A		
842	GR-M (S)	V		
843	GR-P (S)	X		
844	GR-I	A		
846	Saniprene	A	All %; 150° max.	BIPTSDF
848	Silastic	A	(A.S.T.M. D–543–43)	BIPTSDF
849	Superflexite	A	All %; 150° max.	BIPTSDF
853	Thiokol	A	150° max.	T
854	Thiokol	A	150° max.	PTSCDF
855	Triflex	A	All %; 150° max.	BIPTSDF
856	Vistanex	F	Dry salt; aqueous sol.; rm.	
913	Redwood	A	pH 2; rm.; pitch-L; acid plating sol.	T
		A	5%; 36°–170°; 2 yr.; chemical plant	T
		A	1–5% + 1% H_2SO_4; 36°–170°; chemical plant	T
		A	0.1% + 0.1% arsenic acid; 36°–90°	T

Fatty Acids

No.	Material	Ratings	Exposure conditions	Applications
3	Admiralty	A		
4	Admiralty	A	....	CH
6	Adnic	A		
10–17	Aluminum	A	Rm., boil.; moisture inhibits its attack	PTCH
19	Aloyco	A	150–550°	BT
22	Ambraloy	A	....	CH

Table 1. Chemical Resistance of Constructional Materials—(*Continued*)

No.	Material	Ratings	Exposure conditions	Applications
23	Ambraloy	A	....	CH
24	Ambraloy	A	....	CH
29–40	Ampco	AF	Depends on type and presence of oxygen	BIVPTCHDF
54	Brass	F	....	CH
61	Brass	F	....	VCH
63	Brass	A	....	PCH
66	Bronze	A		
73	Bronze	A	....	VT
74	Bronze	A	....	V
75	Bronze	A	....	VP
76	Bronze	A	....	Instruments
77	Bronze	F	....	PCH
86A	Cast iron	X	Sometimes used on crude products	
87	Causul	A	Hot or cold	B
88	Chlorimet	A	All % and temp.	BIVCHF
89	Chlorimet	A	All % and temp.	BIVCHF
111	Copper	A		
114	Copper	A	....	PTCHR
117	Copper	V	....	P
118	Copper	A		
123	Cupro-Ni	A	....	CH
124	Cupro-Ni	A		
139	Durco	A	All % and temp.	BIVHF
140	Durichlor	A	All % and temp.	BIVPCHFR
141	Durimet	A	All % and temp.	BIVHF
142	Durimet	A	All % and temp.	BIVHF
143	Duriron	A	All % and temp.	BIVPHFR
148	Everdur	A	....	BIVR
149	Everdur	A	....	PTCHDR
150	Everdur	A	....	PCH
156	Gold	A		
159	Hastelloy	A	All % and temp.	BIVPTCH
160	Hastelloy	A	All % and temp.	BIVPTCH
161	Hastelloy	A	All % and temp.	BIVPTCH
162	Hastelloy	A	All % and temp.	BIVP
163	Stellite	A	All % and temp.	BIV
165	Stellite	A	All % and temp.	BIV
184	Inconel	A	Boil.	BIVPTCHR
185	Inc-clad	A	....	BTSCH
196, 200, 266	Lead	V	212° max. in absence of oxygen	PTH
216	Monel	A	Boil.	BIVPTCH
217	Monel-clad	A	....	BTSCH
219	Muntz	F	....	CH
224	Nickel	A	Boil.	PTH
225	Ni-clad	A	....	BTSCH
226	Ni-silver	A	....	Instruments
231	Ni-Resist	A	Boil.	BIVPH
234	Olympic	A		
235	Olympic	A		
236	Palladium	A		
240	Platinum	A		
242	Ir-Platinum	A		
244	Rh-Platinum	A		
268	Silver	A		
275	St. 301	A	....	BIVTSCDF
276	St. 302	A	....	BIVTSCDF
278	St. 303	A	....	IVF
280	St. 308	A	....	BIVPTSCDF
281	St. 309	A	....	BIVTSCDF
282	St. 310	A	....	BIVPTSCHDF
283	St. 316	A	....	BIVPTSCHDF
284	St. 317	A	....	BIVPTSCHDF
285	St. 321	A	....	BIVPTSCH
286	St. 347	A	....	BIVPTSCDF
287	St. 403	A	....	IV
288	St. 405	A	....	BIVTSC
290	St. 410	A	....	BIVPTSCHDF
291	St. 414	A	....	IV
292	St. 416	A	....	IV
293	St. 418	A		
294	St. 420	A		
295	St. 430	A	....	BIVPTSCHDF
296	St. 430F	A	....	IV
297	St. 431	A	....	IV
298	St. 440A	A		
300	St. 440C	A	....	IV
301	St. 442	A	....	BIVTSC
303	St. 446	A	....	BIVPTSC
313	St. CF-7	A	Boil.	
319	St. CF-7M	A	Boil.	
360A	Steel	X	Sometimes used on crude products	
367	Super Ni	A	....	PTCH
368	Tantalum	A	....	Not used commercially.
369	Telnic	A		
390	Worthite	A	All %; any temp.	BIV
392	Wyndaloy	AV	68°	
401	Karbate	A	All %; to boil.	BIVPTCHDF
402	Karbate	A	All %; to boil.	BIVPTCHDF
403	Kempruf	A	All %; any temp.	PTDR
404	Acheson	A	All %; any temp.	PTDR
500	Sul. cement	F	100° max.	T
501	Sul. cement	F	100° max.	T
502	Furan cement	A	360° max.	PT
503	Furan cement	A	360° max.	PT
504	Phen. cement	A	360° max.	PT
505	Phen. cement	A	360° max.	PT
506	Silicate cement	F	1600° max.	T
507	Silicate cement	F	1600° max.	T
515	Basolit	A	200° max.	T
517	Carboline	A		
518	C-Basolit	A	200° max.	T
523	Duralon	A	350° max.	BIPTCHDF
524	Durisite	A	350° max.	BIPTCHDF
532	Lumnite	A	1% max.; 90°	TD
535	Nukem	A	350° max.	T
536	Nukem	A	....	T
542	Permanite	A	....	TD
544	Plastite	F	175° max.	TPSD
545	P-Basolit	A	200° max.	T
554	Silastic	A	(A.S.T.M. D-543–43)	
556	Staminite	A	All %	T
559	Thiokol	A	150° max.	
600	Acid brick	A	....	TDR
603, 612, 615	Glass	A	....	THDR
604	Ceratherm	A	200–400°, depending on design	BIPTSCHDFR
606	Stoneware	A	All conditions	BIVPTCHDFR
607	Glass-L	A	600° max.	PTCH
610	Stoneware	A	All %; any temp.	BIVPTCHDR
611	Porcelain	A	All % and temp.	BIPTDR
614	Glass-L	A	Under 302°; agitation	BTCH
616	Pyrex	A		
617	Stoneware	A	140–160°	BIPTSCHDFR
618	Vitreo	A	1000° max.	
621	Vycor	A		
700	Ace Saran	A	122°	P
703	Compar	A	200° max.	TD
704	DC Silicone	A	(A.S.T.M. D-543–43)	HD
706	Formica	A	All %; 212° max.	
711	Haveg	A	All conditions	BIVPTDFR
713	Haveg	A	All conditions	BIVPTDFR
715	Heresite	A	200° max.	
716	Heresite	A	200° max.	
718	Koroseal	F	All %; 100° max.	TD
720	Lamicoid	A		
726	Nukemite	V	....	PTSDF
733	Permanite	A	360° max.	BVPDT
736	Pyroflex	V	....	PTCHD
737	Resilon	F	175° max.	PTSD
738	Resistoflex	A	Conc.; 275° max.	P
739	Resistoflex	A	Conc.; 140° max.	IVP
741	Sealon	A	160° max.	BPTDF
742	Teflon	A	Rm.	VP
743	Textolite	A	Dil.; rm.	T
744	Textolite	F	Dil.; rm.	BI
745	Textolite	F	Dil.; rm.	BT
746	Tygon	A	180° max.	BIPTSCDF
801	Acidseal	F	Any %; 100° max.	BIPTDF
805	Butyl	F	Rm.	
809	Fairprene	V		
814	G.E. Silicone	A	Dil. rm.	VPTDF
817	Heresite	A	Sat.; 212° max.	
835	Perbunan	A	Rm.	
836	Natural (S)	V		
837	Natural (H)	V		
838	GR-S (S)	V		
839	GR-S (H)	V		
840	GR-A (S)	V		
841	GR-A (H)	V		
842	GR-M (S)	X		
843	GR-P (S)	X		
844	GR-I	F		
846	Saniprene	F	Any %; 100° max.	BIPTDF
848	Silastic	A	(A.S.T.M. D-543–43)	
849	Superflexite	F	Any %; 100° max.	BIPTDF
853	Thiokol	A	150° max.	T
854	Thiokol	A	150° max.	PTSCDF
856	Vistanex	F	Rm.	
913	Redwood	F	40–220° max.; +20% NaCl; soap plant	
		A	75% + Dil. H_2SO_4; 160° max.; reduction plant	
		A	75% + Dil. H_2SO_4 + Dil. NaOH; 160° max.; 10 yr.	
		A	5%; 40–185°; 32 yrs. packing plant	

Table 1. Chemical Resistance of Constructional Materials—(*Continued*)

Ferric Chloride

No.	Material	Ratings	Exposure conditions	Applications
3	Admiralty	X		
4	Admiralty	X		
6	Adnic	X	Severe attack	
10–17	Aluminum	X		
22	Ambraloy	X		
23	Ambraloy	X		
24	Ambraloy	X		
29	Ampco	X		
42	Antaciron	A	Cold	
51	Berylco	X	Hot	
54	Brass	X		
61	Brass	X		
63	Brass	X		
66	Bronze	X		
73	Bronze	X		
74	Bronze	X		
75	Bronze	X		
76	Bronze	X		
77	Bronze	X		
86A	Cast iron	X		
88	Chlorimet	X		
89	Chlorimet	V	All % and temp.	BIVCHF
111	Copper	X		
114	Copper	X		
118	Copper	X		
119	Corrosiron	F	48%; rm.; quiet; c.p.	BIP
123	Cupro-Ni	X		
124	Cupro-Ni	X		
139	Durco	X		
140	Durichlor	V	....	BIVPCHFR
141	Durimet	X		
142	Durimet	X		
143	Duriron	V		
148	Everdur	X		
149	Everdur	X		
150	Everdur	X		
159	Hastelloy	X	All % and temp.	
160	Hastelloy	X	All % and temp.	
161	Hastelloy	A	10% to 160°	BIVPTCH
		A	45%; rm. only	BIVPTCH
162	Hastelloy	X	All % and temp.	
163	Stellite	A	All %; rm. only	BIV
165	Stellite	X	All % and temp.	
184	Inconel	X	10%; 70°	
216	Monel	X	10%; 70°	
219	Muntz	X		
224	Nickel	X	10%; 70°	
226	Ni-silver	X		
231	Ni-Resist	X	10%; 70°	
234	Olympic	X		
235	Olympic	X		
245	Pyrasteel	X		
268	Silver	F		
275	St. 301	X		
276	St. 302	X		
		X	Dil.; 70°	
278	St. 303	X		
279	St. 304	X		
280	St. 308	X		
281	St. 309	X		
282	St. 310	X		
283	St. 316	A	....	BIPT
		X	Dil.; 70°	
		X		
284	St. 317	A	....	PT
285	St. 321	X		
286	St. 347	X		
287	St. 403	X		
288	St. 405	X		
290	St. 410	X		
		X	Dil.; 70°	
291	St. 414	X		
292	St. 416	X		
293	St. 418	X		
294	St. 420	X		
295	St. 430	X		
		X	Dil.; 70°	
296	St. 430F	X		
297	St. 431	X		
298	St. 440A	X		
300	St. 440C	X		
301	St. 442	X		
303	St. 446	X		
360A	Steel	X		
364	Stoody	A	10%; boil.; rm.	
		A	25%; rm.	
		F	25%; boil.	
365	Stoody	A	10%; rm.	
		F	10%; boil.	
		F	25%; rm.	
		X	25%; boil.	
367	Super Ni	X		
368	Tantalum	A	200° max.	H
369	Telnic	X		
390	Worthite	X		
401	Karbate	A	All %; to boil.	BIVP
402	Karbate	A	All %; to boil.	BIVP
403	Kempruf	V		T
404	Acheson	V	....	T
500	Sul. cement	A	200° max.	PT
501	Sul. cement	A	200° max.	PT
502	Furan cement	A	360° max.	PT
503	Furan cement	A	360° max.	PT
504	Phen. cement	A	360° max.	PT
505	Phen. cement	A	360° max.	PT
506	Silicate cement	F	1600° max.	T
507	Silicate cement	A	1600° max.	PT
508	Acichlor	A	Dilute and conc.; 300°	
510	Acitite	X		
513	Asplit	A	All %; 350° max.	
514	Asplit-F	A	All %; 350° max.	
515	Basolit	A	80%	T
517	Carboline	A		
518	C-Basolit	A	80%	T
521	Causplit	A	All %; 350° max.	
523	Duralon	A	350° max.	BIPTCHDF
524	Durisite	A	350° max.	BIPTCHDF
534	N-series	A	25%; rm.	BIVPTCDFR
535	Nukem	A	80%; 350° max.	T
536	Nukem	A	80%	T
537	Pecomastic	F	Dilute and conc.; 300°	
538	Penchlor	A	All %; 750° max.	
539	Penchlor	A	All %; 750° max.	
540	Penchlor	A	All %; 500–2000°	
541	Pennsalt	A	All %; 350° max.	
542	Permanite	A	360° max.	TD
544	Plastite	A	175° max.	PTSD
545	P-Basolit	A	80%	T
554	Silastic	A	(A.S.T.M. D–543–43)	
556	Staminite	F	....	T
559	Thiokol	A	150° max.	
600	Acid brick	A	....	TDR
603, 612, 615	Glass	A	....	THDR
604	Ceratherm	A	200–400°, depending on design	BIPTSCHDFR
606	Stoneware	A	All conditions	BIVPTCHDFR
607	Glass-L	A	600° max.	PTCH
610	Stoneware	A	All %; any temp.	BIVPTCHDR
611	Porcelain	A	All %; any temp.	BIPTDR
614	Glass-L	A	All %; under 302°; agitation	BTCH
616	Pyrex	A		
617	Stoneware	A	140–160°	BIPTSCHDFR
618	Vitreo	A	1000° max.	
619	Vitreosil	A	Concentrating solutions	
621	Vycor	A		
700	Ace Saran	A	....	P
704	DC Silicone	A	....	HD
711	Haveg	A	All conditions	BIVPTDFR
713	Haveg	A	All conditions	BIVPTDFR
715	Heresite	A	200° max.	
716	Heresite	A	200° max.	
720	Lamicoid	A		
726	Nukemite	A	....	PTSDF
733	Permanite	A	360° max.	BVPDT
736	Pyroflex	V	....	PTCHD
737	Resilon	A	175° max.	PTSD
740	Saran	F	77°	
741	Sealon	A	160° max.	BPTDF
742	Teflon	A	Rm.	VP
746	Tygon	A	180° max.	BIPTSCHDF
800	Ace Hd. Rub.	A	....	BIVPTDF
801	Acidseal	A	All %; 150° max.	BIPTSDF
805	Butyl	A	Dry salt; aqueous sol.; rm.	
809	Fairprene	A		
817	Heresite	A	Sat.; 212° max.	
835	Perbunan	A	Dry salt; aqueous sol.; rm.	
836	Natural (S)	A		
837	Natural (H)	A		
838	GR-S (S)	A		
839	GR-S (H)	A		
840	GR-A (S)	A		
841	GR-A (H)	A		
842	GR-M (S)	A		
843	GR-P (S)	X		
844	GR-I	A		
846	Saniprene	A	All %; 150° max.	BIPTSDF

Table 1. Chemical Resistance of Constructional Materials—(*Continued*)

No.	Material	Ratings	Exposure conditions	Applications
848	Silastic	A	(A.S.T.M. D–543–43)	
853	Thiokol	A	150° max.	T
854	Thiokol	A	150° max.	PTSCDF
855	Triflex	A	All %; 150° max.	BIPTSDF
856	Vistanex	A	Dry salt; aqueous sol.; rm.	
913	Redwood	F	20% + 5% HCl; 100–180°; 2 years; smelter	T

Ferric Sulfate

No.	Material	Ratings	Exposure conditions	Applications
3	Admiralty	X		
4	Admiralty	X		
6	Adnic	X	Severe attack	
10–17	Aluminum	F	Rm.	
22	Ambraloy	X		
23	Ambraloy	X		
24	Ambraloy	X		
29–40	Ampco	F	Depends upon temperature and oxidizing agents	BIVPTCHDF
51	Berylco	X	Hot	
54	Brass	X		
61	Brass	X		
63	Brass	X		
66	Bronze	X		
73	Bronze	X		
74	Bronze	X		
75	Bronze	X		
76	Bronze	X		
77	Bronze	X		
86A	Cast iron	X		
88	Chlorimet	X		
89	Chlorimet	A	All % and temp.	BIVCHF
111	Copper	X		
114	Copper	X		
117	Copper	V	....	P
118	Copper	X		
119	Corrosiron	A	50%; rm.; quiet; tech. grade	BIP
123	Cupro-Ni	X		
124	Cupro-Ni	X		
139	Durco	A	All % and temp.	BIVHF
140	Durichlor	A	All % and temp.	BIVPCHFR
141, 142	Durimet	A	All % and temp.	BIVHF
143	Duriron	A	All % and temp.	BIVPLFR
148	Everdur	X		
149	Everdur	X		
150	Everdur	X		
159	Hastelloy	X	All % and temp.	
160	Hastelloy	X	All % and temp.	
161	Hastelloy	A	All %; rm. only	BIVPTCH
162	Hastelloy	X	All % and temp.	
163	Stellite	A	10% to boil.	BIV
165	Stellite	A	10% to boil.	BIV
184	Inconel	A	5%; 65°	
193, 196, 200, 266	Lead	A	Various % and temp.	BIPTH
216	Monel	X	5%; 65°	
219	Muntz	X		
224	Nickel	F	5%; 65°	
226	Ni-silver	X		
231	Ni-Resist	X	5%; 65°	
234	Olympic	X		
235	Clympic	X		
245	Pyrasteel	A	....	BIVPTSCHF
270	304-clad	A	Under 5%; 70°	BTSCH
271	316-clad	A	Under 5%; 70°	BTSCH
274	430-clad	A	Under 5%; 70°	BTSCH
275	St. 301	A	....	BIVTSCH
276	St. 302	A	....	BIVTSCH
		A	10%; 70°; boil.	
		A	5% max.; 70°	VTSCDF
278	St. 303	A	....	IV
279	St. 304	A	....	BIVPTSCH
		A	5% max.; 70°	VTSCDF
280	St. 308	A	....	BIVPTSCH
281	St. 309	A	....	BIVTSCH
282	St. 310	A	....	BIVPTSCH
283	St. 316	A	....	BIVPTSCH
		A	5% max.; 70°	VTCHDF
284	St. 317	A	....	BIVPTSCH
285	St. 321	A	....	BIVPTSCH
287	St. 403	A	....	BIVPTSCHDF
295	St. 430	A	....	BIVPTSCH
		A	5% max.; 70°	VTSCHDF
296	St. 430F	A	....	IV
297	St. 431	A	....	IV
298	St. 440A	A		
300	St. 440C	A	....	IV
301	St. 442	A	....	BIVTSCH
303	St. 446	A	....	BIVPTSCH
308	St. CB-30	A	Conc.; rm.	
309	St. CC-35	A	Conc.; boil.	
313	St. CF-7	A	Conc.; boil.	
316	St. CF-20	A	....	BIV
319	St. CF-7M	A	Conc.; boil.	
360A	Steel	X		
364	Stoody	A	10%; rm.; boil.	
365	Stoody	A	10%; rm.; boil.	
367	Supe Ni	X		
368	Tantalum	A	No corrosion up to 392°	Not used commercially
369	Telnic	X		
390	Worthite	A	With H_2SO_4	BIV
401	Karbate	A	All %; to boil.	P
402	Karbate	A	All %; to boil.	P
403	Kempruf	F	....	T
404	Acheson	F	....	T
500	Sul. cement	A	200° max.	PT
501	Sul. cement	A	200° max.	PT
502	Furan cement	A	360° max.	PT
503	Furan cement	A	360° max.	PT
504	Phen. cement	A	360° max.	PT
505	Phen. cement	A	360° max.	PT
506	Silicate cement	F	1600° max.	T
507	Silicate cement	F	1600° max.	T
513	Asplit	A	All %; 350° max.	
514	Asplit-F	A	All %; 350° max.	
515	Basolit	A	200° max.	T
517	Carboline	A		
518	C-Basolit	A	200° max.	T
521	Causplit	A	All %; 350° max.	
523	Duralon	A	350° max.	BIPTCHDF
524	Durisite	A	350° max.	BIPTCHDF
532	Lumnite	A	1% max.; 90° max.	TD
535	Nukem	A	350° max.	T
536	Nukem	A	....	T
537	Penchlor	A	All %; 750° max.	
538	Penchlor	A	All %; 750° max.	
540	Penchlor	A	All %; 500–2000°	
541	Pennsalt	A	All %; 350° max.	
542	Permanite	A	360° max.	TD
544	Plastite	A	175° max.	PTSD
545	P-Basolit	A	200° max.	T
554	Silastic	A	(A.S.T.M. D–543–43)	
556	Staminite	F	....	T
559	Thiokol	A	150° max.	
600	Acid brick	A	....	TDR
603, 612, 615	Glass	A	....	THDR
604	Ceratherm	A	200–400°, depending on design	BIPTSCHDFR
606	Stoneware	A	All conditions	BIVPTCHDFR
607	Glass-L	A	600° max.	PTCH
610	Stoneware	A	All %; any temp.	BIVPTCHDR
611	Porcelain	A	All % and temp.	BIPTDR
614	Glass-L	A	All %; under 302°; agitation	BTCH
616	Pyrex	A		
617	Stoneware	A	140–160°	BIPTSCHDFR
618	Vitreo	A	1000° max.	
621	Vycor	A		
700	Ace Saran	A	....	P
704	DC Silicone	A	....	HD
715	Heresite	A	200° max.	
716	Heresite	A	200° max.	
718	Koroseal	A	All %; 150° max.	TD
720	Lamicoid	A		
726	Nukemite	A	....	PTSDF
733	Permanite	A	360° max.	BVPDT
736	Pyroflex	A	....	PTCHD
737	Resilon	A	175° max.	PTSD
741	Sealon	A	160° max.	BPTDF
742	Teflon	A	Rm.	VP
746	Tygon	A	180° max.	BIPTSCHDF
800	Ace Hd. Rub.	A	....	BIVPTDF
801	Acidseal	A	All %; 150° max.	BIPTSDF
802	Acidseal	A	All %; 150° max.	BIPTSDF
805	Butyl	A	Dry salt; aqueous sol.; rm.	
809	Fairprene	A		
817	Heresite	A	Sat.; 212° max.	
835	Perbunan	A	Dry salt; aqueous sol.; rm.	
836	Natural (S)	A		
837	Natural (H)	A		
838	GR-S (S)	A		
839	GR-S (H)	A		
840	GR-A (S)	A		
841	GR-A (H)	A		
842	GR-M (S)	A		
843	GR-P (S)	X		
844	GR-I	A		

Table 1. Chemical Resistance of Constructional Materials—(*Continued*)

No.	Material	Ratings	Exposure conditions	Applications
846	Saniprene	A	All %; 150° max.	BIPTSDF
848	Silastic	A	(A.S.T.M. D-543-43)	
849	Superflexite	A	All %; 150° max.	BIPTSDF
853	Thiokol	A	150° max.	T
854	Thiokol	A	150° max.	PTSCDF
855	Triflex	A	All %; 150° max.	BIPTSDF
856	Vistanex	A	Dry salt; aqueous sol.; rm.	

Ferrous Sulfate

No.	Material	Ratings	Exposure conditions	Applications
3	Admiralty	V		
4	Admiralty	F	......	CH
6	Adnic	V		
10–17	Aluminum	A	10% max.; rm.	
22	Ambraloy	F	......	CH
23	Ambraloy	F	......	CH
24	Ambraloy	F	......	CH
29–40	Ampco	F	......	BIVPTCHDF
54	Brass	X		
61	Brass	X		
63	Brass	VF	......	PCH
66	Bronze	V		
73	Bronze	F	......	VT
74	Bronze	F	......	V
75	Bronze	F	......	VP
76	Bronze	F		
77	Bronze	X		
85	Cast iron	A		
86A	Cast iron	F		
88	Chlorimet	A	All % and temp.	BIVCHF
89	Chlorimet	A	All % and temp.	BIVCHF
111	Copper	V		
114	Copper	F	......	PTCHR
117	Copper	V	......	P
118	Copper	V		
123	Cupro-Ni	F	......	CH
124	Cupro-Ni	V		
139	Durco	A	All % and temp.	BIVHF
140	Durichlor	A	All % and temp.	BIVPCHFR
141, 142	Durimet	A	......	BIVHF
143	Duriron	A	All % and temp.	BIVPHFR
148	Everdur	F	......	BIVR
149	Everdur	F	......	PTCHDR
150	Everdur	F	......	PCH
159	Hastelloy	A	All % to 180°	BIVPTCH
160	Hastelloy	A	All % to boil.	BIVPTCH
161	Hastelloy	A	All % to boil.	BIVPTCH
162	Hastelloy	A	All % to boil.	BIVP
163	Stellite	A	All %; rm. only	BIV
165	Stellite	A	All %; rm. only	BIV
184	Inconel	A	5%; 150°; unaerated	
193, 196, 200, 266	Lead	A	Various % and temp.	PTH
216	Monel	A	5%; 150°; unaerated	VPT
217	Monel-clad	A	Dil. sol.; 70°	BTSCH
219	Muntz	X		
224	Nickel	A	5%; 150°; unaerated	
225	Ni-clad	A	Dil. sol.; 70°	BTSCH
226	Ni-silver	F	......	Instruments
231	Ni-Resist	A	5%; 150°; unaerated	BIV
234	Olympic	V		
235	Olympic	V		
249	Resistac	F	......	BIVCHF
270	304-clad	A	Dilute; 70°	BTSCH
271	316-clad	A	Dilute; 70°	BTSCH
274	430-clad	A	Dilute; 70°	BTSCH
275	St. 301	A	......	BIVTSCH
276	St. 302	A	......	BIVTSCH
		A	10%; 70°; boil.	
		A	Dil.; 70°	VTSCDF
278	St. 303	A	......	IV
279	St. 304	A	......	BIVPTSCH
		A	Dil.; 70°	VTSCDF
280	St. 308	A	......	BIVPTSCH
281	St. 309	A	......	BIVPTSCH
282	St. 310	A	......	BIVPTSCHD
283	St. 316	A	......	BIVPTSCH
		A	10%; 70°; boil.	
		A	Dil.; 70°	VTCHDF
284	St. 317	A	......	BIVPTSCH
285	St. 321	A	......	BIVPTSCH
286	St. 347	A	......	BIVPTSCHDF
287	St. 403	A	......	IV
288	St. 405	A	......	BIVTSCH
290	St. 410	A	......	BIVPTSCH
291	St. 414	A	......	IV
292	St. 416	A	......	IV
293	St. 418	A		
294	St. 420	A		
295	St. 430	A	......	BIVPTSCH
		A	10%; 70°	
		A	Dil.; 70°	VTSCHDF
296	St. 430F	A	......	IV
297	St. 431	A	......	IV
298	St. 440A	A		
300	St. 440C	A	......	IV
301	St. 442	A	......	BIVTSCH
303	St. 446	A	......	BIVPTSCH
313	St. CF-7	A	Conc.; rm.	
316	St. CF-20	A	......	BIV
319	St. CF-7M	A	Sat.; boil.	
360A	Steel	F		
367	Super-Ni	F	......	PTCH
368	Tantalum	A	No corrosion up to 392°	Not used commercially
369	Telnic	V		
390	Worthite	A	180°; with or without H_2SO_4	BIV
401	Karbate	A	All %; to boil.	P
402	Karbate	A	All %; to boil.	P
403	Kempruf	A	All %; any temp.	
404	Acheson	A	All %; any temp.	
500	Sul. cement	A	200° max.	PT
501	Sul. cement	A	200° max.	PT
502	Furan cement	A	360° max.	PT
503	Furan cement	A	360° max.	PT
504	Phen. cement	A	360° max.	PT
505	Phen. cement	A	360° max.	PT
506	Silicate cement	F	1600° max.	T
507	Silicate cement	F	1600° max.	T
513	Asplit	A	All %; 350° max.	
514	Asplit-F	A	All %; 350° max.	
515	Basolit	A	200° max.	T
518	C-Basolit	A	200° max.	T
521	Causplit	A	All %; 350° max.	
523	Duralon	A	350° max.	BIPTCHDF
524	Durisite	A	350° max.	BIPTCHDF
532	Lumnite	A	1% max.; 90° max.	TD
535	Nukem	A	350° max.	T
536	Nukem	A	......	T
538	Penchlor	A	All %; 750° max.	
540	Penchlor	A	All %; 500–2000°	
541	Pennsalt	A	All %; 350° max.	
542	Permanite	A	360° max.	TD
544	Plastite	A	175° max.	PTSD
545	P-Basolit	A	200° max.	T
554	Silastic	A	(A.S.T.M. D-543-43)	
559	Thiokol	A	150° max.	
600	Acid brick	A	......	TDR
603, 612, 615	Glass	A	......	THDR
604	Ceratherm	A	200–400°, depending on design	BIPTSCHDFR
606	Stoneware	A	All conditions	BIVPTCHDFR
607	Glass-L	A	600° max.	PTCH
610	Stoneware	A	All %; any temp.	BIVPTCHDR
611	Porcelain	A	All % and temp.	BIPTDR
614	Glass-L	A	All %; under 302°; agitation	BTCH
616	Pyrex	A		
617	Stoneware	A	140–160°	BIPTSCHDFR
618	Vitreo	A	1000° max.	
621	Vycor	A		
700	Ace Saran	A	......	P
704	DC Silicone	A	......	HD
711	Haveg	A	All conditions	BIVPTDFR
713	Haveg	A	All conditions	BIVPTDFR
715	Heresite	A	200° max.	
716	Heresite	A	200° max.	
718	Koroseal	A	All %; 150° max.	TD
720	Lamicoid	A		
726	Nukemite	A	......	PTSDF
733	Permanite	A	360° max.	BVPTD
736	Pyroflex	A	......	PTCHD
737	Resilon	A	175° max.	PTSD
740	Saran	F	77°	
741	Sealon	A	160° max.	BPTDF
742	Teflon	A	Rm.	VP
746	Tygon	A	180° max.	BIPTSCHDF
800	Ace Hd. Rub.	A	......	BIVPTDF
801	Acidseal	A	All %; 150° max.	BIPTSDF
802	Acidseal	A	All %; 150° max.	BIPTSDF
805	Butyl	A	Dry salt; aqueous sol.; rm.	
809	Fairprene	A		
817	Heresite	A	Sat.; 212°	
835	Perbunan	A	Dry salt; aqueous sol.; rm.	
836	Natural (S)	A		
837	Natural (H)	A		
838	GR-S (S)	A		
839	GR-S (H)	A		

Table 1. Chemical Resistance of Constructional Materials—(*Continued*)

No.	Material	Ratings	Exposure conditions	Applications
840	GR-A (S)	A		
841	GR-A (H)	A		
842	GR-M (S)	A		
843	GR-P (S)	X		
844	GR-I	A		
846	Saniprene	A	All %; 150° max.	BIPTSDF
848	Silastic	A	(A.S.T.M. D-543-43)	
849	Superflexite	A	All %; 150° max.	BIPTSDF
853	Thiokol	A	150° max.	T
854	Thiokol	A	150° max.	PTSCDF
855	Triflex	A	All %; 150° max.	BIPTSDF
856	Vistanex	A	Dry salt; aqueous sol.; rm.	
913	Redwood	A	Dil.; 72–100°; 23 years; explosives mfgr.	T
		A	Dil.; 72–180°; 4 years; explosives mfr.	T

Formaldehyde

No.	Material	Ratings	Exposure conditions	Applications
3	Admiralty	A		
4	Admiralty	A	....	CH
6	Adnic	A		
10–17	Aluminum	AF	38%; rm.; no free formic acid	BIPTSCHDF
22	Ambraloy	A	....	CH
23	Ambraloy	A	....	CH
24	Ambraloy	A	....	CH
29–40	Ampco	A	....	BIVPTCHDF
54	Brass	V	....	CH
61	Brass	V	....	CHV
63	Brass	A	....	PCH
66	Bronze	A		
73	Bronze	A	....	VT
74	Bronze	A	....	V
75	Bronze	A	....	VP
76	Bronze	A	....	Instruments
77	Bronze	V	....	PCH
86	Cast iron	V		
86A	Cast iron	F	Used where discoloration is not objectionable	
88	Chlorimet	A	All % and temp.	BIVCHF
89	Chlorimet	A	All % and temp.	BIVCHF
111	Copper	A		
114	Copper	A	....	PTCHR
117	Copper	F	....	P
118	Copper	A		
123	Cupro-Ni	A	....	CH
124	Cupro-Ni	A		
127–133	Dowmetal	V	50%	
139	Durco	A	All % and temp.	BIVHF
140	Durichlor	A	All % and temp.	BIVPCHFR
141, 142	Durimet	A	All % and temp.	BIVHF
143	Duriron	A	All % and temp.	BIVPHFR
148	Everdur	A	....	BIVR
149	Everdur	A	....	PTCHDR
150	Everdur	A	....	PCH
159	Hastelloy	A	All % and temp.	
160	Hastelloy	A	All % and temp.	
161	Hastelloy	A	All % and temp.	
162	Hastelloy	A	All % and temp.	
163	Stellite	A	All % and temp.	BIV
165	Stellite	A	All % and temp.	BIV
184	Inconel	A	37%; rm.; storage	
185	Inc-clad	A	37% in container	BTSCH
216	Monel	A	37%; rm.; storage	BIVPTD
217	Monel-clad	A	37% in container	BTSCH
219	Muntz	V	....	CH
224	Nickel	A	37%; rm.; storage	VPTH
225	Ni-clad	A	37% in container	BTSCH
226	Ni-silver	A	....	Instruments
231	Ni-Resist	A	37%; rm.; storage	BIV
234	Olympic	A		
235	Olympic	A		
245	Pyrasteel	A	....	BIVPTSCHF
270	304-clad	A	40%	BTSCH
271	316-clad	A	40%	BTSCH
274	430-clad	A	40%	BTSCH
275	St. 301	V		
276	St. 302	V		
		A	40%; 70°; boils	
278	St. 303	V		
279	St. 304	V		
280	St. 308	V		
281	St. 309	V		
282	St. 310	V		
283	St. 316	V		
		A	40%; 70°; boils	
284	St. 317	V		
285	St. 321	V		
286	St. 347	V		
287	St. 403	V		
288	St. 405	V		
290	St. 410	V		
291	St. 414	V		
292	St. 416	V		
293	St. 418	V		
294	St. 420	V		
295	St. 430	V		
		A	40%; 70°	
296	St. 430F	V		
297	St. 431	V		
298	St. 440A	V		
300	St. 440C	V		
301	St. 442	V		
303	St. 446	V		
308	St. CB-30	A	All %	
309	St. CC-35	A	All %	
313	St. CF-7	A	All %	
316	St. CF-20	A	....	BIV
360A	Steel	F	Used where discoloration is not objectionable	
367	Super Ni	A	....	PTCH
368	Tantalum	A	Extreme limits not determined	CH
369	Telnic	A		
390	Worthite	A	175°	BIV
401	Karbate	A	All %; to boil.	
402	Karbate	A	All %; to boil.	
403	Kempruf	A	All %; any temp.	H
404	Acheson	A	All %; any temp.	H
500	Sul. cement	A	200° max.	PT
501	Sul. cement	A	200° max.	PT
502	Furan cement	A	360° max.	PT
503	Furan cement	A	360° max.	PT
504	Phen. cement	A	360° max.	PT
505	Phen. cement	A	360° max.	PT
506	Silicate cement	A	1600° max.	PT
507	Silicate cement	A	1600° max.	PT
513	Asplit	A	All %; 350° max.	
514	Asplit-F	A	All %; 350° max.	
515	Basolit	A	....	T
518	C-Basolit	A	....	T
521	Causplit	A	All %; 350° max.	
523	Duralon	A	350° max.	BIPTCHDF
524	Durisite	A	350° max.	BIPTCHDF
534	N-series	F	Under 40% aq. sol.; 82°	BIVPTCDFR
535	Nukem	A	....	T
536	Nukem	A	....	T
538	Penchlor	A	All %; 750° max.	
539	Penchlor	A	All %; 750° max.	
540	Penchlor	A	All %; 500–2000°	
541	Pennsalt	A	All %; 350° max.	
542	Permanite	A	360° max.	TD
544	Plastite	F	175° max.	PTSD
545	P-Basolit	A	....	T
554	Silastic	X	(A.S.T.M. D-543-43)	
559	Thiokol	A	150° max.	
600	Acid brick	A	....	TDR
603, 612, 615	Glass	A	....	THDR
604	Ceratherm	A	200–400°, depending on design	BIPTSCHDFR
606	Stoneware	A	All conditions	BIVPTCHDFR
610	Stoneware	A	All %; any temp.	BIVPTCHDR
611	Porcelain	A	All % and temp.	BIPTDR
614	Glass-L	A	Under 302°; agitation	BTCH
616	Pyrex	A		
617	Stoneware	A	140–160°	BIPTSCHDFR
618	Vitreo	A	1000° max.	
619	Vitreosil	A		
621	Vycor	A		
700	Ace Saran	A	37%	P
704	DC Silicone	X		
713	Haveg	A	All conditions	BIVPTDFR
715	Heresite	A		
716	Heresite	A		
720	Lamicoid	A		
726	Nukemite	X		
733	Permanite	A	360° max.	BVPTD
736	Pyroflex	V	....	PTCHD
737	Resilon	F	175° max.	PTSD
740	Saran	A	77°	
741	Sealon	A	160° max.	BPTDF
742	Teflon	A	Rm.	VP
746	Tygon	V	180° max.	BIPTSCHDF
800	Ace Hd. Rub.	A	40%	BIVPTDF
801	Acidseal	A	40% max.; 150° max.	BIPTSDF
809	Fairprene	A		
817	Heresite	A	50%; 212° max.	
836	Natural (S)	V		

Table 1. Chemical Resistance of Constructional Materials—(*Continued*)

No.	Material	Ratings	Exposure conditions	Applications
837	Natural (H)	A		
838	GR-S (S)	V		
839	GR-S (H)	A		
840	GR-A (S)	V		
841	GR-A (H)	A		
842	GR-M (S)	X		
843	GR-P (S)	X		
844	GR-I	V		
846	Saniprene	A	40% max.; 150° max.	BIPTSDF
848	Silastic	X	(A.S.T.M. D-543-43)	
849	Superflexite	A	40% max.; 150° max.	BIPTSDF
853	Thiokol	A	150° max.	T
854	Thiokol	A	150° max.	PTSCDF
855	Triflex	A	40% max.; 150° max.	BIPTSDF

Formic Acid

No.	Material	Ratings	Exposure conditions	Applications
3	Admiralty	A		
4	Admiralty	A	....	CH
6	Adnic	A		
10–17	Aluminum	X		
19	Alloyco	A	5%; 150°	BT
		A	10–100%; boiling	BT
22	Ambraloy	A	....	CH
23	Ambraloy	A	....	CH
24	Ambraloy	A	....	CH
29–40	Ampco	F	No oxygen	BITPVCHDF
54	Brass	X		
61	Brass	X		
63	Brass	A	....	PCH
66	Bronze	A	....	
73	Bronze	A	....	VT
74	Bronze	A	....	V
75	Bronze	A	....	VP
76	Bronze	A	....	Instruments
77	Bronze	X	....	
86A	Cast iron	X		
88	Chlorimet	A	All % and temp.	BIVCHF
89	Chlorimet	A	All % and temp.	BIVCHF
111	Copper	X		
114	Copper	A	....	PTCHR
117	Copper	V	....	P
118	Copper	A	....	
123	Cupro-Ni	A	....	CH
124	Cupro-Ni	A	....	
139	Durco	A	All % and temp.	BIVHF
140	Durichlor	A	All % and temp.	BIVPCHFR
141, 142	Durimet	A	All % and temp.	BIVHF
143	Duriron	A	All % and temp.	BIVPHFR
148	Everdur	A	....	BIVP
149	Everdur	A	....	PTCHDR
150	Everdur	A	....	PCH
159	Hastelloy	A	All % and temp.	
160	Hastelloy	A	All % and temp.	
161	Hastelloy	A	All % and temp.	BIVPTCH
162	Hastelloy	A	All % and temp.	
163	Stellite	A	All % and temp.	BIV
165	Stellite	A	All % and temp.	BIV
184	Inconel	A	90%; storage; rm.	
		F	90%; 212°; distillation column	
185	Inc-clad	A	....	BTSCH
216	Monel	F	90%; storage; rm.	T
		F	90%; 212°; distillation column	
217	Monel-clad	A	....	BTSCH
219	Muntz	X		
224	Nickel	A	90%; storage; rm.	
		F	90%; 212°; distillation column	
226	Ni-silver	A	....	Instruments
234	Olympic	A		
235	Olympic	A		
270	304-clad	A	Under 5%; under 150°	BTSCH
271	316-clad	A	Under 5%; under 150°	BTSCH
275	St. 301	V		
276	St. 302	V		
		A	5% max.; 150° max.	VTSCDF
278	St. 303	V		
279	St. 304	V		
		A	5% max.; 150° max.	VTSCDF
280	St. 308	V		
281	St. 309	V		
282	St. 310	V		
283	St. 316	A	....	BIVPT
		A	5% max.; 150° max.	VTCHDF
284	St. 317	A	....	PT
285	St. 321	V		
286	St. 347	V		
287	St. 403	X		
288	St. 405	X		
290	St. 410	X		
291	St. 414	X		
292	St. 416	X		
293	St. 418	X		
294	St. 420	X		
295	St 430	X		
		X	5% max.; 150° max.	
296	St. 430F	X		
297	St. 431	X		
298	St. 440A	X		
300	St. 440C	X		
301	St. 442	X		
303	St. 446	X		
313	St. CF-7	A	Conc.; rm.	
319	St. CF-7M	A	Conc.; boil.	
360A	Steel	X		
367	Super Ni	A		PTCH
368	Tantalum	A	392° max.; crude and c.p.	H
369	Telnic	A		
390	Worthite	A	175°	BIV
401	Karbate	A	All %; to boil.	H
402	Karbate	A	All %; to boil.	H
403	Kempruf	A	All %; any temp.	
404	Acheson	A	All %; any temp.	
500	Sul. cement	A	200° max.	PTD
501	Sul. cement	A	200° max.	PTD
502	Furan cement	A	360° max.	PTD
503	Furan cement	A	360° max.	PTD
504	Phen. cement	A	360° max.	PTD
505	Phen. cement	A	360° max.	PTD
506	Silicate cement	A	1600° max.	PTD
507	Silicate cement	A	1600° max.	PTD
513	Asplit	A	All %; 350° max.	
514	Asplit-F	A	All %; 350° max.	
515	Basolit	A	200° max.	T
517	Carboline	A		
518	C-Basolit	A	200° max.	T
521	Causplit	A	All %; 350° max.	
523	Duralon	A	350° max.	BIPTCHDF
524	Durisite	A	350° max.	BIPTCHDF
535	Nukem	A	350° max.	T
536	Nukem	A	....	T
538	Penchlor	A	All %; 750° max.	
539	Penchlor	A	All %; 750° max.	
540	Penchlor	A	All %; 500–2000°	
541	Pennsalt	A	All %; 350° max.	
542	Permanite	A	360° max.	TD
544	Plastite	A	175° max.	PTSD
545	P-Basolit	A	200° max.	T
559	Thiokol	X		
600	Acid brick	A	....	TDR
603, 612, 615	Glass	A	....	THDR
604	Ceratherm	A	200–400°, depending on design	BIPTSCHDFR
606	Stoneware	A	All conditions	BIVPTCHDFR
607	Glass-L	A	600° max.	PTCH
610	Stoneware	A	All %; any temp.	BIVPTCHDR
611	Porcelain	A	All % and temp.	BIPTDR
614	Glass-L	A	All %; under 302°; agitation	BTCH
616	Pyrex	A		
617	Stoneware	A	140–160°	BIPTSCHDFR
618	Vitreo	A	1000° max.	
619	Vitreosil	A		
621	Vycor	A		
703	Compar	X		
711	Haveg	A	All conditions	BIVPTDF
713	Haveg	A	All conditions	BIVPTDF
715	Heresite	A	200° max.	
716	Heresite	A	200° max.	
720	Lamicoid	A		
727	Nylon	X		
728	Nylon	X		
729–732	Nylon	X		
733	Permanite	A	360° max.	BVPTD
736	Pyroflex	A	....	PTCHD
737	Resilon	A	175° max.	PTSD
738	Resistoflex	X		
739	Resistoflex	X		
741	Sealon	A	160° max.	BPTDF
742	Teflon	A	Rm.	VP
743	Textolite	A	Rm.	T
744	Textolite	F	Rm.	BI
745	Textolite	F	Rm.	BI
746	Tygon	A	180° max.	BIPTSCHDF
800	Ace Hd. Rub.	A	....	BIVPTDF
801	Acidseal	A	All %; 150° max.	BIPTSDF
805	Butyl	A	Rm.	

Table 1. Chemical Resistance of Constructional Materials—(*Continued*)

No.	Material	Ratings	Exposure conditions	Applications
809	Fairprene	A		
814	G.E. Silicone	V	...	VPTDF
817	Heresite	A	Sat.; 77° max.	
836	Natural (S)	V		
837	Natural (H)	A		
838	GR-S (S)	V		
839	GR-S (H)	A		
840	GR-A (S)	V		
841	GR-A (H)	A		
842	GR-M (S)	X		
843	GR-P (S)	X		
844	GR-I	V		
846	Saniprene	A	All %; 150° max.	BIPTSDF
848	Silastic	A	(A.S.T.M. D–543–43)	
849	Superflexite	A		
853	Thiokol	X		
854	Thiokol	X		
856	Vistanex	A	Rm.	

Hydrocarbons (Aliphatic)

No.	Material	Ratings	Exposure conditions	Applications
3	Admiralty	A		
4	Admiralty	A	...	CH
6	Adnic	A		
10–17	Aluminum	A	...	BIPTSCHDF
22	Ambraloy	A	...	CH
23	Ambraloy	A	...	CH
24	Ambraloy	A	...	CH
29–40	Ampco	A	Depends on halogen acid content	BIVPTCHDF
51	Berylco	A		
54	Brass	A	...	CH
61	Brass	A	...	VCH
63	Brass	A	...	PCH
66	Bronze	A		
67	Bronze	A	...	PIV
73	Bronze	A	...	VT
74	Bronze	A	...	V
75	Bronze	A	...	VP
76	Bronze	A	...	Instruments
77	Bronze	A	...	PCH
86A	Cast iron	A	Standard material of construction	
87	Causul	A	...	B
88	Chlorimet	A	All % and temp.	BIVCHF
89	Chlorimet	A	All % and temp.	BIVCHF
111	Copper	A		
114	Copper	A	...	PTCHR
117	Copper	A	...	P
118	Copper	A		
123	Cupro-Ni	A	...	CH
124	Cupro-Ni	A		
127–133	Dowmetal	A	Dry	
139	Durco	A	All % and temp.	BIVHF
140	Durichlor	A	All % and temp.	BIVPCHFR
141, 142	Durimet	A	All % and temp.	BIVHF
143	Duriron	A	All % and temp.	BIVPHFR
148	Everdur	A	...	BIVR
149	Everdur	A	...	PTCHDR
150	Everdur	A	...	PCH
156	Gold	A		
159	Hastelloy	A	All % and temp.	
160	Hastelloy	A	All % and temp.	
161	Hastelloy	A	All % and temp.	BIVPTCH
162	Hastelloy	A	All % and temp.	
163	Stellite	A	All % and temp.	BIV
165	Stellite	A	All % and temp.	BIV
184	Inconel	A	Boil.	
216	Monel	A	Boil.	BVPTCH
219	Muntz	A	...	CH
224	Nickel	A	Boil.	
226	Ni-silver	A	...	Instruments
231	Ni-Resist	A	Boil.	BIVP
233	NS-5	A	If low in sulfur	
234	Olympic	A		
235	Olympic	A		
236	Palladium	A		
240	Platinum	A		
242	Ir-Platinum	A		
244	Rh-Platinum	A		
268	Silver	A		
275	St. 301	A	...	BIVTSCHDF
276	St. 302	A	...	BIVTSCDF
278	St. 303	A	...	IV
279	St. 304	A	...	BIVPTSCDF
280	St. 308	A	...	BIVPTSCDF
281	St. 309	A	...	BIVTSCDF
282	St. 310	A	...	BIVPTSCDF
283	St. 316	A	...	BIVPTSHD
284	St. 317	A	...	BIVPTSCDF
285	St. 321	A	...	BIVPTSCDF
286	St. 347	A	...	BIVPTSCDF
287	St. 403	A	...	IV
288	St. 405	A	...	BIVTSC
290	St. 410	A	...	BIVPTSCDF
291	St. 414	A	...	IV
292	St. 416	A	...	IV
293	St. 418	A		
294	St. 420	A		
295	St. 430	A	...	BIVPTSCDF
296	St. 430F	A	...	IV
297	St. 431	A	...	IV
298	St. 440A	A		
300	St. 440C	A	...	IV
301	St. 442	A	...	BIVTSCH
303	St. 446	A	...	BIVPTSCDF
306	St. CA-15	A	High temp.	
350	St. HH	A	High temp.	
360A	Steel	A	Standard material of construction	
367	Super Ni	A	...	PTCH
368	Tantalum	A	Good for catalytic and chlorination reactions in liquid phase	CH
369	Telnic	A		
401	Karbate	A	All %; 338° max.	BIVP
402	Karbate	A	All %; 338° max.	BIVP
403	Kempruf	A	All %; any temp.	T
404	Acheson	A	All %; any temp.	T
500	Sul. cement	X		
501	Sul. cement	X		
502	Furan cement	A	360° max.	PTD
503	Furan cement	A	360° max.	PTD
504	Phen. cement	A	360° max.	PTD
505	Phen. cement	A	360° max.	PTD
506	Silicate cement	F	1600° max.	PTD
507	Silicate cement	F	1600° max.	PTD
508	Acichlor	X		
510	Acitite	A	Dilute and conc.; 250°	
515	Basolit	X		
517	Carboline	A		
518	C-Basolit	X		
523	Duralon	A	350° max.	BIPTCHDF
524	Durisite	A	350° max.	BIPTCHDF
534	N-series	V	Rm.	BIVPTCDFR
535	Nukem	A	...	T
536	Nukem	A	...	T
537	Pecomastic	X		
542	Permanite	A	360° max.	TD
544	Plastite	X		
545	P-Basolit	X		
554	Silastic	X	(A.S.T.M. D–543–43)	
559	Thiokol	A	150° max.	
600	Acid brick	A	...	TDR
603, 612, 615	Glass	A	...	THDR
604	Ceratherm	A	200–400°, depending on design	BIPTSCHDFR
606	Stoneware	A	Not commonly used	
607	Glass-L	A	600° max.	PTCH
610	Stoneware	A	All %; any temp.	BIVPTCHDR
611	Porcelain	A	All % and temp.	BIPTDR
614	Glass-L	A	Under 302°; agitation	BTCH
616	Pyrex	A		
617	Stoneware	A	140–160°	BIPTSCHDFR
618	Vitreo	A	1000° max.	
619	Vitreosil	A		
621	Vycor	A		
700	Ace Saran	V	...	P
703	Compar	A	Conc.; −40 to 200°	TD
704	DC Silicone	V	...	HD
706	Formica	A	All %; 212° max.	ID
707	Formica	A	All %; 212° max.	I
708	Formica	A	Ordinary conditions; 122° max.	
711	Haveg	A	All conditions	BIVPTDFR
713	Haveg	A	All conditions	BIVPTDFR
715	Heresite	A	200° max.	
716	Heresite	A	200° max.	
718	Koroseal	V	All %; 100° max.	TD
720	Lamicoid	A	...	T
723	Nixon	F	77°	S
724	Nixon	A	77°	S
725	Nixon	X	77°	
726	Nukemite	A	...	PTSDF
733	Permanite	A	360° max.	BVPTD
735	Polythene	F	Rm.; swells slightly	
		X	176°	
736	Pyroflex	V	...	PTCHD
737	Resilon	X		

Table 1. Chemical Resistance of Constructional Materials—(*Continued*)

No.	Material	Ratings	Exposure conditions	Applications
738	Resistoflex	A	Conc.; −40 to 275°	P
739	Resistoflex	A	Conc.; −40 to 275°	PIV
741	Sealon	V	160° max.	BPTDF
742	Teflon	A	Rm.	VP
744	Textolite	A	77–122°	BI
745	Textolite	A	....	BI
746	Tygon	V	180° max.	BIPTSCHDF
801	Acidseal	V	All %; 150° max.	BIPTSDF
805	Butyl	X		
809	Fairprene	A		
814	G.E. Silicone	V	Rm.	VPTDF
817	Heresite	A	Sat.; 86° max.	
835	Perbunan	A	To 300°; type hydrocarbon governs max. temp.	
836	Natural (S)	X		
837	Natural (H)	V		
838	GR-S (S)	X		
839	GR-S (H)	V		
840	GR-A (S)	X		
841	GR-A (H)	A		
842	GR-M (S)	A		
843	GR-P (S)	A		
844	GR-I	X		
846	Saniprene	V	All %; 150° max.	BIPTSDF
848	Silastic	X	(A.S.T.M. D-543-43)	
849	Superflexite	V	All %; 150° max.	BIPTSDF
853	Thiokol	A	150° max.	T
854	Thiokol	A	150° max.	PTSCDF
856	Vistanex	X		

Hydrocarbons (Aromatic)

No.	Material	Ratings	Exposure conditions	Applications
3	Admiralty	A		
4	Admiralty	A		CH
6	Adnic	A		
10–17	Aluminum	A	....	BIPTSCHDF
22	Ambraloy	A	....	CH
23	Ambraloy	A	....	CH
24	Ambraloy	A		CH
29–40	Ampco	A	....	BIVPTCHDF
54	Brass	A		CH
61	Brass	A	....	VCH
63	Brass	A	....	PCH
66	Bronze	A		
67	Bronze	A	....	PIV
73	Bronze	A	....	VT
74	Bronze	A	....	V
75	Bronze	A	....	VP
76	Bronze	A	....	Instruments
77	Bronze	A	....	PCH
86A	Cast iron	A	Standard material of construction	
87	Causul	A	Hot or cold	B
88	Chlorimet	A	All % and temp.	BIVCHF
89	Chlorimet	A	All % and temp.	BIVCHF
111	Copper	A		
114	Copper	A	....	PTCHR
117	Copper	A	....	P
118	Copper	A		
123	Cupro-Ni	A	....	CH
124	Cupro-Ni	A		
127–133	Dowmetal	A	Dry	
139	Durco	A	All % and temp.	BIVHF
140	Durichlor	A	All % and temp.	BIVPCHFR
141, 142	Durimet	A	All % and temp.	BIVHF
143	Duriron	A	All % and temp.	BIVPHFR
148	Everdur	A	....	BIVR
149	Everdur	A	....	PTCHDR
150	Everdur	A	....	PCH
156	Gold	A		
159	Hastelloy	A	All % and temp.	BIVPTCH
160	Hastelloy	A	All % and temp.	BIVPTCH
161	Hastelloy	A	All % and temp.	BIVPTCH
162	Hastelloy	A	All % and temp.	BIVP
163	Stellite	A	All % and temp.	BIV
165	Stellite	A	All % and temp.	BIV
184	Inconel	A	Boil.	
216	Monel	A	Boil.	BVPTCH
219	Muntz	A	....	CH
224	Nickel	A	Boil.	
226	Ni-silver	A	....	Instruments
231	Ni-Resist	A	Boil.	BIVP
233	NS-5	A	If low in sulfur	
234	Olympic	A		
235	Olympic	A		
236	Palladium	A		
240	Platinum	A		
242	Ir-Platinum	A		
244	Rh-Platinum	A		
268	Silver	A		
275	St. 301	A	....	BIVPTSCDF
276	St. 302	A	....	BIVTSCDF
278	St. 303	A	....	IV
279	St. 304	A	....	BIVPTSCDF
280	St. 308	A	....	BIVPTSCDF
281	St. 309	A	....	BIVTSCDF
282	St. 310	A	....	BIVPTSCDF
283	St. 316	A	....	BIVPTSHD
284	St. 317	A	....	BIVPTSCDF
285	St. 321	A	....	BIVPTSCDF
286	St. 347	A	....	BIVPTSCDF
287	St. 403	A	....	IV
288	St. 405	A	....	BIVTSC
290	St. 410	A	....	BIVPTSCHDF
291	St. 414	A	....	IV
292	St. 416	A	....	IV
293	St. 418	A		
294	St. 420	A		
295	St. 430	A	....	BIVPTSCDF
296	St. 430F	A	....	IV
297	St. 431	A	....	IV
298	St. 440A	A		
300	St. 440C	A	....	IV
301	St. 442	A	....	BIVTSCH
303	St. 446	A	....	BIVPTSCDF
306	St. CA-15	A	High temp.	
350	St. HH	A	High temp.	
360A	Steel	A	Standard material of construction	
367	SuperNi	A	....	PTCH
368	Tantalum	A	Good for catalytic and chlorination reactions in liquid phase	CH
369	Telnic	A		
401	Karbate	A	All %; 338° max.	BIVP
402	Karbate	A	All %; 338° max.	BIVP
403	Kempruf	A	All %; any temp.	T
404	Acheson	A	All %; any temp.	T
500	Sul. cement	X		
501	Sul. cement	X		
502	Furan cement	A	360° max.	PTD
503	Furan cement	A	360° max.	PTD
504	Phen. cement	A	360° max.	PTD
505	Phen. cement	A	360° max.	PTD
506	Silicate cement	F	1600° max.	TD
507	Silicate cement	F	1600° max.	TD
508	Acichlor	X		
510	Acitite	A	Dilute and conc.; 250°	
515	Basolit	X		
517	Carboline	A		
518	C-Basolit	X		
523	Duralon	A	350° max.	BIPTCHDF
524	Durisite	A	350° max.	BIPTCHDF
534	N-series	X		
535	Nukem	A	....	T
536	Nukem	A	....	T
537	Pecomastic	X		
542	Permanite	A	360° max.	TD
544	Plastite	X		
545	P-Basolit	X		
554	Silastic	X	(A.S.T.M. D-543-43)	
559	Thiokol	F	150° max.	
600	Acid brick	A	....	TDR
603, 612, 615	Glass	A	....	THDR
606	Stoneware	A	Not commonly used	
607	Glass-L	A	600° max.	PTCH
610	Stoneware	A	All %; any temp.	BIVPTCHDR
611	Porcelain	A	All % and temp.	BIPTDR
614	Glass-L	A	Under 302°; agitation	BTCH
616	Pyrex	A		
617	Stoneware	A	140–160°	BIPTCHDFR
618	Vitreo	A	1000° max.	
619	Vitreosil	A		
621	Vycor	A		
700	Ace Saran	V		
703	Compar	A	Conc.; −40 to 200°	TD
704	DC Silicone	X		
706	Formica	A	All %; 212° max.	I
707	Formica	A	All %; 212° max.	I
708	Formica	A	Ordinary conditions; 122° max.	
713	Haveg	A	All conditions	BIVPTDFR
715	Heresite	A	200° max.	
716	Heresite	A	200° max.	
718	Koroseal	X		
720	Lamicoid	A	....	T
723	Nixon	F	77°	S
724	Nixon	A	77°	S
725	Nixon	X		
726	Nukemite	X		

Table 1. Chemical Resistance of Constructional Materials—(*Continued*)

No.	Material	Ratings	Exposure conditions	Applications
733	Permanite	A	360° max.	BVPTD
735	Polythene	V	Rm.; swells slightly	
		X	176°	
736	Pyroflex	V	...	PTCHD
737	Resilon	X		
738	Resistoflex	A	Conc.; −40 to 275°	P
739	Resistoflex	A	Conc.; −40 to 275°	IVP
741	Sealon	V	160° max.	BPTDF
742	Teflon	A	Rm.	VP
744	Textolite	A	77–122°	BI
745	Textolite	A	77–122°	BI
746	Tygon	X		
801	Acidseal	X		
802	Acidseal	X		
805	Butyl	X		
809	Fairprene	X		
814	G.E. Silicone	V	Rm.	VPTDF
817	Heresite	V	Sat.; 86°	
835	Perbunan	A	Type of hydrocarbon governs max. temp.	
836	Natural (S)	X		
837	Natural (H)	X		
838	GR-S (S)	X		
839	GR-S (H)	X		
840	GR-A (S)	X		
841	GR-A (H)	V		
842	GR-M (S)	X		
843	GR-P (S)	A		
844	GR-I	X		
846	Saniprene	X		
848	Silastic	X	(A.S.T.M. D-543-43)	
849	Superflexite	X		
853	Thiokol	F	150° max.	T
854	Thiokol	V	150° max.	PTSCDF
856	Vistanex	X		

Hydrochloric Acid

No.	Material	Ratings	Exposure conditions	Applications
3	Admiralty	V		
4	Admiralty	V	...	CH
6	Adnic	V		
10–17	Aluminum	X		
22	Ambraloy	V	...	CH
23	Ambraloy	V	...	CH
24	Ambraloy	V	...	CH
29–40	Ampco	V	...	BIVPTCHDF
42	Antaciron	A	All %; cold	PIV
51	Berylco	V		
54	Brass	X		
61	Brass	X		
63	Brass	V	...	PCH
66	Bronze	V		
73	Bronze	V	...	VT
74	Bronze	V	...	V
75	Bronze	V	...	VP
76	Bronze	V		
77	Bronze	X		
81	CA-FA20	V	1, 5% aerated and agitated; 70°	BIVP
		V	1%; boil.	BIVP
82	CA-MM	F	1, 5% aerated and agitated; 70°	BIVP
		X	1%; boil.	
86A	Cast iron	X		
87	Causul	F	Dil; cold	
88	Chlorimet	A	All % and temp.	BIVCHF
89	Chlorimet	A	All % to 170°	BIVCHF
111	Copper	V		
114	Copper	V	...	PTCHR
118	Copper	V		
119	Corrosiron	F	5%; rm.; unagitated; c.p.	BIPF
		V	25%; rm.; unagitated; c.p.	BIPF
		V	37%; rm.; unagitated; c.p.	BIPF
123	Cupro-Ni	V	...	CH
124	Cupro-Ni	V		
139	Durco	A	Dil.; rm.	BIVHF
140	Durichlor	A	All % and temp.	BIVPCHFR
141, 142	Durimet	A	Dil.; rm.	BIVHF
143	Duriron	V	Low % and temp. (140 preferred)	BIVPHFR
148	Everdur	V	...	BIVR
149	Everdur	V	...	PTCHDR
150	Everdur	V	...	PCH
156	Gold	A	In absence of oxidizing material	
159	Hastelloy	A	All % to 160°	BIVPTSCHDF
160	Hastelloy	A	All % to boil.	BIVPTSCHDF
161	Hastelloy	A	All % to 125°	BIVPTCHDF
162	Hastelloy	A	All %; rm. only	BIV
163	Stellite	A	All %; rm. only	BIV
165	Stellite	X	All %; rm.	
181	Hytensl	F	10% max.; cold	BIVHF
184	Inconel	X	0.5%; boil.	
		A	5% unaerated; 86°	
		F	5% aerated; 86°	
		X	5% unaerated; 158°	
		F	20% unaerated; 86°	
185	Inc-clad	A	5% sol. sat. with N; 86°	BTSCH
216	Monel	A	0.5%; boil.	BIVPTCHDFR
		A	5% unaerated; 86, 158°	
		F	5% aerated; 86°	
		F	20% unaerated; 86°	
217	Monel-clad	A	5% sol. sat. with N; 86°	BTSCH
219	Muntz	X		
224	Nickel	X	0.5%; boil.	
		A	5, 20% unaerated; 86°	BIVPTCH
		F	5% aerated; 86°	
		X	5% unaerated; 158°	
225	Ni-clad	A	5% sol. sat. with N; 86°	BTSCH
226	Ni-silver	V		
231	Ni-Resist	A	5% unaerated; 86°	BIVR
		X	5% aerated; 86°	
		X	5% unaerated; 158°	
		F	20% unaerated; 86°	
233	NS-5	F	Low %; cold	V
234	Olympic	V		
235	Olympic	V		
236	Palladium	A	In absence of oxidizing material	
240	Platinum	A	In absence of oxidizing material	
242	Ir-Platinum	A	In absence of oxidizing material	
244	Rh-Platinum	A	In absence of oxidizing material	
245	Pyrasteel	X		
268	Silver	A	In absence of oxidizing material	
275	St. 301	X		
276	St. 302	X		
		X	Diluted 1:85; boil.	
		A	Diluted 1:85; 70°	
		F	Diluted 1:10; 70°	
		X	Diluted 1:10; boil.	
		A	Vapors; 70°	
		X	Vapors; 212°; 425°	
		X	All %; any temp.	
278	St. 303	X		
279	St. 304	X		
		X	All %; any temp.	
		X	1%; aerated and agitated; 70°	
		X	1%; boil.	
280	St. 308	X		
281	St. 309	X		
282	St. 310	X		
283	St. 316	X		
		A	Diluted 1:85; 70°	
		X	Diluted 1:85; boil.	
		A	Diluted 1:10; 70°	
		X	Diluted 1:10; boil.	
		A	Vapors; 70°	
		X	Vapors; 212°; 425°	
		X	All %; any temp.	
		A	1%; aerated and agitated; 70°	BIVP
		X	1%; boil.	
284	St. 317	X		
285	St 321	X		
286	St. 347	X		
287	St. 403	X		
288	St. 405	X		
290	St. 410	X		
		X	Diluted 1:85; boil.	
		X	Diluted 1:10; boil.	
		X	1%; aerated and agitated; 70°	
291	St. 414	X		
292	St. 416	X		
		X	All %	
293	St. 418	X		
294	St. 420	X		
		X	All %	
295	St. 430	X		
		X	Diluted 1:85; boil.	
		X	Diluted; 1:10; boil.	
		X	All %; any temp.	
296	St. 430F	X		
297	St. 431	X		
298	St. 440A	X		

Table 1. Chemical Resistance of Constructional Materials—(*Continued*)

No.	Material	Ratings	Exposure conditions	Applications
300	St. 440C	X		
301	St. 442	X		
303	St. 446	X		
360A	Steel	X		
364	Stoody	A	10%; rm.	
		F	10%; boil.; conc.	
365	Stoody	F	10%; rm.; boil.	
		X	Conc.; boil.	
367	Super Ni	V	...	PTCH
368	Tantalum	A	662° max.	BPCH
369	Telnic	V		
387	Stainless	X	All %	
388	Stainless	X	All %	
389	Stainless	A	Dilute	
		X	Strong	
390	Worthite	F	15%; 100°; without metal salts	
		X	20% min.	
392	Wyndaloy	V	Very dilute; 68°	
		X	Moderate dilution; 68°	
		X	Conc.; 68°	
401	Karbate	A	All %; to boil.	BIVPTCHD
402	Karbate	A	All %; to boil.	BIVPTCHD
403	Kempruf	A	All %; any temp.	PTDR
404	Acheson	A	All %; any temp.	PTDR
500	Sul. cement	A	200° max.	PTD
501	Sul. cement	A	200° max.	PTD
502	Furan cement	A	360° max.	PTD
503	Furan cement	A	360° max.	PTD
504	Phen. cement	A	360° max.	PTD
505	Phen. cement	A	360° max.	PTD
506	Silicate cement	A	1600° max.	PTD
507	Silicate cement	A	1600° max.	PTD
508	Acichlor	A	Dilute and conc. 300°	
510	Acitite	A	Dilute and conc. 250°	
513	Asplit	A	All %; 350° max.	
514	Asplit-F	A	All %; 350° max.	
515	Basolit	A	200° max.	T
517	Carboline	A		
518	C-Basolit	A	200° max.	T
521	Causplit	A	All %; 350° max.	
523	Duralon	A	350° max.	BIPTCHDF
524	Durisite	A	350° max.	BIPTCHDF
534	N-series	V	...	BIVPTCDFR
		X	Conc.	
535	Nukem	A	350° max.	T
536	Nukem	A	...	T
537	Pecomastic	A	Dilute and conc.; 300°	
538	Penchlor	A	All %; 750° max.	
539	Penchlor	A	All %; 750° max.	
540	Penchlor	A	All %; 500–2000°	
541	Pennsalt	A	All %; 350° max.	
542	Permanite	A	360° max.	TD
544	Plastite	A	60% max; 175° max.	PTSD
545	P-Basolit	A	200° max.	T
554	Silastic	A	10%; (A.S.T.M. D-543-43)	
		X	Conc.; (A.S.T.M. D-543-43)	
556	Staminite	A	All %	T
559	Thiokol	X		
600	Acid brick	A	...	TDR
603, 612, 615	Glass	A	...	THDR
604	Ceratherm	A	200–400°, depending on design	BIPTSCHDFR
606	Stoneware	A	All conditions	BIVPTCHDFR
607	Glass-L	A	600° max.	PTCH
610	Stoneware	A	All %; any temp.	BIPTVCHDR
611	Porcelain	A	All % and temp.	BIPTDR
614	Glass-L	A	All %; under 302°; agitation	BTCH
616	Pyrex	A		
617	Stoneware	A	140–160°	BIPTSCHDFR
618	Vitreo	A	1000° max.	
619	Vitreosil	A	...	PCH
621	Vycor	A		
700	Ace Saran	A	...	P
703	Compar	X		
704	DC Silicone	A	...	HD
706	Formica	F	All %; 212° max.	TD
711	Haveg	A	All conditions	BIVPTDFR
713	Haveg	A	All conditions	BIVPTDFR
715	Heresite	A	200° max.	
716	Heresite	A	200° max.	
720	Lamicoid	X		
723	Nixon	X	10%; 77°	
724	Nixon	A	10%; 77°	
725	Nixon	A	10%; 77°	S
726	Nukemite	A	150° max.	PTSDF
727	Nylon	X	Conc.	
728	Nylon	X	Conc.	
729–732	Nylon	X		
733	Permanite	A	360° max.	BVPTD
735	Polythene	V	Rm.	
736	Pyroflex	A	...	PTCHD
737	Resilon	A	60% max.; 175° max.	PTSD
738	Resistoflex	X		
739	Resistoflex	X		
740	Saran	A	77°	
741	Sealon	A	160° max.	BPTDF
742	Teflon	A	Rm.	VP
743	Textolite	V		
744	Textolite	V	...	BI
745	Textolite	V	...	BI
746	Tygon	A	All %; 180° max.	BIPTSCHDF
800	Ace Hd. Rub.	A	...	BIVPTDF
801	Acidseal	A	All %; 150° max.	BIPTSDF
805	Butyl	F	Rm.	
809	Fairprene	A		
814	G.E. Silicone	X		
817	Heresite	A	Conc.; 122° max.	
836	Natural (S)	A		
837	Natural (H)	A		
838	GR-S (S)	V		
839	GR-S (H)	V		
840	GR-A (S)	V		
841	GR-A (H)	V		
842	GR-M (S)	V		
843	GR-P (S)	X		
844	GR-I	A		
846	Saniprene	A	All %; 150° max.	BIPTSDF
848	Silastic	A	10%; (A.S.T.M. D-543-43)	
		A	Conc.; (A.S.T.M. D-543-43)	
849	Superflexite	A	All %; 150° max.	BIPTSDF
853	Thiokol	X		
854	Thiokol	F	150° max.	PTSCDF
855	Triflex	A	All %; 150° max.	BIPTSDF
856	Vistanex	F	Rm.	
913	Redwood	X	18–20° Bé.; 60–120°; Failed in 3 months	T
		A	18° Bé.; 60–120° asphalt + tar-L; 12 years; chem. plant	T
		A	15%; 36–180°; 5 years; steel pickling	T
		F	5%; 36–90°; 3 years; petroleum by-products plant	T
		A	1½% – 2% H_2SO_4; 40–180°; 21 years; chem. plant	T
		A	0.5% + glycerin; 40–150°; 2 years; soap plant	T

Hydrofluoric Acid

No.	Material	Ratings	Exposure conditions	Applications
3	Admiralty	V		
4	Admiralty	V	...	CH
6	Adnic	V		
10–17	Aluminum	X		
22	Ambraloy	V	...	CH
23	Ambraloy	V	...	CH
24	Ambraloy	V	...	CH
29–40	Ampco	A	18% max.; cold; anhydrous	BIVPTCHDF
54	Brass	X		
61	Brass	X		
63	Brass	V	...	PCH
66	Bronze	V		
73	Bronze	V	...	VT
74	Bronze	V	...	V
75	Bronze	V	...	VP
76	Bronze	V		
77	Bronze	X		
81	CA-FA20	A	48%; 70°	BIVP
		X	48%; 176°	
82	CA-MM	A	48%; 70°	BIVP
		A	48%; 176°	BIVP
86A	Cast iron	X		
88	Chlorimet	A	All % at rm.	BIVCHF
89	Chlorimet	A	All % and temp.	BIVCHF
111	Copper	V		
114	Copper	V	...	PTCHR
118	Copper	V		
123	Cupro-Ni	V	...	CH
124	Cupro-Ni	V		
127–133	Dowmetal	A	Over 5%	
139	Durco	F	Low %; rm.	BIVHF
140	Durichlor	X		

Table 1. Chemical Resistance of Constructional Materials—(*Continued*)

No.	Material	Ratings	Exposure conditions	Applications
141, 142	Durimet	F	Low %; rm.	BIVHF
143	Duriron	X		
148	Everdur	V	....	BIVR
149	Everdur	V	....	PTCHDR
150	Everdur	V	....	PCH
156	Gold	A		
159	Hastelloy	A	All %; rm. only	
160	Hastelloy	A	All %; to boil.	
161	Hastelloy	A	All %; to boil.	
162	Hastelloy	X	All %; to boil.	
163	Stellite	A	All %; rm. only	BIV
165	Stellite	A	All %; rm. only	BIV
181	Hytensl	A	50% max.; cold	BIVHF
184	Inconel	A	10%; 70°	
		X	6%; 170°	
		X	38%; 230°	
		A	100% to 900°	
216	Monel	A	10%; 70°	BIVPTCHDFR
		A	6%; 170°	
		A	38%; 230°	
		A	100%; to 1100°	
217	Monel-clad	A	60%; rm.	BTSCH
219	Muntz	X		
224	Nickel	A	10%; 70°	VPTC
		X	6%; 170°	
		X	38%; 230°	
		A	100% to 1,2	
226	Ni-silver	V		
231	Ni-Resist	A	10%; 70°	BIVP
		X	6%; 170°	
		X	38%; 230°	
233	NS-5	A	Dil. or conc.	V
234	Olympic	V		
235	Olympic	V		
236	Palladium	A		
240	Platinum	A		
242	Ir-Platinum	A		
244	Rh-Platinum	A		
245	Pyrasteel	X		
268	Silver	A		
275	St. 301	X		
276	St. 302	X		
		A	Vapors; 212°	
		X	All %; any temp.	
278	St. 303	X		
279	St. 304	X		
		X	All %; any temp.	
280	St. 308	X		
281	St. 309	X		
282	St. 310	X		
283	St. 316	X		
		A	Vapors; 212°	
		X	All %; any temp.	
284	St. 317	X		
285	St. 321	X		
286	St. 347	X		
287	St. 403	X		
288	St. 405	X		
290	St. 410	X		
291	St. 414	X		
292	St. 416	X		
293	St. 418	X		
294	St. 420	X		
295	St. 430	X		
296	St. 430F	X		
297	St. 431	X		
298	St. 440A	X		
300	St. 440C	X		
301	St. 442	X		
303	St. 446	X		
360A	Steel	A	Over 75%; low temp.; unaerated	
367	SuperNi	F	....	PTCH
368	Tantalum	X		
369	Telnic	V		
390	Worthite	A	Cold, 0.5% in 4% H_3PO_4	BIV
		F	175°, 0.5% in 4% H_3PO_4	BIV
392	Wyndaloy	V	68°	
401	Karbate	A	60% max.; 185° max.	BIVPCHD
402	Karbate	A	60% max.; 185° max.	BIVPCHD
403	Kempruf	A	All %; to boil.	PTR
404	Acheson	A	All %; to boil.	PTR
500	Sul. cement	X		
501	Sul. cement	A	200° max.	PTD
502	Furan cement	X		
503	Furan cement	A	360° max.	PTD
504	Phen. cement	X		
505	Phen. cement	A	360° max.	PTD
506	Silicate cement	X		
507	Silicate cement	X		
513	Asplit	X		

No.	Material	Ratings	Exposure conditions	Applications
514	Asplit-F	A	All %; 350° max.	
515	Basolit	X		
517	Carboline	A		
518	C-Basolit	A	200° max.	T
521	Causplit	X		
523	Duralon	V	350° max.	BIPTCHDF
524	Durisite	V	350° max.	BIPTCHDF
535	Nukem	A	50% max.	T
536	Nukem	X		
538	Penchlor	X		
539	Penchlor	X		
540	Penchlor	X		
541	Pennsalt	A	All %; 350° max.	
542	Permanite	A	360° max.	TD
544	Plastite	V	175° max.	PTSD
545	P-Basolit	X		
556	Staminite	X		
559	Thiokol	X		
600	Acid brick	X		
603, 612, 615	Glass	X	All % and temp.	
604	Ceratherm	X	All %	
606	Stoneware	X		
607	Glass-L	X		
610	Stoneware	X		
611	Porcelain	X		
614	Glass-L	X		
616	Pyrex	X		
617	Stoneware	X	All %	BIPTSCHDFR
618	Vitreo	X		
619	Vitreosil	X		
621	Vycor	X		
700	Ace Saran	A	....	P
703	Compar	X		
712	Haveg	A	All conditions	BIVPTDFR
714	Haveg	A	All conditions	BIVPTDFR
715	Heresite	A	20%; 70°	
716	Heresite	A	20%; 70°	
718	Koroseal	A	60%; 100°	TD
720	Lamicoid	X		
726	Nukemite	A	150° max.	PTSDF
727	Nylon	X	Conc.	
728	Nylon	X	Conc.	
729–732	Nylon	X		
733	Permanite	A	360° max.	BVPTD
735	Polythene	A	48%; rm.	
736	Pyroflex	V	....	PTCHD
737	Resilon	V	175° max.	PTSD
738	Resistoflex	X		
739	Resistoflex	X		
740	Saran	A	77°	
741	Sealon	A	160° max.	BPTDF
742	Teflon	A	Rm.	VP
746	Tygon	F	Dil.; 180° max.	BIPTSCHDF
800	Ace Hd. Rub.	A	....	BIVPTDF
802	Acidseal	A	50% max.; 100°	BIPTSDF
805	Butyl	A		
809	Fairprene	A		
814	G.E. Silicone	X		
817	Heresite	A	48%; 77° max.	
836	Natural (S)	V		
837	Natural (H)	V		
838	GR-S (S)	V		
839	GR-S (H)	V		
840	GR-A (S)	V		
841	GR-A (H)	V		
842	GR-M (S)	A		
843	GR-P (S)	X		
844	GR-I	V		
853	Thiokol	X		
854	Thiokol	F	150° max.	PTSCDF
856	Vistanex	A	Rm.	
913	Redwood	A	3½–5%; 36–90°; pitch + asphalt-L; airplane mfr.	T
		A	3%; 36–90°; 5 years; steel plant	T

Hydrogen Peroxide

No.	Material	Ratings	Exposure conditions	Applications
3	Admiralty	V		
4	Admiralty	A	....	CH
6	Adnic	F	Attacked	
10–17	Aluminum	A	30–95%; rm. to boil.; 99.6% Al, Al-Mg, and Al-Si alloys cause min. decomp.	BIPTSCHD
19	Aloyco	A	Boil.	BT
22	Ambraloy	A	....	CH
23	Ambraloy	A	....	CH

Table 1. Chemical Resistance of Constructional Materials—(*Continued*)

No.	Material	Ratings	Exposure conditions	Applications
24	Ambraloy	A	....	CH
29–40	Ampco	A	If oxygen content is not vital	BITVPCHDF
54	Brass	V	....	CH
61	Brass	V	....	VCH
63	Brass	AV	....	PCH
66	Bronze	V	....	
73	Bronze	A	....	VT
74	Bronze	A	....	V
75	Bronze	A	....	VP
76	Bronze	A	....	Instruments
77	Bronze	V	....	PCH
86A	Cast iron	F		
88	Chlorimet	X		
89	Chlorimet	A	All % and temp.	BIVHF
111	Copper	V		
114	Copper	A	....	PTCHR
117	Copper	V	....	P
118	Copper	V		
123	Cupro-Ni	A	....	CH
124	Cupro-Ni	V		
139	Durco	A	All % and temp.	BIVHF
140	Durichlor	A	All % and temp.	BIVPCHFR
141, 142	Durimet	A	All % and temp.	BIVHF
143	Duriron	A	All % and temp.	BIVPHFR
148	Everdur	A	....	BIVR
149	Everdur	A	....	PTCHDR
150	Everdur	A	....	PCH
156	Gold	A		
159	Hastelloy	F	All %; to boil.	
160	Hastelloy	F	All %; to boil.	
161	Hastelloy	A	All %; to boil.	
162	Hastelloy	F	All %; to boil.	
163	Stellite	A	All %; to boil.	
165	Stellite	A	All %; to boil.	
184	Inconel	A	30%; 86°	
185	Inc-clad	V		
216	Monel	A	30%; 86°	VPTH
217	Monel-clad	V		
219	Muntz	V	....	CH
224	Nickel	A	30%; 86°	PTH
225	Ni-clad	V		
226	Ni-silver	A	....	Instruments
234	Olympic	V		
235	Olympic	V		
236	Palladium	A		
240	Platinum	A		
242	Ir-Platinum	A		
244	Rh-Platinum	A		
245	Pyrasteel	A	....	BIVPTSCHF
268	Silver	A		
270	304-clad	A	All %, any temp.	BTSCH
271	316-clad	A	All %; any temp.	BTSCH
274	430-clad	A	All %; any temp.	BTSCH
275	St. 301	V		
276	St. 302	V		
		A	No influence through catalysis; 70°	
		A	All %; any temp.	VTSCDF
278	St. 303	V		
279	St. 304	V		
		A	All %; any temp.	VTSCDF
280	St. 308	V		
281	St. 309	V		
282	St. 310	V		
283	St. 316	V		
		A	No influence through catalysis; 70°	
		A	All %; any temp.	VTCHDF
284	St. 317	V		
285	St. 321	V		
286	St. 347	V		
287	St. 403	V		
288	St. 405	V		
290	St. 410	V		
291	St. 414	V		
292	St. 416	V		
293	St. 418	V		
294	St. 420	V		
295	St. 430	V		
		X	No influence through catalysis; 70°	
		A	All %; any temp.	VTSCHDF
296	St. 430F	V		
297	St. 431	V		
298	St. 440A	V		
300	St. 440C	V		
301	St. 442	V		
303	St. 446	V		
360A	Steel	F		
367	Super Ni	A	....	PTCH
368	Tantalum	A	All conc. to decomposition temp.	CH
369	Telnic	V		
390	Worthite	A	Warm 30%	BIV
401	Karbate	A	3% max.; rm.	BIVH
400	Karbate	A	3%; max. rm.	BIVH
500	Sul. cement	F	Dilute sol.; low temp. only	T
501	Sul. cement	F	Dilute sol.; low temp. only	T
502	Furan cement	F	Dilute sol.; low temp. only	T
503	Furan cement	F	Dilute sol.; low temp. only	T
504	Phen. cement	F	Dilute sol.; low temp. only	T
505	Phen. cement	F	Dilute sol.; low temp. only	T
506	Silicate cement	F	All %	T
507	Silicate cement	F	All %	T
513	Asplit	A	All %; 350° max.	
514	Asplit-F	A	All %; 350° max.	
515	Basolit	X		
518	C-Basolit	X		
521	Causplit	A	All %; 350° max.	
523	Duralon	A	350° max.	BIPTCHDF
524	Durisite	A	350° max.	BIPTCHDF
534	N-series	X		
535	Nukem	X		
536	Nukem	X		
538	Penchlor	A	All %; 750° max.	
539	Penchlor	A	All %; 750° max.	
540	Penchlor	A	All %; 500–2000°	
541	Pennsalt	A	All %; 350° max.	
542	Permanite	X		
544	Plastite	F	Dil.; 175° max.	PTSD
545	P-Basolit	X		
554	Silastic	A	3%; (A.S.T.M. D–543–43)	
559	Thiokol	V	150° max.	
600	Acid brick	A	....	TDR
603, 612, 615	Glass	A	....	THDR
604	Ceratherm	A	200–400°, depending on design	BIPTSCHDFR
606	Stoneware	A	All conditions	BIVPTCHDFR
607	Glass-L	A	600° max.	PTCH
610	Stoneware	A	All %; any temp.	BIVPTCHDR
611	Porcelain	A	All % and temp.	BIPTDR
614	Glass-L	A	Under 212°	BTCH
616	Pyrex	A		
617	Stoneware	A	140–160°	BIPTSCHDFR
618	Vitreo	A	1000° max.	
619	Vitreosil	A		
621	Vycor	A		
700	Ace Saran	A	....	P
704	DC Silicone	A	....	HD
711	Haveg	A	All conditions	BIVPTDFR
715	Heresite	A	200° max.	
716	Heresite	A	200° max.	
718	Koroseal	A	30% max.; 100° max.	TD
720	Lamicoid	A		
723	Nixon	A	3%; 77°	S
724	Nixon	A	3%; 77°	S
725	Nixon	A	3%; 77°	S
726	Nukemite	A	....	PTSDF
733	Permanite	X		
736	Pyroflex	V	....	PTCHD
737	Resilon	F	Dil.; 175° max.	PTSD
740	Saran	A	77°	
741	Sealon	A	160° max.	BPTDF
742	Teflon	A	Rm.	VP
746	Tygon	F	180° max.	BIPTSCHDF
805	Butyl	V		
809	Fairprene	A		
817	Heresite	A	30%; 212° max.	
836	Natural (S)	V		
837	Natural (H)	V		
838	GR-S (S)	V		
839	GR-S (H)	V		
840	GR-A (S)	V		
841	GR-A (H)	V		
842	GR-M (S)	V		
843	GR-P (S)	X		
844	GR-I	A		
848	Silastic	A	3%; (A.S.T.M. D–543–43)	
853	Thiokol	V	150° max.	T
854	Thiokol	F	150° max.	PTSCDF
856	Vistanex	V		

Iodine

No.	Material	Ratings	Exposure conditions	Applications
3	Admiralty	X		
6	Adnic	A		
10–17	Aluminum	X		
29–40	Ampco	X		

Table 1. Chemical Resistance of Constructional Materials—(*Continued*)

No.	Material	Ratings	Exposure conditions	Applications
63	Brass	X		
66	Bronze	X		
86A	Cast iron	X		
89	Chlorimet	A	All % and temp.	BIVHF
111	Copper	X		
118	Copper	X		
124	Cupro-Ni	X		
127–133	Dowmetal	V	Crystals	
139	Durco	X		
140	Durichlor	X		
141, 142	Durimet	X		
143	Duriron	X		
159	Hastelloy	X	All % and temp.	
160	Hastelloy	X	All % and temp.	
161	Hastelloy	A	Wet gas; all %; rm. only	BIVPTDF
162	Hastelloy	X	All % and temp.	
163	Stellite	A	Wet gas; all %; rm. only	BIV
165	Stellite	X	All % and temp.	
216	Monel	X	5% by vol. U.S.P. tincture 70°	
234	Olympic	X		
235	Olympic	X		
275	St. 301	X		
276	St. 302	X		
		A	Dry; 70°	
		X	Moist; 70°	
278	St. 303	X		
279	St. 304	X		
280	St. 308	X		
281	St. 309	X		
282	St. 310	X		
283	St. 316	A	Dry; 70°	
		X	Moist; 70°	
284	St. 317	A	....	PT
285	St. 321	X		
286	St. 347	X		
287	St. 403	X		
288	St. 405	X		
290	St. 410	X		
291	St. 414	X		
292	St. 416	X		
293	St. 418	X		
294	St. 420	X		
295	St. 430	X	Moist; 70°	
296	St. 430F	X		
297	St. 431	X		
298	St. 440A	X		
300	St. 440C	X		
301	St. 442	X		
303	St. 446	X		
360A	Steel	X		
368	Tantalum	A	....	Not used commercially
369	Telnic	X		
513	Asplit	X		
514	Asplit-F	X		
515	Basolit	A	....	T
518	C-Basolit	A	....	T
521	Causplit	X		
523	Duralon	F	350° max.	BIPTCHDF
524	Durisite	F	350° max.	BIPTCHDF
535	Nukem	X		
536	Nukem	A	....	T
538	Penchlor	A	All %; 750° max.	
539	Penchlor	A	All %; 750° max.	
540	Penchlor	A	All %; 500–2000°	
541	Pennsalt	X		
542	Permanite	X		
544	Plastite	X		
545	P-Basolit	A	....	T
559	Thiokol	X		
600	Acid brick	A	....	TDR
603, 612, 615	Glass	A	....	THDR
604	Ceratherm	A	200–400°, depending on design	BIPTSCHDFR
606	Stoneware	A	All conditions	BIVPTCHDFR
607	Glass-L	A	600° max.	PTCH
610	Stoneware	A	All %; any temp.	BIVPTCHDR
611	Porcelain	A	All % and temp.	BIPTDR
614	Glass-L	A	Under 212°; small capacity units	BTCH
616	Pyrex	A		
617	Stoneware	A	140–160°	BIPTSCHDFR
618	Vitreo	A	1000° max.	
619	Vitreosil	A		
621	Vycor	A		
700	Ace Saran	A	Crystals	P
711	Haveg	X		
715	Heresite	A	200° max.	
716	Heresite	A	200° max.	
720	Lamicoid	A		
726	Nukemite	A	....	PTSDF
733	Permanite	X		
736	Pyroflex	V	....	PTCHD
737	Resilon	X		
740	Saran	F	77°	
741	Sealon	A	160° max.	BPTDF
742	Teflon	A	Rm.	VP
809	Fairprene	X		
817	Heresite	F	Sat.; 77° max.	
836	Natural (S)	V		
837	Natural (H)	V		
838	GR-S (S)	V		
839	GR-S (H)	V		
840	GR-A (S)	V		
841	GR-A (H)	V		
842	GR-M (S)	V		
843	GR-P (S)	X		
844	GR-I	V		
853	Thiokol	X		
854	Thiokol	F	150° max.	PTSCDF

Lactic Acid

No.	Material	Ratings	Exposure conditions	Applications
3	Admiralty	A		
4	Admiralty	A	....	CH
6	Adnic	A		
10–17	Aluminum	AF	All %; rm.	PT
19	Aloyco	A	5%; 150°	
		A	10%; conc.; boil.	
22	Ambraloy	A	....	CH
23	Ambraloy	A	....	CH
24	Ambraloy	A	....	CH
29–40	Ampco	A	....	BIVPTCHDF
51	Berylco	A		
54	Brass	X		
61	Brass	X		
63	Brass	A	....	PCH
66	Bronze	A		
73	Bronze	A	....	VT
74	Bronze	A	....	V
75	Bronze	A	....	VP
76	Bronze	A	....	Instruments
77	Bronze	X		
81	CA-FA20	A	5%; 70°	BIVP
82	CA-MM	A	5%; 70°	BIVP
86A	Cast iron	X		
88	Chlorimet	A	All % and temp.	BIVCHF
89	Chlorimet	A	All % and temp.	BIVHF
111	Copper	A		
114	Copper	A	....	PTCHR
117	Copper	V	....	P
118	Copper	A		
119	Corrosiron	A	25%; rm.; unagitated; technical	BIP
123	Cupro-Ni	A	....	CH
124	Cupro-Ni	A		
139	Durco	A	All % and temp.	BIVHF
140	Durichlor	A	All % and temp.	BIVPCHFR
141, 142	Durimet	A	All % and temp.	BIVHF
143	Duriron	A	All % and temp.	BIVPHFR
148	Everdur	A	....	BIVR
149	Everdur	A	....	PTCHR
150	Everdur	A	....	PCH
156	Gold	A	All temp.	
159	Hastelloy	A	All % to 160°	BIVPTCH
160	Hastelloy	A	All % and temp.	BIVPTCHDF
161	Hastelloy	A	All % and temp.	BIVPTCH
162	Hastelloy	A	All % and temp.	BIVP
163	Stellite	A	All % and temp.	BIV
165	Stellite	A	All % and temp.	BIV
184	Inconel	A	5%; 70°; unaerated	
		A	45%; storage; rm.	
		X	30–60%; evaporator; boil. at 115°	
185	Inc-clad	A	45%; aerated; rm.	BTSCH
216	Monel	A	5%; 70°; unaerated	
		A	45%; storage; rm.	
		X	30–60%; evaporator; boil. at 115°	
217	Monel-clad	A	45%; aerated; rm.	BTSCH
219	Muntz	X		
224	Nickel	A	5%; 70°; unaerated	
		A	45%; storage; rm.	
		X	30–60%; evaporator; boil. at 115°	
225	Ni-clad	A	45%; aerated; rm.	BTSCH
226	Ni-silver	A	....	Instruments
231	Ni-Resist	A	5%; 70°; unaerated	

Table 1. Chemical Resistance of Constructional Materials—(*Continued*)

No.	Material	Ratings	Exposure conditions	Applications
234	Olympic	A		
235	Olympic	A		
236	Palladium	A	All temp.	
240	Platinum	A	All temp.	
242	Ir-Platinum	A	All temp.	
244	Rh-Platinum	A	All temp.	
268	Silver	A	Rm.	
		F	Boil.	
270	304-clad	A	Under 5%; under 150°	BTSCH
271	316-clad	A	Under 5%; under 150°	BTSCH
		A	10%; boil.	
274	430-clad	A	Under 5%; under 150°	BTSCH
275	St. 301	A	...	BIVTSCH
276	St. 302	A	...	BIVTSCH
		A	Sp. gr. 0.96; 70°; boil.	
		A	5% max.; 150° max.	VTSCDF
		X	10%; boil.	
278	St. 303	A	...	IV
279	St. 304	A	5% max.; 150° max.	VTSCDF
		X	10%; boil.	
		F	5%; 70°	BIVP
280	St. 308	A	...	BIVPTSCH
281	St. 309	A	...	BIVTSCH
282	St. 310	A	...	BIVPTSCH
283	St. 316	A	Sp. gr. 0.96; 70°; boil.	
		A	5% max.; 150° max.	VTCHDF
		A	10%; boil.	VTCHDF
		A	5%; 70°	BIVP
284	St. 317	A	...	BIVPTSCH
285	St. 321	A	...	BIVPTSCH
286	St. 347	A	...	BIVPTSCH
287	St. 403	F	Rm.	
288	St. 405	F	Rm.	
290	St. 410	F	Rm.	
		A	Sp. gr. 0.96; 70°; boil.	
		X	5%; 70°	
291	St. 414	F	Rm.	
292	St. 416	F	Rm.	
293	St. 418	F	Rm.	
294	St. 420	F	Rm.	
295	St. 430	A	...	BIVPTSCH
		F	Sp. gr. 0.96; 70°; boil.	
		A	5% max.; 150° max.	VTSCHDF
296	St. 430F	A	...	IV
297	St. 431	A	...	IV
298	St. 440A	A		
300	St. 440C	A	...	IV
301	St. 442	A	...	BIVTSCH
303	St. 446	A	...	BIVPTSCH
360A	Steel	X		
367	SuperNi	A	...	PTCH
368	Tantalum	A	...	Not used commercially
369	Telnic	A		
390	Worthite	A	All %; any temp.	BIV
401	Karbate	A	All %; to boil.	CH
402	Karbate	A	All %; to boil.	CH
403	Kempruf	A	All %; any temp.	T
404	Acheson	A	All %; any temp.	T
500	Sul. cement	A	200° max.	PT
501	Sul. cement	A	200° max.	PT
502	Furan cement	A	360° max.	PT
503	Furan cement	A	360° max.	PT
504	Phen. cement	A	360° max.	PT
505	Phen. cement	A	360° max.	PT
506	Silicate cement	F	1600° max.	T
507	Silicate cement	F	1600° max.	T
508	Acichlor	A	Dilute and conc. 300°	
510	Acitite	A	Dilute and conc. 250°	
513	Asplit	X		
514	Asplit-F	A	All %; 350° max.	
515	Basolit	A	200° max.	T
517	Carboline	A		
518	C-Basolit	A	200° max.	T
521	Causplit	X		
523	Duralon	A	350° max.	BIPTCHDF
524	Durisite	A	350° max.	BIPTCHDF
532	Lumnite	A	1% max.; 90° max.	TD
535	Nukem	A	350° max.	T
536	Nukem	A	...	T
537	Pecomastic	F	Dilute and conc.; 300°	
538	Penchlor	X		
539	Penchlor	X		
540	Penchlor	X		
541	Pennsalt	A	All %; 350° max.	
542	Permanite	A	360° max.	TD
544	Plastite	A	175° max.	PTSD
545	P-Basolit	A	200° max.	T
554	Silastic	A	(A.S.T.M. D-543-43)	
559	Thiokol	A	150° max.	
600	Acid brick	A	...	TDR
603, 612, 615	Glass	A	...	THDR
604	Ceratherm	A	200-400°, depending on design	BIPTSCHDFR
606	Stoneware	A	All conditions	BIVPTCHDFR
607	Glass-L	A	600° max.	PTCH
610	Stoneware	A	All %; any temp.	BIVPTCHDR
611	Porcelain	A	All % and temp.	BIPTDR
614	Glass-L	A	All %; under 302°; agitation	BTCH
616	Pyrex	A		
617	Stoneware	A	140-160°	BIPTSCHDFR
618	Vitreo	A	1000° max.	
619	Vitreosil	A		
621	Vycor	A		
703	Compar	X		
704	DC Silicone	A	...	HD
706	Formica	A	All %; 122° max.	PT
707	Formica	V	All %	
708	Formica	F	All %; rm.	
711	Haveg	A	All conditions	BIVPTDFR
713	Haveg	A	All conditions	BIVPTDFR
715	Heresite	A	200° max.	
716	Heresite	A	200° max.	
718	Koroseal	A	All %; 150° max.	TD
720	Lamicoid	A		
724	Nixon	A	77°	S
726	Nukemite	A	150° max.	PTSDF
733	Permanite	A	360° max.	BVPTD
736	Pyroflex	A	...	PTCHD
737	Resilon	A	175° max.	PTSD
738	Resistoflex	X		
739	Resistoflex	X		
741	Sealon	A	160° max.	BPTDF
742	Teflon	A	Rm.	VP
744, 745	Textolite	V	...	BI
746	Tygon	A	180° max.	BIPTSCHDF
800	Ace Hd. Rub.	A	...	PVBITDF
801	Acidseal	A	All %; 150° max.	BIPTSDF
805	Butyl	F		
809	Fairprene	A		
814	G.E. Silicone	V	...	VPTDF
817	Heresite	A	Sat.; 212° max.	
836	Natural (S)	V		
837	Natural (H)	A		
838	GR-S (S)	V		
839	GR-S (H)	A		
840	GR-A (S)	V		
841	GR-A (H)	A		
842	GR-M (S)	V		
843	GR-P (S)	X		
844	GR-I	V		
846	Saniprene	A	All %; 150° max.	BIPTSDF
848	Silastic	A	(A.S.T.M. D-543-43)	
849	Superflexite	A	All %; 150° max.	BIPTSDF
853	Thiokol	A	150° max.	T
854	Thiokol	A	150° max.	PTSCDF
856	Vistanex	F		
913	Redwood	A	25-27%; 150° max.; 3 years; packing corp.	T

Magnesium Chloride

No.	Material	Ratings	Exposure conditions	Applications
3	Admiralty	A		
4	Admiralty	F	...	CH
6	Adnic	A		
10-17	Aluminum	A	Dry powders	
		F	Sol.; rm.	
19	Aloyco	A	1%; 5%; quiescent, hot	BV
22	Ambraloy	F	...	CH
23	Ambraloy	F	...	CH
24	Ambraloy	F	...	CH
29-40	Ampco	A	...	BIVPTCHDF
54	Brass	X		
61	Brass	X		
63	Brass	AF	...	PCH
66	Bronze	A		
73	Bronze	F	...	VT
74	Bronze	F	...	V
75	Bronze	F	...	VP
76	Bronze	F		
77	Bronze	X		
81	CA-FA20	A	5%; 70°	BIVP
82	CA-MM	A	5%; 70°	BIVP
86	Cast iron	V		
86A	Cast iron	F	Frequently used	
87	Causul	A	...	B
88	Chlorimet	A	All %; and temp.	BIVCHF
89	Chlorimet	A	All %; and temp.	BIVHF

Table 1. Chemical Resistance of Constructional Materials—(*Continued*)

No.	Material	Ratings	Exposure conditions	Applications
111	Copper	A		
114	Copper	F	....	PTCHR
117	Copper	V	....	P
118	Copper	A		
119	Corrosiron	A	25%; rm.; unagitated; technical	BIP
123	Cupro-Ni	F	....	CH
124	Cupro-Ni	A		
139	Durco	A	All % and temp.	BIVHF
140	Durichlor	A	All % and temp.	BIVPCHFR
141, 142	Durimet	A	All % and temp.	BIVHF
143	Duriron	A	140 preferred	BIVPHFR
148	Everdur	F	....	BIVR
149	Everdur	F	....	PTCHDR
150	Everdur	F	....	PCH
159	Hastelloy	A	All % to 160°	BIVPTCH
160	Hastelloy	A	All % to boil.	BIVPTCH
161	Hastelloy	A	All % to 125°	BIVPTCH
162	Hastelloy	A	All %; rm. only	BIV
163	Stellite	A	All %; rm. only	BIV
165	Stellite	A	All %; rm. only	BIV
184	Inconel	A	48%; boil. at 330°	BIVPTCHD
216	Monel	A	48%; boil. at 330°	BIVPTH
217	Monel-clad	A	....	BTSCH
219	Muntz	X		
224	Nickel	A	48%; boil. at 330°	PTH
225	Ni-clad	A	....	BTSCH
226	Ni-silver	F	....	Instruments
231	Ni-Resist	A	48%; boil. at 330°	BIVP
234	Olympic	A		
235	Olympic	A		
245	Pyrasteel	V		
268	Silver	A		
270	304-clad	A	5%; 70°	BTSCH
271	316-clad	A	5%; 70°	BTSCH
275	St. 301	V		
276	St. 302	V		
		F	Solution; hot	
		A	1–5%; 70°	VTSCDF
		X	Hot	
278	St. 303	V		
279	St. 304	V		
		A	1–5%; 70°	VTSCDF
		X	Hot	
		F	5%; 70°	BIVP
280	St. 308	V		
281	St. 309	V		
282	St. 310	V		
283	St. 316	A	....	BIVPTSCH
		F	Solution; hot	
		A	1-5%; 70°	VTCHDF
		X	Hot	
		A	5%; 70°	BIVP
284	St. 317	A	....	BIVPTSCH
285	St. 321	V		
286	St. 347	V		
287	St. 403	V		
288	St. 405	V		
290	St. 410	V		
		X	5%; 70°	
291	St. 414	V		
292	St. 416	V		
293	St. 418	V		
294	St. 420	V		
295	St. 430	V		
		F	Solution; hot	
296	St. 430F	V		
297	St. 431	V		
298	St. 440A	V		
300	St. 440C	V		
301	St. 442	V		
303	St. 446	V		
316	St. CF-20	A	....	BIV
360A	Steel	F	Frequently used	
367	Super Ni	F	....	PTCH
368	Tantalum	A	....	Not used commercially
369	Telnic	A		
390	Worthite	A	274°; 40%	BIV
392	Wyndaloy	A	Very dilute; 68°	
		AV	Moderate dilution; 68°	
		AV	Conc.; 68°	
401	Karbate	A	All % to boil.	PHD
402	Karbate	A	All %; to boil.	PHD
403	Kempruf	A	All %; 1472° max.	PCHD
404	Acheson	A	All %; 1472° max.	PCHD
500	Sul. cement	A	200° max.	PTD
501	Sul. cement	A	200° max.	PTD
502	Furan cement	A	360° max.	PTD
503	Furan cement	A	360° max.	PTD
504	Phen. cement	A	360° max.	PTD
505	Phen. cement	A	360° max.	PTD
506	Silicate cement	F	1600° max.	TD
507	Silicate cement	F	1600° max.	TD
513	Asplit	A	All %	
514	Asplit-F	A		
515	Basolit	A	200° max.	T
518	C-Basolit	A	200° max.	T
521	Causplit	A	All %; 350° max.	
523	Duralon	A	350° max.	BIPTCHDF
524	Durisite	A	350° max.	BIPTCHDF
535	Nukem	A	350° max.	T
536	Nukem	F	....	T
538	Penchlor	A	All %; 750° max.	
539	Penchlor	A	All %; 750° max.	
540	Penchlor	A	All %; 500–2000°	
541	Pennsalt	A	All %; 350° max.	
542	Permanite	A	360° max.	TD
544	Plastite	A	175° max.	PTSD
545	P-Basolit	A	200° max.	T
554	Silastic	A	(A.S.T.M. D–543–43)	
559	Thiokol	A	150° max.	
600	Acid brick	A	....	TDR
603, 612, 615	Glass	A	....	THDR
604	Ceratherm	A	200–400°, depending on design	BIPTSCHDFR
606	Stoneware	A	Not commonly used	
607	Glass-L	V	....	PTCH
610	Stoneware	A	All %; any temp.	BIVPTCHDR
611	Porcelain	A	All %; and temp.	BIPTDR
614	Glass-L	A	All %; 302° max.; agitation	BTCH
616	Pyrex	A		
617	Stoneware	A	140–160°	BIPTSCHDFR
618	Vitreo	A	1000° max.	
621	Vycor	A		
700	Ace Saran	A	....	P
704	DC Silicone	A		
711	Haveg	A	All %; 300° max.	BIVPTDFR
713	Haveg	A	All %; 300° max.	BIVPTDFR
715	Heresite	A	200° max.	
716	Heresite	A	200° max.	
718	Koroseal	A	All %; 150° max.	TD
720	Lamicoid	A		
726	Nukemite	A	....	PTSDF
733	Permanite	A	360° max.	BVPTD
736	Pyroflex	A	....	PTCHD
737	Resilon	A	175° max.	PTSD
741	Sealon	A	160° max.	BPTDF
742	Teflon	A	Rm.	VP
746	Tygon	A	180° max.	BIPTSCHDF
800	Ace Hd. Rub.	A	....	BIVPTDF
801	Acidseal	A	All %; 150° max.	BIPTSDF
802	Acidseal	A	All %; 150° max.	BIPTSDF
805	Butyl	A	Dry salt; aqueous sol.; rm.	
809	Fairprene	A		
817	Heresite	A	Sat.; 212° max.	
835	Perbunan	A	Dry salt; aqueous sol.; rm.	
836	Natural (S)	A		
837	Natural (H)	A		
838	GR-S (S)	A		
839	GR-S (H)	A		
840	GR-A (S)	A		
841	GR-A (H)	A		
842	GR-M (S)	A		
843	GR-P (S)	X		
844	GR-I	A		
846	Saniprene	A	All %; 150° max.	BIPTSDF
848	Silastic	A	(A.S.T.M. D–543–43)	
849	Superflexite	A	All %; 150° max.	BIPTSDF
853	Thiokol	A	150° max.	T
854	Thiokol	A	150° max.	PTSCDF
855	Triflex	A	All %; 150° max.	BIPTSDF
856	Vistanex	A	Dry salt; aqueous sol.; rm.	
913	Redwood	A	30° Bé. bittern; 36–80°; 15 years; insulation factory	T

Mixed Acid

No.	Material	Ratings	Exposure conditions	Applications
3	Admiralty	V		
4	Admiralty	X		
6	Adnic	F	Attacked	
10–17	Aluminum	X		

Table 1. Chemical Resistance of Constructional Materials—(*Continued*)

No.	Material	Ratings	Exposure conditions	Applications
			H_2SO_4 % HNO_3 %	
19	Aloyco	A	50 50 200°	BV
		A	50 50 boil.	BV
		A	75 25 200°	BV
		A	75 25 boil.	BV
		A	70 10 200°	BV
		A	70 10 boil.	BV
		A	30 5 boil.	BV
		A	30 5 200°	BV
		A	15 5 200°	BV
		A	15 5 boil.	BV
		A	90% acetic, 2% H_2SO_4, bal. H_2O; 150°	BV
22	Ambraloy	X		
23	Ambraloy	X		
24	Ambraloy	X		
29–40	Ampco	X		
51	Berylco	X		
54	Brass	X		
61	Brass	X		
63	Brass	VX		
66	Bronze	V		
73	Bronze	X		
74	Bronze	X		
75	Bronze	X		
76	Bronze	X		
77	Bronze	X		
81	CA-FA20	A	57 28 176°	BIVP
82	Ca-MM	X	57 28 176°	
86	Cast iron	A	Nitrating acids; water 20% max.; sulfuric 15% min.	
86A	Cast iron	A	Under 20% water; over 15% sulfuric	
88	Chlorimet	V	....	BIVCHF
89	Chlorimet	A	All % and temp.	BIVHF
111	Copper	V		
114	Copper	X		
118	Copper	V		
123	Cupro-Ni	X		
124	Cupro-Ni	V		
139	Durco	A	All % and temp.	BIVHF
140	Durichlor	A	All % and temp.	BIVPCHFR
141	Durimet	A	All % and temp.	BIVHF
142	Durimet	A	All % to 200 F.	BIVHF
143	Duriron	A	All % and temp.	BIVPHFR
148	Everdur	X		
149	Everdur	X		
150	Everdur	X		
156	Gold	A	To boil; all proportions	
159	Hastelloy	X	All % and temp.	
160	Hastelloy	X	All % and temp.	
161	Hastelloy	A	All % to 150°	BIVP
162	Hastelloy	X	All % and temp.	
163	Stellite	A	All % to 150°	BIV
165	Stellite	A	All % to 150°	BIV
219	Muntz	X		
226	Ni-silver	X		
234	Olympic	V		
235	Olympic	V		
240	Platinum	A	To boil.; all proportions	
242	Ir-Platinum	A	To boil.; all proportions	
244	Rh-Platinum	A	To boil.; all proportions	
			H_2SO_4 % HNO_3 %	
275	St. 301	V		
276	St. 302	V		
		A	50 50 140°; 200°	
		F	50 50 250° (boil.)	
		A	75 25 140°; 200°	
		F	75 25 315° (boil.)	
		A	70 10 140°; 200°	
		X	70 10 335° (boil.)	
		A	30 5 140°; 200°; 230° (boil.)	
		A	15 5 140°; 200°; 220° (boil.)	
278	St. 303	V		
279	St. 304	V		
		A	57 28 176°	BIVP
280	St. 308	V		
281	St. 309	V		
282	St. 310	V		
283	St. 316	V		
		A	50 50 140°; 200°	
		F	50 50 250° (boil.)	
		A	75 25 140°; 200°	
		F	75 25 315° (boil.)	
		A	70 10 140°; 200°	
		X	70 10 335° (boil.)	
		A	30 5 140°; 200°; 230° (boil.)	
		A	15 5 140°; 200°; 220° (boil.)	
		A	57 28 176°	BIVP
284	St. 317	V		
285	St. 321	V		
286	St. 347	V		
287	St. 403	V		
288	St. 405	V		
290	St. 410	V		
		F	57 28 176°	BIVP
291	St. 414	V		
292	St. 416	V		
293	St. 418	V		
294	St. 420	V		
295	St. 430	V		
296	St. 430F	V	Sp. gr. 1.42; all %; rm.	
297	St. 431	V		
298	St. 440A	V		
300	St. 440C	V		
301	St. 442	V		
303	St. 446	V		
360A	Steel	A	Under 20% water; over 15% sulfuric	
367	Super Ni	X		
368	Tantalum	V	Suitable at temp. up to 338° if SO_3 is absent	CH
369	Telnic	V		
390	Worthite	A	175°; 61% sulfuric + 36% nitric	BIV
401	Karbate	X		
402	Karbate	X		
403	Kempruf	X		
404	Acheson	X		
513	Asplit	X		
514	Asplit-F	X		
515	Basolit	X		
518	C-Basolit	X		
521	Causplit	X		
523	Duralon	X		
524	Durisite	X		
535	Nukem	X		
536	Nukem	V	....	T
538	Penchlor	A	All %; 750° max.	
539	Penchlor	A	All %; 750° max.	
540	Penchlor	A	All %; 500–2000°	
541	Pennsalt	X		
542	Permanite	V	Dil.; cold	TD
		X	Conc.; hot	
544	Plastite	X		
545	P-Basolit	X		
556	Staminite	V	Except HF mixtures	T
559	Thiokol	X		
600	Acid brick	A	....	TDR
603, 612, 615	Glass	A	....	THDR
604	Ceratherm	A	200–400°, depending on design	BIPTSCHDFR
606	Stoneware	A	All conditions	BIVPTCHDFR
607	Glass-L	A	600° max.	PTCH
610	Stoneware	A	All %; any temp.	BIVPTCHDR
611	Porcelain	A	All % and temp.	BIPTDR
614	Glass-L	A	Under 302°; agitation	BTCH
616	Pyrex	A		
617	Stoneware	A	140–160°	BIPTSCHDFR
618	Vitreo	A	1000° max.	
619	Vitreosil	A	....	PCH
621	Vycor	A		
703	Compar	X		
711	Haveg	V	Unsuitable for strong oxidizing acids	BIVPTDFR
713	Haveg	V	Unsuitable for strong oxidizing acids	BIVPTDFR
715	Heresite	A	200° max.	

Table 1. Chemical Resistance of Constructional Materials—(*Continued*)

No.	Material	Ratings	Exposure conditions	Appications
716	Heresite	A	200° max.	
718	Koroseal	V	All %; 150° max.	TD
726	Nukemite	X		
727	Nylon	X	Conc.	
728	Nylon	X	Conc.	
729–732	Nylon	X		
733	Permanite	V	Dil.; cold	BVPDT
		X	Conc.; hot	
736	Pyroflex	V	....	PTCHD
737	Resilon	X		
738	Resistoflex	X		
739	Resistoflex	X		
741	Sealon	A	160° max.	BPTDF
742	Teflon	A	Rm.	VP
744, 745	Textolite	X	Conc.	
746	Tygon	V	180° max.	BIPTSCHDF
800	Ace Hd. Rub.	V	....	BIVPTDF
801	Acidseal	V	All %; 150° max.	BIPTSDF
802	Acidseal	V	All %; 150° max.	BIPTSDF
814	G.E. Silicone	X		
817	Heresite	X		
836	Natural (S)	X		
837	Natural (H)	X		
838	GR-S (S)	X		
839	GR-S (H)	X		
840	GR-A (S)	X		
841	GR-A (H)	X		
842	GR-M (S)	X		
843	GR-P (S)	X		
844	GR-I	V		
846	Saniprene	V	All %; 150° max.	BIPTSDF
849	Superflexite	V	All %; 150° max.	BIPTSDF
853	Thiokol	X	150° max.	
854	Thiokol	V	150° max.	PTSCDF
855	Triflex	V	All %; 150° max.	BIPTSDF
856	Vistanex	V		

Nitric Acid

No.	Material	Ratings	Exposure conditions	Applications
3	Admiralty	X		
4	Admiralty	X		
6	Adnic	X	Severe attack	
10–17	Aluminum	A	Over 80%	BIPTSCHDF
		FX	Under 80%	
22	Ambraloy	X		
23	Ambraloy	X		
24	Ambraloy	X		
29–40	Ampco	X		
51	Berylco	X		
54	Brass	X		
61	Brass	X		
63	Brass	X		
66	Bronze	X		
73	Bronze	X		
74	Bronze	X		
75	Bronze	X		
76	Bronze	X		
77	Bronze	X		
81	CA-FA20	A	65%; boil.	BIVP
82	CA-MM	X	65%; boil.	
86A	Cast iron	X		
88	Chlorimet	X		
89	Chlorimet	V	....	BIVHF
111	Copper	X		
114	Copper	X		
118	Copper	X		
119	Corrosiron	A	10, 25, 70%; rm.; unagitated; c.p.	BIVPTCHDFR
123	Cupro-Ni	X		
124	Cupro-Ni	X		
139	Durco	V	All % (except strong) and temp.	BIVHF
140	Durichlor	A	All % and temp.	BIVPCHFR
141, 142	Durimet	A	All % and temp.	BIVHF
143	Duriron	A	All % and temp.	BIVPHFR
148	Everdur	X		
149	Everdur	X		
150	Everdur	X		
156	Gold	A	To boil.	
159	Hastelloy	X	All % and temp.	
160	Hastelloy	X	All % and temp.	
161	Hastelloy	A	All % to 150°	BIV
162	Hastelloy	X	All % and temp.	
163	Stellite	A	All % to 150°	BIV
165	Stellite	A	All % to 150°	BIV
184	Inconel	F	5%; 70°; aerated	
		A	65%; 70°	
216	Monel	X	5%; 70°; aerated	
219	Muntz	X		
224	Nickel	X	5%; 70°; aerated	
226	Ni-silver	X		
231	Ni-Resist	X	5%; 70°; aerated	
234	Olympic	X		
235	Olympic	X		
240	Platinum	A	To boil.	
242	Ir-Platinum	A	To boil.	
244	Rh-Platinum	A	To boil.	
245	Pyrasteel	V		
275	St. 301	A	All %; rm.	BIVTSCHDF
		A	65%; boil.	BIVTSCHDF
276	St. 302	A	All %; rm.	BIVTSCHDF
		A	65%; boil.	BIVTSCHDF
		A	Diluted 1:10; 70°; boil.	
		A	Diluted 1:1; 70° boil.	
		A	Sp. gr. 1.40; conc.; 70°; boil.	
		A	Sp. gr. 1.52; conc.; fuming 70°; boil.	
		A	50% max.; boil.	VTSCDF
		A	65%; boil.	VTSCDF
		A	All %; 70°	VTSCDF
		X	Conc.; boil.	
		A	All %	
278	St. 303	A	All %; rm.	IVF
		A	65%; boil.	IVF
279	St. 304	A	All %; rm.	BIVPTSCHDF
		A	65%; boil.	BIVPTSCHDF
		A	50% max.; boil.	VTSCDF
		A	All %; 70°	VTSCDF
		X	Conc.; boil.	DF
		A	65%; boil.	BIVP
280	St. 308	A	All %; rm.	BIVPTSCH
		A	65%; boil.	BIVPTSCH
281	St. 309	A	All %; rm.	BIVTSCHDF
		A	65%; boil.	BIVTSCHDF
282	St. 310	A	All %; rm.	BIVPTSCHDF
		A	65%; boil.	BIVPTSCHDF
283	St. 316	A	All %; rm.	BIVPTSCHDF
		A	65%; boil.	BIVPTSCHDF
		A	Diluted 1:10; 70°; boil.	
		A	Diluted 1:1; 70°; boil.	
		A	Sp. gr. 1.40; conc.; 70°; boil.	
		A	Sp. gr. 1.52; conc. fuming; 70°	
		F	Sp. gr. 1.52; conc. fuming; boil.	
		A	50% max.; boil.	VTCHDF
		A	65%; boil.	VTCHDF
		A	All %; boil.	VTCHDF
		X	Conc.; boil.	
		F	65%; boil.	BIVP
284	St. 317	A	All %; rm.	BIVPTSCHDF
		A	65%; boil.	
285	St. 321	A	All %; rm.	BIVPTSCHDF
		A	65%; boil.	BIVPTSCHDF
286	St. 347	A	All %; rm.	BIVPTSCHDF
		A	65%; boil.	BIVPTSCHDF
287	St. 403	V	All %; rm.	
288	St. 405	V	All %; rm.	
290	St. 410	V	All %; rm.	
		A	Diluted 1:10; 70°	
		A	Diluted 1:1; 70°	
		A	All %	
		X	65%; boil.	
291	St. 414	V	All %; rm.	
292	St. 416	V	All %; rm.	
		A	All %	
293	St. 418	V	All %; rm.	
294	St. 420	V	All %; rm.	
		A	All %	
295	St. 430	V	All %; 70°	BIVPTSCHDF
		A	Diluted 1:10; 70°; boil.	
		A	Diluted 1:1; 70°; boil.	
		A	Sp. gr. 1.40; conc.; 70°	
		F	Sp. gr. 1.40; conc.; boil.	
		A	50% max.; boil.	VTSCHDF
		X	65%; boil.	
		A	All %; boil.	VTSCHDF
		X	Conc.; boil.	
		A	All %	
296	St. 430F	V		
297	St. 431	V	Sp. gr. 1.42; all %; 70°	
		V	65%; boil.	
298	St. 440A	V	Sp. gr. 1.42; all %; rm.	
300	St. 440C	V	Sp. gr. 1.42; all %; rm.	
301	St. 442	V	Sp. gr. 1.42; all %; rm.; 65%; boil.	BIVTSCH
303	St. 446	V	Sp. gr. 1.42; all %; rm.; 65%; boil.	BIVPTSCHDF

Table 1. Chemical Resistance of Constructional Materials—(*Continued*)

No.	Material	Ratings	Exposure conditions	Applications
313	St. CF-7	A	All %; boil.	
316	St. CF-20	A	....	BIV
319	St. CF-7M	A	All %	
360A	Steel	X		
364	Stoody	A	10%; boil.	
		F	Conc.; boil.	
365	Stoody	A	10%; boil.	
		F	Conc.; boil.	
367	Super Ni	X		
368	Tantalum	A	All %; 392° max.	BCH
369	Telnic	V		
387	Stainless	A	All %	
388	Stainless	X	All %	
389	Stainless	A	All %	
390	Worthite	A	All %; 175°	BIV
392	Wyndaloy	V	Very dilute; 68°	
		X	Moderate dilution; 68°	
		X	Conc.; 68°	
401	Karbate	V	40% max.	BIVPTCH
402	Karbate	V	40% max.	BIVPTCH
403	Kempruf	V	40% max.	T
404	Acheson	V	40% max.	T
500	Sul. cement	A	30% max.; 200° max.	PTD
501	Sul. cement	A	30% max.; 200° max.	PTD
502	Furan cement	X		
503	Furan cement	X		
504	Phen. cement	X		
505	Phen. cement	X		
506	Silicate cement	A	1600° max.	PTD
507	Silicate cement	A	1600° max.	PTD
508	Acichlor	F	Dilute and conc.; 300°	
510	Acitite	F	Dilute and conc.; 250°	
513	Asplit	X		
514	Asplit-F	X		
515	Basolit	A	....	T
518	C-Basolit	A	....	T
521	Causplit	X		
523	Duralon	X		
524	Durisite	X		
534	N-series	X		
535	Nukem	X		
536	Nukem	A	....	T
537	Pecomastic	V	Dilute and conc.; 300°	
538	Penchlor	A	All %; 750° max.	
539	Penchlor	A	All %; 750° max.	
540	Penchlor	A	All %; 500–2000°	
541	Pennsalt	X		
542	Permanite	X		
544	Plastite	F	Dil.; 125° max.	PTSD
545	P-Basolit	A	....	T
554	Silastic	A	10%;(A.S.T.M. D-543-43)	
		X	Conc.; (A.S.T.M. D-543-43)	
556	Staminite	A	All %	T
559	Thiokol	X		
600	Acid brick	A	....	TDR
603, 612, 615	Glass	A	....	THDR
604	Ceratherm	A	200–400°, depending on design	BIPTSCHDFR
606	Stoneware	A	All conditions	BIVPTCHDFR
607	Glass-L	A	600° max.	PTCH
610	Stoneware	A	All %; any temp.	BIVPTCHDR
611	Porcelain	A	All % and temp.	BIPTDR
614	Glass-L	A	All %; under 302°; agitation	BTCH
616	Pyrex	A		
617	Stoneware	A	140–160°	BIPTSCHDFR
618	Vitreo	A	1000° max.	
619	Vitreosil	A	....	PCH
621	Vycor	A		
700	Ace Saran	A	77°	P
703	Compar	X		
704	DC Silicone	A	10%	
707	Formica	F	5% max.; rm.	
708	Formica	X	All %	
711	Haveg	V	5%; 125° max.	BIVPTDFR
715	Heresite	A	200° max.	
718	Koroseal	A	35% max.; 70° max.	TD
		A	10% max.; 150°	TD
720	Lamicoid	X		
723	Nixon	X	10%; 77°	
724	Nixon	A	10%; 77°	S
725	Nixon	A	10%; 77°	S
733	Permanite	X		
735	Polythene	X	Conc.; hot	
736	Pyroflex	V	....	PTCHD
737	Resilon	F	Dil.; 125° max.	PTSD
738	Resistoflex	X		
739	Resistoflex	X		
740	Saran	F	77°	
741	Sealon	A	160° max.	BPTDF
742	Teflon	A	Rm.	VP
744, 745	Textolite	V	....	BI
746	Tygon	F	25% max.; 140° max.	BIPTSCHDF
800	Ace Hd. Rub.	A	16° Bé.; rm.	BIVPTDF
801	Acidseal	A	20% max.; 70° max.	BIPTSDF
805	Butyl	V	35%; rm.	
809	Fairprene	X		
814	G.E. Silicone	X	Conc.	VPTDF
817	Heresite	A	20% max.; 77° max.	
836	Natural (S)	X		
837	Natural (H)	V		
838	GR-S (S)	X		
839	GR-S (H)	V		
840	GR-A (S)	X		
841	GR-A (H)	V		
842	GR-M (S)	X		
843	GR-P (S)	X		
844	GR-I	A		
846	Saniprene	A	20% max.; 70° max.	BIPTSDF
848	Silastic	A	10%;(A.S.T.M. D-543-43)	
		A	Conc.; (A.S.T.M. D-543-43)	
849	Superflexite	A	20% max.; 70° max.	BIPTSDF
853	Thiokol	X		
854	Thiokol	V	150° max.	PTSCDF
856	Vistanex	V	35%; rm.	
913	Redwood	A	5–8% + dil. H_2SO_4; asphalt-L; 21 years; explosives mfr.	T
		A	1%; 40–90°; acid res. brick-L; chemical plant	T
		A	Very dil.; 70–80°; 19 years; explosives mfr.	T

Oleic Acid

No.	Material	Ratings	Exposure conditions	Applications
3	Admiralty	A		
4	Admiralty	A	....	CH
6	Adnic	A		
10–17	Aluminum	A	Rm.	BIPTSCHDF
19	Aloyco	A	300°, 400°	BV
22	Ambraloy	A	....	CH
23	Ambraloy	A	....	CH
24	Ambraloy	A	....	CH
29–40	Ampco	A	....	BIVPTCHDF
54	Brass	F	....	CH
61	Brass	F	....	VCH
63	Brass	A	....	PCH
66	Bronze	A		
73	Bronze	A	....	VT
74	Bronze	A	....	V
75	Bronze	A	....	VP
76	Bronze	A	....	Instruments
77	Bronze	F	....	PCH
86	Cast iron	V		
86A	Cast iron	F		
87	Causul	A	Any temp.	B
88	Chlorimet	A	All % and temp.	BIVCHF
89	Chlorimet	A	All % and temp.	BIVHF
111	Copper	A		
114	Copper	A	....	PTCHR
117	Copper	V	....	P
118	Copper	A		
123	Cupro-Ni	A	....	CH
124	Cupro-Ni	A		
139	Durco	A	All % and temp.	BIVHF
140	Durichlor	A	All % and temp.	BIVPCHFR
141, 142	Durimet	A	All % and temp.	BIVHF
143	Duriron	A	All % and temp.	BIVPHFR
148	Everdur	A	....	BIVR
149	Everdur	A	....	PTCHDR
150	Everdur	A	....	PCH
156	Gold	A	To boil.	
159	Hastelloy	A	All % and temp.	
160	Hastelloy	A	All % and temp.	
161	Hastelloy	A	All % and temp.	
162	Hastelloy	A	All % and temp.	
163	Stellite	A	All % and temp.	BIV
165	Stellite	A	All % and temp.	BIV
184	Inconel	A	Boil.	BIVPTCHR
185	Inc-clad	A	70°	BTSCH
191	Lead	A	Conc.; about 200° max.; agitated; aerated	PTH
216	Monel	A	Boil.	BIVPTCH
217	Monel-clad	A	70°	BTSCH
219	Muntz	F	....	CH
224	Nickel	A	Boil.	PTH
225	Ni-clad	A	70°	BTSCH

Table 1. Chemical Resistance of Constructional Materials—(*Continued*)

No.	Material	Ratings	Exposure conditions	Applications
226	Ni-silver	A		Instruments
231	Ni-Resist	A	Boil.	BIVP
234	Olympic	A		
235	Olympic	A		
236	Palladium	A	To boil.	
240	Platinum	A	To boil.	
242	Ir-Platinum	A	To boil.	
244	Rh-Platinum	A	To boil.	
245	Pyrasteel	A		BIVPTSCHF
249	Resistac	F		BIVCHF
268	Silver	A	To boil.	
270	304-clad	A	70°	BTSCH
271	316-clad	A	70°	BTSCH
274	430-clad	A	70°	BTSCH
275	St. 301	A		BIVTSCDF
276	St. 302	A		BIVTSC
		A	Raw; 300°; 400°	
		A	70°	VTSCDF
278	St. 303	A		IVF
279	St. 304	A		BIVPTSCHDF
		A	70°	VTSCDF
280	St. 308	A		BIVPTSCD
281	St. 309	A		BIVTSCDF
282	St. 310	A		BIVPTSCDF
283	St. 316	A		BIVPTSC
		A	Raw; 300°; 400°	
		A	70°	VTCHDF
284	St. 317	A		BIVPTSCDF
285	St. 321	A		BIVPTSCDF
286	St. 347	A		BIVPTSCDF
287	St. 403	A		IV
288	St. 405	A		BIVTSC
290	St. 410	A		BIVPTSCDF
291	St. 414	A		IV
292	St. 416	A		IV
293	St. 418	A		
294	St. 420	A		
295	St. 430	A		BIVPTSCDF
		A	Raw; 300°	
		A	70°	VTSCHDF
296	St. 430F	A		IV
297	St. 431	A		IV
298	St. 440A	A		
300	St. 440C	A		IV
301	St. 442	A		BIVTSC
303	St. 446	A		BIVPTSC
316	St. CF-20	A		BIV
360A	Steel	F		
367	Super Ni	A		PTCH
368	Tantalum	A		Not used commercially
369	Telnic	A		
390	Worthite	A	All %; any temp.	BIV
401	Karbate	A	All %; to boil.	PTD
402	Karbate	A	All %; to boil.	PTD
403	Kempruf	A	All %; any temp.	BIVPTCHDF
404	Acheson	A	All %; any temp.	BIVPTCHDF
500	Sul. cement	F	Low temp. only	T
501	Sul. cement	F	Low temp. only	T
502	Furan cement	A	360° max.	PT
503	Furan cement	A	360° max.	PT
504	Phen. cement	A	360° max.	PT
505	Phen. cement	A	360° max.	PT
506	Silicate cement	F	1600° max.	T
507	Silicate cement	F	1600° max.	T
513	Asplit	A	All %; 350° max.	
514	Asplit-F	A	All %; 350° max.	
515	Basolit	X		
518	C-Basolit	X		
521	Causplit	A	All %; 350° max.	
523	Duralon	A	350° max.	BIPTCHDF
524	Durisite	A	350° max.	BIPTCHDF
534	N-series	A	212° max.	BIVPTCDFR
535	Nukem	A	350° max.	T
536	Nukem	A		T
538	Penchlor	A	All %; 750° max.	
539	Penchlor	A	All %; 750° max.	
540	Penchlor	A	All %; 500–2000°	
541	Pennsalt	A	All %; 350° max.	
542	Permanite	A	360° max.	TD
544	Plastite	A	175° max.	PTSD
545	P-Basolit	X		
554	Silastic		(A.S.T.M. D-543-43)	
556	Staminite	A	All %	T
559	Thiokol	A	150° max.	
600	Acid brick	A		TDR
603, 612, 615	Glass	A		THDR
604	Ceratherm	A	200–400°, depending on design	BIPTSCHDFR
606	Stoneware	A	All conditions	BIVPTCHDFR
607	Glass-L	A	600° max.	PTCH
610	Stoneware	A	All %; any temp.	BIVPTCHDR
611	Porcelain	A	All % and temp.	BIPTDR
614	Glass-L	A	Under 302°; agitation	BTCH
616	Pyrex	A		
617	Stoneware	A	140–160°	BIPTSCHDFR
618	Vitreo	A	1000° max.	
621	Vycor	A		
700	Ace Saran	A		P
703	Compar	A	200° max.	T
704	DC Silicone	A		
706	Formica	A	All %; 122° max.	PT
707	Formica	V	All %	
708	Formica	F	All %; rm.	
711	Haveg	A	All conditions	BIVPTDFR
713	Haveg	A	All conditions	BIVPTDFR
715	Heresite	A	200° max.	
718	Koroseal	A	Any %; 100° max.	TD
720	Lamicoid	A		T
723	Nixon	A		S
724	Nixon	A		S
726	Nukemite	A	150° max.	PTSDF
727	Nylon	A	Dilute	
728	Nylon	A	Dilute	
729–732	Nylon	X		
733	Permanite	A	360° max.	PVBTD
736	Pyroflex	V		PTCHD
737	Resilon	A	175° max.	PTSD
738	Resistoflex	A	Conc.; 275° max.	P
739	Resistoflex	A	Conc.; 140° max.	IVP
740	Saran	A		
741	Sealon	A	160° max.	BPTDF
742	Teflon	A	Rm.	VP
744, 745	Textolite	F	Dil.	BI
746	Tygon	A	180° max.	BIPTSCHDF
801	Acidseal	F	All %; 100° max.	BIPTSDF
805	Butyl	F	Rm.	
809	Fairprene	V		
814	G.E. Silicone	V		VPTDF
817	Heresite	A	Sat.; 212° max.	
835	Perbunan	A	Rm.	
836	Natural (S)	X		
837	Natural (H)	V		
838	GR-S (S)	X		
839	GR-S (H)	V		
840	GR-A (S)	V		
841	GR-A (H)	V		
842	GR-M (S)	X		
843	GR-P (S)	V		
844	GR-I	X		
846	Saniprene	F	All %; 100° max.	BIPTSDF
848	Silastic	A	(A.S.T.M. D-543-43)	
849	Superflexite	F	All %; 100° max.	BIPTSDF
853	Thiokol	A	150° max.	T
854	Thiokol	A	150° max.	PTSCDF
856	Vistanex	F	Rm.	

Oxalic Acid

No.	Material	Ratings	Exposure conditions	Applications
3	Admiralty	A		
4	Admiralty	A		CH
6	Adnic	A		
10–17	Aluminum	V		
		A	Dry powders	
22	Ambraloy	A		CH
23	Ambraloy	A		CH
24	Ambraloy	A		CH
29–40	Ampco	A		BIVPTCHDF
51	Berylco	A	Cold	
54	Brass	X		
61	Brass	X		
63	Brass	A		PCH
66	Bronze	A		
73	Bronze	A		VT
74	Bronze	A		V
75	Bronze	A		VP
76	Bronze	A		Instruments
77	Bronze	X		
86	Cast iron	V		
86A	Cast iron	F		
89	Chlorimet	A		BIVHF
111	Copper	A		
114	Copper	A		PTCHR
117	Copper	V		P
118	Copper	A		
119	Corrosiron	A	2.1, 7.9%; rm.; unagitated; c.p.	BIPF
123	Cupro-Ni	A		CH

Table 1. Chemical Resistance of Constructional Materials—(*Continued*)

No.	Material	Ratings	Exposure conditions	Applications
124	Cupro-Ni	A		
139	Durco	A	All % and temp.	BIVHF
140	Durichlor	A	All % and temp.	BIVPCHFR
141, 142	Durimet	A	All % and temp.	BIVHF
143	Duriron	A	All % and temp.	BIVPHFR
148	Everdur	A	....	BIVR
149	Everdur	A	....	PTCHDR
150	Everdur	A	....	PCH
156	Gold	A		
159	Hastelloy	A	All % and temp.	
160	Hastelloy	A	All % and temp.	
161	Hastelloy	A	All % and temp.	
162	Hastelloy	A	All % and temp.	
163	Stellite	A	All % and temp.	BIV
165	Stellite	A	All % and temp.	BIV
184	Inconel	A	20–50%; 100–175°	
216	Monel	A	20–50%; 100–175°	VPT
217	Monel-clad	A	5%; any temp.	BTSCH
219	Muntz	X		
224	Nickel	F	20–50%; 100–175°	
226	Ni-silver	A	....	Instruments
231	Ni-Resist	X	20–50%; 100–175°	
234	Olympic	A		
235	Olympic	A		
236	Palladium	A		
240	Platinum	A		
242	Ir-Platinum	A		
244	Rh-Platinum	A		
270	304-clad	A	10%; 70°	BTSCH
		A	5%; any temp.	BTSCH
271	316-clad	A	10%; 70°	BTSCH
		A	5%; any temp.	BTSCH
274	430-clad	A	5%; any temp.	BTSCH
275	St. 301	V		
276	St. 302	A	10%; 70°	
		F	10%; boil.	
		F	25%; boil.	
		F	50%; boil.	
		A	5%; any temp.	VTSCDF
		A	10%; 70°	VTSCDF
		X	10%; boil.	
278	St. 303	V		
279	St. 304	V		
		A	5%; any temp.	VTSCDF
		A	10%; 70°	VTSCDF
		X	10%; boil.	
280	St. 308	V		
281	St. 309	V		
282	St. 310	V		
283	St. 316	V		
		A	10%; 70°	
		F	10%; boil.	
		F	25%; boil.	
		F	50%; boil.	
		A	5%; any temp.	VTCHDF
		A	10%; 70°	VTCHDF
		X	10%; boil.	
284	St. 317	V		
285	St. 321	V		
286	St. 347	V		
287	St. 403	V		
288	St. 405	V		
290	St. 410	V		
291	St. 414	V		
292	St. 416	V		
293	St. 418	V		
294	St. 420	V		
295	St. 430	V		
		A	5%; any temp.	VTSCHDF
		A	10%; 70°	VTSCHDF
		X	10%; boil.	
296	St. 430F	V		
297	St. 431	V		
298	St. 440A	V		
300	St. 440C	V		
301	St. 442	V		
303	St. 446	V		
319	St. CF-7M	A	10%; 212°	
360A	Steel	F		
367	Super Ni	A	....	PTCH
368	Tantalum	A	....	Not used commercially
369	Telnic	A		
390	Worthite	A	175°	BIV
401	Karbate	A	All %; to boil.	PS
402	Karbate	A	All %; to boil.	PS
403	Kempruf	A	All %; any temp.	
404	Acheson	A	All %; any temp.	
500	Sul. cement	A	200° max.	PT
501	Sul. cement	A	200° max.	PT
502	Furan cement	A	360° max.	PT
503	Furan cement	A	360° max.	PT
504	Phen. cement	A	360° max.	PT
505	Phen. cement	A	360° max.	PT
506	Silicate cement	A	1600° max.	PT
507	Silicate cement	A	1600° max.	PT
508	Acichlor	A	Dilute and conc.; 300°	
510	Acitite	A	Dilute and conc.; 250°	
513	Asplit	A	All %; 350° max.	
514	Asplit-F	A	All %; 350° max.	
515	Basolit	A	200° max.	T
517	Carboline	A		
518	C-Basolit	A	200° max.	T
521	Causplit	A	All %; 350° max.	
523	Duralon	A	350° max.	BIPTCHDF
524	Durisite	A	350° max.	BIPTCHDF
535	Nukem	A	350° max.	T
536	Nukem	A	....	T
537	Pecomastic	X		
538	Penchlor	A	All %; 750° max.	
539	Penchlor	A	All %; 750° max.	
540	Penchlor	A	All %; 500–2000°	
541	Pennsalt	A	All %; 350° max.	
542	Permanite	A	360° max.	TD
544	Plastite	A	175° max.	PTSD
545	P-Basolit	A	200° max.	T
554	Silastic	A	(A.S.T.M. D-543-43)	
556	Staminite	A	All %	T
559	Thiokol	A	150° max.	
600	Acid brick	A	....	TDR
603, 612, 615	Glass	A	....	THDR
604	Ceratherm	A	200–400°, depending on design	BIPTSCHDFR
606	Stoneware	A	All conditions	BIVPTCHDFR
607	Glass-L	A	600° max.	PTCH
610	Stoneware	A	All %; any temp.	BIVPTCHDR
611	Porcelain	A	All % and temp.	BIPTDR
614	Glass-L	A	All %; 77° max.; agitation	BTCH
616	Pyrex	A		
617	Stoneware	A	140–160°	BIPTSCHDFR
618	Vitreo	A	1000° max.	
619	Vitreosil	A		
621	Vycor	A		
700	Ace Saran	A	....	P
704	DC Silicone	A		
706	Formica	A	All %; 122° max.	PT
707	Formica	V	All %	
708	Formica	F	All %; rm.	
711	Haveg	A	All conditions	BIVPTDFR
713	Haveg	A	All conditions	BIVPTDFR
715	Heresite	A	200° max.	
718	Koroseal	..	All %; 100° max.	TD
720	Lamicoid	A	....	T
726	Nukemite	A	150° max.	PTSDF
727	Nylon	A	Dilute	
728	Nylon	A	Dilute	
729–732	Nylon	X		
733	Permanite	A	360° max.	PVBTD
736	Pyroflex	V	....	PTCHD
737	Resilon	A	175° max.	PTSD
741	Sealon	A	160° max.	BPTDF
742	Teflon	A	Rm.	VP
744, 745	Textolite	V	....	BI
746	Tygon	A	180° max.	BIPTSCHDF
800	Ace Hd. Rub.	A	....	BIVPTDF
801	Acidseal	..	All %; 100° max.	BIPTDF
809	Fairprene	V		
814	G.E. Silicone	V	Rm.	VPTDF
817	Heresite	A	Sat.; 212°	
836	Natural (S)	V		
837	Natural (H)	V		
838	GR-S (S)	V		
839	GR-S (H)	V		
840	GR-A (S)	V		
841	GR-A (H)	V		
842	GR-M (S)	V		
843	GR-P (S)	X		
844	GR-I	V		
846	Saniprene	..	All %; 100° max.	BIPTDF
848	Silastic	..	(A.S.T.M. D-543-43)	
849	Superflexite	..	All %; 100° max.	BIPTDF
853	Thiokol	A	150° max.	T
854	Thiokol	A	150° max.	PTSCDF

Phenol

No.	Material	Ratings	Exposure conditions	Applications
3	Admiralty	V		
4	Admiralty	F	....	CH
6	Adnic	A		
10–17	Aluminum	A	Rm.–212°; moisture inhibits action	VPFSCHD

Table 1. Chemical Resistance of Constructional Materials—(*Continued*)

No.	Material	Ratings	Exposure conditions	Applications
22	Ambraloy	F	...	CH
23	Ambraloy	F	...	CH
24	Ambraloy	F	...	CH
29–40	Ampco	A	If contamination is no factor	BIVPTCHDF
54	Brass	F	...	CH
61	Brass	F	...	VCH
63	Brass	FV	...	PCH
66	Bronze	V		
73	Bronze	F	...	VT
74	Bronze	F	...	V
75	Bronze	F	...	VP
76	Bronze	F		
77	Bronze	F	...	PCH
85	Cast iron	A	Conc.	
86A	Cast iron	A	Where color is not important; do not use with c.p. acid	
88	Chlorimet	A	All % and temp.	BIVCHF
89	Chlorimet	A	All % and temp.	BIVHF
111	Copper	V		
114	Copper	F	...	PTCHR
117	Copper	V	...	P
118	Copper	V		
123	Cupro-Ni	F	...	CH
124	Cupro-Ni	V		
127–133	Dowmetal	A	248°	
		X	356°	
139	Durco	A	All % and temp.	BIVHF
140	Durichlor	A	All % and temp.	BIVPCHFR
141, 142	Durimet	A	All % and temp.	BIVHF
143	Duriron	A	All % and temp.	BIVPHFR
148	Everdur	F	...	BIVR
149	Everdur	F	...	PTCHDR
150	Everdur	F	...	PCH
156	Gold	A		
159	Hastelloy	A	All % and temp.	
160	Hastelloy	A	All % and temp.	
161	Hastelloy	A	All % and temp.	
162	Hastelloy	A	All % and temp.	
163	Stellite	A	All % and temp.	BIV
165	Stellite	A	All % and temp.	BIV
184	Inconel	A	90%; boil.	PTCH
185	Inc-clad	A	All %; any temp.	BTSCH
216	Monel	A	90%; boil.	VPT
217	Monel-clad	A	All %; any temp.	BTSCH
219	Muntz	F	...	CH
224	Nickel	A	90%; boil.	BIVPTSCH
225	Ni-clad	A	All %; any temp.	BTSCH
226	Ni-silver	F	...	Instruments
231	Ni-Resist	A	90%; boil.	BIVP
234	Olympic	V		
235	Olympic	V		
236	Palladium	A		
240	Platinum	A		
242	Ir-Platinum	A		
244	Rh-Platinum	A		
268	Silver	A		
270	304-clad	A	All %; any temp.	BTSCH
271	316-clad	A	All %; any temp.	BTSCH
275	St. 301	V		
276	St. 302	V		
		A	C.p.; boil.	
		A	Raw; boil.; 212°	
278	St. 303	V		
279	St. 304	V		
280	St. 308	V		
281	St. 309	V		
282	St. 310	V		
283	St. 316	A	...	VP
		A	C.p.; boil.	
		A	Raw; boil.; 212°	
284	St. 317	A	...	PT
285	St. 321	V		
286	St. 347	V		
287	St. 403	V		
288	St. 405	V		
290	St. 410	V		
291	St. 414	V		
292	St. 416	V		
293	St. 418	V		
294	St. 420	V		
295	St. 430	V		
296	St. 430F	V		
297	St. 431	V		
298	St. 440A	V		
300	St. 440C	V		
301	St. 442	V		
303	St. 446	V		
306	St. CA-15	A	Conc.; boil.	
313	St. CF-7	A	Conc.; boil.	

No.	Material	Ratings	Exposure conditions	Applications
360A	Steel	A	Where color is not important; do not use with c.p. acid	
367	Super Ni	F	...	PTCH
368	Tantalum	A	...	Not used commercially
369	Telnic	V		
390	Worthite	A	370°	BIV
401	Karbate	A	All %; to boil.	PH
402	Karbate	A	All %; to boil.	PH
403	Kempruf	A	All %; any temp.	
404	Acheson	A	All %; any temp.	
500	Sul. cement	A	Dilute sol.; low temp. only	PT
501	Sul. cement	A	Dilute sol.; low temp. only	PT
502	Furan cement	A	Phenol sol.; 220° max.	PT
503	Furan cement	A	Phenol sol.; 220° max.	PT
504	Phen. cement	A	Phenol sol.; 220° max.	PT
505	Phen. cement	A	Phenol sol.; 220° max.	PT
506	Silicate cement	A	1600° max.	PT
507	Silicate cement	A	1600° max.	PT
503	Asplit	X		
514	Asplit-F	X		
515	Basolit	A	200° max.	T
517	Carboline	A		
518	C-Basolit	A	200° max.	T
521	Causplit	A	All %; 350° max.	
523	Duralon	A	350° max.	BIPTCHDF
524	Durisite	A	350° max.	BIPTCHDF
534	N-series	X	Above rm.	
535	Nukem	A	350° max.	T
536	Nukem	A	...	T
538	Penchlor	A	All %; 750° max.	
539	Penchlor	A	All %; 750° max.	
540	Penchlor	A	All %; 500–2000°	
541	Pennsalt	A	All %; 350° max.	
542	Permanite	A	360° max.	TD
544	Plastite	F	175° max.	PTSD
545	P-Basolit	A	200° max.	T
554	Silastic	X	(A.S.T.M. D-543-43)	
559	Thiokol	X		
600	Acid brick	A	...	TDR
603, 612, 615	Glass	A	...	THDR
604	Ceratherm	A	200–400°, depending on design	BIPTSCHDFR
606	Stoneware	A	Not commonly used	
607	Glass-L	A	600° max.	PTCH
610	Stoneware	A	All %; any temp.	BIVPTCHDR
611	Porcelain	A	All % and temp.	BIPTDR
614	Glass-L	A	Under 302°; agitation	BTCH
616	Pyrex	A		
617	Stoneware	A	140–160°	BIPTSCHDFR
618	Vitreo	A	1000° max.	
621	Vycor	A		
700	Ace Saran	A	Crystals; 77°	P
703	Compar	V		
704	DC Silicone	X		
713	Haveg	V	Satisfactory for water sol.	BIVPTDFR
715	Heresite	A	200° max.	
718	Koroseal	X		
720	Lamicoid	A	...	T
726	Nukemite	A	150° max.	PTSDF
727	Nylon	X		
728	Nylon	X		
729–732	Nylon	X		
733	Permanite	A	360° max.	PVBTD
736	Pyroflex	A	...	PTCHD
737	Resilon	F	175° max.	PTSD
738	Resistoflex	V		
739	Resistoflex	V		
740	Saran	F	77°	
741	Sealon	A	160° max.	BPTDF
742	Teflon	A	Rm.	VP
746	Tygon	V	180° max.	BIPTSCHDF
801	Acidseal	X		
802	Acidseal	X		
805	Butyl	A	Rm.	
809	Fairprene	F		
817	Heresite	A	Sat.; 122° max.	
835	Perbunan	F	Rm.	
836	Natural (S)	X		
837	Natural (H)	V		
838	GR-S (S)	X		
839	GR-S (H)	V		
840	GR-A (S)	V		
841	GR-A (H)	V		
842	GR-M (S)	X		
843	GR-P (S)	X		
844	GR-I	X		
846	Saniprene	X		
848	Silastic	X	(A.S.T.M. D-543-43)	

Table 1. Chemical Resistance of Constructional Materials—(*Continued*)

No.	Material	Ratings	Exposure conditions	Applications
849	Superflexite	X		
853	Thiokol	X		
854	Thiokol	A	150° max.	PTSCDF
855	Triflex	X		
856	Vistanex	A	Rm.	

Phosphoric Acid

No.	Material	Ratings	Exposure conditions	Applications
3	Admiralty	V		
4	Admiralty	V	......	CH
6	Adnic	F	Attacked	
10–17	Aluminum	X		
19	Aloyco	A	25%; boil.	BV
		A	45%; 70°	BV
22	Ambraloy	V	......	CH
23	Ambraloy	V	......	CH
24	Ambraloy	V		CH
29–40	Ampco	F	Up to 85%; pure	BIVPTCHDF
54	Brass	X		
61	Brass	X		
63	Brass	V	......	PCH
66	Bronze	V		
73	Bronze	V	......	VT
74	Bronze	V	......	V
75	Bronze	V	......	VP
76	Bronze	V		
77	Bronze	X		
81	CA-FA20	A	85% aerated and agitated; 70°	BIVP
		A	85%; boil.	BIVP
82	CA-MM	A	85% aerated and agitated; 70°	BIVP
		F	85%; boil.	BIVP
86	Cast iron	V	Crude	
86A	Cast iron	F	Crude only over 70%	
88	Chlorimet	A	All % to 176°	BIVCHF
89	Chlorimet	A	All % to 176°	BIVHF
111	Copper	V		
114	Copper	V	......	PTCHR
117	Copper	V	Dilute	P
118	Copper	V		
119	Corrosiron	A	10, 25, 87%; rm.; unagitated; c.p.	BIVPF
123	Cupro-Ni	V	......	CH
124	Cupro-Ni	V		
139	Durco	A	All % and temp. except conc. boil.	BIVHF
140	Durichlor	A	All % and temp. (except crude)	BIVPCHFR
141, 142	Durimet	A	All % to 176°	BIVHF
143	Duriron	A	All % and temp. (except crude)	BIVPHFR
148	Everdur	V	......	BIVR
149	Everdur	V	......	PTCHDR
150	Everdur	V	......	PCH
156	Gold	A	All %; any temp.	
159	Hastelloy	A	All % to 180°	BIVPTCHDF
160	Hastelloy	A	All % to boil.	BIVPTCHDF
161	Hastelloy	A	All % to boil.	BIVPTCHDF
162	Hastelloy	A	All % to boil.	BIVP
163	Stellite	A	All % to boil.	BIV
165	Stellite	A	All% to boil.	BIV
181	Hytensl	A	10% max.; cold	BIVF
184	Inconel	A	12%; 212°; unaerated	
		A	57%; 70°	
		X	90%; 220°	
196, 200, 266	Lead	A	80% max.; 212° max.; 85% max. with crude	BIPTSCH
216	Monel	A	12%; 212°; unaerated	BIVPTCHDF
		A	57%; 70°	
		A	90%; 220°	
217	Monel-clad	A	5% max.; 70°	BTSCH
219	Muntz	X		
224	Nickel	A	12%; 212°; unaerated	BVPT
		X	57%; 70°	
		F	90%; 220°	
225	Ni-clad	A	5% max.; 70°	BTSCH
226	Ni-silver	V		
231	Ni-Resist	F	12%; 212°; unaerated	BIVP
		F	57%; 70°	
		X	90%; 220°	
234	Olympic	V		
235	Olympic	V		
240	Platinum	A	All %; any temp.	
242	Ir-Platinum	A	All %; any temp.	
244	Rh-Platinum	A	All %; any temp.	
245	Pyrasteel	V		
268	Silver	A	Under 365°	
		F	Over 365°	
270	304-clad	A	Under 5%; 70°	BTSCH
271	316-clad	A	10%	BTSCH
274	430-clad	A	Under 5%; 70°	BTSCH
275	St. 301	A	......	BIVTSC
276	St. 302	A	......	BIVTSCH
		A	1%; 70°; boil.	
		A	1% at 45 lb./sq. in. press.; 275°	
		A	10%; boil.	
		A	45%; boil.	
		A	80%; 140°	
		X	80%; 230°	
		A	5%; max.; 70°	VTSCDF
		X	10%; 70°	
		A	Weak or medium	
		X	Boil.; Dilute	
		A	Cold; Conc.	
278	St. 303	A	......	IV
279	St. 304	A	......	BIVPTSCH
		A	5% max.; 70°	VTSCDF
		X	10%; 70°	
		A	85%; aerated and agitated; 70°	BIVP
280	St. 308	A	......	BIVPTSCH
281	St. 309	A	......	BIVTSCH
282	St. 310	A	......	BIVPTSC
283	St. 316	A	......	BIVPTSCH
		A	1%; 70°; boil.	
		A	1% at 45 lb./sq. in. press.; 275°	
		A	10%; boil.	
		A	45%; boil.	
		A	80%; 140°	
		F	80%; 230°	
		A	5% max.; 70°	VTCHDF
		A	10%; 70°	VTCHDF
		A	85%; aerated and agitated; 70°	BIVP
284	St. 317	A	......	BIVPTSCH
285	St. 321	A	......	BIVPTSCH
286	St. 347	A	......	BIVPTSCH
287	St. 403	V		
288	St. 405	V		
290	St. 410	V		
		A	1%; 70°	
		X	All %	
		A	85% aerated and agitated; 70°	BIVP
291	St. 414	V		
292	St. 416	V		
		X	All %	
293	St. 418	V		
294	St. 420	X	Medium and strong	
		V		
		A	Dilute	
295	St. 430	V		
		A	1%; 70°	
		A	5% max.; 70°	VTSCHDF
		X	10%; 70°	
		X	All %	
296	St. 430F	V		
297	St. 431	V		
298	St. 440A	V		
300	St. 440C	V		
301	St. 442	V		
303	St. 446	V		
309	St. CC-35	A	85%; 130°	
316	St. CF-20	A	......	BIV
360A	Steel	F	Crude only over 70%	
367	Super Ni	V	......	PTCH
368	Tantalum	V	All % (fluorine below 10 p.p.m.), to 392°	CH
369	Telnic	V		
387	Stainless	A	Dilute	
		X	Medium and strong	
388	Stainless	X	All %	
389	Stainless	A	All %	
390	Worthite	A	5%; boil.	BIV
		A	70%; 175°; with slurry and HF to 0.5%	BIV
		A	85%; 250°	BIV
392	Wyndaloy	AV	68°	
401	Karbate	A	85% max.; to boil.	BIVPTH
402	Karbate	A	85% max.; to boil.	BIVPTH
403	Kempruf	A	All %; any temp.	PTDR
404	Acheson	A	All %; any temp.	PTDR
500	Sul. cement	A	85% max.; 200° max.	PT
501	Sul. cement	A	85% max.; 200° max.	PT
502	Furan cement	A	85% max.; 360° max.	PT
503	Furan cement	A	85% max.; 360° max.	PT
504	Phen. cement	A	85% max.; 360° max.	PT
505	Phen. cement	A	85% max.; 360° max.	PT
506	Silicate cement	A	85% max.; 1600° max.	PT
507	Silicate cement	A	85% max.; 1600° max.	PT
513	Asplit	A	All %; 350° max.	

Table 1. Chemical Resistance of Constructional Materials—(*Continued*)

No.	Material	Ratings	Exposure conditions	Applications
514	Asplit-F	A	All %; 350° max.	
515	Basolit	A	70%	T
517	Carboline	A		
518	C-Basolit	A	70%	T
521	Causplit	A	All %; 350° max.	
523	Duralon	A	350° max.	BIPTCHDF
524	Durisite	A	350° max.	BIPTCHDF
532	Lumnite	A	1% max.; 90° max.	TD
535	Nukem	A	70%; 350° max.	T
536	Nukem	A	70%	T
538	Penchlor	A	All %; 750° max.	
539	Penchlor	X		
540	Penchlor	X		
541	Pennsalt	A	All %; 350° max.	
542	Permanite	A	360° max.	TD
544	Plastite	A	175° max.	PTSD
545	P-Basolit	A	70%	T
554	Silastic	A	(A.S.T.M. D-543-43)	
556	Staminite	V	Cold	T
559	Thiokol	V	150° max.	
600	Acid brick	A	Dil.; cold	TDR
		X	Conc.; hot	
603, 612, 615	Glass	V	Better at low % and temp.	THDR
604	Ceratherm	A	200–400°, depending on design	BIPTSCHDFR
606	Stoneware	F	Under 176°	
607	Glass-L	V	....	PTCH
610	Stoneware	A	All %; any temp.	BIVPTCHDR
611	Porcelain	V	Dil.; cold	BIPTDR
614	Glass-L	A	40% max.; 212° max.; agitation	BTCH
616	Pyrex	X	Glacial	
617	Stoneware	A	140–160°	BIPTSCHDFR
618	Vitreo	V	1000° max.	
619	Vitreosil	V		
621	Vycor	X	Glacial	
700	Ace Saran	A	....	P
703	Compar	X		
704	DC Silicone	A		
707	Formica	F	All %; 122°	
711	Haveg	A	All conditions	BIVPTDFR
715	Heresite	A	200° max.	
718	Koroseal	A	All %; 150° max.	TD
720	Lamicoid	A	....	T
726	Nukemite	A	150° max.	PTSDF
727	Nylon	X	Conc.	
728	Nylon	X	Conc.	
729–732	Nylon	X		
733	Permanite	A	360° max.	BVPTD
736	Pyroflex	A	....	PTCHD
737	Resilon	A	175° max.	PTSD
738	Resistoflex	X		
739	Resistoflex	X		
740	Saran	A	77°	
741	Sealon	A	160° max.	BPTDF
742	Teflon	A	Rm.	VP
744–745	Textolite	V	....	BI
746	Tygon	A	180°	BIPTSCHDF
800	Ace Hd. Rub.	A	85%	BIVPTDF
801	Acidseal	A	All %; 150° max.	BIPTSDF
802	Acidseal	A	All %; 150° max.	BIPTSDF
805	Butyl	A	85%; rm.	
809	Fairprene	A		
814	G.E. Silicone	V	....	VPTDF
817	Heresite	A	Sat.; 104° max.	
836	Natural (S)	A		
837	Natural (H)	A		
838	GR-S (S)	A		
839	GR-S (H)	A		
840	GR-A (S)	A		
841	GR-A (H)	A		
842	GR-M (S)	A		
843	GR-P (S)	X		
844	GR-I	A		
846	Saniprene	A	All %; 150° max.	BIPTSDF
848	Silastic	A	(A.S.T.M. D-543-43)	
849	Superflexite	A	All %; 150° max.	BIPTSDF
853	Thiokol	V	150° max.	T
854	Thiokol	A	150° max.	PTSCDF
855	Triflex	A	All %; 150° max.	BIPTSDF
856	Vistanex	A	85%; rm.	
913	Redwood	V	30%; 36–90°; 1–3 years; airplane manufacture	T

Sodium Bisulfate

No.	Material	Ratings	Exposure conditions	Applications
3	Admiralty	A		
4	Admiralty	F	....	CH
6	Adnic	A		
10–17	Aluminum	V	Sol.	
		A	Dry powders	
22	Ambraloy	F	....	CH
23	Ambraloy	F	....	CH
24	Ambraloy	F	....	CH
29–40	Ampco	A		
51	Berylco	A	Hot and cold	
54	Brass	X		
61	Brass	X		
63	Brass	AF	....	PCH
66	Bronze	A		
73	Bronze	F	....	VT
74	Bronze	F	....	V
75	Bronze	F	....	VP
76	Bronze	F		
77	Bronze	X		
81	CA-FA20	A	10%; 70°	BIPT
82	CA-MM	A	10%; 70°	BIPT
86A	Cast iron	X		
88	Chlorimet	A	All % and temp.	BIVCHF
89	Chlorimet	A	All % and temp.	BIVHF
111	Copper	A		
114	Copper	F	....	PTCHR
118	Copper	A		
123	Cupro-Ni	A	....	CH
124	Cupro-Ni	A		
139	Durco	A	All % and temp.	BIVHF
140	Durichlor	A	All % and temp.	BIVPCHFR
141, 142	Durimet	A	All % and temp.	BIVHF
143	Duriron	A	All % and temp.	BIVPHFR
148	Everdur	F	....	BIVR
149	Everdur	F	....	PTCHDR
150	Everdur	F	....	PCH
159	Hastelloy	A	All % to 180°	BIVPTCH
160	Hastelloy	A	All % to boil.	BIVPTCH
161	Hastelloy	A	All % to boil.	BIVPTCH
162	Hastelloy	A	All % to boil.	BIVP
163	Stellite	A	All %; rm. only	BIV
165	Stellite	A	All %; rm. only	BIV
196, 200, 266	Lead	A	All %	BIPT
219	Muntz	X		
226	Ni-silver	A	....	Instruments
234	Olympic	A		
235	Olympic	A		
240	Platinum	A	All temp.	
242	Ir-Platinum	A	All temp.	
244	Rh-Platinum	A	All temp.	
245	Pyrasteel	V		
249	Resistac	A	....	BIVCHF
275	St. 301	A	....	BIVTSCH
276	St. 302	A	....	BIVTSCH
		A	10%; 70°; boil.	
278	St. 303	A	....	IV
279	St. 304	A	....	BIVPTSCH
		A	10%; 70°	BIVP
280	St. 308	A	....	BIVPTSCH
281	St. 309	A	....	BIVTSCH
282	St. 310	A	....	BIVPTSCH
283	St. 316	A	....	BIVPTSCH
		A	10%; 70°; boil.	
		A	10%; 70°	BIVP
284	St. 317	A	....	BIVPTSCH
285	St. 321	A	....	BIVPTSCH
286	St. 347	A	....	BIVPTSCH
290	St. 410	F	10%; 70°	BIVP
316	St. CF-20	A	....	BIV
360A	Steel	X		
367	Super Ni	A	....	PTCH
368	Tantalum	V	Water sol. only; attacked by molten salt	H
369	Telnic	A		
390	Worthite	A	All %; any temp.	BIV
401	Karbate	A	All %; to boil.	
402	Karbate	A	All %; to boil.	
403	Kempruf	A	All %; any temp.	
404	Acheson	A	All %; any temp.	
500	Sul. cement	A	200° max.	PT
501	Sul. cement	A	200° max.	PT
502	Furan cement	A	360° max.	PT
503	Furan cement	A	360° max.	PT
504	Phen. cement	A	360° max.	PT
505	Phen. cement	A	360° max.	PT
506	Silicate cement	F	1600° max.	T
507	Silicate cement	F	1600° max.	T
513	Asplit	A	All %; 350° max.	
514	Asplit-F	A	All %; 350° max.	
515	Basolit	A	....	T
518	C-Basolit	A	....	T
521	Causplit	A	All %; 350° max.	
523	Duralon	A	350° max.	BIPTCHDF
524	Durisite	A	350° max.	BIPTCHDF

Table 1. Chemical Resistance of Constructional Materials—(*Continued*)

No.	Material	Ratings	Exposure conditions	Applications
535	Nukem	A	350° max.	T
536	Nukem	A	...	T
538	Penchlor	A	All %; 750° max.	
539	Penchlor	A	All %; 750° max.	
540	Penchlor	A	All %; 500–2000°	
541	Pennsalt	A	All %; 350° max.	
542	Permanite	A	360° max.	TD
544	Plastite	A	175° max.	PTSD
545	P-Basolit	A	...	T
554	Silastic	A	(A.S.T.M. D-543-43)	
559	Thiokol	A	150° max.	
600	Acid brick	A	...	TDR
603, 612, 615	Glass	A	...	THDR
604	Ceratherm	A	200–400°, depending on design	BIPTSCHDFR
606	Stoneware	A	All conditions	BIVPTSCHDFR
607	Glass-L	A	600° max.	PTCH
610	Stoneware	A	All %; any temp.	BIVPTSCHDR
611	Porcelain	V	Dil.; cold	BIPTDR
614	Glass-L	A	All %; 302° max.; agitation	BTCH
616	Pyrex	A		
617	Stoneware	A	140–160°	BIPTSCHDFR
618	Vitreo	A	1000° max.	
621	Vycor	A		
700	Ace Saran	A	...	P
704	DC Silicone	A		
711	Haveg	A	All conditions	BIVPTDFR
713	Haveg	A	All conditions	BIVPTDFR
715	Heresite	A	200° max.	
718	Koroseal	A	Any %; 150° max.	TD
720	Lamicoid	A	...	T
726	Nukemite	A	...	PTSDF
733	Permanite	A	360° max.	BVPTD
736	Pyroflex	A	...	PTCHD
737	Resilon	A	175° max.	PTSD
740	Saran	A	77°	
741	Sealon	A	160° max.	BPTDF
742	Teflon	A	Rm.	VP
746	Tygon	A	180° max.	BIPTSCHDF
800	Ace Hd. Rub.	A	...	BIVPTDF
801	Acidseal	A	All %; 150° max.	BIPTDFS
802	Acidseal	A	All %; 150° max.	BIPTSDF
805	Butyl	F		
809	Fairprene	A		
817	Heresite	A	Sat.; 212° max.	
835	Perbunan	F		
836	Natural (S)	A		
837	Natural (H)	A		
838	GR-S (S)	A		
839	GR-S (H)	A		
840	GR-A (S)	A		
841	GR-A (H)	A		
842	GR-M (S)	A		
843	GR-P (S)	X		
844	GR-I	A		
846	Saniprene	A	All %; 150° max.	BIPTSDF
848	Silastic	A	(A.S.T.M. D-543-43)	
849	Superflexite	A	All %; 150° max.	BIPTSDF
853	Thiokol	A	150° max.	T
854	Thiokol	A	150° max.	PTSCDF
855	Triflex	A	All %; 150° max.	BIPTSDF
856	Vistanex	F		

Sodium Carbonate

No.	Material	Ratings	Exposure conditions	Applications
3	Admiralty	A		
4	Admiralty	F	...	CH
6	Adnic	A		
10–17	Aluminum	A	0.1% max.	
		FX	Over 0.1%; Al-Mg alloys best	
22	Ambraloy	F	...	CH
23	Ambraloy	F	...	CH
24	Ambraloy	F	...	CH
29–40	Ampco	A		
54	Brass	V	...	CH
61	Brass	V	...	VCH
63	Brass	AF	...	PCH
66	Bronze	A		
73	Bronze	F	...	VT
74	Bronze	F	...	V
75	Bronze	F	...	VP
76	Bronze	F		
77	Bronze	V	...	PCH
85	Cast iron	A		
86	Cast iron	A		
86A	Cast iron	A		
87	Causul	A	All %; any temp.; sol.	B
88	Chlorimet	A	All % and temp.	BIVCHF
89	Chlorimet	A	All % and temp.	BIVHF
111	Copper	A		
114	Copper	F	...	PTCHR
117	Copper	V		
118	Copper	A		
123	Cupro-Ni	A	...	CH
124	Cupro-Ni	A		
127–133	Dowmetal	X	25%; 212°	
139	Durco	A	All % and temp.	BIVHF
140	Durichlor	A	All % and temp.	BIVPCHFR
141, 142	Durimet	A	All % and temp.	BIVHF
143	Duriron	A	All % and temp.	BIVPHFR
148	Everdur	F	...	BIVR
149	Everdur	F	...	PTCHDR
150	Everdur	F	...	PCH
156	Gold	A		
159	Hastelloy	A	All % to boil.	
160	Hastelloy	A	All % to boil.	
161	Hastelloy	A	All % to boil.	
162	Hastelloy	A	All % to boil.	
163	Stellite	A	All % to boil.	BIV
165	Stellite	A	All % to boil.	BIV
184	Inconel	A	10%; 60°	
216	Monel	A	10%; 60°	BVPTH
217	Monel-clad	A	5%; any temp.	BTSCH
219	Muntz	V		CH
224	Nickel	A	10%; 60°	H
225	Ni-clad	A	5%; any temp.	BTSCH
226	Ni-silver	A	...	Instruments
231	Ni-Resist	A	10%; 60°	BIVP
233	Olympic	A		
234	Olympic	A		
236	Palladium	A		
240	Platinum	A		
242	Ir-Platinum	A		
244	Rh-Platinum	A		
245	Pyrasteel	A	...	BIVPTSCHF
268	Silver	A		
270	304-clad	A	5%; any temp.	BTSCH
271	316-clad	A	5%; any temp.	BTSCH
274	430-clad	A	5%; any temp.	BTSCH
275	St. 301	A	...	BIVTSCH
276	St. 302	A	...	BIVTSCH
		A	5%; boil.	
		A	50%; boil.	
		X	Melting; 1650°	
		A	5%; any temp.	VTSCDF
278	St. 303	A	...	IV
279	St. 304	A	...	BIVPTSCH
		A	5%; any temp.	VTSCDF
280	St. 308	A	...	BIVPTSCH
281	St. 309	A	...	BIVTSCH
282	St. 310	A	...	BIVPTSCH
283	St. 316	A	...	BIVPTSCH
		A	5%; boil.	
		A	50%; boil.	
		X	Melting; 1650°	
		A	5%; any temp.	VTCHDF
284	St. 317	A	...	BIVPTSCH
285	St. 321	A	...	BIVPTSCH
286	St. 347	A	...	BIVPTSCH
287	St. 403	A	...	IV
288	St. 405	A	...	BIVTSCH
290	St. 410	A	...	BIVPTSCH
		A	5%; boil.	
291	St. 414	A	...	IV
292	St. 416	A	...	IV
293	St. 418	A		
294	St. 420	A		
295	St. 430	A	...	BIVPTSCH
		A	5%; boil.	
		A	5%; any temp.	VTSCHDF
296	St. 430F	A	...	IV
297	St. 431	A	...	IV
298	St. 440A	A	...	
300	St. 440C	A	...	IV
301	St. 442	A	...	BIVTSCH
303	St. 446	A	...	BIVPTSCH
316	St. CF-20	A	...	BIV
319	St. CF-7M	A	50%; boil.	
360A	Steel	A		
367	Super Ni	A	...	PTCH
368	Tantalum	V	Water sol. only; 212° max.	Not used commercially
369	Telnic	A		
390	Worthite	A	All %; any temp.	BIV
401	Karbate	A	All %; to boil.	H
402	Karbate	A	All %; to boil.	H
403	Kempruf	A	700° max.	R
404	Acheson	A	700° max.	R
500	Sul. cement	F	100° max.	

Table 1. Chemical Resistance of Constructional Materials—(*Continued*)

No.	Material	Ratings	Exposure conditions	Applications
501	Sul. cement	F	100° max.	
502	Furan cement	A	360° max.	PTD
503	Furan cement	A	360° max.	PTD
504	Phen. cement	A	360° max.	PTD
505	Phen. cement	A	360° max.	PTD
506	Silicate cement	X		
507	Silicate cement	X		
513	Asplit	X		
514	Asplit-F	X		
515	Basolit	X		
517	Carboline	A		
518	C-Basolit	X		
521	Causplit	A	All %; 350° max.	
523	Duralon	A	350° max.	BIPTCHDF
524	Durisite	A	350° max.	BIPTCHDF
535	Nukem	A	350° max.	T
536	Nukem	X		
538	Penchlor	X		
539	Penchlor	X		
540	Penchlor	X		
541	Pennsalt	A	All %; 350° max.	
542	Permanite	A	360° max.	TD
544	Plastite	A	175° max.	PTSI
545	P-Basolit	X		
554	Silastic	A	(A.S.T.M. D-543-43)	
556	Staminite	X		
559	Thiokol	A	150° max.	
600	Acid brick	A	....	TDR
603, 612, 615	Glass	V	All % and temp.	THDR
604	Ceratherm	A	200–400°, depending on design	BIPTSCHDFR
606	Stoneware	A	Not commonly used	
607	Glass-L	V	....	PTCH
610	Stoneware	A	All %; any temp.	BIVPTCHDR
611	Porcelain	V	Dil.; cold	BIPTDR
614	Glass-L	V	All %; 77° max.; agitation	BTCH
617	Stoneware	A	140–160°	BIPTSCHDFR
618	Vitreo	A	1000° max.	
700	Ace Saran	A	....	P
704	DC Silicone	A		
713	Haveg	A	....	BIVPTDFR
715	Heresite	A	200° max.	
718	Koroseal	A	Any %; 150° max.	TD
720	Lamicoid	A	....	T
723	Nixon	AF	2%	S
724	Nixon	A	2%	S
726	Nukemite	A	....	PTSDF
727	Nylon	A		
728	Nylon	A		
729–732	Nylon	A		
733	Permanite	A	360° max.	BVPTD
736	Pyroflex	A	....	PTCHD
737	Resilon	A	175° max.	PTSD
741	Sealon	A	160° max.	BIPTDF
742	Teflon	A	Rm.	VP
746	Tygon	A	180° max.	BIPTSCHDF
800	Ace Hd. Rub.	A	....	BIVPTDF
801	Acidseal	A	All %; 150° max.	BIPTSDF
802	Acidseal	A	All %; 150° max.	BIPTSDF
805	Butyl	A	Dry salt; aqueous sol.; rm.	
809	Fairprene	A		
817	Heresite	A	Sat.; 212° max.	
835	Perbunan	A	Dry salt; aqueous sol.; rm.	
836	Natural (S)	A		
837	Natural (H)	A		
838	GR-S (S)	A		
839	GR-S (H)	A		
840	GR-A (S)	A		
841	GR-A (H)	A		
842	GR-M (S)	A		
843	GR-P (S)	X		
844	GR-I	A		
846	Saniprene	A	All %; 150° max.	BIPTSDF
848	Silastic	A	(A.S.T.M. D-543-43)	
849	Superflexite	A	All %; 150° max.	BIPTSDF
853	Thiokol	A	150° max.	T
854	Thiokol	A	150° max.	PTSCDF
855	Triflex	A	All %; 150° max.	BIPTSDF
856	Vistanex	A	Dry salt; aqueous sol.; rm.	
913	Redwood	A	5–6%; 36–180°; 13 years; insulation prods.	T
		A	11%; 60–212°; 3 years; oil refinery	T

Sodium Chloride

No.	Material	Ratings	Exposure conditions	Applications
3	Admiralty	A		
4	Admiralty	A	....	CH
6	Adnic	A		
10–17	Aluminum	AF	Rm.–boil.; Alclad alloys best	PTSCH
		A	Dry powders	
22	Ambraloy	A	....	CH
23	Ambraloy	A	....	CH
24	Ambraloy	A	....	CH
29–40	Ampco	A		
51	Berylco	A		
54	Brass	V	....	CH
61	Brass	V	....	VCH
63	Brass	AF	....	PCH
66	Bronze	A		
67	Bronze	A	Sea water	BIV
71	Bronze	A	Sea water	BIV
72	Bronze	A	Sea water	BIV
73	Bronze	F	....	VT
74	Bronze	F	....	V
75	Bronze	A	....	VP
76	Bronze	F		
77	Bronze	V	....	PCH
85	Cast iron	A		
86	Cast iron	A		
86A	Cast iron	F	Frequently used	
87	Causul	A	....	B
88	Chlorimet	A	All % and temp.	BIVCHF
89	Chlorimet	A	All % and temp.	BIVHF
111	Copper	A		
114	Copper	F	....	PTCHR
117	Copper	V	....	P
118	Copper	A		
119	Corrosiron	A	25%; rm.; unagitated; c.p.	BIVP
123	Cupro-Ni	A	....	CH
124	Cupro-Ni	A		
139	Durco	A	All % and temp.	BIVHF
140	Durichlor	A	All % and temp.	BIVPCHFR
141, 142	Durimet	A	All % and temp.	BIVHF
143	Duriron	A	All % and temp. (140 preferred)	BIVPHFR
148	Everdur	F	....	BIVR
149	Everdur	F	....	PTCHDR
150	Everdur	F	....	PCH
157	Gun metal	A	Sea water	BIV
159	Hastelloy	A	All % to boil.	
160	Hastelloy	A	All % to boil.	
161	Hastelloy	A	All % to boil.	BIVPH
162	Hastelloy	A	All % to boil.	
163	Stellite	A	All %; rm.	BIV
165	Stellite	A	All %; rm.	BIV
181	Hytensl	A	....	BIVHF
184	Inconel	A	Sat.; 200°	
185	Inc-clad	A	Sat. sol. mixed with steam and air; 200°	BTSCH
193, 196, 200, 266	Lead	A	Dil. sol.; sea water; brine	BIPT
216	Monel	A	Sat.; 200°	BIVPTCHDF
217	Monel-clad	A	Sat. sol. mixed with steam and air; 200°	BTSCH
219	Muntz	V	....	CH
224	Nickel	A	Sat.; 200°	BV
225	Ni-clad	A	Sat. sol. mixed with steam and air; 200°	BTSCH
226	Ni-silver	A	....	Instruments
231	Ni-Resist	A	Sat.; 200°	BIVP
233	NS-5	A	All % and temp.	V
234	Olympic	A		
235	Olympic	A		
245	Pyrasteel	V		
249	Resistac	A	....	BIVCHF
270	304-clad	A	Sat.; boil.	BTSCH
271	316-clad	A	5%; 150°	BTSCH
275	St. 301	V		
276	St. 302	V		
		V	Cold sat. sol.; 70°	
		A	Hot sat. at 212°	
		X	5%; 150°	
		X	Sat.; boil.	
278	St. 303	V		
279	St. 304	V		
		X	5%; 150°	
		X	Sat.; boil.	
280	St. 308	V		
281	St. 309	V		
282	St. 310	V		
283	St. 316	V		
		V	Cold sat. sol.; 70°	
		A	Hot sat. at 212°	
		A	5%; 150°	VTCHDF
		A	Sat.; boil.	VTCHDF
284	St. 317	V		
285	St. 321	V		
286	St. 347	V		

Table 1. Chemical Resistance of Constructional Materials—(*Continued*)

No.	Material	Ratings	Exposure conditions	Applications
287	St. 403	V		
288	St. 405	V		
290	St. 410	V		
291	St. 414	V		
292	St. 416	V		
293	St. 418	V		
294	St. 420	V		
295	St. 430	V		
		X	5%; 150°	
296	St. 430F	V		
297	St. 431	V		
298	St. 440A	V		
300	St. 440C	V		
301	St. 442	V		
303	St. 446	V		
316	St. CF-20	A	....	BIV
360A	Steel	F	Frequently used	
367	Super Ni	A	....	PTCH
368	Tantalum	V	Water sol. only; 212° max.	Not used commercially
369	Telnic	A		
390	Worthite	A	All %; any temp.	BIV
392	Wyndaloy	AV	Moderate dilution; 68°; sea water and brines	
		AV	Conc.; 68°; sea water and brines	
401	Karbate	A	All %; to boil.	BIVPH
402	Karbate	A	All %; to boil.	BIVPH
403	Kempruf	A	All %; any temp.	
404	Acheson	A	All %; any temp.	
500	Sul. cement	A	200° max.	PTD
501	Sul. cement	A	200° max.	PTD
502	Furan cement	A	360° max.	PTD
503	Furan cement	A	360° max.	PTD
504	Phen. cement	A	360° max.	PTD
505	Phen. cement	A	360° max.	PTD
506	Silicate cement	X		
507	Silicate cement	V	1600° max.	TD
600	Acid brick	A	....	TDR
603, 612, 615	Glass	A	....	THDR
604	Ceratherm	A	200–400°, depending on design	BIPTSCHDFR
606	Stoneware	A	Not commonly used	
607	Glass-L	A	600° max.	PTCH
610	Stoneware	A	All %; any temp.	BIVPTCHR
611	Porcelain	A	All % and temp.	BIPTDR
614	Glass-L	A	All %; 302° max.; agitation	BTCH
616	Pyrex	A		
617	Stoneware	A	140–160°	BIPTSCHDFR
618	Vitreo	A	1000° max.	
621	Vycor	A		
700	Ace Saran	A	....	P
704	DC Silicone	A		
706	Formica	A	All %; 212° max.	T
707	Formica	A	All %; 212° max.	
708	Formica	F	All conc.	
711	Haveg	A	All conditions	BIVPTFDR
713	Haveg	A	All conditions	BIVPTDFR
715	Heresite	A	200° max.	
718	Koroseal	A	Any %; 150° max.	TD
720	Lamicoid	A	....	T
723	Nixon	A	10%	S
724	Nixon	A	10%	S
726	Nukemite	A	....	PTSDF
733	Permanite	A	360° max.	PVBDT
736	Pyroflex	A	....	PTCHD
737	Resilon	A	175° max.	PTSD
741	Sealon	A	160° max.	BPTDF
742	Teflon	A	Rm.	VP
746	Tygon	A	180° max.	BIPTSCHDF
800	Ace Hd. Rub.	A	....	BIVPTDF
801	Acidseal	A	All %; 150° max.	BIPTSDF
802	Acidseal	A	All %; 150° max.	BIPTSDF
805	Butyl	A	Dry salt; aqueous sol.; rm.	
809	Fairprene	A		
817	Heresite	A	Sat.; 212° max.	
835	Perbunan	A	Dry salt; aqueous sol.; rm.	
836	Natural (S)	A		
837	Natural (H)	A		
838	GR-S (S)	A		
839	GR-S (H)	A		
840	GR-A (S)	A		
841	GR-A (H)	A		
842	GR-M (S)	A		
843	GR-P (S)	A		
844	GR-I	A		
846	Saniprene	A	All %; 150° max.	BIPTSDF
848	Silastic	A	(A.S.T.M. D–543–43)	
849	Superflexite	A	All %; 150° max.	BIPTSDF
853	Thiokol	A	150° max.	T
854	Thiokol	A	150° max.	PTSCDF
855	Triflex	A	All %; 150° max.	BIPTSDF
856	Vistanex	A	Dry salt; aqueous sol.; rm.	
913	Redwood	A	25%; 36–76°; 22 yr.; tanks always full; chem. plant	T
		F	20%; 40–220°; 5 years; soap plant	T
		A	2½%; 130–200°; packing plant	T
		A	36,000 p.p.m.; 40–100°; 4 years; petroleum plant	T
		A	1–2%; 75°; 18 years; olive cannery	T
		A	8½% + 0.5% acetic acid; 36–110°; paraffin-L; olives	T
		A	Sea water; 36–76°; 28 years; tannery	T
		A	Sea water; 70°; 8 years; fish cannery	T

Sodium Hydroxide

No.	Material	Ratings	Exposure conditions	Applications
3	Admiralty	V		
4	Admiralty	F	....	CH
6	Adnic	A		
10–17	Aluminum	X		
19	Aloyco	F	35%; boil.	BV
		F	Molten; 600°	BV
22	Ambraloy	F	....	CH
23	Ambraloy	F	....	CH
24	Ambraloy	F	....	CH
29–40	Ampco	F	0.5% sol. max.	
54	Brass	V	....	CH
61	Brass	V	....	VCH
63	Brass	VF	....	PCH
66	Bronze	V		
73	Bronze	F	....	VT
74	Bronze	F	....	V
75	Bronze	F	....	VP
76	Bronze	F		
77	Bronze	V	....	PCH
81	CA-FA20	A	Under 20%; boil.	BIVP
		F	Molten; 600°	BIVP
82	CA-MM	A	Under 20%; boil.	BIVP
		A	Molten; 600°	BIVP
85	Cast iron	A		
86	Cast iron	A		
86A	Cast iron	A	Under 70%; under 200°; low velocities	
87	Causul	A	40% max.; 230°	B
88	Chlorimet	A	All % and temp.	BIVCHF
89	Chlorimet	A	All % and temp.	BIVHF
111	Copper	V		
114	Copper	F	....	PTCHR
117	Copper	V	....	P
118	Copper	V		
123	Cupro-Ni	A	....	CH
124	Cupro-Ni	A		
127–133	Dowmetal	V	3%	
139	Durco	A	All % and temp.	BIVHF
140	Durichlor	A	All % and temp. (except boil.)	BIVPCHFR
141, 142	Durimet	A	All % and temp.	BIVCH
143	Duriron	A	All % and temp. (except boil.)	BIVPHFR
148	Everdur	F	....	BIVR
149	Everdur	F	....	PTCHDR
150	Everdur	F	....	PCH
156	Gold	A	All %	
159	Hastelloy	A	All % to boil.	BIV
160	Hastelloy	A	All % to boil.	BIV
161	Hastelloy	A	All % to boil.	BIV
162	Hastelloy	A	All % to boil.	
163	Stellite	A	All % to boil.	BIV
165	Stellite	A	All % to boil.	BIV
184	Inconel	A	50%; 180°	
		A	75%; 275°	
185	Inc-clad	A	....	BTSCH
216	Monel	A	50%; 180°	BIVPTCHDF
		A	75%; 275°	
217	Monel-clad	A	....	BTSCH
219	Muntz	V	....	CH
224	Nickel	A	50%; 180°	BIVPTSCHDF
		A	75%; 275°	
225	Ni-clad	A	70% max.	BTSCH
226	Ni-silver	A	....	Instruments
231	Ni-Resist	A	50%; 180°	BIVPH
		A	75%; 275°	
233	NS-5	A	All % and temp.	

Table 1. Chemical Resistance of Constructional Materials—*(Continued)*

No.	Material	Ratings	Exposure conditions	Applications
234	Olympic	V		
235	Olympic	V		
236	Palladium	A	All %	
240	Platinum	A	Any temp.	
242	Ir-Platinum	A	Any temp.	
244	Rh-Platinum	A	Any temp.	
245	Pyrasteel	V		
249	Resistac	A	....	BIVCHF
268	Silver	A	Any temp.	
270	304-clad	A	20% max.; 70°	BTSCH
271	316-clad	A	20% max.; 70°	BTSCH
274	430-clad	A	20% max.; 70°	BTSCH
275	St. 301	A	Solutions	BIVTSCH
276	St. 302	A	Solutions	BIVTSCH
		A	20%; 230°	
		A	Melting; 610°	
		A	20% max.; 70°	VTSCDF
		A	Dilute	
		X	Conc.	
278	St. 303	A	Solutions	IV
279	St. 304	A	Solutions	BIVPTSCH
		A	20% max.; 70°	VTSCDF
		A	Under 20%; boil.	BIPT
		X	Molten; 600°	
280	St. 308	A	Solutions	BIVPTSCH
281	St. 309	A	Solutions	BIVTSCH
282	St. 310	A	Solutions	BIVPTSCH
283	St. 316	A	Solutions	BIVPTSCH
		A	Melting; 610°	
		A	20% max.; 70°	VTCHDF
		A	Under 20%; boil.	BIPT
		X	Molten; 600°	
284	St. 317	A	Solutions	BIVPTSCH
285	St. 321	A	Solutions	BIVPTSCH
286	St. 347	A	Solutions	BIVPTSCH
287	St. 403	A	....	IV
288	St. 405	A	....	BIVTSCH
290	St. 410	A	....	BIVPTSCH
		X	Cold; strong	
		X	All %; boil.	
		A	Cold; weak or medium	
		A	Under 20%; boil.	BIVP
		X	Molten; 600°	
291	St. 414	A	....	IV
292	St. 416	A	....	IV
		A	Cold; Dilute	
		X	Cold; Conc.	
		X	All %; boil.	
293	St. 418	A		
294	St. 420	A		
		A	Weak and medium	
		X	Strong	
295	St. 430	A	....	BIVPTSCH
		A	20% max.; 70°	VTSCHDF
		A	Dilute	
		X	Conc.	
296	430F	A	....	IV
297	St. 431	A	....	IV
298	St. 440A	A		
300	St. 440C	A	....	IV
301	St. 442	A	....	BIVTSCH
303	St. 446	A	....	BIVPTSCH
313	St. CF-7	A	50%; boil.	
316	St. CF-20	A	....	BIV
319	St. CF-7M	A	70%; boil.	
360A	Steel	A	Under 70%; under 200°; low velocities	
364	Stoody	A	25%; boil.	
365	Stoody	F	25%; boil.	
367	Super Ni	A	....	PTCH
368	Tantalum	X		
369	Telnic	V		
387	Stainless	A	Dilute	
		X	Conc.	
388	Stainless	A	Weak or medium; strong, cold	
		X	Strong, boil.	
389	Stainless	A	Dilute, cold	
		X	Conc.; medium, boil.	
390	Worthite	A	50%; 260°	BIV
		A	70%; 200°	BIV
392	Wyndaloy	A	Very dilute; 68°	
		AV	Moderate dilution; 68°	
		AV	Conc.; 68°	
401	Karbate	A	All %; to boil.	BIVPH
402	Karbate	A	All %; to boil.	BIVPH
403	Kempruf	A	All %; any temp.	TR
404	Acheson	A	All %; any temp.	TR
500	Sul. cement	X		
501	Sul. cement	X		
502	Furan cement	X		
503	Furan cement	A	360° max.	PT

No.	Material	Ratings	Exposure conditions	Applications
504	Phen. cement	X		
505	Phen. cement	X		
506	Silicate cement	X		
507	Silicate cement	X		
508	Acichlor	F	Dilute and conc.; 300°	
510	Acitite	X	Dilute and conc.; 250°	
513	Asplit	X		
514	Asplit	X		
515	Basolit	X		
518	C-Basolit	X		
521	Causplit	A	All %; 350° max.	
523	Duralon	A	350° max.	BIPTCHDF
524	Durisite	A	350° max.	BIPTCHDF
534	N-series	A	70% max. at 160°	BIVPTCDFR
535	Nukem	A	50%	T
536	Nukem	X		
538	Penchlor	X		
539	Penchlor	X		
540	Penchlor	X		
541	Pennsalt	A	All %; 350° max.	
542	Permanite	A	360° max.	TD
544	Plastite	A	175° max.	PTSD
545	P-Basolit	X		
554	Silastic	F	10%;(A.S.T.M. D-543-43)	
		A	50%;(A.S.T.M. D-543-43)	
556	Staminite	X		
559	Thiokol	F	150° max.	
600	Acid brick	A	Dil.	TDR
		V	Conc.	
603, 612, 615	Glass	V	Better at low % and temp.	
604	Ceratherm	X	Conc.; hot	
606	Stoneware	V	Under 10% at 77°	BIVPTDFR
607	Glass-L	V	....	PTCH
610	Stoneware	X		
611	Porcelain	V	Dil.; cold	BIPTDR
614	Glass-L	V	Not recommended	BTCH
616	Pyrex	X	Conc.	
617	Stoneware	X	Conc.; hot	
618	Vitreo	X		
621	Vycor	X	Conc.	
700	Ace Saran	A	50%; 77°	P
703	Compar	X		
704	DC Silicone	A		
706	Formica	F	10%; rm.	F
		F	5%; 176°	F
707	Formica	F	20%; boil.	T
		F	50%; rm.	T
708	Formica	A	10% max.; 122°	P
713	Haveg	A	All conditions	
715	Heresite	V	5%; 70°	
718	Koroseal	A	35%; 90°	TD
		A	10%; 150°	TD
720	Lamicoid	X		
723	Nixon	X		
724	Nixon	X		
726	Nukemite	A	50%; rm.	PTSDF
727	Nylon	A		
728	Nylon	A		
729–732	Nylon	A		
733	Permanite	A	360° max.	BVPTD
735	Polythene	A	50%	
736	Pyroflex	A	....	PTCHD
737	Resilon	A	175° max.	PTSD
738	Resistoflex	X		
739	Resistoflex	X		
740	Saran	F	77°	
741	Sealon	A	160° max.	BPTDF
742	Teflon	A	Rm.	VP
744, 745	Textolite	V	....	BI
746	Tygon	A	180° max.	BIPTSCHDF
800	Ace Hd. Rub.	A	....	BIVPTDF
801	Acidseal	A	All %; 150° max.	BIPTSDF
802	Acidseal	A	All %; 150° max.	BIPTSDF
805	Butyl	A	Rm.	
809	Fairprene	A		
814	G.E. Silicone	V	....	VPTDF
817	Heresite	A	76%; 158° max.	
836	Natural (S)	A		
837	Natural (H)	A		
838	GR-S (S)	A		
839	GR-S (H)	A		
840	GR-A (S)	A		
841	GR-A (H)	A		
842	GR-M (S)	A		
843	GR-P (S)	X		
844	GR-I	A		
846	Saniprene	A	All %; 150° max.	BIPTSDF
848	Silastic	A	10%;(A.S.T.M. D-543-43)	
		A	50%;(A.S.T.M. D-543-43)	
849	Superflexite	A	All %; 150° max.	BIPTSDF
853	Thiokol	F	150° max.	T

Table 1. Chemical Resistance of Constructional Materials—(*Continued*)

No.	Material	Ratings	Exposure conditions	Applications
854	Thiokol	A	150° max.	PTSCDF
855	Triflex	A	All %; 150° max.	BIPTSDF
856	Vistanex	A	Rm.	
913	Redwood	A	3½%; 75°; 15 years; olive cannery	
		A	1–2% + dil. H_2SO_4; 40–212°; soap plant	

Sodium Hypochlorite

No.	Material	Ratings	Exposure conditions	Applications
3	Admiralty	V		
4	Admiralty	F	...	CH
6	Adnic	V	Dil. sol.	
10–17	Aluminum	X		
22	Ambraloy	F	...	CH
23	Ambraloy	F	...	CH
24	Ambraloy	F	...	CH
29–40	Ampco	F	Results may be erratic	
42	Antaciron	A	15%; warm	
54	Brass	X		
61	Brass	X		
63	Brass	VF	...	PCH
66	Bronze	V		
70	Bronze	A	Hot sol. such as sulfite liquors	BIV
73	Bronze	F	...	VT
74	Bronze	F	...	V
75	Bronze	F	...	VP
76	Bronze	F		
77	Bronze	X		
86A	Cast iron	X		
88	Chlorimet	X		
89	Chlorimet	A	All %; moderate temp.	BIVHF
111	Copper	V		
114	Copper	F	...	PTCHR
117	Copper	V	...	P
118	Copper	V		
123	Cupro-Ni	F	...	CH
124	Cupro-Ni	V		
139	Durco	F	Dil.; moderate temp.	BIVHF
140	Durichlor	A	All % and temp.	BIVPCHFR
141, 142	Durimet	F	All % at moderate temp.	BIVHF
143	Duriron	A	All % and temp. (140 preferred)	BIVPHFR
148	Everdur	F	...	BIVR
149	Everdur	F	...	PTCHDR
150	Everdur	F	...	PCH
156	Gold	A	All %	
159	Hastelloy	X	All % to boil.	
160	Hastelloy	X	All % to boil.	
161	Hastelloy	A	All %; rm. only	
162	Hastelloy	X	All % to boil.	
163	Stellite	A	All %; rm. only	
165	Stellite	X	All % to boil.	
184	Inconel		Same as calcium hypochlorite	
216	Monel		Same as calcium hypochlorite	
219	Muntz	X		
224	Nickel		Same as calcium hypochlorite	
226	Ni-silver	F		
231	Ni-Resist		Same as calcium hypochlorite	
233	NS-5	A	Bleach liquids; cold	V
234	Olympic	V		
235	Olympic	V		
236	Palladium	A	All %	
240	Platinum	A	All %	
242	Ir-Platinum	A	All %	
244	Rh-Platinum	A	All %	
245	Pyrasteel	V		
268	Silver	A	All %	
275	St. 301	V		
276	St. 302	V		
		A	Sp. gr. 1.21; aqueous sol.; 70°	
		A	5%	VTSCDF
		A	Weak	
		X	Strong	
278	St. 303	V		
279	St. 304	V		
		A	5%	VTSCDF
280	St. 308	V		
281	St. 309	V		
282	St. 310	V		
283	St. 316	A	...	PT
		A	Sp. gr. 1.21; aqueous sol.; 70°	
		A	5%	VTCHDF
284	St. 317	A	...	PT
285	St. 321	V		
286	St. 347	V		
287	St. 403	X		
288	St. 405	X		
290	St. 410	X		
291	St. 414	X		
292	St. 416	X		
293	St. 418	X		
294	St. 420	X		
295	St. 430	X		
		F	Sp. gr. 1.21; aqueous sol.; 70°	
		X	5%	
		A	Weak	
		X	Strong	
296	St. 430F	X		
297	St. 431	X		
298	St. 440A	X		
300	St. 440C	X		
301	St. 442	X		
303	St. 446	X		
332	St. CH-10	A	10% free Cl_2; boil.	
338	St. CK-25	A	10% free Cl_2; boil.	
360A	Steel	X		
367	Super Ni	F	...	PTCH
368	Tantalum	V	Corroded if sol. is alkaline	H
369	Telnic	V		
387	Stainless	A	Dilute	
		X	Conc.	
388	Stainless	A	Dilute	
		X	Conc.	
389	Stainless	A	Dilute	
		X	Conc.	
390	Worthite	A	Cold; 3% available Cl_2	BIV
401	Karbate	V	...	BIVPH
402	Karbate	V	...	BIVPH
403	Kempruf	V		
404	Acheson	V		
500	Sul. cement	X		
501	Sul. cement	X		
502	Furan cement	X		
503	Furan cement	X		
504	Phen. cement	X		
505	Phen. cement	F	Very dilute sol.; low temp. only	PT
506	Silicate cement	X		
507	Silicate cement	X		
513	Asplit	X		
514	Asplit-F	X		
515	Basolit	X		
518	C-Basolit	X		
521	Causplit	X		
523	Duralon	V	350° max.	BIPTCHDF
524	Durisite	V	350° max.	BIPTCHDF
535	Nukem	A	18.6% free Cl_2; 68°	T
536	Nukem	X		
538	Penchlor	X		
539	Penchlor	X		
540	Penchlor	X		
541	Pennsalt	X		
542	Permanite	V	360° max.	TD
544	Plastite	F	175° max.	PTSD
545	P-Basolit	X		
554	Silastic	A	(A.S.T.M. D-543-43)	
559	Thiokol	X		
600	Acid brick	A	...	TDR
603, 612, 615	Glass	A	...	THDR
604	Ceratherm	A	200–400° depending on design	BIPTSCHDFR
606	Stoneware	A	All conditions	BIVPTCHDFR
607	Glass-L	V	...	PTCH
610	Stoneware	A	All %; any temp.	BIVPTCHDR
611	Porcelain	A	All % and temp.	BIPTDR
614	Glass-L	V	Under 104°	BTCH
616	Pyrex	A		
617	Stoneware	A	140–160°	BIPTSCHDFR
618	Vitreo	A	1000° max.	
621	Vycor	A		
700	Ace Saran	A	5.5% avail. Cl_2; 122°	P
704	DC Silicone	A		
706	Formica	X	5%; rm.	
708	Formica	X	Over 1%	
711	Haveg	X		
713	Haveg	X		
715	Heresite	A	200° max.	
718	Koroseal	A	All %; 150° max.	BIPTSDF
720	Lamicoid	X		
726	Nukemite	A	...	PTSDF
733	Permanite	V	360° max.	BVPTD
736	Pyroflex	A	...	PTCHD

Table 1. Chemical Resistance of Constructional Materials—(*Continued*)

No.	Material	Ratings	Exposure conditions	Applications
737	Resilon	F	175° max.	PTSD
741	Sealon	A	160° max.	BPTDF
742	Teflon	A	Rm.	VP
746	Tygon	V	180° max.	BIPTSCHDF
800	Ace Hd. Rub.	A	....	BIVPTDF
801	Acidseal	A	All %; 150° max.	BIPTSDF
809	Fairprene	V		
817	Heresite	A	5% max.; 212° max.	
836	Natural (S)	A		
837	Natural (H)	V		
838	GR-S (S)	A		
839	GR-S (H)	V		
840	GR-A (S)	V		
841	GR-A (H)	V		
842	GR-M (S)	V		
843	GR-P (S)	X		
844	GR-I	V		
846	Saniprene	A	All %; 150° max.	BIPTSDF
848	Silastic	A	(A.S.T.M. D-543-43)	
849	Superflexite	A	All %; 150° max.	BIPTSDF
853	Thiokol	X		
854	Thiokol	F	150° max.	PTSCDF

Sodium Nitrate

No.	Material	Ratings	Exposure conditions	Applications
3	Admiralty	A		
4	Admiralty	A	....	CH
6	Adnic	A		
10-17	Aluminum	A	Sol.	BIPTSDF
		A	Dry powders	
22	Ambraloy	A	....	CH
23	Ambraloy	A	....	CH
24	Ambraloy	A	....	CH
29-40	Ampco	V		
54	Brass	V	....	CH
61	Brass	V	....	VCH
63	Brass	A	....	PCH
66	Bronze	A		
73	Bronze	A	....	VT
74	Bronze	A	....	V
75	Bronze	A	....	VP
76	Bronze	A	....	Instruments
77	Bronze	V	....	PCH
85	Cast iron	A		
86	Cast iron	A		
86A	Cast iron	A		
89	Chlorimet	A	All % and temp.	BIVHF
111	Copper	A		
114	Copper	A	....	PTCHR
117	Copper	A	....	P
118	Copper	A		
123	Cupro-Ni	A	....	CH
124	Cupro-Ni	A		
139	Durco	A	All % and temp.	BIVHF
140	Durichlor	A	All % and temp.	BIVPCHFR
141, 142	Durimet	A	All % and temp.	BIVHF
143	Duriron	A	All % and temp.	BIVPHFR
148	Everdur	A	....	BIVR
149	Everdur	A	....	PTCHDR
150	Everdur	A	....	PCH
156	Gold	A	Sol.	
159	Hastelloy	X	All % to boil.	
160	Hastelloy	X	All % to boil.	
161	Hastelloy	A	All % to 150°	BIVPCH
162	Hastelloy	X	All % to boil.	
163	Stellite	A	All % to 150°	BIV
165	Stellite	A	All % to 150°	BIV
184	Inconel	A	Molten, 950-1100°	
216	Monel	A	27%; 122°	BVH
217	Monel-clad	A	....	BTSCH
219	Muntz	V	....	CH
224	Nickel	A	Molten, 950-1100°	
226	Ni-silver	A	....	Instruments
231	Ni-Resist	A	Molten, 950-1100°	BIVP
234	Olympic	A		
235	Olympic	A		
236	Palladium	A	Sol.	
240	Platinum	A	For sol. and melts	
242	Ir-Platinum	A	For sol. and melts	
244	Rh-Platinum	A	For sol. and melts	
245	Pyrasteel	A	....	BIVPTSCHF
268	Silver	A	Sol.	
270	304-clad	A	....	BTSCH
271	316-clad	A	Fused	BTSCH
275	St. 301	A	....	BIVTSCH
276	St. 302	A	....	BIVTSCH
		A	Solution; hot	
		X	Fused	
278	St. 303	A	....	IV
279	St. 304	A	....	BIVPTSCH
		X	Fused	
280	St. 308	A	....	BIVPTSCH
281	St. 309	A	....	BIVTSCH
282	St. 310	A	....	BIVPTSCH
283	St. 316	A	....	BIVPTSCH
		A	Solution; hot	
		A	Fused	VTCHDF
284	St. 317	A	....	BIVPTSCH
285	St. 321	A	....	BIVPTSCH
286	St. 347	A	....	BIVPTSCH
287	St. 403	A	....	IV
288	St. 405	A	....	BIVTSCH
290	St. 410	A	....	BIVPTSCH
291	St. 414	A	....	IV
292	St. 416	A	....	IV
293	St. 418	A		
294	St. 420	A		
295	St. 430	A	....	BIVPTSCH
		X	Fused	
296	St. 430F	A	....	IV
297	St. 431	A	....	IV
298	St. 440A	A		
300	St. 440C	A	....	IV
301	St. 442	A	....	BIVTSCH
303	St. 446	A	....	BIVPTSCH
316	St. CF-20	A	....	BIV
360A	Steel	A		
367	SuperNi	A	....	PTCH
368	Tantalum	V	Water sol. only	Not used commercially
		A		
369	Telnic	A		
390	Worthite	A	All %; any temp.	BIV
401	Karbate	A	All %; to boil.	BIVP
402	Karbate	A	All %; to boil.	BIVP
403	Kempruf	A	All %; any temp.	
404	Acheson	A	All %; any temp.	
500	Sul. cement	A	200° max.	PT
501	Sul. cement	A	200° max.	PT
502	Furan cement	V	....	PT
503	Furan cement	V	....	PT
504	Phen. cement	V	....	PT
505	Phen. cement	V	....	PT
506	Silicate cement	X		
507	Silicate cement	X		
513	Asplit	X		
514	Asplit-F	X		
521	Causplit	X		
523	Duralon	V	350° max.	BIPTCHDF
524	Durisite	V	350° max.	BIPTCHDF
535	Nukem	A	350° max.	T
538	Penchlor	A	All %; 750° max.	
539	Penchlor	A	All %; 750° max.	
540	Penchlor	A	All %; 500-2000°	
541	Pennsalt	X		
542	Permanite	A	360° max.	TD
544	Plastite	V	175° max.	PTSD
554	Silastic	A	(A.S.T.M. D-543-43)	
559	Thiokol	A	150 max.	
600	Acid brick	A	....	TDR
603, 612, 615	Glass	A	....	THDR
604	Ceratherm	A	200-400°, depending on design	BIPTSCHDFR
606	Stoneware	A	All conditions	BIVPTCHDFR
607	Glass-L	V	....	PTCH
610	Stoneware	A	All %; any temp.	BIVPTCHDR
611	Porcelain	A	All % and temp.	BIPTDR
614	Glass-L	A	All %; 302° max.; agitation	BTCH
616	Pyrex	A		
617	Stoneware	A	140-160°	BIPTSCHDFR
618	Vitreo	A	1000° max.	
621	Vycor	A		
700	Ace Saran	A	....	P
704	DC Silicone	A		
711	Haveg	A	All conditions	BIVPTFDR
713	Haveg	A	All conditions	BIPTDFR
715	Heresite	A	200° max.	
718	Koroseal	A	All %; 150° max.	TD
720	Lamicoid	A	....	T
726	Nukemite	A	....	PTSDF
733	Permanite	A	360° max.	BVPDT
736	Pyroflex	A	....	PTCHD
737	Resilon	V	175° max.	PTSD
741	Sealon	A	160° max.	BPTDF
742	Teflon	A	Rm.	VP
746	Tygon	A	180° max.	BIPTSCHDF
800	Ace Hd. Rub.	A	....	BIVPTDF
801	Acidseal	A	All %; 150° max.	BIPTDFS
802	Acidseal	A	All %; 150° max.	BIPTDFS
805	Butyl	A	Dry salt; aqueous sol.; rm.	
809	Fairprene	A		

Table 1. Chemical Resistance of Constructional Materials—(*Continued*)

No.	Material	Ratings	Exposure conditions	Applications
817	Heresite	A	Sat.; 212° max.	
835	Perbunan	A	Dry salt; aqueous sol.; rm.	
836	Natural (S)	A		
837	Natural (H)	A		
838	GR-S (S)	A		
839	GR-S (H)	A		
840	GR-A (S)	A		
841	GR-A (H)	A		
842	GR-M (S)	A		
843	GR-P (S)	X		
844	GR-I	A		
846	Saniprene	A	All %; 150° max.	BIPTSDF
848	Silastic	A	(A.S.T.M. D-543-43)	
849	Superflexite	A	All %; 150° max.	BIPTSDF
853	Thiokol	A	150° max.	T
854	Thiokol	A	150° max.	PTSCDF
855	Triflex	A	All %; 150° max.	BIPTSDF
856	Vistanex	A	Dry salt; aqueous sol.; rm.	

Sodium Sulfate

No.	Material	Ratings	Exposure conditions	Applications
3	Admiralty	A		
4	Admiralty	A	...	CH
6	Adnic	A		
10–17	Aluminum	A	Sol.	PTSDF
		A	Dry powders	
22	Ambraloy	A	...	CH
23	Ambraloy	A	...	CH
24	Ambraloy	A	...	CH
29–40	Ampco	A	...	BIVPTCHDF
54	Brass	F	...	CH
61	Brass	F	...	VCH
63	Brass	A	...	PCH
66	Bronze	A		
73	Bronze	A	...	VT
74	Bronze	A	...	V
75	Bronze	A	...	VP
76	Bronze	A	...	Instruments
77	Bronze	F	...	PCH
85	Cast iron	A		
86	Cast iron	A		
86A	Cast iron	A		
88	Chlorimet	A	All % and temp.	BIVCHF
89	Chlorimet	A	All % and temp.	BIVHF
111	Copper	A		
114	Copper	A	...	PTCHR
117	Copper	A	...	P
118	Copper	A		
119	Corrosiron	A	10%; rm. unagitated; c.p.	BIP
123	Cupro-Ni	A	...	CH
124	Cupro-Ni	A		
139	Durco	A	All % and temp.	BIVHF
140	Durichlor	A	All % and temp.	BIVPCHFR
141, 142	Durimet	A	All % and temp.	BIVHF
143	Duriron	A	All % and temp.	BIVPHFR
148	Everdur	A	...	BIVR
149	Everdur	A	...	PTCHDR
150	Everdur	A	...	PCH
159	Hastelloy	A	All % to 180°	BIVPTCH
160	Hastelloy	A	All % to boil.	BIVPTCH
161	Hastelloy	A	All % to boil.	BIVPTCH
162	Hastelloy	A	All % to boil.	BIVPTCH
163	Stellite	A	All %; rm. only	BIV
165	Stellite	A	All %; rm. only	BIV
184	Inconel	A	Sat.; 200°	
185	Inc-clad	A	5%; 70°	BTSCH
193, 196, 200, 266	Lead	A	10% max.; boil.	BIPT
216	Monel	A	Sat.; 200°	VT
217	Monel-clad	A	5%; 70°	BTSCH
219	Muntz	F	...	CH
224	Nickel	A	Sat.; 200°	BIVHT
225	Ni-clad	A	5%; 70°	BTSCH
226	Ni-silver	A	...	Instruments
231	Ni-Resist	A	Sat.; 200°	BIVP
234	Olympic	A		
235	Olympic	A		
245	Pyrasteel	V		
268	Silver	A		
270	304-clad	A	All %; 70°	BTSCH
271	316-clad	A	All %; 70°	BTSCH
274	430-clad	A	5%; 70°	BTSCH
275	St. 301	A	...	BIVTSCH
276	St. 302	A	...	BIVTSCH
		A	Solution; hot	
		A	All %; 70°	VTSCDF
278	St. 303	A	...	IV
279	St. 304	A	...	BIVPTSCH
		A	All %; 70°	VTSCDF
280	St. 308	A	...	BIVPTSCH
281	St. 309	A	...	BIVTSCH
282	St. 310	A	...	BIVPTSCH
283	St. 316	A	...	BIVPTSCH
		A	Solution; hot	
		A	All %; 70°	VTCHDF
284	St. 317	A	...	BIVPTSCH
285	St. 321	A	...	BIVPTSCH
286	St. 347	A	...	BIVPTSCH
287	St. 403	A	...	IV
288	St. 405	A	...	BIVTSCH
290	St. 410	A	...	BIVPTSCH
291	St. 414	A	...	IV
292	St. 416	A	...	IV
293	St. 418	A		
294	St. 420	A		
295	St. 430	A	...	BIVPTSCH
		F	Solution; hot	
		A	5%; 70°	VTSCHDF
		X	All %; 70°	
296	St. 430F	A	...	IV
297	St. 431	A	...	IV
298	St. 440A	A		
300	St. 440C	A	...	IV
301	St. 442	A	...	BIVTSCH
303	St. 446	A	...	BIVPTSCH
313	St. CF-7	A	Sat.; boil.	
316	St. CF-20	A	...	BIV
319	St. CF-7M	A	Sat.; boil.	
360A	Steel	A		
367	Super Ni	A	...	PTCH
368	Tantalum	V	Water sol. only	Not used commercially
369	Telnic	A		
390	Worthite	A	All %; any temp.	BIV
401	Karbate	A	All %; to boil.	PH
402	Karbate	A	All %; to boil.	PH
403	Kempruf	A	All %; any temp.	T
404	Acheson	A	All %; any temp.	T
500	Sul. cement	A	200° max.	PTD
501	Sul. cement	A	200° max.	PTD
502	Furan cement	A	360° max.	PTD
503	Furan cement	A	360° max.	PTD
504	Phen. cement	A	360° max.	PTD
505	Phen. cement	A	360° max.	PTD
506	Silicate cement	X		
507	Silicate cement	V	1600° max.	TD
513	Asplit	A	All %; 350° max.	
514	Asplit-F	A	All %; 350° max.	
515	Basolit	A	20%; 77°	T
517	Caroboline	A		
518	C-Basolit	A	20%; 77°	T
521	Causplit	A	All %; 350° max.	
523	Duralon	A	350° max.	BIPTCHDF
524	Durisite	A	350° max.	BIPTSEHDF
535	Nukem	A	20%; 77°	T
536	Nukem	X		
538	Penchlor	A	All %; 750° max.	
539	Penchlor	A	All %; 750° max.	
540	Penchlor	A	All %; 500–2000°	
541	Pennsalt	A	All %; 350° max.	
542	Permanite	A	360° max.	TD
544	Plastite	F	175° max.	PTSD
545	P-Basolit	A	20%; 77°	T
554	Silastic	A	(A.S.T.M. D-543-43)	
556	Staminite	A	All %	T
559	Thiokol	A	150° max.	
600	Acid brick	A	...	TDR
603, 612, 615	Glass	A	...	THDR
604	Ceratherm	A	200–400°, depending on design	BIPTSCHDFR
606	Stoneware	A	All conditions	BIVPTCHDFR
607	Glass-L	V	...	PTCH
610	Stoneware	A	All %; any temp.	BIVPTCHDR
611	Porcelain	A	Dil.; cold	BIPTDR
614	Glass-L	A	All %; 302° max.; agitation	BTCH
616	Pyrex	A		
617	Stoneware	A	140–160°	BIPTSCHDFR
618	Vitreo	A	1000° max.	
621	Vycor	A		
700	Ace Saran	A	...	P
704	DC Silicate	A		
706	Formica	A	All %; 212° max.	PS
711	Haveg	A	All conditions	BIVPTDFR
713	Haveg	A	All conditions	BIVPTDFR
715	Heresite	A	200° max.	
718	Koroseal	A	All %; 150° max.	TD
720	Lamicoid	A	...	T
726	Nukemite	A	...	PTSDF
733	Permanite	A	360° max.	BVPTD
736	Pyroflex	A	...	PTCHD

Table 1. Chemical Resistance of Constructional Materials—(*Continued*)

No.	Material	Ratings	Exposure conditions	Applications
737	Resilon	F	175° max.	PTSD
740	Saran	A	77°	
741	Sealon	A	160° max.	BPTDF
742	Teflon	A	Rm.	VP
746	Tygon	A	180° max.	BIPTSCHDF
800	Ace Hd. Rub.	A	....	BIVPTDF
801	Acidseal	A	All %; 150° max.	BIPTSDF
802	Acidseal	A	All %; 150° max.	BIPTSDF
805	Butyl	A	Dry salt; aqueous sol.; rm.	
809	Fairprene	A		
817	Heresite	A	Sat.; 212° max.	
835	Perbunan	A	Dry salt; aqueous sol.; rm.	
836	Natural (S)	A		
837	Natural (H)	A		
838	GR-S (S)	A		
839	GR-S (H)	A		
840	GR-A (S)	A		
841	GR-A (H)	A		
842	GR-M (S)	A		
843	GR-P (S)	X		
844	GR-I	A		
846	Saniprene	A	All %; 150° max.	BIPTSDF
848	Silastic	A	(A.S.T.M. D-543-43)	
849	Superflexite	A	All %; 150° max.	BIPTSDF
853	Thiokol	A	150° max.	T
854	Thiokol	A	150° max.	PTSCDF
855	Triflex	A	All %; 150° max.	BIPTSDF
856	Vistanex	A	Dry salt; aqueous sol.; rm.	
913	Redwood	F	20%; 40-110°; 8 years; chemical plant	T

Sodium Sulfide

No.	Material	Ratings	Exposure conditions	Applications
3	Admiralty	V		
4	Admiralty	F	....	CH
6	Adnic	F		
10-17	Aluminum	X	Sol.	
		A	Dry powders	
22	Ambraloy	V	....	CH
23	Ambraloy	V	....	CH
24	Ambraloy	F	....	CH
54	Brass	F	....	CH
61	Brass	F	....	VCH
63	Brass	V	....	PCH
66	Bronze	V		
73	Bronze	V	....	VT
74	Bronze	V	....	V
75	Bronze	V	....	VP
76	Bronze	V		
77	Bronze	F	....	PCH
85	Cast iron	A		
86	Cast iron	A		
86A	Cast iron	A		
88	Chlorimet	A	All % and temp.	BIVCHF
89	Chlorimet	A	All % and temp.	BIVHF
111	Copper	V		
114	Copper	V	....	PTCHR
118	Copper	A		
123	Cupro-Ni	V	....	CH
124	Cupro-Ni	A		
139	Durco	A	All % and temp.	BIVHF
140	Durichlor	A	All % and temp.	BIVPCHFR
141, 142	Durimet	A	All % and temp.	BIVHF
143	Duriron	A	All % and temp.	BIVPHFR
148	Everdur	V	....	BIVR
149	Everdur	V	....	PTCHDR
150	Everdur	V	....	PCH
159	Hastelloy	A	All % to boil.	
160	Hastelloy	A	All % to boil.	
161	Hastelloy	A	All % to boil.	BIVPTCH
162	Hastelloy	A	All % to boil.	BIV
163	Stellite	A	All % to boil.	BIV
165	Stellite	A	All % to boil.	BIV
184	Inconel	A	50%; 320°	TH
185	Inc-clad	A	....	BTSCH
216	Monel	A	50%; 320°	TH
217	Monel-clad	A	....	BTSCH
219	Muntz	F	....	CH
224	Nickel	A	50%; 320°	BIVPTH
225	Ni-clad	A	....	BTSCH
226	Ni-silver	V		
234	Olympic	A		
235	Olympic	A		
245	Pyrasteel	A	....	BIVPTSCHF
268	Silver	X		
270	304-clad	A	....	BTSCH
271	316-clad	A	Sat.	BTSCH
275	St. 301	A	....	BIVTSCH
276	St. 302	A	....	BIVTSCH
		A	50%; boil.	
		X	Sat.	
278	St. 303	A	....	IV
279	St. 304	A	....	BIVPTSCH
		X	Sat.	
280	St. 308	A	....	BIVPTSCH
281	St. 309	A	....	BIVTSCH
282	St. 310	A	....	BIVPTSCH
283	St. 316	A	....	BIVPTSCH
		A	50%; boil.	
		A	Sat.	
284	St. 317	A	....	BIVPTSCH
285	St. 321	A	....	BIVPTSCH
286	St. 347	A	....	BIVPTSCH
287	St. 403	A	....	IV
288	St. 405	A	....	BIVTSCH
290	St. 410	A	....	BIVPTSCH
291	St. 414	A	....	IV
292	St. 416	A	....	IV
293	St. 418	A		
294	St. 420	A		
295	St. 430	A	....	BIVPTSCH
		V	50%; boil.	
		A	Sat.	
296	St. 430F	A	....	IV
297	St. 431	A	....	IV
298	St. 440A	A		
300	St. 440C	A	....	IV
301	St. 442	A	....	BIVTSCH
303	St. 446	A	....	BIVPTSCH
316	St. CF-20	A	....	BIV
360A	Steel	A		
367	SuperNi	V	....	PTCH
368	Tantalum	V	Water sol. only	Not used commercially
369	Telnic	A		
390	Worthite	A	All %; any temp.	BIVP
401	Karbate	A	All %; any temp.	PCH
402	Karbate	A	All %; any temp.	PCH
403	Kempruf	A	All %; any temp.	T
404	Acheson	A	All %; any temp.	T
500	Sul. cement	X		
501	Sul. cement	X		
502	Furan cement	X		
503	Furan cement	A	360° max.	PT
504	Phen. cement	X		
505	Phen. cement	A	360° max.	PT
506	Silicate cement	X		
507	Silicate cement	X		
513	Asplit	X		
514	Asplit-F	X		
515	Basolit	A	20%; 70°	T
517	Carboline	A		
518	C-Basolit	A	20%; 70°	T
521	Causplit	A	All %; 350° max.	
523	Duralon	A	350° max.	BIPTCHDF
524	Durisite	A	350° max.	BIPTCHDF
534	N-series	A	All %; 160°	BIVPTCDFR
535	Nukem	A	20%; 77°	T
536	Nukem	X		
538	Penchlor	X		
539	Penchlor	X		
540	Penchlor	X		
541	Pennsalt	A	All %; 350° max.	
542	Permanite	A	360° max.	TD
544	Plastite	A	175° max.	PTSD
545	P-Basolit	A	20%; 77°	T
554	Silastic	F	(A.S.T.M. D-543-43)	
559	Thiokol	X		
600	Acid brick	A	....	TDR
603, 612, 615	Glass	A	....	THDR
604	Ceratherm	A	200-400°, depending on design	BIPTSCHDFR
606	Stoneware	A	All conditions	BIVPTCHDFR
607	Glass-L	V	....	PTCH
610	Stoneware	A	All %; any temp.	BIVPTCHDR
611	Porcelain	A	All % and temp.	BIPTDR
614	Glass-L	V	Not recommended	BTCH
616	Pyrex	A		
617	Stoneware	A	140-160°	BIPTSCHDFR
618	Vitreo	A	1000° max.	
621	Vycor	A		
700	Ace Saran	A	....	P
704	DC Silicone	F		
713	Haveg	A	All conditions	BIVPTDFR
715	Heresite	A	200° max.	
718	Koroseal	A	All %; 150° max.	TD
720	Lamicoid	A	....	T
726	Nukemite	A	....	PTSDF
733	Permanite	A	360° max.	VPTDB

Table 1. Chemical Resistance of Constructional Materials—(*Continued*)

No.	Material	Ratings	Exposure conditions	Applications
736	Pyroflex	A	....	PTCHD
737	Resilon	A	175° max.	PTSD
741	Sealon	A	160° max.	BPTDF
742	Teflon	A	Rm.	VP
746	Tygon	A	180° max.	BIPTSCHDF
800	Ace Hd. Rub.	A	....	BIVPTDF
801	Acidseal	A	All %; 150° max.	BIPTSDF
802	Acidseal	A	All %; 150° max.	BIPTSDF
809	Fairprene	A		
817	Heresite	A	Sat.; 212° max.	
836	Natural (S)	A		
837	Natural (H)	A		
838	GR-S (S)	A		
839	GR-S (H)	A		
840	GR-A (S)	A		
841	GR-A (H)	A		
842	GR-M (S)	A		
843	GR-P (S)	X		
844	GR-I	A		
846	Saniprene	A	All %; 150° max.	BIPTSDF
848	Silastic	F	(A.S.T.M. D-543-43)	
849	Superflexite	A	All %; 150° max.	BIPTSDF
853	Thiokol	X		
854	Thiokol	X		
855	Triflex	A	All %; 150° max.	BIPTSDF

Sodium Sulfite

No.	Material	Ratings	Exposure conditions	Applications
3	Admiralty	A		
4	Admiralty	F	....	CH
6	Adnic	V		
10-17	Aluminum	AF	Sol.	
		A	Dry powders	
22	Ambraloy	F	....	CH
23	Ambraloy	F	....	CH
24	Ambraloy	F	....	CH
29-40	Ampco	F	....	BIVPTCHDF
54	Brass	X		
61	Brass	X		
63	Brass	AF	....	PCH
66	Bronze	A		
73	Bronze	F	....	VT
74	Bronze	F	....	V
75	Bronze	F	....	VP
76	Bronze	F		
77	Bronze	X		
81	CA-FA20	A	5%; 70°	BIVP
82	CA-MM	A	5%; 70°	BIVP
86A	Cast iron	A		
88	Chlorimet	A	All % and temp.	BIVCHF
89	Chlorimet	A	All % and temp.	BIVHF
111	Copper	A		
114	Copper	F	....	PTCHR
117	Copper	V	....	P
118	Copper	A		
123	Cupro-Ni	F	....	CH
124	Cupro-Ni	A		
139	Durco	A	All % and temp.	BIVHF
140	Durichlor	X		
141, 142	Durimet	A	All % and temp.	BIVCHF
143	Duriron	X		
148	Everdur	F	....	BIVR
149	Everdur	F	....	PTCHDR
150	Everdur	F	....	PCH
159	Hastelloy	X	All % to boil.	
160	Hastelloy	X	All % to boil.	
161	Hastelloy	A	All % and temp.	
162	Hastelloy	X	All % and temp.	
163	Stellite	A	All % and temp.	
165	Stellite	A	All % and temp.	
184	Inconel	A	7.5% plus 2% $NaHCO_3$; 75°	
216	Monel	A	7.5% plus 2% $NaHCO_3$; 75°	T
217	Monel-clad	A	....	BTSCH
219	Muntz	X		
224	Nickel	A	7.5% plus 2% $NaHCO_3$; 75°	
226	Ni-silver	F		
231	Ni-Resist	A	7.5% plus 2% $NaHCO_3$; 75°	
234	Olympic	A		
235	Olympic	A		
270	304-clad	A	5%; 70°	BTSCH
		A	10%; 150°	BTSCH
271	316-clad	A	5%; 70°	BTSCH
		A	10%; 150°	BTSCH
275	St. 301	A	Rm.	BIVTSCH
276	St. 302	A	Rm.	BIVTSCH
		A	50%; boil.	
		A	5%; 70°	VTSCDF
		A	10%; 150°	VTSCDF
278	St. 303	A	Rm.	IV
279	St. 304	A	Rm.	BIVPTSCH
		A	5%; 70°	VTSCDF
		A	10%; 70°	VTSCDF
		A	5%; 70°	BIPT
280	St. 308	A	Rm.	BIVPTSCH
281	St. 309	A	Rm.	BIVTSCH
282	St. 310	A	Rm.	BIVPTSCH
283	St. 316	A	Solutions in pulp and paper industry	BIVPTSCH
		A	50%; boil.	
		A	5%; 70°	VTCHDF
		A	10%; 70°	VTCHDF
		A	5%; 70°	BIVP
284	St. 317	A	Solutions	BIVPTSCH
285	St. 321	A	Rm.	BIVPTSCH
286	St. 347	A	Rm.	BIVPTSCH
287	St. 403	V		
288	St. 405	V		
290	St. 410	V		
		X	5%; 70°	
291	St. 414	V		
292	St. 416	V		
293	St. 418	V		
294	St. 420	V		
295	St. 430	V		
		X	5%; 70°	
296	St. 430F	V		
297	St. 431	V		
298	St. 440A	V		
300	St. 440C	V		
301	St. 442	V		
303	St. 446	V		
313	St. CF-7	A	50%; 70°	
316	St. CF-20	A	....	BIV
319	St. CF-7M	A	50%; 70°	
360A	Steel	A		
367	SuperNi	F	....	PTCH
368	Tantalum	V	Water sol. only	Not used commercially
369	Telnic	A		
390	Worthite	A	All %; any temp.	BIVPT
401	Karbate	A	All %; to boil.	BIVPH
402	Karbate	A	All %; to boil.	BIVPH
403	Kempruf	A	All %; any temp.	T
404	Acheson	A	All %; any temp.	T
500	Sul. cement	A	200° max.	PT
501	Sul. cement	A	200° max.	PT
502	Furan cement	A	360° max.	PT
503	Furan cement	A	360° max.	PT
504	Phen. cement	A	360° max.	PT
505	Phen. cement	A	360° max.	PT
506	Silicate cement	F	1600° max.	PT
507	Silicate cement	F	1600° max.	PT
513	Asplit	A	All %; 350° max.	
514	Asplit-F	A	All %; 350° max.	
515	Basolit	A	....	T
517	Carboline	A		
518	C-Basolit	A	....	T
521	Causplit	A	All %; 350° max.	
523	Duralon	A	350° max.	BIPTCHDF
524	Durisite	A	350° max.	BIPTCHDF
534	N-series	A	Sat.; 160°	BIVPTCDFR
535	Nukem	A	....	T
536	Nukem	A	In sol.	T
538	Penchlor	A	All %; 750° max.	
539	Penchlor	A	All %; 750° max.	
540	Penchlor	A	All %; 500-2000°	
541	Pennsalt	A	All %; 350° max.	
542	Permanite	A	360° max.	TD
544	Plastite	V	175° max.	PTSD
545	P-Basolit	A	....	T
554	Silastic	F	(A.S.T.M. D-543-43)	
559	Thiokol	A	150° max.	
600	Acid brick	A	....	TDR
603, 612, 615	Glass	A	....	THDR
604	Ceratherm	A	200-400°, depending on design	BIPTSCHDFR
606	Stoneware	A	All conditions	BIVPTCHDFR
610	Stoneware	A	All %; any temp.	BIVPTCHDR
611	Porcelain	A	All % and temp.	BIPTDR
614	Glass-L	A	All %; 302° max.; agitation	BTCH
616	Pyrex	A		
617	Stoneware	A	140-160°	BIPTSCHDFR
618	Vitreo	A	1000° max.	
621	Vycor	A		

Table 1. Chemical Resistance of Constructional Materials—(*Continued*)

No.	Material	Ratings	Exposure conditions	Applications
700	Ace Saran	A		P
704	DC Silicone	F		
711	Haveg	A	All conditions	BIVPTDFR
713	Haveg	A	All conditions	BIVPTDFR
715	Heresite	A	200° max.	
718	Koroseal	A	All %; 150° max.	TD
720	Lamicoid	A		T
726	Nukemite	A		PTSDF
733	Permanite	A	360° max.	PVDTB
736	Pyroflex	A		PTCHD
737	Resilon	V	175° max.	PTSD
741	Sealon	A	160° max.	BPTDF
742	Teflon	A	Rm.	VP
746	Tygon	A	180° max.	BIPTSCHDF
800	Ace Hd. Rub.	A		BIVPTDF
801	Acidseal	A	All %; 150° max.	BIPTSDF
802	Acidseal	A	All %; 150° max.	BIPTSDF
809	Fairprene	V		
817	Heresite	A	Sat.; 212° max.	
836	Natural (S)	A		
837	Natural (H)	A		
838	GR-S (S)	A		
839	GR-S (H)	A		
840	GR-A (S)	A		
841	GR-A (H)	A		
842	GR-M (S)	A		
843	GR-P (S)	X		
844	GR-I	A		
846	Saniprene	A	All %; 150° max.	BIPTSDF
848	Silastic	F	(A.S.T.M. D-543-43)	
849	Superflexite	A	All %; 150° max.	BIPTSDF
853	Thiokol	A	150° max.	T
854	Thiokol	A	150° max.	PTSCDF
855	Triflex	A	All %; 150° max.	BIPTSDF
913	Redwood	F	20%; 36-180°; 4 years; photo film mfr.	T
		A	15%; 40-100°; 10 years; chem. plant	T
		A	3%; 36-80°; 5 years; film mfr.	T

Sodium Thiosulfate (Hypo)

No.	Material	Ratings	Exposure conditions	Applications
3	Admiralty	V		
4	Admiralty	F		CH
6	Adnic	V		
10-17	Aluminum	A	1-5% sol.; rm.	S
		A	Dry powders	
22	Ambraloy	V		CH
23	Ambraloy	V		CH
24	Ambraloy	F		CH
29-40	Ampco	X		
54	Brass	F		CH
61	Brass	F		VCH
63	Brass	V		PCH
66	Bronze	V		
73	Bronze	V		VT
74	Bronze	V		V
75	Bronze	V		VP
76	Bronze	V		
77	Bronze	F		PCH
86	Cast iron	V		
86A	Cast iron	A	Do not use if iron contamination is not permissible	
89	Chlorimet	A	All % and temp.	BIVHF
111	Copper	V		
114	Copper	V		PTCHR
118	Copper	V		
123	Cupro-Ni	V		CH
124	Cupro-Ni	V		
139	Durco	A	All % and temp.	BIVHF
140	Durichlor	A	All % and temp.	BIVPCHFR
141, 142	Durimet	A	All % and temp.	BIVHF
143	Duriron	A	All % and temp.	BIVPHFR
148	Everdur	V		BIVR
149	Everdur	V		PTCHDR
150	Everdur	V		PCH
159	Hastelloy	A	All % and temp.	BIVPTHS
160	Hastelloy	A	All % and temp.	BIVPTSCH
161	Hastelloy	A	All % and temp.	BIVPTSCH
162	Hastelloy	A	All % and temp.	
163	Stellite	A	All % and temp.	BIV
165	Stellite	A	All % and temp.	BIV
185	Inc-clad	A		BTSCH
193, 196, 200, 266	Lead	A	General photographic use	PT
217	Monel-clad	A		BTSCH
219	Muntz	F		CH
225	Ni-clad	A		BTSCH
226	Ni-silver	V		
234	Olympic	V		
235	Olympic	V		
245	Pyrasteel	A		BIVPTSCHF
268	Silver	X		
275	St. 301	A		BIVTSCH
276	St. 302	A		BIVTSC
		A	25%; 70° boil.	
278	St. 303	A		IV
279	St. 304	A		BIVPTSC
280	St. 308	A		BIVPTSC
281	St. 309	A		BIVTSC
282	St. 310	A		BIVPTSC
283	St. 316	A		BIVPTSC
		A	25%; 70°; boil.	
284	St. 317	A		BIVPTSC
285	St. 321	A		BIVPTSC
286	St. 347	A		BIVPTSC
287	St. 403	A		IV
288	St. 405	A		BIVTSCH
290	St. 410	A		BIVPTSC
291	St. 414	A		IV
292	St. 416	A		IV
293	St. 418	A		
294	St. 420	A		
295	St. 430	A		BIVPTSC
296	St. 430F	A		IV
297	St. 431	A		IV
298	St. 440A	A		
300	St. 440C	A		IV
301	St. 442	A		BIVTSC
303	St. 446	A		BIVPTSC
313	St. CF-7	A	50%; 70°	
316	St. CF-20	A		BIV
319	St. CF-7M	A	50%; 70°	
360A	Steel	A	Do not use if iron contamination is not permissible	
367	Super Ni	V		PTCH
368	Tantalum	A		Not used commercially
369	Telnic	V		
390	Worthite	A	All %; any temp.	BIV
401	Karbate	A	All %; to boil.	
402	Karbate	A	All %; to boil.	
403	Kempruf	A	All %; any temp.	
404	Acheson	A	All %; any temp.	
500	Sul. cement	A	200° max.	PT
501	Sul. cement	A	200° max.	PT
502	Furan cement	A	360° max.	PT
503	Furan cement	A	360° max.	PT
504	Phen. cement	A	360° max.	PT
505	Phen. cement	A	360° max.	PT
506	Silicate cement	X		
507	Silicate cement	F	1600° max.	PT
513	Asplit	A	All %; 350° max.	
514	Asplit-F	A	All %; 350° max.	
515	Basolit	A		T
518	C-Basolit	A		T
521	Causplit	A	All %; 350° max.	
523	Duralon	A	350° max.	BIPTCHDF
524	Durisite	A	350° max.	BIPTCHDF
535	Nukem	A		T
536	Nukem	A		T
538	Penchlor	A	All %; 750° max.	
539	Penchlor	A	All %; 750° max.	
540	Penchlor	A	All %; 500-2000°	
541	Pennsalt	A	All %; 350° max.	
542	Permanite	A	360° max.	TD
544	Plastite	A	175° max.	PTSD
545	P-Basolit	A		T
554	Silastic	A	(A.S.T.M. D-543-43)	
559	Thiokol	A	150° max.	
600	Acid brick	A		TDR
603, 612, 615	Glass	A		THDR
604	Ceratherm	A	200-400°, depending on design	BIPTSCHDFR
606	Stoneware	A	All conditions	BIVPTCHDFR
607	Glass-L	V		PTCH
610	Stoneware	A	All %; any temp.	BIVPTCHDR
611	Porcelain	A	All % and temp.	BIPTDR
614	Glass-L	A	All %; 302° max.; agitation	BTCH
616	Pyrex	A		
617	Stoneware	A	140-160°	BIPTSCHDFR
618	Vitreo	A	1000° max.	
621	Vycor	A		
700	Ace Saran	A		P
704	DC Silicone	A		
706	Formica	F	All %; 212° max.	T
707	Formica	A	All %; 122° max.	

Table 1. Chemical Resistance of Constructional Materials—(*Continued*)

No.	Material	Ratings	Exposure conditions	Applications
711	Haveg	A	All conditions	BIVPTDFR
713	Haveg	A	All conditions	BIVPTDFR
715	Heresite	A	200° max.	
718	Koroseal	A	All %; 150° max.	TD
720	Lamicoid	A	...	T
726	Nukemite	A	...	PTSDF
733	Permanite	A	360° max.	BVPTD
736	Pyroflex	A	...	PTCHD
737	Resilon	A	175° max.	PTSD
740	Saran	A	77°	
741	Sealon	A	160° max.	BPTDF
742	Teflon	A	Rm.	VP
746	Tygon	A	180° max.	BIPTSCHDF
800	Ace Hd. Rub.	A	...	BIVPTDF
801	Acidseal	A	All %; 150° max.	BIPTSDF
802	Acidseal	A	All %; 150° max.	BIPTSDF
809	Fairprene	V		
817	Heresite	A	Sat.; 212° max.	
836	Natural (S)	A		
837	Natural (H)	A		
838	GR-S (S)	A		
839	GR-S (H)	A		
840	GR-A (S)	A		
841	GR-A (H)	A		
842	GR-M (S)	A		
843	GR-P (S)	X		
844	GR-I	A		
846	Saniprene	A	All %; 150° max.	BIPTSDF
848	Silastic	A	(A.S.T.M. D-543-43)	
849	Superflexite	A	All %; 150° max.	BIPTSDF
853	Thiokol	A	150° max.	T
854	Thiokol	F	150° max.	PTDSCF
855	Triflex	A	All %; 150° max.	BIPTSDF

Stearic Acid

No.	Material	Ratings	Exposure conditions	Applications
3	Admiralty	A		
4	Admiralty	A	...	CH
6	Adnic	A		
10–17	Aluminum	A	Rm.-boil.; moisture inhibits attack	PTSCHDF
19	Aloyco	A	300°, 400°	BV
22	Ambraloy	A	...	CH
23	Ambraloy	A	...	CH
24	Ambraloy	A	...	CH
29–40	Ampco	A	...	BIVPTCHDF
54	Brass	F	...	CH
61	Brass	F	...	VCH
63	Brass	A	...	PCH
66	Bronze	A		
73	Bronze	A	...	VT
74	Bronze	A	...	V
75	Bronze	A	...	VP
76	Bronze	A	...	Instruments
77	Bronze	F	...	PCH
86	Cast iron	A		
86A	Cast iron	F		
87	Causul	A	Any temp.	
88	Chlorimet	A	All % and temp.	
89	Chlorimet	A	All % and temp.	BIVHF
111	Copper	A		
114	Copper	A	...	PTCHR
118	Copper	A		
123	Cupro-Ni	A	...	CH
124	Cupro-Ni	A		
139	Durco	A	All % and temp.	BIVHF
140	Durichlor	A	All % and temp.	BIVPCHFR
141, 142	Durimet	A	All % and temp.	BIVHF
143	Duriron	A	All % and temp.	BIVPHFR
148	Everdur	A	...	BIVR
149	Everdur	A	...	PTCHDR
150	Everdur	A	...	PCH
159	Hastelloy	A	All % and temp.	
160	Hastelloy	A	All % and temp.	
161	Hastelloy	A	All % and temp.	BIVPTSCHF
162	Hastelloy	A	All % and temp.	
163	Stellite	A	All % and temp.	BIV
165	Stellite	A	All % and temp.	BIV
184	Inconel	A	Boil.	BIVPTCH
185	Inc-clad	A	...	BTSCH
216	Monel	A	Boil.	BIVPTCH
217	Monel-clad	A	...	BTSCH
219	Muntz	F	...	CH
224	Nickel	A	Boil.	PTH
225	Ni-clad	A	...	BTSCH
226	Ni-silver	A	...	Instruments
231	Ni-Resist	A	Boil.	BIVP
233	NS-5	A	Hot or cold	
234	Olympic	A		
235	Olympic	A		
245	Pyrasteel	A	...	BIVPTSCHF
249	Resistac	F	...	BIVCHF
268	Silver	A	...	
270	304-clad	A	...	BTSCH
271	316-clad	A	...	BTSCH
274	430-clad	A	...	BTSCH
275	St. 301	A	...	BIVTSCDF
276	St. 302	A	...	BIVTSCDF
278	St. 303	A	...	IV
279	St. 304	A	...	BIVPTSCDF
280	St. 308	A	...	BIVPTSCDF
281	St. 309	A	...	BIVTSCDF
282	St. 310	A	...	BIVPTSCDF
283	St. 316	A	...	BIVPTSCDF
		A	...	VTCHDF
284	St. 317	A	...	BIVPTSCDF
285	St. 321	A	...	BIVPTSCDF
286	St. 347	A	...	BIVPTSCDF
287	St. 403	A	...	IV
288	St. 405	A	...	BIVTSC
290	St. 410	A	...	BIVPTSC
291	St. 414	A	...	IV
292	St. 416	A	...	IV
293	St. 418	A		
294	St. 420	A		
295	St. 430	A	...	BIVPTSCHDF
296	St. 430F	A	...	IV
297	St. 431	A	...	IV
298	St. 440A	A		
300	St. 440C	A	...	IV
301	St. 442	A	...	BIVTSCH
303	St. 446	A	...	BIVPTSCDF
313	St. CF-7	A	212°; c.p.	
		F	212°; commercial	
316	St. CF-20	A	...	BIV
360A	Steel	F		
367	Super Ni	A	...	PTCH
368	Tantalum	A	...	Not used commercially
369	Telnic	A		
390	Worthite	A	All %; any temp.	BIV
401	Karbate	A	All %; to boil.	BIVPHDF
402	Karbate	A	All %; to boil.	BIVPHDF
403	Kempruf	A	All %; any temp.	TR
404	Acheson	A	All %; any temp.	TR
500	Sul. cement	X		
501	Sul. cement	X		
502	Furan cement	A	360° max.	PT
503	Furan cement	A	360° max.	PT
504	Phen. cement	A	360° max.	PT
505	Phen. cement	A	360° max.	PT
506	Silicate cement	F	...	PT
507	Silicate cement	F	...	PT
513	Asplit	A	All %; 350° max.	
514	Asplit-F	A	All %; 350° max.	
515	Basolit	A	200° max.	T
518	C-Basolit	A	200° max.	T
521	Causplit	A	All %; 350° max.	
523	Duralon	A	350° max.	BIPTCHDF
524	Durisite	A	350° max.	BIPTCHDF
535	Nukem	A	350° max.	T
536	Nukem	A	...	T
538	Penchlor	A	All %; 750° max.	
539	Penchlor	A	All %; 750° max.	
540	Penchlor	A	All %; 500–2000°	
541	Pennsalt	A	All %; 350° max.	
542	Permanite	A	360° max.	TD
544	Plastite	A	175° max.	PTSD
545	P-Basolit	A	200° max.	T
554	Silastic	A	(A.S.T.M. D-543-43)	
556	Staminite	A	All %	T
559	Thiokol	A	150° max.	
600	Acid brick	A	...	TDR
603, 612, 615	Glass	A	...	THDR
604	Ceratherm	A	200–400°, depending on design	BIPTSCHDFR
606	Stoneware	A	All conditions	BIVPTCHDFR
607	Glass-L	A	600° max.	PTCH
610	Stoneware	A	All %; any temp.	BIVPTCHDR
611	Porcelain	A	All % and temp.	BIPTDR
614	Glass-L	A	302° max.; agitation	BTCH
616	Pyrex	A		
617	Stoneware	A	140–160°	BIPTSCHDFR
618	Vitreo	A	1000° max.	
621	Vycor	A		
703	Compar	A	200° max.	T
704	DC Silicone	A		
706	Formica	A	All %; 122° max.	PT
707	Formica	V	All %	
708	Formica	F	All %; rm.	
711	Haveg	A	300° max.	BIVPTDFR

Table 1. Chemical Resistance of Constructional Materials—(*Continued*)

No.	Material	Ratings	Exposure conditions	Applications
713	Haveg	A	300° max.	BIVPTDFR
715	Heresite	A	200° max.	
718	Koroseal	F	All %; 100° max.	TD
720	Lamicoid	A	...	T
726	Nukemite	A	150° max.	PTSDF
727	Nylon	A	Weak	
728	Nylon	A	Weak	
729–732	Nylon	X		
733	Permanite	A	360° max.	BVPTD
736	Pyroflex	V	...	PTCHD
737	Resilon	A	175° max.	PTSD
738	Resistoflex	A	Conc.; 275° max.	P
739	Resistoflex	A	Conc.; 140° max.	IVP
741	Sealon	A	160° max.	BPTDF
742	Teflon	A	Rm.	VP
744, 745	Textolite	V		
746	Tygon	A	180° max.	BIPTSCHDF
801	Acidseal	F	All %; 100° max.	BIPTDF
805	Butyl	F		
809	Fairprene	V		
814	G.E. Silicone	V	...	VPTDF
817	Heresite	A	Sat.; 212° max.	
835	Perbunan	A		
836	Natural (S)	V		
837	Natural (H)	V		
838	GR-S (S)	V		
839	GR-S (H)	V		
840	GR-A (S)	V		
841	GR-A (H)	V		
842	GR-M (S)	V		
843	GR-P (S)	X		
844	GR-I	V		
846	Saniprene	F	All %; 100° max.	BIPTDF
848	Silastic	A	(A.S.T.M. D-543-43)	
849	Superflexite	F	All %; 100° max.	BIPTDF
853	Thiokol	A	150° max.	T
854	Thiokol	A	150° max.	PTSCDF
856	Vistanex	F		

Sulfur

No.	Material	Ratings	Exposure conditions	Applications
3	Admiralty	V		
4	Admiralty	A	Solid	
6	Adnic	A	Dry	
10–17	Aluminum	A	...	PTSHDF
22	Ambraloy	F	Solid	
23	Ambraloy	F	Solid	
24	Ambraloy	A	Solid	
29–40	Ampco	F	...	BIVPTCHDF
54	Brass	A	Solid	
61	Brass	A	Solid	
63	Brass	VF	Solid	
66	Bronze	V		
73	Bronze	F	Solid	
74	Bronze	F	Solid	
75	Bronze	F	Solid	
76	Bronze	F	Solid	
77	Bronze	A	Solid	
85	Cast iron	A	In water	
		A	Molten	
86	Cast iron	A		
86A	Cast iron	A		
88	Chlorimet	A	Except high temp.	BIVCHF
89	Chlorimet	A	Moderate temp.	BIVCHF
111	Copper	V		
114	Copper	F	Solid	
118	Copper	V		
123	Cupro-Ni	F	Solid	
124	Cupro-Ni	V		
127–133	Dowmetal	A		
139	Durco	A	All % and temp.	BIVHF
140	Durichlor	A	All % and temp.	BIVPCHFR
141, 142	Durimet	A	All % and temp.	BIVHF
143	Duriron	A	All % and temp.	BIVPHFR
148	Everdur	F	Solid	
149	Everdur	F	Solid	
150	Everdur	F	Solid	
159	Hastelloy	X		
160	Hastelloy	X		
161	Hastelloy	X		
162	Hastelloy	X		
163	Stellite	X		
165	Stellite	X		
184	Inconel	A	Molten, 260°	
		A	Molten, 500°	
216	Monel	A	Molten, 260°	
		F	Molten, 500°	
217	Monel-clad	A	Dry; molten	BTSCH
219	Muntz	A		
224	Nickel	A	Molten, 260°	
		F	Molten, 500°	
226	Ni-silver	F		
231	Ni-Resist	A	Molten, 260°	
		F	Molten, 500°	
234	Olympic	V		
235	Olympic	V		
245	Pyrasteel	V		
270	304-clad	A	Dry; molten	BTSCH
271	316-clad	A	Dry; molten	BTSCH
274	430-clad	A	Dry; molten	BTSCH
275	St. 301	V		
276	St. 302	V		
		A	Fused; 265°	
		F	Boil; 830°	
		A	Dry; molten	VTSCDF
		X	Wet	
278	St. 303	V		
279	St. 304	V		
		A	Dry; molten	VTSCDF
		X	Wet	
280	St. 308	V		
281	St. 309	V		
282	St. 310	V		
283	St. 316	V		
		F	Boil.; 830°	
		A	Dry; molten	VTCHDF
		X	Wet	
284	St. 317	V		
285	St. 321	V		
286	St. 347	V		
287	St. 403	V		
288	St. 405	V		
290	St. 410	V		
291	St. 414	V		
292	St. 416	V		
293	St. 418	V		
294	St. 420	V		
295	St. 430	V		
		A	Dry; molten	VTSCHDF
		X	Wet	
296	St. 430F	V		
297	St. 431	V		
298	St. 440A	V		
300	St. 440C	V		
301	St. 442	V		
303	St. 446	V		
360A	Steel	A		
367	Super Ni	F	Solid	
368	Tantalum	A	392° max.	Not used commercially
369	Telnic	V		
401	Karbate	A	338° max.	D
402	Karbate	A	338° max.	D
403	Kempruf	A	932–1292°	PT
404	Acheson	A	932–1292°	PT
513	Asplit	A	All %; 350° max.	
514	Asplit-F	A	All %; 350° max.	
515	Basolit	A	...	T
518	C-Basolit	A	...	T
521	Causplit	A	All %; 350° max.	
523	Duralon	V	350° max.	BIPTCHDF
524	Durisite	V	350° max.	BIPTCHDF
532	Lumnite	A	10% max.; 90° max.	TD
535	Nukem	A	...	T
536	Nukem	A	...	T
538	Penchlor	A	All %; 750° max.	
539	Penchlor	A	All %; 750° max.	
540	Penchlor	A	All %; 500–2000°	
541	Pennsalt	A	All %; 350° max.	
542	Permanite	A	360° max.	TD
544	Plastite	V	175° max.	PTSD
545	P-Basolit	A	...	T
559	Thiokol	V	150° max.	
600	Acid brick	A	...	TDR
603, 612, 615	Glass	A	...	THDR
604	Ceratherm	A	200–400°, depending on design	BIPTSCHDFR
606	Stoneware	A	Not commonly used	
607	Glass-L	A	600° max.	PTCH
610	Stoneware	A	All %; any temp.	BIVPTCHDR
611	Porcelain	A	All % and temp.	BIPTDR
614	Glass-L	V		
616	Pyrex	A		
617	Stoneware	A	140–160°	BIPTSCHDFR
618	Vitreo	A	1000° max.	
619	Vitreosil	A		
621	Vycor	A		
711	Haveg	A	300° max.	BIVPTFRD
713	Haveg	A	300° max.	BIVPTFDR
715	Heresite	A	200° max.	

Table 1. Chemical Resistance of Constructional Materials—(*Continued*)

No.	Material	Ratings	Exposure conditions	Applications
720	Lamicoid	A	....	T
726	Nukemite	A	....	PTSDF
733	Permanite	A	360° max.	BVPTD
736	Pyroflex	V	....	PTCHD
737	Resilon	V	175° max.	PTSD
741	Sealon	A	160° max.	BPTDF
742	Teflon	A	Rm.	VP
746	Tygon	A	180° max.	BIPTSCHDF
809	Fairprene	A		
817	Heresite	A	212° max.	
836	Natural (S)	V		
837	Natural (H)	V		
838	Gr-S (S)	V		
839	GR-S (H)	V		
840	GR-A (S)	V		
841	GR-A (H)	V		
842	GR-M (S)	V		
843	GR-P (S)	V		
844	GR-I	V		
853	Thiokol	V	150° max.	T
854	Thiokol	V	150° max.	PTSCDF

Sulfur Dioxide

No.	Material	Ratings	Exposure conditions	Applications
3	Admiralty	A	Dry; moist	
4	Admiralty	A	Dry	CH
		F	Moist	
6	Adnic	A		
10–17	Aluminum	A	....	PTCHDF
19	Aloyco	A	70°; moist	PV
22	Ambraloy	A	Dry	CH
		F	Moist	
23	Ambraloy	A	Dry	CH
		F	Moist	
24	Ambraloy	A	Dry	CH
		F	Moist	
29–40	Ampco	F	Dry	BIVPTCHDF
54	Brass	A	Dry	CH
		X	Moist	
61	Brass	A	Dry	VCH
		X	Moist	
63	Brass	A	Dry;	PCH
		AF	Moist	
66	Bronze	A	Dry; moist	
73	Bronze	A	Dry	VT
		F	Moist	
74	Bronze	A	Dry	V
		F	Moist	
75	Bronze	A	Dry	VP
		F	Moist	
76	Bronze	A	Dry	Instruments
		F	Moist	
77	Bronze	A	Dry	PCH
		X	Moist	
81	CA-FA20	A	Moist gas; 70°	BIVP
82	CA-MM	F	Moist gas; 70°	BIVP
86	Cast iron	A		
86A	Cast iron	A	Dry gas only	
88	Chlorimet	A	All %; except high temp.	BIVCHF
89	Chlorimet	A	All % and temp.	BIVHF
111	Copper	A	Dry; moist	
114	Copper	A	Dry	PTCHR
		F	Moist	
118	Copper	A	Dry; moist	
123	Cupro-Ni	A	Dry	CH
		V	Moist	
124	Cupro-Ni	A	Dry; moist	
127–133	Dowmetal	A	Dry	
139	Durco	A	All % and temp.	BIVHF
140	Durichlor	X		
141, 142	Durimet	A	All % and temp.	BIVHF
143	Duriron	X		
148	Everdur	A	Dry	BIVR
		F	Moist	
149	Everdur	A	Dry	PTCHDR
		F	Moist	
150	Everdur	A	Dry	PCH
		F	Moist	
159	Hastelloy	F	Wet gas	
160	Hastelloy	F	Wet gas	
161	Hastelloy	A	Wet gas	BIVPTSCH
162	Hastelloy	F	Wet gas	
163	Stellite	A	Wet gas	BIV
165	Stellite	A	Wet gas	BIV
184	Inconel	A	Dry	
		F	Wet; 70°	
185	Inc-clad	A	Dry; 575°	BTSCH
216	Monel	A	Dry	V
		X	Wet; 70°	
217	Monel-clad	A	Dry; 575°	BTSCH
219	Muntz	A	Dry	CH
		X	Moist	
224	Nickel	A	Dry	
		X	Wet; 70°	
225	Ni-clad	A	Dry; 575°	BTSCH
226	Ni-silver	A	Dry	Instruments
		V	Moist	
231	Ni-Resist	A	Dry	
234	Olympic	A	Dry; moist	
235	Olympic	A	Dry; moist	
245	Pyrasteel	A	....	BIVPTSCHF
249	Resistac	A	....	BIVCHF
270	304-clad	A	Dry; 575°	BTSCH
		A	Moist; 70°	
271	316-clad	A	Dry; 575°	BTSCH
		A	Moist; 70°	
274	430-clad	A	Dry; 575°	BTSCH
275	St. 301	V		
276	St. 302	V		
		A	Gas; 70°, 575°	VTSCDF
278	St. 303	V		
279	St. 304	V		
		A	Gas; 575°	VTSCDF
		A	Gas; 70°, 575°	BIPT
280	St. 308	V		
281	St. 309	V		
282	St. 310	V		
283	St. 316	V		
		A	Gas; 70°, 575°	VTCHDF
		A	Gas; 70°, 575°	BIPT
284	St. 317	V		
285	St. 321	V		
286	St. 347	V		
287	St. 403	V		
288	St. 405	V		
290	St. 410	V		
		F	Gas; 70°	BIVP
		A	Gas; 575°	BIVP
291	St. 414	V		
292	St. 416	V		
293	St. 418	V		
294	St. 420	V		
295	St. 430	V		
		F	Moist gas; 70°	
		A	Gas; 575°	VTSCHDF
296	St. 430F	V		
297	St. 431	V		
298	St. 440A	V		
300	St. 440C	V		
301	St. 442	V		
303	St. 446	V		
316	St. CF-20	A	....	BIV
319	St. CF-7M	A	Dry gas; 500°	
360A	Steel	A	Dry gas only	
367	Super Ni	A	Dry	PTCH
		V	Moist	
368	Tantalum	A	392° max.; wet or dry	CH
369	Telnic	A	Dry; moist	
390	Worthite	A	All %; any temp.	BIVP
401	Karbate	A	330° max.	PH
402	Karbate	A	330° max.	PH
403	Kempruf	A	All %; any temp.	PR
404	Acheson	A	All %; any temp.	PR
500	Sul. cement	A	200° max.	PTD
501	Sul. cement	A	200° max.	PTD
502	Furan cement	A	360° max.	PTD
503	Furan cement	A	360° max.	PTD
504	Phen. cement	A	360° max.	PTD
505	Phen. cement	A	360° max.	PTD
506	Silicate cement	V	1600° max.	PTD
507	Silicate cement	V	1600° max.	PTD
513	Asplit	A	All %; 350° max.	
514	Asplit-F	A	All %; 350° max.	
515	Basolit	A	....	T
518	C-Basolit	A	....	T
521	Causplit	A	All %; 350° max.	
523	Duralon	V	350° max.	BIPTCHDF
524	Durisite	V	350° max.	BIPTCHDF
532	Lumnite	A	10% max.; 1000° max.	TD
535	Nukem	A	....	T
536	Nukem	A	....	T
538	Penchlor	A	All %; 750° max.	
539	Penchlor	A	All %; 750° max.	
540	Penchlor	A	All %; 500–2000°	
541	Pennsalt	A	All %; 350° max.	
542	Permanite	A	360° max.	TD
544	Plastite	F	175° max.	PTSD
545	P-Basolit	A	....	T
559	Thiokol	A	150° max.	
600	Acid brick	A	....	TDR
603, 612, 615	Glass	A	....	THDR

Table 1. Chemical Resistance of Constructional Materials—(*Continued*)

No.	Material	Ratings	Exposure conditions	Applications
604	Ceratherm	A	200–400°, depending on design	BIPTSCHDFR
606	Stoneware	A	All conditions	BIVPTCHDFR
607	Glass-L	A	600° max.	PTCH
610	Stoneware	A	All %; any temp.	BIVPTCHDR
611	Porcelain	A	All % and temp.	BIPTDR
614	Glass-L	AV	....	BTCH
616	Pyrex	A		
617	Stoneware	A	140–160°	BIPTSCHDFR
618	Vitreo	A	1000° max.	
619	Vitreosil	A		
621	Vycor	A		
711	Haveg	A	300° max.	BIVPTDFR
713	Haveg	A	300° max.	BIVPTDFR
720	Lamicoid	A	....	T
726	Nukemite	A	....	PTSDF
733	Permanite	A	360° max.	BVPTD
736	Pyroflex	A	....	PTCHD
737	Resilon	F	175° max.	PTSD
740	Saran	F	77°	
741	Sealon	A	160° max.	BPTDF
742	Teflon	A	Rm.	VP
746	Tygon	F	180° max.	BIPTSCHDF
801	Acidseal	A	Sat.; 150° max.; wet	BIPTDF
809	Fairprene	F		
817	Heresite	A	Vapor	
836	Natural (S)	V		
837	Natural (H)	V		
838	GR-S (S)	V		
839	GR-S (H)	V		
840	GR-A (S)	V		
841	GR-A (H)	V		
842	GR-M (S)	V		
843	GR-P (S)	V		
844	GR-I	V		
846	Saniprene	A	Sat.; 150° max.; wet	BIPTDF
849	Superflexite	A	Sat.; 150° max.; wet	BIPTDF
853	Thiokol	A	150° max.	T
854	Thiokol	A	150° max.	PTSCDF

Sulfuric Acid

No.	Material	Ratings	Exposure conditions	Applications
3	Admiralty	V		
4	Admiralty	V	....	CH
6	Adnic	V	0.5–2.5%, 50%, 95%	
10–17	Aluminum	A	Fuming; rm.	S
		FX	Dil.; rm.	
19	Aloyco	A	All %; 70°	BV
		A	50% max.; boiling	BV
		A	50–93%; 150°	BV
		A	93% sulfuric to 65% oleum; 150°	BV
22	Ambraloy	V	....	CH
23	Ambraloy	V	....	CH
24	Ambraloy	V	....	CH
29–40	Ampco	A	To 10%; 212° max.	BIVPTCHDF
		V	10–75%; depends upon conditions	BIVPTCHDF
42	Antaciron	A	All conditions, with slurry, reducing or oxidizing	BIV
51	Berylco	A	Cold	
54	Brass	X		
61	Brass	X		
63	Brass	V	....	PCH
66	Bronze	V		
67	Bronze	A	Dil.; hot and cold	BIV
73	Bronze	V	....	VT
74	Bronze	V	....	V
75	Bronze	V	....	VP
76	Bronze	V		
77	Bronze	X		
81	CA-FA20	A	5% aerated and agitated; 70°	
		A	5% aerated and agitated; 176°	
		A	78% aerated and agitated; 176°	
		A	93% aerated and agitated; 70°	
82	CA-MM	A	5% aerated and agitated; 70°	
		F	5% aerated and agitated; 176°	
		X	73% aerated and agitated; 176°	
		F	93% aerated and agitated; 70°	
85	Cast iron	A	Over 77%	
86	Cast iron	A	75–95%, 98%-fuming	
		X	To 75%	
86A	Cast iron	A	Over 90%	
87	Causul	A	Cold or hot to 60° Bé.; cold above 60° Bé.	B
88	Chlorimet	A	All % and temp. except hot over 80%	BIVCHF
89	Chlorimet	A	Dil. all temp.; conc. moderate temp.	BIVHF
		F	Intermediate %	BIVHF
111	Copper	V		
114	Copper	V	....	PTCHR
118	Copper	V	95%	
119	Corrosiron	A	25, 87, 95%; rm.; unagitated; c.p.	BIVPTCHDFR
123	Cupro-Ni	V	....	CH
124	Cupro-Ni	V	95%	
139	Durco	A	All % to 176°	BIVHF
		A	Under 45% to boil.	BIVHF
140	Durichlor	A	All % and temp.	BIVPCHFR
141	Durimet	A	All % to 176°	BIVCHF
		A	Under 25% boil.	BIVHF
		F	78% hot	BIVHF
142	Durimet	A	Under 10% boil; all % to 176° except near 60° Bé. (141 preferred)	BIVHF
143	Duriron	A	All % and temp.	BIVPHFR
148	Everdur	V	....	BIVR
149	Everdur	V	....	PTCHDR
150	Everdur	V	....	PCH
156	Gold	A	To boil.	
159	Hastelloy	A	Under 50%; to boil.	BIVPTCHDF
		A	Over 50%; to 160°	BIVPTCHDF
160	Hastelloy	A	Under 60%; to boil.	BIVPTCHDF
		A	Over 60%; to 160°	BIVPTCHDF
161	Hastelloy	A	Under 50%; to boil.	BIVPTCHDF
		A	Over 50%; to boil.	BIVPTCHDF
162	Hastelloy	A	All % to boil.	BIVP
163	Stellite	A	All %; rm. only	BIV
165	Stellite	A	All %; rm. only	BIV
184	Inconel	A	5%; 86°; unaerated	
		X	5%; 86°; aerated	
		X	19%; boil.	
185	Inc-clad	A	Conc.; 70°	BTSCH
191	Lead	A	....	PTH
193, 196, 200, 266	Lead	A	Under 96% to 60°; under 85% to 428° (No. 193, 248° max.)	BIPTSCHDF
216	Monel	A	5%; 86°; unaerated	BIVPTCHDF
		F	5%; 86°; aerated	
		A	19%; boil.	
		A	45, 60%; 140°; unaerated	
		A	80%; 86°	
217	Monel-clad	A	Conc.; 70°	BTSCH
219	Muntz	X		
224	Nickel	A	5%; 86°; unaerated	VPTH
		X	5%; 86°; aerated	
		X	19%; boil.	
225	Ni-clad	A	Conc.; 70°	BTSCH
226	Ni-silver	V		
231	Ni-Resist	A	5%; 86°; unaerated	BIVP
		X	5%; 86°; aerated	
		X	19%; boil.	
233	NS-5	A	Under 60° Bé.; hot or cold	V
234	Olympic	V	95%	
235	Olympic	V	95%	
240	Platinum	A	To boil.	
242	Ir-Platinum	A	To boil.	
244	Rh-Platinum	A	To boil.	
245	Pyrasteel	V		
249	Resistac	A	66° Bé. max.	BIVCHF
270	304-clad	A	Conc.; 70°	BTSCH
271	316-clad	A	5%; 10%; fuming; 70°	BTSCH
274	430-clad	A	Conc.; 70°	BTSCH
275	St. 301	V	Conc.; rm.	
276	St. 302	V	Conc.; rm.; 60° Bé. min.; 180° max.	
		A	Diluted 1:20; 70°	
		X	Diluted 1:20; boil.	
		A	Diluted 1:10; 70°	
		X	Diluted 1:10; boil.	
		A	Diluted 1:1; 70°	
		X	Diluted 1:1; boil.	
		A	Conc.; 1:0; 70°	
		F	Conc.; 1:1; 212°	
		X	Conc.; 1:0; 300°	
		A	Fuming, 11% free SO_3; 212°	
		A	Fuming, 60% free SO_3; 70°, 160°	

Table 1. Chemical Resistance of Constructional Materials—(*Continued*)

No.	Material	Ratings	Exposure conditions	Applications
278	St. 303	V	Conc.; rm.; 60° Bé. min.; 180° max.	
279	St. 304	V	Conc.; rm.; 60° Bé. min.; 180° max.	
		A	5%; aerated and agitated; 70°	BIVP
		X	5%; aerated and agitated; 176°	
		X	78%; aerated and agitated; 176°	
		A	93%; aerated and agitated; 70°	BIVT
280	St. 308	V	Conc.; rm.; 60° Bé. min.; 180° max.	
281	St. 309	V	Conc.; rm.; 60° Bé. min.; 180° max.	
282	St. 310	V	Conc.; rm.; 60° Bé. min.; 180° max.	
283	St. 316	V	Conc. or 15° Bé.; rm.	
		V	5% at 120° max.; 60° Bé. min. 200° max.	
		A	Diluted 1:20; 70°	
		F	Diluted 1:20; boil.	
		A	Diluted 1:10; 70°	
		X	Diluted 1:10; boil.	
		A	Diluted 1:1; 70°	
		X	Diluted 1:1; boil.	
		A	Conc., 1:0; 70°	
		F	Conc., 1:0; 212°	
		X	Conc., 1:0; 300°	
		A	Fuming 11% free SO_3; 212°	
		A	Fuming 60% free SO_3; 70°, 160°	
		A	5%; aerated and agitated, 176°,	BIVP
284	St. 317	V	Over 15° Bé.; rm.	
285	St. 321	V	Conc.; rm.	
		V	Over 60° Bé.; 180°F. max.	
286	St. 347	V	Conc.; rm.	
		V	Over 60° Bé.; 180°F. max.	
287	St. 403	V	Conc.; rm.	
288	St. 405	V	Conc.; rm.	
290	St. 410	V	Conc.; rm.	
		X	5%; aerated and agitated; 176°	
		X	78%; aerated and agitated; 176°	
		A	93%; aerated and agitated; 70°	BIVP
291	St. 414	V	Conc.; rm.	
292	St. 416	V	Conc.; rm.	
293	St. 418	V	Conc.; rm.	
294	St. 420	V	Conc.; rm.	
295	St. 430	V	Conc.; rm.	
		F	Diluted 1:20, 70°	
		X	Diluted 1:20, boil.	
		F	Diluted 1:10, 70°	
		X	Diluted 1:10, boil.	
		A	Conc. 1:0, 70°	
		F	Conc. 1:0, 212°	
		X	Conc. 1:0, 300°	
296	St. 430F	V	Conc.; rm.	
297	St. 431	V	Conc.; rm.	
298	St. 440A	V	Conc.; rm.	
300	St. 440C	V	Conc.; rm.	
301	St. 442	V	Conc.; rm.	
303	St. 446	V	Conc.; rm.	
316	St. CF-20	A	....	BIV
360A	Steel	A	Over 90%	
364	Stoody	A	10%; rm.; boil.	
365	Stoody	A	10%; rm.	
		X	10%; boil.	
367	Super Ni	V		PTCH
368	Tantalum	V	98% max.; 347° max. at 98%; avoid free SO_3	H
369	Telnic	V	95% and under	
387	Stainless	X	All %	
388	Stainless	X	All %	
389	Stainless	A	Dilute; strong	
		X	Medium	
390	Worthite	A	All %; 125°	BIV
		A	1–50%, aerated; 175°	BIV
		A	96–100%, aerated; 175°	BIV
		A	100–110%; 200°	BIV
		F	50–93%, aerated; 140°	BIV
392	Wyndaloy	A	Very dilute	
		V	Other %	
401	Karbate	V	96% max.; 338° max.	BIVPTCHD
402	Karbate	V	96% max.; 338° max.	BIVPTCHD
403	Kempruf	A	100% max. to boil.	PTDR
		A	115% max. to 158°	PTDR
404	Acheson	A	50% max.; to boil.	PTDR
500	Sul. cement	A	50% max.; 200° max.	PTD
501	Sul. cement	A	50% max.; 200° max.	PTD
502	Furan cement	A	50% max.; 360° max.	PTD
503	Furan cement	A	50% max.; 360° max.	PTD
504	Phen. cement	A	50% max.; 360° max.	PTD
505	Phen. cement	A	50% max.; 360° max.	PTD
506	Silicate cement	A	Strong sol. only; 1600° max.	PTD
507	Silicate cement	A	Strong sol. only; 1600° max.	PTD
508	Acichlor	V	Dilute and conc.; 300°	
510	Acitite	A	Dilute and conc.; 250°	
513	Asplit	A	All %; 350° max.	
514	Asplit-F	A	All %; 350° max.	
515	Basolit	A	Diluted; 200° max.	T
517	Carboline	A	50% max.	
518	C-Basolit	A	Diluted; 200° max.	T
521	Causplit	A	All %; 350° max.	
523	Duralon	F	Sp. gr. 1.5; 350° max.	BIPTCHDF
524	Durisite	F	Sp. gr. 1.5; 350° max.	BIPTCHDF
532	Lumnite	A	1% max.; 90° max.	TD
535	Nukem	A	....	T
536	Nukem	A		T
537	Pecomastic	A	Dilute and conc.; 300°	
538	Penchlor	A	All %; 750° max.	
539	Penchlor	A	All %; 750° max.	
540	Penchlor	A	All %; 500–2000°	
541	Pennsalt	A	All %; 350° max.	
542	Permanite	V	Dil.; cold	TD
		X	Conc.; hot	
544	Plastite	F	175° max.	PTSD
545	P-Basolit	A	Dil.; 200° max.	T
554	Silastic	A	10–30%; (A.S.T.M. D-543-43)	
		X	Conc.; (A.S.T.M. D-543-43)	
556	Staminite	A	All %	T
559	Thiokol	X		
600	Acid brick	A	....	TDR
603, 612, 615	Glass	A	....	THDR
604	Ceratherm	A	200–400°, depending on design	BIPTSCHDFR
606	Stoneware	A	All conditions	BIVPTCHDFR
607	Glass-L	A	600° max.	PTCH
610	Stoneware	A	All %; any temp.	BIVPTCHDR
611	Porcelain	A	All % and temp.	BIPTDR
614	Glass-L	A	All %; 302° max.; agitation	BTCH
616	Pyrex	A		
617	Stoneware	A	140–160°	BIPTSCHDFR
618	Vitreo	A	1000° max.	
619	Vitreosil	A	....	PCH
621	Vycor	A		
700	Ace Saran	A	60%; 122°	P
704	DC Silicone	V		
706	Formica	F	25% max.; rm.	
707	Formica	F	25% max.; rm.	T
708	Formica	X	All %	
711	Haveg	V	75%; cold	BIVPTFDR
		V	50%; 300° max.	BIVPTFDR
718	Koroseal	A	50%; 150°	TD
		A	60%; 100°	TD
720	Lamicoid	X		
723	Nixon	A	3%	S
		F	30%	S
724	Nixon	A	3–30%	S
726	Nukemite	A	40%; rm.	PTSDF
727	Nylon	X	Conc.	
728	Nylon	X	Conc.	
729–732	Nylon	X		
733	Permanite	V	Dil.; cold	PVBDT
		X	Conc.; hot	
735	Polythene	X	Conc.; hot	
		V	Rm.	
736	Pyroflex	V	....	PTDCH
737	Resilon	F	175° max.	PTSD
740	Saran	F	77°	
741	Sealon	A	160° max.	BTPDF
742	Teflon	A	Rm.	VP
744, 745	Textolite	V		BI
746	Tygon	V	Dil. to med. conc.; '80° max.	BIPTSCHDF
800	Ace Hd. Rub.	A	50%	BIVPTDF
801	Acidseal	A	50%; 150° max.	BIPTSDF
		A	60%; 100° max.	BIPTSDF
802	Acidseal	A	50%; 150° max.	BIPTSDF
		A	60%; 100° max.	BIPTSDF
805	Butyl	V	70%; 140° max.	
809	Fairprene	V		
814	G.E. Silicone	V	....	VPTDF
817	Heresite	A	50%; 122° max.	

Table 1. Chemical Resistance of Constructional Materials—(*Continued*)

No.	Material	Ratings	Exposure conditions	Applications
836	Natural (S)	V		
837	Natural (H)	V		
838	GR-S (S)	V		
839	GR-S (H)	V		
840	GR-A (S)	V		
841	GR-A (H)	V		
842	GR-M (S)	V		
843	GR-P (S)	X		
844	GR-I	V		
846	Saniprene	A	50%; 150° max.	BIPTSDF
		A	60%; 100° max.	BIPTSDF
848	Silastic	A	10–30%; (A.S.T.M. D–543–43)	
		X	Conc.; (A.S.T.M. D–543–43)	
849	Superflexite	A	50%; 150° max.	BIPTSDF
		A	60%; 100° max.	BIPTSDF
853	Thiokol	X	150° max.	
854	Thiokol	F	150° max.	PTSCDF
855	Triflex	A	50%; 150° max.	BIPTSDF
		A	60%; 100° max.	BIPTSDF
856	Vistanex	V	70%; 140° max.	
913	Redwood	A	5–12%; 36–190°; Falsewood-L; steel pickle	T
		A	5–8% + dil. HNO_3; 36–90°; asphalt-L; explosives mfr.	T
		A	pH4.1; 36–80°; 17 years; plating works	T
		A	pH5.8; 36–80°; 20 years; plating works	T
		A	4–5%; 36–140°; 37 years; armor plate pickling	T
		A	3%; 36–90°; 7 years; plating works	T
		A	2% + 1.5% HCl; 40–180°; 21 years; chemical plant	T
		A	1.5%; 36–120°; 27 years; tannery	T
		A	1% + 1–5% $CuSO_4$; 36–170°; 13 years chemical plant	T
		A	0.5% + dil. NaOH; 40–115°; 2 years; soap specialties	T
		A	Dil. + dil. HCl + dil. lead acetate; 40–150°; 13 years; paint	T
		A	Dil.; 60-200°; 2 years; petroleum refinery	T

Sulfurous Acid

No.	Material	Ratings	Exposure conditions	Applications
3	Admiralty	A		
4	Admiralty	F	....	CH
6	Adnic	A		
10–17	Aluminum	F		
19	Aloyco	A	70%	BV
		A	Sat.; 60 lb./sq. in.; 250°	BV
		A	Sat.; 125 lb./sq. in.; 310°	BV
		A	Sat.; 150 lb./sq. in.; 375°	BV
22	Ambraloy	F	....	CH
23	Ambraloy	F	....	CH
24	Ambraloy	F	....	CH
29–40	Ampco	A	Cold	BIVPTCHDF
		V	Hot	
54	Brass	X		
61	Brass	X		
63	Brass	FA	....	PCH
66	Bronze	A		
73	Bronze	F	....	VT
74	Bronze	F	....	V
75	Bronze	F	....	VP
76	Bronze	F		
77	Bronze	X		
81	CA-FA20	A	Sat.; 70°	BIPT
82	CA-MM	F	Sat.; 70°	BIPT
86A	Cast iron	X		
89	Chlorimet	A	All % and temp.	BIVHF
111	Copper	A		
114	Copper	F	....	PTCHR
118	Copper	A		
123	Cupro-Ni	V	....	CH
124	Cupro-Ni	A		
139	Durco	A	All % and temp.	BIVHF
140	Durichlor	X		
141, 142	Durimet	A	All % and temp.	BIVHF
143	Duriron	X		
148	Everdur	F	....	BIVR
149	Everdur	F	....	PTCHDR
150	Everdur	F	....	PCH
159	Hastelloy	X	All % to boil.	
160	Hastelloy	X	All % to boil.	
161	Hastelloy	A	All % to boil.	BIVPTCHDF
162	Hastelloy	X	All % to boil.	
163	Stellite	A	All % to boil	BIV
165	Stellite	A	All % to boil.	BIV
184	Inconel	F	1%; 68°	
185	Inc-clad	A	....	BTSCH
196, 200, 266	Lead	A	428° max.	BIPTH
216	Monel	F	1%; 68°	
217	Monel-clad	A	....	BTSCH
219	Muntz	X		
224	Nickel	F	1%; 68°	
225	Ni-clad	A	....	BTSCH
226	Ni-silver	V		
234	Olympic	A		
235	Olympic	A		
245	Pyrasteel	A	....	BIVPTSCHF
275	St. 301	V		
276	St. 302	V		
		A	Sat. cold; 60 lb./sq. in.; 70°, 275°	
		F	Sat. cold; 70–125 lb./sq. in.; 320°	
		F	Sat. cold; 150 lb./sq. in.; 350°	
		F	Sat. cold; 200 lb./sq. in.; 400°	
		F	Sat. cold; 300 lb./sq. in.	
278	St. 303	V		
279	St. 304	V		
		F	Sat.; 70°	BIVP
280	St. 308	V	....	PT
281	St. 309	V		
282	St. 310	V		
283	St. 316	V		
		A	Sat. cold; 60 lb./sq. in.; 70°, 275°	
		A	Sat. cold; 75-125 lb./sq. in.; 320°	
		A	Sat. cold; 150 lb./sq. in.; 350°	
		A	Sat. cold; 200, 300 lb./sq. in.; 400°	
		A	Sat.; 70°	BIVP
284	St. 317	V		
285	St. 321	V		
286	St. 347	V	....	T
287	St. 403	F	Rm.	
288	St. 405	F	Rm.	
290	St. 410	F	Rm.	
		X	Sat.; 70°	
291	St. 414	F	Rm.	
292	St. 416	F	Rm.	
293	St. 418	F	Rm.	
294	St. 420	F	Rm.	
295	St. 430	F	Rm.	
		F	Sat. cold; 75–125 lb./sq. in.; 320°	
		F	Sat. cold; 150 lb./sq. in.; 350°	
296	St. 430F	F	Rm.	
297	St. 431	F	Rm.	
298	St. 440A	F	Rm.	
300	St. 440C	F	Rm.	
301	St. 442	F	Rm.	
303	St. 446	F	Rm.	
316	St. CF-20	A	....	BIV
360A	Steel	X		
362	Still Metal	A	....	BIVF
367	Super Ni	V	....	PTCH
368	Tantalum	A	All %; 392° max.	Not used commercially
369	Telnic	A		
390	Worthite	A	All %; any temp.	BIVP
401	Karbate	A	All %; to boil.	BIVPCH
402	Karbate	A	All %; to boil.	BIVPCH
403	Kempruf	A	All %; any temp.	DR
404	Acheson	A	All %; any temp.	DR
500	Sul. cement	A	200° max.	PTD
501	Sul. cement	A	200° max.	PTD
502	Furan cement	A	360° max.	PTD
503	Furan cement	A	360° max.	PTD
504	Phen. cement	A	360° max.	PTD
505	Phen. cement	A	360° max.	PTD
506	Silicate cement	F	1600° max.	TD
507	Silicate cement	F	1600° max.	TD
600	Acid brick	A	....	TDR
603, 612, 615	Glass	A	....	THDR
604	Ceratherm	A	200–400°, depending on design	BIPTSCHDFR
606	Stoneware	A	All conditions	BIVPTCHDFR

Table 1. Chemical Resistance of Constructional Materials—(*Continued*)

No.	Material	Ratings	Exposure conditions	Applications
607	Glass-L	A	600° max.	PTCH
610	Stoneware	A	All %; any temp.	BIVPTCHDR
611	Porcelain	A	All % and temp.	BIPTDR
614	Glass-L	A	302° max; agitation	BTCH
616	Pyrex	A		
617	Stoneware	A	140–160°	BIPTSCHDFR
618	Vitreo	A	1000° max.	
619	Vitreosil	A		
621	Vycor	A		
700	Ace Saran	A	122°	P
711	Haveg	A	All %	BIVPTDFR
713	Haveg	A	All %	BIVPTDFR
720	Lamicoid	A	...	T
726	Nukemite	A	150° max.	PTSDF
733	Permanite	A	360° max.	BVPTD
736	Pyroflex	V	...	PTCHD
737	Resilon	A	175° max.	PTSD
740	Saran	F	77°	
741	Sealon	A	160° max.	BPTDF
742	Teflon	A	Rm.	VP
744, 745	Textolite	V	...	BI
746	Tygon	V	Dil. to med. conc.; 180° max.	BIPTSCHDF
800	Ace Hd. Rub.	A	...	BIVPTDF
801	Acidseal	A	Sat.; 150° max.	BIPTDF
809	Fairprene	A		
814	G.E. Silicone	V	...	VPTDF
817	Heresite	A	Sat.; 212° max.	
836	Natural (S)	V		
837	Natural (H)	V		
838	GR-S (S)	V		
839	GR-S (H)	V		
840	GR-A (S)	V		
841	GR-A (H)	V		
842	GR-M (S)	V		
843	GR-P (S)	X		
844	GR-I	V		
846	Saniprene	A	Sat.; 150° max.	BIPTDF
849	Superflexite	A	Sat.; 150° max.	BIPTDF
853	Thiokol	X		
854	Thiokol	F	150° max.	PTSCDF

Trichlorethylene

No.	Material	Ratings	Exposure conditions	Applications
3	Admiralty	A		
4	Admiralty	A	Dry	CH
		F	Moist	
6	Adnic	A		
10–17	Aluminum	A	Dry; rm. boil.	
22	Ambraloy	A	Dry	CH
		F	Wet	
23	Ambraloy	A	Dry	CH
		F	Wet	
24	Ambraloy	A	Dry	CH
		F	Wet	
29–40	Ampco	F	In absence of hydrolysis and water or steam	BIVPTCHDF
54	Brass	A	Dry	CH
		V	Moist	
61	Brass	A	Dry	VCH
		V	Moist	
63	Brass	A	Dry	PCH
		F	Moist	
66	Bronze	A		
73	Bronze	A	Dry	VT
		F	Moist	
74	Bronze	A	Dry	V
		F	Moist	
75	Bronze	A	Dry	VP
		F	Moist	
76	Bronze	A	Dry	Instruments
		F	Moist	
77	Bronze	A	Dry	PCH
		V	Moist	
85	Cast iron	A		
86	Cast iron	V		
86A	Cast iron	A	Dry only	
88	Chlorimet	A	All % and temp.	BIVCHF
89	Chlorimet	A	All % and temp.	BIVHF
111	Copper	A		
114	Copper	A	Dry	PTCHR
		F	Moist	
117	Copper	F	Absence of moisture	P
118	Copper	A		
123	Cupro-Ni	A	Dry	CH
		F	Moist	
124	Cupro-Ni	A		
127–133	Dowmetal	A	Dry	
139	Durco	A	All % and temp.	BIVHF
140	Durichlor	A	All % and temp.	BIVPCHFR
141, 142	Durimet	A	All % and temp.	BIVHF
143	Duriron	A	All % and temp.	BIVPHFR
148	Everdur	A	Dry	BIVR
		F	Moist	
149	Everdur	A	Dry	PTCHDR
		F	Moist	
150	Everdur	A	Dry	PCH
		V	Moist	
159	Hastelloy	A	All % to boil.	BIVPTCH
160	Hastelloy	A	All % to boil.	BIVPTCH
161	Hastelloy	A	All % to boil.	BIVTCH
162	Hastelloy	A	All % to boil.	BIV
163	Stellite	A	All % and temp.	BIV
165	Stellite	A	All % and temp.	BIV
184	Inconel	A	Boil.	
216	Monel	A	Boil.	BIVPTCHDF
217	Monel-clad	A	...	BTSCH
219	Muntz	A	...	CH
224	Nickel	A	Boil.	TCH
225	Ni-clad	A	...	BTSCH
226	Ni-silver	A	...	Instruments
231	Ni-Resist	A	Boil.	BIV
234	Olympic	A		
235	Olympic	A		
245	Pyrasteel	V		
275	St. 301	V		
276	St. 302	V		
		A	Boil.	
279	St. 304	V		
283	St. 316	A	...	PT
284	St. 317	A	...	PT
285	St. 321	V		
286	St. 347	V		
287	St. 403	X		
288	St. 405	X		
290	St. 410	X		
291	St. 414	X		
292	St. 416	X		
293	St. 418	X		
294	St. 420	X		
295	St. 430	X		
296	St. 430F	X		
297	St. 431	X		
298	St. 440A	X		
300	St. 440C	X		
301	St. 442	X		
303	St. 446	X		
360A	Steel	A	Dry only	
367	Super Ni	A	Dry	PTCH
		F	Moist	
368	Tantalum	A	...	Not used commercially
369	Telnic	A		
390	Worthite	A	175°	
401	Karbate	A	All %; to boil.	BIVPH
402	Karbate	A	All %; to boil.	BIVPH
403	Kempruf	A	All %; any temp.	
404	Acheson	A	All %; any temp.	
500	Sul. cement	X		
501	Sul. cement	X		
502	Furan cement	A	200° max.	PTD
503	Furan cement	A	220° max.	PTD
504	Phen. cement	A	220° max.	PTD
505	Phen. cement	A	220° max.	PTD
506	Silicate cement	X		
507	Silicate cement	V	1600° max.	PTD
600	Acid brick	A	...	TDR
603, 612, 615	Glass	A	...	THR
604	Ceratherm	A	200–400°, depending on design	BIPTSCHDFR
606	Stoneware	A	All conditions	BIVPTCHDFR
607	Glass-L	A	600° max.	PTCH
610	Stoneware	A	All %; any temp.	BIVPTCHDR
611	Porcelain	A	All % and temp.	BIPTDR
614	Glass-L	A	302° max.; agitation	BTCH
616	Pyrex	A		
617	Stoneware	A	140–160°	BIPTSCHDFR
618	Vitreo	A	1000° max.	
619	Vitreosil	A		
621	Vycor	A		
703	Compar	A		
713	Haveg	A	All conditions	BIVPTDFR
718	Koroseal	X		
720	Lamicoid	A	...	T
726	Nukemite	X		
729–732	Nylon	X		
733	Permanite	A	360° max.	BPVTD
735	Polythene	V	Rm.; swells slightly	
		X	176°	
736	Pyroflex	V	...	PTCHD
737	Resilon	X		

Table 1. Chemical Resistance of Constructional Materials—(*Concluded*)

No.	Material	Ratings	Exposure conditions	Applications
738	Resistoflex	A		
739	Resistoflex	A		
740	Saran	V	77°	
741	Sealon	V	160° max.	BPTDF
742	Teflon	A	Rm.	VP
746	Tygon	X		
800	Ace Hd. Rub.	X		
801	Acidseal	X		
802	Acidseal	X		
805	Butyl	X		
809	Fairprene	X		
817	Heresite	X		
836	Natural (S)	X		
837	Natural (H)	X		
838	GR-S (S)	X		
839	GR-S (H)	X		
840	GR-A (S)	X		
841	GR-A (H)	X		
842	GR-M (S)	X		
843	GR-P (S)	V		
844	GR-I	X		
846	Saniprene	X		
848	Silastic	A	(A.S.T.M. D-543-43)	
849	Superflexite	X		
853	Thiokol	V	150° max.	
854	Thiokol	X		
855	Triflex	X		
856	Vistanex	X		

Table 2. Resistance of Metals and Alloys to Corrosive Gases*

Part I. Typical Composition of Alloys for Resistance to Corrosive Gases (in Presence of Moisture)

Composition	Representative Trade Names or Type Number (See Tables 3 and 5)
Moist SO_2[a]	
14–18 Cr, bal. Fe, 0.12 max. C	Stainless type 430[f]
23–30 Cr, bal. Fe, 0.35 max. C	Stainless type 446
17.5–19 Cr, 8–9 Ni, bal. Fe, 0.08 max. C	Stainless type 304[g]
18–20 Cr, 9–10 Ni, bal. Fe, 0.08 max. C	Stainless type 304
22–26 Cr, 12–14 Ni, bal. Fe, 0.20 max. C	Stainless type 309
17–19 Cr, 7–9 Ni, 1–1.5 Cu, 1–1.5 Mo, bal. Fe, 0.15 max. C	Stainless type 315
18–20 Cr, 14 max. Ni, 3–4 Mo, bal. Fe, 0.10 max. C	Stainless type 317
19 Cr, 22 Ni, 3.5 Mo, 1.25 Si, 0.07 max. C, bal. Fe	Durimet T
14.5 Si, 0.8 C, bal. Fe	Duriron
14 Cr, 58 Ni, 17 Mo, 5 W, 6 Fe	Hastelloy C
99.93 Pb, 0.06 Cu	Chemical lead
94 Pb, 6 Sb	Antimonial lead
100 Ta	Tantalum
Moist Nitrous Gases[b]	
14–18 Cr, bal. Fe, 0.12 max. C	Stainless type 430[f]
23–30 Cr, bal. Fe, 0.35 max. C	Stainless type 446
17.5–19 Cr, 8–9 Ni, bal. Fe, 0.08 max. C	Stainless type 304[g]
18–20 Cr, 9–10 Ni, bal. Fe, 0.08 max. C	Stainless type 304
22–26 Cr, 12–14 Ni, bal. Fe, 0.20 max. C	Stainless type 309
17–19 Cr, 7–9.5 Ni, 1–1.5 Cu, 1–1.5 Mo, bal. Fe, 0.15 max. C	Stainless type 315
18–20 Cr, 14 max. Ni, 3–4 Mo, bal. Fe, 0.10 max. C	Stainless type 317
14 Cr, 58 Ni, 17 Mo, 5 W, 6 Fe	Hastelloy C
55–60 Ni, 18–24 Cr, 5–7 Mo, 4–8 Fe, 1–8 Cu	Illium G
14.5 Si, 0.8 C, bal. Fe	Duriron
100 Ta	Tantalum
Moist Hydrochloric Acid Vapors[c]	
62 Ni, 30 Mo, 5 Fe	Hastelloy B
14 Cr, 58 Ni, 17 Mo, 5 W, 6 Fe	Hastelloy C[h]
14.5 Si, 3 Mo, bal. Fe	Durichlor
100 Ta	Tantalum
Moist Chlorine[d]	
14 Cr, 58 Ni, 17 Mo, 5 W, 6 Fe	Hastelloy C[h]
100 Ta	Tantalum
Moist Hydrofluoric Acid Vapors, Cold[h]	
55–60 Ni, 18–24 Cr, 5–7 Mo, 4–8 Fe, 1–8 Cu	Illium G
99.93 Pb, 0.06 Cu	Chemical lead
86 Cu, 10 Al, 4 Fe	Ampco

* Based on Chilton and Huey [*Ind. Eng. Chem.*, **24**, 125 (1932)]; "Tables of Chemical Compositions, Physical and Mechanical Properties and Corrosion-resistant Properties of Corrosion-resistant and Heat-resistant Alloys" [*Proc. Am. Soc. Testing Materials*, **30**, Part I, Suppl. (1930)]; "Modern Alloys" [*Chem. & Met. Eng.*, **39**, 497 (1932)]; and other sources.

[a] Alloys from column headed Moist Sulfurous Atmosphere, A.S.T.M. Tables, 1930.

[b] Alloys taken from column headed Nitric Acid, recommendations for all concentrations at room and some higher temperature.

[c] Alloys taken from column headed Hydrochloric Acid, A.S.T.M. Tables, 1930; same recommendations.

[d] Alloys taken from column headed Chlorine in Aqueous Solution; same recommendations.

[e] Alloys taken from column headed Hydrofluoric Acid, A.S.T.M. Tables, 1930, and are the only ones unqualifiedly recommended by the manufacturers; none are recommended for hot solutions.

[f] Nearest type number to compositions listed, A.S.T.M. Tables, 1930.

[g] Columbium-modified compositions also applicable.

[h] Low temperatures only.

Part II. Typical Composition of Alloys for Resistance to Direct Gaseous Attack (in Absence of Moisture)

Max. temp. recommended, °C.	Composition	Representative trade names or type number (see Tables 3 and 5)
	Attack by Air and Oxidizing Fuel Gases[a]	
500	70 Cu, 29 Ni, 1 Sn	Adnic
500[b,c]	67 Ni, 30 Cu, 1.4 Fe, 1 Mn	Monel
540	14 Cr, 0.35 C, bal. Fe	Stainless type 420[d]
595	13 Cr, 2 Ni, 0.12 C, bal. Fe	Stainless type 414
700	72 Ni, 18 Co, 6.5 Fe, 2.5 Ti, 0.5 Al	Konel
700[b,c]	99.4 Ni	Nickel
700[b,c]	79.5 Ni, 13 Cr, 6.5 Fe	Inconel
760[b]	18–20 Cr, 9–10 Ni, 0.2 C, bal. Fe	Stainless Type 302
800[b]	58 Ni, 20 Cr, 20 Fe, 2 Mn	Hastelloy A
800[b]	85 Ni, 10 Si, 3 Cu, 2 Al	Hastelloy D
815	14–18 Cr, bal. Fe	Stainless type 440
815[b,c]	22–26 Cr, 12–14 Ni	Stainless type 309
815[b,c]	36 Ni, 11–15 Cr, bal. Fe	ATV-1
925[b]	23–30 Cr, 0.35 max. C, bal. Fe	Stainless type 446
950[b,c]	26.5 Ni, 14 Cr, 3.5 W, bal. Fe	ATV-3
1000	50 Co, 30 Cr, 15.5 W, bal. Fe	Stellite No. 1
1000	65 Co, 30 Cr, 4 W, bal. Fe	Stellite No. 6
1000	60 Co, 30 Cr, 8 W, bal. Fe	Stellite No. 12
1000[b,c]	58 Ni, 17 Mo, 14 Cr, 5 W, 6 Fe	Hastelloy C
1150[c]	80 Ni, 20 Cr	Nichrome V
	Attack by Sulfur Gases[e]	
540	18–20 Cr, 9–10 Ni, 0.2 C, bal. Fe	Stainless type 302
540	14 Cr, 0.35 C, bal. Fe	Stainless type 420[d]
815	14–18 Cr, bal. Fe	Stainless type 440
925	23–30 Cr, 0.35 max. C, bal. Fe	Stainless type 446
	Attack by Hydrogen, Nitrogen and Ammonia[f]	
480	18–20 Cr, 9–10 Ni, 0.2 C, bal. Fe	Stainless type 302
540	22–26 Cr, 12–14 Ni	Stainless type 309
705	18–20 Cr, 14 max. Ni, 3–4 Mo, bal. Fe, 0.10 max. C	Stainless type 317
785	20 Ni, 8 Cr, 1 Si, 0.4 C, bal. Fe	Cyclops No. 17
900	35 Ni, 18 Cr, bal. Fe	Chromax
1000	45 Ni, 55 Cu	Advance

* Based on Chilton and Huey [*Ind. Eng. Chem.*, 24, 125 (1932)]; "Tables of Chemical Compositions, Physical and Mechanical Properties and Corrosion-resistant Properties of Corrosion-resistant and Heat-resistant Alloys" [*Proc. Am. Soc. Testing Materials*, **30**, Part I, Suppl. (1930)]; "Modern Alloys," [*Chem. & Met. Eng.*, **39**, 497 (1932)]; and other sources.

[a] Alloys are listed under the lowest maximum operating temperature recommended by any manufacturer for material of substantially the same composition (A.S.T.M. Tables, 1930). Manufacturers vary widely in their recommendations; some recommend alloys of the same or even lower alloy content for higher temperatures.

[b] Recommended for oxidizing fuel gases at the same temperature (A.S.T.M. Tables, 1930).

[c] Nickel alloys recommended only in absence of sulfur-containing gases.

[d] Nearest type number to compositions listed, A.S.T.M. Tables, 1930.

[e] Only those alloys recommended for H_2S, SO_2, and SO_3 are included; the temperature is the lowest recommended by any manufacturer for substantially the same composition (A.S.T.M. Tables, 1930).

[f] Taken from column with this heading in A.S.T.M. Tables, 1930; the temperatures given are the lowest recommended by any manufacturer for any of the three gases for substantially the same alloy composition. Note that the metals should not be used at high pressure at these temperatures; see Maxwell [*Trans. Am. Soc. Metals*, 24, 213 (1936)].

Table 3. Directory of Materials for the Construction of Chemical Equipment*

Metals and Alloys

No.	Material	Manufacturer	Essential Nominal Chemical Composition, %	Forms Available	Primarily for
1	Acipco stainless	Amer. Cast Iron Pipe Co., Birmingham, Ala.	Various standard stainless steels; see Nos. 275–360	C, T (centrifugal)	
2	Admiralty	Generally available†	70 Cu, 1 Sn; 29 Zn	CR, D, E, P, S, T, R, W	C
3	Admiralty, antimonial	Chase Brass & Copper Co., Waterbury, Conn.	71 Cu; 27.95 Zn; 1 Sn; 0.05 Sb	T, W	C
4	Admiralty, arsenical	Amer. Brass Co., Waterbury, Conn.	70 Cu; 28.95 Zn; 1 Sn; 0.05 As	T	C
5	Admiralty, phosphorized	Scovill Mfg. Co., Waterbury, Conn.	70 Cu; 29 Zn; 1 Sn; 0.03 P	....	C
6	Adnic	Scovill Mfg. Co., Waterbury, Conn.	70 Cu; 29 Ni; 1 Sn	S	C, A
7	Advance	Driver-Harris Co., Harrison, N.J.	55 Cu; 45 Ni	B, CR, HR, P, R, S, W	H
8	Alchrome 3	Wilbur B. Driver Co., Newark, N.J.	Fe; 20 Cr; 3 Al	R, W	H
9	Alchrome 6	Wilbur B. Driver Co., Newark, N.J.	Fe; 20 Cr; 6 Al	R, W	H
10	Alcoa Alclad 35	Aluminum Co. of Amer., Pittsburgh, Pa.	Surface layer of Al alloy which is anodic to core and therefore protects it electrolytically	CR, HR, P, S, T	C
11	Alcoa Alloy 43	Aluminum Co. of Amer., Pittsburgh, Pa.	Al; 5 Si	C	C
12	Alcoa Alloy B214	Aluminum Co. of Amer., Pittsburgh, Pa.	Al; 1.8 Si; 3.8 Mg	C	C
13	Alcoa Alloy 356-T4	Aluminum Co. of Amer., Pittsburgh, Pa.	Al; 7 Si	C	C
14	Alcoa 2S	Aluminum Co. of Amer., Pittsburgh, Pa.	99 min. Al	B, CR, HR, D, E, F, P, S, R, T, W	C
15	Alcoa 3S	Aluminum Co. of Amer., Pittsburgh, Pa.	Al; 1.2 Mn	CR, HR, D, E, P, S, T	C
16	Alcoa 52S	Aluminum Co. of Amer., Pittsburgh, Pa.	Al; 2.5 Mg; 0.25 Cr	D, E, P, S, T	C
17	Alcoa 61S	Aluminum Co. of Amer., Pittsburgh, Pa.	Al; 1 Mg; 0.6 Si; 0.25 Cr; 0.25 Cu	B, CR, HR, D, E, F, P, S, R, T, W	C
18	Allegheny metal	Allegheny Ludlum Steel Corp., Brackenridge, Pa.	Various standard stainless steels; see Nos. 275–360	B, C, CR, P, S, R, T, W	C, H, A
19	Aloyco-20	Alloy Steel Products Co., Linden, N.J.	Fe; 19–21 Cr; 28–30 Ni: 4.0–4.5 Cu; 2.5–3.0 Mo; 1.5 max. Si; 0.65–0.85 Mn; 0.07 max. C	C	C
20	Alray D	Alloy Metal Wire Co., Prospect Park, Pa.	Fe; 35 Cr; 15 Ni	W	H
21	Alumel	Hoskins Mfg. Co., Detroit, Mich.	94 Ni; 4 Al; 1 Si; 1 Mn	CR, W	H
22	Ambraloy 901	Amer. Brass Co., Waterbury, Conn.	95 Cu; 9 Al	B, P, S, R, T, W	C
23	Ambraloy 917	Amer. Brass Co., Waterbury, Conn.	82 Cu; 9½ Al; 5 Ni; 2½ Fe; 1 Mn	B, P, R	C
24	Ambraloy 927	Amer. Brass Co., Waterbury, Conn.	76 Cu; 21.95 Zn; 2 Al; 0.05 As	T	C
25	American stainless	Amer. Steel Castings Co., Newark, N.J.	Various standard stainless steels; see Nos. 275–360	C	C, H
26	American AW	Amer. Steel Castings Co., Newark, N.J.	Fe; 23.5 Ni; 19 Cr; 0.07 max. C	C	C
27	A metal	Midvale Co., Philadelphia, Pa.	Fe; 19 Cr; 35 Ni; 0.35 C; 1 Si; 0.5 Mn	C	H
28	AMF	Midvale Co., Philadelphia, Pa.	Fe; 46–50 Ni; 0.1–0.2 C; 1–2 Mn	C	C
29	Ampco 8	Ampco Metal, Inc., Milwaukee, Wis.	Cu; 7½ Al; 2–2½ Fe	B, CR, HR, D, P, S, R, T, W	C, A
30	Ampco 12	Ampco Metal, Inc., Milwaukee, Wis.	Cu; 8.5–9.3 Al; 2.5–3.25 Fe	C	C, A
31	Ampco 15	Ampco Metal, Inc., Milwaukee, Wis.	Cu; 9–10 Al; 2.75–3.75 Fe	B, CR, HR, D, P, S, R, T	C, A
32	Ampco 16	Ampco Metal Inc., Milwaukee, Wis.	Cu; 9.6–10.3 Al; 3–4 Fe	C	C, A
33	Ampco 18	Ampco Metal, Inc., Milwaukee, Wis.	Cu; 10.3–11 Al; 3.0–4.25 Fe	B, C, D, R, T	C, A
34	Ampcoloy 40	Ampco Metal, Inc., Milwaukee, Wis.	Cu; 9.5–10.5 Al; 0.20 max. Fe	C	C, A
35	Ampcoloy 45	Ampco Metal, Inc., Milwaukee, Wis.	Cu; 9.7–10.9 Al; 2.0–3.5 Fe; 4.5–5.5 Ni; 1.5 max. Mn	B, C, D, R, T	C, A
36	Ampcoloy 49	Ampco Metal, Inc., Milwaukee, Wis.	Cu; 6½–9 Al; 4 max. Fe; 2 max. Mn	B, D, R, T, W	C, A
37	Ampcoloy 62	Ampco Metal, Inc., Milwaukee, Wis.	55–60 Cu; 0.5–1.5 Al; 0.4–2.0 Fe; 1.5 max. Mn; 1.0 max. Sn; 0.4 max. Pb; 0.5 max. Ni; bal. Zn	C	C, A
38	Ampcoloy 66	Ampco Metal, Inc., Milwaukee, Wis.	60–68 Cu; 3–7 Al; 2–4 Fe; 2.5–5 Mn; 0.5 max. Sn; 0.2 max. Pb; 0.5 max. Ni; bal. Zn	C	C, A,
39	Ampcoloy A3	Ampco Metal, Inc., Milwaukee, Wis.	Cu; 9–10 Al; 1.25 max. Fe	B, C, D, R, T	C, A
40	Ampcoloy E123	Ampco Metal, Inc., Milwaukee, Wis.	Cu; 9.5–11 Al; 1.5 max. Fe	C	C, A
41	Amsco Alloys	Amer. Manganese Steel Div., Amer. Brake Shoe Co., Chicago Heights, Ill.	Now produced under trade name Thermalloy by Electro Alloys Div.		
42	Antaciron	Worthington Pump & Machinery Corp., Harrison, N.J.	Fe; 14.5 Si	C	CA
43	Armco and Rustless stainless	Amer. Rolling Mill Co., Middletown, Ohio	Various standard stainless steels; see Nos. 275–360	B, CR, HR, D, P, S, W, strip	C, H, A
44	Armco aluminized steel	Amer. Rolling Mill Co., Middletown, Ohio	Hot-dip Al coating on mild or Cu-bearing steel	S, strip	C, H
45	Armco galvanized Paintgrip	Amer. Rolling Mill Co., Middletown, Ohio	Bonderized galvanized sheets	S, strip	C
46	Armco galvanized steel or ingot iron	Amer. Rolling Mill Co., Middletown, Ohio	Hot-dip galvanized on mild or Cu-bearing steel or Armco Ingot Iron	S, strip	C
47	Armco Zincgrip	Amer. Rolling Mill Co., Middletown, Ohio	Galvanized steel	S, strip	
48	Armco Zincgrip Paintgrip	Amer. Rolling Mill Co., Middletown, Ohio	Bonderized galvanized steel	S, strip	
49	Asarco Lead	Amer. Smelting & Ref. Co., New York, N.Y.	Pb; 0.06 Cu; 0.02 Bi	C, CR, R, S, T, W	C
50	Beraloy	Wilbur B. Driver Co., Newark, N.J.	Cu; 1.9 Be.; 0.5 max. Co; 0.5 max. Ni	B, CR, D, HR	
51	Berylco 25	Beryllium Corp., Reading, Pa.	Cu; 1.9–2.15 Be; 05.0–0.4 Co; 0.1 max. Fe; 0.15 max. Si	B, C, CR, HR, P, S, R, T, W	C
52	Borod	Stoody Co., Whittier, Calif.	60 tungsten carbide; 40 steel	Welding rod	A
53	Brass, aluminum	Generally available†	76 Cu; 21.5–22 Zn; 2–2.5 Al	CR, D, T	

* This portion of *Chemical Engineering's* Twelfth Report on Materials of Construction provides a quick means of identifying materials available for the construction of chemical plant equipment. Generous cooperation on the part of the manufacturers has made it possible to include nearly 600 items. Materials are arranged alphabetically within each of seven classes of material. Numbers in the first column are used to identify materials appearing in Table 1.

† "Generally available" copper alloys may be obtained from such companies as the following: American Brass Co., Waterbury, Conn.; Bridgeport Brass Co., Bridgeport, Conn.; Bristol Brass Co., Bristol, Conn.; Chase Brass & Copper Co., Waterbury, Conn.; Mueller Brass Co., Huron, Mich.; New England Brass Co., Taunton, Mass.; Phelps Dodge Copper Prod. Corp., New York, N.Y.; Revere Copper & Brass, Inc., New York, N.Y.; Riverside Metal Co., Riverside, N.J.; Scovill Mfg. Co., Waterbury, Conn.; Seymour Mfg. Co., Seymour, Conn.; Wolverine Tube Co., Detroit, Mich.

NOTE: Primarily for: C, corrosion; H, heat; A, abrasion. Forms available: B, bars; C, castings; CR, cold rolled; D drawn; E, extrusions; F, forgings; HR, hot rolled; P, plates; R, rods; S, sheets; T, tubes W, wire.

Table 3. Directory of Materials for the Construction of Chemical Equipment—(*Continued*)

No.	Material	Manufacturer	Essential Nominal Chemical Composition, %	Forms Available	Primarily for
54	Brass, cartridge	Generally available* (Corrosion data by Amer. Brass)	66.5–69.5 Cu; 30.5–33.5 Zn	B, P, S, R, T, W	C
55	Brass, forging	Scovill Mfg. Co., Waterbury, Conn.	59 Cu; 2 Pb; bal. Zn	R, W	C
56	Brass, free cutting	Scovill Mfg. Co., Waterbury, Conn.	61.5 Cu; 3 Pb; bal. Zn	R, W	C
57	Brass, high	Generally available*	66 Cu; 34 Zn	....	C
58	Brass, low leaded	Scovill Mfg. Co., Waterbury, Conn.	65 Cu; 0.5 Pb; bal. Zn	S, T, W	C
59	Brass, med. leaded	Scovill Mfg. Co., Waterbury, Conn.	65 Cu; 0.9 Pb; bal. Zn	S, R, W	C
60	Brass, high leaded	Scovill Mfg. Co., Waterbury, Conn.	63 Cu; 1.75 Pb; bal. Zn	S, R, W	C
61	Brass, naval	Generally available* (Corrosion data by Amer. Brass)	60 Cu; 39.25 Zn; 0.75 Sn	B, P, S, R, W, D, T	C
62	Brass, naval, low leaded	Scovill Mfg. Co., Waterbury, Conn.	61 Cu; 0.75 Sn; 0.75 Pb; bal. Zn (corrosion resistance same as 61)	R, W	C
63	Brass, red	Generally available* (Corrosion data by Chase, Amer. Brass)	85 Cu; 15 Zn	B, S, R, T, W, P	C
65	Bronze, aluminum	Generally available*	82–95 Cu; 5–10 Al; Fe; Mn; Ni; Sn	C, CR, F, HR, P, R, S, T, W	C
66	Bronze, commercial	Chase Brass & Copper Co., Waterbury, Conn.	90 Cu; 10 Zn	P, S, T, W	C
67	Bronze, 600 forgeable bearing	Mueller Brass Co., Port Huron, Mich.	58 Cu; 2.5 Mn; 1.5 Al; 0.7 Si; bal. Zn	B, D, R, F	C, A
68	Bronze, gilding 95%	Scovill Mfg. Co., Waterbury, Conn.	95 Cu; 5 Zn (corrosion resistance same as commercial bronze)	S, R, T, W	C
69	Bronze, hardware	Scovill Mfg. Co., Waterbury, Conn.	89 Cu; 8 Zn; 2 Pb; 1 Ni	D, R, W	C
70	Bronze, high leaded tin	Mueller Brass Co., Port Huron, Mich.	80 Cu; 10 Pb; 10 Sn	C	C
71	Bronze, manganese	Mueller Brass Co., Port Huron, Mich.	58 Cu; 1 Sn; 1.4 Fe; 0.25 Mn; bal. Zn	B, D, R, F	C
72	Bronze, high strength manganese	Mueller Brass Co., Port Huron, Mich.	65 Cu; 3 Fe; 4 Mn; 5 Al; bal. Zn		C
73	Bronze, phosphor, 5% A	Generally available* (Corrosion data by Amer. Brass)	94.8–95.5 Cu; 4.3–5 Sn; P	B, S, R, W, P, CR, D, T	C
74	Bronze, phosphor, 8% C	Generally available* (Corrosion data by Amer. Brass)	Cu; 7–9 Sn; 0.03–0.25 P	B, S, R, T, W, P, CR, D	C
75	Bronze, phosphor, 10% D	Generally available* (Corrosion data by Amer. Brass)	89.5–90 Cu; 10–10.5 Sn; P	B, S, R, T, W, P, CR, D	C
76	Bronze, phosphor, special free cutting	Amer. Brass Co., Waterbury, Conn.	88 Cu; 4 Zn; 4 Sn; 4 Pb	B, S, R, P	C
77	Bronze, Tobin	Amer. Brass Co., Waterbury, Conn.	60 Cu; 39.25 Zn; 0.75 Sn	B, S, R, T, W, P, D, HR, CR	C
78	Bronze, Tobin	Mueller Brass Co., Port Huron, Mich.	60 Cu; 1 max. Sn; bal. Zn	....	C
79	Buflokast gray iron	Buflovak Equipment Div., Blaw-Knox Co., Buffalo, N.Y.	Fe; 3.2–3.6 C; 2 max. Ni; 1–2 Si; 0.6–0.9 Mn	C	
80	CA stainless	Cooper Alloy Fdry. Co., Hillside, N.J.	Various standard stainless steels; see Nos. 275–360	C	C
81	CA-FA20	Cooper Alloy Fdry. Co., Hillside, N.J.	Fe; 19–21 Cr; 28–30 Ni; 3.5 Mo; 4–4½ Cu; 0.07 max. C	C	C
82	CA-MM	Cooper Alloy Fdry. Co., Hillside, N.J.	67 Ni; 30 Cu; 1.4 Fe; 0.1 Si; 0.15 C	C	C
83	Calite A	Calorizing Co., Wilkinsburg, Pa.	Fe; 35 Ni; 15 Cr	B, C, P, S, R, W	H
84	Calite B28	Calorizing Co., Wilkinsburg, Pa.	Fe; 25 Cr; 10 Ni; 1 Mo	C	H
85	Cast iron	Corrosion data from *Standards of Hydraulic Institute*, Sec. G, 1941, Hydraulic Inst., New York	All-cast-iron pumps		
86	Cast iron	Corrosion data from *Circ.* 312, 1939, Crane Co., Chicago	Cast-iron valves		
86A	Cast iron	Corrosion data from A. W. Spitz, Eng. Dept., Amer. Cyanamid Co., New York	Ordinary unalloyed cast iron		
87	Causul metal	Lukenheimer Co., Cincinnati, Ohio	Fe; 19 Ni; 2.2–2.8 C; 4 Cu; 1.5 Cr	C	C
88	Chlorimet 2	Duriron Co., Dayton, Ohio	63 Ni; 32 Mo; 3 max. Fe; 0.15 max. C; 1 Si; 1 Mn	C	C
89	Chlorimet 3	Duriron Co., Dayton, Ohio	60 Ni; 18 Mo; 18 Cr; 2 Fe; 0.07 max. C; 1 Si; 1 Mn	C	C
90	Chromax	Driver Harris Co., Harrison, N.J.	Fe; 19 Cr; 35 Ni	B, CR, HR, P, R, S, W	H
91	Chromel A	Hoskins Mfg. Co., Detroit, Mich.	80 Ni; 20 Cr	B, CR, HR, D, R, W	H
92	Chromel C and D	Hoskins Mfg. Co., Detroit, Mich.	35–61 Ni; 16–18½ Cr; 23–46½ Fe		
93	Chromel P	Hoskins Mfg. Co., Detroit, Mich.	90 Ni; 10 Cr	CR, W	H
94	Circle L stainless	Lebanon Steel Fdry., Lebanon Pa.	Various standard stainless steels; see Nos. 275–360	C	C, A
95	Circle L 13	Lebanon Steel Fdry., Lebanon, Pa.	Fe; 13 Cr; 0.25 max. C; 0.75 Mn; 0.75 max. Ni; 0.4 Mo	C	C, A
96	Circle L 21	Lebanon Steel Fdry., Lebanon, Pa.	Fe; 19 Cr; 9 Ni; 0.07 max. C; 0.75 Mn; 0.75 Cb	C	C
97	Circle L 22M	Lebanon Steel Fdry., Lebanon, Pa.	Fe; 19 Cr; 9 Ni; 0.07 max. C; 0.75 Mn; 1.75 Mo; 0.25 Se	C	C
98	Circle L 22XM	Lebanon Steel Fdry., Lebanon, Pa.	Fe; 20 Cr; 10 Ni; 0.07 max. C; 0.75 Mn; 3 Mo	C	C
99	Circle L 23XM	Lebanon Steel Fdry., Lebanon, Pa.	Fe; 19 Cr; 9 Ni; 0.2 max. C; 0.75 Mn; 3 Mo	C	C
100	Circle L 24	Lebanon Steel Fdry., Lebanon, Pa.	Fe; 10 Cr; 20 Ni; 0.2 max. C; 0.75 Mn	C	C
101	Circle L 30H	Lebanon Steel Fdry., Lebanon, Pa.	Fe; 24 Cr; 12 Ni; 0.5 max. C; 0.75 Mn	C	H
102	Circle L 31	Lebanon Steel Fdry., Lebanon Pa.	Fe; 29 Cr; 9 Ni; 0.3 max. C; 0.75 Mn	C	C
103	Circle L 31H	Lebanon Steel Fdry., Lebanon, Pa.	Fe; 29 Cr; 9 Ni; 0.5 max. C; 0.75 Mn	C	H
104	Circle L 32	Lebanon Steel Fdry., Lebanon, Pa.	Fe; 15 Cr; 35 Ni; 0.5 max. C; 0.75 Mn	C	H
105	Circle L 32XMC	Lebanon Steel Fdry., Lebanon, Pa.	Fe; 15 Cr; 35 Ni; 0.07 max. C; 3.25 Mo; 2.25 Cu; 0.75 Mn	C	C
106	Circle L 34	Lebanon Steel Fdry., Lebanon, Pa.	Fe; 21 Cr; 29 Ni; 0.07 max. C; 3.25 Mo; 0.75 Mn	C	C
107	Circle L 41	Lebanon Steel Fdry., Lebanon, Pa.	Fe; 15 Cr; 65 Ni; 0.5 max. C; 0.5 Mn	C	H
108	Colomony	Wall-Colomony Corp., Detroit, Mich.	68–80 Ni; 7–19 Cr; 2–4 B; Fe; Si	B, C, R	C, H, A
109	Colonial 610	Vanadium Alloys Steel Co., Latrobe, Pa.	Fe; 16–18 Cr; 1 Ni; 0.12 max. C; S (optional)	B, D, HR, P, R, S, W	C
110	Copel	Hoskins Mfg. Co., Detroit, Mich.	55 Cu; 45 Ni	B, CR, HR, D, R, W	C, H
111	Copper	Generally available* (Corrosion data by Chase)	99.9 + Cu	B, P, S, R, T, W	C
112	Copper, beryllium	Generally available*	97.5 Cu; 2.15 Be.; 0.35 Ni	B, C, CR, D, HR, P, R, S, W	C

* See p. 1527, † footnote.

Table 3. Directory of Materials for the Construction of Chemical Equipment—(*Continued*)

No.	Material	Manufacturer	Essential Nominal Chemical Composition, %	Forms Available	Primarily for
113	Copper, cadmium	Phelps Dodge Copper Prod. Corp., New York, N.Y.	99 Cu; 1 Cd	B, CR, D, HR, R, W	
114	Copper, deoxidized	Generally available* (Corrosion data by Amer. Brass)	99.9+ Cu; 0.01–0.03 P	B, C, CR, D, HR, P, R, S, T	C
115	Copper, O.F.H.C.	Amer. Metal Co., New York, N.Y.	99.9+ Cu	B, C, CR, D, HR, P, R, S, T, W	C
116	Copper, P.D.C.P.	Phelps Dodge Copper Prod. Corp., New York, N.Y.	99.9+ Cu	B, C, D, CR, HR, R, T, W	C
117	Copper, phosphorized	Mueller Brass Co., Port Huron, Mich.	99.9 Cu; trace P	T	C
118	Copper, tellurium	Chase Brass & Copper Co., Waterbury, Conn.	99.5 Cu; 0.5 Te	B, R, T, W	C
119	Corrosiron	Pacific Fdry. Co. Ltd., San Francisco, Calif.	Fe; 14.5 Si	C	C
120	Croloy stainless	Babcock & Wilcox Tube Co., Beaver Falls, Pa.	Various standard stainless steels; see Nos. 275–360	T	C, H
121	Cupaloy	Westinghouse Electric Corp., Pittsburgh, Pa.	99.4 Cu; 0.1 Ag; 0.5 Cr	B, C, R	C
122	Cupron	Wilbur B. Driver Co., Newark, N.J.	Cu; 45 Ni	B, CR, D, HR	
123	Cupro-nickel, 20%	Generally available* (Corrosion data by Amer. Brass)	70 Cu; 20 Ni	B, P, S, R, T, W	C
124	Cupro-nickel, 30%	Generally available* (Corrosion data by Chase)	70 Cu; 30 Ni	B, C, CR, D, HR, P, R, S, T, W	C
125	Disston stainless	Henry Disston & Sons Inc., Philadelphia, Pa.	Various standard stainless steels; see Nos. 275–360	B, HR, P	C, H, A
126	Dopploy 30	Sowers Mfg. Co., Buffalo, N.Y.	Fe; 18.5 Ni; 2.35 Cr; 2.85 C; 1 Mn	C	C
127	Dowmetal C	Dow Chemical Co., Midland, Mich.	Mg; 9 Al; 2 Zn; 0.1 Mn	C (perm. mold)	C
128	Dowmetal FS-1	Dow Chemical Co., Midland, Mich.	Mg; 3 Al; 1 Zn; 0.3 Mn	B, E, P, S, R, T	C
129	Dowmetal H	Dow Chemical Co., Midland, Mich.	Mg; 6 Al; 3 Zn; 0.2 Mn	C (sand)	C
130	Dowmetal J-1	Dow Chemical Co., Midland, Mich.	Mg; 6.5 Al; 1 Zn; 0.2 Mn	B, E, P, S, R, T, F	C
131	Dowmetal M	Dow Chemical Co., Midland, Mich.	Mg; 1.5 Mn	B, E, P, S, R, T, F	C
132	Dowmetal O-1	Dow Chemical Co., Midland, Mich.	Mg; 8.5 Al; 0.5 Zn; 0.2 Mn	B, E, R, F	C
133	Dowmetal R	Dow Chemical Co., Midland, Mich.	Mg; 9 Al; 0.6 Zn; 0.2 Mn	C (die)	C
134	Duraloy stainless	Duraloy Co., Scottdale, Pa.	Various standard stainless steels; see Nos. 275–360	C, T (centrifugal)	C, H, A
135	Duraloy	Duraloy Co., Scottdale, Pa.	Fe; 26–30 Cr; 3 Ni; 2.75 C	C, T (centrifugal)	A
136	Duraloy HCA	Duraloy Co., Scottdale, Pa.	Fe; 26–30 Cr; 1 C	C, T (centrifugal)	C, A
137	Duraloy HCN	Duraloy Co., Scottdale, Pa.	Fe; 25 Cr; 12 Ni; 1 C	C, T (centrifugal)	C, H, A
138	Durco stainless	Duriron Co., Dayton, Ohio	Various standard stainless steels; see Nos. 275–360	C	C, H, A
139	Durco D-10	Duriron Co., Dayton, Ohio	57 Ni; 23 Cr; 8 Cu; 4 Mo; 2 W; 1 Mn	C	C
140	Durichlor	Duriron Co., Dayton, Ohio	Fe; 0.85 C; 14.5 Si; 3 Mo; 0.35 Mn	C	C
141	Durimet 20	Duriron Co., Dayton, Ohio	Fe; 20 Cr; 29 Ni; 0.07 max. C; 2 Mo; 4 Cu; 1 Si	C	C
142	Durimet T	Duriron Co., Dayton, Ohio	Fe; 19 Cr; 22 Ni; 0.07 max. C; 2 Mo; 1 Cu; 1 Si	C, HR, B	C
143	Duriron	Duriron Co., Dayton, Ohio	Fe; 0.80 C; 14.5 Si; 0.35 Mn	C	C
144	Elverite A	Babcock & Wilcox Co., New York, N.Y.	Fe; 3–3.5 C; 0.35 Mn; 0.25–1 Si	C	A
145	Elverites B and C	Babcock & Wilcox Co., New York, N.Y.	Fe; 3–3.5 C; 1–1.8 Cr; 3.75–4.75 Ni; 0.25–1 Si	C	A
147	Enduro stainless	Republic Steel Corp., Cleveland, Ohio	Various standard stainless steels; see Nos. 275–360	B, HR, D, CR, S, C, P, R, T, W, strip	C, H
148	Everdur 1000	Amer. Brass Co., Waterbury, Conn.	94.9 Cu; 4 Si; 1.1 Mn	Casting ingots	C
149	Everdur 1010	Amer. Brass Co., Waterbury, Conn.	95.8 Cu; 3.1 Si; 1.1 Mn	B, P, S, R, T, W	C
150	Everdur 1015	Amer. Brass Co., Waterbury, Conn.	98.25 Cu; 1.5 Si; 0.25 Mn	B, P, S, R, T, W	C
151	Fahrite stainless	Ohio Steel Fdry. Co., Cincinnati, Ohio	Various standard stainless steels; see Nos. 275–360	C	C, H, A
152	Frontier 5	Frontier Bronze Corp., Niagara Falls, N.Y.	89 Cu; 10 Al; 1 Fe	C	C
153	Frontier 11	Frontier Bronze Corp., Niagara Falls, N.Y.	88 Cu; 5 Ni; 5 Sn; 2 Zn	C	C
154	Frontier 40	Frontier Bronze Corp., Niagara Falls, N.Y.	Cu; 5 Zn; 0.5 Mg; 0.5 Cr; 0.2 Ti	C	C
155	Genesee stainless	Symington-Gould Corp., Rochester, N.Y.	Various standard stainless steels; see Nos. 275–360	C	C, H, A
156	Gold	Baker & Co. Inc., Newark, N.J.	99.99 Au	B, C, CR, HR, D, P, S, R, T, W	C
157	Gun metal	Mueller Brass Co., Port Huron, Mich.	88 Cu; 10 Sn; 2 Zn	C	C
158	Hascrome	Haynes Stellite Co., Kokomo, Ind.	Fe; 10–14 Cr; 0.8–1.2 C; 3–5 Mn	C, R (rolled, extruded)	H, A
159	Hastelloy A	Haynes Stellite Co., Kokomo, Ind.	Ni; 17–21 Mo; 17–21 Fe	B, C, CR, HR, D, P, S, R, T, W	C, H
160	Hastelloy B	Haynes Stellite Co., Kokomo, Ind.	Ni; 24–32 Mo; 3–7 Fe; 0.02–0.12 C	B, C, CR, HR, D, P, S, R, T, W	C, H
161	Hastelloy C	Haynes Stellite Co., Kokomo, Ind.	Ni; 14–19 Mo; 4–8 Fe; 0.04–0.15 C; 12–16 Cr; 3–5.5 W	B, C, CR, HR, D, P, S, R, T, W	C, H, A
162	Hastelloy D	Haynes Stellite Co., Kokomo, Ind.	Ni; 8–11 Si; 2–5 Cu; 1 max. Al	C, R	C, H, A
163	Haynes Stellite 1	Haynes Stellite Co., Kokomo, Ind.	Co; 28–34 Cr; 11–15 W	R (cast)	C, H, A
164	Haynes Stellite 3	Haynes Stellite Co., Kokomo, Ind.	. . . .	C	C, H, A
165	Haynes Stellite 6	Haynes Stellite Co., Kokomo, Ind.	Co; 25–31 Cr; 3–6 W	C, S, R (cast)	C, H, A
166	Haynes Stellite 12	Haynes Stellite Co., Kokomo, Ind.	Co; 26–32 Cr; 6–10 W	C, R (cast)	C, H, A
167	Haynes Stellite 21	Haynes Stellite Co., Kokomo, Ind.	Co; 25–30 Cr; 4.5–6.5 Mo; 1.5–3.5 Ni; 2 max. Fe; 0.2–0.35 C	C, S	C, H
168	Haynes Stellite 23	Haynes Stellite Co., Kokomo, Ind.	Co; 0.35–0.5 C; 23–29 Cr; 1.5 max. Ni; 4–7 W; 2 max. Fe	C, S	C, H
169	Haynes Stellite 27	Haynes Stellite Co., Kokomo, Ind.	Ni; 0.35–0.5 C; 23–29 Cr; 5–7 Mo; 2 max. Fe; 30 min. Co	C, S	C, H
170	Haynes Stellite 30	Haynes Stellite Co., Kokomo, Ind.	Co; 3.5–5 C; 23–29 Cr; 13–17 Ni; 5–7 Mo; 2 max. Fe	C, S	C, H
171	Haynes Stellite 31	Haynes Stellite Co., Kokomo, Ind.	Co; 0.45–0.6 C; 23–28 Cr; 9–12 Ni; 6–9 W; 1.5 max. Fe	C, S	C, H
173	Haynes Stellite 93	Haynes Stellite Co., Kokomo, Ind.	Fe; 15–19 Cr; 4–7 Co; 0.5–3 V; 13–17 Mo	C, R	H, A
174	Haynes Stellite 98M2	Haynes Stellite Co., Kokomo, Ind.	Co; 27–31 Cr; 16–19 W	C	C, H, A

* See p. 1527, † footnote.

Table 3. Directory of Materials for the Construction of Chemical Equipment—(*Continued*)

No.	Material	Manufacturer	Essential Nominal Chemical Composition, %	Forms Available	Primarily for
175	Haynes Stellite Multimet	Haynes Stellite Co., Kokomo, Ind.	20 Cr; 20 Co; 20 Ni; 3 Mo; 2 W; 1 Cb; 0.14 N; 0.1–0.35 C	B, C, CR, HR, D, P, S, R, T, W	C, H
176	Haynes Stellite stainless	Haynes Stellite Co., Kokomo, Ind.	Various standard stainless steels; see Nos. 275–360	C	C, H, A
177	Haynes Stellite Star J	Haynes Stellite Co., Kokomo, Ind.	Co; 29–34 Cr; 15–19 W	C	C, H, A
178	Haystellite, hard grade	Haynes Stellite Co., Kokomo, Ind.	90–96 W; 0.5–2 Co; 3.5–4.5 C	R, crushed grains	A
179	Haystellite, tough grade	Haynes Stellite Co., Kokomo, Ind.	79–85 W; 8–11 Co; 3.5–4.5 C	Cast inserts, crushed grains	A
180	Herculoy A and B	Revere Copper and Brass, New York, N.Y.	96–98 Cu; 1.75–3 Si; 0.25–0.5 Sn	B, CR, D, HR, P, R, S, T, W	C
181	Hytensl	Amer. Manganese Bronze Co., Philadelphia, Pa.	63 Cu; 23 Zn; 4 Al; 3 Fe; 3 Mn	B, C, HR, R	C, A
182	Illium G	Burgess-Parr Co., Freeport, Ill.	56 Ni; 22 Cr; 6 Mo; 6 Fe; 6 Cu; Mn; Si; C	C	
183	Illium R	Burgess-Parr Co., Freeport, Ill.	55–60 Ni; 18–24 Cr; 5–8 Mo; 5–8 Fe; 2–6 Cu; 0.5–1.75 Mn; Si; C	B, C, CR, P, R, S, T, W	
184	Inconel	International Nickel Co., New York, N.Y.	79.5 Ni; 13 Cr; 6.5 Fe; 0.08 C; 0.2 Cu; 0.25 Mn	B, C, CR, HR, D, P, S, R, T, W	C, H
185	Inconel-clad steel	Lukens Steel Co., Coatesville, Pa.	Steel sheet clad on one or both sides	HR, P, heads	C
186	Indium metal	Amer. Smelting and Ref. Co., New York, N.Y.	In	B, C, CR, R, S, T, W	C
187	Ingersoll stainless	Ingersoll Steel Div., Borg-Warner Corp., Chicago, Ill.	Various standard stainless steels; see Nos. 275–360	P, S	C, H
188	Iso cast stainless	Empire Steel Castings, Inc., Reading, Pa.	Various standard stainless steels; see Nos. 275–360	C	C, H, A
189	K42B	Westinghouse Electric Corp., Pittsburgh, Pa.	42 Ni; 22 Co; 18 Cr; 14 Fe; 2.2 Ti	B, CR, R, W, F	H
190	Kanthal alloys	C. O. Jelliff Mfg. Corp., Southport, Conn.	Fe; 20–25 Cr; Co; Al	W, ribbon	H
191	Lead	Eagle-Picher Co., Cincinnati Ohio	99.9+ Pb	B, C, CR, S, R, T, W	C
192	Lead, antimonial	Amer. Smelting & Ref. Co., New York, N.Y.	94 Pb; 6 Sb	C, CR, D, R, S, T, W	C
193	Lead, antimonial	National Lead Co., New York, N.Y.	Pb; 4–12 Sb	B, C, CR, S, R, T, W, pipe	C
194	Lead, antimonial	Northwest Lead Co., Seattle, Wash.	93.45 Pb; 6.5 Sb; 0.04–0.08 Cu	C, S, T, W	C
195	Lead, chemical	Amer. Smelting & Ref. Co., New York, N.Y.	99.93 Pb; 0.06 Cu	B, C, CR, R, S, T, W	C
196	Lead, chemical	National Lead Co., New York, N.Y.	99.93 Pb; 0.06 Cu	B, C, CR, S, R, T, W, pipe	C
197	Lead, chemical	Northwest Lead Co., Seattle, Wash.	99.95 Pb; 0.04–0.08 Cu	C, S, T, W	C
198	Lead, chemical tellurium	Northwest Lead Co., Seattle, Wash.	99.9 Pb; 0.02–0.06 Te; 0.04–0.08 Cu	C, S, T, W	C
199	Lead, tellurium	Amer. Smelting & Ref. Co., New York, N.Y.	99.88 Pb; 0.045 Te; 0.06 Cu	B, C, CR, R, S, T, W	C
200	Lead, tellurium	National Lead Co., New York, N.Y.	99.88 Pb; 0.045 Te; 0.06 Cu	B, C, CR, R, S, T, W, pipe	C
201	Lukens CrCuNi steel	Lukens Steel Co., Coatesville, Pa.	Fe; 0.1 max. C; 0.4–0.6 Mn; 0.4–0.6 Cr; 0.4–0.6 Cu; 0.4–0.8 Ni	HR, P, heads	C
202	Lukens CrNiMo steel	Lukens Steel Co., Coatesville, Pa.	Fe; 0.25 max. C; 0.7–0.9 Mn; 0.8–1 Ni; 0.3–0.5 Mo	HR, P, heads	H, A
203	Lukens CrMn steel	Lukens Steel Co., Coatesville, Pa.	Fe; 0.45 max. 0.8–1 Mn; 0.3–0.5 Cr	HR, P, heads	H, A
204	Lukens CrMo steel	Lukens Steel Co., Coatesville, Pa.	Fe; 0.45 max. C; 0.3–0.6 Mn; 0.4–0.6 Cr; 0.3–0.6 Mo	HR, heads	H
205	Lukens MnMo steel	Lukens Steel Co., Coatesville, Pa.	Fe; 0.2 max. C; 1.2–1.5 Mn; 0.4–0.6 Mo	HR, P, heads	H, A
206	Lukens MnV steel	Lukens Steel Co., Coatesville, Pa.	Fe; 0.2 max. C; 1–1.4 Mn; 0.12 max. V	HR, P, heads	H
207	Lukens ½Mo steel	Lukens Steel Co., Coatesville, Pa.	Fe; 0.2 max. C; 0.6–0.9 Mn; 0.4–0.6 Mo	P, heads	H
208	Lukens 8½ Ni steel	Lukens Steel Co., Coatesville, Pa.	Fe; 0.12 max. C; 0.5–0.8 Mn; 8–9 Ni	HR, P, heads	C
209	Lukens NiCr steel	Lukens Steel Co., Coatesville, Pa.	Fe; 0.2 max. C; 0.4–0.6 Mn; 1–1.2 Ni; 0.4–0.6 Cr	HR, P, heads	H, A
210	Mayari R	Bethelehem Steel Co., Bethlehem, Pa.	Fe; 0.12 max. C; 0.2–1 Cr; 0.25–0.75 Ni; 0.5–0.7 Cu; 0.5–1 Mn	B, CR, HR, P, R, S, T, W	A
211	Michiana stainless	Michiana Products Corp., Michigan City, Ind.	Various standard stainless steels; see Nos. 275–360	C	H
212	Midvaloy stainless	Midvale Co., Philadelphia, Pa.	Various standard stainless steels; see Nos. 275–360		
213	Milwaukee stainless	Milwaukee Steel Fdry., Milwaukee, Wis.	Various standard stainless steels; see Nos. 275–360	C	C, H, A
214	Misco stainless	Michigan Steel Casting Co., Detroit, Mich.	Various standard stainless steels; see Nos. 275–360	C	C, H, A
215	Miscrome stainless	Michigan Steel Casting Co., Detroit, Mich.	Various standard stainless steels; see Nos. 275–360	C	C, H, A
216	Monel	International Nickel Co., New York, N.Y.	67 Ni; 30 Cu; 1.4 Fe; 0.1 Si; 0.15 C	B, C, CR, D, HR, P, R, S, T, W	C
217	Monel-clad	Lukens Steel Co., Coatesville, Pa.	Steel sheet clad on one or both sides	HR, P, heads	C
218	Mueller 85-5-5-5	Mueller Brass Co., Port Huron, Mich.	85 Cu; 5 Zn; 5 Sn; 5 Pb	C	C
219	Muntz metal	Generally available* (corrosion data by Amer. Brass)	60 Cu; 40 Zn	B, P, S, R, T, W	C
220	National Al alloys	National Smelting Co., Cleveland, Ohio	Al; 0–4 Cu; 0–4 Mg; 0–7.5 Si; 0–0.6 Mn; 0–1.5 Ni	C	C
221	Niag	Mueller Brass Co., Port Huron, Mich.	46 Cu; 10 Ni; 2.5 Pb; bal. Zn	B, D, R, F	C
222	Nichrome	Driver Harris Co., Harrison, N.J.	60 Ni; 15 Cr; Fe	B, CR, HR, P, R, S, T, W	H
223	Nichrome V	Driver Harris Co., Newark, N.J.	80 Ni; 20 Cr	B, CR, D, HR, P, R, S, T, W	H
224	Nickel	International Nickel Co., New York, N.Y	99.4 Ni; 0.2 Mn; 0.1 Cu; 0.15 Fe; 0.05 Si	B, C, CR, HR, D, P, R, S, T, W	C, H
225	Nickel-clad	Lukens Steel Co., Coatesville, Pa.	Steel sheet clad on one or both sides	HR, P, heads	C
226	Nickel silver, 18% A	Generally available* (corrosion data by Amer. Brass)	65 Cu; 18 Ni; 17 Zn	B, C, CR, D, P, R, S, T, W	C
227	Nickel silver, 18% B	Generally available*	55 Cu; 18 Ni; 27 Zn	B, C, CR, D, P, R, S	C
228	Nicloy	Babcock & Wilcox Tube Co., Beaver Falls, Pa.	Fe; 3½, 5, or 9 Ni	T (seamless)	C
229	Ni-Hard	International Nickel Co., New York, N.Y.	Fe; 3.4 C; 1.5 Cr; 4.5 Ni; 0.6 Si	C	A
230	Nilstain	Wilbur B. Driver Co., Newark, N.J.	Fe; 18–20 Cr; 8–10 Ni; 0.2 max. C; 2 max. Mn	D, HR, R, W	C
231	Ni-Resist	International Nickel Co., New York, N.Y.	Fe; 2.8 C; 14 or 20 Ni; 6 Cu (optional); 2 Cr; 2 Si	C	C, H

* See p. 1527, † footnote.

Table 3. Directory of Materials for the Construction of Chemical Equipment—(*Continued*)

No.	Material	Manufacturer	Essential Nominal Chemical Composition, %	Forms Available	Primarily for
232	Nirex	Driver-Harris Co., Harrison, N.J.	80 Ni; 14 Cr; 6 Fe	B, CR, HR, P, R, S, W	H
233	NS-5 Alloy	Lunkenheimer Co., Cincinnati, Ohio	50 Ni; 46 Cu; 2.4 Si; 1.6 Mn	C (valve seats, disks)	C, H, A
234	Olympic bronze, type A	Chase Brass & Copper Co., Waterbury, Conn.	96 Cu; 3 Si; 1 Zn	P, S, R, T, W	C
235	Olympic bronze, type B	Chase Brass & Copper Co., Waterbury, Conn.	97.5 Cu; 1.5 Si; 1 Zn	P, S, R	C
236	Palladium	Baker & Co., Newark, N.J.	99.991 Pd	B, C, CR, HR D, P, S, R, T, W	C, H
237	Pennalloy	Pennsylvania Elec. Steel Casting Co., Hamburg, Pa.		C	A
238	Permite Al alloys	Aluminum Industries, Cincinnati, Ohio	Al; 0–5 Cu; 1.5–7.5 Si; 0–1 Fe; 0–0.4 Mg	C	C
239	Pioneer	Pioneer Alloy Products Co., Cleveland, Ohio	65 Ni; Cr; Mo; Fe	C	C
240	Platinum	Baker & Co., Newark, N.J.	99.99 Pt	B, C, CR, HR, D, P, S, R, T, W	C, H
241	Platinum	J. Bishop & Co., Malvern, Pa.	99.95 Pt	B, C, CR, HR, D, P, S, R, T, W	C, H
242	Platinum, iridio	Baker & Co, Newark, N.J.	Pt; 5–30 Ir	B, CR, HR, C, D, P, S, R, T, W	C, H
243	Platinum, iridium	J. Bishop & Co., Malvern, Pa.	Pt; 10–30 Ir	B, CR, HR, C, D, P, S, R, T, W	C, H
244	Platinum, rhodio	Baker & Co., Newark, N.J.	Pt; 5–40 Rh	B, C, CR, HR, D, P, S, R, T, W	C, H
245	Pyrasteel	Chicago Steel Fdry. Co., Chicago, Ill.	Fe; 25–27 Cr; 12–14 Ni; 0.1–0.35 C; Mo, Cb, Se (optional)	C	C, H
246	Pyrocast	Pacific Fdry. Co. Ltd., San Francisco, Calif.	Fe; 22–30 Cr	C	H
247	Pyrocast stainless	Pacific Fdry. Co. Ltd., San Francisco, Calif.	Various standard stainless steels; see Nos. 275–360	C	H
249	Resistac	Amer. Manganese Bronze Co., Philadelphia, Pa.	88 Cu; 10 Al; 2 Fe	B, C, HR, R	C, H
250	Reynolds 2S	Reynolds Metals Co., Louisville, Ky.	99 Al; 1 max. Fe + Si; 0.2 max. Cu	B, R, W	C
251	Reynolds 3S	Reynolds Metals Co., Louisville, Ky.	Al; 1–1.5 Mn; 0.7 max. Fe; 0.6 max. Si; 0.2 Cu	B, R, W	C
252	Reynolds 14S	Reynolds Metals Co., Louisville, Ky.	Al; 3.9–5 Cu; 0.5–1.2 Si; 1 max. Fe; 0.4–1.2 Mn; 0.2–0.8 Mg	B, R, W	C
253	Reynolds 17S	Reynolds Metals Co., Louisville, Ky.	Al; 3.5–4.5 Cu; 0.8 max. Si; 1 max. Fe, 0.4–1 Mn; 0.2–0.8 Mg	B, R, W	C
254	Reynolds 18S	Reynolds Metals Co., Louisville, Ky.	Al; 3.5–4.5 Cu; 1.7–2.3 Ni; 0.9 max. Si; 1 max. Fe; 0.45–0.9 Mg	B, R, W	C
255	Reynolds 24S	Reynolds Metals Co., Louisville, Ky.	Al; 3.8–4.9 Cu; 1.2–1.8 Mg; 0.5 max. Si	B, R, W	C
256	Reynolds 24S Pureclad	Reynolds Metals Co., Louisville, Ky.	Al 24S core covered with Al of high purity	S, P	C
257	Reynolds 25S	Reynolds Metals Co., Louisville, Ky.	Al; 3.9–5 Cu; 0.5–1.2 Si; 1 max. Fe	B, R, W	C
258	Reynolds 32S	Reynolds Metals Co., Louisville, Ky.	Al; 11–13.5 Si; 0.5–1.3 Cu; 1.0 max. Fe; 0.8–1.3 Mg; 0.5–1.3 Ni	B, R, W	C
259	Reynolds A51S	Reynolds Metals Co., Louisville, Ky.	Al; 0.6–1.2 Si; 1 max. Fe; 0.35 Cu; 0.45–0.8 Mg	B, R, W	C
260	Reynolds 52S	Reynolds Metals Co., Louisville, Ky.	Al; 2.2–2.8 Mg.; 0.45 max. Fe + Si; 0.1 max. Cu	B, R, W	C
261	Reynolds 53S	Reynolds Metals Co., Louisville, Ky.	Al; 0.35 max. Fe; 1.1–1.4 Mg. 0.15–0.35 Cr	B, R, W	C
262	Reynolds 61S	Reynolds Metals Co., Louisville, Ky.	Al; 0.4–0.8 Si; 0.7 max. Fe; 0.8–1.2 Mg	B, R, W	C
263	Reynolds R301	Reynolds Metals Co., Louisville, Ky.	High strength Al alloy core clad with corrosion-resistant Al alloy of intermediate strength	S, P	C
264	Rezistal stainless	Crucible Steel Co. of Amer., New York, N.Y.	Various standard stainless steels; see Nos. 275–360	B, CR, HR, D, P, S, R, W	C, H, A
265	Roofloy	Amer. Smelting & Ref. Co., New York, N.Y.	Pb; 0.25 Sn; 0.02 Mg; 0.02 Bi	CR, R, T	C
266	St. Joe lead	St. Joseph Lead Co., New York, N.Y.	99.90+ Pb; 0.04–0.08 Cu; 0.002–0.02 Ag; 0.002 max. Fe; 0.001 max. Zn; 0.002 max. As, Sb, Sn	Pigs	C
267	Silfram	Stoody Co., Whittier, Calif.	Fe; 30 Cr; 10 Ni	C, welding rod	A
268	Silver	Baker & Co. Inc., Newark, N.J.	99.9+ Ag	B, C, CR, HR, D, P, S, R, T, W	C
269	Silver	Handy & Harman, New York, N.Y.	99.9+ Ag	B, C, CR, D P, R, S, T, W	C
270	Stainless-clad 304	Lukens Steel Co., Coatesville, Pa.	Sheet steel clad on one or both sides	HR, P, heads	C
271	Stainless-clad 316	Lukens Steel Co., Coatesville, Pa.	Sheet steel clad on one or both sides	HR, P, heads	C
272	Stainless-clad 347	Lukens Steel Co., Coatesville, Pa.	Sheet steel clad on one or both sides	HR, P, heads	C
273	Stainless-clad 410	Lukens Steel Co., Coatesville, Pa.	Sheet steel clad on one or both sides	HR, P, heads	C
274	Stainless-clad 430	Lukens Steel Co., Coatesville, Pa.	Sheet steel clad on one or both sides	HR, P, heads	C
275	Stainless type 301		Fe; 16–18 Cr; 6–8 Ni; 0.08–0.15 C	B, CR, D, HR, P, S, W	C
276	Stainless type 302	NOTE: Listed below are the producers of standard stainless steels (Nos. 275–305) for which type analyses have been established by the Amer. Iron and Steel Inst. The designations are intended to apply only to wrought products but are so commonly applied to castings as well that they have been adopted by many foundries to identify comparable casting alloys. Types marked † are no longer in force, but occasional reference is still made to them. Numbers in ( ) are the number of standard A.I.S.I. alloys for which the manufacturer has provided corrosion data; they are solely to acknowledge cooperation in the preparation of these tables and have no reference to the number of steels produced by any company.	Fe; 17–19 Cr; 8–10 Ni; 0.08–0.15 C	B, CR, D, HR, P, S, W, T, R	C
277	Stainless type 302B		Fe; 17–19 Cr; 8–10 Ni; 0.08–0.15 C; 2–3 Si	B, CR, D, HR, P, S, W	C, H
278	Stainless type 303		Fe; 17–19 Cr; 8–10 Ni; 0.15 max. C; 0.07 min. P, S, Se; 0.6 max. Zr, Mo; 2 max. Mn	B, CR, D, HR, P, S, W	C
279	Stainless type 304		Fe; 18–20 Cr; 8–11 Ni; 0.08 max. C; 2 max. Mn	B, CR, D, HR, P, R, S, T, W	C
280	Stainless type 308		Fe; 19–21 Cr; 10–12 Ni; 0.08 max. C	B, CR, D, HR, P, R, S, T, W	C
281	Stainless type 309		Fe; 22–24 Cr; 12–15 Ni; 0.2 max. C	B, CR, D, HR, P, R, S, W	H, C
282	Stainless type 310		Fe; 24–26 Cr; 19–22 Ni; 0.25 max. C	B, CR, D, HR, P, R, S, T, W	C, H
283	Stainless type 316		Fe; 16–18 Cr; 10–14 Ni; 0.1 max. C; 1.75–2.75 Mo	B, CR, D, HR, P, R, S, T, W	C, H
284	Stainless type 317†		Fe; 17.5–20 Cr; 10–14 Ni; 0.1 max. C; 3–4 Mo	B, CR, HR, P, R, S, T, W	C, H
285	Stainless type 321	Allegheny Ludlum Steel Corp., Pittsburgh, Pa. (25)	Fe; 17–19 Cr; 8–11 Ni; Ti, 5xC min.	B, CR, D, HR, P, R, S, T, W	C, H

Table 3. Directory of Materials for the Construction of Chemical Equipment—(*Continued*)

No.	Material	Manufacturer	Essential Nominal Chemical Composition, %	Forms Available	Primarily for
286	Stainless type 347	Alloy Metal Wire Co., Prospect Park, Pa.	Fe; 17–19 Cr; 9–12 Ni; Cb, 10xC min.	B, CR, D, HR, P, R, S, T, W	C, H
		Amer. Chain & Cable Co., Bridgeport, Conn.			
		Amer. Rolling Mill Co., Middletown, Ohio (4)			
		Babcock & Wilcox Tube Co., Beaver Falls, Pa. (7)			
		Bethlehem Steel Co., Bethlehem, Pa.			
287	Stainless type 403	A. M. Byers Co., Pittsburgh, Pa.	Fe; 11.5–13 Cr; 0.15 max. C; (turbine quality)	B, CR, D, HR, P, R, S, T, W	C, A
		Carnegie-Illinois Steel Co., Pittsburgh, Pa. (4)			
		Carpenter Steel Co., Reading, Pa.			
288	Stainless type 405	Cooper Alloy Fdry. Co., Elizabeth, N.J. (3)	Fe; 11.5–13.5 Cr; 0.08 max. C; 0.1–0.3 Al	B, CR, D, HR, P, S, W	C, H
289	Stainless type 406	Copperweld Steel Co., Warren, Ohio	Fe; 12–14 Cr; 0.15 max. C; 3.5–4.5 Al	B, CR, D, HR, P, S, W	C
290	Stainless type 410	Crucible Steel Co. of America, New York, N.Y. (4)	Fe; 11.5–13.5 Cr; 0.15 max. C	B, CR, D, HR, P, R, S, T, W	C, A, H
		Wilbur B. Driver Co., Newark, N.J.			
291	Stainless type 414	Eastern Stainless Steel Corp., Baltimore, Md.	Fe; 11.5–13.5 Cr; 1.25–2.5 Ni; 0.15 max. C	B, D, HR, P, R, S, W	C, A
292	Stainless type 416	Firth Sterling Steel Co., McKeesport, Pa.	Fe; 12–14 Cr; 0.15 max. C; 0.07 min. P, S, Se; 0.6 max. Zr, Mo	B, CR, D, HR, P, R, S, T, W	C, A
		Globe Steel Tubes Co., Milwaukee, Wis.			
		Henry Disston & Sons, Philadelphia, Pa.			
293	Stainless type 418†	Ingersoll Steel Div., Borg-Warner Corp., New Castle, Ind.	Fe; 12–14 Cr; 0.15 max. C; 2.5–3.5 W	....	C, H
294	Stainless type 420		Fe; 12–14 Cr; 0.15 max. C	B, CR, D, HR, P, R, S, W	C, A
		Jessop Steel Co., Washington, Pa.			
295	Stainless type 430	Joslyn Mfg. & Supply Co., Chicago, Ill.	Fe; 14–18 Cr; 0.12 max. C	B, CR, HR, P, R, S, T, W	C, H
296	Stainless type 430F	Latrobe Electric Steel Co., Latrobe, Pa.	Fe; 14–18 Cr; 0.12 max. C; 0.07 min. P, S, Se; 0.6 max. Zr, Mo	B, CR, D, HR, R, W	C, H
		McLouth Steel Corp., Detroit, Mich.			
		Michiana Products Corp., Michigan City, Ind.			
297	Stainless type 431	Midvale Co., Philadelphia, Pa.	Fe; 15–17 Cr; 1.25–2.5 Ni; 0.2 max. C	B, CR, D, HR, P, S, W	C, A, H
298	Stainless type 440A	Pittsburgh Steel Co., Pittsburgh, Pa.	Fe; 16–18 Cr; 0.6–0.75 C; 0.75 max. Mo	B, CR, HR, D, P, S, R, W	C, A
		Timken Roller Bearing Co., Canton, Ohio			
299	Stainless type 440B	Republic Steel Corp., Cleveland, Ohio	Fe; 16–18 Cr; 0.75–0.95 C; 0.75 max. Mo	B, CR, HR, D, P, S, R, W	C, A
300	Stainless type 440C	Rotary Electric Steel Corp., Detroit, Mich.	Fe; 16–18 Cr; 0.95–1.2 C; 0.75 max. Mo	B, CR, HR, D, P, S, R, W	C, A
301	Stainless type 442	Sharon Steel Corp., Sharon, Pa.	Fe; 18–23 Cr; 0.25 max. C	B, CR, D, HR, P, R, S	H, C
		Stanley Works, New Britain, Conn.			
302	Stainless type 443	Summerill Tubing Co., Bridgeport, Pa.	Fe; 18–23 Cr; 0.2 max. C; 0.9–1.25 Cu	B, CR, D, HR, P, R, S, T, W	C
		Superior Steel Corp., Carnegie, Pa.			
303	Stainless type 446	Universal Cyclops Steel Corp., Bridgeville, Pa.	Fe; 23–27 Cr; 0.35 max. C; 0.25 max. N	B, CR, D, HR, P, R, S, T, W	C, H
		Vanadium Alloys Steel Co., Latrobe, Pa. (5)			
		Wallingford Steel Co., Wallingford, Conn.			
304	Stainless type 501		Fe; 4–6 Cr; 0.1 min. C	B, CR, D, HR, P, R, S, T, W	C
305	Stainless type 502		4–6 Cr; 0.1 max. C	B, CR, D, HR, P, R, S, T, W	C
306	Stainless type CA-15	NOTE: Listed below are the producers of standard *cast* stainless steel (Nos. 306–360) for which type analyses have been established by the Alloy Casting Institute. Designations marked † are no longer in force, but occasional reference is still made to them. Numbers in ( ) are the number of standard A.C.I. alloys for which the company has provided corrosion data; they are solely to acknowledge cooperation in the preparation of these tables and have no reference to the number of steels produced by any company.	Fe; 11–14 Cr; 1 max. Ni; 0.15 max. C	C	C
307	Stainless type CA-40		Fe; 11–14 Cr; 1 max. Ni; 0.2–0.4 C	C	C
308	Stainless type CB-30		Fe; 18–22 Cr; 2 max. Ni; 0.3 max. C	C	C
309	Stainless type CC-35†		Fe; 26–30 Cr; 4 max. Ni; 0.35 max. C	C	C
310	Stainless type CC-50		Fe; 26–30 Cr; 4 max. Ni; 0.5 max. C	C	C
311	Stainless type CD-10M†		Fe; 27–30 Cr; 3–6 Ni; 0.1 max. C; 2 max. Mo	C	C
312	Stainless type CE-30		Fe; 26–30 Cr; 8–11 Ni; 0.3 max. C	C	C
313	Stainless type CF-7		Fe; 18–20 Cr; 8–10 Ni; 0.07 max. C	C	C
314	Stainless type CF-10		Fe; 18–20 Cr; 8–10 Ni; 0.1 max. C	C	C
315	Stainless type CF-16		Fe; 18–20 Cr; 8–10 Ni; 0.16 max. C	C	C
316	Stainless type CF-20		Fe; 18–20 Cr; 8–10 Ni; 0.2 max. C	C	C
317	Stainless type CF-7Se	Allegheny Ludlum Steel Corp., Pittsburgh, Pa.	Fe; 18–20 Cr; 8–10 Ni; 0.07 max. C; 0.2–0.35 Se	C	C
318	Stainless type CF-7C	Amer. Cast Iron Pipe Co., Birmingham, Ala.	Fe; 18–20 Cr; 8–10 Ni; 0.07 max. C; 8xC-1 Cb	C	C
319	Stainless type CF-7M	American Manganese Steel Div., American Brake Shoe & Fdry. Co., Chicago Heights, Ill.	Fe; 18–20 Cr; 8–10 Ni; 0.07 max. C; 1.5–3.5 Mo	C	C
320	Stainless type CF-10M	American Steel Castings Co., Newark, N.J.	Fe; 18–20 Cr; 8–10 Ni; 0.1 max. C; 1.5–3.5 Mo	C	C
321	Stainless type CF-16M	Atlas Fdry. Co., Irvington, N.J.	Fe; 18–20 Cr; 8–10 Ni; 0.16 max. C; 1.5–3.5 Mo	C	C
322	Stainless type CF-7MC†	Babcock & Wilcox Co., Barberton, Ohio	Fe; 18–20 Cr; 8–10 Ni; 0.07 max. C; 1.5–3.5 Mo; 8xC-1 Cb	C	C
		Calorizing Co., Wilkinsburg, Pa.			
		Chicago Steel Fdry., Chicago, Ill.			
		Copper Alloy Fdry. Co., Elizabeth, N.J.			
		Crane Co., Chicago, Ill.			
323	Stainless type CG-7	Driver-Harris Co., Harrison, N.J.	Fe; 20–22 Cr; 10–12 Ni; 0.07 max. C	C	C
324	Stainless type CG-10	Duraloy Co., Scottdale, Pa. (9)	Fe; 20–22 Cr; 10–12 Ni; 0.1 max. C	C	C
325	Stainless type CG-16	Duriron Co., Inc., Dayton, Ohio	Fe; 20–22 Cr; 10–12 Ni; 0.16 max. C	C	C
326	Stainless type CG-16Se	Electric Steel Fdry. Co., Portland, Ore.	Fe; 20–22 Cr; 10–12 Ni; 0.16 max. C; 0.2–0.35 Se	C	C
327	Stainless type CG-7C	Electro-Alloys Div. Amer. Brake Shoe Co., Elyria, Ohio	Fe; 20–22 Cr; 10–12 Ni; 0.07 max. C; 8xC-1 Cb	C	C
		Empire Steel Castings, Inc., Reading, Pa.			
328	Stainless type CG-7M	General Alloys Co., Boston, Mass.	Fe; 20–22 Cr; 10–12 Ni; 0.07 max. C; 1.5–3.5 Mo	C	C
		General Metals Corp., Oakland, Calif.			
329	Stainless type CG-10M	Grede Foundries, Milwaukee, Wis.	Fe; 20–22 Cr; 10–12 Ni; 0.1 max. C; 1.5–3.5 Mo	C	C
330	Stainless type CG-16M		Fe; 20–22 Cr; 10–12 Ni; 0.16 max. C; 1.5–3.5 Mo	C	C

Table 3. Directory of Materials for the Construction of Chemical Equipment—(*Continued*)

No.	Material	Manufacture	Essential Nominal Chemical Composition, %	Forms Available	Primarily for
331	Stainless type CG-7MC†	Haynes Stellite Co., Kokomo, Ind. Hoskins Mfg. Co., Detroit, Mich. Key Co., East St. Louis, Ill.	Fe; 20–22 Cr; 10–12 Ni; 0.07 max. C; 1.5–3.5 Mo; 8xC-1 Cb	C	C
332	Stainless type CH-10	Lebanon Steel Fdry., Lebanon, Pa.	Fe; 22–26 Cr; 12–15 Ni; 0.1 max. C	C	C
333	Stainless type CH-20	Michiana Products Corp., Michigan City, Ind.	Fe; 22–26 Cr; 12–15 Ni; 0.2 max. C	C	C
334	Stainless type CH-10C	Michigan Steel Casting Co., Detroit, Mich.	Fe; 22–26 Cr; 12–15 Ni; 0.1 max. C; 8xC-1 Cb	C	C
335	Stainless type CH-10M	Midvale Co., Philadelphia, Pa. Milwaukee Steel Fdry., Milwaukee, Wis.	Fe; 22–26 Cr; 12–15 Ni; 0.1 max. C; 1.5–3.5 Mo	C	C
336	Stainless type CH-20M†	National Alloy Div. Blaw-Knox Co., Blaw-Knox, Pa.	Fe; 22–26 Cr; 12–15 Ni; 0.2 max. C; 1.5–3.5 Mo	C	C
337	Stainless type CH-10MC†	Ohio Steel Fdry. Co., Cincinnati, Ohio Otis Elevator Co., Buffalo, N.Y.	Fe; 22–26 Cr; 12–15 Ni; 0.1 max. C; 1.5–3.5 Mo; 8xC-1 Cb	C	C
338	Stainless type CK-25	Pacific Fdry. Co., Ltd., San Francisco, Calif.	Fe; 23–27 Cr; 19–22 Ni; 0.25 max. C	C	C
339	Stainless type CM-25	Shawingan Chemicals, Ltd., Montreal, Que.	Fe; 8–11 Cr; 19–22 Ni; 0.25 max. C	C	C
340	Stainless type CN-7	Sivyer Steel Casting Co., Milwaukee, Wis. Standard Alloy Co., Cleveland, Ohio Sterling Alloys Inc., Woburn, Mass.	Fe; 18–22 Cr; 20–30 Ni; 0.07 max. C; may contain others	C	C
341	Stainless type CN-25†	Symington-Gould Corp., Rochester, N. Y. (1) Taylor Wharton Iron & Steel Co., High Bridge, N.J.	Fe; 18–22 Cr; 20–30 Ni; 0.25 max. C; may contain others	C	C
342	Stainless type CS-25†	Utility Electric Steel Fdry., Los Angeles, Calif.	Fe; 8–12 Cr; 29–32 Ni; 0.25 max. C	C	C
343	Stainless type CT-7	Warman Steel Casting Co., Huntington Park, Calif.	Fe; 13–17 Cr; 34–37 Ni; 0.07 max. C; may contain others	C	C
344	Stainless type CT-25†		Fe; 13–17 Cr; 34–37 Ni; 0.25 max. C; may contain others	C	C
345	Stainless type HB		Fe; 18–22 Cr; 2 max. Ni	C	H
346	Stainless type HC		Fe; 26–30 Cr; 4 max. Ni	C	H
347	Stainless type HD		Fe; 26–30 Cr; 3–6 Ni	C	H
348	Stainless type HE		Fe; 26–30 Cr; 8–11 Ni	C	H
349	Stainless type HF		Fe; 18–23 Cr; 8–11 Ni	C	H
350	Stainless type HH		Fe; 23–27 Cr; 11–14 Ni	C	H
351	Stainless type HI		Fe; 26–30 Cr; 14–17 Ni	C	H
352	Stainless type HK		Fe; 23–27 Cr; 19–22 Ni	C	H
353	Stainless type HL		Fe; 28–32 Cr; 19–22 Ni	C	H
354	Stainless type HN		Fe; 18–22 Cr; 23–26 Ni	C	H
355	Stainless type HP		Fe; 28–32 Cr; 29–31 Ni	C	H
356	Stainless type HS		Fe; 8–12 Cr; 29–32 Ni	C	H
357	Stainless type HT		Fe; 13–17 Cr; 33–37 Ni	C	H
358	Stainless type HU		Fe; 17–21 Cr; 37–41 Ni	C	H
359	Stainless type HW		Fe; 10–14 Cr; 58–62 Ni	C	H
360	Stainless type HX		Fe; 15–19 Cr; 64–68 Ni	C	H
360A	Steel	Corrosion data from A. W. Spitz, Eng. Dept., Amer. Cyanamid Co., New York	Plain carbon steel		
362	Still Metal	Amer. Manganese Bronze Co., Philadelphia, Pa.	Cu; Sn	C	C
363	Stoodite	Stoody Co., Whittier, Calif.	Fe; 33 Cr; 4.5 Mn; 2 Si; 4 C	Welding rod	A
364	Stoody 1	Stoody Co., Whittier, Calif.	Co; 25 Cr; 13 W; 2 C	C, welding rod	C, H, A
365	Stoody 6	Stoody Co., Whittier, Calif.	Co; 25 Cr; 5.5 W; 1 C	C, welding rod	C, H, A
366	Stoody Self Hardening	Stoody Co., Whittier, Calif.	Fe; 6 Cr; 2.5 Mn; 1.5 Si; 1 C	Welding rod	A
367	Super Nickel	Amer. Brass Co., Waterbury, Conn.	70 Cu; 30 Ni	B, S, R, T, W, P	C
368	Tantalum	Fansteel Metallurgical Corp., North Chicago, Ill.	99.9+ Ta	B, CR, D, S, R, T, W	C
369	Telnic Bronze	Chase Brass and Copper Co., Waterbury, Conn.	98.3 Cu; 1 Ni; 0.2 P; 0.5 Te	R	C
370	Thermalloy Stainless	Electro Alloys Div., Amer. Brake Shoe Co., Elyria, Ohio	Various standard stainless steels; see Nos. 275–360	C	H
371	Timken Stainless	Steel and Tube Div., Timken Roller Bearing Co., Canton, Ohio	Various standard stainless steels; see Nos. 275–360	B, CR, HR, R, T, W	C, H
372	Tisco 150 Alloy	Taylor-Wharton Iron and Steel Co., Easton, Pa.	Fe; 2.5–3 C; 2 Si; 1–1.5 Ni; 28–32 Cr	C	A, C
373	Tisco Timang MnNi Steel	Taylor-Wharton Iron and Steel Co., Easton, Pa.	Fe; 0.6–0.8 C; 13–15 Mn; 3 Ni	P, R	A
374	Tisco Mn Steel	Taylor-Wharton Iron and Steel Co., Easton, Pa.	Fe; 12 Mn (Hadfield type)	C	A
375	Toncon Cu Mo Iron	Republic Steel Corp., Cleveland, Ohio	....	B, CR, HR, P, S	C
376	Tophet A	Wilbur B. Driver Co., Newark, N.J.	80 Ni; 20 Cr	D, HR, R, W	H
377	Tophet C	Wilbur B. Driver Co., Newark, N.J.	Fe; 60 Ni; 15 Cr	D, HR, R, W	H
378	Tophet D	Wilbur B. Driver Co., Newark, N.J.	Fe; 35 Ni; 18.5 Cr	B, CR, D, HR, R, W	H
379	Tube Borium	Stoody Co., Whittier, Calif.	60 tungsten carbide; 40 steel	Welding rod	A
380	Tuf-Stuf	Mueller Brass Co., Port Huron, Mich.	86.9 Cu; 10 Al; 3 Fe; 0.1 Mn	B, D, R, F	C, H, A
381	U. S. S. Stainless	Carnegie-Illinois Steel Corp., Pittsburgh, Pa.	Various standard stainless steels; see Nos. 275–360	B, CR, HR, P, S, R, strip	C, H
382	U. S. S. Cor-Ten	Carnegie-Illinois Steel Corp., Pittsburgh, Pa.	Fe; 0.12 max. C; 0.25–0.55 Cu; 0.5–1.25 Cr; 0.65 max. Ni	B, HR, P, S, R, strip	C, A
383	U. S. S. Man-Ten	Carnegie-Illinois Steel Corp., Pittsburgh, Pa.	Fe; 0.25 max. C; 1.1–1.6 Mn; 0.2 min. Cu	B, HR, P, S, R, strip	C, A
384	U. S. S. A-R	Carnegie-Illinois Steel Corp., Pittsburgh, Pa.	Fe; 0.35–0.5 C; 1.5–2 Mn; 0.15–0.3 Si	B, HR, P, S, strip	A
385	Utiloy Stainless	Utility Electric Steel Fdry., Los Angeles, Calif.	Various standard stainless steels; see Nos. 275–360	C	C, H
386	Vanadium Stainless	Vanadium Alloys Steel Co., Latrobe, Pa.	Various standard stainless steels; see Nos. 275–360	B, HR, D, P, R, W	C, H, A

Table 3. Directory of Materials for the Construction of Chemical Equipment—(*Continued*)

No	Material	Manufacturer	Essential Nominal Chemical Composition, %	Forms Available	Primarily for
387	Vanadium Stainless B	Vanadium Alloys Steel Co., Latrobe, Pa.	Fe; 14–18 Cr; 0.12 min. C	B, HR, D, P, R, W	C, A
388	Vanadium Stainless 795	Vanadium Alloys Steel Co., Latrobe, Pa.	Fe; 14–18 Cr; 2 max. Ni; 0.15 min. C	B, HR, P, R, W	C, A
389	Vanadium Stainless U	Vanadium Alloys Steel Co., Latrobe, Pa.	Fe; 17–19 Cr; 7–9.5 Ni; 0.15 max. C; 1–1.5 Cu; 1–1.5 Mo	B, HR, P, R, W	C, H
390	Worthite	Worthington Pump and Machinery Corp., Harrison, N.J.	Fe; 20 Cr; 24 Ni; 0.07 max. C; 3.25 Si; 3 Mo; 1.75 Cu; 0.5 Mn	B, C, HR, R, W	C
391	Wrought Iron, Genuine	A. M. Byers Co., Pittsburgh, Pa.	Fe; 0.02 C; 0.03 Mn; 0.12 P; 0.15 Si; 0.02 S	B, HR, P, R, S, T	C
392	Wyndaloy	Wyndale Mfg. Corp., Indianapolis, Ind.	60 Cu; 20 Ni; 20 Mn	B, CR, HR, D, P, R, T, F	C, A

Carbon and Graphite

No.	Material	Manufacturer	Description
400	International	International Graphite and Electrode Corp., St. Mary's, Pa.	Graphite. Electrodes and various shapes
401	Karbate (carbon)	National Carbon Co., Cleveland, Ohio	Impervious carbon } Pipe, fittings, pumps, valves, towers and auxiliary parts, tanks, heat exchangers, brick, raschig rings, plates, rods, etc.
402	Karbate (graphite)	National Carbon Co., Cleveland, Ohio	Impervious graphite
403	National Kempruf	National Carbon Co., Cleveland, Ohio	Carbon
404	National Acheson	National Carbon Co., Cleveland, Ohio	Graphite
405	Speer	Speer Carbon Co., St. Mary's, Pa.	Carbon and graphite. Brick, plates, blocks, tubes, cylinders, bushings, shapes
406	Stackpole	Stackpole Carbon Co., St. Mary's, Pa.	Carbon and graphite. Tubes, pipe, rods, plates, bearings, seal rings, crucibles, heat-treating boxes and molds Impervious graphite. Pipe, bearings, seal rings, special shapes (injector body and nozzles)

Cement, Mortar, Putty

No.	Material	Manufacturer	Description
500	Sul. cement	Generally available	Sulfur cement, silica aggregate, plasticized with Thiokol
501	Sul. cement	Generally available	Sulfur cement, carbon aggregate, plasticized with Thiokol
502	Furan cement	Generally available	Furan derivative, silica aggregate, sets by chemical reaction
503	Furan cement	Generally available	Furan derivative, carbon aggregate, sets by chemical reaction
504	Phen. cement	Generally available	Phenol-formaldehyde, silica aggregate, sets by chemical reaction
505	Phen. cement	Generally available	Phenol-formaldehyde, carbon aggregate, sets by chemical reaction
506	Silicate cement	Generally available	Sodium silicate, silica aggregate, slow setting
507	Silicate cement	Generally available	Sodium silicate, silica aggregate, sets by chemical reaction
508	Acichlor	Pecora Paint Co., Philadelphia, Pa.	Firm putty
509	Acidol	The Sullivan Co., Memphis, Tenn.	Pouring cements
510	Acitite	Pecora Paint Co., Philadelphia, Pa.	Ready mixed cement; troweling
511	Alkor	Atlas Mineral Products Co., Mertztown, Pa.	Furan derivative; carbon aggregate
512	Alkor-S	Atlas Mineral Products Co., Mertztown, Pa.	Furan derivative; silica aggregate
513	Asplit	Pennsylvania Salt Mfg. Co., Philadelphia, Pa.	Chemical hardening resin cement
514	Asplit F	Pennsylvania Salt Mfg. Co., Philadelphia, Pa.	Chemical hardening resin cement
515	Basolit	Nukem Products Corp., Buffalo, N.Y.	Sulfur-silicate cement
516	Carbo Korez	Atlas Mineral Products Co., Mertztown, Pa.	Phenol-formaldehyde; carbon aggregate
517	Carboline	Carboline Co., St. Louis, Mo.	Carbonaceous resin cement
518	Carbon Basolit	Nukem Products Corp., Buffalo, N.Y.	Sulfur-carbon cement
519	Carbo Vitrobond	Atlas Mineral Products Co., Mertztown, Pa.	Sulfur cement; carbon aggregate
520	Carolina	Charlotte Chemical Laboratories, Charlotte, N.C.	Acidproof cement
521	Causplit	Pennsylvania Salt Mfg. Co., Philadelphia, Pa.	Chemical hardening resin cement
522	Charlab	Charlotte Chemical Laboratories, Charlotte, N.C.	Chemical putty
523	Duralon	U.S. Stoneware Co., Akron, Ohio	Furan base resin cements
524	Durasite	U.S. Stoneware Co., Akron, Ohio	Furan base resin cements
525	Fairprene	E. I. du Pont de Nemours & Co., Wilmington, Del.	Synthetic elastomer cement
526	Filtros	Filtros, Inc., East Rochester, N.Y.	Acidproof cement
527	Haveg 41-G	Haveg Corp., Newark, Del.	Quick setting, phenolic resin grouting cement
528	Haveg 41-R	Haveg Corp., Newark, Del.	Quick setting phenolic resin cement for Haveg 41
529	Haveg 43-HF	Haveg Corp., Newark, Del.	Quick setting phenolic resin cement for Haveg 43
530	Haveg 60-R	Haveg Corp., Newark, Del.	Quick setting furane resin cement for Haveg 60
531	Korez	Atlas Mineral Products Co., Mertztown, Pa.	Phenol-formaldehyde; silica aggregate
532	Lumnite	Lumnite Div., Universal Atlas Cement Co., New York	Hydraulic cement; powder for mixing with aggregate and water to make concrete
533	N-380 Silicate	Philadelphia Quartz Co., Philadelphia, Pa.	Sodium silicate; to be combined with special proprietary quick-setting cements at time of use
534	N-series	Union Bay State Chemical Co., Cambridge, Mass.	Neoprene base cements to be used as lining for tanks, etc.
535	Nukem All-Purpose	Nukem Products Corp., Buffalo, N.Y.	Resinous cement
536	Nukem silicate	Nukem Products Corp., Buffalo, N.Y.	Silicate cement
537	Pecomastic	Pecora Paint Co., Philadelphia, Pa.	Putty; troweling; brushing
538	Penchlor acidproof	Pennsylvania Salt Mfg. Co., Philadelphia, Pa.	Quick setting silicate cement
539	Penchlor acidproof S25	Pennsylvania Salt Mfg. Co., Philadelphia, Pa.	Quick setting silicate cement

NOTE: Nos. 500–507 are the eight standard types of cement, or mortar, that predominate in chemical construction. Each has been rated against 55 chemicals by C. R. Payne, President of Electro Chemical and Eng. Co., who, rather than speak only for one company's products, has consented to provide this comprehensive treatment of cements according to type. His ratings are on the conservative side.

Table 3. Directory of Materials for the Construction of Chemical Equipment—*(Continued)*

No.	Material	Manufacturer	Description
540	Penchlor fireproof	Pennsylvania Salt Mfg. Co., Philadelphia, Pa.	Quick setting silicate cement
541	Pennsalt PRF	Pennsylvania Salt Mfg. Co., Philadelphia, Pa.	Chemical hardening resin cement
542	Permanite	Maurice A. Knight, Akron, Ohio	Furan base resin cement
543	Plastikon	B. F. Goodrich Rubber Co., Akron, Ohio	Rubber-base putty
544	Plastite	U.S. Stoneware Co., Akron, Ohio	Calking putty
545	Plasul Basolit	Nukem Products Corp., Buffalo, N.Y.	Plasticized sulfur-silicate cement
546	Porox	Patterson Fdry. & Machine Co., East Liverpool, Ohio	Silicate cement
547	Quigley	Quigley Co., New York, N.Y.	Acidproof cements
548	Reardon	Reardon Industries, Cincinnati, Ohio	Acidproof cement
549	Redux	Resinous Products & Chemicals Co., Philadelphia, Pa.	Metal-to-metal resin adhesive
550	Sauereisen 31	Sauereisen Cements Co., Pittsburgh, Pa.	Acidproof cement
551	Sauereisen 44	Sauereisen Cements Co., Pittsburgh, Pa.	Plastic; acid tank sealer
552	Sauereisen 48	Sauereisen Cements Co., Pittsburgh, Pa.	Joint compound
553	Silicate	Philadelphia Quartz Co., Philadelphia, Pa.	Sodium silicate; to be used with silica for acid-resistant cements
554	Silastic 121	Dow Corning Corp., Midland, Mich.	Silicone elastomers in putty form for calking
555	Stackpole	Stackpole Carbon Co., St. Mary's, Pa.	Carbonaceous cements for carbon, graphite, and carbon-to-metal joints
556	Staminite	Robinson Clay Product Co., Akron, Ohio	Acidproof cement
557	Sulsilo	The Sullivan Co., Memphis, Tenn.	Premixed silicate cement
558	Tegul Vitrobond	Atlas Mineral Products Co., Mertztown, Pa.	Sulfur cement; silica aggregate
559	Thiokol	Thiokol Corp., Trenton, N.J.	Liquid (100% solids) rubber polymers
560	Vitrex	Atlas Mineral Products Co., Mertztown, Pa.	Sodium silicate; sets by chemical reaction

Ceramics

No.	Material	Manufacturer	Description
600	Acidproof brick	Generally available*	Brick, stone, tile, rings, plates, cylinders, tower packing, tower linings, etc.
601	Alsop glass lining	Alsop Engineering Co., Milldale, Conn.	Glass-lined tanks, mixers, and filters
602	Amersil fused silica	Amersil Co., Inc., Hillside, N.J.	Fused silica ware. Pans, pipes, gas coolers, absorbers, insulators, tubes, plates
603	Carrara	Pittsburgh Plate Glass Co., Pittsburgh, Pa.	Structural glass, flat or bent
604	Ceratherm	U.S. Stoneware Co., Akron, Ohio	Ceramic ware. Process equipment
605	Fiberglas	Owens-Corning Fiberglas Corp., Toledo, Ohio	Fibrous glass cloth and mat. Air filters, thermal insulation, tower packing, filter cloth
606	"General" chemical stoneware	General Ceramics & Steatite Corp., Keasbey, N.J.	Chemical stoneware. Tanks, kettles, pipe, fittings, valves, pumps, coils, filters, etc.
607	Glascote glass lining	Glascote Products, Inc., Cleveland, Ohio	Glass-lined tanks and processing equipment
608	Hanovia fused quartz	Hanovia Chemical & Mfg. Co., Newark, N.J.	Transparent fused quartz in all shapes
609	Illinois chemical porcelain	Illinois Electric Porcelain Co., Macomb, Ill.	Ceramic chemical porcelain. Pipes, valves, fittings
610	Knight-Ware	Maurice A. Knight, Akron, Ohio	Chemical stoneware. Tanks, kettles, pipe, fittings, valves, pumps, coils, filters, etc.
611	Lapp porcelain	Lapp Insulator Co., Le Roy, N.Y.	Chemical porcelain. Pipe, fittings, valves, plug cocks, towers, tower packing special shapes
612	Pennvernon	Pittsburgh Plate Glass Co., Pittsburgh, Pa.	Window glass, flat
613	Permaglas	A. O. Smith Corp, Milwaukee, Wis.	Glass-lined steel equipment
614	Pfaudler glass lining	The Pfaudler Co., Rochester, N.Y.	Glass-lined steel equipment
615	Pittsburgh plate glass	Pittsburgh Plate Glass Co., Pittsburgh, Pa.	Polished plate glass, flat or bent
616	Pyrex	Corning Glass Works, Corning, N.Y.	Glass. Pipe, fittings, sight glasses, gage glasses, heat-resistant plate
617	"U.S." chemical stoneware	U.S. Stoneware Co., Akron, Ohio	Chemical stoneware. Tanks, kettles, pipe, fittings, valves, pumps, coils, filters, etc.
618	Vitreo	Vitreous Steel Products Co., Cleveland, Ohio	Acid-resisting porcelain enameled steel. Drying and evaporating trays, miscellaneous parts
619	Vitreosil	The Thermal Syndicate Ltd., New York, N.Y.	Vitreous silica. Pipes, tubes, rods, plates, dishes, retorts, stills, HCl acid plant, and apparatus for absorption, cooling, and condensing
620	Vitreous enamel	Vitreous Enameling & Stamping Co., New York, N.Y.	Enameled tanks and specialties
621	Vycor	Corning Glass Works, Corning, N.Y.	96% silica glass. Tubing, rods, flat ware, various shapes

* Acidproof brick and stone products are available from the following (trade names, where they differ from name of company, are in parentheses): Acme Brick, Co., Fort Worth, Texas (Acitex, Everlast, La Perla); Alabama Clay Products Co., Birmingham, Ala; Alberene Stone Corp. of Va., New York, N.Y.; Atlas Mineral Products Co., Mertztown, Pa.; Belden Brick Co., Canton, Ohio; Charlotte Chemical Labs., Charlotte, N.C. (Carolina); Claycraft Co., Columbus, Ohio; Custodis Construction Co., New York, N.Y. (Aco); Electro-Chemical Supply & Engineering Co., Paoli, Pa.; Filtros Inc., East Rochester, N.Y.; General Refractories Co., Philadelphia, Pa. (Acido); Harbinson-Walker Refractories Co., Pittsburgh, Pa.; B. Mifflin Hood Co., Daisy, Tenn.; Keagler Brick Co., Steubenville, Ohio; Kewaunee Mfg. Co., Kewaunee, Wis. (Karcite, Kemrock); Metropolitan Paving Brick Co., Canton, Ohio; Parker-Russell Mining & Mfg. Co., St. Louis, Mo.; Patterson Fdry. & Machine Co., East Liverpool, Ohio; Quigley Co., New York, N.Y.; Robinson Clay Product Co., Akron, Ohio; Southern Clay Mfg. Co., Chattanooga Tenn.; Thornton Firebrick Co., Clarksburg, W. Va.

Plastics

No.	Material	Manufacturer	Description
700	Ace Saran	Amer. Hard Rubber Co., New York, N.Y.	Vinylidene chloride. Pipe, fittings, tubing
701	Celcon	Celanese Plastics Corp., New York, N.Y.	Ethyl cellulose
702	Celluloid	Celanese Plastics Corp., New York, N.Y.	Cellulose nitrate
703	Compar	Resistoflex Corp., Belleville, N.J.	Solution of compounded polyvinyl alcohol. Solution for dipping or painting
704	DC 801–804	Dow Corning Corp., Midiand, Mich.	Solvent dispersed silicone resins for coatings
706	Formica CHN-5	Formica Insulation Co., Cincinnati, Ohio	Reinforced phenol-formaldehyde. Sheet, tube, rod
707	Formica CN-22	Formica Insulation Co., Cincinnati, Ohio	Reinforced phenol-formaldehyde. Sheet and rod
708	Formica LN-41	Formica Insulation Co., Cincinnati, Ohio	Reinforced melamine. Sheet, tube, rod
709	Forticel	Celanese Plastics Corp., New York, N.Y.	Cellulose propionate
710	Geon	B. F. Goodrich Chemical Co., Cleveland, Ohio	Polyvinyl chloride plastics } Pipe, pumps, tanks, blowers, agitators, valves, towers, coolers, fume duct, hoods
711	Haveg 41	Haveg Corp., Newark, Del.	Phenolic-asbestos plastic }
712	Haveg 43	Haveg Corp., Newark, Del.	Phenolic-graphite plastic }
713	Haveg 60	Haveg Corp., Newark, Del.	Furan-asbestos plastic }
714	Haveg 63	Haveg Corp., Newark, Del.	Furan-graphite plastic }

Table 3. Directory of Materials for the Construction of Chemical Equipment—(*Continued*)

No.	Material	Manufacturer	Description
715	Heresite M66	Heresite and Chemical Co., Manitowoc, Wis.	Transparent molding powder
716	Heresite MF 66	Heresite and Chemical Co., Manitowoc, Wis.	Black molding powder
717	Insurok	Richardson Co., Melrose Park, Ill.	Molded and laminated plastics. Sheet, rod, tube, fabricated parts, special moldings
718	Koroseal	B. F. Goodrich Co., Akron, Ohio	Plasticized polyvinyl chloride. Sheet lining for tanks and fume ducts
719	Kriston	B. F. Goodrich Chemical Co., Cleveland, Ohio	Allyl ester thermosetting materials
720	Lamicoid	Mica Insulator Co., Schenectady, N.Y.	Phenolic laminated, fabric base. Sheets
721	Lumarith	Celanese Plastics Corp., New York, N.Y.	Cellulose acetate
722	Micarta	Westinghouse Electric Corp., Pittsburgh, Pa.	Laminated plastics, fabric or kraft base. Sheets, chanels, angles, molded shapes
723	Nixon CA (Nixonite)	Nixon Nitration Works, Nixon, N.J.	Cellulose acetate. Sheets, rods, molding powders
724	Nixon CN (Nixonoid)	Nixon Nitration Works, Nixon, N.J.	Cellulose nitrate. Sheets, rods, tubes
725	Nixon EC	Nixon Nitration Works, Nixon, N.J.	Ethyl cellulose. Molding powders
726	Nukemite	Nukem Products Corp., Buffalo, N.Y.	Synthetic resin sheet and coating
727	Nylon FM-1	E. I. du Pont de Nemours & Co., Wilmington, Del.	Injection moldings. Heat-resistant injection molded parts
728	Nylon FM-3	E. I. du Pont de Nemours & Co., Wilmington, Del.	Injection and extrusion moldings. Wire covering
729	Nylon FM-4	E. I. du Pont de Nemours & Co., Wilmington, Del.	Injection, compression, and extrusion moldings
730	Nylon FM-100	E. I. du Pont de Nemours & Co., Wilmington, Del.	Injection and extrusion moldings
731	Nylon FM-101	E. I. du Pont de Nemours & Co., Wilmington, Del.	Injection, compression, and extrusion moldings. Tubing, sheeting, wire covering, gasketing
732	Nylon FM-102	E. I. du Pont de Nemours & Co., Wilmington, Del.	Injection and extrusion moldings. Wire covering
733	Permanite	Maurice A. Knight, Akron, Ohio	Furan base resin. Laminates, fabricated shapes
734	Phenolite	National Vulcanized Fibre Co., Wilmington, Del.	Laminated phenolics. Sheet, tube, rod
735	Polythene	E. I. du Pont de Nemours & Co., Wilmington, Del.	Polyethylene. Sheet, rod, tube, molding powders, film, filament. Pipe, pipe lining, gaskets, etc.
736	Pyroflex Construction	Maurice A. Knight, Akron, Ohio	Thermoplastic resin plus steel, brick, etc., to make a functional unit
737	Resilon	U.S. Stoneware Co., Akron, Ohio	Bituminous linings and membranes
738	Resistoflex	Resistoflex Corp., Belleville, N.J.	Compounded polyvinyl alcohol extruded tubing. Fiber or wire braided hose
739	Resistoflex Compar	Resistoflex Corp., Belleville, N.J.	Compounded polyvinyl alcohol molded. Molded shapes; abrasion resistant gaskets, pump diaphragms, sand blasting mats, rollers, etc.
740	Saran	Dow Chemical Co., Midland Mich.	Vinyl chloride-vinylidene chloride copolymer. Pipe, pipe fittings, tube, tube fittings
741	Sealon	Maurice A. Knight, Akron, Ohio	Polyvinyl chloride elastomer. Tank linings, sheets, gaskets
742	Teflon	E. I. du Pont de Nemours & Co., Wilmington, Del.	Polymerized tetrafluoroethylene. Rods, tubes, sheets, beading, gaskets, thin tapes
743	Textolite 1422	General Electric Co., Pittsfield, Mass.	Rods and plates
744	Textolite 2001	General Electric Co., Pittsfield, Mass.	Graphited phenolic laminate. Sheet, rod, tube
745	Textolite 2013	General Electric Co., Pittsfield, Mass.	Fabric base, phenolic laminate. Sheet, rod, tube
746	Tygon	U.S. Stoneware Co., Akron, Ohio	Synthetic compounds. Linings, tubing, protective coatings, etc.
747	Vimlite	Celanese Plastics Corp., New York, N.Y.	Mesh reinforced plastic
748	Vulcanized Fibre	National Vulcanized Fibre Co., Wilmington, Del.	Sheets, tubes, rod
749	Zerok	Atlas Mineral Products Co., Mertztown, Pa.	Polyvinyl chloride. Tank linings and coatings

Rubber

No.	Material	Manufacturer	Description
800	Ace hard rubber	Amer. Hard Rubber Co., New York, N.Y.	Vulcanized rubber. Rod, sheet, tube, molded parts, linings, pipe, fittings, etc.
801	Acidseal E	B. F. Goodrich Co., Akron, Ohio	Hard rubber. Sheet lining for tanks and miscel. metal parts*
802	Acidseal MA and PA	B. F. Goodrich Co., Akron, Ohio	Soft rubber. Sheet lining for tanks and miscel. metal parts*
803	Armstrong	Armstrong Cork Co., Lancaster, Pa.	Gaskets
804	Boston	Boston Woven Hose & Rubber Co., Boston, Mass.	Conveyor and transmission belt, hose, mechanical goods
805	Butyl (GR-I)	Stanco Distributors, Inc., New York, N.Y.	Solid copolymer of isobutylene and isoprene
806	Crane	Crane Packing Co., Chicago, Ill.	Packing and mechanical seals
807	Custoplast	Custodis Construction Co., New York, N.Y.	Soft rubber and neoprene tank linings
808	Dayton	Dayton Rubber Mfg. Co., Dayton, Ohio	Oilproof rubber belt, transmission belt
809	Fairprene	E. I. du Pont de Nemours & Co., Wilmington, Del.	Sheet and coated fabric made from vulcanized neoprene. Linings, diaphragms, gaskets, packing
810	Firestone	Firestone Tire & Rubber Co., Akron, Ohio	Vibration dampeners, adhesives, sheet, tape, hose, belting, fabric, moldings, extrusions
812	Garlock	Garlock Packing Co., Palmyra, N.Y.	Gaskets, packings, moldings, Klozure oil seals
813	Gates	Gates Rubber Co., Denver, Colo.	Belts, hose, moldings
814	G. E. Silicone rubber	General Electric Co., Pittsfield, Mass.	Sheets, extruded shapes, molded parts, coating pastes
816	Greene-Tweed	Greene, Tweed & Co., New York, N.Y.	Packing, gaskets, sheet
817	Heresite rubber	Heresite & Chemical Co., Manitowoc, Wis.	Synthetic rubber coatings
818	Hewitt	Hewitt Rubber Corp., Buffalo, N.Y.	Hose, belting, packing, moldings, extrusions
820	Hycar (GR-A)	B. F. Goodrich Chemical Co., Cleveland, Ohio	Nitrile type synthetic rubber
821	Jenkins	Jenkins Bros. Rubber Div., Bridgeport, Conn.	Mechanical goods, packing, valve discs, tape, moldings, extrusions
822	Johns-Manville	Johns-Manville Sales Corp., New York, N.Y.	Gaskets
825	Linear	Linear Packing & Rubber Co., Philadelphia, Pa.	Packing
826	Luzerne	Luzerne Rubber Co., Trenton, N.J.	Hard rubber pipe, fittings, valves, shapes, tanks, rayon equipment
827	Manhattan	Manhattan Rubber Mfg. Div., Passaic, N.J.	Belting, blocks, hose, pipe, rolls
828	Neobon	Atlas Mineral Products Co., Mertztown, Pa.	Neoprene-base lining for tanks, fans, fume ducts, etc.

* In addition to the lining material named, this company produces most of the following: hose, belt, packing, gaskets, moldings, extrusions, vibration dampeners, rubber-metal bonded products, hard rubber pipe and fittings.

Table 3. Directory of Materials for the Construction of Chemical Equipment—(*Concluded*)

No.	Material	Manufacturer	Description
829	Neoprene	E. I. du Pont de Nemours & Co., Wilmington, Del.	Polymer of chloroprene. Crude neoprene for compounding and curing
830	Parakote	Paramount Rubber Co., Detroit, Mich.	Plating rack insulation for racks to be coated at Paramount plant
831	Paramount	Paramount Rubber Co., Detroit, Mich.	Neoprene, buna-S, natural rubber. Sheets for tank lining
834	Parlon	Hercules Powder Co., Wilmington, Del.	Chlorinated rubbers. Used as a base for concrete paints
835	Perbunan (GR-A)	Stanco Distributors, Inc., New York, N.Y.	Copolymer of butadiene and acrylonitrile. Solid sheets
836	Permobond Natural (soft)	U.S. Rubber Co., New York, N.Y.	Linings for tanks, pipes, fittings, valves
837	Permobond Natural (hard)	U.S. Rubber Co., New York, N.Y.	
838	Permobond GR-S (soft)	U.S. Rubber Co., New York, N.Y.	
839	Permobond GR-S (hard)	U.S. Rubber Co., New York, N.Y.	
840	Permobond GR-A (soft)	U.S. Rubber Co., New York, N.Y.	
841	Permobond GR-A (hard)	U.S. Rubber Co., New York, N.Y.	
842	Permobond GR-M (soft)	U.S. Rubber Co., New York, N.Y.	
843	Permobond GR-P (soft)	U.S. Rubber Co., New York, N.Y.	
844	Permobond GR-I	U.S. Rubber Co., New York, N.Y.	
845	Plioweld	Goodyear Tire & Rubber Co., Akron, Ohio	Rubber-lined tanks, pipe, etc.*
846	Saniprene	B. F. Goodrich Co., Akron, Ohio	Hard rubber. Sheet lining for tanks and miscellaneous metal parts*
847	Self Vulcanizing	Self Vulcanizing Rubber Co., Chicago, Ill.	Linings and coatings
848	Silastic 181	Dow Corning Corp., Midland, Mich.	Silicone rubber in form of crepes for molding and extruding
849	Superflexite	B. F. Goodrich Co., Akron, Ohio	Hard rubber. Sheet lining for tanks and miscellaneous metal parts*
850	Stokes	Jos. Stokes Rubber Co., Trenton, N.J.	Hard rubber pipes, valves, fittings, miscellaneous moldings and extrusions
851	Tensilgrip	Amer. Wringer Co., Woonsocket, R.I.	Natural- and synthetic-rubber-lined tanks, pipe, pumps, ducts, etc.
852	Thermoid	Thermoid Rubber Div., Trenton, N.J.	Belting, hose, packing
853	Thiokol (GR-P)	Thiokol Corp., Trenton, N.J.	Solid form
854	Thiokol	Thiokol Corp., Trenton, N.J.	Water dispersions for linings
855	Triflex	B. F. Goodrich Co., Akron, Ohio	Hard rubber layer sandwiched between two soft rubber layers. Sheet lining for tanks and miscellaneous metal parts*
856	Vistanex	Stanco Distributors, Inc., New York., N.Y.	Polymerized isobutylene. Viscous liquid to solid, depending on molecular weight
857	Vulcanized	Vulcanized Rubber Co., New York, N.Y.	Hard and semihard moldings

* In addition to the lining material named, this company produces most of the following: hose, belt, packing, gaskets, moldings, extrusions, vibration dampeners, rubber-metal bonded products, hard rubber pipe and fittings.

Wood

No.	Material
900	Eastern red cedar
901	Port Orford cedar
902	Western red cedar
903	White cedar
904	Southern tidewater cypress
905	Douglas fir (coast type)
906	Hard maple
907	White oak
908	Southern longleaf pine
909	Southern shortleaf pine
910	Northern white pine
911	Red pine
912	Yellow poplar
913	California redwood
914	Spruce

Following is a list of producers of wooden tanks, towers, pipe, and culverts. Companies in italic are members of the National Wood Tank Institute, Chicago, whose executive director, S. E. Chaney, provided all data on wood's resistance to chemicals.

ALABAMA: Hightower Box & Tank Co., Birmingham. ARKANSAS: Fordyce-Crossett Sales Co., Fordyce; Leird Lumber Co., Little Rock; Smith Fabricating Shop, Hot Springs. CALIFORNIA: Acme Tank Mfg. Co., Los Angeles; C. F. Braun Co., Los Angeles; Fluor Corp., Los Angeles; *W. D. Hall Co., El Cajon;* Hammond Lumber Co., San Francisco; *Industrial Manufacturers Ltd., Los Angeles;* Inman Tank, Pipe & Crossarm Co., San Leandro; *Pacific Tank & Pipe Co., Oakland;* Pacific Wood Tank Corp., San Francisco; Pope & Talbot Inc., San Francisco; Redwood Manufacturers Co., San Francisco; San Mateo Planing Mill, San Mateo; Union Lumber Co., San Francisco; *George Windeler Co., San Francisco.* COLORADO: Plattner Co., Denver; Stearns-Roger Mfg. Co., Denver. CONNECTICUT: G. H. Manville Pattern & Model Co., Waterbury. FLORIDA: G. M. Davis & Sons, Palatka; Timber Fabrications, Miami. GEORGIA: McCarr-Turner Co. ILLINOIS: *Batavia Metal Products Co., Batavia;* Benson Cooling Tower Co., Chicago; Binks Mfg. Co., Chicago; California Redwood Distributors, Chicago; *U.S. Challenge Co., Batavia;* Chicago Wooden Tank Co., Chicago; J. P. Devine Mfg. Co., Mt. Vernon; Eagle Tank Co., Chicago; *Johnson & Carlson, Chicago;* C. Jacobson & Co., Chicago; Lord & Bushnell Lumber Co., Chicago; McKeown Bros. Co., Chicago; E. W. Schmeling & Sons, Chicago; Technical Plywoods, Chicago; *Wendnagel & Co., Chicago;* Whyte-Coleman Co., Chicago. IOWA: Beckman Bros., Des Moines; *Dultmeter Tank Co., Manning;* Iowa Wind Mill & Pump Co., Cedar Rapids; Kretchmer Mfg. Co., Council Bluffs; Storm Lake Tank & Silo Co., Storm Lake; Wheeler Lumber, Bridge & Supply Co., Des Moines. KANSAS: Perdue Tank Co., Wichita; J. F. Pritchard & Co., Kansas City; Scherer Mfg. Co., Kansas City; Souder Tank Co., Madison; Stevens Tank Co., Wichita. KENTUCKY: *W. E. Caldwell Co., Louisville.* LOUISIANA: Lincoln Tank Co., Shreveport; McGuffin Tank Co., Shreveport; Moran Tank Co., Shreveport. MAINE: Stevens Tank & Tower Co., Auburn. MARYLAND: *Baltimore Cooperage Tank & Tower Co., Baltimore;* Economy Silo & Mfg. Co., Frederick; John Eppler Co., Baltimore; Maryland Engineering Co., Pikesville. MASSACHUSETTS: Rodney Hunt Machine Co., Orange; James Hunter Machine Co., North Adams; E. D. Jones & Sons Co., Pittsfield; New England Tank & Tower Co., Everett; Plymold Corp., Lawrence; Riggs & Lombard Inc., Lowell: J. C. Roy Lumber Co., Chicopee; *A. T. Stearns Lumber Co., Boston.* MICHIGAN: Kalamazoo Tank & Silo Co., Kalamazoo; Michigan Pipe Co., Bay City. MINNESOTA: Midway Lumber Co., St. Paul; Terminal Mfg. Co., St. Paul; Twin City Tank, Silo, & Specialty Co., Minneapolis. MISSOURI: Lillie-Hoffman Cooling Towers Inc., St. Louis; Schubert-Christy Corp., Afton; Water Cooling Equipment Corp., St. Louis. NEBRASKA: Nebraska Bridge Supply & Lumber Co., Omaha. NEW HAMPSHIRE: Improved Paper Machinery Corp., Nashua; Nashua Milling Corp., Nashua. NEW JERSEY: Acme Tank Co., Jersey City; *Atlantic Tank Corp., North Bergen;* A. J. Corcoran Inc., Jersey City; *General Tank Works Inc., Kearney;* Hanson-Van Winkle-Munning Co., Matawan. NEW YORK: *Arrow Tank Co., Buffalo;* M. C. Bascom & Co., Bolivar; Carley Heater Co., Olean; J. Holland & Sons, Brooklyn; Howard Wood Tank Co., Brooklyn; F. E. Hudson & Sons, Buffalo; Hydro & Chemical Tank Co., New York; David Isseks & Sons, Brooklyn; *O. G. Kelley & Co., New York; Mayer Tank Mfg. Co., Brooklyn;* Market Mfg. Co., Syracuse; Noble & Wood Machine Co., Hoosick Falls; Peerless Tank & Tower Co., New York; Phillips Cooling Tower Co., New York; Sandy Hill Iron & Brass Works, Hudson Falls; U.S. Plywood Corp., New York; Wenneis Tank Co., New York; *Wilcox-Johnson Tank Co., Victor;* A. Wyckoff & Son Co., Elmira. OHIO: Black-Clawson Co., Hamilton; Brown Lunber Co., Massilon; *Hauser-Stander Tank Co., Cincinnati;* Harvey Loehr Lumber Co., Canton; Shartle Bros. Machine Co., Middleton. OKLAHOMA: Black, Sivalls & Bryson Inc., Oklahoma City; National Tank Co., Tulsa; Parkersburg Rig & Reel Co., Tulsa; Producers Tank Co., Seminole. OREGON: Beall Pipe & Tank Corp., Portland; Cottage Grove Lumber Co., Cottage Grove; National Pipe & Tank Co., Portland. PENNSYLVANIA: Downingtown Mfg. Co., Downingtown; Eastern Wood Products Co., Williamsport; Everett Forest Products Co., Everett; Amos H. Hall & Sons, Philadelphia; H. K. Porter Co., Pittsburgh; E. F. Schlichter Co., Philadelphia; C. H. Wheeler Mfg. Co., Philadelphia; Woolford Wood Tanks, Darby. TENNESSEE: *O. G. Kelley & Co., Johnson City;* James E. Stark Co., Memphis. TEXAS: Axtell Co., Fort Worth; Cowser & Co., Dallas; Drane Tank Co., Fort Worth; Federal Tank Co., Midland; Hayward Tank Co., Greggton; Hudson Engineering Corp., Houston; Martin Tank Co., Corsicana; M & V Tank Co., Wichita Falls; *Alexander Schroeder Hardwood Lumber Co., Houston;* Tex. Well Equipment Mfg. Co., Fort Worth; Well Machinery & Supply Co., Fort Worth; Wilborne Bros. Co., Amarillo. WASHINGTON: American Wood Pipe Co., Tacoma; Brooks Lumber Co., Belingham; Brooks Tank Co., Everett; Cascade Pipe & Flume Co., Seattle; Federal Pipe & Tank Co., Seattle; Horizontal Stave Tank Co., Seattle; Robinson Wood Tank Co., Bellingham; Weyerhaeuser Sales Co., Tacoma; Whatcom Falls Mill Co., Bellingham. WEST VIRGINIA: Parkersburg Rig & Reel Co., Parkersburg. WISCONSIN: Beloit Iron Works, Beloit; *Dunck Tank Works, Milwaukee;* Frank Hamechek Machine Co., Kewannee; Nekoosa Foundry & Machine Works, Nekoosa; Charles H. Stehling Co., Milwaukee; Stoelting Mfg. Co., Kiel.

Refractory Materials. Representative Makers of Refractories and High Temperature Mortars*

Manufacturer	Principal Types
Acme Brick Co., Ft. Worth, Tex.	Firebrick and clay, high-temperature cements, plastic refractories
Alberene Stone Corp. of Va., New York........	Refractory linings
American Crucible Co., Shelton, Conn.	Graphite crucibles, silica and mullite refractories
Armstrong Cork Co., Lancaster, Pa.	Refractory insulating brick, high-temperature mortars, 1800° insulating block and cement
Atlas Lumnite Cement Co., New York.	Cement for refractory, heat-resisting, and insulating concrete
Babcock & Wilcox Co., New York.	Glass plant refractories, high-temperature mortars, plastic refractories, insulating and kaolin refractories
Bartley Crucible & Refr. Co., Trenton, N.J.	Graphite crucibles, firebrick, magnesite refractories
Betson Plastic Fire Brick Co., Buffalo.	High-temperature mortars, plastic refractories
Botfield Refractories Co., Philadelphia.	Chrome, firebrick, plastic refractories, high-temperaure mortars
Philip Carey Co., Lockland, Ohio.	High-temperature mortars, insulations
Carborundum Co., Perth Amboy, N.J.	Silicon carbide, aluminum oxide, mullite and fused cast refractories and high-temperature mortars
Champion Spark Plug Co., Detroit.	Sillimanite plastic refractories, electric furnace refractories
Corhart Refractories Co., Louisville, Ky.	High-temperature mortars, electrocast mullite refractories
Corundite Refractories, Inc., Massillon, Ohio.	Firebrick, high-temperature mortars, plastic refractories, alumina, silica and mullite refractories
Denver Fire Clay Co., Denver, Colo.	Firebrick, diaspore and sillimanite refractories, high-temperature mortars, plastic refractories, fireclays
W. S. Dickey Clay Mfg. Co., Kansas City, Mo.	Fireclay refractories
Joseph Dixon Crucible Co., Jersey City, N.J..	Graphite crucibles
Ehret Magnesia Mfg. Co., Valley Forge, Pa......	High-temperature mortar
Electro Refrs. & Alloys Corp., Buffalo.	Mullite, fused alumina, silicon carbide and magnesia refractories
Emsco Refractories Co., Vernon, Calif.	Firebrick, glass plant refractories, high-temperature mortars
The Exolon Co., Blasdell, N.Y.................	Silicon carbide, alumina refractories
General Abrasive Co., Niagara Falls, N.Y......	Alumina and silicon carbide
General Ceramics Co., New York...........	Special refractories
General Refractories Co., Philadelphia.	Fired and unfired chrome and magnesite, firebrick, high-temperature mortars, plastic and silica refractories
Gladding, McBean & Co., Los Angeles.	Firebrick, insulating brick and plastics, high-temperature mortars, plastic refractories, fireclays
A. P. Green Fire Brick Co., Mexico, Mo.	Firebrick, insulating firebrick, high-temperature mortars, plastic and castable refractories, fireclays
Harbison-Walker Refrs. Co., Pittsburgh.	Refractories of most types including regular and super fireclay, high-alumina, silica, chrome, magnesite, Fosterite brick; clays; insulating firebrick and mortars; high-temperature mortars

Manufacturer	Principle Types
Haws Refractories Co., Johnstown, Pa.	Firebrick of all kinds, silica brick, fireclays
Illinois Clay Products Co., Joliet, Ill.	Firebrick, high-temperature mortars, insulating cements, coatings and brick
Ironton Fire Brick Co., Ironton, Ohio.	Fireclay refractories, refractory cements, fireclays
Johns-Manville, New York.	Bonding mortars, castables, ramming mixtures, plastic and insulating refractories
Laclede-Christy Clay Prod. Co., St. Louis.	Firebrick, high-temperature mortars, plastic refractories, glass plant refractories, fireclays
E. J. Lavino & Co., Philadelphia.	Chrome and magnesite refractories, high-temperature mortars, silica refractories, fireclays
Massillon Refractories Co., Massillon, Ohio.	Firebrick, high-temperature mortars, plastic refractories, special compositions
McLain Fire Brick Co., Pittsburgh............	Various
McLeod & Henry Co., Troy, N.Y.	Firebrick, high temperature mortars, plastic-refractories, fireclays
Mullite Refractories Co., Shelton, Conn.	High-temperature mortars, plastic refractories, mullite refractories
National Carbon Co., Inc., Cleveland............	Carbon refractories
Niles Fire Brick Co., Niles, Ohio.........	Firebrick, insulating refractories
North American Refrs. Co., Cleveland.	Fireclay, super, insulating and silica brick, high-temperature mortars, plastic refractories, fireclays
Norton Co., Worcester, Mass.	High-temperature mortars, silicon carbide, fused alumina and magnesia, raw materials, cements, refractory shapes
Pacific Clay Products Co., Los Angeles..........	Plastic refractories
Pyro Clay Products Co., Oak Hill, Ohio........	Glass plant refractories
Quigley Co., Inc., New York.	Firebrick, insulating refractories, super firebrick, high-temperature mortars, plastic refractories
Ramtite Co., Chicago...	High-temperature mortars
Refractory & Insulation Corp., New York.....	High-temperature mortars
Robinson Clay Product Co. of N.Y., New York.	High-temperature mortars, firebrick and clay, insulating refractories
Ross Tacony Crucible Co., Philadelphia.	Graphite crucibles and stopper heads, magnesite refractories
St. Louis Fire Brick & Insulation Co., Huntington Park, Calif........	Various
Seaboard Refrs. Co., Perth Amboy, N.J.	Firebrick, high-temperature mortars, plastic and insulating refractories, silicon carbide and mullite refractories
Chas. Taylor Sons Co., Cincinnati.	Firebrick, glass plant and insulating refractories, sillimanite
The United States Stoneware Co., Akron, Ohio.	Plastic and castable refractories
M. D. Valentine & Bro. Co., Woodbridge, N.J.	Firebrick
Vitrefrax Corp., Los Angeles.	Glass plant refractories, firebrick, high-temperature mortars, plastic refractories, fireclays

Table 4a. Materials of Construction—Base Metals. Physical Properties and Methods of Fabrication*

Material	Form for which tensile properties are recorded	Tensile strength, 1000 lb./sq. in.	Yield point, 1000 lb./sq. in.	Elongation, % in 2 in.	Reduction of area, %	Elastic modulus, lb./sq. in. (multiply by 10^6)	Brinell hardness	Machining qualities	Methods of fabrication†
Ferrous Alloys									
Abrasion Resisting	Plates	110	60	20	40	...	225	Fair	R
Allegheny Metal 18-8	Casting	70–87	36–46	50–75	45–60	28–30		...	R, W
Allegheny Metal 18-8 S	Casting	70–87	35–46	50–75	45–65	28–30		...	R, W
Allegheny Metal 20-10 S	Casting	68–80	40–60	50–75	45–65	28–30		...	R, W
Allegheny Metal 25-12	Casting	75–90	40–60	25–40	25–45	28–30		...	R, W
Allegheny Metal 25-12 S	Casting	75–90	40–55	30–50	30–50	28–30		...	R, W
Allegheny Metal 25-20	Casting	60–75	40–55	40–60	40–60	28–30		...	R, W
Allegheny Metal 25-20 S	Casting	60–75	40–55	40–60	40–60	28–30		...	R, W
Allegheny Metal 20-25 S	Casting	60–75	40–55	40–60	40–60	28–30		...	R, W
Allegheny Metal 20-25 SM	Casting	60–75	50–70	40–60	40–60	28–30		...	R, W
Allegheny Metal 29-9	Casting	80–95	45–70	20–30	20–35	28–30		...	R, W
Allegheny Metal 29-9 S	Casting	75–95	36–46	20–40	20–45	28–30		...	R, W
Allegheny Metal 18-8 M	Casting	73–85	35–46	45–60	45–60	28–30		...	R, W
Allegheny Metal 18-8 SM	Casting	73–85	35–45	45–60	45–60	28–30		...	R, W
Allegheny Metal 19-10 M	Casting	70–80	35–45	45–60	45–60	28–30		...	R, W
Allegheny Metal 19-10 SM	Casting	70–80	40 60	45–60	45–65	38–30		...	R, W
Allegheny Metal 28-4	Casting	70–90	35–50	10–25	10–25	28–30		...	R, W
Allegheny Metal 15-35	Casting	50–75	40–56	40–60	40–60	28–30		...	R, W
Allegheny Metal 18-8 C	Casting	78–90	40–55	35–50	40–50	28–30		...	R, W
Allegheny Metal 18-11 C	Casting	75–90	...	35–50	40–50	28–30		...	R, W
Allegheny Metal 12 Cr.	Casting	Depends on heat-treatment	...	...	...	28–30		...	R, W
Allegheny Metal 17 Cr.	Casting	Depends on heat-treatment	...	...	...	28–30		...	R, W
Alray C	Annealed	95–175	35	35–40	60	...	...	...	W, B
Alray D	Annealed	75 160	...	35 40	60	...	...	...	R, W, B
Amsco F-1	...	50–70	40–50	4–8	4–8	26	150–180	Fair	W
Amsco F-3	As cast	40–60	30–45	0–2	...	...	170	Fair	W
Amsco F-5	As cast	50–70	35–45	24–40	...	...	180–200	Fair	W
Amsco F-6	...	...	...	...	...	...	...	...	W
Amsco F-8	As cast	70–90	25–45	40–70	45–75	29	150–180	Good	W
Amsco F-10	As cast	80-100	40–55	30–50	30–50	23.5	160–190	Fair	W
Amsco manganese steel	Water quenched from 1850°F.	120–140	50–60	25–35	25–35	...	185–200	...	R, W
Amsco Ni-Mn welding rod									
B & W 5150							450–750		
Bethlehem 235	...	...	...	...	...	...	...	Fair	
Bethlehem 300	...	...	...	...	...	...	...	Fair	
Buflokast gray iron	Bar	30–50					150–250	Good	
Calmar 18-8	Annealed casting	80–90	38–45	55–65	...	...		Fair	
Calmar 18-8 Cb	Annealed casting	80–85	40–45	40–45	...	...		Fair	
Calmar 18-8 M	Annealed casting	85–95	40–50	25–30	...	...	150	Fair	
Caloxo 18	Cast, heat-treated	65–75	45–52	20–25	35–40	...		Fair	
Caloxo 15–35	...	55–65	35–40	10–15	...	...		Good	
Caloxo 25-12	Annealed casting	75–85	40–45	30–35					
Caloxo 25–20	Annealed casting	72–80	35–42	35–40	...	...	135–145	Fair	
Caloxo 28–10	Annealed casting	70–80	40–45	20–30					
Carbon-molybdenum steel	Plates	65–87	...	...	...	...	145	Good	F, R, W
Causul metal	...	32	...	...	...	...	140–180		
Chromax	Casting	60	40	2	3	...	...	Good	
Chromax	Wrought	70–150	30–40	30	...	...	...	Good	DD, F, R, W, B
Chromel C	...	105	60	25–35	55	31	...	Good	F, R, W, B
4-6 chrome	Drawn	66	39	45	...	30	143	Good	F, W
4-6 chrome	Annealed	67	26	39	81	30	128	Good	F, W
Chrome-copper-nickel steel	...	65	35	30	50	...	145	Fair	F, R, W
Chrome-manganese steel	Plate	100	55	25.2	45	...	200	Fair	F, R, W
Cimet	Casting	65	...	0	...	...	...	Good	
Circle L 1	Cast, normalized and drawn	90	60	25	50	...	180	Good	W
Circle L 2	Cast, oil quenched and drawn	125	95	15	35	...	250	Fair	W
Circle L 3	Cast, oil quenched and drawn	160	130	10	15	...	350	Fair	W
Circle L 5	Cast, normalized and drawn	100	70	24	50	...	190	Fair	W
Circle L 10	Cast, normalized and drawn	100	75	18	30	...	215	Good	W
Circle L 11	Cast, normalized and drawn	100	75	7.5	8.5	...	200	Fair	W
Circle L 12	Cast, normalized and drawn	85	55	22	40	...	170	Good	W
Circle L 13	Cast, oil quenched and drawn	200	150	5	10	...	500 max.	Good	W
Circle L 19	Cast, normalized and drawn	70	40	30	50	...	150	Good	W
Circle L 21	Cast, water quenched	70	32	45	50	...	150	Fair	W
Circle L 22	Cast, water quenched	70	32	45	50	...	150	Fair	W
Circle L 22 XM	Cast, water quenched	75	35	45	50	...	160	Fair	W
Circle L 23	Cast, water quenched	75	35	40	45	...	160	Fair	W
Circle L 30	Cast, water quenched	80	40	30	30	...	170	Fair	W
Circle L 31	Cast, water quenched	90	50	25	25	...	180	Fair	W
Circle L 34	Cast, water quenched	65	30	30	30	...	135	Fair	W
Colonial 610	Annealed	85	54	29.9	54	28	187	Fair	W, B
Cooper 14	Annealed cast bar	75	50	20	30	27	180	Good	DD, F, R, W, B
Cooper 16	Annealed cast bar	75	50	20	30	29	180	Good	DD, F, R, W, B
Cooper 17	Annealed cast bar	80	40	47	50	28	155	Good	DD, F, R, W, B
Cooper 17 S	Annealed cast bar	80	40	47	50	28	150	Good	DD, F, R, W, B
Cooper 17 S Mo	Annealed cast bar	85	43	47	50	28	160	Good	DD, F, R, W, B
Cooper 17 S Cb	Annealed cast bar	80	40	47	50	28	150	Good	DD, F, R, W
Cooper 18	Annealed cast bar	79	46.5	23	24	28	140	Good	DD, F, R, W, B
Cooper 19 A	As cast bar	70	35	7	8	28	190	Good	DD, F, R, W, B

† B = brazing; DD = deep drawing; F = flanging; R = riveting; W = welding.

Table 4a. Materials of Construction—Base Metals. Physical Properties and Methods of Fabrication— (*Continued*)

Material	Form for which tensile properties are recorded	Tensile strength, 1000 lb./sq. in.	Yield point, 1000 lb./sq. in.	Elongation, % in 2 in.	Reduction of area, %	Elastic modulus, lb./sq. in. (multiply by 10^6)	Brinell hardness	Machining qualities	Methods of fabrication†
Cooper 19	As cast bar	60	35	3	4	28	170	Good	DD, F, R, W, B
Cooper 20	As cast bar	65	40	7	8	28	165	Good	
Cooper 21	Cast bar	60	45	3	2	28	170	Good	
Cooper 21 A-B-C	Cast bar	70	35	10	12	28	150	Good	
Cooper 22	Cast bar	90	60	30	35	28	180	Good	DD, F, R, W, B
Cooper 22 P-M	Cast bar	90	60	30	35	28	180	Good	F, R, W
Cooper KNC 3	Cast bar	65	50	15	12	28	180	Good	W
Corrosiron	Casting	16	....	0	0	....	275–325	Fair	
Crane 18-8 Mo	Annealed cast bar	85	48	53	55	28.6	135–150	Poor	W, B
Crane 5 Cr-Mo cast steel	Cast, heat-treated bar	110	80	18	30	....	220–245	Fair	W
Croloy 2	Annealed seamless tube	60 min.	25 min.	30	....	....	163 max.	Fair	DD, F, R, W
Croloy 5	Seamless tubing	60 min.	25 min.	30	....	29	163 max.	Fair	DD, F, R, W
Croloy 9	Seamless tubing	60 min.	30 min.	30 min.	....	29	170 max.	Fair	DD, F, R, W
Croloy 12	Seamless tubing	65–78	35–45	25–35	....	28	135–160	Fair	DD, F, R, W
Croloy 16-13-3	Seamless tubing	75 min.	30 min.	35	....	28	135–175	Fair	DD, F, R, W
Croloy 18	Seamless tubing	70–85	40–60	27–38	....	29	140–180		
Croloy 18-8	Seamless tubing	75 min.	30 min.	50	....	28	150	Fair	DD, R, F, W
Croloy 25-20	Seamless tubing	75 min.	35 min.	35	....	29.7	115–175	Fair	DD, F, R, W
Croloy 27	Seamless tubing	75–90	45–60	15–30	....	29	160–195	Fair	F, W
Dopploy	Cast	35	....	....	....	....	....	Good	
Duraloy 18-8 S	Heat-treated casting	65	35	35	40	....	180	Good	
Duraloy 18-8 S Mo	Heat-treated casting	65	35	35	40	....	180	Good	
Duraloy 35-15	As cast	60	35	5	6	24	170	Good	
Duraloy 25–20 Mo	Heat-treated casting	65	45	20	20	....	185	Fair	
Duraloy A	Casting	50	40	1	....	....	....	Good	
Duraloy B	Casting	90	55	5	5	....	....	Good	
Duraloy N	As cast	80	50	20	15	....	190	Good	
Duraloy NS Mo	Heat-treated casting	75	35	25	30	....	195	Fair	
Durco D-18	Sand casting	90	55	5	5	....	180	Fair	W
Durco KA 2S	Sand casting	80–90	30–40	40–55	40–55	....	130–150	Fair	W
Durco KA2S Mo	Sand casting	85–95	45–55	40–50	45–55	....	140–160	Fair	W
Durichlor	Bar	17	....	....	....	....	....	Poor	W
Durimet	Sand casting	65–75	30–40	35–50	40–55	....	130–150	Fair	W
Duriron	Bar	16–18	....	....	....	....	....	Poor	W
Economet	Casting	60	30	7	10	....	180	Good	
Economy Hardface-Self Hardening	....	....	....	....	....	....	450–550	Poor	W
Elverite A	....	....	....	....	....	....	500		
Elverite B	....	....	....	....	....	....	650–700		
Elverite C	....	....	....	....	....	....	650–700		
Empire 8	Annealed casting	105	60	20–25	25–30	....	225	Fair	W
Empire 11	Annealed casting	105	65	15–20	30–35	....	250	Fair	W
Empire 12	Annealed casting	85	50	15–25	30–35	....	200	Fair	W
Empire 14	Annealed casting	95	55	20–25	45–50	....	225	Fair	W
Empire 15	....	....	....	....	....	....	....	....	W
Empire 16	Annealed casting	75	40	50–55	60–65	....	165	Fair	W
Empire 17	Annealed casting	55	35	1–2	3–4	....	180	Fair	W
Empire 20	Annealed casting	75	40	35–40	40–50	....	175	Fair	W
Empire 23	Annealed casting	60	30	30–35	35–40	....	180	Fair	W
Esco 45	....	95–100	55	40	40	30	217	Good	
Fahrite C 7	....	84	40	25	24	25	160	Fair	
Fahrite C 7 M	....	85	42	27	26.5	25	179	Fair	
Fahrite C 8 A	....	80	36	50	60	28	160	Fair	
Fahrite C 8 A M	....	90	45	50	60	28	170	Fair	
Fahrite C 12	....	100	45	20	35	29	179–400	Fair	
Fahrite C 73	....	80	45	10	10	29	200	Fair	
Genesee 212	....	100	80	25	50	....	200	Good	
Genesee 255	....	110	80	20–25	50–55		180–200	Fair	
Genesee 304	....	....	....	....	....	....	....	Fair	
Genesee 305	....	....	....	....	....	....	....	Fair	
Genesee 412	Cast	80–140	....	25–35	30–40		170–200	Poor	
Hard facing rod 217	....	....	....	....	....	....	600–700	Poor	
Hard facing welding rod 459	....	....	....	....	....	....	500–600	Poor	
Hoskins 502									
(*a*) Hot rolled	Hot rolled	110	....	....	....	....	....	Good	
(*b*) Casting	Casting	70							
Ing Aclad stainless clad steel 304	Sheets and plates	60–70	40–50	30–50	50–60	29	145–160	Fair	DD, F, R, W, B
Ing Aclad stainless clad steel 309	Sheets and plates	60–70	40–50	30–50	50–60	29	150–200	Poor	DD, F, R, W, B
Ing Aclad stainless clad steel 317	Sheets and plates	60–70	40–50	....	....	29	145–175	Poor	DD, F, R, W, B
Ing Aclad stainless clad steel 347	Sheets and plates	60–70	40–50	30–50	50–60	29	145–160	Poor	DD, F, R, W, B
Manganese-molybdenum steel	Plates	95	60	25	45	....	200	Fair	F, R, W
Mayari R	Hot rolled	70 min.	50 min.	22	50	28–30	150	Good	F, R, W, B
Midvaloy 13	Heat-treated wrought bars	80–200	35–170	7–30	11–70	28	140–400	....	DD, F, R, W, B
Midvaloy 18-8	Annealed wrought and cast bars	70–145	30–110	20–60	30–70	28	130–280	Fair	
Midvaloy 18-8 Se	Annealed wrought bars	85	35	45	56	28	170	Good	
Midvaloy 1700	Annealed wrought and cast bars	65–90	45–55	20–32	40–60	29	145–210	Fair	
Midvaloy 17-35	Castings	55–65	40–45	5–12	10–20	....	150–170	Fair	
Midvaloy 25-12	Bars and casting	70–115	40–60	9–45	9–50	28	140–230	Fair	
Midvaloy 25-20	Wrought and cast bars	85–100	40–50	25–40	24–50	....	170–200	Fair	
Midvaloy 2802	Castings	75–100	60–80	20–30	40–60	29	150–200	Fair	
Midvaloy ATV 3	Wrought bars	100–112	45–70	24–33	40–45	....	185–238	Good	

† B = brazing; DD = deep drawing; F = flanging; R = riveting; W = welding.

Table 4a. Materials of Construction—Base Metals. Physical Properties and Methods of Fabrication— (*Continued*)

Material	Form for which tensile properties are recorded	Tensile strength, 1000 lb./sq. in.	Yield point, 1000 lb./sq. in.	Elongation, % in 2 in.	Reduction of area, %	Elastic modulus, lb./sq. in. (multiply by 10^6)	Brinell hardness	Machining qualities	Methods of fabrication†
Misco 18–8	Room temperature, quench annealed	70	30	40	46	26	197 max.	Fair	
Misco C	Room temperature, as cast	85	50	15–25	15–25	21.9	229 max.	Fair	
Misco HN-1	...	60	30	3	3	...	217 max.	Good	
Misco HN-2	Room temperature, as cast	60	30	3	3	22	217 max.	Good	
Misco Metal‡	Room temperature	60	35	4	4	23.9	217 max.	Fair	
Miscrome 1	Room temperature Normalized and tempered	95	60	17	20	30	241 max.	Good	
Miscrome 4	Air quenched and tempered, room temperature	80	50	20	30	...	217 max.	Good	
Nichrome	Wrought	95–175	60	35	50	...	...	Good	DD, F, R, W, B
Nichrome	Casting	65	...	2	3.5	...	...	Good	
2% nickel steel	Plates	...	...	...	...	...	140	Poor	F, R, W
Ni-Hard	Casting	30–40	...	0	0	...	575–750	Fair	
Nilstain	Annealed wire	85–350	35–300	5–60	0–75	27	140–425	Good	W
Ni-Resist	Casting	25–40	...	...	...	12–17	130–175	Good	
Pyrocast	Casting	60–70	60–70	0	0	...	300–600	Good	
Q Alloy A	Casting	68	34	5	6	24.9	180	Good	W
Q Alloy B	Casting	66	33	4.5	5.5	23.6	176	Good	W
Q Alloy CN-1	Casting	85	43	50	60	26	190	Good	
Q Alloy CN-1-H	Casting	80	40	16	17	23	177	Fair	W
Q Alloy CN-1-Mo	Casting	85	43	50	60	26	190	Fair	W
Q Alloy CN-2	Casting	75	37	50	60	35	180	Fair	W
Q Alloy CN-2 Mo	Casting	75	37	50	60	35	180	Good	DD, F, W
O-4	Cast bar	100–110	65–80	18 min.	30 min.	30	190–230	Good	
O-12	Cast bar, heat-treated	75–130	39–90	25–10	55–25	30	140–250	Good	
O-16	Heat-treated casting	70–80	39–43	20–15	30–20	28	140–175	Good	
O-18	Casting water quenched, 2000°F.	78–85	39–42	60–65	60–70	28.6	135–160	Fair	
O-24	Heat-treated casting	80–85	40–43	50–45	55–45	29	140–150	Fair	
O-30	Casting, stress relieved	45–70	35–60	2–3	2–3	29	170–190	Good	
Sivyer 5% Cr	Heat-treated casting	100–110	75–90	18–22	45–55	...	229	Good	
Sivyer 60	Heat-treated casting	70–80	35–45	45–55	45–60	...	156	Fair	
Sivyer 62	Casting	75–85	35–45	35–45	35–45	...	163	Fair	
Sivyer 64	Casting	85–95	40–50	20–30	20–30	...	197	Fair	
Sivyer 66	Heat-treated casting	75–150	45–125	8–25	15–50	...	175–325	Good	
Sivyer 70	Casting	60–70	...	2–7	2–8	...	200–225	Fair	
Spang Chalfant 1	Annealed	66.4	40.6	36.4–40.3	74.1	30	131	Good	F, W
Spang Chalfant 2	Hot finished	62.4	42	36	67.5	30	126	Good	F, W
Spang Chalfant 3	Normalized and drawn	64.1	32.5	37	62.5	30	126	Good	F, W
Stainless steel 301	Annealed sheet	90	45	55	...	29	143–170	Fair	DD, F, R, W, B
Stainless steel 302	Annealed sheet	90	45	55–60	65–55	29	130–162	Fair	DD, F, R, W, B
Stainless steel 302 B	Annealed sheet	80–100	40–50	60–50	70–60	...	150–180	Fair	
Stainless steel 303	Annealed sheet	80–90	35–45	60–50	70–55	...	130–150	Fair	W, B
Stainless steel 304	Annealed bar, plate	80–95	35–45	55–60	55–65	29	135–185	Good	DD, F, R, W
Stainless steel 308	Annealed bar	80 90	30–40	55–65	60–70	29	135–185	Poor	DD, F, R, W
Stainless steel 309	Annealed sheet	95	45	45	65–50	29	170	Fair	DD, F, R, W, B
Stainless steel 309 S	Annealed	85	35	40	...	29	150		
Stainless steel 310	Annealed sheet	90–110	40–60	55–45	60–50	...	146–210	Fair	DD, F, R, W
Stainless steel 311	Annealed	90–110	45–50	30–40	35–45	29	175	Good	DD, W, R, F
Stainless steel 315	As rolled	101	56	40	62	28.5	196	Fair	W, B
Stainless steel 316	Annealed sheet	85	40	45–55	75–60	28	135–185	Fair	DD, F, R, W, B
Stainless steel 317	Annealed sheet	85	40	40–50	...	28	150–200	Fair	DD, F, R, W, B
Stainless steel 321	Annealed sheet	80–90	35–45	60–55	65–55	...	135–185	Fair	DD, F, R, W
Stainless steel 325	Annealed	85–95	45–55	30–40	45–55	28–30	160–190	...	W, R
Stainless steel 329	Annealed	100	75	20	...	29		Good	DD, F, R, W, B
Stainless steel 330	Rolled bars	60–110	40–50	40	50	...	190–220	...	F, R, B
Stainless steel 347	Annealed sheet	85	40	45–55	65–55	29	140–180	Fair	DD, F, R, W, B
Stainless steel 403	Annealed	65–85	35–45	25–35	60–65	28	135–165	Fair	DD, F, R, W
Stainless steel 405	Annealed	75	48	31	65	...	143		
Stainless steel 406									
Stainless steel 410	Annealed sheet	65	35	30	65–60	29	135–160	Fair	DD, F, R, W, B
Stainless steel 414	Quenched in oil from 1750°F.	128	113	20	53	28	269	...	DD, F, W, B
Stainless steel 416	Annealed sheet	70–85	40–60	35–25	65–60	...	145–185	Good	Welding, forging
Stainless steel 418	Annealed	87	64	29	64				
Stainless steel 420									
(*a*) Annealed	Annealed	105	55	23	55	29.5	207	Fair	R, W, B, DD, F
(*b*) Tempered	Tempered	225	185	8	20		430		
Stainless steel 420 F	Quenched in oil	134	76	6.5	7.6	...	278	Good	R, W, B
Stainless steel 430	Annealed sheet	75	45	27	55–40	29	146	Fair	DD, F, R, W, B
Stainless steel 430 F	Annealed sheet	70–90	40–55	30 20	55–40	28	145–185	Good	R, W, B
Stainless steel 431	Oil quenched 1750°F.	195	165	17	57	...	418		
Stainless steel 440									
(*a*) Annealed	Annealed	100	60	23	40	29	197	Fair	R, W, B
(*b*) Tempered	Tempered	250	190	3	6		495		
Stainless steel 441	Annealed	235	186	6	14	29.5	430	...	W, B
Stainless steel 442	Annealed	80	55	35	55	...	171	Fair	DD, F, R, W, B
Stainless steel 446	Annealed sheet	80	50	23	50–40	29	175	Fair	F, R, W, B
Stainless steel 501	Seamless tubes	60	25	30	...	29	163	Fair	DD, F, R, W
Stainless steel 502	Seamless tubing	60	25	30	...	29	163	Fair	DD, F, R, W
Standard alloy H R 4	...	...	...	...	...	...	...	Fair	DD, F, R, W
Standard alloy H R 3	...	...	...	...	...	...	...	Fair	R, W

† B = brazing; DD = deep drawing; F = flanging; R = riveting, W = welding.
‡ Composition given is for castings.

Table 4a. Materials of Construction—Base Metals. Physical Properties and Methods of Fabrication— (*Continued*)

Material	Form for which tensile properties are recorded	Tensile strength, 1000 lb./sq. in.	Yield point, 1000 lb./sq. in.	Elongation, % in 2 in.	Reduction of area, %	Elastic modulus, lb./sq. in. (multiply by 10^6)	Brinell hardness	Machining qualities	Methods of fabrication†
Standard alloy H R 6	......	90	50	50	50	30	200	Fair	R, W
Symington Nirosta K A 2	......	65	35	35–40	55–65	......	145–160	Good	
Thermalloy A	......	......	......	......	......	......	170–240		
Thermalloy B	......	......	......	......	......	......	170–240		
Thermalloy E	......	......	......	......	......	......	170–240		
Thermalloy 72	......	......	......	......	......	......	170–240		
Timken 16-13-3	Annealed	80 min.	30 min.	40 min.	......	......	200	Fair	DD, F, R, W
Timken 2512	Annealed tubing	90 min.	50 min.	30 min.	......	......	163 max.	Good	
Timken 2% Cr-0.5% Mo	Annealed	65.9	40.6	40.3	74.1	......	137	Good	F, W
Timken 5% Cr-Mo	Annealed	66.6	26.3	39	80	......	146	Good	F, W
Timken DM	Annealed	66.5	35.2	36.5	72.7	......	149	Good	F, W
Timken Sicromo 5S	......	82.2	45	38	76	......	163	Fair	F, W
Timken Sicromo 7	Annealed	60 min.	25 min.	30 min.	......	......	179	Good	F, DD, R, W
Timken Sicromo 7M	Annealed	60 min.	25 min.	30	......	......	179 max.	Good	DD, F, R, W
Timken Sicromo 9M	Annealed	60 min.	25 min.	30 min.	......	......	179	Good	DD, F, R, W
Tisco 41	Casting	115–130	60–80	6–12	8–20	......	220–235	......	W
Tisco 53	Casting	130–140	115–120	10–15	30–35	......	280–300	Good	W
Tisco 80	Casting	......	......	......	......	......	300–600	Fair	
Tisco 102	Casting	65–70	25–30	50–60	45–60	......	......	Poor	
Tisco 108	Casting	80–90	35–45	15–25	15–25	......	160–180	Fair	
Tisco 130	Casting	40–60	30–40	0–3	0–5	......	200–220	Good	
Tisco 131	Casting	75–100	70–90	5–12	6–15	......	190–210	Good	
Tisco 160	Casting	20–35	......	......	......	......	120–170	Good	
Tophet C	Annealed wire	110	60	25					
Warman 5M	Casting	100	65	20–25	60–65	......	......	Good	W
Warman 13	Heat-treated casting	85–95	45–50	25–30	50–60	......	......	Good	
Worthite	Rolled	97	48	45	57	30	170	Good	W
	Sand casting	72	33	40	40	30	150		
X-7	Casting	75.4	41.1	17	17.5	24.2	172	Fair	W
X-ITE	Casting	62.5	35.8	5.5	7.2	25.9	169	Good	W
X-ITE-B	Casting	60	34	5.8	7.6	25	165	Good	W
Non-ferrous Alloys									
Admiralty	Sheet, soft to hard	52–85	......	6–60	......	15	......	Fair	DD, F, R, W, B
Adnic	Annealed strip	55	25	50	......	21	......	Fair	DD, F, R, W, B
Advance	Wrought	60–100	25–75	30–35	50–65	......	......	Good	DD, F, R, W, B
Alcoa 2 S§	A. S. T. M. specimen	13–24	5–21	15–45	60–80	10.3	......	Good	DD, F, R, W, B
Alcoa 3 S	A. S. T. M. specimen	16–29	6–25	10–40	45–75	10.3	......	Good	DD, F, R, W, B
Alcoa 52 S	A. S. T. M. specimen	29–41	14–36	8–30	40–65	10.3	......	Good	DD, F, R, W, B
Alcoa 53 S	A. S. T. M. specimen	16–39	7–33	20–35	50–70	10.3	......	Good	DD, F, R, W, B
Alcoa Alclad (72 S) 3 S	A. S. T. M. specimen	16–29	6–25	4–30	......	......	......	Good	DD, F, R, W
Alcoa 43	A. S. T. M. specimen	19–29	9–13	6	......	10.3	......	Fair	R, W, B
Alcoa B 214	A. S. T. M. specimen	17–19	12–13	......	......	10.3	......	Good	R, W
Alcoa 220-T 4	A. S. T. M. specimen	45	25	14	......	10.3	......	Good	R, W
Alcoa 356	A. S. T. M. specimen	25–32	16–22	2–6	......	10.3	......	Good	R, W, B
Alcoa 406	A. S. T. M. specimen	16	......	6.5	......	10.3	......	Good	R, W, B
Alcumite	Sand castings	65–70	23–25	30–35	30–35	16.5	120–140	Good	W, B
40 Alloy	Heat-treated	30–40	15–35	1–10	......	10	50–110	Good	
Alray A	Soft annealed	80–200	......	35–40	60	......	......	Poor	F, R, W, B
Aluminum 99+	Casting	12	......	40	......	......	18	Poor	
Aluminum 98–99	Casting	12	......	......	......	......	18	Poor	
Aluminum bronze	Sheet, soft to hard	55–105	......	7–65	......	......	65–200	Fair	DD, F, R, B
Aluminum bronze	As cast, bar	70–80	32–35	17–22	19–25	......	120–140	Good	
Aluminum bronze 5%	......	57–71	10–50	30–65	......	16	......	Fair	
Aluminum brass	Annealed tube	52–100	15–76	10–70	......	15	......	Fair	DD, F, R, B
Ambrac 850	Sheet and rod, soft to hard	50–82	......	5–40	......	19	58–145	Fair	DD, F, R, W, B
Ampco 18	Cast bar	77–84	35–42	10–14	6–10	14.4	159–183	Good	R, W, B
Ampcoloy	Cast bar	70–80	38–45	12–20	14–20	15	131–143	Good	R, W, B
Antimonial Admiralty	Annealed tube	50	......	70	......	......	......	Fair	
Antimonial lead	Rolled sheet	4.5	2.2	50	......	2	7.5	Good	DD, F, W
Antimonial lead	Rolled	4	3	39	......	2	9	Good	F, W
Arsenical copper	Annealed tube	35	12	40	......	15	......	Fair	F, R, W, B
Asarco acid lead	Cold rolled	2.3	......	40	100	2	4.5	Fair	DD, F, W
Beryllium copper	Heat-treated sheet	175	......	5	......	19	350	Fair	DD, F
Beryllium copper	Cold rolled annealed	60–80	......	3.5	......	18–19	......	Fair	DD, F, R, W, B
Cadmium-copper	Rod	35–70	20–60	16–70	70–85	20	......	......	DD, F, B
Chemical lead	Rolled sheet	2.3	1.2	50	......	1.5	4.7	Good	DD, F, W
Chemical lead	Rolled	1.9–2.3	0.7	50	......	1–2	4.5–5.5	Good	F, W
Chromel A	Annealed	110	63	25–35	55	31	......	Good	F, R, W, B
Colmonoy 6	......	42	......	......	......	......	587	Poor	W, B
Chromium copper	Heat-treated rod	72	61	25	......	......	125	Fair	DD, F
Commercial bronze F. C.	Rod half hard	45	40	30	......	17	......	Good	
20% Cupro-nickel	Annealed tube	55	......	45	......	......	......	Fair	DD, F, R, W, B
30% Cupro-nickel	Sheet, soft to hard	55–77	22–70	5–40	......	20	70–140	Fair	DD, F, R, W, B
Deoxidized copper	Tube	31–55	12–44	6–56	60–72	16	42–111	Tough	DD, F, R, W, B
Durco D-10	Sand castings	65	45	5	5	......	175	Fair	
Everdur-1000	Castings	50	20	25	25	15	80	Fair	W, B
Everdur-1015	Sheets, soft to hard	40–65	15–50	8–46	......	15	45–125	Fair	DD, F, R, W, B
Gold	Annealed	14	......	45	......	11.3	35	Poor	DD, F, R, W, B
Hastelloy A	Rolled annealed	110–120	47–52	40–48	40–54	27	200–215	Good	DD, F, W
Hastelloy B	Rolled annealed	130–140	60–65	40–45	40–45	30.7	210–235	Fair	DD, F, W
Hastelloy C	Casting	72–80	48–48	10–15	11–16	28.5	175–215	Fair	DD, F, W
Hastelloy D	Casting	36–40.5	......	0	0	28.8	......	Poor	W

† B = brazing; DD = deep drawing; F = flanging; R = riveting; W = welding.
§ In cast form known as Alcoa 100.

Table 4a. Materials of Construction—Base Metals. Physical Properties and Methods of Fabrication— (*Concluded*)

Material	Form for which tensile properties are recorded	Tensile strength, 1000 lb./sq. in.	Yield point, 1000 lb./sq. in.	Elongation, % in 2 in.	Reduction of area, %	Elastic modulus, lb./sq. in. (multiply by 10^6)	Brinell hardness	Machining qualities	Methods of fabrication†
Hardware bronze	Annealed	38	16	35	...	15	...	Good	R, B
Herculoy A	Hard to soft sheet strip	67–113	30.113	5–60	...	16	...	Fair	DD, F, R, W, B
Herculoy B	Hard to soft rod,	42–88	14–60	20–70	75–85	15	...	Fair	W, B
High brass	Sheet	46–90	15–55	3–64					
High silicon bronze	Rod	44–95	13–74	16–70	61–80	15	...	Fair	DD, F, B
Ilium G	Casting	60–70	50–60	...	...	...	160–210	Fair	W
Ilium R	Rolled annealed	100	50	30–45	...	27.5	175–240	Good	DD, R, W
Inconel	Hot-rolled rods to cold-drawn wire	80–185	35–165	2–45	...	31	...	Fair	DD, F, R, W, B
Inconel-clad steel	Plate	55	30	30	50	...	120	Fair	F, R, W
Indium metal	Cold-rolled sheet	0.5	...	...	100	...	1.0	Poor	DD, F, W
Iridio platinum	Annealed	40–160	...	...	...	...	90–280	Good	DD, F, R, W, B
10% iridium platinum	Cold-drawn annealed wire	65	...	27	...	...	140	Fair	DD, F, R, W, B
30% iridium platinum	Cold-drawn annealed wire	145	...	...	...	...	240	Good	DD, F, R, W, B
40 M alloy	Casting	26	13	9	...	...	50	Fair	
Molybdenum	Drawn wire	390	...	...	65	51.4	147	Poor	B
Monel	Hot-rolled rods to cold-drawn wire	80–170	40–130	2–45	...	26	...	Fair	DD, F, R, W, B
Monel-clad steel	Plate	55	30	30	50	...	...	Fair	F, R, W
Muntz	Sheet	50–80	...	9.5–48	...	12.8	76–172	Good	DD, F, B
Naval brass	Annealed wire	55	23	40	60	...	...	Fair	
Nichrome V	Wrought	100–200	60	35	55	...	...	Good	DD, F, R, W, B
Nickel	Hot-rolled rods to cold-drawn wire	65–165	20–130	2–45	...	30	...	Fair	DD, F, R, W, B
Nickel-clad steel	Plates	55	30	30	50	...	110	Fair	F, R, W
Nickel silver 18% (Cu 65, Ni 18, Zn 17)	Sheet soft to hard	58–85	...	4–40	...	18	75–130	Fair	DD, F, R, W, B
Nickel silver 18% (Cu 55, Zn 27, Ni 18)	Sheet soft to hard	60 99	...	4 45	...	18	75–175	Fair	DD, F, R, W, B
Nirex	Wrought	80–175	40	40	50	...	...	Good	DD, F, R, W, B
O F H C	Rod and wire	32–65	12–45	5–50	60–90	18	30–120	Fair	DD, F, R, W, B
Olympic bronze	Sheet	55–110	20–87	5–65	...	...	...	...	DD, F, R, W, B
P D C P copper	Rod or wire	30–65	15–45	5–50	60–90	16–18	30–120	Fair	DD, F, R, W, B
Palladium	Annealed	20	...	40	...	13.8	46	Poor	DD, F, R, W, B
Phosphor bronze A	Sheet, soft to hard	48–80	...	8–50	...	15	60–140	Fair	DD, F, R, W, B
Phosphor bronze D	Sheet, soft to hard	66–102	...	12–65	...	15	75–190	Fair	DD, F, R, W, B
Phosphor bronze F. C.	Rod	60	...	20	...	15	120	Good	B
Pioneer	Casting	74	36.5	42	...	...	150	Good	
Platinum	Annealed wire	17	...	44	...	21.4	42	Poor	DD, F, R, W, B
Platinum	Annealed wire	20	...	40	...	2.5	40	Poor	DD, F, R, W, B
Red X 10	Heat-treated	33–48	30–45	1	...	10	100–140	Good	
Red brass	Sheet	40–83	14–58	3–47	...	...	50–120	Fair	DD, F, R, W, B
Rhodio platinum	Annealed	30–100	...	...	...	...	70–150	Good	DD, F, R, W, B
Roofloy	Cold-rolled sheet	4	...	25	100	2	8.5	Fair	
70 S alloy	Heat treated	28–40	26–38	1–5	...	...	70–90	Fair	
Silver, fine‖	...	...	...	...	...	10.3			
Silver, fine¶	Annealed sheet 800°F	18–23	12	50	...	10.3	25–35	Fair	DD, F, R, W, B
Stellite 1	Weld deposit	47	...	0	...	34.9	...	Poor	W, B
Stellite 6	Casting	105	...	1	...	40.3	...	Poor	W, B
Super nickel-clad steel	Plate	55	30	30	50	...	...	Fair	F, R, W
Tantalum	Sheet, wire	42–178	...	...	...	27	75–125	Poor	DD, F, W
Tellurium lead	Rolled sheet	3.2	1.6	40	...	1.5	6.0	Good	DD, F, W
Tophet A	Annealed wire	120	70	25					
Tungsten	Drawn wire	590	...	...	65	51.4	360	Poor	B
Wolverine commercial bronze	Tube	36–63	14–48	8–60	25-60	15	...	Fair	DD
Y alloy	Casting	30–50	28–45	0–3	...	10	80–150	Good	

† See footnote, p. 1539.
‖ Coin, sterling, and clad are also available.
¶ Coin and sterling are also available.

Table 4b. American Iron and Steel Institute Standard Steels. Chemical Compositions*

Introduction. The ever-growing variety of chemical compositions and quality requirements of steel specifications has been a matter of concern in the steel industry for years. That condition prompted the general technical committee of the American Iron and Steel Institute to make a study of it with a view to simplifying the problems of both consumer and producer. Having in mind the potential benefits of simplification and standardization, their survey was first directed toward determining which of the many grades being currently specified were the ones in most common demand; also the feasibility of combining specifications having like requirements

That survey showed that, exclusive of stainless steels and tool steels, steels with several thousand different combinations of chemical elements were being manufactured to meet the individual demands of purchasers of the primary rolled and forged steel products used in American industry.

The facts thus developed have convinced steel producers that, if their efforts could be concentrated upon a limited number of standardized grades, deliveries could be expedited and service improved. Moreover, there would be better opportunity to achieve advances in manufacturing practices and quality and to study more fully the possibilities of application inherent in those grades.

The general technical committee selected the chemical compositions and physical test properties originally listed as representing steels of proved merit and in extensive use for a wide variety of purposes. Such steels are called *standard steels.*

In keeping with the intention of the general technical committee to review the standard steels at frequent intervals, an intention that was nullified because of the war, these revised lists of standard steels are presented and should be substituted for the corresponding lists in the latest editions of the product manual sections designated here.

* Reprinted from American Iron and Steel Institute. Manufacturers' Standard Practice, July 1, 1946.

Table 4b. American Iron and Steel Institute Standard Steels. Chemical Compositions—(*Continued*)

Carbon Steel Semifinished Products

Basic Open-hearth Carbon Steels for Semifinished Products
Subject to Permissible Variations for Check Analysis

A.I.S.I. designation	C	Mn	P, max.	S, max.
C 1008	0.10 max.	0.30/0.50	0.040	0.050
C 1010	0.08/0.13	0.30/0.60	.040	.050
C 1012	0.10/0.15	0.30/0.60	.040	.050
C 1015	0.13/0.18	0.30/0.60	.040	.050
C 1016	0.13/0.18	0.60/0.90	.040	.050
C 1017	0.15/0.20	0.30/0.60	.040	.050
C 1019	0.15/0.20	0.70/1.00	.040	.050
C 1020	0.18/0.23	0.30/0.60	.040	.050
C 1022	0.18/0.23	0.70/1.00	.040	.050
C 1023	0.20/0.25	0.30/0.60	.040	.050
C 1025	0.22/0.28	0.30/0.60	.040	.050
C 1030	0.28/0.34	0.60/0.90	.040	.050
C 1035	0.32/0.38	0.60/0.90	.040	.050
C 1040	0.37/0.44	0.60/0.90	.040	.050
C 1043	0.40/0.47	0.70/1.00	.040	.050
C 1045	0.43/0.50	0.60/0.90	.040	.050
C 1050	0.48/0.55	0.60/0.90	.040	.050
C 1055	0.50/0.60	0.60/0.90	.040	.050
C 1085	0.80/0.93	0.70/1.00	.040	.050

A.I.S.I. designation	C	Mn	P	S
C 1109	0.08/0.13	0.60/0.90	0.045	0.08/0.13
C 1115	0.13/0.18	0.70/1.00	.045	0.08/0.13
C 1118	0.14/0.20	1.30/1.60	.045	0.08/0.13
C 1120	0.18/0.23	0.70/1.00	.045	0.08/0.13
C 1137	0.32/0.39	1.35/1.65	.045	0.08/0.13
C 1145	0.42/0.49	0.70/1.00	.045	0.04/0.07
C 1151	0.48/0.55	0.70/1.00	.045	0.08/0.13

Hot-rolled Carbon Steel Bars

Basic Open-hearth and Acid Bessemer Carbon Steels
for Hot-rolled Bars
Subject to Permissible Variations for Check Analysis

A.I.S.I. number	Chemical composition limits, %			
	C	Mn	P, Max.	S, max.
C 1008	0.10 max.	0.30/0.50	0.040	0.050
C 1010	0.08/0.13	0.30/0.60	.040	.050
C 1012	0.10/0.15	0.30/0.60	.040	.050
C 1015	0.13/0.18	0.30/0.60	.040	.050
C 1016	0.13/0.18	0.60/0.90	.040	.050
C 1017	0.15/0.20	0.30/0.60	.040	.050
C 1019	0.15/0.20	0.70/1.00	.040	.050
C 1020	0.18/0.23	0.30/0.60	.040	.050
C 1022	0.18/0.23	0.70/1.00	.040	.050
C 1023	0.20/0.25	0.30/0.60	.040	.050
C 1025	0.22/0.28	0.30/0.60	.040	.050
C 1030	0.28/0.34	0.60/0.90	.040	.050
C 1035	0.32/0.38	0.60/0.90	.040	.050
C 1040	0.37/0.44	0.60/0.90	.040	.050
C 1043	0.40/0.47	0.70/1.00	.040	.050
C 1045	0.43/0.50	0.60/0.90	.040	.050
C 1050	0.48/0.55	0.60/0.90	.040	.050
C 1055	0.50/0.60	0.60/0.90	.040	.050
C 1060	0.55/0.65	0.60/0.90	.040	.050
C 1065	0.60/0.70	0.60/0.90	.040	.050
C 1070	0.65/0.75	0.60/0.90	.040	.050
C 1078	0.72/0.85	0.30/0.60	.040	.050
C 1080	0.75/0.88	0.60/0.90	.040	.050
C 1085	0.80/0.93	0.70/1.00	.040	.050
C 1095	0.90/1.05	0.30/0.50	.040	.050
B 1010	0.13 max.	0.30/0.60	0.07/0.12	.600

Note 1: When silicon is specified in standard basic open-hearth steels, silicon may be ordered only as 0.10 per cent maximum; 0.10 to 0.20 per cent; or 0.15 to 0.30 per cent. In the case of many grades of basic open-hearth steel, special practice is necessary in order to comply with a specification including silicon.

Note 2: Acid bessemer steel is not furnished with specified silicon content.

Basic Open-hearth Sulfurized Carbon Steels for Hot-rolled Bars
Subject to Permissible Variations for Check Analysis
(See Note)

A.I.S.I. number	Chemical composition limits, %			
	C	Mn	P, Max.	S
C 1109	0.08/0.13	0.60/0.90	0.045	0.08/0.13
C 1112	0.10/0.16	1.00/1.30	.045	0.08/0.13
C 1115	0.13/0.18	0.70/1.00	.045	0.08/0.13
C 1116	0.14/0.20	1.10/1.40	.045	0.16/0.23
C 1117	0.14/0.20	1.00/1.30	.045	0.08/0.13
C 1118	0.14/0.20	1.30/1.60	.045	0.08/0.13
C 1120	0.18/0.23	0.70/1.00	.045	0.08/0.13
C 1137	0.32/0.39	1.35/1.65	.045	0.08/0.13
C 1141	0.37/0.45	1.35/1.65	.045	0.08/0.13
C 1144	0.40/0.48	1.35/1.65	.045	0.24/0.33
C 1145	0.42/0.49	0.70/1.00	.045	0.04/0.07
C 1151	0.48/0.55	0.70/1.00	.045	0.08/0.13

Note: Sulfurized steel is not subject to check analysis for sulfur. Phosphorized steel is not subject to check analysis for phosphorus.

Acid Bessemer Sulfurized Carbon Steels for Hot-rolled Bars
Subject to Permissible Variations for Check Analysis
(See Note 1)

A.I.S.I. number	Chemical composition limits, %			
	C	Mn	P	S
B 1111	0.08/0.13	0.70/1.00	0.07/0.12	0.10/0.15
B 1112	0.08/0.13	0.70/1.00	0.07/0.12	0.16/0.23
B 1113	0.08/0.13	0.70/1.00	0.07/0.12	0.24/0.33

Note 1: Sulfurized steel is not subject to check analysis for sulfur.
Note 2: Acid bessemer steel is not furnished with specified silicon content

Basic Open-hearth and Acid Bessemer Carbon Steels
for Cold-finished Bars
Subject to Standard Permissible Variations for Check Analysis

A.I.S.I. designation	Chemical composition limits, %			
	C	Mn	P, max.	S, max.
C 1008	0.10 max.	0.30/0.50	0.040	0.050
C 1010	0.08/0.13	0.30/0.60	.040	.050
C 1015	0.13/0.18	0.30/0.60	.040	.050
C 1016	0.13/0.18	0.60/0.90	.040	.050
C 1019	0.15/0.20	0.70/1.00	.040	.050
C 1020	0.18/0.23	0.30/0.60	.040	.050
C 1022	0.18/0.23	0.70/1.00	.040	.050
C 1025	0.22/0.28	0.30/0.60	.040	.050
C 1030	0.28/0.34	0.60/0.90	.040	.050
C 1035	0.32/0.38	0.60/0.90	.040	.050
C 1040	0.37/0.44	0.60/0.90	.040	.050
C 1045	0.43/0.50	0.60/0.90	.040	.050
C 1050	0.48/0.55	0.60/0.90	.040	.050
C 1095	0.90/1.05	0.30/0.50	.040	.050
B 1010	0.13 max.	0.30/0.60	0.07/0.12	.060

Note: When silicon is specified in standard basic open-hearth steels, silicon may be ordered only as 0.10 per cent maximum; 0.10 to 0.20 per cent; or 0.15 to 0.30 per cent. In the case of many grades of basic open-hearth steel, special practice is necessary in order to comply with a specification including silicon.

Basic Open-hearth Sulfurized Carbon Steels for Cold-finished Bars
Subject to Standard Permissible Variations for Check Analysis
(See Note 2)

A.I.S.I. designation	Chemical composition limits, %			
	C	Mn	P, max.	S
C 1115	0.13/0.18	0.70/1.00	0.045	0.08/0.13
C 1116	0.14/0.20	1.10/1.40	.045	0.16/0.23
C 1117	0.14/0.20	1.00/1.30	.045	0.08/0.13
C 1118	0.14/0.20	1.30/1.60	.045	0.08/0.13
C 1137	0.32/0.39	1.35/1.65	.045	0.08/0.13
C 1141	0.37/0.45	1.35/1.65	.045	0.08/0.13
C 1144	0.40/0.48	1.35/1.65	.045	0.24/0.33
C 1151	0.48/0.55	0.70/1.00	.045	0.08/0.13

Note 1: When silicon is specified in standard basic open-hearth steels, silicon may be ordered only as 0.10 per cent maximum; 0.10 to 0.20 per cent; or 0.15 to 0.30 per cent. In the case of many grades of basic open-hearth steel, special practice is necessary in order to comply with a specification including silicon.

Note 2: Sulfurized steel is not subject to check analysis for sulfur. Phosphorized steel is not subject to check analysis for phosphorus.

Table 4b. American Iron and Steel Institute Standard Steels. Chemical Compositions—(*Continued*)

Acid Bessemer Sulfurized Carbon Steels for Cold-finished Bars Subject to Standard Permissible Variations for Check Analysis (See Note 1)

A.I.S.I. designation	Chemical composition limits, %			
	C	Mn	P	S
B 1111	0.08/0.13	0.70/1.00	0.07/0.12	0.10/0.15
B 1112	0.08/0.13	0.70/1.00	0.07/0.12	0.16/0.23
B 1113	0.08/0.13	0.70/1.00	0.07/0.12	0.24/0.33

NOTE 1: Sulfurized steel is not subject to check analysis for sulfur.
NOTE 2: Acid bessemer steel is not furnished with specified silicon content.

Basic Open-hearth and Acid Bessemer Carbon Steels for Rods Subject to Standard Permissible Variations for Check Analysis

A.I.S.I. designation	Chemical composition limits, %			
	C	Mn	P, max.	S, max.
C 1006	0.08 max.	0.25/0.40	0.040	0.050
C 1008	0.10 max.	0.30/0.50	.040	.050
C 1010	0.08/0.13	0.30/0.60	.040	.050
C 1012	0.10/0.15	0.30/0.60	.040	.050
C 1013	0.11/0.16	0.60/0.90	.040	.050
C 1015	0.13/0.18	0.30/0.60	.040	.050
C 1016	0.13/0.18	0.60/0.90	.040	.050
C 1017	0.15/0.20	0.30/0.60	.040	.050
C 1019	0.15/0.20	0.70/1.00	.040	.050
C 1020	0.18/0.23	0.30/0.60	.040	.050
C 1022	0.18/0.23	0.70/1.00	.040	.050
C 1025	0.22/0.28	0.30/0.60	.040	.050
C 1030	0.28/0.34	0.60/0.90	.040	.050
C 1034	0.32/0.38	0.50/0.80	.040	.050
C 1038	0.35/0.42	0.60/0.90	.040	.050
C 1040	0.37/0.44	0.60/0.90	.040	.050
C 1041	0.36/0.44	1.35/1.65	.040	.050
C 1051	0.45/0.56	0.85/1.15	.040	.050
C 1054	0.50/0.60	0.50/0.80	.040	.050
C 1057	0.50/0.61	0.85/1.15	.040	.050
C 1059	0.55/0.65	0.50/0.80	.040	.050
C 1060	0.55/0.65	0.60/0.90	.040	.050
C 1062	0.54/0.65	0.85/1.15	.040	.050
C 1064	0.60/0.70	0.50/0.80	.040	.050
C 1066	0.60/0.71	0.85/1.15	.040	.050
C 1069	0.65/0.75	0.40/0.70	.040	.050
C 1075	0.70/0.80	0.40/0.70	.040	.050
C 1078	0.72/0.85	0.30/0.60	.040	.050
C 1095	0.90/1.05	0.30/0.50	.040	.050
B 1006	0.08 max.	0.45 max.	0.07/0.12	.060
B 1010	0.13 max.	0.30/0.60	0.07/0.12	.060

NOTE 1: When silicon is specified in standard basic open-hearth steels, silicon may be ordered only as 0.10 per cent maximum; 0.10 to 0.20 per cent; or 0.15 to 0.30 per cent. In the case of many grades of basic open-hearth steel, special practices are required in order to comply with a specification including silicon.
NOTE 2: Acid bessemer steel is not furnished with specified silicon content.

Basic Open-hearth Sulfurized Carbon Steels for Rods Subject to Standard Permissible Variations for Check Analysis (See Note 2)

A.I.S.I. designation	Chemical composition limits, %			
	C	Mn	P, max.	S
C 1108	0.08/0.13	0.50/0.80	0.045	0.07/0.12
C 1109	0.08/0.13	0.60/0.90	.045	0.08/0.13
C 1111	0.08/0.13	0.60/0.90	.054	0.16/0.23
C 1112	0.10/0.16	1.00/1.30	.045	0.08/0.13
C 1115	0.13/0.18	0.70/1.00	.045	0.08/0.13
C 1117	0.14/0.20	1.00/1.30	.045	0.08/0.13
C 1118	0.14/0.20	1.30/1.60	.045	0.08/0.13
C 1120	0.18/0.23	0.70/1.00	.045	0.08/0.13
C 1137	0.32/0.39	1.35/1.65	.045	0.08/0.13
C 1141	0.37/0.45	1.35/1.65	.045	0.08/0.13

NOTE 1: Special practice is necessary to comply with a specification of a minimum silicon content in a sulfurized steel when the maximum limit of sulfur is over 0.055 per cent.
NOTE 2: Sulfurized steel is not subject to check analysis for sulfur.

Acid Bessemer Sulfurized Carbon Steels for Rods Subject to Standard Permissible Variations for Check Analysis (See Note 1)

A.I.S.I. designation	Chemical composition limits, %			
	C	Mn	P	S
B 1111	0.08/0.13	0.70/1.00	0.07/0.12	0.10/0.15
B 1112	0.08/0.13	0.70/1.00	0.07/0.12	0.16/0.23
B 1113	0.08/0.13	0.70/1.00	0.07/0.12	0.24/0.33

NOTE 1: Sulfurized steel is not subject to check analysis for sulfur.
NOTE 2: Acid bessemer steel is not furnished with specified silicon content.

Acid Open-hearth Carbon Steels for Rods Subject to Standard Permissible Variations for Check Analysis

A.I.S.I. designation	Chemical composition limits, %			
	C	Mn	P, max.	S, max.
D 1049	0.43/0.50	0.50/0.80	0.050	0.050
D 1054	0.50/0.60	0.50/0.80	.050	.050
D 1059	0.55/0.65	0.50/0.80	.050	.050
D 1064	0.60/0.70	0.50/0.80	.050	.050
D 1069	0.65/0.75	0.40/0.70	.050	.050
D 1075	0.70/0.80	0.40/0.70	.050	.050

NOTE: The standard steels listed here may be ordered with a silicon content of 0.10 to 0.25 per cent; or 0.15 to 0.35 per cent.

Basic Open-hearth and Acid Bessemer Carbon Steels for Wire Subject to Standard Permissible Variations for Check Analysis

A.I.S.I. designation	Chemical composition limits, %			
	C	Mn	P, max.	S, max.
C 1006	0.08 max.	0.25/0.40	0.040	0.050
C 1008	0.10 max.	0.30/0.50	.040	.050
C 1010	0.08/0.13	0.30/0.60	.040	.050
C 1012	0.10/0.15	0.30/0.60	.040	.050
C 1013	0.11/0.16	0.60/0.90	.040	.050
C 1015	0.13/0.18	0.30/0.60	.040	.050
C 1016	0.13/0.18	0.60/0.90	.040	.050
C 1017	0.15/0.20	0.30/0.60	.040	.050
C 1019	0.15/0.20	0.70/1.00	.040	.050
C 1020	0.18/0.23	0.30/0.60	.040	.050
C 1022	0.18/0.23	0.70/1.00	.040	.050
C 1025	0.22/0.28	0.30/0.60	.040	.050
C 1030	0.28/0.34	0.60/0.90	.040	.050
C 1034	0.32/0.38	0.50/0.80	.040	.050
C 1038	0.35/0.42	0.60/0.90	.040	.050
C 1040	0.37/0.44	0.60/0.90	.040	.050
C 1041	0.36/0.44	1.35/1.65	.040	.050
B 1006	0.08 max.	0.45 max.	0.07/0.12	.060
B 1010	0.13 max.	0.30/0.60	0.07/0.12	.060

NOTE 1: When silicon is specified in standard open-hearth steels, silicon may be ordered only as 0.10 per cent maximum; 0.10 to 0.20 per cent; or 0.15 to 0.30 per cent. In the case of many grades of basic open-hearth steel, special practice is necessary in order to comply with a specification including silicon.
NOTE 2: Acid bessemer steel is not furnished with specified silicon content.

Acid Open-hearth Carbon Steels for Wire Subject to Standard Permissible Variations for Check Analysis

A.I.S.I. designation	Chemical composition limits, %			
	C	Mn	P, max.	S, max.
D 1049	0.43/0.50	0.50/0.80	0.050	0.050
D 1054	0.50/0.60	0.50/0.80	.050	.050
D 1059	0.55/0.65	0.50/0.80	.050	.050
D 1064	0.60/0.70	0.50/0.80	.050	.050
D 1069	0.65/0.75	0.40/0.70	.050	.050
D 1075	0.70/0.80	0.40/0.70	.050	.050

NOTE: The standard steels listed here may be ordered with a silicon content of 0.10 to 0.25 per cent; or 0.15 to 0.35 per cent.

Table 4b. American Iron and Steel Institute Standard Steels. Chemical Compositions—(*Continued*)

Alloy Steel

Open-hearth and Electric Furnace Alloy Steels
Bars, Bar-strip, Billets, Blooms, and Slabs
Subject to Permissible Variations for Check Analysis
(See Note 3)

A.I.S.I. number	Chemical composition limits, %								
	C	Mn	P	S	Si	Ni	Cr	Mo	V
A 1320	0.18/0.23	1.60/1.90	0.040	0.040	0.20/0.35				
A 1330	0.28/0.33	1.60/1.90	.040	.040	0.20/0.35				
A 1335	0.33/0.38	1.60/1.90	.040	.040	0.20/0.35				
A 1340	0.38/0.43	1.60/1.90	.040	.040	0.20/0.35				
A 2317	0.15/0.20	0.40/0.60	.040	.040	0.20/0.35	3.25/3.75			
A 2330	0.28/0.33	0.60/0.80	.040	.040	0.20/0.35	3.25/3.75			
A 2335	0.33/0.38	0.60/0.80	.040	.040	0.20/0.35	3.25/3.75			
A 2340	0.38/0.43	0.70/0.90	.040	.040	0.20/0.35	3.25/3.75			
A 2345	0.43/0.48	0.70/0.90	.040	.040	0.20/0.35	3.25/3.75			
E 2512	0.09/0.14	0.45/0.60	.025	.025	0.20/0.35	4.75/5.25			
A 2515	0.12/0.17	0.40/0.60	.040	.040	0.20/0.35	4.75/5.25			
E 2517	0.15/0.20	0.45/0.60	.025	.025	0.20/0.35	4.75/5.25			
A 3115	0.13/0.18	0.40/0.60	.040	.040	0.20/0.35	1.10/1.40	0.55/0.75		
A 3120	0.17/0.22	0.60/0.80	.040	.040	0.20/0.35	1.10/1.40	0.55/0.75		
A 3130	0.28/0.33	0.60/0.80	.040	.040	0.20/0.35	1.10/1.40	0.55/0.75		
A 3135	0.33/0.38	0.60/0.80	.040	.040	0.20/0.35	1.10/1.40	0.55/0.75		
A 3140	0.38/0.43	0.70/0.90	.040	.040	0.20/0.35	1.10/1.40	0.55/0.75		
A 3141	0.38/0.43	0.70/0.90	.040	.040	0.20/0.35	1.10/1.40	0.70/0.90		
A 3145	0.43/0.48	0.70/0.90	.040	.040	0.20/0.35	1.10/1.40	0.70/0.90		
A 3150	0.48/0.53	0.70/0.90	.040	.040	0.20/0.35	1.10/1.40	0.70/0.90		
E 3310	0.08/0.13	0.45/0.60	.025	.025	0.20/0.35	3.25/3.75	1.40/1.75		
E 3316	0.14/0.19	0.45/0.60	.025	.025	0.20/0.35	3.25/3.75	1.40/1.75		
A 4023	0.20/0.25	0.70/0.90	.040	.040	0.20/0.35	....	....	0.20/0.30	
A 4024	0.20/0.25	0.70/0.90	.040	.035/0.050	0.20/0.35	....	....	0.20/0.30	
A 4027	0.25/0.30	0.70/0.90	.040	.040	0.20/0.35	....	....	0.20/0.30	
A 4028	0.25/0.30	0.70/0.90	.040	.035/0.050	0.20/0.35	....	....	0.20/0.30	
A 4032	0.30/0.35	0.70/0.90	.040	.040	0.20/0.35	....	....	0.20/0.30	
A 4037	0.35/0.40	0.70/0.90	.040	.040	0.20/0.35	....	....	0.20/0.30	
A 4042	0.40/0.45	0.70/0.90	.040	.040	0.20/0.35	....	....	0.20/0.30	
A 4047	0.45/0.50	0.70/0.90	.040	.040	0.20/0.35	....	....	0.20/0.30	
A 4063	0.60/0.67	0.75/1.00	.040	.040	0.20/0.35	....	....	0.20/0.30	
A 4068	0.64/0.72	0.75/1.00	.040	.040	0.20/0.35	....	....	0.20/0.30	
A 4130	0.28/0.33	0.40/0.60	.040	.040	0.20/0.35	....	0.80/1.10	0.15/0.25	
E 4132	0.30/0.35	0.40/0.60	.025	.025	0.20/0.35	....	0.80/1.10	0.18/0.25	
E 4135	0.33/0.38	0.70/0.90	.025	.025	0.20/0.35	....	0.80/1.10	0.18/0.25	
A 4137	0.35/0.40	0.70/0.90	.040	.040	0.20/0.35	....	0.80/1.10	0.15/0.25	
E 4137	0.35/0.40	0.70/0.90	.025	.025	0.20/0.35	....	0.80/1.10	0.18/0.25	
A 4140	0.38/0.43	0.75/1.00	.040	.040	0.20/0.35	....	0.80/1.10	0.15/0.25	
A 4142	0.40/0.45	0.75/1.00	.040	.040	0.20/0.35	....	0.80/1.10	0.15/0.25	
A 4145	0.43/0.48	0.75/1.00	.040	.040	0.20/0.35	....	0.80/1.10	0.15/0.25	
A 4147	0.45/0.50	0.75/1.00	.040	.040	0.20/0.35	....	0.80/1.10	0.15/0.25	
A 4150	0.48/0.53	0.75/1.00	.040	.040	0.20/0.35	....	0.80/1.10	0.15/0.25	
A 4317	0.15/0.20	0.45/0.65	.040	.040	0.20/0.35	1.65/2.00	0.40/0.60	0.20/0.30	
A 4320	0.17/0.22	0.45/0.65	.040	.040	0.20/0.35	1.65/2.00	0.40/0.60	0.20/0.30	
E 4337	0.35/0.40	0.60/0.80	.025	.025	0.20/0.35	1.65/2.00	0.40/0.60	0.20/0.30	
A 4340	0.38/0.43	0.60/0.80	.040	.040	0.20/0.35	1.65/2.00	0.70/0.90	0.20/0.30	
E 4340	0.38/0.43	0.60/0.80	.025	.025	0.20/0.35	1.65/2.00	0.70/0.90	0.20/0.30	
A 4608	0.06/0.11	0.40 Max	.040	.040	0.25 Max	1.40/1.75	....	0.15/0.25	
A 4615	0.13/0.18	0.45/0.65	.040	.040	0.20/0.35	1.65/2.00	....	0.20/0.30	
E 4617	0.15/0.20	0.45/0.65	.025	.025	0.20/0.35	1.65/2.00	....	0.20/0.27	
A 4620	0.17/0.22	0.45/0.65	.040	.040	0.20/0.35	1.65/2.00	....	0.20/0.30	
E 4620	0.17/0.22	0.45/0.65	.025	.025	0.20/0.35	1.65/2.00	....	0.20/0.27	
A 4621	0.18/0.23	0.70/0.90	.040	.040	0.20/0.35	1.65/2.00	....	0.20/0.30	
A 4640	0.38/0.43	0.60/0.80	.040	.040	0.20/0.35	1.65/2.00	....	0.20/0.30	
E 4640	0.38/0.43	0.60/0.80	.025	.025	0.20/0.35	1.65/2.00	....	0.20/0.27	
A 4812	0.10/0.15	0.40/0.60	.040	.040	0.20/0.35	3.25/3.75	....	0.20/0.30	
A 4815	0.13/0.18	0.40/0.60	.040	.040	0.20/0.35	3.25/3.75	....	0.20/0.30	
A 4817	0.15/0.20	0.40/0.60	.040	.040	0.20/0.35	3.25/3.75	....	0.20/0.30	
A 4820	0.18/0.23	0.50/0.70	.040	.040	0.20/0.35	3.25/3.75	....	0.20/0.30	
A 5120	0.17/0.22	0.70/0.90	.040	.040	0.20/0.35	....	0.70/0.90		
A 5130	0.28/0.33	0.70/0.90	.040	.040	0.20/0.35	....	0.80/1.10		
A 5140	0.38/0.43	0.70/0.90	.040	.040	0.20/0.35	....	0.70/0.90		
A 5145	0.43/0.48	0.70/0.90	.040	.040	0.20/0.35	....	0.70/0.90		
A 5150	0.48/0.53	0.70/0.90	.040	.040	0.20/0.35	....	0.70/0.90		
A 5152	0.48/0.55	0.70/0.90	.040	.040	0.20/0.35	....	0.90/1.20		
E 52095	0.95/1.10	0.25/0.45	.025	.025	0.20/0.35	....	0.40/0.60		
E 52098	0.95/1.10	0.25/0.45	.025	.025	0.20/0.35	....	0.90/1.15		
E 52100	0.95/1.10	0.25/0.45	.025	.025	0.20/0.35	....	1.20/1.50		
E 52101	0.95/1.10	0.25/0.45	.025	.025	0.20/0.35	....	1.30/1.60		
A 6120	0.17/0.22	0.70/0.90	.040	.040	0.20/0.35	....	0.70/0.90	....	0.10 min.
A 6145	0.43/0.48	0.70/0.90	.040	.040	0.20/0.35	....	0.80/1.10	....	0.15 min.
A 6150	0.48/0.53	0.65/0.90	.040	.040	0.20/0.35	....	0.80/1.10	....	0.15 min.

Table 4b. American Iron and Steel Institute Standard Steels. Chemical Compositions—(*Concluded*)

Alloy Steel—(*Concluded*)

A.I.S.I. number	Chemical composition limits, %								
	C	Mn	P	S	Si	Ni	Cr	Mo	V
E 6150	0.47/0.53	0.70/0.90	0.025	0.025	0.20/0.35	....	0.80/1.10	....	0.15 min.
A 6152	0.48/0.55	0.70/0.90	.040	.040	0.20/0.35	....	0.80/1.10	....	0.10 min.
A 8615	0.13/0.18	0.70/0.90	.040	.040	0.20/0.35	0.40/0.70	0.40/0.60	0.15/0.25	
A 8617	0.15/0.20	0.70/0.90	.040	.040	0.20/0.35	0.40/0.70	0.40/0.60	0.15/0.25	
A 8620	0.18/0.23	0.70/0.90	.040	.040	0.20/0.35	0.40/0.70	0.40/0.60	0.15/0.25	
A 8622	0.20/0.25	0.70/0.90	.040	.040	0.20/0.35	0.40/0.70	0.40/0.60	0.15/0.25	
A 8625	0.23/0.28	0.70/0.90	.040	.040	0.20/0.35	0.40/0.70	0.40/0.60	0.15/0.25	
A 8627	0.25/0.30	0.70/0.90	.040	.040	0.20/0.35	0.40/0.70	0.40/0.60	0.15/0.25	
A 8630	0.28/0.33	0.70/0.90	.040	.040	0.20/0.35	0.40/0.70	0.40/0.60	0.15/0.25	
A 8632	0.30/0.35	0.70/0.90	.040	.040	0.20/0.35	0.40/0.70	0.40/0.60	0.15/0.25	
A 8635	0.33/0.38	0.75/1.00	.040	.040	0.20/0.35	0.40/0.70	0.40/0.60	0.15/0.25	
A 8637	0.35/0.40	0.75/1.00	.040	.040	0.20/0.35	0.40/0.70	0.40/0.60	0.15/0.25	
A 8640	0.38/0.43	0.75/1.00	.040	.040	0.20/0.35	0.40/0.70	0.40/0.60	0.15/0.25	
A 8642	0.40/0.45	0.75/1.00	.040	.040	0.20/0.35	0.40/0.70	0.40/0.60	0.15/0.25	
A 8645	0.43/0.48	0.75/1.00	.040	.040	0.20/0.35	0.40/0.70	0.40/0.60	0.15/0.25	
A 8647	0.45/0.50	0.75/1.00	.040	.040	0.20/0.35	0.40/0.70	0.40/0.60	0.15/0.25	
A 8650	0.48/0.53	0.75/1.00	.040	.040	0.20/0.35	0.40/0.70	0.40/0.60	0.15/0.25	
A 8655	0.50/0.60	0.75/1.00	.040	.040	0.20/0.25	0.40/0.70	0.40/0.60	0.15/0.25	
A 8660	0.55/0.65	0.75/1.00	.040	.040	0.20/0.35	0.40/0.70	0.40/0.60	0.15/0.25	
A 8720	0.18/0.23	0.70/0.90	.040	.040	0.20/0.35	0.40/0.70	0.40/0.60	0.20/0.30	
A 8735	0.33/0.38	0.75/1.00	.040	.040	0.20/0.35	0.40/0.70	0.40/0.60	0.20/0.30	
A 8740	0.38/0.43	0.75/1.00	.040	.040	0.20/0.35	0.40/0.70	0.40/0.60	0.20/0.30	
A 8742	0.40/0.45	0.75/1.00	.040	.040	0.20/0.35	0.40/0.70	0.40/0.60	0.20/0.30	
A 8745	0.43/0.48	0.75/1.00	.040	.040	0.20/0.35	0.40/0.70	0.40/0.60	0.20/0.30	
A 8747	0.45/0.50	0.75/1.00	.040	.040	0.20/0.35	0.40/0.70	0.40/0.60	0.20/0.30	
A 8750	0.48/0.53	0.75/1.00	.040	.040	0.20/0.35	0.40/0.70	0.40/0.60	0.20/0.30	
A 9255	0.50/0.60	0.70/0.95	.040	.040	1.80/2.20				
A 9260	0.55/0.65	0.70/1.00	.040	.040	1.80/2.20				
A 9261	0.55/0.65	0.75/1.00	.040	.040	1.80/2.20	....	0.10/0.25		
A 9262	0.55/0.65	0.75/1.00	.040	.040	1.80/2.20	....	0.25/0.40		
E 9310	0.08/0.13	0.45/0.65	.025	.025	0.20/0.35	3.00/3.50	1.00/1.40	0.08/0.15	
E 9315	0.13/0.18	0.45/0.65	.025	.025	0.20/0.35	3.00/3.50	1.00/1.40	0.08/0.15	
E 9317	0.15/0.20	0.45/0.65	.025	.025	0.20/0.35	3.00/3.50	1.00/1.40	0.08/0.15	
A 9437	0.35/0.40	0.90/1.20	.040	.040	0.20/0.35	0.30/0.60	0.30/0.50	0.08/0.15	
A 9440	0.38/0.43	0.90/1.20	.040	.040	0.20/0.35	0.30/0.60	0.30/0.50	0.08/0.15	
A 9442	0.40/0.45	1.00/1.30	.040	.040	0.20/0.25	0.30/0.60	0.50/0.50	0.08/0.15	
A 9445	0.43/0.48	1.00/1.30	.040	.040	0.20/0.35	0.30/0.60	0.30/0.50	0.08/0.15	
A 9747	0.45/0.40	0.50/0.80	.040	.040	0.20/0.35	0.40/0.70	0.10/0.25	0.15/0.25	
A 9763	0.60/0.67	0.50/0.80	.040	.040	0.20/0.35	0.40/0.70	0.10/0.25	0.15/0.25	
A 9840	0.38/0.43	0.70/0.90	.040	.040	0.20/0.35	0.85/1.15	0.70/0.90	0.15/0.25	
A 9845	0.43/0.48	0.70/0.90	.040	.040	0.20/0.35	0.85/1.15	0.70/0.90	0.20/0.30	
A 9850	0.48/0.53	0.70/0.90	.040	.040	0.20/0.35	0.85/1.15	0.70/0.90	0.20/0.30	

NOTE 1: The lowest standard maximum phosphorus or sulfur limit that may be specified for acid open-hearth or acid electric furnace steel is 0.05 per cent.

NOTE 2: Lowest minimum silicon limit to be specified for acid open-hearth or acid electric furnace alloy steel is 0.15 per cent.

NOTE 3: Phosphorus and sulfur on basic open-hearth product shall be 0.040 max. Phosphorus and sulfur on basic electric furnace product shall be 0.025 max.

NOTE 4: Small quantities of certain elements may be found in alloy steel which are not specified or required. These elements are to be considered as incidental and acceptable to the following maximum amounts: copper, 0.35 per cent; nickel, 0.25 per cent; chromium, 0.20 per cent; molybdenum, 0.06 per cent.

NOTE 5: When a grade is specified to a method of manufacture (electric furnace or open-hearth) not listed in the above table the phosphorus and sulfur limits shall be, unless otherwise specified, as follows: basic electric furnace—0.025 max., basic open-hearth—0.040 max., acid open-hearth or electric—0.050 max.

Enameled ware to a considerable degree avoids the objectionable mechanical features of glassware but at the expense of a reduction in chemical resistance. The glass used for coating iron or steel is less resistant to chemical action than the best grade of chemical glass, which because of its coefficient of expansion and melting point, however, cannot be used for enameling. In general, enameled ware may be used successfully only under moderate process temperatures and under relatively mild conditions of mechanical and temperature shock.

Stoneware largely avoids the mechanical fragility of glass and silica but substitutes a problem in heat transfer, the thickness of the material ordinarily used causing this to become a serious problem when considerable quantities of heat are to be supplied or removed.

Other ceramic materials present problems, in general, similar to those already mentioned, the design problems accompanying the use of porcelain or stoneware, tile or brick, of cementlike material in general, being accompanied by limitations as to temperature, pressure, heat transfer, size of equipment, etc., which are not present when working with the metals or alloys. Enameled equipment is the only type of this group in which the working surface is non-metallic, and the physical strength is supplied by a metal. In tile lining, difficulty is also encountered in obtaining satisfactory cement of resistance equal to the tile.

Hard and soft rubber are frequently used for the handling of acids, especially dilute aqueous solutions. The resistance properties of the two are similar, both being unattacked by dilute aqueous solutions except those of oxidizing agents and both being swelled by organic solvents. An important difference between the two lies in the fact that the soft rubber is used as a lining, usually for steel, while the hard rubber can be used alone.

Synthetic rubberlike polymers are becoming of increasing importance as engineering raw materials because of the superiority to natural rubber in many important properties such as resistance to oxidation, solvents, oil, and many chemicals.

(Continued on p. 1558.)

Table 5. Precious Metals and Their Alloys*

Metals and alloys having an annealed hardness not exceeding 150 Brinell hardness number generally are available in all forms; above this hardness, consult with the manufacturer.

Most of the materials listed are available from American Platinum Works; Baker & Co.; J. Bishop Platinum Works; Siegmund Cohn; and others. Handy & Harman supply principally silver and gold alloys; clad materials can be had from the above and from General Plate Co.; D. E. Makepiece; P. R. Mallory; I. Stern & Co.; H. A. Wilson, etc. Certain alloys are supplied by manufacturers of dental alloys.

Material	Chem. equip.	High temp.	Anodes	Spinnerets	Catalysts	Jewelry	Contacts	Dental	Solder	Density, gm./cm³	Liquidus °C. Solidus °C.	Properties (annealed) hardness B.h.n.-V.h.n.	Annealed tensile lb./sq. in.	Annealed elongation %	Elec. trical resistivity 20°C., microhm-cm.	Temp. coeff. of resistivity 20–100°C.	Expansion per °C. at 20°C ×10⁻⁶
Platinum		X			X			X		21.45	1773.5	40	18,000	35	10.6	0.0036	8.9
Platinum (commercial)	X	X	X		X		X	X		21.45	approx. 1773.5	44	22,000	35	11.0	.0036	8.9
El. deposited Pt							X			21.4	....	500–600					
3.5% Rh 96.5% Pt	X	X	X							20.9	1800	60	25,000	35	16.6	.0022	
10% Rh 90% Pt		X		X	X					20.0	1850	90	45,000	35	19.2	.0017	
20% Rh 80% Pt		X								18.7	1900	120	70,000	40	20.8	.0014	
5% Ir 95% Pt	X		X			X				21.5	1780	90	40,000	25	19.0	.002	
10% Ir 90% Pt	X		X			X	X	X		21.5	1800	130	55,000	20	25.0	.0013	8.9
25% Ir 75% Pt							X			21.7	1870	240	125,000	20	33.0	.0006	
5% Ru 95% Pt	X		X			X	X			20.7	1810	130	60,000	34	31.5	.0009	
10% Ru 90% Pt			X				X			19.9	1840	190	85,000	30	43.0	.0008	
20% Pd 80% Pt	X						X	X		18.5	1730	95	50,000	35	24.5	.001	
10% Au 90% Pt	X	X					X			21.2	1660–1730	140	77,000	25	24.0	.002	
5% Ni 95% Pt	X	X					X			20.0	1700–1740	135	65,000	20	23.3	.002	
4% W 96% Pt	X						X			21.3	1800–1825	150	75,000	25	36.9		
Pt-clad Cu, Ni, Ag	X		X				X										
Palladium		X			X	X	X			12.0	1554	38	22,000	25	10.8	.0037	11.8
Pd electroplate						X	X			.....	....	150–400					
5% Ru 95% Pd						X	X			12.0	1600	90	55,000	50	23.5	.00002	
40% Ag 60% Pd						X	X	X		11.3	1330–1390	100	51,000	47	42.0	.00002	
30% Au 70% Pd								X	X	13.5	1500–1520	80	45,000	40	23.0	.001	
40% Cu 60% Pd							X			10.6	1160–1180	120	75,000	25	25.0	.0003	
16% Cu 40% Ag 44% Pd							X	X		10.8	1050–1120	200‡	100,000	..	26		
15% Pt 25% Au 60% Pd							X	X		14.8	1300	230	145,000	10			
Pd-clad Cu, Ni, Ag						X	X	X									
Iridium†		X								22.5	2454	270	....	..	5.3	.0036	6.8
Rhodium†		X								12.4	1966	135	73,000	..	4.1	.0042	8.3
Rh electroplate						X	X	X		.....	....	600					
Ruthenium†					X					12.4	2450	360	....	..	8.0	......	9.1
Osmium†										22.7	2700	400	....	..	9.5	.0042	5.0
Gold (fine)	X							X	X	19.3	1063	30	18,000	30	2.35	.0039	14.2
Gold electroplate						X			X	19.2	1063						
30% Pd 70% Au	X						X	X	X	16.3	1420	72	48,000				
30% Pt 70% Au	X			X						19.8	1200–1410	110‡	68,000	22	25		
30% Ag 70% Au							X		X	15.4	1020–1040	33	28,000	40	9.9		
6% Pt 25% Ag 69% Au							X			16.1	1110–1030	70	55,000	35	14.9		
12% Cu 13% Ag 75% Au						X	X	X	X	15.4	890–910	115‡	70,000	40			
10% Pd 20% Pt 70% Au	X			X						18.5	....	100‡	50,000				
13% Cu 16% Ag 14% Pt 57% Au								X		15.3	920–1300	210‡	104,000	13			
15% Cu 18% Ag 17% Pd 50% Au							X	X		13.4	975–1050	170‡	84,000	34			
Silver (commercial)	X					X	X			10.5	960	30	26,000	50	1.62	.004	17
Ag electroplate	X				X	X	X			10.5	960	60–130			1.7–2.2		
3% Pt 97% Ag	X						X			10.7	965–975	40	28,000	40	3.7		
10% Pd 90% Ag	X						X			10.6	1045–1055	35	27,000	45			
7.5% Cu 92.5 Ag						X	X			10.4	760–800	50‡	42,000	42	2.1		
10% Cu 90% Ag						X	X			10.3	778–895	55‡	42,000	40	2.2		
15% Zn 25% Cu 60 Ag									X	9.5	680–720	105	64,000				
16% Zn 34% Cu 50% Ag									X	9.4	690–775	115	50,000				
16% Zn 18% Cd 16% Cu 50% Ag									X	9.5	627–635	......	50,000				
Ag Clad Ni, Fe, Cu	X					X	X										

* By C. M. Wise, platinum metals specialist, International Nickel Co.
† These metals are almost wholly employed as alloying elements with other platinum metals.
‡ These alloys can be effectively age-hardened.

Table 6a. Glass, Glass-lined, and Fused Silica Equipment

Physical Properties of Low-expansion Glasses, Fused Quartz, and Fused Silica

Material	Specific gravity	Specific volume, cu. in./lb.	Tensile strength, lb./sq. in.	Modulus of elasticity, lb./sq. in. (multiply by 10⁶)	Hardness*	Thermal expansion per °C. (multiply by 10⁻⁵)	Thermal cond., cal./sec., cm.², °C., cm. (multiply by 10⁴)	Specific heat cal./°C., gm.	Softening point, °F.	Breakdown voltage, 60 cycles, volts/mil	Dielectric constant, 60 cycles	Refractive index, nAD	Transparency†	Forms available‡
Borosilicate glass	2.23	12.4	10,000	98		0.32	24.5	0.20	1505	3200 (0.1 in.)	4.6	1.47	T, TL	S, R, T, other
96% silica glass	2.18	12.7	......	......		.080	...	....	2750 ± 90	3000	4.0	1.458	T	R, T, other
Fused quartz	2.20	12.6	4000	105–126	4.9	.054	35	.25	2600	500 (¼ in.)	Approx. 3.8	1.459	T	S, R, T, other
Fused silica	2.07	13.4	400–800	94–114	...	.054	25	....	2600	250 (¼ in.)	3.7	.....	TL, O	S, R, T, other

* Hardness: 2.5 mm. ball, 25 kg. load, depth in 1/200 mm.
† T = transparent; TL = translucent; O = opaque.
‡ S = sheets; R = rods; T = tubes.

Table 6b. Glass-lined and Enameled Steel*

Coefficients of heat transfer for typical chemical enamels:

The P, S, M, L, and LL series of process units wherein the glass is applied by the dry dust method for very severe chemical service (Pfaudler No. 24)	
Steam to water being heated	U = 60– 70
Steam to boiling water	U = 70– 80
Cooling, water to water	U = 40– 50
The R series of reaction units for severe chemical service (Pfaudler No. 42)	
Steam to water being heated	U = 80– 90
Steam to boiling water	U = 85– 95
Cooling, water to water	U = 50– 60
Standard units for organic acids and mild chemical conditions (Pfaudler No. 48)	
Steam to water being heated	U = 90–100
Steam to boiling water	U = 95–105
Cooling, water to water	U = 50– 60

NOTE: All values based upon agitation of contents of unit.

Pressures, temperatures, corrosion conditions, and any thermal shock anticipated for any set of conditions should be definitely ascertained and this information supplied to the manufacturer.

* Courtesy of The Pfaudler Co.

Table 7. Chemical Stoneware, Porcelain, and Cements

Physical Properties of Chemical Stoneware and Porcelainware

Chemical Stoneware*

Specific gravity	2.2
Hardness, scleroscope	100
Ultimate tensile strength, lb./sq. in.	2000–3000
Ultimate compressive strength, lb./sq. in.	80,000
Modulus of rupture, lb./sq. in.	5000–13,000
Modulus of elasticity, lb./sq. in.	8,000,000
Specific heat	0.2
Thermal conductivity, B.t.u./hr., sq. ft., °F., in.	10–35
Linear thermal expansion/°F	0.0000020
Water absorption, %	0–2

Chemical Porcelainware†

Specific gravity	2.41
Ultimate tensile strength, lb./sq. in.	5000–8000
Ultimate compressive strength, lb./sq. in.	100,000
Modulus of rupture, lb./sq. in.	12,000–15,000
Modulus of elasticity, lb./sq. in.	10,400,000
Specific heat	0.2
Thermal conductivity, B.t.u./hr., sq. ft., °F., in.	8.4
Linear thermal expansion/°F	0.0000023
Water absorption, %	0

* This table, which has been prepared by the General Ceramics Co., gives the physical properties of an average grade of chemical stoneware. It should be emphasized here that "chemical stoneware" is not the name of a definite material, such as an alloy, but a generic term applied to a wide variety of ceramic compositions, and hence that, in any particular composition designed to give optimum properties in one respect, it will ordinarily be impossible to secure optimum properties in all other respects.

† Data supplied by Lapp Insulator Co.

Table 8. Refractory Materials

Physical Properties of Refractory Materials

Complete revision of earlier *Chem. & Met. Eng.* data, compiled by L. J. Trostel, General Refractories Co., Baltimore, with additional material on kaolin super duty and insulating refractories supplied by the Babcock & Wilcox Co., New York.

Type of brick	Silica	High heat duty (No. 1) fireclay	Super duty fireclay	Super duty kaolin	Alumina-diaspore, 70% Al_2O_3	Sillimanite (mullite)	Chrome	Unburned chrome[1]	Magnesite	Unburned magnesite[1]	Bonded silicon carbide (grade A)	Bonded fused alumina	Kaolin insul. refr. (2600 °F.)
Typical composition, %:													
SiO_2	96	50–57	52	52	22–26	35	6	5	3	5	7–9	8–10	57.7
Fe_2O_3	1	1.5–2.5	1	0.6	1–1.5	0.5			6	8.5	0.3–1	1–1.5	2.4
FeO							15	12					
Al_2O_3	1	36–42	43	45.4	68–72	62	23	18	2	7.5	2–4	85–90	36.8
TiO_2		1.5–2.5	2	1.7	3.5	1.5					1	1.5–2.2	1.5
CaO	2			0.1					3	2			0.6
MgO				0.2			17	32	86	64			0.5
Cr_2O_3							38	30		10			
SiC											85–90		
Flux[2]		1–3.5	2		1–1.5	0.5					1.5	0.8–1.3	
P.C.E. (with approx. equivalent temp., °F.)[3]	31–32 (3056–3092°)	31–33 (3056–3173°)	33–34 (3173–3200°)	34 (3200°)	36 (3290°)	37–38 (3308–3335°)	41+ (3578°+)	41+ (3578°+)	41+ (3578°+)	41+ (3578°+)	39 (3389°)	39+ (3389°+)	29–30 (2984–3002°)
Deformation under load,[4] % (at lb./sq. in. and temp., °F., shown)	Shears (25 p.s.i. 2900°)	2.5–10* (25 p.s.i. 2460°)	2–4† (25 p.s.i. 2640°)	0.5† (25 p.s.i. 2640°)	1–4† (25 p.s.i. 2640°)	0.0–0.5† (25 p.s.i. 2640°)	Shears (28 p.s.i. 2740°)	Shears (28 p.s.i. 2955°)	Shears (28 p.s.i. 2765°)	Shears (28 p.s.i. 2940°)	0–1 (50 p.s.i. 2730°)	1 (50 p.s.i. 2730°)	0.3 (10 p.s.i. 2200°)
Resistance to spalling, % loss in appropriate A.S.T.M. panel test	Poor	5–20	0–4	No loss	No loss	No loss	Poor	Fair	Poor	Fair	Good	Good	Good
Permanent linear change on reheating[5] (after 5 hr. at temp., °F., shown)	(+)0.5–.8 (2640°)	(±)0–1.5 (2550°)	(±)0–1.5 (2910°)	(−)0.75–1 (2910°)	(−)2–4 (2910°)	(−)0–0.8 (2910°)	(−)0.5–1 (3000°)	(−)0.5–1.0 (3000°)	(−)1–2 (3000°)	(−)0.5–1.5 (3000°)	(+)2[6] (2910°)	(+)0.5 (2910°)	(−)0.2 (2600°)
Porosity (as open pores), %	20–30	15–25	12–15	18	34–38	20–25	20–26	10–12	20–26	10–12	13–28	20–26	75
Weight per brick (std. 9 in. straight), lb.	6.5	7.5	8.5	7.7	7.5	8.5	11	11.3	10	10.7	8–9.3	9–10.6	2.25
Specific heat (60–1200°F.)	0.23	0.23	0.23	0.22	0.23	0.23	0.20	0.21	0.27	0.26	0.20	0.20	0.22
Relative slag resistance:[7]													
Acid steel slag	Good	Fair	Fair	Fair	Good	Good	Poor	Poor	Poor	Poor	Good	Good	Poor
Basic steel slag	Poor	Poor	Poor	Poor	Fair	Fair	Good	Good	Good	Good	Good	Good	Poor
Mill scale	Fair	Poor	Fair	Good	Fair	Fair	Good	Good	Good	Good	Fair	Fair	Good
Coal ash slag	Poor	Fair	Fair	Fair	Fair	Fair	Fair	Fair	Good	Good	Good	Good	Poor

1 Made by hydraulic pressing.

2 Includes CaO + MgO + alkalies.

3 Pyrometric cone equivalent; terms "fusion," "softening," "deformation," and melting points heretofore loosely used.

4 Data marked (*) are from A.S.T.M. test C 16–36 with high heat duty time-temperature schedule; those marked (†) are from same test with super duty time-temperature schedule; others determined by other commonly used tests.

5 (+) means expansion; (−) means shrinkage.

6 Oxidizing atmosphere.

7 Ratings affected somewhat by varying temperature and type of atmosphere prevailing. Resistance to coal ash slag affected by furnace temperature as well as analysis and fusion point of slag.

Table 9. Structural Carbon and Graphite

Physical Characteristics of Carbon and Graphite Products

Material and form	Apparent sp. gr.	Weight, lb./cu. ft.	Strength, lb./sq. in. Tensile	Strength, lb./sq. in. Compressive	Strength, lb./sq. in. Transverse	Elastic, modulus, lb./sq. in. (multiply by 10^5)	Specific resistance, ohms./in.^{3}	K (thermal expansion) (see Note)	Thermal conductivity, B.t.u./hr./sq. ft./°F./ft.
Carbon cylinders:									
8 in. dia.	1.54	96.0	660	2,920	1,320	5.5	0.0013	13	6.0
10–14 in. dia. inc.	1.525	95.0	470	2,120	950	5.4	.0013	12	6.0
17–24 in. dia. inc.	1.54	96.0	400	2,200	790	5.4	.0014	13	6.0
30–40 in. dia. inc.	1.54	96.0	400	1,910	810	4.3	.0026	12	6.0
Carbon blocks:									
4 × 4 in. to 6 × 6 in. inc.	1.57	97.8	840	4,100	1,670	9.4	.0018	14	4.0
6 × 6 in. to 20 × 20 in. inc.	1.55	96.7	500	2,140	990	7.1	.0016	15	4.0
15 × 30, 24 × 30 and 24 in. sq.	1.54	96.0	400	1,910	810	4.3	.0026	12	4.0
Carbon tubes:									
½–4 in. i. d. inc.	1.51	94.2	885	10,200	2,700	21.0	.0014	15	3.0
5–10 in. i. d. inc.	1.49	93.0	980	8,140	2,550	17.0	.0016	21	3.0
Carbon brick:									
Dependent on application	1.56	96.7–97.8	970–1,530	5,340–8,320	1,950–3,070	8.9–10.3	0.0015–0.0016	13–14	3.0
Graphite cylinders:									
To 5⅛ in. dia. inc.	1.56	97.3	760	3,050	1,750	8.8	.00036	5–12	84.0
6–12 in. dia. inc.	1.55	96.7	610	3,420	1,810	8.0	.00037	6–12	79.0
14 in. dia.	1.53	95.3	580	3,180	1,490	6.7	.00039	8–12	70.0
16 and 18 in. dia.	1.53	95.3	500	3,180	1,490	6.7	.00040	8–12	70.0
20 in. dia.	1.53	95.3	440	3,180	1,490	6.7	.00040	8–12	70.0
Graphite squares and slabs:									
To 5 in. thick inc.	1.56	97.3	700	3,050	1,750	8.8	.00036	5–12	94.0
6 in. thick to 144 sq. in.	1.55	96.7	700	3,420	1,810	8.0	.00037	6–12	84.0
Over 144 sq. in. section	1.53	95.3	570	3,180	1,490	6.7	.00039	8–12	79.0
Graphite tubes:									
½–4 in. i. d. inc.	1.68	104.7	780	4,550	2,820	14.0	.0003	12	94.0
5–10 in. i. d. inc.	1.67	104.0	870	5,100	2,980	13.0	.0003	12	84.0
Graphite brick, standard sizes	1.56	97.3	700	3,050	1,750	8.8	.00036	5–12	84.0
Karbate No. 1 (impervious carbon):									
Tubes ½–2 in. i. d. inc.	1.77	110.0	1,700	10,500	4,170	29.0	.00164	27	3.0
Over 2 in. i. d.	1.76	110.0	2,000	10,500	4,640	26.0	.0016	33	2.8
Karbate No. 2 (impervious graphite):									
Tubes ½–2 in. i. d. inc.	1.86	116.0	2,600	8,900	4,650	23.0	.00034	23	85.0
Over 2 in. i. d.	1.91	119.0	2,350	10,500	4,980	21.0	.00033	24	75.0
Carbocell (porous carbon):*									
Grade C (finest)	1.34	84.0	500	1,530	2,700	>1.2	.0020	6	3.0
Grade 60	1.05	69.0	190	600	850	>1.2	.0070	27	1.5
Grade 50	1.05	69.0	180	500	830	>1.2	.0070	27	1.4
Grade 40	1.04	69.0	120	320	900	>1.2	.0057	27	1.0
Grade 30	1.04	69.0	100	250	770	>1.2	.0070	27	1.0
Grade 20	1.03	68.0	90	240	700	>1.2	.0070	27	1.0
Grade 10	1.03	68.0	80	160	300	>1.2	.0080	27	1.0
Graphicell (porous graphite):*									
Grade C (finest)	1.35	84.0	600	1,080	1,680	....	.00045	6	60.0
Grade 60	1.05	69.0	110	250	500	....	.0012	21	50.0
Grade 50	1.05	69.0	110	250	500	....	.0012	21	45.0
Grade 40	1.04	69.0	100	190	500	....	.0013	21	45.0
Grade 30	1.04	69.0	80	200	520	....	.0017	21	40.0
Grade 20	1.03	68.0	60	140	310	....	.0020	21	30.0
Grade 10	1.03	68.0	50	140	270	....	.0020	22	20.0

NOTE: Coefficient of thermal expansion per degree: To temperature t°F. = [K + 0.0039t (°F.)] 10^{-7}; to temperature t°C. = [1.8 K + 0.007t (°C.)] 10^{-7}. Carbon graphite products are resistant to most acids and alkalies.

Physical Properties of Porous Carbon and Porous Graphite

Material and form	Porosity, %	Average pore diameter, Inches	Average pore diameter, Microns	Filter action, minimum diameter of particle retained, in.	Average water* permeability at 5 lb./sq. in. pressure, gal./sq. ft./min.	Average air† permeability at 2 in. H_2O pressure, cu. ft./sq. ft./min.	Material and form	Porosity, %	Average pore diameter, Inches	Average pore diameter, Microns	Filter action, minimum diameter of particle retained, in.	Average water* permeability at 5 lb./sq. in. pressure, gal./sq. ft./min.	Average air† permeability at 2 in. H_2O pressure, cu. ft./sq. ft./min.
Carbocell							Graphicell						
Grade C	36	0.0002	5	......	0.30		Grade C	36	0.0002	5	......	0.30	
Grade 60	48	.0013	33	0.00047	14.0		Grade 60	48	.0013	33	0.00047	14.0	
Grade 50	48	.0019	48	.00079	30.0		Grade 50	48	.0019	48	.00079	30.0	
Grade 40	48	.0027	69	.00098	45.0	4.0	Grade 40	48	.0027	69	.00098	45.0	4.0
Grade 30	48	.0039	99	.00173	80.0	8.5	Grade 30	48	.0039	99	.00173	80.0	8.5
Grade 20	48	.0055	140	.00300	120.0	17.0	Grade 20	48	.0055	140	.00300	120.0	17.0
Grade 10	48	.0075	190	.00590	175.0	33.0	Grade 10	48	.0075	190	.00590	175.0	33.0

Note: Carbocell can be treated so as to be wettable for use in caustic filtration. Both are resistant to most acids and alkalies.

* Water at 70°F., 1-in.-thick plate.

† Air at 70°F. and 760 mm. Hg. pressure, 15 per cent relative humidity, 1-in.-thick plate.

Table 10. Physical Properties of Synthetic and Natural Rubbers

Material	Specific gravity of base material	Tensile strength, lb./sq. in.	Hardness, Shore durometer	Maximum temp. for use, °F.*	Dielectric strength, volts/mm.	Effect of heat	Abrasion resistance	Effect of sunlight†	Effect of aging	Machining qualities
Chemigum, oil-resistant	1.0 –1.5	800–4,000	30–90	300	...	Stiffens	Excellent	Equal to rubber	Stiffens	Can be ground
Chemigum, tire	1.0 –1.15	1,000–4,000	50–65	450	...	Stiffens	Good	None	Better than rubber	
GR-I (Butyl)	0.91	500–3,000	15–90	250–300	25,000	Stiffens slightly	Excellent	None	Highly resistant	Can be ground
GR-M (Neoprene)	1.25–1.30	1,000–4,500	10–95	300	...	Stiffens slightly	Excellent	None	Highly resistant	Can be ground
GR-N (Perbunan)	0.96	500–5,000	30–90	300	...	Stiffens	Excellent	Slight	Highly resistant	Can be ground
GR-P (Thiokol FA)	1.34	1,400	25–90	200	...	Hardens slightly	Fairly good	None	None	Excellent
GR-P (Thiokol ST)	1.27	500–2,000	30–90	250–300	...	Hardens slightly	Good	None	None	Excellent
GR-S (Buna S), hard	0.94	4,000–11,000	70–95[1]	220	...	...	...	...	Highly resistant	Excellent
GR-S (Buna S), soft	0.94	500–3,000	25–95	300	...	Stiffens	Excellent	Deteriorates	Highly resistant	Can be ground
Hycar OR-15, soft	1.00	500–4,000	20–95	300	...	Stiffens	Excellent	Slightly better than natural rubber	Highly resistant	Can be ground
Hycar OR-25, soft	0.99	500–3,000	20–95	300	...	Stiffens	Excellent	Slightly better than natural rubber	Highly resistant	Can be ground
Hycar OR-15, hard	1.00	4,000–11,000	70–95‡	275	...	...	...	...	Highly resistant	Excellent
Hycar OS-10, soft	0.98	500–3,500	20–95	300	...	Stiffens	Excellent	Deteriorates	Highly resistant	Can be ground
Koroseal, soft	1.40	500–2,500	30–80	190	15,000–30,000	Softens	Good	None	Highly resistant	Can be ground
Koroseal, hard	1.40	2,000–9,000	80–100	212	30,000–50,000	Softens	Excellent	None	Highly resistant	Good
Pliolite, No. 40	1.06	4,000–5,000	...	160–248	...	Softens	...	None	None	
Resistoflex	1.26	2,000–5,000	55–95	250	6,000–10,000	Softens	Good	None	None	
Tygon T	1.33–1.36	9,000	...	175	35,000–50,000	Softens	Good	Nonc	...	Excellent
Vistanex, medium	0.9	200	...	...	...	...	...	None	Better than rubber	Cannot be machined
Vistanex, high	0.9	550	...	...	...	...	...	None	Better than rubber	Cannot be machined
Natural rubber, hard	0.93	4,000–11,000	70–95‡	220	...	...	...	...	Highly resistant	Excellent
Natural rubber, soft	0.93	500–5,000	20–95	300	...	Softens	Excellent	Deteriorates	Moderately resistant	Can be ground

* Maximum temperature suitable for service depends greatly upon the exact service conditions. Maximum temperature for use as a packing can be much higher than the maximum temperature suitable for tank lining. Individual cases should be referred to the supplier for recommendations.

† Effect of exposure to sunlight under tension.

‡ Type D.

Table 11. Vulcanized Fiber*

Marketed forms	Sheets, rods, and tubes
Thickness, $\frac{1}{1000}$ in.	4 up
Forming properties	Good
Machining qualities	Good
Colors	Red, brown, white, gray, black, olive
Effect of heat	Stable to charring
Effect of water	Swells and softens. Warps on drying. Not recommended
Effect of mineral, animal, and vegetable oils	Very slight absorption
Specific gravity	1.2–1.5
Specific volume, cu. in./lb	23–18.5
Tensile strength, lb./sq. in.	8,000–16,000
Breakdown voltage, 60 cycle ($\frac{1}{16}$-in. sample), volts/mil	175–500

* From *Chem. & Met. Eng.*, December, 1932.

Table 12. Wood for Chemical Equipment
Properties of Woods*

Wood (common name)	Specific gravity, oven-dry, 12% moisture	Weight/cu. ft., oven-dry, 12% moisture	Hardness, load to embed a 0.444-in.-diam. ball to ½ its diam., lb.		Static bending			Impact bending, height of drop of 50-lb. hammer to cause complete failure, in.¶	Compression			Ultimate tensile strength perpendicular to grain, lb./sq. in.	Maximum shearing strength parallel to grain, lb./sq. in.	Cleavage, load to cause splitting, lb./in. of width	Workability‖	Paint holding‖	Nailing‖	Resistance to decay‖
			End	Side	Proportional limit, lb./sq. in.	Modulus of rupture,§ lb./sq. in.	Modulus of elasticity 10^{-6} lb./sq. in.		Proportional limit parallel to grain, lb./sq. in.	Ultimate strength parallel to grain, lb./sq. in.	Proportional limit perpendicular to grain, lb./sq. in.							
Softwoods:																		
Bald cypress	0.46	32	660	510	7,200	10,600	1.44	24	4,470	6,360	900	270	1,000	170	Fair	Very good	Splits rarely	Very good
Douglas fir	.48	34	760	670	8,100	11,700	1.92	30	6,450	7,420	910	300	1,140	180	Poor	Poor	Splits easily	Fair to good
Hemlock, western	.42	29	940	580	6,800	10,100	1.49	26	5,340	6,210	680	310	1,170	200	Fair	Fair	....	Poor
Larch, western	.52	36	1,110	760	7,900	11,900	1.71	32	5,950	7,490	1,080	310	1,360	160	Poor	Poor	Splits; use blunt nails	Fair
Pine, ponderosa	.4	28	550	450	6,300	9,200	1.26	17	4,060	5,270	740	400	1,160	220	Good	Fair	Does not split readily	Fair
Redwood, virgin	.4	28	790	480	6,900	10,000	1.34	19	4,560	6,150	860	240	940	150	Fair	Very good	....	Very good
Redwood, second growth	.34	24	710	400	5,500	8,300	1.12	16	3,750	5,240	640	280	930	160	Fair	Very good	....	Very good
Spruce, Sitka	.4	28	760	510	6,700	10,200	1.57	25	4,780	5,610	710	370	1,150	210	Good	Fair	Little tendency to split	Fair
White-cedar, Port Orford	.42	29	730	560	7,700	11,300	1.73	28	5,890	6,470	760	400	1,080	220	Fair	Excellent	Splits rarely	Very good
Hardwoods:																		
Ash, commercial white†	.6	42	1,720	1,320	8,900	15,400	1.77	43	5,790	7,410	1,410	940	1,950	480	Poor	Good	Tendency to split	Poor
Aspen‡	.38	26	510	350	5,600	8,400	1.18	21	3,040	4,250	460	260	850	210	Good	Fair	Does not split	Very poor
Balsa‡	.15	11	.....	.....	.....	2,500	....	..	.....	.....	.....	.....	.....	...	....	....	....	Poor
Basswood,‡ American	.37	26	520	410	5,900	8,700	1.46	16	3,800	4,730	450	350	990	230	Good	Very good	Does not split	Very poor
Beech, American	.64	45	1,590	1,300	8,700	14,900	1.72	41	4,880	7,300	1,250	1,010	2,010	490	Poor	Good	Tendency to split	Poor
Birch, yellow, sweet	.62	43	1,480	1,260	10,100	16,600	2.01	55	6,130	8,170	1,190	920	1,880	520	Poor	Good	Tendency to split	Poor
Cherry, black	.5	35	1,470	950	9,000	12,300	1.49	29	5,960	7,110	850	560	1,700	350	Fair	Good	Tendency to split	Good
Chestnut, American	.43	30	720	540	6,100	8,600	1.23	19	3,780	5,320	760	460	1,080	250	Good	Good	Splits frequently	Very good
Cottonwood,‡ eastern	.4	28	580	430	5,700	8,500	1.37	20	3,490	4,910	470	580	930	270	Good	Fair	Does not split	Very poor
Elm, American	.5	35	1,110	830	7,600	11,800	1.34	39	4,030	5,520	850	600	1,510	...	Fair	Good	Difficult to split	Fair
Elm, rock	.63	44	1,510	1,320	8,000	14,800	1.54	56	4,700	7,050	1,520	.....	1,920	...	Fair	Good	Difficult to split	Fair
Hickory, shagbark	.72	50	.....	.....	10,700	20,200	2.16	67	.....	9,210	2,170	.....	2,430	...	Very poor	Good	Splits readily	Poor
Khaya (African mahogany)	.45	32	1,080	790	7,890	10,700	1.48	22	.....	5,680	980	.....	1,340	...	Good	Good	Tendency to split	Good
Magnolia, southern	.5	35	1,280	1,020	6,800	11,200	1.4	29	3,420	5,460	1,060	740	1,530	430	Fair	Good	Tendency to split	Poor
Mahogany	.7	49	1,350	1,330	7,100	9,600	1.2	..	.....	6,240	1,770	.....	1,480	...	Good	Good	Tendency to split	Good
Maple, sugar (hard)	.63	44	1,840	1,450	9,500	15,800	1.83	39	5,390	7,830	1,810	.....	2,330	...	Fair to poor	Good to fair	Tendency to split	Poor
Maple, silver (soft)	.47	33	1,140	700	6,200	8,900	1.14	25	4,360	5,220	910	500	1,480	340	Fair to poor	Good to fair	Tendency to split	Poor
Oak, commercial red	.63	44	1,580	1,290	8,500	14,300	1.82	43	4,580	6,760	1,250	800	1,780	410	Poor	Good	Tendency to split	Good
Oak, commercial white	.68	48	1,520	1,360	8,200	15,200	1.78	37	4,760	7,440	1,320	800	2,000	450	Poor	Good	Tendency to split	Good
Pecan	.66	46	1,930	1,820	9,100	13,700	1.73	44	5,180	7,850	2,130	.....	2,080	660				
Sycamore	.49	34	920	770	6,400	10,000	1.42	26	3,710	5,380	860	720	1,470	400	Poor	Good	Fair	Poor
Sweetgum	.49	34	950	690	8,100	11,900	1.49	32	4,700	5,800	860	800	1,610	380	Poor	Good	Tendency to split	Fair
Tupelo, water (tupelo gum)	.5	35	1,200	880	7,200	9,600	1.26	23	4,280	5,920	1,070	700	1,590	360	Poor	Good	Tendency to split	Fair
Tupelo, black (black gum)	.5	35	1,240	810	7,300	9,600	1.2	22	3,470	5,520	1,150	500	1,340	340	Poor	Good	Tendency to split	Fair
Walnut, black	.55	38	1,050	1,010	10,500	14,600	1.68	34	5,780	7,580	1,250	690	1,370	320	Fair	Stains well	....	Very good
Yellow poplar‡	.4	28	560	450	6,100	9,200	1.5	20	3,550	5,290	580	520	1,100	280	Good	....	Does not split readily	Poor

* Markwardt and Wilson, Strength and Related Properties of Woods Grown in the United States, *Tech. Bull.* 479, U. S. Department of Agriculture, September, 1935.
† Includes white, blue, and green.
‡ Soft species of hardwoods.
§ Assumed to be equal to maximum tensile strength parallel to the grain, which is developed seldom in actual use; although tensile strength is actually higher than modulus of rupture for straight-grained material, it is reduced more seriously by steeper slopes of grain.
‖ These properties of plywoods will be different from those of normal wood, depending upon the adhesive used and the type of construction.
¶ Standard test specimen, 2 by 2 by 30 in. (A.S.T.M. D143-27).

Table 12. Wood for Chemical Equipment—(*Concluded*)

Condition of Woods after 31 Days Immersion in Cold Solutions*

Examined after 7 days drying

	Fir	Oak	Oregon pine	Yellow pine	Spruce	Redwood	Maple	Cypress
Hydrochloric acid, 5%	NAC	NAC	NAC	SS	SS	SS	NAC	NAC
Hydrochloric acid, 10%	NAC	NAC	NAC	SS	SS	SS	NAC	NAC
Hydrochloric acid, 50%	SS,SB,SWF	SS,WF	S,WF	S,WF	S,WF	S,WF	S,WF	S,WF
Sulfuric acid, 1%	NAC	NAC	NAC	SS	SS	NAC	NAC	SS,SB
Sulfuric acid, 5%	SS	SS	SS	SS	SS,SB	SS,SB	NAC	SS,SB
Sulfuric acid, 10%	S,FSD	S,FSD	S,FSD	S,FSD	S,FSD	S,FSD	S,FSD	S,FSD
Sulfuric acid, 25%	SSp,FSD	SSp,FSD	SSp,FSD	SSp,FSD	SSp,FSD	SSp,FSD	SSp,FSD	SSp,FSD
Caustic soda, 5%	S,NAC	MSh,SWp	SS	SS,FSD	SSp,FSD	SSp,FSD	MSh	SSp,FSD
Caustic soda, 10%	S,FSD	MSh,WF,Horny	SS	SS,SB,FSD	SS,SB,FSD	SS,SB,FSD	MSh	S,SB,FSD
Alum, 13%	NAC	NAC	NAC	NAC	NAC	NAC	NAC	NAC
Sodium carbonate, 10%	SB,GC	NAC	GC	SB,GC	SB,GC	SB,GC	GC	SB,GC
Calcium chloride, 25%	NAC	NAC	NAC	NAC	NAC	NAC	NAC	NAC
Common salt, 25%	NAC	NAC	NAC	SS,GC	SS,GC	SS,GC	NAC	NAC
Water	NAC	NAC	NAC	NAC	NAC	NAC	NAC	NAC
Sodium sulfide	SS,SB	MSh,WF	SB	SB	SB	SB	MSh,FSD	FSD

Condition of Woods after 8 Hr. Boiling in Solutions*

Examined after 7 days drying

	Fir	Oak	Oregon pine	Yellow pine	Spruce	Redwood	Maple	Cypress
Hydrochloric acid, 10%	SB,S	FSD	FSD	FSD	FSD	FSD	FSD	FSD
Hydrochloric acid, 50%	FD,Ch,B,S,NG	FD,Ch,B,S,NG	FD,Ch,B,S,NG	FD,Ch,B,S,NG	FD,Ch,B,S,NG	FD,Ch,B,S,NG	FD,Ch,B,S,NG	FD,Ch,B,S,NG
Sulfuric acid, 4%	SB,GC	SB,GC	SB,GC	SB,GC	SB,GC	SB,GC	SB,GC	SB,GC
Sulfuric acid, 5%	SS,GC	SB,GC	SB,GC	SB,GC	SB,FSD	SB,GC	SB,GC	SB,FSD
Sulfuric acid, 10%	SS,GC	BFD,Wpd,NG	Sp,FD,NG	B,Sp,FD,NG	B,Sp,FD,NG	SB,FSD	SB,FSD	B,FD
Caustic soda, 5%	SS	MSh	S	GC	S,GC	S,GC	Sh	SSp
Alum, 13%	SB,GC	NAC	NAC	SB,GC	SB,GC	SB,GC	NAC	SB,GC
Sodium carbonate, 10%	SB,GC	GC	GC	GC	GC	GC	GC	SB,GC
Calcium chloride, 25%	SB,GC	SB,SS,GC	NAC	SB,GC	SB,GC	NAC	NAC	SB,GC
Common salt, 25%	NAC	NAC	NAC	SB,GC	NAC	SB,GC	NAC	NAC
Water	NAC	NAC.	NAC	SB,GC	NAC	NAC	NAC	NAC

* The two tables describing the condition of eight varieties of woods used for tanks and other chemical-resistant uses are based on a report of James K. Stewart, consulting chemist, to the Mountain Copper Co., Martinez, Calif. Tests were conducted on samples 1 by 4 by ¼ in. in size, seasoned and chosen so as to be as nearly as possible in the same physical condition as the woods would be when used for equipment construction. Results of the tests are described by terms explained in the following key:

Abbreviation key:

B —Brittle
Ch —Charred
FD —Fiber disintegrated
FSD—Fiber slightly disintegrated
GC —Good condition
MSh—Much shrunk
NAC—No apparent change
NG —No good
S —Softer
SB —Slightly brittle
Sh —Shrunk
Sp —Spongy
SS —Slightly softer
SSp —Slightly spongy
SWF—Slightly weakened fiber
SWp—Slightly warped
WF —Weakened fiber
Wpd—Warped

Representative Makers of Wood Tanks and Pipe for Chemical Applications

Acme Tank Co., New York
Alert Pipe & Supply Co., Bay City, Mich.
Atlantic Tank Corp., North Bergen, N.J.
Axtell Co., Fort Worth, Tex.
Baltimore Cooperage Tank & Tower Co., Baltimore
Black, Sivalls & Bryson, Inc., Oklahoma City, Okla.
C. F. Braun & Co., Alhambra, Calif.
W. E. Caldwell Co., Louisville, Ky.
Caspar Lumber Co., San Francisco
Challenge Co., Batavia, Ill.
A. J. Corcoran, Inc., Jersey City, N.J.
Cypress Tank Co., Shreveport, La.
Dempster Mill Mfg. Co., Beatrice, Neb.
Drane Tank Co., Fort Worth, Tex.
Drummond Mfg. Co., Louisville, Ky.
Dunck Tank Works, Inc., Milwaukee
G. Elias & Bro., Buffalo
Engle Tank Co., Chicago
Federal Pipe & Tank Co., Seattle, Wash.
Fibre Conduit Co., Orangeburg, N.Y.
Fleming Tank Co., Pittsburgh
Fluor Corp., Ltd., Los Angeles
Foster-Wheeler Corp., New York
General Tank Corp., Kearny, N.J.
Amos H. Hall & Sons, Philadelphia
Hammond & Little River Redwood Co., Samoa, Calif.
Harry Cooling & Equipment Co., Doylestown, Pa.
Hauser-Stander Tank Co., Cincinnati
Henderson Bros. Co., Waterbury, Conn.
R. R. Howell & Co., Minneapolis
James Hunter Machine Co., North Adams, Mass.
Johnson & Carlson, Chicago
Kalamazoo Tank & Silo Co., Kalamazoo, Mich.
Lille-Hoffman Cooling Towers, Inc., St. Louis
Lincoln Tank Co., Shreveport, La.
Marley Co., Kansas City, Kans.
Michigan Pipe Co., Bay City, Mich. (Pipe)
National Tank Co., Tulsa, Okla.
National Tank & Pipe Co., Portland, Ore.
New England Tank & Tower Co., Everett, Mass.
Pacific Cooperage Co., Portland, Ore.
Pacific Tank & Pipe Co., Oakland, Calif.
Pacific Wood Tank Corp., San Francisco
Parkersburg Rig & Reel Co., Parkersburg, W. Va
Fred C. Pfeil, Inc., Buffalo, N.Y.
J. F. Pritchard & Co., Kansas City, Mo.
Redwood Mfrs. Co., San Francisco
Wm. B. Scaife & Sons Co., Oakmont, Pa.
Schubert-Christy Corp., St. Louis
A. T. Stearns Lumber Co., Boston
Treadwell Construction Co., Midland, Pa.
Union Lumber Co., San Francisco
U.S. Wind Engine & Pump Co., Batavia, Ill.
Wendnagel Co., Chicago
C. H. Wheeler Mfg. Co., North Phila., Pa.
G. Woolford Wood Tank Co., Darby, Pa.
A. Wyckoff & Son Co., Elmira, N.Y.

Table 13. Chemical Resistance of Gasket Materials

Resistance ratings[1] A = Good F = Fair C = Depends on conditions X = Unsuitable

CAUTION: Do not use table without reading footnotes and text.

Chemical	Metals[2]: Lead	Copper	Aluminum	Monel	Nickel	Iron and steel	Stainless 304	Stainless 316	Stainless 347	Others with good resistance[3]	Asbestos: White (comp. or woven)[4]	Blue (comp. or woven)[4]	Compressed sheet	Comp., rubber bonded: White (Buna-S)[5]	White (Neoprene)[5]	Blue (butyl)[5]	Woven, rubber frictioned: Blue (Neoprene)[5]	White (Neoprene)[5]	Blue (butyl)[5]	Rubber: GR-S	Gr-S[5]	GR-S	Neoprene	Neoprene[5]	Neoprene	Buna-N	Buna-N[5]	Buna-N	Butyl	Butyl	Thiokol	Natural	Silicone[6]	Miscellaneous: Glass fabric and silicone elastomer[7]	Glass fabric and synthetic rubber[7]	Cork composition	Plant-fiber sheet	Plant-fiber sheet	Teflon[8]	Teflon[5]
Source of data[10]	G	G	G	G	G	G	G	G	G	G	J	J	U	P	P	P	P	P	P	A	P	U	A	P	U	A	P	U	S	U	U	U	D	C	C	A	A	U	E	P
Acetic acid, crude	X	F	F	F	F	X	..	F	..	Ag	C	A	C	A	A	A	A	A	A	C	A	C	C	A	F	C	A	F	A	F	F	C	..	A	A	C	X	F	A	A
Pure	F	F	A	A	F	X	F	F	..	Ag	C	A	C	A	A	A	A	A	A	C	A	C	C	A	F	C	A	F	A	F	F	C	A	A	A	C	X	F	A	A
Vapors	X	F	C	F	F	X	F	F	..	Ag	C	A	C	A	A	A	A	A	A	A	A	C	A	A	F	A	A	F	A	F	F	C	..	A	A	C	X	F	A	A
150 lb./sq. in., 400°F	X	F	C	F	F	X	..	F	..	Ag	C	A	F	..	..	..	..	..	..	X	..	X	X	..	F	X	..	F	..	X	X	X	..	A	X	X	X	X	A	..
Acetic anhydride	X	X	A	A	..	C	F	F	F	Ag	C	A	..	A	A	A	A	A	A	C	A	..	A	A	..	C	A	..	A	..	..	..	..	C	C	X	X	..	A	A
Acetone	A	A	A	A	A	A	A	A	A	...	A	..	A	X	X	X	X	X	X	A	X	A	A	X	C	X	X	X	A	A	C	F	F	A	A	A	A	C	A	A
Acetylene	A	C	A	A	..	A	A	A	..	...	A	..	A	..	..	..	..	..	..	A	..	A	A	..	C	A	..	A	..	A	C	A	..	X	X	A	A	C	A	..
Air	A	A	A	A	..	A	A	..	A	...	A	..	A	A	A	A	A	A	A	A	A	A	A	A	A	A	A	A	A	A	A	A	A	A	A	A	A	A	A	A
Aluminum chloride	X	F	X	A	..	F	X	X	F	...	A	..	A	A	A	A	A	A	A	A	A	A	A	A	A	A	A	A	A	A	..	A	..	F	A	F	C	..	A	A
Aluminum fluoride	..	..	X	..	..	..	C	C	C	...	A	..	..	..	..	..	..	..	..	A	..	..	A	..	..	A	..	..	A	..	..	..	..	F	A	F	C	..	A	..
Aluminum sulfate	A	F	C	F	C	X	F	F	F	Ag	A	..	..	A	A	A	A	A	A	A	A	A	A	A	A	A	A	A	A	A	..	A	..	A	A	F	F	..	A	A
Alums	A	F	C	F	C	X	F	F	F	Ag	A	..	A	A	A	A	C	C	C	A	C	A	A	C	A	A	C	A	A	A	..	A	..	A	A	F	F	..	A	A
Ammonia gas, cold	A	C	A	A	..	A	A	A	..	...	A	..	A	A	A	A	A	A	A	A	F	A	A	F	A	A	F	A	..	A	A	A	..	A	A	X	X	A	A	A
Hot	X	X	..	C	..	C	C	..	..	...	A	..	A	..	..	..	..	..	..	C	..	C	C	..	A	C	..	C	..	C	X	C	..	A	A	X	X	C	A	A
Ammonium chloride	A	X	X	F	F	C	F	F	..	...	C	A	A	A	A	A	A	A	A	A	A	..	A	A	A	C	A	..	A	A	C	A	..	F	A	F	F	X	A	A
Ammonium hydroxide	A	X	F	C	..	A	A	A	A	...	A	..	A	A	A	A	A	A	A	F	F	C	F	F	C	F	F	C	A	A	X	C	A	A	A	X	X	F	A	A
Ammonium nitrate	X	X	F	C	..	A	A	A	A	...	C	A	..	F	F	F	F	F	F	A	C	A	A	C	A	A	C	A	A	A	..	A	..	..	A	F	F	..	A	A
Ammonium phosphate, monobasic	A	F	X	C	..	X	A	A	..	...	C	A	..	A	A	A	A	A	A	A	A	A	A	A	A	A	A	A	A	A	..	A	..	..	A	F	F	..	A	A
Dibasic	A	F	F	A	..	F	A	A	..	...	A	..	..	A	A	A	A	A	A	A	A	A	A	A	A	A	A	A	A	A	..	A	..	..	A	F	F	..	A	A
Tribasic	A	F	F	A	A	A	A	A	..	...	A	..	..	A	A	A	A	A	A	A	A	A	A	A	A	A	A	A	A	A	..	A	..	..	A	F	F	..	A	A
Ammonium sulfate	A	F	..	A	..	A	A	A	A	...	C	A	A	A	A	A	A	A	A	A	A	A	A	A	A	A	A	A	A	A	..	A	A	..	A	F	F	..	A	A
Amyl acetate	..	F	F	A	..	C	A	A	A	...	A	..	A	C	C	C	C	C	C	X	X	X	X	X	X	X	X	X	..	..	C	F	..	X	C	A	A	C	A	A
Amyl alcohol	..	A	..	A	..	..	..	..	..	...	A	..	A	C	C	C	C	C	C	A	X	A	A	X	A	A	X	A	A	A	A	A	..	X	A	A	F	A	A	A
Aniline, aniline oil	..	X	X	A	..	A	A	A	A	...	A	..	A	..	..	..	..	..	..	A	..	X	X	..	F	F	..	X	A	A	A	X	..	X	C	X	A	..	A	A
Aniline dyes	..	..	..	A	..	..	A	..	..	...	A	..	A	..	..	..	..	..	..	..	..	F	..	..	F	..	..	F	..	A	C	F	..	X	C	C	C	..	A	A
Asphalt	..	A	..	A	..	A	A	..	..	...	A	..	F	..	..	..	..	..	..	X	..	X	A	..	F	A	..	F	X	X	A	X	A	C	C	A	A	F	A	A
Barium chloride	..	..	X	..	A	..	F	A	..	...	C	A	A	A	A	A	A	A	A	A	A	A	A	A	A	A	A	A	A	A	..	A	A	..	A	F	F	..	A	A
Barium hydroxide	X	X	X	..	A	..	A	..	..	...	A	..	A	A	A	A	A	A	A	A	A	A	A	A	A	A	A	A	A	A	A	A	A	A	A	F	F	F	A	A
Barium sulfide	A	X	..	A	..	..	A	A	A	...	A	..	..	..	..	..	..	..	..	A	..	A	A	..	A	A	..	A	A	A	A	A	..	..	A	F	F	..	A	A
Beer	..	A	A	A	..	A	A	A	..	...	A	..	A	A	A	A	X	X	X	A	A	A	A	A	A	A	..	X	A	A	X	A	..	A	A	A	A	A	A	A
Beet sugar liquors	..	A	A	A	..	A	A	A	..	...	A	..	A	A	A	A	A	A	A	A	A	A	A	A	A	A	A	A	..	A	X	A	..	A	A	A	A	A	A	A
Benzene, benzol	A	A	A	A	..	A	A	A	..	...	A	..	C	X	X	X	X	X	X	X	X	X	X	X	X	X	X	X	X	X	C	X	X	X	X	A	A	A	A	A
Benzine, petroleum ether, naphtha	A	A	A	A	..	A	A	A	..	...	A	..	C	X	X	X	X	X	X	X	X	X	F	X	X	A	X	X	X	X	C	X	X	X	C	A	A	A	A	A
Black sulfate liquor	..	F	..	A	..	A	A	..	..	...	A	..	A	..	..	..	..	..	..	A	..	A	A	..	A	A	..	A	..	A	X	A	..	..	..	A	A	X	A	..
Blast furnace gas	X	X	..	C	..	A	C	..	A	...	A	..	C	..	..	..	..	..	..	A	..	C	A	..	C	A	..	C	..	C	X	C	..	F	C	C	C	X	A	..
Borax	..	F	F	A	A	A	A	A	A	...	A	..	A	A	A	A	A	A	A	A	A	A	A	A	C	A	A	F	A	A	X	A	A	A	A	A	A	A	A	A
Boric acid	A	F	A	A	A	X	A	A	A	...	A	..	..	A	A	A	A	A	A	C	A	A	C	A	A	C	A	A	A	A	X	A	A	A	A	A	A	..	A	A
Bromine	X	X	..	..	..	X	X	X	X	Ta, Zr	X	C	X	X	X	X	X	X	X	X	X	X	X	X	X	X	X	X	X	X	F	X	..	X	X	X	X	X	A	A
Butane	..	..	A	A	..	A	..	A	..	...	A	..	A	..	..	..	..	..	..	X	..	X	A	..	F	C	..	C	X	..	C	X	..	X	C	A	A	X	A	..
Butyl acetate	..	A	A	..	..	..	A	A	A	...	A	..	A	C	C	C	X	X	X	X	X	X	X	X	X	X	X	X	F	F	C	F	..	X	A	A	A	C	A	A
Butyl alcohol, butanol	..	A	..	A	..	A	..	..	..	...	A	..	A	X	C	X	C	C	X	A	X	A	A	C	A	A	X	A	A	A	C	A	..	F	A	A	F	X	A	A
Calcium bisulfite	A	X	C	X	..	X	..	A	A	...	C	A	C	A	A	A	A	A	A	A	A	C	A	A	..	A	A	..	A	C	..	C	..	F	A	F	F	..	A	A
Calcium chloride	X	A	C	F	..	A	C	C	..	...	C	A	A	A	A	A	A	A	A	A	A	A	A	A	A	A	A	A	A	A	A	A	A	F	A	F	F	A	A	A
Calcium hydroxide	C	..	..	A	A	A	F	F	..	...	A	..	A	A	A	A	A	A	A	A	A	A	C	A	A	C	A	A	A	A	C	A	A	F	A	F	C	F	A	A
Calcium hypochlorite	X	C	X	C	..	C	C	C	C	Ta, Zr	C	A	A	A	A	A	A	A	A	F	A	C	F	A	C	F	A	C	..	A	..	C	A	F	F	X	X	F	A	A
Caliche liquors	..	..	..	A	..	A	A	..	..	...	C	A	..	..	..	..	..	..	..	..	..	A	..	..	A	..	..	A	..	A	X	A	..	..	A	..	..	..	A	..
Cane sugar liquors	A	A	A	A	..	A	A	A	..	...	A	..	A	A	A	A	A	A	A	A	A	A	A	A	A	A	A	A	..	A	X	A	..	A	A	F	A	F	A	A
Carbolic acid, phenol	A	X	A	A	..	C	A	A	..	...	A	..	A	C	C	C	C	C	C	C	C	X	C	C	C	X	C	X	..	X	..	X	..	F	C	A	A	..	A	A
Carbon dioxide, dry	A	A	A	A	..	A	A	A	..	...	A	..	A	C	C	C	..	..	..	A	..	A	A	..	A	A	..	A	A	A	A	A	A	A	A	X	A	X	A	A
Wet	X	F	F	A	..	F	A	A	..	...	A	..	A	C	C	C	..	..	..	A	..	A	A	..	A	A	..	A	A	A	A	A	A	A	A	X	A	X	A	A
Carbon bisulfide	..	X	A	A	..	A	A	A	..	...	A	..	A	..	..	..	..	..	..	X	..	X	X	..	X	X	..	X	X	X	C	X	..	X	X	A	A	X	A	..

Carbon monoxide, hot	..	X	..	..	..	A	A	A	A	....	A	..	A	..	..	..	..	..	..	A	..	C	A	..	C	A	..	C	A	C	X	C	..	A	C	X	A	C	A	..
Carbon tetrachloride	F	C	C	A	..	C	C	C	C	Ag	A	..	F	C	C	C	X	X	X	X	X	X	X	X	X	A	X	X	X	X	C	X	X	X	X	A	A	C	A	A
Castor oil	..	..	A	A	..	A	A	..	..	....	A	..	A	C	C	C	X	X	X	A	X	A	A	X	A	A	X	A	A	A	X	A	..	C	C	A	A	A	A	A
China wood oil, tung oil	..	..	A	A	..	A	A	..	..	....	A	..	A	C	C	C	X	X	X	X	X	X	A	X	A	A	X	F	..	..	C	X	..	X	C	A	A	C	A	A
Chlorinated solvents, dry	A	A	A	A	..	A	A	..	..	....	A	..	C	C	C	C	X	X	X	X	X	X	X	X	X	X	X	C	X	X	X	X	X	X	X	A	A	X	A	A
Wet	F	X	X	A	..	X	C	..	..	....	A	..	C	C	C	C	X	X	X	X	X	X	X	X	X	X	X	C	X	X	C	X	X	X	X	A	X	X	A	A
Chlorine, dry	A	A	A	A	..	A	A	A	..	....	X	..	F	X	X	C	C	X	C	X	X	F	X	C	F	F	X	F	F	F	C	A	..	X	X	X	X	X	A	A
Wet	F	X	X	X	..	X	X	C	..	Ag	X	A	F	X	X	C	C	X	C	X	X	F	X	C	F	F	X	F	F	F	X	F	..	X	X	X	X	X	A	A
Chloroacetic acid	X	X	X	..	F	X	X	X	X	Ta, Zr	X	A	..	C	C	C	X	X	X	F	X	..	F	X	..	F	X	..	..	..	..	..	..	X	C	X	X	..	A	A
Chlorosulfonic acid	F	..	..	F	..	..	F	..	..	....	X	A	C	C	C	C	X	X	X	C	X	X	C	X	X	C	X	X	..	X	X	X	..	X	F	X	X	X	A	A
Chromic acid	A	X	X	F	..	..	..	A	..	....	X	A	C	C	C	C	X	X	X	X	X	X	X	X	X	X	X	X	C	C	X	X	..	X	F	F	X	C	A	A
Citric acid	A	A	A	A	..	X	A	A	..	....	C	A	C	A	A	A	A	A	A	A	A	F	A	A	A	A	A	F	A	F	X	A	..	A	A	A	A	F	A	A
Coke oven gas	..	A	..	A	..	A	..	A	..	....	A	..	A	..	..	..	..	..	..	C	..	C	C	..	C	C	..	C	C	..	X	C	..	A	C	X	A	X	A	..
Copper chloride	A	C	X	F	..	F	X	X	..	....	C	A	A	A	A	A	A	A	A	A	A	A	A	A	F	A	A	A	A	A	X	A	A	A	A	A	F	..	A	A
Copper sulfate	A	C	X	A	..	X	A	A	..	....	C	A	A	A	A	A	A	A	A	A	A	A	A	A	A	A	A	A	A	F	C	A	A	A	A	A	F	A	A	A
Corn oil	A	..	A	A	..	A	A	A	..	....	A	..	A	C	C	C	X	X	X	X	X	X	F	X	F	A	X	A	A	A	C	X	..	C	C	A	A	C	A	A
Cottonseed oil	A	C	A	A	..	A	A	A	..	....	A	..	A	C	C	C	X	X	X	X	X	X	F	X	F	A	X	A	A	A	C	X	..	C	C	A	A	X	A	A
Creosote, coal tar	..	A	A	A	..	A	A	A	..	....	A	..	A	..	..	..	..	..	..	X	..	X	F	..	F	A	..	A	X	X	C	X	..	C	C	A	A	X	A	A
Wood	..	A	A	A	..	A	A	A	..	....	A	..	C	C	C	C	X	X	X	X	X	X	A	X	F	A	X	A	X	X	C	X	..	C	C	A	A	X	A	A
Creosols, cresylic acid	..	A	A	F	..	F	..	A	..	....	A	..	..	C	C	C	X	X	X	X	X	X	X	X	X	C	X	F	X	X	..	X	..	C	C	A	A	..	A	A
Dowtherm, A	..	X	A	..	..	A	..	..	..	....	A	..	C	..	..	..	X	X	X	X	X	X	X	X	X	X	X	X	..	X	X	X	..	..	..	C	C	X	A	..
E	..	A	X	..	..	A	..	..	..	....	A	..	C	..	..	..	X	X	X	X	X	X	X	X	X	X	X	X	..	X	X	X	..	..	..	C	C	X	A	..
Ethers	A	A	A	A	..	A	..	..	..	....	A	..	A	C	C	C	X	X	X	X	X	C	X	X	C	C	X	C	A	C	X	C	X	C	..	A	A	X	A	A
Ethyl acetate	..	A	F	A	..	A	A	A	..	....	A	..	A	C	C	C	X	X	X	X	X	F	X	X	X	X	X	X	A	F	F	F	..	F	F	A	A	A	A	A
Ethyl cellulose	..	..	..	A	A	..	..	..	..	....	A	..	F	..	..	..	..	..	..	A	..	F	A	..	F	A	..	F	..	F	C	F	..	F	F	A	A	X	A	A
Ethyl chloride	..	A	..	A	A	A	A	A	..	....	A	..	A	..	..	..	..	..	..	X	..	A	X	..	A	X	..	A	A	A	C	A	X	F	F	A	A	C	A	A
Ethylene glycol	A	A	A	A	..	A	A	A	A	....	A	..	A	..	..	..	..	..	..	A	..	A	A	..	A	A	..	A	A	A	C	A	..	A	A	A	A	A	A	A
Ferric chloride	X	X	X	X	X	X	X	X	X	Ta, Zr	C	A	..	C	C	C	C	C	C	A	C	A	A	C	A	A	C	A	A	A	..	A	..	F	A	A	F	..	A	A
Ferric sulfate	A	X	X	X	X	X	F	A	..	....	C	A	..	A	A	A	C	C	C	A	A	A	A	A	A	A	A	A	A	A	..	A	..	F	A	A	F	..	A	A
Formaldehyde	X	F	F	A	..	F	A	A	..	Ag	A	..	C	A	A	A	C	C	C	A	A	A	A	A	C	A	A	A	A	A	A	A	..	C	A	A	A	F	A	A
Formic acid	X	F	X	C	C	X	F	F	..	Ag	X	A	..	A	A	A	C	C	C	A	A	A	A	A	X	F	A	X	..	A	..	A	A	X	A	F	F	..	A	A
Freon	A	A	A	A	..	C	C	C	..	....	A	..	C	..	..	..	..	..	..	C	..	X	A	..	F	A	..	F	..	F	C	X	..	X	X	A	A	C	A	..
Fuel oil	A	A	..	A	..	A	A	..	..	....	A	..	A	..	..	..	..	..	..	X	..	X	F	..	F	A	..	A	X	X	C	X	X	A	A	A	A	X	A	A
Fuel oil, acid	A	C	..	A	..	C	C	..	..	....	A	..	..	..	..	..	..	..	..	X	..	..	F	..	..	A	..	..	X	..	..	..	..	A	A	A	A	..	A	A
Furfural	..	A	A	A	..	A	A	A	..	....	A	..	C	..	..	..	..	..	..	A	..	X	A	..	X	X	..	X	..	..	C	X	..	C	A	A	A	X	A	A
Gasoline, sour	A	C	C	A	..	C	C	A	..	....	A	..	..	..	..	..	..	..	..	X	..	X	F	..	A	A	..	A	X	X	C	X	X	X	A	A	A	..	A	A
Refined	A	A	A	A	..	A	A	A	..	....	A	..	C	..	..	..	..	..	..	X	..	X	F	..	A	A	..	A	X	X	C	X	X	X	A	A	A	X	A	A
Gelatin	..	..	A	A	..	..	A	A	..	....	A	..	A	A	A	A	A	A	A	A	A	A	A	A	A	A	A	A	A	A	X	A	..	A	A	A	A	A	A	A
Glucose	..	A	A	A	..	A	A	A	..	....	A	..	A	A	A	A	A	A	A	A	A	A	A	A	A	A	A	A	A	A	X	A	..	A	A	A	A	A	A	A
Glue	..	..	A	A	..	A	A	A	..	....	A	..	A	A	A	A	A	A	A	A	A	F	A	A	A	A	A	A	A	F	X	F	..	A	A	A	A	C	A	A
Glycerin, glycerol	A	F	A	A	..	A	A	A	..	....	A	..	A	C	C	C	C	C	C	A	X	A	A	X	A	A	X	A	A	A	C	A	..	A	A	A	A	A	A	A
Green sulfate liquor	..	..	..	A	..	A	..	..	..	....	A	..	A	..	..	..	..	..	..	A	..	A	A	..	A	A	..	A	..	A	X	A	..	..	..	F	F	X	A	A
Hydrobromic acid	C	C	X	C	C	X	X	X	..	....	C	A	..	..	..	..	..	..	..	C	..	X	C	..	..	C	..	..	A	..	..	A	..	X	F	X	X	..	A	..
Hydrochloric acid, <150°F	F	X	X	C	C	X	X	X	..	Ag	C	A	F	A	A	A	A	A	A	F	A	F	F	A	F	F	A	A	C	A	X	A	A	X	F	X	X	..	A	A
>150°F	X	X	X	C	C	X	X	X	..	H-B, Ta, Zr	C	A	X	F	F	F	C	C	C	X	C	X	X	C	F	X	C	X	C	..	X	F	..	X	X	X	X	..	A	A
Hydrocyanic acid	..	..	..	A	..	C	A	A	..	....	A	..	..	..	..	..	..	..	..	F	..	..	F	..	..	F	..	..	..	..	..	..	..	X	F	X	X	..	A	..
Hydrofluoric acid, cold, <65%	F	C	X	F	X	X	X	X	X	Cu-Ni	X	A	X	..	..	..	..	..	..	C	..	F	A	..	A	C	..	..	C	F	C	F	..	X	F	X	X	X	A	..
>65%	C	F	X	A	..	F	X	X	X	Ag	X	A	X	..	..	..	..	..	..	X	..	X	X	..	X	X	..	X	C	C	X	X	..	..	..	X	X	X	A	..
Hot, <65%	X	X	X	C	X	X	X	X	X	....	X	A	X	..	..	..	..	..	..	X	..	C	X	..	A	X	..	..	C	C	X	C	..	..	..	X	X	X	A	..
>65%	X	F	X	A	..	C	X	X	X	Ag	X	A	X	..	..	..	..	..	..	X	..	C	X	..	C	X	..	..	C	C	X	C	..	..	..	X	X	X	A	..
Hydrofluosilicic acid	A	..	..	C	..	X	X	X	X	....	X	A	A	..	..	..	..	..	..	F	..	A	F	..	F	F	..	A	..	A	C	A	..	X	..	X	X	X	A	..
Hydrogen gas, cold	A	A	A	A	..	A	A	A	..	....	A	..	A	..	..	..	..	..	..	A	..	F	A	..	A	A	..	A	A	A	C	F	A	A	A	A	A	C	A	..
Hot	X	X	X	A	..	A	A	A	A	....	A	..	A	..	..	..	..	..	..	A	..	F	A	..	A	A	..	A	A	A	C	F	..	A	A	X	A	C	A	..
Hydrogen peroxide	F	X	A	F	F	X	A	A	..	....	A	..	..	C	C	C	X	X	X	A	X	F	A	X	A	A	X	..	..	F	C	F	A	A	A	X	X	..	A	..
Hydrogen sulfide, dry, cold	..	X	A	A	A	A	A	A	A	....	A	..	..	C	C	C	X	X	X	A	C	C	A	C	F	A	C	C	A	A	A	C	..	C	A	X	A	..	A	A
Hot	..	X	A	X	X	X	C	C	A	502	A	..	..	C	C	C	X	X	X	C	C	C	C	C	F	C	C	C	A	A	A	C	..	C	A	X	A	..	A	A
Wet, cold	..	X	A	A	A	C	A	A	A	....	A	..	..	C	C	C	X	X	X	F	C	C	F	C	F	F	C	C	A	A	A	C	..	C	A	X	A	..	A	A
Hot	..	X	A	X	X	X	C	C	A	....	A	..	..	C	C	C	X	X	X	C	X	C	C	X	F	C	X	C	A	A	A	C	..	C	A	X	A	..	A	A
Kerosene	A	A	..	A	..	A	A	A	..	....	A	..	A	C	C	C	X	X	X	X	X	C	F	X	F	A	X	A	X	X	C	C	X	F	A	A	A	C	A	A
Lacquers	..	C	A	A	..	C	A	A	..	....	A	..	C	C	C	C	X	X	X	X	X	X	X	X	X	X	X	X	C	X	C	X	..	X	A	A	A	X	A	A
Lacquer solvents	..	C	A	A	..	C	A	A	..	....	A	..	C	C	C	C	X	X	X	X	X	X	X	X	X	X	X	X	C	X	C	X	..	X	A	A	A	X	A	A
Lactic acid, cold	..	C	C	A	A	X	C	F	F	Sn	A	..	A	A	A	A	A	A	A	A	A	C	A	A	C	A	A	C	A	C	X	C	..	A	A	A	F	C	A	A
Hot	..	C	X	C	C	X	C	C	C	Ag (fair)	A	A	A	A	A	A	A	A	A	C	A	C	C	A	C	C	A	C	A	C	X	C	..	A	A	X	X	C	A	A
Linseed oil	A	A	A	A	..	A	A	A	..	....	A	..	A	C	C	C	C	C	C	X	X	C	F	X	F	A	X	A	A	A	C	C	..	X	C	A	A	C	A	A
Lubricating oils, sour	A	C	C	A	..	C	C	..	A	....	A	..	A	C	C	C	C	C	C	X	X	F	F	X	A	A	X	A	X	X	C	F	..	C	A	A	A	C	A	A
Refined	A	A	A	A	..	A	A	..	A	....	A	..	A	C	C	C	C	C	C	X	X	F	F	X	A	A	X	A	X	X	C	F	A	C	A	A	A	C	A	A
Magnesium chloride	X	F	X	F	F	F	F	F	..	Ag	C	A	A	A	A	A	C	C	C	A	A	A	A	A	A	A	A	A	A	A	C	A	..	A	A	X	F	C	A	A
Magnesium hydroxide	..	X	X	A	A	A	A	A	..	....	A	..	A	A	A	A	C	C	C	F	A	A	F	A	A	F	A	A	A	A	C	A	A	A	A	A	F	C	A	A
Magnesium sulfate	..	A	C	A	..	A	A	A	..	....	C	A	..	A	A	A	C	C	C	A	A	A	A	A	A	A	A	A	A	A	..	A	A	A	A	A	A	..	A	A
Mercuric chloride	..	X	X	X	X	C	X	X	X	....	A	A	..	A	A	A	C	C	C	A	A	..	A	A	..	A	A	..	A	..	..	..	..	A	A	A	A	..	A	A
Mercury	..	X	X	A	..	A	A	A	..	....	A	..	C	..	..	..	..	..	..	A	..	..	A	..	..	A	..	..	A	..	..	..	..	A	A	A	A	..	A	..

Table 13. Chemical Resistance of Gasket Materials—(*Concluded*)

Chemical	Metals[2]										Asbestos									Rubber														Miscellaneous						
	Lead	Copper	Aluminum	Monel	Nickel	Iron and steel	Stainless 304	Stainless 316	Stainless 347	Others with good resistance[3]	White (comp. or woven)[4]	Blue (comp. or woven)[4]	Compressed sheet	Comp., rubber bonded: White (Buna-S)[5]	Comp., rubber bonded: White (Neoprene)[5]	Comp., rubber bonded: Blue (butyl)[5]	Woven, rubber frictioned: Blue (Neoprene)[5]	Woven, rubber frictioned: White (Neoprene)[5]	Woven, rubber frictioned: Blue (butyl)[5]	GR-S	GR-S[5]	Gr-S	Neoprene	Neoprene[5]	Neoprene	Buna-N	Buna-N[5]	Buna-N	Butyl	Butyl	Thiokol	Natural	Silicone[6]	Glass fabric and silicone elastomer[7]	Glass fabric and synthetic rubber[7]	Cork composition	Plant-fiber sheet	Plant-fiber sheet	Teflon[8]	Teflon[5]
Source of data[10]	G	G	G	G	G	G	G	G	G	G	J	J	U	P	P	P	P	P	P	A	P	U	A	P	U	A	P	U	S	U	U	U	D	C	C	A	A	U	E	P
Methyl alcohol, methanol	A	A	A	A	..	A	A	A	..		A	..	A	F	F	F	C	C	C	A	F	A	A	F	A	A	F	A	A	A	C	A	..	F	A	A	F	C	A	A
Methyl chloride	A	A	..	A	..	A	..	..	..		A	..	C	C	C	C	..	..	..	X	X	C	X	X	C	X	X	C	A	C	X	C	..	C	A	C	C	..	A	A
Milk	..	..	A	A	A	A	..	A	..		A	..	..	A	A	A	X	X	X	A	A	A	A	A	A	A	A	A	A	A	X	A	A	A	A	A	A	..	A	A
Mineral oils	A	A	A	A	..	A	A	A	..		A	..	A	C	C	C	X	X	X	X	X	C	F	X	F	A	X	A	X	X	C	C	A	A	A	A	A	C	A	A
Natural gas	A	C	A	A	..	A	A	A	..		A	..	A	..	..	..	..	..	..	F	..	F	A	..	A	A	..	A	X	X	C	F	..	C	A	A	A	..	A	..
Nickel chloride	..	X	X	C	..	..	F	F	..		C	A	..	A	A	A	A	A	A	A	A	A	A	A	A	A	A	A	A	A	..	A	..	A	A	F	F	..	A	A
Nickel sulfate	..	X	X	C	..	..	A	A	..		C	A	..	A	A	A	A	A	A	A	A	A	A	A	A	A	A	A	A	A	..	A	..	A	A	F	F	..	A	A
Nitric acid, crude	X	X	C	X	X	X	C	C	..	Pt	X	A	C	X	X	X	X	X	X	X	X	X	X	X	X	X	X	X	C	C	..	X	..	A	A	X	X	..	A	A
Diluted	X	X	X	X	X	X	A	A	..	Pt	X	A	C	C	C	C	X	X	X	X	X	X	X	X	X	X	X	X	C	A	..	X	A	F	A	X	X	..	A	A
Concentrated	X	X	A	X	X	X	F	F	..	Pt	X	A	X	X	X	X	X	X	X	X	X	X	X	X	X	X	X	X	C	C	..	X	A	A	X	X	X	..	A	A
Nitrobenzene	..	F	..	..	..	A	..	A	..		C	A	..	..	..	..	..	..	..	X	..	X	X	..	X	F	..	C	X	X	..	X	..	X	C	F	X	..	A	A
Oleic acid	X	X	A	A	A	..	A	A	..		A	..	A	..	..	..	..	..	..	F	..	F	C	..	A	F	..	A	A	..	..	F	..	X	X	A	A	..	A	A
Oleum spirits	..	A	..	A	..	A	..	..	..		X	A	..	C	C	C	X	X	X	F	X	..	A	X	..	A	X	..	..	..	..	..	..	X	X	A	A	..	A	A
Oxalic acid	X	C	A	A	..	C	C	C	..	Ag	X	A	A	A	A	A	A	A	A	F	A	F	F	A	..	F	A	..	A	..	..	F	..	C	A	A	F	C	A	A
Oxygen, cold	A	A	A	A	..	A	A	A	..		A	..	A	..	..	..	..	..	..	C	..	C	A	..	A	C	..	C	A	..	C	C	A	X	X	C	A	C	A	..
<500°F	X	A	A	A	..	A	A	A	..		A	..	C	..	..	..	..	..	..	X	..	C	X	..	C	X	..	C	C	..	..	C	A	X	X	X	X	X	X	..
500–1000°F	X	X	C	A	..	A	A	A	A		A	..	..	..	..	..	..	..	..	X	..	..	X	..	..	X	..	..	..	..	..	X	X	X	X	X	X	..	X	..
>1000°F	X	X	X	X	A	X	X	X	A	Inc., 502	X	..	X	..	..	..	..	..	..	X	..	X	X	..	X	X	..	X	..	X	X	X	..	X	X	X	X	X	X	..
Ozone	..	..	..	..	..	..	..	..	..		..	..	..	..	..	..	..	..	..	..	..	..	..	..	..	..	..	..	..	..	..	..	A	..	..	..	..	..	..	..
Palmitic acid	A	A	A	A	..	A	A	A	..		A	..	A	..	..	..	..	..	..	A	..	F	A	..	A	A	..	A	A	F	C	F	..	X	C	A	A	C	A	..
Petroleum oils, crude, <500°F	C	C	A	C	C	A	A	A	..	410	A	..	A	C	C	C	X	X	X	X	X	C	F	X	A	A	X	A	X	X	C	C	..	F	A	A	C	X	A	A
>500°F	X	X	A	X	X	A	A	A	..	410, 502	A	..	C	..	..	..	X	X	X	X	X	C	X	X	X	X	X	X	X	X	X	C	..	X	X	X	X	X	X	X
>1000°F	X	X	X	X	X	X	C	C	A	410, 502	C	..	X	..	..	..	X	X	X	X	X	X	X	X	X	X	X	X	X	X	X	X	..	X	X	X	X	X	X	X
Phosphoric acid, crude	C	X	X	X	X	C	C	C	..		X	A	A	A	A	A	C	C	C	F	A	C	F	A	C	F	A	C	A	C	..	C	..	X	A	X	X	..	A	A
Pure, <45%	A	F	C	F	C	X	A	A	..		X	A	A	A	A	A	C	C	C	F	A	C	F	A	C	F	A	C	A	C	..	C	..	X	F	X	X	..	A	A
>45%, cold	A	F	X	F	C	X	A	A	..		X	A	A	A	A	A	C	C	C	F	A	C	F	A	C	F	A	C	A	C	..	C	..	X	X	X	X	..	A	A
Hot	X	C	X	C	..	X	X	C	..		X	A	..	F	F	F	C	C	C	X	C	C	X	C	C	X	C	C	A	C	..	C	..	X	X	X	X	..	A	A
Picric acid, molten	X	X	F	X	X	A	A	A	..		C	A	A	..	..	..	..	..	..	..	..	C	..	..	C	..	..	C	..	C	X	C	..	X	X	X	X	C	A	..
Water solution	X	X	X	C	X	C	A	A	..		X	A	A	C	C	C	C	C	C	F	C	C	F	C	A	F	C	A	..	A	X	A	..	X	A	X	X	A	A	A
Potassium chloride	A	A	C	A	A	A	A	A	..		A	..	..	A	A	A	A	A	A	A	A	A	A	A	A	A	A	A	A	A	..	A	A	A	A	A	F	..	A	A
Potassium cyanide	X	X	X	A	..	A	A	A	..		A	..	..	..	..	..	..	..	..	A	..	A	A	..	A	A	..	A	A	A	..	A	..	A	A	A	F	..	A	..
Potassium hydroxide	X	X	X	A	A	C	F	F	..		A	..	..	C	C	C	C	C	C	C	C	A	C	C	A	C	C	A	A	A	..	A	A	A	A	X	X	..	A	A
Potassium sulfate	A	A	A	A	A	A	F	F	..		A	..	..	A	A	A	A	A	A	A	A	A	A	A	A	A	A	A	A	A	..	A	A	A	A	A	X	..	A	A
Producer gas	A	C	A	A	..	A	..	..	..		A	..	A	..	..	..	..	..	..	X	..	..	F	..	..	A	..	..	..	..	C	..	..	A	A	C	A	..	A	..
Propane	A	..	..	A	..	A	A	A	..		A	..	A	..	..	..	..	..	..	..	..	..	A	..	F	A	..	A	X	..	A	..	..	C	A	A	A	..	A	..
Sewage	A	C	F	A	..	F	F	F	..		A	..	A	..	..	..	..	..	..	A	..	C	A	..	F	A	..	A	..	X	C	C	A	A	A	X	F	C	A	..
Soap solutions	A	C	..	A	..	A	A	..	..		A	..	C	A	A	A	A	A	A	A	A	C	A	A	F	A	A	A	A	..	C	C	A	A	A	F	F	C	A	A
Soda ash, sodium carbonate	A	C	X	A	..	A	A	A	..		A	..	A	A	A	A	A	A	A	A	A	A	A	A	A	A	A	A	A	A	C	A	A	A	A	X	X	A	A	A
Sodium bicarbonate, baking soda	A	C	X	A	A	C	A	A	..		A	..	A	A	A	A	A	A	A	A	A	A	A	A	A	A	A	A	A	A	C	A	..	A	A	A	F	A	A	A
Sodium bisulfate	A	F	C	A	A	X	C	C	..		C	A	A	A	A	A	A	A	A	A	A	A	A	A	A	A	A	A	A	A	C	A	..	A	A	A	F	A	A	A
Sodium chloride	A	F	X	A	A	A	F	A	..		A	..	A	A	A	A	A	A	A	A	A	A	A	A	A	A	A	A	A	A	C	A	A	A	A	A	A	A	A	A
Sodium cyanide	X	X	X	F	..	A	..	A	..		A	..	..	..	..	..	..	..	..	A	..	A	A	..	A	A	..	A	A	A	..	A	..	A	A	A	A	..	A	A
Sodium hydroxide	F	X	X	A	A	A	F	F	..		A	..	A	A	A	A	A	A	A	C	A	C	C	A	C	C	A	C	A	A	C	C	A	A	A	X	X	C	A	A
Sodium hypochlorite	X	C	X	C	C	X	X	X	..	Pt	C	A	A	A	A	A	A	A	A	F	A	C	F	A	C	F	A	C	A	C	..	C	A	A	A	X	X	C	A	A
Sodium metaphosphate	A	C	A	A	A	..	A	..	..		C	A	A	A	A	A	A	A	A	A	A	A	A	A	C	A	A	A	A	A	..	A	..	A	A	F	F	A	A	A
Sodium nitrate	A	F	A	A	A	A	F	A	..		C	A	C	A	A	A	A	A	A	A	A	C	A	A	C	A	A	C	A	A	..	C	..	A	A	A	A	..	A	A
Sodium perborate	..	..	A	A	A	C	A	A	..		A	..	C	..	..	..	..	..	..	A	..	C	A	..	C	A	..	C	A	A	..	C	..	A	A	X	X	..	A	A
Sodium peroxide	..	..	A	A	A	C	A	A	..		A	..	C	..	..	..	..	..	..	F	..	C	F	..	C	F	..	C	A	A	..	C	..	C	A	X	X	..	A	A
Sodium phosphate, monobasic	..	C	A	A	A	C	..	A	..		C	A	A	A	A	A	A	A	A	F	A	A	F	A	C	F	A	A	A	A	..	A	..	C	A	F	F	A	A	A
Dibasic	..	A	A	A	A	C	..	A	..		A	..	A	A	A	A	A	A	A	F	A	A	F	A	C	F	A	A	A	A	..	A	..	C	A	F	F	A	A	A
Tribasic	A	X	X	A	A	A	..	A	..		A	..	A	A	A	A	A	A	A	F	A	A	F	A	C	F	A	A	A	A	..	A	..	C	A	F	F	A	A	A
Sodium silicate	X	C	X	A	A	A	..	A	..		A	..	A	A	A	A	A	A	A	A	A	A	A	A	A	A	A	A	A	A	..	A	A	A	A	F	X	..	A	A
Sodium sulfate	A	A	C	A	A	A	A	A	A		A	..	A	A	A	A	A	A	A	A	A	A	A	A	A	A	A	A	A	A	..	A	A	A	A	A	F	..	A	A

Sodium sulfide	A	X	X	F	F	A	A	A	A	...	A	..	A	..	..	..	..	..	..	A	..	A	A	..	A	A	..	A	A	A	..	A	..	A	A	X	X	..	A	A
Sodium thiosulfate, "hypo"	A	X	X	C	..	C	A	A	..	...	A	..	A	A	A	A	A	A	A	A	A	A	A	A	..	A	A	A	A	A	..	A	A	A	A	A	F	A	A	A
Soybean oil	..	..	..	..	..	..	A	A	..	...	A	..	A	..	..	..	..	..	..	X	..	C	A	..	A	A	..	A	A	A	..	C	..	X	C	A	A	C	A	A
Stannic chloride	..	X	X	X	X	..	C	C	..	Ta, Zr	X	A	A	A	A	A	A	A	A	A	A	A	A	A	A	A	A	A	A	F	..	A	..	C	A	C	X	C	A	A
Steam, <500°F	C	A	A	A	A	A	A	A	A	...	A	..	A	..	..	..	..	..	..	F	..	A	F	..	C	F	..	A	C	..	..	C	A	F	A	X	X	X	A	..
500–1000°F	X	C	C	C	C	A	A	A	A	502	A	..	C	..	..	..	..	..	..	X	..	X	X	..	X	X	..	X	..	..	..	X	..	X	X	X	X	X	X	..
>1000°F	..	X	X	X	X	X	A	A	A	Inc. 502	C	..	X	..	..	..	..	..	..	X	..	X	X	..	X	X	..	X	..	..	..	X	..	X	X	X	X	X	X	..
Stearic acid	A	C	C	A	A	C	A	A	..	Ag	A	..	A	C	C	C	X	X	X	A	X	C	A	X	A	A	X	A	A	C	..	C	A	X[9]	F[9]	A	A	C	A	A
Sulfur	..	X	A	X	X	A	F	F	..	...	A	..	A	..	..	..	..	..	..	A	..	F	A	..	A	A	..	F	A	F	A	F	..	A	C	A	A	C	A	..
Sulfur chloride	A	X	..	C	..	C	..	C	..	...	X	A	C	..	..	..	..	..	..	C	..	X	C	..	F	C	..	F	X	X	X	X	..	X	C	X	X	X	A	..
Sulfur dioxide, dry	A	A	A	A	A	A	A	A	..	...	A	..	..	C	C	C	..	..	..	C	C	C	C	C	C	C	C	C	A	C	..	C	..	F	C	F	X	C	A	A
Sulfur trioxide, dry	A	A	A	A	..	A	A	..	..	...	A	..	..	C	C	C	..	..	..	C	C	C	C	C	C	C	C	C	A	C	..	A	..	X	C	X	X	..	A	A
Sulfuric acid, <10%, cold	A	C	C	C	C	X	F	F	..	Pt	F	A	A	A	A	A	A	A	A	A	A	A	A	A	A	A	A	A	A	A	..	A	A	A	A	X	X	..	A	A
Hot	A	X	C	C	X	X	X	F	..	Pt	F	A	A	C	C	C	C	X	C	F	F	A	F	F	A	F	F	A	A	A	..	A	..	F	A	X	X	..	A	A
10–75%, cold	A	X	C	C	C	X	X	F	..	Pt	X	A	F	X	X	X	X	X	X	C	X	C	C	X	C	C	X	C	C	A	..	C	A	A	A	X	X	..	A	A
Hot	A	X	X	C	X	X	X	X	..	Pt	X	A	F	X	X	X	X	X	X	X	X	C	X	X	C	X	X	C	C	A	..	C	..	F	A	X	X	..	A	A
75–95%, cold	A	X	C	C	C	C	A	A	..	Pt	X	A	C	X	X	X	X	X	X	C	X	C	C	X	C	C	X	C	C	A	..	C	X	X	F	X	X	X	A	A
Hot	A	X	X	C	X	F	X	X	..	Pt	X	A	C	X	X	X	X	X	X	X	X	X	X	X	X	X	X	X	C	C	..	X	..	X	F	X	X	X	A	A
Fuming	A	X	C	X	X	C	C	F	..	Pt	X	A	X	X	X	X	X	X	X	X	X	X	X	X	X	X	X	X	X	X	..	X	..	X	X	X	X	..	A	A
Sulfurous acid	A	C	C	X	X	A	X	C	..	Pt	X	A	A	A	A	A	A	A	A	F	A	C	F	A	C	F	A	C	A	C	..	A	..	F	A	X	X	..	A	A
Tannic acid	X	A	X	A	A	..	F	F	..	...	A	..	..	A	A	A	A	A	A	F	A	C	F	A	C	F	A	C	A	C	..	A	..	A	A	A	A	..	A	A
Tar	..	..	A	..	..	A	A	..	..	...	A	..	A	..	..	..	..	..	..	C	..	C	A	..	C	A	..	C	X	X	..	C	..	C	A	A	A	X	A	..
Tartaric acid	A	..	A	C	C	X	C	A	..	...	A	..	..	A	A	A	A	A	A	A	A	C	A	A	C	A	A	C	A	C	..	A	..	A	A	A	F	..	A	A
Toluene	A	..	A	A	..	A	..	..	..	...	A	..	A	C	C	C	X	X	X	X	X	X	X	X	F	F	X	A	X	X	A	X	X	X	C	A	A	X	A	A
Trichloroethylene	F	C	C	A	..	C	C	C	..	...	A	..	C	C	C	C	X	X	X	X	X	X	X	X	C	X	X	C	X	X	X	X	X	X	X	A	A	X	A	A
Turpentine	A	C	A	A	..	..	A	A	..	...	A	..	A	C	C	C	X	X	X	X	X	C	C	X	A	A	X	A	X	X	C	C	..	C	C	A	A	X	A	A
Vinegar	..	..	C	A	..	C	F	A	..	...	A	..	A	A	A	A	A	A	A	A	A	C	A	A	C	A	A	C	A	A	..	C	A	A	A	A	F	X	A	A
Water, acid mine, containing oxidizing salts	C	C	C	X	X	X	A	A	..	...	A	..	A	A	A	A	A	A	A	A	A	A	A	A	C	A	A	A	A	A	X	A	..	A	A	X	F	C	A	A
No oxidizing salts	..	..	A	A	..	C	..	X	..	...	A	..	A	A	A	A	A	A	A	A	A	A	A	A	C	A	A	A	A	A	X	A	..	A	A	X	F	C	A	A
Water, fresh (tap, boiler feed, etc.)	A	A	A	A	..	C	A	A	..	...	A	..	A	A	A	A	A	A	A	A	A	A	A	A	C	A	A	A	A	A	C	A	A	A	A	F	A	A	A	A
Distilled laboratory grade	X	X	A	C	A	X	A	A	..	...	A	..	A	X	X	X	X	X	X	A	X	A	A	X	C	A	X	A	A	A	C	A	A	A	A	X	A	A	A	A
Return condensate	A	A	A	A	..	A	A	A	..	...	A	..	A	C	C	C	C	C	C	A	C	A	A	C	C	A	C	A	A	A	C	A	A	A	A	F	A	A	A	A
Water, sea water	A	C	X	A	..	C	F	F	..	...	A	..	A	A	A	A	A	A	A	A	A	A	A	A	C	A	A	A	A	A	C	A	A	A	A	F	F	A	A	A
Whisky and wines	..	A	..	A	..	X	F	A	..	...	A	..	A	C	C	C	X	X	X	A	C	A	A	C	A	A	C	A	A	A	X	A	..	A	A	A	A	A	A	A
Zinc chloride	A	X	X	A	..	C	X	X	..	...	X	A	C	A	A	A	A	A	A	A	A	C	A	A	C	A	A	C	A	A	..	C	..	A	A	F	X	A	A	A
Zinc sulfate	..	X	C	A	..	C	A	A	..	...	X	A	A	A	A	A	A	A	A	A	A	A	A	A	A	A	A	A	A	A	..	A	..	A	A	F	F	A	A	A

As explained in the text, at least eight factors must be considered when a gasket is chosen for corrosive service. Any table of this type is therefore of questionable value in the selection of a gasket for a critical installation, since it is obviously impossible to take into account the effects of all these variables, singly or in combination. However, the table will be useful in indicating the degree of safety with which the various materials may be used in the service shown.

[1] Those materials known to be suitably resistant have been rated as A. Those whose resistance is only fair, but not so low as to be dangerous, have been given an F rating. An X denotes that the material is totally unsuitable. The C rating means that the use of this material is dependent upon specific service conditions and should not be selected unless carefully investigated. The blank spaces for the most part represent absence of data. However, in some instances, in the data on metals, a space has been left blank because some other metal listed is entirely suitable and less costly.

[2] Unless otherwise noted, the corrosion resistance of the metals has been rated on the basis of exposure to a pure reagent in the temperature range from 50°F. to boiling.

[3] Abbreviations in this column stand for: cupro-nickel, Hastelloy-B, Inconel, platinum, silver, stainless steels types 410 and 502, tantalum, and zirconium.

[4] The generally accepted temperature limit for a good grade compressed asbestos sheet, also called asbestos composition sheet, is 750°F. However, some grades are successfully used at considerably higher temperatures. This type of sheet is used for smooth flanges. For rough flanges, gaskets cut from asbestos-metallic sheet or formed by folding asbestos-metallic cloth are preferred. The latter, and gaskets cut from felted asbestos sheet, are indicated for flanges when bolt pressures are necessarily limited because of the type of flange material.

[5] Data from the Pfaudler Co. are given from the special point of view of the suitability of the gasket material for use with glass-lined steel equipment.

[6] Data in this column apply specifically to Silastic 181, a special silicone rubber for use in gasketing produced by Dow-Corning Corp.

[7] Fiberglas fabric filled with Silastic silicone rubber (polysiloxane elastomer) has a usable compressibility of about 20 per cent and shows the chemical resistance cited here over the temperature range from −85 to 392°F. For Fiberglas fabric filled with chemically resistant synthetic rubber the temperature range is approximately −40 to 257°F. Both the silicone rubber and the ordinary synthetic rubber are available as gasket materials in which the reinforcing fabric is a metal cloth (brass, aluminum, iron, stainless steel). The chemical properties of these constructions are the same as those given here for the Fiberglas-reinforced material, with the properties of the metal in the cloth imposed upon them. The metal-cloth construction provides for increased mechanical strength and electrical conductivity.

[8] Teflon is the Du Pont trade name for polymerized tetrafluoroethylene. It is completely inert in the presence of all known chemicals. It is not affected by any known solvent or combination of solvents. It is chemically stable up to 617°F., but, being a plastic, it is not recommended for gasket applications above 392°F. or for high pressures unless confined in a tongue-and-groove or similar joint.

[9] Hot.

[10] Sources of data: A, Armstrong Cork Co.; C, Connecticut Hard Rubber Co.; D, Dow-Corning Corp.; E, E. I. du Pont de Nemours & Co.; G, Goetze Gasket and Packing Co.; J, Johns-Manville Corp.; P, The Pfaudler Co.; S, Stanco Distributors, Inc.; U, United States Rubber Co.

Polymers such as neoprene and Perbunan can be compounded and cured to yield vulcanizates similar to natural rubber in their stress-strain and elastic properties. Other polymers such as Thiokol are swollen even less than neoprene or Perbunan by aliphatic hydrocarbons but have lower elongations and do not retain their properties as well at elevated temperatures. Plasticized polyvinyl chloride such as Koroseal can be used where less rubberlike properties are required.

Wood, while fairly inert chemically, is readily dehydrated by concentrated solutions and hence shrinks badly when subjected to the action of such solutions. It is also slowly hydrolyzed by acids and alkalies, especially when hot. In tank construction, if sufficient shrinkage once takes place to allow crystals to form between the staves, it becomes very difficult to make the tank tight again.

Structural carbon is available for use as tank linings, tower packing, absorption systems, and piping. It is suitable for use with most chemical materials except strong oxidizing agents. An impervious form, Karbate, is available; Karbate No. 2, impervious graphite, has good thermal conductivity and can be used in heat-transfer equipment.

CHEMICAL RESISTANCE OF GASKET MATERIALS

By E. C. Fetter and H. H. Dunkle

All the gasket materials commonly employed in the chemical industry will be found in the column headings of Table 13. Each is rated against some 200 chemical mediums. Ratings were assigned, by the manufacturers indicated, with gasket service specifically in mind. It is not safe to assume that a material suitable for general process equipment will also be suitable for gaskets in the same service because: (1) Many materials, *e.g.*, alloy cast irons, do not lend themselves to gasket fabrication. (2) Metal or metal-jacketed gaskets are usually thin, highly stressed, and exposed to turbulence; and, under these conditions, they may be attacked by an agent not normally considered corrosive.

Several materials in the table are rated by more than one manufacturer. Where disagreement occurs, it reflects differences in the gasket material or the exposure conditions. Significantly, these differences are not discernible in the table—warning enough that the table is not an automatic selector. The following variables, none of which can be completely or adequately taken into account in a tabulation such as this, must all be carefully considered in selecting a gasket material:

1. *Concentration of Corrosive Agent.* Dilute solutions are not necessarily less corrosive than concentrated, and, as in the familiar case of sulfuric acid and iron, the reverse is frequently true.

2. *Purity of Corrosive Agent.* Contaminants and minor constituents often influence material selection as much as the primary chemical. Fluorides in phosphoric acid, oxygen in boiler feed water, or carry-over acid in any acid-derived salt solution are examples.

3. *Temperature.* It affects the mechanical properties of the gasket as well as the aggressiveness of the corrosive agent. Non-metallic gaskets and those incorporating low-melting metals are, with a few exceptions, limited to service temperatures below 250°F. Metal-jacketed asbestos constructions may be used up to 850°F. Above 850°F. only all-metal gaskets are satisfactory.

4. *Location of Gasket.* Attack may be accelerated by aeration or condensing water vapor, which frequently accompany intermittent or liquid-level exposures. Galvanic contact with a more noble metal should be avoided.

5. *Type of Gasket Construction.* Materials abnormally subject to stress corrosion should not be used in a gasket that relies on high localized stress to form a seal. An imperfectly resistant metal will frequently do for a solid metal gasket where it would quickly fail as the thin jacket of a metal-clad gasket.

6. *Frequency of Opening Joint.* If the joint will be opened frequently anyway, a less resistant material may be satisfactory, since impending failure will be noted, and replacement can be made with no extra trouble.

7. *Relative Cost of Materials.* Gaskets are generally such a small item that it is poor economy to flirt with lost production or high labor charges for replacing a cheap gasket.

8. *Scrap Value.* Sometimes the scrap value of a rare-metal gasket, together with its longer service, will make it more economical in the long run than an initially less expensive material.

For mechanical considerations in gasket selection, see Sec. 5.

Table 14. Concrete for Chemical-tank Construction*

Solutions	Applications	Treatment
Ammonium hydroxide	Storing strong ammoniacal liquors	Inside coating of tar
Bleach liquors	Bleach tanks, stuff chests, and bleachers in paper plants	No special treatment necessary
Brines	Tanks for holding brine solutions in paper manufacturing, salt works, and vegetable-pickling plants	No special treatment necessary
Calcium chloride	Containers	No special treatment necessary
Coconut oil	Storage tanks	No special treatment necessary
Cottonseed oil	Storage tanks	No special treatment necessary
Fish oil	Storage tanks	No special treatment necessary
Fluosilicates	Depositing tanks in electrolytic refining of lead	Asphalt coating used
Glycerin	Storage tanks	Coating of rich mortar
Hemlock liquors	Storage tanks	Treated on inside with a brush coat of neat cement
Hydrochloric acid	Chemical-sediment tanks (dilute acid)	Lined inside and outside with special coating
Leaching bark	Storing and handling of leaching-bark solutions	Inner surface plastered with cement mortar
Mineral oils	Storage tanks	Coating of rich mortar
Molasses	Storage tanks	No special treatment necessary
Peanut oil	Storage tanks	No special treatment necessary
Quebracho extract	Handling of quebracho extracts	Inside plaster coat of 1:2 cement mortar
Sodium silicate	Storage tanks	
Soybean oil	Storage tanks	No special treatment necessary
Sulfuric acid	Digester tanks for boiling wood chips	Lined with brick laid in resin cement
	Depositing tank in electrolytic refining of zinc	Asphalt coating used
	Storing of fuming acid	Lined with lead
	Storing 5% sulfuric acid	Lined with mastic
Tanning liquors	Vats for beam-house work in a tannery	No special treatment necessary
Water	Storage tanks	Coating of rich mortar
Zinc chloride	Storing concentrated solutions of zinc chloride	

* From *Chem. & Met. Eng.*, September, 1929.

SECTION 22

FUELS

BY

Harry A. Curtis, Ph.D., Director, Tennessee Valley Authority; Member, American Chemical Society, American Institute of Chemical Engineers, American Society for Engineering Education, American Association for Advancement of Science. (Solid Fuels)

H. M. Weir, B.Ch.E., Ph.D., Consulting Engineer, Philadelphia, Pa.; Member, American Chemical Society, American Institute of Chemical Engineers; American Petroleum Institute, American Association for Advancement of Science. (Liquid Fuels)

W. A. Myers, B.S.E., M.S.E., Assistant Manager, Research and Development Department, Atlantic Refining Company; Member, American Institute of Chemical Engineers, American Chemical Society. (Liquid Fuels)

Wilbert J. Huff, Ph.D., D.Sc., Chairman, Department of Chemical Engineering, University of Maryland; Member, American Chemical Society, American Institute of Chemical Engineers, American Gas Association, American Association for Advancement of Science. (Gaseous Fuels)

Harlan W. Nelson, Ph.D., Supervisor, Fuels Research, Battelle Memorial Institute; Member, American Chemical Society, American Coke and Coal Chemicals Institute, American Association for Advancement of Science. (Solid Fuels)

CONTENTS

SOLID FUELS

PAGE

Coal . . . 1560
- Consumption of Bituminous Coal and Lignite in the United States . . . 1560
- Formation of Coal . . . 1560
- Analysis and Testing of Coal . . . 1560
  - Proximate Analysis . . . 1560
  - Ultimate Analysis . . . 1561
  - Heating Value . . . 1561
  - Ash-softening Temperature . . . 1561
- Physical Properties . . . 1561
  - Banded Constituents in Coal . . . 1562
  - Mineral Matter in Coal . . . 1562
  - Classification of Coal . . . 1562
  - Spontaneous Combustion of Coal . . . 1564
  - Storage of Coal . . . 1564
  - Briquetting . . . 1564
- Carbonization of Coal . . . 1564
  - Thermal Decomposition of Coal . . . 1564
  - The Plastic Stage in Coal Carbonization . . . 1564
  - The Process of Carbonization in By-product Ovens . . . 1565
  - Products of Coal Carbonization . . . 1565
  - Test Methods of Interest in Carbonization Practice . . . 1565
  - Properties of Coke . . . 1566
- Combustion of Coal . . . 1566
  - The Process of Combustion . . . 1566
  - Clinker Formation . . . 1566
  - Mechanical Stokers . . . 1567
  - Pulverized Coal . . . 1567

Miscellaneous Solid Fuels . . . 1568
- Petroleum Coke . . . 1568
- Gas Coke . . . 1568
- Peat . . . 1568
- Wood . . . 1568
- Charcoal . . . 1568
- Straw . . . 1568
- Tanbark . . . 1568
- Bagasse . . . 1568

LIQUID FUELS

Purchase of Fuel Oils . . . 1569
- Units of Sale . . . 1569

Checking Receipts and Testing Quality of Fuel Oils . . . 1569
- Sampling the Oil and Determining Receipts . . . 1569
- Specific Gravity . . . 1569
- Specifications for Fuel Oils . . . 1570
- Viscosity . . . 1570
- Flash Point . . . 1570

Fuel Oil vs. Coal . . . 1570
- Fuel Costs . . . 1570
- Labor . . . 1570
- Operating Cost . . . 1571
- Repairs, Maintenance, and Depreciation . . . 1571
- Investment Costs . . . 1571

Properties of Fuel Oils . . . 1571

Details of Fuel-oil Burning Equipment . . . 1572
- General . . . 1572
- Storage Tanks . . . 1573
- Pumps—Piping . . . 1573
- Fuel-oil Heaters . . . 1573
- Burners . . . 1573

Miscellaneous Liquid Fuels . . . 1575
- Coal Tar . . . 1575
- Tar Oil . . . 1575
- Gasoline . . . 1575
- Kerosene . . . 1575
- Alcohol and Benzol . . . 1575

GASEOUS FUELS

Description of Various Fuel Gases . . . 1576

Impurities and By-products . . . 1577
- Data on Gas Manufacture . . . 1579
- The Manufacture of Producer Gas . . . 1579
- The Manufacture of Water Gas . . . 1579
- Carbureted Water Gas . . . 1580
- The Cracking of Oil in Gas Making . . . 1581
- Oil Gas. Pacific Coast Methods . . . 1581
- Single-shell Oil Gas Methods . . . 1581
- References on Pacific Coast Oil-gas Methods . . . 1581
- The Dayton Process . . . 1582
- Recent Developments in Oil Gas Manufacture . . . 1582
- Coal Gas . . . 1582

Combustion Data . . . 1582
- Ignition Temperatures . . . 1582
- Limits of Inflammability of Mixtures of Inflammable Gases and Vapors . . . 1583
- Speed of the Propagation of Flame . . . 1587
- Checking Flame Propagation . . . 1587
- Specific Flame Output or Intensity of Combustion . . . 1587
- Flame Temperatures . . . 1588
- Combustion Calculations from Fuel-gas Analysis . . . 1589

The Flow of Gas through Pipe Lines . . . 1589

The Design of an Atmospheric Gas Burner . . . 1590

Gas as an Industrial Fuel . . . 1591
- The Atmospheric System . . . 1592
- Pressure Systems . . . 1593

SOLID FUELS

BY HARRY A. CURTIS

REVISED BY HARLAN W. NELSON

General References: Lowry *et al.*, "Chemistry of Coal Utilization," 2 vols., John Wiley & Sons, Inc., New York, 1945. Haslam and Russell, "Fuels and Their Combustion," McGraw-Hill, New York, 1925. A.S.T.M. Standards on Coal and Coke. Moore, "Coal," John Wiley & Sons, Inc., New York, 1940. Marks, "Mechanical Engineers Handbook," 4th ed., McGraw-Hill, New York, 1941. Bulletins, Technical Papers, Reports of Investigations, and published papers by U.S. Bureau of Mines.

COAL

Of the solid fuels, only coal and the coke derived from it are of prime importance industrially. Within limited areas, a few other solid fuels, such as cordwood, sawmill waste, bagasse, and bark, are used in considerable quantities.

Consumption of Bituminous Coal and Lignite in the United States, 1938–1947. Table 1, from the 1947 "Minerals Yearbook," U.S. Bureau of Mines, indicates the general trend of bituminous coal and lignite consumption in this country during the period 1938–1947.

During this same period the annual production of Pennsylvania anthracite amounted to about 50 million tons until 1941, when production increased to 56 million tons. Further increases resulted in an output of 63,701,-363 tons during 1944.

Formation of Coal. Vast deposits of vegetable material, accumulated during past geologic ages, constitute the parent material of coal. By processes involving biochemical action, submersion in water, heat, and pressure, the more resistant portions of this material, together with degradation products, have been transformed into coal. The initial step in this process, that of the collection of vegetable matter, its partial decomposition, and its transformation into peat principally through biochemical action, is taking place in present-day peat bogs.

The next stage in the coal-forming process resulted from the deposition of sedimentary material over the deposit of peat, during a period of subsidence of the land surface. In following geologic periods, the pressure of this overburden compacted the peat material, and lignitic coals were formed. The transformation into the higher rank bituminous coals and anthracite has occurred in those regions where intense thrust pressures have been developed in the earth's crust. The gradual metamorphism of coal through the higher ranks has been accomplished by increased heat and pressure extending through long but varied geologic intervals of time.

The term **rank** is used to differentiate coals with respect to their degree of metamorphism, this being evidenced by progressive changes in the content of fixed carbon. There is no definite line of demarcation among the several ranks; and the analyses, properties, and characteristics of one group merge into those of the next group. It is generally assumed that differences in rank are not caused by differences in original source materials. Within the several ranks of coal are various types or **varieties** which exhibit typical properties by reason of differences in source material.

Analysis and Testing of Coal. The **proximate analysis** consists of determinations of moisture, volatile matter, fixed carbon, and ash. The analytical procedures followed are purely arbitrary, and standard methods adopted by the A.S.T.M. and reported in Standards for Coal and Coke must therefore be followed carefully.

Moisture is determined by drying the coal under standard conditions for 1 hr. at 104 to 110°C. Values obtained in this way represent the sum of the surface moisture, from mine waters, rain, or preparation plant, and the bed or inherent moisture. The latter is characteristic of various coals and ranges from 2 to 4 per cent for anthracites and bituminous coals in Pennsylvania and West Virginia; from 7 to 15 per cent for the bituminous coals in Illinois, Indiana, and western Kentucky; and from 30 to 40 per cent for lignites. Additional moisture may be released by heating to higher temperatures, but such moisture is considered to be an integral part of the coal and is held by physical or chemical forces. Under comparable conditions of exposure, the amount of surface moisture progressively increases as the size of the coal particles is decreased.

The **volatile matter** is determined by heating a sample of coal in a covered crucible for 7 min. at 950°C. The loss in weight, minus the moisture, represents the amount

Table 1. Consumption of Bituminous Coal and Lignite, by Consumer Class, with Retail Deliveries in the United States, 1938–1947

In thousands of net tons

Year	Colliery fuel	Electric power utilities*	Bunker, foreign trade†	Railroads‡ (class I)	Coke plants		Steel and rolling mills	Cement mills§	Other industrials‖	Retail dealer deliveries‖	Total of classes shown
					Beehive	Oven					
1938	2,493	38,245	1,352	73,921	1,360	45,266	8,412	4,483	94,034	68,520	338,086
1939	2,565	43,979	1,477	79,072	2,298	61,216	9,808	5,274	100,514	71,570	377,773
1940	2,443	50,973	1,426	85,130	4,803	76,583	10,040	5,633	108,026	87,700	432,757
1941	2,489	61,861	1,643	97,384	10,529	82,609	10,902	6,832	122,379	97,460	494,088
1942	2,708	65,636	1,585	115,410	12,876	87,974	10,434	7,570	133,271	104,750	542,214
1943	2,702	76,403	1,647	130,283	12,441	90,019	11,238	5,851	142,816	122,764	596,164
1944	2,712	78,887	1,559	132,049	10,858	94,438	10,734	3,789	131,898	124,906	591,830
1945	2,442	71,603	1,785	125,120	8,135	87,214	10,084	4,215	127,164	121,805	559,567
1946	1,951	68,743	1,381	110,166	7,167	76,121	8,603	7,009	118,659	100,586	500,386
1947¶	2,489	86,003	1,689	109,296	10,142	94,522	10,048	7,872	124,459	99,163	545,683

* Federal Power Commission.
† Bureau of Census, U.S. Department of Commerce.
‡ Association of American Railroads.
§ Includes small amount of anthracite.
‖ Estimates based upon reports collected from a selected list of representative manufacturing plants and retail dealers.
¶ Subject to revision.

of gaseous constituents produced by the decomposition of the coal substance. The **ash** is the inorganic residue resulting from the burning of the coal. It consists principally of silica, alumina, ferric oxide, and lime, together with smaller amounts of magnesia, titanium oxide, alkali compounds, and sulfur compounds. These compounds are derived largely from the clay, shale, slate, pyrite, and other mineral constituents in the coal. **Fixed carbon** is determined by subtracting from 100 the percentages of moisture, volatile matter, and ash. It represents the coke residue, minus the ash.

The percentage of sulfur, heating value, and ash-softening temperature are commonly reported with the proximate analysis but are separate determinations.

The proximate analysis is the most widely used procedure for evaluating coal, particularly when the general characteristics of other coals from the same district are known. It falls far short, however, of being a complete criterion for evaluation for utilization in specific equipment or for a specific process.

For the **ultimate analysis,** the percentages of carbon, hydrogen, nitrogen, and sulfur are determined by direct analytical methods. Ash is determined as in the proximate analysis. Since there is no satisfactory method for the direct determination of oxygen, it is found by subtracting the sum of the other five components from 100. The percentage of oxygen found in this way is subject to the errors incurred in the other determinations, and especially by the change in weight of the ash-forming mineral constituents upon ignition. Because the air-dried samples used for these determinations contain moisture, the oxygen and hydrogen in this moisture are included in the analysis. When the moisture content is known, the results may be calculated to the "dry" basis.

Table 2 lists the results of proximate and ultimate analyses of selected American coals of various ranks, as reported by the U.S. Bureau of Mines.

The **heating value** is obtained by the complete combustion of a unit quantity of coal in an oxygen-bomb calorimeter under carefully defined conditions. The "gross" or "high" heating value is obtained by this method, as the latent heat of moisture in the combustion products is recovered. The results may be expressed on the "as received," "dry," or "dry and ash-free" basis.

If the ultimate analysis is available, the heating value of most high-rank coals may be calculated with an accuracy within about 2 or 3 per cent by means of the **Dulong formula:**

$$\text{B.t.u./lb.} = 14{,}544\ C + 62{,}028\left(H - \frac{O}{8}\right) + 4050\ S$$

where C, H, O, and S are expressed as fractional weights obtained from the analysis.

The **ash-softening temperature** of coal ash is determined by a standard method adopted by the A.S.T.M. Ashes that fuse in the range 1900 to 2200°F. are considered **low fusing**; those in the range 2200 to 2600°F. **medium fusing**; and those above 2600°F. **high fusing.** In general, coal ashes having low softening temperatures are likely to form clinkers, but the chemical composition of the ash, combustion conditions, and other factors affect the clinker formation. Table 3 shows the composition and softening temperatures of the ashes from several selected coals.

Physical Properties

The A.S.T.M. has adopted standard procedures for determining the **true** and **apparent specific gravities** of coal and coke. It is necessary to distinguish between the apparent specific gravity of a lump of porous material, such as coke, and the true specific gravity of the substance forming the lump.

Coals differ considerably in specific heats, depending upon the kind of coal, its ash and moisture content, etc. The range is from about 0.25 to 0.38. For metallurgical coke of 5 per cent ash, the specific heats shown in Table 5 have been determined.

Table 2. Analyses of Selected Coals of Various Ranks
"As received" basis

Rank*	State	Seam	Proximate analysis, %				Ultimate analysis, %					Heating value, B.t.u./lb.
			Moisture	Volatile matter	Fixed carbon	Ash	Carbon	Hydrogen	Oxygen	Sulfur	Nitrogen	
Anthracitic:												
Meta-anthracite	Rhode Island	Uncorrelated	1.0	4.0	66.7	28.3	...	...	...	...	...	9,620
Anthracite	Pennsylvania	Mammoth	2.3	3.1	87.7	6.9	86.7	2.2	2.9	0.5	0.8	13,540
Anthracite	Pennsylvania	Big Lykens	2.1	7.5	80.3	10.1	80.9	3.3	4.2	0.5	1.0	13,480
Semianthracite	Virginia	Merrimac	2.2	12.4	67.4	18.0	72.4	3.6	4.7	0.5	0.8	12,270
Bituminous:												
Low-volatile	West Virginia	Pocahontas No. 3	3.5	18.2	74.4	3.9	84.0	4.8	5.6	0.6	1.1	14,550
Medium-volatile	West Virginia	Sewell	3.1	25.0	66.8	5.1	...	...	...	1.3	...	14,290
High-volatile A	Pennsylvania	Pittsburgh	2.6	30.0	58.3	9.1	76.6	5.2	6.2	1.3	1.6	13,610
High-volatile A	Kentucky	Elkhorn	3.1	35.0	58.9	3.0	79.2	5.7	10.0	0.6	1.5	14,290
High-volatile B	Ohio	Middle Kittanning	8.2	36.1	48.7	7.0	68.4	5.6	16.4	1.2	1.4	12,160
High-volatile B	Kentucky	No. 6	7.2	39.8	48.8	4.2	71.5	5.8	14.3	2.6	1.6	12,950
High-volatile C	Illinois	No. 2	12.1	40.2	39.1	8.6	62.8	5.9	17.4	4.3	1.0	11,480
High-volatile C	Indiana	No. 6	12.4	36.6	42.3	8.7	63.4	5.7	18.6	2.3	1.3	11,420
Subbituminous:												
Subbituminous A or high-volatile bituminous C	Wyoming	Uncorrelated	16.5	34.2	38.1	11.2	...	...	...	2.1	...	9,740
Subbituminous B	Wyoming	Monarch	23.2	33.3	39.7	3.8	54.6	6.4	33.8	0.4	1.0	9,420
Subbituminous C	Wyoming	Uncorrelated	24.6	27.7	39.9	7.8	....	...	....	1.1	...	8,610
Lignitic:												
Lignite	North Dakota	Beulah	34.8	28.2	30.8	6.2	42.4	6.7	43.3	0.7	1.7	7,210

* According to A.S.T.M. method of classification.

Table 3. Ash-softening Temperatures and Ash Composition of Selected Coals*

Sample	Softening temperature, °F.	Analysis of ash, %							
		SiO_2	Al_2O_3	Fe_2O_3	TiO_2	CaO	MgO	$Na_2O + K_2O$	SO_3
Montana subbituminous	2060	30.7	19.6	18.9	1.1	11.3	3.7	2.4	12.2
Illinois bituminous	2320	46.2	22.9	7.7	1.0	10.1	1.6	1.5	8.9
Pennsylvania bituminous	2500	49.7	26.8	11.4	1.2	4.2	0.8	2.9	2.5
West Virginia semibituminous	2730	51.0	30.9	10.7	1.9	2.1	0.9	1.4	0.6
Kentucky bituminous	+2900	58.5	30.6	4.2	1.8	2.0	0.4	1.6	0.9

* *U.S. Bur. Mines Bull.* 209.

Table 4. Typical Specific Gravities

Fuel	Specific gravity		Pores, %
	True	Apparent	
Bituminous coal	1.25–1.45		
By-product coke	1.75–2.00	0.75–1.1	40–60
Low-temperature coke	1.50–1.75	0.5 –1.1*	30–70
Charcoal	1.4 –1.7	0.3 –0.6	65–80
Anthracite	1.45–1.7		
Wood	0.5 –1.1		

* By unusual procedures in preparing coal previous to carbonization, and control of carbonization conditions, an apparent specfic gravity as high as 1.4 may be obtained.

Table 5. Specific Heat of Coke

Temperature range, °C	20–260	20–538	20–815	20–1093
Average specific heat	0.240	0.303	0.338	0.363

The **bulk density** is a measure of the weight per cubic foot of broken coal, including the void space between particles. It is of interest in the storage of coal and in the preparation of coal to be charged in by-product coke ovens. Factors affecting the bulk density of coal are specific gravity, size consist, moisture content, degree of compaction, and use of oil in preparing the coal. Typical values for the bulk density of coals are as follows: subbituminous and high-volatile bituminous, 42 to 56 lb./cu. ft.; low- and medium-volatile bituminous, 49 to 57 lb./cu. ft.; anthracite, 50 to 58 lb./cu. ft.

Size stability refers to the ability of coal to withstand breakage during handling and shipping. It is determined by twice dropping a 50-lb. sample of coal from a height of 6 ft. onto a steel plate. From the size distribution before and after the test, the size stability is reported as a percentage factor (A.S.T.M. D440–37T). The **friability** test measures the tendency of a coal to break during handling. It is actually the complement of size stability and is determined by the standard tumbler test (A.S.T.M. D441–37T). In general, the high-volatile bituminous coals are less friable than the low-volatile coals.

The **grindability index** is a term used to indicate the relative ease of pulverizing a coal. The A.S.T.M. has adopted both the Hardgrove-machine method and the ball-mill method as tentative standard procedures. The Hardgrove-machine method (A.S.T.M. D409–37T), through use of a grinding machine of special design, furnishes a relative grindability index based on a coal that is assumed to have an index of 100. The ball-mill method (A.S.T.M. D408–37T) determines the relative amounts of energy required to pulverize different coals by placing a sample of coal in a ball mill and finding the number of revolutions required to grind it so that 80 per cent of the sample passes a No. 200 sieve. Approximate conversions of ball-mill grindability indexes into Hardgrove grindability indexes may be made as follows:

Ball-mill grindability index, %	20	30	40	50	60	70	80	90	100
Hardgrove grindability index, %	39	43	56	68	80	90	100	110	118

Banded Constituents in Coal. If a small lump of bituminous coal is examined, it will usually be found to consist of laminae. Across the faces of the lump at right angles to the original bedding plane of the coal, the cross sections of these laminae appear as bands. The four ingredients of coal giving rise to these bands have been given distinctive names, although much confusion has been occasioned in attempts to correlate the terms used by different investigators, particularly those in England and the United States. The brightest of the ingredients has been called **vitrain** by British investigators and is to be correlated with the anthraxylon discussed by American writers. It is derived from the woody plant material and contains the coking ingredient of coal. The **clarain** of British writers is also a bright coal, although not so bright as vitrain, and it forms the slightly duller bands noted in American coals. It is an attritus of plant debris and has been called **transparent attritus** by some American investigators. The dullest ingredient in British banded coals has been called **durain.** It is also an attritus but is characterized by its opacity. It does not ordinarily occur in typical American banded coals but is the main ingredient of American splint coal. The fourth macroscopic ingredient of coals is mineral charcoal, or **fusain.**

Table 6. Nomenclature of Banded Constituents of Coal

Macroscopic description	British nomenclature	German nomenclature	American nomenclature	
			Coal types	Constituents
Uniform brilliant black bands.	Vitrain	Vitrit	....	Anthraxylon. Term used to include the uniform brilliant bands in all coals
Charcoal-like layers and fragments that readily soil the fingers	Fusain	Fusit	.....	Fusain
Bright coal; clearly laminated; composed of innumerable brilliant fragments and bands with some duller material.	Clarain	Clarit	Bright coal	Translucent attritus. Contains anthraxylon, spores, cuticles, resins, together with some opaque attritus and fusain
			Semi-splint coal	Intermediate in properties and constitution between bright and splint coal
Dull coal; dull and non-reflecting in the hand specimen; lamination poor or absent.	Durain	Durit	Splint coal	Very largely opaque and semitranslucent attritus with spores, cuticles, resins, and a little anthraxylon

Mineral Matter in Coal. The ash-forming mineral constituents in coal have originated from (1) original plant ash, (2) sedimentary deposition, (3) deposits from percolating ground waters, and (4) material from the roof or floor of the mine added during the mining process. The principal ingredients present are slate, clay, sandstone, shale, carbonates, pyrite, and gypsum. These and other minor constituents give rise to the ash residue following combustion of the coal. The ash ordinarily weighs less than the original mineral matter. During the combustion process, clay and shale lose their water of hydration; the carbonates are decomposed, freeing carbon dioxide; and pyrite is oxidized to ferric oxide, giving off sulfur dioxide. The relative amounts of mineral constituents present determine the characteristics of the ash and impose certain limitations on the utilization of the coal.

The **sulfur** in coal occurs in three forms: (1) pyritic sulfur in the form of pyrite or marcasite, (2) organic sulfur, which consists of sulfur chemically combined with the coal substance, and (3) sulfate sulfur, which appears as iron or calcium sulfate. There is no evidence that sulfur occurs in the coal in its free state. The relative proportions of the sulfur found in coal vary widely, although more than small amounts of the sulfate form are not present in most freshly mined coals. During combustion of a coal, from 70 to 90 per cent of the sulfur appears in the combustion products as sulfur dioxide. About half of the sulfur in the coal carbonized in by-product ovens is retained in the coke.

Classification of Coal. Because of their variable and complex nature, attempts to classify coals have been based on almost every physical, chemical, or use characteristic, but few have found wide acceptance. The A.S.T.M. has adopted standard methods for classifying coals according to rank, variety, grade, and size.

In classification according to **rank,** coals are grouped according to their degree of metamorphism or progressive alteration by dynamochemical processes in the series

Table 7. Classification of Coals by Rank*

F.C. = fixed carbon; V.M. = volatile matter

Class	Group	Limits of fixed carbon or B.t.u. mineral-matter-free basis	Requisite physical properties
I. Anthracitic	1. Meta-anthracite	Dry F.C., 98% or more (dry V.M., 2% or less)	
	2. Anthracite	Dry F.C., 92% or more and less than 98% (dry V.M., 8% or less and more than 2%)	
	3. Semianthracite	Dry F.C., 86% or more and less than 92% (dry V.M., 14% or less and more than 8%)	Non-agglomerating†
II. Bituminous§	1. Low-volatile bituminous coal	Dry F.C., 78% or more and less than 86% (dry V.M., 22% or less and more than 14%)	
	2. Medium-volatile bituminous coal	Dry F.C., 69% or more and less than 78% (dry V.M., 31% or less and more than 22%)	
	3. High-volatile *A* bituminous coal	Dry F.C., less than 69% (dry V.M., more than 31%); and moist‡ B.t.u., 14,000‖ or more	
	4. High-volatile *B* bituminous coal	Moist‡ B.t.u., 13,000 or more and less than 14,000‖	
	5. High-volatile *C* bituminous coal	Moist B.t.u., 11,000 or more and less than 13,000‖	Either agglomerating or non-weathering¶
III. Subbituminous	1. Subbituminous *A* coal	Moist B.t.u., 11,000 or more and less than 13,000‖	Both weathering and non-agglomerating
	2. Subbituminous *B* coal	Moist B.t.u., 9500 or more and less than 11,000‖	
	3. Subbituminous *C* coal	Moist B.t.u., 8300 or more and less than 9500‖	
IV. Lignitic	1. Lignite	Moist B.t.u., less than 8300	Consolidated
	2. Brown coal	Moist B.t.u., less than 8300	Unconsolidated

* This classification does not include a few coals that have unusual physical and chemical properties and come within the limits of fixed carbon or B.t.u. of the high-volatile bituminous and subbituminous ranks. All these coals either contain less than 48 per cent dry mineral-matter-free fixed carbon or have more than 15,500 moist mineral-matter-free B.t.u.

† If agglomerating, classify in low-volatile group of the bituminous class.

‡ Moist B.t.u. refers to coal containing its natural bed moisture but not including visible water on the surface of the coal.

§ It is recognized that there may be non-caking varieties in each group of the bituminous class.

‖ Coals having 69 per cent or more fixed carbon on the dry mineral-matter-free basis shall be classified according to fixed carbon, regardless of B.t.u.

¶ There are three varieties of coal in the high-volatile C bituminous coal group, namely, variety 1, agglomerating and non-weathering; variety 2, agglomerating and weathering; variety 3, non-agglomerating and non-weathering.

lignite to anthracite. The A.S.T.M. method (D388–38) is based upon the fixed carbon and heating value on a mineral-matter-free basis. A few requisite physical properties are used to differentiate among lower rank coals. Table 7 shows this method of classification.

The dry fixed carbon and moist B.t.u. are calculated to the **mineral-matter-free basis** by the following formulas:

$$\text{Dry mm-free F.C.} = \frac{\text{F.C.} - 0.15\text{S}}{100 - (\text{M} + 1.08\text{A} + 0.55\text{S})} \times 100 \quad (1)$$

$$\text{Dry mm-free V.M.} = 100 - \text{dry mm-free F.C.} \quad (2)$$

$$\text{Moist mm-free B.t.u.} = \frac{\text{B.t.u.} - 50\text{S}}{100 - (1.08\text{A} + 0.55\text{S})} \times 100 \quad (3)$$

where mm = mineral matter.
F.C. = percentage of fixed carbon.
V.M. = percentage of volatile matter.
M = percentage of moisture.
A = percentage of ash.
S = percentage of sulfur.

Moist refers to coal containing natural bed moisture but not including visible water on the surface of the coal.

Coals are also classified according to **variety,** *i.e.,* according to the nature and biochemical alteration of the original plant ingredients. Common varieties are defined by the A.S.T.M. (D493–38T) as:

1. *Common Banded Coal.* The common variety of bituminous and subbituminous coal. It consists of a sequence of irregularly alternating layers or lenses of (1) homogeneous black material having a brilliant vitreous luster; (2) grayish-black, less brilliant, striated material usually of silky luster; and (3) generally thinner bands or lenses of soft, powdery, and fibrous particles of mineral charcoal. The difference in luster of the bands is greater in bituminous than in subbituminous coal.

2. *Splint Coal.* A variety of bituminous or subbituminous coal, commonly having a dull luster and grayish-black color, of compact structure, often containing a few thin irregular bands with vitreous luster. When struck, it is resonant. It is hard and tough and breaks with an irregular, rough, sometimes splintery fracture. It is free-burning and does not swell on heating.

3. *Cannel Coal.* A variety of bituminous or subbituminous coal of uniform and compact fine-grained texture with a general absence of banded structure. It is dark gray to black in color, has a greasy luster, and is noticeably of conchoidal or shell-like fracture. It is non-caking, yields a high percentage of volatile matter, ignites easily, and burns with a luminous smoky flame.

4. *Boghead Coal.* A variety of bituminous or subbituminous coal resembling cannel coal in appearance and behavior during combustion. It is characterized by a high percentage of algal remains and volatile matter. Upon distillation, it gives exceptionally high yields of tar and oil.

The A.S.T.M. classification according to **grade,** or quality, is actually a shorthand system for the designation of coal properties with respect to size designation and inorganic impurities (A.S.T.M. D389–37). For the classification of bituminous coals by **size,** see A.S.T.M. D431–38 in "Standards on Coal and Coke." Commercially, coals are referred to by such terms as run of mine, which is unscreened broken coal from the mine; slack coal, which is all the coal passing through screen of a given size, such as 1- or 2-in. slack; and double-screened sizes such as egg, stove, nut, pea, and stoker. Stoker coal for industrial use may also be a slack size.

Table 8 shows the standard sizing of Pennsylvania anthracite as approved and adopted by the Anthracite Institute.

Table 8. Standard Anthracite Specifications According to Size, Breaker Standard

Round-mesh screen

Name	Test mesh, in.		Oversize max., %	Undersize, %		Max. impurities, %	
	Through	Over		Max.	Min.	Slate	Bone
Broken	4 3/8	3 1/4	5	15	7 1/2	1 1/2	2
Egg	3 1/4	2 7/16	5	15	7 1/2	1 1/2	2
Stove	2 7/16	1 5/8	5	15	7 1/2	2	3
Chestnut	1 5/8	13/16	5	15	7 1/2	3	4
Pea	13/16	9/16	10	15	7 1/2	5	5
No. 1 buckwheat	9/16	5/16	10	15	7 1/2		
No. 2 buckwheat (rice)	5/16	3/16	10	15	7 1/2		
No. 3 buckwheat (barley)	3/16	3/32	10	20	10		

Spontaneous Combustion of Coal. The primary cause of spontaneous combustion of coal is oxidation of the coal substance itself. Moisture and pyrites in the coal are contributing factors, as are many other items, such as size of the coal and particularly the segregation of fines in the coal pile, and exposure to the wind, sun, steam pipes, or any external source of heat. Oxidation of the coal substance occurs immediately when the coal at the face of the seam is broken and exposed to the air. The rate of oxidation in most coals increases only very slowly with temperature up to about 50°C. If conditions for oxidation are particularly favorable and conditions for heat dissipation are poor, the temperature rises above this point, and more rapid oxidation occurs with still further increase in temperature until finally the kindling point of the coal is reached. For most bituminous coals, the kindling temperature is above 150°C. Anthracite is not liable to spontaneous heating.

Storage of Coal. Underwater storage avoids weathering and danger of spontaneous combustion and has been used with good results. For outdoor storage, the danger of spontaneous combustion and inconvenience and loss therefrom are lessened by observing the following rules:

1. Do not pile in such a way as to permit the segregation of lump and fine coal, for this increases the likelihood of fires started in the segregated fine coal. Layer piling is best.
2. Do not pile too high, for this tends to prevent the escape of heat from regions of local heat development in the pile.
3. Keep storage piles away from external sources of heat.
4. When coal is being placed in storage, compact the pile with a road roller or tractor. For long-term storage, sealing the pile with oil or tar has been done successfully.
5. Take the temperature of the pile regularly. A temperature above 50°C. at any point is a danger signal, and coal in that section should be removed.
6. Avoid storing freshly mined coal, since most coals, if prone to spontaneous heating, are especially so during the first few weeks after being mined.

Briquetting. Roll presses are used almost exclusively in American practice. Coal-tar pitch and petroleum asphalt are the two binders used most extensively. Pitch of about 170°F. melting point is usually specified, or asphalt of 50 to 60 penetration test. It requires 8 to 10 per cent pitch or 6 to 8 per cent asphalt to give a satisfactory briquet. "Sulfite pitch," obtained by evaporating the black liquor from paper mills using the sulfite process, is used to a small extent as a binder, usually in admixture with asphalt or other binders. Starch is used as a binder for charcoal briquets. An emulsion of asphalt in starch paste (Hite binder) has been used successfully in two plants making anthracite culm briquets. The asphalt renders the briquet water-resistant; and, by replacing a part of the normally required asphalt with starch, a less smoky briquet is obtained. Coals differ considerably in the ease with which they can be briquetted, and this is particularly true of bituminous coals.

"Packaged fuel" is a trade name applied to a combination of briquetting and packaging of coal screenings compressed into 3- or 4-in. cubes. Six or eight of these cubes are wrapped in strong paper and sealed with gummed tape. Production of packaged fuel increased rapidly from about 25,000 tons in 1935, to 284,000 tons in 1940, but it has decreased since that time.

Carbonization of Coal

The carbonization of coal is accomplished by heating a suitable coal, or a blend of coals, in an oven or retort in the absence of air. The complex coal substance is broken down, with the evolution of combustible gases and condensible tars and oils, leaving a residue of coke. The ability to form a strong coke is found only in the bituminous class of coals, and relatively few of these are coking coals suitable for use in by-product ovens. The coal must form a strong coherent coke, should not expand in the oven during the heating process, and should not contain over 1.5 per cent sulfur or 9 per cent ash. For good yields of gas, the dry and ash-free volatile matter should be over 35 per cent. Coal charged into by-product ovens is usually a blend of high-volatile coal with 10 to 50 per cent low-volatile coal. Blending is done to improve the quality of the coke or to avoid damage to the oven caused by expansion of the charge.

Thermal Decomposition of Coal. When the temperature throughout a small mass of crushed bituminous coal is raised slowly, gases are evolved, slowly at first. They consist principally of carbon dioxide, carbon monoxide, and water vapor. As the temperature rises, the composition of the evolved gases changes continually. Traces of gaseous hydrocarbons, particularly methane, appear early in the process. Upon further heating, liquefiable hydrocarbons appear in the volatile products, but the rate of evolution of gases and vapors continues low until the **softening temperature** of the particular coal is reached. A coal does not fuse at a well-defined temperature, but for each coal there is a short temperature range within which enough liquid products form to cause the whole mass of coal particles to coalesce more or less completely. The degree of fusion varies widely for different coals.

The rate of evolution of volatile products formed by decomposition of the coal rapidly increases as the temperature rises above the softening point. If the coal is one that fuses sufficiently to become plastic, the evolved gases form bubbles that work their way out of the plastic mass, causing it to become spongy. As decomposition of the coal substance progresses, with continued evolution of volatile products, the plasticity of the coking mass becomes less and less, until finally the bubbles are trapped and the mass takes on a fairly rigid cellular structure. Further evolution of gas can then occur only by diffusion through the cell walls or through cracks in the structure.

As the temperature is carried still higher, several processes occur simultaneously. Volatile products continue to be evolved, and the coke pieces shrink in volume, these two processes being opposite in their effect so far as the apparent specific gravity is concerned. Depending upon the nature of the original coal, either may predominate, *i.e.*, the apparent specific gravity may either increase or decrease. The true specific gravity of the coke substance, however, always increases, going from about 1.5 at 500°C. to 1.9 or higher at 1000°C. The reactivity of the coke decreases steadily during the devolatilization process, and for most cokes the hardness and strength of the coke increase with the devolatilization.

The Plastic Stage in Coal Carbonization. The temperature range during which fusing coals are in a plastic condition is relatively small, seldom more than 100°C. and often less than 50°C. Since most fusing coals soften below 410°C., it follows that in most cases the end of the plastic range lies below 500°C. The degree of fluidity, the rate of gas evolution, the rate of temperature increase, and probably several other factors, combine to determine the lump or apparent specific gravity of cellular coke which forms at the end of the plastic range.

Those coals which exhibit fusion when heated do not have sharply defined fusing points, but for each coal there is a characteristic temperature within a few degrees of which the coal will soften. The method of Foxwell [*Fuel*, **3**, 122 (1924)], as modified by Layng [*Ind. Eng.*

Chem., **17**, 165 (1925)], and later by Ball and Curtis [*Ind. Eng. Chem.*, **22**, 137 (1930)], for determining the plastic range of temperature, is based on the fact that, if a stream of inert gas is passed at a constant rate through a column of crushed coal while the temperature of the coal is slowly raised, a marked increase of the resistance to gas flow through the column may be observed when the coal softens.

Dilatometer methods have been used by Agde [*Brennstoff-Chem.*, **10**, (1929)] and by Damm [*Fuel*, **8**, 163 (1928)]. J. D. Davis of the U.S. Bureau of Mines has devised a successful plastometer, and Giesler [*Glückauf*, **70**, 178 (1939)] has also described a plastometer method which has been investigated by Soth and Russell [*Trans. Am. Inst. Mining Met. Engrs.*, **157**, 281 (1944)].

The Process of Carbonization in By-product Ovens. Modern by-product coke ovens are commonly about 40 ft. long, 12 to 14 ft. high, and 12 to 18 in. in average width, with an outward taper of about 1.5 to 4 in. from pushing to discharge end. The ovens are separated by heating flues and are arranged side to side in "batteries" of as many as 80 ovens. The heating flues are usually vertical, and coal gas or producer gas is used as a fuel. Coal is charged from the top; and, at the end of the coking period, doors at each end of the oven are opened, and the coke is pushed from the oven onto a wharf or into special quenching cars. The volatile products evolved are collected in mains; and tars, light oils, fuel gas, and other primary by-products are recovered.

When the coal is charged into a hot oven, the layer of coal adjacent to the heated walls is quickly decomposed. A plastic layer is formed, which moves slowly toward the center of the oven as carbonization proceeds. The gases and primary tar vapors, evolved by decomposition of the coal substance in the plastic zone and the semicoke residue, travel outward toward the wall of the oven and upward through the coke and semicoke. Because of the resistance of the plastic layer to the passage of gases, the travel is outward rather than inward. Coal is an extremely poor conductor of heat, and the center of the charge remains at a low temperature for several hours after being charged. Figure 1 shows the temperature gradient in a cross section of an 18-in. coke oven being heated on a 22-hr. schedule. The plastic zone is about 0.5 in. thick, and the drop in temperature from the heated side to the inner side may be from 700 to 900°F. The two plastic layers finally meet in the center at the end of the coking process. The average coking rate is about 1 in./hr., and the final carbonizing temperature is from 900 to 1000°C.

Products of Coal Carbonization. In addition to about 1400 lb. of coke, the complex volatile constituents yield the following by-products per ton of coal charged:

1. About 11,000 cu. ft. of fuel gas of approximately 550 B.t.u./cu. ft.
2. Two to four gallons of crude light oil removed from the fuel gas by scrubbing with oil.
3. An aqueous liquor containing ammonia, ammonium salts, and various other water-soluble compounds. This liquor is usually heated with a slurry lime to remove all ammonia and is then discarded.
4. Ammonia, recovered as indicated above and also scrubbed out of the fuel gas. In most coke plants, all the ammonia is converted to ammonium sulfate. About 20 lb. are recovered.
5. Tar, amounting to about 10 gal., is collected from various points in the by-product recovery system. The tar is composed of a large number of liquids and of crystalline compounds, such as naphthalene, anthracene, and others, dissolved in the complex liquid mixture.

In the ordinary coke oven and gas retort, the volatile products liberated from the coal are subjected to subsequent **cracking** by the hot coke and hot walls, and the volatile products listed above are therefore not those initially liberated from the coal. In the various **low-temperature carbonization** processes, there is much less cracking of the initially liberated products, and the gaseous and liquid products finally collected are therefore different both in character and in relative amounts.

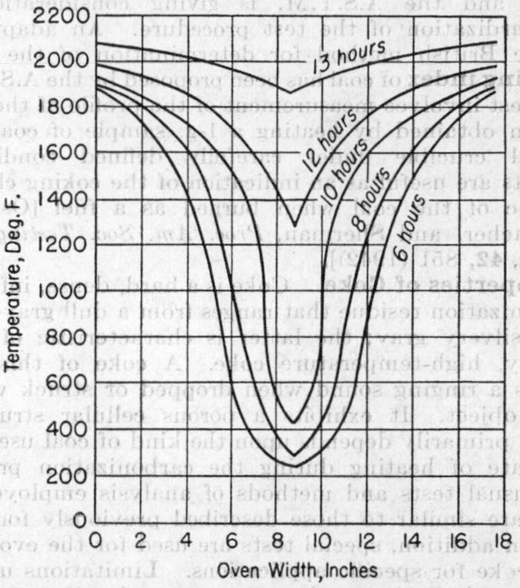

FIG. 1. Temperature gradients across a coke oven at different intervals from the start of the coking period. (*Haslam and Russell, "Fuels and Their Combustion," McGraw-Hill, New York,* 1925.)

Typical yields for a low-temperature carbonization process operating on a high-rank coking coal having 35 per cent volatile matter are:

Semicoke (12% volatile matter)	1500 lb.
Tar	30 gal.
Ammonia	2 lb.
Gas, 950 B.t.u./cu. ft.	3000 cu. ft.

Test Methods of Interest in Carbonization Practice. In addition to the plasticity tests previously mentioned, numerous tests have been proposed for purposes of evaluating the suitability of coals for carbonization or determining by-product yields and properties. Because of differences in temperature, rates of heating, mass effects, and dimensional factors, **laboratory tests** are not completely satisfactory for these purposes, but some may be correlated successfully with commercial practice. For estimating by-product yields, the U.S. Steel Corporation test and the British Gray-King carbonization assay test are of limited usefulness when correlated with the results of a selected coal in full-scale tests. The **Bureau of Mines–American Gas Association test** is carried on in an apparatus of somewhat larger capacity, with either 90 or 180 lb. of coal charged in a cylindrical welded-steel retort. A by-product collection system patterned after commercial practice furnishes data regarding by-product recovery. It is useful in comparing the carbonizing properties of various coals; the effects of variations in size of charge, rate of heating, final coking temperature, bulk density of charge; and other factors. The Fischer low-temperature assay test is useful in indicating the theoretical yields of liquid by-products on low-temperature (500°C.) distillation of coal.

Following the pioneering work of Altieri in 1935 [*Proc. Am. Gas. Assoc.*, **17**, 812 (1935)], tests for measuring relative **expansion pressures** as exerted against the walls of coke ovens have been devised by several investigators [Auvil, Davis, and McCartney, *U.S. Bur. Mines Rept. Investigations*, 3644 (1944); Fuchs *et al.*, *Penn. State College Min. Ind. Exp. Sta. Bull.* 34, (1941); Russell, *Am. Inst. Mining Met. Engrs. Tech. Pub.* 1118 (1939); Brown, *Proc. Am. Gas Assoc.*, **20**, 640 (1938)]. The tests are used in many commercial coke plants, as an aid in blending coals and avoiding damaging pressures in the oven, and the A.S.T.M. is giving consideration to standardization of the test procedure. An adaptation of the British method for determination of the **free-swelling index** of coal has been proposed by the A.S.T.M. The test involves measurement of the profile of the coke button obtained by heating a 1-g. sample of coal in a special crucible under carefully defined conditions. Results are useful as an indication of the coking characteristic of the coal when burned as a fuel [Ostborg, Limbacher, and Sherman, *Proc. Am. Soc. Testing Materials*, **42**, 851 (1942)].

Properties of Coke. Coke is a hard, dense, infusible carbonization residue that ranges from a dull gray-black to a silvery gray; the latter is characteristic of good quality, high-temperature coke. A coke of this type makes a ringing sound when dropped or struck with a hard object. It exhibits a porous cellular structure, which primarily depends upon the kind of coal used and the rate of heating during the carbonization process. The usual tests and methods of analysis employed for coke are similar to those described previously for coal; and, in addition, special tests are used for the evolution of a coke for specific applications. Limitations usually are placed on the maximum content of sulfur (1.0 and 1.3 per cent, respectively, for foundry and blast-furnace coke), phosphorus, and ash.

Table 9. Reactivity by Coal Research Laboratory Method. T_{15} and T_{75} for Various Coals and Cokes

Fuel	T_{15}, °C.	T_{75}, °C.
Illinois No. 6 seam coal	180	225
Pittsburgh seam coal	230	285
Pocahontas No. 3 seam coal	245	295
Anthracite	310	425
Semicoke	205	260
Foundry coke:		
Undercarbonized	270	390
Well-carbonized	500	565
Blast-furnace coke:		
A	455	520
B	500	565
C	550	600

Cokes having a combination of specific properties are desired for utilization in blast furnaces, foundry cupolas, gasification sets, or furnaces. The relative evaluation of coke is difficult. The **reactivity** of coke toward air or oxygen, carbon dioxide, and steam has been studied by numerous investigators. Some confusion has been brought about by the introduction of the terms "reactivity" for the reactions of carbon dioxide or steam, "combustibility" for tests employing air or oxygen, and "ignition point" for those tests in which air or oxygen is used and the extent of the reaction is observed by temperature measurement. Mayers ("Chemistry of Coal Utilization," vol. 1, p. 897, John Wiley & Sons, Inc., New York, 1945) points out that such differentiation appears unnecessary, because all these are simply devices for measuring the same property, and that the results of adequately performed tests may be convertible, at least in theory, into any of the others. The method of the Coal Research Laboratory, Carnegie Institute of Technology, based on the determination of the ignition temperature by the crossing point method, characterizes cokes by the temperatures at which the reaction rates become great enough to produce fixed rates of temperature rise. Two values of this quantity, T_{15} and T_{75}, define both the magnitude of the reaction rate at a given temperature and its rate of change with temperature. Values for these quantities, T_{15} and T_{75}, observed by Sebastian and Mayers [*Ind. Eng. Chem.*, **29**, 1118 (1937)] are shown in Table 9.

Combustion of Coal

Combustion of coal is commonly carried on in fuel beds supported by a grate, and air to support the combustion process is blown through the bed. An exception to this is coal burned in the pulverized form.

The three basic types of fuel beds, as given by Nicholls [*Am. Inst. Mining Met. Engrs., Coal Div. Tech. Pub.* 629 (1935)], are illustrated in Fig. 2. The fuel beds in all furnaces, ranging from hand-fired to stoker-fired installations, may be defined in terms of one or more of these types.

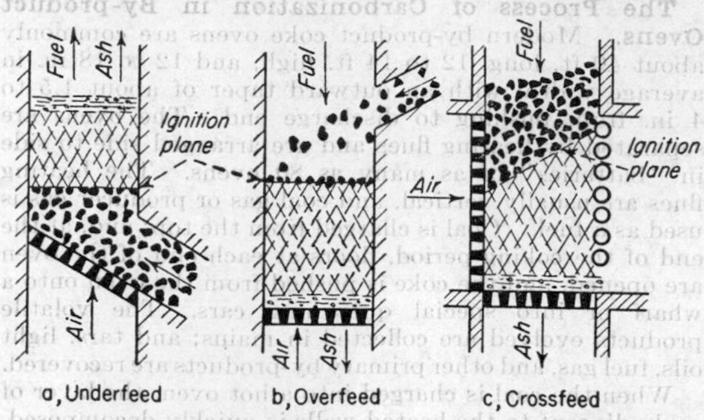

FIG. 2. The three primary types of fuel beds. (*Nicholls.*)

The Process of Combustion. The over-all result of the combustion process is the oxidation of the carbon and hydrogen in the fuel to carbon dioxide and water vapor, with the liberation of heat. In the usual overfeed fuel bed, the fuel bed may be divided into zones to illustrate the processes taking place: (1) the ash zone, immediately above the grates, which serves to protect the grates from excessive temperatures and, to a certain extent, to preheat the incoming air; (2) the oxidation zone, where the exothermic reaction $C + O_2 = CO_2$ takes place within a vertical distance of only a few inches; (3) the reduction zone, where part of the carbon dioxide produced in the oxidation zone is reduced to carbon monoxide by the endothermic reaction $C + CO_2 = 2CO$; and (4) the distillation zone, where volatile matter is distilled from the fresh coal placed on the fuel bed, and the coke residue is preheated. The volatile combustible gases from the distillation zone, and carbon monoxide from the reduction of carbon dioxide, are burned above the fuel bed, provided that (1) secondary air is provided, (2) the air and combustible gases are thoroughly mixed, and (3) a temperature above the ignition temperature of the gas-air mixture is maintained in the combustion chamber. Lack of observance of any one of these requirements will result in lowered combustion efficiency and the emission of smoke from bituminous coals.

The rate of burning is determined by the rate of admission of **primary air** from below the grate; the efficiency of the combustion process is determined by control of admission of **secondary air.** The rate of formation of the carbon dioxide in the oxidation zone is instantaneous as far as the rate of reaction between carbon and oxygen is concerned. The limiting factor in the reaction of the two components is the rate of diffusion of the oxygen in the air to the surface of the coke particle. Thick fuel beds produce greater quantities of carbon monoxide, with a consequent increase in secondary air requirements.

Clinker Formation. The amount and type of clinkers formed in fuel beds during the combustion process are often the limiting factor in determining the selection of coals for specific applications. Because of the basic complexities and interdependence of the numerous factors affecting their formation, the relative evaluation of coals in this respect is difficult. Ash is a nonhomogeneous mixture consisting of refractory and fluxing oxides in varying proportions, and its fusion characteristics alone furnish only an indication of the probable clinkering tendencies of a coal. Other factors, such as the nature of the atmosphere surrounding the ash particles, whether oxidizing or reducing, coking properties of the coal, size of the original coal particles, type of burning equipment in use, and firing practice, are also of inherent importance in determining the degree and type of clinker formation. The degree of importance of combinations of these factors varies with the use application of the fuel; and requirements for gas producers, water-gas generators, and stoker-fired installations may be quite different. For a review of the fundamental aspects of ash fusibility and clinkering, see Chap. 15, by E. P. Barrett, in "The Chemistry of Coal Utilization," Vol. 1.

Mechanical Stokers. Most coal used as an industrial fuel is burned either on mechanically operated grates and stokers or in pulverized form. Hand firing of furnaces, except in some locomotives and older or very small installations, is obsolete. Figure 3 shows diagrammatically the burning of coal on a hand-fired grate and in stoker-fired units of the overfeed, spreader, traveling-grate, and underfeed types. Description of these stokers is given in Sec. 24, 1637; and Sec. 20, pp. 1378*ff*.

A summary of information on stokers is presented in Table 10, adapted from Haslam and Russell ("Fuels and Their Combustion," McGraw-Hill, New York, 1925).

Pulverized Coal. For reasons of economy, efficiency, and flexibility, most large steam boiler installations built since 1930 have been equipped to burn pulverized coal. Because the unit cost of pulverizing coal becomes increasingly higher as the capacity of the plant is decreased, installations of less than about 30,000 lb. steam/hr. are stoker-fired.

Pulverized-coal installations may be classified according to the method of pulverizing and supplying the fuel to the burner. The two systems are:

1. The unit system, in which the coal is ground, mixed with part of the air required for combustion, and blown directly into the combustion chamber. Most of the recent installations are of this type, because first costs are lower and the layout is less extensive and requires less floor space.

2. The storage system, where the coal is dried, ground, and sent to storage, from which it is subsequently removed as required, mixed with air, and delivered to the combustion chamber. Originally used as an insurance against outages and to give greater flexibility in primary-air–coal ratios, capital and operating costs are higher than for the unit system.

The coal is pulverized in mills of the roller, impact, or ball type. Air is forced through the mill. The finer sizes are carried to the burner, and larger particles are returned to the mill by a cyclone separator or classifier

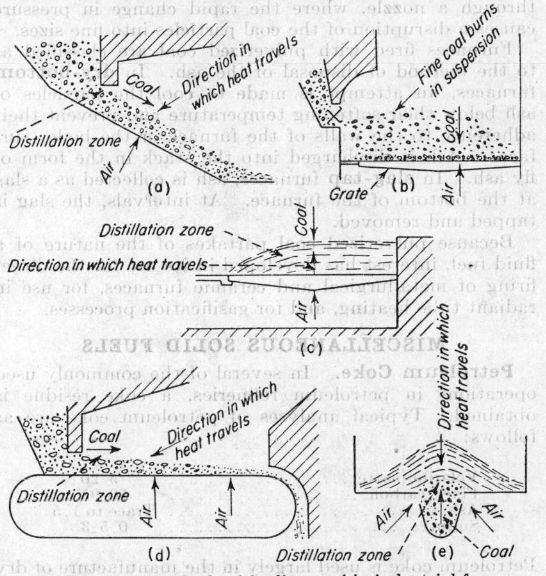

FIG. 3. Principal methods of feeding coal in industrial furnaces. (*a*) Overfeed stoker. (*b*) Spreader stoker. (*c*) Hand-fired furnace. (*d*) Traveling grate stoker. (*e*) Underfeed stoker. (*Haslam and Russell, "Fuels and Their Combustion," McGraw-Hill, New York,* 1925.)

Table 10. Summary of Information on Stokers

	Traveling grate	Overfeed stokers	Spreader stokers	Underfeed stokers
Description	Fuel from hopper is carried horizontally into furnace on a continuous web. The web is cooled coming out through the ash pit and returns to coal hopper	Grate extends from hopper into furnace at a rather steep incline, although coal flows down only if a rocking or plunger movement is imparted to grate bars	Coal is projected onto the grate by revolving paddles or a stream of air. Fine coal burns in suspension; larger pieces fall to the grate and are burned. Stationary, dumping, or traveling grates used	Raw coal is pushed up through the fuel bed and cinder falls off onto cinder plate. Gases are distilled in an oxidizing atmosphere. Capable of high overloads
Fuel used	1. Coke breeze, steam sizes of anthracite, or 2. High-volatile Midwestern coals. In general, any non-coking, clinkering coal may be used, but preferably no mixtures	All coals can be used. Good all-round stoker. Is mainly used on Midwest fuels. Coking coals can be used. Will burn refuse fuels	Both coking and non-coking coals can be used	High-volatile coals, coking coals, and slack or fines may be burned. Ash must not be easily fusible
Draft	Natural: 0.25–0.60 in. water Forced: 1–2 in. water pressure with coke 1–4 in. water pressure with Illinois and similar coals	Natural: 0.25–0.6 in. water Forced: 1–3 in. water pressure	Forced draft usually not over 2 in. water. Close air-fuel ratios required	All forced draft Normal: 2–4 in. water pressure in wind box Maximum: 5–7 in. water pressure in wind box
Rate of combustion, lb./sq. ft./hr.	Avg.: 30–35 10 lb./0.1 in. water, for natural drafts	Avg.: 25–35	Avg.: 35 to 45	Avg.: 30–40 10 lb. per 1 in. water pressure
Means of regulation	1. Height of coal gate 2. Speed of grate 3. Amount and distribution of air	1. Rate of plunger feeder 2. Rate of ash removal 3. Amount and distribution of air	1. Rate of coal feed 2. Amount and distribution of air	1. Rate of feed 2. Amount and distribution of blast
Miscellaneous	Watch for live coals going over end of grate	Fireman needed for ash removal at times, and fire must be cleaned	Close regulation of air is necessary. Critical factors are smoke and fly ash	Air is admitted over clinker plate to burn out cinder. Fuel bed 12–24 in. deep

for regrinding. The several types of mills used to pulverize coal are described in the section on Size Reduction and Size Enlargement. pp. 1154*ff*. Recently, considerable interest has developed in the *coal atomizer* as a means of pulverizing and drying coal in one operation [Yellott and Singh, *Power Plant Eng.*, **49,** 82 (December, 1945)]. Coal and steam or air at moderate pressures are passed through a nozzle, where the rapid change in pressure causes a disruption of the coal particles into fine sizes.

Furnaces fired with pulverized coal are classified as to the method of disposal of the ash. In **dry-bottom** furnaces, an attempt is made to cool the particles of ash below their softening temperature and prevent their adherence to the walls of the furnace or the boiler surfaces. Ash is discharged into the stack in the form of fly ash. In **slag-tap** furnaces, ash is collected as a slag at the bottom of the furnace. At intervals, the slag is tapped and removed.

Because pulverized coal partakes of the nature of a fluid fuel, interest has developed in its use for the direct firing of metallurgical and ceramic furnaces, for use in radiant tube heating, and for gasification processes.

MISCELLANEOUS SOLID FUELS

Petroleum Coke. In several of the commonly used operations in petroleum refineries, a coke residue is obtained. Typical analyses of petroleum coke are as follows:

	%
Volatile matter	5–20
Fixed carbon	80–95
Ash	Trace to 1.5
Sulfur	0.5–3

Petroleum coke is used largely in the manufacture of dry cells, carbon electrodes, and electric-furnace resistance elements.

Gas Coke. Gas coke is the residue remaining from the manufacture of coal gas. It is very soft and easily friable and is not suitable for use in metal foundries or furnaces. It is used mainly in the manufacture of water gas.

Peat. Peat is commercially unimportant as a fuel in the United States; no sales of peat for use as fuel were reported to the U.S. Bureau of Mines in 1939.

Typical analyses of peat are shown in Table 11.

Wood. The fuel values of different woods (except resinous woods) are nearly proportional to their weights on a dry basis. The resinous woods possess higher heating values than the non-resinous ones. Table 12 gives pertinent data for various woods.

Charcoal. Charcoal is the residue remaining after the destructive distillation of wood. It absorbs moisture readily, often containing as much as 10 to 15 per cent water. In addition it usually contains about 2 to 3 per cent ash and 0.5 to 1.0 per cent hydrogen. The heating value of charcoal is about 12,000 to 13,000 B.t.u./lb.

Straw. Depending upon its moisture content, straw has a heating value of from 5000 to 6500 B.t.u./lb.

Table 11. Analyses and Calorific Values of Air-dried Peat*

Kind of peat	Locality	Water, %	Ash, %	Sulfur, %	Calorific value of air-dried peat, B.t.u./lb.
Brown, fibrous	Fremont, N.H.	6.34	7.93	0.69	9290
Brown, fibrous	Hamburg, Mich.	7.50	6.55	.28	9090
Light brown, fibrous	Rochester, N.H.	11.64	4.06	.22	9083
Dark brown	Westport, Conn.	12.70	4.12	.24	8590
Brown, fibrous	Westport, Conn.	19.69	3.23	.19	7691
Brown	Kent, Conn.	12.10	7.22	.83	7684
Brown, fibrous	Cicero, N.Y.	14.57	7.42	.25	7576
Brown	Black Lake, N.Y.	8.68	16.61	.99	7522
Brown, fibrous	La Martine, Wis.	9.95	16.77	.79	7468
Salt marsh	Kittery, Me.	13.50	12.04	1.94	7319
Black	Greenland, N.H.	6.62	24.11	1.01	7186
Brown, fibrous	Madison, Wis.	6.99	18.77	0.38	6943
Brown, sandy	Kent, Conn.	9.06	36.06	1.46	5924

* U.S. Bureau of Mines.

Table 12. Approximate Weights and Heating Values per Cord of Fuel Woods*

Variety of wood	Weight per cord, lb.		Available heat units per cord, million B.t.u.		Equivalent in heat value to tons of coal†	
	Green	Air dry	Green	Air dry	Green	Air dry
Ash, white	4300	3800	19.9	20.5	0.77	0.79
Beech	5000	3900	19.7	20.9	.76	.80
Birch, yellow	5100	4000	19.4	20.9	.75	.80
Chestnut	4900	2700	12.9	15.6	.50	.60
Cottonwood	4200	2500	12.7	15.0	.49	.58
Elm, white	4400	3100	15.8	17.7	.61	.68
Hickory	5700	4600	23.1	24.8	.89	.95
Maple, sugar	5000	3900	20.4	21.8	.78	.84
Maple, red	4700	3200	17.6	19.1	.68	.73
Oak, red	5800	3900	19.6	21.7	.75	.83
Oak, white	5600	4300	22.4	23.9	.86	.92
Pine, yellow	....	....	21.1	22.0	.81	.85
Pine, white	....	....	12.9	14.2	.50	.55
Walnut, black	....	....	18.6	20.8	.72	.80
Willow	4600	2300	10.9	13.5	.42	.52

* The Use of Wood for Fuel, *U.S. Dept. Agr. Bull.* 753.
† Short ton (2000 lb.) of coal having a heating value of 13,000 B.t.u./lb.

Tanbark. Tanbark is the residue remaining after the bark has been used in tanning operations. It usually contains from 60 to 70 per cent water and has a heating value of 2500 to 3000 B.t.u./lb. In the use of tanbark as a fuel, a very large combustion space is required.

Bagasse. Bagasse is the solid residue remaining after sugar cane has been crushed by pressure rolls. It usually contains from 40 to 50 per cent water. The dry bagasse has a heating value of 8000 to 9000 B.t.u./lb.

LIQUID FUELS

BY H. M. WEIR AND W. A. MYERS

REFERENCES: Haslam and Russell, "Fuels and Their Combustion," McGraw-Hill, New York, 1925. Performance of Pressure Type Oil Burners, *Iowa State College Eng. Exp. Sta. Bull.* 151, 1941. Steiner, Viscosity Compensation for Oil Burners, *Heating, Piping, Air Conditioning,* **14,** 272 (1942). Steiner and Ravnsbeck, "Oil Burner Service Manual," McGraw-Hill, New York, 1942. Oil Piping, Its Design and Application, *Heating & Ventilating,* 70–74 (September, 1941); 86–91 (October, 1941); 82–87 (November, 1941). For Information on Domestic Heating: Automatic Mechanical Draft Oil Burners Designed for Domestic Installations, *U.S. Bur. Standards C.S.* 75, 1942. American Society for Testing Materials, A.S.T.M. Standards on Petroleum Products and Lubricants.

Introduction. Except in unusual and relatively unimportant circumstances, the only commercial liquid fuels sufficiently cheap for power generation for industrial heating are certain fractions of petroleum oil. The term fuel oil, as used hereafter, will refer to these materials only.

The petroleum refiner uses crude oil as his raw material. This material consists of a whole series of hydrocarbons varying from dissolved fixed gases to heavy, nearly solid compounds. Certain fractions of this crude petroleum which may be separated by simple distillation will have the necessary properties for use as a fuel oil. The petroleum refiner also practices forms of destructive distillation which are called either thermal or catalytic cracking. In these processes some hydrocarbons suitable for fuel oil are also produced. The commercial fuel oil that the refiner sells is usually a blend in various proportions of some of the materials separated by primary distillation with some of the products of destructive distillation, and it represents one of the end products of his operations on crude petroleum.

Since fuel oil is made by blending various fractions, it is possible for the refiner to make a variety of products to suit various kinds of burners. Below are given the specifications by which such fuel oils are sold. If these specifications are met, satisfactory operation of suitable types of burning equipment is ensured irrespective of the previous history of the liquid.

The American Petroleum Institute issues a *Statistical Bulletin of the Industry*, based upon reports of the Bureau of Mines and of The Bureau of Foreign and Domestic Commerce. These data indicate a domestic and export consumption of the following amounts of oils, expressed in 42-gal. barrels. The figures, in million bbl./day, apply to both 1947 and 1948 with reasonable approximation.

Crude oil, natural gasoline, benzol, and refined petroleum products	6.0
Motor fuels	2.5
Gas oil and distillate fuels	1.0
Residual fuel oil	1.3

The trend of the figures has been continually upward for many years. It is obvious that production and consumption of fuel oils is a major feature of United States industrial activity.

Purchase of Fuel Oils

Units of Sale. Fuel oil is sold in the United States in multiples of the 42-gal. barrel, the contents being measured at 60°F. Measurements made at other temperatures must be corrected to the standard, using expansion coefficients or the tables cited below under Quality of Material.

Though fuel oil is purchased for its available heat, this factor is never specified nor is it usually determined. As a matter of fact, the heat of combustion per unit weight varies only to a small extent. However, since fuel oil is invariably purchased by volume, the differences in weight, hence in heating value, should be allowed for in considering price.

Checking Receipts and Testing Quality of Fuel Oils

In drawing contracts and in making acceptance tests, it is advisable to refer to Petroleum Products and Lubricants, *Am. Soc. Testing Materials Rept. Comm. D2.* This report is issued annually and contains standard methods for determination of any physical property. The descriptions of tests given below are intended only to indicate the nature of the tests. Actual determinations of these empirical factors should be carried out precisely as described in the above publication.

Sampling the Oil and Determining Receipts. On receipt in consumers' tanks, the volume of oil should be measured and its average temperature (°F.) obtained immediately. Any recognized method of sampling for laboratory inspection may be used under conditions dictated by the fact that heavy fuel oil, especially, may not be homogeneous but may contain considerable water and salts in suspension or at the bottom of tanks.

Specific Gravity. This determination complements that of temperature of shipments in checking the volume of receipts. Determination can be made by a hydrometer graduated in terms of specific gravity, but it is preferably made with a hydrometer carrying an arbitrary scale termed **Degrees A.P.I.** The latter is defined by

$$\text{Degrees A.P.I.} = \frac{141.5}{\text{specific gravity } 60°\text{F.}/60°\text{F.}} - 131.5$$

If the measured temperature of the oil is 60 ± 30°F., the true volume at 60°F. can be determined with sufficient accuracy for most cases from the following coefficients of cubical expansion:

Up to 35°A.P.I. = 0.0004 per degree
35 to 50°A.P.I. = 0.0005 per degree

Table 13. Detailed Requirements for Fuel Oils—A.S.T.M. D396–48T(1948) *

Grade of fuel oil†	Flash point, deg. F.	Pour point, deg. F.	Water and sediment, % by volume	Carbon residue on 10% bottoms, %	Ash, % by weight	Distillation temperature, deg. F. 10% point	10% point	End point	Saybolt viscosity, sec. Universal at 100°F.		Furol at 122°F.		Kinematic viscosity, centistokes At 100°F.		At 122°F.		Gravity, deg. A.P.I.	Corrosion, copper strip, 3 hr. at 122°F.§
	Min.	Max.	Max.	Max.	Max.	Max.	Max.	Max.	Max.	Min.	Max.	Min.	Max.	Min.	Max.	Min.	Min.	
No. 1. A distillate oil intended for vaporizing pot-type burners and other burners requiring this grade of fuel	100 or legal	0	trace	0.15	....	420	...	625	...	...	...	..	2.2	1.4	...	...	35	pass
No. 2. A distillate oil for general-purpose domestic heating in burners not requiring No. 1 fuel oil	100 or legal	20‡	0.10	0.35	....	‖	675	...	40	...	...	..	(4.3)	...	...	..	26	
No. 4. An oil for burner installations not equipped with preheating facilities	130 or legal	20	0.50	....	0.10	...	...	...	125	45	...	..	(26.4)	(5.8)				
No. 5. A residual-type oil for burner installations equipped with preheating facilities	130 or legal	....	1.00	....	0.10	...	...	...	...	150	40	..	...	(32.1)	(81)			
No. 6. An oil for use in burners equipped with preheaters permitting a high-viscosity fuel	150	....	2.00¶	....	....	...	...	...	...	...	300	45	...	...	(638)	(92)		

* Recognizing the necessity for low-sulfur fuel oils used in connection with heat-treatment, non-ferrous metal, glass, and ceramic furnaces, and other special uses, a sulfur requirement may be specified in accordance with the following table:

Grade of Fuel Oil	Sulfur, Max., Per Cent
No. 1	0.5
No. 2	1.0
No. 4	no limit
No. 5	no limit
No. 6	no limit

Other sulfur limits may be specified only by mutual agreement between the purchaser and the seller.

† It is the intent of these classifications that failure to meet any requirement of a given grade does not automatically place an oil in the next lower grade unless in fact it meets all requirements of the lower grade.

‡ Lower or higher pour points may be specified whenever required by conditions of storage or use. However, these specifications shall not require a pour point lower than 0°F. under any conditions.

§ The exposed copper strip shall show no gray or black deposit.

‖ The 10 per cent point may be specified at 440°F. maximum for use in other than atomizing burners.

¶ The amount of water by distillation plus the sediment by extraction shall not exceed 2.00 per cent. The amount of sediment by extraction shall not exceed 0.50 per cent. A deduction in quantity shall be made for all water and sediment in excess of 1.0 per cent.

Abridged volume-correction tables of greater actual and recognized legal accuracy will be found in the publication named above, the same being based upon a complete table in *Nat. Bur. Standards Circ.* 154.

Specifications for Fuel Oils. The specifications described in Table 13 are those promulgated as Tentative Specifications for Fuel Oils by the A.S.T.M., as revised for 1948.

Water and Sediment. (For Grades 1 to 5, Inclusive.) Standard Method of Test for Water and Sediment in Petroleum Products by Means of Centrifuge, A.S.T.M. Designation: D96.

Water by Distillation. (For Grade 6.) Standard Method of Test for Water in Petroleum Products and Other Bituminous Materials, A.S.T.M. Designation: D95.

Sediment by Extraction. (For Grade 6.) Sediment in Fuel Oil by Extraction, A.S.T.M. Designation: D473.

Pour Point. Standard Method of Test for Cloud and Pour Points, A.S.T.M. Designation: D97.

Carbon Residue. Standard Method of Test for Carbon Residue of Petroleum Products (Ramsbottom Carbon Residue), A.S.T.M. Designation: D524.

Ash. Procedure for Determination of Ash as Described in the Standard Methods of Analysis of Oils, A.S.T.M. Designation: D482. Sample shall be thoroughly mixed to ensure that the portion for ash determination is representative of the sample.

Distillation. Distillation of Grade 1 oil shall be made in accordance with the Standard Method of Test for Distillation of Gasoline, Naphtha, Kerosene, and Similar Petroleum Products, A.S.T.M. Designation: D86 and of Grade 2 in accordance with the Standard Methods of Testing Gas Oils, A.S.T.M. Designation: D158.

Viscosity. The standard test method designated as D88–44 calls for the use of a Saybolt Furol or Saybolt Universal viscosimeter. The reading obtained is the time, in seconds, required for 60 cc. of oil at constant temperature to flow through the Furol or Universal orifice under its own (continually decreasing) head. On the other hand, if only relative values of viscosity are required, the expensive Saybolt instrument may be dispensed with and a pipette-type viscosimeter such as described by Ferris [*Ind. Eng. Chem.*, **20**, 974 (1928)] may be used.

The Saybolt Furol orifice is to be used for viscous oils only and at a single standard temperature of 122°F. If the oil shows less than 25 sec. efflux time at this temperature, its viscosity is to be measured at 100°F. with the Saybolt Universal orifice.

Viscosity of Grade No. 1 shall be determined in accordance with the Tentative Method of Test for Kinematic Viscosity (A.S.T.M. Designation: D445) and of Grades 2, 4, 5, and 6 in accordance with the Standard Method for Viscosity by Means of the Saybolt Viscosimeter (A.S.T.M. Designation: D88). The application of viscosity values to the design and operation of plant fuel-burning equipment is discussed on p. 1572.

Flash Point. Special equipment is required for determining the flash point of fuel oil, according to the American standard method D93 using the A.S.T.M. Pensky-Martens tester. The flash point has scarcely any connection with the manner in which an oil behaves in an oil burner. However, it does bear some relation to safety in storage even though it be obscure. In general, the lower the flash point, the greater the possibility of fire in storage tanks due to a spark or flame. The small consumer will not find it necessary to check the flash point of his fuel oils, but its value should be obtained from the refiner and provision made to keep storage tanks well below this temperature.

Fuel Oil vs. Coal

Fuel Costs. Figure 4 will be of assistance in making this basic calculation with any given set of data. The example indicates that 1 ton of 9500 B.t.u./lb. coal burned in a furnace with 80 per cent efficiency is equivalent to 3.5 bbl. of 148,000 B.t.u./gal. oil burned in a furnace with 70 per cent efficiency. Whereas the utilization of 80 per cent of the heating value of coal is quite exceptional, this ratio of heat into steam vs. heat in the fuel is not uncommonly obtained with oil-burning equipment. The average coal-burning boiler probably operates at something less than 65 per cent efficiency. On the other hand, an over-all boiler-furnace efficiency of 75 per cent should always be obtainable with fuel oil, provided that the equipment is in reasonably good operating condition and firemen are alert and passably well informed.

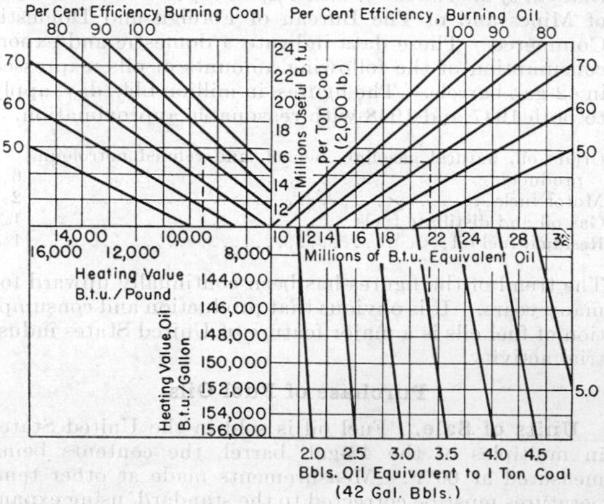

FIG. 4. Conversion chart. Number of barrels of fuel oil equivalent to 1 ton of coal.

The higher average efficiency of oil-burning equipment reflects the relatively lower percentages of excess air necessary for combustion and the absence of ash-pit losses. Nevertheless, there is an unavoidable loss in fuel-oil firing which may be considered to correspond to ash-pit loss in coal firing.

Complete combustion of oil involves the formation of approximately an equal weight of water—since oil is about one-eighth hydrogen. The flue gases thus carry out the stack not only the sensible heat in the other gases but also all the latent-heat equivalent of the water (steam) formed. Depending on the oil used and the stack temperature, this loss may vary from 7 to 9 per cent, whereas the corresponding "hydrogen-steam" loss from anthracite coal is of the order of 2.5 per cent and from bituminous coal up to 4.5 per cent.

Labor. The oil-fired plant shows a great saving in labor cost over the hand-fired coal-burning plant, chiefly because of the elimination of coal passing and ash handling. Even the most modern stoker and ash-handling equipment, or powdered-coal-burning equipment, as installed in larger plants, requires more labor than oil firing. A few firemen can control a large number of oil burners efficiently, and the labor connected with cleaning fires is obviated. If the oil burners are properly controlled, it is not necessary to clean flues so often as with equally well operated coal-burning equipment. Good-class firing labor can be retained for oil-fired installations because of the pleasant working conditions. As a matter of fact, only conscientious and intelligent firemen should be employed, for, though fuel-oil firing is capable of exact control, poorly informed or careless operators can easily waste several times their daily wage in the price of oil inexpertly burned.

Operating Cost. The cost of atomizing fuel oil depends more on the primary cost of power, steam, or air than it does on the size of the oil-burning equipment itself. This follows from the fact that large installations consist of a number of small atomizing units.

The steam required for atomization of oil is, under the best circumstances, somewhat less than 1.5 per cent of the steam generated by the boiler, but many good burners require 2 per cent, a figure which may be taken as the average of good practice. Poor burners poorly operated may consume up to 5 per cent of the steam generated with otherwise efficient boiler equipment.

The power used in pulverizing coal for powdered coal installations is 1.5 to 2 per cent of the output of the boiler equipment and hence is comparable to that required for oil burners using steam atomization. Power consumption for low-pressure air or mechanical atomization is comparable to that for steam atomization. The power consumption for high-pressure air atomization is two to three times as high as this. In general, the use of high-pressure air atomization burners can be justified only if special conditions obtain, such as on shipboard, where high-capacity burners are imperative and boiler feed-water makeup is expensive.

Repairs, Maintenance, and Depreciation. The repair and maintenance cost of the oil-burning equipment itself is invariably small since the only moving parts in many installations are the pumps. Five to seven per cent of the total investment is usually ample allowance for repairs, and fixed-rate depreciation of 10 per cent per annum is considered good practice.

Oil firing is synonymous with high-temperature combustion zones; hence high-grade refractories must be used if the repair cost of furnace linings is to be kept within reason. However, the same applies to powdered-coal- or stoker-fired-coal-burning furnaces. The expansion and contraction of furnace linings, due to opening firing doors in hand firing of coal, are obviated by any of the above alternatives. Proportional decreases in maintenance cost partly, if not wholly, offset the effect of inherently high temperature operations.

Investment Costs. The investment required for oil-burning equipment is influenced largely by items of tankage and piping. Since the kind of tankage and its location are frequently prescribed by local ordinance, these factors are not entirely under the control of the designer. The oil-storage volume advisable per unit of oil-burning equipment is also contingent on regularity of delivery, which in turn usually reflects the distance from source of oil supply.

As a rough general figure, the investment may be assumed to vary from about \$3000 per 1000 gal. of oil burned per day for small installations (150 to 200 boiler hp.) to about \$1200 per 1000 gal. of oil per day for 5000 boiler-hp. plants. These costs include tankage, piping, heaters, pumps, meters, etc., set up ready to operate.*

The comparative installation costs of coal-fired equipment scarcely admit of any generalization. These costs may vary from almost negligible figures in the case of hand-fired coal furnaces to the very substantial investment entailed by overhead bunkers, coal hoists or elevators, stokers, ash-handling devices, etc. However, two items of investment, the fuel inventory and the ground area required, are almost always in favor of oil-burning equipment, even in comparison with hand-fired coal installations.

The wide distribution of fuel-oil supply points usually makes it feasible to carry a smaller oil inventory than coal. Even if large volume reserves of oil must be carried, the fact is important that oil does not deteriorate in storage and is not subject to spontaneous combustion or slow oxidation which reduces the heating value of stored coal.

Another advantage accruing to the use of oil fuels is that the fuel storage may be placed at a considerable distance from the point of use without materially increasing handling costs. Where ground space is at a premium, this, together with the fact that oil requires only about 60 to 70 per cent of the volume of coal of equal heating value may present real advantages.

Properties of Fuel Oils

Under Purchase of Fuel Oils (p. 1569), those properties which form the subject of buyers' and sellers' contracts are discussed. In this section, properties of fuel oils of interest to the engineers designing equipment or to the users of fuel oil are presented. These include heat content of oils, air required for combustion, flue-gas analysis as influenced by excess air, and oil viscosity as a function of the temperature.

The high heating value of an oil (hydrogen burned to liquid water) in terms of both B.t.u. per gallon and B.t.u. per pound is given as a function of the A.P.I. gravity of the oil in Fig. 5. The example, illustrated by

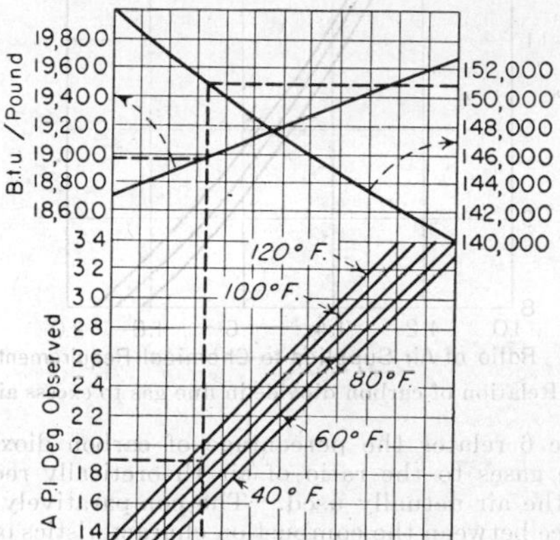

FIG. 5. Heating value of fuel oil.

dotted lines on Fig. 5, shows that an oil which indicates 19° A.P.I. gravity at 100°F. has a true A.P.I. gravity (measured at 60°F.) of 16.8. Furthermore, its high heating value is 151,000 B.t.u./gal. or 18,900 B.t.u./lb.

The air required for the complete combustion of an oil may be accurately calculated from the data of an ultimate organic analysis, using the formula

$$\text{Pounds of air per pound of oil} = 0.1152\ C + 0.3456\left(H - \frac{O}{8}\right) + 0.04328$$

A fairly good approximation of the air required for combustion can, however, be made without recourse to an elementary analysis. This circumstance is due to the fact that fuel oils do not differ greatly in their elementary

* These investment figures represent 1939–1940 conditions. A factor should be applied to compensate for increased costs of labor and material.

composition. As a matter of fact, such oils range from 83 to 87 per cent carbon content, together with 11 to 13 per cent by weight of hydrogen. The remainder consists, for the most part, of sulfur and water dissolved or mechanically mixed with the oil. Despite the similarity in the ultimate composition of such oils, their chemical constitution is very complex and includes rather large molecular weight hydrocarbons of the saturated, unsaturated, and cyclic types. Aside from these hydrocarbons, a few per cent of other organic compounds containing sulfur, together with occasional small amounts of oxygen or nitrogen, are usually present. The varied arrangement of these few elements in the molecules of the mixture is not reflected in large differences in heat content or air required for combustion. Accordingly, the volume of air (cubic feet at 60°F., 29.92 in. Hg) required to burn a unit (weight or volume) of oil, allowing no excess air, may be calculated within a very few per cent by dividing the high B.t.u. heating value by 100.

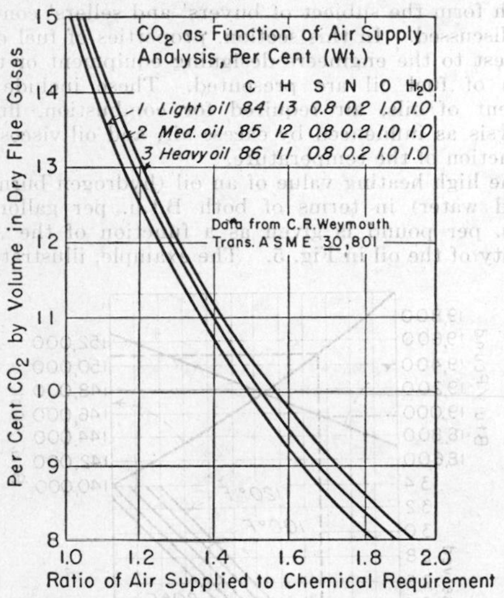

FIG. 6. Relation of carbon dioxide in flue gas to excess air used.

Figure 6 relates the percentage of carbon dioxide in dry flue gases to the ratio of air theoretically required and to the air actually used. The comparatively small difference between the combustion characteristics of light and heavy oil is illustrated by the chart, which is useful in orienting an observer on the meaning of CO_2 determinations in flue gases from oil-fired equipment. (Note that the usual Orsat gas analysis yields per cent of constituents approximately on the basis of dry gas irrespective of the water content of sample.)

The viscosity of fuel oil at the point of injection to the burner is of considerable importance from the standpoint of efficient operation. This statement is equivalent to stressing the importance of the temperature of the oil since the desired viscosity is obtained with power-plant fuels by raising the temperature. Different oils of the same A.P.I. gravity, or specific gravity, do not necessarily have the same viscosity at any given temperature. This, together with the fact that routine laboratory determinations of viscosity are never carried out at the temperature at which the oil is used, forms a problem conveniently solved by reference to the temperature-viscosity chart for petroleum oils available from the Amer. Soc. for Testing Materials, Philadelphia. A straight line drawn on this chart between points representing two viscosity determinations at different temperatures enables the observer to read the viscosity of the particular oil at any other temperature. (The chart is equally applicable to the determination of the viscosity-temperature relationship of lubricating oils.)

If only one viscosity determination is available a rough approximation to the change in viscosity with temperature can be determined by drawing a line through the point parallel to a line drawn on the chart to represent an oil of similar type, *i.e.*, depending on whether the oil is made from a paraffinic, mixed, or naphthenic, base crude. As a general rule the type of the fuel oil is not well known by the user; hence the method is of limited application. In any case, if other than the roughest approximation is desired, the actual determination of viscosity at two temperatures is necessary.

The minimum temperature to which an oil must be raised for successful burning is conditioned not only by the oil chosen but also by the burner used. To get best results with mechanical burners, the viscosity should not exceed 150 sec. Saybolt Universal. A higher viscosity oil (lower temperature) can usually be handled efficiently by steam atomizing burners. Since the oil must be brought to steam-atomizing burners at pressures from 10 to 70 lb./sq. in. and to mechanical burners at from 100 up to perhaps 250 lb./sq. in., the effect of viscosity on pumping costs requires consideration. Although it is advantageous to locate the oil-feed pumps near the burners, this is not always possible. In such cases the choice of oil, the viscosity to which it must be reduced, and the kind of burners present a problem of economics involving the laws of fluid flow. The proper size of pump can be shown by an analysis of the situation in the light of the discussion on pp. 1414*ff*. No oil burner operates with the regularity inherently possible if it is necessary to heat the oil much above the point where it begins to vaporize. Hence the feed pumps, piping, etc., should be chosen of such dimensions that the required volume and pressure of the oil supply do not demand viscosity reductions corresponding to excessive temperatures.

Details of Fuel-oil Burning Equipment

General. The National Board of Fire Underwriters publishes a pamphlet "Regulations . . . for the Installation of Oil-burning Equipments." In order to make certain that the lowest insurance rates will apply, equipment should be designed to conform to these regulations, which are those recommended by the National Fire Protection Association. Although these regulations ensure physically "safe" equipment, it sometimes happens that municipalities or other local districts have certain additional rules governing such installations. It is the part of wisdom, therefore, to obtain the approval of the local fire authorities before actually installing any proposed equipment. The details that are particularly subject to regulation are location of tanks (with respect to property lines), strength of materials, arrangement of drains or gravity cross connections from one storage tank to another, etc.

The principal units necessary for oil-burning equipment are indicated diagrammatically in Fig. 7. These are the storage tanks, pumps, oil heaters, and burners. Probably the most generally satisfactory procedure to adopt in providing equipment for fuel-oil burners is to purchase a complete "fuel-oil set." A number of manufacturers market these sets consisting of strainers, pump, steam heater, and pressure regulators for oil passing to the burners. All the equipment is frequently mounted on one base plate. Nevertheless it sometimes happens that purchase or manufacture of the equipment in parts rather than as a whole may be desirable. Hence, a few notes may be of value.

Storage Tanks.* Division of storage tankage into at least two parts facilitates cleaning and simplifies gaging of receipts. If the latter advantage is to be fully realized, each tank should hold at least one unit of delivery. A 13,000-gal. tank will hold the contents of any tank car; tank trucks ordinarily deliver less than 3000 gal. The maximum size of tanks is a matter of judgment predicated on the regularity of supply and influenced by local regulation. Ten days' to 3 weeks' storage capacity should be provided. Storage tanks for heavy oils should be provided with steam coils to keep the oil fluid for the pump suction. A bottom drawoff or, in the case of underground tanks, a sump with separate suction line should be installed for drawing off water and sediment. If possible, tanks should be arranged for filling by gravity from tank car or truck. A steam hose connection near the inlet to the system will be advantageous, especially for cold-weather deliveries.

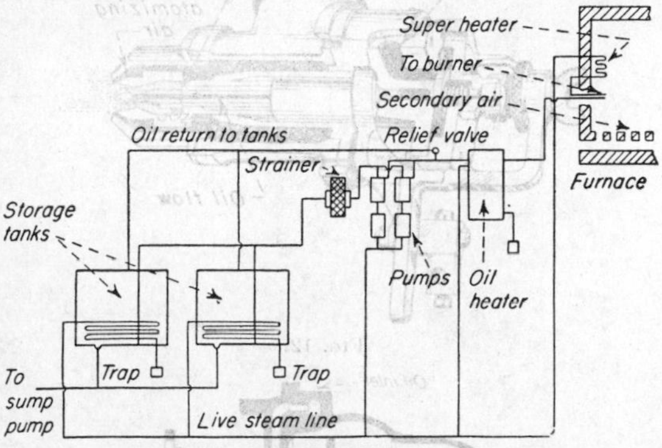

FIG. 7. Fuel-oil firing system.

A simple gage pole generally meets all measuring requirements for fuel-oil storage. Gage glasses are troublesome with heavy-oil installations and are not to be generally recommended. If a continuous indicator is desired, one of several float devices on the market may be purchased. However, it is doubtful if any of these will serve the purpose better for outside tanks above ground than the following homemade device. Pass a $\frac{1}{10}$-in. steel wire through a gland on the top of the tank and over an old bicycle rim and wheel arranged as a pulley turning on its original ball-bearing axle. Connect one end of the wire to a 2-ft. square wooden float in the tank and the other to a weighted indicator guided by, and registering over, a scale on the outside of the tank. If the gland is lightly packed with lead wool, there will be no danger of accidentally ignited gases striking back through the opening, and all insurance regulations in this regard will be met. The device will operate over a period of years without attention.

Pumps—Piping. Almost any type of pump of the proper size and capable of producing the requisite pressure may be used. A horizontal duplex steam pump provided with an air chamber to reduce the pulsation on the outlet oil is the most generally satisfactory, though motor-driven rotary pumps are gaining in favor for this service. Strainers should be inserted before the pump and a two-compartment-type strainer is advisable, since cleaning can be done without affecting operation. Pumps should be installed in duplicate, particularly where fire-protection equipment is dependent on the continuity of fuel-oil (steam) supply. The pumps should be placed as close to the storage tanks and at as low a level as possible. Advantage should be taken of every possible opportunity to reduce resistance in the pump suction line.

* See Sec. 20.

Piping from the pumps to the oil heater and from the heater to the burners should be designed for oil velocities of the order of 2 ft./sec. or lower.

Fuel-oil Heaters. Design or selection of the oil heater should be made with the requirement of easy cleaning in mind. Well-designed tubular units are most advantageous from this standpoint alone, but the standard double-pipe unit is often attractive because it makes it possible to heat the oil almost to the burner tip itself. Elegance in design and most efficient heating units can be obtained only by applying heat-transfer formulas such as discussed in detail in Sec. 6. Nevertheless, heaters for fuel-oil-burning equipment are generally so small that the investment savings from elaborate design scarcely compensate for the time consumed. Such heaters may be designed on the assumption that the heat-transfer coefficient from steam to oil is 80 B.t.u./(sq. ft.)(°F.)(hr.), with an oil velocity of 3 ft./sec.; or 40 B.t.u./(sq. ft.)(hr.)(°F.), with an oil velocity of 1 ft./sec. The specific heat of the oil may be taken as 0.5. Enough surface should be provided to reduce the viscosity of the oil to the proper point for the particular burner when the surfaces are fouled sufficiently to reduce heat-transfer rates to about one-third the above values.

Burners. There are five types of fuel-oil burners in use at the present time, namely: (1) steam atomizing; (2) high-pressure air atomizing; (3) low-pressure air atomizing; (4) spray nozzle (also called "mechanical atomizing"); (5) rotary mechanical atomizing.

There are a large number of companies in the United States manufacturing fuel-oil burners. Furthermore, experience indicates that it is almost impossible for a mechanically minded individual to observe the operation of fuel-oil burners for any length of time without feeling the urge to design his own improvement thereof. As a result of this situation, there are a large number of satisfactory designs in each of the above types of burners as well as a host of very medicore appliances. The figures that follow illustrate a few of the good burners that are on the market, although the choice of a burner for illustration does not necessarily mean it is considered as the most efficient in its class.

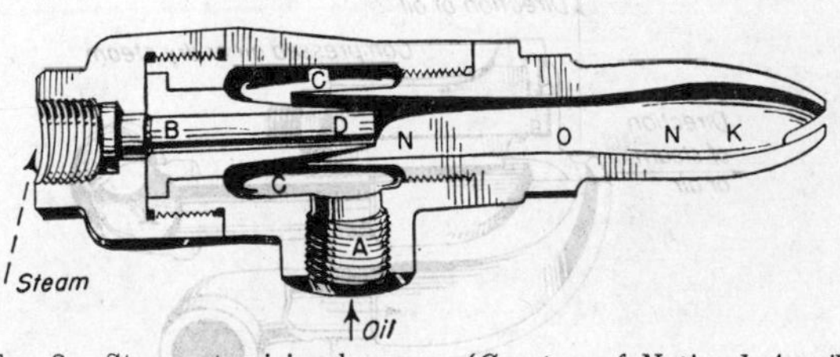

FIG. 8. Steam-atomizing burner. *(Courtesy of National Aeroil Burner Co.)*

Figure 8 shows an example of the inside mixing type of steam-atomizing burner. The slot in the burner tip illustrated produces a flat flame but various forms of tips are interchangeable to give the form of flame most suited to the combustion volume. When colloidal fuels, consisting of a suspension of coal or coke in fuel oil, or pitch, tar, or heavy fuel which may contain dirt or coke in suspension is to be burned a more positive commingling of steam and oil is desirable and a mixing "core" such as is illustrated in Fig. 9 is preferable. Various forms of both core and tips are available to fit the needs of a particular installation.

The common feature of these burners is that the oil and steam for atomization are mixed within the burner itself and they are representative of this form of steam-atomizing burner.

Outside-mixing-type burners are also on the market,

one type being illustrated in Fig. 10. In this form the steam disintegrates the oil stream into fine droplets after the oil leaves the confining channel.

With either of these burners, the primary air is drawn in around the burner, and secondary air is usually supplied through a checkerwork in the base of the furnace. The steam consumption of these burners may be as low as 1 per cent of that generated in the boiler. However, about 1½ to 2 per cent is more usual, and with poor operation it is possible to use as much as 5 per cent of the total steam. Two pounds of steam per gallon of oil burned corresponds roughly to 1.5 per cent of the steam generated in a boiler running at 80 per cent efficiency.

Proper adjustment of the steam quantities can be made by observing the character of the flame. A white

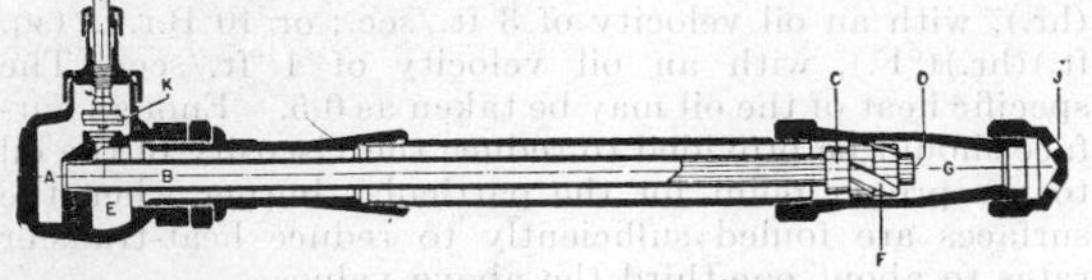

FIG. 9. Steam-atomizing burner. (*Courtesy of National Aeroil Burner Co.*)

flame indicates too much steam, a smoky one too little, and a bright orange to yellow flame indicates the proper mixture. All these observations are predicated on the assumption that sufficient air for combustion is present.

It is hardly necessary to state that the only function of the steam is to break up the oil into fine droplets so that it can quickly and thoroughly come in contact with air for combustion. Having effected this atomization, dry steam has practically no further effect on the combustion. However, if the steam is wet, combustion is slowed down, burners operate in an irregular fashion, and a portion of the heat of the oil is required to supply the latent heat of vaporization of water. To obtain the most effective atomization of the oil, provision should be made so that only dry steam can reach the burner. A small coil in the

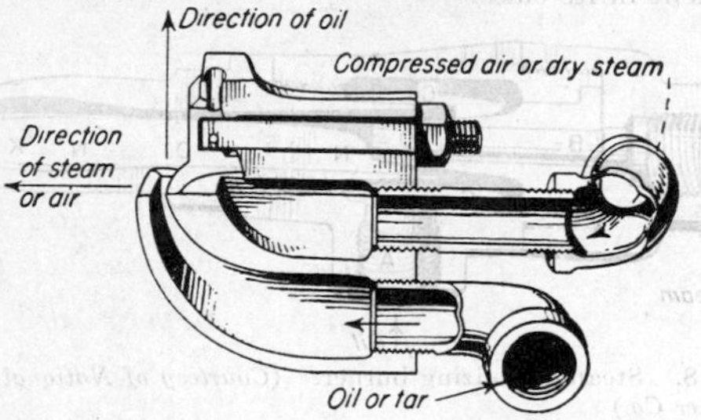

FIG. 10. Steam-atomizing burner. (*Courtesy of W. N. Best Corp.*)

fire box of the furnace can be used with advantage to superheat the steam before it is passed to the burners.

Figure 11 shows an example of the high-pressure air-atomizing burner. These burners differ little in design from the steam-atomizing burner and, in fact, are often used interchangeably. A drip tank should be installed on the air lines ahead of compressed-air-atomizing burners, since occasional slugs of water may extinguish the flame, particularly before the combustion chamber has attained full operating temperature.

Figures 12 and 13 show two examples of low-pressure air-atomizing burners. Burners of this type require air supply at pressures from a few ounces up to 4 to 5 lb./sq. in. Such burners may require atomizing air equal to 50 or 60 per cent of the total necessary for complete

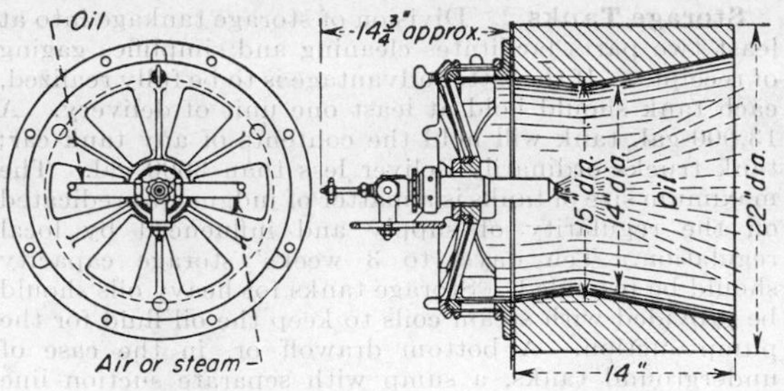

FIG. 11. High-pressure air (or steam-) atomizing burner. (*Courtesy of Schutte and Koerting Co.*)

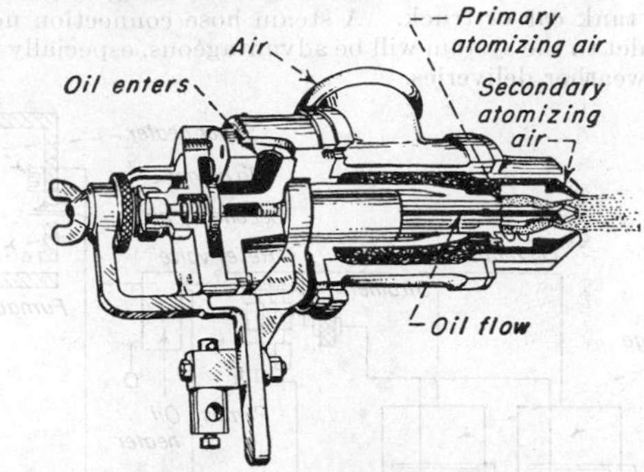

FIG. 12.

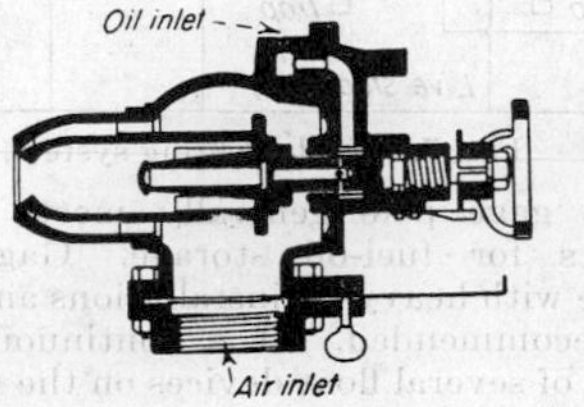

FIG. 13. Low-pressure air-atomizing burner. (*Courtesy of Schutte and Koerting Co.*)

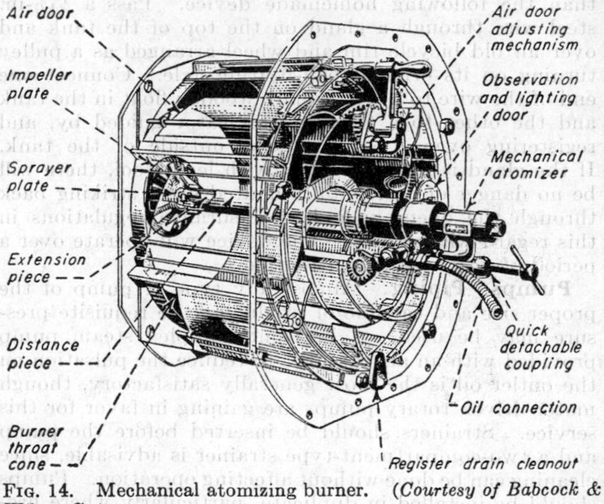

FIG. 14. Mechanical atomizing burner. (*Courtesy of Babcock & Wilcox Co.*)

combustion. Motor-driven blowers and less frequently steam-turbine-driven blowers, are used to supply the air.

The mechanical or spray-nozzle type of burner (Fig. 14) depends upon special-shaped orifices to atomize the oil forced to the burner under a pressure of 100 to 200 lb./sq. in. Some burners will operate fairly satisfactorily at lower pressure, but this type is essentially a high-pressure, low-viscosity oil burner. Heavy fuel oil may of course be burned with these appliances, but the temperature must be raised so that the viscosity of the oil ranges from 100 to 125 sec. Saybolt Universal for best results.

When mechanical spray burners are used, it is practically necessary to arrange to supply all air for combustion through openings around the atomizer. The best results can be secured only if the air is guided in proper channels with respect to the pattern of the oil spray. Air registers such as are shown in the illustrations are therefore always supplied with these burners. The size of air ports in these registers should be easily and accurately adjustable.

Mechanical spray burners are very popular for marine boilers where elimination of steam consumption and saving of space and power for air blowers or compressors are of importance. The two principal disadvantages of this type of burner are (1) the necessity of removing practically all suspended matter from the oil, else the small burner orifices will clog up; and (2) interdependence of quantity of oil burned and character of the atomization. As a matter of fact, variation of capacity over wide ranges can be secured only by changing the nozzles in these burners. Small changes in capacity can be effected by changing the oil pressure, though the procedure is less effective with small nozzles than with larger sizes.

The last of the five classes of burners, the rotary mechanical type, is finding extended application for household heating boilers. The invasion of the power-plant field by this type of burner is improbable because of the difficulty of handling heavy fuel oil.

The selection of the proper type of burner involves many factors of purely local significance. Thus, if very heavy oil is to be burned under boilers upon which the load varies greatly, the mechanical spray burner is at a serious disadvantage and the tendency of steam jet burners to "blow out" with low fires must be considered. If medium- or light-gravity oil is to be used and the load is steady, either mechanical burners or steam burners will operate satisfactorily. The use of high-pressure air burners is becoming obsolete for boiler plants, but they are still used to advantage in some types of metallurgical furnaces.

Relative efficiencies of the various types of burners also admit of no general statement. Proper size of burners for the particular duty, and skill in firing, probably weigh more in the efficiency obtained than any differences of design.

A new fuel-oil installation should at first be under the scrutiny of an informed operator or engineer who can interpret flue-gas analyses and who can direct work to determine the optimum oil temperatures, burner adjustment, draft conditions, secondary air regulation, etc. By following this procedure and by occasionally checking up intelligent and conscientious firemen with results of flue-gas analyses, the high efficiency inherent to oil-burning equipment can be established and maintained.

Miscellaneous Liquid Fuels

A discussion of liquid fuels can hardly ignore certain fuels other than fuel oil. Gasoline and kerosene are used to a considerable extent for stoves in isolated localities, and gasoline is commonly used as the fuel for plumbers' torches and similar devices. Alcohol and benzol are also used to a limited extent, usually in small and highly specialized devices. The devices that use any one of these higher priced fuels are usually made as a unit by the manufacturer, who supplies detailed instructions as to quality of fuel suitable and method for using most satisfactorily.

Industrial use of such liquid fuels as shale oil, coal tar, tar oil, and distillates or residues from low-temperature carbonization of coal is sometimes justified. Suitable methods for using such materials are to be learned only by trial, and little in the way of generalization can be offered.

Coal Tar. Coal tar is a by-product of the manufacture of coke and of coal gas. This tar is a viscous mixture consisting, for the most part, of aromatic compounds. Its heating value varies from 15,000 to 16,500 B.t.u./lb. In order to burn coal tar in regular fuel-oil burners, the tar must be filtered and preheated to such a temperature that its viscosity is reduced to that of the oils for which the particular burner is designed.

Tar Oil. Tar oil is obtained by the distillation of coal tar and consists of so-called "creosote oil," "anthracene oil" and other materials. Its heating value is about the same as that of coal tar.

Gasoline. The composition of an average gasoline is: carbon, 83.5 to 85 per cent; hydrogen, 15.0 to 15.8 per cent; nitrogen plus sulfur plus oxygen, 0 to 1 per cent. The heating value is about 20,000 B.t.u./lb. Gasoline containing tetraethyl lead should not be used for heating equipment.

Kerosene. The average composition of kerosene is carbon 84 per cent, hydrogen 16 per cent. Sulfur should not exceed 0.125 per cent (United States government specification). The heating value varies from 20,000 to 21,000 B.t.u./lb. The government specifications require a distillation end point of 625°F. maximum and a flash point of 115°F. minimum.

Alcohol and Benzol. Table 14 gives pertinent combustion data for various alcohol and benzol fuels.

Table 14. Combustion Data for Various Alcohol and Benzol Fuels

Fuel	O_2 required for combustion, lb./lb.	Air required for combustion, lb./lb.	Products of combustion, lb./lb.			Approximate higher heating value, B.t.u./lb.
			CO_2	H_2O	N_2	
Ethyl alcohol (C_2H_6O)	2.08	9.04	1.91	1.17	6.95	12,780
Methyl alcohol (CH_4O)	1.5	6.50	1.38	1.12	5.0	9,550
Benzol (C_6H_6)	3.1	13.32	3.39	0.69	0.24	18,000
Denatured alcohol	1.81	7.83	1.66	1.15	6.02	11,600
50% mixture of alcohol and benzol	2.45	10.60	2.53	0.92	8.16	14,200

GASEOUS FUELS

BY WILBERT J. HUFF

General References: Meade, "The New Modern Gasworks Practice," vol. 1, Eyre and Spottiswoode, Ltd., London; "Modern Gas Works Practice," Van Nostrand, New York, 1921. Morgan, "American Gas Practice," 2 vols. (privately published by author, Columbia University). "Gludd's International Handbook of the By-product Coke Industry," American ed. by Jacobson, Chemical Catalog, 1932. Lowery *et al.* (National Research Council Committee), "Chemistry of Coal Utilization," John Wiley & Sons, Inc., New York, 1945. Lunge, "Coal Tar and Ammonia," Van Nostrand, New York, 1916. Warnes, "Coal Tar Distillation and Working Up of Tar Products," Benn, London, 1923. *Proceedings American Gas Association*, New York, particularly the technical and natural gas sections. "Fuel-flue Gases" [new edition (1948) entitled "Gaseous Fuels"], American

Gas Association. "Combustion," American Gas Association. "Gas Chemist's Handbook," 3rd ed., American Gas Association. Altieri, "Light Oils and Light Oil Products," American Gas Association. Altieri, "Gas Analysis and the Testing of Gaseous Materials," American Gas Association. Seil, "Dry Box Purification," American Gas Association. Bulletins, Technical Papers, and Reports of Investigations, U.S. Bureau of Mines. Circulars, Technologic Papers, Scientific Papers, and Research Papers and Miscellaneous Publications, Bureau of Standards, U.S. Department of Commerce.

DESCRIPTION OF VARIOUS FUEL GASES*

Acetylene. The use of acetylene as a fuel and illuminant is generally limited to cutting and welding operations requiring high flame temperature, to small isolated lighting plants, and to single "carbide" lights. It is made from calcium carbide and water. To avoid a dangerous rise in temperature, sufficient water (about ½ gal./lb. carbide) should be present in the generator. The crude gas contains as impurities: ammonia, hydrogen sulfide, and phosphine, which must be removed before the gas can be used for indoor illumination. Acetylene forms explosive acetylides, particularly with copper, has wide explosive limits when mixed with air, and is explosive per se at pressures of 5.9 lb./sq. in. gage or greater and at 15 lb./sq. in. requires only 540°C. to set it off. It ignites at 635°C. at atmospheric pressure. Its use as a liquid is therefore prohibited, and it is ordinarily dissolved in acetone under pressure. (Fulweiler, chapter on Industrial Gases, in Rogers, "Manual of Industrial Chemistry," 5th ed., Van Nostrand, New York, 1931. Vogel, "Das Acetylen," Spamer, Leipzig, 1923. Leeds and Butterfield, "Acetylene. The Principles of Its Generation and Use," Griffin, London, 1910. Nieuwland and Vogt, "The Chemistry of Acetylene," Reinhold, New York. *Cf.* also Jones *et al.*, *U.S. Bur. Mines Rept. Investigations*, 3567, 3755, 3809, 3826.)

Blast-furnace gas is a by-product from the smelting of iron ore with coke and preheated air in the blast furnace. About one-third of the exit gases from the top of the furnace is used for heating the blast stoves, and the remainder may be burned. The gas is carefully cleaned and washed before use. The low B.t.u. value requires regenerative preheating, as with producer gas. [Camp and Francis, chapter on Blast Furnace Gas, in Bacon and Hamor, "American Fuels," McGraw-Hill, New York, 1922. Wagner, "The Cleaning of Blast Furnace Gases," McGraw-Hill, New York, 1914. *Blast Furnace and Steel Plant*, **17**, 1048–1052 (1929).]

Blue water gas (see pp. 1579–1580) is the product obtained by the interaction of steam and a highly heated solid carbonaceous fuel. The fuel is brought to a high temperature by blasting with air, after which the air supply is cut off and steam is injected. The blast of air is again admitted to restore the temperature, after which another steam run is made. [Fulweiler, chapter on City Gas, *op. cit.* Meade, "Modern Gas Works Practice," *op. cit.* Travers, *Trans. Inst. Chem. Engrs. (London)*, **2**, 65 (1924). Morgan, "American Gas Practice" (privately printed), vol. 1, Chap. 15, 1931. Fulweiler, *Proc. 1st Intern. Conf. Bituminous Coal*, p. 472; *U.S. Bur. Mines Bull.* 203; *Tech. Papers* 246, 274, 284, 335; *Rept. Investigations* 2183. Pettyjohn, *Am. Gas Assoc. Proc.*, 1930, p. 1535. Lowry *et al.*, "Chemistry of Coal Utilization," Chap. 37, John Wiley & Sons, Inc., New York, 1945.]

Carbureted Water Gas (see pp. 1580–1581). As the thermal content of blue water gas is too low to meet present public requirements, the gas is carbureted with oil gas which is formed by the thermal decomposition (see p. 1581) of the oil usually in supplementary shells connected in series to the generator shell of the water gas machine.

Coal gas (retort) (see p. 1582) is obtained by the destructive distillation of bituminous coal of suitable characteristics, usually designated as a gas coal, in a closed, highly heated retort of fire clay or silica. (Fulweiler, chapter on City Gas, in Rogers, "Manual of Industrial Chemistry," 5th ed., Van Nostrand, New York, 1931. Meade, *loc. cit.* Morgan, *loc. cit.* Also publications of the American Gas Institute and the American Gas Association, particularly the Carbonization Committee of the latter. See also references under Coke-oven Gas.)

Coke-oven gas is a coal gas derived from the distillation of a bituminous coal generally known as a "coking coal," which is somewhat lower in volatile content than the usual gas coal. The carbonizing chamber is very much larger than the coal-gas retort and is built up from silica forms. (Sperr, chapter on the Technology of Coke, in Bacon and Hamor, "American Fuels," McGraw-Hill, New York, 1922. See references under Coal Gas; also Haslam and Russell, "Fuels and Their Combustion" chapter on the Carbonization of Coal, with bibliography, McGraw-Hill, New York, 1925. Gluud, American ed. by Jacobson. *Cf.* also Lowry *et al.*, *loc. cit.*

Hydrogen. The use of hydrogen as a fuel is limited to certain special industrial purposes, such as certain welding and cutting operations. One of the methods of production at present favored involves the catalytic oxidation of the carbon monoxide in blue water gas to carbon dioxide with steam. Electrolysis and the low-temperature fractionation of coal gas are also used. (Taylor, "Industrial Hydrogen," Reinhold, New York, 1921. Greenwood, "Industrial Gases," Sec. VI, Bailliere, Tindall and Cox, 1920. Pincass, "Die Industrielle Herstellung von Wasserstoff," Steinkopf, Dresden, 1933. See also "Fixed Nitrogen," Reinhold, New York, 1932.)

Natural gas is obtained by drilling through overlying strata to tap porous rocks, generally in sandstone of open texture or broken limestone, known as sands, above which a relatively impervious compact shale or cap rock has been folded in the form of a dome or inverted container. (Cross, "Handbook of Petroleum, Asphalt and Natural Gas," 1928 revision, Kansas City Testing Laboratory. Diehl, "Natural Gas Handbook," Metric Metal Works, Erie, Pa., 1927. *U.S. Bur. Mines Bull.*, particularly 65, 82, 134, 163, 232; and other publications of the same bureau. Bacon and Hamor, "American Fuels," McGraw-Hill, New York, 1922. Dunstan *et al.*, "The Science of Petroleum," Oxford University Press, London, 1938.)

Oil Gas. The oil gas manufactured for public utility distribution on the Pacific coast is made by the gasification of oil with steam in a chamber containing hot checker brick. The heat is obtained by burning oil in the same chamber, and the process is a cyclic one, as in the water gas process. [Pike and West, *Ind. Eng. Chem.*. **21**, 104–109 (1929). Morgan, *op. cit.*]

There are a number of other oil gas processes, none of which are, however, widely used in the United States. The gases from oil refineries, particularly from the operation of various cracking processes for the production of gasoline from higher boiling oils, are utilized to some extent, and this utilization for a time tended to increase. Recent developments in the production of synthetic gasoline and synthetic rubber have provided other outlets. The recent widespread supplanting of manufactured gas by natural gas from long-distance transmission mains has stimulated great interest in standby substitute high-B.t.u. oil-gas processes, and development of such processes continues.

Petroleum Gases. The fractionation of gasolines, particularly those recovered from natural gas, has made available a large quantity of light hydrocarbon liquids and vapors, chiefly propane and butane. These are available commercially for shipment under pressure. As fuel gases they possess certain advantages, as in isolated situations where only a small quantity of gas is required, which may be serviced from pressure containers. In non-inflammable mixtures with air, the vapors, particularly butane, are distributed by pipe line from a central point for the servicing of small communities. Propane may be distributed directly without admixture. (Thomas and Setrum, *Proc. Am. Gas Assoc.*, 1928, pp. 1284*ff*. Odell, *U.S. Bur. Mines Rept. Investigations* 2840, November, 1927. Oberfell, *Gas Age-Record*, Feb. 2, 1929. Odell, *U.S. Bur. Mines Bull.* 294. *Repts. Am. Gas Assoc., Water-gas Committee.* Dunstan *et al.*, *loc. cit.* Morgan, *op. cit.*, Chap. 20.) The general distribution of natural gas has lead to greatly increased use of liquefied petroleum gases to assist in meeting peak loads, and for substitute gas enrichment. Developments have afforded catalytic re-forming processes by which liquefied petroleum gases may quickly be converted to manufactured gas substitutes also. (See pp. 1577 and 1581.)

Producer gas (see p. 1579) is generated by blasting a deep hot bed of coal or coke continuously with a mixture of air and steam. Because of the large percentage of nitrogen in the gas thus obtained, its heating value is low. Solid fuels of widely different characteristics, including wood waste, may be employed in the process, and the variations in the thermal value of the resulting gas with the various volatile contents of the solid fuels are correspondingly large. Gas having as much as 180 B.t.u./cu. ft. is made with high-volatile coals, while low-volatile coke with poor operations may give 110 B.t.u./cu. ft. or lower. (Rambush, "Modern Gas Producers," Benn, London, 1923. *U.S. Bur. Mines Bull.* 7 and 13. Haslam and Russell, chapter on Producer Gas with bibliography in "Fuels and Their Combustion," McGraw-Hill, New York, 1925. Reports, Car-

* Analysis of typical industrial fuel gases will be found in Tables 15 and 16.

Table 15. Analyses of Fuel Gases*

Gas	Constituents of gas, % by volume							Illuminants		Sp. gr. air = .000	Cu. ft. of air required for combustion of 1 cu. ft. gas	B.t.u./cu. ft. gross	B.t.u./cu. ft. net	Products of combustion, cu. ft./cu. ft. gas				Ultimate CO_2, %	Net B.t.u./cu. ft. of the products of combustion	Flame temperature corrected for dissociation
	CO	CO_2	H_2	N_2	O_2	CH_4	C_2H_6	C_2H_4	C_6H_6					CO_2	H_2O	N_2	Total (dry)			
Natural gas, Texarkana	....	0.80	....	3.20	...	96.00	.....	...	...	0.57	9.17	967	873	0.97	1.92	7.29	8.26	11.7	80.2	3580
Natural gas, Cleveland	....	....	....	1.30	...	80.50	18.20	...	...	.65	10.70	1131	1025	1.17	2.16	8.50	9.67	12.1	81.1	3600
Natural gas, Oil City, Pa	....	....	....	1.10	...	67.60	31.30	...	...	.71	11.70	1232	1120	1.30	2.29	9.26	10.56	12.3	81.7	3620
Retort coal gas (horizontal)	8.6	1.5	52.5	3.5	0.3	31.4	.....	1.1	1.1	.42	5.00	575	510	0.50	1.21	3.99	4.49	11.2	83.5	3665
Coke-oven gas	6.3	1.8	53.0	3.4	.2	31.6	.....	2.7	1.0	.42	5.19	588	521	.51	1.25	4.13	4.64	11.0	82.7	3660
Coke-oven gas, Koppers ovens	6.8	2.2	47.3	6.0	.3	33.9	.....	2.6	0.9	.44	5.23	591	525	.54	1.23	4.19	4.73	11.4	82.3	3650
Carbureted water gas	33.4	3.9	34.6	7.9	.9	10.4	.....	6.7	2.2	.65	4.37	536	496	.74	0.75	3.54	4.28	17.2	88.5	3815
Blue water gas	42.8	3.0	49.9	3.3	.5	0.5	.....	...	...	.53	2.26	308	281	.46	.51	1.82	2.28	22.3	89.7	3800
Theoretical water gas	50.0	....	50.0	....	...	....	.....	...	...	.52	2.39	325	298	.50	.50	1.89	2.39	20.9	90.3	3830
Anthracite producer gas	24.0	7.5	16.5	50.2	.6	1.2	.....	...	...	.85	1.05	134	124	.33	.19	1.36	1.69	19.5	65.6	3000
Bituminous producer gas	27.0	4.5	14.0	50.9	.6	3.0	.....	...	...	.86	1.24	150	140	.35	.19	1.49	1.84	19.0	69.2	3160
Blast-furnace gas	27.5	10.0	3.0	58.0	1.0	0.5	.....	...	...	1.00	0.78	102	100	.38	.04	1.21	1.59	23.9	61.0	2800
Oil gas (Protero 1920)	6.8	1.0	59.2	2.7	0.1	25.4	.....	3.8	1.0	0.35	4.91	575	510	.47	1.21	3.91	4.38	10.7	84.2	3725

* From "Combustion," 2d ed., Table 20, p. 34, American Gas Association, 1926. Reproduced by permission.

Table 16. Industrial Fuel Gases Commonly Available*

Component gases	CO	H_2	CH_4	C_2H_6	Illuminants	CO_2	O_2	N_2	Blue gases	Hydrocarbons	Inerts	B.t.u./cu. ft.	Sp. gr.
Producer gas	25.3	13.2	0.4	....	....	5.4	0.5	55.2	38.5	0.4	61.1	129	0.878
Blue gas	38.3	52.8	0.4	....	....	5.5	.1	2.9	91.1	0.4	8.5	295	.521
Blue gas, 250 B.t.u./cu. ft	34.8	42.0	0.4	....	....	5.5	.2	17.1	76.8	0.4	22.8	250	.618
Coal gas	5.9	53.2	29.6	....	2.7	1.4	.7	6.5	59.1	32.3	8.6	548	.375
Coal gas — producer gas	7.3	51.3	26.6	....	3.2	2.1	.9	8.6	58.6	29.8	11.6	527	.431
Carb. water gas, no blow run	32.6	38.2	8.0	2.2	8.6	3.4	.5	6.5	70.8	18.8	10.4	530	.643
Carb. water gas, with blow run	24.1	32.5	9.0	2.2	10.3	4.6	.6	16.7	56.6	21.5	21.9	530	.703
Carb. water gas, low gravity	21.9	49.6	10.9	2.5	6.1	3.6	.4	5.0	71.5	19.5	9.0	536	.539
Carb. water gas, no blow run	25.7	32.6	8.6	7.2	13.1	6.1	.6	6.1	58.3	28.9	12.8	603	.675
Reformed gas, refinery	14.3	50.9	15.9	5.0	2.4	2.2	.9	8.4	65.2	23.3	11.5	530	.497
Reformed gas, natural	14.5	43.5	29.7	1.7	1.0	2.4	.2	7.0	58.0	32.4	9.6	530	.472
Refinery gas, high paraffin	1.2	6.1	4.4	72.5	15.0	...	.2	0.6	7.3	91.9	0.8	1644	.990
Refinery gas, high olefin	1.2	13.1	23.3	21.7	39.6	0.1	1.0	...	14.3	84.6	1.1	1468	.890
Natural gas	....	....	78.8	14.0	....	.4	....	6.8	....	92.8	7.2	1013	.635
Natural gas	....	....	82.8	16.3	0.8	.1	....	....	....	99.9	0.1	1145	.650
Butane, commercial	....	....	....	....	....	....	....	....	....	100.0	....	3200	1.950
Propane, commercial	....	....	....	....	....	....	....	....	....	100.0	....	2550	1.523

* From Perry, "Gas Mixtures," *Proc. Am. Gas Assoc.*, p. 809, 1931.

bonization Comm. and Subcomm. on Producer Gas, *Am. Gas. Assoc., Proc.*, 1927, 1928, 1929, 1930.)

Re-formed Gas. Although applicable to any gas transformed by suitable treatment, the term "re-formed gas" is ordinarily applied to lower thermal value gases obtained by the pyrolysis and steam decomposition of high thermal value gases, such as natural gas or oil refinery gas. The steam minimizes carbon loss and possesses other advantages. Catalytic re-forming to meet peak loads has recently become important. (*Cf.* Odell, *U.S. Bur. Mines Bull.* 301; *Tech. Paper* 483; *Rept. Investigations* 2973 and 2991. Also Schlegel, *Am. Gas Assoc. Proc.*, 1930. p. 1466. Morgan, *op. cit.*, Chap. 19.) Milbourne, Catalytic Cracking Plants for Relieving Gas Utility Peak Loads, Paper 48-A-101 *Am. Soc. Mech. Engrs.*, Fuel Division, Annual Meeting, Nov. 28–Dec. 3, 1948.

Impurities and By-products

Sulfur Impurities in Gaseous Fuels. The chief sulfur impurity in gaseous fuels is **hydrogen sulfide.** Certain natural and petroleum gases are found free from this undesirable impurity, but its presence may otherwise be expected rather universally in raw fuel gases in amounts which may range from approximately 100 gr./100 cu. ft. in blue and carbureted water gas to several hundred grains per 100 cu. ft. in coal and coke-oven gases. In refinery gases from sulfur crudes and natural gases from sulfur-bearing regions the concentration may be several thousand grains per 100 cu. ft.

Another important sulfur impurity is **carbon disulfide,** which may contribute as much as 80 per cent of the organic sulfur present in manufactured fuel gases. In such gases, however, the total organic sulfur is relatively small, usually much less than the 30 gr./100 cu. ft. of gas limit permitted by most states.

Other sulfur compounds, which may be present in small amounts, are the thiophenes, carbon oxysulfide, mercaptans, thioesters, and organic sulfides. [Repts. Am. Gas Assoc. Purification Committee. Morgan, "American Manufactured Gas Practice," 1931 ed., Chap. 23. "International Handbook of the By-Product Coke Industry," by W. Gluud, American ed. by Jacobson, Chap. 9, Reinhold, New York, 1932. Huff and Milbourne, *Am. Gas Assoc. Proc.*, 1930, p. 856; *Ind. Eng. Chem.*, **22**, 1213 (1930). Dotterweich and Huff, *Am. Gas Assoc. Proc.*, 1938, p. 699. Huff and Holtz, *Am. Gas Assoc. Proc.*, 1927, pp. 1431, 1436; *Ind. Eng. Chem.*, **19**, 1268 (1927); **22**, 639 (1930). Denig and Powell, Liquid Purification, *Am. Gas Assoc. Proc*, 1933, p. 913. Denig, *Am. Gas Assoc. Proc.*, 1933, p. 903. Sperr, *Am. Gas Assoc. Proc., Tech. Sec.*, 1921, p. 282; 1923, p. 1200. Huff, *Proc. 2d Intern. Conf. Bituminous Coal* II, 1928, p. 814. Bottoms, *Ind. Eng. Chem.*, **23**, 501 (1931). Lowry *et al.*, *op. cit.*, Chap. 26. *Cf.* Reports of Organic Sulfur Committee, American Gas Association.]

In the **iron oxide process** for the removal of hydrogen sulfide, as generally practiced in America, the oxide, which is ordinarily rusted iron borings, bog-iron ore, or a by-product from the refining of bauxite, is mixed with wood shavings. Ground corncobs also make an excellent support, although wood is to be preferred. About 25 lb. oxide are used per bushel of shavings. With boring oxides, the rusting may take place on the shavings, and a prepared purifying material may be offered. The shavings serve as a fluffing agent, permitting the ready passage of gas through the material and serving to expose the oxide to the gas stream. Oxides vary enormously in their purifying properties, and this property is dependent not upon the iron content but upon the chemical and physical state of the oxide. Humidity effects play a most important part in the reaction. Data developed by Huff and Milbourne (*loc. cit.*) indicate that the fouling reaction is best maintained at a humidity of about 65 per cent. The fouled oxide is revivified by the action of oxygen in air in the presence of moisture; the study above mentioned showed that high humidities, just sufficiently under saturation to avoid the precipitation of liquid moisture, are best. Spent oxide may be revivified *in situ* simultaneously with fouling by admitting a small quantity of air with the gas, about 0.5 per cent more oxygen (on total volume basis) than is required for the reaction $2H_2S + O_2 \rightarrow 2S + 2H_2O$. The reaction is, however, more complex than this equation indicates. The revivification process is relatively slow and, when practiced simultaneously, humidity conditions favorable to it must prevail. The oxide may be revivified by removal from the box or by by-passing the box and passing air through it, but care must be taken to avoid overheating and explosions.

The oxide is placed in cast-iron or steel boxes of large cross section in order to effect contact with a minimum of pressure drop. Wooden trays are placed in the boxes, and the oxide-shavings mixture is placed in two or three layers usually from 18 to 30 in. deep. If only a single layer is used, it may be about 4 ft. deep. The oxide is distributed over at least three or four boxes in series, and suitable valve arrangements make it possible to change the order to permit the placing of a badly fouled box out of the line or in a position where it may not receive much hydrogen sulfide and so be revivified.

A formula that allows for the several factors in calculating the area of the boxes is known as the Steere formula [*Bull.* 37, Steere Engineering Co., Detroit, Mich. or *Gas Age*, **43**, 227 (1919)]. This is

$$G = \frac{3000 \times (D + C) \times A}{S}$$

or

$$A = \frac{G \times S}{3000 \times (D + C)}$$

where G is the maximum amount of gas in cubic feet corrected to 60°F., to be purified per hour; D is the total depth in feet of oxide through which the gas passes consecutively. When a single catch box is used for two or more parallel sets of boxes, disregard the catch box in obtaining D; A is the cross-sectional area in square feet of oxide through which the gas passes in any one layer in the series; C is the factor: 4 for a two-box, 8 for a three-box, and 10 for a four-box series. Where a single catch box is used for two two-box sets, $C = 6$; S is the factor for grains hydrogen sulfide per 100 cu. ft. gas entering the purifiers as follows:

Grains H_2S per 100 Cu. Ft. Unpurified Gas	Factor S
1000 or more	720
900	700
800	675
700	640
600	600
500	560
400	525
300	500
200 or less	480

By-products in Gas Manufacture. In the carbonization of coal for the production of gas and coke, there are obtained as by-products: ammonia, light oils of the aromatic-hydrocarbon series, and tar. Other by-products, such as cyanides, are sometimes obtained.

Ammonia. The ammonia yields may vary from 20 to 26 lb. ammonium sulfate equivalent per ton of coal carbonized. The *free ammonia* in the liquor is chiefly carbonates and sulfides. All these decompose with heat and ammonia may be recovered by steam distillation. The *fixed ammonium salts* must be decomposed by the action of a base, such as lime.

The ammonia that remains in the gas is recovered by water scrubbing or by washing with a dilute solution of sulfuric acid in a lead-lined saturator. Today, the greater part of the ammonia recovered from the carbonization of coal is fixed as the sulfate. Free-ammonia and aqueous-ammonium solutions of commerce are more conveniently prepared in high purity by synthetic-ammonia processes. (Lunge, "Coal Tar and Ammonia," Van Nostrand, New York, 1916. Calvert, "The Manufacture of Sulfate of Ammonia and Crude Ammonia," Benn, London, 1917. Parrish, "Design and Working of Ammonia Stills," Van Nostrand, New York, 1924. Meade, "Modern Gas Works Practice," Benn, London, 1921. Gluud, "International Handbook of the By-Product Coke Industry," American ed. by Jacobson, Chap. 10, 1932. Morgan, *op. cit.*, Chap. 22. Lowry *et al.*, *op. cit.*, Chaps. 27 and 32.)

The Recovery of Light Oils. The light oils, chiefly benzene, toluene, and xylene, existing in certain industrial gases, particularly coal gas and carbureted water gas, may be recovered by scrubbing these gases with wash oil and subsequently distilling the oil solution to volatilize the light oil dissolved therein. *Cf.* Sperr, *Trans. Am. Inst. Chem. Engrs.*, **9**, 169 (1917); *Gas Age*, **41**, 393–397. Gluud-Jacobson, *op. cit.*, Chap. 11. Fisher, *U.S. Bur. Mines Bull.* 412, Lowry *et al.*, *op. cit.*, Chap. 28.

Tar. An examination of coal tar on a large scale has been made by Weiss and Downs [*Ind. Eng. Chem.*, **15**, 1022 (1923)]. Water gas tar has been examined by Downs and Dean [*Ind. Eng. Chem.*, **6**, 366 (1914)]. A recent extensive study of the composition of coal tar and light oil has been made by Fisher (*U.S. Bur. Mines Bull.* 412). A bibliography of 255 references is appended. A new procedure for the characterization of water gas tars was developed at Penn State and applied to 13 tars. (*Research Bulletin* 2, Gas Production Research Committee, Am. Gas Assoc., September, 1949.)

Further sources: Warnes, "Coal Tar Distillation," 3d ed., Benn, London, 1923. Lunge, "Coal Tar and Ammonia," Van Nostrand, New York, 1916. Huff, *Chem. & Met. Eng.*, **26**, 113 (1922). Weiss, "Recent Progress in Science in Relation to the Gas Industry," Chap. 8, Am. Gas Assoc., 1926. Gluud, *op. cit.*, Chap. 12. The viscosity of some coke-oven tars is discussed by Huff, *Ind. Eng. Chem.*, **15**, 1026 (1923).

Compressibility of Natural Gas at High Pressures. Deviations from the simple gas laws may be important in many engineering operations involving fuel gases. (Burrell and Robertson, *U.S. Bur. Mines Tech. Papers* 131, 158. Johnson and Berwald, *U.S. Bur. Mines Tech. Paper* 539, U.S. Bur. Mines Monograph 6. American Gas Assoc., Gas Measurement Committee, Report 1, Natural Gas Dept.). As natural gas is more compressible under usual high pressures at ordinary temperatures than is called for by the simple gas laws, gas purchased at an elevated pressure gives a greater volume when the pressure is reduced than it would if the gas were ideal. Burrell and Robertson (*op. cit.*, *Tech. Paper* 158) give the following formula for calculating the compressibility of natural gas:

$$D = aP_1 + bP_2 + cP_3 + dP_4 +, \text{etc.}$$

where D expresses the percentage deviation from Boyle's law, P_1; P_2; P_3; P_4; etc., represent the partial pressures of the respective constituents expressed in atmospheres under the conditions in question, and a, b, c, d, e, and f are characteristic factors for constituents of natural gas having the following values:

Methane	a	0.228
Ethane	b	0.90
Propane	c	1.9
Carbon dioxide	d	0.67
Nitrogen	e	0.01
Air	f	0.05

The work of Johnson and Berwald (*loc. cit.*) extends the study and gives deviation curves for characteristic natural gases. Contrary to the assumption implicit in the work of Burrell and Robertson, Johnson and Berwald found that the deviation curves were not always straight lines. For details, their work should be consulted.

Gross and Net Heating Values. The **gross heating value** is the maximum utilizable heat in the products of combustion and is obtained only under such conditions that the steam is actually condensed to water. This is possible only if the products are cooled to the starting temperature, and even then there may be loss as the entering air may have been partly dry, and the flue gases must leave saturated.

In the usual applications, the excess air and the failure to cool the products of combustion generally render none of this latent heat available, and the reduced value obtained is designated the **net heating value** [*cf.* Lichty and Brown, *Ind. Eng. Chem.*, **23**, 1419 (1931); and Porter, *Ind. Eng. Chem.*, **23**, 1433 (1931)]. Osborne, Stimson, and Fiock (Mech. Engr., March 1935, p. 63) give 18,919 B.t.u./lb.-mole water condensed. (*Cf.* also *Bur. Standards Research Paper* 209.)

The density of moist gases is treated under Humidification (p. 758*ff.*) and Drying (p. 810*ff.*).

Data on Gas Manufacture

The Manufacture of Producer Gas. Producer gas is made by blowing humidified air into a deep ignited bed of solid fuel, usually coal or coke. The primary reaction is the combustion of the fuel, first giving CO_2 and N_2. As the gases progress, the CO_2 first formed is reduced to CO, and the water vapor is partly decomposed yielding CO, CO_2, and H_2. The thermochemistry of the various producer-gas reactions may be expressed as follows:

$$C + O_2 = CO_2 + 174{,}600 \text{ B.t.u.} \quad (1)$$

$$CO_2 + C = 2CO - 70{,}200 \text{ B.t.u.} \quad (2)$$

$$C + H_2O = CO + H_2 - 70{,}900 \text{ B.t.u.} \quad (3)$$

$$C + 2H_2O = CO_2 + 2H_2 - 71{,}600 \text{ B.t.u.} \quad (4)$$

$$CO + H_2O = CO_2 + H_2 - 700 \text{ B.t.u.} \quad (5)$$

The thermochemical values stated are based on products and reactants at 60°F. and liquid water.

There is some divergence of opinion concerning the exact mechanism of the reactions and their relative velocities, which cannot be considered here in detail. There is, however, evidence to show that the thin oxidation zone, immediately above the ash zone, gives predominating reaction (1). As the partial pressure of the free oxygen reaches a low value or is practically gone, carbon monoxide appears and rapidly increases. Shortly above this point, steam decomposition begins and free hydrogen appears rapidly at first and then increases only very slowly. This zone is sometimes designated the *primary reduction zone* and is relatively thin. In the higher or *secondary reduction zone*, no reduction of steam by carbon occurs. This secondary reduction zone serves chiefly as a heat interchanger, the hot gas serving to heat the incoming fuel. Above the secondary reduction zone is the *distillation* zone, which is relatively unimportant with low-volatile fuels such as coke and anthracite, but which may contribute a considerable quantity of thermal energy in the form of gaseous hydrocarbons from high-volatile fuels. Indeed, it may represent as much as 40 per cent of the heating value of the gas.

Above the fuel bed, the heating value of the gas may drop somewhat because of the Neumann reversion [*Stahl u. Eisen*, **33**, 394 (1913) and *Z. Ver. deut. Ing.*, **58**, 1481, 1501 (1914)],

$$2CO \rightarrow CO_2 + C$$

and to the leakage of CO_2 and steam around the edges of the fuel bed and through blowholes.

The capacity of a producer is a variable quantity, depending chiefly upon the quality of the fuel supplied, the method of operation, the design of the producer, and the character of the demand for gas with respect to quality and quantity. With hand-firing conditions, gasification rates for short periods as high as 15 lb. fuel per square foot grate area per hour have been attained with gas coals having a low percentage of high-fusing ash. About 8 to 9 lb./sq. ft./hr. is obtained with good-quality bituminous coal, and with lower grade fuels the gasification rate may not exceed 6 to 7 lb. The limitations are imposed by the clinker conditions and the necessity of avoiding both the blowing over the fuel and the extension of the oxidizing and primary reducing zones to such heights that the relatively slow reduction of carbon dioxide to the monoxide does not have time to proceed sufficiently, and the gas discharged is consequently too high in CO_2 and too low in heating value. Although this last disadvantage may be overcome to some extent by the use of a deeper fuel bed, and the excessive oxidizing temperatures may be reduced somewhat by the addition of more steam to the air blast, the blowing over of fuel presents a practical limitation on velocity conditions for any given grade of fuel. To avoid this with high gasification rates it is necessary to use mechanical operations, which ensures uniform fuel-bed condition free from zones of excessive gas velocity. The fuel should also be carefully sized. Observing suitable precautions, American gas producers may average gasification rates of 15 lb.; and 25 lb. are possible with good coals. It is claimed that, with mechanical poking and continuous ash removal, gasification rates as high as 50 lb. have been attained by some producers.

Increasing the depth of the fuel bed raises the temperature of the primary reduction zone and affords a longer time of contact. This gives better decomposition of the steam. According to Haslam [*Ind. Eng. Chem.*, **16**, 782 (1924)], the constituents of the producer gas come to an apparent equilibrium constant dependent upon the thickness of the fuel bed alone, and independent of gas velocity (rate of firing), ratio of pounds of steam to pounds of coal, or temperature of the exit gas. In this

$$K' = \frac{(CO_2)(H_2)}{(CO)(H_2O)} = 0.096L$$

in which L is the depth of the fuel bed in feet.

Excessive amounts of steam should be avoided, as these, in accordance with equilibrium considerations, raise the CO_2 concentration and produce steam in the exit gases, an inert diluent which removes heat from the fuel bed. On the other hand, the use of some steam is desirable, as its endothermic decomposition prevents excessive clinker formation, and serves to convert some of the sensible heat developed from the carbon-oxygen combustion into potential energy in the form of CO and H_2. The most desirable value is usually about 0.4 lb. water in the air blast per pound of coal gasified.

The importance of a satisfactory size and uniform space conditions in the producer bed has already been indicated. The coal should be spread uniformly, the fuel bed should be constantly poked, and the ash should be continuously removed to avoid channeling and attendant unequal temperature and contact conditions.

Blauvelt [*Trans. Am. Inst. Mining Engrs.*, **18**, 614 (1890)], discussing soft-coal producer-gas practice, emphasizes the importance of placing the producer so as to lose as little as possible of the sensible heat of the gas and to prevent the condensation of the hydrocarbon vapors. He recommends a high fuel bed, keeping the producer cool on top, thereby preventing the breaking down of the hydrocarbons and the deposit of soot and the formation of CO_2. He advises the use of as much steam with air as will maintain incandescence.

For many purposes a clean producer gas must be used. As it is difficult to clean a hot gas, it is necessary to cool and scrub. The sensible-heat losses can, however, be greatly diminished by the use of water-jacketed side walls and waste-heat boilers, if the capacity and demand will warrant the necessary investment.

In addition to precautions indicated in the foregoing, it is important to avoid mixing the ash with hot coal in the poking; a good distribution of the entering air steam should be secured, and the steam control should be such that a constant air-steam ratio is maintained under constant load.

Practically any solid fuel can be used in producers, provided that the density-surface conditions are such that the fuel will form a satisfactory combustion bed under reasonable velocities. Economic considerations, therefore, play a primary part in the choice of the fuel. For the higher B.t.u. producer gases it is necessary, of course, to use a fuel high in volatile combustible, and it should preferably be closely sized, non-coking, low in ash, with the ash non-clinkering, and low in sulfur. The fuel should also be low in moisture.

It is impossible to deal extensively with the action of different types of fuels, and different types of producers. The reader is referred to texts such as Haslam and Russell on "Fuels and Their Combustion," McGraw-Hill, New York, 1925, and Rambush, "Modern Gas Producers," Van Nostrand, New York, 1923. Considerable valuable data are contained in publications of the *U.S. Bur. Mines*, such as *Bulls.* 7 and 13. The use of the net hydrogen-volatile-matter ratio of coals in certain gas-producer calculations is discussed in *Ind. Eng. Chem.*, **20**, 1371 (1928). Such calculations are, however, limited by difficulties in securing true average gas samples.

Developments in Germany, made public by the postwar work of the Technical Industrial Intelligence Committee and various military and naval missions have demonstrated important possibilities for producer gas and blue water gas manufactured with low-cost high-tonnage oxygen. (*Cf.* Newman, *Am. Inst. Min. Met. Engrs., Tech. Pub.* 2116 (1946); *idem, Am. Gas Assoc. Proc.*, 1946, p. 442; Rushton and Downs, *Am. Gas Assoc. Proc.* 1947, p. 717.) See also references to use of oxygen in water-gas manufacture below.

The Manufacture of Water Gas. For public service distribution the water gas is generally carbureted with oil gas and is known as *carbureted gas*, the use of blue water gas being confined to certain industrial conditions and as a diluent for peak-load demands on natural gas, oil gas, or coal-gas distributing systems.

Since the manufacture of carbureted water gas involves necessarily the manufacture of blue gas, consideration of blue gas logically comes first. The blue-gas machine consists primarily of a generator which is a steel shell, lined with insulation and fire brick, and may vary in free internal diameter from about 3 to about 10 ft. This generator is equipped with an iron charging door at the top, ground to prevent leakage, and suitable clean-out doors at the side near the bottom. Connections at the top are made to lead away the hot gases; and, to admit steam for down run, air-blast, steam, and gas-outlet connections are provided at the bottom of the generator. The fuel is ordinarily charged to a depth of from 7 to 9 ft.

Solid fuels, properly sized, are employed. Anthracite coal and, later, coke have been favorite fuels and are still widely used; particularly, coke. Shortages of those fuels made the use of certain bituminous coals necessary, and this has led to an extensive development with attending economies. Not all bituminous coals are satisfactory. Certain chemical and physical properties that must exist are, however, as yet not well defined. A non-coking coal that will not shatter on heating has obvious advantages.

The fire is lighted and blown with an air blast until the fuel bed attains satisfactory incandescence. The chemical reactions of the heated bed under blasting correspond to those of an air-blown producer. However, no water vapor, beyond that ordinarily present in the air, is admitted, and the velocity-contact conditions established seek to store the maximum amount of heat in the generator at the highest possible rate, while the combustion products are maintained high in carbon dioxide. Carbon monoxide in the blast products may represent an energy that cannot be economically used. The blast products may be used to heat checkerwork for oil cracking and for raising steam in a waste-heat boiler, or they are discharged to the atmosphere through a stack valve.

When the bed has attained a satisfactory temperature, and before the carbon monoxide production becomes excessive, the air supply is shut off, and steam is admitted to produce the desired blue gas. The steam reacts with the carbonaceous fuel to produce carbon monoxide and hydrogen with some carbon dioxide, but the velocity, contact, and temperature conditions are so chosen that only low percentages of the dioxide are obtained. Under these conditions some small amounts of nitrogen, methane, and organic sulfur compounds are also present.

According to the reactions given for the gas producer, the endothermic decomposition of the steam rapidly reduces the fuel-bed temperature, which reduction favors the production of carbon dioxide rather than the desired monoxide; therefore it is necessary to discontinue the steam admission and subject the bed to air blasting until a satisfactory temperature is again attained. The gas making and blasting thus go on in cyclic fashion, with pauses for fuel charging and clinkering. The gas-making time lost for these two later operations was, however, greatly reduced, in fact almost eliminated, by the development of automatic grates and chargers.

It will be seen that blue-gas manufacture involves the balancing of the temperatures of the fuel bed, which must not be allowed to get too hot, thus giving excessive carbon monoxide losses in the blast; nor too low, involving carbon dioxide losses in the gas-making period.

If the steam is admitted solely from the bottom of the fuel bed, this rapidly grows too cold and may not again ignite when blasted. Moreover, the top of the bed grows excessively hot, involving higher sensible-heat losses. To overcome these difficulties, the steam run is divided, part up and part down through the bed. To prevent explosions, the steam flow just preceding and immediately after the air blast is upward, thus avoiding the mixing of air and combustible gas. To secure fires that are readily cleaned, it is customary to use a considerable excess of steam. Often 50 per cent of the steam used may not be decomposed.

The efficiency of blue water gas manufacture is necessarily low. Some of the heat losses include:

1. Losses as sensible heat in the gases and combustion products that leave the generator at high temperature.
2. Losses as sensible heat in clinker and ashes, and unconsumed fuel that may be removed from the generator.
3. Radiation and convection losses from the generator.
4. Losses due to the vaporization of water in the fuel.
6. Stand-by losses when the machine is under heat but idle.
6. Combustion losses during fire-cleaning periods.
7. Losses due to the use of excessive amounts of steam during the steam run.

The preparation of a heat and material balance upon a blue water gas machine is a difficult matter, and no standard practice has been developed. Some of the shortcomings of some published tests have been discussed by Travers [*Trans. Inst. Chem. Engrs.*, **2**, 65 (1924)]. The Water Gas Committee of the American Gas Association has developed a test code for public utility use with certain carbureted gases (*Am. Gas Assoc. Proc.*, p. 1507, 1930).

Some conception of heat and material balances that may be secured in blue water gas manufacture may be obtained from the data of Morris (*Am. Gas Assoc. Proc.*, pp. 39–45, 1922).

New developments affording low-cost tonnage oxygen give promise of the manufacture of blue water gas by continuous processes in which the solid carbonaceous fuels are treated simultaneously with oxygen and steam. A symposium on Synthesis Gas (for the manufacture of synthetic liquid fuels) reported in *Ind. Eng. Chem.* **40**, 558 (1948), reviews the state of this art and includes bibliographies citing important sources of information. The field is under active experimentation and development but at the time of writing has not been brought to commercial utilization in the United States. Powell (*Am. Gas Assoc. Proc.*, p. 631, 1947) discusses Future Possibilities in Methods of Gas Manufacture, citing 39 references. (*Cf.* references on oxygen producer gas.)

Carbureted Water Gas. Carbureted water gas consists largely of a mixture of a rich oil gas and blue water gas. The production of the latter has already been discussed, and from the principles mentioned it was shown that the gases leaving the generator at the end of the blast must of necessity carry a considerable quantity of potential heat as carbon monoxide, in addition to the sensible heat due to the high temperatures at the top of the fuel bed. This heat is utilized in part by the maintenance of a satisfactory "cracking" temperature in two succeeding shells of approximately the same diameter as the generator. The shells are designated the **carburetor** and **superheater** in the order in which the blue gas passes through. The first, or carburetor, is about as high as the generator, but the superheater is much higher in order to maintain a satisfactory natural draft, thus permitting generator charging from above.

Ordinarily the carburetor and superheater are filled with checker brick in order to provide extensive heat-transfer surfaces, thus assisting in the cracking of the oil.

In the normal operating cycle, the generator and checker chambers are first brought to the requisite temperatures by air-producer operation with combustion of the producer gas by admitting secondary air between the generator and carburetor. The machine is then placed on the run, *i.e.*, during part of the cycle, steam is admitted and blue gas formed. This passes over to the top of the carburetor and downward, here meeting a fine spray of oil which is carried through the checker chambers and cracked to an oil gas and some tar. As the fire cools and the steam decomposition falls off, the generator is again blasted with air and the carburetor ignited with the secondary air blast until satisfactory temperatures are again attained, when the steam run is again begun. With automatic opening and closing of the valves, these cycles follow one another after intervals of only a few minutes.

When extremely heavy petroleum oils are used, the quantity of checker surface in the carburetor is greatly reduced in order to avoid stopping the interstices with heavy carbon deposits. In addition, some oil may be thrown directly on the top of the fuel bed during the downstream run. (*Cf.* Dashiell, *Am. Gas Assoc. Proc.*, p. 886, 1930. Also see Gas Production Committee Report at 1932 Convention of the Am. Gas Assoc., "Use of Heavy Oil for Carburetion" by Hartzel and Lueders.)

The carbureted water gas apparatus as well as the Pacific coast oil gas apparatus has been used for **reforming natural gas and oil refinery gas** (for references *cf.* Description, p. 1577). Carbureted water gas apparatus has also been converted for use in the manufacture of oil gas by replacing the generator fire with a refractory screen. Oil alone, suitably treated with air and steam, is used as the fuel in this process (Johnson, *Am. Gas Assoc. Proc.*, p. 892, 1932).

Following the introduction of bituminous coal as a fuel in water gas manufacture, important modifications were introduced into the fundamental water gas cycle. One of these is the **back run** in which the down-run steam is admitted at the top of the superheater, whence it passes backward to the lower part of the carburetor, thence to the top of the fuel bed. The blue gas produced is led directly to the wash box by a by-pass or back-run pipe. This arrangement does away with the hot valve between the carburetor and lower portion of the generator and uses instead a valve in the wash box. The process cools the top of the superheater and preheats the down-run steam and avoids unduly high temperatures in the checkers, thereby minimizing an objection often encountered in soft-coal operation due to the combustion of volatile matter during the blow. The gas leaving the bottom of the generator has a relatively low temperature because of heat interchange with the ash and clinker cooled by the blast. By passing the checkers and going directly to the wash box, it decreases waste-heat losses from the checkers. Important thermal economies are claimed for it. The carbureting oil is added during the up run. The process is discussed in the *Annual Reports* of the Water Gas Committee of the American Gas Association for 1924 and 1925. A similar process, save that the steam is admitted at the top of the generator, is the **Chrisman down-run process.** This is described in the 1925 *Report.*

To bring the fuel bed to temperature in bituminous-coal operation while avoiding the overheating of the checkers, the **blow-run** operation is sometimes used. In this, during the latter part of the blow, the carburetor air is cut off and the stack valve through which the combustion products are wasted is closed, thus forcing the air-producer gas from the generator into the holder.

The **reversed air-blast process** is related to the blow run and the back run. The reverse blast is admitted to the top of the superheater at the portion of the cycle generally used for the blow run and follows the path of the back run. It is described by Howard [*Am. Gas Assoc. Monthly,* **7,** 579–584 (1925)].

American practice in water gas making has tended toward automatic grate operations which greatly reduce the labor charges and stand-by losses for cleaning and clinkering. The first of these to be successfully used was the A.B.C. grate of the Western Gas Construction Co. described by Ramsburg at the First International Coal Conference (*Proc.*, 1926, pp. 514*ff.*). Automatic grates are now offered by the Semet-Solvay Co., the United Gas Improvement Co., and the Gas Machinery Co., and the operations of some of these have been discussed in *Reports* of the Water Gas Committee of the American Gas Association (*Am. Gas Assoc. Proc.*, p. 1245, 1929; p. 1573, 1930; p. 1172, 1931).

The Cracking of Oil in Gas Making. The cracking of oil to hydrocarbons of low molecular weight is of considerable importance to the gas manufacturer who relies upon oil gas or carbureted water gas. The literature relating thereto is extensive as cracking is an old art. The manufacture of gas from oil was discussed in England as early as 1792. The considerations involved are many and complex and have been dealt with at some length in texts on petroleum technology. The art has been greatly advanced by the introduction of catalytic processes. *Cf.* Milbourne Catalytic Cracking Plants for Relieving Gas Utility Peak Loads, Paper 48-A-101, Fuel Division, Am. Soc. Mech. Engrs. Annual Meeting, Nov. 28–Dec. 3, 1948, and paper by Horsfield presented at the 1948 Production and Chemical Conference of the Am. Gas Assoc. Experimental data directly applicable to problems of the carbureted water gas maker will be found in a paper by Downing and Pohlman (*Am. Gas Inst. Proc.*, p. 587, 1916).

Temperature and time of contact are important variables in the cracking and are, in certain measure, interdependent. Actually, the operator of the water gas machine is usually obliged by load considerations to maintain adequate capacity, and consequently the most important variable at his command is the temperature of the cracking chambers. In *U.S. Bur. Mines Bull.* 203 are summarized various prior investigations on temperature as follows:

1. The candles per gallon reach a maximum between 1300 and 1350°F.
2. The percentage of illuminants reaches a maximum between 1300 and 1400°F.
3. The percentages by volume of methane and hydrogen increase with increasing temperature.
4. The volume of gas increases with the temperature.
5. The percentage of carbon formed from the cracking of the oil increases with the temperature.
6. The percentage of tar formed decreases with increasing temperature.
7. The B.t.u./gal. oil increases with the temperatures and quite often the maximum is not reached under 1500°F.

The above conclusions apply, in general, to temperatures up to 1600°F., which is about the upper limit of the experiments made, so far as data are available.

The control of the process of cracking involves not only thermodynamics but also chemical kinetics, and equilibrium is not reached in practice.

The theory and art of oil cracking have advanced markedly in recent years. In the theory the more notable advances have occurred in developments involving the conception of free radicals and chain reaction kinetics.

The chemical composition of the oil used for carbureting has, of course, some considerable importance, but commercial oils are in general so complex chemically that the evaluation from its chemical composition is not convenient or certain. In general, a saturated paraffin may be said to rank highest in carbureting efficiency, with olefins somewhat less efficient. Naphthenes are better than aromatics, and asphaltics have been given the lowest value. Others place the asphaltics above the aromatics and place unsaturates at the bottom of the list. The value of a hydrocarbon fraction increases with its boiling point. Holmes [*Ind. Eng. Chem.*, **24**, 325 (1932)] has proposed a method of evaluating the carbureting power of a gas oil, based on the specific gravity, average boiling point, etc. Chief reliance in oil testing for carbureted water gas use is ordinarily placed upon small-scale cracking tests such as was described by Dick before the 1933 Production and Chemical Conference of the American Gas Assoc. [*Am. Gas J.*, **138** (No. 6), 32–34] or in the "Gas Chemists' Handbook" (1929 ed., p. 66). Analytical methods and interpretation have been considered, conspicuously by Mighill (*Am. Gas Assoc. Proc.*, pp. 1093, 1454, 1927).

A low content of sulfur is desirable because this simplifies the purification problem.

Modern carbureted water gas practice has grown so involved (embracing anthracite, coke, and bituminous coal fuels, gas oil, and heavy oil operations, reforming, with back run or Chrisman down run or reversed air blast, and with blow run) that material and heat-balance considerations have become exceedingly complex, and a truly typical carbureted water gas operation can scarcely be said to exist. Valuable data on various combinations of the elements that may enter into carbureted water gas operation may be found in reports of committees of the American Gas Association and in the annual proceedings of that association. An informative and complete analysis of a carbureted water gas operation in Great Britain is presented in the 7th Report of the Research Sub-committee of the Institute of Gas Engineers of Great Britain [*Gas J.*, **158**, 800–827 (1922)].

Where gas oil is employed for carbureting, oil is not thrown on the fire, and re-forming is not practiced, it is possible to follow the oil efficiencies obtained by a procedure known as the Providence modification of the Pacific coast method.

This is described by the Subcommittee on Uniform Oil Efficiency of the Water Gas Committee of the American Gas Association in the 1926 and 1927 Proceedings of that association.

Oil Gas. Pacific Coast Methods. On the Pacific coast, except in the state of Washington, manufactured gas for city distribution has generally been made from California residuum oils. The processes used are somewhat akin to the water gas process. There is a heating period in which oil is burned to bring the generator to temperature, and the products of combustion are wasted. This is followed by a making period in which heated steam, carbon, and oil interact to yield stable gas together with some tar and lampblack. The make is followed by the blow in which carbon deposits are burned off by air. The length of the cycle varies but is ordinarily much longer than the water gas cycle.

One of the major units is the large two-shell **Jones oil-gas generator.** Heat and material balances on such a generator have been reported by Pike and West [*Ind. Eng. Chem.*, **21**, 104 (1929)]. Because of the widespread distribution of natural gas, oil gas has become of relatively less importance on the West coast, and the introduction of a development of the Jones process, in which the lampblack was gasified, was halted (*Am. Gas Assoc. Proc.*, 1930, p. 1349).

Single-shell Oil Gas Methods. For smaller loads recourse is had to the single-shell oil gas machines, which may be further classified, according to methods of operation, in the **heat up and make down** or **straight shot** and the **heat and make down** types. One drawback possessed by these machines is the relatively large amount of lampblack produced.

References on Pacific Coast Oil-gas Methods. Further information on the above methods will be found in the treatise on the Production of Oil Gas by the Educational Committee of the Pacific Service Employees' Association under the direction of Cowles, Henderson, and Yard (1922); Yard in the *Am. Gas Assoc. Monthly*, **7**, 741–743 (1925), and in the section on Oil Gas Machines in the 1925 Report of the Water Gas Committee, the American Gas Association (*Am. Gas Assoc. Proc.*, 1925, p. 1278).

The Dayton Process. Although the foregoing oil gas methods have been very successful on the Pacific coast, owing no doubt in large measure to the ready availability of suitable oil supplies, they have not been widely used elsewhere. For small plants in other parts of the country the Dayton process has been proposed. This utilizes small cast-iron retorts which are first brought to temperature by combustion of oil in the refractory setting about the retort. The supply of combustion oil is then greatly decreased, and a mixture of oil and air is admitted to the retort through an atomizer. Partial combustion and cracking follow.

This gas possesses so high a gravity that it cannot conveniently be used for stand-by or peak equipment in systems distributing mixed coal and water gas.

The original Dayton process has been improved to increase the capacity and eliminate the provisions for external heating, but with means for heat exchange. Such generators may have capacities of 1,000,000 cu. ft. of gas a day. Heating values greater than 500 B.t.u. cannot be made from oil, and the usual output may have values from 300 to 500 B.t.u. Such gas can be mixed with natural gas to replace a high specific gravity, high B.t.u. manufactured gas and may be used to some advantage to meet peak demands on natural gas systems.

A modification of the Dayton process, known as the **Faber** process, uses as raw material natural gas, refinery gas, or liquefied hydrocarbon gases obtained from natural gas. This is essentially a gas re-forming process.

Further information will be found in an article by Binnall [*Ind. Eng. Chem.*, **13**, 242 (1921) and in *Am. Gas Assoc. Proc.*, pp. 1361, 1482, 1930].

Table 17. Comparison of Principal Operating Results, Coal-gas Tests*

	Utica Koppers gas ovens	Rochester U.G.I. verticals	Lowell horizontals	Stamford Glover-West continuous vertical through retorts
Coal:				
Proximate analysis (dry basis):				
Volatile	35.2	35.8	35.1	34.3
Fixed carbon	57.7	57.6	58.4	58.4
Ash	7.1	6.6	6.5	7.3
Sulfur	0.9	1.0	0.9	0.9
Moisture as charged (wet basis)	2.1	2.5	1.67	2.2
Coal gas:				
Cu. ft. at 30 in., 60° satur., per ton dry coal	11,315	11,347	12,110	14,300
Cu. ft. at 30 in., 60° satur., per lb. dry coal	5.66	5.67	6.06	7.15
Heating value, B.t.u./cu. ft.	573	540	542	530
B.t.u. in gas, per lb. dry coal	3,245	3,065	3,280	3,790
Average analysis, %:				
CO_2	2.2	1.7	2.40	3.0
Illuminants	4.0	3.1	3.05	2.8
O_2	0.8	0.5	0.75	0.2
H_2	46.5	49.7	47.95	54.5
CO	6.3	6.9	7.35	10.9
CH_4	32.1	29.9	27.15	24.2
N_2	8.1	8.2	11.35	4.4
Dry coke produced as % dry coal carbonized	70.2	70.7	68.6	70.0
Gal. dry tar per ton dry coal	12.4	11.4	11.6	14.6
Ammonia per ton dry coal, lb.	6.4	5.9	5.0	3.4
Producer fuel, lb./ton coal	292	318	229	275
Producer fuel, % coal carbonized	14.6	15.9	11.5	13.7
Waste-heat steam from and at 212°F./lb. producer fuel, lb	0.59	1.96	1.14	5.74
Steaming based on dry coal, %	....	1.0	....	12.0

* *Am. Gas Assoc. Proc.*, 1928, p. 1118.

Recent Developments in Oil Gas Manufacture. Shortages of oil and coal for residential heating encountered during the war and price increases on these fuels have favored the expansion of house heating by gas, thus contributing to the development of unprecedented peak demands on gas utilities. To meet this need and to provide quickly a high B.t.u. substitute in the event of unexpected natural-gas transmission failure, active experimentation on new gasification processes for low-grade high-carbon residue oils have been going on apace. These use the oil, with little or no solid fuel, for combustion as well as carburetion. This permits quick startup. Regenerative preheating of steam and air gives good economy and high capacity. Several processes are under examination, but at the time of writing no preferred process appears to have received general adoption. [(*Cf.* The Hall High B.t.u. Oil Gas Process, *Am. Gas Assoc. Monthly*, **30**, 11, p. 27 (1948); The Twin Generator Oil-Gas Process, idem, **31**, 2, p. 21 (1949); Chambers, The Twin Generator Oil-Gas Process, *Am. Gas Assoc. Proc.*, p. 760, 1947.]

Coal Gas. Coal gas is produced by the pyrogenetic decomposition of suitable coals in externally heated refractory retorts. As the potential heat of the gas made usually represents only about 20 per cent of the total heat input, the economic success of this means of gas production is largely determined by the form value of the other products, particularly the coke, which usually represents some 50 or 60 per cent of the heat input. Considerable attention is therefore given to the selection of proper coals in order to secure cokes of salable quality. Although in certain localities very important domestic markets have been developed, the great markets lie in the metallurgical industries. Consequently, recent coal-gas production has tended to follow by-product metallurgical practice.

In general, three types of carbonizing processes may be said to be in operation in the American gas industry: (1) horizontal and (2) vertical retorts; and (3) coke ovens. Comparative heat and material balances have been made upon certain units representing such processes by a committee of the American Gas Assoc. (*Proc.*, 1928, p. 1115). The operating results shown in Table 17 have been taken from this report.

Combustion Data

Ignition Temperatures. The term "ignition temperature" is sometimes used as a characteristic property of a substance. Any such implication is, however, in error. A system ignites when the rate of gain of heat due to the oxidation reaction is greater than the rate of loss of heat. It follows therefore that ignition temperatures are dependent upon the properties of the particular system in question and are not characteristic of the igniting substance alone. The term as ordinarily applied is used to signify the temperature at which rapid combustion occurs in ordinary air.

In ignition temperature measurements two methods are used to avoid error due to slow combustion prior to ignition: (1) by exploding the gaseous mixture by adiabatic compression and (2) by preheating independently each stream of gas and air before mixing.

Limits of inflammability are dependent in part upon the conditions of determination. When a weak source of ignition is used, certain mixtures near the limits may not inflame. The limit is lower for upward propagation of flame, and the limits are wider as the tube diameter is increased, although usually not markedly wider above diameters of 5 cm. except in certain cases, such as acetylene whose upper limit is much increased by enlarging the tube diameter above 5 cm. In closed tubes the length of the tube may also affect the results.

Normal variations of atmospheric pressure do not appreciably affect the limits of inflammability. The effect of larger variations in pressure is neither simple nor uniform but is specific for each inflammable mixture. Reduction of pressure below 760 mm. narrows the range of flammability by raising the lower limit and decreasing the higher limit. At a suitably low limit the limits coincide; below this point there is no propagation of flame. Increase of pressure above atmospheric does not always widen the limits. In some mixtures this increase

Table 18. Gas Combustion Constants

No.	Gas	Formula	Sp. gr. air = 1.000*	Heat of combustion† B.t.u./cu. ft. Gross	Net‡	B.t.u./lb. Gross	Net‡	Experimental error in heat of combustion % + or −
1	Carbon	C	. . . .	. . . .	. . . . . .	14,093¶	14,093¶	0.012
2	Hydrogen	H_2	0.06959	325.0	275.0	61,100	51,623	.015
3	Oxygen	O_2	1.1053					
4	Nitrogen (atms.)	N_2	0.9718§					
5	Carbon monoxide	CO	0.9672	321.8	321.8	4,347	4,347	.045
6	Carbon dioxide	CO_2	1.5282					
	Paraffin series $C_nH_{2n} + 2$							
7	Methane	CH_4	0.5543	1,013.2	913.1	23,879	21,520	.033
8	Ethane	C_2H_6	1.04882§	1,792	1,641	22,320	20,432	.030
9	Propane	C_3H_8	1.5617§	2,590	2,385	21,661	19,944	.023
10	n-Butane	C_4H_{10}	2.06654§	3,370	3,118	21,308	19,680	.022
11	Isobutane	C_4H_{10}	2.06654§	3,363	3,105	21,257	19,629	.019
12	n-Pentane	C_5H_{12}	2.4872§	4,016	3,709	21,091	19,517	.025
13	Isopentane	C_5H_{12}	2.4872§	4,008	3,716	21,025	19,478	.071
14	Neopentane	C_5H_{12}	2.4872§	3,993	3,693	20,970	19,396	.11
15	n-Hexane	C_6H_{14}	2.9704§	4,762	4,412	20,940	19,403	.05
	Olefin series C_nH_{2n}							
16	Ethylene	C_2H_4	0.9740	1,613.8	1,513.2	21,644	20,295	.021
17	Propylene	C_3H_6	1.4504§	2,336	2,186	21,041	19,691	.031
18	n-Butene (Butylene)	C_4H_8	1.9936§	3,084	2,885	20,840	19,496	.031
19	Isobutene	C_4H_8	1.9336§	3,068	2,869	20,730	19,382	.031
20	n-Petene	C_5H_{10}	2.4190§	3,836	3,586	20,712	19,363	.037
	Aromatic series $C_nH_{2n} + 6$							
21	Benzene	C_6H_6	2.6920§	3,751	3,601	18,210	17,480	.12
22	Toluene	C_7H_8	3.1760§	4,484	4,284	18,440	17,620	.21
23	Xylene	C_8H_{10}	3.6618§	5,230	4,980	18,650	17,760	36
	Miscellaneous gases							
24	Acetylene	C_2H_2	0.9107	1,449	1,448	21,500	20,776	.16
25	Naphthalene	$C_{10}H_8$	4.4208§	5,854**	5,654**	17,298**	16,708**	**
26	Methyl alcohol	CH_3OH	1.1052§	867.9	768.0	10,259	9,078	.027
27	Ethyl alcohol	C_2H_3OH	1.5890§	1,600.3	1,450.5	13,161	11,929	.030
28	Ammonia	NH_3	0.5961§	441.1	365.1	9,668	8,001	.088
29	Sulfur	S	. . . .	. . . . .	. . . . .	3,983	3,983	.071
30	Hydrogen sulfide	H_2S	1.1898§	647	596	7,100	6,545	.30
31	Sulfur dioxide	SO_2	2.264					
32	Water vapor	H_2O	0.6215§					
33	Air	. . . .	1.0000					

From "Fuel-Flue Gases," published by the American Gas Association, New York, and reproduced by permission.

All gas volumes corrected to 60°F. and 30 in. Hg dry. For gases saturated with water at 60°F., 1.73% of the B.t.u. value must be deducted.

* Densities calculated from values given in grams per liter at 0°C. and 760 mm. in the "International Critical Tables" allowing for the known deviations from the gas laws. Where the coefficient of expansion was not available, the assumed value was taken as 0.0037/°C. Compare this with 0.003662, which is the coefficient for a perfect gas. Where no densities were available, the volume of the mole was taken as 22.41151. Some of the materials cannot exist as gases at 60°F. and 30 in. Hg pressure, in which case the values are theoretical ones given for ease of calculation of gas problems. Under the actual concentrations in which these materials are present, their partial pressure is low enough to keep them as gases.

† Converted to mean B.t.u. per pound (1/180 of the heat per pound of water from 32 to 212°F.) from data by Frederick O. Rossini, National Bureau of Standards, letter of Apr. 10, 1937, except as noted.

‡ Deduction from gross to net heating value determined by deducting 18,919 B.t.u./lb.-mole of water in the products of combustion. Osborne, Stimson, and Ginnings, *Mech. Eng.*, p. 163, March, 1935, and Osborne, Stimson, and Fiock, *Bur. Standards* (U. S.) *Research Paper* 209.

§ From 3d ed. of Combustion, published by American Gas Association.

¶ See data for Carbon, *Bur. Standards Research Paper* 1141.

** The asterisks denote that either the density or the coefficient of expansion has been assumed.

narrows the limits, and a mixture that can propagate flame at atmospheric pressure may not be able to do so at higher pressures. The flammability of hydrogen-air and carbon monoxide-air mixtures (downward propagation of flame) is narrowed at both limits by a moderate increase in pressure above atmospheric. Under analogous conditions the range of each gas in the paraffin series is narrowed on the lower limit side and widened at the upper limit side.

Limits of Inflammability of Mixtures of Inflammable Gases and Vapors (from *U.S. Bur. Mines Bull.* 279, 1938 ed., pp. 7*ff.*).

A simple formula, of additive character, was advanced by Le Chatelier to connect the lower limits of single gases with the lower limit of any mixture of them. It is

$$\frac{n_1}{N_1} + \frac{n_2}{N_2} = 1$$

in which N_1 and N_2 are the lower limits in air for each combustible gas separately and n_1 and n_2 are the percentages of each gas in any lower-limit mixture of the two in air.

The formula expresses the fact that, for example, a mixture of air, carbon monoxide, and hydrogen, which contains one-quarter of the amount of carbon monoxide and three-quarters of the amount of hydrogen necessary to form a lower-limit mixture, will be a lower-limit mixture. If the formula expresses experimental facts, the lower limits of inflammability form a series of inflammability equivalents for the individual gases of a mixture.

The formula also leads to the deduction that lower-limit mixtures if mixed in any proportions give rise to mixtures that are also at their lower limits; or, vice versa, the formula may be deduced from the latter statements as a postulate.

The formula may be generalized to apply to any number of combustible gases, thus:

$$\frac{n_1}{N_1} + \frac{n_2}{N_2} + \frac{n_3}{N_3} + \cdot \cdot \cdot = 1$$

and, so far as it expresses experimental results truly, may be applied to high-limit mixtures, with the appropriate rewording of the definition of n_1 . . . *and* N_1 . . . .

A small algebraic transformation gives a more useful formula for calculating the limits of any mixture of combustible gases which obeys it, as follows:

$$L = \frac{100}{\frac{p_1}{N_1} + \frac{p_2}{N_2} + \frac{p_3}{N_3} + \cdot \cdot \cdot}$$

Table 19. Minimum Ignition Temperatures and Flash Points of Combustible Liquids, Gases, and Vapors*

Name	Formula	Ignition temp.		Flash point	
		°F.	°C.	°F.	°C.
Acetal	$C_6H_{14}O_2$	446	230		
Acetaldehyde	C_2H_4O	527	275	−17	−27
Acetanilide	C_8H_9NO	...	...	345	174
Acetic acid	$C_2H_4O_2$	1022	550	107	42
Acetic anhydride	$C_4H_6O_3$	738	392	127	53
Acetone	C_3H_6O	1042	561	0	−18
Acetophenone	C_8H_8O	...	...	221	105
Acetyl chloride	C_2H_3OCl	...	...	40	4
Acetylene	C_2H_2	581	305	Gas	Gas
Acrolein	C_3H_4O	532	278		
Acrylonitrile	C_3H_3N	898	481	23	−5
Aldol	$C_4H_8O_2$	...	...	181	83
Allyl alcohol	C_3H_6O	712	378	70	21
Allyl chloride	C_3H_5Cl	909	487		
Ammonia	NH_3	1204	651	Gas	Gas
Amyl acetate	$C_7H_{14}O_2$	750	399	77	25
Isoamyl acetate	$C_7H_{14}O_2$	714	379	92	33
Amyl alcohol	$C_5H_{12}O$	801	427	100	38
p-Iso-amyl alcohol	$C_5H_{12}O$	650	343	114	45
s-Iso-amyl alcohol	$C_5H_{12}O$	...	...	103	39
tert-Amyl alcohol	$C_5H_{12}O$	...	...	67	19
Amyl benzene	$C_{11}H_{16}$	491†	255†		
Amyl chloride	$C_5H_{11}Cl$	498	259		
tert-Amyl chloride	$C_5H_{11}Cl$	649	343		
Amylene	C_5H_{10}	523	273		
Isoamyl ether	$C_{10}H_{22}O$	802	428		
Amyl propionate	$C_8H_{16}O_2$	...	...	106	41
Aniline	C_6H_7N	986	530	160	71
Anthracene	$C_{14}H_{10}$	882	472		
Benzaldehyde	C_7H_6O	377	192	148	64
Benzene	C_6H_6	1076	580	12	−11
Benzoic acid	$C_7H_6O_2$	1063†	573†	250	121
Benzyl acetate	$C_9H_{10}O_2$	860	460	216	102
Benzyl alcohol	C_7H_8O	802	428	213	101
Benzyl chloride	C_7H_7Cl	1161	627	140	60
Bromo-benzene	C_6H_5Br	1270	688	149	65
Butadiene	C_4H_6	804	429		
Butane	C_4H_{10}	826	441	Gas	Gas
Isobutane	C_4H_{10}	1010	543	Gas	Gas
Di-chloro butane	$C_4H_8Cl_2$	543	284	0	−18
Butyl acetate	$C_6H_{12}O_2$	790	421	84	29
Butyl alcohol	$C_4H_{10}O$	653	345	100	38
Iso-butyl alcohol	$C_4H_{10}O$	813	434	82	28
tert-Butyl alcohol	$C_4H_{10}O$	892	478	52	11
s-Butyl alcohol	$C_4H_{10}O$	777	414	70	21
s-Butyl benzene	$C_{10}H_{14}$	...	...	126	52
Butyl bromide	C_4H_9Br	901	483		
Butyl carbitol	$C_8H_{10}O_3$	442	228	172	78
Butyl cellosolve	$C_6H_{14}O_2$	472	244	141	61
Butyl chloride	C_4H_9Cl	860	460		
Butylene	C_4H_8	829	443	Gas	Gas
Butyl formate	$C_5H_{10}O_2$	635	335	64	18
Butyl propionate	$C_7H_{14}O_2$	799	426	90	32
Butyraldehyde	C_4H_8O	446	230	20	−7
Butyric acid	$C_4H_8O_2$	1026	552	170	77
Carbon disulfide	CS_2	248	120	−22	−30
Carbon monoxide	CO	1128	609	Gas	Gas
Cellosolve	$C_4H_{10}O_2$	460	238	104	40
Cellosolve acetate	$C_6H_{12}O_3$	715	379	124	51
Chlorobenzene	C_6H_5Cl	...	...	85	29
Chloro-ethyl acetate	$C_4H_7O_2Cl$	...	...	129	54
Creosote	...	637	356	165	74
o-Cresol	C_7H_8O	1110	599	178	81
m-Cresol	C_7H_8O	1159	626	187	86
Croton aldehyde	C_4H_6O	450	232	55	13
Cyanamide	CN_2H_2	...	...	285	141
Cyanogen	C_2N_2	1562	850	Gas	Gas
Cyclohexane	C_6H_{12}	565	296	1	−17
Cyclohexanol	$C_6H_{12}O$	...	...	154	68
Cyclohexanone	C_6H_8O	...	...	93	34
Cyclopropane	C_3H_6	928	498	Gas	Gas
Cymene	$C_{10}H_{14}$	871	466	117	47
Decalin	$C_{10}H_{18}$	504	262	136	58
Decane	$C_{10}H_{22}$	482	250	115	46
Di-chloro-ethyl ether	$C_4H_8OCl_2$	696	369	131	55
Di-chloro-ethylene	$C_2H_2Cl_2$	856	441	57	14
Di-ethanol amine	$C_4H_{11}NO_2$	1224	662	280	133
Di-ethylene glycol	$C_4H_{10}O_3$	775	413	255	184
Di-methyl aniline	$C_8H_{11}N$	700	371	145	63
Dioxan	$C_4H_8O_2$	511	266	54	12
Di-phenyl	$C_{12}H_{10}$	...	...	235	113
Di-phenyl amine	$C_{12}H_{11}N$	...	...	307	153
Di-phenyl methane	$C_{13}H_{12}$	...	...	266	130
Di-phenyl oxide	$C_{12}H_{10}O$	...	...	239	115
Di-vinyl ether	C_4H_6O	680	360	−22	−30
Dodecane	$C_{12}H_{26}$	993	534	165	74
Ethane	C_2H_6	882	472	Gas	Gas
Ethyl acetate	C_4H_8O	903	484	28	−2
Ethyl alcohol	C_2H_6O	738	392	54	12
Ethyl benzene	C_8H_{10}	1027	553	59	15
Ethyl bromide	C_2H_5Br	952	511		
Ethyl butyrate	$C_6H_{12}O_2$	865	463	78	26
Ethyl chloride	C_2H_5Cl	963	517	−58	−50
Ethylene	C_2H_4	914	490	Gas	Gas
Ethylene chlorohydrin	C_2H_5OCl	797	425	140	60
Ethylene dichloride	$C_2H_4Cl_2$	775	413	56	13
Ethylene glycol	$C_2H_6O_2$	775	413	232	111
Ethylene oxide	C_2H_4O	804	429	Gas	Gas
Ethyl ether	$C_4H_{10}O$	379	193	−49	−45
Ethyl formate	C_3H_6O	1071	577	−4	−20
Ethyl mercaptan	C_2H_6S	570	299		
Ethyl propionate	$C_5H_{10}O_2$	889	476	54	12
Furfural	$C_5H_4O_2$	736	391	140	60
Gasoline	Regular	536	280	−47	−44
Gasoline	73 octane	570	299		
Gasoline	92 octane	734	390		
Gasoline	100 octane	804	429		
Glycerin	$C_3H_8O_3$	739	393	320	160
Heptane	C_7H_{16}	451	233	25	−4
Hexane	C_6H_{14}	478	248	−15	−26
Isohexane	C_6H_{14}	543	284		
Hexyl alcohol	$C_6H_{14}O$	572	300	137	58
Hydrocyanic acid	HCN	1000	538	Gas	Gas
Hydrogen	H_2	1065	574	Gas	Gas
Hydrogen sulfide	H_2S	558	292	Gas	Gas
Isoprene	C_5H_8	824	440		
Kerosene	...	491	255	100–165	38–74
Methane	CH_4	1170	632	Gas	Gas
Methyl acetate	$C_3H_6O_2$	936	502	14	−10
Methyl alcohol	CH_4O	878	470	52	11
Methyl bromide	CH_3Br	999	537		
Methyl butyl ketone	$C_6H_{12}O$	991	533	95	35
Methyl butyrate	$C_5H_{10}O_2$	...	...	57	14
Methyl cellosolve	$C_3H_8O_2$	551	288	107	42
Methyl chloride	CH_3Cl	1170	632	Gas	Gas
Methyl cyclo-hexane	C_7H_{14}	545	285	25	−4
Methyl cyclo-hexanol	$C_7H_{14}O$	...	...	154	68
Methyl cyclo-pentane	C_6H_{12}	624	329		
Methyl ethyl ketone	C_4H_8O	957	514	30	−1
Methyl formate	$C_2H_4O_2$	456	236	−2	−19
Methyl propyl ketone	$C_5H_{10}O$	941	505	60	16
Methyl salicylate	$C_8H_8O_3$	850	454	219	104
Methylene chloride	CH_2Cl_2	1224	662		
Naphtha	...	450–531	232–277	20–110	−7–43
Naphthalene	$C_{10}H_8$	1038	559	176	80
Nicotine	$C_{10}H_{14}N_2$	471	244		
Nitrobenzene	$C_6H_5NO_2$	900	482	198	92
Nonane	C_9H_{20}	545	285	88	31
Octane	C_8H_{18}	446	230	56	13
Oil:					
Cable insulating	...	738	392		
Castor	...	840	449	445	229
Coconut	...	...	...	420	216
Cod liver	...	662†	350†		
Corn	...	...	...	490	254
Cottonseed	...	650	343	486	252
Creosote	...	637	336	165	74
Fish	...	531†	277†	420	216
Flax	...	772†	411†		
Gas	...	640	338	150	66
Lard	...	650	343	363	184
Linseed, raw	...	650	343	432	222
Linseed, boiled	...	650	343	403	206
Lubricating	...	711	377		
Lubricating, cylinder	...	783	417	535	279
Lubricating, spindle	...	778	414	169	76
Lubricating, turbine	...	700	371	400	204
Menhaden	...	828	442	435	224
Mineral seal	...	...	...	170	77
Neatsfoot	...	828	442	470	243
Olive	...	650	343	437	225
Palm	...	600	315	323	162
Paraffin	...	...	...	444	229
Peanut	...	833	445	540	282
Pine	...	...	...	172	78
Pine tar	...	...	...	144	62
Rape	...	836	447	464	240
Rosin	...	648	342	266	130
Signal	...	...	...	200	93
Soybean	...	...	...	540	282
Sperm	...	586	308	428	220
Straw	...	...	...	315	157
Tallow	...	...	...	492	256
Transformer	...	...	...	295	146
Tung	...	855	457	552	289
Turkey red	...	833	445	476	247

* Compiled by G. W. Jones, U.S. Bureau of Mines.

† Determinations made in oxygen.

Table 19. Minimum Ignition Temperatures and Flash Points of Combustible Liquids, Gases, and Vapors*—(*Concluded*)

Name	Formula	Ignition temp.		Flash point	
		°F.	°C.	°F.	°C.
Oil (*Continued*)					
Whale	....	878	470	476	246
Paraffin	....	473	245	390	199
Paraldehyde	$C_6H_{12}O_3$	460	238	63	17
Pentane	C_5H_{12}	527	275	−40	−40
Petroleum ether	....	624	329	−69	−56
Phenol	C_6H_6O	1319	715	175	79
Pinene	$C_{10}H_{16}$	527	275		
Propane	C_3H_8	898	481	Gas	Gas
Propyl acetate	$C_5H_{10}O_2$	......	......	58	14
Isopropyl acetate	$C_5H_{10}O_2$	860	460	40	4
Propyl alcohol	C_3H_8O	822	439	59	15
Isopropyl alcohol	C_3H_8O	853	456	53	12
Propyl benzene	C_9H_{12}	......	......	87	31
Isopropyl benzene	C_9H_{12}	......	......	102	39
Propyl bromide	C_3H_7Br	914	490		
Propyl chloride	C_3H_7Cl	968	520		
Propylene	C_3H_6	856	458	Gas	Gas
Propylene chlorhydrin	C_3H_7OCl	......	......	125	52
Propylene dichloride	$C_3H_6Cl_2$	1035	557	60	16
Propyl ether	$C_6H_{14}O$	372	189		
Isopropyl ether	$C_6H_{14}O$	830	443	−18	−28
Propylene oxide	C_3H_6O	......	......	−20	−29
Propyl formate	C_4H_8O	......	......	27	−3
Isopropyl formate	C_4H_8O	......	......	22	−6
Propylene glycol	$C_3H_8O_2$	790	421	207	97
Pyridine	C_5H_5N	900	482	74	23
Stearic acid	$C_{18}H_{36}O_2$	743	395	385	196
Styrene	C_8H_8	914	490	86	30
Tetra-decane	$C_{14}H_{30}$	......	......	212	100
o-Toluidine	C_7H_9N	900	482	202	94
p-Toluidine	C_7H_9N	900	482	188	87
Toluene	C_7H_8	1026	552	40	4
Trichloro ethylene	C_2HCl_3	865	463		
Tri-ethylene glycol	$C_6H_{14}O_4$	700	371	313	156
Tri-methyl benzene	C_9H_{12}	948	509		
Turpentine	$C_{10}H_{16}$	464	240	95	35
Valeric acid	$C_5H_{10}O_2$	1274	690		
Vinyl ether	C_4H_6O	680	360		
Vinyl acetate	$C_4H_6O_2$	800	427	18	−8
o-Xylene	C_8H_{10}	925	496	63	17
o-Xylidine	$C_8H_{11}N$	......	......	206	97

* Compiled by G. W. Jones, U.S. Bureau of Mines.
† Determinations made in oxygen.

Table 20. Limits of Flammability of Gases and Vapors in Air*

Type	Name	Formula	Limits of inflammability, % by volume		% by volume combustibles in air. Mixture for theoretical complete combustion	Ratio of lower limit to P.C.C.,† column (4) + (6)	Ratio of upper limit to P.C.C.,† column (5) + (6)
			Lower	Upper			
(1)	(2)	(3)	(4)	(5)	(6)	(7)	(8)
Paraffin hydrocarbons	Methane	CH_4	5.00	15.00	9.47	0.53	1.58
	Ethane	C_2H_6	3.10	12.45	5.64	.55	2.21
	Propane	C_3H_8	2.10	10.10	4.02	.52	2.51
	Butane	C_4H_{10}	1.86	8.41	3.12	.60	2.70
	Isobutane	C_4H_{10}	1.80	8.44	3.12	.58	2.71
	Pentane	C_5H_{12}	1.40	7.80	2.55	.55	3.06
	Isopentane	C_5H_{12}	1.32	......	2.55	.52	
	Hexane	C_6H_{14}	1.25	6.90	2.16	.58	3.19
	Heptane	C_7H_{16}	1.00	6.00	1.87	.53	3.21
	Octane	C_8H_{18}	0.95	3.20‡	1.65	.58	1.94
	Nonane	C_9H_{20}	.83	2.90‡	1.47	.56	1.97
	Decane	$C_{10}H_{22}$	.67‡	2.60‡	1.33	.50	1.95
	Dodecane	$C_{12}H_{26}$	.60‡	......	1.12	.54	
	Tetradecane	$C_{14}H_{30}$	.50	......	0.96	.52	
Olefines	Ethylene	C_2H_4	2.75	28.60	6.52	.42	4.39
	Propylene	C_3H_6	2.00	11.10	4.44	.45	2.50
	Butadiene	C_4H_6	2.00	11.50	3.67	.54	3.13
	Butylene	C_4H_8	1.98	9.65	3.37	.59	2.86
	Amylene	C_5H_{10}	1.65	7.70	2.72	.61	2.84
Acetylenes	Acetylene	C_2H_2	2.50	80.00	7.72	.32	10.36
	Allylene	C_3H_4	1.74	......	4.97	.35	
Aromatics	Benzene	C_6H_6	1.35	6.75	2.72	.50	2.49
	Toluene	C_7H_8	1.27	6.75‡	2.27	.56	2.97
	Styrene	C_8H_8	1.10‡	6.10‡	2.05	.54	2.98
	o-Xylene	C_8H_{10}	1.00	6.00‡	1.95	.51	3.08
	Naphthalene	$C_{10}H_8$	0.90‡	......	1.71	.53	
	Anthracene	$C_{14}H_{10}$	.63‡	......	1.25	.50	
Cyclic hydrocarbons	Cyclopropane	C_3H_6	2.45	10.45	4.44	.55	2.25
	Cyclohexene	C_6H_{10}	1.22‡	4.81‡	2.40	.51	2.00
	Cyclohexane	C_6H_{12}	1.33	8.35‡	2.27	.59	3.68
	Methyl cyclohexane	C_7H_{14}	1.15	......	1.95	.59	
Terpenes	Turpentine	$C_{10}H_{16}$	0.80‡	......	1.47	.54	
Alcohols	Methyl alcohol	CH_4O	6.72	36.50‡	12.24	.55	2.98
	Ethyl alcohol	C_2H_6O	3.28	18.95‡	6.52	.50	2.91
	Allyl alcohol	C_3H_6O	2.52	18.00‡	4.97	.51	3.62
	Propyl alcohol	C_3H_8O	2.15	13.50	4.44	.48	3.04
	Isopropyl alcohol	C_3H_8O	2.02‡	......	4.44	.45	
	Propylene glycol	$C_3H_8O_2$	2.62‡	12.55‡	4.97	.53	2.53
	Butyl alcohol	$C_4H_{10}O$	1.70	......	3.37	.50	
	Isobutyl alcohol	$C_4H_{10}O$	1.68	......	3.37	.50	
	Amyl alcohol	$C_5H_{12}O$	1.19	......	2.72	.44	
	Isoamyl alcohol	$C_5H_{12}O$	1.20	......	2.72	.44	
	Triethylene glycol	$C_6H_{14}O_4$	0.89‡	9.20‡	2.72	.33	3.39
Aldehydes	Acetaldehyde	C_2H_4O	3.97	57.00†	7.72	.51	7.38
	Croton aldehyde	C_4H_6O	2.12	15.50‡	4.02	.53	3.86
	Butyraldehyde	C_4H_8O	2.47	......	3.67	.67	
	Furfural	$C_5H_4O_2$	2.10‡	......	4.02	.52	
	Paraldehyde	$C_6H_{12}O_3$	1.30	......	2.72	.48	

* Compiled by G. W. Jones, U.S. Bureau of Mines.
† P.C.C., per cent combustible in air. Mixture for theoretical complete combustion.
‡ Determinations made at elevated temperature.

Table 20. Limits of Flammability of Gases and Vapors in Air*—*(Concluded)*

Type	Name	Formula	Limits of inflammability, % by volume		% by volume combustibles in air. Mixture for theoretical complete combustion	Ratio of lower limit to P.C.C.,† column (4) + (6)	Ratio of upper limit to P.C.C.,† column (5) + (6)
			Lower	Upper			
(1)	(2)	(3)	(4)	(5)	(6)	(7)	(8)
Ethers	Methyl ethyl ether	C_3H_8O	2.00	10.10	4.44	.45	2.27
	Di-ethyl ether	$C_4H_{10}O$	1.85	36.50	3.37	.55	10.83
	Di-vinyl ether	C_4H_6O	1.70	27.00	4.02	.42	6.72
Ketones	Acetone	C_3H_6O	2.55	12.80	4.97	.51	2.58
	Methyl ethyl ketone	C_4H_8O	1.81	9.50‡	3.67	.49	2.59
	Methyl propyl ketone	$C_5H_{10}O$	1.55	8.15‡	2.90	.53	2.81
	Methyl butyl ketone	$C_6H_{12}O$	1.22	8.00‡	2.40	.51	3.33
	Methyl-iso-butyl ketone	$C_6H_{12}O$	1.35	7.60‡	2.40	.56	3.16
Acids	Acetic acid	$C_2H_4O_2$	4.05	......	9.47	.43	
	Hydrocyanic acid	HCN	5.60	40.00	14.34	.39	2.79
Esters	Methyl formate	$C_2H_4O_2$	5.05	22.70‡	9.47	.53	2.40
	Ethyl formate	$C_3H_6O_2$	2.75	16.40‡	5.64	.49	2.91
	Methyl acetate	$C_3H_6O_2$	3.15	15.60‡	5.64	.56	2.77
	Ethyl acetate	$C_4H_8O_2$	2.18	11.40‡	4.02	.54	2.84
	Propyl acetate	$C_5H_{10}O_2$	1.77	8.00‡	3.12	.57	2.56
	Isopropyl acetate	$C_5H_{10}O_2$	1.78	7.80‡	3.12	.57	2.50
	Cellosolve acetate	$C_6H_{12}O_3$	1.71	......	2.72	.63	
	Butyl acetate	$C_6H_{12}O_2$	1.39	7.55‡	2.55	.55	2.96
	Amyl acetate	$C_7H_{14}O_2$	1.12‡	......	2.16	.52	
	Ethyl nitrate	$C_2H_5NO_3$	3.80	......	10.68	.36	
	Ethyl nitrite	$C_2H_5NO_2$	3.01	50.00	8.51	.35	5.88
Hydrogen	Hydrogen	H_2	4.00	74.20	29.50	.14	2.52
	Deuterium	De	4.90	75.00	29.50	.17	2.54
Oxides	Carbon monoxide	CO	12.50	74.20	29.50	.42	2.52
	Ethylene oxide	C_2H_4O	3.00	80.00	7.72	.39	10.36
	Propylene oxide	C_3H_6O	2.00	22.00	4.97	.40	4.43
	Acetic anhydride	$C_4H_6O_3$	2.67‡	10.10‡	4.97	.54	2.03
	Dioxan	$C_4H_8O_2$	1.97‡	22.25‡	4.02	.49	5.53
	Di-ethyl peroxide	$C_4H_{10}O_2$	2.34	......	3.67	.64	
	Acetal	$C_6H_{14}O_2$	1.65‡	......	2.40	.69	
Nitrogen compounds	Ammonia	NH_3	15.50	26.60	21.82	.71	1.22
	Cyanogen	C_2N_2	6.60	42.60	9.47	.70	4.50
	Acrylonitrile	C_3H_3N	3.05	17.00‡	5.29	.58	3.21
	Pyridine	C_5H_5N	1.81	12.40‡	3.24	.56	3.83
	Quinoline	C_9H_7N	1.21‡	......	1.91	.63	
	Nicotine	$C_{10}H_{14}N_2$	0.75‡	4.00‡	1.53	.49	2.61
Sulfur compounds	Carbon disulfide	CS_2	1.25	50.00	6.52	.19	7.67
	Hydrogen sulfide	H_2S	4.30	45.50	12.24	.35	3.72
	Carbon oxysulfide	COS	11.90	28.50	12.24	.97	2.33
	Ethyl mercaptan	C_2H_6S	2.80	18.20	4.44	.63	4.10
Chlorides	Methyl chloride	CH_3Cl	8.25	18.70	12.24	.67	1.53
	Vinyl chloride	C_2H_3Cl	4.00	21.70	7.72	.52	2.81
	Ethyl chloride	C_2H_5Cl	4.00	14.80	6.52	.61	2.27
	Propyl chloride	C_3H_7Cl	2.60	11.10‡	4.44	.59	2.50
	Allyl chloride	C_3H_5Cl	3.28	11.15‡	4.97	.66	2.24
	Chlorobutene	C_4H_7Cl	2.02	9.25‡	3.67	.55	2.52
	Butyl chloride	C_4H_9Cl	1.85	10.10‡	3.37	.55	3.00
	Isobutyl chloride	C_4H_9Cl	2.05	8.75‡	3.37	.61	2.60
	Amyl chloride	$C_5H_{11}Cl$	1.60	8.63‡	2.72	.59	3.18
	Mono-chloro-benzene	C_6H_5Cl	1.35‡	7.05‡	2.90	.47	2.43
	Benzyl chloride	C_7H_7Cl	1.10	......	2.40	.46	
	Di-chloro-ethylene	$C_2H_2Cl_2$	7.00	19.50‡	9.47	.74	2.06
	Ethylene dichloride	$C_2H_4Cl_2$	6.20	15.90‡	7.72	.80	2.06
	Propylene dichloride	$C_3H_6Cl_2$	3.40	14.50‡	4.97	.68	2.92
Bromides	Methyl bromide	CH_3Br	13.50	14.50	12.24	1.10	1.18
	Ethyl bromide	C_2H_5Br	6.75	11.25	6.52	1.04	1.73
Gases	Blast furnace gas	....	35.00	73.50			
	Coal gas	....	6.50	36.00			
	Coal gas	....	5.30	33.00			
	Natural gas	....	4.30	13.50			
	Natural gas	....	4.90	15.00			
	Oil gas	....	4.75	32.50			
	Producer gas	....	20.70	73.70			
	Water gas	....	6.00	70.00			
Petroleum products	Gasoline, regular	....	1.40	7.50			
	Gasoline, 73 octane	....	1.50	7.40			
	Gasoline, 92 octane	....	1.50	7.60			
	Gasoline, 100 octane	....	1.45	7.50			
	Naphtha	....	1.10	6.00			
Metallo-organic compounds	Tetra-methyl lead	$C_4H_{12}Pb$	1.80	......	2.72	0.66	
	Tetra-methyl tin	$C_4H_{12}Sn$	1.90	......	2.72	.70	

* Compiled by G. W. Jones, U.S. Bureau of Mines.
† P.C.C., per cent combustible in air. Mixture for theoretical complete combustion.
‡ Determinations made at elevated temperatures.

in which p_1, p_2, p_3 are the proportions of each combustible gas present in the original mixture, free from air and inert gases, so that

$$\frac{p_1}{N_1} + \frac{p_2}{N_2} + \frac{p_3}{N_3} + \cdots = 100$$

An example of the use of the formula will make its application clear: To calculate the lower limit of a "natural gas" of the composition

Methane.	80 per cent (lower limit, 5.3 per cent)
Ethane.	15 per cent (lower limit, 3.22 per cent)
Propane.	4 per cent (lower limit, 2.37 per cent)
Butane.	1 per cent (lower limit, 1.86 per cent)

gives

$$L = \frac{100}{\frac{80}{5.3} + \frac{15}{3.22} + \frac{4}{2.37} + \frac{1}{1.86}} = 4.55 \text{ per cent}$$

The accuracy of the formula has been tested carefully for many mixtures. In general, it may be said that, although the formula is often correct or very nearly so, there are

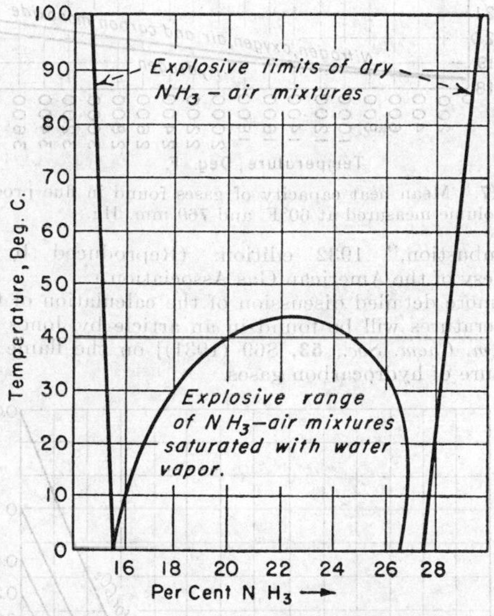

FIG. 15. Limits of inflammability of ammonia-air mixtures.

some marked exceptions. It seems that the limits (lower and higher) of mixtures of hydrogen, carbon monoxide, and methane taken two at a time or all together and of water gas and coal gas may be calculated with approximate accuracy. The same is true for mixtures of the simpler paraffin hydrocarbons, including "natural gas." Sometimes, however, the differences between the calculated and observed values are very large; . . . Many of the greater discrepancies are found with upward-propagating flames, especially when one of the constituents is a vapor, such as ether or acetone, capable of giving rise to the phenomenon known as a "cool flame." Le Chatelier's law is useful when its applicability has been proved, but it must not be applied indiscriminately.

A brief account of the method of calculating limits of complex industrial gases, as those just mentioned, will be found in *U.S. Bur. Mines Tech. Paper* 450.

Speed of the Propagation of Flame. Flame speeds are of importance in certain problems of gaseous combustion. Thus in questions of interchangeability it is usually difficult to substitute a major quantity of a slow-burning gas for a fast-burning gas. The same holds for problems in burner design where flash back and blowoffs must be considered. In heat transfer, questions of combustion intensity or flame output are related to flame speeds. Mixture compositions below or above which flame speeds are zero determine the lower or upper limits of inflammability.

Payman has proposed the use of the Le Chatelier mixture calculation for the estimation of certain flame speeds in certain complex mixtures. Whenever the law holds true, the percentage amount of a mixed gas of known composition which gives a certain speed of flame can be calculated from the observed amounts of constituent gases which, separately, give the same flame speeds. The formula is the same as that used for calculating the dilution limits. There are many marked exceptions, and the application cannot be deemed general enough to constitute a law. Natural gas mixtures, however, give results in fairly close agreement with those calculated (*U.S. Bur. Mines Tech. Paper* 427).

Checking Flame Propagation. Flame propagation is impossible in a **coal gas + air** mixture when the tubes are smaller than 0.08 in. or 2 mm. in inside diameter. **Methane + air** mixtures will not propagate flame through tubes whose internal diameter is smaller than 0.142 in. or 3.6 mm. **Hydrogen,** however, has been found to flame through tubes whose internal diameter was 0.035 in. or 0.9 mm.

Specific Flame Output or Intensity of Combustion. A conception known as specific flame output, or

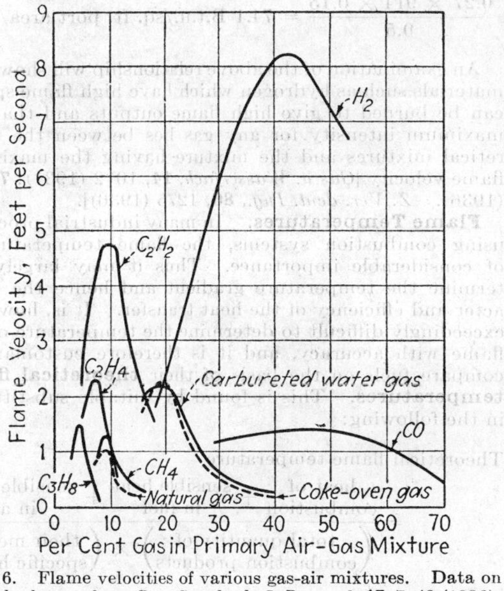

FIG. 16. Flame velocities of various gas-air mixtures. Data on individual gases from *Bur. Standards J. Research,* **17,** 7–43 (1936). Data on natural gas, coke-oven gas, and carbureted water gas from "Combustion," Am. Gas Assoc., 1932 (as compiled by Elliott and Denues of U.S. Bureau of Mines for Marks, "Mechanical Engineers' Handbook," McGraw-Hill, New York, 1941).

intensity of combustion, is sometimes used in comparing gaseous fuels. This is expressed by the relation $J = (u \times H)/K$, where J = specific flame output, B.t.u. per square foot of port area per second; u = flame velocity in feet per second; H = heat developed per cubic foot of air-gas mixture, as will be further explained; and K = ratio of the port area to the area of the inner cone. The result may be expressed as primary flame output when based on the heat developed in the primary combustion or as total flame output when based upon the total heat developed in both primary and secondary combustion.

The calculation of primary flame output involves the relationship

$$H = \frac{hx_t(1 - x)}{(1 - x_t)}$$

where h = net B.t.u. value of the gas.
x = decimal fraction of combustible gas in the primary mixture.
x_t = decimal fraction of combustible gas in the stoichiometric mixture for complete combustion.

Assume that the primary air-gas mixture contains 15 per cent methane whose net B.t.u. at 60°F. and 30 in. Hg. = 914 and the stoichiometric mixture contains 9.5 per cent methane. From Fig. 16 the flame speed of a 15 per cent mixture is 0.27 ft./sec. Therefore

$$\frac{914 \times 0.095(0.85)}{0.905} = 81.4$$

Assume a value of 0.5 for K, and

$$J = \frac{0.27 \times 81.4}{0.5} = 43.9 \text{ B.t.u./sq. ft. port area/sec.}$$

The evaluation of the total flame output is made similarly except that $H = hx$. The total flame output would be

$$\frac{0.27 \times 914 \times 0.15}{0.5} = 74.1 \text{ B.t.u./sq. ft. port area/sec.}$$

An examination of the above relationship will show that materials such as hydrogen which have high flame speeds can be burned to give high flame outputs and that the maximum intensity for any gas lies between the theoretical mixtures and the mixture having the maximum flame velocity [*Gas u. Wasserfach*, **14**, 1012 (1931); **79**, 17 (1936). *Z. Ver. deut. Ing.*, **80**, 1275 (1936)].

Flame Temperatures. In many industrial processes using combustion systems, the flame temperature is of considerable importance. Thus it may largely determine the temperature gradient and hence the character and efficiency of the heat transfer. It is, however, exceedingly difficult to determine the temperature of the flame with accuracy, and it is therefore customary to compare fuels on the basis of their **theoretical flame temperatures.** This is found by suitable substitution in the following:

Theoretical flame temperature

$$= \frac{\begin{matrix}\text{heat of} \\ \text{combustion}\end{matrix} + \begin{matrix}\text{sensible heat} \\ \text{in fuel}\end{matrix} + \begin{matrix}\text{sensible heat} \\ \text{in air}\end{matrix}}{\begin{pmatrix}\text{total quantity of} \\ \text{combustion products}\end{pmatrix} \times \begin{pmatrix}\text{their mean} \\ \text{specific heat}\end{pmatrix}}$$

The calculation of theoretical flame temperatures is most simply made on a trial-and-error basis, *i.e.*, a probable temperature is assumed and corresponding mean specific heat values are chosen. If the sum of the heats divided by the chosen heat capacities gives a temperature higher than that assumed, it is necessary to make a new trial by assuming a new and higher temperature and choosing again a corresponding specific heat value, and so on, until values of flame temperature and mean specific heat are found to correspond. It is always necessary to use the low or net heating value of the gas as the latent heat of condensation of the water formed is not available to raise the temperature of the products of combustion in the flame.

The flame temperatures thus found are the uncorrected theoretical temperatures. If these temperatures are low, no further correction is necessary, but at temperatures approximating 3000°F. or higher the dissociation of water vapor and of carbon dioxide sets in with a corresponding lowering of the flame temperature.

For convenience in the solution of such problems there is presented a curve showing mean specific heats per cubic foot and a curve showing the dissociation of carbon dioxide and water at high temperatures from the booklet

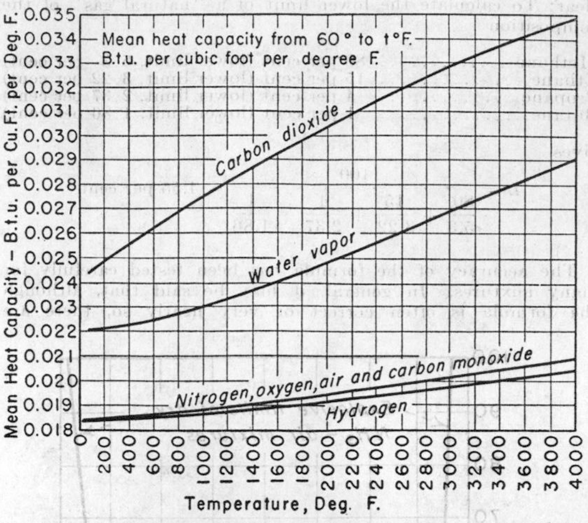

FIG. 17. Mean heat capacity of gases found in flue products. Gas volume measured at 60°F. and 760 mm. Hg.

"Combustion," 1932 edition. (Reproduced by the courtesy of the American Gas Association.)

A more detailed discussion of the calculation of flame temperatures will be found in an article by Jones *et al.* [*J. Am. Chem. Soc.*, **53**, 869 (1931)] on the flame temperature of hydrocarbon gases.

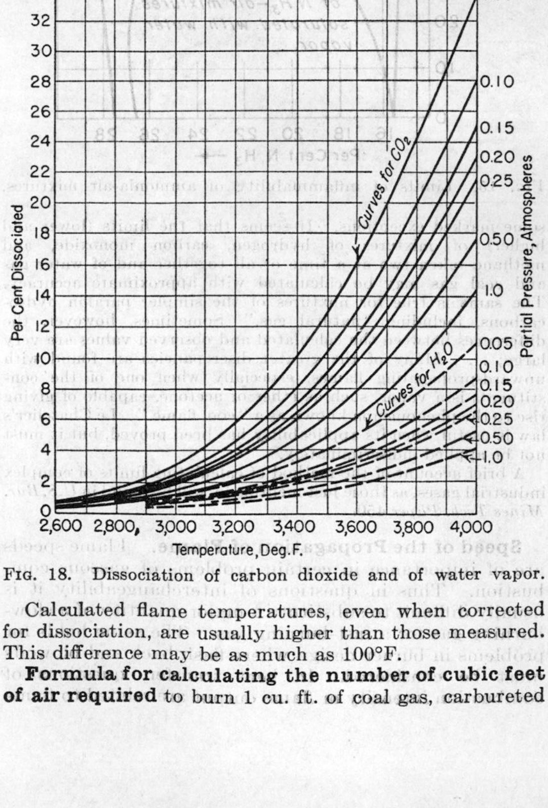

FIG. 18. Dissociation of carbon dioxide and of water vapor.

Calculated flame temperatures, even when corrected for dissociation, are usually higher than those measured. This difference may be as much as 100°F.

Formula for calculating the number of cubic feet of air required to burn 1 cu. ft. of coal gas, carbureted water gas, and natural gas, a convenient approximation is

$$\frac{(\text{Gross B.t.u.}) - 50}{100}$$

Combustion Calculations from Fuel-gas Analysis. In combustion calculations involving industrial gases, there are two possible sources of error arising from uncertainties concerning the chemical composition of (1) the illuminants and (2) the paraffin hydrocarbons. Repeated analyses accompanied by calorimetric tests usually make it possible to assign approximate composition and heating values to these components. In isolated analyses, such a procedure is not available. Watson and

Table 21. Flame Temperatures

Computed theoretical (maximum) flame temperatures of various gases			Experimentally determined flame temp. in air		
	1	2	(Sodium-line reversal method)		
	Theoretical flame temp. (computed max. temp.)[a] °F.	Computed maximum flame temp., with allowance for radiation loss[b] °F.	% combustible	Flame temp., °F.	Ref.
Carbon monoxide (CO)	4475	3092	32.0	3812	*c*
Hydrogen (H_2)	4010	3488	31.6	3713	*d*
Methane (CH_4)	3750	3038	10.0	3407	*d,e*
Ethane (C_2H_6)	3820	....	5.8	3443	*e*
Propane (C_3H_8)	3840	....	4.15	3497	*e,f*
Butane (C_4H_{10})	3870	....	3.2	3443	*e*
Methyl ether $(CH_3)_2O$					
Ethylene (C_2H_4)	4250	3452	7.0	3587	*e*
Propylene (C_3H_6)	4090	....	4.5	3515	*e*
Butylene (C_4H_8)	4030	....	3.4	3506	*e*
Acetylene (C_2H_2)	4770	4208	9.0	4217	*d*
Blue water gas (310 B.t.u./cu. ft.)	4167				
Carbureted water gas (578 B.t.u./cu. ft.)	4090				
Natural gas (1047 B.t.u./cu. ft.)	3740				
Producer gas (136 B.t.u./cu. ft.)	3050				

In column 1 are shown the theoretical (calculated) flame temperatures of various gases, including that of carbon monoxide, as given by Haslam (Haslam, "Fuels and Their Combustion," p. 279, McGraw-Hill, New York, 1925), without allowance for radiation. Similar computations, according to Bone and Townend (Bone and Townend, "Flame and Combustion in Gases," p. 199, London, 1927), using allowances for radiation according to Helmholtz and Callendar's experiments and mean specific heats after Holborn and Henning, are shown in column 2.

[a] Values according to Haslam.

[b] Values according to Bone and Townend.

[c] Lewis and von Elbe, "Combustion, Flames and Explosions of Gases" Macmillan, New York, 1938.

[d] Jones, Lewis, and Seaman, *J. Am. Chem. Soc.*, **53**, 3992 (1931).

[e] Jones, Lewis, Friauf, and Perrott, *Ind. Eng. Chem.*, **53**, 869 (1931).

[f] Loomis and Perrott, *Ind. Eng. Chem.*, **20**, 1004 (1928).

Ceaglske [*Ind. Eng. Chem.*, anal. ed. **4**, 70 (1932)] use a "triple combustion" to estimate the average compositions and heating values of both illuminants and paraffins by extending the Bureau of Mines Orsat analysis (*cf. U.S. Bur. Mines Bull.* 197), which includes a hydrogen-carbon monoxide slow combustion over copper oxide and a paraffin combustion over a platinum spiral, to include an additional total combustion with oxygen. The method is not adapted to gases containing acetylene.

The Flow of Gas through Pipe Lines*

Where conditions permit the application and the necessary data are available, the flow of gas in pipe lines can be most logically dealt with in accordance with Reynolds criterion, as discussed in Sec. 6 of this handbook. The application has been limited because quantitative values of the viscosity relationships in commercial gas mixtures are generally not available. Another difficulty arises because gas flowing in long lines under high pressure may undergo a marked change in density.

* See Sec. 5.

The problem presented by the application of Reynolds criterion to flow problems such as are handled by simple gas engineering flow formulas has been considered in a paper by Huff and Logan (*Am. Gas Assoc., Proc.*, 1935, p. 687).

These authors show that at ordinary temperatures a function of the specific gravity of a gas referred to air can be substituted for the absolute viscosity of common fuel gases. The density viscosity relationship is $13.8s^{0.129} = z \times 10^3$, where s is the specific gravity of the gas referred to air as 1 and z is the absolute viscosity of the gas in centipoises. With this simplification for low-pressure flows, in which the change in density of the gas is not important and in which the flow is turbulent, it is possible to derive a flow formula resembling former formulas, but resting upon the firm foundation of Reynolds criterion and consequently applicable to a wide variety of conditions. This formula is

$$Q = \frac{1281h^{0.543}d^{2.631}}{S^{0.468}(L')^{0.543}} \text{ cu. ft./hr.}$$

where h = the pressure drop, in. of water.
d = the diameter of the pipe, in.
S = specific gravity of the gas at the temperature and pressure in question relative to air at room temperature and 30 in. of mercury.
L' = the length of the pipe, yd.

If it is desired to express this formula in feet of pipe L, the relationship becomes

$$Q = \frac{2331h^{0.543}d^{2.631}}{S^{0.468}L^{0.543}} \text{ cu. ft./hr.}$$ See Fig. 19.

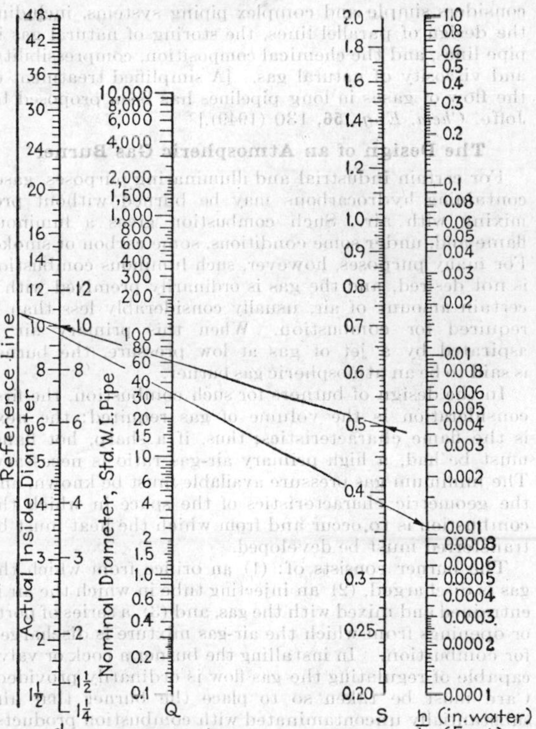

FIG. 19. Flow of gas in commercial pipes (M Cu. Ft. = 1000 cu. ft.)

The U.S. Bureau of Mines in cooperation with the Natural Gas Department of the American Gas Association has studied the flow of natural gas through **high-pressure** transmission lines. The reported results obtained favor for the use of a formula developed by Weymouth [*Trans. Am. Soc. Mech. Engrs.*, **34,** 1091–1104 (1912)] (*cf.* Berwald and Johnson, *U.S. Bur. Mines Rept. Investigations* 3153, December, 1931). This is

$$Q_c = 18.062 \frac{T_0 + t_1}{P_0} \sqrt{\frac{(P_1^2 - P_2^2)d^{5\frac{1}{3}}}{GTL}}$$

where Q_c = number of cubic feet of gas flowing per hour, corrected to 60°F., and base pressure (14.4 lb./sq. in.)
P_1 = absolute inlet-line pressure, lb./sq. in.
P_2 = absolute outlet-line pressure, lb./sq. in.
T = absolute flowing temperature, °F., *i.e.* $(T_0 + t_2)$.
L = length of line, miles.
T_0 = absolute temperature, 460°F.
P_0 = absolute pressure base (14.4 lb./sq. in., as generally adopted in natural-gas engineering practice).
t_1 = 60°F.
t_2 = observed flowing temperature, °F.
G = specific gravity of gas (air = 1).

A nomograph for the graphical solution of the Weymouth formula has been developed by Mathis (*Gas Age-Record*, Apr. 30, 1932, pp. 536–538).

Extensive studies of Johnson and Berwald on the flow of natural gas have been compiled in Monograph No. 6 issued jointly by the U.S. Bureau of Mines and the American Gas Association, entitled "Flow of Natural Gas through High Pressure Transmission Lines." This considers simple and complex piping systems, including the design of parallel lines, the storing of natural gas in pipe lines, and the chemical composition, compressibility, and viscosity of natural gas. [A simplified treatment of the flow of gases in long pipelines has been proposed by Joffe, *Chem. Eng.*, **56,** 130 (1949).]

The Design of an Atmospheric Gas Burner*

For certain industrial and illuminating purposes, gases containing hydrocarbons may be burned without premixing with air. Such combustion gives a luminous flame and, under some conditions, some carbon or smoke. For many purposes, however, such luminous combustion is not desired, and the gas is ordinarily premixed with a certain amount of air, usually considerably less than is required for combustion. When this primary air is aspirated by a jet of gas at low pressure, the burner is said to be an atmospheric gas burner.

In the design of burners for such combustion, the first consideration is the volume of gas required; the next is the flame characteristics; thus, if a sharp, hot flame must be had, a high primary air-gas ratio is necessary. The minimum gas pressure available must be known, and the geometric characteristics of the space in which the combustion is to occur and from which the heat must be transferred must be developed.

The burner consists of: (1) an orifice from which the gas is discharged, (2) an injecting tube in which the air is entrained and mixed with the gas, and (3) a series of ports or openings from which the air-gas mixture is discharged for combustion. In installing the burner a cock or valve capable of regulating the gas flow is ordinarily provided. Care must be taken so to place the burner that air, substantially uncontaminated with combustion products, is freely available at the air-shutter opening of the mixing tube under conditions of maximum demand, and provision must also be made to permit the ready flow of the requisite secondary air to the ports and to permit the free discharge of the combustion products. The inner cone of the flame must never impinge an object, as the incomplete combustion resulting may produce dangerous amounts of carbon monoxide.

* An excellent extended discussion on the design of gas burners is contained in Vandaveer, "Gaseous Fuels," Chap. VII, Am. Gas Assoc., 1948.

Orifice. In selecting an orifice, a high coefficient of discharge is desirable, as entrainment of more air may be obtained. The rate of flow of gas from a sharp-edged orifice may be computed by means of the formula

$$Q = 1658.5AK \sqrt{\frac{H}{d}}$$

where Q = the quantity of gas flowing, cu. ft./hr.
A = the area of the orifice, sq. in.
K = the orifice constant or discharge coefficient.
H = the orifice pressure, in. of water (above the atmosphere).
d = the specific gravity of the gas as it approaches the orifice (air = 1).

Up to an orifice pressure of 10 in. of water, no serious error due to gravity changes because of increased pressure at the orifice is introduced.

The various values of K for different types of sharp-edged orifices are shown in Fig. 20 (*U.S. Bur. Standards Tech. Paper* 193; *Sci. Paper* 359).

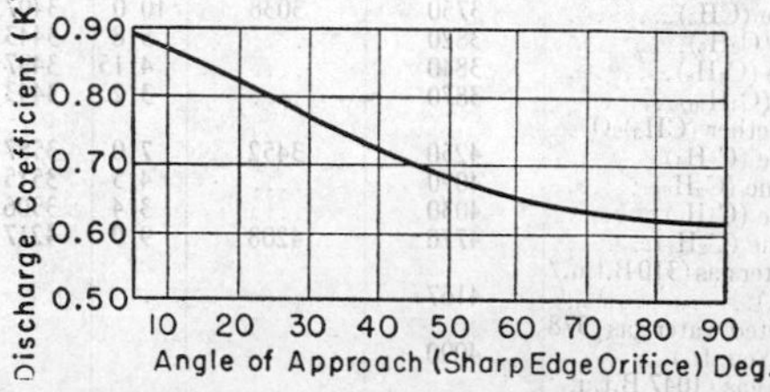

FIG. 20. Relation between orifice constant K and angle of approach for sharp-edged orifice. Commercial orifices are nearly all of channel type (fixed-spud orifices) to hold the discharge coefficient as nearly constant as possible with variations in drill sizes.

Air-gas Ratio. For flexible atmospheric burner operation it is desirable to operate with a primary air-gas ratio in the mixing tube of such a character that ordinary changes in composition will not be critical. In older manufactured gas appliances the primary air-gas ratio is from 1½ to 2, *i.e.*, from 30 to 40 per cent of the total air required. With advance in the technique of burner design, it has been possible to secure much higher percentages of aeration in the primary mixture. Where access of secondary air to the ports is somewhat restricted, as in some radiant-type heaters fitted with ceramic grilles, the percentage primary aeration may be as much as 80 per cent. Such high primary air-gas ratios are, however, more critical and are avoided where flexibility is important.

If the primary air-gas ratio is too low, the flame lengthens unduly and may produce carbon monoxide if impingement results. The tips may turn yellow and the combustion may become smoky. Too high primary air-gas ratio may lead to strike backs with carbon monoxide production, or blowoffs which are equally undesirable.

The Injecting Tube. To secure maximum air entrainment with a minimum gas pressure, the injecting or air-mixing tube should be in the form of a venturi (refer to Sec. 5).

In *Bur. Standards Tech. Paper* 193 the following are given as some items of importance in regard to the injector tube:

a. The change of the lines of approach of the inlet to the outlet should be gradual.

b. The approach should follow approximately a curvature which should not be less than 3 in. radius for a ⅝-in. throat, and other sizes should be in proportion. For this size tube, the orifice should discharge about 1.5 in. from this throat.

c. The outlet angle should be about 2 deg. The outlet tube should not be too short. Six throat diameters is a minimum length for good service. A satisfactory area of the throat can be determined by multiplying the area of the ports by 0.43.

The Bureau of Standards (*Circ.* 394) develops the relationship between the burner design and the momentum of the gas stream with particular reference to the entrainment of the air. The product of the mass and velocity of the jet of gas, as it leaves the orifice, bears a sufficiently constant ratio to the momentum of the mixture of gas and primary air as it passes any definite cross section of the burner to permit the use of a constant in most problems of design, *i.e.*, $MV/mv = C$. The value of C is the same for burners of different sizes but of the same geometric design. The considerations are approximate and subject to some modification as the burner heats up and with increasing pressure and low gravity gases. According to this, geometrically similar burners have a characteristic injecting constant

$$k = \frac{Q}{q}\sqrt{\frac{aD}{d}}$$

where k = burner constant.
q = quantity of gas flowing in unit time.
Q = volume of the air-gas mixture flowing in unit time.
a = area of the orifice.
D = specific gravity of the air-gas mixture, air = 1.
d = specific gravity of the gas, air = 1.

The value of k should be determined experimentally but should not be assumed to be greater than $0.8\sqrt{P}$, where P is the total port area, although well-designed burners, properly aligned with a sharp-edged orifice, have shown values as great as $1.2\sqrt{P}$.

A number of useful relationships may be derived from this treatment of the burner constant.

It has been recommended that free area of the mixer head, through which the air is admitted, be fixed approximately at 1.25 times the port area.

The Burner Head. Considerations of importance relating to the burner head are the port area, the depth and coefficient of discharge of the ports, the individual port area, the spacing of the ports, the burner volume, and the distribution of secondary air.

For natural gas, butane, and propane 12,000 to 15,000 B.t.u./hr. can be burned per square inch of port area; for manufactured gases, from 22,000 to 26,000 B.t.u./sq. in. of port area. For universal burners for use on either fuel, around 20,000 B.t.u./sq. in. should be selected. In general, the total port area should be made as large as possible. The figures above are for minimum areas.

Drilled ports are preferred. No gauze should be placed in the burner. For small iron or steel burners the burner head, tube, and injector should be cast in one piece and not cemented.

Individual port areas are important factors in stability and blowoff or strike back. According to the Bureau of Standards, the following areas should not be exceeded:

For manufactured gas: No. 40 M.D.S. (0.0075 sq. in.).
For natural gas: No. 30 M.D.S. (0.013 sq. in.).
For propane and butane: No. 32 M.D.S. (0.0106 sq. in.).

Universal burners require small ports with total areas approaching the recommendations for natural gas.

The port relations must be such that the flame will not strike back owing to too low a velocity of air-gas flow nor must it blow off owing to too high a velocity. To secure high air-gas ratios without strike-back, ample momentum must be imparted to the air-gas mixture to ensure adequate pressure within the burner head to cause a proper velocity of flow through deep burner ports.

Kowalke and Ceaglske (*Proc. Am. Gas Assoc.*, 1929, pp. 662–686) derive certain relationships between the air-gas ratios and the ratio of throat area to port area and orifice area to throat area. The work was further developed (*Proc. Wis. Utilities Assoc.*, 1930). The equations obtained in this work have been rearranged ("Combustion," 1932 ed., published by American Gas Association) as follows:

$$A_P = \frac{0.85}{C} \times (R + 1) \times \frac{d + R}{d} \times A_o$$

where A_P = port area, sq. in.
A_o = area of the orifice, sq. in.
C = coefficient of discharge of the ports. Values may range from 0.45 to 1.00; 0.60 is sometimes taken.
R = primary air-gas ratio.
d = density of the gas, air = 1.

The value 0.85 is characteristic of the burner and is applicable only to a well-designed burner, such as those using injecting tubes meeting the suggestions of the Bureau of Standards. Values from 0.35 to 0.7 have been found.

The general form of the above equation is as follows:

$$A_P = \frac{K^2 A_o}{CF} \times \frac{(R + 1)(R + d)}{d}$$

where K is the discharge coefficient of the gas orifice (values ranging from 0.75 to 0.85) and F is the injector characteristic.

Arrangement of ports in more than two rows without access of secondary air between them should be avoided. Burner ports should not be drilled too close to the injector tube, and the air-gas stream should not be deflected against certain ports. Care should be taken to secure substantially equal pressure on all ports. It is difficult to estimate the coefficient of discharge of the ports because of the various ways in which ports may be made. For rough approximations, values from 0.6 to 0.8 may be used, although values from 0.45 to 1.00 have been found.

Wills [*Western Gas*, August, 1931 (this author gives another approach to the problem of burner design)] recommends that the free cross-sectional area of the burner head be three times the area of the ports to be supplied. If an even distribution of heat is required, the difference in velocity of the air-gas mixture between the first and last ports must not be too great.

Air-shutter Opening. The Bureau of Standards recommends that the possible air-shutter opening should be made so large that the linear velocity through the opening does not exceed 4 or 5 ft./sec.

GAS AS AN INDUSTRIAL FUEL

Furnace Atmospheres. Three classes of furnace atmospheres have been recognized, namely, the oxidizing, the neutral, and the reducing atmospheres. The American Gas Association characterizes these as follows:

Name of atmosphere	Oxygen, %	Carbon monoxide and hydrogen
Oxidizing	Over 0.05	Not over 0.5% total $CO + H_2$
Neutral	Not over 0.05	Not over 0.5% total $CO + H_2$
Reducing	Not over 0.05	Over 0.5% total $CO + H_2$

This classification is for purposes of definition and clarification only and does not rest upon the behavior of the furnace atmosphere toward specific materials such as metals at elevated temperatures.

Carbon dioxide, water vapor, and oxygen will scale metals, while CO and H_2 are reducing. Neutral atmospheres produced by ideal combustion conditions are actually scaling. A better classification of furnace atmospheres is based upon the effect upon the product as scaling, carburizing, decarburizing, nitriding, bright annealing, etc. Much of the information on furnace atmospheres is empirical in nature and based upon imperfect analytical procedures.

In furnaces in which atmospheres must be controlled, two classes may be recognized:

1. Furnaces in which the work is in contact with the combustion atmosphere.
2. Furnaces in which the heating is indirect and the work is subject to a controlled or controllable atmosphere, as in muffle furnaces.

Certain remarks and precautions about the choice of industrial burner systems are apropos here. The systems will be described later. Thus atmospheric burners should not be used for the production of reducing atmospheres as the low burner-head pressure may permit the admission of pockets of excess air and consequent explosions. Pressure burners are to be preferred over atmospheric burners where wide flexibility is desired. Diffusion-flame burners possess certain advantages where low flame temperatures and high rates of heat transfer are desired. These burners allow precipitations of carbon and a blanket of reducing gas. Nozzle-mixing burners do not allow satisfactory control of atmospheres on turn-down conditions. Premixed systems are very successful in the production of controlled atmospheres.

In the so-called protective atmospheres water vapor is frequently found very deleterious and must be removed. Furnace technique for special work has now become so developed that it is possible to remove undesirable gases, such as water vapor and carbon dioxide, from combustion gases almost completely.

It will be obvious to the chemical engineer that the action of furnace atmospheres and their effective control involve a consideration of the thermodynamics of the various oxidizing and reducing actions which the product may undergo and the thermodynamics of the furnace atmosphere. Thus, in bright annealing, the constituents in the furnace atmosphere must not dissociate to give available oxygen in partial pressures sufficient to repress the dissociation of the oxide of the product, as otherwise the product would tarnish or scale. It is obviously impossible to discuss the many applications of gas to controlled furnace atmospheres and the precautions that must be observed. The American Gas Association through appropriate committees on industrial gas research has and is engaged upon studies that are helpful, and the reports of such committees and booklets of that association, particularly "Combustion" and "Fuel-flue Gases"* may be consulted.

Industrial Combustion Systems. Combustion systems have been classified by the American Gas Association† as follows:

I. Atmospheric.
II. Pressure.

* New edition (1948), entitled "Gaseous Fuels."

† In the booklet "Combustion" 2d ed., 1926, prepared under the direction of the Industrial Gas section of the association. Permission to use material contained therein is gratefully acknowledged.

A. Air compressed
 (1) without automatic proportioning.
 (2) with automatic proportioning.

B. Gas compressed
 (1) without automatic proportioning.
 (2) with automatic proportioning.

C. Mixture compressed
 (1) without automatic proportioning.
 (2) with automatic proportioning.

It will be impossible to describe in detail the various types of apparatus available for these systems; therefore merely the general principles involved will be given, somewhat along the lines adopted by the Gas Association.

The object of the various combustion systems is to provide accurate regulation of the air-gas ratio in the mixture supplied to the burner ports and an accurate control of the rate of gas supply.

Diffusion-flame Burners. Since the above classification was prepared, the industrial use of burners in which there is no premixing of air with gas has extended greatly.

These burners depend upon the diffusion of all the air required for combustion. They are known as diffusion-flame or luminous-flame burners. They are characterized by a luminous flame of high radiant emissivity and a low flame temperature. They will be referred to in the concluding paragraphs (p. 1595).

I. The Atmospheric System

For operations requiring temperatures up to 1000°F., and where an ample supply of secondary air to the combustion chamber is assured, atmospheric burners can be applied successfully. Properly designed burners will operate without flashing back when the gas flow is reduced to one-tenth the rated consumption, thus permitting thermostatic control. The simplicity of the atmospheric system causes it to be widely used, but the necessity for a positive draft and the long time required to attain furnace temperatures from 500 to 1000°F. often causes it to be rejected in favor of a special combustion system.

Some of the considerations involved in the design have already been set forth in the division entitled The Design of an Atmospheric Gas Burner.

Additional considerations follow:

1. Each individual burner port should receive its supply of secondary air by drilling the ports in raised teats when this is possible, and the ports should be so arranged that secondary air is readily accessible to every individual port.
2. The burner body should be as small as possible consistent with good operation in order to minimize flash-back objections.
3. No cement joints should be used, as these may crack under heat and leak gas.
4. Burners should be made of cast iron.
5. Air mixers should have flat faces and be capable of being rigidly fixed at any desired adjustment.
6. Burners and manifolds should be rigidly connected to their appliance.
7. Standard A.G.A. orifice spuds and gas cocks are available for gas consumptions up to 80 cu. ft./hr. and should be used when possible.
8. The use of wire gauzes in burners should be avoided.
9. The burner should, where possible, be placed close to the appliance without causing the cone to impinge and without causing floating from the banking of burned gases.
10. Free exit of burned gases from under the heated

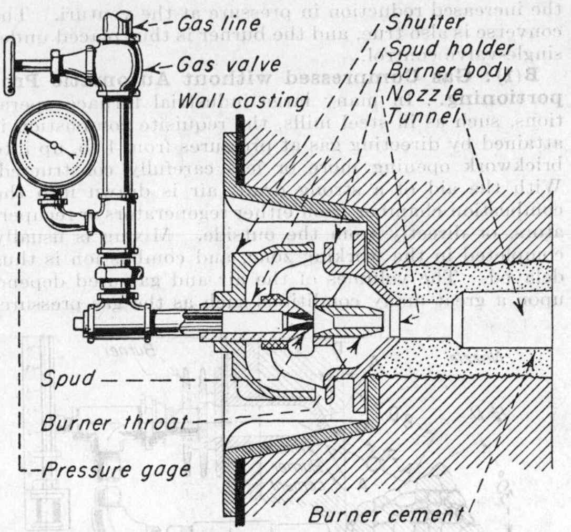

FIG. 21. Atmospheric industrial gas-burner installation with individual air control to each burner.

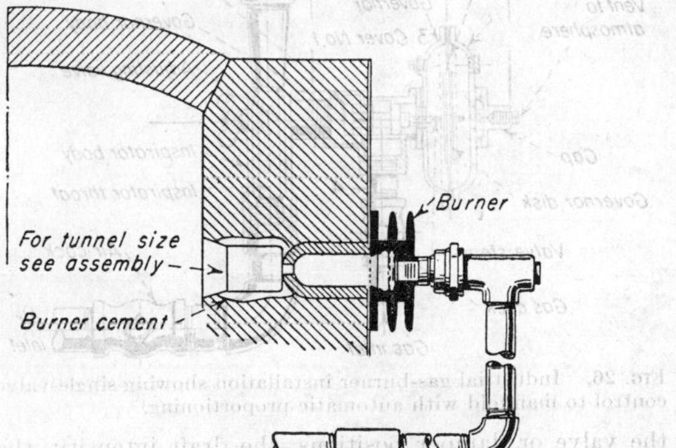

FIG. 22. Atmospheric industrial gas burner with manifold mixing.

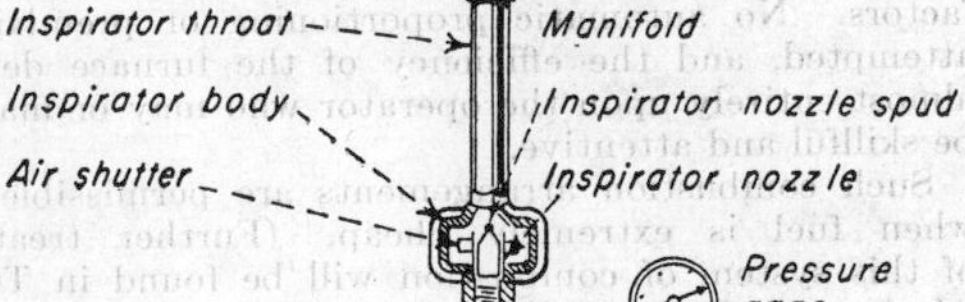

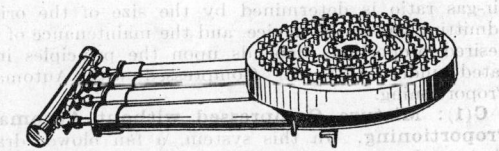

FIG. 23. Atmospheric industrial gas burner, circular, four-range capacity.

surface must be provided by placing the burner sufficiently low.

11. A draft inducing an excess of secondary air should be reduced by a permanent restriction in the flue. No dampers should be used.

Figures 21, 22, 23, and 24 illustrate some of the many forms of atmospheric burners.

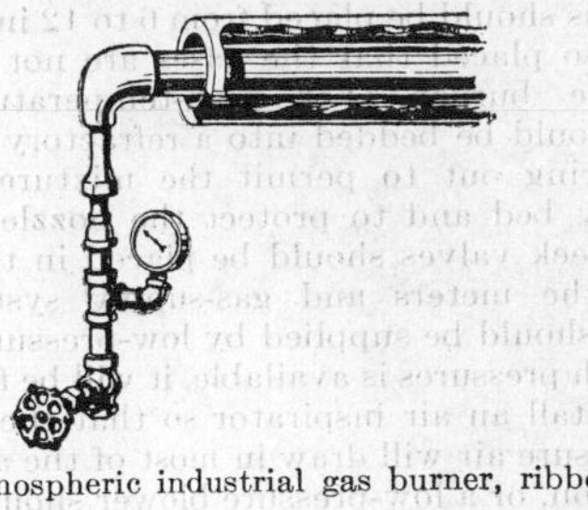

FIG. 24. Atmospheric industrial gas burner, ribbon type.

II. Pressure Systems

A(1): Air Compressed without Automatic Proportioning (Two-valve Control). Next to the atmospheric system, the simplest combustion system consists of independently controlled air and gas lines meeting in a mixing tee. The air, which constitutes the bulk of the mixture, is passed straight through the tee and inspirates gas from the side outlet. The proportions of the mixture should be correct for perfect combustion, without secondary air. From the mixing tee, the combustible mixture is passed to the burner.

Air is usually supplied at pressures of 1 to 2 lb./sq. in., but the system may operate satisfactorily at air pressures as low as 6 in. water, if the supply lines are large and no important friction loss occurs in these when working at capacity. The gas is supplied at the service pressure.

The intimate mixture of air and gas before entering the furnace makes it possible to attain high temperatures, and good temperature control may be obtained by the system; but it possesses the disadvantage that furnace atmospheres cannot be readily duplicated and controlled owing to possible independent variations in the two supply pressures.

The mixing tees have a slight restriction at the air-supply end to create a suction at the point where the gas enters the mixture. Many mixing tees are made from a simple pipe fitting without an inspirating device.

This system develops burner pressures from ½ to 8 in. water, and the same type burner may therefore be used as those that are employed in other blast-combustion systems.

Ordinarily, in any burner of any system the flame may be held near the burner port by a glowing refractory. When the mixture issuing from the port contains sufficient air to support combustion, and flash back must consequently be prevented, the design of the port must be such that such flash back is extinguished. This can be accomplished by designing the burner port to give the issuing mixture a velocity higher than the velocity at which flame will propagate, thus blowing the ignited mixture to the mouth of the port. Since the limiting velocity is dependent upon the diameter of the tube through which the flame may propagate, it is very desirable to use small ports of considerable depth, since such ports permit the use of low-mixture velocities without the danger of flash back. If the flame is not held by a glowing refractory, it is important to limit the velocity at the head of the burner so that the flame will stick and not blow off. This may involve the placing of the ports in such a manner that the issuing mixtures from several adjacent ports mingle to give a low resultant velocity. Since in blast systems no secondary air access need be provided, this desirable condition may be attained by hooding the mouth of the burner or by directing port streams against each other to interfere, blend, and slow the resultant streams. Pilot flames are also used to hold the flame on the burner.

When the mixture is blown against a refractory material, this should be placed from 6 to 12 in. from the nozzle and so placed that the gases are not reflected back upon the burner. For high-temperature work the nozzle should be bedded into a refractory wall with side walls flaring out to permit the mixture to reach the refractory bed and to protect the nozzle from burning out. Check valves should be placed in the gas lines to protect the meters and gas-supply system. The air pressure should be supplied by low-pressure blowers. If air at high pressures is available, it will be found economical to install an air inspirator so that the energy in the high-pressure air will draw in most of the air required for combustion, or a low-pressure blower should be installed.

The pressure systems without automatic proportioning are not recommended for anything except installations so small that the expenditure for more elaborate systems is not warranted.

A(2): Air Compressed with Automatic Proportioning (Single-valve Control). The automatic proportioning referred to in combustion is applied to the automatic control of the ratio of fuel to air without attention by the operator. Because of fluctuating supply pressures and changing flow rates, separate valve controls for fuel and air cannot be manipulated in such a manner as to ensure the maintenance of a given furnace atmosphere at all times. Consequently, various devices have been developed to regulate the amount of mixture admitted to a burner or a set of burners. A homogeneous constant-composition mixture is thus ensured. The composition can, of course, be set as desired.

A thoroughly mixed and correctly proportioned gas-air blend suitable for complete combustion is the ideal mixture to discharge from the burner nozzle. A minimum of combustion space is required because combination occurs immediately upon ignition.

Figures 25 and 26 illustrate arrangements for automatic proportioning.

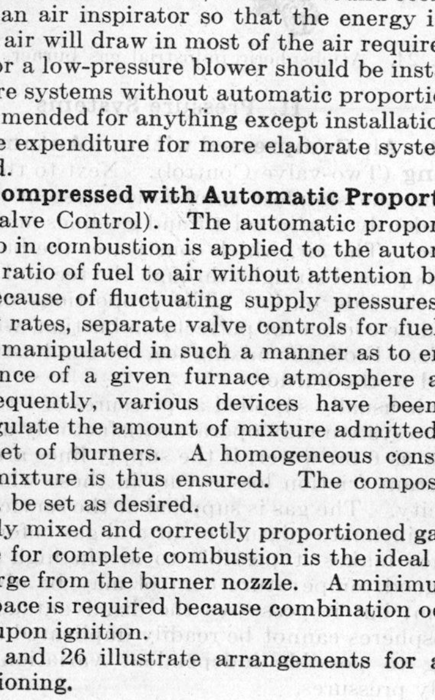

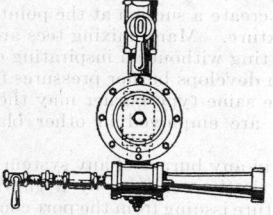

FIG. 25. Gas burner with automatic proportioning.

The system referred to in this section operates by utilizing the energy in the air to inspirate gas which has been reduced to atmospheric pressure by a diaphragm regulator. The air is supplied at from 1 to 2 lb./sq. in. At the narrow part of the venturi, the pressure is below atmospheric, and gas at atmospheric pressure is drawn through an orifice into this region of reduced pressure. By proper construction of the venturi and gas orifice, it is possible to make both fluids follow the expression $V = \sqrt{2gh}$, in which V is the stream velocity; g is the acceleration due to gravity = 32.16 ft./sec.; and h is the head or pressure difference causing the flow. A given air flow will inspirate a given gas flow; with increased air flow, the amount of gas inspirated will be increased with the increased reduction in pressure at the venturi. The converse is also true, and the burner is thus placed under single-valve control.

B(1): Gas Compressed without Automatic Proportioning. In many large industrial furnace operations, such as in steel mills, the requisite combustion is attained by directing gas at pressures from 1 lb. up into brickwork opening more or less carefully constructed. With the aid of a strong draft, air is drawn into the combustion chamber from either regenerators or recuperators, or directly from the outside. Mixing is usually completed in the working zone, and combustion is thus delayed. The volumes of the air and gas used depend upon a great many conditions such as the gas pressure,

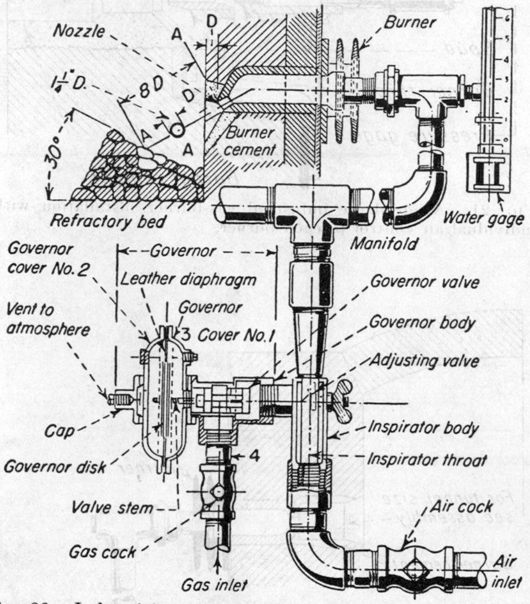

FIG. 26. Industrial gas-burner installation showing single-valve control to manifold with automatic proportioning.

the valve or damper positions, the draft intensity, the character of the air or gas passages or both, and other factors. No automatic proportioning or premixing is attempted, and the efficiency of the furnace depends almost entirely upon the operator who may or may not be skillful and attentive.

Such combustion arrangements are permissible only when fuel is extremely cheap. (Further treatment of this system of combustion will be found in Trinks, "Industrial Furnaces," Wiley, New York, vol. I, 1923, vol. II, 1925.)

B(2): Gas Compressed with Automatic Proportioning. This system is generally termed the *high-pressure* system. The compressed gas is carried to the point of application without the use of fire screens, fire checks, etc. All the air for combustion is injected by the gas through the use of patented inspirators which proportion the amounts of air and gas at the appliance. The air-gas ratio is determined by the size of the orifice admitting gas to the device, and the maintenance of the desired proportions depends upon the principles indicated under $A(2)$, Air Compressed with Automatic Proportioning.

C(1): Mixture Compressed without Automatic Proportioning. In this system, a fan blower draws air from the atmosphere and gas from a supply at normal service pressure into the fan casing where a homogeneous mixture is at once formed. A double valve turns the air and gas up and down together. This system is designed for constant heat requirements.

This type of system can be used chiefly for single-burner nozzle applications and for work that requires little or no

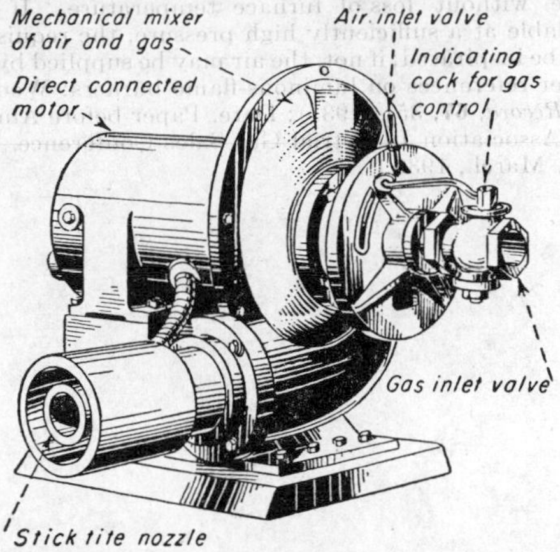

FIG. 27. Industrial gas burner and mixer, the mixture being compressed without automatic proportioning.

turn down and will permit the fan to be placed close to the furnace without overheating. Multiple burner use, involving long manifold piping must be avoided, as flash backs may occur with variable burner pressures. Figure 27 shows such an installation, and Fig. 28 shows the type of nozzle used.

C(2): Mixture Compressed with Automatic Proportioning. The fan blower described in *C*(1) can be greatly improved in reference to automatic proportioning by the addition of a gas governor. If this is done, and suitable admission ports are chosen for the gas and the air, a certain degree of automatic proportioning is obtained. A fan blower so equipped is shown in Fig. 29.

A much more elaborate type, in which all the air for combustion mingles with the gas homogeneously is that of the Kemp Co.; it is illustrated in Fig. 30. Admission valves for air and gas with a number of different settings are provided to give such ratios of air to gas as may be desired. Once set, the ratio remains constant.

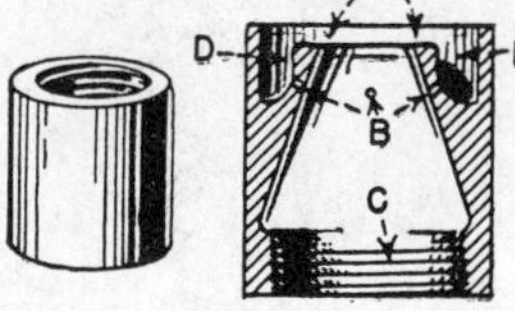

FIG. 28. Burner to hold the flame on fan-type mixture compression system.

Since the system handles a combustible mixture, various precautions are used to prevent backfiring, including "backfire checks" and specially designed burners with screens. Explosion relief heads are also furnished, but these are added precautions very rarely, if ever, needed.

Another subdivision is used to classify **mixing compressors drawing in only a portion of the air required,** usually less than that necessary to support combustion. The air-gas ratio is usually 2:1. This mixture injects the remaining air at the burners. The amount injected here will of course depend somewhat upon the furnace conditions and burner adjustments, so the system is not completely automatic in proportioning. Flash back may occur in these burners if the mixture flow is reduced to one-third normal operations. Automatic heat control of the gradual shutdown type is therefore applicable only when the low comsumption is greater than one-third the normal.

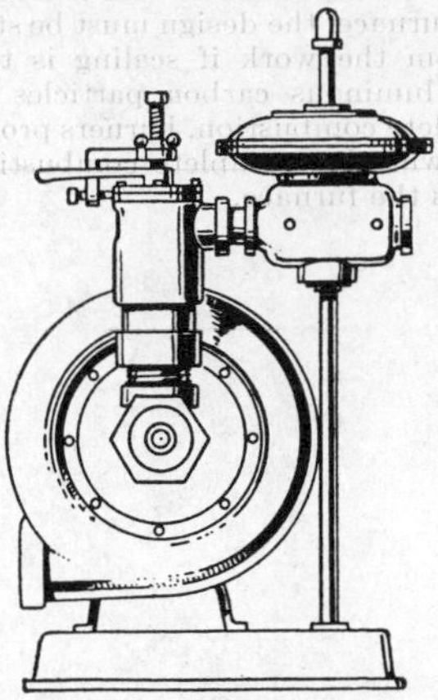

FIG. 29. Mixture compressed with some automatic proportioning—fan type.

Diffusion-flame or Luminous-flame Burners. It is claimed that diffusion-flame or luminous-flame burners offer the following advantages:

1. An increased rate of heat transfer.
2. More uniform temperature control.
3. Less maintenance of furnace refractories.
4. Greater degree of turn down of burners.
5. Less scale is formed, and such scale as may be formed is more easily removed.

Diffusion- or luminous-flame burners admit little or no primary air to the burner. The air and gas streams emitted from the burner nozzle should be free of turbulence and should travel along the furnace at the same speed, burning taking place only at the surfaces of the individual streams of gas and air as diffusion of one into the other occurs. By controlling the velocity of air and gas streams or speed of mixing, the length of flame as well as the luminosity may be controlled within desired limits. A mathematical analysis of the factors affecting the shape of such flames has been made by Burke and Schumann [*Ind. Eng. Chem.*, **20**, 998 (1928)].

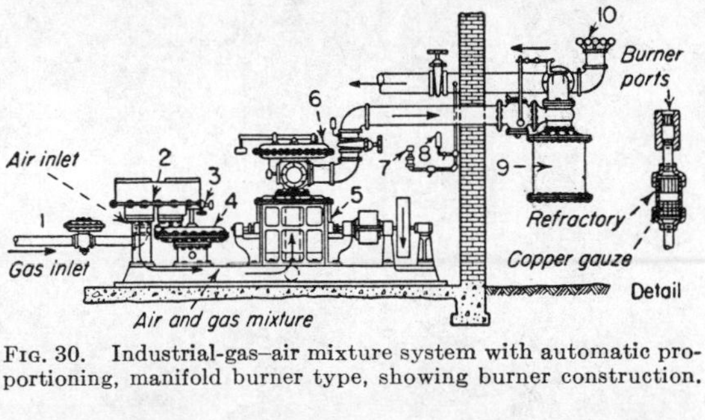

FIG. 30. Industrial-gas–air mixture system with automatic proportioning, manifold burner type, showing burner construction.

The radiating power of a luminous flame is from four to six times as great as that of a non-luminous flame; thus for certain high temperatures a faster rate of heat transfer can be obtained from the luminous flame than from non-luminous flames, even though the flame temperature of the luminous flame is several hundred degrees lower.

In applying luminous flames to industrial installations, it is desirable to secure the services of a manufacturer of gas-burner equipment experienced in such work. A number have given considerable time and attention to luminous-flame burners that can be applied to furnaces for various types of operations. As there is uncombined oxygen in the furnace, the design must be such that this is kept away from the work if scaling is to be avoided. Although the luminous carbon particles indicate temporary incomplete combustion, burners properly designed and operated will give complete combustion of the fuel before it leaves the furnace.

Luminous-flame burners can be supplied in a design such that, if the gas is purchased on an off-peak basis, oil can be burned in them when the gas supply is interrupted. Gas varying from 350 to 1500 B.t.u. can be used in the same burner, and the change-over can sometimes be made without loss of furnace temperature. If gas is available at a sufficiently high pressure, the requisite air may be inspirated; if not, the air may be supplied by a fan. [Other references on luminous-flame burners: Wood, *Gas Age-Record*, **67**, 351 (1931); Dare, Paper before American Gas Association Industrial Gas Sales Conference, Cleveland, March, 1939.]

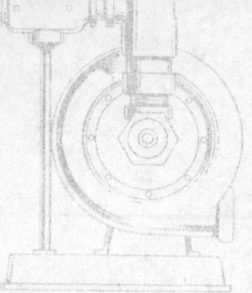

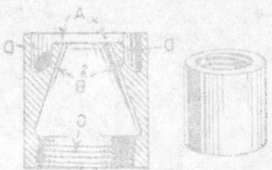

SECTION 23

FURNACES AND KILNS

BY

Percy H. Royster, A.B., A.M., Technical Advisor, Metallurgical Division, Bureau of Mines, U.S. Department of the Interior.

C. H. Evans, B.S., Chemical Engineer, E. I. du Pont de Nemours & Co., Inc.; Member, American Institute of Chemical Engineers, American Chemical Society, American Association for the Advancement of Science.

A. C. Mueller, Ph.D., Chemical Engineer, E. I. du Pont de Nemours & Co., Inc., Member, American Institute of Chemical Engineers, American Society of Mechanical Engineers, American Chemical Society, American Association for the Advancement of Science.

W. Trinks, Professor Emeritus, Carnegie Institute of Technology; Consulting Engineer.

CONTENTS

CLASSIFICATION OF FURNACES AND KILNS

	Page
According to Name	1598
According to Use	1598
According to Movement of Charge	1598
According to Fuel or Other Source of Heat	1598
According to Method of Heat Salvage	1598
According to Method of Heat Application	1598
According to Design or Shape	1598
According to a Special Name	1599

TEMPERATURES IN FURNACES AND KILNS

Control of Temperatures	1599
Temperature Ranges of Process Furnaces	1600

INDUSTRIAL FURNACES

Design Factors	1600
Fuels Used in Industrial Furnaces	1600
Heat Distribution	1600
Heat Transfer and Heat Release	1600
Furnace Construction	1601
Effect of Furnace Size and Shape on Heat Transfer	1602
Uses and Modifications	1605
Boiler Furnaces	1605
Cyclone Furnaces	1605
Petroleum Furnaces	1605
Metallurgical Furnaces	1605
Ceramic Furnaces	1606

PROCESS FURNACES

Rotary Kilns	1608
Uses of Rotary Kilns	1608
Description of Rotary Kilns	1608
Time of Passage in Rotary Kilns	1609
Charge Volume in Rotary Kilns	1609
Heat Transfer in Rotary Kilns	1609
Capacity of Rotary Kilns	1609
Thermal Efficiency of Rotary Kilns	1611
Size Segregation in Kilns	1612
Shaft Furnaces	1612
Uses of Shaft Furnaces	1612
General Description of Shaft Furnaces	1612
Shaft Furnaces for Lime Production	1612
Factors Affecting Performance of Shaft Furnaces	1613
Heat Transfer and Efficiency in Shaft Furnaces	1614
Blast Furnaces	1614
Bessemer Converters	1616
Gas Producers and Water-gas Sets	1617
Pebble Heater	1617
Thermofor Cracking Units	1618
Fluidized-solids Reactors	1618

	Page
FluoSolids Furnaces	1619
Reverberatory Furnaces	1619
Uses of Reverberatory Furnaces	1619
Description of Reverberatory Furnaces	1619
Open-hearth Steel Furnace	1620
Glass Furnace	1621
Special Designs of Process Furnaces	1621
By-product Coke Ovens	1621
Mannheim Furnaces	1621
Multiple-hearth Furnaces	1622
Flash Roasters	1623

WASTE-HEAT RECOVERY

General	1623
Checkerbrick Recuperators	1623
Blast-furnace Stoves	1623
Open-hearth and Glass-tank Regenerators	1624
Coke-oven Regenerators	1624
The Pebble Stove	1625

THERMAL TREATMENT OF UNSTABLE GASES

Thermal Treatment of Unstable Gases	1625

GAS REACTION AT ELEVATED TEMPERATURES

Gas Reaction at Elevated Temperatures	1626

REFERENCES: Trinks, "Industrial Furnaces," vol. 1, 3d ed., 1934; vol. 2, 2d ed., 1942, Wiley, New York. De Lorenzi, "Combustion Engineering," Combustion Engineering-Superheater, Inc., New York, 1948. Griswold, "Fuels, Combustion, and Furnaces," McGraw-Hill, New York, 1946. Haslam and Russell, "Fuels and Their Combustion," McGraw-Hill, New York, 1925. Sweetser, "Blast Furnace Practice," McGraw-Hill, New York, 1938. Etherington, "Modern Furnace Technology," Lippincott, Philadelphia, 1938. Martin, "Chemical Engineering and Thermodynamics Applied to the Cement Rotary Kiln," Technical Press, London, 1932. Gilbert, "Heat Transmission in Rotary Kilns," series of 12 parts plus appendix, "Cement and Cement Manufacture," Concrete Publications, Ltd., London, December, 1932; March, June, August, October, December, 1933; January, May, July, October, 1934; April, 1935; January, February, 1936. Buell, "The Open Hearth Furnace," 3 vols. (1936–1939), Penton, Cleveland. Eckel, "Cement, Limes, and Plasters," Wiley, New York, 1928. McNamara, "Ceramics," vol. 3, The Pennsylvania State College, Pennsylvania, 1944. Rosenholtz and Oesterle, "Elements of Ferrous Metallurgy," 2d ed., Wiley, New York, 1933. Bell, "American Petroleum Refining," 3d ed., Van Nostrand, New York, 1945. Nelson, "Petroleum Refinery Engineering," 3d ed., McGraw-Hill, New York, 1949. Francis, "The Making, Shaping, and Treating of Steel," 5th ed., United States Steel Corp., Pittsburgh, 1940. Mawhinney, "The Heating of Steel," Reinhold, New York, 1945. Yearbook of the American Iron and Steel Institute, New York. "Metals Handbook," 1948 ed., American Society for Metals, Cleveland.

CLASSIFICATION OF FURNACES AND KILNS

This section describes the commercial apparatus and structures used for the generation and application of high temperatures. Some of these designs are used with low temperatures to carry out drying operations as described in Sec. 13. Units heated by electricity are not described here but are included in Sec. 27.

A large number of widely varying designs are used in high-temperature work, many being custom-built for a specific job. Thus a number of different classifications must be applied to cover the field as an aid in identifying which of the existing types would have features suitable for application to a processing problem.

1. According to Name. The terms "furnace" and "kiln" have a wide variance in their meaning as commonly used in different industries. The use of adjectives does not always clarify this situation, *e.g.*, a "lime kiln" may be a vertical shaft or a rotating cylinder slightly inclined from the horizontal position. Consequently care must be exercised in the use of these and other terms in this field.

In a strict sense, a *furnace* is a chamber made of a refractory material in which heat is generated at a high-temperature level, its design involving no consideration of the use of this heat. A *kiln* is an apparatus or chamber of a refractory material in which heat is used, but also it commonly includes the generation of the heat. Moreover, this term generally implies the use of the heat for drying or degasifying as well as for the subsequent further heating of the charge to a high temperature.

An *oven* is an apparatus or chamber of a refractory material in which heat is used at the lower levels of the high-temperature range, say below 1000°F. However, a *Dutch oven* is a chamber for generating heat which is separate from the one in which the heat is used.

Other general terms are stove, muffle, retort, calciner, roaster, etc. All these terms tend to have different meanings in different industries, and quite often they are used interchangeably.

2. According to Use. Furnaces, kilns, etc., are often identified by their uses, such as cement kilns, lime kilns, gas producers, coke ovens, blast furnaces, glass tanks, zinc retorts, muriatic acid furnaces, sulphur burners, black ash kilns, and many others. All these units have different designs as specifically dictated by the physical and chemical properties of the charge being processed; consequently these are classified as *processing furnaces*. There is another category in which the nature of use merely requires a heated space into which the charge can be placed. This charge might consist of individual items such as billets or ceramic articles, or of solids or fluids confined in vessels or pipes which are positioned in the heated space. It includes heat-treating ovens, melt pots, ceramic kilns, japanning ovens, petroleum pipe stills, and pipe cracking furnaces, steam boiler furnaces, etc., and these are classified as *industrial furnaces*.

3. According to Movement of Charge. The charge movement may be either *periodic* or *continuous*. In the periodic or batch type, there is no movement of the charge during the heating process. It includes heat-treating furnaces for very large items, pot furnaces, salt baths, low-capacity muffles, etc. In the continuous type, the charge moves constantly. It enters the apparatus at one point and leaves at another. The charge may be moved by a pusher, on a conveyor, in cars, by gravity, by movement of the hearth, or in a fluid stream. Pushers or conveyors are used in billet-heating furnaces. Cars are used in tunnel kilns. Gravity is used in shaft furnaces and in rotary kilns. Rotating annular hearths are used for heat-treating. In multideck (Herreshoff) furnaces, the charge is alternately moved inward and outward by rakes with gravity discharge from deck to deck. In fluidized-solids units, the charge is moved in a gas stream.

4. According to Fuel or Other Sources of Heat. Furnaces may be fired with solid fuels on a grate; or with liquid, gaseous, or powdered fuels in an air stream through burner nozzles. They may be heated by electricity, including induction, conduction, arc, and radiant types. Heat also may be generated within the charge by an exothermic reaction or by mixing a fuel with the charge.

5. According to Method of Heat Salvage. Fuel is saved when the outgoing products of combustion preheat the combustion air or the fuel or both or the incoming ware or charge. In *regenerative furnaces*, combustion air (or combustion air and fuel) and products of combustion flow alternately through the same checkerwork, and the flame travels in alternately opposite directions through the furnace. In a *recuperative furnace*, products of combustion and combustion air flow on opposite sides of a heat-transferring partition wall, and the flame travels through the furnace always in the same direction. If the charge or ware leaves the furnace at or near room temperature, the combustion air or the incoming ware can be preheated by the outgoing ware. This type of furnace is called the *compensating type*.

6. According to Method of Heat Application. These include the *direct furnaces* and the *indirect heaters*. The former includes those furnaces where the combustion gases come into direct contact with the charge, while the latter includes those where these gases are separated from the charge by a wall through which the heat must pass. The direct furnaces are brick-firing (and similar) tunnels and kilns, rotary kilns, blast furnaces, open hearths, reverberatory furnaces, shaft furnaces, Thermofor kilns, etc. The indirect heater types are muffles, radiant tubes, zinc retorts, steam boilers, petroleum cracking and pipe still furnaces, melting pots, crucibles, etc. A convection furnace is a modification of the direct type since the charge is heated by the hot combustion gases although it is shielded from a direct view of the flame. Indirect heaters are adaptable where special atmospheres are needed.

7. According to Design or Shape. *Vertical furnaces* include cupolas, blast furnaces, shaft kilns, gas producers, pebble heaters, etc. *Horizontal furnaces* include tunnels, rotary kilns, ceramic kilns, heat-treating furnaces, etc. *Tank furnaces* include reverberatory, open hearth, glass, fusion, or melting pots, etc. *Space furnaces* include ore roasters, sulphur burners, rotary kilns, pipe stills, and steam boiler furnaces, particularly those fired with gaseous, liquid, or powdered fuels. *Circular furnaces* include those for muriatic acid, billet heating, chromite ore, etc. Multideck furnaces may be classed as both vertical and circular. Fluidized-solids furnaces may be classed as vertical units and sometimes as space furnaces.

8. According to a Special Name. Furnaces are often designated (but not classified) with a proper name such as that of the inventor, of the manufacturing company, of the location of first use, or with a trade name. This includes Lepol kiln, Herreshoff furnace, Trail roaster, Mannheim furnace, Nichols-Freeman roaster, Thermofor kiln, etc.

TEMPERATURES IN FURNACES AND KILNS

The temperature to be carried in a furnace or kiln depends not only on the material that is to be heated but also on the manufacturing process to which the material is to be subjected after having been heated Thus steel requires temperatures from 600° to 3100°F. for different operations. It is impossible to tabulate temperatures for all operations, because new processes are continually introduced. Table 1 contains temperature data on a number of industrial-furnace operations.

Table 1. Temperatures in Various Heating Furnaces

Degrees Fahrenheit

Annealing:	
Aluminum	750
Brass and copper	930
Cold-rolled strip sheets, steel	1250–1400
German silver	1200
Glass	1000–1150
High-carbon steel	1500
Malleable iron, long-cycle	1600
Malleable iron, short-cycle	1800
Manganese steel castings	1900
Monel or nickel, sheets or wire	1470
Pack-rolled steel sheets	1600
Steel, castings or forgings	1650
Bisque firing porcelain	2250
Bluing steel	500
Burning:	
Common brick	1790–1860
Firebrick	2400–2750
Red earthenware	1640–1900
Sewer pipe	2100–2400
Stoneware	2250–2300
Terra cotta	1650–2300
Vitrified brick or floor tile	2200–2430
Calorizing (heating in aluminum powder)	1700
Carburizing	1750
China manufacture, vitrified	2420
China manufacture, glost	2200–2300
Cyaniding	1800
Drawing, after quenching steel	1200
Drying lacquer	300
Enameling, wet process	1200
Glazing porcelain	1830
Hardening high-speed steel	2200
Heating:	
Brass for rolling	1400
Copper for rolling	1560
Mild steel for rolling and forging	2275
Preformed skelp for tube welding	2550
Stainless steel for rolling	1750–2100
Steel for drop forging or extrusion	2370–2400
Tool steel for rolling	1900
Heat-treating medium-carbon steel	1550
Japanning	180–450
Nitriding steel	950
Normalizing:	
Steel pipes	1650
Steel sheets	1750
Stainless steel	1700–2000
Patenting wire	700
Porcelain decorating	1400
Strain relieving	500 min.
Tempering in oil	500
Tempering high-speed steels	630
Vitreous enameling of steel sheets	1600
Vitreous enameling of castings	1850

1. Control of Temperatures. The instrumentation for maintaining the furnace temperature at a desired level is described in Sec. 19. It generally consists of one or more devices for measuring the temperature of either the furnace or the product, the variation in this measurement being used to operate automatically the fuel valves or feeders, draft dampers, etc. Other auxiliaries must be provided for (1) carrying the furnace temperature at a value lower than flame temperature and (2) attaining the high temperatures required by some processes.

Table 2. Ideal Flame Temperatures for Various Fuels Burned in Theoretical Quantity of Cold Air*

Fuel (and its lower heating value, B.t.u./cu. ft.) disregarding dissociation	Flame Temperature, °F.
Natural gas (935)	3780
By-product coke-oven gas (428)	3730
Clean producer gas (128)	2930
Blast-furnace gas (92)	2600
Blue water gas (279)	4100
Carbureted water gas (506)	3810
Town gas [mixed-retort water gases (518)]	3800
Commercial butane (2977)	3640
Commercial propane (2371)	3660
Coal (14,080 B.t.u./lb.)	4020
Coal (12,400 B.t.u./lb.)	3560
Lignite, dry (9900 B.t.u./lb.)	3740
Lignite, wet (6917 B.t.u./lb.)	3430
Coal tar (15,827 B.t.u./lb.)	4050
Grade 2 fuel oil (18,410 B.t.u./lb.)	3900
Grade 6 fuel oil (17,410 B.t.u./lb.)	3820
Petroleum coke (15,820 B.t.u./lb.)	3840

* For actual temperatures, see Tables 15 and 21, Sec. 22, Fuels.

Object 1 is attained by (*a*) a small flame or a plurality of small flames in a large furnace, such as in the oil-cracking furnace; (*b*) delayed combustion which produces a long sluggish flame; (*c*) dilution of products of combustion with cold air; (*d*) dilution of hot products of combustion with colder products of combustion which have already passed over the charge; (*e*) cooling of hot products of combustion by their imparting heat to thin walls before they enter the heating chamber, a method which is wasteful of fuel; and (*f*) use of muffles.

Table 3. Temperatures in Various Process Furnaces

Degrees Fahrenheit

Alumina	1800–2000
Barytes to BaS	2400–2700
Blast furnace (pig iron)	2700–3300
Blue-gas manufacture, direct-fired	3000
Boron carbide, arc manufacture	5000
Burning:	
Limestone	1700–2600
Natural cement	1800–2200
Portland cement	2800–2900
Calcium carbide manufacture	3000
Coal carbonization, high-temperature	1800–2200
Coal carbonization, low-temperature	900–1400
Copper smelter	2800
Cracking petroleum to gasoline	750–850
Cracking petroleum to acetylene	2000–2500
Distillation of zinc	1350–1700
Distilling hardwood	700–800
Glass melting	2400–2600
Hydrogen chloride from NaCl and H_2SO_4	1000–2200
Hydrogen manufacture from H_2O and Fe	1200
Lead smelter	2200
Nitrogen fixation, arc process	5400
Open-hearth steel furnace	2850–3100
Roasting iron pyrites	1830
Silicon tetrachloride manufacture	950–1800
Water-gas manufacture, approx	1800
Zinc chloride manufacture, from Cl_2 and Zn	950

Object 2 requires different methods. The upper attainable temperature is a fraction of the ideal flame temperature which would be obtained with adiabatic, complete combustion. Ideal flame temperatures are given in Table 2. Since combustion is never ideal, heat is given off by the flame during combustion, and the actual rise in temperature is frequently only 60 to 90 per cent of the ideal rise.

Another limiting factor, noticeable only at very high temperatures, is dissociation of CO_2 and H_2O, which lowers combustion temperature. Combustion temperature is increased by preheating combustion air or fuel, or both. Since specific heats of gases increase with temperature, the increase of flame temperature is always smaller than the increase of preheat temperature. Hydrocarbon fuels (oil, natural gas, coke-oven gas, etc.) are seldom preheated to high temperatures, unless they have been diluted, because they decompose and choke the furnace passages with carbon. Flame temperatures may be increased by enriching the combustion air with oxygen. The highest temperatures are obtained when the heat is generated directly within the charge. This can be accomplished by (1) exothermic reaction of the charge; (2) combustion of fuel mixed with the charge, including the use of oxygen; (3) direct contact of the flame with the charge, particularly in a space furnace with a fluidized charge, and also where a pulverized charge is mixed with a fuel gas prior to combustion; and (4) the use of electricity for induction, conduction, or arc heating.

2. Temperature Ranges of Process Furnaces. The upper range of temperatures which have been attained in various furnaces for different processes is shown in Table 3. The use of oxygen to increase the temperature level of large commercial operations is currently gaining some acceptance because of the recent decrease in its cost resulting from wartime developments in its production methods. At present, the attainment of high temperatures is limited by the inadequacies of the commercially available refractories. New refractories, now in the development phase of commercial production, offer some promise of satisfactory performance at higher levels than the present types.

INDUSTRIAL FURNACES

DESIGN FACTORS

In the design or selection of industrial furnaces, many factors must be considered in order to obtain the desired performance. Some of these factors are temperature required; fuel; distribution of heat; heat-transfer rate; combustion volume; flame size and velocity; refractories; furnace size and shape; vents, ports, and stacks; construction details; recovery of waste heat; and economy of the system.

Fuels Used in Industrial Furnaces. Furnaces in which the combustion chamber is used also for abstracting some of the heat being generated, such as water-cooled walls in boiler furnaces, are fired by solid, liquid, or gaseous fuels and may have combustion chambers as large as 70,000 cu. ft.

Furnaces using solid fuels on grates may be hand-fired, or the fuel may be mechanically introduced by means of underfeed stokers, reciprocating grates, spreader stokers, chain- and traveling-grate stokers. (Descriptions of these stokers, firing rates, and fuel limitations are given in Sec. 24, Power Generation. Table 10 in Sec. 22, Fuels, also summarizes data on mechanical stokers.)

Solid fuels may be pulverized and conveyed into the furnace through burners by means of a portion of the air needed for combustion, the remaining air for combustion being added at the burner. Two basic types of furnaces for powdered coal depend on whether the ash is removed as a powder or as a molten slag. Furnaces may be fired horizontally, vertically, or tangentially. The usual range of heat release is given in Table 4. In certain new designs, it is possible to obtain very high rates of heat release in the primary combustion chamber, but these rates are found to be more nearly normal when the secondary combustion chamber is included. Rate of flame propagation is dependent on the volatile matter in the fuel and the ratio of fuel to air. Velocities as high as 45 ft./sec. can be obtained. For further data, see Haslam and Russell, "Fuels and Their Combustion."

Table 4. Usual Heat-release Rates in Furnaces
B.t.u./(hr.)(cu. ft.)

	Refractory lined	Water walls
Hand-fired coal	5,000–10,000	
Stoker-fired coal	20,000–35,000	40,000–55,000
Pulverized coal	10,000–18,000	12,000–35,000
Fuel oil, at furnace temperatures of		
600°–1500°F.	1,000–15,000	
1500°–1800°F.	15,000–22,000	
1800°–2000°F.	22,000–35,000	30,000–50,000
2000°–3000°F.	30,000–40,000	25,000–65,000
Marine service		50,000–80,000
Gas	15,000–20,000	20,000–30,000

Furnaces fired by liquid fuels may use various devices for atomizing the fuel, and a discussion of typical burners is given under Fuels, Sec. 22. Also, see Power Generation, Sec. 24. In order to be atomized properly, the oil must have low viscosity. Heaters are sometimes needed to reduce the viscosity of the oil before it enters the burners.

Gas-fired furnaces may operate with large flames filling the combustion chamber, or the gas may be premixed with air and burned in small refractory cups which radiate the heat to the charge. High rates of radiant-heat transfer can be obtained in the latter case.

Heat Distribution. The heat generated by the combustion of fuel is distributed as follows: (1) heat usefully imparted to the charge; (2) heat stored in furnace walls, in slag, or in containers and carriers of the charge; (3) heat lost through the furnace walls; (4) heat radiated through openings; (5) heat imparted to cooling water; (6) unconsumed combustible in ash; (7) sensible heat in exhaust gases; (8) unconsumed combustible in the exhaust gases; and (9) heat conducted outside the furnace by metal extending through the walls.

Useful heat imparted to the charge includes the heats for evaporating hydroscopic and crystallizing water as well as the sensible heat in the charge and the heat of reaction of the charge.

Approximate wall losses (referred to mean area between inside and outside) can be taken from Fig. 1. During a closed heating cycle, the sum of wall losses plus heat storage in walls is practically constant. Hence the chart for wall losses approximately covers both items. The heat radiated through openings is given in Fig. 2. Figure 3 gives the factor by which these values must be multiplied in order to obtain correct values for actual conditions. The heat that escapes through a steel bar that protrudes from the furnace equals approximately 15 per cent of the heat that could be radiated from the furnace through an opening having the same cross section as the bar for usual rates of heating or 30 per cent after a steady state is reached.

Heat Transfer and Heat Release. The most important way of transferring heat in furnaces is by radiation. Convective heat transfer is relatively small. A detailed theoretical analysis of furnace heat transfer is seldom possible or reliable. Therefore, most calculations are based on simplifying assumptions and on semi-empirical equations. For treatment of the combustion-space heat transfer, see Heat Transmission, Sec. 6. The rates of heat transfer into the charge are governed by the properties and dimensions of the charge. The

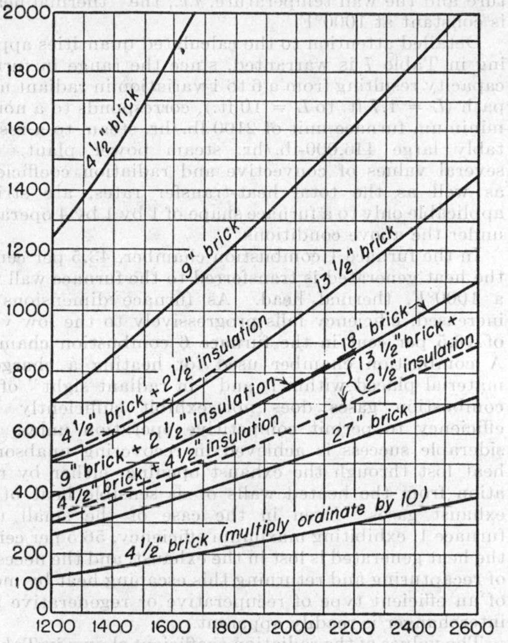

FIG. 1. Heat losses through furnace walls (for atmospheric pressure in furnace). (*Trinks, Industrial Furnaces, vol.* 1, *p.* 88, *Wiley, New York*, 1934.)

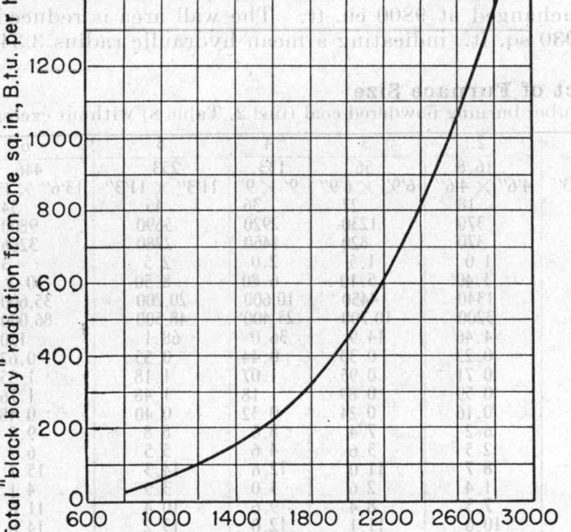

FIG. 2. Heat radiation through openings (black body). (*Trinks, op. cit., pp.* 112–113.)

rates of heat conduction can be estimated from charts given in Heat Transmission, Sec. 6.

The combustion volume of furnaces is determined by the fuel, the method of firing, the temperature and life limitations of the refractory, or by the methods used to prevent overheating the walls, such as water screens. Transpiration cooling by passing a cooling fluid through a porous wall countercurrent to the heat flow has been used and is being developed further. Sometimes, water-cooled plates are placed within the walls to help reduce refractory temperatures. Table 4 lists the usual ranges of heat release for various types of firing, but much higher rates are obtained in special designs. An alternate method of determining furnace sizes is based on the rate of heat absorption in the walls rather than on heat release in the combustion space, thus using surface as a basis rather than volume. When calculated by this method, present-day boiler furnaces absorb heat at rates of 50,000 to 100,000 B.t.u./(hr.)(sq. ft.).

Further information on furnaces for power generation is given in Sec. 24. Also, data are given there for the design of stacks.

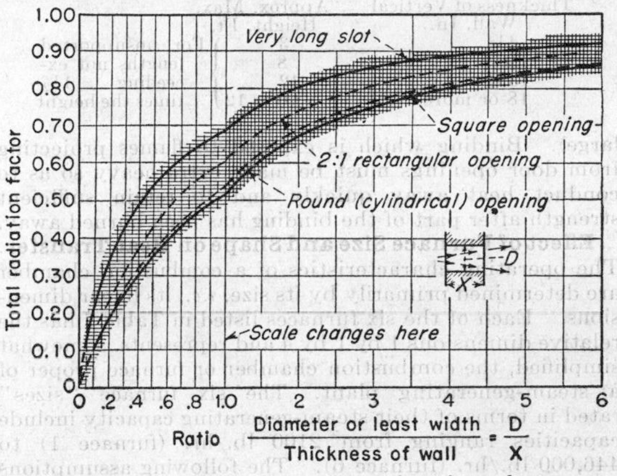

FIG. 3. Radiation through openings of various shapes, as a fraction of the radiation from freely exposed surface of same area as cross section of opening. Multiplying factor for Fig. 2 above. (*Trinks, op. cit., pp.* 112–113.)

Furnace Construction. Retorts or ovens at temperatures of 1000°F. or less present no serious problems in construction or maintenance. In order to reduce heat losses, as few as possible metallic connectors are run through the wall. In furnaces and kilns, difficulties arise that grow with the furnace temperature and with the chemical affinity between the furnace lining and the charge (stock or ware). The commonly used construction material is firebrick composed of silica and alumina plus impurities. For temperatures above 2750°F., either the silica content or the alumina content is greatly increased. The result is silica brick or high-alumina brick (sillimanite, mullite, or fused corundum). Silica brick is used as roof and side-wall material in open-hearth furnaces.

Metallic construction materials in the interior of furnaces and kilns must be adapted to the furnace temperature. Unalloyed steel oxidizes rapidly at temperatures above 1000°F. The rate of oxidation grows exponentially with the temperature. For temperatures above 1000°F., either cast iron is used or else iron is alloyed with chromium and nickel so as to produce heat-resisting metals. The higher the furnace temperature, the less iron is permissible in the metal. For temperatures above 1600°F., metallic materials are not recommended, except for electric resistors.

Expansion and contraction alter the thrust of sprung arches and cause furnaces to collapse, unless the walls are restrained by structural material, such as I beams, channels, rails, tie rods, and cast buckstays. Expansion allowances for various materials are given in Table 5, and height-thickness data in Table 6. The binding must be strong enough to resist being spread by the thrust of sprung arches and to push the brickwork of a cooling furnace back into place against the frictional resistance of overlying bricks. Binding is always arranged near the corners of a furnace (and elsewhere if the furnace is large). Binding which is exposed to flames projecting from door openings must be made extra heavy so as to conduct heat away quickly and to retain sufficient strength after part of the binding has been burned away.

Table 5. Expansion Allowances for Refractories

Material	Expansion Allowance, In./Ft.
Fire clay and 50% alumina	1/16–3/32
60–80% alumina	3/32–1/8
Chrome	5/32
Silica:	
Handmade	1/8–3/16
Power-pressed	3/16
Forsterite	1/4
Magnesite	1/4

Table 6. Height-thickness Data for Vertical Walls

Thickness of Vertical Wall, In.	Approx. Max. Height, Ft.	
4½	3	For unsupported lengths not exceeding 1½ times the height
9	8	
13½	12	
18 or more	Over 12	

Effect of Furnace Size and Shape on Heat Transfer. The operating characteristics of a combustion chamber are determined primarily by its size, *i.e.*, its linear dimensions. Each of the six furnaces listed in Table 7 has the relative dimensions 1 by 1 by 4 and represents, somewhat simplified, the combustion chamber or furnace proper of a steam-generating plant. The six furnace "sizes" rated in terms of their steam-generating capacity include capacities ranging from 2100 lb./hr. (furnace 1) to 446,000 lb./hr. (furnace 6). The following assumptions were made to calculate the comparative performances shown in Table 7. The heat release is maintained constant at 50,000 B.t.u./(hr.)(cu. ft.). The powdered coal used as fuel has the analysis shown in Table 8, column 2. In each furnace, the air is preheated sufficiently to maintain the temperature of the combustion products at 3500°F. The area of the "cold" surface, *i.e.*, the exhaust opening plus the exposed surface of the liquid-cooled elements, in each case is adjusted to control the temperature of the refractory walls at 2500°F. In these circumstances, the difference between the gas temperature and the wall temperature, *i.e.*, the "thermal head," is constant at 1000°F.

Detailed attention to the calculated quantities appearing in Table 7 is warranted, since the range in furnace capacity resulting from a 6 to 1 variation in radiant mean path (L = 1.7 ft. to L = 10 ft.), corresponds to a normal minimum furnace unit of 2100 lb./hr. steam to a respectably large 446,000-lb./hr. steam power plant. The several values of convective and radiation coefficients, as well as the total heat-transfer rates, are strictly applicable only to a furnace shape of 1 by 1 by 4 operating under the above conditions.

In the furnace 1 combustion chamber, 43.5 per cent of the heat generated is transferred to the furnace wall with a 1000°F. thermal head. As furnace dimensions are increased, efficiency falls progressively to the low value of 16.5 per cent in the furnace 6 combustion chamber. A combustion chamber used for heating a charge of material placed within it and "in radiant sight" of the combustion gases does not exhibit sufficiently high efficiency to permit competitive operation unless considerable success is achieved in recovering unabsorbed heat lost through the exhaust opening, either by radiation from the heated walls or as sensible heat of the exhaust gases. Even in the case of the small unit, furnace 1, exhibiting maximum efficiency, 56.5 per cent of the heat generated is lost in the exhaust, and the necessity of recapturing and returning this escaping heat by means of an efficient type of recuperative or regenerative heat interchanger is readily apparent.

The values of the radiation coefficient shown in Table 7, although properly applicable only to a parallelepiped combustion chamber with the relative dimensions 1 by 1 by 4, may be used for other furnace shapes. For example, if furnace 6 is constructed as a cylinder with a 15 ft. 2 in. diameter and 54-ft. axial length, the volume remains unchanged at 9800 cu. ft. The wall area is reduced to 2930 sq. ft., indicating a mean hydraulic radius 3.34 ft.

Table 7. Calculated Effect of Furnace Size

Heat transfer from gas at 3500°F. to wall at 2500°F., in combustion chamber burning powdered coal (fuel 2, Table 8) without excess air

Furnace	1	2	3	4	5	6
1. Nominal rating, 1000 lb. steam/hr.*	2.1	16.8	56.	133.	253.	446.
2. Horizontal section, ft.	2′3″ × 2′3″	4′6″ × 4′6″	6′9″ × 6′9″	9′ × 9′	11′3″ × 11′3″	13′6″ × 13′6″
3. Height, ft.	9	18	27	36	45	54
4. Volume V, cu. ft.	46	370	1230	2920	5690	9800
5. Wall area A_w, sq. ft.	91	370	820	1460	2280	3266
6. Mean hydraulic radius, ft.	0.5	1.0	1.5	2.0	2.5	3
7. Radiant mean path L, ft.	1.70	3.40	5.10	6.80	8.50	10.2
8. Coal rate, lb./hr.	167	1340	4450	10,600	20,200	35,600
9. Draft air, standard cu. ft./min.	394	3200	10,700	25,400	48,500	86,000
10. Combustion products, lb./sec.	0.547	4.46	14.9	36.0	68.1	120
11. Specific mass flow G, lb./(sec.)(sq. ft.)	0.108	0.22	0.33	0.44	0.55	0.67
12. h_v (Nusselt), B.t.u./(hr.)(sq. ft.)(°F.)	0.35	0.71	0.95	1.07	1.18	1.35
13. $P_c L$, ft.-atm.	0.30	0.59	0.89	1.18	1.48	1.76
14. $P_w L$, ft.-atm.	0.08	0.16	0.24	0.32	0.40	0.48
15. e_c emissivity (CO_2) 3500°F., %	4.7	6.2	7.4	8.0	8.8	9.5
16. e_w emissivity (H_2O) 3500°F., %	1.3	2.5	3.6	4.6	5.5	6.2
17. e (uncorrected) $e_c + e_w$ 3500°F. %	6.0	8.7	11.0	12.6	14.3	15.7
18. Δe, correction, 3500°F. %	0.2	1.4	2.6	3.0	3.9	4.1
19. e (corrected) 3500°F., %	5.8	7.3	8.4	9.6	10.4	11.6
20. e_c, 2500°F., %	7.7	10.0	11.1	12.0	13.2	14.0
21. e_w, 2500°F., %	2.5	4.5	6.0	7.5	8.7	9.8
22. e (uncorrected) $e_c + e_w$ 2500°F., %	10.2	14.5	17.1	19.5	21.9	23.8
23. Δe, correction, 2500°F., %	0.2	1.4	2.6	3.0	3.9	4.1
24. e (corrected) 2500°F., %	10.0	13.1	14.5	16.5	18.0	19.7
25. q_r', B.t.u./(hr.)(sq. ft.) 3500°F.	24,600	31,000	35,800	40,800	44,100	49,500
26. q_r'' B.t.u./(hr.)(sq. ft.) 2500°F.	14,000	17,300	19,200	21,800	24,000	26,000
27. q_r net interchange, B.t.u./(hr.)(sq.ft.)	10,600	13,700	16,600	19,000	20,100	23,500
28. h_r (q_r/1000°F.)	10.6	13.7	16.7	19.0	20.1	23.5
29. $h_r + h_v = h$ (total)	10.9	14.4	17.8	20.1	21.3	24.8
30. hA_w, million B.t.u./hr.	1.00	5.35	14.6	29.4	48.5	81.0
31. Total heat release, million B.t.u./hr.	2.30	18.5	61.5	146	278	490
32. Furnace efficiency (30)/(31), %	43.5	28.8	23.8	20.1	17.5	16.5

* Based on generating steam at 600 lb./sq. in. and 800°F. with an 85 per cent boiler efficiency.

and a radiant mean path of 11.4 ft. This causes an increase of 12 per cent in P_cL and P_wL (radiant path length times partial pressure of CO_2 or H_2O, see Sec. 6), thus increasing the value of the total gas emissivity from 11.6 to 12.75 per cent at 3500°F. Concurrently, the emissivity at 2500°F. is increased from 19.4 to 20.2 per cent.

If the furnace is made cubical with the length of the edges of the cube 21 ft. 5 in., the volume still remains 9800 cu. ft. but the wall area is decreased to 2470 sq. ft. The value of the radiant mean path is increased 19 per cent above the value 10.2 for furnace 6, *i.e.*, $L = 12.1$. Since, of all rectangular-shaped furnaces, the cube exhibits a maximum value of L, the variation in a furnace of fixed volume due to the changes in relative dimensions does not exceed 20 per cent, and within this range of variation, the value of the radiation coefficient may be taken to vary linearly with L.

Table 9 shows the coefficient of radiant-heat transfer for the furnace 6 combustion chamber (Table 7) for each of the seven fuels listed in Table 8. It is seen that the high-grade coal (fuel 2) which was used for illustration in Table 7 exhibits the smallest value of the radiation coefficient for gas at 3500°F. and wall at 2500°F. The two high-hydrogen fuels (coke-oven gas and natural gas) result in coefficients respectively 73 and 63 per cent higher than for coal. High values of the radiation coefficient, however, are not necessarily associated with high hydrogen, since tar (fuel 3) with only 7.12 per cent H_2O in its combustion products is in third place with the radiation coefficient only 12 per cent less than for natural gas. It is not possible to draw general conclusions with respect to the dependence of the radiation coefficient on the composition of the fuel or the analysis of the combustion products. In any particular case, there is no method at hand which is simpler and

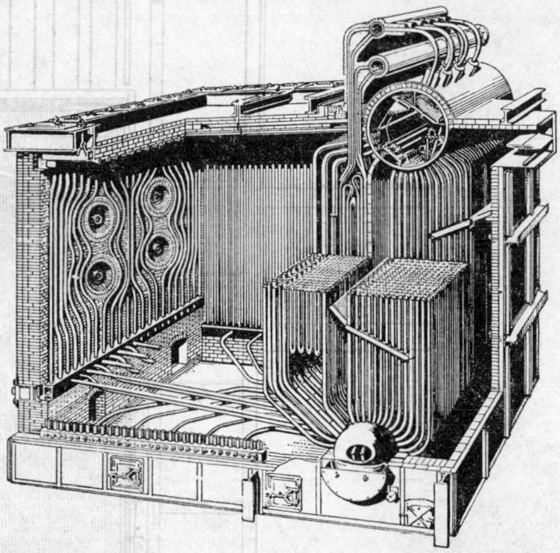

FIG. 4*A*. C-E vertical unit boiler, type VU-50. Standardized steam-generating unit built in capacities from 30,000 to 350,000 lb./hr. and pressures up to 1000 lb./sq. in. Suitable for firing by pulverized coal in any type of stoker, oil, or gas. (*Combustion Engineering-Superheater, Inc.*)

Table 8. Combustion of Industrial Fuels

Fuel:
1. Blast-furnace gas: (dry basis) % by volume, CO_2 14.22, CO 25.61, H_2 2.13, N_2 58.04, CH_4 tr.; saturated with H_2O at 80°F.
2. High-grade bituminous coal: (natural basis) % by weight, C 79.35, H 5.14, O 5.82, N 1.68, S 1.26, ash 4.15, moisture 2.00
3. Coke-oven tar: (dry basis), % by weight, C 90.45, H 5.55, O 2.35, N 1.05, S 0.60
4. Low-grade bituminous coal: (natural basis) % by weight, C 59.47, H 5.72, O 12.00, N 1.50, S 1.70, ash 11.92, moisture 8.50
5. Fuel oil: (dry basis) % by weight, C 86.48, H 11.63, O 0.18, N 0.74, S 0.97
6. Coke-oven gas: (dry basis) % by volume, CO_2 1.35, CO 7.22, H_2 52.70, N_2 3.17, CH_4 30.62, C_2H_6 0.65, C_2H_4 3.51, H_2S 0.48; saturated with H_2O at 80°F.
7. Natural gas: (dry basis) % by volume, CH_4 93.45, C_2H_6 3.67, C_3H_8 1.78, C_4H_{10} 0.65, C_2H_4 0.45; saturated with H_2O at 80°F.

Fuel	1	2	3	4	5	6	7
1. Gross H.V., B.t.u./lb.	......	14,260	16,390	10,900	19,300		
2. Gross H.V., B.t.u./cu. ft.	89.9	......	......	......	......	575	1090.4
3. Net H.V., B.t.u./lb.	......	13,820	15,800	10,400	18,250		
4. Net H.V., B.t.u./cu. ft.	88.8	......	......	......	......	513	984.3
5. Required air, cu. ft./lb.	......	142.9	161.6	110.1	183.9		
6. Required air, cu. ft./cu. ft.	0.6684	......	......	......	......	5.01	12.40
7. Combustion products, cu. ft./lb.	......	143.6	168.1	118.5	194.9		
8. Combustion products, cu. ft./cu. ft.	1.566	......	......	......	......	5.74	13.57
Analysis of combustion products, % by volume:							
9. CO_2	25.53	17.38	16.86	15.80	14.17	8.30	8.04
10. H_2O	3.95	4.68	7.12	11.32	12.34	22.92	19.95
11. N_2	70.52	77.94	76.02	72.88	73.49	68.88	72.01
12. Orsat CO_2	26.6	18.2	18.2	17.8	16.2	10.8	10.1
13. SO_2, % by volume	nil	0.103	0.042	0.170	0.059	0.084	tr.

Lines 1–4: Heating value (H.V.), gross and net per unit of fuel, lb. or cu. ft., dry, 60°F., 30 in. Hg.
Lines 5–8: Air required for complete combustion per unit of fuel: cu. ft., 60°F., 30 in. Hg, 3.5 gr. moisture/cu. ft.
Lines 9–11: Analysis of combustion products including H_2O, sulphur-free basis, with theoretical air.
Line 12: CO_2 by volume as determined by Orsat analysis (H_2O free).

Table 9. Combustion of Fuels in a Large Boiler Furnace

Based on furnace 6, Table 7, heat release 50,000 B.t.u./(hr.)(cu. ft.)
Calculated for combustion without excess air

Fuel	Blast-furnace gas 1	Bituminous coal, high H.V. 2	Tar 3	Bituminous coal, low H.V. 4	Fuel oil 5	Coke-oven gas 6	Natural gas 7
1. Fuel rate:							
Lb./hr.	......	35,300	31,000	47,100	26,800		
Cu. ft./min.	92,000					15,900	8,350
2. Required air, cu. ft./min.	61,500	83,600	83,500	81,700	82,600	80,000	104,000
3. P_cL	2.56	1.74	1.69	1.58	1.39	0.83	0.80
4. P_wL	0.40	0.47	0.71	1.13	1.22	2.29	1.99
5. e (corrected) at 3500°F.	13.3	11.3	14.3	14.9	14.6	17.9	17.1
6. e (corrected) at 2500°F.	20.6	19.7	21.6	25.4	25.0	29.0	27.3
7. h_r, B.t.u./(hr.)(sq. ft.)(°F.)	29.3	21.9	31.8	29.5	28.9	38.0	36.4

H.V. = heating value.

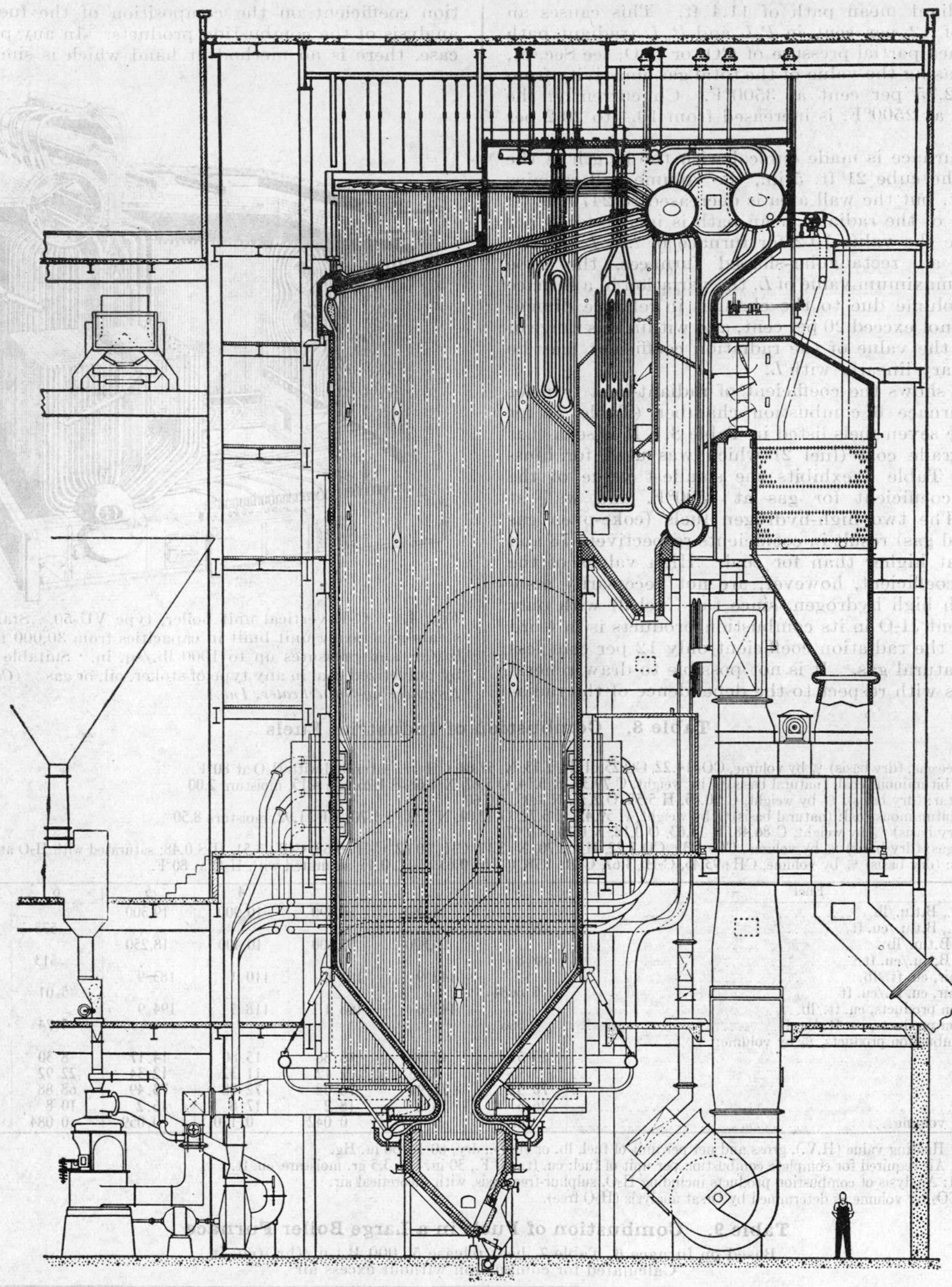

FIG. 4B. Large boiler at Port Jefferson station of Long Island Lighting Co. Operating pressure, 1350 lb./sq. in., with maximum continuous capacity of 425,000 lb. of steam per hour at steam temperature of 955°F. *(Combustion Engineering–Superheater, Inc.)*

more direct than carrying out the calculations shown in Tables 7 and 9.

It should be emphasized that the values of the radiation coefficient in Table 7 refer only to heat transfer due to radiation from the non-luminous combustion products CO_2 and H_2O and do not include any radiation which may be due to a luminous flame which might be encountered in burning any of the fuels in Table 8, except blast-furnace gas. It should be emphasized further that combustion was calculated without excess air whereas in normal practice 10 to 30 per cent excess air is present. Furthermore, the data in Tables 7 and 9 do not represent the performance of any existing furnaces but are calculated values to demonstrate the effect of furnace size. Furnace temperatures of 3500°F. are seldom maintained in the presence of heat-absorption surfaces and will be affected by the relative volume to cooled surface area of the combustion space.

USES AND MODIFICATIONS

Boiler Furnaces. Boilers for pressures of 250 lb./sq. in. or less may be of the fire-tube, water-tube, or double-shell construction. In the fire-tube boiler, the combustion gases pass through the tube; and in the water-tube boiler, the steam and water are inside the tubes. For the smaller sizes, *i.e.*, evaporation rates less than 15,000 lb./hr., there are many standard boilers available. These boilers are usually oil- or gas-fired and have all the control equipment mounted on the same frame as the boiler. To install, it is only necessary to connect the water, steam, gas, or oil, and electric lines to the boiler and to add a stack. Efficiencies may range up to 70 per cent. For pressures above 50 lb./sq. in., the water-tube boiler is used. The arrangement of tubes takes many forms with tubes within, as well as outside, the combustion spaces, the tubes being bent, straight, vertical, or inclined. Standard boilers can be obtained in sizes up to 40,000 lb./hr. of steam. The larger boilers may range up to 1,000,000 lb./hr. of steam and are usually individually designed, although there is some tendency to standardization even in the larger boilers. Figure 4 illustrates a standard vertical-type boiler and a large central-power-station boiler. Because of more extensive heat recovery by means of air preheaters and feed-water heaters, the efficiency is 80 to 85 per cent and may exceed 90 per cent in exceptional cases.

In some processes, waste heat is recovered in boilers which have no combustion space, and the design becomes structurally simpler. The gases may pass through or across the tubes, and in some cases the tubes may have external fins to improve the rate of heat removal from the gases.

Dowtherm is vaporized in natural circulation boilers, units having been built in sizes ranging from 50,000 to 30,000,000 B.t.u./hr.

High rates of heat release and heat transfer are being obtained in new furnace designs using oil or gas firing where the combustion chamber is a large cooled tube or where the cylindrical chamber is formed from bent tubes. In these furnaces, the flame has a cyclonic motion. Heat release is high, but secondary combustion chambers are used; and, when this volume is added to the cyclone chamber, the heat-release rate is greatly reduced.

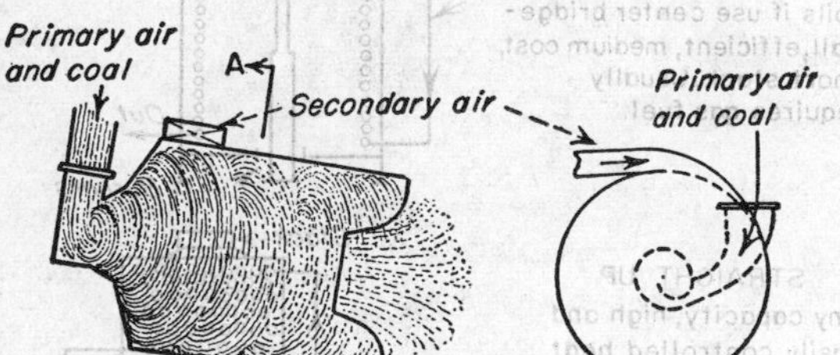

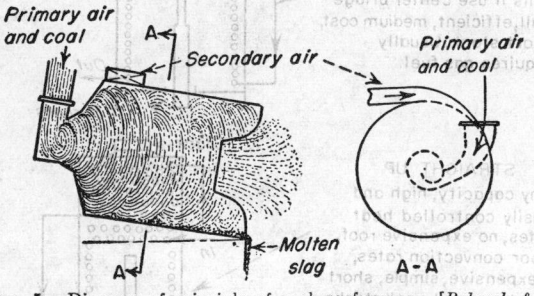

FIG. 5. Diagram of principle of cyclone furnace. [*Babcock & Wilcox Co.; Trans. Am. Soc. Mech. Engrs.*, **69**, 613 (1947).]

A **cyclone furnace**, as shown in Fig. 5 for crushed-coal firing, has been developed by Babcock & Wilcox Co. (*Trans. Am. Inst. Mech. Engrs.*, **69**, 613, August, 1947). Heat-release rates as high as 545,000 B.t.u./(hr.)(cu. ft.) have been attained in two powerhouses. However, when the primary and secondary furnace volumes are included, the over-all rate is reduced to the more conventional figure of 32,800 B.t.u./(hr.)(cu. ft.) (*Power*, March, 1949, p. 72). Very rapid and complete combustion of the coal in this furnace is claimed to result from its special design. The ash from the coal forms a molten slag layer which covers the interior surface of the cyclone. The fresh incoming coal is thrown by centrifugal force on to this slag layer, where it is held. The air stream at extremely high velocity scrubs this coal and produces complete combustion. The ash is withdrawn from the bottom as a molten slag. Improvements are claimed in the reduction of slag deposition on the heat-absorption surfaces, less fly ash, low crushing costs of the coal, and potentially smaller furnace volumes.

Petroleum Furnaces. Petroleum furnaces are used to heat and vaporize feed to cracking units and distillation columns. Several designs are available, as shown in Fig. 6. One design (Petrochem) is a vertical, cylindrical chamber with the vertical pipes in a single layer near the walls. The upper section of the pipes which are in a convective zone may have fins on them. Other pipe-still designs have a more conventional combustion space with tubes in the radiant section near the walls and preheating pipes in a convection zone in the stack. Efficiencies are about 80 per cent and the radiant section heat-transfer rate ranges from 4800 to 9500 B.t.u./(hr.)(sq. ft. outside tube surface). Furnace volumes are about 4 cu. ft./sq. ft. radiant tube surface. The heat-release rate is 20,000 to 30,000 B.t.u./(hr.)(cu. ft.), about 50 per cent of the heat being absorbed in the radiant sections.

Metallurgical Furnaces. Furnaces are used in the heat-treatment of metals such as annealing, normalizing, and "drawing" (tempering). Many specialized furnaces are designed for these purposes and may be either batch or continuous operation. The continuous furnaces may be arranged either in straight line or with circular rotating hearths. Batch furnaces may be of conventional design or specialized designs, such as rotating hearths, removable bottoms, stationary bottom, or muffle. Continuous furnaces are classified according to method of moving material, such as roller, pusher, conveyor, walking beam, and tunnel types.

For small temperature differences within the metal, a number of practical rules for iron and steel are: (1) The heat penetrates at rate of 16 to 20 min./in. diameter for low-carbon steel and 30 to 35 min./in. diameter for high-carbon or alloy steel. (2) Steel can be heated in a

Table 10. Fuel Consumption of Typical Furnaces

	B.t.u./1000 Brick
Brick kilns:	
Hoffman continuous kiln	3,920,000–4,760,000
Ruebon kiln	5,880,000–9,240,000
Grates and trough	4,760,000–6,300,000
Staffordshire kiln	4,760,000–6,300,000
Buhrer continuous tunnel	3,220,000–3,900,000
Metal melting:	B.t.u./Ton
Iron and steel:	
Puddling	14,000,000
Open-hearth	3,000,000–6,500,000
Air furnace for malleable castings	8,600,000
Copper, reverberatory	4,200,000
Tin, smelting	19,200,000
Rolling-mill furnaces:	
Busheling	5,600,000–7,000,000
Billet heating	2,240,000–2,520,000
Billet heating (4 by 4)	2,800,000–4,200,000
Continuous furnace (cold billets)	1,500,000–2,250,000
Continuous-bloom furnace	1,400,000–2,100,000
Reheating cold blooms	1,960,000–2,800,000
Soaking pits (hot to cold ingots)	250,000–2,000,000
Tire furnaces	4,500,000
Wheel furnaces	8,400,000
Sheet annealing	2,800,000
Tin-plate mill	2,500,000
Rivet making:	
Rivet-making furnaces	1,260,000
Forging:	
Drop forging or bolt heating	6,000,000–7,000,000
Ingot forging	4,000,000–5,000,000
Ingot forging (regeneration furnace)	3,500,000–4,000,000
Brass annealing	1,500,000–2,000,000
Heat-treatment:	
Casehardening	4,500,000–5,000,000
Quenching	2,000,000–3,000,000
Tempering	1,500,000–2,000,000

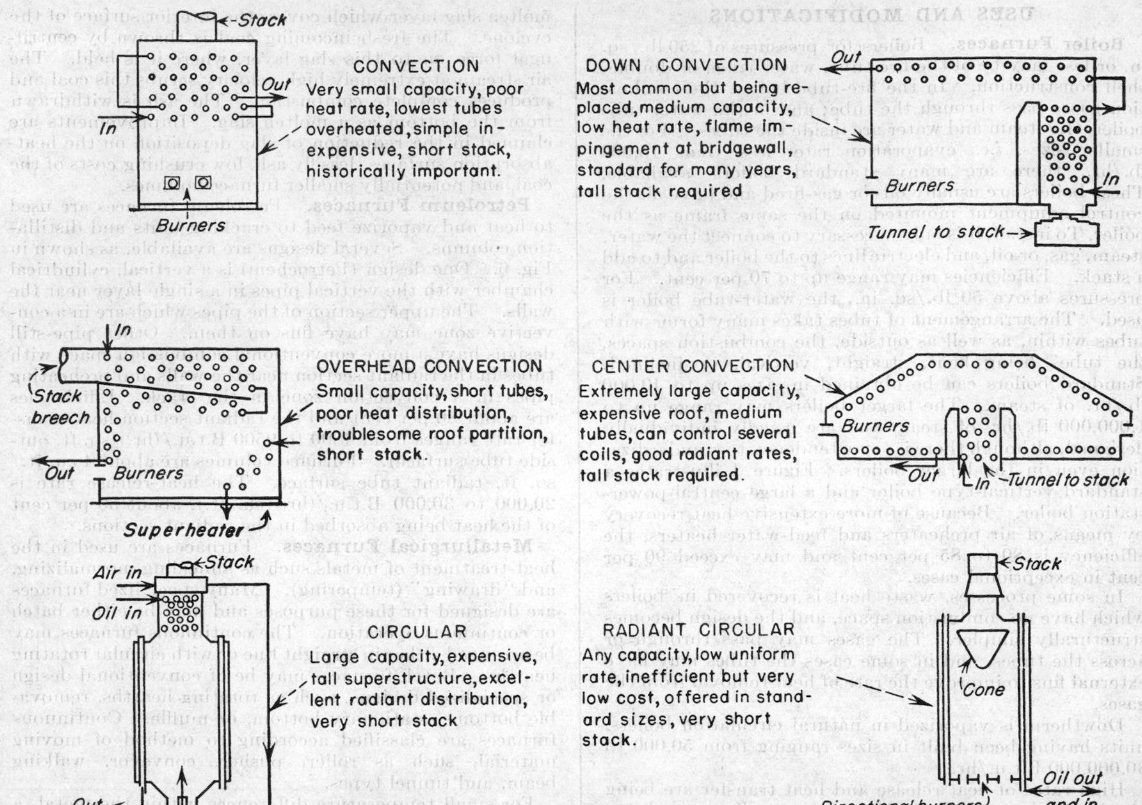

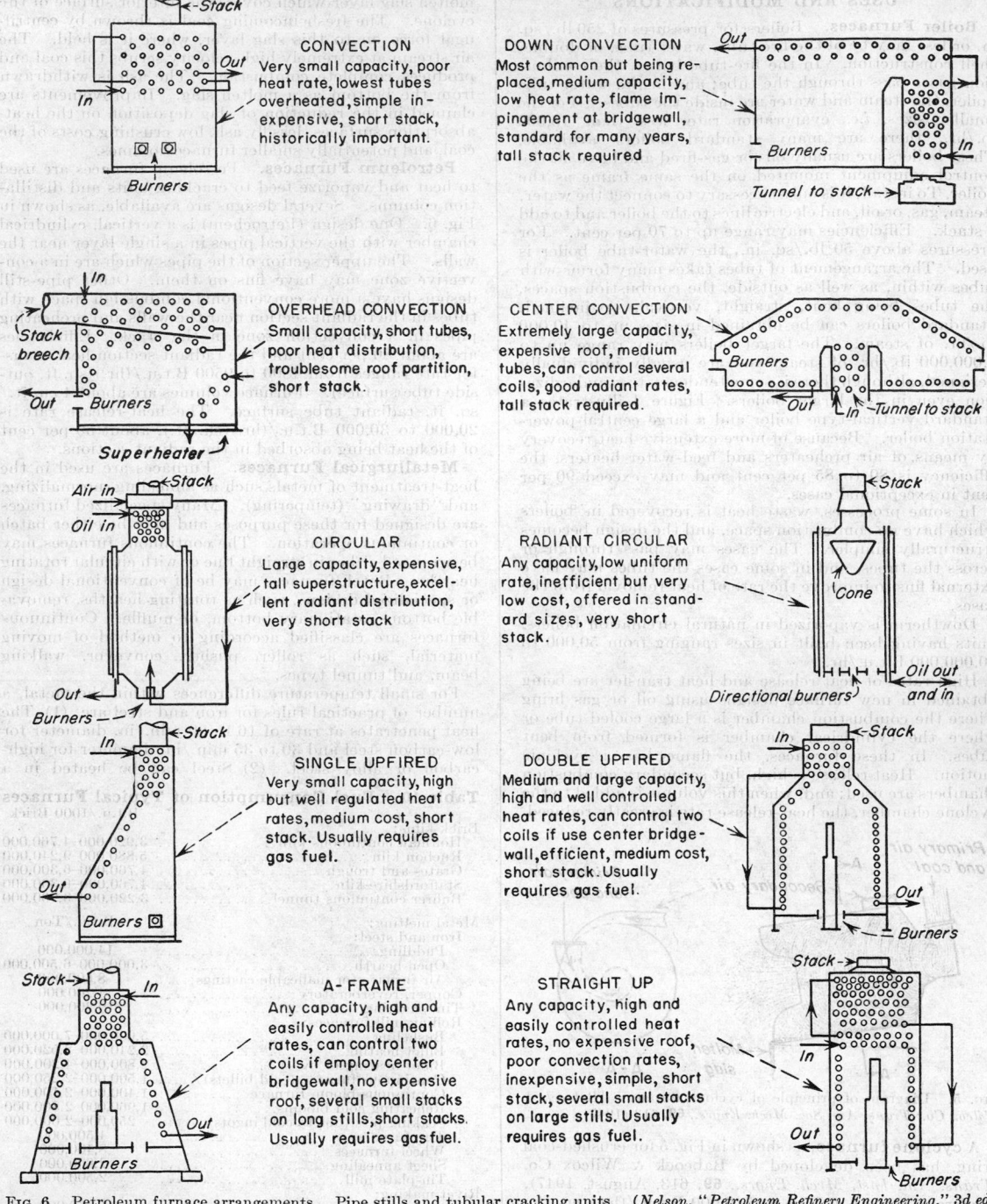

FIG. 6. Petroleum furnace arrangements. Pipe stills and tubular cracking units. (*Nelson, "Petroleum Refinery Engineering," 3d ed., p. 526, McGraw-Hill, New York,* 1949.)

furnace at rates of 30 lb./(hr.)(sq. ft. hearth area) for heat-treating, 60 lb./(hr.)(sq. ft.) for annealing, and 110 to 130 lb./(hr.)(sq. ft.) for work where temperature gradients of 40° to 50°F./in. are allowable. (3) In continuous furnaces, steel is heated at rates of 50 to 100 lb./(hr.)(sq. ft. hearth area). Fuel consumptions of typical furnaces are listed in Table 10.

Ceramic Furnaces. The ceramic furnaces are generally called kilns and include the operations of dehydration, oxidation, calcination, and vitrification. The

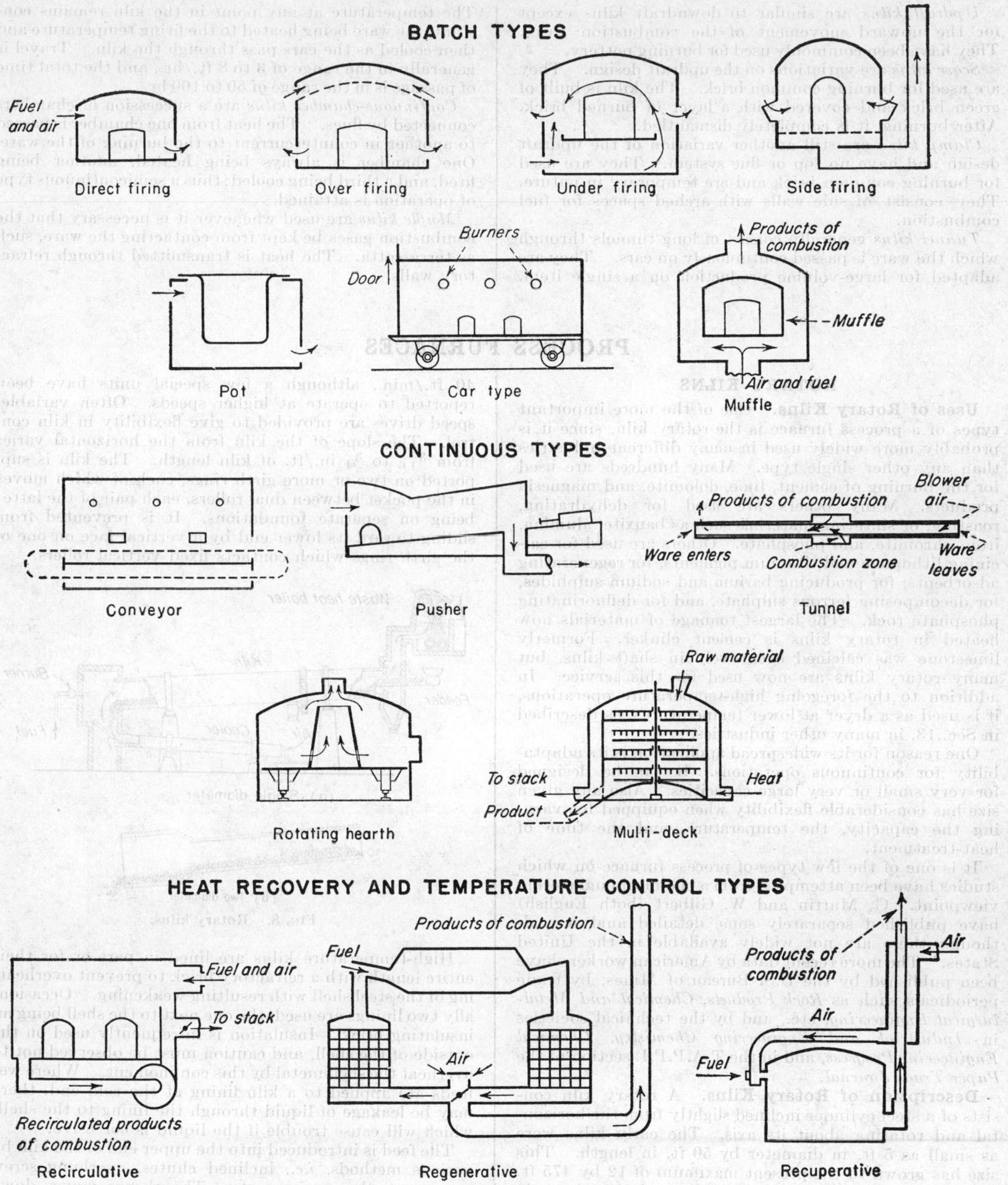

FIG. 7. Diagrams of various industrial furnaces.

kilns are of various types, horizontal and space burners and gaseous, liquid, and solid fuels being used. Where product quality is not important, the ceramic ware will be exposed to the combustion gases and sometimes directly to the flame. Dutch ovens are frequently used for the generation of heat. Various designs are shown in Fig. 7 and some are described briefly below. See Table 10 for fuel consumptions.

Downdraft kilns operate as batch units and are the most common type, being used for brick, sewer pipe, tile, and stoneware. The name is derived from the direction of the passage of the combustion gases in contacting the charge. The gases then go up inside the walls to the top of the kiln and to a chimney. The charge is "set" in the kiln and heating started. The temperature is raised on a definite time schedule until the "firing" temperature is reached. Then the furnace is cooled slowly and unloaded.

Updraft kilns are similar to downdraft kilns except for the upward movement of the combustion gases. They have been commonly used for burning pottery.

Scove kilns are variations on the updraft design. They are used for burning common brick. The kiln is built of green brick and covered with a layer of burned brick. After burning, it is completely dismantled.

Clamp kilns are still another variation of the updraft design and have no top or flue system. They are used for burning common brick and are temporary in nature. They consist of side walls with arched spaces for fuel combustion.

Tunnel kilns generally consist of long tunnels through which the ware is passed continuously on cars. They are adapted for large-volume production on a single item. The temperature at any point in the kiln remains constant, the ware being heated to the firing temperature and then cooled as the cars pass through the kiln. Travel is generally in the range of 3 to 8 ft./hr., and the total time of passage is in the range of 50 to 100 hr.

Continuous-chamber kilns are a succession of chambers connected by flues. The heat from one chamber is passed to another in countercurrent to the burning of the ware. One chamber is always being heated, another being fired, and a third being cooled; thus a semicontinuous type of operation is attained.

Muffle kilns are used whenever it is necessary that the combustion gases be kept from contacting the ware, such as terra cotta. The heat is transmitted through refractory walls.

PROCESS FURNACES

ROTARY KILNS

Uses of Rotary Kilns. One of the more important types of a process furnace is the rotary kiln, since it is probably more widely used in many different industries than any other single type. Many hundreds are used for the burning of cement, lime, dolomite, and magnesia products. Many others are used for dehydrating, roasting, or sintering materials such as bauxite, alumina, iron, chromite, and phosphate. Others are used for calcining lithopone and titanium pigments, for regenerating adsorbents, for producing barium and sodium sulphides, for decomposing ferrous sulphate, and for defluorinating phosphate rock. The largest tonnage of materials now heated in rotary kilns is cement clinker. Formerly limestone was calcined exclusively in shaft kilns, but many rotary kilns are now used for this service. In addition to the foregoing high-temperature operations, it is used as a dryer at lower temperatures, as described in Sec. 13, in many other industries.

One reason for its widespread application is its adaptability for continuous operations. It can be designed for very small or very large capacities. Also any given size has considerable flexibility when equipped for varying the capacity, the temperature, and the time of heat-treatment.

It is one of the few types of process furnace on which studies have been attempted from a chemical-engineering viewpoint. G. Martin and W. Gilbert (both English) have published separately some detailed analyses, although these are not widely available in the United States. The more recent data by American workers have been published by the U.S. Bureau of Mines, by trade periodicals such as *Rock Products*, *Chemical and Metallurgical Engineering*, etc., and by the technical societies in *Industrial and Engineering Chemistry*, *Chemical Engineering Progress*, and in the T.A.P.P.I. section of the *Paper Trade Journal*.

Description of Rotary Kilns. A rotary kiln consists of a steel cylinder inclined slightly from the horizontal and rotating about its axis. The early kilns were as small as 5 ft. in diameter by 50 ft. in length. This size has grown to the present maximum of 12 by 475 ft. The dimensions of a kiln are generally such that the ratio of kiln length to the square of the shell diameter is usually between 1.5 and 5. Some kilns have two diameters, part of its length being one diameter, say 9 ft., the remainder being of another diameter, say 12 ft. It is claimed that this enlargement increases the kiln capacity, decreases the fuel consumption, and improves the product quality. Figure 8 shows both types of kilns.

Rotation is accomplished by a girth gear which gives peripheral speeds of the kiln shell in the range of 25 to 40 ft./min., although a few special units have been reported to operate at higher speeds. Often variable-speed drives are provided to give flexibility in kiln control. The slope of the kiln from the horizontal varies from $\frac{3}{16}$ to $\frac{3}{4}$ in./ft. of kiln length. The kiln is supported on two or more girth rings, each of which moves in the pocket between dual rollers, each pair of the latter being on separate foundations. It is prevented from sliding toward its lower end by a vertical face on one of the girth rings which contacts fixed vertical rollers.

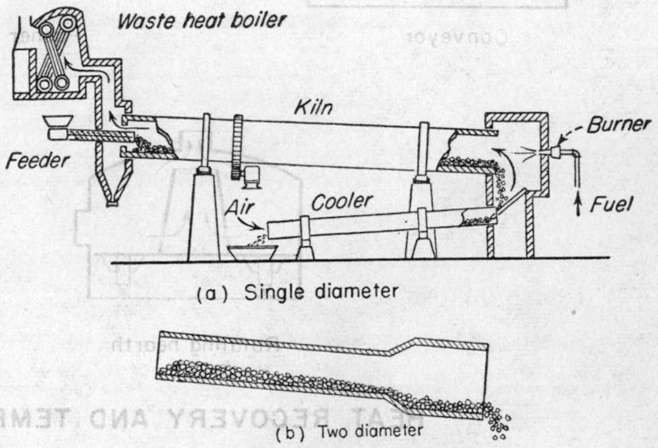

Fig. 8. Rotary kilns.

High-temperature kilns are lined in part or for their entire length with a refractory brick to prevent overheating of the steel shell with resulting weakening. Occasionally two linings are used, the one next to the shell being an insulating brick. Insulation is infrequently used on the outside of the shell, and caution must be observed not to overheat the shell metal by this confinement. Where wet feeds are applied to a kiln lining at the cold end, there may be leakage of liquid through the lining to the shell, which will cause trouble if the liquid is corrosive.

The feed is introduced into the upper end of the kiln by various methods, *i.e.*, inclined chutes, overhung screw conveyors, slurry pipes, etc. The charge moves down the kiln during rotation, being lifted on the rising side, then falling back, thus pursuing a zigzag path. Sometimes ring dams or chokes of a refractory material are installed within the kiln to build a deeper bed at one or more points, thus changing the flow pattern. The hot product is discharged from the lower end of the kiln into quench tanks, onto conveyors, or into cooling devices which may or may not recover its heat content. These cooling and heat-recovery devices include rotating inclined cylinders, inclined slow-moving grates, shaking grates, etc.

Firing may be accomplished at either end, depending on whether concurrent or countercurrent flow of the charge and gases is desired. Sometimes a solid fuel is mixed with the charge and burned as it moves down the kiln. Gaseous, liquid, or powdered fuels may be used. The burner may be installed directly at the end of the kiln with combustion occurring inside of it. In this case, the discharge-end housing usually consists of a fixed or movable kiln hood through which the fuel pipe enters the kiln. A center position for the fuel pipe is used when the flame is wanted off the charge. Some users prefer an off-center position toward the trough between the charge chord and the descending kiln lining. In some cases, the fuel is burned in a separate combustion chamber, with only hot gases entering the kiln. The kiln and the hood (combustion chamber) have open ends which coincide with each other. The gap may be closed by a sliding seal or enclosed by a bustle chamber for the introduction of secondary combustion air. Sometimes secondary air is admitted also through pipe ports distributed along the length of the kiln in connection with delayed combustion, the kiln being operated under a negative pressure.

The hot exhaust gases are generally discharged from the kiln into dust and fume knockdown equipment to avoid contamination of the atmosphere. Sometimes dust recovery is desired because the recovered material has an economic value. Gas-cleaning equipment includes cyclones, settling chambers, scrubbing towers, and electrical precipitators. Recently developments have been started on the use of supersonic energy for collecting the dust particles. Sometimes heat-recovery devices are utilized on these gases both within and outside the kiln. These result in an increase in kiln capacity, or a decrease in fuel consumption, or both. Waste-heat boilers, grates, coil systems, chains, and flights are used for this purpose.

It is necessary generally to make some provision at the upper end of the kiln for catching spill back of feed material. This often occurs as a result of overloading beyond capacity, insufficient slope of kiln, rotation being too slow, or the feed not being placed sufficiently far into the kiln.

When installed, instrumentation gives an excellent control of the temperature within the kiln.

A general discussion of kilns and their structural design has been published by Dickie [*Chem. & Met. Eng.*, **46** (5), 326 (1939)].

Time of Passage in Rotary Kilns. The time of passage of the material through a rotary kiln can be estimated from the following equation developed from studies of Sullivan, Maier, and Ralston [*U.S. Bur. Mines Tech. Paper* 384 (1927)].

$$\theta = \frac{0.19L}{NDS}$$

where θ = time of passage in minutes, L = kiln length in feet, N = rate of rotation in r.p.m., D = kiln diameter in feet, and S = slope of kiln in feet per foot. This equation is suitable for kilns without interior restrictions to the movements of the solids. Other equations must be used when dams are used within or at the end of the kiln, as presented in the form of nomographs by Bayard [*Chem. & Met. Eng.*, **52**(3), 100–102 (1945)].

Charge Volume in Rotary Kilns. The charge volume in rotary kilns varies in practice from 3 to 12 per cent of the kiln volume. This quantity can be increased by the use of dams, which results in a higher kiln capacity by allowing the feed rate to be increased. As implied in the preceding paragraph, the use of faster rotation would decrease the charge volume held in the kiln.

Heat Transfer in Rotary Kilns. The major source of heat transfer within the kiln is due to radiation from the flame and hot gases. This occurs directly to both the charge surface and the wall, and from the latter it is transferred to the charge by reradiation as well as by contact conduction during rotation. A small amount of heat is transferred directly by convection to both the charge and the wall.

Generally, a dry-feed kiln will have three zones and a wet-feed kiln will have four zones of heating.

1. *Drying zone* at feed end where moisture is driven off.
2. *Heating zone* in which the charge is heated to the reaction temperature, *i.e.*, decomposition temperature for limestone or "burning" temperature for cement.
3. *Reaction zone* where the charge is burned, decomposed, oxidized, reduced, etc.
4. *Final treatment zone* where the reacted charge is superheated, or "soaked" at a specified temperature, or if desired it is cooled before discharge.

The rates of heat transfer in each zone will be different, and the factors affecting these rates will also differ in degree. No adequate theory has been developed which would predict accurately the heat transfer and heat efficiency for a kiln of different size from the one for which data are available. Thus it is necessary to use experience and available operating data for several materials in order to determine an approximate size of kiln for a different charge material. The detailed method of analyzing the heat transfer of a cement kiln as published by Gilbert can be used as a guide for estimating the performance of a kiln on other materials with reasonably satisfactory results.

Rotary kilns operate at various temperatures throughout their length. A graph of approximate gas and charge temperatures for a cement kiln is shown in Fig. 9. The

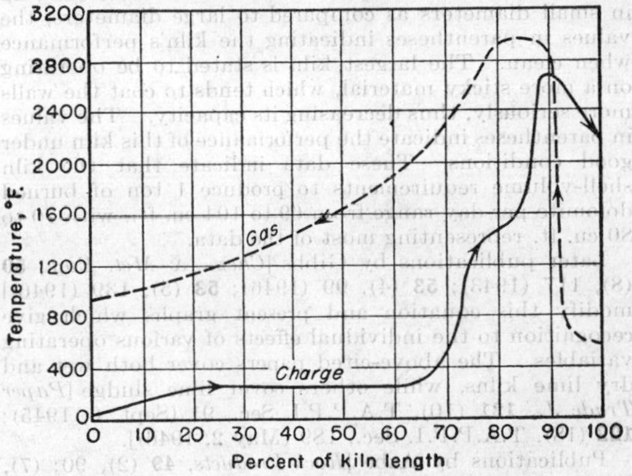

FIG. 9. Temperatures in rotary kiln on wet-process cement.

maximum temperatures for the charge are about 2600° to 2800°F. and for the gases are about 2800° to 3000°F. Over-all heat-transfer rates have been estimated to be in the range of 2500 to 6000 B.t.u./(hr.)(cu. ft.) based on total volume inside the kiln without deduction for the charge volume. Long kilns give a better heat economy than short kilns for high-temperature work. However an equally high heat economy can be obtained in nearly all cases by the application of suitable heat-recovery equipment and by the adoption of proper operating practices.

Capacity of Rotary Kilns. The capacity of any specific rotary kiln depends on the physical and chemical properties of the charge material; its time of passage, temperature, and heat requirements; and on the temperature and amount of hot gases supplied for heating.

Thus a specific kiln will have different capacity ratings for different materials. However, in practice it is found that the range of capacities obtained on the same type of material from the same size kiln may be just as large as the difference in capacity found on two different materials handled in the same size of kiln.

Seveıal short-cut methods for estimating kiln capacity have been published but all have limitations in use. The following equation offered by Gibbs [*Rock Products*, **45** (11), 58 (1942)] gives an approximation of commercial practice for single-diameter kilns.

$$\text{Tons of product/day} = \frac{kLD^2}{100}$$

This equation relates product capacity to kiln volume, where L is kiln length in feet, D is shell diameter in feet, and k is a factor whose magnitude depends on the charge material and other variables. It was formulated in connection with the observation of dolomite kilns, for which the data are given in Table 11. The very high ratio of

Table 11. Dolomite Kiln Data

Size of kiln		Average production, tons/day	Secondary air	Shell volume/capacity ratio, cu. ft./(ton/day)	k
$D \times L$, ft.	Shell volume, cu. ft.				
6 × 110	3,110	30(40)	Hot	104(78)	0.76(1.01)
7 × 115	4,420	60	Hot	74	1.06
8 × 125	6,280	90	Hot	70	1.12
8 × 125	6,280	80*	Cold	79	1.00
9 × 250	15,880	210(230)	Hot	76(69)	1.04(1.15)
Average value of k					1.00
Value of k under favorable conditions					1.15

* Note the difference in production from the same kiln which results from using either hot or cold secondary air.

shell volume to capacity for the smallest kiln is believed to be due to the larger percentage effect of ring build-up in small diameters as compared to large diameters, the values in parentheses indicating the kiln's performance when clean. The largest kiln is stated to be operating on a more sticky material, which tends to coat the walls more seriously, thus decreasing its capacity. The values in parentheses indicate the performance of this kiln under good conditions. These data indicate that the kiln shell-volume requirements to produce 1 ton of burned dolomite per day range from 69 to 104 cu. ft., with 70 to 80 cu. ft. representing most of the data.

Later publications by Gibbs [*Chem. & Met. Eng.*, **50** (8), 117 (1943); **53** (4), 99 (1946); **53** (5), 139 (1946)] modify this equation and present graphs which give recognition to the individual effects of various operating variables. The above-cited papers cover both wet and dry lime kilns, while others cover lime sludge [*Paper Trade J.*, **121** (10), T.A.P.P.I. Sec., 91 (Sept. 6, 1945); **122** (18), T.A.P.P.I. Sec., 189 (May 2, 1946)].

Publications by Azbe [*Rock Products*, **49** (2), 90; (7), 80; (8), 90 (1946)] give data on the performance of rotary lime kilns. Worthwhile graphs on the effect of various operating factors are included. His observations indicate that a kiln of standard design should produce 1 ton of high calcium lime for each 35 cu. ft. of kiln volume, measured within the lining. He rates kiln volume requirements in the range of 30 to 40 cu. ft./(ton/day) of lime as "very good," and those of 40 to 50 cu. ft. as "good." Based on a heat requirement for burning lime of 2249 B.t.u./lb. of CaO, the "very good" rating would represent an effective over-all heat transfer rate of 6300 to 4700 B.t.u./(hr.)(cu. ft.).

The use of kiln volume measured inside of the lining is preferred over the use of shell volume. The difference between these two volumes is substantial. The space occupied by a single lining will vary from about 12 per cent of the shell volume for a large-diameter kiln to about 35 per cent for a small-diameter kiln. Also, double linings are used occasionally. Furthermore, the definition of D as the diameter inside of the lining is preferred. The modification of Gibbs' equation for this change of definition makes it more applicable for short-cut estimates but its use is still limited.

Table 12. Cement Kiln Data

Size of kiln		Dry process		Wet process	
Shell, $D \times L$, ft.	Volume inside of lining, cu. ft.	Production, bbl./day	Volume/capacity ratio, cu. ft./(bbl./day)	Production, bbl./day	Volume/capacity ratio, cu. ft./(bbl./day)
6 × 60	1,100	200	5.5	140	7.9
8 × 125	4,600	700	6.6	500	9.2
10 × 150	9,200	....	...	1000	9.2
10 × 200	12,200	1700	7.2	1200	10.2
11.5 × 475	40,500			3500	11.5
12 × 475	43,800	5000	8.8	3600	12.1

Some commercial performance data for cement kilns are shown in Table 12. Cement clinker weighs 367 lb./bbl. A heat and material balance for a cement kiln has been published by Lacey and Woods [*Ind. & Eng. Chem.*, **27** (4), 379 (1935)].

The modern design of cement kilns is based on the ratio of effective heat-release volume to capacity, with consideration being given to the effect of kiln length on heat economy. The effective heat-release volume is defined as being the same as the kiln volume measured inside of the lining, the volume of the charge being disregarded. A high ratio of kiln length to effective kiln cross-section area gives a greater heat economy than small ratios. Figure 10 shows performance data for

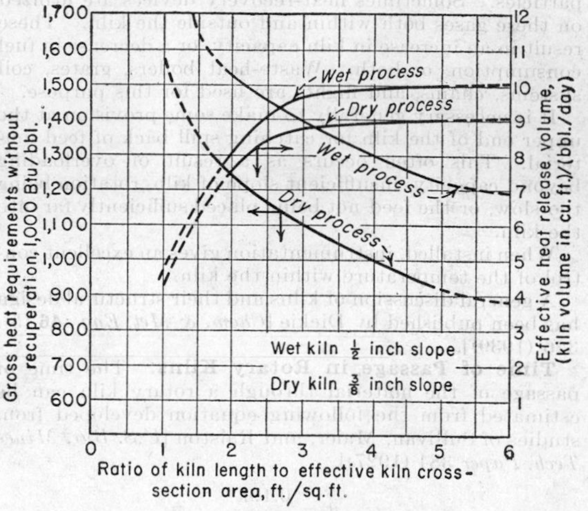

Fig. 10. Kiln performance data for cement clinker of average burnability. (*Fuller Co.*)

kilns operating on cement clinker of average burnability. The dashed extensions of the solid lines are estimated extrapolations. This graph can be used to relate the capacity of different-sized kilns on cement or it can be used to select the approximate size of a proposed cement kiln to meet a desired performance in capacity and heat economy.

In designing a rotary kiln for a new charge material for which no actual kiln data are available, the new material should be compared on the basis of detailed estimates of heat requirements with some material for which kiln data are available. The very detailed calculations on cement kiln capacity and heat transfer by Gilbert as published in "Cement and Cement Manufacture" are valuable as a guide and as a reference. Less detailed calculations covering a wet-process cement kiln are given by Griswold in "Fuels, Combustion, and Furnaces."

Short-cut estimates of kiln capacity can be made by using the data already presented here as well as that given in the next table. Table 13 gives some approximate values for commercial kilns on several materials. The ratios of volume to capacity are based on the volume measured inside of the lining and on the daily production. The spread of the values in this table indicates a large variation in commercial-kiln performance. This variation is caused primarily by two factors: (1) extent of facilities within the kiln or auxiliary to it for improving the heat transfer and the efficiency of utilizing the heat, and (2) operating procedures which may be rigid or loose.

Table 13. Short-cut Factors for Kiln Capacity

	Volume/capacity ratio, cu. ft./(ton/day)	Heat requirements, B.t.u./lb.
Alumina	45–60	
Cement—dry process	29–66	1515
Cement—wet process*	46–103	3255
Dolomite—dead burned	55–76	2525
Dolomite lime	......	2185
Lime	35–70	2214
Pigments	15–220	
Barytes (based on BaS product)	40–90	904

* Feed to kiln, 43.6 per cent water, 56.4 per cent solids.

When it is desired to increase the capacity of an existing kiln installation, consideration should be given to the following changes:

1. Increase charge volume held in kiln.
2. Increase temperature and quantity of combustion gases.
3. Decrease quantity of air in excess of combustion needs.
4. Increase speed of rotation of kiln.
5. Install ring dams at intermediate and discharge points.
6. Increase capacity of feeding and discharge mechanisms.
7. Decrease moisture content of feed material.
8. Increase temperature of feed material.
9. Install chains or flights, etc., in feed end.
10. Preheat all combustion air.
11. Reduce leakage of cold air into kiln at hot end.
12. Increase stack draft by increasing height or by use of jets.
13. Install instrumentation to control the kiln at maximum capacity conditions.

Thermal Efficiency of Rotary Kilns. The thermal efficiencies of rotary kilns vary widely, depending on the extent of heat utilization or recovery facilities and on the economy of the operating practices. Kiln length is a major factor, and kilns with a high ratio of length to diameter have a greater thermal efficiency than those with a low ratio. However, the forcing of a kiln to produce a capacity beyond its optimum will reduce its thermal efficiency. Under such circumstances, the best performance from both product-capacity and thermal-efficiency viewpoints can be obtained by the use of instrumentation to control automatically in a coordinated manner the kiln-operating variables of feed quantity, fuel quantity, air quantity, product discharge temperature, and an intermediate-product temperature. The use of flights or chains inside of the kiln and of heat-recovery equipment on the gases and product leaving the kiln can increase substantially the thermal efficiency of a kiln installation. Efficiencies ranging from 45 per cent to more than 80 per cent have been reported. A reasonably satisfactory range based on present fuel prices and construction costs would be 65 to 75 per cent utilization and recovery of the heat content of the fuel plus any heat of reaction of the charge.

The foregoing discussion makes no distinction from an efficiency calculation standpoint between the heat utilized in the kiln and that recovered (or utilized) outside of the kiln. With countercurrent flow of the combustion gases and the charge material, an exceptionally long kiln will give high efficiencies within itself. However, good economics may dictate that a shorter kiln be installed with a waste-heat boiler on the hot gases to obtain an equivalent thermal efficiency at a lower investment. The heat in the hot product usually is recovered as preheat in the combustion air. When the heat consumption to produce a unit weight of product is calculated on an over-all installation including not only the kiln but also the heat recovery facilities, it is possible to obtain values which are less than the normally calculated heat requirements as listed in Table 13. Thus, mathematically, it is possible to obtain efficiencies higher than 100 per cent. This can occur because the values given in this table represent the required heat input, from which no allowance could be made for an unknown amount of heat recovery. The only heat unavailable for recovery is that required for an endothermic chemical or physical change in the product. The economics of the value of the wasted heat compared with the investment for recovery facilities decides the extent of the latter to be installed.

For cement clinker by the dry process, fuel consumptions varying from 800,000 to 1,300,000 B.t.u./bbl. have been reported in the United States. Values of 600,000 B.t.u./bbl. have been reported from Europe but no details are known about the installation. Consumptions of this low magnitude indicate circumstances of a special nature which may not be justified economically in other locations. The figure of 800,000 B.t.u./bbl. of dry-process clinker represents an efficiency of approximately 70 per cent in the over-all utilization of heat, based on Table 13. It is understood that there are three kilns of a special design at two different locations in the United States which are operating in this range [*Rock Products*, **48** (12), 81 (1945)]. A Swedish installation, for which heat economy was very important, reports 850,000 B.t.u./bbl. of wet-process clinker, but the ratio of kiln volume to capacity is quite high, being 19 cu. ft./(bbl./day), in addition to extensive recovery facilities.

In connection with lime burning, the U.S. Bureau of Mines made an over-all industry survey of fuel consumption, which is reported in *Information Circular* 7174, Lime-Fuel Ratios of Commercial Lime Plants in 1939. This study included both small and large operators, but excluded the captive units where the product was used within the producing company. Also, the necessary use of average values narrowed the spread between the maximum and minimum fuel consumptions. Bituminous coal was used for 57 per cent of the lime produced in rotary kilns and the fuel consumption ranged from 5200 to 2890 B.t.u./lb. of lime, the latter value being a thermal efficiency of 76.5 per cent. Producer gas was used for 21 per cent of the lime and the fuel consumption varied from 8670 to 3090 B.t.u./lb., the latter being 71.5 per cent efficiency. Fuel oil was used for 14 per cent of the lime and the fuel consumption ranged from 6000 to 3750 B.t.u./lb., the latter being 59.0 per cent efficiency. Natural gas was used for 8 per cent of the lime and the fuel consumption varied from 5000 to 3120 B.t.u./lb., the latter being 71.0 per cent efficiency. This survey covered also the pot and shaft kilns. According to the average commercial practice in 1939, the ranking of these kilns in decreasing order of maximum thermal efficiencies was shaft, pot, and rotary. The highest efficiency of all combinations was coke fuel in a shaft kiln, although only a very small amount of lime was produced in this manner. These data show the tremendous variation which exists in commercial practice and indicate a large field for the profitable application of chemical engineering.

Size Segregation in Kilns. When an assemblage of solid particles, not very closely screened, is rotated within a cylinder, the solids assume a lunar shape, as shown in Fig. 11. This causes serious size segregation. The finest sizes remain at the bottom, in contact with the hot brick. The coarser particles form the upper layer of the agitated mass. As the kiln completes a revolution, the exposed brick, in an upper position, absorbs radiant heat from the gas mass. As the heated brick completes its circuit, it passes under, and in conductive contact with, the fine particles. These fines are thus effectively heated by direct solid-to-solid transfer. The larger particles are heated by direct radiation from gas and brick, and become adequately calcined. The particles of size intermediate between the fine and coarse remain throughout a complete revolution "sandwiched" between the coarse and fine layers and are protected from heat by the excellent insulated properties of these layers, thus perhaps escaping complete calcination.

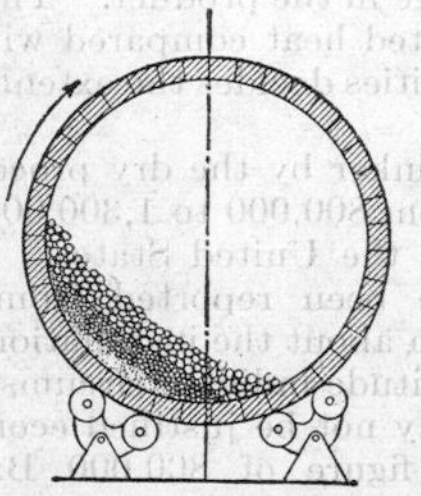

FIG. 11. Size segregation of solids in rotary kiln.

This factor of segregation is offset by some kiln operators who classify or screen the kiln feed so that only a narrow range of particle size is fed at one time. Also faster kiln speeds which give a better agitation of the charge are used.

SHAFT FURNACES

Uses of Shaft Furnaces. The oldest and most important use of a shaft furnace is the *blast furnace* used for the production of pig iron. It has been used also for the production of ferro-alloys and of non-ferrous metals such as copper and lead. A new use is the manufacture of phosphorus, by the reduction of phosphate rock. A *cupola* is a shaft-type furnace for melting iron for casting. Formerly lime was calcined exclusively in this type of furnace, but many of these have been replaced by rotary kilns. *Gas producers* are also of this type, and there are many widely varying designs. Some chemicals are made in shaft furnaces, being fed with briquetted mixtures of the reacting materials. The *Pebble Heater*, marketed by the Babcock & Wilcox Co. of New York is a special design of a shaft furnace for supplying heat at a temperature level above that available from metallic surface units. The *Thermofor* kiln developed by the Socony-Vacuum Oil Co. and the "fluidized-solids" reactor developed by the Standard Oil Development Co., New York, fall into this general category and are currently used almost exclusively for petroleum cracking. The *FluoSolids* reactor being developed by the Dorr Co., New York, for inorganic chemical reactions is a type of fluidized-solids unit. Its use for calcining lime, roasting ores, etc., has recently been announced.

It is believed that there will be further new applications of this general type of furnace in the next few years, particularly in the chemical-manufacturing field. It has outstanding merit as a gas-solids contactor. Frequently, shaft furnaces are called *shaft kilns*.

General Description of Shaft Furnaces. A shaft furnace is a vertical, refractory-lined cylinder in which a stationary or a descending column of solids is maintained, and through which an ascending stream of heated gas is propelled. Three types of shaft furnaces may be distinguished as to the method of fuel application: (1) one in which solid fuel is charged as discrete particles, alone or admixed with other solid charge material; (2) another, in which a fuel (solid, liquid, or gaseous) is burned in a separate combustion chamber (Dutch oven), the hot products of combustion being blown into the shaft at a low level of the column; and (3) another, in which the fuel (gaseous, liquid, or pulverized solids) is introduced through nozzles and burned in the bottom of the shaft. These differences affect the manner in which heat transfer must be considered. In the first modification, the heat is generated directly within the charge, and it is impossible to separate generation from utilization of the heat. In the second modification, the generation of heat is separated physically from its utilization, and the hot gases serve as a transferring medium. The principles of heat transfer as described under Industrial Furnaces apply here. The third modification falls between the other two in its manner of transferring heat but approaches the first modification more closely since the generation of heat takes place in the immediate vicinity or in direct contact with the lower portion of the charge bed. In all cases, the maximum economy of fuel utilization depends on recovering the exhaust heat from the discharged solids and gases for preheating the incoming charge or fuel-air blast.

Feeding of the charge solids into the top of the shaft may be continuous from a bucket elevator, or intermittent from a skip hoist or crane. The downward movement of these solids within the shaft is controlled by the method and timing of the discharge at the bottom. The discharge may be through manually or mechanically operated gates on a definite time schedule, or it may be continuous over a vibrating or reciprocating grate. Gases are forced through the solids by a blower, or by induced draft from a stack or discharge fan. Heat-recovery auxiliaries may be used.

A shaft furnace was the earliest and today is still an outstanding example of a continuous countercurrent device for heating solid particles by means of a heated gas. In operation, the downward flow of material across each horizontal section, and the upward flow of gas is invariant with time, thus satisfying the ideal requirements for "steady-state" conditions. The full advantages of this type of design as a gas-solids contactor have begun to be recognized in recent years, and its use for chemical operations is being considered more widely, as indicated in the discussion of its uses.

Shaft Furnaces for Lime Production. A shaft furnace used for burning lime is generally known as a shaft kiln. A study of such units has been reported by Victor J. Azbe (*Rock Products*, August, September, October, November, 1945, and January, 1946). Figure 12 shows a shaft furnace for lime production.

The early designs of shaft kilns made about 500 lb. of lime/(day)(sq. ft. of shaft area). Modern kilns make about 3000 lb./sq. ft. A rate of 4000 lb./sq. ft. appears feasible when all parts are adequately designed. Shaft heights range from 30 to 80 ft. One set of shaft kilns is being fed with 5-in. lump limestone. The active height is 37.5 ft., and the shaft area is 55 sq. ft. The draft required is 5 in. of water pressure, which gives a requirement of 0.13 in. of draft/ft. of bed height. The combustion gas rate is 73 cu. ft./(sq. ft. of shaft area)(min.).

Approximately 80 tons of lime are burned per day in each kiln, which is equivalent to 2900 lb./(day)(sq. ft. of shaft area) or equivalent to 77.5 lb./(day)(cu. ft. of active shaft volume).

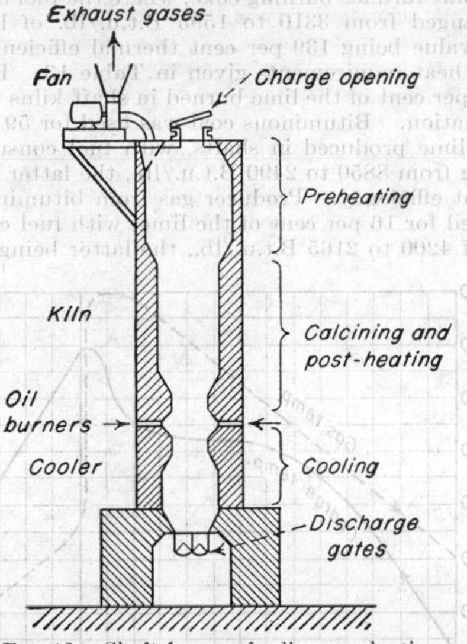

FIG. 12. Shaft furnace for lime production.

Factors Affecting Performance of Shaft Furnaces. The major factors affecting the performance of shaft kilns are:

1. Heat requirements of charge material.
2. Physical characteristics of the charge material, *i.e.*, particle size and range of size distribution, shape of particle, nature of particle surface, random packing characteristics, and changes in these physical properties which occur during their passage through the furnace.
3. Features of kiln design, *i.e.*, active shaft height and cross-sectional area of bed, amount of total draft applied, whether natural or forced, to cause gas flow through the charge bed, temperature of hot gases entering the bed, amount of heat recovery from hot discharge solids in cooler, nature of discharge and control of retention time in the shaft, etc.

The heat required for burning limestone has been estimated by Azbe, as shown in Table 14.

Table 14. Heat Requirements for Calcining Limestone*

1.785 lb. of $CaCO_3$/1.0 lb. of CaO

	B.t.u./lb. CaO
Preheat of limestone to 1648°F.	862
Heat of dissociation of limestone	1212
Postheating of lime to 2400°F.	175
Total heat put into lime in kiln	2249
Heat removed from lime in cooler	557

* Azbe, *Rock Products*, **48** (10), 102 (1945).

The size and shape of the charge particles control the amount of surface over which heat may be transmitted to the particle and also the depth of penetration through which the heat must pass to reach the center of each particle. Also, this size and shape control the nature of the random packing in the shaft and the extent of voids for gas passage. As particle size is decreased, the surface area of the particles increases and, at the same time, the depth of heat penetration decreases. Both these factors tend to improve furnace performance. With small particle size, however, the charge column presents an objectionably high resistance to the passage of gas, resulting in a greater increase in the power required for maintaining gas flow.

With closely screened material, the percentage of voids (usually 37 per cent) is independent of particle size. With unscreened particles showing a wide variation in size, the void volume is decreased, irregularity in gas flow results, and heat transfer is impaired. Since the time required for heat penetration varies with particle size although the passage time is the same, small particles are overburned, while the large particles are left with uncalcined cores. In lime burning, large particles (4 to 8 in.) are preferably heated in a shaft kiln, while sizes below 3 in. are calcined in a rotary kiln. If the smaller sizes are treated in a shaft furnace at normal rates, the draft power becomes excessive. If resort is had to slow blowing, the small particles tend to become overburned.

It has been found that there is a large difference between the total surface of the particles (as determined by their size and shape) and the "effective surface" actually exposed to the passing gas stream. In practice, it has been estimated that as little as 10 to 25 per cent of the total surface is effective in heat transfer when unscreened particles are treated, because of decreased void volume, channeling, gas pockets, irregular gas distribution, and "flow shadows."

Irregular-shaped particles exhibit greater surface area than regular-shaped cubes and spheres, the amount of this increase being possibly 25 per cent. The effect of particle size and size distribution on effective surface is shown in Fig. 13, taken from Azbe. Curve *A* shows

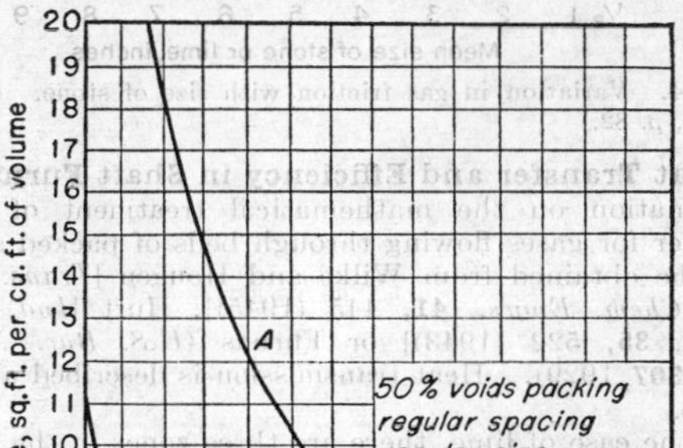

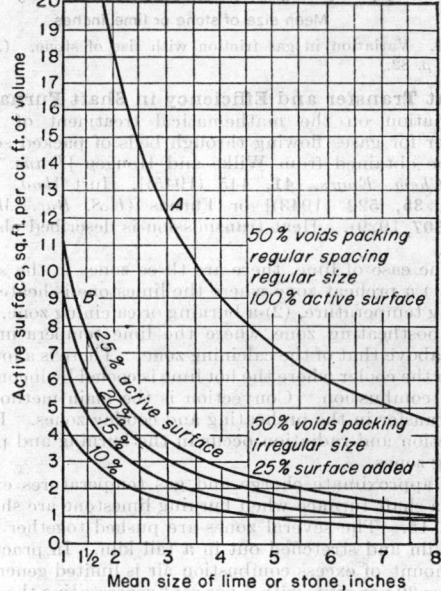

FIG. 13. Curve *A* shows surface variation with stone size, 100 per cent active surface. Curves in group *B* show effect of irregular stone size. [*Azbe, Rock Products*, **48**, p. 81 (*Sept.*, 1945).]

the calculated surface based on an assumed 50 per cent void volume and cubical-shaped particles. The *B* set of curves applies to such unscreened, irregular-shaped particles as are usually encountered in practice.

The laws governing the flow of fluids through packed beds given in Sec. 5 are applicable to shaft furnaces, although difficulty may be experienced in applying these laws directly to an actual furnace charged with material of unknown physical characteristics, unless pressure and flow tests have been made to establish an adequate "base performance." Since the pressure drop in a bed is affected by the size and shape of the interstitial voids, the horizontal and vertical non-uniformity of the bed, the changes in gas composition during passage, and other operating factors, test data for a given material are necessary to proper design. When several reliable base points have been determined, however, it is frequently possible to estimate the effect of minor deviations from base performance with satisfactory accuracy.

In the case of limestone, Fig. 14, from Azbe, shows the effect of particle size on the gas-flow friction through the bed, assuming that the friction varies as the square of the gas mass velocity and inversely with the particle size, and utilizing base points established during actual kiln operations.

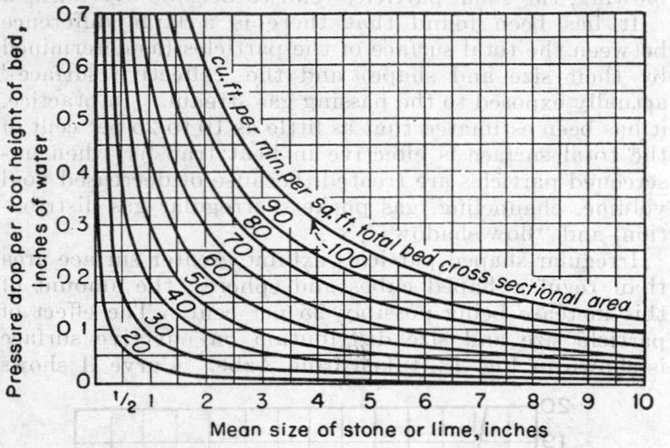

FIG. 14. Variation in gas friction with size of stone. (*Azbe, op. cit.*, p. 82.)

Heat Transfer and Efficiency in Shaft Furnaces. Information on the mathematical treatment of heat transfer for gases flowing through beds of packed solids may be obtained from Wilke and Hougen [*Trans. Am. Inst. Chem. Engrs.*, **41**, 445 (1945)]; Hurt [*Ind. Eng. Chem.*, **35**, 522 (1943)] or Furnas (*U.S. Bur. Mines Bull.* **307**, 1929). Heat transmission is described also in Sec. 6.

In the case of lime, there are three zones in the shaft kiln: (1) a preheat zone where the limestone is heated to burning temperature, (2) a burning or calcining zone, and (3) a postheating zone where the lime temperature is raised above that of the calcining zone. There is a fourth zone in the cooler where the hot lime is cooled by incoming air for combustion. Convection is the main method of heat transfer in the preheating and cooling zones. Both convection and radiation occur in the burning and postheating zones.

The approximate charge and gas temperatures existing in a shaft furnace when burning limestone are shown in Fig. 15. The several zones are pushed together in a short kiln and stretched out in a tall kiln. In practice, the amount of excess combustion air is limited generally to below 20 per cent, with 5 per cent representing the best performance.

The thermal efficiencies for inadequately equipped and poorly operated shaft kilns will be in the range of 40 to 60 per cent, whereas well-equipped and well-operated ones will exceed 80 per cent. The principles of operation of a shaft kiln, wherein cold solids are intimately mixed with hot gases, facilitates the attainment of high efficiencies within the shaft. This factor makes the shaft kiln inherently better than the rotary kiln, although the latter may approach the former in efficiency when it is fully equipped with heat-utilization and -recovery facilities. Increasing the height of a shaft increases its ability for greater efficiencies.

In the U.S. Bureau of Mines *Information Circular* 7174, mentioned under Rotary Kilns, there are given the results of an over-all industry survey of lime to fuel ratios for 1939. The highest thermal efficiency was reported for a shaft furnace burning coke, where the fuel consumption ranged from 3310 to 1595 B.t.u./lb. of lime, the latter value being 139 per cent thermal efficiency based on the heat requirements given in Table 13. However, only 2 per cent of the lime burned in shaft kilns used this combination. Bituminous coal was used for 59 per cent of the lime produced in shafts, with fuel consumptions varying from 8850 to 2490 B.t.u./lb., the latter being 89 per cent efficiency. Producer gas from bituminous coal was used for 16 per cent of the lime, with fuel consumptions of 4200 to 2165 B.t.u./lb., the latter being 102 per cent efficiency. Wood was used for 11 per cent of the lime, with fuel consumption varying from 6600 to 2065 B.t.u./lb., the latter being 107 per cent efficiency. Natural gas was used for 10 per cent of the lime, with fuel consumptions of 3880 to 2120 B.t.u./lb., the latter being 104 per cent efficiency. Fuel oil was used for 1.3 per cent of the lime, with fuel consumptions ranging from 4000 to 2550 B.t.u./lb., the latter being 87 per cent efficiency. This latter consumption corresponds to 35.7 gal. of fuel oil per ton of lime. These values should be compared with those shown under Rotary Kilns and FluoSolids Furnaces.

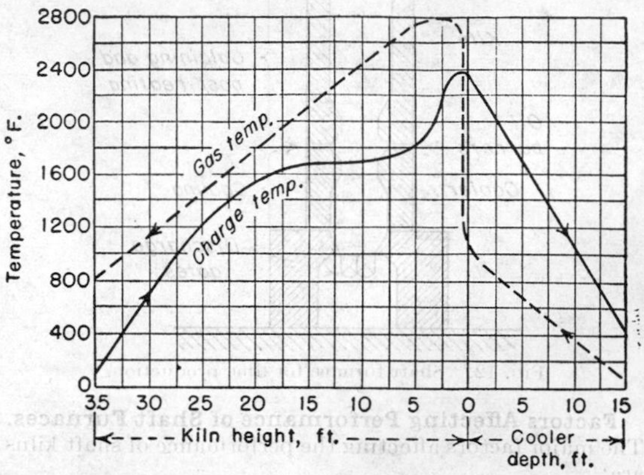

FIG. 15. Approximate temperatures in a shaft furnace for burning limestone.

Blast Furnaces. The blast furnace was developed for the conversion of iron ore into molten pig iron. It is a shaft furnace, in which a continually descending column of iron ore, coke, and limestone is blasted with preheated air. In its descent through the shaft, all the Fe_2O_3 (or Fe_3O_4) of the ore is reduced to FeO, by countercurrent reaction with the CO (34.6 per cent) in the ascending gas stream. A portion (35 to 65 per cent) of the resultant FeO is reduced to carburized metallic iron, the remaining FeO being reduced to Fe by carbon in the hearth. The gangue material of the ore and the ash of the coke (largely Al_2O_3 and SiO_2) are fluxed with the CaO of the limestone, calcined in the shaft, to form a fluid slag. The molten iron and slag, as it accumulates in the bottom of the furnace ("hearth"), forms immiscible liquid layers, the slag floating on the iron. Slag is "flushed" from the hearth through a water-cooled copper nozzle (cinder notch) at intervals of 1 to 2 hr. Molten pig iron is withdrawn ("tapped") from a hole (iron notch) drilled through the refractory lining of the hearth at 4- to 6-hr. intervals.

The height of the charge column in the furnace (stock line) is maintained by introducing "rounds" of ore, coke, and stone at 8- to 20-min. intervals through a rotating bell-and-hopper distributor. The charge materials are conveyed to the top of the furnace by means of a skip car running on a steeply inclined skip track. Furnace gas exhausting from the charge column (250° to 325°F.) is withdrawn through gas offtakes (downcomers). This gas, containing from 25 to 28 per cent CO, 1 to 4 per cent H_2, after cleaning is used for preheating the blast, generating steam for power, and when mixed with coke-oven gas for firing coke ovens and open-hearth furnaces.

Atmospheric air, preheated to 900° to 1200°F. at 13 to 24 lb./sq. in. gage pressure, is forced into the furnace at a level 6 to 8 ft. above the iron notch through 8 to 18 copper nozzles (tuyères) distributed symmetrically about the perimeter of the hearth. In present-day American practice, blast volumes of 44,000 to 100,000 cu. ft./min. are encountered. The dimensions of an average modern blast furnace are shown in Fig. 16, from Sweetser. This furnace has a hearth diameter of 25 ft., a maximum diameter of 28.25 ft., a stock-line diameter of 19 ft., with an active volume of 34,900 cu. ft. The largest units thus far constructed in this country have 27.5, 30.0, and 20.0 for these three diameters. The maximum sustained daily production from these largest units is 1750 tons.

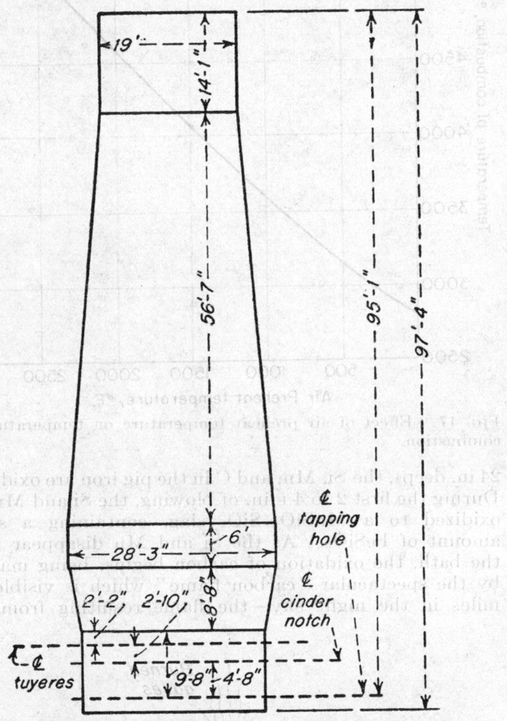

FIG. 16. Dimensions of typical large blast furnace. (*Sweetser, "Blast Furnace Practice," p. 12, McGraw-Hill, New York,* 1938.)

The capacity of a blast furnace is determined by the rate of blowing (cubic feet per minute blast volume) and the coke consumption (pounds of coke per ton of pig iron). In practice, coke consumption varies widely from furnace to furnace, the average being close to 1700 lb./ton of iron, ranging from 1110 to 1880 for steelmaking grades of iron. Furnace capacity varies directly with the rate of blowing and inversely as the coke consumption, the average ratio of blast to coke being 52.5 cu. ft. of air/lb. of stock-line coke. Thus, 65,000 cu. ft./min. of blast will produce 1,040 tons of pig iron per day at average coke consumption. With low coke operation (0.72 tons per ton) and with large blast volumes (95,000 cu. ft./min.), a daily production of 1780 tons is realized. Although operating costs and overhead charges per ton of metal are somewhat reduced by increased capacity, the economic success of a furnace operation is determined primarily by the cost of fuel, *i.e.*, the coke consumption.

A material balance for a typical 1000-ton/day blast-furnace operation is shown in Table 15. It is observed that the air blast constitutes slightly more than one-half of the total furnace charge.

Table 15. Typical Blast-furnace Material Balance

Material Charged	Tons/Day
Iron ore	1625
Iron and steel scrap	125
Sintered flue dust	150
Open-hearth slag	50
Limestone	450
Coke	850
Air blast	3410
Total	6660
Material Produced	
Pig iron	1000
Slag	575
Flue dust	170
Gas	4915
Total	6660

Furnace performance relative to reducing coke consumption or increasing furnace capacity can be improved by employing:

1. Blast temperatures in excess of present-day maximum limit of 1200° to 1450°F.
2. Removing moisture from the blast.
3. Low-sulfur coke (well below 1 per cent), permitting use of less limestone, higher grade ore and concentrates, lower metal and slag temperatures, and reduced volume of slag.
4. Ore with improved uniformity of particle size, realized by screening, crushing the oversize, and agglomerating the fines by sintering, briquetting, nodulizing, or pelletizing, thus decreasing the FeO reduced by carbon in the hearth.
5. Increased blast volumes, *e.g.*, in excess of the present-day maximum of 100,000 cu. ft./min.
6. Increased stock-line diameters, thereby decreasing gas velocity and avoiding excess flue-dust production.
7. Higher blast pressures (up to 40 lb./sq. in. gage pressure) with normal or higher blast volumes and with a restriction at the furnace top to maintain a higher static pressure at the stock line.

Pressure operation of two furnaces by the Republic Steel Corp. [Slater, *Steel*, **120** (23), 102 (June 9, 1947)] indicates that iron production is increased, coke consumption is reduced, flue dust is decreased, and heat requirements per ton of pig iron are reduced. Top pressures in the range of 10 lb./sq. in. gage were used and blast volumes were increased by 10,000 to 15,000 cu. ft./min. Coke consumption was decreased by 130 to 255 lb./ton of iron. Furnace capacity was increased by 11 to 20 per cent. The use of pressure permits higher blast volumes with lower gas velocities at the stock line. Test operation at top pressures of 20 to 25 lb./sq. in. is being arranged.

It has frequently been suggested that the blast be enriched with oxygen. This increases the temperature of combustion in the tuyère region and has an effect similar to an increase in blast temperature or a decrease in blast moisture. With increased oxygen content, the heat supplied to the furnace is decreased, since the nitrogen passing through the furnace enters at blast temperature (900° to 1200°F.) and exhausts from the stock line at a much lower temperature (250° to 325°F.). The nitrogen although chemically inert, acts as a heat carrier to increase the supply of heat to the furnace operation.

Because the air quantity used in a blast furnace is so large, the moisture carried by this air has a substantial effect on flame temperatures and on the heat requirements. This moisture is converted to hydrogen and carbon monoxide with a residual moisture content of only 0.00041 per cent. The heat absorbed by this reaction is 4685.9 B.t.u./lb. of carbon. The decrease in the combustion-zone temperature resulting from varying concentrations of moisture in the air is shown in Table 16.

Table 16. Effect of Variations in Atmospheric Moisture on the Combustion Temperature in an Ideal Adiabatic Gas Producer

Atmospheric condition	Dry	Winter	Normal	Summer humid	Extreme heat and humidity
Atmospheric temperature..	60	32	60	90	100
% Humidity............	0	100	60	90	100
% H_2O in wet air, vol.....	0.00	0.61	1.05	4.26	6.38
Combustion temperature, °F..................	2755	2680	2630	2250	2016
Decrease in temperature due to blast moisture...	0	75	125	505	745

When air is enriched with oxygen, the ratio of nitrogen to oxygen is decreased. The quantity of oxygen required to react with the carbon remains the same, as also does the amount of heat generated. Therefore the temperature of the combustion is increased and the amount of available heat is increased because of the smaller quantity of diluent nitrogen. The rise in combustion temperature for varying amounts of added oxygen is shown in Table 17.

Table 17. Effect of Oxygen Enrichment
Combustion Temperature in an Ideal Adiabatic Gas Producer Blown with an Air-oxygen Blast

1. % enrichment*........	0	10	20	30	40
2. % O_2 in moisture..........	20.97	23.06	25.16	27.27	29.37
3. Cu. ft. blast/lb. carbon......	74.87	68.25	62.50	57.55	53.51
4. Combustion temperature, °F..	2755	2995	3250	3610	3980
5. Bosh, gas, % CO...........	34.68	37.58	40.25	43.12	44.43
Bosh gas, % N_2............	65.32	62.42	59.75	56.88	55.57
6. Equivalent hot blast, °F.†......	60	320	680	1050	1450

* % enrichment: cu. ft. of O_2 added to each 100 cu. ft. of atmospheric oxygen.

† Equivalent hot-blast temperature: preheat of atmospheric blast required to give the same combustion temperature as is attained with oxygen enrichment.

The effect of preheating the air is also large. As the blast is preheated from 60° to 3000°F. (the maximum attainable in any type of blast-furnace stove with firebrick refractories), the combustion temperature is progressively raised from 2755° to 5750°F. The combustion temperature for various preheats is shown in Fig. 17.

The heat loss through the brick lining of the bosh and hearth of a 1400 ton/day blast furnace was measured hourly for more than 1 year. With changes in operating conditions, it varied from a minimum of 200 B.t.u. to a maximum of 310 B.t.u./lb. of carbon oxidized, the average being 262 B.t.u. This heat loss reduces the temperature of combustion by 145°F.

Bessemer Converters. The Bessemer bottom-blown converter (Fig. 18) is a pear-shaped, vertical furnace, in which air is blown through a bath of molten pig iron to produce molten steel. It is mounted on trunnions to permit tilting the vessel from a vertical to a horizontal position. One of the trunnions is hollow, through which air at 20 to 30 lb./sq. in. gage pressure is supplied to a wind box located at the bottom of the vessel. The converter shell (vessel) is constructed of steel plate, lined with 12 to 24 in. of mica schist, sandstone, or ganister (SiO_2). It is reported that the removable bottom piece averages about 40 blows and the lining about 2450 blows.

When tilted into its horizontal position, molten pig iron is charged into the vessel, from 11 to 35 tons being the limit of present-day U.S. converter practice. Air from the wind box at sonic velocity passes through 150 to 350 cylindrical holes, ½ to ⅝ in. in diameter and 18 in. long, forming the tuyères in the furnace bottom. As the air discharges upwardly through the molten bath (10 to

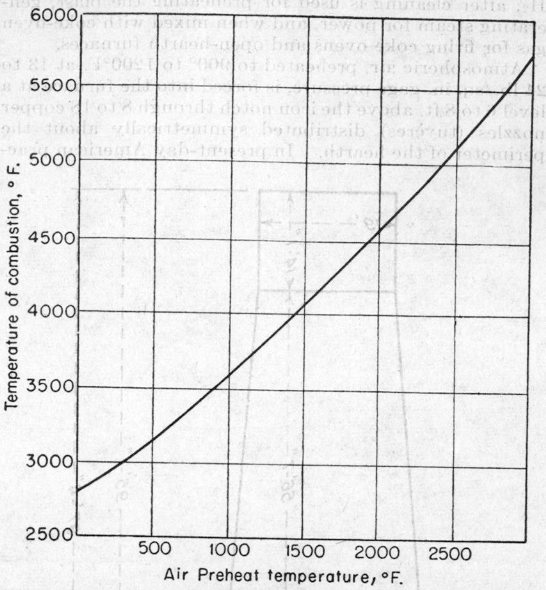

Fig. 17. Effect of air preheat temperature on temperature of combustion.

24 in. deep), the Si, Mn, and C in the pig iron are oxidized. During the first 2 to 4 min. of blowing, the Si and Mn are oxidized to a $MnSiO_3$–SiO_2 slag, containing a small amount of $FeSiO_3$. As the Si and Mn disappear from the bath, the oxidation of carbon begins, being marked by the spectacular "carbon flame" which is visible for miles in the night sky—the flame resulting from the

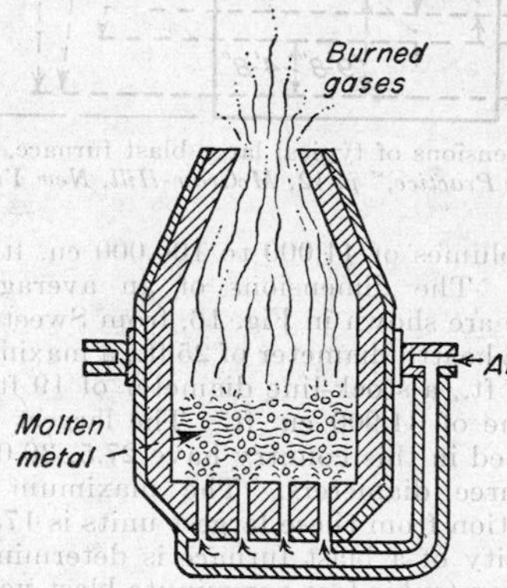

Fig. 18. Bessemer converter.

burning of the CO produced in the vessel with the oxygen in the atmosphere after its discharge from the converter. In 10 to 20 min., at the "drop" of the carbon flame, the "blow" is terminated, the "blown metal" and slag (vessel cinder) being poured from the tilted vessel into a ladle.

In an "acid-lined" converter (SiO_2) without CaO

additions to the charge, none of the phosphorus of the pig iron is oxidized. Thus Bessemer steel exhibits a usual content of 0.09 to 0.12 per cent phosphorus and is generally regarded as inferior to basic open-hearth or electric-furnace steel carrying from 0.015 to 0.035 per cent of phosphorus.

In the basic Bessemer process (Thomas converter), the vessel is lined with dead-burned magnesite. Limestone or dolomite and limestone are charged with the pig iron, thus producing a "basic slag." In this process, phosphorus is removed as $Ca_3P_2O_8$ dissolved in a calcium silicate slag. Although the Thomas converter has been extensively used in England and on the continent continuously during the past 93 years, no commercial Thomas converter plant has operated in the United States since the introduction of the Bessemer process in 1856.

Gas Producers and Water-gas Sets. Gas producers are shaft furnaces designed to produce a low-B.t.u. fuel gas from bituminous coal or other solid fuels by blasting it with a mixture of steam and air. There are many design modifications of gas producers, such as Wellman, Wood, Morgan, Chapman, and Wollaston.

Water-gas generators are shaft furnaces designed for the production of a high B.t.u. fuel gas from coke or other solid fuels by blasting it alternately with air and with steam.

Both operations have been described in Sec. 22, Fuels. Haslam and Russell include extensive descriptions of both types.

Pebble Heater. The pebble heater was developed in the early 1940's by the Babcock & Wilcox Co. for the purpose of heating steam to temperatures higher than could be obtained in metallic units. It can be used for heating air, hydrogen, methane, steam, etc., for use in various industrial operations. Also it can be used for recovering the heat from hot gases discharged from another process. Furthermore, the lower section of the heater can be used as a reactor for vapor-phase operations at high temperature. It is a special form of shaft furnace.

Its construction and operation have been described by Norton [*Chem. & Met. Eng.*, **53** (7), 116 (1946)]. Figure 19 shows its general arrangement, (*a*) vertical cross section of a unit for heating air, (*b*) horizontal section through the combustion chamber, and (*c*) variable feeder which regulates the flow of the pebbles through the system. A bucket elevator (not shown) lifts the cold pebbles from the bottom and charges them in at the top to complete their recycle. The pebbles are heated in the upper chamber by direct contact with combustion gases and then are passed through a throat to the lower chamber where their heat is transmitted to air, steam, or other gases. The throat increases the resistance to flow of gases between the two chambers. In operation, the two chambers are kept at the same pressure so that there is no flow. A damper in the stack from the upper chamber may be used to cause a small gas flow in either direction through this throat. An average cycle on the pebbles is 30 to 50 min.

Table 18. Pebble Heater Typical Operating Data*

	Unit 1	Unit 2
Gas being heated	Air	Steam
Exit gas temperature, °F.	2180	1800
Inlet gas temperature, °F.	100	230
Weight of gas heated, lb./hr.	1800	4400
Fuel heat, B.t.u./hr	1,760,000	5,550,000
Weight of flue gas, lb./hr.	2240	8000
Pebble circulation, lb./hr.	2150	9500†
Flue gas exit temperature, °F.	750	805
Lower bed diameter, in.	19	27

* Norton, *Chem. & Met. Eng.*, **53** (7), 117 (1946).
† Computed.

The maximum temperature to which gases can be heated is restricted by refractory limitations and obtainable combustion temperatures. The pebbles have a mullite composition (72 per cent Al_2O_3, 28 per cent SiO_2). Other suitable compositions are kaolin, an 85 per cent Al_2O_3 mix, and 99.9 per cent Al_2O_3. Temperatures of 2300°F. have been attained. Typical operating data are shown in Table 18.

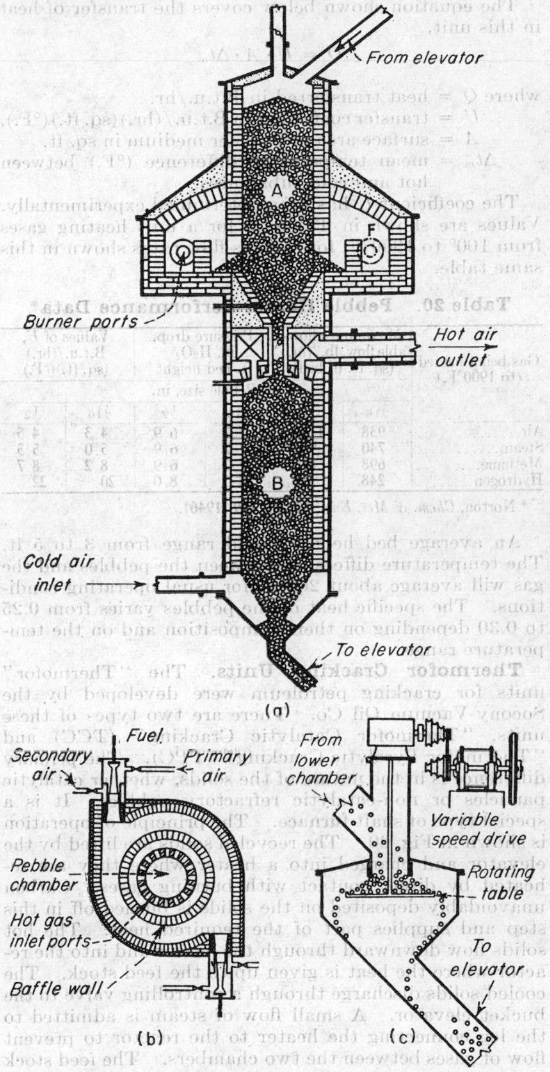

FIG. 19. Pebble heater. (*a*) Vertical section; *A*, pebble heating chamber; *B*, heat-transfer chamber. (*b*) Horizontal section, combustion chamber. (*c*) Variable feeder. [*Norton, Chem. & Met. Eng.*, **53**, 117 (1946).]

The large surface of these pebbles available for transmitting heat is a major advantage of this unit. The pebble surface per cubic foot in the furnace is compared with other types of surface in Table 19.

Table 19. Heat-transfer Surface per Cubic Foot of Volume*

	Sq. Ft.
Pebbles, 5⁄16 in. diameter	135.5
Pebbles, ½ in. diameter	86.5
Steel tubes, 2½ in. o.d. (3¼ by 3¼ in. on centers)	9.0
Regenerator checkerbrick	3.0–8.0

* Norton, *J. Am. Ceram. Soc.*, **29** (7), 189 (1946).

The limiting factor which governs the capacity of the heater is the gas velocity through the bed, which will cause pebble lifting or stoppage of pebble flow through the throat. Table 20 shows the maximum allowable flow of various gases per square foot of bed cross-sectional area with different sizes of pebbles when the gas is being heated to 1900°F. This table shows also the pressure drop per foot of bed height at these maximum flow rates.

The equation shown below covers the transfer of heat in this unit.

$$Q = U \cdot A \cdot \Delta t_m$$

where Q = heat transferred in B.t.u./hr.
U = transfer coefficient in B.t.u./(hr.)(sq. ft.)(°F.).
A = surface area of transfer medium in sq. ft.
Δt_m = mean temperature difference (°F.) between hot and cold substances.

The coefficient U has been determined experimentally. Values are shown in Table 20 for a unit heating gases from 100° to 1900°F. for the gas-flow rates shown in this same table.

Table 20. Pebble Heater Performance Data*

Gas being heated (to 1900°F.)	Maximum allowable flow, lb./(hr.)(sq. ft. of bed)		Pressure drop, in. H_2O/ft. bed height		Values of U, B.t.u./(hr.)(sq. ft.)(°F.)	
	Pebble size, in.					
	5/16	1/2	5/16	1/2	5/16	1/2
Air	938	1180	7.7	6.9	4.3	4.5
Steam	740	933	7.7	6.9	5.0	5.3
Methane	698	880	7.7	6.9	8.2	8.7
Hydrogen	248	312	10.5	8.0	20.	22.

* Norton, *Chem. & Met. Eng.*, **53** (7), 119 (1946).

An average bed height would range from 3 to 5 ft. The temperature difference between the pebbles and the gas will average about 200°F. for usual operating conditions. The specific heat of the pebbles varies from 0.25 to 0.30 depending on their composition and on the temperature range.

Thermofor Cracking Units. The "Thermofor" units for cracking petroleum were developed by the Socony-Vacuum Oil Co. There are two types of these units, "Thermofor Catalytic Cracking" (TCC) and "Thermofor Pyrolytic Cracking" (TPC). The primary difference is in the nature of the solids, whether catalytic particles or non-catalytic refractory pebbles. It is a special type of shaft furnace. The principle of operation is shown in Fig. 20. The recycled solids are lifted by the elevator and charged into a heater where they are reheated by direct contact with burning gases. Carbon unavoidably deposited on the solids is burned off in this step and supplies part of the required heat. The hot solids flow downward through the heater and into the reactor where the heat is given up to the feed stock. The cooled solids discharge through a controlling valve to the bucket elevator. A small flow of steam is admitted to the leg connecting the heater to the reactor to prevent flow of gases between the two chambers. The feed stock is admitted into the bottom of the reactor, and the products are withdrawn at its top. These units are being used to produce olefins, gasolines, and aromatic compounds. Temperatures of 1400° to 1800°F. are employed. Contact times of the feed stock with the solids at high temperatures in the reactor are controlled in the range of 0.1 to 3.0 sec. Heat-transfer rates as high as 15,000 B.t.u./(hr.)(°F.)(cu. ft. of pebbles) have been obtained in the reactor. Each unit must be specifically designed for its proposed service.

Fluidized-solids Reactors. The use of fluidized solids for carrying out high-temperature reactions was developed by the Standard Oil Development Co. early in World War II for the catalytic cracking of various petroleum feed stocks to give high yields of gasoline. Some forty units were installed during the war to supply military needs. Numerous additional units have since been built both in the United States and abroad. The fluidized-solids techniques and equipment are now being tried in other fields. The M. W. Kellogg Co. has undertaken to develop its use further in the catalytic and hydrocarbon fields. The Dorr Co. has engaged in similar activities for non-catalytic purposes (see FluoSolids).

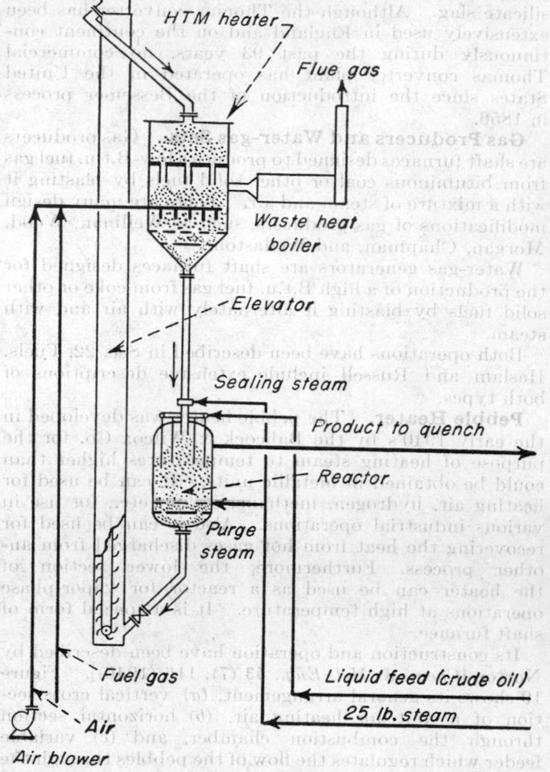

FIG. 20. Thermofor pyrolytic cracking unit. [*Socony-Vacuum Oil Co., Inc.; Chem. Eng.*, **55,** 113 (1948).]

The "Fluid Catalytic Cracking" units consist primarily of a reactor and a regenerator with auxiliaries. For gasoline production, this equipment is followed by the necessary refining equipment, as shown in Fig. 21. The reactors are vertical, cylindrical steel vessels containing a mass of finely divided catalyst suspended by the flow of vaporized feed stock. Gas velocities through the reactor are low, in the order of 0.5 to 2 ft./sec., based on the empty cross section, but sufficiently high to maintain a fluidized bed of solids having many of the properties of a liquid. The feed stock is vaporized before it enters the reactor by contact with hot, regenerated catalyst (1000° to 1150°F.). The vaporized feed entrains the regenerated catalyst and carries it into the reactor through a distributing grid. The reactor temperature is maintained at 800° to 1000°F. The cracked feed stock passes from the catalyst bed through cyclone dust collectors installed in the upper part of the reactor and thence to the fractionating system.

Carbon unavoidably deposited on the catalyst during the reaction must be removed periodically by being carried through the following cycle. From the reactor bed, spent catalyst flows by gravity through a steam-stripping section and discharges through a standpipe into a heated air stream. This air flow entrains the catalyst and conveys it up through a grid into a regenerator where the carbon is burned off, at temperatures of 1000° to 1150°F. The regenerator is similar in construction to the reactor but normally is lined with a refractory to protect the metal walls. Internal metal parts, including the cyclones, are fabricated from temperature-resistant alloys. Hot regenerated catalyst flows by gravity through a standpipe to be entrained by the feed entering the reactor, thus completing the catalyst cycle. The hot flue gas from the regenerator passes through a waste-heat boiler, a Cottrell precipitator, and thence to a stack. On the newer units, a microspheroidal catalyst

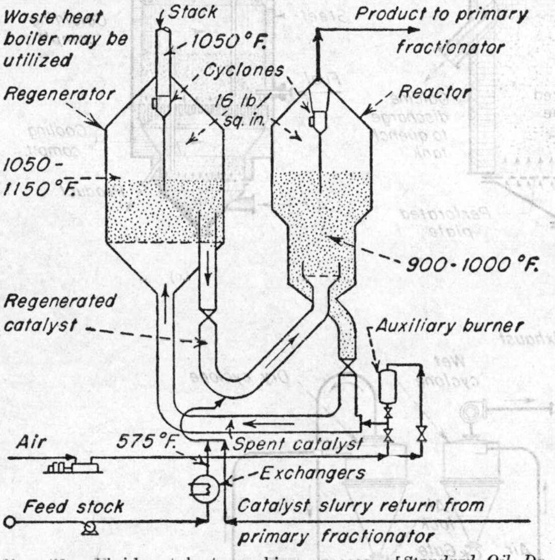

FIG. 21. Fluid catalyst cracking process. [*Standard Oil Development Co.; Petroleum Refiner*, **27**(9), 89 (1948).]

is employed which can be satisfactorily recovered from the gas stream without using a Cottrell unit. There is some loss of catalyst from attrition, and makeup must be added continuously.

A rapid flow of catalyst is maintained between the reactor and regenerator, in order to maintain constant catalyst activity and uniform bed temperatures. For example, a 15,000 bbl./day unit has been reported to include a reactor 25 ft. diameter by 50 ft. high, containing 75 to 100 tons of catalyst. In this unit, 8 to 50 tons/min. of catalyst were circulated to the regenerator, where the holdup was 150 to 300 tons. The heat transfer from the regenerator to the reactor will range above 150 million B.t.u./hr. Units have been installed with capacities as high as 41,000 bbl./day.

The fluidized-solids techniques appear to have special application for gas-solids contacting processes, particularly those requiring close control of temperature. The solid particles may be catalytic or non-catalytic, may participate as a direct reactant, or serve merely as a heat carrier. The use of these techniques in fields other than petroleum is experimental, but progress is being reported. Every unit must be designed especially for its proposed purpose.

FluoSolids Furnaces. This furnace is a vertical shaft, but the single or multiple beds are fluidized (boiling movement) by the flow of gases upward through them. It is an adaptation of the fluidized-solids principle by the Dorr Co. to non-catalytic operations. Figure 22 shows the cross sections of several modifications of FluoSolids furnaces.

The Dorr Co. reports (*Chem. Eng.*, December, 1947, pp. 112–115) the use of these furnaces for (1) roasting (1700°F.) an arseno-pyrite gold ore at 10 to 15 tons/day, (2) calcination (1700° to 1750°F.) of crystalline high-calcium limestone at a rate of 11 tons/day of burned lime and (3) calcination (1850°F.) of a calcium carbonate sludge to lime in a pilot plant at 6 tons/day. A unit for calcining limestone to lime at a rate of 100 tons/day is being built. The oil consumption in the 11 ton/day unit was 43 gal./ton of lime at a calcining temperature of 1700° to 1750°F. with two heat-recovery compartments. The lime contained less than 1 per cent CO_2 and had a chemical availability of 98 per cent of the possible maximum. Calcination at 1600°F. gave an oil consumption of 40 gal./ton and still lower consumptions are expected in the 100 ton/day unit which will have four heat-recovery compartments. These oil consumptions should be compared with those shown under Rotary Kilns and Shaft Kilns.

It is claimed that operations of the following types can be carried out advantageously by fluidization: (1) calcination of materials such as limestone, dolomite, magnesite, metal hydrates, and certain pigments; (2) partial calcination such as that employed in reducing the magnesium carbonate in dolomite to the oxide with only slight alteration of the calcium carbonate; (3) oxidation, particularly when it can be accomplished with air at elevated temperatures; (4) reduction of metallic oxides such as the reduction of hematite to magnetite; (5) roasting of arseno-pyrites and sulphides of zinc, copper, and iron; (6) chlorination, including volatilization of metallic chlorides; (7) sulphatization such as the conversion of copper oxide to copper sulphate; and (8) heat transfer from solid to gas phase or the reverse.

REVERBERATORY FURNACES

Uses of Reverberatory Furnaces. A reverberatory furnace, when used in the production of the following products, is conventionally called: (steel) open-hearth furnace, (wrought iron) puddling furnace, (cast iron) air furnace, (copper) copper reverberatory, and (glass) glass tank. Their design has been developed particularly for melting solids and for refining and heating the resulting liquids.

Description of Reverberatory Furnaces. A reverberatory furnace consists of a comparatively shallow, rectangular, refractory hearth for holding a molten charge. Fuel combustion is maintained directly above the bath, which is enclosed by vertical side walls and covered with a low-arched, silica-brick roof. The walls and roof receive radiant heat from the hot combustion products and reradiate this heat to the surface of the bath. The temperature of the walls in steel furnaces approaches closely to 3000°F. If this temperature limit is exceeded, however, failure of the roof refractories may result. Although the conventional placing of the roof near the surface of the bath tends to increase gas velocity and improve convective heat transfer, the resulting decrease in the radiant mean path decreases the radiant-heat transfer with an over-all loss in furnace efficiency.

Puddling furnaces, air furnaces, and copper reverberatories are not customarily provided with either regenerators or recuperators. Open-hearth furnaces and glass tanks are invariably equipped with Siemens-type, checkerbrick regenerators, in order (1) to recover sensible heat of the exhaust gases and return it to the furnace in the preheated air, and (2) to permit ready attainment of the desired elevated furnace temperature. Two regenerators are usually provided at each end of the furnace. When a thermally stable fuel gas is used

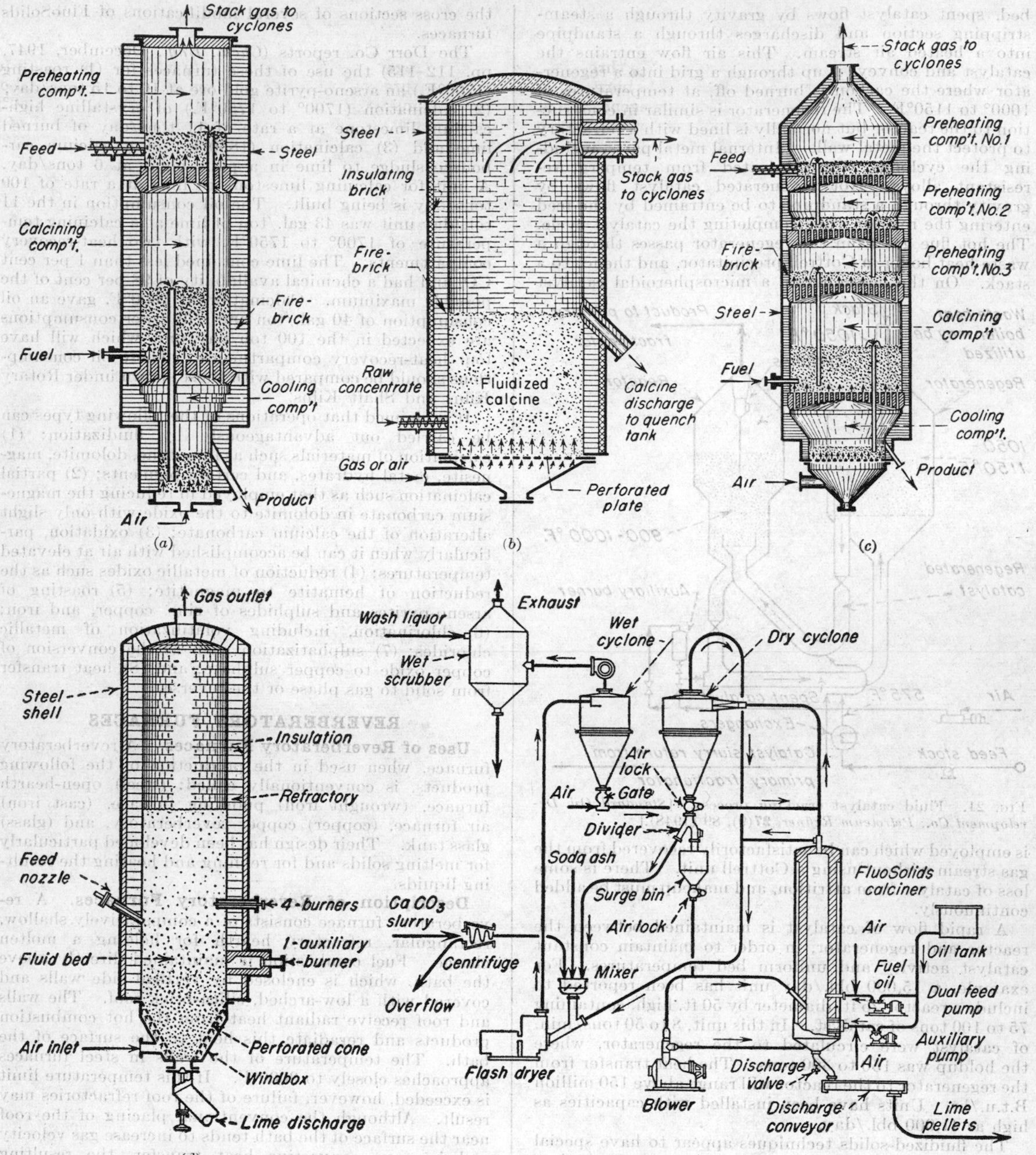

FIG. 22. FluoSolids furnaces. (*a*) FluoSolids pilot kiln with three superposed fluid beds. (*b*) Adaptation of dense phase fluidization to roasting concentrates. (*c*) Type of design which will be used for commercial lime kilns. (*d*) Calciner design being developed for lime sludge. (*e*) Flow diagram for lime-sludge calcination for use in such industries as sulphate pulp, beet sugar, and water softening. [*The Dorr Company; Chem. Eng.*, **54**, 112 (1947).]

(blast-furnace gas, producer gas), one regenerator is used to preheat the fuel and the other to preheat the combustion air. When unstable fuels are used (oil, tar, natural gas, etc.), only the combustion air is preheated, the two regenerators at each end being operated in parallel. The direction of flow of combustion products over the surface of the bath is reversed at regular intervals (about 20 min.), this reversal being effected by the valve arrangement shown in Fig. 7. Figures 23 and 24 show vertical cross sections of the width and length of a typical 150-ton open-hearth furnace, the latter view showing the regenerators.

Open-hearth Steel Furnace. In 1948, the 96 steel plants in the United States operated 954 open-hearth furnaces to produce 83.6 million tons of steel ingots and castings—88 per cent of the total steel production. In the manufacture of open-hearth steel, unheated steel scrap and pig iron (as a cold solid or as a molten liquid) form the charge. Iron ore is introduced to oxidize the silicon, manganese, carbon, and phosphorus in the metal. Limestone (or lime) is added to control the sulphur and to flux the resultant slag.

The hearth dimensions of the typical 150-ton open-hearth furnace in Figs. 23 and 24 are 15 by 40 ft. with a bath depth of 15 to 24 in. The charge is introduced through refractory-lined doors in the front wall. The finished steel and slag are removed through a tap hole

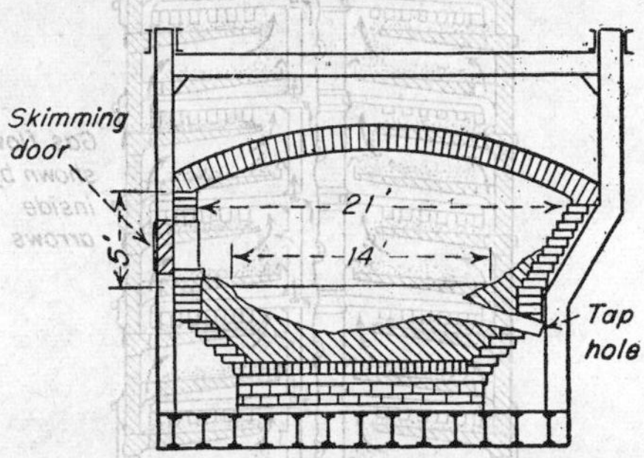

FIG. 23. Cross section of open-hearth steel furnace.

located in the back wall. Inlet ports for fuel and air are located at the ends, the regenerators being placed below the furnace level and extending under the charge floor. Combustion takes place with a sluggish flame directed somewhat downwardly to impinge on the bath surface, the slow combustion taking place throughout the greater part of the furnace length.

The tonnage in a "heat" (batch) is determined largely by the area of the bath surface, altered somewhat by the depth of the metal bath, which may vary from 15 to 30 in. Open-hearth furnaces in present-day steelmaking are rated from 25 to 500 tons per heat. The time required to complete a heat, charged with 50 per cent cold steel

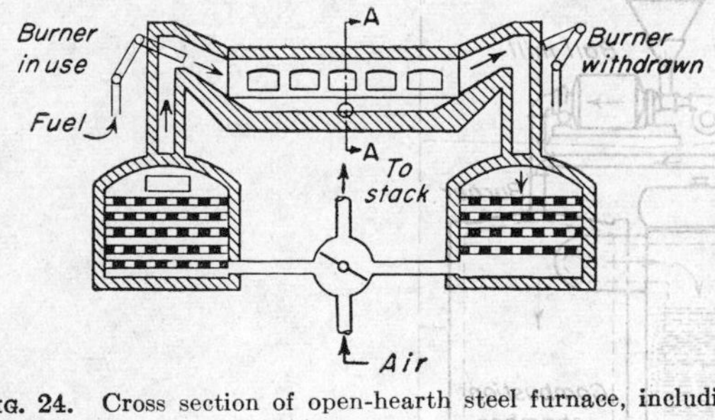

FIG. 24. Cross section of open-hearth steel furnace, including regenerators.

scrap and 50 per cent molten iron, will range from 9 to 13 hr.

The combustion chamber of the furnace shown in Figs. 23 and 24 has a width of 21 ft. and is 46 ft. from port to port. The vertical cross-sectional area of the furnace will be 137 sq. ft. with a combustion volume of 6300 cu. ft. The inside area of the refractories including the molten metal surface will be 2900 sq. ft. When burning fuel tar with 35 per cent excess air, the fuel consumption would be 3450 lb./hr. and the combustion air would be 12,820 cu. ft./min.

The gas and wall temperatures will be 3130°F. and 2920°F., respectively, and the heat of combustion 49 million B.t.u./hr. In addition, the incoming preheated air would return about 41 million B.t.u./hr. for a total heat through-put of 90 million B.t.u./hr. The total heat-transfer coefficient would be 25.8 B.t.u./(hr.)(sq. ft.)(°F. of temperature difference) which is equivalent to 5420 B.t.u./(hr.)(sq. ft.). About 98.5 per cent of this heat transfer would be due to radiation. The total transfer within the furnace would be 15.7 million B.t.u./hr. About 7 million B.t.u./hr., or 7.8 per cent, is radiated through the exhaust ports. The release of heat of combustion within the furnace is 7800 B.t.u./(hr.)(cu. ft.), while the total heat through-put is 14,300 B.t.u./(hr.)(cu. ft.). Both these values are low when compared with the rate of 25,000 to 50,000 B.t.u./(hr.)(cu. ft.) found in modern steam boilers.

The heat sent to the regenerators would be 67.3 million B.t.u./hr., of which 41 million are recovered and returned to the furnace. This indicates an over-all 61 per cent recovery in the regenerators.

Glass Furnace. A glass melting furnace (tank) is generally similar to a steel furnace except that gas flow is crosswise of the long dimension and the hearth is divided into two parts by a partly submerged bridge and a perforate wall. Further, the gas flow is only in the first section so that the second section is heated from the first by radiation through the perforate wall. A typical glass tank would be 36 ft. wide by 100 ft. long. The bottom is flat and the side walls are vertical. The regenerators are similar to those of a steel furnace.

The charge is fed continuously at one end of the melting section across the width of the tank. It melts and gradually flows toward the other end. The submerged bridge holds back unmelted material and gas bubbles floating near the surface. The molten glass flows under the bridge into the refining section. It is discharged at the far end. The manner of discharge is governed by the purpose for which the glass is used.

The depth of the glass tank and the respective sizes of the melting and refining sections depend on the type of glass being made and the time needed for melting, release of bubbles, etc. The dividing wall prevents contamination of the glass in the refining section with dust, etc., carried in the combustion gas stream.

The wall and bath temperatures are in the range of 2500° to 2650°F. Glass furnaces are built to larger dimensions than steel furnaces and refractories such as mullite are used.

SPECIAL DESIGNS OF PROCESS FURNACES

By-product Coke Ovens. Coke ovens (retorts) are a special design of a process furnace wherein it is desired to heat coal to temperatures of 1800° to 2200°F. without contact with the combustion gases. Many designs have been used in the past, but the present style has an oven with a very narrow rectangular, slightly tapering base (42 ft. long by 14.5 in. at one end and 17.5 in. at the other end) with a moderate height (12 ft.). These oven spaces filled with coal are alternated with narrow flues for the hot combustion gases, as shown in Fig. 30. Heat is transmitted from each flue through its side walls to the adjacent oven spaces. A detailed description is given in Sec. 22, Fuels.

Mannheim Furnaces. These furnaces are a special design of a circular muffle in which sulphuric acid is reacted with salt to produce salt cake and hydrochloric acid. It consists of a refractory hearth up to 18 ft. in diameter with a silicon carbide arch or cover. Hot flue gases are circulated around this muffle. The major portion of the heat is transmitted through the arch and radiated to the charge on the hearth. The feed materials are mixed and charged continuously to the center of the hearth where they are pushed around by rabble arms, underdriven.

The charge is gradually worked toward the periphery as the reaction generates hydrogen chloride gas. This gas is withdrawn through a separate duct to an absorption system. The salt cake is discharged at the periphery. Figure 25 shows a diagrammatic cross section of a Mannheim furnace. Combustion-chamber temperatures in the range of 2200°F. are used for heating. The salt cake is discharged from the hearth at about 1000°F.

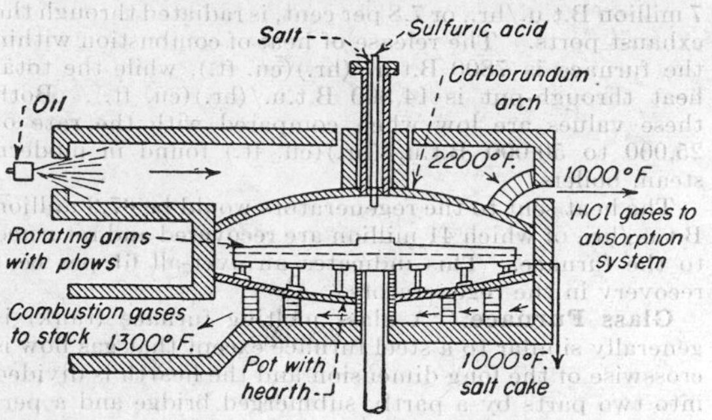

FIG. 25. Mannheim-type mechanical hydrochloric acid furnace.

Multiple-hearth Furnaces. Multiple-hearth furnaces have been in use for many years for ore processing. Modifications are known under various names, the Herreshoff, McDougall, Wedge, Nichols, etc. Figure 26 shows the general design. It consists of a number of annular-shaped hearths mounted above each other. There are rabble arms on each hearth driven from a common center shaft. The feed is charged at the periphery of the upper hearth. The arms move the charge inward to the center where it falls to the next hearth. Here it is moved to the periphery from which it falls to the next hearth. This continues down the furnace. The hollow center shaft is cooled internally by forced air circulation. Burners may be mounted at any of the hearths, and the circulated air is used for combustion. These furnaces handle granular materials and give a long countercurrent path between the flue gases and the charge material. Industrial sizes are built from 6 to 22 ft. in diameter and include 4 to 16 hearths. Total hearth areas range from 70 to over 3000 sq. ft. They are used for roasting ores; drying and calcining lime, magnesite, and carbonate

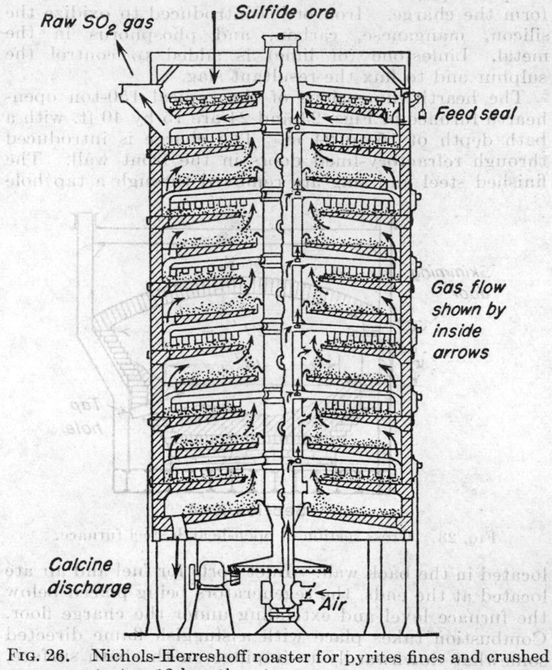

FIG. 26. Nichols-Herreshoff roaster for pyrites fines and crushed ore concentrates, 12 hearths.

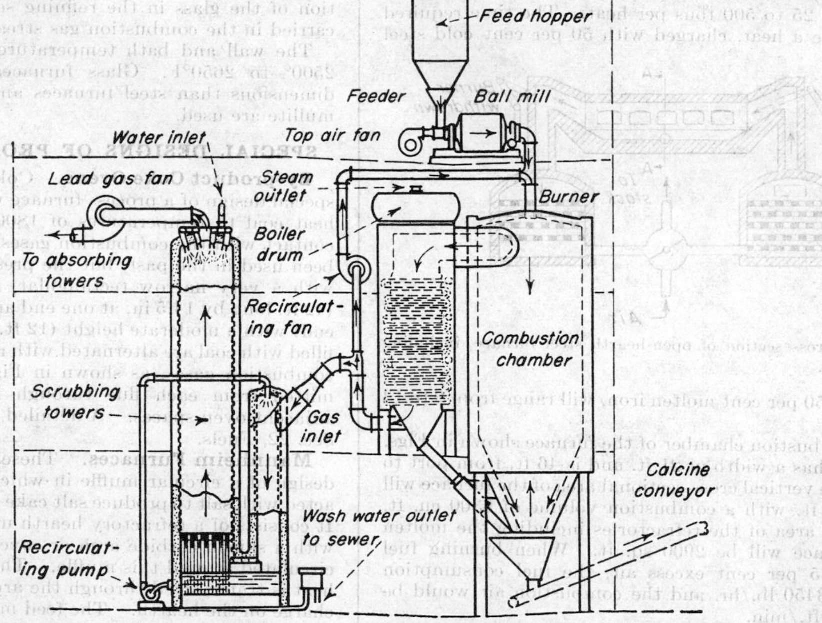

FIG. 27. Nichols-Freeman flash roasting system for making sulphur dioxide gas. *(Griswold, "Fluids, Combustion, and Furnaces," p. 425, McGraw-Hill, New York, 1946.)*

sludges; reactivation of decolorizing earths; and burning of sulphides to produce sulphur dioxide.

Flash Roasters. Flash roasters are a special type of fluidized-solid reactor and may be classified as a space burner. They are especially suitable for very fine particles such as sulphides of zinc, iron, copper, and nickel which react with oxygen in the air with the generation of heat. Sulphur dioxide and the metallic oxide will be formed. The sulphur dioxide may be recovered as such or be made into sulphuric acid. The use of oxygen instead of air gives a more concentrated sulphur dioxide gas, and one company is reported to be installing an oxygen plant. The metal oxide is recovered for the production of the metal. Figure 27 shows a system of this type. The fine charge particles are carried by a stream of air into the combustion chamber where burning occurs. A tangential take-off gives a swirling motion to the gases so that the larger particles settle in the hopper bottom of the combustion chamber. The hot gases are passed through waste-heat boilers where more solids settle and are collected. The gases are finally scrubbed and sent to sulphur dioxide recovery. Temperatures of about 1800°F. are maintained in the combustion chambers.

WASTE-HEAT RECOVERY

General. Combustion, in every fuel-fired furnace, generates thermal energy which appears initially as the sensible heat of the products of combustion. Since a furnace is, by definition, a heat "source" as distinguished from a thermal "sink," 100 per cent furnace efficiency is realized if (1) combustion is complete, (2) no heat is lost through the furnace walls, and (3) any incombustible residue from the fuel is removed from the furnace at 60°F. With liquid and gaseous fuels, in furnaces of large capacity, little difficulty is encountered in realizing 96 to 99 per cent efficiency in practice. Complete combustion of a fluid fuel is a matter merely of proper burner design. Reducing loss of heat by wall conduction can be attained by providing sufficient wall thickness. No material incombustible residue results from the combustion of liquids and gases. With solid fuels, loss of energy results from unconsumed carbon in the ash and in the unrecovered sensible heat of the ash when it is not discharged at atmospheric temperatures. Furnace efficiencies in excess of 90 per cent with solid fuels, however, are encountered.

If the definition of a "furnace" is extended to include the sink as well as the source, *i.e.*, if the structural elements provided for the absorption and utilization of the heat generated are included in the term furnace, serious loss of over-all efficiency will result when combustion products are discharged as incompletely cooled stack gases. In the case of liquid heaters and liquid evaporating furnaces, *e.g.*, pipe still or steam boiler, it is in theory possible, by providing sufficient tube area, to cool the stack gases to any desired approach to atmospheric temperature. In actual plant construction, of course, the cost of installing and maintaining an excessively large tube area imposes an economic limit on the extent to which this method of attaining furnace efficiency can be carried out. In usual design, it is found desirable arbitrarily to limit the fraction of heat transferred to the liquid-cooled elements, and to transfer the function of further heat utilization to some auxiliary element. In boiler practice, where insufficiently cooled exhaust heat is transferred by a recuperator to the combustion air for return to the furnace, the recuperator is termed a preheater. When exhaust heat is captured and returned as preheated feed water, the recuperator is termed an economizer. In almost every modern pipe-still or steam-boiler installation, some form of recuperative waste-heat recovery is conveniently employed. Since the temperature of the exhaust gases in furnaces of this type does not ordinarily exceed the maximum tolerance of structural metals, the refractory heat interchangers have not found extensive application. Wherever chemical and metallurgical furnaces discharge exhaust gases at somewhat higher temperatures, refractory, recuperative thermal interchangers appear essential.

Checkerbrick Recuperators. The earliest industrial example of the recapture of exhaust heat and its return to the furnace operation as preheated air for combustion was the application by Nielson of a recuperative stove to the iron blast furnace in Scotland in 1812. Difficulty immediately experienced in maintaining the cast-iron pipes in this hot-blast stove and the limitation in blast temperature, imposed by this type of recuperator, led to the development, again in Scotland, by Cowper in 1832, of the refractory regenerative, checkerbrick interchanger, *i.e.*, the present-day blast-furnace hot-blast stove. A century of rigid adherence to this original checkerbrick design has led to the identification of the terms, "blast-furnace stove," "checkerbrick interchanger," and "Cowper" as synonyms. Preheating combustion air in open-hearth furnaces, ingot-soaking pits, glass-melting tanks, by-product coke ovens, heat-treating furnaces, and the like have been universally carried out in regenerators constructed of fire-clay, chrome, or silica brick shapes. Although many geometric arrangements have been used in practice, the so-called basket-weave design shown in Fig. 28*a* and *b* has been adopted in many applications and is typical enough of current checkerbrick design.

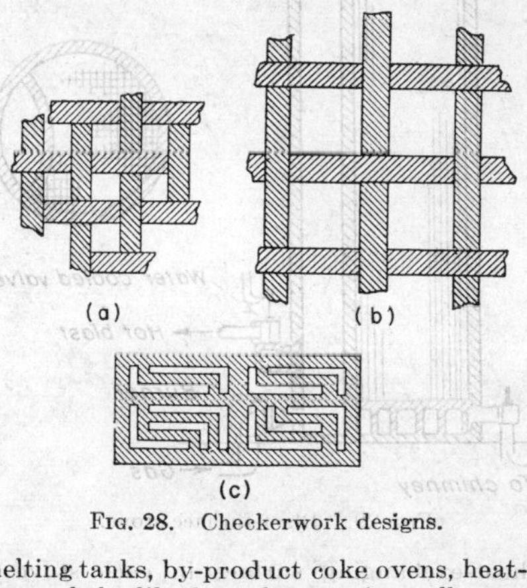

FIG. 28. Checkerwork designs.

In blast-furnace stove construction, standard 9 by 4.5 by 2.5 in. firebrick, assembled in basket-weave design, form square flues 3.25 by 3.25 in. (Fig. 28*a*). In open-hearth regenerators, 18 by 6 by 3 in. tiles form flues 7.5 by 7.5 in. (Fig. 28*b*). Special shapes have been devised for more complicated, if frequently less rugged, heat-absorbing elements, *e.g.*, coke-oven tiles (Fig. 28*c*). Standard firebrick are cheaper than special shapes, and this fact has tended to confine regenerator design to the readily available and less expensive standard refractories.

Blast-furnace Stoves. A modern blast furnace, producing 1650 tons pig iron/day, will be blown with 100,000 standard cu. ft./min. of atmospheric air, preheated to temperatures ranging in normal practice from 900° to

1200°F., with 1000°F. close to an average. To preheat this blast volume, a set of four stoves is usually provided. A vertical and horizontal section of one such stove is shown in Fig. 29. Each stove consists of a vertical steel cylinder 24 ft. in diameter, 110 ft. high, topped with a spherical dome. The shell has an 18-in. firebrick lining. A side combustion chamber is separated by a bridge wall with a lens-shaped horizontal cross section. The remaining volume is filled with heat-absorbing checkerwork having a horizontal cross section of 280 sq. ft. The height of the checkerwork is 94 ft. and its volume 26,400 cu. ft. The volume of the flues is 32 per cent of the total checker volume. This arrangement of a refractory forms 1220 flues, each 94 ft. long for a total flue length of 21.7 miles in each stove (86.8 miles of flue in the set of four stoves).

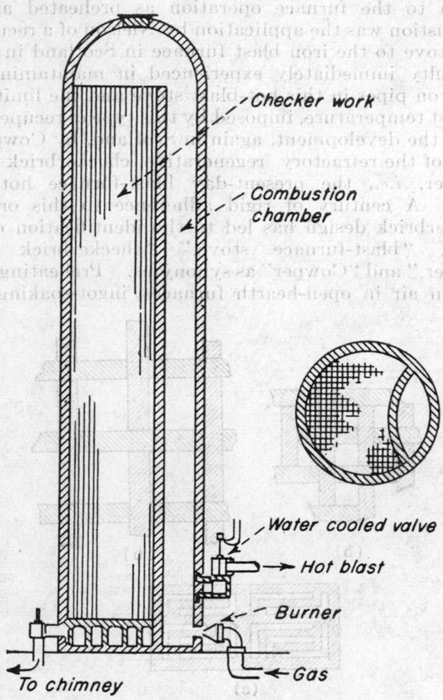

Fig. 29. Blast-furnace stove.

The heat-exchanging surface in each stove is just under 3 acres (124,000 sq. ft.), indicating 4.7 sq. ft./cu. ft. of checker volume. In operation, each stove is carried through a two-step, 4-hr. cycle. In one 3-hr. "on-gas" step, the checkers are heated by the combustion of blast-furnace gas. In the alternating "on-wind" 1-hr. step, the checkers are cooled by the passage of cold air through the stove. At any given time, three stoves are simultaneously on gas, while a single stove is on wind.

After 3 hr. of "on gas," an "on-wind" step is initiated. At the start, about one-half of the air, entering at 200°F. (blower discharge temperature at 15 psig.) passes through the checkers, the other half being by-passed around the stove through the cold-blast mixer valve. The gas passing through the stove exhausts initially at 2000°F. Mixing this with the unheated air produces a blast temperature of 1000°F. The temperature of the heated air from the stove falls rapidly, minute by minute, throughout the "on-wind" step. The fraction of total air volume by-passed through the mixer valve is continually decreased by progressive closing of this valve, its operation being automatically regulated under control of a thermocouple and potentiometer. At the end of 60 min. of "on-wind" operation, in usual practice, the cold-blast mixer valve is practically closed, the entire blast then passing through the checkers.

Satisfactory approach to uniform blast temperature can readily be realized by this automatic control of the mixer valve, provided that the uniform blast temperature does not greatly exceed one-half of the combustion temperature in the preceding on-gas step. The rapid decrease in temperature exhibited by the air discharging from the checkers is a characteristic feature of classical checkerwork heat transfer. The thickness of the refractory flue walls retards the flow of heat by thermal diffusion into the central portions of the brick. Although the heat removed in an "on-wind" step is less than 5 per cent of the total sensible heat stored in the stove refractories, the introduction and removal of heat from such large-dimensioned refractory elements is sluggish, with the result that the in-and-out movement of heat is largely a skin effect confined closely to the refractory surface.

Open-hearth and Glass-tank Regenerators. Because of the higher working temperatures, more drastic thermal shock, and dirtier gases encountered in open-hearth and glass-tank regenerators, checkerwork construction in these furnace units, while somewhat similar to that employed in blast-furnace stoves, requires considerable modification. The vertical height of the flues is limited by the elevation of the furnace above plant level. Short flues from 10 to 16 ft. are common in contrast to the 85 to 95 ft. flue lengths in blast-furnace stoves. Larger brick shapes (Fig. 28*b*) form flue cross sections five times as large as the stove flues, and the percentage of voids in the checkerwork is 51 per cent in contrast with the stove 32 per cent voids. In a typical open-hearth (Figs. 23 and 24), a checker volume of 7500 cu. ft. contains 810 flues having a heat-transfer area of 3.4 sq. ft./cu. ft., for a total of 25,000 sq. ft. for each of the two regenerators. The total flue length in the two regenerators is 3.7 miles, only 4.2 per cent of the flue length in the blast-furnace stoves, although the gas to be heated and cooled is 12.5 per cent of the air heated in its blast-furnace counterpart.

As a result of the larger dimensions of flue and the restricted surface per unit gas passed, regenerators employed with this type of reverberatory furnace exhibit much lower efficiency than would be realized with smaller flue dimensions. In view, however, of the large amount of iron oxide contained in the open-hearth exhaust gas and the alkali fume present in glass-tank stack gases, resort to smaller checker dimensions has appeared impractical.

Coke-oven Regenerators. In the by-product coke oven, waste-heat recovery is effected in the standard Siemens manner, although, as seen in Fig. 30, the dimensions of the upstream and downstream regenerators show little outward resemblance either to the blast-furnace stove or to the reverberatory-furnace regenerators. From structural necessity, the coke-oven regenerator is located under the oven itself and must assume the dimensions of an extremely narrow parallelepiped. Fortunately, the design problem is simplified because of the absence of fume and dust in the flue system. Special regenerator blocks are commonly employed, a typical design being shown in Fig. 28*c*. An oven 40 ft. by 12 ft. by 16 in., carbonizing 24 tons/day coal to produce 17 tons/day coke, will be provided with a pair of regenerators having a horizontal cross section of 70 sq. ft. containing 210 flues ("slots"), and an over-all volume of 300 cu. ft. Because fuels used in underfiring are either cleaned coke-oven gas, clean blast-furnace gas, or mixtures of the two, difficulty with dirt and fume accumulation in the flues is not encountered, and because of the lower working

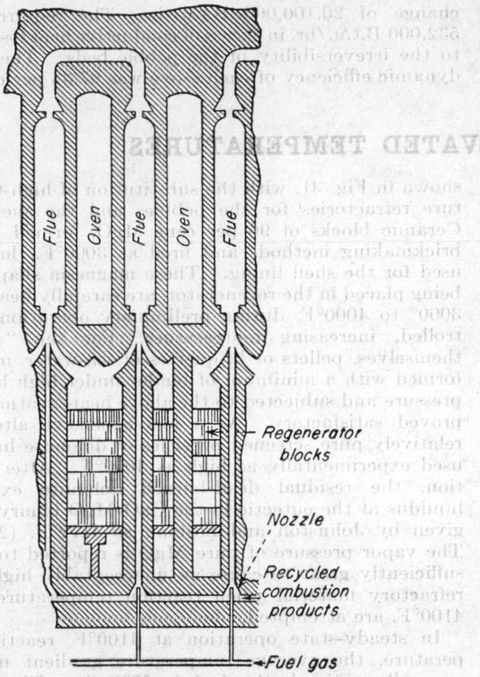

FIG. 30. By-product coke oven and regenerators.

temperatures in coking, this intricate type of flue has been found satisfactory. Attempts to duplicate coke-oven regenerator construction in other Siemens units have not been successful.

The Pebble Stove. Although considerable ingenuity has been applied to the design of the Cowper checkerwork, involving variations in bricklaying patterns, and special ceramic shapes, the checkerbrick regenerators still exhibit five engineering defects: (1) high initial cost of construction; (2) unsatisfactory thermal efficiency with clean heat-exchanging surfaces; (3) tendency to lose efficiency with dust- and fume-coated surfaces inevitable with dirty gas; (4) inaccessibility of surface causing lost time and high labor cost in any cleaning operation; and (5) danger of fusion, slagging, and spalling when subjected to high maximum temperatures and rapid temperature changes. It is possible that one or even all of these difficulties are inherent in the Cowper checkerbrick system, as such. No serious attempt, however, to alter 1832 construction appears to have been made prior to 1929, when the Department of Agriculture undertook to substitute the so-called "pebble bed" heat-exchanging structure for the classical checkerwork in order to provide higher air preheat in the blast-furnace smelting of phosphate rock.

In this departure from regenerator precedent, a mass of small refractory particles, enclosed in a brick-lined steel shell, was substituted as a functional equivalent of standard checkerwork (see Fig. 31). In the operation of an experimental phosphate blast furnace, blast temperatures as high as 2000°F. were readily obtained. In later operation of a 25 ton/day blast furnace producing pig iron, ferromanganese, ferrochromium, and ferrosilicon, air preheat temperatures of 2800°F. were attained and maintained. In connection with the process of converting air into NO [Daniels and Gilbert, *Ind. Eng. Chem.*, **40**, 1719 (1948)] air was preheated to 3600°F. in magnesia refractory pebble stoves.

Aside from the engineering value of the elevated temperatures attained with pebble-stove regenerators, extremely high thermal efficiencies are observed to be an inherent characteristic of this type of heat interchanger and indicate its importance in processes where gases are subjected to a specified and restricted time-temperature history.

THERMAL TREATMENT OF UNSTABLE GASES

Usually a furnace is interposed between two regenerators, but with the regenerators exhibiting efficiencies in excess of 95 per cent, the central furnace itself frequently becomes reduced in function and may be eliminated in

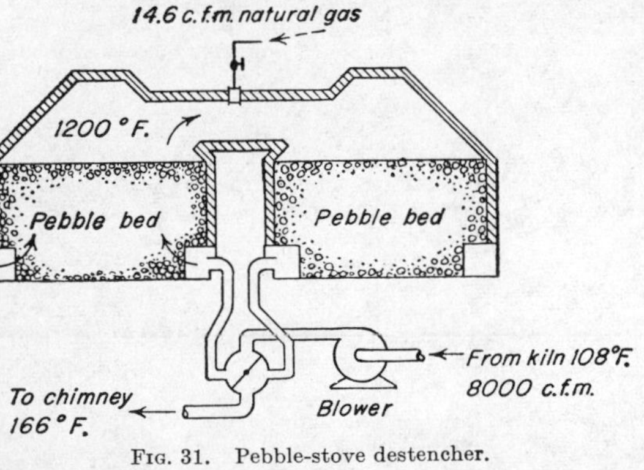

FIG. 31. Pebble-stove destencher.

practice, leaving as a structural residue merely two regenerators in series. When a gas is passed through a pair of pebble-type regenerators connected back to back, it is heated in the upstream regenerator and is immediately cooled in the downstream regenerator. This back-to-back (Fig. 31) arrangement permits a thermally unstable gas to be subjected to a controlled time-temperature history of any selected character, with the result that any required degree of thermal decomposition in the gas can be effected. Examples of such decomposition are encountered in the vapor-phase cracking of petroleum and other hydrocarbons.

Decomposition of thermally unstable gas in its simplest industrial application may be illustrated by the destenching of chimney gases from a sewage-disposal kiln. Unpleasant odors produced in the gas heating of the kiln caused a serious neighborhood nuisance, although the concentration of the noxious constituents was small and not identified. Heating the chimney gas to 1200°F., however, was found to remove all detectable odor.

The destencher proposed by Cottrell and developed by the Research Corp., New York, consists of two regenerators, 15 ft. in diameter, connected in series as shown in Fig. 31. Each stove contains 1230 cu. ft. of 1.2-in. sandstone pebbles having a surface in each bed of 44,700 sq. ft.

If the thermodynamic reversibility of the beds were complete, and if the stoves were adiabatic, no fuel whatever would be required to heat the chimney gas from 108° to 1200°F. and recool, since the sensible heat recovered in the downstream stove in 1 hr. would be returned to the incoming gas during the subsequent 1 hr. of the stove's operation in the upstream position.

Gas consumption over a period of years corresponded to a thermal input of 925,000 B.t.u./hr., a figure 532,000

B.t.u./hr. greater than the conductive heat losses. In passage through the two stoves in series, the heat exchange in the upstream stove was 10,800,000 and 9,300,000 B.t.u./hr. in the downstream stove, a total heat interchange of 20,100,000 B.t.u./hr. The requirement of 532,000 B.t.u./hr. in excess of conductive heat losses is due to the irreversibility of the pebble beds. The thermodynamic efficiency of each stove was 97.34 per cent.

GAS REACTIONS AT ELEVATED TEMPERATURES

When a reactive gas is forced through two regenerators in series, with reversal of gas flow in the Siemens manner, the gas may be heated to reaction temperature and promptly cooled with almost complete recovery of sensible heat. This type of heat-treatment of a reactive gas is of technical value in many industrial processes involving gas reactions, both in synthesis and in thermal decomposition. An example of such an operation is the fixation of nitrogen according to the equation

$$N_2 + O_2 \rightleftharpoons 2NO$$

When air is heated to 4000°F. and above, equilibrium concentrations of NO of more than 2 per cent result. At such elevated temperatures, the rate of NO formation is rapid, and the gas need not remain at reaction temperature for more than a fraction of a second to attain equilibrium. As the gas is cooled, however, decomposition of NO is also rapid. Unless the cooling rate is quite fast ("shock quenching"), no NO remains in the cooled gas. In heating and recooling a gas by passage through pebble beds connected back to back, it is possible not only to recover the greater part of the sensible heat of the reacted gas but to carry out the cooling at extremely rapid rates.

Acceptable operating costs in this process are made possible because (1) the construction cost of pebble-type regenerators, per unit volume of gas, is low; (2) regenerator reversibility is high; and (3) quenching rates in the downstream regenerator may be made rapid enough to prevent excessive amounts of NO dissociating during cooling. Technical problems involved in this Cottrell process are concerned only with providing refractories capable of operating at the required reaction temperatures.

Regenerators for NO formation are similar to those shown in Fig. 31, with the substitution of high-temperature refractories for the pebbles and the shell lining. Ceramic blocks of 96 per cent MgO, formed by usual brickmaking methods and fired at 3000°F., have been used for the shell lining. These magnesia shapes, after being placed in the regenerator, are carefully heated from 3000° to 4000°F. during preliminary operation at controlled, increasing temperature. For the "pebbles" themselves, pellets or briquettes of sea-water magnesia, formed with a minimum of binder under high hydraulic pressure and subjected to the above heat-treatment have proved satisfactory. As a less expensive alternative, relatively pure screened particles of dolomite have been used experimentally as high as 4000°F. After calcination, the residual dead-burned dolomite exhibits a liquidus at the eutectic in the CaO-MgO binary system, given by Johnston and Sosman as 4172°F. (2300°C.). The vapor pressure of pure MgO is reported to become sufficiently great to cause an unacceptably high loss of refractory material when reaction temperatures above 4100°F. are attempted.

In steady-state operation at 4100°F. reaction temperature, the average temperature gradient measured upwardly through the bed is 67°F./in. The average velocity of gas flow through the interstitial voids, at 4100°F. and 1.9 psig. pressure is 800 in./sec., resulting in a cooling rate of 53,600°F./sec. (67 × 800). Although experiments have failed to determine the rate of decomposition of NO, at temperatures between 3000° and 4100°F., it has been found that with a gas-quenching rate of 53,600°F./sec., less than a 0.2 per cent NO loss is experienced in gas passage through the downstream regenerator. This result is sufficient to indicate that no more than 1 per cent of the NO formed is decomposed in 0.001 sec. at 4100°F.

SECTION 24

POWER GENERATION AND MECHANICAL POWER TRANSMISSION

BY

Theodore Baumeister, B. S., M. E., Stevens Professor of Mechanical Engineering, Columbia University; Consulting Engineer, American Gas & Electric Service Corp., and E. I. du Pont de Nemours & Co. (Power and Power Machinery)

William Staniar, M. E., Consulting engineer. (Mechanical Power Transmission)

CONTENTS

POWER AND POWER MACHINERY

	PAGE
Systems and Sources of Power	1628
Loads and Load Curves	1630
Costs of Service	1631
Investment Costs	1631
Production Costs	1631
Steam Plants	1633
Cycles and Performance	1633
Fuel- and Ash-handling Systems	1637
Coal Handling	1637
Oil Handling	1637
Ash Handling	1637
Boilers	1637
Furnaces	1639
Gas Firing	1640
Oil Firing	1640
Coal Firing	1640
Furnace Walls	1641
Waste Fuels	1641
Waste Heat	1641
Superheaters	1641
Economizers and Air Preheaters	1642
Draft Systems	1642
Boiler Accessories	1643
Steam Prime Movers	1644
Steam Engines	1644
Steam Turbines	1645
Condensers and Auxiliaries	1647
Feed-water Heaters	1647
Boiler Feed Pumps	1647
Feed-water Treatment	1647
Piping	1649
Instruments	1652
Internal-combustion Plants	1652
Water Power Plants	1655
Plant Operation and Scheduling of Units	1657
Heat Pumps	1657

MECHANICAL POWER TRANSMISSION

	PAGE
Belting	1660
Leather Belting	1660
Rubber Belting	1662
Stitched Canvas Belting	1663
Hair Belting	1663
Belt Joining	1663
Belt Drives	1664
Pulleys	1665
Cast-iron Pulleys	1665
Steel Pulleys	1665
Wooden Pulleys	1665
Gearing	1666
Bearings	1666
Shafting	1667
Couplings	1667
Gear Reduction Units	1669
Variable-speed Drives	1670
Short-center Drives	1670
Clutches	1671
Chain Driving	1672
Lubrication of Power-transmission Equipment	1673
High-starting Torque Control	1673

POWER AND POWER MACHINERY

BY THEODORE BAUMEISTER

REFERENCES: *General:* Kent, "Mechanical Engineer's Handbook," "Power," Wiley, New York, 1936. Marks, "Mechanical Engineers' Handbook," 4th ed., McGraw-Hill, New York, 1941. Justin and Mervine, "Power Supply Economics," Wiley, 1934. Myers, "Reducing Industrial Power Costs," McGraw-Hill, 1935. Faires, "Applied Thermodynamics," Macmillan, New York, 1948. Steinberg and Smith, "Economy Loading of Power Plants," Wiley, 1943.

Fuels: Haslam and Russell, "Fuels and Their Combustion," McGraw-Hill, New York, 1925. Spiers, Technical Data on Fuel, British World Power Conference, 1937.

Steam plants: Barnard, Ellenwood, and Hirshfeld, "Heat Power Engineering," Wiley, New York, 1933. Gaffert, "Steam Power Stations," 3d ed., McGraw-Hill, New York, 1946. Skrotzki and Vopat, "Applied Energy Conversion," McGraw-Hill, New York, 1945.

Steam plant equipment: Am. Soc. Mech. Engrs., Boiler Construction Code, 1948. Baumeister, "Fans," McGraw-Hill, New York, 1935. Betz and Betz, "Handbook of Industrial Water Conditioning," Philadelphia, 1945. Keenan and Keyes, "Thermodynamic Properties of Steam," Wiley, New York, 1936. Kristal and Annett, "Pumps," McGraw-Hill, New York, 1940. McAdams, "Heat Transmission," 2d ed., McGraw-Hill, New York, 1942. Newman, "Modern Turbines," Wiley, New York, 1944. de Lorenzi, "Combustion Engineering," Combustion Engr. Co., New York, 1947. Crocker, "Piping Handbook," 4th ed., McGraw-Hill, New York, 1945.

Water power plants: Addison, "Applied Hydraulics," Wiley, New York, 1944. Barrows, "Water Power Engineering," 3d ed., McGraw-Hill, New York, 1943. Creager and Justin, "Hydroelectric Handbook," Wiley, New York, 1927.

Internal-combustion plants: Am. Soc. Mech. Engrs., Annual Reports on Oil Engine Power Costs. Boyer, "Diesel and Gas-engine Power Plants," McGraw-Hill, New York, 1943. Jennings and Obert, "Internal Combustion Engines," International Textbook, Scranton, Pa., 1944. Standard Practices, Diesel Engine Manufacturers Association, 1935.

Heat pumps: Sporn, Ambrose, and Baumeister, "Heat Pumps," Wiley. Sporn, Baumeister, and Ambrose, Industrial Applications of the Heat Pump, *Chem. & Met. Eng.*, June, 1946. Latham, Compression Distillation, *Mech. Eng.*, March, 1946. Baumann and Marples, Some Applications of the Heat Pump as a Heating Machine, *Brown Boveri Rev.*, July–August, 1943. Economy of Concentrating Plants, With and Without Heat Pumps, *Sulzer Tech. Rev. (Switz.)*, 1945. Peter, Heat Pumps for Evaporation, *Escher Wyss Bul.* 25007 (e). Faber, The Value of Heat with Special Reference to the Heat Pump, *Proc. Inst. Mech. Engrs. (London)*, 1946. Ambrose, Kemler, Smith, and Holladay, Progress Report on the Heat Pump, *Heating & Ventilating*, Dec., 1946.

Miscellaneous: Power Test Codes, Am. Soc. Mech. Engrs. Standard Specifications, Am. Soc. Testing Materials. Standards of the Heat Transfer Institute. Standards of the Hydraulic Institute.

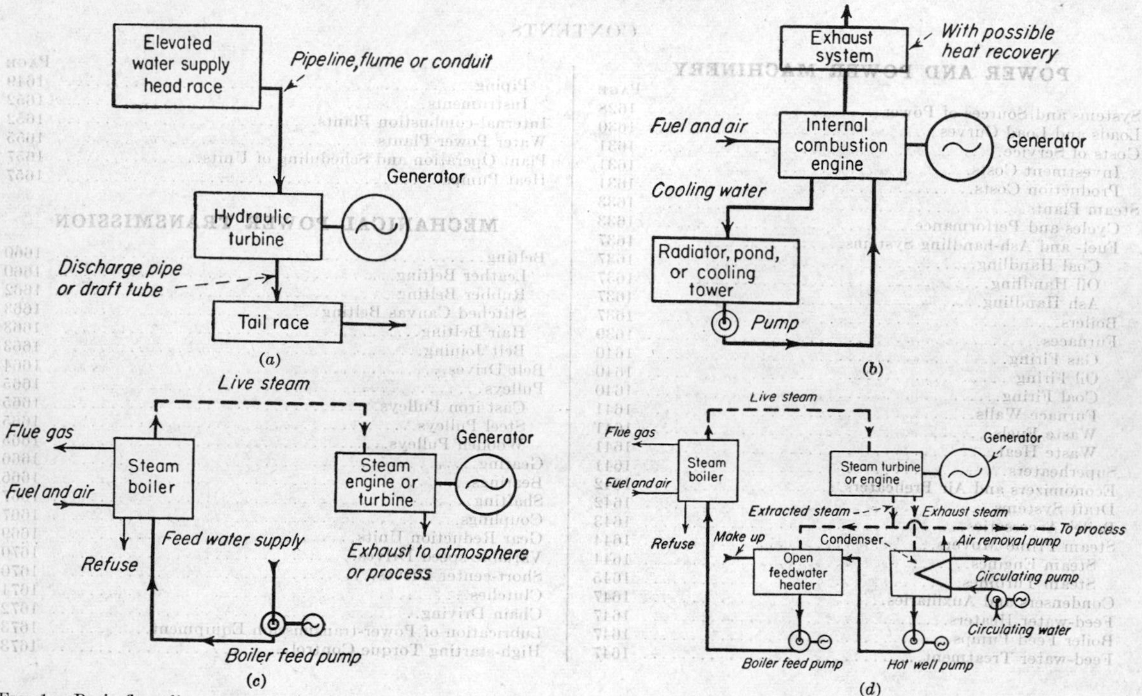

FIG. 1. Basic flow diagrams of power-plant cycles. (*a*) Hydroelectric. (*b*) Internal-combustion. (*c*) Non-condensing steam. (*d*) By-product, condensing steam.

SYSTEMS AND SOURCES OF POWER

The power plant in the chemical industry must provide a variety of services, *i.e.*, (1) mechanical or electric power; (2) process heat or steam; (3) drinking, service, and process water; (4) refrigeration; (5) compressed air and gases; and (6) heating, ventilation, and air conditioning. The power plant is thus multipurpose, and many of the services are delivered as a "by-product" of the others. The greater the number of services to be provided, the greater is the number of alternate power cycles and systems that can be projected.

Table 1. Allocation of Prime Mover Capacity in the United States*

	10^6 Hp.
Electric central stations	45
Industrial power plants	20
Electric railways	3
Isolated non-industrial	2
Mines and quarries	3
Agricultural prime movers	73
Automotive	965
Airplanes	4
Locomotives	88
Marine	30
Total	1233

* Derived from National Resources Board, 1936.

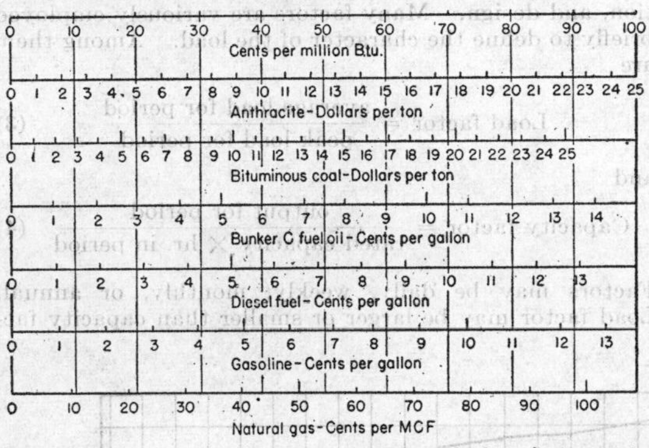

FIG. 2. Alignment chart: comparative fuel prices, cents/million B.t.u.

Commercial power plants start with raw energy in the form of (1) elevated water supply or (2) heat of combustion of fuel. Basic elementary flow sheets of hydroelectric, steam, and internal-combustion plants are shown in Fig. 1. Fuel-burning plants are of greater commercial significance than hydroelectric plants (see Table 1). In the former, fuel constitutes the largest item of operating expense. Representative fuel analyses and costs are shown in Tables 2 and 3 and in Fig. 2. Further details on fuels are to be found in Sec. 22. The cents per million B.t.u. basis is best for direct comparison of fuel costs. Such prices can be applied to thermal performance to determine the fuel production cost. Thus the heat rate on a power plant is defined as

$$\text{Heat rate, B.t.u./kw.-hr.} = \frac{\text{heat supplied in fuel, B.t.u.}}{\text{energy generated, kw.-hr.}} \quad (1)$$

and

$$\text{Fuel production cost, cents/kw.-hr.} = \frac{\left(\begin{array}{c}\text{heat rate,}\\ \text{B.t.u./kw.-hr.}\end{array}\right)\left(\begin{array}{c}\text{fuel price,}\\ \text{cents/million B.t.u.}\end{array}\right)}{10^6} \quad (2)$$

Table 3. Weight and Space Requirements of Some Typical Fuels

Fuel	Density, lb./ft.³	High heating value, B.t.u./lb. or/ft.³	Weight/million B.t.u., lb.	Space/million B.t.u., ft.³
Bituminous coal	52	14,000	71	1.4
Anthracite	57	12,500	80	1.4
Coke	30	12,500	80	2.7
Bunker C fuel oil	60	18,500	54	0.9
Diesel fuel	55	19,500	51	0.93
Gasoline	45	20,000	50	1.1
Natural gas	0.04 N.T.P.	1,000	36	910 N.T.P.
Manufactured, city, gas	0.04 N.T.P.	550	72	1820 N.T.P.

Table 2. Analyses of Some Typical Fuels

Coals—Gravimetric Basis, as Fired*

Fuel	Volatile matter	Fixed carbon	Moisture	Ash	H_2	C	$O_2 + N_2 + S$	High heating value, B.t.u./lb.	Chemically necessary air, lb./lb. fuel
Lignite	17	35	39	9	2	36	14	6,500	4.8
Subbituminous	31	38	22	9	3	54	12	8,800	7.2
Low-rank bituminous	37	42	11	10	4	65	10	11,600	8.8
Middle-rank bituminous	37	50	5	8	4	75	8	12,800	10.0
High-rank bituminous	29	60	3	7	5	77	8	14,000	10.6
Low-rank semibituminous	22	59	3	6	5	79	7	14,600	10.8
High-rank semibituminous	13	76	5	6	4	79	5	14,500	10.5
Semianthracite	9	77	6	8	3	79	3	13,700	10.2
Anthracite	2	85	3	10	2	82	3	13,000	10.2
Coke	..	..	..	13		86	1	12,500	10.0

* Many commercial coals as found in the present markets are poorer than these by 1000 to 2000 B.t.u./lb., because of higher moisture and ash.

Liquids—Gravimetric Basis, as Fired

Fuel	H_2	C	$O_2 + N_2 + S$	High heating value, B.t.u./lb.	Sp. gr., deg. A.P.I.	High heating value, B.t.u./gal.	Chemically necessary air, lb./lb. fuel
Bunker C	11.5	85	3.5	18,500	10–18	150,000	13.7
Diesel fuel	13	85	2.0	19,200	22–28	145,000	14.3
Furnace oil	13.5	85	1.5	19,500	30–35	140,000	14.4
Kerosene	14	85	1.0	19,800	45	133,000	14.6
Gasoline	15	85	0	20,300	55	130,000	14.9

Gases—Volumetric Basis

Fuel	H_2	CO	CH_4	Heavy hydrocarbons	$O_2 + N_2 + CO_2$	High heating value, B.t.u./ft.³ N.T.P.	Sp. gr. (air = 1)	Chemically necessary air, ft.³/ft.³ gas
Natural gas	..	..	96	2	3	1100	0.6	10
Coke oven gas	50	6	32	5	7	600	0.4	5.2
Producer gas	15	26	2	1	56	150	0.85	1.2
Blast-furnace gas	5	23			72	90	1.0	0.67

The heat rates of several representative power plant services are given in Table 4. Among the fuel-burning systems the internal-combustion unit gives minimum heat rate, 11,000 B.t.u./kw.-hr., with the smallest sizes (less than 2000 kw.). The best steam plant heat rates obtain only with the largest units (200,000 kw. maximum); high pressure and temperature (2300 lb./sq. in.

Table 4. Typical Over-all Thermal Performance of Fuel-burning Power Plants

Type of plant	Plant heat rate, B.t.u./kw.-hr.	Plant thermal efficiency
All stationary steam plants, average	25,000	0.14
Central station steam plants, average	16,000	.21
Best record, large central station steam plant	10,100	.34
Small non-condensing industrial steam plant	35,000	.10
Small condensing industrial steam plant	20,000	.17
"By-product" power steam plant	4,500–5,000	.75
Diesel plant	11,500	.30
Natural gas engine plant	14,000	.24
Gasoline engine plant	16,000	.21
Producer gas engine plant	18,000	.19

and 1050°F. maximum); condensing (1 in. Hg abs.); multistage extraction feed heating (eight stages maximum). The absolute minimum of heat rates will obtain with multipurpose plants in which all exhaust heat is charged to process. This would be the mechanical equivalent of heat (3413 B.t.u./kw.-hr.) divided by generator efficiency (0.9+), the boiler efficiency (0.8+), auxiliary power allowance (0.9+), and realization ratio (0.9+), or an over-all heat rate for power generation of 4500 to 5000 B.t.u./kw.-hr.

LOADS AND LOAD CURVES

Power must be generated at the instant of its use, because there is no practical way of storing, in appreciable amounts, mechanical energy, electrical energy, steam, heat, or compressed air. Proper design and proper operation can be effected only if loads and their fluctuations, as to nature and magnitude, are accurately known. Graphical presentation of load data, on the daily, monthly, seasonal, or annual basis prevails. Integrated readings over a period of 15 min., 30 min., or 1 hr. are employed rather than instantaneous readings. Flywheel and storage effects must handle momentary fluctuations. Some representative load curves are given in Figs. 3 and 4. Figures 3*b* and 4*b* show the load duration curves as calculated from Figs. 3*a* and 4*a* by integration of the area under the load curve. These curves, together with energy-time and energy-utilization curves, are all useful in studies involving plant loading, operation, and design. Many factors are variously employed briefly to define the character of the load. Among them are

$$\text{Load factor} = \frac{\text{average load for period}}{\text{peak load for period}} \tag{3}$$

and

$$\text{Capacity factor} = \frac{\text{output for period}}{\text{rated capacity} \times \text{hr. in period}} \tag{4}$$

Factors may be daily, weekly, monthly, or annual. Load factor may be larger or smaller than capacity fac-

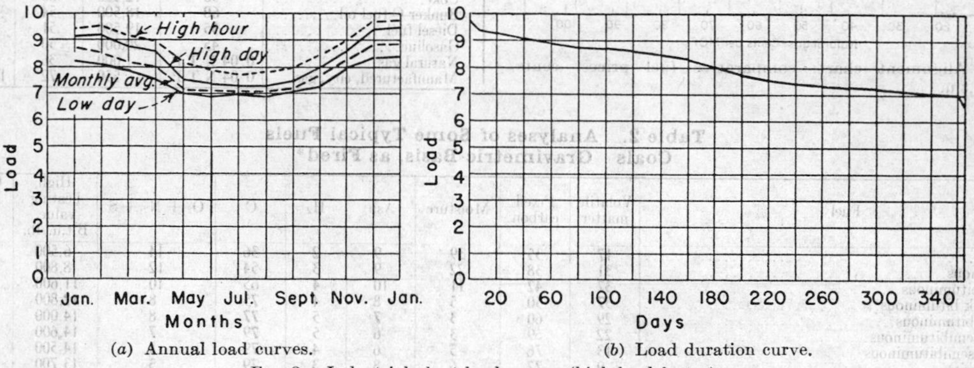

(*a*) Annual load curves. (*b*) Load duration curve.

FIG. 3. Industrial-plant load curves (high load factor).

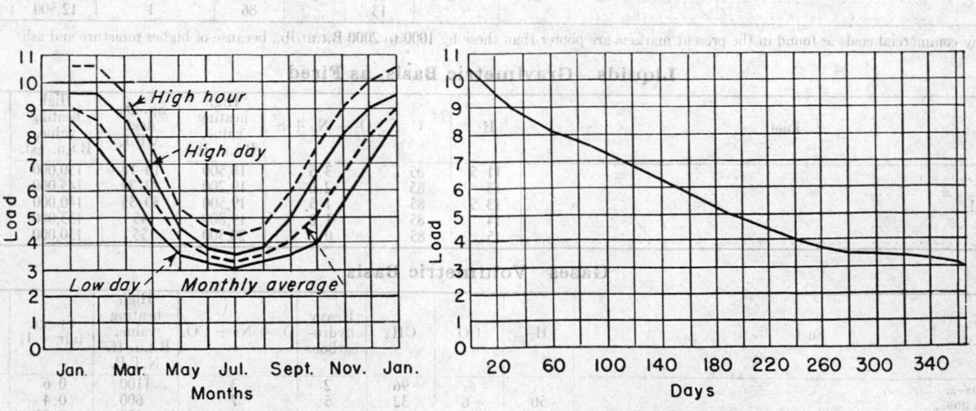

(*a*) Annual load curves. (*b*) Load duration curve.

FIG. 4. Industrial-plant load curves (low load factor).

tor. Each can be applied to any service offered by, or any equipment included in, a power plant.

In the design of an industrial plant typically to deliver electric power and process steam, the load curves for each of the services must be considered separately and in combination. If the power is to be generated (see Fig. 1*d*), as a by-product of the process steam, using an arrangement in which the boiler delivers steam at high pressure to a turbine or engine, with exhaust delivered to the process header, the demand for one service will limit the availability of the other. Consider, as in Fig. 5, *a*, the electric-demand, and *b*, the process-steam-demand curves. If all the power is to be generated as a by-product of the process steam, the available electric generation will be as shown in curve *c*. A comparison of this curve with the kilowatt-demand curve *a* discloses periods when there is a deficiency of electric energy and others when there is an excess. Deficiencies can be met only by purchase of electric power or by use of condensing steam equipment or hydroelectric or internal-combustion

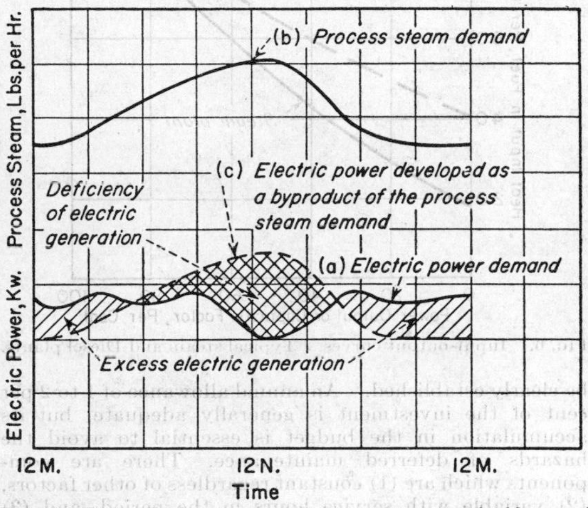

FIG. 5. Daily load curves to demonstrate non-coincidence of process steam and electric-power load demands.

engines. Excess electric generation entails some disposal to some user or the installation of a reducing valve to by-pass the prime mover. The simple, thermodynamically desirable, by-product power plant thus becomes a complicated reality. The non-coincidence of the power and process heat demands requires some power to be generated at a rate less favorable than 4500 B.t.u./kw.-hr.

The loads and load-duration curves assist in fixing the propriety of equipment sizes, operating schedules, and maintenance programs. The peaks, troughs, averages, and non-coincidences of the loads all have their influence. Spare power equipment is ordinarily omitted in modern industrial practice. Availability factors, defined as

$$\text{Availability factor} = \frac{\text{hr. operated or operable}}{\text{total hr. in period}} \quad (5)$$

are high, being in excess of 90 or 95 per cent for fuel-burning systems and in excess of 99 per cent for hydroelectric equipment. Scheduled outage can be effected during lower load periods if equipment sizes are correct. The severity of forced, or breakdown, outages can likewise be minimized. Larger sizes favor lower unit investment and labor costs. The fuel economy of steam equipment improves with larger size. Internal-combustion and hydroelectric plant efficiencies, however, are not much affected by size. Thus, considering all factors, the best sizes for boilers, turbines, and auxiliaries would not be the same for two different load types, illustrated by Figs. 3 and 4, even though the peaks and troughs might be equal. One of the most practical and economical ways of meeting loads with wide variations is to use a multiplicity of smaller units rather than to have wide range governing at sustained efficiency on a larger unit.

COSTS OF SERVICE

The range of power plant **investment** is illustrated by the data of Table 5. These reflect not only the price paid for materials and labor but an additional 10 to 50 per cent for land, engineering, drafting, purchasing, inspection, legal services, construction equipment and services, contractors' fee, interest during construction, and preliminary operation. Labor approximates one-fourth to one-third of the total investment in a power plant.

Table 5. Range of Power Plant Investments

Type of Plant	
Condensing steam power plant	\$100–\$200 per kw.-hr.
Diesel engine power plant	\$125–\$250 per kw.-hr.
Hydroelectric power plant	\$125–\$500 per kw.-hr.
Non-condensing steam power plant	\$125–\$300 per kw.-hr.
Steam boiler plant	\$3–\$8 per lb./hr. steam-making capacity
Refrigeration plant	\$500–\$1500 per ton
Manufactured gas plant	\$2–\$6 per ft.³/hr. gas-making capacity
Service water, pumping plant	\$10–\$30 per gal./min. capacity
Potable water supply plant	\$200–\$500 per gal./min. capacity
Compressed air plant	\$20–\$50 per cu. ft./min. capacity

Costs may be expressed on (1) the unit basis, *e.g.*, mills per kilowatt-hour, or cents per thousand pounds of steam, or (2) the total basis, *e.g.*, dollars per month or year. The cost of service at the point of use is all-important. It must include those items generally classified as (1) a fixed group, which is a function of investment, and (2) an operating or production group, which is a function of output. The **fixed charge** group, expressed as an annual percentage of the investment, embraces (1) cost of money (3 to 10 per cent), (2) insurance and taxes (1 to 4 per cent), and (3) depreciation (2 to 10 per cent).

Cost of money reflects bond interest, mortgage and underwriting discounts, and dividends on stock. Insurance covers such matters as fire protection, public liability, boiler explosion, and property damage. Taxes include real estate, franchise, business, and workmen's compensation. Depreciation is both physical and functional and is made up of allowances for wear and tear, obsolescence, change in use, and inadequacy. Depreciation may be computed on the straight line, the sinking fund, the secondhand, the retirement, or the observed basis. Statistical data, as in Table 6, on useful life are needed for these calculations. If the enterprise is short-lived, depletion of resources or current income may be substituted for depreciation allowances. The aggregate of all fixed charges runs typically 10 per cent for hydroelectric, 12 per cent for steam, and 13 per cent for internal-combustion plants.

Production or **operating costs** include charges for (1) fuel, (2) operating labor and supervision, (3) maintenance materials and labor, and (4) supplies and expense. All these costs have (1) a constant component, which is independent of load or output, and (2) a variable component, which is a function of output. The constant component will depend upon the equipment installed and upon the equipment in service. **Fuel charge** is based on the price at the power plant. In the case of coal, this is the mine price, plus a terminal charge plus a haulage charge. Efficiency of fuel-burning and heat power equipment varies with load, and the unit production costs for fuel will therefore vary similarly. There is one load on any fuel-burning system where the efficiency will be a maximum and the fuel cost a minimum. The input-output curves are particularly useful for showing the over-all fuel performance (see Fig. 6). These are lines that are essentially straight or concave upward. They show an intercept typically (1) on steam plants of 10 to 20 per cent and (2) on internal-combustion plants of 20 to 40 per cent, of the full load heat input.

Table 6. Probable Life of Structures and Apparatus

Steam Power Plants

Item	Years
Accumulators	15
Boilers, fire tube	15
Boilers, water tube	20
Boiler accessories	20
Breechings, steel	20
Buildings, steel frame, brick	40
Wood frame	25
Cables and feeders	25
Coal and ash handling	20
Compressors, air	20
Condensers, steam	20
Cranes	30
Economizers and air heaters	15
Electric generators	25
Electric motors	25
Engines, small steam	20
Engines, large steam	30
Evaporators	20
Fans and blowers	20
Feed-water heaters	20
Foundations	Same as equipment supported
Fuel-burning equipment	20
Fuel-oil-handling systems	20
Furniture and fixtures	15
Pipe and pipe covering	20
Pumps, centrifugal and propeller	20
Pumps, direct-acting	25
Stacks, brick- or concrete-lined	30
Steam turbines	20
Superheaters	20
Switchboards	20
Tools and shop machinery	15
Transformers	15

Hydroelectric Power Plants

Item	Years
Concrete structures	75
Cranes and hoists	60
Dams, concrete	75
Dams, earth	75
Dams, timber	10
Flumes, concrete	20
Flumes, steel	30
Flumes, wood	10
Gates	50
Generators	50
Miscellaneous powerhouse equipment	25
Pipe lines, concrete	40
Pipe lines, steel	20
Pipe lines, wood	20
Powerhouse substructure	75
Powerhouse superstructure	50
Screens and racks	50
Steel structures, exposed	40
Transformers	20
Waterwheels	50

Internal-combustion Power Plants

Item	Years
Automobiles	5
Batteries, storage, stationary	10
Batteries, storage, automobile	2
Cooling towers	15
Compressors, portable	5
Compressors, stationary	20
Diesel engines, slow-speed	20
Diesel engines, high-speed	10
Fuel-oil systems	20
Gas engines	20
Gas producers	15
Gas supply systems	20
Gas turbines	5
Gasoline engines, automotive	5
Heat exchangers	20
Tanks, steel	25

Operating labor costs are best estimated by development of the pay roll for the project at hand. Some average unit prices, like $1.25 per man-hour, may conveniently be employed. Labor varies somewhat with load, but ordinarily it is unwise to reduce or expand the staff with plant output. Good economy requires the continuance of operating skill and experience. This is particularly true for steam plants where the hiring of good firemen is always justified. Some variation of labor cost with output is reasonable, and from one-half to three-fourths of the total labor cost may be considered constant.

Maintenance materials and labor constitute a most elusive element of cost. Standards and schedules must be clearly established. An annual allowance of 1 to 2 per cent of the investment is generally adequate, but its accumulation in the budget is essential to avoid the hazards of deferred maintenance. There are components which are (1) constant regardless of other factors, (2) variable with service hours in the period, and (3) variable with output for the period. As an approximate rule, from one-half to three-fourths of the cost should be considered constant with the remainder variable. The cost of maintenance usually aggregates 50 per cent for materials and 50 per cent for labor, even though the tendency is toward a higher proportion for labor.

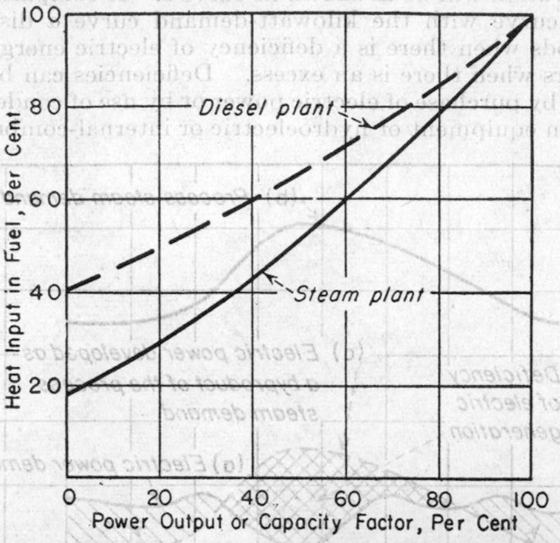

FIG. 6. Input-output curves. Typical steam and Diesel plants.

Supplies and expense cover numerous miscellaneous minor items such as lubricants, water, waste, and chemicals, many of which become large with poor maintenance. They are higher on internal-combustion plants than on steam plants. They should be allocated similarly to maintenenace and with steam plants may run from 10 to 50 per cent of the maintenance costs.

Total cost of service, for a selected steam plant, is shown in Fig. 7 as a function of capacity factor. The data are all given as straight lines, which is not rigorously correct for actual plants. With a multiplicity of units in the plant, the curves would be stepped whenever another unit is put into service. The fuel line would also be curved upward because of varying efficiency. Costs for any product of a power plant, whether electric, steam, water, air, or refrigeration, would follow an equation that reflects a demand charge (constant portion) and an energy charge (variable portion). These components may variously be subdivided to reflect capacity, peak prepared for, readiness to serve, energy delivered, and service hours charges. Rate schedules are based on such considerations. The following are some typical cost equations:

Wholesale power, \$/year/kw.

$$= \$15.00 + 0.25 \text{ per month} + 0.003 \times 8760 \times \text{capacity factor} \quad (6)$$

where the terms on the right are capacity, peak prepared for, and energy charges, respectively.

Live steam, \$/year

$$= \$1.00 \times \text{max. demand, lb./hr.,} + 60¢ \times \text{steam used million lb.} \quad (7)$$

Cost equations and data like Fig. 7 can be resolved into unit cost figures where Tables 7*a* and 7*b* are pertinent.

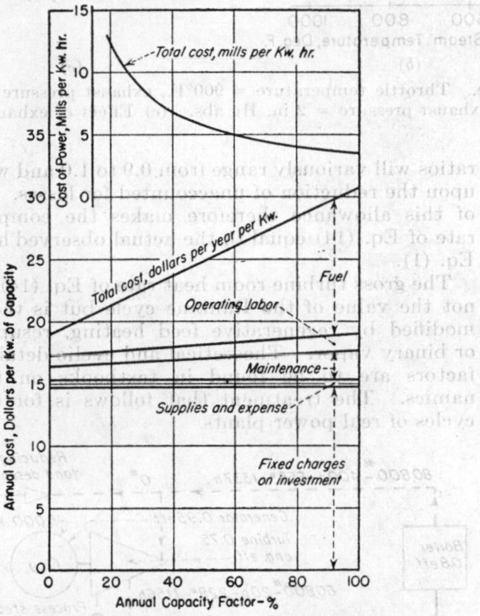

FIG. 7. Annual cost of generation, typical large steam plant.

Table 7a. Selected Power Plant Production and Fixed Costs

Plant	A	B	C	D
Plant type	Steam central station	Steam industrial	Diesel	Hydro
Plant capacity, kw	100,000	5,000	3,000	100,000
Capacity factor, %	70	50	30	65
Load factor, %	75	60	40	75
Annual energy output, 10⁶ kw.-hr.	614	21.9	7.9	570
Investment, \$/kw	123	135	160	220
Fuel cost, ¢/million B.t.u.	12	20	35	
Plant heat rate, B.t.u./kw.-hr.	12,000	25,000	13,000	
Fixed charges:				
Annual %	12	12	13	10
Mills/kw.-hr.	2.41	3.7	7.9	3.87
Production costs:				
Fuel, mills/kw.-hr.	1.44	5.0	4.5	
Labor and supervision, mills/kw.-hr.	0.20	2.1	2.4	0.1
Maintenance m and l, mills/kw.-hr.	0.30	0.8	0.6	0.2
Supplies and expense, mills/kw.-hr.	0.10	0.3	0.7	0.1
Total production cost, mills/kw.-hr.	2.04	8.2	8.2	0.4
Total cost of power, mills/kw.-hr.	4.45	11.9	16.1	4.27

STEAM PLANTS

Cycles and Performance. The performance of the basic steam plant of Fig. 1*c* is best measured by the Rankine cycle standard, which is represented on the *pv* diagram by Fig. 8, where there are two constant-pressure phases (4-1) and (2-3) connected by an isentropic phase (1-2). The work of the cycle is evaluated by thermal methods using the properties of steam as given in Sec. 3, Tables 241, 242, 243, and in the Mollier chart of Sec. 4, Fig. 21, p. 333. The work of the prime mover, from the general energy equation, is

Table 7b. Production and Fixed Costs in a Steam Boiler Plant

Size of plant, lb./hr.	100,000
Capacity factor, %	60
Load factor, %	65
Steam output, lb./year	525 × 10⁶
Heat added to steam, B.t.u./lb.	1,100
Boiler efficiency, %	82
Heat in fuel, B.t.u./year	705 × 10⁹
Fuel price, ¢/million B.t.u	20
Investment, \$/lb./hr. capacity	5
Fixed charges:	
Annual, %	12
¢/thousand lb. steam	11.4
Production cost:	
Fuel, ¢/thousand lb. steam	26.9
Labor and supervision, ¢/thousand lb. steam	3.8
Maintenance m and l, ¢/thousand lb. steam	1.1
Supplies and expense, ¢/thousand lb. steam	0.2
Total production cost, ¢/thousand lb. steam	32.0
Total cost of steam, ¢/thousand lb. steam	43.4

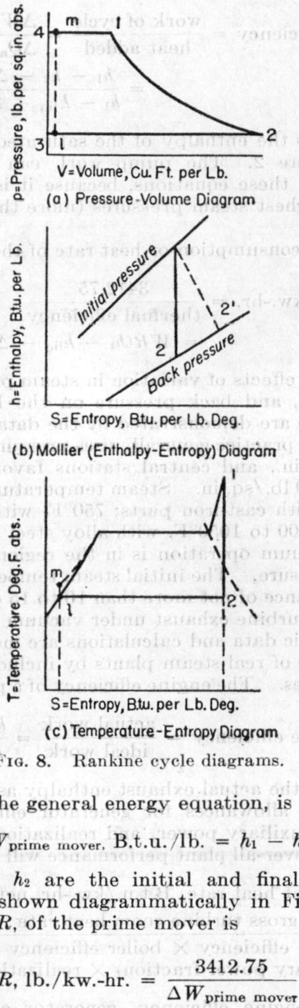

FIG. 8. Rankine cycle diagrams.

$$\Delta W_{\text{prime mover}}, \text{ B.t.u./lb.} = h_1 - h_2 \quad (8)$$

where h_1 and h_2 are the initial and final enthalpies, B.t.u./lb., as shown diagrammatically in Fig. 8*b*. The water rate, *WR*, of the prime mover is

$$WR, \text{ lb./kw.-hr.} = \frac{3412.75}{\Delta W_{\text{prime mover}}} \quad (9)$$

The work of the cycle is the work of the prime mover less the work of the feed pump (area 3*lm*4, Fig. 8*a*), or

$$\Delta W_{\text{cycle}}, \text{ B.t.u./lb.} = h_1 - h_2 - \Delta W_{\text{pump}} \quad (10)$$

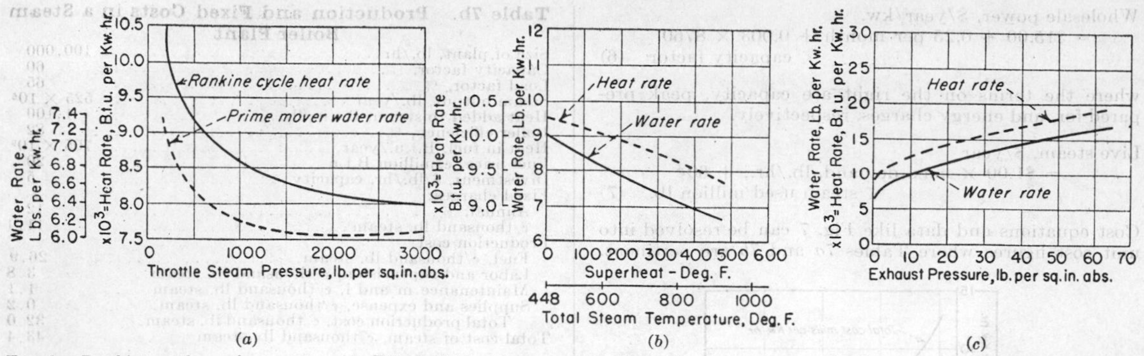

FIG. 9. Rankine cycle performance. (a) Effect of throttle pressure. Throttle temperature = 900°F., exhaust pressure = 2 in. Hg abs. (b) Effect of superheat. Throttle pressure = 400 lb./sq. in., exhaust pressure = 2 in. Hg abs. (c) Effect of exhaust pressure. Throttle conditions = 400 lb./sq. in., 800°F., 1415.8 B.t.u./lb.

The thermal efficiency of the cycle is

$$\text{Thermal efficiency} = \frac{\text{work of cycle}}{\text{heat added}} = \frac{\Delta W_{\text{cycle}}}{\Delta Q_{\text{added}}}$$

$$= \frac{h_1 - h_2 - \Delta W_{\text{pump}}}{h_1 - h_{\text{liq}} - \Delta W_{\text{pump}}} \quad (11)$$

where h_{liq} is the enthalpy of the saturated liquid at the back pressure 2. The pump work can ordinarily be neglected in these equations, because it is small except with the highest steam pressures (more than 900 lb./sq. in.).

The heat consumption or heat rate of the cycle is

$$HR, \text{ B.t.u./kw.-hr.} = \frac{3412.75}{\text{thermal efficiency}}$$

$$= WR(h_1 - h_{\text{liq}} - \Delta W_{\text{pump}}) \quad (12)$$

Some of the effects of variation in steam pressure, steam temperature, and back pressure on the Rankine cycle performance are demonstrated by the data of Fig. 9.

Industrial practice generally uses pressures of less than 600 lb./sq. in., and central stations favor pressures in excess of 900 lb./sq. in. Steam temperatures are limited to 450°F. with cast-iron parts; 750°F. with carbon steel parts; and 900 to 1050°F. with alloy steel (Cr, Mo, etc.) parts. Vacuum operation is in the region of 1 to 3 in. Hg abs. pressure. The initial steam temperature is fixed by maintenance of not more than 10 to 15 per cent moisture in the turbine exhaust under vacuum operation.

These basic data and calculations are modified to give performance of real steam plants by including inefficiencies and losses. The engine efficiency of a prime mover is

$$\text{Engine efficiency} = \frac{\text{actual work}}{\text{ideal work}} = \frac{h_1 - h_2'}{h_1 - h_2} \quad (13)$$

where h_2' is the actual exhaust enthalpy as illustrated in Fig. 8b. If allowances for generator efficiency, boiler efficiency, auxiliary power, and realization ratio are included, the over-all plant performance will result:

Over-all plant heat rate, B.t.u./kw.-hr. net output

$$= \frac{\text{gross turbine room heat rate}}{\text{generator efficiency} \times \text{boiler efficiency} \times (1 - \text{auxiliary power fraction}) \times \text{realization ratio}} \quad (14)$$

Values of engine efficiency, generator efficiency, and boiler efficiency can be estimated from the data given below. Auxiliary power will range from 3 to 12 per cent of the gross output, being higher for non-condensing plants, and increasing with steam pressure. Realization ratios will variously range from 0.9 to 1.0 and will depend upon the reduction of unaccounted for losses. Inclusion of this allowance therefore makes the computed heat rate of Eq. (14) equal to the actual observed heat rate of Eq. (1).

The gross turbine room heat rate of Eq. (14) is usually not the value of the Rankine cycle but is the value as modified by regenerative feed heating, resuperheating, or binary vapor. Theoretical and cyclic details of these factors are to be found in textbooks on thermodynamics. The treatment that follows is for the actual cycles of real power plants.

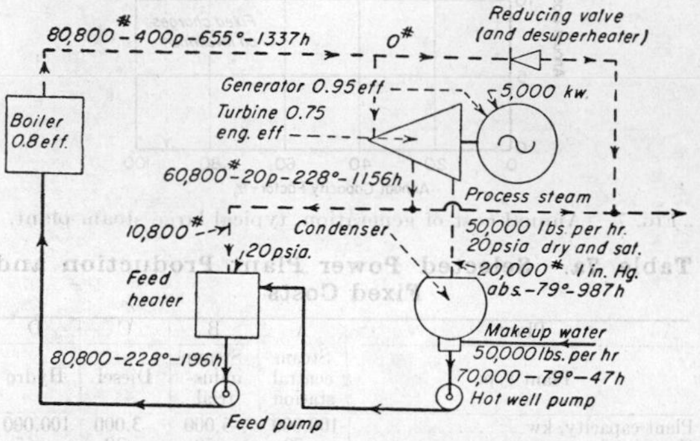

FIG. 10. Steam-plant cycle diagram showing by-product generation of electric power and process steam. Heat supplied in steam = 92 × 10⁶ B.t.u./hr. Heat supplied in fuel = 115 × 10⁶ B.t.u./hr. Net plant electric send-out = 4700 kw.

A typical heat balance and its calculations are shown in Figs. 10 and 11 and Table 8. This is the cycle of an industrial power plant that requires the delivery of 5000 kw. of electric energy and 50,000 lb./hr. of dry saturated process steam at 20 lb./sq. in. abs. Because all power cannot here be generated as a by-product of the process steam flow, a condensing element is added to the turbine and operated at 1 in. Hg abs. pressure. The steps in the calculation can be followed in Table 8.

Figure 12 shows the results of several similar heat balance calculations for different throttle pressures and temperatures. The same load (5000 kw.) is imposed on the generator terminals, and the same process steam output (50,000 lb./hr. at 20 lb./sq. in. abs., dry and saturated) prevails in all cases. The curves show that, for increasing steam pressures, there is an increase in the necessary throttle temperature, a decrease in the throttle flow, an increase in the by-product power, and a decrease in the fuel consumption and cost. Such analyses serve to guide the selection of optimum steam pressures and temperatures even though many other factors must also be taken into account.

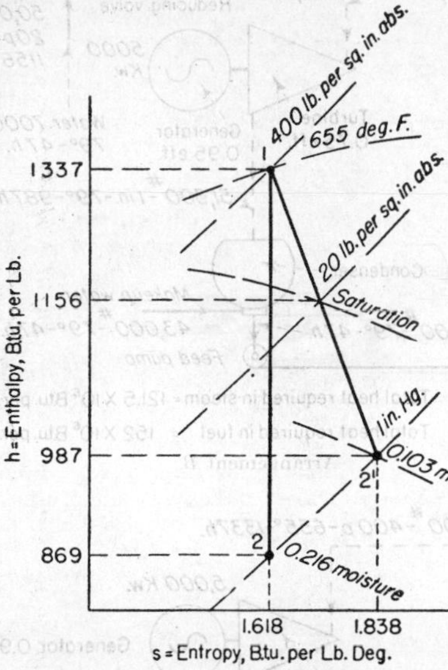

FIG. 11. Turbine expansion line on Mollier diagram (used in example of Fig. 10 and Table 8).

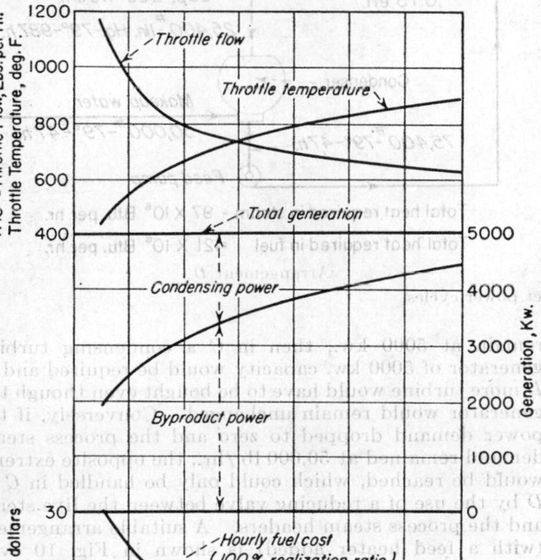

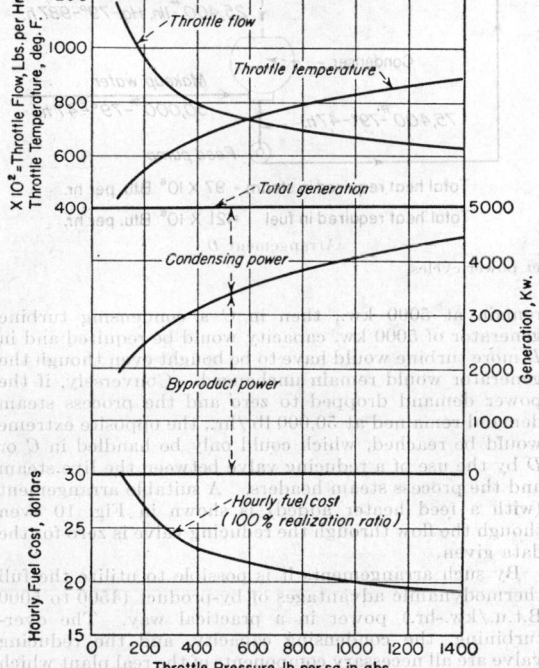

FIG. 12. Performance of by-product power cycle as affected by throttle pressure (see Fig. 10 and Table 8).

The heat balance of Fig. 10 is for one set of load conditions. Load curves, as discussed above, show not only variations but also non-coincidence of the several service demands. The steps in the reasoning that lead to the selection of an economical cycle for a suitable, adequate, and reliable power plant are as follows, all predicated on the hypothesis that 5000 kw. and 50,000 lb./hr. of process steam are required.

Table 8. Heat Balance Calculations, By-product Power Cycle*

Requirements: 5000 kw. and 50,000 lb./hr. process steam at 20 lb./sq. in. abs., dry and saturated; throttle conditions, 400 lb./sq. in. abs., 655°F.; exhaust conditions: 1 in. Hg abs.

Item	Performance
Condenser:	
Steam flow, lb./hr	20,000
Pressure, in. Hg abs	1
Hot-well temperature, °F	79
Hot-well enthalpy, B.t.u./lb	47
Make-up water flow, lb./hr	50,000
Water flow out, lb./hr	70,000
Feed heater:	
Heater pressure, lb./sq. in. abs	20
Heater saturation temperature, °F	230
Enthalpy, water in, B.t.u./lb	47
Enthalpy, water out, B.t.u./lb	196
Heat added to water, B.t.u./lb	149
Heat added to water, 10^6 B.t.u./hr	10.4
Bleed enthalpy, B.t.u./lb	1,156
Heat available from bleed, B.t.u./lb	960
Extraction required, lb./hr	10,800
Boiler feed, lb./hr	80,800
Turbine generator:	
Enthalpy at throttle, B.t.u./lb	1,337
Enthalpy at bleed point, B.t.u./lb	1,156
Enthalpy drop to bleed point, B.t.u./lb	181
Extraction, lb./hr	60,800
Internal generation by extracted steam, kw	3,220
Enthalpy at exhaust, B.t.u./lb	987
Enthalpy drop to exhaust, B.t.u./lb	350
Condenser flow, lb./hr	20,000
Internal generation by condenser flow, kw	2,050
Total internal generation, kw	5,270
Generator efficiency	0.95
Generator electrical and mechanical losses, kw	270
Gross load, generator terminals, kw	5,000
Throttle water rate, lb./kw.-hr	16.2
Condenser water rate, lb./kw.-hr	4.0
Boiler:	
Enthalpy at superheater outlet, B.t.u./lb	1,337
Enthalpy at feed, B.t.u./lb	196
Heat added to steam, B.t.u./hr	1,141
Heat added to steam, 10^6 B.t.u./hr	92
Boiler efficiency	0.8
Heat supplied in fuel, 10^6 B.t.u./hr	115
Plant performance:	
Auxiliary power	0.06
Auxiliary power, kw	300
Net plant electric sendout, kw	4,700
Plant realization ratio	0.95
Net plant heat supplied, 10^6 B.t.u./hr	121
Fuel price, ¢/million B.t.u	20
Fuel cost/hr	$24.20
Allocation of heat charges:	
To process:	
Steam flow, lb./hr	50,000
Enthalpy of steam, B.t.u./lb	1,156
Enthalpy of make-up, B.t.u./lb	47
Heat to process, B.t.u./lb	1,109
Heat to process, 10^6 B.t.u./hr	55.4
To power (balance), 10^6 B.t.u./hr	36.6
Fraction chargeable to process	0.60
Fraction chargeable to power	0.40
Fuel cost, process steam, per hr	$14.50
Fuel cost, power, per hr	$9.70
Fuel cost, process steam, ¢/thousand lb	29
Fuel cost, power, net, mills/kw.-hr	2.06
Net plant heat rate, chargeable to power, B.t.u./kw.-hr	10,300

* *Cf.* Figs. 10, 11, and 12.

Arrangement A (Fig. 13). For complete independence of power and process steam output, arrangement *A* would require two boilers, one of suitable pressure and size serving the condensing turbine generator, and, the other of suitable pressure and size to serve the process steam demand. Although maximum flexibility is obtained by this arrangement, the space requirements, investment, fuel, labor, and maintenance would all be a maximum.

Arrangement B (Fig. 13) combines the two boilers of arrangement *A* and adds a reducing valve and desuperheater. The complete independence of the power and process steam supplies is maintained. No by-product

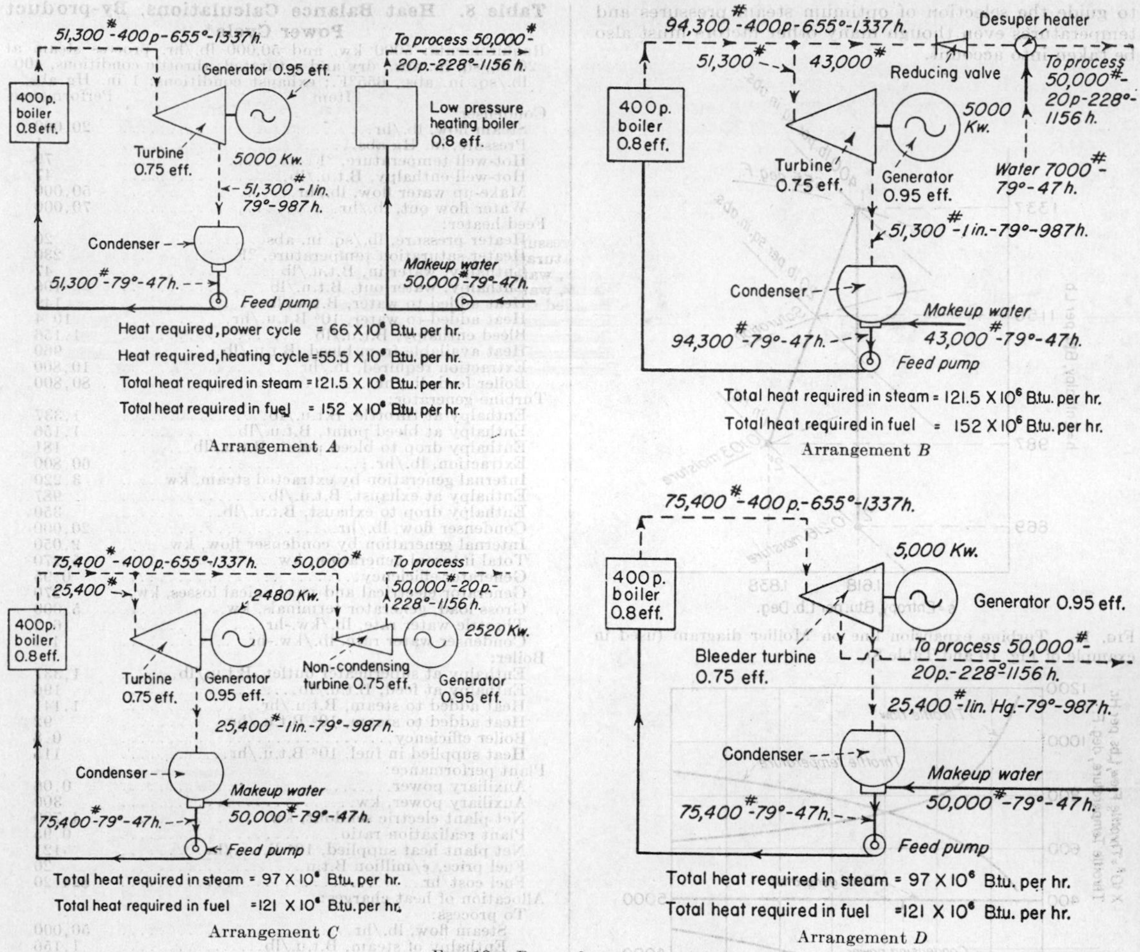

FIG. 13. By-product power cycles.

power is generated, since all process steam passes through the reducing valve. The investment is still high, and the pump work and operating costs are a maximum.

Arrangement C (Fig. 13) substitutes a non-condensing turbine generator for the reducing valve of arrangement *B*. By-product power (4500 to 5000 B.t.u./kw.-hr. heat rate) can be generated in the amount dictated by the process steam flow. Additional generation is supplied by the condensing unit at a higher heat rate (15,000 B.t.u./kw.-hr.). Boiler size is reduced, thus lowering investment at the same time that fuel costs are reduced. Complete independence of the power and process steam demands is lost. All by-product power, which is delivered by the back-pressure unit, must be accepted. The non-condensing unit must be operated on base load, and all power load swings must be carried on the condensing turbine.

Arrangement D (Fig. 13) is essentially the same as arrangement *C* except that a single constant-pressure extraction turbine generator is substituted for the two units. This lowers the investment but does not avoid the dependence of power generation on process steam demand.

If, in arrangements *C* or *D*, the process steam demand were to drop to zero but the power demand were to remain at 5000 kw., then in *C* a condensing turbine generator of 5000 kw. capacity would be required and in *D* more turbine would have to be bought even though the generator would remain unchanged. Conversely, if the power demand dropped to zero and the process steam demand remained at 50,000 lb./hr., the opposite extreme would be reached, which could only be handled in *C* or *D* by the use of a reducing valve between the live steam and the process steam headers. A suitable arrangement (with a feed heater added) is shown in Fig. 10 even though the flow through the reducing valve is zero for the data given.

By such arrangements it is possible to utilize the full thermodynamic advantages of by-product (4500 to 5000 B.t.u./kw.-hr.) power in a practical way. The overturbining, the condensing capacity, and the reducing valve are all necessary components of the real plant which add to the basic investment cost. The fundamental thermodynamic operating gain is consequently compromised. The alternative would be to operate during some periods on a straight non-condensing power cycle with its prohibitive heat rates of 25,000 to 50,000 B.t.u./kw.-hr.

The many possible combinations of expected power and process steam loads cannot usually be predicted, nor can the burden of their detailed analysis be borne by the design. A reasonable practical compromise is to design for the usual expected load coincidences but to be prepared, as above, to meet the maximum power peak and the minimum process steam demand simultaneously, and vice versa. The whole problem may be restated in rate schedule terms, *i.e.*, "capacity cannot be traded for energy, and energy cannot be traded for capacity." Alternate solutions to the problem might of course include consideration of the use of purchased power, internal-combustion engines, hydroelectric power, gas turbines, and heat pumps simultaneously to meet the variety of demands imposed on the industrial power plant.

The thermodynamic advantages of regenerative feed heating, as typified by the cycle of Fig. 10, are analogous to those of by-product power. If steam can be used for feed heating instead of being exhausted to atmosphere or to a condenser, the energy generated, by expansion from throttle pressure to heater pressure, is at the low heat rate of mechanical equivalence of by-product power. The aggregate heat savings on all the power generated are a function of the final feed temperature and the number of stages of heating, as shown by the data of Fig. 14. Any

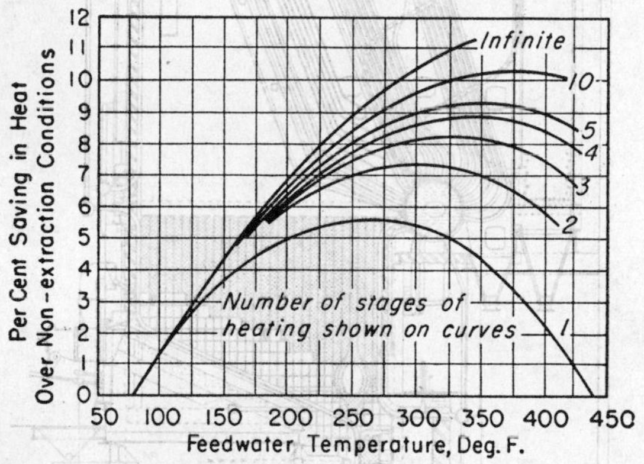

FIG. 14. Feed-water heat saving by extraction. Steam conditions, 400 lb./sq. in., 700°F., to 1 in. abs. back pressure.

source of low-pressure steam may be used, but the greatest gain requires maximum generation of work by the steam prior to its use in a feed heater. High engine efficiency is desired, and main unit bleed is consequently preferred to auxiliary exhaust. Final feed temperatures are determined from data like Fig. 14 plus considerations of pump types and locations in cycle, feed heater designs, deaerator location, economizers, air heater, flue gas temperatures, and furnace characteristics. Central stations use four to eight stages of heating with final temperatures between 350 and 475°F. Industrial plants seldom exceed three stages of heating, or 300 to 350°F., unless plants are of exceptional size.

Fuel- and Ash-handling Systems. Adequate fuel storage is needed to assure (1) continuity of supply regardless of strikes, frozen waterways, or bad weather; and (2) the advantage of favorable markets for purchase. Coal or oil storage may range from one month's to one year's supply. Gas, in holders, is seldom stored for more than a few days. Inexpensive outdoor storage is best for the bulk of the fuel. Bituminous coal piles, subject to spontaneous combustion, should not be too deep (10 to 20 ft.) and may be placed under water. Otherwise any convenient height for storing and reclaiming is used. Oil tanks may be underground or aboveground but must comply with underwriters' standards. Access for cleaning and blowdown is essential. Fuel shipments should not be limited to a single route. Plant location is influenced by accessibility to rail, water, and truck deliveries. Bulk storage, external to plant, is for long periods; and transient storage, within plants, is to accommodate daily, weekend, and overnight loads. Coal-handling equipment is operated on a single day shift basis during 5 days each week. Basic design data are given in Table 9.

Table 9. Design Data, Fuel and Ash Handling

Angle of repose of coal,* measured from horizontal:	
Over 1-in. size, deg.	50
Pulverized	60
Bulk (Apparent) Densities	Lb./Ft.^{3}
Anthracite	55–60
Semibituminous coal	52–57
Bituminous coal	50–55
Subbituminous coal	45–50
Fuel oil	55–60
Diesel fuel	52–57
Gasoline	45–50
Hogged fuel	20–27
Bagasse	10
Sawdust	17–20
Pulverized coal, from pulverizer	30
Pulverized coal, after deaeration	35
Ashes, fly ash	35–45
Ashes, pulverized fuel, wet bottom, water-quenched	90
Ashes, pulverized fuel, dry bottom, dry	50–65
Ashes, pulverized fuel, dry bottom, after wet handling	80
Ashes, stoker	50–60

* May run to 90 deg. or bridge over with sufficient moisture or when compacted.

Practically every type of conveying, hoisting, and dredging machinery is used for **coal handling**. Reference should be made to Sec. 20 and 21 for details of available equipment, which include unloading towers; trestles; car dumps; locomotives; gantry cranes and shovels; drag-line scrapers; belt, bucket, screw, flight and apron conveyors; skip hoists and elevators; hand trucks; wheel barrows; and lorries.

A crusher to reduce run-of-mine coal to a size not less than ½ in. is desirable. Coal is fed to stokers or mills from (1) overhead steel or concrete bunkers, (2) silos, or (3) external pit with elevating conveyor. Tramp iron remover, automatic weighing scales and samplers, and belt trippers may be included. Coal spouts of stainless steel avoid plugging and corrosion.

Reciprocating, rotary, or screw pumps are used in conjunction with strainers and steam or electric heaters to remove **fuel oil** from storage and deliver to day or transfer tanks, or to burners. Bunker C or No. 6 residue fuel oil, which is standard petroleum boiler fuel, can be pumped when the viscosity is reduced by heating, with curves of Fig. 15 as typical.

Coal and **ash handling** may be combined in varying degrees by the use of the same equipment. Manual methods prevail on small plants (less than 25,000 to 50,000 lb. steam/hr.). Hand cars; wheelbarrows; and mechanical, pneumatic, or hydraulic conveyors may deliver refuse to an elevated silo or directly to the ultimate disposal point. Hydraulic conveyors are preferred in large plants, where ash sluicing is by service water through nozzles, under each boiler, to move refuse through a covered cast-iron floor trench to an external settling basin where water is decanted and ash dredged or pumped to an overhead storage silo or to ultimate discard. Nozzles are spaced about 50 ft. apart and require 200 to 300 gal./min. of water under 100 lb./sq. in. pressure. Sluicing is effected once or twice a shift or a day with ashes moved at a typical rate of 1 ton/min.

Boilers. Steam generators for burning fuel and delivering steam variously use, in addition to a furnace and a boiler, superheater, economizer, air preheater, water walls, draft system, setting and casing, accessories, and instruments and controls. All pressure parts must comply with the law. The A.S.M.E. Boiler Code is the generally accepted American standard. Operators, in turn, are licensed under local ordinances. Boiler sizes range from 5000 to 1,000,000 lb. steam/hr. with over-all peak efficiency from 60 to 90 per cent, the higher values obtaining on larger units equipped with heat traps and designed for use with expensive fuels and high load

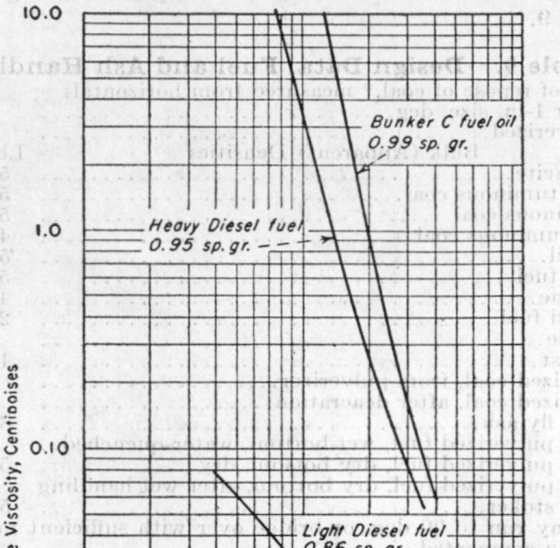

FIG. 15. Liquid fuel viscosities.

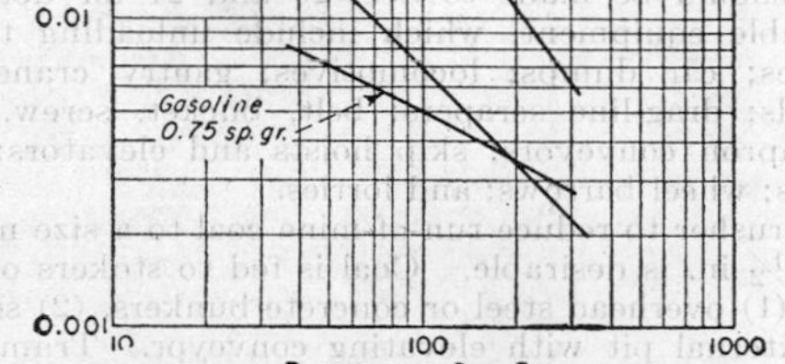

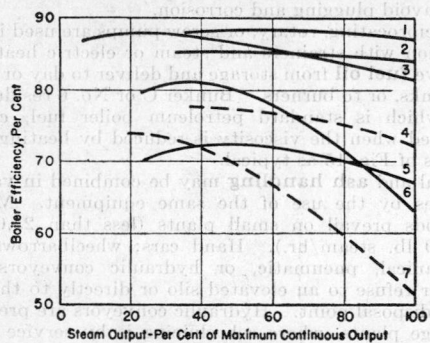

FIG. 16. Some boiler efficiency curves. 1. Coal-fired drum boiler, economizer, and air heater. 2. Gas-fired drum boiler, economizer, and air heater. 3. Oil-fired naval boiler (destroyer). 4. Coal-fired industrial boiler, no air heater. 5. No. 3 buckwheat-fired header boiler, no economizer or air heater. 6. Bituminous-fired four-drum boiler, economizer, no air heater. 7. Coal-fired locomotive boiler.

factors. Figure 16 shows some typical over-all performances of efficiency vs. output. The output was formerly expressed in boiler horsepower, but the actual pounds of steam are preferred in present practice. A developed boiler horsepower is defined as the addition of 33,479 B.t.u./hr. to the steam. This is equivalent to the generation of 34.5 lb. steam/hr. from and at 212°F. Such evaporation could formerly be effected on 10 sq. ft. of heating surface, but modern boilers will operate at many times this rate (1000 per cent on Navy boilers, 300 to 500 per cent on industrial installations, 100 per cent is condition for banking). Boiler horsepower is thus obsolete and should not be used.

Boiler surface exposed to the fire may transfer heat at a rate of 100,000 B.t.u./hr./sq. ft. Water circulation is essential to keep the surface wet and thus prevent burning. High capacity per square foot of surface is sought to keep down the investment. High capacity requires maximum temperature difference between the gases and water. High efficiency for best fuel economy requires that gases be cooled to the lowest possible exit tempera-

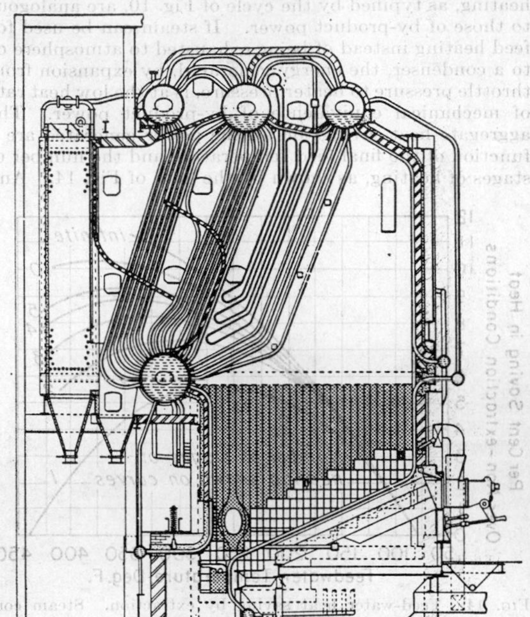

FIG. 17*A*. Four-drum Stirling-type boiler with underfeed stoker. Type suitable for sizes from 30,000 to 300,000 lb./hr., pressures from 150 to 1000 lb./sq. in. Pulverized-coal-, oil-, or gas- or stoker-fired. (*Babcock & Wilcox Co.*)

ture with resulting low temperature differences. Maximum capacity per square foot and maximum efficiency cannot consequently be expected in the same boiler.

Boilers are (1) fire tube or (2) water tube (Figs. 17*A*, 17*B*, 17*C*). **Fire tube boilers** have gases inside the tubes and water outside. They may be vertical or horizontal, externally or internally fired. They are characterized by low pressure (less than 200 lb./sq. in.), low efficiency (less than 70 per cent), low ratings (less than 5 lb. steam/hr./sq. ft.), and slow steaming. They are inexpensive, easy to install, require low headroom, and are particularly suitable for hand or spreader stoker firing with coal.

Water tube boilers use tubes, less than 4 in. diameter, with water inside and hot gases circulating outside through baffled passages. They may be (1) straight tube with box or sectional headers into which tubes are

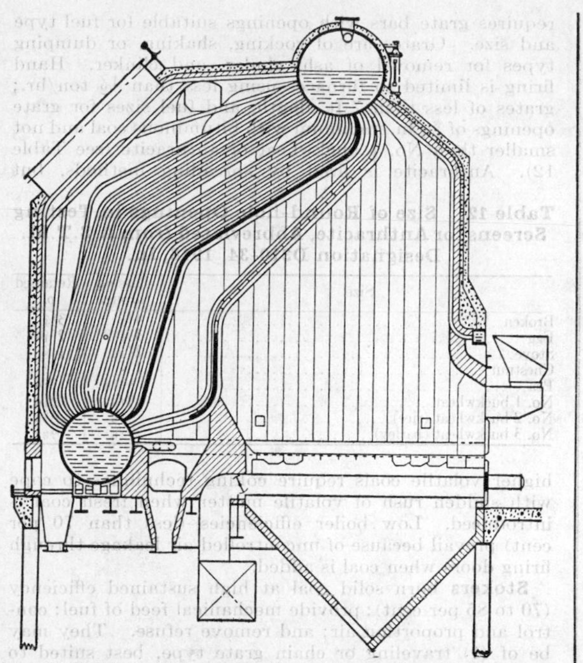

FIG. 17*B*. Integral furnace boiler with spreader stoker. Type suitable for sizes from 5000 to 50,000 lb./hr., pressures from 160 to 600 lb./sq. in.; temperatures from saturation to 700°F. Oil-, gas-, or stoker-fired. (*Babcock & Wilcox Co.*)

rolled and accessibility is by handholes opposite the tube ends, or (2) bent-tube type (Figs. 17*A*, *B*, and *C*), where tubes are rolled into walls of drums.

Improved suitability, greater flexibility of design, and lower costs have resulted in practical elimination of straight-tube types in modern industrial practice. Water tube boilers contrasted with fire tube boilers are more costly initially (as much as 2 to 1); built in larger sizes (up to 1,000,000 lb. steam/hr.); suitable for highest pressures (2500 lb./sq. in.) and highest temperatures (1000°F.); built for highest efficiencies (85 to 90 per cent), highest ratings, and rapid steaming; and they can be designed to accommodate any desired fuel or method of firing.

Furnaces. Furnaces should be designed for two alternate fuels wherever possible to take advantage of best market conditions. Firing may be by gas, oil, coal in solid form, pulverized coal, waste fuel, or waste heat. The standard furnace for industrial boilers is a refractory-lined structure with insulating brick and an enclosing casing of metal, brick, tile, or plastic. Furnace size is determined by the usual allowable performance, as evidenced by the data of Table 10 showing volumetric heat release rates and excess air values from the equation

Table 10. Typical Furnace Firing Rates for Industrial Boilers

Fuel	Heat liberation rate, B.t.u./hr./ft.³	Excess air, %
Natural gas	20,000–40,000	10– 25
Fuel oil	20,000–50,000	15– 30
Pulverized coal	15,000–35,000	15– 30
Stokers, large	20,000–40,000	20– 40
Stokers, small	15,000–25,000	30– 60
Hand-fired	5,000–10,000	50–150

$$\text{Per cent excess air} = \frac{3.78(O_2 - CO/2)}{N_2 - 3.78(O_2 - CO/2)} \times 100 \quad (15)$$

where O_2, N_2, and CO are the volumetric percentages of these components in the dry flue gas. Despite the presence of free oxygen in the products, carbon monoxide or hydrogen will appear as unburned combustible. Excess air as a function of carbon dioxide is shown in Fig. 18.

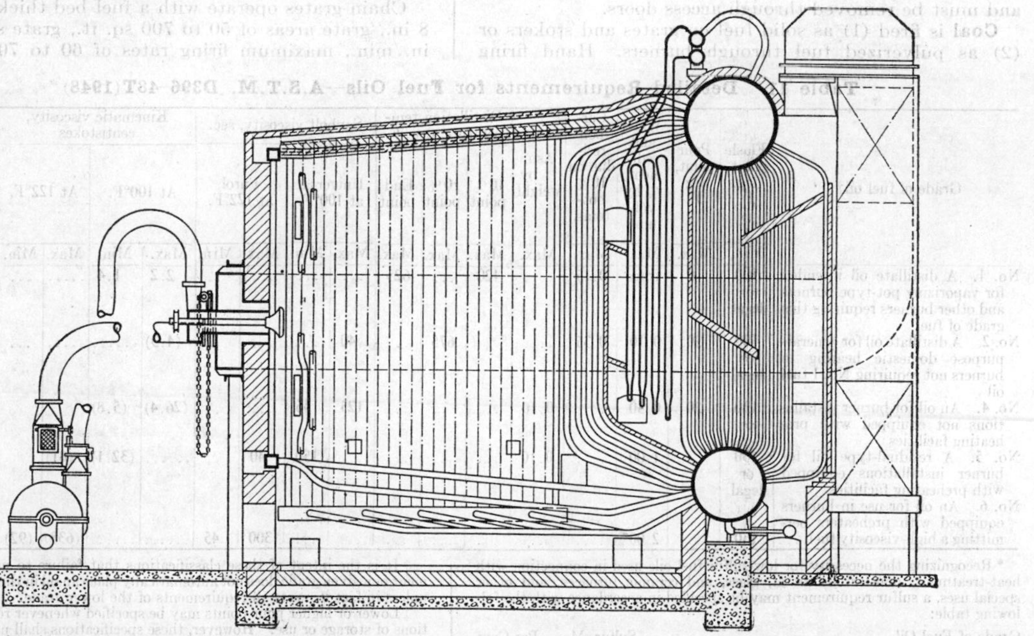

FIG. 17*C*. Pulverized-coal- and natural-gas-fired boiler. Type suitable for sizes from 20,000 to 200,000 lb./hr., pressures from 160 to 1000 lb./sq. in., temperatures from saturation to 900°F. Pulverized-coal-, oil-, or gas-fired. (*Combustion Engineering Corp.*)

Gas firing is the simplest method of firing. Burners are feeders and mixers. Flame length is reduced by greater premixing with primary air. Simple refractory walls generally suffice, even though water-cooled walls may be employed.

Oil firing uses mechanical, steam, or air atomizing burners (see Sec. 21, p. 1573, and Table 11), the first being most prevalent with the oil delivered to the burner tip

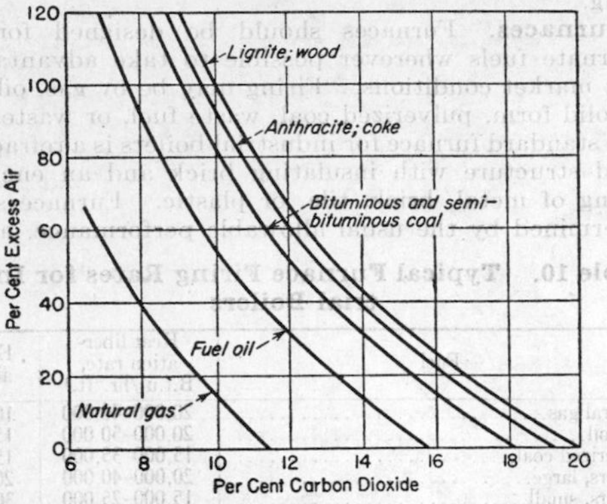

FIG. 18. Excess air and carbon dioxide in products of combustion of typical fuels.

under pressures of 100 to 300 lb./sq. in. A single nozzle will pass from 10 to 600 gal./hr. with a 2 to 1 range of capacity on a single tip or sprayer plate. Change of tips and a multiplicity of burners give the necessary capacity range on installation. Steam atomizing burners will use 1 per cent of the boiler output for atomizing alone; oil pressures range from 10 to 100 lb./sq. in.; and firing ranges of 7 to 1 are obtainable on a single burner. Each burner is provided with adjustable louvers for control and guidance of air for most effective mixing. Furnace shape and proportions contribute to mixing. Burner tips carbonize and must be cleaned. Refuse forms as a hard mass and must be removed through access doors.

Coal is fired (1) as solid fuel on grates and stokers or (2) as pulverized fuel through burners. Hand firing requires grate bars with openings suitable for fuel type and size. Grates are of rocking, shaking, or dumping types for removal of ash, cinder, and clinker. Hand firing is limited to furnaces using less than ½ ton/hr.; grates of less than 7 ft. depth, and fuel sizes for grate openings of ¾ in. minimum with bituminous coal and not smaller than No. 1 buckwheat or anthracite (see Table 12). Anthracite is fired by spreading methods, but

Table 12. Size of Round-hole Openings in Testing Screens for Anthracite, Abbreviated from A.S.T.M. Designation D310-34, 1948, In.

Size	Passing through	Retained on
Broken	4⅜	3¼
Egg	3¼–3	2⁷⁄₁₆
Stove	2⁷⁄₁₆	1⅝
Chestnut	1⅝	¹³⁄₁₆
Pea	¹³⁄₁₆	⁹⁄₁₆
No. 1 buckwheat	⁹⁄₁₆	⁵⁄₁₆
No. 2 buckwheat (rice)	⁵⁄₁₆	³⁄₁₆
No. 3 buckwheat (barley)	³⁄₁₆	³⁄₃₂

higher volatile coals require coking techniques to cope with sudden rush of volatile matter when fresh coal is introduced. Low boiler efficiencies (less than 70 per cent) prevail because of uncontrolled air leakage through firing doors when coal is added.

Stokers burn solid coal at high sustained efficiency (70 to 85 per cent); provide mechanical feed of fuel; control and proportion air; and remove refuse. They may be of (1) traveling or chain grate type, best suited to anthracite or non-coking bituminous, where coal is introduced at the front end, burns as the grate moves toward the rear where the refuse is dumped; (2) spreader type (Fig. 17*B*), which scatters non-coking bituminous coal to the horizontal grate surface and from which ash is removed by shaking or dumping; (3) underfeed type (Fig. 17*A*), best suited to caking or coking coals and in which fresh fuel is introduced from beneath, in troughs between inclined tuyères, the ram action serving to break up the cake. The ash is dumped at the rear through grinders.

Chain grates operate with a fuel bed thickness of 4 to 8 in., grate areas of 50 to 700 sq. ft., grate speeds of 12 in./min., maximum firing rates of 60 to 70 lb./hr./sq.

Table 11. Detailed Requirements for Fuel Oils—A.S.T.M. D396–48T(1948)*

Grade of fuel oil†	Flash point, deg. F.	Pour point, deg. F.	Water and sediment, % by volume	Carbon residue on 10% bottoms, %	Ash, % by weight	Distillation temperature, deg. F. 10% point	10% point	End point	Saybolt viscosity, sec. Universal at 100°F.		Furol at 122°F.		Kinematic viscosity, centistokes At 100°F.		At 122°F.		Gravity, deg. A.P.I	Corrosion, copper strip, 3 hr. at 122°Fe
	Min.	Max.	Max.	Max.	Max.	Max.	Max.	Max.	Max.	Min.	Max.	Min.	Max.	Min.	Max.	Min.	Min.	
No. 1. A distillate oil intended for vaporizing pot-type burners and other burners requiring this grade of fuel	100 or legal	0	trace	0.15	....	420	...	625	...	...	...	..	2.2	1.4	.....	...	35	pass
No. 2. A distillate oil for general-purpose domestic heating in burners not requiring No. 1 fuel oil	100 or legal	20‡	0.10	0.35	....	‖	675	...	40	...	...	..	(4.3)	......	.....	...	26	
No. 4. An oil for burner installations not equipped with preheating facilities	130 or legal	20	0.50	....	0.10	...	...	...	125	45	...	..	(26.4)	(5.8)				
No. 5. A residual-type oil for burner installations equipped with preheating facilities	130 or legal	...	1.00	....	0.10	...	...	...	...	150	40	..	......	(32.1)	(81)			
No. 6. An oil for use in burners equipped with preheaters permitting a high-viscosity fuel	150	...	2.00¶	....	...	...	...	...	...	...	300	45	.....	.....	(638)	(92)		

* Recognizing the necessity for low-sulfur fuel oils used in connection with heat-treatment, non-ferrous metal, glass, and ceramic furnaces, and other special uses, a sulfur requirement may be specified in accordance with the following table:

Grade of Fuel Oil	Sulfur, Max., Per Cent
No. 1	0.5
No. 2	1.0
No. 4	no limit
No. 5	no limit
No. 6	no limit

Other sulfur limits may be specified only by mutual agreement between the purchaser and the seller.

† It is the intent of these classifications that failure to meet any requirement of a given grade does not automatically place an oil in the next lower grade unless in fact it meets all requirements of the lower grade.

‡ Lower or higher pour points may be specified whenever required by conditions of storage or use. However, these specifications shall not require a pour point lower than 0°F. under any conditions.

§ The exposed copper strip shall show no gray or black deposit.

‖ The 10 per cent point may be specified at 440°F. maximum for use in other than atomizing burners.

¶ The amount of water by distillation plus the sediment by extraction shall not exceed 2.00 per cent. The amount of sediment by extraction shall not exceed 0.50 per cent. A deduction in quantity shall be made for all water and sediment in excess of 1.0 per cent.

ft., and best operating rates of 30 to 40 lb./hr./sq. ft.; and they are available for boiler sizes up to 250,000 lb./hr.

Spreader stokers (Fig. 17*B*) operate with a fuel bed not over 3 or 4 in. thick and a grate length less than 10 ft. with dumping grates; burn fuel at rates of 25 to 50 lb./hr./sq. ft. of grate area; and are available in sizes for boilers generally delivering less than 50,000 lb. steam/hr. and burning non-caking bituminous coal, particularly of low fusion temperature.

Underfeed stokers (Fig. 17*A*) single or multiple retort, operate with fuel beds as thick as 3 ft., furnace lengths of 6 to 18 ft., firing rates as high as 75 lb./hr./sq. ft. of projected area with best performance at 30 to 50 lb./hr./sq. ft., built in sizes for boilers ranging from 10,000 lb. steam/hr. (single retort) to 300,000 lb./hr. (inclined multiple retort).

Table 13a. Square-hole, Woven-wire Sieve Sizes for Tests of Fineness of Powdered Coal, Abbreviated from A.S.T.M. Designation D197-30 and E11-39, 1948

Size or sieve designation		Sieve opening, mm.	Wire diameter, mm.
Microns	No.		
1190	16	1.19	0.50 -0.70
590	30	0.59	0.29 -0.42
297	50	0.297	0.170-0.253
149	100	0.149	0.096-0.125
74	200	0.074	0.045-0.061

Table 13b. Square-Hole, Woven-wire Sieve Sizes for Analysis of Crushed Bituminous Coal, Abbreviated from A.S.T.M. Designation D311-30 and E11-39, 1948

Size or sieve designation	Sieve opening, mm.	Wire diameter, mm.
1.06 in.	26.9	3.43-4.50
¾ in.	19.1	3.10-3.91
0.530 in.	13.4	2.39-3.10
⅜ in.	9.52	2.11-2.59
0.265 in.	6.73	1.60-2.11
4760 microns (No. 4)	4.76	1.14-1.68
3360 microns (No. 6)	3.36	0.87-1.32

Pulverized coal (Fig. 17*C*) offers the greatest latitude in coal selection, shows the best over-all economy for steam raising in boiler sizes greater than 25,000 to 50,000 lb. steam/hr. (1 to 2 tons coal/hr.), and is used in units as large as 1,000,000 lb. steam/hr. (50 tons coal/hr.). Grinding is in mills of the ball, roller, impact, or attrition types to a fineness where 75 to 80 per cent will pass through a 200 mesh screen (see Table 13). Power consumption is from 10 to 20 kw.-hr./ton and is primarily determined by grindability as measured on the Hardgrove index, anthracite having a value of 40, Indiana bituminous 60, Pittsburgh bituminous 70, and Pocahontas 100. The power requirement of the mill in kilowatts is substantially constant regardless of the load on the mill. Grinding parts of mills need renewal after 30,000 to 60,000 tons output. Maintenance costs aggregate about 5¢/ton of coal.

Air-swept mills are used either for storage or direct firing systems. In the former the fuel is pulverized and stored in bulk in a bin or silo from which it is drawn and delivered to burners as dictated by boiler load. In the latter the fuel is pulverized as needed by the load and delivered directly to the burners. This entails wide ranges of operation for the mills and/or a multiplicity of mills. Mills are built in capacities of 0.5 to 20 tons/hr. The storage system, because of its reservoir of powdered fuel, allows mills to run at a steady load and is often preferred for industrial plants even though the investment may be nearly double.

Many designs of burners are in use which must be accommodated to the furnace in order to give the requisite degree of turbulence and firing rates, as shown in Table 10. Furnaces are of (1) wet or (2) dry types, dependent upon the method of ash removal. A low-fusion-temperature ash (1800 to 2400°F.) is needed in the former; and, if the furnace is kept hot enough at all loads, by reduction of black surface, the refuse can be tapped in a molten condition. The ash can alternately be removed in a dry powdered form with high-fusion-temperature ash (2200 to 2800°F.) and with adequate cooling of the furnace at all loads. Furnaces must be designed for one or the other method of ash removal, since the two cannot be combined. At best not more than 75 per cent of the ash is taken out in the ash pit with either method, and often the removal does not exceed 20 per cent. Furnace proportions are not only a matter of volumetric combustion rates but are also a matter of the time needed to burn out a fuel particle. This may require a flame travel of 50 to 100 ft. Ignition is by oil torch, supplementary oil or gas burners. Combination burners, using alternate fuels, are a recommended design procedure (see Fig. 17*C*).

Furnace walls may be plain refractory, water-cooled, or steam-cooled. Air-cooled walls are no longer used. Fuel and combustion conditions dictate the extent to which the various surfaces are used. Radiant heat transfer to cold surface will alter the furnace temperature and contribute to or interfere with the fusion of the ash. Water-cooled surfaces may be plain 3- or 4-in.-diameter tubes, longitudinally finned or stud tubes, or bare or refractory-covered tubes. Superheater or reheater tubes need attention to prevent overheating or burning, particularly during starting periods. Water wall surfaces are most effective for steam raising (up to 150 lb./hr./sq. ft.), but they are the most expensive form of boiler surface ($25 per sq. ft.). The extent of water wall surfaces ranges from 10 to 25 per cent of the boiler water heating surface, being a minimum with underfeed or spreader stokers and a maximum with oil or pulverized coal firing. Surrounding the furnace is a layer of insulation and a brick and steel or plastic casing.

Waste fuels are available in many industries and include such items as hogged fuel, bagasse, sawdust, tan bark, petroleum coke, blast furnace gas, sewage gas, black liquor, sulfur, garbage, and trash. Each waste fuel offers its own problems. Often the calorific power is low and much moisture a natural accompaniment. Combustion must be completed before any attempt is made to absorb the heat in steam raising. This may require the use of a supplementary fuel if incineration is a primary objective. Reverberatory Dutch oven constructions prevail for furnaces with no cooling of walls so as to give maximum furnace temperatures. Each fuel is best handled in a furnace of appropriate design.

Waste heat is available in gases discharged from many high-temperature industrial operations such as smelting and refining furnaces, cement kilns, ovens, roasters, heat-treating, ceramics, and glassmaking. Heat transfer rates are improved by high mass flow on the gas side, which in turn dictates high draft loss. If the available gas temperatures are not high and if attempt is made to generate steam at any considerable pressures, the mean temperature difference will be low and much surface will be required. Investment charges for boiler surface must therefore be balanced against operating charges for fan power. The economic break-even point on waste heat for steam raising is in the neighborhood of 1000°F. For higher gas temperatures, the use of waste heat will be favored; and, for lower gas temperatures, the use of fuel is likely to be indicated.

Superheaters. Superheaters with carbon steel parts are suitable for final steam temperatures of 750°F., but chrome steel parts are needed for 1000°F. Convection superheaters (see Figs. 17*A* and 17*C*) are placed in the gas path while radiant superheaters are placed in furnace walls. Resulting steam-temperature curves are shown in Fig. 19, where the drooping characteristic is representative of the radiant type and the rising characteristic representative of the convection type. Constancy of steam temperature is variously accomplished by the proper superheater position in the gas path, *i.e.*, (1) interdeck (Fig. 17*A*); (2) convection and radiant elements in series; (3) countercurrent and parallel-flow adaptations; (4) desuperheaters and atemperators; (5) internal dampers, baffles, and by-passes for hot gases; and (6) separately fired zones in the boiler furnace. Superheater elements must be protected against burning during starting up periods, by use of screening, flooding, or pendant arrangements.

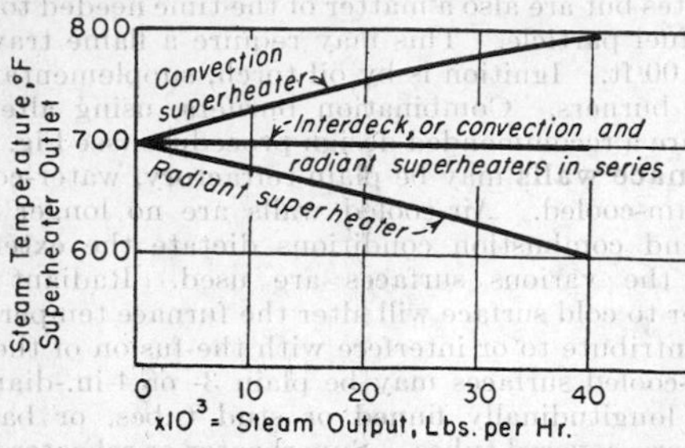

Fig. 19. Superheater characteristics.

Economizers and Air Preheaters. Economizers and air preheaters are used to improve boiler efficiency by reduction of flue gas temperatures and by betterment of combustion conditions. Final flue gas temperatures are fixed by the dew point (125 to 150°F.). Flue gas temperature drops with load. To assure adequate safety margin above the dew point at all partial loads, it is therefore general practice to use 300°F. as minimum full-load gas temperature. **Economizers** are pressure surface and consequently more expensive per square foot than air heaters. Economizers are built of 2- to 4-in.-diameter steel tubes when used with deaerated feed water. Cast-iron tubes were formerly used for low pressures without degasification of feed. Finned tubes may be used because the controlling thermal resistance is on the gas side. Economizers may be of the steaming or non-steaming type. If entering feed water temperature is more than 100°F. below the boiler drum saturation temperature, the non-steaming type is indicated. Economics of economizers vs. preheaters must be balanced on any proposed installation. Heat transfer rates, for clean surfaces, are approximated by

$$U = 2.0 + 0.0015m \qquad (16)$$

where U equals B.t.u./hr./sq. ft./°F. and m is the mass velocity, gas side, lb./hr./sq. ft. flow area. The maximum mass velocity is between 5000 and 10,000 lb./hr./sq. ft., and water velocities are below 10 ft./sec. Economizer surfaces are cleaned by soot blowing or washing.

Air preheaters are of the recuperative or regenerative types. The former is (1) a plate construction using a multiplicity of flat sheet-metal surfaces, spaced 1 or 2 in. apart, with hot gases and cold air blowing in alternate spaces; or (2) a tubular construction with gas inside and air outside 4-in.-diameter tubes arranged between tube sheets. The regenerative heater is a drum filled with corrugated sheet metal, which slowly rotates (2 to 3 r.p.m.), in a horizontal or vertical plane, so that it is in the hot gas stream for approximately one-half a revolution and in the cold air stream for the remaining period. The regenerative heater, because of its moving element, must be provided with locks and gas seals. It is more compact than the recuperative types and is particularly favored for larger boilers. Air heaters are used alone or in series with an economizer. The amount of air-heater surface needed is predicated on the allowable degree of preheat that can be tolerated in the furnace, *i.e.*, 250°F. maximum with stokers; and 600°F. maximum with gas, oil, and powdered coal. Heat transfer rates are less than the values prevailing for economizers, because of the presence of the two layers of gas that are involved in the heat flow path; and they may be estimated by the equation

$$U = 0.5 + 0.0005m, \qquad (17)$$

where U is the heat transfer rate, B.t.u./hr./sq. ft./°F., and m is the mass velocity, lb./hr./sq. ft. flow area.

Draft Systems. Forced, induced, or natural draft, separately or in combination, may be employed. The most favored is the combination called "balanced draft," in which the furnace pressure is maintained at a slightly negative value (less than 0.1 in. water). The simplest, most reliable form of draft is the induced draft of a stack, which is theoretically given as

$$\text{Theoretical draft, in. water} = \frac{H(\bar{w}_{ca} - \bar{w}_{hg})}{\bar{w}_w} \times 12 \qquad (18)$$

where H = height of stack, ft.
$\bar{w}_{ca}$ = density of cold air, lb./cu. ft.
$\bar{w}_{hg}$ = density of hot gas in the stack, lb./cu. ft.
$\bar{w}_w$ = density of water, lb./cu. ft.

Chimney friction and velocity head must be estimated by the methods of Sec. 6 and deducted from the theoretical draft to give the net available draft. Usual velocities are given in Table 14.

Table 14. Usual Allowable Velocities, Duct and Piping Systems

Item	Velocity, Ft./Min.
Forced-draft ducts	2,500–3,500
Induced-draft flues and breeching	2,000–3,000
Chimneys and stacks	2,000
Water lines	600
High-pressure steam lines	10,000
Low-pressure steam lines	12,000–15,000
Vacuum steam lines	25,000
Compressed-air lines	2,000
Refrigerant vapor lines:	
High pressure	1,000–3,000
Low pressure	2,000–5,000
Refrigerant, liquid	200
Brine lines	400
Ventilating ducts	1,200–3,000
Register grilles	500

Stacks are built of brick, concrete, plain or lined steel; may be guyed for structural strength; and are carried on separate foundations or on the building steel. Stacks are practically limited to drafts of less than 3 in., and heights seldom exceed 200 to 300 ft. above ground level or 100 to 200 ft. above the roof of the plant. Actual draft requirements may range as high as 30 in. of water and compel the use of mechanical draft equipment, *i.e.*, jets and fans. Jets, using steam, are limited to the smallest installations and are very inefficient (less than 1 per cent thermal efficiency). Fans and blowers of the centrifugal type prevail. Specification and selection originate with estimation of requisite capacity and head. Table 15 should help on the latter, and combustion calculations are basic to the former. After best estimates are prepared, the values should be raised to include allowances for extreme furnace and setting conditions, leakage, dirtiness, and wear of fan. Allowances can aggregate 25 per cent on both head and capacity. Allocation among separate blowers is dependent upon selection of unit system or the common duct system for a battery of boilers. The system-head curve will ultimately be developed as in Fig. 20. This curve must be balanced against the fan characteristics, which are the inherent exact performance of centrifugal and propeller machinery.

Table 15. System Resistance Factors, Forced- and Induced-draft Systems

Head loss α (volume rate of flow)^n

Item	Exponent n
Ducts	1.8–2
Boilers and economizers	1.8–2.5
Superheaters	1.6–1.8
Air preheaters	2.0–2.3
Stokers and solid fuel beds	1.0
Forced-draft system, preheater, air ducts and stoker	1.0–1.3
Induced-draft system for boiler, superheater, economizer, preheater, flues, and breeching	2.0–2.3

The nature of the fan characteristic will depend upon design detail of **centrifugal fans,** as illustrated by Fig. 21, which shows a rising head characteristic for forward-curved blades, a flat characteristic for radial-tip blades,

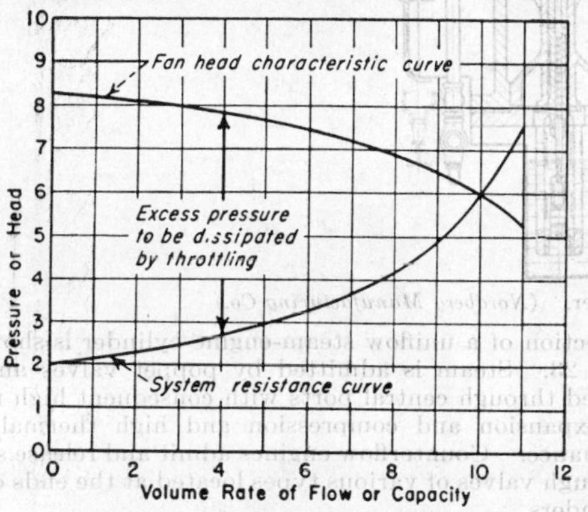

FIG. 20. Fan and system characteristics.

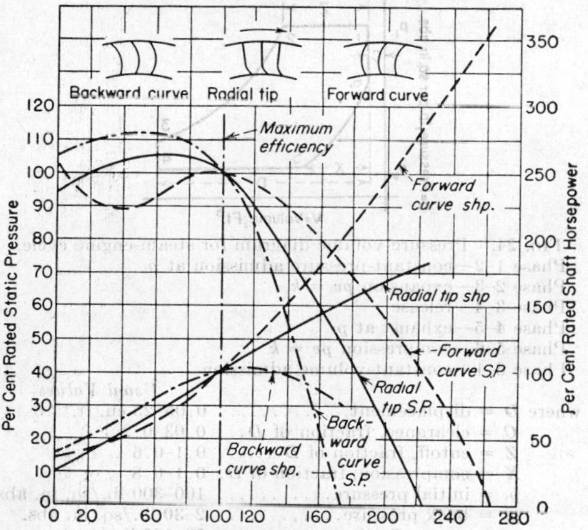

FIG. 21. Comparative characteristics for centrifugal fans. Percentage basis.

and a falling characteristic for backward-curved blades. The shaft-horsepower curves are, respectively, concave upward, straight inclined, and concave downward. The backward-curved blade fan is favored for forced draft, because it is high speed and has maximum reserve head and self-limiting horsepower. The radial-tip or forward-curved fans are preferred for induced draft, because they are slower speed and handle maximum volumes with a given wheel size. Constant-speed drive is generally preferred. This requires that all excess head between the system and the fan characteristic curves be throttled out by damper. This arrangement gives minimum investment but maximum operating expense for power. Variable speed will save power, because power is proportional to the cube of the speed at the same efficiency point. This entails additional investment for wound rotor motors, magnetic couplings, hydraulic couplings, or steam turbines. In many cases, adjustable vanes in the fan inlet offer a reasonable economic compromise with constant-speed operation. Axial-flow fans are limited to forced-draft service and to small capacities because of their high speed and characteristics, as shown in Fig. 22.

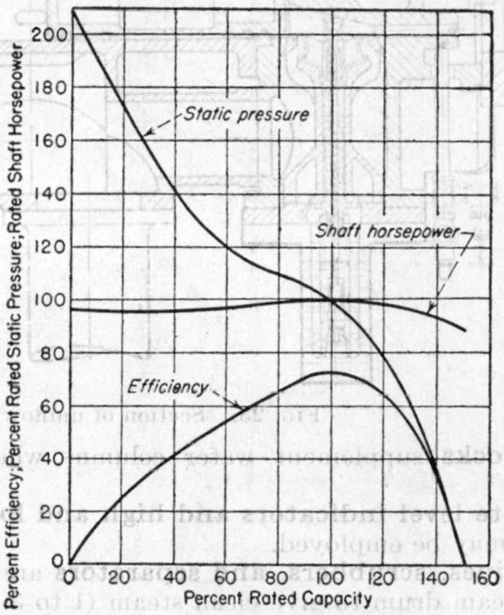

FIG. 22. Percentage characteristics—axial-flow fan.

Breeching and duct work are generally of plate with external insulation. An internal cement lining is sometimes included to protect against erosion by dust-laden gases. Dampers and cross connections are essential for sectionalizing and for basic control purposes. Splitter vanes are used in sharp radius turns for good aerodynamic design.

Cinder traps and **dust catchers** are indicated when fly ash must be eliminated. They are generally bulky, heavy, and expensive and often of little direct economic advantage. If installed, they should be placed ahead of the induced-draft fan to reduce erosion on this equipment.

Powdered-coal stack gases may contain 80 per cent of the ash in the fuel and the bulk less than 50 μ size. Powdered-coal dust loadings range from 3 to 20 oz./1000 cu. ft. Settling chambers and mechanical separators will remove larger particles; many will show 60 per cent efficiency even with the finest particles; but electrostatic precipitators are needed for high efficiency (more than 90 per cent) on the finest particles (less than 40 μ). Particles smaller than 20 μ are invisible to the naked eye.

Boiler Accessories. Pressure gages are required on all pressure parts and are of the Bourdon tube type.

Safety valves are required and are of the spring-loaded poppet type usually less than 3 or 4 in. diameter, a lift less than 1 in., and a relieving capacity less than 50,000 lb./hr./valve. A multiplicity of valves is employed, each set for a different relieving pressure, the first to open being on the superheater outlet. Power control valves may supplement, but not supersede, safety valves.

Water columns for direct indication of water level are a legal requirement. They must be so installed as to give true level, and must be well illuminated and readily visible to the operator.

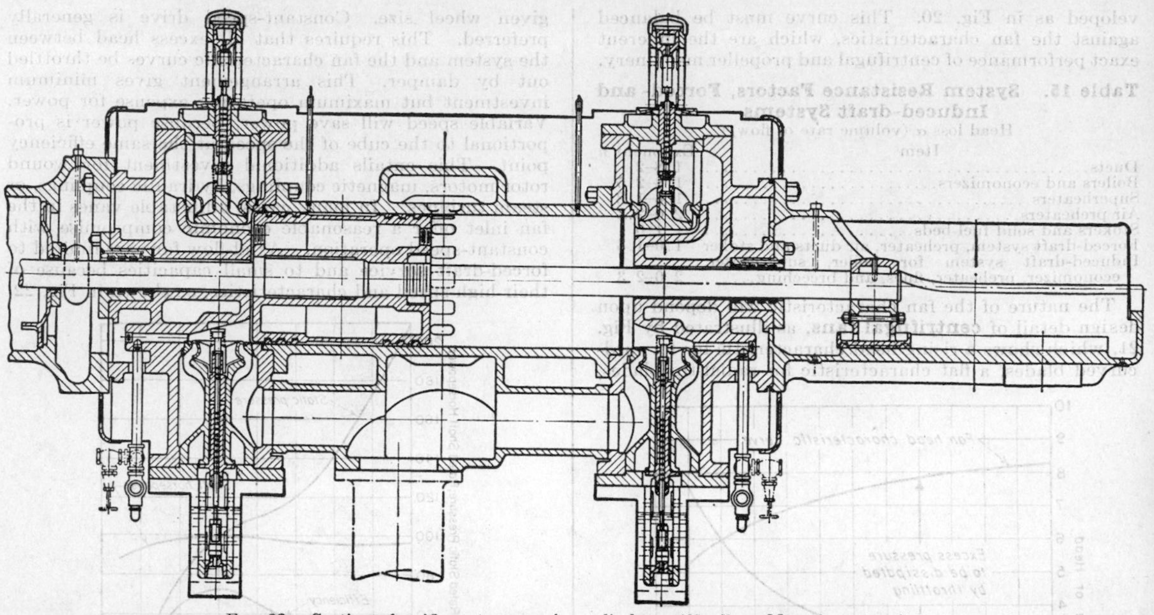

FIG. 23. Section of uniflow steam-engine cylinder. (*Nordberg Manufacturing Co.*)

Try cocks supplement water columns when out of service.

Remote level indicators and high and low water alarms may be employed.

Dry pipes, scrubbers, and separators are included in the steam drum to give clean steam (1 to 3 p.p.m. of solids). Disengagement rates range from 500 to 5000 lb. steam/hr./sq. ft. of disengaging surface.

Blowdown system will be (1) surface to remove floating matter and (2) bottom to remove heavier solids.

Soot blowers, generally steam or compressed air, are operated once or twice a shift to clean surfaces.

Instruments of indicating, recording, or integrating types are used to give pressures, temperatures, water flow, steam flow, draft, carbon dioxide, and fuel weights.

Combustion control may be made fully automatic and should be superimposed only on an adequate basic manual control system.

Steam Prime Movers. Steam prime movers may be reciprocating engines or turbines. Advantages of the former are slow speed, high efficiency in small sizes, starting torque, and foolproof operation. Advantages of the latter are high speed, maximum sizes (200,000 kw.), minimum floor space and bulk, maximum efficiencies (80 to 90 per cent engine efficiency in larger sizes), suitability for highest pressures (2300 lb./sq. in.), highest temperatures (1050°F.), and highest vacuums (1 in. Hg abs.). Engines and turbines both have utility, *e.g.*, engines to drive reciprocating pumps and compressors, turbines to drive electric generators, centrifugal pumps, and turbocompressors.

Steam engines may be of the following types:

Single or multiple cylinder.
Horizontal, vertical, or inclined.
Simple or multiple expansion.
High or low speed.
Condensing or non-condensing.
Counterflow or uniflow.
Single or multiple valve.
Slide, plug, or poppet valve.
Flyball or flywheel governed.
Throttle or cutoff governed.
Reversible or non-reversible.

A section of a uniflow steam-engine cylinder is shown in Fig. 23. Steam is admitted by poppet valves and released through central ports with consequent high ratios of expansion and compression and high thermal performance. Counterflow engines admit and release steam through valves of various types located at the ends of the cylinders.

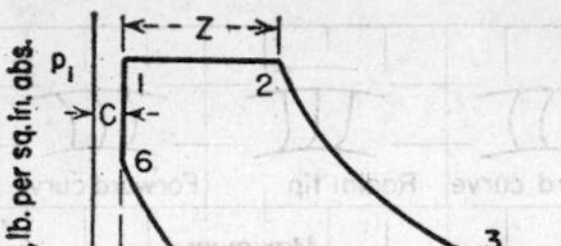

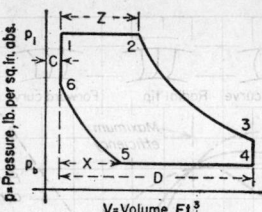

FIG. 24. Pressure-volume diagram for steam-engine cycle.
Phase 1–2—constant-pressure admission at p_i
Phase 2–3—expansion $pv = k$
Phase 3–4—release
Phase 4–5—exhaust at p_b
Phase 5–6—compression $pv = k$
Phase 6–1—constant-volume admission

		Usual Values
where D =	displacement	0.05–25 cu. ft.
C =	clearance, fraction of D	0.03–0.2
Z =	cutoff, fraction of D	0.1–0.6
X =	compression, fraction of D	0.1–0.8
p_i =	initial pressure	100–300 lb./sq. in. abs.
p_b =	back pressure	2–30 lb./sq. in. abs.
p_m =	mean effective pressure	50–125 lb./sq. in.

The basic thermodynamic cycle is illustrated by Fig. 24. The hypothetical mean effective pressure, hyp. p_m, is computed by the following formula:

$$\text{hyp. } p_m, \text{ lb./sq. in.} = p_i\left[Z + (Z + C)\log_e\frac{1 + C}{Z + C}\right] - p_b\left[(1 - X) + (X + C)\log_e\frac{X + C}{C}\right] \quad (19)$$

where the terms are as defined in Fig. 24.

The actual indicated mean pressure (ind. p_m) from a real indicator card is related to the hypothetical value by the diagram factor (Table 16), where

$$\text{ind. } p_m = \text{hyp. } p_m \times \text{diagram factor} \tag{20}$$

Table 16. Steam Engine Diagram Factors at Rated Load

Over-all range	0.50–0.95
Usual range	0.75–0.95
Simple slide-valve engine, small size	0.50–0.80
Simple single-valve automatic engine, well-maintained	0.80–0.90
Simple four-valve engine	0.85–0.95
Compound four-valve engine	0.75–0.85
Uniflow engine	0.85–0.95

Usual values of ind. p_m range from 50 to 125 lb./sq. in. Horsepower is given by

$$\text{HP.} = \frac{p_m L a n}{33{,}000} \tag{21}$$

where L = length of stroke, ft.
a = area of piston, sq. in.
n = number of cycles completed per min.

Dimensions are limited in practice, as given in Fig. 24, and stroke = (1 to 2) × bore; piston speed = 500 to 750 ft./min.; 300 r.p.m. with poppet or slide valves, 150 r.p.m. with Corliss or plug valves.

Water rates are computed by means of the same basic data:

Water rate, lb./hp.-hr.

$$= \frac{13{,}750}{p_m}\left[(Z + C) - (X + C)\frac{p_b}{p_i}\right]\bar{w}_i \tag{22}$$

where $\bar{w}_i$ = initial steam density, lb./cu. ft.

Hypothetical or indicated values of p_m may be used in Eq. (22) to give two different water rates. The actual measured water rate will be from 25 to 100 per cent higher than the indicated water rate because of cylinder condensation and internal leakage. Compound or uniflow design reduces condensation as does superheat. The latter shows 1 per cent gain for each 10°F. of superheat. Steam temperature for running clearances is limited to 600 to 700°F. (200°F. of superheat).

Friction mean pressure of 3 to 10 lb./sq. in. or mechanical efficiency of 0.8 to 0.95 bring values of power and steam consumption to the shaft of the engine. Engines may be governed by throttling the initial pressure p_i or by varying the cutoff Z. The latter gives better economy at part load (see Fig. 25). Both water rates and total steam (Willans') lines are shown.

Steam turbines are essentially a series of calibrated nozzles in which kinetic energy is developed by expansion of the fluid and the energy is delivered at the rotating shaft. Impulse turbines contain stationary nozzles, followed by a row or rows of moving vanes mounted on the periphery of a wheel or wheels. Reaction turbines contain moving nozzles. Vanes or buckets are arranged in an annulus about the drum so that passages between adjacent buckets contain a throat section for fluid acceleration. Both impulse and reaction turbines have stationary nozzles to bring the fluid to the moving blades. Full circumferential admission is required on reaction turbines because of pressure drop across the moving buckets. Partial circumferential admission may be used on impulse turbines. Hence the reaction principle is limited to large turbines (more than 2000 kw.) or to the vacuum end only.

Velocity for free expansion of steam is given by

$$\text{Jet velocity, ft./sec.} = 223.7\sqrt{\Delta h} \tag{23}$$

where Δh is the enthalpy drop, B.t.u./lb., between upstream and downstream pressures and as evaluated by the Mollier chart [see Eq. (8) and Fig. 17, p. 332, Sec. 3] for isentropic expansion in the ideal limit. Jet velocities range from 500 to 5000 ft./sec. Practical bucket speeds are limited to 600 ft./sec. (maximum 1200 ft./sec.). Single-velocity-stage impulse blading runs at 0.45 jet

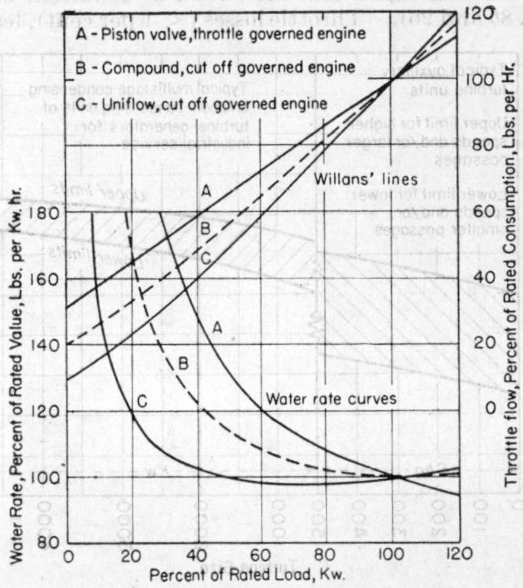

Fig. 25. Willans' lines and water-rate curves for typical steam engines.

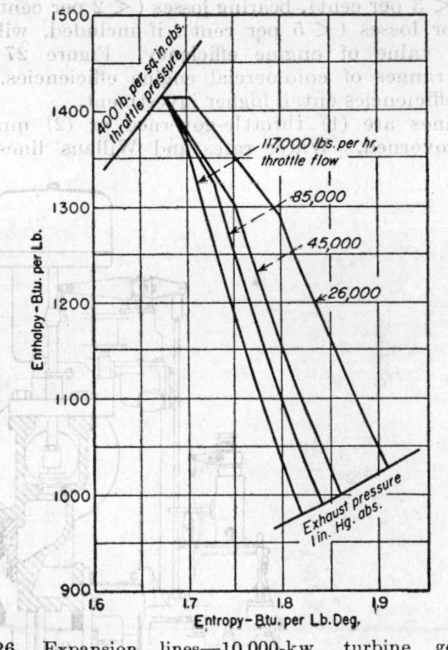

Fig. 26. Expansion lines—10,000-kw. turbine generator. Throttle, 440 lb./sq. in. abs., 800°F. Exhaust, 1 in. Hg abs.

speed and reaction blading at 0.9 jet speed for usual 10 to 20 deg. exit vane angles. Staging is needed to give acceptable speeds and efficiencies. Impulse turbines may be (1) velocity staged by reentry on a single wheel; (2) Curtis, two- or three-row velocity; or (3) Rateau pressure staged with a single wheel following each row of nozzles, in a cellular construction. Reaction turbines use more stages than do impulse turbines.

Internal component losses in turbines include nozzle

(2 to 10 per cent), blade (10 to 50 per cent), disk and fanning (2 to 5 per cent), and leakage (1 to 5 per cent). The aggregate effect is expressed in engine efficiency, where [see Eq. (8)]

$$\text{Engine efficiency} = \text{actual work/Rankine cycle work} \quad (24)$$

The resulting expansions are shown as condition curves (Figs. 8*b* and 26). Throttle losses (< 1 per cent), leaving

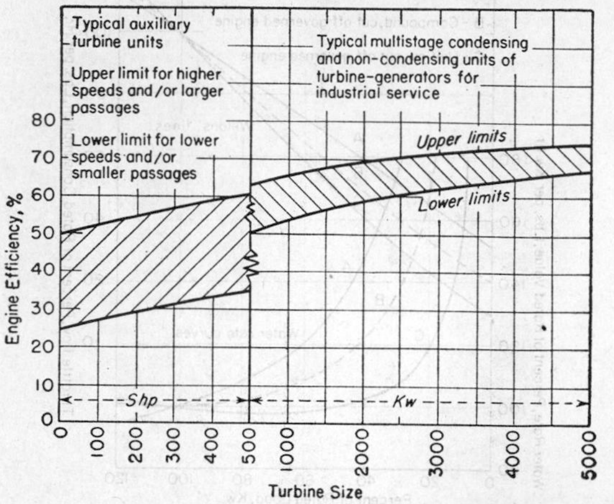

FIG. 27. Representative engine efficiencies for small steam turbines.

losses (< 3 per cent), bearing losses (< 2 per cent), and generator losses (< 5 per cent), if included, will give over-all value of engine efficiency. Figure 27 gives typical ranges of commercial engine efficiencies. The higher efficiencies entail higher investment.

Turbines are (1) throttle-governed or (2) multiple-nozzle-governed. Water rates and Willans' lines similar to those for steam engines (Fig. 25) obtain. Multiple-nozzle governing is more expensive initially but gives improved part-load economy.

Multistage turbines are particularly adaptable to extraction operation. The pressure on any stage varies directly with flow as typified by Fig. 28. For industrial process steam applications the turbine may contain an internal variable orifice or grid valve (Fig. 29) to alter the opening below the bleed point and thus to maintain con-

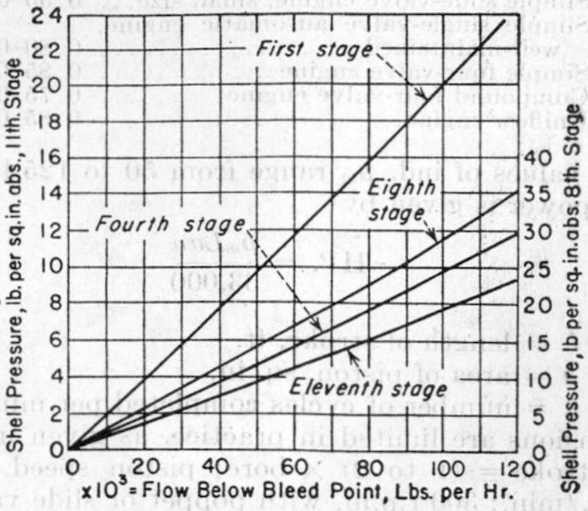

FIG. 28. Turbine shell pressures—10,000-kw. unit, 3,600 r.p.m., 14 stages. Throttle, 440 lb./sq. in. abs., 800°F. Exhaust, 1 in. Hg abs.

stancy of extraction pressure. Resulting performance is typified by the data of Fig. 30. The grid valve is under control of the extraction pressure governor, and the first stage valves are under the control of the speed governor. Turbines are available (1) with two automatic constant extraction pressures and (2) for mixed-pressure operation where low-pressure steam may be introduced at an intermediate stage.

Auxiliary turbine equipment includes lubricating system; labyrinth stuffing boxes, glands, and sealing system; stop valves and check valves; casing drains;

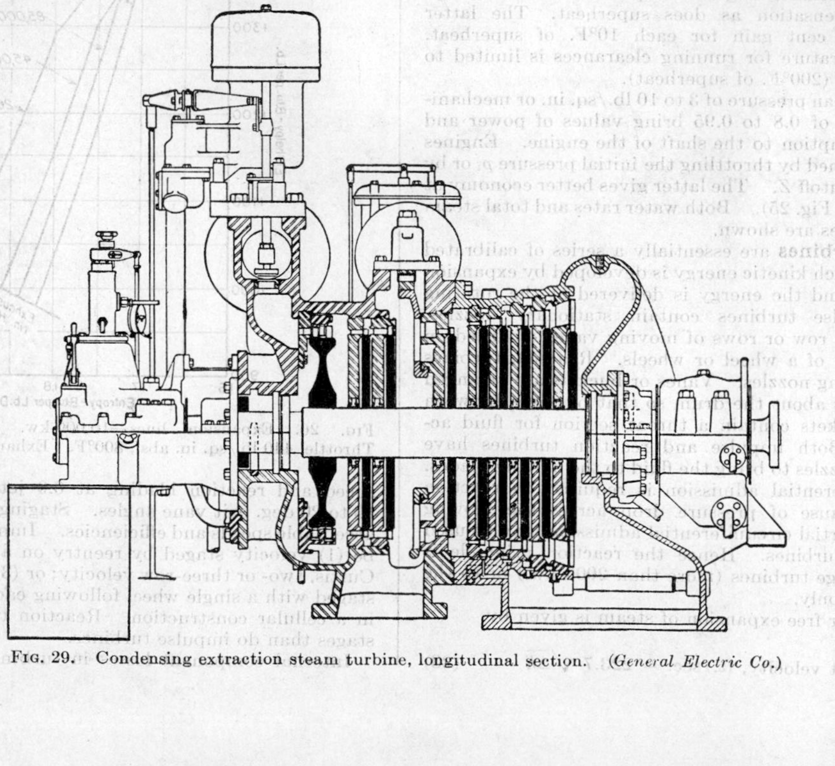

FIG. 29. Condensing extraction steam turbine, longitudinal section. (*General Electric Co.*)

insulation blanket; vacuum breaker; atmospheric relief valve or rupturing diaphragm; and operating and supervisory instruments.

Condensers and Auxiliaries. Condensers are of (1) the surface type (Fig. 31) or (2) the barometric type (Fig. 32). The former is a horizontal nest of ⅝- to 1-in.-diameter copper alloy tubes between tube sheets, the assembly placed in a steel or cast-iron shell, with water circulating through the tubes in a one-, two-, three- or four-pass arrangement, and steam condensing on the outside. The latter is a mixing chamber where circulating water is injected into the steam space, condensing the steam, the mixture being withdrawn by a pump or barometric tail pipe. Injection condensers are of minor importance, being limited to small installations where pure feed water is of no concern and poor vacuums (3 or 4 in. Hg abs.) are acceptable. Surface condensers prevail, conserve pure water, and give the highest vacuums (less than 1 in. Hg abs.). The value of vacuum is illustrated by the data of Fig. 9*c*, but turbines alone can effectively utilize the highest vacuums because of the prohibitive volumes encountered.

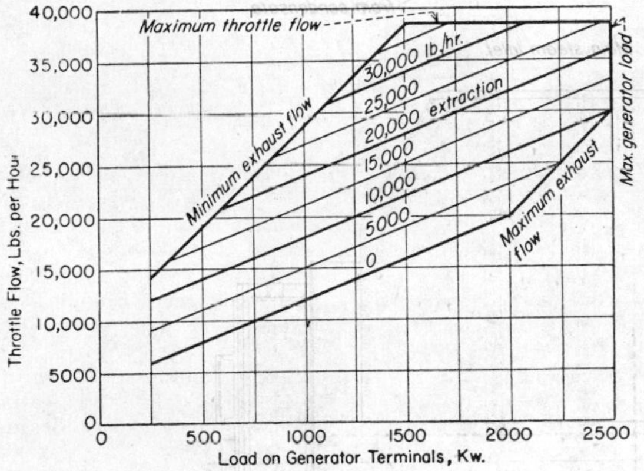

FIG. 30. Willans lines for an extraction turbine generator with single constant pressure extraction point.

Circulating water requirements are large (see Fig. 33) and may range to 50 or 100 lb./lb. steam condensed. Surface requirements can be estimated from Fig. 34. Cleanliness factors of 60 to 90 per cent are usual to allow for contamination. Reverse flow and divided water boxes are desirable. There should be no refrigeration of the condensate, and oxygen content can be held below 0.01 cc./l. with a deaerating hot well. Pressure drop between the exhaust hood and the hot well should be negligibly small.

Circulating pumps are of the centrifugal or axial-flow types with characteristics as in Fig. 35. Condenser elevation and siphon loop limit head requirements to friction and kinetic energy in the condenser passages and intake and discharge piping. Circulating water intake and discharge systems must prevent mixing of hot and cold water, must contain screens to eliminate solids, and may include chlorination equipment to reduce organic growth.

The source of water is usually a river or pond; but, where supply is limited, a reclamation system utilizing evaporative cooling in **spray ponds** or **towers** is used. The loss by evaporation will at least equal the water rate of the prime mover. With spray ponds the loss may be doubled because of physical carry-over by wind. With cooling towers, these losses are less. Induced-draft towers are generally preferred to forced or natural draft. Cooling to within 10 or 20°F. of the entering wet-bulb air temperature is good design.

Air removal pumps of the steam-jet type, two or three stage, prevail. Typical operating capacity for steam-turbine service and surface condensers serving steam turbines is 2 to 5 cu. ft./min. free air for sizes up to 100,000 lb. steam/hr. Values must be at least double for engines and surface condensers. Capacity must be many times larger on injection condensers. Multistage jets are provided with inter and after condensers for thermal economy and require typically 10 lb. steam at 100 to 200 lb./sq. in. to move 1 lb. dry air (see Sec. 20).

Hot-well pumps, designed for a head of 100 to 200 ft. may be positive-displacement but are more likely to be multistage centrifugal. Figure 36 gives typical performance and shows the critical significance of suction conditions.

Feed-water Heaters. Feed-water heaters are of (1) open, contact type or (2) closed, non-contact type. In open heaters steam is supplied to a steel or cast-iron shell, and water is introduced into the same space as spray through nozzles or over trays. The heat transfer space requirement is given by Eq. (25) for the deaerating type of heater.

$$\text{Contact volume, cu. ft.} = \frac{\text{B.t.u./hr.}}{30{,}000 \text{ to } 40{,}000} \tag{25}$$

Final water temperature is the saturated value for the operating pressure. In closed heaters, water is circulated at velocities less than 10 ft./sec. through non-ferrous tubes (0.5 to 1.0 in. diameter) arranged in a bundle and contained in a cylindrical steel shell. Heat-transfer data are given by Fig. 37, and usual terminal temperature differences are 3 to 10°F. at full load. Closed heaters are of a multipass construction with vertical or horizontal arrangement. Straight or U-tubes are preferred to curved, corrugated, or helical tubes, particularly with distilled water. Drainage is by traps or pumps. Open heaters are not suitable for operation on engine exhaust steam because of oil contamination. When open heaters are arranged in a series for stage heating, a pump is needed under each heater. Open heaters are most conveniently vented under positive pressure. A standard arrangement for a feed heating cycle is to use such a heater to give a 215°F.+ source of hot water and to provide 10 or 15 min. storage of such water under the heater.

Boiler Feed Pumps. Continuity of feed-water supply, under pressure, is essential for safe operation of boilers. Multiplicity of feed pumps, with alternate types of drive, prevails. Pumps may be of the injector type or the reciprocating slow-speed type for capacities below 25,000 lb./hr. but the centrifugal types are more common, particularly for all larger plants. System-resistance curves and pump characteristics are shown typically in Fig. 38. With constant-speed operation, excess head at part loads must be throttled. With variable speed, however, there is a large operating saving because power is proportional to the cube of the speed; but investment charges are high for speed control on motors, steam turbines, or hydraulic couplings. Suction conditions must comply with data of Fig. 39 because of possible flashing and cavitation. Specific speeds must comply with experience limits as set in Sec. 20. Head characteristics shall be continuously steep, as in Fig. 38, to assure good parallel operation of pumps.

Feed-water Treatment. Feed water must be pure to reduce or eliminate (1) scaling of surfaces and possible burning of boiler tubes, (2) corrosion and pitting of surfaces, and (3) foaming and priming with consequent carry-over of solids. The sources of contamination include (1) floating or suspended solids, (2) dissolved solids or hardness, (3) dissolved gases, (4) acidity or

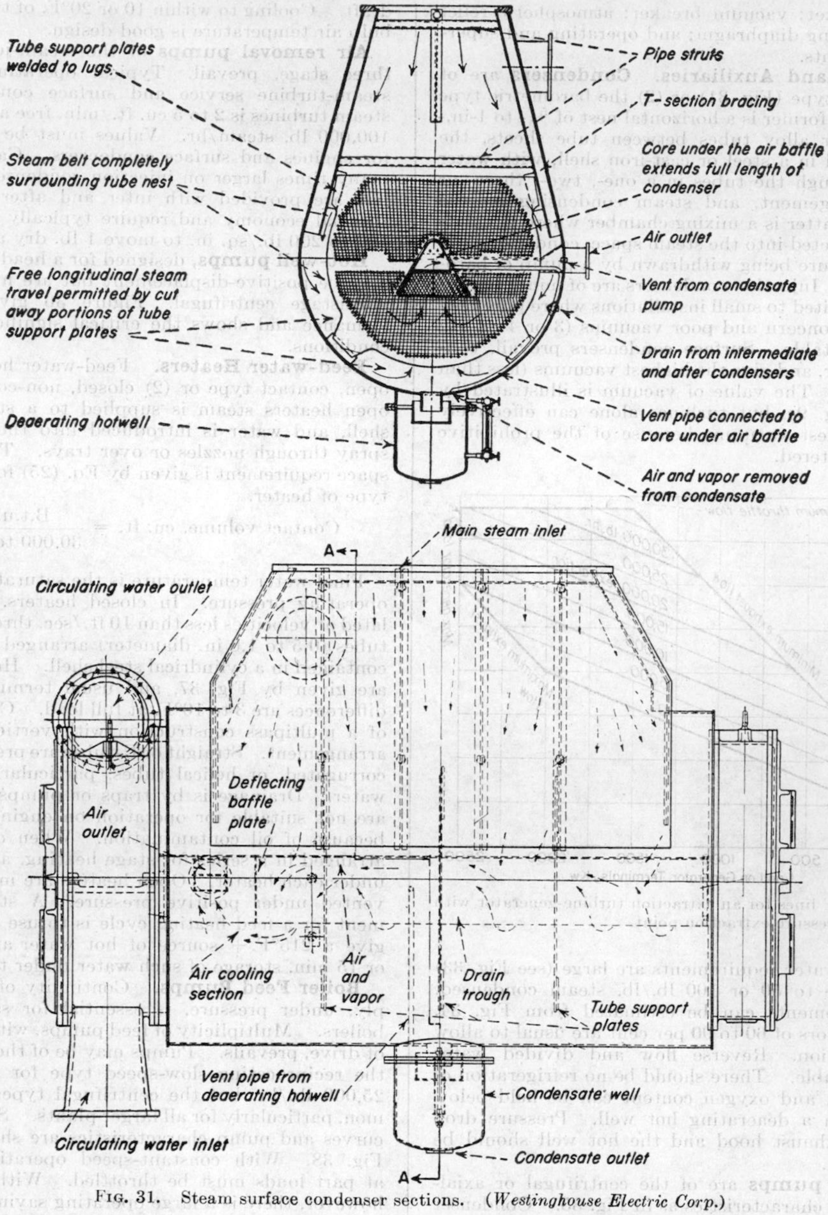

FIG. 31. Steam surface condenser sections. (*Westinghouse Electric Corp.*)

alkalinity, and (5) suspended or dissolved liquids and greases.

Among the prominent methods employed for controlling these items to make available waters suitable for boiler feed are:

1. Screens and racks.
2. Filtration, sedimentation, coagulation.
3. Chemical treatment, including zeolites, sodium, and acid; lime soda; and demineralization.
4. Steam evaporators.
5. Deaerators.
6. Chemical residual scavengers and supplementary treatment.
7. Separators, eliminators, scrubbers.

These methods, separately or in combination, give the control of feed water to less than 300 p.p.m. total solids and pH of 9. This should deliver clean steam (1 to 3 p.p.m. of solids) from boiler water of pH 10 to 11 and solids of less than 2000 p.p.m., maintained by intermittent or continuous boiler blowdown.

Evaporators may be single- or multiple-effect. On steam plants with no process requirements the make-up will run from 1 to 5 per cent. With process steam the make-up may run to 100 per cent when the choice is (1) chemical treatment or (2) single-effect evaporators using the prime mover exhaust as a heat source to deliver vapor directly to process. If the make-up amounts to more than 10 to 20 per cent the use of an evaporator-distiller combination is precluded because of inadequate heat-absorbing capacity in the feed system. An open preheater serves to degasify the water and to eliminate some hardness before introduction into the evaporator shell itself. Surfaces can be made self-scaling by use of bowed or helical tubes. Economical selection of evaporators is based on 30 to 40°F. mean temperature difference.

Deaerators are of the steam-scrubbing type and will deliver water with less than 0.005 cc./l. oxygen. Open

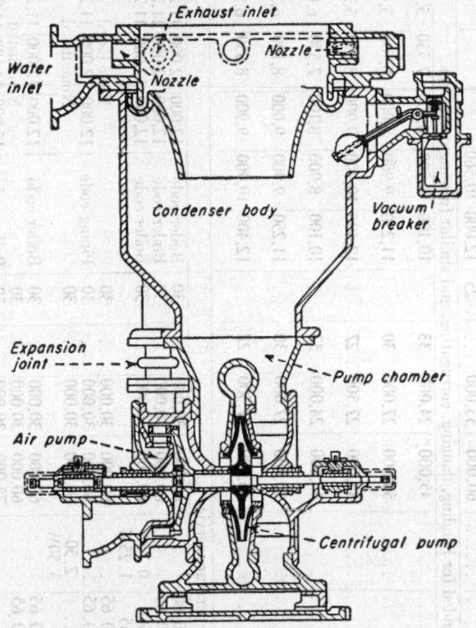

Fig. 32. Multijet injection condenser section.

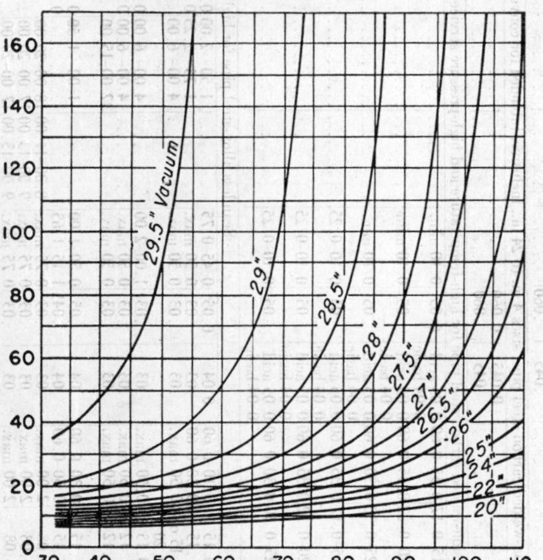

Fig. 33. Hypothetical condenser circulating-water requirements (29.92 in. Hg barometer).

heaters and condenser hot wells will ordinarily deliver water with not less than 0.03 cc./l. oxygen. A 30°F. temperature rise is the usual specification for good deaerator performance. Positive-pressure operation is preferable to vacuum operation because of simplification of the air-removal system. Vent condensers preserve heat and water, but low pH (4 to 5) makes material selection for tubes and sheets difficult. Removal of ammonia and carbon dioxide is much more troublesome than removal of oxygen. Residual scavenging of oxygen is accomplished by chemical introduction such as sodium sulfite with 10 to 15 p.p.m. residual in boiler water.

Piping. The piping services in a power plant include feed water, live steam, process steam, exhaust steam,

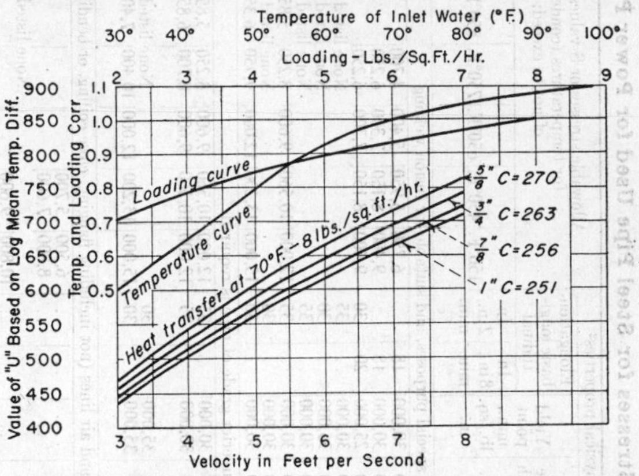

Fig. 34. Steam surface-condenser heat-transfer rates. (*Heat Exchange Institute.*)

Heat transfer based on $C\sqrt{\text{velocity}}$. Applicable to Muntz, Admiralty, red brass, aluminum-brass, copper, and arsenical-copper tubes. For copper-nickel tubes, use 90 per cent of above values.

Loading curve based on $\sqrt[4]{\dfrac{\text{lb./sq. ft./hr.}}{8}}$

Curves apply to condensers serving turbines; for engine service use 65 per cent of above values.

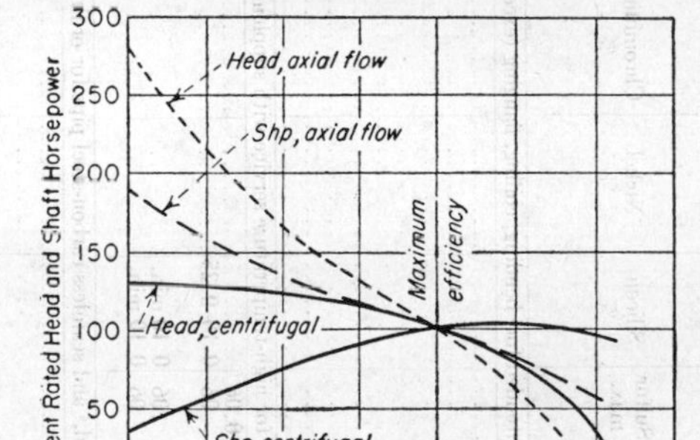

Fig. 35. Comparative characteristics of centrifugal- and axial-flow circulating water pumps—percentage basis.

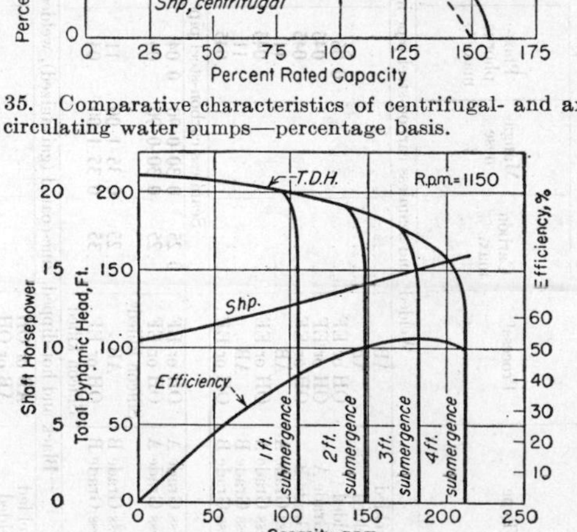

Fig. 36. Two-stage hot-well pump characteristics showing influence of suction head submergence.

Table 17. Consolidated Table of Chemical Compositions, Physical Properties, and Allowable Stresses for Steel Pipe Used for Power Piping*

A.S.T.M. serial designation	Type	Process[1]	Chemical composition: Carbon max. %	Manganese %	Phosphorus max. %	Sulfur max. %	Silicon %	Nickel %	Chromium %	Molybdenum %	Tungsten or vanadium %	Physical properties[2]: Tensile strength min., lb./sq. in.	Yield point min., lb./sq. in.	Elongation, basic longitudinal: In 8 in., min. %	Elongation: In 2 in., min. %	Allowable stresses or S values, lb./sq. in. for temperatures (condensed)[3] of not to exceed: 150°F.	450°F.	650°F.	750°F.	850°F.	950°F.	1000°F.
A53	Welded[4] and seamless carbon-steel pipe intended for[5] bending, coiling, flanging (except for butt-welded), and other special purposes, and suitable for fusion welding																					
	Butt-welded	AB										50,000	30,000	18	..	6,750	5,950	5,400	4,700			
	Lap-welded	AB										50,000	30,000	18	..	9,400	8,150	7,300	6,250			
	Lap-welded	OH or EF			0.06							45,000	25,000	20	30	9,400	8,150	7,300	6,250			
	ERW Grade A	OH or EF			.045							48,000	30,000	..	35				None listed			
	ERW Grade B	OH or EF			.045							60,000	35,000	..	30				None listed			
	Seamless Grade A	AB			.11							48,000	30,000	..	35				None listed			
	Seamless Grade A	OH or EF			.045							48,000	30,000	..	35	12,000	10,560	9,600	8,250	5,850	2,600	1,350
	Seamless Grade B	AB			.11							60,000	30,000	..	30				None listed			
	Seamless Grade B	OH or EF			.045							60,000	30,000	..	30	15,000	13,200	12,000	9,950	6,350	2,600	1,350
A106	Seamless carbon-steel pipe for high-temperature service with supplementary requirements where a superior grade of pipe is required[5]																					
	Seamless Grade A	OH or EF	0.25	0.30–0.90	0.04	0.06						48,000	30,000	..	35	12,000	10,560	9,600	8,250	5,850	2,600	1,350
	Seamless Grade A	OH or EF Silicon killed[6]	.25	0.30–0.90	.04	.06	0.10–0.25					48,000	30,000	..	35	12,000	10,560	9,600	8,700	6,850	3,800	2,000
	Seamless Grade B	AB	.25	0.35–1.00	.11	.06	0.10 min.					60,000	35,000	..	30				None listed			
	Seamless Grade B	OH or EF Silicon killed	.35	0.35–1.00	.04	.06	0.10 min.					60,000	35,000	..	30	15,000	13,200	12,000	10,400	7,400	3,800	2,000
A120	Black and hot-dipped, zinc-coated (galvanized), welded,[4] and seamless carbon-steel pipe for ordinary uses in steam, water, gas, and air lines (not including flanging, close coiling, or bending)																					
	Butt-welded	AB or OH												..	..	6,500	5,700					
	Lap-welded	AB or OH												..	..	8,800	7,600					
	Electric-welded	AB, OH, or EF												..	..				None listed			
	Seamless	AB, OH, or EF												..	..	10,800	9,600					
A135	Electric-resistance-welded carbon-steel pipe[7] intended for conveying liquid, gas, or vapor. Grade A only is adapted for flanging and bending																					
	ERW Grade A	OH or EF			0.045	0.060						48,000	30,000	..	35	10,200	8,950	8,150	7,000	4,950	2,200	1,150
	ERW Grade B	OH or EF			.045	.060						60,000	35,000	..	30	12,750	11,200	10,200	8,450	5,400	2,200	1,150
A139	Electric-fusion-welded carbon steel pipe, sizes 4 in. to 24 in., inclusive, intended for conveying liquid, gas, or vapor. Grade A only is adapted for flanging and bending																					
	EFW Grade A	OH or EF			0.045	0.060						48,000	30,000	..	30	9,600	8,400					
	EFW Grade B	OH or EF			.045	.060						60,000	35,000	..	25	12,000	10,560					
A155	Electric-fusion-welded carbon steel pipe for high-temperature and high-pressure service intended for bending, flanging, corrugating, and similar forming operations																					
	EFW Grade A	OH or EF	0.15–0.17	0.35–0.60	0.05 acid 0.04 basic	0.05	0.10 min.					45,000	24,000	33	..	10,100	8,900	8,100	7,550	5,150	2,350	1,200
	EFW Grade B	OH or EF	0.20–0.22	0.35–0.60	0.05 acid 0.04 basic	.05	0.10 min.					50,000	27,000	30	..	11,250	9,900	9,000	8,100	5,400	2,350	1,200
	EFW Grade C	OH or EF	0.25–0.30	0.30–0.60	0.05 acid 0.04 basic	.05	0.10 min.					55,000	27,500	27	..	12,400	10,900	9,900	8,550	5,700	2,350	1,200
	EFW Grade A	OH or EF Silicon killed[6]	0.15–0.17	0.35–0.60	0.05 acid 0.04 basic	.05	0.10–0.25					45,000	24,000	33	..	10,100	8,900	8,100	7,550	6,000	3,400	1,800
	EFW Grade B	OH or EF Silicon killed[6]	0.20–0.22	0.35–0.60	0.05 acid 0.04 basic	.05	0.10–0.25					50,000	27,000	30	..	11,250	9,900	9,000	8,100	6,250	3,400	1,800
	EFW Grade C	OH or EF Silicon killed[6]	0.25–0.30	0.30–0.60	0.05 acid 0.04 basic	.05	0.10–0.25					55,000	27,500	27	..	12,400	10,900	9,900	8,550	6,500	3,400	1,800
A158	Seamless alloy-steel pipe for high-temperature service																					
	Ferritic P3a	OH or EF	0.15	0.10–0.60	0.04	0.05	0.45–0.75		1.50–2.00	0.60–0.80		60,000	30,000	..	30	Boiler code		12,000	12,000	11,200	8,000	5,850
	Ferritic P3b	OH or EF	.15	0.30–0.60	.03	.03	0.50 max.		1.75–2.25	0.45–0.65		60,000	30,000	..	30	Boiler code		12,000	12,000	11,200	8,000	5,850
	Ferritic P5a	OH or EF	0.15 or 0.20	0.50 max.	.03	.03	0.50 max.		4.00–6.00	0.45[8]–0.65	0.75[8]–1.25W	60,000	30,000	..	30	Boiler code		12,000	12,000	11,200	8,000	5,850
	Ferritic P5b	OH or EF	0.15	0.50 max.	.03	.03	1.00–2.00		4.00–6.00	0.45–0.65		60,000	30,000	..	30				None listed			
	Ferritic P5c[9]	OH or EF	.12	0.50 max.	.03	.03	0.50 max.		4.00–6.00	0.45–0.65		60,000	30,000	..	30	Piping code		12,000	12,000	11,200	8,000	5,850
	Ferritic P6	OH or EF	.12	0.50 max.	.03	.03	0.50 max.		12.00–15.00		2.50–3.50W	60,000	30,000	..	30				None listed			
	Ferritic P11	OH or EF	.15	0.30–0.60	.04	.05	0.50–1.00		1.00–1.50	0.45–0.65		60,000	30,000	..	30	Boiler code		12,000	12,000	11,200	8,000	5,850
	Ferritic P15	OH or EF	.15	0.30–0.60	.04	.045	1.15–1.65			0.45–0.65		60,000	30,000	..	30				None listed			
	Austenitic P8a	OH or EF	.08	2.00 max.	.03	.03	0.75 max.	8.00–11.00	18.00–20.00			75,000	30,000	..	50	Boiler code		15,000	14,600	14,000	12,300	10,000
	Austenitic P8b[10]	OH or EF	.08	2.00 max.	.03	.03	0.75 max.	9.00–13.00	17.00–20.00			75,000	30,000	..	50	Boiler code		15,000	14,600	14,000	12,300	10,000
	Austenitic P8d[11]	OH or EF	.08	2.00 max.	.03	.03	0.75 max.	9.00–13.00	17.00–20.00			75,000	30,000	..	50				None listed			

* By Sabin Crocker, *Heating, Piping, Air Conditioning,* January, 1946. Reproduced by permission.

Table 17. Consolidated Table of Chemical Compositions, Physical Properties, and Allowable Stresses for Steel Pipe Used for Power Piping—(*Concluded*)

A.S.T.M. serial designation	Type	Process[1]	Chemical composition: Carbon max. %	Manganese %	Phosphorus max. %	Sulfur max. %	Silicon %	Nickel %	Chromium %	Molybdenum %	Tungsten or vanadium %	Physical properties[2]: Tensile strength min., lb./sq. in.	Yield point min., lb./sq. in.	Elongation, basic longitudinal: In 8 in., min. %	In 2 in., min. %	Allowable stresses or S values, lb./sq. in. for temperatures (condensed)[3] of not to exceed: 150°F.	450°F.	650°F.	750°F.	850°F.	950°F.	1000°F.
A206	Seamless carbon molybdenum alloy steel pipe for high-temperature service suitable for bending, flanging, and similar forming operations and for fusion welding																					
	Ferritic P1	OH or EF	0.10–0.20	0.30–0.60	0.04	0.05	0.10–0.50	....	....	0.45–0.65	.....	55,000	30,000	..	30	Boiler code or piping code		11,000	11,000	10,500	8,000	5,000
A213	Seamless alloy steel boiler and superheater tubes (these materials are sometimes specified for pipe for high-temperature service and made to nominal pipe size)																					
	Ferritic T3	EF	0.15	0.40–0.60	0.03	0.03	0.45–0.75	....	1.50–2.00	0.60–0.80	.....	60,000	25,000	..	30	Boiler code		12,000	12,000	11,200	8,000	5,850
	Ferritic T5	EF	.15	0.50 max.	.03	.03	0.50 max.	....	4.00–6.00	0.45–0.65	.....	60,000	25,000	..	30	Boiler code		12,000	12,000	11,200	8,000	5,850
	Ferritic T11	EF	.15	0.30–0.60	.03	.03	0.50–1.00	....	1.00–1.50	0.45–0.65	.....	60,000	25,000	..	30	Boiler code		12,000	12,000	11,200	8,000	5,850
	Ferritic T12	EF	.15	0.30–0.60	.04	.04	0.30 max.	....	0.80–1.10	0.45–0.65	.....	60,000	25,000	..	30	Boiler code		12,000	12,000	11,200	8,000	5,850
	Ferritic T13	EF	.15	0.50 max.	.03	.03	1.00–2.00	....	4.00–6.00	0.45–0.65	.....	60,000	25,000	..	30	Boiler code		12,000	12,000	11,000	6,000	4,200
	Ferritic T14	EF	.15	0.30–0.60	.03	.03	0.50 max.	....	1.75–2.25	0.45–0.65	.....	60,000	25,000	..	30	Boiler code		12,000	12,000	11,200	8,000	5,850
	Ferritic T16[12]	EF	.12	0.50	.03	.03	0.50 max.	....	4.00–6.00	0.45–0.65	.....	60,000	25,000	..	30	Boiler code		11,000	11,000	10,800	8,000	5,850
	Ferritic T17	EF	0.15–0.25	0.30–0.60	.04	.045	0.15–0.30	....	0.80–1.10	....	0.15 min. V	60,000	25,000	..	30	Boiler code		12,000				
	Ferritic T21	EF	0.15	0.30–0.60	.03	.03	0.50 max.	....	2.75–3.25	0.80–1.00	.....	60,000	25,000	..	30	Boiler code		12,000	12,000	11,200	8,250	6,250
	Ferritic T22	EF	.15	0.30–0.60	.03	.03	0.50 max.	....	2.00–2.50	0.90–1.10	.....	60,000	25,000	..	30	Boiler code		12,000	12,000	11,200	8,250	6,250
	Austenitic TP-304	EF	.08	2.00	.03	.03	0.75 max.	8.00–11.00	18.00–20.00	....	.....	75,000	30,000	..	35	Boiler code		15,000	14,600	14,000	12,300	10,000
	Austenitic TP-316	EF	.08	2.00	.03	.03	0.75 max.	11.00–14.00	16.00–18.00	2.00–3.00	.....	75,000	30,000	..	35	Boiler code		15,000	14,600	14,000	12,600	11,200
	Austenitic TP-321[13]	EF	.08	2.00	.03	.03	0.75 max.	9.00–13.00	17.00–20.00	....	.....	75,000	30,000	..	35	Boiler code		15,000	14,600	14,000	12,300	10,000
	Austenitic TP-347[14]	EF	.08	2.00	.03	.03	0.75 max.	9.00–13.00	17.00–20.00	....	.....	75,000	30,000	..	35	Boiler code		15,000	14,600	14,000	12,300	10,000
A280	Seamless chrome molybdenum alloy steel pipe for high-temperature service suitable for bending, flanging, and similar forming operations and for Fusion welding																					
	Seamless	OH or EF	0.10–0.20	0.30–0.60	0.04	0.05	0.10–0.30	....	0.40–0.60	0.45–0.65	.....	55,000	30,000	..	30	Boiler code		11,000	11,000	10,500	8,000	5,000

[1] Abbreviations: *AB*, acid bessemer; *EF*, electric furnace; *EFW*, electric-fusion-welded; *ERW*, electric-resistance-welded; *OH*, open hearth.

[2] In addition to the tests specified, each length of pipe shall be given a hydrostatic test and other physical tests as prescribed in the respective specification.

[3] For intermediate *S* values, see American Standard Code for Pressure Piping, ASA B31, or A.S.M.E. Boiler Construction Code. The boiler code *S* values below 650°F. are the same as at 650°F. in all cases, whereas the piping code values are as tabulated. Allowable stresses for A53 and A135 pipe above 750°F. are for the boiler code only; likewise those for non-silicon-killed A155 pipe at 1000°F. The boiler code also contains silicon-killed varieties of the following items for which there is no equivalent in the piping code: (*a*) A53 seamless pipe in both grades A and B to which it assigns the same *S* values as shown for silicon-killed A106 pipe in this table; and (*b*) A135 resistance-welded pipe.

[4] Welded pipe 4 in. and under in nominal diameter may be butt-welded. Welded pipe over 4 in. in nominal diameter shall be lap-welded or electric-welded.

[5] Grade A pipe should be used for close coiling, cold bending, or forge welding.

[6] Not shown in the A.S.T.M. specification but can be obtained with this silicon range.

[7] For applications where the temperature is below 650°F., *S* values equal to the corresponding seamless grades may be used if the pipe is subjected to supplemental tests and/or heat-treatments which demonstrate the strength of the weld to be equal to the minimum tensile strength specified for the pipe.

[8] Either molybdenum or tungsten shall be used.

[9] Grade P5c shall have a titanium content of not less than four times the carbon content and not more than 0.70 per cent; or a columbium content of eight to ten times the carbon content.

[10] Grade P8b shall have a titanium content of not less than five times the carbon content and not more than 0.60 per cent.

[11] Grade P8d shall have a columbium content of not less than ten times the carbon content and not more than 1.0 per cent.

[12] Grade T16 shall have a titanium content of not less than four times the carbon content and not more than 0.70 per cent.

[13] Grade TP-321 shall have a titanium content of not less than five times the carbon content and not more than 0.60 per cent.

[14] Grade TP-347 shall have a columbium content of not less than ten times the carbon content and not more than 1.00 per cent.

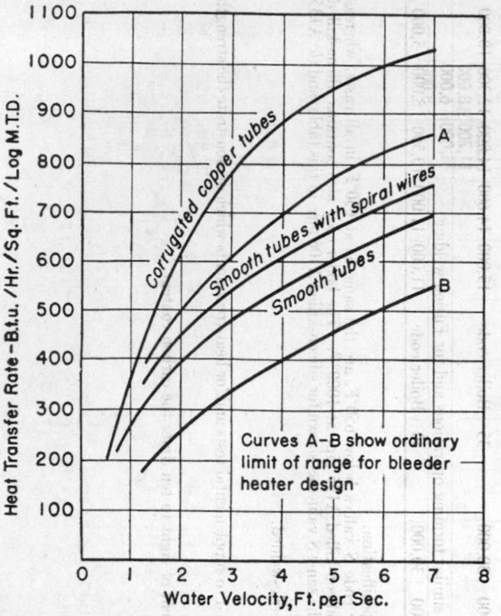

FIG. 37. Heat-transfer rates for feedwater heaters, 5/8 in. and 3/4 in. o.d. No. 18 or 20 B.W.G. tubes.

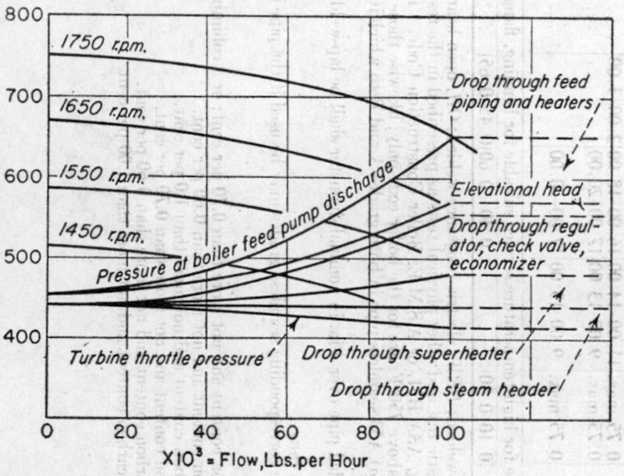

FIG. 38. Boiler-feed system resistance curve and pump characteristics.

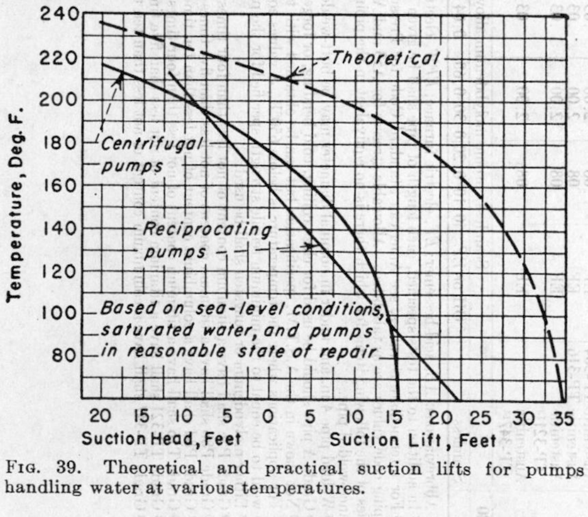

FIG. 39. Theoretical and practical suction lifts for pumps handling water at various temperatures.

vacuum, compressed air, refrigeration, gas, fuel, lubricant, service water, circulating water, safety valve, vent, blowdown, and sewage. Piping materials must be selected for the service and include carbon steel; alloy steel; black, gray, or galvanized iron; brass; bronze; copper; vitreous; clay; and wood stave. The best specifications are set by the standards of the A.S.T.M., where the data of Table 17 are representative of steel pipe for power services. Types include seamless tubing, lap-welded, butt-welded, spiral-welded or riveted, and cast. Commercial sizes are fixed by the A.S.A. standards (see Sec. 5). Joints may be welded, threaded, flanged, or bell-and-spigot. Selection of the type and size of pipe requires consideration of service, operating pressure, operating temperature, velocity, and investment and production economics. Friction loss is estimated by the methods of Sec. 6. Table 14 gives useful velocity data.

Piping systems may be the unit type, with or without interconnections, or the loop type. Needs for sectionalization for repairs must be recognized. Valves should be so installed on steam lines that there are always two valves that can be closed between the pressure source and the point of maintenance. Standard valves and fittings are used with prevalent types, as detailed in Sec. 6. Large valves are motor-operated.

Expansion provisions are met by slip joints and flexible bellows if pressures and temperatures are low (less than 100 lb./sq. in. and 250°F.). Otherwise expansion loops and bends are substituted, the design of which is a major problem in stress analysis. Anchorage must prohibit thrust from reaching connected machine members. Hangers, supports, and cradles are needed to hold piping in place. Separators, drainers, traps, and reducing valves are important accessories. Thermal **insulation** is typically:

Air-cell covering (three- or four-layer) for low-pressure steam (5 lb./sq. in.).

Diatomaceous earth and magnesia, canvas- or sheet-metal-covered, for high-pressure, high-temperature steam and water (1 to 3 in.).

Cork, felt, hair, waterproofed, for refrigeration and cold-water lines (1 to 2 in.).

Ground cork in waterproof paint for subatmospheric temperature water lines, subject to sweating.

The economical thickness of insulation is determined as given in Sec. 6. Color schemes may be used for identification; but accessibility of valves, a well-trained operating personnel, and stenciled directions at valve locations are better practice.

Instruments. Operating instruments are installed for (1) protection of equipment and (2) improved economy. Commercial, not laboratory, accuracy is to be expected. Adequate maintenance personnel must be available to assure continued useful functioning. Records should be limited to those which are used and understood by the operating personnel. Instruments of the indicating, recording, or integrating types include clocks, synchroscopes, and frequency meters; pressure and vacuum gages, barometers; flowmeters; tachometers and speed indicators; volt, ampere, watt, and watt-hour meters; CO_2 meters; salinity meters, pH instruments; weightometers; level indicators and recorders; and turbine supervisory instruments. See pp. 1263*ff*.

INTERNAL-COMBUSTION PLANTS

Internal-combustion **engines** are of the reciprocating type except in the few instances of gas turbines. The former operate on either the Otto or the Diesel cycle and the latter on the Brayton cycle (Fig. 40). Thermodynamically these cycles give higher thermal efficiencies with higher ratios of compression $R_v = V_1/V_2$ (phase 1-2). For the Otto and Brayton cycles this is computed by

$$\text{Thermal efficiency} = 1 - \frac{1}{R_v^{k-1}} \quad (26)$$

where k is the ratio of specific heats, c_p/c_v. For the Diesel cycle, the efficiency is increased also by shortening the heat addition line (phase 2-3) so that

$$\text{Thermal efficiency} = 1 - \frac{1}{R_v^{k-1}}\left[\frac{(R_c)^k - 1}{k(R_c - 1)}\right] \quad (27)$$

where R_c = cutoff ratio, V_3/V_2.

Practical engines develop approximately one-half of these ideal air card standard values. They are limited to liquid or gaseous fuels. Representative performance of real engines is given in Table 18 and Fig. 41.

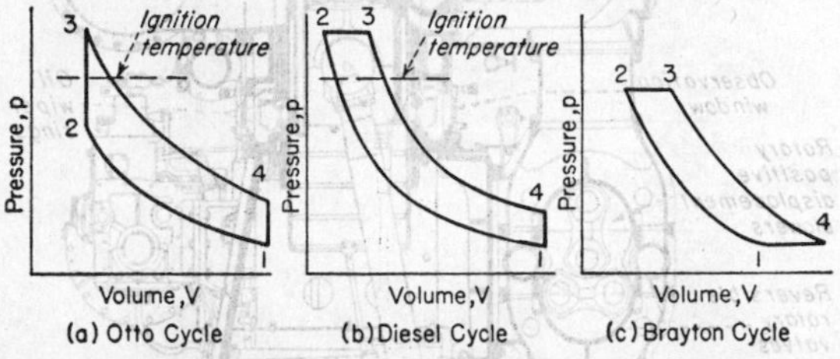

FIG. 40. Ideal indicator cards (pressure-volume diagrams) for internal-combustion cycles.
Phase 1-2—isentropic compression; compression ratio $R_v = V_1/V_2$
Phase 2-3—heat addition at constant pressure or volume
Phase 3-4—isentropic expansion
Phase 4-1—heat rejection at constant pressure or volume

Otto engines are essentially mixture engines in which an explosive fuel-air mixture is externally made in a carburetor or mixing valve and compression (phase 1-2, Fig. 40) is limited to a value below the fuel ignition temperature. Ignition is usually by electric spark. Diesel engines are of the injection type where air alone is compressed and fuel is sprayed into the combustion chamber toward the end of the compression stroke (1-2). Compression temperature must exceed the ignition temperature of the fuel. Engines may be (1) four cycle, using four strokes (two revolutions) to complete a cycle, suction, compression, expansion, and exhaust; or (2) two-cycle, using two strokes (one revolution) to complete a cycle, compression and expansion. Figure 42 shows a representative (*a*) actual four-cycle Otto engine indicator card and (*b*) actual two-cycle Diesel engine indicator card.

The four-stroke principle is favored with multicylinder, automative-type engines; and the two-stroke principle is preferred with the larger and lower speed injection engines. A typical four-cycle gas engine is shown in Fig. 43 and a typical two-cycle Diesel engine in Fig. 44. The two-cycle principle gives twice the power strokes for a given size and speed, thus reducing engine weight and cost. Scavenging requires some 1.2 times displacement, and mixture-type engines therefore show poorer economy with two-cycle as opposed to four-cycle. Scavenging with two-cycle Diesels is with air at 5 lb./sq. in. obtained by crankcase compression; front end compression; or separate rotary, reciprocating, or centrifugal scavenging pump.

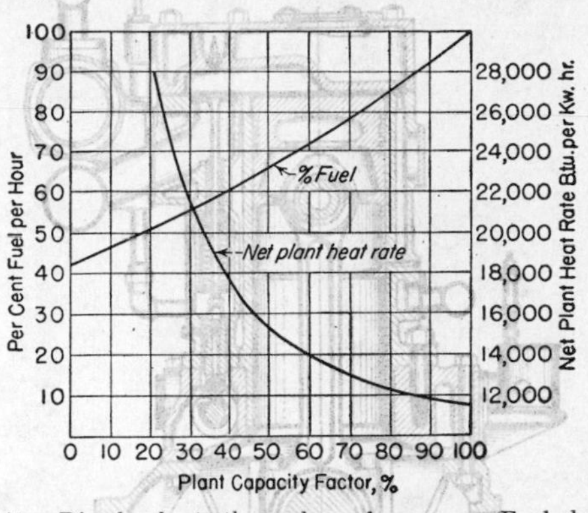

FIG. 41. Diesel plant thermal performance. Fuel heating value = 19,000 B.t.u./lb. Sp. gr. = 20° A.P.I. 148,000 B.t.u./gal. Auxiliary power = 5 per cent at 60 per cent capacity factor.

Types of engines that are available in the market and have merit for some particular service include Otto or Diesel, mixture or injection, two-cycle or four-cycle, single-acting or double-acting, gaseous or liquid fuel, volatile or non-volatile liquid fuel, high-speed or low-speed, horizontal or vertical, solid or air injection, and cylinders in line or radial.

Diesel fuel specifications are shown in Table 19. Supercharging by means of a rotary or turbocompressor

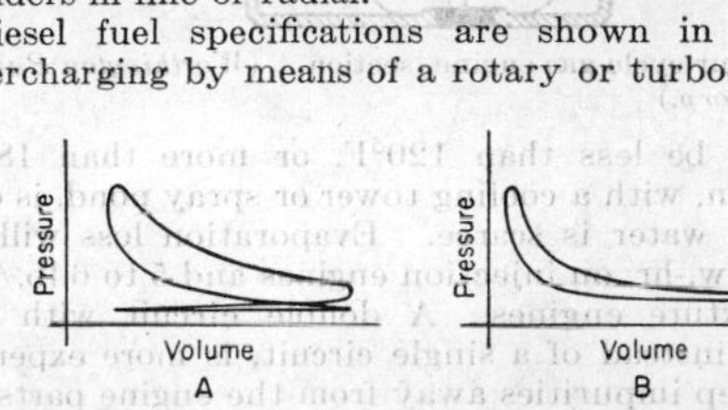

FIG. 42. Actual internal-combustion engine indicator cards. *A* = four-cycle mixture engine, Otto cycle. *B* = two-cycle injection engine, Diesel cycle.

to an initial pressure as high as 15 lb./sq. in. increases the weight charge and the mean pressure in the power cylinder, resulting in lower weights per horsepower.

Starting is usually by storage batteries with electric motor for small size and for automotive engines (less than 100 to 200 hp.). Larger engines use compressed air from bottles (200 lb./sq. in.) introduced through the cylinder head into the combustion chamber. If the

Table 18. Performance of Some Typical Internal-combustion Engines

Type	Fuel	Bhp.	Compression ratio	Brake m.e.p.	Piston speed ft./min.	Weight, lb./cu. in. piston displacement	Weight, lb./bhp.	Over-all heat rate, B.t.u./kw.-hr.
Mixture engines:								
Automotive engines	Gasoline	10– 200	4.5– 7.5	50– 100	800–1600	3 –6	10– 50	15,000
	Kerosene	10– 200	3.5– 4.5	40– 75	800–1600	3 –6	15– 55	15,000
Stationary gas engines	Natural gas	150– 800	4 – 7	50– 70	600–1200	4 –8	50–140	14,000
Injection engines:								
Solid injection, spark ignition	Diesel fuel	25– 100	5 – 7	50– 80	800–1200	3±	12– 15	14,000
Air injection Diesel	Diesel fuel	300–5000	12 –15	50– 75	600–1000	4 –8	25–200	11,000
Solid injection, compression ignition								
High-speed	Diesel fuel	20–1200	12 –17	50–110	900–1500	3.5–8	15–100	13,000
Medium-speed	Diesel fuel	50– 750	12 –15	40– 75	800–1500	3 –6	20–100	12,000
Low-speed	Diesel fuel	100–5000	13 –14	40– 75	600–1000	4 –8	25–100	11,000

engine drives an electric generator, the latter may be used as a motor for starting.

Water **cooling** prevails on stationary engines. Water must be clean and pure to prevent clogging and corrosion of the jacket system. The jacket water temperature

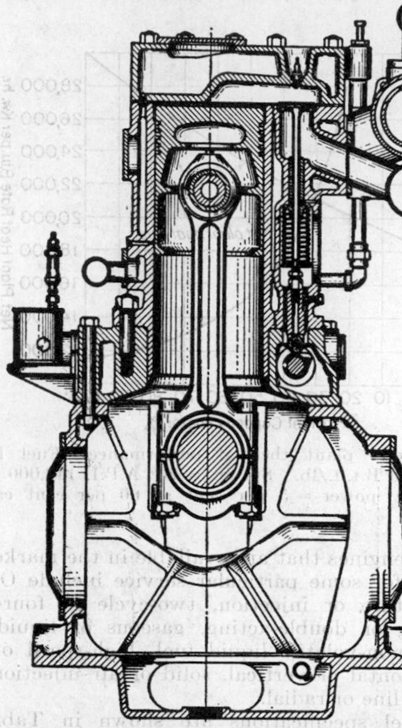

FIG 43. Four-cycle gas engine, section. (*Worthington Pump & Machinery Corp.*)

should not be less than 120°F. or more than 180°F. Reclamation, with a cooling tower or spray pond, is common where water is scarce. Evaporation loss will run 3 to 4 lb./kw.-hr. on injection engines and 5 to 6 lb./kw.-hr. on mixture engines. A double circuit with heat exchanger, instead of a single circuit, is more expensive but will keep impurities away from the engine parts and will reduce water-treating expense.

Lubrication costs may become the single largest operating expense. They are a function of engine design, type, and size; maintenance practices; operating temperatures; service hours; and load. Over-all lubricant consumption on Diesel engines ranges from 200 to

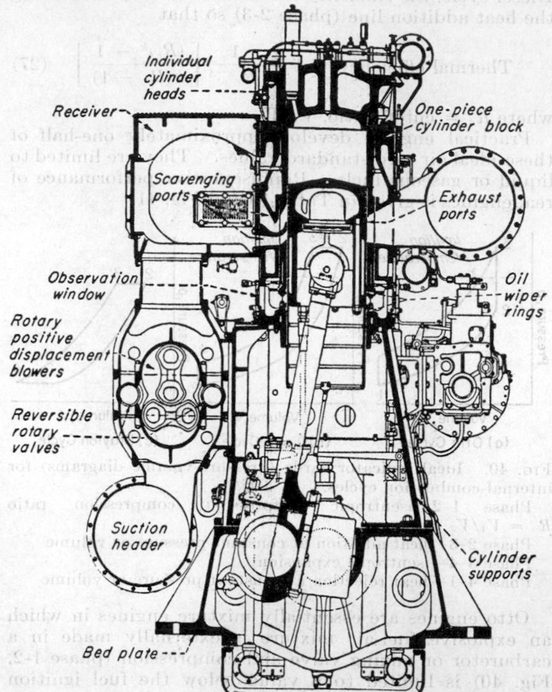

FIG. 44. Two-cycle Diesel engine, section. (*Busch-Sulzer Bros. and Power.*)

3000 kw.-hr./gal., with 1000 to 2000 kw.-hr./gal. as representative.

The potentialities for **waste heat** utilization on internal-combustion engines would, superficially, appear to meet the demands of industrial power plants for by-product power generation. Mixture engine heat input is typi-

Table 19. Tentative Specifications for Diesel Fuel Oils, Abbreviated from ASTM Designation D975-48T, 1948[a]

Grade of Diesel fuel oil	Flash point, °F., min.	Pour point, °F., max.	Water and sediment, % by volume, max.	Carbon residue on 10% residuum, %, max.	Ash, % by weight, max.	Distillation temperatures, °F. 90% point, max.	Distillation temperatures, °F. End point, max.	Viscosity at 100°F. Seconds Saybolt Universal or Kinematic Min.	Viscosity at 100°F. Seconds Saybolt Universal or Kinematic Max.	Sulfur, % by weight, max.	Copper strip corrosion, max.	Cetane number, min.
No. 1-*D*—a volatile distillate fuel oil for engines in service requiring frequent speed and load changes	100 or legal	20[b]	Trace	0.15	0.01	...	625	1.4[d]	....	0.50	Negative[e]	40[c]
No. 2-*D*—a distillate fuel oil of low volatility for engines in industrial and heavy mobile service	100 or legal	20[b]	0.10	0.35	0.02	675	...	......	45 sec. (5.8)[d]	1.0	Negative[e]	40[c]
No. 4-*D*—a fuel oil for low- and medium-speed engines	130 or legal	20[b]	0.50	....	0.10	...	...	45 sec. (5.8)[d]	125 sec. (26.4)[d]	2.0	....	25[c]

[a] The numerical values contained in the table are similar to those of the corresponding grades of commercially obtainable oil-burner fuels except for addition of ash content, sulfur limits, and cetane number. To meet special operating conditions, modification of individual fuel properties become necessary as outlined below in footnotes *b* and *c*.

[b] For cold-weather operation the pour point should be specified 10°F. below the temperature at which the engine is to be operated.

[c] Low-atmospheric temperatures as well as engine operation at high altitudes may require use of fuels with higher cetane ratings.

[d] Kinematic viscosity in centistokes.

[e] The exposed copper strip shall show no gray or black deposit.

Note. Lighter or heavier grades of distillates and heavier grades of residual fuel oils may be procured by mutual agreement between the purchaser and the seller.

cally allocated: one-fourth to power, three-eighths to cooling system, and three-eighths to exhaust. Injection engine heat distribution is roughly one-third each to power, cooling water, and exhaust. The heat in cooling water is limited to levels below 180°F. and is of doubtful worth. Exhaust heat is at higher levels (see Fig. 45). The exhaust gas temperatures of injection engines are lower than those of mixture engines because of less heat to exhaust, lower engine heat rate, and higher excess air. At part loads, the heat is at such a low level as to limit waste heat boilers to the lowest operating pressures. Likewise, low heat transfer rates require excessive boiler

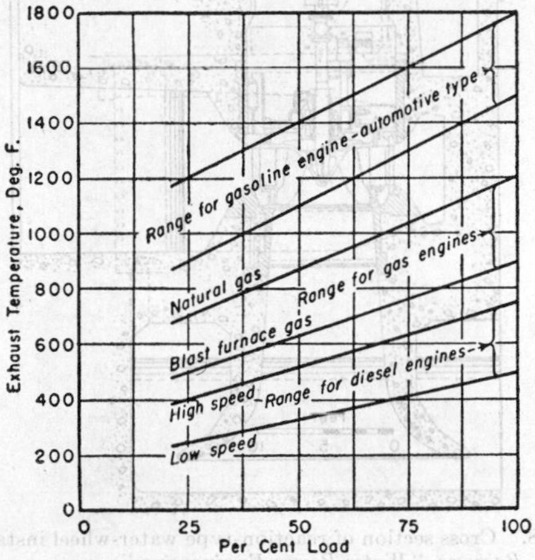

FIG. 45. Typical exhaust gas temperatures for internal-combustion engines.

surface with resultant poor economics. Waste heat opportunities of Diesel installations may be estimated by the equation

Heat recoverable from exhaust gases, B.t.u./hr.

$$= \text{shp.} \times C \times \frac{\Delta T}{4} \qquad (28)$$

where shp. = shaft horsepower at load.

ΔT = temperature drop of exhaust gases through heater, °F. (usually 50 to 100°F.).

C = 10 for four-cycle engines.

C = 24 for two-cycle engines.

Exhaust temperatures for four-cycle Diesels range from 600° to 750°F. at full load and for two-cycle Diesels from 350 to 550°F. at full load.

Foundations must be designed for shock loads and to minimize vibration problems. Exhaust silencers must be true wave traps to eliminate low-frequency vibrations and noise. Filters on air and fuel supply assure reliable operation by admitting only the cleanest air and fuel to the cylinder. Instruments include exhaust gas pyrometers, smoke indicators, jacket thermometers, lubrication gages, and injection nozzle testers.

Gas turbines operate on the Brayton cycle (Fig. 40) or a modification using intercooling on compression, reheating on expansion, and regenerative heating and cooling to improve thermal performance. Basically, the gas turbine plant is composed of a compressor, a combustion chamber, a turbine, and a generator, as shown in Fig. 46. Air is the working fluid and is generally compressed in an axial-flow unit. Fuel is burned with high excess air (500 per cent) to limit temperatures (less than 1400°F.) entering the turbine. The electric generator is used as a starting motor. The output of the unit is given by

$$\Delta W_T \times \eta_T - \frac{\Delta W_c}{\eta_c} = \text{net work of unit} \qquad (29)$$

where ΔW_T = ideal work of turbine.

ΔW_c = ideal work of compressor.

η_T = engine efficiency of turbine.

η_c = compression efficiency of compressor.

The net output, for given turbine and compressor values of work, must be high to make the unit economically attractive. High engine and compression efficiencies are imperative. This requirement has lead to the perfection of centrifugal, turbo, and rotary compressors. Engine and compression efficiencies of 80 per cent are obtainable in current practice. The influence of compression ratio and combustion temperature on performance is typified by the data of Fig. 46.

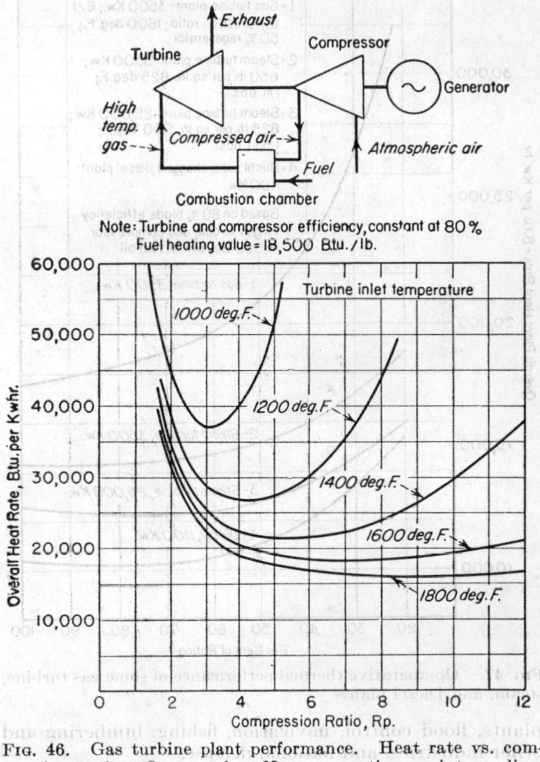

FIG. 46. Gas turbine plant performance. Heat rate vs. compression ratio. Open cycle. No regeneration or intercooling. (*After Salisbury.*)

Gas turbines are light, compact, high-speed machines with the minimum water requirements. They are limited to liquid or gaseous fuels, need high temperatures, intercooling and regeneration for best thermal performance, and offer scant opportunity for by-product power generation. The gas turbine art is, however, developing so rapidly that the use of closed cycles, regeneration, pressure combustion, and solid fuels must be anticipated. Data of Fig. 47 show some comparative performances.

WATER POWER PLANTS

Water power, today, is synonymous with hydroelectric power. Where process heat is needed, this entails the use of electric furnaces, electric boilers, or heat pumps. The power available at a hydrosite is theoretically given by

$$\text{Water horsepower, whp.} = \frac{Qh}{8.8} \qquad (30)$$

$$\text{Water kilowatts, wkw.} = \frac{Qh}{11.8} \qquad (31)$$

where Q = water flow rate, cu. ft./sec.
h = head on site, ft.

Hydrology is basic to an understanding of available water flow because runoff varies widely with climatic conditions. Hydrographic data are available from government sources for most watersheds. Reservoirs reduce the extremes of flow variation by providing storage. The economic factors affecting the capacity to be installed include load requirements, head, stream flow, cost of development, value of output, auxiliary power

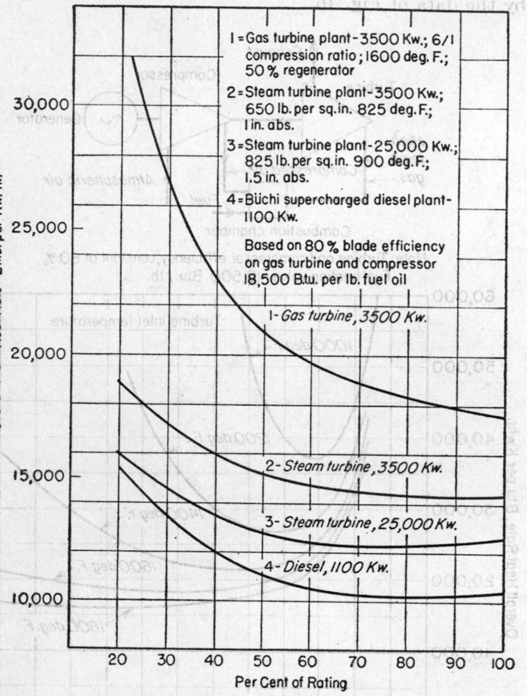

FIG 47. Comparative thermal performance of some gas turbine, steam, and Diesel plants.

plants, flood control, navigation, fishing, lumbering and other industries, and national defense.

Prime capacity is that continuously available while firm capacity will be much larger if the hydroplant is supplemented by fuel-burning units and if the load curves show swings that permit variable-capacity operations. Studies leading to the extent of development and capacity to be installed are extremely complicated and time-consuming. Base load, peak load, run of river, and pumped storage plants variously result in efforts to fit installed capacity, runoff, and storage to the load curve. Capacity (kilowatts) must be clearly distinguished from energy (kilowatt-hours). The complexity of the problem stems from the fact that the bulk of the investment is for structures and works with only a small incremental investment in wheels and generators, usually less than \$25 per kilowatt. Thus a prime capacity installation might cost \$1000 per kilowatt; but, with greater wheel capacity installed with the same structures to give firm capacity, suitable for the load curve, the cost might be reduced to \$200 per kilowatt.

Water-wheel generator sets offer efficiencies of the order of 90 per cent and are available in sizes up to 100,000 kw. They will be of the (1) impulse, Pelton type if heads are over 500 ft.; (2) the reaction, Francis, inward-flow type (Fig. 48) for heads of 50 to 500 ft.; or (3)

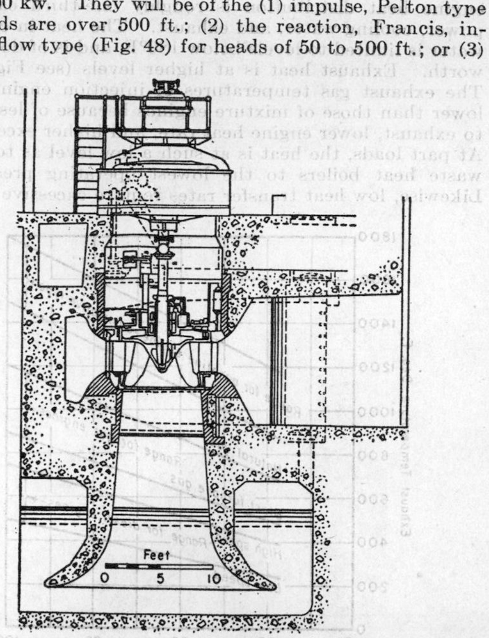

FIG. 48. Cross section of reaction-type water-wheel installation. (*H. K. Barrows, "Water Power Engineering.*")

the reaction, propeller, or Kaplan, axial-flow type for heads of 10 to 100 ft. The features and proportions of different runners are illustrated in Fig. 49, and the efficiency characteristics by Fig. 50. In Fig. 50, the efficiency curve for the propeller unit shows a sharp peak,

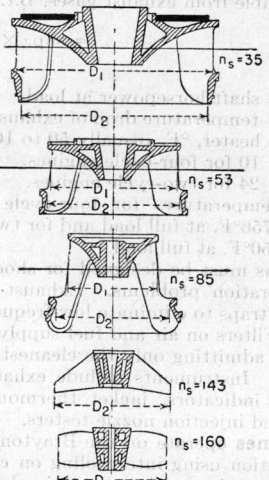

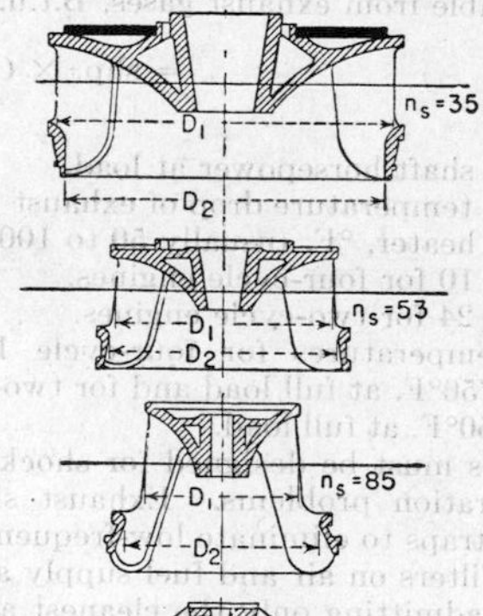

FIG. 49. Comparison of reaction runners of equal power but different specific speeds (n_s).

whereas the Kaplan unit, with its adjustable blades, gives high sustained efficiency at higher investment. The selection of water wheels must comply with the experience curves of Fig. 51, which gives the allowable specific speed, n_s, as a function of site head, where specific speed is defined as

$$n_s = \frac{\text{r.p.m.} \times \text{shp.}^{0.5}}{\text{head}^{1.25}} \tag{32}$$

where r.p.m. = revolutions per minute.
shp. = shaft horsepower.
head = head on unit, ft.

This and cavitation phenomena are the criteria for judging the suitability of a water wheel for given service conditions.

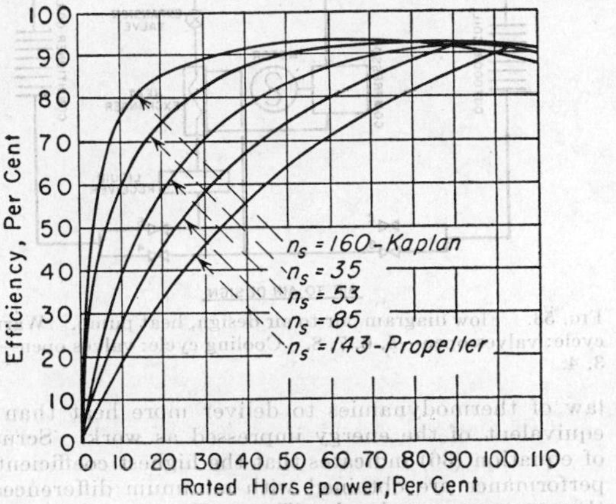

FIG. 50. Efficiency characteristics of selected hydraulic turbine types.

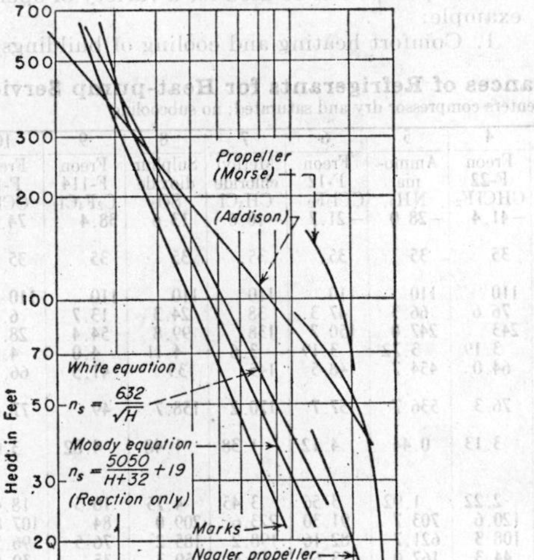

FIG. 51. Water-wheel experience data; specific speed vs. head.

The reaction-type wheels run submerged, full of water, and in a continuous uninterrupted water column from headrace to tailrace. The **draft tube** permits the maintenance of this unbroken liquid column, the setting of the wheel safely above tail water, and the utilization of the head between exit of the wheel and tail-water level. The wheel is thus operated at a negative pressure on the underside, and the draft tube serves to regain most of the exit kinetic energy leaving the wheel while the unit is placed in a protected position above flood level in the tailrace.

Structures include dams of reinforced concrete for heights up to 500 ft., or earth filled for heights up to 300 ft. Head works include diversion canals, control gates, trash racks, log booms, flash boards, and spillways. Penstocks are of steel for high heads (5000 ft. max.), wood stave pipe for heads less than 300 ft., and rock tunnels or open flumes. Surge tanks, standpipes, and relief valves all supplement governors to give adequate regulation without water hammer. Logways, fishways, and navigation locks may be required in supplementary structures.

PLANT OPERATION AND SCHEDULING OF UNITS

After a plant is placed in service the operator can reduce expense only by reducing production costs. Investment charges are beyond his control. Labor is essentially determined by design. Deferred maintenance

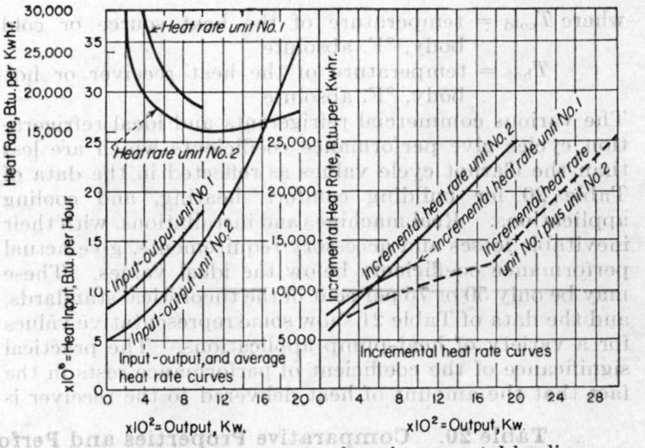

FIG. 52. Performance curves to show incremental loading procedures for power generating units.

offers an irrational choice. Practically, therefore, reduction in cost resolves into reduction of fuel expense. Thermal loading for economy prevails, in which (1) the necessary equipment must be in operation to assure reliability of supply; (2) equipment selected shall give best over-all economy; and (3) load shall be allocated among selected units for lowest thermal expense. Maximum economy is effected by incremental loading. Increment curves are obtained from the input-output curves by using a step-by-step approximation for d (input)/d (output), as in Fig. 52. Incremental rate curves are drawn for each unit separately and then combined in all possible combinations (see Fig. 52) on the basic principle that best economy obtains when the load is divided among units so that the incremental rates are equal. These principles are applicable to all equipment in the plant whether turbines, engines, boilers, pumps, fans, or other auxiliaries. A corollary to incremental loading principles requires the equal division of load among units that are identical. Average rates must not be confused with increment rates, because there are practically no fundamental relations between them, and the profits of any enterprise accrue on the averages and not on the increments.

HEAT PUMPS

Heat pumps are advantageously used where heat has to be transferred from one temperature level to a second, slightly or moderately higher, temperature level. This may require the removal of heat from subatmospheric temperature levels, as in refrigeration plants, or the entire operation may be carried out above atmospheric temperature levels for the purposes of warming. Thermodynamically, the operations are the same and the efficiency of the process is expressed as a coefficient of performance, c.p. (the equivalent of the reciprocal of thermal efficiency on heat engines), as follows:

$$\text{c.p.} = \frac{\text{Refrigeration}}{\text{Work}} \text{ as a cooling machine} \qquad (33)$$

$$\text{c.p.} = \frac{\text{Heat delivered}}{\text{Work}} \text{ as a warming machine} \qquad (34)$$

The theoretical maximum coefficient of performance is obtained on the Carnot cycle, where it is defined as

$$\text{c.p.} = \frac{T_{\text{cold}}}{T_{\text{hot}} - T_{\text{cold}}} \text{ as a cooling machine} \qquad (35)$$

$$\text{c.p.} = \frac{T_{\text{hot}}}{T_{\text{hot}} - T_{\text{cold}}} \text{ as a warming machine} \qquad (36)$$

where T_{cold} = temperature of the heat source or cold body, °F. absolute
T_{hot} = temperature of the heat receiver or hot body, °F. absolute

The various commercial refrigerants and ideal refrigeration cycles give performance coefficients which are less than the Carnot cycle values as reflected in the data of Table 20 for building comfort, heating, and cooling applications. Real machines and installations, with their inevitable losses and accessory requirements, give actual performance coefficients below the ideal values. These may be only 50 or 75 per cent of the theoretical standards, and the data of Table 21 show some representative values for a variety of heat-pump applications. The practical significance of the coefficient of performance rests in the fact that the amount of heat delivered to the receiver is equal to the mechanical equivalent of the power supplied to drive the unit, multiplied by the coefficient. Thus the values in Table 21, for an electrically driven unit, reflect the possibility of delivering more than the mechanical equivalent of a kw.-hr. (3413 B.t.u.). With a coefficient ranging from 3 to 17 the heat delivered per kw.-hr. impressed on the motor terminals becomes from 10,000 to 60,000 B.t.u. The heat pump thus utilizes the second

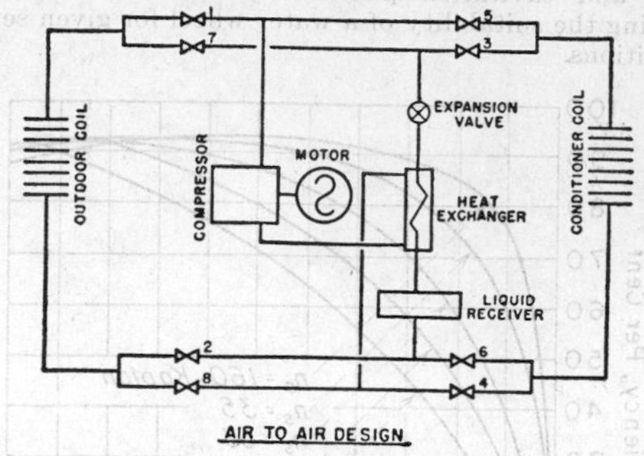

Fig. 53. Flow diagram, air to air design, heat pump. Warming cycle: valves open—5, 6, 7, 8. Cooling cycle: valves open—1, 2, 3, 4.

law of thermodynamics to deliver more heat than the equivalent of the energy impressed as work. Scrutiny of equation (36) indicates that the highest coefficients of performance are obtained with minimum differences in the temperature levels ($T_{\text{hot}} - T_{\text{cold}}$) and maximum absolute values for the receiver temperature, T_{hot}.

Heat pumps can be used for a variety of operations, for example:

1. Comfort heating and cooling of buildings.

Table 20. Comparative Properties and Performances of Refrigerants for Heat-pump Service

Note: Ideal cycle values used throughout: refrigerant enters compressor dry and saturated; no subcooling

Item	Unit	1	2	3	4	5	6	7	8	9	10	11
Refrigerant		Carnot	Free air	Dense air	Freon F-22	Ammonia	Freon F-12	Methyl chloride	Sulphur dioxide	Freon F-114	Freon F-11	Steam
Chemical formula		.....	$O_2 + N_2$	$O_2 + N_2$	$CHClF_2$	NH_3	CCl_2F_2	CH_3Cl	SO_2	$C_2F_4Cl_2$	CCl_3F	H_2O
Boiling point	°F.	.....	.....	.....	−41.4	−28.0	−21.7	−10.6	13.6	38.4	74.7	212
Suction temperature (saturation)	°F.	35	35	35	35	35	35	35	35	35	35	35
Head temperature (saturation)	°F.	110	110	110	110	110	110	110	110	110	110	110
Suction pressure	lb./sq. in. abs.		14.7	50	76.6	66.3	47.3	38	24.3	13.7	6.2	0.10
Head pressure	lb./sq. in. abs.	.....	73.5	250	243	247.0	150.7	138	99.8	54.4	28.1	1.275
Ratio of compression	Rp	.....	5.0	5.0	3.19	3.72	3.19	3.6	4.11	4.0	4.53	12.75
Refrigerating effect, Q_c	B.t.u./lb.	.....	32.4	32.4	64.0	454.7	48.5	145	135	41.5	66.1	999
Heat delivered, as a heating machine, Q_h	B.t.u./lb.	.....	50.9	50.9	76.3	536.7	57.7	170.2	158.7	49	77.4	1192
Refrigerant weight per ton as a cooling machine	Lb./min./ton	.....	6.18	6.18	3.13	0.44	4.12	1.38	1.48	4.82	3.03	0.2
Refrigerant volume per ton at suction conditions as a cooling machine	Ft.³/min./ton	.....	77.0	22.7	2.22	1.92	3.56	3.45	4.75	10.5	18.4	590
Enthalpy, h_h	B.t.u./lb.	.....	.....	.....	120.6	703.7	91.30	223.6	209.0	84	107.8	1270
Enthalpy, h_c	B.t.u./lb.	.....	.....	.....	108.3	621.7	82.16	198.2	185.2	76.5	96.5	1077.1
Enthalpy, h_{liq}	B.t.u./lb.	.....	.....	.....	44.3	167.0	33.65	53.4	50.3	35	30.4	78.0
Ideal compressor work, W(net)	B.t.u./lb.	.....	18.5	18.5	12.3	82.0	9.14	25.4	23.8	7.5	11.3	193
Horsepower, as a cooling machine	hp./ton		2.69	2.69	0.908	0.85	0.888	0.826	0.832	0.853	0.808	0.91
Horsepower, as a heating machine	hp. per 1000 B.t.u./min.		8.57	8.57	3.80	3.60	3.74	3.52	3.53	3.61	3.44	3.82
c.p. as a cooling machine	c.p.	6.6	1.75	1.75	5.20	5.55	5.31	5.71	5.67	5.53	5.85	5.18
c.p. as a heating machine	c.p.	7.6	2.75	2.75	6.20	6.55	6.30	6.7	6.67	6.53	6.85	6.18
Efficiency as a cooling machine		1.00	0.265	0.265	0.788	0.841	0.805	0.865	0.859	0.838	0.885	0.785
Efficiency as a heating machine		1.00	0.362	0.362	0.816	0.862	0.829	0.882	0.878	0.859	0.90	0.814
Toxicity		.....	Min.	Min.	Low	High	Low	Medium	High	Low	Low	Low
Underwriters Laboratory classification on toxicity		.....	.....	.....	5	2	6	4	1	6	5	.....
Poisonous decomposition products		.....	No	No	Yes	.....	Yes	Yes	.....	Yes	Yes	No
Flammability		.....	Non	Non	Non	Yes	Non	Yes	Non	Non	Non	Non
Odor			No	No	No	Yes	No	No	Yes	No	No	No
Corrosiveness			No	No	No	Yes	No	Yes	Yes	No	No	No

Table 21. Actual Performance of Heat-pump Installations

Operating Results in Evaporation and Distillation Installations

Process	Kw.-hr. input	lb./hr. evaporated capacity	Lb. water evaporated per kw.-hr.	Approx. c.p.	Evaporator, °F.	Suction pressure, lb./sq. in. abs.
Evaporating plant handling milk products....	73	2,200	30.1	8.9	120	
Evaporating plant handling milk products and unfermented fruit juices..........	240	6,600	27.1	8.0	120	
Evaporating plant in chemical works......	94	1,540	16.3	4.8	...	0.86
Water evaporating plant for distillation of drinking water..........	75	2,750	36.6	10.6	212	14.65
Water evaporating plant for distillation of drinking water from sea water............	0.02	1	50.0	14.7	213	14.75
Estimated performance of makeup evaporator for a power plant	315	20,000	63	17	293	60

Operating Results on Comfort Heating Installations

Installation	Date installed	Heat source	Volume of conditioned space, cu. ft.	Capacity, tons	Coefficient of performance, c.p.
Atlantic City Electric Co., Salem, N.J......	1934	Well water	76,800	20	3.3
Ohio Power Co., Steubenville, Ohio..	1936	Outside air	170,000	40	3.4
Mangel, Riverside, Calif.	1937	City water	20,000	2	5.1
Southern California Edison Co., Montebello, Calif..........	1938	Outside air	50,000	10	3.8
United Illuminating Co., New Haven, Conn....	1940	Well water	1,570,000	320	2.9
Norwich Corp., Electricity Dept., Norwich, England...........	1945	River water	500,000	...	3.2
Webber, Indianapolis, Ind..........	1945	Earth	7,000	3	3.6
Equitable Building, Portland, Ore.......	1948	Well water and waste	2,275,000		7.8

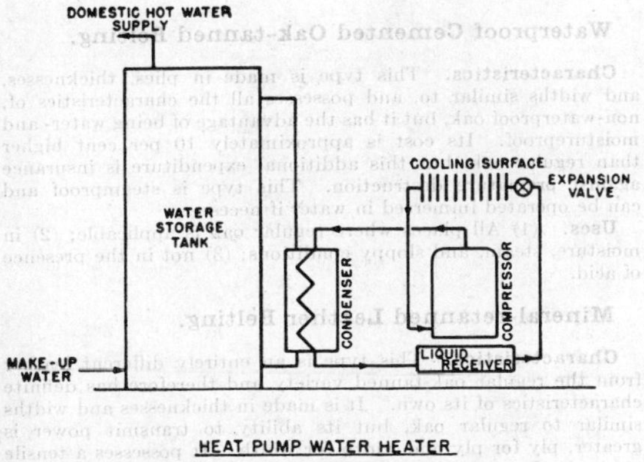

FIG. 54. Hot-water heat pump.

2. Controlling atmospheric temperature and humidity in industrial plants.
3. Domestic and industrial heating of water and other liquids.
4. Concentrating, evaporating, and distilling.
5. Desiccation.
6. Simultaneous heating and refrigerating.

In the fields (1) of comfort heating and (2) the control of atmospheric temperature and humidity, the source of heat can be the outdoor air, surface water, well water,

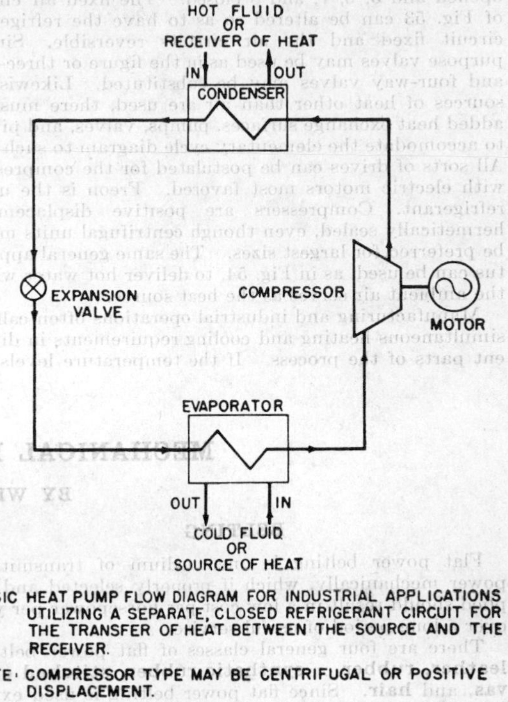

FIG. 55. Heat-pump flow diagram for simultaneous heating and cooling operations.

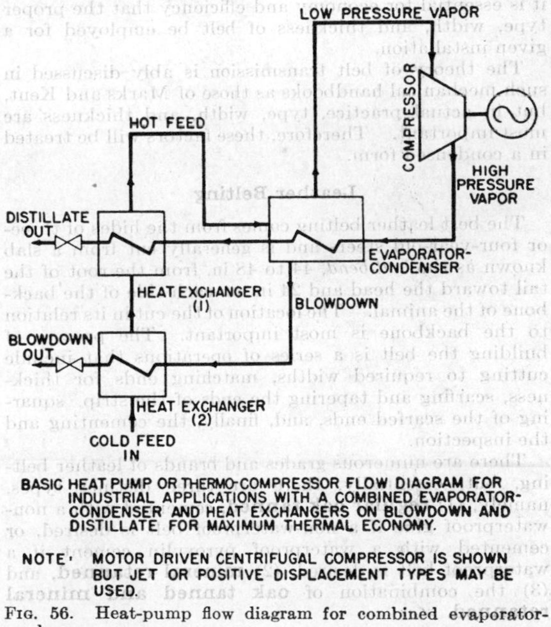

FIG. 56. Heat-pump flow diagram for combined evaporator-condenser.

the ground itself, or industrial waste water and gases. The basic flow diagram, using air as the heat source, is shown in Fig. 53. Here the warming cycle requires that valves numbers 1, 2, 3, and 4 be closed, and valves numbers 5, 6, 7, and 8 be opened. For the cooling cycle the process is reversed, with valves numbers 1, 2, 3 and 4 opened and 5, 6, 7, and 8 closed. The fixed air circuit of Fig. 53 can be altered so as to have the refrigerant circuit fixed and the air circuits reversible. Single-purpose valves may be used as in the figure or three-way and four-way valves may be substituted. Likewise, if sources of heat other than air are used, there must be added heat exchange surfaces, pumps, valves, and piping to accomodate the elementary cycle diagram to such use. All sorts of drives can be postulated for the compressor, with electric motors most favored. Freon is the usual refrigerant. Compressers are positive displacement, hermetically sealed, even though centrifugal units might be preferred for largest sizes. The same general apparatus can be used, as in Fig. 54, to deliver hot water where the ambient air serves as the heat source.

Manufacturing and industrial operations often call for simultaneous heating and cooling requirements in different parts of the process. If the temperature levels are sufficiently close together a cycle, as illustrated in Fig. 55, is attractive. If the objective is distillation, evaporation, concentration, or drying, the heat pump, generally called a thermo compressor, can be used as illustrated in Fig. 56. Here the vapor from the evaporating space enters the compressor suction, its pressure is raised in the compressor, and higher pressure vapor is delivered to the condensing coil. This coil is submerged in the evaporator so that the latent heat is thus transferred from suction to discharge and back again, with minimum outside energy addition through the compressor shaft. This arrangement avoids the use of a separate refrigerant. The compressor handles the evaporated vapor, usually steam, directly. For temperatures in the range of 50° to 200°F. the vapor pressures are low and the specific volumes high. Centrifugal, axial flow, and jet compressors are consequently well suited to the service. It is entirely reasonable to find thermo compressors, as reflected in the data of Table 21, with high over-all performance coefficients, superior to multiple-effect evaporators.

MECHANICAL POWER TRANSMISSION

BY WILLIAM STANIAR

BELTING

Flat power belting is one medium of transmitting power mechanically, which if properly selected and applied should result in a low cost per horsepower per year over a long period of useful service.

There are four general classes of flat power belting: **leather, rubber** or **synthetic rubber, stitched canvas,** and **hair.** Since flat power belting is used extensively for transmitting power in a wide variety of industrial applications under various atmospheric conditions, it is essential for economy and efficiency that the proper type, width, and thickness of belt be employed for a given installation.

The theory of belt transmission is ably discussed in such mechanical handbooks as those of Marks and Kent, but in actual practice, type, width, and thickness are most important. Therefore, these factors will be treated in a condensed form.

Leather Belting

The best leather belting comes from the hides of three- or four-year-old steers and is generally cut from a slab known as the *butt bend,* 44 to 48 in. from the root of the tail toward the head and 24 in. on each side of the backbone of the animal. The location of the cut in its relation to the backbone is most important. The process of building the belt is a series of operations that include cutting to required widths, matching ends for thickness, scarfing and tapering the ends of the strip, squaring of the scarfed ends, and, finally, the cementing and the inspection.

There are numerous grades and brands of leather belting, but, essentially, there are three distinct types, namely, (1) **regular oak tanned,** cemented with a non-waterproof glue if a non-waterproof belt is desired, or cemented with a waterproof pyroxylin cement if a waterproof belt is desired, (2) **mineral retanned,** and (3) the combination of **oak tanned and mineral retanned.**

Regular Oak Tanned.

Characteristics. Made in 1 ply, 2 ply, and 3 ply. Known as singles, doubles, and triples. 4 ply is very rare.

Thickness of singles, $^{10}/_{64}$ to $^{14}/_{64}$ in.; doubles, $\frac{1}{4}$ to $\frac{3}{8}$ in.; triples, $^{15}/_{32}$ to $^{36}/_{64}$ in.

Widths of singles, $\frac{1}{8}$ to 10 in.; doubles, 1 to 72 in.; triples, 24 to 72 in.

Fabricated with a hot non-water-resisting glue. It does not possess flexibility when new. Ultimate tensile strength, 4000 to 6500 lb./sq. in. It has low stretch. Coefficient of friction: new oak, 0.27 to 0.45; well-worn oak, 0.35 to 0.60. It will take any type of metallic fastener and can be made endless at the point of application if necessary.

Uses. Singles: (1) Small power applications: (2) light machine-tool driving; (3) on pulleys 3 and 4 in. diameter and over; (4) on light shifting drives; (5) where moisture or acid is not present.

Doubles: (1) medium and heavy power applications; (2) all manner of machine-tool driving; (3) if light double, on pulleys 6 in. diameter and over; (4) if regular double, on pulleys 10 in. diameter and over; (5) on shifting drives; (6) where moisture or acid is not present.

Waterproof Cemented Oak-tanned Belting.

Characteristics. This type is made in plies, thicknesses, and widths similar to, and possesses all the characteristics of, non-waterproof oak, but it has the advantage of being water- and moistureproof. Its cost is approximately 10 per cent higher than regular oak, but this additional expenditure is insurance against premature destruction. This type is steamproof and can be operated immersed in water if necessary.

Uses. (1) All places where regular oak is applicable; (2) in moisture, steam, and sloppy conditions; (3) not in the presence of acid.

Mineral-retanned Leather Belting.

Characteristics. This type is an entirely different leather from the regular oak-tanned variety and therefore has definite characteristics of its own. It is made in thicknesses and widths similar to regular oak, but its ability to transmit power is greater, ply for ply and width for width. It possesses a tensile strength of approximately 6000 lb./sq. in. and has a coefficient of friction of 0.60 to 1.10. It possesses low "stretch." It has extreme flexibility and is therefore efficient on high-speed, small pulley work. It is fabricated with a waterproof cement and is therefore impervious to the action of moisture. It is moisture-, steam-, and heatproof and will resist the action of corrosive acids for a considerable period of time. It will take any type of metallic fastener and can be made endless at the point of application. It can be repaired, cut down, rebuilt, and salvaged in a manner similar to oak-tanned leather.

Uses. **Singles:** (1) High-speed motor drives, up to 7½ hp., 1200 to 1800 r.p.m.; (2) high-speed spindle drives of various machine tools such as internal and external grinders; (3) mule-pulley installations on small single- and multiple-drill presses; (4) on pulleys as small as 1¼ in. diameter; (5) pivoted motor-base drives when made endless.

Doubles: (1) High-speed motor drives, 10 hp. and up, 720 to 1200 r.p.m.; (2) high-speed pulleys 6 in. diameter and over; (3) high-speed spindle drives of the larger machine tools; (4) automatic idler and pivoted motor-base drives when made endless.

Combination Oak- and Mineral-retanned Leather.

Characteristics. This type of leather belting is made, as the name implies, of a combination of the regular oak-tanned and mineral-retanned types joined flesh side to flesh side with a cement insoluble in water. It is made in doubles only. It is made in similar widths to regular oak. It is specially manufactured for shifting, step-cone, and flanged-pulley drives because the oak tannage resists the transverse crumpling action on the flexible retanned type of leather caused by the action of shifter forks and pulley flanges. It is moisture- and steamproof but will not resist the action of corrosive acids. It will take any style of metallic fastener and can be made endless at point of application. Its repair and salvage values are similar to those of all types of leather.

Uses. (1) Heavy slow-speed loads, such as 200 to 500 ft./min.; (2) shifting and loose pulley drives; (3) step-cone and flanged-pulley drives; (4) pulleys should not be under 8 in. diameter; (5) heavy-duty milling machines or lathe work.

Formulas for Leather Belting. The power-transmitting capacity of leather belting may be calculated by the following formula:

$$\text{Horsepower} = \frac{T_e WS}{33{,}000}$$

where T_e = effective belt tension, lb./in. of width.
W = belt width, in.
S = belt speed, ft./min.

The allowable effective belt tension for various grades and thicknesses of leather belting is as follows:

Grade of leather	Allowable effective tension T_e, lb./in. width	
	Single ply (3/16 to 7/32 in. thick)	Double ply (5/16 to 11/32 in. thick)
Oak-tanned	45	65
Mineral-retanned	50	70
Combination oak- and mineral-retanned	...	70

Atmospheric Effects on Leather Belting. **Regular Oak Non-waterproof Cemented:** An atmosphere charged with steam or moisture causes separation of the lap and plies because of its action on the cement or glue. An atmosphere charged with the mists or vapors of the corrosive acid attacks the leather, causing rapid deterioration.

Regular Oak Waterproof Cemented: An atmosphere charged with steam or moisture has no effect, because in this type the leather is dressed and the pyroxylin cement is impervious to the action of moisture. An atmosphere charged with the mists or vapors of the corrosive acids attacks the leather regardless of the dressing, causing rapid deterioration.

Special Tannage Waterproof Cemented: An atmosphere charged with steam or moisture has no effect on this type because of the special tanning of the leather and the pyroxylin cement. An atmosphere charged with the mists or vapors of the corrosive acids attacks this type but is slow in its action.

Combination of Oak and Special Tannage: This type is waterproof because of the pyroxylin cement, but the oak ply will not resist the action of the mists and vapors of the corrosive acids.

Leather-belting Dressings. There are numerous cheap and low-grade stick and semiliquid dressings on the market most of which contain resin, tar, and graphite. All such ingredients are injurious to leather and also to its power-transmitting ability; caution should therefore be exercised in their use. Belting of any type does not pull its load by being stuck to the pulleys. It functions through its own natural frictional grip.

Leather belting requires periodical lubrication or dressing, because the heat generated by the transmission of power is absorbed by the belt, resulting in a drying out of the cod oil and beef tallow used in the currying process. A dressing should contain similar ingredients so that penetration and lubrication of the fibrous structure will result. Reputable leather-belting manufacturers supply dressings containing the above currying materials which can always be employed to an advantage on oak leather belting whether waterproof or non-waterproof. Special tannage leather requires a special dressing, which is also supplied by the respective manufacturers.

Under normal operating conditions leather belting should be dressed every 2 or 3 weeks, and the most effective results are obtained by applying the dressing to the outer ply. This allows the material to work its way through the leather and therefore prevents an accumulation on the pulley side of the belt.

The Effect of Centrifugal Force. As a belt moves in a curved path around a pulley, each particle of the belt is subjected to a centrifugal force that acts radially from the pulley. These forces create a centrifugal tension throughout the length of the belt entirely apart from the initial belt tension. A diagram illustrating the action of centrifugal force on a horizontal belt drive is shown by Fig. 57.

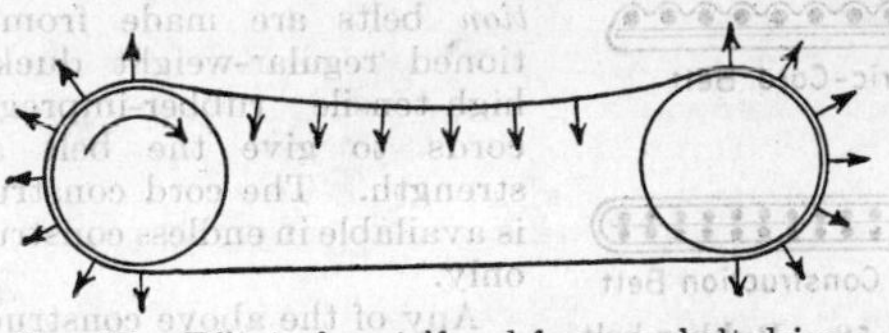

FIG. 57. Effect of centrifugal force on belt drive.

Centrifugal tension decreases initial tension. On heavy belting operating at high speeds on fixed centers, the centrifugal tension may equal the initial tension, with the result that power transmission is not possible. The following formula may be used to ascertain the centrifugal tension in a belt:

$$T_c = \frac{WS^2}{116{,}000}$$

where T_c is centrifugal tension, lb.; W, weight of belt, lb./ft.; and S = speed, ft./min.

Horsepower Losses Due to Centrifugal Tension for Leather Belting

Speed of Belt, Ft./Min.	Percentage off Rated Hp.
1,000	1
2,000	4
3,000	8
4,000	15
5,000	23
6,000	34
7,000	46
8,000	60
9,000	76
10,000	95

Notes: At 10,000 ft./min. centrifugal tension almost equals effective tension or $T_1 - T_2$.
These losses are applicable to belts operating on fixed centers.

Practical Belt Speeds. Belt speeds of 1000 to 3000 ft./min. are common, but the most economical and efficient belt velocities range from 3500 to 4500 ft./min. Speeds higher than 5000 ft./min. cause excessive strain and wear on the belt.

Rubber Belting

Construction. Friction surface rubber belting is made by vulcanizing together two or more thicknesses or plies of cotton duck, impregnated and frictioned with a tough, slow-aging rubber (natural or synthetic) compound. For most rubber belting, 28- to 32-oz. duck is used.

Manufacture. Manufacture of the belt begins with cutting of the frictioned duck to the desired width, then folding or plying the strips in such a manner as to give the width and thickness of belt required. The raw belt is then placed in a large steam-heated horizontal vulcanizing press to cure the rubber compound. After vulcanization the belt is ready for use.

Types of Rubber Belting. *Regular folded construction* consists of one or more envelopes of duck or separate duck plies encased in an envelope with a binder strip to form a round edge belt.

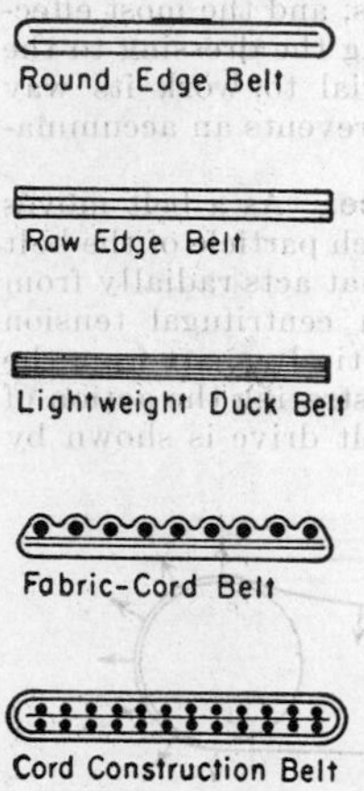

FIG. 58. Rubber-belt constructions.

Raw Edge Construction. The plies of duck are laid one on top of the other with no envelope or binder strip. *Lightweight* (8 to 10 oz.) *frictioned duck* plies are laid one on top of the other with no envelope or binder strip. For the same thickness belt, more plies are possible than with regular heavyweight duck.

Fabric-cord and cord-construction belts are made from frictioned regular-weight duck and high-tensile rubber-impregnated cords to give the belt added strength. The cord construction is available in endless construction only.

Any of the above constructions is normally available frictioned with natural or synthetic rubber compounds. The most widely used synthetic rubber types are Neoprene (for oil or chemical resistance) and Buna S (GR-S). Any of these constructions can be made "staticproof" by the use of conductive rubber compounds for the rubber friction.

Characteristics. Natural or synthetic rubber belting is made in 2, 3, 4, 5, 6, 7, 8, 9, 10, and 12 plies in thickness and from 1 to 84 in. in width. The number of plies increases in proportion to width, pulley diameters, and power requirements. The tensile strength depends upon weight and quality of the duck employed. It will hold any style of metallic fastener and also can be rawhide or wire laced. It can be made endless, but this should be done by the manufacturer. It has no salvage value.

Uses. (1) High-speed motor drives, based on correct number of plies for pulley diameters; (2) crossover drives from line- to countershafts of machine tools when controlled by tight and loose pulleys; (3) on shifting drives; (4) on step-cone and flanged pulleys; (5) in mild chemical atmospheres since synthetic rubber frictioned belts are quite moisture- and steamproof and are quite resistant to mineral oils and corrosive acid fumes.

Dressings for Rubber Belting. It is impossible actually to dress or lubricate the fibers of a friction surface rubber belt, because the rubber friction compound between and in the duck plies cannot be penetrated. The immediate use of vegetable castor oil or boiled linseed oil on new rubber belting will remove the bloom of the rubber and give a non-dangerous adhesiveness to the belt.

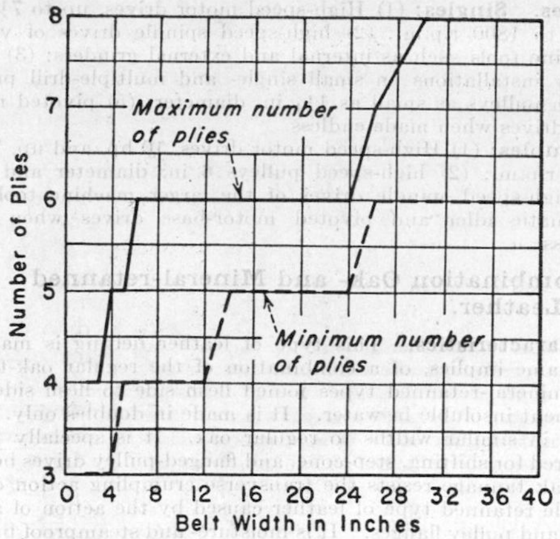

FIG. 59. Relation of number of plies and belt width for rubber belting.

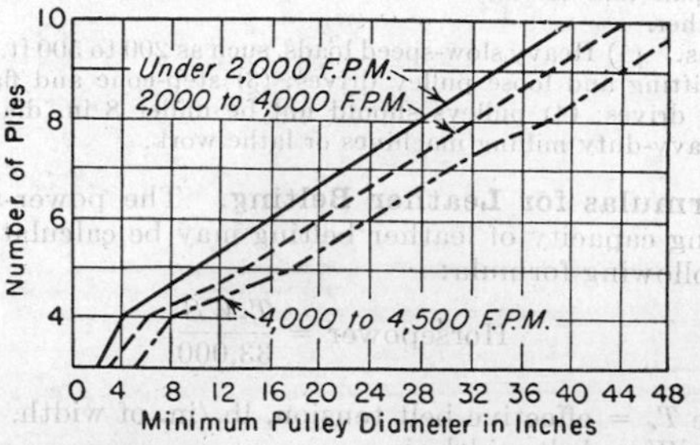

FIG. 60. Relation between number of plies to pulley diameters and speeds for rubber belting.

Formula for Rubber Belting. Rubber-belting manufacturers use the following formula:

$$\text{Hp.} = \frac{WNS(T_1 - T_c)}{33{,}000}\left(1 - \frac{1}{R}\right)$$

where Hp. = horsepower capacity of belt.
W = width, in.
N = number of plies.
S = belt speed, ft./min.
T_1 = tight-side tension, lb./in. width.
T_c = centrifugal tension, lb./in. width.
$\left(1 - \frac{1}{R}\right)$ = correction factor for arc of contact (see table).

Arc of Contact Correction Factors

Arc of contact, deg.	$1 - \frac{1}{R}$	Arc of contact, deg.	$1 - \frac{1}{R}$
90	0.29	190	0.51
100	.32	200	.53
110	.35	210	.55
120	.37	220	.57
130	.39	230	.58
140	.41	240	.60
150	.44		
160	.46		
170	.48		
180	.50		

For good quality rubber belting 32-oz. duck is generally used. Based on an average ultimate tensile strength for this weight of 440 lb., the allowable effective tension is 25 lb./in. width of duck or a factor of safety of

17½. $(T_1 - T_c)$ can therefore be calculated as 25 lb. for a 32-oz. duck belt and 20 lb. for 28-oz. duck.

Stitched Canvas Belting

Construction. There are two types of construction used in the manufacture of this class of belting, one known as the round edge and the other as the folded edge. The difference in the two methods of construction is in the manner in which the duck is folded before stitching and in the building up of the various numbers of duck plies.

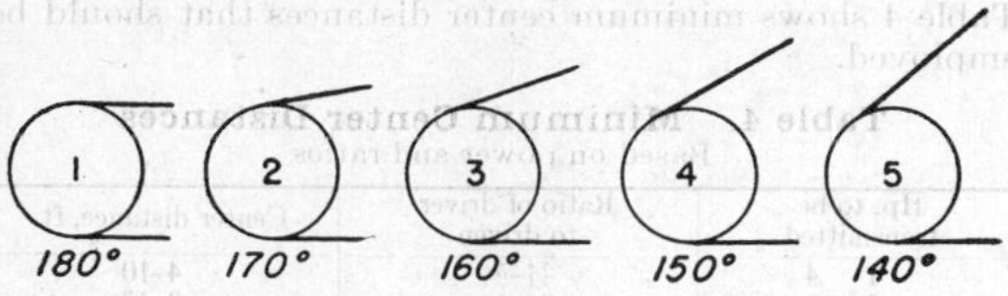

FIG. 61. Loss of contact on driving pulley by high ratios.

The round-edge construction gives a smoother surface to the action of shifter forks and prevents premature cracking of the edges of the belt. The better grades of this class of belt are usually made from 36- to 37½-oz. duck, which possesses a tensile strength of from 550 to 600 lb./in. width, therefore giving the belt high tensile strength.

After folding for either the round- or folded-edge construction, the belt is stitched and is then thoroughly impregnated with certain compounds by either the immersion or the pressure process, thus rendering the belt impervious to mechanical and atmospheric actions. After the belt is thoroughly impregnated it is stretched and is then ready for the transmission of power.

Characteristics. It has a high coefficient of friction, extreme flexibility, resistance to ply separation, and is elastic. It will hold any type of metallic fastener but is difficult to make endless. It has extreme uniformity and therefore results in a true running belt. It has no salvage value.

Uses. (1) High-speed motor drives, based on correct number of plies for pulley diameters; (2) crossover drives from line- to countershafts of machine tools, preferably when clutch controlled; (3) shifting drives, preferably when the belt is of 6-ply construction; (4) it will resist the action of water, steam, heat, and mineral oil but will not resist the action of the corrosive acids; (5) all manner of slow speed, heavy driving, when not subjected to the atmospheric conditions stated.

Formula for Stitched Canvas Belting. As in rubber belting, the allowable effective pull is based on so many pounds per inch ply of the belt and varies according to the weight of the duck employed. The average good grade of stitched canvas belting is made up of 36-oz. duck, 28 lb./in. width allowable effective pull on tight side.

The formula for 36-oz. duck at 28 lb. allowable effective pull per inch is

$$\text{Horsepower} = \frac{WNS(T_1 - T_c)\left(1 - \frac{1}{R}\right)}{33{,}000}$$

Use 28 lb. for the value of $(T_1 - T_c)$ in the above formula.

Hair Belting

Construction. Belting of this type is made of hair yarn of various kinds and mixtures of hair and wool.

In weaving, the hair yarn becomes the warp of the belt while cotton yarn is used as the filler and also a binder to hold the several layers of hair together in a compact and homogeneous fabric. The warp or lengthwise hair strands form the portion of the belt that gives the traction and takes the load pull. The cotton yarn that binds the several layers of camel's-hair yarn together acts as a check on the hair and restrains its stretch.

In some types of hair belting the edges are made of cotton or leather known as friction edges. These edges are designed to protect the belt against the action of shifter forks.

After weaving, the belt is impregnated with a combination of heavy oils which preserve the entire fabric, lubricate it, and aid the frictional qualities.

Characteristics. It is highly elastic and therefore should be placed on the pulleys extremely tight. It has no laps or plies. It will remain soft and pliable during its life. It possesses an average ultimate tensile strength of 6300 lb./sq. in. It has extreme flexibility and therefore can be run on high-speed small pulleys. It will hold any style of metallic fastener but cannot be made endless. It has no salvage value.

Uses. (1) High-speed motor drives, based on correct thickness for pulley diameters; (2) intermittent loads where shocks must be absorbed by the belt; (3) shifting and cross drives; (4) heavy hard-pull drives; (5) on crossover drives from line to counter of machine tools, whether clutch or tight- and loose-pulley control; (6) it is not affected by mineral oils unless the amount is excessive. It will stand heat to 300°F. and is not affected by extreme cold. It will stand for long periods the action of corrosive acids but should not be used where caustic is present. It is moisture- and steamproof; (7) all manner of driving when not subjected to atmospheric conditions stated.

Formula for Hair Belting. The following allowable effective tensions may be used in calculating the power transmitting capacity of hair belting:

Light or single	3/16 in. thick	= 40 lb./in. of width
Light or double	1/4 in. thick	= 50 lb./in. of width
Double	9/32 to 3/8 in. thick	= 70 lb./in. of width
Extra heavy	9/16 in. thick	= 90 lb./in. of width

Formula:

$$\text{Horsepower} = \frac{TWS}{33{,}000}$$

where T is the effective pull lb./in. width; W, the belt width, in.; and S, the belt velocity, ft./min.

Table 1. Recommended Minimum Pulley Diameters for Hair Belting

Weights of hair belting	Thickness, in.	Min. pulley diameter, in.	
		Under 2000 ft./min.	2000 to 4000 ft./min.
Light	3/16	3	5
Light double	1/4	5	7
Double	9/32–3/8	10	14
Extra heavy	9/16	16	24

Belt Joining

Improper and careless joining is one of the principal factors of belting maintenance costs and one of the causes of short belting life.

The fastener must not unduly strain the belt; must not cut or weaken the belt; must resist wear caused by pulley contact; must conform to the pulley circumference. It should allow the belt ends to fit closely and prevent breakage at joint. It must be quick of application.

The Endless Method. Joining a belt so that the result is an endless band is the most efficient method, because the joint is integral with the belt and therefore eliminates the use of any foreign material. All types of belting have stretch, a factor that necessitates the use of take-up facilities where the endless belt is employed; otherwise considerable trouble will be experienced.

Leather belting offers the greatest possibilities for the endless or cemented joint, because in reality the belt itself is composed of cemented joints. To make a leather belt endless the ends should be scarfed or scraped down to a thin edge, 3, 4, or 6 in. back, depending on the width, and the cement or glue applied to the scarfed surfaces. These surfaces are then placed together and subjected to continuous pressure for 2 or 3 hr. When joining with pyroxylin cement as on a waterproof leather belt, the scraped surfaces are first sized with the cement and allowed to dry. After drying, another application of the cement is made and the surfaces placed together. Pressure is then applied for at least 3 or 4 hr.

The endless joint should be employed on all drives where center adjustment is possible—on high-speed spindle driving of machine tools, on large engine installations, and on automatic idler and pivoted motor-base drives.

The endless joint should not be used on drives where center adjustment is not possible—on vertical drives without idlers and on quarter-turn drives.

Endless Rubber Belting. Rubber belting of the ply type can be made endless by cutting and stepping the plies back from the ends so that when placed together they coincide. When set in this position, vulcanization is necessary. This can be done at the point of application by the use of small portable vulcanizing machines. Rubber belting of the cord type is now manufactured endless, with no joint or splice.

The Laced Joint. Rawhide is the most common material employed for making the laced belting joint and if properly made results in a good joint. It should always be employed where hand shifting is necessary because of the safety feature. Rawhide lacing is being done away with on general driving because of the time required to apply it and the advent of the quicker and efficient metallic methods.

Metallic Fasteners. A number of metallic fasteners are on the market, and they are all used extensively because of their efficiency and time-saving factor of application. The wire hook and pressed-steel-pin type, the plate type, and the staple or prong type are the most popular.

The wire hook type can be used on single, light double, and regular double leather belting up to 8 in. wide and on all types of rubber and fabric belting from 2- to 6-ply up to 8 in. wide, on drives where unfastening must be done quickly, on high-speed work where the endless joint is not practicable, where an insert is necessary, and on shifting drives. Hooks made from ordinary carbon steel, stainless steel, bronze, or monel metal are commercially available to suit various corrosive atmospheric conditions.

The wire hook should not be used on hand-shifted drives, on heavy double leather, or on rubber or fabric belting over 6 plies in thickness, on automatic idler drives.

Table 2. Minimum Pulley Diameters and Belt Thicknesses for Wire Hooks

Hook size	Thickness of belt	Minimum pulley diameter, in.
2	2 or 3-ply fabric	2½ and 3
4	4-ply fabric and single leather	6
6	6-ply fabric and regular double leather	12

Table 3. Plate Sizes for Various Belt Widths and Minimum Pulley Diameters

Belt width, in.	Number and size of plates				
	4-in.-diam. pulley and larger	6-in.-diam. pulley and larger	9-in.-diam. pulley and larger	12-in.-diam. pulley and larger	24-in.-diam. pulley and larger
2	....	1-No. 607	1-No. 63		
4	....	2-No. 607	2-No. 63		
6	4-No. 66	4-No. 67	2-No. 103	2-No. 149	
8	....	2-No. 147	4-No. 63	2-No. 189	
10	....	....	4-No. 83	4-No. 109	
12	....	....	4-No. 103	4-No. 149	4-No. 1611
14	....	....	4-No. 123	4-No. 1409	4-No. 1611
18	....	....	....	6-No. 149	6-No. 1611
24	....	....	....	....	6-No. 2211

The plate fastener can be used on single, light, and regular double leather from 1 in. to any width required, and on all types of fabric belting from 3 to 12 ply and in widths from 1 in. to any width required, on open and crossed drives, on hard-pull slow-speed driving, and on shifting drives.

The plate fastener should not be used on any type of idler installation, on high-speed small pulley work, on hand-shifted belting, or on special tannage leather belting.

Belt Drives

Minimum Centers. Certain minimum center distances must be used for fixed center flat belt drives to ensure sufficient tension for trouble-free operation. Table 4 shows minimum center distances that should be employed.

Table 4. Minimum Center Distances
Based on power and ratios

Hp. to be transmitted	Ratio of driver to driven	Center distance, ft.
1– 4	2/1–5/1	4–10
5– 9	2/1–5/1	8–12
10– 14	2/1–5/1	8–14
15– 24	2/1–5/1	9–15
25– 39	2/1–5/1	10–16
40– 49	2/1–5/1	12–17
50– 74	2/1–5/1	13–18
75– 99	2/1–5/1	18–24
100–124	2/1–5/1	20–26
125–149	2/1–5/1	22–28
150–200	2/1–5/1	26–32

Much shorter center distances may be employed on a motor drive equipped with a pivoted motor base (see Short-center Drives, p. 1670).

Vertical Drives. A belt drive in a vertical position without take-up facilities, although necessary at times, is always troublesome. The natural stretch of the belting substance causes sag of the belt away from the bottom pulley, resulting in power loss and excessive maintenance. When a vertical drive must be employed without take-up provisions, use a center distance as short as possible, because this provides less accumulated stretch.

Crossed-belt Drives. Belting is crossed for the purpose of reversing direction of rotation and arc of contact increase, but consideration must be given to pulley diameters, belting width, and type of fastener employed if good results are to be obtained. The pulleys should be of reasonable diameter and as near equal in this dimension as possible. Belting over 8 in. wide should not be crossed because of the large area of belt in contact. A poor joint or projecting metallic fastener will quickly cause the destruction of a crossed belt.

Quarter-twist Drives. Quarter-twist drives are employed for transmitting power by belting at right angles, either vertically or horizontally. Leather belting gives the best results, because it can be especially constructed to suit this type of drive. An empirical rule for this installation follows:

RULE: *The center of the face of the loose side of the driver must line with the center of the face of the tight side of the driven.*

Mule or Guide-pulley Drives. This type of driving makes it possible to transmit power by belting up over or around a corner. It is accomplished by the aid of a guide, or mule pulley as it is generally termed, operating on independent spindles or shafts. Machine tools, such as single- and multiple-drill presses, are frequently driven in this manner. This type of drive is difficult to maintain and should not be used unless absolutely necessary. Best results are obtained from a narrow belt width, not in excess of 6 in.

Cone-pulley Driving. Tapered cone-pulley drives are employed when variation in speed is required; but, unless a properly constructed belt is used, constant trouble will result. Crossed belting operates to a better advantage on cone pulleys and especially where the taper is extreme. Cone-pulley belts are as a rule shifted frequently, and lateral stiffness of more than one ply is necessary. This is accomplished by making the outer ply wider and of a stiffer material than that which is against the pulley.

The Crowning of Pulleys. Belting connecting the pulleys of parallel shafting tends to run toward that part of the pulley which is largest in diameter. It is for this reason that pulleys are higher at the center of the face or rim than they are on the edges. This raise is termed the *crown*. Crown-faced pulleys should be employed on all belted installations with the exception of the tight and loose pulley drive. The twin, or tight- and loose-driven, pulley should be crowned, but the driver, or drum pulley, should be flat or straight faced so as to permit easy sliding of the belt from the loose to tight pulley and vice versa.

PULLEYS

There are three general classes of pulleys for the transmission of power by belting, namely, **cast iron, steel,** and **wood.**

Cast-iron Pulleys

The cast-iron pulley is made in the following types:

1. Single arm, solid or split.
2. Double arm, solid or split.
3. Keyless or interchangeable bushed.

Construction. Cast-iron pulleys are usually designed and built to withstand the strains under which the respective belt can be expected to operate. The cast-iron pulley is of rigid construction and therefore runs true if properly set and fastened to the shaft. Because of its rigidity, sudden shock, loads, or falls are liable to cause cracking or fracture of the iron.

Speeds. Standard pulleys are made for rim speeds up to 5000 ft./min. Rim speeds up to 6000 ft./min. are permissible but such pulleys should be dynamically balanced.

Applications. The cast-iron pulley will stand heat, moisture, steam, and acid fumes; it therefore has a wide range of usefulness. A cast-iron pulley will last for years if operated according to its capacity.

Steel Pulleys

Types. The pressed-steel pulley is made in three distinct types: the one-piece rolled edge, crowned rim, split; the one-piece crowned rim, without edge roll, split; and the two-piece crowned rim, edge roll, split.

All have respective advantages and disadvantages; therefore, as to their usage, it is difficult to say which is the best, because for the average industrial service the three types are efficient.

Construction. The steel pulley is a fabricated structure; therefore, there exists a wide difference in design. Within working limits of belting, however, strength of joints and the greatest possible rigidity have been secured. The difference in type particularly pertains to the construction of the rim, which has been noted in the above paragraph.

Speeds. The pressed-steel pulley can be operated with safety at high rim speeds, 4000 to 4500 ft./min. being common practice. On tests, this type of pulley has been run at 14,000 ft./min. for periods of 30 hr. without signs of distress. This would not be practicable in power transmission because 4500 ft./min. is the most economical belt speed.

Applications. Steel pulleys can be operated in similar places to those where cast-iron ones are used, with the exception that they should not be used in the presence of corrosive acids or in severe moisture and steam.

This pulley is generally of the split type and is furnished with interchangeable metallic bushings which make it possible to use either clamp, key, or set screw.

Wooden Pulleys

Types. Under normal conditions the modern wooden pulley will perform in every way as efficiently as the cast-iron or steel pulley.

There are three distinct types made, namely, the heavy-duty type of split design with standard bore into which the required wooden bushing is placed, the wooden-rim iron-center type, and the solid-wood type.

Construction. The wooden pulley is usually of the split type and for average driving is clamped to the shaft. It is made from hard maple and finished to all standard sizes.

Speeds. The wooden pulley is safe to any speed up to 5000 ft./min. and has long life if operated under correct mechanical and atmospheric conditions.

Applications. Wooden pulleys are satisfactory for all ordinary driving and in operations such as crushers, stamp mills, oil-well rigs, and in hazardous operations such as explosives manufacturing. Wooden pulleys should not be used where water, dampness, or steam is present.

Selection. In selecting pulleys the following points should be borne in mind: (1) The surface of the rim should have a high coefficient of friction, yet it must not be rough because belt slippage cannot be entirely eliminated and, when a belt slips on a rough pulley, wear is rapid. (2) A pulley should combine strength with light weight. Unnecessary weight loads up the shaft and, generally, increases the friction of the drive. (3) The pulley should be able to withstand severe service and resist breaking in case of sudden overload or shock. (4) A pulley should be, to a certain extent, a heat conductor since belt slip generates heat. (5) A split pulley has great advantages over a solid pulley because it can be mounted on the shaft with the minimum amount of shutdown and labor. (6) A pulley should be exactly round and of such material and construction that it is unaffected by atmospheric conditions. (7) A pulley should be designed so that it will rotate with the least amount of air resistance.

Ordering Pulleys. When ordering pulleys, careful observance of the following rules will avoid errors and save time:

1. **Service:** Give horsepower, r.p.m., and character of service.
2. **Description:** State whether solid, split, clamp hub, tight and loose, flanged or special.
3. **Diameter:** Specify diameter in inches. This should be the first dimension given. If exact diameter is required, mention this fact and state whether measurement shall be made at crown or edge of rim. Unless otherwise specified, the diameter of the pulley is the diameter at the top of the crown.
4. **Face:** Specify face in inches. This should be the second dimension given and should be specified in accordance with the width of the belt unless an exact width of face is desired in which case the fact should be noted by having the word *exact* follow dimension of face.
5. **Bore:** Specify exact diameter of shaft in inches. This should be the third dimension given.
6. **Crown or Straight Face:** After specifying dimensions of pulleys, state whether crown or straight face. If neither is specified, crown face is generally furnished. Drum pulleys for shifting belts have straight faces. Each pulley of a pair of tight and loose pulleys should have crown face.
7. **Key-seat or Set-screwed:** State whether key-seated or set-screwed, or both.

Belt Width and Pulley Face. The face of a pulley is generally determined by the width of belt to be employed. As an example, if a 6-in. double leather belt is to operate over a 24-in.-diameter pulley, the pulley should be specified as a 24 by 6 in. However, the actual width of the face of the pulley would be approximately $6\frac{1}{2}$ in., because it is a standard with pulley manufacturers to make the face of the pulley at least ½ in. wider than the belt to be used, so as to overcome the possibility of the belt's running off the pulley because of weave. Pulleys to accommodate belts wider than 12 to 14 in. up are usually made 1 to 1½ in. wider in face than the width of the belt.

GEARING

The general subject of gearing is so well treated in Marks' "Mechanical Engineers' Handbook" and similar handbooks that merely the various types will be mentioned here. The various types of power gears are as follows: **spur, miter, bevel, mortise, worm, herringbone,** and **helical.**

BEARINGS

Both plain and antifriction types of bearings are used to support power transmission shafting and accessories. Plain or sleeve-type bearings may be grouped as follows:

Plain or Sleeve Bearings

Description	Uses
One-piece cast-iron babbitted pillow block (Fig. 62)	Rough, slow-speed applications or for experimental work where adjustment is not required
Split cast-iron babbitted pillow block (Fig. 63)	Rough, slow-speed applications where shaft removal is necessary without disturbing base of bearing
Self-lubricating rigid split babbitted pillow block (Fig. 64)	Heavy line, counter, or jack shafting for speeds up to 400 r.p.m. When adjacent to gears or heavy chain drives, use turned liner type
Self-lubricating ball-and-socket babbitted pillow block (Fig. 65)	Light and medium line, counter and jack shafting for speeds up to 350 r.p.m. can be used in inverted position
Self-lubricating ball-and-socket hanger bearing (Fig. 66)	Light- and medium-duty line, counter, and jack shafting for speeds up to 350 r.p.m. Normally supported from overhead steel or timber in inverted position as shown. Should not be used for heavy chain or gear drives. Standard distances from base to center of shaft from 8 to 36 in.

Bearing Metals. Plain bearings are usually lined with metals such as babbitt, bronze, or brass, which are known as "bearing metals." The use of bearing metals provides a quick and inexpensive means for replacement of worn bearings. These metals are usually specified to be of a composition different from that of the supported moving part. Babbitt metal is used almost universally in connection with steel shafts for power transmission work. The original Babbitt metal formula calls for 89.3 per cent tin, 3.6 per cent copper, and 7.1 per cent antimony. Other satisfactory babbitt metals having a lower tin content in combination with lead, zinc, and/or nickel have been developed.

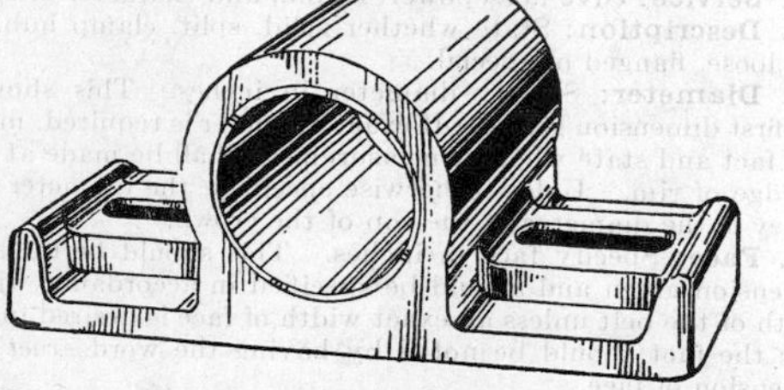

FIG. 62. One-piece cast-iron babbitted pillow block.

Lubrication of Plain or Sleeve Bearings. Lubricant selection for industrial bearings depends upon relative speeds of the rubbing surfaces, the pressure on the rubbing surfaces, and the temperature to which the bear-

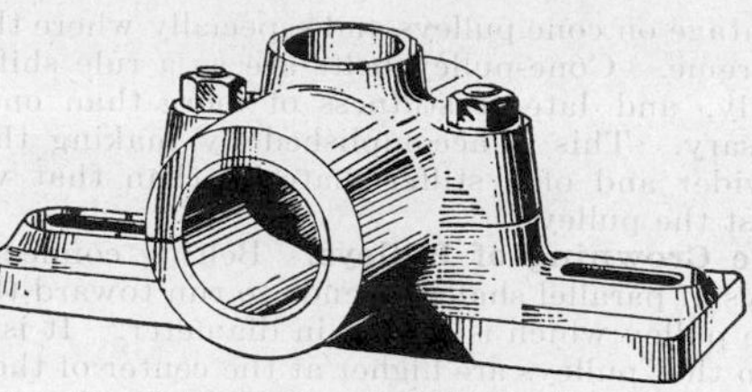

FIG. 63. Split cast-iron babbitted pillow block.

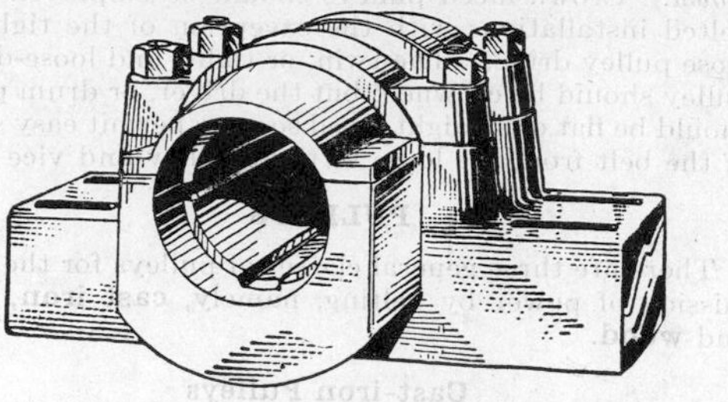

FIG. 64. Self-lubricating rigid split babbitted pillow block.

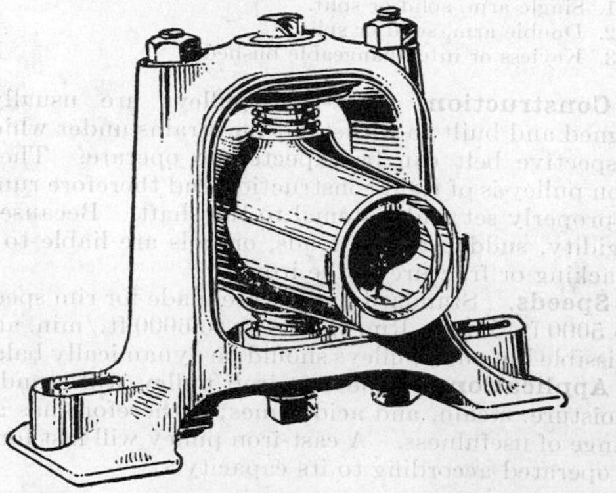

FIG. 65. Self-lubricating ball-and-socket babbitted pillow block.

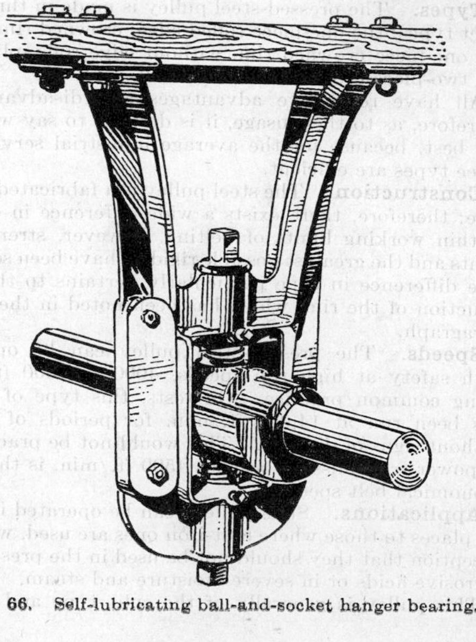

FIG. 66. Self-lubricating ball-and-socket hanger bearing.

ing is subjected. Since bearing friction losses are approximately proportional to the lubricant viscosity under ordinary operating conditions, it is advisable to use oil lubrication whenever possible or practicable. In those cases where oil leakage presents a housekeeping problem or where grease will act as a seal against entrance of abrasive dust to the bearing, grease lubrication is usually recommended.

Load-carrying Capacity of Plain Babbitted Bearings. Bearing pressures for average industrial head, line, and countershaft installations should not exceed 150 lb./sq. in. of projected area (bearing length times shaft diameter).

Bearing Clearance. The diameter of a shaft must be less than the inside diameter of the bearings. The difference in diameters is termed "bearing clearance." It is usual practice to allow approximately 0.001 in. clearance per inch of shaft diameter.

Oilless Bearings. The oilless or self-lubricating bearings are those that contain within themselves sufficient lubrication to assure continuous service. The general types are as follows:

The bronze and graphite type, which is constructed of high-grade phosphor bronze into which are cast symmetrical grooves for graphite, varying in design according to the service for which the bushing is intended; the hardwood, lubricant-impregnated type, which is made of hard seasoned wood, thoroughly impregnated with a specially prepared lubricating compound (no additional lubricant is ever required); and the metaline type, which is constructed of either phosphor-bronze or gun-metal-bronze and into which are inserted small-diameter metaline plugs. The metaline plugs are a mixture of graphite and antifriction metals. The oilless bearing is particularly adaptable for high-speed spindles and in the cotton and textile industries, generally, where cleanliness is a feature.

Spacing of Bearings. The number of bearings used on a shaft, and their spacing, has a direct influence on the dead-load capacity of the shaft. Deflection of shafting should not exceed 0.01 of 1 in./ft. Beyond this, binding in the bearings occurs, causing increased friction load and the possibility of burned-out bearings. On line shafting, 8-ft. centers are good practice where the pulleys can be set close to the bearings.

Bearings for Power Shafting. Based on the numerous designs of bearings manufactured, the selection of a type is at times difficult. The factors governing bearing selection are: (1) diameter and speed of shaft; (2) power and dead load; (3) support; (4) lubricant and lubrication method; (5) space limitations; (6) operating conditions; (7) initial and maintenance costs.

Antifriction Bearings. With the present trend toward unit drives with enclosed gears, the use of ball or roller bearings in such transmission units has been adopted almost universally. The tabulation in Fig. 67 is intended to assist the engineer in identifying the various antifriction-bearing types and to present a brief summary of the characteristics of each type.

SHAFTING

Cold-finished Steel Shafting. Because of its low cost and adequate strength for the average power-transmission requirements of industry, cold-finished steel shafting is widely employed.

It frequently occurs, however, that stronger steels are required for shafting because of slow speed, heavy loads, and shocks. This makes necessary the use of the S.A.E. alloy steels.

Specification Numbers for S.A.E. Steels. A numeral index system is used for the specification of S.A.E. steels, which facilitates the specifying of these steels. Such numerals are partly descriptive of the quality of material covered by these numbers. The first numeral of the number indicates the class to which the steel belongs; thus the numeral 1 indicates a carbon steel, 2 a nickel steel, and 3 a nickel chromium steel. In the case of the alloy steels the second numeral of the number generally indicates the approximate percentage of the predominant alloying element. Usually the last two or three numerals of the number indicate the average carbon content in "points" or hundredths of 1 per cent. Thus 2340 indicates a nickel steel of approximately 3 per cent nickel (3.25 to 3.75) and 0.40 per cent carbon (0.35 to 0.45); 71360 indicates a tungsten steel of about 13 per cent tungsten and 0.60 per cent carbon. The basic numerals for the various qualities of steels specified are as follows:

Steel	Basic Numeral
Carbon	1
Nickel	2
Nickel-chromium	3
Molybdenum	4
Chromium	5
Chromium-vanadium	6
Tungsten	7
Silicomanganese	9

Physical Properties of Cold-finished Steel Shafting

Type	S.A.E. numbers				
	Tensile strength, lb./sq. in.	Yield point, lb./sq. in.	Elongation, in 2 in. %	Reduction of area, %	Size diameter, in.
Hot-rolled soft steel or screw stock	50/60,000	30/35,000	20/30	35/45	1 to 2
Cold-finished screw stock S.A.E. 1112 or 1120	70/100,000	55/90,000	10/20	35/60	Up to 1
Cold-finished screw stock S.A.E. 1112 or 1120	50/60,000	30/40,000	20/35	30/45	2¾ to 4
Cold-finished 35/45 carbon S.A.E. 1040	70/90,000	35/50,000	20/30	35/50	1½ to 2
Cold-finished annealed 35/45 carbon alloy—S.A.E. 3140, 2340, 6140, etc.	90/110,000	60/70,000	20/30	55/65	Up to 1½

Shafting Horsepower Formulas

Cold-finished steel

HEAD SHAFTING (First Shaft from Prime Mover):

Heavy strains and loads .... horsepower $= D^3R/125$
Average loading .... horsepower $= D^3R/110$
Light loading .... horsepower $= D^3R/100$

INTERMEDIATE SHAFTING (Such as Jacks and Counters):
Horsepower $= D^3R/90$

LINE SHAFTING:

Heavy load, with bearings about 8 ft., center-to-center .... horsepower $= D^3R/100$
Medium load, with bearings about 8 ft., center-to-center .... horsepower $= D^3R/90$
Light load, with bearings about 6 ft. center-to-center .... horsepower $= D^3R/75$

D = diameter of shaft.
R = r.p.m. of shaft.

CONSTANTS

125 = 2800 lb. fiber stress
110 = 3000 lb. fiber stress
100 = 3200 lb. fiber stress
90 = 3400 lb. fiber stress
75 = 4000 lb. fiber stress

COUPLINGS

Couplings, of either the rigid or the flexible type, are used to connect power transmitting shafts together. Rigid couplings are used where it is desired to maintain permanent alignment on a permanent type of installations. Flexible couplings are used where permanent and perfect alignment is not possible.

Rigid Couplings. The following general rigid coupling types are used for power transmission work.

1. Flanged with plain-finished faces.
2. Flanged with male and female faces.
3. Flanged with separator plate.
4. Ribbed-sleeve type.

	Type	Description		Type	Description
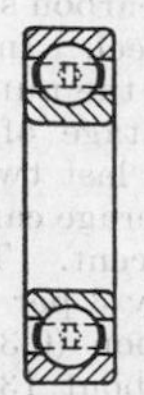 (a)	Deep-groove ball bearing	Available in single or double row types. Primarily designed for radial loads only but it will carry some thrust load. Sometimes made for heavy radial loads with filling notch on inner race to permit use of more balls than standard construction permits.	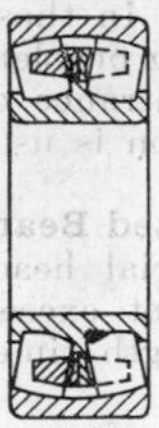 (e)	Spherical roller bearing	Self-aligning roller bearings have two rows of barrel shaped rollers rolling on a common spherical raceway. Used for heavy-load applications to compensate for misalignment and shaft deflection. Made in both straight bore and taper bore for adaptor sleeves. The thrust capacity is quite high with respect to radial capacity.
 (b)	Separable ball bearing (magneto bearing)	Single-row ball bearing with a shallower groove in the inner ring than standard ball bearing and only one shoulder forming outer race to permit bearing separation. Made in small sizes only for instruments, magnetos, etc. Chief advantage is simplicity in mounting. Usually mounted opposed to each other with some axial play.	 (f)	Cylindrical roller bearing	Cylindrical roller bearings are designed to carry high radial loads with minimum deflection. For the same size bearing the cylindrical roller has a greater radial load capacity than a ball bearing. This type is available in single or double row types.
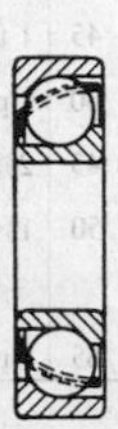 (c)	Angular contact ball bearing	Angular contact bearings are of grooved type designed so that direction of load through the balls forms an angle with the radial plane of the bearing Used instead of deep groove ball bearings for applications where thrust load predominates. Usually mounted in opposed pairs.	 (g)	Needle bearing	Bearing is made from a hardened steel shell into which is fitted a full complement of small-diameter hardened steel rollers. The inner race may be omitted if installed on hardened shaft. Designed primarily for oscillating applications or for continuous service where load is intermittent or variable.
 (d)	Self-aligning ball bearing	Self-aligning ball bearings have two rows of balls rolling on a common spherical raceway. Used to compensate for misalignment and shaft deflection. Radial load capacity is approximate equal to equivalent size single-row deep-groove ball bearing and the wide series bearings have considerable thrust capacity. Most sizes are available with taper bore for adaptor sleeves.	 (h)	Tapered roller bearing	Tapered-roller bearings are of the separable type, usually mounted in pairs opposed to each other. They have a high load-carrying capacity both radially and axially in one direction.

Fig. 67. Antifriction bearing types.

The flanged type with male and female faces facilitates permanency of alignment, but one of the shafts must be backed off to permit removal.

The flanged type with separator plate permits disconnecting the shaft without removal of the coupling.

Flexible Couplings. It is desirable to employ flexible couplings between motors and gear reduction units and similar applications, because initial alignment is perfect at the time of installation and still more rarely remains perfect. To avoid bending stresses in the shafting and excessive loads on supporting bearings, the general types of flexible or self-aligning couplings shown in Table 5 are used.

Any of the flexible couplings listed will accommodate nominal amounts of angular and parallel misalignment. However, the torsionally resilient types should be used on drives subject to severe shock loads or loads of a vibrating character.

The Flexible Floating Shaft. This transmission device is essentially an elongated flexible coupling employed where the driving shaft must be remote from the driven shaft. A flexible floating shaft generally consists of an independent shaft of the required length and diameter having the rigid half of a coupling keyed to each end. To the driving and driven shafts are keyed halves of a flexible coupling. Applications of flexible shafts are usually made for reasons of excessive misalignment. The maximum length of floating shafts is controlled only by the critical speed of the shaft.

Table 5

Design	Torsionally resilient	Supplemental lubrication required
Pin and rubber or fiber bushing	Limited	No
Laminated pin and bushing	Limited	No
Laminated spring pack	Yes	Yes
Spring cylindrical grid	Yes	Yes
Floating center block	No	Yes
Double engagement spline	No*	Yes

* Although the double engagement spline type is classed as being nonresilient, it is capable of absorbing severe shocks and vibration because the sliding metallic splines are protected by an oil film.

GEAR REDUCTION UNITS

Since the source of power for most unit process applications is a high-speed electric motor, gear reducers must be utilized to obtain the required speed reduction to avoid a multiplicity of chains, gears, shafting, and belting for ratios above 7 to 1. Modern speed reduction units utilize five general types of gearing alone or in combination, namely, (1) the worm and worm wheel; (2) the spur; (3) the helical gear; (4) the herringbone; and (5) the spiral bevel gear. These various types of gear units have been fairly well standardized so that suitable sizes and types are commercially available for almost any power requirement. Trade catalogues of gear reduction unit manufacturers should be consulted for specific information on available standard ratios and power capacities.

1. Conventional Worm Gear Reduction Units. *Types and Worm Arrangements.* Modern worm gear reducers are available in the following general constructions and worm arrangements:

1. The bottom driven single reduction unit.
2. The top driven single reduction unit.
3. The vertical single reduction unit with the output shaft extending either above or below the unit.
4. The combination helical and worm gear unit.
5. The compound or double reduction unit.

Construction. The modern worm gear is constructed with as steep a helix angle as possible (up to 42 deg.) to permit operation at maximum efficiency.

The worm of a modern reducer is usually integral with an alloy steel shaft with the threads hardened and ground. The worm is rigidly supported on antifriction bearings.

The worm gear usually consists of a chilled cast bronze rim shrunk on, and pinned to a high-grade cast-iron center. The worm wheel is usually mounted on heavy-duty antifriction bearings. The entire worm and worm wheel assembly is encased in a rugged oiltight cast-iron or welded-steel housing.

Ratios and Speeds. Single reduction worm gear units are usually employed for ratios ranging from 4 to 1 up to 80 to 1. It is possible to build worm gear units having higher reduction ratios, but their low efficiency beyond that point precludes their economic application. (The efficiency of a single reduction worm gear unit expressed in per cent is approximately $100 - \frac{1}{2}$ the ratio.)

To obtain higher reduction ratios and still retain the right-angle feature of the single reduction unit, combination helical and worm gear reducers are manufactured. The primary reduction in these units is by means of relatively small helical gears, and the secondary reduction is by means of a worm gear.

It is possible to obtain very high reductions by means of a compound or double reduction units.

Modern worm reduction units can be operated at worm speeds up to 4000 r.p.m. and are therefore suitable for direct connection to high-speed motors and turbines.

Applications. The worm gear reduction unit is essentially a right-angle drive, and the housings and gear arrangements are furnished by manufacturers to suit this arrangement to industrial requirements.

The worm gear unit can be applied to almost any type of drive where the input speed is high and the output speed is low.

2. Air-cooled Worm Gear Reduction Units. The capacity of conventional worm gear units to transmit mechanical power is limited by the strength of their component parts and by their ability to dissipate frictional heat. For high input speeds, it is frequently necessary to use an oversize worm gear unit, because a unit of the proper mechanical rating would operate at temperatures so excessive that lubrication of the gears and bearings would be difficult.

Several manufacturers now market air-cooled worm gear units which can, at high speeds, transmit more power than conventional units. By utilizing a double-wall construction of the gear housing, and mounting a fan on one end of the worm shaft, the unit is cooled by an air stream drawn through the space between the housing walls in much the same manner as a totally enclosed fan-cooled motor. The high-velocity air stream over the outer surface of the oil reservoir greatly increases the rate of heat transfer. It is therefore possible to transmit more power for a given temperature rise with the air-cooled design than with the conventional worm gear unit. However, the air-cooled unit has no advantage over the conventional unit if the input speed is low (less than 580 r.p.m.) or if the unit is operated intermittently.

Air-cooled units are at least as efficient as a correspondingly larger conventional unit having the same thermal rating because the power consumed by the fan is offset by the additional friction losses encountered in oversize conventional units. For a given power capacity, the air-cooled unit occupies less floor space and is lower in initial cost than conventional units.

At the present time, air-cooled units have been developed only in the single reduction types.

3. Helical Gear Reduction Units. Helical gears are used in gear reduction units (usually motorized) of the following general types:

1. Single, double, and triple reduction units designed for direct connection to a motor.
2. Single reduction planetary gear reduction unit designed for direct connection to a motor.
3. Double reduction planetary gear unit with primary reduction by helical gears and secondary reduction by stub spur gears.

Helical gearing is used for high-speed medium-power-capacity industrial gear units because of relatively silent operation, high efficiency, and long life.

Ratios and Power Capacities. The power capacities of present standard units are similar for both the helical-in-train and the planetary gear assemblies. Higher ratios per reduction are possible, however, with the planetary assembly.

Table 6

Gear assembly	Reduction	Standard unit, hp.	Ratios available	Motor r.p.m.
Helical-in-train	Single	½–75	1:2– 8:1	580–1750
Helical-in-train	Double	½–75	8:1– 40:1	580–1750
Helical-in-train	Triple	½–75	40:1–288:1	580–1750
Planetary	Single	½–75	3.6:1– 10:1	580–1750
Planetary	Double	½–75	10:1– 87:1	580–1750

Efficiency. The efficiency of single reduction units at full load is approximately 97 per cent, double reduction units about 95 per cent, and triple reduction units about 92 per cent.

4. Herringbone Gear Units. Herringbone gear reduction units are available commercially as single, double, and triple reduction units. In single and triple reduction, the input and output shafts are parallel but offset. In the double reduction type, high- and low-speed shafts in line or offset are available. Choice of shaft arrangement is governed entirely by installation convenience rather than by superiority of design.

The advantages of the herringbone gear unit over other types are: (1) continuous, smooth meshing of gears (two teeth always having two contact points); (2) greater strength; (3) absence of end thrust; (4) high permissible peripheral speeds; and (5) ability to withstand shock loads and loads of a vibrating nature because of very low backlash.

Because of high permissible peripheral speeds, single reduction units are frequently applied as speed increasers.

The herringbone gear is usually made from a heat-treated high-carbon-steel casting or forging. The pinion is usually forged from a high carbon or alloy steel, heat-treated, with the gear teeth cut integral with the shaft.

Table 7

Reduction	Ratios	Capacity, hp.	Permissible input speeds
Single	1.7:1–10:1	1½–700	Up to 1800 r.p.m.
Double	10:1–70:1	3–1300	Up to 1800 r.p.m.
Triple	70:1–320:1	½–500	Up to 1800 r.p.m.

Efficiency. The single reduction herringbone continuous-tooth type gear unit is approximately 97 per cent efficient at full load, the double reduction unit about 95 per cent, and the triple about 92 per cent.

5. The Spiral Bevel Gear Unit. Spiral bevel gears are usually employed in gear units of the following types:

1. The single reduction spiral bevel gear unit.
2. In combination with the herringbone or helical gear unit as the primary reduction to permit right-angle drive arrangement.

The advantages of spiral bevel gears over conventional bevel gears for driving shafts mounted at right angles (either horizontal or vertical) are relatively more silent operation, higher efficiency, and longer life.

The maximum ratio available in commercial units is about 7 to 1. Input speeds up to 2000 r.p.m. are possible with a maximum power capacity of about 500 hp.

6. The Motor Reducer. Motorized gear reduction units with the motor either built integral or mounted on the gear housing are available with the following general gear arrangements:

1. Single reduction planetary type with helical gears.
2. Double reduction planetary type with helical gears for primary reduction and stub tooth spur gears for second reduction.
3. Single reduction helical gear train.
4. Double reduction helical gear train.
5. Triple reduction helical gear train.
6. Motorized single reduction worm.

The first five types are available in all N.E.M.A. standard motor sizes from ¾ up to 75 hp. The motorized worm gear reducer is available only in fractional horsepowers up to about 3 hp.

Advantages. The advantages of the motor reducer over a separate motor and reducer assembly are: (1) elimination of base plate and coupling; (2) compactness; (3) permanent alignment of motor and reducer; (4) ideal for driving vertical shafts on mixers, because it can be mounted directly on the top of the vessel; and (5) lower initial cost.

Disadvantages. Motor trouble results in complete shutdown of unit until integral motor can be repaired. With separate motor and gear unit under similar circumstances, a spare motor can be substituted.

VARIABLE-SPEED DRIVES

Many modern forms of apparatus require minute variations in applying the speed. Frequently, when such variation is desired, the control must be rapidly and easily accessible to the operator.

For many operations of this character the ideal installation is the variable-speed motors, but it is difficult to obtain fine adjustment with this method. Therefore, there are two distinct types of variable-speed control—electric and mechanical, the mechanical device being most widely used for graduated speed reduction or increase.

Mechanical Infinitely Variable Speed Units. There are two types that appear to be the most popular; namely, the Reeves, which utilizes smooth-surfaced cone disks with their apices facing inward, between which operates a specially constructed rubber and wood block belt; and the P.I.V. unit, which utilizes radially toothed cone disks with their apices facing inward, between which operates a laminated toothed chain. The former is available in ratios to 16 to 1 and with horsepower capacities from fractional to 150, whereas the latter is available in ratios to 6 to 1 with horsepower capacities from 1 to 15. Selection depends upon installation space available and the refinement of infinite variation required. Both types can be had in either horizontal or vertical construction, and both can be supplied motorized.

Applications. These units can be used wherever variable speed is desired. They are therefore invaluable in canning, textile, machine, mining, paper, and cement industries.

Advantages. The advantages of utilizing the mechanical speed-variation unit are: (1) It provides an infinite number of speeds. (2) The transmission of power is positive at all speeds. There is practically no slippage. (3) It automatically locks in place at any desired speed. (4) It may be mounted in any position; therefore power may be applied easily by belting, chain, or gearing. (5) It may be driven in either direction. (6) The speeds may be changed while the machine is running. (7) Ratios may be obtained as high as 16 to 1.

Control. These units can be controlled either by hand or remote electric. When hand controlled, a handwheel is provided directly on the machine. If the machine is hung on the ceiling, a sprocket is provided on the shifting screw over which passes a chain to the hand-controlled sprocket below. When the unit is placed at a distance from the operator's position at the driven machine, the electric or remote control has been designed in order that the unit may be used to its full advantage.

SHORT-CENTER DRIVES

To permit driving a machine direct from a motor for speed ratios of 7 to 1 or less, a number of methods have been devised to overcome the defects of ordinary open flat belting operated on short centers. The methods in common use are:

1. The conventional V-belt drive.
2. The V-flat drive.
3. The detachable V-belt drive.
4. The pivoted motor base flat belt drive.
5. The silent or roller chain drive.

1. The Conventional V-belt Drive. The V-belt drive consists of one or more V-shaped belts composed of cotton cords and fabric, impregnated with and embedded in rubber operating between V-grooved cast-iron or steel sheaves. The cross-sectional sizes of V belts have been standardized as shown in Fig. 68 and Table 8.

Standard-pitch-length belts covering a wide range are available to suit almost any sheave-diameter and center-distance combination.

Standard construction belts are fabricated for use under normal atmospheric conditions. Special heat-resistant belts are available on special order for temperatures above 140°F. For operation in the presence of mineral oil, corrosive chemicals, etc., Neoprene synthetic rubber frictioned belts are available. For use in hazardous locations, "static-proof" belts frictioned with conductive rubber are available.

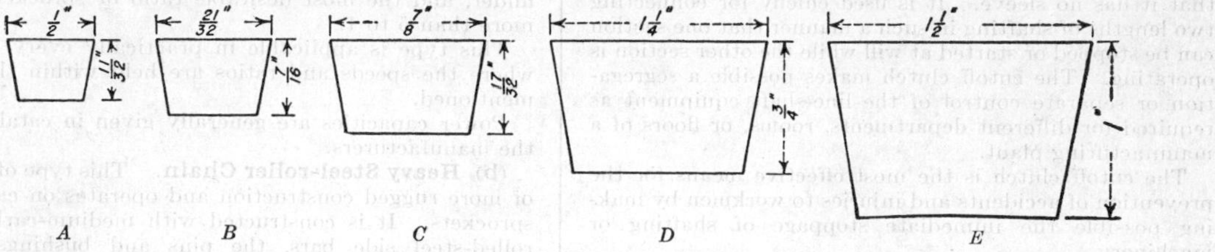

FIG. 68. Standard V-belt cross sections.

The approximate horsepower capacity of a single V belt operating over minimum recommended diameter sheave with 180 deg. arc of contact at various speeds is shown in Table 8.

Table 8. Power Capacity of V Belts for Various Belt Speeds

Section	Min. sheave diam., in.	Hp. capacity per belt, velocity, ft./min.			
		1000	2000	3000	4000–5000
A (½ × 11⁄32 in.)	3.0	0.7	1.3	1.6	
B (21⁄32 × 7⁄16 in.)	5.4	1.4	2.6	3.5	4.0
C (⅞ × 17⁄32 in.)	9.0	2.8	5.5	7.6	9.0
D (1¼ × ¾ in.)	13.0	5.1	9.7	13.3	15.5
E (1½ × 1 in.)	21.0	8.5	16.2	22.7	27

Operation. The V-belt drive will operate in either direction with the tight strands on either the top or bottom. The centers may be horizontal, inclined on an angle, or vertical. Provision for a slight adjustment between centers should be made for installing and removing belts and to take up for stretch.

2. The V-flat Drive. When the velocity ratio between the driving and driven shaft exceeds 3 to 1 and the center distance is equal to or less than the diameter of the large wheel, grooving of the large wheel may be omitted without sacrifice of power capacity or efficiency. A V-flat drive consists of a grooved driving sheave, a set of V-flat belts, and a flat-faced driven pulley. The bottom or narrow face of the V belt contacts the flat face of the pulley. This method is economical both on new installations and on change-over from existing flat-belt drives, since a flat-faced pulley costs less than a grooved pulley of equal size and since, on change-overs from flat-belt drives, it is usually necessary only to machine off the crown of the pulley.

3. The Detachable V-belt Drive. For V-belt drives, which must operate on fixed centers without idlers, detachable V belts have been developed to permit take-up for stretch. Both riveted and bolted construction are available.

Detachable V belting is also useful for compressor or three-bearing motor drives where it is necessary to remove a bearing to permit installation of conventional endless V belts.

4. The Pivoted Motor Base Flat-belt Drive. Pivoted motor bases designed to maintain the correct tension for short-center flat-belt drives are available in either the gravity or the reaction-torque types.

The gravity type consists of a cast iron or welded steel base so designed that the overhung weight of the motor beyond the pivot shaft provides the necessary belt tension. By increasing or decreasing the distance from the pivot point to the motor, the belt tension can be adjusted to suit the load requirements. Once set, the belt tension remains substantially constant regardless of belt stretch. Best results are obtained from the gravity-type motor base if the tight side of the belt is nearest the pivot.

The reaction torque motor base is of welded-steel construction with a cradle supported on a pivot arm so located that the reaction torque tends to rotate the motor about the pivot in a direction opposite from the pulley direction of rotation. This drive must be mounted, however, so that the reaction torque will tighten, not loosen the belt, which means that the tight side must be nearest the pivot axis.

5. Chain Drives. See Chain Driving, pp. 1672 and 1673.

CLUTCHES

The problem of connecting full-speed power to dead load is always present in industrial operations. In many instances it is solved by the tight-and-loose pulley, but where starting loads are severe and in cases of chain connection the clutch must be employed.

The clutch that controls the load by friction mediums is the one most universally used. Numerous types and designs of such clutches are on the market, each possessing some individual characteristic as to the methods employed for engaging the friction-resistance portions. However, in the final analysis, although their mechanisms may differ, the result is the same.

Friction Clutches. Types. A number of friction clutches that are on the market may be classified as follows:

1. Toggle and wood friction-block clutch.
2. The lined-disk and self-oiling sleeve clutch.
3. High-speed clutches.
4. The worm-and-gear-controlled clutch.
5. The friction pick-up and positive-lock clutch.
6. The coil-spring clutch.
7. The air-controlled clutch.

Construction. The modern friction clutch consists generally of three principal parts, namely, the sleeve and friction-material drum or ring which operates loose on the shaft and to which is keyed the load-connecting medium such as pulley, gear, sprocket, or sheave; the clamp disk or part that engages the friction ring or blocks of the sleeve drum, this part being keyed to the power-input shaft; and the engaging mechanism, which is controlled by a sliding yoke and collar.

Application. The *friction clutch* is applicable to almost any type of driving and is indispensable to the driving of certain machine tools and where the power transmitted by chain is intermittent.

The *split clutch* should be used on a line shaft or on any shaft where it is impossible to slide the clutch from the end of the shaft.

The *solid clutch* should be used at, or near the end of, a line shaft where it can be readily removed without disturbing other elements of transmission.

When using a split clutch the driving member on the

sleeve such as pulley, gearing, or sheave should be split also.

Speeds. As far as speed is concerned friction clutches are designed in two classes, namely, high- and low-speed clutches.

The low-speed clutch can be operated up to 500 r.p.m., while the high-speed clutch can be safely operated at speeds as high as 2000 r.p.m.

Cutoff Friction Clutches. This mechanism, sometimes termed a *friction cutoff coupling*, is similar in construction to the regular friction clutch with the exception that it has no sleeve. It is used chiefly for connecting two lengths of shafting in such a manner that one section can be stopped or started at will while the other section is operating. The cutoff clutch makes possible a segregation or separate control of the line-shaft equipment as required for different departments, rooms, or floors of a manufacturing plant.

The cutoff clutch is the most effective means for the prevention of accidents and injuries to workmen by making possible the immediate stoppage of shafting or machinery.

Jaw Clutches. The jaw clutch is a convenient method for connecting or disconnecting a load where shock is of no consequence and the speed is low. Therefore it is usually employed for very rough driving. It should not be considered for the average industrial transmission system and should not be used on speeds exceeding 60 r.p.m.

It is constructed with either square or spiral jaws, and either type may be equipped with a sleeve to hold the driving member. This type of clutch may also be used as a cutoff coupling.

CHAIN DRIVING

Selection. The transmission of power by chain has progressed rapidly, so that at the present time there is a type and design for almost any installation. The selection of type, therefore, is of the utmost importance because each style of chain manufactured is for a specific field and when misapplied is costly.

Advantages. The general advantages of chain driving are: (1) Certain types are approximately 98 per cent efficient. (2) Relatively high speeds can be obtained. (3) Not affected by heat, cold, or moisture. (4) Certain types will transmit within reason any amount of power. (5) Give a positive velocity ratio. (6) Can be used on short or long centers. (7) Chains cannot slip.

Types. There are three distinct types of power-transmission chains, namely, (1) malleable-iron detachable, (2) steel roller, and (3) the silent chain. Each type has its own construction and applications, and they will therefore be discussed separately.

1. Malleable-iron Detachable Chain. Construction. This type is manufactured in approximately 25 stock sizes designated by number. Special sizes and designs can be obtained, but they are generally for purposes other than the transmission of power. This type is composed of individual malleable-iron links so designed as to allow ease of assembly.

Ratios and Speeds. The most desirable speed for this type is 400 ft./min. and under, and the most desirable ratio of sprockets is not more than 5 to 1. Higher ratios have a tendency to shorten the life of the chain.

Application and Capacities. This type finds application in practically every form of industry where the speeds and ratios are held within the limits mentioned.

The power capacities of this chain as to tensile strength and horsepowers are usually given in catalogues of the various chain manufacturers.

2. Steel-roller Chains. Types. This chain is manufactured in three styles: (*a*) the light-roller, (*b*) the heavy-steel-roller, and (*c*) the finished steel-roller power chains.

(a) **Light Steel-roller Chain.** This type of chain possesses three times the tensile strength of the malleable-iron type and operates on cast tooth sprockets. It is constructed with medium-carbon hot-rolled-steel side bars, the pins and bushings are of hardened steel, and the rollers are of either malleable iron or hardened steel.

The most desirable speed for this type is 700 ft./min. or under, and the most desirable ratio of sprockets is not more than 5 to 1.

This type is applicable in practically every industry where the speeds and ratios are held within the limits mentioned.

Power capacities are generally given in catalogues of the manufacturers.

(b) **Heavy Steel-roller Chain.** This type of chain is of more rugged construction and operates on cast tooth sprockets. It is constructed with medium-carbon hot-rolled-steel side bars, the pins and bushings are of hardened steel, and the rollers are of either malleable iron or hardened steel.

The most desirable speeds for this type of chain is 700 ft./min. or under, and the most desirable ratio of sprockets is not more than 5 to 1.

This type is applicable in practically every industry where the speeds and ratios are held within the limits mentioned.

The power capacities can be obtained from catalogues of chain manufacturers.

(c) **Finished Steel-roller Chain.** This type is intended for general power-transmission purposes and is designed to operate on accurately cut sprocket wheels. It is much more expensive, but it has demonstrated its economy compared with the less costly chains which operate on cast tooth wheels.

It is constructed with special rolled-steel heat-treated side bars, the pins are made from alloy steel hardened, and the bushings are steel, casehardened. The rollers are high-carbon steel, heat-treated.

It is possible to operate this type of chain at speeds as high as 1000 ft./min. However, it is more desirable from an economical and long-life standpoint to limit its operating speed to approximately 700 ft./min. This type allows a wide range of ratios and can be operated successfully as high as 10 to 1. However, the average commercial-ratio requirements vary up to 7 to 1.

Based on its strength and accuracy of design, this type of chain is applicable to all manner of industrial driving within the speeds and ratios mentioned. This includes driving by motor.

The ultimate or breaking tensile strengths of finished steel-roller chains are given in catalogues of the various chain manufacturers.

3. Silent Chain. This is a distinct type of power-transmission chain and is manufactured for almost any reasonable power capacity with a sustained efficiency of approximately 98.5 per cent.

It is manufactured in several types, but the difference of design is almost entirely in the construction of the pin or joint connection of the links.

There are two types of what may be termed rocker pin construction—one-pin and segmental-bushing joint, and a number of single-pin bushless joints.

All types of silent chains operate on cut-tooth-steel or cast-iron sprockets, and the permissible speeds and ratios are similar for all.

The most desirable speed for the silent chain is 1200 to 1500 ft./min. At speeds higher than 1500 ft./min., the silent chain should be encased to assure proper lubrication. The most desirable ratios for silent-chain driving ranges from 1 to 1 to 7 to 1. Ratios as high as 15 to 1, however, have been successfully employed.

The power capacities for silent chains are thoroughly given by tables, etc., in data books furnished by the various manufacturers of this type of chain.

Lubrication of Power Chains. The severe conditions under which most chains operate make exacting demands not only on the manufacturer of the chain but on the user. The former produces the best chain possible, and the consumer must also do his best to keep the chain clean and well lubricated.

Plastic lubrication such as grease and non-fluid oil should be used on the malleable-iron and unfinished-steel-roller chains. Fluid lubrication such as mineral oil should be used on finished-steel-roller and silent chains. Best results are obtained when these types are encased in what is known as an oil-retaining casing. These casings are supplied by the manufacturer of the chain.

LUBRICATION OF POWER-TRANSMISSION EQUIPMENT

Whether bearings, gears, or other relatively moving mechanical elements are being dealt with, complete lubrication implies the separation of the opposed working parts by an oil film which transmits forces from one part to another and prevents contact of metal on metal. Correct lubrication implies further that the fluid film consists of a correctly selected oil of a quality suited to the service, whether the service be one in which the oil performs its duty but once and goes to waste, or one in which it must perform its duty repeatedly over long periods of time.

The factors that are essential to the building of the oil film are: (1) the presence of a wedge-shaped clearance space, free from grooving such as might interfere with formation of an oil film; (2) speed or relative motion of the parts by which the carrying action is accomplished; (3) adhesiveness of the oil to assure that it will follow the moving surfaces; (4) a certain degree of oil body or viscosity—a characteristic that resists the too rapid escape of the oil from the film; (5) lubricating value of the oil which largely determines the strength of the oil film under pressure.

There are also conditions which tend to break down the oil film, chief of which are (1) heavy bearing loads, (2) high temperatures. Heavy bearing loads create intense pressures. If the carrying effect of speed or the adhesiveness or body of the oil is insufficient, a heavy load will prevent the formation or maintenance of a complete oil film.

The effect of increased temperature on all oils is to decrease their body or viscosity. With any given oil there is therefore a temperature beyond which the satisfactory lubrication of a given bearing becomes impossible. The maintenance of lubrication therefore depends on securing a satisfactory balance between the film-forming factors of oil wedge, speed, adhesiveness, oil body, and lubricating value, and the film-destroying influences of pressure and temperature.

Efficient lubrication is created and maintained by the action of an oil wedge, and this wedge must be maintained by a regular oil supply. In modern transmission equipment there are three ways in which the oil is supplied:

(1) By the ring, chain, collar, capillary, splash systems, circulating and pressure systems, when the same oil is to serve repeatedly; (2) by the automatic system such as drop-feed, wick-feed, and bottle oiler, when the oil loss must be compensated for; (3) by the hand-application system, when the need is intermittent, periodic, or irregular. When the parts cannot be lubricated by any available method of oil application, it may be necessary to use grease. Grease cups can at times be applied where an oiling device cannot.

HIGH-STARTING TORQUE CONTROL

Direct-connecting a motor to a load, particularly when high ratios are involved, generally necessitates rigid connecting methods. In view of the absence of mechanical slip with rigid connections, the motor is required to come to full speed under the applied load of the driven apparatus. The torque required to start a heavy applied load sets up destructive strains in both the electrical and the mechanical equipment, unless relieved by either electrical or mechanical devices. Economically it is desirable to use standard induction motors with "across the line" start, since such a start permits the motor to come to full speed in a few seconds, provided that the load is moderate or light. If the load is heavy, severe starting torque strains are produced. This condition can be relieved by special motors and electric starting equipment or by mechanical devices interposed between motor and load.

Electrical Methods. *A.c. Slip-ring Motor.* This type accelerates gradually under load in view of the use of collector rings and adjustable rheostats. The acceleration in speed is procured by wasting energy in resistance.

A.c. High-starting-torque Motor. This type differs from the standard induction motor in that it has two windings, one of which controls the resistance during starting and the other the reactance during running. The starting torque of the standard induction motor is approximately 175 per cent of the full-load torque, whereas for a light starting torque motor it is approximately 250 per cent.

D.c. Motor. This type possesses large starting torque capacity. The series motor can be used when the load is always connected to the motor and when the constancy of speed with variations in load is not especially desired. The series motor has a rapid rise in speed at high load; *i.e.*, if the load is not connected, the motor will race.

D.c. Compound Wound Motor. The advantage of this type is that at starting, when the current through the armature and series field winding is large, the total field excitation is large; hence there is an increase in torque capacity. Another advantage over the series motor is that it will not speed up indefinitely when released from load. A disadvantage is that its running speed decreases considerably with load increase.

Mechanical Methods. During the past few years, mechanical methods for overcoming high starting-torque resistance have been perfected. The number in actual use is relatively small when compared with the special motor equipment available. However, in view of many advantages, the mechanical is becoming a competitor of the electrical.

Advantages. (1) Elimination of expensive electrical starting equipment, (2) minimum size of mechanical transmission equipment, (3) economy in current where starting and stopping are frequent, (4) possibility of using standard induction across-the-line start motors, (5) the motor horsepower size can be based on the running instead of the starting load, (6) reduction in starting shock, and (7) overload release capacity.

Self-actuating mechanisms for high starting-torque service are of two types: those that serve as a coupling (combining flexibility) between motor and load and those that act integrally with belt pulleys, chain sprockets, gears, and V-belt sheaves. The former are known as high starting-torque couplings, and the latter as pulley-, sprocket-, or gear-type starters.

The Mechanical Slip-ring Starter. This is a unit which by slipping friction automatically starts machinery smoothly and with uniform acceleration from across-the-line start motors regardless of the starting torque involved. It consists of a spider, two friction bands, and a drum. The spider or driving element is keyed to the motor shaft, and the drum or driven element is keyed to the shaft of equipment to be driven. The weighted bands which lie inside the drum are secured at one end to the spider. The opposite or trailing ends of the bands are free. As the motor speeds up, centrifugal force causes the bands to expand and exert a frictional drag on the inside of the drum. The torque thus transmitted is dependent upon motor speed, diameter of drum, and the weight of the bands.

Performance. The mechanical slip-ring starter is designed to transmit any required torque at a specified speed in r.p.m. For any application the full-load speed of the motor and the maximum horsepower to be transmitted must be known. The device is usually proportioned to transmit a maximum of 125 per cent of the rated horsepower of the motor. The device can, under normal conditions, remain in a stalled condition, *i.e.*, with the motor revolving at full speed and the drum or driven element held stationary for a few minutes without injury to the device or motor. When the driving shaft is at rest, it is physically disconnected from the driven shaft. When power is applied, and as the motor increases in speed, centrifugal force acts on the bands and therefore develops very gradually a connection between the driving and driven shafts. When the driving shaft reaches full speed, the device is transmitting its maximum predetermined torque to the driven shaft. The driven shaft eventually reaches its full speed. During the period of acceleration, a gradually diminishing amount of slippage is occurring in the device, but, when the driven shaft reaches full speed, all slippage ceases, and the functioning is that of an ordinary flexible coupling, with the additional protection of slippage, in the event of overload. The device is inherently a limit-torque coupling, since it will not transmit to the driven machine or impose on the motor more than the maximum load for which it is designed.

General Applications.

1. High inertia starting loads, such as are encountered in centrifugals, extractors, etc., where the power required to start is frequently five or six times the actual running load.

2. Where starting is difficult owing to excessive bearing friction, the device assists any motor by permitting almost full speed before the application of the load. This allows the use of motors with low starting-torque characteristics.

3. The torque-limiting feature of the device is useful in many cases where sudden running overloads or complete stalling of the motor is likely. It is possible to build the device to transmit a maximum of 150 to 200 per cent of the motor rating, so that, in the event of the driven machinery jamming, the device will slip and allow the motor to continue at full speed until the overload relay takes the motor off the line.

4. In the driving of ball, pebble, and tube mills, where the maximum running load can be definitely determined, improved starting and operation can be obtained by selecting a motor that is just equal to the running conditions and by applying a mechanical slip-ring starter of approximately 15 per cent more horsepower capacity. Such a combination produces jerkless starting with across-the-line control and permits the size of the mechanical power equipment to be proportioned to the running instead of the starting load.

Horsepower and Speeds for Mechanical Slip-ring Starters, Coupling and Integral Types

Size of starter	R.p.m.						
	495	600	685	750	870	1160	1750
5½ × 2½	0.16	0.3	0.45	......	0.91	2.15	7.5
6 × 3	.42	0.75	1.13	1.47	2.3	5.5	19.1
8 × 2½	.95	1.63	2.5	3.27	5.1	12.1	42.2
9 × 3	1.92	3.4	5.1	6.68	10.4	24.7	86.5
10 × 3½	3.27	5.8	8.7	11.4	17.7	42.0	147.0
10 × 4	3.75	6.65	10.0	13.1	20.3	48.0	169.0
11 × 5	7.5	13.4	20.0	36.2	40.7	95.7	338.0
12 × 3½	6.8	12.2	18.2	24.0	37.5	89.0	310.0
12 × 4½	8.6	15.2	22.8	30.0	46.5	111.0	377.0
12 × 6	12.4	22.0	33.0	43.0	67.0	159.0	555.0
14 × 5	19.6	34.8	52.2	68.2	106.0	253.0	
15 × 5	24.2	43.0	64.5	84.0	131.0	312.0	
16 × 6	41.0	73.0	109.0	143.0	223.0	530.0	
18 × 6	59.5	105.0	158.0	207.0	322.0	765.0	
18 × 9	89.5	158.0	237.0	310.0	485.0	1150.0	
18 × 12	119.0	210.0	315.0	413.0	645.0	1530.0	
20 × 6	85.0	150.0	225.0	295.0	460.0	1090.0	
20 × 9	127.0	225.0	338.0	445.0	690.0	1640.0	
20 × 12	170.0	302.0	452.0	590.0	920.0	2180.0	
24 × 12	258.0	460.0	690.0	900.0	1410.0	3300.0	
26 × 12	438.0	780.0	1160.0	1520.0	2370.0	5620.0	
28 × 10	750.0	1330.0	2000.0	2520.0	4080.0	9700.0	
28 × 12	900.0	1610.0	2400.0	3150.0	4900.0	11600.0	

Note. Direction of rotation of driving shaft must always be given.

5. The device makes possible the use of two-speed a.c. motors for very gradual starting of heavy loads instead of d.c. or complicated a.c. arrangements. It takes care of the acceleration up to the low-speed setting of the motor. The motor control then handles the load from low to high speed.

6. With heavy inertia loads it is at times impossible to use synchronous motors because of their inability to "pull in." This results in overmotoring. In such cases the mechanical slip-ring starter, by eliminating the inertia factor WR^2 of the driven machine, enables a motor of the correct size for the running load to pull in with ease.

Permissible Speeds. On many low-speed motors the mechanical slip-ring starter is practicable, but its proportions become large because of low centrifugal force. The most adaptable speeds are from 600 to 1800 r.p.m., although 3600 r.p.m. is possible.

Power Capacity. Horsepower capacities range from fractional to any practical requirement. Mechanically there is no limit.

SECTION 25
REFRIGERATION

BY

Hans C. Duus, Ph. D., Member, American Institute of Chemical Engineers, American Chemical Society. Chemical Engineer, E. I. du Pont de Nemours & Co. (Refrigeration)

Barnett F. Dodge, Sc. D., Professor of Chemical Engineering, Yale University. Member, American Institute of Chemical Engineers, American Chemical Society. (Low-temperature Processes)

CONTENTS

REFRIGERATION

	Page
Definitions	1676
Units of Refrigeration	1676
The Dense-air Machine	1677
Description and Theory of Operation of Vapor-compression and Absorption Refrigeration Systems	1677
The Simple Vapor-compression Refrigerating Machine	1677
Sample Calculation of Performance Coefficient	1678
Indicator Card and Volumetric Efficiency	1679
Dual Compression	1680
Wet Compression	1680
Multistage Compression	1680
Split-stage Compression or Binary Cycles	1681
Compressors	1681
Steam Jets	1682
Lubricants	1682
Compressor Capacity	1682
Continuous Absorption Systems	1683
Absorption Systems with Other Refrigerants	1684
Platen-Munters Continuous Absorption System	1684
Intermittent Absorption Machine	1685
Refrigerants Used in Vapor-compression and Absorption Systems	1686
Condensers	1687
Types of Condensers	1688
Submerged-coil Condensers	1688
Atmospheric Condensers	1688
Bottom-inlet Atmospheric Condensers	1688
Countercurrent Atmospheric Condensers	1688
Double-pipe Condensers	1688
Shell-and-tube Condensers	1688
Horizontal Multipass Condenser	1689
Heat Transfer in Ammonia Condensers	1689
Condenser Water	1690
Cooling-water Towers	1690
Corrosion in Condenser Systems	1690
Silicate Treatment of Condenser Water	1690
Cooling Systems	1690
Evaporation or Expansion Coils	1691
Heat Transfer in Evaporators	1692
Liquid Controls	1592
Brine Piping	1692
Copper Piping	1692
Refrigeration Pipe Lines	1692
Insulation Values of Building Materials	1693
Brines	1694
Small-scale Refrigeration	1697
Natural Refrigeration Processes	1698
Ice Manufacture	1698
Flakice	1699
Pak-Ice	1700
Eutectic Salt-ice Mixtures	1700
Manufacture of Solid Carbon Dioxide	1700
The Heat-pump Principle	1701

LOW-TEMPERATURE PROCESSES

Fundamental Methods of Producing Refrigeration at Low Temperatures	1702
Work of Gas Liquefaction and Separation	1703
Analysis to Determine Power Requirement and Distribution of Losses	1703
Second Law	1703
Reversibility and Irreversibility	1704
Analysis of Losses	1704
Results of Thermodynamic Analysis	1705
Gas Liquefaction	1707
Rectification at Low Temperatures	1708
Various Low-temperature Processes	1709
Expanders	1712
Compressors	1713
Heat Exchangers	1713
Purification	1714
Low-cost Tonnage Oxygen	1715
Miscellaneous Processes	1716
Low-temperature Insulation	1717

REFRIGERATION

REFERENCES: Goodenough, "Principles of Thermodynamics," Holt, New York, 1920. Lorenz-Heinel, "Neuere Kühlmaschinen," Oldenbourg, Munich and Berlin, 1913. Macintire, "Handbook of Mechanical Refrigeration," Wiley, New York, 1928. Morrison, "Power's Practical Refrigeration," McGraw-Hill, New York, 1928. Moyer and Fittz, "Refrigeration," McGraw-Hill, New York, 1932. Dodge, "Chemical Engineering Thermodynamics," Chap. X, Refrigeration, McGraw-Hill, New York, 1944. Plank, "Amerikanische Kältetechnik," V.D.I. Verlag, Berlin, 1929. Siebel, "Compendium of Mechanical Refrigeration," Nickerson and Collins, Chicago, 1918. Instructions for Operation, Care and Repair of Refrigerating Plants, Navy Dept., Bur. Eng., Government Printing Office, Washington, 1918. Crocker, "Piping Handbook," Chap. XVI, McGraw-Hill, New York, 1945. Quinn and Jones, "Carbon Dioxide," Reinhold, New York, 1936. Periodicals: *Refrig. Eng.; Ice and Refrigeration.*

Refrigeration as considered in the following may be broadly defined as the art of producing cold, referring particularly to cooling below atmospheric temperature. The means most commonly employed for such cooling is to induce a change of phase in a heat-abstracting body such as is involved in the vaporization of liquid ammonia or the melting of ice. In the production of liquid air, however, cooling is brought by expanding the gas through a nozzle (Joule-Thomson effect) or by causing it to do work against a brake (Claude system). Other physical changes, such as the contraction of stretched rubber or the extension of a steel spring, the passage of an electric current through a bimetallic junction, and, in fact, any reversible physical change involving the expenditure of work, are capable of producing cold. Radiation of heat from the surface of the earth to interstellar space may be used in certain localities for freezing water, and cool ground waters have from time immemorial been used for preserving foodstuffs.

In recent years, many new applications of refrigeration have developed, and great expansion has taken place in established applications. Among the new uses may be mentioned the concentration of penicillin and blood plasma, which involves the evaporation of water at very low temperature (−20°C.). The dehydration of food at low temperature also shows promise, because taste may be preserved. In the fabrication of aluminum parts, chilling is employed to control age-hardening, and in electric welding, refrigerated electrodes have been found to give longer service because less molten metal adheres to the tips. In established applications, the great expansion in the use of frozen foods may be mentioned, with corresponding developments in storage and shipment, notably in marine transportation. In purely chemical industries, such as synthetic rubber manufacture, a great deal of refrigeration is used. The use of dry air in blast-furnace operation, although an old process, has been stimulated by increased demands for steel. Production of pure oxygen for various operations has been extensively developed in Russia and is being followed closely in other countries. The present discussion will not consider details of these numerous applications and will be limited to the general principles of how heat is transferred from a lower to a higher temperature, which is the distinguishing characteristic of refrigeration.

Definitions. A **refrigerant** in its broadest sense may be defined as any material which is used for abstracting heat. In a narrower, but more commonly used, sense the term refers only to those materials which are used in **mechanical refrigeration.**

Mechanical refrigeration includes those processes in which the refrigerant is recovered and recirculated, as distinguished from those in which the spent refrigerant is wasted (ice refrigeration, ice-cream freezing by salt-ice mixtures, cold ground waters in spring houses). Mechanical refrigeration falls into two general groupings, depending on the method used in recirculating the refrigerant.

In the **compression** system (dense air or vapor-compression machines) a compressor is used, which may have either a positive-displacement mechanism (reciprocating or rotary compressor) or an impeller (centrifugal compressor). Although the thermodynamic cycle is the same for both types of compression, the kinematic considerations, particularly with regard to the refrigerant used, are markedly different. The dense-air machine operates on the Carnot cycle and the vapor-compression machine on the reversed Rankine (steam-engine) cycle.

The **absorption system** differs essentially from the compression system in requiring no positive work input to produce cold through generally requiring some accessory power. Circulation is effected by absorption of the refrigerant in appropriate liquids or solids which are regenerated by heat in another part of the refrigerating system. Absorption machines are classified as **continuous or intermittent,** the latter corresponding quite closely to batch processes. All continuous machines use liquid absorbents because of the great practical difficulties in moving a solid absorbent about from the absorption side to the generator. When continuous refrigeration is required with intermittent units, two or more machines must be installed.

This classification is neither absolute nor all-inclusive. Combinations of absorption and compression systems have been developed (Westinghouse-Leblanc). The processes for manufacturing liquid air or solid carbon dioxide do not fall into the above scheme since they employ compression but do not recirculate the refrigerant. The distinctions are useful, however, in considering the operations involved.

Practically, mechanical refrigeration is the most important branch of the art, and developments have centered almost wholly about the vapor-compression machine.

Absorption machines still find considerable industrial application where large quantities of waste heat are available. Interest has also been revived in absorption systems tied in with heating arrangements for year-round air conditioning, involving summer cooling and winter heating. The main investigations have been in the direction of developing refrigerant-solvent combinations which can produce refrigeration with a minimum heat expenditure. [Zellhoefer, *Refrig. Eng.*, **33,** 317 (1937).]

Units of Refrigeration. The unit of refrigeration in the United States is the **standard ton** of 288,000 B.t.u. which is very nearly equal to the heat of fusion of 2000 lb. of ice at 32°F. The **standard commercial ton of refrigeration** is at the rate of 200 B.t.u./min., 12,000 B.t.u./hr., or 288,000 B.t.u./24 hr. Note that the standard ton has the dimensions of heat, while the standard commercial ton has the dimensions of heat divided by time. The **standard rating of a refrigerating**

machine, which applies only to compression and absorption systems using a condensable vapor, is the number of commercial tons of refrigeration it performs under certain prescribed conditions. These conditions are:

1. Nothing but liquid shall enter the expansion valve, and nothing but vapor shall enter the compressor cylinder (impeller in a centrifugal compressor) of the compression refrigerating system or the absorber of the absorption system.
2. There shall be 9°F. (5°C.) subcooling of the liquid entering the expansion valve and 9°F. (5°C.) superheating of the vapor entering the compression cylinder or the absorber. The points at which subcooling and superheating are determined must be within 10 ft. of the cylinder or absorber.
3. The inlet pressure is that which corresponds to a saturation temperature of 5°F. (−15°C.).
4. The outlet pressure from the compressor cylinder or generator is that which corresponds to a saturation temperature of 86°F. (30°C.).

The American Society of Refrigerating Engineers has for a number of years been sponsoring a compilation of standard methods for rating refrigeration equipment such as compressors, condensers, evaporators, and brine coolers. Various circulars dealing with these subjects are being issued from time to time as completed.

The **British unit of refrigeration** is based on a rate of cooling of 1 kg.-cal./sec. or 237.6 B.t.u./min., with inlet pressures corresponding to a saturation temperature of 23°F. (−5°C.) and outlet pressures corresponding to a saturation temperature of 59°F. (15°C.).

The Dense-air Machine. From a thermodynamic standpoint, the dense-air machine operating on the Carnot cycle is capable of the highest efficiency. In practice, the efficiency is low because of the large volumes of air which must be handled for a relatively small effect. The safety of air as regards toxicity, flammability, and odor was an important factor in its favor some years ago, but in recent years refrigerants have been developed which are practically as safe and more efficient. For operating details, the reader may consult U.S. Navy, "Manual of Engineering Instructions," Chap. 17, Government Printing Office, Washington.

DESCRIPTION AND THEORY OF OPERATION OF VAPOR-COMPRESSION AND ABSORPTION REFRIGERATION SYSTEMS

The basic principle involved in refrigerating system is that of transferring heat from an environment at low temperature to one at a higher temperature, by causing a volatile liquid (the refrigerant) to absorb heat at the low temperature by vaporization and to dissipate this heat at the high temperature by condensation. Vaporization and condensation are respectively induced by maintaining a lower or higher pressure than the saturation pressures of the refrigerant at the lower and higher temperatures. Both systems employ an **evaporator** (expansion coil) and a **condenser.** The apparatus used for transferring the refrigerant from the low- to the high-pressure side is technically the **refrigerating machine,** but the term is not always used in strict conformity with this definition. In the vapor-compression system this machine is actuated by a prime mover and in the absorption machine by heat. Economy requires that a given amount of refrigeration be produced with a minimum expenditure of work or heat.

The Simple Vapor-compression Refrigerating Machine. A diagrammatic representation of a simple compression system of refrigeration is shown in Fig. 1. Refrigerant vapor is drawn from the expansion coil A at the pressure p_1 by the compressor B; forced into the condenser C at a pressure p_2, dependent on the temperature of the cooling water, where the vapor liquefies; and collected in the receiver D from which it returns through the expansion valve E, under the pressure difference $p_2 - p_1$, into the expansion coil.

In this process the following heat quantities are involved, which may be taken from the tables of thermodynamic properties (Sec. 3):

H_b = heat content of vapor leaving evaporator (expansion coil).
H_c = heat content of vapor leaving compressor.
h_e = heat content of liquid entering evaporator (expansion coil).

If the compression is adiabatic (isentropic) the work expended is

$$H_c - H_b$$

The net heat abstracted is

$$H_b - h_e$$

Note that this is the heat of evaporation less the heat required to cool the liquid refrigerant from the temperature of the receiver to the temperature of the expansion coil. The ratio of the net heat abstracted to the work expended is known as the performance coefficient β and may be written

$$\beta = \frac{H_b - h_e}{H_c - H_b} \quad (1)$$

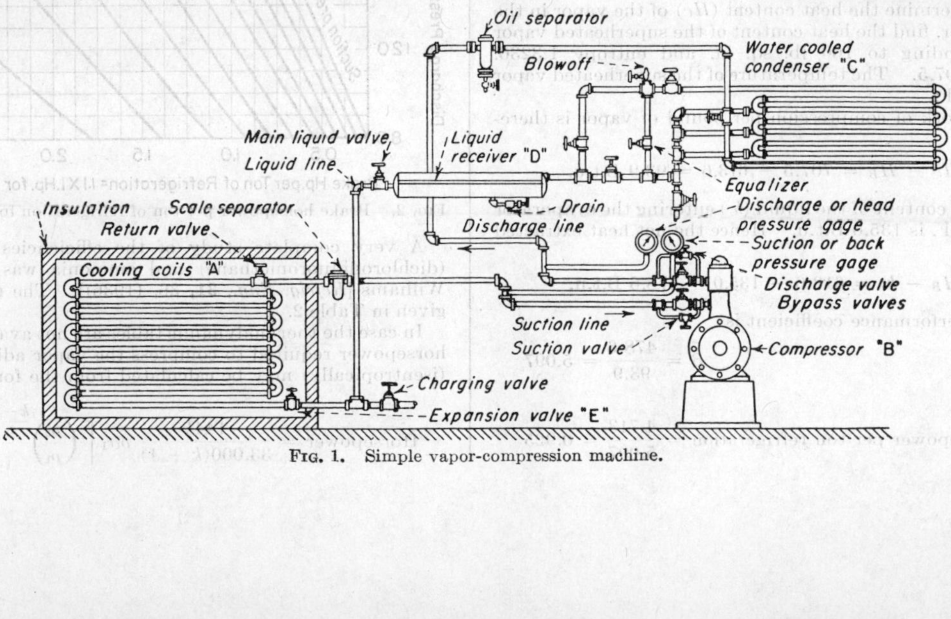

FIG. 1. Simple vapor-compression machine.

The inverse of this ratio is the efficiency coefficient E of the simple steam engine or turbine.

The horsepower and the kilowatts required per standard commercial ton of refrigeration follow directly from the definition of β and the conversion factors:

$$1 \text{ horsepower} = 2{,}546 \text{ B.t.u./hr.}$$
$$1 \text{ kilowatt} = 3{,}415 \text{ B.t.u./hr.}$$
$$1 \text{ standard commercial ton} = 12{,}000 \text{ B.t.u./hr.}$$

This gives

$$\text{Horsepower per ton} = \frac{12{,}000}{2{,}546\beta} = \frac{4.713}{\beta} \qquad (2a)$$

$$\text{Kilowatt per ton} = \frac{12{,}000}{3{,}415\beta} = \frac{3.514}{\beta} \qquad (2b)$$

which represent the **theoretical horsepower** or **kilowatts** per standard commercial ton of refrigeration.

The theoretical horsepower can be realized only in a frictionless machine with a weightless piston. In general it will be very nearly equal to the **indicated horsepower** shown by the actual card of an indicator, which reveals the imperfections in the operation of the valves and the departure from strictly adiabatic compression. The **brake horsepower** is the indicated horsepower plus the power necessary to overcome the friction in the cylinder walls and bearings and the inertia of the piston. The brake horsepower will be 10 to 20 per cent more than the theoretical horsepower. The difference between brake horsepower and indicated horsepower is sometimes designated as **friction horsepower.** The friction horsepower is relatively larger in small than in large machines.

Since refrigeration involves a conversion of work into heat it might appear that the performance coefficient is dependent only on the temperature difference and independent of the refrigerant used. This is only approximately true, as shown by Table 1, carbon dioxide being a marked exception. The temperature difference is, however, important; and, consequently, a machine tested according to the standard American temperature interval will show much better performance when tested on the British standard.

Sample Calculation of Performance Coefficient. Assume that the refrigerant is ammonia; the absolute evaporator pressure (p_1) 35 lb./sq. in. corresponding to 5.89°F., and the absolute condenser pressure (p_2) 160 lb./sq. in. corresponding to 82.64°F.

The heat content (H_B) of the saturated vapor in the evaporator is 613.6 B.t.u./lb., and the entropy (S) is 1.3236 B.t.u./lb./°F.

To determine the heat content (H_C) of the vapor in the condenser, find the heat content of the superheated vapor corresponding to 160 lb./sq. in. and entropy 1.3236. This is 707.5. The temperature of the superheated vapor is 199.5°F.

The work of compression per pound of vapor is therefore

$$H_C - H_B = 707.5 - 613.6 = 93.9 \text{ B.t.u.}$$

The heat content of the liquid (h_e) entering the evaporator at 82.64°F. is 135.0 B.t.u. Hence the net heat extracted is

$$H_B - h_e = 613.6 - 135.0 = 478.6 \text{ B.t.u.}$$

The performance coefficient is

$$\beta = \frac{478.6}{93.9} = 5.097$$

and

$$\text{Horsepower per ton refrigeration} = \frac{4.713}{5.097} = 0.925$$

Table 1 gives the theoretical performance coefficient for ammonia, which will be found to hold approximately for all other refrigerants except those boiling much below ammonia. Carbon dioxide will require a greater power

Table 1. Comparison of Refrigerants*

One-ton refrigeration, 5 to 86°F.

Cycle	Weight, lb./min.	Volume, cu. ft./min.	Ratio of compression	Coefficient of performance	Horsepower per ton	Relative efficiencies, %
Ideal	......	......	.....	5.74	0.8214	100
Ammonia	0.4214	3.44	4.93	4.85	0.973	84.5
Propane	1.396	3.35	3.64	4.88	0.9668	85
Carbon dioxide	3.74	0.999	3.11	2.56	1.843	44.6
Sulfur dioxide	1.388	9.24	5.63	4.735	0.995	82.5
Ethyl ether	1.555	60.8	7.12	4.86	0.971	84.6
Dichloroethylene	1.768	108.4	8.23	5.14	0.918	89.4
Trichloroethylene	2.137	513	10.84	5.085	0.928	88.5
Water	0.1996	1972	21.9	4.1	1.15	71.5

* Carrier and Waterfill, *Refrig. Eng.*, **10**, 415 (1923).

Table 2. Large Enclosed York V. S. A. Compressors with Ammonia and Freon

Evaporator temperature, 35°F., condenser temperature, 110°F.

Size	No. of cylinder	Speed, r.p.m.	Capacity, tons	Brake horsepower per ton
Ammonia				
7½ × 7½	2	360	55.2	1.10
8 × 8	2	360	68.4	1.10
9 × 9	2	300	84.3	1.07
10 × 10	2	300	118.7	1.07
11 × 13	2	277	173.0	1.07
12½ × 14½	2	257	234.0	1.06
Freon				
9 × 7½	2	360	43.7	1.27
11 × 9	2	300	66.5	1.23
12½ × 10	2	300	95.4	1.215
14 × 10	2	300	119.6	1.20
13½ × 13	2	277	133.7	1.20
15 × 13	2	277	165.1	1.19

expenditure. The brake horsepower, arbitrarily taken as 10 per cent greater than the theoretical is shown in Figs. 2 and 3.

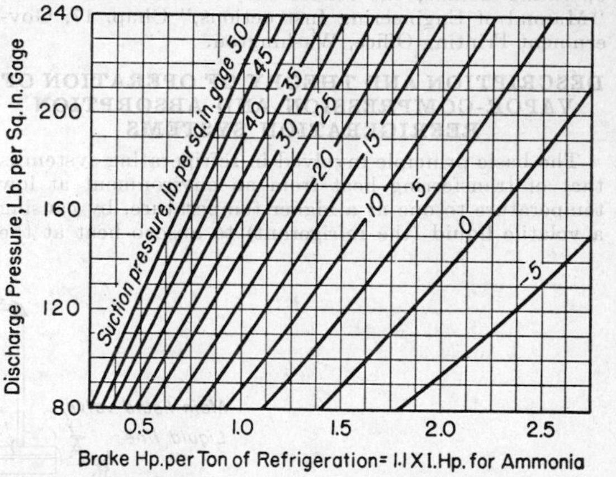

FIG. 2. Brake horsepower per ton of refrigeration for ammonia.

A very complete study of the efficiencies of Freon (dichlorodifluoromethane) and ammonia was made by Williams [*Refrig. Eng.*, **31**, 36, (1936)]. The results are given in Table 2.

In case the thermodynamic tables are not available, the horsepower required to compress the vapor adiabatically (isentropically) may be calculated from the formula

$$\text{Horsepower} = \frac{144k}{33{,}000(k-1)}\, p_1 v_1 \left[\left(\frac{p_2}{p_1}\right)^{\frac{k-1}{k}} - 1\right]$$

where p_1 = absolute intake pressure (35 lb. in above example).
p_2 = absolute discharge pressure (160 lb. in above example).
v_1 = volume compressed, cu. ft./min.
k = ratio of specific heat of the vapor, c_p/c_v.

The volume v_1 may usually be found with sufficient accuracy from the general gas law equation expressed in the appropriate units.

The net heat extracted is similarly the difference between the heat of evaporation at the evaporating temperature and the sensible heat of the liquid between the condenser temperature and the evaporator temperature (specific heat times temperature difference).

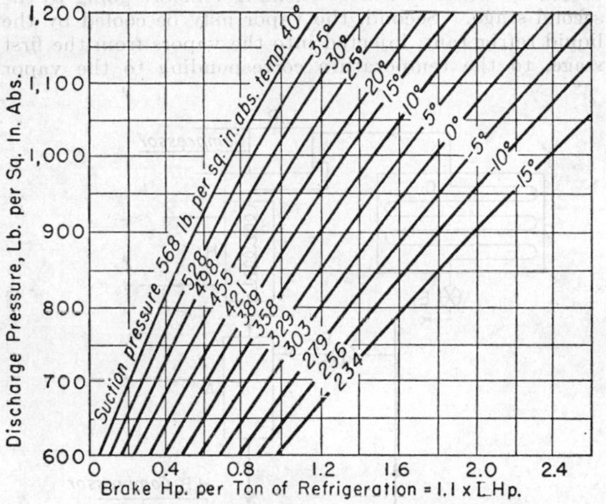

FIG. 3. Brake horsepower per ton of refrigeration for carbon dioxide.

While the performance coefficients of refrigerants having critical temperatures well above that of the available cooling water will be approximately equal for a given temperature interval, the degrees of superheating will vary greatly. In an adiabatic compression the following relation holds:

$$\frac{T_2}{T_1} = \left(\frac{p_2}{p_1}\right)^{\frac{k-1}{k}}$$

where T_1 and p_1 = absolute suction temperature and pressure.
T_2 and p_2 = absolute discharge temperature and pressure.
k = ratio of specific heats.

The compression ratio p_2/p_1, which is determined by the molal latent heats, will be nearly the same for all refrigerants (Trouton's law) while the specific heat ratio will be lowest for those refrigerants having high molal heat capacities. The refrigerants having complex molecules will therefore have a much smaller temperature rise in adiabatic compression than those having simple molecules. In the case of ethyl ether ($C_4H_{10}O$) there is actually no superheating but supercooling and liquefaction, while ammonia (NH_3) and sulfur dioxide (SO_2) give the greatest superheating on compression.

Indicator Card and Volumetric Efficiency. The work of compression is shown by the indicator diagram (Fig. 4). In the ideal case the piston at the beginning of the suction stroke touches the cylinder head and the volume is zero. As the piston moves out, vapor is drawn in and the volume increases at the constant pressure p_1 along the line *AB*. The vapor is now compressed adiabatically along the line *BC* until the pressure equals the condenser pressure p_2, after which it is forced into the condenser and the volume decreases along the line *CH*. The area *ABCH* is then a measure of the compression work.

In an actual compressor it is necessary to provide a small clearance at the end of the compression stroke, represented by the segment *EH*. Before suction can actually start, the vapor enclosed in this space expands adiabatically along the line *EF*. The work in an actual compressor is therefore represented by the area *FBCE*, which is smaller than the area *ABCH*. However, the volume of gas handled by an actual compressor is less than by an ideal compressor, the ratio being given by the ratio of the lines *FB/AB*. The work represented by area *EF'F* is the excess of work required by an actual compressor handling a volume of gas corresponding to the line *FB*. This excess will become a smaller and smaller fraction, as the clearance represented by the line *EF* is decreased.

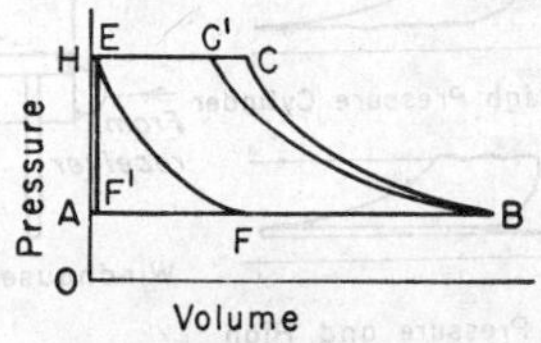

FIG. 4. Ideal indicator diagram for vapor-compression machine.

If the compression were isothermal instead of adiabatic, it would follow a line *BC'* lying below the adiabatic *BC*. The work requirement would therefore be less. In practice this cannot be attained completely, but by compressing the vapor in two or more stages and cooling the superheated vapor the isothermal line may be approached (see Stage Compression, p. 1439*ff*.).

The **volumetric efficiency** is the ratio of vapor volume handled by an actual compressor to the volume handled by an ideal compressor of the same piston displacement and would therefore be equal to the ratio *FB/AB*. In practice the ratio will be lower because of superheating of the vapor entering the cylinder, leakage, etc., but reliable manufacturers will guarantee above 80 per cent. Volumetric efficiencies of ammonia compressors are shown in Table 3.

Table 3. Volumetric Efficiency of Ammonia Compressors

Condenser pressure, lb./sq. in. gage		Suction pressure, lb./sq. in., gage				
		0	10	20	30	40
120	*A*	0.77	0.83	0.87	0.89	0.91
	B	.60	.70	.77	.81	.84
	C	.52	.65	.72	.77	.80
160	*A*	.74	.80	.83	.86	.88
	B	.54	.65	.72	.76	.80
	C	.44	.58	.66	.72	.75
200	*A*	.71	.77	.81	.84	.86
	B	.49	.61	.68	.72	.75
	C	.37	.52	.62	.67	.71

A, no clearance.
B, 4 per cent clearance.
C, 6 per cent clearance.

For a centrifugal compressor the compressor work and volumetric efficiency do not admit a graphical representation as does the reciprocating compressor. A centrifugal compressor has no clearance and therefore a 100 per cent volumetric efficiency. However, there are slippage losses around the impeller vanes which correspond roughly to clearance losses. The operation of a reciprocating compressor may be compared to a series of batch processes, while the operation of a centrifugal compressor is more like a continuous process.

Dual Compression. When two refrigerating temperatures are required, the **dual-compression** system may be employed. The general principle of this design is to draw in the vapor from the low-pressure (low-temperature) expansion coils, and, at or near the end of the stroke, vapor from the higher pressure coils will be admitted. The whole mixture will then be compressed to the saturation pressure corresponding to the temperature of the condenser water. The advantages of this system are the use of one condenser for the two cooling systems and a markedly higher capacity, since the low-pressure cooling unit requires much larger piston displacement than the higher pressure unit. The chief disadvantage is the difficulty of securing continuous adjustment between the loads on the two systems. Two schemes of dual

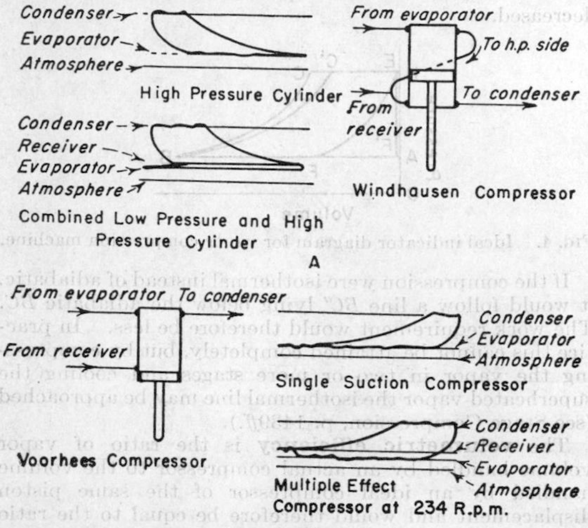

FIG. 5. Indicator diagrams for Voorhees and Windhausen systems of dual compression. (*Reprinted by permission from Macintire, "Handbook of Mechanical Refrigeration," Wiley, New York*, 1928.)

compression (Voorhees and Windhausen) with the corcorresponding indicator cards are shown in Fig. 5. In some household units, the dual-compression principle is employed to maintain two temperatures, one for freezing ice cubes and a higher one at the optimum value for food preservation.

Wet Compression. In wet compression liquid refrigerant is admitted with the refrigerant vapor from the expansion coils. This liquid evaporates during adiabatic compression and thereby makes the compression approach more nearly to the isothermal line. The vapor leaves the compressor with a much smaller degree of superheat, which also lessens the load on the condenser. Although the scheme is sound in theory, it is difficult in practice to adjust the amount of liquid added, and the capacity is very much reduced owing to the increased cylinder volume required by the liquid after vaporization. The scheme can naturally not be used with a refrigerant like ethyl ether, which liquefies on adiabatic compression, and it would not be particularly advantageous for those refrigerants which superheat slightly.

Multistage Compression. The most promising developments toward diminishing the power requirements of refrigeration are along the lines of multistage compression. This principle has long been used in air compression, but in refrigeration the problem is more complicated. In general it will prove economical with ammonia only where suction pressures lower than 5 lb./sq. in. (gage) are encountered. As in the case of wet compression, it can be applied with best advantage to those refrigerants which superheat markedly on adiabatic compression (ammonia, sulfur dioxide, and carbon dioxide but not ethyl ether, dichlorodifluoromethane, propane, and butane). Multistage compression is frequently used with the latter type of refrigerant, but intercoolers are of much less importance than with ammonia, and the power savings are less marked.

Several possibilities in multiple compression are presented. First, the vapor from the first stage may be cooled by water in a heat exchanger before going to the second stage. Second, the vapor may be cooled by the liquid refrigerant, injected into the vapors from the first stage, to the temperature corresponding to the vapor

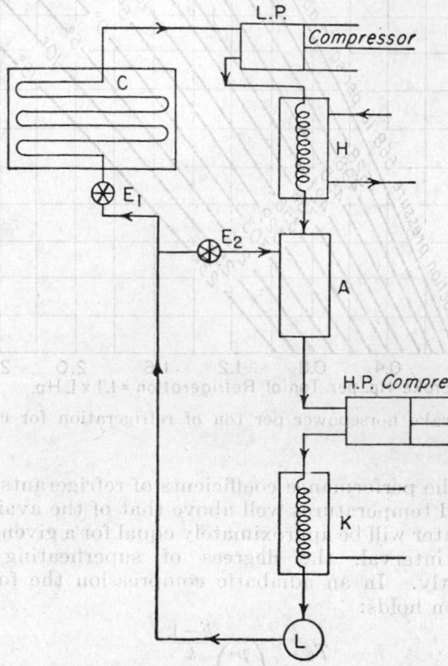

FIG. 6. Multiple-stage (two-stage) compression. (*Macintire, "Handbook of Mechanical Refrigeration," Wiley, New York*, 1928.)

pressure of the refrigerant equal to the pressure at the beginning of the second stage. Third, a combination of both may be used. A diagrammatic sketch of this third scheme is shown in Fig. 6.

Vapor is pumped from the expansion coils C by the low-pressure compressor LP into the heat exchanger H where it is cooled by water. From H it passes into the accumulator A where it meets liquid refrigerant from the liquid receiver L which cools the vapor further. The high-pressure compressor then compresses the vapor to the saturation pressure of the condenser K in which it liquefies and is collected by the liquid receiver L. The liquid refrigerant then returns to the expansion coils through the valve E_1 and to the accumulator through the valve E_2.

When ammonia is used for low-temperature cooling, a booster compressor is generally used to raise the low-pressure vapor to some intermediate pressure. The vapors are cooled before going to the high-pressure compressor, partly to save power, partly to take out the superheat which would continue to build up in the second compressor. A further advantage is the saving in cylinder volume.

The power saving and the correct intermediate pressures are shown in Figs. 7 and 8. An extended analysis of the power savings in obtaining low temperatures by multistage compression has been made by Sloan [*Refrig. Eng.*, **45,** 419 (1943)].

Split-stage Compression or Binary Cycles (Two or More Refrigerants). When very low temperatures

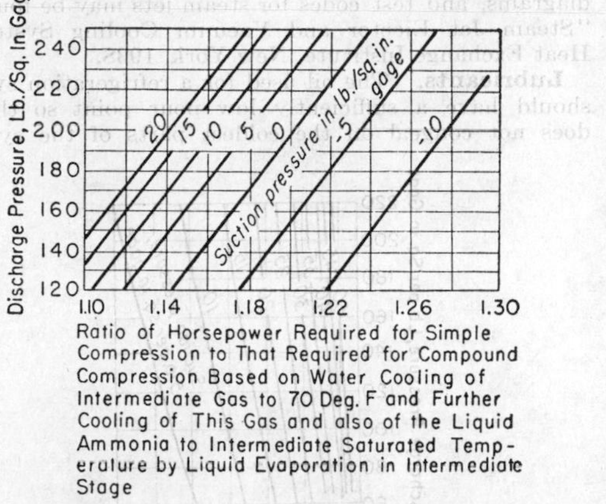

FIG. 7. Power saving by multiple-stage compression, compared with single-stage compression and evaporation.

are required, it is sometimes advantageous to employ two or more refrigerants in order to maintain pressures within reasonable limits. Thus a refrigerant of relatively high boiling point will have such a low vapor pressure at low temperatures that excessive cylinder volume will be needed, whereas a refrigerant of low boiling point will have excessive pressures at the available condenser water temperatures. It may also have a critical temperature below that of the available cooling water and will there-

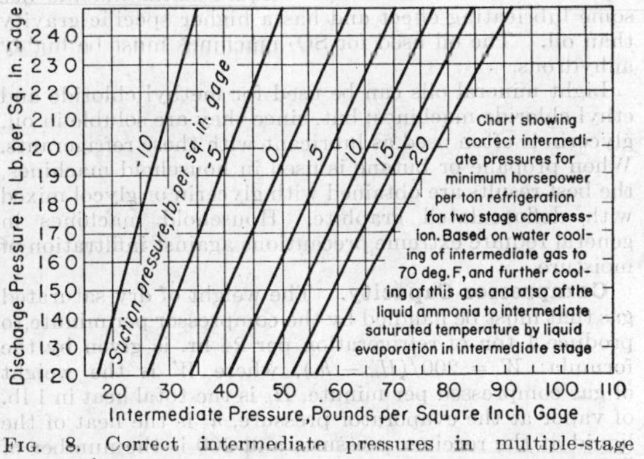

FIG. 8. Correct intermediate pressures in multiple-stage compression.

fore fail to liquefy. In such cases, a low boiling refrigerant may be used to produce the low temperature, but the condenser is cooled by a high boiling refrigerant instead of by water.

A system using carbon dioxide for the low-temperature refrigerant and ammonia for cooling the condenser was described by Kitzmiller [*Power*, **75,** 92 (1932)]. Since ammonia and carbon dioxide react vigorously, great care must be taken in sealing joints. Recently developed refrigerants of the fluorinated and chlorinated hydrocarbons (Freon) permit combinations of refrigerants that are non-active, and may often find applications in low-temperature problems.

Systems using three or more refrigerants but employing the same general principles are frequently designated as **cascade** cycles.

Compressors. Two general types of compressors are used for refrigeration: (1) compressors with positive displacement and (2) centrifugal compressors. Of those with positive displacement, the most common type is the reciprocating compressor, which may be either **horizontal** or **vertical.** The vertical type is largely displacing the horizontal type, the latter being found principally in very large refrigerating systems. In the United States two-cylinder compressors are common, but three or four cylinders are also used. Development at the present time tends largely toward the design of very high-speed machines, since high speed gives greater capacity for a given cylinder volume. Electric drive, which is used in the greatest number of installations, is also more readily adapted to high speeds, 1200 to 1800 r.p.m. Oil- and steam-engine drives function more satisfactorily with low-speed machines (250 r.p.m.). High-speed machines require more careful design with respect to lubrication, balancing, and foundations; demand higher priced materials of construction; and have greater power losses because of friction and inertia in the piston. Water jackets are sometimes employed, chiefly to prevent the metal parts from overheating. They thus increase the volumetric efficiency but provide hardly enough cooling to make the compression isothermal. The poppet valve operated with springs is used on low-speed compressors, but high-speed machines often use lightweight plate or ribbon valves. Lubrication may be of the **splash** type if the compressor is enclosed, but for the open-frame construction or higher speed machines, **forced-feed** lubrication is desirable. An oil separator is generally provided near the condenser to return oil carried over by the refrigerant vapors.

Carbon dioxide compressors do not differ essentially from ammonia compressors except that heavier construction is required and stuffing boxes are more tightly packed. The cylinder diameter is smaller and the stroke longer.

Freon compressors handle a vapor of very high density, and the kinetic energy losses are therefore relatively high. To overcome these losses, ample gas passages must be provided and "streamlined." Suction valves should be 80 per cent larger than with ammonia and discharge valves about 100 per cent larger. However, water jackets are not needed with Freon because the gas superheats very little on compression. An extensive discussion of Freon compressor design is given by Williams [*Refrig. Eng.*, **31,** 36 (1936)].

Rotary compressors with positive displacement may be of the **blade** type operating on an eccentric, the blade being kept in contact with the casing by centrifugal force; or of the **pendulum** type. Rotary compressors are well adapted to high rotative speeds and direct connection to a high-speed electric motor; valves are not needed. They have been built only in small units, principally for household use, which show lower efficiencies than reciprocating compressors, but small units are in general less efficient than large ones. A seven-cylinder rotary compressor with automatic cutoff of separate cylinders under variable load has recently been developed for air-conditioning work by the Airtemp Division of the Chrysler Corp.

In the **centrifugal compressor** low-pressure vapor is drawn into impellers which impart a high velocity to the vapors. The high-speed vapors emerge through discharge vanes, and their kinetic energy is then converted into pressure energy. The speed is usually 3000 to 6000 r.p.m. Several stages are necessary to raise the vapor to the discharge velocity (see Fig. 9). Since the kinetic energy is a function of mass, refrigerants of high gas density are preferred. Slippage losses are less with low-pressure refrigerants than with high-pressure refrigerants; and, since only high-molecular-weight refrigerants can give high densities at low pressures, a special class of such refrigerants has been developed for these machines. In general, centrifugal compressors cannot be used advantageously below certain load limits characteristic of each refrigerant. Zwicke [*Refrig. Eng.*, **45**, 179 (1943)] gives the following lower limits in tons refrigeration: ammonia, 1400; methyl chloride, 780; Freon 12, 810; sulfur dioxide, 561; Freon 114, 260; Freon 21, 260; Freon 11, 150; methylene chloride, 83; and Freon 113, 62.

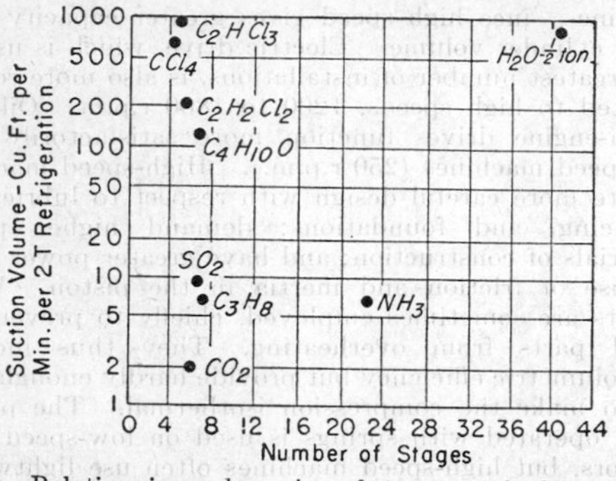

FIG. 9. Relative size and number of stages required for various refrigerants (5° to 86°F.). [*Carrier and Waterfill, Refrigerating Eng.*, **10**, 423 (1924).]

Steam Jets. Low-pressure water vapor may be compressed by high-pressure steam in a steam jet, the operation involving conversion of kinetic energy into pressure-volume energy. In this way a vacuum can be created over water with resultant evaporation and cooling. This method is frequently very useful where moderate cooling is needed, such as in chilling water for air conditioning or other uses. Temperatures from 35°F. up can generally be attained.

Steam jets have been greatly improved in recent years, and various operating details have appeared in the literature. Steam requirements are given in Table 3*a*.

Table 3a. Steam and Water Requirements for Steam Jet Vacuum Cooling Units*

Lb. steam/hr./ton refrigeration
Gal./min./ton refrigeration

Booster condenser pressure, in. Hg abs.	Chilled water temp., °F				
	40	45	50	55	60
1.5	23.4	20.7	17.6	16.1	12.9
2.0	31.4	27.0	22.5	20.8	16.5
2.5	41.2	32.8	27.5	23.6	20.6
Condensing water (75°F.):					
1.5	6.51	6.05	5.44	5.16	4.51
2.0	4.45	3.98	3.52	3.34	2.90
Condensing water (85°F.):					
2.0	8.45	7.56	6.67	6.34	5.50
2.5	6.16	5.80	4.57	4.10	3.75

* Stinson, *Refrig. Eng.*, **46**, 316 (1943).

Condensers used with steam jets may be of three kinds: (1) Surface condensers with steam condensed on the outside of the tubes, in which condensed vapor and motive steam from the booster are pumped from the high vacuum with recovery of motive steam condensate; (2) low-level jet condensers in which steam and vapor are mixed directly with the cooling water; and (3) barometric condensers, in which steam and vapor are also mixed directly with the cooling water but are removed by gravity without an auxiliary pump.

For variable load, several jets are required with provision for cutting out as the load varies. Steam at 2 lb. gage pressure is said to be suitable for steam jet refrigeration [Stevens, *Refrig. Eng.*, **40**, 146, (1940)]. As the motive steam pressure decreases a point is reached at which the compressor "breaks back" or suddenly drops all the load.

A full discussion of cycles, standards, thermodynamic diagrams, and test codes for steam jets may be found in "Steam Jet Ejector and Vacuum Cooling Systems," Heat Exchange Institute, New York, 1938.

Lubricants. The oil used for a refrigeration system should have a sufficiently low pour point so that it does not congeal on the coldest parts of the system.

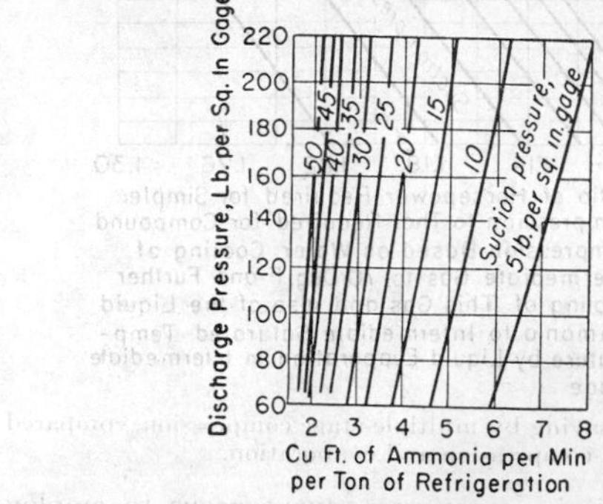

FIG. 10. Volume of ammonia compressed per ton of refrigeration. (*Reprinted by permission from Macintire "Handbook of Mechanical Refrigeration," Wiley, New York, 1928.*)

Most of the oil companies prepare such a grade of mineral oil, the usual specification being a pour point of −20 to −30°F. Such oils may be used for ammonia, carbon dioxide, and sulfur dioxide. Propane dissolves to a very appreciable extent, and the oil when removed from the oil separator foams extensively. Liquid sulfur dioxide has some lubricating effect and has a higher specific gravity than oil. The oil used for SO_2 machines must be highly anhydrous.

Light mineral oils can be used for methyl chloride and ethyl chloride machines; but, since they are soluble in oil, glycerin is often used as lubricant with these refrigerants. When propane or butane is used in household machines, the best results are obtained with glycerin or glycol mixed with deflocculated graphite. Household machines in general require extreme precautions against infiltration of moisture.

Compressor Capacity. The weight of dry saturated gas that must be handled by the compressor per minute to produce 1 ton of refrigeration per 24 hr. is given by the formula: $W = 200/(H_b - h_e)$, where W is the weight of gas compressed per minute, H_b is the total heat in 1 lb. of vapor at the evaporator pressure, h_e is the heat of the liquid at the receiver pressure, and 200 is the number of B.t.u. removed per minute to equal 1 ton of refrigeration per 24 hr. (Table 4 and Figs. 10 and 11).

The formula for the theoretical capacity, in tons of refrigeration per 24 hr., of a double-acting compressor, taking into consideration the specified pressures in the condenser and the refrigerator, is

$$T = \frac{d^2 \times 0.7854 \times L \times 2 \times N \times h}{1728 \times 200}$$

where T is the tons refrigeration in 24 hr.; d is the diameter of compressor cylinder in inches; L is the stroke in inches; N is the number of revolutions per minute; h is the refrigeration effect of 1 cu. ft. of ammonia vapor at evaporator pressure.

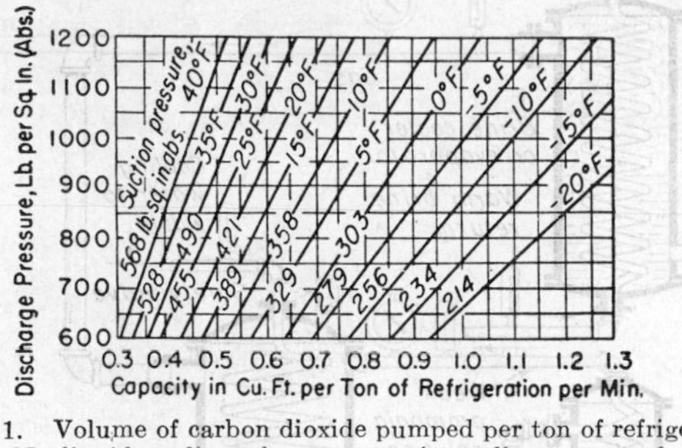

FIG. 11. Volume of carbon dioxide pumped per ton of refrigeration. No liquid cooling; dry compression; allowance made for volumetric efficiency. *(Reprinted by permission from Macintire "Handbook of Mechanical Refrigeration," Wiley, New York,* 1928.)

The capacity of vertical single-acting (V. S. A.) enclosed machines is shown in Table 5.

Table 4. Cubic Feet of Ammonia Gas to Be Pumped per Minute to Produce 1 Ton of Refrigeration per Day of 24 Hr.

Evaporator		Condenser								
P	P	103	115	127	139	153	168	185	200	218
	T	65°	70°	75°	80°	85°	90°	95°	100°	105°
4	−20°	5.84	5.90	5.96	6.03	6.09	6.16	6.23	6.30	6.43
6	−15°	5.35	5.40	5.46	5.52	5.58	5.64	5.70	5.77	5.83
9	−10°	4.66	4.73	4.76	4.81	4.86	4.91	4.97	5.05	5.08
13	− 5°	4.09	4.12	4.17	4.21	4.25	4.30	4.35	4.40	4.44
16	0°	3.59	3.63	3.66	3.70	3.74	3.78	3.83	3.87	3.91
20	5°	3.20	3.24	3.27	3.30	3.34	3.38	3.41	3.45	3.49
24	10°	2.87	2.90	2.93	2.96	2.99	3.02	3.06	3.09	3.12
28	15°	2.59	2.61	2.65	2.68	2.71	2.73	2.76	2.80	2.82
33	20°	2.31	2.34	2.36	2.38	2.41	2 44	2.46	2.49	2.51
39	25°	2.06	2.08	2.10	2.12	2.15	2.17	2.20	2.22	2.24
45	30°	1.85	1.87	1.89	1.91	1.93	1.95	1.97	2.00	2.01
51	35°	1.70	1.72	1.74	1.76	1.77	1.79	1.81	1.83	1.85

The values in this table are calculated theoretical values. In practice, allowance must be made for losses. Temperatures are in degrees Fahrenheit.

CONTINUOUS ABSORPTION SYSTEMS

In the conventional continuous absorption system (Fig. 12) the refrigerant, usually ammonia, is drawn from the expansion coils F at a given pressure p_1 and dissolved in the absorber G where the combined pressure of solvent and refrigerant must be less than p_1. The refrigerant is recovered by applying heat to the generator A; separated from entrained water by the analyzer or rectifier B; liquefied in the condenser C; collected in a liquid receiver D (not shown); and returned to the expansion coils F through the expansion valve E. Since the absorber is at lower pressure than the generator, it is necessary to insert between them a pump H to transfer the ammonia solution usually known as strong aqua. The generator

Table 6. Factors for Capacities at Different Condenser Pressures

Condenser pressure, lb./sq. in. gage	Ammonia suction pressure							
	6 lb.	9 lb.	12 lb.	16 lb.	19 lb.	24 lb.	28 lb.	33 lb.
200	0.975	0.969	0.975	0.977	0.978	0.977	0.976	0.977
168	1.029	1.028	1.028	1.028	1.02	1.027	1.026	1.022
153	1.046	1.049	1.05	1.05	1.043	1.05	1.042	1.042

does not remove all the ammonia from the water, and this weak aqua solution is returned to the absorber through the heat exchanger I. Considerable heat is evolved in the absorption of ammonia vapor, and cooling water must be supplied to keep down the temperature and pressure. In the absorption system two complete cycles are in operation: (1) the refrigerant cycle, and (2) the absorbent cycle. Just as the absorbent contains some ammonia, so the refrigerant contains a small quantity of water.

The heat requirements of an absorption system may be approximately calculated by the following method (Glazebrook, "Dictionary of Applied Physics," Article on Refrigeration, Macmillan & Co., Ltd., London, 1921). Assume that a quantity of heat Q_2 is available at an absolute temperature T_2, and is transferred to the generator at a lower temperature T_1. The work W available from this heat in a Carnot cycle is given by the relation

$$W = Q_2 \frac{(T_2 - T_1)}{T_2}$$

Next assume that a quantity of heat Q_1 is to be transferred from the evaporator at a temperature T_4 to the condenser at the temperature T_3. The work required to effect this transfer is

$$W = Q_1 \frac{(T_3 - T_4)}{T_3}$$

The work available may now be set equal to the work required, from which it follows that

$$\frac{Q_2}{Q_1} = \frac{T_2}{T_3}\left(\frac{T_3 - T_4}{T_2 - T_1}\right)$$

If Q_1 is regarded as refrigeration, the heat requirement per unit of refrigeration is given by the ratio Q_2/Q_1. The value of this ratio calculated by the above method is necessarily a minimum, since there are various losses and irreversible effects that will add to the heat requirements.

Calculation of pressures, temperatures, and heat effects is facilitated by the use of an enthalpy concentration table for water-ammonia solution of which an abbreviated form is given in Table 7.

Results of various performances under practical operating conditions are given in Tables 8 and 9.

Table 5. Capacities of V. S. A. Enclosed-type Machines

Cylinder			Piston speed per min.		Displacement per min.		Capacity in tons/24 hr., 185 lb./sq. in. condensing pressure								
Number	Bore	Stroke	R.p.m.	Ft.	Cu. in.	Cu. ft.	6 lb. −15°F., refg.	9 lb. −10°F., refg.	12 lb. −5°F., refg.	16 lb., 0°F. Ice	16 lb., 0°F. Refg.	19 lb. 5°F., refg.	24 lb. 10°F., refg.	28 lb. 15°F., refg.	33 lb. 20°F., refg.
1	3	3	300	150	6,362	3.68	0.47	0.51	0.65	0.47	0.75	0.85	1.96	1.07	1.24
1	4	4	275	183	13,828	7.99	0.98	1.12	1.35	1.0	1.6	1.78	2.0	2.2	2.6
2	4	4	275	183	27,646	15.98	1.96	2.24	2.7	2.0	3.2	3.5	4.0	4.4	5.2
2	5	5	240	200	47,124	27.3	3.34	3.84	4.6	3.3	5.3	5.97	6.8	7.7	8.85
2	6	6	220	220	74,644	43.2	5.5	6.6	7.6	5.5	8.8	10.0	11.4	12.6	14.5
2	7	7	210	245	113,146	65.6	8.5	10.2	11.9	8.5	13.6	15.5	17.4	19.6	22.4
2	8	8	200	266	160,848	93.08	12.2	14.6	16.9	12.1	19.5	22.1	24.9	28.0	32.1
2	9	9	190	285	217,570	125.8	16.6	19.2	22.9	16.6	26.4	29.8	33.8	37.9	43.5
2	10	10	180	300	282,744	163.62	22.1	26.1	30.4	21.9	35.0	38.8	44.9	50.4	57.8
2	12	12	170	340	461,448	267.04	35.1	41.9	48.6	35.0	56.0	63.4	71.6	80.3	92.4

Multiply above capacities by factors in Table 6 to obtain capacities at different condenser pressures.

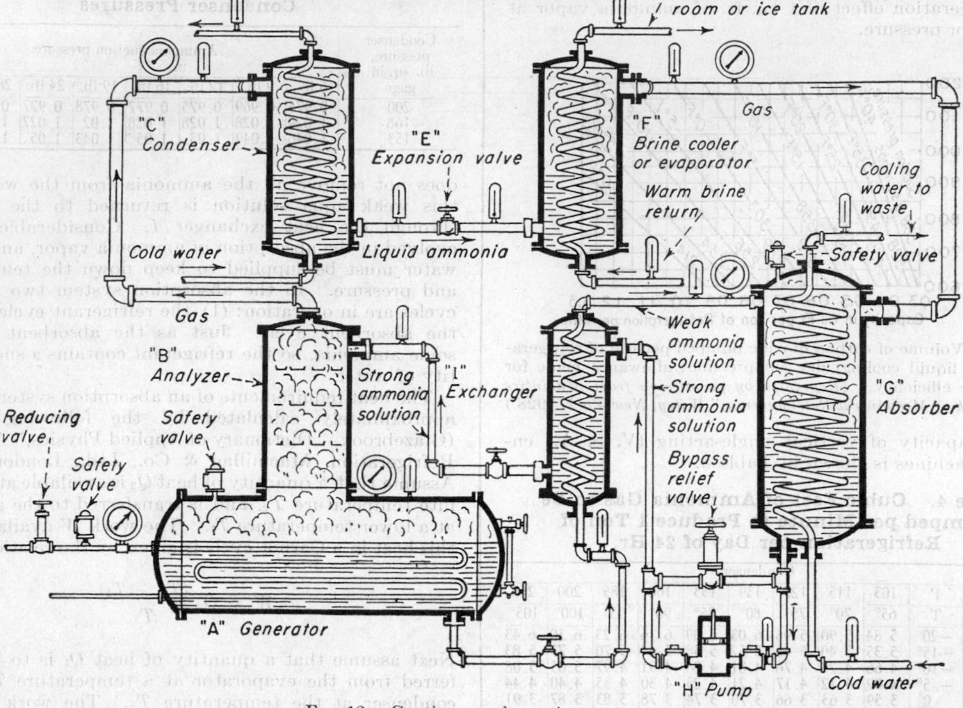

FIG. 12. Continuous absorption system.

Absorption units in large sizes are still being installed in custom-built units. Bubble cap rectifying towers are used to separate ammonia so that it reaches the evaporator in very high purity. Shell and tube exchangers with interval baffles are used to recover heat. Installations are bulky; but, where waste heat is available, economies over mechanical refrigeration may be obtained.

Absorption Systems with Other Refrigerants. Interest in the use of absorption systems for air conditioning has been revived recently [Zellhoefer, *Refrig. Eng.*, May, 1937; Hainsworth, *ibid.*, **48**, 97 (1944)]. Because of its odor, ammonia cannot be used in this application, and considerable research has been devoted toward finding other combinations of refrigerants and solvents that will be as efficient as the ammonia water system. The prime requirement is that the refrigerant shall be highly soluble to minimize the heat waste in warming the solvent. Various cycles that may be used with such units are discussed by Taylor [*Refrig. Eng.*, **49**, 188 (1945)].

Platen-Munters Continuous Absorption System. In the Planten-Munters system (**Electrolux Household Refrigerator**), circulation of refrigerant and absorbent is effected by the operation of hydrostatic forces developed within the system itself, and consequently there are no mechanical moving parts. In order to produce these hydrostatic forces, it is necessary to add a third component, which must differ in density (molecular weight) from the refrigerant and be capable of separation from the refrigerant. The usual combination is ammonia, water, and hydrogen.

The essential features are shown in Fig. 13. Strong ammonia solution is heated in the lower portion of the generator *G*, the weak liquor flowing back to the absorber *A*, while the ammonia-water vapors are separated in the rectifier *R* and the ammonia vapors liquefied in the condenser *C*. The liquid ammonia passes through a heat exchanger *D* into the evaporator *E*. In the evaporator it flows into an atmosphere of hydrogen and is vaporized, though the process is more analogous to humidification. The ammonia-hydrogen mixture then moves back to the absorber *A* through the heat exchanger, and the ammonia is dissolved leaving substantially pure hydrogen.

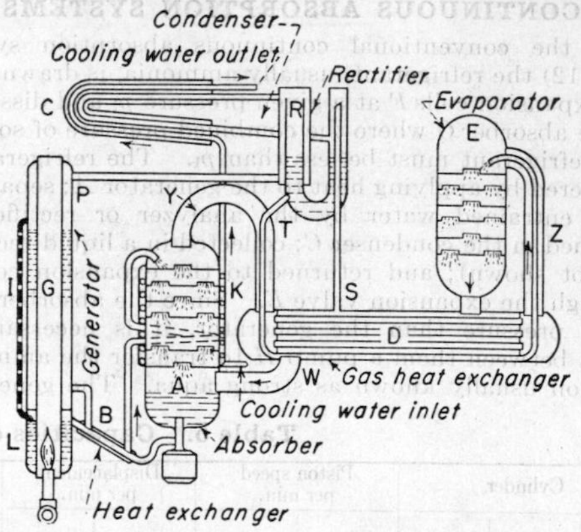

FIG. 13. Electrolux gas-heated refrigerating unit.

Circulation is produced because the column of hydrogen in the absorber is opposed by a heavier column of hydrogen-ammonia vapors in the evaporator, although the total gas pressure throughout the unit is equal. A difference in hydrostatic pressure, obtained partly by heat, partly by change in liquid density, also prevails between the generator and absorber.

The Platen-Munters principle has as yet found application only in the small household refrigerator. To

Table 7. Properties of Aqua-ammonia Solutions*

Liquid concentration, weight per cent ammonia

t = °F.; h_f = enthalpy of liquid (heat content); h_v = enthalpy of vapor; x_v = vapor concentration (weight per cent ammonia)

Pressure, lb./sq. in.		0	10	20	30	40	50	60	70	80	90	100
1	t	101.8	49.6	12.4	−16.3	−40.3	−64.0	−83.4	−95.1	−99.6	−102.6	−105.0
	h_f	69.7	−16.6	−85.7	−141.8	−182.2	−208.6	−219.8	−217.6	−200.9	−176.7	−146.0
	h_v	1105.0	678.0	586.2	551.0	539.0	522.0	510.9	505.5	501.2	496.2	490.7
	x_v	0.00	83.00	96.00	99.50	99.99	100.00	100.00	100.00	100.00	100.00	100.00
10	t	193.2	137.9	99.1	65.3	34.9	7.8	−13.3	−25.8	−33.4	−38.7	−41.3
	h_f	161.1	73.2	3.3	−58.1	−105.2	−134.5	−149.2	−146.0	−130.6	−108.0	−79.3
	h_v	1143.0	743.3	630.2	583.9	558.9	545.5	534.2	527.5	523.5	520.6	519.2
	x_v	0.00	74.18	92.49	98.15	99.75	99.90	99.98	100.00	100.00	100.00	100.00
20	t	228.0	173.2	132.0	97.2	65.0	36.5	14.5	1.4	−7.3	−13.2	−16.6
	h_f	196.0	109.1	37.4	−24.8	−73.8	−104.6	−120.2	−117.4	−102.9	−80.8	−52.9
	h_v	1156.0	781.4	656.0	602.5	575.1	557.3	545.8	538.4	533.6	530.9	528.3
	x_v	0.00	69.99	90.52	97.39	99.44	99.87	99.96	99.99	99.99	100.00	100.00
30	t	250.3	195.3	153.3	117.3	84.4	54.9	32.5	18.7	9.3	3.4	−0.6
	h_f	218.7	131.8	59.6	−3.3	−53.0	−85.1	−101.3	−99.0	−85.1	−62.8	−35.6
	h_v	1163.7	805.1	674.8	615.3	585.0	565.7	553.0	545.0	539.5	536.0	533.7
	x_v	0.00	67.35	88.80	96.70	99.17	99.84	99.93	99.97	99.99	99.99	100.00
50	t	281.1	226.1	182.5	145.5	111.3	80.6	57.8	43.0	33.1	26.2	21.7
	h_f	250.0	163.5	90.4	27.3	−24.1	−57.7	−74.2	−72.8	−59.4	−38.2	−11.4
	h_v	1173.5	838.8	699.7	634.0	598.9	576.4	562.6	553.7	547.5	543.1	540.3
	x_v	0.00	63.50	86.60	95.52	98.69	99.75	99.89	99.95	99.99	99.99	100.00
60	t	292.7	237.7	193.7	156.2	121.4	90.6	67.6	52.3	42.2	34.8	30.2
	h_f	262.0	175.5	102.0	39.1	−12.9	−47.0	−63.6	−62.8	−49.5	−28.7	−2.0
	h_v	1177.0	851.6	709.8	641.1	604.2	580.6	566.3	556.7	550.3	545.6	542.6
	x_v	0.00	61.98	85.63	95.07	98.43	99.68	99.87	99.95	99.98	99.99	100.00
80	t	312.0	257.9	212.8	174.0	138.6	107.7	83.2	67.7	57.2	49.2	44.4
	h_f	218.9	195.8	122.3	58.7	6.2	−28.4	−46.6	−46.1	−32.9	12.8	13.8
	h_v	1182.4	873.7	727.2	653.3	613.7	588.1	571.7	561.8	554.7	459.4	546.1
	x_v	0.00	59.34	83.94	94.20	97.96	99.48	99.83	99.93	99.97	99.99	100.00
100	t	327.8	273.7	228.0	188.7	152.7	121.0	96.3	80.5	69.5	61.4	56.0
	h_f	298.3	212.7	138.6	75.2	22.0	−13.5	−32.0	−31.7	−19.0	0.5	26.8
	h_v	1186.6	892.6	742.2	663.9	622.0	594.4	576.2	565.4	558.0	552.5	548.6
	x_v	0.00	56.97	82.37	93.23	97.49	99.19	99.80	99.93	99.96	99.98	100.00
120	t	341.3	287.5	241.5	201.4	164.9	132.5	107.6	91.4	80.0	71.6	66.0
	h_f	312.4	227.1	153.1	89.6	36.0	−0.6	−19.2	−19.5	−7.2	11.6	38.1
	h_v	1189.8	908.9	754.4	673.4	628.8	599.7	580.5	568.6	560.8	554.7	550.5
	x_v	0.00	54.91	81.11	92.44	97.08	98.93	99.70	99.89	99.94	99.98	100.00
140	t	353.0	299.5	253.4	212.4	175.8	142.7	117.4	101.2	89.4	80.8	74.8
	h_f	324.7	240.0	166.1	102.2	48.6	10.9	−8.0	−8.6	3.5	22.3	48.1
	h_v	1192.4	921.5	765.9	682.4	635.4	604.5	584.2	572.0	563.1	556.7	552.0
	x_v	0.00	53.18	79.92	91.78	96.66	98.72	99.54	99.79	99.89	99.96	100.00
160	t	363.6	310.2	264.0	222.5	185.4	151.8	126.2	109.8	97.9	89.0	82.6
	h_f	335.9	251.2	177.5	113.7	59.8	21.2	2.0	1.3	13.3	31.9	57.1
	h_v	1194.5	933.0	776.0	690.1	641.2	608.6	587.5	574.6	565.2	558.3	553.2
	x_v	0.00	51.70	78.85	91.12	96.26	98.51	99.40	99.71	99.87	99.95	100.00
180	t	373.0	319.9	273.2	231.6	193.8	160.2	134.4	117.7	105.5	96.5	89.8
	h_f	346.0	261.6	187.5	124.0	69.9	31.0	11.3	10.5	22.2	40.4	65.4
	h_v	1196.3	943.0	785.2	696.9	646.6	612.6	590.5	576.8	566.9	559.6	554.1
	x_v	0.00	50.41	77.83	90.49	95.87	98.30	99.28	99.66	99.86	99.94	100.00
200	t	381.8	328.8	281.7	239.8	202.0	167.7	142.2	124.9	112.6	103.2	96.3
	h_f	355.3	271.4	196.8	133.5	79.7	39.8	20.3	18.8	30.4	48.2	73.0
	h_v	1197.8	952.0	793.7	703.7	651.5	616.0	593.1	578.8	568.4	560.6	554.8
	x_v	0.00	49.25	76.86	89.88	95.52	98.09	99.20	99.63	99.84	99.94	100.00
250	t	401.0	348.2	300.8	257.7	219.2	184.5	158.4	140.8	127.6	118.2	110.8
	h_f	376.0	293.0	217.5	155.2	100.7	59.5	39.6	37.5	48.2	65.8	90.1
	h_v	1200.5	970.1	812.9	717.5	661.8	623.4	598.4	582.4	570.9	562.5	555.9
	x_v	0.00	46.94	74.70	88.64	94.75	97.67	99.02	99.58	99.81	99.93	100.00
300	t	417.3	365.3	316.9	273.2	234.6	198.9	172.4	154.4	141.6	130.1	123.2
	h_f	393.9	311.7	235.3	172.4	119.8	77.1	56.1	53.4	64.6	81.0	104.1
	h_v	1202.4	984.5	829.3	730.3	671.2	629.6	602.9	585.8	573.6	536.6	556.1
	x_v	0.00	45.17	72.70	87.44	94.00	97.43	98.84	99.52	99.78	99.92	100.00

* Jennings and Shannon, Lehigh University Studies, Science and Technology Series 1, 1938. Data obtained through courtesy of Prof. Jennings. Original paper contains much more complete data.

function properly it is necessary that leakage be avoided, and this entails excessive construction costs on large units. The efficiencies and capacities of a small unit are shown in Fig. 14, efficiency being based on the ratio of refrigeration produced to heat supplied.

The Platen-Munters system employs hydrogen, ammonia, and water and is therefore known as a **three-fluid cycle** as distinguished from a **two-fluid** as in large absorption units. The third fluid is necessary to obtain circulation without a pump.

Intermittent Absorption Machine. In the intermittent absorption machine the absorber and generator, and the condenser and the evaporator, are combined. During the cooling period the refrigerant is taken up by the absorbent, which may be water, silica gel, activated charcoal, or chlorides of the alkaline earth metals. When the absorbent has become saturated it is regenerated by heating and the vapors condensed in the evaporator which is cooled by running water. These machines have not proved very successful for two reasons: (1) the gen-

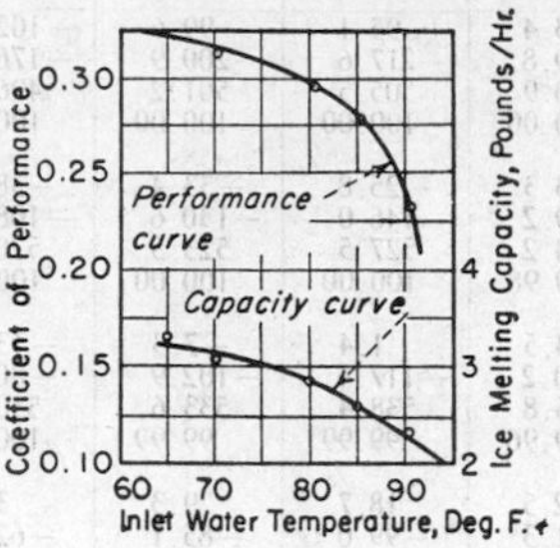

FIG. 14. Effect of varying cooling-water temperature. Gas rate, 2.5 cu. ft./hr.; room temperature, 70°F.; quantity of water, 7 gal./hr.

eral desirability of maintaining continuous refrigeration and (2) the great explosion hazard in overheating during regeneration.

Table 8. Test Results at Quincy Market Cold Storage and Warehouse Co.

Machine	Capacity of machine, tons	Capacity during test, tons	Suction temperature, °F.	Discharge pressure, lb. gage	Steam conditions: Pressure, lb./sq. in.	Steam conditions: Superheat, °F.	Steam conditions: Vacuum, in. Hg	Compressor i.hp./ton of refrigeration	Steam per ton, lb.
1	1000	750	+10	115	150	100	28	0.9	228
2	400	400	−10	115	150	100	28	1.25	390
2	400	400	+10	115	150	100	28	0.9	282
3	500	288	−27	139	140	125	28	1.34	426
4	500	370	−20	130	...	...	..	1.64	
5	150	150	−10	115	125	125	28	1.28	430
5	150	150	+10	115	125	125	28	0.92	376
6	100	100	−10	115	125	125	28	1.32	459
6	100	100	+10	115	125	125	28	0.94	389
7	225	225	−10	115	...	...	..	....	342.2
7	225	225	+10	115	...	...	..	....	316.8

1 and 2, cross-compound Corliss engines; 3, two-stage feather-valve compressor driven by Uniflow engine; 4, two-stage electrically driven compressor; 5 and 6, tandem compound engines; 7, absorption machine.

The discharge pressures in these tests are exceptionally favorable and result in low power consumption.

The absorption machine was more economical than the 400-ton compression machine with the low-temperature suction gas, and less economical than the same machine with high-temperature suction gas.

Station operation results show that absorption machines require 25 to 35 lb. of steam per ton of refrigeration with 0° brine and 150 lb./sq. in. condenser pressure. Makers' guarantees are from 55 to 60 lb. of live steam per hour per ton of ice.

Table 9. Performance of Absorption Machine with Different Condenser and Suction Pressures*

	Condenser pressure, lb./sq. in. abs.								
	155			185			215		
	Suction pressure, lb./sq. in. abs.								
	15	30	45	15	30	45	15	30	45
S.L., %	24.0	35.0	42.0	22.0	32.0	38.0	18.0	28.0	36.0
W.L., %	13.1	25.8	33.7	10.9	22.3	29.2	6.3	17.7	26.9
S.G., lb.	30.1	27.9	22.9	41.3	30.9	26.2	48.7	34.1	27.9
S.P., lb.	31.8	29.5	24.3	43.4	32.8	28.0	51.1	36.4	30.1
Relative capacities:									
Absorption	0.97	1.03	1.09	0.94	1.00	1.05	0.91	0.97	1.03
Compression	0.46	1.05	1.62	0.43	1.00	1.56	0.41	0.95	1.49

S.L. is strong liquor.
W.L. is weak liquor.
S.G. is the steam consumption of the generator.
S.P. is the steam consumption of the pump.
* Voorhees, "Refrigerating Machines."

REFRIGERANTS USED IN VAPOR-COMPRESSION AND ABSORPTION SYSTEMS

The choice of a refrigerant is always determined by the specific conditions of refrigeration to be met. A refrigerant cannot be used in a compression machine if the temperature of the available cooling water lies above the critical temperature of the refrigerant. The temperature of cooling waters in America is usually too high to use carbon dioxide with advantage (see Fig. 15). For absorption machines, it is necessary that the absorbent be capable of taking up large quantities of the refrigerant. In centrifugal compression machines it is necessary that the vapor pressures be as low as possible so as to reduce the number of stages and that the molecular weight be as high as possible so as to reduce the peripheral speed of the rotor.

In vapor-compression machines of the reciprocating type, refrigerants of high vapor pressure are desired since these give low piston displacement and smaller friction losses. A comparison of the vapor pressures of various materials is shown in Fig. 16, where the logarithm of the vapor pressure in atmospheres is plotted against the reciprocal of the absolute centigrade temperature. The slopes of all the lines are very nearly parallel (Trouton's law), water being the notable exception, and the compression ratios for any given temperature interval are very nearly equal.

Other properties of importance are: chemical stability, inertness to metals (corrosion), ease of detection in case of leaks, and behavior with lubricants. Cheapness of the refrigerant is a very important item in larger units since it is not practicable to prevent leakage altogether.

Ammonia is the most important refrigerant used in industrial work, but the irritating and toxic character of the vapor excludes it from certain applications, such as marine refrigeration or air conditioning, where carbon dioxide and dichlorodifluoromethane (Freon 12) are preferred. Although the thermodynamic efficiency of ammonia is not very different from other refrigerants, its low gas density is a distinct practical advantage in reducing power losses in refrigerant piping. In certain chemical processes, notably those in which free chlorine may be present, ammonia is also undesirable. Propane having nearly the same pressures may be substituted in machines designed for ammonia. In household refrigerators, the use of ammonia is limited to the gas-fired absorption unit of the Platen-Munters type. Automatic regulation is more difficult with ammonia because of its high latent heat per pound, which means that a small departure from the average flow upsets the temperature a great deal more than with a refrigerant of low latent heat.

The most important refrigerants used in household machines are sulfur dioxide, methyl chloride, and difluorodichloromethane. The last-named refrigerant has been developed in response to the demand for a non-toxic, nonflammable material for household machines [Midgeley and Henne, *Ind. Eng. Chem.*, **22**, 542 (1930)]. Dimethyl ether, first proposed by Linde at the beginning of the modern period of refrigeration developments, was used in some German units. The low-pressure, high-molecular-weight refrigerants—methylene chloride and the more complex fluoro-chloro-methanes—are used in centrifugal compressors for air-conditioning work.

Binary mixtures have been proposed as refrigerating fluids, but they have been unsuccessful because one of the components is always likely to leak out more rapidly than the other and when this happens it becomes impossible to interpret indicated pressures satisfactorily. They also require a greater work expenditure if any rectification occurs in the evaporator. If two refrigerants have nearly identical vapor pressures, such as methyl chloride and dimethyl ether, or carbon dioxide and nitrous oxide, they cannot be separated and would act as a one-component fluid. Two refrigerants may also form constant-boiling mixtures which behave similarly, but the vapor pressures of such binaries have not been deter-

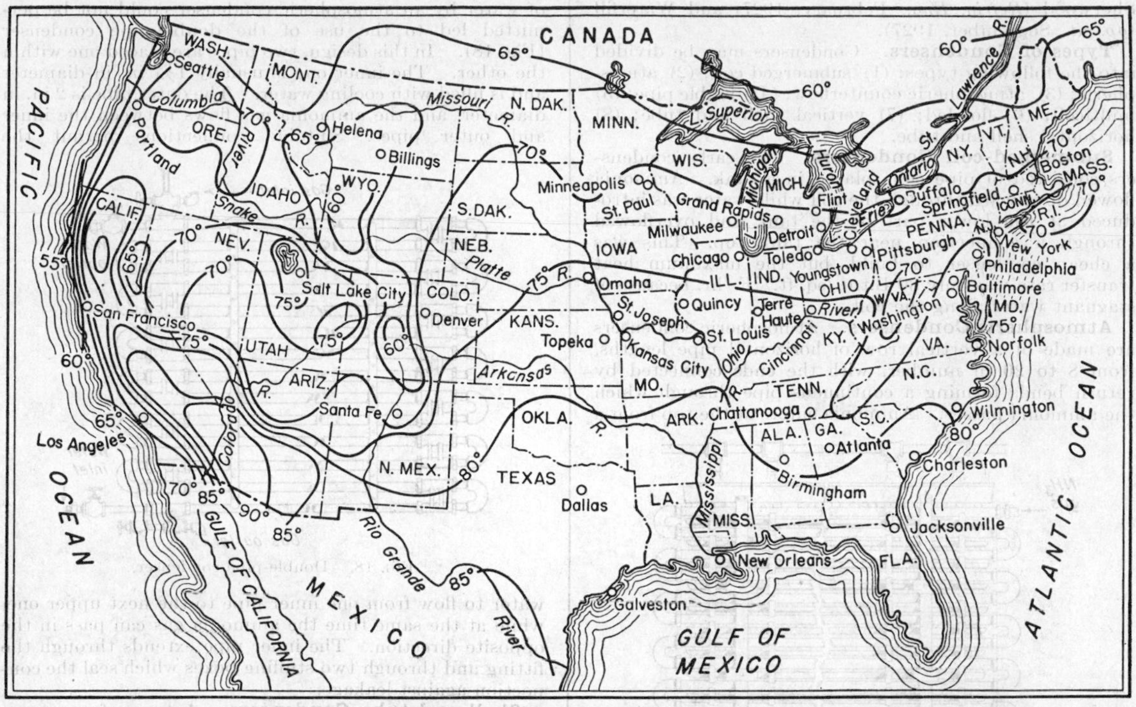

FIG. 15. Surface-water temperatures (during summer months) in the United States. (*Reprinted by permission from Macintire, "Handbook of Mechanical Refrigeration," Wiley, New York, 1928.*)

mined. There are no essential advantages in binary mixtures except that the freezing points might be lowered.

Small admixtures of warning or detecting agents are sometimes added. Thus peppermint may be added to

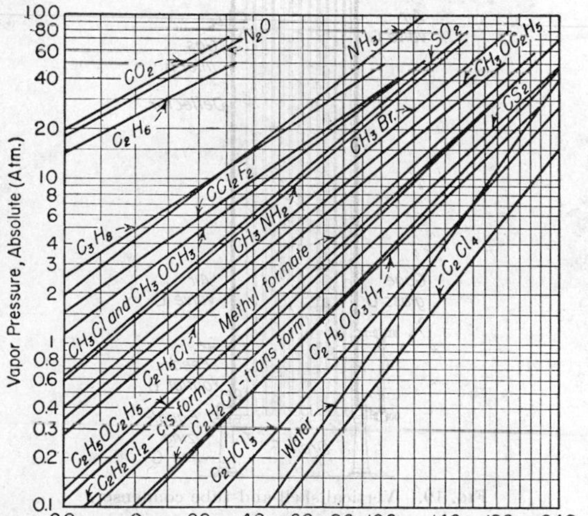

FIG. 16. Vapor pressures of refrigerants.

carbon dioxide, acrolein to methyl chloride (both of which have characteristic odors), or ethyl nitrite for the detection of pinholes by discoloration of starch-potassium iodide paper. With ammonia, a sulfur candle, a piece of litmus paper, or a glass rod dipped in acid serves as means of detection. Refrigerants containing chlorine can usually be detected by lighting a taper which gives a bluish-green flame in the escaping vapors, but this procedure is to be used with care in the case of methyl chloride, which may form an explosive mixture with air.

The heat-transfer characteristics of refrigerants, in both the liquid and gaseous phases, are also important, but very few comparative data have been obtained so far. According to R. C. Doremus (Eng. Sec., *Elec. Refrigeration News*, Sept. 23, 1931) the rate of heat transfer for sulfur dioxide is about 85 per cent as great as for methyl chloride under the same conditions of evaporation.

Thermodynamic Properties of Refrigerants. The thermodynamic properties of various refrigerants are given in Sec. 3.

CONDENSERS

The primary function of the condenser is to transfer to the environment (usually water or air) the heat extracted at the lower temperature. If c is the specific heat of water (or air), W the weight of water sent through the condenser per minute, Δt the temperature rise in degrees Fahrenheit, and n the number of tons refrigeration produced, then

$$W = \frac{200n}{c\Delta t}$$

Usually Δt will not be more than 30°F., from which it follows that W is approximately 7 lb./min./ton refrigeration or about 1100 to 1200 gal./day/ton refrigeration.

In considering any refrigeration installation it is always important to determine whether the condensing water requirements can be met from the available sources. In large metropolitan centers this may often be a serious problem [Gardner, *Refrig. Eng.*, **41**, 17 (1941)]. The question of cooling-water economy was discussed by

Sherwood (*Refrig. Eng.*, February, 1927) and Waterfill (*op. cit.*, September, 1927).

Types of Condensers. Condensers may be divided into the following types: (1) submerged coil; (2) atmospheric; (3) atmospheric counterflow; (4) double pipe; (5) multicoil; (6) flooded; (7) vertical shell and tube; (8) horizontal shell and tube.

Submerged-coil Condensers. The early condensers consisted of pipe coils placed in a tank. Ammonia flowed downward through the coil while water was introduced at the lower part of the tank and overflowed through a connection near the tank top. This was a cheap condenser to build, but the maximum heat transfer rate was only 30 B.t.u./sq. ft./°F./hr. because of stagnant water along the coil.

Atmospheric Condensers. Atmospheric condensers are made of a vertical row of horizontal pipe lengths, from 8 to 20 in number, with the ends connected by return bends forming a continuous pipe through which the ammonia passes. Ammonia enters at the top from a

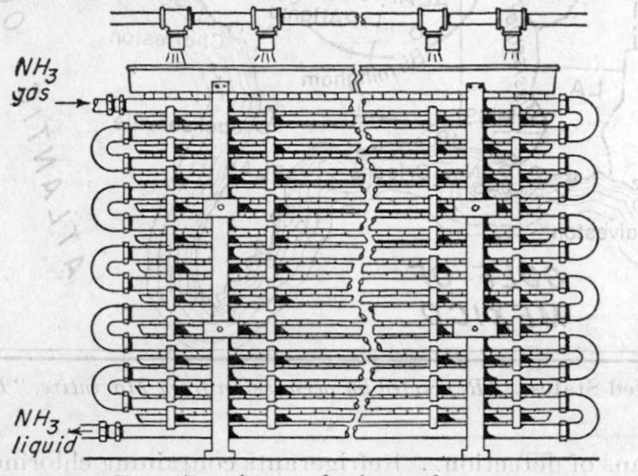

FIG. 17. Atmospheric condenser.

header connecting two or more of the rows of pipes and flows downward, losing its heat to the water. Liquid ammonia settles in the bottom coils and passes off to the receiver or storage tank. Water is fed into a trough placed over the top pipe and, overflowing, drops into the top pipe and thence successively flows over each pipe into a basin beneath the coil. This type is illustrated in Fig. 17.

Bottom-inlet Atmospheric Condensers. In bottom-inlet condensers the flow is made countercurrent by feeding the vapor at the bottom and by taking liquid off at the top, but, unless bleeder connections are taken off every few coils, the liquid tends to flow back toward the bottom. The condition prevails in those condensers where, after passing upward through the two lower pipes, the ammonia vapor is conveyed by an outside connection to the top pipe and then flows downward through the remainder of the bank. Both designs are erratic in operation and are subject to flooding and slugging.

Countercurrent Atmospheric Condensers. As stated above, the coldest water in the ordinary atmospheric condenser is at the top in contact with the hottest gases, while the hottest water at the bottom is in contact with the coldest ammonia. To reduce the ammonia temperature on leaving the condenser, condensers are designed with the gas inlet at the bottom where it meets the hottest water and the liquid discharge higher where the water is cooler. In its passage upward through the coils the condensing ammonia will have a tendency to trickle downward against the entering vapor. To prevent the bottom coils from filling up with liquid, trap drains are provided at several points.

Double-pipe Condensers. The desirability of having a condenser suitable for locations where the splashing of water by an atmospheric condenser could not be permitted led to the use of the double-pipe condenser (Fig. 18). In this design, two pipes are placed one within the other. The inner one is usually 1¼ in. in diameter and is filled with cooling water. The outer pipe is 2 in. in diameter, and the ammonia gas flows between the inner and outer pipes. Suitable connections permit the

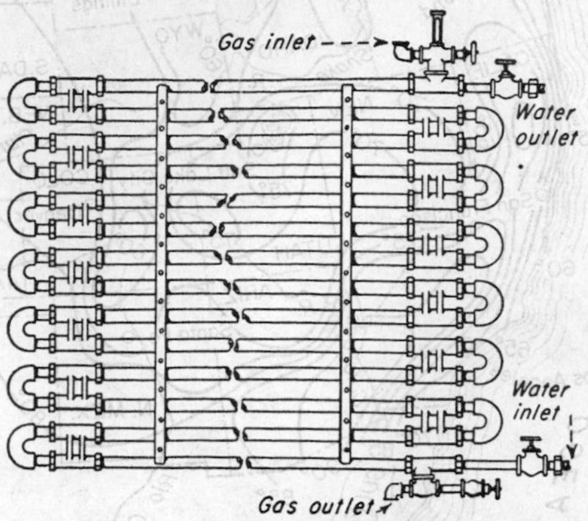

FIG. 18. Double-pipe condenser.

water to flow from one inner pipe to the next upper one, while at the same time the ammonia gas can pass in the opposite direction. The inner pipe extends through the fitting and through two stuffing boxes which seal the connection against leakage.

Shell-and-tube Condensers. A type of construction of the vertical single-pass multitube condenser is

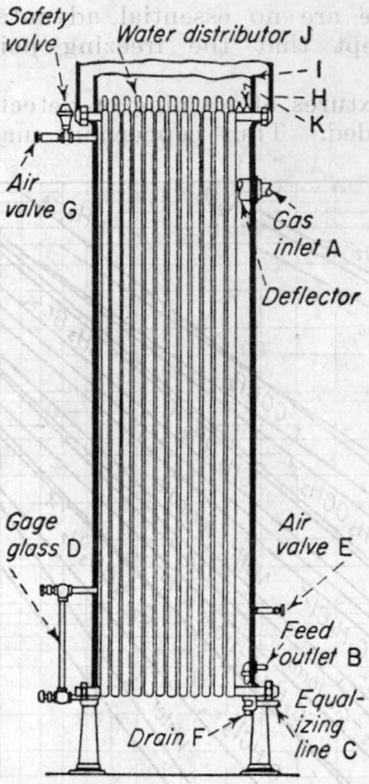

FIG. 19. Vertical shell-and-tube condenser.

illustrated in Fig. 19. This consists of a welded shell with flared ends riveted to heavy tube sheets and a number of charcoal-iron tubes. The ammonia gas inlet is at the point *A*. In order to provide for a more even distribution of the gas at the inlet, a special deflector is attached to the inlet nozzle as shown. The ammonia is condensed on the tubes, collected in the bottom part of the shell, and is drained off at the feed outlet *B*. An equalizing line is provided at C. A drain is installed at F, while purge connections may be made at points E and F. A circular water box H is attached to the top sheet, and a special water baffle I, having serrated edges K, is placed within the water box as shown. Water-distributing devices J are placed on the top end of each tube. These distributors are made of cast iron and are hollow and with spiral grooves, and cause the water to flow in a corkscrew motion down the tube. A certain amount of air is drawn in through the hollow cores of the device. The vertical shell-and-tube condenser has the advantage

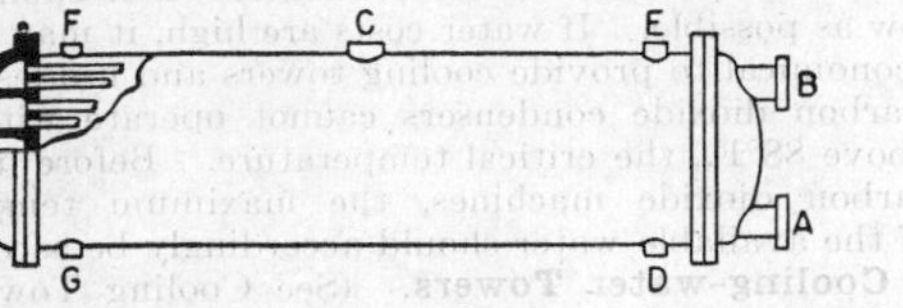

FIG. 20. Horizontal multipass condenser. (*Courtesy Struthers Wells.*)

that a greater amount of cooling surface can be obtained for a given floor space than with any other type.

Horizontal Multipass Condenser. A construction of a horizontal multipass condenser is illustrated in Fig. 20. Water enters at the inlet A and passes through a bank of 1¼-in. tubes in the lower part of the shell, then flows backward and forward several times through the tubes in the upper part and finally out through the exit B. Refrigerant vapors enter at the inlet C placed on the top of the shell near the middle. The vapors condense, and the condensate is removed at D where it meets

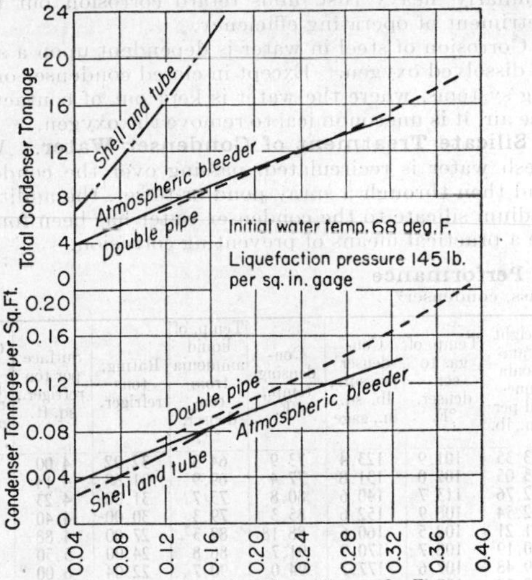

FIG. 21. Comparative performances of various condensers.

the coldest cooling water. A purge for non-condensable gases is provided at E. A safety valve is installed at F, and a drain valve at G. The number of passes depends on the diameter of the shell, being 2 to 6 on a 12-in. shell and 8 to 12 on a 24-in. shell. The total number of tubes varies from about 30 on a 12-in. shell to 160 on a 24-in. shell.

Heat Transfer in Ammonia Condensers (See Sec. 6, Heat Transmission, pp. 455*ff*.). Exhaustive investigations on heat transfer in ammonia condensers were carried on by the University of Illinois; the results appear in the University of Illinois Bulletin 25. In these investigations tests were run on shell-and-tube, atmospheric-bleeder, and double-pipe condensers, and both the unit condenser tonnage and the total condenser tonnage have been plotted against the unit water rate for initial water temperature of 68°F., as shown in Fig. 21. When the total tonnage developed is taken into consideration the shell-and-tube condenser shows the greatest capacity. The total area of the bleeder condenser exposed to saturated ammonia is 105 sq. ft., while that of the double-pipe condenser is 92 sq. ft. and in the case of the shell-and-tube condenser is 251 sq. ft.

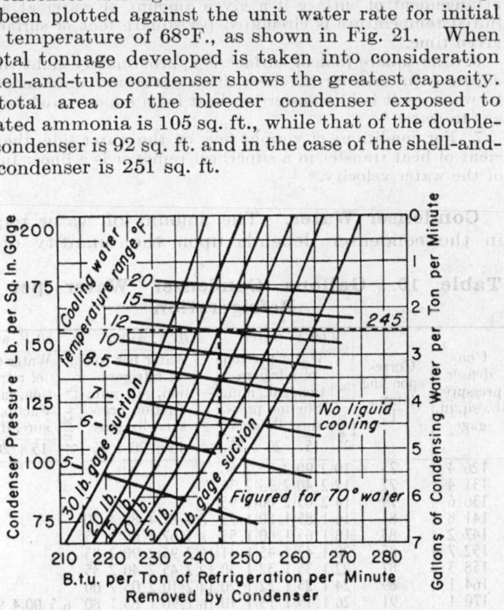

FIG. 22. Condenser water requirements. (*Reprinted by permission from Macintire, "Handbook of Mechanical Refrigeration," Wiley, New York,* 1928.)

As reported by the University of Illinois Bulletin 186, the heat removed by a 16 ft. vertical shell-and-tube condenser having thirty 2-in. tubes is found to vary with the water velocity, or water flow, as shown in Fig. 22 and Table 10. The variations in heat transfer with varying water flows and tonnages are shown in Fig. 23.

The conclusions reached in these tests were as follows:

1. The thickness of the layer of liquid ammonia adhering to the tubes of the vertical shell-and-tube condenser materially affects the rate of heat transfer per unit of surface.

2. At a constant water rate the coefficient of heat transfer decreases with increasing values of the mean temperature differ-

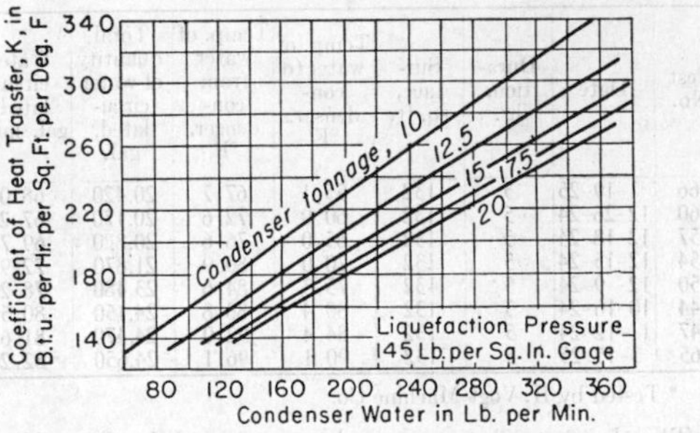

FIG. 23. Variation of heat transfer with water rate for shell-and-tube condensers.

ence between the ammonia and the water in the vertical shell-and-tube condenser.

3. Irrespective of the proportions of the vertical shell-and-tube condenser no appreciable subcooling of the liquid occurs.

4. The condenser tonnage developed per square foot of surface in the shell-and-tube condenser is independent of the size or proportions of the condenser and is a function only of the initial temperature of the water and the amount circulated per square foot of surface per unit of time.

5. The condenser tonnage developed per square foot of surface in the condensers when the effective surface is altered by raising

the liquid level or by plugging pumps, as was done in the investigation, is approximately independent of the proportions or arrangement of surface if a given amount of water at a given initial temperature is circulated per square foot of surface in a given time.

6. One square foot of surface is approximately eight times as effective in transferring heat from saturated ammonia vapor to water as it is in transferring heat from superheated ammonia gas to water.*

7. For conditions of viscous flow on the water side, the coefficient of heat transfer in a superheat remover is a linear function of the water velocity.*

Condenser Water. The amount of water required in the condenser depends upon the quantity of heat

Table 10. Gallons Condenser Water per Ton Refrigeration

Condenser pressure, lb./sq. in. gage	Corresponding temp., °F.	60°F. water: Range °F.	60°F. water: Water per ton of refrigeration, gal./min. Suction pressure, lb. gage 15	20	25	70°F. water: Range °F.	70°F. water: Water per ton of refrigeration, gal./min. Suction pressure, lb. gage 15	20	25	80°F. water: Range °F.	80°F. water: Water per ton of refrigeration, gal./min. Suction pressure, lb. gage 15	20	25
126.4	75	10	2.90	2.85	2.80								
131.4	77	12	2.40	2.35	2.30								
136.6	79	14	2.05	2.00	1.95								
141.8	81	16	1.85	1.80	1.75	6	4.87	4.80	4.70				
147.2	83	18	1.65	1.60	1.55	8	3.67	3.60	3.55				
152.7	85	20	1.47	1.45	1.41	10	2.95	2.90	2.85				
158.3	87	22	1.35	1.32	1.30	12	2.45	2.40	2.35				
164.1	89	24	1.25	1.22	1.20	14	2.10	2.05	2.00				
170.1	91	26	1.15	1.13	1.10	16	1.90	1.85	1.80	6	5.00	4.90	4.80
176.2	93	28	1.08	1.05	1.03	18	1.68	1.63	1.60	8	3.75	3.70	3.65
182.6	95	30	1.00	0.99	0.97	20	1.50	1.48	1.45	10	3.00	2.95	2.90
189.1	97	32	0.95	0.93	0.90	22	1.38	1.35	1.32	12	2.52	2.48	2.43
195.7	99	...	...	...	...	24	1.28	1.25	1.22	14	2.13	2.10	2.06
202.5	101	...	...	...	...	26	1.17	1.15	1.13	16	1.92	1.85	1.83
209.5	103	...	...	...	...	28	1.10	1.07	1.05	18	1.70	1.67	1.63
216.5	105	...	...	...	...	30	1.02	1.00	0.98	20	1.53	1.50	1.48
223.7	107	...	...	...	...	32	0.95	0.93	0.92	22	1.40	1.37	1.35
231.1	109	...	...	...	...	...	...	...	...	24	1.30	1.28	1.25
238.7	111	...	...	...	...	...	...	...	...	26	1.20	1.16	1.15
246.5	113	...	...	...	...	...	...	...	...	28	1.12	1.08	1.06
254.5	115	...	...	...	...	...	...	...	...	30	1.05	1.02	1.00
262.7	117	...	...	...	...	...	...	...	...	32	1.00	0.98	0.95

which must be removed to liquefy the ammonia. This may be calculated as follows: first, find the weight of ammonia to be circulated per minute per ton.

The heat to be removed per pound is $Q = H_s - h_l$, where Q is the B.t.u. to be removed by the cooling water; H_s is the total heat in 1 lb. of discharge gas; h_l is the heat of ammonia liquid at discharge pressure. The pounds of water required per pound of ammonia is found by the formula

$$W = \frac{Q}{(t_o - t_i)c_w}$$

where t_o = temperature of outlet water.
t_i = temperature of inlet water.
c_w = specific heat of water.

The condensing pressure depends on the temperatures and quantity of the cooling water. With 60°F. water in large quantities, it is possible to get a condensing pressure as low as 145 lb./sq. in.; with 75°F. water, 165 lb./sq. in.; with 85°F. water, 185 lb./sq. in. In ordinary practice the pressure may be somewhat higher.

Table 11 gives the performances obtained on a multipass condenser.

The temperature of the condenser water should be as low as possible. If water costs are high, it may be more economical to provide cooling towers and use less water. Carbon dioxide condensers cannot operate with water above 88°F., the critical temperature. Before installing carbon dioxide machines, the maximum temperature of the available water should accordingly be ascertained.

Cooling-water Towers. (See Cooling Towers Sec. 12.) Cooling towers or cooling ponds may be used to recover waste water from condensers.

Towers are usually figured for a small range of water cooling, about 5°F., and are designed to handle approximately 6 gal. water per ton of refrigeration and cool this to within 5°F. of the wet-bulb temperature with an assumed air velocity of 5 m.p.h. See Fig. 56, p. 793.

Corrosion in Condenser Systems. Ammonia condensers present a severe corrosion problem. A large area of iron or steel is exposed to a continuous flow of water, which is frequently quite corrosive. To obtain efficient heat transfer it is desirable to have the metal bare, and ordinary methods of protection, such as painting, are used at the expense of condensing efficiency. Similarly heavy rust films retard corrosion but to the detriment of operating efficiency.

Corrosion of steel in water is dependent upon a supply of dissolved oxygen. Except in closed condenser or cooling systems, where the water is kept out of contact with the air, it is uneconomical to remove the oxygen.

Silicate Treatment of Condenser Water. Where fresh water is recirculated, passing over the condensers and then through a spray pond or tower, the addition of sodium silicate to the condenser water has been found to be a practical means of preventing corrosion.

Table 11. Condenser Performance

8-in. multitube, seven-pass, condenser*

Test No.	Date	Duration, hr.	Surface, sq. ft.	Temp. of water to condenser, °F.	Temp. of water from condenser, °F.	Total quantity of water circulated, gal.	Water circulated, gal./min.	Total weight of ammonia liquefied, lb.	Weight of ammonia liquefied per min., lb.	Temp. of gas to condenser, °F.	Condenser pressure, lb./sq. in., gage	Condensing temp., °F.	Temp. of liquid ammonia from condenser, °F.	Rating, tons refriger.	Surface per ton refriger., sq. ft.	Cooling water per ton refriger., gal./min.
66	1-19-25	5	132	55.1	67.7	20,420	68.07	4,004	13.35	101.9	123.4	73.9	64.3	32.92	4.00	2.06
60	12-26-24	5	132	60.0	72.6	20,170	67.23	3,915	13.05	102.0	131.8	77.4	69.9	31.88	4.14	2.10
57	12-18-24	5	132	65.0	76.6	20,820	69.70	3,829	12.76	113.7	140.6	80.8	75.7	31.00	4.23	2.25
54	12-15-24	5	132	70.0	80.9	21,870	72.90	3,763	12.54	108.9	152.6	85.3	79.3	30.00	4.40	2.43
50	12- 9-24	5	132	75.3	84.6	23,480	78.26	3,363	11.21	102.5	160.6	88.18	82.3	27.00	4.88	2.90
44	10-16-24	5	132	80.4	88.8	24,150	80.50	2,412	10.19	108.7	170.9	91.7	86.8	24.00	5.50	3.35
47	11-12-24	5	132	84.4	92.0	24,470	81.60	2,843	9.48	107.6	177.7	94.0	90.7	22.04	6.00	3.70
65	1-15-25	5	132	90.1	96.1	24,650	82.20	2,490	8.30	107.9	193.3	98.8	93.8	19.40	6.72	4.24

* Tested by H. Vogt Machine Co.

COOLING SYSTEMS

Cooling Systems. The term **cooling system** refers to that portion of a refrigeration unit in which cold is applied. Two types of cooling system are in use: (1) **direct-expansion** and (2) **indirect-brine** systems.

With the **direct-expansion** system, where ammonia is allowed to boil in the cold-storage rooms, there is danger from ammonia leakage at all times. The pipe lines may become corroded or may split because of imperfect welding, or fittings may be broken accidentally with resulting damage to life or commodities. In a large system with long supply and return pipes, or extensive refrigerating

* If the superheated vapor is condensing, the rate of heat transfer is probably as good as for saturated vapor. The above conclusions apply to the case in which superheated vapor goes to a lower degree of superheat.

piping, the amount of the initial charge has to be very large, and constant care to maintain the piping tight at all times is required.

In the **brine system of indirect refrigeration,** the high-pressure side is the same as in the direct-expansion system. The low-pressure side consists of a brine cooler, usually of the shell-and-tube type, similar in construction to a steam condenser. The brine is a non-freezing solution of sodium chloride or calcium chloride, of such concentration as will not freeze at the temperature carried in the cooling system. The brine system then is really an additional unit, in which the brine is kept cool by boiling ammonia and the cold-storage rooms or other refrigerating applications are kept cold by the brine. Appreciable amounts of refrigeration may be stored up in the brine to take up peak loads or provide reserve refrigeration for closing down periods.

In the direct-expansion system one less heat transfer is necessary, and consequently the expansion coils can be maintained at a higher temperature and pressure so that the compression work is less. Moreover, the refrigerant extracts heat by evaporation (latent heat) so it can be distributed at room temperatures, whereas brine extracts heat only by being colder than the surroundings (sensible heat) and must be distributed cold in well-insulated piping. The direct-expansion system is therefore much more efficient in theory, and the only obstacle to its general adoption is the difficulty of constructing leakproof systems.

Evaporation or Expansion Coils.* Just as the function of the condenser is to dissipate heat, so the function of the evaporating coil is to collect heat at the lower temperature of the refrigerating system.

It is quite possible to inject the liquid refrigerant into the material to be cooled and then cause it to evaporate, but this is in general undesirable for two reasons: first, the refrigerant will have a deleterious effect on the substance cooled, and, second, its vapor pressure will be lower because of dilution and the power expenditure in compression will be greater. However, in case the refrigerant can also be used as a solvent, this system may prove very satisfactory, and one notable application is found in the dewaxing of petroleum oils by liquid propane [Beiter, *Refrig. Eng.*, **40**, 293 (1940)] and in the Edeleanu process for extracting asphaltic materials from mineral oils by sulfur dioxide.

The simplest method of cooling brine is to thrust a pipe coil into the brine tank with one end connected to the liquid refrigerant supply and the other to the compressor. An older form of construction with a series of continuous coils attached to headers is shown in Fig. 24. Liquid

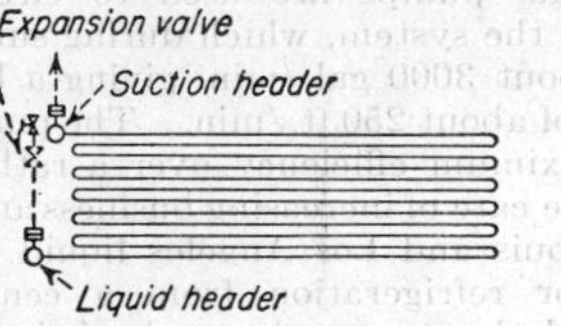

FIG. 24. Continuous-coil evaporator.

ammonia is fed into the bottom and drawn off on top. Under these conditions liquid ammonia is present on the inside surface of the coils, and if this is accidentally drawn into the compressor it may fill up the clearance space in the cylinder and blow off the head on the compression stroke. This difficulty is overcome by placing an accumulator (shown by dotted lines in Fig. 25) between the expansion coil and the compressor. The accumulator may serve the further purpose of precooling the liquid ammonia returning to the expansion coil. This arrangement of coils with accumulator is known as the **flooded system.** In order to facilitate transfer of heat, brine tanks may be built with partitions and bulkheads so arranged that the brine may be agitated and swirled around and through the coils.

A modern short evaporator coil is shown in Fig. 26. This is sometimes made with three sets of pipes between

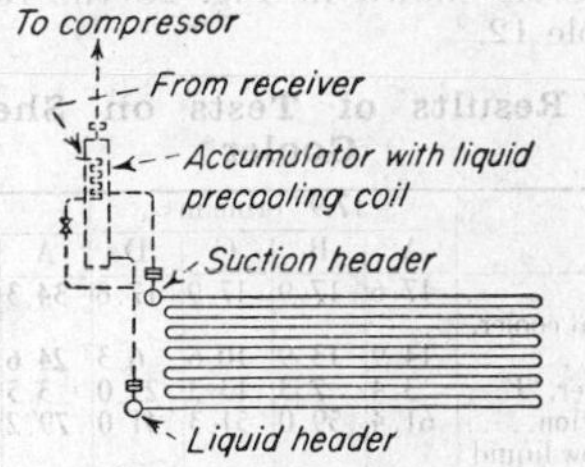

FIG. 25. Evaporator coil with accumulator.

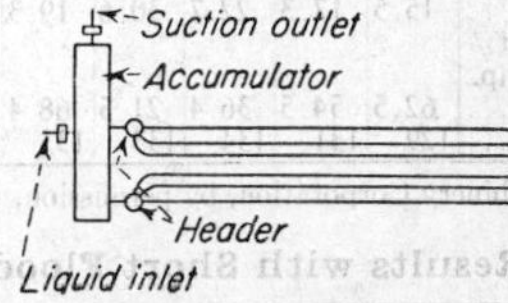

FIG. 26. Short evaporator coil.

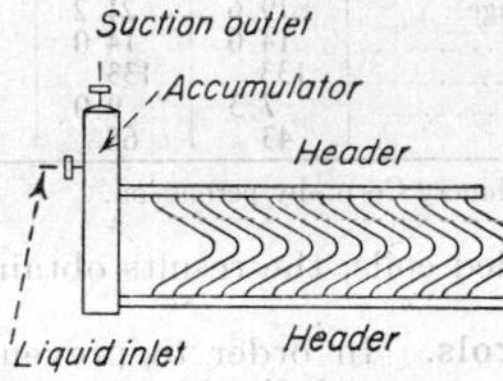

FIG. 27. Herringbone coil.

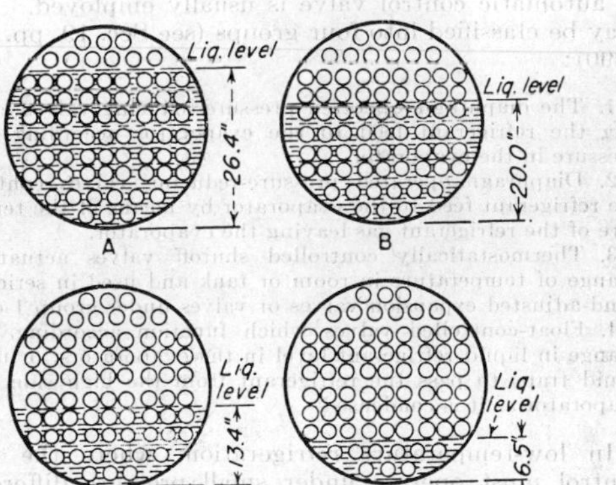

FIG. 28. Liquid levels used in shell-and-tube coolers. Shell diameter about 32.9 inches.

the upper and lower headers and placed in a covered brine compartment. The herringbone coil, shown in Fig. 27, can be placed in a trough through which the brine is circulated at high velocity.

Besides the *submerged* types of coils, which are used chiefly in ice making, shell-and-tube brine coolers, and double-pipe brine coolers with brine flowing through the inner tube are also used. Shell-and-tube coolers are largely displacing other types in more recent installations (Fig. 28). Brine flows through the tubes, and a surface

* For a more general discussion see Consley, Heat Transfer in Ammonia Shell-and-tube Brine Coolers as Affected by Operating Conditions, *Refrig. Eng.*, **35**, 409 (1938).

of 8 to 15 sq. ft. is usually allowed per ton of refrigeration. They may be placed either upright or horizontal and require small floor space. A float valve can be placed in the liquid receiver to regulate the amount of ammonia fed to the expansion coils.

Heat Transfer in Evaporators. (See Sec. 7, pp. 500–505.) The York Ice Machinery Corp. conducted a series of tests on heat transfer in various evaporators. For shell-and-tube evaporators with tube arrangements and liquid levels shown in Fig. 28 the results were as given in Table 12.

Table 12. Results of Tests on Shell-and-Tube Cooler*

	17.9° brine in				34° brine in			
	A	B	C	D	A	B	C	D
Brine temp., °F	17.6	17.9	17.9	17.8	34.3	34.0	33.9	33.6
Suction pressure at cooler, lb. gage	14.9	13.9	10.6	6.3	24.6	22.7	18.0	10.5
Superheat at cooler, °F	3.4	7.3	13.3	23.0	3.5	7.8	14.5	28.7
Tons of refrigeration	61.4	59.0	51.3	41.0	79.2	75.5	66.2	49.8
Tube surface below liquid level, %	87.5	62.5	37.5	12.5	87.5	62.5	37.5	12.5
Height of ammonia, in	26.4	20.0	13.4	6.5	26.4	20.0	13.4	6.5
Mean temp. difference	15.5	17.3	22.7	30.6	19.3	21.3	27.5	38.9
B.t.u. transfer/sq. ft./hr./deg. mean temp. difference	62.5	54.5	36.4	21.5	68.4	56.8	38.6	20.5
Brine velocity, ft./min	129	131	134	133	133	133	135	133

* The York Ice Machinery Corporation, by permission.

Table 13. Results with Short Flooded Coils*

	A	B	C	D
1¼-in. pipe, ft	500	500	500	500
Suction pressure, lb. gage	19.6	21.2	18.9	21.5
Brine temperature, °F	14.0	14.0	14.0	14.0
Brine velocity, ft./min	133	133	133	133
Tons refrigeration	7.3	9.0	12.0	15.7
B.t.u. transfer	43	68	66	120

* The York Ice Machinery Corp., by permission.

For short flooded coils, the results obtained are shown in Table 13.

Liquid Controls. In order to prevent evaporating coils from overflowing with liquid refrigerant, some form of automatic control valve is usually employed. These may be classified into four groups (see Sec. 19, pp. 1289–1290):

1. The diaphragm-operated pressure-reducing valves controlling the refrigerant feed to the evaporator according to the pressure in the evaporator.
2. Diaphragm-operated pressure-reducing valves controlling the refrigerant feed to the evaporator by means of the temperature of the refrigerant gas leaving the evaporator.
3. Thermostatically controlled shutoff valves actuated by change of temperature in room or tank and used in series with hand-adjusted expansion valves or valves under group 1 or 2.
4. Float-controlled valves which function according to the change in liquid refrigerant level in the evaporators, or used as liquid traps to pass the refrigerant from the high side to the evaporator as it is condensed.

In low-temperature refrigeration, where the liquid control must operate under small pressure differences, special thermostatic valves have been developed [Carter, *Refrig. Eng.*, **47**, 96 (1944); **50**, 39 (1945)].

Brine Piping. Brine is seldom allowed to warm more than 20°F. in performing refrigeration, and the heat absorbed per pound is much smaller than that absorbed by the evaporation of an equal weight of refrigerant. Consequently brine piping is larger than refrigerant piping. Standard-weight piping is sufficient for all ordinary purposes.

The same rule for threads applies as in any threaded pipe work. Threads should not be shouldered and should be entered three or four by hand. A 5-in. pipe ought to enter the fitting about ¾ or ⅞ in. and have two or three threads sticking out to make certain that it is not shouldered.

Red lead is a very good heavy lubricant to be used for screwing the pipe together, and it helps a little to seal joints. White lead will make a better seal but sets so hard that pipes are separated with difficulty and is little used. A little asphalt mixed with red lead makes a better joint in case of imperfections in the threads. (Private information from Torrance, Carbondale Ice Machine Co.)

Copper Piping. Copper piping is extensively used for handling refrigerants in direct-expansion systems. Joints can be sweated together, and the piping can readily be bent around obstructions. However, ammonia attacks copper, and such tubing can therefore not be used except with the Freon type of refrigerants or methyl and ethyl chlorides. Small commercial installations of 1 to 5 tons capacity are tending to go over to copper piping with these refrigerants because of the ease of installation.

Refrigeration Pipe Lines. There are two general systems of pipe-line refrigeration for distributing refrigeration for large central plants—those circulating ammonia and those circulating brine solutions as refrigerating mediums. The latter seems to be the preferred method. A brine system operated by the Merchants Refrigerating Co., New York, serving the dairy products, poultry, and produce trades, carries 400 refrigerator boxes, varying in size from 200 cu. ft. to 10,000 cu. ft. aggregating a total of 2,750,000 cu. ft. refrigerated space (Oakley, Am. Soc. Refrigerating Eng., annual meeting, 1924).

Brine is cooled to 0°F. by two horizontal multipass shell-and-tube brine coolers and is circulated through the system by means of four 12-in. mains, making a dual system on the trunk lines so that, in the event of a break in one pair of trunk lines, the other pair will carry the load. These mains are carried under the street at a depth of 4 to 7 ft. and are properly supported and insulated with three or more layers of 1-in. hair felt. The mains are full-weight wrought-iron pipe with flanged joints and ring gaskets, alternate joints being fitted with heavy cast-iron sleeves and calked with lead. Expansion joints of the corrugated type are introduced to take up the initial movement of the pipe. Great care must be exercised in construction to keep the losses at a minimum. The normal brine losses average about ½ bbl. a day, while the accidental losses over which the distributor has no control are about twice this amount. The specific gravity of the brine is maintained at about 1.23. The solution is tested several times a year and kept slightly alkaline. No trouble is experienced with deterioration of the mains, some of which have been in service over 30 years.

Centrifugal pumps are used to circulate the brine throughout the system, which during summer conditions requires about 3000 gal./min. giving a brine velocity in the mains of about 250 ft./min. The pumps are designed to give maximum efficiency over a rather wide range of head to take care of increasing business and changing load.

In St. Louis and Los Angeles liquid ammonia is distributed for refrigeration from a central compressor station, and the vapors drawn back by the compressor suction. The total length of pipe line is about 15 miles in each city.

In a brine-circulating system there are power losses which are absent in a direct-expansion system, such as the lower operating back pressure to the compressor, the cost of brine pumping, and the consequent heat delivered to the brine through pumping. However, this is offset by features of safety, simplicity, and a more uniform suction pressure to the compressor by having the ammonia confined to bring coolers under immediate control.

The general application of pipe-line refrigeration is most successful in market districts where the demand for

Table 14. Piping for General Storage

Cu. ft. of space cooled per linear foot of 1¼-in. and of 2-in. pipe. 15.67 lb. suction pressure. Temperature of ammonia expansion, 0°F. Mean temperature of brine, 10°F.

Size of room, cu. ft.	Room temp., 40°F. Tons refriger.	Direct expansion, in. 1¼	Direct expansion, in. 2	Brine, in. 1¼	Brine, in. 2	Room temp., 36°F. Tons refriger.	Direct expansion, in. 1¼	Direct expansion, in. 2	Brine, in. 1¼	Brine, in. 2
1,000	0.7	8	12	6	9	0.75	7	10	4.5	6
2,000	1.1	10	15	7	11	1.3	8	12	7	7
3,000	1.5	11	16	8	12	1.6	9	14	6	8
4,000	1.8	12	18	9	13	2.0	10	15	6.5	10
5,000	2.2	13	19	10	14	2.4	11	16	7	11
7,000	2.8	14	20	11	15	3.0	12	17	8	13
10,000	3.6	16	23	12	17	3.9	13	19	9	14
15,000	4.8	18	26	13	19	5.2	15	21	10	15
20,000	6.0	20	28	15	21	6.5	16	25	11	16
30,000	7.5	23	32	17	24	8.0	19	28	13	20
40,000	9.3	25	35	19	26	10.0	21	30	15	21
60,000	12.8	27	37	20	28	13.8	22	32	15.5	22
80,000	16.6	28	39	21	29	17.9	23	33	16	23
100,000	20.3	29	40	22	30	22.0	24	34	17	24

Size of room, cu. ft.	Room temp., 32°F. Tons refriger.	Direct expansion, in. 1¼	Direct expansion, in. 2	Brine, in. 1¼	Brine, in. 2	Room temp., 28°F. Tons refriger.	Direct expansion, in. 1¼	Direct expansion, in. 2	Brine, in. 1¼	Brine, in. 2
1,000	0.8	6	8	4	5	1.1	4	5	2	3
2,000	1.4	7	9	5	6	1.9	4.5	6	2.5	3.5
3,000	1.7	8	12	5.5	7	2.3	5	7	3	4
4,000	2.2	8.5	12.5	6	8	3.0	5.5	7.5	3.5	4.5
5,000	2.6	9	14	6.5	9	3.5	6	8	4	5
7,000	3.2	10	16	7	11	4.3	6.5	9	4.5	5.5
10,000	4.2	11	18	7.5	12	5.7	7	10	5	6
15,000	5.6	12	19	8	13	7.5	8	11	5.5	7
20,000	7.0	13	23	9	15	9.5	9	12	6	8
30,000	8.7	15	25	10	17	11.7	10	15	6.5	9
40,000	10.8	17	26	11	17.5	14.6	11	16	7	10
60,000	15.0	18	27	12	18	20.2	12	17	7.5	10.5
80,000	19.4	18.5	27.5	12.5	18.5	26.2	12.5	17.5	8	11
100,000	23.8	19	28	13	19	32.1	13	18	8.5	11.5

For other ammonia and brine temperatures multiply the number of feet of pipe found from the above table by the constants in the following table.

Suction pressure abs.	Ammonia temp., °F.	Brine temp., °F.	Room temp., °F. Direct expansion 40	Direct expansion 36	Direct expansion 32	Direct expansion 28	Brine 40	Brine 36	Brine 32	Brine 28
19.46	5	15	1.2	1.2	1.2	1.2	1.2	1.2	1.3	1.4
23.64	10	20	1.3	1.4	1.5	1.6	1.5	1.6	1.8	2.3
28.24	15	25	1.6	1.7	1.9	2.2	2.0	2.4	3.0	
32.25	20	30	2.0	2.3	2.7	3.5	3.0	4.3		

service is in confined territory. Power is the greatest item of expense, and, therefore, the cost per ton of refrigeration delivered to the system should be kept to a minimum (Fig. 29).

Pipe-line systems should be equipped with instruments to obtain data properly, and records should be kept of such items as tons of refrigeration delivered to the system, cost of power to deliver a ton, etc.

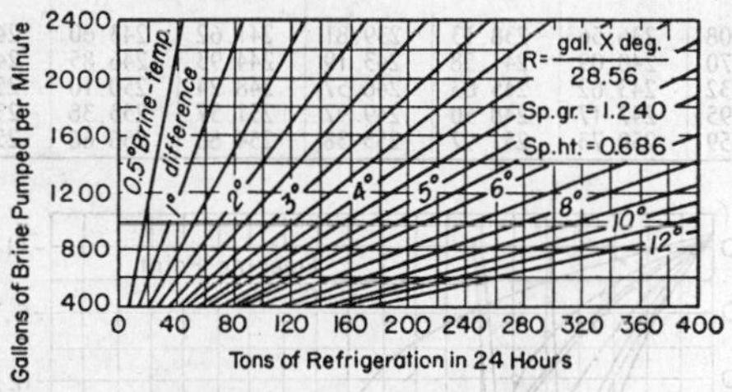

FIG. 29. Brine flow required for varying tonnages of refrigeration.

Insulation Values of Building Materials (see Sec. 6, Tables 4 to 7, pp. 457*ff*., and p. 479).

An important factor in insulation is the perfection of construction, which is subject to wide variations. Unless the joints are carefully made and airtight, moisture will get in and freeze, causing the ultimate deterioration of the insulation. This is probably the factor most responsible for the divergence in laboratory experiments and practical results. A liberal factor of safety should be allowed to ensure good operating results.

It is customary to assume an average maximum temperature during the period of peak refrigerating loads. The average temperatures in the United States may be obtained from government reports. The average temperature experienced during a 24-hr. period for conditions that are likely to prevail for a week's time should be used in making estimates. The actual choice of the thickness of insulation for any particular case will be decided by the relative costs of the insulation itself and the book value of a ton of refrigeration.

The **standard insulation of the American Warehouse Association** is: For the walls, ceilings, floors, partitions, etc., of cold-storage buildings, the insulation

Table 15. Conductivity and Density of Various Insulating Materials*

U.S. Bureau of Standards

Material	Thermal conductivity, B.t.u./(sq. ft.) (in.) (hr./°F.)	Density, lb./cu. ft.	Description of material
Air	0.175	0.08	Ideal air space
Air cell, ½ in.	0.458	8.80	Asbestos paper and air spaces
Air cell, 1 in.	0.500	8.80	Asbestos paper and air spaces
Asbestos mill board	0.830	61.0	Pressed asbestos
Asbestos wood	3.700	123.0	Asbestos and cement
Balsawood	0.350	7.5	Light and soft across grain
Calorax	0.221	4.0	Fluffy, finely divided mineral matter
Cork	0.337	5.3	Granulated ⅛ to 3/16 in.
Cork	0.330	10.0	Regranulated 1/16 to ⅛ in.
Cork board	0.279	6.9	No artificial binder—low density
Cork board	0.308	11.3	No artificial binder—high density
Cotton wool	0.292	......	Loosely packed
Fibrofelt	0.329	11.3	Felted vegetable fibers
Fire-felt wool	0.625	43.0	Asbestos sheet coated with cement
Fire-felt sheet	0.583	26.0	Soft, flexible asbestos sheet
Flaxlinum	0.329	11.3	Felted vegetable fibers
Hair felt	0.246	17.0	
Hard maple wood	1.125	44.0	Across grain
Infusorial earth	0.583	43.0	Natural blocks
Insulite	0.296	11.9	Pressed wool pulp—rigid
Kapok	0.238	0.88	Vegetable fiber—loosely packed
Keystone hair	0.271	19.0	Hair felt combined with building paper
Linofelt	0.300	11.3	Vegetable fiber combined with paper
Lithboard	0.379	12.5	Mineral wool and vegetable fibers
Mineral wool	0.275	12.5	Medium packed
Mineral wool	0.288	18.0	Felted in blocks
Oak wood	1.000	38.0	Across grain
Planer shavings	0.417	8.8	Various
Pulp board	0.458	......	Stiff pasteboard
Pure wool	0.263	5.0	
Rock cork	0.346	21.0	Mineral wool and binder—rigid
Slag wool	0.750	15.0	
Tar roofing	0.707	55.0	
Virginia pine wood	0.958	34.0	Across grain
White pine wood	0.791	32.0	Across grain
Wool felt	0.363	21.0	Flexible paper stock

* See Physical and Chemical Data, Sec. 3, pp. 194*ff*.

consists of corkboard of good quality and of medium density with both outer surfaces sealed by either dipping or coating with an asphaltic mastic, or by applying a waterproof Portland cement plaster as a finish. An additional ½ in. of waterproof Portland cement is used

Table 16. Heat Content of Calcium Chloride Brines, B.t.u./Lb.

From value at 32°F., where it equals 200 B.t.u. minus the heat of solution

Temp., °F.	Specific gravity, 60°/60°F.												
	1.08	1.10	1.12	1.14	1.16	1.18	1.20	1.22	1.24	1.26	1.28	1.30	1.303
	Freezing point, °F.												
	22.5	19.1	15.1	10.6	5.4	−0.4	−6.4	−13.2	−20.7	−29.6	−41.0	−56.8	−59.8
	Lb. $CaCl_2$/100 lb. brine												
	9.2	11.4	13.6	15.7	17.8	19.8	21.8	23.7	25.6	27.5	29.4	31.2	31.4
−85	33.63	40.58	47.37	54.02	60.54	66.89	73.23	79.39	85.44	91.31	97.13	102.83	103.55
−80	35.47	42.40	49.17	55.80	62.30	68.63	74.95	81.09	87.12	92.97	98.77	104.46	105.18
−75	37.34	44.24	50.99	57.60	64.08	70.39	76.69	82.81	88.82	94.65	100.43	106.10	106.82
−70	39.23	46.11	52.83	59.42	65.87	72.16	78.44	84.54	90.54	96.35	102.11	107.76	108.47
−65	41.14	48.00	54.70	61.26	67.69	73.96	80.22	86.30	92.27	98.06	103.80	109.44	110.14
−60	43.07	49.91	56.59	63.13	69.54	75.78	82.01	88.07	94.02	99.79	105.52	111.13	111.83
−59.8°	43.15	49.98	56.65	63.19	69.61	75.85	82.08	88.13	94.08	99.85	105.58	111.20	111.90
−59.8°	60.60	71.56	82.30	92.82	103.13	113.19	123.16	132.88	142.43	151.73	160.93	169.94	171.10
−55	63.34	74.49	85.40	96.08	106.57	116.79	126.93	136.81	146.52	155.97	165.30	173.73	174.10
−50	66.25	77.60	88.67	99.56	110.23	120.63	130.94	140.99	150.87	160.49	169.97	176.86	177.23
−45	69.31	80.89	92.20	103.28	114.18	124.79	135.30	145.56	155.64	165.45	175.13	179.98	180.36
−40	72.57	84.40	95.96	107.32	118.44	129.30	140.04	150.53	160.82	170.86	180.24	183.12	183.50
−35	76.01	88.15	100.02	111.66	123.08	134.23	145.25	155.97	166.58	176.89	183.41	186.27	186.63
−30	79.71	92.21	104.43	116.39	128.15	139.62	150.96	162.04	172.90	183.49	186.59	189.41	189.77
−25	83.72	96.62	109.24	121.61	133.77	145.62	157.34	168.79	180.01	187.01	189.77	192.55	192.91
−20	88.07	101.50	114.63	127.49	140.13	152.45	164.64	176.55	187.57	190.24	192.95	195.71	196.05
−15	93.01	107.06	120.76	134.21	147.41	160.30	173.04	185.48	190.87	193.49	196.15	198.86	199.20
−10	98.64	113.44	127.91	142.11	156.04	169.63	183.06	191.60	194.19	196.73	199.35	202.01	202.35
− 5	105.07	120.88	136.36	151.53	166.42	180.94	192.45	195.00	197.51	199.99	202.55	205.18	205.51
0	112.60	129.62	146.29	162.62	178.67	193.48	195.95	198.40	200.84	203.26	205.77	208.34	208.67
5	121.63	140.24	158.44	176.31	193.87	197.08	199.45	201.82	204.18	206.54	208.99	211.52	211.84
10	133.07	153.84	174.13	194.07	198.38	200.69	202.97	205.24	207.53	209.83	212.21	214.70	215.01
15	149.39	173.51	197.04	199.88	202.11	204.32	206.49	208.68	210.89	213.12	215.45	217.88	218.18
20	176.58	199.43	201.63	203.75	205.86	207.95	210.03	212.13	214.27	216.42	218.70	221.07	221.37
25	201.49	203.58	205.63	207.62	209.62	211.60	213.58	215.59	217.75	219.74	221.94	224.27	224.56
30	205.79	207.74	209.65	211.51	213.38	215.26	217.14	219.06	221.04	223.06	225.20	227.47	227.75
35	210.12	211.90	213.67	215.41	217.15	218.93	220.70	222.54	224.44	226.39	228.46	230.69	230.96
40	214.44	216.08	217.70	219.31	220.94	222.60	224.28	226.03	227.75	229.73	231.74	233.90	234.17
45	218.78	220.26	221.74	223.22	224.74	226.29	227.87	229.53	231.27	233.08	235.02	237.13	237.39
50	223.11	224.44	225.79	227.14	228.54	229.98	231.47	233.04	234.69	236.44	238.31	240.36	240.62
55	227.46	228.64	229.85	231.07	232.35	233.69	235.08	236.56	238.13	239.81	241.62	243.60	243.85
60	231.82	232.86	233.91	235.00	236.17	237.40	238.70	240.09	241.58	243.19	244.93	246.85	247.09
65	236.18	237.05	237.98	238.95	240.00	241.12	242.32	243.62	245.03	246.57	248.24	250.10	250.35
70	240.55	241.27	242.07	242.90	243.83	244.85	245.95	247.17	248.50	249.97	251.57	253.38	253.61
75	244.92	245.50	246.17	246.86	247.67	248.58	249.59	250.73	251.97	253.38	254.86	256.66	256.88

between the layers of the cork to seal the voids between the cork granules against atmospheric or other moisture. The standard thickness for temperatures down to 32°F. is two layers of 2-in. corkboard; to this is added 1 in. for each 15°F. below 32°F. Piping, fittings, etc., should be covered with molded cork covering, which should have its outer and inner surfaces sealed with a rubber or asphaltic mastic. Standard brine-pipe covering for temperatures down to 0°F. varies from 2 to 3 in. thickness with the diameter of the pipe. Table 14 (taken from Marks, "Mechanical Engineers' Handbook," 4th ed., McGraw-Hill, New York, 1941, p. 2170) shows the cubic feet of space cooled per linear foot of pipe in general storage. Table 15 gives pertinent data on the thermal conductivities and densities of various insulating materials.

BRINES

A *brine* may be defined as a liquid of low freezing point used in the transmission of refrigeration without change of state (see Sec. 3, Tables 52 to 121, pp. 177–183). The brines commonly employed in refrigeration are calcium and sodium chlorides. The latter is cheaper but cannot be used below its eutectic point of −6.03°F. Calcium chloride of commercial grade does not operate very satisfactorily below −40°F. The specific gravity, freezing point, composition by weight, and heat content per pound of solution are given in Tables 16 and 17, which have been compiled by Jessup [*Refrig. Eng.*, **22**, 168–169 (1931)].

Although these brines have the great advantage of low cost, they have the disadvantage of being extremely corrosive, and the calcium brines have the disadvantage of throwing down insoluble precipitates with untreated

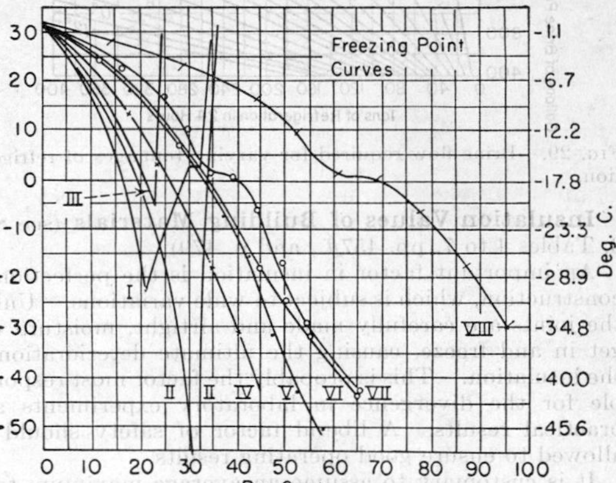

FIG. 30. Compositions and freezing points of solutions.
I. $MgCl_2$, per cent by weight
II. $CaCl_2$, per cent by weight
III. NaCl, per cent by weight
IV. Methanol, per cent by volume
V. Ethylene glycol, per cent by volume
VI. Denatured alcohol, per cent by volume
VII. Glycerol U.S.P. 96.5%, per cent by volume
VIII. Glycerol U.S.P. 60%, per cent by volume

Table 17. Heat Content of Sodium Chloride Brines, B.t.u./Lb.

Temp., °F.	Specific gravity, 60°/60°F.															
	1.03	1.04	1.05	1.06	1.07	1.08	1.09	1.10	1.11	1.12	1.13	1.14	1.15	1.16	1.17	1.178
	Freezing point, °F.															
	27.6	26.0	24.4	22.7	20.8	19.0	17.0	14.9	12.7	10.4	8.0	5.4	2.7	−0.3	−3.4	−6.0
	%, NaCl															
	4.14	5.51	6.86	8.21	9.56	10.88	12.20	13.51	14.81	16.09	17.36	18.63	19.87	21.11	22.33	23.30
−6.03	62.26	70.26	78.21	86.12	93.97	101.77	109.50	117.16	124.76	132.29	139.75	147.11	154.42	161.65	168.81	174.45
−4	64.12	72.42	80.67	88.88	97.03	105.12	113.14	121.09	128.97	136.79	144.53	152.17	159.75	167.26	174.69	176.03
−2	66.03	74.65	83.21	91.74	100.20	108.60	116.93	125.19	133.37	141.48	149.52	157.45	165.33	173.13	177.49	177.59
0	68.07	77.05	85.98	94.86	103.67	112.43	121.11	129.71	138.24	146.70	155.07	163.34	171.55	178.98	179.07	179.16
2	70.27	79.66	88.99	98.28	107.50	116.67	125.73	134.74	143.65	152.49	161.25	169.90	178.48	180.56	180.64	180.73
4	72.64	82.50	92.31	102.06	111.74	121.35	130.86	140.33	149.70	158.98	168.18	177.25	182.03	182.15	182.22	182.29
6	75.22	85.61	95.95	106.22	116.43	126.57	136.61	146.57	156.44	166.23	175.90	183.47	183.64	183.74	183.80	183.86
8	78.08	89.09	100.04	110.95	121.75	132.50	143.15	153.73	164.19	174.56	184.83	185.09	185.24	185.33	185.38	185.43
10	81.32	93.10	104.81	116.45	128.00	139.48	150.88	162.16	173.33	184.41	186.56	186.71	186.84	186.92	186.96	187.00
12	85.04	97.72	110.32	122.90	135.35	147.70	159.98	172.13	184.21	188.00	188.20	188.33	188.45	188.51	188.54	188.58
14	89.36	103.14	116.84	130.47	144.00	157.44	170.76	183.96	189.44	189.66	189.84	189.96	190.06	190.11	190.12	190.15
16	94.46	109.60	124.65	139.62	154.48	169.24	183.87	190.85	191.11	191.31	191.48	191.58	191.67	191.71	191.71	191.72
18	100.74	117.62	134.40	151.09	167.68	184.13	192.26	192.54	192.78	192.97	193.12	193.21	193.28	193.30	193.29	193.30
20	108.77	127.97	147.06	166.04	184.90	193.64	193.98	194.24	194.46	194.63	194.76	194.83	194.89	194.90	194.88	194.88
22	119.54	141.96	164.26	186.43	195.03	195.39	195.70	195.94	196.14	196.29	196.40	196.46	196.50	196.50	196.46	196.45
24	135.41	162.72	189.89	196.38	196.80	197.14	197.42	197.63	197.81	197.94	198.04	198.09	198.12	198.10	198.05	198.03
26	160.83	196.19	197.70	198.18	198.57	198.88	199.14	199.33	199.49	199.61	199.69	199.72	199.73	199.70	199.64	199.61
28	198.38	199.00	199.53	199.98	200.34	200.63	200.86	201.03	201.17	201.27	201.33	201.34	201.34	201.30	201.23	201.19
30	200.27	200.85	201.35	201.77	202.11	202.38	202.59	202.74	202.86	202.94	202.98	202.98	202.96	202.91	202.82	202.77
32	202.16	202.71	203.18	203.57	203.88	204.12	204.31	204.44	204.54	204.60	204.63	204.61	204.58	204.51	204.41	204.35
34	204.05	204.57	205.01	205.37	205.65	205.87	206.04	206.14	206.22	206.26	206.28	206.24	206.20	206.11	206.00	205.93
36	205.94	206.43	206.84	207.17	207.43	207.62	207.76	207.85	207.91	207.93	207.93	207.88	207.82	207.72	207.60	207.51
38	207.84	208.29	208.67	208.97	209.20	209.37	209.49	209.56	209.60	209.60	209.58	209.51	209.44	209.33	209.19	209.10
40	209.73	210.15	210.50	210.78	210.98	211.12	211.22	211.26	211.28	211.27	211.23	211.15	211.06	210.94	210.79	210.68
42	211.62	212.01	212.33	212.57	212.75	212.87	212.95	212.97	212.97	212.94	212.89	212.79	212.68	212.54	212.38	212.27
44	213.52	213.87	214.16	214.38	214.53	214.62	214.68	214.68	214.66	214.61	214.54	214.43	214.31	214.16	213.98	213.86
46	215.41	215.73	215.99	216.18	216.31	216.38	216.41	216.39	216.35	216.29	216.20	216.07	215.93	215.77	215.58	215.44
48	217.30	217.59	217.82	217.99	218.09	218.14	218.15	218.11	218.05	217.96	217.85	217.71	217.56	217.38	217.17	217.03
50	219.20	219.46	219.66	219.80	219.87	219.89	219.88	219.82	219.74	219.63	219.51	219.34	219.18	218.99	218.77	218.62
52	221.09	221.32	221.49	221.60	221.65	221.65	221.62	221.53	221.43	221.31	221.17	220.99	220.81	220.60	220.37	220.20
54	222.99	223.19	223.33	223.41	223.44	223.41	223.35	223.25	223.13	222.99	222.83	222.63	222.44	222.22	221.97	221.79
56	224.88	225.04	225.16	225.22	225.22	225.17	225.09	224.96	224.83	224.67	224.49	224.27	224.06	223.82	223.57	223.38
58	226.78	226.92	227.00	227.03	227.00	226.93	226.83	226.68	226.52	226.34	226.15	225.92	225.69	225.44	225.17	224.97
60	228.68	228.79	228.84	228.84	228.79	228.69	228.57	228.40	228.22	228.02	227.81	227.56	227.32	227.05	226.77	226.56
62	230.80	230.65	230.68	230.65	230.57	230.46	230.31	230.12	229.92	229.70	229.47	229.20	228.95	228.67	228.37	228.15
64	232.47	232.51	232.51	232.46	232.36	232.22	232.05	231.84	231.62	231.38	231.13	230.85	230.58	230.29	229.98	229.74
66	234.37	234.38	234.35	234.27	234.15	233.98	233.79	233.56	233.32	233.07	232.80	232.50	232.21	231.90	231.58	231.33
68	236.27	236.25	236.19	236.09	235.94	235.74	235.53	235.28	235.02	234.75	234.46	234.14	233.84	233.52	233.18	232.92
70	238.17	238.12	238.03	237.90	237.72	237.51	237.27	237.00	236.72	236.43	236.13	235.80	235.47	235.13	234.79	234.51
72	240.07	239.99	239.87	239.71	239.51	239.27	293.01	238.72	238.42	238.11	237.79	237.44	237.10	236.75	236.39	236.10
74	241.97	241.86	241.72	241.53	241.30	241.04	240.76	240.44	240.12	239.79	239.45	239.09	238.74	238.37	237.99	237.70
76	243.87	243.74	243.06	243.34	243.09	242.81	242.50	242.16	241.82	241.47	241.12	240.74	240.37	239.99	239.60	239.29
78	245.77	245.61	245.40	245.15	244.88	244.57	244.25	243.89	243.53	243.16	242.79	242.39	242.00	240.61	241.21	240.88
80	247.67	247.47	247.24	246.98	246.68	246.34	245.99	245.61	245.23	244.84	244.45	244.04	243.64	243.23	242.81	242.47

waters. Corrosion in closed systems can be largely overcome by the addition of sodium dichromate, $Na_2Cr_2O_7.2H_2O$, 100 lb. and 200 lb. being required per 1000 cu. ft. of calcium and sodium brines, respectively. Enough caustic is added to make the brines slightly alkaline. In open systems using sodium brines, disodium phosphate, $Na_2HPO_4.12H_2O$, may be used at the rate of 100 lb./1000 cu. ft., but the solutions should be colorless to phenolphthalein. Open calcium-brine systems may be protected by adding zinc dust at the rate of 60 lb./1000 cu. ft., a little at a time.

Magnesium chloride brines are used to some extent, but their eutectic temperature is not low enough to give them much advantage over sodium chloride.

Other materials that may be used are methanol, denatured alcohol, ethylene glycol, and glycerin. These have been much used for protection of automobile radiators and are no more corrosive than water. Their main disadvantage is cost, but they have been used in place of brines for household refrigerators. Methanol and denatured alcohol solutions are not inflammable at refrigeration temperatures, but at higher temperatures they may give off inflammable vapors. A comparison of freezing points is shown in Fig. 30. The eutectic for the alcohols lies well below that of calcium chloride. The alcohols will give no difficulty from crystallization; but, if the concentrations of the sodium and calcium brines are greater than 23 or 33 per cent, respectively, solid salt may deposit out. When solutions begin to freeze, their fluidity persists until some lower temperature is reached. The temperature at which the slush of ice crystals and liquid ceases to flow through a ¼-in. pipe has arbitrarily been defined as the **flow point** [Olsen, Brunjes, and Olsen, *Ind. Eng. Chem.*, **22**, 1316 (1930)]. A comparison of freezing points and flow points is given in Table 18.

Nessler's Solution. To determine ammonia leakage into the brine Nessler's solution is used, which is prepared as follows: Dissolve 17 g. mercuric chloride in about 300 cc. distilled water; dissolve 35 g. potassium iodide in 100 cc. water. Add the former solution to the latter, with constant stirring, until a slight permanent red precipitate is formed. Next dissolve 120 g. potassium hydrate in about 200 cc. water; allow the solution to cool and then add to the previous solution and make up with water to 1 l. Add mercuric chloride solution until a permanent precipitate again forms. Allow to stand till settled and decant off the clear solution for use. Store in glass-stoppered blue bottles in a dark place.

Table 18. Freezing Points and Flow Points of Ethylene Glycol, Methanol, Denatured Alcohol, and Glycerol*

Composition		Freezing point		Flow point	
Weight, %	Volume, %	°C.	°F.	°C.	°F.
		Ethylene glycol			
10	9.10	− 3.3	26.1	− 4.3	24.3
20	18.39	− 7.8	18.0	− 8.6	16.6
30	27.86	−13.5	7.7	−14.0	+ 6.8
35	32.67	−17.1	1.2	−19.1	− 2.4
40	37.53	−22.1	− 7.8	−24.5	−12.1
45	42.44	−26.7	−16	−34.5	−30.1
50	47.41	−35.4	−31.7	−43.9	−47.4
55	52.42	−41.7	−43.0	−47	−52.6
60	57.48	−46	−50.8	−53	−63.4
		Synthetic methanol			
10	12.24	− 6.8	19.8	− 8.5	16.7
15	18.13	−10.5	13.1	−13.9	7.0
20	23.89	−16.2	2.8	−18.4	− 1.1
25	29.50	−21.2	− 6.2	−24	−11.2
30	34.98	−28	−18.4	−34	−19.2
35	40.33	−35.2	−31.4	−40	−40
40	45.56	−42	−43.6	−49.5	−57.1
		Denatured alcohol			
10	11.9	− 4.3	24.26	− 6	21.2
15	17.75	− 6.9	19.58	− 8.5	16.7
20	23.37	−10	14	−14	6.8
25	28.25	−14	6.8	−16	3.2
30	34.33	−18.1	− 0.2	−21.4	− 6.5
35	39.65	−23.5	−10.3	−25.5	−13.9
40	44.85	−28	−18.4	−34.5	−30.1
45	49.95	−31.6	−24.9	−39.9	−39.9
50	54.95	−35.5	−31.9	−42.4	−44.3
60	64.66	−42.0	−43.6	−46.4	−52.5

Glycerol

96.5%	100%	96.5%	60%	Freezing point °C.	Freezing point °F.	Flow point °C.	Flow point °F.
10	9.65	8.14	14.23	− 2.2	28.0	− 3	26.6
20	19.30	16.62	29.09	− 5.3	22.46	− 9.0	15.8
30	28.95	25.46	44.65	− 8.8	16.3	−15.8	3.5
35	33.78	30.03	52.71	−12.4	9.5	−18.9	− 2.0
40	38.60	34.70	60.96	−17.2	1.04	−19.5	− 3.1
45	43.43	39.47	69.40	−18.0	0.4	−20.5	− 4.9
50	48.25	44.36	78.03	−21.4	− 6.5	−28	−18.4
55	53.08	49.35	86.91	−27.5	−17.5	−35.9	−32.6
60	57.90	54.46	95.98	−34.0	−29.2	−41.9	−43.6
70	67.55	65.03	.....	−41.5	−42.7		

* From *Ind. Eng. Chem.*, **22**, 1316 (1930).

Table 19. Specific Heats (Extrapolated) of Non-freezing Solutions

Weight, %	+23°F. −5°C.	+14°F. −10°C.	+5°F. −15°C.	−4°F. −20°C.	−13°F. −25°C.	−22°F. −30°C.	−31°F. −35°C.	−40°F. −40°C.
	Methanol [Bose, *Z. phys. Chem.*, **58**, 585 (1907)]							
10	1.02							
20	0.98	0.98	0.98					
30	0.92	.92	.92	0.91	0.91			
40	0.87	.87	.86	.86	.85	0.85	0.84	0.83
50	0.81	.80	.79	.79	.78	.77	.76	.76
60	0.76	.76	.75	.74	.73	.73	.72	.71
70	0.71	.70	.69	.69	.68	.67	.66	.66
80	0.66	.66	.65	.64	.64	.63	.62	.62
90	0.62	.61	.60	.60	.59	.58	.58	.57
100	0.57	.56	.56	.55	.55	.54	.54	.53
	Ethanol [Bose, *Z. phys. Chem.*, **58**, 585 (1907)]							
10	1.04							
20	1.04	1.04						
30	0.99	0.99	0.99	0.99				
40	0.93	0.92	.92	.91	0.91	0.90		
50	0.86	0.86	.85	.84	.84	.83	0.82	
60	0.80	0.79	.78	.77	.76	.75	.75	0.74
70	0.73	0.72	.71	.70	.69	.68	.67	.66
80	0.66	0.64	.63	.62	.61	.60	.59	.57
90	0.59	0.58	.57	.56	.54	.53	.52	.51
100	0.54	0.52	.51	.50	.49	.48	.47	.46
	Ethylene Glycol (*Tech. Papers*, Carbide and Carbon Chemicals Corp.)							
10	1.00							
20	0.94	0.94						
30	0.87	.87	0.86	0.86				
40	0.84	.84	.83	.83	0.82	0.82		
50	0.78	.78	.77	.76	.75	.75	0.74	0.74
60	0.73	.72	.72	.71	.71	.70	.70	.69
70	0.70	.70	.69	.69	.68	.68	.67	.67
80	0.68	.68	.67	.67	.66	.66	.65	.65
90	0.65	.65	.64	.64	.63			
100	0.52	.50						
	Glycerin ("International Critical Tables" and Landolt-Börnstein Tables, 5th ed.)							
10	0.96							
20	.93	0.93						
30	.89	.89	0.89					
40	.85	.85	.85	0.85				
50	.79	.79	.79	.79	0.79			
60	.76	.77	.77	.77	.77	0.77		
70	.71	.71	.71	.72	.72	.72	0.72	0.72

Table 20. Viscosities (Extrapolated) of Refrigerating Solutions (Centipoises)

Weight, %	0°C. 32°F.	−5°C. 23°F.	−10°C. 14°F.	−15°C. 5°F.	−20°C. −4°F.	−25°C. −13°F.	−30°C. −22°F.	−35°C. −31°F.	−40°C. −40°F.	Freezing point
	Sodium chloride [Jessup, *Refrig. Eng.*, **12**, 171 (1925)]									
10.5	2.1	2.4			....	....	....	....	....	+20°F.
16.8	2.4	2.8	3.2	....	....	....	....	....	....	+10°F.
21.0	2.7	3.1	3.6	4.2	....	....	....	....	....	0°F.
	Calcium chloride [Jessup, *Refrig. Eng.*, **12**, 171 (1925)]									
11.0	2.1	2.4			....	....	....	....	....	+20°F.
16.0	2.6	2.9	3.3		....	....	....	....	....	+10°F.
20.0	3.1	3.5	4.0	4.5	....	....	....	....	....	0°F.
22.8	3.6	4.1	4.6	5.2	5.9	....	....	....	....	−10°F.
25.2	4.0	4.6	5.1	5.8	6.6	7.4	....	....	....	−20°F.
27.2	4.6	5.2	5.8	6.6	7.5	8.4	9.6	....	....	−30°F.
29.0	5.1	5.8	6.6	7.4	8.4	9.5	10.8	12.3	14.1	−40°F.
	Methanol ("International Critical Tables" and Landolt-Börnstein Tables, 5th ed.)									
10.0	2.6	3.2	....		....	....	....	....	....	+20°F.
16.8	3.0	3.7	4.5		....	....	....	....	....	+10°F.
22.0	3.4	4.2	5.0	6.1	....	....	....	....	....	0°F.
26.2	3.6	4.4	5.3	6.5	7.9	....	....	....	....	−10°F.
30.4	3.7	4.5	5.5	6.8	8.3	10.3	....	....	....	−20°F.
34.2	3.6	4.4	5.3	6.5	7.9	9.7	12.0	....	....	−30°F.
38.4	3.4	4.2	5.0	6.1	7.3	8.9	11.0	13.3	16.5	−40°F.
	Ethanol ("International Critical Tables" and Landolt-Börnstein Tables, 5th ed.)									
15.0	4.1	5.2			....	....	....	....	....	+20°F.
22.8	5.7	7.4	9.5		....	....	....	....	....	+10°F.
29.0	6.9	9.0	11.8	15.5	....	....	....	....	....	0°F.
35.0	7.2	9.5	12.4	16.4	22.0	....	....	....	....	−10°F.
41.2	7.0	8.9	11.6	15.0	19.7	26.0	....	....	....	−20°F.
48.8	6.7	8.4	10.8	13.8	17.7	23.0	30.0	....	....	−30°F.
56.8	6.1	7.6	9.5	12.1	15.4	19.8	25.5	32.5	42.5	−40°F.
	Glycerol [Green and Parke, *J. Soc. Chem. Ind.*, **58**, 319 (1939)]									
30	....	6.5	....	....	....	....	....	....	....	+15°F.
40	....	10.3	14.4	....	....	....	....	....	....	+ 4.3°F.
50	....	18.8	24.4	....	48.1	....	....	....	....	− 9.4°F.
60	....	41.6	59.1	....	108.0		244.0	....	....	−30.5°F.

In calcium chloride brine, Nessler's solution will form a yellow precipitate, but if no ammonia is present the precipitate will be almost white. In water or brine the precipitate will be yellow if there is but a trace of ammonia present, and a reddish brown if there is considerable ammonia in the sample.

Pressure Drop and Power Required in Brine Circulation. The pressure drop in brine pipes is determined by the rate of flow, the internal diameter and length of the pipe, and the viscosity of the brine. The method for finding the pressure drop with water has already been discussed in Sec. 5, pp. 377*ff.* and FIG. 21, p. 379. The pressure drop for brines may be found by multiplying the values for water by the ratio of the kinematic viscosities of brine and water. This factor f is given by the formula

$$f = \frac{\text{viscosity of brine}}{\text{viscosity of water}} \times \frac{\text{density of water}}{\text{density of brine}}$$

The viscosities of most solutions have not been experimentally determined at low temperatures. However, the curve obtained by plotting the logarithm of the absolute viscosity against the logarithm of the absolute temperature gives a straight line [Genereaux, *Ind. Eng. Chem.*, **22**, 1382 (1930)]. Extrapolation of this plot has given the viscosities in Table 20. Thermal conductivities of brines are given in Table 21.

SMALL-SCALE REFRIGERATION

The construction of fractional-tonnage refrigeration machines equipped with condenser and cooling system has become an important industry. They fall normally into two classes:

1. **Commercial units** from ¼ to 5 tons for use in retail stores, butcher shops, clubs, etc. 2. **Household units,** with ice-melting capacities of from 50 to 500 lb./day ($\frac{1}{40}$ to ¼ ton).

The commercial units are nearly all compression systems, usually operated with electric power and using a water-cooled condenser. Intermittent-absorption systems of this size were popular a few years ago, but a number of disastrous explosions, mainly due to inexperienced operation, have rendered them practically obsolete.

Table 21. Thermal Conductivities of Refrigerating Brines

Brines	Weight, %	Temp., °C.	Conductivity,‡ cal./cm./sec./°C.
*NaCl	12.5	32	0.001403
	25.0	32	.001141
*$CaCl_2$	15.0	32	.001383
	30.0	32	.001315
*$MgCl_2$	11.0	32	.001376
	14.5	32	.001329
	22.0	32	.001290
	29.0	32	.001238
†CH_3OH	0	19	.00141
	25	19	.00107
	50	19	.00078
	75	19	.00061
	100	19	.00050
†C_2H_5OH	0	11	.00149
	25	10	.00104
	50	11	.00079
	75	12	.00059
	100	10	.00047
†Glycerin	0	20	.00140
	25	20	.00119
	50	20	.00101
	75	20	.00081
	100	20	.00070
*Ethylene glycol	100	0	.00064

* Landolt-Börnstein Tables, 5th ed.
† Lees, *Phil. Trans. Roy. Sos.*, (A) **191**, 399 (1898).
‡ For conversion factors see pp. 41 and 456.

Household machines must be automatic in operation, which, with the compression system, necessitates electric drive. In these systems the condenser is usually air cooled, by either a fan or a combined condenser radiator on top of the cabinet. The refrigerants commonly used are sulfur dioxide, methyl chloride, ethyl chloride, and Freon (F-12). Ammonia, however, is seldom used in machines with an air condenser. Among the household refrigerators, there is only one important unit employing the absorption system, namely, the Electrolux (*cf.* Continuous Absorption Systems, p. 1684). This unit is

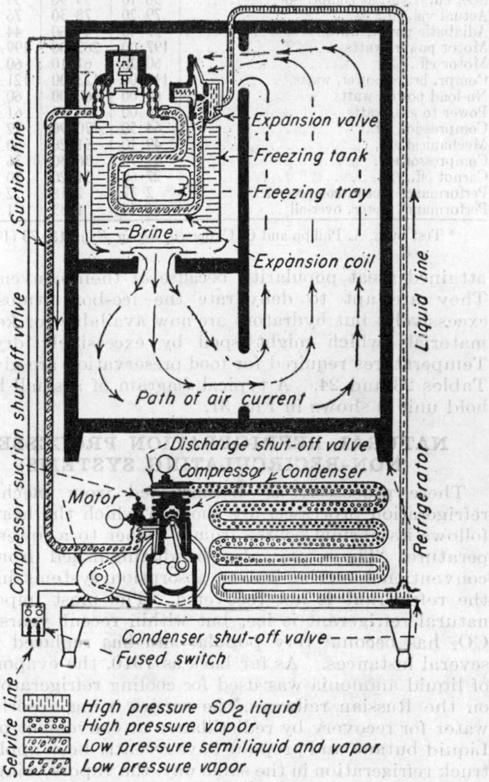

FIG. 31. Household refrigeration unit (Kelvinator, dry system).

noiseless, since it has no moving mechanical parts, and gives no interference with radio reception, since it has no motor.

The automatic control of small machines is effected by a thermostat which turns off the power at any desired temperature and by an automatic pressure-reducing valve for maintaining a constant suction pressure. In addition to the automatic control many commercial units are equipped with brine tanks for storing up refrigeration. Alcohol may be used in place of brine to provide a non-freezing solution and has the advantage over brine of being non-corrosive.

In designing household machines it is imperative to prevent all leakage of refrigerant. This would not only be dangerous to health but would seriously interfere with the operation of the machine.

Small machines are usually less efficient than large ones (see Table 22), but their operating costs are not excessive (1 to 2 kw.-hr./day for compression units or 50 cu. ft. gas/day for absorption units) and they have

Table 22. Tests on SO_2 Household Machine*

Data and results at constant room temperature, suction pressure of 15.22 lb.

	Number of test											
	1	2	3	4	5	6	7	8	9	10	11	12
Discharge pressure, lb. abs	49.80	62.30	62.80	69.80	77.30	84.80	92.80	104.80	114.80	124.30	133.80	139.30
Suction pressure, lb. abs	15.22	15.22	15.22	15.22	15.22	15.22	15.22	15.22	15.22	15.22	15.22	15.22
SO_2, lb./hr	6.34	6.16	6.03	5.92	5.86	6.61	5.28	4.98	4.70	4.38	3.95	3.91
Sat. temp. and pressure, i″	184.20	184.20	184.20	184.20	184.20	184.20	184.20	184.20	184.20	184.20	184.20	184.20
Exp. valve liq. temp., °F	60.50	67.50	71.80	73.80	75.80	74.80	75.80	74.80	75.80	77.70	76.70	74.50
Capacity ice, lb./24 hr	159.20	152.50	148.00	144.30	142.10	136.20	128.10	121.00	114.00	106.00	95.80	95.30
Compressor, r.p.m	339.00	341.00	342.00	340.00	339.00	340.00	338.00	338.00	337.00	337.00	338.00	337.00
Suction temp., compressor, °F	61.50	71.30	73.80	77.30	78.80	78.80	83.70	83.70	85.50	88.00	91.00	87.40
Discharge temp., °F	127.50	146.80	149.70	157.00	162.00	163.20	172.50	173.00	174.10	175.50	182.00	184.50
Sat. SO_2, cu. ft./hr	31.90	31.00	30.40	29.80	29.50	28.30	26.60	25.10	23.70	22.10	19.90	19.70
Piston displ., cu. ft./hr	45.60	45.90	46.00	45.70	45.60	45.70	45.50	45.50	45.40	45.40	45.40	45.40
Apparent vol. eff., %	70.00	67.60	66.10	65.20	64.60	61.80	58.50	55.20	52.20	48.60	43.70	43.40
SO_2, cu. ft./hr. at compressor	36.10	35.90	35.30	34.90	34.70	33.20	31.60	29.80	28.20	26.40	24.00	23.60
Actual vol. eff., %	79.20	78.30	76.70	76.30	76.10	72.80	69.50	65.50	62.10	58.20	52.80	52.00
Adiabatic power, watts	37.40	44.60	44.00	47.00	50.10	50.70	51.00	52.10	51.70	50.60	47.90	48.50
Motor power, watts	197.00	202.00	199.00	203.00	204.00	208.00	213.00	220.00	224.00	232.00	236.00	233.00
Motor eff., %	60.20	61.10	60.80	61.20	61.40	62.00	63.00	64.00	64.50	65.70	66.00	65.80
Compr. brake power, watts	118.00	123.00	121.00	124.00	125.00	129.00	134.00	141.00	144.00	152.00	156.00	153.00
No-load power, watts	60.00	60.00	60.00	60.00	60.00	60.00	60.00	60.00	60.00	60.00	60.00	60.00
Power to gas, watts	58.00	63.00	61.00	64.00	65.00	69.00	74.00	81.00	84.00	92.00	96.00	93.00
Compression eff., %	64.50	70.90	72.20	73.40	77.10	73.50	68.90	64.40	61.60	55.20	49.90	52.20
Mechanical eff., %	49.10	51.20	50.40	51.60	52.00	53.50	55.20	57.40	58.20	60.50	61.50	60.80
Compressor eff., %	31.70	36.30	36.40	37.90	40.10	39.30	38.10	37.00	35.90	33.30	30.70	31.70
Carnot eff., %	27.60	31.20	30.90	31.80	33.60	33.40	32.10	31.10	30.00	28.10	25.80	26.90
Performance factor, compressor	2.37	2.18	2.15	2.04	2.00	1.86	1.68	1.51	1.39	1.22	1.08	1.09
Performance factor, over-all	1.42	1.33	1.30	1.25	1.22	1.15	1.06	0.97	0.89	0.80	0.71	0.72

* Test by L. A. Philipp and C. C. Spreen, *Refrig. Eng.*, **13**, 75 (1927).

attained great popularity because of their convenience. They are apt to dehydrate the ice-box atmosphere excessively, but hydrators are now available for keeping materials which might spoil by excessively dry air. Temperatures required for food preservation are given in Tables 23 and 24. A typical diagram of a small household unit is shown in Fig. 31.

NATURAL REFRIGERATION PROCESSES (NON-RECIRCULATING SYSTEMS)

These processes, as distinguished from mechanical refrigeration processes, are those in which the heat flow follows its normal course from a higher to a lower temperature. They may also be distinguished from the conventional mechanical or absorption systems in that the refrigerant is not recovered. The most important natural refrigerant is ice, but within recent years solid CO_2 has become very popular and has replaced ice in several instances. As far back as 1910, the evaporation of liquid ammonia was used for cooling refrigerator cars on the Russian railways, the vapors being absorbed in water for recovery by redistillation at convenient points. Liquid butane and propane have recently been applied to truck refrigeration in the same way, the vapors being used as fuel in the motor [Schlumbohm, *Refrig. Eng.*, **42**, 14 (1941)]. Chemical refrigeration methods in which a solid is dissolved in a liquid or two solids melt and go into solution are other examples.

ICE MANUFACTURE

Manufactured ice is made by two methods: (1) the **plate** system and (2) the **can** system. In the **plate system** a plate at 0°F. or lower is immersed in a tank of water. In a week's time a plate of ice about 1 ft. thick is formed which is removed from the plate by permitting hot gas from the compressor to run through the hollow part of the plate. Although the ice is of excellent quality, it is non-uniform in thickness and has not found much favor in the retail trade.

In the **can system,** a can containing 300 or 400 lb. water is immersed in brine at such a temperature that the water will freeze in about 44 hr. time. If distilled water is used the freezing is straightforward; but, if raw hard water saturated with air is used, two effects must be overcome. Unless the water is agitated, air bubbles will be set free and freeze into the ice giving it a marblelike appearance and poor strength. Agitation with low- or high-pressure air is the usual method of overcoming this difficulty. The dissolved salts in the water will also precipitate on freezing and deposit on the ice surface to cause discoloration. Air agitation will also prevent this effect by keeping the particles in suspension, but at the end of the freezing period a core of turbid water remains. This is removed by a **core sucker** and replaced with fresh water. The air used for agitation must be dehumidified at a temperature below 32°F. to avoid freezing of the pipes. Air agitation is a rather expensive operation but is preferable to the use of distilled water such as was used

Table 23. Cold-storage Temperatures

Food articles	°F.	Food articles	°F.
Fruit		Liquids	
Apples	32–36	Beer, ale, porter, etc	33
Bananas	34	Cider	30
Berries, fresh	36	Ginger ale	36
Cranberries	33–36	Wines	40–45
Cantaloupes	40	Flour and Meal	
Dates, figs, etc	50–55	Buckwheat flour	36–40
Fruits, dried	35–40	Cornmeal	36–40
Grapes	34–36	Oatmeal	36–40
Lemons	34–36	Wheat flour	36–40
Oranges	34–36	Vegetables	
Peaches	34–36	Asparagus	34–35
Pears, watermelons	34–36	Cabbage	34–35
Meats		Carrots	34–35
Brined	38	Celery	34-35
Beef, fresh	33	Dried beans	32–40
Beef, dried	36–40	Dried corn	35
Calves	32–33	Dried peas	35–40
Hams, ribs, shoulders (not brined)	20	Onions	36
Hogs	29–32	Parsnips	34–36
Lard	38	Potatoes	36–40
Livers	20–30	Sauerkraut	35
Sheep, lambs	32	Miscellaneous	
Oxtails	30	Cigars, tobacco	35
Sausage casings	20	Furs, woolens, etc	35
Tenderloin, butts, etc	33	Honey	45
Fish		Hops	40
Fresh fish	20	Maple sirup, sugar	40–45
Dried fish	36	Oils	35
Oysters in shell	30–35	Poultry, dressed, iced	28–30
Oysters in tubs	25	Poultry, dry picked	26–28
Canned Goods		Poultry, scalded	20
Sardines	35–40	Game, to freeze	15–18
Fruits	35–40	Game, after frozen	25–28
Meats	35–40	Poultry, to freeze	15–18
Butter, Eggs, Etc.		Poultry, after frozen	25–28
Butter	18–20	Nuts, in shell	35–40
Butterine	18–20	Chestnuts	33
Cheese	34		
Eggs	31		

Table 24. Favorable Conditions of Temperature and Humidity Artificially Created and Maintained in Various Food-manufacturing Processes and Storage*

	Temp., °F.	Relative humidity, %
Bakeries:		
Dough rooms	80	80
Proof box	90–95	80–90
Bread cooling	70	65
Prepared powdered beverages and crisp cereals	75	35–40
Chewing gum:		
Rolling and scoring chicle	75	50
Wrapping and packing	70	45
Confectionery:		
Enrobing and hand dipping	60–65	55
Hard-candy manufacturing	70	40
Starch room	75–85	50
Packing	65	50
Dairy products:		
Butter manufacturing	60	60
Chill room	40	60
Fruits:		
Apple storage	31–34	80–85
Avocado packing	40	50
Bananas:		
Holding ripe fruit	56	70–75
Holding green fruit	58	70–75
Slow ripening	60–62	90
Normal ripening	64–68	90
Fast ripening	70–72	90
Danger of chilling	Below 49	
Meat products:		
Butter substitutes:		
Churn room	70	60
Print room	60	60
Chill room	30	60
Cooler	55	60
Bacon slicing	60	48
Egg candling	60	48
Sugar storage	80	35

* This table is offered only to demonstrate the wide variance of conditions that may be demanded even within a single plant. *Food Industries, May*, 1931.

in the older installations. The brine used for freezing is vigorously stirred to promote heat transfer. At the completion of freezing the cans are immersed in water at room temperature or above to loosen the cake, which is then removed from the can by dumping. The cake next goes to the **scoring machine** which cuts grooves at the points where it must be split up into the 25- and 50-lb. lots for the retail trade. Cakes that have been checked or cracked are used for crushed ice.

Ice for refrigerator cars is manufactured in plants at strategic points along the railroad lines, with icing platforms sufficiently long to handle a whole train of cars at once. The time of icing a vegetable or fruit car with cake ice is 1 to 2½ min. A car in transit from the Rio Grande Valley to Chicago will use about 8 to 12 tons of ice. Many attempts have been made to refrigerate cars by mechanical means, but they have not been successful in this country though they are used to some extent abroad. The chief objection seems to be that no systems are wholly reliable without costly supervision.

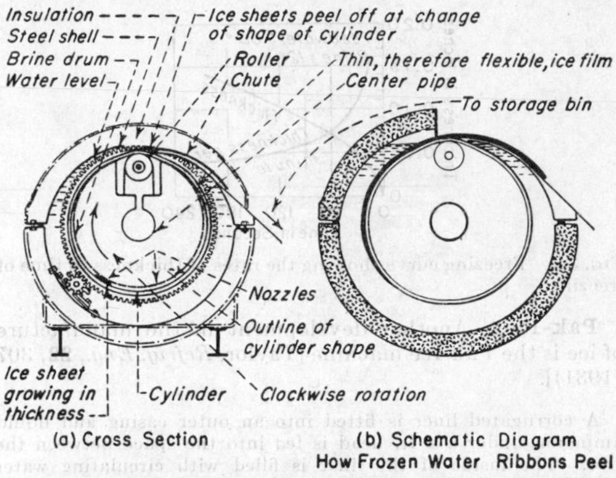

Fig. 32. Flakice machine.

Flakice. For numerous applications ice cakes have to be cracked up before they can be used. Field has developed a machine for freezing water directly in small chips or flakes [*Refrig. Eng.*, **31**, 95 (1936)]. The machine consists of a flexible cylinder with several metal panels separated by rubber strips. This cylinder is cooled internally while revolving partly submerged in water. A thin layer of ice freezes upon the metal panels, which is discharged from the panels as the freezing edge emerges from the water.

The method of discharge is unique. The freezing cylinder is slightly flexible. Inside the cylinder at the point of discharge is a deflecting roller which distorts the circular form. The ice sheet, being rigid, leaves the cylinder at a tangent and rolls onto a chute where it cracks up and drops into a bin. The relation of thickness to time of freezing is shown in Fig. 34.

The operation of the machine is indicated by Figs. 32 and 33.

The width of the strips depends on the width of the panels, since no ice forms on the rubber separators be-

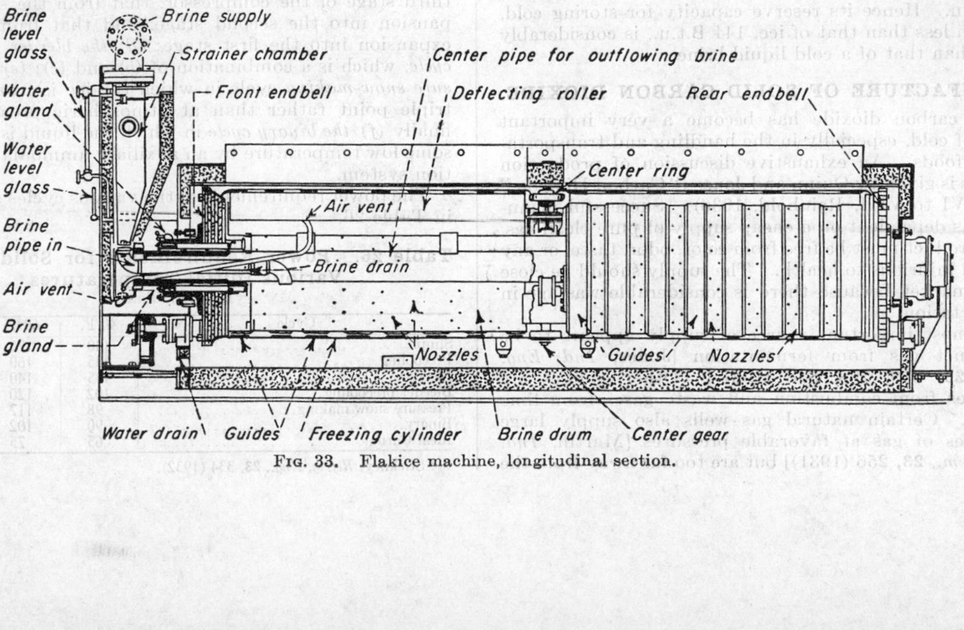

Fig. 33. Flakice machine, longitudinal section.

cause of their low thermal conductivity. The thickness is about ⅛ in. and the length not over 2 ft. In storage, about 20 to 30 per cent of the space is void. The chips are usually frozen to a temperature 10 to 12° below 32°F. and are therefore crisp and dry. Because of their large exposed surface, the chips cool water about six times as rapidly as does crushed ice.

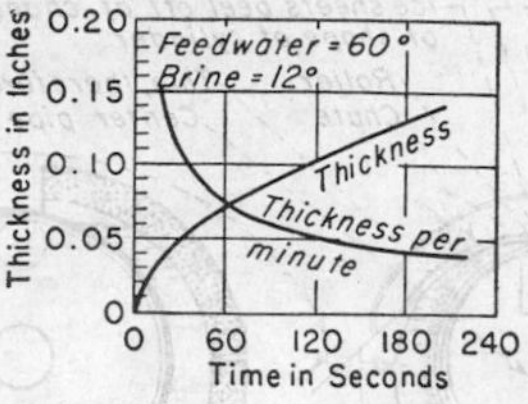

FIG. 34. Freezing curve showing the rates of thickness to time of freezing.

Pak-Ice. Another development in the manufacture of ice is the Pak-Ice machine [Taylor, *Refrig. Eng.*, **22**, 307 (1931)].

A corrugated liner is fitted into an outer casing and liquid ammonia under a 4 ft. head is fed into the space between the two. The inside of the liner is filled with circulating water (18 ft./sec.) which freezes rapidly on the liner surface to a thickness not over 0.008 in. and is constantly removed by tool scrapers.

The ice scraped off the sides is driven toward the center and carried out of the machine in a stream of water and transferred to a bin when the water is drained off.

Both the Flakice and the Pak-Ice systems dispense with the items of air agitation, brine circulation, filling, and dumping cans, which, as noted above, add considerably to the cost of ice. They do not, however, produce the cakes of clear ice demanded by the retail trade.

Eutectic Salt-ice Mixtures. Where temperatures lower than the freezing point of ice are desired, it is possible to freeze a salt solution to the eutectic temperature −6°F. Such mixtures have been used in sealed cans and carried about for the servicing of ice-cream cabinets in isolated localities. When the brine melts, the can is returned for freezing. Both Flakice and Pak-Ice machines can be used for freezing such mixtures. "Ice-cream bullets," another form of eutectic salt-ice mixture, have been manufactured by one of the larger metropolitan ice producers as a substitute for salt and cracked ice in making homemade ice cream.

The latent heat of fusion of 1 lb. of eutectic mixture is 101 B.t.u. Hence its reserve capacity for storing cold, although less than that of ice, 144 B.t.u., is considerably better than that of a cold liquid brine.

MANUFACTURE OF SOLID CARBON DIOXIDE

Solid carbon dioxide has become a very important source of cold, especially in the handling and transportation of foods. An exhaustive discussion of production and uses is given by Quinn and Jones ("Carbon Dioxide," Chaps. VI to VIII, Reinhold, 1936). Successful manufacture is dependent on a cheap supply of pure clean gas, as the product must be free from color, odor, taste, or any material injurious to health. The supply should be close to the market because there is considerable wastage in transportation.

The most important source of supply appears to be by-product gas from fermentation [Jones, *Ind. Eng. Chem.*, **23**, 519, 798, 848 (1931)]. Other sources are flue gases from combustion and waste gases from lime burning. Certain natural gas wells also supply large quantities of gas at favorable pressures [Martin, *Ind. Eng. Chem.*, **23**, 256 (1931)] but are too far away from the market. The same may be said of other industrial sources, such as the production of CO_2 as a by-product of the manufacture of hydrogen by the water-gas reaction.

Recovery of carbon dioxide is generally accomplished by absorption in cold sodium or potassium carbonate solutions. On heating, the resulting bicarbonate liberates CO_2 and reverts to the carbonate. The equilibrium concentrations for a $2N$ K_2CO_3 solution (12 per cent) for a gas mixture containing 15 per cent CO_2 are given by Quinn and Jones (Table 25).

Table 25. Percentages of Carbonate and Bicarbonate at Varying Temperatures in a $2N$ K_2CO_3 Solution with 15 Per Cent CO_2 at Atmospheric Pressure

Temp., °C.	% K as K_2CO_3	% K as $KHCO_3$
20	9.5	90.5
40	14.2	85.8
60	22.2	77.8
80	34.0	66.0
100	47.9	52.1

Absorption is usually effected in coke towers and desorption in steam-heated lye boilers. Water vapor carried out with the gas is removed by condensers.

The CO_2 gas is next purified by chemical reagents such as permanganate or dichromate, or by adsorption on activated charcoal, silica, or aluminum gel. Traces of water vapor are removed by calcium chloride, by refrigeration, or by adsorption.

Following purification, the gas is compressed in three- or four-stage compressors and is condensed. Oil filters or separators are necessary to prevent oil contamination. The liquid carbon dioxide is the starting point for manufacture of the solid; but, if there is a demand for the liquid itself, it may be filled into cylinders.

Liquid carbon dioxide is transformed into solid carbon dioxide by the cooling effect of its own evaporation. At a temperature of 84°F. and 70 atm., the yield of solid on expansion to atmospheric pressure is 0.23 lb./lb. of liquid, and 0.77 lb. of evaporated gas must be recompressed in a three- or four-stage compressor. Several cycles have been devised for carrying out this recompression most economically. These are: (*a*) *the simple cycle* in which the expanded gas is simply recompressed after expansion; (*b*) *the precooling cycle* in which the expanded gas is allowed to cool the liquid before it is expanded; (*c*) *the bleeder cycle* in which the liquid is expanded in three stages corresponding to the three stages of the compressor, the gas flashing in the first expansion being sucked into the third stage of the compressor, that from the second expansion into the second stage, and that from the last expansion into the first stage; (*d*) *the bleeder precooling cycle*, which is a combination of (*b*) and (*c*); (*e*) the *pressure snow-making* cycle in which the ice is frozen at the triple point rather than at atmospheric pressure; and finally (*f*) *the binary cycle* in which the liquid is cooled to some low temperature by an auxiliary ammonia refrigeration system.

The power requirements of the various cycles are shown in Table 26.

Table 26. Power Requirements for Solid CO_2 for Various Initial Temperatures*

Kw.-hr./Ton

Cycle	40°F.	60°F.	80°F.
Simple	140	175	240
Precooling	135	160	200
Bleeder	115	140	155
Bleeder precooling	102	120	145
Pressure snow making	98	117	142
Binary	90	102	120
Ideal Carnot	65	75	90

* Stickney, *Refrig. Eng.*, **23**, 334 (1932).

The simple cycle obviously requires less capital investment than the more complicated cycles and may therefore be most satisfactory for small plants. The pressure snow-making cycle is used by most large plants. For further comparison see Rabe and Duevel, *Refrig. Eng.*, **22**, 18, 90, 260, 388 (1931).

The snow formed by any of the above cycles is in a light fluffy condition and is squeezed into solid blocks (10 by 10 by 10 in.) in a hydraulic press at 2000 lb. pressure. The transfer from the snow-making equipment to the press is effected by rabble arms.

Numerous other processes have been developed. Mention may be made of the Carba process in which the solid block is frozen directly without auxiliary pressure. Capital expenditures are low, but the regulation is delicate. Somewhat similar in principle are the Linde-Sürth and Agefko processes. The Maiuri process (*Cold Storage and Produce Rev.*, Sept. 21, 1931) uses an ammonia absorption system to produce temperatures of −76 to −94°F., which are used to cool an alcohol water bath surrounding freezing cans containing liquid carbon dioxide under 80 to 100 lb. pressure.

Solid carbon dioxide in 50-lb. blocks wrapped in paper are shipped in specially built refrigerator cars or trucks. When removed from the truck, it will not evaporate completely in 24 hours' time on exposure to the atmosphere, which is a decided advantage in handling as compared with water ice. Comparison of water ice and solid carbon dioxide is shown in Table 27.

Table 27. Physical Properties of Ice and Solid CO_2

Physical property	Solid CO_2	Water ice
Specific gravity	1.53	0.90
Melting point, °F	−109.6	32
Latent heat of fusion, B.t.u./lb.	82.0	144
Latent heat of sublimation, B.t.u./lb.	240.6	
Sensible heat of gas to 32°F., B.t.u./lb.	34.4	
Net refrigerating effect, B.t.u./lb.	275	144

In many applications solid carbon dioxide may be used directly, but in some applications the vapors are detrimental, and the condensation of moisture may produce excessive dryness in the refrigerated space. The usual method of avoiding these effects is to use the solid indirectly through an intermediate low-freezing liquid, such as aqueous solutions of alcohol or methanol, the flow of which is regulated automatically to meet the refrigerating requirements. Calcium chloride brines cannot be used as their minimum freezing point is −60°F. (eutectic temperature), but 75 per cent solutions of ethanol or methanol in water have eutectic temperatures of −200°F. and are never in danger of freezing up by solid CO_2. Systems employing auxiliary circulating liquids are used quite extensively in truck refrigeration.

THE HEAT-PUMP PRINCIPLE

In a mechanical refrigeration system, heat is abstracted in one part of the cycle at a low temperature and transferred in another part of the cycle to a higher temperature. Although mechanical refrigeration is primarily designed to produce a temperature lower than that of the environment, it can also be used to produce a higher temperature. The first proposal of this sort was made by Lord Kelvin, who showed that a more efficient utilization of heat for warming could be obtained if it were applied to a steam engine operating between steam temperature and outside atmospheric temperature, driving a refrigerating machine operating between outside atmospheric temperature and inside room temperature. This proposal has acquired considerable interest in the last few years [Stevenson, Faust, and Roessler, *Refrig. Eng.*, **23**, 83 (1932)]. To illustrate the principle, a gas engine may be considered. This has a normal over-all efficiency of about 30 per cent in the conversion of **heat into work.** A refrigerating machine operating between an outside freezing temperature (32°F.) and normal room temperature (68°) has an over-all efficiency of about 900 per cent for the conversion of **work into heat.** Accordingly 3000 B.t.u. in the gas engine will produce 900 B.t.u. of work, which will raise

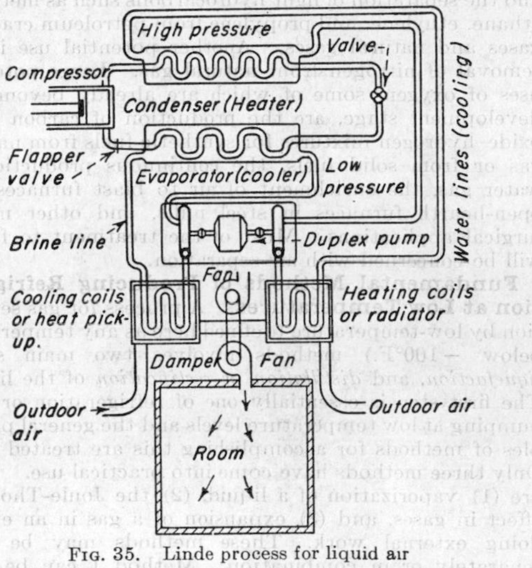

FIG. 35. Linde process for liquid air

8100 B.t.u. of heat from 32° to 68°F. If the outside temperatures are at 0°F., the refrigeration efficiency will drop to about half. Moreover, if heat is distributed by the ordinary type of radiator, it will have to be raised to a higher temperature level than 68°F. Nevertheless, assuming electric power at 2¢ per kw.-hr. for driving the refrigeration unit and gas to be used in a gas furnace at 75¢ per 1000 cu. ft., the cost of heating by the heat pump is only 4 per cent greater in New Orleans, Charleston, S.C., and Los Angeles, while it becomes 30 per cent greater in Washington, D.C. (Stevenson *et al.*, *loc. cit.*). With lowered costs of electric power and the advantage that it can be used in summer for air conditioning, the Kelvin warming engine may find considerable development in the future. An installation that has been tried in practice is shown in Fig. 35 [Chawner, *Electric West*, **66**, 177 (1931)] abstracted in *Mech. Eng.*, **53**, 686 (1931).

A further application of the heat-pump principle is found in the vapor recompression systems of evaporation (see Sec. 7, p. 519). Also see Sec. 24, pp. 1657*ff*.

LOW-TEMPERATURE PROCESSES

BY BARNETT F. DODGE

REFERENCES: Ruhemann, "The Separation of Gases," Oxford University Press, New York, 1940. Wien and Harms, "Handbuch der Experimental Physik," Leipzig, 1929, vol 9, p. 47, section by Lenz, Gasverflüssigung und ihre thermodynamischen Grundlagen. Dodge, "Chemical Engineering Thermodynamics," Chap. X, McGraw-Hill, New York, 1944. Ruhemann and Ruhemann, "Low Temperature Physics," Cambridge University Press, London, 1937. Bliss and Dodge, Oxygen Manufacture, Thermodynamic Analyses of Processes Depending on Low Temperature Distillation of Air, *Chem. Eng. Progress*, **45**, 51, 129 (1949). Oxygen, Past, Present and Prospects, *Chem. Eng.*, p. 123 (Jan., 1947). Downs and Rushton, Tonnage Oxygen, *Chem.*

Eng. Progress, **1**, 12 (1947). Air Separation Principles and Technology, *Chem. Eng.*, p. 126 (March, 1947). Clark, Large Scale Production of Oxygen and Atmospheric Gases, BIOS Final Report 591, P.B. 41229 (1946). Lobo, Technical Data Pertaining to Air; Its Liquefaction and Distillation, P.B. 8900 (1945). Carlsmith, Large Scale Production of Oxygen, FIAT Final Report 1120, P.B. 88840 (1947).

The field of low-temperature processes is of growing importance. Present applications include the separation of air into its component gases, particularly oxygen for the welding and cutting of metals, nitrogen as an inert gas and as a raw material for synthetic ammonia, argon for filling lamps, and the rare gases for display signs; the separation of helium from natural gases; the liquefaction and storage of natural gas and of liquid oxygen for explosives and for rocket engines; the separation of hydrogen and other gases such as ethylene from coke-oven and similar gases; and the separation of light hydrocarbons such as methane, ethane, ethylene, and propylene from petroleum cracking gases and natural gases. Another potential use is the removal of nitrogen from natural gas. Large potential uses of oxygen, some of which are already beyond the development stage, are the production of carbon monoxide–hydrogen mixtures for synthetic fuels from natural gas or from solid fuels, the continuous production of water gas, the enrichment of air to blast furnaces and open-hearth furnaces in steel mills, and other metallurgical applications. Most of the treatment to follow will be concerned with air separation.

Fundamental Methods of Producing Refrigeration at Low Temperatures. A process for gas separation by low-temperature (defined here as any temperature below −100°F.) methods involves two main steps, *liquefaction*, and *distillation* or *rectification* of the liquid. The first step is essentially one of refrigeration or heat pumping at low temperature levels and the general principles of methods for accomplishing this are treated next. Only three methods have come into practical use. They are (1) vaporization of a liquid, (2) the Joule-Thomson effect in gases, and (3) expansion of a gas in an engine doing external work. These methods may be used separately or in combination. Method 1 can be used to reach liquid air temperatures by employing a series of liquids, the one of lowest boiling point absorbing heat from the system to be refrigerated and delivering this heat to the next higher boiling fluid in a condenser-boiler combination. Finally the fluid of highest boiling point is condensed by air or water cooling and thus discharges the heat to the atmosphere. Such a process is called a cascade, and it has been applied to the liquefaction of air and of natural gas (see Fig. 47) and to the separation of air into its components. The lowest temperature that can be produced by this method is 63°K., the triple point of nitrogen, since there is no more volatile fluid that will condense at this temperature.

Although the self-cooling produced on expanding a gas such as air from a high pressure to 1 atm. through a throttle is relatively small, by using an efficient heat exchanger to make it accumulative (see Fig. 36) temperatures within a few degrees of absolute zero can be reached. The refrigerating effect in such a scheme is the isothermal enthalpy difference between streams 1 and 3. The fraction of a gas that can be liquefied on passage though an apparatus using this process is readily obtained from an enthalpy balance, which leads to the equation

$$x = \frac{H_3 - H_1 - q_L}{H_3 - H_2} \tag{1}$$

where x = fraction liquefied, H_1, H_2, and H_3 are the enthalpies of unit mass of fluid at the points designated by the numbers, and q_L = heat leaking in from the surroundings per unit of entering gas. The maximum possible degree of liquefaction occurs when $q_L = 0$ and H_3 and H_1 refer to the same temperature.

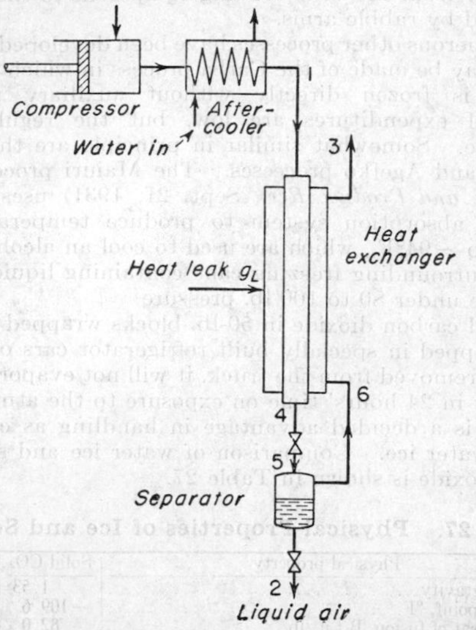

FIG. 36. Simple air liquefaction process.

Method 3 is probably the most important of the three methods, since it offers the best possibility for an economical process on a large scale. It utilizes expansion with external work (isentropic expansion at the limit) instead of isenthalpic expansion, and the amount of cooling for a given pressure difference is much greater. This is illustrated diagrammatically in Fig. 37, where $T_A - T_B$ is the isentropic cooling and $T_A - T_C$ the isenthalpic cooling for adiabatic expansions between the same pressure limits.

Although expansion in an engine can theoretically be made to approach complete reversibility and hence should be more efficient than a Joule-Thomson expansion, nevertheless it is possible to devise a low-temperature, air-separation process using the latter type expansion which has substantially the same power requirement as a process using an engine for refrigeration. The reason for this lies in two facts, (1) the refrigerating effect from a Joule-Thomson effect is considerably increased by

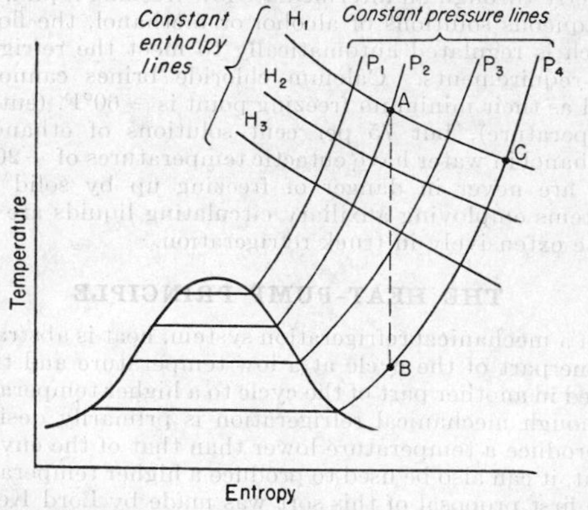

FIG. 37. Comparison of the cooling effects in isentropic and isenthalpic expansions.

increasing the pressure and precooling the gas by an auxiliary liquid-vapor refrigeration cycle, and (2) only a relatively small fraction of the power input to the compressor of an air-separation system is needed to supply the refrigeration, the major portion being necessary to carry out the separation of the gases. For details on this point reference may be made to the paper by Bliss and Dodge (*loc. cit.*).

The great problem in the practical realization of expansion with external work has been the development of a suitable engine. This was originally solved by the French engineer Georges Claude who developed a reciprocating engine that, though not very efficient, was at least better than a throttle. Recently both reciprocating and turbine engines of high efficiency (80 per cent and over) have been developed, and the development of compact efficient turboexpanders makes possible the separation of air on a very large scale in a low-pressure (and consequently low-power) compact system.

Work of Gas Liquefaction and Separation. Cost of power is an important consideration in any large-scale gas liquefaction and separation process, and hence it becomes important to know the thermodynamic efficiency of any proposed process. This is defined as the ratio of the minimum reversible work for the given process to the actual work. This is the product of two other efficiencies defined below:

$$\text{Cycle efficiency} = \frac{\text{reversible work for the process}}{\text{theoretical work for ideal operation of the cycle}} \tag{2}$$

$$\text{Practical efficiency} = \frac{\text{theoretical work for ideal operation of the cycle}}{\text{actual work for the cycle}} \tag{3}$$

The reversible work is the least possible amount of work to effect the change in question and is calculated from the expression

$$-W = \Delta H - T_0 \Delta S \tag{4}$$

where T_0 is the lowest temperature on the absolute scale at which large amounts of heat may be rejected. The reversible work clearly depends only on the initial and final states of the system and is entirely independent of the cycle or the mechanisms used. On the other hand, certain cycles may have inherent irreversible effects such as throttle expansions or unavoidable temperature differences but can, in imagination, be operated in a frictionless or ideal manner, and the calculated work is used to give the cycle efficiency. This furnishes a measure of how good the cycle is without regard to the imperfections of the equipment used to realize the cycle. The practical efficiency is a measure of the approach to perfection of the equipment as distinct from that of the process.

The minimum or reversible work is readily calculated from Eq. (4) and a table or diagram of thermodynamic properties. Some figures for a few processes are given herewith, based on T_0 = 300°K. (80°F.).

Process	Reversible Work, Kw.-hr.	
1. Liquefaction of oxygen starting with O_2 gas at 80°F. and 1 atm. and producing liquid at the normal boiling point	2.56	(per lb.-mole of air)
2. Complete separation of air at 80°F. and 1 atm. into oxygen and nitrogen (including argon) gases at the same pressure and temperature	0.1615	(per lb.-mole of air)
	0.767	(per lb.-mole of O_2)
	1.95	(per 1000 cu. ft. of O_2)
3. Separation of air at 80°F. and 1 atm. into gaseous nitrogen (plus argon, etc.) at 80°F. and 1 atm. and liquid O_2 at the normal boiling point	0.700	(per lb.-mole of air)
	3.33	(per lb.-mole of O_2)

4. Separation of air (assumed a binary system) into gaseous oxygen and nitrogen.

Oxygen purity % O_2 in oxygen product	Nitrogen purity % O_2 in the nitrogen product	Hp. hr./lb. mole of pure O_2 in the oxygen product
100	0	1.033
95	0	.950
95	0.01	.910
70	0	.663
50	0	.447
30	0	.174
70	0.01	.615

Even the best low-temperature processes have a relatively low thermodynamic efficiency (15 to 20 per cent) because of the multiplication of a number of factors each resulting from a given irreversible effect, such as friction in engines, fluid friction in pipes and through equipment, heat leak, throttling, temperature differences across heat exchangers, and mass transfer in rectifying columns between fluids not in phase equilibrium. The last named effect is generally overlooked, but it is probably the greatest single contributor to loss in efficiency. An ideally operated adiabatic rectifying column has an efficiency of 67 per cent, and under practical operating conditions its efficiency is of the order of 35 to 40 per cent.

Analysis to Determine Power Requirement and Distribution of Losses. The first and second laws of thermodynamics are the fundamental tools for low-temperature-process analysis. Neglecting kinetic energy and potential energy due to position in the gravitational field, the first law may be written

$$\Delta H = Q - W_{sh} \tag{5}$$

where ΔH is the summation of the enthalpy differences for all fluids entering and leaving the section in question, Q designates the algebraic summation of all heat exchanges between the section and its surroundings including auxiliary cycles, and W_{sh} is the shaft work. There will, in general, be as many first-law equations as there are units of equipment such as heat exchangers, expanders, columns, or points of mixing of streams. If the first law is applied to the whole, which is customary, one of these equations is dependent. There will, of course, be considerably more unknowns, so it will be necessary to hold some of them as parameters. While the choice of these parameters is arbitrary, convenience in the solution of the system of simultaneous equations will in most cases serve as a suitable guide. Solution will yield all the necessary quantities, compositions, enthalpies, and temperatures. Obviously different solutions will result from different choices of variables to be fixed. Solutions obtained may be unworkable, however, because of hidden second-law violations, which will be discussed presently.

If one is to obtain a practical, in contrast to a theoretical, figure for the power requirement from such an analysis, it is clear that one must make due allowance for the inevitable irreversible effects that always accompany the operation of any equipment in actual practice. This means that reasonable estimates, based on practice, must be made for such things as heat leak, compressor and expander efficiencies, temperature differences in heat exchangers, pressure drop due to fluid flow, and excess reflux above the minimum for rectification.

Second Law. The part played by the second law in these analyses is not so straightforward. It leads to no particular set of orderly equations, but only to the observation that all driving forces must be positive. For example, applied to heat transfer, it means that the fluid giving up heat must be higher in temperature than the fluid receiving heat at all points in the exchanger. This seems so obvious as not to require discussion, but the difficulty arises from the fact that fixing the terminal temperature differences of exchangers does not insure absence of negative Δt's at intermediate points. Such virtual second-law violations are likely to occur when one of the fluids undergoes a phase change or even in the absence of phase changes, when one of the fluids is near its critical state. The place where a virtual second-law violation is most likely to occur is in a liquefier. This is illustrated in Fig. 38, which shows the temperature-

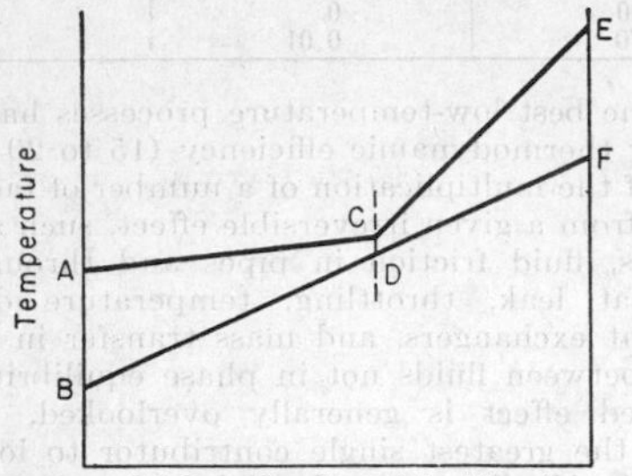

FIG. 38. Examination of liquefier in light of the second law.

enthalpy relation of two streams in an exchanger when liquefaction occurs in one of the fluids. At the point of incipient liquefaction C, the Δt may be much less than at the two terminals. It is easy to see that reduction in the terminal Δt's could readily lead to an impossible case of zero or negative Δt at the "pinch."

Reversibility and Irreversibility. The measure of the general quality of a process is its reversibility. A process will approach the minimum power requirement as it approaches the reversible one. On the other hand, any item of equipment will approach minimum size as it departs from reversibility—most easily seen with heat exchangers.

The temperature difference is a measure of reversibility in heat exchangers, *i.e.*, when it is small, reversibility is approached, and vice versa. The absolute temperature at which temperature differences occur is also important. Thus a 5°C. temperature difference at ambient levels is not badly irreversible, but the same at 77°K. is considerably more so. The entropy increase in a heat exchanger, of course, weighs both these effects, and the smaller such increase is, the closer the approach to reversibility. A good guide to heat exchangers in process planning is always to exchange sensible heat with sensible heat and latent with latent. If latent and sensible heats are exchanged it is impossible to avoid excessive temperature differences. Processes have been proposed in which gaseous helium was to be used as a cooling medium for reflux with a compound-type column. Such a provision leads to great temperature difference at the lowest temperature of the process and hence to considerable irreversibility.

In general, the entropy increase is the best guide to the degree of irreversibility of any process. According to the second law of thermodynamics, the entropy of an isolated system must either increase or remain constant. If one imagines a *virtual* change in such a system and finds $\Delta S < 0$ for the entire system, then one can say at once that the change is not possible. If all changes occur reversibly, $\Delta S = 0$, and if there is any irreversible effect of any kind occurring in the system, $\Delta S > 0$. One must be careful to avoid making the erroneous assumption that entropy always increases in any irreversible process. The statement that $\Delta S \geqq 0$ applies to an *isolated* system, *i.e.*, the particular system plus its surroundings. One must likewise avoid assuming that any virtual process is possible if it involves an increase in entropy of the isolated system. For example, it is easily possible to imagine a heat exchanger in which the net $\Delta S > 0$ but the assumed conditions would be impossible because of negative temperature differences.

Analysis of Losses. Values of the thermodynamic efficiency for low-temperature-liquefaction and gas-separation processes generally fall in the range of 3 to 30 per cent. From the standpoint of process improvement it is important to account for the 97 to 70 per cent of the work done and be able to allocate it to different steps of the process in order to know where the greatest improvement in efficiency can be made. The inefficiency is due, of course, to various irreversible effects which are well known, and tracking them down and assigning to each a definite fraction of the total work done is what is meant by the term analysis of losses.

The fundamental equation for such an analysis is:

$$W = W_{\text{rev}} + T_o \Sigma M \Delta S \tag{6}$$

which simply states that the total work is the sum of the reversible work and a series of terms each representing the loss in availability for a certain step in the process. For example, an algebraic summation of the entropies of the fluids entering and leaving a heat exchanger, each multiplied by the corresponding mass of the fluid, gives $\Sigma M \Delta S$ for the exchanger; and when this is multiplied by the absolute ambient temperature, one obtains the work that was required as a result of the irreversibilities in this exchanger.

In the application of this method to actual processes there are several details that require further elucidation which can best be done by numerical examples. Consider the simple air-liquefaction process sketched in Fig. 36. Conditions for the process are indicated in Table 1. For

Table 1. States and Values of Properties for Process of Fig. 36*

Properties based on Kellogg Temperature-Entropy diagram

Point	Pressure, atm.	Temperature, °F.	Mass, lb.	H, B.t.u./lb.	S, B.t.u./(lb.)(°R.)
1	100	80	41	121.0	0.583
2	1	−318	1	−53.0	0
3	1	70	40	127.4	0.911
4	100	−165	41	32.0	.355
5	1	−318	41	32.0	.583
6	1	−318	40	35.0	.598

* Lobo, Technical Data Pertaining to Air; Its Liquefaction and Distillation, P.B. 8900 (1945).

a heat leak of 2.0 B.t.u./lb. of entering air, the fraction of air liquefied is 0.0244, as calculated by Eq. (1). The total work requirement for the process assuming three-stage adiabatic compression and 75 per cent compression efficiency is 11,650 B.t.u./lb. of liquid air. The reversible work is 312 and the problem is to account for the difference. Using the data in the table, the following calculations are made:

For the exchanger,

$$T_o \Sigma M \Delta S = 540[40(0.911 - 0.598) - 41(0.583 - 0.355)] = 1720 \text{ B.t.u.}$$

Similarly for the throttle valve,

$$T_o \Sigma M \Delta S = 5040$$

The compression of the air is irreversible not only due to frictional losses but also to the fact that adiabatic compression followed by constant pressure cooling is inherently irreversible. Determination of the loss could be made from ΔS values, just as was done for the low-temperature parts of the process, but since the work of compression was obtained algebraically, it was more convenient to calculate the work of reversible isothermal compression and take the difference. The isothermal work is 6980 B.t.u. and hence the loss in the compression itself is 11,650 − 6980 = 4670 B.t.u.

One may now set up a balance sheet of work requirements and/or losses as follows:

	B.t.u.	Per cent
1. Reversible work	312	2.7
2. Loss in compression	4,670	39.8
3. Loss in exchanger	1,720	14.6
4. Loss due to throttling	5,040	42.9
Total	11,742	100.0

The discrepancy of about 1 per cent in the total work is well within the error of the calculation.

This analysis shows at once where the greatest gains in efficiency can be made. There is, however, one irreversible effect, namely, heat leak, whose effect does not appear directly in such an analysis because it is distributed over all the steps of the process. An insight into the effect of heat leak can, however, be readily obtained by making a first-law or enthalpy balance for zero heat leak and recalculating the work per pound of liquid produced. The total work obtained in this way is 8020 B.t.u. Therefore, we may conclude that heat leak alone increased the work by 3630 B.t.u., or 31 per cent of the work is attributable directly to this cause.

As a second illustration an air-separation process will be chosen, as this introduces some new problems. In order not to complicate the calculation and obscure the principles the very simple process shown in Fig. 39 has been chosen.

In the preparation of the *TS* diagrams for air, oxygen, and nitrogen, arbitrary base values of *H* and *S* are chosen and when the three diagrams are to be used together there are certain relations between the *H* and *S* of a mixture and those of the pure components which must be taken into account. If we assume that the various gases and mixtures are ideal at low pressure (1 atm., *e.g.*) the following relations apply:

$$H_{\text{mix}} = \Sigma x_i H_i \quad (7)$$

$$S_{\text{mix}} = \Sigma x_i S_i - R\Sigma x_i \ln x_i \quad (8)$$

the *H*'s and *S*'s for the individual gases being taken at the same temperature and total pressure. Using these equations at the state $p = 1$ atm. and $t = 80°F.$, one can readily derive the following relations for placing values for air read from the Kellogg *TS* diagram on the basis of the Millar and Sullivan diagrams [*U.S. Bur. Mines Tech. Paper* 424 (1928)]:

$$\frac{H_K \times 29}{1.8} + 872 = H_{M \text{ and } S}$$

$$S_K \times 29 + 1.518 = S_{M \text{ and } S}$$

To obtain *H* and *S* for any other mixture than air Eqs. (7) and (8) are used directly, but it is to be noted that a difficulty arises in the case of a saturated vapor as at point 5 in Fig. 39. Oxygen does not exist as a vapor at the temperature and total pressure of the mixture. This situation can be handled in various ways and the very simple one has been chosen of extrapolating into the unstable region. This is a very slight extrapolation and cannot introduce serious error.

Conditions chosen for the process are indicated in Table 2 along with the various properties calculated by the methods previously outlined or illustrated. In addition it is to be noted that a heat leak of 50 p.c.u./lb.-mole of air was assumed and it was distributed 70 per cent to the column and 30 per cent to the exchanger. It was further assumed that the column would yield 65 per cent of the oxygen in the air as pure oxygen.

Proceeding as in the previous example, losses may be distributed:

	p.c.u./lb.-mole air	Per cent
Reversible work	150	3.8
Loss in compression	1480	37.7
Loss in rectifying column	1790	45.5
Loss in heat exchanger	510	13.0
Total	3930	100.0

It is of interest to note that the greatest loss is in the column and next greatest in the compression. Again, of course, the effect of heat leak does not appear directly but can be calculated. Presumably the column losses could be further localized to show the losses in the throttle valve, boiler, etc., if desirable.

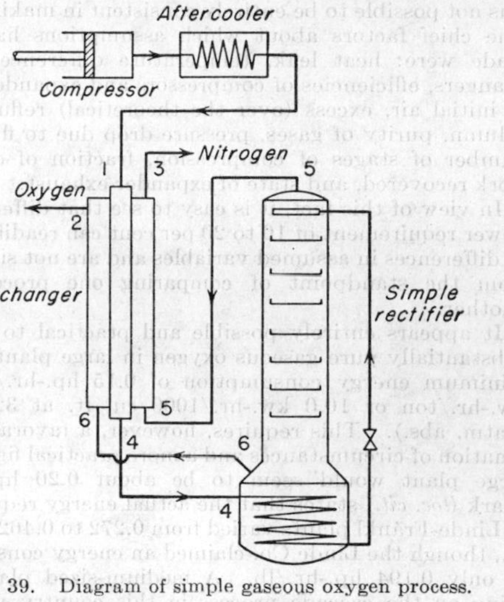

FIG. 39. Diagram of simple gaseous oxygen process.

It is believed that the method for analysis of the losses has been sufficiently demonstrated by means of simple examples so that it can readily be applied to any process for liquefaction or separation at low temperatures and such an analysis should prove valuable as a starting point for process improvement.

Table 2. States and Values of Properties for Process of Fig. 39*

Basis: 1 lb.-mole of entering air

Point	*P*, atm.	*t*, °F.	*T*, °K.	Composition	State	*H*, p.c.u./lb.-mole	*S*, p.c.u./(lb.-mole)(°K.)
1	54.5	80	300	air	Superheated	2877	19.82
2	1.0	71	295	O_2	Superheated	3215	28.30
3	1.0	71	295	8.52% O_2	Superheated	2884	27.31
4	54.5	−196	147	air	Superheated	1397	12.44
5	1.0	.....	79.3	8.52% O_2	Sat. vapor	1379	18.11
6	1.0	.....	90.2	O_2	Sat. vapor	1788	19.99

* Properties on basis of Millar and Sullivan, *U.S. Bur. Mines Tech. Paper* 424 (1928).

Results of Thermodynamic Analyses. Bliss and Dodge (*loc. cit.*) present results of the analysis of 17 liquid oxygen and 12 gaseous oxygen processes. Some of the conclusions from this study are summarized below. In order to follow the results conveniently a classification of processes on the basis of the method used for producing refrigeration is first presented as follows:

1. Joule-Thompson effect only
2. Joule-Thompson effect plus auxiliary refrigeration with an ordinary liquid-vapor cycle at moderate or high-temperature levels, *i.e.*, relative to liquid-air temperature
3. Joule-Thomson effect plus approximately reversible expansion of the air or products in an expander
4. Refrigeration essentially due only to approximately reversible expansions of auxiliary fluid or fluids such as helium through expanders, *i.e.*, processes in which the fluid remains entirely in the gas phase
5. Refrigeration essentially due only to auxiliary fluid or fluids operating in liquid-vapor cycles, *i.e.*, the cascade process
6. Refrigeration essentially due only to approximately reversible expansion of air or products in an expander, *i.e.*, low-pressure processes
7. Processes using an auxiliary nitrogen-liquefaction cycle

It was very difficult, if not impossible, to put all processes on a strictly comparable basis. Many assumptions have to be made in the course of such calculations and it was not possible to be entirely consistent in making them. The chief factors about which assumptions had to be made were: heat leak, temperature differences in exchangers, efficiencies of compressors and expanders, state of initial air, excess (over the theoretical) reflux in the column, purity of gases, pressure drop due to fluid flow, number of stages of compression, fraction of expander work recovered, and state of expander exhaust.

In view of this fact, it is easy to see that differences in power requirement of 10 to 20 per cent can readily be due to differences in assumed variables and are not significant from the standpoint of comparing one process with another.

It appears entirely possible and practical to produce substantially pure gaseous oxygen in large plants with a minimum energy consumption of 0.15 hp.-hr./lb. (224 kw.-hr./ton or 10.0 kw.-hr./1000 cu. ft. at 32°F. and 1 atm. abs.). This requires, however, a favorable combination of circumstances and a more practical figure for a large plant would seem to be about 0.20 hp.-hr./lb. Clark (*loc. cit.*) states that the actual energy requirement in Linde-Fränkl plants varied from 0.272 to 0.462 hp.-hr./lb., though the Linde Co. claimed an energy consumption of only 0.194 hp.-hr./lb. A medium-sized plant operating on the cascade process in this country yielded a figure of 0.22 hp.-hr./lb. It seems equally possible and practicable to produce essentially pure liquid oxygen with a minimum energy consumption of 0.5 hp.-hr./lb., although 0.6 would probably represent a more practical figure.

Processes of various types including ones classified as 2, 3, 5, and 6 are comparable from the standpoint of energy requirements and the choice between them must be based on other considerations. Processes 1 and 4 are definitely inferior from this standpoint; process 1 because it utilizes only the Joule-Thomson effect in air for refrigeration and process 4 because of the large inherent Δt in the reflux condenser of the column. So little work has been done with process 7 that no general conclusion can be drawn about it, but on the basis of the one type analyzed for gaseous oxygen, it would appear to require somewhat more work than the four processes first mentioned. This is believed to be reasonable because the separate nitrogen cycle has its own irreversible effects and the total gas handled in the air and nitrogen circuits is considerably greater than in a process such as 6 where the air is its own refrigerant.

On the other hand, process 7 has certain advantages which may offset the somewhat greater power requirement. Among these may be mentioned: (1) the fact that the expander always operates on highly purified gas (2) since the air to be separated is at substantially atmospheric pressure, its temperature as it leaves the exchanger will be considerably lower than air at 4 to 5 atm. and hence it will carry less impurities into the column, and (3) the compound column is simpler to control and requires fewer plates than a double column. However, these advantages are partly balanced by certain disadvantages such as the fact that the heat-exchange and purification system is probably bulkier than in process 6, due to the low pressure of the air and the added N_2 system. Furthermore the moisture content of the air is so much greater that external removal must be used.

It has been claimed that the cascade process for gas liquefaction is much more efficient than the other well-known methods and this, by implication, leads one to expect that this process for air separation should have the least energy requirement. This is not borne out by the calculations. Furthermore it should be noted that the cascade process is more complex than the others and since it has no power advantage, it is not likely to be considered for the production of oxygen on a tonnage scale. However, it should be noted that it involves no cold moving parts.

It may be noted that the higher yield attainable in double columns is very beneficial in reducing the energy requirement of gaseous oxygen processes, but it is much less so in the case of liquid oxygen. The predominant portion of the work is required for the liquefaction, and when the yield is increased, the liquefaction work is increased proportionately.

There is an important point in regard to internal purification (discussed later) which warrants discussion here since it has a bearing on the thermodynamics of the process. In the simplest example of processes of class 6, where the air is compressed to 4 to 6 atm. abs., cooled in regenerators or reversing exchangers by separated gases, and a portion expanded for refrigeration, the separated gases returning from the column will cool the entering air down to the saturation point. Thus, (1) the turbine will operate in the wet region, which probably lowers its efficiency, and (2) the cold-end Δt of the exchangers is too large to permit complete removal of carbon dioxide and deriming must be more frequent. Methods for overcoming these difficulties are discussed under the heading of Heat Exchangers.

A general study by means of over-all balances only was made on processes of type 6 with the object of investigating the effect on power requirement of (1) heat leak, (2) expander efficiency, (3) warm-end Δt, and (4) material leak. These effects under comparable conditions were as follows:

1. A heat leak of 100 B.t.u./lb.-mole of air increased the energy requirement about 20 per cent over that for no heat leak.
2. An expander efficiency of 40 per cent increased the energy requirement about 20 per cent over that with 100 per cent efficiency.
3. A warm-end temperature difference of 20°F. increased the energy requirement about 40 per cent over that with zero difference.
4. A material leak of 5 per cent of the cold air at the expander inlet increased the energy requirement by about 25 per cent.

The energy or thermodynamic efficiency of the gaseous oxygen processes analyzed varied from 9 to 23 per cent and the liquid oxygen ones from 6 to 29 per cent. At first thought this would seem to allow considerable leeway for marked improvement. However, this is probably not the case as an analysis of the losses will show. In general, the irreversible effects are of two kinds, (1) those inherent in the particular process such as a Joule-Thomson expansion or an adiabatic column, and (2) those which can be reduced to zero at the limit. Those of class (1) may be avoided by modifying the process but usually this means greater complexity of equipment. Any attempt to reduce the irreversible effects of class (2) will inevitably lead to smaller driving forces and larger equipment. From a consideration of the economic balance between power cost and fixed charges on the investment in equipment, it appears very doubtful if significant increases in over-all efficiency over the maxima cited above will be achieved in the near future.

Gas Liquefaction. Low-temperature gas-liquefaction processes may be based either on the Joule-Thomson expansion or engine expansion. Processes based on the first are usually called Linde processes after the German engineer who pioneered in the field. Processes based on the use of engine expansion are commonly given the name Claude after the French engineer who did much of the original development work. It might be noted that an engine-expansion process also generally uses a Joule-Thomson expansion for the final refrigeration in order to avoid formation of liquid in the engine. Equation (1) may be used to calculate the percentage liquefaction in a simple Linde process such as that shown in Fig. 36. It should be noted that this percentage is determined by conditions at the warm end of the heat exchanger which is just ahead of the expansion valve. Under certain conditions, notably with hydrogen and helium at room temperature, $H_1 > H_3$ and no liquefaction could be obtained. Assuming H_3 to be fixed at the lowest possible pressure and the highest possible temperature (when $t_3 = t_1$), the maximum degree of liquefaction occurs when H_1 is a minimum. The criterion for this is

$$\left(\frac{\partial H}{\partial p_1}\right)_{T_1} = T_1\left(\frac{\partial V_1}{\partial T_1}\right)_{p_1} - V_1 = 0 \qquad (9)$$

or

$$T_1\left(\frac{\partial V_1}{\partial T_1}\right)_{p_1} = V_1 \qquad (10)$$

This is the equation of the Joule-Thomson inversion and hence it is concluded that maximum liquefaction occurs when the initial pressure for a given temperature of gas entering the exchanger is the inversion temperature. Referring to Fig. 40, which shows a generalized Joule-

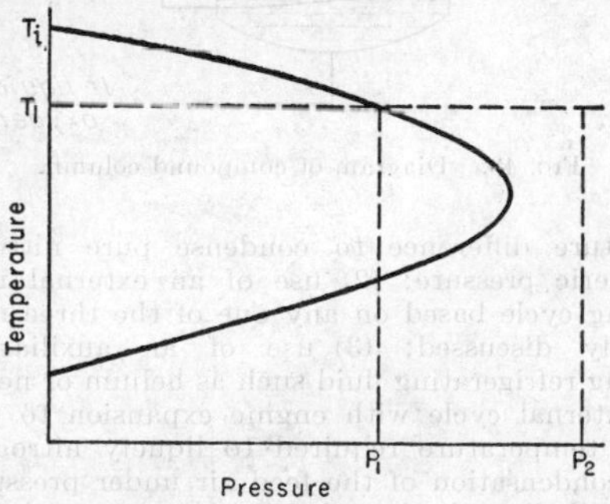

FIG. 40. Relation of degree of liquefaction to the Joule-Thomson inversion.

Thomson inversion curve, no liquefaction could be obtained at any pressure if the entering temperature were above T_i, which is known as the Joule-Thomson inversion temperature. For air this temperature is about 325°C., for hydrogen about 200°K., and for helium about 24°K. In general, as the temperature of the gas entering the main heat exchanger is lowered, the value of $H_3 - H_1$ increases and the degree of liquefaction increases. For example if air at 200 atm. is precooled to −100°F. by a Freon refrigeration cycle, the yield of liquid air is about 2.2 times as great as for water cooling to 80°F. Referring to Fig. 40 again, if the entering temperature were T_1, Eq. (10) shows that the maximum degree of liquefaction would occur when the pressure is p_1. Higher pressures, such as p_2, would actually give less liquid.

Figure 41 shows a diagram of a process for liquefying hydrogen using precooling with liquid air.

One of the problems in liquefying a gas such as hydrogen or helium is the presence of impurities such as oxygen and nitrogen which are solids at the low temperatures involved. Oxygen can be removed by chemical means but nitrogen is particularly difficult to remove. The Kapitza scheme shown in Fig. 41 solves this in an ingenious manner. Ordinary cylinder hydrogen throttled to low pressure is the feed gas. After precooling in exchangers I and II by exchange of heat with return gas and in the two liquid air (or nitrogen) containers, it passes

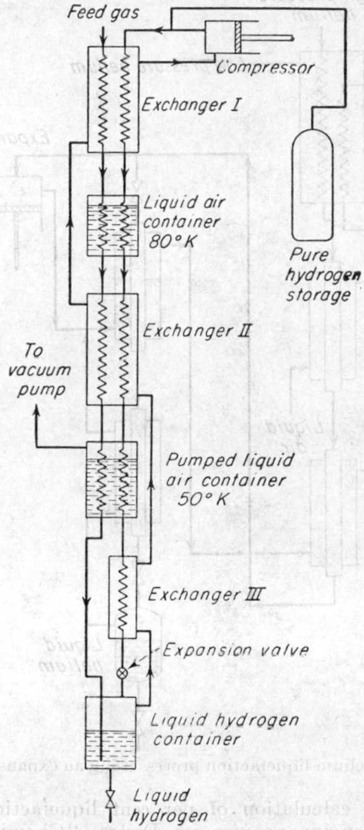

FIG. 41. Simplified diagram of the Kapitza scheme for hydrogen liquefaction.

directly to the liquid-hydrogen container, where it condenses by evaporating an equivalent amount of hydrogen. At the same time the impurities are condensed as solids and drawn off with the liquid product. The gaseous hydrogen, now completely purified, returns through the exchangers and is then compressed to about 150 atm. and is sent down through the various exchangers and liquid-air containers to the expansion valve, where it liquefies by the Joule-Thomson effect. This system must be started the first time with an independent source of purified hydrogen, but once started the supply of pure hydrogen in the storage system can be replenished from the hydrogen circulating in the closed system. Further details on plants using this scheme are given by Blanchard and Bittner [*Rev. Sci. Instruments*, **13**, 394 (1942)] and Huffman [*Chem. Rev.*, **40**, 1 (1947)].

To liquefy helium by the Joule-Thomson process requires precooling with liquid hydrogen, since the inversion temperature is well below the temperature of liquid air. Both hydrogen and helium can be liquefied by an expansion-engine process without the necessity of using any precooling by other fluids. Figure 42 shows a diagram of the process used at the Physics Laboratory of Yale University for the liquefaction of helium. This is the expansion-engine process developed by Kapitza with the addition of precooling by means of liquid air. Arthur D. Little, Inc., produces a helium liquefier and cryostat, developed by Dr. S. C. Collins, which uses two reciprocating expanders and no auxiliary cooling system. The helium is continuously recycled in a system especially designed to avoid contamination of the gas.

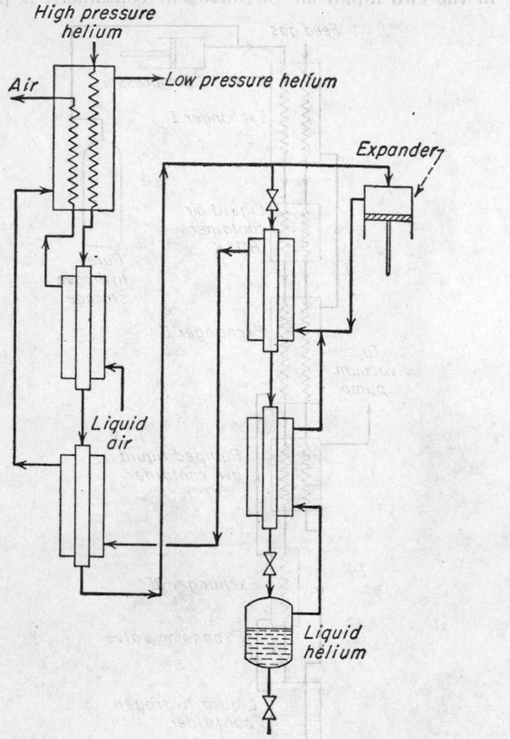

Fig. 42. Helium liquefaction process using an expansion engine.

For the calculation of per cent liquefaction by an expansion-engine process see Dodge, "Chemical Engineering Thermodynamics," McGraw-Hill, New York, 1944. Further information on the liquefaction of hydrogen and helium is available in the following papers: Helium—Lane, *Rev. Sci. Instruments,* **12,** 326 (1941); Hydrogen—Starr, *Rev. Sci. Instruments,* **12,** 193 (1941); Keyes, Gerry, and Hicks, *J. Am. Chem. Sco.,* **59,** 1426 (1937); Johnston, Begman, and Hood, P.B. 31880 (1946).

Rectification at Low Temperatures. A low-temperature gas-separation process is essentially a liquefaction cycle plus a rectification column. The latter does not differ in principle or materially in construction from those used above room temperature, but some conditions peculiar to low-temperature columns are worth mentioning. In the following discussion, attention will be focused on columns for rectification of liquid air.

Just as at ordinary temperatures, low-temperature columns must have boilers and condensers and may consist of a single or exhausting column or may have both exhausting and enriching sections as in double columns or in compound columns. It is important to note that the heat that is to be added in the boiler must always be obtained by removing it from some other cold part of the system and never from any outside source. Thus, in the single column or in the lower section of a double column, the heat for boiling is obtained by condensation of the air to produce the liquid feed. In a single column only about 70 per cent of the oxygen in the air can be recovered and pure nitrogen cannot be made, because the vapor leaving the top plate will be in a phase equilibrium with liquid air and will contain about 7 per cent oxygen.

In order to recover a high percentage of the oxygen and/or make pure nitrogen, a reflux of liquid nitrogen must be produced. This can be effected in several possible ways. Briefly, these are (1) evaporation of the liquid oxygen in a combination boiler-condenser at a reduced pressure such that there will be a sufficient

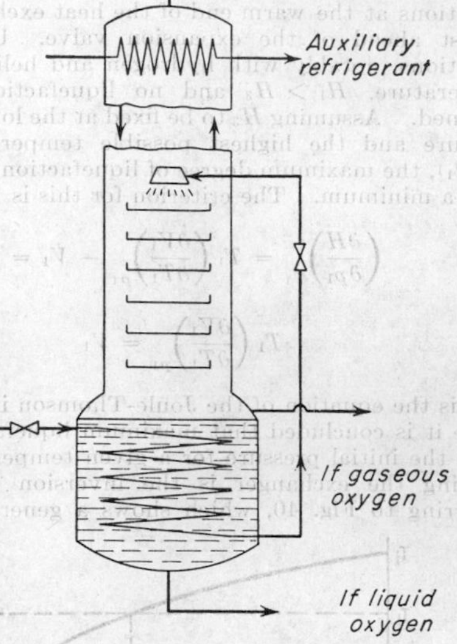

Fig. 43. Diagram of compound column.

temperature difference to condense pure nitrogen at atmospheric pressure; (2) use of an external nitrogen liquefying cycle based on any one of the three methods previously discussed; (3) use of an auxiliary non-liquefying refrigerating fluid such as helium or neon used in an external cycle with engine expansion to produce the low temperature required to liquefy nitrogen; (4) partial condensation of the feed air under pressure in a condenser-boiler, the heat being given up to the boiling oxygen with means provided for separating the condensate into two fractions, one of which is rich in nitrogen; and (5) use of a double column, one section operating at atmospheric pressure and the other at a pressure (4 to 5 atm.) sufficient to liquefy nitrogen at the oxygen boiling point.

Methods 1 and 3 are of minor importance at present. Method 2 is used in at least one large plant and is likely to increase in importance. Method 4 is the basis of the so-called "backward return" condenser of the Claude process, and method 5 is the method in most common use.

A column with both enriching and exhausting sections at substantially the same pressure and provided with refrigeration below the boiling point of nitrogen at the column pressure, say by method (1), (2), or (3) discussed above, will be known as a compound column and is illustrated in Fig. 43. Another form of compound column uses a nitrogen-liquefaction cycle in which nitrogen is condensed under pressure by the boiling oxygen and then throttled into the top of the column as the reflux (see Fig. 50).

Figure 44 shows a Linde double column, which consists of a column operating at elevated pressure surmounted by an atmospheric-pressure column. The boiler of the upper column is at the same time the reflux condenser for both columns. Gaseous air plus enough liquid to take care of heat leak into the column (more liquid, of course, if liquid-oxygen product is withdrawn) enters the exchanger at the base of the lower column and condenses, giving up heat to the boiling liquid and thus supplying the vapor flow for this column. The liquid air enters an intermediate point in this column, as shown. The vapors

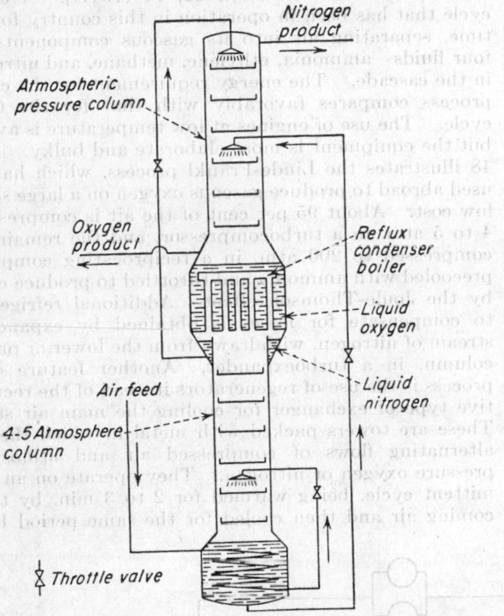

Fig. 44. Linde double column for air separation.

rising in this column are partially condensed to form the reflux, and the uncondensed vapor passes to an outer row of tubes and is totally condensed, the liquid nitrogen collecting in an annulus, as shown. If this column is operated at 4 to 5 atm. the liquid oxygen boiling at 1 atm. is cold enough to condense pure nitrogen. The liquid that collects in the bottom of the lower column contains about 45 per cent O_2 and forms the feed for the upper column. Such a double column can produce a very pure oxygen with high oxygen recovery or a very pure nitrogen with oxygen of moderate purity. It cannot produce both products simultaneously in a high state of purity, owing primarily to the fact that air is really a ternary mixture and the argon must be taken off with one of the products.

The determination of number of plates is made by the usual procedure for rectifying columns; but it is particularly to be noted that it is not satisfactory to treat air as a binary solution of oxygen and nitrogen when a pure oxygen (99 per cent or better) is to be made, because most of the plates are required to separate the oxygen-argon binary. This is a much more difficult separation than the oxygen-nitrogen binary; and, if the argon is not separated, only 95 per cent oxygen would be produced.

In analyzing a low-temperature, gas-separation process, the rectifying column must be examined for possible second-law violations. In other words, one must check to find out if there is sufficient reflux available to permit the attainment of the desired degree of separation. Details of methods for doing this for all three types of columns are given in the paper by Bliss and Dodge (*loc. cit.*). They are based on the use of enthalpy-concentration diagrams. Such an analysis also reveals the fact that there is appreciably more than the minimum amount of reflux necessary for separating the liquid feed, available in the upper section of a double column separating air into gaseous products. This makes possible either one of two procedures:

1. Some of the nitrogen may be withdrawn from the condenser-boiler as a vapor instead of being condensed and used as reflux and this nitrogen under pressure may be expanded in a turbine to produce some refrigeration.

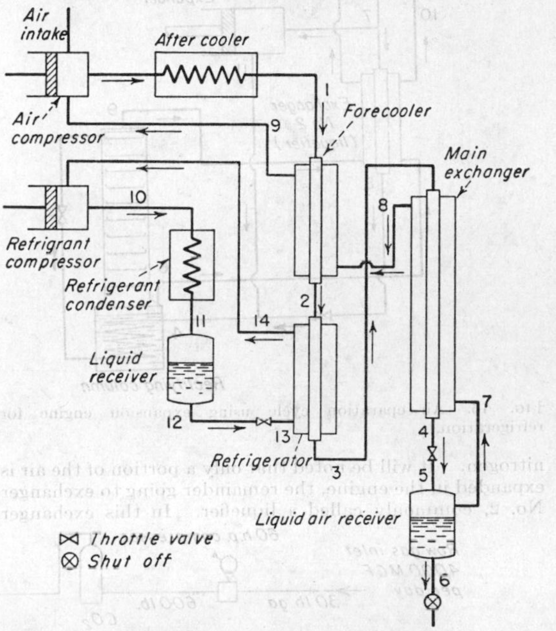

Fig. 45. Liquid-air cycle using Joule-Thomson effect supplemented by an auxiliary refrigerator.

This is used in the Linde-Fränkl process, to be described presently.

2. Extra air can be added to the upper column as a saturated vapor. Since this air requires very little compression, the net result is a greater yield of oxygen with no increase in power requirement. This scheme is usually called the Lachmann procedure. It poses a problem of purification of the extra air and of course reduces the driving force in the column and hence requires more plates. Further details on this procedure are given in the Bliss and Dodge paper.

Various Low-temperature Processes. There are many possible cycles for gas liquefaction and separation. Six of them are illustrated by flow sheets in Figs. 45, 46, 47, 48, 49, and 50. Figure 45 shows a Linde cycle for air liquefaction using two methods for obtaining the refrigeration, namely, precooling with a refrigerant such as ammonia or one of the Freons and the Joule-Thomson effect. The air is compressed to 100 to 200 atm., this high a pressure being necessary to obtain sufficient refrigeration by the Joule-Thomson effect. The precooling would generally lower the temperature of the compressed air to about −40°F., though lower temperatures could be used to advantage. Such a process is used for small liquid air, nitrogen, or oxygen plants.

Figure 46 shows an expansion-engine or Claude cycle for producing pure gaseous oxygen and an impure gaseous

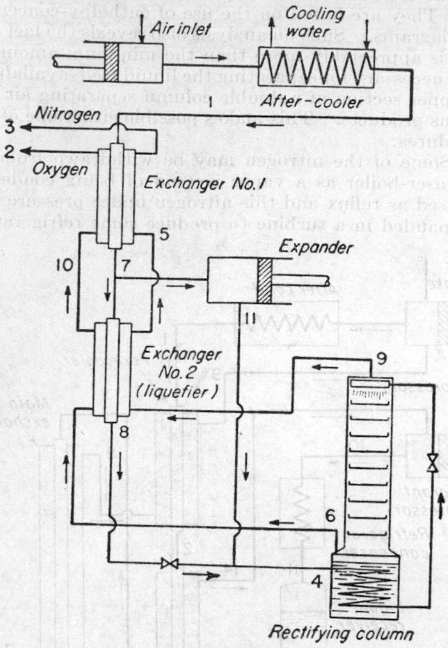

FIG. 46. Air-separation cycle using expansion engine for refrigeration.

nitrogen. It will be noted that only a portion of the air is expanded in the engine, the remainder going to exchanger No. 2, commonly called a liquefier. In this exchanger enough of the air is liquefied to compensate for heat leak into the column. For a liquid oxygen product, the portion to the liquefier would, of course, have to be greater. There are many possible variations of this cycle, but the one shown has been in common use in medium-sized plants for the production of both gaseous and liquid oxygen. With reciprocating expanders the cycle producing gaseous oxygen has operated with an air pressure of about 250 lb./sq. in., but this of course depends on the expander efficiency and size of the plant. Recently plants using a somewhat similar cycle but with efficient turboexpanders have operated at pressures as low as 70 lb./sq. in. gage.

Figure 47 shows a cascade cycle using the two fluids, ammonia and ethylene, for the liquefaction of natural gas. For further details on this process, see Miller and Clark [*Chem. & Met. Eng.*, **48**, 74 (1941)]. A cascade cycle that has been in operation in this country for some time, separating air into its gaseous components, uses four fluids—ammonia, ethylene, methane, and nitrogen—in the cascade. The energy requirement for the cascade process compares favorably with that for the Claude cycle. The use of engines at low temperature is avoided, but the equipment is more elaborate and bulky. Figure 48 illustrates the Linde-Fränkl process, which has been used abroad to produce gaseous oxygen on a large scale at low cost. About 95 per cent of the air is compressed to 4 to 5 atm. in a turbocompressor, and the remainder is compressed to 200 atm. in a reciprocating compressor, precooled with ammonia and throttled to produce cooling by the Joule-Thomson effect. Additional refrigeration to compensate for losses is obtained by expanding a stream of nitrogen, withdrawn from the lower or pressure column, in a turboexpander. Another feature of the process is the use of regenerators instead of the recuperative type of exchanger for cooling the main air stream. These are towers packed with metal in which there are alternating flows of compressed air and atmospheric pressure oxygen or nitrogen. They operate on an intermittent cycle, being warmed for 2 to 3 min. by the incoming air and then cooled for the same period by the

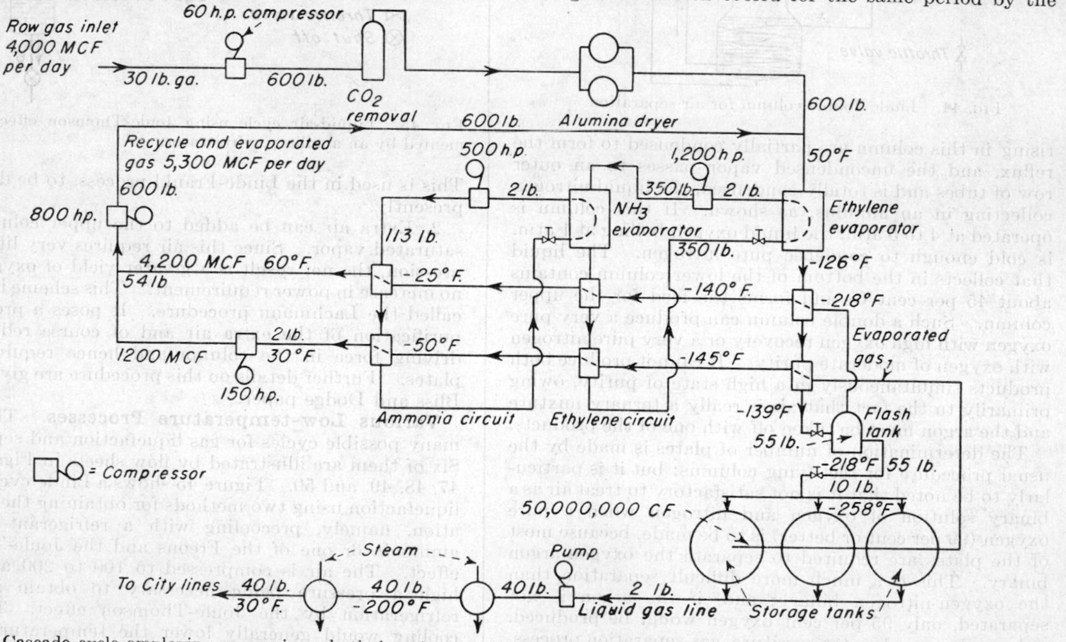

FIG. 47. Cascade cycle employing ammonia and ethylene for liquefaction of natural gas. [*Miller and Clark, Chem. Met. Eng.*, **48**, 74 (1941).]

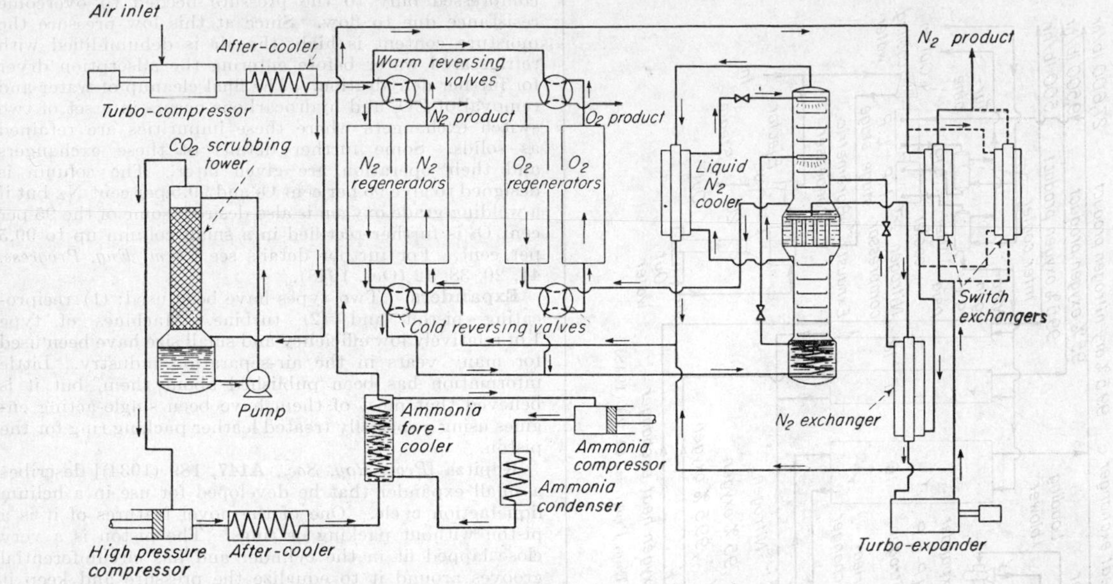

FIG. 48. Linde-Fränkl air-separation cycle.

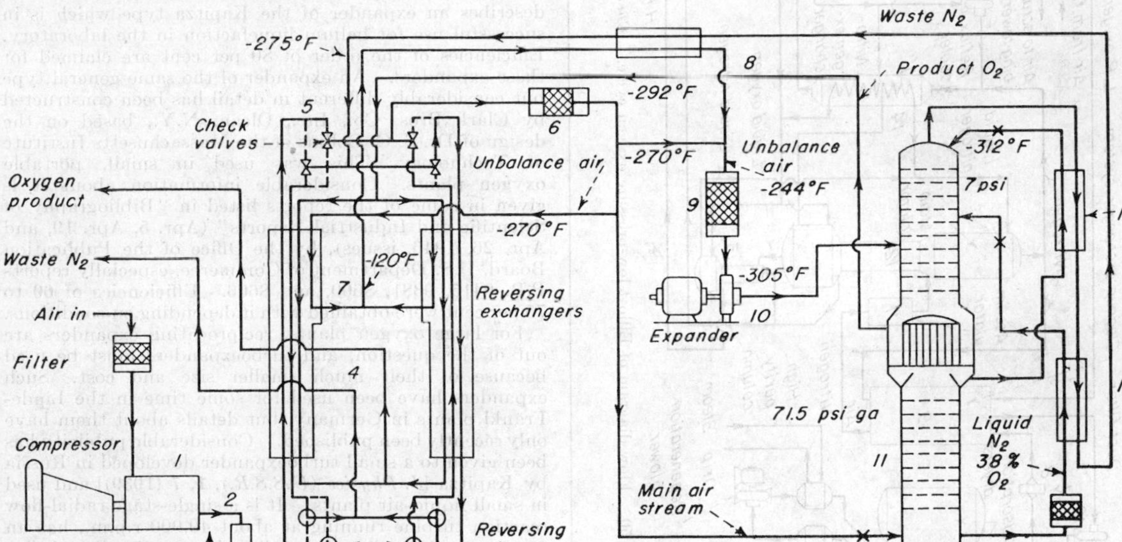

FIG. 49. M. W. Kellogg low-pressure cycle for gaseous oxygen. [*Chem. Eng.*, *March*, 1947, p. 134.]

separated gases leaving the column. In addition to giving efficient heat exchange and low pressure drop, the water vapor and CO_2 in the air are alternately deposited and reevaporated so that they act as purifiers as well as heat exchangers.

The M. W. Kellogg process for gaseous oxygen on a tonnage scale is shown in Fig. 49. This is a low-pressure (about 80 lb./sq. in. gage) expansion-engine process using a conventional double column and reversing exchangers for air purification. The oxygen stream does not reverse so that this process can produce high-purity oxygen if desired. The mass-unbalance system (to be explained later) is used to reduce the cold-end Δt of the exchangers. Extra, low-pressure air is added to the upper section of the column. Referring to the diagram, 2 is a catalytic oxidation unit to remove hydrocarbons, 5 shows the check valves for automatic control of flows at the cold end, 6 is an active-carbon filter, 7 indicates the high-pressure unbalance channel, and 9 is a silica-gel dryer for starting.

The Elliott tonnage oxygen process is shown diagrammatically in Fig. 50. The main features of this process are the following: The refrigeration and column reflux are supplied by a nitrogen liquefaction cycle in which nitrogen is compressed to about 80 lb./sq. in. gage, cooled in exchangers by returning nitrogen and by the oxygen product, and then a portion expanded in a turbine to supply the refrigeration needed while the remainder is liquefied and supplied to the column as reflux. The air is

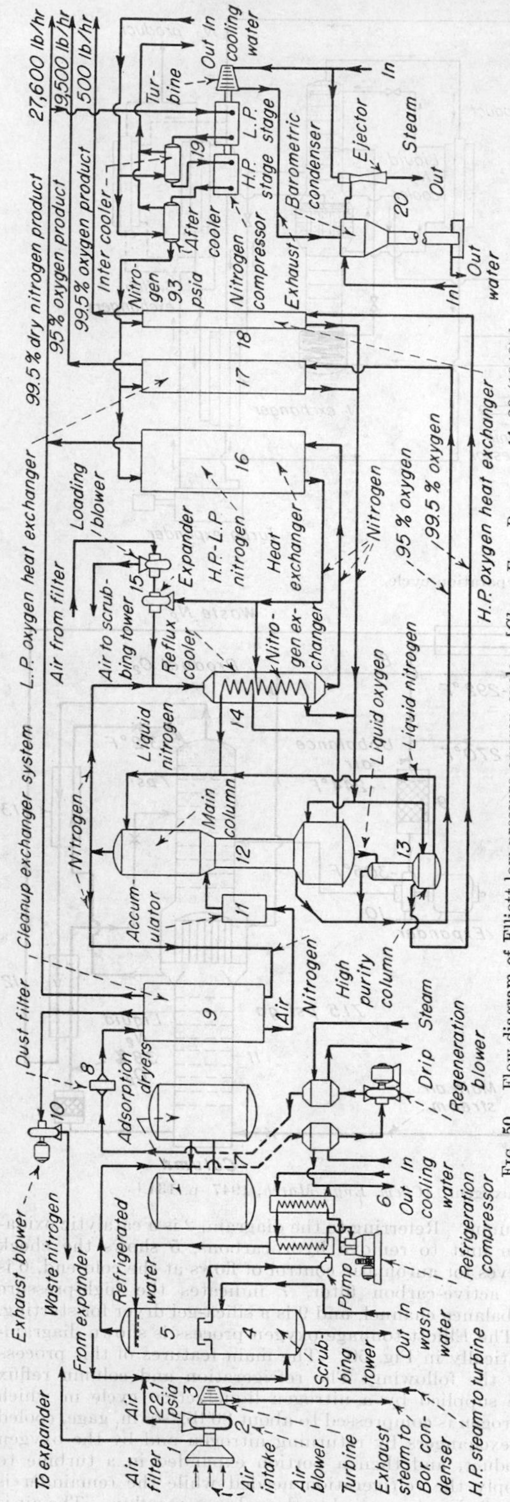

FIG. 50. Flow diagram of Elliott low-pressure oxygen cycle. [*Chem. Eng. Progress*, **44**, 20 (1948).]

compressed only to the pressure needed to overcome resistance due to flow. Since at this low pressure the moisture content is high, the air is dehumidified with refrigerated water before entering the adsorption dryer for further dehydration. The final cleanup of water and removal of CO_2 and hydrocarbons occurs in a set of two switch exchangers where these impurities are retained as solids. Some further details of these exchangers and their operation are given later. The column is designed to give 95 per cent O_2 and 99.5 per cent N_2, but if a welding-grade oxygen is also desired, some of the 95 per cent O_2 is further rectified in a small column up to 99.5 per cent. For further details see *Chem. Eng. Progress*, **44**, 20, 38, 42 (Oct. 1948).

Expanders. Two types have been used: (1) reciprocating piston and (2) turbine. Machines of type 1 of relatively low efficiency and small size have been used for many years in the air-separation industry. Little information has been published about them, but it is believed that most of them have been single-acting engines using a specially treated leather packing ring for the piston.

Kapitza [*Proc. Roy. Soc.*, **A147**, 189 (1934)] describes a small expander that he developed for use in a helium liquefaction cycle. One of the novel features of it is a piston without packing or rings. The piston is a very close lapped fit in the cylinder and has circumferential grooves around it to equalize the pressure and keep it centered. Lane [*Rev. Sci. Instruments*, **12**, 326 (1941)] describes an expander of the Kapitza type which is in successful use for helium liquefaction in the laboratory. Efficiencies of the order of 80 per cent are claimed for these expanders. An expander of the same general type but considerably different in detail has been constructed by Clark Bros. Co., Inc., Olean, N.Y., based on the design of Dr. S. C. Collins of the Massachusetts Institute of Technology. This was used in small, portable oxygen plants. Considerable information about it is given in some of the reports listed in "Bibliography of Scientific and Industrial Reports" (Apr. 5, Apr. 19, and Apr. 26, 1946, issues), by the Office of the Publication Board, U.S. Department of Commerce, especially reports P.B. 9415, 9381, 8600, and 8606. Efficiencies of 60 to 75 per cent were obtained with it depending on conditions.

For large oxygen plants, reciprocating expanders are out of the question, and turboexpanders must be used because of their much smaller size and cost. Such expanders have been used for some time in the Linde-Fränkl plants in Germany, but details about them have only recently been published. Considerable publicity has been given to a small turboexpander developed in Russia by Kapitza [*J. Physics* (*U.S.S.R.*), **1**, 7 (1939)] and used in small liquid-air plants. It is a single-stage radial-flow reaction turbine running at about 40,000 r.p.m., has an 8-cm. wheel, and can handle about 1250 lb. air/hr. Kapitza describes the principles underlying its design in some detail in his paper. A single-stage turbine of the same general type has been built by the Elliott Company and the Sharples Corporation in collaboration, and some details on its design and performance are available in the government reports referred to above. It has a 6⅞-in. wheel running at 22,000 r.p.m., is capable of an efficiency of 80 per cent with an expansion ratio of 5 to 1, and has a capacity of 7000 lb. air/hr. at operating temperature.

Some further details on the Elliott-Sharples expander are given in a paper by Swearingen [*Trans. Am. Inst. Chem. Engrs.*, **43**, 85 (1947)]. This paper gives the basis for the design and illustrates many of the mechanical details. It also gives some mechanical details and performance data on the German turboexpanders used in Linde plants as early as 1936. The Elliott Co. has since built larger expanders for tonnage oxygen plants but no details are available. Clark Bros. Co. has also built turboexpanders for large plants.

Compressors. In small air-separation plants the compressor has always been of the reciprocating type but in very large plants this type of compressor becomes so bulky and expensive that it is being replaced by the turbocompressor. This type is characterized by high speed, large capacity for a given bulk, and relatively low pressure ratio per stage. There is no theoretical limit to the pressure that may be obtained, however, if one uses enough stages or wheels in series. There is a limit to the pressure ratio in any one stage set by the velocity of sound in the gas. It is generally considered not to be good practice to approach too closely to the sonic velocity. Allowing a maximum velocity, say about 70 per cent of the sonic for the particular conditions, one can calculate at least approximately the pressure rise obtainable from a wheel by conversion of the kinetic energy to the energy associated with static pressure. To reach pressures of the order of 100 lb./sq. in., which are required for the tonnage oxygen processes, the compressor would probably be in at least two sections of about 4 or 5 stages each with intercooling between them.

Turbocompressors are of two general types, *axial flow*, and *radial flow* or *centrifugal*. Both are capable of handling very large volumes with high adiabatic efficiency. Efficiencies up to 88 per cent are claimed for the axial-flow type and somewhat lower for the centrifugal. In general the centrifugal machine is used for the higher pressures and it is also somewhat more flexible in handling varying capacities. Because of its high speed the best drive for the turbocompressor is a steam turbine. Centrifugal compressors have been built for pressures as high as 900 lb./sq. in. Pressures considerably higher than this are used in some low-temperature processes, and in those cases reciprocating compressors would still be used, as well as in all small plants where the volume does not lend itself to use of the turbocompressor. A recent series of papers by Karassik [*Chem. Eng.*, Oct., 110, Nov., 132, Dec., 126 (1947); Jan., 118, Feb., 134 (1948)] gives useful information on centrifugal compressors from the viewpoint of the process engineer.

Heat Exchangers. Efficient heat exchangers are essential parts of any low-temperature process since for good performance the gases returning from the low-temperature regions must be brought up to a close temperature approach with the entering gas. As long as plants were relatively small and pressures high enough so that pressure drop was not very critical, it was not difficult to build efficient, compact heat exchangers of a double-pipe or shell-and-tube type which were satisfactory. But the advent of large-size oxygen plants using low pressures of the order of 75 to 100 lb./sq. in. called for new types of heat exchangers which would combine large surface per unit of volume with low pressure drop. One of the important developments along this line was the Fränkl regenerator, which has been used extensively in oxygen plants in Europe. It consists of a cylindrical vessel packed with cylindrical disks, or "pancakes," of metal made by rolling a corrugated strip of aluminum or copper in a spiral. Figure 51 shows one of these spiral elements. The surface area per unit of volume is of the order of 500 to 1000 sq. ft./cu. ft., the pressure drop is of the order of a few lb./sq. in., and warm-end temperature approaches of 2° to 3°F. are common.

In an air-separation process, two regenerators are used for heat exchange between air and nitrogen and two for air and oxygen. Their operation is intermittent, as follows: in one-half of the cycle regenerator No. 1 is storing up heat taken from air passing through it and No. 2 is giving up heat to oxygen or nitrogen passing in a direction counter to that of the air. At the end of the half-cycle the flows are switched and No. 2 now becomes the storer of heat and No. 1 gives up heat to returning gas. In order not to have too great a temperature swing the half cycles must be quite short—of the order of 2 to 3 min. The switching is generally done by air-operated valves at the warm end of the regenerators, the cold-end flows being controlled by check valves. In addition to its heat-exchange function a regenerator performs an equally important function of air purification, which will be discussed under this heading. A recent paper [Lund and Dodge, *Ind. Eng. Chem.*, **40**, 1019 (1948)] gives performance data on small Fränkl regenerators.

Another significant development is the reversing exchanger. In its simplest form it consists of two adjacent finned passages through which air and oxygen or nitrogen pass in countercurrent flow. At the end of a short period of 2 to 3 min. the flows are reversed, *i.e.*, the air flows in the passage which formerly contained the pure separated gas and in the opposite direction. The action of a

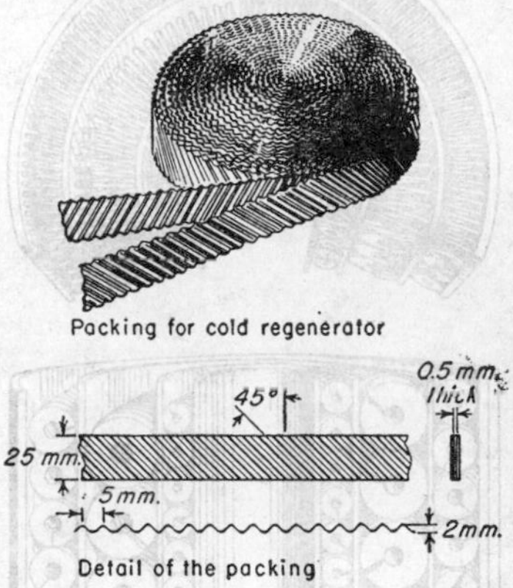

FIG. 51. Packing element of a Fränkl regenerator. (*FIAT Final Report* 840, *July*, 1946.)

reversing exchanger is recuperative rather than regenerative and the only reason for reversing the flows is that the exchanger also acts as a purifier of the air and frequent and regular reversal is necessary to prevent accumulation of the deposited solids. Extended surface obtained by the use of fins is necessary not only to obtain a large transfer surface per unit of volume but also to aid in retaining the impurities deposited as solids.

The idea of the reversing exchanger is not new, having been patented more than 20 years ago, but little use of it was made until World War II. Then reversing exchangers based on the work of S. C. Collins were extensively used in small, portable oxygen plants and they are now being installed in some of the large plants for tonnage oxygen.

The type of finning developed by Collins is shown in Fig. 52. This shows a three-annulus tube for three fluids, each annulus being packed with an edge-wound helix of copper ribbon solder-bonded to the tube walls. Performance data on exchangers of this type are given in a paper by Trumpler and Dodge [*Trans. Am. Inst. Chem. Eng.*, **43**, 75 (1947)]. The method of manufacture and the application to small oxygen plants is discussed by Collins [*Chem. Eng.*, **53**, 106 (Dec., 1946)]. Exchangers for large plants have been constructed by assembling in parallel with suitable manifolds a large number of the relatively small-diameter (about 3 in.) tubes of the general Collins type but with a different type of finning, details of which have not been disclosed.

Another type of extended-surface exchanger suitable for large plants has been disclosed by the Elliott Company. A model of a section of this exchanger is shown in Fig. 53 along with a view of the fin stampings, which are solder-bonded to the walls of the rectangular heat-exchange passages. Such an exchanger has a surface area of about 300 sq. ft./cu. ft., which is comparable to that in a Fränkl regenerator. This particular exchanger is used as a "switch" exchanger for air purification, as will be described later.

Purification. Little has been said about purification, but this is a very important operation in any low-temperature apparatus. In small plants, water vapor has been removed by activated alumina or other drying agents or deposited as ice in "switch" exchangers, a set of two exchangers in one of which the water is deposited as ice and in the other it is removed by thawing with warm air. After a suitable period of hours, the two exchangers are switched. Carbon dioxide is removed by

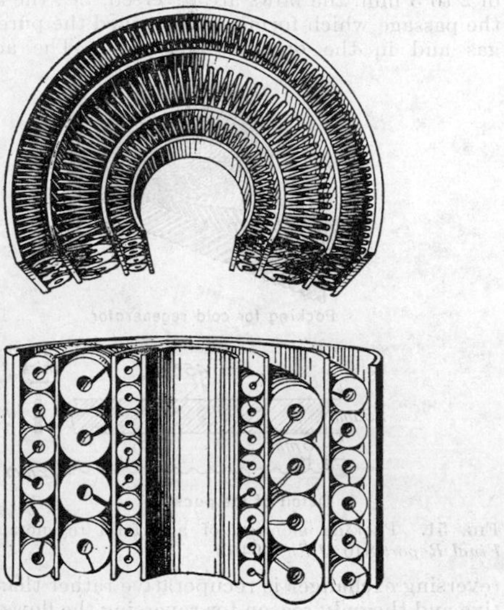

Fig. 52. Photograph of sections through ribbon-packed exchangers. [*Chem. Eng. Progress*, **43**, 77 (1947).]

solid KOH or by scrubbing in packed towers with caustic soda solution. For large-scale production of cheap oxygen the cost of chemicals rules out these methods, and the purely physical method of purification by regenerators or reversing exchangers is preferred.

The principle on which the purifying action of regenerators and reversing exchangers depends is as follows. When air flows through the passages and cools, water and carbon dioxide deposit as solids at the places where the partial pressure of vapor in the air stream exceeds the vapor pressure of the deposited solids. Because of the effect of the large surface and the fact that the gas is at all times in close proximity to a colder surface, the solids are retained at the point of deposition. When the flow is reversed the returning gas is somewhat colder and hence the solid deposit at any point will also be a little colder. Consequently it will have a lesser tendency to evaporate but counteracting this is the fact that the total pressure is lower, and hence at any level in the exchanger the partial pressure of vapor will be considerably less than it was during the air-cooling half of the cycle. There must be some value of Δt for a given pressure difference at which the two opposing tendencies are just balanced. At greater values of Δt (the value varying with temperature level in the exchanger) the deposits will not completely evaporate in the gas-warming phase of the cycle and hence will accumulate; at lower values of Δt the exchanger should be entirely self-cleaning. Further details on the purifying action of reversing exchangers are given in a paper by Lobo and Skaperdas [*Trans. Am. Inst. Chem. Engrs.*, **43**, 69 (1945)]. The purifying action is treated quantitatively in this paper and it is shown that the critical job is CO_2 removal, the allowable Δt for water being generally greater than the actual Δt, at least for all temperatures above -100°F. On the other hand the maximum allowable Δt for CO_2 removal is of the order of 10°F. and with a balanced mass flow in the exchanger the cold-end Δt will be greater than this due to difference in specific heats and to heat leak. To get around this difficulty Trumpler devised the "unbalanced-flow" system. There are several variants of this but the general principle of all of them is to unbalance the flow in such a way as to increase the heat capacity of the returning low pressure gases relative to that of the air. One way to

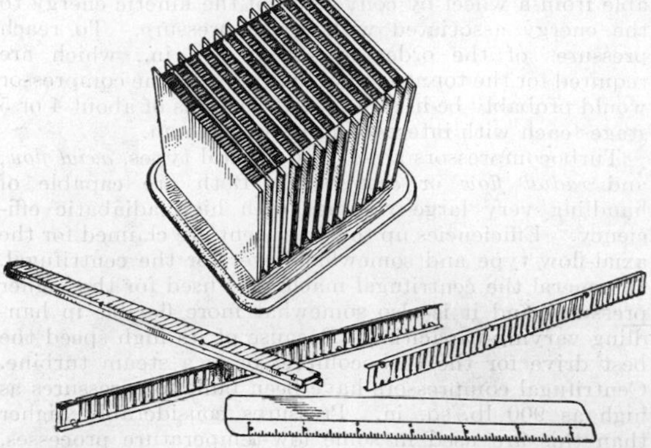

Fig. 53. Elements of the Elliott heat exchanger. (*Courtesy of Elliott Co.*)

accomplish this is to bypass a portion of the air around the cold end of the exchanger and send it to the expander or, better still, return a portion of the compressed air that has passed through the exchanger back through it part of the way in a separate non-reversing channel and then to the expander. This second procedure, used in the process illustrated in Fig. 49, is equivalent to the first as far as mass unbalance is concerned but is obviously better from a purification standpoint. Another method is to split the returning nitrogen into two portions, one of which goes through the regular purifying channel and the other (smaller) portion goes through a special channel known as the unbalance passage. This portion of the N_2 is then recirculated back to the main N_2 passage so that it passes twice through the cold end of the exchanger and thus increases the mass of the returning gas in relation to that of the air.

A third method of mechanical (also called "internal") purification by deposition in exchangers is the use of switch exchangers. In this case two exchangers alternate on- and off-stream. During the on-stream period the solid impurities accumulate on the transfer surfaces as in the other two types of exchangers but the period is much longer. In the Elliott process the exchangers are switched every 4 hr. When off-stream, the accumulated solids are removed by evaporation into a stream of the product nitrogen which has been heated a little. By passing the stream of clean nitrogen through the exchanger first in the nitrogen passage from the warm to the cold end and then back through the plugged air passage in the opposite direction, temperature levels throughout the exchanger are raised only a few degrees and the scavenging operation is accomplished without much loss of cold.

All large oxygen plants now building or planned apparently intend to use one of the three methods of internal purification for CO_2 and H_2O removal, but in one process a preliminary dehydration is accomplished in alumina dryers. Chemical removal of CO_2 is apparently not in the running, though it would seem to offer some advantages from the standpoint of trouble-free operation. The cost for chemicals could be reduced materially by recausticizing the carbonate with lime.

Hydrocarbons in the form of oil spray or of oil vapor and gases such as methane, ethylene, and acetylene are impurities in air that can cause trouble. Oil spray can be removed by filters and oil vapor by adsorbents, but the lower hydrocarbons tend to accumulate in the liquid oxygen and are generally removed by regular purging. Acetylene in particular must be prevented from accumulating, since it constitutes an explosion hazard.

The conditions under which acetylene will accumulate in a low-temperature system can be calculated at least approximately from available vapor-pressure data on acetylene. Such calculations show that it would accumulate in the boiler of an oxygen column if the concentration in the entering air was more than about 0.05 p.p.m. This calculation is subject to several uncertainties and should be relied on only as a general guide. One scheme for avoiding acetylene accumulation is to locate the air intake as far as possible from sources of contamination. Then a regular determination of the acetylene content of the liquid oxygen should be routine and if dangerous accumulation is indicated, *i.e.*, approaching the solubility limit, some of the liquid can be purged to the atmosphere. Another effective means of controlling acetylene is to provide a silica gel adsorber in the air line just before the rectifying column. The amount adsorbed is so small that reactivation is not a problem. In passing, it might be mentioned that the temperature at the cold end of regenerators or reversing exchangers is not low enough for effective removal. However, the switch exchangers of the Elliott process operate on air at a much lower pressure so that the cold-end temperature is 30° to 40°F. lower and it is claimed that they are quite effective in acetylene removal. In spite of this, the precaution of providing a silica-gel adsorber is taken.

Low-cost Tonnage Oxygen. This had been considered to be "just around the corner" for more than 30 years but is at last a reality. This is probably due both to developments in oxygen manufacture during the war and to a better understanding of the potentialities of low-cost oxygen and hence a more insistent demand. The latter is probably the more important reason as the principles and techniques of large-scale oxygen manufacture were well known before the war. The uses for oxygen on a tonnage scale that are now either actually existent or considered likely to be developed soon may be listed as follows:

1. Complete gasification of coal or other solid fuels by continuous processes. The objectives here are either to produce a gas of high thermal value and which can, if desired, be transported long distances, or to produce a gas to be used for the synthesis of ammonia, methanol, or liquid fuels.
2. Reaction with natural gas to yield a synthesis gas for producing hydrocarbons for liquid fuels and oxygenated organic compounds.
3. Oxidation of ammonia to nitric oxides, SO_2 to SO_3, and hydrocarbons to oxygenated derivatives.
4. Roasting of sulphide ores so as to permit better thermal efficiency, easier recovery of SO_2, and utilization of lower grade ores.
5. Smelting of iron ore in the blast furnace.
6. Refining of iron to produce steel by the Bessemer or open-hearth processes.
7. Production of very high temperatures by combustion for special reactions or melting operations that require such temperatures.

The advantages of oxygen in all these cases are, of course, due to the elimination of nitrogen and for the most part have been recognized for a long time but have awaited the availability of oxygen at a price of the order of \$3/ton (12.5¢/1000 cu. ft. at 80°F. and 1 atm.). Some of these uses are still only in the talking stage or in the pilot-plant stage but plants are actually operating in this country or are in the process of design and construction for many of these uses.

The majority of steel companies are actively experimenting on the use of oxygen in open-hearth furnaces and Bessemer converters and some with oxygen-enriched air for blast furnaces. The first two uses have been pretty well established as practical and economic but insufficient work has been done on the blast-furnace use. For considerable detail on these applications along with test data see a paper by Strassburger [*Steel*, **123**, 16, 148 (1948)].

No cost figures are available from actual installations but many estimates have been made which indicate that oxygen can be produced in a very large plant at a cost approaching or even bettering the figure given above. The data in Tables 3 and 4 are taken from Downs and Rushton (*loc. cit.*). By combining oxygen manufacture with synthesis of liquid fuels from natural gas and utilizing the waste heat from the synthesis to produce steam for turbines to drive the air compressors, it is estimated [Keith, *Am. Gas. Assoc.* Monthly, **28**, 253, 296 (1946)] that oxygen can be made in an 850-ton/day unit for \$1.15/ton (4.7¢/1000 cu. ft.).

Table 3. Estimated Total Plant Costs, October, 1946, for Producing 95 Per Cent Oxygen

Plant capacity:					
Tons/day	120	240	360	480	1,000
Tons/hr.	5	10	15	20	42
Millions of cu. ft. of 95 per cent O_2/day	2.88	5.76	8.64	10.52	24.00
Approximate plant cost	\$900,000	\$1,300,000	\$1,700,000	\$2,000,000	\$3,400,000

Table 4. Estimated Operating Charges, October, 1946, for Producing 95 Per Cent Oxygen

	Plant capacity, tons/day				
	120	240	360	480	1,000
Total utility cost/ton*	\$1.29	\$1.27	\$1.25	\$1.22	\$1.20
Operating labor/ton†	0.68	0.34	0.23	0.17	0.08
Maintenance/ton‡	0.54	0.39	0.34	0.30	0.25
Fixed charges/ton §	2.70	1.95	1.70	1.50	1.22
Total operating cost/ton	\$5.21	\$3.95	\$3.52	\$3.19	\$2.75
Total operating cost/1000 cu. ft.	21.7¢	16.5¢	14.7¢	13.3¢	11.5¢
Kw.-hr./1000 cu. ft. for compression	13.4	13.1	12.8	12.7	12.5
Kw.-hr./1000 cu. ft. for auxiliaries	2.2	2.2	2.2	2.2	2.2
Total kw.-hr./1000 cu. ft.	15.6	15.3	15.0	14.9	14.7

* This figure includes charges for a negligible amount of make-up water at 1¢/1000 gal. electric power for pumps, cooling towers, and auxiliaries at 0.4¢/kw.-hr.

† Operating labor is taken at \$80/day including labor overhead.

‡ 2½ per cent of plant cost.

§ 12½ per cent of plant cost.

Oxygen of 90 to 95 per cent purity is entirely satisfactory for most of these uses and is cheaper to make than the 99.5+ per cent grade required for welding and cutting. The lower purity grade permits the use of regenerators which cannot make the high-purity grade because of the inevitable mixing of some air with the oxygen when the flows are reversed. Although the 99.5 per cent oxygen can be made with reversing exchangers, they require a third passage for the oxygen in heat-exchange relationship with the other two but in which the flow is not reversed. This complicates the exchangers and increases their cost. The lower purity oxygen also requires appreciably less reflux and fewer plates, since the difficult argon-oxygen separation is avoided. Several processes have been proposed for tonnage oxygen manufacture and three of them were described briefly in a previous section. At the present writing several plants with capacities from 150 to 400 tons of oxygen per day are in actual operation and one plant is now in construction for a capacity of 1700 tons/day in two units. No data on any of these plants have been released. It is perhaps desirable to recall that up to the present time no tonnage oxygen process except the Linde-Fränkl has any considerable operating experience behind it. About 75 Linde-Fränkl plants have been built in Europe, the largest having a capacity of almost 135 tons/day. For further information on large oxygen plants in this country and on estimated costs and potential uses, reference is made to Downs, *Chem. Eng.*, 113; 121 (Aug., 1948).

Air enriched in oxygen content to a value of 25 to 50 per cent may find large-scale uses, for example in blowing blast furnaces, to cite one potential use. Enriched air can be made in either of two ways: by a direct elimination of some of the nitrogen by a low-temperature process, or by dilution of 90 to 95 per cent oxygen with air. The ideal or reversible work to make, by a separation process, enriched air with 40 per cent O_2 is only 55 per cent of that required to make it by dilution, per unit of contained oxygen, and the ratio is still more favorable for lower enrichment. However, much of the theoretical advantage in power requirement is wiped out when actual processes with their unavoidable irreversible effects are considered, and it is the judgment of some of the experts who have studied the question that enriched air can best be made by dilution. There is a considerable practical advantage to the latter process in that the low-temperature system would handle much less air. For some further discussions of enriched air processes see Hochgesand, *Mitt. Forsch. Anst. GHH Konzern*, **4**, 14 (1935); Shuftan, Conference on Developments in Industrial Production and Use of Gaseous Oxygen, Inst. of Chem. Engrs. and Physical Society (Low Temperature Group) (London), Feb. 13, 1948; FIAT final report No. 1120 (P.B. 88840).

Miscellaneous Processes. The separation of hydrogen from coke-oven gas by low-temperature methods has been developed to a process of some importance in Europe but it has never found any application in this country. For this reason it will not be discussed but reference is made to Ruhemann (*loc. cit.*) for details.

For some years the U.S. Bureau of Mines has operated plants for the extraction of helium from natural gas and during the last war the production was greatly expanded. The history of the development up to 1937 is reviewed by Seibel [*Trans. Am. Soc. Mech. Engrs.*, **59**, 55 (1937)]. The helium content of the natural gases processed in the various plants varies from 1 to 8 per cent, the nitrogen content from 12 to 80 per cent, with the balance chiefly methane but with small amounts of higher hydrocarbons also present.

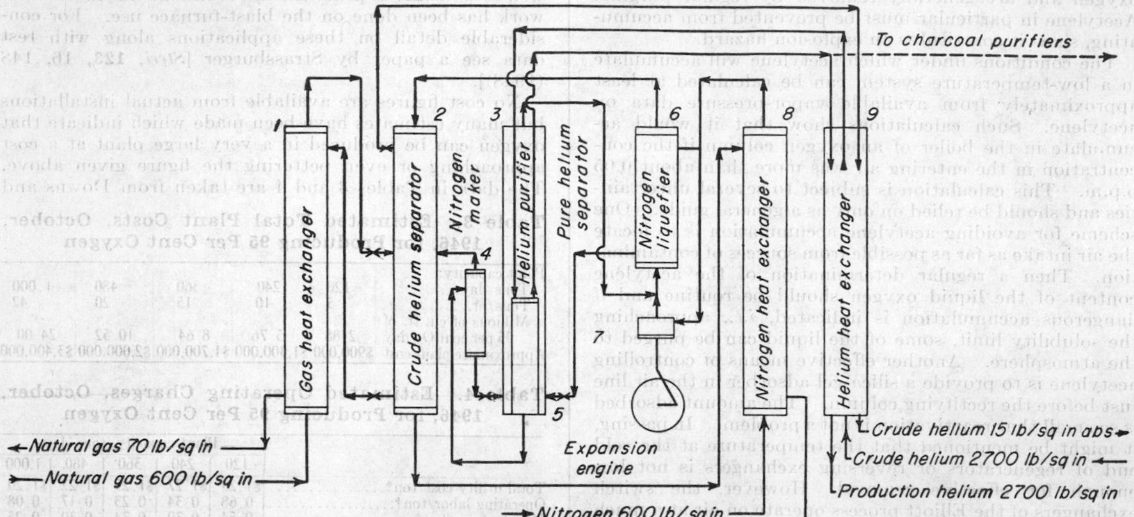

FIG. 54. Simplified diagram of Bureau of Mines helium separation process. [*Chem. Eng. Progress*, **44**, 567 (1948).]

Figure 54 shows a flow diagram of the Bureau of Mines helium separation process. Refrigeration is obtained from two sources (1) Joule-Thomson expansion of the gas from about 600 lb./sq. in. gage, and (2) an auxiliary nitrogen refrigeration cycle involving engine expansion from 600 to 15 lb./sq. in. abs. No rectification is used since the difference between the boiling points of nitrogen and helium is so great that a single partial condensation step will effect a considerable enrichment to yield a crude helium of 60 per cent content and a second partial condensation at higher pressure yields a helium of 98.5 per cent. A further purification to 99.7 per cent or higher can be achieved by adsorption on activated carbon at about −275°F. Rectification would require liquefaction of the helium and this would require refrigeration to much lower temperatures.

Referring to the diagram, the gas at 600 lb./sq. in. after treatment to remove CO_2, H_2S, and water vapor is cooled and almost completely condensed in exchanger 1 by returning low-pressure gas and then throttled into separator 2 at a pressure of 200 to 300 lb./sq. in., where both phases are further cooled by nitrogen vapor from the auxiliary cycle. The gas phase withdrawn from the separator is the crude helium product of 60 per cent helium and 40 per cent nitrogen. It is brought to atmospheric temperature in exchanger 9 by exchange of heat with itself after compression to 2700 lb./sq. in. on its way to further purification. The liquid phase from separator 2 is throttled and passed back through exchanger 1 and is brought to atmospheric temperature and finally recompressed to pipe-line pressure. The second partial condensation is carried out at 2700 lb./sq. in. in vessel 5 surrounded by liquid nitrogen in 3. The gas phase from this condensation, which is 98.5 per cent helium, is brought to atmospheric temperature in 9. The liquid phase from the partial condensation in 5 which is mostly nitrogen is throttled to 4 to flash off dissolved helium which joins the crude helium stream in 2. The liquid phase from 4 passes to vessel 3 where it serves as the cooling agent for the helium purification step. The rest of the diagram shows the nitrogen liquefaction cycle for supplying the liquid which maintains the bath in 3 around the partial condenser 5. For further details on the process and on the plant equipment reference is made to a paper by Mullins [*Chem. Eng. Progress*, **44**, 567 (1948)] from which the diagram in Fig. 54 was taken. This paper also gives some enthalpy data and phase equilibrium data for the natural gas processed at the Otis, Kansas, plant of the Bureau of Mines.

Removal of nitrogen from natural gases of high nitrogen content by low-temperature rectification is a process of potential importance. There are large quantities of gas that have nitrogen contents ranging from 1 to 34 per cent and appreciable amounts with a much higher nitrogen content. Removal of the inert nitrogen would have obvious economic advantages due to the increase in heating value of the gas. This is discussed in some detail in a paper by Mullins and Wilson, Prospective Benefits from Removing Excess Nitrogen from Natural Gas, published by the American Gas Association (1948). This paper also contains an extensive bibliography on gas separation. Processes for nitrogen removal are treated in considerable detail in a paper by Deschner and Bodle [*Oil Gas J.*, Apr. 15 and 29 (1948)]. Flow diagrams, material balances, heat balances, and power requirements are given for three processes.

The separation of light hydrocarbons in a more or less pure state from mixtures such as petroleum refinery gases and cracked natural gases is an important operation at the present time, conducted primarily to recover the olefins for use in chemical syntheses. A typical process for producing 95 per cent ethylene is described by Pratt and Foskett [*Trans. Am. Inst. Chem. Engrs.*, **42**, 149 (1946)] as follows:

The olefin-bearing gas from the cracking furnace or other source is compressed in three or four stages to a pressure of 600 lb./sq. in. gage. The gas is cooled between stages and small quantities of condensate and water removed in interstage separators. The gas leaving the last stage of compression is cooled to an economical approach to the cooling water temperature, then subcooled to approximately 70°F. by high-level refrigeration. The cracked gas, after removal of water and hydrocarbon condensate, flows to the dehydrators. These units remove the remaining moisture from the gas by adsorption on activated bauxite or alumina. The dried gas is then cooled to 0°F. and introduced to the demethanizer tower. Hydrogen and methane are taken overhead in this tower and released to the plant fuel system. Reflux is produced by a runback-type condenser refrigerated by the evaporation of liquid ethylene at 5 lb./sq. in. gage. The reflux temperature is approximately −130°F. A small quantity of ethylene is lost in the overhead product in order to satisfy the dew point requirements at the top of the tower of −130°F. and 575 lb./sq. in. gage. The bottoms from the methane tower, essentially free of hydrogen and methane, flow to the ethylene tower, where ethylene is removed as an overhead product. Reflux is produced by a condenser refrigerated by propane or ammonia evaporating at −15°F.

The subsequent processing of the ethane and heavier components which comprise the ethylene tower bottoms depends on the number and the purity of the products desired. An ethane tower and propylene tower are required if a propylene-propane product is required. On the other hand, if ethylene is the only primary product, the ethane and C_3's may be recycled to the cracking furnace; in which case, a single tower would suffice to separate the recycle stream from the C_4's and heavier components.

Table 5 shows the compositions of the gas streams at various points. Low-level refrigeration is obtained from a vapor-compression refrigeration system using ethylene which is evaporated at −150°F. and condensed by propane or ammonia, which in turn discharges the heat to cooling water. The paper referred to gives further details on operating conditions, gas compression, power requirement, purification of the feed gas, materials of construction, and instrumentation.

Table 5. Stream Compositions Mole Per Cent

Constituent	Feed	Methane tower		Ethylene tower		Ethane tower		Propylene tower	
		Net overhead	Bottoms	Net overhead	Bottoms	Net-overhead	Bottoms	Net overhead	Bottoms
Hydrogen	11.5	40.0							
Methane	16.5	56.0	0.5	2.0					
Ethylene	19.4	4.0	25.6	95.0	0.9	2.8			
Ethane	17.7		24.9	3.0	32.6	95.0	1.4	1.5	
Propylene	8.9		12.5		17.0	2.2	24.3	25.7	
Propane	24.0		33.7		45.7		68.6	72.2	5.0
C_4's	1.5		2.1		2.9		4.3	0.6	69.0
C_5's	0.5		0.7		0.9		1.4		26.0
Total	100.0	100.0	100.0	100.0	100.0	100.0	100.0	100.0	100.0
Moles/100 moles of feed	100	28.8	71.2	18.7	52.5	17.5	35.0	33.1	1.9

Low-temperature Insulation. The minimization of heat leakage from the surroundings into apparatus operating at low temperatures is a very important factor in any low-temperature process but lack of space precludes any extensive treatment of the subject. Fortunately a symposium on the subject has recently been published in *Chem. Eng. Progress* for 1948 and attention is called to the following papers in this symposium:

White, Problems and Developments, **44,** 647; Palmer and Taylor, Glass-fiber Insulation at Liquid-air Temperatures, **44,** 652; Stone and Bradley, Thermal Insulation for Low-temperature Application, **44,** 723; McIntire and Kennedy, Styrofoam for Low-temperature Insulation, **44,** 727; Buskirk and Surland, A New Plastic Low-temperature Insulation, **44,** 803; Cellular-glass Insulation for Low-temperature Equipment, **44,** 804.

SECTION 26

PLANT LOCATION*

BY

Robert S. Aries, M. Ch. E., M.A. (Econ.), M. Sc., D. Ch. E., Consulting Engineer and Economist; Adjunct Professor of Chemical Engineering, Polytechnic Institute of Brooklyn, N.Y.; Technical Director, Northeastern Wood Utilization Council; Member American Chemical Society, National Society of Professional Engineers, American Economic Association, Technical Association of The Pulp and Paper Industry, American Statistical Association, Society for the Advancement of Management, American Marketing Association, American Association for the Advancement of Science, Chemical Market Research Association, American Management Association, American Society of Mechanical Engineers, American Institute of Chemists, Association of Consulting Chemists and Chemical Engineers.

CONTENTS

	Page
Aim and Method of Plant Location Studies	1720
Main Determinants of Plant Locations	1720
Raw Materials	1720
Fuels and Power	1721
Labor Supply	1722
Geographic Factors	1723
Water Resources	1723
Transportation Facilities	1724
Markets	1724
Laws and Public Practices	1725
Company Policies	1725
Possible Other Factors	1725
General Sources of Information	1725
Aids in Plant Location Analysis	1726
Maps	1726
Raw-material Surveys	1726
Labor Surveys	1727
Two Steps to Determine Plant Location	1727
Methods Used in Making Preliminary Selection	1727
Apparent Association	1727
Questionnaires	1727
Relative Cost Magnitudes	1727
Sieve Procedure	1728
Final Comparative Evaluation of Sites	1728
Alternative Method of Weighted Scores	1729
The Role of National Defense in Choosing a Plant Location	1729

REFERENCES: Hoover, The Location of Economic Activity, Chap. 25 in "The Growth of The American Economy" by Williamson, Prentice-Hall, New York, 1944. Hoover, "Location Theory and the Shoe and Leather Industries," Harvard University Press, 1937. Weber, "Theory of the Location of Industries," University of Chicago Press, Chicago, 1929. Losch, "Die Raumliche Ordnung der Wirtschaft," 2d ed., Gustav Fischer Verlag, Jena, 1944; University Microfilms, Ann Arbor, Mich., 1942. McLaughlin, "Growth of American Manufacturing Area," University of Pittsburgh, 1938. Fischer, What to Consider before Relocating a Plant, *Iron Age*, **129,** 1007 (1932). Wood, Where to Locate the Plant, *Factory Management*, **95,** 23 (1938). Kirkpatrick, Why These New Chemical Industries Went South, *Chem. & Met. Eng.*, **41,** 400 (1934). Elbourne, Selection of Works Site, *Engineering*, **139,** 288 (1935). Garver, Boddy, and Nixon, "The Location of Manufacturers in the U.S. 1899–1929," University of Minnesota Press, Minneapolis, 1933. Mears, Strategy in Industrial Location, *Harvard Bus. Rev.*, **17,** 1 (1938). Warner, Significance of Location upon Production Costs, *Trans. Am. Inst. Chem. Engrs.*, **32,** 193 (1936). "Power Requirements in Electrochemical, Electrometallurgical and Allied Industries," Federal Trade Commission, 1938. "Maps of Selected Industries Reported in the Census of Manufacturers 1937," Bureau of the Census, 1941. U.S. National Resources Committee, "The Structure of American Economy," Washington, D.C., 1939. U.S. Planning Board, "Industrial Location and National Resources," Washington, D.C., 1942. Plant Location with an Eye to the Future, *Chem. & Met. Eng.*, **48,** 80 (1941) (supplemented by Plant Location bibliography, 1931–1941). West, Selecting the Site for a Factory, *Chem. Ind.*, **40,** 369 (1937). Raisz, Geographical Distribution of the Mineral Industry of the United States, *Mining Met.*, March, 1941. Report on the Location of Industry, *Political and Economic Planning*, London, 1939. Perry and Cuno, Economic and Technical Factors in Chemical Plant Location, and New Bibliography on Plant Location, *Chem. & Met. Exp.*, **41,** 434 (1934). U.S. Census Bureau, "Location of Manufactures 1899–1929," Washington, D.C., 1933. Dennison, "Location of Industry and the Depressed Areas," Oxford, New York, 1940. Taylor, Discussion on the Geographical Distribution of Industry, *Geog. Jour.*, **42,** 22 (1938). *Monthly Labor Review* (Department of Labor); "Minerals Yearbook" (Bureau of Mines); Census Reports (Bureau of the Census); various studies of the Bureau of Foreign and Domestic Commerce; also see Sec. 26 of the previous editions of this handbook for a treatment by Cuno entitled Economic Factors in Chemical Plant Location. Aries, "Techniques & Theory of Plant Location with Special Reference to the Chemical Process Industries," Overland Commercial Corp., Brooklyn, N.Y., 1949. Papers presented to the 41st Annual Meeting of the American Institute of Chemical Engineers, New York, 1948, Dr. R. S. Aries presiding: Aries, Introductory Remarks; Boyd, Mineral Resources of the United States and Their Relation to Industry Location; Eskew, Some Agricultural Resources of the United States and Their Relation to Industry Location; McLaughlin, Location of Industries in the South; Dimmitt, National Security Factors in Plant Location; Powell and von Lossberg, The Relation of Water Supply to Chemical Plant Location;† Aries and Othmer, Theory and Methods of Determining Plant Location;† Sommers, Location Factors in Heavy Chemical Plants;† Hoyer, Location Factors of the Plastics Industry;† Bell, Factors in the Location of Fertilizer Plants; Faith, Plant Location in the Agricultural Process Industries;† Sittenfield, The Economics of Petroleum Chemical Plant Location.† Hoover, "The Location of Economic Activity," McGraw-Hill, New York, 1948. McLaughlin, "Why Industries Move South," National Planning Assn., 1949.

† Published in *Chem. Eng. Progress*, May, 1949.

* Acknowledgment. In any field so long and widely cultivated as that dealt with here, any sudden large accessions of new knowledge are scarcely expected, and striving for originality could readily be carried beyond the bounds of practical wisdom. The single aim of the present treatment has been to try to provide the utmost helpfulness to practicing engineers in need of guidance or reminder on problems of plant location. We have therefore thought it proper, and indeed necessary, to draw freely upon the thought and language of the comprehensive presentation of this subject by Dr. Charles W. Cuno, which appeared as Sec. 26 of the second edition of this handbook, and appreciative acknowledgment is hereby made accordingly.

PLANT LOCATION

AIM AND METHOD OF PLANT LOCATION STUDIES

In any problem of plant location, the aim must be *to select the place that will enable the necessary materials to be assembled, the manufacturing processes to be carried out, and the product to be delivered to the customers, at the lowest total cost.*

This, of course, is simply a statement in explicit terms of the general problem of plant location; but in this statement will be found the master key for the solution of almost all such problems. Although the ultimate administrative choice of a location may in some cases be influenced by intangible considerations, not capable of being assigned any definite weight in the comparison of costs, yet the function of the engineer will generally consist in pointing out and measuring in dollars all the elements of the particular problems that are capable of at least approximate evaluation, and in making the necessary final comparison of indicated total costs among all the potential sites of sufficient promise to justify their inclusion. In doing this, he will naturally seek to make his analysis as comprehensive as practicable and will not fail to include in it some mention of the imponderables known to him or discernible through his study but not assigned any definite weight in his figures.

Some chemical industries enjoy little freedom of location, and in others the range of choice will be found rather narrowly limited, for more or less obvious reasons. For example, since ethylene, a by-product of oil refining, cannot be economically transported by any method thus far developed, plants designed for using it must be located near the oil refineries. Again, for products of very bulky or highly perishable character, or those burdened by high transportation rates for their distribution, any location not in close proximity to the market is generally out of the question. Dry ice, for example, is distributed over only a small area from any one plant; and industrial explosives are usually produced close to the fields of use.

With respect to such limitations, or obviously controlling factors, one may generalize to some extent. In the so-called *basic industries*, using previously unprocessed materials and selling their products to other manufacturers for further processing, the locational pull toward cheap raw materials and low-cost power is strong, especially where the primary material is perishable or where the weight or bulk is greatly reduced in the processing. Thus, a mineral extractive plant will generally be located with a view to minimizing the cost of providing the ore; and beet-sugar-extraction plants are located in the best growing areas. Producers of industrial alcohol from molasses, depending mainly on foreign sources for this raw material, are attracted to seaboard locations, not entirely because of the loss of weight in processing, but for a combination of reasons, including promptness and convenience of delivery from tankers and nearness to major markets. Most electrochemical and electrometallurgical processes, such as aluminum reduction and the manufacture of abrasives, are established of necessity where power is available in its cheapest form, typically in the vicinity of favorable hydroelectric sites. *Intermediate industries*, using already processed materials for further manufacture, tend to settle in well-developed industrial areas, which offer advantages for both the procurement of the necessary materials and the distribution of their products. *Tributary industries*, depending on favorable freight differentials, or *complementary industries* (*e.g.*, the production of coke for the pig-iron industry) are for obvious reasons economically restricted to the districts, or to proximity to the consuming industries, on which they respectively depend for their existence.

More commonly, however, many different factors will be found important in determining the best location for a chemical plant, providing a wide field for study and judgment. Moreover, prospective as well as current conditions must be considered; the engineer's responsibility is to cover the foreseeable future. Here such questions are involved as the development of new raw-material resources, shifts in population and markets, and the advent of new transportation facilities and of new chemical processes and products. In looking ahead, the continuing availability and adequacy of the necessary supplies, as well as their probable future cost, must of course be considered. A knowledge of trends is often essential for the solution of plant location problems; and an economic background is desirable for the engineer making studies of this character. Market research is sometimes an essential part of a successful plant location analysis.

A plant uneconomically located may be hopelessly handicapped in the unending struggle against competitors. Moreover, a change of location after the plant is established is likely to entail the sacrifice of a major part of the investment in site, buildings, and equipment, together with serious interruption of manufacture, loss of customers, sacrifice of valuable employees, and heavy expense for moving. Brief reflection upon these facts will make apparent the cardinal necessity of deliberate, painstaking, and, as nearly as possible, exhaustive study before any important question of plant location is decided.

MAIN DETERMINANTS OF PLANT LOCATIONS

The factors decisive of problems of plant location will be found to relate mainly to (1) *production costs*, with which we shall include plant construction costs as reflected in the yearly charges for depreciation and in interest on the investment; and (2) *transfer costs*, under which head we shall group for convenience the costs of transporting both incoming materials and outgoing shipments of products.

The necessary examination of comparative production costs will require separate consideration mainly of (1) raw materials, (2) fuels and purchased power, (3) labor supply, (4) geographic factors, and (5) water resources. Comparative analysis of transfer costs brings the consideration of (6) transportation facilities and rates and (7) markets. Affecting both production and transfer costs will be (8) laws and established public practices. In addition to all these basic factors, it will often be necessary also to take account of (9) special company or industry policies, and (10) possible other tangible or intangible considerations.

From the present general viewpoint, each of these 10 individual factors requires brief separation consideration.

Raw Materials. Under this head may be conveniently included the necessary mill supplies, and containers for the product, as well as the raw materials consumed in the manufacturing process. These last, it should be noted, may be either entirely unprocessed natural materials or, as is more commonly the case, materials that have already undergone, in other hands, more or less manufacture. From the present viewpoint the distinction between entirely raw and semimanufactured materials is of little importance; anything that is to be brought into the plant from an outside source for processing may be regarded as a raw material. Mill supplies, which are comparable with raw materials so far as plant location is concerned, include such items as lubricants, abrasives, machine parts, tools, furnace linings, and paint for maintenance. Containers, which are obviously on the same basis, may be bought complete and ready for use or fabricated on the plant from purchased components, which then become simply the raw materials for one of its manufacturing operations.

The cash outlay for materials is the largest single item of cost represented in the gross value of products for nearly all industry groups, accounting for about 50 per cent of the value of chemicals and allied products, 55 per cent of the value of rubber products, and 77 per cent of the value of petroleum and coal products, according to the 1939 figures of the Bureau of the Census.

Manufacturing industries differ widely in the variety of materials that must be brought together for their operations. The locational influence of a particular material may be counteracted by an equally strong attraction of a second required material in another area. In general, the greater the number of materials required in important quantity, the less influential any one of them is likely to be in determining the plant location. In a case of approximate balance among materials, other factors may decide the issue. It is quite possible, of course, that various raw-material factors, and other factors also, instead of working against one another, may be mutually reinforcing in favor of a particular area. For example, the occurrence in Alabama of iron ore, coal, and limestone in proximity to one another and to Southern markets at some distance from other steel centers multiplies the advantages of that area as a steel-plant location. Such combinations, however, are the exception rather than the rule. Wood-pulp mills, representing a compromise between sources of cellulose and of fuel, power, and water, would illustrate the more usual sort of situation.

Depending on the nature of his process, the manufacturer may have more or less latitude in the choice of his materials. For the production of ethyl alcohol by fermentation, for example, blackstrap molasses may be used, indicating a seaboard location, or, under certain conditions, cheap grains or possibly other farm products, suggesting quite different locations. Usually the manufacturer can choose among two or more fuels the one most economical and satisfactory for his purpose. On the other hand, in the production of high-grade carbon black, for example, under the present commercial processes, natural gas is indispensable, and the choice of location is accordingly restricted.

If the raw-materials cost is a large part of the value of the finished product, and especially if there is a great loss of weight in the processing, sources of raw materials have a compelling importance in the location of manufacture, unless the materials are so plentiful that they may be obtained almost anywhere at about equal cost. Few materials, however, are so wholly available; and, as pointed out above, there is a strong tendency among industries with large expenditures for crude commodities to locate near the sources of their raw materials.

Industries located near their markets but with a high proportion of expenditure for raw materials are likely to be those that use materials in which there is little or no weight loss in the processing operations. For example, in the petroleum-refining industry many refineries are located close to the consumer market and at distances of more than 1000 miles from the source of crude oil. In this industry the primary material, crude oil, is almost fully converted to marketable products. The pull of the market in such cases will therefore exert a much greater influence relatively, because, assuming equal or approximately equal freight rates, it makes little difference whether the crude materials or the finished products are transported. This factor has been further accentuated by greater flexibility in refining methods, which permits adjusting the products to the market.

The fertilizer industry provides an interesting case of weight changes, space requirements, and perishability. It uses a long list of materials. The fertilizer-manufacturing establishments are in about 350 counties in the United States. For the industry as a whole, the establishments are usually drawn toward their sources of materials. Since a variety of materials are used in particular establishments, the processing plants are near many different sources that produce the principal materials utilized. Thus there are fertilizer establishments near such places as meat-packing centers, fishing areas, deposits of phosphate rock, and garbage-collecting centers.

Usually the fertilizer establishments may be classified as of two types—the organic refuse and the inorganic phosphorus-potash-nitrogen plants. The first type use perishable materials that undergo great loss of weight in processing. The inorganic type tend to conduct their activities in large central chemical factories whose highly concentrated products are shipped to local mixing plants where carriers, chiefly inert and alkali ingredients, are added to dilute the concentrates to usable proportions.

Fertilizer manufacturing is more important in the South than anywhere else in the United States. Sulfur and phosphate rock mining, sulfuric acid manufacturing, and the production of cottonseed meal are closely interrelated with fertilizer manufacturing in that area. Most of the Southern establishments use large quantities of sulfuric acid to process fertilizers based on the phosphate rock mined in Florida, Tennessee, and Virginia. Both the loss of weight and bulk in processing the materials and the perishable nature of some of the materials encourage the fertilizer plants to locate near their principal sources of materials.

Alternative sources for the essential material must not be overlooked. For example, it has been said that for sulfuric acid manufacture it is necessary to establish only one fact, *i.e.*, can a given consumer territory use the minimum practicable production unit of 50 tons per day? If sulfur from Texas is considered as the only source of raw material, this statement is undoubtedly true; but there are also sources of sulfur dioxide, such as zinc works, pyrite, and other sulfides. Examples are the Atlantic ports using foreign pyrite and plants in and around St. Louis using sulfur dioxide from zinc refineries of Montana, Colorado, Utah, Tennessee, and New Jersey.

Fuels and Power. The influence of energy resources on industrial location is, in general, similar to that of raw materials, but several distinctive features are peculiar to fuels and power.

All the energy resources except coal are transported in whole or in part by special methods peculiar to themselves. Pipe lines for oil and gas, and high-tension transmission lines for electric energy, follow a geographical pattern differing from the network of railways, waterways, and highways over which most raw materials are carried; and this results in an unusual structure of transportation costs and charges. The peculiarities are particularly striking in the transmission of electricity, where geographical variations in rates are based less upon the combined costs of generation and transmission than upon the nature of the load and the rate policies of particular utility systems.

The several energy resources often compete with one another and may be substituted one for another. For general heating purposes, coal, crude or fuel oil, and natural gas are all widely employed; for carefully controlled heating, electricity and gas (natural or manufactured) each has its special advantages; for generating electricity, water power, steam (raised from coal, oil, or gas); and internal-combustion engines (gasoline or Diesel) all play a part.

Fuel and power are wholly consumed in the process of manufacture, and consequently, unlike most raw materials, do not enter into the weight of the manufactured product. Moreover, transportation costs of coal, oil, and natural gas constitute a relatively high proportion of the delivered price, whereas zones of low-priced electric power are comparatively narrow; and geographical variations in fuel and power costs are therefore very large. Although other factors must often be considered, heavy consumers of fuel or power tend to be locationally drawn toward low-cost power or fuel areas.

Industrial consumption of anthracite has fallen steadily, as better markets have been found in heating and combustion equipment developed for utilization of the smaller sizes. Plant-owned water power is no longer a considerable fraction of the total, although the development of private hydroelectric sites by electrochemical industries has recently increased the share of this energy source. The most profound alternation in the last thirty years is the decline of the proportions of industrial energy supplied by bituminous coal and its replacement by natural gas, fuel oil, and purchased electric energy.

It should be noted that, in the case of coal, costs are at their lowest at the mines and increase rapidly with increasing distance from the coal fields by rail, but less rapidly at points served by water transportation. In general, costs of 10¢ or less per 1,000,000 B.t.u. are found only at the mine mouth; points from 11 through 18¢ occur either close to the mines or on developed waterways (inland or coastal); and points above 18¢ are at considerable distances from the coal fields and accessible only by rail.

By contrast to oil, natural gas is heavily consumed in the immediate producing areas and shows a marked cost increase at distant points. In the Southwestern fields, where this resource is most abundant relative to the market, and where billions of cubic feet are annually blown into the air, natural gas is obtainable for industrial purposes at 3 to 6¢ per 1,000,000 B.t.u. Noteworthy among these points are the Texas Panhandle, northern Oklahoma, and Jackson, Miss. At other gas-consuming points, costs run as high as 16 to 21¢ per 1,000,000 B.t.u.; and in some of the gas fields this fuel is priced too high to be used for industrial purposes.

The somewhat peculiar pattern of gas costs is a result of the location of pipe lines and of enormous geographical differences in marketability. The average value of gas at the point of consumption in 1938 was 16.1¢ per 1,000,000 B.t.u. for general industrial use, compared with 74.2¢ for domestic use, 47.1¢ for commercial use, 4.3¢ for consumption in the oil and gas fields, and 0.9¢ for use in the carbon-black industry. Although industrial sales are more lucrative than the "dump" consumption in the field or in making carbon black, they rank far behind domestic and commercial use in attractiveness as markets. The Appalachian production, therefore, goes mostly into these uses. Only in the Southwest area are reserves so great that large-scale industrial consumption can be expected for a considerable period in the future. As additional pipe lines are brought out from this area, the scope of industrial consumption can be broadened to some extent. Beyond a distance of a few hundred miles, however, pipe-line construction appears uneconomical unless a large share of the transported gas can be sold for domestic and commercial consumption.

Over recent decades, interfuel competition has favored natural gas and oil at the expense of coal and has undoubtedly contributed substantially to the noticeable southward drift of heavy fuel-consuming industries. For direct heating or steam raising, the three major fuels are readily substituted for one another, and the selection of fuel at a given location is in general simply a matter of delivered price plus costs of handling, burning, and waste disposal. Even at the same cost, however, the greater convenience and cleanliness of the petroleum fuels and their greater ease of heat control in special thermal processing may make them preferable.

Labor Supply. Study of the supply and cost of labor will require examination into such individual factors as the quantity of labor available in each community considered, its uniformity or diversity, national origins, average intelligence, background of industrial or mechanical experience, existing wage scales, and indicated general efficiency. The alert investigator will not fail to look into questions of adequate housing; local transportation; schools; churches; recreational facilities; and other community conditions, experience, and habits that may make for unrest or tend toward general contentment.

In general, wages are lower in country districts and small towns than in large cities. Similarly, and at least in part for reasons obviously related to the cost of living and therefore presumably permanent, wages tend to be somewhat lower in the Southern states than in the North Atlantic and Middle West industrial districts though in recent years this differential has been diminishing. In both cases, union pressures are being exerted in the direction of eliminating these differences in wage rates.

Nevertheless, farseeing industrial policy would seem now to call for centrifugal rather than centripetal patterns in further plant location—for some dissemination rather than additional congestion of our great manufacturing centers. Too much concentration of industry and of population imposes needless hardships on employees. There are indications that in our progress from pioneer conditions we may already have become somewhat too urban a nation; and there are signs also that the industrial tide is now turning outward. Nationally, it is beyond question that we can profit from a more widespread distribution of manufacturing activities and from the partial industrialization of areas now exclusively agricultural. And, other things being equal, wise management will prefer new plant locations apart from the crowded industrial centers and in areas where more normal and natural conditions of life prevail.

The necessity, pointed out above, of looking ahead so far as is possible has special force with respect to the labor-cost factor. A case in point is the effect of wage and hour legislation on the labor factor in plant location. Prior to its enactment, low wages and long hours were considered industrial advantages in highly competitive industries. The labor differential, for example, was a major factor in the movement of cotton mills from New England to Southern rural districts, where unskilled workers in turn found the wages and hours in the mills preferable to what they had on the farm or in other rural occupations. There may still be a differential after the establishment of minimum wages and maximum hours, but it is much smaller and may be no longer a major factor in the location of cotton mills. Unionization, which is also advancing rapidly in the South, tends to have a similar equalizing effect.

In the more basic branches of the chemical industry, labor costs are a relatively small part of the total cost of production, and labor is not likely to be the controlling factor in the problem of plant location. In plants of this sort, the salary roll for technical and research men may exceed the wage roll.

For convenience, we shall here include technical and managerial personnel with labor. A large manufacturing unit in the chemical industry, especially where the processes are complicated, may require highly trained managerial ability at the top and in addition a whole series of junior technicians and executive personnel to direct and integrate various operations that are not so clearly differentiated in a smaller concern. Production control absorbs a large part of the internal efforts of such a group; and relatively large staffs may be needed either because the business is growing rapidly or because frequent changes are being made in production techniques and in plant organization. The reluctance of such men to leave their established professional associations may prove an obstacle to decentralization. One reason given for the concentration of steel plants in the Pittsburgh area, for example, is the ease with which experienced personnel can be recruited there; and it is said that competent steel men can more readily be attracted to new undertakings if they can continue to reside in that district.

On the other hand, it may be said, in general, that for an ambitious man the opportunity for successful accomplishment, and for advancement, apparent to him in the new enterprise should outweigh all thoughts of its location. Nevertheless, this natural tendency of men of kindred and perhaps relatively narrow interests to flock together, and the natural reluctance to leave established associations, cannot always be disregarded in problems of plant location. Doubtless the incentive can always be made great enough to cause a sufficient staff of such men to move to a new plant site. In some instances of plant removals, it has been found advisable to pay the costs of moving families and of advancing junior personnel to more responsible positions. If enough keymen can be transferred, it will generally be possible to train the required additional supervisors and foremen at the new location.

Geographic Factors. In addition to determining the availability of minerals and non-minerals, physical features are important for other reasons in plant location analysis. Rough terrain, for example, may make difficult or preclude mining operations. Weather affects the cost of manufacturing. A too vigorous climate may be unfavorable to operations that must be outdoors or that require large building space or special protection from the elements. It may necessitate the erection of enclosed plants where they are not otherwise necessary for the process.

Land contour, atmospheric temperatures, humidity, precipitation, waterways, and type of soil have locational repercussions. Approximately two-fifths of the total land area of continental United States is characterized by rough mountainous or hilly terrain, presenting obstacles to location directly or as barriers to transportation.

Again, the exact location of a site may depend on subsurface features, *e.g.*, the strength of the underlying strata for supporting the weight of buildings and equipment or for preventing excessive vibration of machinery. This is an important element, especially in hydroelectric plant location.

Water Resources. Water is used extensively in a direct manner in processing, washing, cooling, and humidification. Much of the water used by industry is not obtained from public water systems. If a plant uses large amounts of water, it will be located at or near a body of water or where abundant supplies are available from wells. For example, all large electric power plants, except Diesel plants, must be on rivers or fresh-water lakes. Thus the options for location are greatly reduced. Even where there is a supply of water, location of plants may be vetoed because of restrictions on its use, for sanitary, recreational, or other reasons, or because of qualitative deficiencies of the water or competitive use by other plants. A good example of an abundance of water not available at many sites is in Chicago, where virtually all the lakefront is publicly owned.

In general, the western part of the country is short of water, and industries requiring large volumes of water can be located in relatively few districts in that region.

Water can be obtained from either surface or ground sources. Usually it is cheaper to tap the surface; yet in some places the construction of upstream storage dams may be needed to regulate the flow to industrial plants. The chemical engineer should carefully estimate the amount and type of water available during all seasons in view of both the immediate and potential requirements of the proposed plant. In some districts and for some processes, surface water has to be treated for contamination and discoloration. In many areas, ground water is the more dependable source; in some places, it is the only adequate source. Its provision, however, raises problems. Since ground-water moves very slowly, it often contains a higher proportion of dissolved minerals, which makes it unsuitable for cooking, laundering, and some industrial purposes unless treated. The higher proportion of carbonic acid sometimes found in ground water is undesirable, because it causes corrosion of metal equipment.

For general manufacturing, the water should be pure enough to avoid the rapid formation of boiler scale. Steam is of such general significance in manufacturing that, if the water supply cannot be made satisfactory for the boiler feed except at very high cost, the locality is clearly unfavorable as an industrial location.

In many chemical industries, water enters directly into the solutions containing the materials to be processed. In these cases, abundant local supplies of pure water are indispensable to successful operation and a determining factor in plant location. In certain districts, for example, the color of the natural water prevents the manufacture of bond paper. The same is true of sugar refining, rayon, hides, and *process* industries in general. The lack of adequate pure water has been a handicap to the expansion of rayon, cellophane, or textile plants. In some areas, the production of glucose sugar from corn products has been found impractical because the sulfur in the local water renders the product cloudy. River waters are usually lower in mineral analysis but higher in bacterial content and turbidity than well waters. The latter are usually high in carbonate and bicarbonate.

The *temperature* of the water sometimes plays a dominant role. Some industries (such as pulp mills) have to install expensive refrigerating equipment to bring the temperature of river water down to requirements, entailing considerable costs. Many fermentation and other chemical processes require water that is not only free of impurities but at a definite temperature. It should be noted that high summer temperatures are not limited to the Southern states. On the other hand, areas with wide annual extremes of heat and cold may be unfavorable to the location of industries.

Wherever large amounts of water are required for cooling and where the supply is not adequate, it is necessary to resort to expensive artificial means of recirculating and lowering the temperature, by the use of large reservoirs, ponds with spraying devices, and cooling towers.

In steam-electric plants and in industries using steam directly for power, large amounts of *condensing water* are necessary. Lack of adequate condensing water is the main reason why cheap power usually cannot be produced in the neighborhood of coal mines. In the newer power plants, condensers require about 800 tons of water per ton of coal consumed. Water for this purpose does not have to be pure, since it is usually applied externally in surface condensers to lower the temperature. Thus water obtained either from surface sources or from wells can be used usually without treatment. On the other hand, at some places river water is so acid that it corrodes the boiler and condenser tubes; and, under such circumstances, the water must be treated or tubes of noncorrosive metal installed.

An indirect use of water important in locational analysis is that it is a *carrier of industrial wastes*. If the cost of installing a waste-treatment process is much in excess of the value of the materials recovered, the industry should seek a site on a stream into which it is allowed to discharge. The problem of waste liquors is large in process plants, especially in view of recent government agitation on the subject, and is a major factor in the studies of plant location.

Transportation Facilities. Transportation is the most important single factor in location analysis. The location of production will depend on the relative costs of transporting materials in their unprocessed and processed states. The attraction of the raw material will be greater, the greater the weight loss in processing. Recent legislation on basing point and uniform pricing systems will have an important bearing on transportation costs.

Speed is an important qualitative aspect of transportation, including frequency and dependability of schedules. Goods in transit frequently represent a considerable absorption of working capital. In fact, it is customary for some materials, such as zinc and electrolytic copper, to be sold at delivered prices that include interest on the value of the product while it is in transit. Speed in transportation also reduces storage requirements. Prompt service to customers, of course, is vital.

In considering the relative advantages of two different sources or markets, not only their distance from the place of manufacture but also the related but not always proportionate costs of carriage along the best route must be taken into account. A route that provides cheaper service, such as a navigable waterway, may be said to bring the points it serves "economically closer." Thus iron ore from the upper Great Lakes ports is shipped several hundred miles by water to lower ports at a much lower cost than it could be shipped over a relatively short land route. The importance of ocean, lake, or river shipping facilities should not be neglected, since water transportation is usually the cheapest method.

Where a location has more than one railroad or, better still, railroads and barge or steamship lines, the freight competition makes for lower rates, higher efficiency, and better service. A careful rate study should be made by the engineer locating a new plant and an application for special rates (or a sounding out) should be made if necessary.

Carriers typically charge rates that rise with the progress of fabrication from raw materials to final products; in some cases this has been done at the instance of the Interstate Commerce Commission. The result is a tendency for manufacturing to be located away from the source of raw materials and close to market, unless the weight loss in fabrication is sufficiently great to offset this difference in rates.

The practice of carriers of increasing their charges with distance, but at a diminishing rate of increase as distance lengthens, causes total transportation charges to be lower if manufacture can be located at one end of the line from the material source to the market, since the resultant single long-haul charge will be less than the sum of the equivalent two short-haul charges. Unless, then, processing costs are sufficiently lower at some other place to offset its disadvantage in transportation costs, fabrication should occur either at the source of the material or at the market. Railroads, however, frequently grant fabrication-in-transit privileges, permitting a raw material to be stopped for fabrication at some place intermediate between its source and ultimate market, and the product then to be carried on to the market, for substantially the same total charges as if the haul had not been interrupted. The in-transit privilege has been of great importance in the establishment of processing plants in places where they could not otherwise be profitably operated.

Typical examples of the influence of transportation costs on plant location can be found in sulfuric acid and caustic soda. About 32 tons of sulfur (low freight rate) will make 100 tons of sulfuric acid (higher freight rate). Plainly, then, the acid from any producing unit must be sold in a limited area, and its manufacture should be near the point of consumption. On the other hand, 1.5 tons of salt are needed to produce 1 ton of caustic soda, while the freight rates on salt and caustic are approximately the same. Thus, in the location of caustic soda manufacture, there is no particular pull toward the market, but the nearness of adequate salt supplies is an important consideration.

It remains to point out—what would seem obvious but is sometimes overlooked—that in any problem of plant location the factor of freight-on-product is of equal importance whether the product is to be sold "delivered" or "f.o.b. plant." It makes no difference that under the terms of sale the customers will pay the freight. The customer can always add—and will add—freight charges to purchase price where the transportation cost is not prepaid. The significant thing from his viewpoint, and from any sound viewpoint of plant location, is always the total delivered cost, including all transportation charges.

Markets. In recent years, proximity to markets has been of rising importance. This is perhaps partly due to increasing consumer-consciousness on the part of American business, dating largely from the "buyers' strikes" of the 1930's; and partly to the general intensification of competitive forces and the speeding up of our business life. A major reason, to view the matter more broadly, is the desire to give better service to customers; and, for chemical producers, selling largely to manufacturing consumers, this consideration is likely to apply with maximum force. Generally, however, this advantage is to be weighed against possible definite economies such as may be had from nearness to raw materials, power-cost savings, and the economics of large-scale production.

Nowadays, however, there appears to be less tendency to stress the gains from large-scale manufacture and from concentration in areas of similar industrial production. Especially this appears to be the trend in some of the larger concerns. The chief reasons given are such as: (1) to avoid a dominant position in any one industrial community, (2) to reduce the variety and volume of top-management decisions, (3) to reduce the operating disadvantages of bigness, (4) to have a more wholesome environment for the working forces, (5) to deal more effectively with personnel problems, and (6) to satisfy the needs of industrial consumers who would be disadvantageously located with respect to a single plant and consequently more disposed to buy from competitors.

Finally, it may be noted that, in addition to the delivery-cost advantage and the actual service advantage of proximity to important markets, there may be a sentimental value in such location, depending apparently in part on the local pride and sense of self-importance of the buyers. There is an element of rooting for the home team. Thus the disposition of the automotive industry to give a preference to suppliers located in or near the Detroit district and to press their important suppliers to establish themselves there would appear to go beyond any demonstrable advantages of time and distance.

Laws and Public Practices. The *Federal government* has never played an entirely neutral part in the determination of industry patterns, and political developments of the past two decades would appear to threaten the further interposition of artificial forces from that quarter. A protective tariff may permit industrial locations that would not be possible under free trade; excise taxes raise prices and may thus affect location; and Federal expenditures may provide a stimulus to certain types of industries by fostering an extra demand. The engineer, though in some cases he cannot disregard these factors, should bear in mind their factitious character and their relative liability to change and uncertainty of duration in comparison with the normal operation of economic forces.

State governments, through their power to tax and regulate the use of their highways, have sometimes erected barriers to the intelligent location of plants. They thus in effect levy special taxes on out-of-state goods (though this is, in principle, unconstitutional), give preferences to local producers, apply embargoes (*e.g.*, on alcoholic liquors), and discourage the entry of trucks from other states.

For the most part, however, state action influencing plant location is more constructively directed to the collection and dissemination of industrially significant facts, through the means of industrial surveys and information services. These activities are described below under the head of General Sources of Information; and for convenience we include also at that point a list of state agencies charged with more actively inducing the location of industry within their boundaries.

In 1939, 23 state governments, 21 of which were in the South and East, supplemented fact-finding and fact-disseminating with special tax concessions. In these states commissions had the power to examine firms requesting such concessions and to decide which should be thus favored. The Mississippi Industrial Commission went further and supervised municipal promotional efforts, deciding whether or not to allow a municipality to subsidize new plants with public funds. In considering the advantage of any such financial aid, the engineer should, of course, not overlook the obligations that its acceptance will naturally impose.

Local interests often offer special concessions in order to develop industry within their localities. These can be gifts of cash, property, and services, loans and stock subscriptions, tax exemptions, and promises of labor "loyalty." Here the same caution about reciprocal obligations, of course, applies as in the case of state subsidies.

Public utilities also may influence a particular location. Railroads maintain industrial agents for the purpose of promoting the establishment of industries along their rights of way. Rate concessions may be offered if approved by the appropriate government body, and detailed surveys of prospective plant sites may be offered.

Municipal restrictions, such as nuisance laws on fumes and waste disposal, may have a special importance for a chemical plant; and the existence of such regulations, or the likelihood of their enactment, must always be looked into.

Company Policies. The individual policies of the particular business concern, or of the industry of which it is a part, may have an important and perhaps decisive influence in the selection of a plant location. Policies of production and of distribution, financial policies, and firmly held practices of management and of pricing may all have a bearing.

The advantages of concentrating production in large plants or in highly developed industrial centers (already considered from the viewpoint of labor relations above), as against the gains to be had from decentralization, will vary from industry to industry. In many cases, a compromise must be sought between (1) the economies of greater specialization of operations, mass purchasing, centralized service, and availability of skilled labor; and (2) the wastes due to the complexity and inflexibility of large enterprises, the disadvantages and perils of the great concentration of labor, and the high transport costs and rents likely to be associated with the centralization of operations. Where the advantages of centralization are great, the locational pattern is likely to consist of a few large clusters of plants; whereas, if transportation costs are dominant, plants are likely to be distributed regionally.

The organization of chemical and other production involves inter-product, inter-plant, and inter-firm relationships that should be taken into account by the engineer. Such are the problems of substituting materials or fuels, by-product utilization, dovetailing of fluctuating labor or equipment requirements, and similar considerations.

Pricing policies have an indirect but often important influence on location. For example, by the establishment of a definite geographic system of delivered prices, producers may be able to locate at points that would not be economical under a system of prices established at the plant. Such arbitrary setups, perhaps contravening economic law, are liable to government attack and should not be too firmly relied on.

Sometimes small-scale chemical and other operations are found in non-urban non-industrial areas because they are designed to tap small pools of labor; however, such operations are also peculiarly adapted for manufacture in the major centers of industrial activity.

Possible Other Factors. Beyond this point, it is difficult to generalize. Every question of where to locate a plant is a distinct individual problem that may involve special factors not included in the above enumeration. It can hardly be too much stressed, however, that the responsibility of the engineer entrusted with finding the solution of any such problem is exhaustive and that it includes the discovery and pointing out of every factor that may have an important bearing on the success of the location chosen, as well as the evaluation of each such factor when identified.

GENERAL SOURCES OF INFORMATION

The trouble and expense of making an exhaustive survey to determine the best location for a new plant are likely to prove substantial, and short cuts are sometimes sought. One easy but perilous procedure is to invite proposals from local chambers of commerce, railroads, power companies, and real estate agents and to base the decision on a comparison of these proposals. This has the disadvantage of not being thorough and of leading to confusion, misunderstanding, and unfair comparisons. It is at least conceivable that biased or incomplete information may be worse than none at all for the comparison of locations, and the difficulty of checking the reliability and comparability of data from different sources is often formidable. Nevertheless this procedure appeals to some executives as easy, inexpensive, and adequate for their purposes. Small manufacturers in particular have been known to use the inquiry about a proposed new location as a means of getting free advertising and offers of financial assistance. Such tactics, however, are a means of avoiding rather than solving the problem of plant location as we are here considering it.

The essential information required for locating a plant is determined largely by the requirements of the industry in question. It will generally consist primarily of facts about the supply and cost of materials, labor, power, transportation, and markets and will also naturally include information on the character and cost of sites, taxation, public improvements, housing, and community conditions affecting the welfare and contentment of labor. For particular industries, important information may include data on climate, topography, population composition and trends, income patterns, and other matters of special interest.

The necessary information thus covers a great variety of subjects and will usually be found widely scattered among many public and private sources. It will vary in scope from material covering the whole United States to exclusively local data. Since no single body, public or private, has the primary responsibility of compiling information useful to industrial location, it falls on the manufacturer or his consultant to determine what is needed and where to find it.

Impartial information valuable to location studies is provided by many Federal, state and local government departments and special agencies. For example, the U.S. Department of Labor compiles and publishes wage and hour statistics and cost of living and price data; the U.S. Department of Commerce conducts regular censuses of population, manufacturing, distribution, and business and issues special industrial and commercial bulletins; while the Department of the Interior prepares data and maps on topographic features, resources, etc. Among special agencies handling information useful to industrial location are the Federal Trade Commission, the Federal Power Commission, the Interstate Commerce Commission, the Employment Service of the Federal Security Agency, and the War Assets Corporation. Maps of the Coast and Geodetic Survey, Geological Survey, Bureau of Plant Industry and Soils, Public Roads Administration, and Weather Bureau are particularly helpful.

Regional and state agencies furnish details that supplement information compiled by the Federal government and cover additional subjects as well. State planning boards, for example, have assembled information on resources and industrial opportunities in their areas and have made special studies of unemployment, skilled and unskilled labor supply, wage scales, cost and location of factory sites, and location of vacant buildings available for industrial occupancy. Taxation and laws affecting industrial operations are among the items covered particularly by local units of government.

State informational functions sometimes consist chiefly of the publication of bulletins that are supplied to those businessmen who request information concerning investment opportunities. Most states, however, advertise in national publications, and some add field officers to their central informational staffs. The success of state advertising and other informational efforts probably varies widely; some states pronounce themselves well satisfied with the results achieved, although, in general, experience seems to cast doubt upon the effectiveness of such advertising as a means of influencing industrial location.

Two types of industrial promotion have been provided for in state law. One provides for publicity only, and the other type involves active and aggressive promotion either by the state or by local communities.

State agencies established to handle advertising to attract industry (generally located at the state capitols) have included:

Connecticut Development Commission.
Illinois Development Council.
Indiana Division of Publicity.
Kansas Industrial Development Bureau.
Kentucky Progress Commission.
Maine Development Commission.
Massachusetts Industrial and Development Commission.
New England Council and Southern Governors Council (regional organizations).
New Hampshire Planning and Development Commission.
New Jersey Council.
North Carolina Department of Conservation and Development.
Pennsylvania Department of Commerce.
Vermont Department of Conservation and Development.
Washington State Progress Commission.
Wyoming Department of Commerce and Industry.

State agencies responsible for actively inducing the location of industry within their boundaries have included:

Alabama Industrial Authority.
Arkansas Agricultural and Industrial Commission.
Louisiana State Board of Commerce and Industry.
Mississippi Industrial Commission.
North Carolina Department of Conservation and Industry.
South Carolina Board of Promotion of External Trade.

In addition, 43 states have created planning boards, which carry on research activities but usually do not have directly promotional functions.

Private agencies handling information important to location studies include railroad and power companies, which benefit directly from the location of industries on their lines and maintain departments to aid industries seeking location for new plants or expansion of existing facilities. The railroads are specially qualified to answer questions about freight rates, shipping schedules, and sites adjacent to railroad lines, just as the power companies are specially equipped to answer questions about power rates and conditions under which the use of purchased power is economical. Both the railroads and the power companies, through their close contacts with industry, have valuable funds of information that extend beyond the range of their immediate activities, and both are in a position to give useful advice about plant location within their own territories, though their advice may not be impartial with respect to places served by competing companies.

Other private sources of information include trade associations and trade union organizations, and other special interest groups; research organizations such as the National Industrial Conference Board and Standard & Poor's Corp. and civic agencies such as the New England Council and various local chambers of commerce. Many of these private sources publish valuable information, and many will answer questions on particular industry problems.

AIDS IN PLANT LOCATION ANALYSIS

Maps. It will readily be seen that a conclusive location study requires comprehensive searches for facts of various sorts and a considerable work of analysis of the data collected. To facilitate this analysis and assure effective presentation of the results of it, charts and diagrams will be found quite helpful; and these should be used even when the "sieve procedure" described below is not resorted to. Maps are especially helpful in visualizing raw-material sources, location of competitive producers, competitive freight-rate areas, consumer territories, etc.

Raw-material Surveys. Findings as to raw-material resources can readily be expressed in the form of maps; and, when the locations of the principal plants using the material in question are added, these charts not only will give a broad picture of the industry as a whole—its sources of raw material and what sources are not being utilized—but will often bring to light districts of large consuming and buying power that are not being adequately served by local plants. These "virgin consumer territories," of course, may or may not offer logical opportunities for new plants, depending on whether or not they answer favorably to the other major factors of plant location.

Labor Surveys. The country is replete with labor surveys—all more or less worthless, since labor conditions are continually changing. The entry of a single large industry into a district may change the labor situation entirely.

Comparative labor costs should be figured always on the basis of cost per unit of output and not merely by a comparison of hourly or daily wage rates. Wide variations will often be encountered in the efficiency of labor in different districts.

Inquiry into the labor conditions of a particular area should include such items as quantity of labor available; nationality, with percentage of foreign-born and of Negroes, and whether the foreign-born are predominantly from northern Europe or from the Mediterranean countries; union affiliations and practices; and the habitual local conditions as to contentment or unrest and the current experience with respect to labor turnover. As elsewhere indicated, the labor survey should also cover community conditions with respect to housing, water supply, trolley or bus service, local newspapers, schools, churches, and facilities for recreation.

TWO STEPS TO DETERMINE PLANT LOCATION

Ordinarily the choice of a plant location will be accomplished, and its superiority proved, in two steps, as follows:

1. First, by selecting for close final comparisons the several geographic areas and specific locations that preliminary investigation shows to be the most promising.
2. Next, by determining, from a detailed study of comparative costs, just which location can reasonably be expected to prove the most advantageous.

METHODS USED IN MAKING PRELIMINARY SELECTION

For the necessary first step of selecting for final close comparisons a small number that appear most promising out of all the possible locations, the usual starting point is to determine the paramount locational factor of the industry, or the several dominant factors, along with some approximate measure of their relative importance. Such factors would include (1) *cost of assembling materials*, which will usually be the paramount consideration for industries using heavy, bulky, or perishable raw materials and making relatively light, compact, or non-perishable products, as in the case of cement manufacture, ore smelting, or vegetable-oil extraction; (2) *cost of fuel or electric energy*, as in heat-treating and many chemical operations such as the production of caustic soda and chlorine, or of alloys, aluminum, magnesium, etc.; (3) *availability and cost of labor*, as in textiles, machine tools, and the synthetic fiber branch of the chemical industry; (4) *delivery costs*, which may be paramount in the case of heavy, bulky, or perishable products or those involving special risks in transportation, such as building materials or explosives; (5) *intangible factors*, such as the location of a parent plant or administrative office, or the availability of a suitable site. Each of the above categories may suggest an approach directed to the paramount factor and give the engineer helpful guidance in his preliminary choice among possible locations.

Here, however, a word of caution is in order against uselessly restricting the range of choice. The variability of production techniques in the chemical industry, with respect to both materials and processes, must not be lost sight of. In a great many cases, processes can be varied so as to use less of those materials and services which are relatively expensive at a particular location and more of the relatively cheap ones; and there will often be a choice among radically different processes that may involve entirely new ranges in the choice of location. It is the responsibility of the engineer charged with locating a plant to overlook no such possibility of variation which may enlarge the scope of the choice and perhaps bring important cost benefits.

This may necessitate preliminary cost computations for two or more different processes. For example, in the production of caustic soda, either the electrolytic or the lime-soda process may be used, depending on the relative availability of materials, fuel, and power, and the demand for products. In an area with cheap electric power and good access to sources of salt, which is close to a market for chlorine, the basis of site selection would be the requirements of the electrolytic process. In certain other areas, the requirements of the lime-soda process would govern. Similarly, in the production of phenol, the Raschig process is less costly than the one involving the sulfonation of benzene, but the latter is economical in localities where its by-products can be disposed economically. This versatility in processes and products contributes to the competitive structure of the chemical industries.

To determine the dominant locational factors of the industry in question, or otherwise to narrow the range of possible locations to which final intensive study must be given, the following methods are often employed:

1. Apparent Association. A guide commonly adopted by outside consultants rather than directly by the manufacturing firms seeking plant locations themselves is that of "apparent association." The analyst examines the characteristic habitat of the industry in question, notes its chief features, and points to them as dominant locational influences. For instance, if a certain type of chemical industry is usually found in the immediate proximity of oil refineries, one might conclude without knowing anything further about the industry that nearness to oil refineries was the dominant locational requirement. The risks and limitations of such a method are evident, however, even where it can be made to yield a definite and plausible statement. In most cases, the real reasons for location cannot be safely deduced merely from the location of established plants. Some further insight into the actual character of the industry in question is necessary.

2. Questionnaires. A somewhat more penetrating method, also favored by consultants, is the use of questionnaires to the firms in the industry, asking which requirements they deem most important in determining desirability of location. This method is also subject to grave defects. In the first place, there is no clear indication on such questionnaires as to the meaning of the term "importance." Some respondents, for instance, may indicate as "most important" those items regarded as absolutely indispensable, like water supply, although by others these might be taken for granted and omitted altogether. Some may indicate the items that bulk largest in total cost, regardless of whether or not they can be influenced by a change of location (*e.g.*, fuel costs), while others will stress factors like rent or taxes, which vary locationally within an entirely different and much smaller geographic range of reference. Finally, there is at least the possibility that some of the respondents will be particularly conscious of certain locational factors that they view with resentment (*e.g.*, taxes), and may exaggerate their importance.

3. Relative Cost Magnitudes. A more direct approach in seeking to gage the relative importance of the various locational factors is to compute the relative magnitude of different items in the cost bill of the industry. Thus, in analyzing the manufacture of cement, we can make the following tabulation from data in the 1939 Census of Manufactures:

Cost item	Total expenditure	Percentage of total value of product
Materials and supplies	$ 34,034,358	17.7
Wages	31,588,404	16.4
Fuel	24,164,457	12.5
Purchased electricity	9,908,469	5.1
Salaries	7,408,199	3.8
Other items, including interest, legal and other professional services, taxes, profits, etc.	85,507,417	44.5
Value of products	$192,611,304	100.0

Such a tabulation as this can be useful as a basis for provisional elimination of unimportant factors and for further analysis of important ones. However, it cannot stand by itself as a complete statement of relative weights of factors, because it ignores the differences in savings potential of the several cost components. To illustrate, we might conclude, from the figures given, that materials and supplies were rather more important than wages in determining location and that electric power rates were considerably less than half as important as fuel prices. This would be misleading to say the least, since it implies that we have as equally available alternatives the saving of say 5 per cent on materials expenditure or 5 per cent on labor costs or 5 per cent on electric power charges. In actual fact, some of the expenditures may be highly variable according to location and others not so; and it is impossible to say that one requirement is more important locationally than another until we take this into account. In the alcoholic-beverage and tobacco industries, for example, a major cost item is the Federal tax; yet so far as choice of location within the United States is concerned, its influence is absolutely nil. However, local, city, state, and county taxes play a very important part in locating the chemical plant. Another defect of the breakdown of product value is that it cannot readily be extended to cover the requirements of access to market.

4. Sieve Procedure. One helpful method of narrowing down the range of possible locations to which final intensive study must be given, especially suitable for large firms or public agencies, is the so-called "sieve procedure." This involves the use of a number of outline maps of the United States (or other regions if the possible choice is otherwise) and consists essentially in first blocking out on individual maps the areas decided to be unsuitable for successive particular reasons, and finally combining in one composite map all the blacked-out areas of the individual maps. The white areas thus "sifted out" will, of course, represent on each individual map the districts found unobjectionable with respect to the particular reason to which that map relates; and on the composite map, the districts finally found unobjectionable for any reason. Thus attention can be focused on a relatively small expanse of presumably practicable locations, and the problem is greatly simplified.

For example, regions generally unsuitable because of topography can be sifted out on the first map; those with inadequate access to transport routes on the second; those with too sparse or too dense population on the third and fourth; areas of highly productive agricultural land on the fifth; etc. The adaptability of this method to the particular requirements of a wide variety of locational problems will be readily apparent. Most of the necessary work of collecting the basic data will be found, appropriately enough, already to have been done at public expense; and the results have a wide usefulness not only to planning agencies but also to individual manufacturers seeking plant locations.

Such a broad survey is, in effect, a rough method of preliminary zoning for industrial development; it does not remove the need for intensive studies of the resources in specific areas and of the needs of specific industries or plants, but it does simplify the problem by first eliminating great numbers of unsuitable locations. By narrowing consideration to a small fraction of the total area, this procedure can greatly reduce the number of intensive surveys needed.

FINAL COMPARATIVE EVALUATION OF SITES

The range being narrowed to a conveniently small number of the very best available sites, it remains to make the final choice among them. This, apart from intangible considerations, will regularly depend on (1) the comparative amounts of investment necessary for each of the locations finally considered and (2) the probable operating costs respectively attributed to each location. The relative importance of these two elements will naturally depend on the financial circumstances and policies of the operating company. In any event, it will be logical to include in operating costs a reasonable rate of allowance for interest on the necessary investment. The elements ordinarily to be included in each cost group are shown in the following diagram, though in case of any individual problem its comprehensiveness should be checked to make sure that no important essential is omitted.

Investment Cost

Engineering	Real estate	Plant	Facilities
Studies	Options	Grading	Housing
Investigations	Title searches	Buildings	Power
Reports	Purchases	Material and labor	Transportation
Surveys	Rights of way	Equipment	Water
Designs	Leases		Waste disposal
Estimates	Permits		
Layouts	Privileges		
Inspection			

Operating Cost

Transportation	Manufacturing	Maintenance	Overhead
Raw materials (inbound)	Raw materials	Repairs	Supervision
	Labor	Renewals	Office expense
Finished product (outbound)	Power, light, heat	Upkeep	Welfare
	Water		Taxes
Employees	Storage		Insurance
	Supplies		Depreciation
	Sanitation		Interest

As a necessary starting point, a round figure for the expected yearly output of the plant to be located (with reasonable allowance of reserve capacity) is taken. The necessary investment and probable operating costs are calculated for that volume. Finally, the corresponding totals and per unit costs are computed and compared. The method is illustrated in the following tabulation, in which a yearly output of 30,000 tons of product is assumed and locations *A*, *B*, *C*, and *D* are finally compared.

From the above figures, it will be noted that there are wide differences in the cost of labor, power, and transportation. Site *C* shows a substantial advantage in production cost and would appear plainly to be the selection indicated, although the required investment would be considerably lower for *D*.

An important limitation of this type of analysis, suggested above, is that cold figures may fail to tell the whole story, since some intangible considerations cannot be translated into costs. However, this sort of comprehensive determination of all the measurable elements of the problems should always at least afford the responsible administrative officers of the company a firm footing and definite background from which finally to consider and appraise the intangible factors.

A further limitation, also foreshadowed above, is the difficulty of projecting cost estimates forward over long periods. Here about all that can be added to what has already been said is that the engineer's responsibility extends only to all discernible trends and all reasonably foreseeable changes; he is not expected to be clairvoyant, and the rest must simply remain part of the inscrutable future.

Investment Costs

	A	*B*	*C*	*D*
Real estate	$ 3,750	$ 18,000	$ 60,000	
Rights of ways, permits, etc.	300	2,000	4,700	$ 800
Grading site	2,540	4,440	6,460	1,500
Plant buildings	1.131,800	1,407,200	1,011,100	1,000,000
Equipment installed	718,920	761,000	680,060	701,100
Power	134,610	149,770	105,000	149,900
Transportation	22,000	22,630	104,330	16,000
Water	88,450	78,850	113,400	69,220
Waste disposal	54,400	59,010	32,700	41,570
Miscellaneous	23,500	31,200	38,000	22,200
Total	$2,180,270	$2,534,100	$2,155,750	$2,002,290
Per ton	$ 72.68	$ 84.47	$ 71.86	$ 66.74

Operating Costs

	A	*B*	*C*	*D*
Freight on raw materials	$ 48,750	$ 47,000	$ 45,750	$ 25,500
Freight on finished product	110,000	67,500	60,000	105,000
Raw materials	97,300	97,300	82,110	97,300
Labor	589,680	708,400	582,800	718,080
Power (light, heat, etc.)	45,000	75,420	62,010	100,000
Water	30,460	51,830	25,460	41,692
Refrigeration	3,200	18,890		
Storing, packing, etc	9,460	11,090	7,600	12,000
Supplies	3,060	3,000	2,940	2,700
Sanitation	....	11,600	8,840	9,660
Repairs, renewals, upkeep	11,940	14,100	9,060	9,800
Supervision and office expenses	62,060	78,640	56,810	76,100
Welfare	34,500	25,000	27,750	10,000
Taxes, insurance, etc	22,210	24,900	23,200	23,900
Depreciation	43,500	50,000	40,610	40,000
Interest on investment	130,800	152,040	129,360	120,120
Miscellaneous	6,000	6,100	5,000	6,700
Total yearly	$1,250,920	$1,422,810	$1,169,300	$1,398,550
Per ton	$ 41.70	$ 47.43	$ 38.98	$ 46.62

ALTERNATIVE METHOD OF WEIGHTED SCORES

A logically parallel but somewhat different method of comparative evaluation of locations is that of "weighted scores." Thus, in a particular case, labor supply may be weighted as having an importance of 250, fuel supply 330, and power 100, with other factors bringing the total weight to 1000. The appraisal of locations involves scoring each place with regard to each factor on a percentage basis, with 100 per cent representing perfection. These percentages are then multiplied into the weights, and the total is added up to each location, the place scoring highest being presumably preferred. This method permits intangibles to be evaluated (as can, in fact, also be done under the cost-comparison method, through a similar exercise of judgment) and yet give an answer in definite numerical form.

It should be recognized, however, that this device of weighted scores shares all the limitations of the above-described cost-comparison procedure and has at least one additional drawback. The assignment of weights and of percentage ratings is of course based on the informed judgment of the investigator, and the result will be as trustworthy as his judgement but no more so. For example, if two locations have approximately equal wage rates but there is the prospect of a more rapid labor turnover at one of the locations, the scoring procedure would call for grading that location down a certain number of points to take account of the probable loss in efficiency resulting from the higher turnover. On the other hand, an investigator using the cost-comparison method would have to estimate the probable effect of the higher turnover in terms of labor costs, spoilage of materials, and perhaps various overhead items. In the hands of any given investigator, both methods ought, of course, to give the same result. The only difference is that in one procedure the estimates are made in dollars and cents and in the other case they are made in points. One is guesswork to the same extent as the other. This being the case, the dollar-and-cents comparison would probably appeal to most people as involving less hocus-pocus and as attempting to express the differences in locational advantage in terms of the direct effect upon particular cost elements. Either method is equally well adapted for taking account of imponderables or uncertainties by expert guesswork.

The more important disadvantage of the weighted-score method is that its weights imply that the proportions in which the various materials and services are required are uniform as between different locations. If, for instance, we give labor costs a weight of 200 and various overhead items 150, we are ignoring the possibility that at a high-labor-cost location it might be feasible to use a somewhat more mechanized method of production and cut down on labor requirements at the expense of some additional machine overhead and power costs. The procedure of cost comparison on the other hand does not suffer from this inflexibility. We can compare the estimated costs of production, delivery, etc., for various locations on the assumption that at each location the setup most appropriate to local conditions would be used.

THE ROLE OF NATIONAL DEFENSE IN CHOOSING A PLANT LOCATION

The war potential of the chemical industry of the country makes dispersion of plants an important plant-location factor. The problem is especially urgent, even at the present time, since the large plants now being built would be difficult to move and reestablish, should war seem near at some future time.

The new factor to be considered is the atomic bomb. This weapon, together with long-range bombers and guided missiles, for the first time exposes the entire United States to effective enemy attack. The only known defense against such weapons is *space*, and dispersion of plants protects them by virtue of space. Space is a defense because it prevents the enemy from easily and simultaneously destroying vital industries in one surprise blow. The government has estimated that a single atomic bomb has an effective destructive radius of 3 miles. It is not considered likely that an enemy would use such weapons against areas of industrial concentration measuring less than 5 square miles, or areas with less than 50,000 inhabitants, unless such areas included a war plant of especially great importance.

The National Security Resources Board has listed the objectives which the enemy will primarily attempt to destroy: (1) highly important individual plants, (2) explosive or flammable installations, the destruction of which would endanger other important installations, (3) utilities, (4) transport facilities, (5) armed-forces bases, (6) dams and bridges, (7) air bases and supply centers, and (8) concentrations of plants producing important industrial items. An index of the strategic importance of a location can be estimated by enumerating and weighting all installations within a 3-mile radius which fall in the foregoing classifications.

Although the possibility of building underground plants has been often considered, the expense is too great to be borne by the manufacturer alone, unless the government considers the installation vital enough to contribute to the cost.

In addition to considering the factors already enumerated, the location should also be examined in terms of the future. The probable industrial and population growth should be considered. The terrain is also important. Hills tend to limit the effectiveness of the atomic-bomb explosion.

At the present time security measures rarely dictate plant location. Often location in a non-strategic area adds too greatly to the initial or operational plant cost. Unless the government is able to assist financially, private industry must disregard security considerations in favor of meeting competition and serving the available markets.

SECTION 27

ELECTRICITY AND ELECTRICAL ENGINEERING

BY

R. G. Warner, Ph. B., E. E., Vice President, United Illuminating Co., New Haven, Conn.; Fellow, American Institute of Electrical Engineers.

F. T. McNamara, Ph. B., E. E., Associate Professor of Electrical Engineering, Yale University; Associate Member, Institute of Radio Engineers; Associate Member, American Institute of Electrical Engineers.

CONTENTS

	Page
Definitions—Units	1732
Electrical Circuits	1733
Magnetic Circuits	1736
Electric and Magnetic Relations	1737
Electrical Measurements	1737
Photo-, Piezo-, Thermoelectricity	1742
Thermionic Tubes	1742
Conduction of Electricity through Gases	1747
Gaseous-discharge Tubes	1748
Rectifiers	1750
Rheostats	1752
Magnets	1752
Electrical Wiring	1753
Cost of Power	1755
Illumination	1755
Electric Heating	1759
Transformers	1762
Generators	1763
Synchronous Converters	1765
Motors	1766
Summary of Motor Applications	1770

ELECTRICITY AND ELECTRICAL ENGINEERING

DEFINITIONS—UNITS*

The **electron** is the elementary charge, or quantity of negative electricity.

Any ultimate carrier of electricity is called an **ion.** It may be an electron, or a charged atom, molecule, or group of molecules.

Current (I, i) is the rate of flow of electricity. The unit, the **ampere** (amp.) is equal to a flow of 1 coulomb (6.285×10^{18} electrons) per second. The fundamental measurement of the ampere is determined by the deposition of silver from a silver nitrate solution (see p. 1772).

Potential difference (E, e, or V, v) or **electromotive force** (e.m.f.) is the force or pressure causing current to flow in a circuit. The unit, the **volt,** is based on a standard primary cell. The "normal" Weston cadmium cell has a potential of 1.0183 volts at 20°C. (see p. 1772). Electromagnetically, 1 volt is induced in a conductor cutting magnetic flux at the rate of 10^8 lines per second.

Resistance (R, r) is the property of a circuit (or substance) opposing the flow of current and causing heat when current flows. The unit, the **ohm** (Ω), is the resistance which limits the current, caused by 1 volt, to 1 amp. One ampere flowing through a 1-ohm resistance produces heat at the rate of 1 watt (10^7 ergs/sec.).

Resistivity (ρ) is the specific resistance of a substance, usually expressed in ohms per centimeter cube, or ohms per circular mil-foot. A **circular mil-foot** is a conductor having a length of 1 ft. and a uniform cross section which is 0.001 in. in diam. The resistance of 1 cir. mil-ft. of copper at 20°C. is 10.371 ohms, and at 60°C., 12.0 ohms.

Conductivity is the reciprocal of resistivity.

Conductance (G, g) is a property of a circuit which permits the flow of electricity. The unit, the **mho** is the reciprocal of the ohm. In direct-current circuits the conductance equals $1/R$; in alternating-current circuits the conductance equals $R/(R^2 + X^2)$.

Inductance (L) is a circuit property expressing the ratio of *flux linkages* (total flux times number of turns) to the current producing the flux. The unit, the **henry,** equals 10^8 flux linkages per ampere. *Self-inductance* is this ratio referring to flux linkages with the circuit producing the flux; *mutual inductance* is the ratio referring to flux linkages with one circuit caused by current in a neighboring circuit.

Capacitance (C) expresses the ability of a circuit to hold or store electric charges. The unit, the **farad,** is the capacitance which will store 1 coulomb for each volt. Commonly the **microfarad** is used, which is one-millionth of a farad.

Dielectric constant (ϵ) is the specific inductive capacity of a material. It is equal to the ratio of the capacitances of two condensers of identical size, one using the particular dielectric, the other using air or a vacuum as the dielectric.

Dielectric strength (S) is the rupturing strength of an insulating material when subjected to voltage stress. It is usually expressed in volts per mil or volts per millimeter or multiples thereof. Actually the breakdown varies with the shape of the electrodes and does not increase in proportion to the thickness of the material.

* For more complete definitions see "American Standard Definitions of Electrical Terms," American Standards Association, C42-1941.

Frequency (f) is the number of cycles of alternating current per second.

Reactance (X) is the property of alternating-current circuits opposing the flow of current due to inductance or capacitance. The unit is the **ohm.** The reactance is equal to $2\pi fL$ for *inductive circuits* and to $1/(2\pi fC)$ for *capacitive circuits.* When inductive reactance, X_L, and capacitive reactance, X_C, are both present, their effects oppose one another and, in a series circuit, the total $X = X_L - X_C$.

Susceptance (B, b) is a property of a circuit which permits the flow of the reactive component of current. The susceptance in mhos is equal to $X/(R^2 + X^2)$.

Impedance (Z) is the property of alternating-current circuits opposing the flow of current, combining the effects of resistance and reactance. The unit, the **ohm,** is the impedance requiring 1 volt to cause 1 amp. to flow: $Z = \sqrt{R^2 + X^2}$.

Admittance (Y) is the property of a circuit which permits current to flow, including both the conductance and susceptance of the circuit. The admittance in mhos is equal to $\sqrt{G^2 + B^2}$.

Magnetic flux (ϕ) is the magnetic flow or lines of magnetic force through a magnetic circuit. The unit, the **maxwell,** is one line.

Flux density (B) is the flux per unit area. The unit, the **gauss,** is 1 maxwell per square centimeter. Flux density is also expressed in lines per square inch.

Magnetomotive force (m.m.f.) is the force which causes flux in a magnetic circuit. The unit, the **gilbert,** is equal to 1.257 amp.-turns. Frequently magnetomotive force is expressed in ampere-turns.

Reluctance ($\mathcal{R}$) is the magnetic resistance to flux. The unit reluctance limits the flux to 1 maxwell with a m.m.f. of 1 gilbert.

Magnetic field intensity (H) is the m.m.f. per unit length of path usually expressed in gilberts (or ampere-turns) per centimeter (or per inch).

Permeability (μ) is the magnetic conductivity of a material. The permeability of air and many other substances is unity.

Power (P, p) is the rate of supplying or developing energy. The unit, the **watt,** is the power delivered by 1 volt with 1 amp. flowing (unity power factor). One horsepower equals 746 watts; 1 kw. = 1.34 hp.

Energy (W, w) is the capacity for doing work. The unit, the **watt-hour** is the energy supplied or developed by 1 watt in 1 hr. The **kilowatt-hour,** 1000 watt-hours, is the common unit for billing energy (occasionally but erroneously this is abbreviated to the form kilowatt).

Volt-amperes (Va), frequently called the **apparent power,** is the product of volts and amperes in a circuit. At unity power factor, the volt-amperes equal the watts.

Power factor (P.F.) is the ratio of the power to the volt-amperes and hence must be equal to, or less than 1.0. The volt-amperes in an alternating-current circuit are frequently greater than the watts due to the effect of inductance or capacity: P.F. $= W/Va = \cos\theta$, where θ is the angular phase difference between voltage and current.

Reactive volt-amperes (*vars*) is the component of the volt-amperes not producing power: $Vars = EI \sin\theta$ and the total $Va = \sqrt{w^2 + vars^2}$.

Table 1. Electrical and Magnetic Units

Quantity	Symbol	Practical unit	m*	S†
Charge	q	Coulomb	10^{-1}	3×10^{9}
Current	I, i	Ampere	10^{-1}	3×10^{9}
Electromotive force	E, e	Volt	10^{8}	$\frac{1}{3} \times 10^{-2}$
Resistance	R, r	Ohm	10^{9}	$\frac{1}{9} \times 10^{-11}$
Inductance	L	Henry	10^{9}	$\frac{1}{9} \times 10^{-11}$
Capacitance	C	Farad	10^{9}	9×10^{11}
Magnetic flux	ϕ	Maxwell	1	$\frac{1}{3} \times 10^{-10}$
Flux density	B	Gauss	1	$\frac{1}{3} \times 10^{-10}$
Power‡	P, p	Watt	10^{7}	10^{7}
Energy‡	W, w	Watt-second	10^{7}	10^{7}
		Watt-hour	36×10^{9}	36×10^{9}

* For a given quantity, m is the ratio of the number of electromagnetic units to the number of practical units, or m is the ratio of the magnitude or size of a practical unit to the magnitude of an electromagnetic unit.

† For a given quantity, S is the ratio of the number of electrostatic units to the number of practical units, or S is the ratio of the magnitude or size of a practical unit to the magnitude of an electrostatic unit.

‡ For more complete equivalents of units of power and energy, see Table 18 in Sec. 1.

ELECTRICAL CIRCUITS

Each circuit to carry current must be completed from source through apparatus and back to source.

Direct current or continuous current is unidirectional and practically non-pulsating, although there may be some variation due to the commutators of generators.

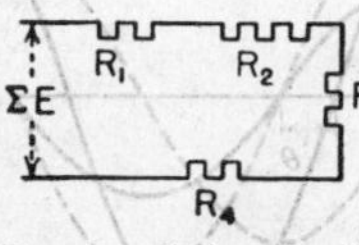

FIG. 1. Series circuit.

Ohm's law for any part or parts of a circuit:

$$E = IR \qquad I = \frac{E}{R} \qquad R = \frac{E}{I}$$

In a **series circuit** (Fig. 1):

$$I_1 = I_2 = I_3, \text{ etc.}$$
$$\Sigma E = E_1 + E_2 + E_3 + \cdots E_n$$
$$\Sigma R = R_1 + R_2 + R_3 + \cdots R_n$$

In a **parallel circuit** (Fig. 2):

$$E_1 = E_2 = E_3, \text{ etc.}$$
$$\Sigma I = I_1 + I_2 + I_3 + \cdots I_n$$
$$\sum \frac{1}{R} = \frac{1}{R_1} + \frac{1}{R_2} + \frac{1}{R_3} + \cdots \frac{1}{R_n}$$

In **combination circuits,** compute equivalent values for each group and then combine group values.

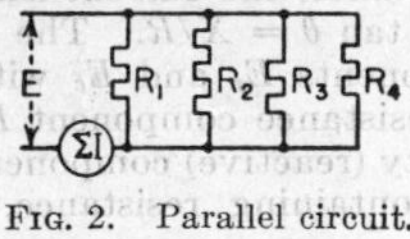

FIG. 2. Parallel circuit.

Kirchhoff's laws: (1) The algebraic sum of the currents at any point must equal zero; or the currents coming into a point must equal the currents leaving. (2) The algebraic sum of the voltages about a complete loop in a circuit must equal zero; or between any two points the voltage must be the same as determined over each path.

Resistances:

$$R = \frac{\rho l}{a}$$

where l is length and a is area, in the same units as resistivity. ρ is nearly constant for any given material except for temperature changes. For different materials resistivity ranges from microhms to megohms per centimeter cube (Tables 2 and 3):

$$\rho_t = \rho_{20}[1 + \alpha(t - t_{20})]$$

or

$$\rho_t = \rho_0[1 + \alpha(t - t_0)]$$

where ρ_t is the resistivity at any temperature t; ρ_{20} is the resistivity at 20°C.; α is the temperature coefficient of resistance and is equal to 0.00393 for copper at 20°C. (Table 2); and ρ_0 is the resistivity at 0°C.

The **effective resistance** in alternating-current circuits increases at higher frequencies because of nonuniform distribution of currents and because of losses.

Table 2. Conductors and Resistances

Material	Resistivity, microhms/cm. cube, ρ_{20}	Temperature coefficient, α_{20}	Max. working temperature, °C.
Aluminum	2.82	0.00421	
Carbon (filament)	4050	−0.00054	
Graphite	1365	−0.00028	
Copper	1.72	0.00393	
Iron	9.78	0.00634	
Lead	20.63	0.00367	
Mercury	95.9	0.00089	
Nickel	7.24	0.00491	500
Platinum	10.6	0.00369	
Silver	1.62	0.00361	
Tin	11.4	0.0040	
Tungsten	5.75	0.00454	
Zinc	5.92	0.00325	
Alloy:			
Advance	48–49	0.00001 to 0.00002	500
Alumel	33.3	0.0012	
Brass	5.7	0.0014 to 0.002	
Bronze	13–18	0.0005	
Calido	110	0.00012	1000
Cast iron	57–114		
Chromel	70–110	0.00011 to 0.00054	
Climax	87	0.00067	600
Constantan	47–51	−0.00004 to +0.00001	
Duralumin	3.35		
German silver	17–41	0.00004 to 0.00038	
Ideal	49	0.000005	500
Invar	75	0.002	
Karma	103	0.0001	1150
Lucero	46	0.00076	500
Manganin	34–100	−0.00003 to −0.00002	100
Monel	42.5–45	0.00002 to 0.002	900
Nichrome	110	0.0004 to 0.00603	1000
Nichrome II	109–111	0.00015	
Nichrome III	90–97	0.00005 to 0.00019	1100
Nichrome IV	98–103	0.00018	1150
Phosphor bronze	2–12	0.003 to 0.004	
Swedish iron	20	0.0012	
Welding iron	18	0.006	

Note. The resistivity and temperature coefficients are based on data in the "International Critical Tables," vol. 6, p. 159, McGraw-Hill, and are subject to variation with the commercial composition of the materials and with the heat-treatment.

Power in a direct-current circuit is equal to the product of volts and amperes:

$$W = EI = I^2R = \frac{E^2}{R}$$

Energy is the power for a given time or $I^2Rt = EIt$.

Alternating current varies periodically in magnitude and direction. Commercial alternating current approximates the **sine wave,** and in this discussion sine waves will be assumed.

$$e = E_{\max} \sin 2\pi ft$$

where e is the instantaneous value, f is frequency, t is time in seconds and $2\pi ft$ is the angle in radians.

The **effective value** of alternating current (or voltage) gives the same heating effect in a resistance circuit as the corresponding direct current. Inasmuch as the heating is proportional to the squared value, the effective value of alternating current is obtained by taking the square root of the average of the instantaneous squared values. This is referred to as the **root-mean-square (r.m.s.) value** and is designated as E or I. For sine waves, $E = 0.707E_{\max}$ and $I = 0.707I_{\max}$.

Table 3. Dielectric Properties

Resistivity ρ, dielectric strength S for samples of given thickness and the dielectric constant ϵ of materials used as insulators.

Material	ρ, ohms/cm. cube	S, kv./mm.	Thickness of sample, mm.	ϵ
Asbestos paper	2×10^5	4	1	2.7
Cellulose	10^9	......	......	3.9–7.5
Enamel	10^{14}	20–25	0.02	
Fiber, vulcanized	5 to 20×10^9	8–18	1	5–7.5
		3–6	12	
Fish paper	......	10–15	0.1–1.2	
Glass	10^{10} to 10^{14}	230	1.4	5–8.5
Pyrex	10^{14}	134	6.35	4.8
Maple, paraffined	3×10^{10}	4.5	15	4.1
Marble	10^9 to 10^{11}	2–4	25	8.3
Mica	10^{15} to 10^{17}	80–200	0.05	2.15–2.5
Oil, transformer	2 to 15×10^{13}	10–25	3.81	
Paraffin	10^{16} to 5×10^{18}	15–50	......	1.9–2.3
Petrolatum	2 to 10×10^{12}	20	2.5	2.2
Phenolic insulating materials, molded	10^{11} to 10^{12}	9–280	......	5.0–7.5
Porcelain	10^{12} to 10^{15}	12–28	5	4.4–6
Pressboard	10^9	12–5	0.2–3.0	
Rubber, hard	3×10^{13}	10–38	1–0.5	3
Rubber, soft vulcanized	1 to 15×10^{15}	......	......	2.7
Silica, fused	5×10^{17}	......	......	3.5
Slate	10^8	0.2–0.4	25	6.0–7.5
Soapstone	6×10^8	1.0	25	
Varnished cambric	......	60–45	0.1–0.4	3.5–5.5
Woods, hard, dry	10^{10} to 10^{13}	0.4–0.6	25	3.0

Note. These values are based on data in the "International Critical Tables," vol. 2, p. 310, McGraw-Hill, and are subject to considerable variation with temperature and moisture conditions. The dielectric strength decreases with thicker pieces of material.

Average value of alternating current or voltage over a complete cycle is zero. The average value of a half cycle of a sine wave is $2/\pi$ times the maximum, hence $E_{\text{avg}} = 0.636E_{\max}$. This value has little importance. It is

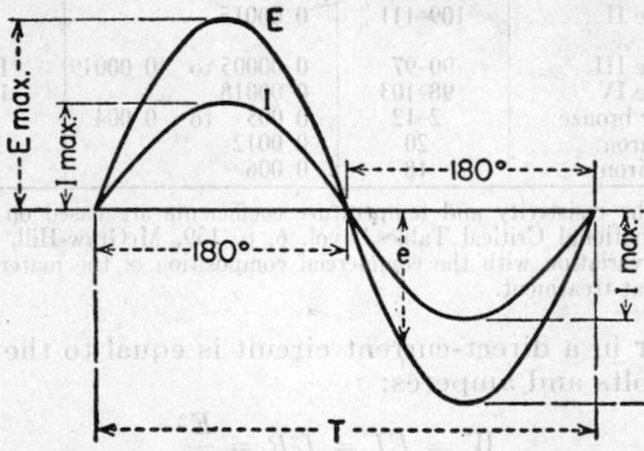

Fig. 3. Sine waves; non-inductive circuit.

used with the effective value as a ratio to designate the form factor. Form factor is $E_{\text{eff}}/E_{\text{avg}}$. For a sine wave the form factor is 1.11.

In non-inductive circuits the current is in phase with the voltage (Fig. 3), *i.e.*, it reaches a maximum value at the same time as the voltage and passes through zero with the voltage. When the circuit (or portion of the circuit) contains resistance only, the current is in phase with the voltage, and the following relations hold:

$$e = iR \qquad E_{\max} = I_{\max}R \qquad E = IR$$

Inductive circuits require additional voltage; since the alternating current produces alternating flux and this change of flux induces a voltage, it is necessary to impress additional voltage (for a given current) to overcome the induced voltage. For a circuit containing inductance only, the current is out of phase with the voltage, lagging it by a quarter cycle or 90 deg. If

$$i = I_{\max} \sin 2\pi ft$$

then

$$e = 2\pi fLI_{\max} \cos 2\pi ft = E_{\max} \cos 2\pi ft$$

where $2\pi fL$ is the inductive reactance (X), and $XI_{\max} = E_{\max}$ and $XI = E$.

In a circuit containing resistance and inductance, the current and voltage are not in phase and the current lags the voltage by an angle θ, where $\theta = \tan^{-1} X/R$. The e.m.f. may be divided into components E_r and E_l, where the current is in phase with the resistance component E_r and lags by 90 deg. the inductance (reactive) component E_l (Fig. 4).

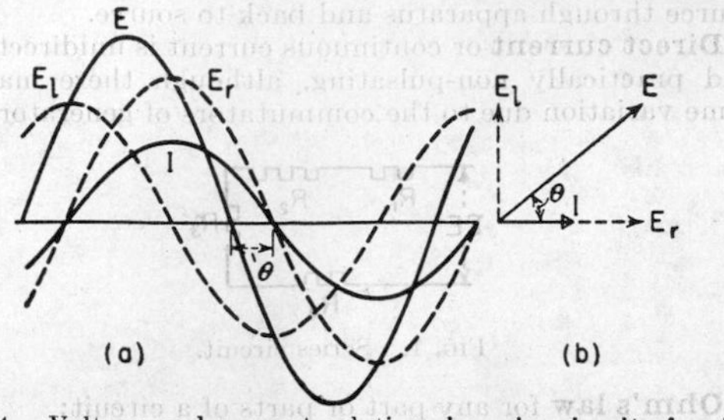

Fig. 4. Voltage and current relations in series circuit of resistance and inductance. (*a*) Sine waves; (*b*) vectors.

Capacitive circuits also have a phase difference between the voltage and current. Inasmuch as capacitance absorbs coulombs in proportion to the voltage, the rate of absorbing (the current) is proportional to the rate of change of voltage. Hence the current is maximum when the voltage is changing fastest (at zero) and the current leads the voltage by 90 deg. for a circuit containing capacitance only. If

$$i = I_{\max} \sin 2\pi ft$$

then

$$e = -\frac{1}{(2\pi fC)} I_{\max} \cos 2\pi ft = -XI_{\max} \cos 2\pi ft$$

where $1/(2\pi fC)$ is the capacity reactance and $XI_{\max} = E_{\max}$ and $XI = E$. In a circuit containing resistance and capacity reactance, the current leads the voltage by an angle θ where $\tan\theta = X/R$. The e.m.f. may be divided into components E_r and E_c with the current in phase with the resistance component E_r and leading by 90 deg. the capacity (reactive) component E_c (Fig. 5).

In a circuit containing resistance, inductance, and capacitance,

$$E = \sqrt{(E_r)^2 + (E_l - E_c)^2} = \sqrt{(IR)^2 + (IX_l - IX_c)^2} = IZ$$

$$Z = \sqrt{R^2 + (X_l - X_c)^2}$$

$$\tan\theta = \frac{(X_l - X_c)}{R} \qquad \cos\theta = \frac{R}{Z}$$

Resonance in a circuit containing inductance and capacity exists when the reactance due to each is equal; the current is limited only by resistance in the series circuit, and the voltages across the reactances are equal and opposite and may be very large. Under this condition $f = 1/(2\pi\sqrt{LC})$ and is the natural frequency of the circuit. This is the frequency at which such a circuit tends to oscillate. By changing the value of L and/or C, the natural frequency may be adjusted to any desired value as in radio sending and receiving sets.

Vector representation is commonly used for indicating magnitude and phase position of alternating currents and voltages. Each vector is the expression of a sine wave; it rotates counterclockwise, one revolution per cycle; hence only one frequency can be shown on a

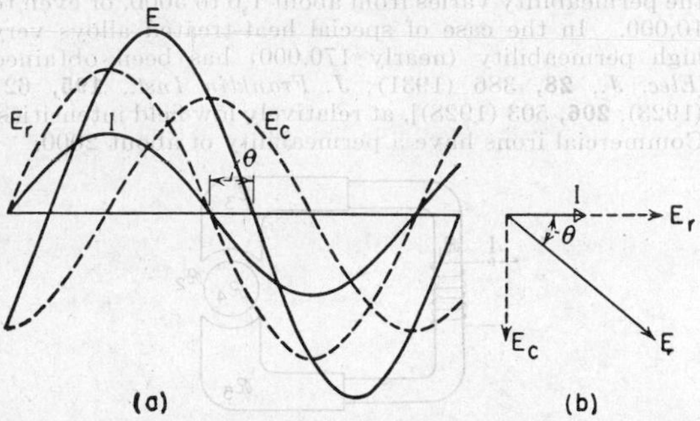

FIG. 5. Voltage and current relations in series circuit of resistance and capacitance. (*a*) Sine waves; (*b*) vectors.

diagram. The vectors are pictured in an instantaneous position and may be added or subtracted to obtain the same results as by adding or subtracting sine waves. Figures 4*b* and 5*b* show the same relations as the sine waves of Figs. 4*a* and 5*a*, respectively. The lengths of vectors are usually shown as effective values to an arbitrary scale.

The **energy** in an alternating-current circuit is the integrated product of the instantaneous voltage and

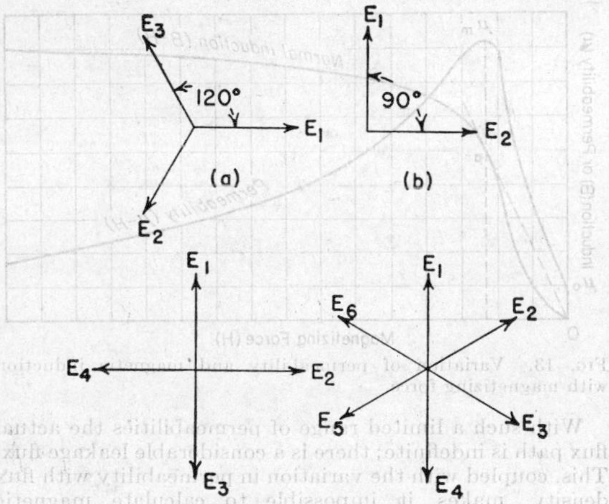

FIG. 6. Polyphase systems voltage vectors. (*a*) Three-phase; (*b*) two-phase; (*c*) four-phase; (*d*) six-phase.

current. The energy delivered to inductance or to capacitance is stored during a portion of the cycle and then returned; hence it does not affect the average energy. The energy for a period t is $EIt \cos \theta$. The angle θ is the phase displacement between E and I, and $\cos \theta$ is called the **power factor.**

Power in an alternating-current circuit is the instantaneous product of the voltage and current. Since e and i are sine waves, the product of the instantaneous values is a sine wave of double frequency; hence the power in a single-phase circuit is pulsating. The average value of power is

$$W = I^2R = \frac{EIR}{Z} = EI \cos \theta$$

Polyphase systems are commonly used because of better economy in generation and transmission and because they permit more rugged motors with a uniform delivered torque. Such a system has two or more

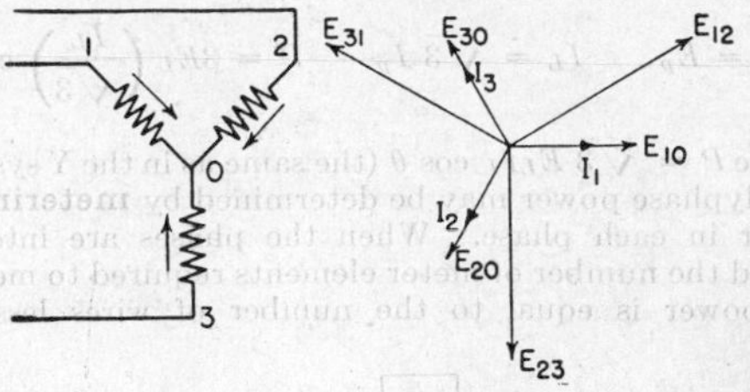

FIG. 7. Three-phase system, Y-connection and corresponding vectors of voltage and current for 1.0 power factor.

e.m.fs. displaced in phase by a definite amount with respect to one another. The common polyphase systems are three-phase, two-phase, and six-phase. Figure 6 shows the displaced voltage vectors of these systems.

A balanced polyphase system has equal loads in all the phases. Although in each phase the power is pulsating, the total power in a balanced polyphase system is constant.

Calculations for a polyphase system are normally made for each phase and then the results combined.

Three-phase Y system (Fig. 7). The line voltages are equal to the vector differences of the phase voltages. Using vector notation,

$$\dot{E}_{12} = \dot{E}_{10} - \dot{E}_{20}$$
$$\dot{E}_{23} = \dot{E}_{20} - \dot{E}_{30}$$
$$\dot{E}_{31} = \dot{E}_{30} - \dot{E}_{10}$$

For balanced phase voltages, $E_L = \sqrt{3}\, E_p$, E_L = line voltage and E_p = phase voltage. The line currents are the phase currents.

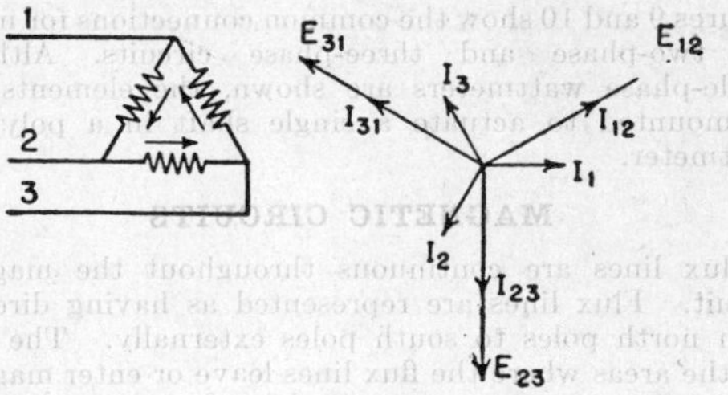

FIG. 8. Three-phase system, delta-connection and corresponding vectors of voltage and current for 1.0 power factor.

Three-phase Delta System (Fig. 8). The line voltages are the phase voltages, $E_L = E_p$. The line currents equal the vector difference of the phase currents:

$$\dot{I}_1 = \dot{I}_{12} - \dot{I}_{31}$$
$$\dot{I}_2 = \dot{I}_{28} - \dot{I}_{12}$$
$$\dot{I}_3 = \dot{I}_{31} - \dot{I}_{23}$$

For balanced loads, $I_L = \sqrt{3}\, I_p$, I_L = line current and I_p = phase current.

Polyphase power is the sum of the power in the individual phases. For the three-phase balanced Y system

$$P = 3E_pI_p \cos \theta$$

where θ is the angle between E_p and I_p but

$$E_L = \sqrt{3}E_p \qquad I_L = I_p$$

$$P = 3\left(\frac{E_L}{\sqrt{3}}\right) I_L \cos\theta = \sqrt{3}\,E_L I_L \cos\theta$$

and for a balanced Δ system

$$P = 3E_p I_p \cos\theta$$

but

$$E_L = E_p \qquad I_L = \sqrt{3}\,I_p \qquad P = 3E_L\left(\frac{I_L}{\sqrt{3}}\right)\cos\theta$$

Hence $P = \sqrt{3}\,E_L I_L \cos\theta$ (the same as in the Y system).

Polyphase power may be determined by **metering** the power in each phase. When the phases are interconnected the number of meter elements required to measure the power is equal to the number of wires less one.

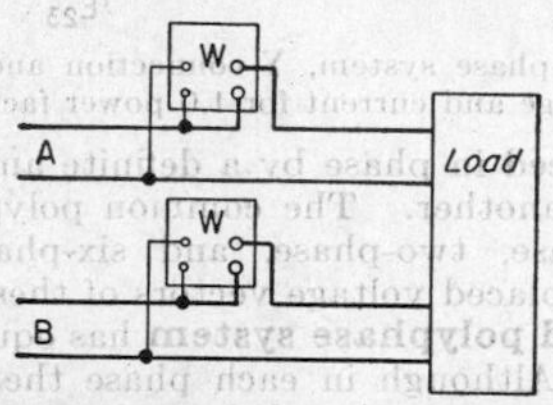

FIG. 9. Power measurement in two-phase four-wire circuit.

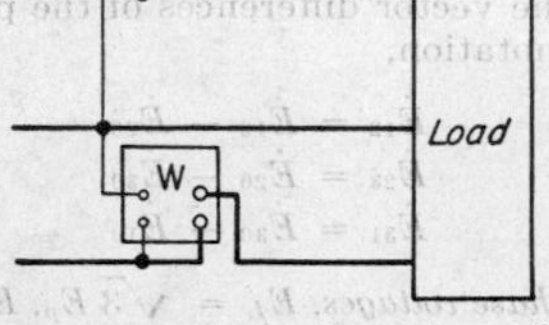

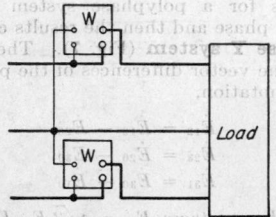

FIG. 10. Power measurement in three-phase three-wire circuit or two-phase three-wire circuit.

Figures 9 and 10 show the common connections for metering two-phase and three-phase circuits. Although single-phase wattmeters are shown, the elements may be mounted to actuate a single shaft in a polyphase wattmeter.

MAGNETIC CIRCUITS

Flux lines are continuous throughout the magnetic circuit. Flux lines are represented as having direction from north poles to south poles externally. The poles are the areas where the flux lines leave or enter magnetic materials.

Rowland's law (*the magnetic Ohm's law*):

$$\text{m.m.f.} = \phi\mathcal{R} \qquad \phi = \frac{\text{m.m.f.}}{\mathcal{R}}$$

In **series circuits** (Fig. 11):

$$\phi_1 = \phi_2 = \phi_3 = \cdots$$
$$\Sigma\text{m.m.f.} = \text{m.m.f.}_1 + \text{m.m.f.}_2 + \cdots$$
$$\Sigma\mathcal{R} = \mathcal{R}_1 + \mathcal{R}_2 + \mathcal{R}_3 + \mathcal{R}_4 + \cdots$$

In **parallel circuits** (Fig. 12):

$$\text{m.m.f.}_1 = \text{m.m.f.}_2 = \cdots$$
$$\Sigma\phi = \phi_1 + \phi_2 + \cdots$$
$$\sum\frac{1}{\mathcal{R}} = \frac{1}{\mathcal{R}_1} + \frac{1}{\mathcal{R}_2} + \cdots$$

In **combination circuits**, compute equivalent values for each group and then combine group values.

Reluctance.

$$\mathcal{R} = \frac{l}{\mu a}$$

where l is the length, a is the area, and μ is the permeability.

The permeability varies greatly in magnetic materials with the flux density (Fig. 13). In different materials the permeability varies from about 1.0 to 5000, or even to 10,000. In the case of special heat-treated alloys very high permeability (nearly 170,000) has been obtained [*Elec. J.*, **28**, 386 (1931); *J. Franklin Inst.*, **195**, 621 (1923), **206**, 503 (1928)], at relatively low field intensities. Commercial irons have a permeability of about 2000.

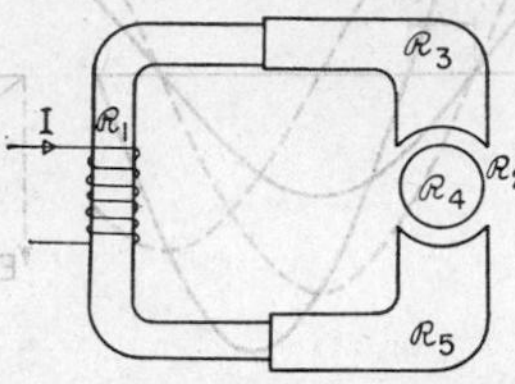

FIG. 11. Series magnetic circuit.

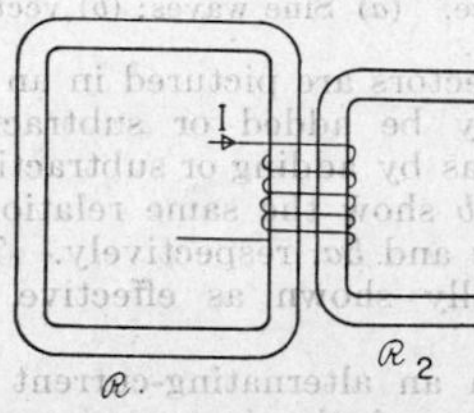

FIG. 12. Parallel magnetic circuit.

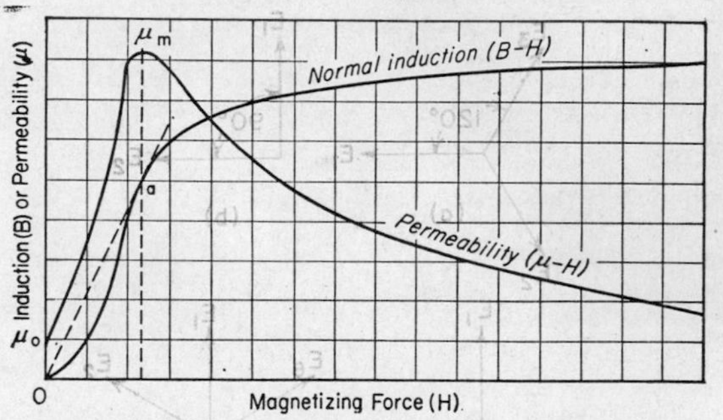

FIG. 13. Variation of permeability and magnetic induction with magnetizing force.

With such a limited range of permeabilities the actual flux path is indefinite; there is a considerable leakage flux. This, coupled with the variation in permeability with flux density, makes it impossible to calculate magnetic circuits with the same precision as electric circuits. Most calculations involving magnetic materials use (*B-H*) magnetization curves directly rather than permeability (see Fig. 14). In electrical apparatus the iron is normally magnetized to a point near the knee (where the curve bends rapidly) of the magnetization curve.

Energy is stored *in the magnetic field;* it must be supplied when the field is established and is returned when the field is destroyed. Energy is not necessary to maintain a magnetic field, although, if the field is produced electrically, energy is required to supply the copper loss of the magnetizing coil.

Hysteresis loss occurs in iron with a changing magnetic flux and is attributed to a molecular friction. With an alternating flux, this loss in watts is $P_h = \eta f B_{max}^{1.6} 10^{-7}$ per unit volume, where η is the hysteresis coefficient, f is the frequency, and B_{max} is the maximum flux density. Silicon steels are used to reduce the hysteresis loss.

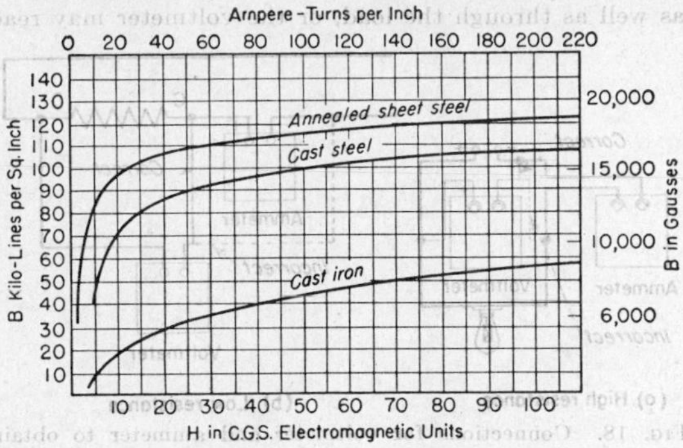

FIG. 14. Normal magnetization characteristics.

Table 4. Values of Hysteretic Constant*

Material	η
Hard tungsten steel	0.058
Hard nickel	.039
Hard cast steel	.025
Cast iron	.013
Soft nickel	.013
Cast steel	.012
Cobalt	.012
Electrolytic iron	.009
Annealed cast steel	.008
Ordinary sheet iron	.004
Annealed iron sheet	.002
Silicon steel	.0009
Best silicon steel	.0006

* Data from *J. Franklin Inst.*, **170**, 1 (1910).

Eddy currents are caused by voltages induced in the iron by a changing flux and tend to oppose the change of the flux. Eddy-current losses in iron subjected to a pulsating or alternating flux are reduced by laminating the circuit; this increases the length of path of the secondary currents and hence increases the resistance of their paths. With an alternating flux, the eddy-current loss per unit volume is

$$P_e = ks^2f^2B^2_{max}$$

where s is the thickness of the laminations and k is a constant varying with the resistivity of the iron, the wave shape of the induced voltages, and the flux distribution in the iron.

ELECTRIC AND MAGNETIC RELATIONS

Magnetic flux from electric current (Fig. 15):

$$\phi = \frac{\text{m.m.f.}}{\mathcal{R}} = \frac{1.257NI}{\mathcal{R}}$$

where NI = ampere-turns, $\mathcal{R}$ = reluctance, and ϕ = flux.

Direction of flux lines is reversed if direction of current is reversed (Fig. 15).

Electromotive force (Fig. 16) is induced in a conductor in a magnetic field (*a*) when there is relative physical motion of conductor with respect to the field, or (*b*) when there is change in magnitude of the field interlinking the conductor.

$$E = \frac{Z\phi \times 10^{-8}}{t} \qquad e = \left(\frac{d\phi}{dt}\right) N \times 10^{-8}$$

where Z is the number of conductors cutting the flux ϕ in t sec.; N is the number of turns. The direction of induced e.m.f. is reversed if in (*a*) the direction of motion is reversed or if the field is in the opposite direction, and in (*b*) if the field is increasing or if the field is in the opposite direction and decreasing. The direction of induced e.m.f. is such as to cause a current tending to oppose the change.

Mechanical force (Fig. 17) is exerted on a conductor carrying current when in a magnetic field, tending to move the conductor perpendicularly with respect to both the conductor and the field.

$$F = kBlI$$

$k = 8.83 \times 10^{-8}$ when F is the force in pounds, l the effective length of conductor in inches, B the com-

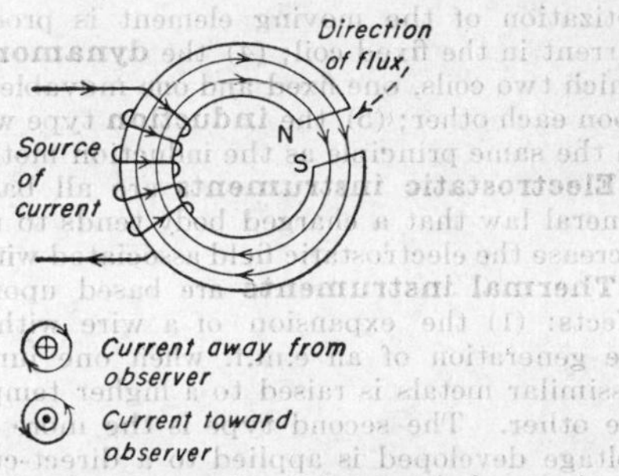

FIG. 15. Magnetic flux produced by electrical current

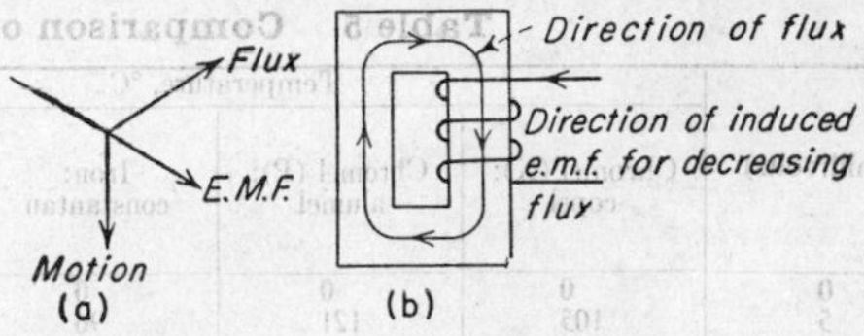

FIG. 16. Direction of induced e.m.fs. (*a*) For maximum e.m.f., flux, motion, and conductor are mutually perpendicular; (*b*) e.m.f. induced in coil by changing of flux linking with it.

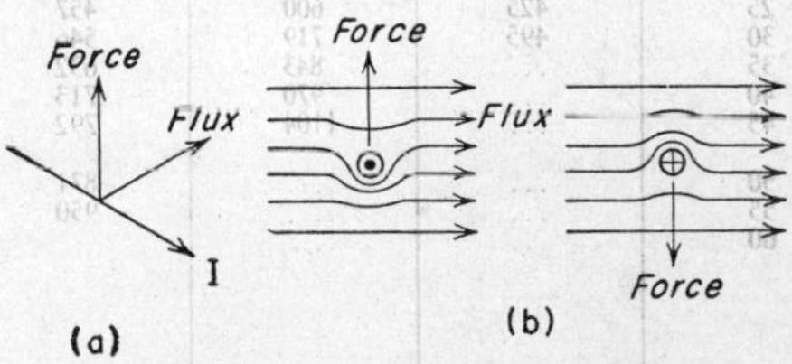

FIG. 17. Force on conductor carrying current in magnetic field. (*a*) Direction of force, perpendicular to flux and current; (*b*) distortion of flux lines due to current.

ponent of flux density perpendicular to the conductor in lines per square inch, and I the current in amperes. $k = 10.2 \times 10^{-8}$ when F is in kilograms, l in centimeters, and B in gausses.

Electrical machines consist of interlinked magnetic and electric circuits, usually fairly simple in nature, although in some cases the interactions become quite involved. *Iron* is the usual magnetic material and *copper* the electrical material; hence these together with various forms of *insulation* are the principal materials used in construction.

ELECTRICAL MEASUREMENTS*

The measurement of the magnitude of any phenomenon is usually based upon some effect produced by it. The effects utilized in electrical measuring devices are (1) *electromagnetic*, (2) *electrostatic*, (3) *thermal*, and (4)

* See Sec. 19, Process Control.

chemical. Electrical measuring devices which merely indicate, such as ammeters and voltmeters, are called *instruments;* devices which totalize with time, such as watt-hour meters and ampere-hour meters, are called *meters.*

Electromagnetic instruments are all based upon the fact that the current-carrying circuit has a force exerted upon it when located in a magnetic field. These instruments may be classified by types as follows: (1) The **moving-magnet** type in which a permanent magnet is caused to move under the influence of an electric current; (2) The **moving-coil** type in which the stationary and movable elements of (1) are interchanged; (3) the **moving-iron** type which differs from (1) in that the magnetization of the moving element is produced by the current in the fixed coil; (4) the **dynamometer** type in which two coils, one fixed and one movable, exert a force upon each other; (5) the **induction** type which operates on the same principle as the induction motor.

Electrostatic instruments are all based upon the general law that a charged body tends to move so as to increase the electrostatic field associated with it.

Thermal instruments are based upon one of two effects: (1) the expansion of a wire with heat, or (2) the generation of an e.m.f. when one junction of two dissimilar metals is raised to a higher temperature than the other. The second type is the more common; the voltage developed is applied to a direct-current instrument (see Thermoelectricity, pp. 1272*ff.*).

across the terminals, and the scale may therefore be calibrated in volts.

When both an ammeter and a voltmeter are used simultaneously, one of two *errors* is inevitable; the ammeter may read the current through the voltmeter as well as through the load, or the voltmeter may read

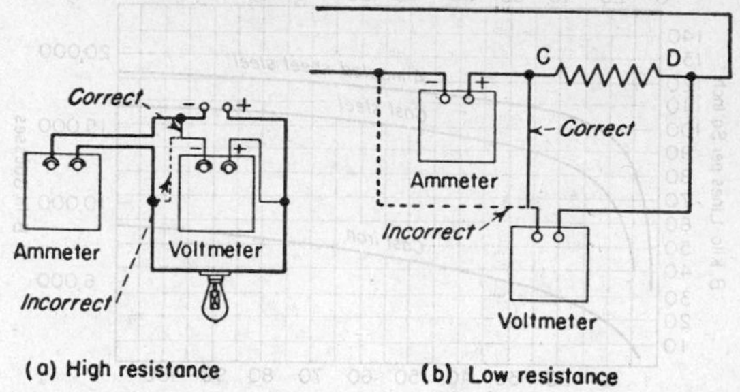

FIG. 18. Connections for voltmeter and ammeter to obtain minimum effect of power taken by instruments.

the potential drop across the ammeter as well as across the load. The magnitude of the load resistance is the factor that decides which method shall be used. Figure 18 shows the correct and incorrect way of making a measurement in the case of a high and also a low resistance.

A **wattmeter** is made by having the current in one ele-

Table 5. Comparison of the More Common Thermocouples*

E (millivolts)	Temperature, °C. Chromel (X): copel	Chromel (P): alumel	Iron: constantan	Platin-rhodium† gold-palladium
0	0	0	0	0
5	105	121	96	131
10	195	244	186	237
15	277	365	277	335
20	353	483	367	429
25	425	600	457	513
30	495	719	546	607
35	...	843	632	694
40	...	970	713	779
45	...	1104	792	866
50	...	....	871	954
55	...	....	950	1044
60	...	....	...	1136

E (millivolts)	Temperature, °C. Platinum: platin-rhodium (Heraeus)	Platinum: platin-rhodium (Johnston-Matthey)	Copper: constantan
0	0	0	0
1	147	146	25
2	265	260	49
3	374	364	72
4	478	461	94
5	578	553	115
6	675	641	136
7	769	725	156
8	861	806	176
9	950	884	195
10	1037	959	213
11	1122	1032	232
12	1206	1103	250
13	1289	1173	268
14	1372	1242	285
15	1455	1311	302
16	1537	1379	320
17	1620	1447	336
18	1704	1515	353

* "International Critical Tables," vol. 1, p. 59, McGraw-Hill.
† 10 per cent Rh; 40 per cent Pd.

Chemical instruments are based on Faraday's law of electrolysis.

Electromagnetic instruments constitute the great bulk of commercial meters. Electrostatic instruments are used mainly on high voltages or as precise laboratory instruments. The chief importance of thermal instruments is for high-frequency measurements, and chemical instruments are seldom used except as standards. Accordingly, the discussion that follows will be limited to instruments of the first class.

All **electromagnetic instruments** are essentially ammeters in that the deflection of the needle is an effect produced by the flow of current. When used as a voltmeter, a large impedance (normally a pure resistance) is inserted in the current-carrying circuit. With constant impedance, the current is proportional to the voltage

ment proportional to the load current and the current in the other element proportional to the load voltage. The two currents react upon each other to produce a deflection proportional to the average product of voltage and current. The wiring diagram of a single-phase wattmeter is shown in Fig. 19.

When *measuring three-phase power,* it is customary to use the **two-wattmeter method,** the wiring diagram of which is shown in Fig. 10. When using this method the power factor of the load may be determined from the two wattmeter readings, as indicated by the nomogram of Fig. 20. A **polyphase wattmeter** is merely a two-wattmeter arrangement under a single cover, both elements acting on a single shaft.

The range of any instrument may be extended by the use of a multiplier or shunt in the case of direct current or by an instrument transformer in the case of alternating current. A **multiplier** is a large resistance inserted in series with a voltage element so that only a small fraction of the total voltage will appear across the instrument terminals; similarly, a **shunt** is a low resistance used to by-pass a large portion of a given current around an ammeter. In self-contained instruments the multiplier or

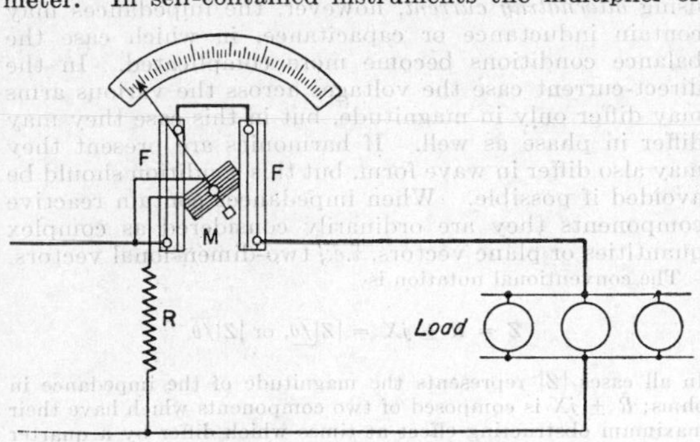

FIG. 19. Internal and external connections of wattmeter.

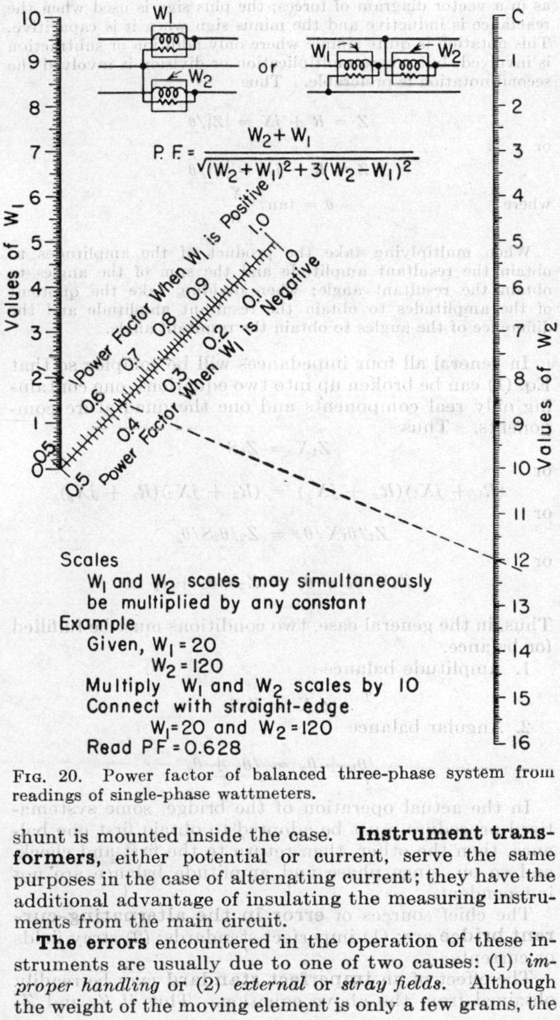

FIG. 20. Power factor of balanced three-phase system from readings of single-phase wattmeters.

shunt is mounted inside the case. **Instrument transformers,** either potential or current, serve the same purposes in the case of alternating current; they have the additional advantage of insulating the measuring instruments from the main circuit.

The errors encountered in the operation of these instruments are usually due to one of two causes: (1) *improper handling* or (2) *external* or *stray fields.* Although the weight of the moving element is only a few grams, the pressure on the pivot may vary from 1 to 100 tons/sq. in.; consequently great care should be exercised to prevent shock or impact; otherwise the pivot will mushroom or the bearing jewel crack, thus causing large and erratic errors. Stray or external fields are objectionable because they modify the fields existing due to the quantity measured. Most modern instruments are enclosed in a steel case to eliminate these effects; nevertheless it is well to avoid proximity to all known sources of such disturbance.

Certain instruments, *e.g.*, the **power-factor meter** or the **synchroscope,** are made *without a control spring.* In such instruments the moving element comes to rest when its field at maximum value coincides with a rotating field similar to that in a polyphase induction motor.

Recording instruments are based on the same general principles as the indicating instruments. A clockwork mechanism is added which draws a sheet of ruled paper under an indicator equipped with a pen and inkwell so that a permanent record is obtained (see Sec. 19, Process Control).

Integrating meters, such as the watt-hour meter, are essentially small motors of very slow speed so adjusted that the speed of rotation is proportional to the power absorbed. The moving element drives a gear train which integrates the number of revolutions made and so the energy utilized.

Besides these instruments and meters, however, there are several **comparison instruments** which enable two quantities of the same or related kinds to be compared. Such for example are: (1) the **potentiometer** for comparing e.m.fs. with the e.m.f of a standard cell, and (2) the **Wheatstone bridge** for comparing two resistances, or more generally, the alternating-current bridge, for comparing impedances, inductances, capacitances, etc.

The **potentiometer** or **opposition principle** is that, if two equal and opposing e.m.fs. are inserted in a circuit,

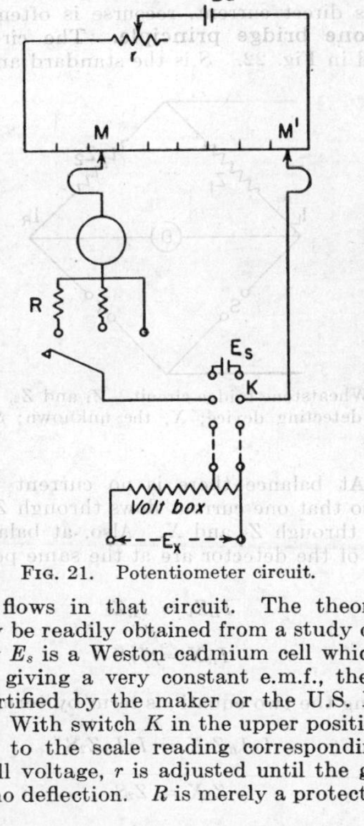

FIG. 21. Potentiometer circuit.

no current flows in that circuit. The theory of the method may be readily obtained from a study of Fig. 21. The battery E_s is a Weston cadmium cell which has the property of giving a very constant e.m.f., the value of which is certified by the maker or the U.S. Bureau of Standards. With switch K in the upper position and M and M' set to the scale reading corresponding to the standard cell voltage, r is adjusted until the galvanometer shows no deflection. R is merely a protective resistance for the galvanometer. This fixes the value of the working current, and in commercial instruments the scale will now read volts directly. Switch K is now thrown to the lower position and M and M' readjusted until the galvanometer again shows no deflection. Then the unknown voltage is equal to the scale reading between M and M'.

The **volt box** is used merely to intercept a definite fraction of the unknown voltage in order to bring it within the scale of the instrument. It is apparent that the unknown voltage supplies no current to the potentiometer when it is being measured; that the measurement is independent of galvanometer calibration; that the source of working current (Ba) must be very constant; and that the ratio of the various resistances must be exactly as indicated on the scale. Current may be determined by measuring the voltage drop across a standard resistance. (For details see *Leeds and Northrup, Bull.* 705.)

The principle enumerated above may be applied to alternating-current measurements also, but with much less accuracy because (1) there is no source of alternating potential comparable to the standard cell for direct potential and (2) alternating potentials may differ from each other in (*a*) magnitude, (*b*) phase, (*c*) frequency, and (*d*) wave form, whereas direct potentials may differ only in magnitude. Additional difficulties with this type of measurement will appear in the discussion of the Wheatstone's-bridge principle. [For information concerning alternating-current potentiometers, see Larsen, *Elektrotech. Z.*, (V) **41**, 1039 (1910); Drysdale, *Electrician*, **75**, 157 (1915); Drysdale, *J. Inst. Elec. Eng.*, **68**, 339 (1930).]

The simplest method of **determining the resistance or impedance** of a given element is to send a known current through the element and measure the drop of potential across its terminals. The ratio of volts to amperes gives the impedance in ohms of the element.

For precision measurements using alternating current as well as direct current, recourse is often had to the **Wheatstone bridge principle.** The circuit used is illustrated in Fig. 22. S is the standard and X the unknown.

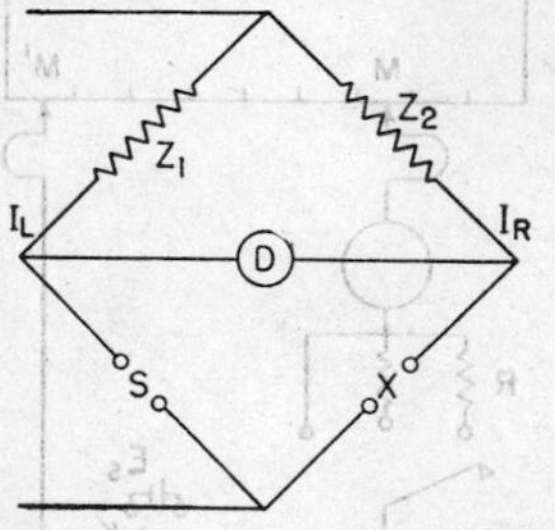

FIG. 22. Wheatstone-bridge circuit. Z_1 and Z_2, known impedances; D, detecting device; X, the unknown; S, a standard impedance.

At balance there is no current through the detector so that one current flows through Z_1 and S and the other through Z_2 and X. Also, at balance the two terminals of the detector are at the same potential and, therefore,

$$I_L Z_1 = I_R Z_2$$

and

$$I_R X = I_L S$$

Multiplying the two equations term by term,

$$I_L I_R Z_1 X = I_L I_R Z_2 S$$

or

$$Z_1 X = Z_2 S \quad (1)$$

or

$$X = \frac{Z_2 S}{Z_1}$$

When using *direct current*, all the impedances are pure resistances, and Eq. (1) becomes $R_1X = R_2S$. When using *alternating current*, however, the impedances may contain inductance or capacitance, in which case the balance conditions become more complicated. In the direct-current case the voltages across the various arms may differ only in magnitude, but in this case they may differ in phase as well. If harmonics are present they may also differ in wave form, but this condition should be avoided if possible. When impedances contain reactive components they are ordinarily considered as complex quantities or plane vectors, *i.e.*, two-dimensional vectors. The conventional notation is

$$Z = R \pm jX = |Z|\underline{/\theta}, \text{ or } |Z|\overline{\backslash\theta}$$

In all cases $|Z|$ represents the magnitude of the impedance in ohms; $R \pm jX$ is composed of two components which have their maximum obstructing effect at times which differ by a quarter of a cycle; their effective impedance is therefore obtained by combining them at right angles and obtaining the resultant as in a vector diagram of forces; the plus sign is used when the reactance is inductive and the minus sign when it is capacitive. This notation is quite simple where only addition or subtraction is involved, but where multiplication or division is involved the second notation is preferable. Thus

$$Z = R + jX = |Z|\underline{/\theta}$$

or

$$Z = R - jX = |Z|\overline{\backslash\theta}$$

where

$$\theta = \tan^{-1}\frac{X}{R}.$$

When multiplying take the product of the amplitudes to obtain the resultant amplitude and the sum of the angles to obtain the resultant angle; when dividing take the quotient of the amplitudes to obtain the resultant amplitude and the difference of the angles to obtain the resultant angle.

In general all four impedances will be complex so that Eq. (1) can be broken up into two equations, one containing only real components and one the quadrature components. Thus

$$Z_1X = Z_2S$$

or

$$(R_1 + jX_1)(R_x + jX_x) = (R_2 + jX_2)(R_s + jX_s)$$

or

$$Z_1\underline{/\theta_1}X\underline{/\theta_x} = Z_2\underline{/\theta_2}S\underline{/\theta_s}$$

or

$$Z_1X\underline{/\theta_1 + \theta_x} = Z_2S\underline{/\theta_2 + \theta_s}$$

Thus, in the general case, two conditions must be fulfilled for balance.

1. Amplitude balance

$$Z_1X = Z_2S$$

2. Angular balance

$$\underline{/\theta_1 + \theta_x} = \underline{/\theta_2 + \theta_s}$$

In the actual operation of the bridge, some systematized procedure must be adopted to obtain first one balance, then the other, then return to the first and check, and so on, since phase and amplitude balance are not independent.

The chief sources of **error in the alternating-current bridge** are: (1) imperfect standards; (2) stray fields or currents.

The effect of an **imperfect standard** may be readily obtained from the above equation. Thus if Z_1 and Z_2

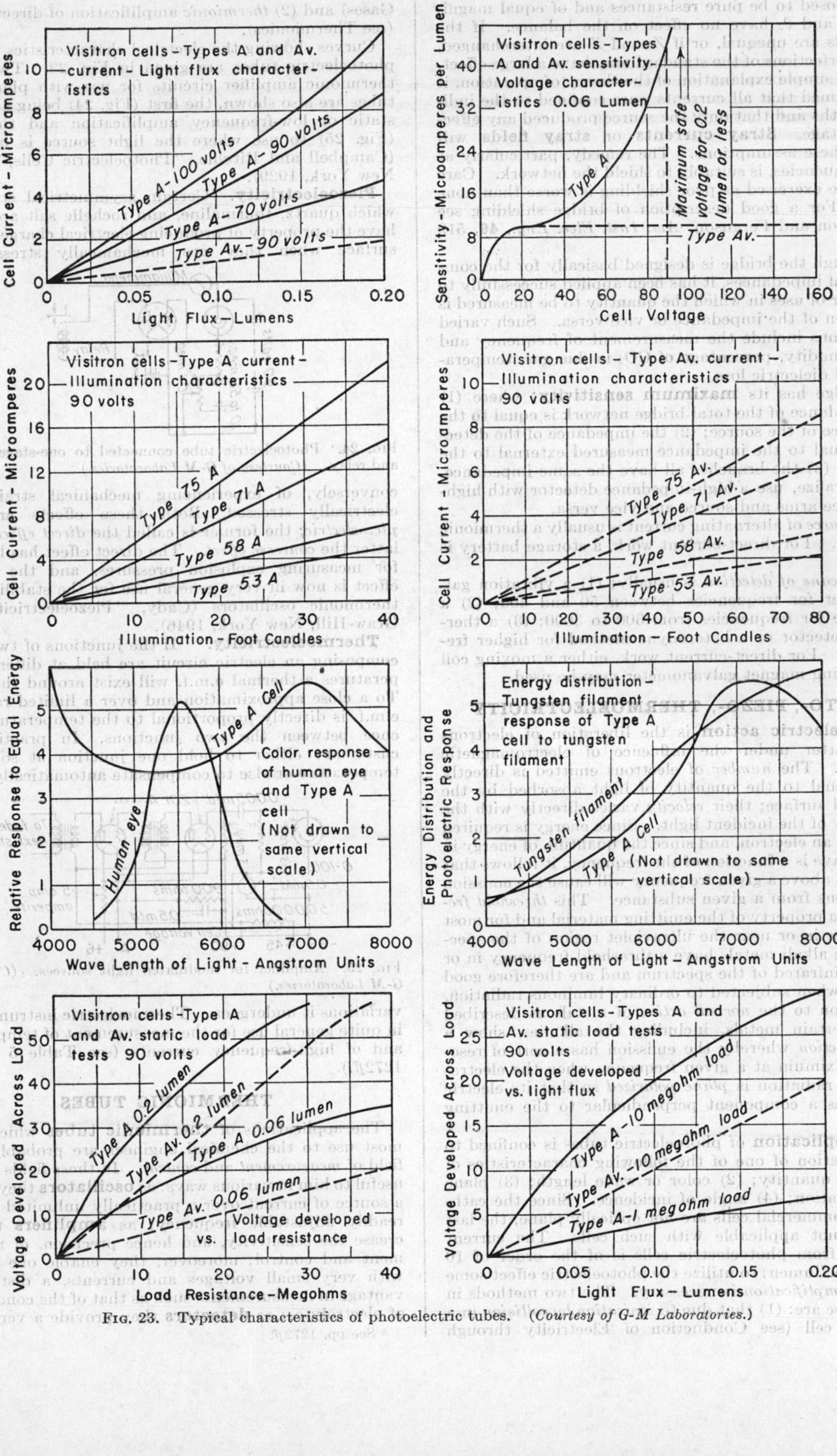

FIG. 23. Typical characteristics of photoelectric tubes. (*Courtesy of G-M Laboratories.*)

are supposed to be pure resistances and of equal magnitude, θ_1 and θ_2 have no effect on the balance. If the standards are unequal, or if Z_2 and X are interchanged, the imperfections of the standard may have a large effect.

In the simple explanation of the theory of operation, it was assumed that all currents were confined to the indicated paths and that only the source produced any effective voltage. **Stray currents** or **stray fields** will vitiate these assumptions. The remedy, particularly at high frequencies, is suitably to shield the network. Care should be exercised as poor shielding is worse than none at all. For a good description of bridge shielding see Shackelton and Ferguson, *Am. Inst. Elec. Eng.*, **46**, 519 (1927).

Although the bridge is designed basically for the comparison of impedances, it has been applied successfully to a number of uses in which the quantity to be measured is a function of the impedance or vice versa. Such varied applications include the measurement of frequency and time, humidity, percentage of CO_2 in flue gas, temperature, and dielectric loss.

A bridge has its **maximum sensitivity:** where (1) the impedance of the total bridge network is equal to the impedance of the source; (2) the impedance of the detector is equal to the impedance measured external to the detector; (3) the branches all have the same impedance. To generalize, use a high-impedance detector with high-impedance arms and source and vice versa.

The *source* of alternating current is usually a thermionic oscillator. For direct-current work, a storage battery is used.

The *means of detection* is usually: (1) a vibration galvanometer for frequencies between 50 and 300; (2) a telephone for frequencies from 300 to 3000; (3) a thermionic detector or a heterodyne method for higher frequencies. For direct-current work, either a moving coil or a moving magnet galvanometer may be used.

PHOTO-, PIEZO-, THERMOELECTRICITY

Photoelectric action is the liberation of electrons from matter under the influence of electromagnetic radiation. The *number* of electrons emitted is directly proportional to the quantity of light absorbed by the irradiated surface; their *velocity* varies directly with the frequency of the incident light. Since energy is required to release an electron, and since the quantum of energy in a light wave is a function of the frequency, it follows that only light above a given frequency will cause the emission of electrons from a given substance. This *threshold frequency* is a property of the emitting material and for most metals lies in or near the ultraviolet region of the spectrum; the alkali metals have a threshold frequency in or near the infrared of the spectrum and are therefore good emitters when subjected to ordinary luminous radiation. In addition to the *normal action* of metals as described above, certain metals, including the alkalies, show a *selective action* whereby the emission has a sort of resonance maximum at a given frequency when the electromagnetic radiation is *plane polarized* so that its electric vector has a component perpendicular to the emitting surface.

The **application** of photoelectric tubes is confined to the utilization of one of the following characteristics of light: (1) quantity; (2) color or wave length; (3) plane of polarization; (4) angle of incidence. Since the cathodes of commercial cells are not optically plane, the last two are not applicable with such cells. The current obtained from photoelectric cells is of the order of 10 microamp./lumen; to utilize the photoelectric effect some form of *amplification* is necessary. The two methods in general use are: (1) that due to *ionization by collision* in a gas-filled cell (see Conduction of Electricity through Gases) and (2) *thermionic* amplification of direct currents (see Thermionics).

Curves showing the electrical characteristics of typical photoelectric tubes are given in Fig. 23. Two typical thermionic amplifier circuits for use with photoelectric tubes are also shown, the first (Fig. 24) being adapted to static or low-frequency amplification and the second (Fig. 25) to use where the light source is modulated (Campbell and Ritchie, "Photoelectric Cells," Pitman, New York, 1929).

Piezoelectricity. Certain asymmetrical crystals, of which quartz, tourmaline, and rochelle salt are typical, have the property of producing electrical charges on their surface when they are mechanically stressed and,

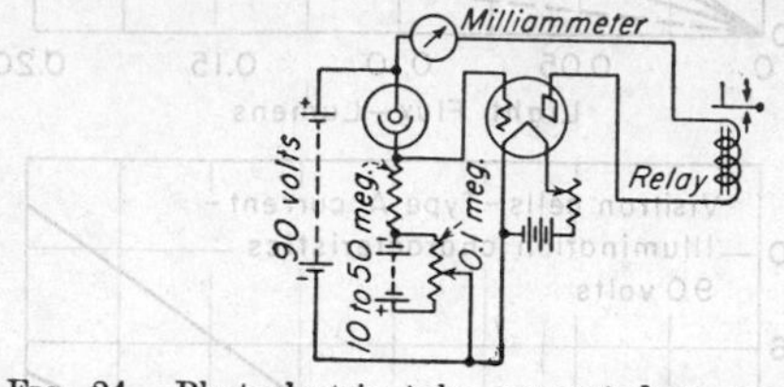

FIG. 24. Photoelectric tube connected to one-stage amplifier and relay. (*Courtesy of G-M Laboratories.*)

conversely, of experiencing mechanical strain where electrically stressed. Both these effects are called *piezoelectric;* the former is called the *direct effect* and the latter the *converse effect.* The direct effect has been used for measuring explosion pressures, and the converse effect is now in very general use for the stabilization of thermionic oscillators (Cady, "Piezoelectricity," McGraw-Hill, New York, 1946).

Thermoelectricity.* If the junctions of two metals composing an electric circuit are held at different temperatures a thermal e.m.f. will exist around the circuit. To a close approximation and over a limited range this e.m.f. is directly proportional to the temperature difference between the two junctions. In practice it is customary either to hold one junction at some fixed temperature or else to compensate automatically for any

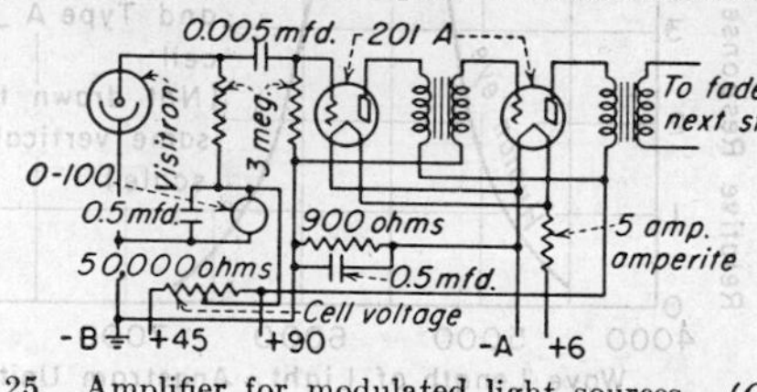

FIG. 25. Amplifier for modulated light sources. (*Courtesy of G-M Laboratories.*)

variations it undergoes. Thermoelectric instruments are in quite general use for the measurement of temperatures and of high-frequency currents (see Table 5 and pp. 1272*ff*.).

THERMIONIC TUBES

The applications of **thermionic tubes** which are of most use to the chemical engineer are probably in the field of *measurement* and *control.* In these fields they are useful to him in various ways: as **oscillators** they provide a source of current over a practically unlimited range of readily adjustable frequency; as **amplifiers** they increase the sensitivity, and hence precision, of measurement and control; moreover, they enable one to work with very small voltages and currents, a distinct advantage in measurements such as that of the conductance of electrolytes; as **detectors** they provide a very versa-

* See pp. 1272*ff*.

the electrical indicator and in the heterodyne arrangement enable one to apply audible methods to the range of supersonic frequencies.

The outstanding **advantages** of thermionic tubes as measurement or control instruments are: (1) the extremely *small influence* they exert upon the circuit which is being tested; (2) the *wide range* of frequency and amplitude over which observations may be made; and (3) the *high sensitivity* which they make possible. Their outstanding **disadvantage** is the *variability* of tube and associated apparatus with time, making it necessary to use substitution methods or frequent calibrations where high accuracy is desired.

The **fundamental phenomena** of thermionic tubes have been well described by Irving Langmuir in an article entitled The Pure Electron Discharge, *Gen. Elec. Rev.*, **18**, p. 327 (1915). Good reference books on thermionic tubes and their applications are Terman, "Radio Engineers' Handbook," McGraw-Hill, New York, 1943, and Reich, "Theory and Applications of Electron Tubes," 2d ed., McGraw-Hill, New York, 1944. For our purposes it is sufficient to note that, although a vacuum is ordinarily an almost perfect insulator, it may be made a fair conductor by releasing free electrons within the enclosure. In the thermionic tube this is accomplished by heating the filament or cathode; the mechanism of this emission is closely analogous to the evaporation of water upon the application of heat. Under working conditions not all the "evaporated" electrons are drawn to the plate or anode, which is held at a relatively high positive potential with respect to the cathode by means of an external B battery. In consequence a cloud of free electrons called the *space charge* gathers in the region between anode and cathode and exerts a repulsive or resistive effect on other electrons trying to reach the anode. In the triode or three-electrode tube a metallic grid or mesh, usually held at a small negative potential with respect to the cathode by means of a C battery, is inserted into this space-charge region and, by its tendency to neutralize or reinforce the space-charge effect as its potential is varied, gives a powerful and sensitive means of controlling the current between cathode and anode. It should be noted that the grid acts as a valve which may be turned or controlled by a very small amount of power, whereas the amount of power which it in turn controls, derived from the B battery, may be very great.

The **characteristics** of a thermionic tube are commonly presented in graphic form similar to the curves of Fig 26. The three constants of greatest importance are: the **amplification constant** μ, which indicates the relative effect of small changes of grid and plate voltages on the flow of plate current; the **mutual conductance** g_m, which indicates the change in plate current produced by a small change in grid voltage, and the **internal** or **plate resistance** r_p, which is the ratio of a small change in plate voltage to the resultant change in plate current. Mathematically these become

$$\mu = -\frac{\Delta e_p}{\Delta e_g}\,(I_p = \text{const.}) \qquad r_p = \frac{\Delta e_p}{\Delta i_p}\,(E_g = \text{const.})$$

$$g_m = \frac{\mu}{r_p} = \frac{\Delta i_p}{\Delta e_g}\,(E_p = \text{const.})$$

From the curves in Fig. 26 it will be seen that these are true constants over only a limited range.

Tube manufacturers publish booklets giving the characteristics of the more commonly used receiving and transmitting tubes, photoelectric cells, cathode-ray tubes, etc. For further information write to the tube manufacturer: RCA Manufacturing Co., Harrison, N.J.; General Electric Co., Schenectady, N.Y.; Westinghouse Electric Corp., Pittsburgh, Pa.; Raytheon Manufacturing Co., Waltham, Mass.; Sylvania Tube Co., Emporium, Pa., etc.

Thermionic amplifiers for high frequencies, where the second harmonic is remote from the fundamental and may be eliminated by means of a tuned circuit, are sometimes operated over the complete tube characteristic, but for most purposes amplifiers are required to be distortionless and operation is limited to the linear portion of the tube characteristic. In radio work *distortionless* means that the second harmonic is not more than 5 per cent of the fundamental. In such applications the circuits would be adjusted to operate about some such point as *A* (Fig. 26). With a sinusoidal input voltage,

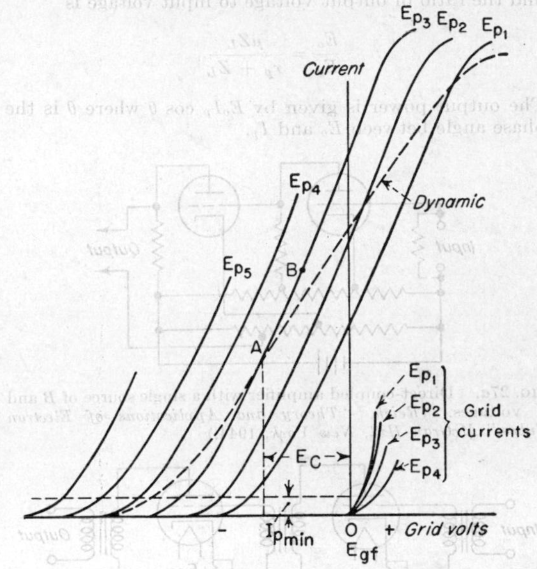

FIG. 26. Thermionic-tube characteristics.

the grid potential would just reach zero during the positive half of the cycle for the same value of input which would reduce the plate current to the minimum I_p on the negative half of the cycle. These represent the two limits for *Class A*, or *linear amplifiers*. If the current goes below the minimum I_p, the output is no longer a true reproduction of the input due to the curvature of the characteristic, and, if the grid acquires a positive potential, it will draw a grid current, as indicated by the curves emanating from $E_g = 0$ (Fig. 26). This current flowing through a high impedance will distort the input voltage. Actually the *dynamic* or operating characteristic will differ from the *static* characteristic due to the impedance drop in the load, as indicated by the dotted curve through *A*, so that the actual potentials may differ somewhat from those indicated by the above simple explanation.

For Class *A* or linear amplifiers the prediction of operating results is relatively simple. An alternating voltage E_g, limited in magnitude by the range of linearity of the tube characteristic, is assumed to be impressed on the grid. The three tube constants μ, r_p, and g_m, are obtained from the tube manufacturer's data, or from the published tube characteristics, for the desired operating point. The alternating component of plate current is then obtained by means of the equivalent plate generator theorem, which states that any alternating voltage E_g impressed on the grid electrode acts as though it were a larger alternating voltage (μE_g) inserted in the plate circuit. This equivalent plate voltage will cause an alternating plate current I_p to flow through the internal resistance of the tube r_p and the external load Z_L. The magnitude of this current is given by

$$I_p = \frac{\mu E_g}{r_p + Z_L}$$

Z_L is calculated by the rules given under Electrical Circuits (pp. 1733*ff*.). This alternating current flowing through Z_L causes an output voltage

$$E_o = I_p Z_L = \frac{\mu E_g Z_L}{r_p + Z_L}$$

and the ratio of output voltage to input voltage is

$$\frac{E_o}{E_g} = \frac{\mu Z_L}{r_p + Z_L}$$

The output power is given by $E_o I_p \cos \theta$ where θ is the phase angle between E_o and I_p.

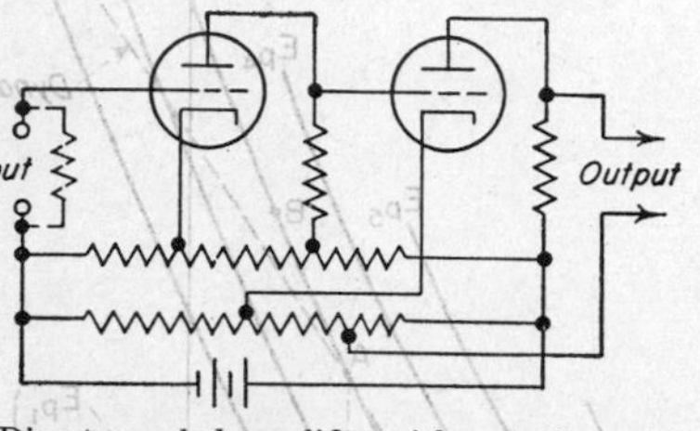

FIG. 27*a*. Direct-coupled amplifier with a single source of *B* and *C* voltages. (*Reich, "Theory and Applications of Electron Tubes," McGraw-Hill, New York,* 1944.)

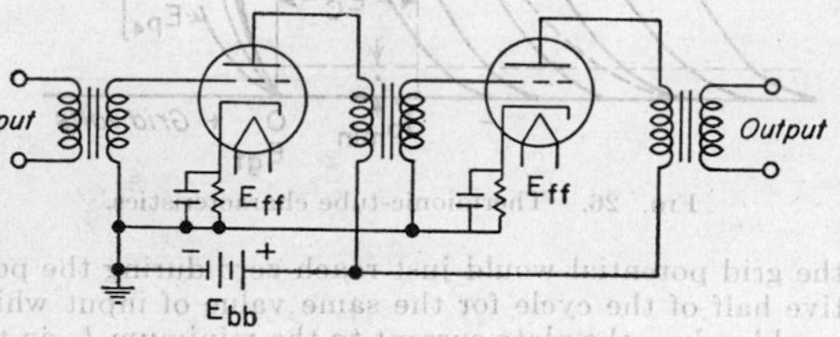

FIG. 27*b*. Untuned transformer-coupled amplifier. (*Reich, "Theory and Applications of Electron Tubes," McGraw-Hill, New York,* 1944.)

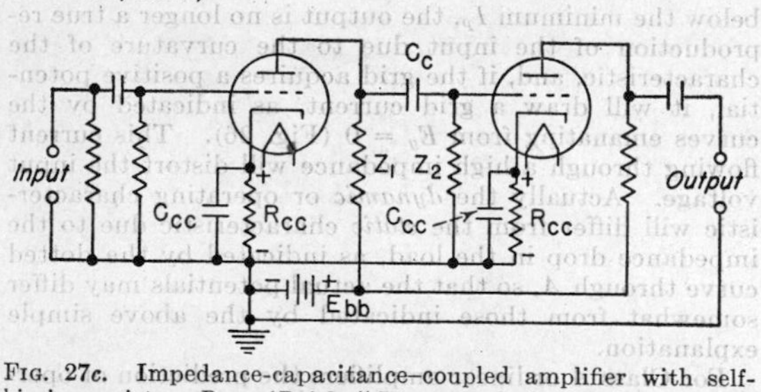

FIG. 27*c*. Impedance-capacitance–coupled amplifier with self-biasing resistors R_{cc}. (*Reich, "Theory and Applications of Electron Tubes," McGraw-Hill, New York,* 1944.)

It should be noted that these calculations apply only to the effects caused by a signal E_g, which was assumed to be alternating. In most cases the constant, or d.c., values are not required to be calculated. In particular cases this may not be true. Attention is also called to the fact that in many cases not all the output voltage of one tube is transferred to the input of the next tube, some may be lost in a connecting impedance.

If the **output** of such a thermionic amplifier is to govern another thermionic tube or any other potentially operated device, the amplifier is adjusted for **maximum voltage amplification** which occurs when Z, the external impedance, is as large as is feasible. Under this condition the *effective amplification* is $\mu Z/(r_p + Z)$. Note that this is always less than the nominal amplification, though it may be made greater by inserting a transformer. By using a transformer very *high amplifications* may be obtained over a limited range of frequency; but, where a wide band is to be covered, as in the audio amplifier of a radio receiving set, transformers having a ratio of 3 to 1 or less should be used.

If the **maximum power output** for a given input voltage is desired, the external impedance is made just equal to the internal impedance of the tube. If the **maximum undistorted power output** which the tube can handle is desired, the external impedance is made equal to twice the tube impedance for triodes and the input voltage is raised to its limiting value. The optimum load impedance for tetrodes and pentodes should be obtained from the tube manufacturer. In no case should the input voltage be made so large that appreciable grid current flows. Where power output is desired, the reactance in the external circuit should be reduced to a minimum.

The **precautions** to be observed in operating such amplifiers are numerous and complex. Good apparatus and good electrical connections are of course essential. A common trouble, especially at high frequencies, is that of keeping the electrical quantities in the assigned paths and keeping all disturbances out. This involves the difficult and complex problem of **shielding**. A peculiar type of such disturbance occurs by means of the *feedback* of energy from the external plate circuit to the external grid circuit through the grid-to-plate capacitance of the tube itself. This may be eliminated by means of a four-element tube (screen grid), or it may be neutralized by means of an auxiliary balancing circuit as in the **neutrodyne** (Fig. 28). No high impedance should be

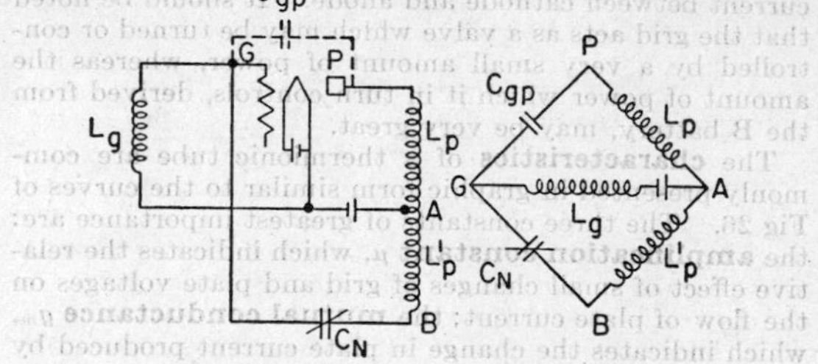

FIG. 28. Neutrodyne circuit.

common to two or more stages of an amplifier except that specifically inserted for coupling purposes. The output, voltage or power, is not uniform over a *range of frequencies* if the external impedance is a function of frequency; at high frequencies this condition always holds owing to the shunting effect of the plate-to-filament capacitance of the tube itself. The *grid current* cannot be neglected if the input circuit has a high impedance. With very low-impedance input circuits, the tube may be operated about some such point as *B* (Fig. 26) and the grid allowed to swing to a positive potential. Such an arrangement will greatly increase the power output of the tube, but care must be exercised that the grid current has no deleterious effect.

Where **larger outputs** are desired than can be obtained with a given tube, several possibilities present themselves: (1) Larger tubes may be used. (2) Two or more tubes may be connected in parallel, giving twice the output current, and the same output voltage, for a given input voltage. Note that the effective internal impedance then becomes r_p/n, where n represents the number of tubes in parallel. (3) Two tubes may be connected in push pull or back to back (Fig. 30). This arrangement will handle twice the previous input voltage and give twice the output voltage with the same current. Since at any instant the distortion produced by one tube tends to neutralize that produced by the other, this arrangement is capable of giving somewhat more than twice the undistorted output of a single tube.

Two other classes of amplifier are worthy of mention. The *Class B amplifier* is normally a push-pull arrangement

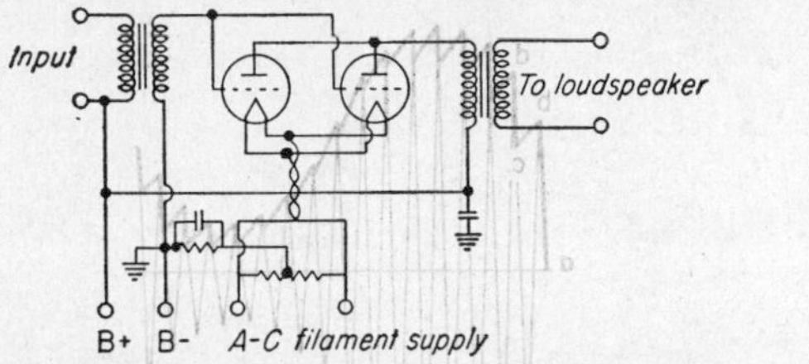

Fig. 29. Parallel-connected amplifier. (*By permission, Radio Corporation of America, from RCA Technical Series Bull. RC 14.*)

with the operating point fixed at or near cutoff (Fig. 31). Each tube carries current for one-half cycle, and the output is essentially a true reproduction of the input. It is used at high frequencies to amplify modulated waves and at audio-frequencies when high output and efficiency are required. Special care should be taken that the grid current does not cause distortion. The *Class C amplifier* is used only at high frequencies, and the external circuit is always tuned to a single frequency which it is desired to amplify. The tube is given a high negative bias so that current flows during only a small fraction of a cycle. In this way the external circuit is periodically excited by an impulse of current, but the resonant action of the circuit tends to minimize harmonics. This arrangement gives very high efficiency but relatively low power amplification because of the large amount of power required to excite the grid.

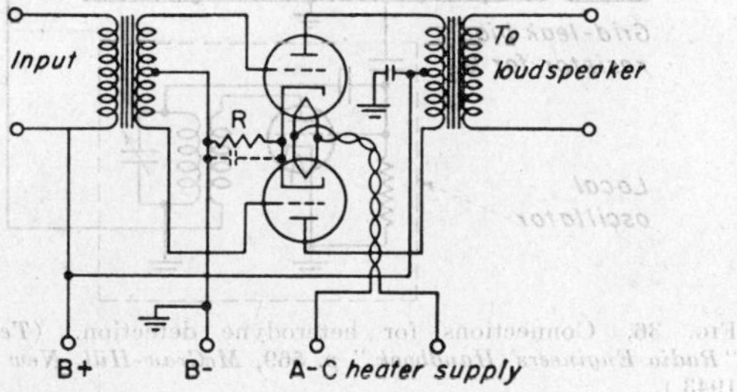

Fig. 30. Push-pull amplifier. (*By permission, Radio Corporation of America, from RCA Technical Series Bull. RC 14.*)

Oscillators are essentially amplifiers with some means of feeding part of the output energy back into the grid circuit for the purpose of maintaining the electrical vibrations. The oscillations are started by some sort of an electrical disturbance such as the closing of a switch; the *frequency* of the oscillations is given by $f = 1/(2\pi\sqrt{LC})$, where L and C are, respectively, the inductance and capacitance of the oscillating circuit. Figure 32 illustrates a so-called *Hartley oscillator*. The circuit between A and B represents the output impedance of the tube and L_g furnishes the necessary feedback. $L = L_p + L_g + 2M$, where M is the mutual inductance between L_p and L_g. In order to have good *wave form*, the output impedance should be large and the feedback small. If possible the oscillator should be separated from the measuring circuit so that load variations will not affect the frequency or amplitude of the oscillator output.

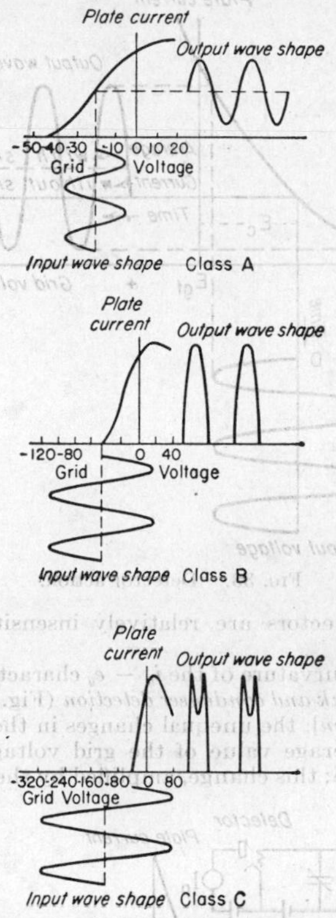

Fig. 31. Typical amplifier characteristics.

With this arrangement the oscillator may be worked at high efficiency, similar to Class C amplifiers and the wave form corrected in the intermediate circuits.

The operation of the amplifier is limited, for the most part, to the linear portion of the tube characteristic; the oscillator (and classes B and C amplifiers) swing over the entire tube characteristic; the operation of the **detector**

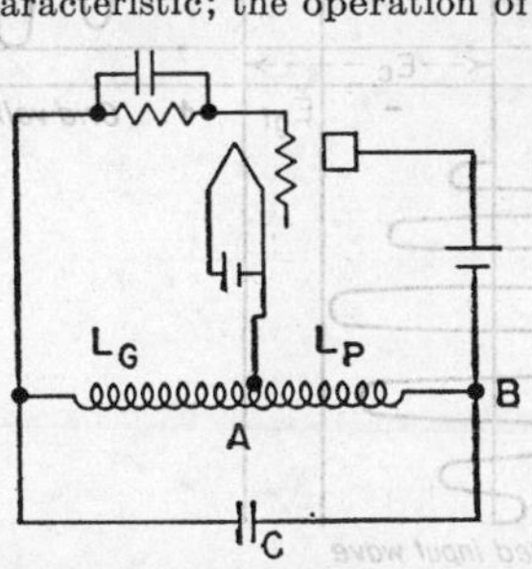

Fig. 32. Hartley oscillator.

is limited, for the most part, to the curvature of a tube characteristic. Either the $i_p - e_g$ or the $i_g - e_g$ characteristic may be used. From the diagram (Fig. 33), it is seen that a sinusoidal input voltage causes unequal increase and decrease of the resultant current. The increase (in this case) of the average current is a measure of the **rectification effect.** Mathematical analysis shows that it is proportional to the square of the input voltage for small inputs, though by proper arrangements it may be made directly proportional to the input voltage for large voltages. It follows from what has just been said that detectors are relatively insensitive to small voltages.

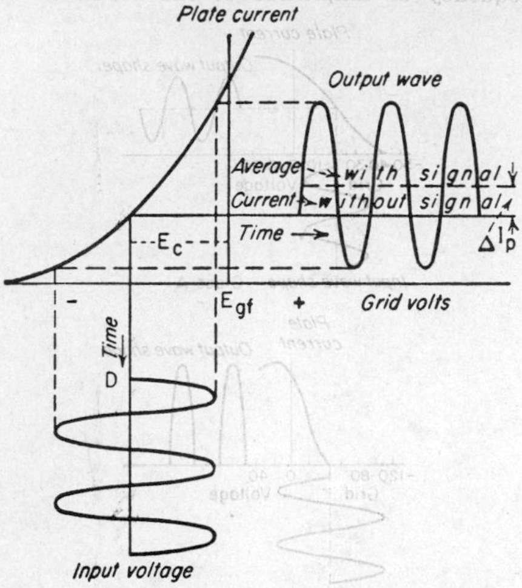

FIG. 33. Detector action.

When the curvature of the $i_g - e_g$ characteristic is used [called *grid-leak and condenser detection* (Fig. 35*A*), or *grid-circuit detection*], the unequal changes in the grid current cause the average value of the grid voltage to become more negative; this change, amplified by the action of the tube, causes a decrease in the average value of plate current which may be observed on a direct-current meter. Because of the amplifying action of the tube this method *is more sensitive* for small voltages than that described below, but it has two compensating disadvantages: first, that grid current flows and therefore power is consumed in the input circuit; and second, that the effective input voltage varies with frequency owing to the effect of the condenser shunting the grid leak. With due care these disadvantages may be minimized for any given installation.

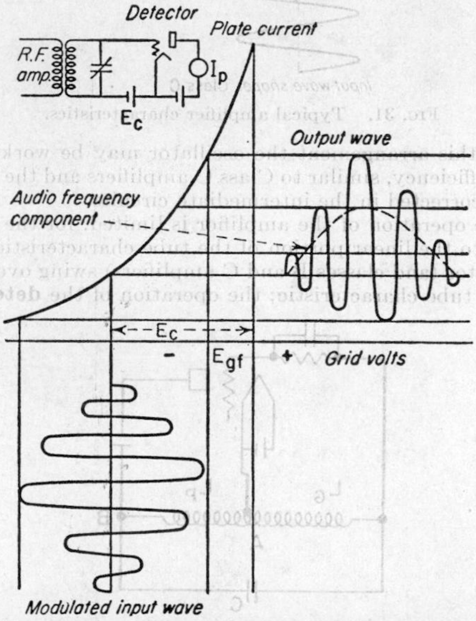

FIG. 34. Grid-bias detection of a modulated wave.

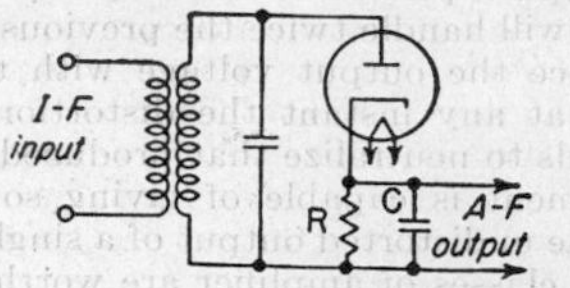

FIG. 35*a*. Diode detector. (*By permission, Radio Corporation of America, from RCA Technical Series Bull. RC 14.*)

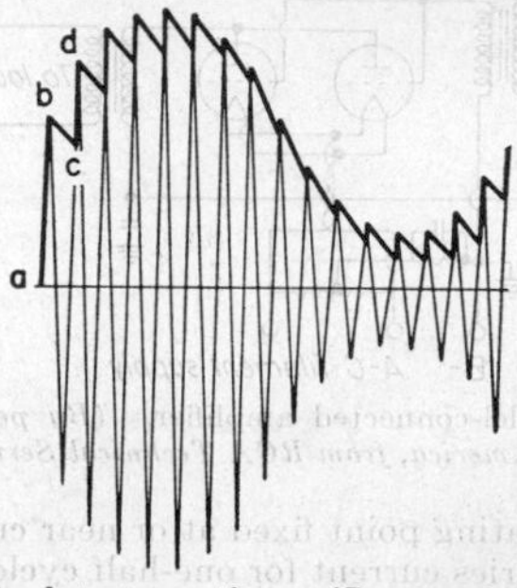

FIG. 35*b*. Diode detector action. (*By permission, Radio Corporation of America, from RCA Technical Series Bull. RC 14.*)

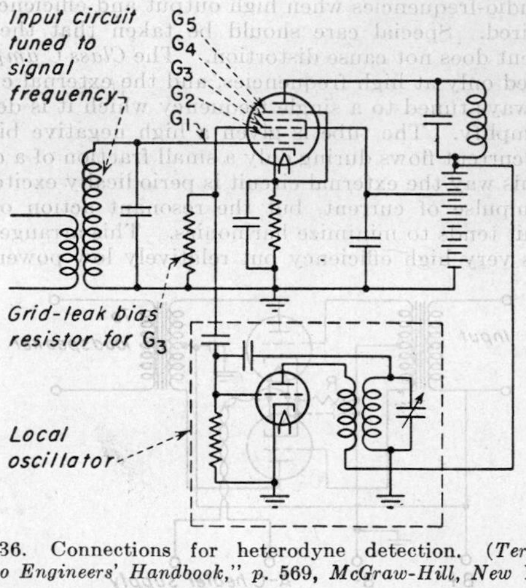

FIG. 36. Connections for heterodyne detection. (*Terman, "Radio Engineers' Handbook," p. 569, McGraw-Hill, New York, 1943.*)

When the curvature of the $i_p - e_g$ characteristic is used (called *plate-circuit detection*) the C battery is adjusted so that operation is on the point of greatest curvature. This method will handle *larger signals* than the grid-leak and condenser method but is less sensitive. It consumes a minimum of power in the input circuit and has no frequency error until the input capacitance begins to have an effect (Fig. 34).

A simple rectifier, being a unidirectional conductor, may also be used as a detector. This method of operation, called **diode detection,** is in very common use in radio receiving circuits where the signal is a modulated wave. The detector gives an output voltage proportional to the amplitude of the signal voltage and so follows the envelope of the signal from cycle to cycle. In this field it has the advantage of producing less distortion than other detectors because its dynamic characteristic is more nearly linear than is that of other detectors. Like grid-current detection this method draws current from the input circuit and so reduces selectivity. It is readily applicable to large signals.

The **heterodyne method** (Fig. 36) of detection is used to detect very small inputs. In this method an auxiliary voltage of slightly different frequency (it is called the **homodyne method** if the two frequencies are the same) is added to the voltage to be detected in the input circuit. The output is then directly proportional to the unknown voltage and of a frequency equal to the difference between the unknown and auxiliary voltages. This method makes it possible to use audible methods of detection at *supersonic frequencies*. When the frequency of the auxiliary voltage is known, it also enables one to determine the frequency of the unknown voltage (see also Rectifiers).

CONDUCTION OF ELECTRICITY THROUGH GASES

A gas in its normal state is a very **poor conductor** of electricity because of absence of ions, or carriers of electricity. Any ultimate carrier of electric charge, be it molecule, atom, or electron, is called an **ion**. These carriers are present in small quantities under all conditions, and a plentiful supply may be engendered in a variety of ways. The *source* of these carriers may be the gas itself or its container or the electrodes of the discharge; usually all these sources contribute to the supply.

There are three major methods for releasing ions, usually electrons, **from solid bodies**. These are: *heat* (thermionics), *light* (photoelectrics), and *positive-ion bombardment*. The mechanism of the first two is discussed elsewhere; that of the third is somewhat obscure but seems to be similar in effect to a mechanical impact. The major cause of the production of **ions in gases** is *electron impact* with an atom or molecule.

If voltage is applied between two electrodes in a gas at a pressure, of 10 to 1000 mm. Hg, the resultant current-voltage curve is given by Fig. 37. At very low potentials

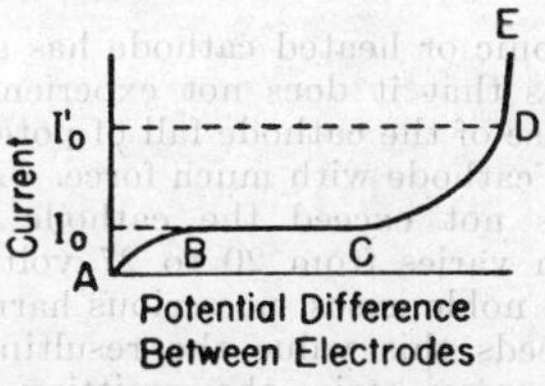

FIG. 37. Gaseous-discharge characteristics.

the characteristic is practically linear (Ohm's law). The carriers in this case are the few ions which seem to be omnipresent. Between B and C the current is saturated because all these available carriers are being drawn to the electrodes and no new carriers are being generated. At C these carriers are being so accelerated by the electric field that the electrons possess sufficient kinetic energy to release new electrons by impact with the gas molecules. The current consequently begins to increase. This increase is accumulative until at D the positive ions, which resulted from the release of free electrons from the gas molecules, acquire sufficient kinetic energy to ionize and the current increases very rapidly.

The phenomena that occur in this region depend upon many factors, chief of which are the shape, size, and material of the electrodes and the voltage regulation of the source of energy. With sources having poor voltage regulation, the voltage will drop with increase of current as shown in Fig. 38. Hence a higher voltage is required to start the discharge than to maintain it. This type of response is made use of in gaseous glow tubes as voltage regulators. With sources having very poor regulation, the discharge may be instantaneous in the form of a spark. If the voltage is maintained so that the electrodes become heated due to ionic bombardment, electrons being emitted from the heated cathode, the discharge is self-sustaining and is called an *arc*.

It is found that for a given distance d, between electrodes, there is a critical pressure P_0, at which the spark will most readily pass. This potential is called the **minimum sparking potential** for the gas and electrode material in question. It varies from 60 to 400 volts, the lower values applying when electrodes of the electropositive metals are used. For any given distance between

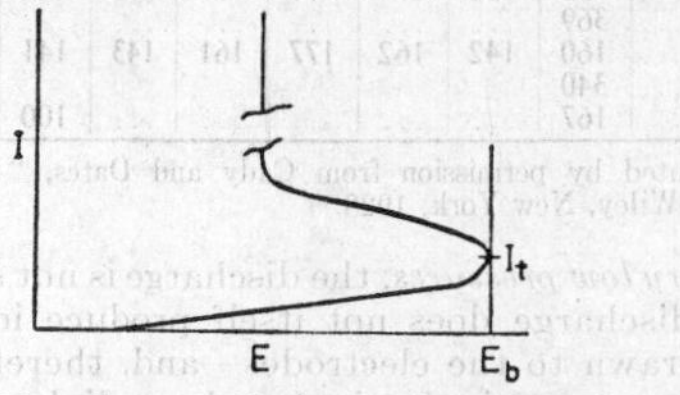

FIG. 38. Gaseous-discharge characteristic.

plates it is always possible to find some pressure at which a spark passes at this minimum potential. At higher pressures the mean free path of the gas is so short that higher potentials are required to accelerate the electrons sufficiently to cause ionization; at lower pressures the probabilities of an electron striking a neutral gas molecule are reduced. It is found that P_0d is a constant for any gas and that above P_0, if the electric field is uniform, the sparking potential is proportional to Pd. In other words, the sparking potential is proportional to the mass of gas between the electrodes. This is called **Paschen's law** (Fig. 39). It should be noted that under the proper conditions a short length of gas makes a better insulator than a long one.

As the *pressure is reduced*, the discharge gradually loses its spark characteristic and spreads out toward the walls

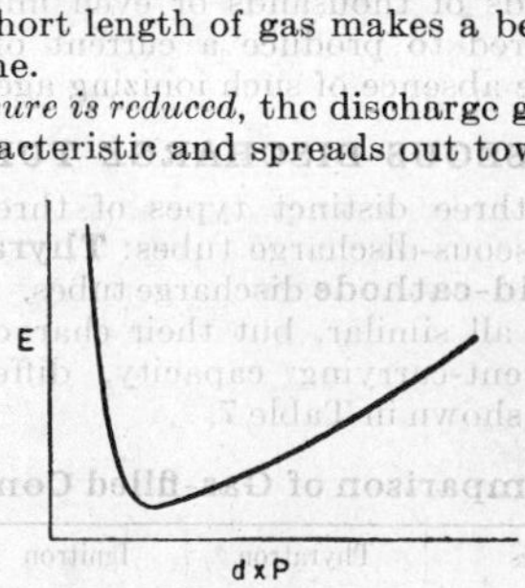

FIG. 39. Minimum sparking potential related to gas pressure P and distance d between electrodes.

of the tube and takes on the appearance of a **glow discharge**. At about 1 mm. Hg pressure we have the familiar Geissler tube effects (Fig. 40). Here the discharge has actually broken up and, going from anode to cathode, has a positive column (glow), a Faraday dark space, a negative glow, a cathode dark space, and a cathode glow. In some cases the positive column is *striated*, especially if the gas is contaminated by a small amount of another gas. With further reduction in pressure the negative glow and cathode dark space increase in length while the positive column recedes to the anode. Finally the whole tube becomes dark and we have what is known as the **dark discharge.**

At about 1 mm. Hg pressure the discharge is characterized by its **cathode fall of potential,** which is that sharp drop of potential which occurs within a few millimeters of the cathode. When the cathode fall is *normal* its value is equal to the minimum sparking potential, and the cathode glow covers only a portion of the cathode. The current density is then about 0.4 milliamp./sq. cm. and varies directly as the pressure. When the current increases so that the glow extends over the entire cathode, the current density increases and the cathode fall of potential increases to *abnormal* values, sometimes reaching thousands of volts.

Table 6. Minimum Sparking Potentials for Various Atmospheres and Electrode Materials*

Gas	Potentials in volts for electrodes of									
	Pt	Hg	Ag	Cu	Fe	Zn	Al	Mg	Na	K
H_2	300	...	295	280	230	213	190	168	185	172
N_2	232	226	...	...	...	...	...	207	178	170
O_2	369	...	...	...	...	...	...	310		
He	160	142	162	177	161	143	141	125	80	69
Air	340									
A	167	...	...	...	...	...	100			

* Reprinted by permission from Cady and Dates, "Illuminating Engineering," Wiley, New York, 1928.

At *very low pressures*, the discharge is not self-sustaining—the discharge does not itself produce ions to replace those drawn to the electrodes—and, therefore, some external source of ionization, as heat, light, or X rays, is

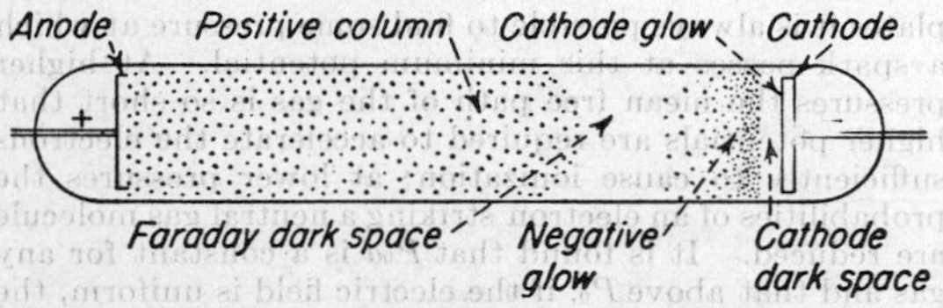

FIG. 40. Typical glow discharge in a gas at about 1 mm. pressure.

required to maintain the discharge. With modern X-ray tubes, hundreds of thousands or even millions of volts may be required to produce a current of a few milliamperes in the absence of such ionizing agents.

GASEOUS-DISCHARGE TUBES

There are three distinct types of three-element, or controlled, gaseous-discharge tubes: **Thyratrons, Ignitrons,** and **cold-cathode** discharge tubes. Functionally the tubes are all similar, but their characteristics, particularly current-carrying capacity, differ greatly in magnitude, as shown in Table 7.

Table 7. Comparison of Gas-filled Control Tubes*†

Characteristics	Thyratron	Ignitron	Cold-cathode tube
Current capacity	Up to 100 amp	5–10,000 amp.	Up to 100 ma.
Deionization time	10^{-4} sec.	10^{-4} sec.	10^{-2} sec.
Ionization time	10^{-6} sec.	10^{-6} sec.	10^{-4} sec.
Cathode heating time	Finite	0	0
Deterioration in stand-by service	Yes	No	No
Accuracy of characteristics	± 2 volts	Variable	± 10 volts
Sustaining voltage	15 volts	15 volts	75 volts

* Taken from article by Ingram in *Elec. Eng.*, July, 1939, p. 345.
† All values are approximate only.

The phenomena in gas-filled tubes are more complex, are less well understood, and are less precisely controlled than are those in vacuum tubes. Nevertheless such tubes have assumed a wide and growing importance in the field of electrical engineering.

The gases most commonly used are the noble gases, argon, neon, helium, krypton and xenon, and mercury vapor. The noble gases possess the advantage that the gas concentration is not dependent on temperature. With mercury vapor, however, the equilibrium vapor pressure, and hence gas concentration, is dependent upon the temperature of the condensed mercury. When the ambient temperature varies or when the heat generated in the tube because of the passage of current causes an increase in the temperature of the condensed mercury, the gas concentration and hence the tube operating characteristics change also. At the equilibrium pressure the vapor must be condensed as rapidly as it is produced, hence requiring air cooling of glass-walled tubes or water cooling of metal-walled tubes or else limitation of the current to the capacity of the available cooling means. The noble gases have the disadvantage that they "clear up" or are absorbed by the electrodes and enclosing walls with the passage of time and so reduce the gas concentration.

The gas in a tube may fulfill one or several of the following functions:

1. It may reduce the vaporization of the cathode material. One of the earliest applications of this phenomenon was to the Mazda-C or nitrogen-filled lamps. Besides other functions the high gas pressure served to prevent the evaporation of the tungsten filament and so permitted its operation at a higher temperature, resulting in higher luminous efficiency, with a given life. In present-day high-pressure mercury-vapor lamps the same principle is used to protect the (unheated) oxide-coated cathode. It is also used to protect the heated cathodes of commercial battery charging tubes known as Tungars and Rectigons.

2. It may be used to increase the current flow by the process of ionization by collision. Ions so produced may have a secondary effect of producing a greater emission from the cathode, but they are also used to obtain greater current flow, without any secondary phenomena, as in the gas-filled photoelectric tubes.

3. It may be used to control the breakdown voltage in a tube (see Paschen's law).

4. By far its most important use is to neutralize the "space charge" of the electrons in a tube, by means of the positive ions produced by ionization by collision and thus to reduce the voltage drop across the tube. With reduced voltage drop less heat is generated in the tube for a given current flow and so the load-carrying capacity is increased. Moreover a lower voltage drop means better voltage regulation. Gas-filled tubes may use two different types of cathode in addition to the heated cathode normally used in thermionic vacuum tubes. These are the cold cathode and the mercury pool cathode.

The thermionic or heated cathode has a limitation in gas-filled tubes that it does not experience in vacuum tubes. Because of the cathode fall of potential, positive ions strike the cathode with much force. As long as this potential does not exceed the cathode disintegration voltage, which varies from 20 to 27 volts for mercury vapor and the noble gases, no serious harm is done. If the drop exceeds this value the resulting positive-ion bombardment may strip the emitting surface from thoriated or oxide-coated cathodes. As long as the current flow does not exceed the thermionic emission in a high vacuum the cathode fall of potential does not exceed these permissible values. Hence the peak current rating of a gas tube with a coated thermionic cathode is limited to the value obtainable in vacuum. The load must be fixed with this in mind, since the tube will not protect itself.

Cold-cathode tubes are essentially small current tubes (usually less than 100 milliamp.). Such tubes are characterized by a cathode fall of potential of the order of the minimum sparking potential (see Table 6). Typical tubes of this type are RCA-0A4 tube and the Western Electric Co. 346A tube. Since the minimum sparking potential is reduced by using coated cathodes both these tubes use such an arrangement. They are, however, subject to greatly reduced life as the current drain is increased. Thus the Western Electric 346A tube has the life expectancies shown in Table 8.

Table 8. Tube Life Expectancy with Current

Current, milliamp.	Life Expectancy, hr.
35	100
20	1000
15	5000

Mercury-pool cathode tubes are essentially arc discharge devices characterized by a cathode fall of potential of the order of the minimum ionization potential of the vapor. The cathode fall of potential is of the order of 10 volts, and the total drop across the tube is of the order of 15 to 20 volts. The allowable cathode emission is unlimited, except for tube heating, since the cathode restores itself by the condensation of the mercury vapor.

To summarize: For very small current a cold cathode may be used at the expense of a relatively high voltage drop across the tube.

For medium current a thermionic cathode may be used, but the total current carried may not exceed the thermionic emission without endangering the surface of coated cathodes. For very large currents and for operations where the tube may be subjected to severe or prolonged overloads the mercury-pool cathode is the choice.

Gas Diodes. With the differences noted above, the operation of gas diodes is similar to that of vacuum-tube diodes. The tube conducts current whenever the voltage of the anode exceeds that of the cathode by some small value called the firing voltage. At other times the tube should not conduct. The firing voltage is a few volts for hot-cathode mercury-vapor tubes and may be several hundred volts for cold-cathode tubes. The mercury-pool-type tube is not self-starting by means of its anode voltage and so normally includes some sort of starting device, such as an auxiliary electrode and also excitation anodes to keep the mercury vaporized during the non-conducting portion of the cycle.

Inverse Voltage. Because of the presence of the gas or vapor the tube will "break down," or fail as an insulator, at a much lower voltage than would a similar vacuum tube. During the non-conducting cycle the tube is subjected to an inverse voltage equal to the sum of the load voltage and the applied anode voltage. The peak value of this inverse voltage may be equal to twice the peak value of the applied anode voltage. Hence the tube construction and gas pressure must be designed to withstand this inverse voltage without breakdown. This problem is particularly acute in mercury-vapor rectifiers, especially those in which a number of anodes are supplied by a common cathode. Modern mercury-arc rectifiers are supplied with insulated metallic baffles around the anode, and in some cases the open end of this baffle is covered by a screen connected to a negative potential. These devices shorten the tube deionization time and lengthen the arc path between anodes, both of which reduce the tendency to backfire. In addition some mercury-arc rectifiers have anode heaters to prevent the condensation of mercury on the anodes during the light load period and so to prevent the formation of a cathode spot on the anode.

The correlation of a mercury-vapor diode characteristic with condensed mercury temperature is shown in Fig. 41*a* taken from Steiner, Gable, and Maser [Engineering Features of Gas-filled Tubes, *Elec. Engineering*, **51**, 313 (1932)]. The vapor pressure, and hence gas concentration, varies rapidly with temperature, doubling for about a 10°C. rise within the normal operating region. At very low temperatures the gas concentration is low and the positive ions produced may not be sufficient to neutralize the space charge of normal current flow; and, if the cathode is a coated surface, this may lead to cathode disintegration. If the pressure is high the inverse voltage breakdown is low, as shown in Fig. 41*a*. Hence mercury-vapor tubes must be operated within a restricted range of temperature; at the low-temperature end because of current-carrying limitations and at the high-temperature end because of voltage breakdown limitations. Hence these tubes normally require some sort of cooling system, especially in large sizes.

Controlled Discharge. It is not possible to control the flow of current in gaseous-discharge tubes in the same sense that it is possible to control the flow of current in vacuum tubes. Once current starts flowing, or the gas is ionized, the tube is filled with a cloud of positive and negative ions. The positive ions move toward negative electrodes, and the negative ions move toward positive electrodes, forming ion sheaths around them and protecting the remainder of the discharge path from the electrostatic effects of these electrodes. The thicknesses of

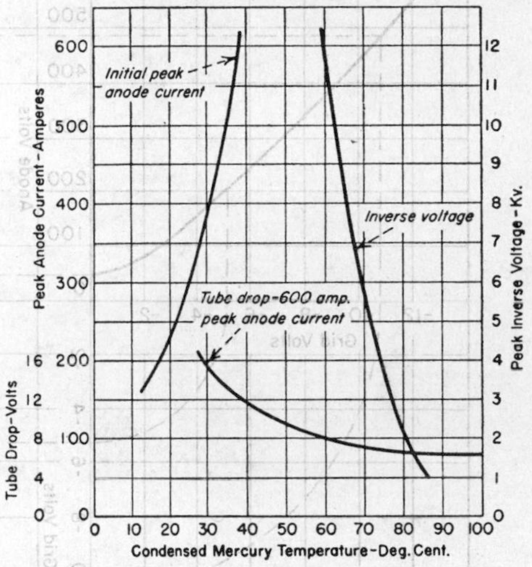

FIG. 41*a*. Tube voltage drop, arc-drop potential, and initial anode current at start of conduction, for a typical two-element mercury-vapor tube. [*Elec. Eng.*, **51**, 313 (1932), by permission of the American Institute of Electrical Engineers.]

these sheaths adjust themselves so that the electrostatic effect is completely neutralized; usually this requires only a very thin sheath so that the greater portion of the discharge path is in no way influenced by the presence of these electrodes.

Before current starts or before the gas is ionized, these electrodes exert the same electrostatic effect that they do in vacuum tubes. Hence a negative third electrode, or "grid," can prevent electron flow from the cathode and so can prevent ionization by collision and the initiation of the gaseous discharge. Thus control in such tubes is limited to the starting or initiation of the discharge. Once the discharge is under way the third electrode loses all control. The four generally used methods of extinguishing the discharge are (1) use of an alternating voltage, (2) opening the anode circuit by a switch or relay, (3) application of a surge of negative voltage to the anode through a capacitor, and (4) overshooting of the voltage due to the inductance of the circuit and the dynamic characteristic of the tube when a capacitor is discharged through the tube.

Figure 41*b* shows a typical control characteristic of a three-electrode thermionic cathode gaseous-discharge tube, called a **Thyratron.** The region to the right and above the curve is the conducting region; that to the left and below the curve is the non-conducting region. The curve represents the crossover or critical adjustment of voltages. By the construction shown in Fig. 41*b* the critical grid voltage at any instant during the cycle of an assumed applied alternating voltage may be determined. Whenever the voltage on the third electrode is more positive than the critical grid voltage, and the anode voltage is positive, the tube will conduct. It will continue to conduct until the end of the positive half cycle or until the anode reaches the minimum sustaining voltage (sometimes called extinction voltage).

As an example, if a sinusoidal voltage equal to 60 per cent of the peak critical grid voltage and lagging the anode voltage by 90 deg. is applied to the grid, the tube will start conducting at point *X* and will continue to approximately the end of the positive half cycle. The tube deionizes during the negative half cycle of anode voltage, and the third electrode again controls the instant of starting during the next positive half cycle. With a pure resistance load the current is directly proportioned to the anode voltage (neglecting tube drop), and the crosshatched area in the diagram is proportional to the average current flow. A study of this diagram will show how the average current varies or may be controlled by varying the phase of the applied grid voltage, by varying the amplitude of the applied grid voltage, or by varying the constant voltage bias about which the alternating grid voltage is impressed, or by combination of these methods. Control by phase variation is the most flexible and is most generally used.

Another method of controlling the initiation of the arc is employed in **Ignitron** tubes. These are mercury-pool tubes with a third electrode of refractory material such as silicon carbide, immersed in the pool. When a heavy current, say 5 amp., is sent through this electrode, a sufficient voltage gradient is produced near the mercury surface to start the arc. By controlling the time of current flow through this igniter rod the time of starting the arc may be controlled. Because of the heavy currents required it is frequently desirable to make this a two-step process. The discharge in a Thyratron is controlled as outlined in the preceding paragraph, and the Thyratron is discharged through the igniter rod and so controls the initiation of the arc in the mercury-pool tube. Thus very large currents may be controlled with practically no energy drain on the controlling source.

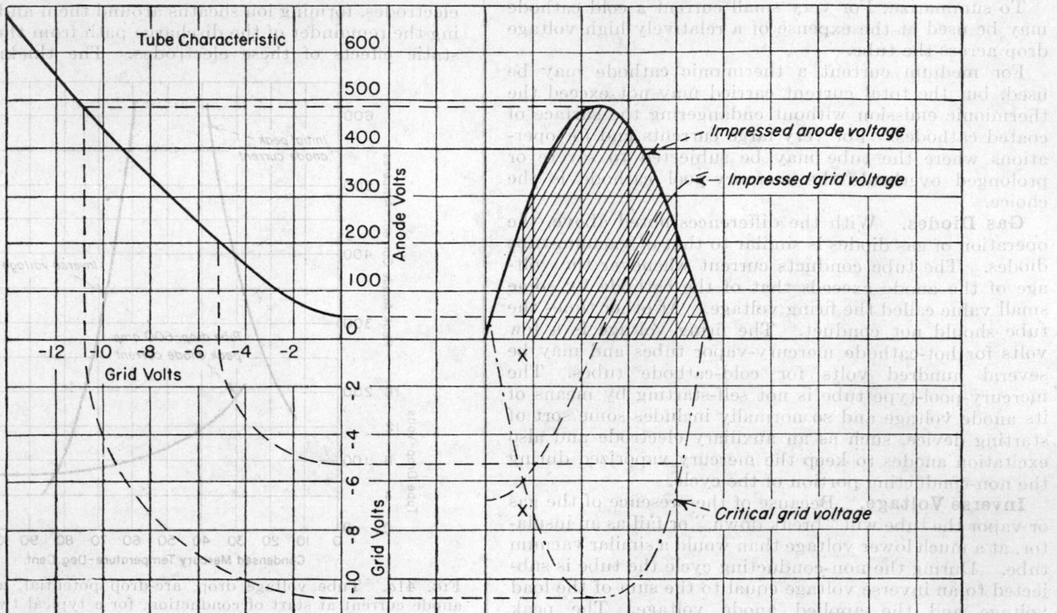

Fig. 41*b*. Thyratron control characteristics.

RECTIFIERS

Rectifiers are devices for converting alternating to direct current. They are made in such varied and diverse types that it will be possible to mention only a few of the more important here.

Large power rectifiers commonly use a mercury-pool cathode and are known as **mercury-arc rectifiers.** In all such rectifiers some auxiliary provision must be made for starting the discharge. In some cases auxiliary anodes, known as excitation anodes, are required to maintain ionization during the non-conducting portion of the cycle. The glass-bulb types are made in sizes up to 500 amp. at 500 volts continuous load. They are normally permanently evacuated, *i.e.*, they require no auxiliary pumping equipment. Because of gas leakage around the electrode seal it is more difficult to use permanent evacuation on steel-tank types, although such rectifiers have been made in sizes up to 750 amp. at 600 volts. Generally, however, auxiliary pumping equipment is used in the larger sizes to keep the pressure of foreign gases at a suitably low value. Operation of such devices at 20,000 volts output is quite feasible.

Figure 41*c*, taken from "Applied Electronics," by the Electrical Engineering Staff, Massachusetts Institute of Technology, p. 375, Wiley, New York, 1944, shows a comparison of efficiencies of mercury-arc rectifiers (including all auxiliaries) and other conversion equipment for 3000-kw. units operating at the prescribed voltages 600, 1500, and 3000.

Since commercial power-supply systems normally

furnish large blocks of power by polyphase systems (*e.g.*, three-phase), the large power rectifiers normally utilize polyphase circuits. The rectifiers themselves may consist of a number of anodes that operate from a common cathode or of a number of single-anode rectifiers with their cathodes connected together externally. The increase in output voltage for a given input voltage as the

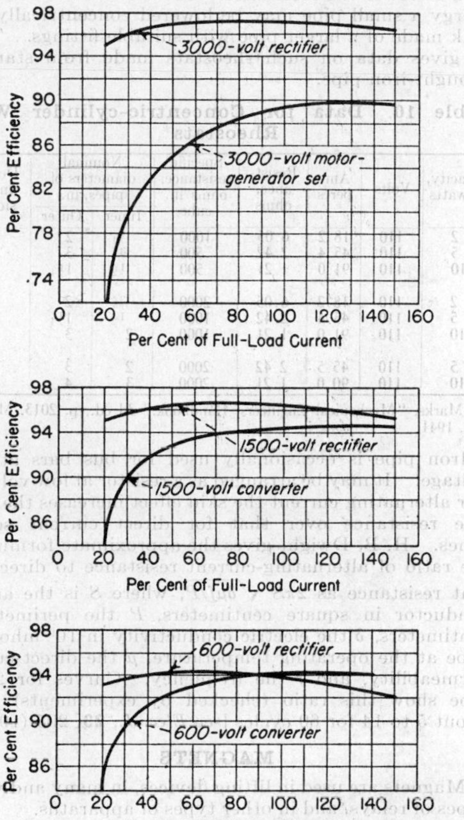

FIG. 41c. Comparison of efficiencies of mercury-arc rectifiers and other conversion equipment. (*By permission, from "Applied Electronics," by the Electrical Engineering Staff, Massachusetts Institute of Technology, Wiley, New York,* 1944.)

number of phases is increased is shown in the following table.

p	2	3	4	6	12	∞
Ed/Es	0.9	1.17	1.27	1.35	1.40	$\sqrt{2}$

p = number of phases. Ed = output voltage (including tube drop), and Es = r.m.s. anode-to-neutral voltage.

Since the load is driven or pulsed at least p times each cycle of the a.c. supply it may be shown that the ripple voltage in the output will be integer multiples n of p times the frequency of the supply. The magnitude of such ripple voltages is shown below.

n	2	3	4	6	8	9	12	18	24
$\sqrt{2}\,En/Ed$	0.667	0.250	0.133	0.057	0.032	0.025	0.014	0.006	0.0035

En = r.m.s. value of ripple voltage.

This indicates the effectiveness of a large number of phases in reducing the ripple voltage. Moreover, as the frequency of the ripple voltage is increased, the problem of further reducing this ripple by auxiliary devices is greatly simplified.

On the other hand, the polyphase rectifier, in addition to requiring more tubes, makes poorer utilization of the supply transformers. Thus if P_L is the required rating of the transformer secondary windings and P is the d.c. output power (including the loss in the rectifier tubes) then $U = \dfrac{P}{P_L}$ is given by the following table for star-connected transformers.

p	2	3	4	6	12	∞
U	0.636	0.675	0.636	0.551	0.399	0

With special transformer connections, *e.g.*, the delta double Y, and the use of interphase reactors the utilization factor may be greatly improved.

The higher pulsing frequency of polyphase connections makes the transformer leakage reactance more effective, and the regulation of polyphase rectifiers therefore increases as the number of phases is increased.

The **Ignitron** mercury-arc rectifier is being used in considerable numbers in electrochemical industries. The operating principle has been described under gaseous-discharge tubes.

Of the **separately maintained rectifiers,** *i.e.*, those using some auxiliary means to maintain the supply of free electrons, three in quite general use have externally heated cathodes as the sources of electrons. These three are the thermionic tubes and gaseous-discharge tubes of high-voltage and low-voltage types.

Thermionic tubes have unilateral conductivity and so may be used as rectifiers. Owing to the absence of gas they may be used on very high voltages but can pass only a very small current, limited by the saturation current of the cathode, and they have a high internal voltage drop.

Hot-cathode, gaseous-discharge tubes are of two different types. The *high-voltage type*, of which the 866 tubes or **Phanotron** is representative, contains only a small amount of mercury vapor so that the tube will still sustain relatively high voltages, 5000 to 7500 volts as a peak value, and in addition the internal voltage drop is less than 15 volts. The *current*, however, must be limited to the emission current of the cathode, since with greater currents the cathode fall of potential rises and the positive-ion bombardment is severe enough to strip the emitting surface from the cathode.

The *low-voltage type*, of which the **Tungar** and **Rectigon** are typical, contain argon at 3 to 8 mm. pressure and will pass 10 or 12 amp. at very low voltages. The maximum permissible voltage is only 100 to 200 volts. The current flow may be greater than the emission current from the cathode, since the high gas pressure protects the cathode from excessive positive-ion bombardment.

Several other types of rectifiers are used occasionally but only for small currents and small voltages. Among these are *mechanical vibrators, copper oxide rectifiers,* and *point-to-plate rectifiers.*

The point-to-plate rectifier supplies small unidirectional currents without any accessory apparatus, such as heating transformers, etc. The pointed electrode has a high potential gradient around it and so will produce the ionization necessary for the maintenance of the discharge. When it is the anode, the fast-moving electrons are easily able to reach it; when it is the cathode, the electrons are impeded by the gas molecules from reaching the anode. Highly insulated transformers are required with such tubes to prevent breakdown due to surges.

Copper oxide rectifiers are, as their name implies, copper disks on which an oxide coating has been formed by sudden quenching. When used with another electrode, such as lead, they pass current unidirectionally, provided that the current density is low. They are used extensively in conjunction with d.c. meters, to measure small high-frequency currents.

Selenium rectifiers, consisting essentially of a metal-plated steel or aluminum disk on one face of which a thin layer of selenium is applied, have characteristics similar to copper oxide rectifiers. They are, however, slightly smaller and more efficient and have a wider permissible temperature range than copper oxide rectifiers. Their economic range of application appears to be limited to loads of less than 15 kw. At low voltages, say a few hundred volts at the most but mainly at voltages less than 100, and high currents they have gained wide acceptance. The individual disks, which may be connected in series and in parallel, are made in sizes up to about 8 in. square, have current capacities up to about 20 amp., and will withstand inverse voltages of about 15 r.m.s. volts. Because of their high self-capacitance they are not suitable for high-frequency applications. Table 9 [extracted from Richards, Characteristics and Applications of Selenium Rectifiers, *J. Inst. Elec. Engrs.*, **88**, (III), 238 (1941)] indicates the operating range of such rectifiers.

Table 9. Selenium Rectifier Data*

Max. voltage and current loadings for each size of disk for continuous operation in ambient temperatures up to 35°C.

	Inverse voltage						
Diam. of disk, mm	18	25	35	45	67	84	112
Applied d.c. volts	15	15	15	15	15	12	12
Applied a.c. volts	18	18	18	18	18	16	15

	D.c. output current (amp. mean) per disk							With additional cooling			
Diam. of disk, mm	18	25	35	45	67	84	112	45	67	84	112
D.c.	0.06	0.12	0.23	0.47	0.9	1.8	3.1	0.78	1.5	4.5	7.5
A.c. single phase:											
Half wave	0.04	0.075	0.15	0.3	0.6	1.2	2.0	0.5	1	3	5
Bridge or push pull	0.075	0.15	0.3	0.6	1.2	2.4	4.0	1.0	2	6	10
Voltage doubler	0.03	0.06	0.12	0.25							
A.c. three phase:											
Half wave	0.1	0.2	0.4	0.8	1.6	3.2	5.3	1.3	2.6	8	13
Bridge	0.11	0.22	0.45	0.9	1.8	3.6	6.0	1.5	3.0	9	15
Push pull	0.13	0.27	0.55	1.1	2.2	4.5	7.5	1.85	3.7	11	18

* Reprinted by permission of The Institution of Electrical Engineers from Richards, Characteristics and Applications of Selenium Rectifiers, *J. Inst. Elec. Engrs.*, **88** (III), 238 (1941).

Filters should be used with all the above types of rectifiers if steady output voltage is desired. An electric filter consists essentially of electrical reservoirs (condensers) on the input and output ends to take up the fluctuations in source and load, and a choke coil to prevent these fluctuations from reaching from input to output and vice versa. Frequently the choke coil is split and a third condenser added at its mid-point.

RHEOSTATS

Rheostats are adjustable resistances for controlling various circuits. Wire- or strip-wound resistance units with insulating supports arranged with suitable switches are commonly used as field rheostats, starting boxes, and for other purposes.

Carbon rheostats of the compression type are useful for fine adjustment. The resistance varies nearly inversely as the pressure, permitting a wide range of values. The resistance decreases with temperature increase, but this may not be objectionable for laboratory work.

Slide-wire rheostats are convenient for a variety of laboratory uses. These are obtainable in sizes from a fraction of an ohm to several thousand ohms, usually for use in circuits of less than 250 volts.

Water-cooled rheostats can be used to absorb large amounts of energy. Galvanized iron wire submerged in running water makes a suitable rheostat, where the various turns are kept apart. This is not satisfactory for continuous operation on direct current because of slime forming on the wire. The water should be kept below the boiling temperature.

Liquid rheostats are built in a variety of forms. A plate partly lowered into a water tank or barrel permits a variation in resistance. Pure water is not suitable for voltages below 1000; for lower voltages, salt or acid is used to lower the resistance. For moderate amounts of energy a small pipe may be lowered concentrically in a tank made of a larger pipe with suitable fittings. Table 10 gives data on such rheostats made from standard wrought-iron pipe.

Table 10. Data for Concentric-cylinder Water Rheostats*

Capacity, kilowatts	Volts	Amperes	Resistance, ohms	Specific resistance, ohms/in. cube	Nominal diameters of pipes, in.		Depth of immersion, in.
					Inner	Outer	
2	110	18.2	6.05	1000	¼	2	35
5	110	45.4	2.42	500	¾	3	35
10	110	91.0	1.21	500	½	1¼	33
2	110	18.2	6.05	2000	¾	2	36
5	110	45.4	2.42	1000	½	1¼	33
10	110	91.0	1.21	1000	2	3	34
5	110	45.5	2.42	2000	2	3	34
10	110	90.0	1.21	2000	3	4	37

* Marks, "Mechanical Engineers' Handbook," 3d ed., p. 2013, McGraw-Hill, 1941.

Iron pipe is occasionally used for bus bars at high voltage. It may be arranged as a resistor at low voltages. For alternating current the skin effect increases the effective resistance over that for direct current several times. H. B. Dwight gives the approximate formula for the ratio of alternating-current resistance to direct-current resistance as $2\pi S \sqrt{\delta\mu f/P}$, where S is the area of conductor in square centimeters, P the perimeter in centimeters, δ the electric conductivity in 10^9 mhos/cm. cube at the operating temperature, μ the direct-current permeability, and f the frequency. Curves for 2.5-in. pipe show this ratio (checked by experiments) to be about 5 to 14 for 60 cycles [see *Elec. J.*, **23**, 295 (1926)].

MAGNETS

Magnets are used in lifting devices, in many and varied types of relays, and in other types of apparatus.

Permanent magnets rely on residual magnetism for their usefulness. The m.m.f. retained is approximately constant. The resultant constant field is particularly desirable for magnetos, instruments, and meters. The ability of steel to retain its magnetism depends on the kind of steel, its heat-treatment, and the shape of the particular piece. Short magnets tend to become demagnetized unless made glass hard. Steels with tungsten, chromium, or cobalt are used for permanent magnets.

The steel to be used as a permanent magnet is usually magnetized electrically, the particular method varying with the shape of the magnet. In order that the field may be constant the magnet is usually overmagnetized and then artificially aged. Aging may be hastened by temperature, mechanical shock, vibration, or demagnetizing fields.

Alloys of iron, nickel, and aluminum have been developed which have very high coercive force and high residual induction. These qualities permit short magnets of small volume which are little affected by vibration, stray fields, or temperature up to 500°F. These alloys are difficult to machine and are usually cast so that a minimum of grinding to size is necessary. Comparative data of various materials used for permanent magnets is given by Williams [*Elec. Eng.*, **55**, 19 (1936)];

Webb [*Inst. Elec. Eng.*, **82**, 303 (1938)]; Adams [*Gen. Elec. Rev.*, **41**, 518 (1938)]; and Bozorth [*Rev. of Modern Physics*, **19**, 29 (1947)].

Electromagnets depend on electric currents in coils for their m.m.f. and are usually made of soft iron so that the flux will vary with the magnitude of the current in the coil.

The pull of a magnet, where the air gap between the stop and the armature is small, may be expressed by the equation

$$F = kB^2$$

where F is in dynes per square centimeter, $k = 0.0398$, and B in gausses. For B in lines per square inch and F in pounds per square inch,

$$k = 1.39 \times 10^{-8}$$

With direct-current magnets the current in the solenoid is limited by the resistance of the winding, giving a constant m.m.f. regardless of position of the armature. As the air gap is reduced, the flux density increases greatly increasing the pull. In order to increase the pull and reduce the time of moving an armature, the coil is frequently wound for a lower voltage and is provided with a resistance which is automatically inserted in the coil circuit as the armature comes to its final position.

Solenoids are a very common form of electromagnet, used where a rod or plunger is to be moved through some distance. The coil is wound on a tube into which the plunger is drawn. The pull increases as the plunger approaches the center of the coil (Fig. 42), at which time the pull F in pounds is

$$F = \frac{CANI}{l}$$

where l is the length of the solenoid in inches; C is the unit pull per ampere-turn per inch of the coil and depends on the proportions of the coil and the properties and the length of the plunger; A is the cross section of the plunger in square inches; and NI is the ampere-turns of the coil (see Table 11). The pull may be increased at the end of the stroke by adding a magnetic stop (Fig. 42).

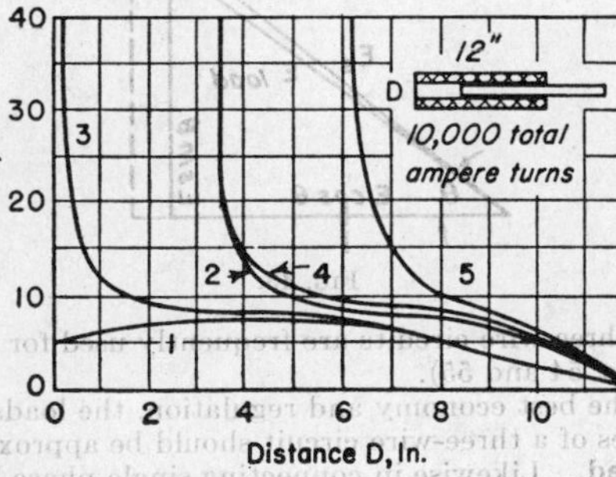

FIG. 42. Pull of solenoid on plunger. 1, coil and plunger; 2, coil and plunger with stop; 3, iron-clad coil and plunger; 4 and 5, same as 3 with different length stop. (*Marks, "Mechanical Engineers' Handbook," McGraw-Hill, New York*, 1941.)

Solenoids must be designed with sufficient radiating surface so that the temperature does not become excessive. The design must be sufficiently liberal so that with the increased resistance of the coil at its maximum temperature the required current will flow. A solenoid may be rewound for another voltage by changing the size of wire. If a solenoid operates satisfactorily with E volts when wound with a wire of bare diameter D, it may be rewound with wire of bare diameter d to operate on voltage e, where $D^2E = d^2e$. The new winding will occupy the same space as the old except for the space occupied by the insulation [see Brooks, *Ind. Eng.*, **87**, 349 (1929)].

Magnets are built into plates to form *chucks* for holding steel while it is being ground. *Magnetic separators* of many forms are built for separating magnetic material from other materials (see pp. 1091*ff*.).

Table 11. Maximum Pull per Square Inch of Core for Solenoids with Open Magnetic Circuit*

Length of coil, in.	Length of plunger, in.	Area of core, sq. in.	Total amp.-turns	Max. pull, lb./sq. in.	1000 × C
6	Long	1	15,000	22.4	9.0
9	Long	1	11,330	11.5	9.1
9	Long	1	14,200	14.6	9.2
10	10	2.76	40,000	40.2	10.0
10	10	2.76	60,000	61.6	10.3
10	10	2.76	80,000	80.8	10.1
12	Long	1	11,200	8.75	9.4
12	Long	1	20,500	16.75	9.8
18	36	1	18,200	9.8	9.7
18	36	1	41,000	22.5	9.8
18	18	1	18,200	9.8	9.7
18	18	1	41,000	22.5	9.8

* "Standard Handbook for Electrical Engineers," 6th ed., McGraw-Hill, New York, 1933.

Lifting magnets are particularly useful in handling magnetic materials such as scrap iron, billets, pigs, and castings.

Alternating-current magnets inherently have considerable reactance, which limits the current and the flux. The relations are essentially the same as in a transformer thus

$$\phi_{max} = \frac{E \times 10^8}{(4.44fN)}$$

where E is the impressed voltage, f is the frequency, and N is the number of turns. This indicates that the flux and consequently the pull are nearly the same regardless of the armature position instead of increasing rapidly with decreased air gaps as in the d.c. magnet.

Because the flux is continuously changing, the iron of the a.c. magnet must be laminated. The pull, being reduced to zero twice each cycle, causes vibration of the armature and consequently considerable noise. This is reduced by a shading coil consisting of a short-circuited coil about a portion of the pole, so that there is some pull on the armature at all times (see also Underhill, "Magnets," McGraw-Hill, 1924).

ELECTRICAL WIRING

Electrical wiring should always be installed with care and neatness; joints should be mechanically secure, well soldered, and taped. Care should be exercised to avoid the two most common faults, (1) incomplete circuits and (2) short circuits. All installations should comply with the National Electrical Code and with the National Electrical Safety Code.

The **minimum size of wire** for a particular installation must be large enough (1) to provide *carrying capacity* as indicated in the Electrical Code, (2) to have *sufficient strength* on spans as indicated in Safety Code, and (3) so that the resistance and reactance drop permit *good regulation*. **Regulation** is the rise in voltage from full load to no load expressed as a percentage of the full-load voltage. Good regulation varies with the particular applications; and, while 3 per cent may be a maximum variation for lighting, 10 per cent may be reasonable for motors.

The regulation for a **direct-current circuit** may be easily calculated, as the voltage change from full load to no load is the IR drop of the circuit. The resistance of the circuit (two wires) may be obtained from Table 12; the current is determined from the load.

Table 12. Wire Table

Size, American wire gage	Area, cir. mils	Diameter, in.	Weight bare, lb./1000 ft.	Resistance, ohms/1000 ft. at 20°C.	Reactance per wire per 1000 ft. at 60 cycles—distance between centers						Allowable carrying capacity, amp., with rubber insulation
					1 in.	6 in.	1 ft.	2 ft.	4 ft.	8 ft.	
	1,000,000	1.152	3,090	0.0108	.....	0.063	0.079	0.095	0.111	0.127	455
	800,000	1.031	2,470	.0135	.....	.066	.082	.098	.113	.129	410
	600,000	0.893	1,850	.0180	0.027	.069	.085	.101	.117	.133	355
	500,000	.815	1,540	.0216	.030	.071	.087	.103	.119	.135	320
	400,000	.728	1,240	.0270	.032	.073	.089	.105	.121	.137	280
	300,000	.630	926	.0361	.035	.076	.092	.108	.124	.140	240
	250,000	.575	772	.0433	.038	.079	.095	.111	.127	.143	215
0000	211,600	.460	653	.0490	.040	.081	.097	.113	.129	.145	195
000	167,800	.410	518	.0618	.042	.083	.099	.115	.131	.147	165
00	133,100	.365	411	.0779	.045	.086	.102	.118	.134	.150	145
0	105,500	.325	326	.0983	.047	.089	.105	.121	.137	.153	125
1	83,690	.289	258	.124	.050	.091	.107	.123	.139	.155	110
2	66,370	.258	205	.156	.053	.094	.110	.126	.142	.158	95
4	41,740	.204	126	.248	.058	.099	.115	.131	.147	.163	70
6	26,250	.162	79.5	.395	.063	.104	.120	.136	.152	.168	55
8	16,510	.128	50.0	.628	.069	.110	.126	.142	158	.174	40
10	10,380	.102	31.4	.999							30
12	6,530	.081	19.8	1.59							20
14	4,107	.064	12.4	2.52							15
16	2,583	.051	7.8	4.02							7
18	1,624	.040	4.9	6.38							5

Note. Values for wire sizes larger than 0000 are for stranded wire. Allowable current-carrying capacity is for code rubber insulation without more than three conductors in raceway, as shown in the 1947 National Electrical Code. Greater current is allowed for other insulations and for conductors in open air.

Example. To find the regulation of a line supplying a 50-kw. d.c. load at 220 volts. The line is 200 ft. long and consists of two 0000 wires.

$$\frac{50{,}000 \text{ watts}}{220 \text{ volts}} = 227.3 \text{ amp.}$$

Resistance of 1000 ft. 0000 wire = 0.049 ohm
Resistance of 400 ft. 0000 wire = 0.0196 ohm
The drop = 227.3 amp. × 0.0196 ohm = 4.46 volts

$$\text{Regulation} = \frac{(4.46 \times 100)}{220} = 2 \text{ per cent}$$

In an **alternating-current circuit,** the power factor of the load and reactance of the circuit must be taken into account.

Example. To find the regulation of a three-phase line supplying 440 volts to a 90-kva. load of 0.8 power factor. The line is 2000 ft. long and consists of three 0 wires with 24-in. spacing of wires.

From Table 12 resistance of 0 wire is 0.0983 per 1000 ft.
From Table 12, reactance of 0 wire is 0.121 per 1000 ft. at the given spacing
Resistance of 2000 ft. = 0.197
Reactance of 2000 ft. = 0.242
Calculations should be per phase. Assume Y connection.
Kilovolt-amperes per phase = 30
Volts per phase, $440/\sqrt{3} = 254$
Amperes per phase, 30,000/254 = 118
IR drop per phase, 0.196 × 118 = 23.2
IX drop per phase, 0.240 × 118 = 28.3
Referring to vector diagram $E \cos\theta = 203.2$ and $E \sin\theta = 152.4$.
203.2 + 23.2 = 226.4
152.4 + 28.3 = 180.7

$$\text{Sending voltage per phase} = \sqrt{(226.4)^2 + (180.7)^2} = 289.5$$

$$\text{Sending voltage} = 289.5\sqrt{3} = 501.4$$

$$\text{Regulation} = \frac{(501.4 - 440)100}{440} = 14 \text{ per cent}$$

The **most economical size of wire** is that which gives the lowest annual cost. The annual cost consists of the fixed or capital costs (interest on investment, depreciation, and taxes) and the cost of energy (copper losses) in the wires. In general, the *minimum annual cost* will be found for that size of wire for which the fixed cost is equal to the cost of energy loss in the wire. This wire size may be larger than the permissible minimum. It is desirable when installing electrical wiring to anticipate future increase of loads.

The *higher the voltage* the less the copper required for a given load. Since the load is equal to EI, if the voltage is doubled the current is halved and the copper cross section may be halved for the same current density or reduced to one-fourth for the same percentage voltage drop. When voltages are increased, the insulation problems and hazards to employees working on live circuits are increased. Hence, for industrial motors, 440 (460) volts is the common maximum, although larger motors are operated at 2300, 6600, and at 13,200 volts. For lighting, 115-volt lamps are more rugged than 230-volt lamps.

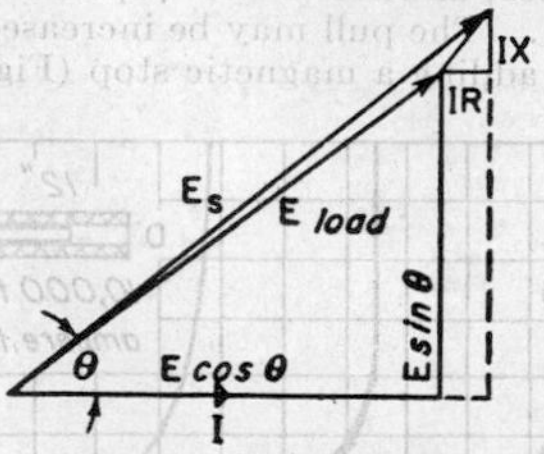

FIG. 43.

Hence three-wire circuits are frequently used for lighting (see Figs. 54 and 55).

For the best economy and regulation, the loads on the two sides of a three-wire circuit should be approximately **balanced.** Likewise in connecting single-phase loads to a polyphase circuit, the phase loads should at all times be approximately balanced. Polyphase motors normally draw balanced currents tending to decrease the effect of other loads which may not be balanced. The maximum permissible unbalance depends on the particular installation and would, in general, be about 10 to 15 per cent of the full-load value.

Approximate Values of Wire. No. 10 wire has an approximate diameter of 0.1 in., a cross section of 10,000 cir. mils, and a resistance of about 1 ohm/1000 ft. The cross section doubles for every third size of wire (smaller size number), and the resistance is reduced to one-half. Again, the cross section increases tenfold for 10 sizes of wire, and the resistance is one-tenth.

COST OF POWER

The **rates** for energy from a utility to an industrial consumer must include items to cover the *fixed charges* and *operating costs* of the utility. These costs are similar to those of a privately owned plant.

The **demand charge** in the rate is to cover a portion of the fixed charges. This demand charge is usually based on the maximum 15-min. load, inasmuch as the utility must have available sufficient generating and distributing equipment to supply that load. Since the peak load of all industries will not occur during the same period, the demand charge is less than the fixed charges for an equivalent private plant. Since the apparatus to supply a given amount of power is determined by the voltage and the current, the required rating is greater if the power factor is low, and, therefore, the demand charge is frequently based on kilovolt-amperes rather than kilowatts.

The **energy charge** in the rate must cover the operating costs including coal, labor, distribution losses, etc., and frequently includes some of the fixed charges. This charge, per kilowatt-hour, may be decreased with increase of kilowatt-hours used.

Owing to better efficiency and lower first cost of large units, and frequently better operating conditions and diversity of loads, power from a utility costs less than that from an industrial power plant, except in the case of unusual local conditions, such as large steam requirements for process work.

The true cost of power in an industry should be valued in terms of increased efficiency of labor, better product, etc. In most industries the cost of power is less than 4 per cent of the value added by the manufacture of the product. It may therefore be economical to add a large percentage to the cost of apparatus or power, provided that it will increase the value of the product by a small amount.

ILLUMINATION

Correct lighting has proved its ability to *increase production* economically, to give *greater comfort* to workmen, and to *reduce accidents*. Because of the proved value of increased illumination, lighting standards have been gradually raised. **Good illumination** requires light of *sufficient intensity* in the *correct place, without glare,* and usually with only *soft shadows.*

The **intensities,** expressed in **foot-candles,** found by experience to be suitable for various applications are given in Table 13. These values should be considered minimum, and, for seeing under difficult conditions, higher values should be used. **Glare** is present when the light intensity in the line of vision is extremely high, particularly in comparison with the light intensity on near-by objects. Glare interferes with seeing and causes discomfort. Lighting installations should be designed to minimize glare from either the light source directly or from its reflected image. Reflectors and lighting glassware are used to *diffuse the light.* The intensity of brightness is reduced by increasing the size of reflector or enclosing globe. The quantity of light is expressed in lumens. A **lumen** is the amount of light which will give a light intensity of 1 ft.-candle on 1 sq. ft.

Table 13. Recommended Levels of Illumination for Industrial Interiors*

	Foot-candles measured on the work
Aisles, stairways, passageways	5–10
Assembling:	
Rough	10
Medium	20
Fine	50–100
Extra fine	100 up
Chemical works:	
Hand furnaces, boiling tanks, stationary dryers, stationary or gravity crystallizing	5
Mechanical furnaces, generators and stills, mechanical dryers, evaporators, filtration, mechanical crystallizing, bleaching	10
Tanks for cooking, extractors, percolators, nitrators, electrolytic cells	20
Clay products and cements:	
Grinding, filter presses, kiln rooms	5
Molding, pressing, cleaning, and trimming	20
Enameling	30
Color and glazing	30
Elevator—freight and passenger	10
Engraving	100 up
Forge shops and welding	10
Foundries:	
Charging floor, tumbling, cleaning, pouring, and shaking out	10
Rough molding and core making	10
Fine molding and core making	20
Garage—automobiles:	
Storage—dead	2
live	10
Repair department and washing	30–50
Glass works:	
Mix and furnace rooms, pressing and lehr, glass-blowing machines	10
Grinding, cutting glass to size, silvering	30
Fine grinding, polishing, beveling, inspection, etching, and decorating	50–100
Machine shops:	
Rough bench and machine work	20
Medium bench and machine work	30
Fine bench and machine work	50–100
Offices:	
Private and general	25–50
Offices (*Continued*):	
Close work	30–50
No close work	20
Distribution of mail	30
Drafting room	30–50
Packing and boxing	10
Paint manufacturing	20
Paint shops:	
Dipping, spraying, firing	20
Rubbing, ordinary hand painting and finishing	20
Fine hand painting and finishing	50
Store and stockrooms:	
Rough and bulky material	5
Medium or fine material requiring care	10–20
Structural steel fabrication	10
Sugar grading	30
Testing:	
Rough	20
Fine	30
Extra fine instruments, scales, etc	100 up
Textile mills:	
Cotton:	
Opening and lapping, carding, drawing	10
Spooling, spinning, drawing-in, warping, weaving, quilling, inspecting, knitting, slashing (over beam end)	20
Silk:	
Winding, throwing, dyeing	30
Quilling, warping, weaving, and finishing	30–50
Woolen:	
Carding, picking, washing, and combing	10
Twisting and dyeing	10
Drawing-in, warping:	
Light goods	20
Dark goods	50
Weaving:	
Light goods	20
Dark goods	100
Knitting machines	20
Toilet and washrooms	10
Upholstering:	
Automobile, coach, and furniture	30
Warehouse	5

These foot-candle values represent order of magnitude rather than exact level of illumination. Where 30 or more foot-candles are recommended specialized lighting is necessary to supplement the general illumination. The combination systems must provide proper direction of light, diffusion, and minimizing of reflected glare.

* By permission of the Illuminating Engineering Society; from "Illuminating Engineering Society Lighting Handbook," 1947.

Hence to obtain an intensity of 10 ft.-candles on an area of 100 sq. ft. would require 1000 lumens projected on the surface.

Artificial illumination is usually obtained from incandescent lamps, fluorescent lamps, arc lamps, or gaseous-discharge lamps. For lighting of highways and large areas, lighting units combining two types of lamps are sometimes used.

Table 14. Mounting Heights of Lighting Units*

Direct lighting units				Semi-indirect and indirect lighting	
Actual spacing between units, ft.	Distance of units from floor, ft., not less than ft.	Desirable mounting height in industrial interiors	Desirable mounting height in commercial interiors	Actual spacing between units, ft.	Recommended suspension length (top of bowl to ceiling), ft.
7	8	12 ft. above floor if possible—to avoid glare, and still be within reach from step ladder for cleaning. Where units are to be mounted much more than 12 ft. it is usually desirable to mount the units at ceiling or on roof trusses	The actual hanging height should be governed largely by general appearance, but, particularly in offices and drafting rooms, the minimum values shown should not be violated	7	1–3
8	8½			8	1–3
9	9			9	1–3
10	10			10	1½–3
11	10½			11	2–3
12	11			12	2–3
14	12½			14	2½–4
16	14			16	3–4
18	15			18	3–4
20	16			20	4–5
22	18			22	4–5
24	20			24	4–6
26	21			26	4–6
28	22			28	5–7
30	24			30	5–7

Note. Based upon the assumption that the plane of work is 30 in. above the floor.

* Tables 14, 15, 16 reprinted by permission from *Westinghouse Lamp Co., Bull. E-108.*

The **incandescent lamp** produces light from a heated filament of tungsten (early filaments were of carbon and tantalum) enclosed in a vacuum or an inert gas. Through continuous research the efficiency has been gradually increased from about 1.5 lumens/watt in 1881, to 10 lumens/watt for 25-watt lamps and 20 lumens/watt for 1000-watt lamps (see Table 17).

The **arc light** is the oldest type of electric light; it derives the light from an arc maintained between two electrodes. The electrodes may be of a variety of materials, although carbon electrodes have been most common. The d.c. magnetite arc produces a brilliant white light with an efficiency of about 18 lumens/watt between a cathode of magnetic oxide of iron and titanium and an anode of copper. Many arc lights have been superseded by series incandescent lamps; these are as efficient on alternating current as on direct, and the costs of replacing electrodes are eliminated.

Electric-discharge lamps produce luminescence from an arc or discharge in a vacuum or a rarefied gas. There are several types of these lights, and they are unstable except when provided with a special transformer, or reactor, or resistor to limit the current. Only the mercury-vapor lamp, the neon-type lamp, and the fluorescent lamps will be briefly discussed here.

The **mercury-vapor lamp** is used in many industrial plants because of its high efficiency (30 to 40 lumens/watt) and its freedom from sharp shadows. The high actinic value of this lamp makes it useful for photographic work. In the early mercury-vapor lamps the arc was struck by tilting the tube. The present-day lamp is provided with a transformer which serves as a stabilizing ballast and has sufficient voltage to establish the arc without external starting mechanism. See Table 19.

Because of color and the time required for starting, mercury-vapor lights are frequently used with incandescent lighting units. As these units have high light intensity, the mounting height should not be less than 20 ft.

Table 15. Room Index for Narrow and Average Rooms

For indirect lighting, use ceiling height		Ft.						
		9 to 9½	10 to 11½	12 to 13½	14 to 16½	17 to 20	21 to 24	25 to 30
For direct lighting, use mounting height		Ft.						
		7 and 7½	8 and 8½	9 and 9½	10 to 11½	12 to 13½	14 to 16½	17 to 20
Room width, ft.	Room length, ft.	Room index						
9 (8½–9½)	8–10	1.0	0.8	0.6	0.6			
	10–14	1.0	0.8	0.8	0.6			
	14–20	1.2	1.0	0.8	0.6	0.6		
	20–30	1.2	1.2	1.0	0.8	0.6	0.6	
	30–42	1.5	1.2	1.0	0.8	0.6	0.6	0.6
	42 up	2.0	1.5	1.2	1.0	0.8	0.6	0.6
10 (9½–10½)	10–14	1.2	1.0	0.8	0.6	0.6		
	14–20	1.2	1.0	0.8	0.6	0.6	0.6	
	20–30	1.5	1.2	1.0	0.8	0.6	0.6	
	30–42	1.5	1.2	1.2	1.0	0.8	0.6	0.6
	42–60	2.0	1.5	1.2	1.0	0.8	0.6	0.6
	60 up	2.0	1.5	1.5	1.0	1.0	0.8	0.6
12 (11–12½)	10–14	1.2	1.0	0.8	0.8	0.6	0.6	
	14–20	1.5	1.2	1.0	0.8	0.6	0.6	
	20–30	1.5	1.2	1.2	1.0	0.8	0.6	0.6
	30–42	2.0	1.5	1.2	1.0	0.8	0.6	0.6
	42–60	2.0	1.5	1.5	1.2	1.0	0.8	0.6
	60 up	2.0	2.0	1.5	1.2	1.0	0.8	0.6
14 (13–15½)	14–20	1.5	1.2	1.0	1.0	0.8	0.6	0.6
	20–30	2.0	1.5	1.2	1.0	0.8	0.6	0.6
	30–42	2.0	1.5	1.5	1.2	1.0	0.8	0.6
	42–60	2.0	2.0	1.5	1.5	1.0	0.8	0.6
	60–90	2.5	2.0	2.0	1.5	1.2	1.0	0.6
	90 up	2.5	2.0	2.0	1.5	1.5	1.2	0.8
17 (16–18½)	14–20	2.0	1.5	1.2	1.0	0.8	0.6	0.6
	20–30	2.0	1.5	1.5	1.2	1.0	0.8	0.6
	30–42	2.5	2.0	1.5	1.2	1.0	1.0	0.6
	42–60	2.5	2.0	2.0	1.5	1.2	1.2	0.8
	60–110	2.5	2.0	2.0	1.5	1.2	1.2	0.8
	100 up	3.0	2.5	2.0	2.0	1.5	1.2	1.0
20 (19–21½)	20–30	2.5	2.0	1.5	1.2	1.0	0.8	0.6
	30–42	2.5	2.0	2.0	1.5	1.2	1.0	0.8
	42–60	2.5	2.5	2.0	2.0	1.5	1.2	0.8
	60–90	3.0	2.5	2.0	2.0	1.5	1.2	1.0
	90–140	3.0	2.5	2.5	2.0	1.5	1.5	1.0
	140 up	3.0	2.5	2.5	2.0	1.5	1.5	1.0
24 (22–26)	20–30	2.5	2.0	2.0	1.5	1.2	1.0	0.8
	30–42	3.0	2.5	2.0	1.5	1.2	1.2	0.8
	42–60	3.0	2.5	2.5	2.0	1.5	1.2	1.0
	60–90	3.0	2.5	2.5	2.0	1.5	1.5	1.0
	90–140	3.0	3.0	2.5	2.0	2.0	1.5	1.2
	140 up	3.0	3.0	2.5	2.0	2.0	1.5	1.2
30 (27–33)	30–42	3.0	2.5	2.5	2.0	1.5	1.2	1.0
	42–60	3.0	3.0	2.5	2.5	1.5	1.5	1.0
	60–90	4.0	3.0	3.0	2.5	2.0	1.5	1.2
	90–140	4.0	3.0	3.0	2.5	2.0	2.0	1.5
	140–180	4.0	3.0	3.0	2.5	2.0	2.0	1.5
	180 up	4.0	3.0	3.0	2.5	2.0	2.0	1.5
36 (34–39)	30–42	4.0	3.0	2.5	2.0	1.5	1.5	1.0
	42–60	4.0	3.0	3.0	2.5	2.0	1.5	1.2
	60–90	5.0	3.0	3.0	3.0	2.0	2.0	1.5
	90–140	5.0	4.0	3.0	3.0	2.5	2.0	1.5
	140–200	5.0	4.0	3.0	3.0	2.5	2.0	1.5
	200 up	5.0	4.0	3.0	3.0	2.5	2.0	1.5
40 or more	42–60	5.0	4.0	3.0				
	60–90	5.0	4.0	4.0				
	90–140	5.0	4.0	4.0				
	140–200	5.0	5.0	4.0				
	200 up	5.0	5.0	4.0				

The **neon lamps**, used extensively in signs, require relatively high voltage varying with the length of tubing. Because of its color it is not desirable for ordinary illumination.

The **fluorescent lamps** obtain a high lumen output by using salts deposited on the inside of the tube, called

Table 16. A Guide to the Selection of Reflecting Equipment and Coefficients of Utilization

Lighting unit		Efficiency based upon: Illumination on horizontal	Efficiency based upon: Illumination on vertical	Appearance of lighted room	Direct glare	Reflected glare	Shadows	Maintenance	Depreciation factor (*D*): Clean conditions	Depreciation factor (*D*): Average conditions	Depreciation factor (*D*): Dirty conditions	Ceiling / Walls / Room index	Very light (70%): Fairly light (50%)	Very light (70%): Fairly dark (30%)	Very light (70%): Very dark (10%)	Fairly light (50%): Fairly light (50%)	Fairly light (50%): Fairly dark (30%)	Fairly light (50%): Very dark (10%)	Fairly dark (30%): Fairly dark (30%)	Fairly dark (30%): Very dark (10%)
													Coefficients of utilization (*U*)							
Direct lighting—general industrial reflectors												Calculation data—general units								
1 RLM Dome White bowl lamp 90–180°, 0% 0–90°, 66%	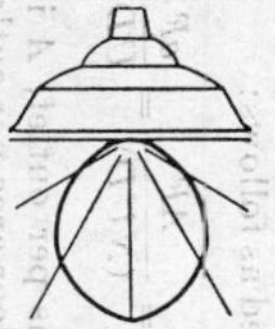	*A* Excellent	*B* Good	*B* Good	*B*+ Very good	*B* Good	*B*+ Very good	*A*− Very good	0.80	0.75	0.65	0.6	0.32	0.28	0.25	0.32	0.28	0.25	0.27	0.25
												0.8	.40	.36	.34	.39	.35	.33	.35	.33
												1.0	.43	.39	.37	.42	.39	.37	.39	.37
												1.2	.46	.43	.41	.45	.43	.41	.43	.41
												1.5	.48	.45	.43	.47	.45	.43	.45	.43
												2.0	.52	.50	.48	.51	.49	.47	.49	.47
												2.5	.56	.54	.52	.55	.53	.51	.53	.51
												3.0	.57	.55	.53	.56	.54	.52	.54	.52
												4.0	.60	.58	.56	.59	.57	.55	.57	.55
												5.0	.61	.59	.57	.60	.58	.57	.58	.56
2 Glassteel diffuser Clear lamp 90–180°, 7% 0–90°, 60%	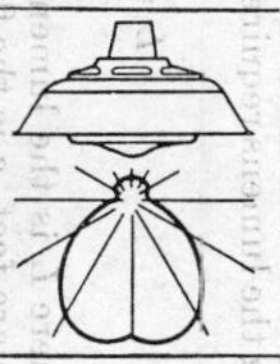	*A*− Very good	*B* Good	*A*− Very good	*A*− Very good	*B*+ Very good	*A* Excellent	*B* Good	0.75	0.70	0.60	0.6	.29	.25	.21	.28	.24	.21	.23	.21
												0.8	.36	.32	.29	.35	.31	.28	.31	.28
												1.0	.39	.36	.33	.38	.35	.33	.34	.32
												1.2	.42	.39	.36	.41	.38	.36	.37	.35
												1.5	.45	.42	.39	.43	.40	.38	.39	.38
												2.0	.49	.46	.43	.48	.45	.43	.44	.42
												2.5	.53	.50	.47	.51	.49	.47	.47	.46
												3.0	.54	.52	.49	.52	.50	.49	.49	.47
												4.0	.57	.55	.53	.55	.53	.51	.51	.50
												5.0	.58	.56	.54	.56	.54	.53	.52	.51
Store and general utility units																				
3 White glass enclosing globe 90–180°, 35% 0–90°, 45%	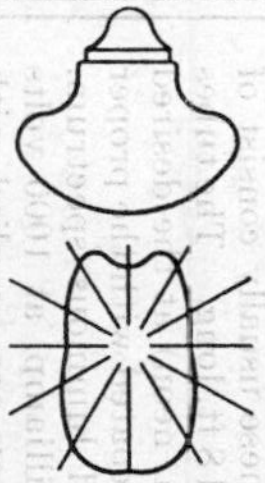	*B*+ Very good	*B*+ Very good	*A* Excellent	*B* Good	*B*+ Very good	*A*− Very good	*B*+ Very good	0.80	0.75	0.65	0.6	.22	.17	.14	.20	.16	.13	.14	.12
												0.8	.27	.22	.19	.25	.21	.18	.19	.17
												1.0	.31	.26	.23	.28	.24	.21	.22	.19
												1.2	.35	.30	.26	.31	.27	.24	.25	.22
												1.5	.38	.33	.29	.34	.30	.27	.27	.24
												2.0	.42	.38	.33	.38	.34	.31	.31	.28
												2.5	.46	.41	.37	.41	.37	.34	.34	.31
												3.0	.49	.45	.40	.43	.39	.36	.36	.33
												4.0	.53	.48	.44	.47	.43	.40	.38	.36
												5.0	.55	.51	.47	.49	.45	.42	.40	.38
4 Enclosed semi-indirect cased-glass bottom etched top 90–180°, 51% 0–90°, 21%	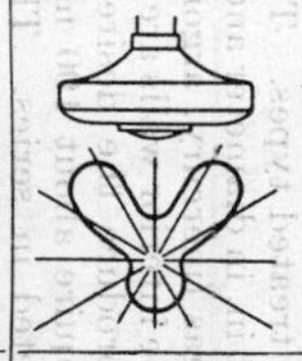	*C*+ Very fair	*C*+ Very fair	*A* Excellent	*A* Excellent	*A* Excellent	*A* Excellent	*B*− Very fair	0.75	0.65	....	0.6	.16	.12	.10	.13	.10	.08	.08	.07
												0.8	.20	.16	.14	.17	.14	.11	.11	.09
												1.0	.23	.19	.17	.19	.16	.14	.13	.11
												1.2	.26	.22	.19	.22	.18	.16	.14	.13
												1.5	.29	.25	.21	.24	.20	.19	.16	.14
												2.0	.32	.28	.25	.27	.23	.21	.18	.17
												2.5	.35	.31	.28	.29	.26	.24	.20	.19
												3.0	.38	.34	.31	.31	.28	.26	.22	.21
												4.0	.41	.38	.35	.34	.31	.29	.24	.23
												5.0	.43	.39	.37	.36	.33	.31	.26	.24
5 Open indirect 90–180°, 80% 0–90°, 0%		*C*+ Very fair	*C* Fair	*B*+ Very good	*A*+ Excellent	*A*+ Excellent	*A*+ Excellent	*C* Fair	0.70	0.60	....	0.6	.15	.12	.10	.11	.09	.07	.05	.04
												0.8	.18	.15	.13	.13	.11	.09	.07	.06
												1.0	.22	.19	.16	.15	.13	.11	.08	.07
												1.2	.25	.22	.19	.18	.15	.13	.09	.08
												1.5	.27	.24	.21	.20	.17	.15	.10	.09
												2.0	.30	.27	.25	.22	.19	.17	.11	.10
												2.5	.34	.31	.28	.24	.22	.20	.13	.12
												3.0	.36	.33	.30	.26	.24	.22	.14	.13
												4.0	.40	.37	.34	.28	.26	.24	.15	.14
												5.0	.42	.39	.37	.30	.28	.26	.17	.15

phosphors, which glow when activated by ultraviolet rays (see Table 18). Various colors are obtained by using different combinations of phosphors. The arc or discharge is struck between two heated tungsten filaments, one at each end of the tube, which have been treated to produce high electronic emission. The auxiliary control equipment limits the current to a predetermined value and is not interchangeable with that for lamps of a different size. The losses in the auxiliaries are 20 to 30 per cent of the nominal watts of the lamp with which they are designed to operate. The power factor of many of these lamps with auxiliaries is about 50 to 60 per cent; however, this can be improved with condensers. Newer control devices give an over-all power factor of nearly unity. The lower temperature, higher efficiency, and available color qualities give it a large field of application. A useful life of nearly 2500 hr. may be expected.

Table 17. Incandescent Lamps—Light Output of Inside-frosted Lamps*

115–125-volt standard		115–125-volt daylight		230–250-volt standard	
Size, watts	Lumen output	Size, watts	Lumen output	Size, watts	Lumen output
15	141	60	540	50	480
25	260	100	1,060	100	1,250
40	465	150	1,600	200	3,050
60	835	200	2,350	300	4,850
100	1,630	200	3,650	500	8,800
150	2,600	500	6,400	750	13,600
200	3,700			1,000	19,100
300	5,650				
500	9,950				
750	15,500				
1,000	21,500				
1,500	33,000				

* Data from "General Electric Catalog," 1948.

In general fluorescent lamps give appreciably more lumens per watt than do incandescent lamps even though the fluorescent lamps have losses in their auxiliaries while incandescent lamps have no such losses. It is difficult to compare the lives of the two types, since the life of the incandescent unit is determined primarily by the number of burning hours whereas that of the fluorescent unit, at least the hot-cathode type, seems to be dependent largely on the number of times it is switched on and off. The first cost of bulbs and fixtures is definitely higher for the fluorescent units than for the incandescent. With higher first costs and lower operating costs it would be expected that the fluorescent type would find its chief field of usefulness where it is operated for long periods per day. The difference in the luminous spectra of the two types sometimes dictates a choice. The "daylight" fluorescent tube gives much lower intensity on the yellow portion of the visible spectrum than does the incandescent lamp, and this may cause an unpleasant visual sensation with certain objects or in some surroundings. In some cases a blending of the two types gives the most satisfactory results. If higher luminous intensities are required with the fluorescent units, a large portion of their economic advantage is lost.

Fluorescent lamps are also made employing cold cathodes of the treated types. These usually consist of tubes less than 1 in. in diameter and 8 ft. long. The tubes are filled with gas (mercury, argon, neon) at the desired pressure, and the interior walls are coated with the proper phosphors to produce the desired luminous spectrum. These tubes require about 100 milliamp. at 1000 volts and are connected in series. They are supplied with cutouts in case of the failure of an individual tube and do not appear to be subject to shortened life span by repeated switching. At present the first cost and annual operating cost are somewhat higher than for the hot-cathode type of fluorescent lighting. For further discussion see Inman and Thayer [*Trans. Am. Inst. Elec. Engrs.*, **57**, 723 (1938)]; Sabatine, Cold Cathode Fluorescents at War, *Illum. Eng.*, **38** (4), pp. 171, 279, 280 (1943)]; Bradley and Lee, Fluorescent Sources at Work, *Illum. Eng.*, **39** (1), (1944)]; and Smith, Cold Cathode at War, *Illum. Eng.*, **40** (8), 558 (1945)].

In planning industrial lighting the illumination sources should be uniformly spaced, and, for incandescent lamps, the distance between units should approximate the height of the ceiling. A convenient way to *locate outlets* is to divide the room into equal squares or rectangles locating the outlets at the center of each space. The distance between outlet and wall is thus one-half the distance between outlets. The appearance of the resulting fixtures and the location of machines must be considered also in locating outlets. The recommended mounting height of units is given in Table 14. For high lighting intensity and for contrast, local lighting units supplementing the general illumination are necessary.

Table 18. Data on Fluorescent Lamps*

Description	Nominal watts (add auxiliary watts for total)				
	15	20	30	40	100
	Lumen output				
Preheat-starting:					
Daylight	555	800	1350	1960	3900
White	622	940	1485	2310	4200
4500° white	585	860	1380	2110	4000
Blue	315	460	780		
Green	900	1300	2250		
Red	45	60	120		
Instant-starting:					
Daylight	...	....	....	1920	
White	...	....	....	2300	
4500° white	...	....	....	2100	

Life of lamps: 3 burning hours per start, 2500 hours.
6 burning hours per start, 4000 hours.
12 burning hours per start, 6000 hours.

* Data from "Illuminating Engineering Society Lighting Handbook," 1947, and "General Electric Catalog," 1948.

Table 19. Mercury-vapor Discharge Lamps Used in General Lighting*

Nominal size, watts†	Starting time, minutes	Average life with 5 burning hours per start, hours	Lumen output
100	3	1000	3,000
250	4	2000	10,000
400	7	4000	16,000

* Data from "Illuminating Engineering Society Lighting Handbook," 1947.
† Add 20 to 40 watts for transformer.

The **size of lamp** for each unit is selected to give the necessary light, expressed in lumens. *Additional lumens* at the lamp sources are required, because all the light falling on the walls and ceiling is not reflected; with dark surfaces a considerable amount of light is absorbed. This is taken into account by a **coefficient of utilization** based on a room index (Table 15) for the dimensions of the room, the type of unit, and the color of the walls. The lumens from a unit are decreased by the light absorbed by the lighting unit itself and by dust accumulating on the unit. These factors are considered in calculating the lumens required as follows:

$$L = \frac{AF}{(NUD)} = \frac{aF}{UD}$$

where L is the lumens per outlet, A is the floor area in square feet, a is the floor area per outlet in square feet, N is the number of outlets, F is the desired foot-candles based on Table 13, U is the coefficient of utilization from Table 16, and D is the expected average efficiency of the lighting unit from Table 16. The size of lamp is selected according to the required lumens from Tables 17, 18 and 19.

Wiring should be installed to provide not only for modern lighting but for possible future uses of the structure. It is recommended that the wiring provide for a capacity of about 4 to 6 watts/sq. ft. of floor area for offices, drafting rooms, and factories, 4 watts/sq. ft. for stores and school rooms, 3 watts/sq. ft. for neighborhood stores and storage areas in factories, and 1 watt/sq. ft. for storage areas in garages and unimportant basements. The wire size selected for each branch should allow not more than 2 volts drop to the farthest outlet. For branch circuits a minimum wire size of No. 12 is recommended; and for runs from panel board to first outlet from 50 to 100 ft., No. 10 wire is the smallest that should be used; runs exceeding 100 ft. to first outlet should be avoided.

Additional data on illumination standards and calculations may be obtained from "Illuminating Engineering Society Lighting Handbook," 1947.

ELECTRIC HEATING

Heating is produced (1) when current is forced through the resistance of a conductor and (2) in some cases by an arc. The first is by far the more common, giving light in incandescent lamps and heat in a variety of devices. The induction-type heaters are a modification of (1), utilizing the material to be heated as the resistance.

The **advantages of electric heat** are the convenience of the control, the facility in producing the heat where desired, and the elimination of combustion and its products. This permits heating under unusual conditions as with hydrogen surrounding the object being heated.

The **rate** at which **energy** is converted into heat in a resistance is equal to I^2R. The **ultimate temperature** of a device is that which will cause the heat energy to be dissipated at the same rate as it is produced. Heating devices utilize conductors of much higher specific resistance than copper and which will not deteriorate rapidly at the required temperature. Many of these resistance materials are alloys which have a small temperature coefficient of resistance (Table 2).

The **efficiency** of a heating device may be expressed as the ratio of the energy absorbed by the material being heated to the total energy consumed. This efficiency is relatively high in electric heaters, as ventilation to remove products of combustion is unnecessaty. Further, the heater element can usually be placed in close proximity to the material being heated, and occasionally it is possible to use either the material itself or its container as the heater element. For a discussion of the latter method see Adam, *Trans. Electrochem Soc.*, **70**, 143 (1936).

Insulation for electric conductors is usually heat insulation also. Consequently, solid insulation about the conductors of heating ovens is used sparingly and in many cases is replaced by air.

Resistance furnaces and ovens are constructed in a variety of forms to be most useful and efficient for the work to be performed. The more common forms include (1) the box type, with single-door opening and with modifications, such as the car type where the material is carried on trucks or cars with and without heaters in the cars, the pusher type where one piece (or container) is used to push the next, and the conveyor type for continuous process; (2) the rotary hearth furnace used for continuous heating; and (3) the crucible or pot furnaces with resistors surrounding the pot or with the heating elements submerged in the bath itself.

Box-type ovens for temperatures up to 600 and even 700°F. are listed by electrical manufacturers for connected loads from about 1.5 up to nearly 100 kw. These ovens are suitable for drying, baking, evaporating, japanning, and low-temperature heat-treatment. **Box-type furnaces** for temperatures up to 1850°F. are listed with connected loads approximately the same as for the ovens. These furnaces are suitable for hardening, annealing, tempering, and carburizing.

Commercial **pot-type furnaces** suitable for melting babbitt and other soft metals have connected loads from 4.5 to 22 kw. holding from 150 to 1500 lb. of metal. To expedite the handling of the metal the furnace may be equipped for bottom pouring or for tilting. Other furnaces of the same general type use a lead, salt, or cyanide bath for hardening or tempering tools or machine parts. In large sizes as for continuous hardening or drawing of wire or strip, or for heating large parts, standard furnaces have capacities up to 15,000 cu. in. and have a connected load of 75 kw. Such a furnace with lead bath will average in a 9-hr. day a production of 1200 lb. steel/hr. raised to 1200°F. with 800 kw.-hr./day (see Catalogue 250, Westinghouse Electric Corp.). Certain pot-type furnaces for heat-treatment use a conducting salt: one electrode is immersed in the salt; the containing vessel is used for the other. An arc is produced to start the melting of the salt which is non-conducting in its solid state [see Hall, *Trans. Am. Soc. Steel Treating*, p. 399 (1929)].

Many resistor-type furnaces utilize **heating units** which are easily replaced; these are connected by steel bus bars as copper oxidizes rapidly at furnace temperature. In designing an oven or other heating device, such units may be applied. Certain units use nickel-chromium ribbon wound on porcelain insulators. Another form is a flat heater element protected by a steel casing; still others are designed for immersion in the liquid to be heated. The cartridge-type heater is used for local heating in a portion of a machine, the cartridge fitting closely in a hole drilled in the machine, so that the heat may be conducted away.

Table 20. Steel-sheathed Strip Heaters*
(Dimensions ⅜ in. thick by 1½ in. wide)

Watts	Volts	Max. length, in.
500	115	24
350	115	18
250	115	12
150	115	7

Note. Heaters for 230 volts have the same dimensions and price. These may be connected in series for 440-volt circuits.
* Data from General Electric Supply Catalogue.

In the **design**† of a **heating element,** materials must be selected that will withstand the operating temperature of the element without deterioration. Nickel-chromium

Table 21. Electric Heater Units*

Kw.	Volts	Length, in.	Width, in.	Height, in.
1.75	110	21½	6¼	10
2.5	110	21½	6¼	10
3.5	220	36½	6¼	10
5.0	220	36½	6¼	10

* Westinghouse catalogue 247.

Table 22. Electric Cartridge Heaters*

Watts	Volts	Outside diam., in.	Length, in.	Terminal connection
75	110	0.496	3²²⁄₃₂	Monel wire
100	110	.496	4⁷⁄₃₂	Monel wire
150	110	.621	2⅝	Monel wire
250	110	.621	4⅛	Monel wire
200	110	.746	4⁹⁄₃₂	Monel wire
150	110	.933	3⅝	Brass terminal
250	110	.933	5⅛	Brass terminal
400	110	.933	5⅝	Brass terminal
300	110	1.306	4¹³⁄₁₆	Brass terminal
600	110	1.306	5³⁄₁₆	Brass terminal

* Westinghouse catalogue 247.

† By Hood Worthington.

alloys are capable of withstanding an operating temperature of 1150°C., and certain less expensive alloys operate satisfactorily at lower temperatures. Figure 44 gives the allowable heat flux at the surface of nichrome IV wire or ribbon as a function of the temperature of the material being heated and the method of applying the heating element. The empirical design curve for exposed radiating strips or rods is given by Fleischmann [*Trans. Am. Inst. Elec. Engrs.*, **48,** 1196 (1929)] as useful because it gives long-lived elements even under conditions of close spacing and double banking.

Woodson's curve for confined radiant coils [*J. Am. Inst. Elec. Engrs.*, **44,** 1354 (1925)] checks closely with Driver-Harris Company's recommendation for range units and radiant heaters made of coiled wire.

The two curves marked **wire-wound** and **ribbon-wound insulated tube** give the allowable heat flux at the surface of a wire wound with the turns one diameter apart and of a ribbon wound in such a way that the surface of the ribbon is twice the surface on which it is wound, respectively. Both windings are assumed to be placed over ⅛-in. asbestos insulation and to be delivering heat to a metal tube having the temperature given on the ordinate. If the ribbon or wire is differently spaced, or the thickness or thermal conductivity of the insulation is changed, the allowable heat flux will be altered. The factor of safety enters the last two curves only in that the maximum temperature which the winding can stand is taken as 900°C. instead of 1150°C.

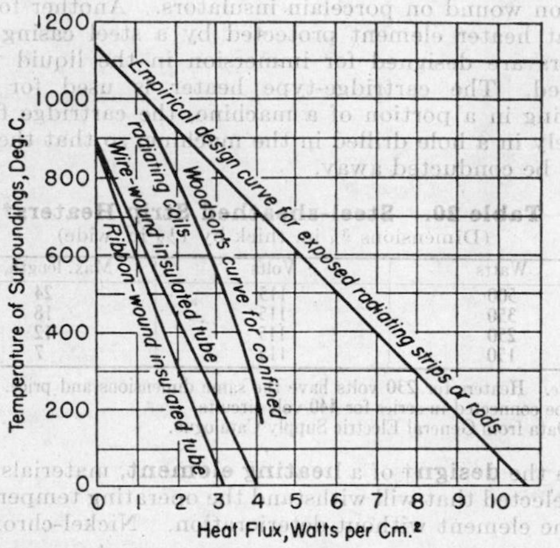

Fig. 44. Permissible heat flux from nichrome IV.

From these curves it will be apparent that the method of embedding a winding in an insulating medium should be avoided if it is possible to use the method of supporting the winding so that it is free to give up its heat by radiation. However, the use of a winding over an insulated tube has the advantage of simplicity of construction.

The use of Figs. 45 and 46 will be made plain by an example. The position of the circle in Fig. 45 indicates that, if the winding is such that it permits a heat flux of 1.9 watts/sq. cm., a 5-kw. heater operating on 110 volts may be constructed from 73 ft. of No. 7 B. & S. gage nichrome IV **wire.** For a 6-kw. heater on 220 volts (Fig. 46), 120 ft. of No. 10 wire should be used. (Note the slightly higher heat flux used to get an even wire size.)

When **ribbon** is used there are more possibilities. Suppose that the allowable heat flux in a 20-kw. heater to be used on a 220-volt line is 3.7 watts/sq. cm. Moving horizontally at 3.7 watts/sq. cm. across the lower left-hand field of Fig. 47 to the line marked 20 kw. at 220 volts, and then vertically to the upper left-hand field, one finds that there are four sizes of ribbon that may be used. Then moving horizontally to intersect the 3.7 watts/sq. cm. line in the right-hand field and reading

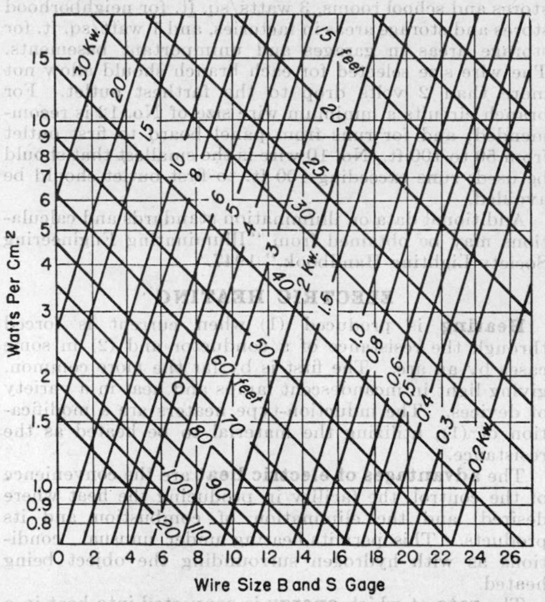

Fig. 45. Dimensions of nichrome IV wire for 110-volt heaters, 0.2 to 30 kw.

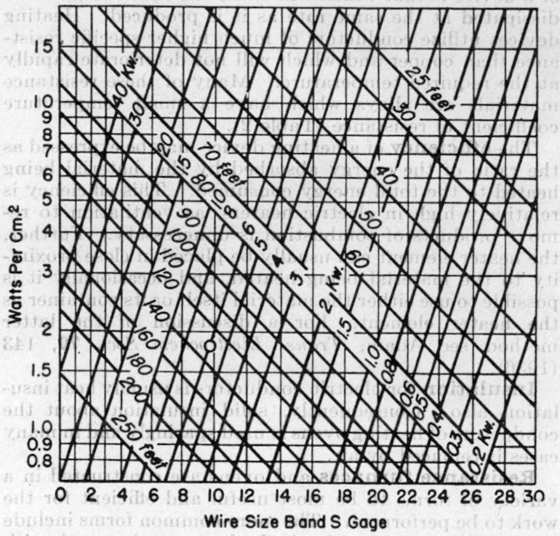

Fig. 46. Dimensions of nichrome IV wire for 220-volt heaters, 0.2 to 40 kw.

from the lengths at 220 volts, one finds the following solutions: 37 ft. of 1 in. × 32 gage; 51 ft. of ¾ in. × 26 gage; 70 ft. of ½ in. × 20 gage; or 86 ft. of ⅜ in. × 16 gage.

Because of their superior resistance to oxidation, the ½ in. × 20 gage or the ⅜ in. × 16 gage are preferable. If, however, the ribbon is too thick it loses its advantage in ease of winding and uniform covering of the winding area, and wire may be used with equal satisfaction.

For certain heaters the **control** may be manual; for

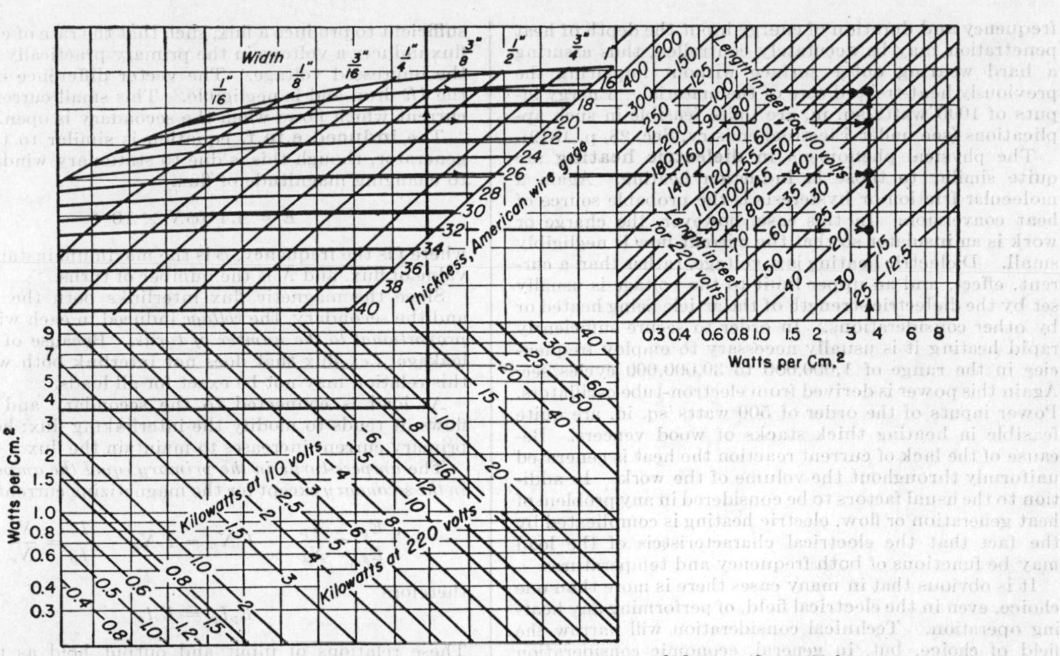

FIG. 47. Dimensions of nichrome IV ribbon for 110- and 220-volt heaters, 0.4 to 60 kw.

others the temperatures are controlled automatically by thermostats or pyrometers. The **energy** delivered to a heating device may be varied by (1) adjusting the applied voltage as from transorfmer taps, (2) changing the connections so that the voltage applied to the heating elements is varied, (3) rheostatic control in series with the element, (4) resistance change of the heater due to the temperature coefficient of the heater material, and (5) interrupting the supply to part or all of the heater elements.

Example. On a single-phase or direct-current heater with two elements, three values of heating may be obtained by (*a*) both elements in parallel, (*b*) one element giving half value, and (*c*) the two elements in series giving one-fourth value. Rheostatic control will give finer regulation. A three-phase heater may have switches provided to connect the three elements in Δ for full heat or in Y for one-third value.

Within recent years three additional methods of electric heating have gained widespread adoption. These are radiant, or infrared heating, and the two methods of high-frequency, or electronic, heating known as induction heating, applicable to conductors, and dielectric heating, applicable to insulators or poor conductors.

Radiant heating finds its greatest usefulness where only the surface of the work needs be heated, as in drying painted surfaces. In such applications the maximum temperature is determined by the effect of temperature on the chemical process which occurs during the drying operation. Within limits a higher temperature means a shorter drying period.

For such work there are two sources of heat in general use, lamps and exposed resistor units (see p. 1759 *et seq.*). The lamp has the advantage that the glass globe prevents the loss of heat by convection and that its high operating temperature permits a large energy output to be obtained from a small source. By using reflectors this high energy output can be concentrated at the desired spot. Because of the low thermal capacity the lamp reaches its operating temperature rapidly and is very efficient for intermittent operation. The maximum electrical input to such lamps is about 1 watt/sq. cm. in commercial practice, and the output depends on the lamp and reflector assembly as a source of radiation. The over-all efficiency, which is usually between 60 and 70 per cent, depends on the absorptive power of the articles heated. The time necessary to attain a given temperature is proportional to the heat capacity per unit area of the receiving surface.

Intensities at least ten times as high as from lamps may be obtained from exposed resistor elements (see Figs. 44 and 45). These higher intensities are necessary if the rapid heating of "massive" articles is to be achieved. Articles are classified as "light" or "massive" according to whether their heat capacity per unit area of surface is small or large. Exposed heater elements suffer losses by convection and possess thermal inertia and so are slower to reach a final operating temperature.

For many years the **induction heating** principle has been used in electric furnaces, particularly for melting non-ferrous arcs. These furnaces usually operated at power system frequencies, except for the Ajax-Northrup induction furnace, which operates at frequencies of 10,000 or 12,000 cycles/sec. derived from the oscillatory discharge of a bank of condensers. The induction-type furnace utilizes the material to be heated as the secondary of a transformer; the primary coil is connected to the supply. Certain furnaces provide a magnetic structure for the flux path interlinking the primary and secondary. These uses of induction heating persist, and in addition motor-generator sets or mercury-arc rectifiers are used as frequency converters; for such application the former up to 10,000 cycles/sec. and the latter up to 1000 cycles/sec. In all such applications the purpose is to heat the charge or load throughout its mass, usually up to its melting point.

The so-called electronic, or **high-frequency induction** furnace is designed to heat only the shell of the material as on casehardening of gears and other parts subjected to heavy wear. It utilizes frequencies from 100,000 to 500,000 cycles/sec., or even higher, derived from electron-tube oscillators. By proper choice of frequency and duration of energy input the depth of heat penetration may be accurately controlled, thus assuring a hard wearing outer surface without impairing the previously heat-treated core of the material. Energy inputs of 1000 watts/sq. in. are quite feasible in such applications (see under Electrochemistry, Sec. 28, p. 1812).

The physical phenomena in **dielectric heating** are quite similar to those in induction heating. Again a molecular friction or hysteresis is the probable source of heat conversion. In this case, however, the charge or work is an insulator so that the current flow is negligibly small. Dielectric heating is a voltage, rather than a current, effect, and an upper limit to the voltage is usually set by the dielectric strength of the article being heated or by other considerations. In order to secure sufficiently rapid heating it is usually necessary to employ frequencies in the range of 1,000,000 to 30,000,000 cycles/sec. Again this power is derived from electron-tube oscillators. Power inputs of the order of 500 watts/sq. in. are quite feasible in heating thick stacks of wood veneers. Because of the lack of current reaction the heat is generated uniformly throughout the volume of the work. In addition to the usual factors to be considered in any problem of heat generation or flow, electric heating is complicated by the fact that the electrical characteristcis of the load may be functions of both frequency and temperatures.

It is obvious that in many cases there is more than one choice, even in the electrical field, of performing any heating operation. Technical consideration will narrow the field of choice, but, in general, economic consideration will determine the final choice. Some of these considerations are capital costs, labor costs, power costs, maintenance, processing time, and floor space required. It is not possible to generalize on these factors, since each application is a separate problem and should be reviewed by a competent authority. The large electrical manufacturers usually supply all types of electrical heating units and are in a position to make recommendations for any particular application.

Thermostats indicate (or record) the furnace temperature and, by means of two adjustable contacts for the "high" and "low" temperature limits, close the control circuits at high for shutting off the supply to the heaters and at low for turning on the power to the heating elements. The thermostat is actuated by heating of a bulb containing mercury, a liquid, or a gas, causing a pressure change, which is transmitted through a tube to the indicating and controlling mechanism. The maximum temperature to be controlled by thermostats is 1000°F.

For temperature above 1000°F., **pyrometers** are used. The pyrometer is actuated by the voltage from a thermocouple placed in the furnace or oven. The pyrometer is a sensitive instrument and should be mounted in a location free from vibration. The calibration is for a particular type of thermocouple and may vary with length of wires between the couple and the instrument. By means of two adjustable contacts the power to the heater is controlled similarly to the thermostat control. See Section 19.

TRANSFORMERS

A transformer consists of two or more *electric circuits* interlinked by a *magnetic circuit.* The electric circuits are usually insulated from each other but may be electrically connected as in the autotransformer. This device permits the transfer of energy from a circuit of one alternating voltage to that of another voltage and serves to insulate the two circuits from one another. The primary winding is the one connected to the supply; the winding connected to the receiver circuit is the secondary.

The **magnetizing current** of a transformer is just sufficient to produce a flux, such that the rate of change of flux induces a voltage in the primary practically equal to the impressed voltage. The vector difference is due to the IR drop and is negligible. This small current is the current which flows when the secondary is open.

The **induced e.m.f.** equation is similar to that in a generator, though this is due to stationary windings and to changing magnitude of flux.

$$E = 4.44 f \phi N \times 10^{-8}$$

where f is the frequency, ϕ is the maximum instantaneous value of flux, and N is the number of turns.

Since the magnetic flux interlinks both the primary and the secondary, the *voltage* induced in each winding *is proportional to the number of turns.* Because of a slight leakage, *i.e.*, flux that does not interlink both windings, this relation may not be exact for all loads.

As load is connected to the secondary and current flows, it tends to modify the interlinking flux; hence the primary current increases to maintain the flux.

The *ampere-turns in the primary equal the ampere-turns in the secondary* except for the magnetizing current:

$$\frac{E_p}{E_s} = \frac{N_p}{N_s} \qquad I_pN_p = I_sN_s \qquad \frac{I_s}{I_p} = \frac{N_p}{N_s}$$

therefore

$$E_pI_p = E_sI_s$$

These relations of input and output hold as to volts, amperes, power, and power factor, except as modified slightly by the losses and by the magnetizing current. Subscripts refer to primary or secondary.

The **losses** in a transformer consist of *iron losses* (hysteresis and eddy-current losses) and *copper losses* (primary and secondary). These losses are small so that the full-load efficiency is about 98 per cent and is higher for large transformers. The iron losses of transformers which are continuously on the supply line will reduce the daily average efficiency.

In **polyphase circuits,** transformers are used to change from one system to another as well as to change the voltage. Any polyphase system may be secured from any other polyphase system with transformers provided with suitable windings. A few of the more common *transformer connections* are shown diagrammatically in Fig. 48. The desired voltage for both systems may be obtained from transformers with the proper voltage ratio. A single-phase load cannot be balanced on a polyphase supply by any combination of transformers; neither can a polyphase load be taken from a single-phase supply without energy-storage devices.

Autotransformers (Fig. 49) have a portion of the winding in common for the primary and secondary. In this portion of the winding, the current is the difference between the primary and secondary currents resulting in a higher utilization of materials, particularly where the transformer ratio is near unity. For example, with a ratio 3 to 4, the capacity as an autotransformer would be four times the capacity used as an ordinary transformer. Since the primary and secondary are interconnected the autotransformer does not insulate one section of the circuit from the other.

The **induction regulator** is an autotransformer or booster transformer with a variable ratio. In the *single-phase regulator*, the flux common to the primary and secondary is varied by turning the primary with respect to the secondary. In the *three-phase regulator*, a constant voltage is added vectorially; the angle is varied by the position of the rotor so that the resultant voltage is regulated to the value desired. Regulators are commonly used to control distribution voltages; by means of the line-drop compensators the point of regulation may be at a distance from the regulator.

The **constant current transformer** utilizes the repulsion between the primary and secondary coils to separate the coils when the current starts to increase. The greater separation increases the leakage flux, reducing

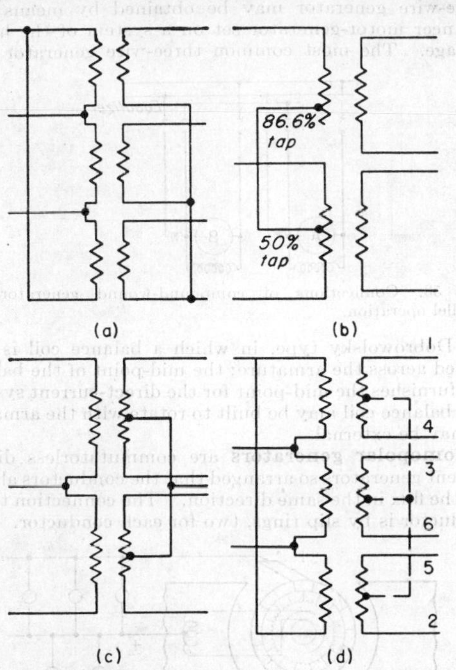

FIG. 48. Transformer connections in polyphase circuits. (*a*) Three-phase delta to three-phase Y; (*b*) three-phase to two-phase (Scott connection); (*c*) two-phase three-wire to two-phase five-wire; (*d*) three-phase delta to six-phase diametrical.

the voltage until the current is the same. If the current starts to decrease, the coils come closer together so that the voltage increases and the current also. By properly counterbalancing the moving coil, the current can be maintained practically constant.

Instrument transformers are used to reduce voltage (potential transformers) and current (current trans-

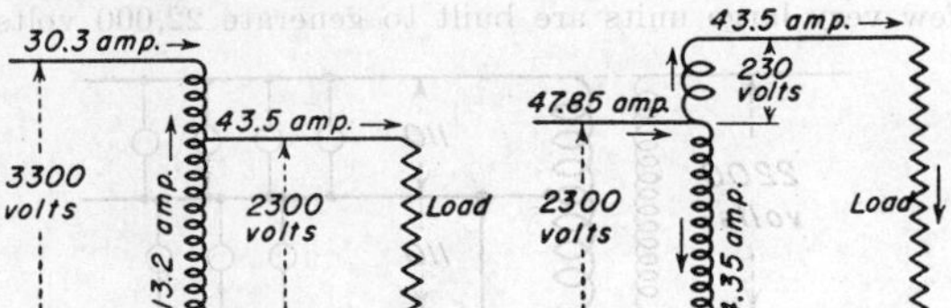

FIG. 49. Autotransformer connections. (*a*) To reduce voltage at load; (*b*) to increase voltage at load.

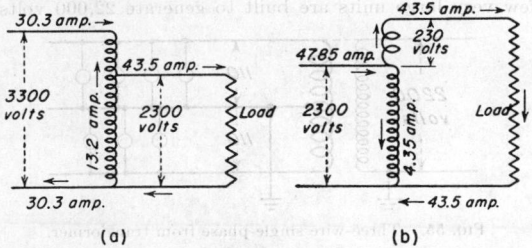

formers) for applying to instruments and relays. This permits arrangement of main leads independent of the switchboard location by running small wires to the instruments; it permits standardizing on 5-amp. 110-volt instruments with proper scales according to the ratio of the transformers, and it permits insulating the instruments from high-potential circuits.

Table 23. Approximate Cost of Small Oil-filled Transformers

Kva.	Single-phase, $/kva.	Three-phase, $/kva.
5	22.00	
9	.....	23.60
15	14.50	22.00
75	8.50	12.00

GENERATORS

Direct-current Generators. Direct-current generators consist of a stationary field and yoke and a rotating armature with commutator. The **armature** is laminated to reduce the eddy-current losses and has slots for the armature windings to reduce the reluctance of the magnetic path. Inasmuch as the conductors pass alternately under north and south poles the e.m.f. induced is alternating. The armature conductors are connected through a switching device, the commutator and brushes, to the external circuit so that the e.m.f. delivered is unidirectional.

The armature winding is continuous in order that the commutator may function; hence there are two or more paths for current through the armature:

$$E = \left(\frac{Z}{\text{paths}}\right) \phi p \left(\frac{\text{r.p.m.}}{60}\right) \times 10^{-8}$$

where Z is the total number of conductors on armature; paths refer to number of parallel paths through armature; ϕ is total flux per pole actually entering the armature (because of leakage this is less than the total flux per pole); p is the number of poles; and r.p.m. is the speed of the armature in revolutions per minute. For a given machine this may be written $E = k'\phi$ r.p.m.

The speed of the generator is usually fixed by the driving engine or motor; and the number of conductors, paths, and poles by the designer. Hence any *adjustment of e.m.f.* must be by field control; this may be by a hand-controlled rheostat or by automatic devices.

Types of generators are designated by the field connections. When the field circuit is connected across the armature, it is a **shunt** machine; if connected in series

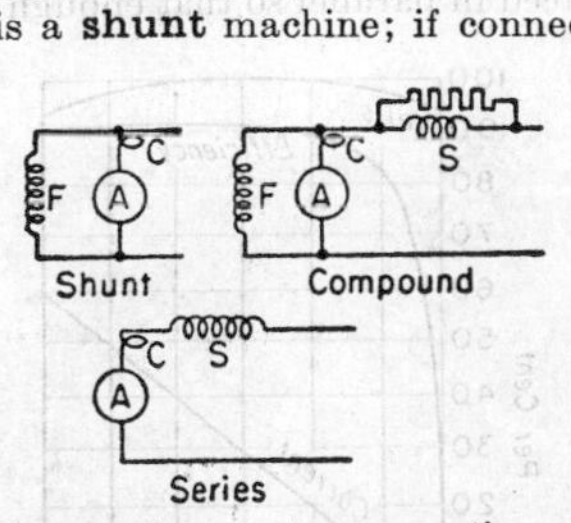

FIG. 50. Direct-current generator connections. *A*, armature; *F*, shunt field; *C*, commutating field; *S*, series field. Note that diverter resistance is shown in parallel with series field of compound-wound generator.

with the armature, it is a **series** machine; if there are both shunt and series field coils, it becomes a **compound** machine; the field may be connected to some other source making it a **separately excited** machine.

Armature reaction is the distortion of the main field due to the current in the armature. This distortion tends to induce e.m.fs. in the commutated coils and may also cause a slight weakening of the field.

Commutation is the reversing of current in the coil under the brush. The brush covers two or more commutator segments; hence one or more armature coils are short-circuited; if there is an e.m.f. induced while it is short-circuited, current will flow and sparking will result The self-inductance of the coil delays the reversal of the current, and this also causes sparking. Carbon brushes with higher specific resistance decrease the sparking. In older machines, brushes were shifted with load to change the flux in the commutated coil. Modern machines usually have **commutating poles** connected in series with the armature, which furnish sufficient flux to aid the reversal of current and neutralize the effect of armature reaction at the point of commutation. The commutating pole makes it unnecessary to shift brushes with load, but the proper position for the brushes is critical, and the manufacturer's setting should not be disturbed.

Regulation of direct-current generators, *i.e.*, change of voltage with load, depends on the type of generator (Fig. 51). The voltage of the separately excited gen-

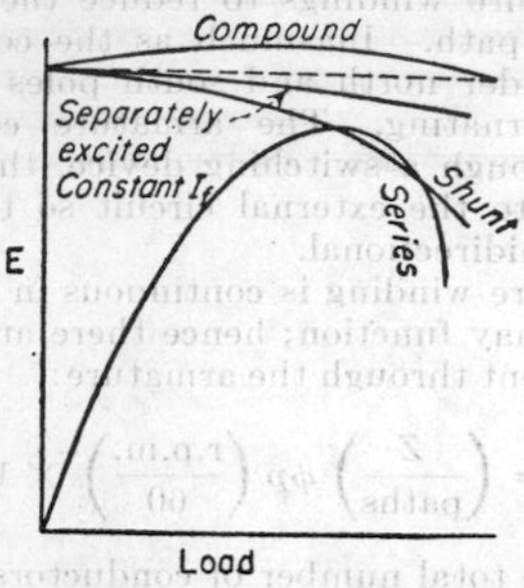

FIG. 51. Voltage characteristics of direct-current generators.

erator decreases with load because of *IR* drop in the armature circuit and the effect of armature reaction. The shunt machine has a further reduction in voltage with load because the lower voltage is impressed on the field. In the compound-wound machine, the series-field coil increases the m.m.f. with load. This may be adjusted by the number of turns on the series field so that the full-load voltage is just equal to the no-load voltage (flat compounded) or so that it is more or less as desired. On commercial generators, excess turns are usually put in the series field and then diverters (resistances) are placed in parallel so that enough of the current

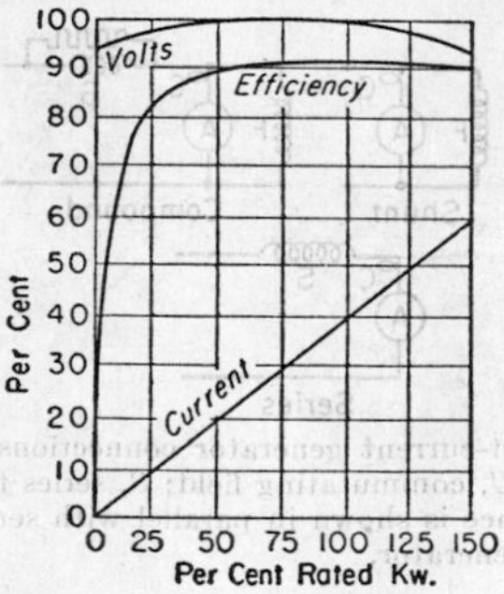

FIG. 52. Typical characteristics of compound-wound generator.

is diverted to give the desired characteristic. The series generator is seldom used.

Paralleling of generators is frequently desirable in order that more than one generator may be used to supply a load. **Shunt machines** may be paralleled with little difficulty, care being taken that the same terminal voltages exist and the same polarities are connected together. For proper distribution of load, all generators operating in parallel should have the same regulation. **Compound-wound generators** must be provided with an equalizer connected in order that two or more machines be stable when paralleled. The *equalizer* is a low-resistance conductor connecting the series fields in parallel. With the equalizer bus compound machines may be adjusted to divide the loads properly.

Three-wire generators are used to obtain the advantage of double the load voltage for distribution. Common voltages are 115 or 230 volts. Motor loads are usually connected at the higher voltage. The equivalent of the three-wire generator may be obtained by means of a balancer motor-generator set on a system of the higher voltage. The most common three-wire generator is of

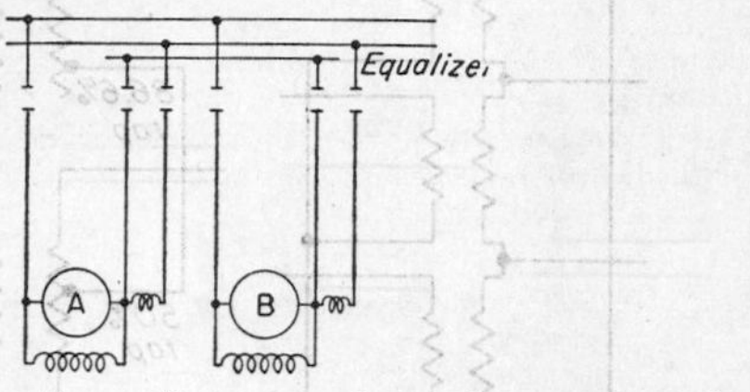

FIG. 53. Connections of compound-wound generators for parallel operation.

the Dobrowolsky type, in which a balance coil is connected across the armature; the mid-point of the balance coil furnishes the mid-point for the direct-current system. The balance coil may be built to rotate with the armature or may be external.

Homopolar generators are commutatorless direct-current generators so arranged that the conductors always cut the flux in the same direction. The connection to the conductor is by slip rings, two for each conductor. The

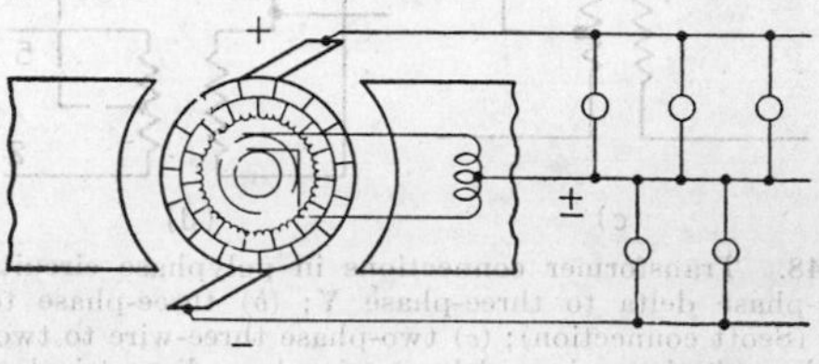

FIG. 54. Three-wire direct-current generator.

machine is not standard, is of low voltage, heavy, and expensive.

Alternating-current Generators. Alternating-current generators are built with rotating fields and stationary armatures, permitting a very rugged construction; the only moving contacts are for the field current at low voltage. Commercial alternators generate at 2300 volts in small sizes, 13,500 for medium and large sizes, and a few very large units are built to generate 22,000 volts.

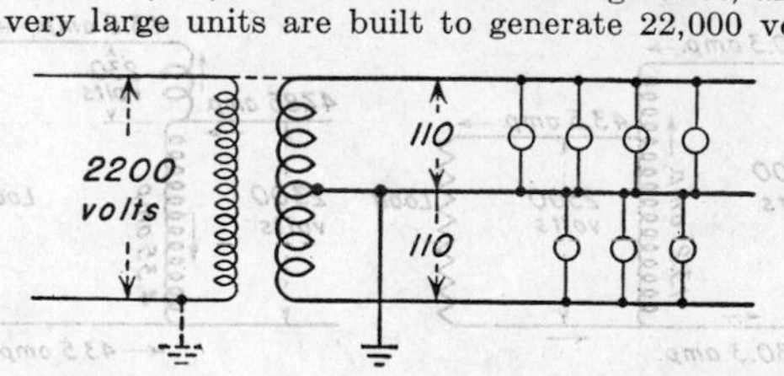

FIG. 55. Three-wire single-phase from transformer.

Because of the large capacity and high voltage the dissipation of the heat due to the losses is a major problem requiring enormous amounts of air for cooling.

Three-phase alternators are supplied for the usual installation. Two phase is frequently obtained by means of transformers from a three-phase generator; by adding damper windings on the field-pole structure, heavy single-phase loads may be supplied. Three groups of windings are wound 60 electrical degrees between centers (360 electrical degrees to each pair of poles). The e.m.fs.

of these windings would be 60 electrical degrees apart (Fig. 56*a*), and by reversing the middle winding and consequently its e.m.f. (Fig. 56*b*), the three e.m.fs. are 120° apart. The windings can then be connected in Δ or in *Y*, the latter being more common.

A large range of **excitation** (field-circuit current) is necessary to provide constant external voltage, with variation in load and power factor (Fig. 57). As individual exciters are frequently furnished with each alternator, a portion of this control can be provided by the rheostat in the exciter field circuit, reducing the I^2R losses otherwise present in the alternator field rheostat. Certain types of automatic regulators utilize the exciter field circuit entirely for control of alternator voltage.

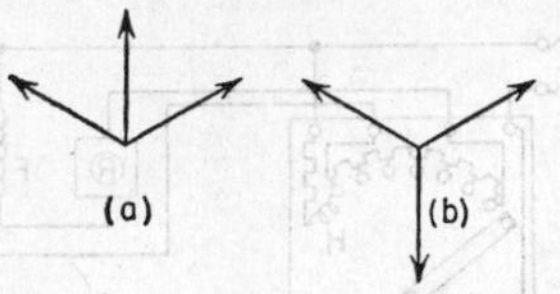

FIG. 56. Three-phase voltage vectors. (*a*) Voltages 60 deg. apart; (*b*) voltages 120 deg. apart.

Parallel operation of commercial alternators is quite satisfactory. To connect an alternator to an energized bus, it is necessary to synchronize by hand or automatically. *Synchronizing* is adjusting the time of the voltage wave, by changing the speed of the incoming alternator, to obtain the correct frequency and the correct instantaneous polarity; lamps may be used, although the synchroscope is the instrument usually used to indicate this. In addition it is necessary to have correct voltage and phase rotation.

Load division of paralleled alternators is dependent on the speed (governor) adjustments of the prime movers. When it is desired to increase the output of an alternator, the throttle is opened more, tending to increase its speed, but actually changing the phase position of its

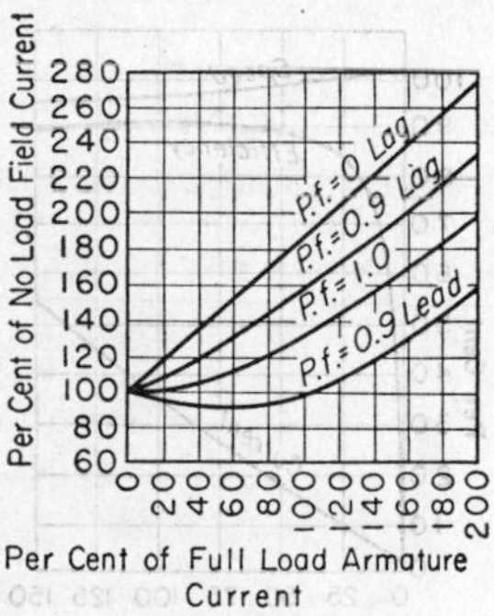

FIG. 57. Excitation curves of a turboalternator.

generated e.m.f. with respect to that of the other alternators. The frequency of all alternators in parallel must be the same, which thus fixes their speed.

Excitation change of paralleled alternators tends to change their **reactive current.** Since the total reactive current (at a given voltage) is fixed by the load, if the excitation of one alternator is increased, it furnishes a larger proportion of the total reactive current.

An induction motor may be driven slightly above synchronous speed; it will then deliver energy to the supply line. This **induction generator** receives its excitation from the supply line and hence ceases to generate if the supply is disconnected. The power factor of the generator is fixed by the voltage, load, and design of the machine; it cannot be adjusted.

Synchronous or **rotary converters** (Figs. 58 and 59) are used to convert alternating current to direct current. Alternating current is applied to, and direct current taken from, a common armature winding rotating within a salient pole structure. The converter operates as a synchronous motor and is started in like manner.

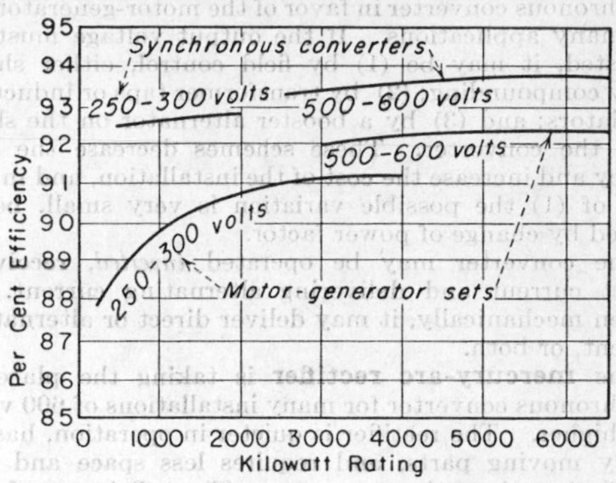

FIG. 58. Efficiency of synchronous converters and motor-generator sets.

The power factor may be adjusted by field change as in the motor, but the efficiency and output of the converter are reduced at decreased power factor. The characteristics as a direct-current generator are limited, since the speed and voltage are fixed by the frequency and the alternating voltage.

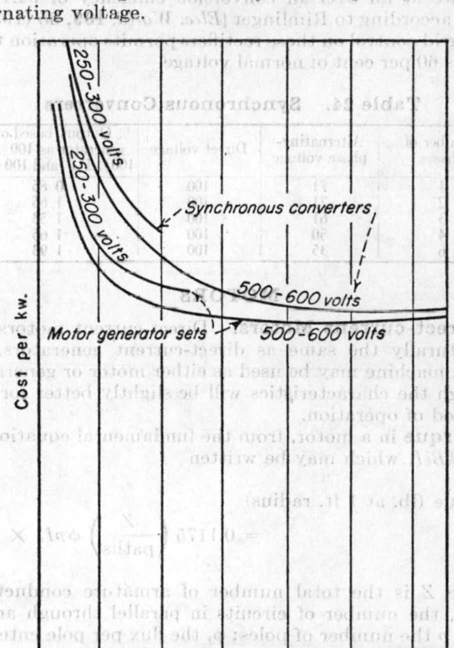

FIG. 59. Relative prices of synchronous converters and motor-generator sets.

To obtain a standard direct voltage from a standard alternating voltage, it is necessary to use transformers, because the *conversion ratio is fixed.* Since the efficiency and capacity increase with more phases, and the transformers are in use for proper voltage, six-phase converters are common. In large sizes the converter has the advantage over a motor-generator set of *higher efficiency, lower first cost,* and *smaller space.*

The **disadvantage** of the synchronous converter is that the direct voltage is directly proportional to the alternating voltage. This difficulty disqualifies the synchronous converter in favor of the motor-generator set for many applications. If the output voltage must be adjusted, it may be (1) by field control, either shunt or by compounding; (2) by transformer taps or induction regulators; and (3) by a booster alternator on the shaft with the converter. These schemes decrease the efficiency and increase the cost of the installation, and in the case of (1) the possible variation is very small, being caused by change of power factor.

The converter may be operated *inverted,* receiving direct current and delivering alternating current. If driven mechanically, it may deliver direct or alternating current, or both.

The **mercury-arc rectifier** is taking the place of synchronous converter for many installations of 600 volts and higher. The rectifier is quieter in operation, has no heavy moving parts, and requires less space and less foundation than the converter. The efficiency of the rectifier is high over a wide range of load and increases with the voltage. (See also pp. 1750, 1751.)

The application of mercury-arc rectifiers is extended by introduction of the grid, permitting control of voltage, current, or load. Grid-controlled rectifiers, rated at 2000 and 3000 kw. to supply 500 volts to chlorine cells, operate at an over-all conversion efficiency of 94.1 per cent, according to Rimlinger [*Elec. World,* **109,** 35 (1938)]. The grid control on these rectifiers permits operation to as low as 60 per cent of normal voltage.

Table 24. Synchronous Converters

Number of phases	Alternating-phase voltage	Direct voltage	% Output, based on d.c. generator as 100%, at 100% P.F. and 100% Eff.
1	71	100	0.85
2	71	100	1.65
3	61	100	1.33
4	50	100	1.65
6	35	100	1.93

MOTORS

Direct-current Motors. Direct-current motors are structurally the same as direct-current generators. A given machine may be used as either motor or generator, though the characteristics will be slightly better for one method of operation.

Torque in a motor, from the fundamental equation, is $F = kBlI$, which may be written

$$\text{Torque (lb. at 1 ft. radius)} = 0.1175\left(\frac{Z}{\text{paths}}\right)\phi p I_a \times 10^{-8}$$

where Z is the total number of armature conductors; *paths,* the number of circuits in parallel through armature; p the number of poles; ϕ, the flux per pole entering the armature; and I_a, the current to the armature.

It is evident that all factors are fixed by the design of the motor, except ϕ and I_a; or, $T = K\phi I_a$ for a given motor.

Speed in a motor is dependent on the voltage and field as follows:

$$V = E + I_aR_a$$

where V is the impressed voltage, E is the induced voltage in armature, and I_aR_a is resistance drop in armature circuit.

Since

$$E = k'\phi\ \text{r.p.m.} = V - I_aR_a$$

$$\text{R.p.m.} = \left(\frac{V - I_aR_a}{k'\phi}\right)$$

For commercial motors, under normal operation, I_aR_a is less than 5 per cent of V. Hence the speed is approximately proportional to V/ϕ.

Starting of direct-current motors requires a rheostat in the armature circuit, except for fractional-horsepower motors, because of the low inherent resistance of the armature. This rheostat, usually built into a starting box, increases the resistance of the armature circuit so that, during starting I_aR_a is the larger part of V.

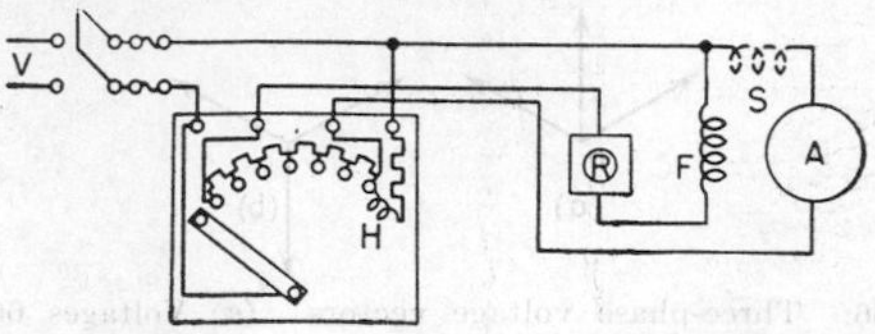

FIG. 60. Starting box and shunt (or compound) motor connections. *R,* rheostat; *F,* shunt field; *A,* armature; *S,* series field.

Shunt motors are the most common type of direct-current motor, furnishing nearly constant speed (within 5 per cent) from no load to full load. The speed of commercial motors may be increased nearly 25 per cent by field-circuit (rheostat) control; for small motors 50 per cent speed range is available (see Fig. 61 for characteristic curves). Figure 60 shows circuits for shunt motor with internal connections of starting box.

Adjustable speed shunt motors are standard for 2 to 1, 3 to 1, and 4 to 1 speed range at increase in price. For greater speed range in a motor it is necessary to resort to other means than field-rheostat control, largely because of armature reaction effects. One scheme used successfully is to shift the armature axially, giving a flux

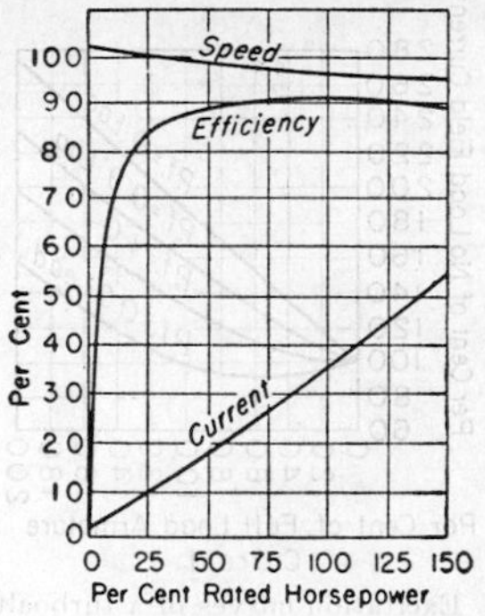

FIG. 61. Typical characteristics of shunt motor.

change through the armature. In this type of motor the commutating poles are offset from the center line of the main poles, so that, with the weak main field and hence greater armature reaction, the effect of the commutating poles is increased, giving good commutation over a large speed range. One of the best arrangements for increased speed range and good speed regulation is to provide independent voltage supplies for the field and armature. With constant voltage applied to the field, the motor speed will vary directly with the voltage applied to the armature. This arrangement requires a direct-current generator for each motor but it is quite common for applications requiring a wide speed range, such as rolling mills and elevators.

Armature circuit resistance may be used to *reduce the speed* of the motor with the load. Aside from the cost of the rheostat, the losses are relatively large, the regulation poor, and the speed decrease at light load is negligible. For a given decrease in speed, the rheostat loss is the same proportion of the input; *i.e.*, with 40 per cent speed reduction, the loss is 40 per cent of the input.

Potentiometer control is very convenient for obtaining reduced speed from small motors where the accompanying losses are not excessive. Referring to Fig. 63, the resistance R_1 and R_2 are proportioned to give the desired voltage to the armature. This reduced voltage will vary somewhat with load on the motor; the

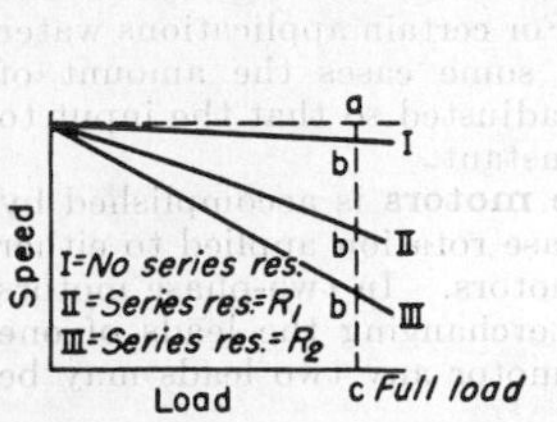

FIG. 62. Speed-torque characteristics of shunt motor with armature resistance.

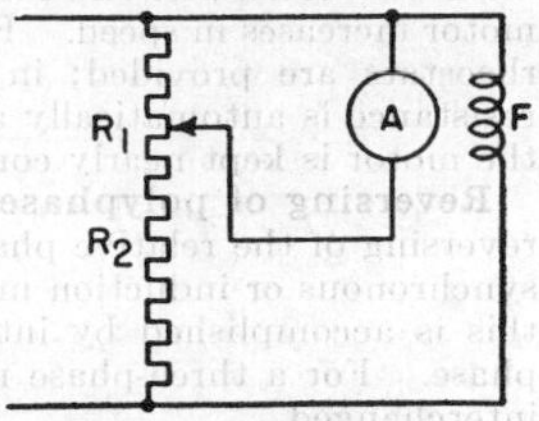

FIG. 63. Potentiometer control to operate small motor at low speed.

larger the current in R_1 relative to that in A, the more constant will be the voltage across A.

Series motors are used for loads requiring heavy starting torque, and where variable speed is permissible. The armature current is also the field current; hence the torque is approximately proportional to the current squared. The speed varies nearly inversely as the load current. Because of the excessive speed at no load, a series motor should not be connected to its load by a belt. The series motor is used for streetcars because of the heavy starting torque.

Compound-wound motors combine the characteristics of the shunt and series motors. The starting

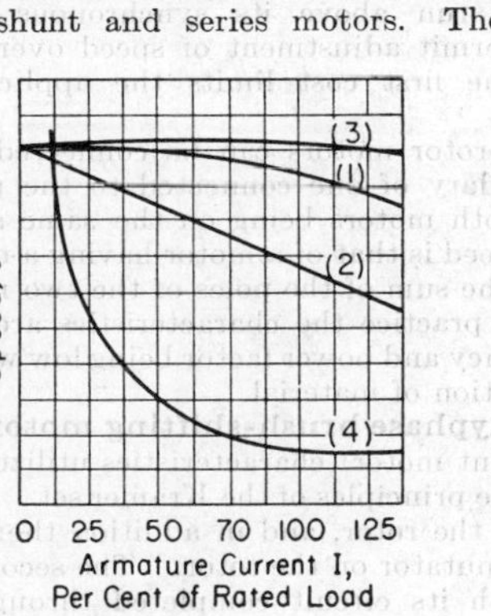

FIG. 64. Speed-load characteristics of direct-current motors. 1, shunt; 2, compound; 3, differential compound; 4, series.

torque is good with speed regulation intermediate between that of the shunt and series motors. This type of motor is frequently used with a flywheel for heavy duty of short duration, as on a punch press.

To **reverse** the direction of rotation of a direct-current motor, it is necessary to reverse the direction of current through the armature circuit with respect to the field circuits. Commutating poles are a part of the armature circuit and should always be connected to the armature in the same way. Any series-field turns must be connected so that current through them is not reversed with respect to that in the shunt field.

Alternating-current Motors. These motors may be classed as *polyphase* or *single phase* and further classed by type of connection or characteristics as *synchronous, induction, brush shifting,* etc.

The **polyphase synchronous motor** is constructed similarly to an alternator. An alternator deprived of its driving torque will continue to operate as a synchronous motor. The *speed* is fixed only by the frequency of the supply and the number of poles.

$$\text{R.p.m.} = 120\frac{f}{p}$$

where f is the frequency and p the number of poles.

Starting is facilitated by providing an amortisseur or damper winding in the pole faces of the synchronous motor. The field circuit is short-circuited through a small resistance, and low-voltage alternating current, from an autotransformer, is applied to the armature. The motor starts as an induction motor approaching

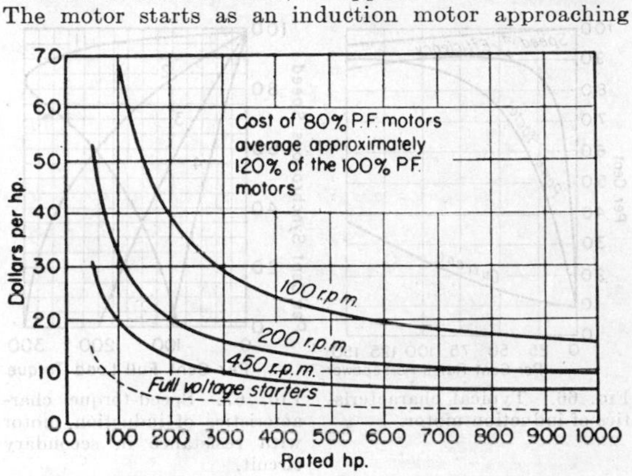

FIG. 65. Approximate prices of low-speed 100 per cent power factor synchronous motors.

synchronous speed. Direct current is applied to the field as the motor pulls in to synchronous speed, and full voltage is then applied to the armature. Modern synchronous motors have fair starting torque, and with built-in clutch any reasonable load may be started.

The **excitation adjustments** by means of a field rheostat give the most important characteristics of the synchronous motor. By increasing the field current the power factor of the motor may be made leading, and by decreasing the field current the power factor may be made lagging. Synchronous motors are frequently operated with leading power factor to compensate for the lagging current of induction motors. Purchased power usually costs less if high power factor is maintained.

Table 25. Synchronous-motor Speeds

Poles	Revolutions per minute for 60 cycles	Revolutions per minute for 25 cycles
2	3600	1500
4	1800	750
6	1200	500
8	900	375
10	720	300
12	600	250
14	514.3	214.3
16	450	187.5
18	400	166.7
20	360	150

Synchronous condensers are used to regulate voltage by means of power-factor control. The synchronous machine without mechanical loading is called a "synchronous condenser" or "synchronous phase modifier."

The **polyphase induction motor** is most useful because of its rugged construction. The **squirrel-cage** type has no moving contacts; the stator consists of a winding similar to that of the synchronous-motor armature; the rotor winding resembles a squirrel cage, bars secured to end rings. The rotor bars are sometimes bolted and soldered to the end connecting rings, sometimes welded and sometimes cast with the end rings.

The *speed characteristics* are similar to those of the direct-current shunt motor decreasing about 2 to 5 per cent from no load to full load; the no-load speed is the synchronous speed as defined under Synchronous Motors. The available speeds are limited, and there is no speed adjustment. Characteristic curves are shown in Fig. 66. It will be noted that the slip (decrease in speed as a percentage of synchronous speed) increases almost directly with the load for normal loading. Hence from a measurement of the slip and knowledge of the slip at full

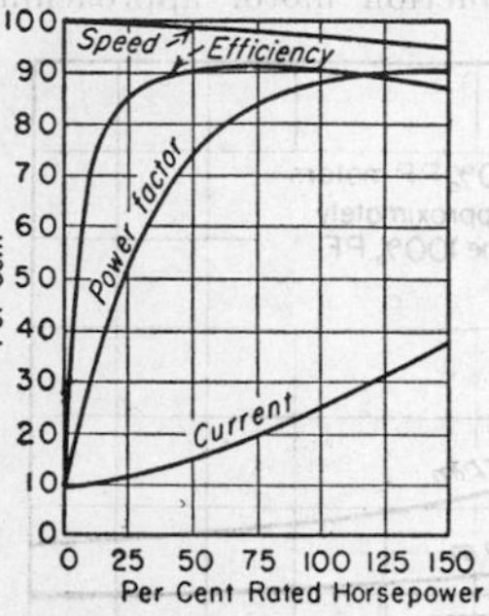

Fig. 66. Typical characteristics of induction motor.

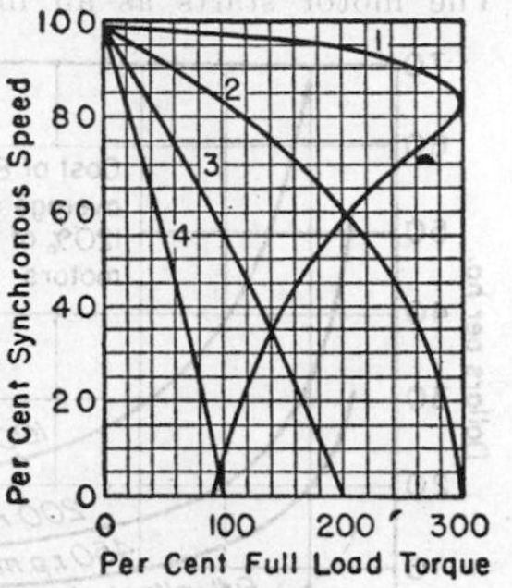

Fig. 67. Speed-torque characteristics of induction motor with resistance in secondary circuit.

load (name-plate data), the approximate load may be determined.

To reduce the current on *starting*, low voltage is applied from a starter that contains an autotransformer. The starting torque with normal voltage is not large and is reduced with voltage ($T = KE^2$ at constant speed). The starting voltage is usually about two-thirds of the normal. Small motors, below 5 hp. (the largest permissible motor depends on the system) are started on full voltage. The stalling or breakdown torque of an induction motor is usually three times that of full load.

Multispeed motors of the squirrel-cage type are obtained by reconnecting the stator windings for different numbers of poles. A single winding may be reconnected to give half the number of poles and hence twice the speed. By arranging two windings, each for two sets of poles, it is possible to have four speeds. For example, with one winding for 4 and 8 poles, the other for 6 and 12 poles, on 60 cycles the motor would have synchronous speeds of 600, 900, 1200, and 1800 r.p.m.

Special Characteristics. By making the rotor with high resistance, a motor with high starting torque and lower starting current is obtained. The regulation and efficiency are poor. Because of heating in the rotor, this type of motor is usually applied on intermittent service such as hoisting.

Low starting current may be obtained by placing the rotor bars in deep slots. This reduces the maximum torque as well.

By arranging *two squirrel-cage windings*, one of high reactance, the other of high resistance, it is possible to obtain high starting torque, low starting current, and good speed regulation.

Wound rotors or **phase-wound rotors** for induction motors permit varying the resistance of the rotor circuit. Increased resistance of the rotor circuit reduces the starting current and improves the power factor at starting, and, up to a given amount, the starting torque is increased. If the resistance is left in the circuit, the regulation is poor. However, this permits speed reduction with load at reduced efficiency. This method of operation is exactly analogous to the operation of the shunt motor with resistance in the armature circuit. By removing the resistance, the speed is increased giving a characteristic similar to that of the squirrel-cage motor. The controller provides for starting the motor with the maximum resistance and then reducing the amount as the motor increases in speed. For certain applications water rheostats are provided; in some cases the amount of resistance is automatically adjusted so that the input to the motor is kept nearly constant.

Reversing of polyphase motors is accomplished by reversing of the relative phase rotation applied to either synchronous or induction motors. In two-phase motors this is accomplished by interchanging the leads of one phase. For a three-phase motor any two leads may be interchanged.

Speed Control. Inherently, an induction motor is a constant-speed motor. Its speed is dependent on the frequency. Voltage change will affect the speed a little, particularly in a motor with a high-resistance rotor; however, the effect is in general so small (nil at light loads) that it is not used. Pole changing and the wound-rotor motor with resistance have already been discussed.

The energy consumed in the resistor of the wound-rotor motor reduces the efficiency because it is wasted. This energy can be supplied to an alternating-current commutator motor (or converter), helping to drive the main motor (Kramer system) or to drive an induction generator feeding back to the electrical supply (Scherbius system). By suitable control devices, this commutating machine can also be used as a generator, causing the main motor to run above its synchronous speed. These devices permit adjustment of speed over a wide range, though the first cost limits the application to large motors.

Wound-rotor motors can be connected in series, *i.e.*, the secondary of one connected to the primary of the second, both motors being on the same shaft. The resultant speed is that of a motor having a number of poles equal to the sum of the poles of the two motors actually used. In practice the characteristics are not desirable, the efficiency and power factor being low with uneconomical utilization of material.

The **polyphase brush-shifting motor** with *constant-speed* (shunt motor) characteristics utilizes in one motor some of the principles of the Kramer set. The primary is placed on the rotor, and in addition there is a winding with commutator on the rotor. The secondary is on the stator with its circuit completed through two sets of brushes on the commutator. When the brushes are together, the operation is similar to an induction motor with short-circuited secondary. As the brushes are spread apart, an e.m.f. is introduced in the secondary corresponding to an IR drop; however, this e.m.f. is constant, *i.e.*, independent of the current flowing. With this e.m.f. introduced, the speed is reduced at no load as well as with load. Increasing the brush shift increases the speed reduction. Reversing the brush shift reverses the e.m.f. and permits operation above synchronous speed.

A polyphase brush-shifting motor with *series-motor characteristics* is also available. The stator windings are connected in series with the brushes on the commutating winding. By shifting the brushes the speed can be adjusted.

Single-phase motors are ordinarily used in fractional-horsepower sizes for many laborsaving devices. In larger sizes they are used only where a polyphase power supply is not conveniently available. Installations of single-phase motors of larger than 7½ hp. are unusual.

The **single-phase induction motor** has no torque at standstill; the starting torque must be supplied by some auxiliary device. A polyphase induction motor, after being started, will continue to run with only one phase connected to the supply. The available output is reduced, and hence the utilization of material is poorer; yet the power factor and efficiency for the smaller output are nearly as high as for the polyphase machine.

Split-phase starting utilizes an auxiliary winding which is disconnected after starting, usually by a centrifugal switch. This auxiliary circuit has the phase of

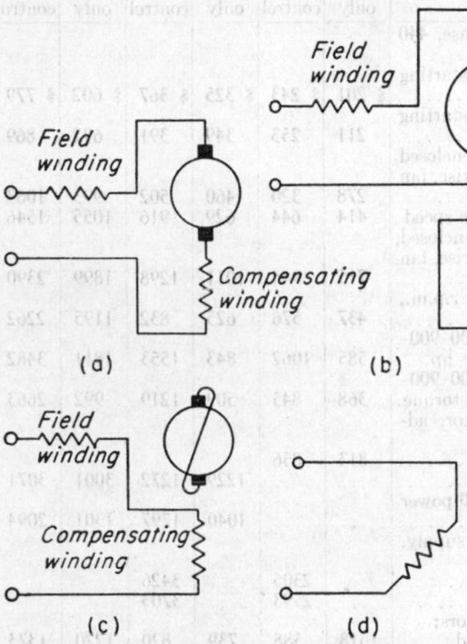

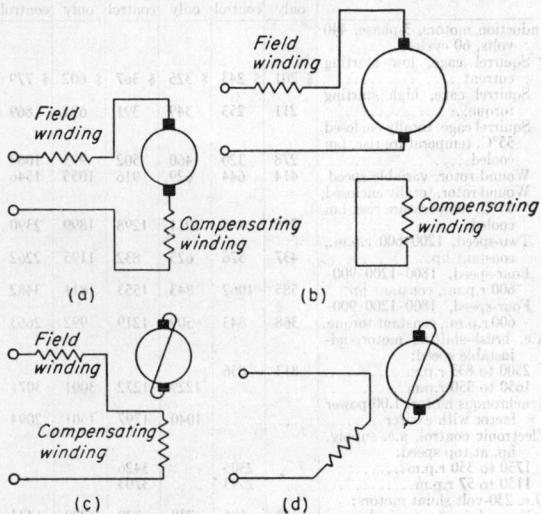

FIG. 68. Single-phase motors with series characteristics. (*a*) Series motor conductively compensated; (*b*) series motor inductively compensated; (*c*) and (*d*) repulsion motors.

its current displaced from that in the main winding by a series resistance or a condenser. These displaced currents are similar to those in a polyphase motor producing a small torque.

The **repulsion-start** induction motor utilizes a commutating winding on the armature with the brushes short-circuited giving torque as a repulsion motor. As the motor accelerates, a centrifugal device operates at about three-fourths speed to short-circuit the entire commutator. This makes the armature, in effect, a squirrel-cage rotor and the motor operates as an induction motor.

Shading poles are used for starting small induction motors such as fans. A section of the pole area is enclosed by a low-resistance band. As the flux starts to change, an e.m.f. is induced in this shading band, the resulting current retarding the change of flux. Hence the flux in the shaded portion of the pole occurs after that in the unshaded portion; this phase displacement is sufficient to react on the squirrel-cage rotor and give a starting torque.

Series motors for operation on single-phase circuits require a completely laminated magnetic structure. To reduce sparking, only one armature turn is connected between commutator segments, resulting in a large number of segments. To reduce the reactance of the armature circuit a compensating winding is placed in slots in the pole faces; this winding may be short-circuited (inductive) (Fig. 68*b*), or in series with the armature (conductive) (Fig. 68*a*). To reduce the reactance of the field, fewer field turns are used compared with a normal direct-current motor, and more turns are needed on the armature. As a result the poles appear shorter and the armature larger than for the direct-current motor. Preventive resistance leads are frequently connected between the armature coils and the commutator segments to reduce sparking. The single-phase series motor is quite satisfactory on direct current, and this type of motor is used as a universal motor. In the larger sizes the voltage applied when operated on direct current should be lower than when operated on alternating current.

Table 26. Approximate Full-load Current for Motors*

Horsepower	Single-phase, 220 volts	Three-phase				Direct current, 230 volts
		Squirrel cage		Wound rotor		
		220 volts	2200 volts	220 volts	2200 volts	
⅙	1.67					
¼	2.4					
½	3.5	2.5	..	.....	....	2.3
¾	4.7	2.8	..	.....	....	3.3
1	5.5	3.3	..	3.9	....	4.2
1½	7.6	4.7	..	.....	....	6.3
2	10	6	..	7.2	....	8.3
3	14	9	..	10	....	12.3
5	23	15	..	15	....	19.8
7½	34	22	..	25	....	28.7
10	43	27	..	28	....	38
15	.....	38	..	45	....	56
20	.....	52	..	56	....	74
25	.....	64	7	67	7.5	92
30	.....	77	8	82	9	110
40	.....	101	10	106	11	146
50	.....	125	13	128	14	180
60	.....	149	15	150	16	215
75	.....	180	19	188	19	268
100	.....	246	25	246	25	357
125	.....	310	32	310	32	443
150	.....	360	36	364	37	
200	.....	480	49	490	52	

Note. For motors of other voltages the current is inversely proportional to the voltage. Current for two-phase motors (four-wire) is about 87 per cent of the three-phase value.

* Compiled from the National Electrical Code.

In the series motor (Fig. 68*b* and *c*) the current in the compensating winding was produced by transformer action from the armature. In a similar manner the current in the armature may be induced by transformer action from the compensating field as in Fig. 68*c*. This type of motor is the **repulsion motor,** and a more suitable voltage may be obtained for the armature through the transformer action. A more common connection for the repulsion motor is shown in Fig. 68*d*, where the main field and compensating winding are combined into one. By changing the brush position, the relative amount of field and compensation component may be changed. This method of operation is common for starting single-phase induction motors.

Modifications for the series and repulsion motors have been developed with power supplied in various ways to the armature and field, some using auxiliary brushes; certain combinations give improved power factor, speed modifications, etc. The **repulsion-induction motor** is provided with two sets of windings on the armature, one with commutator giving repulsion characteristics, and, at the same time, the other, a squirrel-cage, giving induction-motor characteristics.

Reversing rotation of single-phase motors of the induction type is accomplished through the starting device. With a split-phase motor the connections to the starting winding may be reversed. In the repulsion motor the brushes are shifted. Shading-pole motors are built for one rotation only. Series motors are reversed by reversing the armature with respect to the field.

Condenser-type induction motors approximate polyphase characteristics in a motor operated from a single-phase supply. The motor is essentially a two-phase motor, one phase being supplied directly from the line, the other phase having a condenser in series with it, thus producing a split-phase effect. By selecting the proper condenser this circuit may be used throughout the operating range of the motor, with resulting good characteristics. The starting torque is small; for larger starting torque, additional capacitance may be connected and then disconnected as the motor comes up to speed.

SUMMARY OF MOTOR APPLICATIONS

The available power supply is the first consideration. The commercial supply of power is usually three phase, 60 cycles, and the alternating-current data given apply to

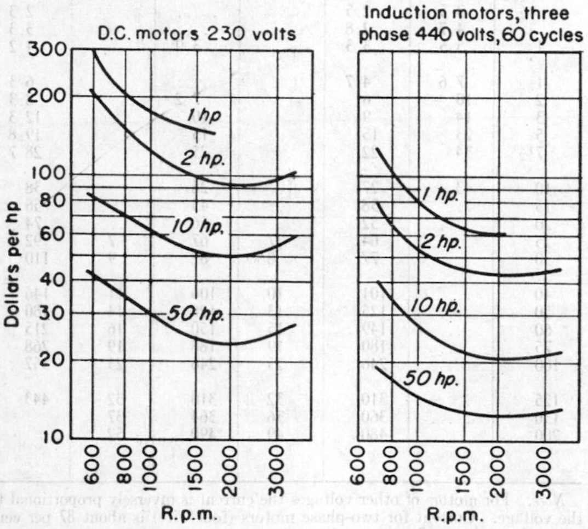

FIG. 69. Approximate prices of motors with manual or minimum starting devices.

this type of equipment; two-phase motors have similar characteristics and costs. Direct current is desirable for many drives because of the convenience of adjusting speeds. However, with an alternating-current supply the cost of converting to direct current (especially for one or two motors) may be so great as to make direct-current motors impractical. With an alternating-current supply of 2300 volts, motors larger than 75 hp. may be operated at the supply voltage to eliminate cost of transformers.

Single-phase motors are used in small sizes where direct current or polyphase power is not conveniently available.

For **constant speed** the direct-current shunt motor, the induction motor, or the synchronous motor may be used. For *large starting torque*, the induction motor should have either a high-resistance squirrel-cage rotor (which causes more change in speed with load) or a wound rotor with resistance control. The synchronous-motor speed is the only one that does not change slightly with load. The direct-current shunt-motor speed may be kept constant by field control.

For **adjustable speed** the direct-current shunt motor is preferable because its change of speed with load is small, and the speed may be changed, at will, over a wide range. Similar characteristics are obtainable from the polyphase brush-shifting motor, though the change in speed with load is somewhat larger than with the direct-current motor. Where two or more fixed speeds are desired, the multispeed induction motor can be applied. For fairly constant load the wound-rotor induction motor with resistance can be utilized to give reduced speed.

For **heavy starting torque** or **intermittent heavy loads** the series direct-current motor may be used; if less speed variation is desired the compound direct-current motor should be applied. The wound-rotor motor with

Table 27. Typical Prices of Motors without and with Common Starting or Control Devices
(Rated speeds about 1150 r.p.m.)

Type	10 hp.		20 hp.		50 hp.	
	Motor only	With control	Motor only	With control	Motor only	With control
Induction motors, 3-phase, 440 volts, 60 cycles:						
Squirrel cage, low starting current	$ 201	$ 243	$ 325	$ 367	$ 602	$ 779
Squirrel cage, high starting torque	211	253	349	391	692	869
Squirrel cage, totally enclosed, 55°C. temperature rise, fan cooled	278	320	460	502	993	1063
Wound rotor, variable speed	414	644	629	916	1055	1546
Wound rotor, totally enclosed, 55°C. temperature rise, fan cooled	715	945	1011	1298	1899	2390
Two-speed, 1200–600 r.p.m., constant hp	437	576	625	832	1195	2262
Four-speed, 1800–1200–900–600 r.p.m., constant hp	585	1062	843	1553	1811	3482
Four-speed, 1800–1200–900–600 r.p.m., constant torque	368	845	509	1219	992	2663
A.c. brush-shifting motor, adjustable speed:						
2500 to 833 r.p.m	813	856				
1650 to 550 r.p.m	....	....	1229	1272	3001	3071
Synchronous motor, 1.00 power factor with exciter	....	....	1040	1797	1301	2094
Electronic control, a.c. supply, hp. at top speed:						
1750 to 350 r.p.m	....	2305	....	3426		
1150 to 57 r.p.m	....	2593	....	3703		
D.c. 230-volt shunt motors:						
Normal constant speed	518	588	739	820	1220	1373
Armature and field control, 1400 to 650 r.p.m	518	741	739	1038		
Constant speed, totally enclosed, 55°C. temperature rise, fan cooled	870	940	1242	1323	2157	2310
Field control, 1500 to 500 r.p.m., totally enclosed, 55°C. temperature rise, fan cooled (hp. at 750 r.p.m.)	1620	1821	2340	2552	4070	4454

resistance can be similarly applied, though the losses will be greater. The double-rotor squirrel-cage motor supplies heavy starting torque, but for heavy intermittent loads the demand on the system would be great.

The **price** of a motor and its control equipment should always be considered in terms of the usefulness of the particular characteristics. For example, assume a motor delivering 10 hp. 2500 hr. per year. The cost of energy would be about $400. The motor cost might be only $200 if constant speed were permissible, or $700 if close adjustment of speed over a wide range were desired. The fixed charges on this difference in cost taken at 20 per cent would be $100, 25 per cent of the cost of the energy. Assuming, however, that the power cost was about 4 per cent of the total manufacturing cost, this additional cost of the adjustable-speed motor would be economical if it reduced the total manufacturing cost or increased the value of product by more than 1 per cent. It must be assumed further that suitable electrical supply is available for either motor. In general, the *first cost* of suitable control in electrical apparatus must be considered in terms of operating value.

SECTION 28

ELECTROCHEMISTRY

BY

C. L. Mantell, Ph. D., Consulting Chemical Engineer, New York; Member American Institute of Mining and Metallurgical Engineers, American Institute of Chemical Engineers; Author of "Industrial Electrochemistry."

CONTENTS

	Page
Electrochemical Industries	1772
Electrochemical Units	1772
Standard Cells	1772
Electrochemical Laws	1772
Electrochemical Equivalents	1775
Coulometers	1775
Electrolytic Dissociation	1780
Conductivity	1782
Conversion Data	1781
Electrochemical Energy	1782
Polarization, Overvoltage, and Passivity	1786
Polarization	1786
Electrolytic Rectifiers	1787
Overvoltage	1787
Passivity	1788
Superimposed Alternating Current on Direct Current	1788
Primary and Secondary Cells	1788
Dry Cells	1790
Secondary Cells	1792
Electrolysis and pH Measurement	1793
Electroanalysis	1793
pH and Its Measurement	1793
Electrolytic Hydrogen and Oxygen Production	1795
Hydrogen and Oxygen Production	1795
Oxidation and Reduction Products	1796
Electroplating and Electrotyping	1796
Electroplating on Aluminum	1798
Plating 3S and Strong Alloys	1799
Plating Castings	1799
Electrodeposition of Rubber	1799
Electrotyping	1800
Electrolytic Cleaning and Pickling	1800
Electrolytic Polishing	1801
Electrolytic Refining	1801
Copper	1801
Lead	1802
Nickel	1803
Tin	1804
Silver	1804
Gold	1805
Bismuth	1805
Electrowinning	1805
Copper	1805
Zinc	1805
Cadmium Recovery	1806
Alkaline Chloride Electrolysis	1806
Aqueous Electrolysis—Chlorine and Caustic	1806
Mercury Cells	1806
Diaphragm Cells	1807
Operating Data	1810
Fused Electrolytes	1810
Aluminum	1810
Magnesium	1811
Beryllium	1811
Calcium	1811
Cerium	1811
Lithium	1811
Sodium	1812
Potassium	1812
Barium	1812
Electrothermics	1812
Electric Furnaces	1812
Electric Steel Furnaces	1813
Ferroalloys	1816
Electric Furnaces for Non-ferrous Metals	1817
Electric-furnace Products	1818
Calcium Carbide	1818
Cyanamid	1818
Silicon Carbide	1818
Boron Carbide	1818
Graphite	1819
Fused Alumina	1819
Silicon	1819
Fused Silica	1819
Fused Quartz	1819
Carbon Bisulfide	1819
Phosphorus	1820
Zinc	1820
Gaseous Electrothermics	1820
Nitrogen Fixation	1820
Ozone	1820
Materials of Construction for Electrochemical Processes	1820
Operating Characteristics	1821
Amorphous Carbon and Graphite Electrodes	1822
Power for Electrochemical Processes	1824
Energy Consumption of Electrochemical Products	1825
Power Costs	1825

ELECTROCHEMISTRY

References: For general references to electrochemical processes and engineering, the reader is referred to Mantell, "Industrial Electrochemistry," 3d ed., McGraw-Hill, New York, 1950. For theoretical electrochemistry, reference should be made to Dole, "Principles of Experimental and Theoretical Electrochemistry," McGraw-Hill, New York, 1935; Foerster, "Elektrochemie wässeriger Lösungen," 4th ed., Barth, Leipzig, 1923; Walden, "Elektrochemie nichtwässeriger Lösungen," Barth, Leipzig, 1924; and Lorenz, "Elektrochemie geschmolzener Salze," Barth, Leipzig, 1909. None of the latter books is available in English. For plating, reference should be made to Blum and Hogaboom, "Principles of Electroplating and Electroforming," 3d ed., McGraw-Hill, New York, 1949; for calcium carbide to Taussig, "Die Industrie des Kalziumkarbides," Knapp, Halle, 1930. For the individual metals, reference should be made to the recent metallurgical works on this particular topic.

Electrochemical Industries. Electrochemistry has been defined as the science which treats of the chemical changes produced by the electric current and of the production of electricity from the energy of chemical reactions. Theoretically the two branches are of equal importance. Industrially, however, the chemical and physical changes produced by the use of the electric current are by far the more important.

Electrochemical engineering is primarily a branch of chemical engineering to which portions and viewpoints of electrical engineering and metallurgy have been joined. Electrochemical engineering deals not only with all the electrochemical theories, processes, and operations but also with the furnishing and utilization of electrical power to the industries; the design, construction, and operation of the equipment, machinery, and plants employed to produce the electrochemical products; the economic considerations involved in the competition of chemical and electrochemical methods for the preparation of the same or similar products, as well as the sale, distribution, and consumption of the materials produced.

Electrochemical industries may be roughly divided into several classes or groups, as shown in Fig. 1. Each group and subdivision has problems and applications of the chemical and electrical engineering unit processes but with the emphasis placed differently in the different groups.

The scope of the electrochemical engineering industries is very wide. To illustrate the breadth of the industries, a list of products of the electric furnace and electrolytic cell, at least as far as the major materials are concerned, is given (see Table 1).

Electrochemical processes of an endothermic (absorption of energy) nature have frequently supplanted purely chemical processes and in some cases have allowed the production of new products that could hardly be obtained in any other way. Thus copper is now almost entirely refined by electrochemical means. All the chlorine used for water purification, sanitation, and bleaching is the product of electrolytic cells. Aluminum can be commercially made only by fused salt electrolysis. Calcium carbide and the synthetic abrasives of the silicon carbide or fused alumina type are not possible by other than electrothermal methods.

Chemical reactions often are made to take place by a series of steps because the most direct processes cannot be used and because of the difficulty of conversion of thermal into chemical energy. In electrochemical processes the needed energy is introduced in an electrical rather than a thermal form. When the electric current is used for heating, it may be applied at the point where it is desired. Electrochemical processes are more direct than the corresponding chemical ones. Sometimes the electrochemical method is more expensive but is preferred, inasmuch as purer products are produced.

Electrochemical processes generally operate satisfactorily only under constant conditions. They must be as simple as possible. Raw materials should be as pure as can be obtained or manufactured within the economic limitations of the process. Accumulation of impurities causes rapid decreases in efficiency.

Electrochemical Units. The **ampere** is defined electrochemically as the unvarying electric current which will deposit silver at the rate of 0.00111800 g./sec. from a solution of $AgNO_3$ in water under a given set of conditions. The current in amperes flowing across any point, divided by the cross-sectional area of the conductor, is called the **current density** (c.d.) (amperes per square foot or amperes per centimeter square, etc.). So-called **c.d.** meters used for control purposes in electroplating consist of cathode surfaces of definite area, connected in series with an ammeter, the mounting and connections being so arranged that the meter appears above the level of the bath and the whole apparatus can be hung on the cathode rod. The meter thus indicates amperes per unit of area.

The **coulomb** or **ampere-second** is that quantity of electricity which will deposit 0.00111800 g. silver from a solution of $AgNO_3$.

The **volt** in practical use is that of the International System of Electrical Units. In this system the volt cannot be easily produced as defined, because of the definition of the ampere. The e.m.f. of a voltaic cell, however, can be determined against the international ohm and the international ampere, and such a cell is used as a medium for realizing the international volt.

Standard cells have provided the means for making comparisons of e.m.f. The cells so used may be divided into primary standards, or normal cells, and secondary standards. The first are those by means of which the value of the volt is maintained, as at the National Bureau of Standards or elsewhere. The second are those suitable for general laboratory use.

The Weston standard cell consists of an amalgam of cadmium, a solution of cadmium sulfate having a concentration corresponding to that of a solution saturated at 4°C., and pure mercury overlaid with Hg_2SO_4. This combination has a very low temperature coefficient and is constant when properly made. However, it is not reproducible to the degree required in a primary standard. The "normal Weston" or "normal cadmium" cell, having an excess of cadmium sulfate, is therefore the standard maintained at the various government laboratories upon which rests the duty of establishing the volt. Its e.m.f. is taken as 1.0183 volts at 20°C. by international agreement. These cells are reproducible to better than 10 microvolts.

Electrochemical Laws. Conductors of electricity may be sharply divided into three classes. The first class, the *metallic* or *electronic* conductors, consists of the metals, alloys, and a few other substances such as carbon.

Table 1. Products of Electric Furnace and Electrolytic Cell

Product	Raw material	Applications of product
Alumina, fused	Bauxite (natural aluminum oxide)	Abrasives and refractories
Alumina, pure	Bauxite	Insulating material; aluminum metal
Aluminum metal	Bauxite	Electric power transmission cable; lightweight alloys for airplanes, automobiles, and trucks; deoxidizing agent for steel; alumino-thermic reactions; ammonal (explosives); acid containers; cooking utensils
Aluminum, pure	Aluminum metal	Corrosion-resistant coatings
Ammonium persulfate	Ammonium bisulfate solution	Oxidizing and bleaching agent
Anthraquinone	Oxidation of anthracene	Chemicals; dyestuffs
Antimony	Antimony ores	Alloying agent
Barium	Fused barium chloride	Alloys; electron emission
Beryllium	Beryl	Light alloys
Bismuth	Lead refining slimes	Alloys
Boron carbide	Anhydrous boric acid and coke	Abrasives
Cadmium	Zinc electrowinning slimes	Alloys; plating
Calcium	Calcium chloride	Alloying agent; radio tubes; lamps; special uses
Calcium carbide	Lime and coke	Acetylene for welding, cutting, and lighting; acetone, acetic acid
Calcium cyanamid	Calcium carbide (nitrogen of the air)	Fertilizer; ammonia; cyanides
Carbon bisulfide	Coke and sulfur	Solvent; insecticide; carbon tetrachloride; artificial silk (viscose)
Caustic	Water, salt	Soap; paper industry; explosives
Cerium metals	Rare-earth chlorides	Pyrophoric alloys; automatic lighters; tracer bullets and shells
Chlorine gas	Water, salt	Bleaching; gas warfare; mustard gas, phosgene, chlorpicrin, silicon tetrachloride; explosives; chlorbenzol; water purification; surgery (Dakin solution); detinning; artificial plastics; hydrochloric acid; aluminum chloride for oil refining; sanitation
Chrome yellow	Lead	Paint pigment
Chromic acid	Oxidation of chromium sulfate solution	Chromium plating; chemicals
Chromium	Complex chromium ores	Alloying agents
Chromium	Chromic acid and sulfuric acid solution	Plating; alloys
Cobalt	Complex cobalt ores	Alloying agent
Copper, pure	Copper ore; crude copper	Electrical and brass industries
Cuprous oxide	Copper	Paint pigment
Deuterium	Water	Chemical for scientific use
Ferrochrome	Chrome ore	Special and high-speed steels; armor plate; projectiles
Ferrocolumbium	Residues from tin ores	Alloys; addition agents to steel; stainless steels
Ferromanganese	Manganese ore and coke	Steel; permanganates
Ferromolybdenum	Molybdenum ore	Special steels
Ferrosilicon	Iron, silica rock, coke	Steel manufacture; hydrogen production
Ferrosilicon-titanium	Bauxite	Steel deoxidizer
Ferrotitanium	Titanium ore	Scavenger in steel manufacture
Ferrotungsten	Tungsten ore	Special and high-speed steels
Ferrovanadium	Iron vanadate	Special steels; automobile steels
Fluorine	Fused potassium acid fluoride; anhydrous HF	Chemical reactant
Gold	Copper refining slimes	Jewelry; coinage; industrial alloys
Graphite	Coal and coke	Electrodes; lubricants; paints
Hydrogen	Water, sodium hydroxide	Ballooning; hydrogenated fats
Hydrogen peroxide	Sulfuric acid solution	Chemicals; antiseptic; bleaching and oxidizing agent
Hypochlorite	Water, salt	Disinfectants; bleaches
Indium	Indium oxide in sulfuric acid solution	Nontarnishable silver alloy; jewelry; television
Iodoform	Alcohol, potassium iodide, sodium carbonate solution	Antiseptic; disinfectant
Iron, electrolytic	Ferrous ammonium sulfate solution	Electromagnetic purposes, powder metallurgy
Iron, pig	Iron ore	Steel industry
Iron, pure or "Swedish"	Pyrrhotite	Tubes and special steels
Iron powder	Iron ores or ferrous sulfate	Powder metallurgy
Lead refined	Crude lead	Alloys; fittings; acid chambers
Lithium metal	Lepidolite, lithium salts	Light alloys
Magnesium metal	Magnesium chloride	Flash-light powders; lightweight alloys; tracer bullets and flares
Manganese, electrolytic	Mn ore	Alloying agent; deoxidizer
Mercuric oxide	Mercury	Catalyst and chemical
Nickel, refined	Crude nickel	Alloys; plating industry; dairy equipment; utensils
Nickel powder	Nickel	Powder metallurgy
Nitric acid	Air	Explosives; fertilizers
Oxygen	Water, sodium hydroxide	Oxy-welding; oxy-cutting
Ozone	Air	Sterilization of water; sanitation
Palladium	Nickel refining slimes	Industrial alloys
Para-aminophenol	Reduction of nitrobenzene	Chemicals; photographic developer; dyes
Perborates	Borax	Bleaching agents for textiles
Perchloric acid	Hydrochloric acid	Salts of perchloric acid
Phosphoric acid	Phosphate rock, coke, and sand	Acid phosphates; cleaners; food products
Phosphorus	Phosphate rock, coke, and sand	Matches; phosphorus compounds; phosphor bronze; smoke screens
Platinum	Copper refining slimes	Electrical uses
Platinum	Nickel refining slimes	Catalysts; jewelry; industrial alloys
Potassium chlorate	Potassium chloride	Primers; matches; dyeing
Potassium hydroxide	Potassium chloride	Soap; chemicals; mercerizing cotton; electroplating
Potassium perchlorate	Sodium chlorate solution and converted to potassium salt	Oxidizing agent; explosives; medicine
Potassium persulfate	Potassium sulfate solution	Oxidizing agent, bleaching

Table 1. Products of Electric Furnace and Electrolytic Cell—(*Concluded*)

Product	Raw material	Application of product
Quartz, fused	Quartz rock	Silica tubes; heat-resisting materials; optical uses; lenses
Rhodium	Nickel refining slimes	Industrial alloys
Rubber	Latex containing the usual components for compounding rubber	Household and industrial rubber goods
Silicon	Sand and coke	Silicon steel; hydrogen for balloons; resistance units; silicides; silicon tetrachloride
Silicon carbide	Sand, sawdust, and coke	Abrasives and refractories
Silver	Copper refining slimes	Jewelry; coinage; industrial alloys
Solder	Impure tin-lead alloys	Metal joining
Sodium bichromate	Chromium salts	Dyeing; tanning
Sodium metal	Caustic (Castner)	Peroxides; cyanides; bleaching; mining
Sodium metal	Salt	Alloys; tetraethyl lead; organic synthesis
Sodium perchlorate	Sodium salts, $NaClO_3$	Fireworks
Sorbitol	Glucose	Humectants
Tantalum	Potassium tantalum fluoride	Acid-resistant lining for chemical equipment
Thorium	Fused potassium thorium fluoride	Electron emission; X-ray targets
Tin, refined	Impure tin, tin dross	Tin-plate industry, bronzes
Tungsten	Sodium tungstate solution	Alloys; plating; electric bulbs
Uranium	Fused potassium uranous fluoride	Alloys
White lead	Lead	Paint pigment
Zinc, pure	Zinc ore	Brass; galvanizing
Zinc metal, pure	Zinc ore	Brass industry

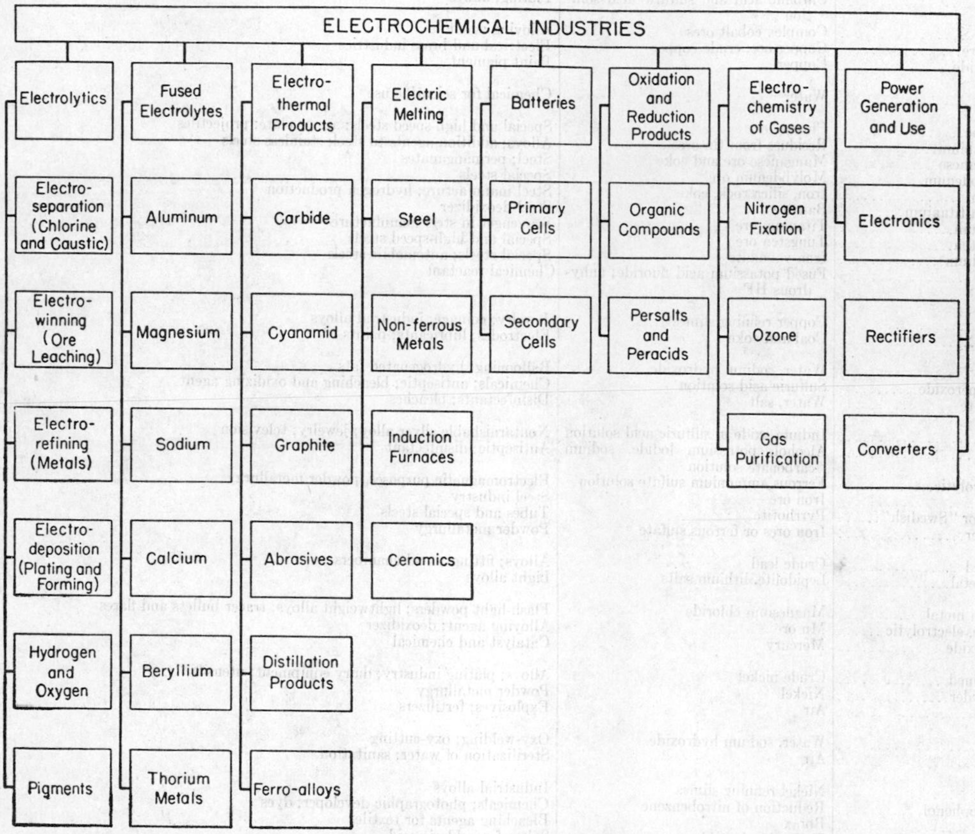

FIG. 1. Classification of the electrochemical industries.

The current passes through these without the accompaniment of any quantity of matter. Those of the second class are termed *electrolytic* conductors. They embrace in general the solutions of acids, bases, and salts, fused salts, some solid substances, and hot gases. In electrolytic conduction the passage of the current is always connected with a movement of matter. When the current leaves the electrolyte, it cannot take the matter with it; the latter is consequently set free. Chemical effects are produced. These mark the chief distinction between metallic or electronic and electrolytic conduction. In a third class with *mixed* conductors, the current passes partly in a metallic and partly in an electrolytic manner. Examples of these are the β form of silver sulfide and the solutions of the alkali and alkaline-earth metals in liquid ammonia. The greater part of electrochemistry is concerned with the chemical effects resulting from electrolytic conduction, and the corresponding electrical effects necessary for their production.

If two carbon plates are dipped into a dilute solution of HCl and connected with a source of direct current such as a battery, electrolysis will take place. At the negative carbon plate, hydrogen gas is given off; while at the positive plate, chlorine is evolved. The gases are produced only at the carbon plates. The plates are termed *electrodes*, the negative the *cathode* and the positive one the *anode*. If a $AgNO_3$ solution be electrolyzed, silver is deposited on the cathode and oxygen evolved at the anode. Whatever the solution, chemical action takes place only at the electrodes.

Michael Faraday (1791–1867) discovered the quantitative relations between the amount of electricity that passes through a solution and the quantity of matter separated at the electrodes. His first law is: *The quantities of substances set free at the electrodes are directly proportional to the quantity of electricity which passes through the solution.* A second law expresses the fundamental relation between quantities of different substances liberated at the electrodes by the same quantity of electricity. It is: *The same quantity of electricity sets free the same number of equivalents of substances at the electrodes.* In other words, the quantity of silver liberated at the cathode by the passage of 20 coulombs electricity through a solution of a silver salt is double that which would be obtained by the passage of 10 coulombs. The passage of the same quantity of electricity through solutions of a silver salt, a copper salt, an iron salt, a zinc salt, and an acid will set free quantities of silver, copper, iron, zinc, and hydrogen which are proportional to their equivalent weights.

The neutral dissolved molecules of an electrolyte consist of two oppositely charged parts, called *ions*. Those which move toward the cathode are called *cations*, and those toward the anode, *anions*. When the current passes, the positive ions are attracted toward the negatively charged cathode where their charge is neutralized and they are set free. In a similar manner, the anions move to, and are discharged at, the anode.

The **electrochemical equivalent** of an element or a group of elements is the number of grams of that substance set free by the passage of 1 coulomb of electricity through an electrolyte. Electrochemical equivalents are proportional to chemical equivalents.

One coulomb of electricity (by definition) sets free 0.00111800 g. silver from a solution of a silver salt. If the gram equivalent of silver (its gram-atomic weight divided by its valence) is divided by the electrochemical equivalent of silver, we find that

$$107.88 \div 0.00111800 = 96{,}494 \text{ coulombs}$$

This quantity of electricity is called a **faraday** (F). It is taken as 96,500 coulombs or 26.8 amp.-hr. If 1 faraday is passed through an electrolytic conductor, 1 g. equivalent of some substance will be liberated at each electrode. If 1 faraday is passed through a solution containing several electrolytes, the summation of the quantities of the different products set free at each electrode, when expressed in equivalents, will be unity.

In practice more than 1 faraday is needed for the liberation of a gram equivalent of a substance. This is due not to the failure of Faraday's laws but to other causes. Side reactions may take place. The products of the electrolysis may suffer mechanical loss. Secondary reactions may take place at the electrodes. In addition there may be current leaks, short circuits, and losses in the form of heat. The ratio of the theoretical to the actual quantity of current used is the **current efficiency.** In a similar manner, the actual amount of a product formed from a definite amount of current, divided by the theoretical amount, also gives us the current efficiency. In commercial practice, current efficiencies may vary from as low as 25 to 30 per cent in the decomposition of certain fused salts and in chromium plating to as high as 92 to 95 per cent in copper refining and 100 per cent in the electrolytic oxidation of anthracene to anthraquinone.

The **current concentration** is the current divided by the volume upon which the current acts. If a high concentration is to be produced of some compound subject to chemical decomposition, a high current concentration is desirable.

Electrochemical Equivalents

The electrochemical equivalents of the elements are tabulated. The use of Table 2 has been simplified in that not only the ordinary valences of the elements are included but also the changes of valence which occur in oxidation-reduction reactions. For example, while iron exhibits the valence of 2 or 3, in the reduction of ferric to ferrous iron there is a valence change of 1. In Table 2, valence changes of 1, when they are not a normal valence or normal valence change of the element, are indicated as *n*. The calculations in Table 2 are based on the 1945 atomic weights. Mass and energy conversion factors allowing greater ease of use of the electrochemical equivalent tables are given in Tables 4, 5, and 6.

Coulometers. Galvanometers are used for the *detection* of current and occasionally for the *measurement* of small currents. For large currents, ammeters are generally used and time recorded. Coulometers find application for the measurement of current (in coulombs) in small-scale experimental work. Measurement is made of the electrode products due to the passage of the current. From Faraday's laws, the quantity of electricity passed through can be calculated. Coulometers depend upon operation under 100 per cent current efficiency. Side reactions must be eliminated.

Coulometers are of several classes. In the **weight coulometer** the gain in weight of the cathode of an electrolytic cell, due to the deposition of metal from a solution of its salt, is measured. **Volume coulometers** are those in which the volume of a gas liberated as the result of electrolysis, or the volume of mercury set free during electrolysis of a suitable mercury salt, is measured. In **titration coulometers** the change in concentration or the amount of a substance set free at one of the electrodes is determined by analytical methods.

The errors of coulometers are those inherent in the measurement of weight and volume or in titration, and also those due to imperfections in the coulometer itself. The silver-weight and the iodine-titration coulometers are the most accurate, partly because of the high equivalent weights of iodine and silver (1 g. Ag corresponds to 894.53, 1 g. I to 760.33 coulombs).

The silver coulometer has been studied. On the accuracy of the measurement of quantities of electricity depend not only the value of the faraday and the definition of the ampere but also the value of the e.m.f. of the normal Weston cell which is employed as a universal standard of e.m.f. The errors in the silver coulometer have been so completely eliminated that the results obtained for the e.m.f. of this cell by investigators in several countries, using three types of the coulometer, agree to about 1 part in 100,000. [See Rosa and Vinal, *Bur. Standards Bull.*, **13,** 479 (1916); *Nat. Bur. Standards Sci. Paper* 285.]

Table 2. Electrochemical Equivalents of the Elements

1 Element	Symbol	2 Atomic weight	3 Val. or chg. val.	4 Mg./coulomb	5 Coulombs/mg.	6 G./amp.-hr.	7 Amp.-hr./g.	8 Lb./1000 amp.-hr.	9 Amp.-hr./lb.
Actinium	Ac	**227***	3	0.78 4$\overline{11}$	1.27 $\overline{533}$	2.82 $\overline{280}$	0.35 4$\overline{26}$	6.22 $\overline{320}$	160.$\overline{689}$
			n	2.35 $\overline{233}$	0.42 5$\overline{11}$	8.46 $\overline{839}$	0.11 8$\overline{09}$	18.6$\overline{6}$ $\overline{961}$	53.$\overline{563}$
Alabamine	Ab	**221***	7	0.32 7$\overline{17}$	3.05 $\overline{656}$	1.17 $\overline{779}$	0.84 9$\overline{04}$	2.59 $\overline{656}$	385.$\overline{120}$
			5	0.45 $\overline{803}$	2.18 $\overline{326}$	1.64 $\overline{891}$	0.60 6$\overline{46}$	3.63 $\overline{523}$	275.$\overline{086}$
			3	0.76 $\overline{339}$	1.30 $\overline{995}$	2.74 $\overline{819}$	0.36 $\overline{388}$	6.05 $\overline{871}$	165.$\overline{052}$
			2	1.14 $\overline{508}$	0.87 3$\overline{30}$	4.12 $\overline{228}$	0.24 2$\overline{58}$	9.08 $\overline{807}$	110.$\overline{034}$
			1	2.29 $\overline{016}$	0.43 6$\overline{65}$	8.24 $\overline{456}$	0.12 1$\overline{29}$	18.1$\overline{7}$ $\overline{614}$	55.0$\overline{17}$
Aluminum	Al	**26.97**	3	0.09 316	10.73 $\overline{415}$	0.33 53$\overline{8}$	2.98 1$\overline{71}$	0.73 93$\overline{8}$	1352.$\overline{480}$
			n	0.27 94$\overline{8}$	3.57 8$\overline{05}$	1.00 6$\overline{13}$	0.99 39$\overline{0}$	2.21 8$\overline{15}$	450.82$\overline{7}$
Antimony	Sb	**121.76**	5	0.25 235	3.96 27$\overline{2}$	0.90 847	1.10 07$\overline{5}$	2.00 28$\overline{3}$	499.29$\overline{4}$
			3	0.42 059	2.37 76$\overline{3}$	1.51 41$\overline{1}$	0.66 045	3.33 80$\overline{5}$	299.57$\overline{6}$
			2	0.63 088	1.58 50$\overline{9}$	2.27 11$\overline{7}$	0.44 030	5.00 70$\overline{7}$	199.71$\overline{7}$
			n	1.26 17$\overline{6}$	0.79 254	4.54 23$\overline{4}$	0.22 015	10.01 4$\overline{15}$	99.858
Argon	A	**39.944**	n	0.41 393	2.41 58$\overline{8}$	1.49 01$\overline{4}$	0.67 108	3.28 51$\overline{9}$	304.39$\overline{6}$
Arsenic	As	**74.91**	5	0.15 25$\overline{4}$	6.44 1$\overline{06}$	0.55 89$\overline{1}$	1.78 9$\overline{18}$	1.23 2$\overline{19}$	811.5$\overline{60}$
			3	0.25 87$\overline{6}$	3.86 4$\overline{64}$	0.93 15$\overline{2}$	1.07 3$\overline{51}$	2.05 3$\overline{66}$	486.9$\overline{36}$
			2	0.38 81$\overline{3}$	2.57 6$\overline{42}$	1.39 7$\overline{29}$	0.71 56$\overline{7}$	3.08 0$\overline{49}$	324.6$\overline{24}$
			n	0.77 62$\overline{7}$	1.28 8$\overline{21}$	2.79 4$\overline{57}$	0.35 78$\overline{4}$	6.16 0$\overline{97}$	162.3$\overline{12}$
Barium	Ba	**137.36**	2	0.71 171	1.40 50$\overline{7}$	2.56 21$\overline{6}$	0.39 030	5.64 85$\overline{8}$	177.03$\overline{5}$
			n	1.42 34$\overline{2}$	0.70 253	5.12 43$\overline{1}$	0.19 515	11.29 71$\overline{7}$	88.518
Beryllium	Be	**9.02**	2	0.04 67$\overline{4}$	21.3$\overline{9}$ $\overline{688}$	0.16 8$\overline{25}$	5.94 $\overline{358}$	0.37 0$\overline{92}$	269$\overline{5}$.$\overline{963}$
			n	0.09 34$\overline{7}$	10.6$\overline{9}$ $\overline{844}$	0.33 6$\overline{50}$	2.97 $\overline{179}$	0.74 1$\overline{85}$	134$\overline{7}$.$\overline{981}$
Bismuth	Bi	**209.00**	5	0.43 316	2.30 86$\overline{1}$	1.55 93$\overline{8}$	0.64 128	3.43 78$\overline{4}$	290.88$\overline{0}$
			3	0.72 193	1.38 51$\overline{7}$	2.59 89$\overline{6}$	0.38 477	5.72 97$\overline{3}$	174.52$\overline{8}$
			2	1.08 29$\overline{0}$	0.92 345	3.89 84$\overline{5}$	0.25 651	8.59 46$\overline{0}$	116.35$\overline{2}$
			n	2.16 58$\overline{0}$	0.46 172	7.79 68$\overline{9}$	0.12 826	17.18 9$\overline{20}$	58.176
Boron	B	**10.82**	5	0.02 242	44.59 $\overline{337}$	0.08 073	12.38 $\overline{704}$	0.17 79$\overline{8}$	5618.$\overline{669}$
			3	0.03 737	26.75 $\overline{602}$	0.13 45$\overline{5}$	7.43 2$\overline{23}$	0.29 66$\overline{2}$	3371.$\overline{201}$
			2	0.05 606	17.83 $\overline{735}$	0.20 18$\overline{2}$	4.95 4$\overline{82}$	0.44 49$\overline{5}$	2247.$\overline{467}$
			n	0.11 21$\overline{2}$	8.91 8$\overline{67}$	0.40 36$\overline{5}$	2.47 7$\overline{41}$	0.88 98$\overline{9}$	1123.$\overline{734}$
Bromine	Br	**79.916**	7	0.11 831	8.45 26$\overline{3}$	0.42 590	2.34 79$\overline{5}$	0.93 896	1065.0$\overline{13}$
			6	0.13 802	7.24 51$\overline{1}$	0.49 689	2.01 25$\overline{2}$	1.09 54$\overline{5}$	912.86$\overline{8}$
			5	0.16 563	6.03 75$\overline{9}$	0.59 626	1.67 70$\overline{8}$	1.31 45$\overline{4}$	760.72$\overline{4}$
			4	0.20 704	4.83 00$\overline{7}$	0.74 533	1.34 16$\overline{9}$	1.64 31$\overline{7}$	608.57$\overline{9}$
			3	0.27 605	3.62 25$\overline{5}$	0.99 377	1.00 62$\overline{6}$	2.19 09$\overline{0}$	456.43$\overline{4}$
			2	0.41 407	2.41 50$\overline{4}$	1.49 06$\overline{6}$	0.67 084	3.28 63$\overline{4}$	304.28$\overline{9}$
			1	0.82 815	1.20 75$\overline{2}$	2.98 13$\overline{2}$	0.33 542	6.57 26$\overline{9}$	152.14$\overline{5}$
Cadmium	Cd	**112.41**	2	0.58 244	1.71 69$\overline{3}$	2.09 67$\overline{7}$	0.47 692	4.62 25$\overline{8}$	216.32$\overline{9}$
			n	1.16 48$\overline{7}$	0.85 846	4.19 35$\overline{3}$	0.23 846	9.24 51$\overline{6}$	108.16$\overline{5}$
Calcium	Ca	**40.08**	2	0.20 76$\overline{7}$	4.81 5$\overline{37}$	0.74 76$\overline{1}$	1.33 7$\overline{60}$	1.64 8$\overline{19}$	606.7$\overline{26}$
			n	0.41 53$\overline{4}$	2.40 7$\overline{68}$	1.49 5$\overline{21}$	0.66 88$\overline{0}$	3.29 6$\overline{38}$	303.3$\overline{63}$
Carbon	C	**12.010**	4	0.03 111	32.13 9$\overline{89}$	0.11 201	8.92 77$\overline{5}$	0.24 694	4049.5$\overline{58}$
			2	0.06 223	16.06 9$\overline{94}$	0.22 402	4.46 38$\overline{7}$	0.49 388	2024.7$\overline{79}$
			n	0.12 446	8.03 49$\overline{7}$	0.44 804	2.23 19$\overline{4}$	0.98 776	1012.3$\overline{90}$
Cerium	Ce	**140.13**	4	0.36 303	2.75 45$\overline{9}$	1.30 69$\overline{1}$	0.76 516	2.88 12$\overline{5}$	347.07$\overline{2}$
			3	0.48 404	2.06 59$\overline{4}$	1.74 25$\overline{5}$	0.57 387	3.84 16$\overline{6}$	260.30$\overline{4}$
			1	1.45 21$\overline{2}$	0.68 865	5.22 76$\overline{5}$	0.19 129	11.52 4$\overline{99}$	86.768
Cesium	Cs	**132.91**	1	1.37 73$\overline{1}$	0.72 606	4.95 83$\overline{0}$	0.20 168	10.93 1$\overline{18}$	91.481
Chlorine	Cl	**35.457**	7	0.05 248 9	19.05 1$\overline{24}$	0.18 896	5.29 20$\overline{1}$	0.41 659	2400.41$\overline{7}$
			6	0.06 123 8	16.32 9$\overline{64}$	0.22 046	4.53 60$\overline{1}$	0.48 603	2057.5$\overline{00}$
			5	0.07 348 6	13.60 8$\overline{03}$	0.26 455	3.78 00$\overline{1}$	0.58 323	1714.5$\overline{83}$
			4	0.09 185 8	10.88 6$\overline{42}$	0.33 069	3.02 40$\overline{1}$	0.72 904	1373.66$\overline{7}$
			3	0.12 248	8.16 48$\overline{2}$	0.44 092	2.26 80$\overline{1}$	0.97 205	1028.7$\overline{50}$
			2	0.18 372	5.44 32$\overline{1}$	0.66 137	1.51 20$\overline{0}$	1.45 80$\overline{8}$	685.83$\overline{3}$
			1	0.36 743	2.72 16$\overline{1}$	1.32 27$\overline{5}$	0.75 600	2.91 61$\overline{6}$	342.91$\overline{7}$
Chromium	Cr	**52.01**	6	0.08 983	10.13 $\overline{247}$	0.32 33$\overline{8}$	3.09 2$\overline{35}$	0.71 29$\overline{3}$	1402.$\overline{668}$
			4	0.13 47$\overline{4}$	7.42 1$\overline{65}$	0.48 50$\overline{7}$	2.06 1$\overline{57}$	1.06 9$\overline{39}$	935.1$\overline{12}$
			3	0.17 96$\overline{5}$	5.56 6$\overline{24}$	0.64 67$\overline{6}$	1.54 6$\overline{18}$	1.42 5$\overline{85}$	701.3$\overline{34}$
			2	0.26 94$\overline{8}$	3.71 0$\overline{82}$	0.97 01$\overline{3}$	1.03 0$\overline{78}$	2.13 8$\overline{78}$	467.5$\overline{56}$
			1	0.53 89$\overline{6}$	1.85 5$\overline{41}$	1.94 0$\overline{27}$	0.51 53$\overline{9}$	4.27 7$\overline{56}$	233.7$\overline{78}$

Table 2. Electrochemical Equivalents of the Elements—*(Continued)*

1 Element	Symbol	2 Atomic weight	3 Val. or chg. val.	4 Mg./coulomb	5 Coulombs/mg.	6 G./amp.-hr.	7 Amp.-hr./g.	8 Lb./1000 amp.-hr.	9 Amp.-hr./lb.
Cobalt	Co	58.94	3	0.20 35$\overline{9}$	4.91 17$\overline{7}$	0.73 28$\overline{7}$	1.36 45$\overline{0}$	1.61 57$\overline{0}$	618.92$\overline{5}$
			2	0.30 53$\overline{9}$	3.27 45$\overline{2}$	1.09 93$\overline{1}$	0.90 96$\overline{6}$	2.42 35$\overline{6}$	412.61$\overline{7}$
			1	0.61 07$\overline{8}$	1.63 72$\overline{6}$	2.19 86$\overline{1}$	0.45 48$\overline{3}$	4.84 71$\overline{1}$	206.30$\overline{8}$
Columbium	Cb	**92.91**	5	0.19 25$\overline{6}$	5.19 32$\overline{0}$	0.69 32$\overline{1}$	1.44 25$\overline{5}$	1.52 82$\overline{8}$	654.33$\overline{2}$
			4	0.24 07$\overline{0}$	4.15 45$\overline{6}$	0.86 65$\overline{2}$	1.15 40$\overline{4}$	1.91 03$\overline{5}$	523.46$\overline{6}$
			3	0.32 09$\overline{3}$	3.11 59$\overline{2}$	1.14 70$\overline{2}$	0.86 55$\overline{3}$	2.54 71$\overline{3}$	392.59$\overline{9}$
			2	0.48 14$\overline{0}$	2.07 72$\overline{8}$	1.73 30$\overline{4}$	0.57 70$\overline{2}$	3.82 06$\overline{9}$	261.73$\overline{3}$
			1	0.96 28$\overline{0}$	1.03 86$\overline{4}$	3.46 60$\overline{7}$	0.28 85$\overline{1}$	7.64 13$\overline{8}$	130.86$\overline{6}$
Copper	Cu	63.57	2	0.32 93$\overline{8}$	3.03 60$\overline{2}$	1.18 57$\overline{6}$	0.84 33$\overline{4}$	2.61 41$\overline{6}$	382.53$\overline{2}$
			1	0.65 87$\overline{6}$	1.51 80$\overline{1}$	2.37 15$\overline{2}$	0.42 16$\overline{7}$	5.22 83$\overline{1}$	191.26$\overline{6}$
Dysprosium	Dy	162.46	3	0.56 069	1.78 35$\overline{1}$	2.01 84$\overline{9}$	0.49 542	4.44 90$\overline{1}$	224.88$\overline{6}$
			n	1.68 20$\overline{7}$	0.59 450	6.05 54$\overline{7}$	0.16 514	13.34 00$\overline{2}$	74.962
Erbium	Er	167.2	3	0.57 75$\overline{5}$	1.73 14$\overline{5}$	2.07 91$\overline{7}$	0.48 09$\overline{6}$	4.58 37$\overline{8}$	218.16$\overline{0}$
			n	1.73 26$\overline{4}$	0.57 71$\overline{5}$	6.23 75$\overline{1}$	0.16 03$\overline{2}$	13.75 13$\overline{5}$	72.72$\overline{0}$
Europium	Eu	152.0	3	0.52 50$\overline{4}$	1.90 46$\overline{1}$	1.89 01$\overline{6}$	0.52 90$\overline{6}$	4.16 70$\overline{8}$	241.91$\overline{1}$
			n	1.57 51$\overline{3}$	0.63 48$\overline{7}$	5.67 04$\overline{7}$	0.17 63$\overline{5}$	12.40 12$\overline{4}$	80.63$\overline{7}$
Fluorine	F	19.00	1	0.19 68$\overline{9}$	5.07 89$\overline{5}$	0.70 88$\overline{1}$	1.41 08$\overline{2}$	1.56 26$\overline{5}$	639.93$\overline{7}$
Gadolinium	Gd	**156.9**	3	0.54 17$\overline{9}$	1.84 57$\overline{2}$	1.95 04$\overline{6}$	0.51 27$\overline{0}$	4.30 00$\overline{2}$	232.73$\overline{7}$
			n	1.62 538	0.61 52$\overline{4}$	5.85 13$\overline{7}$	0.17 09$\overline{0}$	12.90 00$\overline{7}$	77.57$\overline{9}$
Gallium	Ga	69.72	3	0.24 08$\overline{3}$	4.15 23$\overline{2}$	0.86 69$\overline{8}$	1.15 34$\overline{2}$	1.91 13$\overline{7}$	523.18$\overline{4}$
			2	0.36 12$\overline{4}$	2.76 82$\overline{2}$	1.30 04$\overline{8}$	0.76 89$\overline{5}$	2.86 70$\overline{6}$	348.78$\overline{9}$
			1	0.72 24$\overline{9}$	1.38 41$\overline{1}$	2.60 09$\overline{5}$	0.38 44$\overline{7}$	5.73 41$\overline{2}$	174.39$\overline{5}$
Germanium	Ge	**72.60**	4	0.18 80$\overline{8}$	5.31 68$\overline{0}$	0.67 71$\overline{0}$	1.47 68$\overline{9}$	1.49 27$\overline{5}$	669.90$\overline{6}$
			2	0.37 61$\overline{7}$	2.65 84$\overline{0}$	1.35 42$\overline{0}$	0.73 84$\overline{5}$	2.98 54$\overline{9}$	334.95$\overline{3}$
			n	0.75 23$\overline{3}$	1.32 92$\overline{0}$	2.70 83$\overline{9}$	0.36 92$\overline{2}$	5.97 09$\overline{9}$	167.47$\overline{6}$
Gold	Au	197.2	3	0.68 11$\overline{7}$	1.46 80$\overline{5}$	2.45 22$\overline{3}$	0.40 77$\overline{9}$	5.40 62$\overline{4}$	184.97$\overline{2}$
			2	1.02 17$\overline{6}$	0.97 87$\overline{0}$	3.67 83$\overline{4}$	0.27 18$\overline{6}$	8.10 93$\overline{6}$	123.31$\overline{4}$
			1	2.04 35$\overline{2}$	0.48 93$\overline{5}$	7.35 66$\overline{8}$	0.13 59$\overline{3}$	16.21 87$\overline{1}$	61.65$\overline{7}$
Hafnium	Hf	178.6	4	0.46 26$\overline{9}$	2.16 12$\overline{5}$	1.66 57$\overline{0}$	0.60 03$\overline{5}$	3.67 22$\overline{3}$	272.31$\overline{3}$
			n	1.85 07$\overline{8}$	0.54 03$\overline{1}$	6.66 28$\overline{0}$	0.15 00$\overline{9}$	14.68 89$\overline{5}$	68.07$\overline{8}$
Helium	He	4.003	*n*	0.04 148 $\overline{2}$	24.10 69$\overline{2}$	0.14 93$\overline{3}$	6.69 63$\overline{7}$	0.32 92$\overline{3}$	3037.42$\overline{1}$
Holmium	Ho	163.5	3	0.56 47$\overline{7}$	1.77 06$\overline{4}$	2.03 31$\overline{6}$	0.49 18$\overline{5}$	4.48 23$\overline{5}$	223.09$\overline{7}$
			n	1.69 43$\overline{0}$	0.59 02$\overline{1}$	6.09 94$\overline{8}$	0.16 39$\overline{5}$	13.44 70$\overline{5}$	74.36$\overline{6}$
Hydrogen	H	1.0081	1	0.01 044 7	95.72 46$\overline{5}$	0.03 760 8	26.59 01$\overline{8}$	0.08 291	12061.10$\overline{2}$
Deuterium†	D	**2.01471**	1	0.02 087 8	47.89 77$\overline{1}$	0.07 516 0	13.30 49$\overline{2}$	0.16 570	6035.01$\overline{1}$
Illinium	Il	**146***	3	0.50 43$\overline{2}$	1.98 28$\overline{8}$	2.72 33$\overline{2}$	0.55 08$\overline{0}$	3.90 25$\overline{6}$	272.53$\overline{9}$
			n	1.51 29$\overline{5}$	0.66 09$\overline{6}$	5.44 66$\overline{3}$	0.18 36$\overline{0}$	11.00 76$\overline{0}$	90.84$\overline{6}$
Indium	In	**114.76**	3	0.39 641	2.52 26$\overline{6}$	1.42 70$\overline{7}$	0.70 074	3.14 61$\overline{4}$	317.84$\overline{9}$
			n	1.18 92$\overline{2}$	0.84 089	4.28 12$\overline{0}$	0.23 358	9.43 84$\overline{3}$	105.95$\overline{0}$
Iodine	I	**126.92**	7	0.18 790	5.32 22$\overline{5}$	0.67 641	1.47 84$\overline{0}$	1.49 12$\overline{2}$	670.59$\overline{3}$
			6	0.21 921	4.56 19$\overline{3}$	0.78 914	1.26 72$\overline{0}$	1.73 97$\overline{6}$	574.79$\overline{4}$
			5	0.26 305	3.80 16$\overline{1}$	0.94 697	1.05 60$\overline{0}$	2.08 77$\overline{1}$	478.99$\overline{5}$
			4	0.32 881	3.04 12$\overline{9}$	1.18 37$\overline{1}$	0.84 480	2.60 96$\overline{3}$	383.19$\overline{6}$
			3	0.43 841	2.28 09$\overline{6}$	1.57 82$\overline{8}$	0.63 360	3.47 95$\overline{1}$	287.39$\overline{7}$
			2	0.65 762	1.52 06$\overline{4}$	2.36 74$\overline{2}$	0.42 240	5.21 92$\overline{7}$	191.59$\overline{8}$
			1	1.31 52$\overline{3}$	0.76 032	4.73 48$\overline{4}$	0.21 120	10.43 85$\overline{3}$	95.799
Iridium	Ir	193.1	4	0.50 02$\overline{6}$	1.99 89$\overline{6}$	1.80 09$\overline{5}$	0.55 54$\overline{6}$	3.97 03$\overline{8}$	251.86$\overline{5}$
			3	0.66 70$\overline{1}$	1.49 92$\overline{2}$	2.40 12$\overline{4}$	0.41 65$\overline{9}$	5.29 38$\overline{4}$	188.89$\overline{9}$
			1	2.00 10$\overline{4}$	0.49 97$\overline{4}$	7.20 37$\overline{3}$	0.13 88$\overline{6}$	15.88 15$\overline{1}$	62.96$\overline{6}$
Iron	Fe	55.84	3	0.19 28$\overline{8}$	5.18 44$\overline{6}$	0.69 43$\overline{8}$	1.44 01$\overline{3}$	1.53 08$\overline{5}$	653.23$\overline{0}$
			2	0.28 93$\overline{3}$	3.45 63$\overline{0}$	1.04 15$\overline{8}$	0.96 00$\overline{8}$	2.29 62$\overline{8}$	435.48$\overline{7}$
			1	0.57 86$\overline{5}$	1.72 81$\overline{5}$	2.08 31$\overline{5}$	0.48 00$\overline{4}$	4.59 25$\overline{6}$	217.77$\overline{4}$
Krypton	Kr	**83.7**	*n*	0.86 73$\overline{6}$	1.15 29$\overline{3}$	3.12 24$\overline{9}$	0.32 02$\overline{6}$	6.88 39$\overline{0}$	145.26$\overline{6}$
Lanthanum	La	**138.92**	3	0.47 986	2.08 39$\overline{3}$	1.72 75$\overline{0}$	0.57 887	3.80 84$\overline{9}$	262.57$\overline{1}$
			n	1.43 95$\overline{9}$	0.69 464	5.18 25$\overline{1}$	0.19 296	11.42 54$\overline{7}$	87.524

Table 2. Electrochemical Equivalents of the Elements—(*Continued*)

1 Element	Symbol	2 Atomic weight	3 Val. or chg. val.	4 Mg./coulomb	5 Coulombs/mg.	6 G./amp.-hr.	7 Amp.-hr./g.	8 Lb./1000 amp.-hr.	9 Amp.-hr./lb.
Lead	Pb	207.21	4	0.53 681	1.86 284	1.93 253	0.51 746	4.26 050	234.715
			2	1.07 363	0.93 142	3.86 506	0.25 873	8.52 099	117.357
			1	2.14 725	0.46 571	7.73 011	0.12 936	17.04 198	58.679
Lithium	Li	6.940	1	0.07 192	13.90 490	0.25 890	3.86 247	0.57 078	1751.988
Lutecium	Lu	175.0	3	0.60 449	1.65 429	3.26 425	0.45 952	7.34 874	208.437
			n	1.81 347	0.55 143	6.52 849	0.15 317	14.39 287	69.479
Magnesium	Mg	24.32	2	0.12 601	7.93 586	0.45 364	2.20 440	1.00 010	999.901
			n	0.25 202	3.96 793	0.90 727	1.10 220	2.00 020	499.951
Manganese	Mn	54.93	7	0.08 132	12.29 772	0.29 274	3.41 603	0.64 537	1549.487
			6	0.09 487	10.54 090	0.34 153	2.92 803	0.75 294	1328.132
			5	0.11 384	8.78 409	0.50 983	2.44 002	0.90 352	1106.777
			4	0.14 230	7.02 727	0.51 229	1.95 202	1.12 941	885.421
			3	0.18 974	5.27 045	0.68 305	1.46 401	1.50 587	664.066
			2	0.28 461	3.51 363	1.02 458	0.97 601	2.25 881	442.711
			1	0.56 921	1.75 682	2.04 916	0.48 800	4.51 762	221.355
Masurium	Ma	97.8*	7	0.14 478	6.90 695	0.52 121	1.91 860	1.14 908	870.262
			n	1.01 347	0.98 671	3.64 850	0.27 409	8.04 356	124.323
Mercury	Hg	200.61	2	1.03 943	0.96 207	3.74 195	0.26 724	8.24 958	121.218
			n	2.07 886	0.48 103	7.48 390	0.13 362	16.49 917	60.609
Molybdenum	Mo	95.95	6	0.16 572	6.03 439	0.59 658	1.67 622	1.31 523	760.321
			5	0.19 886	5.02 866	0.71 590	1.39 685	1.57 828	633.601
			4	0.24 858	4.02 293	0.89 487	1.11 748	1.97 285	506.881
			3	0.33 143	3.01 720	1.19 316	0.83 811	2.63 047	380.160
			2	0.49 715	2.01 146	1.78 974	0.55 874	3.94 570	253.440
			1	0.99 430	1.00 573	3.57 948	0.27 937	7.89 141	126.720
Neodymium	Nd	144.27	3	0.49 834	2.00 665	1.79 403	0.55 740	3.95 516	252.834
			n	1.49 503	0.66 888	5.38 209	0.18 580	11.86 548	84.278
Neon	Ne	20.183	*n*	0.20 915	4.78 125	0.75 294	1.32 813	1.65 995	602.428
Nickel	Ni	58.69	3	0.20 273	4.93 270	0.72 982	1.37 019	1.60 899	621.509
			2	0.30 409	3.28 846	1.09 474	0.91 346	2.41 348	414.340
			1	0.60 819	1.64 423	2.18 947	0.45 673	4.82 696	207.170
Nitrogen	N	14.008	5	0.02 903 2	34.44 446	0.10 452	9.56 795	0.23 042	438.995
			4	0.03 629 0	27.55 568	0.13 064	7.65 436	0.28 802	347.196
			3	0.04 838 7	20.66 676	0.17 419	5.74 077	0.38 403	260.397
			2	0.07 258 0	13.77 784	0.26 129	3.82 718	0.57 604	173.598
			1	0.14 516	6.88 892	0.52 258	1.91 359	1.15 209	86.799
Osmium	Os	190.2	8	0.24 637	4.05 889	0.88 694	1.12 747	1.95 537	447.485
			6	0.32 850	3.04 416	1.18 259	0.84 560	2.60 717	383.558
			5	0.39 420	2.53 680	1.41 911	0.70 467	3.12 860	319.632
			4	0.49 275	2.02 944	1.77 389	0.56 373	3.91 075	255.706
			3	0.65 700	1.52 208	2.36 518	0.42 280	5.21 433	191.779
			2	0.98 549	1.01 472	3.54 777	0.28 187	7.82 150	127.853
			1	1.97 098	0.50 736	7.09 554	0.14 093	15.64 299	63.926
Oxygen	O	16.0000	2	0.08 290 2	12.06 250	0.29 845	3.35 069	0.65 796	1519.850
			n	0.16 580	6.03 125	0.59 689	1.67 535	1.31 592	759.925
Palladium	Pd	106.7	4	0.27 642	3.61 762	0.99 513	1.00 489	2.19 388	455.812
			3	0.36 857	2.71 332	1.32 684	0.75 367	2.92 518	340.859
			2	0.55 285	1.80 881	1.99 026	0.50 245	4.38 777	227.906
			n	1.10 570	0.90 440	3.98 052	0.25 122	8.77 554	113.953
Phosphorus	P	31.02	5	0.06 429	15.55 448	0.23 144	4.32 069	0.51 025	1959.832
			3	0.10 715	9.33 269	0.38 574	2.59 241	0.85 041	1175.899
			2	0.16 073	6.22 179	0.57 861	1.72 828	1.27 562	783.933
			n	0.32 145	3.11 090	1.15 722	0.86 414	2.55 124	391.966
Platinum	Pt	195.23	4	0.50 578	1.97 716	1.82 080	0.54 921	4.01 417	249.117
			2	1.01 155	0.98 858	3.64 160	0.27 460	8.02 834	124.559
			n	2.02 311	0.49 429	7.28 319	0.13 730	16.05 669	62.279
Polonium	Po	210*	6	0.36 269	2.75 714	1.30 570	0.76 587	2.87 857	347.394
			4	0.54 404	1.83 810	1.95 855	0.51 058	4.31 786	231.596
			2	1.08 808	0.91 905	3.91 710	0.25 529	8.63 572	115.798
			n	2.17 617	0.45 952	7.83 420	0.12 765	17.27 145	57.899

Table 2. Electrochemical Equivalents of the Elements—(*Continued*)

1 Element	Symbol	2 Atomic weight	3 Val. or chg. val.	4 Mg./coulomb	5 Coulombs/mg.	6 G./amp.-hr.	7 Amp.-hr./g.	8 Lb./1000 amp.-hr.	9 Amp.-hr./lb.
Potassium	K	39.096	1	0.40 514	2.46 82$\overline{8}$	1.45 85$\overline{0}$	0.68 563	3.21 54$\overline{5}$	310.99$\overline{8}$
Praseodymium	Pr	140.92	3	0.48 677	2.05 43$\overline{6}$	1.75 23$\overline{7}$	0.57 065	3.86 33$\overline{2}$	258.84$\overline{5}$
			n	1.46 03$\overline{1}$	0.68 479	5.25 71$\overline{2}$	0.19 022	11.58 99$\overline{6}$	86.282
Protoactinium	Pa	**231**	5	0.59 84$\overline{5}$	2.08 $\overline{874}$	1.72 $\overline{352}$	0.58 0$\overline{21}$	3.79 $\overline{972}$	263.$\overline{177}$
			3	0.79 $\overline{793}$	1.25 $\overline{325}$	2.87 $\overline{254}$	0.34 8$\overline{12}$	6.33 $\overline{286}$	157.$\overline{906}$
			2	1.19 $\overline{689}$	0.83 $\overline{550}$	4.30 $\overline{881}$	0.23 $\overline{208}$	9.49 $\overline{929}$	105.$\overline{271}$
			n	2.39 $\overline{378}$	0.41 $\overline{775}$	8.61 $\overline{762}$	0.11 6$\overline{04}$	18.9$\overline{9}$ $\overline{859}$	52.6$\overline{35}$
Radium	Ra	**226.05**	2	1.17 12$\overline{4}$	0.85 379	4.21 64$\overline{8}$	0.23 716	9.29 57$\overline{4}$	107.57$\overline{6}$
			n	2.34 24$\overline{9}$	0.42 690	8.43 29$\overline{5}$	0.11 858	18.59 14$\overline{8}$	53.788
Radon	Rn	**222**	*n*	2.30 $\overline{052}$	0.43 $\overline{468}$	8.28 $\overline{187}$	0.12 0$\overline{75}$	18.2$\overline{5}$ $\overline{839}$	54.7$\overline{69}$
Rhenium	Re	**186.31**	7	0.27 581	3.62 56$\overline{8}$	0.99 292	1.00 71$\overline{3}$	2.18 90$\overline{1}$	456.82$\overline{8}$
			6	0.32 178	3.10 77$\overline{2}$	1.15 84$\overline{0}$	0.86 326	2.55 38$\overline{4}$	391.56$\overline{7}$
			5	0.38 613	2.58 97$\overline{7}$	1.39 00$\overline{8}$	0.71 938	3.06 46$\overline{1}$	326.30$\overline{6}$
			4	0.48 267	2.07 18$\overline{2}$	1.73 76$\overline{1}$	0.57 550	3.83 07$\overline{7}$	261.04$\overline{4}$
			3	0.64 356	1.55 38$\overline{6}$	2.31 68$\overline{1}$	0.43 163	5.10 76$\overline{9}$	195.78$\overline{3}$
			2	0.96 537	1.03 59$\overline{1}$	3.47 52$\overline{1}$	0.28 775	7.66 15$\overline{3}$	130.52$\overline{2}$
			1	1.93 06$\overline{7}$	0.51 795	6.95 04$\overline{2}$	0.14 388	15.32 30$\overline{6}$	65.261
Rhodium	Rh	102.91	4	0.26 661	3.75 08$\overline{5}$	0.95 978	1.04 19$\overline{0}$	2.11 59$\overline{6}$	472.59$\overline{9}$
			3	0.35 547	2.81 31$\overline{4}$	1.27 97$\overline{1}$	0.78 143	2.82 12$\overline{8}$	354.44$\overline{9}$
			2	0.53 321	1.87 54$\overline{3}$	1.91 95$\overline{6}$	0.52 095	4.23 19$\overline{2}$	236.30$\overline{0}$
			1	1.06 64$\overline{2}$	0.93 771	3.83 91$\overline{3}$	0.26 048	8.46 38$\overline{3}$	118.15$\overline{0}$
Rubidium	Rb	**85.48**	1	0.88 580	1.12 89$\overline{2}$	3.18 88$\overline{9}$	0.31 35$\overline{9}$	7.03 03$\overline{0}$	142.24$\overline{1}$
Ruthenium	Ru	101.7	8	0.13 17$\overline{4}$	7.59 09$\overline{5}$	0.47 42$\overline{5}$	2.10 86$\overline{0}$	1.04 55$\overline{4}$	856.44$\overline{4}$
			6	0.17 56$\overline{5}$	5.69 $\overline{322}$	0.63 23$\overline{3}$	1.58 14$\overline{5}$	1.39 40$\overline{5}$	717.3$\overline{33}$
			5	0.21 07$\overline{8}$	4.74 43$\overline{5}$	0.75 88$\overline{0}$	1.31 78$\overline{7}$	1.67 28$\overline{6}$	597.77$\overline{8}$
			4	0.26 34$\overline{7}$	3.79 54$\overline{8}$	0.94 85$\overline{0}$	1.05 43$\overline{0}$	2.09 10$\overline{8}$	478.22$\overline{2}$
			3	0.35 13$\overline{0}$	2.84 66$\overline{1}$	1.26 46$\overline{6}$	0.79 07$\overline{2}$	2.78 81$\overline{0}$	358.66$\overline{7}$
			2	0.52 69$\overline{4}$	1.89 77$\overline{4}$	1.89 69$\overline{9}$	0.52 71$\overline{5}$	4.18 21$\overline{6}$	239.1$\overline{11}$
			1	1.05 38$\overline{9}$	0.94 88$\overline{7}$	3.79 39$\overline{9}$	0.26 35$\overline{7}$	8.36 43$\overline{1}$	119.55$\overline{6}$
Samarium	Sm	**150.43**	3	0.51 962	1.92 44$\overline{8}$	1.87 06$\overline{3}$	0.53 458	4.12 40$\overline{4}$	242.48$\overline{1}$
			n	1.55 88$\overline{6}$	0.64 149	5.61 19$\overline{0}$	0.17 819	12.37 2$\overline{11}$	80.827
Scandium	Sc	45.10	3	0.15 57$\overline{9}$	6.41 90$\overline{7}$	0.56 08$\overline{3}$	1.78 30$\overline{7}$	1.23 64$\overline{2}$	808.78$\overline{9}$
			n	0.46 73$\overline{6}$	2.13 96$\overline{9}$	1.68 24$\overline{9}$	0.59 43$\overline{6}$	3.70 92$\overline{5}$	269.59$\overline{6}$
Selenium	Se	**78.9**	6	0.13 63$\overline{7}$	7.33 $\overline{283}$	0.49 09$\overline{4}$	2.03 $\overline{690}$	1.08 $\overline{234}$	923.9$\overline{21}$
			4	0.20 45$\overline{6}$	4.88 $\overline{855}$	0.73 64$\overline{1}$	1.35 793	1.62 35$\overline{2}$	615.94$\overline{7}$
			2	0.40 91$\overline{2}$	2.44 $\overline{428}$	1.47 $\overline{283}$	0.67 89$\overline{7}$	3.24 $\overline{703}$	307.9$\overline{74}$
			n	0.81 82$\overline{4}$	1.22 $\overline{214}$	2.94 $\overline{566}$	0.33 94$\overline{8}$	6.49 $\overline{406}$	153.98$\overline{7}$
Silicon	Si	28.06	4	0.07 269	13.75 $\overline{624}$	0.26 17$\overline{0}$	3.82 1$\overline{18}$	0.57 69$\overline{5}$	1733.$\overline{257}$
			n	0.29 078	3.43 $\overline{906}$	1.04 $\overline{680}$	0.95 52$\overline{9}$	2.30 7$\overline{79}$	433.3$\overline{14}$
Silver	Ag	107.880	1	1.11 793‡	0.89 451	4.02 454	0.24 848	8.87 259	112.707
Sodium	Na	22.997	1	0.23 831	4.19 62$\overline{0}$	0.85 792	1.16 56$\overline{1}$	1.89 13$\overline{9}$	528.71$\overline{2}$
Strontium	Sr	87.63	2	0.45 40$\overline{4}$	2.20 24$\overline{4}$	1.63 45$\overline{5}$	0.61 17$\overline{9}$	3.60 35$\overline{6}$	277.50$\overline{3}$
			n	0.90 80$\overline{8}$	1.10 1$\overline{22}$	3.26 91$\overline{0}$	0.30 58$\overline{9}$	7.20 71$\overline{3}$	138.75$\overline{2}$
Sulfur	S	32.06	7	0.04 746	21.06 $\overline{987}$	0.17 08$\overline{6}$	5.85 27$\overline{4}$	0.37 66$\overline{8}$	2654.$\overline{759}$
			6	0.05 537	18.05 $\overline{989}$	0.19 93$\overline{4}$	5.01 6$\overline{64}$	0.43 94$\overline{6}$	2275.$\overline{508}$
			5	0.06 645	15.04 $\overline{991}$	0.23 92$\overline{0}$	4.18 0$\overline{53}$	0.52 73$\overline{6}$	1896.$\overline{257}$
			4	0.08 306	12.03 $\overline{993}$	0.29 90$\overline{1}$	3.34 4$\overline{42}$	0.65 91$\overline{9}$	1517.$\overline{005}$
			3	0.11 07$\overline{4}$	9.02 $\overline{994}$	0.39 86$\overline{7}$	2.50 8$\overline{32}$	0.87 89$\overline{2}$	1137.$\overline{754}$
			2	0.16 61$\overline{1}$	6.01 $\overline{996}$	0.59 80$\overline{1}$	1.67 22$\overline{1}$	1.31 83$\overline{9}$	758.$\overline{503}$
			1	0.33 22$\overline{3}$	3.00 998	1.19 60$\overline{2}$	0.83 61$\overline{1}$	2.63 67$\overline{7}$	379.2$\overline{51}$
Tantalum	Ta	**180.88**	5	0.37 488	2.66 75$\overline{1}$	1.34 95$\overline{7}$	0.74 098	2.97 52$\overline{9}$	336.10$\overline{1}$
			4	0.46 860	2.13 40$\overline{1}$	1.68 69$\overline{6}$	0.59 278	3.71 91$\overline{2}$	268.88$\overline{1}$
			3	0.62 480	1.60 05$\overline{1}$	2.24 928	0.44 459	4.95 88$\overline{2}$	201.66$\overline{1}$
			2	0.93 720	1.06 70$\overline{1}$	3.37 39$\overline{3}$	0.29 639	7.43 82$\overline{4}$	134.44$\overline{0}$
			1	1.87 44$\overline{0}$	0.53 350	6.74 78$\overline{5}$	0.14 820	14.87 64$\overline{7}$	67.220
Tellurium	Te	**127.61**	6	0.22 040	4.53 72$\overline{6}$	0.79 343	1.26 03$\overline{7}$	1.74 92$\overline{1}$	571.68$\overline{6}$
			4	0.33 060	3.02 48$\overline{4}$	1.19 01$\overline{5}$	0.84 023	2.62 38$\overline{2}$	381.12$\overline{4}$
			2	0.66 119	1.51 24$\overline{2}$	2.38 02$\overline{9}$	0.42 012	5.24 76$\overline{4}$	190.56$\overline{2}$
			n	1.32 23$\overline{8}$	0.75 621	4.76 05$\overline{8}$	0.21 006	10.49 52$\overline{8}$	95.281

Table 2. Electrochemical Equivalents of the Elements—(*Concluded*)

1 Element	Symbol	2 Atomic weight	3 Val. or chg. val.	4 Mg./coulomb	5 Coulombs/mg.	6 G./amp.-hr.	7 Amp.-hr./g.	8 Lb./1000 amp.-hr.	9 Amp.-hr./lb.
Terbium	Tb	159.2	3	0.54 99$\overline{1}$	1.81 8$\overline{47}$	1.97 9$\overline{69}$	0.50 51$\overline{3}$	4.36 4$\overline{47}$	229.1$\overline{13}$
			n	1.64 9$\overline{74}$	0.60 61$\overline{6}$	5.93 9$\overline{07}$	0.16 83$\overline{8}$	13.09 $\overline{340}$	76.37$\overline{1}$
Thallium	Tl	204.39	3	0.70 601	1.41 64$\overline{1}$	2.54 16$\overline{4}$	0.39 345	5.59 0$\overline{02}$	178.57$\overline{1}$
			1	2.11 80$\overline{3}$	0.47 214	7.62 49$\overline{1}$	0.13 115	16.80 0$\overline{05}$	59.524
Thorium	Th	232.12	4	0.60 135	1.66 29$\overline{3}$	2.16 48$\overline{5}$	0.46 193	4.77 26$\overline{8}$	209.52$\overline{6}$
			n	2.40 53$\overline{9}$	0.41 573	8.65 9$\overline{40}$	0.11 548	19.09 0$\overline{70}$	52.382
Thulium	Tm	169.4	3	0.58 51$\overline{5}$	1.70 8$\overline{97}$	2.10 6$\overline{53}$	0.47 47$\overline{1}$	4.64 4$\overline{10}$	215.3$\overline{27}$
			n	1.75 5$\overline{44}$	0.56 96$\overline{6}$	6.31 959	0.15 82$\overline{4}$	13.93 $\overline{230}$	71.7$\overline{76}$
Tin	Sn	118.70	4	0.30 751	3.25 19$\overline{0}$	1.10 70$\overline{5}$	0.90 330	2.44 06$\overline{2}$	409.73$\overline{2}$
			2	0.61 503	1.62 59$\overline{5}$	2.21 40$\overline{9}$	0.45 165	4.88 12$\overline{4}$	204.86$\overline{6}$
			n	1.23 00$\overline{5}$	0.81 297	4.42 81$\overline{9}$	0.22 583	9.76 24$\overline{8}$	102.43$\overline{3}$
Titanium	Ti	**47.90**	4	0.12 40$\overline{9}$	8.05 8$\overline{46}$	0.44 67$\overline{4}$	2.23 8$\overline{46}$	0.98 48$\overline{8}$	1015.$\overline{348}$
			3	0.16 54$\overline{6}$	6.04 $\overline{384}$	0.59 56$\overline{5}$	1.67 8$\overline{84}$	1.31 3$\overline{18}$	761.5$\overline{11}$
			1	0.49 63$\overline{7}$	2.01 4$\overline{61}$	1.78 $\overline{694}$	0.55 96$\overline{1}$	3.93 9$\overline{53}$	253.$\overline{837}$
Tungsten	W	**183.92**	6	0.31 765	3.14 81$\overline{1}$	1.14 35$\overline{4}$	0.87 447	2.52 10$\overline{8}$	396.65$\overline{5}$
			5	0.38 118	2.62 34$\overline{2}$	1.37 22$\overline{5}$	0.72 873	3.02 53$\overline{0}$	330.54$\overline{6}$
			4	0.47 648	2.09 87$\overline{4}$	1.71 53$\overline{2}$	0.58 298	3.78 16$\overline{2}$	264.43$\overline{7}$
			3	0.63 530	1.57 40$\overline{5}$	2.28 70$\overline{9}$	0.43 724	5.04 21$\overline{7}$	198.32$\overline{7}$
			2	0.95 295	1.04 93$\overline{7}$	3.43 06$\overline{3}$	0.29 149	7.56 32$\overline{5}$	132.21$\overline{8}$
			1	1.90 59$\overline{1}$	0.52 468	6.86 12$\overline{6}$	0.14 575	15.12 6$\overline{50}$	66.109
Uranium	U	**238.07**	6	0.41 117	2.43 20$\overline{6}$	1.48 02$\overline{3}$	0.67 557	3.26 16$\overline{8}$	306.59$\overline{1}$
			5	0.49 341	2.02 67$\overline{1}$	1.77 62$\overline{7}$	0.56 298	3.91 60$\overline{1}$	255.49$\overline{2}$
			4	0.61 676	1.62 13$\overline{7}$	2.22 03$\overline{4}$	0.45 038	4.89 25$\overline{2}$	204.39$\overline{4}$
			3	0.82 235	1.21 60$\overline{3}$	2.96 04$\overline{6}$	0.33 779	6.52 33$\overline{5}$	153.29$\overline{5}$
			2	1.23 35$\overline{2}$	0.81 069	4.44 06$\overline{8}$	0.22 519	9.78 50$\overline{3}$	102.19$\overline{7}$
			1	2.46 70$\overline{5}$	0.40 534	8.88 13$\overline{7}$	0.11 260	19.57 0$\overline{06}$	51.098
Vanadium	V	**50.95**	5	0.10 56$\overline{0}$	9.47 0$\overline{07}$	0.38 01$\overline{5}$	2.63 05$\overline{7}$	0.83 80$\overline{8}$	1193.$\overline{209}$
			4	0.13 19$\overline{9}$	7.57 6$\overline{06}$	0.47 51$\overline{8}$	2.10 4$\overline{46}$	1.04 7$\overline{60}$	954.56$\overline{7}$
			3	0.17 59$\overline{9}$	5.68 $\overline{204}$	0.63 35$\overline{8}$	1.57 8$\overline{34}$	1.39 6$\overline{79}$	715.9$\overline{25}$
			2	0.26 39$\overline{9}$	3.78 8$\overline{03}$	0.95 03$\overline{6}$	1.05 2$\overline{23}$	2.09 5$\overline{19}$	477.2$\overline{84}$
			1	0.52 7$\overline{98}$	1.89 4$\overline{01}$	1.90 0$\overline{73}$	0.52 61$\overline{1}$	4.19 0$\overline{38}$	238.6$\overline{42}$
Virginium	Vi	**224***	1	2.32 $\overline{124}$	0.43 0$\overline{80}$	8.35 $\overline{648}$	0.11 96$\overline{7}$	18.42 $\overline{288}$	54.2$\overline{80}$
Xenon	Xe	**131.3**	*n*	1.36 06$\overline{2}$	0.73 49$\overline{6}$	4.89 8$\overline{24}$	0.20 41$\overline{6}$	10.79 8$\overline{77}$	92.60$\overline{3}$
Ytterbium	Yb	**173.04**	3	0.59 772	1.67 30$\overline{2}$	2.15 1$\overline{79}$	0.46 473	4.74 38$\overline{9}$	210.79$\overline{7}$
			n	1.79 31$\overline{6}$	0.55 767	6.45 5$\overline{38}$	0.15 491	14.23 1$\overline{67}$	70.266
Yttrium	Y	88.92	3	0.30 71$\overline{5}$	3.25 5$\overline{74}$	1.10 5$\overline{74}$	0.90 43$\overline{7}$	2.43 7$\overline{74}$	410.2$\overline{16}$
			n	0.92 14$\overline{5}$	1.08 5$\overline{25}$	3.31 7$\overline{22}$	0.30 14$\overline{6}$	7.31 3$\overline{22}$	136.739
Zinc	Zn	65.38	2	0.33 87$\overline{6}$	2.95 1$\overline{97}$	1.21 9$\overline{52}$	0.81 9$\overline{99}$	2.68 8$\overline{59}$	371.9$\overline{42}$
			n	0.67 75$\overline{1}$	1.47 599	2.43 905	0.41 0$\overline{00}$	5.37 7$\overline{18}$	185.9$\overline{71}$
Zirconium	Zr	91.22	4	0.23 63$\overline{2}$	4.23 1$\overline{53}$	0.85 07$\overline{6}$	1.17 54$\overline{2}$	1.87 5$\overline{60}$	533.1$\overline{64}$
			n	0.94 52$\overline{8}$	1.05 $\overline{788}$	3.40 $\overline{303}$	0.29 38$\overline{6}$	7.50 $\overline{239}$	133.2$\overline{91}$

NOTE. Atomic weights in boldface type indicate those in which changes have been made since the last revision of this table in 1929, or new additions to the list since that time.

Digits overscored may, if desired, be dropped from the values, rounding them off to the nearest preceding digit; such digits have been carried as a matter of convenience and uniformity in calculating and tabulating but are in excess of the number of significant figures in the primary data and hence do not add to the true accuracy of the results.

* Best value known; not included in the official list.

† This is the second isotope of hydrogen and is the only isotope included in the table, as no others have as yet been isolated to a sufficient degree to have their atomic weights determined.

‡ This value varies from the basic figure of 1.1180 mg. because of the rounding off of the value of the faraday to 96,500 coulombs; other values also differ in the same proportion.

Electrolytic Dissociation

In dilute solutions of nonelectrolytes the gas law equation

$$PV = nRT$$

applies to osmotic pressure. There, P is the osmotic pressure, V the volume of the solution, n the number of moles in the solution, R the gas constant, and T the absolute temperature. $R = 0.0821$ l.-atm./°K. (the absolute temperature scale) or 1.985 cal./deg. For electrolytes, P is always greater than the value calculated. For electrolytes the expression becomes

$$PV = inRT$$

where i, always greater than unity, represents a degree of abnormality.

The presence of a solute causes a depression of the freezing point of a solvent. For non-electrolytes the extent

Table 3. Atomic Numbers, Atomic Weights, and Isotopes of the Elements

Atomic No.	Symbol	Mean atomic weight	Known isotopes	Atomic weight of isotopes
1	H	1.0078	3	1-2-3
2	He	4.002	2	4
3	Li	6.94	2	6-7
4	Be	9.02	2	(8)-9
5	B	10.82	2	10-11
6	C	12.01	2	12-13
7	N	14.008	2	14-15
8	O	16.0000	3	16-17-18
9	F	19.00	1	19
10	Ne	20.183	3	20-21-22
11	Na	22.997	3	23(1L-1G)
12	Mg	24.32	3	24-25-26
13	Al	26.97	1	27
14	Si	28.06	3	28-29-30
15	P	31.02	1	31
16	S	32.06	3	32-33-34
17	Cl	35.497	3	35-37-39
18	A	39.944	3	36-38-40
19	K	39.096	3	39-40-41
20	Ca	40.08	6	40-42-43-44-46-48
21	Sc	45.10	1	45
22	Ti	47.90	5	46-47-48-49-50
23	V	50.95	1	51
24	Cr	52.01	4	50-52-53-54
25	Mn	54.93	7	55(3L-3G)
26	Fe	55.84	4	54-56-57-58
27	Co	58.94	2	57-59
28	Ni	58.69	5	58-60-62-64
29	Cu	63.57	2	63-65
30	Zn	65.38	5	64-66-67-68-70
31	Ga	69.72	2	69-71
32	Ge	72.60	5	70-72-73-74-76
33	As	74.91	1	75
34	Se	78.96	6	74-76-77-78-80-82
35	Br	79.916	2	79-81
36	Kr	83.7	6	78-80-82-83-84-86
37	Rb	85.48	2	85-87
38	Sr	87.63	4	84-86-87-88
39	Y	88.92	1	89
40	Zr	91.22	5	90-91-92-94-96
41	Cb	92.91	1	93
42	Mo	96.0	8	92-94-95-96-97-98-100-(102)
43	Ma	97.8		
44	Ru	101.7	7	96-99-100-101-102-104
45	Rh	102.91	2	101-103
46	Pd	106.7	6	102-104-105-106-108-110
47	Ag	107.880	2	107-109
48	Cd	112.41	9	106-108-110-111-112-113-114-116

Atomic No.	Symbol	Mean atomic weight	Known isotopes	Atomic weight of isotopes
49	In	114.76	2	113-115
50	Sn	118.70	10	112-114-115-116-117-118-119-120-(121)-122-124
51	Sb	121.76	2	121-123
52	Te	127.61	8	120-122-123-124-125-126-128-130
53	I	126.92	1	127
54	Xe	131.3	9	124-126-128-129-130-131-132-134-136
55	Cs	132.91	5	133(2L-2G)
56	Ba	137.36	7	130-132-134-135-136-137-138
57	La	138.92	1	139
58	Ce	140.13	4	136-138-140-142
59	Pr	140.92	1	141
60	Nd	144.27	5	142-143-144-145-146
61	Il	146.		
62	Sm	150.43	7	144-147-148-149-150-152-154
63	Eu	152.0	2	151-153
64	Gd	156.9	5	155-156-157-158-160
65	Tb	159.2	1	159
66	Dy	162.46	4	161-162-163-164
67	Ho	163.5	1	165
68	Er	167.64	4	166-167-168-170
69	Tm	169.4	1	169
70	Yb	173.04	5	171-172-173-174-176
71	Lu	175.0	1	175
72	Hf	178.6	5	176-177-178-179-180
73	Ta	180.88	1	181
74	W	184.0	4	182-183-184-186
75	Re	186.31	2	185-187
76	Os	191.5	6	186-187-188-189-190-192
77	Ir	193.1	2	191-193
78	Pt	195.23	5	192-194-195-196-198
79	Au	197.2		
80	Hg	200.61	8	196-198-199-200-201-202-203-204
81	Tl	204.39	8	201-203-205-207-209-211-213-215
82	Pb	207.21	16	201-202-203-204-205-206-207-208-209-210-211-212-213-214-215-216
83	Bi	209.00	14	205-206-207-208-209-210-211-212-213-214-215-216-217-219
84	Po	210.		
85	Ab	221.		
86	Rn	222.		
87	Vi	224.		
88	Ra	226.05	4	226-228-230-232
89	Ac	227.		
90	Th	232.12	8	229-230-231-232-233-234-235-236
91	Pa	231.		
92	U	238.07	8	233-234-235-236-237-238-239-240

NOTE. Parentheses indicate an uncertain value.

Table 4. Conversion Data for Metric and Avoirdupois Units

	1 Mg./sec. ↓ ÷	2 G./hr. ↓ ÷	3 Kg./day ↓ ÷	4 Metric tons/year ↓ ÷	5 Oz./hr. ↓ ÷	6 Lb./hr. ↓ ÷	7 Lb./day ↓ ÷	8 Net tons/year ↓ ÷	
1. Mg./sec. → ×	1	3.6	0.0864	0.031557	0.12699	0.0079367	0.19048	0.034786	÷ ← Sec./mg.
2. G./hr. → ×	0.27778	1.	0.024	0.0087658	0.035274	0.0022046	0.052911	0.0096626	÷ ← Hr./g.
3. Kg./day → ×	11.57407	41.66667	1.	0.36524	1.46975	0.091860	2.20462	0.40261	÷ ← Days/kg.
4. Metric tons/year → ×	31.68877	114.07955	2.73791	1.	4.02404	0.25150	6.03606	1.10231	÷ ← Years/metric ton
5. Oz./hr. → ×	7.87487	28.34953	0.68039	0.24851	1.	0.0625	1.5	0.27393	÷ ← Hr./oz.
6. Lb./hr. → ×	125.99790	453.59243	10.88622	3.97611	16.	1.	24.	4.38291	÷ ← Hr./lb.
7. Lb./day → ×	5.24991	18.89969	0.45359	0.16567	0.66667	0.041667	1.	0.18262	÷ ← Days/lb.
8. Net tons/year → ×	28.74767	103.49124	2.48379	0.90719	3.65051	0.22816	5.47582	1.	÷ ← Years/net ton
	Sec./mg. × ↑	Hr./g. × ↑	Days/kg. × ↑	Years/metric ton × ↑	Hr./oz. × ↑	Hr./lb. × ↑	Days/lb. × ↑	Years/net ton × ↑	

Table 5. Conversion Data for Metric and Troy Units

	1 Mg./sec. ↓ ÷	2 G./hr. ↓ ÷	3 Kg./day ↓ ÷	4 Metric tons/year ↓ ÷	5 Oz./hr. ↓ ÷	6 Oz./day ↓ ÷	7 1000 oz./year ↓ ÷	
1. Mg./sec. → ×	1.	3.6	0.0864	0.031557	0.11574	2.77778	1.01458	÷ ← Sec./mg.
2. G./hr. → ×	0.27778	1.	0.024	0.008766	0.032151	0.77162	0.28183	÷ ← Hr./g.
3. Kg./day → ×	11.57407	41.66667	1.	0.36524	1.33941	32.15072	11.74281	÷ ← Days/kg.
4. Metric tons/year → ×	31.68877	114.07955	2.73791	1.	3.66774	88.02578	32.15072	÷ ← Years/metric ton
5. Oz./hr. → ×	8.63986	31.10359	0.74649	0.27265	1.	24.	8.76581	÷ ← Hr./oz.
6. Oz./day → ×	0.36	1.29598	0.031104	0.011360	0.041667	1.	0.365224	÷ ← Days/oz.
7. 1000 oz./year → ×	0.98563	3.54823	0.085157	0.031104	0.11408	2.73791	1.	÷ ← Years/1000 oz.
	Sec./mg. × ↑	Hr./g. × ↑	Days/kg. × ↑	Years/metric ton × ↑	Hr./oz. × ↑	Days/oz. × ↑	Years/1000 oz. × ↑	

of this lowering is given by the expression

$$\Delta = K\frac{n}{N}$$

where Δ is the lowering of the freezing point, K is a constant, n is the number of formula weights of solute present in N formula weights of solvent. The value becomes 1.858 °K./g.-mole solute/liter. Electrolytes always give values that exceed 1.858.

Table 6. Energy Conversion Data

$$\text{Kg./hp.-hr.} = \frac{\text{g./amp.-hr.} \times 0.7465}{\text{voltage of the reaction}}$$
$$\text{Hp.-hr./kg.} = \text{amp.-hr./g.} \times 1.3411 \times \text{voltage}$$
$$\text{Lb./hp.-hr.} = \frac{\text{g./amp.-hr.} \times 1.6457}{\text{voltage of the reaction}}$$
$$\text{Hp.-hr./lb.} = \text{amp.-hr./g.} \times 0.60786 \times \text{voltage}$$
$$\text{Kg./hp.-yr.} = \frac{\text{g./amp.-hr.} \times 6543.8}{\text{voltage of the reaction}}$$
$$\text{Lb./hp.-yr.} = \frac{\text{g./amp.-hr.} \times 14{,}426.5}{\text{voltage of the reaction}}$$

In order to explain this behavior of electrolytes, Arrhenius in 1887 formulated a theory of **electrolytic dissociation.** It was assumed that the molecules of electrolytes break up into equivalent quantities of positively and negatively electrified particles or ions when dissolved in water. The solutions of electrolytes are electrically neutral. The abnormal osmotic effects produced by electrolytes may then be accounted for by the increase in the number of particles of solute present in a solution. The theory does not assume that all the molecules in solution are dissociated. Let α equal the degree of dissociation or the fraction of each formula weight dissociated into ions, and n the number of ions into which each molecule dissociates, then

$$i = (1 - \alpha) + n\alpha$$
$$= 1 + (n - 1)\alpha$$
$$\alpha = \frac{(i - 1)}{(n - 1)}$$

Thus α may be calculated from osmotic pressure or from freezing-point determinations.

When n is the valence of the ion, the quantity of electricity carried by any gram ion is nF. One gram ion contains Avogadro's number of ions, which is 6.06×10^{23}. A single ion must carry the charge equivalent to the amount carried by the gram ion divided by the number of ions present, or a simple multiple n of this quantity if the ion has a valence of more than 1. This ultimate quantity of negative electricity is called the **electron.** It amounts to

$$\frac{96{,}500}{(6.06 \times 10^{23})} = 1.59 \times 10^{-19} \text{ coulomb}$$

Solutions of electrolytes in solvents other than water conduct the electric current. It may be inferred that electrolytic dissociation takes place in these solvents. Substances that show conduction of the electric current in non-aqueous solutions are not necessarily dissociated in water. This solvent, however, is more effective in bringing about dissociation than almost all others. Molten salts exhibit the same phenomena as solutions of electrolytes.

Solvents with high dielectric constants, like water, possess a high dissociating power, while those with low dielectric constants dissociate dissolved material to a less degree. The attraction of electric charges for each other is inversely proportional to the dielectric constant of the surrounding medium.

In the newer theories, the electric charges upon the ions are assumed to set up electrostatic fields which do not allow the ions to behave independently as demanded by the gas laws. The anomaly of strong electrolytes is to be attributed entirely to these interionic attractions.

Conductivity. In industrial operations, owing to the resistance of the electrolyte, certain amounts of electrical energy are converted into heat. From a practical viewpoint the resistance of the electrolyte is important in that it represents one of the ways in which electrical energy is consumed.

The resistance of any conductor of uniform cross section is

$$R = r\left(\frac{l}{a}\right)$$

where l is the length, a the cross-sectional area, r the specific resistance, and R the total resistance. The **specific resistance** is defined as the resistance of a unit cube of a conductor of a given material, expressed in ohms per centimeter cube or ohms per inch cube. The reciprocal of resistance is conductivity, and the reciprocal of specific resistance is the **specific conductivity,** denoted by K, which may be defined by the expression

$$K = \frac{1}{r} = \frac{l}{Ra} = \frac{lI}{Ea}$$

where I is the current and E the potential drop between the electrodes.

Conductivity of a solution is a function of the nature of the electrolyte, the solvent, the concentration, and the temperature. Specific conductivity varies directly with concentration up to a maximum point, after which there is a decrease. The $K - t$ relation

$$K_t = K_{18}[1 + b(t - 18)]$$

is almost linear, where b is 0.02 to 0.025 for salts and bases, and 0.01 to 0.016 for acids.

The **equivalent conductivity** (Λ) equals KV, where V is the volume of the solution containing 1 g. equivalent of the solute. Λ varies directly with V. K decreases with dilution except in very concentrated solutions, while Λ increases. At first the change in value is rapid, gradually diminishing until, at sufficiently high dilutions, a practically constant maximum termed Λ_∞, or equivalent conductivity at infinite dilution, is reached. It can be shown that $\Lambda_v/\Lambda_\infty = \alpha$, giving an electrical method for determination of the degree of ionization at any dilution. α increases with dilution, approaching 1 as the limit. Equivalent conductivities of a number of inorganic acids, bases, and salts are shown in Table 7. The values for the concentrations of $0.001N$ are in many cases equal to or approximately the same as Λ_∞, so that α values may be calculated from the tables. From the relations $\Lambda_v = KV$ and $K = 1/r$, $\Lambda_v = V/r$, and $r = V/\Lambda_v$, the resistance of electrolytes at various concentrations may be calculated from the tables. V can be calculated from the normalities or concentrations given. If, instead of normalities, concentrations C in mil-equivalents per liter are used, the normalities must be multiplied by 10^3 and $\Lambda_v = 10^6\ K/C$, from which specific conductivities and resistances may be calculated. The values in Table 7 are reciprocal ohms. This table has been condensed from those of Washburn and Klemenc and Parker and Klemenc in the "International Critical Tables," vol. 6, by permission.

Electrochemical Energy

Differences of potential are electrochemically produced by voltaic or galvanic cells, which may be defined as any arrangement by which the energy of chemical reactions or

Table 7. Equivalent Conductivities of Salts, Acids, and Bases

Salts	°C.	0.001*N*	0.01*N*	0.1*N*	0.2*N*	0.5*N*	1*N*	2*N*	3*N*	4*N*	5*N*	6*N*	7*N*	10*N*	20*N*	Ref.
$AgNO_3$	..	113.0	107.62	94.2	87.8	77.3	67.5	55.9	48.4	42.3	...	...	...	...	...	1,2,3
Ag_2SO_4	..	116.8	103.3	...	...	...	...	...	...	...	...	...	...	...	...	4,5
$AlCl_3$	..	...	...	...	...	65.0	56.2	44.5	34.2	27.1	...	...	...	...	...	1
$Al(NO_3)_3$	25	125.	108.	87.9	80.5	69.6	...	...	...	...	...	...	...	...	...	6
$Al_2(SO_4)_3$	25	107.0	67.3	...	...	...	...	...	...	...	...	...	...	...	...	7
$BaCl_2$	..	115.44	106.52	90.65	85.23	77.18	70.04	60.5	51.5	...	...	...	...	...	...	2,8
$Ba(NO_3)_2$	..	111.56	100.82	78.83	70.08	56.52	...	...	...	...	...	...	...	...	...	2,3,9
$CaCl_2$	..	111.8	103.23	88.07	82.68	74.82	67.45	58.0	49.7	42.4	35.6	...	23.5	...	...	2
$Ca(NO_3)_2$	..	108.34	99.39	82.37	75.84	65.61	55.79	42.7	33.5	26.5	...	14.2	...	5.	...	2,10
$CdCl_2$	..	104.8	82.9	50.0	41.0	29.6	21.6	14.1	10.	7.	4.8	...	3.	1.2	...	11,12,13
$Cd(NO_3)_2$	..	108.2	96.2	79.7	73.5	63.4	53.9	40.8	31.4	24.2	18.4	13.7	11.	...	...	12,13
$CdSO_4$	..	97.58	70.23	42.15	35.84	28.7	23.56	17.9	14.	10.8	8.2	...	...	...	...	2,12
$CoCl_2$	..	...	...	...	...	...	60.8	49.2	39.7	32.1	...	...	12.	...	...	1,14
$Co(NO_3)_2$	..	...	...	...	...	66.2	58.0	46.8	37.6	29.9	23.3	...	...	...	...	1
$CoSO_4$	25	113.4	82.5	51.4	43.9	35.3	29.3	...	...	...	...	...	...	...	...	6
$Cr(NO_3)_3$	..	...	108.6	80.2	74.0	64.8	55.9	43.2	33.3	25.5	19.	...	...	...	...	1,15
$CuCl_2$	..	...	...	83.2	77.1	67.0	56.9	43.4	33.6	26.0	...	...	...	...	...	8,16
$Cu(NO_3)_2$	..	...	...	81.9	76.0	67.2	57.8	45.5	36.3	28.4	...	...	...	4.6	...	16
$CuSO_4$	..	98.42	71.64	43.8	37.6	30.7	25.74	20.	16.2	...	...	...	...	...	...	2,10
$FeCl_2$	..	...	...	...	...	69.4	60.6	48.1	38.8	30.8	24.0	18.1	...	...	...	1
$FeSO_4$	..	...	...	...	...	30.8	25.8	19.5	15.3	11.9	...	...	...	...	...	17
$Fe(NO_3)_3$	..	...	...	...	...	75.9	63.8	48.5	37.5	28.5	...	...	...	...	...	1
$HgCl_2$	..	...	2.6	1.5	...	1.0	...	...	...	...	...	...	...	...	...	12
KBr	..	129.1	124.13	113.98	110.17	105.06	101.2	96.5	91.9	86.9						8,18,19,20 21
KCl	..	127.07	122.18	111.79	107.74	102.25	98.08	92.3	88.1							22,23,24
$KClO_3$	..	116.7	111.5	99.0	93.6											21,25
KF	..	108.45	103.85	93.63	89.1	82.1	75.7									9,16,26
KNO_3	..	123.37	117.93	104.56	98.53	89.12	80.33	69.0	61.3	...	...	...	...	...	...	9,22,25
K_2SO_4	..	126.7	115.6	94.8	87.8	78.3	71.5									2,3,9,17,27
LiBr	..	...	...	84.3	80.8	74.1	67.2	57.7	50.5	44.2						19
LiCl	..	96.2	91.84	82.76	77.69	70.5	63.18	53.0	45.2	...	33.3	...	...	11.3	...	22
$LiNO_3$	..	92.58	88.33	78.95	74.79	67.8	60.61	...	...	...	...	...	...	...	...	9,22
$MgCl_2$	0	...	65.7	55.2	51.8	46.0	40.6	...	...	...	...	...	...	...	...	6,28,29,30,31
$Mg(NO_3)_2$	..	102.5	94.52	80.4	75.2	67.2	59.5	47.9	39.2	32.1	...	...	...	...	...	5,8,9
$MgSO_4$	..	99.9	76.0	49.57	43.0	34.8	28.99	21.4	16.1	12.0	8.8	...	...	...	...	2,3,32,33
$MnCl_2$	..	...	...	...	...	66.4	57.4	45.3	35.7	28.1	21.6	16.3	...	...	...	1
$Mn(NO_3)_2$	..	...	...	...	...	66.8	58.9	47.5	37.9	30.3	23.5	...	...	...	...	1
$MnSO_4$	..	...	...	...	...	30.0	24.7	18.5	14.2	10.8	8.1	5.8	3.8	...	...	17
$Na_2B_4O_7$	25	86.	78.6													34
NaBr	..	...	...	95.9	91.7	84.5	78.0	69.1	60.5	53.0	...	...	...	...	...	19
NaCl	..	106.27	101.72	91.82	87.53	80.76	74.19	64.6	56.4	42.6	...	...	...	...	...	22
$NaClO_3$	..	95.5	91.0	80.8	76.4	69.1	61.8	51.7	43.5	36.1	...	...	...	...	...	8,16,35
Na_2CO_3	..	112.	96.1	72.8	65.	54.4	45.4	...	...	...	...	...	...	...	...	36,37
Na_2CrO_4	..	...	...	82.1	75.8	66.3	57.8	46.6	37.8	30.6	...	...	...	...	...	16,38
NaF	..	87.65	83.33	72.9	67.8	59.8	51.8	...	...	...	...	...	...	...	...	9,21
$NaNO_3$	..	102.60	97.93	87.04	82.09	73.88	65.72	54.6	46.1	39.2	...	...	...	...	...	8,9,22
Na_3PO_4	25	...	118.	...	...	...	...	...	...	...	...	...	...	...	...	34
Na_2SO_4	..	105.8	96.1	77.6	70.4	59.4	50.3	39.6	...	...	...	...	...	...	...	1,2,9,17,20,26,27,36,37
NH_4F	..	...	...	89.9	83.4	74.3	65.6	...	...	...	...	...	...	...	...	16
$NiCl_2$	..	...	...	...	...	70.8	62.1	50.6	41.0	33.3	...	...	...	...	...	1
$Ni(NO_3)_2$	..	...	...	...	...	66.9	57.6	46.0	37.2	29.3	22.7	16.	...	...	...	1

1 Heydweiller, *Z. anorg. allgem. Chem.*, **116**, 42 (1921).
2 Kohlrausch and Grüneisen, *Sitzb. preuss. Akad. Wiss.*, p. 1215 (1904).
3 Noyes and Melcher, *Carnegie Inst. Wash. Pub.* 63, p. 71 (1907).
4 Drucker, *Z. physik. Chem.*, **96**, 381 (1920).
5 Hunt, *J. Am. Chem. Soc.*, **33**, 795 (1911).
6 Jones *et al.*, *Carnegie Inst. Wash. Pub.* 170 (1912).
7 Walden, *Z. physik. Chem.*, **2**, 49 (1888).
8 Clausen, *Ann. Physik*, **37**, 51 (1912).
9 Noyes and Falk, *J. Am. Chem. Soc.*, **34**, 454 (1912).
10 Kohlrausch and Holborn, "Das Leitvermögen der Elektrolyte," Teubner, Leipzig, 1916.
11 Grotrian, *Ann. physik. Chem.*, **151**, 378 (1874).
12 Grotrian, *Ann. physik. Chem.*, **18**, 177 (1883).
13 Wershoven, *Z. physik. Chem.*, **5**, 481 (1890).
14 Mazzetti, *Gazz. chim. ital.*, **54**, 891 (1924).
15 Bjerrum, *Z. physik. Chem.*, **59**, 336 (1907).
16 Heydweiller, *Ann. Physik*, **37**, 739 (1912).
17 Klein, *Ann. physik. Chem.*, **27**, 151 (1886).
18 Gropp, *Dissertation*, Rostock (1915).
19 Heydweiller, *Ann. Physik*, **30**, 873 (1909).
20 Heydweiller, "Gesammelte Abhandlungen von Fr. Kohlrausch," vol. 2, Barth, Leipzig, 1911
21 Kohlrausch and Steinwehr, *Sitzb. preuss. Akad. Wiss.*, p. 581 (1902).
22 Kohlrausch and Maltby, *Sitzb. preuss. Akad. Wiss.*, p. 665 (1899).
23 Parker and Parker, *J. Am. Chem. Soc.*, **46**, 312 (1924).
24 Weiland, *J. Am. Chem. Soc.*, **40**, 131 (1918).
25 Walden and Ulich, *Z. physik. Chem.*, **106**, 49 (1923).
26 Kohlrausch, *Sitzb. preuss. Akad. Wiss.*, p. 1002 (1900).
27 Sherrill, *J. Am. Chem. Soc.*, **32**, 741 (1910).
28 Jones *et al.*, *Carnegie Inst. Wash. Pub.* 180 (1913).
29 Kohlrausch and Grotrian, *Nachr. kgl. Ges. Wiss. Göttingen*, p. 405 (1874).
30 Kohlrausch and Grotrian, *Ann. physik. Chem.*, **154**, 1 (1875).
31 Kohlrausch and Grotrian, *Ann. physik. Chem.*, **154**, 215 (1875).
32 Foster, *Phys. Rev.*, **8** 257 (1899).
33 Harkins and Paine, *J. Am. Chem. Soc.*, **41** 1155 (1919).
34 Walden, *Z. physik. Chem.*, **1**, 529 (1887).
35 Flügel, *Z. physik. Chem.*, **79**, 577 (1912).
36 Kohlrausch, *Ann. physik. Chem.*, **6**, 145 (1879).
37 Kohlrausch, *Ann. physik. Chem.*, **26**, 161 (1885).
38 Claussen, *Dissertation*, Rostock (1911).

Table 7. Equivalent Conductivities of Salts, Acids, and Bases—(*Concluded*)

Salts	°C.	0.001*N*	0.01*N*	0.1*N*	0.2*N*	0.5*N*	1*N*	2*N*	3*N*	4*N*	5*N*	6*N*	7*N*	10*N*	20*N*	Ref.
$NiSO_4$	. .	94.4	69.8	43.8	37.9	30.6	25.4	19.3	15.1	....	....	....	....	....	....	10,17,39,40
$PbCl_2$	. .	118.98	102.0	....	....	....	....	....	....	....	....	....	....	....	....	5,9
$Pb(NO_3)_2$	. .	116.0	103.44	77.18	67.29	53.15	41.97	30.6	....		....	....	....	....	....	2,9,41
$SrCl_2$	. .	114.3	105.3	90.4	85.1	76.1	67.9	57.5	49.7	....	....	....	....	....	....	2,29,30,31
$Sr(NO_3)_2$	. .	108.16	98.90	80.82	73.70	62.64	52.00	38.4	28.9	21.1	16.4	....	....	....	....	2,9
$ZnCl_2$	. .	....	....	86.5	79.3	68.7	56.2	39.5	29.6	22.9	18.4	15.2	12.9	8.	0.7	16,41
$Zn(NO_3)_2$	. .			80.5	75.2	67.2	59.2	47.9	38.8	31.0	....	....	....	....	....	16
$ZnSO_4$	. .	98.5	72.8	45.4	39.1	31.4	26.0	20.0	15.6	11.9	8.9	....	....	....	....	2,10
$FeCl_3$	. .	....	....	....	....	66.5	53.1	37.5	27.8	20.9	15.9	12.4	....	....	....	1
Acids:																
HBr	. .	....	....	355.2	347.7	328.4	301.1	....	....	....	....	....	....	....	....	16
HCl	. .	377.	369.3	350.1	341.5	326.6	300.5	253.8	214.7	182.	152.	....	....	64.3	....	10,37,42,43,44
$HClO_3$	. .	....	....	343.6	334.6	316.6	291.4	....	....	....	....	....	....	....	....	8,16,19
$HClO_4$	25	....	405.	384.	373.	359.	342.	291.	226.	185.	150.	118.	91.	41.	....	46,47,48
H_2CrO_4	25	....	193.	186.	....	....	....	....	....	....	....	....	....	....	....	7
HNO_3	. .	372.	364.	345.7	336.	321.	305.5	259.	220.	184.	154.	129.1	108.6	64.9	5.5	10,31,12,49,50
H_3PO_4	. .	....	202.7	96.4	....	....	....	24.	....	....	....	14.	13.	11.4	6.4	49
H_2SO_3	. .	....	130.	80.	....	....	....	24.	....	....	....	....	....	....	....	51
H_2SO_4	. .	360.	308.6	232.9	....	....	198.6	182.7	166.5	150.5	134.6	119.1	105.	70.	12.	5,49,52,53,54
Bases:																
$Ba(OH)_2$	. .	217.	207.	179.7	....	....	....	....	....	....	....	....	....	....	....	36,49
$Ca(OH)_2$	25	....	220.	....	....	....	....	....	....	....	....	....	....	....	....	47
KOH	. .	234.	228.	213.	207.	197.	184.	160.2	140.4	122.0	105.6	90.6	77.1	39.6	....	36,37,55
LiOH	. .	....	....	180.5	172.2	156.9	139.	113.3	94.4	78.7	65.3	....	....	....	....	8,16,19
NaOH	. .	....	208.4	195.3	189.0	175.5	158.4	129.4	104.2	86.7	68.9	54.2	43.	20.5	4.4(19*N*)	36,37,55,56,57,58

[39] Pfanhauser, *Z. Elektrochem.*, **7**, 698 (1901).
[40] Murata, *Chem. Soc. Japan, Bull.*, **3**, 47 (1928
[41] Long, *Ann. physik. Chem.*, **11**, 37 (1880).
[42] Goodwin and Haskell, *Phys. Rev.*, **19**, 369 (1904).
[43] Lorenz and Ostwald, *Z. anorg. allgem. Chem.*, **114**, 209 (1920).
[44] Noyes and Cooper, *Carnegie Inst. Wash. Pub.* 63, p. 115 (1907).
[45] Brownson and Cray, *J. Chem. Soc., London*, **127**, 2923 (1925).
[46] Linde, *Z. Elektrochem.*, **30**, 255 (1924).
[47] Ostwald, "Lehrbuch der allgemeinen Chemie," Engelmann, Leipzig, 1891–1903.
[48] Smith, *J. Am. Chem. Soc.*, **45**, 360 (1923).
[49] Noyes and Eastman, *Carnegie Inst. Wash. Pub.* 63, 239 (1907).
[50] Veley and Manley, *Phil. Trans. Roy. Soc. (London)*, **A**, **191**, 365 (1898).
[51] Lindner, *Monatsh.*, **33**, 613 (1912).
[52] Gibson and Gibson, *Proc. Roy. Soc.* (Edinburgh), **30**, 254 (1909).
[53] Kohlrausch, *Ann. physik. Chem.*, **159**, 233 (1876).
[54] Kohlrausch, *Ann. physik. Chem.*, **17**, 69 (1882).
[55] Heydweiller, *Ann. Physik*, **48**, 681 (1915).
[56] Bousfield and Lowry, *Phil. Trans. Roy. Soc. (London)*, **A**, **204**, 253 (1905).
[57] Noyes and Kato, *Carnegie Inst. Wash. Pub.* 63, 151 (1907).
[58] Raikes, Yorke, and Ewart, *J. Chem. Soc. London*, p. 630 (1926).

of certain physical processes, such as diffusion, is converted into electrical energy.

If the reaction involved in a voltaic cell is written completely as a thermochemical equation, and on an equivalent basis for the components involved, and the thermal energy is converted into electrical units, the electrical energy (in watts) divided by the faraday (in coulombs), the quotient of watts per equivalent divided by coulombs per equivalent will be volts. This will be the rough calculation of the theoretical decomposition voltage of the reaction of the cell under the conditions and limitations of the reaction.

The effective voltage of a primary cell is a function of the amount of current drawn from the cell. The greater the current, the lower will be the voltage across the terminals. When the current flowing is infinitely small, the voltage will have its maximum value, a figure which is termed the e.m.f. of the cell, or voltage on open circuit. Conversely, the greater the current forced into the cell, the higher the necessary voltage that must be applied across the terminals. With an infinitely small current the minimum value of applied voltage approaches the e.m.f. of the cell as a limit.

The maximum amount of electrical energy can be developed only when a cell operates isothermally and its reactions are completely reversible.

In a reversible process involving the conversion of energy, $U = A - Q$, where Q is the heat absorbed by the system (following the custom of thermodynamics, heat developed by a reaction is taken as negative), and A the external work done by the system when its total energy decreases by U. Electrical energy converted from chemical energy has a maximum value equal to A. In electrical units $A = nFE$, where n is the number of equivalents involved. If Q is negligible,

$$U = nFE$$

and

$$E = \frac{U}{nF}$$

The e.m.f. of a cell can be calculated from either thermochemical data or the theoretical decomposition voltage of a compound.

The relation between electrical energy of a system and the heat of reaction is given by the Gibbs-Helmholtz equation

$$A - U = \frac{TdA}{dT}$$

in which T is the absolute temperature. In chemical reaction systems, the external work is small. Since A approaches zero as a limit, $U = -Q$. By substitution in the Gibbs-Helmholtz equation for A and U, we obtain

$$nFE + Q = \frac{TnF\,dE}{dT}$$

whence

$$E = \frac{-Q}{nF} + \frac{T\,dE}{dT}$$

If E is expressed in volts and Q in calories, then

$$E = \frac{-JQ}{nF} + \frac{T\,dE}{dT}$$

Table 8. Electrode Potentials

Reaction	Volts	Ref.
$Li^+ + e = Li$	−2.959	1
$Rb^+ + e = Rb$	−2.925	1
$K^+ + e = K$	−2.924	1
$Ca^{++} + 2e = Ca$	−2.76	2
$Na^+ + e$	−2.714	1
$Zn^{++} + 2e = Zn$	−0.761	1
$Cr^{++} + 2e = Cr$	−0.557	3
$Fe^{++} + 2e = Fe$	−0.44	1
$Cd^{++} + 2e = Cd$	−0.401	1
$Ni^{++} + 2e = Ni$	−0.23	4
$Sn^{++} + 2e = Sn$	−0.136	1
$Pb^{++} + 2e = Pb$	−0.122	1
$H^+ + e = \frac{1}{2}H_2$	0.000	
$Cu^{++} + 2e = Cu$	0.344	1
$Ag^+ + e = Ag$	0.797	1
$Hg_2^{++} + 2e = 2Hg$	0.798	1
$Au^{+++} + 3e = Au$	1.36	5
$Ba^{++} + 2e = BaHg$	−1.570	6
$NaCl + e = NaHg(+Cl^-$ in sat. NaCl)	−1.837	7
$Sr^{++} + 2e = SrHg$	−1.793	6
$Tl^+ + e = TlHg$	−0.336	1
$Sb_2O_3 + 6H + 6e = 2Sb + 3H_2O$	0.144	8
$BiOCl + 2H^+ + 3e = Bi + Cl^- + H_2O$	0.158	9,10
$As_2O_3 + 6H^+ + 6e = 2As + 3H_2O$	0.234	11

[1] Lewis and Randall, "Thermodynamics and the Free Energy of Chemical Substances," McGraw-Hill, New York, 1923.
[2] Drucker and Luft, *Z. physik. Chem.*, **121**, 307 (1926).
[3] Grube and Breitinger, *Z. Elektrochem.*, **33**, 112 (1927).
[4] Haring and Van den Bosche, *J. Phys. Chem.*, **33**, 161 (1929).
[5] Jirsa and Jellinek, *Chem. Listy*, **18**, 1 (1924); *Z. Elektrochem.*, **30**, 286 (1924).
[6] Danner, *J. Am. Chem. Soc.*, **46**, 2385 (1924).
[7] Danner, *J. Am. Chem. Soc.*, **44**, 2832 (1922).
[8] Schuhmann, *J. Am. Chem. Soc.*, **46**, 52 (1924).
[9] Jellinek and Kühn, *Z. physik. Chem.*, **105**, 337 (1923).
[10] Noyes and Chow, *J. Am. Chem. Soc.*, **40**, 739 (1918).
[11] Schuhmann, *J. Am. Chem. Soc.*, **46**, 1444 (1924).

where $J = 4.182$ is the electrical equivalent of heat. It will be observed that, when dE/dT is positive, the e.m.f. of a reversible voltaic cell increases with rise in temperature; when zero, the electrical energy is equal to the chemical energy. By the use of this equation, heats of reaction may be determined by e.m.f. measurements. The equation also allows the determination of the theoretical voltage of a cell for the calculation of the theoretical energy and energy efficiency.

The existence of a definite tendency toward the passage from the atomic to the ionic state is designated as electrolytic solution pressure (p). In general, the values decrease in the same order as the increase of the electropositive character of the metal. The non-metals are also assumed to have electrolytic solution pressures, the order in the case of the halogens and sulfur being fluorine, chlorine, bromine, iodine, and sulfur. If a bar of zinc is placed in water, some zinc atoms give up two electrons each to the bar of metal, passing into the water as positively charged zinc ions. An electric double layer is thus formed at the interface of the metal and the liquid. The electrostatic attraction of the negative charges accumulating on the metal surface opposes the passage of atoms to the ionic state. If a piece of zinc is immersed in a solution of a copper salt, copper ions will discharge and deposit as copper atoms on the zinc. The negative charges on the zinc will be reduced, and more zinc atoms will be able to assume the ionic state. The amount of zinc entering the solution will be equivalent to the amount of copper deposited. When any metal A is immersed in a solution of a salt of another metal B having a lower electrolytic solution pressure, B is deposited and A enters solution. Electrolytic solution explains the displacement of metals by others from solution, and the solution of metals in acids (displacement of hydrogen). When a metal of a high electrolytic solution pressure is in contact with a solution of its own ions, the tendency of the atoms to pass into the ionic state is opposed by the osmotic pressure P of the metal ions in solution. If $p > P$, the metal will dissolve; if $P > p$, the metal will deposit.

The **single-electrode potential** is the difference of potential between the electrode and the solution around it. The e.m.f. of a cell is equal to the difference of the single-electrode potentials of the electrodes of the cell.

$$E = e_1 - e_2$$

where e_1 and e_2 are single-electrode potentials. The normal hydrogen electrode is the standard, and a single potential of zero is assigned to it. A normal hydrogen electrode consists of a platinized-platinum plate, half immersed in a normal H^+ solution and half surrounded by pure hydrogen gas which is bubbled through the solution. The hydrogen is dissolved in the platinized platinum and behaves like a metal. A definite and reproducible potential difference is set up between the electrode and the H^+ solution. The single potential of an electrode can be obtained by measuring the e.m.f. of a cell in which the hydrogen electrode is combined with the electrode in question. The sign of this single potential will be positive or negative according as the electrode in question is the positive or negative pole of the cell. Such values are single-electrode potentials on the hydrogen scale.

In order to indicate the direction of the polarity between a metal and a solution, the sign of the charge on the metal is placed before the potential difference between the two phases—the so-called potential of the metal or the electrode potential.

This convention, according to Bancroft [*Trans. Am. Electrochem. Soc.*, **33**, 79 (1918)] the only one that can be adopted universally, has been adopted officially by the Bunsen Gesellschaft, the American Electrochemical Society, and the National Bureau of Standards and is employed by most European electrochemists and largely in this country. Another convention, advocated by Lewis [*J. Am. Chem. Soc.*, **35**, 1 (1913)], is extensively employed in this country. According to this, the potential difference of a metal-solution junction is considered as positive when there is a tendency for positive electricity to flow from left to right through the junction as written, and as negative when there is a tendency for positive electricity to flow from right to left.

In solutions containing their own ions, noble metals (*e.g.*, with electrolytic solution pressures lower than that of hydrogen) acquire a positive potential, while base metals (*e.g.*, with electrolytic solution pressures greater than that of hydrogen) acquire a negative potential. In accordance with this convention, the potential of a metal in contact with a solution containing its own ions is positive when $p < P$, and negative when $p > P$.

A short arrow will sometimes be placed above a metal-solution junction to indicate the direction in which the positive current tends to flow. When the arrow points toward the metal, its potential is positive; when it points away from it, its potential is negative.

The magnitude of a single-electrode potential will be a function of p and P. For reactions of the reversible type,

$$M^n\oplus \rightleftarrows M + n\oplus$$

Nernst developed the equation

$$neF = -RT \log_e \frac{p}{P}$$

on the assumption that the gas laws are valid in respect to ions in strong electrolytes, where p is the electrolytic solution pressure of metal M; P the osmotic pressure of the M ions in solution; e the single-electrode potential corresponding to the equilibrium; R the gas constant; and T the absolute temperature. For dilute solutions, P is a function of c, the concentration, and $P = kc$. Then

$$neF = -RT \log_e \frac{p}{kc}$$

$$e = -\frac{RT}{nF} \log_e \frac{p}{k} + \frac{RT}{nF} \log_e c$$

For a pure metal at a given temperature

$$-\frac{RT}{nF} \log_e \frac{p}{k} = e_o$$

or the normal electrode potential of the specific equilibrium. If $T = 291$ (18°C.), $R = 8.32$ joules and, using Briggs' logarithms for the cation,

$$e_c = e_o + \frac{(0.058 \log c)}{n}$$

and for the anion,

$$e_a = e_o - \frac{(0.058 \log c)}{n}$$

if $c = 1$, $e = e_0$. The potential of an electrode equilibrium is then the potential difference of the electrode material and a molar solution of the ion involved.

A table of electrode potentials will furnish data as to the quantitative aspect of electrode equilibriums in a concise form. A table of the more important single potentials is given, the values being taken from Gerke, "International Critical Tables," vol. 6, McGraw-Hill, by permission.

In addition to electrode reactions between metal electrodes and metal ions, gas electrodes exist in that hydrogen, the halogens, and oxygen also are known to ionize. Nitrogen, however, does not ionize. The gases are bubbled against the surface of a platinized-platinum electrode in an electrolyte containing the ion concerned. The gases are dissolved or sorbed by the metal and can ionize. The P of a gas dissolved in platinum is a function of the concentration of the gas in the metal and the gas pressure on the entire system.

Many reactions, when resolved into their constituent oxidation and reduction processes, give electrode reactions that involve transference of electricity from ion to ion. Cells involving these reactions are oxidation-reduction cells, and the electrode systems oxidation-reduction electrodes. The essential reactions are the changes in the amount of electricity associated with the substances. For a reaction of the type $M^x \rightleftarrows M^y + (x - y) \oplus$, the potential difference varies with ionic concentration according to the relation

$$e = e_0 + \frac{0.058}{n} \log \frac{c}{c'}$$

where c = concentration of M^x ions and c' = concentration of M^y ions. When $c = c' = 1$, $\log \frac{c}{c'} = 0$. Table 9 gives representative values of oxidation-reduction potentials for some common cells.

Table 9. Reduction Reactions

	Volts	Ref.
$Cr^{+++} + e = Cr^{++}$	−0.40	1.2
$Cu_2O + H_2O + 2e = 2Cu + 2OH^-$	−0.34	3
$CuS + 2H^+ + e = Cu + H_2S$	−0.259	4
$SbO + 2H^+ + 3e = Sb + H_2O$	−0.212	5
$PbS + 2H^+ + 2e = Pb + H_2S$	0.07	4
$PbO_2 + H_2O + e = PbO + 2OH^-$	0.27	6
$Ti^{+++} + e = Ti^{++}$	0.37	7
$Cu^{++} + 2Cl^- + e = CuCl_2^-$	0.455	8
$K_3Fe(CN)_6 + K^+ + e = K_4Fe(CN)_6$	0.486	9
$H_3AsO_4 + 2H^+ + 2e = H_3AsO_3 + H_2O$	0.57	10
$MnO_4^- + e = MnO_4^{--}$	0.664	11
$Fe^{+++} + e = Fe^{++}$	0.747	12,13
$Tl^{+++} + 2e = Tl^+$	1.21	14
$Sn^{++++} + 2e = Sn^{++}$	1.25	15
$MnO_2 + 4H^+ + 2e = Mn^{++} + 2H_2O$	1.33	16
$Ce^{++++} + e = Ce^{+++}$	1.55	17
$MnO_4^- + 4H^+ + 3e = MnO_2 + 2H_2O$	1.58	18
$Co^{+++} + e = Co^{++}$	1.81	19.20

[1] Forbes and Richter, *J. Am. Chem. Soc.*, **39**, 1140 (1917).
[2] Grube and Breitinger, *Z. Elektrochem.*, **33**, 112 (1927).
[3] Allmand, *J. Chem. Soc., London*, **95**, 2151 (1909).
[4] Knox, *Trans. Faraday Soc.*, **4**, 29 (1908).
[5] Schuhmann, *J. Am. Chem. Soc.*, **46**, 52 (1924).
[6] Glasstone, *J. Chem. Soc., London*, **121**, 1456 (1922).
[7] Forbes and Hall, *J. Am. Chem. Soc.*, **46**, 385 (1924).
[8] Carter and Lea, *J. Chem. Soc., London*, **127**, 499 (1925).
[9] Lewis and Sargent, *J. Am. Chem. Soc.*, **31**, 355 (1909).
[10] Foerster and Pressprich, *Z. Elektrochem.*, **33**, 176 (1927).
[11] Sackur and Taegener, *Z. Elektrochem.*, **18**, 718 (1912).
[12] Noyes and Brann, *J. Am. Chem. Soc.*, **34**, 1016 (1912).
[13] Popoff and Kunz, *J. Am. Chem. Soc.*, **51**, 382 (1929).
[14] Grube and Hermann, *Z. Elektrochem.*, **26**, 291 (1920).
[15] Forbes and Bartlett, *J. Am. Chem. Soc.*, **36**, 2030 (1914).
[16] Tower, *Z. physik. Chem.*, **32**, 566 (1900).
[17] Baur and Glaessner, *Z. Elektrochem.*, **9**, 534 (1903).
[18] Brown and Tefft, *J. Am. Chem. Soc.*, **48**, 1128 (1926).
[19] Jahn, *Z. anorg. allgem. Chem.*, **60**, 292 (1908).
[20] Lamb and Larson, *J. Am. Chem. Soc.*, **42**, 2024 (1920).

In concentration cells, potential differences are set up between electrodes in contact with different concentrations of the same electrolyte. The e.m.f. of a concentration cell is a function of the ratio of the ionic concentrations at the two electrodes. Only when the concentration ratio is large are the e.m.f. values large.

Polarization, Overvoltage, and Passivity

1. Polarization. If current from an external source is sent into a cell, the voltage of the cell terminals is raised above its static value. The cell is said to be polarized. When, however, an appreciable current is passing across the boundary between an electrode and a solution, the value of the potential difference between the two is changed from its equilibrium value as given by the Nernst equation. The difference between these two values is called **polarization.** Complete or partial removal of this difference is termed **depolarization.** Any agent that does this work is called a **depolarizer.**

A primary cell becomes polarized when discharged so that its e.m.f. falls below its static value. When the potential of an electrode is raised, the electrode is anodically polarized; when the potential is lowered or becomes more negative, the electrode is cathodically polarized. The amount of polarization is the difference between the actual and the equilibrium values of the electrode potential.

Owing to irreversibility of electrode processes and ohmic resistances in the cell at various points, the working voltage of a cell always exceeds the theoretical decomposition voltage. The percentage ratio between the theoretical quantity of energy necessary for the production of a given amount of a substance and the actual quantity of energy, the latter always being the larger, is termed the **energy efficiency** of the process. The energy efficiency may also be described as the product of the **current efficiency** and the percentage ratio of the theoretical to the actual voltage.

Electrode processes occur essentially at the surface of the electrode, in contact with the electrolyte or in thin films adjacent to these electrodes. The bulk of the electrolyte may be considered merely as a reservoir for ions and as a conducting medium. At the cathode the discharge of an ion would tend to reduce the concentration of the ions in the cathode film. As a result, the single potential of the electrode tends to become more negative and cathodic polarization begins. The higher the current density, the greater will be the tendency for the concentration of the ions in the cathode film to be reduced. This tendency is opposed by diffusion of the ions from the region of higher concentration in the electrolyte to that of lower concentration around the cathode. Convection currents, mechanical agitation, and other compensating processes will tend to reduce this form of concentration polarization. A similar set of conditions holds true for the anode, with the exception that the single potential of the anode hence becomes more positive.

If the current become so large that the concentration of metal ions on the cathode surface is reduced practically to zero, no larger current can pass, however great the potential difference is made, unless some other ions begin to deposit. This value is termed the **limiting** or **maximum current.** It depends upon the concentration, temperature, and rate of stirring of the solution.

If the polarization occurring at ordinary current density is not greater than that accounted for by concentration polarization, the process is considered as reversible. Otherwise the process is considered to be irreversible.

Polarization may be caused by factors that interfere with the main electrode process. Films of non-conducting substances may form on the electrode surfaces, and the current density at the uncoated sections of the electrodes will increase. Enormous polarization may set in at electrode surfaces completely covered with films.

Commercial utilization of this type of effect is found in anodic oxidation of aluminum and its alloys, tantalum, and other metals in rectifiers, electrolytic lightning arresters, etc.

Electrolytic Rectifiers (see Electrical Section). Electrodes which, as anodes, have become covered with an insulating film show different behaviors when employed as cathodes. With some the insulating layer is easily removed, but in the case of metals having difficultly reducible oxides (Al and Ta) the film continues for some time although H^+ ions may be discharged through it. Such electrodes are **electrolytic valves.** They possess the property of permitting current passage in only one direction. A cell with an aluminum plate as cathode and a lead sheet as anode in a suitable electrolyte may be placed in series with and between a generator and a storage cell battery. The interposed cell does not interfere with the charging current to the cells but prevents the storage cells from "feeding back" to the generator when that machine is shut down.

Electrolytic valves may be employed as rectifiers for a.c.-d.c. conversion. A typical arrangement is shown in Fig. 2. Four rectifier cells, each containing an aluminum electrode and one of carbon, iron, or platinum, are connected as shown. The film formation on aluminum allows the passage of appreciable current only when the electrode is used as cathode. From the figure it can be verified that the current produced on the right-hand circuit will be direct current, and that both positive and negative waves of the alternating current are utilized. The energy efficiency is about 60 per cent.

It so happens that in some cases polarization is definitely necessary for the production of the desired electrode process. In chromium deposition, the cathodic hydrogen film is an important factor in the production of the chromium metal. Polarization effects may markedly affect the type of deposits obtained. When these cannot be eliminated by variation of current and voltage, depolarizers are employed. In high-quality nickel plating, hydrogen adsorbed by the nickel affects the ductility of the metal. Oxidizers like hydrogen peroxide are added to convert the hydrogen codeposited with the nickel into water. Chlorides in various plating baths and in refining solutions find application in overcoming anode polarization by increasing the rate of anode corrosion.

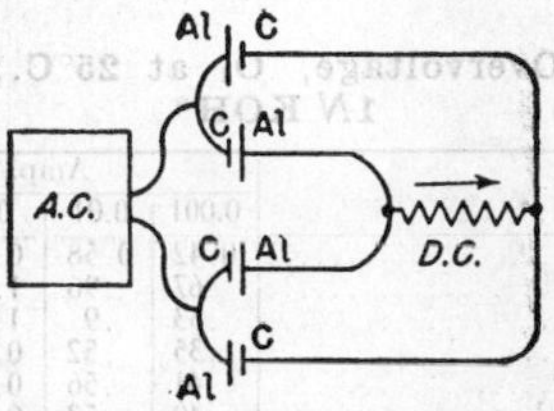

FIG. 2. Electrolytic rectifier.

2. Overvoltage. Hydrogen Overvoltage. Even at low current density, some polarization is usually needed to cause cathodic hydrogen evolution. At higher current density, these polarizations may be of considerable magnitude. The polarization voltage needed for the evolution of a gas at an electrode material is known as the overvoltage of the material for the gas under the conditions stated. The hydrogen overvoltage of an electrode is the difference between the (actual) cathode potential for hydrogen evolution and the equilibrium (theoretical) potential of hydrogen in the same electrolyte. Hydrogen overvoltage of an electrode material may be obtained from the current-density-electrode-potential curves for hydrogen evolution at a cathode of that material. The values vary directly with the current density, the extent of this increase usually being greater than accountable by concentration changes in the electrolyte. Hydrogen overvoltages at very high current density for nearly all substances approach limiting values of about 1.3 volts. A table of hydrogen overvoltages for various metals (Table 10) is given below. This is condensed from Knobel's data in "International Critical Tables," McGraw-Hill, by permission.

Hydrogen overvoltage values for most cathode materials are of practical importance even at moderate current density. The highest hydrogen overvoltages are shown by the soft metals of low melting point, while those of high melting point have relatively small values.

Overvoltage depends upon a large number of factors such as condition of the electrode surface, purity of the electrode, temperature, impurities and colloidal substances in the electrolyte, external pressure, imposed alternating current, current density, and time of electrolysis.

Overvoltage necessitates increased energy expenditure in the separation of electrolytic gases. In addition it may actually change the nature of an electrode process as the result of the change of electrode potential.

Table 10. Overvoltage, H_2 at 25°C.; Electrolyte $2N$ H_2SO_4*

	Amp./sq. cm.				
	0.001	0.01	0.1	0.5	1.0
Aluminum	0.56	0.83	1.0	1.24	1.29
Bismuth	.78	1.05	1.14	1.21	1.23
Brass	.50	0.65	0.91	1.23	1.25
Cadmium	.98	1.13	1.22	1.25	1.25
Carbon	.....	0.70	0.9	1.1	1.17
Copper	.48	0.58	0.8	1.19	1.25
Duriron	.20	0.29	0.61	0.86	1.02
Gold	.24	0.39	0.59	0.77	0.80
Graphite	.60	0.78	0.98	1.17	1.22
Iron	.40	0.56	0.82	1.26	1.29
Lead	.52	1.09	1.18	1.24	1.26
Mercury	.9	1.04	1.07	1.1	1.12
Monel	.28	0.38	0.62	0.86	1.07
Nickel	.56	0.75	1.05	1.21	1.24
Palladium	.12	0.3	0.7	1.	1.
Platinum, platinized	.015	0.03	0.04	0.05	0.05
Platinum, smooth	.024	0.07	0.29	0.57	0.68
Silver	.47	0.76	0.88	1.03	1.09
Tellurium	.4	0.45	0.48	0.54	0.6
Tin	.86	1.08	1.22	1.24	1.23
Zinc	.72	0.75	1.06	1.20	1.23

* Knobel, *J. Am. Chem. Soc.*, **46**, 2613, 2751 (1924). Knobel, Caplan, and Eiseman, *Trans. Am. Electrochem. Soc.*, **43**, 55 (1923). Knobel and Joy, *Trans. Am. Electrochem. Soc.*, **44**, 443 (1923). Newbery, *J. Chem. Soc., London*, **105**, 2419 (1914); *Trans. Faraday Soc.*, **15 I.**, 126 (1919).

Table 11. Overvoltage, O_2 at 25°C.; Electrolyte $1N$ KOH*

	Amp./sq. cm.				
	0.001	0.01	0.1	0.5	1.0
Copper	0.42	0.58	0.66	0.74	0.79
Gold	.67	.96	1.24	1.53	1.63
Graphite	.53	.9	1.09	1.19	1.24
Nickel, smooth	.35	.52	0.73	0.82	0.85
Nickel, spongy	.41	.56	0.69	0.74	0.76
Platinum, platinized	.40	.52	0.64	0.71	0.77
Platinum, smooth	.72	.85	1.28	1.43	1.49
Silver	.58	.73	0.98	1.08	1.13

* References are same as for Table 10.

Oxygen Overvoltage. In a manner similar to hydrogen overvoltage, oxygen overvoltage exists in the course of anodic evolution of oxygen. Only the noble metals, and a few others like those of the iron group which can be "ennobled," can be satisfactorily employed for the evolution of oxygen. A table of values is given in Table 11 (condensed from Knobel's tables in "International Critical Tables").

Halogen Overvoltage. Overvoltages for the halogens exist of a nature comparable to hydrogen and oxygen overvoltages. On continued electrolysis these frequently attain very large values. Values for chlorine, bromine, and iodine overvoltages are given in Table 12 (condensed from Knobel's tables, in "International Critical Tables," McGraw-Hill, by permission).

Table 12. Overvoltage*

Cl_2 at 25°C.; Electrolyte Saturated Solution of NaCl or KCl

	Amp./sq. cm.				
	0.001	0.01	0.1	0.5	1.0
Graphite	.....	.....	0.25	0.42	0.5
Platinum, platinized	0.006	0.016	.026	.05	.08
Platinum, smooth	.008	.03	.054	.16	.24

Br_2 at 25°C.; Electrolyte Saturated Solution of NaBr or KBr

	Amp./sq. cm.			
	0.01	0.1	0.5	1.0
Graphite	0.002	0.027	0.16	0.33
Platinum, platinized	.002	.012	.07	.2
Platinum, smooth	.002	.....	.26	.4

I_2 at 25°C.; Electrolyte Saturated Solution of NaI or KI

	Amp./sq. cm.			
	0.01	0.1	0.5	1.0
Graphite	0.013	0.1	0.4	0.8
Platinum, platinized	.006	.03	.09	.2
Platinum, smooth	.004	.03	.12	.22

* References are same as for Table 10.

3. Passivity. In the case of the iron-group metals and chromium, solution of these metals as anodes may be displaced by other reactions without the formation of visible films on the anode. Normally the anodic solution of this group of metals requires a considerable polarization, but at higher current density, oxygen evolution sets in, the metal is "passivated" or ennobled. Reduction of current density does not immediately eliminate the passive state of the metal. Similar phenomena occur with other metals, particularly with the noble ones whose reversible potentials and oxygen-evolution potentials are close together. In general, halogen ions interfere with passivity while oxidizing ions favor its inception. Chromium and its alloys are easily rendered passive and are thus resistant to anodic solution or corrosion. Passivity may markedly affect the type of anode reactions taking place.

The passive state in metals when used as anodes in electrolytic cells is analogous to chemical passivity, a state that is of considerable importance in the corrosion resistance of metals when used as structural materials.

Many theories of anodic metal solution and passivity have been proposed. In general, the anodic polarization required for the solution of certain metals is attributed to the formation of metal-oxygen complexes at the electrode surface which make the single potential of the electrode more positive and hinder its solution. The value of this quantity may become so great that other anode reactions set in. The passivation of a metal, whether anodic or chemical, will be favored by those conditions which tend to produce a high oxygen concentration in the surface layers of the metal.

Superimposed Alternating Current on Direct Current. Superimposing an alternating current on a direct current in electrolysis causes a decrease in electrode potential, reduces any irreversibility of the reaction, acts as a depolarizer, reduces hydrogen overvoltage, affects oxygen evolution at anodes, lowers chlorine overvoltage, aids anodic solution of metals, and allows higher current density of direct current at an electrode.

The electrochemical effects are a function of the direct current, and the ratio of alternating current to direct current and the frequency of the alternating current determine the magnitude of its effects. The modified Wohlwill gold-refining process is the only industrial use of superimposed alternating current on direct current.

Primary and Secondary Cells

Batteries. The conversion of chemical into electrical energy is the function of primary cells or batteries. The term *batteries* is usually applied to an assembly of identical units or cells but is often loosely used for designating a single unit. In a **primary battery** the chemically reacting parts require renewal or replacement, while in a **secondary** or **storage battery,** reactions being reversible to a high degree, the chemical conditions are restored after partial or complete discharge by reversal of the current flow, *i.e.*, by sending electric current into the cell. The high cost of primary batteries makes the production of electricity in large quantities from them impractical, so that they are largely used for services of an intermittent nature, or for those demanding electric current for short times.

From the table of single-electrode potentials (Table 8), we can determine the e.m.f. of a primary cell resulting from the combination of two electrode systems, one with a high positive or oxidizing and the other with a strongly negative or reducing potential. The e.m.f. of the resulting cell is then the difference between the two potentials. A large number of systems may be theoretically set up,

Table 13. Primary Batteries

Type	System	E.m.f.
Bunsen	HgZn $\mid$ 1 part H_2SO_4 + 12 parts $H_2O \mid$ concentrated $HNO_3 \mid$ C	1.94
Bichromate or Poggendorf	HgZn $\mid$ 1 part H_2SO_4 + 12 parts $H_2O \mid$ concentrated solution $Na_2Cr_2O_7$ + $H_2SO_4 \mid$ C	2.00
Daniell or gravity	HgZn $\mid$ 5% solution $ZnSO_4.6H_2O \mid$ saturated solution $CuSO_4.5H_2O \mid$ Cu	1.07
Féry	Zn $\mid$ 12% NH_4Cl solution $\mid$ depolarizing C	1.2
Grove	HgZn $\mid$ 1 part H_2SO_4 + 12 parts $H_2O \mid$ fuming $HNO_3 \mid$ Pt	1.66
Leclanché	HgZn $\mid$ 20% NH_4Cl solution—$MnO_2 \mid$ C	1.5
Lalande	Zn $\mid$ 18–19% NaOH solution-oxides of Cu $\mid$ Cu	0.95
Dry cell	Zn $\mid$ NH_4Cl—$ZnCl_2$—$MnO_2 \mid$ C	1.53
Le Carbone	Zn $\mid$ 20% NaOH $\mid$ special C	1.4
Air cell	Zn $\mid$ solution NaOH $\mid$ special C	1.25

but in the case of most of them they will be found unsuited for practical use because of chemical activity of the anode material, costliness, or tendency to passivity, as well as other reasons. In practice zinc is almost invariably the soluble anode in a primary cell.

On the assumption that the electrode system and its resulting reactions are reversible, the e.m.f. of the cell may be calculated from the Gibbs-Helmholtz equation. If the cell has a zero temperature coefficient, the theoretical e.m.f. may be calculated from the heats of reaction of the materials of the cell. In practice, however, electrode systems never behave absolutely reversibly. At times the degree of irreversibility is considerable. Inasmuch as soluble anode materials are commonly used and oxygen evolution does not occur, passivity is the only irreversible effect to be feared. This is eliminated by the proper choice of metal and electrolyte. At the cathode where the process is a discharge of a metallic ion to metal, the reaction takes place very nearly reversibly. If, however, the cathode system is an oxidation-reduction electrode consisting of an oxidizing agent in contact with an indifferent electrode and serving to depolarize the discharge of hydrogen ions, a considerable overvoltage may be needed for the hydrogen discharge which will in turn lower the e.m.f. of the cell. If depolarizers are used, the e.m.f. of the cell is increased above the value corresponding to reversible cathodic hydrogen discharge. The reaction occurs so quickly between the depolarizer and the discharged hydrogen that the electrode never becomes saturated with the gas, and hydrogen discharge takes place at a less negative cathode potential. The more effective and rapid the action of the depolarizer and hydrogen, the lower is the hydrogen concentration in the electrode, the more positive the cathode potential, and the more nearly it reaches its equilibrium value.

Liquid depolarizers act more rapidly than do solids but tend to diffuse toward the anode which they may strongly attack. This may necessitate a diaphragm between the anode and the cathode, which increases the internal resistance of the cell. For satisfactory use solid depolarizers, in addition to reacting quickly with the discharged hydrogen, should have high electrical conductivity and make good contact. This may be obtained, as in the case of the dry cells, by adding graphite to the MnO_2 depolarizer.

Internal resistances in the cell lower the e.m.f. of the unit. These may be due to the type of construction, diaphragms, concentration changes, electrolyte resistances, as well as those of a film or polarization nature not taken care of by depolarizers.

Scores of different primary batteries were on the market a generation ago, but now only a few of commercial importance. At the present time the list includes the **Daniell** cell, the **Lalande**, and the **Leclanché**, of which the dry cell and the **Féry** cell are modifications. Of the entire group, the dry cell is by far the most important, followed by the Lalande, often termed the caustic soda primary battery. Table 13 lists laboratory and commercial cell systems.

The Daniell cell consists of the system

$$Zn \mid ZnSO_4 \mid CuSO_4 \mid Cu$$

and the chemical reaction is

$$Zn + CuSO_4 = ZnSO_4 + Cu$$

The cell e.m.f. is a function of the concentration of the $ZnSO_4$ solution, having a maximum value of about 1.14 volts. If the $ZnSO_4$ solution is not acid with H_2SO_4, the voltage is 1.07. Inasmuch as the chemical reactions between the constituents of the cell continue whether the cell be used or not, the setup is not adapted to stand on open circuit.

A modification of the Daniell cell is known as the **gravity battery,** shown diagrammatically in Fig. 3.

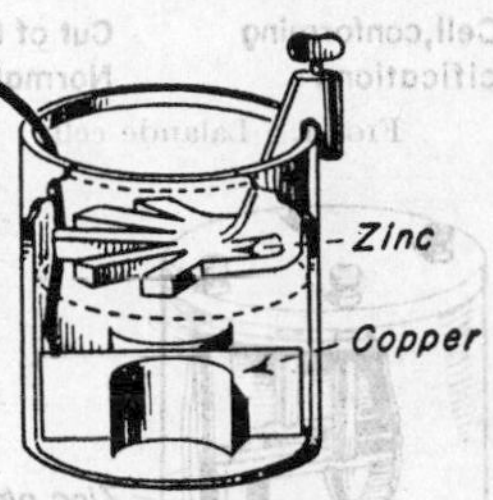

FIG. 3. Gravity cell.

The heavier $CuSO_4$ solution is placed at the bottom of the cell in contact with a spread-out copper-sheet electrode to which a rubber-covered wire connection is made. The zinc electrode is frequently in the form of a cast crowfoot suspended at the top of the jar, surrounded by $ZnSO_4$ solution which has been carefully poured on top of the $CuSO_4$ solution. The edges of the jars are ordinarily coated with paraffin to prevent creepage of $ZnSO_4$ crystals over the top, and evaporation is reduced by covering the solution with a thin layer of mineral oil. It is estimated that under average conditions only about 30 per cent of the zinc is electrochemically utilized.

The **Lalande** cell has a soluble zinc anode, an alkaline electrolyte, and a solid cathodic depolarizer. The system is Zn | alkali solution-oxides of copper | Cu. The electrolyte is usually 18 to 19 per cent NaOH. The construction of a typical cell is shown in Fig. 4. The CuO acts as a depolarizer and is made either in a compressed or in a loose form. In the manufacture of the compressed form, the CuO is first made in a very fine powder, mixed with a binder, and compressed under heavy pressures, then given a baking treatment, after which the outer surface is metallized by partial reduction to lower the resistance, inasmuch as CuO alone is a very poor conductor. The compressed oxide element is usually a flat plate or a hollow cylinder. In commercial cells the zinc electrode is usually cast in a cylindrical form and amalgamated with mercury, so that the electrode contains as much as 2.5 per cent of the latter metal. The CuO and zinc electrode are mounted in a glass jar made of heavy construction to withstand the corrosive action of the caustic solution, and the electrolyte covered with a layer of mineral oil to prevent absorption of CO_2 from the atmosphere by the solution as well as the evaporation of the solution. Purity of materials is quite important for the successful operation of the cell. In the **Edison** type, the electrodes are flat plates, the compressed copper oxide being in the center of the cell on each side of which is a zinc plate cast with ribs. The construction of the cell is shown in Fig. 5.

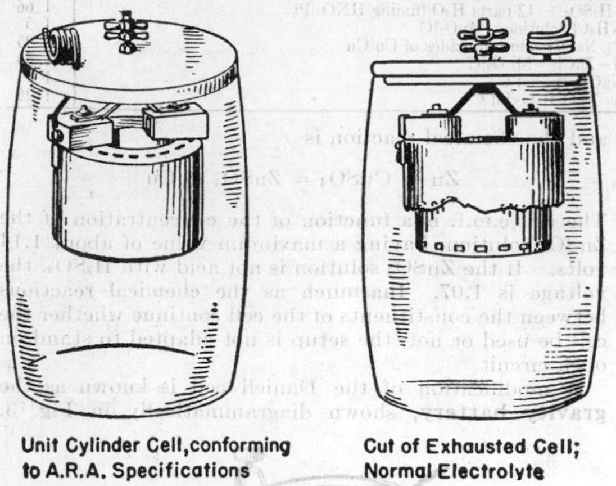
Unit Cylinder Cell, conforming to A.R.A. Specifications
Cut of Exhausted Cell; Normal Electrolyte

FIG. 4. Lalande cell.

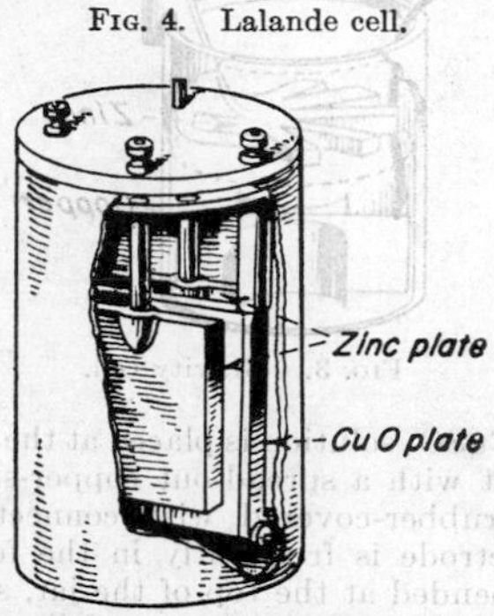

FIG. 5. Edison cell.

The chemical reaction of the cell may be given as

$$Zn + 2NaOH + CuO = Na_2ZnO_2 + H_2O + Cu$$

The e.m.f. of the cell is about 0.95, but the terminal voltage drops to less than two-thirds of this when heavy currents are furnished. The internal resistance of the cell is low. It is adapted to both closed- and open-circuit work. Because of its simple construction, ready working, relative cheapness, and despite its low voltage, it has found extensive use for the operation of signal systems and for railway work. Batteries are commercially manufactured in sizes from 75 amp.-hr. up to cells with a rated capacity of 1000 amp.-hr. When cells are run down, fresh zinc plates are added, and the oxide electrode, which has been largely reduced to copper, is washed and reoxidized by heating at 150°C.

The **Leclanché** cell is of the system

$$Zn|NH_4Cl - MnO_2|C$$

The electrolyte is a strong NH_4Cl solution, usually about 20 per cent, to which various hygroscopic substances, such as glycerin, $ZnCl_2$, or at times $CaCl_2$, may be added to lessen the tendency of the cell to lose water. The reaction of the cell is given as

$$Zn + 2NH_4Cl + 2MnO_2 = Zn(NH_3)_2Cl_2 + H_2O + Mn_2O_3$$

In the original form, the carbon rod was contained in a porous cup filled with crushed carbon and MnO_2, the mixture being tamped to obtain intimate contact. In later forms the MnO_2 and carbon were molded by the use of a binder into a cylindrical form, the zinc rod being suspended centrally as in Fig. 6. The cell e.m.f. is about 1.5 volts, but the terminal voltage drops rapidly when heavy currents are drawn from the cell, showing values of 1.1 to 1.2 volts at currents as low as 0.1 to 0.2 amp. for an ordinary cell. The cell is suitable for open- or closed-circuit work if large currents are drawn intermittently and only for a short time. The dry-battery modification of the Leclanché cell has largely replaced this unit in commercial work.

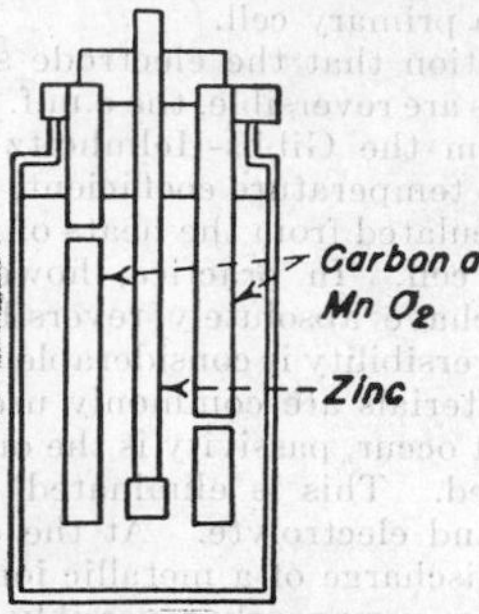

FIG. 6. Leclanché cell.

Dry Cells. The dry cell is so built that its electrolyte is contained in an absorbent material which prevents its spilling out with the cell in any position. The cell, however, is not dry, as one of the essential requirements in its construction is to be sufficiently wet under all ordinary conditions. The zinc electrode is made the container for the cell, the electrolyte consisting of a water solution of NH_4Cl and $ZnCl_2$, held partly in an adsorbent material that lines the zinc container and partly in the mixture of ground carbon or graphite and MnO_2 which serves as a depolarizer. The latter is bulky and occupies most of the interior of the cell. Often the electrolyte is made into a jelly by the use of colloidal materials such as gum tragacanth, agar-agar, flour, or, more commonly, starch. In American practice the cell is completely sealed at the top, the newer types having a vent for the escape of gas.

The cells are made in two general sizes, the larger ones for ignition, signaling, and miscellaneous intermittent use, the smaller ones for flashlights, radio batteries and, similar purposes. In the case of the larger cells the zinc container is made of sheet with the bottom soldered with lap seams, but in the smaller cells the containers are zinc stampings. The zinc, for electrochemical reasons, must have a high degree of purity but to withstand the manufacturing operations must also have good mechanical properties, *i.e.*, high tensile strength and elongation as well as stiffness. The use of pure zinc avoids galvanic couples and local corrosion of the metal.

The electrical conductivity of the MnO_2 depolarizer is very low, so that granulated carbon, more or less completely graphitized, is added to increase the conductivity of the mixture. As regards the NH_4Cl electrolyte, it is desirable that it be free of metals such as Cu, Pb, Fe, As, Ni, Co, and Sb, which may be plated out by the zinc, as well as free from negative radicals, such as sulfates, which form compounds less soluble than the chlorides. Insulation and sealing compounds are usually resin, sealing wax, or bituminous pitches with fillers such as ground silica, fibrous talc, and coloring matter added. The entire cell is ordinarily insulated by a paper container or carton surrounding it.

Two general methods of manufacture are employed. In the first, used for the large cells, the zinc container is lined with a sulfite- and ground-wood-pulp board into which the MnO_2—NH_4Cl—$ZnCl_2$ mixture is tamped around the central carbon electrode, after which the cell is sealed with a sealing compound. In the manufacture

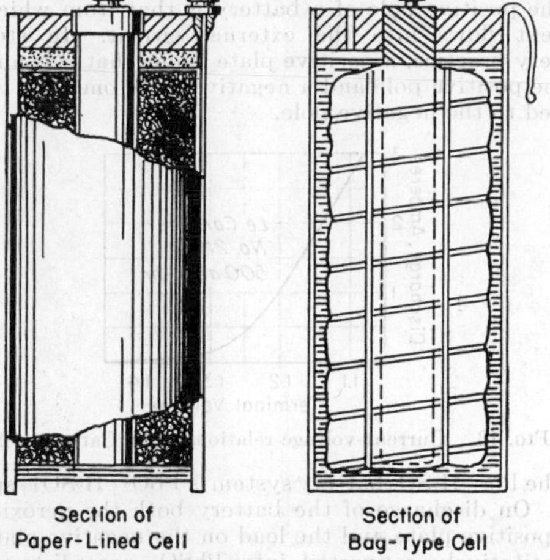

FIG. 7. Dry cells.

of cells of the flash light type, the carbon rod with its surrounding mixture of depolarizer and electrolyte is wrapped in a muslin bag and tied with a string, forming a unit which is placed in the zinc can, leaving sufficient space between the two for the electrolyte in the form of a paste. A solution of ammonium chloride and zinc chloride used for the electrolyte is thickened with flour or

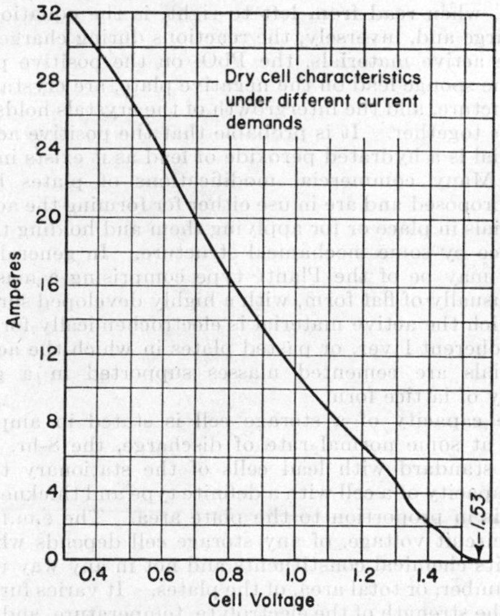

FIG. 8. Dry-cell characteristics.

starch. The **desiccated cell** is manufactured dry, being either of the paper-lined or of the bag type, the cell being provided with an opening in the seal or the center of the carbon rod, through which the water necessary to make the cell active may be introduced. The energy of a 90-g. *D*-cell is equal to almost 13,000 watt-sec. Such a unit will deliver about 2.73 amp.-hr. before its voltage falls to 1.13 volts on a discharge through 83.3 ohms for 4 hr. a day. These conditions correspond to radio use. The average voltage during this discharge is 1.3. Table 14 gives the commercially important dry-cell sizes and voltages. Figure 8 shows dry-cell characteristics, and

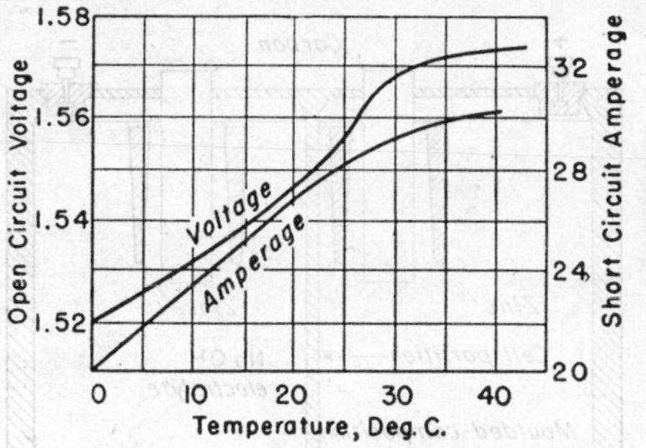

FIG. 9. Temperature-current-voltage relations of No. 6 dry cells.

Fig. 9 shows the effect of temperature on the voltage and amperage of No. 6 dry cells.

Table 14. Dry Cells

Type	Sizes				Minimum voltage
	Diameter, in.	Height, in.	Diameter, mm.	Height, mm.	
A	5/8	1 7/8	16	48	1.47
B	3/4	2 1/8	19	54	1.48
C	15/16	1 13/16	24	46	1.49
D	1 1/4	2 1/4	32	57	1.50
E	1 1/4	2 7/8	32	73	1.50
F	1 1/4	3 7/16	32	87	1.50
6	2 1/2	6	63	152	1.50

The **Féry** cell is a modification of the Leclanché type for use where very low currents are required, designed to avoid the use of cathodic depolarizers completely, relying on air dissolved in the carbon for depolarizing the hydrogen discharge. The type used on the French railways consists of a glass jar at the bottom of which a zinc-plate cathode rests. External connection is made by a copper wire insulated from the electrolyte. The anode plate carries a cross-shaped insulator of synthetic plastic or ebonite on top of which the hollow carbon-cylinder cathode rests. The cathode is about half the diameter of the jar and is pierced with holes. It projects several inches above the surface of the electrolyte, which is 12 per cent NH_4Cl. Oxygen for depolarization is adsorbed from the atmosphere by the carbon. A cell weighing 2.1 kg. has a 90 amp.-hr. capacity and an open-circuit voltage of about 1.2. In practice the cells are discharged at currents of 20 to 50 milliamp. and have a life of about 6 months.

The most recent development in primary cells has been the manufacture of so-called **breather batteries** or air cells of the system Zn | 10 to 20 per cent NaOH | C. No depolarizers are employed, inasmuch as the carbon is of an adsorbent type which takes up oxygen from the air for depolarization. Two-cell units are manufactured for operation of 2-volt radio tubes. The cell shows an open-circuit voltage of about 1.25 and a closed-circuit voltage of 1.12; a closed-circuit amperage of 0.38. The batteries are said to be good for about a year of average service when used in connection with radio reception. They were developed for rural districts not served by power lines. Figure 10 shows the construction of the air cell of the National Carbon Co. Figure 11 shows the voltage characteristics of this battery.

Primary batteries of the adsorbent-carbon depolarizing type are also built for semaphore, highway flashing systems, lighthouses, railway-signal and similar work. A typical one is the **Le Carbone** cell, the construction of which is shown in Fig. 12, and the characteristics in Fig. 13. A cell whose cylindrical glass jar is approximately 10 in. high and 7 in. in diameter is rated at 500 amp.-hr., with initial closed-circuit voltage of over 1 volt at rates of discharge not greater than 3 amp. The positive or carbon element has a capacity of 2000 to 2500 amp.-hr., while the zinc circular element weighing about 825 g. lasts 500 amp.-hr., after which it is renewed. The cell holds about 4 l. of a 20 per cent NaOH electrolyte, which is renewed every 500 amp.-hr. The open-circuit voltage of the cell is 1.4 to 1.5 volts.

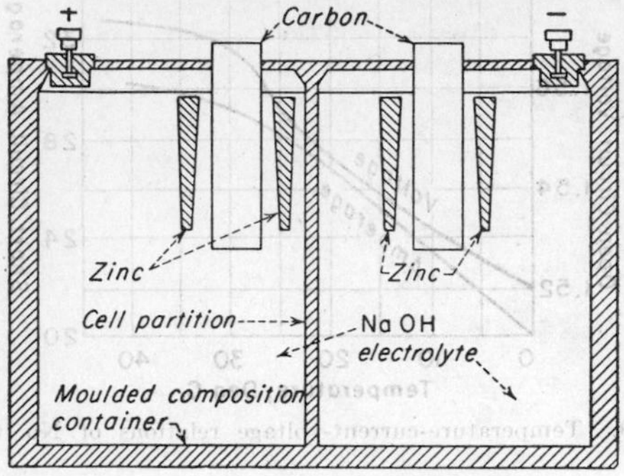

FIG. 10. Air cell, two-cell battery.

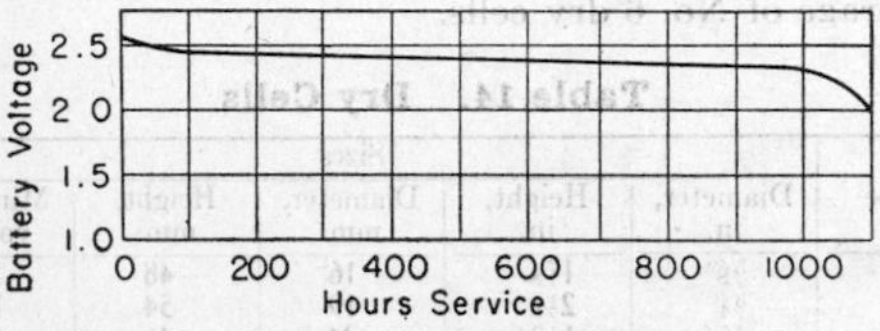

FIG. 11. Voltage characteristics of air cell.

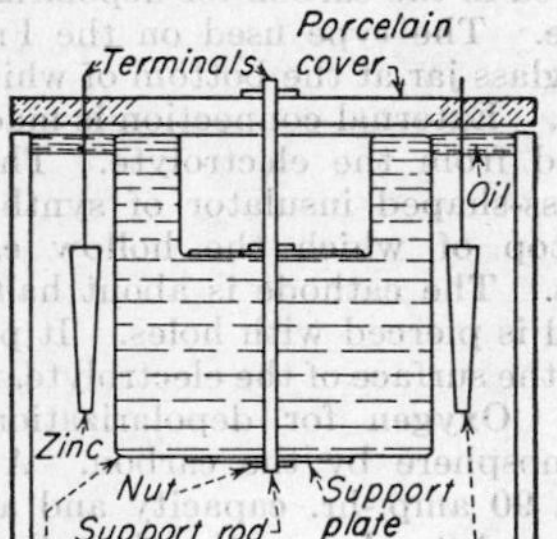

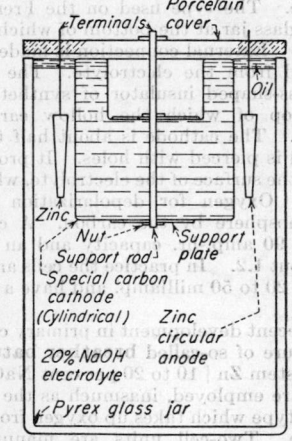

FIG. 12. Le Carbone caustic cell.

A number of specialized cell systems were developed during World War II for short-time use in either high-amperage-discharge applications or high-voltage use.

Secondary Cells. Cells that are reversible to a high degree, in that the chemical conditions may be restored by causing current to flow into the cell on charge, are used as storage batteries or electric accumulators. The Daniell or gravity cell as well as the Lalande is reversible to a high degree, but these cells have practical disadvantages preventing their use as storage batteries. The form in widest commercial use is the Pb—H_2SO_4 type. The only other form of any prominence is the Ni-Fe-caustic cell, often termed the **Edison** battery.

The positive pole of a battery is that from which the current flows into the external circuit. In storage-battery practice, a positive plate is one that is connected to the positive pole and a negative plate one that is connected to the negative pole.

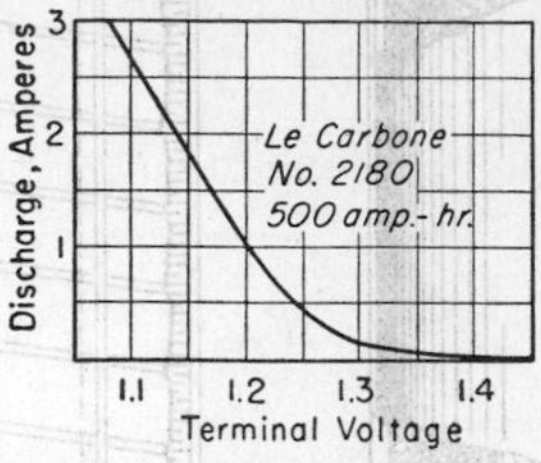

FIG. 13. Current-voltage relations of Le Carbone cell.

The Pb—H_2SO_4 battery system is $PbO_2 \mid H_2SO_4 \mid$ sponge Pb. On discharge of the battery both the peroxide on the positive plate and the lead on the negative plate are quantitatively converted into $PbSO_4$ according to the reactions

$$PbO_2 + H_2SO_4 = PbSO_4 + H_2O + O$$

and

$$Pb + H_2SO_4 = PbSO_4 + 2H$$

which may be combined into the reaction

$$PbO_2 + Pb + 2H_2SO_4 = 2PbSO_4 + 2H_2O$$

which, when read from left to right, is the equation of discharge and, inversely, the reactions during charge.

The active materials, the PbO_2 on the positive plate and the sponge lead on the negative plate, are crystalline in structure, and the intergrowth of the crystals holds the masses together. It is probable that the positive active material is a hydrated peroxide of lead as it exists in the cell. Many commercial modifications of plates have been proposed and are in use either for forming the active materials in place or for applying them and holding them in place by some mechanical structure. In general the plates may be of the Planté type comprising a mass of lead, usually of flat form, with a highly developed surface on which the active material is electrochemically formed as a coherent layer, or pasted plates in which the active materials are cemented masses supported in a grid, usually of lattice form.

The capacity of a storage cell is stated in ampere-hours at some normal rate of discharge, the 8-hr. rate being standard with lead cells of the stationary type. The capacity of a cell with a definite type and thickness of plate is in proportion to the plate area. The e.m.f., or open-circuit voltage, of any storage cell depends wholly upon its chemical constituents and not in any way upon the number, or total area, of the plates. It varies further with the strength of the electrolyte, temperature, and to a minor extent with the state of charge of the plates, internal resistance of the cell, polarization, and acid-concentration effects. Per ampere-hour of discharge, the amount of active material converted into $PbSO_4$ is 0.135 oz. sponge lead and 0.156 oz. PbO_2, independent of the rate of discharge. The amount of active material actually present in the plate is some three to six times that which under normal discharge of the cell is converted into $PbSO_4$. Part of this excess is present to give long life to the plates.

The open-circuit voltage of a lead cell varies from 2.06 to 2.14 according to the strength of the electrolyte and the temperature and may be calculated from the formula

$$E = 1.850 + 0.917(G - g)$$

where G is the specific gravity of the electrolyte and g the specific gravity of water at the cell temperature.

Typical charge and discharge curves for the stationary type of lead cell are given in Fig. 14. It is often desirable to determine the relative performance of positive and negative plates in a cell. This may be done by taking the voltage between either group and a reference electrode such as zinc, sponge lead, or preferably cadmium. Cadmium curves are included in the diagram.

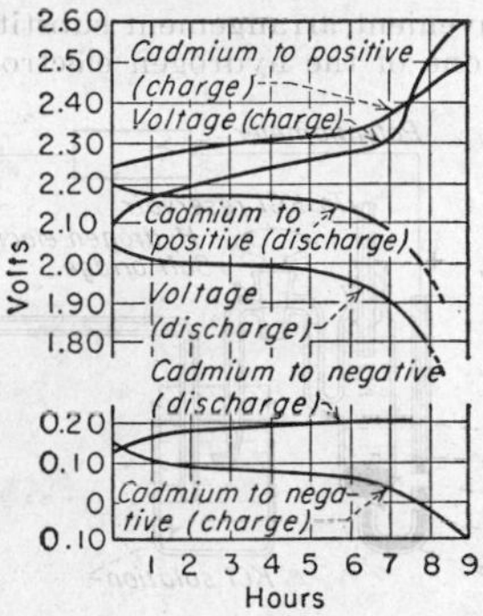

FIG. 14. Typical charge-discharge curves of lead storage battery.

Storage batteries are used for stand-by electric power services where the pasted-plate type is preferred; for electric vehicles; train lighting; gasoline-engine and automobile starting, lighting, and ignition; signaling and control work, as well as for many other applications.

Two alkaline storage batteries are in commercial use, the first being the **Hubbell** consisting of the system Ni threads and Ni oxide | KOH | Fe, used in miners' lamps, and the more important **Edison** battery consisting of the following system: finely divided Ni + Ni peroxide | 21 per cent KOH | finely divided Fe. The active materials of the Edison battery consist of nickel peroxide for the positive plate and finely divided iron for the negative. Small amounts of LiOH are added to the electrolyte, and certain amounts of mercury are incorporated with the iron of the negative plate to overcome the passivity of the iron, while layers of flake nickel are added to the positive plate nickel oxide to increase its conductivity. The reactions of the cell are:

$$8OH + 3Fe = Fe_3O_4 + 4H_2O$$
$$8K + 6NiO_2 = 2Ni_3O_4 + 4K_2O$$
$$8KOH + 6NiO_2 + 3Fe = Fe_3O_4 + 2Ni_3O_4 + 8KOH$$

or

$$6NiO_2 + 3Fe = Fe_3O_4 + 2Ni_3O_4$$

which, when read from left to right, are the reactions of discharge and, inversely, those of charge. It is probable that the iron and nickel oxides are both hydrated. In charging the battery, the electrolyte density does not change as in the lead storage cell, but concentration changes of the electrolyte in the pores of the active materials do occur with perhaps the formation of higher oxides of nickel.

The positive plate consists of a nickel-plated steel frame into which are pressed perforated tubes filled with alternate layers of nickel hydrate and metallic nickel in very thin flakes. The tube is made from a thin sheet of steel, nickel-plated and perforated, and has a spirally lapped joint. The negative plate consists of a grid of nickel steel with oblong openings into which perforated steel boxes containing finely divided iron with mercury are placed.

Characteristic normal charge, and discharge, curves for the Edison battery are given (Fig. 15). The average voltage on discharge is approximately 1.2, the initial open-circuit voltage 1.5, and the final voltage at the end of discharge a little less than 1.

Edison cells are used for ignition and lighting of gasoline motor cars but because of their high internal resistance are not used for motor starting. They find application in electric vehicles, storage-battery streetcars, mining locomotives, and industrial trucks. In contradistinction

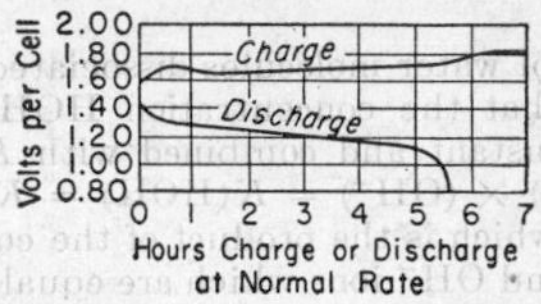

FIG. 15. Charge-discharge curves for Edison alkaline battery.

to lead storage batteries, they are not used for load regulation in power systems because of their heavy voltage drop at high discharge rates. The commercial cells show ampere-hour efficiencies of 82 per cent and watt-hour efficiencies of 60 per cent, and an average capacity of about 13 watt-hr./lb. of cell. The Edison cell is especially sensitive to reduction of electrolyte temperature, showing a critical point about 50°F. below which the capacity falls off very rapidly. Its performance at low temperatures, such as are met in the northern parts of the United States, is therefore unsatisfactory.

Electrolysis and pH Measurement

Electroanalysis. Electrochemical methods are employed to a considerable extent in analytical work for the determination of metals from solution, the separation of these metals one from the other by deposition at controlled voltages. Gauze electrodes and rotating anodes are employed so that high current densities may be used to shorten the necessary time for analysis. In other cases mercury cathodes, which adsorb metals of the salt solution being electrolyzed, are advantageous. For specific methods, reference should be made to the standard works on analytical chemistry.

pH and Its Measurement. This section is a condensation of the discussion of pH in various publications of the Leeds and Northrup Co. and is used by their permission.

The concentration of H^+ ion in any solution may be expressed in terms of the normal H^+ ion solution containing 1 g. H^+ ion per liter. In most solutions the H^+ ion concentration is only a small fraction of that in a normal solution and is expressed as powers of 10 to avoid decimals with a large number of ciphers, or fractions with large denominators. A concentration of 0.0001 g. H^+ ion per liter thus becomes $1/10^4$ or 1×10^{-4}. The symbol pH is used to designate the logarithm of the reciprocal of the H^+ ion concentration, or the negative logarithm of the H^+ ion concentration. Thus a 0.0001 N solution of H^+ ion equals 1×10^{-4}, or $-\log(H^+) = 4 = \log(1/H^+)$; therefore pH = 4. Concentrations that are uneven decimal fractions of normal can also be expressed in pH units. Thus for a solution in which the concentration is 2.73×10^{-4} N, the pH number is 3.566. This can be proved by the use of a logarithm table, which shows that $\log 2.73 = +0.434$ and $\log 10^{-4} = -4.000$. Since the numbers are to be multiplied, the logarithms are added, so $-4.000 + 0.434 = -3.566$. Therefore $\log (H^+) = -3.566$, so $-\log (H^+) = 3.566$, and pH = 3.566.

If it is desired to know the actual figure for the H^+ ion concentration when only the pH value is given, it can be found by the reverse of the calculation just given. Thus, 9.63 pH means that $(H^+) = 1 \times 10^{-9.63}$. The exponent $-9.63 = -10 + 0.37$; hence $10^{-9.63} = 10^{-10} \times 10^{+0.37}$. The logarithm table shows that the exponent 0.37 corresponds to the number 2.34. Therefore $(H^+) = 10^{-9.63}$ is the same as $(H^+) = 2.34 \times 10^{-10}$.

In a similar manner the alkalinity of a base may be expressed as $pOH = \log 1/(OH^-)$. In an acid-base equilibrium water is formed and neutrality is reached when

$$\frac{(H^+) \times (OH^-)}{(HOH)} = K$$

The number of water molecules dissociated is so small in comparison that the concentration HOH may be considered as constant and combined with K, making the equation $(H^+) \times (OH^-) = K(HOH) = Kw$. At 25°C., Kw is 10^{-14}, which is the product of the concentration of the H^+ ions and OH^- ions which are equal to each other, so that the concentration of each is 10^{-7} and in water, pH = 7. In an acid solution the concentration of H^+ ions is greater than that of water, because of the dissociation of the acid and according to its dissociation constant. In an analogous manner in basic solutions the OH^- ion concentration is greater than that of water. But, since water is present, there must be some OH^- ions in any acid solution and some H^+ ions in any basic solution, the concentration of one of these varying inversely with the concentration of the other and the product of the two concentrations being equal to 10^{-14}. Then $(H^+)(OH^-) = 10^{-14}$ and $\log (1/H^+) + \log (1/OH^-) = 14$, and pH + pOH = 14.

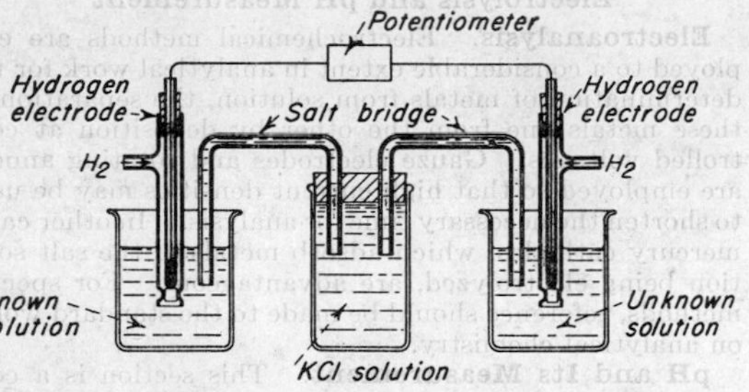

FIG. 16. Theoretical hydrogen-ion concentration setup. *(Leeds and Northrup.)*

It is evident that, if the concentration of either ion is known, that of the other can be computed so that the reaction of any solution, whether acid, alkaline, or neutral, can be expressed in terms of pH. An acid-alkaline scale can be set up in terms of H^+ ion concentrations with a pH = 0 at one end representing a normal H^+ ion solution, and a pH = 14 at the other end representing the H^+ ion concentration of a normal OH^- ion solution. Any pH number from zero to 7 thus indicates acidity with decreasing acidity as the number increases. pH 7 indicates neutrality, and any pH number between 7 and 14 indicates an alkaline solution with increasing alkalinity (or decreasing acidity) as the number increases.

H^+ ion concentration in a water solution can be determined electrically by the use of two hydrogen electrodes in contact with the solution and the voltage developed by the two electrodes determined by the use of a potentiometer. Such an arrangement is shown in Fig. 16. From the following formulas the H^+ ion concentration or pH can be calculated.

$$\frac{V}{0.0001983T} = \log \frac{C_n}{C}$$

where T is the absolute temperature, V the voltage, C_n the known concentration, and C the unknown. At 25°C. for the temperature of the two solutions, the equation becomes

$$\frac{V}{0.0591} = \log \frac{C_n}{C}$$

If the known solution is a normal H^+ ion solution, then $C_n = 1$ and $\log 1/C = $ pH, and the equation becomes

$$\frac{V}{0.0591} = \text{pH}$$

A more convenient arrangement substitutes a calomel electrode for one of the hydrogen electrodes as a result

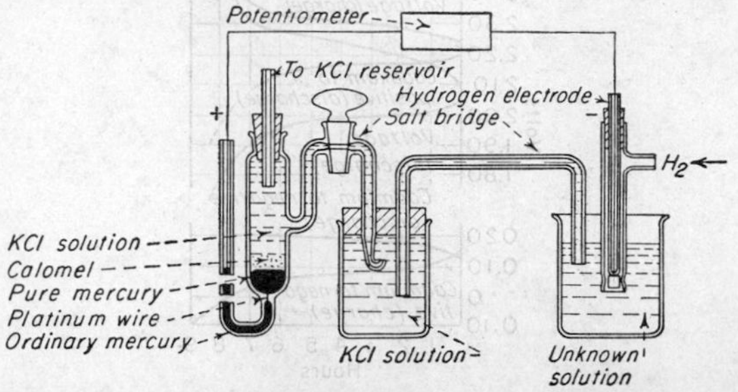

FIG. 17. Calomel electrode setup for hydrogen-ion measurement. *(Leeds and Northrup.)*

of which the arrangement becomes that of Fig. 17. The formula then becomes

$$\frac{V - v}{0.0591} = \log \frac{1}{C} = \text{pH}$$

The potential of the calomel electrode varies with its temperature as well as with the KCl concentration as tabulated.

Table 15. Variation of Calomel-electrode Potential with Temperature and KCl Concentration

Calomel electrode	Potential at 20°C.	Potential at 25°C.	Potential at 30°C.
Tenth normal	0.3379	0.3376	0.3371
Normal	.2860	.2848	.2835
Saturated	.2496	.2458	.2420

For many practical purposes in industry the quinhydrone electrode is substituted for the hydrogen electrode. The basis of this electrode is a piece of platinum or gold exactly the same as used for the hydrogen electrode, but the surface of the metal is not platinized and it is not supplied with gaseous hydrogen. Instead, a small quantity of quinhydrone (benzoquinhydrone) is dissolved in the solution, and, within certain pH ranges, the electrode in the solution acquires a potential that is definitely related to the H^+ ion concentration of the solution. The potential is measured against that of a calomel electrode, and the pH value is found from the measured voltage in the same manner as with a hydrogen electrode. Quinhydrone is so slightly soluble in acid that only small quantities are necessary to saturate the solution. In solutions more alkaline than pH = 8, quinhydrone is more soluble, dissociates, and becomes oxidized. The quinhydrone electrode potential is altered to such an extent that the measured voltage is no longer a linear function of pH. The quinhydrone electrode is therefore not suitable for use in highly alkaline solutions.

The relation of voltage to pH is given by the expression

$$\mathrm{pH} = \frac{0.7177 - 0.00074t}{0.0001983T} \frac{V - v}{}$$

where t is the temperature of the solution and the electrodes, V the measured voltage, v the correction factor for

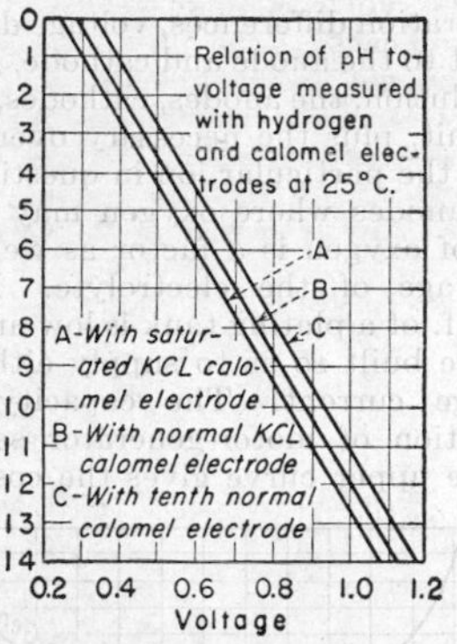

FIG. 18. pH-voltage relation of hydrogen electrode. (*Leeds and Northrup.*)

the potential of the calomel electrode, and T the absolute temperature. With a saturated calomel electrode at 25°C., $v = 0.2458$ and the equation becomes

$$\frac{0.453 - V}{0.0591} = \log \frac{1}{\mathrm{H}^+} = \mathrm{pH}$$

The pH-voltage curves of the hydrogen electrode, calomel electrode arrangement and the quinhydrone electrode, saturated KCl calomel electrode, are given in Figs. 18 and 19.

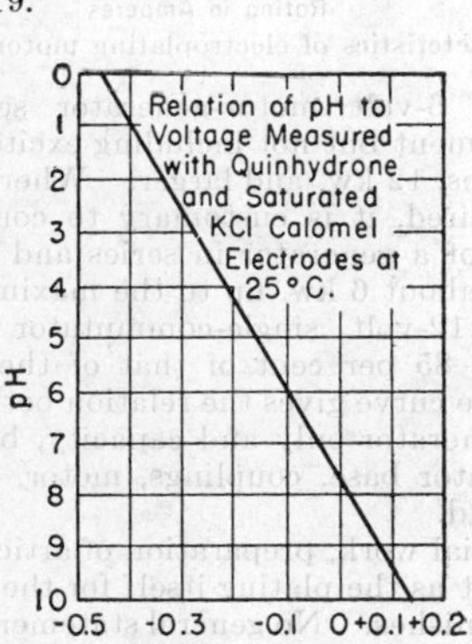

FIG. 19. pH-voltage relation of quinhydrone electrode. (*Leeds and Northrup.*)

Glass electrode is a Na_2O, CaO, SiO_2 glass bulb, thin walled, blown on the end of a glass tube and filled with pH 1 HCl. A quinhydrone electrode in this solution is the electrode terminal. Voltage measurements through the glass wall are made with a calomel reference electrode in the solution under test.

Antimony electrode, used for industrial controls, consists of prepared Sb with its end immersed in the solution whose pH is being measured.

Electrolytic Hydrogen and Oxygen Production

Hydrogen and Oxygen Production. Commercial hydrogen and oxygen cells are built entirely of iron or steel, insulating materials, and asbestos cloth for diaphragms. Every effort is made to have simple and inexpensive construction. Great care is taken to reduce all contact voltages to a minimum, in order to obtain the lowest possible cell voltages. Present-day cells are so well designed that per unit of current almost theoretical yields of the gases are obtained. The electrolyte ordinarily employed is an alkaline solution of either 15 per cent NaOH or its equivalent KOH. Sulfuric acid electrolytes are no longer used. Theoretical data for electrolytic hydrogen production are tabulated, and operating data for a typical cell are plotted in Fig. 20.

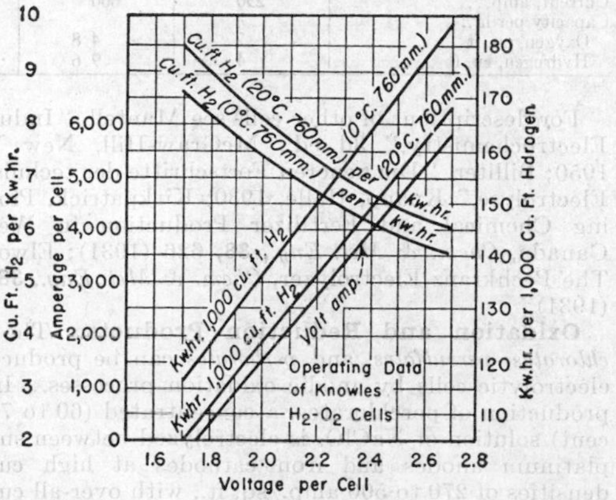

FIG. 20. Current, voltage, and gas-volume relations for hydrogen-oxygen cells.

In general the cells use iron electrodes, the anode of which is ordinarily nickel plated to reduce the oxygen overvoltage. In the **Levin** cell the cathode is plated with cobalt to reduce the hydrogen overvoltage.

Two typical cells will be described. The **Knowles** cell consists of an outer tank containing a number of sheet-iron, or steel, gas-collecting hoods, rectangular in cross section and of considerably greater length than breadth. They alternately contain anodes and cathodes, each held in its hood by two conducting bolts which are insulated from the bells. Alternate hoods carry asbestos extensions which serve as diaphragms.

The Levin cell consists of a thin vertical tank divided by two asbestos diaphragms into three compartments, the two outer ones being the anode and the inner the cathode compartment. The gases produced pass upward through sight indicators to collecting manifolds. The entire tank is closed. For laboratory use, multiple or filter-press types of cells are used in which the plates act as bipolar electrodes. They are separated by asbestos diaphragms

Table 16. Electrolytic Hydrogen-cell Constants
(Knowles)

Quantity of hydrogen liberated per 1000 amp.-hr.		Amp.-hr. required to liberate unit mass or volume of hydrogen, per	
Grams	37.65	Gram	26.56
Pounds	0.0830	Pound	12,050
Cubic feet at 20°C. and 760 mm. pressure saturated with water vapor	16.25	Cubic foot at 20°C. and 760 mm. pressure saturated with water vapor	61.55
Cubic feet at 0°C. and 760 mm. pressure dry	14.79	Cubic foot at 0°C. and 760 mm. pressure dry	67.61
Cubic meters at 20°C. and 760 mm. pressure saturated with water vapor	0.4604	Cubic meter at 20°C. and 760 mm. pressure saturated with water vapor	2171
Cubic meters at 0°C. and 760 mm. pressure dry	0.4189	Cubic meter at 0°C. and 760 mm. pressure dry	2387

with rubber packings at the point where they insulate adjacent plates. Each plate except the end ones has three holes, one on each side at the top to lead off the gases and one at the bottom for the electrolyte, which may be a 15 to 30 per cent NaOH. They are designed with sufficient plates to allow operation on 110-volt circuits. Hydrogen purer than 99 per cent can be obtained if means are provided for balancing the pressures on opposite sides of the diaphragms. Similar large-sized industrial units are now in operation.

Table 17. Levin Cell Characteristics

	Type *A*	Type *B*	Type M-1250
Dimensions, in.	30 × 25 × 6¼	43 × 37 × 8½	
Weight, lb.	145	325	
Current, amp.	250	600	1250
Capacity per hr.:			
Oxygen, cu. ft.	2	4.8	10
Hydrogen, cu. ft.	4	9.6	20

For descriptions of other cells see Mantell, "Industrial Electrochemistry," 3d ed., McGraw-Hill, New York, 1950; Billiter, "Die neueren Fortschritte der technischen Electrolyse," Knapp, Halle, 1930; Kirkpatrick, Pioneering Chemical and Fertilizer Production in Western Canada, *Chem. & Met. Eng.*, **38**, 626 (1931); Elworthy, The Pechkranz Electrolyzer, *Chem. & Met. Eng.*, **38**, 714 (1931).

Oxidation and Reduction Products. The *perchlorates*, *persulfates*, and *perborates* can be produced in electrolytic cells by anodic oxidation processes. In the production of perchlorates, a concentrated (60 to 70 per cent) solution of $NaClO_3$ is electrolyzed between smooth platinum anodes and iron cathodes at high current densities of 270 to 500 amp./sq. ft., with over-all current efficiencies of 70 to 85 per cent and an energy consumption of 1.5 to 1.8 kw.-hr./lb. $NaClO_4$. This salt is deliquescent, and the potassium and ammonium salts are formed by double decomposition from the corresponding chlorides. These are used in fireworks and the explosives industries. The persulfates are made by the use of $(NH_4)_2SO_4$ solution in an anodic compartment with H_2SO_4 in the cathodic compartment separated by a diaphragm. The electrolyte is kept at about 15°C., the anode platinum operated at a high current density, and the cathode lead of a much larger surface. The current efficiency exceeds 70 per cent. Simpler processes avoid the use of diaphragms by the addition of about 0.2 per cent K_2CrO_4 to prevent cathodic reduction, or else employ $KHSO_4$ and the persulfate product $K_2S_2O_8$ is precipitated. Current densities are of the order of 400 to 600 amp./sq. ft. and energy consumptions for $(NH_4)_2S_2O_8$ about 1 to 1.2 kw.-hr./lb. **Sodium perborate** is prepared either by the interaction of borax and H_2O_2 or by the electrolysis of sodium borate containing Na_2CO_3 at anodic current densities of 200 to 350 amp./sq. ft., between platinum anodes and iron cathodes, with the addition of chromate to the electrolyte. Energy consumption is about 3 kw.-hr./lb. sodium perborate.

A small number of **organic compounds** are made electrolytically by cathodic reduction or anodic oxidation. Among these are bromoform, iodoform, chloral, and paraaminophenol.

Sorbitol and **mannitol** [hexahydric alcohols $C_6H_6(OH)_6$] are the electrolytic reduction products of glucose (U. S. Patents 1,612,361; 1,653,004; 1,712,951–2; 1,990,582) and are produced by batch processing by cathodic reduction in alkaline glucose solutions containing Na_2SO_4 with Pb anodes and Pb-Hg cathodes.

White lead was made by the Sperry process by the electrolytic corrosion of lead anodes in a sodium acetate electrolyte containing Na_2CO_3.

Electroplating and Electrotyping

Electroplating is concerned with the production of metallic coatings on metal objects and, to a limited extent, on non-metallic objects. In general, the metal plated is dissolved from an anode of the material and deposited on the object to be plated as cathode, although this arrangement may be modified and the metal plated may be obtained from a dissolved salt in the plating bath. Where soluble anodes are used, electrochemical action at the cathode is the reverse of that at the anode. The e.m.f. applied at the terminals of the cell is used to overcome concentration differences, voltage drops through the films adjacent to the anode and cathode, the ohmic resistance of the solution, the anodes, cathodes, connectors, and external circuit, plus the necessary overvoltages for the deposition of the particular ion in question. In the case of insoluble anodes where oxygen may be evolved, the overvoltage of oxygen is a factor as well as the decomposition voltage of the electrolyte. In general, the terminal e.m.f. of a plating tank is low and electroplating generators are built so as to supply either 6- or 12-volt high-amperage current. The capacity and cost per kilowatt relation of motor-generator sets are given in Fig. 21. The upper curve gives the cost to the user of

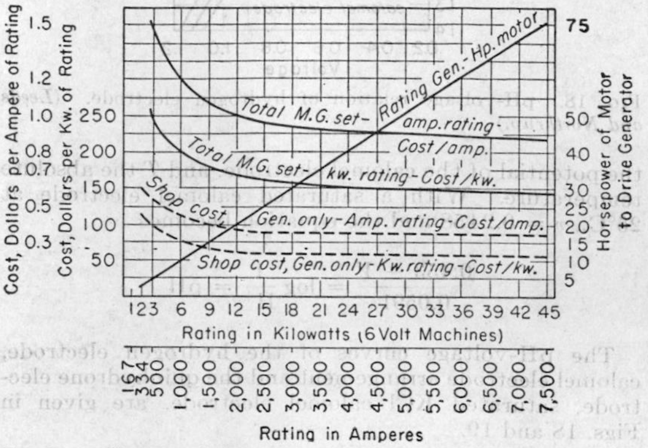

FIG. 21. Characteristics of electroplating motor-generator sets.

standard size 6-volt motor-generator sets, including starting equipment but not including exciters commonly used on all sizes, 12 kw. and larger. Where more than 6 volts are required, it is customary to connect the two commutators of a generator in series and operate at 12 volts. From about 6 kw. up to the maximum sizes, the cost of the 12-volt single-commutator machines is approximately 85 per cent of that of the 6-volt units. The dotted-line curve gives the relation between the shop cost of the generator only and capacity, but it does not include generator base, couplings, motor, and manufacturing overhead.

In commercial work, preparation of articles for plating is as important as the plating itself for the production of high-quality finishes. No general statement can be made as to methods used for preparing articles for plating, in that the material of the object, its condition, shape, size, composition, kind of surface, and other factors must be considered. Inasmuch as the plating will follow exactly the contours of the object to be plated, smooth finishes can be produced only on smooth-surface objects. In general, rough surfaces may be cut down or made smoother by sandblasting, mechanical cleaning methods, grinding, and buffing with abrasives; or the object may be pickled, generally in H_2SO_4 or occasionally in HCl solutions. Removal of dirt, grease, and lacquers is accomplished by some form of alkaline cleaner, often containing oxide solvents such as cyanides, and detergents such as Na_3PO_4, as well as other materials. These may be operated alone or with the aid of the electric current in which the object is suspended from a rod and is alternately made anode and cathode in the circuit. At other times grease and dirt may be removed by organic solvents, usually of the petroleum type. In some cases after cleaning, the object may be further pickled or else may receive additional mechanical surfacing treatment for the production of preliminary polishes. In the case of non-ferrous articles, pickling is ordinarily done in so-called "bright" acid dips consisting of mixtures of concentrated H_2SO_4 and HNO_3 with the addition of small amounts of salt. Depending upon the object and the type of surface to be produced on the article, the preliminary flow sheet may vary widely in different industries. In order to avoid contamination of solutions as a result of the material being carried over when articles are transferred, and the introduction of acids into cyanide plating baths or vice versa, all operations, irrespective of their nature, are followed by washing or rinsing.

Plating baths may be considered to consist of a number of different parts: (1) the salt containing the metallic ion or radical; (2) an additional salt whose function is to increase the conductivity of the bath; (3) a material to effect the anode corrosion and prevent passivity of the anode; (4) a so-called addition agent to effect the type of deposit produced; and (5) a buffer material to maintain the proper pH of the solution. Some baths may contain all these while others may not. For example, a common nickel-plating solution will have the metal ion in the form of $NiSO_4.7H_2O$ (105 g./l.), NH_4Cl or $(NH_4)_2SO_4$ (15 g./l.) to increase the conductivity of the bath, $NiCl_2.6H_2O$ (15 g./l.) to assist anode corrosion, and H_3BO_3 (15 g./l.), which acts as a buffer to maintain the pH of the solution at 5.3. In the case of a tin bath, the tin salt would be furnished by Na_2SnO_3, the conducting salt by NaOH, which also assists anode corrosion, the addition agent to effect the deposit being glucose or other organic material.

X-ray examination shows that all electrodeposited metals have crystalline structures, the differences in their physical properties and appearances being caused by differences in the size and shape of the crystals. Plating baths are operated within definite current-density ranges beyond which poor types of deposits result. In general, for the production of fine grained or so-called amorphous deposits, which are readily polishable, plating baths show low ionic metallic concentration produced not by dilute solutions but either by the use of salts showing low ionization or salts whose ionization is depressed by the addition of another salt having a common ion, or by the employment of compounds producing the metallic ion not by primary but by secondary ionization, or by reduction at the cathode. This explains the widespread use of double salts such as $NiSO_4.(NH_4)_2SO_4.6H_2O$, complex compounds such as the double cyanides, and, in the specific case of chromium, chromic acid. Deposition potentials in general are lower than the decomposition voltage of water, but in the case of chromium and alloys, such as brass, the deposition potential is above that of the decomposition of water, and hydrogen is evolved at the cathode. This latter condition is essential for the deposition of chromium, and about 70 per cent of the current is consumed in hydrogen evolution. In general, however, current efficiencies of plating baths are high, usually being above 90 per cent. A large number of metals are commercially plated. The general characteristics of the baths employed are given in Table 18. After plating and drying, the objects may be polished, usually with mild abrasives such as tripoli and rouge, on high-speed wheels.

Mechanical equipment for plating assumes many different forms. Large quantities of small objects are plated in barrels which rotate in the plating solutions. Larger objects are generally wired or hung on rods which in turn connect with cathode bars. In recent years there has been a tremendous development of automatic plating, in which the successive operations involved in the entire production flow sheet are mechanized by having the objects hung on rods suspended from conveyors with automatic mechanisms causing the object to pass through different baths such as cleaners, dips, plating solution, rising out of the tank at one end and dipping into the next one by a tripping-and-raising action connected to the conveyor. The plan view of a machine of the type widely employed in the automotive industries for plating, where the objects receive successive finishes of copper, nickel, and chromium, is shown in Fig. 22.

Table 18. Electroplating Baths
(Mantell)

Metal	Bath type	Anode	Cathode c.d. amp./sq. ft.	Temp., °C.
Copper	Acid sulfate	Cu	15–40	25–50
	Alkaline cyanide	Cu	3–14	35–40
Nickel	Single sulfate	Ni	5–20	20–30
	Double sulfate	Ni	3–6	20–30
	Sulfate-chloride	Ni	14–50	50–60
Iron	Chloride	Fe	100–180	90–110
	Double sulfate	Fe	20–30	20–30
Cobalt	Sulfate	Co	30–165	20–30
Chromium	Chromic acid	Pb	100–300	40–50
Zinc	Sulfate	Zn	12–30	20–30
	Sulfate (hot)	Zn	80 100	50–60
	Chloride	Zn	40–100	20–40
	Alkaline cyanide	Zn	8–20	40–50
Cadmium	Alkaline cyanide	Cd	10–50	20–30
Lead	Fluoborate	Pb	5–20	20–30
	Perchlorate	Pb	20–30	20–30
Tin	Alkaline stannite	Sn	10	50
	Alkaline stannate	Sn	5–15	60
Silver	Cyanide	Ag	3–8	15–25
Gold	Cyanide	Au	1–5	60–80
Platinum	Phosphate	Pt	1	70
Brass	Cyanide	Cu-Zn alloy	2–3	32–45

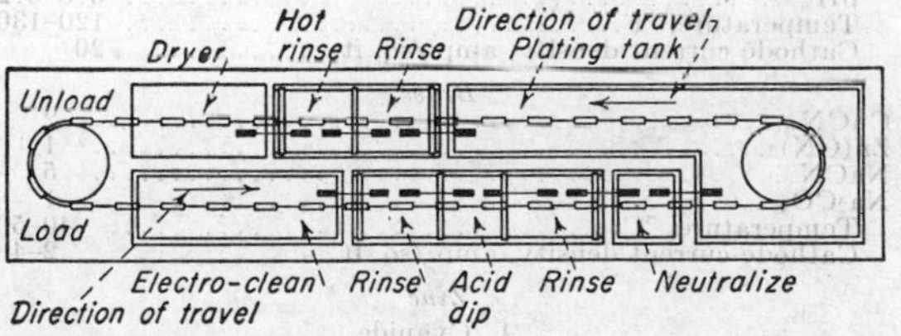

FIG. 22. Diagram shows plan view of return-type plating machine, containing relatively few operations. This type of machine may be expanded to embrace almost any number of operations. Or two separate and distinct operating routines may be built into one of these machines, thus serving the purpose of two units; *i.e.*, rack may be loaded on conveyor at one end, pass through the various operations and be removed at other end for color buffing, then again racked and replaced on conveyor for return to starting point through the prescribed processes. One side of machine may be used for cleaning and copper plating and the other side for the preparing and applying of finishing nickel coat, or for final chromium deposit. The adaptability of this machine is almost unlimited.

Commercial Thickness of Electroplates

Ni, rarely over 0.001 in.; usually 0.0005 in. or under
Cr, generally 0.00002–0.00004 in.
Zn, usually over 0.0005 in.; frequently around 0.001 in.
Cd, usually 0.0002–0.0003 in.; rarely over 0.0005 in.
Au, usually less than 0.00005 in.

Composite plates:

Cu − Ni − Cr, 0.0003, 0.0005, 0.00002 in.
Ni − Cu − Ni − Cr, 0.0002, 0.0003, 0.0003, 0.00002 in.

Conversion Factors for Electroplaters

Amp./sq. dm. × 9.3 = amp./sq. ft.
Amp./sq. ft. × 0.108 = amp./sq. dm.
G./l. × 0.134 = oz./gal.
Oz./gal. × 7.5 = g./l.

For thickness of 0.001 in. for any metal,

$$\text{Weight plate in oz./sq. ft.} = \frac{\text{sp. gr. metal}}{12}$$

Typical Plating-bath Formulas

(Concentrations in avoirdupois ounces per gallon)

Chromium

Chromic acid, CrO_3	33
Sulfuric acid, H_2SO_4	0.34
Temperature, °F	113
Cathode current density, amp./sq. ft.	150–200

Copper

A. Cyanide

CuCN	3
NaCN	4
Na_2CO_3	2
$Na_2S_2O_3.5H_2O$	⅛ (optional for brightener)
Temperature, °C	40–50
Cathode current density, amp./sq. ft.	3–6

B. Acid

$CuSO_4.5H_2O$	24
H_2SO_4	6
Glue	⅛ (optional for brightener)
Temperature	room
Cathode current density, amp./sq. ft.	25–50

Nickel

$NiSO_4.7H_2O$	32
$NiCl_2.6H_2O$	2
H_3BO_3	4
pH	5.8–6.2
Temperature, °F	120–130
Cathode current density, amp./sq. ft.	20

Brass

CuCN	2
$Zn(CN)_2$	1
NaCN	5
Na_2CO_3	1
Temperature, °C	40–50
Cathode current density, amp./sq. ft.	2–4

Zinc

A. Cyanide

(1)		(2)	
$Zn(CN)_2$	8	ZnO	6
NaCN	3	NaCN	10
NaOH	7	NaOH	5
Temperature			room
Cathode current density, amp./sq. ft.			10–20

B. Acid Sulfate

$ZnSO_4.7H_2O$	48
NH_4Cl	4
$NaC_2H_3O_2.3H_2O$	2
H_2SO_4	0.4
Temperature	room
Cathode current density, amp./sq. ft.	15–30

Cadmium

	(1)	(2)*
CdO	3	6
NaCN	7	1.6
NaOH	2	4
$NiSO_4.7H_2O$		0.13
Goulac		1.5
Temperature	Room	Room
Cathode current density, amp./sq. ft.	10	25

* U.S. Patent 1,681,509.

Lead

Basic lead carbonate	20
Hydrofluoric acid, 50%	32
Boric acid	14
Glue (optional)	0.025
Temperature	Room
Cathode current density, amp./sq. ft.	10–20

Tin

Sodium stannate	12
Sodium hydroxide	2
Sodium acetate crystals	2
Temperature, °C	60–80
Cathode current density, amp./sq. ft.	10–30

Gold

Gold fulminate, dwt	0.3 or 5
NaCN	2
$Na_2HPO_4.12H_2O$	0.5
Temperature, °C	60–80
Cathode current density, amp./sq. ft.	1

Silver

	Strike	Plate
NaCN	16	5
AgCN	⅔	4
Temperature	Room	Room
Cathode current density, amp./sq. ft.	5+	3–5

The relation of time of plating, thickness of deposit, and current in commercial baths for nickel is given in Fig. 23, for copper in Fig. 24, for zinc in Fig. 25, and for cadmium in Fig. 26.

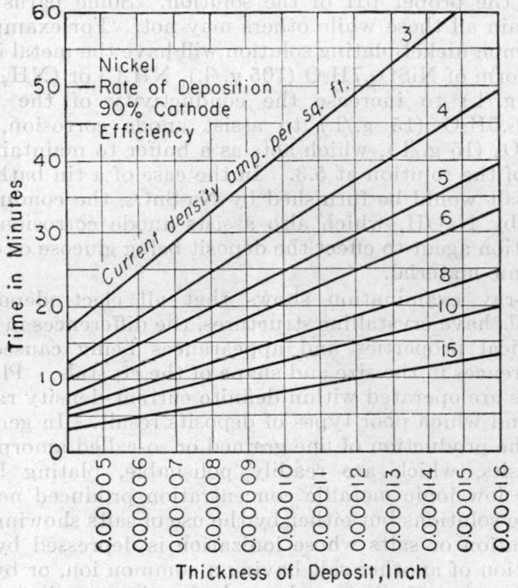

FIG. 23. Plating characteristics of nickel bath.

Electroplating on Aluminum. In any application of electroplating on aluminum, the type of plate should be selected for the particular job. It is possible to apply a flash coating of zinc to aluminum and then plate with nickel, copper, or other metals. Unfortunately the corrosion resistance of such a plate is poor and suited only for indoor service. The zinc alone does not suffer this disadvantage and is useful for special purposes. Chromium may be applied directly to aluminum, but it has not yet been possible to obtain quite as satisfactory luster as is possible when chromium is applied over nickel. There is a possibility, however, that such chromium plates will have application for special corrosion jobs, particularly against mild alkalies. For general use it is advised that nickel be first applied to the aluminum after a surface roughening, and then the desired metal should be applied to the nickel. The aluminum alloys are divided into three classes, each requiring a somewhat different plating procedure, as follows:

Plating 2*S* Aluminum.

1. Remove grease.
2. Rinse in clear cold water.
3. Dip for from 10 to 15 sec. in 1 part of 48 to 52 per cent of hydrofluoric acid to 9 parts of water.
4. Rinse in clear cold water.

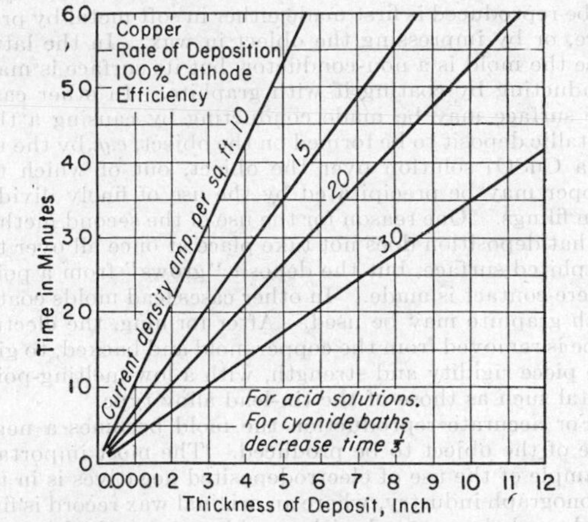

FIG. 24. Plating characteristics of copper bath.

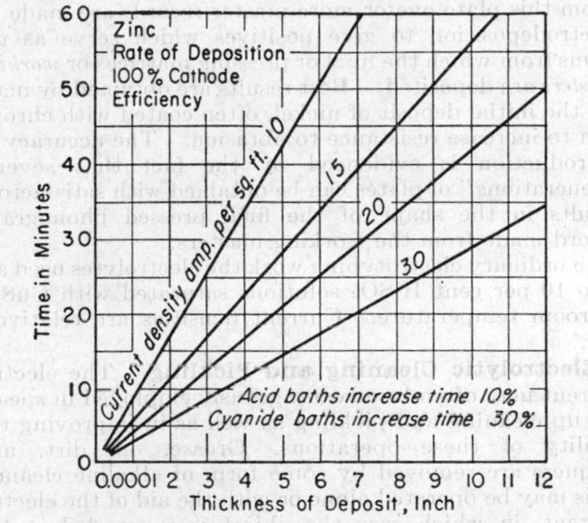

FIG. 25. Plating characteristics of zinc bath.

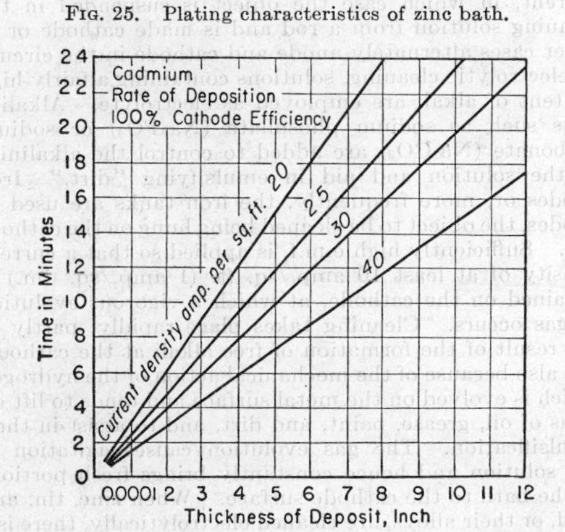

FIG. 26. Plating characteristics of cadmium bath.

5. Dip in the following solution: nickel chloride 37 oz., HCl (sp. gr. 1.18) 0.2 gal., water 1.0 gal., temperature 75 to 80°F. The time of dip must be carefully determined as this is the important step in the procedure. This is accomplished by plating a series of specimens after different times in the dip and then bending or breaking them. A 35-sec. dip usually gives good results, although such factors as temperature, purity of the chemicals in the dip, and the temper and polish of the metal plated on affect the time required. The speed of the action of the dip may be regulated by the amount of acid and the temperature. Generally a time of dip of 15 to 40 sec. is desirable in production work. The actual composition of the nickel dip is not so important as the resultant adhesion of the electroplate.
6. Rinse in clear cold water.
7. Plate at 15 amp./sq. ft. in the following nickel bath: single nickel salts 19 oz./gal., magnesium sulfate 10 oz./gal., ammonium chloride 2 oz./gal., boric acid 2 oz./gal. The time of plating should be determined to give the desired corrosion resistance. A time of plating of 1 hr. is suggested, but this can be reduced later if it is found that a thinner plate meets the service requirements.
8. Rinse in cold water and dry by means of a hot-water dip.
9. Polish nickel.
10. Clean and plate with any desired metal.

Plating 3*S* and Strong Alloys. For this group of alloys a different etching process is needed; otherwise the plating procedure is unchanged. Step 5 (as given above) will therefore be rewritten for these alloys for substitution in the above procedure.

5. Dip in the following solution: HCl (sp. gr. 1.18) ½ gal., water ½ gal., manganous sulfate ($MnSO_4.2H_2O$) ½ oz., temperature 75 to 80°F. The time of dip must be carefully determined by experiment as this is the most important step in the procedure. It will generally be found somewhere between 10 and 60 sec.

This group of alloys presents several peculiarities. In the first place, the heat-treated metal gives a plated product that is more resistant to corrosion than the untreated metal. There are also differences between the alloys. 51*S* is most readily plated. 17*ST* is readily plated, as screw-machine products, but often gives trouble in sheet form due to metal streaks. 25*S* may give trouble in sheet form.

Plating Castings.* Castings require a still different etching process. In addition, it is possible to omit steps 3 and 4 of the procedure for 2*S*. For the etching, step 5 of the 2*S* procedure, the following method should be substituted:

5. Dip in the following solution: HNO_3 (sp. gr. 1.42) 3 parts, hydrofluoric acid (48 to 52 per cent) 1 part, temperature 75 to 80°F.

The proper timing of this dip must be carefully determined by experiment, as it is the most important step in the plating procedure. Each alloy composition will have a dipping time most suited to it, although the differences between alloys are slight, as the dipping range for each alloy is wide. Usually the proper time of dip lies between 15 and 30 sec. for die castings, and 60 to 120 sec. for sand castings.

The container for the dip should be lead lined. It should be painted with a mixture of 1 part of beeswax to 4 parts of paraffin. This is particularly necessary above the solution line, as the greatest attack of the lead takes place here.

The **electrodeposition of rubber** was invented in

* 195 alloy presents special difficulties. Even a 3-min. etch is not sufficient to secure adhesion, although the corrosion resistance appears fairly good. The procedure for the strong alloys may be applied to secure good adhesion, but it is sometimes difficult to secure uniform results.

1921 by Sheppard and Eberlin in the United States (U. S. Patent 1,476,374, Dec. 4, 1923) and about the same time by Klein and Szegvari in Budapest (British Patent 223,189). It has entered the technical field not as a competitor of mechanical methods but rather as a supplementary process and as a means of reducing costs.

Electrolytic deposits can be made from artificial rubber dispersions in aqueous solutions of slightly alkaline characteristics, as well as from natural latex preserved with ammonia. The rubber particles, existing as colloidal suspensions, are negatively charged and on the passage of current move toward the anode. If the current is reversed, the particles will also reverse their direction. With a steady flow of current in one direction, the rubber will deposit on the anode in a solid uniform layer. It has been found that in the Sheppard process sulfur, pigments, and accelerators can be mixed with the latex or rubber dispersion and a rubber compound deposited by electrolysis on metal or any other substance. This compound will possess approximately the same percentage composition as the total solids in the bath. It is washed and dried and can then be vulcanized.

In general a current density of 0.25 to 0.33 amp./sq. in. proved satisfactory. The anode metal varies over a wide range, including lead, cadmium, zinc, tin, antimony, alloys of these metals, and iron and steel under certain conditions. Copper and copper alloys, however, proved objectionable. A typical plating bath follows:

Solids	Concentration, g./100 cc.	Percentage of total solids
Rubber from latex	8.00	53.3
Sulfur	0.30	2.0
Zinc oxide	1.50	10.0
Whiting	4.50	30.0
Carbon black	0.30	2.0
Paraffin wax	0.30	2.0
Tuads (accelerator)	0.03	0.2
Gum arabic	0.075	0.5
Total solids	15.00	100.0
Ammonia	0.16	

Ordinary fine sulfur, colloidal sulfur, or polysulfide sulfur may be used. Among the pigments and fillers, litharge, lithopone, white lead, magnesium carbonate, zinc, oxide, titanox, clay, silica, certain carbon blacks, and a few others may be compounded without great difficulty. Softening agents and such materials as glue, dextrins, or gums for facilitating the dispersion of the added substances are readily incorporated. Accelerators insoluble in water may be employed, since they will be colloidally suspended in the solution. It has been found advantageous to homogenize the ingredients before mixing with the rubber emulsion.

Thickness of deposit is somewhat limited but not to so narrow a margin as might be expected from the nonconducting characteristics of rubber. Layers varying from 0.01 to 0.15 in. have been produced. The bath has sufficient throwing powers to effect a uniform coating on edges, corners, etc. Adequate stirring is of course important for smoothness of deposit.

Rubber and rubber compounds may be deposited not only as a coating on metals or other materials, but they may also be plated on a mold and later stripped off. Among the applications of the process described are the production of coated fabrics for automobiles; waterproofed textiles; coated metal wire for insulation and protective purposes; coated metal apparatus such as rubber-lined tanks, etc.; the direct production of rubber gloves, inner tubes for tires, and the like, as well as rubber sheets and rubber slabs for stock in the rubber industry.

The economic advantage of electrodeposition, as compared to mechanical milling and spreading, lies in the reduction of power and labor. Mixing and compounding operations are cheaper if done by high-speed mills. Furthermore, electrodeposited rubber will generally have a greater strength than the mechanically milled product, since it has not been degraded by friction.

Electrotyping. Electrotyping has as its object the reproduction of the printer's type setup, of engravings and medals, while electroforming produces or reproduces an object by electrodeposition. The mold of the object to be reproduced is first made either in soft metal by pressure, or by impressing the object in wax. In the latter case the mold is a non-conductor, but its surface is made conducting by coating it with graphite. In other cases the surface may be made conducting by causing a thin metallic deposit to be formed on the object, *e.g.* by the use of a $CuSO_4$ solution over the object, out of which the copper may be precipitated by the use of finely divided iron filings. One reason for the use of the second method is that deposition does not take place at once all over the graphited surface, but the deposit "grows" from a point where contact is made. In other cases lead molds coated with graphite may be used. After forming, the electrotype is removed from the copper mold and backed, to give the piece rigidity and strength, with a low-melting-point metal such as those of the tin-lead alloy type.

For accurate reproduction the mold becomes a negative of the object to be produced. The most important example of the use of electrodeposited negatives is in the phonograph industry, where an original wax record is first made and then coated with graphite, upon which copper is deposited to form a *master matrix* which is a negative. From this plate one or more master records are made by electrodeposition to give positives which serve as the forms from which the final or pressing matrices or *working masters* are deposited. Best results are obtained by making the initial deposit of nickel, often coated with chromium to increase resistance to abrasion. The accuracy of reproduction is evidenced by the fact that several "generations" of plates can be obtained with satisfactory results in the shape of the final pressed phonograph record made from the working masters.

In ordinary electrotyping work the electrolytes used are 8 to 10 per cent H_2SO_4 solutions saturated with $CuSO_4$ at room temperature. Current densities are relatively low.

Electrolytic Cleaning and Pickling. The electric current may often be advantageously employed in speeding up cleaning and pickling as well as in improving the quality of these operations. Grease, oil, dirt, and lacquers are removed by some form of alkaline cleaner. This may be operated alone or with the aid of the electric current, in which case the object is suspended in the cleaning solution from a rod and is made cathode or in other cases alternately anode and cathode in the circuit. In electrolytic cleaning, solutions containing a fairly high content of alkali are employed as electrolyte. Alkaline salts such as sodium phosphate (Na_3PO_4) or sodium carbonate (Na_2CO_3) are added to control the alkalinity of the solution and aid in emulsifying "dirt." Iron anodes or, more frequently, the iron tanks are used as anodes, the object to be cleaned being hung on the cathode bar. Sufficiently high e.m.f. is applied so that a current density of at least 10 amp./sq. ft. (1 amp./sq. dm.) is obtained on the cathode, at which a vigorous evolution of gas occurs. Cleaning takes place rapidly, partly as the result of the formation of free alkali at the cathode, but also because of the mechanical action of the hydrogen which is evolved on the metal surface and tends to lift off films of oil, grease, paint, and dirt, and to assist in their emulsification. The gas evolution causes agitation of the solution and hence constantly brings fresh portions of the bath to the cathode surface. When zinc, tin, and lead, or their alloys, are cleaned electrolytically, there is a tendency for them to dissolve in the alkaline solution and be redeposited in small amounts in thin films. These films may prevent adherence of electroplatings. Under such conditions the redeposited metal is removed by causing the piece to function as anode for a short time. Electrolytic cleaning has been applied industrially in operations not connected with electroplating, such as the removal of paint from paint cans so that they may be reused, cleaning of steel barrels, drums, and shipping containers.

It is commercially desirable to get the most effective use of the electrical energy employed. The cleaning bath should be of such composition and concentration as to have high conductivity and resultant low bath resistance and low voltage drop across the terminals of the tank. The current supplied to the bath should be such as to give the most effective gas-generation rate at the anode and cathode. A current density of at least 10 amp./sq. ft. is necessary; but the current density may be pushed to such a high figure that the gas bubbles evolved at the electrodes may blanket the electrode surfaces and cause high voltage drops through the gas films formed adjacent to the electrode surface. Too high a current density will increase the voltage drop across the bath, with resultant increase of electrical energy consumption without increased efficiency. Actually, too high a current density and too rapid a generation of gas bubbles on the object being cleaned may result in decreased efficiency of cleaning and higher power costs than necessary.

Electrolytic pickling has recently found commercial employment, particularly for the removal of oxides on machined parts that have been heat-treated. The operation is analogous to electrolytic cleaning in that the object to be pickled is hung on the cathode rod and suspended in an acid electrolyte containing chiefly H_2SO_4. Operating temperatures vary from room temperatures to those near the boiling point, depending upon the object treated. Sufficiently high current density is applied so that hydrogen is copiously evolved at the cathode. The hydrogen in expanding pries off the scale, bit by bit, and bubbles to the surface. Lead anodes are commonly used. If the electrolyte contains chlorides as the result of addition of either HCl or NaCl, the anode will be attacked and a thin layer of lead deposited on the object being pickled. Experimental tests do not indicate any striking advantage of the electrolytic over chemical pickling methods as far as the amount of acid used is concerned, but it is possible to employ lower temperatures and obtain better surfaces with electrolytic pickling than by chemical methods. In the case of steel, the amount of iron dissolved by the two methods varies considerably with different kinds of scale. Electrolytic pickling in general is more rapid than chemical pickling.

Scale-removing methods have been devised which consist first of electrolytic pickling followed by electrolytic cleaning. In one method the object is hung as cathode in a lead- or rubber-lined tank with anodes of metals which form protective films on the object, such as lead which is commonly used, or tin or zinc. Electrolytic pickling is done in an electrolyte consisting of dilute H_2SO_4 and HCl and some NaCl at 65°C. Because of the presence of chlorides in the electrolyte and resultant anode corrosion, thin adherent metal films of lead are deposited on the object being pickled. This action continues until all the scale has been removed and all the surface coated. At this time the object is removed from the acid bath and placed as anode in an electrolytic cleaner containing soda ash, caustic soda, and a small quantity of Na_3PO_4 at 90 to 100°C. In this bath the lead coating deposited in the acid bath is removed, after which the object is withdrawn, washed, and dried. The process is stated to be free from the difficulties of pickling and hydrogen embrittlement. It need not be very carefully controlled, inasmuch as in the acid bath, after the first layer has been deposited, further action results in the deposition of porous spongy deposits without detriment to the object.

In the Bullard Dunn process (U. S. Patent 1,775,671, Sept. 16, 1930) applied to many objects such as heat-treated castings, forgings, etc., in the descaling bath lead or tin is used as anode, the electrolyte being (with Pb) 2.2 oz. salt, 11.0 fluid oz. 66°Bé. H_2SO_4, 3.5 fluid oz. 20°Bé. HCl, and 113.5 fluid oz. of H_2O. Of the anode surface 75 per cent is duriron and 25 per cent is lead. Current density is 60 amp./sq. ft. and anode-cathode spacing is 8 to 10 in. at 6 volts across the bath. Deleading is done in an alkaline cleaner. In descaling, the object under treatment is the cathode; in deleading, it is the anode.

Electrolytic Polishing. Strip steel, rods and sheet, particularly of the nickel chromium or "stainless" alloys, are polished by anodic treatment in phosphoric acid solutions as a competitive method to mechanical surfacing.

Anodic oxidation of aluminum has received considerable study and attention. Films of oxide are produced on aluminum and its alloys for protection against corrosion, for electrical insulation, for decorative effects, and for other uses. Protective adherent oxide films can be produced anodically on aluminum and its alloys. These can be dyed, painted, oiled, or otherwise treated. A number of different methods have been devised. Chubb (U. S. Patents 999,749, 1,068,410, and 1,068,411) uses sodium silicate, ammonium borate, or other electrolytes; Mershon (U. S. Patent 1,065,704), borax; Presser (U. S. Patent 1,117,240), sodium carbonate; Abernathy (U. S. Patent 1,323,236), potassium permanganate and H_2SO_4; Flick (U. S. Patent 1,526,127) ammonia and ammonium sulfide, the article after treatment being dyed by acid dyes; Kujirai (U. S. Patent 1,735,286), oxalic acid or oxalates; Bengough and Stuart (U. S. Patent 1,771,-910), 3 per cent chromic acid solution. Other processes employing H_2SO_4 electrolytes have also been developed. Airplane and dirigible parts of aluminum alloys are often anodically treated in a 3 per cent chromic acid solution in which the voltage across the bath is raised from 0 to 40 volts in 15 min., kept at 40 volts for 35 min., held at 40 to 50 volts for 10 min., and at 50 volts for 5 min., after which they are washed dried, and oiled or greased. The coatings are very resistant to corrosion.

Electrolytic Refining

Copper. The art of electrorefining of copper was invented by James Elkington, in 1865, shortly after the discovery of the dynamo. The invention of the dynamo made possible the electrolytic refining of copper, and the pure metal in turn opened up the great electrical industry of today, which is the largest customer of the copper plants. In electrolytic refining of metals the starting material is a highly concentrated alloy. The object of refining is to remove the last impurities, to recover not only the principal metal in very pure form but also the foreign constituents, especially the precious metals. The impure metal is made the anode at which it dissolves in the electrolyte, to be deposited at the cathode in pure form, inasmuch as the impurities either do not dissolve and fall to the bottom of the tank as slime or mud, or, if they are dissolved, are precipitated by constituents of the electrolyte, or else can remain in the electrolyte without being deposited on the cathode. In general, in the case of copper refining, the anode material is comparatively pure, running 98 per cent copper or better and in most cases well above 99. During the smelting of the copper and lead ores, the base metals act as carriers for the precious metals so that the latter are found in the crude copper or lead. About 75 per cent of the total silver production of the world is recovered as a by-product of copper, lead, zinc, and nickel refining. In addition a very large portion of the platinum produced is a by-product of nickel refining, while considerable palladium is obtained from the same source as well as from copper slimes. Selenium and tellurium are produced almost entirely as by-products of copper refining.

The crude copper from the smelter, termed blister, comes to the refinery in the form of slabs which are sampled very carefully for their precious metal content and then cast into anodes for the refining cells. Cathodes are so-called starting sheets or thin sections of copper deposited on blanks employed as cathodes in starting-sheet cells, from which the deposited copper is stripped off as a sheet. Connectors are then attached and the starting sheet becomes the cathode in the refining cell. The cathode copper is exceedingly pure, usually running considerably better than 99.96 per cent copper in its commercial remelted forms. In order to have high electrical conductivity, the copper must be free of impurities, particularly the metalloids such as arsenic and antimony, and in order to avoid brittleness it must be free from tellurium and lead. The electric conductivity is a very delicate measure of copper purity. It is given according to the Matthiessen standard in which the resistance of 1 mg. pure soft copper at 0°C. is 0.14172 international ohm.

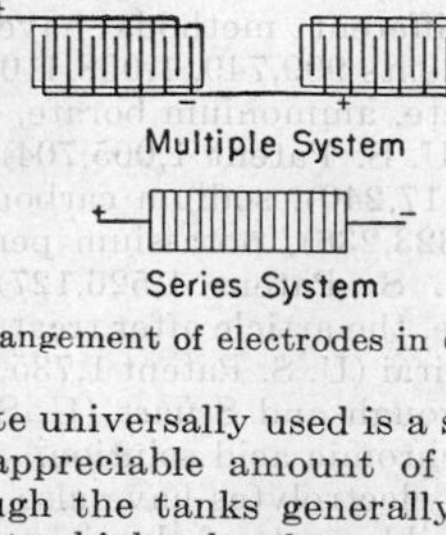

FIG. 27. Arrangement of electrodes in copper refining.

The electrolyte universally used is a solution of $CuSO_4$ containing an appreciable amount of H_2SO_4, which is circulated through the tanks generally in cascade, *i.e.*, from one tank at a higher level to a succeeding one at a lower level. The use of addition agents such as glue and goulac to cause better cathode deposits, and reduce nodules and crystals, is general. Current densities in American refineries vary according to local conditions, power costs, and other items. Constant attention is directed toward reduction of voltage across the tanks to reduce energy cost. Every effort is made to eliminate contact resistances and unnecessary voltage drops through the tanks.

Soluble sulfates of impurities in the anode pass into the electrolyte, portions of which are purified at intervals. Nickel sulfate from such purification becomes a by-product of copper refining and finds extensive employment in nickel electroplating. In some plants copper builds up in the electrolyte to too great an extent, and portions of the electrolyte are run into depositing out tanks in which insoluble anodes are used, so that part of the copper of the electrolyte is precipitated.

Two different systems of electrode arrangement are in use. In the parallel, or multiple, system all cathodes are in parallel and all anodes are in parallel, the tanks being in series. In the series system, only the first and last electrodes are connected to the electric circuit, so that all the electrodes in the circuit, with the exception of the first and the last, are bipolar, one side functioning as cathode and the other as anode. The arrangements of the two systems are illustrated diagrammatically in Fig. 27. The multiple system finds much wider application in that in the United States only one plant and part of another employ the series system.

In the multiple tank, close attention must be paid to the contacts. The series tank has relatively no contacts or conducting bars, and the electrodes are very close together, the anodes being thin, even plates. To produce such anodes they must either be rolled or specially hand cast, and the material used must be of good quality. The interest on the metal tied up in process and the investment in plant are less in the series system. The series system requires no starting sheets, but much closer supervision is needed to keep up the quality of the cathodes. Since lead-lined tanks cannot be used in series work because of the relatively high voltages used, tank maintenance becomes an important item.

Average data for copper refining are shown in Table 19. For greater detail, see Mantell, "Industrial Electrochemistry," 3d ed., McGraw-Hill, New York, 1950.

Table 19. Copper Refining

Electrolyte:	
% Cu	3–3.5
G. Cu/l.	35–42
% H_2SO_4	15–18
Current:	
Amp./sq. ft. cathode	15–20
Voltage per tank	0.2–0.4
% current efficiency	90–95
Lb. Cu/kw.-day	200–250
Anodes:	
Composition	99+ % Cu
Weight, lb.	500–700
Cathodes:	
Original weight, lb.	7–11
Finished weight, lb.	145–300

Lead. The successful electrolytic lead-refining method, known as the Betts process, is used in the large refineries at Trail, B.C.; East Chicago, Ind.; Omaha, Neb.; and at an English plant in Newcastle-on-Tyne.

Anodes of lead bullion and cathodes of electrolytic lead in sheet form are connected in multiple. They are supported on copper bars across a tank containing lead fluosilicate ($PbSiF_6$) and free hydrofluosilicic acid (H_2SiF_6) as electrolyte. The anode mud clings to the uncorroded anode and contains practically all the impurities. It is collected from the anode and from the bottom of the tank and treated for the recovery of Sb, As, Bi, Cu, and Se. Any tellurium that may be present is discarded. The lead deposit on the cathode is melted together with the starting sheet and cast into bars.

During electrolysis, Pb, Sn, Zn, Fe, Ni, and Co go into solution while Cu, Sb, As, Bi, Cd, Ag, Au, Se, and Te remain at the anode. In the first group, tin will invariably deposit with the lead, since these two metals are so close together in the electrolytic series. It is therefore necessary to subject the lead bullion to a softening treatment before it is cast into anodes. The other soluble metals will not precipitate with the lead. Of the insoluble metals, small amounts of antimony may pass to the cathode, especially when a current density as high as 17 to 18 amp./sq. ft. and a temperature of 37 to 38°C. are employed. This may be removed by poling the cathode lead with air in a kettle, which at the same time will separate any tin that may be present.

The electrolyte contains 7 to 10 per cent lead and 8 to 12 per cent total H_2SiF_6, the free H_2SiF_6 content being 3 to 5 per cent. Multiplying the lead value by 0.7 will give an approximate figure for the lead combined with H_2SiF_6. To ensure a solid cathode deposit, a hot strong solution of glue is added daily, to the amount of about 0.013 per cent of the weight of the electrolyte. In making the electrolyte at the plant, H_2SiF_6 is formed by the action of HF on SiO_2, then allowed to combine with lead, PbO, or white lead to yield $PbSiF_6$.

Tanks are arranged in double cascades. If impure anodes are used, the rate of circulation of the electrolyte is 3 to 4 gal./min., while with pure anodes it becomes 7 gal./min. The solution is raised from the sump by means of copper centrifugal pumps equipped with bronze shafts. An electrolyte loss of 5 to 10 lb. H_2SiF_6 per ton of normal lead bullion is attributed in part to the dissociation of the acid. If the anodes contain appreciable quantities of impurities, the loss will approach the higher figure.

Ordinarily, impurities in lead bullion do not total more than 2 per cent, 1 to 1.25 per cent being antimony. Anodes containing more than 2.25 per cent of foreign metals cause difficulties in electrolysis.

Table 20. Electrolytic Lead

Electrolyte:	
% Pb	7–10
% Total H_2SiF_6	12–15
Temperature, °C	35–40
Current:	
Current density, amp./sq. ft.	16–18
Voltage per tank	0.4–0.6
Anodes:	
% Pb	98+
Weight, lb	375–400
Cathodes:	
Original weight, lb	9–12
Finished weight, lb	130–150

Data on refining practice are given in Table 20. Cathodes are cast on sloping cast-iron tables fed by a trough at the upper end. The electrode spacing is 1¾ to 2½ in., the usual range being 2½ to 2¼ in. center to center and 1⅝ in. face to face. Rectangular, flat-bottomed tanks, similar to those used in copper refining, are employed in the electrolytic lead-refining plants. These tanks are coated with petroleum-residue asphalt not exceeding ¼ in. in thickness.

To ensure a full surface for slime particles, about 20 to 28 per cent of the anode is left unattacked, returning to the anode kettles as scrap. The anode mud which may drop to the bottom of the tank is likely not only to cause short circuits but to set up a chemical reaction between its components and the free acid of the electrolyte. A 10-day cleanup and renewal of anodes are the usual plant practice. The presence of bismuth, tellurium, and selenium is the determining factor in the treatment of the mud. If these metals are absent, the process is a simple one; but, with bismuth to be recovered, the treatment becomes complicated.

Nickel. The largest share of the world nickel production is obtained from the Sudbury district of Ontario, Can. The ores are copper-nickel sulfides containing cobalt, iron, and precious metals. The ores are smelted to low-grade mattes which are then blown in basic converters for the removal of iron. The product of these converters, termed *bessemer matte*, is shipped to the nickel refineries where the process is a combination thermal and electrolytic one.

The process is based on the fact that, in a molten system containing nickel sulfide, copper sulfide, and sodium sulfide, in general two liquid layers are formed, the upper carrying the bulk of the sodium and copper sulfides and the lower the bulk of the nickel sulfide. A separation is made of these two layers.

The bessemer matte, containing approximately 54 per cent Ni, 26 per cent Cu, 20 per cent S, and 0.30 per cent Fe, is smeltered in a water-jacketed blast furnace with coke and soda flux. On solidification in pots a **top** and a **bottom** are formed. About 90 per cent of the total copper in the form of top is transferred to converters, where it is blown to blister copper ready for market. The separation between top and bottom is marked by a clean line of cleavage.

The bottom, containing the nickel, is again smelted with soda flux, poured, and allowed to solidify into two layers. The second bottom is nickel sulfide (Ni, 70 per cent; Cu, 0.90 per cent; Fe, 0.25 per cent), which is ground, washed free of soda leached with dilute H_2SO_4 to remove iron, given a chloridizing roast, and leached to remove the copper; then follows a second roast with soda ash to remove all fractional remaining impurities. The resulting black oxide of nickel of the following analysis, Ni, 77.60 per cent; Cu, 0.10 per cent; Fe, 0.25 per cent; Si, 0.10 per cent; and S, 0.015 per cent, is ready for the market or reduction to metal in the open-hearth furnace. It is tapped from the furnace into pig nickel, nickel shot, and nickel anodes. The nickel anodes go to the electrolytic refinery where they are refined under the conditions tabulated.

The electrolyte, purified of iron and other metals, enters the catholyte box at an acidity equivalent to pH of 5.2. During refining, the electrolyte decreases in nickel content leaving the tank at a pH of 4.0. The electrolyte is purified of iron and precipitable metals by passage over finely divided nickel, using the cementation method. Further removal of iron is obtained by adjusting the pH of the solution to 5.2 and heating it to approximately 135°F., at which point ferric sulfates are precipitated. The electrolyte is then cooled down to 125°F. and returned to the cells through rubber-lined or hard-rubber piping. Every precaution is taken to keep the electrolyte out of contact with metals, the liquid coming in contact only with hard rubber, mastic, and, in a single case, with lead at the discharge-box end of the cell, where there is a lead apron.

Owing to the use of the closely woven special canvas diaphragms and the resultant voltage drop across them, the energy efficiency will be low in comparison with copper refining. The anodes are just slightly larger than the cathodes. A 3-day plating gives a 12-lb. starting sheet. Top buckle strips are made from 1½-day starting sheets being fastened together in the same manner as is common practice in copper plants.

Chlorides are not added for anode corrosion, but they are present from various sources to the extent of 75 mg./l. The sulfur content of the electrolytically refined nickel is negligible.

The precious metal content of the original anodes is ¾ oz./ton, being mostly platinum, palladium, and some rhodium. Slimes from primary anodes will run 12 oz./ton. These slimes contain a large amount of nickel oxide,

Table 21. Data on Electrolytic Refining of Nickel

Electrolyte:	
% Ni	40 g./l.
% Boric acid	20 g./l.
Temperature, °C	52–57
Circulation apparatus	Hard-rubber pumps 10 rotaries, each 1100 kva.
Current:	
Amp./sq. ft. cathode	11.5–12
Voltage per tank	2.4–2.5
% current efficiency	93–94
Kw.-hr./lb. Ni	1.1
Anodes:	
Composition	Precious metals ¾ oz./ton
Length, width, thickness, in	36 × 27 × 2
Weight, lb	415–425
Mode of suspension	Cast lugs
Life, days	32–33
% Scrap	36–40
Cathodes:	
Starting sheet blanks	Aluminum, Na_2S dipped, 30 lb. each
Size starting sheet, length by width, in	36 × 28
Weight, lb	11–12
Mode of suspension	Nickel loops
Replaced after how many days	14
Weight, lb	125
Deposition vats, length by width by depth	16 ft. 9½ in. × 2 ft. 10½ in. × 5 ft. 2 in.
Number of anodes, cathodes	29, 30
Tank material	Concrete, mastic, or gilsonite lined
Anode mud:	
Composition	
From primary anodes	Precious metals 12 oz./ton, Ni 30–40%
From secondary anodes	Precious metals 500 oz./ton

the nickel content being as high as 30 to 40 per cent. The slime goes back for reworking and remelting into so-called secondary anodes which have an original precious-metal concentration of 30 oz./ton, yielding slimes running 500 oz./ton. These slimes are concentrated by further operations to a product containing 5000 oz./ton for shipment.

Electrolytic nickel-dissolving cells are used for restoring the metal content of the electrolyte.

Cathode nickel is not remelted, but the cathodes are cut into squares either 2, 4, 6, 8, or 9 in. as specified by the customer. These are packed in barrels for shipment.

Tin. During the First World War the electrolytic refining of tin was developed in the United States as a means of dealing with the metal obtained from the smelting of complex or impure Bolivian ores from which straight dry thermal methods of refining could produce only a poor grade of metal.

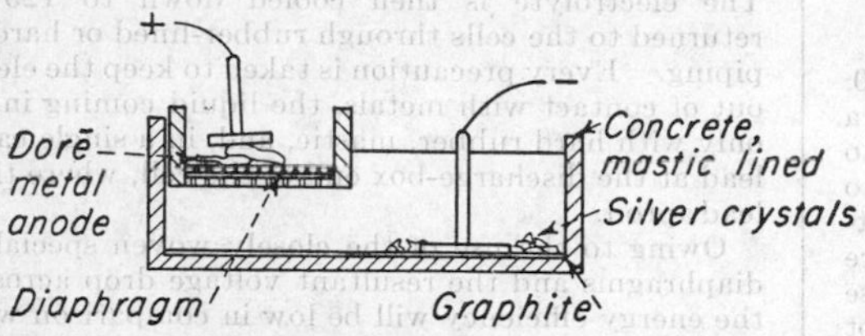

Fig. 28*a*. Thum cell.

The bath employed commercially was 8 per cent H_2SO_4, 4 per cent cresolphenol sulfonic acid, and 3 per cent Sn. Practically, lead is the only impurity that dissolves; consequently the electrolyte must contain a radical that will form an insoluble compound with lead, such as a sulfate, chromate, fluoride, etc. The other metals occurring as impurities in tin (As, Sb, Bi, Cu, etc.) are not dissolved and remain in the anode slimes.

The operating data are given in Table 22. The total acid in the electrolyte is calculated as H_2SO_4, although it was partly H_2SO_4 and partly cresol sulfonic acid. Glue-cresylic acid emulsions were used as addition agents.

Table 22. General Data on Electrolytic Tin
(Mantell)

Electrolyte:	
% Sn	3
% total acid (as H_2SO_4)	10.2
Temperature, °C	35
Circulation, gal./min	5
Circulation apparatus	Vertical centrifugal pumps
Current:	
Amp./sq. ft. cathode	8–10
Voltage per tank	0.3–0.35
Current, kw. per generator	4500
% current efficiency	85
Kw.-hr./lb. Sn	0.085
Anodes:	
% composition	Sn, 96.0; Bi, 1.0; Sb, 0.25; As, 0.15; Cu, 0.25; Pb, 1.0
Length, width, thickness, in	33 × 36 × 1¼
Weight, lb	350
Mode of suspension	Cast lugs
Life, days	21
% scrap	25
Cathode:	
Size starting sheet, length by width by thickness, in	34 × 37 × 0.03
Weight, lb	8 to 10
Mode of suspension	Wrapped
Replaced after how many days	7
Weight, lb	100
Deposition tanks:	
Length, width, depth	12 ft. 11 in. × 3 ft. 5 in. × 3 ft. 6 in.
Number anodes, cathodes	26, 27
Electric connection	Walker
Material of construction	Wood, lead lined
Anode mud:	
% anode	5
% composition	Pb, 20; Cu, 5; As, 3; Sb, 5; Sn, 30; Bi, 20

The addition agent consumed was of the order of ⅓ to 3 lb. glue and 8 to 16 lb. cresylic acid per ton of refined tin. Refinery starting sheets were made by pouring molten electrolytic tin over an inclined steel table of the size and shape of sheet desired, a method similar to that used for making starting sheets for the electrolytic refining of lead. The electrolytic tin analyzed better than 99.98 per cent Sn.

Silver. The raw material from which pure silver is produced is in the main concentrates from anode slimes from copper refining as well as those of other non-ferrous metals such as lead, nickel, and zinc; silver concentrates resulting from the desilverization of lead by the Parkes process; and silver-gold bullion of various compositions including secondary metal and scrap. From copper-refining slimes the crude silver may contain 95 per cent Ag, 1 to 3 per cent Au, the remaining 2 per cent being Cu, Bi, Pb, Te, Fe, Ni, Pt, and small amounts of other

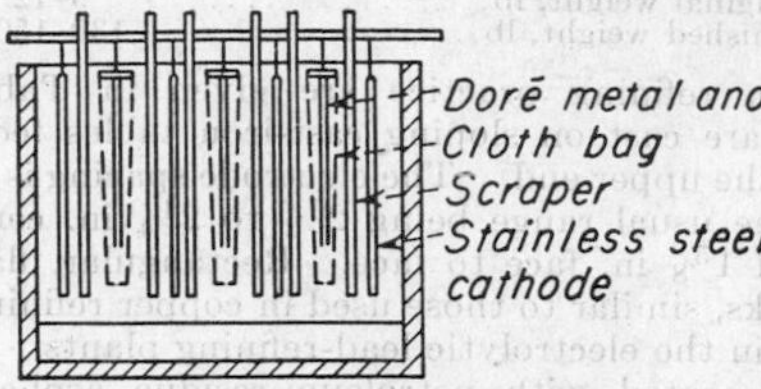

Fig. 28*b*. Moebius cell.

metals; while from the desilverization of lead the crude silver may contain 98 per cent Ag; 0.5 per cent Au; 1.5 per cent being Cu, Bi, Pb, and Zn. Such material is cast into anodes and refined electrolytically in a $AgNO_3$-$Cu(NO_3)_2$ electrolyte.

In the electrolytic refining of copper, all the precious metals originally in the anodes drop to the bottom of the tanks during the process of electrolysis and are known as *slime* or *anode mud.* When the scrap anodes are lifted from the tanks after electrolysis, the slime is sluiced from the bottom of the tank and pumped to the silver refinery. On reaching the silver refinery, the slime is allowed to settle in large tanks, filtered, and given a light roast (300°C.). This converts all the copper into copper oxide, readily leachable with 10 per cent H_2SO_4. The leached slime, containing less than 1 per cent copper, is melted and refined in small oil-fired reverberatory *doré furnaces.* During the refining process, the lead forms a lead slag which is skimmed off and sent to the lead refinery for further treatment. Antimony is volatilized and recovered from the flue dust. Selenium and tellurium are partly volatilized but are mainly removed by the addition of alkaline fluxes. The doré silver is cast into anodes and electrolyzed for the parting of the silver from the gold, using both the Thum and the Moebius systems (Fig. 28*a* and Fig. 28*b*).

In the Thum system the electrolytic cells consist of shallow, glazed, porcelain tanks, or mastic-lined concrete tanks, the bottom of which, lined with graphite slabs, forms the cathode. The doré silver anodes are arranged horizontally in wooden frames above the cathode. These frames are covered with canvas which acts as a diaphragm and prevents the gold slime from mixing with the cathode deposit. The silver is deposited in a loose crystalline form which is scraped from the bottom of the tank at regular intervals. In the Moebius system, the anodes and cathodes are arranged vertically in much the same manner as in copper refining, the anodes being enclosed in canvas bags in order to catch the gold slime. The cathodes are sheets of stainless steel. Mechanically operated wooden scrapers are used to remove the silver crystals from the cathodes into trays which are periodically withdrawn and emptied. The cathode deposit of silver crystals is very pure. After washing, it is melted in large graphite retorts into standard 1000-oz. bars, assaying over 999 in fineness.

The gold slime resulting from the silver electrolysis contains all the gold, platinum, and palladium which were originally present in the blister copper. It is treated with H_2SO_4 to remove the excess of silver present and is then melted in graphite crucibles and cast into gold anodes to be electrolyzed using the Wohlwill process.

The silver electrolyte is silver and copper nitrates (60 g. Ag/l., 40 g. Cu/l.) operated at a current density of 40 to 50 amp./sq. ft.

Gold. The process employed for gold refining is due to Wohlwill. The method, which is used for the refining of gold in all the United States mints, consists in electrolyzing gold anodes in a hot acid solution of 7 to 8 per cent gold chloride and 10 per cent HCl. In order to maintain low interest charges on the gold tie-up, a high current density of 110 amp./sq. ft. of cathode surface is used. This necessitates a thorough stirring of the electrolyte, which is accomplished by means of a small air lift.

In gold electrolysis, the platinum and palladium are totally soluble and are allowed to accumulate in the electrolyte until present in sufficient quantity to recover. The amount of platinum and palladium recovered is small, averaging only about 1 oz. of the combined metals per 1,000,000 lb. copper refined.

The electrolysis cells are constructed of glazed porcelain and chemical stoneware. They are relatively small in size, in that the electrolyte is an expensive one. The cathodes, having the same area as the anodes, are of thin gold foil or thin rolled-gold sheet. In copper refineries the anodes are fairly thin. In governmental refineries or mints where interest on metal tied up is not a factor, the anodes may be fairly thick. At periodic intervals, because of the accumulation of impurities, part of the electrolyte is drawn off for purification and replaced by a strong $AuCl_3$ solution.

For the treatment of anodes of high silver content, Wohlwill modified his process by using an unsymmetrical alternating current in which an alternating current of low frequency and greater current density is superimposed on the direct current. The quantities of gold dissolved and deposited are functions only of the direct current. The alternating current opposes passivation which would develop in a high-silver-content anode as the AgCl layer forms over the gold anode, lessening the active area and increasing the effective current density. The alternating current prevents the formation of an insulating film over the anode and causes the AgCl to flake off readily.

The modified Wohlwill process employing alternating current is not used in copper refineries, but the method finds employment in the government mints and assay offices.

Bismuth. Bismuth-rich slimes are produced as a by-product of the Betts method of lead refining. The slimes are washed, dried, and smelted under oxidizing conditions with Na_2CO_3 and NaOH to produce alkaline slags which carry off the arsenic and antimony as sodium compounds and the copper as oxide. The residue is crude bismuth containing mainly silver and lead. This is refined electrolytically in an acid $BiCl_3$ solution containing somewhat more than 100 g. HCl/l. and 3 to 4 g. of bismuth in a Thum cell, the same type as is used for silver refining, at 50 to 60°C. The lead from the crude bismuth anode goes into solution as $PbCl_2$ but does not deposit with the bismuth. It concentrates in the electrolyte, portions of which are periodically removed and the $PbCl_2$ crystallized out.

Electrowinning

Copper. In several places in the world, copper-containing minerals are treated for the recovery of the copper by hydrometallurgical methods, in which the mineral is leached or dissolved by a solution which in turn becomes an electrolyte and is stripped of a portion of its metal values by electrolytic precipitation, the stripped or spent electrolyte being returned for leaching fresh ore in a cyclic process. The metallic values of ores treated by hydrometallurgical or electrowinning methods are usually very low, being of the order of 1.5 to 1.75 per cent copper (or less). At the plant of the Chile Copper Co. at Chuquicamata, Chile, the principal copper minerals are chalcanthite ($CuSO_4.5H_2O$), brochantite [$CuSO_4$.-$3Cu(OH)_2$], and atacamite [$CuCl_2.3Cu(OH)_2$], and the leach contains H_2SO_4. In the electrolytic deposition insoluble copper silicide anodes are used. Power consumption per pound of copper is much higher than in copper refining because of the greater voltages across the cells.

In the plant of the Inspiration Consolidated Copper Co. at Inspiration, Ariz., the ore consisted of mixed oxides and sulfides in which the total copper of the ore was of the order of 1.1 to 1.2 per cent. The leach was a H_2SO_4-$Fe_2(SO_4)_3$ solution, and the insoluble anodes were 8 per cent antimonial lead.

In general, the ores were leached in large concrete tanks. At Chile the tanks are of reinforced concrete lined with mastic, while at Inspiration they were reinforced concrete lined with lead. In general, despite the fact that hydrometallurgical plants are operating with extremely low grade ores, the copper recovery is of the order of 80 to 90 per cent and the production costs are considerably lower than in pyrometallurgical plants operating smelters. These smelters, in addition, have the advantage of using high-copper-content ores.

Operating data for different plants vary widely depending upon ores treated, leaching solutions used, impurities encountered, and methods of purification. For further data see Mantell, "Industrial Electrochemistry," 3d ed., McGraw-Hill, New York, 1950. Cell electrolytes are 20 to 35 g. Cu./l., 20 to 45 g. H_2SO_4/l. Current densities average 10 to 12, although as low as 5 amp./sq. ft. are used. Energy consumptions are 1 to 1.5 kw.-hr./lb. Cu; tank voltages are 1.8 to 2.5, and current efficiencies 65 to 90 per cent.

Zinc. Vast quantities of complex zinc ores are available which are not readily amenable to pyrometallurgical methods, but in the treatment of which leaching processes, coupled with electrolytic deposition, have particular advantages. In addition, the subsidiary metals can be completely recovered.

Electrolytic zinc plants are now operating on a very large scale in a number of places in the world. In general, all the processes employed involve H_2SO_4 and the production of $ZnSO_4$ electrolytes, which can be divided into those termed low-acid, low-current density, and the high-acid, high-current density, also known as the Tainton process.

The ores treated are of the complex zinc-lead-copper-iron sulfide type. These ores, averaged after roasting, contain 33 per cent or higher Zn, 4 to 4.5 per cent S, and a soluble zinc content of the order of 82 to 86 per cent, 10 per cent Pb, 21 to 22.5 per cent Fe, and 1.7 to 2 per cent Cu, with about 7 to 7.5 per cent of the ore being insoluble in acids. In the roasting of concentrate containing 5 per cent iron (or less), practically all the iron combines with zinc as $ZnO.Fe_2O_3$ (ferrite), which is nearly insoluble in dilute H_2SO_4. As the iron content increases above 5 per cent, the combination of zinc and iron oxides can be partly controlled by temperature regulation during roasting. The roasting furnaces are operated to convert the zinc sulfide in the ore to acid-soluble oxides and sulfates and, at the same time, to produce SO_2 which is used for the manufacture of H_2SO_4 employed in leaching.

In general, the procedure is cyclic. Electrolyte from the cells after the greater portion of the zinc content has been deposited is used to leach roasted ore, termed calcine, the solids and liquids being separated, the $ZnSO_4$ solution being purified by means of metallic zinc. Spent electrolyte from deposition tanks contains 10 to 11.5 per cent H_2SO_4 and about 2.5 per cent zinc. After the leaching treatment, the liquor contains about 0.5 per cent H_2SO_4 and 10 per cent zinc.

The high-acid, high-current-density method is employed in the electrolytic zinc plants erected to treat the zinc-bearing ores of the Coeur d'Alene district. These ores cannot be classed as particularly favorable for electrolytic treatment, inasmuch as they tend to form an unusually large amount of insoluble zinc ferrite in the roast and yield appreciable quantities of gelatinous silica in the leaching operation. Furthermore, they contain relatively large amounts of cobalt, which is one of the most troublesome impurities from the standpoint of electrolytic zinc treatment. In the Tainton high-acid process the return electrolyte used for leaching carries 28 to 30 per cent free acid, and the electrolysis is carried out at a current density of 100 amp./sq. ft., both these amounts being about three times as great as the corresponding figures in ordinary electrolytic zinc practice.

Table 23. Electrolytic Zinc

	Low acid, low c.d.	High acid, high c.d.
Electrolyte:		
Zn to cells, g./l.	100–150	200–220
H_2SO_4, cell discharge, g./l.	100–130	280
Current:		
Density, amp./sq. ft.	30–40	100
Current efficiency, %	90–94	87–92
Voltage per tank	3.7	3.2–3.5
Anodes:		
Material	Lead	Lead alloy
Construction	Solid	40–50% perforated
Cathodes:		
Sheets	Aluminum	Aluminum
Zn weight on cathode, lb.	8–11	22–24
Tanks	Wood, lead lined; or lined concrete	Wood, lead lined
Tank mud	MnO_2	MnO_2

In the Tainton process, zinc ferrite is separated magnetically from the calcine and in the leaching process is first added to a strongly acid solution resulting from stripping the electrolyte in the deposition cells, the neutralization of the acid being then completed by the use of the ZnO portion of the calcine free from ferrite. With the allowable production of zinc ferrite in the calcine, less rigid control on roasting is necessary than is needed for the usual type of leaching of zinc ores.

Cadmium Recovery. Cadmium is found in very small quantities in practically all zinc ores. It can be profitably produced only as a by-product in the manufacture of some other metal. Important sources of raw material are bag-house condensation products from lead and copper furnaces, zinc ores, "blue" powder, and particularly cadmium residues from electrolytic zinc plants. These materials are leached, purified by chemical treatment and by the precipitation of heavy metals with zinc, to produce liquors containing only zinc, cadmium, and free H_2SO_4. These liquors are electrolyzed between insoluble lead anodes and aluminum cathodes. In earlier plants, rotating cathodes were used, but in present practice they are stationary. Operating data are given in Table 24. Cadmium has found widespread application in electroplating and rustproofing. Demand for the metal has at times exceeded the supply and markedly increased the price.

Table 24. Electrolytic Recovery of Cadmium

Raw material	Precipitate from Zn dust purification of Zn electrolyte
Electrolyte:	
Cd to cells, g./l.	100–120
Zn to cells, g./l.	80
Concentration H_2SO_4, cell discharge, g./l.	70–80
Temperature, °C	35
Current per cell:	
Amp./sq. ft.	4.25
Current efficiency, %	80–90
Kw.-hr./lb.	0.82
Voltage per tank	2.6
Anodes:	
Composition	Lead
Anode-cathode spacing, in.	3½
Cathodes:	
Composition	Aluminum
Removed	24 hr.
Weight, lb.	100
Tanks:	
Number anodes and cathodes	27,26
Material of construction	Wood, lead lined

Manganese is leached from reduced manganese ores by sulfate solutions which are electrolyzed in multicompartment cells to produce pure metal.

Electrowinning procedure have been proposed for antimony, chromium, and cobalt.

Alkaline Chloride Electrolysis

Aqueous Electrolytes—Chlorine and Caustic. Chlorine and NaOH are made by the electrolysis of a solution of an alkaline chloride, generally NaCl. The sodium ion travels to the cathode. The cathodic products are a solution of NaOH and hydrogen gas evolved at the cathode. The chlorine ion travels to the anode, gives up its charge, becomes a chlorine molecule, Cl_2, and to the greater extent is evolved as gaseous chlorine, although some dissolves in the solution. The anodes almost universally in use are those non-attackable by chlorine, consisting of graphite. In every commercial cell, precautions are taken to keep the anodic and cathodic products separate so that they do not interact to form hypochlorites. The products of electrolysis are removed as rapidly as possible from the sphere of electrolytic action. Cells in commercial use fall, with only minor exceptions, into those in which the hydroxyl ions are not liberated inside the electrolysis chamber (as in the mercury cells) and diaphragm cells in which mixing of the anolyte and catholyte is prevented by a porous diaphragm. These may be subdivided into those rectangular in shape and those circular or cylindrical in form. The bell-jar cell is an example of the type which does not use a diaphragm, but in which mixing of the anodic and cathodic products is prevented by regulation of the electrolyte fed to the cell.

The fundamental principles of construction of the more important cells used commercially are shown in Fig. 29.

The **mercury** cell employs a large iron-grid cathode with graphite anodes and an intermediate electrode consisting of mercury. In one portion or chamber of the cell a solution of NaCl is electrolyzed to produce chlorine at the anode and sodium metal at the mercury cathode. The sodium metal alloys with the mercury to form an amalgam. Circulation of the mercury by mechanical means, such as pumps, screws, or rocking, causes the mercury from the NaCl compartment to pass into another compartment where it functions as an anode in a NaOH electrolyte, with the production of hydrogen at the iron cathode. The second section is often termed the denuder. The sodium from the amalgam is discharged with the formation of caustic solution. A number of modifications of a mechanical nature for the circulation of the mercury have been proposed. In general, mercury cells produce caustic liquors of much higher concentration than those of other types.

In the **Castner** cell the mercury is circulated from two outside electrolysis compartments to a central denuder section. In the **Sorensen** cell the mercury is mechan-

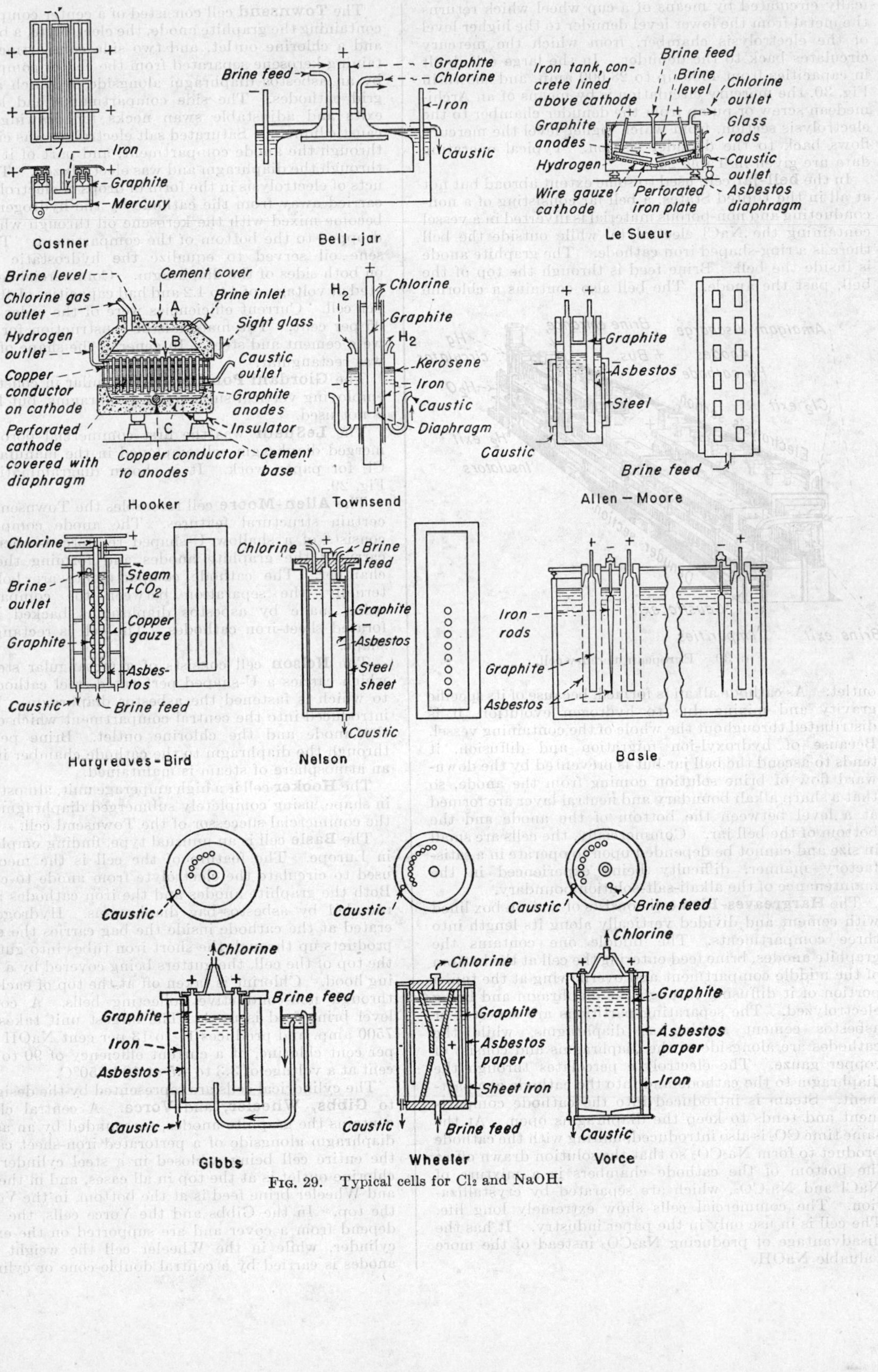

FIG. 29. Typical cells for Cl_2 and NaOH.

ically circulated by means of a cup wheel which returns the metal from the lower level denuder to the higher level of the electrolysis chamber, from which the mercury circulates back to the denuder. In the large cell, built in capacities from 4000 up to 28,000 amp. and shown in Fig. 30, the mercury circulation is by means of an Archimedean screw or pump from the denuder chamber to the electrolysis section, from which higher level the mercury flows back to the denuder section. Typical operating data are given in Table 25.

In the **bell-jar** cell, used to some extent abroad but not at all in the United States, a bell jar consisting of a nonconducting and non-porous material is inverted in a vessel containing the NaCl electrolyte, while outside the bell there is a ring-shaped iron cathode. The graphite anode is inside the bell. Brine feed is through the top of the bell, past the anode. The bell also contains a chlorine

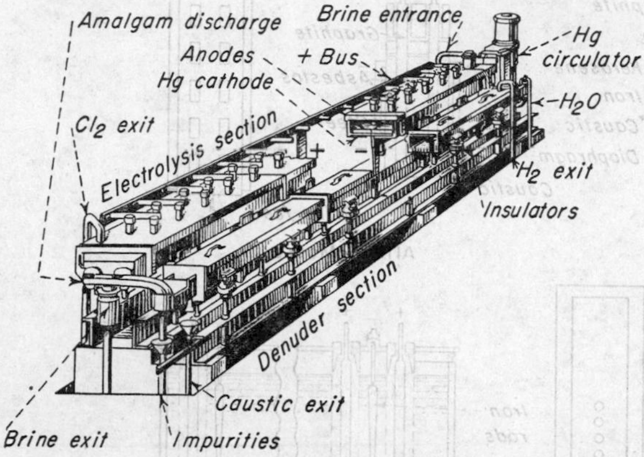

FIG. 30. European mercury cell.

outlet. As cathode alkali is formed, because of its specific gravity and mixing due to hydrogen evolution, it is distributed throughout the whole of the containing vessel. Because of hydroxyl-ion migration and diffusion, it tends to ascend the bell jar but is prevented by the downward flow of brine solution coming from the anode, so that a sharp alkali boundary and neutral layer are formed at a level between the bottom of the anode and the bottom of the bell jar. Commercially, the cells are small in size and cannot be depended upon to operate in a satisfactory manner, difficulty being experienced in the maintenance of the alkali-salt-solution boundary.

The **Hargreaves-Bird** cell consists of an iron box lined with cement and divided vertically along its length into three compartments. The middle one contains the graphite anodes, brine feed entering the cell at the bottom of the middle compartment and overflowing at the top, a portion of it diffusing through the diaphragm and being electrolyzed. The separating partitions are asbestos or asbestos cement serving as diaphragms, while the cathodes are alongside of the diaphragms and consist of copper gauze. The electrolyte percolates through the diaphragm to the cathode and into the cathode compartment. Steam is introduced into the cathode compartment and tends to keep the diaphragms open. At the same time CO_2 is also introduced, uniting with the cathode product to form Na_2CO_3 so that the solution drawn off at the bottom of the cathode chambers is a mixture of NaCl and Na_2CO_3, which are separated by crystallization. The commercial cells show extremely long life. The cell is in use only in the paper industry. It has the disadvantage of producing Na_2CO_3 instead of the more valuable NaOH.

The **Townsend** cell consisted of a center compartment containing the graphite anode, the electrolyte, a brine feed and a chlorine outlet, and two side compartments containing kerosene separated from the center compartment by an asbestos diaphragm alongside of which are iron-grid cathodes. The side compartments had hydrogen exits and adjustable swan necks for drawing off the caustic liquors. Saturated salt electrolyte was circulated through the anode compartment, and part of it diffused through the diaphragm and was electrolyzed. The products of electrolysis in the form of drops of electrolyte were carried away from the cathode by the hydrogen gas, to become mixed with the kerosene oil through which they dropped to the bottom of the compartment. The kerosene oil served to equalize the hydrostatic pressure on both sides of the diaphragm. Commercial cells operated at voltages of 4 to 4.2 and had capacities of 4000 amp. per cell. Current efficiencies were of the order of 90 to 95 per cent. The materials of construction for the cell were cement and steel. In general, the shape of the cell was rectangular.

The **Giordani Pomilio** cell is similar in construction, employing vertical submerged diaphragms, but kerosene is not used.

The **LeSueur** was the first commercial porous, submerged diaphragm cell employed in the manufacture of Cl_2 for paper work. It is shown diagrammatically in Fig. 29.

The **Allen-Moore** cell resembles the Townsend cell in certain structural features. The anode compartment consists of a shallow U-shaped reinforced-concrete pot carrying the graphite anodes and forming the anode chamber. The cathode compartments are bolted externally, the separation between the compartments being made by asbestos diaphragms backed by perforated sheet-iron cathodes. The cell is rectangular in shape.

The **Nelson** cell consists of a rectangular steel tank which carries a U-shaped perforated-steel cathode plate to which is fastened the asbestos diaphragm. Brine is introduced into the central compartment which contains the anode and the chlorine outlet. Brine percolates through the diaphragm to the cathode chamber in which an atmosphere of steam is maintained.

The **Hooker** cell is a high amperage unit, almost cubical in shape, using completely submerged diaphragms, being the commercial successor of the Townsend cell.

The **Basle** cell is an unusual type, finding employment in Europe. The feature of the cell is the mechanism used to circulate the electrolyte from anode to cathode. Both the graphite anodes and the iron cathodes are surrounded by asbestos-bag diaphragms. Hydrogen generated at the cathode inside the bag carries the cathode products up through the short iron tubes into gutters on the top of the cell, the gutters being covered by a collecting hood. Chlorine is taken off at the top of each anode through non-conductive collecting bells. A constant-level brine feed is used. The largest unit takes about 7500 amp. and produces 11 to 13 per cent NaOH and 98 per cent chlorine, at a current efficiency of 90 to 92 per cent at a voltage of 3.3 to 4.5 at 40 to 50°C.

The cylindrical cells are represented by the designs due to **Gibbs, Wheeler,** and **Vorce.** A central chamber contains the graphite anodes, surrounded by an asbestos diaphragm alongside of a perforated iron-sheet cathode, the entire cell being enclosed in a steel cylinder. The chlorine outlet is at the top in all cases, and in the Gibbs and Wheeler brine feed is at the bottom, in the Vorce at the top. In the Gibbs and the Vorce cells, the anodes depend from a cover and are supported on the external cylinder, while in the Wheeler cell the weight of the anodes is carried by a central double-cone or cylindrical

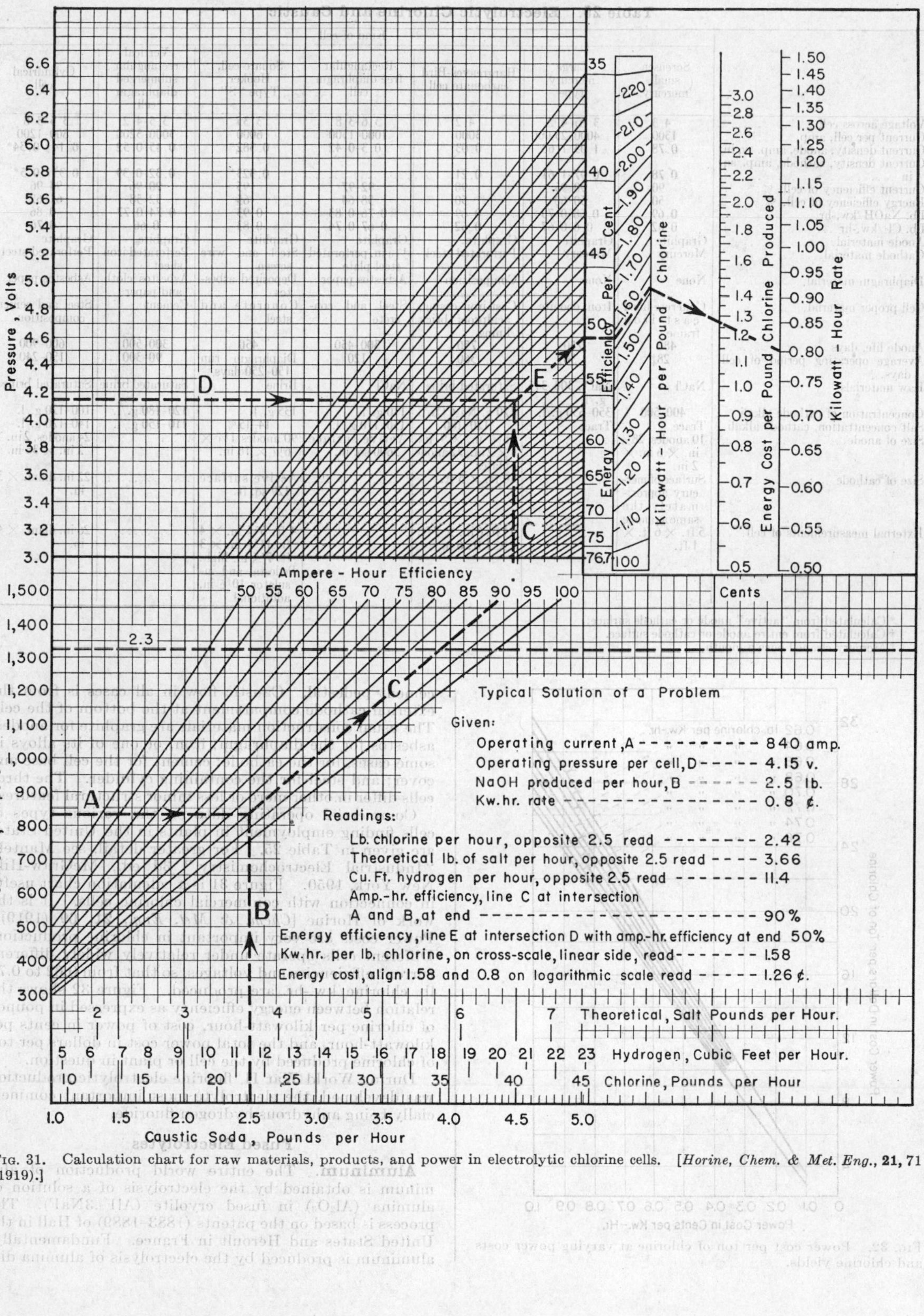

FIG. 31. Calculation chart for raw materials, products, and power in electrolytic chlorine cells. [*Horine, Chem. & Met. Eng.*, **21**, 71 (1919).]

Table 25. Electrolytic Chlorine and Caustic

	Kind of cell						
	Sorensen small mercury	Large mercury type	Hargreaves-Bird carbonate cell	Rectangular free diaphragm cell	Square cell, Hooker Type "S"	Vertical rectangular submerged diaphragm cell	Cylindrical cell
Voltage across cell	4.3	3.85–4.2	4.2	3.6–3.8	3.35	3.5–4.2	3.5–3.6
Current per cell, amp	1300	4000–25,000	3000	1000–1500	6000	3000–3200	800–1200
Current density, anode, amp./sq in.	0.78	1.03–1.68	0.03	0.3–0.42	0.382*	0.45–0.52	0.14† 0.34*
Current density, cathode, amp./sq. in.	0.78	0.97–1.61	0.21	....	0.323*	0.32–0.39	0.35† 0.5*
Current efficiency of cell, %	90	90–95	90	92–97	95	90–96	94–96
Energy efficiency of cell, %	50	50–60	50	58–64	65	52–58	61–62
Lb. NaOH/kw.-hr	0.69	0.68–0.75	0.69	0.76–0.83	0.93	0.74–0.77	0.86
Lb. Cl_2/kw.-hr	0.62	0.6–0.67	0.62	0.67–0.74	0.83	0.66	0.76
Anode material	Graphite	Graphite	Graphite	Graphite	Graphite	Graphite	Graphite
Cathode material	Mercury	Mercury	Perforated steel	1/16 in. perforated steel plate	Steel and wire screen	Perforated iron sheet	Perforated steel
Diaphragm material	None	None	Composition	Asbestos paper	Deposited asbestos	Asbestos cloth and paper	Asbestos paper
Cell proper material	Concrete-lined cast-iron frame	Iron, ebonite	Cast-iron casing, acidproof brick lining	Steel and concrete	Concrete and steel	Cement	Steel and cement composition
Anode life, days	480	750	750	300–450	450	300–500	600–900
Average operating period of cell, days.	28‡	30‡	360	120	Diaphragm run 150–250 days	90–300	150–240
Raw materials	NaCl	NaCl, 300–315 g./l.	Saturated brine	NaCl	Brine	Saturated brine	Saturated brine
Concentration of cathode alkali	400–500	350–650 g./l.	170 g. Na_2CO_3/l.	110 g./l.	135 g./l.	120 180 g./l.	100 120 g./l.
Salt concentration, cathode alkali	Trace	Trace	120–180	150–170 g./l.	14–15%	110–150 g./l.	140–170 g./l.
Size of anode	10 anodes, 28½ in. × 6 in. × 2 in.	....	16½ in. × 9 in. × 2 in., 72 anode plates	Active surface 3600 sq. in.	90 anodes 1¼ × 6¼ × 18 in.	....	24 anodes, 2 in. × 2 in. × 36 in.
Size of cathode	Surface of mercury approximately the same as anode	....	10 ft. × 5 ft.	....	Active surface 129 sq. ft.	....	22 in. diam. × 34 in.
External measurements of cell	5 ft. × 6 ft. × 1 ft.	....	11 ft. 6 in. × 7 ft. × 1 ft. 6 in.	....	4 ft. 5⅝ in. × 4 ft. 11½ in. × 3 ft. 7⅝ in. high. Pedestal and insulator 10½ in. additional	....	26 in. diam. × 42 in.

* Calculated from "active" anode or cathode surface.
† Calculated from entire anode or cathode surface.
‡ Mercury cleaned once a month.

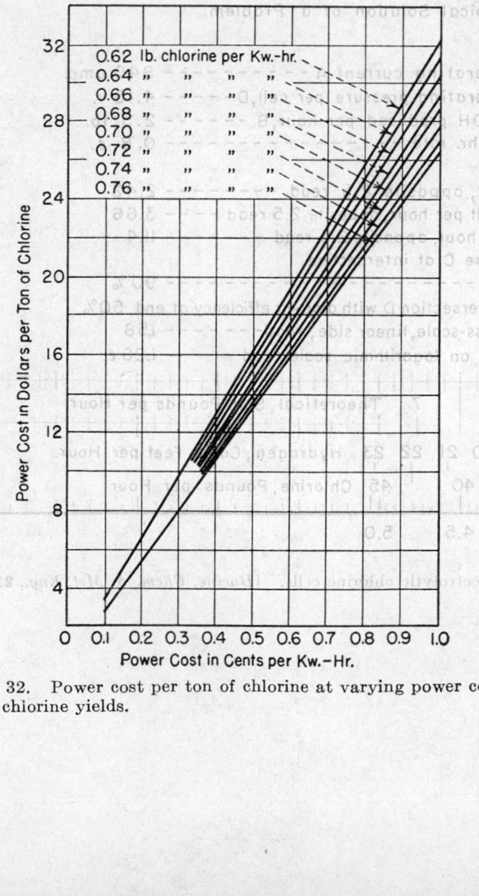

FIG. 32. Power cost per ton of chlorine at varying power costs and chlorine yields.

cement pedestal. Caustic flow in all cases is from the external cathode compartment at the bottom of the cell. The usual construction materials are graphite for anodes; asbestos for the diaphragm; iron, or one of its alloys in some cases, for the cathode; cement for the cell base and cover; and steel for the containing cylinder. The three cells differ in other more or less minor structural features.

Commercial operating data on the various types of cells finding employment in plants in the United States are given in Table 25. For greater detail, see Mantell, "Industrial Electrochemistry," 3d ed., McGraw-Hill, New York, 1950. Figure 31 is a calculation chart useful in connection with commercial chlorine cells. It is the work of Horine [*Chem. & Met. Eng.*, **21**, 69 (1919)]. Power costs are very important in chlorine production. Different cells operate under relatively widely different current efficiencies and voltages, so that from 0.62 to 0.76 lb. chlorine/kw.-hr. are produced. Figure 32 shows the relation between energy efficiency as expressed in pounds of chlorine per kilowatt-hour, cost of power in cents per kilowatt-hour, and the total power cost in dollars per ton of chlorine produced by the cell or plant in question.

During World War II, fluorine electrolytic production was developed, the electrolyte most important commercially being anhydrous hydrogen fluoride.

Fused Electrolytes

Aluminum. The entire world production of aluminum is obtained by the electrolysis of a solution of alumina (Al_2O_3) in fused cryolite ($AlF_3.3NaF$). The process is based on the patents (1883–1889) of Hall in the United States and Heroult in France. Fundamentally, aluminum is produced by the electrolysis of alumina dissolved in a bath of aluminum fluoride and the fluoride of one or more metals more electropositive than aluminum, such as sodium, potassium, or calcium. Cryolite melts at about 1000°C. and at temperatures slightly above its melting point is able to dissolve as much as 10 to 20 per cent of its weight of alumina with resultant decrease in its melting point. As the result of electrolysis alumina is decomposed, aluminum metal being deposited at the cathode in a molten condition (melting point about 660°C.) and oxygen at the carbon anode with which the oxygen reacts to form the probable primary product of CO_2. This is believed to be subsequently reduced to CO by the hot carbon. The thermal effect of the oxidation of the carbon anodes is to reduce the amount of electrical energy required to maintain the bath in a fused state. The bath itself is not appreciably decomposed by the current.

Table 26. Operating Data for Aluminum and Magnesium

Factors	Aluminum	Magnesium
Raw material	Purified Al_2O_3	Anhydrous $MgCl_2$
Melting point material, °C.	Cryolite 1000	$MgCl_2$ 708
Melting point metal, °C.	Al 660 (about)	Mg 651
Bath material	Cryolite + Na, Al, Ca fluorides	$MgCl_2$ + NaCl + $CaCl_2$
Furnace:		
Shape	Rectangular	Rectangular 6 ft. × 5 ft. × 11 ft.
Shell	1–2-in. steel	Steel
Anode arrangement	Suspended vertically	Suspended vertically
Anode material	Carbon	Graphite
Cathode material	Rammed carbon lining	Steel pot
Voltage across cell	5.5–7	6.3
Amperage of cell	20,000–40,000	30,000–70,000
Current density, amp./sq. in.	....	16–35
Operating temperature, °C.	900–1000	670–730
Concentration of raw material in bath	2–5% Al_2O_3	
Theoretical decomposition voltage	2.0	
Carbon consumption	0.6–0.85 lb./lb. Al	
Raw material consumption	2–2.2 lb. Al_2O_3/lb. Al	4–5 lb. $MgCl_2$/lb. Mg
Current efficiency, %	70–90	75–80
Energy efficiency, %	25–40	20–30
Energy consumption	10 kw.-hr./lb. Al	8–9 kw.-hr./lb. Mg
Specific gravity molten Al	2.29 (1000°C.)	
Specific gravity molten cryolite	2.095 (1000°C.)	

At the operating temperature the fused cryolite (or modified bath) and molten aluminum are very reactive and destructive of containers. Hall solved the problem by the use of an iron crucible or a box lined with carbon. To date no better materials of construction have been found.

The reduction cell or "pot" is a strong steel box, usually rectangular in shape, provided with a carbon lining 6 to 10 in. or more in thickness. Figure 33 shows the

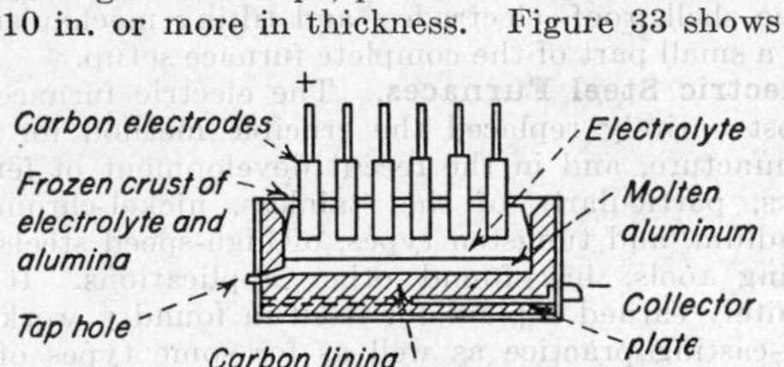

Fig. 33. Electrolytic aluminum cell.

essential features of such a cell. External and internal dimensions of the cell and lining vary considerably with current capacity. In commercial practice the smallest cells are 8000 amp. and the largest of the order of 40,000 amp. Generally the more current that can be used in a cell, the lower will be the producing cost of 1 lb. aluminum. Aluminum production is approximately proportional to the current, while it takes just about as much labor to run a small cell as a large one. The limit of current capacity is set by increase in difficulties involved in changing anodes, breaking the frozen top crust, and operating the larger cells. For multiple-electrode pots the practical limit is reached at about 50,000 amp. Operating data are given in Table 26.

Aluminum Refining. Aluminum is refined electrolytically in an *inverted* cell in a fused electrolyte composed of cryolite, AlF_3, and BaF_2 nearly saturated with Al_2O_3 [Frary, Electrolytic Refining of Aluminum, *Trans. Am. Electrochem. Soc.*, **47**, 275 (1925)]. Heavier aluminum alloys at the bottom of the cell are in contrast with the carbon anode, and the refined, purified metal rises through the electrolyte to make contact with graphite cathodes suspended vertically through a top frozen crust of electrolyte which contains a large concentration of alumina.

Magnesium. Magnesium is prepared in large quantities by the chloride process, in which the raw material and electrolyte are $MgCl_2$ to which some NaCl may be added to lower the melting point. Figure 34

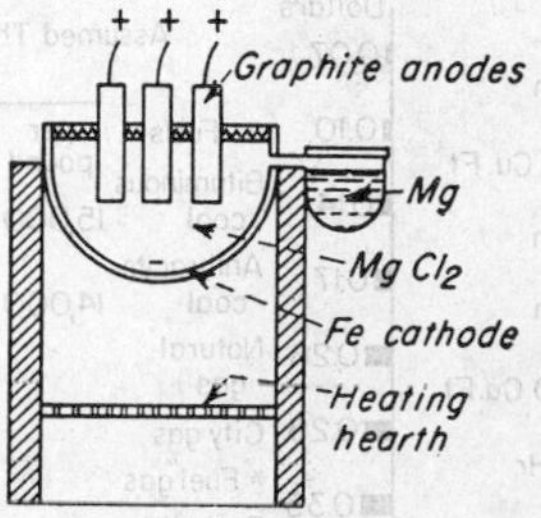

Fig. 34. Magnesium cell.

shows a typical cell. Operating data are given in Table 26. For manufacturing processes involving preparation, purification, and drying of the $MgCl_2$ and its electrolysis, see Gann, The Magnesium Industry, *Trans. Am. Inst. Chem. Engrs.*, **24**, 206 (1930); Mantell, "Industrial Electrochemistry," 3d ed., McGraw-Hill, 1950; or Hunter, *Trans. Electrochem. Soc.*, **86**, 21–31 (1944).

Beryllium. Beryllium is produced from beryl by chemical concentration, producing chlorides or fluorides. Mixed fused salts are employed as electrolyte.

Calcium. Calcium is made by fused salt electrolysis according to data in Table 27 (see Mantell and Hardy, "Calcium Metallurgy and Technology," Am. Chem. Soc. Monograph 100, Reinhold, New York, 1945).

Table 27. Calcium Production

Raw material	Anhydrous $CaCl_2$
Melting point $CaCl_2$, °C.	780
Melting point Ca, °C.	800
Specific conductivity $CaCl_2$	1.9 at 800°C.
Type of cell	Contact electrode
Cell temperature, °C.	780–800
Current density, amp./sq. in.	600–650
Energy consumption:	
Kw.-hr./ton	50,000
Kw.-hr./lb.	22–24
Energy efficiency, %	10

Cerium. Metallic cerium is produced from fused $CeCl_3$ plus alkali chlorides in cast-iron pot cells with graphite anodes, at 5.5 to 6.5 amp./sq. in. at voltages of 10 to 15. The metal is used for pyrophoric alloys.

Lithium. Lithium is produced in small amounts by the electrolysis of fused halides of potassium and lithium which are produced from lithium minerals such as lepidolite, spodumene, and triphylite. The electrolytes used are either LiCl and KCl, or these with some LiBr. The metal is useful as an alloying agent for non-ferrous metals such as copper. Its specific gravity is 0.534, melting point 186°C.

Sodium. Sodium is produced on a large scale from the electrolysis of fused NaOH or NaCl. In the older method (NaOH) the Castner vertical cell is employed in which iron anodes and cathodes are used, with external heating of the cell. In the newer method (NaCl) graphite anodes are used and the cell assumes a number of different forms. In both cases sodium metal comes to the top of the electrolyte.

Potassium is prepared electrolytically in a manner similar to that employed for sodium, from either KOH or halide salts.

Barium has been prepared from a fused fluoride electrolyte.

Electrothermics

The manufacture of many electrochemical and electrometallurgical products requires temperatures higher than

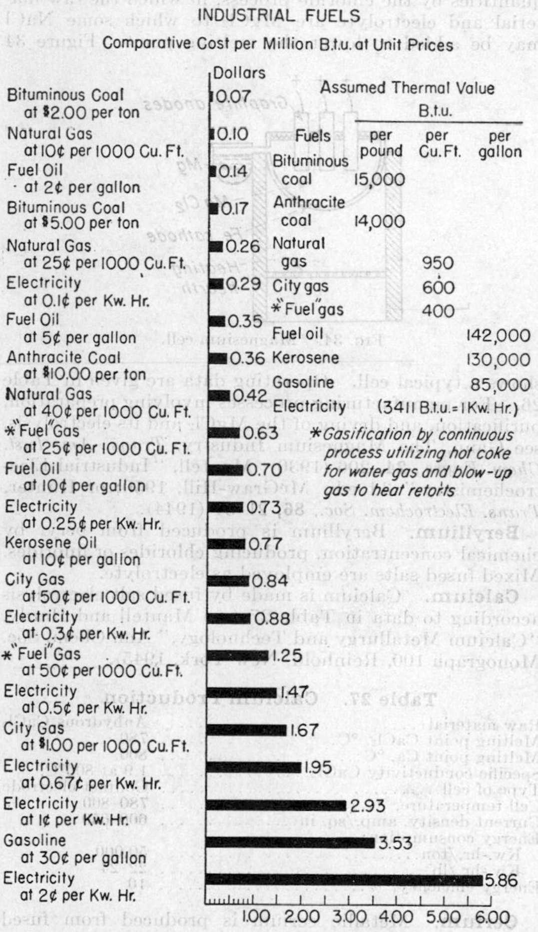

FIG. 35. Comparative cost of industrial fuels.

those obtainable by ordinary combustion methods. In this particular field there is no competition with heating obtainable by the use of fuels. In general, electric heating shows greater flexibility of application than do competitive methods of gas or solid-fuel combustion. Electric heat can usually be developed at, or adjacent to, the point of use more rapidly than fuel heat. In electric heating there is a lower temperature gradient between the heat source and the point of use. In contradistinction, flame temperatures are often 1000 to 2000°F. higher than the working temperatures required. Electric heating shows higher relative efficiency, but unless power rates are very low it is more costly (see relations in Fig. 35).

The field of application of electric furnaces can be subdivided into (1) those employed for iron and steel; (2) those commonly employed in the production of ferroalloys; and (3) those especially adapted to the melting of non-ferrous metals and alloys. To these must be added (4) the discussion of non-metallic products of the electric furnace of the arc and resistance types, such as CaC_2, silicon carbide (SiC), and graphite, as well as distillation products such as CS_2.

The electric arc is the simplest and most practical of the general methods of electric heating. It has found the widest application in the steel industry. In general, the arc is formed between two carbon electrodes, but in order to strike the arc the electrodes are brought in contact, after which the gap may be gradually produced and the arc still maintained. If the gap becomes too wide, the arc will break. Hence in arc furnaces some means of regulating the distance between the electrodes must be provided. Arc phenomena are explained on the assumption that some of the electrode material is vaporized by the heat of the arc and that these vapors serve as the conductor. The arc gives us the highest temperatures yet obtained. The limit in commercial practice is set by the materials forming the electrodes and the materials of construction of the furnace body. The intensity of the heat of the arc may be appreciated from the fact that carbon vaporizes at 3500 to 3600°C.

Electric Furnaces

Modern electric furnaces are somewhat complicated mechanisms. The most widely used electric furnace is of the three-phase arc type. The complete electrical equipment of a furnace installation includes:

1. A transformer, usually of the multiple type, to step down the voltage of the power supply system to the voltage or voltages needed for the furnace.
2. A secondary bus line and supporting structures between the transformer and the furnace.
3. Reactors, usually in the primary circuit of the transformer, to maintain arc stability and limit current fluctuations to the desired value.
4. Switching, instrument, and meter equipment.
5. Small d.c. electrode motors with automatic regulators.
6. Except in small hand-tilting furnaces, a tilting motor for the operation of the tilting mechanism.

It will thus be seen that the furnace proper, consisting of the shell, roof, electrodes, and tilting mechanism, is only a small part of the complete furnace setup.

Electric Steel Furnaces. The electric furnace has almost entirely replaced the crucible method for steel manufacture, and in the recent development of ferrous alloys, particularly of the stainless, nickel-chromium, vanadium, and tungsten types, of high-speed steels and cutting tools, has found other applications. It has definitely earned a place for itself in foundry work and steel-casting practice as well as for some types of rail steels. In general, the only part the electric current and electric furnace play in the manufacture of steel is in the production of heat. For the development of this heat only two successful types of furnace, the arc type and the induction type, are available. The advantages of electric-furnace application to steel metallurgy are the quick availability of the heat produced, the unusually high temperatures, the ease of regulation, the steady maintenance of any desired temperature, the cleanliness of the furnace and the method of heating, the non-production of harmful gases, and the ease of control of the furnace atmosphere for the production of oxidizing, reducing, or neutral conditions at the will of the operator.

In arc furnaces the heating of the charge is chiefly by radiation from an arc or arcs situated immediately above the bath. Many different arrangements have been proposed by various inventors, as well as utilization of single-phase, two-phase, and three-phase circuits. In the majority of arc furnaces the arcs play between the electrodes in the bath itself. Such furnaces differ in the manner in which the current is introduced into the furnace and let out of it. This can be done entirely by electrodes entering through the roof or top, in which case there are as many arcs as electrodes and never less than two of each. In the Heroult and 'Lectromelt furnaces as examples, the arcs are in series. In another type the bottom of the furnace may be built of such materials as to allow it to be used as an electrode, giving a direct-arc free-hearth-electrode furnace, an example of which is the Girod. In a modification of this type the metallic or conductor-material hearth electrode may be separated from the fluid charge by the refractory which forms the hearth lining. Although at room temperatures these refractories are insulators, at the operating temperatures of the steel-refining furnace they conduct fairly freely. Under such circumstances considerable currents can pass through the hearth. Such a construction results in the direct-arc buried-hearth-electrode type of furnace such as the Greaves-Etchells and Electrometals. In addition, other furnace designs may combine electrode arrangements of two or more types, giving mixed types of furnaces such as the Newkirk, the Booth-Hall, and the Nathusius. Instead of being inserted through the roof of the furnace, electrodes may be introduced through the side, or through the sides and top, giving arcs between the electrodes, the heat being reflected down from the furnace roof to the bath or material under treatment. These are the indirect-arc type of furnace represented by the Stassano and the Rennerfelt as applied to steel melting and the Detroit rocking furnace used in non-ferrous melting. The different classes of furnaces are illustrated diagrammatically in Fig. 36.

The **Heroult** furnace heads the list of electric furnaces used in the manufacture of steel. It consists essentially of a shallow hearth lined with calcined dolomite or magnesite, roofed with silica brick. The hearth slopes up in front toward a pouring lip. The melting chamber is surrounded by poorly conducting material. The whole furnace is enclosed in a shell or casing of steel sheet. Present-day Heroult furnaces are usually three electrode, three phase. The electrodes are supported above the roof of the furnace through which they project to within an inch of the surface of the slag over the metal. Electrodes are so far separated from each other as to prevent arcs between them. Heating occurs through arcs between the electrode and the slag, the slag and the metal underneath, the current carried thus in series to the next electrode but passing through a layer of slag and forming a second arc as it jumps between the slag and the electrode. Practically all the heat is formed by the arcs above the slag which acts as a shield to the metal to protect it from the carbon vapors thrown off at the bottom of the electrode and from the very high temperatures at this point. Portions of this heat are transmitted to the metal through the slag and may be distributed to all parts of the bath by conduction and convection. Heroult furnaces are designed in ½-, 1-, 1½-, 2-, 3-, 7-, 10-, 15-, 25-, 40-, and 100-ton capacities for open-top, chute, or machine charges. In the larger sizes, six electrodes may be employed.

It is considered good practice to obtain a power consumption of approximately 500 kw.-hr./net ton in the making of acid steel castings in a 3-ton Heroult furnace with an electrode consumption of 10 to 12 lb. carbon/net ton. In the making of alloy steels by the basic process in a 25-ton Heroult furnace, the power consumption is approximately 525 kw.-hr./gross ton good ingots.

The **Moore 'Lectromelt** furnace is similar in type to the Heroult furnace, being of the direct-arc series type employing three electrodes operating on a three-phase circuit. While the Heroult has been used largely in steel refining, the 'Lectromelt furnace has found its place in steel and iron foundry work. It is estimated that over 80 per cent of the electric-furnace foundry capacity is produced in 'Lectromelts. The furnaces are of the quick-melting type, generally using acid linings. In foundry practice, the electric furnace shows a 75 per cent thermal efficiency as compared with a 35 per cent value for a well-operated foundry cupola. The reducing atmosphere of the electric furnace makes possible the utilization of light scrap, turnings, borings, and similar low-cost materials not ordinarily suitable for open-hearth and cupola melting.

A typical energy distribution during the melting period in acid hearth practice shows:

	%
Useful energy	67
Heat flow through enclosure	16
Heat flow through openings	7
Escaping gases	5
Cooling-water loss	5
Sources of energy:	
Electric power	93.4
Charge	0.46
Reactions	2.26
Electrodes	3.88

Typical energy expenditures during melting and refining in basic practice with a 15-ton furnace are as shown in the table below.

Energy Distribution in Basic Practice

	%
Steel	45.80
Slag	8.50
Electrodes	10.60
Gases	3.20
Open-door loss	3.70
Cooling water	4.80
Roof surface loss	14.20
Side surface loss	5.20
Bottom surface loss	4.00

The **Swindell** furnace is a three-electrode three-phase direct series arc furnace with a balanced load on each phase. No bottom furnace connections are used. In general, the construction and electrode locations are similar to the Heroult. One of the features of the standard line of Swindell furnaces that distinguishes it from all other makes is that the electrodes do not tilt with the furnace when tapping but are raised in a vertical position until they clear the furnace, after which the ports are covered and the furnace tilted. In this furnace the electrode masts are separate from the furnace shell.

A multiple system of electrode melting furnaces has been developed for the continuous pouring of castings. Two complete furnaces are mounted on a revolting platform by which means the same transformer and set of electrodes are used for each furnace alternately. Each furnace contains the same features of design as described and is complete in itself.

The **Greene** furnace may be classified as a three-electrode three-phase direct-series-arc furnace, similar to the Heroult save for special mechanical and electrical features designed for specific installations and locations. The Greene furnace shell is of heavy reinforced-steel construction mounted to roll over on the horizontal axis of the rolling cylinder shell in order to pour the metal.

The **Fiat** furnace is essentially an Italian modification of the Heroult design. The hearth is hemispherical,

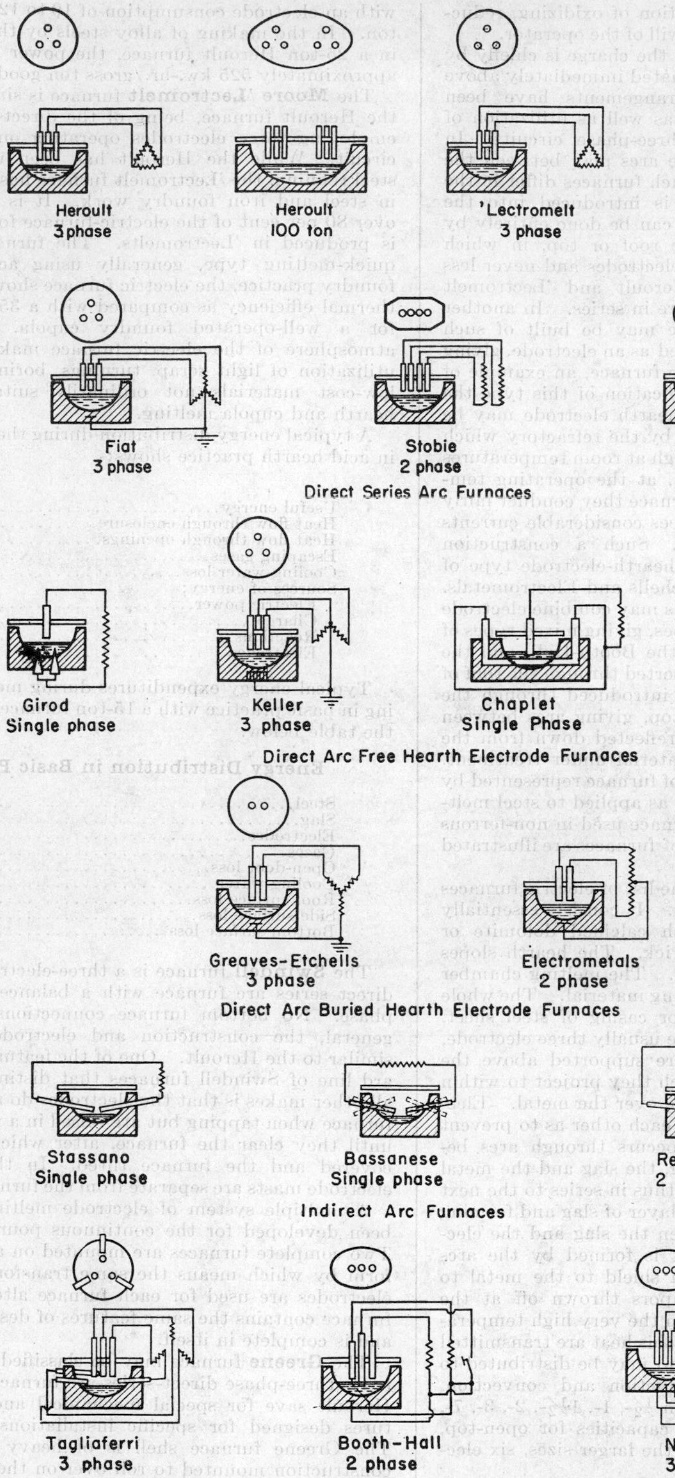

FIG. 36. Types of arc furnaces.

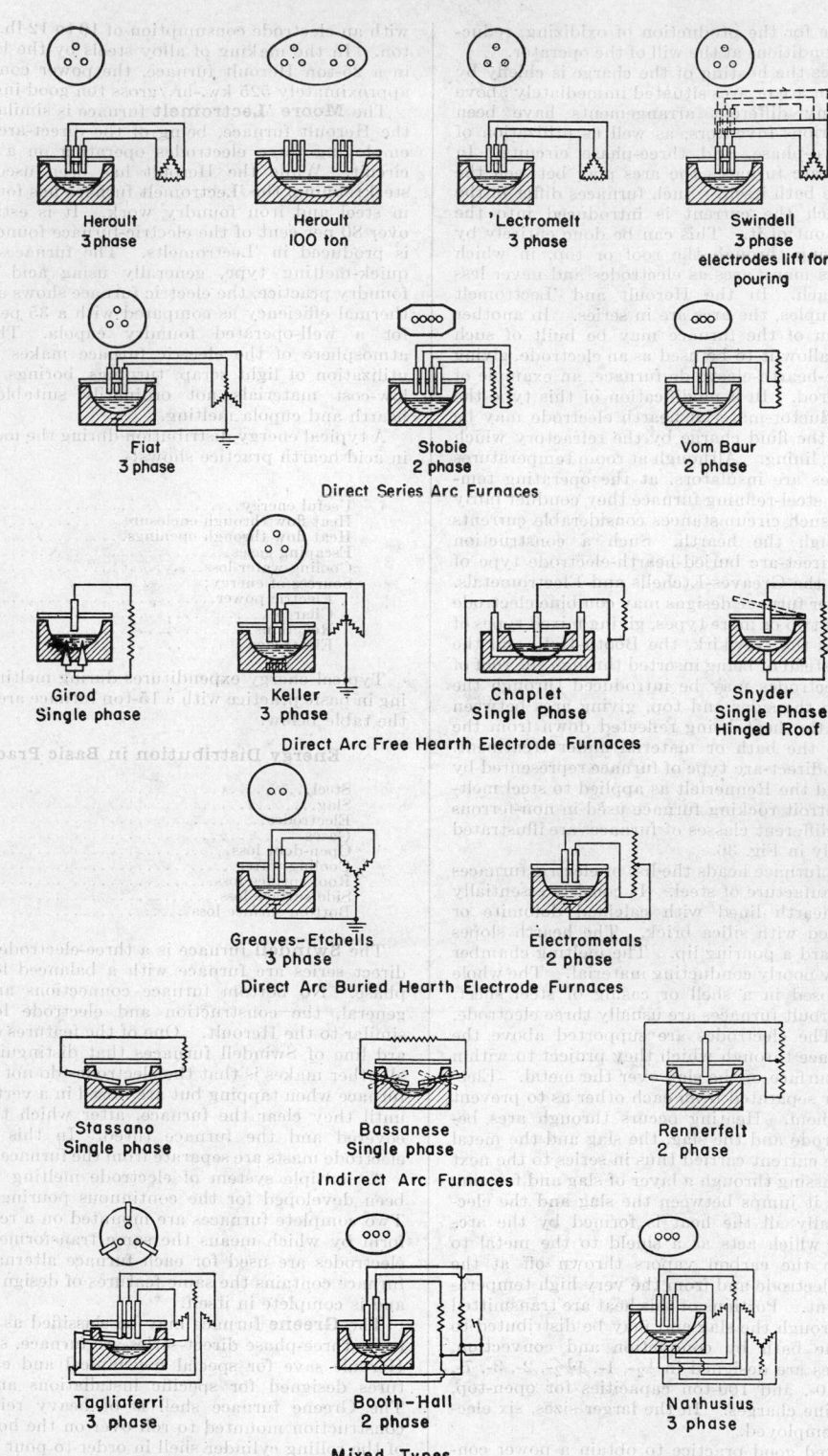

mounted on two circular shoes of cast steel. Electrodes are graphite, arranged at the points of a triangle and provided with special economizers. The hearth is slightly conductive and in practice is connected on the secondary side and grounded. When phase loads are balanced, very little current flows to ground.

The **Stobie** furnace is fed by a single-, two-, or three-phase current, whichever is available. For capacities larger than 6 tons, the furnace is of the four-electrode type supplied by a two-phase Scott transformer, the phases being independent of each other. For more than 25 tons, a design having six electrodes and three-phase current, each phase with its own return, is used. The feature of the furnace is the electrode economizers attached to the furnace electrode gear. The power factor ranges from 0.85 to 0.93.

The **Girod** furnace resembles the Heroult as far as the major constructional parts are concerned. Current enters through an electrode or electrodes passing through the roof, forming an arc between the bottom of the electrode and the slag, passing through the steel bath, and finally leaving by a number of soft-steel electrodes set in the hearth, which in turn are connected with a copper conductor plate under the base of the furnace. When more than one electrode is used, the arcs are in parallel. In operation the steel electrodes melt down somewhat, and when tapping takes place this molten part runs out but is replaced by the next charge. The furnaces use one or more electrodes, being single phase in smaller sizes and three phase in larger ones.

The **Keller** hearth-electrode furnace is characterized by its composite hearth made of a number of iron rods or a grid of iron bars fastened to a metal base plate from which the bars project vertically upward. The space between the bars is filled with a mixture of burned magnesite and tar rammed in place. The hearth is water jacketed and connected to the source of current. Under furnace-operating conditions the refractory in the spaces between the hearth bars becomes conductive and carries some of the current.

The **Chaplet** furnace is a special modification of the hearth-electrode type which avoids water cooling by the use of a hearth channel which leaves the hearth sideways and connects with a block or blocks of iron in subsidiary chambers at the side of the furnace. Transformer connections are made to these vertical blocks. The steel in the channel of the hearth electrode partially melts during operation.

The **Snyder** furnace uses a hearth electrode in order that high-voltage currents may be employed as far as possible. Commercial refractories at high temperatures become good conductors and thus limit the voltage values possible because of power loss by current leaks through the refractory. In the original type of Snyder furnace the shell was of steel with a thick refractory lining. It tilted for pouring. A single small electrode entered through the roof. From this the current passed in a long arc to the slag and left by a single water-cooled steel conductor set in the hearth, connecting with the metal bath in the furnace. The special feature of this furnace is the hinged roof which permits charging through the top rather than through the door.

The **Greaves-Etchells** furnace consists of a rectangular steel shell with a low-domed, silica-brick-lined roof. The walls are of chrome brick and the hearth a dolomite-magnesite mixture tamped in place with a pitch or tar binder on to a copper plate in the bottom of the shell. Three-phase currents are used, two phases being connected to vertical electrodes entering through the roof and the third phase to ground, to which the hearth is also connected. When the hearth is hot, its resistance is less than that of the arcs. Special transformer connections are supplied to permit of operation of the furnace in a number of different ways. Power can be put through the top electrodes only and the furnace employed as a direct-series-arc type. By proper switching, power can be introduced through the top electrodes and caused to pass to the hearth, as in the Girod furnace, with the exception that in the Greaves-Etchells furnace the hearth is of the buried type instead of being free.

The **Electrometals** furnace has been developed chiefly in Great Britain. In the United States it is known as the **Grönwall-Dixon** furnace. It uses a three-phase, high-voltage current from Scott transformers so that two phases are employed on the furnace. There are two roof electrodes and a buried hearth electrode which acts as a neutral return. The loads on the roof electrodes are balanced, and very little power goes through the hearth, which has a negligible resistance. The silica-brick-lined roof of the furnace is detachable. The hearth electrode is formed of steel conductors embedded in the lowest part, contact with the dolomite hearth being made by a graphite-rich bonding material.

In the **Stassano** furnaces the electrodes enter the furnace shell through the side, there being two electrodes for single-phase furnaces and three or six for three-phase units. The usual type of Stassano furnace is circular in plan, rotating slowly about an axis inclined from the vertical. The roof is domed and lined with magnesite brick, while the melting chamber is lined with dolomite, the brick being insulated from the steel furnace shell. The electrodes are of small diameter and symmetrically arranged around the furnace, three-phase transformer connections being delta.

The **Bassanese** furnace is a modification of the Stassano, having its electrodes mounted on mechanisms that make possible the use of direct as well as indirect action of the arc on the bath. The electrodes are inclined to a variable angle so that the arc may be made free burning or directly projected on the metal.

In the **Angelini** modification of the Stassano, the electrodes are arranged so that arcs may be formed between movable electrodes above the metal surface as well as between each electrode and the metal. This arrangement permits the use of single-phase currents at higher voltages than those employed in other designs.

The **Rennerfelt** furnace is of the indirect-arc type but differs from the Stassano in that it employs two-phase current obtained from a three-phase supply by a Scott transformer as well as using three electrodes, two entering through the side of the furnace and one through the roof. The roof electrode operates as a common or neutral return. Arcs are sprung between the electrodes above the surface of the furnace charge, but the unit may be operated with arcs in contact with the charge for short periods, to increase the activity and fluidity of slags. In the operation of the furnace the arcs burn regardless of whether the furnace is charged or not, so that the hearth may be sintered by electric heat and kept hot for any length of time.

The **Tagliaferri** furnace combines the indirect-arc heating of the Stassano with the direct-series-arc principle of the Heroult. Two- or three-phase, low-voltage current, generally the latter, is used. The three-phase unit has three roof electrodes and during the refining period operates as a direct-series-arc furnace. Three auxiliary electrodes enter through the side of the furnace. During the melting period, arcs are sprung between the roof and corresponding side electrodes.

In the **Booth-Hall** furnace there are one, two, or three roof electrodes for single-, two-, or three-phase circuits, as well as hearth electrodes composed of conductors buried in the hearth. With single- and two-phase power the current passes through the hearth electrodes during the refining period, while with three-phase supply the hearth acts as a neutral point. There is an individual hearth electrode for each roof electrode but located on the opposite side of the furnace instead of being directly underneath the bottom of the roof electrode. An auxiliary roof electrode is provided, connected with the return to the transformer in parallel with the hearth electrode.

The **Nathusius** furnace combines the principles of the direct-series arc, the buried-hearth type, and the resistance idea. As in the Greaves-Etchells, the furnace charge is heated from above by arcs and from below by currents through the hearth electrodes. In the Nathusius furnace the latter type of heating can be independently regulated through star-transformer connections to the hearth electrodes independent of those through the roof. The hearth as well as the whole of the charge may be included in the circuit as an ohmic instead of an inductive resistance. Three-phase currents are used, the transformer secondaries being star connected to the roof electrodes, and their separate returns to the corresponding ones in the hearth. The metal bath in the furnace thus becomes the neutral point of the transformer. A feature of the furnace is its great flexibility.

Induction Furnaces for Iron and Steel. In the United States the use of the induction furnace for ferrous metallurgy has been very limited, finding almost no application; but in the form of the **Ajax-Wyatt** and **Ajax-Northrup** induction furnaces, it has been extensively employed in non-ferrous metal melting. In recent years the Ajax-Northrup high-frequency furnace has found greater application in steel and ferrous alloy manufacture, particularly on a small scale and for special materials.

An induction furnace may be described as a step-down transformer in which the metal under treatment is the short-circuited secondary coil or coils. The induction furnace is subject to the magnetic and electrical losses due to hysteresis in the iron core in the same manner as transformers; but on account of the design of the furnace, these losses are usually greater than those in transformers, with resulting lower power factor. In the induction steel furnace there is only one turn in the secondary winding, consisting of the steel to be refined in a ring-shaped channel through the center of which the primary coil and the core pass.

One of the large number of arrangements of primary and secondary which have been proposed, only two are used.

The **Kjellin** furnace is shown diagrammatically in Fig. 37. The core is built up of laminated sheets of soft iron

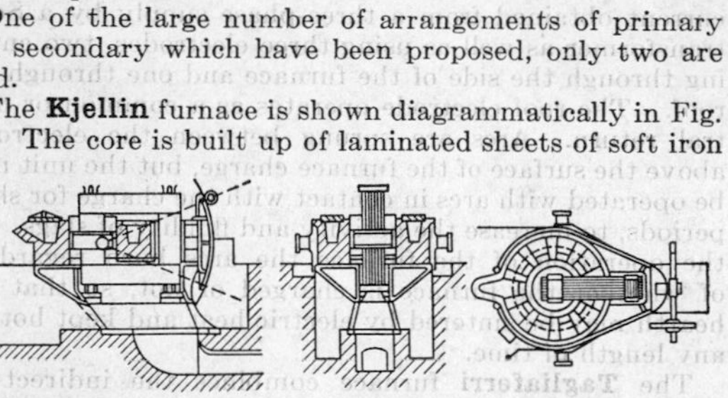

Fig. 37. Diagrammatic sketch of the Kjellin induction furnace.

and, like an ordinary transformer core, forms a closed circuit. The primary is made of insulated copper wire and air cooled, or in other designs made of water-cooled copper tube. The secondary is the ring-shaped refractory-lined channel for the steel, insulated by brickwork. The entire furnace is enclosed in a steel shell. Charging and tapping doors are provided in both the fixed and the tilting types. Some of the molten material must always be left in the furnace so that currents can pass on recharging. Furnaces have been built up to 8½ ton capacity, operating on low frequencies of the order of five cycles. Operating data for a 2-ton furnace show a consumption of about 170 kw. at 3000 volts in the primary and 30,000 amp. in the secondary. Power consumptions are of the order of 750 to 850 kw.-hr./ton of steel with cold charges of pig iron or pig iron and scrap.

The **Röchling-Rodenhauser** furnace uses a combination induction-and-resistance principle. The unit consists of a soft-iron core with two primary windings, while the secondary has the shape of an 8, the middle part of which is a comparatively broad hearth to allow working of the bath and handling of the slags. Auxiliary secondary windings consisting of a few turns of strip copper are provided to heat the central hearth. These are separated from the primaries by a small air gap and connected with hearth electrodes or pole plates at opposite sides of the central hearth, through which currents may pass from the pole plates. These pole plates are made of corrugated cast steel embedded in the furnace wall behind a layer of refractory which at furnace-operating temperatures becomes conducting.

Single-phase furnaces have been built up to a rated capacity of 8½ tons, consuming 700 to 750 kw. with a primary voltage of 4000 to 5000. Three-ton furnaces show power factors of 0.7 to 0.8 at a frequency of 25 cycles. With molten charges, about 120 to 160 kw.-hr./ton of steel is consumed when Bessemer steel is refined to open-hearth quality, while for reduction of the sulfur and phosphorus to 0.01 per cent, 200 to 300 kw.-hr. are required. On cold scrap, the power consumption is of the order of 900 kw.-hr./ton.

Ferroalloys. In the development of the steel industry the production of alloy steels brought forth a demand for alloying agents in a form readily usable, as well as refining materials. These are the substances in general termed the ferroalloys, including ferrosilicon, ferromanganese, ferrochromium, ferromolybdenum, ferrotungsten, ferrovanadium, and silicomanganese. A large part of these are made in the electric furnace, generally of the arc type.

Ferrosilicon is made either in the blast furnace, in which case the percentage Si seldom exceeds 15 per cent,

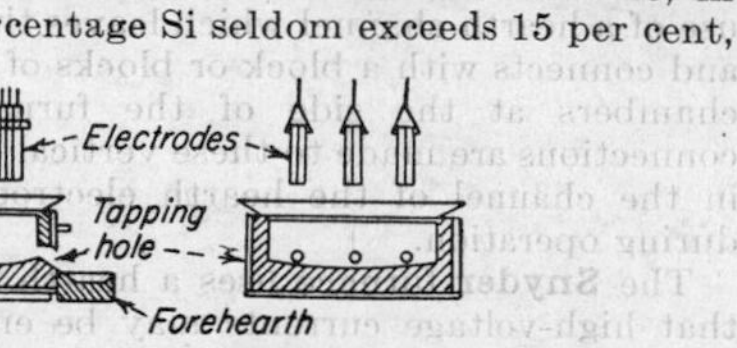

Fig. 38. Ferrosilicon furnace.

or in the electric furnace, where higher percentage Si alloys can be made. The raw materials are silica, carbon, and iron, the process being essentially a carbon-reduction one taking place at high temperatures. All ferrosilicon furnaces are of the mixed arc-resistance type.

Ferromanganese containing up to about 20 per cent Mn is made in the blast furnace almost exclusively, but there has been considerable electric-furnace manufacture of higher percentage Mn alloys. The furnaces used are low shafts working with open tops, in some cases having a hearth electrode but more usually with a neutral electrode in three-phase furnaces.

Ferrochromium, largely employed for chromium steels, armor plate, projectiles, cutting tools, as well as stainless steels, is made entirely in the electric furnace from chromite, carbon, and slag-forming materials. In the case of the low-carbon-content ferrochromium, silicon may be the reducing agent. The reduction of carbon content of high-carbon ferrochromium is accomplished in furnaces of the same type as used for steel refining, in which the material is remelted by the use of fluxes such as lime and fluorspar, with the addition of chromite.

Ferrotungsten is produced by electric-furnace smelting of a tungsten mineral [wolframite ($FeMnWO_4$), ferberite ($FeWO_4$), or scheelite ($CaWO_4$)], a carbonaceous fuel, and slagging materials, with the addition of iron ore or scrap iron if the iron content of the original tungsten mineral be insufficient.

Ferrovanadium is prepared from vanadium ores which are smelted with iron and fluxes, with reducing agents such as silicon 90 per cent or ferrosilicon, 50 per cent Si, or coke.

Ferrosilicon-titanium was produced in considerable quantities as a by-product of the electric-furnace refining of bauxite, in which the iron, silicon, and titanium content of the bauxite is reduced by carbon to form the ferroalloy. This settles to the bottom of the furnace and is tapped out, while the alumina of the bauxite becomes the slag which is drawn off as a purified alumina, granulated, and worked up for use in the manufacture of aluminum. Ferrosilicon-titanium finds use as a steel deoxidizer.

In general, furnaces for **ferromolybdenum** and **ferrotungsten** are small of 200 to 750 kw. capacity, and are single phase. Ferromolybdenum meets competition from calcium molybdate. Ferrochromium and ferrovanadium furnaces are 2000 to 5000 kw., usually three phase, and like all the other ferroalloy furnaces, are of the arc-resistance type. The ferrosilicon, ferromanganese, and silicomanganese furnaces are larger, being 3500 to 12,000 kw., three phase. Electrode current densities are 30 to 60 amp./sq. in. cross section. Power consumptions are of the order of 1.2 to 3 kw.-hr./lb. ferroalloy, being highest for ferromolybdenum.

Electric Furnaces for Non-ferrous Metals. For non-ferrous melting, three furnaces are of industrial importance in the United States and find wide application. The first to be considered is the *Detroit rocking furnace* in which it is estimated that about 85 per cent of the foundry output of the United States is melted and 20 to 25 per cent of the rolling-mill tonnage of brass and Cu-Zn alloys. It is estimated that 75 to 80 per cent of the rolling-mill production of brass in the United States is melted in *Ajax-Wyatt induction furnaces* which also find some application in foundries. The third furnace is the *Ajax-Northrup high-frequency induction unit.*

The **Detroit rocking furnace** consists of a steel cylinder mounted horizontally on cog gearings whereby it can be rocked by means of a small motor. The steel

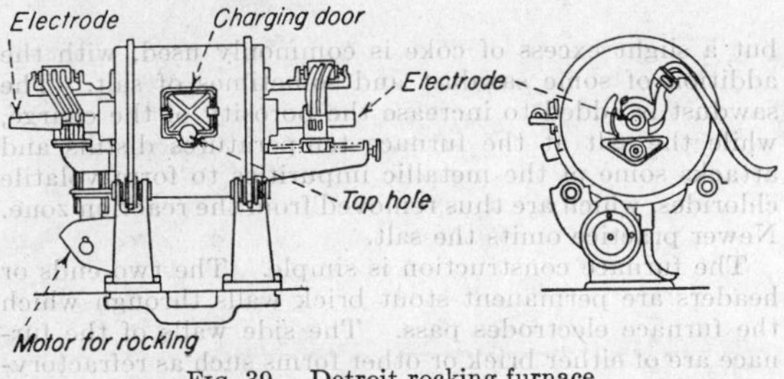

FIG. 39. Detroit rocking furnace.

cylinder is lined with insulating brick which are in turn covered by refractories. The charging door is on the upper side of the drum, the spout for pouring being located directly below it. Graphite electrodes are introduced horizontally through centers at the ends of the drum. They can be adjusted for length of arc and power input, usually by hand. Single-phase alternating current is employed. Heating of the metal is not by direct contact with the arc but by reflected or radiated heat. During charging, the electrodes can be withdrawn until their ends are flush with the refractory lining of the furnace to avoid breakage. During operation, as soon as superficial melting of the charge is started, rocking of the furnace is caused to take place. The furnace has the advantages of rapid melting, a totally enclosed body so that volatilization losses, particularly of zinc, are cut down, as well as control of the furnace atmosphere which, because of the presence of the electrodes, is ordinarily reducing. Inasmuch as each charge can be completely poured from the furnace, it shows great flexibility when different alloys have to be melted. The furnace has been applied to non-ferrous alloys, particularly brass and bronze, and to a smaller extent to gray iron, special iron-alloy work, and the duplexing of cupola-melted iron. On brass it shows a power consumption of 250 to 300 kw.-hr./ton; 540 kw.-hr./ton on melting cast-iron borings; 600 kw.-hr./ton on synthetic gray iron, and 100 kw.-hr./ton on duplexing cupola iron.

The **Ajax-Wyatt** induction furnace is illustrated in Fig. 40. In operation, a single-phase alternating current of

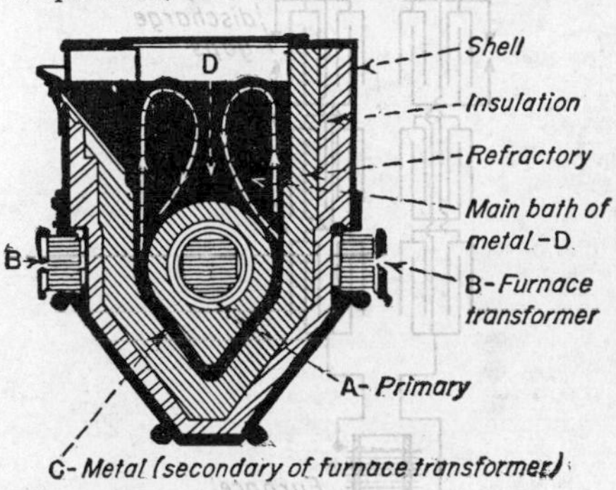

FIG. 40. Ajax-Wyatt induction furnace.

220, 440, or 550 volts is fed to the primary coil A in the center of the furnace, which in turn energizes the furnace transformer B. This in turn induces a voltage in the V-shaped channel C which is always filled with molten metal or alloy and acts as a secondary of the furnace transformer. A current of relatively high value flows in this V-shaped channel and, owing to the resistance of the molten metal, the channel is heated at a constant rate. At the same time the magnetic fields about this channel set up electromagnetic forces which eject the hot metal out of the channel, so that it is in turn replaced by colder metal from the main bath above the melting channel. The circulation follows the path shown in the figure. Thus, with a furnace stationary throughout the melting cycle, the metal bath is thoroughly mixed by an internal automatic stirring action, while the electric energy is converted to heat directly within the metal. The furnace has the disadvantage that it must always be partly filled in order to operate and cannot stop with cold charges. It cannot, therefore, be used with charges of a varying nature without some trouble. It has the advantages, however, of a simple mechanical design, great steadiness in working, small metal loss, small wear and tear on the lining, and in continuous operation a low energy consumption. This is of the order of 250 to 270 kw.-hr./ton of red brass (85 per cent Cu, 5 per cent Pb, 5 per cent Zn, 5 per cent Sn); 200 kw.-hr./ton of yellow brass (75 per cent Cu, 2 per cent Sn, 3 per cent Pb, 20 per cent Zn) for plumbing fixtures; 275 kw.-hr./ton on nickel-silicon; 285 kw.-hr./ton of 4 to 5 per cent tin bronzes, 310 kw.-hr./ton of copper, and 90 kw.-hr./ton of zinc.

In the high-frequency **Ajax-Northrup** induction furnace, the material to be heated is not in the shape of a ring as in the ordinary induction furnace but is held in a crucible placed in the field of a high-frequency coil. The heating is produced by eddy currents generated in the material to be melted. If this is a non-conductor, a conducting crucible is used. Since the eddy currents increase as the square of the frequency, the reason for the use of high frequency is evident. The frequencies are of the order of 10,000 to 12,000 cycles/sec. It has been found most convenient and satisfactory to use the oscillatory discharge of a bank of condensers as a source of high-frequency current. The desired frequency is obtained by the proper proportioning of the capacity and inductances of the oscillatory circuit. Figure 41 shows

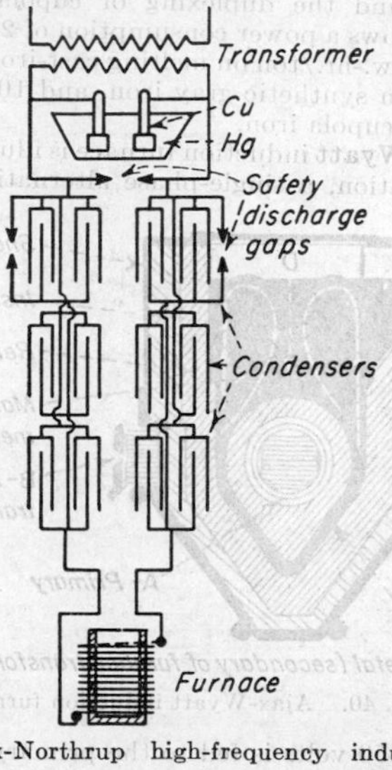

FIG. 41. Ajax-Northrup high-frequency induction furnace (small sizes).

the circuits of a single-phase Ajax-Northrup furnace. The spark gap consists of water-cooled copper over mercury in a hydrogen atmosphere. With the two gaps, in whichever direction current from the secondary of the transformer is flowing, positive current must always leave a mercury surface. When the voltage between the mercury and the graphite reaches a certain minimum value, the mercury opens the circuit completely with great suddenness, causing very rapid and regular oscillations. The power input is controlled by regulation of the spark gaps. The induction coil around the furnace is a water-cooled flattened copper tube. The furnaces have been applied for nickel-silver where the energy consumption is 0.17 kw.-hr./lb.; special steels produced at 660 kw.-hr./short ton; silver and silver alloys, precious metals, special alloys, stainless steels, and a variety of other products.

Electric-furnace Products

Calcium carbide is made in the electric furnace according to the reaction

$$CaO + 3C = CaC_2 + CO - 125{,}000 \text{ cal.}$$

It does not seem probable that temperatures of 2000°C. are exceeded in the furnaces. The commercial material is dark colored and crystalline, but if pure it is colorless and transparent. The 80 per cent pure product melts in the neighborhood of 1800°C. The raw materials are carbon in the form of charcoal, low-ash anthracite, low-ash coke (or sometimes petroleum coke), and lime.

The furnaces in use are very large, being of the vertical-arc type. Ingot furnaces, in which the lower electrode is a small car which can be removed when filled, were formerly used, but hearth-electrode tapping furnaces of either the non-continuous or continuous type are generally found in all carbide plants. In general, the furnaces are three electrode three phase. In the newest furnaces, single electrodes of sectional construction are found with capacities of 40,000 to 300,000 amp. The furnaces are single phase. Electrode consumption is of the order of 2 to 4 per cent, current density on the electrodes being 35 to 50 amp./sq. in., while energy consumptions may be from as low as 1.5 to 3 kw.-hr./lb., with perhaps an average energy efficiency of 60 per cent or better.

The annual United States production of CaC_2 is of the order of a quarter million tons. For specific data on furnaces, capacities, operation, and design, see Mantell, "Industrial Electrochemistry," 3d ed., McGraw-Hill, New York, 1950; and Taussig, "Die Industrie des Kalziumkarbides," Knapp, Halle, 1930.

Cyanamid. Finely divided CaC_2 absorbs nitrogen at 1000°C., giving cyanamid according to the equation

$$CaC_2 + N_2 = CaCN_2 + C$$

The carbide is heated in furnaces, and nitrogen is passed over the heated material. The nitrogen is generally made by fractional distillation of liquid air. Starting with a carbide containing 75 to 80 per cent CaC_2 80 to 90 per cent of the theoretical amount of nitrogen will be absorbed, resulting in a product containing about 20 to 22 per cent nitrogen with lime and carbide as impurities. If cyanamid is treated with superheated steam, ammonia is evolved according to the equation

$$CaCN_2 + 3H_2O = CaCO_3 + 2NH_3$$

Cyanamid may be converted into cyanides by the fusion of a mixture of salt, cyanamid, and carbide, and a rapid chilling of the melt, the equation being

$$CaCN_2 + C + 2NaCl = CaCl_2 + 2NaCN$$

Silicon carbide (trade names *Carborundum, Crystolon,* etc.) is made by heating a suitable charge of carbon and silica sand in a horizontal resistance furnace. The carbon used is either high-grade anthracite or good-quality coke such as petroleum coke. The proportions roughly correspond to the equation

$$SiO_2 + 3C = SiC + 2CO$$

but a slight excess of coke is commonly used, with the addition of some sawdust and sometimes of salt. The sawdust is added to increase the porosity of the charge, while the salt at the furnace temperatures distills and attacks some of the metallic impurities to form volatile chlorides, which are thus removed from the reaction zone. Newer practice omits the salt.

The furnace construction is simple. The two ends or headers are permanent stout brick walls through which the furnace electrodes pass. The side walls of the furnace are of either brick or other forms such as refractory-faced castings to hold the charge in place. The bottom of the furnace is usually made of insulating firebrick or gannister. Through the center of the furnace a heating core connects the electrodes, the charge surrounding the heating core (see Fig. 42). Operating data are given in Table 28.

Boron carbide, B_4C, is made in a resistance furnace from B_2O_3 and coke [Ridgway, *Trans. Electrochem. Soc.,* **66,** 117 (1934)].

Table 28. Products of Resistance Furnaces

Factors	Graphite	Silicon carbide	Fused alumina
Raw materials	Low-ash anthracite or petroleum coke	Coke, 98% silica sand	Bauxite, coke, scrap iron
Additions		Sawdust and salt	
Furnace:			
Type	Resistance	Resistance	Vertical arc resistance
Size	1000 hp.	2000 hp.	550 kw.
Length, ft	30	30	
Cross section, ft		10 × 10	
Cross section of charge, ft. diam	2	3	
Walls	Refractory brick or concrete blocks	Refractory brick, cast-iron or steel supports	Steel, water cooled
Initial voltage	200	230	100–110
Final voltage	80	75	
Initial current	...	6000	2500
Maximum current at 200 volts, amp	3700		
Final current, amp	9000	20,000	
Current density across furnace charge, amp./sq. ft	900–2250	650–2200	
Core temperature, °C	...	2350	
Furnace temperature, °C	2200	1820–2220	2000–2200
Length of run, hr	24	36	
% Conversion of material	90–100	70–80	95–100
Energy consumption, kw.-hr./lb	1.5	3.2–3.85	1.0–1.5
Energy efficiency, %	25–30	55–70	

Graphite is made in a horizontal resistance furnace similar in construction to the SiC furnace. Graphite is produced either as a powder or as shaped articles such as electrodes, brushes, etc. When loose graphite or powder is made, low-ash anthracites or petroleum cokes are employed as raw materials. One theory of graphite formation is that the ash content of the raw material, containing iron, aluminum, and other oxides, functions as a catalyst for the conversion of amorphous carbon into graphite at the furnace temperature through the intermediary formation of carbides, which dissociate to give

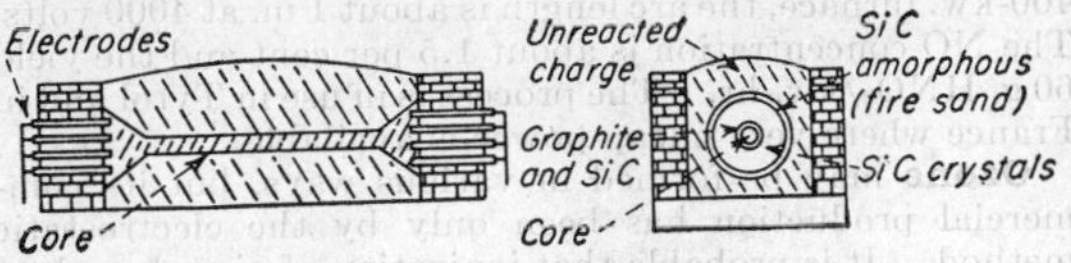

FIG. 42. Silicon carbide furnace.

graphite. In the preparation of graphite articles the materials are formed and baked into an amorphous-carbon article, after which they are graphitized in the electric furnace. The ash content of graphite is very low, in that at the furnace temperature most of the metallic oxides are volatilized. A sketch of a typical furnace is given, as well as operating data. The ready machinability of graphite is well known, as well as its resistance to most forms of chemical attack. It is the standard anodic material for electrolysis of NaCl to produce chlorine and caustic.

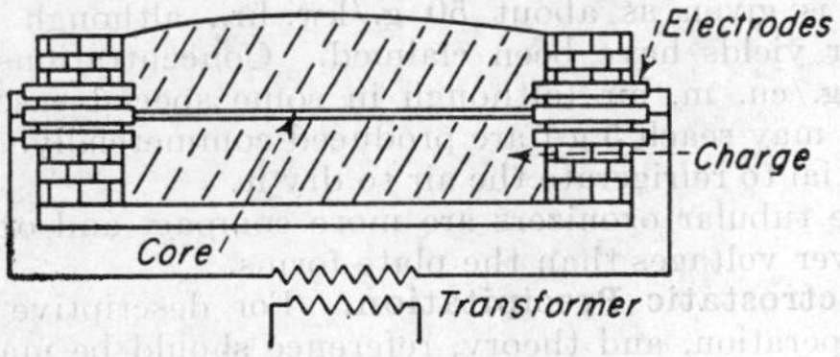

FIG. 43. Graphite furnace.

Fused alumina is made in the electric furnace. The product is extensively used as an abrasive, particularly in grinding wheels, and as a refractory. The raw material is a red bauxite containing a few per cent of oxides other than alumina (iron and silicon) which give tougher products than those made from purer bauxite. The material is first calcined to remove water, and the fusion done in vertical arc furnaces with a bottom electrode. The ore is fed into the furnace, melted in the arc between the top and bottom electrodes until the furnace is filled, after which the entire shell of the furnace is removed, the block of fused material allowed to cool, taken away from the furnace bottom, and broken up into commercial products. Operating data are given in Table 28.

Silicon is made in three-phase arc furnaces with bottom electrodes by the reduction of silica with coke, a 95 per cent silicon being produced. It may also be made from cheap SiC and silica sand. The energy efficiency of the furnace is given at about 50 to 55 per cent. The material is used for high-grade silicon steels and as a reducing agent for the production of low-carbon ferrovanadium and ferrochromium, as well as silicon-aluminum and silicon-copper alloys.

Fused silica (vitreosil, etc.) is made from silica sand by passing a current through carbon rods or plates embedded in the material. When sufficient sand has been fused, the carbon resistor is withdrawn. By utilizing the hole left in the interior of the melt, the latter is blown and molded to the desired form. The crude articles thus obtained are trimmed and polished by the use of the oxyhydrogen flame, sandblast, and abrasives.

Fused Quartz. In recent years clear fused quartz has been made under pressure in electric furnaces. The clean natural crystals of small size are packed as densely as possible in a graphite or carbon crucible, so that in the inevitable cracking of the crystals as the temperature is raised the parts cannot separate and permit gas, which may be present in small quantities, to enter the crevices and form bubbles. In a modified vacuum furnace, the quartz is heated to melting (about 1800°C.) as quickly as possible, usually in 45 min. or less, while the pressure is kept as low as possible. The resultant transparent slugs, containing a few small bubbles, are placed in another graphite crucible suspended in a vertical carbon-tube furnace, with a graphite piston closely fitting the crucible and weighted. The slugs are heated to fusion, the bubbles mostly collapsed by the weight which also extrudes the quartz through the bottom of the crucible in rods, tubes, and other desired forms. When large blocks are to be made, a vacuum furnace is used which is capable of withstanding very high pressure. As soon as the quartz is fused, the vacuum valve is closed and the pressure raised. Thus are produced very large blocks of quartz more free from bubbles than many kinds of the best optical glass.

Carbon bisulfide is made by one company in an arc resistance furnace from charcoal and sulfur. Coke cannot be used. The process and furnace design are to be credited to Taylor (*Trans. Am. Electrochem. Soc.*, vols. 1 and 2). The CS_2 distills out of the furnace and is condensed in external condensers. Data are given in Table 29. Most of the CS_2 production is by thermal methods, not of the electric furnace type.

Phosphorus is electrothermally produced from phosphate rock and reducing materials such as carbon, with sand as a slag-forming product, in arc furnaces. The phosphorus distills out of the furnace and is collected. Phosphoric acid is also produced in a somewhat similar manner, but as the phosphorus distills from the furnace it is allowed to oxidize to form phosphoric acid. Data are given in Table 29.

Table 29. Distillation Products of Electric Furnaces

Factors	Phosphorus	CS_2
Raw materials	Bone ash or phosphate rock, coal, and SiO_2	Charcoal, sulfur
Furnace:		
Type	Arc, carbon lining, vertical electrode	Resistance, two phase, a.c., shaft feed
Capacity, kw		240–330
Current, amp		4000–6000
Voltage		60
Temperature, °C	1150–1450	
Production per 24 hr., lb		14,000
Energy consumption, kw.-hr./lb	4.0–5.5	0.4–0.5
% thermal efficiency		30–45
% P recovered	80–90	

Zinc is produced by electrothermal processes whereby zinc ore is reduced to metal, volatilized out of the furnace, and condensed.

Gaseous Electrothermics

Nitrogen Fixation. In the United States there has been very little application of the fixation of nitrogen from the air in electric furnaces either of the arc or of the spark type. Atmospheric air is a mixture of 80 per cent by volume of nitrogen and 20 per cent of the oxygen. The simplest fixation process would be one that would combine these, converting them into nitrogen oxides which in turn could be made into HNO_3 and nitrates. This can be done by the processes of Bradley and Lovejoy, Birkeland and Eyde, Schoenherr, and Pauling. All of these employ electric discharges through air to produce a very high temperature. The spark or arc is merely a heating means to reach the high temperature necessary for the combination of the nitrogen and oxygen. The mixture of air and oxides of nitrogen thus produced is treated with water or with an alkaline solution to give HNO_3 or nitrates, or a mixture of nitrates and nitrites.

The higher the temperature the more rapidly does the conversion of nitrogen and oxygen into nitrogen oxides take place. But the decomposition reaction proceeds in the same order. It is therefore necessary to have a very high temperature for efficient reaction, with a quick removal of the reaction products from the temperature zone so that they may be cooled down as rapidly as possible to prevent decomposition.

In the **Bradley-Lovejoy** process, now no longer in operation, a wheel carrying a set of electrodes was rotated so that the electrodes passed opposite and by a stationary set of electrodes so as to make and break continuous sparks, at the rate of about 6900 sparks per second, in the space through which the air was passed. The units were limited in size, and the desire to have larger ones led to the construction of arc furnaces.

In the **Birkeland-Eyde** process, employed in Norway where cheap power is available, the arc is deviated magnetically by means of a single-phase magnet field until the arc breaks; then a new arc is formed and the cyclic process continued. The high-voltage flame or arc is formed between two water-cooled copper electrodes by the use of a 50-cycle current at 500 volts. In ordinary operation of the furnace, a flame is formed at each reversal of the current every 0.02 sec. The furnace consists of a narrow iron chamber lined with firebrick. Air enters through the walls and leaves the furnace at a temperature of 1000°C., containing 1 per cent NO. The gases pass through the steam boiler in which they are cooled to 200°C., then through a cooling apparatus where the temperature is reduced to 50°C., and then into an acidproof, brick-lined oxidation chamber where the reaction $NO + \frac{1}{2}O_2 = NO_2$ is completed. The NO_2 is then absorbed in water to form HNO_3 which is concentrated. The net yield is stated to be 62.5 g. pure HNO_3/kw.-hr.

The **Schoenherr** furnace has as its characteristic feature the use of a very long a.c. arc around which the air moves in a helical path. The process was used successfully for a time by the Badische Co. in Germany and has been operated on a commercial scale in Norway. In the latter plant the arc was 5 m. long in a 447-kw. furnace and 7 m. long in a 746-kw. furnace. It is estimated that 3 per cent of the power is used in the formation of the NO, 40 per cent is recovered in the form of hot water, 17 per cent is lost by radiation, 30 per cent is used in the steam boiler in which the gases are cooled, and 10 per cent is removed by water cooling after the gases have passed through the steam boiler.

The **Pauling** process uses a fan-shaped arc or an electric discharge quite similar to that obtained in a horn lightning arrester. The arc is lighted where the electrodes are nearest together and is blown upward by the hot air rising between the electrodes. The arc is broken every half period of the alternating current. In a 400-kw. furnace, the arc length is about 1 m. at 4000 volts. The NO concentration is about 1.5 per cent and the yield 60 g. HNO_3/kw.-hr. The process is in use in Tyrol and in France where very cheap power is available.

Ozone may be formed in various ways, but its commercial production has been only by the electrostatic method. It is probable that ionization of air takes place with consequent dissociation of the oxygen, which recombines to form ozone. Within working limits in commercial ozone generators, the production is roughly proportional to the electrostatic intensity, and with alternating current to the frequency. Ozone generators have been made in various forms, but in general two or more discharging surfaces are placed in juxtaposition so as to form a condenser with an air gap which may or may not be furnished with a dielectric element. Most successful ozone generators have smooth electrodes and dielectrics and assume either the cylindrical or the plate form. They are operated on voltages ranging from 5000 to 25,000 and frequencies from 50 to 500 cycles. An average yield is given as about 50 g./kw.-hr., although much higher yields have been claimed. Concentrations of 1 to 3 g./cu. m. air (although in some special cases the figure may reach 5 g.) are produced commercially. It is essential to refrigerate the air to dry it.

The tubular ozonizers are more compact and operate at lower voltages than the plate forms.

Electrostatic Precipitation. For descriptive matter, operation, and theory, reference should be made to Sec. 15, pp. 1039–1045.

Materials of Construction for Electrochemical Processes

Materials of construction for electrochemical apparatus may in general be subdivided into three classes: (1) those employed in electrolytic cells in which aqueous electrolytes are used, (2) those finding application in fused electrolyte work, and (3) those in electrothermic operations. These may be further subdivided as in Table 30.

Table 30. Materials of Construction

1. Electrolytic cells—aqueous electrolytes
   - *a*. Anodes
   - *b*. Cathodes
   - *c*. Diaphragms
   - *d*. Tank materials
   - *e*. Pipe lines, circulation systems
   - *f*. Floors and buildings
   - *g*. Bus bars and power lines

2. Electrolytic cells—fused salt electrolytes
   - *a*. Anodes
   - *b*. Cathodes
   - *c*. Diaphragms
   - *d*. Container or tank materials
   - *e*. Circulation systems, stirrers
   - *f*. Floors and buildings
   - *g*. Bus bars and power lines

3. Electric furnaces
   - *a*. Electrodes and their operation
   - *b*. Furnace linings
   - *c*. Charging mechanisms
   - *d*. Gas and dust-collecting systems
   - *e*. Furnace body and shell
   - *f*. Auxiliary equipment
   - *g*. Bus bar and power lines

Material	Use	Characteristics	Remarks
		1*a*. Aqueous Electrolytes—Anodes	
Carbon, amorphous	In some early forms of cells	Appreciably attacked by chlorine	No longer used save in unusual cases, in NaCl electrolysis
Chilex, mainly copper silicide	Copper electrowinning	Does not introduce Fe into solution. Films formed on surface catalyze anodic reactions	Used at Chile Copper Co., Chuquicamata, Chile
Graphite, artificial	Chlorine and chlorate cells	Only slightly attacked by Cl_2; long life which can be lengthened by impregnation with Co salts, linseed oil, and synthetic resins, etc.	Standard for NaCl electrolysis. Can be machined to almost any shape. Not useful in cells containing H_2SO_4 or acid solutions; disintegrates in these
Iron	Electrolytic H_2-O_2 cells	Not attacked in alkaline solution. Can be passivated in certain alkaline solutions. Attacked by solutions containing chlorides, nitrates, sulfates, or acid solutions. Iron salts in solution are undesirable, causing lowered energy efficiency	Standard anode for electrolytic H_2 and O_2—generally coated with Ni to reduce O_2 overvoltage. Cheapest material of construction, universally available
Iron-silicon alloys, duriron	Formerly in copper electrowinning	Attacked by chlorides and nitrates in sulfate solutions. Introduced Fe into solution	Used only in isolated cases
Lead (lead-antimony alloy, chemical lead).	Electrodeposition and electrorefining of metals; electrowinning	Not attacked by sulfates in absence of chlorides. Often becomes coated, and then acts like a lead peroxide electrode	One of the standard insoluble or nonattackable electrodes
Lead-silver alloys	Zn electrodeposition	See Pb	Have been proposed as anode in certain per cent for NaCl electrolysis. Form protective films resistant to anodic corrosion. Films have low O_2 overvoltage
Magnetite, fused	Early chlorine cells; copper electrowinning	Used as hollow castings. Good resistance to attack but introduces Fe into solutions	Castings were fragile
Platinum	Electrochemical analysis; persalts, peracids	Excellent insoluble electrode. Low O_2 overvoltage. Resistant to most forms of attack but affected by chlorides. Generally used in form to give greatest surface per unit of weight, as screen, gauze, thin sheet, or even foil or plating over other metals	High cost is disadvantage. Used only where other materials fail and Pt is needed. Platinized Pt electrodes give lower O_2 overvoltages
		1*b*. Aqueous Electrolytes—Cathodes	
Aluminum	Zn electrowinning; Ni refining	Used to form starting sheets	High overvoltage of Al allows deposition of metal. In Ni refining, Al blanks given a Na_2S dip so that starting sheet is easily pried off
Copper	Starting sheet blanks for Cu refining	Not soluble in electrolyte, when used as cathode with current on cell	Competes with Pb blanks
Iron	Chlorine and caustic	....	Universally used
Lead and lead alloys	Starting blanks for Cu refining and electrowinning	Not attacked in sulfate solutions	Usually oiled or coated to allow easy removal of formed starting sheet
Stainless steel	Electrolytic Mn and Fe	Resistant to sulfate solutions	Allows stripping of deposited metal
		1*c*. Aqueous Electrolytes—Diaphragms	
Alundum	Electrolytic organic preparations	Ordinarily high voltage drop	
Asbestos	Chlorine and caustic cells	Resistant to almost all forms of attack. Can be woven so as to maintain porous structure	Almost universal. Can be built to show very low voltage drop or pressure loss through diaphragm
Asbestos-$BaSO_4$ and/or iron oxides, and plastics.	Special Cl_2-NaOH cells, Townsend	....	Offers mechanical difficulty in some cases. Is the so-called "Baekeland diaphragm"
Clay ware	Laboratory operations	Ordinarily high voltage drop	
Metal screens	Filter press and bipolar H_2-O_2 generators	Act as moderately effective diaphragms	Are mechanically strong but costly
Silica, "filtros"	Electrolytic organic preparations	Ordinarily high voltage drop	
Textiles (duck, linen, muslin) plastics, vinyon.	Cathode boxes in Ni refining, Ag refining, diaphragms	Acts largely as filter or separator	Keeps anode slimes away from cathode deposit. Low voltage drop
		1*d*. Tank Materials	
Carbon-sulfur-sand mixtures	General use	Properly made mixes are very acid resistant	Easily fabricated, cast, or built up
Ceramics or clay ware	Silver refining		
Concrete, lined	Electrorefining and electrowinning; chlorine cells	Linings of mastic, asphalt, bitumen, or pitch determine character of tank	Concrete should be of slow-setting type
Plastics (bakelite, celluloid, etc.)	Electroplating	In plating barrels, etc., is satisfactory	
Soapstone	Nelson chlorine cell	Shows appreciable resistance to salt corrosion	Easily fabricated
Steel	Chlorine cells, H_2-O_2 cells, chlorate cells	Steels and irons are alkali-resistant	Standard satisfactory tank material
Steel, rubber lined	Electroplating and similar operations	Character depends upon type of rubber and kind of construction used for tank	Steel tanks welded; several rubber-lining processes. Tanks stand temperatures to boiling point of H_2O
Wood (cypress, red wood, etc.)	Electroplating, electrorefining, electrowinning	Kind of wood selected depends upon electrolyte, concentration, constitution, temperature, etc.	Wood protected by acid-resistant paints of bitumen or asphalt type
Wood, lead lined	Electrorefining (Cu, etc.), electrowinning	Used where solutions are free of chlorides	Chemical lead, 6–8% Sb, used for stiffness. Sheets "burned" together

Table 30. Materials of Construction—(*Concluded*)

Material	Use	Characteristics	Remarks
		2a. Fused Salt Electrolytes—Anodes	
Carbon, amorphous	Aluminum, cerium, etc.	Oxidizes around 500–550°C. characteristics in Table 31	Cheap; easily made. See standard sizes, Tables 32, 33, 34
Graphite	Na, Mg, Be electrolysis, Pb alloys	Resists oxidation better than amorphous carbon. Better anode for Cl_2 liberation	Often too expensive. See standard sizes, Table 35
Iron	Castner sodium cell	Sometimes iron alloys or Ni-Fe used	Anode is in form of casting, is cheap, and has relatively long life
		2b. Fused Salt Electrolytes—Cathodes	
Carbon	Aluminum	Carbon in plastic mass form, rammed into place in steel box or pot and baked *in situ*	Raw materials are ground coke, petroleum coke, tars, and pitches
Iron	Na from NaCl, Na from NaOH, Be, lead alloys, Ca from $CaCl_2$	Is satisfactory cathode for Na deposition; does not alloy with Na	In Be electrolysis, cathode is special water-cooled casting. In Na from NaOH, cathode is cylindrical. In Pb alloy manufacture, iron pot holds molten Pb which is cathode
Steel	Magnesium	Cathode is steel pot container	Cathode pot is externally heated
		2c. Fused Salt Electrolytes—Diaphragms	
Iron or iron-alloy wire gauze	Na from NaOH, Na from NaCl	Diaphragm prevents passage of Na metal particles to anode. Diaphragm material resistant to electrolyte	Openings in mesh must not be so fine as to clog readily, but still small enough to serve as diaphragm
		2d. Fused Salt Electrolytes—Containers	
Steel or cast iron	Al, Mg, Ca, Na, Be, etc.	Common container is made of steel plate or cast in one or more pieces of ordinary or special cast iron	Frozen layers of electrolyte adjacent to walls often protect shells or containers if they are unlined
		3a. Electric Furnaces—Electrodes	
Carbon	General electric furnace work; iron and steel furnaces; carbide, etc.	Characteristics are a function of raw materials (coal, petroleum coke, etc.) and method of manufacture (extruded, molded, etc.). Can be turned and machined, but operations are more difficult than with graphite	For standard sizes, connecting pins, carrying capacities, etc., see Tables 32, 33, 34 Cheaper than graphite. Lower current density must be used than for graphite
Carbon, sectional of prebaked shapes, "Miguet."	Carbide and other furnaces	Allows continuous feed; blocks of electrodes joined to frame	For details, see *Trans. Am. Electrochem. Soc.*, **52**, 335 (1927)
Carbon, baked in place, "Söderberg."	Iron and steel furnaces, carbide	Carbonaceous paste is fed to electrode shell which burns off as electrode is consumed. Electrode baked in place. Electrode is continuous	
Graphite	General electric furnace work; Detroit rocking furnace, non-ferrous melting, cyanamid	See Table 31 for electrical and physical values. Can be readily machined, joined, and cut to unusual shapes	For standard sizes, connecting pins, carrying capacities, see Tables 35, 36, 37. More expensive than carbon
		3b. Electric Furnaces—Linings	
Carbon, preformed block	Ferrovanadium and other furnaces	Useful only in reducing atmosphere	Can stand higher temperatures than any common refractory
Chromite	Basic electric furnace practice	Has limits of use and special applications	Expensive; has high melting point
Clay brick	CS_2, phosphorus, etc.	Firebrick is suitable	Brick serves as structural material as well as to resist action of furnace charges and gases
Dolomite	Basic electric furnace practice	Has limits of use and special application	Is cheap; widely used
Ground brick	Base of resistance furnaces, as SiC and graphite	Electrical insulator	Cheap
Layer of material under treatment	Fused alumina	Unfused layer of material adjacent to shell protects it	
Magnesite	Basic electric furnace practice	Has limits of use and special application	Must be carefully burned to prevent shrinkage
Silica-type refractory	Acid electric furnace practice		

Table 31. Comparison of Amorphous and Graphitic Carbon Electrodes

	Acheson graphite electrodes	Amorphous carbon electrodes*	Gas-baked amorphous carbon electrodes†	Large electro-thermal electrodes	Copper	Aluminum	Iron
Specific resistance, ohms/in. cube	0.00032	0.00124	0.00161	0.00220	0.00000065	0.00000120	0.00000380
Specific resistance, ohms/cm. cube	0.000813	0.00325	0.00400	0.00550			
Comparative section area for same voltage drop	1.	3.8	4.4	6.8			
Weight, lb./cu. in.	0.0574	0.0564	0.0560	0.058	0.320	0.090	0.280
Weight, lb./cu. ft.	99.0	97.5	97.0	100.0	554	155	484
Apparent density, g./cc.	1.585	1.558	1.55	1.60			
Tensile strength, lb./sq. in.							
Lengthwise	800–1000	1000–1500	1000–1500	....	20,000–30,000	24,000–30,000	30,000–50,000
Crosswise	500–600	600–900	600–900				
Temperature of oxidation in air, °C.	640	500	500	500			

* Very good electric baked electrode (small).
† Typical small electrode.

In fused salt electrolysis, the containers are almost universally steel, either bare or lined with carbon. When the containers are unlined, an effective lining is usually formed by a fused layer of the electrolyte or the material being melted. Such is the case in fused alumina furnaces where the furnace consists of a steel shell which is protected by a layer of the alumina adjacent to the steel. In electric-furnace work the container of the furnace is always sheet steel or cast iron refractory lined, the type of refractory used depending upon the furnace operations. The refractory may be acid, in which case it is silica in the form of silica brick or ground material mixed with a binder and tamped into place; or basic, when it is either magnesite, dolomite, or in some cases chromite. In addition various other refractories may be used, such as those of the clay type or the aluminous refractories and in special cases silicon carbide materials. For preventing large heat losses, insulating refractories may be used, but they are never exposed to the furnace bath. They may be fire clay, diatomaceous earth, kieselguhr, magnesia, or asbestos compositions, as well as other special products. In some unusual cases where other materials fail as the result of high temperatures, the temperatures exceeding the point where the refractory gives satisfactory load service or else the melting point of the refractory being exceeded and reducing conditions maintained in the furnace, carbon refractories find application. They are used in the form of preshaped blocks, or they may be rammed in place. For the base of resistance furnaces, particularly of the graphite, silicon carbide, and electrode baking furnaces, ground brick grog or gannister is employed. For the side walls and heads of such furnaces, high-quality clay refractory brick serves. For insulating materials in such furnaces the usual substances cannot be employed, in that they would contaminate the charge of the furnace. Insulation is therefore obtained by the use of relatively thick layers of finely ground calcined carbonaceous materials such as those made from petroleum coke or low-ash anthracite.

Table 32. Standard Amorphous-carbon Electrode Data
(National Carbon Co., Inc.)

Diam., in.	Area, sq. in.	Weight per 60 in. length threaded, lb.	General limits current-carrying capacity, amp.	Current density, amp./sq. in.
Round electrodes				
6	28	92	1,200– 1,700	40–60
8	50	158	2,000– 3,000	40–60
10	79	250	3,000– 4,800	40–60
12	113	360	4,500– 6,800	40–60
14	154	510	5,400– 8,500	35–55
16	201	590	7,000–11,000	35–55
17	227	730	7,900–12,500	35–55
20	314	971	11,000–17,300	35–55
24	452	1437	15,000–25,000	35–55
Square electrodes				
6	36	125	1,500– 2,200	40–60
8	64	224	2,500– 3,800	40–60
10	100	333	4,000– 6,000	40–60
12	144	500	5,500– 8,500	40–60
14	196	647	6,800–10,800	35–55
16	256	864	9,000–14,000	35–55
20	400	1320	14,000–20,000	35–55

Carbon electrodes for use in electric furnaces are made from either calcined petroleum coke or low-ash anthracite coal, or mixtures of these. Materials before molding into the electrode are carefully ground and screened and a composite aggregate made of such screen sizes as to give mechanically strong electrodes. The aggregate is mixed with a binder, molded under relatively high pressures, baked, and machined for continuous feeding of the electrodes.

Furnace manufacturers have eliminated electrode designs which necessitate shutdowns for electrode changes. Continuous feed of electrode has become standard practice. The manner of joining electrodes has been reduced to two methods. The first method employs electrodes turned down and threaded at one end, and drilled and tapped at the other end. These can be fitted together as male and female joints. The second method calls for both ends of the electrode to be drilled and tapped; a small threaded pin serves as the connecting medium.

Table 33. Amorphous Carbon—Weights of Unthreaded and Threaded Electrodes
(National Carbon Co., Inc.)

Size, in.	Unthreaded	Threaded
4 × 40	30	
5⅛ × 48	63	
6 × 6 × 48	100	
6 × 48	80	77
6 × 60	95	92
7 × 48	105	98
7 × 60	133	126
8 × 48	134	127
8 × 60	165	158
10 × 28 × 28	492	
10 × 48	222	209
10 × 60	263	250
12 × 48	292	270
12 × 12 × 60	500	
12 × 60	382	360
14 × 60 extruded	525	493
14 × 60 molded	542	510
14 × 72 extruded	622	590
14 × 72 molded	642	610
16 × 60	647	590
16 × 72	781	724
16 × 16 × 80	1152	
16 × 16 × 84	1210	
16 × 16 × 110	1584	
17 × 60 extruded	743	686
17 × 60 molded	787	730
17 × 72 extruded	871	814
17 × 72 molded	949	892
18 × 72	1027	970
20 × 72 extruded	1223	1130
20 × 72 molded	1285	1192
20 × 20 × 80	1760	
20 × 20 × 84	1848	
20 × 84 extruded	1403	1310
20 × 84 molded	1515	1422
20 × 20 × 90	1980	
20 × 90 molded	1638	1545
20 × 20 × 100	2200	
20 × 20 × 120	2640	
20 × 180	3033	2940
24 × 72 extruded	1760	1630
24 × 72 molded	1890	1760
24 × 84 extruded	2130	2000
24 × 84 molded	2226	2096
24 × 110 extruded	2750	2620

Table 34. Amorphous Carbon—Standard Connecting-pin Data
(National Carbon Co., Inc.)

Size of pin (pitch diam. by length of thread), in.	Diam. of electrode used for, in.	Max. diam. top of threads, in.	Max. length over all, in.	Average weight, lb. oz.	Pins per package
3 × 6	6	3.28	6⅝	2 3	24
3 × 6	7	3.28	6⅝	2 3	24
4 × 8	8	4.41	9⅛	6 0	12
5 × 10	10	5.41	11⅛	11 0	6
6 × 12	12	6.41	13⅛	19 0	3
7 × 14	14	7.59	14	28 0	Crate as desired
8½ × 17	16	9.13	17	51 0	Crate as desired
8½ × 17	17	9.13	17	51 0	Crate as desired
8½ × 17	18	9.13	17	51 0	Crate as desired
10 × 20	20	10.63	20	85 0	Crate as desired
10⅞ × 24	24	11.69	24	120 0	Crate as desired
12 × 24	24	12.81	24	150 0	Crate as desired

Present-day dowel pins are turned out of stock just large enough to ensure good threads. Electrodes are now drilled and threaded in special self-centering lathes. Mechanical details on the dowel pins such as an end boss, chamfering the first thread, and reduction of thread tolerances have done away with electrode overhangs. Joint trouble has been reduced to a minimum.

Desire for a continuous feed of electrode as it is consumed has brought the development of the Söderberg electrode. In this method, electrode paste (the "green" mix) is tamped into a shell which is a vertical continuation of the electrode holder. As the electrode is consumed, the tamped mixture passes through zones of increasing temperature up to a heating ring where baking takes place. The electrode is continuous in that the form is kept filled up with electrode paste. It is jointless, baked, and renewed right in the furnace.

Table 35. Graphite Electrodes
(Acheson Graphite Corp.)

Size, diam., in.	Approximate weight per piece, lb.	Size rectangular, in.	Approximate weight per piece, lb.
Cylindrical			
1/16 × 12	0.002	1¼ × 12 × 12	2.10
⅛ × 12	0.008	½ × 4 × 24	2.94
¼ × 12	0.035	½ × 6 × 24	4.50
⅜ × 24	0.15	½ × 12 × 12	4.05
7/16 × 24	0.21	¾ × 3½ × 19½	3.13
½ × 24	0.26	¾ × 5 × 18	4.21
⅝ × 24	0.40	¾ × 12 × 12	6.75
¾ × 24	0.60	⅞ × 2 × 21¼	2.20
⅞ × 24	0.82	1 × 4 × 30	7.50
1 × 24	1.00	1 × 6 × 30	11.30
1 × 48	2.00	1 × 12 × 12	8.35
1⅛ × 24	1.4	1¼ × 3 × 36	7.88
1¼ × 24	1.75	1¼ × 5 × 30	11.50
1½ × 24	2.60	2 × 4 × 30	14.50
Square			
2 × 2 × 30	7.50	2 × 4 × 40	19.00
2 × 2 × 36	8.90	2 × 6 × 28½	20.00
4 × 4 × 13	12.00	2 × 7 × 30	25.75
4 × 4 × 17	15.75	2 × 7 × 40	32.48
4 × 4 × 40	37.00	3 × 6 × 30	33.25
6 × 6 × 40	84.50	4 × 8¾ × 15	31.45
6 × 6 × 48	102.00	2 × 8 × 48	44.00
8 × 8 × 48	179.00	3 × 9 × 48	75.20
		3 × 11 × 48	92.00
Tubular			
2¼ o.d. × 1 1/16 i.d. × 24	4.6	4 × 10 × 48	112.00

Average practice allows a current density on the electrodes of 20 to 40 amp./sq. in. cross section.

Furnace electrodes made of Acheson graphite are employed when very high currents are necessary, and a balancing of all factors shows that graphite electrodes, although costing more per unit, would allow more economical operation.

Power for Electrochemical Processes

It is estimated that over 200,000 kw. capacity represents the annual purchase of d.c. machines for electrolytic service. Direct-current power can be obtained in several ways:

1. Purchased alternating current converted to direct current by:
   a. Synchronous converters.
   b. Motor-generator sets.
   c. Rectifiers.
2. Generated power:
   a. Alternating-current generation and conversion as for purchased power.
   b. Direct-current generation:
      (1) Geared steam turbines.
      (2) Steam engines.
      (3) Diesel engines.

The bulk of d.c. power is obtained from synchronous converters or motor-generator sets, while a few geared steam turbines are in operation. Recently the mercury-arc rectifier has found wide application.

At voltages above 600, the mercury-arc rectifier with no moving parts is favored over rotary converters or motor-generator sets for continuous electrochemical loads (for construction, etc., see Sec. 27).

The Cu-CuO or oxide type of rectifier has found application in low-voltage (6 to 15) practice such as electroplating, cleaning, etc. The selenium rectifier is competitive.

Table 36. Standard Graphite Electric-furnace Electrodes
(Acheson Graphite Corp.)

Round electrode, diameter by length, in.	Approximate weight per piece, lb.	Sectional area, Sq. in.	Sectional area, Sq. cm.	Approximate carrying capacity, amp.
2 × 24 (50 × 610 mm.)	4.50	3.1416	20.266	600– 800
2½ × 30 (63 × 760 mm.)	8.40	4.9087	31.661	800– 1,200
3 × 40 (75 × 1000 mm.)	16.60	7.0686	45.595	1,200– 1,800
4 × 40 (100 × 1000 mm.)	30.00	12.566	81.025	1,800– 2,300
5⅛ × 40 (130 × 1000 mm.)	48.00	20.629	133.084	2,300– 3,000
6 × 48 or 60 (150 × 1220 or 1500 mm.)	75.00 or 94.00	28.274	182.369	3,000– 4,000
7 × 48 or 60 (175 × 1220 or 1500 mm.)	107.00 or 134.00	38.485	248.234	4,000– 5,500
8 × 48 or 60 (200 × 1220 or 1500 mm.)	140.00 or 175.00	50.265	324.227	5,500– 6,500
9 × 48 or 60 (225 × 1220 or 1500 mm.)	176.00 or 220.00	63.617	410.348	6,500– 8,500
10 × 48 or 60 (250 × 1220 or 1500 mm.)	230.00 or 285.00	78.640	506.662	8,500–10,000
12 × 48 or 60 (300 × 1220 or 1500 mm.)	320.00 or 400.00	113.100	729.608	10,000–15,000
14 × 60 (350 × 1500 mm.)	500.00	153.94	993.169	15,000–20,000
16 × 60 (400 × 1500 mm.)	700.00	201.06	1297.239	20,000–26,000
18 × 60 (450 × 1500 mm.)	900.00	254.47	1641.839	26,000–30,000

Table 37. Graphite Electrode Current Density
(Acheson Graphite Corp.)
Amp./sq. in. cross section

Diam. of electrode	Cross section sq. in.	Circumference, in.	Amp. 1000	1500	2000	2500	3000	3500	4000	4500	5000	5500	6000	7000	8000	9000	10,000	12,000	15,000	18,000	20,000
2	3.14	6.28	318.3	447.3																	
2½	4.90	7.85	203.6	305.4	401.2																
3	7.07	9.50	141.5	212.2	283	353	424														
4	12.6	12.6	80	120	160	200	240	280	320												
5⅛	20.6	16.1	48	73	97	121	146	169	194	218	242										
6	28.3	18.85	35	53	70	88	106	124	141	159	177	194	212	247							
7	38.5	22.0	26	39	52	65	78	91	104	118	130	143	156	182	208	234					
8	50.3	25.13	20	30	40	50	60	70	80	90	100	110	120	140	160	180	200	240	300		
9	63.6	28.27	16	24	31	39	47	55	63	71	79	86	94	110	126	141	187	188	236	283	
10	78.6	31.42	13	19	25	32	38	44	51	57	64	70	76	89	102	114	127	153	191	230	254
12	113.1	37.70	9	13	18	22	26	31	35	40	44	48	53	62	70	79	88	106	133	159	177
14	153.9	43.98	7	10	13	16	20	23	26	29	33	36	39	46	52	59	65	78	97	117	130
17	227.0	53.41	4	7	9	11	13	15	18	20	22	24	26	31	35	40	44	53	66	80	88
24	452.4	75.40	2	3	4	6	7	8	9	10	11	12	13	15	18	20	22	26	33	40	44

Table 38. Energy Consumption of Electrochemical Products
(Mantell)

Product	Kw.-hr./lb.	Lb./kw.-hr.	Voltage per tank. cell. or furnace
Alumina, fused	1 - 1.5	0.67- 1	100 -110
Aluminum	10 -12	0.08- 0.1	5.5 - 7
Cadmium	0.8	1.25	2.6
Calcium	22 -24	0.04	
Calcium carbide	1.3 - 1.4	0.71- 0.77	
Carbon bisulfide	0.4 - 0.5	2 - 2.5	60
Caustic, diaphragm cells	1.16 - 1.43	0.68- 0.86	3.4 - 4.2
Caustic, mercury cells	1.45	0.69	4.1 - 4.3
Chlorine, diaphragm cells	1.3 - 1.6	0.62- 0.76	3.4 - 4.2
Chlorine, mercury cells	1.6	0.62	4.1 - 4.3
Copper, electrorefining:			
Multiple system	0.09 - 0.16	6.3 -11	0.18- 0.4
Series system	0.074	13.5	16 - 18
Copper, electrowinning	1 - 1.5	0.67- 1	1.9 - 2.4
Ferrochromium, 70%	2 - 3	0.33- 0.5	90 -120
Ferromanganese, 80%	1.5 - 3	0.33- 0.67	90 -115
Ferromolybdenum, 50%	3 - 4	0.25- 0.33	50 -150
Ferrosilicon, 50%	2 - 3.5	0.28- 0.5	75 -150
Ferrotungsten, 70%	1.5 - 2	0.5 - 0.67	90 -120
Ferrovanadium	2 - 3.5	0.28- 0.5	150 -250
Gold	0.15	6.6	1.3 - 1.6
Graphite	1.5 - 2.0	0.67	80 -200
Iron	1.8 - 2.0	0.5 - 0.55	4 - 4.4
Lead	0.04 - 0.05	24	0.35- 0.6
Magnesium, chloride process	8	0.125	6 - 7
Manganese	5 - 6	0.16- 0.2	4 - 5
Nickel	1.1	0.9	2.4 - 2.5
Phosphorus	4 - 5.5	0.18- 0.25	
Silico-manganese	2 - 3	0.33- 0.5	90 -120
Silicon carbide	3.2 - 3.85	0.26- 0.31	75 -230
Silver, Moebius	0.31	3.2	2.7
Silver, Thum	0.41	2.4	3 - 3.5
Sodium	7.1 - 7.3	0.13- 0.14	
Steel, cold charge	0.25 - 0.4	2.5 - 4	75 -150
Steel, hot charge	0.05 - 0.2	5 -20	
Tin	0.085	11.8	0.3 - 0.35
Zinc	1.4 - 1.56	0.64- 0.7	3.5 - 3.7

Table 39. Typical Power Costs

	¢ per Kw.-hr.
New York harbor (steam)	0.67-1.3
Norway	0.1 -1.5
Sweden	0.1 -1.5
Switzerland	0.3
French Alps	0.17-0.3
England	0.3 -0.5
Scotland	0.3 -0.5
Germany (brown coal)	0.38
Niagara Falls	0.3 up
Massena, N.Y.	0.36
Alabama (steam)	0.38-0.6
Tennessee (steam)	0.38-0.6
California (steam)	0.38-0.6
Ontario, Can.	0.15-0.4
Shawinigan, Que., Can.	0.15-0.3
Arvida, Que., Can.	0.1
Bonneville	0.1 -0.2
Alcoa, Tenn.	0.3 -0.37

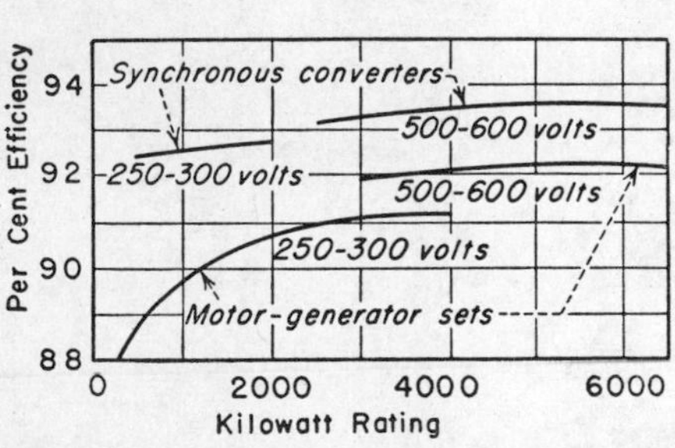

FIG. 44. Efficiency of synchronous converters and motor-generator sets.

Converter efficiency includes losses of transformer and induction regulator or booster and 0.3 per cent a.c. lead loss. Motor-generator set efficiency is based on 100 per cent power factor for motors. All losses are calculated or measured in accordance with the A.I.E.E. rules. The resulting efficiencies are close to the actual efficiencies.

Requirements of the electrolytic circuits are usually pretty definitely known. As a rule a voltage range of 10 per cent plus and minus will be ample to meet all operating requirements. This range is easily obtained on the ordinary self-excited generator.

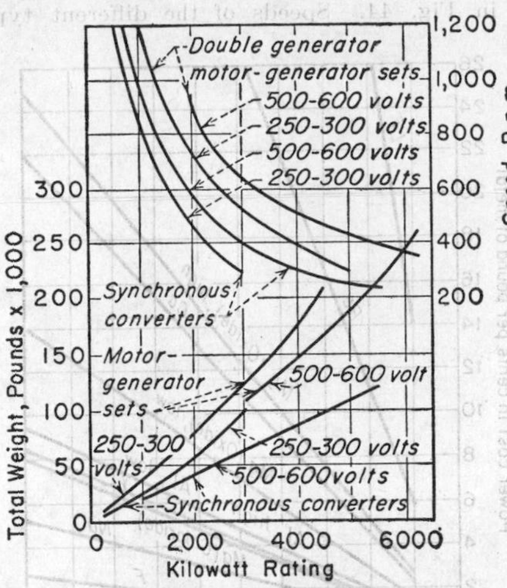

FIG. 45. Speed of synchronous converters and motor-generator sets. (*Westinghouse Electric Corp.*)

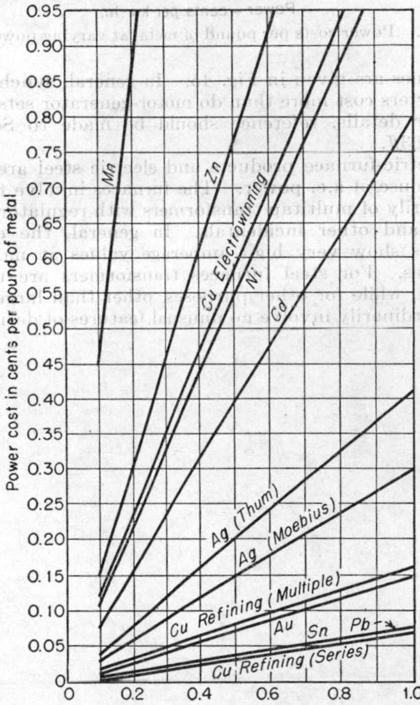

FIG. 46. Power costs for metal production.

A comparison of a.c. generation followed by conversion to direct current, and d.c. generation, shows that the latter has the advantages of lower first cost, slightly lower fuel cost, lower labor cost, with a possibility of lower maintenance cost, but the disadvantage of lack of flexibility when interconnected with other power systems, or in the conversion of the plant to other than electrolytic use, or when major changes are made in electrolytic circuits.

In general, synchronous converters show higher efficiencies than do motor-generator sets. This relation is shown in Fig. 44. Speeds of the different types of machines are given in Fig. 45. In general, synchronous converters cost more than do motor-generator sets. For further details, reference should be made to Sec. 27, pp. 1763*ff*.

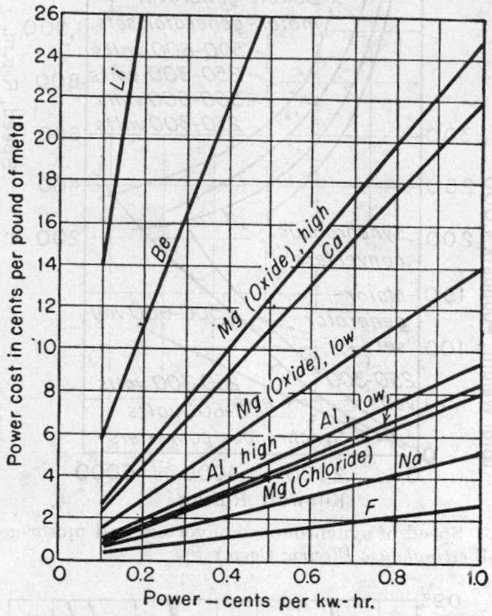

Fig. 47. Power costs per pound of metal at varying power rates.

Electric-furnace products and electric steel are made by the use of a.c. power. The furnaces involve the use ordinarily of multitap transformers with regulators, controls, and other incidentals. In general, the electric circuits show very high amperage values at moderate voltages. For steel furnaces transformers are usually special, while for other purposes, other than large sizes, they ordinarily involve no unusual features of design.

Power cost is an important factor in all electrolytic or electrothermal products. Figure 46 shows the power cost in cents per pound of metal at varying power costs in cents per kilowatt-hour for the electrolytic and hydrometallurgical industries, while Fig. 47 gives similar data for the fused electrolytes and Fig. 48 gives similar data for electric-furnace products. The curves are plotted from the power consumptions for these products given in Table 38.

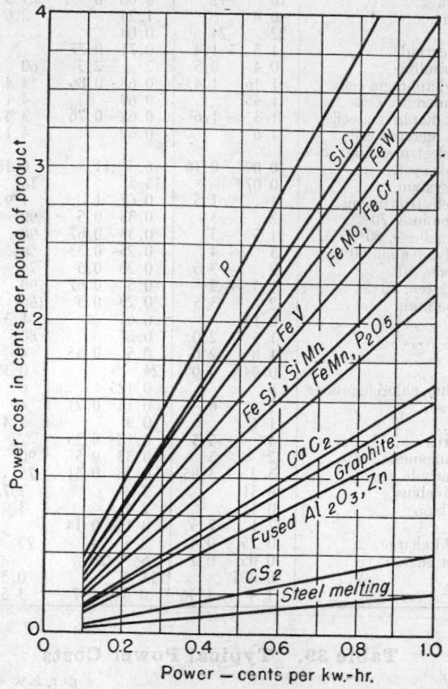

Fig. 48. Power costs per pound of product from electric furnaces with fused electrolytes at varying power rates.

Some typical power costs at various localities are given in Table 39. Unless otherwise stated the generation of power is in hydroelectric plants.

SECTION 29

ACCOUNTING AND COST FINDING

BY

George A. Prochazka, Jr., E. M., Management Consultant; formerly Chemical Economist, Ammonia Department, E. I. du Pont de Nemours & Co.; formerly Manager, Central Dyestuff & Co.; Member, American Institute of Chemical Engineers, American Institute of Mining and Metallurgical Engineers; Licensed Professional Engineer, N.Y. State; Author, "Accounting and Cost Finding for the Chemical Industries," McGraw-Hill, New York, 1928.

CONTENTS

	Page
References	1828
General Accounting	1828
Analyzing Financial Statements	1829
Fixed-property Accounting	1831
Depreciation	1831
Cost-finding Systems	1833
Pay-roll Accounting and Wage Control	1834
Budgeting and Expense Accounting	1835
Expense Distributions	1837
Service Department Costs	1838
General Overhead Costs	1839
Production Department Costs	1839
Product Costs, Variances, and Yields	1840
Material Accounting and Control	1841
Production Control and Production Lines	1843
Cost Estimating	1844

ACCOUNTING AND COST FINDING

REFERENCES: "Accountants' Handbook," 3d ed., Ronald, New York, 1943. Esquerre, "Applied Theory of Accounts," Ronald, New York, 1923. Foulke, "Practical Financial Statement Analysis," McGraw-Hill, New York, 1945. "Cost and Production Handbook," Ronald, New York, 1934. Knoeppel and Seybold, "Managing for Profit," McGraw-Hill, New York, 1937. Gardner, "Variable Budget Control," McGraw-Hill, New York, 1940. Williams, "The Flexible Budget," McGraw-Hill, New York, 1934. Leith, "Mineral Valuations of the Future," Am. Inst. Min. & Met. Engrs., 1938. Income Tax Depreciation and Obsolescence, *Bull. F*, Superintendent of Documents, Washington, D.C., 1942. Marston and Agg, "Engineering Valuation," McGraw-Hill, New York, 1936. Grant and Norton, "Depreciation," Ronald, New York, 1949. Camman, "Basic Standard Costs," American Institute Publishing Co., New York, 1932. Presgrave, "The Dynamics of Time Study," McGraw-Hill, 1945. "Systems of Wage Payment," National Industrial Conference Board, Inc., New York, 1930. Carroll, "Timestudy for Cost Control," McGraw-Hill, New York, 1943. Lowry, Maynard, and Stegemerten, "Time and Motion Study," 3d ed., McGraw-Hill, New York, 1940. Kress, "How to Rate Jobs and Men," Factory Management & Maintenance Library, McGraw-Hill, New York. Benge, Burk, and Hay, "Manual of Job Evaluation," Harper, New York, 1941. Halsey, "Making and Using Industrial Service Ratings," Harper, New York, 1944. Prochazka, "Accounting and Cost Finding for the Chemical Industries," McGraw-Hill, New York, 1928. Lawrence, "Cost Accounting for War Production," Prentice-Hall, New York, 1942. Clark, "The Gantt Chart," Ronald, New York, 1922. Muther, "Production Line Technique," McGraw-Hill, New York, 1944. Tyler, "Chemical Engineering Economics," 3d ed., McGraw-Hill, New York, 1947. Bliss, The Costs of Process Equipment and Accessories, *Trans. Am. Inst. Chem. Engrs.*, **37** (1941). Walker, "The Building Estimator's Reference Book," 9th ed., F. R. Walker Co., Chicago, 1940. Pulver, "Construction Estimates and Costs," McGraw-Hill, New York, 1940. Raymond, "Quantity and Economy in Manufacture," McGraw-Hill, New York, 1931. Blocker, "Cost Accounting," McGraw-Hill, New York, 1940. Churchill, "Pricing for Profit," Macmillan, New York, 1932.

GENERAL ACCOUNTING

Importance of Accounting. There is no quicker approach to the economics of an enterprise than through an analysis of its accounting records. Knowledge of accounting is essential to cost determination and cost control.

Debits and Credits. Every business transaction has two phases, and these give rise to the debits and credits of bookkeeping. Goods are bought, shipped, and billed by the vendor; an invoice is received and its dollar amount is debited to the account, Raw Material, and credited to Accounts Payable. Every business transaction is recorded in this manner by two entries, one a debit by custom placed on the left side of the ledger and the other a credit, placed on the right side of the ledger.

Journals and Ledgers. The accountant uses two types of book, each of which serves a different purpose. The **journal** is the book of original entry; debits and credits are first entered there. From the journal the entries are posted to the ledgers. The purpose of this posting is to classify the transactions into accounts. Those dealing with Raw Material are placed in one account, those dealing with Accounts Payable in another, etc.

Controlling Accounts. The labor of posting transactions to ledgers has been greatly reduced by means of controlling accounts, subsidiary ledgers, and classified journals. Under this pattern, the invoices are entered in a specially ruled journal with columns for Accounts Payable, Raw Material, Supplies, Expenses, etc.; and at the end of the month only the column totals are posted to Raw Material, to Accounts Payable, etc., in the ledger. When necessary, details of the amounts owed each vendor are kept in a subsidiary Accounts Payable Ledger, which has an account for each vendor. The subsidiary ledgers are controlled by accounts of like name in the general ledger. This simply means that, when the figures in the secondary book are added, they must agree with the total of the controlling account. The method requires a subdivision of the ledgers and journals. A typical subdivision of the journals is invoice register, sales book, pay roll, cash book, and general journal.

Reserve Accounts. The true nature of a reserve account is not always understood. Commonly it is believed that officials have set aside money for use during some future lean period, but that is not the purpose of a reserve account. It is established to provide for future losses or payments that at the moment can only be estimated. Thus there are reserves for depreciation, obsolescence, depletion, bad debts, accrued taxes, and other contingencies. A reserve is rarely actually funded through a setting aside of money. The usual pattern is to add the estimated loss to the costs and thus to withhold the amount from the profits. An equal sum is credited to a reserve, but no provision is made for the future payment of the loss out of the general business funds. For this reason it is good accounting practice in the preparation of a financial statement to deduct a reserve account from the appropriate asset, thus:

Machinery and equipment	$500,000
Less reserve for depreciation	100,000
Depreciated value	$400,000

Bookkeeping Machines. Bookkeeping machines are essentially writing-calculating devices arranged to produce two, three, or more accounting records simultaneously. With a single writing, for instance, it is possible to enter a vendor's invoice (1) in a voucher register, (2) in an accounts payable ledger, and (3) on a remittance statement. There are similar combinations for sales, pay rolls, etc., but in each case the purpose is to make a journal entry, a ledger posting, and an auxiliary notation in one writing.

Tabulating Machines. When a great mass of statistical material must be analyzed from several viewpoints, the easiest procedure is to use tabulating equipment. Operation of tabulating equipment follows the pattern: (1) an original written record, (2) a card punched to prepare the data for the machine operations, (3) a selective sorting of the cards to obtain those containing the desired information, and (4) a tabulation of the selected cards to summarize the desired information. Various mathematical operations can also be performed in conjunction with the sorting and tabulating.

Classification of Accounts. Regardless of whether mechanical equipment or hand posting is used, the success of the accounting depends to no small degree on the care with which ledger accounts are selected. The formation of transactions into suitable accounts requires attention to volume of activity as well as to kind. It is the function of the accounts to tell what is going on in the business. This purpose is best served when there are neither too many nor too few accounts. For practical bookkeeping purposes, accounts may be divided into four classes: (1) expense, (2) income, (3) asset, and (4) liability.

Chart of Accounts. To facilitate day-to-day accounting, it is customary to prepare a formal list of accounts with codes for conveniently charging items to specific accounts. The list includes all the classifications that have been found desirable for a complete financial analysis of the activities of the enterprise. Because titles are not always adequately descriptive, the chart of accounts often also carries explanatory comment that defines the scope of the various accounts.

Trial Balances. At the end of an accounting period, before data are taken from the books for use in preparing financial statements, it is customary to take a trial balance. The procedure is to list the open balances of each account, placing the debits in one column and the credits in another column of the trial balance. If the two columns are equal in amount, the books are in balance, and the task of preparing financial statements can be safely undertaken.

Financial Statements. Two statements are required to explain the affairs of a business. The balance sheet shows the asset, liability, and net worth values existing as of a particular moment. The profit and loss statement shows the effect of the business activity, *i.e.*, what caused the change from the previous financial condition. In order to make financial statements more informative, they are usually presented in comparative form; *i.e.*, the results of the previous year are stated alongside the values of the current year. Typical comparative statements are shown in Tables 1 and 2.

Table 1. Comparative Balance Sheet XYZ Company as of December 31, 1948

	1948	1947	Increase − Decrease
STATEMENT OF ASSETS			
Cash	$ 2,855,600	$ 1,751,810	$1,103,790
Notes and trade acceptances receivable	215,500	386,400	170,900
Accounts receivable, less reserves	1,482,350	1,579,250	−96,900
Inventories at cost	2,450,115	2,293,810	151,305
Total current assets	$ 7,003,565	$ 6,016,270	$ 987,295
Land buildings, and machinery	$15,380,610	$14,181,900	$1,198,710
Less reserves for depreciation	7,589,005	7,037,000	552,005
Depreciated fixed assets	$ 7,791,605	$ 7,144,900	$ 646,705
Investments	$ 1,258,000	$ 680,000	$ 578,000
Deferred charges	214,800	179,980	34,820
TOTAL ASSETS	$16,267,970	$14,021,150	$2,246,820
STATEMENT OF LIABILITIES			
Accounts payable	$ 1,080,950	$ 1,135,410	$ −54,460
Accrued charges	56,720	65,400	−8,680
Reserved for Federal taxes	598,600	586,900	11,700
Interest accrued	26,500		26,500
Dividends payable	250,000	200,000	50,000
Total current liabilities	$ 2,012,770	$ 1,987,710	$ 25,060
Ten-year 3¾ % notes	1,500,000		1,500,000
Capital stock 500,000 shares	5,000,000	5,000,000	
Surplus	7,755,200	7,033,440	721,760
TOTAL LIABILITIES	$16,267,970	$14,021,150	$2,246,820

Working Capital, etc. Usage of financial terms tends to be loose; some of the more usual terms and their relation to the balance sheet equation are presented in Fig. 1. Current assets are sometimes referred to as liquid assets. Net worth is the capital and surplus and thus represents the book value of the owners' interest in the business. Working capital is the difference between the current assets and the current liabilities.

Table 2. Comparative Profit and Loss Statement XYZ Company for Year Ending December 31, 1948

	1948	1947	Increase − Decrease
Sales, less returns, credits, etc.	$12,751,813	$11,885,901	$865,912
Less cost of goods sold	9,855,198	9,082,300	772,898
Gross profit	$ 2,896,615	$ 2,803,601	$ 93,014
Provision for depreciation and obsolescence	552,005	516,211	35,794
General overhead expense	788,550	759,280	29,270
Profit from operations	$ 1,556,060	$ 1,528,110	$ 27,950
Add income from investments	40,800	34,000	6,800
Total income	$ 1,596,860	$ 1,562,110	$ 34,750
Deduct Federal taxes	598,600	586,900	11,700
Accrued interest on notes payable	26,500		26,500
NET PROFITS ADDED TO SURPLUS	$ 971,760	$ 975,210	$−3,450
SURPLUS ACCOUNT			
Surplus as of January 1	$ 7,033,440	$ 6,258,230	$775,210
Net profits added to surplus	971,760	975,210	−3,450
Less dividends paid	250,000	200,000	50,000
Surplus as of December 31	$ 7,755,200	$ 7,033,440	$721,760

ANALYZING FINANCIAL STATEMENTS

Balance-sheet Analysis. Financial statements do not reveal to the casual reader all the information contained in them, and it is consequently common practice to analyze a series of statements to determine the trend of the business and to establish the nature of its shortcomings. Four methods are used for analysis, namely, (1) percentage method, (2) ratio method, (3) trend method, and (4) disposition and source of income method.

Percentage Method. This is a simple method and essentially seeks to establish the proportions of the balance-sheet items by calculating each as a percentage of the total. During this study a few general facts are also obtained such as:

1. Solvency—excess of assets over liabilities.
2. Cash on hand—if relatively small.
3. Receivables—if too large.
4. Inventories—if excessive.
5. Effect of deducting intangibles from surplus.
6. Depreciation—amount taken.
7. Improper balance-sheet combinations that hide facts.
8. Possible tampering with assets through appraisals or marking up of investments.

Ratio Method. Information about the affairs of a company can be obtained from both the balance sheet and the profit and loss statement by comparing the items with financial practice. Nine ratios are commonly used:

1. Fixed assets per $100 of net sales.
2. Inventory per $100 of net sales.
3. Working capital per $100 of net sales.
4. Net worth per $100 of net sales.
5. Receivables per $100 of net sales.
6. Fixed assets per $100 of net worth.
7. Liabilities per $100 of net worth.
8. Current ratio; the current assets divided by the current liabilities should be 2 or higher.
9. Acid test; the quick assets divided by the current liabilities should be 1 or higher.

Trend Method. The rate of increase or decrease of balance-sheet items over a period of years clearly shows the direction in which the enterprise is moving. The method requires a grouping of the balance-sheets items so that the values can be carried along without encountering items that do not appear each year. The total assets of 1 year are then taken as a reference point and called 100. The values of all other years are referred to this index number base and then are plotted on semilogarithmic chart paper. The graph gives a better picture of the trend than a tabulation.

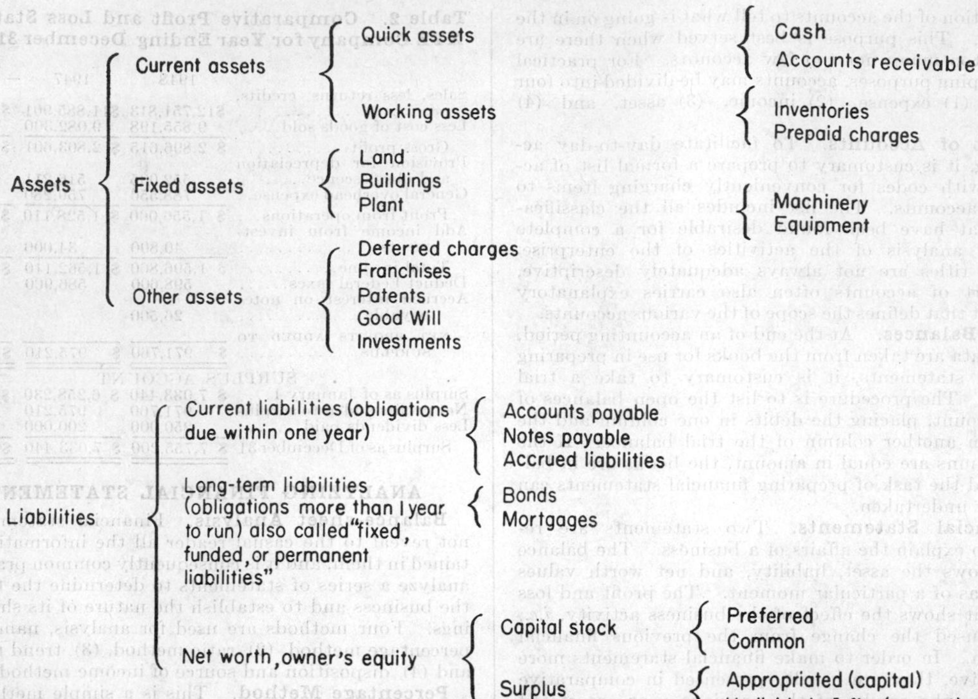

FIG. 1. Balance-sheet terms.

Disposition and Source of Income. Of the four methods for analyzing financial statements, the most informative is the one that examines the disposition and source of income. The analysis is based on a rearrangement of the balance-sheet equation, assets equals liabilities, into increase of assets plus decrease of liabilities, or **disposition of funds,** equals increase of liabilities plus decrease of assets, or **source of funds.** The analysis can be made from a comparative balance sheet such as that given in Table 1 by tabulating the increases and the decreases as shown in Table 3.

Table 3. Disposition and Source of Income

Source of Income*

Income from profit and loss statement	$ 971,760
Appropriated for depreciation	552,005
Reduction in receivables	267,800
Increase in liabilities	88,200
Borrowed on 10-year notes	1,500,000
Total funds available	$3,379,765

Disposition of Funds

Cash increased	$1,103,790
Inventories increased	151,305
Plant additions	1,198,710
Investments increased	578,000
Deferred charges increased	34,820
Accounts payable, etc., decreased	63,140
Dividends paid	250,000
Total disposition of funds	$3,379,765

* Compiled from Tables 1 and 2.

Profit and Loss Analysis. Comments on profit and loss analysis appear in the discussion of the Ratio Method. In the preparation of profit and loss statements it is common practice to subdivide the statement on a supporting schedule so as to disclose the source of the profits by departments, lines, or products. Even though these data may have been carefully compiled from cost records, it does not follow that elimination of the unprofitable items will increase profits. Greater losses may actually result from this procedure. Many of the expenses are fixed and must be carried by other departments. It is necessary, therefore, to recast the statement, reapportioning all fixed expenses that cannot be eliminated simultaneously with the unprofitable products. The

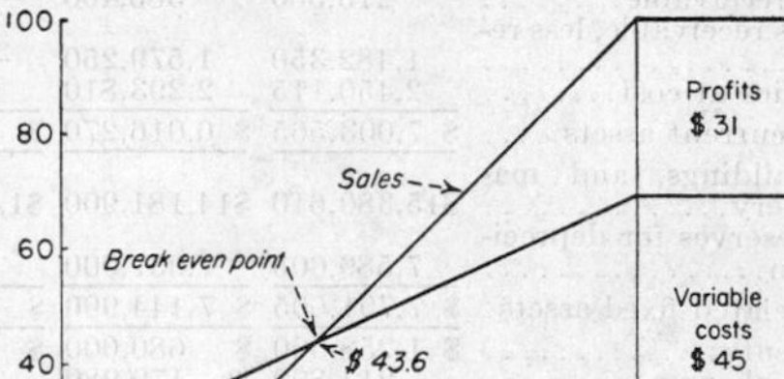

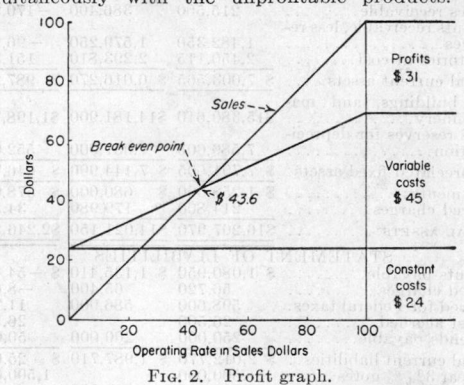

FIG. 2. Profit graph.

usual requirement is not to eliminate, but to substitute more profitable items.

Minimum Profitable Rate of Operations. The no-profit point relationship that exists between volume and cost can be studied advantageously by means of the profit graph shown in Fig. 2. An alternate plan for drawing the chart which places the constant costs above the variable costs and parallel to them is shown in Knoeppel and Seybold, "Managing for Profit," McGraw-Hill, New York, 1937. The horizontal scale in Fig. 2 expresses the operating rate in dollars of sales and the vertical scale the costs, profits, and sales in dollars. The chart can be applied to the profits of the business as a whole or to its component parts. The rate at which profits increase or decrease above and below the break-even point and the location of this point in terms of sales dollars can be calculated as follows:

$$\text{Profit variation ratio} = \frac{\text{constant costs} + \text{profits}}{\text{sales}} \quad (1)$$

$$\text{Break-even point} = \frac{\text{constant costs}}{\text{profit variation ratio}} \quad (2)$$

FIXED-PROPERTY ACCOUNTING

Fixed Charges. Expenses that continue unchanged regardless of the volume of business activity are spoken of as fixed charges. To the cost accountant fixed charges connote depreciation, local taxes, fire insurance, and other constant costs. Many of these charges are related to property values and cannot be apportioned equitably unless there is available a detailed system of property records.

Property Classifications. There are many separate items in a fixed property inventory. To make the records easily manageable it is important to have well-devised property classifications. Six property characteristics usually must be considered in the development of a classification system, namely, (1) type of property, (2) construction elements, (3) location of asset, (4) departmental ownership, (5) process use, and (6) useful life. Some of the information made available through these classifications is required for insurance purposes, some for cost allocation, and some for cost determination. With well-devised classifications it is usually possible to handle all these requirements with two sets of records, (1) a detailed accounting that serves essentially for insurance purposes and (2) an abridged or summary record that is used for depreciation and cost allocations.

Land. Under the caption land is placed the original acreage cost and all subsequent improvements thereto. Such things as pavements, wells, and railroad sidings are logically included because at time of liquidation they cannot be separated. Idle land should be kept on the records apart from occupied land as should also items that are depreciable or involve depletion. Mineral rights are valued separately from land. The subject is informatively discussed by Leith, "Mineral Valuations of the Future," Am. Inst. Min. & Met. Engrs., 1938.

Buildings and Structures. The principal accounting problem with buildings is what to do with the mechanical equipment such as air conditioning, elevators, heating plant, and electric wiring. Usually this is considered part of the building because separate values for these units are not available. A composite depreciation rate is used varying from 1.5 to 3.5 per cent. Ordinary repairs are written off as operating cost; but replacements are charged against depreciation reserves. A charge against the reserve reduces the net amount of depreciation taken and thus in effect increases the length of time during which the building can be depreciated. When a building has a long useful life and contains much mechanical equipment this simple composite depreciation method gets into difficulties because of changes in prices. The elevator replacement, for instance, may cost twice as much as the original installation, and thus too much depreciation is canceled out. In such cases it is preferable to separate the mechanical equipment from other building costs (see also Depreciation Accountancy).

Machinery and Equipment. Machinery should be distinguished on the records as to type so that it can be grouped according to life expectancy. To charge departments for depreciation it is necessary to know departmental ownership of the machinery and also its process use if expenses are apportioned on the basis of production centers. Equipment usually brings accounting difficulties because it exists in endless variety and usually has a small item value. Much serves as accessories to larger machines, and frequently most of the equipment is considered part of the machine. When replacements are numerous, this plan raises an accounting problem similar to that of the mechanical equipment of buildings.

The Spare-equipment Problem. Items like electric motors, which from time to time are removed from service and overhauled, present a difficult property accounting problem. Motor replacements tend to confuse the fixed-property records when they occur in substantial number. Frequently the problem can be handled more advantageously by accumulating all motor costs in one account and distributing the charges in proportion to the electric-motor horsepower assigned to various departmental locations on a list kept up to date by the maintenance department. This group treatment of equipment for depreciation and other maintenance costs can also be applied to instruments, pumps, piping, belting, etc.

Depreciation

Depreciation Accountancy. Classification of property into suitable accounts greatly simplifies the problem of depreciation accountancy. Four types of depreciable property accounts are recognized by the Bureau of Internal Revenue, namely, (1) composite, (2) classified, (3) group, and (4) item or unit accounts. The composite account has no subdivisions and requires the use of an over-all rate, as discussed under Buildings and Structures. The classified accounts provide broad subdivisions of the depreciable property such as buildings, machinery and equipment, furniture and fixtures, and transportation equipment. The group account uses classifications as discussed under The Spare-equipment Problem, namely, assets similar in kind and of approximately the same useful life. Item accounts require the depreciation of each unit separately, and this involves considerable accounting detail which often is not warranted. Both the Bureau of Internal Revenue and the Interstate Commerce Commission favor group accounts for depreciation.

In modern depreciation accountancy the first-cost asset value is kept intact on the ledgers, and all depreciation charges are accumulated in separate reserve accounts. Subsequent expenditures that do not affect the life of the asset are considered repairs and are written off as an expense. Small replacements or renewals of machine parts, under the income tax regulations, can be charged to the depreciation reserves, reducing the accumulated depreciation. The effect is to increase the period over which depreciation can be taken; but, as noted under Buildings and Structures, complications creep in when there is a substantial difference between the first cost and the replacement cost. Under such conditions or when the value of the replacement is substantial, the regulations require that the cost of the original part be removed from the asset account and the replacement cost be entered as a capital expenditure. The nature of these charges and the depreciation adjustments involved are shown in Table 5.

Amortization, Depletion, Depreciation, and Obsolescence. These four terms are all concerned with the recovery of capital values through a calculated bookkeeping charge. Amortization relates to bond values, leaseholds, etc.; depletion to mineral or natural resources; depreciation to industrial facilities; and obsolescence to loss caused by chemical engineering and economic progress. Depreciation has been defined through the income tax procedures as a reasonable allowance for the exhaustion, wear, and tear of property used in trade or business, including a reasonable allowance for obsolescence.

Depreciation Rates and Methods. No specific method or depreciation rate is prescribed by the Bureau of Internal Revenue. The fundamental requirement is that the rate be reasonable and the method consistently applied. If a taxpayer's experience with depreciation and obsolescence exceeds the usual concept of useful life, extra allowances can be obtained if the facts can be demonstrated. For guidance as to the general trend in useful life, the Bureau of Internal Revenue has published a pamphlet, Income Tax Depreciation and Obsolescence, *Bull. F.*, Revised January, 1942, obtainable from the Superintendent of Documents at Washington, D.C. From this have been tabulated data related to the chemical industries as shown in Table 4. The base for applying rates usually is first cost less estimated salvage value. Four methods for applying depreciation rates are commonly used, and each of these has a special field of usefulness. The methods are (1) straight-line, (2) reducing-balance, (3) sinking-fund, and (4) production-unit depreciation.

Straight-line Depreciation. This is the most widely used of all the depreciation methods. It is extremely simple to apply. The machinery cost $200,000 and has a useful life of 10 years and an estimated salvage value of $2000. The straight-line depreciation charge is therefore $19,800 annually. Sometimes it is argued that the straight-line method does not fairly state the asset values during the life of the property, but there are so many other variables to the situation that this question is largely academic.

Reducing-balance Depreciation. The particular advantage of the reducing-balance method lies in the fact that, when extraordinary obsolescence is a factor, capital values can be quickly reduced to levels at which sudden obsolescence has only a nominal effect. The rapid reduction in capital values is indicated by the charges for a machine costing $1000 and having a salvage value of $10 and a useful life of 10 years. The depreci-

Table 4. Useful Life for Calculating Depreciation and Obsolescence*

Buildings

With mechanical equipment included, depreciation for a factory building, depending on the construction, ranges 2¼ to 3% per annum. With mechanical equipment excluded, the useful life is factory building 50 years, office 67 years, and warehouse 75 years.

General Equipment

	Years		Years
Air compressors, pumps	20–25	Motors, electric	20
Air conditioning	10–20	Office equipment, composite	15
Automobiles	3–5	Office equipment, mechanical	8
Carpenter shop equipment	20–25	Pipes, brass, copper	Life of building
Construction equipment	10	Pipes, iron	20–25
Cooling towers, spray ponds	15	Plumbing	25
Cranes	20–25	Power, hydroelectric	30–40
Elevators, freight	25	Power, steam-electric	20–25
Elevators, passenger	20	Sewers, sprinklers	Life of building
Fire equipment	20	Switchboards	25
Furniture and fixtures	20	Trailers	10
Heating, boilers, burners, tanks	20	Trucks, gas, outside	4–8
Heating, radiators	25	Trucks, inside	15
Locomotives	30–35	Tank cars	30
Machine-shop equipment	20–25	Water-supply systems	67
Meters	25–30	Wiring systems, electric	20

Machinery, Composite Life in Various Processing Industries

	Years		Years
Acids	15	Glass, container	15
Alkalies	22	Glass, window	17–20
Aniline dyes	20	Laundry	14
Atmospheric nitrogen	15	Oxygen	18
Brewery	20	Packinghouse	17–20
Bottling equipment	13	Paint and varnish	20
Carbide and carbon	15	Petroleum refining	25
Carbonic gas products	16	Pharmaceuticals	20
Cement	20–25	Pulp, soda	20
Chromium products	15	Pulp, sulfate	17
Clay	15–20	Pulp, sulfite	20
Coal-tar products	20	Rayon manufacturing	16
Cottonseed oil	25	Rubber	17
Distilling	15–20	Soap	20
Electrochemical	17	Sugar refining	28–30

Machinery, Chemical Type

	Years		Years
Autoclaves, rubber	10	Filters, wheel	20
Blow cases	2–5	Grinding mills	20
Briquetting machines	18	Heat exchanger	20
Bucket elevators	20	Hydraulic presses	25
Burners, sulfur	10	Ice machines	25
Calendars	20–25	Kettles, nitrating, reducing, jacketed	6
Centrifugals	15–25	Kilns, lime	22
Centrifuges	18	Laboratory equipment	10
Conveyors, belt	15	Mixers, dough	15–20
Conveyors, flight or gravity	20	Packing machinery	15–20
Conveyors, screw	10	Piping, process	2–20
Crushers, gyratory, jaw, roll	12–17	Pulverizers	12
Dryers, vacuum	20	Pumps	10–20
Dust collectors	15	Scales	10–25
Electrolytic cells	15	Screens and sifters	12
Evaporators	15–17	Speed reducers	12
Evaporators, multiple	25–35	Storage, acid	12–20
Filter press	17	Vacuum pans	15

* Tabulated from U.S. Treasury Department, Bureau of Internal Revenue, *Bull. F*, revised January, 1942.

ation is first year \$369, second \$233, third \$147, fourth \$93, fifth \$59, sixth \$37, seventh \$23, eighth \$14, ninth \$9, and tenth \$6. It is not generally realized that the straight-line percentages are hopelessly inadequate for reducing balance depreciation. Instead of using 10 per cent of the first cost for a 10-year life, it is necessary to take 36.9 per cent of the reduced balance. The equal annual factor is calculated for n years of useful life by the following formula:

$$\text{Annual reducing-balance factor} = 1 - \sqrt[n]{\frac{\text{salvage value}}{\text{first cost of asset}}}$$

Sinking-fund Depreciation. This method does not, as might be implied from its title, pay money into a sinking fund but merely uses a sinking-fund formula to calculate the depreciation charge. Only the equal annual sinking-fund payment is charged to costs. The method and its variants are used in the public utility field where the question of a fair return on the fair value of the property depreciated is an important issue. A full discussion of the subject appears in Marston and Agg, "Engineering Valuation," McGraw-Hill, New York, 1936.

Production-unit Depreciation. This method for depreciating property is used mostly by companies that derive their earnings from natural resources. Exhaustion of values in such cases depends not on the passage of time but on the rate at which the resources are worked. There are so many tons of ore in the ground, and each one taken out reduces the value a proportionate amount. The method also applies to tools and dies, which usually have a productive life based on the number of times the tool is used. Here obsolescence is also a factor, however, and the value of the tools item by item often is so small as to raise practical difficulties in applying the production-unit depreciation method.

Retirements. Depreciation rates are only estimates, and the actual life of property may exceed or fall short of the expectancy. When the life exceeds expectancy, the depreciation rate must be reduced. An asset with an estimated life of 20 years, for instance, is found at the end of 15 to have a useful life of at least 25 years, and the rate is therefore reduced from 5 to 2.5 per cent. When a unit is retired before the end of its estimated life because of fire damage or natural cause, the loss must be calculated as shown by Table 5. Similar calculations are required to correct the first cost and the depreciation reserve for replacements or improvements made in a machine, though sometimes it is more practical to deduct the replacement cost from the depreciation reserve and thus in effect extend its useful life. The engineering economy of retiring old equipment is discussed by Grant and Norton, "Depreciation," Ronald, New York, 1949.

Table 5. Loss Calculation for a Scrapped-plant Unit

First cost of unit		\$10,987.58
Depreciation reserves, 4½ years at 15 per cent to be deducted		7,416.62
Depreciated value		3,570.96
Renewals to be added		563.08
Present book value		4,134.04
Value to be realized:		
Salvage	\$ 350.00	
Fixed property for sale	1,925.00	
Total to be deducted		2,275.00
Value to be scrapped		1,859.04
Liquidation expense		750.00
Loss on scrapped-plant unit		\$ 2,609.04

COST-FINDING SYSTEMS

The Cost Problem. Through accounting methods it is not possible to find the exact cost of a product except when the factory output is limited to a single item. Apportionment of the joint costs of manufacture among several products often must follow accounting convenience instead of realities because there is no practical method for precisely ascertaining the facts or for utilizing them when they are available. Contemplation of these difficulties brings the belief that the prime purpose of a cost-finding system should be cost reduction and cost control rather than cost determination. There is also a very practical objective in cost finding, and that is determination of the cost of goods sold, information that is required to close the books of account so that the profit for the period can be established.

Types of Cost Systems. A confusing variety of cost systems is found in the literature of cost accounting. These patterns result from detailed differences in the methods that have been used to handle two fundamental problems of cost accounting. One is a question of finding the most suitable methods for collecting the costs of individual products from among the countless factory charges. The other is a question of what price level and what rate of operations to use as a basis for cost determination. To each of these problems there are two general answers, and thus there are four general cost concepts to which every cost system should be referred in order to establish its salient characteristics. A cost system collects its costs on either a job or a process basis and figures its values on either a standard or an actual basis. These four fundamental conditions, which are varyingly interpreted, define a cost system as to type. A unit-cost system is not a separate type but merely reports costs in greater detail by subdividing either job or process costs into their component operations.

Job-cost Systems. One of the distinguishing characteristics of a job-cost system is the use of serially numbered production orders that authorize performance of the task and also supply a serial number as a means for collecting the cost of the task. Bill Smith works 8 hr. on Job 28892, and the timekeeper notes the fact on the cost ticket. The storekeeper issues 500 lb. of raw material against Job 28892 and sends the requisition to the office where all the factory vouchers against Job 28892 are collected, priced, and listed on a cost sheet so that at conclusion of the work it can be figured that the 18,975 lb. of product made on Job 28892 cost \$0.1783 per pound.

Although the job-cost procedure is usually thought of as an actual method, the work can be priced at either actual or standard cost. The principal objection to the job-cost method stems from the fact that vouchers are lost and charged to the wrong accounts. Also it involves an enormous amount of paperwork. Various interpretations of job-cost procedures seek to overcome these difficulties and often do so successfully. In general, the job-cost procedure belongs in the field of intermittent manufacture.

Process-cost Systems. Continuous manufacturing operations are most conveniently evaluated by means of process-cost systems. Thus, if the manufacture is sulfuric acid, all incoming sulfur is charged to the process and all deliveries of acid are credited to the process. Better control over labor and expense can be maintained if the process is divided into its units, such as burners, towers, and chambers, and the costs for the unit operations are collected separately. With a process-cost system, countless requisitions for insignificant amounts of material or services are avoided. Everything is charged to the process when the voucher or the invoice is rendered.

The simplicity of the process-cost system has led to its adoption for intermittent manufacture in departments producing many products. It is not well suited to these conditions, however, because the problem of allocating charges to products and the need for detailed opening and closing inventories leads to a system with extensive internal records. Then in reality a secondary job-costing system exists within the outer framework of a process-cost system. Additional comments on process costs appear under Product Costs and Variances. Either actual or standard costs can be used in process costing.

Standard-cost Methods. A standard cost is really nothing more than a precise estimate of the values attainable in practice. Frequently, a practical basis for solving many cost problems can be obtained only through a standard-cost approach. There are too many variables to obtain an answer without first assuming a standard cost for some of the unknowns. This is the case with the service departments such as maintenance, power, and building space. The maintenance men occupy building space, do work for the power plant, and also repair buildings. It is a vicious circle that can be broken only by using the standard-cost assumption. Then there are the goods such as dye intermediates, which are used to make other intermediates and also to make dyestuffs. Here again are complications that are easily disposed of by the standard-cost assumption. When a company makes 15,000 or more products, cost determination becomes so burdensome and so subject to mechanical error that standard costs are essential to a practical solution of the problem. See Camman, "Basic Standard Costs," American Institute Publishing Co., 1932. Essentially the argument for standard costs is convenience and a better basis for cost control.

Actual-cost Methods. One of the great advantages of actual cost is the close adherence to the facts that is required by the method. A number of leading chemical companies report costs on an actual basis. They are opposed to standard costs. Some standard-cost procedures go far afield and level the hills and valleys of production on such a long-term base that the costs are essentially meaningless. Actual costs are always to be preferred; but in their development the basic principle can be too strictly interpreted. For instance, if a department has a total processing cost of $10,000 monthly and processes only 200 batches instead of the 250 required to justify the department's economic existence, the significant fact is that there is a loss of $40 per batch from the deficiency of 50 batches in the work load. Actual article costs of practical value in solving marketing problems are not obtained by scattering this $2000 over the 200,000 lb. actually produced through 200 batches. The procedure loads the inventories with a $2000 loss that should be immediately accepted and not deferred. Also the existence of a controllable loss is hidden in the cost figures. The standard-cost method takes a more analytic approach to the problem, but it too has its difficulties. Other aspects of the subject are discussed under Plant Capacity and Burden Rates.

Selection and Design of a Cost System. Each business should have a cost system specially designed to meet local conditions. Before any forms or procedures are designed it is important to study the opportunities in the plant for obtaining accurate information. The facilities for weighing, measuring, and recording the flow of product need to be carefully considered. There are seven major phases in cost accounting, and in each of these an important stage of the accounting is brought to a conclusion. The seven focal points are:

1. Cost charges—collection of classified charges that constitute the total cost exclusive of materials.
2. Departmental costs—distribution of cost charges to departments or processing units, including labor apportionment.
3. Output of departments—collection of data required to credit departments for performance.
4. Raw-material costs—receipt, disbursement, consumption, and inventory of raw material with prices.
5. Work-in-process costs—charging of material, labor, and expense by process or product and crediting of finished work.
6. Finished stock—receipt and shipment of finished stock and inventory with prices.
7. Cost of goods sold—evaluation of sales at cost to obtain data for profit and loss statement.

PAY-ROLL ACCOUNTING AND WAGE CONTROL

Labor-cost Control. In chemicals manufacture labor often is only a small element of cost; but sometimes it is large, and then there is need for cost controls such as are used for material and expense. The problem, however, is more complex. It is generally approached by establishing scientifically the magnitude of a fair labor performance. This requires task analysis, work simplification, and time studies. The available techniques for controlling pay rolls and labor performance are discussed in this section with textbook references.

Organization for Pay-roll Control. A complete program for pay-roll control involves five activities, namely, (1) timekeeping, (2) pay-roll detail, (3) labor performance, (4) labor costs, and (5) personnel relations. In the larger plants these functions may be organized as five departments, timekeeping, pay roll, standards, cost, and personnel. It is common practice to have the personnel and standards department under authority of the works manager and the pay roll, timekeeping, and cost departments under authority of the controller. A small plant is not likely to have a standards department and usually has the other four functions handled by various people in the business office.

Pay-roll Records. A pay-roll accounting system requires five basic records, (1) "in and out" time card, (2) pay-roll record, (3) employee's pay statement, (4) employee's earnings record, and (5) job reports for charging costs. To show that the Fair Labor Standards Act is being observed, an employer must preserve pay-roll records for about 4 years. Details of the various government record requirements are available in publications issued by the Wage and Hour Division, U.S. Department of Labor. Data commonly required are (*a*) name in full, (*b*) home address, (*c*) date of birth if under nineteen, (*d*) occupation in which employed, (*e*) time of day and name of day on which employee's work week begins, (*f*) regular hourly rate of pay, (*g*) basis on which wages are paid, (*h*) hours worked each workday and hours worked each week, (*i*) total daily or weekly straight-time earnings, (*j*) total weekly overtime excess compensation, (*k*) total additions to or deductions from wages paid, (*l*) total wages paid each pay period, and (*m*) date of payment and pay period covered by the payment.

Timekeeping. There are two timekeeping activities, namely, (1) the clocking of an employee's in and out time and (2) the clocking of time spent on individual jobs. The "in and out" time card is generally used for wage computations and as the medium for posting pay-roll records. If a factory uses wage incentives additional information from sources other than the "in and out" time card is required to make up the pay roll. An incentive plan usually requires a detailed record of the individual's daily and weekly performance, and this record is generally convenient as a source for the supplementary wage payment.

Charging Pay Rolls to Costs. Three general patterns are used for translating pay rolls into departmental and product costs, (1) by classification, (2) by job ticket, and (3) by measured work. Combinations of these methods are commonly used. Under the classification method, the pay roll is grouped by departments so that subtotals can be submitted to the cost department, which may allocate to products empirically. There are disadvantages in that transfers of employees often are made in the factory without advising the pay-roll department. The second method uses timekeepers and job tickets in the factory to obtain the time spent on each job. These reports are priced separately at individual wage rates and are reconciled with the pay-roll totals. Disadvantages are (1) the difficulty of obtaining reliable information even when the data are time-stamped and (2) the large amount of paperwork. A simplified version uses the factory labor reports to build a percentage scale for pay-roll allocation to products or jobs. The third method uses the measured time of an incentive plan to charge pay rolls to jobs. A disadvantage is the large amount of effort required to determine the standards for task performance. When labor costs are small the first method is preferable, and when they are proportionately large the third is desirable.

Labor Productivity. Study shows that the output of individuals varies from 50 to 140 per cent of the so-called "normal worker." The performance of an individual also varies during the day to reach a morning and afternoon peak. This subject is fully discussed by Presgrave, "The Dynamics of Time Study," 2d ed., McGraw-Hill, New York, 1945.

Incentive Methods. The great difficulty of controlling the productivity of workers has led to the design of numerous incentive systems. Details of many plans are discussed informatively in "Systems of Wage Payment," National Industrial Conference Board, Inc., New York, 1930. The general pattern of a modern wage plan is carefully to determine task performance and set it as an allowed performance time. A good worker may produce 10 hr. of measured work in an 8-hr. day and is paid at his base rate for 10 hr. work. It has been found that an incentive ranging from 20 to 30 per cent or more is required to bring the average productivity up to desirable levels. An incentive of 25 per cent is commonly offered. Sometimes it is overlooked that workers also must be trained to task performance.

Group and Supervisor Incentives. Group incentives are not desirable when workers who make no real contribution to productivity are included within the group. Supervisor incentives require study of the phases of departmental activity such as cost, output, and quality, which reflect supervisory effort. It is essential that the rewards be based on measurable quantities and not merely on qualities that can be judged only on the basis of opinions. In the chemical field, workers cleaning filter presses or doing similar labor can be placed on ordinary group incentives based on output. Operators responsible for process performance, however, require a reward patterned after a supervisor's incentive.

Task Determination. The key to incentive methods is task determination. Efforts at task determination often are immediately confronted with the fact that manufacturing methods require extensive revision. This usually is so because productive operations were turned over to the workman without adequate study of task requirements. Frequently the plant layout is bad. Sometimes as much as 40 per cent of a man's time is devoted to needless shifting of material. All obvious crudities must be removed before task determination can be undertaken. The four major phases of task determination are (1) analysis of task elements, (2) improvement of methods, (3) stop-watch readings, and (4) adjustment of the watch readings to those for a normal operator. It should be remembered that task determination is not merely a question of output but also requires attention to quality and economy of material consumption.

Time and Motion Study. The purpose of time study is to find out how long it takes to do a job the right way, and this involves study of the elements of the task. These usually are thought of in terms such as get, place, process, and delay. Two types of watches are used, one reading to 0.01 min. and the other to 0.0001 hr., or approximately ⅓ sec. When motions are too fast to be caught by eye, motion pictures are taken and studied frame by frame. Motions are of five types, finger, wrist, forearm, upper arm, and body; and the object is to get the task done with the fewest possible motions. Effort rating, or appraising the speed at which men work, is one of the important phases of time study. This is done on the trained judgment of the observer, who reduces the watch readings to the level of a normal operator. These normal values also include full allowance for fatigue and for the difficulties inherent to working conditions. Two standard texts on the subject are Carroll, "Timestudy for Cost Control," 2d ed., McGraw-Hill, New York, 1943, and Lowry, Maynard, and Stegemerten, "Time and Motion Study," 3d ed., McGraw-Hill, New York, 1940.

Job Specifications. Instruction of employees in their duties often is overlooked. Every job should have written specifications telling what the job is and what kind of man is required. Usually these specifications are set up in conjunction with a program of evaluation devised to fix equitable wage rates. It is convenient, therefore, to set up the personnel requirements in terms of the factors discussed under the next caption. A letter-sized form is commonly used, one side being reserved for job title, department, rate, functions, and the more detailed duties. The other side is used for the personnel requirements and the evaluation factors.

Job Evaluation for Wage Control. The older practice of paying the market for workers and granting them periodic increases leads to unbalanced wage scales. It is becoming standard practice to rate jobs on a wage scale graduated in proportion to the difficulty of performance judged in terms of such factors as skill, mental and physical effort, responsibility, and job conditions. These factors usually are weighted about 50 per cent for skill, 20 per cent for responsibility, and 15 per cent each for effort and for conditions. The scales vary from 500 to 2000 difficulty points, and each point is appraised at an appropriate number of cents per hour. To simplify the wage structure, the scale from lowest to highest pay level is grouped into 10 or more labor grades so that, for example, all jobs falling within the range 880 to 940 difficulty points will carry a rate of 90¢ an hour. A fundamental requirement in job evaluation is uniformity in rating. This can be attained only by persistently checking the value of skill in one job as compared with another. The pamphlet by Kress, "How to Rate Jobs and Men," Factory Management and Maintenance Library, McGraw-Hill, is informative as is also the text by Benge, Burk, and Hay, "Manual of Job Evaluation," Harper, New York, 1941.

Merit Rating. Job evaluation appraises the task and merit rating the man's performance of his tasks. Two methods are used to rate men. One judges personal qualities and the other actual task performance; but the latter is difficult and therefore not much used. Factors used for merit rating include the following: job knowledge, quantity of work, quality of work, adaptability, dependability, and attitude. Merit rating requires training in rating; otherwise the program becomes too much a matter of opinion. A good text on the subject is Halsey, "Making and Using Industrial Service Ratings," Harper, New York, 1944.

Labor Reports to Management. Well-selected labor statistics submitted weekly are helpful in judging the tenor of labor conditions. Data commonly used for the purpose are number on pay roll, number present, lateness, absenteeism, disciplinary action, overtime, work on incentive, productive hours, new employees, employees discharged, employees leaving, labor turnover, sickness, leaves, vacations, and accidents. A departmental basis is desirable in preparing these statistics.

BUDGETING AND EXPENSE ACCOUNTING

Budgetary Principles. A budget is an accounting device used to control accomplishment in relation to a forecast of performance. Although the concept is concerned primarily with expenditures or receipts, it has been applied to many other phases of business. Sales, cash, profit, and production have all come within the field of budgetary control, though often there seems little reason for departing from the older concept of sales quotas, cash requirements, and production schedules. The general pattern of budgetary control is quite simple. It requires (1) a forecast of performance, (2) a record of accomplishment, and (3) a periodic comparison of forecast with accomplishment.

Difficulties often arise when there is a grave disparity between the operating rates used in the forecasts and the operating rates achieved in performance. It permits the department head to observe that the budgets were developed for an operating rate of 65 per cent and that his department certainly needs more money running at 90 per cent of capacity. The flexible budget seeks to close this gap in control by studying the relationship between expenses and operating rates. If the characteristics of the fixed and the variable expenses have been fully established it is possible to tell the department head what his expenses should have been at 90 per cent of capacity. The problem is discussed in detail by Williams, "The Flexible Budget," McGraw-Hill, New York, 1934, and by Gardner, "Variable Budget Control," McGraw-Hill, New York, 1940.

The Budget Forecast, Accomplishment, and Comparison. Budgetary control requires a subdivision of the operations by departments and by expense types. The tools for budgeting thus are (1) an organization chart, (2) a chart of accounts, (3) an expense ledger, (4) a forecast schedule, and (5) a comparison sheet. The forecast can be made on an annual basis, though a quarterly period is preferable because it does not so heavily discount the future. Monthly values must be developed from the forecasts to permit those short-term comparisons which are so essential for the control of expenditures. Figure 3 shows a budget comparison sheet used to advise department heads. The data pertaining to accomplishment are obtained from an expense ledger and the standards from the forecast schedules. Suggestions for suitable accounts are available in the discussions of departmental costs. A system of departmental cost accounting is actually the first step in budgetary control.

Expense Classifications. For the purposes of expense accounting in relation to budgetary control, the word "expense" embraces all departmental outgo except

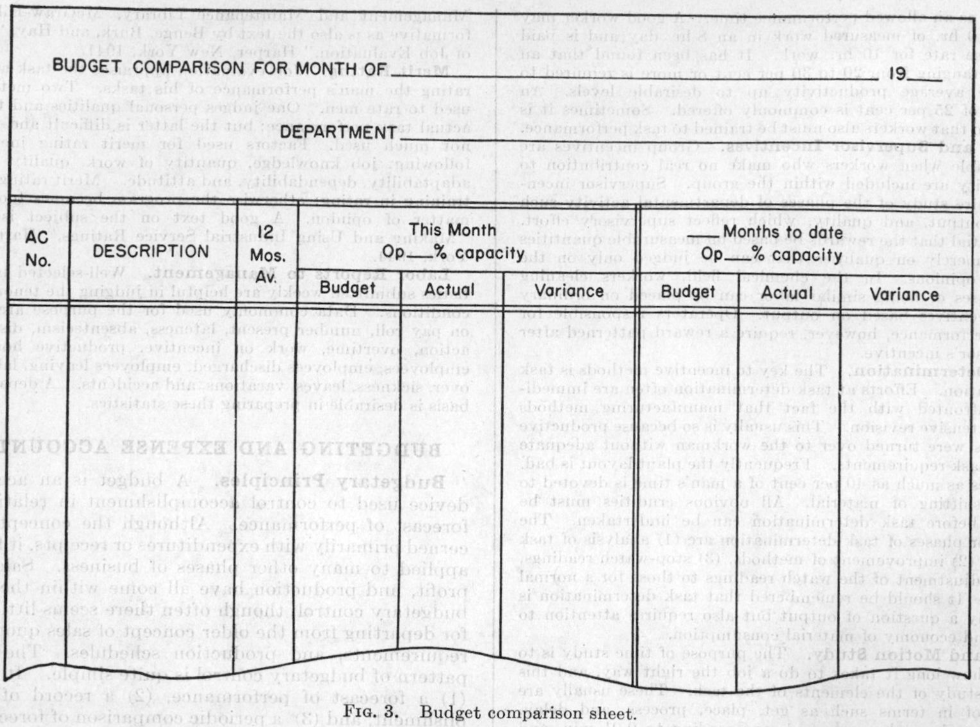
BUDGET COMPARISON FOR MONTH OF ______ 19___
DEPARTMENT ______
AC No.
DESCRIPTION
12 Mos. Av.
This Month Op.___% capacity
Budget
Actual
Variance
___ Months to date Op.___% capacity
Budget
Actual
Variance

FIG. 3. Budget comparison sheet.

materials charged directly to products. Often the pattern of the cost-finding system requires productive labor to be excluded from the expense category. Expenses charged to departments may be classified according to their source as (1) processing cost elements or charges obtained from the general books of account and (2) distributed charges obtained from the operations of service departments. The charges obtained from the general books can be subdivided into four groups, (1) depreciation, (2) expense, (3) supplies, and (4) pay rolls. Frequently it is desirable to divide these further as, for instance, expense into taxes, insurance and other expense or supplies into general supplies, metered supplies, and fuel. A classification of the service charges appears later under that caption.

Depreciation as a Cost Element. The details of depreciation charges and their allocation on the basis of capital values are discussed under Fixed-property Accounting. Today there is a marked trend toward the exclusion of depreciation charges from operating costs. Depreciation is not an out-of-pocket expense but is an estimate of the current cost of a loss that will be sustained when at some future date the equipment can no longer be used because it is worn out. Instead of injecting these estimates into the product costs, some accountants prefer to provide for them by withholding adequate sums for depreciation from the gross profits obtained in profit and loss accountancy. When this procedure is followed it is desirable to evaluate depreciation per unit of product so that provision can be made in the selling prices for depreciation. It is a big item, especially in the chemical industries.

Expenses. Sometimes expenses can be charged directly to departments benefited, but often allocations on the basis of capital or pay-roll values are necessary. Local taxes usually are charged to buildings on a first-cost basis, except the personal component, which may be charged to departments in accordance with machinery values or perhaps to factory overhead and then on a pay roll dollar to departments. Social security and unemployment insurance, usually classified as taxes, may be similarly distributed with other labor overhead charges in ratio of the pay-roll dollar. Insurance has many subdivisions; and some, like boiler, can be charged direct whereas others, like fire, are allocated on a capital-value basis. Many other expenses, such as administrative, office, and selling, can be charged directly to departments.

Supplies. Requisitions are generally used to accumulate supply costs; but this requires considerable paperwork, which sometimes is poorly handled. There are omissions and pricing errors. It is preferable to charge supplies directly to departments insofar as possible by vendor's invoice. The metered supplies, electricity, gas, and water are also called utilities. Distribution of these charges is discussed with the service department charges. Better control over supplies can be maintained through a subsidiary ledger as discussed under Departmental Expense Accounting.

Pay Roll. For the control of departmental costs a single item, pay roll, is inadequate. A subsidiary ledger as discussed under Departmental Expense Accounting can be used to analyze the payroll month by month for each department on the basis of unit operations. Procedures for obtaining these charges are discussed under Pay-roll Accounting and Wage Control.

Departmental Expense Accounting. Three general methods are used for accumulating departmental costs, namely, (1) by voucher register, (2) by cost-sheet distribution, and (3) by expense ledger. The small company finds the voucher-register method convenient. This book of original entry for invoices is ruled so as to provide columns for the various departments, and charges as they occur are posted to the departmental account. At the end of the month the column total gives a departmental cost if all expenditures have been routed through the record. A secondary analysis is required to classify departmental costs by expense types.

The cost-sheet distribution shown in Fig. 4 is desirable because of the completeness with which it presents the interrelations of charges to departments. The method used to prepare the distribution is discussed in Prochazka, "Accounting and Cost Finding for the Chemical Industries," McGraw-Hill, New York, 1928.

new products, and (4) unfruitful effort. The segregation of expenditures so that these four end results can be correctly established requires the extensive use of distribution principles. Table 6 presents distribution methods that are applicable to various departmental activities. Discussions of the procedures appear later under the corresponding captions.

Direct-labor Distributions. Because of its extreme simplicity, distribution of expense on a direct-labor

COST DISTRIBUTION SHEET — MONTH OF JANUARY, 1946

	TOTALS		COST ELEMENTS				SERVICE DISTRIBUTIONS						
	Departmental costs	Total costs	Depreciation	Expense	Supplies	Pay roll	Labor overhead	Factory maintenance	Building space	Power	Transportation	Technical service	General supervision
Production Department A	$14,887		$1,520	$ 380	$ 715	$ 8,562	$ 510	$ 130	$ 780	$ 855	$ 155	$ 880	$ 400
Production Department B	9,659		1,010	216	111	6,222	370	140	300	510	190	290	300
Production Department C	8,920		780	185	92	5,660	340	218	330	460	125	380	350
Production Department D	11,504		1,855	278	1,008	4,490	270	320	980	1,200	203	600	300
Total production		$44,970											
Administrative	$ 4,953		$ 123	$ 1,895	$ 95	$ 2,630	$ 135	......	$ 75	......	......	......	......
Sales	9,580		160	4,560	290	3,830	190		120		$ 180	$ 250	
Office	3,827		96	1,720	130	1,605	96		180				
Warehouse	1,776		73		350	550	33		300	......	320	......	$ 150
Total general overhead		20,136											
Plant replacements		592						$ 592					
Labor overhead	$ 2,281		$ 58	$ 1,588	$ 55	$ 580	......	......	......	......	......	......	......
Factory maintenance	1,435		172	192	95	921	$ 55	......	......	......	......	......	......
Building space	3,551		981	1,522	110	722	43			$ 75	$ 98	......	......
Power, steam, and electric	3,066		585	125	1,390	615	36	......	$ 155	......	110	$ 50	......
Transportation	1,311		122	118	235	789	47	......		......			
Technical service	2,444		194	290	185	1,510	90	......	175	......	......	......	......
General supervision	1,584		28	380	28	980	58	......	110	......	......	......	......
Service distributions							$2,273	$1,400	$3,505	$3,100	$1,381	$2,450	$1,500
Total service costs							2,281	1,435	3,551	3,066	1,311	2,444	1,584
Service variances		63					$ 8	$ 35	$ 46	$ −34	$ −70	$ −6	$ 84
Total costs		$65,761	$7,757	$13,449	$4,889	$39,666							

FIG. 4. Cost-distribution sheet.

The expense ledger accumulates the costs currently as in the voucher register, but a separate page in a ledger is used for each department. The ruling and the captions are similar to those shown in Fig. 4. It is a desirable method for budgetary control because classified accountings can be rendered promptly. The expense ledger is made more informative by supporting it with subsidiary ledgers in which charges for supplies, pay roll, etc., can be explained through classified columnar entries. Thus a pay-roll entry of $48,550 in the expense ledger for Department A would appear in the subsidiary ledger as Supervision $5100, Sulfonation $9850, Precipitation $2280, Centrifuging $4590, Mechanics in the Department $4800, General Labor $4300, etc.

EXPENSE DISTRIBUTIONS

Distribution Principles. Expense distribution requires selection of a suitable divisor for reducing total costs to a unit basis. Ten fundamental quantities commonly used are money, time, task, capacity, pieces, weight, volume, area, length, and power. To make the expense distribution, a record of the total departmental activity is required in terms of the factor and a report on the number of factor units chargeable to each beneficiary of the activity. For instance, there are 100,000 sq. ft. of building space, the total cost is $20,000 or 20¢ a square foot, and Department A is using 5000 sq. ft., Department B 9000, etc. That is the general procedure regardless of the activity or the factor selected.

Distribution in Cost Accounting. The money spent to operate a business produces four end results, (1) a burden of overhead retrieved through sales, (2) new facilities or processes, (3) a processing of materials into

basis is widely used. All that is required is to go over the pay roll and say Smith, Brown, Williams, etc., are productive because they are actually engaged in the manufacture of products. The total of their pay is $5000, departmental expenses are $7500, and thus the burden

Table 6. Expense-distribution Factors

Production departments	To products manufactured
	1. Direct labor dollars or hours
	2. Output, machine-hour, or other capacity factor
General overhead	To products sold per dollar of sales
Administrative	
Sales	
Office	
Warehouse	
Service departments	To other departments
Labor maintenance	Per dollar of pay roll or per man
Factory overhead	Per hour of service, man-hours combined with machine time or separately
Upkeep of grounds	Per square foot of building area, ground, or floor
Building space	Per square foot of space occupied
Power	Per 1000 lb. of steam
	Per kilowatt-hour of electricity
Transportation	Per hour of service, man-hours combined with equipment time or separately
Technical service	Per engineer- or chemist-hour, -day, or -week
General supervision	Departmental capacity, empirically weighted

rate is 150 per cent. Departmental records charge Product X with $1500 of productive labor, and its share of the expense is therefore $2250. There are refinements to the procedure such as using direct-labor hours instead of dollars. The method is not desirable when there is difficulty in defining direct labor and its detailed tasks or if the ratio of expense to labor is high, since this tends to magnify the inherent errors. There is a further difficulty in that many expenses in the chemical industry are not proportional to direct labor; but this can be surmounted by the use of selective burden rates.

Output or Machine-hour Distributions. Departmental expense can be allocated on the basis of factors such as pounds, gallons, batch, or machine-hour. These are all functions of capacity rated at a normal or an attained value. Application of the charges is based on operating statistics and differs from the direct-labor method in that both labor and expense are included in the processing charged to product. If several products are made in the same department, selective burden rates are desirable, since the average unit cost is likely to vary substantially from the cost of processing individual products.

Selective or Multiple Burden Rates. A separate burden rate should be calculated for each dissimilar product manufactured in a department. If the products are very numerous, it may be preferable to divide them into classes and have a separate rate for each class. Each burden rate should take into consideration the extent to which departmental facilities are used in the manufacture of a product. The method applies to either direct labor or output allocations of expense. The purpose is to see that the products are charged correctly for their respective shares of depreciation, expense, supplies, labor, and the various services such as steam, electricity, technical, and supervision. Procedures for controlling multiple burden rates are discussed under Control of Processing Credits.

Plant Capacity and Burden Rates. Selection of an operating rate as a base for burden charges raises debatable questions. Three general solutions for the problem are available: (1) use of the actual operating rate, (2) use of a long-term average or normal operating rate, and (3) use of a short-term predetermined operating rate. Each of these has various advantages and disadvantages, and thus the selection of a suitable base depends to a large extent on the nature of the cost problems.

Use of the actual operating rate as a burden base holds the advantage that there are no variances that must be considered in connection with the costs. The use of an actual rate is to be preferred if the cost problem is of a relatively simple pattern that permits adequate management control on an actual basis. When there are a great number of products each with a great number of cost components, adequate control is not readily maintained except by trying to hold operations to a predetermined rate. This is more easily accomplished by selecting a short-term base than by selecting a long-term normal rate. When the repricing of standards is a burdensome task, however, the long-term base is preferable even though it is less useful for management control of costs. It should be noted that a prime purpose in using predetermined burden rates is to call to management's attention the existence of controllable variances, and the basis for rates should be fixed correspondingly. Thus, if capacity has been set at 85 per cent of the one-shift machine ratings and unusually good business permits operations at close to 100 per cent on a two- or three-shift basis, the burden rates should take such facts into account.

SERVICE DEPARTMENT COSTS

Service Charges. As noted in the discussion Standard-cost Methods, it is necessary to make arbitrary rules for the distribution of service charges in order to end the wheel-within-a-wheel accounting that seeks with perfect justice to charge maintenance to power and power to maintenance and thus on ad infinitum through all the service charges. Service classifications should be selected so as to obtain rational unit charges such as per square foot of building space, because this makes the department heads realize that they are being charged for service rendered and not just being loaded with an incomprehensible 20 or 30 per cent of overhead. The following are representative service classifications:

Labor overhead	Power
Factory maintenance	Transportation
Upkeep of grounds	Technical service
Building space	General supervision

Labor Overhead. The cost of providing and maintaining a staff of employees should be determined, and this can be done on the basis of two units, (1) per worker, and (2) per dollar of pay roll. Under modern industrial conditions these costs are considerable. Among the usual costs are the following:

1. Staff to hire workers and keep records.
2. Equipment, supplies, stationary, and manuals.
3. Space for office, lunchroom, lockers, and first aid.
4. Employees' compensation insurance.
5. Unemployment insurance.
6. Social security.
7. Group insurance for life, health, etc.
8. Pension and retirement plans.
9. Vacations.
10. Medical, examinations, and first aid.
11. State and municipal pay-roll taxes.

Factory Maintenance. In the chemical industry maintenance reaches large proportions and may include a machine shop with power tools, carpenter shop, electrical, pipe, sheet metal, and paint. If these services are not very large in scope it is preferable to use one average maintenance rate that includes power and minor supply items. The statistics reported on maintenance work are not always very accurate, even when a time stamp is used to check a man in and out. A maintenance man usually has little interest in accurately reporting his time to particular jobs, and overrefinement in values is therefore hardly warranted.

Upkeep of Grounds. In an enterprise with little factory maintenance, upkeep of grounds and buildings usually is considered part of factory maintenance. Guards, watchmen, sweepers, porters, and elevator operators are all charged to maintenance. In a large undertaking it is preferable to separate these charges and use the three services, factory maintenance, building space, and upkeep of grounds. Items such as care of lawns, fences, highways, and sewers; snow removal; yard illumination; refuse removal; fire equipment; and watchmen and guards belong under upkeep of grounds; and the total cost can be allocated to building space on a floor-area basis.

Building Space. It is preferable to develop the cost of building space complete with all services, such as janitor, elevator, air conditioning, water, and grounds upkeep, included. The distribution basis is per square foot occupied by a department, and in calculating the charge it is customary to exclude stairways, halls, and general utility aisles. Items entering the cost of providing 1 sq. ft. of building space are as follows:

Depreciation, taxes, insurance	Power—heat, light, water, power for elevator, and air conditioning
Expense—repairs, painting, mortgage interest	Transportation—refuse removal
Supplies—janitor	Factory-maintenance service
Pay roll—janitor, elevator	Upkeep of grounds
Labor overhead	Supervision

Power. Of all the service charges, power is often the most difficult, because many of the facts of its consumption are unknown, and accumulation of data in regard to consumption leads to a maze of detail. In addition to steam and electricity, utilities found in a factory include items such as air conditioning, compressed air, city gas, acetylene, carbon dioxide, refrigeration, hydraulic power, vacuum, and water. The water itself may be available hot, chilled, city water, and well water. Through water meters it is usually possible to obtain data on the consumption of water in various areas and then subdivide the consumption empirically. City gas generally has limited consumption. Most of the other utilities often serve few departments and need not be considered as distributable entities.

Costs of Steam and Electricity. Cost determination for a powerhouse generating by-product electricity raises complex issues. A heat balance is useful in coming to a conclusion as to the fairness of contemplated allocations; but even then there are difficulties in that a B.t.u. of high temperature is worth more than one of low temperature. In calculating the cost of by-product electricity it should be noted that a calculation that charges less than 3413 B.t.u./kw.-hr. is incorrect, this being the mechanical equivalent. The loss in heat content of steam multiplied by the water rate of the turbine gives the number of B.t.u. chargeable to the generation of electric power, and the result is likely to be 4000 to 4500 B.t.u./kw.-hr., depending on the efficiency of the turbine.

Steam Distribution. Although steam is a very costly commodity, ranging as it does from below 30¢ per 1000 lb. to more than $1 depending on the cost of fuel and the efficiency of the powerhouse, usually little provision is made for recording its consumption. Some steam meters may be available, but even then the problem becomes one of making theoretical calculations as to the possibilities for consumption from studies of pipe flow and heat requirements for process use. Considerable steam is required for heating buildings. With a 7½-month heating season and a climate comparable to New York, it can be assumed that about 1 lb. of coal is required to heat 1 cu. ft. of building space, exclusive of hot water. Another figure is 5 gal. oil/sq. ft. radiator/heating season, or about 1 gal./10 cu. ft. Efficiency of the heating plant and construction of the building can throw these estimates wide of the mark.

Electric Distributions. A reasonable apportionment of electric-power costs can be made on a horsepower-hour basis, though a disturbing factor is lack of knowledge of motor loading and hours of running time. In the discussion of the spare-equipment problem under Fixed-Property Accounting, it was noted that there is advantage in accumulating all electrical costs, including replacement charges, and distributing them on a horsepower basis. One of the difficulties in electric-power distribution is consumption for welding and for electric furnaces or processes. If these charges are large, the consumption should be metered. Lighting in a rough way can be figured as 1 kw.-hr./1000 sq. ft. of area/hr. of illumination.

Transportation. A satisfactory basis for distributing transportation costs can be developed from daily reports covering the activities of the department, which may include items such as unloading material, fuel handling, transfer of in-process materials, refuse removal, ash removal, snow removal, maintenance of grounds, and loading shipments. Whether labor should be charged at a differential rate depends on the size of the operation and the type of mechanical equipment. A single hourly rate usually is entirely comparable with the accuracy of the transportation service reports. For both internal and external transportation it is desirable to study routing and procedures, since a great deal of time often is wasted in a planless order of work. Trucking costs involve items such as depreciation, taxes, insurance, drivers, helpers, repairs, tires and tubes, gasoline, oil and grease, garage expense, and supervision.

Technical Service. Subdivision of technical service into chemical laboratory and engineering probably will prove desirable. Both activities are likely to have two general purposes, (1) creation of something new and (2) control of routine operations. Usually it is preferable to segregate these activities. Thus technical service may exist as quite a number of departments. Often it is desirable to make routine control a direct operating department responsibility. Charges for technical service should be placed on an hourly, daily, or weekly basis using the time of productive workers as in any other cost procedure.

General Supervision. Whenever possible supervision should be charged directly as a departmental cost. A certain amount of supervision cannot be treated this way because of its general nature. Properly included with supervision are activities like production planning and control, material control, and general factory overhead. Factory accounting sometimes is also included. Allocation of supervision can be based on a calculated-requirements basis. Some departments, because of the intricate nature of the manufacture, require a disproportionate amount of supervision. These needs can be established by analysis and then used regardless of the departmental operating rate. If a department is idle or runs below capacity it is not equitable to charge other departments for the deficiency. On the other hand, however, when a department runs above capacity it is likely to require a great deal of extra supervision. The problem is quite complex, and some prefer to solve it by charging general supervision in proportion to output (see also Plant Capacity and Burden Rates).

GENERAL OVERHEAD COSTS

General Overhead. Two general-overhead burdens are encountered in a manufacturing plant. One is essentially commercial in nature, and the other is productive. A line of demarcation between these two sometimes is drawn with difficulty. The two differ in that one is deducted from the gross profits derived through the sale of goods and the other is charged to costs and added to the values that are laid aside as inventory. The general character of the productive overhead is discussed under Service Department Costs and also under Expense Distribution. Commercial overhead is usually evaluated per dollar of sales and is divided departmentally as noted below.

Administrative, Sales, Office, and Warehouse. These four general overhead activities sometimes are very large and require subdivision, as, for instance, sales, which may include advertising, sales analysis and research, sales, deliveries, etc. The costs for these departments can be expressed in terms of the same cost elements and service charges discussed under Expense Classifications. The warehouse is sometimes partly productive, various operations of a factory type being conveniently performed there. In metalworking the warehouse cuts steel to production size; in wool combing wool is sorted into grades for processing in the warehouse; and in chemical operations bottling, packaging, and blending are often warehouse assignments. It may at times be desirable to segregate these costs and charge them to products in ratio to the material values; but, as a rule, it is preferable to consider warehousing in all its phases as a commercial operation deductible from the gross profits achieved through the sale of goods.

PRODUCTION DEPARTMENT COSTS

Selection of Cost Units. Although the goal in cost determination is a unit price such as 13¢ per lb. of finished product, determinants other than pounds produced sometimes need consideration. Unit cost depends in part on a relatively constant cost for making ready to produce, in part on processing charges that vary with lot size, and in part on the quality or chemical characteristics of the material placed into process. Production of copper from copper matte, for instance, requires the removal of sulfur, and the cost of the operation depends to no small degree on the amount of sulfur in the matte. A similar situation exists in a drying operation. However, for cost-accounting purposes, generally it is preferable to use a simple convenient unit such as per pound produced. More effective recognition can be given to these other factors by using them to place into production only the most favorable combinations of lot size, quality, etc. A formula for calculating economic lot size is available under Cost Estimating.

Batch Processing. In chemicals manufacture, the batch often is a logical processing unit. This fixes the material placed into process and permits close control of the yield of finished product. The cost of processing a batch or a series of batches should not be based on average departmental costs when dissimilar products are made; for this procedure does not give proper recognition to the amount of depreciation, expense, supplies, labor, steam, electricity, technical service, supervision, etc., required to process each product. A separate batch standard should be developed for each product or product class. Subdivision of batch costs into unit operations usually is not required except when the in-process material is sent to other departments for various finishing operations such as drying or grinding.

Continuous Processing. Control over continuous operations generally can best be maintained by studying separately the cost of the various unit operations such as filtration, evaporation, and centrifuging. Each unit operation should be costed on the basis of the most rational factor, which may be per pound of solid removed, per pound of water evaporated, or per pound of salt finished. When such a cost basis is used for control purposes, it is important to check the accuracy of the data against the process formula. If this is not done, it may be discovered that the cost of evaporation is low only because it is based on more water than is available in the process.

Production Department Costs. There are two fundamental patterns for accumulating production department costs. One includes productive labor, and the other excludes it because the accounting procedure requires that the productive labor be charged directly to products and be used as a basis for charging departmental expense to product. When this method is used the departmental cost elements shown in Fig. 4 should include only the indirect labor, *i.e.*, the portion not directly chargeable to manufacture of product.

Control of Processing Credits. By using a processing ledger of the same format as the expense ledger discussed under Departmental Expense Accounting it is possible to control closely the value of the constituents of the burden rates such as depreciation, expense, supplies, pay roll, maintenance, technical service, power, and supervision. This is particularly important when selective or multiple burden rates are used. Through the process ledger it is possible to obtain in a department a variance for each cost element and each service charge and by an analysis of these values to arrive at a detailed understanding of the true costs of production. Procedures for a process ledger of this type are discussed in Prochazka, "Accounting and Cost Finding for the Chemical Industries," McGraw-Hill, New York, 1928.

PRODUCT COSTS, VARIANCES, AND YIELDS

Charges to Work in Process. Product costs customarily are accumulated on work-in-process cost sheets, and these usually are columnar forms suitable for analysis of the charges made monthly for items such as material, labor, and expense. Figure 9 suggests the general pattern. Ordinarily a separate cost sheet is used for each product or job; but sometimes it is preferable to use separate sheets for the material phase, as in foundry work, where the natural basis is material for a class of products and production costs by process or product. From this viewpoint there are four fundamental classifications of work-in-process cost sheets:

1. Job costs—a separate sheet for each product or job, and each sheet has all charges for material, labor, and expense.
2. Process costs—a separate sheet for each process, and each sheet has all charges for material, labor, and expense.
3. Class costs for materials with separate sheets for labor and expense on a product basis.
4. Class costs for materials with separate sheets for labor and expense on a process basis.

Credits to Work in Process. Credits to work in process can be made on either an actual- or a standard-cost basis and need not be on the same basis as that used for charging costs. One procedure throws the variances into raw material or process and the other into finished-product yields. The decision rests on whether it is desirable to throw the variances in one or the other direction or whether any variances at all are desired. Credits to work in process are given for one or more of the following items: finished stock, component materials, class products, by-products, spoiled work, seconds, reworks, and processing services or variances.

Finished Stock. This term is generally used to denote goods that are ready for market. If the finished items are held after removal from process as a warehouse inventory, a ledger sheet covering each product is required; and, if there are not too many items, this can be patterned after Fig. 9, using appropriate entries. Determination of cost of goods sold is an important objective in cost accounting. Details of the cost of goods sold can be developed in the finished stock ledger or can be tabulated separately as desired.

Component Materials. Some articles are manufactured not for market but for use in the manufacture of other products. From an accounting viewpoint they partake of the characteristics of both raw and finished materials, and it may be desirable to bracket them with either of these groups or to give them a separate classification titled component parts, component materials, or intermediates.

Class Products. The distinguishing feature of class products lies in the fact that they are related through a voluntary association. The producer can choose the particular ceramics to be fired in a kiln or the castings poured on a single day or the stampings plated in the same bath. It is not necessary that class products be like in kind, though in cost accounting products of similar characteristics often are bracketed as a class. In class costing, material, labor, and expense may be joint or only the material or only the processing. The usual distribution principles are used for allocations. Class costing often is desirable because paperwork is thus greatly reduced.

By-products. If a manufacturing process causes a splitting up of raw material into several products, the incidental items are called by-products, and the major revenue producers are joint or coproducts. The refining of petroleum and the slaughtering of animals are two well-known by-product manufacturers. Sometimes within a limited degree it is possible to vary the ratios of the end products.

By-product accounting is a question of the economics of the marketplace. The problem is one of pricing so that the group as a whole will yield the greatest net profit, and that usually is attained by converting each item into its most valuable form. For instance, cottonseed oil placed in the soap kettle has a low value, and this may be enhanced by converting the oil into a cooking fat. Unit process costs are helpful in determining whether it is profitable to dispose of a by-product at the splitting off stage as a salable waste or to carry it on to a higher end point. The various methods of accounting for by-products are discussed at length by Lawrence, "Cost Accounting," Prentice-Hall, New York, 1940. It is desirable when a by-product is split off to carry its costs separately and include a fair portion of the joint cost as determined by the earning capacity of the by-product at various advantageous price levels.

Second-quality Products and Spoiled Work. Goods of second quality can be either sold or reworked. If a proper market can be found, disposal as second quality is preferable as a rule; but often there is no option other than to rework the product. Costs of rework should be charged to the product costs as the true cost of 1000 lb. is the cost of 900 lb. first quality, plus the cost of making and reworking 100 lb. The problem of pricing second-quality goods is akin to that of pricing by-products, the determinant being the greatest possible return from the first and second quality considered as a unit. Spoiled work should be credited at cost and should not be ignored, because otherwise its value will appear as a large and unexplained work-in-process variance or as unnecessarily high actual cost.

Variance Accounting. There are three general procedures for disposing of cost variances obtained when standard cost procedures are used, (1) charge them against cost of goods sold, (2) charge them against profits, and (3) attempt to adjust the costs so as to eliminate the variances. The more usual procedure is to charge them against the cost of goods sold, because it is argued that the variances are due to manufacturing operations and should be borne by them. It is important to clear variances out of all accounts promptly, inasmuch as their effect is to conceal profits or losses. Variances are due to differences in price, rate of operations, efficiency, quantities used, quality, and other causes. They are found in the material and processing accounts, and their causes should be traced in detail because this leads to close operating control. The sources of variances in manufacturing costs have been diagrammed in detail by Camman, "Basic Standard Costs," American Institute Publishing Co., 1932.

Yields. One of the most important considerations in the manufacture of chemicals is the yield of finished product obtained from a specified amount of raw material. Yield can be calculated as either a percentage of the raw material placed in process or as a percentage of the amount that theoretically can be produced from a given amount of raw material. The latter method is preferable; because, when yield is referred to a molecular weight, adjustments are made for the quality of the raw and the final product, and the entire calculation is placed on a scientific basis. Accurate yield determination requires exact factory measurements. This is difficult if too short a factory run is used as a yield base.

MATERIAL ACCOUNTING AND CONTROL

Materials Defined. To be classified as a material, the item under consideration should add an appreciable amount to the cost of the finished product, and its consumption should be in direct proportion to the rate of production. It is not essential that the material itself be present in the finished goods. From this viewpoint it is possible to include as direct materials some of the expensive supply items and to exclude minor raw materials.

Material Control. Modern material-accounting procedures require a centralized material control, and usually the group in charge of this activity is associated with production planning and control. It is a function of the group to verify usage of materials against a standard-material bill. This cannot be done if weights and measures are not carefully tended from receipt to shipment. In addition to the many technical controls for quality, materials may be quantitatively controlled in connection with any of the following 10 transactions:

1. By the seller when he ships goods.
2. By the warehouse when goods are received.
3. By the warehouse when goods are delivered for process use.
4. By the production department when goods are received.
5. By the production department when goods are put into process.
6. By the production department when goods are removed from process.
7. By the production department when goods are delivered to the warehouse.
8. By the warehouse when goods are received from a production department.
9. By the warehouse when goods are shipped.
10. By the customer when he receives goods.

Stock Sheets. From a controlling viewpoint there are two kinds of materials, (1) package goods and (2) bulk goods. Figures 5 and 6 show two types of stock sheets, one for package the other for bulk goods. A much closer check is possible with package goods, because each container, being serially numbered, can be recorded as having been shipped to a particular customer or as having been processed in a particular batch. The delivery can be noted on the same horizontal line as the receipt. For bulk goods a similar control, but on a larger scale, can be effectuated by segregating shipments.

Raw-material Consumption. Requisitions used for the withdrawal of materials from the warehouse are frequently used as vouchers for raw-material consumption; nevertheless they are inadequate for the verification of material usage, which requires a rather detailed analysis. Figure 7 shows a long-established method for tabulating the consumption of raw materials on a batch-process basis. Figure 8 shows what is required to recon-

STOCK RECORD																	
RECEIVED								DELIVERED									
Date	Identification Nos.			From	Gross	Tare	Net	Date	Order No.	To	Gross	Tare	Net	Error		Remarks

FIG. 5. Bulk-goods stock sheet.

STOCK RECORD										
RECEIVED					BALANCE ON HAND	DELIVERED				Remarks
Date	Identification Nos.		From	Quantity		Date	Order #	To	Quantity	

FIG. 6. Bulk-goods stock sheet.

TABULATION OF RAW MATERIAL CONSUMPTION — Article No. 1

Date	Lot No.	RM 1	RM 5	RM 11	RM 3	RM 7	RM 18	RM 20	RM 15	RM 6	RM 12	RM 4	RM 19	RM 5		OUTPUT FINISHED STOCK
12-1426	103	200	825	65	277	150	225	323	281	4600	1800	22	307	950		216
16	104	200	827	65	277	150	225	323	185	4600	1800	21	415	959		205
19	105	200	855	65	277	157	225	323	95	4800	1800	22	394	1909		421
														1188		219
21	106	200	745	65	277	140	225	323	100	4800	1800	65	1116	1138		205
23	107	200	846	65	277	125	225	323	661	18800	7200	21	345			
									45	4800	1800	21	380	1232		206
27	108	200	836	65	277	140	225	1615	175	5400	1800	18	340	1072		196
29	109	200	854	65	277	145	1350	323								
1231-26		1400	5788	455	1939	1007	225	323	201	5400	1800	23	375	1125		207
1-3-27	110	200	757	65	257	140	225	323	63	5400	1800	18	355	1137		189
5	111	200	873	65	277	140	225	323	130	5400	1800	18	450	1070		200
7	112	200	807	65	277	135	225	323	160	5600	1800	20	385	1090		203
10	113	200	824	65	277	135	225	323	135	5500	1800	21	400	1085		212
12	114	200	846	65	277	135	225	323	185	5700	1800	19	380	1035		201
14	115	200	805	65	277	135	225	323	165	5700	1800	21	395	1046		212
17	116	200	820	65	277	145	225	323	169	5700	1800	20	360	1032		198
19	117	200	832	65	277	140	225	323	155	5700	1800	22	410	1011		214
		1600	6564	520	2196	1105	2025	3230	1583	60300	19800	242	4575	14261		2662
Cost		3000	12352	975	4135	2112	3375	4845	2244	79100	27000	307	5691	16170		3083
21	118	200	854	65	277	140	225	323	165	5700	1800	21	409	0		0
												21	409			
24	119	200	769	65	249	140	225	323	115	5700	1800					
26	120	200	850	65	277	140	225	323	130	5400	3600					
									410	16800						
28	121	200	818	65	277	145	225	323								
31	121	200	790	65	257	140	900	1292								
131-27		1000	4081	325	1337	705										

FIG. 7. Tabulating raw-material consumption.

RAW MATERIAL CONSUMPTION DATA — *Raw Material No. 1*

Req'd by	Date	Rec'd.	Error+	Inven'y	Total	Inven'y	Error-	Returns	Used	Used	Work in Process	Price	Amount
Dept. C	*1-3-37*	*500*		*179*		*560*	*19*			*1600*	*Article No. 1-110-117*	*25*	*400 00*
	7	*500*							*2600*	*1000*	*" " 1 118-122*	*25*	*250 00*
	12	*500*											
	17	*500*											
	21	*500*											
	26	*500*											
Dept. D	*1-3-27*	*1000*					*12*	*38*		*850*	*Article No. 32-570-571*	*25*	*212 50*
	10	*1000*								*1400*	*" " 33-308-311*	*25*	*350 00*
	17	*1000*								*200*	*" " 40-79-80*	*25*	*50 00*
									2950	*500*	*" " 51-792-801*	*25*	*125 00*
		6000		*179*	*6179*	*560*	*31*	*38*	*5550*	*5550*	*Total R.M. No. 1*		*1387 50*

FIG. 8. Checking raw-material consumption.

cile the quantities withdrawn from the warehouse on requisitions with the quantities that the tabulation indicates have been put into process.

Raw-material Pricing. If standard costs are used for pricing materials the transition from actual to standard can be made either when the materials are received or when they are put into process. The latter procedure is

RAW MATERIAL LEDGER

		JANUARY			FEB
		Quantity	Price	Amount	Quan
Debits					
Inventory, initial	1	2179	.25	544 75	
Purchase invoices	2	10000		2450 00	
" "	2				
Factory handling	3			11 70	
Received without invoice	4				
In transit previous month	5				
Accounting transfer	6				
Surplus warehouse	7				
" factory	7				
" miscellaneous	7				
" unit price	7			38 30	
" price adjust.	7				
Totals	8	12179		3044 75	
Credits					
Work in process consum	9	5550	.25	1387 50	
" " " "	9				
Used non-productively	10				
Cost of material sold	11				
Purchase credits	11				
No invoice previous month	12				
In transit	13				
Accounting transfer	14				
Deficit, warehouse	15				
" factory	15	31	.25	7 75	
" miscellaneous	15				
" unit price	15				
" price adjust.	15				
Inventory, final	16	6598	.25	1649 50	
Totals	8	12179		3044 75	

FIG. 9. Material-ledger entries.

preferable if the materials are not too numerous in that it does not unnecessarily anticipate profits or losses. If the task of recasting product-cost standards is not excessive, the raw-material-cost standards can be based on relatively short-term price movements. The price base should include all transportations charges and handling costs to the stage laid down in the warehouse ready for withdrawal. All costs from that point on usually are charged to current production. The price base also should allow for such shrinkage and spoilage as normally attend the use of a raw material.

Material Ledgers and Records. A ledger form convenient for a monthly accounting of each raw material is illustrated by Fig. 9. In addition to the stock, tabulating, and check sheets previously shown, many other material records are required. Receipt of materials usually is reported on serially numbered tickets with copies for accounting, procurement, material control, and the warehouse. This ticket can also be used to clear the goods through inspection, or a separate form can be used for laboratory control. Procurement procedures usually involve a purchase requisition, an inquiry form, a purchase order, and a purchase change order. Withdrawal of material for processing generally is based on a requisition and returns to the warehouse on a material-credit slip.

PRODUCTION CONTROL AND PRODUCTION LINES

Production Planning and Control. Substantial operating economies can be obtained through a well-designed system for production planning and control. The major purpose is to get customers' orders out on time, and this may involve working to special orders or to shelf-inventory requirements. Systems are specially designed, because the production control needs vary from plant to plant. In a large factory the system may reach formidable proportions; nevertheless, only five fundamentals are involved:

1. Planning or determining what is to be done and how.
2. Scheduling or setting the timetables for production.
3. Dispatching or sending off operations on schedule.
4. Follow-up or searching out laggard operations that require special attention.
5. Records or posting of performance conditions and results.

Production-control Records. Extensive and varied record systems are used for production control. Some of the forms commonly used are production orders, schedules, route sheets, operation sheets, move tickets, display charts, and control boards. A tray file often is used to show the progress of factory operations, either through a progressive movement of the tickets or by keeping the tickets stationary in serial order and posting operation status daily. Two types of graphic charts are convenient for production control. Figure 10 illustrates a chart that is desirable for recording equipment loads and also for other purposes discussed by Clark, "The Gantt Chart," Ronald, New York, 1922. Figure 11 illustrates the cumulative chart which is so convenient for controlling production against forecasts, sales, and inventories, both warehouse and in process. Each upward step indicates a new batch put into process.

Production Lines. For package-filling operations and other tasks that require considerable handling, the production-line procedure often permits substantial economies. This type of

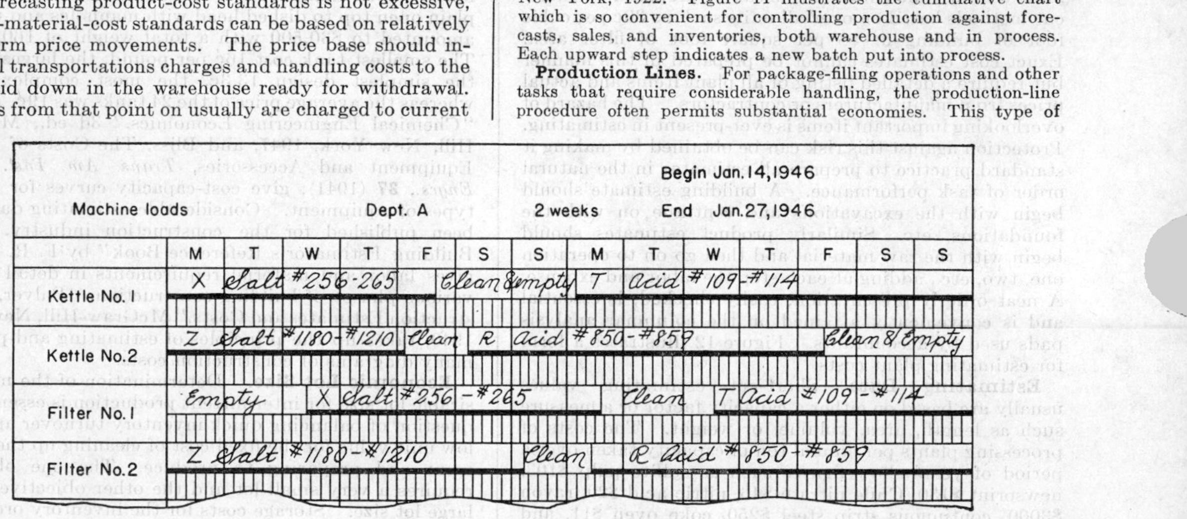

FIG. 10. Gantt chart.

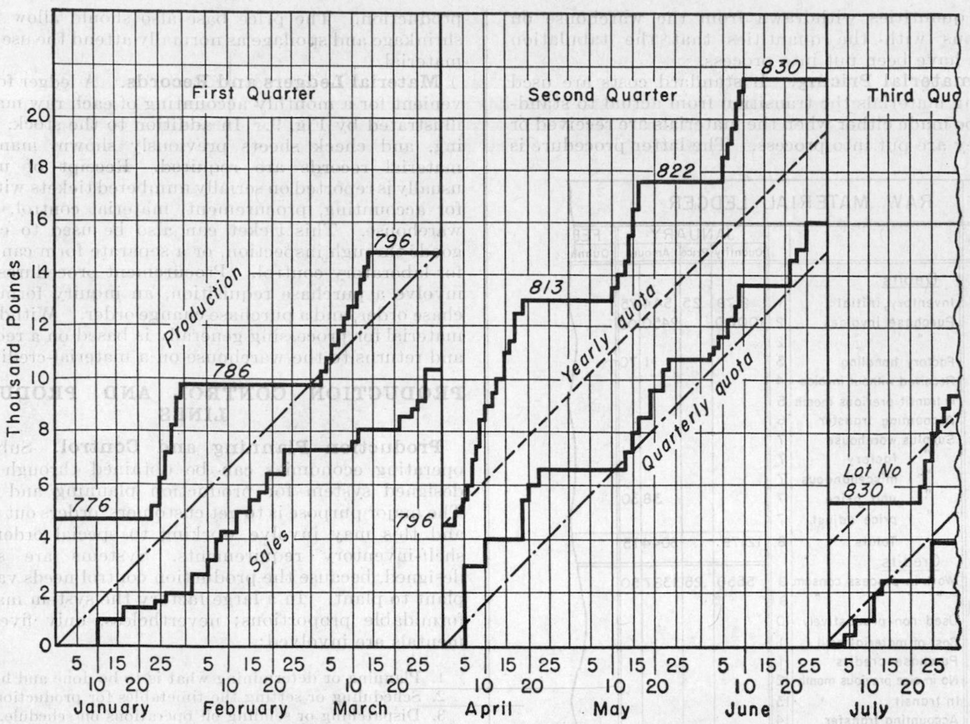

FIG. 11. Cumulative production control chart.

work has been highly publicized through automobile assembly lines; but it has other applications fully discussed by Muther, "Production Line Technique," McGraw-Hill, New York, 1944. The fundamental principle is to determine the work load by time study and to balance the tasks among sufficient operators to secure the desired volume. Movement of the material can be either intermittent by hand or continuous by conveyors.

COST ESTIMATING

Estimating Principles. Cost estimates are constantly required in business. Sometimes an approximate cost is sufficient, and this can be prepared by short-cut methods based on over-all units such as 30¢ per cubic foot of building or $8 per square foot of filter area. Exact cost estimates cannot be prepared in this manner but require a detailed listing of all plant items and actual prices from manufacturers or contractors. The hazard of overlooking important items is ever-present in estimating. Protection against this risk can be obtained by making it standard practice to prepare all estimates in the natural order of task performance. A building estimate should begin with the excavations and continue on with the foundations, etc. Similarly product estimates should begin with the raw material and then go on to operation one, two, etc., adding at each step the labor and expense. A neat orderly presentation, easily checked, is essential and is conveniently attained on the columnar analysis pads used by accountants. Figure 12 illustrates a form for estimating plant costs.

Estimating Data. Short-cut estimating values usually are based on either a capacity factor or a measure such as length, area, volume, or weight. The costs of processing plants per annual ton of capacity taken over a period of years show figures such as sulfite pulp $103, newsprint $130, Chile nitrate $46, nitric acid $20, rayon $3000, continuous strip steel $250, coke oven $11, and steam-electric power $100 per kilowatt of capacity. Various weight figures are heavy processing machinery 25¢ to 50¢ per pound, precision machinery $1 to $2 per pound, automobile 20¢, refrigerator 40¢, structural steel fabricated 8¢. These values give a rough idea of capital requirements, but they are dangerous because of great variations among jobs. In a single year, for instance, building costs ranged from 16¢ to $1 per cubic foot for various types of construction. A tank order placed early in 1949 for 24 tanks welded from 3/8-in. and 1/2-in. steel plate, ranging in size from 2 ft. diameter by 3 ft. long to 7 ft. diameter by 30 ft. long and varying from plain open top to dished head with manholes and nozzles, amounted to $30,500 with a total weight of 160,000 lb. The smallest tank cost 16¢ per pound, the largest 15.3¢, the simplest design 13.3¢, the most complex 29.3¢, whereas the average price of the 24 tanks was 19¢. Tyler, "Chemical Engineering Economics," 3d ed., McGraw-Hill, New York, 1947, and Bliss, The Costs of Process Equipment and Accessories, *Trans. Am. Inst. Chem. Engrs.*, **37** (1941), give cost-capacity curves for various types of equipment. Considerable estimating data have been published for the construction industry. "The Building Estimator's Reference Book" by F. R. Walker gives labor and material requirements in detail for the various phases of building construction. Pulver, "Construction Estimates and Costs," McGraw-Hill, New York, 1940, takes up the principles of estimating and presents many diagrams of construction costs.

Economic Lot Size. Determination of the most desirable lot size for intermittent production is essentially a question of balancing quick inventory turnover against a low unit value for the fixed cost of cleaning up the equipment and preparing to produce. The one objective requires a very small lot and the other objective a very large lot size. Storage costs for the inventory ordinarily are not a factor in the computation. The subject is fully

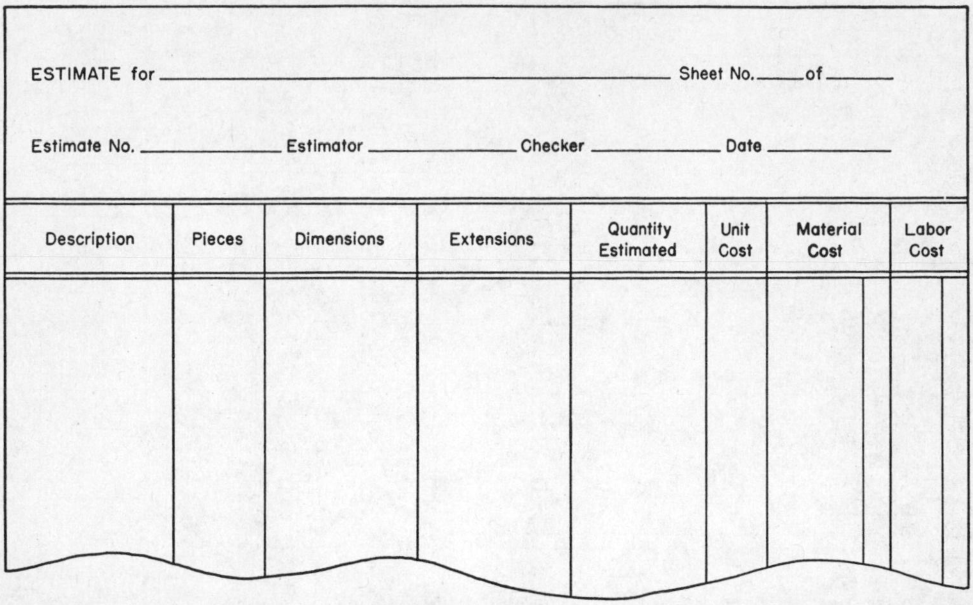

ESTIMATE for Sheet No. of
Estimate No. Estimator Checker Date
Description
Pieces
Dimensions
Extensions
Quantity Estimated
Unit Cost
Material Cost
Labor Cost

FIG. 12. Estimating form.

discussed by Raymond, "Quantity and Economy in Manufacture," McGraw-Hill, New York, 1931, with details of all the factors that sometimes need be considered. Ordinarily the following formula is adequate:

$$\text{Economic lot size, lb.} = \frac{i}{R}\sqrt{\frac{2SA}{Ui}}$$

where i is the annual interest rate in decimals, R is the desired capital return in decimals, S is the constant cost in dollars for setting up the production lot, A is the annual sales in pounds, and U is the unit-product cost in dollars.

Profit Margins. In estimating profits it is desirable to calculate the return on both total cost and processing cost in order to ensure adherence to a sound economic position. A profit of 25 per cent on a product having a total cost of $1 yields a profit of 50 per cent on the processing charges when the processing cost is 50¢ and a profit of 100 per cent when the processing cost is only 25¢. If these material-processing cost relationships are not studied, there may be a tendency to under- or overprice a product when abnormally low or high material values are encountered (see Churchill, "Pricing for Profit," Macmillan, New York, 1932). Profit margins often are figured gross, *i.e.*, prior to deduction of the general overhead burden of commercial charges such as administrative, sales, office, and warehouse.

Turnover on Capital. The return on investment can be figured as a function of turnover on capital by using the following formulas:

Cost of goods sold = factory cost (material, labor, expense) (1)

Gross profit = sales − cost of goods sold (2)

Net profit = gross profit − general overhead (administrative, sales, office, warehouse) (3)

Per cent profit on sales = (net profit ÷ sales) × 100 (4)

Total investment = fixed investment (land, buildings, plant) + working capital (inventories, accounts payable, cash) (5)

Turnover = sales ÷ total investment (6)

Return on investment = turnover × profit on sales (7)

Interest and Contingencies. Interest on capital is not an item to be included in a cost estimate except when the interest covers the charge for capital borrowed during a construction period. A man who uses capital for manufacturing receives profits, not interest. This concept is discussed from several viewpoints by Blocker, "Cost Accounting," McGraw-Hill, New York, 1940. No provision for unforeseen contingencies should be included in a competitive estimate. An estimator is supposed to provide for all job requirements through itemized lists; and, when all details have been considered, there should be no remaining contingencies that need coverage through an arbitrary percentage added to the estimate

SECTION 30

SAFETY AND FIRE PROTECTION*

BY

H. L. Miner, Manager, Safety and Fire Protection Division, Service Department, E. I. du Pont de Nemours & Co.; Past President, Member of Board of Directors, National Fire Protection Association; Past Vice-President, Member of Executive Board of National Safety Council, Past General Chairman, Chemical Section of National Safety Council; Member American Society of Mechanical Engineers, Army Ordnance Association, International Association of Fire Chiefs, Executive Committee of National Fire Waste Council, Safety Code Correlating Committee and its Executive Committee of American Standards Association, American Society of Safety Engineers and Veterans of Safety, Board of Directors of Delaware Safety Council; Chairman or Member of national association committees related to safety and fire protection.

E. J. Meyers, E. E., Assistant Manager, Safety and Fire Protection Division, Service Department, E. I. du Pont de Nemours & Co.; Member Board of Directors of Delaware Safety Council; Chairman of Committee on Static Electricity, National Fire Protection Association; Member, Committee on Protection against Lightning, National Fire Protection Association; Member, Committee on Special Extinguishing Systems, National Fire Protection Association.

C. R. Groves, Ch. E., Engineer, Safety and Fire Protection Division, Service Department, E. I. du Pont de Nemours & Co.; Member, American Society of Safety Engineers.

CONTENTS

	Page
Introduction	1848
Legislation	1848
The Industrial Plant	1848
Location	1848
Arrangement of Buildings	1848
Water Supplies	1849
Fire Mains	1849
Hydrants and Hose	1850
Fire-alarm Equipment	1851
Buildings	1851
Type of Construction	1851
Outside Walls	1851
Horizontal Cutoffs	1851
Vertical Cutoffs	1851
Elevators	1851
Exits	1852
Lighting	1852
Color in Industry	1853
Ventilation	1853
Sanitation	1853
Automatic Sprinkler Systems	1854
Inside Hose	1854
First-aid Fire Extinguishers	1854
Equipment	1854
Type	1854
Location	1854
Guarding	1855
Processes	1864
Closed Systems	1864
Flammable Liquids	1864
Combustion	1866
Apparent Ignition Temperature	1866
Flash Point	1866
Flash Fire	1868
Explosive Range	1868
Vapor or Dust Explosion	1868
Explosion or Detonation	1868
Pressure Rupture	1868
Spontaneous Combustion	1868
Decomposition	1868
Location and Arrangement of Storage Tanks for Flammable Liquids	1868
Testing Devices for Flammable Vapors	1869
Static Electricity	1869
Compressed Gases in Cylinders	1869
Dust Explosions	1869
Special Safety Protection Equipment	1870
Goggles and Helmets	1870
Respiratory Protection	1870
Special Fire Protection Equipment	1870
Foam (Two-solution System)	1870
Foam (Dry-powder Generators)	1870
Carbon Dioxide	1870
Steam	1871
Auxiliary Sprinkler Systems	1871
Other Inert Gases	1871
Organization and Work	1871
Safety and Fire Protection Organization	1871
Employee Selection and Training	1871
Investigation and Classification of Industrial Injuries	1871
Housekeeping	1872
Maintenance	1872
Watchman Service	1872
Plant Fire Brigade	1872
Mechanical Tables and Data for the Safety and Fire Protection Engineer	1873
First Aid	1875
Shock	1875
Sunstroke	1875
Fainting	1875
Sprains	1875
Wounds	1875
Burns and Scalds	1875
Acid and Alkali Burns	1875
Gas Poisoning	1875
Poison Antidotes	1876
Artificial Respiration	1878
Arterial Bleeding	1878
Bibliography	1879

* This section was originally prepared by H. L. Miner, P. V. Tilden, and G. H. Miller, all then of the Safety and Fire Protection Division, Service Department, E. I. du Pont de Nemours & Co.

SAFETY AND FIRE PROTECTION

INTRODUCTION

Accident prevention and fire protection have come to be so much a part of good business, particularly in recent years, that at least certain fundamental phases of these subjects have a logical place in a handbook of this nature. It is recognized, however, that space will not permit more than a brief treatment of them, and students of safety and fire protection are referred to the two national organizations: the National Safety Council, 20 North Wacker Drive, Chicago, Ill., and the National Fire Protection Association, 60 Batterymarch Street, Boston, Mass., as well as to other organizations and societies, some of which are included in the Bibliography.

Safety and fire protection begin with the design of plants—buildings, equipment, and processes. Because careful attention has been given these matters in the past, the majority of present-day accidents and fires result from what is commonly known as human causes. Plans should always be examined by someone competent to decide upon their safety and fire protection aspects. The object is to remove, as far as possible, the human factor from the hazards by making protection permanent or automatic.

For references to more detailed information on items related to safety and fire protection, refer to pp. 1879 to 1884.

LEGISLATION

Most states and most cities have regulations and ordinances governing safety and fire protection measures for buildings, equipment, and processes; and it is always necessary to examine these to make certain that their provisions are being followed. If the information desired cannot be obtained from the local authorities, the two national organizations previously mentioned might be of some assistance in procuring copies of these regulations and ordinances. Many states now have compensation laws which require that compensation insurance be carried and further require certain reports when accidents do occur. These provisions must of course be followed.

THE INDUSTRIAL PLANT

Location. Many factors, such as raw-material supplies, transportation facilities, and labor supply, must be considered in locating industrial plants. From the standpoint of safety and fire protection, it is important, particularly where corrosive, poisonous, or explosive materials are to be handled or produced, to make certain that the location of the operation does not constitute undue danger to adjacent life and property. State laws and municipal ordinances should be consulted and satisfied.

Arrangement of Buildings. Buildings will first of all be arranged to permit economic production, since it is usually possible to provide safety and fire protection under any given arrangement through type of construction, by fire walls, barricades, etc. However, an effort should always be made to limit the height of factory buildings, to provide good separation—50 ft. or more between important units—and most important, to minimize conflagration areas by definite fire stops. Buildings housing hazardous occupancies should be segregated and isolated or barricaded to prevent life loss and/or extensive property damage. Buildings in which explosives are handled or stored should be separated and barricaded in accordance with the "American Table of Distances," published by The Institute of Makers of Explosives, 103 Park Avenue, New York, N.Y. See Table 1.

The distance to be maintained between regular plant-operating buildings depends largely upon the type of construction; the area and height; importance of the unit

Table 1. American Table of Distances*

Blasting and electric blasting caps		Other explosives		Inhabited building barricaded,† ft.	Public railway barricaded,† ft.	Public highway barricaded,† ft.
No. over	No. not over	Lb. over	Lb. not over			
1,000	5,000	....	....	15	10	5
5,000	10,000	....	....	30	20	10
10,000	20,000	....	....	60	35	18
20,000	25,000	....	50	73	45	23
25,000	50,000	50	100	120	70	35
50,000	100,000	100	200	180	110	55
100,000	150,000	200	300	260	155	75
150,000	200,000	300	400	320	190	95
200,000	250,000	400	500	360	215	110
250,000	300,000	500	600	400	240	120
300,000	350,000	600	700	430	260	130
350,000	400,000	700	800	460	275	140
400,000	450,000	800	900	490	295	150
450,000	500,000	900	1,000	510	305	155
500,000	750,000	1,000	1,500	530	320	160
750,000	1,000,000	1,500	2,000	600	360	180
1,000,000	1,500,000	2,000	3,000	650	390	195
1,500,000	2,000,000	3,000	4,000	710	425	210
2,000,000	2,500,000	4,000	5,000	750	450	225
2,500,000	3,000,000	5,000	6,000	780	470	235
3,000,000	3,500,000	6,000	7,000	805	485	245
3,500,000	4,000,000	7,000	8,000	830	500	250
4,000,000	4,500,000	8,000	9,000	850	510	255
4,500,000	5,000,000	9,000	10,000	870	520	260
5,000,000	7,500,000	10,000	15,000	890	535	265
7,500,000	10,000,000	15,000	20,000	975	585	290
10,000,000	12,500,000	20,000	25,000	1,055	635	315
12,500,000	15,000,000	25,000	30,000	1,130	680	340
15,000,000	17,500,000	30,000	35,000	1,205	725	360
17,500,000	20,000,000	35,000	40,000	1,275	765	380
		40,000	45,000	1,340	805	400
		45,000	50,000	1,400	840	420
		50,000	55,000	1,460	875	440
		55,000	60,000	1,515	910	455
		60,000	65,000	1,565	940	470
		65,000	70,000	1,610	970	485
		70,000	75,000	1,655	995	500
		75,000	80,000	1,695	1,020	510
		80,000	85,000	1,730	1,040	520
		85,000	90,000	1,760	1,060	530
		90,000	95,000	1,790	1,075	540
		95,000	100,000	1,815	1,090	545
		100,000	125,000	1,835	1,100	550
		125,000	150,000	1,900	1,140	570
		150,000	175,000	1,965	1,180	590
		175,000	200,000	2,030	1,220	610
		200,000	225,000	2,095	1,260	630
		225,000	250,000	2,155	1,295	650
		250,000	275,000	2,215	1,330	670
		275,000	300,000	2,275	1,365	690
		300,000	325,000	2,335	1,400	705
		325,000	350,000	2,390	1,435	720
		350,000	375,000	2,445	1,470	735
		375,000	400,000	2,500	1,500	750
		400,000	425,000	2,555	1,530	765
		425,000	450,000	2,605	1,560	780
		450,000	475,000	2,655	1,590	795
		475,000	500,000	2,705	1,620	810

* By permission of The Institute of Makers of Explosives.

† Barricaded, as here used, signifies that the building containing explosives is screened from other buildings, railways, or from highways by either natural or artificial barriers. Where such barriers do not exist, the distances should be doubled.

to continued plant production; hazard of the processes; the land available on the chosen site for present manufacture and for future expansion; values involved and the available protection. In addition, the extent of the exposure from outside properties not under the control of the plant will affect the distance to be decided upon between that property and the nearest plant buildings.

The following minimum *overhead clearances* should be provided for footways, roads, and railroads:

Table 2. Overhead Clearances

Projects	Clearance	Measured above
Floors, platforms, stairs, and other regular routes of foot travel	6′ 6″	Footway
Roadways—main thoroughfares	15′ 0″	Road
Other roadways	12′ 0″	Road
Broad-gage tracks	22′ 0″	Top of rails
Narrow-gage tracks outside buildings	10′ 6″	Top of rails
Narrow-gage tracks where passing through barricades	6′ 6″	Top of rails
Narrow-gage tracks inside buildings	6′ 6″	Platform of cars, lorries, and other conveyances

Where present installations of pipe lines, barricades, doorways, valve handles, belt-shifter levers, etc., cannot be rearranged to secure this clearance, their presence should be clearly indicated from both directions of approach.

Side clearance, railroads, and trams are measured from center to center of tracks, or from center of tracks to nearest point of structure.

Table 3. Side Clearances

Projects	Broad gage	Narrow gage with rolling stock, 6 ft. wide	Narrow gage with rolling stock, 4 ft. wide
Between parallel lines of track of same gage*	13′ 0″	8′ 6″	6′ 6″
Between parallel lines of broad- and narrow-gage track*	....	11′ 0″	10′ 0″
From tracks to buildings and structures other than loading and unloading platforms	7′ 6″	5′ 6″	4′ 6″

* Where tracks are on a curve, allowance should be made for projection of car bodies over track, so that a clearance equivalent to the above is maintained under all conditions of use.

Side clearances for loading and unloading platforms where locomotives are operated are to be measured from center of track to nearest point of platform or structure:

Broad-gage track	6′ 6″
Narrow-gage track with rolling stock, 6 ft. wide	4′ 6″
Narrow-gage track with rolling stock, 4 ft. wide	3′ 6″

On narrow-gage tracks where cars are pushed by hand, it is permissible to bring platforms for loading or unloading, raised transfer tracks, etc., as close as possible to car platforms. As far as practicable, this is to be avoided where cars are operated by locomotives of any description. In all such cases, clearance between car platform and structure, if below the standard, is to be as small as possible and is not to be over 2 in., unless varying width of cars makes this impossible. If platform is above floor of car, front of platform is to be sheathed down to 6 in. below car-floor level. If side clearance from structures, etc., is less than the standard on tracks on which locomotives are operated, signs, "Caution—Insufficient Clearance" (18 by 12 in., yellow with black letters) should be posted 72 in. above the rails (and at the standard side clearance) at each end of the obstruction or group of obstructions. The distance away from the obstruction at which these signs are to be posted is to be governed by speed conditions. For medium speeds a distance of 30 ft. is recommended.

Table 4. Minimum Width for Footways, Doors, Runways, Etc.

Regular routes of foot travel, in	36
Passage between machines, in	24
Safety exits and doorways, in	28
Gangplanks and runways, in	30
Runways for two-wheel hand trucks (allowing two trucks to pass), in	60

Water supplies for plant fire protection should be available from at least two independent sources. The **primary supply** which maintains pressure on the mains should be automatic and of good pressure but may be of limited capacity. One or more connections from a reliable public water system or a gravity tank of 100,000 gal. capacity elevated on a 125-ft. trestle provides a good primary supply. Pressure should be adequate to give at least 15 lb./sq. in. flowing pressure on highest sprinkler heads, or in their absence at least 65 lb./sq. in. on yard hydrants. This should be augmented by a **secondary supply.** One or more fire pumps located in an isolated or properly protected fire-resistive pump house with an unlimited suction supply forms a most satisfactory secondary source. Where rivers, ponds, lakes or similar natural water supply are not available, concrete reservoirs or suction tanks of at least 100,000 gal. capacity for each 1000 gal./min. pumping capacity are necessary. With such reservoirs, adequate filling connections should be provided. A reliable power supply for operating the pumps, whether steam or electric, is of utmost importance and should be given serious consideration, particularly with the view of preventing its interruption during a serious fire. Where a public fire department is available, steamer connections are provided on the plant mains so that city pumpers or fire boats can discharge water at high pressure into plant mains.

Fire Mains. Where the plant is of sufficient size, an independent system of underground fire mains supplying sprinkler systems and yard hydrants should be provided. These mains should not supply water for domestic purposes except in the event of an emergency. The size of the mains should depend upon the size of the system; but, in general, mains should not be smaller than 8 in. where hydrants as well as sprinkler systems are supplied. The branches to individual hydrants should be not smaller than 6 in. (For friction losses in mains, see Flow of Fluids section.) Because of its much longer life in underground service, cast-iron pipe Class C or D, American Water Works Association Standard (130 or 173 lb.) or class 250 (250 lb. working pressure) centrifugal cast-iron pipe (see A.S.A. Standard A-21.2–1939) should be used. Where the ground contains corrosive substances, it is necessary to protect the outside of the pipe, including the joints, by a special treatment. The pipe should first be thoroughly cleaned of all loose and fixed mill scale and should be cleaned from all grease and foreign matter. A primary coat of pipe-line enamel is then applied with a brush. Next, one or more coats of hot coal-tar pitch are applied according to the manufacturer's directions. Immediately following the application of the last hot coat and before it has set up, the pipe shall be spirally wound with asbestos pipe-line felt or fabric. The edges of the felt must be overlapped, and hot coating shall be used to seal the laps as needed. All joints must be carefully protected in a similar manner. Care must be exercised in laying underground pipe so that foreign materials such as stones or pieces of wood are not allowed to enter the pipe. The depth of the mains below the ground surface should be sufficient to prevent the water in them from freezing in the coldest weather experienced in the locality. Further information can be obtained from the "Regulation for Outside Protection" published by the National Board of Fire Underwriters, 85 John St., New York, N. Y.

In order to provide flexibility and to facilitate repairs, sectional control valves are installed in the mains. Check valves are necessary to prevent backflow of water where pressures of the different supply lines vary. State laws must be complied with, and special precautions must be taken where there is a cross connection between public water supplies and a fire-main system which is supplied by non-potable water.

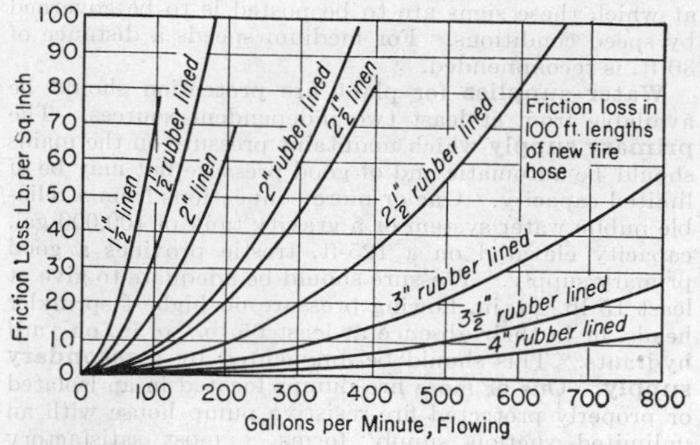

FIG. 1. Friction loss in fire hose. Curves based on the following assumptions: (1) 14 lb. friction loss per 100 ft. of 2.5-in. cotton rubber-lined hose with 250 gal./min. flowing (smooth rubber lining). Where lining is rough or old, friction losses may be 50 to 100 per cent higher. (2) Friction loss varies directly with the square of the quantity flowing. (3) Friction loss varies inversely as the fifth power of the diameters. (4) Friction loss in linen hose is taken from data in "Fire Engine Tests and Fire Stream Tables," published by the National Board of Fire Underwriters.

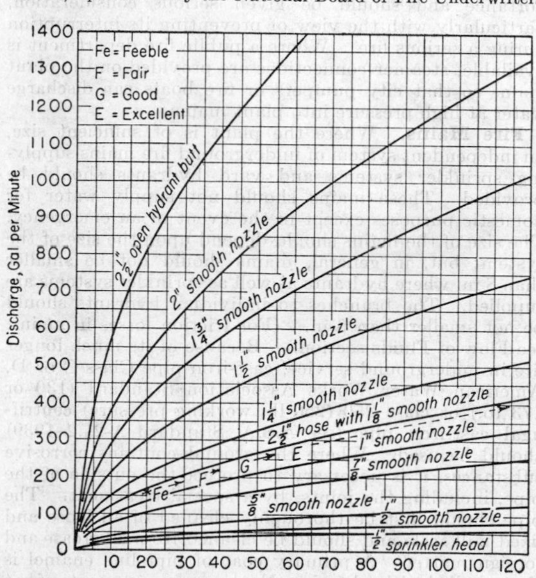

FIG. 2. Discharge of water through smooth nozzles and sprinkler heads. Assumed coefficients of discharge (0.99 to 0.997) based on experiments by Freeman, *Trans. Am. Soc. Civil Eng.*, **21, 24** (1889). Sprinkler-head data from tests by Underwriters' Laboratories. The 2½-in. open-hydrant butt-curve data from Crosby, Fiske, Forster, "Handbook of Fire Protection," p. 713, Table 18, Van Nostrand, 1924—two-way Mathews Hydrant. The 2-, 1¾-, 1½-, 1¼-, 1⅛-, and 1-in. nozzle curve data from Crosby, Fiske, Forster, p. 698, Table 12. The ⅞-, ⅝-, and ½-in. nozzle curve data from "Fire Stream Tables" (pamphlet by Factory Mutuals, Boston). The ½-in. sprinkler-head curve data from Crosby, Fiske, Forster, p. 717.

Hydrants and Hose. Hydrants are specially designed to be substantial, to provide an unobstructed waterway from the mains to the outlets, and are easily operated and arranged so that when the hydrant is shut a drain at the base opens to allow the water in the hydrant barrel to drain, thus preventing freezing. Hydrants are usually located so that at least two streams are available for each building without the use of more than 150 ft. of hose. This generally necessitates locating hydrants not more than 300 ft. apart; but where areas are congested, it may be necessary to place them much nearer together. Hydrants are placed not closer than 40 ft. to the buildings which are to be protected. The threads on the hydrant outlets and the threads on the hose should be uniform throughout the plant and should conform to the public standard in the locality. Where possible, threads conforming to the National (American) Standard Fire Hose Coupling Screw Threads should be used (see National Fire Protection Association publication "National Standard Fire Hose Couplings").

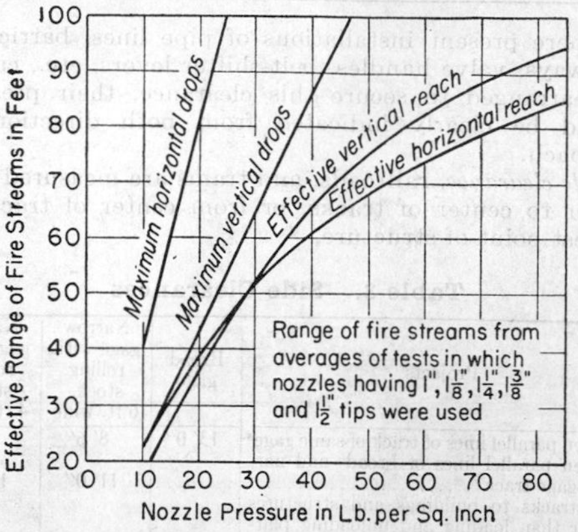

FIG. 3. Range of fire streams. Nozzle pressures as indicated by pitot tube. Data based on experiments by Freeman, *Trans. Am. Soc. Civil Eng.*, **21**, 303 (1889).

Standard single-jacketed, cotton, rubber-lined mill-yard hose is preferred for outside fire protection on industrial plants, because of its flexibility and relatively low friction loss; 2½-in. hose provided with 1⅛-in. nozzles is considered minimum for outside protection. See the tables on Friction Loss in Fire Hose; Discharge of Water through Smooth Nozzles and Sprinkler Heads; and Range of Fire Streams.

To save valuable time at the start of a fire, the hose is ordinarily connected directly to the hydrant and is kept in hose houses or in hose boxes at the hydrants. These also contain miscellaneous equipment. This hose is supplemented by a reserve supply on hose reels conveniently located throughout the plant yard.

Table 5. Minimum Safe Distance between Hose Nozzles and High-tension Wires

Voltage	Safe distance 1⅛-in. nozzle, ft.		Safe distance 1½-in. nozzle, ft.	
	Fresh water	Salt water	Fresh water	Salt water
1,100	6	25	9	30
2,200	11	25	16	30
6,600	19	30	29	35
11,000	20	30	30	35
22,000	25	30	33	40
33,000	30	35	40	45

Table 5* shows what is considered to be the minimum distance between hose nozzles and high-tension wires, under which there may be danger of passage of current to employees holding the hose line.

* By permission of the National Fire Protection Association.

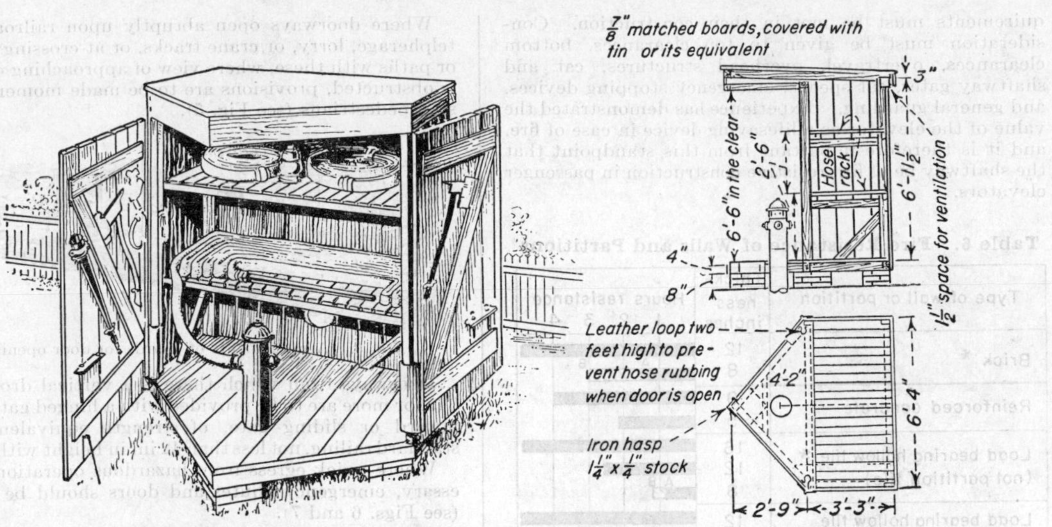

FIG. 4. Underwriters' hose house and equipment. (*National Fire Protection Association.*)

Fire-alarm Equipment. Private fire-alarm systems are necessary on practically all plants for promptly notifying the fire brigade and for exit of employees. These systems should operate on normally closed, electrically supervised circuits. Fire-alarm boxes are of three types:

1. Interfering.
2. Positive non-interfering.
3. Positive non-interfering and successive.

The **interfering** type of box has no ability to control its circuit once it has been started in operation; and hence, if more than one box is pulled at or about the same time, a mixed, a confused, or a totally multilated signal will result.

The **positive non-interfering** type of box is so arranged that should two or more be pulled at the same time a definite, accurate signal will be received from one of the stations.

The **positive non-interfering and successive** type of box is non-interfering as just described but has the added feature that after the box selecting control of the circuit has completed its signal the other box or boxes will automatically take control, one after the other; hence the term successive.

Fire-alarm boxes should be installed in conspicuous locations and mounted so that accidental operation will not be caused by vibration or jarring. When provided inside buildings they should preferably be located near exits. Boxes located outside the buildings should be well detached from the buildings, and the location should be designated at night by a red light. Gongs may be located throughout manufacturing buildings, automatically coded so that they ring the number of the box pulled. The plant whistle may be automatically or manually operated. Recording punch registers provide visual as well as permanent records of all signals passing over the fire alarm system. It is good practice to operate a fire-alarm system from storage batteries installed in duplicate sets, one set operating the system while the other set is in reserve or is being charged. The system should be tested daily by actually pulling at least one box. Every box on the system should be operated once a month.

BUILDINGS

Type of Construction. Permanent factory buildings should not be of frame construction except where necessary to meet manufacturing conditions. Noncombustible construction or fire-resistive construction is to be preferred. For further details of construction, consult the "Building Code" recommended by the National Board of Fire Underwriters.

Outside Walls (Buildings Other Than of Frame Construction). Where distance between buildings is limited by uncontrollable factors, and the various buildings expose one another, (1) Door openings should be protected with standard automatic fire doors. (2) Windows should be of wired glass in approved steel sash without ventilating sections or with these sections arranged to close automatically at time of fire. Under severe exposure, fire shutters or cornice sprinklers are advisable. (3) Fire walls should be parapeted to prevent spread of fire through roofs.

Horizontal Cutoffs. Where floor areas are large, or where combustible construction is involved, fire-wall cutoffs should be provided. These cutoffs should be standard fire walls of at least 13 in. of brick or of equal construction, with all door openings properly protected with single or double automatic fire doors. All other necessary openings through fire walls, such as belt, shaft, or conveyor openings, should be suitably guarded to prevent spread of fire. The fire walls, where roofs are combustible, should extend through the roofs and should be parapeted to a height of 3 ft. Where side walls of the building are combustible, the fire walls should be extended out several feet beyond the exterior walls.

All new interior partitions should preferably be fire-resistive or at least of non-combustible construction.

Table 6 shows the relative fire resistances of various types of wall and partition construction.

Vertical Cutoffs. In multiple-story buildings, each floor should preferably be entirely cut off from the other floors. Where it is necessary for stairways, elevators, or dumbwaiters to extend through floors, enclosures should be provided at least equivalent to the existing floor construction of the building. All openings in these enclosures should be equipped with automatic fire doors.

All floors should be maintained in a good state of repair, and when interior changes are made, which in any way affect vertical cutoffs, proper regard for the spread of fire should be considered.

Elevators. Elevators are a frequent source of accidents, and experience has shown that certain safety requirements must be met in their construction. Consideration must be given to top clearances, bottom clearances, overtravel, overhead structures, car and shaftway gates, car speeds, emergency stopping devices, and general guarding. Experience has demonstrated the value of the elevator as a lifesaving device in case of fire, and it is therefore important from this standpoint that the shaftway be of fire-resistive construction in passenger elevators.

Table 6. Fire Resistance of Walls and Partitions*

Type of wall or partition	Thickness inches	Hours resistance 1 2 3 4
Brick *	12 8	 A B
Reinforced concrete	9 5	
Load bearing hollow tile * (not partition tile)	16 12 8	 A B
Load bearing hollow tile plastered both sides	12 8	
Concrete block	8	C D E
Cement plaster, metal lath and frame	2	
Gypsum block	4	F G
Cement plaster, metal lath both sides, wood frame	7/8	
Cement plaster, metal lath one side wood frame	3/4	

* Three-quarter-inch cement or gypsum plastering on one side of a brick or tile wall will increase the fire resistance by 1 hr.

Fire-resistance Limits (see chart above):

A. When combustible members come into wall.
B. When framing is noncombustible.
C. For bearing walls due to characteristic splitting along center of wall.
D. For non-bearing wall.
E. For cinder concrete, load-bearing, or non-load-bearing wall.
F. For unreinforced partition.
G. For partition reinforced with steel rods in joints.

(*Compiled from test data of Bureau of Standards and Underwriters' Laboratories.*)

Safety requirements for elevators, hoists, dumbwaiters, and escalators are fully covered in the "Safety Code for Elevators," A.S.A. Standard A17.1, to which reference should be made for detailed information.

Exits. Although the following covers the subject in a general way, reference should be made to the "Building Exit Code," the National Fire Protection Association, for information in greater detail.

It is important that exits be of ample capacity to accommodate safely and properly the employees in case of fire and panic and to enable them to reach a place of safety outside of the building in which they are working. Every factory or working space should have at least two means of egress from each story. These should be so located and constructed that the occupants will use them if a fire occurs, as a matter of course—preferentially.

In planning exits for new and old buildings, such factors as the following should be thoroughly considered: (*a*) number and location; (*b*) types; (*c*) construction characteristics; (*d*) maintenance; (*e*) drills.

As far as practicable, doors should not be placed so as to open directly upon moving machinery, electrical equipment, open containers of hot or corrosive liquids, pits, railroad or tram tracks.

Where doorways open abruptly upon railroad, tram, telpherage, lorry, or crane tracks, or at crossing of routes or paths with these, where view of approaching cars, etc., is obstructed, provisions are to be made momentarily to stop pedestrians (see Fig. 5).

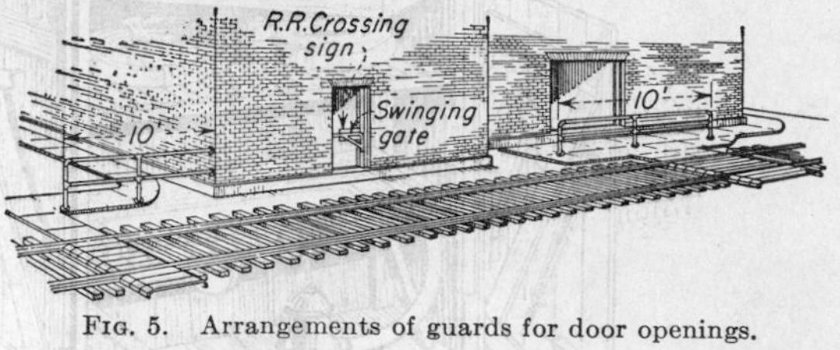

FIG. 5. Arrangements of guards for door openings.

Doorways from which there is a vertical drop of 4 ft. 0 in. or more are to be provided with a hinged gate to open inward or sliding gate, of strength equivalent to the standard railing, not less than 42 in. in height with midrail.

Where quick egress from hazardous operations is necessary, emergency chutes and doors should be installed (see Figs. 6 and 7).

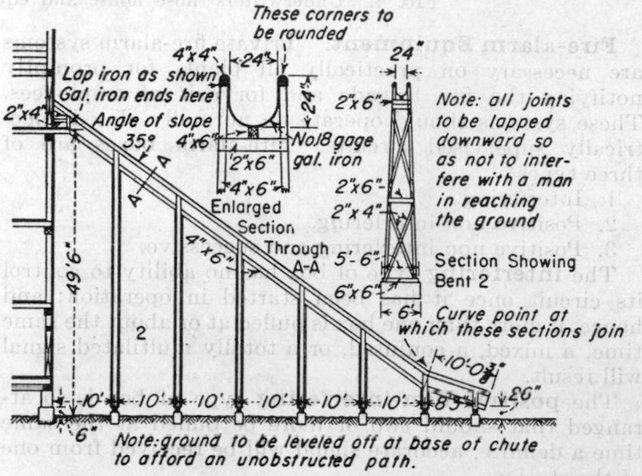

FIG. 6. Emergency exit chute.

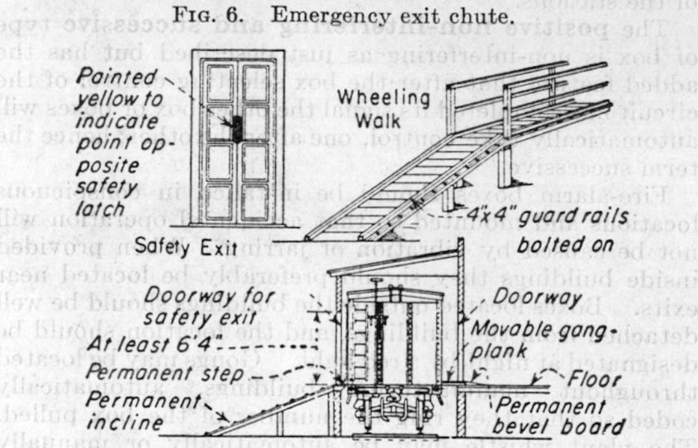

FIG. 7. Arrangements for exits, gangplanks, and wheeling walks.

Lighting (see also pp. 1755*ff.*). Certain factors which have been found to affect the eyes must be considered in the design and installation of lighting sources. Among these are reflection, absorption, distribution, diffusion, intensity, steadiness, color, and glare. Among the controlling practical factors are height, spacing, and size of lamps and type of reflectors or globes.

Adequate daylight illumination, properly applied, is the ideal light. Light from above is generally better than light from side windows only. Skylights and monitor windows should be provided wherever practicable. Large window areas, equipped when necessary with awnings, window shades or blinds, and diffusive or refractive glass, together with light interiors, are desirable.

Artificial lighting, which may be divided into four systems, direct, indirect, semi-indirect, and local, requires investigation as to the nature of the work for which lighting is to be provided. First of all, the intensity of illumination, which is expressed in foot-candles, for the various types of work and for the various locations in buildings, should be determined.

Table 7. Recommended Intensities of Illumination

	Recommended Ft.-candles Measured 30 In. above the Floor or Ground
Roadways and yard thoroughfares	0.2
Storage space; aisles, stairways, and passageways	5
Where discrimination of detail is not essential	10
Where slight discrimination of detail is essential	20
Where moderate discrimination of detail is essential	30
Where close discrimination of detail is essential	50
Where discrimination of minute detail is essential	100

Where work involving discrimination of fine details is involved a combination of at least 20 ft.-candles of general lighting plus specialized supplementary lighting is necessary. The supplementary lighting will vary with the conditions of contrast in the materials being handled or inspected. For complete information and recommendations for lighting in various industries, see the A.S.A. Standard A-11-1942 prepared by the Illuminating Engineering Society.

Color in Industry. The proper application of color in painting walls, floors, ceilings, machines and machine parts, piping, etc., has a definite and useful function in modern industry. Interior lighting can be materially improved, for instance, by the proper use of various colors, and eyestrain can be reduced by scientific blending of backgrounds. Hazardous machine parts, dangerous materials, traffic hazards, warning signs, and protective equipment should all be designated by a standard Safety Color Code. Several leading paint manufacturers have adopted the color code and can offer assistance in industrial painting and color identification. The following table outlines the basic colors and their application, as recognized by leading safety authorities, including the National Safety Council:

Color—Name	Application
High-visibility yellow	Curbs, aisles, low beams, stairway approaches, floor elevation changes, etc.
Alert orange	Interior surfaces of electric switch boxes, fuse boxes, machine guards, etc. Exposed parts of pulleys, gears rollers, cutting devices, etc.
Safety green	To identify first-aid rooms, stretchers, cabinets for gas masks, safety showers, etc.
Fire protection red	Fire hydrants and valves, sprinkler system piping, fire alarms, location of fire extinguishers and firefighting equipment
Precaution blue	To identify equipment or apparatus which should not be used, moved, or started, such as electrical controls, railroad cars, valves, elevator gates, ovens, tanks, and dryers
Traffic white (including gray and black)	Traffic control, aisle markings, storage areas, corners, waste receptacles, etc.

For further information and details on the standard color codes recommended for identification of piping systems, see American Standard A-13.

Ventilation. In considering the subject of ventilation, the following factors must be taken into account: amount and distribution of air supply, temperature, humidity, motion, velocity, odors, dust, bacteria, and toxic or flammable vapors.

Blower and exhaust systems may constitute fire hazards in themselves or may introduce hazards contributing to the causes and the spread of fires, even though they are properly located, installed, and safeguarded. Where practicable the ducts of such systems should not pierce floors, fire walls, etc., which would aid in the spread of fire. If such installations are necessary, proper protective devices should be provided at these cutoffs.

The two main divisions of the ventilating problem are: first, proper ventilation for a specified number of people in a confined space; second, the removal of harmful or dangerous substances in the air.

A deficiency of oxygen or high humidity in a room where people are working will result in excessive fatigue on the part of the occupants. This can be overcome by supplying the proper amount of fresh air. The following recommendations will serve as examples:

Table 8. Air Requirements

Project	Renewals of Air per Hour
Work shops	10–15
General offices	4– 5
Engine rooms	10–20
Boiler rooms	10–20
Stairways and halls	1– 2
Mills (light work)	4– 6

Whether the air is supplied by windows, a forced-draft system, a suction system, or a combination of these, depends on the conditions involved.

If harmful or flammable dusts or vapors are present in the air, it is essential to keep them below the harmful or hazardous concentration for the particular dust or vapor in question. State and Federal health codes should be consulted where they apply.

The means of ventilating an enclosure where harmful or flammable dusts or vapors are present will usually consist of a hood with a suction fan placed in the vortex constructed over the source of the dusts or vapors, to keep them from escaping into the room. Where this is not possible, ventilating ducts are distributed throughout the room, usually with the suction duct being near the floor when the vapors are heavier than air and near the ceiling when lighter than air. See "Regulations for the Installation of Blower and Exhaust Systems for Dust, Stock and Vapor Removal," National Board of Fire Underwriters, also *Safe Practices Pamphlets* 32 and 37, National Safety Council.

Sanitation. Change houses with lockers, washrooms, lunchrooms, and toilet facilities adequate for all employees should be provided. Hygiene among workers is an important factor in combating health hazards and makes for efficiency. Change rooms should be separated or located away from the working operations. Shower baths should be provided for occupations where workmen necessarily become dirty or are exposed to poisonous dust, fumes, or vapors. In some operations it is necessary to provide special working clothes and also to supervise and to provide washing facilities for this clothing.

To ensure comfortable and desirable conditions, it is essential to have all locker rooms, lunchrooms, showers, and toilets adequately heated, ventilated, and illuminated. Cleanliness is necessary in these places, and attention should be given to their maintenance.

A drinking-water system with modern bubbler fountains located at convenient places should be installed throughout the plant. When water is taken from a city drinking-water supply, it can generally be assumed that the water is satisfactory and will meet health codes; but if taken from other sources it is essential to make frequent chemical and bacteriological analyses. The water should be kept at a constant temperature, not too cold, by use of individual refrigerated fountains. These will be found to be more satisfactory and economical in the long run than a central refrigerating system. Refrigerating equipment, if used in hazardous locations where flammable vapors or dusts may be encountered, should be approved by the Underwriters' Laboratories, Inc., for Class I or Class II locations.

Automatic Sprinkler Systems. The automatic sprinkler is the best single safeguard against loss of life or property by fire. It is a device for automatically discharging water on a fire, and with proper water supplies it will either extinguish the fire or hold it in check. The automatic feature is secured through the heat from the fire which melts a soft solder compound and allows the head to operate. In the case of the bulb type of sprinkler, operation is secured through bursting of the bulb owing to expansion of the liquid it contains. The water is supplied by a piping system, the sprinklers being at intervals along the pipe. Sprinkler-system design requires expert knowledge to secure proper spacing and location of heads, the proper pipe sizes, control valves, and adequate water supplies. All portions of the building should be protected. Sprinkler installations should follow present-day standards. (See "Regulations for the Installation of Sprinkler Equipments," National Board of Fire Underwriters.) These regulations require adequate water supplies, standard and approved material, heads properly spaced and located under various types of ceiling construction, and piping properly supported.

Automatic sprinklers are designed to operate at various fixed temperatures, their selection depending upon the normal temperature of the space to be protected as shown in Table 9.

Table 9. Operating Temperatures of Automatic Sprinklers

Rating	Service temp. range, °F.	Color	Rating of sprinkler heads, °F.	
			Solder type	Quartzoid bulb
Ordinary	Under 100	Uncolored	150–165	135
Intermediate	100–150	White	212	175
Hard	150–225	Blue	286	250
Extra hard	225–300	Red	360	325

When corrosive vapors are present, special heads, such as chromium-plated, lead-plated, or wax-covered (corroproof) heads, or a combination of these or the bulb type should be employed. Where the heads are subject to mechanical injury they are provided with guards.

When the space to be protected is heated, a wet system with water in the pipe lines under pressure to the sprinkler heads at all times is installed. When installed in unheated spaces, a dry system is used with water held in check by a dry valve located in a small heated enclosure, the sprinkler piping being filled with air. When a sprinkler head opens, the air pressure is reduced. The dry valve then operates and water flows into the system.

Unsealed sprinkler heads (ordinary heads with struts or links removed) may be used in conjunction with quick-opening or deluge valves to discharge large quantities of water over an entire area or at certain critical points. **Cornice or window sprinklers** may be used for producing a water curtain on the outside of buildings to prevent the spread of fire. Unsealed heads, and cornice or window sprinklers, may be used in addition to standard automatic sprinkler systems. For quantities of water discharged from automatic sprinkler heads, see Fig. 2.

Inside Hose. Linen hose is usually used for such equipment, although in buildings containing acid fumes, or where there is excessive moisture in the air, rubber fire hose may be necessary. Small hand hose can be attached to wet pipe sprinkler systems under conditions specified in "Regulations Governing the Installation of Sprinkler Equipments," National Board of Fire Underwriters. In most cases inside hose is either 1½ or 1¼ in. with ½-in. brass nozzles and stored on semiautomatic hose racks or so located that all parts of each story of the building are within 20 ft. of a nozzle for a first-aid stream when attached to not more than 75 ft. of hose. For quantities of water discharged from inside hose, see Fig. 2.

First-aid Fire Extinguishers. These are designed for incipient fires and should be installed in addition to other equipment, such as sprinklers and inside hose. Incipient fires are divided into three groups:

Class A fires are those in ordinary combustible material, where the quenching and cooling effects of quantities of water or solutions containing a large percentage of water are of primary importance.

Class B fires are those in oils, greases, flammable liquids, etc., where the blanketing or smothering effect of the extinguishing agent is of greatest importance.

Class C fires are incipient fires in electrical equipment where the non-conducting property of the extinguishing agent is of prime importance.

A unit has been adopted for convenience in measuring fire protection afforded by first-aid fire appliances. This unit is composed of from one to five appliances, depending upon the extinguishing value of the kind of appliance comprising the unit. For ordinary combustible occupancies and extra hazardous occupancies at least one unit should be provided for every 2500 sq. ft. of floor area or not over 50 ft. of travel from any point to reach the nearest unit. For light hazard occupancies one unit should be provided for every 5000 sq. ft. or not over 100 ft. of travel from any point to reach the nearest unit.

Extinguishers are designed for certain types of fires, as shown in Table 10. This table also shows: (*a*) the number of devices required to form one unit; (*b*) whether the device needs protection against freezing; and (*c*) the necessary frequency of recharging. For further details, see "Regulations Governing First Aid Fire Appliances," National Board of Fire Underwriters.

EQUIPMENT

Type. In considering the type of equipment from the standpoint of safety, particularly common equipment or that which is not peculiar to certain processes, the principal considerations are that it be properly guarded and free from projecting shafts, exposed moving parts, and projecting setscrews. Machines that are automatically fed should be given preference, and the point of operation should in all cases be adequately guarded. Rolls and other such equipment where there is the possibility of the employees being injured between moving parts should be provided with quick-stopping devices either automatically or manually controlled. Remote controls are frequently used for machines in which explosive substances are handled. In all cases suitable means should be provided for adequate cleaning without undue hazard to the employees. Pressure vessels should be installed in accordance with the "Boiler Construction Code," American Society of Mechanical Engineers. Further information on this subject will be found in Sec. 24, pp. 1628 to 1660, and in the *Safe Practices Pamphlets*, National Safety Council.

Location. Equipment should be located to provide adequate working areas, aisles, and light. It is advisable to locate machines upon which work requiring good light is performed along the walls near windows where daylight may be secured.

Table 10. First-aid Fire Appliances Classification

Kind of extinguisher	Classification and number of devices	Protection from freezing required	Recharging
1¼ gal. soda-acid	A-2	Yes	Renew yearly
1½ gal. soda-acid	A-2	Yes	Renew yearly
2½ gal. soda-acid	A-1	Yes	Renew yearly
17 gal. soda-acid (wheeled)	A-1	Yes	Renew yearly
30 gal. soda-acid (wheeled)	A-1	Yes	Renew yearly
33 gal. soda-acid (wheeled)	A-1	Yes	Renew yearly
40 gal. soda-acid (wheeled)	A-1	Yes	Renew yearly
60 gal. soda-acid (wheeled)	A-1	Yes	Renew yearly
80 gal. soda-acid (wheeled)	A-1	Yes	Renew yearly
1¼ gal. foam	A-2; B-2	Yes	Renew yearly
1½ gal. foam	A-2; B-2	Yes	Renew yearly
2½ gal. foam	A-1; B-1	Yes	Renew yearly
5 gal. foam	A-1; B-1	Yes	Renew yearly
17 gal. foam (wheeled)	B-1	Yes	Renew yearly
33 gal. foam (wheeled)	B-1	Yes	Renew yearly
2½ gal. calcium chloride	A-1	No	Renew yearly
2½ gal. pump tank	A-2	Yes	Keep full
5 gal. pump tank	A-1	Yes	Keep full
22 gal. bucket tank containing 5 standard fire pails	A-1	Yes	Keep full
50 gal. cask with 3 standard fire pails	A-1	Yes	Keep full
12 qt. standard fire pail	A-5	Yes	Keep full
12 qt. foam fire pail	A-5; B-5	Yes	Renew yearly
12 qt. sand fire pail	A-5; B-5	No	Keep full
1 gal. loaded stream	A-2; B-4	No	Renew yearly
1¾ gal. loaded stream	A-1; B-2	No	Renew yearly
1 qt. carbon tetrachloride	B-2; C-2	No	Keep full
1¼ qt. carbon tetrachloride	B-2; C-2	No	Keep full
1½ qt. carbon tetrachloride	B-2; C-2	No	Keep full
2 qt. carbon tetrachloride	B-2; C-2	No	Keep full
1 gal. carbon tetrachloride	B-2; C-1 or B-2; C-2	No	Keep full
2 gal. carbon tetrachloride	B-2; C-1 or B-2; C-2	No	Keep full
3 gal. carbon tetrachloride	B-2; C-1 or B-2; C-2	No	Keep full
7½ lb. carbon dioxide cylinder	B-2; C-1	No	Weigh semiannually
10 lb. carbon dioxide cylinder	B-2; C-1	No	Weigh semiannually
15 lb. carbon dioxide cylinder	B-1; C-1	No	Weigh semiannually
20 lb. carbon dioxide cylinder	B-1; C-1	No	Weigh semiannually
50 lb. carbon dioxide cylinder (wheeled)	B-1	No	Weigh semiannually
100 lb. carbon dioxide cylinder (wheeled)	B	No	Weigh semiannually
12 lb. dry chemical	B-1; C-1	No	Check yearly

Guarding. The following section divided into two parts, guarding of general machinery, and protection against falling and slipping, covers the subject in a general way. Further information may be obtained from the *Safe Practices Pamphlets*, National Safety Council; and "Safety Code for Mechanical Power Transmission Apparatus," American Standards Association Standard B-15.

Guarding of General Machinery

Alteration or Replacement of Existing Protective Appliances. These recommendations are not to be construed as requiring alteration or replacement of present protective appliances except under the following conditions:

a. Where the protection at present afforded is not adequate and is clearly less than that afforded by the standard.

b. Where present protective appliances fall below the requirements of the state.

c. Where present protective appliances are in need of general repair or replacement.

Operation without Protection. New and existing machinery is not to be put into operation until adequate permanent or temporary protective appliances are in place. If necessary to operate machinery for adjustment temporarily without guards, warning signs are to be displayed. Defective guards are to have warning signs displayed and to be repaired or replaced as soon as possible.

Strength. To be of substantial and durable construction. No guard or railing to be installed or kept in use which can be broken, collapsed, or displaced by the weight of a man's body lurching violently against it.

Material. To be of metal, except under the following conditions when wood may be employed:

a. Existing wooden guards that are substantial and otherwise satisfactory.

b. Temporary conditions, where, on account of the uncertain future of the operations, impossibility of immediately applying metal guards, etc.

c. In buildings containing explosives, etc., which are readily ignited or exploded by spark, and in which the addition of metal is to be avoided.

d. In situations where, on account of the peculiar nature of the process, iron is objectionable or is subject to rapid deterioration.

Wood is not to be used for protection against belt breakage where likely to be splintered and thrown, or for any but temporary installation in buildings where its use materially increases the general fire risk, particularly in forming a flammable connecting link by which flame may travel from the floor to a flammable roof or ceiling.

Provisions for Oiling and Greasing. Guards to be applied so that it will be unnecessary to enter, open, or remove them for routine oiling. Except where unavoidable, closed guards shall not enclose bearings, and as far as practicable outboard bearings on shafts should be outside filled railings. Extension oil pipes, preferably not smaller than ½ in. standard pipe, may be installed in order to reach bearings, etc., which cannot be satisfactorily oiled otherwise. Modern systems employ to a great extent the pressure grease systems which are fed through small lines and do not require access to the bearing.

Where necessary for the lubrication of gears, an opening not over 6 in. square may be left in the guard away from the inrunning side of the point of mesh, if provided with an attached shutter or cover closing by gravity or springs.

Under exceptional conditions, where access to oil and grease cups, etc., while machinery is in motion, is essential and can be secured in no other way, closed guards may be hinged; or railings and closed guards may be provided with a hinged section. In either case these are to be held closed by a latch. If this provides access to a moving part, the opening should be ample in size and should be situated at the point of travel where least danger is introduced. In this case, covers or gates should not be self-closing. No gates or removable section for employees to enter railed enclosures to do routine oiling while machinery is in motion are permissible. Gates provided for repairs and adjustments must be locked.

Oiling Walkways and Platforms. The oiling of overhead shafting by means of portable ladders is to be avoided as far as practicable by the installation of platforms or walkways.

Platforms and walkways should be at least 30 in. in width and are not to be less than 24 in., protected on open or exposed sides with standard railings. As a general rule, the best location for oiling walkways and platforms is 24 to 48 in. below the level of the shafting.

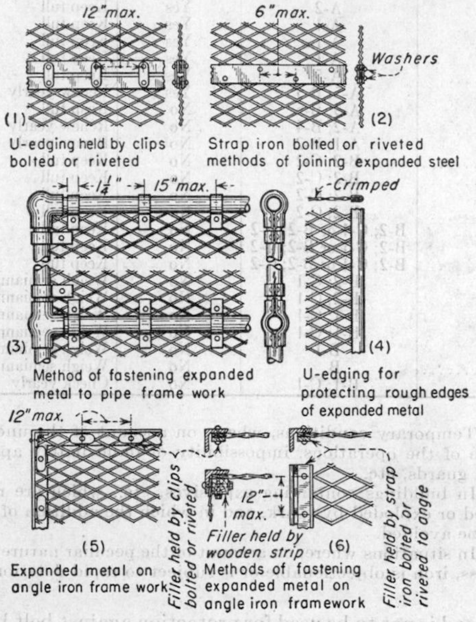

FIG. 8. Framework for expanded metal guards.

Oiling walkways and platforms are not to be located more than 24 in. (vertical distance) above the shafting for which they are provided or more than 30 in. (horizontal distance) to one side—for best distances see Table 17.

Material and Construction. Sheet or Perforated Metal Guards. Not recommended for very large guards or filling in railings on account of obstructing view of machinery. Should be galvanized or painted.

a. Gage. In no case, less than No. 22 (0.0313 in.) to be used.

When Largest Side of Guard Does Not Exceed 4 *Sq. Ft. in Area.* (1) Not lighter than No. 22 gage to be used if constructed on supporting framework. (2) Not lighter than No. 16 gage (0.0625 in.) to be used if not constructed on supporting framework.

When Largest Side of Guard is 4 *Sq. Ft. in Area or Greater.* (1) To be constructed on supporting framework. (2) Not lighter than No. 18 gage (0.05 in.) to be used. (All specifications for U.S. Standard Gage.)

b. Seams and Joints. To be welded, riveted, or bolted.

c. Edges and Corners. No exposed sharp edges or corners.

d. Framework. To be constructed of strap iron not less than 1 by 3/16 in., or angle iron not less than 1 by 1 by 1/8 in. made up by welding, rivets, or bolts. If bolts are used, corners of large guards to be braced with angles or gussets. Strength of framework to be proportioned to size of guard.

Expanded Metal Guards. Expanded metal is recommended for constructing closed guards for pulleys and large gears and for filling in railings.

a. Gage and Mesh. (1) Guards within 4 in. of nearest moving part: No. 18 gage, 3/4-in. mesh. (2) Guards not included in (1): No. 13 gage, 1½-in. mesh.

b. Seams and Joints. See Fig. 8 for permissible methods.

c. Edges and Corners. No exposed edges or corners where metal has been cut to be left unprotected (see Fig. 8, for permissible methods). Roughest side of metal to be placed inward. All metal to be dipped or painted.

d. Framework. All guards to have supporting framework of not lighter than 1¼-in. iron pipe; not lighter than 1 by 1 by 1/8 in. angle iron; not lighter than 1 by 3/16 in. strap iron. Angle and strap-iron frames to be made up by welding or with rivets or bolts. If bolts are used, corners of large guards to be braced with angles or gussets. The strength of framework to be proportioned to size of guard.

See Fig. 8, (3), (5), and (6) for methods of fastening expanded metal to framework.

e. Clearance from Moving Parts. Although 2 in. is shown as minimum in Tables 14, 15, and 16, 4 in. should be used if possible.

Cast-iron Guards. This type of guard is frequently provided on manufactured machinery and is satisfactory if providing complete protection.

Woven Wire-mesh Guards. Where used, the following specifications to be observed:

a. Guards within 4 in. of nearest moving part: ½-in. mesh, wire not smaller than 0.063 in. diameter.

b. Guards not included in (*a*): 1-in. mesh, wire not smaller than 0.105 in. diameter.

Should be galvanized or painted. General construction to conform with that specified for expanded metal.

Wooden Guards. *a. Material.*

Distance of Guard from Nearest Moving Point, In.	Construction
6 or less	Solid panels, not less than 3/4 in. thick.
Over 6	Solid panels not less than 3/4 in. thick or slats not less than 7/8 by 2 in. with not over 1½ in. between slats and not over 4 ft. 0 in. unsupported length.

As far as practicable to be of sound, straight-grained stock, free from knots and cracks, and all exposed surfaces dressed.

b. Construction. No nails to be used for securing removable permanent guards to structure or machine. Where explosive dusts are present, tongue-and-groove boards should not be used and all seams and cracks should be filled.

c. Edges and Corners. Edges and corners to be beveled.

d. Framework. The use of a framework is recommended for paneled guards and is necessary for slatted guards. Stock should be not less than 1½ by 1½ in. or equivalent for small guards. Where metal is not objectionable, the use of angle braces for reinforcement and for securing to structure or machine is recommended.

e. Painting. Permanent wooden guards exposed to acid fumes should receive at least two coats of acid-resisting paint.

Method of Securing to Structures, Machines, Etc. Guards protecting parts that must be frequently repaired or adjusted to be applied so that they are readily removable and readily replaceable. Such guards to be strong enough to retain their shape when removed. If sheet-metal or built-up metal guards for this purpose have their edges fastened to the structure, machine, etc., these edges must be reinforced with angle or strap iron for their full length.

Nails or staples not to be used for attaching metal guards. Use bolts, screws, or sockets. No unreinforced sheet metal is to be used for supporting guard unless washers are placed under heads of bolts and screws.

Table 11. Design of Railing Guards. Material and Sizes

Project	Minimum requirements		
	Pipe railings	Angle-iron railings	Wood railings
Height of top rail above floor	42 in.	42 in.	42 in.
Height of center of mid-rail above floor	21 in.	21 in.	21 in.
Spacing of posts or stanchions	8 ft.	8 ft.	8 ft.
Size of posts	1¼-in. pipe	2 × 2 × ¼ in.	2 × 4 in.
Size of top rail	1¼-in. pipe	2 × 2 × ¼ in.	2 × 4 in.
Size of mid-rail	1¼-in. pipe	1¼ × ⅜ in. (strap iron)	1 × 6 in. (full) or 2 × 6 in.

The use of railings without mid-rail is not permissible.

Toeboards. Iron toeboards to be ¼ by 4 in. on edge; wood toeboards to be 1 by 4 in. on edge; concrete toeboards (curbing) to be at least 4 in. wide by 4 in. high, bonded in.

Where toeboards are objectionable from their tendency to collect snow or ingredients, they may be raised a maximum of 1 in. above the platform.

b. Pipe Railings. Screwed fittings or welding to be used. Where subject to rough usage, as on trestles, posts should be 2 in.

c. Angle-iron Railings. All joints riveted, bolted, or welded. The use of angle iron where expanded metal is used as filler is recommended as preferable to iron pipe.

d. Wood Railings. Material to be sound. Hand and mid-rails to be dressed on all four sides, and edges of handrails beveled. Handrails to be set with longer dimension horizontal. Posts to be set with longer dimension perpendicular to handrail. Where subject to rough usage or where iron pipe or cable is used for hand or mid-rails, posts to be at least 4 by 4 in.

Method of Securing to Structures. Pipe and angle-iron railings to be secured as follows:

a. To wood by flanges or angles secured by lag or through-bolts.
 by sockets at least 4 in. deep, securely fastened down by flanges.
 by socketing in holes bored through timbers not less than 4 in. thick, provided that a flange or similar device is employed to keep the post from slipping down.

b. To concrete by setting in the concrete at least 6 in. deep when poured.
 by drilling at least 6 in. deep and grouting with cement grit.
 by flanges secured through multiplex floors by bolts.
 by sockets at least 6 in. deep set in or grouted into the concrete.

c. To brickwork by flanges secured by through-bolts.
 by drilling to a sufficient depth and setting in pipe or flange with anchor bolts and concrete.

d. To metal work by flanges with through-bolts or rivets.
 by sockets at least 4 in. deep, bolted or riveted.

The Following Methods Are Not to Be Employed. Flattening pipe, drilling, and bolting; fastening flanges with nails or wood screws, with bolts set in lead, with screws set in wooden plugs, or with expansion bolts.

Wooden railings are to be secured as follows:

By nailing only where at least 12 in. bearing surface on cross timbers may be secured. See Fig. 10.

By nailing where at least 4 in. bearing surface on cross timbers may be secured together with one diagonal bracing on each post. See Fig. 10.

By toe nailing only where post may be rigidly braced. Post bracing must not project into footways.

Fillers. Expanded metal, sheet and perforated metal to be used for filling in railings. Method of attachment

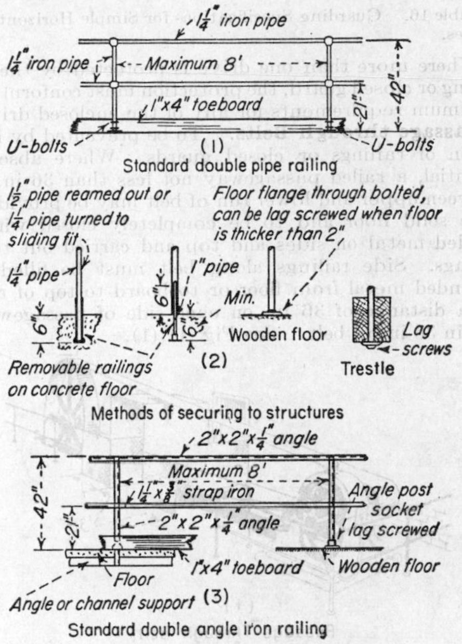

FIG. 9. Methods of securing railing to structures.

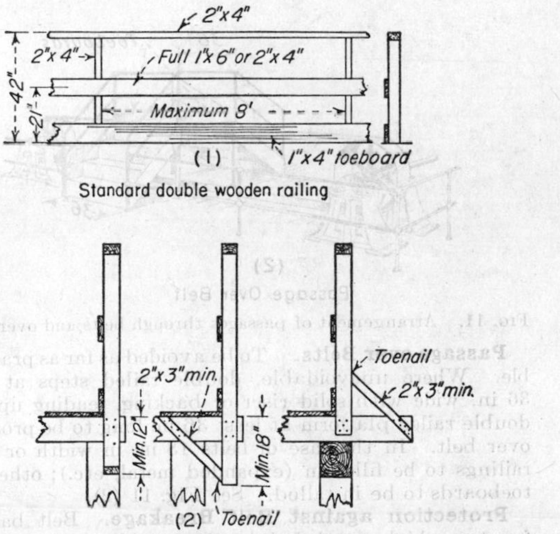

FIG. 10. Methods of securing wood railings to structures.

is shown in Fig. 8 (3), (5), and (6). U edging is recommended for binding the edges of expanded metal.

Toeboards (as protection for machinery). To be used where a slip or projection of the foot under the railing would bring about a close proximity to moving machinery and around the edges of belt pits and belt openings.

Guards for Belts and Pulleys

General Protection. All belts and pulleys within 7 ft. 0 in. of ground, floors, platforms, stairs, and walkways to be protected against contact.

Minimum requirements for guarding are given in:

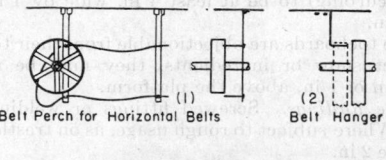

Table 14. Guarding Specifications for Simple Vertical Belt Drives.
Table 15. Guarding Specifications for Simple Inclined Belt Drives.
Table 16. Guarding Specifications for Simple Horizontal Belt Drives.

Where more than one drive is protected by the same railing or closed guard, the protection must conform to the minimum requirements for any of the enclosed drives.

Passage through Belts. To be prevented by installation of railings or closed guards. Where absolutely essential, a railed passageway not less than 36 in. wide between upper and lower run of belt may be provided, to have solid floor and to be completely closed with expanded metal on sides and top and carried out to side railings. Side railings along belt must be filled with expanded metal from floor or toeboard to top of railing for a distance of 36 in. on each side of passageway, if within 36 in. of belt. See Fig. 11 (1).

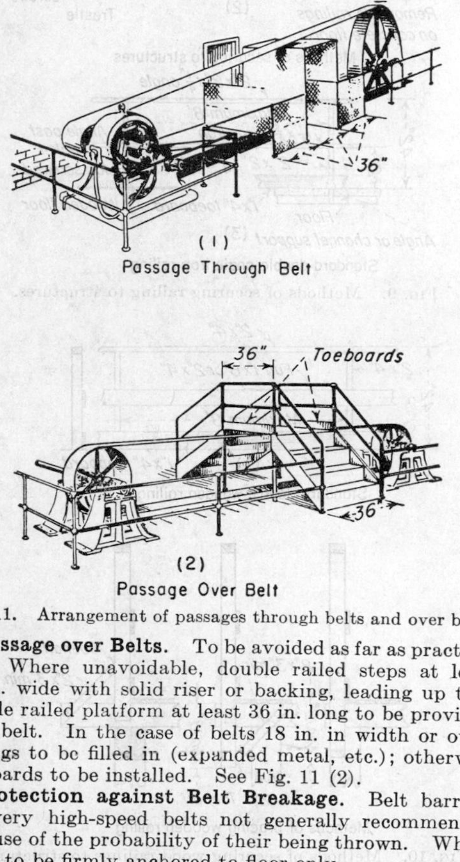

Fig. 11. Arrangement of passages through belts and over belts.

Passage over Belts. To be avoided as far as practicable. Where unavoidable, double railed steps at least 36 in. wide with solid riser or backing, leading up to a double railed platform at least 36 in. long to be provided over belt. In the case of belts 18 in. in width or over, railings to be filled in (expanded metal, etc.); otherwise toeboards to be installed. See Fig. 11 (2).

Protection against Belt Breakage. Belt barriers for very high-speed belts not generally recommended because of the probability of their being thrown. Where used, to be firmly anchored to floor only.

Routes* and stations† as well as important control centers, such as switchboards and grouped steam valves, and, in general, any equipment the breaking of which would cause a serious accident should not be located under or in direct line with heavy high-speed belts. Where this is unavoidable, the following protection to be applied:

Belt width	Belt speed	Overhead protection	End protection
Up to and including 8 in. or over 18 in..........	Any	None	None
8 to 12 in., inclusive..........	3000 ft./min. and over	Continuous belt trough	Belt barriers
Over 12 in., and up to and including 18 in..........	Any	Continuous trough	Belt barriers

Belt barriers to consist of structural steel with vertical members extending from floor to not less than 24 in. above run of belt. Maximum height of barrier, 84 in. Barriers to be securely braced or stayed.

Belt Clearance. A clearance exceeding the width of the belt by one-fifth to be provided between edge of pulley and structures and between edge of pulley and bearings or stationary objects. Overhead belts that are frequently removed from the pulleys and allowed to hang on the shafting should be provided with belt perches. See Fig. 12 (1) and (2).

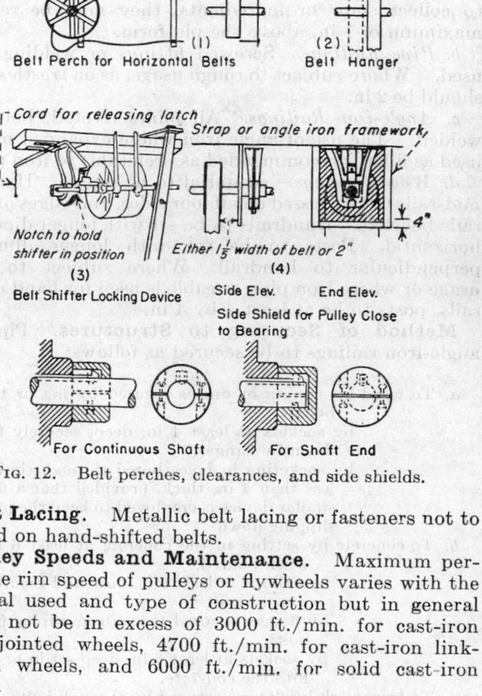

Fig. 12. Belt perches, clearances, and side shields.

Belt Lacing. Metallic belt lacing or fasteners not to be used on hand-shifted belts.

Pulley Speeds and Maintenance. Maximum permissible rim speed of pulleys or flywheels varies with the material used and type of construction but in general should not be in excess of 3000 ft./min. for cast-iron flange-jointed wheels, 4700 ft./min. for cast-iron link-jointed wheels, and 6000 ft./min. for solid cast-iron wheels.

Laminated wood pulleys not to be installed or run where continually subjected to the action of moisture.

Pulleys not regularly in use to be removed from shafting, etc., which is in continued operation.

The use of short, heavy drives with adjustable belt tightener pulleys to be avoided wherever practicable. V-belt drives with automatic tension feature are satisfactory (see Sec. 24, p. 1670).

Belt guides of any character (except special pulleys designed for the purpose) not to be installed for the purpose of keeping belts on pulleys or keeping belts from working off into contact with other objects.

* Refers to general paths, passages, walkways, exits, stairs, ladders, or aisles used frequently by building employees and others.

† Refers to regular working stations of employees where they are apt to remain for some time—including work benches, desks, lunch tables, lavatories, bulletin boards, and similar congregating places.

Belt Shifters and Clutches

Provision. If in a group drive it is necessary to stop the individual units in the course of normal operation or for frequent adjustment, sufficient tight and loose pulleys with belt shifters, clutches, etc., to be installed to obviate the necessity of throwing belts on or off moving pulleys by hand.

Arrangement. Control levers, etc., to be accessible from a safe position outside railings or guards and located where not readily exposed to accidental contact. All control levers, etc., to automatically lock themselves in the off position and to be arranged for permanent locking in this position by means of a padlock. See Figs. 12 (3) and 14.

Tight and loose pulleys to be arranged so that the loose pulley is readily accessible for lubrication.

Belt shifters and clutches controlling the transmission of power to rooms other than those in which they are located to have extension controls in such rooms, arranged for padlocking in the off position.

Belt shifters, clutches, etc., controlling the motion of such machines as mixing rolls, calenders, and pickers, which, while in motion, are fed by hand from a point close to the intake, to have control within easy reach of the operator's working position. In the case of group drives, group control to be within easy reach of operators of all machines in the group.

Guarding. Friction clutches, unless of the safety type (no revolving projections whatsoever), to be completely enclosed. See Fig. 13 (6).

Gear and Friction Drives. Guarding. All power driven or hand-operated gears or friction drives up to 6 ft. 0 in. in diameter, wherever located, to be completely enclosed as specified under "Use and Design of Closed Guards." Provisions for oiling and greasing to be made. Gears over 6 ft. 0 in. in diameter constitute special cases which must be considered on their own merits. Where it is impracticable to totally enclose them, pinion and point of mesh should be covered and operators or oilers protected from contact with gear teeth and spokes.

Where frequent adjustments are necessary with the machinery in motion the installation of a railing or general guard, protecting other moving parts as well as the gears, does not obviate the necessity of separately enclosing the gears.

Chain and Sprocket Drives. Guarding. All hand-operated or power-driven chain and sprocket drives to be protected as specified for belts and pulleys.

Sprocket chains located over 7 ft. above routes and stations, which in breaking would endanger those beneath, to be provided with troughs or netting.

Provisions for oiling and greasing to be made.

Rope and Cable Drives. Guarding. To be protected as specified for belts and pulleys.

Shafting

Revolving Projections. No projecting setscrews, bolts, keys, etc., on shafting pulleys, gears, etc., unless protected by closed guards (not railings). See Fig. 13 (1) and (5). **Safety couplings** (flanged) and **safety collars** without guards are permissible if bolts and setscrews do not project beyond flanges.

Ends of shafts and arbors which project beyond bearings, etc., to be cut off flush or completely enclosed. See Fig. 13 (7 and 8). Absolutely smooth shafts over 2 in. in diameter without keyways are allowed an unguarded projection beyond bearing of not over 1 in.

Guarding Vertical and Inclined Shafts. All vertical or inclined shafts (regardless of speed) which are exposed to contact to be protected in either of the following ways:

a. By sheet or expanded metal enclosure covering all exposed portions up to 6 ft. above ground, floor, platform, walkway, etc.

b. Protection on all exposed sides by standard 42-in. double railing located not less than 36 in. away from shaft.

Guarding Horizontal Shafts. All horizontal shafts within 7 ft. 0 in. of ground, floor, platform, walkway, etc., to be protected against contact. Minimum requirements for guarding are given in Table 17 and Fig. 28.

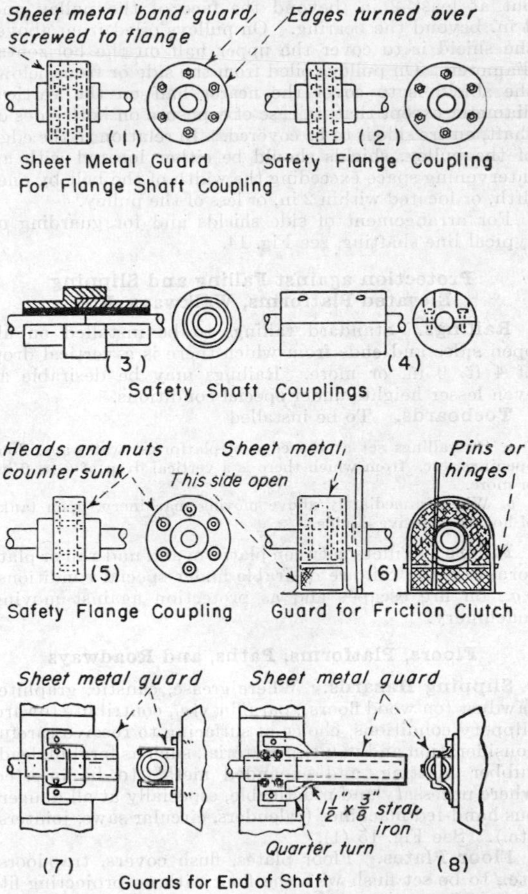

FIG. 13. Guarding clutches, couplings, and shaft ends.

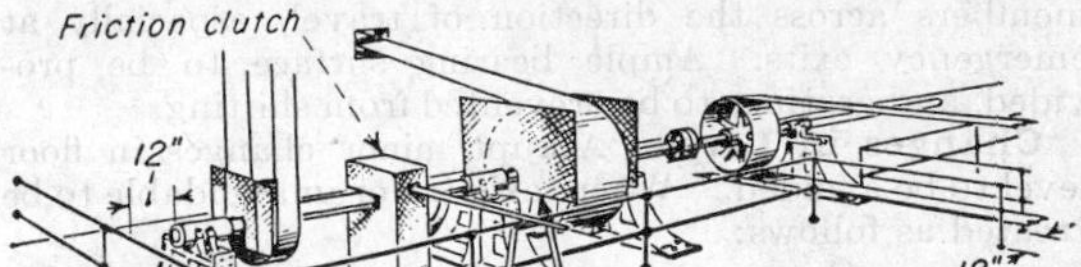

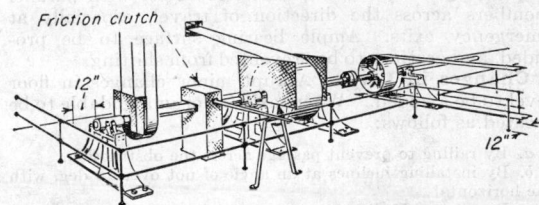

FIG. 14. Guards for line shafting and pulleys.

Guarding Pulleys on Horizontal Shafting. Protection to be applied in accordance with Tables 14, 15, and 16 combined with Table 17.

In the case of line shafting protected by open railings, it is permissible for the face of large pulleys to be within 6 in. (horizontal distance) of railing, provided that sides of pulleys and especially the onrunning points of the belts are properly protected.

Where the sides of pulleys on shafts from 24 in. below the floor or platform to 66 in. above are closer than 12 in. (horizontal distance) to the bearing, side shields must be installed. Where the sides of pulleys on shafts more than 66 in. above the floor or platform are closer than 18 in. to the bearing, side shields must be installed. See Fig. 12 (4).

Side shields must cover at least one-half (180 deg.) the side of pulleys up to 48 in. diameter and should extend out at least 2 in. beyond the face of the pulley and 4 in. beyond the bearing. On pulleys oiled from above, the shield is to cover the upper half on the horizontal diameter. On pulleys oiled from the side or from below, the shield is to cover the nearer half on the vertical diameter except that, in case of exposure on both sides of shaft, entire side is to be covered. In relation to the edge of the pulley, shields should be either located with an intervening space exceeding the width of the belt by one-fifth, or located within 2 in. or less of the pulley.

For arrangement of side shields and for guarding of typical line shafting, see Fig. 14.

Protection against Falling and Slipping Elevated Platforms, Walkways, Etc.

Railings. Standard railings to be installed on all open sides and ends from which there is a vertical drop of 4 ft. 0 in. or more. Railings may be desirable at even lesser heights under special conditions.

Toeboards. To be installed

a. On railings set at the edge of platforms, walkways, floor openings, etc., from which there is a vertical drop of 6 ft. 0 in. or more.

b. When immediately above moving machinery, open tanks of hot or corrosive liquids.

Fillers. Fillers between platform and mid-rail or platform and top rail are desirable under special conditions, *e.g.*, on fire escapes and as protection against moving machinery.

Floors, Platforms, Paths, and Roadways

Slipping Hazards. Where grease, caustic, graphite, sawdust (on wood floors), paraffin, etc., contribute toward slippery conditions, choice of surfacing to receive careful consideration and non-slip material, such as feralun, lead, rubber matting, and expanded metal, to be inserted where necessary and practicable, especially at all dangerous hand-fed machines (calenders, circular saws, jointers, etc.). See Fig. 15 (1).

Floor Plates. Floor plates, flush covers, trapdoors, etc., to be set flush with the floor with no projecting fittings. Smooth metal not to be used for this purpose.

Gratings. As far as possible to be set with parallel members across the direction of travel, especially at emergency exits. Ample bearing surface to be provided, and gratings to be prevented from shifting.

Changes in Level. Abrupt minor changes in floor level to be avoided. Where existing or unavoidable to be treated as follows:

a. By railing to prevent passage across the obstruction.

b. By installing inclines at an angle of not over 15 deg. with the horizontal.

c. By steps. If difference in level is between 6 and 10 in., which may be considered as one step, a 1¼-in. projection or "nosing" should be provided along its edge.

Trenches and Excavations. Floor openings, trenches, excavations, etc., of a temporary nature to be clearly indicated, if necessary by railing or roping off during the day with the addition of temporary lights after dark if area is not thoroughly illuminated. All permanent openings of this character inside buildings or across regular routes outside to be covered or fenced off.

Traps and Hatches. To be provided with flush doors or covers, having swinging guards with stops to hold them open in a position somewhat beyond the vertical [see Fig. 16 (1)] or to be protected on at least three sides by railings with toeboards, etc. See Fig. 16 (2). Traps and hatches set flush in platforms, floors, or other footways to be protected by fixed or hinged grids with strong bars not over 6 in. apart, in addition to the trapdoors or covers, unless standard railing protection is provided. Where materials are discharged through such grids, bars may be placed farther apart, but should be as near 6 in. as practicable.

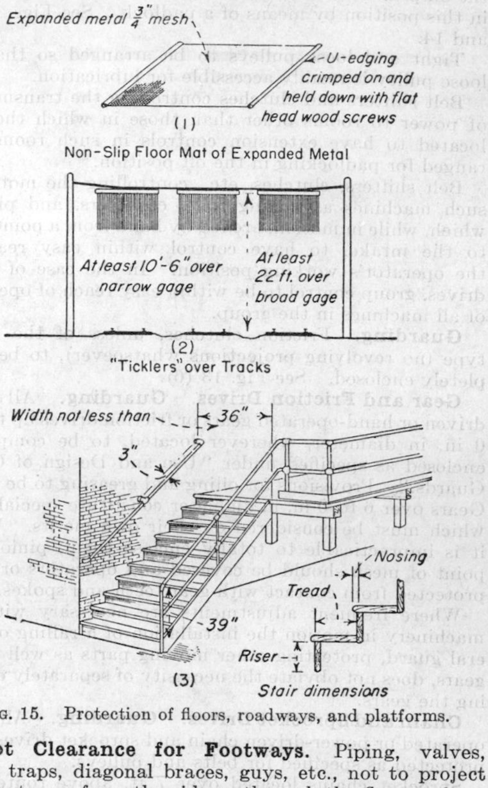

FIG. 15. Protection of floors, roadways, and platforms.

Foot Clearance for Footways. Piping, valves, steam traps, diagonal braces, guys, etc., not to project into footways from the side or through the floor to form tripping hazards. Where unavoidable, their presence to be indicated by railings or by fender boards or markers, painted yellow and at least 3 ft. 6 in. high. See Fig. 16 (4). The boundaries of aisles in shops and manufacturing buildings, the floors of which may be congested, should be denoted by painting 4-in. stripes of yellow paint on the floor; materials and trucks not in actual use should be kept back from the areas so indicated.

Piping on Floors. Piping should not be installed across routes either on or just above the floor. Where unavoidable, steps or inclines are to be provided as shown in Fig. 17 (2), (3), (4).

Projections into Footways. Equipment to be arranged to avoid the projection into footways of such obstacles as clutch and belt-shifter levers, post indicator and hydrant wrenches, valve stems, etc. Where unavoidable, they are to be guarded by railings or indicated by fender boards, etc. Valve wheels to be used instead of inverted L- or T-extension valve handles.

Lighting. Ample natural or artificial lighting to be provided and maintained for routes and footways, stairs, exits, and where any of the foregoing obstacles to safe

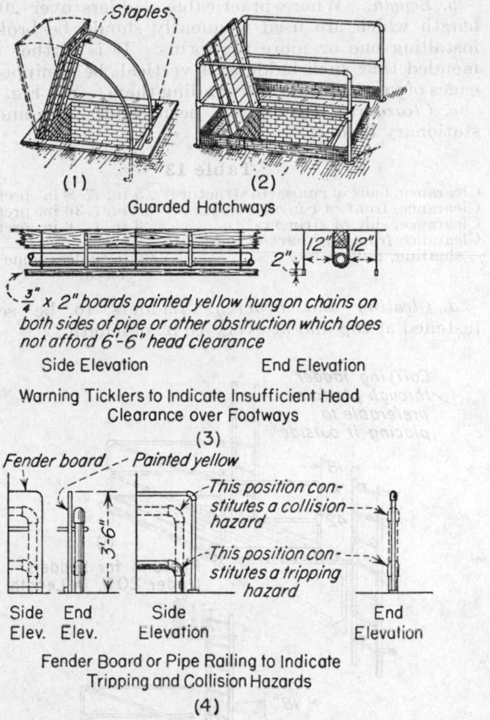

FIG. 16. Hatchway protection and guards for tripping hazards.

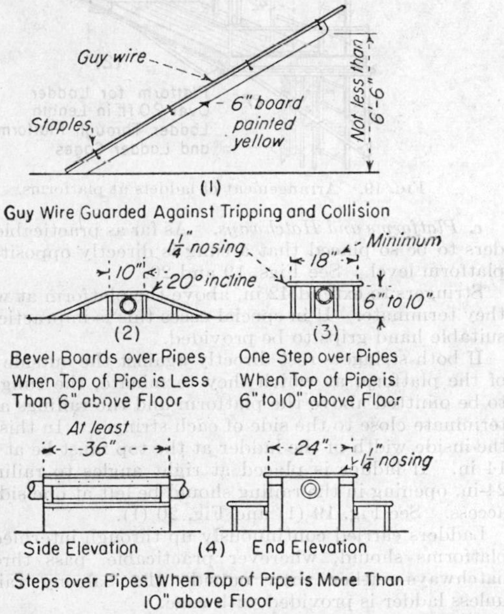

FIG. 17. Protection for some hazards of tripping.

travel are present. Such lights to be so arranged that deep shadows and glare are minimized.

Stairs*

Angle. Angle with the horizontal should not exceed 45 deg.; best angle, 35 deg.

* For additional details see the "Building Exits Code" of the National Fire Protection Association.

Risers and Treads. Height of riser [see Fig. 15 (3)] should not exceed 8 in. and be preferably 7 in. **All risers must be equal,** including first step above and below floors. Width of tread should be not less than 8 in. and preferably 10 in. Nosing to be not less than ¾ in. or more than 1¼ in. **All treads and nosings must be equal.** Best practice gives sum of riser and tread (exclusive of nosing) as 17½ in.

Table 12. Angles for Various Risers and Treads

Riser, in.	Tread, in.	Angle of stairs with horizontal
6½	9	35° 50′
	10	33° 2′
	11	30° 38′
	12	28° 28′
7	9	37° 53′
	10	35° 0′
	11	32° 30′
	12	30° 15′
7½	9	39° 42′
	10	36° 53′
	11	34° 18′
	12	32° 0′
8	9	41° 36′
	10	38° 40′
	11	36° 1′
	12	33° 42′
8½	9	43° 21′
	10	40° 22′
	11	37° 42′
	12	35° 18′
9	9	45° 0′
	10	41° 59′
	11	39° 17′
	12	36° 52′

Width of Stairs. Width of stairs consisting of over three steps should be not less than 36 in. Special stairways for infrequent use (oiling walkways, etc.) may be narrower but should not be less than 30 in.

Landings. Length and width to be not less than width of stairs. Spiral stairs and winders (steps set at an angle to the main flight) not permissible on new construction. Landing platform or clear space at least 36 in. deep to be provided at the foot of all stairs.

Doors. No doorways with doors hung to swing toward stairs except where stair platform at least 36 in. wide is provided. Door sills should be on same level as platform. Doors to be arranged to swing back so as not to block platform or stairs extending to upper landings.

Material and Construction. *a. Wood.* Treads not less than 1-in. board, clear sound stock, securely nailed. Treads to be supported directly by side members and not by cleats nailed to same or to wall. Front edges of treads to be slightly rounded. No cleats or other projections on treads are permissible.

b. Metal. Treads of smooth metal or parallel rods not permissible. Metal stairs with narrow treads (less than 9 in.) are classed as ladders and should not be used except at angles less than 65 deg. with the horizontal. Except in special cases (step ladders, engine-room ladders, etc.) they are objectionable, and round-rung ladders are preferred.

c. Concrete. Treads should be roughened, preferably by mixing in abrasive materials, unless (as is desirable on much used stairways) patented non-slip treads are installed.

Railings. Handrails with mid-rails to be provided as specified for platforms and walkways. **Standard height of railings: 42 in. above center of tread.** Lower posts should be set at extreme bottom of stairs with handrail and mid-rail terminated at this point, thus avoiding dangerous projections. Stairways built against the wall to have outside railing and if 4 ft. or more in width to have inside handrail with 3 in. clearance from wall. Completely enclosed stairways to have at least one handrail and if over 4 ft. in width, handrails on both sides. If a space is left between side of stairs and wall, complete double railing to be provided. See Fig. 15 (3).

If materials are piled around stair wells, railings should be entirely filled or stairwell enclosed.

Ladders

General Use. For continuous operation or frequent attendance, stairs should be installed in preference to ladders wherever practicable. Ladders should not be used for emergency exit.

Specifications for Iron or Steel Ladders (Fixed Ladders). *a. Stringers.* Not less than 2 by ½ in. or equivalent. Iron pipe should not be used for stringers. Width between stringers not less than 14 in. Sharp edges, burrs, etc., to be removed.

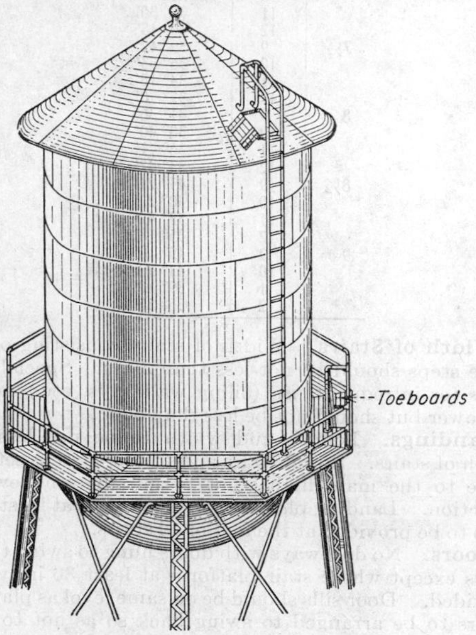

FIG. 18. Arrangement of ladders on elevated tanks.

b. Rungs. Not less than ¾-in.-diameter round bar to be used. Rungs to be spaced uniformly not less than 12 in. or more than 15 in. center to center. To be smooth and secured against turning.

c. Splice Plates. Cross section equivalent to stringers. To be double riveted or bolted to stringers.

d. Brackets. Cross section of supporting brackets to be at least equivalent to cross section of stringers. To be riveted or bolted to stringers and structure. Brackets to be spaced not more than 10 ft. center to center, except in the case of ladders 14 ft. or less in length.

e. Stack and Elevated Tank Ladders. U-bars without stringers are permissible for stack ladders but are not recommended for use elsewhere. Diagonal bracing (lattice) on structural steel of elevated tanks, etc., not to be accepted as equivalent to a ladder. Elevated tank ladders not to be carried up outside of balcony but through a hatchway provided for the purpose (Fig. 18).

f. Ladders with Flat Tread. Not to be used except on special work such as engine rooms. Vertical distance between treads should not exceed 10 in., and width of treads should not be less than 4½ in. Angle of ladder should not exceed 65 deg. Handrails to be provided.

Placing of Permanently Fixed Ladders. *a. Angle.* Ladders should not be set at less than 70 deg. with the horizontal.

b. Length. Where practicable, ladders over 20 ft. in length which are used frequently should be broken by installing one or more platforms. It is further recommended that such ladders, if vertical, be equipped with cages of vertical slats surrounding them. See Fig. 19 (2).

c. Clearance. Minimum permissible clearance for stationary ladders:

Table 13

Clearance, back of rungs (to structure)	5 in. (7–8 in. preferable)
Clearance, front of rungs	30 in. (36 in. preferable)
Clearance, side of stringers	3 in. (6 in. preferable)
Clearance from any part of ladder to shafting, pulleys, etc.	36 in. unless same are enclosed.

d. Cleating and Bracing. Ladders to be securely fastened at top and at every 10 ft. of length.

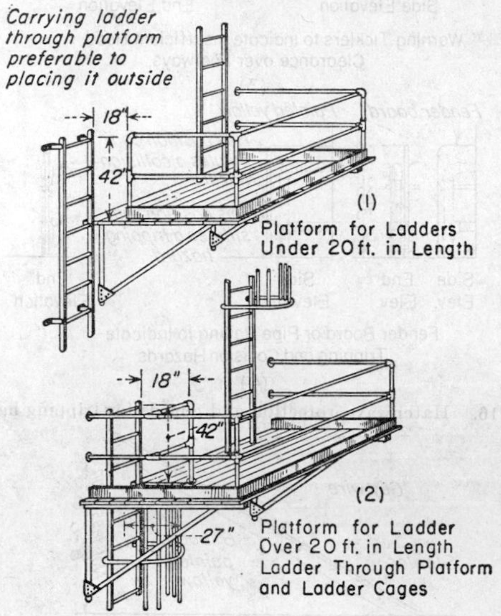

FIG. 19. Arrangement of ladders at platforms.

e. Platforms and Hatchways. As far as practicable ladders to be so placed that a rung is directly opposite the platform level. See Figs. 19 and 20.

Stringers to extend 42 in. above the platform at which they terminate. If in special cases this is impracticable, suitable hand grips to be provided.

If both stringers rest directly against the outside edge of the platform at which they terminate, the rungs are to be omitted above the platform and the railings are to terminate close to the side of each stringer. In this case the inside width of the ladder at the top must be at least 14 in. If ladder is placed at right angles to railing, a 24-in. opening in the railing should be left at one side for access. See Fig. 19 (1) and Fig. 20 (1).

Ladders carried continuously up through intermediate platforms should, wherever practicable, pass through hatchways rather than outside the balcony railing, unless ladder is provided with cage.

All open-ladder hatchways to be protected on three sides with standard railing and toeboard. The opening in the railing is to be on the less exposed of the two sides next the ladder. See Fig. 20 (2).

Portable Ladders (Straight Side, Round-rung Wooden Ladders). All wooden ladders should be equal or superior to the American Standard Association's Safety Code for Ladders Standard A-14.

a. Length. Maximum length 30 ft.

b. Non-slip Feet. All portable ladders to be shod with spikes, sharpened points, or safety feet securely fastened. The Underwriters' Laboratories Lists of Inspected Appliances Relating to Accident Hazard shows satisfactory approved ladder feet. Where ladders are used

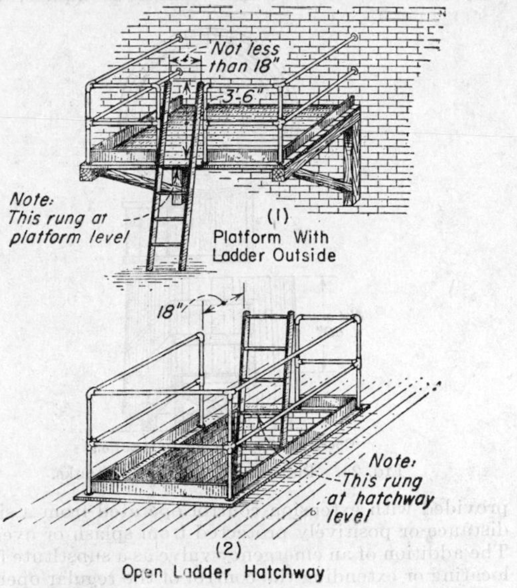

FIG. 20. Arrangement of ladders at hatchways and platforms.

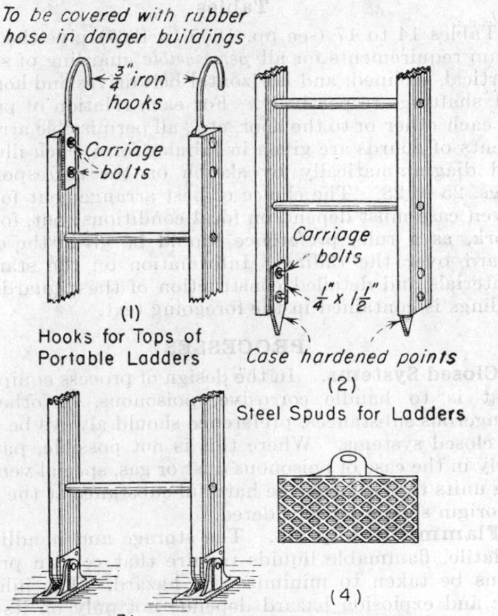

FIG. 21. Types of safety feet for ladders.

exclusively on shafting at a uniform height, metal shafting hooks may be substituted. See Fig. 21 (1), (2), (3), and (4).

NOTE. **Portable ladders are unsafe when resting on metal flooring.** They should be secured or held.

c. Portable ladders for permanent use in certain buildings should be clearly marked with the name of the building.

Step Ladders. Not to exceed 15 ft. in length, to be of substantial construction and are to be provided with rigid automatic braces between both pairs of front and back rails. For every foot of length they should be at least 1 in. wider at the bottom than at the top.

Inclines, Gangplanks, and Wheeling Walks. Design and Construction. *a.* Width to be not less than 30 in. and recommended angle with horizontal 7½ deg., not to exceed 15 deg. maximum.

b. Permanent or temporarily fixed inclines or wheeling walks to be railed as provided for platforms. Where absolutely necessary for unloading bales, etc., removable rails, gates, or chains may be introduced. Where trucks or barrows are used, inclines, etc., to be at least 36 in. in width with guard rails at least 4 in. high securely fastened to flooring; floor boards are to be properly secured against spreading. See Fig. 7 (3). For double trucking a width of at least 60 in. should be provided.

c. Movable gangplanks to be cleated or otherwise secured to prevent shifting. Smooth metal not to be used for gangplanks.

Skids. Permanent skids used for receiving or storage of drums or barrels, which are elevated above the ground or floor, are to be planked solid between each pair of rails. Between each pair of skids a 12-in. running board should be provided, and the clearance between the ends of drums at this point should not be less than 12 in. See Fig. 22 (1).

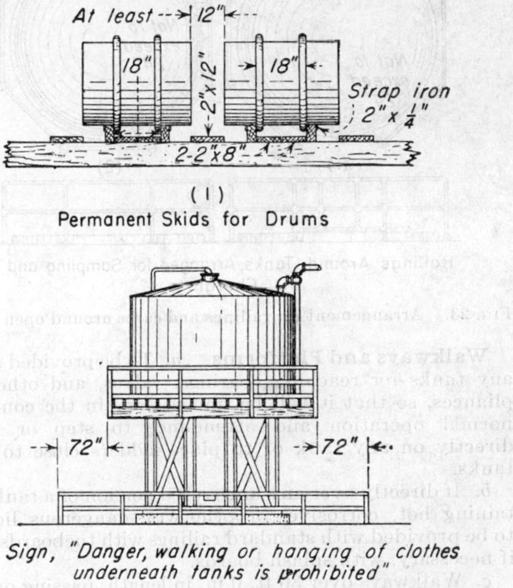

FIG. 22. Arrangement of drum skids and railings under tanks.

Tanks, Vats, and Pans. Location. At least 36 in. separation to be provided from open containers of hot, corrosive, or otherwise dangerous liquids to routes passing alongside and 72 in. separation to stations. Passageways between two rows of open containers or between a row of open containers and walls, railing, etc., to be at least 72 in. wide.

Protection of Routes and Stations below Tanks. Routes or stations should not be situated directly below open containers of hot, corrosive, or otherwise dangerous liquids, or where subject to their leakage or overflow.

Where this cannot be avoided, they are to be protected by roofs, tight floors, pans, etc., properly drained to discharge to a safe point or by the provision of adequate overflow pipes or gutters or, if necessary, by both.

Unused spaces directly below any tanks containing hot, corrosive, or otherwise dangerous liquids to be railed off and standard danger signs posted. Such railings should be located 72 in. out from outermost edges of tanks or platform overhead. See Fig. 22 (2).

Railings and Curbs. *a.* No open tanks, vats, or pans to be installed with upper edge less than 6 in. above the adjacent floor level.

b. Open tanks, vats, or pans to have upper edges at least 36 in. above adjacent floor level or to be protected on all accessible sides by standard railings, covers, or hoods. Where operating conditions permit, the upper edges of such tanks not otherwise protected should be at least 3 ft. 6 in. above the adjacent floor level.

c. Where necessary for filling, sampling, etc., such railings may be broken with offsets or inclined sections, provided that distance between top of tank and mid-rail and between mid-rail and top rail is not greater than 24 in. Gates or chains may be introduced into railings only where absolutely necessary and then are to be permanently secured so as to be movable but not removable (see Fig. 23).

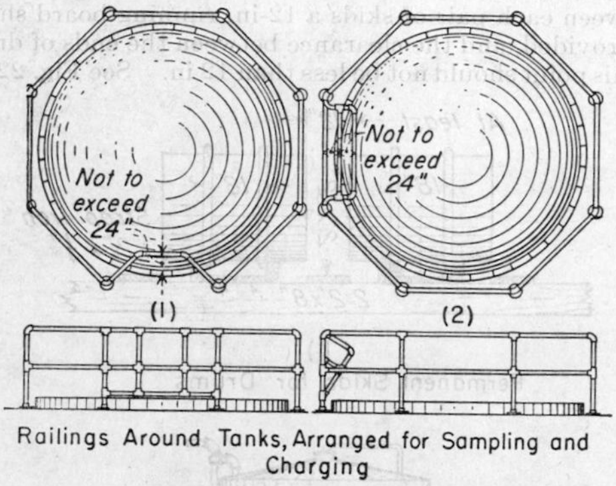

FIG. 23. Arrangement for railings and curbs around open tanks.

Walkways and Platforms. *a.* To be provided above any tanks for reaching bearings, valves, and other appliances, so that it will be unnecessary in the course of normal operation and attendance to step or stand directly on any tank or to place ladders close to open tanks.

b. If directly over any uncovered portion of a tank containing hot, corrosive, or otherwise dangerous liquids, to be provided with standard railings with toeboards, and, if necessary, with splash boards.

c. Walkways over 30 ft. 0 in. in length, passing over or close to tanks containing hot, corrosive, or otherwise dangerous liquids should be provided with facilities for escape at both ends. Wherever practicable, steps, rather than ladders, should be installed for this purpose and are to be located where not subjected to splash, leakage, or overflow.

d. Permanent ladders are not to be located close to the sides of tanks containing hot, corrosive, or otherwise dangerous liquids for purposes of observation or attendance. For such uses or for the operation of isolated acid and caustic valves frequently operated, small platforms at least 36 in. wide should be installed provided with ladder or preferably with stairs leading away from side of tank or valves. See Fig. 24.

e. Where repairs, cleaning, etc., are of frequent occurrence, suitable permanent platforms or walkways to be provided for this purpose.

Location of Controlling Devices. Valves, belt shifters, clutch levers, and other controlling devices to be located at least 36 in. away from edges of tanks containing hot, corrosive, or otherwise dangerous liquids, unless

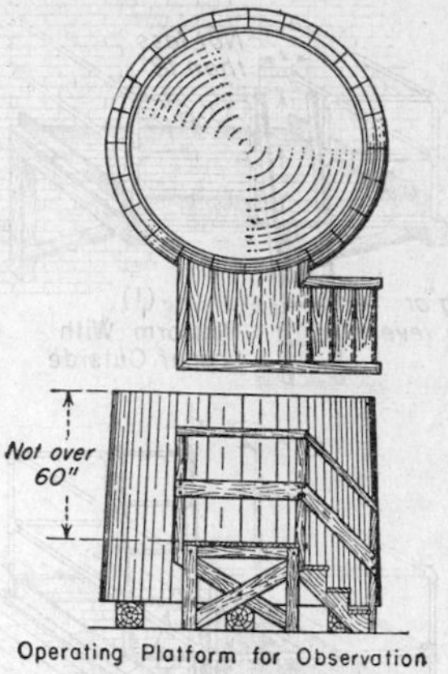

FIG. 24. Platforms for observing tanks.

provided with extension control operated from a similar distance or positively protected from splash or overflow. The addition of an emergency valve as a substitute for relocating or extending the control of the regular operating valve is not permissible.

Tables

Tables 14 to 17 (see pp. 1865 to 1867) give the minimum requirements for all *permissible* guarding of simple vertical, inclined, and horizontal belt drives and horizontal shafting, respectively. For each relation of pulleys to each other or to the floor, etc., all permissible arrangements of guards are given in tabular form, each illustrated diagrammatically by sketch on the corresponding Figs. 25 to 28. The choice of best arrangement for any given case must depend on local conditions; but, for new work, as a rule, preference should be given the closed guard over the railing. Information on the standard materials and detailed construction of these guards and railings is contained in the foregoing text.

PROCESSES

Closed Systems. In the design of process equipment that is to handle corrosive, poisonous, or otherwise dangerous substances, preference should always be given to closed systems. Where this is not possible, particularly in the case of poisonous dust or gas, special ventilating units that remove the harmful substance at the point of origin should be considered.

Flammable Liquids. The storage and handling of volatile, flammable liquids require that certain precautions be taken to minimize the hazard. The inherent fire and explosion hazard depends not only on its flash point, but also upon the apparent ignition temperature, explosive range, vapor density with respect to air diffusibility of the vapor in air, etc. Although data on these properties are incomplete at the present time, and those published by different authorities vary, the following table indicates the relative hazard of certain liquids (see also pp. 1868 to 1870).

Any volatile liquid that gives off flammable vapors at or below ordinary room temperature is considered hazardous from the fire prevention standpoint and warrants special consideration. The following definitions

Table 14. Guarding Specifications for Simple Vertical Belt Drives

The following table gives minimum requirements for guarding. Each exposed side of drive must be considered independently and guarding designed to meet worst condition of exposure.

Height of center of shaft above floor: Upper pulley, in.	Height of center of shaft above floor: Lower pulley, in.	Type of guarding	Minimum horizontal distance from guard to edge of pulley, in.	Height of top of guard above floor, in.	Filler (expanded metal, etc.)	Toeboard	Top of guard	Bottom of guard	Fig. 25, sketch number
0–24	0 or below floor	Railing	36	42	None	Around floor opening	......	......	1
		Railing	18	42	Complete	None	......	......	2
		Belt enclosure	2	Above top of pulley	Complete	None	Closed	Carried to floor	3
24–72	0 or below floor	Railing	36	42	None	Around floor opening	......	......	4
		Belt enclosure	2	72	Complete	None	Open	Carried to floor	5
		Belt enclosure	2	Above top of pulley	Complete	None	Closed	Carried to floor	6
24–72	0–72	Railing	36	42	None	Around floor opening	......	......	7
		Belt enclosure	2	72	Complete	None	Open	See Note A	9
		Belt enclosure	2	Above top of pulley	Complete	None	Closed	See Note A	10
72 or over	0 or below floor	Railing	18	42	None	Around floor opening	......	......	12
		Belt enclosure	2	6 above center of pulley	Complete	None	Open	Carried to floor	13
72 or over	0–24	Railing	36	42	None	Around floor opening	......	......	14
		Railing	18	42	Complete	None	......	......	15
		Belt enclosure	2	6 above center of pulley	Complete	None	Open	See Note A	17
72 or over	24–72	Railing	36	42	None	None			18
		Belt enclosure	2	6 above center of pulley	Complete	None	Open	See Note A	19
Over 72	Over 72 but bottom of pulley less than 84 above floor	Railing	18	42	None	None			20
		Belt enclosure	2	6 above center of pulley	Complete	None	Open	See Note A	21

NOTE A. Bottom of guard to be closed unless lower edge is 6 in. or less above floor and at least 2 in. below bottom of pulley.

Table 15. Guarding Specifications for Simple Inclined Belt Drives

The following table gives minimum requirements for guarding. Each exposed side of drive must be considered independently, with guarding designed and railings spaced to meet worst conditions.

Height of center of shaft above floor: Upper pulley, in.	Height of center of shaft above floor: Lower pulley, in.	Type of guarding	Minimum horizontal distance from guard to: Belt edges. (For spacing of side guards. Dimension D), in.	Face of lower pulley. (For spacing of end guards. Dimension E), in.	Face of upper pulley. (For spacing of end guard. Dimension F), in.	Height of top of guard above floor, in.	Filler (expanded metal, etc.)	Toeboard	Top of guard	Bottom of guard	Fig. 26, sketch number
0–24	0 or below floor	Railing	36	36-Note C	36	42	None	Around floor opening	......	......	1
		Railing	18	18-Note C	18	42	Complete	None	......	Carried to floor	2
		Belt enclosure	2	2	2	Above top of pulley	Complete	None	Closed	Carried to floor	3
24–72	0 or below floor	Railing	36	36-Note C	36	42	None	Around floor opening	......	......	4
		Belt enclosure	2	2	2	72	Complete	None	Open	Carried to floor	5
		Belt enclosure	2	2	2	Above top of pulley	Complete	None	Closed	Carried to floor	6
24–72	0–72	Railing	36	36	36	42	None	Around floor opening	......	......	7
		Belt enclosure	2	2	2	72	Complete	None	Open	See Note A	9
		Belt enclosure	2	2	2	Above top of pulley	Complete	None	Closed	See Note A	10
72 or over	0 or below floor	Railing	36	36-Note C	18-Note B	42	None	Around floor opening	......	......	12
		Belt enclosure	2	2	2	78	Complete	None	Open	Carried to floor	13
72 or over	0–24	Railing	36	36	18-Note B	42	None	Around floor opening	......	......	14
		Railing	18	18	18-Note B	42	Complete	None	......	Carried to floor	15
		Belt enclosure	2	2	2	78	Complete	None	Open	See Note A	17
72 or over	24–72	Railing	36	36	18-Note B	42	None	None			18
		Belt enclosure	2	2	2	78	Complete	None	Open	See Note A	19
Over 72	Over 72, but bottom of pulley less than 84 above floor	Belt enclosure	2	2	....	See Note D	Complete	None	Open	Closed	20

NOTE A. Bottom of guard to be closed unless lower edge is 6 in. or less above floor at least 2 in. below bottom of pulley.
NOTE B. If center of shaft of upper pulley is higher than 72 in. above floor this dimension to be taken to nearest part of belt at a height of 72 in.
NOTE C. If lower pulley is below floor, one-half this distance may be used as spacing of end railing from belt at floor level.
NOTE D. Guard to be carried up 6 in. above center of pulley.

Table 16. Guarding Specifications for Simple Horizontal Belt Drives

The following table gives minimum requirements for guarding on sides of drives. Each exposed side of each drive must be considered independently and guarding designed to meet worst condition of exposure.

Height of pulley above floor: Upper edge, in.	Height of pulley above floor: Lower edge, in.	Type of guarding	Minimum horizontal distance from guard to edge of pulley, in.	Height of top of guard above floor, in.	Height of bottom of guard above floor, in.	Filler (expanded metal, etc.)	Toeboard	Top of guard	Bottom of guard	Fig. 27, sketch number
0–36	....	Railing	36	42	....	None	Around floor opening*	....	....	2
		Railing	6	42†	....	Complete†	None	....	Carried to floor	3
		Belt enclosure	2	Above top of pulley	0	Complete	None	Closed	Carried to floor	6
		Belt enclosure	2	Above top of pulley	Below belt	Complete	None	Closed	Closed	7
36–66	....	Railing	36	42	....	None	Around floor opening*	....	....	9
		Railing	6	At least 6 above pulley†	See Note A	Complete†	None	....	See Note A	10
		Belt enclosure	2	Above top of pulley	0	Complete	None	Closed	Carried to floor	11
		Belt enclosure	2	Above top of pulley	Below belt	Complete	None	Closed	Closed	12
66 and over	Under 66	Railing	36	42	....	None	Around floor opening*	....	....	13
		Railing	6	78†	See Note A	Complete†	None	....	See Note A	14
		Belt enclosure	2	78	0	Complete	None	Open	Carried to floor	15
		Belt enclosure	2	78	Below belt	Complete	None	Open	Closed	16
66 and over	66–84	Railing	18	42	....	None	None	....	....	17
		Belt enclosure	2	84	Below belt	Complete	None	Open	Closed	19
		Belt enclosure	2	Above top of pulley if below 84	Below belt	Complete	None	Closed	Closed	20

NOTE A. If filler extends 2 in. below bottom of pulley it may be terminated not more than 6 in. above floor, provided that, if part of belt or pulley is below floor, toeboards are provided around floor opening.

* If part of belt or pulley is below floor.

† In case of long horizontal drives guarded with filled railings located not closer than 12 in. to the belt, the "intermediate space" (see abjacent sketch) between the pulleys may be protected by *standard 42-in. railings without filler*. The "intermediate space" is considered as extending horizontally to within 36 in. of the center of the shaft of pulleys less than 60 in. in diameter and to within 6 in. of edge of pulleys 60 in. or more in diameter.

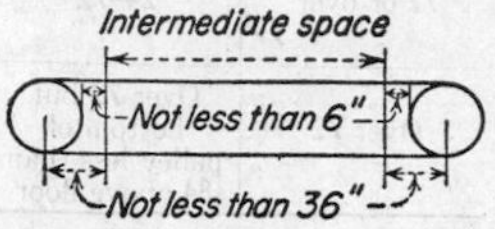

Table 17. Guarding Specifications for Horizontal Shafting

The following table gives minimum requirements for protection. In designing shafting guards minimum requirements for protection of belts and pulleys (see Tables 14, 15, and 16) and facilities for oiling shafting while in motion must be considered.

Shafting accessible from:	Vertical distance from shaft to floor etc., in.	Type of guarding	Horizontal distance from guard to shaft, in.: Best practice	Min.	Max.	Height of top of guard, in.	Height of bottom of guard	Filler (expanded metal, etc.)	Toeboard	Top of guard	Bottom of guard	Fig. 28, sketch number
Floor level platform or walkway	0–30 above	Railing§	24	24	30	42	....	None	Along railing	....	....	1, 2, 3
		Railing	12	6	24	42	6 in. below shaft or carried to floor	Complete*	None†	....	....	4, 5, 6, 7
		Shaft enclosure	2	..	..	Above top of shaft		Complete‡	....	Closed	Open	10, 11, 12
		Shaft enclosure	2	..	..	Above top of shaft	Below shaft	Complete	....	Closed	Closed	13
Floor level platform	30–66 above	Railing§	24	24	30	42	....	None	None†	....	....	8
		Shaft enclosure	2	..	..	Above top of shaft	Below shaft	Complete	....	Closed	Closed	14
Floor level platform or walkway	66–84 above	Railing§	18	18	24	42	....	None	None†	....	....	9
		Shaft enclosure	2	..	..	Above top of shaft	Below shaft	Complete	....	Closed	Closed	15
		Shaft enclosure	2	..	..	6 in. above top of Shaft	Below shaft	Complete	....	Open	Closed	16
Floor level (by portable ladder)	Over 84 above	None										
Elevated platform or walkway	0–12 below	Railing	18	18	24	42	....	None	Along railing	....	....	17 and 18
Elevated platform or walkway	12–24 below	Railing	12	8	12	42	....	None	Along railing	....	....	19

* If shaft is between 0 and 12 in. above floor, filler between mid-rail and top rail may be omitted. If shaft is between 24 and 30 in. above floor, filler between floor and mid-rail may be omitted.

† Not required as guard for machinery, but may be otherwise necessary. See Part II of "Safety Standards."

‡ Not necessary to extend enclosure to protect inaccessible side of shaft which is away from platform or walkway.

§ If end of shaft is accessible, railing to be extended around it and located not less than 12 in. (horizontal distance) from end bearing.

may be helpful when analyzing situations that involve the use of such materials:

Combustion,[1] generally speaking, implies some form of chemical change accompanied by the evolution of both heat and light. Although there are examples of true combustion in the entire absence of oxygen, the term usually means oxidation. There are two types of combustion; namely, slow combustion which is not accompanied by light such as the oxidation of iron; and rapid combustion which we know as "fire," due to the presence of flame. The term "fire" includes both combustion and flame, and the results of both.

Apparent ignition temperature of an element or compound, whether solid, liquid, or gaseous, is the temperature required to initiate or cause oxidation sufficiently rapid to be self-sustained when the heating or heated element is removed.

Flash point of a liquid is the temperature at which it gives off vapor sufficient to form an ignitible mixture with the air contained in the vessel used. It does not mean that no evaporation takes place below that temperature,

[1] Miner, *S.A.E. Journal*, **21,** No. 6 (1927).

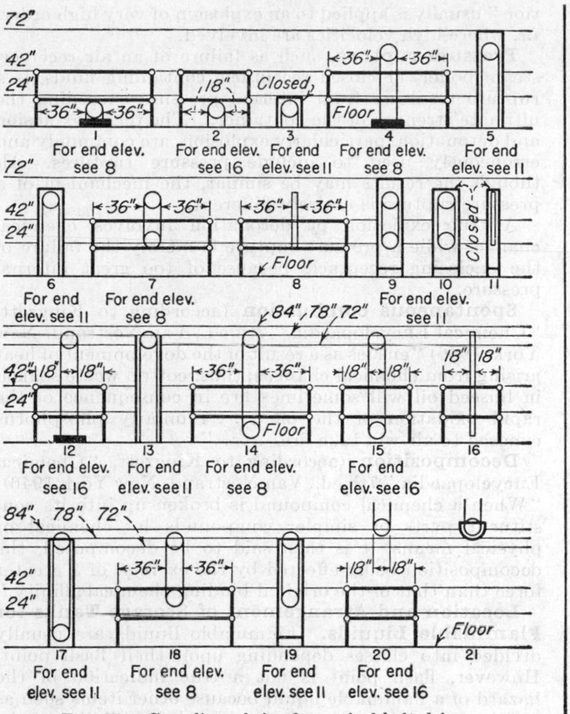

FIG. 25. Guarding of simple vertical belt drives.

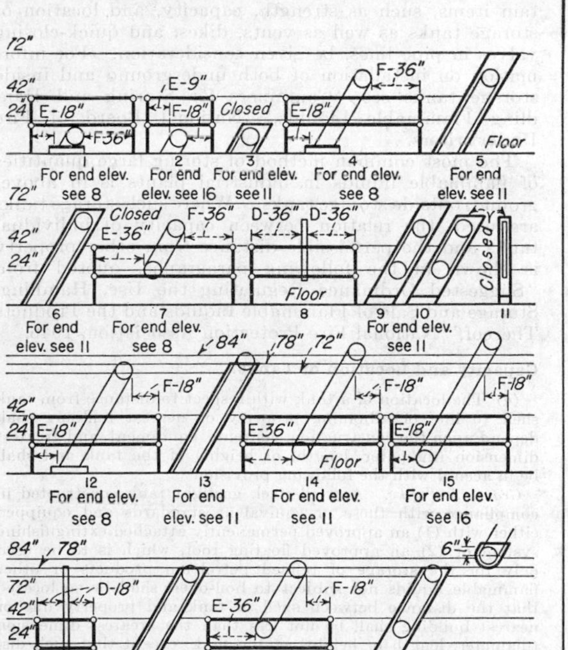

FIG. 26. Guarding of simple inclined belt drives.

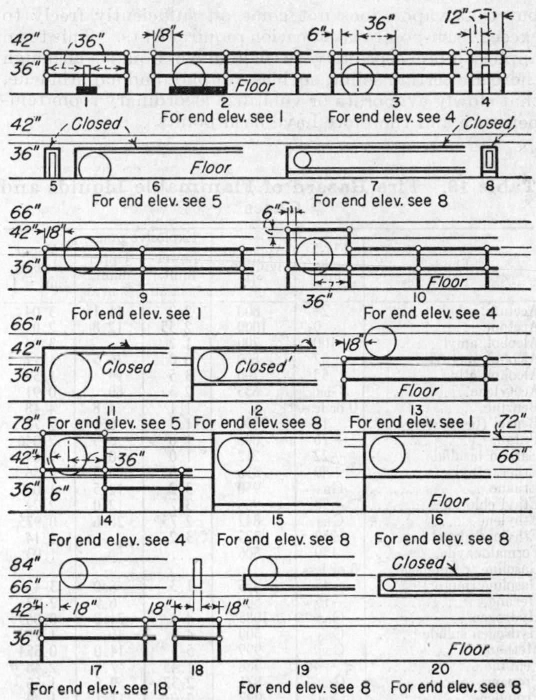

FIG. 27. Guarding of simple horizontal belt drives.

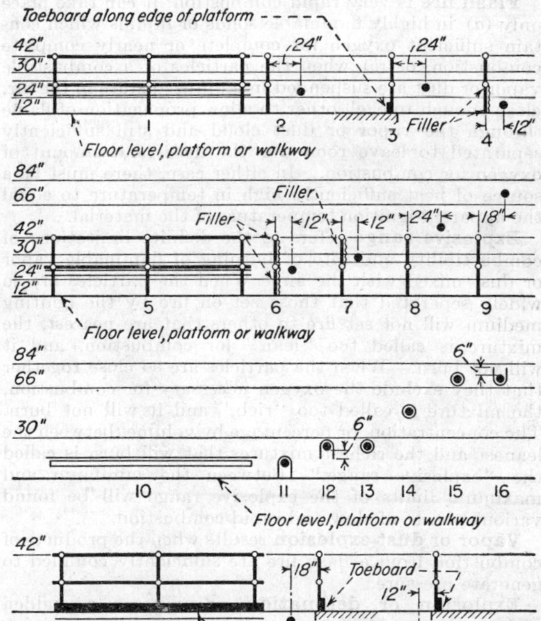

FIG. 28. Guarding of horizontal shafting.

but that vapor does not come off sufficiently freely to exceed flash-point classification requirements. This term applies more especially to flammable liquids, although there are certain solids, such as camphor and naphthalene, that slowly evaporate or volatilize at ordinary room temperature and therefore have flash points.

Table 18. Fire Hazard of Flammable Liquids and Gases*

Fluid	Flash point, °F.	Apparent ignition, °F.	Explosive range		Vapor density (air = 1)
			Lower limit, % in air	Upper limit, % in air	
Acetate, ethyl	24	800	2.2	11.15	3.04
Acetone	0	1000	2.55	12.8	2.0
Alcohol, amyl	100	700	1.2	....	3.04
Alcohol, methyl	52	800	6.0	36.5	1.11
Alcohol, ethyl	55	700	3.5	19	1.59
Acetylene	Gas	635	2.5	80	0.91
Benzine	0 or less	....	1.1	4.8	4.48
Benzol (benzene)	12	1000	1.5	8.0	2.77
Butane	−76	806	1.6	8.5	2.046
Carbon bisulfide	−22	212	1.0	50	2.64
Ether, ethyl	−49	356	1.85	36.5	2.56
Ethane	Gas	950	3.2	12.5	1.049
Ethyl chloride	−58	966	3.7	12.0	2.22
Ethylene	Gas	842	2.75	28.6	0.975
Ethylene glycol	232	775	3.2	....	2.14
Formaldehyde	130	806	....	.....	1.03
Gasoline (casinghead)	0 or less				
Gasoline (motor)	−45	495	1.3	6.0	3–4
Hexane	−15	500	1.2	6.9	2.97
Hydrogen	Gas	1085	4.1	74.2	0.069
Hydrogen sulfide	Gas	500	4.3	46	1.189
Methane	Gas	999	5.3	14.0	0.554
Pentane	< −40	588	1.45	7.5	2.48
Propane	Gas	871	2.37	9.5	1.56
Propylene	Gas	927	2.0	11.1	1.49
Pyridine	68	900	1.8	12.5	2.73
Toluol	40	1026	1.27	7.0	3.14

* Based on "Fire Hazard Properties of Certain Flammable Liquids, Gases, and Volatile Solids," publication of the National Fire Protection Association.

Flash fire is very rapid combustion; it can take place only (*a*) in highly flammable solids or liquids which contain sufficient oxygen for complete or nearly complete combustion, or (*b*) when the particles of a combustible vapor or dust are suspended in a diffused state in the air, close enough to each other to allow propagation of flame through the vapor or dust cloud and still sufficiently separated to leave room for the necessary amount of oxygen for combustion. In either case, there must be a source of heat sufficiently high in temperature to equal the apparent ignition temperature of the material.

Explosive range refers to the definite limitations of combustibility and rate of burning of flammable vapor or dust mixed with the air. When the particles are so widely separated that those set on fire by the igniting medium will not set fire to others that are nearest, the mixture is called too "lean" for combustion, and it will not burn. When the particles are so close together that they exclude the oxygen necessary for combustion, the mixture is called too "rich," and it will not burn. The concentration, or percentage by volume, between the leanest and the richest mixtures that will burn is called the "explosive range." Between the minimum and maximum limits of the explosive range will be found various phases of slow and rapid combustion.

Vapor or dust explosion results when the products of combustion from a flash fire are sufficiently confined to generate pressure.

Explosion or detonation presupposes a sudden violent change of pressure, characteristically involving the liberation and expansion of a large volume of gas due to high temperature. The change taking place is a progressive one proceeding from one part of the exploding material to the next adjoining part. The term "detonation" usually is applied to an explosion of very high order, *i.e.* where high velocities are involved.

Pressure rupture, such as failure of an air receiver, steam boiler, or closed receptacle containing fluids, is a rupture resulting from internal pressure exceeding the ultimate strength of the container. The terms explosion and detonation, particularly explosion, are commonly and erroneously used to include pressure ruptures. Although the results may be similar, the mechanism of a pressure rupture is entirely different.

A true explosion or detonation involves chemical change, while a pressure rupture is merely the failure of the enclosing receptacle because of too great internal pressure.

Spontaneous combustion (according to Kingzett, "Chemical Encyclopaedia," 7th ed., Van Nostrand, New York, 1946) "ensues as a result of the development of heat arising from chemical changes; thus, cotton waste soaked in linseed oil will sometimes fire in consequence of the rapid oxidation of the oil. . . . Similarly phosphorus exposed to air will take fire. . . . "

Decomposition (according to Kingzett, "Chemical Encyclopaedia," 7th ed., Van Nostrand, New York,1946). "When a chemical compound is broken up into its constituent parts or simpler compounds by chemical or physical means, it is then said to be decomposed, the decomposition being effected by the exercise of a greater force than that of the original binding chemical affinity."

Location and Arrangement of Storage Tanks for Flammable Liquids. Flammable liquids are usually divided into classes depending upon their flash point. However, flash point is not a true indication of the hazard of a flammable liquid because other items such as ignition temperature, explosive range, diffusibility of the vapors, and density of the vapors also affect the hazard. The potential hazard involved in the storage and handling of flammable liquids in industrial plants requires that certain items, such as strength, capacity, and location of storage tanks as well as vents, dikes, and quick-closing valves in pipe lines, be given consideration. For information on installation of both underground and inside storage tanks see "Containers for Storing and Handling Flammable Liquids," National Board of Fire Underwriters.

The most common method of storing large quantities of flammable liquids in industrial plants is in aboveground, outside storage tanks. Where such storage tanks are used, the relation between capacity of individual tanks and the permissible distance from other property is shown in the following paragraphs, quoted from "Suggested Ordinance Regulating the Use, Handling, Storage and Sale of Flammable Liquids and the Products Thereof," National Fire Protection Association, 1946.

Capacity and Location of Tanks.

(*a*) The location of a tank with respect to distance from tank shell to line of adjoining property or nearest building shall depend upon the construction, contents, equipment, and greatest dimension (diameter, length, or height) of the tank and shall be in accord with the following provisions:

Group A Tanks. Any all-steel, gastight tank constructed in compliance with these or equivalent standards and equipped either with (1) an approved permanently attached extinguishing system or (2) an approved floating roof, which is to be used only for the storage of refined petroleum products or other flammable liquids not subject to boil-over, shall be so located that the distance between shell of tank and property line or nearest building shall be not less than the greatest dimension (diameter, length or height) of the tank, except that such distance need not exceed 120 ft.

Group B Tanks. Any all-steel, gastight tank constructed in compliance with these or equivalent standards but not equipped either with (1) an approved permanently attached extinguishing system or (2) an approved floating roof, which is to be used only for the storage of refined petroleum products or other flammable liquids not subject to boil-over, shall be so located that the distance between shell of tank and property line or nearest building shall be not less than 1½ times the greatest dimension (diameter, length, or height) of the tank, except that such distance need not exceed 175 ft.

Group C Tanks. Any all-steel, gastight tank constructed in compliance with these or equivalent standards and equipped either with (1) an approved permanently attached extinguishing system or (2) an approved floating roof, which is to be used for the storage of crude petroleum or other flammable liquid subject to boil-over, shall be so located that the distance between shell of tank and property line or nearest building shall be not less than twice the greatest dimension (diameter, length, or height) of the tank except that such distance shall be not less than 20 ft. and need not exceed 175 ft.

Group D Tanks. Any all-steel, gastight tank constructed in compliance with these or equivalent standards and not equipped either with (1) an approved permanently attached extinguishing system or (2) an approved floating roof, which is to be used for the storage of crude petroleum or other flammable liquid subject to boil-over, shall be so located that the distance between shell of tank and property line or nearest building shall be not less than three times the greatest dimension (diameter, length, or height) of the tank except that such distance shall be not less than 20 ft. and need not exceed 350 ft.

(*b*) The minimum distance between shells of any two all-steel, gastight tanks shall be not less than one-half the greatest dimension (diameter, length, or height) of smaller tank except that such distance shall not be less than 3 ft., and for tanks of 18,000 gal. or less the distance need not exceed 3 ft.

(*c*) Tanks shall be so located as to avoid possible danger from high water.

(*d*) When tanks are located on a stream without tide they shall, where possible, be downstream from burnable property.

Three methods of handling these liquids are in common use—by pumping, by gravity, and by compressed air or inert gas. (The use of compressed air for this purpose is extremely dangerous and should not be allowed.) Pumping is preferable to gravity discharge and should be used wherever possible.

Testing Devices for Flammable Vapors. Flammable-vapor indicators, which read in terms of the lower limit of the explosive range, are now available for determining the approximate amount of flammable vapor in air. Most of the instruments will operate satisfactorily, not only in any given solvent vapor but also in any mixture of vapors even though the composition may be unknown.

Static Electricity. Static electricity may develop from handling materials—solids, liquids, or gases—or from the operation of equipment such as belts. The handling of dry granular substances, such as sulfur in metal chutes, bins, or even through relatively dry air, will generally result in the generation of static electricity. Similarly, static electricity may be generated by the flow of certain liquids through pipe lines.

The discharge of static electricity in the presence of flammable dusts or vapors is considered hazardous; and, where this may occur, a proper system of grounding, humidification of the air in the room, or the use of certain static eliminators which are available today is usually warranted. For additional information on the hazards of static electricity, consult "Static Electricity," a publication of the National Fire Protection Association.

Compressed Gases in Cylinders. The Interstate Commerce Commission has established definite regulations that must be complied with in **filling and transporting** cylinders of compressed gases. These rules also include specifications for the construction of the containers. In **using** these cylinders, a few of the more important precautions to be taken are listed below; additional safeguards are outlined in the following regulations, published by the National Board of Fire Underwriters: Acetylene Equipment for Lighting, Heating and Cooking, 1930; Gas Systems for Welding and Cutting, 1947; Liquified Petroleum Gases, 1949.

1. Extreme care should be exercised in handling so that cylinders are neither dropped nor permitted to strike against each other.
2. Cylinders should be protected against mechanical injury when in use.
3. Precautions should be taken that the safety devices with which the cylinders are equipped are not tampered with.
4. When cylinders are not in use, outlet valves should be kept tightly closed even though cylinders may be considered empty.
5. When the cylinders have been exhausted, the discharge valves should be closed and protecting caps screwed securely into position.
6. Before attaching the required pressure regulator to the cylinder, the valve should be slightly opened and then closed to be sure it is in proper working condition.
7. Only the special wrenches and tools provided by the manufacturer should be used on valves.
8. Make sure that the thread connections on regulators correspond to those on the cylinder outlet as the threads may vary somewhat for different gases.
9. Regulators and pressure gages provided for use for a particular gas must not be used on cylinders containing other gases.
10. All cylinders should be protected against excessive rise in temperature and against extremes of weather. They should not be exposed to continued dampness.
11. In order to avoid confusion, full cylinders should be stored apart from the empty ones.
12. Cylinders of compressed gas should not be stored near highly flammable materials.
13. In welding or cutting, no sparks should be allowed to strike the cylinders.
14. Acetylene is shipped in cylinders completely filled with a porous material which is charged with a suitable solvent (usually acetone).
15. Every possible precaution should be taken to prevent oxygen from coming in contact with oil or grease.
16. The cylinders received by the plant should bear a conspicuous standard label indicating the kind of gas. The color of the label shows whether the gas is flammable, corrosive, or inert.
17. Oxygen should never be used as a substitute for compressed air.

Dust Explosions (see definition under Flammable Liquids). Dust explosions can occur in any combustible dust or any material that will burn or oxidize, which includes some metallic dusts.

When there is a proper mixture for rapid combustion, it is said to be within the explosive range, which includes all percentages of dust and air from the lower to the upper limit. Below the lower limit, flame propagation will not take place and, above the upper limit, there may be burning, but an explosion or flash fire is not likely to occur in apparatus or enclosed spaces unless air is admitted. The lower and the upper limit may be widely separated, depending upon the character of the dust.

The degree of flammability of a dust depends upon its moisture content and upon its fineness. The drier and the finer a dust is, the greater its flammability.

Even though the intensity of the explosion may vary, according to the percentage of dust and air, any dust combination within or above the explosive range is dangerous and should not be permitted. The safeguarding of this hazard, as with flammable gases, involves the dilution below the lower limit of the explosive range and immediate removal while in the diluted state.

In most disastrous dust explosions there are usually two explosions—a primary and a secondary. The primary involves a relatively small quantity of dust and is local in character, but of sufficient intensity to dislodge dust which has collected in buildings—on ledges, side walls, floors, or which may be released due to rupture of equipment or otherwise thrown into the air. If ignited, a more general explosion, the secondary, may occur which may result in major structural damage. Precautions against dust explosions consist of keeping the premises scrupulously clean and eliminating sources of ignition.

SPECIAL SAFETY PROTECTION EQUIPMENT

The use of protective equipment is a secondary measure against injury and health hazards which cannot be eliminated by the design and arrangement of equipment, and by the installation of adequate ventilation. Since the human factor enters into the wearing of such equipment, its dependability is variable and at times uncertain.

For the protection of the head and eyes, such equipment as goggles, shields, helmets, hoods, masks, and hats should be worn. The breathing of poisonous gases and fumes, or of an atmosphere in which there is a deficiency of oxygen, must be guarded against by the use of the proper respiratory protective devices. The hands, arms, legs, and feet may be protected by the use of rubber or canvas gloves, rubber boots and aprons, leggings, shoes, and other clothing.

Protective equipment should not be transferred from one employee to another unless it is sterilized. See American Standard Safety Code for the Protection of Heads, Eyes, and Respiratory Organs for additional information.

Goggles and Helmets. Goggles may be divided into approximately seven distinct types. Table 19 will serve as a guide in selecting the general type for the work to be done.

Table 19. Recommended Head and Eye Protection

Conditions	Protection
Chipping, riveting, calking, etc.	Types 2 and 6
Scaling, grinding, etc.	Type 1, 2, or 6; preferably Type 2
Exposure to dust and wind	Type 1, 2, 5, or 7; depending on conditions
Babbiting	Babbiting mask
Handling corrosive chemicals	Types 4 and 5, or face masks, or hoods; depending on conditions
Sandblasting	Sandblast cabinets or helmets
Sunlight, snow, and similar sources of glare	Type 1, 2, 3, or 5 goggles; the shade of lenses to be determined by conditions
Oxyacetylene welding and furnace work	Helmets, shields, or type 3 goggles; shade 6 lenses
Electric welding	Helmets or shields with shade 10 to 12 windows
Nitrometer work	Special nitrometer shield

Type 1. Non-shatterable lenses in strong frames with no side shields.
Type 2. Non-shatterable lenses in strong frames with side shields.
Type 3. Oxyacetylene welding.
Type 4. Close-fitting rubber, little or no ventilation, non-shatterable lenses.
Type 5. Cup type, close-fitting, wide vision, little or no ventilation, non-shatterable lenses.
Type 6. Cup type, close-fitting, wide vision, with non-shatterable lenses, ventilated.
Type 7. Cup type, tight-fitting, non-shatterable lenses.

Respiratory Protection. Devices for respiratory protection are divided into the following general classes by the Bureau of Mines, and the uses for each class are indicated:

1. Gas Mask—Canister Type. For use as protection against special gases in concentrations not exceeding 2 per cent in air. Some canisters are available with an extra filter to give better protection against dust, smoke, etc. Should be used only when nature of gas is known and sufficient oxygen supply is available.
2. Gas Mask—Canister Universal Type. For protection against all gases, including carbon monoxide, not exceeding 2 per cent concentration in air. Should be used when not certain of nature of gas, or as protection against carbon monoxide. Do not use in atmosphere containing less than 16 per cent oxygen.
3. Self-contained Oxygen Breathing Apparatus. Self-contained units for supplying pure oxygen from high-pressure cylinders. Apparatus is equipped with reducing valve, face mask, etc. Can be used for complete respiratory protection in any atmosphere.
4. Supplied-air Respirators. Four types are available; hose masks with supplied air from blower with unit, special hose masks without blower, air-line respirators and abrasives masks for use with compressed-air supply. Can be used for complete respiratory protection, but movement of individual restricted by air supply line, and in some cases use of life line and rescue harness advisable.
5. Dispersoid (Dust, Fume, and Mist) Respirators. Mechanical filter units for removing dusts and fumes from otherwise breathable air. Several approved units are available for various types of toxic and non-toxic dusts and fumes.
6. Non-emergency Gas Respirators (Chemical Cartridge Respirators). A respirator with chemical filter for respiratory protection in atmospheres not immediately dangerous to life or containing not more than 0.1 per cent of certain vapors by volume.

Table 20. Standard Colors for Gas-mask Canisters

Gases, Vapors, Smoke, Etc.	Color
Acid	White
Organic vapor	Black
Ammonia	Green
Acid and organic vapors	Yellow
Hydrocyanic acid	White with green stripe at bottom of canister
Acid, organic and ammonia vapors	Brown
Universal (all gases)	Red

The various canisters can be obtained provided with an additional filter to give better protection against dust, solid particles of smoke, etc. These canisters will have the same color scheme with a stripe in a contrasting color.

SPECIAL FIRE PROTECTION EQUIPMENT

Special fire protection equipment is receiving an increasing amount of attention. New or unusual processes should be carefully analyzed, the hazards determined, and the best methods of fire extinguishing provided. In some cases more protection is required than that afforded by the standard system of automatic sprinklers or by first-aid extinguishing appliances. The following are some of the more common systems in use.

Foam (Two-solution System). Foam is generated by mixing a solution of aluminum sulfate, which is known as A solution, with a solution of bicarbonate of soda and a foam stabilizer, which is known as the B solution. The solutions are stored in separate tanks until required for use when they are pumped to a special mixing device, which discharges foam. Foam is particularly useful in class B fires, which are those in oils, greases, flammable liquids, etc., where the blanketing or smothering effect of the extinguishing agent is of greatest importance.

Foam (Dry-powder Generators). Foam is generated in a special device by mixing water with dry chemicals. The discharge is continuous as long as the water is supplied and the chemicals fed into the device. The foam is piped through permanent piping or hose to the location where needed. The smaller sizes of generators are readily portable.

Carbon Dioxide. Carbon dioxide may be stored in the shipping containers under pressure and piped to the apparatus or location where required. At this point it may be discharged either automatically or manually through suitable nozzles, thereby smothering the fire. There is a slight cooling effect, although the characteristic of carbon dioxide extinguishing is the smothering effect. The automatic discharge is secured through fixed temperature or through temperature rate-of-rise devices. The former operate when a predetermined temperature has been reached, the latter when the increase in temperature exceeds a predetermined rate of rise.

Steam is sometimes used for special locations or in certain equipment. It is not generally recommended.

Auxiliary Sprinkler Systems. Automatic sprinklers, or in some cases sprinklers with the fusible links removed, may be supplied with water controlled by a quick opening valve or by a deluge valve which in turn may be actuated thermostatically or manually by remote pull handles. This arrangement is used where it is desirable to discharge water over extra-hazardous equipment or areas.

Other inert gases, such as flue gas or nitrogen, is sometimes piped to the interior of grinding, pulverizing, mixing, or conveying equipment, where extra-hazardous products are handled or where there may be acute hazards due to presence of flammable dust or gases.

ORGANIZATION AND WORK

Safety and Fire Protection Organization. This organization should be essentially the same as the production organization. There should be a central committee composed of the plant manager or the assistant plant manager as permanent chairman and the safety and fire protection supervisor, if there be one, as permanent secretary. The members of this committee should be the remainder of the plant executive staff.

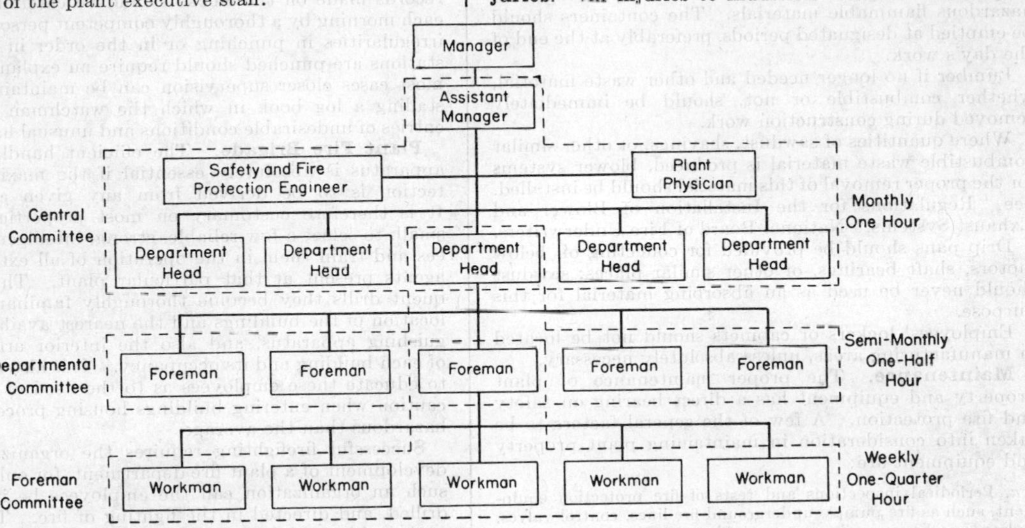

FIG. 29. Chart of a typical safety and fire protection organization.

There should also be a departmental committee in each department, or such subdivision, composed of the head of the department as permanent chairman and the foremen as members. In order to carry this organization to the workmen, each foreman should have a committee made up of himself as permanent chairman and his men as members, the latter being rotated if there are too many to serve at one time.

The central safety and fire protection committee will take care of general policies regarding safety and fire protection. The departmental committee will be responsible for safety and fire protection in their departments, and the foreman's committees in their particular areas.

Figure 29 shows an outline of a typical plant safety and fire protection organization.

Employee Selection and Training. During the Second World War, the importance of the selection and training of individual employees for specific industrial jobs was clearly demonstrated. The problems of quickly selecting and training millions of men and women with no prior experience presented an excellent opportunity for the application on a large scale of methods developed through the combined efforts of United States industries and Federal agencies. The results obtained proved the importance of the two primary factors required for developing efficient and safe employees; first, preemployment physical examination and selection of employees properly equipped physically and mentally for the job to which they will be ultimately assigned; and, second, training the employees in the proper and safe way to perform the individual job or operation. Job Instruction Training, Job Methods Training, and Job Relations Training are some of the titles of available methods of proved instruction. For detailed information the Training Within Industry service of the Federal government should be consulted. Not only should employees be thoroughly examined by a competent physician prior to employment, but annual reexamination is advisable, with more frequent checks in industries handling toxic materials. All employee selection, training, and physical check programs should be supervised or at least coordinated by the safety supervisor or the individual assigned to handle the safety and personnel programs.

Investigation and Classification of Industrial Injuries. All injuries to industrial workers occurring in the course of and arising out of employment, including industrial disease, should be promptly investigated and reported to the proper authorities. A standard system of reporting and classifying injuries should be used, in order that results obtained will be uniform and comparable with those in general use. The American Standard Z16.1–1945 Method of Compiling Industrial Injury Rates should be followed in reporting injuries to the National Safety Council and for record purposes. It will be necessary to use special forms in states requiring reports, but all statistics used for comparison should be based on the Standard Z16.1 method.

Although the primary objective of every industrial safety organization is the prevention of injuries by the elimination of the cause, a careful analysis of accidents can be used to excellent advantage in the elimination of hazards that may cause other injuries. A study of the plant injury experience and that of the industry in general may be used to evaluate the need for further accident prevention activities on the plant and in different departments. Uniform records may thus be used for determining the relative effectiveness of the safety program as compared with similar establishments or industries and the progress made in accident prevention within an establishment or industry.

Housekeeping. Housekeeping has a direct bearing on the number of accidents and fires that occur, as well as on the plant operating costs. It is essential for the safety and fire protection engineer or for the employee who is charged with this work to pay particular attention to this very important item. The rule, "a place for everything and everything in its place," should be enforced. Keen interest in housekeeping should be evidenced by the plant management, who should require his subordinates, and they in turn the workmen, to be directly responsible for the housekeeping in their immediate vicinity. The maximum result from the minimum effort expended comes from daily or even more frequent sweeping and collection of refuse. Employees should be trained to minimize the cleaning required by depositing in receptacles provided for the purpose all dirty waste and general refuse, discarded packing materials, and any other waste materials as such and those used for maintenance and operation as they are produced.

Suitable metal containers with deep-lipped covers should be provided for oil-soaked waste, cleaning rags, and hazardous flammable materials. The containers should be emptied at designated periods, preferably at the end of the day's work.

Lumber if no longer needed and other waste material, whether combustible or not, should be immediately removed during construction work.

Where quantities of sawdust, shavings, or other similar combustible waste material is produced, blower systems for the proper removal of this material should be installed. See "Regulations for the Installation of Blower and Exhaust Systems," National Board of Fire Underwriters.

Drip pans should be provided for collecting oil below motors, shaft bearings, or other similar places; sawdust should never be used as an absorbing material for this purpose.

Employees' lockers or cabinets should not be located in manufacturing areas, unless absolutely necessary.

Maintenance. The proper maintenance of plant property and equipment has a direct bearing on safety and fire protection. A few of the general factors to be taken into consideration in maintaining plant property and equipment are:

a. Periodical inspections and tests of fire protection equipment, such as fire pumps, underground fire lines, control valves, hydrants and hose equipment, sprinkler systems, etc.

b. Periodical inspections and tests of safety equipment, such as respirators, gas masks, goggles, guards, railings, emergency showers, ladders, etc.

c. Regular inspections of buildings, platforms, tanks, pressure vessels, machinery, etc.

d. Prompt repairs of any equipment in case it should be needed.

Watchman Service. The plant watchman holds a position of trust and is an important factor in the protection of property against fire and theft. His responsibility is great since he is often in sole charge over half of the daily 24 hr. Therefore, care should be taken.

To Select the Right Man. One with unquestionable character, habits, and reliability; courage, keen intelligence; physically fit and able-bodied and with sufficient mechanical ability.

To instruct him fully in the duties necessary for efficient service. This should be the duty of a responsible person, preferably the plant management who is familiar with all plant conditions. The watchman should familiarize himself thoroughly with the plant, and in particular stairways, elevators, fire doors, pipe lines, sprinkler systems, hand hose, first-aid extinguishing equipment, and the fire-alarm system. He should know the location of sprinkler control valves. He should have the necessary knowledge to start and run the fire pumps. He should know how to use the telephone, and such telephone numbers as would enable him to call assistance should be posted near the telephone. He should have sufficient instruction regarding the electrical equipment to enable him to manipulate switches to control the lighting of the plant where and when necessary, or to shut off the current in case of accident.

To Support Him Thoroughly. The watchman should report undesirable conditions and unusual happenings to his superior, action upon which should be taken at once to give the watchman the confidence to report further undesirable conditions. If decision is unfavorable on his recommendations, the reasons should be fully explained to him.

In order to ensure that the watchman visits all the important points in the plant and as evidence that he has properly performed his duty, a watchclock system should be provided with stations well located throughout the property. The portable-clock system is a very reliable and usually an inexpensive system to install. The records made on the clocks should be carefully checked each morning by a thoroughly competent person, and any irregularities in punching or in the order in which the stations are punched should require an explanation. In most cases closer supervision can be maintained by installing a log book in which the watchman can make entries of undesirable conditions and unusual happenings.

Plant Fire Brigade. The efficient handling of fire apparatus is absolutely essential if the maximum protection is to be derived from any given equipment. It is therefore customary on most properties, even if small, to select a few reliable, strong, intelligent employees and train them in the operation of all extinguishing agents present at that particular plant. Through frequent drills they become thoroughly familiar with the location of the buildings and the nearest available extinguishing apparatus, and also the interior arrangement of each building and its occupancy. It is also customary to educate these employees as to the need of using due caution when entering buildings housing processes more hazardous than the average.

Successful firefighting requires the organization and development of a plant fire department, for only through such an organization can the employees be instructed, drilled, and directed in the fighting of fire. This phase of the subject might be summarized as follows:

a. For the prompt extinguishing of fire the employees must know the location and use of the first-aid fire protection equipment, and the minor extinguishing apparatus, such as chemical extinguishers, pails, small hand hose, etc.

b. To obtain maximum benefit from the major equipment, *i.e.,* outside hydrants, large fire hose, fire pumps, chemical engines, etc., certain employees must know the location and be trained in the proper method of handling and directing this apparatus, as well as knowing its capacity.

c. To prevent confusion and delay, a definite line of authority must be established.

d. To check successfully the spread of fire, those in authority must have knowledge as to building construction, arrangement, nature of contents, and inter-exposure between buildings.

For additional information on fire brigades, consult the "Suggestions for the Organization, Drilling and Equipment of Private Fire Brigades," National Board of Fire Underwriters.

MECHANICAL TABLES AND DATA FOR THE SAFETY AND FIRE PROTECTION ENGINEER

Table 21. Average Ultimate Strength of Common Metals

Lb./sq. in.

Material	Tension	Compression	Shear	Modulus of elasticity
Aluminum	15,000	12,000	12,000	11,000,000
Brass, cast	24,000	30,000	36,000	9,000,000
Bronze, gunmetal	32,000	20,000	......	10,000,000
Bronze, manganese	60,000	120,000		
Bronze, phosphor	50,000	......	......	14,000,000
Copper, cast	24,000	40,000	30,000	10,000,000
Copper wire, annealed	36,000	......	......	15,000,000
Copper wire, unannealed	60,000	......	......	18,000,000
Iron, cast	15,000	80,000	18,000	12,000,000
Iron wire, annealed	60,000	......	......	15,000,000
Iron wire, unannealed	80,000	......	......	25,000,000
Iron, wrought	48,000	46,000	40,000	27,000,000
Lead, cast	2,000	......	......	1,000,000
Steel castings	70,000	70,000	60,000	30,000,000
Steel, plow	270,000			
Steel, structural	60,000	60,000	50,000	29,000,000
Steel wire, annealed	80,000	......	......	29,000,000
Steel wire, unannealed	120,000	......	......	30,000,000
Steel wire, crucible	180,000	......	......	30,000,000
Steel wire, suspension bridge	200,000	......	......	30,000,000
Steel wire, piano	300,000			
Tin, cast	3,500	6,000	......	4,000,000
Zinc, cast	5,000	20,000	......	13,000,000

"Machinery's Handbook."

Table 22. Tensile and Compressive Strength of Different Kinds of Steel

The ultimate strength of steel in tension and compression is practically the same and may, for different kinds of steel, be assumed as follows:

Kind of steel	Ultimate strength lb./sq. in.	Kind of steel	Ultimate strength lb./sq. in.
Structural steel for rivets	55,000	Machine steel	75,000
Structural steel for beams	60,000	Gun steel	90,000
Boiler steel for rivets	50,000	Axle steel	100,000
Boiler steel for plates	60,000	Spring steel	125,000

"Machinery's Handbook."

Table 23. Average Ultimate Strength of Common Materials Other than Metals

Lb./sq. in.

Material	Compression	Tension
Bricks, best hard	12,000	400
Bricks, light red	1,000	40
Brickwork, common	1,000	50
Brickwork, best	2,000	300
Cement, Portland, 1 month old* (4½ gal. water per sack of cement)	4,000	400
Cement, Portland, 1 year old* (4½ gal. water per sack of cement)	6,000	600
Concrete, Portland cement* (7½ gal. water per sack of cement)	2,000	200
Concrete, Portland, 1 year old* (7½ gal. water per sack of cement)	3,500	350
Hemlock	4,000	6,000
Pine, shortleaf yellow	6,000	9,000
Pine, Georgia (longleaf)	8,000	12,000
Pine, white	5,500	7,000
White oak	7,000	10,000

"Machinery's Handbook" and (*) Portland Cement Association.

Table 24. Influence of Temperature on the Strength of Metals

Material	Degrees Fahrenheit 210	400	570	750	930	1100	1300	1475
	Strength of metal in % of the strength at 70°F.							
Wrought iron	104	112	116	96	76	42	25	15
Cast iron	...	100	99	92	76	42		
Cast steel	109	125	121	97	57			
Structural steel	103	132	122	86	49	28		
Copper	95	85	73	59	42			
Bronze	101	94	57	26	18			

"Machinery's Handbook."

Table 25. Strength of U. S. Standard Bolts from 1/4 to 3 In. in Diameter

Bolt: Diam. of bolt, in.	No. of threads per in.	Areas: Full bolt, sq. in.	Bottom of thread, sq. in.	Tensile strength, lb.: At 10,000 lb./sq. in.	At 12,500 lb./sq. in.	At 17,500 lb./sq. in.	Shearing strength, lb.: Full bolt, At 7,500 lb./sq. in.	Full bolt, At 10,000 lb./sq. in.	Bottom of thread, At 7,500 lb./sq. in.	Bottom of thread, At 10,000 lb./sq. in.
1/4	20	0.049	0.027	270	340	470	380	490	200	270
5/16	18	.077	.045	450	570	790	580	770	340	450
3/8	16	.110	.068	680	850	1,190	830	1,100	510	680
7/16	14	.150	.093	930	1,170	1,630	1,130	1,500	700	930
1/2	13	.196	.126	1,260	1,570	2,200	1,470	1,960	940	1,260
9/16	12	.248	.162	1,620	2,030	2,840	1,860	2,480	1,220	1,620
5/8	11	.307	.202	2,020	2,520	3,530	2,300	3,070	1,510	2,020
3/4	10	.442	.302	3,020	3,770	5,290	3,310	4,420	2,270	3,020
7/8	9	.601	.419	4,190	5,240	7,340	4,510	6,010	3,150	4,190
1	8	.785	.551	5,510	6,890	9,640	5,890	7,850	4,130	5,510
1 1/8	7	.994	.693	8,930	8,660	12,130	7,450	9,940	5,200	6,930
1 1/4	7	1.227	.890	9,890	11,120	15,570	9,200	12,270	6,670	8,900
1 3/8	6	1.485	1.054	10,540	13,180	18,450	11,140	14,850	7,910	10,540
1 1/2	6	1.767	1.294	12,940	16,170	22,640	13,250	17,670	9,700	12,940
1 5/8	5 1/2	2.074	1.515	15,150	18,940	26,510	15,550	20,740	11,360	15,150
1 3/4	5	2.405	1.745	17,450	21,800	30,520	18,040	24,050	13,080	17,440
1 7/8	5	2.761	2.049	20,490	25,610	35,860	20,710	27,610	15,370	20,490
2	4 1/2	3.142	2.300	23,000	28,750	40,250	23,560	31,420	17,250	23,000
2 1/4	4 1/2	3.976	3.021	30,210	37,770	52,870	29,820	39,760	22,660	30,210
2 1/2	4	4.909	3.716	37,160	46,450	65,040	36,820	49,090	27,870	37,160
2 3/4	4	5.940	4.620	46,200	57,750	80,840	44,580	59,400	34,650	46,200
3	3 1/2	7.069	5.428	54,280	67,850	94,990	53,020	70,690	40,710	54.280

Marks, "Mechanical Engineers' Handbook," McGraw-Hill.

Table 26. Safe Load for Ropes and Chains*

Caution: When handling molten metal, wire ropes and chains should be 25 per cent stronger than indicated in table. Manila rope should not be used for this purpose.

NOTE: If three-leg slings are used, the safe load given for double (two-leg) slings may be increased 50% and four-leg slings 100%		When used for a straight pull	Double (two-leg) sling		
			When used at 60° from the horizontal	When used at 45° from the horizontal	When used at 30° from the horizontal
	Diam., in.	Lb.	Lb.	Lb.	Lb.
	¼	1,060	1,835	1,500	1,060
Plow steel	⅜	2,300	3,980	3,250	2,300
	½	4,000	6,920	5,650	4,000
Wire rope	⅝	6,200	10,730	8,770	6,200
	¾	9,200	15,900	13,000	9,200
(6 strands of 19 wires)	⅞	11,600	20,000	16,400	11,600
	1	15,200	26,300	21,500	15,200
	1⅛	18,800	32,500	26,600	18,800
For crucible steel rope, reduce loads one-fifth.	1¼	24,000	41,500	33,980	24,000
	1⅜	28,000	48,500	39,600	28,000
	1½	32,000	55,400	45,250	32,000
	Diam. of iron, in.	Lb.	Lb.	Lb.	Lb.
	¼	1,060	1,835	1,500	1,060
	⅜	2,385	4,130	3,370	2,385
	½	4,240	7,345	6,000	4,240
Crane chain	⅝	6,630	11,485	9,375	6,630
	¾	9,540	16,525	13,500	9,540
(Best grade of wrought iron, handmade, tested short-link chain)	⅞	12,960	22,450	18,325	12,960
	1	16,950	29,350	23,975	16,950
	1⅛	20,040	34,700	28,350	20,040
	1¼	24,750	42,875	35,000	24,750
	1⅜	29,910	51,800	42,300	29,910
	1½	35,600	61,650	50,350	35,600
	Diam., in.	Lb.	Lb.	Lb.	Lb.
	⅜	120	200	170	120
	½	250	420	350	250
	⅝	360	600	500	360
	¾	520	880	720	520
	⅞	620	1,040	840	620
Manila rope	1	750	1,250	1,050	750
	1⅛	1,000	1,700	1,400	1,000
(Best long-fiber grade)	1¼	1,200	2,050	1,700	1,200
	1½	1,600	2,700	2,200	1,600
	1¾	2,100	3,600	3,000	2,100
	2	2,800	4,800	4,000	2,800
	2½	4,000	6,800	5,600	4,000
	3½	6,000	10,200	8,400	6,000

* Based on table prepared by National Founders' Association.

Definitions, Cautions, and Instructions Governing the Purchase and Use of Chain

Issued by United States Chain and Forging Company, Pittsburgh, now the McKay Co.

Proof Test. The *proof test* is applied to chain for the purpose of detecting defects in material or manufacture.

The test is applied in a standard chain testing machine and shows the load in pounds which the chain, in the condition and at the time it left the factory, has withstood, under a test in which the load has been applied in direct tension to a straight length of chain with a uniform rate of speed represented by a separation of the tester heads not exceeding 10 in./min.

Average Ultimate Load. The *average ultimate load* is the load in pounds at which the chain, in the condition and at the time it left the factory, has been found by experience to break, under a test in which the load is applied in direct tension to a straight length of chain with a uniform rate of load represented by a separation of the tester heads not exceeding 10 in./min.

Safe Working Load. The *safe working load* is the maximum load in pounds which, at any time or under any condition, should ever be applied to the chain, even when the chain is in the same condition it was when it left the factory, and when the load is applied in direct tension to a straight length of chain.

Cautions. The above terms "proof test," "average ultimate load," and "safe working load," contain no implication of what load the chain will safely withstand if any of the above factors are changed.

Any change in the above factors, such as twisting of the chain, deterioration of the chain by strain, by usage, by weathering, or by lapse of time, or acceleration in the rate of application of the load, or variation in the angle of the load to some sharper angle resulting from the configuration or structure of the material constituting the load, will lessen the load that the chain will safely withstand.

Instructions Regarding Attachments. Where attachments, such as hooks or rings, are desired for use with chain in sustaining loads, care should be taken to select the attachments of the type, grade, and size recommended herein for use with the type, grade, and size chain with which such attachments are to be used.

When this is done, the recommended safe working load of such chain, in the sense that the term is defined above, will apply to the chain and attachments thereon when used together.

When chain and attachments thereon are used together, it is impossible to state what load will be sustained, if the attachments thereon are not of the type, grade, and size recommended for use with the particular type, grade, and size of chain used.

Purchasers should note that all the cautions above set forth apply not only to the use of chain but also to the use of attachments thereon.

SUGGESTIONS

Take up slack and then start load slowly.
Keep chains free from twists, knots, and kinks.
Lift from center of hooks, never from the point.
Distribute the load evenly on all legs.
Inspect your sling chains regularly.

Table 28. Identification of Piping Systems

Class of Material	Color
Fire protection	Red
Dangerous	Yellow
Safe	Green
Protective	Blue
Extra valuable	Purple

For subdivisions of these classifications see Standard A-13 "Scheme for the Identification of Piping Systems," American Standards Association, New York, N. Y.

Table 27. Maximum Free Air That Can Be Supplied in Cubic Feet per Minute to the Pressure Vessel for Different Sizes of Safety Valves at Stated Pressures*

Diam. of valve, in.	Gage pressure, lb./sq. in.															
	50	100	150	200	250	300	350	400	500	600	800	1000	1200	1600	2000	2400
¼	...	...	...	...	...	...	...	53	61	70	84	97	109	128	147	160
½	20	32	42	51	59	67	74	111	129	147	177	205	230	270	304	330
¾	37	59	78	96	112	127	141	176	224	232	242	346	386	423	474	518
1	58	94	124	152	178	202	224	248	286	324	390	450	500	586		
1¼	84	135	180	221	259	293	325	352	400	443	509					
1½	114	186	248	302	354	400	444	478	528	568	634					
2	189	306	410	501	592	668	741									
2½	282	457	613	750	880	998	1114									
3	393	638	856	1050	1230	1398	1557									

The foregoing table is based on the following formulas:

$$Q = 28\,PDl \text{ for 45-deg. bevel-seat valves}$$
$$Q = 40\,PDl \text{ for flat-seat valves}$$

where Q = discharge in cu. ft. free air/min.
P = absolute pressure at which the safety valve opens (gage pressure + 14.7 lb. at sea level).
D = diam., in., of the inside edge of the bearing surface between the disk and seat.
l = vertical lift of the safety-valve disk from its seat, in., representing lift for minimum discharge capacity for satisfactory operation of the valve.

* From A.S.M.E. Boiler Construction Code. By permission of American Society of Mechanical Engineers.

Table 29. Coeffiients of Expansion*

The following coefficients of expansion, per degree Fahrenheit, of the principal flammable liquids shall be used in determining outages:

Acetone	0.00085	Gasoline (°A.P.I.*)	
Amyl acetate	.00068	50 -55	0.00055
Benzol (benzene)	.00071	55.1-60	.00060
Carbon bisulfide	.00070	60.1-65	.00065
Ether	.00098	65.1-70	.00070
Ethyl acetate	.00079	70.1-75	.00075
Ethyl (grain) alcohol	.00062	75.1-80	.00080
Methyl (wood) alcohol	.00072	80.1-85	.00085
Toluol (toluene)	.00063	85.1-90	.00090

1°A.P.I. (American Petroleum Institute), according to the following formula:

$$°\text{A.P.I.} = \frac{141.5}{\text{specific gravity}} - 131.5$$

* By permission, from regulations of the Interstate Commerce Commission, Washington, D. C.

Table 30. Outage Chart for Flammable Liquids

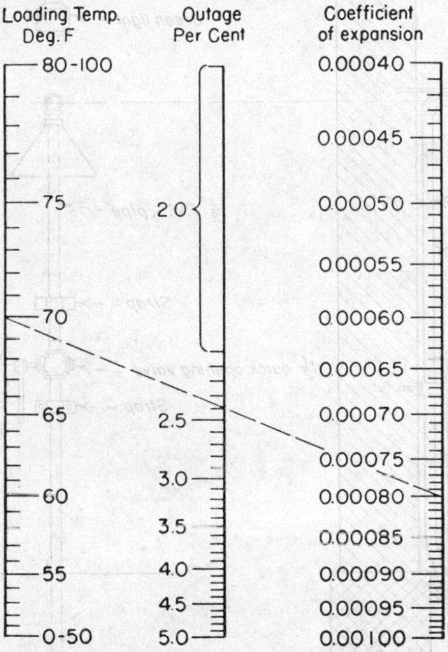

Example: Suppose the temperature of the liquid at time of loading is 70°F. and its coefficient of expansion is 0.00080. Lay a ruler on the chart running from 70°F. to 0.00080 as shown by the dotted line and the required outage is 2.4 per cent where the ruler crosses the outage scale. "Outage" is the percentage of space in the interior of the container which must not be filled with liquid, in order to allow for the expansion of the liquid and the maximum temperature to which it will be subjected in transit. (From regulations of the Interstate Commerce Commission, Washington, D.C., by permission.)

FIRST AID

Every workplace should provide reasonable first aid in order that injured persons may be properly cared for in emergencies and until further treatment can be secured. Where full-time plant physicians are not employed, small industrial first-aid units may be provided, which should be placed under the care of competent persons who should be available at all times and who should have received first-aid training in a standard first-aid course given under the direction of either the American Red Cross or the U. S. Bureau of Mines.

The following outline is recommended merely as **first aid.** In all cases a physician should be summoned at once.

Treatment for Shock. Keep the patient warm. Keep head lower than feet. Cover with blankets or coats. Put hot-water bottle to feet. Give stimulants, such as hot coffee, hot tea, or teaspoonful of aromatic spirits of ammonia in a half glass of water.

Treatment for Sunstroke and Heat Exhaustion. First take the temperature with a clinical thermometer. Sunstroke, or thermic fever, is characterized by rise in body temperature, and the immediate treatment is to reduce the fever by the application of something cold to the head, such as ice bag or cold compresses (cloths wrung out in cold water). Put the patient into a tub of cold water (after first removing the clothing), or sponge the body with cold water.

Heat exhaustion is the same as shock, and the body temperature is below normal. Keep the patient covered with blankets and put a hot-water bag to the feet. Give stimulants to drink, such as strong tea or coffee or aromatic spirits of ammonia.

Treatment for Fainting. Place patient in lying-down position with head lower than the rest of the body so that the brain will receive more blood. Loosen clothing around neck—see that there is plenty of cold air Sprinkle face and chest with cold water. Put smelling salts or ammonia to nose. Rub limbs toward body. Give stimulants when patient can swallow.

Treatment for Sprains. Absolute rest until doctor arrives so as not to do more damage. Do not allow joint to be used—elevate it if possible and apply cold. Wring cloths in cold water and apply to joint or shower the joint with very cold water.

Treatment of Ordinary Wounds. Do not allow the wound to be touched—exposure to the air is much safer than the application of dressings which are not surgically clean. Ordinary water is dangerous as it may contain infectious organisms. Strong antiseptics, such as bichloride of mercury or carbolic acid, should not be used. Peroxide of hydrogen is not sufficiently strong to kill germs; but cover the wound with several layers of sterile gauze and bandage loosely; then place patient under the care of a physician as soon as possible.

Burns and Scalds. In the treatment of burns, the main object is to exclude the air as quickly and completely as possible from the injured part. A 5 per cent tannic acid solution (freshly prepared) is to be preferred, although if this is not available a thin paste made with water and baking soda may be used as a temporary application. Place several layers of sterile gauze over the burned area; saturate with 5 per cent tannic acid solution and bandage loosely. In case of serious burns on the body, remove all clothing and treat as recommended above. If signs of collapse are shown, keep the patient warm and give hot stimulants if not unconscious.

Acid and Alkali Burns. With either, wash off, as quickly as possible, with a large quantity of clean water. For this purpose, emergency showers should be provided. See Figs. 30, 31, and 32. The treatment following this for either acid or alkali burns is similar to that for ordinary burns.

When emergency showers are not available, water from a tap or from an ordinary drinking fountain may be used; the latter is particularly advantageous in either acid or alkali burns of the eyes.

Gas Poisoning. In all cases of gas poisoning, particularly from nitrous fumes, hydrogen sulfide, carbon monoxide, etc., even though the persons show no apparent effects, they should be placed under the care of a physician. These cases frequently do not appear serious until several hours (even 24 to 48 hr.) after the exposure.

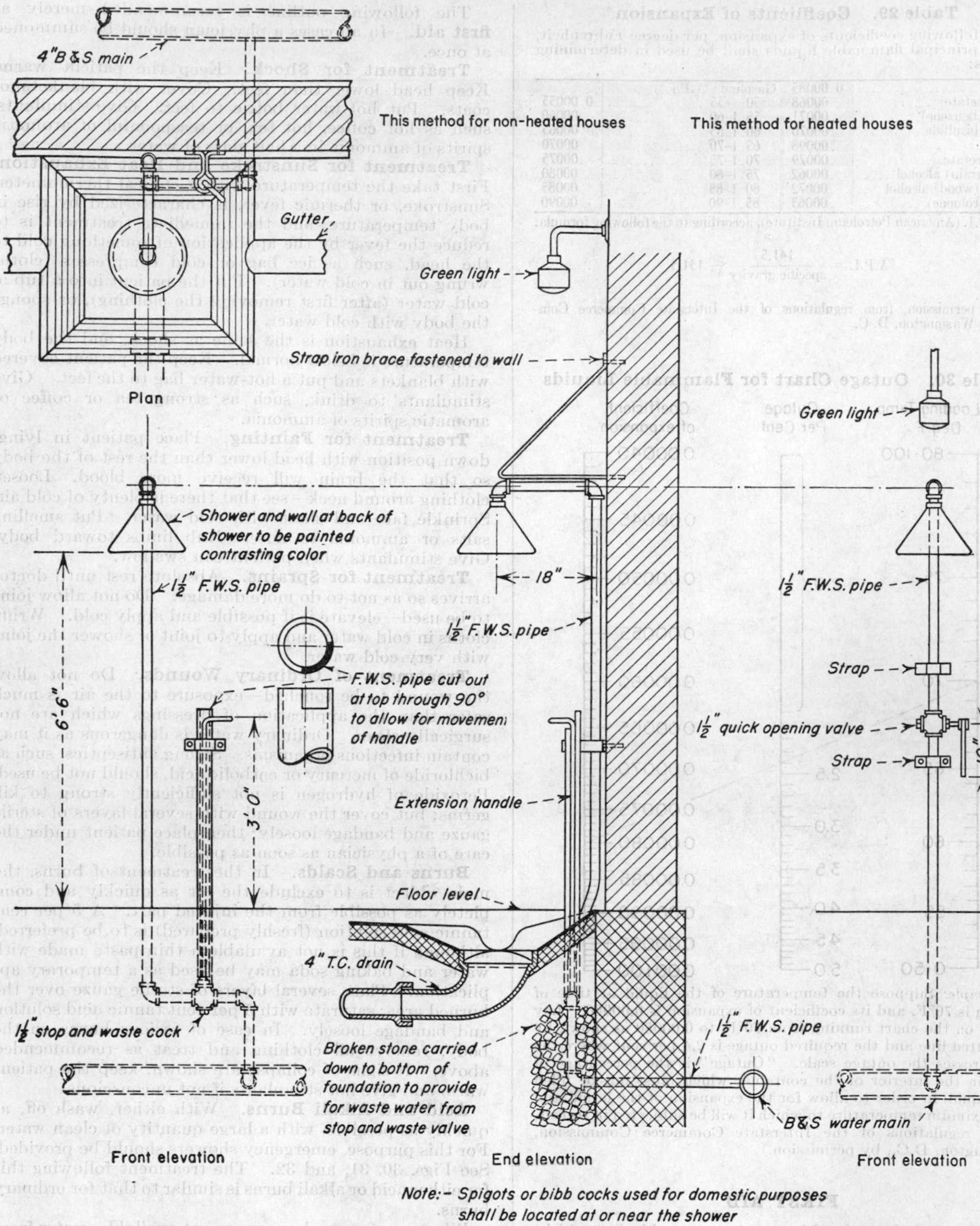

FIG. 30. Arrangement for emergency showers.

From the time of the exposure, until a physician is secured, persons affected by gas poisoning should avoid any unnecessary activity and should be kept as quiet and composed as possible—preferably put to bed. When a person is rendered unconscious by exposure to a gas and breathing has ceased, artificial respiration must be used immediately. Remove the patient to fresh air and begin artificial respiration at once and send for a physician.

Poison Antidotes—First-aid Measures

General Instructions

Call Physician Immediately. In order that the physician may bring appropriate remedies, always state clearly what has happened and the geographical location of the patient.

The following *general instructions* are recommended as strictly *first-aid measures*, and no attempt is made to outline full medical treatment, which should be determined by the physician.

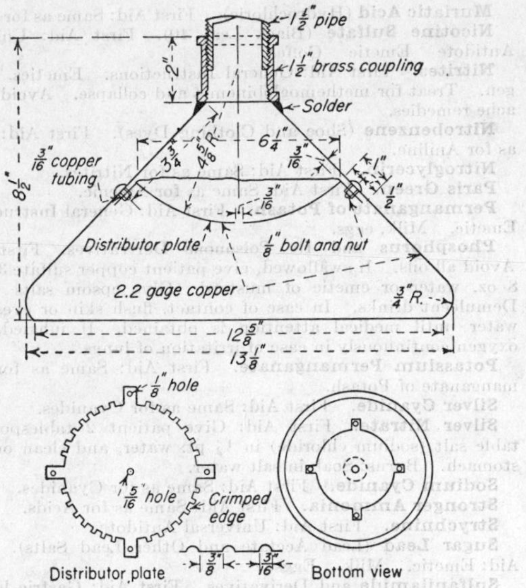

FIG. 31. Head for emergency water shower (local fabrication).

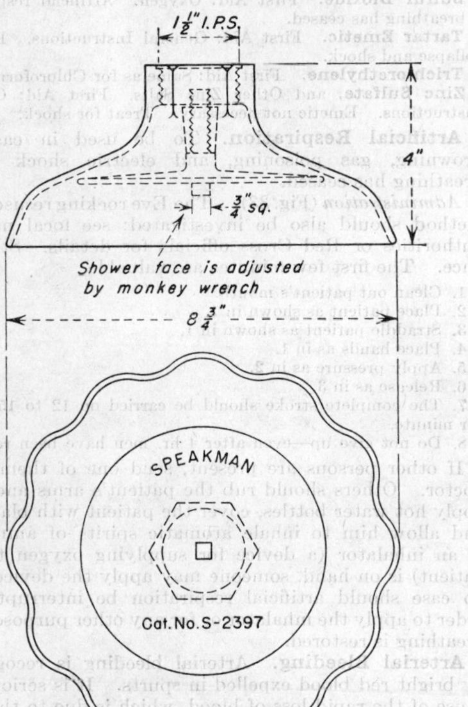

FIG. 32. Emergency shower head. (*Speakman Company.*)

I. Dilute the poison by giving 4 to 7 glassfuls of any one of the following emetics:

a. Salt and lukewarm water (1 cupful of common salt and 1 qt. of water).

b. Soapy warm water.

II. Wash out the poison (empty the stomach) by having the patient stick his finger down his throat to produce vomiting. *Warning:* Only a physician should use the stomach tube.

III. Repeat the above procedure at least three times.

IV. Give antidote if known.

V. After the stomach has been emptied and cleaned of the poisonous materials, give soothing (demulcent) drink, such as milk, raw eggs, flour or starch and water.

VI. If the patient feels weak or faint, and to prevent collapse and shock, he should lie down, without a pillow, and should be kept quiet and warm and given strong coffee or tea. Smelling salts or aromatic spirits of ammonia may be inhaled.

Universal Antidote in Case Poison Is Unknown. When the poison is not known, give in a glass of water a heaping teaspoonful of two parts of activated charcoal, one part of magnesium oxide, one part of Fuller's earth, and one part of tannic acid. If the mixture is not at hand, give an emetic and induce vomiting as described under General Instructions I.

If the Patient Is Unconscious.

1. *Never give unconscious patient anything by mouth or attempt to cause vomiting.*

2. Lay the patient down, preferably on the left side, with the head low. Remove any false teeth, chewing gum, tobacco, or other foreign objects that may be in the mouth.

3. If the patient is not breathing, place in the Schaefer prone position and apply artificial respiration.

4. Keep the patient warm until physician arrives.

5. In case of shallow breathing or cyanosis (blueness of the skin, lips, ears, fingernail beds), give oxygen with carbon dioxide or commercial oxygen. Administer artificial respiration if needed.

6. In case the poison is unknown, give the *universal antidote* only after patient regains consciousness.

Collapse or Shock. To prevent collapse or shock, especially in severe cases, keep the patient warm, quiet, lying down, head low.

Stimulants. If conscious, give strong coffee or tea; smelling salts or aromatic spirits of ammonia may be inhaled.

If Poison Is Inhaled. If poison is inhaled as a vapor or gas, remove immediately from exposure to fresh air.

Artificial respiration is lifesaving if breathing has stopped.

Oxygen may be used together with artificial respiration in all cases where breathing has stopped.

Oxygen should be used in all cases of cyanosis or following the inhalation of irritant gases.

If Poison Contacts Skin or Eyes. Wash off with large amounts of water. Keep wet until seen by a physician. Remove contaminated clothing and wash before reusing.

Acetanilid. First Aid: General Instructions. Delayed effects such as cyanosis should receive oxygen immediately. Force fluids by mouth. Treat for methemoglobinemia. *Antidote:* Picrotoxin (use with caution). Nikethamide or Coramine contraindicated.

Acids. First Aid: *Internal. Do not attempt to produce vomiting.* Give large quantities of water or milk. Avoid emetics and stomach tube. *External.* In case of contact with the skin or eyes, flush repeatedly with large quantities of water and keep the area wet until physician arrives, or for at least 2 hr. following exposure. *For minor burns of skin:* Apply ice cold alcohol or paste of baking soda and water.

Do not apply grease or oil!

Aconite (Its Preparations or Derivatives). First Aid: Cause vomiting. Give coffee if patient is conscious. Keep patient in horizontal position. Maintain warmth. Artificial respiration if breathing has stopped.

Alcohol, Ethyl. First Aid: Cause vomiting if conscious. Following General Instructions.

Alcohol, Wood (Methyl Alcohol, Methanol). First Aid: Same as for Ethyl Alcohol.

Alkaloidal Poisons. First Aid: General Instructions. Universal Antidote.

Ammonia. First Aid: Same as for Acids.

Aniline. First Aid: If absorbed, cleanse skin thoroughly with soapy water and scrub brush. Oxygen for cyanosis or headache. Treat methemoglobinemia. If swallowed, wash from stomach immediately with salt solution (1 tablespoonful table salt to quart of water).

Antimony and Its Preparations (Tartar Emetic). First Aid: General Instructions. Treat for shock and collapse.

Antipyrine. First Aid: Same as for Acetanilid.

Antiseptic Tablets (Corrosive Sublimate). First Aid: Same as for Bichloride of Mercury.

Arsenic, its compounds. Fowler's solution, Paris Green, weed killers, Rough on Rats, Ant Paste, and other insecticides containing arsenic. First Aid: Same as for Antimony.

Barbituric Acid Derivatives. First Aid: Same as for Acetanilid. Artificial respiration. Be sure breathing is not obstructed. Remove mucus by suction. If conscious, empty stomach.

Barium. First Aid: Give epsom or glauber salt (sodium sulfate), 2 tablespoonsful (30 g.) in 1 pt. (500 cc.) of water. Cause vomiting.

Benzene (Benzol). First Aid: If swallowed, General Instructions. If inhaled, oxygen. Treat for anoxia.

Bichloride of Mercury. First Aid: *Give at once* 15 g. (1 teaspoonful) sodium thiosulfate or sodium formaldehyde sulfoxylate in 500 cc. (1 pt.) water, by mouth, or glass of milk with 4 tablespoonsful charcoal. Then empty stomach and repeat several times. Give demulcent drinks of milk and raw eggs.

Bitter Almonds, Oil of. First Aid: (1) If inhaled, oxygen. (2) If swallowed, emetic of mustard (1 teaspoonful of dry mustard in glass of warm water), starch followed by magnesia. (3) For skin burns, bathe freely with water and keep the area wet until medical attention is obtained, or for at least 2 hr. following exposure.

Calcium Cyanide. First Aid: Same as for Cyanides.

Camphor. First Aid: General Instructions. Give artificial respiration. Treat for shock. Avoid alcohol and oils.

Carbon Disulfide. First Aid: Oxygen 93 per cent, carbon dioxide 7 per cent or O_2. Artificial respiration if breathing has stopped.

Chloroform. First Aid: Oxygen. Artificial respiration. General Instructions.

Cyanides or Hydrocyanic Acid. First Aid: Carry patient to fresh air. Have him lie down. Remove contaminated clothing but keep patient warm. Start the following first-aid treatment immediately and *call a physician.*

If patient is conscious and breathing: (1) Break an amyl nitrite pearl in a cloth and hold lightly over the nose for not more than 15 to 20 sec. Repeat every 5 min. for 25 min. if recovery is not forthcoming. (2) If this product has been *swallowed,* give patient 1 pt. of 1 per cent sodium thiosulfate solution (or soapy water or mustard water) by mouth every 15 min. until vomiting occurs. (Eliminate this for liquid HCN.)

If patient has stopped breathing: Give artificial respiration until breathing starts. Break an amyl nitrite pearl in a cloth and hold lightly over nose for not more than 20 sec. repeating every 5 min. for 25 min. or until breathing starts.

If patient is unconscious but breathing: Break an amyl nitrite pearl in a cloth and hold lightly over nose for not more than 20 sec. repeating every 5 min. for 25 min. if recovery is not forthcoming. Give oxygen from an inhalator.

Never give anything by mouth to an unconscious person. In all cases keep patient quiet and warm until a physician arrives.

Denatured Alcohol. First Aid: Same as for Alcohol.

Dinitro-ortho-cresol. First Aid: General Instructions.

Fluorides (Their Compounds or Derivatives). First Aid: Give at once large draughts of lime water or milk of magnesia or paste of chalk.

In case of contact with hydrofluoric acid:

1. Immediately flush skin (until whiteness disappears) and eyes (for at least 15 min.) with plenty of water; pay particular attention to skin under nails.
2. Immerse in mixture of ice and alcohol.
3. *Antidote:* Magnesium oxide or calcium gluconate.
4. Avoid use of greasy ointments.
5. Remove and wash clothing before reuse.

Formaldehyde (Formalines). First Aid: Give milk freely or water with baking soda, 1 teaspoonful in glass of water. Then induce vomiting.

Fowler's Solution. First Aid: Same as for Arsenic.

Hydrochloric Acid. First Aid: Same as for Acids.

Hydrocyanic Acid or Cyanides. First Aid: If inhaled, same as for Cyanides. If swallowed, same as for Cyanides and following antidote: Sodium thiosulfate 15 g. (1 teaspoonful) in ½ pt. (250 cc.) water.

Hydrofluoric Acid. First Aid: Same as for Fluorides.

Iodine. First Aid: Give patient thin starch-water paste. Cause vomiting. Antidote: Sodium thiosulfate 15 g. (1 teaspoonful) in ½ pt. (250 cc.) water.

Mercury. First Aid: Same as for Bichloride of Mercury.

Methyl Chloride (Refrigerants). First Aid: Oxygen. Artificial respiration. For skin contact: Cold water. Avoid oils or grease.

Muriatic Acid (Hydrochloric). First Aid: Same as for Acids.

Nicotine Sulfate (Black Leaf 40). First Aid: Universal Antidote. Emetic. Coffee.

Nitrites. First Aid: General Instructions. Emetics. Oxygen. Treat for methemoglobinemia and collapse. Avoid headache remedies.

Nitrobenzene (Shoe and Clothing Dyes). First Aid: Same as for Aniline.

Nitroglycerin. First Aid: Same as for Nitrites.

Paris Green. First Aid: Same as for Arsenic.

Permanganate of Potash. First Aid: General Instructions. Emetic. Milk, eggs.

Phosphorus and Its Poisonous Derivatives. First Aid: Avoid all oils. If swallowed, give patient copper sulfate 3 gr. to 8 oz. water or emetic of mustard. Give epsom salts ½ oz. Demulcent drinks. In case of contact, flush skin or eyes with water until medical attention is obtained. If inhaled, give oxygen continuously in case of irritation of lungs.

Potassium Permanganate. First Aid: Same as for Permanganate of Potash.

Silver Cyanide. First Aid: Same as for Cyanides.

Silver Nitrate. First Aid: Give patient 2 tablespoonsful table salt (sodium chloride) in ½ pt. water, and clean out the stomach. Burns: Soak in salt water.

Sodium Cyanide. First Aid: Same as for Cyanides.

Stronger Ammonia. First Aid: Same as for Acids.

Strychnine. First Aid: Universal Antidote.

Sugar Lead (Lead Acetate and Other Lead Salts). First Aid: Emetic. Milk. Eggs.

Sulfanilamide and Derivatives. First Aid: Gastric lavage. Warm water. Sodium bicarbonate, 30 gr. Force water by mouth.

Sulfur Dioxide. First Aid: Oxygen. Artificial respiration if breathing has ceased.

Tartar Emetic. First Aid: General Instructions. Prevent collapse and shock.

Trichlorethylene. First Aid: Same as for Chloroform.

Zinc Sulfate, and Other Zinc Salts. First Aid: General Instructions. Emetic not necessary. Treat for shock.

Artificial Respiration. To be used in cases of drowning, gas poisoning, and electric shock where breathing has ceased.

Administration (Fig. 33). The Eve rocking resuscitator method should also be investigated; see local medical authorities or Red Cross officials for details. Start at once. The first few minutes are valuable.

1. Clean out patient's mouth.
2. Place patient as shown in 1.
3. Straddle patient as shown in 1.
4. Place hands as in 1.
5. Apply pressure as in 2.
6. Release as in 3.
7. The complete stroke should be carried on 12 to 15 times per minute.
8. Do not give up—even after 4 hr. men have been revived.

If other persons are present, send one of them for a doctor. Others should rub the patient's arms and legs, apply hot water bottles, cover the patient with blankets, and allow him to inhale aromatic spirits of ammonia. If an inhalator (a device for supplying oxygen to the patient) is on hand, someone may apply the device. In no case should artificial respiration be interrupted in order to apply the inhalator or for any other purpose until breathing is restored.

Arterial Bleeding. Arterial bleeding is recognized by bright red blood expelled in spurts. It is serious because of the rapid loss of blood, which is due to the fact that there is too much pressure at the wound to allow the blood to clot. The flow of blood cannot be stopped by ordinary pressure over the wound. To control arterial bleeding, press with your fingers or thumb on the artery between the bleeding point and the heart. The points where the arteries come close to the surface of the skin are shown in Fig. 34. A tourniquet should be applied wherever possible at these points to control bleeding in the particular artery. The tourniquet should be released for 3 or 4 sec. at 20-min. intervals.

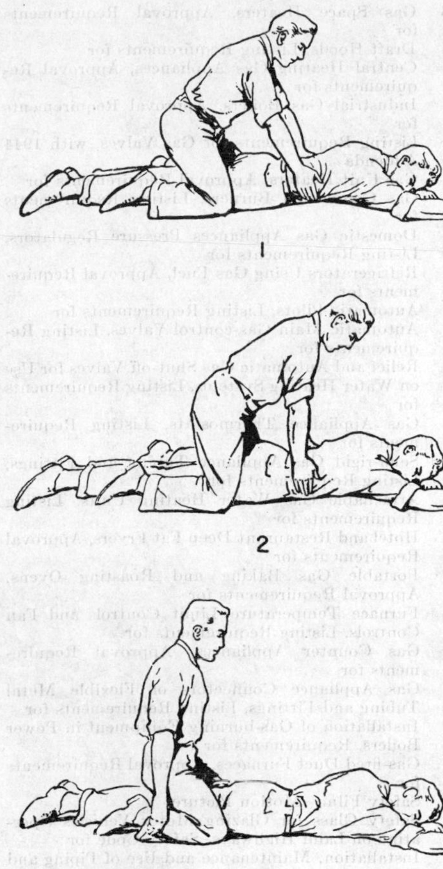

FIG. 33. Positions for administering artificial respiration. (*By permission of The National Electric Light Association.*)

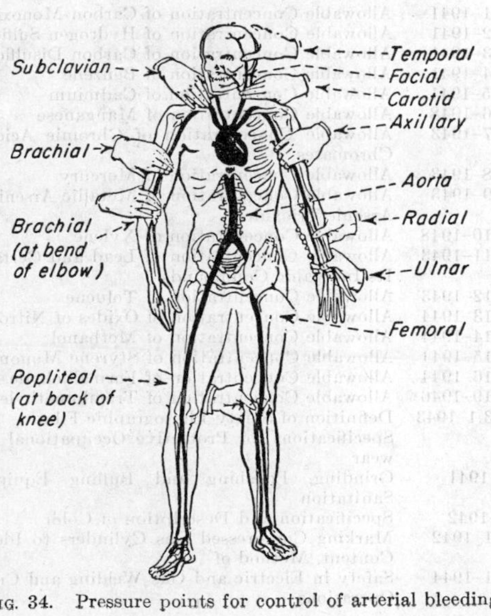

FIG. 34. Pressure points for control of arterial bleeding.

BIBLIOGRAPHY

American Engineering and Industrial Standards

May be secured from The American Standards Association, 70 East 45 St., New York 17, N.Y.

A2.1–1942 Fire Tests of Building Construction and Materials, Methods of
A2.2–1942 Fire Tests of Door Assemblies, Methods of
A9.1–1946 Building Exits Code
A10.1–1939 Manual of Accident Prevention in Construction
A10.2–1944 Building Construction, Safety Code for
A11–1942 Industrial Lighting
A12–1932 Floor and Wall Openings, Railings and Toe Boards, Safety Code for
A13–1928 Scheme for the Identification of Piping Systems
A14–1948 Construction, Care and Use of Ladders, Safety Code for
A17.1–1937 Safety Codes for Elevators, Dumbwaiters, and Escalators, with Supplement
A17.2–1945 Inspection of Elevators
A23.1–1948 School Lighting
A39–1933 Window Cleaning
A40.6–1943 Backflow Preventions in Plumbing Systems
A41.1–1944 Building Code Requirements for Masonry
A55.1–1948 Administrative Requirements for Building Codes
A57.1–1943 Building Code Requirements for Structural Steel
A85–1942 Protective Lighting for Industrial Properties
B5.17–1943 Markings for Grinding Wheels
B7.1–1947 Use, Care and Protection of Abrasive Wheels, Safety Code for the
B8–1932 Protection of Industrial Workers in Foundries, Safety Code for
B9–1939 Mechanical Refrigeration, Safety Code for
B11.1–1948 Power Presses and Foot and Hand Presses, Safety Code for
B13–1924 Logging and Sawmill Safety Code
B15–1927 Mechanical Power-transmission Apparatus, Safety Code for
B19–1938 Compressed-air Machinery and Equipment, Safety Code for
B24–1927 Forging and Hot Metal Stamping, Safety Code for
B26–1925 Fire-hose Coupling Screw Thread
B28a–1927 Rubber Mills and Calendars, Safety Code for
B30.1–1943 Jacks, Safety Code for
B30.2–1943 Cranes, Derricks, and Hoists, Safety Code for
B31.1–1942 Code for Pressure Piping, with Supplement
B33.1–1935 Hose Coupling Screw Threads
B40.1–1939 Indicating Pressure and Vacuum Gages
C1 1946 National Electrical Code
C2.1–1941 Safety Rules for the Installation and Maintenance of Electrical Supply Stations
C2.2–1941 Safety Rules for the Installation and Maintenance of Electric Supply and Communication Lines
C2.3–1941 Safety Rules for the Installation and Maintenance of Electric Utilization Equipment
C2.4–1939 Safety Rules for the Operation of Electric Equipment and Lines
C2.5–1940 Safety Rules for Radio Installations
C5 Code for Protection against Lightning
C74–1942 Machine Tool Electrical Standards
D6.1–1948 Manual on Uniform Traffic Control Devices for Streets and Highways
D7.1–1941 Inspection Requirements for Motor Vehicles
D8.1–1943 Railroad Highway Grade Crossing Protection
D10.1–1942 Adjustable Face Traffic Control Signal Head Standards
K2–1927 Gas Safety Code for Installation and Work in Buildings
K13–1930 Code for Identification of Gasmask Canisters
L1.1–1947 Textile Safety Code
L3.1–1941 Cotton Rubber-lined Fire Hose for Public and Private Fire Department Use, Specifications for
L17 Specifications for Women's Industrial Clothing
L18 Specifications for Protective Occupational Clothing
M2–1926 Installing and Using Electric Equipment in Coal Mines, Safety Rules for
M10–1938 Miscellaneous Outside Coal Handling Equipment
M11–1927 Wire Rope for Mines
M12.1–1946 Ladders and Stairs for Mines, Construction and Maintenance of

M13–1942	Rock-dusting Coal Mines to Prevent Coal Dust Explosions
M14–1930	Use of Explosives in Bituminous Coal Mines
M15–1931	Coal Mine Transportation, Safety Code for
M17–1930	Fire Fighting Equipment in Metal Mines
M24–1932	Safety Rules for Installing and Using Electrical Equipment in Metal Mines
O1.1–1944	Woodworking Machinery, Safety Code for
P1–1936	Paper and Pulp Mills, Safety Code for
Z2–1938	Protection of Heads, Eyes and Respiratory Organs, Safety Code for
Z4.1–1935	Industrial Sanitation in Manufacturing Establishments, Safety Code for
Z4.2–1942	Drinking Fountains, Specifications for
Z4.3–1935	Sanitary Privy
Z7.1–1942	Illuminating Engineering Nomenclature and Photometric Standards
Z8–1941	Laundry Machinery and Operations, Safety Code for
Z9	Fundamentals Relating to the Design and Operation of Exhaust Systems
Z9.1–1941	Safety in Electroplating Operations
Z11.6–1947	Flash and Fire Points by Means of Open Cup, Method of Test for
Z11.7–1947	Flash Point by Means of the Method of the Pensky-Martens Closed Tester, Method of Test for
Z11.24–1936	Flash Point by Means of the Tag Closed Tester, Method of Test for
Z11.42–1940	Stoddard Solvent, Specifications for
Z12.1–1946	Installation of Pulverized Fuel Systems, Safety Code for
Z12.2–1944	Prevention of Dust Explosions in Starch Factories, Safety Code for
Z12.3–1946	Prevention of Dust Explosions in Flour and Feed Mills, Safety Code for
Z12.4–1942	Prevention of Dust Explosions in Terminal Grain Elevators, Safety Code for
Z12.5–1942	Prevention of Dust Explosions in Woodworking Plants, Safety Code for
Z12.6–1946	Pulverizing Systems for Sugar and Cocoa, Safety Code for
Z12.7–1946	Prevention of Dust Explosions in Coal Pneumatic Cleaning Plants, Safety Code for
Z12.8–1946	Prevention of Dust Explosions in Wood Flour Manufacturing Establishments, Safety Code for
Z12.9–1946	Prevention of Dust Ignitions in Spice Grinding Plants, Safety Code for
Z12.10–1943	Use of Inert Gas for Fire and Explosion Prevention, Safety Code for
Z12.11–1946	Prevention of Dust Explosions in the Manufacture of Aluminum Bronze Powder, Safety Code for
Z12.12–1946	Prevention of Sulphur Dust Explosions and Fires, Safety Code for
Z12.13–1946	Prevention of Dust Ignitions in Country Grain Elevators, Safety Code for
Z12.14–1943	Suggested Good Practices for the Application of Suction and Venting for the Control of Dust in Grain Elevators and Storage Units
Z12.15–1946	Explosion and Fire Protection in Plants Producing or Handling Magnesium Powder or Dust, Safety Code for
Z16.1–1945	Method of Compiling Industrial Injury Rates
Z16.2–1941	Compiling Industrial Accident Causes
Z20.1–1941	Portable Steel and Wood Grandstands, Building Code Requirements
Z21.1–1948	Domestic Gas Ranges, Approval Requirements for
Z21.2–1938	Flexible Gas Tubing, Listing Requirements for
Z21.3–1940	Hotel and Restaurant Ranges and Unit Broilers, Approval Requirements for
Z21.4–1932	Private Garage Heaters, Approval Requirements for
Z21.5–1940	Clothes Dryers, Approval Requirements for
Z21.6–1932	Incinerators, Approval Requirements for
Z21.7–1932	Gas Heated Ironers, Approval Requirements for
Z21.8–1948	Installation of Conversion Burners in House Heating and Water Heating Appliances, Requirements for
Z21.9–1948	Hot Plates and Laundry Stoves, Approval Requirements for
Z21.10–1945	Gas Water Heaters, Approval Requirements for
Z21.11–1948	Gas Space Heaters, Approval Requirements for
Z21.12–1937	Draft Hoods, Listing Requirements for
Z21.13–1945	Central Heating Gas Appliances, Approval Requirements for
Z21.14–1934	Industrial Gas Boilers, Approval Requirements for
Z21.15–1944	Listing Requirements for Gas Valves, with 1944 Addenda
Z21.16–1940	Gas Unit Heaters, Approval Requirements for
Z21.17–1948	Gas Conversion Burners, Listing Requirements for
Z21.18–1934	Domestic Gas Appliances Pressure Regulators, Listing Requirements for
Z21.19–1942	Refrigerators Using Gas Fuel, Approval Requirements for
Z21.20–1940	Automatic Pilots, Listing Requirements for
Z21.21–1948	Automatic Main Gas-control Valves, Listing Requirements for
Z21.22–1935	Relief and Automatic Gas Shut-off Valves for Use on Water Heating Systems, Listing Requirements for
Z21.23–1940	Gas Appliance Thermostats, Listing Requirements for
Z21.24–1941	Semi-rigid Gas Appliance Tubing and Fittings, Listing Requirements for
Z21.26–1941	Attachable Gas Water Heating Units, Listing Requirements for
Z21.27–1940	Hotel and Restaurant Deep Fat Fryers, Approval Requirements for
Z21.28–1948	Portable Gas Baking and Roasting Ovens, Approval Requirements for
Z21.29–1941	Furnace Temperature Limit Controls and Fan Controls, Listing Requirements for
Z21.31–1941	Gas Counter Appliances, Approval Requirements for
Z21.32–1942	Gas Appliance Connectors of Flexible Metal Tubing and Fittings, Listing Requirements for
Z21.33–1942	Installation of Gas-burning Equipment in Power Boilers, Requirements for
Z21.34–1942	Gas-fired Duct Furnaces, Approval Requirements for
Z22.31–1946	Safety Film—Motion Picture
Z26.1–1938	Safety Glass for Glazing Motor Vehicles Operating on Land Highways, Safety Code for
Z27–1933	Installation, Maintenance and Use of Piping and Fittings for City Gas
Z33.1–1938	Regulations for the Installation of Blower and Exhaust Systems for Dust, Stock and Vapor Removal
Z35.1–1941	Industrial Accident Prevention Signs, Specifications for
Z37.1–1941	Allowable Concentration of Carbon Monoxide
Z37.2–1941	Allowable Concentration of Hydrogen Sulfide
Z37.3–1941	Allowable Concentration of Carbon Disulfide
Z37.4–1941	Allowable Concentration of Benzene
Z37.5–1941	Allowable Concentration of Cadmium
Z37.6–1948	Allowable Concentration of Manganese
Z37.7–1943	Allowable Concentration of Chromic Acid and Chromates
Z37.8–1943	Allowable Concentration of Mercury
Z37.9–1943	Allowable Concentration of Metallic Arsenic and Arsenic Trioxide
Z37.10–1948	Allowable Concentration of Xylene
Z37.11–1943	Allowable Concentration of Lead and Certain of Its Inorganic Compounds
Z37.12–1943	Allowable Concentration of Toluene
Z37.13–1944	Allowable Concentration of Oxides of Nitrogen
Z37.14–1944	Allowable Concentration of Methanol
Z37.15–1944	Allowable Concentration of Styrene Monomer
Z37.16–1944	Allowable Concentration of Formaldehyde
Z37.19–1946	Allowable Concentration of Trichloroethylene
Z38.3.1–1943	Definition of Safety Photographic Film
Z41	Specifications for Protective Occupational Footwear
Z43–1941	Grinding, Polishing and Buffing Equipment Sanitation
Z44–1942	Specification and Description of Color
Z48.1–1942	Marking Compressed Gas Cylinders to Identify Content, Method of
Z49.1–1944	Safety in Electric and Gas Welding and Cutting Operations

Safe Practices Pamphlets

Published by National Safety Council, 20 North Wacker Drive, Chicago 6, Ill.

General

No.
1 Ladders—1940.
2 Stairs and Ramps—1942.
3 Steam Boilers—1939.
4 Overhead Traveling Cranes—1939.
6 Fiber Rope—1941.
11 Floors and Flooring—1939.
12 Scaffolds—1937.
13 Grinding Wheels—1936.
14 Goggles—1940.
15 Elevators—1935.
16 Protective Clothing—1941.
17 Plant Yards and Grounds—1938.
18 Power Presses, 1929.
19 Exits, Fire Alarms and Fire Drills—1943.
20 Woodworking Machinery and Equipment—1939
21 Industrial Accident Records and Analyses—1946
22 Industrial Shop Lighting—1941.
23 Gas Welding and Flame Cutting—1941.
24 Fire Extinguishment—1939.
25 Acids and Caustics—1941.
26 Wire Rope—1937.
27 Industrial Sanitation (Drinking Water, Wash and Locker Rooms and Toilet Facilities)—1941.
28 Commercial Explosives—1939.
29 Electric Equipment in Industrial Plants—1939.
30 Discontinued.
31 Fire Causes and Prevention—1939.
32 Exhaust Systems—1939.
33 Hoisting Apparatus—1939.
34 Industrial Explosion Hazards (Gases, Vapors and Flammable Liquids)—1941.
35 Conveyors—1939.
36 Fire Brigades—1941.
37 Industrial Ventilation—1939.
38 Safety Posters and Bulletin Boards—1939.
39 Machine Shops—1939.
40 Suggestion Systems—1941.
41 Hand Tools—1939.
42 Organizing a Complete Industrial Safety Program—1939.
43 Discontinued.
44 Cutting Oils and Emulsions—1941.
45 Industrial Housekeeping—1941.
46 Fuel Handling, Storing and Firing—1940.
47 Compressed Air Machinery and Equipment—1939.
48 Railroads in Industrial Plants—1942.
49 Construction and Equipment of Steam Boilers—1939.
50 Reducing Fatigue—1940.
51 Discontinued.
52 Static Electricity—1942.
53 Checking Plans and Specifications for Safety—1938.
54 Handling Material (Hand and Truck)—1938.
55 Industrial Power Trucks and Tractors—1940.
56 Investigation of Industrial Accidents—1939.
57 Discontinued.
59 Warehouses and Shipping Rooms—1941.
60 Chemical Laboratories—1942.
61 Mechanical Refrigeration—1937.
62 Discontinued.
63 Discontinued.
64 Respiratory Protective Equipment—1938.
65 Teaching Safety to New Employees—1941.
66 Discontinued.
67 Maintaining Interest in Safety—1938.
68 Pressure Vessels—Fired and Unfired (Part II) Steam Jacketed Vessels, Digesters, Stills, Blow Cases and Autoclaves—1939.
69 Discontinued.
70 Maintenance and Repair Men—1938.
71 Chlorine—1936.
72 Safety Committees—1939.
73 Foundries—1940.
74 Safety Contests—1937.
75 Safety Inspections—1941.
76 Portable Electric Hand Tools—1940.
77 Safety Meetings—1937.
78 Mathematical Tables and Data for the Safety Engineer—1939.
79 Discontinued.
80 Industrial Safety Rules—Their Formulation and Use—1941.
81 Industrial Accident Prevention Signs—1943.
82 Discontinued.
83 Training for First Aid and Rules for First Aid Contests—1941.
84 The Safety Man in Industry—1939.
85 Forging and Hot Metal Stamping—1934.
86 Discontinued.
87 Safety in the Medium Sized Plant—1942.
88 Identification of Piping Systems—1937.
89 Discontinued.
90 Discontinued.
91 Spray Coating—1940.
92 Discontinued.
93 Topics for Safety Meetings—1940.
94 Discontinued.
95 Compressed Gases—1939.
96 Industrial Power Departments—1938.
97 Pulverized Coal Systems—1940.
98 Use and Care of Hoisting Chains—1943.
99 Falls of Workers—Their Causes and Prevention—1937.
100 Safety Stunts—Part I—1939.
101 Safety Stunts—Part II—1937.
102 Off the Job Accidents—1939.
103 Purchasing for Safety—1939.
104 Dust Explosions—1941.
105 Electric Welding—1941.
106 Conservation of Personal Protective Equipment—1943.
107 Women in Industry—1942.
108 Office Safety—1943.
109 Safety Observation Plan—1947
110 Mechanical Power Transmission Apparatus—1947.

Safe Practices—Special Industries

Au-1 Heat Treating—1931.
Au-2 Motor Block Testing—1937.
CE-1 Mercantile Establishments—1941.
CE-2 Hotels—1939.
CE-3 Hospitals—1939.
Cem-1 Cement Rock Quarrying and Crushing—1931.
Cem-2 Raw and Finished Cement Mill Grinding—1941.
Cem-3 Cement Burning (Including Fuel Handling)—1940.
Cem-4 Cement Mill Shops—1941.
Cem-5 Storing, Packing and Shipping Cement—1940.
Cem-6 Cement Mill Yards and Railroads—1940.
Chem-1 Chemical Pipe Lines and Tanks—1941.
Chem-2 Fume Poisoning from Nitric and Mixed Acids—1940.
Chem-3 Chemical Burns—1937.
Chem-4 Discontinued.
Chem-5 Pyroxylin Lacquer Manufacture—1935.
Chem-6 Cyanide Compounds—1940.
Chem-7 Industrial Waste Disposal—1948
Con-1 Building Construction—1936.
Con-2 Discontinued.
Con-3 Excavation Work—1941.
D-1 Discontinued.
D-2 Selecting Drivers for Commercial Vehicles—1941.
D-3 Training Drivers for Commercial Vehicles—1933.
D-4 Commercial Vehicle Accident Records—1939.
D-5 Preventing Vehicle Accidents—1937.
D-6 Garages and Repair Shops—1941.
F-1 Food Preserving and Canning—1929.
F-2 Candy, Chocolate and Cocoa Manufacture—1930.
F-3 Bakery Operations—1941.
F-4 Milk Bottling Plants—1939.
F-5 Macaroni Plants—1936.
Hy-1 State Highway Employees—1943.
L-1 Dry Cleaning and Dyeing Establishments—1939.
M-1 Discontinued.
M-2 Mine Rescue Work—1937.
Mar-1 Discontinued.
Mar-2 Marine Boilers—1938.
Me-1 Cleaning and Finishing Rooms in Foundries—1936.
Me-2 Blast Furnaces—1938.
Me-3 Discontinued.
Me-4 Rod Mills—1930.
Mun-1 Municipal Employees—1945.
PP-1 Paper and Pulp Mills—1940
PP-2 Paper Box Manufacturing—1936.
PT-1 Leather Tanneries—1937.

Pet-1 Discontinued.
Pet-2 Discontinued.
Pet-3 Discontinued.
Pet-4 Discontinued.
Pet-5 Discontinued.
Pet-6 Discontinued.
Pet-7 Discontinued.
Pet-8 Drums and Barrels (Handling, Cleaning and Filling)—1936.
PU-1 Protecting Public Utility Employees on Streets and Highways—1938.
PU-2 Discontinued.
PU-3 Linemen's Rubber Protective Equipment—1939.
PU-4 Handling of Poles—1939.
Ref-1 Delivery of Ice—1941.
Ref-2 Ice-processing Machines—1941.
RR-1 Use of Motor, Hand and Push Track Cars—1940.
Ru-1 Compounding Materials Used in the Rubber Industry Pt. I—1937.
Ru-2 Vulcanizers and Devulcanizers—1936.
Ru-3 Compounding Materials Used in the Rubber Industry Pt. II—1938.
T-1 Cotton Mills—1938.
VOC-1 Industrial and Vocational Training—1941.
W-1 Wood Furniture Manufacture—1940.
W-2 Operating Commercial Lumber-yards—1932.

Health Practices

1 Chromium—1929.
2 Physical Examinations in Industry—1936.
3 Lead—1942.
4 Industrial Dust—1939.
5 Health Service in Industry—1939.
6 Industrial Eye Hazards—1934.
7 Carbon Monoxide—1941.
8 Caring for Injured Workers—1938.
9 Gases and Vapors—1942.
10 Skin Affections—1942.
11 Nursing Service in Industry—1930.
12 Discontinued.
13 Physical Defects—1931.
14 Benzol—1940.
15 Discontinued.
16 Physical Therapy in Industry—1946.
17 Back Injuries—1936.
18 Lighting and Health—1941.
19 Illness in Industry—1936.

Publications of the National Fire Protection Association on Fires, Fire Prevention, and Fire Protection

(*Corrected to May* 1, 1949)

Copies of the publications listed below will be mailed on application to National Fire Protection Association, 60 Batterymarch St., Boston, Mass.

A. Standard Regulations for Fire Protection and the Safeguarding of Hazards

Published by the National Board of Fire Underwriters

NOTE: The following N.F.P.A. Regulations have been adopted by, and are the official standards of the National Board of Fire Underwriters.

1. **Acetylene Equipment** for Lighting, Heating and Cooking. (1930.)
2. **Air Conditioning, Warm Air Heating, Air Cooling and Ventilating Systems.** (1946.)
3. **Airplane Hangars, Construction and Protection of.** (1943.)
4. **Blower and Exhaust Systems for Dust, Stock and Vapor Removal.** (1947.)
5. **Carbon Dioxide Fire Extinguishing Systems and Inert Gas for Fire and Explosion Prevention.** (1946.)
6. **City Gas**, Installation, Maintenance and Use of Piping and Fittings. (1943.)
7. **Combustible Fibres**, Storage and Handling. (1947.)
8. **Dip Tanks** Containing Flammable Liquids, Including Hardening and Tempering Tanks; Flow Coat Work. (1941.)
9. **Dry Cleaning and Dry Dyeing Plants**, Safeguarding of. (1944.)
10. **Dust Explosions**, Prevention of in Grain Elevators, Flour Mills, etc. (1948.)
11. **Dust Explosions in Industrial Plants**, Fundamental Principles for the Prevention of. (1938.)
12. **Electric Cars and Trolley Buses, Including Houses and Yards.** (1935.)
13. **Electric Wiring and Apparatus** (National Electrical Code). (1947.)
14. **Fire Brigades, Private.** (1937.)
15. **Fire Department Hose Connections for Sprinkler and Standpipe Systems.** (1939.)
16. **Fire Pumps, Centrifugal.** (1948.)
17. **First Aid Fire Appliances** (fire extinguishers), Installation, Maintenance and Use of. (1948.)
18. **Flammable Liquids**, Containers for Storing and Handling. (1941.)
19. **Flammable Liquids and Petroleum**, Pipe Lines and Discharge from Tank Cars. (1932.)
20. **Foam Extinguisher Systems.** (1946.)
21. **Fruits and Vegetables**, Coloring and Ripening of. (1938.)
22. **Fur Storage.** (1947.)
23. **Garages**, Construction and Protection of. (1939.)
24. **Gas Systems for Welding and Cutting.** (1947.)
25. **Gasoline Vapor Gas Machines, Lamps, and Systems.** (1926.)
26. **Heating and Cooking Appliances** (kerosene and fuel oil). (1937.)
27. **Hose Houses for Mill Yards**, Construction and Equipment. (1940.)
28. **Incinerators.** (1938.)
29. **Internal Combustion Engines** (gas, gasoline, kerosene, fuel oil) and **Coal Gas Producers** (pressure and suction systems). (1934.)
30. **Liquefied Petroleum Gases**, Storage and Handling. (1949.)
31. **Liquefied Petroleum Gases at Utility Gas Plants.** (1949.)
32. **Merchandise Vaults and Safes.** (1948.)
33. **Municipal Fire Alarm Systems.** (1941.)
34. **Nitrocellulose Motion Picture Films** (storage and handling). (1939.)
35. **Oil Burners—Installation in Stoves Originally Designed for Solid Fuels.** (1932.)
36. **Oil Burning Equipments**, and the Storage and Use of Oil Fuels in Connection Therewith. (1941.)
37. **Outside Protection**, Underground Piping Systems Supplying Water for Fire Extinguishment. (1931.)
38. **Ovens for Japan, Enamel, and Other Flammable Finishes.** (1931.)
39. **Paint Spraying and Spray Booths.** (1946.)
40. **Photographic and X-ray Nitrocellulose Film.** Storage and Handling. (1930.)
41. **Piers and Wharves**, Construction and Protection. (1935.)
42. **Proprietary, Auxiliary and Local Systems** for Watchman, Fire Alarm and Supervisory Service. (1949.)
43. **Protection of Openings in Walls and Partitions.** (1939.)
44. **Protective Signaling Systems**, Central Station, for Watchman, Fire Alarm and Supervisory Service. (1940.)
45. **Pulverized Fuel Systems**, Installation of. (1935.)
46. **Pyroxylin Plastic**, Storage, Handling, and Use of. (1940.)
47. **Pyroxylin Plastic**, Storage and Sale of in Other Than Plants Manufacturing Articles Therefrom. (1940.)
48. **Sprinkler Equipments**, automatic and open systems. (1940.)
49. **Standpipe and Hose Systems.** (1938.)
50. **Tanks.** (Water, Gravity and Pressure, Towers, etc.) (1941.)
51. **Valves**, Controlling Water Supplies for Fire Protection. (1931.)
52. **Water Spray Nozzles and Extinguishing Systems.** (1947.)
53. **Waterproofing of Floors and Drainage, and Installation of Scuppers.** (1937.)

Standards and Special Committee Reports

Published by the National Fire Protection Association

100. **Anaesthetic Gases and Oxygen.** Construction and Installation of Piping Systems; Oxygen Chambers. (1934.)
100*a*. **Anaesthetics, Combustible and Operating Room Explosions.** (1944.)
103. **Building Exits Code.** Stairs and enclosures, fire escapes, ramps, horizontal exits, doors, aisles, and corridors, elevators,

escalators, slide escapes, alarm systems, fire exit drills, signs and lighting. Requirements for schools, department stores, factories, theatres and places of public assembly, hospitals, sanitariums and corrective institutions, hotels and apartment houses, and office buildings. 9th ed. (1948.)

103*a*. **Control of Water Waste from Private Fire Protective Equipment.** (1938.)

103*b*. **Dust Explosions in Industrial Plants,** Fundamental Principles for the Prevention of. (1948.)

104. **Farm Storage for Flammable Liquids.** (1947.)

107. **Fire Exit Drills and Alarm Systems.** (Reprint from 103.) (1935.)

107*a*. **Flameproofing of Textiles.** (1941.)

108*a*. **Fire Hose,** Care of. (1936.)

108*b*. **Fire Pumps, Steam.** Operation and Maintenance. (1937.)

108*c*. **Fumigation of Grain Storage and Grain Processing Plants.**

108*d*. **Gas Hazards on Vessels to be Repaired,** Regulations Governing the Control of. (1948.)

108*e*. **General Storage Standard.** (1946.)

109. **Fire Fighting Equipment in Metal Mines.** (1930.)

109*a*. **Grandstands, Tents and other Places of Outdoor Assembly.** (1948.)

109*b*. **Hotel Fire Safety Law,** Guide for. (1948.)

109*c*. **Hose and Hose Couplings.** (1937.)

109*d*. **Lacquer Manufacturing Plants.** Suggestions for the Protection of. (1946.)

109*e*. **Hose and Ladder Work.**

109*f*. **Fur Storage.** (1947.)

110. **Lightning,** Protection of Life and Property against. (1937.)

110*b*. **Marinas,** Requirements for the Location, Construction and Operation of.

111*a*. **Motor Craft,** Fire Protection Regulations for. (1948.)

111*b*. **Marine Terminals,** Operation of. (1937.)

112. **Mechanical Refrigeration,** Hazards and Safeguards of. (1931.)

112*a*. **National Standard Fire Hose Couplings and Fire Department Hose Connections for Sprinkler and Standpipe Systems.** (1934.)

114. **Protection of Records.** Consolidated reports of the Committee on Protection of Records 1942–1946. Including Standards on Vaults.

114*a*. **Recommended Fire Defense Training Courses.** (1942.)

114*b*. **Reference List for Firemen's Training.** (1948.)

114*c*. **Rubbish Handling and Incinerators.** (1948.)

115. **Volunteer Fire Departments for Rural and Small Communities.** (1947.)

115*a*. **Salvaging Operations.** (1938.)

116. **Shoe Factories.** Suggestions for Their Improvement as Fire Risks. (1923.)

116*a*. **Spark Arresters for Chimneys and Stacks.** (1936.)

116*b*. **Spontaneous Ignition of Coal and Other Mining Products.** (1936.)

116*c*. **Sprinkler Systems,** Care and Maintenance. (1940.)

117*a*. **Standard Methods of Fire Tests of Building Construction and Materials and Door Assemblies.** (1942.)

117*b*. **Standard Threads for Small Hose Couplings.** (1922.)

118. **Static Electricity.** (1947.)

119*a*. **Summer Homes in Forested Areas.** Fire Protection and Fire Prevention. (1935.)

119*b*. **Table of Material Subject to Spontaneous Heating.** (1947.)

119*c*. **Tank Vehicles for Flammable Liquids.** (1948.)

120. **A Table of Common Hazardous Chemicals.** (1944.)

120*a*. **Trailer Coaches and Trailer Coach Camps.**

120*c*. **Vessels in Course of Construction and During Lay-up,** Prevention of Fire on. (1938.)

121. **The Watchman.** Suggestions for Guidance in Selection. Instruction and Duties. (1925.)

122. **Water Charges.** Report of Committee on Public Water Supplies for Private Fire Protection. (1932.)

123. **Water Systems for Fire Protection on Farms.** (1938.)

N.F.P.A. Standards Published by U.S. Government

These publications may be obtained from the Government Printing Office, Washington, D.C. Remittances should be included with orders, in cash, postal money order, or coupons sold by the Superintendent of Documents, Government Printing Office. As a convenience to members, single copies will be furnished by the Executive Office when ordered with other publications.

DA1678. **Gasoline and Kerosene on the Farm,** Safe use and storage of. 5 cents

DA1590. **Fire Protective Construction on the Farm.**

B. Suggested Municipal Ordinances

1. **Tank Trucks, Tank Trailers and Tank Semi-trailers for the Transportation of Flammable Liquids.** Regulating the Construction and Operation of.

1*a*. **Bureau of Fire Prevention.** To establish, provide officers and define their powers and duties.

1*b*. **Automatic Sprinkler Ordinance.**

3. **Explosives,** Ordinance for Cities.

3*b*. **Fireworks Ordinance,** Suggested.

4. **Flammable Liquids** and the products thereof, to regulate the use, handling, storage and sale of. Including Appendixes on Gasoline Service Stations, and Rooms, Cabinets and Outside Houses for Flammable Liquids.

4*a*. **Fumigation Ordinance, Model.**

4*b*. **Forest Fire Ordinances, County and Municipal.**

5. **Oil Burning Equipments,** and Oil Storage in Connection Therewith; **Stove or Range Oil Burners.**

5*a*. **Oil Burning Equipments.**

5*b*. **Petroleum Wharves.**

6. **Piers and Wharves,** Regulating Construction and Protection of.

9. **Fire Prevention Code for Cities.** Advisory Outline.

C. Reference Books

See also National Fire Codes

CFF. **Crosby-Fiske-Forster Handbook of Fire Protection.** The standard handbook of fire protection engineering, including fundamental reference data for experienced men specializing in fire prevention and fire protection. Also used as a text for beginners. Contains background information and an explanation of the principles underlying N.F.P.A. standards, but does not give the detailed provisions of standards found in the several volumes of National Fire Codes. Includes hydraulic data, charts, tables, illustrations of automatic sprinklers, and other fire protection devices, and much other material not elsewhere available. 10th ed. (1948), 1544 pp. Price $9.50.

C6. **Fire Tests of Building Columns.** An experimental investigation of the resistance of columns, loaded and exposed to fire or to fire and water with record of characteristic effects. Published by Underwriters' Laboratories, Inc., 390 pp.

C9. **A Model Records and Reporting System for Fire Departments.** A comprehensive manual on best procedures, including 27 sample forms, 78 pp., illustrated (1938).

C10. **Fire Service Training Texts.** (Revised 1941.) Compiled by the Department of Trade and Industrial Education, Oklahoma A. and M. College, in cooperation with the fire department instructors of Oklahoma and cooperating states by W. Fred Heisler.

Unit I. Forcible Entry and Minor Extinguishment Practices, 40 pp.
Unit II. Ladder Practices. 48 pp.
Unit III. Hose Practices. 56 pp.
Unit IV. Salvage and Overhaul Practices. 32 pp.
Unit V. Fire Steam Practices. 64 pp.
Unit VI. Fire Apparatus Practices. 84 pp.
Unit VII. Ventilation Practices. 34 pp.
Unit VIII. Rescue Practices. 72 pp.
Unit X. Inspection Practices. 96 pp.

C12. **Municipal Fire Administration.** A training course for fire administrators.

C14. **Industrial Fire Brigades, A Training Manual.** For use in classes where employees are trained in private fire brigade plant protection. 176 pp., 6 by 9 in., illustrated.

C16. **Airplane Crash Fire Fighting Manual.** A comprehensive reference manual prepared by experts, covering every phase of this type of fire fighting. 96 pp., printed in color, paper bound.

D. Popular Educational and Miscellaneous Pamphlets

1*b*. **Artillery Ammunition Manufacture.*** Guise and Slicer.

* Reprint from *Quarterly of the National Fire Protection Association.*

2. **Absorption of Heat by Waterfog.**† Hendricks.
2a. **Airplane Crash Fire Fighting.*** January, 1944. Tryon.
2b. **An Analysis of Watchman Failure.***
2d. **Automatic Sprinkler Tables,** 1945.*
3a. **Canada Views the Sprinkler Tax, How.**
3b. **Bleaching Wood with Hydrogen Peroxide.***
3c. **Building Exits for Fire Safety.**
3d. **Building Codes and Fire Prevention Codes.***
8b. **Definition of Fire Protection Engineering.***
10. **Drying by Infra-red Lamps.*** Richards.
10a. **Dust Explosion Hazards of Powdered Metals.*** Brown.
12a. **Dwelling Inspection by Fire Departments.***
12c. **Employee Organization for Fire Safety.**
12d. **Electrical Inspections by Firemen.**
13. **Explosion Prevention in Army Ordnance Plants.** Field.†
13a. **Fighting Airplane Crash Fires.** Doolin.†
13b. **Fighting Magnesium Fires.***
13c. **Finishing Floors and Bowling Alleys.**
14. **Fire Alarm Central Stations.*** Revised and reprinted. Carroll.
15. **Fire and Building Department Co-ordination.**
15a. **Fire Department Organization.*** Firebrace.
16a. **Fire Department Records.***
16b. **Fire Extinguisher Explosions.***
17. **Fire Effects on Fire-resistive Construction.** Burton.
17a. **Fire Gases.*** Ferguson.
17b. **Firemen's Training.** Bond.
17c. **Fire Hazards of Termite Control.*** Morgan.
17d. **Fire Hazard Tests of Building Materials.*** Steiner.
19. **Fire Prevention Week Handbook.** Suggestions for guidance in planning the observance of Fire Prevention Week. Illustrated.
19a. **Fire Protection in Engineering.*** Moulton.
19b. **Fire Protection in Refineries.** Prepared by American Petroleum Institute, 116 pp., 8 by 10½ in., illustrated.
19c. **Fire Protection Engineering Education.**
20b. **Fire Retardant Treatments for Wood.***
20c. **Fire Safety Education for Employees.***
20d. **Fire Safeguards for Magnesium.***
21. **Fire Test of Brick Joisted Buildings.***
22. **Food Dehydration.** Hallowell.
23. **Industrial Fire Training Films.***
27. **Inspected Electrical Equipment.** Underwriter's Laboratories, Inc.
28. **Inspected Fire Protection Equipment and Materials.** Published by Underwriter's Laboratories, Inc.
29. **Inspected Gas, Oil and Miscellaneous Appliances.** Underwriter's Laboratories, Inc.
29a. **Large Spherical Compressed Gas Containers.**
30a. **Magnesium and Its Alloys.*** Hirst and Guise. (1942.)
31. **Memphis Industrial Fire Brigade.*** Robinson.
32b. **Motion Picture Films,** on Fire Prevention and Protection, List of.
33a. **Municipal Fire Apparatus Budget Programs.*** Bugbee.
34. **New Method of Protecting Conveyor Openings.** Thompson.
35. **New Developments in Chimneys and Flues.***
35a. **New Developments in Carbon Dioxide.**
35b. **Physiological Considerations Relating to Fire Gas Exposure.**† Gray (1948).
36. **Preventing Cutting and Welding Fires.**
36b. **Public Water Systems in the War Emergency.**† (1942.)
36c. **Plastics.***
37. **Organization Plan for a Fire Prevention Committee.**
37b. **Postwar Municipal Fire Protection.***
37c. **Problems in the Use of Public Water Supplies for Private Fire Protection.**†
38. **Preventing Air Conditioning Fires.***
42. **Safeguarding Gas Appliances.*** Vandaveer.

* Reprint from *Quarterly of the National Fire Protection Association.*
† Reprint from *Proceedings, National Fire Protection Association.*

44a. **Selected Demonstrations for Use in Fire Safety Education.**
44b. **Shipyard Fire Protection.**† Clachos.
45a. **Spontaneous Ignition of Chopped Hay.*** Roethe, Bradshaw, and Hoffman.
45c. **Spontaneous Ignition in Mines and Mineral Products,** Bibliography on.
46. **Sprinkler Protection in War Industry.***
46a. **Sprinklers Provide Safety to Life.***
47a. **A Study of Record Container Performance.***
47b. **Synthetic Rubber.*** Hallowell.
48a. **Treatment of Burns with Aniline Dyes.*** Aldrich.
49b. **Venting of Tanks Exposed to Fires.***
50. **Thermostats.**† Hendricks.
50b. **Water Supplies for Rural Fire Protection.*** Charnock
50c. **Water Spray Protection.** (1942.)
51. **Wired Glass Windows in the Fall River Conflagration.** (See also E103.)

Other Reference Books and Pamphlets

FIRE PROTECTION: Assheton, "History of Explosions," Institute of Makers of Explosives, New York. "Fire Tests of Building Columns," Associated Factory Mutual Fire Insurance Companies, Boston. Crosby, Fiske, and Forster, "Hand Book of Fire Protection," Van Nostrand, New York. Dana, "Automatic Sprinkler Protection," Wiley, New York. Dominge and Lincoln, "Fire Insurance Inspection and Underwriting," Spectator Company, New York. Eichel, "Fire Prevention and Protection for Hospitals," Wiley, New York. Gamble, "Outbreaks of Fire, Their Causes and Means of Prevention," Lippincott, Philadelphia. Hough and Lawson, "Fire Prevention Year Book," Hough-Lawson, Baltimore. Hutson, "Fire Prevention and Protection," Spectator Company, New York. Owen, "Notes on Hydraulics," Insurance Press, New York. Phillips, "The Handling of Dangerous Goods," Crosby, Lockwood, London. Shepperd, "Practical Hydraulics for Firemen," Sheppard, New York. Von Schwartz, "Fire and Explosion Risks," Griffin, London. Wallace, "Fire Losses Locomotive Sparks," Barr-Erhardt Press, New York.

GENERAL SAFETY: Chase, "Men and Machines," Macmillan, New York. DeBlois, "Industrial Safety Organization," McGraw-Hill, New York. Fisher, "Mental Causes of Accidents," Houghton Mifflin, Boston. Heinrich, "Industrial Accident Prevention," McGraw-Hill, New York. Lange, "Handbook of Safety and Accident Prevention," McGraw-Hill, New York. Williams, "The Manual of Industrial Safety," McGraw-Hill, New York.

INDUSTRIAL HEALTH HAZARDS AND MEDICAL SERVICE: Hamilton, "Industrial Poisons in the United States," Macmillan, New York. Kober, "Industrial Health," Blakiston, Philadelphia. Rambousek, "Industrial Poisoning From Fumes, Gases and Poisons of Manufacturing Processes," Legge, London. Thompson, "The Occupational Diseases—Their Causation, Symptoms, Treatment and Prevention," Appleton, New York. Underhill, "Toxicology," Philadelphia.

PUBLICATIONS OF VARIOUS BUREAUS: Various organizations and governmental departments have issued bulletins from time to time relating to safety and fire protection. Among these are the bulletins, technical papers, and circulars of the U.S. Department of Labor, U.S. Bureau of Mines, U.S. Public Health Service, U.S. Bureau of Standards, Federal Board for Vocational Education, Interstate Commerce Commission, and others.

PERIODICALS: *Fire Engineering* (Semimonthly), Case, Sheppard, Mann Publishing Corp., New York. *Fire Protection* (Monthly), National Underwriter Co., Indianapolis. *National Safety News* (Monthly), National Safety Council, Chicago. *Safety Engineering* (Monthly), Alfred M. Best Co., New York. *Quarterly*, National Fire Protection Association, Boston.

* Reprint from *Quarterly of the National Fire Protection Association.*
† Reprint from *Proceedings, National Fire Protection Association*

INDEX

A

Abbé rotary cutters, 1161
Abbreviations, mathematical signs, symbols, and abbreviations, 5
symbols of chemical engineering, 3
technical journals, 4
Abietic acid-oleic acid-propane system, 722, 724
Abnormal void content, 394
Abrasives, classification of, 937
grinding of, 936
Absorbent cotton, hygroscopic moisture of, 777
Absorbers, packed, over-all resistance in, 693–698
Absorption, of acetone, 548
from air with water, design of column for, 669
of ammonia in acetic or sulfuric acids, 706
from air by water, H.T.U. for, 693
in water with spray column, 701
batch apparatus, design of, 557
of carbon dioxide in alkaline solutions, 703
in aqueous NaOH solutions, 705
in carbonate solutions, 703
in diethanolamine solutions, 705
in sodium or potassium hydroxide solution, 704
in sugar-lime solutions, 700
in water, 693, 695
of chlorine in sodium hydroxide, 706
in water, 695
columns, limiting velocities for, 680
mass transfer in, 686
plate efficiencies of bubble-tray, 698
pressure drops for, 680
and compression refrigeration system, 1676
equilibrium, prediction of, 532
gas, 667–711
accompanied by chemical reaction, 702
column selection for, 668
columns and packings in, 707
exit strength (or column height) in, 708
H.T.U. of column packing for, 671
plate towers vs. packed towers in, 707
solubility data for, 668
vapor-liquid equilibrium in, 668
velocity (or column diameter in), 708
gas-liquid ratio in, 668
Absorption, of hydrogen chloride in water, 697
liquor-gas ratio in gas, 707
of nitrogen oxides in water and nitric acid, 706
in non-aqueous systems, 698
of olefins and acetylenes in aqueous copper solutions, 706
rate of, in spray towers, 699
in agitated vessels, 702
refrigerating machine, 1677
refrigeration system, 1676, 1683
intermittent, 1685
selection of solvent in, 668
strength of luminous flames, 493
of sulfur dioxide in caustic, 706
from flue gas in ammonium sulfite-bisulfite solution in spray tower, 702
in sodium carbonate solution in cyclone spray tower, 701
in water, 696
in cyclone spray tower, 701
systems, economic design of, 707
refrigerants for, 1686
Absorptivities, 483
Accelator, Infilco, 947
Acceleration of gravity, 45
Accessories, for evaporation, 520
Accounting, cost, 1827*ff*.
Acetic acid, aqueous, vapor pressures of, 170
constructional materials for, 1461
data for ternary systems with, 721, 723, 724, 725, 726
densities of aqueous solutions of, 186, 187
-ethyl ether-water system, 721
extraction from water, by benzene, 751
by isopropyl ether, 750
by methyl isobutyl ketone, 749, 751
specific heats of aqueous solutions of, 234
or sulfuric acids, absorption of ammonia in, 706
-water benzene system, 751
-methyl isobutyl ketone system, 751
-solvents, 723
Acetic anhydride, constructional materials for, 1462*ff*.
Acetone, absorption of, 548
from air by water, design of column for, 669
by silica gel, 912
Acetone-chloroform vapor-liquid equilibriums, 630
Acetylene(s), compressibility factors of, 208
as fuel, 1576
Acetylene(s), and olefins, absorption of, in copper solutions, 706
Acid(s) (*see also under specific names such as* Acetic, Nitric)
constructional materials for, 1461*ff*.
equivalent conductivities of, 1783
fatty, constructional materials for, 1485
indicators, 285
-resisting cements, 453
-treated clays, as adsorbents, 896
Acidity of transformer oil, reduction of, with activated alumina, 911
Acoustic flocculation of dust, 1049
Acoustic velocity, 375
Action, average-position, 1338
Activated alumina, 909
for drying gases, 880
properties of, 905, 910
Activated carbon, for drying gases, 881
pressure drop through beds of, 905, 906
properties and applications of, 905
removal of benzol from, 906
use in Suchar process for sugar refining, 903
use in water filtration, 903
Activated carbon filter, 1049
Activation energy, 323
Activation reagents, 1088
Activity, 307
coefficient(s), 307
binary constants for equations for, 528
of 1-butene in furfural, 646
calculation of, from boiling-point curves, 534
for partly miscible solutions, 534
from total pressure curves, 534
estimation of from azeotropic data, 533
of ethyl alcohol in trichloroethylene, 653
of ethyl alcohol in water, 653
function of composition, 527
of isobutane in furfural, 646
liquid, 526
effect of temperature on, 529
prediction of, 533
of trichloroethylene in ethyl alcohol, 653
of trichloroethylene in water, 653
utilization of, 532
Henry's law, 532
utilization of, prediction of absorption equilibrium, 532
prediction of azeotropes, 532

Activity, coefficient(s), prediction of distribution in liquid-liquid extraction, 533
prediction of heats of solution, 532
prediction of immiscibility, 533
of water in ethyl alcohol, 653
of water in trichloroethylene, 653
Actual-cost methods, 1834
Actual gases, 290
Addition, algebraic, 63
Additive pressures, 290
volumes, 291
Adiabatic compression, 303
expansion, 303
flow of gases, 376, 379
humidification, 759
-saturation lines for air-water vapor, 811
Adjustable-cam cycle controller, 1326
Adjustable orifice, 1284
Adjustment, controller, 1338
Adjust-O-Feeder, 1374
Admittance, 1732
Adsorbent(s) and catalysis, 915
carbons, characteristics of, 901
desulfurization of kerosene by, 913
industrial, 888
(solid), properties and applications of, 905
treatments, pre-centrifuging, 1004
Adsorption, 885–916, 1196
of acetone from air by silica gel, 912
of benzene from air by silica gel, 912
and crystallization, 887
curves for anions and cations, 915
curves, typical, 886
degree of, and pH, 900
Fuller's earth for, 889
from gases, 887
heats of, 301
and hydrogen-ion curves, 900
industrial, 888
liquid-phase, 914
particle-size determination by, 1113
purification of oils by, 889
rates on solid desiccants, 882
of sulfur dioxide from air by silica gel, 912
theory, 886
unit for dehumidification, 786
as a unit operation, 887
Adsorptive efficiency, 904
Aero pulverizer, 1141
Aerofall Mills Limited mills, 1135
Aerotec tube, 1028
Agent, control, 1338
Agents, deflocculating, 1088
Agglomeration tabling, 1080
Agitated batch crystallizer, 1062
Agitated-pan dryer, 856
Agitated vessels, gas absorption rates in, 702
Agitation, by gas dispersion, 1176
Agitation of pulp in filters, 985
Agitation flow sheet, intermediate, 952
Agitator, air-lift, 1207, 1220
gate, 1222
horseshoe, 1222
model designs, 1228
Agitator settling in containers, 1394–1395
Agitators for packing containers, 1394–1395
Agnesi, witch of, 81
Aid, first, 1875
Aids, filter, 969
Air, for combustion, calculation of excess, 338
compressibility factors of, 207
Joule-Thomson effect of, 203
liquefaction process, 1702
moist, enthalpy of, 760
entropy, 760
humidity ratio, 760
thermodynamic properties of, 760
volume of, 760
water vapor pressure of, 760
ratios of specific heats of, at high pressures, 233
removal from evaporators, 511
removal pumps, 1647
requirements for buildings, 1853
requirements for combustion of gases, 1589
sampling of, 1102
saturated, thermodynamic properties of, 250
separation of constituents of, 1703
specific heats of at high pressures, 233
superheated, thermodynamic properties of, 250
Air-acetone absorption by water, design of column for, 669
Air-atomizing burners, 1574
Air-benzene vapor mixtures, humidity chart for, 814
Air-blast process, reversed, 1581
Air-CCl_4 vapor mixtures, humidity chart for, 813
Air cell, 1792
Air classifier and hammer mill in closed circuit, 1118
Air cleaners for bottles, 1398
Air conditioner, unit, 784
Air conditioning, 777
equipment, 779
system using silica gel, 914
Air-cooled worm-gear reduction units, 1669
Air filters, 1013, 1045
Air jig, Conset, 1077
Air lift, 1438
agitator, 1207, 1220
as mixer, 1203
Air preheaters, 1642
Air pumps, 520
Air-separating traps for centrifuges, 1003
Air sparger, 1219
Air-toluene vapor mixtures, humidity chart for, 815
Air-valve relay, non-bleed, 1322
Air velocity, effect of, on drying, 803
Air-water vapor mixtures, humidity chart for, 811
Air-o-xylene vapor mixtures, humidity chart for, 816
Airmat dust filter, 1046, 1047
Air-maze dust filters, 1046
Airplex dust filter, 1046
A.I.S.I. standard steels, chemical composition of, 1543*ff*.
Aisles in factories, operating, 1369
Ajax-Northrup furnace, 1816
Ajax-Northrup induction furnace, 1818
Ajax-Wyatt furnace, 1816
Ajax-Wyatt induction furnace, 1817
Akins classifier, 926
Alcohol (denatured)-water flow points of, 1696
freezing points of, 1696
Alcohol as fuel, 1575
Alcohol recovery by adsorbents, 908
Algebra, 62
binomial theorem, 64
cubic equations, 67
determinants, 68
factoring, 63
infinite series, 65
linear equations, 66
notation, 62
quadratic equations, 66
quartic and higher equations, 68
symbols of, collective, 62
Algebraic approximations, 63
binomial, 64
Newton's method, 67
Algebraic calculations, for extraction, 739, 740, 741
Alignment charts, 100–105
Aliphatic hydrocarbons, constructional materials for, 1492
Alkaline chloride electrolysis, 1806
Allen-Moore cell, 1808
Allen settling cone, 940
Allis-Chalmers crushers, 1120, 1122, 1126, 1127, 1128
Allis-Chalmers mills, 1132, 1133
Allis-Chalmers packer, 1385, 1386
Alloys, composition of, 1526*ff*.
fusible, 454
low-melting-point, 454
melting points of, 281
and metals, corrosion by gases, 1526
properties of, 1460
sampling of, 1101
thermal conductivity of, 456
Aloxite, grinding of, 936
Aloxite filters, 968
Alphabet, Greek, 46
Alpha-Lux gas density balance, 1297
Alternating current, 1733
generators, 1764
magnets, 1753
motors, 1767
precipitators, 1045
Alternators, three-phase, 1764
Alum, constructional materials for, 1463
Alumina, activated, 909
for drying gases, 880
hydrate, drying of, 826
(fused) production, 1819
properties and applications of, 905 910

Aluminum, anodic oxidation of, 1801
chloride, constructional materials for, 1464
electrolytic production of, 1810
electroplating on, 1798
hydroxide, spray-drying of, 845
oxide grains, drying of, 869
pipe, 424
refining, 1811
sulfate, constructional materials for, 1464
continuous countercurrent decantation, 951
densities of aqueous solutions of, 177, 178
spray-drying of, 845
Alundum filters, 968
Amagat's law, 291
Ambient-temperature compensation, 1271
American filter, 981
American Iron and Steel Institute standard steels, chemical composition of, 1543*ff.*
American Pulverizer crushers, 1129
American Ring crusher, 1129
American table of distances (safety and fire protection), 1848
American Warehouse Association, standard insulation of, 1693
Amines, constructional materials for, 1467
Amirglas dust filter, 1046
Ammonia, absorption in acetic or sulfuric acids, 1706
from air by water, H.T.U. for absorption of, 693
-air mixture, limits of inflammability of, 1587
aqua solutions, properties of, 1685
aqueous, densities of solutions, 178
enthalpy-concentration diagram for, 253
specific heats of solutions, 234
vapor pressures of, 172
brake horsepower per ton of refrigeration of, 1678
compressibility factors of, 208
Ammonia compressors, volumetric efficiency of, 1679
condenser, vertical shell-and-tube, heat transfer for, 480
constructional materials for, 1467
in manufactured gas, 1578
partial pressures of, over NH_3, aqueous, 172
as refrigerant, 1678
saturated, thermodynamic properties of, 250
superheated, thermodynamic properties of, 250–252
volume compressed per ton of refrigeration, 1682, 1683
in water, absorption of, 693
H.T.U. data for, 688
with spray column, 701
Ammonium acetate, densities of aqueous solutions of, 178
bichromate, densities of aqueous solutions of, 178
chloride, constructional materials for, 1467
Ammonium acetate, chloride, densities of aqueous solutions of, 178
chromate, densities of aqueous solutions of, 178
nitrate, densities of aqueous solutions of, 178
phosphate, constructional materials for, 1470
sulfate, constructional materials for, 1471
drying of, 833
Amortization, 1831
Amortization of power plant equipment, 1632
Ampere, 1732, 1772
Ampere-second, 1772
Amplification constant, 1743
Amplifier(s), impedance-capacitance-coupled, 1744
linear, 1743
parallel-connected, 1745
push-pull, 1745
thermionic, 1743
transformer-coupled, 1744
typical characteristics, 1745
Amsler polar planimeter, 89
Analysis, of coal, 1560
dimensional, 93–97
applied to theoretical physics, 97
falling bodies, 95
pressure, gas, 95
Rayleigh's problem of heat transfer, 96
fuel gases, 1577
fuels, 1629
particle size, 1017
Analytical geometry, 77–81
circle, 78
conic, 77
ellipse, 78
hyperbola, 79
miscellaneous, 80
parabola, 79
Anderson expeller, 1074
Andrews kinetic elutriator, 1084
Anemometer, 398, 1288
hot wire, 399
Robinson cup, 399
vane, 398
Aneroid barometer, 365
Angelini furnace, 1815
Anger mill, 1147
Angle(s), 53
bisecting, method of, 59
complementary, 53
congruent, 54
contact (of fluids), 363
dihedral, 55, 56
polyhedral, 56
right, inscribed in semicircle, 55
spherical, 58
supplementary, 53
trigonometric functions of, 70
Angle-type industrial thermometers, 1270
Angström unit, 754
Angular measure conversion table and equivalents, 41
Angular measurements, trigonometric, 70
protractor, 70
radian or circular, 70
Angular measurements, sexagesimal, 70
Angular propeller mixer, 1208
Anhydrone, for drying gases, 881
Aniline,-n-heptane-methylcyclohexane system, 722
-methylcyclohexane-n-heptane, 723
selectivity diagram for, 723
ternary system, distribution diagram for, 722
triangular phase diagram for, 721
specific heats of, 235
Anions, adsorption curves for, 915
Annular spaces, heat transfer in, 470
Annulus, area of, 56
construction of, 61
Anodic oxidation of aluminum, 1801
Anomalistic year, 45
Anthracite coal, grinding of, 1156
Anticipatory control, 1337 (*see* Rate action)
Antidotes, poison, 1876
Antifriction bearings, 1667
Antimony electrode, 1309, 1795
Aperture of screens, 955
A.P.I. degrees conversion table and equivalents, 38
and sp. gr. relationship, 1296
gravity, 1569
A.P.I.-A.S.M.E. code of welding, 1243
A.P.I.-A.S.M.E. code of pressure vessel design, 1238
Apothecaries' measure (table), 43
Apparent power, 1732
Apparent reaction order, 324
Apparent viscosity, 1197–1202
Approach, velocity of, 401
Approximations, algebraic, 63
binomial, 64
Newton's method, 67
trigonometric, 72
Apron conveyors, capacities, 1348
power determination, 1348
skirt-board drag, 1348
Apron (conveyor) feeder, 1371
Aqueous solutions, heat capacities of, at infinite dilution of, 530
heats of solution at infinite dilution of, 530
Arc furnaces, 1814
Archimedes principle, 362
Arc(s), circular, bisecting of, 60
lengths of, 33
from angle in degrees, 33
from central angle, 33
from chord and arc height, 33
Area(s), of circles, 29, 55
of circles and squares of equal areas, 33
of circular segments, 32
of cone, 57
of cycloid, 57
of cylinders, 57
of ellipse, 56
of hyperbola, 57
irregular, 57
meters, 408, 1283, 1288
of parabola, 57
of parallelogram, 55
of prism, 57

Area(s), of quadrilateral, 55
of rectangle, 54
of rhombus, 55
of sector of annulus, 56
segments of circles, 32
of spherical triangle, 76
surface conversion table and equivalents, 41
of triangle, 54
Arith-log graph paper, 99
Arithmetic integration, 92–93
Arithmetical progression, 65
Arm mixers, 1204–1208, 1220
Armature reaction, 1763
Aromatic hydrocarbons, constructional materials for, 1493
Arrhenius equation, 323
Arsenic acid, densities of aqueous solutions of, 178
Arterial bleeding, 1878
Artificial respiration, 1878
Asbestos, fiber, hygroscopic moisture of, 780
grinding of, 1151
paper, hygroscopic moisture of, 779
Asbestos-cement pipe, 431
Ash, of coals, 1561
composition of coals, 1561
handling systems, 1637
liquid fuels, 1570
softening temperatures of coals, 1561
A.S.M.E. code for pressure vessels, 1238
A.S.T.M. classification of coal, 1562
distillation criteria of separation, 603
comparison with true-boiling-point method, 607
fuel oil specifications, 1654
Atmosphere, variation of, density with elevation, 361
pressure with elevation, 361
Atmospheric condensers, 1688
drum dryer, 863
pan dryer, 857
pollution, 1017, 1047
Atomic numbers of elements, 1781
volumes, 538
weights of elements, 1781
Atomization, theory of, 840
Atomizers, centrifugal disk, power consumption of, 848
for spray dryers, 839
Atomizing burners, 1573*ff*.
Attrition mills, disk, 1143
double-runner, 1123
performance of, 1144
Auger packers, 1384
Allis-Chalmers, 1385
Autoclaves, 1256, 1257
laboratory, 1257
rocking type, 1257
Autofrettage or "self-hooping," 1241
Autogenous welding, 1243
Automatic control, fundamentals, 1309
loop dryers, 852
mechanisms, 1320
pulverizer, 1142
regulator, 1337
Automatic control, reset, 1322, 1337, 1338
controller, 1323
plus rate action, 1322
scales, 1292, 1293, 1382, 1383, 1384, 1385
shutdown system, pneumatic-electric, 1335
small refrigeration machines, 1697
system applications, 1329
terms, 1337
(*See also* Controllers)
Auto-raise thickener, Hardinge, 944
Autotransformers, 1762
connections, 1763
Auxiliaries, filter, 990
Availability factor of power plant, 1631
Average-position action, 1338
Average value of alterating current, 1734
Avogadro's number, *N*, 47
Avoirdupois and metric units, equivalents of, 1781
Avoirdupois weights and measures (table), 43
Axial-flow fans, 1448
characteristics, 1643
pump, 1420
Axonometric charts, 100
Azeotrope(s), former, 651
heterogeneous, 631
formation of homogeneous, 630
hydrocarbon (table), 635
occurrence in distillation, 630
separating agents for, 643
prediction of, 532
Azeotropic binary mixtures, maximum boiling point (table), 634
minimum boiling point (table), 633
effect of pressure on composition of ethyl alcohol-water azeotrope (table), 631
estimating composition, 631
Azeotropic data, estimation from activity coefficients, 533
Azeotropic distillation, 629, 651
ternary mixtures (table), 634
water-benzene azeotrope composition (table), 632

B

Babcock butterfat-milk tester, 1002
Babcock & Wilcox pulverizer, 1124, 1138
with external air classifier, 1135
Type E, 1155
Back ionization, 1041
Backward-curved-blade fans, 1448
feeding of evaporators, 510
-return condenser, 1708
Baffle-plate mixers, 1203
Baffles, pressure drop across, 393
Bag-closing equipment, 1388
hand sewing, 1390
wire tying, 1390
Bag-filling machine, St. Regis two-tube screw-type-valve, 1388
St. Regis belt, 1388
Bag filters, 1029
house, 1031
Bags, holders, 1389
Dustite, 1390
Swellgrip, 1390
liners, insertion of, 1390
sleeve valve, 1390
Bagasse as fuel, 1568
Bagpak Model A, 1388, 1389
Bagpaker, 1392, 1393
Bailey, automatic hydrometer, 1297
synchronized cam telemeter, 1325
Baker's mixer, 1207
Balance, of centrifuges, 1005
heat, 347
material, 337
mechanical energy, 376
sheets, 1829
terms of, 1830
Balanced polyphase system, 1735
Balanced valve, 1327
Ball & Jewell Improved Patented rotary cutters, 1161
Ball and ring ring-roller mills, 1116
Ball-and-socket joint, 436
Ball mill(s), 1116, 1130
air classifiers with, 1135
closed-circuit, 932
grindability determination with, 1114
mixer, 1212, 1220, 1221
speed of, 1131
steel-ball charge of, 1131
Ball Peb mills, 1132
Ball-type pilot valve, 1321
Band, control, 1338
Banded constituents of coals, 1562
Bantam Mikro-Pulverizer, 1140
Bar screens, 958
Barium, production of, 1812
Barium chloride, densities of aqueous solutions of, 178
Barium hydroxide, dissociation pressures of, 174
Barium oxide, for drying gases, 881
Barium peroxide, dissociation pressures of, 174
Barometer, aneroid, 365
mercury, 365
reduction to standard temperature, 365
standard, 365
Barometric condenser, 520
Barometric legs, 991
Barrel(s), filling and weighing, 1392
liners, insertion of, 1394
mixer, tumbling, 1212
Bartlett & Snow, crushers, 1122, 1129
single-roll crusher, 1128
two-roll crushers, 1128
Barytes, grinding of, 1151
ore, grinding of, 1135
Base(s), equivalent conductivities of, 1783
indicators, 285
Base-metal thermocouples, 1270
Base metals, physical properties of, 1539
Basic slag, grinding of, 1151
Basis, dry, 800
dry-weight, 800

Basis, selection of, in problem, 334
wet-weight, 800
Basket centrifuges, 992, 997, 998
costs of, 1007
Basket-type evaporator, 506
Basle cell, 1808
Bassanese furnace, 1815
Batch apparatus, design of, 557
Batch centrifuge, 992
Batch crystallizer, tank, 1062
Batch distillation, simple, 580
Batch dryers, direct, 814
through-circulation, 320
tunnel, 822
Batch furnaces, 1607
Batch presses, 1073
Batch rectification, 594
Batch steam distillation, 582
Batch temperature control, simple, 1330
Batch vs. continuous evaporators, 512
Batch vs. continuous mixing, 1230
Batching scale, 1294
Batteries, 1788
Bauer Bros. mills, 1144
Bauer Double Disk mills, 1144
Baum jig, 1076
Baumé and sp. gr. relationship, 1296, 1297
Baumé (degrees), °Tw., lb./gal., lb./cu. ft., sp. gr., equivalents, 38
Bauxite, as adsorbent, 896
grinding and drying of, 1119
Beaded-tube rotameter, 1288
Beaker sampling, 1100
Beam lengths for gas radiation, 491
Bearing metals, 1666
Bearings, 1666
Beer, filtration of, 988
Beet-sugar plant filtration, 987
Beiswenger and Piroomov correlations (distillation), 587
Bell-jar cell, 1808
Bell-and-spigot pipe, 436
Bellows, 411, 1280
flowmeter, 1286
for thermometers, 1271
thermostat, 787
Belt(s), 1660
clearances, 1858
compression, 991
conveyors, 1354
drives, 1664
fasteners, 1663
joining, 1663
maximum inclinations, 1356
power chart, 1358
self-aligning idlers, 1356
shifters and clutches, 1859
sliding, 1357
speeds, 1661
trippers, 1356
width and pulley face, 1665
Belt-type packer, 1388
Bends, 389
Benzene, adsorption from air by silica gel, 912
liquid-vapor relations, 594
toluene mixtures, partial pressure-composition diagram for, 577 578
Benzene-water-acetic acid system, 751
Benzene-air vapor mixtures, humidity chart for, 814
Benzoic acid-water-toluene system, extraction rate data for, 750
Benzol, as fuel, 1575
materials of construction for, 1472
removal from activated carbon, 906
Berl saddles, (ceramic), liquor-film H.T.U. for, 692
pressure drop in, 394, 395, 682
loading velocities for, 683
Bernoulli's theorem, 375
Berthoud-Valeton theory of crystallization, 1058
Beryllium production, 1811
Bessemer converters, 1616
Bevel-disk, single-seated valve, 1327
Bevel gear unit, spiral, 1670
Bevel-plug valve, 1328
Beverages, filtration of, 988
Bimetal thermostat, 787
Bimetallic pipe, 430
Bimetallic, spiral, 1276
thermometers, 1270
Bimolecular reactions at constant volume, 322
gases, at constant pressure, 323
Binary constants for equations for activity coefficients, 528
Binary cycles of refrigerants, 1681
Binary mixtures, constant boiling, temperature-composition diagrams for, 578
constant pressure liquid-vapor equilibrium data for, 573–575
maximum boiling-point azeotropic, 634
minimum boiling-point azeotropic, 633
phase diagrams for, 577
and Raoult's law, 525
Binary Margules equations, 527
Binary Van Laar equation, 527
Bindicator solids-level meters, 1290
Bin-Dicators, 1382, 1383
Bin level indicator, 1290
Binomial approximations, 64
theorem, 64–65
Bins, getting material to flow from, 1373
storage in, 1377
Bird-Young filter, 983
Birkeland-Eyde process, 1820
Bismuth, electrolytic refining, 1805
Bituminous coal, 1560
Black body, 483
conditions, 1275
Blacks, grinding of, 1156
Blade paddle mixer, 1204–1208, 1220
Blade rotary compressors, 1681
Blade turbine mixer, 1210
Blake crusher, 1120
Blast furnace, 1612, 1614
dimensions, 1615
gas as fuel, 1576
pipe precipitator, 1042
stoves, 463, 1623
Blaw-Knox contactor, 1037
jet pulverizer, 1146
Bleaching clays, production of, 896
Bleaching of oils, 890
Bleeding, arterial, 1878
Blending and size reduction, 1119
Blending of solids with solids, 1221
power required for, 1221
Block-tin pipe, 430
Blocks, metal, 283
Blood, animal, spray-drying of, 845
Blower(s), 991
turbine as mixer, 1210
Blue Streak Dual Screen pulverizer, 1141
Blue water gas as fuel, 1570
Blueprints, drying of, 869
Board measure, definition of, 43
Boat sampling, 1097
Boghead coal, 1563
Boiler(s), 1637
accessories, 1643
C-E vertical unit, 1603
efficiency curves, 1638
feed pumps, 1647
furnaces, 1605
water-level meter, 1290
Boiling liquids, heat transfer to, 478
Boiling-point, curves, calculation from activity coefficients, 534
effect of external pressure on, 293
elements and inorganic compounds, 110–128
elevation of, effect on multiple-effect evaporator, 500, 509
by non-volatile solutes, 319
estimation of, for concentrated aqueous solutions, 319
initial, 606
normal, of sulfuric acid, aqueous, 168
organic compounds, 129–148
relationships of, 565
Bolted-flange closures, 1249
Bond-paper dryers, 867
Bone black, grinding of, 1156
Bone char, as adsorbent, 897
analyses of, 897
manufacture of, 897
revivification of, 898
Bonnot crushers, 1122
dry pan crusher, 1127
Bookkeeping, 1827*ff.*
machines, 1828
Booster(s), 1439
ejector, 1453
response, 1338
Booth-Hall furnace, 1815
Borda mouthpiece, 407
Boric acid, constructional materials for, 1473
Boron carbide production, 1818
Bosch flowmeter, 1254
Bottle capper, 1400
Bottle-cleaning equipment, 1397
Bottle-closing equipment, 1399
Bottle sampling, 1100
Bound moisture, 800
Bourdon tube, 1271
pressure gage, 1254, 1255
Bowen spray dryer, 843, 844
Bowl, graining, 857

Bowl mill, 1139
grinding coal in, 1155
Bowl ring-roller mills, 1116
Box, skid, 1364
Box-type electric furnaces, **1759**
ovens, 1759
Boyle point, 290
Boyle's law, 290
Brabender plastograph, 1201
Bradley-Hercules 3-roller mill, 1137
Bradley-Lovejoy process, 1820
Brake horsepower, 1678
Brass annealing furnaces, 1605
Brass pipe, 428
Brassert disintegrator, 1037
Brayton cycle, 1653
Brazing solders, 453
Brazing vs. welding, 1244
Bread, hygroscopic moisture of, 779
Breakdown voltage of glasses, 1548
Breather batteries, 1791
Breweries, centrifuges in, 1011
Brick kilns, 1605
Bridge-circuit, telemeter, 1325
Bridge telemeter, self-balancing, 1325
Bridgman, dead-weight piston gage, 1255
electric lead seal, 1254
packing, 1246
"unsupported area" joints, 1246
Briggs filter thickener, 949
Briggsian or common logarithms, 50
Brine(s), 1694
circulation, flow of, 1697
power required in, 1697
pressure drop in, 1697
flow for varying tonnages of refrigeration, 1693
flow point of, 1695, 1696
heat content of calcium chloride, 1694
heat content of sodium chloride, 1695
piping, 1692
system of refrigeration, 1691
thermal conductivities of refrigerating, 1697
Brinell hardness of materials, 1539*ff*.
Briquetting, 1188
of coal, 1564
Bristol, bellows-type flowmeter, 1286
free-vane flapper, 1322
Pyrometer circuit, 1277
British unit of refrigeration, 1677
British weights and measures, 44
Brix, degrees, and sp. gr. relationship, 1296, 1297
Bromine, constructional materials for, 1474
Bronze pipe, 428
Brookfield viscometer, 1201
Brown, continuous-balance potentiometer circuit, 1278
pneumatic transmitter, 1324
radiation pyrometer, 1276
relay valve, 1322
Brown and Sites dialyzer, 756
Brownian motion, 1019
Brush-shifting motor, polyphase, 1768
Bubar fly ash separator, 1030
Bubble-cap columns (extraction), 752
rates for, 753
Bubble-cap plate towers, design of, 597–602
Bubble-cap scrubber, 1037
Bubble-plate column, as mixer, 1217
rectifying towers, 596
Bubble-tray absorption columns, plate efficiencies of, 698
Bubbler mixer, 1217
Bubbler-type liquid-level meter, 1290
Bucket elevators, 1348
Bucket pen for recorders, 1267
Budget comparison, 1836
forecast, 1835
Budgetary principles, 1835
Budgeting and expense accounting, 1835
Buell collectors, 1028, 1030
Buffer action and pH, 1308
Buflovak drum dryer, 863
Buflovak evaporator, heat transfer in, 503
Buhrstone disk mills, 1116
Buhrstone mill, 1124, 1144
Builders-Providence feeder, 1376
Building construction for fire protection, 1851
Building materials, insulation values of, 1693
thermal conductivity of (table), 457–458
Building requirements for pallets, 1369
Building walls, fire resistance of, 1852
Bulged ends of tanks, volume of, 1378
Bulk density of spray-dried materials, 841
Bunkers, storage in, 1377
Buoyancy, center of, 362
curve of, 362
definition of, 362
and flotation, 362
force of, 362
Buoyancy-type, displacers, 1289
level indicator, 1289
Bureau of Mines helium separation process, 1716
Universal gas mask, 908
Burner(s), air-atomizing, 1574
control system, kiln, 1335
diffusion-flame, 1592
luminous-flame, 1595
mechanical atomizing, 1574
steam-atomizing, 1573, 1574
Burt filter, 974
1,3-Butadiene, saturated, thermodynamic properties of, 252
n-Butane, saturated and superheated, thermodynamic properties of, 254
saturated, thermodynamic properties of, 254
Butanol-water vapor-liquid equilibriums, 632
1-Butene, in furfural, activity coefficients of, 646
Butterfly, dampers, 1328
valves, 1328
Butters filter, 976
Buttress threads, 1244
Butyl acetate, dehydration of, by percolation, 911
By-products, casting of, 1840
coke ovens, 1621, 1625
condensing steam power-plant cycle, 1628
of gas manufacture, 1578
ovens, coal carbonization in, 1565
power cycles, 1636
recoveries dust collector, 1023

C

Cachaza, filtration of, 987
Cadmium carbonate, dissociation pressures of, 174
Cadmium cell, 1772
Cadmium electroplating, 1799
Cadmium nitrate, densities of aqueous solutions of, 178
Cadmium recovery, electrolytic, 1806
Cage press, 1073
Cailletet and Mathias' law, 292
Caking of crystals, 1071
prevention of, 777, 1071
Calcium arsenate, drying of, 866
grinding of, 1158
Calcium carbide production, 1818
Calcium carbonate, dissociation pressures of, 174
drying of, 826, 866
Calcium chloride, brines, heat content of, 1694
constructional materials for, 1475
densities of aqueous solutions of, 178
thermal conductivities of, 1697
-water, viscosities of, 1696
Calcium cyanamide, dissociation pressures of, 174
Calcium hydroxide, densities of aqueous solutions of, 178
Calcium hypochlorite, constructional materials for, 1476
densities of aqueous solutions of, 178
Calcium nitrate, densities of aqueous solutions of, 178
Calcium oxalate, dissociation pressures of, 174
Calcium oxide, for drying gases, 881
Calcium production, 1811
Calcium stearate, drying of, 826
Calcium sulfate, for drying gases, 880
Calculation(s), of heats of reaction, technical, 333–358
Calculus, 81–93
graphical, 90–93
tables, 90–91
integral, 85–89
Calder-Fox scrubber, 1023
Calibrating thermocouples, 1274
Calibration of distilling columns, 609
Calibration of gages, 367
Callow settling cone, 940
Calomel electrode, 1794
Calorific value, measurement of, 1301
Calorimeter, gas, 1301
Calorimetric determinations of heats of reaction, 296

Camachine score-cutter, 1162
web slitting, 1162
Can mixer, rotary, 1206
Can system of ice manufacture, 1698
Cane mud, filtration of, 987, 990
Cane-sugar refining, 898
Cane-sugar sirups, filtration of, 986, 990
Cannel coal, 1563
Canvas belting, 1663
Capacitance, 1338, 1732
Capacitance-type liquid-level meter, 1290
Capacitive circuits, 1732, 1734
Capacity, 1338
coefficient (extraction), 743, 744, 749, 751
factor of power plants, 1630
of kilns, 1609ff.
lag, 1338
pumps, 1415
specific inductive, 1732
tanks, horizontal, 1380
of vertical cylindrical, 1381
volume, conversion table and equivalents, 40
Capillarity, 363
and surface tension, 363
"Capillary Conditioner," 1049
Capillary flow, 800
internal moisture gradient, 801
rise, 363
Capper for bottles, rotary, 1400
Caps, shape of, for bubble-cap towers, 599
spacing of, 599
Car sampling, 1097
Car-type furnace, 1607
Carbon(s), (activated), filter, 1049
pressure drop through beds of, 905, 906
properties and applications of, 905
removal of benzol from, 906
adsorbent, characteristics of, 901
plant for, 902
bisulfide production, 1819
black, hygroscopic moisture of, 780
colloidal, use in cement grinding, 1116
decolorizing, 898
grinding of, 1156
effect of, in removing color, 900
electrodes, 1822
filters, 968
gas-adsorbent, 904
graphite equipment, physical properties of, 1550
manufacturers of, 1534
heats of combustion of, 244
medicinal, 909
pipe, 431
products, hygroscopic moisture of, 780
rate of decolorization by, 900
residue of liquid fuels, 1570
rheostats, 1752
steel bars, chemical composition of A.I.S.I. standard, 1544
structural for equipment, 1558
tetrachloride, constructional materials for, 1477
Carbon(s), vegetable, effect of on sugar liquors, 902
Carbon dioxide, absorption in alkaline solutions, 703
absorption in aqueous NaOH solutions, 705
absorption in carbonate solutions, 703
absorption in diethanolamine solutions, 705
absorption in sodium or potassium hydroxide solution, 704
absorption in sugar-lime solution, 700
absorption in water of, 693, 695
brake horsepower per ton of refrigeration of, 1679
compressibility factors of, 206
diethanolamine equilibrium data, 677
emissivity of, 490
radiation of heat from, 490–492
ethanolamine equilibrium data, 677
Joule-Thomson effect of, 203
liquefaction with silica gel purification, 913
-monoethanolamine equilibrium data, 678
-Na_2CO_3-$NaHCO_3$ equilibrium data, 678
purification of, 908
saturated, thermodynamic properties of, 254
solid manufacture, 1700
physical properties of, 1701
-sulfur dioxide, temperature-pressure diagrams for, 572
vapor-liquid equilibrium relations for, 572
superheated, thermodynamic properties of, 255
-triethanolamine equilibrium data, 677
volume pumped per ton of refrigeration, 1683
Carbon monoxide, compressibility factors of, 207
heats of combustion of, 244
Mollier chart for, 258
Carbonates, grinding of, 1150
Carbonic acid, constructional materials for, 1479
Carbonization of coal, 1564
Carborundum, grinding of, 936
production, 1818
Carbrox, characteristics of, 902
Carbureted water gas as fuel, 1576
manufacture of, 1580
Cardan, method, cubic equations, 67
Carrier centrifugal refrigerating machine, 785
Carrier sublimation, 660
Carriers, 1357
horizontal run-around, 1359
pivoted bucket, 1359
Carter Vacuum packer, 1394
Carton folding, forming, sealing, lining, and cartoning equipment, 1404
Cartridge heaters, electric, 1760
Cascade cycle, for natural gas liquefaction, 1710
in refrigeration, 1681
Casein, synthetic, drying of, 837
Cast-iron pipe, 418–423
Caster trailer, 1364
Castings, plating, 1799
Castner chlorine cell, 1806
Catalysis and adsorbents, 915
Catalytic gas reactions, 329
Catchalls, for evaporators, 514
Catgut, hygroscopic moisture of, 780
Cathode fall of potential, 1747
Cationic reagents, 1088
Cations, adsorption curves for, 915
Caustic-chlorine, electrolytic, 1806
Caustic soda, constructional materials for, 1511
continuous countercurrent decantation, 951
filtration of, 988
C.C.D., 950
CCl_4-air vapor mixtures, humidity chart for, 813
Cells, galvanic, 306
primary and secondary, 1788
standard, 1772
cadmium, 1772
Weston, 1772
Cellulose acetate, drying of, 826
hygroscopic moisture of, 779
Cement(s), acid-resisting, 453
metal joint, 453
pipe, cement-lined, 434
clinker, grinding of, 1153
closed-circuit grinding, 935
grinding of, 933
high temperature, manufacturers of, 1538
kiln, 1610
manufacturers of, 1534
materials, grinding of, 1152
sampling of, 1101
slurry, filtration of, 986, 990
grinding, 934
and solders, 453
temperatures in rotary kiln, 1609
use of colloidal carbon as grinding aid in, 1116
Center of buoyancy, 362
Center of pressure, 361
centipoise, 369
Center-slung centrifuges, 998
Centigrade degrees to Fahrenheit degrees, 42
Centigrade temperature scale, 1269
Centipoise, 1197
Central America, weights and measures of, 44
Centralized batching systems, 1294
Centrifugal blowers, 1439
Centrifugal clarifier, 992
Centrifugal collectors, 1023, 1028
Centrifugal compressors, 1681
Centrifugal-discharge elevator, 1348
Centrifugal disks for spray dryers, 839
Centrifugal dryers, 997
Centrifugal fan mixer, 1210
Centrifugal fans, 1448, 1643
Centrifugal filter, 992, 1002

Centrifugal force, 992
effect on belts, 1661
Centrifugal-impeller mixer, 1210
Centrifugal pump as mixer, 1204
Centrifugal pumps, 1417
Centrifugal purifier, 992
Centrifugal refrigerating machine, 786
Centrifugal sedimentation, particle-size determination by, 1112
Centrifugal separator, 992
Centrifugal stills for molecular distillation, 657
Centrifugally cast cast-iron pipe, 423
Centrifuges, 992–1013
apparatus, 997
applications, 1007
basket, 992, 997
batch, 992
centrifugal, 992, 1002
continuous, 992
cost of, 1006
design, 997
energy requirements, 1005
test-tube, 992, 1001
theory, 992
Centrifuging, 887
costs of, 1005
partial, 1004
Centri-Merge scrubber, 1038
Ceramic equipment, manufacturers of, 1535
Ceramic furnaces, 1606
Ceramic grids, costs of, 709
Cereal foods, hygroscopic moisture of, 779
Cereals, milling of, 1147
Cerium production, 1811
C-E vertical unit boiler, 1603
Chain(s), driving, 1672
feeder, 1373
grates for coal, 1640
specifications of, 1874
Chamber scrubber, 1035
Change, demand, 1338
Channels, flow in, 382, 383
Manning formula for, 383
Chaplet furnace, 1815
Characteristic of logarithm, 50
Characterization factor, 565
Characterized valves, 1327
Charcoal, adsorption processes, 907
grinding of, 1156
as fuel, 1568
Charles's law, 290
Charlotte colloid mill, 1169
Chars, metal-adsorbent, 909
Chart(s), alignment, 100–105
line coordinate, 105
comfort, 781
humidity for solvent vapors, 812
air-benzene vapor mixtures, 814
air-CCl_4 vapor mixtures, 813
air-toluene vapor mixtures, 815
air-o-xylene vapor mixtures, 816
psychrometric, 759
recorder, round, 1267
Chaser mill, 1214, 1220, 1227
Chattock manometer, 366
Check- or hand-weighing scale, 1387
Check valves, 449
for high pressures, 1251
Checkerbrick recuperators, 1623
Checkerwork designs for furnaces, 1623
Chemical absorption analyzer, 1303
composition of A.I.S.I. standard steels, 1543ff.
engineering, symbols and nomenclature of, 3
equilibrium, 294
calculation of constants, methods, 350
in heterogeneous system, 348
solution of equilibrium problems, 348
at high pressure, 352
instruments, 1738
and physical principles, 287–358
pumps, 1418
reaction accompanied by gas absorption, 702
kinetics, 321
refrigeration, 1698
resistance of constructional materials, 1461*ff*.
gasket materials, 1558
stoneware equipment, 1549
ware pipe, 434
Chemicals, grinding of, 1157
Cherry-Burrell Viscolizer, 1167
Chézy, coefficients, formula, 377, 383
Manning formula for, 383
Chip-removing device, 924
Chloracetic acid, constructional materials for, 1480
Chlorine, absorption of, in sodium hydroxide, 706
cells, raw materials, chart for, 1809
caustic, 1807
constructional materials for, 1481
feeder, 1375
production, power costs of, 1810
in water, absorption of, 695
Chloroform-acetone vapor-liquid equilibriums, 630
Choice of evaporation type and size, 522
Chrome green, drying of, 837
Chrome yellow, drying of, 819
Chromel-alumel thermocouples, 1273
Chromic acid, constructional materials for, 1482
densities of aqueous solutions of, 178
Chromium alloys, thermal conductivity of, 456
chloride, densities of aqueous solutions of, 179
sulfate, drying of, 866
Chronoflo feeder, 1376
Chutes, emergency, 1852
outflow, 1362
retarding baffles, 1361
spiral, 1362
trickler, 1362
Cinder traps, 1643
Circle(s), 29, 31, 32, 33, 57, 78
analytical problems of, 55, 78
areas of segments, 31
diameters, areas, circumferences, 29
diameters, with sides of squares of equal areas, 33
Circle(s), equation of, 78
geometrical construction for, 60
involute of, construction of, 62
length of circular arcs, 33
segments, 31
Circuit(s), capacitive, 1734
combination, 1733
control, 1338
electrical, 1733
inductive, 1732, 1734
magnetic, 1736
neutrodyne, 1744
non-inductive, 1734
parallel, 1733
potentiometer, 1739
series, 1733
Circular arcs, lengths of, 33
measure table, 43
Circular furnaces, 1598
Circular inch, 43
Circular measure table, 43
Circular mil, 43
Circular mil-foot, 1732
Circular orifice, 1284
Circular percentage graph paper, 99
Circular pipe, velocity traverse, 399
Circular segments of circles, 31, 32
Circulating load, 930
and work of crushing, 931
load limits, in grinding, 932
Circulating mixing systems, 1203, 1204, 1215, 1219, 1221
Circulating pumps, 1647
Circulator, control medium, 1331
Circulators, 1439
Circumferences of circles, 29
Citric acid, constructional materials for, 1483
Clairaut's equation (calculus), 84
Clamp kilns, 1608
Clapeyron equation, 293
for estimation of latent heats of vaporization, 293
Clapeyron-Clausius equation, 293, 564
Clarain in coals, 1562
Clarification (*see also* Filtration)
by centrifugation, 994
capacity, 938
mechanics of, 937
Clarifier(s), 1002
disk-bowl, 999
Dorr, 943
Dorr Duo-clarifier, 944
Dorr multifeed, 947
Dorr S-7, 944
Dorr Sifeed, 943
Dorr Squarex, 943
Dorr Type-S, 947
Hardinge sand-filter, 948
hollow-bowl, 999
Clariflocculator, Dorrco, 947
Clark Bros. Co. expander, 1712
Classification, 922–937
of abrasives, 937
of coals, 1562
of dryers, 813
based on heat transfer (chart), 817
based on materials handled. 872–873
of furnaces and kilns, 1598

Classification, of lithopone, 936
of whiting, 936
Classifier efficiency and cut size, 1118
Classifier(s), Akins, 926
Dorr, 923
Dorr Bowl, 924
Hardinge countercurrent, 926
High-weir, 926
submerged-spiral, 926
types of size, 1118
use of size, 1118
Classifying crystallizer, Krystal, 1068
Claude oxygen process, 1707
Clausius-Clapeyron equation, *see* (Clapeyron-Clausius equation)
Clavarino formula, 1239
Clay(s), acid-treated, as adsorbents, 896
drying of, 837
as adsorbents, 889
analyses of, 889
bleaching, production of, 896
drying of, 826, 838
grinding of, 1150
hygroscopic moisture of, 780
sewer pipe, 435
Cleaning and concentrating, 1119
electrolytic, 1800
equipment for bottles, 1397
of evaporator scale, 513
Clearances, 1849
Clinker formation, 1566
Close-coupled pumps, 1419
Closed-circuit ball mill, 932
grinding, 930
equipment, 931
two-stage, 932
of whiting, 936
tube-mill, 932
Closed-circuit continuous grinding, 1117
Closed- vs. open-circuit grinding, 931
Closed-top separatory cone, 1083
Closing and filling, collapsible-tubes, 1411
small-bag, 1410
Closure(s), bolted-flange, 1249
for high-pressure vessels, 1243
for large vessels, 1249
removable head, 1249
self-sealing, 1249
for small vessels, 1249
vessel, 1248
Vickers-Anderson, 1250
Cloth collectors, 1013, 1029
Clutches, 1671
Coagulation, 1004
Coal(s), 1560–1568
analysis of, 1560, 1629
anthracite, drying of, 833
grinding of, 1156
ash consumption of, 1561
A.S.T.M. classification of, 1561
banded characteristics of, 1562
bituminous, 1560
grinding of, 1154
boghead, 1563
briquetting of, 1564
cannel, 1563
carbonization of, 1564
products of, 1565
clarain in, 1562
Coal(s), classification of, 1562
combustion of, 1564, 1566
consumption of, 1560
conveying by flight conveyors, 1347
density of, 1562
dewatering of, 1361
handling, in elevators, 1349
with hydraulic conveyors, 1361
with skip hoists, 1352
drying of, 333, 838
durain in, 1562
filters, 968
formation of, 1560
free swelling of, index of, 1566
friability of, 1562
as a fuel, 1560
vs. fuel oil, 1570
fusain in, 1562
grindability index of, 1562
Hardgrove, 1562
handling machinery, 1637
heating values of, 1561
mineral matter in, 1562
physical properties of, 1561
pulverized, 1641
rank of, 1560
reactivities of, 1566
sampling of, 1100
softening temperatures of, 1564
specific gravities of, 1561
splint, 1563
stability of, 1562
storage of, 1564
tar, as fuel, 1575
testing of, 1560
screens, 1640, 1641
thermal decomposition of, 1564
transparent attritus in, 1562
vitrain in, 1562
Coal-fired boiler, 1639
Coal gas, manufacture of, 1582
(retort) as fuel, 1576
Coal jig, 1076
Cobalt sulfate, dissociation pressures of, 174
Cochrane tilting-type flowmeter, 1286
Cocks, 451
Cocoa powder, grinding of, 1148
Coconut, drying of, 869
Coefficient(s), activity (*see* Activity coefficient)
of contraction, 401
diffusion, of gases and vapors, 539
in liquids, 540
of discharge, 401
for orifices and nozzles, 401
vs. Reynolds numbers, 404
for venturi tube, 401, 403, 407
viscous flow, 404
columbium tubing, 430
drag, 1017
film, in heat transfer, 478
fugacity, of gases and vapors, 536
heat-transfer, in evaporation, 500, 501
liquid activity, 526
effect of temperature on, 529
of utilization of illumination, 1758
unit, equation for, 537
Coffee, drying of, 866
Coffee extract, spray-drying of, 845
Coil(s), evaporator, 507, 1691
immersed in liquids, heat transfer for, 481
-type unit air conditioner, 784
Coke box, 1029
grinding of, 1156
hygroscopic moisture of, 780
ovens, by-product, 1621, 1625
gas as fuel, 1576
regenerators, 1624
specific heats of, 1562
Colburn flooding data for packed columns (extraction), 753
Cold-cathode discharge tubes, 1748
Cold junction, of thermocouples, 1272
compensation, 1274
Cold storage of foods, 1699
Cold-storage temperatures, 1698
Collapsible-tube filling and closing, 1411
Collection, dust or mist (*see* Dust collection)
Collection equipment, 1021
Collectors, cloth, 1013, 1029
Collectors for froth flotation, 1088
Colloid(s), destabilizing of, 939
mills, 1145, 1169
Charlotte, 1168
mixer, 1212, 1216
Colloidal solutions, 1196
definition of, 939
Color(s), analyzer, 1302
comparator, 1302
drying of, 819
for gas-mask canisters, 1870
grinding of, 1157
in industry, 1853
measurement of, 1302
removal of, by carbon, 900
systems, 891
Colorimetric method of determination of the hydrogen-ion concentration, 283, 285
Column(s), construction of, 710
diameter (or gas velocity) in gas absorption, 708
dimensions and effect of, on packed distillation towers, 619
height (exit-gas strength) in gas absorption, 708
packing, Stedman, 608
and packings in gas absorption, 707
rectifying, types of, 596
Scheibel, agitated for extraction, 748
selection for gas absorption, 668
Combination circuit, 1733
magnetic, 1736
Combinations and permutations, 66
Combustion, of fuels, 1603
coal, 1566
industrial systems, 1592
spontaneous, 1564
gaseous fuels, 1582
in furnaces, 1639
heat of formation from standard calculation, 345
heat of reaction from standard calculation, 344

Combustion, heats of, 244–246
intensity of, 1587
problem, 338
Combustion Engineering classifiers, 1135
Combustion Engineering Co., Raymond Flash dryer, 834, 835, 836
Comfort chart, 781
Commercial dryers, classification by materials handled, 872
Commercial refrigeration units, 1697
Common or Briggsian logarithms, 50
Common logarithms, 6
Commutating poles, 1764
Commutation, 1763
Compacting, 1186
Company policies as plant location factor, 1725
Comparison test, for convergence, 65
for divergence, 65
Compartment dryers, 815
Compartment mills, 1116, 1130
Compeb mill, 1132
grinding cement in, 1153
Compensating furnace, 1598
Compensating lead wires, 1272
Compensation, drift, 1338
reference-junction, 1272
Complementary angles, 53
Component, floating, 1338
Components, definition of in phase rule, 315
Composition of A.I.S.I. standard steels, chemical, 1543*ff*.
Composition, effect of, on packed distillation towers, 619
Composition-temperature diagrams at constant pressure, 578
Compound vessels for high pressure, 1241
Compound-wound generators, voltage characteristics, 1764
Compound-wound motors, 1767
Compounding dials, 1292
Compressed air mixers, 1215
Compressibility, 205–209
acetylene, 208
air, 207
ammonia, 208
carbon dioxide, 206
carbon monoxide, 207
chart, gas mixtures, 353
factors of gases and vapors (chart), 353
Gilliland's method, 354
Kay's method, 354
pure gases, 352
pressure unknown, 352
temperature unknown, 353
volume unknown, 352
ethane, 207
ethylene, 208
hydrogen, 205
hydrogen-nitrogen mixture, 209
liquids, 209
methane, 207
methyl chloride, 208
natural gas, 1578
nitrogen, 206
oxygen, 205
Compressing, 1189
Compression, adiabatic, 303
belts, 991
of data, graphical, 98
drying gases by, 884
dual, 1680
multistage, 1680
split-stage, 1681
wet, 1680
fittings, 442
of gas, 1439
isothermal, 302
multistage, 1258
point of (sedimentation), 938
power saving in multistage, 1681
refrigeration system, 1676
and absorption system, 1676
system, hydraulic, 1260
refrigerants for, 1686
of vapors, work of, 1678
Compressor(s), 1258, 1681, 1713
auxiliaries for, 1260
capacity, 1682
efficiency of, 1260
German, 1260
for high pressures, 1258, 1446
Hofer 1000-atm., 1259
intercoolers for, 1260
load-controllers for, 1260
mercury-piston, 1259, 1260
Michels 2500-atm., 1259
Norwalk 5-stage, 1000-atm., 1259
reciprocating, 1439, 1443
Sulzer, 1258
Concentrating and cleaning, 1119
Concentration, effective-diagram, 723
electrostatic methods of, 1093
of materials, 1072
miscellaneous methods of mechanical, 1072
sink-and-float, 1080
Concentration-enthalpy chart, 1052
Concentration-solvent content diagram, 724
Concentrator bowls, 999
Concentrators, spiral, 1079
Humphreys, 1079
Concentric-cylinder water rheostats, 1752
Concrete for chemical-tank construction, 1558
Concrete pipe, 435
Condensate meter, 1283
Condensate removal from evaporators, 511
Condensation, dropwise film-type, 476, 477
in presence of non-condensable gas, 477
partial, 563
retrograde, 575
processes, ejectors for, 1455
pump, Langmuir, 658
Condenser(s), 1647, 1677, 1687
backward-return, 1708
barometric, 520
calculations, 521
dry, 520
detection, 1746
synchronous, 1767
for evaporators, 520
calculations for, 521
Condenser(s), jet, 520
water requirements for, 521
moisture, 991
parallel-current, 520
low level, 520
performance, 1690
for steam jets, 1682
surface, 520
tubing, 424
conical entrance, 388
vacuum scraper, 855
various types, 520
countercurrent, 520
water, 1690
per ton refrigeration, 1690
wet, 520
Condenser-type induction motors, 1770
Condensing extraction steam turbine, 1646
Condition, desired, 1338
Conductance, electrical, 1732
separator, table, 1094
toboggan, 1094
Conduction of electricity through gases, 1747
heat transmission by, 456, 460
several bodies in parallel, 460
several bodies in series, 460
steady, 456
unsteady, 462
and radiation; drying and, 806
Conductivity, cells, electrical, 1302, 1303
-type hygrometer, 1299
electrical, 1732, 1782
measurement of, 1302
equivalent, 1782
heat (tables), 456–458, 459, 461
specific, 1782
thermal, equivalents (table), 41
of glasses, 1548
measurement of, 1303
Conductors, electronic, 1772
metallic, 1772
and resistance, 1733
Cone(s), area of, 57
volume, 57
closed-top separatory, 1083
crushers, 1116, 1126
joint, 1245
Seger, 1272
separatory, 1082
Conical mills, 1130
Conics, 77
Coning and quartering, 1097
Conjugate hyperbola, 79
Conjugate layers, 719
Conjugate lines for interpolation of tie-line data, 724
Connector, joint-ring, 1252
Connersville vacuum pump, 990
Conservation of mass, law of, 289
Conset air jig, 1077
Consistency, 1197, 1199
definition of, 369
measurement of, 1300
variations of, in mixing, 1202
of various materials, 1540
Constant(s), binary, for equations for activity coefficients, 528
critical, 291

Constant(s), critical, determination of values, 348
equilibrium, 309
estimation of, 291
fundamental, 295
gas, 290
methods of writing, 350
numerical, values of, 295
specific reaction velocity, 321, 322
use in high-pressure calculations, 352–358
use in problems, 350–352
dimensional, 94
gas combustion, 1583
numerical, 46
fundamental physical, 47
Constant-boiling binary mixtures, 630
temperature-composition diagrams for, 578
Constant-current transformers, 1763
Constant-displacement pump, 1282
Constant-energy-of-distortion of steels, 1239
Constant-level fillers for liquids, 1397
Constant rate period of drying, 800, 802
graph, 804
in through-circulation drying, 805
Constant-speed floating action, 1338
Constant-speed paddle wheel viscometer, 1301
Constant-volume feeders, 1134
Constant-weight feeders, 1134, 1373
Construction, of columns, 710
of furnaces, 1601
for high pressures, 1242
materials of, 1457–1558
chemical resistance of, 1461*ff*.
for electrochemical processes, 1820
for evaporators, 515
safety, 1856
Consumption of coal, U.S., 1560
Contact angle, 363
Contact catalytic gas reactions, 329
Contact in extraction, 716
Contact filter plant flow sheet, 894
Contact filtration, 888
Contacting, gas-liquid, 1175
Contactor, liquid vortex, 1037
Contacts, high-low signal, 1268
Containers, agitator settling in, 1394
gas storage in, 1380
handling of, 1362
liner insertion into, 1394
storage in, 1377, 1378, 1382
Content, critical moisture, 800
equilibrium moisture, 800
moisture, 800
Continuous absorption refrigeration systems, 1683
Continuous-bucket elevators, 1349
Continuous centrifuge, 992
Continuous-chamber kilns, 1608
Continuous-coil evaporator, 1691
Continuous countercurrent decantation, 950
Continuous countercurrent dialyzer, 755
Continuous countercurrent differential contact extraction, 717
calculations for, 729, 742
Continuous direct dryers, 814
Continuous equilibrium vaporization, 585
of petroleum fractions, 587
Continuous-flow conveyors, 1351
applications and limitations, 1351
power determinations, 1354–1355
Continuous furnaces, 1607
Continuous liquid sampling, 1100
Continuous press, 1074
Continuous proportioning, 1293
viscometer, 1301
weigh checking, 1294
weigh feeders, 1292
Continuous reactors, 1257
Continuous rectification, 588
Continuous-sheeting direct dryers, 848
Continuous thickening, 937
Continuous thickeners, mechanical, 941
non-mechanical, 940
Continuous through-circulation dryers, 823
Continuous vs. batch mixing, 1211, 1216, 1230
Continuous vs. batch operation of evaporators, 512
Contour-line charts, 100
Contraction, coefficient, 401
loss, 388
Control(s), 1338
agent, 1338
band, 1338
circuit, 1338
corresponding, 1338
in crushing and grinding, size, 932
anticipatory, 1337
automatic, 1337
cycle, 1331
of distillation equipment, 1336
Dowtherm vaporizer, 1335
electric motors for process, 1329
elements, final, 1326
of flow, proportioning pump for ratio, 1326
of fractionation column, 1336
fundamentals of automatic, 1309
of immersion heater, 1331
installation of process, 1328
of level in closed vessel under pressure, 1334
of liquid level, differential-pressure, 1334
mechanisms, automatic, 1320
medium circulator, 1331
piping for process, 1328
pneumatic transmission for remote, 1333
power units for process, 1328
process, 1263–1340
glossary, 1337
programming, 1326
of rate of drying, 1335
simple batch-temperature, 1330
of steam-jet vacuum system, 1333
system applications, automatic, 1329
Control(s), **cycle**, system, dry-bulb, 1335
flow, 1333
for heat exchanger, 1334
kiln-, 1335
pressure-, 1332
ratio flow-, 1333, 1334
recirculating dryer, 1335
saturable-reactor, 1332
temperature-, 1332
wet-bulb, 1335
terms, automatic, 1337
vacuum, 1333
valves, 1326
rotary-stem-motion, 1326
sliding-stem-motion, 1326
damping, 1338
differentiating, 1338
effect, 1338
element, 1338
for evaporators, 520
index setting, 1338
instrument, 1338
level, in drum dryers, 863
of motor speeds, 1768
point, 1338
point setting, 1338
refrigerant liquid, 1692
response, 1338
of small refrigeration machines, 1697
setting, 1338
system, 1338
for tray and compartment dryers. 819
valve for, flow, 1251
packing liquids, 1396
Controlled discharge, 1749
medium, 1338
variable, 1338
Controller(s), 1320
adjustable, 1326
adjustment, 1338
anticipating, 1337
automatic, 1337
automatic "reset," 1323
cycle, 1326
differential pressure, 1333
direct-operated, 1338
electrical, 1323
elements of automatic, 1310
fixed position, 1320
floating, 1320
function, 1338
"hunting" in process, 1329
lag, 1338
Leeds and Northrup air-operated, 1322
maximum flow, 1320
minimum flow, 1320
multiposition, 1320
operation of automatic, 1310
pilot-controlled, 1320
proportional-plus-reset, 1320
proportional position, 1320, 1323
ratio, 1325
relay operated, 1320
reset, 1320
response, 1338
self-operated, 1320
sequence, 1326
simple proportional, 1323

Controller(s), throttling, 1320
two-position (on-and-off), 1320
Controlling means, 1338
Controlling a piston, slide valve, 1321
Convection, forced, heat transfer by (graph), 468
heat transmission by, 456, 463, 468
and heat transfer by, 544
mass transfer by, 541
natural, 462
Convergence, comparison test for (algebra), 65
ratio test for (algebra), 65
Convergent series, 65
ratio test for, 65
Conversion factors, for coefficient of heat transfer, 466
electroplater's, 1798
energy, 1782
for metric and avoirdupois units, 1781
for metric and troy units, 1781
refrigeration, 1678
for thermal conductivity, 456
Conversion response, 1338
tables (*see* Equivalents)
of viscosities, 1198
Converters, Bessemer, 1616
rotary, 1765
synchronous, 1765, 1766
Conveying dryers, pneumatic, 834
Conveying-screen horizontal dryer, 823
Conveyor(s), apron, 1348
belt, 1354
continuous flow, 1351
drag, 922
drives, 1343
flight, 1347
furnace, 1607
grasshopper, 1361
gravity roller, 1358
hydraulic, 1360
pneumatic, 1360
retarding, 1347
scales, totalizing, 1294
screw, 1344
sewing head on slat, 1392
vibrating, 1361
Conveyor-type batching scales, 1294
Cooley jig, 1077
Cooling, crystallizer, 1069
ejectors for, 1456
of gases by expansion, 1702
and heating of fluids, 467
(chart), 446
in isentropic and isenthalpic expansions, 1702
of kettles, 1223–1224
of mixers, 1224
systems, 1690
Cooling-water towers, 1690
Cooke short-columns elutriator, 1085
Cooking oil, use of adsorbents with, 902
Coordinates, 70, 77
line, charts of, 105
polar, 77
graph paper, 99
Copper, alloy pipe, 428
electrolytic refining, 1801
Copper, electroplating, 1799
electrowinning, 1805
piping, 1692
solutions, absorption of olefins and acetylenes in, 706
densities of aqueous solutions of, 179
drying of, 837
specific heats of aqueous solutions of, 234
Copper-constantan thermocouples, 1273
Copper nitrate, densities of aqueous solutions of, 179
Copper oxide rectifiers, 1751
Copper sulfate, constructional materials for, 1484
Copperas, tank crystallization of, 1062
Core sucker for ice manufacture, 1698
Corn sirup, filtration of, 988
Cornell Versator, 1168
Corner taps, 405
Corona, electrical, 1039
Correction, droop, 1338
Corrective action, 1338
Correlation of drying data, 808
Corresponding control, 1338
Corresponding states, theory of, 292
use of, 353–358
Corrosion, in condenser systems, 1690
of filters, 989
tests, 1458*ff*.
and welding and riveting, 1459
Corrosion-resisting construction of filters, 984
Corrosive gases, resistance of metals and alloys to, 1526
Corrugated-roll crushers, 1127
Cost(s), accounting, 1827*ff*.
of centrifuges, 1006
of ceramic grids, 709
estimating, 1844
of evaporators, 507
finding, 1827*ff*.
systems, 1833
of packing materials, 709
of partition rings, 709
or porcelain and stoneware towers, 709
of power generation (curves), 1633
of power-plant services, 1632
of spiral tile, 709
systems, 1833
selection and design of, 1834
of transporting water vapor, 856
Cotton, batting filters, 968
filter mediums, chain weave, 967
duck, 967
fabrics, 967
table felt, 968
twills, 967
Cottonseed refining, 902
Cottrell precipitator, 1041
Couette viscometer, 1201
Coulomb, 1772, 1775
Coulometers, 1775
Countercurrent apparatus, 548–555
Countercurrent atmospheric condenser, 1688
Countercurrent condenser, 520
Countercurrent decantation, continuous, 950
Countercurrent extraction with reflux, 717
calculation of, 733
Countercurrent multistage contact extraction, 717
calculation of, partially miscible systems, 731
immiscible solvents, 740, 741
Coupling, Victaulic, 442
Couplings, 1667
Cox chart, 294, 564
Crackers, hygroscopic moisture of, 779
Cracking process, fluid catalyst, 1619
Cracking units, 1606
thermofor, 1618
Cream, centrifugal separation, 1000
Cream separators, costs of, 1006
Creep of metals, effect in high pressure design, 1240
Critical constants, 291
of elements and inorganic and organic compounds, 204, 205
Critical humidity, 1071
Critical moisture content, 800, 802, 807
values of, 807
Critical pressure, 565
ratio, gases, 402
Critical region, 383
Critical solution temperature, 318, 721
Critical temperature, 565
as function of b. pt. and sp. gr., 566
Critical velocity, 375, 383, 384
cup anemometer, Robinson, 399
in curved pipe, 384
effect of curvature on, 384
for fluids, 375
Crockett and Linney magnetic separator, 1092
Cross-flow heat exchangers, temperature difference in, 465
Crowning of pulleys, 1665
Crusher product size distribution, 1125
Crushers, 1116
classification of, 1116
description of, 1120
Crushing, 1120
circulating load limits, 932
equipment, closed-circuit, 931
selection of, 1117
fine, 932
fineness of, 932
and grinding, 1107, 1110
references, 1107–1110
mediums, 934
work of, 931
and circulating load, 931
Crutcher, soap, 1210
Cryolite, drying of, 826
Cryostat, Collins, 1708
Crystal(s), caking of, 1071
filter dryer, 824
filtration of, 986, 990
formation, 1054
forms, 1050
geometry of, 1057

Crystal(s), growth, 1058
Berthoud-Valeton theory of, 1058
ΔL law, 1059
limitations of, 1060
habit, 1058
invariant, 1058
methods of forming in soln., 1055
prevention of caking of, 779
rate of solution of, 1058
simultaneous growth and formation of, 1061
size and shape, 1050
types of, 1050
ionic, 1050
liquid, 1051
molecular, 1050
valence, 1050
Crystallization, 1050–1072
and adsorption, 887
apparatus, 1061–1070
effect of impurities on, 1057
fractional, 1053
heat of, 1052
Mier's theory of, 1055
seeded and unseeded, 1054
systems, 1051
tanks, 1062
theory of, 1050
yield of, 1051
Crystallizing evaporators, 1065
Crystallographic systems, 1050
Cube(s), of numbers, 23
roots of numbers, 23
by slide rule, 52
surface of, 57
volume of, 57
Cubic equations, 67
Cubic system of crystals, 1051
Cunningham, correction to Stokes's law, 1019
Cupola furnace, 1612
Cuprous chloride, densities of aqueous solutions of, 179
Cuprous salt solutions, solubility of olefins in, 679
Curb press, 1073
Current concentration, 1775
density, 1772
direct, 1733
reactive, 1765
transformers, constant, 1763
efficiency, 1775
meter, 399, 1288
Curvature, effect on Reynolds number, 384
Curve of buoyancy, 362
Curved pipe, friction factor for, 383
Curves, analytical problems, 80
exponential, groups of, 81, 97
hyperbolic, graphs of, 81, 97, 98
logarithmic, 98
miscellaneous, 80
parabolic, graphs of, 98
probability, 81
temperature, vapor-pressure, 98
trigonometric functions of, 80
Cut size, of particle separators, 1026
Cutler-Hammer gas calorimeter, 1301
Cutoff friction clutches, 1672
Cutter, fiber, 1162
Cutters, precision, 1162
knife, 1161
rotary knife, 1160
score, 1162
slitting, 1162
Cutting, 1160
Cyanamid production, 1818
Cyanide pulp, filtration of, 985
Cyanide slime, filtration of, 990
Cycle(s), by-product power, 1636
control, 1331
controllers, 1326
adjustable-cam, 1326
power-plant, 1628
of steam plants, 1633
Cycling, 1338
Cycloid, construction of, 62
area of, 57
length of arc of, 57
Cycloidal meter, 411
Cyclone, furnace, 1605
scrubbers, 1036
separators, 1013, 1023
collection efficiency, 1026
commercial equipment, 1028
cost of, 1028
design factors, 1027
Dutch state mines, 1068, 1083
entrainment collection, 1028
fields of application, 1024
flow pattern, 1024
proportions, 1024
Cyclonic-spray gas scrubber, 700
Cylinder(s), area and volume of, 57
compressed gases in, 1869
drag coefficient for, 1018
target efficiencies of, 1022
dryers, 866
hydraulic power, 1328
pneumatic power, 1328
solid, heating and cooling of, 462
steel, stresses in, 1237, 1238
bursting pressure of, 1239, 1240
compound, for pressure vessel, 1241
stock, decolorization by adsorbents, 892, 895
Cylindrical surfaces, mass transfer rates from, 546
Cylindrical tanks, volume of, 1379

D

Dairy industry, centrifuges in, 1007
Dalton's law, 290
deviations from, 291
use in humidity calculations, 336
Dampers, butterfly, 1328
slide, 1328
Damping, 1338
control, 1338
Daniell cell, 1789
Darco, characteristics of, 902
Darcy equation, 377
Dark discharge, 1747
D'Arsonval galvanometer, 1276
Data, compression of, graphical, 98
drying, methods of correlating, 808
smoothing, 91
Dayton process of oil gas manufacture, 1582
D.D.T., grinding of, 1158
Dead, neutral, 1338
Dead-period lag, 1338
Dead spot, 1338
Dead time, 1338
Dead-weight gage tester, 1280
piston gage, 1255
Bridgman design, 1255
differential-piston, type of, 1256
with lever arm, 1256
Dead zone, 1338
Deaerators, 1649
Debits and credits, 1828
Decantation, continuous countercurrent, 950
and settling, 1003
Decarburization of steels, 1235
Decimals, conversion table and equivalents, fractions, 37
inches to feet, 37
seconds, minutes, degrees, 37
Deck table, Deister-Overstrom, 1079
Decolorizing carbons, 898
applications of, 901
isotherms, 901
manufacturing methods for, 899
Decomposition pressure, 662
Definite and multiple proportions, law of, 289
integrals, 88
table of, 85–87
Deflection, 1338
Deflocculating agents, 1088
Degrees, A.P.I. and sp. gr. relationship, 1296
Baumé, °Tw., lb./gal., lb./cu. ft., sp. gr., °A.P.I. equivalents, 38
Brix and sp. gr. relationship, 1296, 1297
centigrade and Fahrenheit equivalents, 42
of freedom, definition of, 297, 315
of gaseous molecules, 297
to radians, conversion table and equivalents, 36
100ths radian, 36
Twaddell, 1297
equivalents, 38
Dehumidification, 757*ff.*
adsorption unit for, 786
dehumidifiers, 779
methods, miscellaneous, 786
moisture removed in, 786
systems, 783
Dehydrating and size reduction, 1119
Dehydration, definition, 800
of butyl acetate by percolation, 911
of gas with silica gel, 913
of wood by solutions, 1558
Deister-Overstrom deck table, 1079
DeLaval nozzles, 406
Delay, 1338
Delta L (ΔL) law, 1059
Delta system, three-phase, 1735
Demand change, 1338
for electricity, 1755
Demineralizing water with adsorbents, 915
Dense-air refrigerating machine, 1677
Density(ies), of aqueous organic solutions, 186–193
bulk, of spray-dried materials, 841
of coal, 1562

Density(ies), conversion table and equivalents, 40
current, 1772
flux, 1732
of fuels, 1629
of gases, measurements of, 1297
of gases at standard conditions, 176
of glasses, 1548
of inorganic solutions, aqueous, 177–185
of insulating materials, 1693
of liquids, measurement of, 1297
by automatic hydrometer, 1297
by differential-pressure, 1297
by displacement, 1297
expansion factor for, 535
by weight, 1297
measurement of, 1296
of mercury, 176
of mixtures, 289
calculation of, 289
of pure substances, 175, 176
of solids, determination by immersion, 363
of water, 175, 176
Denver equipment feeder, 1374
Denver Mineral jig, 1077
Denver Sub-A flotation cell, 1089, 1091
Departure, 1338
Dephlegmation, 563
Depolarization, 1786
Depolarizer, 1786
Depreciation, 1831
as a cost element, 1836
rates, 1832
Depressers, 1088
Depressor-bar printers for recorders, 1268
Depth of flotation, 362
Derivative action, 1338
calculus, 81, 82
partial, 82
Derris, grinding of, 1158
Desiccants, 854
solid, 877
adsorption rates on, 882
Desiccated cell, 1791
Design, of bubble-cap plate towers, 597
capacities of trays, 600
cap-slot and riser areas, 602
plate spacings, 598
pressure drop and slot opening, 601
reboil heat, 602
reflux supply, 602
shape of caps, 599
spacing of caps, 599
types of liquid flow, 599
vapor velocity in, 597
weirs and downspouts, 600
calculations in sublimation, 662
effect of temperature in, 1240
of gas absorption systems, 707
of gas burners, 1590
of pressure vessels, 1237–42
Desired, condition, 1338
value, 1338
Desorption of oxygen from water in spray absorber, 700
Destabilization of colloids, 939
Destencher, pebble-stove, 1625
Destructive distillation, 563
Desublimation, 660
Desulfurization of kerosene by adsorbents, 913
Detector action, 1746
Detergents, spray-drying of, 845
Determinants, 68–70
multiplication of, 70
properties of, 69
second-order, 68
third-order, 69
Detonation, 1868
Detroit rocking furnace, 1817
Deviation, 1338
Dew point, 759, 811
of complex vapor mixture, calculation of, 586
control of humidifiers, 782
method of determining humidity, 776
recorder, 1299
Dewcel dew-point humidity controller, 1300
Diagram factors, steam engine, 1645
Diagrams, flow, use of, 333
steel, 1633
thermodynamic, 330–333
Dial and chart illumination, 1292
Dialysis, 753–756
applicability of, 755
calculation of dialysis coeffs., 754
equipment, 756
membrane properties, 754
nomenclature, 756
theory, 754
Diameter(s), of circles, hundredths, 29
with squares of equal area, 33
economical pipe size, 384, 385, 386
equivalent, definition of, 375
hydraulic, 375
molecular, of non-electrolytes, 755
rectilinear, law of, 292
of spheres, hundredths, 34
Diaphragm, gas meter, 1283
liquid-level meter, 1290
motor, pneumatic, 1328
valve, direct-acting, 1328
reverse-acting, 1328
meter, 367, 411
orifices, etc., 389
pump, Hardinge, 944
pumps, 990
thermostat, 787
Dichlorodifluoromethane (F-12), saturated, thermodynamic properties of, 261
superheated, thermodynamic properties of, 262
Dichloromonofluoromethane (F-21), saturated, thermodynamic properties of, 263
superheated, thermodynamic properties of, 263, 264
Dielectric constant, 1732
of glasses, 1548
heating, 1762
strength, 1732
Dielectric dryers, 815, 870
Dielectric loss factors in dielectric dryers, 871
Diesel cycle, 1653
engine, two cycle, 1654
-fuel purification, centrifuges in, 1010
plant thermal performance, 1653
plants, input-output curves for, 1632
Diethanolamine–CO_2 equilibrium data, 677
solutions, absorption of CO_2 in, 705
Diethylene glycol, aqueous, vapor pressures of, 170
Differential calculus, 81–85
equations, 84–85
linear, 84
Differential flowmeters, 1285
pressure control of liquid level, 1334
pressure, controllers, 1333
density meter, 1297
manometer-type level meter, 1289
viscometer, 1301
temperature, liquid-level meter, 1290
measurement, 1334
type flowmeters, 1287
Differential gap, 1338
Differential hygrostat, 787
Differential-piston dead-weight gage, 1256
Differentiating control, 1338
Differentiation, formulas, graphical, 90
Diffusion, coefficient (dialysis), 755
of gases and vapors, 539
Fick's law of, 754
in liquid extraction, 743
flame burners, 1592
gaseous, 292
in gases, 538
internal, 800
liquid, 801
in liquids, 540
molecular, theory of, 540
vapor, 801
Diffusional flow and internal moisture gradient, 801
Diffusional operations, general theory of, 523–559
equilibrium relations in, 525
Diffusivity in hydrogen-carbon dioxide mixture-variation with composition, 538
liquid, 809
thermal, of wood, 463
Diffusor-type pumps, 1419
Dihedral angle, 55, 56
Dilatant fluid, 1198
Dilution metering, 412
Dimensional analysis, 93–97
application to theoretical physics, 97
constants, 94
falling bodies, 95
pressure, gas, 95
problems of, 94
Rayleigh, heat transfer, 96
Dimensional constant, g_c, 360
Dings induced-roll separator, 1093
Diodes detector, 1746
gas, 1749

Diphenylhexane-docosane-furfural system, 721
Dipper sampling, 1100
Direct-acting diaphragm valve, 1328
Direct current, 1733
generators, 1763
voltage characteristics, 1764
motors, 1766
Direct dryers, 813
batch, 814
continous, 814
continuous-sheeting, 848–853
rotary, 828
Direct-fire evaporators, 505
Direct-fired kettles (mixers), 1221, 1224
Direct-firing furnaces, 1607
Direct humidifiers, 782
Direct-indirect rotary dryers, 829
Direct-operated controller, 1338
Direction, of flow, indicators, 400
of induced e.m.fs., 1737
Directory of materials of construction, 1527
Directrix, 77
Discharge, coefficients of, 400
controlled, 1749
head of pumps, 1415
liquid-sealed, 1002
Dished ends of tanks, volume of, 1378
Disintegration, explosive, 1166
and pulverizing, 1117
Disintegrator or multiple-cage mill, 1124
Disk attrition mills, 1143
Disk-bowl clarifiers, 999
Disk feeders, 1134
Disk filters, rotary, 971
Disk hammer mills, 1116
Disk-mill principle, 1144
Disk rupture, 1252
Disk series, 396
Disk-type fan, 1448
Disks, centrifugal, for dryers, 839
drag coefficient for, 1018
Disperser, turbine, 1211
Dispersing, 1195, 1196, 1216, 1218, 1229
agents in grinding, 1115
Dispersion, 1196
drying and grinding, 1119
of liquid droplets, 1169
Dispersoids, characteristics of gas, 1019
(*See also* Particles)
Displacement, 1338
Displacement pump, 1439
Dissociation, electrolytic, 1780, 1782
Dissociation pressures, 174
of barium hydroxide, 174
of barium peroxide, 174
of cadmium carbonate, 174
of calcium carbonate, 174
of calcium cyanamide, 174
of calcium oxalate, 174
of cobalt sulfate, 174
of ferrous sulfate, 174
of manganese dioxide, 174
of mercuric oxide, 174
of potassium bicarbonate, 174
of potassium carbonate, 174
Dissociation pressures, of potassium dihydrogen phosphate, 174
of potassium hydride, 174
of silver carbonate, 174
of sodium carbonate, 174
of sodium dihydrogen phosphate, 174
Dissolution, 1196
Dissolving, 1195, 1196, 1218, 1229
Distance-velocity lag, 1338
Distances, American table of, 1848
Distilland layer in molecular distillation, 657
Distillation(s), 561–660, 887
azeotrope occurrence in, 630
azeotropic, 629, 651
batch steam, 582
columns, plate efficiencies for, 610
data and generalizations, 563
definitions, 563
dephlegmation, 563
destructive, 563
equipment, control of, 1336
evaporation, 563
extractive, 629, 634
flash, 585
fractionation, 563
laboratory, 606
liquid fuels, 1570
low-temperature, 609, 1702
molecular, 655–660
partial condensation, 563
petroleum, 602
processes, applications of, 580
methods of calculation, 580
principles, 580
rectification, 563
rectifying columns, 563
reflux ratio, 563
simple batch, 580
steam, effect of liquid water in still on, 583
steam consumption in batch, 584
vaporization efficiency in, 583
towers, packed, 616
true-boiling-point, 607
use of phase rule in, 576
Distillation (batch), ejectors for, 1455
Distillery by-products, spray-drying of, 846
Distilling columns, laboratory, 609
calibration of, 609
equipment, Podbielniak, 610
Distinct roots (calculus), 85
Distribution(s), equilibrium diagrams for, 719–728
fractional, 715, 739
ideal law for, 729
for immiscible solvents, 729
law for, 729
mathematical expression for, 729
in oil-solvent systems for, 728
of partially miscible solvents, 719
of solute between liquid phases, 320
velocity, 385
Distributor, rotary, 1375
Divariant system, 315
Divergence, comparison test for (algebra), 65
Divergent series, 65
Diverging duct, pressure loss in, 388
Division, algebraic, 63
graphic, 54
slide-rule, 51, 52
Dobrowolsky generator, 1764
Docosane-diphenylhexane-furfural system, 721
Dodge crusher, 1120
jaw crushers, performance of, 1125
Dolomite kiln, 1610
Door openings, guards for 1852
Dorr, bowl classifier, 924
air-lift agitator, 1220
clarifier, 943
classifier, 923
Duo-clarifier, 944
multifeed clarifier, 947
S-7 clarifier, 944
Sifeed clarifier, 943
Squarex clarifier, 943
thickener, single-compartment 941
Torq thickener, 941
traction thickener, 942
tray thickener, 942
washing-tray thickener, 953
Dorrco, clariflocculator, 947
Duo-Filter, 944
filter, 979
flocculator, 947
Hydro-Treator, 947
pumps, 945
repulper, 955
suction pump Duplex, 946
Type S clarifiers, 947
V-type pump, 946
Double-cone classifier, 1138
joint, 1247
mixer, 1212, 1221
Double-drum dryer, 863
filter, 980
flaker, 1164
Double-helical mixer, 1210
Double Impeller Crusher, 1129
Double-motion mixer, 1206, 1220
Double Package Maker, 1406
Double-pipe crystallizer, 1063
Double-reaction flow sheet, 951
Double-seat V-port valve, 1327
Double-seated parabolic plug valve, 1327
Double strainers for filters, 984
Double-truck dryer, 818
Dough mixer, 1207, 1208, 1220
Doughy masses, mixing of, 1220
Downdraft kilns, 1607
Downspouts and weirs, 600
Dowtherm A, saturated, thermodynamic properties of, 257
in boilers, 1605
Draft, of flooding objects, 362
gage, 366
vapor in process control, 1335
vaporizer control, 1335
Draft-heated kettles (mixers), 1221, 1223, 1224
Draft system, 1642
tube propeller mixer, 1209
Drag coefficient, 1017, 1018
Drag conveyor, 922
Drag-scraper storage systems, 1376
Draining, 922

Dravo solids-level meter, 1291
Dressings, belt, 1661, 1662
Drierite, for drying gases, 880
Drift, 1338
 compensation, 1338
Drinking-water systems, 1853
Drives, conveyor, 1343
 for belts, 1664
 short-center, 1670
 variable-speed, 1670
Driving mechanisms for centrifuges, 1004
Droop, 1338
 correction, 1338
Drop, surface, size for gas atomization of water, 840
Drop size, 1174
Droplets, dispersion of liquid, 1169
Dropping-mercury electrode, 1307
Drops, evaporation from liquid, 805
 evaporation of small, 540
Dropwise condensation, 477
Drugs, grinding of, 1158
Drum dryer, 863–866
 double-, 863
 finned, 864
 single-, 863
 twin-, 863
 vacuum, 864
Drum feeder, for mills, 1134
 filling and weighing, 1392
 filter, rotary, 971
 flakers, 1164
Dry air, specific volume of, 811
Dry- and wet-bulb thermometry, 557
Dry basis, 800
Dry-bulb control system, 1335
 temperature for air-water vapor, 811
Dry cells, 1790
Dry-cleaners' solvent clarification, 1011
Dry colors, grinding of, 1157
Dry condenser, 520
Dry elutriation, 1084
Dry gas meters, 411
Dry grinding, 1115
Dry measure (table), 43
Dry-pan crusher, 1122
Dry products, filling equipment for, 1400
Dry tables, 1079
 Sutton, Steele, and Steele, 1079
Dry-weight basis, 800
Dry vs. wet grinding, 1118
Dryers, agitated-pan, 856
 auxiliary equipment for, 820
 batch through-circulation, 820
 bond-paper, 867
 centrifugal, 997
 classification of, 813
  based on heat transfer (chart), 817
  based on materials handled, 872–873
 compartment, 815
 continuous through-circulation, 823
 crystal filter, 824
 cylinder, 866
 dielectric, 815, 870
 direct 813
Dryers, direct continuous-sheeting, 848–853
 direct rotary, 828
 double-drum, 863
 drum, 863–866
 efficiency, 800, 814
 festoon, 848
 fields of application, 820
 finned drum, 823, 864
 flash, 834–836
 graining bowl, 857
 high-frequency, 870
 indirect, 814
 indirect-direct rotary, 829
 indirect rotary, 859
 infra-red, 815, 868
 initial comparison of, 874
 Link-Belt Roto-louvre, 825
 loop, 848
 low-temperature vacuum, 854
 Multipass Airlay, 849, 851
 Perkins, 824
 plant, tests on, 809
  correlation of data from, 810
 pneumatic conveying, 834
 preliminary selection of, 874
 print, 849
 relative capacity of various rotary, 861
 Roto-louvre, 824
 screw conveyor, 862
 seals for rotary, 829
 selection of, 874
 single-drum, 863
 spray, 838–848
 Standard National print, 850
 steam-tube rotary, 860
 tenter, 849, 851, 852
 tray, 815
 trough, 862
 tunnel, 821
 twin-drum, 863
 types of, 815
 vacuum drum, 864
 vacuum pan, 857
 vacuum rotary, 858
 vacuum shelf, 853
 Vertex, 834
 vertical turbo-, 821
 vibrating, 824
 vibrating conveyor, 863
 Vissac, 824
 Yankee, 867
Drying, 799–884
 conduction and radiation in, 806
 constant rate period of, 802
 control of rate of, 1335
 critical moisture content in, 807
 data, methods of correlating, 808
 definition of, 800
 effect of air velocity on, 803
 falling rate period of, 806
 gases, 877–884
  by compression, 884
  with desiccants, 878
  methods of, 877
  by refrigeration, 884
 and grinding bauxite, 1119
 liquid drops, 805
 periods of, 800
 psychrometry, application to, 810–813
Drying, reasons for, 800
 of solids, 800–877
 with solvent recovery, 875
 terms used, definition of, 800
 tests, 874
 theory and fundamental concepts of, 800–813
 thixotropic filter cakes, 824
 time, equations for estimating, 808
 vacuum shelf, ejectors for, 1455
Drying gasoline with activated alumina, 911
Dual compression, 1680
Duct, diverging, 388
 thermostat, 787
Dühring's rule, 294, 300, 341
 application to viscosity, 379
 calculations of latent heats, 300
 in evaporation, 500
Dulong formula for heating value of coal, 1561
Dulong and Petit's law, 298
Duo-Clarifier, Dorr, 944
Duo-Filter, Dorrco, 944
Duplex pump, Dorrco suction, 946
Durain in coals, 1562
Durand's rule for integration, 90
Duraseal mechanical seal, 1424
Duriron pipe, 423
Dust, properties (*see* Particles)
Dust catchers, 1643
Dust collection, 1013
 equipment, 1021
  acoustic flocculation, 1049
  air filters, 1045
  bag filter, 1029
  cloth collectors, 1029
  coke box, 1029
  cyclone separators, 1023
  electrical precipitator, 1039
   alternating current, 1045
   applications, 1041
   cost of, 1044
   single-stage, 1041
   theory of, 1039
   two-stage, 1044
  entrainment, for, 1028
  gravity settling chambers, 1021
  impingement separators, 1022
  mechanical centrifugal separators, 1028
  miscellaneous collectors, 1049
  miscellaneous inertial separators, 1028
  packed-bed separators, 1029
  scrubbers, 1034
  thermal precipitation, 1050
Dust explosion, 1868, 1869
Dusting, amount of, in rotary dryers, 832
Dustite bag holder, 1390
Dustop air filter, 1046
Dusts and mists, vapor pressure of, 293
Dutch oven, 1598
Dutch State Mines cyclone separator, 1068, 1083
Dutch weaves as filter mediums, 968
Dyestuffs, grinding of, 1157
Dynamic error, 1338
Dynamics of particles, 1013, 1017
Dynamometer instruments, 1738

E

Ebonite cone joint, for high pressures, 1253
Eccentric circular orifices, 405
orifice, 1284
Economic design of gas absorption systems, 707
Economical pipe diameter, 384
for streamline flow, 385
Economizers, 1642
Eddy currents, magnetic, 1737
Edge filter, 984
Edge runner, 1214, 1220, 1221
Edison battery, 1790
Edwards gas-density balance, 1297
Effect, control, 1338
salting out, 320
of temperature on equilibrium, 311
Effective-concentration diagram, 723
Effective (electrical) resistances, 1733
Effective temperature lines of humidity chart, 777
Effective value of alternating current, 1734
Efficiencies, of extraction towers, 717, 729, 747, 748, 749, 751, 752
of Florida fuller's earth, 895
Efficiency, collection of dust, 1021, 1026
dryer, 800
evaporative, of dryers, 814
of gas liquefaction and separation 1703
grinding, 1115
of heating devices, 1759
of kilns, 1611
of pumps, 1431
target, 1022
volumetric, 1679
Eggs, spray-drying of, 846
Ejector arrangements, 1454
steam-jet, 1453
Elastic-breakdown pressure, relation to bursting pressure, 1240
failure of steels, 1239
follow-up, 1338
modulus of materials, 1539*ff*.
Electric furnaces, 1812
furnace products, 1773, 1818
Electric insulation, 1759
Electric lamps, data on, 1758
Electric lead seals, 1253
Electric leads, for high-pressure apparatus, 1253
Electric measuring instruments, 1276
Electric meters, 1286
Electric motors, for process control, 1329
proportioning, 1329
Electric relays, 1323
Electric telemeters, 1325
Electric wind, 1040
Electrical box-type furnaces, 1759
ovens, 1759
Electrical breakdown voltage of glasses, 1548
Electrical cartridge heaters, 1760
Electrical circuits, 1732
Electrical conductivity, measurement of, 1302
controllers, 1323
Electrical-discharge lamps, 1756
Electrical energy, 1733
Electrical engineering and electricity, 1731–1770
Electrical heating, 1759
units, 1759
Electrical instruments, recording, 1739
Electrical insulation materials, hygroscopic moisture of, 779
Electrical and magnetic relations, 1737
Electrical measurements, 1737
Electrical, methods of liquid-level measurement, 1290
Electrical power, 1733, 1735
cost of, 1755
Electrical precipitators, 1013, 1039
Electrical resistance furnaces, 1759
Electrical resistances, 1733
Electrical units, 1733
Electrical wiring, 1753, 1759
Electrically controlled weighing, 1294
Electrically heated kettles, 1221, 1224
Electricity, conduction through gases, 1747
and electrical engineering, 1731–1770
conduction of electricity through gases, 1747
cost of power, 1755, 1825
definitions, 1732
electric heating, 1759
electric and magnetic relations, 1737
electric and magnetic units, 1732
electric wiring, 1753
electrical circuits, 1733
electrical measurements, 1737
gaseous-discharge tubes, 1748
generators, 1763
illumination, 1755
magnetic circuits, 1736
magnets, 1752
motors, 1766
photo-, piezo-, thermoelectricity, 1742
rectifiers, 1750
rheostats, 1752
synchronous converters, 1766
thermionic tubes, 1742
transformers, 1762
modulus of glasses, 1548
static, 1869
Electroanalysis, 1793
Electrochemical energy, 1782
equivalent, 1775
equivalents of elements, 1776
industries, 1772
classification, 1774
laws, 1772
processes, power for, 1824
reduction reactions, 1786
units, 1772
Electrochemistry, 1771–1826
Electrode(s), measuring, 1309
potentials, 1785
-type liquid-level meter, 1290
Electro-float separator, 1094
Electrolux refrigerator, 1684
Electrolysis, alkaline chloride, 1806
Electrolysis and pH measurement. 1793
Electrolytes, fused, 1810
Electrolytic cell, aluminum, 1811
products, 1773
chlorine, raw materials chart for, 1809
Electrolytic cleaning and pickling, 1800
Electrolytic dissociation, 1780, 1782
Electrolytic hydrogen, 1795
Electrolytic manufactures, 1796
Electrolytic oxygen, 1795
Electrolytic rectifiers, 1787
Electrolytic refining of metals, 1801
Electrolytic valves, 1787
Electromagnetic instruments, 1738
Electromagnets, 1753
Electrometals furnace, 1815
Electrometric method of determination of the hydrogen-ion concentration, 285
Electrometric pH measurement, 1309
Electromotive force, 1732, 1737
Electron, 1732, 1782
Electronic conductors, 1772
liquid-level meters, 1290
relay, 1323
Electroplater's conversion factors, 1798
Electroplates, thickness of, 1798
Electroplating, 1796
of alloys, 1799
baths, 1797
cadmium, 1799
copper, 1799
motor-generator sets, 1796
rubber, 1799
zinc, 1799
Electrostatic concentration methods, 1093
Electrostatic fields, 1039
Electrostatic instruments, 1738
Electrothermics, 1812
gaseous, 1820
Electrotyping, 1796, 1800
Electrowinning of metals, 1805
Elements, atomic numbers of, 1781
atomic weights of, 1781
boiling point of, 110–128
control, 1338
critical constants of, 204, 205
electrochemical equivalents of, 1776
energy conversion data, 1782
final control, 1326, 1338
heat capacities of, 219–224
heats of fusion of, 210–212
heats of vaporization of, 210–212
isotopes of, 1781
melting point of, 110–128
physical properties of, 110–128
refractive index of, 110–128
solid, linear expansion of, 200, 201
solubility of, 110–128
specific gravity of, 110–128
Elevation of boiling point, 282
Elevators, bucket, 1348
centrifugal-discharge, 1348
continuous-bucket, 1349
continuous-flow, 1351

Elevators, gravity-discharge, 1349
perfect-discharge, 1349
and safety, 1851
Elex electrical precipitator, 1043
Elimination of odors, 907
Elliott oxygen process, 1711
Elliott-Sharples expander, 1712
Ellipse, construction of, 78
Ellipsoid, volume of, 58
Elongation of materials, 1539*ff*
of steels, 1235
Elutriation, 1084
dry, 1084
particle-size detn. by, 1112
wet, 1084
Elutriator, Andrews kinetic, 1084
Cooke short-column, 1085
Traxler and Baum, 1084
Emergency showers, 1876
Emission spectrometry, 1306
Emissivity, of carbon dioxide, 490
definition of, 483, 484
of sulfur dioxide, 492
of surfaces for radiant heat, 485–487
of water vapor, 490
Emulsification, 1167
Emulsifier mixer, 1206
Emulsifying, 1195, 1196, 1216, 1229
Emulsions, breaking, pre-centrifuging, 1004
characteristics of, 1167
permanent, 1196
stability of, 1167
temporary, 1196
Enamel frit, drying of, 869
Enameled steel equipment, 1549
Enameled ware, 1547
Energy, balance, mechanical (fluid flow), 375
total (flow of fluids), 375
charge for electricity,
conservation of, 375
kinetic, 375
law of, 295
units of, 295
electrical, 1733
electrochemical, 1782
heat and work conversion table and equivalents, 40
magnetic, 1736
of real gases, variation of, with pressure or volume, 301
requirements of centrifiges, 1005
-temperature diagram, 331
units of, 295
Engine efficiency, 1646
heat, 304
indicator, 1280
Engineering thermodynamic properties, 249–281
England, weights and measures of, 44
Engler gap between distillation fractions, 603
Enlargement loss, 388
Enskog's formula, viscosity of gases, 372
Enthalpy, 296, 344
concentration chart, 1052
$MgSO_4$-H_2O system, 1052
effect of pressure on, 568
of moist air, 760
Entrainment, in evaporators, 514
separation, 514, 1028
Entrance, loss, 388
rounded, 389
sharp-edged, 389
trumpet-shaped, 388
Entropy, 303
change with pressure, volume, and temperature, 304
of formation, in equilibrium-constant calculations, 348
of moist air, 760
of processes, 1704
and second law of thermodynamics, 303
variation of, with temperature volume and pressure, 304
-heat content diagram, 332
-temperature diagram, 331
Eötvös constant, 363
equation, for surface tension, 363
Equal-comfort lines of humidity chart, 777
Equation(s), binary Margules, 527
binary Van Laar, 527
of circle, 78
Clairaut's, 84
Clausius-Clapeyron, 293
cubic (algebra), 67
differential, 84
homogeneous, 84
linear, algebra, 66
calculus, 84
quadratic (algebra), 66
quartic and higher, 68
quaternary Margules, 529
simultaneous, 68
of state, 352
Gilliland's variation of Beattie-Bridgman, for mixtures, 354
at high pressure, 352
summary of thermodynamic, 314
ternary Margules, 528
trigonometric, 72
unknowns, 67
Equilibrium, calculations, 350–352, 355–358
in binary systems, 718
chemical, 294
constants, calculations, 348
determination of values of, 348–352
effect of temperature on (graph), 350
relations between, 350, 351
use in problems, 350–352, 357–358
criteria of, 306
effect of pressure on, 311, 352
effect of temperature on, 311, 350
distribution diagram, 722
free energy change and, 309–311
high-pressure, 352–358
phase, and solutions, 315
selectivity diagrams, 722, 723
in ternary systems, 719
Equilibrium data, CO_2-Na_2CO_3-$NaHCO_3$, 678
constant K for hydrocarbons vs. temp., 568, 569, 570, 571, 572
constant, variations with pressure, 576
Equilibrium data, constants for hydrocarbons, 568
of floating bodies, 362
lines in countercurrent apparatus, 548
moisture content, 800, 810
moisture of desiccants, 878, 880
prediction of absorption, 532
relations in diffusional operations, 525
vapor-liquid, of mixtures, 537
vaporization, continuous, 585
Equinoctial time, 45
Equipment, auxiliary, for tray and compartment dryers, 819
depreciation and obsolescence rates of, 1832
flow through, 390
safety and, 1854
Equivalence of coal and fuel oil, 1570
Equivalent, electrochemical, 1775
Equivalent conductivities of salts, acids, and bases, 1783
Equivalent conductivity, 1782
Equivalents and conversion tables, 36–44
angular measure, 41
area and surface, 41
capacity and volume, 40
conductivity, thermal, 41
decimals, 37
degrees to radians, 36
density, 40
energy, heat, or work, 40, 1782
heat, energy, or work, 40
heat flow, 41
gages, metal and sheet, 39
linear measure, 41
mass, 40
masses per unit length, 40
measure, linear, 41
metal, wire and sheet gages, 39
minutes to radians, 36
power, 41
pressure, 40
radians to degrees and minutes, 36
sheet metal gages, 39
specific gravity, °Bé., °Tw., lb./gal., lb./cu. ft., 38
surface and area, 41
temperature, 42
thermal conductivity, 41
velocity, 41
volume and capacity, 40
weights per unit length, 40
wire and sheet metal gages, 39
work, energy, or heat, 40
Equivalents, of coefficient of heat transfer, 466
for electroplaters, 1798
of elements, electrochemical, 1776
for metric and avoirdupois units, 1781
for metric and troy units, 1781
refrigeration, 1678
thermal conductivity, 456
viscosity, 1198
Erasable-line arithmetic graph paper, 100
Error, 1338
curve of probable, 81
dynamic, 1338

Estimation, of critical constants, 291, 292
of latent heat of vaporization, 299
by Clapeyron equation, 293
of vapor pressure, by Dühring rule, 294
by Othmer method, 294
Ethane, compressibility factors of, 207
saturated, thermodynamic properties of, 256
superheated, thermodynamic properties of, 256, 257
Ethanol-ethyl acetate vapor-liquid equilibriums, 630
Ethanol-water, flow points of, 1696
freezing points of, 1696
specific heats of, 1676
thermal conductivities of, 1697
viscosities, 1696
Ethanolamine-CO_2 equilibrium data, 677
Ether, absorption in H_2SO_4, 676
vapor pressure of, in H_2SO_4 solutions, 676
Ethyl acetate-ethanol vapor-liquid equilibriums, 630
Ethyl alcohol, aqueous, enthalpy-concentration diagram for, 261
densities of aqueous solutions of, 188
specific heats of, 235
and water, mixtures, densities of aqueous solutions of, 189
specific gravity of solutions of, 190
Ethyl alcohol–water, azeotropes, 631
solvent, 723
Ethylamine, saturated, thermodynamic properties of, 257
Ethyl chloride, Joule-Thomson Effect of, 203
saturated, thermodynamic properties of, 258
Ethyl ether-acetic acid-water system, 721
Ethylene, argon mixtures, deviations from Dalton's law, 291
compressibility factors of, 208
glycol, thermal conductivity of, 1697
-water, flow points of, 1696
freezing points of, 1696
-water, specific heats of, 1696
saturated, thermodynamic properties of, 257
superheated, thermodynamic properties of, 259
Eutectic salt-ice mixtures, 1700
Evaporation, 499–522, 1196
(*See also* Evaporator)
coils, 169
Dühring's rule in, 500, 501
effect of feed temperature, 510
heat transfer and coefficient in, 500–507
latent heats of, of paraffin hydrocarbons, 568
from liquid drops, 805
from plane surfaces parallel to wind, 545
vs. power in dryers, 871
Evaporation, rates from spherical surfaces, 546
of small drops, 540
by solar heat, 505
by sulfite pulp liquor, 503
of water, quantity (table), 501
Evaporative efficiency of dryers, 814
Evaporator(s), 500–522, 1648, 1677
accessories, 520
air removal from, 511
basket type, construction of, 506
calculation of required number of effects of, 515
coil, construction of, 505, 507
comparison of types, 507
condensate removal from, 511
condenser calculations for, 521
construction of, 505
continuous vs. batch, 512
controls, 520
cost of, 507
crystallizing, 1065
direct-fire-heated, 505
economical number of effects, 517
ejectors for, 1456
entrainment in, 514
"extra steam" value, 519
feeding of, methods, 510
film coefficients in, 478
foam and entrainment in, 514
forced circulation, construction of, 479, 503, 507
heat-transfer coefficients in, 507
heat transfer in, 478, 500, 501, 1692
heating surface, calculation of, 500
horizontal-tube, construction of, 501
heat-transfer coefficients in, 506
inclined-tube, construction of, 502
heat-transfer coefficients in, 507
kettles, 505
long-tube, construction of, 502, 506
materials of construction, 515
methods of feeding of, 510
miscellaneous, 503
multiple-effect, calculations, 515–519
effect of boiling-point elevation on, 500
temperature distribution in, 509
natural circulation, 479
operating difficulties with, 519
operation of, 509–515
continuous vs. batch, 512
optimum vacuum in operation of, 510
Porrion type, 505
salt removal from, 512
salting in, 514
scale removal from, 513
selection of, 522
single-effect calculations for, 507
solar heat, 505
steam temperature in, 509
temperature drop in, 500
types of, classification, 505
vertical-tube, construction of, 501
heat-transfer coefficients in, 506
Exact weight, bag holder, 1387
check or hand weighing rule, 1387
sacking scale, 1387
Excavations, 1860
Exchangers, heat, cost of, 482
reversing, 1713
Excitation adjustments of motors, 1767
Excitation curves of turboalternator, 1765
Exhausters, 1439
Exit loss, 388
Expanders, 1712
Expansion, adiabatic, 303
coils, 1691
cooling of gases by, 1702
engine in air-separation cycle. 1710
liquefaction by, 1708
factor, nozzles (table), 406
loss, 388
sudden, 388
venturi meter (table), 403
of gases, thermal, coefficients of. 200
of glasses, 1548
isothermal, 302
lines for turbine generator, 1645
of liquids, cubical, 202
of miscellaneous substances, linear, 201
of refractories, 1602
of solids, cubical, 202
Expeller, Anderson, 1074
Expense accounting, budgeting and, 1835
Expense classifications, 1835
Expense distributions, 1837
Explosion, dust, 1869
dust hazard, 1016
Explosions, 1868
Explosive disintegration, 1166
range, 1868
shattering, 1166
Exponential curves, graphs of, 81, 97
Exponential functions, 81
Exponential series, algebra, 64
Exponents, algebraic, 62, 63
integral, positive, 64
Expression, 1072
Extension leads, 1272
Extract, drying of, 866
Extraction, 320, 1196
applications of, 715
calculation methods for, 729–749
-column capacity, 744
equipment for, 747–753
factors of design, 747
liquid-liquid, 747
mixers, 747
phase separation, 747
spray and packed towers, 748
wetted-wall towers, 748
liquid-liquid, prediction of distribution in, 533
operating methods for, 716–718
phase equilibrium in, 718 728
solvent, 713–753
solvent recovery in, 715
Extractive distillation, 629, 634
choice of separating agent, 643
column design of, 645
comparison with azeotropic distillation, 655
effect of solvent concentration (table), 644

Extractive distillation, example of, 650
novel features, 645
separation of benzene from cyclohexane, 634
Extrapolation, 92
Extrusion, 1180–1183
of metals, 1180
of plastics, 1181

F

Fabrication of materials, 1539*ff*.
Fabrics, for dust filters, 1034
permeability of, 1030
resistance factors for, 1031
Factor(s), of plant location, 1720
power, 1732, 1735
wall effect, 394
Factoring, algebraic, 63–64
Factory trucks, 1363
Fahrenheit temperature scale, 1269
Fahrenwald flotation machine, 1091
Fairmount crushers, 1127, 1128
Fairmount single-roll crusher, 1122, 1128
Falling-ball-viscosity, 1197
Falling-film still, 657
Falling-rate period in drying, 800, 806
Falling sphere viscosity, 1199
False body, 1197
Fan(s), 1439, 1447
axial-flow, 1448
backward-curved blade, 1448
centrifugal, 1448, 1643
disk type, 1448
forward-curved blade, 1448
straight blade, 1448
Fan, propeller-type, 1208, 1448
Fan mixer, centrifugal, 1210
Fan nozzles, 1171
Fan and system characteristics, 1643
Fanning equation, 377
friction factors, 381
(chart), 382
Fanno line, 376
Farad, 1732
Faraday, 1775
definition of, 47
laws, 1775
Fatty acids, constructional materials for, 1485
Faxen equation (dialysis membranes), 754
Faxen viscosity, 1199
Feed, drying of, 838
jig, 1077
meal milling, 1147
perforated-pipe, 863
plate, optimum, location (distillation), 624
rate of, and work done in crushing, 931
regulators for centrifuges, 1003
splash, for drum dryer, 864
spray film for drum dryer, 864
stage, location of in countercurrent extraction tower, 735
Feeder(s) and feeder mechanisms, 1370–1376
Adjusto-O-Feeder, 1374
apron (conveyor), 1371
Feeder(s) and feeder mechanisms, Builders-Providence, 1376
carton, 1404
chain, 1373
chlorine, 1375
Chronoflo, 1376
constant-weight, 1373
conveyor (apron), 1371
Denver Equipment, 1374
flight-conveyor, 1372
Galigher, 1374
gas, 1375
for grinding mills, 1134
Geary, 1374
D. W. Haering, 1375
H-O-H, 1375
hopper, oscillating, 1374
lampwick drip, 1374
launder, 1375
lifting-gate, 1370
louvered bins, 1374
Milton Roy, 1374
Omega Machine, 1374
oscillating hopper, 1374
pH meter-controlled, 1376
plunger, 1371
pocket, 1371
proportioning, 1375
reagent, 1374
reciprocating-plate, 1372
revolving-plate, 1371
revolving table, 1373
for rock, 954
roll, 1370
Ross, 1372
rotary, 1375
rotating-vane, 1371
Rotodip meter, 1374
scraper, 1373
screw-conveyor, 1370
shaker, 1372
siphon, 1374
sliding tray, 1374
stirrup, 1370
swinging-plate, 1372
table, 1371
undercut-gate, 1370
vibrating, 1361, 1372
Wallace and Tiernan, 1374, 1375
zinc-dust, 1376
Feeding, methods of, for evaporators, 510
Feeding devices, 1291
Feed-water heat saving by extraction, 1637
heaters, 1647
treatment, 1647
Feinc filter, 979
Feld gas scrubber, 1213
Feldspar, grinding of, 1149
Fenske-Varteressian diagram (extraction), 733
Ferric chloride, constructional materials for, 1487
densities of aqueous solutions of, 179
Ferric nitrate, densities of aqueous solutions of, 179
Ferric sulfate, constructional materials for, 1488
densities of aqueous solutions of, 179
Ferroalloys production, 1816
Ferrochromium production, 1816
Ferromanganese production, 1816
Ferromolybdenum production, 1817
Ferrosilicon production, 1816
Ferrosilicon-titanium production, 1817
Ferrotungsten production, 1817
Ferrous alloys, physical properties of, 1539
pipe, 413
Ferrous sulfate, constructional materials for, 1488
densities of aqueous solutions of, 179
dissociation pressures of, 174
Ferrovanadium production, 1817
Fertilizer industry, 1721
Fertilizer materials, grinding of, 1151*ff*.
Féry cell, 1789
Festoon dryer, 848
Fiat furnace, 1813
Fiber, cutter, Stokes Universal, 1162
drying of, 828
saturation point, 800
vulcanized, physical properties of, 1551
Fiber-plastic pipe, 436
Fibrous-glass filter mediums, 968
Fibrous materials, hygroscopic moisture of, 779
Fick's law of diffusion, 754
Fifth-wheel, trailer, 1364
Field intensity, magnetic, 1732
Figures, significant, rules for, 50
Filament line, 375
Filing machines, 1163
Fillet, area of, 56
Filling and closing, collapsible-tubes, 1411
small-bag, 1410
equipment, 1381
for dry products, 1400
for liquids, 1397
and weighing equipment, 1392
Film coefficients, in heat transfer, 478
liquid (extraction), 743
equations for liquid, in extraction, 743
removal in heated kettles (mixers), 1221, 1222
scrubbers, 1039
spray, feed for drum dryer, 864
theory, application to extraction, 742
Filter(s), aids, 969
air, 1013, 1045
centrifugal, 1002
continuous vacuum, 970
corrosion-resisting construction, 984
cost of, 989
electrical, 1752
enclosed pressure, 970
gravity, 970
Hardinge sand-, 948
industrial application, 985
intermittent vacuum, 970
leaves, 974
makes of, 971–984
mediums, 967–968

Filter(s), nutsch, 970
operation of, 992
papers and pulps, 968
plant (contact) flow sheet, 894
plate-and-frame, 970
power used on, 989
pressure, 970
sizes of, 989
thickeners, 947
Briggs, 949
Oliver-Borden, 948
types of, 970–971
Filtrate pumps, 991
Filtration, 887, 964–992
aids, 969
of barium sulfate, 989
of beet-sugar liquors, 987
of cachaza, 987
of cane-sugar sirups, 986
of cement slurry, 986, 990
clarification, 964
constant-pressure tests, 966
contact, 888
of cyanide pulp, 985
definition of, 964
effect of cake thickness, 966
effect of particle size, 966
effect of pressure, 966
effect of solid content, 966
effect of temperature, 966
effect of temperature and viscosity, 970
effect of type of filter medium, 966
equation, significance of, 965
food products, 988
leaf tests, 968
mediums, 967
paper pulp, 986
petroleum products, 987
phosphoric acid, 988
pigments, 988
pressure tests, 970
purpose of, 964
rate of, 965
salt and crystals, 986
sewage sludge, 986
theory of, 965
titanium pigments, 989
typical plots of data, 967
of water, use of
activated carbon for, 903
European household filter, 904
Final control elements, 1326, 1338
Financial statements, 1829
Fine reduction crushers, 1120
Fine reduction jaw crusher, 1125
Fine Reduction Superior McCully crusher, 1126
Finite time lag, 1338
Finned pipes, heat transfer through, 473
Fire, alarm equipment, 1851
appliances, classification of, 1855
brigade, plant, 1872
extinguishers, first-aid, 1854
hazard of flammable liquids and gases, 1868
hose, friction losses in, 1850
mains, 1849
protection, 1847–1884
(*See also* Safety)
bibliography on, 1879
Fire, protection, data for engineers (table), 1873
equipment for, 1854, 1870
fire brigades, 1872
housekeeping and, 1872
ladders and, 1862
organization as, 1871
shafting and, 1859
stairs and, 1861
watchman's role in, 1872
water supplies for, 1849
resistance of building walls, 1852
streams, range of, 1850
tube boilers, 1638
Firing rates of furnaces, 1639
First aid, 1875
First-aid fire appliances classification, 1855
First-aid fire extinguishers, 1854
First law of thermodynamics, 295, 1703
First-order reaction, 321
at constant pressure, 322
at constant volume, 321
Fisher-Gardner mobilometer, 1200
Fittings, for ammonia, 446
table, 446
cast iron, 445
screwed (table), 446
table, 446
compression, 442
effect on orifice location, 407
flanged, steel (table), 443
for high pressures, 1250
malleable iron, 446
screwed (table), 446
soldered, 442
Fixed carbon in coals, 1561
Fixed charges, 1831
of power plants, 1631
Fixed points for thermometers, 1269
Fixed-position controllers, 1320
Fixed-property accounting, 1831
Flakers, drum, 1164
Flakice, 1699
Flaking, 1164
Flame temperatures, 1588
Flame velocities of gas-air mixtures, 1587
Flames, luminous, "absorption" strength of, 494
emissivity of, 494
of powdered coal, 495
radiation of heat from, 493
Flammable liquids and gases, fire hazard of, 1868
outage chart for, 1875
storage tanks for, 1868
Flammable materials, grinding of, 1119
Flammable vapors, testing devices for, 1868
Flange(s), 442
cast-iron, 445
table, 445
faces for retaining gaskets, 453
pipe joints, 419, 441
standards, 442
steel, 443
table, 443
taps for, 405
orifices, 1284
Flapper(s), 991
Bristol free-vane, 1322
free-vane type, 1323
-type pilot valve, 1320
Flash distillation, 585
Flash dryer, 834
Flash drying and grinding, 1119
Flash fire, 1868
Flash float, 1076
Flash point, 1584, 1866
of liquid fuels, 1570
pulverization, 1146, 1147
roaster, 1622, 1623
vaporization, 585
Flax, hygroscopic moisture of, 778
Flexible couplings, 1668
Flight conveyor(s), 1347
feeder, 1372
Float, flash, 1076
gage, 367
level indicator, 1289
rope, and pulley level indicator, 1289
-type, mercury manometers, 1285
non-indicating liquid-level controller, 1289
Floating action, 1338
average-position, 1338
constant-speed, 1338
Floating-ball viscometer, 1301
controller, 1320
Floating-bell manometers, 1286
Floating bodies, equilibrium of, 362
stability of, 362
Floating shaft, flexible, 1668
Flocculation, 1190
definition of, 939
and mixing, 1196
Flocculator, Dorrco, 947
Jeffrey, 947
Flooding, of extraction columns, 752
Flooding data (Colburn) for packed columns (extraction), 753
Flooding velocities for tower packings, 684
Floor loadings for storage systems, 1379
Floor trucks, 1363
Flotation concentrates, filtration of, 990
cell, Denver Sub-A, 1089, 1091
laboratory, 1090
depth of, 362
and buoyancy, 362
froth, 1085
machines, 1089
Fahrenwald, 1091
Steffensen, 1089
Flour, hygroscopic moisture of, 779
milling, 1147, 1148
Flow, of brine, 1697
capillary, 800
close control of, valve for, 1251
control of, at high-pressure, 1254
control system, 1333
integrators, 1287
measurement, 1282
nozzles, 1283, 1285
proportioning pump for ratio control of, 1326
ratio, 1333, 1334

Flow, diagrams, in dilution problems, 337
in distillation problems, 338
in high-pressure problems, 357
in material balance problems, 338
systems, space-time-yield, 329
use, 333
direction, indicator, 400
of fluids, approx. integration of, 381
differential equation for, 378
laminar, 375
measurement of, 396–412
non-isothermal, liquids, 381
of non-Newtonian fluids, 385
nozzles, 406
in pipes and channels, 377
alignment chart, 379
pulsating, 408
streamline, 375
kinetic energy of, 376
of suspensions, 385
turbulent, 375, 384
viscous, 375
through equipment, 390
of fluids, 359–412
through equipment, 390
in pipes and channels, 377
of gases in pipe lines, 1589
lag, 1338
Master Kom-bi-nator, 1168
mixers, 1197, 1203, 1215
points of brines, 1695, 1696
sheet, double-reaction, 951
intermediate agitation, 952
-stress curve, 1197
of suspensions, 385
turbulent, through smooth pipes, 542, 543
Flowmeter(s), bellows-type, 1286
differential, 1285
integrators, 1287
resistance (electrical), 1286
slack-diaphragm, 1286
for high pressure, 1254
Flue gas(es), sampling of, 1102
water vapor in, 338
Fluid(s), bath, boiling at constant pressure, above zero degrees, 282, 283
boiling at constant pressure, below zero degrees, 282
for thermostatic control, above zero degrees, 283
for thermostatic control, below zero degrees, 282
catalyst cracking process, 1619
catalytic cracking units, 1618
definition of, 369
dilatant, 1198
non-Newtonian, 1197
rheopectic, 1199
thixotropic, 1198
dynamics, terminology in, 374
energy or jet mill, 1116, 1124, 1145
energy reduction mill, 1146
flow of, 359–412
friction, heat transfer, and mass transfer, analogy among, 541
ideal, 369
in motion, 369–395
Fluid(s), perfect, 369
statics, 360–363
Fluidity, 370
of mixtures, 1197
Fluidized-solids reactor, 1612, 1618
Fluorescent lamps, 1758
data on, 1758
Fluorspar, drying of, 826
grinding of, 1151
FluoSolids furnaces, 1619, 1620
reactor, 1612
Flux density, 1732
from electric current, magnetic, 1737
linkages, 1732
magnetic, 1732
Flyball governor, 1287, 1295
Foam in evaporators, 514
Foam production, 1176
"Fog" nozzles, 1172
Follow-up, 1338
elastic, 1338
Food(s), cold storage of, 1699
humidity requirements of, 1699
hygroscopic moisture of cereal, 779
products, filtration of, 988
Force of buoyancy, 362
Formula, Chézy, 377
Darcy, 377
Fanning, 377
Francis, 409
Weymouth, 383
Williams and Hazen, 377
magnetomotive, 1732
on particle, centrifugal, 994
Forced-circulation evaporators, 503, 507
heat transfer in, 479
Forced convection, gases, heat-transfer coefficients for, 467
heat transfer by (graph), 468
Forced-feed lubrication, 1681
Forecast, budget, 1835
Forging furnaces, 1605
Fork truck, 1366
pneumatic-tire, 1366
Formaldehyde, constructional materials for, 1490
Formation, heats and free energies of, 236–243, 298
Formation of coal, 1560
Formed-bell manometer, 1286
Formed-displacer meter, 1286
Formed-tube manometer, 1286
Formula(s), of differential calculus, 81
of integral calculus, 85
Formic acid, constructional materials for, 1491
densities of aqueous solutions of, 186
Forward-curved-blade fans, 1448
Foster Wheeler ball mills, 1132, 1135
Foster Wheeler pulverizers, 1141
Fountain pen for recorders, 1267
bucket pen, 1267
V-pen, 1267
Four-wheel-steer trailer, 1364
Foxboro, Dewcel dew-point humidity controller, 1300
potentiometer circuit, 1278
Fractional calculations, algebraic, 63
crystallization, 1053
of NaCl and $NaNO_3$, 1054
liquid distribution, 715, 739
Fractionation(s), 563
low-temperature, 609
column, control of, 1336
with molecular still, 659
Fractions, decimal equivalents, 37
Francis formula, 409
Fränkl regenerator, 1713
Franz isodynamic magnetic separator, 1093
Free convection, heat transfer by, 463, 474
charts, 472, 475
Free energy(ies), 305
change, standard, 309
effects of nucleation, 1057
equation, as function of temperature, 305
of formation, 236–243
of inorganic compounds, 236–243
of organic compounds, 236–243
function, 314
interfacial, 363
friction head, 377
variation of, with pressure and temperature, 305
Free moisture content, 800
Free-swelling index of coal, 1566
Free-vane, flapper, Bristol type, 1322
Freedom, degrees of, 297, 315
Freezing, ejectors for, 1456
point(s), of brines, 1696
lowering of, 319
of solutions, 1694
of some aqueous solutions, 281
Freon as refrigerant, 1678
Frequency, 1732
Féry radiation pyrometer, 1276
Friability of coals, 1562
Friction, clutches, 1671
coeffs. against steel plates, 1347
factors, for fabricated packings, 396
for in-line tube banks, 392
for non-circular cross sections, 382
for non-isothermal flow, 382
fluid, 377
alignment chart, 382
correlation of data, 377
in pipes, 377
head of liquids, 1415
horsepower, 1678
Friction loss (*see* pressure drop)
Frigidisc grinder, 1144
Frit, enamel, drying of, 869
Froth flotation, 1085
Frothers, 1088
Fruit(s), dried, grinding of, 1148
juice, filtration of, 988
pulp, drying of, 837
Fuel(s), 1559–1596
beds, types of, 1566
charge at power plant, 1631
combustion of, 1603
comparative cost of, 1812
consumption in furnaces, 1605
in kilns, 1611
density of, 1629

Fuel(s), gaseous, 1575–1596
gases, analyses of, 1577*ff.*
corrosion of alloys by, 1526
impurities in, 1577
handling systems, 1637
liquid, 1568–1575
miscellaneous liquid, 1575
miscellaneous types, 1568
oil(s), burning equipment, 1572
heaters, 1573
properties of, 1570, 1571
specifications for, 1654
as plant location factor, 1720
solid, 1560–1568
used in furnaces, 1600
waste, 1641
Fugacity(ies), charts for pure gases, 355
in equilibrium calculations, 355
for gas mixtures, 355
use, 355–356
coefficients (charts) of gases and vapors, 536
of gases, 306
of liquids and solids, 307
use of to calculate effect of gas-law deviations, 535
Fuller's earth, 889
flow sheet for manufacture of, 889
granulation of, 1158
revivification of, 893
Fuller solids-level meter, 1290
Function, controller, 1338
Function(s), expanded into series, 83
exponential, 81
given angles, 71
hyperbolic, 81
implicit, 82
integral, 85–87
transcendental, 87
trigonometric, 70–72
of curves, 80
logarithms, 19
Fundamental constants, 295
Fundamental physical constants, 47
Funicular state, 800
Furfural, activity coeffs. of 1-butene in, 643
activity coeffs. of isobutane in, 643
-docosane-diphenylhexane system, 721
solubility of liquid hydrocarbons in, 647
-soybean oil system, 728
Furnace(s), Ajax-Northrup, 1816, 1818
Ajax-Wyatt, 1816, 1817
Angelini, 1815
arc, 1814
atmospheres, 1591
Bassanese, 1815
batch, 1607
Bessemer, 1616
billet-reheating, heat transfer in, 496
blast, 1614
boiler, 1605
Booth-Hall, 1815
brass annealing, 1605
ceramic, 1606
Chaplet, 1815
Furnace(s), compensating, 1598
construction, 1601
continuous, 1607
cyclone, 1605
Detroit rocking, 1817
Dutch oven, 1598
electric, 1812
box-type, 1759
resistance, 1759
high-frequency, 1761
resistor-type, 1759
steel, 1812
Electrometals, 1815
Fiat, 1813
firing rates, 1639
FluoSolids, 1619
forging, 1605
fuel consumption in, 1605
Girod, 1815
glass, 1621
Greaves-Etchells, 1815
Greene, 1813
Grönwall-Dixon, 1815
heat transmission in combustion chambers of, 493
heat-treatment, 1605
Heroult, 1813
Herreshoff, 1598
horizontal, 1598
industrial, 1598, 1600
iron and steel induction, 1816
Keller, 1815
Kjellin, 1816
and kilns, 1597–1626
Mannheim, 1621
metal-melting, 1605
metallurgical, 1605
multiple-hearth, 1622
Nathusius, 1816
non-ferrous metals, electric, 1817
open-hearth steel, 1620
petroleum, 1605
process, 1598, 1608, 1621
products, electric, 1773
pyrites, 1622
recirculative, 1607
recuperative, 1598, 1607
regenerative, 1598, 1607
Rennerfelt, 1815
reverberatory, 1619
rivet making, 1605
Röchling-Rodenhauser, 1816
Schoenherr, 1820
shaft, 1612
size, effect on performance, 1602
Snyder, 1815
space, 1598
Stabie, 1815
Stassano, 1815
Swindell, 1813
Tagliaferri, 1815
vertical, 1598
wall dimensions, 1602
walls, heat conduction through, 460
Fusain in coals, 1562
Fused electrolytes, 1810
Fused-silica pipe, 440
Fusible alloys, 454
Fusion, as a size enlargement method, 1191
heat of, 299
Fusion, latent heat of, by Othmer's rule, 342, 343
by Trouton's rule, 343
temperatures, 281
of refractories, 281
welding, 1243

G

g_c, gravitational conversion factor, 360
Gage(s), Bourdon, 367, 1254
calibration of, 367
compound, 367
dead-weight piston, 1255
Bridgman design, 1255
differential-piston type of, 1256
with lever-arm, 1256
diaphragm, 367
differential, 364
draft, 366
float, 367
glass, 1289
glasses, 436
hook, 367
inclined pressure, 366
interface, 367
liquid column, 364
mechanical, 367
micro, 366
multiplying, 365
open, 364
point, 367
pressure, 364, 1279
calibration of, 367
mechanical, 367
stick, 1289
tester, dead-weight, 1280
Gages, wire and sheet metal conversion table and equivalents, 39
Galigher feeder, 1374
Galvanic cells, 306
Galvanometer, d'Arsonval, 1276
Galvanometers, 1775
Gantt chart, 1843
Gap, differential, 1338
Gardner mobilometer, 1197, 1200, 1220
Gas(es), absorbers, Murphree plate efficiencies of, 698
turbine, 1211, 1216, 1217
absorption, 667–711
accompanied by chemical reaction, 702
columns and packing in, 707
exit strength (or column height) in, 708
gas velocity (or column diameter in), 708
general design procedure, 668
liquor-gas ratio in, 717
plate towers vs. packed towers in, 707
rates in agitated vessels, 702
in spray towers, 699
systems, economic design of, 707
actual, 290
Joule-Thomson effect, 301, 333
adiabatic flow of, 376
-adsorbent carbons, 904
carbons, applications of, 906
adsorption from, 887

Gas(es), -air mixtures, flame velocities of, 1587
analyzer, thermal conductivity 1302
calorimeter, 1301
-density balance, 1297
density, by static-pressure measurement, 1298
measurement, 1297
measurement of density of, 1297
meter, diaphragm, 1283
water-sealed, 1283
-pressure thermometers, 1270
thermal conductivities of, 1303
thermometer, constant-volume, 1269
-atomizing nozzles, 1173
burners, design of, 1590
calculation of true temperature of, 473
coefficients of thermal expansion of, 200
coke, 1568
combustion constants, 1583
compression, 1439
conduction of electricity through, 1747
constants, 290
cooling, at constant pressure, 337
at constant volume, 337
corrosive, resistance of metals and alloys to, 1526
cost of natural, 1722
in cylinders, compressed, 1869
densities of, 104, 176
diffusion, 292, 538
diodes, 1749
discharge, critical ratio, 402
dispersion, 1175
fluid attrition systems, 1178
mechanical agitators, 1178
precipitation and generation methods, 1178
dispersoids, characteristics of, 1019
drying, 877–884
by compression, 884
with desiccants, 878
methods of, 877
by refrigeration, 884
engine, four-cycle, 1654
equilibrium calculations, 356
feeding of, 1375
flow in pipe lines, 1589
flow through limestone, 1614
fugacity, 306
heat capacity of, 297, 1588
calculations, 340, 357–358
heat transfer by forced convection in, 467
heated refrigeration, 1684
heaters and coolers, design of (chart), 469
holders, 1380
ideal, 289, 296, 335
calculations of, 289
as industrial fuel, 1591
laws, calculations of, 289, 290
lifting power of, 292
liquefaction of, 291, 1702, 1707
and liquid mixing, 1216–1217
liquid ratio in gas absorption, 668, 707

Gas(es), making, oil cracking in, 1581
manufacture, 1578
masks, 908
canisters, standard colors for, 1870
(Universal) of U.S. Bureau of Mines, 908
meters, wet, 410
mixers, propeller as, 1208
turbo-, 1211
mixing, 1215
in closed vessel, 541
mixtures of, 290
critical temperatures of, 291
sensible heat content of, 340
molal volumes of, 205
phase resistance in packed columns, 687
in pipes, heat transfer to, 469
pressure loss in flow of, 381
producers, 1617
radiation, beam lengths for, 491
from non-luminous, 490
ratios of specific heats of, at 1 atm., 233
reactions at high temperature, 1626
contact catalytic, 329
real, variation of energy and heat content with pressure or volume, 301
sampling of, 1102
scrubber, 1213, 1217
separation by low-temperature methods, 1702
solubility of, 317
in non-aqueous liquids, 676
in water, 673
spargers, 1203
specific heats of, 229
storage of, 1380
thermal conductivities of, 461
turbines, 1655
and vapors, diffusion coefficients of, 539
viscosity of, 369
water-vapor content of, 173, 174
Gaseous-discharge tubes, 1748
Gaseous electrothermics, 1820
Gaseous fuels, 1575*ff.*
analyses of, 1629
Gaseous ions, 1747
Gas-filled control tubes, 1748
Gas-fired boiler, 1639
Gaskets, 451
flanges for retaining, 453
for high pressures, 1245
materials of construction for, 1554
Gas-law deviations, calculation of effect of, 535
Gasoline, drying with activated alumina, 911
as fuel, 1575
recovery from natural gas, 906
Gate mixer, 1205, 1222
Gate valves, 447, 1328
table, 448
Gaulin colloid mill, 1169
homogenizer, 1167
Gauss, 1732
Gayco external air classifier, 1135
Gayley dry blast, 884

Gearing, 1666
Gearless gyratory crushers, 1126
Gear-pump meter, 1282
reduction units, 1669
-type rotary pump, 1438
Geary feeder, 1374
Geiger-Müller counter tube, 1304
General Mills Co. vacuum filler, 1402
General overhead costs, 1839
Generation of power, 1627–1660
Generators, 1763
a.c., 1764
Dobrowolsky, 1764
homopolar, 1764
induction, 1765
paralleling of, 1764
sets, efficiency of motor-, 1765
prices of, 1765
three-wire, 1764
voltage characteristics of compound-wound, 1764
direct-current, 1764
Geographic factors of plant location, 1723
Geometric mean, 66
Geometrical constructions, 59–62
mean, 59, 66
progression, 66
Geometry, analytical, 77–81
and mensuration, 53–58
plane, 53–55
theorems of, 53–54
solid, 55–57
theorems of, 55–56
spherical, theorems of, 57–58
German high-pressure compressors, 1260
Gibbs cell, 1808
Gilbert, 1732
Gilliland's correlation of theoretical plates vs. reflux ratio, 622
method for gas mixtures, high-pressure calculations, 354–355
Gilsonite, grinding of, 1156
Giordani Pomilio cell, 1808
Girod furnace, 1815
Glands, smothering, 1423
Glass, electrode, 1309, 1795
manometer, 1267
equipment, 1548
fiber filter mediums, 968
furnace, 1621
pipe, 436
properties of, 1460
Glass-lined pipe, 436
Glass-tank regenerators, 1624
Glauber's salt, costs of crystallization of, 1070
Globe valves, 447
table, 450
Glow discharge, 1747
Glue, drying of, 866, 869
hygroscopic moisture of, 780
Glycerin-water, specific heats of, 1696
thermal conductivities of, 1697
Glycerol, densities of aqueous solutions of, 191
specific heats of, 235
Glycerol-water, flow points of, 1696
freezing points of, 1696
viscosities of, 1696

Gold, electrolytic refining, 1805
 (placer), jigs for, 1077
Governor-type flow integrator, 1287
Grab sampling, 1097
Gradient, moisture, in drying, 800
Grainer, 922
Graining bowl, 857
Granigg magnetic separator, 1092
Granular bed filters, 968
 solids, mass and heat transfer in gases through, 547
Granulation, 1186
 of briquettes, 1187
 of fertilizers, 1187
 by fusion, 1187
 of soft materials, 1158
 by Spheronizing, 1188
 by spray drying, 1188
Graphic recording, 1292
Graphical calculations, for extractions, 729–741
Graphical calculus, 90–93
 tables, 90, 91
Graphical division, 54
 differentiation, 90
 integration, 90
 multiplication, 54
Graphical methods of determining number of plates, 629
Graphical representation of ternary systems, 719
 of thermodynamic functions, 330
Graphite and carbon equipment, physical properties of, 1550
 drying of, 869
 electrodes, 1822
 filters, 968
 grinding of, 1156
 manufacturers of, 1534
 pipe, 431
 production, 1819
Graphs and graph paper, 97–100
Grasshopper conveyor, 1361
Gravitational constant, value, 45, 47
Gravitational conversion factor, g_c, 360
Gravitational displacements of spherical particles, 1020
Gravity, acceleration due to, local, 375
 battery, 1789
 concentrates, filtration of, 990
 filters, 970
 relative, of ingredients of mixture, 1202, 1217
 roller conveyor, 1358
 sedimentation, particle-size determination by, 1112
 settling chamber, 1021
Gravity-discharge elevators, 1349
Gray surface, 484
Great Britain, weights and measures of, 44
Greaves-Etchells furnace, 1815
Greek alphabet, 46
Greene furnace, 1813
Grid-bias detection, 1746
Grid-circuit detection, 1746
Grid-leak detection, 1746
Grid packing, pressure drop in, 396
 hard-rubber pipe, 438
Griffin mill, 1137
Grindability, 1114
 of coals, 1562
 index, Hardgrove, 1114
Grinders, description of, 1130
Grinding, of abrasives, 936
 aids, 1115
 of aloxite, 936
 of carborundum, 936
 dry, 1115
 and drying of bauxite, 1119
 efficiency, 1115
 equipment, classification of, 1116
 selection of, 1116
 of flammable materials, 1120
 of lithopone, 936
 mediums, 1130
 metal powders, 1119
 open-circuit, 930
 continuous, 1117
 safety in, 1119
 wet, 1115
 vs. dry, 1118
 work of, 931
 and circulating load, 931
 (*See also* Crushing, and grinding)
Grinding mills, classification of, 1116
 comparison of, 1149
 description of, 1130
 particle-size classifiers, 1135
 regulating feeders for, 1134
Grizzly, 955
Grönwall-Dixon furnace, 1815
Gross heating value, 1578
Guarding, 1855
Guards for door openings, 1852
Guided-float rotameter, 1288
Gums, grinding of, 1159
Gypsum, drying of, 837
 grinding of, 1154
Gyratory crushers, 1116, 1125, 1126
 sifter, 958

H

Haering feeder, 1375
Hagan Ring Balance Meter, 1254
Hair belting, 1663
 filter mediums, 968
 hygrometer, 1298
 method of determining humidity, 777
Half cell, 1309
Halogen overvoltage, 1788
Hammel-Dahl small-flow valve, 1327
Hammer crushers, 1116, 1122, 1129
Hammer mills, 1116, 1139
 in closed circuit with air classifier, 1118
 for grinding mica, 1151
 with internal air classifiers, 1141
 without internal air classifiers, 1139
 Mikro-pulverizer, 1123
Hancock jig, 1077
Hand electric platform truck, 1369
Hand pallet trucks, 1369
Hand sewing textile bags, 1390
Hand weighing, 1385
 scale, 1387
Hardgrove, grindability detn. with, 1114
 index of coal, 1562
Hardinge, air classifiers, 1135, 1136
 automatic backwash sand-filter, 948
 auto-raise thickener, 944
 ball mills, 1134
 ball and pebble mills, performance of, 1134
 conical ball mills, 1134
 conical mill, 1133
 countercurrent classifier, 926
 diaphragm pump, 945
 direct rotary dryer, 829
 disk feeder, 1134
 double-shell, indirect-direct rotary dryer, 830
 Feedometer, 1134
 mill for coal grinding, 1156
 for wet grinding, 1151
 sand-filter clarifier, 948
 weigh feeder, 1292
Hardness, 1114
 Brinell, of materials, 1539*ff.*
 of glasses, 1548
 Moh scale of, 1114
Hargreaves-Bird cell, 1808
Harmonical mean, 66
 progression, 66
Hartley oscillator, 1745
Harz jig, 1076
Hatches and traps, 1860
Haultain Infrasizer, 1085
Haüy, law of, 1050
Haveg pipe, 437
Hay, grinding of, 1148
Hays oxygen meter, 1303
Head, of fluid, 364
 friction, 377
 pressure, 360, 377
 static, 360, 377
 velocity, 377
Head of liquids, friction and velocity, 1415
Health, dust hazard, 1017
Heat(s), of adsorption, 301
 balance, in ammonia reaction, 347
 use of, in distillation, design, 604
 balance of byproduct power cycle, 1635
 in calculation of theoretical flame temperature, 347
 calculations, 357, 358
 capacities, 296
 of the elements, 219–224
 of gases, 297, 1588
 of infinite dilution of aqueous solutions, 530
 of inorganic compounds, 219–224
 of liquids, 297
 of saturated vapors, 297
 of solids, 298
 of water, 225
 of combustion, 244–246
 conduction of, 456–463
 constant, summation of, 344–345
 content, 296, 565
 of calcium chloride brines, 1694
 chart, high pressure, 357
 use of, 356

Heat(s), content, entropy and, chart of, 332
of gaseous mixtures, 340
of liquid state, 340
of pure gaseous state, 339
of real gases, variation of, with pressure or volume, 301
of sodium chloride brines, 1695
of crystallization, 1052
distribution in furnaces, 1600
effects, calculation of, in flow, 376
in crystallization, 1052
emissivity of (table), 485–486
energy, or work, and conversion table and equivalents, 40
engine, 304
entropy and, diagrams (*see* Mollier diagram)
exchangers, 1713
control system for, 1334
flow through, 390
heat transfer in, 464, 480–482
tubing for, 424*ff*.
flow, conversion table and equivalents, 41
units, equivalents, 40
of formation, 236–243, 298, 335
in calculation of standard heat of reaction, 344
in inorganic compounds, 236–243
of organic compounds, 236–243
of fusion, 299
of elements, 210–212
of inorganic compounds, 210–212
of miscellaneous materials, 219
of organic compounds, 213, 214
humid, 759, 811
integral, of solution, 299
and mass transfer, simultaneous, 557
simultaneous for system liquid water-air-water vapor, 558
simultaneous, under condition of large-concentration driving forces, 559
natural convection of, 463
pumps, 1657, 1701
radiation of, from coal flames, 495
due to carbon dioxide, 490
due to water vapor, 490
from flue gas, 493
from gases, calculation, 490
from luminous flames, 493
from non-luminous flames, 490
from particles, 493
between solids, 484
from sulfur dioxide, 492
between surfaces, 484
between surfaces of solids, 484
of reaction, calculation of, 299, 344
calorimetric determination of, 296
conventions, 344
from heats of combustion, 345
from heats of formation, 344
at specified temperatures, 346
standard, 335
variation with temperature, 299
regenerators, 463
-release rates in furnaces, 1600

Heat(s), requirements for calcining limestone, 1613
of refrigeration absorption system, 1683
saving by extraction, feed-water, 1637
through solids, 460
of solution, 246–248, 299, 341, 344
at infinite dilution of aqueous solutions, 530
of inorganic compounds in water, 246–248
of organic compounds in water, 248
prediction of, 532
of sublimation, 209
transfer of (*see* Heat transfer)
of transition, 299
transmission of, 455–498
by conduction and convection, 456–480
miscellaneous over-all coefficients, 480–482
radiant, 483–498
(*See also* Heat transfer)
units, conversion of, 335
latent heat, 335
standard, 335
types of, 335
of vaporization, 299
of elements, 210–212
estimation, 299
of hydrocarbon, molal, 217
of inorganic compounds, 210–212
of organic compounds, 215, 216
of wetting, 301
silica gel, 884
Heat pumps, 1657, 1701
Heat transfer, for ammonia condensers, 482, 1689
basis of classifying dryers, 817
in blast-furnace stoves, 463
in economizers and preheaters, 1642
in evaporators, 1692
in furnaces, 1600, 1602
and mass transfer, by convection in tubes, 544
and fluid friction, analogy among, 541
in gases through granular solids, 547
in pebble heater, 1617
properties of refrigerants, 1687
rates in boiler surfaces, 1638
in rotary kilns, 1609
and size reduction, 1119
in sublimation, 661
in unsteady state, 462
Heat-transfer coefficients, for coils immersed in liquids, 481
in combustion chamber, 495
in condensers, 476
by convection, 456
conversion factors for, 456, 466
effect of liquor level on, 505
effect of surface conditions on, 505
in evaporating electrolytic caustic, 504

Heat-transfer coefficients, in evaporation, 500, 501
in evaporators, 478, 500–505
film coefficients, 478
over-all coefficients, 481
film, 478
in forced-circulation evaporator, 503–5
to gases, 467
graphic interpretation of over-all coefficients, 470
in heat regenerators, 463
in horizontal-tube evaporators, 501
in inclined-tube evaporators, 502
individual coefficient, 464
for jacketed vessels, 482
to liquids in tubes, 469
in long-tube evaporators, 502
mean temperature difference in, 464
miscellaneous, 480–482
in mixers, 1221–1224, 1229
Heat-treatment furnaces, 1605
Heaters, for centrifuges, 1003
control of immersion, 1331
electric, 1759
cartridge, 1760
efficiency of, 1759
feed-water, 1647
fuel-oil, 1573
petroleum, heat transfer in, 496
Heating and cooling of fluids, 467
(chart), 468
dielectric, 1762
electric, 1759
radiant, 1761
units, electric, 1759
of kettles, 1221–1224
machine, 304
surface, calculation of, in evaporators, 500
values of coals, 1561
of gaseous fuels, 1578
of liquid fuels, 1571
Heavy-media separation processes, 1081
Heights, circular segments, 32
Helical element, 1279
gear reduction units, 1669
mixer, 1210
Helium, liquefaction of, 1707
separation process, 1716
Helix element for thermometers, 1271
Hemp, hygroscopic moisture of, 778
Henry, 1732
Henry's law, 317, 532, 673
deviations from, 317
n-Heptane-aniline-methylcyclohexane system, 722, 723
Herbage, grinding of, 1148
Hermetic separators, 1002
Heroult furnace, 1813
Herreshoff furnace, 1598
as mixer, 1205
Herringbone gear units, 1669
Hersey filter, 1033
Hess, law of, 344, 345, 346
Heterodyne detection, 1746, 1747
Heterogeneous azeotropes, 631
Heterogeneous reactions, kinetics of, 328

H.E.T.P. and H.T.U., for packed distillation towers, 619
experimental values, 620
H.E.T.S., application to continuous differential extraction towers, 746, 752
data for extraction, 748, 752
use in extraction, 729, 746
Hexagon, construction in and about circle, 60, 61
Hexagonal system of crystals, 1051
"High" contact, 1329
High-frequency dryers, 870
High-frequency induction furnace, 1761
High pressure, applications in industry, 1234
calculations, 352
check valve for, 1251
compressors, 1258, 1446
design, 1237
effect of temperature in design of, 1240
intensifiers, 1258
joints, 1243
liquid-level indicator, 1253
measurement, 1254
manometer, 1254
needle valve, 1250
packing for, 1247, 1258
pumps, 1258
reaction vessels, 1256
safety precautions, 1261
sight glass, 1252
quartz window, 1253
technique, 1233–1262
valves, 1250
relief, 1252
vessels, closures, 1243
design code, 1238
design and construction of, 1237, 1242
stresses in walls of, 1237
wall thickness of, 1237, 1240
High-low control, 1339
High-low signal contacts, 1268
High-speed Hoepner sewing head, 1391
High-starting torque control, 1673
High vacuum manometric range, 655
High-weir classifier, 926
Higher order reactions, 323
Hildebrand's rule, 300
Hills-McCanna (metering) pump, 1437
Hindered motion in settling, 996
Hindrance factor (dialysis membranes), 754
Hoepner 02 scale, 1385
sewing head, 1391
Hofer 1000-atm. compressor, 1259
H-O-H feeder, 1375
Holders, bag, 1389
Dustite bag, 1390
gas, 1380
Swellgrip bag, 1390
Holding time, in continuous-flow mixing tanks, 1230
of centrifugal pumps, 1204
Hollow-bowl centrifuges, 997
Hollow-bowl clarifiers, 999
Hollow-cone nozzles, 1171
Homoazeotrope, positive, 630
negative, 630
Homodyne detection, 1747
Homogeneous azeotropes, in distillation, 630
effect of pressure on composition of, 631
Homogeneous equations (calculus), 84
Homogeneous reactions, 321
Homogenization, 1167
Homogenizer, 1167, 1213, 1216
Homogenizer, Gaulin, 1167
Homogenizer, Manton-Gaulin, 1168
Homogenizer, Marco Flow Master, 1168
Homogenizer valve, Split-Flo, 1168
Homopolar generators, 1764
Hook gage, 367
Hooker cell, 1808
Hooke's law, 1256
Hoop stress, 1237
in compound cylinder, 1241
Hoop tension, 362
Hopper feeder, oscillating, 1374
Hopper scale, automatic, 1293
Horix rotary filler for liquids, 1397, 1398, 1399
Horix vacuum filling machine, 1399
Horizontal compressors, 1681
Horizontal conveying-screen dryer, 823
Horizontal crusher, 1129
Horizontal furnaces, 1598
Horizontal pipes, heat loss from, 476
Horizontal tanks, capacities of, 1380
Horizontal-tube evaporator, 501, 506
Horsepower, 40, 41, 44, 1732
brake, 1678
friction, 1678
indicated, 1678
Horsepower losses in belting, 1661
Horseshoe type mixer, 1205, 1222
Hose, flexible metallic, 431
rubber fabric, 439
Hoses, friction losses in fire, 1850
Hot cathode, gaseous-discharge tubes, 1751
Hot junction of thermocouples, 1272
Hot water pumps, 1416
Hot-well pumps, 1647
Hot-wire anemometer, 399
Hot-wire direction indicator, 400
Household refrigeration units, 1697
Kelvinator, 1697
Household water filter, 904
Housekeeping and safety, 1872
Howard crystallizer, 1063
dust chamber, 1021
H.T.U., for absorption of ammonia from air by water, 693
calculation of number required in countercurrent apparatus, 552
of column packing in gas absorption, 671
in countercurrent apparatus, 550
data, for ammonia in water, 688
liquor film for ceramic Berl saddles, 692
use in extraction, 744, 745
H.T.U. and H.E.T.P., experimental values, 620
for packed distillation towers, 619
Hubbell cell, 1793
Hudson belt cleaner, 1357
Human-hair filter mediums, 968
Humid heat, 759, 811
volume, 759
Humidification, 757*ff.*
adiabatic, 759
nozzles for, 1173
Humidifier(s), 779
dew-point humidity control in, 782
direct, 782
heads, 782
indirect, 779
with humidity control, 781
Humidity, 759, 800, 811
absolute measurement of, 1298, 1299
percentage of, 811
automatic control of, 1299
controller, 1299
measurement of, 1298
chart, for air-benzene vapor mixtures, 814
for air-CCl_4 vapor mixtures, 813
for air-toluene vapor mixtures, 815
for air-water vapor mixtures, 811
for air-o-xylene vapor mixtures, 816
for solvent vapors, 812
control for humidifier, 781
controlling devices, 787
critical, 1071
curve of, 811
determination of, 776
dew-point method, 776
hair-hygrometer method, 777
thermal conductivity method, 777
wet-bulb method, 777
effect of, in electrostatic separators, 1095
percentage, 759
ratio, for moist air, 760
relative, 759, 811
percentage absolute, 811
(relative) required for foods, 1699
Humidity-temperature relations in dryers, 812
Hum-mer screen, 957
Humphreys spiral concentrator, 1079
Hunter and Nash, calculating solvent extraction, 729, 732, 741
interpolating tie-line data, 724
oil solvent systems, 728
"Hunting," in process controllers, 1329
Hydrates, transition points of, 283
Hydraulic compression system, 1260
Hydraulic conveyors, 1360, 1361
applications, 1360
booster jets, 1361
sluicing, 1361
Hydraulic efficiency of pumps, 1431
Hydraulic Institute standards for pumps, 1424*ff.*
Hydraulic jolter, Syntron, 1395
Hydraulic pallet hand truck, 1368
Hydraulic power cylinder, 1328
transmissions, 1324

Hydraulic radius, definition of, 375
for various cross sections (table), 378
Hydrazine, densities of aqueous solutions of, 192
Hydriodic acid, partial pressures of, over HI, aqueous, 170
Hydro-Treator, Dorrco, 947
Hydrobromic acid, partial pressures of, over HBr, aqueous, 170
Hydrocarbon(s), aliphatic, constructional materials for, 1492
aromatic, constructional materials for, 1493
azeotropes, 635
heats of combustion of, 244–246
latent heats of evaporation of, 568
light-absorption curves, 1305
liquid, solubility of, in furfural, 647
molal heats of vaporization of, 217
solubility in oil, 676
vapor mixture, calcn. of dew point of, 587
variation of specific gravity with temperature, 536
Hydrochloric acid, constructional materials for, 1494
freezing points of aqueous solutions of, 281
furnace, 1622
partial pressures of, over HCl, aqueous, 167
specific heats of aqueous solutions of, 234
Hydroclone, 1036
Hydroelectric power-plant cycle, 1628
equipment, life of, 1632
Hydrofluoric acid, constructional materials for, 1495
Hydrofluosilicic acid, densities of aqueous solutions of, 179
Hydrogen, from coke-oven gas process, 1716
compressibility factors of, 205
electrode, 1309
electrolytic, 1795
as fuel, 1576
heats of combustion of, 244
liquefaction, Kapitza process for, 1707
and nitrogen, mixture, compressibility factors of, 209
overvoltage, 1787
temperature-entropy diagram for, 266, 267
water vapor content of, 173
Hydrogen bromide, densities of aqueous solutions of, 179
Hydrogen chloride, at 1 atm., aqueous, enthalpy-concentration diagram for, 268, 269
constructional materials for, 1494
densities of aqueous solutions of, 179
in water, absorption of, 697
Hydrogen fluoride, constructional materials for, 1495
densities of aqueous solutions of, 179
Hydrogen-ion, chart, 1308
colorimetric method of determination of, 283, 285
concentration and adsorption curves, 900
electrometric method of determination of, 285
indicator chart, 284
measurement of, 1307
setup, 1794
Hydrogen peroxide, constructional materials for, 1496
densities of aqueous solutions of, 179
Hydrogen sulfide-amine solutions, 677
removal from fuel gases, 1577
Hydrometer, 1297
Hydrostatic head meter, 1289
Hydrostatic paradox, 362
pressure, 360
Hydrostatics, 360
hypodermic needle tubing, 417
ideal fluid, 369
plastic, 369
Hydroxyl-ion concentration chart, 1308
Hygrometer(s), 1298
method of determining humidity, hair, 777
Hygroscopic material, 800
Hygroscopic moisture, carbon products, 780
cereal foods, 779
fibrous materials, 779
inorganic materials, 780
leather, 779
organic materials, 780
papers, 778
rubber, 779
synthetic fibers, 779
various materials, 778
Hygrostat, 786
differential, 786
Hylo Power unit, 1258
Hyperbola, 79
analytical problems, 57
area of, 57
conjugate, 79
construction of, 79
curves, 97, 98
definition, 79
Hyperbolic curves, graphs of, 81
Hyperbolic functions, inverse, 76
trigonometric, 76
Hyperbolic, natural, or Napierian logarithms, 50
Hyperbolic trigonometric functions, 76
Hypo, constructional materials for, 1518
Hypochlorite (sodium), constructional materials for, 1513
Hysteresis loss, 1736
Hysteretic constant values, 1737

I

Ice, manufacture, 1698
physical properties of, 1701
-salt mixtures, eutectic, 1700
Ideal-distribution law (extraction), 729
Ideal gas, 296
expansion and compression of, adiabatic, 303
isothermal, 302
heat capacity of, 297
laws of, 289, 335
pressure-volume relations of, 330
Ideal mixtures, relative volatilities of, 526
Ideal solution, 308
Ignition temperatures, 1582, 1584
apparent, 1866
Ignitrons, 1748, 1750, 1751
Illumination, 1755
of buildings, 1852
intensities of, 1853
utilization coefficient of, 1758
Immersion heater, control of, 1331
specific gravity and volume by, 363
test, 1458
Immiscibility, prediction of, 533
I.M.M. series of sieves, 963
Imp pulverizer, 1141
and grinding D.D.T., 1158
Impact nozzles, 1172
Impact tube, 397
multi-type for flow direction, 400
Impactor hammer crusher, 1129
Impax separator, 1030
Impedance, 1732
-capacitance-coupled amplifier, 1744
determination of, 1740
Impeller mixer, 1210, 1211, 1215, 1218
Impingement of dust
principles, 1022
separator, 1013, 1022
Implicit functions (calculus), 82
Impregnating, ejectors for, 1456
Impurities in gaseous fuels, 1577
Incandescent lamps, data on, 1758
Incentive methods, 1834
Inch, circular, 43
Inclined-tube evaporators, 502, 507
manometer, 1279
Inclined U-tube, 366
Indeterminate forms, 83
Indicated efficiency of pumps, 1431
Indicated horsepower, 1678
Indicating devices, 1267
and recording instruments, 1267, 1268
Indicator(s), acids, bases, and salts, 285
card, 1679
diagrams for dual compression systems, 1680
of flow direction, 400
and hydrogen-ion concentration, 283–285
Indirect-direct rotary dryers, 829
Indirect dryers, 813, 814
batch, 815
continuous, 815
Indirect-fired kettles (mixers), 1221, 1224
Indirect humidifiers, 779
Indirect rotary dryer, 859
Indoor pile storage, 1377
Induced e.m.f., direction of, 1737

Induced-roll magnetic separator, rings, 1093
Inductance, 1732
mutual-, and self-, 1732
Induction furnace, high-frequency, 1761
Induction generator, 1765
Induction motors, condenser-type, 1770
polyphase, 1768
repulsion-start, 1769
single-phase, 1769
Induction regulator, 1762
Induction-type instruments, 1738
Inductive capacity, 1732
circuits, 1732, 1734
Industrial adsorbents, 888
Industrial combustion systems, 1592
Industrial furnaces, 1598, 1600
Industrial-plant load curves, 1630
Industrial trucks, tractors, and trailers, 1362
Industries, classification of electrochemical, 1774
Industries, electrochemical, 1772
Inertia, moment of, for plane figures, 361
Inertial separators, 1022, 1028
scrubbers, 1038
Infilco Accelator, 947
Infinite series (algebra), 65
Inflammability, limits of, 1582, 1585
Information sources on plant location, 1725
Infrared dryers, 815, 868
spectrometry, 1305
Infrasizer, Haultain, 1085
Ingersoll-Rand mechanical seal, 1424
Initial boiling point, 606
comparison of dryers, 874
moisture distribution, 800
Injectors, as mixers, 1203, 1215, 1216
Inorganic compounds, boiling point of, 110–128
critical constants of, 204, 205
heat capacities of, 219–224
heats and free energies of, 236–243
heats of solution of, in water, 246–248
heats of vaporization of, 210–212
melting points of, 110–128
physical properties of, 110–128
refractive indexes of, 110–128
solubilities of, 110–128
in water, 196–199
specific gravities of, 110–128
vapor pressures of, above 1 atm., 149
up to 1 atm., 150–152
Inorganic materials, hygroscopic moisture of, 780
Inorganic salts, elevations of boiling point of aqueous solutions of, 282
spray-drying of, 846
Inorganic solutions, densities of, aqueous, 177–185
Input-output curves for power plants, 1632
Insecticides, grinding of, 1157
Installation, of process control, 1328
valve, 1327
Instant spray dryer, 843
Instrument(s), control, 1338
electromagnetic, 1738
recording electrical, 1739
transformers, 1739
functions of, 1265
terminology, 1266
maintenance of, 1336
for power plants, 1652
service departments, 1337
supervision of, 1336
Instrumentation, process, 1265
Insulating materials, density of, 1693
thermal conductivity of, 1693
thermal conductivity of (tables), 457–458
Insulation, electric, 1759
low-temperature, 1717
materials, hygroscopic moisture of electrical, 779
optimum thickness of, by McMillan chart, 480
thermal conductivity of, 457–458
values of building materials, 1693
Integral calculus, 85–90
Integral exponents, positive, 64
Integral functions, 85–87
Integrals, definite, 88
Integraph, mechanical, 89
Integrating meters, 1739
Integration, arithmetic, 92–93
graphical calculus, 90
single, double, and triple, 88
by substitution, 87, 89
volume determination by, 90
Integrator(s), 89
for circular charts, 89
flow, 1287
Intensifier, pressure, 1258
Intensity of combustion, 1587
Intercoolers for compressors, 1260
Interface, free energy of, 363
location of, gages for, 367
level indicator, 1289
Interfacial tension, 363
(table), 363
Intermediate agitation flow sheet, 952
Intermediate pressures in multistage compression, 1681
Intermittent absorption machine, 1685
settling tanks, 939
thickening, phases of, 938
Internal-combustion power plants, 1652
cycle, 1628
equipment, life of, 1632
Internal diffusion, 800
Internal moisture gradients, 801
and capillary flow, 801
and diffusional flow, 801
Internal resistance, 1743
Interpolation, 92
Interval, timer, 1326
timing, 1326
Invariant crystals, 1058
Inverse hyperbolic functions, 76
Inverse voltage, 1749
Inversion temperature, 333
Inverted plastic, 1198
Inverted plasticity, 1197, 1198, 1201
of starch suspensions, 1201
Investment costs as plant location factors, 1728, 1729
Investments, power plant, 1631
Involute, construction of, 62
Iodine, adsorption, by magnesia, 916
of from toluene solution, 892
constructional materials for, 1497
sublimation of, 663
Ion(s), exchangers, 915
gaseous, 1747
Ionic crystals, 1050
Ionization gage, 1281
gas, 1039
Ireland, weights and measures of, 44
Iron-constantan thermocouples, 1273
Iron ore, sampling of, 1101
Iron oxide(s), grinding of, 1157
process of gas purification, 1577
Iron and steel induction furnaces, 1816
Irreversibility of processes, 1704
Isenthalpic expansion, cooling in, 1702
Isentropic expansion, cooling in, 1702
Isobaric cooling, calculation, 337
Isobars, 330
Isobutane, in furfural, activity coefficients of, 643
saturated, thermodynamic properties of, 264
superheated, thermodynamic properties of, 264, 265
Isochores, 330
Isodynamic magnetic separator, Franz, 1093
Isoelectric point, 909
Isometric charts, 100
Isomorphism, 1050
Isopropyl alcohol, densities of aqueous solutions of, 191
Isotherm, 330
Isothermal expansion and compression, 302
streamline flow, in straight passages, 385
Isotopes of elements, 1781

J

Jack, lift, 1364
trucks, 1364
Jack truck, mechanical, 1364
Jacketed apparatus for evaporation, 505
Jacketed vessels, over-all heat transfer coefficients of, 482
Jar closing equipment, 1399
Jaw clutches, 1672
crushers, 1116, 1120
designs, 1125
performance, 1125
Jay Bee Pulverizer, 1141
Jeffrey, Baum-type jig, 1076
crushers, 1122, 1129
flocculator, 947
jig, 1076
single-roll crusher, 1128
-Steffensen three-drum magnetic separator, 1092
Swing Hammer pulverizer, 1140
weigh feeder, 1292

Jet(s), 1439
condenser, 520
mills, 1116, 1124, 1145
mixers, 1203, 1215, 1216
pipe for poitioning piston, 1321
-pulverizer, Blaw-Knox, 1146
pump, 1439
steam, 1682
-type pilot valve, 1321
velocity of steam, 1645
Jig, Baum, 1076
coal, 1076
Conset air, 1077
Cooley, 1077
Denver Mineral, 1077
feed, 1077
Hancock, 1077
Harz, 1076
Jeffrey, 1076
Jeffrey-Baum, 1077
for placer gold, 1077
Jigging, 1076
Job analysis, for trucks, tractors, and trailers, 1369
Job-cost systems, 1833
Job evaluation for wage control, 1835
Job specifications, 1835
Joint(s), bell-and-spigot, 419
Bridgman "unsupported area," 1246
cone, 1245
double-cone, 1247
flexible, 419
glass pipe, 436
for high-pressure apparatus, 1243
kinematic viscosity, 370
"lens-ring," 1245
metropolitan, 419
narrows siphon, 419
porcelain-pipe, 439
ring, 1246
-ring connector, 1252
tongue and groove, 1245
Vickers-Anderson, 1250
Ward wave ring pressure, 1246
Jolter, Syntron hydraulic, 1395
Jones oil-gas generator, 1581
Joule-Thomson effect, 203, 301, 333, 1706
of air, 203
of carbon dioxide, 203
of ethyl chloride, 203
in liquid-air cycle, 1709
of methane, 203
thermodynamic functions and, 301
inversion and degree of liquefaction, 1707
Joule-Thomson expansion, 1702
Journals, abbreviations of technical, 4
Jumbo swing-hammer mill, 1151
Jute filter mediums, 968
hygroscopic moisture of, 778

K

Kaolin, drying of, 826
grinding of, 1150
hygroscopic moisture of, 780
Kapitza expander, 1712
process of hydrogen liquefaction, 1707
Kay's method for gas mixtures, high-pressure calculations, 354
Kegs, filling and weighing, 1392
Keller furnace, 1815
Kellogg oxygen process, 1711
Kelly filter, 972
Kelvin temperature scale, 1269
Kelvinator refrigerator, 1697
Kennedy air-swept tube-mill, 1155
Kennedy-Van Saun crusher, 1126
Kent Maxecon mill, 1137
Kerosene, decolorization by adsorbents, 891
desulfurization by adsorbents, 913
as fuel, 1575
Kestner evaporator, 502, 506
Kettles, coefficients of heat transfer in, 1221, 1222
direct-fired (mixers), 1221, 1224
Dowtherm-heated (mixers), 1221, 1223, 1224
electrically heated, 1221, 1224
in evaporation, 505
heat transfer in, 1221–1224, 1229
heated, film removal in, 1221, 1222
indirect-fired (mixers), 1221, 1224
mercury-vapor heated, 1221, 1224
oil-heated, 1221, 1224
Kick's law of crushing and grinding, 931, 1114
Kieselguhr, hygroscopic moisture of, 780
Kiln(s), brick, 1605
capacity, 1611
-control system, 1335
downdraft, 1607
and furnaces, 1597–1626
rotary, 828, 831, 1608
size segregation in, 1612
thermal efficiency of, 1611
Kilowatt, 1732
Kilowatt-hour, 1732
Kinetic elutriator, Andrews, 1084
Kinetics, chemical reaction, 321
Kinetics of nucleation, 1057
Kirchhoff's laws, 299, 483, 1733
Kistiakowsky, equation of, 300, 341
Kjellin furnace, 1816
Kneader, 1207, 1208, 1220
vacuum, 1208
Kneading, 1220
Knife cutters, precision, 1161
rotary, 1160
Knit goods, drying of, 869
Knowles cell, 1795
Knudsen-Langmuir equation for evaporation, 658
Kominuter mill, 1133
Kraft paper, drying of, 868
Krystal classifying crystallizer, 1068
Kue-Ken balanced jaw crusher, 1120
Kutter's formula, 383

L

Labeling machines, 1412
Labor-cost control, 1834
Labor productivity, 1834
Labor surveys for plant location analysis, 1726
Laboratory distillations, 606
(*See also* under Molecular distillation)
Laboratory distilling columns, 609
Laboratory flotation cell, 1090
Lacquer clarification, 1011
Lacquer solvents, sampling of, 1102
Lactic acid, constructional materials for, 1498
Ladders, 1862
Lag, capacity, 1338
controller, 1338
dead-period, 1338
distance-velocity, 1338
flow, 1338
time lag, 1338
Lalande cell, 1789
Lamé formula, 1238, 1239
Laminar flow, 375
Laminated phenolic tubing, 438
Laminated vessels, 1243
Lampblack, grinding of, 1156
Lampwick drip feeder, 1374
Langmuir condensation pump, 658
Langmuir-Knudsen equation for evaporation, 658
Lapped joints, 441
Latent heat(s), 210–218
calculation of, 293, 341
of evaporation of paraffin hydrocarbons, 568
of fusion, 293
of solutions, 501
of solvent from solution, 300
of sublimation, 299
of transition, 299
of vaporization, 218, 293
Latex, drying of, 869
Lattices, space, 1050
Launders, feeders for, 1375
Laundry drying, 1011
Laurie cyclone, 1151
Law(s), Amagat's, 291
Boyle's, 289
Cailletet-Mathias, 292
Charles's, 290
conservation of mass, 289
Dalton's, 290
definite proportions, 289
Dulong and Petit's, 298
electrochemical, 1772
Faraday's, 1775
Henry's, 317, 532, 673
Hess, 344, 345, 346
ideal gas, 289, 335
ideal mixtures, 289
Kirchhoff's, 1733
Ohm's, 1733
of particle motion, 1019
Paschen's, 1747
as plant location factors, 1725
Raoult's, 317
Rowland's, 1736
simple distribution, 320
thermodynamics, 303, 313
Lea-Nurse air-permeability method of measuring surface, 1114
Leaching, 715, 929
application of methods in, 718
calculations for, 746
diffusion rates in, 742
equipment for, 746

Leaching, operations, 746
Lead, electrolytic refining, 1802
oxides, grinding of, 1157
pipe, 429
Lead arsenate, drying of, 826
Leaf tests, filtration, 968
Leak detector, 1307
Leather belting, 1660
hygroscopic moisture of, 779
packing rings, 1247
Le Carbone cell, 1792
Leclanché cell, 1789
Leeds & Northrup, air-operated controller, 1322
flyball governor, 1287
humidity controller, 1299
potentiometer circuit, 1278
radiation pyrometer, 1276
Lengths of circular arcs of circle, 33
from angle in degrees, 33
from chord and arc height, 33
Lens-ring joint, 1245
Lessing rings, pressure drop in, 395
Le Sueur cell, 1808
Level in closed vessel under pressure, control of, 1334
control in drum dryers, 863
differential-pressure control of liquid, 1334
Levin cell, 1795
Life of power plants, 1632
Lift, air, 1438
Lift jack, 1364
Lift of pumps, suction, 1415
Lifting-gate feeder, 1370
Lifting magnets, 1753
Lifting power of gas, 292
Light-absorption curves of hydrocarbons, 1305
Light oils in manufactured gas, 1578
Lighting of buildings, 1852
Lignin, spray-drying of, 845
Lignite, 1560
Lime, grinding of, 1154
hydrated, cleaning and concentrating, 1120
kiln, 1610
rock, grinding of, 1151, 1152
shaft furnaces, 1612
Limestone, grinding of, 1150, 1151
Limited mill, 1143
Limiting velocities for absorption columns, 680
Limits of flammability, 1582, 1585
Linde double column, 1709
Linde-Fränkl air-separation cycle, 1711
Linde-Fränkl oxygen plants, 1706
Linde oxygen process, 1707
Line contact, in joints, 1245
Line(s) coordinate charts, 105
equations of, 77
geometrical constructions of, 59
slope of, 77
Linear amplifiers, 1743
equations, 66
calculus, 84
Linear measure, conversion table and equivalents, 41
expansion of miscellaneous substances, 201
Linear measure, expansion of the solid elements, 200
metric system (table), 43
Linen, hygroscopic moisture of, 778
Liner insertion into containers, 1394
Lines of motion, 375
Linkages, flux, 1732
Link-Belt Roto-louvre dryer, 825
Liquefaction, critical constants, 291
of gases, 1702, 1707
and separation of, work of, 1703
Liquid(s), activity coefficients, 526
effect of temperature on, 529
classification of, 534
column manometer, 364
complex, 369
compressibilities of, 209
concentration, effect of, on plate efficiencies of distillation columns, 613
controls in refrigeration, 1692
cubical expansion of, 202
density, expansion factor for, 535
diffusion, 801
coefficients in, 540
diffusivity, 809
droplets, dispersion of, 1169
drops, evaporation from, 805
feeding of, 1374
filling and weighing equipment, 1395*ff*
(*See also* Fluids)
Liquid-crystal phase, 1051
Liquid-film resistance for O_2 desorption from water, 692
flow, types of, in bubble-cap towers, 599
Liquid fuels, 1568*ff*.
analysis of, 1629
vs. coal, 1570
miscellaneous, 1575
purchase of, 1569
requirements for, 1569
specific gravity of, 1569
specifications for, 1570
storage of, 1573
fugacity of, 307
and gas mixing, 1216, 1217
mixing, 1215–1216
and solids mixing, 1220
Liquid-gas ratio, in gas absorption, 668
Liquid-head-type speed indicator, 1296
level, differential-pressure control of, 1334
in tank under-pressure, 1289
measurement, 1289
proportioning, 1293
Liquid-level indicator for high pressures, 1253
-liquid extraction, 714, 718
equipment, 747
prediction of distribution in, 533
Liquid-liquid separation, 996
separators, 1000
Liquid-liquid-solid centrifuges, 1001
Liquid-liquid-solid separation, 997
lowering of freezing point of, by solute, 319
measure (table), 43
miscellaneous, specific heats of, 235
Liquid-liquid-solid separation, mixtures, binary, 319
mutual solubility, 318
natural-convection coefficients for, 474
Newtonian, 369
non-isothermal flow of, 381
non-Newtonian, 369
partially miscible, 318
Liquidometer hydraulic transmission, 1324
Liquid-phase adsorption, 914
resistance in packed columns, 691
solute distribution between, 320
Liquid-piston rotary blower, 1452
pumping of, 1414
rate, effect of, on pressure drop in packed towers, 393
rheostats, 1752
seal depth, effect of, on plate efficiencies of distillation columns, 613
Liquid-sealed bell, 1279
heat capacity of, 297
heat of vaporization, 299
hold-up, 686
ideal, 369
immiscible, 318
Liquid-sealed discharges, 1002
simple, 369
Liquid-solid extraction, 718
Liquid-solid separation, 994
solubility of gases in, 317
solids in, 319
solution of solids, rate of, 329
specific heats of, 228
storage of, 1378
systems ternary for which phase equilibrium data are available, 726, 727
thermal conductivity of, 459
Liquid-vapor equilibrium data in gas absorption, 668
of mixtures, 537
relations, 568
relations for CO_2-SO_2 at constant pressure, 572
vapor pressure of, 293, 299
Liquid-vapor relations, effect of pressure on, 569
viscosity of, 373–374
volume equivalents, 40
vortex contactor, 1037
warming of, inside tubes, 469, 471
outside tubes, 471, 474
water, effect of, in steam distillation, 583
Liquor, level of, effect of, on evaporation, 505
sulfite pulp liquor, evaporation of, 503
Liquor-film H.T.U. for ceramic Berl saddles, 692
Liquor-gas ratio in gas absorption, 707
Lithium chloride, for drying gases, 881
production, 1811
Load(s), circulating, 930
control for compressors, 1260
curves of power plants, 1630

Load(s), factor of power plants, 1630
of power plants, 1630
for ropes and chains, safe, 1874
Lithopone, classification of, 936
drying of, 826
filtration of, 988
grinding of, 936
Little, A. D., helium liquefier, 1708
Loading velocities, for tower packings, 683
Loadings for storage systems, floor, 1379
Local velocity, 397
Location, plant, 1719–1730
Logarithmic curves, 98
Logarithmic graph paper, 98
Logarithmic-reciprocal graph paper, 99
Logarithmic series (algebra), 64
Logarithms, calculations, examples, 50
characteristic of, 50
common Briggsian, 50
five-place, 6
definitions by kinds, 50
mantissa of, 50
natural, hyperbolic or Napierian, 50
trigonometric functions and numbers, 19
trigonometric functions on slide rule, 52
Loglog, duplex slide rule, 50–51
graph paper, 98
Long-tube natural-circulation evaporator, 502, 506
Loop dryer, 848
Loss, hysteresis, 1736
Loss-in-weight, feeders, 1293
hopper, 1292
Louvered bins, 1373
Louvers, 1328
"Low" contact, 1329
Low-Draft-Loss collector, 1030
Lowering of the freezing point, 319
Low-level condenser, 520
Low-lift platform truck, 1365
Low-melting-point alloys, 454
Low-side ring-roller mill, 1121
Low-temperature carbonization of coal, 1565
Low-temperature distillations, 609
Low-temperature insulation, 1717
Low-temperature processes, 1701, 1709
Low temperature, rectification at, 1708
Low-temperature refrigeration, production of, 1702
Low-temperature vacuum dryers, 854
Lubricants for power transmission equipment, 1673
Lubricants for refrigeration systems, 1682
Lubricating-oil purification, centrifuges in, 1009
Lubricating stock, purification by adsorbents, 893
Lubrication, of bearings, 1666
of centrifuges, 1005
of compressors, 1681
of power chains, 1673
Lubrication, of power-transmission equipment, 1673
Lucite tubing, 437
Lumber, hygroscopic moisture of, 780
Luminosity, of flames, 493
Luminous-flame burners, 1595
Lune, definition of, 58
Lurgi filter, 981
Lynch granular dust filter, 1029

M

Macaroni, hygroscopic moisture of, 779
McCabe-Thiele, calculation, 591
application of, 593
diagram, typical, 591
Machine, heating, 304
refrigerating, 304
sewing bags, 1390
Machinery, depreciation and obsolescence rates of, 1832
Machining, 1162
qualities of materials, 1539*ff*.
Maclaurin and Taylor formulas, 83
McLeod gage, 1280
MacMichael viscometer, 1201
McMillan chart, for optimum thickness of insulation, 480
Magnesia, activated, properties of, 916
as an adsorbent, 916
adsorption of iodine by, 916
adsorption of sulfur by, 916
regeneration of, 916
Magnesium, electrolytic production of, 1811
Magnesium carbonate, drying of, 826
spray-drying of, 845
Magnesium chloride, constructional materials for, 1499
densities of aqueous solutions of, 179
spray-drying of, 845
thermal conductivities of, 1697
Magnesium perchlorate, for drying gases, 881
Magnesium sulfate, crystallization of, 1067, 1068
densities of aqueous solutions of, 179
supersaturation curves of, 1056
-water system, enthalpy-concentration chart, 1052
phase diagram, 1053
Magnetic-drag tachometer, 1295
Magnetic and electric relations, 1737
circuits, 1736
conductivity, 1732
eddy currents, 1737
field intensity, 1732
flux, 1732
from electric current, 1737
Ohm's law, 1736
units, 1733
Magnetic reducing valve, 1252
Magnetic roasting, 1093
Magnetic separation, 1091
Magnetic separator, Jeffrey-Steffensen three-drum, 1092
Crockett and Linney, 1092
Franz, 1093
Magnetic separator, Granigg, 1092
ring-and-drum, 1092
rings induced-roll, 1093
Wetherill, 1092
Magnetization, normal, 1737
Magneto tachometer, 1295
Magnetomotive force, 1732
Magnets, 1752
Maintenance of instruments, 1336
of power plant, 1632
and safety, 1872
Maloney-Schubert diagram (extraction), 734
Malt extract, drying of, 866
Malted milk, drying of, 866
Manchurian mica, grinding of, 1151
Manganese, electrowinning, 1806
ore, drying of, 833
Manganese dioxide, dissociation pressures of, 174
Manganese sulfate, spray-drying of, 845
Mannheim furnaces, 1621
Manning formula, 383
average values of n for, 384
theory of overstrain in pressure vessels, 1240
Mannitol production, 1796
Manometer(s), 364, 1279
calibration of, 367
capillary error in, 365
floating bell, 1286
formed bell, 1286
formed tube, 1286
glass, 1267
high pressures, 1254
inclined, 366
liquid column, 364
mechanical pressure, 367
micro, 366
multiplying, 365
technique in use of, 365
tilting, 366, 1286
type liquid-level meter, 1289
(*See also* Gage)
Mantissa of logarithms, 50
Manton-Gaulin Laboratory Homogenizer, 1168
Maps for plant location analysis, 1726
Marco Flow Master Homogenizer, 1168
Marcy ball mill, 1133
grate-type continuous ball mill, 1123
rod mills, performance of, 1134
Margules equations, binary, 527
quaternary, 529
ternary, 528
Markets as plant location factor, 1724
Masonite process, 1166
Mass(es), action law of, 289
conversion table and equivalents, 40
law of conservation, 289
law of material balances, 289
per unit length, conversion table and equivalents, 40
weights and measures, metric system, 40, 43
Mass flow in sublimation, 661

Mass spectrometer, 1306
spectrometry, 1307
spectrum, 1307
Mass transfer, between phases, 549
by convection, 541
effect of pressure on rate of, 544
in extraction, 742
in leaching, 742
rates of, 742–746, 748–752
and heat transfer, by convection in tubes, 544
and fluid friction, analogy among, 541
in gases through granular solids, 547
inside wetted-wall tubes, 543
simultaneous, 557
under condition of large-concentration driving forces, 559
for system liquid water–air–water vapor, 558
over-all resistance in countercurrent apparatus, 549
in packed absorption columns, 686
rate of, 538
from cylindrical surfaces, 546
Mass velocity, mean, 374
superficial, 374
Massco-Adams reagent feeder, 1374
Masticator, 1208, 1220
Material(s), accounting of, 1841
balances, 337
applications to extraction, 729–746
in blast furnace, 1615
in dilution problems, 337
in distillation problems, 338, 625
in lime kilns, 348
in methanol converter, 357
use of, 337
of construction, 1457–1558
directory of, 1527
for electrochemical processes, 1820
for evaporators, 515
for high pressures, 1234–1237
pump, 1424
safety, 1856
handling equipment, 1343
factors in selection of, 1343
recommendations, 1344
miscellaneous, densities of aqueous solutions of, 194, 195
heats of fusion of, 217
specific heats of, 235
storage of, 1376–1381
for valves, 1328
Mathematical signs and symbols, 5
Mathematical tables and weights and measures, 1
Mathematics, 49–105
Mathias, Cailletet and, law of, 292
Mathieson Alkali high-pressure valve, 1251
Maxima and minima, 82
Maximum-boiling-point azeotropic binary mixtures, 634
mixtures, (diagram), 579
Maximum flow controllers, 1320
Maximum useful work or free energy, 305
Maximum work, 305
Maxwell, 1732
Mean free path, 656
Mean, geometric, 59, 66
harmonic, 66
hydraulic radius, 375
mass velocity, 374
temperature difference, 464
velocity, traverse for, 399
Means, controlling, 1338
Measure, angular, equivalents, 41
linear, conversion table and equivalents, 41
Measurement, angular, 70
differential-temperature, 1334
element of, 1266
electrical, 1737
of flow, 396–412, 1282
liquid-level, 1289
particle, 1013, 1017
of process variables, 1266
temperature, 1269
of three-phase power, 1738
of time, 45
Measures and weights, 43–44
English system, 44
metric system, 43
of various systems of different countries, 44
Measuring, electrodes, 1309
element, 1266
instruments, electric, 1276
Mechanical atomizing burner, 1574
Mechanical centrifugal separator, 1028
Mechanical continuous thickeners, 941
Mechanical efficiency of pumps, 1431
Mechanical energy balance, 376
Mechanical filter, 1032
Mechanical jack truck, 1364
Mechanical meter, 412
Mechanical power transmission, 1660–1674
Mechanical pressure gages, 367
Mechanical refrigeration, 1676
Mechanical sampling, 1100
Mechanical scrubbers, 1038
Mechanical seals for pumps, 1423
Mechanical shakers, 964
Mechanical stokers, 1567
Mechanical unloaders for centrifuges, 998
Mechanical vibrators, 991
Mechanically operated relay, 1323
Mechanisms, automatic control, 1320
recording, 1267
Medicinal carbons, 909
Medium, controlled, 1338
grinding, 934
Melting points, alloys, 281
elements and inorganic compounds, 110–128
organic compounds, 129–148
Membrane properties for dialysis, 754
Mensuration, geometry and, 53–62
Mercer-Robinson mills, 1144
unique cutters, 1161
Mercuric oxide, dissociation pressures of, 174
drying of, 826
Mercury-arc rectifiers, 1750, 1766
barometer, 365
chlorine-cell, 1806
density of, 176
in glass thermometers, 1267, 1270
manometer, float-type, 1285
-piston compressor, 1259, 1260
-pressure thermometers, 1270
saturated, thermodynamic properties of, 265
switches, 1324
-vapor heated kettles (mixers), 1221
-vapor lamp, 1756, 1758
data on, 1758
Meridian, 45
Merit rating, 1835
Merkel chart, use in evaporation, 501
Merrill precipitation press, 972
press, 972
Mesh, 955
Metacenter, 362
Metal(s), area, 408, 1288
bellows, 411
blocks, 283
current, 399
cycloidal, 411
diaphragm, 411
dilution, 412
dry gas, 411
electric, 1286
formed-displacer, 1286
impulse wheel, 412
mechanical, 412
melting furnaces, 1605
mixture, 412
molten, sampling of, 1102
nozzle, 400, 406
orifice, 400, 404
piston, 412
power-factor, 1739
prover, 1283
quantity, 410
rate-of-flow, 1283
rotameters, 408
rotary-disk water, 411
stearates, grinding of, 1160
thermal, 412
turbine, 412
V-notch, 954
vane, 412
venturi, 400, 406
weir, 1288
wet gas, 410
Metal(s) **and alloys,** composition of, 1526*ff.*
bearing, 1666
corrosion by corrosive gases, 1526
electrolytic refining of, 180
powders, grinding of, 1119
properties of, 1460, 1539
sampling of, 1101
thermal conductivity (tables), 456
Metal-adsorbent chars, 909
Metal-fabric filter mediums, 968
Metal-joint cements, 453
Metallic conductors, 1772
Metallurgical concentrates, filtration of, 985, 986
furnaces, 1605
Metering, by dilution, 412
fillers for liquids, 1397

Metering, and proportioning pumps, 1437
Methane, compressibility factors of, 207
Joule-Thomson effect of, 203
saturated, thermodynamic properties of, 265
superheated, thermodynamic properties of, 270, 271
Methanol-water flow points of, 1696
freezing points of, 1696
specific heats of, 1696
thermal conductivities of, 1697
viscosities of, 1696
Methods of fabrication of materials, 1539*ff*.
Methods of vaporization, 580
Methods used in plant location, 1727
Methyl alcohol, densities of aqueous solutions of, 187
partial pressures of, over methyl alcohol, 172
specific heats of, 235
Methyl chloride, compressibility factors of, 208
saturated, thermodynamic properties of, 271
superheated, thermodynamic properties of, 271, 272
Methyl formate, saturated, thermodynamic properties of, 272
Methyl isobutyl ketone-acetic acid-water system, 751
Methyl methacylate pipe and tubing, 437
Methylamine, saturated, thermodynamic properties of, 271
Methylcyclohexane-aniline-n-heptane system, 722, 723
Methylene chloride, saturated, thermodynamic properties of, 272
Metric and avoirdupois units, equivalents of, 1781
Metric and troy units, equivalents of, 1781
Metric Gravitometer, 1298
Metric system, weights and measures, 43
Metropolitan pipe joint, 419
Mho, 1732
Mica, grinding of, 1151
Michels 2500 atm. compressor, 1259
Microfarad, 1732
Micromanometers, 366
Micron, definition, 41, 43, 1196
Micronizer, 1145
fluid-energy or jet will, 1124
for graphite grinding, 1156
Microporous rubber filter medium, 968
Microscopic particle-size measurement, 1113
Mier's theory of crystallization, 1055
Mikro-Atomizer, 1141
and grinding D.D.T., 1158
operating principle, 1143
Mikro-Chipper, 1163
Mikro-Collector, 1034
Mikro-Pulverizer, 1140
and grinding D.D.T., 1158
hammer mill, 1123
Mil, circular, 43
Mil-foot, circular, 1732
Milk, drying of, 866
skim, spray-drying of, 846
Mill(s), 930
ball, 1220
colloid, for mixing, 1212
roller, 1074
three-roll sugar, 1074
Mill feeders, 1134
Milling machines, 1163
Millivoltmeters, 1276
Milton Roy feeder, 1374
Mine & Smelter Supply crusher, 1123
mills, 1133
Mineral(s), black, grinding of, 1156
grinding of, 1149
matter in coals, 1562
oils, filtration of, 990
pigments, grinding of, 1157
Miner's inch, definition of, 43
Minima and maxima (calculus), 82
Minimum boiling-point azeotropic binary mixtures, 633
mixtures, (diagram), 579
flow controllers, 1320
plates at total reflux (distillation), 623
reflux ratio, 591, 623
Minutes to radians, conversion tables and equivalents, 36
Mist (*see* Dust)
Mists and dusts, vapor pressure of, 293
Mixco turbine, 1179
Mixed acid, constructional materials for, 1500
Mixed-flow pump, 1422
Mixer(s), air-lift, 1207, 1220
arm, 1204, 1205
baffle-column, 1215
ball or pebble mill, 1212, 1220
blade, 1205
bubbler, 1217
centrifugal fan, 1210
centrifugal pump type, 1203, 1204
circulating, 1203, 1223
coefficient of heat transfer in, 1221
colloid mill, 1212, 1216
comparison of types, 1217
compressed air, 1215
cooling of, 1224
double-cone, 1212
double motion, 1206
dough, 1220
Dowtherm heated, 1224
edge runner, 1214
electrically heated, 1224
emulsifying, 1206
for extraction equipment, 747
Feld gas scrubber, 1213
fitting, to operation, 1214
flow, 1203, 1204
for gas and liquid mixing, 1203, 1208, 1215, 1217
for gas mixing, 1208, 1215
gate, 1205
heat transfer in, 1221, 1223
Herreshoff, 1205
homogenizer, 1213, 1216
horseshoe, 1205
Mixer(s), injector, 1203, 1215
jet, 1203, 1215, 1216
kneader, 1207
for liquid mixing, 1215
for liquid and solid mixing, 1217
masticating, 1208
mercury-vapor-heated, 1224
miscellaneous, 1212
mixing rolls, 1214
mushroom, 1212
nozzle, 1203
oil-heated, 1224
orifice column, 1203
packed towers, 1203
paddle, 1204, 1205, 1206, 1215, 1217
pan, 1214
for paste mixing, 1220
planetary, 1206
for plastics mixing, 1220
pony, 1206
portable propeller, 1209
power input to paddle, 1225
power required, for colloid mill 1216
for double cone, 1212
for paddle, 1215, 1225
for propeller, 1225, 1226
for turbine, 1225, 1226
power transmission, 1226
propeller, 1208–1210, 1215, 1218
pug mill, 1209
pump type, 1216
rake, 1205
revolving-cone, 1213, 1219
ribbon, 1210
rotating pan, 1206
side-entrance propeller, 1209
soap crutcher, 1210
for solid mixing, 1221
for solids and gas mixing, 1221
tower, 1204
transmission of, to mixer, 1226
traveling, 1205, 1219
tumbling, 1212
barrel, 1212
turbine, 1210, 1215, 1218
turbo-, 1217
turbo-disperser, 1211
turbo-gas-absorber, 1211, 1217
turbulence, 1203
turbulent-flow tubes, 1216
types of, 1202 et seq.
vaned disk, 1217
Votator, 1213
whipper, 1206
Mixing, analysis of action in bread dough, 1220
batch vs. continuous, 1230
between thickeners, 954
continuous, 1230
degree of, 1195
effect of, 1195
fundamentals of, 1195
gases with gases, 1203, 1215
heat transfer in, 1221, 1229
immiscible liquids, 1229
liquids with gases, 1216, 1229
liquids with liquids, 1215
liquids with solids, 1217, 1220
miscellaneous functions of, 1229
objectives of, 1195, 1196

Oil(s), petroleum, specific heats of, 567
thermal conductivity of, 460
purification by adsorption, 890
purifiers, costs of, 1006
Oil-filled transformers, 1763
Oil-heated kettles, 1221, 1224
Oiling and greasing, safety in, 1855
Oilless bearings, 1667
Oleates, grinding of, 1160
Olefins and acetylenes, absorption of, in aqueous copper solutions, 706
solubility in cuprous salt solutions, 679
Oleic acid, constructional materials for, 1503
Oleic acid abietic acid-propane system, 722, 724
Oliver-Borden, filter thickener, 948
filter, 976
horizontal filter, 981
pipe-line strainers, 984
precoat filter, 979
pressure filter, 988
top-feed filter, 980
Omega Machine feeder, 1374
Open channels, determination of level in, 367
flow in, 377
hydraulic radius in, 377, 378
Manning formula for, 383
Open-circuit continuous grinding, 1117
Open-circuit grinding, 930
Open-end rod mill, 1133
Open-hearth regenerators, 1624
Open-hearth steel furnace, 1620
Open-impeller centrifugal pumps, 990
Open-kettle temperature-control system, 1329
Open- vs. closed-circuit grinding, 931
Operating aisles in factories, 1369
Operating costs, as plant location factors, 1728, 1729
Operating diagram in sublimation, 662
Operating difficulties in evaporation, location of, 519
Operating of evaporators, 509
Operating lines in countercurrent apparatus, 548
Optical pyrometers, 1270, 1274, 1275
automatic, 1275
Morse, 1275
use for measuring flame radiation, 494
Optimatic pyrometer, 1275
Optimum final temperature difference, in condensers and coolers, 479
in heat exchangers, 479
Optimum velocity in heat exchangers, 479
Order of reaction, 321
apparent, 324
determination, 327
first order, 321
Ore(s), grinding of, 1148
sink-and-float concentration of, 1080
Organic chemicals, spray-drying of, 846
Organic compounds, critical constants of, 204, 205
heats and free energies of, 236–243
heats of fusion of, 213, 214
heats of solution of, in water, 248
heats of vaporization of, 215, 216
melting point of, 129–148
miscellaneous, densities of aqueous solutions of, 192, 193
physical properties of, 129–148
solubility of, 129–148
specific gravity of, 129–148
vapor pressures of, above 1 atm., 165, 166
up to 1 atm., 153–165
Organic liquids, specific heats of, 225–227
Organic materials, drying of, 828
hygroscopic moisture of, 780
Organic solids, specific heats of, 230–232
Organic solutions, densities of, aqueous, 186–193
Orifice(s), 389, 400, 404
accuracy of, 407
columns as mixers, 1203, 1215, 1216
formula for liquid flow, 401
free discharge from, 406
gas burners, 1590
gas flow through, 401
general principles of, 400*ff.*
location relative to fillings, 407
meters, 400
seals and purges for, 1287
taps, 404
nomenclature for formulas, 401
in pipes, 404
plates, 1283
pressure drop across, 388
pressure recovery following, 404
rounded, 407
segmented, 406
sharp-edged, 404
square-edged, 404
theoretical formulas, 401
unsubmerged, operation of, 404
working formula for, 403
(*See also* Nozzles)
Orifice-type viscometer, 1301
Orlon acrylic fiber, 1034
Orsat analyzer, 1303
Orthorhombic system of crystals, 1051
Oscillating hopper feeder, 1374
Oscillating mills, 1130
Oscillating packer, 1383, 1385
Oscillating screens, 957
Oscillating series, algebra, 65
Oscillation, 1338
Oscillators, 1745
Hartley, 1745
Othmer method, 294, 301
for estimation of heats of solution, hydration, dissociation, 342
for estimation of latent heats of fusion, 294, 300
for estimation of latent heats of vaporization, 294, 300
for estimation of vapor pressures, 294
Othmer-Tobias tie-line correlation plot for extraction, 724
Otto cycle, 1653
Outage chart for flammable liquids, 1875
Ovens, 1598
box-type electric, 1759
by-product coke, 1621
Over-all heat-transfer coefficients, 480–482
graphic interpretation of, 470
Over-all temperature difference, 464
Over-firing furnace, 1607
Overflow, 930
Overhead clearances, 1849
Overhead costs, 1839
Overhead storage systems, 1376
Overlapping principle, 1058
Overstrain of steels, 1236–1237
Overstrom mud screen, 957
Overvoltage, 1787
Oxalic acid, constructional materials for, 1504
densities of aqueous solutions of, 187
Oxidation of aluminum, anodic, 1801
Oxidation-reduction potential, measurement of, 1309
Oxygen, from air, 1703
compressibility factors of, 205
desorption from water, liquid-film resistance for, 692
effect on metal corrosion, 1459
electrolytic, 1795
enrichment, effect on gas producer, 1616
hazard due to oil, 1262
low-cost tonnage, 1715
meter, 1303
overvoltage, 1788
plant costs, 1715
process, 1705
Elliott, 1711
M. W. Kellogg, 1711
saturated, thermodynamic properties of, 273
superheated, thermodynamic properties of, 273
from water, desorption of, in spray absorber, 700
Oxygen-nitrogen mixture at 1 atm., enthalpy-concentration diagram for, 272
vapor-liquid equilibrium relations at const. pressure, 575
Oyster shells, grinding of, 1151, 1152
Ozone production, 1820

P

Package wrapping, 1406
"Packaged" fuels, 1564
Packages, and filling equipment, 1381*ff.*
packaging equipment for shelf, 1397
by weight, 1294
Packed absorbers, over-all resistance, 693–698
Packed-bed collectors, 1013
Packed-bed separators, 1029
Packed columns, extraction, flooding data for (Colburn), 753

Packed columns, extraction, gas-phase resistance, 687
liquid-phase resistance, 691
as mixer, 1217
Packed distillation towers, 616
Packed scrubbers, 1038
and spray towers for extraction, 748
estimation of capacity coefficients, 751
rate data for, 750
Packed towers, pressure drop in, 393
flow through, 393
vs. plate towers in gas absorption, 707
rectifying towers, 597
Packed tubes, heat transfer to gases in, 469
Packer(s), auger, 1384, 1385
belt-type, 1388
Carter Vacuum, 1394
oscillating, 1383
Risco, 1385
shifting-tube belt, 1388
Triangle Package Machinery Co., 1403
U.S. Automatic Box Machinery Co., 1403
valve-bag, 1387
Vibrox, 1395
Packing(s), Bridgman, for high pressures, 1246
piston rod, 1247, 1248
Poulter, 1248
for pressure pumps, 1258
rings, leather, 1247
plunger, 1247
for stirrer shafts, 1247
column, Stedman, 608
and columns in gas absorption, 707
dimensions, effect of, on packed distillation towers, 619
dry materials into containers, 1392
element, 1713
liquids, temperature control for, 1396
materials, costs of, 709
for packed distillation towers, 617
piston-rod, 1446
and shaft seals, 1420
supports, 710
Paddle mixers, 1204, 1205, 1215, 1217, 1222
double-motion, 1206
Paint(s), grinding, use of surface-active agents in, 1115
spraying, 1173
standard curves for classifying, 1201
Pak-ice, 1700
Pallet(s), 1367
transportation, trailer-train, 1367
trucks, hand, 1369
Pan crushers, 1116, 1126
Pan dryers, agitated, 856
atmospheric, 857
vacuum, 857
Pan mixer, rotary, 1206, 1214
Paper, drying of, 869
hygroscopic moisture of, 777
pulp, filtration of, 986
Paper, stock, critical moisture content, 867
Paperboard, drying of, 868
Parabola, analytical problems, 79
area of, 57
construction of, 62
interlaced, 91
polar equation, 79
Parabolic curves, graphs of, 97, 98
Parabolic plug valves, 1327
double-seated, 1327
Paraboloid, of revolution, volume of, 58
Paradox, hydrostatic, 362
Paraffin gases, sp. hts. of, 567
Paraffin wax, decolorization of by adsorbents, 892
Parallel-connected amplifier, 1745
circuit, 1733
circuits, magnetic, 1736
current condenser, 520
feeding of evaporators, 510
Parallelogram, 55
Parks-Cramer humidity controller, 1299
Partial centrifuging, 1004
Partial condensation, 563
Partial derivatives, 82
Partial differential equation, 84
Partial molal properties, 308
Partial pressures, CH_3OH over methyl alcohol, aqueous, 172
HBr over HBr, aqueous, 170
HCl over HCl, aqueous, 167
HI over HI, aqueous, 170
HNO_3 over HNO_3, aqueous, 169, 170
H_2O over HBr, aqueous, 170
H_2O over HCl, aqueous, 166
H_2O over HNO_3, aqueous, 169, 170
H_2O over H_2SO_4 aqueous, 169
H_2O over methyl alcohol, aqueous, 172
H_2O over NH_3, aqueous, 171
H_2O over sodium carbonate, aqueous, 172
H_2O over sodium hydroxide, aqueous, 173
H_2O over SO_2, aqueous, 167
of water vapor, 758
H_2SO_4 over H_2SO_4, 169
NH_3 over NH_3, aqueous, 172
SO_2 over SO_2, aqueous, 167
SO_3 over H_2SO_4, fuming, 169
Particle(s), atmospheric, 1017, 1047
centrifugal force on, 992
charging of, 1040
classification, 1016
dynamics, 1013, 1017
hazards, explosion, 1016
health, 1017
measurements, 1013, 1017
mobility, 1040
motion, in electrical field, 1040
in fluids, 1017
paths, 375
properties, 1016, 1019
sampling, 1017
size, 1016, 1019
average, 1113
classifiers, 1135
Particle(s), size, distribution, 955
in atomization, 840
as factor in mixing, 1202
measurement, 1111, 1112, 1113
representation, 1017, 1113
Zsigmondy classification of solid. 939
Partition, law of, 320
Partition rings, costs of, 709
Pascal's principle, 360
Paschen's law, 1747
Passivity, 1786, 1788
Pastes, mixing of, 1220–1221
Pauling process, 1820
Pay roll, 1836
Pay-roll accounting, 1834
records, 1834
Peabody scrubber, 1036
Peak-grid voltage vs. power output, 1743
Pease-Anthony scrubber, 1036
Peat, as fuel, 1568
Pebble heater, 1617
furnace, 1612
Pebble mill(s), 1116, 1130
for grinding refractory siliceous materials, 1150
mixer, 1212, 1220, 1221
Pebble-stove destencher, 1625
Peck carrier, 1358
Pelleting, 1189
Peltier e.m.f., 1272
Pendular state, 800
Pendulum rotary compressors, 1681
Pennsylvania Crusher, 1129
Pens for recorders, 1267
Pentagon, construction in circle, 60
Per cent relative humidity, 811
Percentage, absolute humidity, 811
humidity, 759
precision, 50
relative humidity, 759
Perchloric acid, densities of aqueous solutions of, 181
Percolation, 888
Perfect-discharge elevators, 1349
Perfect fluid, 369
Perforated-pipe feed, 863
Perforated-plate columns (extraction), 752
extraction rates for, 753
Performance coefficients of refrigerants, 1678
screen, 960
Perihelion, 45
Period, constant-rate, 800
falling-rate, 800
Periods of drying, 802–808
constant-rate, 802
falling-rate, 806
Peripheral speeds of rotary dryers, 832
Perkins dryer, 824
Permanent magnets, 1752
Permeability, 1732, 1736
of fabrics, 1030
particle-size detn. by, 1112
Permutations and combinations, 66
Peroxide (hydrogen), constructional materials for, 1496
Petit, Dulong and, law of, 298

Petroleum, criteria of separation, 603
distillation, 602
furnaces, 1605
arrangements, 1606
gases as fuels, 1576
heaters, heat transfer in, 496
measurement and electrolysis, 1793
meter-controlled feeder, 1376
oils, decolorization of, by adsorbents, 892
sp. ht. of, 567
thermal conductivities of, 460
pH, controlling equipment, 1309
electrode assembly, 1309
measurement of, 1307
recording equipment, 1309
products, filtration of, 987
sampling of, 1101
refineries, centrifuges in, 1008
refining, 914
Petroleum coke as fuel, 1568
Phanotron, 1751
Pharmaceuticals, grinding of, 1158
Pharo cyclone, 1151
Phase, change in sublimation, 661
diagram(s), 315
for binary mixtures, 577
for extraction, effect of temperature, 721
$MgSO_4$-H_2O system, 1053
temperature-composition for partly miscible liquids, 318
ternary, with solid components, 725
triangular, 719
types of, 720
for water, 316
equilibriums, in condensed systems (extraction), 718
and solutions, 315
ternary condensed systems, data for, 726, 727
relations on triangular diagram, 719
rule, 315
applications of, 315–320
use in distillation, 576
Phase-wound rotors, 1768
Phases of sedimentation, 938
Phenol, constructional materials for, 1505
Phosphate(s), grinding of, 1151, 1152
pebble, drying of, 833
sand, cleaning and concentrating, 1120
Phosphoric acid, aqueous, vapor pressures of, 167
constructional materials for, 1507
continuous countercurrent decantation, 951
densities of aqueous solutions of, 181
filtration of, 988
specific heats of aqueous solutions of, 234
Phosphorus production, 1820
Photoelectric tubes, characteristics of, 1741
Photoelectricity, 1742
Photographic emulsion, drying of, 869
Physical and chemical constants, numerical fundamental, 47
of the elements and inorganic compounds, 110–128
organic compounds, 129–148
of pure substances, 110–148
Physical and chemical principles, 287–358
Piazza scrubber, 1038
Pickling, electrolytic, 1800
Pie(circular percentage) graph paper, 99
Piezoelectricity, 1742
Piezometer, opening, 396
rings, 397
taps, specifications for, 397
Pig iron, sampling of, 1101
Pigment(s), filtration of, 988, 990
grinding of, 1157
inorganic, drying of, 828
lacquer clarification, 1011
mineral, grinding of, 1157
spray-drying of, 845
white, grinding of, 1157
Pile storage, 1376
Pilot-controlled controller, 1320, 1321
Pipe, 413–450, 1430–1435
aluminum, 424
asbestos cement, 431
bends, 389
bimetallic, 430
block-tin, 430
brass, 428
bronze, 428
carbon, 431
cast iron, 418–423
cement-lined, 434
cements and solders for, 453
chemical ware, 434
clay sewer, 435
concrete, 435
copper, 425–429, 1692
alloy, 429
cost of, 384
diameter, economic (chart), 386
Durichlor, 423
Duriron, 423
economic diameter, 384
ferrous, 413
fittings, 389, 441
flanges, 441
flow in, 377
alignment chart, 379
flow chart, 379
friction in, glass, 436
alignment chart, 379
table, 1430
glass-lined, 436
graphite, 431
Haveg, 437
joints, 441
lead, 429
lead-lined, 429
linear expansion of, 413
lines, refrigeration, 1692
Lucite, 437
metal hose, 431
nickel, 430
alloy, 430
nipples, 442
non-ferrous, 424–431
non-metallic, 431–440
plastic, 436–438
Pipe, porcelain, 438
precipitator, blast-furnace, 1042
rubber, hard, 438
rubber-lined, 438
sampling, 1100
Saran, 438
silica, fused, 440
stainless, 417
standards, 413
steel, 413, 1650
stills, 1606
supports, 453
and tanks (wood), manufacturers of, 1553
taps, 405
tin, 430
unusual cross sections, 385
volumes of, 1379
wood, 440
wood-lined, 440
wrought iron, 413
wrought steel, 413
Pipe-line taps for orifices, 1284
Piping, for brine, 1692
for high pressures, 1250
for liquid fuels, 1573
for power plants, 1649, 1650
for process control, 1328
for refrigerant storage, 1693
systems, color identification of, 1874
Pirani gage, 1281
Piroomov and Beiswenger correlations (distillation), 587
Piston gage, dead-weight, 1255
Piston meters, 412
Piston-rod packing, 1247, 1446
Piston variable area-meter orifice, 1288
Pit-cast cast-iron pipe (tables), 419
Pitch, pulverizing of, 1156
Pitometer, 398
Pitot tubes, 397, 1283, 1289
Pivoted-belt weigh feeders, 1292
Pivoted-bucket carrier, 1359
Plait point, 720
Planck's law, 484, 1274
Plane geometry, theorems 53–54
Plane surfaces, heat transfer to fluids flowing along, 473
Plane triangles, solution of, 72
Plane trigonometry, 70–74
Planimeter, Amsler's polar, 89
Plant dryers, correlation of data from, 810
tests on, 809
Plant location, 1719–1730
aids in analysis of, 1726
aim of, 1720
factors of, 1720
company policies, 1725
fuels, 1721
geography, 1723
information sources on, 1725
labor surveys, 1727
laws, 1725
maps, 1726
markets, 1724
municipal restrictions, 1725
national defense, 1729
power, 1722
public practices, 1725

Plant location, factors of, public utilities, 1725
raw materials, 1720, 1726
transportation, 1724
water, 1723
methods of, 1720
references on, 1719
steps in determining, 1727
Plastic(s), definition of, 369
equipment, manufacturers of, 1535
granules, drying of, 869
inverted, 1198
mixing of, 1220–1221
pipe, 436
solid, 369
stage in coal carbonization, 1564
Plasticity, 1197
inverted, 1197
of mixtures, 1197*ff*.
of starch suspensions, 1201
pseudo-, 1197
Plastico-viscous solid, definition, 369
Plastograph, Brabender, 1201
Plate(s), calculations for multicomponent mixtures, 622, 654
efficiencies of bubble-tray absorption columns, 698
countercurrent apparatus, 550
distillation columns, 610
data on hydrocarbon systems, 615
data on non-hydrocarbon systems, 614
(Murphree) of gas absorbers, 698
fasteners for belts, 1664
precipitator, vertical-flow, 1042
punched metal, 958
resistance, 1743
spacings in bubble-cap towers, 598
system of ice manufacture, 1698
towers vs. packed towers in gas absorption, 707
Platen-Munters continuous absorption refrigerator, 1684
Platen press, 1073
Platform truck, hand electric, 1369
high-lift, 1365, 1366
low-lift, 1365, 1366
Plating baths, 1797
formulas, 1798
Plating-tank control, 1331
Platinum-platinum rhodium thermocouples, 1273
Plug valve, double-seated parabolic, 1327
parabolic, 1327
ratio, 1327
Plunger feeder, 1371
packing rings, 1247
pumps, 990
Plywood sheets, drying of, 869
Pneumatic balance system, 1322
Pneumatic conveying dryers, 834
Pneumatic conveyor, Dracco, 1360
Pneumatic diaphragm motor, 1328
Pneumatic-electric automatic-shutdown system, 1335
Pneumatic power cylinder, 1328
Pneumatic Scale Corp., air cleaner for bottles, 1398
rotary capper, 1400
vacuum filler, 1397
Pneumatic telemeter, Republic, 1324
telemetering, 1324
Pneumatic-tire fork truck, 1366
Pneumatic transmission, 1333
for remote control, 1333
transmitter, Brown, 1324
Republic, 1324
Pneumatics, 360
Pocket feeder, 1371
Podbielniak contactor for extraction, 748
Podbielniak's distilling apparatus, 610
Point, Boyle, 290
of compression (sedimentation), 938
control, 1338
dew, 759
efficiency, relating to over-all Murphree efficiency, 552
gage, 367
Point-to-plate rectifier, 1751
Poise, 1197
Poison antidotes, 1876
Poisson's ratio in designs for high pressures, 1239
Polar coordinate(s), 77
graph paper, 99
Polar planimeter, Amsler's, 89
Polarization, 1786
Polarizing optical pyrometer, 1275
Polarography, 1307
Polishing, electrolytic, 1801
Polls, commutating, 1764
Pollution, atmospheric, 1017, 1047
Polygon, construction, 61
spherical, 58
spherical excess of, 58
Polyhedral angle, 56
Polymer, drying of, 821
Polyphase Duplex Trig slide rule, 51
Polyphase motor, brush-shifting, 1768
induction, 1768
reversing of, 1768
synchronous, 1767
Polyphase power, 1735
Polyphase systems, electrical, 1735
Polyphase wattmeter, 1738
Ponchon, design method for extraction, 729, 730, 731, 734
Pony mixer, 1206, 1220
Porcelain, equipment, 1549
filters, 968
miscellaneous, specific heats of, 235
pipe, 438
joints, 439
towers, costs of, 709
Porous-block filters, 968
Porous carbon and porous graphite, physical properties of, 1550
Porous septa, characteristics, 1177
Porous sparger mixer, 1217
Porrion evaporator, 505
Portable propeller mixer, 1209
Portable scales, 1291
Portland cement, grinding of, 1152
Position action, average-, 1330
Position control, proportional, 1321
Positioner, valve, 1329
Positive-displacement meters, 1282
Positive homoazeotrope, 630
Pot furnace, 1607
Pot press, 1073
Potassium bicarbonate, densities of aqueous solutions of, 181
dissociation pressures of, 174
Potassium bromide, densities of aqueous solutions of, 181
Potassium carbonate, densities of aqueous solutions of, 181
dissociation pressures of, 174
Potassium chlorate, densities of aqueous solutions of, 181
Potassium chloride, densities of aqueous solutions of, 181
specific heats of aqueous solutions of, 234
Potassium chromate, densities of aqueous solutions of, 181
Potassium chrome alum, densities of aqueous solutions of, 182
Potassium dichromate, densities of aqueous solutions of, 182
Potassium dihydrogen phosphate, dissociation pressures of, 174
Potassium hydride, dissociation pressures of, 174
Potassium hydroxide, densities of aqueous solutions of, 182
specific heats of aqueous solutions of, 234
Potassium nitrate, densities of aqueous solutions of, 182
Potassium persulfate, drying of, 837
Potassium production, 1812
Potassium sulfate, densities of aqueous solutions of, 182
Potassium sulfite, densities of aqueous solutions of, 182
Potential, single-electrode, 1785
Potential difference, 1732
minimum sparking, 1747
Potentials, electrode, 1785
Potentiometer, 1276, 1277, 1278, 1739
circuit, 1739
control of motors, 1767
Poulter, high-pressure window, 1253
plunger packing, 1247, 1248
Pounds per gallon—cubic foot equivalents, 38
Pour point of liquid fuels, 1570
Powder metallurgy, 1189
Power(s), algebraic, 63
by logarithms, 50
by slide rule, 52
apparent, 1732
consumption of centrifugal disk atomizers, 848
conversion table and equivalents, 41
cycles, by-product, 1636
heat balance of, 1635
cylinder, hydraulic, 1328
pneumatic, 1328
electric, 1732, 1733, 1735
cost, 1755
factor, 1732, 1735
meter, 1739
measurement, 1736
output vs. peak-grid voltage, 1743

Power(s), electric, polyphase, 1735
for electrochemical processes, 1824
generation of, 1627–1660
annual cost of, 1633
lifting, of gas, 292
machinery, 1628
as plant location factor, 1720
saving in multistage compression, 1681
shafting bearings, 1667
transmission, mechanical, 1660–1674
units for process control, 1328
used on filters, 989
Power plant, cycles, 1628
equipment, life of, 1632
input-output curves for, 1632
investments, 1631
operation, 1657
piping, 1650
thermal performance of, 1630
water, 1655
Power requirements, for brine circulation, 1697
for centrifuges, 1005
for colloid mill mixer, 1216
for crushing and grinding, 936
for double-cone mixer, 1212, 1221
for dough mixer, 1208, 1220
for homogenizer mixer, 1216
input vs. degree of mixing, 1195, 1205
for kneader mixer, 1208, 1220
for liquid mixer, 1215
of low-temperature processes, 1703
for masticator mixer, 1220
for mushroom mixer, 1220
for paddle mixer, 1205, 1207, 1215
for paste mixer, 1208
for propeller mixer, 1208, 1226
for solid CO_2 manufacture, 1700
transmission to mixer, 1226
for tumbling mixer, 1221
for turbine mixer, 1210
Prater pulverizers, 1141
Precious metals and alloys, 1548
Precipitation, 887, 1196
of dust (*see* Dust collection)
Precipitative extraction, 715
Precipitator, electrical, 1013, 1039
Spaulding, 947
Precipitron, 1045
Precision, percentage, 50
Precision cutters, 1162
Precision knife cutters, 1161
Preconditioning for filtration, 970
Prediction of values of relative volatility, 529
Preforming and drying, 823
Preforming materials for drying, methods of, 824
Preheaters, air, 1642
Preliminary selection of dryers, 874
Preliminator mill, 1132
Premier colloid mill, 1169
Presettling tanks for centrifuges, 1003
Press, cage, 1073
continuous, 1074
curb, 1073
platen, 1073
Press, pot, 1073
screw, 1074
Pressed cakes, milling of, 1148
Presses, batch, 1073
Pressure(s), absolute, definition of, 360
additive, 290
use of, in humidity problems, 336
average static, 397
barometric, 365
calculations, 290–291
center of, 361
constant, temperature-composition, diagrams at, 318
control system, 1332
controllers, differential, 1333
relay-operated, 1320
conversion table and equivalents, 40
critical, 565
drop and slot opening in bubble-cap towers, 601
effect on enthalpy, 568
effect of, on equilibrium constant, 576
on homogeneous azeotropes, 631
on liquid-vapor relations, 569
on packed distillation towers, 619
pseudocritical, 567
ratio for gases, 402
reduced, 568
definition of, 360
drop, through activated carbon, 905
absorption columns, 680
across baffles, 396
bends, 389
for Berl saddles, 395, 682
for brine circulation, 1697
in centrifuges, 1006
cloths (fabrics), 1030
contraction, 388
cost of, 384, 385
curved pipe, 383
cyclone separators, 1024
diaphragms, 389
diverging duct, 388
dust layers, 1030
effect of liquor rate, 395
elbows, 390
enlargement, 388
entrance, 388
exits, 388
expansion, 388
in fire hose, 1850
fittings, 389
granular solids, 393
head, 377
heat exchangers, 390
miscellaneous, 387
orifices, 389
packed towers, 393
in pipe, 1430
pipe fittings, 389
for Raschig rings, 681
smooth bends, 389
across tube banks, 390
effect of, on equilibrium, 311
on mutual solubility of liquids, 318
Pressure(s), effect of, on solubility of solid in liquid, 319
on solute distribution between liquid phases, 320
of total, on rate of mass transfer, 544
expression as head of fluid, 360
filters, 970
gage, 360, 364
for power plants, 1643
high (*see* High pressure)
hydrostatic, 360
intensity of, 360
local static, 396
loss (*see also* Pressure, drop)
by fluid friction, gases, 377
by friction, alignment chart inflow, 378, 379
measurement of, 364
in multistage compression, intermediate, 1681
normal, 360
partial, use of in humidity problems, 336, 337
pumps, 990
Dorrco, 945
recovery, following orifices, 404
regulating valve, 1252
resultant, 361
rupture, 1868
spray nozzles, 1170
static, 360, 396
taps, location of, 405
size of, 397
tests for filtration, 969
total, calculation of curves from activity coefficients, 534
variation with position in fluid, 360
water vapor, of moist air, 760
partial, 758
Pressure-filled expansion thermometer, 1271
gages, 1279
measurement, 1279
regulator, self-operated, 1320
taps, location of, 1284
Pressure-temperature diagrams, 331
Pressure-volume diagram, 330
Pressure-volume vs. pressure diagram, 330
Pressure-volume-temperature diagram, 330
Prilling (crystallization), 1069
Primary battery, 1788
cells, 1788
Prime mover capacity in U.S., 1629
Prime movers, steam, 1644
Principle, Archimedes, 362
Principle, Pascal's, 360
Principles, physical and chemical, 287–358
Principles of non-ideal liquid mixtures, 526
Print dryer, 849
Print hammer for recorders, 1268
Print-wheel for recorders, 1268
Printers for recorders, 1267, 1268
Prism, surface area of, 57
volume of, 57
Prismoidal formula for integration, 57, 90
volume, 57

Probability, charts, 100
curves, 81
theory and charts, 100
Problem solutions, general considerations, 333
Process(es), control, 1263*ff.*
electric motors for, 1329
elements of, 1310
glossary, 1337
installation of, 1328
piping for, 1328
power units for, 1328
terms, 1337
three-way valve in, 1330
controllers, "hunting" in, 1329
cost systems, 1833
furnaces, 1598, 1608, 1621
instrumentation, 1265
reversible, 302, 304
safety and, 1864
variables, 1266
Process-gas sampling, 1017
Proctor & Schwartz, through-circulation dryer, 820, 824
standard, 818
three-truck, 818
two-truck, 818
Producer gas as fuel, 1576
manufacture of, 1579
Product costs, variances, and yields, 1840
Production control and production lines, 1843
Production and costs of steam boiler plant, 1633
Profile paper, 100
Profit and loss analysis, 1830
Programming control, 1326
Progressions, 65
Propagation of flame, speed of, 1587
Propane, saturated, thermodynamic properties of, 274
superheated, thermodynamic properties of, 274
Propane–oleic acid–abietic acid system, 722, 724
Propeller mixers, 1208, 1215, 1218, 1222
horsepower consumption of, 1208, 1226
Propeller pumps, 1419
Propeller-type fan, 1208, 1448
Property classifications for accounting, 1831
Proportion, algebra, 63
Proportional controller, simple, 1323
plus-reset controller, 1320
position control, 1321
position controller, 1320, 1323
Proportioning, continuous, 1293
of liquids, 1293
motors, electric, 1329
pump for ratio control of flow, 1326
definite and multiple, laws of, 289
feeders, 1375
and metering pumps, 1437
for mixing, 1220
n-Propyl alcohol, densities of aqueous solutions of, 190
specific heats of, 235
Protractor, 70
Protection equipment, fire, 1870
safety, 1870
Pseudocritical pressure, 567
Pseudoplastic fluid, 1197
Pseudoplasticity, 1197
Psychrometer(s), 1298
sling, 777
Psychrometric chart, 759
Psychrometry, application to drying, 810
Public practices as plant location factors, 1725
Public utilities as plant location factor, 1725
Puffing of materials, 1166
Pug mill mixer, 1209, 1220
Pugging operation, 1209
Pulley(s), 1665
diameters for belts, 1664
safety and, 1858
Pulp, agitation in filters, 985
cyanide, filtration of, 985
hygroscopic moisture of, 778
liquor, sulfite, evaporation of, 503
paper, filtration of, 986, 990
settling-area requirements of, 946
types of in settling, 938
Pulsation dampener, 1280
Pulverization, flash, 1146, 1147
Pulverized coal, 1567, 1641
boiler, 1639
direct-fired system, 1155
radiation from flames of, 495
Pulverizers, 1139
Pulverizing, 1147
circulating load limits, 932
equipment, open-circuit, 931
fine, 932
fineness of, 932
mediums, 934
work of, 931
and circulating load, 931
and disintegration, 1117
Pulverizing Machinery Co., hammer mill, 1123
pulverizer, 1140
Pump(s), air removal, 1647
axial-flow, 1420
boiler feed, 1647
centrifugal, 1417
for centrifuges, 1002
chemical, 1418
circulating, 1647
close-coupled, 1419
diaphragm, 990
diffusor-type, 1419
displacement, 1439
Dorrco, 945
v-type, 947
Duplex, Dorrco suction, 946
efficiency of, 1431
evaporator, 520
air, 520
filtrate, 991
Hardinge diaphragm, 945
heat, 1657
for high pressure, 1258
hot-well, 1647
jet, 1439
for liquid fuels, 1573
materials of construction, 1424
metering and proportioning, 1437
Pump(s), mixed flow, 1422
for molecular distillation, 658
open-impeller centrifugal, 990
plunger, 990
pressure, 990
propeller, 1419
ratio control of flow, proportioning, 1326
rotary, 1437
selection of, 1420
suction lifts for, 1652
sump, 1419
vacuum, 990, 1439
vertical, 1421
volute-type, 1419
Pumping equipment, for liquids, 1414
Pumping of liquids, 1414
equipment, 1414
Punched metal plates, 958
Purges for orifice meters, 1287
Purification of gases for liquefaction, 1714
of soluble crystalline substances, 887
Push-pull amplifier, 1745
Push-pull propeller mixer, 1208
Pusher furnace, 1607
Putty, manufacturers of, 1534
Putty chaser, 1214, 1220
Pyramid, area and volume of, 57
spherical, 58
Pyrator mill, 1133
Pyrethrum, grinding of, 1158
Pyrites roaster, 1622
Pyrolytic cracking unit, 1618
Pyrometer, controller for electric furnace, 1326
optical, 1270, 1274
radiation, 1270, 1274
standardization of, 1269
Pyrometers, 1762
Pyrometry, radiation, 1274

Q

Quadrant, definition, 57, 70
Quadratic equations, algebra, 66
Quadrilateral, 55
Quadruple-effect evaporators, 516
Quaker City grinding mill, 1158
Quantity meters, 410
Quartic and higher equations, 68
Quartz (fused), production, 1819
window, for high pressures, 1253
Quaternary Margules equations, 529
Quinhydrone electrode, 1309

R

R, gas constant, 47, 290
Radiamatic radiation pyrometer, 1276
Radian(s), 70
definition of, 43
to degrees and minutes, conversion table and equivalents, 36
Radiant heating, 1761
Radiation, coefficients of heat transfer, 473
from clouds of particles, 493

Radiation, definition of, 456
from gases, 490–493
of heat, through openings, 1601
transmission, 483–498
from non-luminous gases, 490
pyrometers, 1270
pyrometry, 1274
Radiator, perfect, 483
Radius, determination of, without center, 60
hydraulic, 375, 378
mean hydraulic, definition, 375
taps, 405
Raffinate, definition of in extraction, 715
Rake mixer, 1205
Rake product, 930
Ramsay and Young rule, 294
Ranarex gas-density balance, 1298
Rank of coals, 1560
Rankine cycle for steam, 1633
temperature scale, 1269
Raoult's law, 317, 525
binary mixture, 525
deviations, 317, 534
systems that follow, 526
Raschig rings, pressure drop in, 395, 681
loading velocities for, 683
Rate curves of drying, 802
of drying in constant-rate period (graph), 804
floating, 1338
-of-flow meters, 1283
in mixing, 1197
of mass transfer, 538
from cylindrical surfaces, 546
Rating, merit, 1835
Rating of refrigerating machine, standard, 1676
Ratio, algebraic, 63
test for convergence, 65
control of flow, proportioning pump, 1326
controllers, 1325
flow-control system, 1333, 1334
of heat capacity at constant pressure to that at constant volume, 296, 297
plug valves, 1327
of specific heats of air at high pressures, 233
of gases at 1 atm., 229
Raw materials, as plant location factor, 1720
surveys, 1726
Rayleigh, problem of heat transfer, 96
Raymond, bowl mill, 1155
crushers, 1121
flash dryer, 834, 835, 836
high-side mill, 1137
ring-roller mill for coal grinding, 1156
Imp hammer mill, 1155
power vs. capacity of, 1155
pulverizers, 1140, 1141, 1142
and soap grinding, 1160
ring-roller mill, 1138, 1155
for graphite grinding, 1156
and grinding D.D.T., 1158
and grinding sulfur, 1158
Raymond, Screen Pulverizer, 1140
Vertical Mill, 1141, 1142
and grinding D.D.T., 1158
Rayon, hygroscopic moisture of, 778
Rayotube radiation pyrometers, 1275
Raytheon Sonic Oscillator, 1169
Reactance, 1732
Reactions, apparent order of, 324
concurrent, 324
consecutive, 324
conventions, 344
determination of order of, 327
equation for rate of, 321
first-order, 321
gas, contact catalytic, 329
heats of, at specified temperatures, 346
from heats of combustion, 345
from heats of formation, 344
heterogeneous, kinetics of, 328
higher order, 323
homogeneous, kinetics of, 321
kinetics of, general, 321
non-isothermal, rates of, 327
order of, 321
reversible, 324
second-order, 322
sides, 324
simultaneous, 324
illustrative cases of, 325
standard heats of, 344
temperature effect on velocity of, 323
velocity, constant, specific, 321
of contact catalytic gas reactions, 329
effect of temperature on, 323
equations for first-order reactions, at constant pressure, 322
at constant volume, 321
equations for reversible reactions, 325
equations for second-order reactions at constant pressure, 323
at constant volume, 322
in flow systems, 322, 323, 326, 329
general equations for, 321
in heterogeneous systems, 328
non-isothermal cases, 327
orders of, 321
determination of, 327
promotion by mixing, 1196
simultaneous, 324
of solids with dissolved substances, 329
vessels, 1256
autoclaves, 1256
batch reactors, 1256
continuous reactors, 1256
for high pressures, 1256
Reactive current, 1765
volt-amperes, 1732
Reactivities of coal and coke, 1566
Reactors, continuous, 1257
Reagents, activation, 1088
cationic, 1088
feeders, 1374
for heavy-media separation, 1082
Real solutions, 308
Réaumur temperature scale, 1269
Reboil heat in bubble-cap towers, 602
Receivers, 990
Reciprocals of numbers, 27
by slide rule, 52
Reciprocating compressors, 1439, 1443
Reciprocating-piston meters, 1282
Reciprocating-plate feeder, 1372
Reciprocating pumps, 1426, 1436
Reciprocating screens, 957
Recirculating-dryer control system, 1335
Recirculative furnaces, 1607
Recorder, dew-point, 1299
round chart, 1267
strip chart, 1267
types of, 1267
Recording electrical instruments, 1739
and indicating instruments, 1267, 1268
-mechanisms, 1267
Recovery of waste heat, 1623
Rectangle, circle construction around, 60
length and area of, 54
Rectangular passages, velocity traverse, 399
sections, heat transfer in, 470
tanks, volume of, 1379
Rectification, 563
batch, 594
continuous, 588
at low temperatures, 1702, 1708
Rectifier(s), 1750
electrolytic, 1787
mercury-arc, 1766
Rectifying columns or towers, 563
types of, 596
towers, bubble plate, 596
packed tower, 597
sieve-plate, 596
Rectigon, 1751
Rectilinear diameter, law of, 292
Recuperative furnaces, 1598, 1607
Recuperators, checkerbrick, 1623
Redington, F. B., cartoning machine, 1406
Redler conveyor, 1351
Reduced pressure, 568
temperature, 568
Reducer, motor, 1670
Reducing valve, magnetic, 1252
pilot-operated, 1320
Reduction, of area of materials, 1539*ff*.
engineering mills, 1145
gyratory crusher, Type R, 1122
ratio in size reduction operations, 1113
reactions, electrochemical, 1786
units, gear, 1669
Reductionizer, 1145
Reentrant short tube, 407
Reference-junction compensation, 1272
Refining of cane sugar, 898
of petroleum, 914
Reflectors for infrared dryers, 868
Reflux, 563
calculation methods for, 733
limits of, 736, 737, 738

Reflux, minimum, 717, 736
ratio, 563
effect of, on packed distillation towers, 619
minimum, determination, 623
theoretical minimum, 591
in solvent extraction, 717, 733, 739
supply in bubble-cap towers, 602
total, and extraction, 591, 717, 736, 738
Re-formed gas as fuel, 1577
Re-forming natural and oil refinery gases, 1580
Refractive index, elements and inorganic compounds, 110–128
of glasses, 1548
Refractories, expansion of, 1602
fusion temperature of, 281
Refractory materials, manufacturers of, 1538
physical properties of, 1549
siliceous materials, grinding of, 1150
Refrigerant(s), 1676
comparison of, 1678
for heat pumps, 1658
heat-transfer properties of, 1687
performance coefficients, 1678
for vapor compression and absorption systems, 1686
vapor pressures of, 1687
Refrigerating brines, machine, 304
centrifugal, 786
solutions, viscosities of, 1696
thermal conductivities of, 1697
Refrigeration, 1675–1717
absorption system, 1676, 1683
for air-conditioning, 784
brines for, 1692, 1693, 1694, 1695, 1697
British units, 1677
carbon dioxide, 1679, 1683, 1700, 1701
cascade cycles, 1681
chemical, 1698
compression system, 1676
compressors for, 1677, 1681, 1682, 1687, 1713
condensers, 1677, 1682, 1690, 1708
continuous absorption, 1683
cooling systems, 1690
cooling towers, 1690
definitions, 1676
dense-air machine, 1677
drying gases by, 884
evaporators, 1677, 1691, 1692
expanders, 1712
of foods, 1699
gas liquefaction, 1702, 1707
heat exchangers, 1713
heat pump, 1657, 1701
heat transfer, 1687, 1689, 1692
ice manufacture, 1698
insulation, 1693, 1717
low-temperature, 1701, 1702
lubricants, 1681, 1682
natural, 1698
pipe lines, 1692
Platen-Munters system, 1684
processes, natural, 1698
Refrigeration, refrigerants for, 1676, 1678, 1686, 1687
small-scale, 1697
standard ton of, 1676
steam jets, 1682
system, intermittent absorption, 1685
thermal conductivity for (table), 458
units of, 1676, 1677, 1678
vapor compression systems, 1677
water treatment for, 1686
Refrigerator, Electrolux, 1684
tests on sulfur dioxide, 1698
Regenerative furnaces, 1598, 1607
Regenerators, coke-oven, 1624
glass-tank, 1624
of heat, heat transfer in, 463
open-hearth, 1624
Regulations for oil-burning equipment, 1572
Regulators, automatic, 337
for centrifuges feed, 1003
induction, 1762
self-operated pressure, 1320
temperature, 1320
Relative capacity of various rotary dryers, 861
Relative heat content, total, 344
Relative humidity, measurement of, 1298
percentage, 759, 811
Relative viscosity, 369
Relative volatility, 579
direct prediction of values of, 529
effect of, on plate efficiencies of distillation columns, 613
Relay-operated, controller, 1320
Relay-operated pressure, 1320
Relays, electrical, 1323
electronic, 1323
mechanically operated, 1323
non-bleed air valve, 1322
valve, 1322
Reluctance, 1732, 1736
Remote control, pneumatic transmission for, 1333
Remote transmission, 1324
Rennerfelt furnace, 1815
Representation, graphical, of thermodynamic functions, 330
Republic, carbon dioxide Orsat, 1304
filter, 988
pneumatic, telemeter, 1324
transmitter, 1324
resistance (electrical) flowmeter, 1286
Repulper, Dorrco, 955
Repulsion-induction motor, 1769
Reset, automatic, 1322, 1338
plus rate action, 1322
controller, 1320
automatic, 1323
Resin (synthetic) ion-exchangers, 916
Resins, grinding of, 1159
Resistance(s), and conductors, 1733
of constructional materials, chemical, 1461ff.
electrical, 1732, 1733
furnaces, 1759
factors for fabrics, 1031
for dust layers, 1031
Resistance(s), flowmeter, 1286
internal, 1743
plate, 1743
specific, 1782
thermometers, 1270, 1274
Resistivity, electrical, 1732
Resistor-type furnaces, 1759
Resonance, 1734
Respiration, artificial, 1878
Respirators, 1870
Response, booster, 1338
control, 1338
controller, 1338
conversion, 1338
floating, 1338
Resultant pressure, 361
Retarding conveyors, 1347
Retirements (accounting), 1833
Retrograde condensation in multicomponent hydrocarbon system, 575
Retrograde phenomena, 575
Reverberatory furnaces, 1619
Reverse-acting diaphragm valve, 1328
Reversed-current air classifiers, 1135, 1136
Reversible process, 302, 304
Reversibility of processes, 1704
Reversing exchanger, 1713
Reversing of polyphase motors, 1768
Reversing rotation of motors, 1770
Revolving-cone mixer, 1213, 1219
Revolving-plate feeder, 1371
Revolving table feeder, 1373
Reynolds criterion, 383
Reynolds number, definition of, 375
in design of mixers, 1224, 1225, 1226
effect of curvature on, 384
and orifices, 1284
Rhe, definition of, 370
Rheopectic fluid, 1199
Rheostats, 1752
Rhombus, 55
Ribbon, area of, 56
electric resistance, 1760
mixer, 1210
nichrome, 1760
-packed exchangers, 1714
target efficiencies of, 1022
Richardson, batching scale, 1294
Duo-Screw Feed Bulk Weighing Scale, 1384
Enclosed Dustproof Sacking Scale, 1383
oscillating bag packer, 1385
Riddles, 958
Rietz Disintegrator, 1141
Riffles, 1078
Right spherical triangles, 74
Rigidity, definition of, 369
Riley Atrita pulverizer, 1141
stoker pulverizers, 1141
Ring balance meter, 1254
Ring-and-drum magnetic separator, 1092
Ring hammer mills, 1116
Ring joint, 1246
Ring packings, pressure drop in, 395
Ring piezometer, 397

Ring-roller mills, 1116, 1137
for grinding limestone, 1150
grinding pigments in, 1157
grinding titanium pigment, 1157
low-side, 1121
with internal air classification, 1138
without internal classification, 1137
with internal screen classification, 1137
Risco Packer, 1385
Rise, capillary, 363
Riser areas and cap-slot in bubble-cap towers, 602
Rising-bubble viscometer, 1301
Ritter electrostatic separator, 1095
Rittinger's Law, 931, 1114, 1115
Rivet-making furnaces, 1605
Riveting and corrosion, 1459
Roasting, magnetic, 1093
Robins screen, 957
Robinson attrition mill, operating characteristics of, 1148
Robinson cup anemometer, 399
Röchling-Rodenhauser furnace, 1816
Rocking autoclave, 1257
Rod mills, 1116, 1130
Allis-Chalmers, 1133
grindability tests, 1114
Rod packing, 1218
Roll crushers, 1116
feeder, 1370
Roller mills, 1074
Rolls, compression, 992
mixing, 1214, 1220
Root(s), algebraic, 63
slide-rule, 52
complex, 85
cube, 23
equal, 85
square, 23
Root-mean square (R.M.S.) electrical value of, 1734
Rope, hygroscopic moisture of, 778
Ropes and chains, safe loads for, 1874
Ross feeder, 1372
Rotameters, 408, 1288
armored, 1254
Ro-Tap shaker, 964
Rotary blowers, 1439, 1452
Rotary capper for bottles, 1400
Rotary compressors, 1439, 1452
Rotary converters, 1765
Rotary crushers, 1116, 1122, 1129
Rotary disk filter, 971
Rotary-disk water meter, 411
Rotary distributor, 1375
Rotary drum filter, 971
Rotary dryers, direct, 828
indirect, 859
indirect-direct, 829
relative capacities of, 861
seals for, 829, 830
steam-tube, 860
vacuum, 858
Rotary fillers for liquids, 1397
Rotary kilns, 828, 831, 1608
Rotary knife cutters, 1160
Rotary piston meter, 412
Rotary pumps, 1437
Rotary shears, 1162
Rotary-stem-motion control valves, 1326
Rotary valve, 1328
Rotating can mixer, 1206
Rotating hearth furnace, 1607
Rotating-impeller meter, 1282
Rotating nozzles, 1172
Rotating-paddle solids-level meter, 1290
Rotating pan mixer, 1206, 1220
Rotating-vane feeder, 1371
Rotation of motors, reversing, 1770
Rotex screen, 957
Rotoclone, 1029
Rotodip meter feeder, 1374
Roto-louvre dryer, 824
Rotors, phase-wound, 1768
wound, 1768
Rotor scrubber, Centri-Merge, 1038
Rotor type of electrostatic concentration, 1093
Rotten stone, grinding of, 1156
Roughness factors, Manning formula (table), 384
pipe friction, 382
Round chart recorder, 1267
Rounded entrance, 388
Rounded orifice, 407
Rowland's law, 1736
Rubber, 1547
belting, 1662
electrodeposition of, 1799
equipment, manufacturers of, 1536
fabric, hose, 439
pipe, 438
filter mediums, 968
hard, grinding of, 1159
heating of, 463
hygroscopic moisture of, 779
latex centrifugation, 1012
physical properties of, 1551
Rules, significant figures, 50
Rupture, pressure, 1868
Rupture disk, 1252
Russia, weights and measures of, 44

S

Sacking scale, 1383
Saddles, Berl, pressure drop in, 395, 682
Saran pipe and tubing, 438
S.A.E., steels, specification numbers of, 1667
Safety, bibliography on, 1879
building construction, 1851
and construction materials, 1856
data for engineers (table), 1873
distances of (table), 1848
equipment for, 1854
eye protection for, 1870
and fire protection, 1847–1884
in grinding, 1119
head protection for, 1870
in high-pressure technique, 1261
lighting construction and, 1852
loads for ropes and chains, 1874
organization, 1871
overhead clearances and, 1849
protection equipment, 1870
Safety, role of elevators in, 1851
side clearances and, 1849
in transportation, 1367
valves for power plants, 1643
St. Regis two-tube screw-type-valve bag-filling machine, 1388
St. Regis Valve-bag-filling machines, 1386
Salad oil, use of adsorbents with, 902
Salicylic acid sublimation, 664
Salt(s), equivalent conductivities of, 1783
filtration of, 986, 990
indicators, 285
removal of, from evaporators, 512
Salt-ice mixtures, eutectic, 1700
Salting in evaporators, 514
Salting out, 715
Salting out effect, 320
Samco vacuum filler for liquids, 1397
Sampling, 1095–1102
air, 1102
beaker, 1100
boat, 1097
bottle, 1100
car, 1097
cement, 1101
coal, 1100
coning and quartering, 1097
continuous liquid, 1100
dipper, 1100
dust, 1017
flue gas, 1102
gases, 1102
grab, 1097
iron ore, 1101
lacquer solvents, 1102
mechanical, 1100
metals and alloys, 1101
methods of, 1097
molten metal, 1102
petroleum products, 1101
pig iron, 1101
pipe, 1100
shovel, 1100
steel, 1101
thief, 1100
water, 1102
Sand(s), definition of, 939
drying of, 833
filters, 968
clarifier, Hardinge, 948
Hardinge automatic backwash, 948
Sanitation, 1853
Saturable-reactor control system, 1332
Saturated air, specific volume of, 811
humidity curve, 811
vapors, heat capacity of, 297
volume, 759
Saturation point, fiber, 800
Saunders valve, 1327
Saws, 1163
Sawtooth crusher, 1128
Saybolt, seconds, 370
conversion to viscosity, 370
Scale, deposits, formation of, 513
optimum cycle for evaporators forming, 517, 518
heat transfer coefficients through, 464

Scale, removal of, 513
electrolytic, 1801
Scales, for basic weighing, 1291
temperature, 1269
(*See also* Weighing)
Scalping, 955
Schaffer feeder belt system, 1292
Scheibel extraction column, 748
Schierenbeck method of constructing pressure vessels, 1243
Schmidt number, effect of on analogy between mass transfer and fluid friction in smooth pipes, 543
Schoenherr furnace, 1820
Schütte & Koerting sp. gr. indicator, 1297
Schutz-O'Neill Limited mill, performance of, 1159
Schutz-O'Neill mill, 1143
Scoop feeder for mills, 1134
Score cutters, 1162
Scoring machine for ice manufacture, 1699
Scott evaporator, 506
Scove kilns, 1608
Scraper feeder, 1373
vacuum, condenser, 855
Screen scales, 963, 1640, 1641
Screening, 955–964, 1111
definitions, 955
efficiency, 960
equipment, 955
mechanical variables in, 962
sieve testing in, 962
Screens, 955
arrangement of, 962
bar, 958
capacity of, 960
efficiency of, 960
equipment for, 955
grizzlies, 955
Hum-mer, 957
mechanical shakers for, 964
mechanical variables in, 962
Niagara, 957
oscillating, 957
Overstrom mud, 957
performance of, 960
punched plates, 958
reciprocating, 957
riddles, 958
Robins, 957
Rotex, 957
Selectro, 956
shaking, 956
sifters, 958
silk cloth, 960
testing of, 962
Tyler standard, 964
TY-Rock, 957
vibrating, 956, 1361
wire, 959
Screw conveyors, 1344
dryer, 862
feeder, 1370
Screw feeder for mills, 1134
Screw press, 1074
Screw threads, 1244
Screw-type-valve bag-filling machine, St. Regis, 1388
Screwed pipe joints, 441
Scrubbers, dust, 1013, 1034
Seal(s), Bridgman, 1254
electric lead, 1253
liquid depth, effect of, on plate efficiencies of distillation columns, 613
for orifice meters, 1287
packing and shaft, 1420
for pumps, mechanical, 1423
for rotary dryers, 829
Seamless tubing (table), 417
Second law of thermodynamics, 303, 1703
Second-order reaction, 322
at constant pressure, 323
at constant volume, 322
Secondary battery, 1788
Secondary cells, 1788, 1792
Sector, annulus, area, 56
spherical, 58
Sedberry pulverizers, 1141
Sedimentation, 947–955
phases of, 938
Seed cakes, milling of, 1148
Seed crystals, screen analysis of, 1059
Seeded crystallization, 1054
Seeds, vegetable, drying of, 821
Seger cones, 1272
Segment(s), calculations, 56
of circles, 32
areas of, 32
of spheres, 35
Segmental orifices, 406, 1284
Selden batch sublimation equipment, 663
Selection of dryers, 874
Selection of pumps, 1420
Selectivity diagrams, acetone from water with trichloroethane, 723
acetic acid-water-solvents, 723
equilibrium diagrams, 723
methylcyclohexane-aniline-n-heptane, 723
oleic acid from abietic acid-propane, 724
Selectivity of solvent(s) in extraction, 715
diagrams, 722
Selectro screen, 956
Selenium rectifiers, 1752
Self-balancing bridge telemeter, 1325
"Self-hooping," or autofrettage, 1241
Self-inductance, 1732
Self-lubricating bearings, 1666
Self-operated, controller, 1310, 1320
pressure regulator, 1320
temperature regulator, 1320
Self-sealing closure, 1249
Self-synchronous motors, Selsyn, 1325
Selsyn, motor telemeter, 1325
self-synchronous motors, 1325
Semilive skid and lift jack, 1364
Semilogarithmic graph paper, 99
Sensible heat contents, 339
gaseous mixtures, 340
liquid state, 341
pure gaseous state, 340
Sensitivity, floating, 1338
Sensitivity of measuring instruments, 1266
Separating agents for azeotropes, 643
Separation, heavy-media processes, 1081
and liquefaction, work of, 1703
magnetic, 1091
miscellaneous methods of mechanical, 1072
Separator(s), 990
cream, 1000
dust or mist (*see* Dust collection equipment)
packed-bed, 1029
Dutch State Mines cyclone, 1068, 1083
Electro-float, 1094
impingement, 1013
Jeffrey-Steffensen three-drum magnetic, 1092
table conductance, 1094
toboggan conductance, 1094
Separatory cones, 1082
closed-top, 1083
Sequence controller, 1326
Series, 64, 65
binomial, 64, 65
convergent, 65
divergent, 65
functions expanded into (calculus), 83
infinite, 65
non-convergent, 65
tests for, 65
Series circuit, 1733
magnetic, 1736
Series motors, 1767, 1769
Service departments, instrument, 1337
costs, 1838
Setting, control, 1338
index, 1338
point, 1338
Settlers, in evaporator systems, 512
Settling, 887, 994
Allen cone, 940
area requirements of pulps, 946
Callow cone, 940
chamber, 1021
in containers, agitator, 1394
and decantation, 1003
rate for particles, 994, 995
tanks, intermittent, 939
types of pulps in, 938
velocity (of particles), 1017, 1021
Sewage sludge, centrifugation, 1012
drying of, 837
filter cake, drying of, 838
filtration of, 986, 990
Sewer pipe, clay, 435
Sewing heads, 1391
Sewing unit for bag packers, 1389
Shading poles, 1769
Shaft furnaces, 1612
kilns, 1612
and packing seals, 1420
Shafting, 1667
safety and, 1859
Shaker(s), feeder, 1372
for filling bags, 1411
mechanical, 963
Newark End-shaker, 964
Ro-Tap, 964
Shaking screens, 956

Shale and stone, drying of, 833
Shape factors for non-spherical particles, 394
in particle-size measurement, 1113
Shattering, explosive, 1166
Shaving machines, 1163
Shear, maximum of steels, 1239
in mixing, 1195, 1197, 1198
Shear-strain-energy theory, 1239
Shears, rotary, 1162
Sheepskin, hygroscopic moisture of, 779
Sheet-calipering device, 1295
Sheet metal and wire gages, conversion table and equivalents, 39
Shelf packages, packaging equipment for, 1397
Shelf, vacuum, dryers, 853
Shell-and-tube condensers, 1688
Shifting-tube belt packer, 1388
Short-center drives, 1670
Short-column elutriator, Cooke, 1085
Short tube, reentrant, 407
standard, 407
Shovel sampling, 1100
Showers, emergency, 1876
Shunt, electric, 1739
machines, 1764
motors, 1766
Side clearances, 1849
Side-firing furnaces, 1607
Sidereal time, 45
Sieve-plate, columns (extraction), 752
rates for, 753
rectifying towers, 596
Sieve testing, 962
Sieving, 1111
(*See also* Screening)
Sifeed clarifier, Dorr, 943
Sight-gage glass, 1267
Sight-glass, high-pressure, 1252
Sigma blades for mixers, 1207
Signal contacts, high-low, 1268
Significant figures, rules for, 50
Signs and symbols, mathematical, 5
Silent chains, 1672
Silica, equipment, 1548
filters, 968
fused, production, 1819
gel, 911
adsorption of acetone from air by, 912
adsorption of benzene from air by, 912
adsorption of sulfur dioxide from air by, 912
air-conditioning system, 914
application of use of, 911
dehydration of gas by, 913
drying of, 826
for drying gases, 880
heat of wetting, 884
properties and applications of, 905
purification of carbon dioxide, 913
removal of water from, by air, 913
spray-drying of, 845
use in dehumidification, 786
water content of, as function of temperature, 912
Silica, grinding of, 1149
pipe, 440
ware, properties of, 1460
Silicate treatment of condenser water, 1690
Silicon carbide production, 1818
Silicon production, 1819
Silk, filter mediums, 968
hygroscopic moisture of, **778, 779**
Silo, storage in, 1377
Silver, electrolytic refining, 1804
Silver carbonate, dissociation pressures of, 174
Silver nitrate, drying of, 869
Silver solders, 453
Simple gas law calculations, 335, 336, 337
Simple proportional controller, 1323
Simpson's rule, applied to volumes, 58
for areas, 57, 89
Simpson's rule for integration, 88, 89
Simultaneous equations, 68
Simultaneous reactions, kinetics of, 324
Sine wave of alternating current, 1733
Single-compartment Dorr thickener, 941
Single-drum dryers, 863
Single-effect evaporators, calculations for, 507
Single-electrode potential, 1785
Single-runner attrition mill, 1144
Single-seat V-port valve, 1327
Single-seated bevel-disk valve, 1327
Sink-and-float concentration, 1080
Sinking-fund depreciation, 1833
Siphon feeder, adjustable, 1374
Sirocco collector, 1028, 1029
Sirups, cane-sugar, filtration of, 986
Sisal hemp, hygroscopic moisture of, **778**
Sites of plants, determination of, 1728
Size, analysis of particles, 1017
particle, 1016
classification and size reduction, 1117
classifiers, types of, 1118
use of, 1118
control in crushing and grinding, 932
distributions in batch and continuous closed-circuit grinding, 1117
enlargement, 1179
by fusion, 1191
reduction, 1103*ff.*
and dehydrating, 1119
and drying, 1119
and heat transfer, 1119
miscellaneous methods of, 1160
and mixing and blending, 1119
and size classification, 1117
segregation in kilns, 1612
Sizing, 955
Skid box, 1364
Skid and lift jack, semilive, 1364
Skids, 1363, 1863
Skip hoists, 1352
applications, 1350
Skip hoists, automatic loader, 1351
hoisting unit, 1350
power charts, 1352
slack cable protection, 1351
Slabs, solid, heating and cooling of, 462
Slack diaphragm, 1279
flowmeter, 1286
Slag (basic), grinding of, 1152
Slat conveyor, sewing head on, 1392
Slate, pulverizing of, 1154
Sleeve valve bags, 1390
Slide dampers, 1328
Slide rule, 50–53
description of,
loglog duplex, 50–51, **53**
operations, 50–53
polyphase duplex, 51, **53**
reciprocals by, 52
trigonometric functions, 52–53
types of, 50–51, **53**
Slide valve controlling a piston, 1321
Slide-valve pilot valve, 1320
Slide-wire rheostats, 1752
Sliding belts, 1357
Sliding-stem-motion control valves, 1326
Sliding tray feeder, 1374
Sliding-vane rotary compressor, 1452
Slime-thickening tanks, capacity of, 938
Slimes, definition of, 939
Sling psychrometer, 777, 1299
Slip-ring starter, 1673
Slitters, top, 1162
Slitting cutters, 1162
Slopes of rotary dryers, 832
Sludge, filtration of, 990
Slurry densities, in evaporators, 512
Slurries, pumping of, 512
Small-bag filling and closing, 1410
Small-scale refrigeration, 1697
Smidth mills, 1133
Smith Engineering crusher, 1126
Smoke (*see* Dust)
Smoke detector, 1302
Smooth-roll crushers, 1127
Smoothing data, 91
Smothering glands, 1423
Snow point, 660
Snyder furnace, 1815
Soap(s), crutcher, 1210
grinding of, 1160
hygroscopic moisture of, **780**
Soapstone, grinding of, 1150
Sodium acetate, densities of aqueous solutions of, 182
drying of, 866
Sodium arsenate, densities of aqueous solutions of, 182
Sodium bicarbonate, granulation of, 1158
Sodium bichromate, densities of aqueous solutions of, 182
Sodium bisulfate, constructional materials for, 1508
Sodium bromide, densities of aqueous solutions of, 182
Sodium carbonate, constructional materials for, 1509
densities of aqueous solutions of, 182

Sodium carbonate, dissociation pressures of, 174
freezing points of aqueous solutions of, 281
specific heats of aqueous solutions of, 234
Sodium chlorate, densities of aqueous solutions of, 182
Sodium chloride, brines, heat content of, 1695
thermal conductivities of, 1697
constructional materials for, 1510
densities of aqueous solutions of, 182
Sodium nitrate, fractional crystallization of, 1054
specific heats of aqueous solutions of, 234
and water, viscosities of, 1696
Sodium chromate, densities of aqueous solutions of, 182
Sodium dihydrogen phosphate, dissociation pressures of, 174
Sodium formate, densities of aqueous solutions of, 182
Sodium hydroxide, absorption of chlorine in, 706
constructional materials for, 1511
densities of aqueous solutions of, 182
freezing points of aqueous solutions of, 281
at 1 atm., aqueous, enthalpy-concentration diagram for, 275
specific heats of aqueous solutions of, 234
Sodium hypochlorite, constructional materials for, 1513
Sodium nitrate, constructional materials for, 1514
densities of aqueous solutions of, 182
Sodium nitrite, densities of aqueous solutions of, 183
Sodium perborate production, 1796
Sodium phosphate, crystallization of, 1064
drying of, 866
production, 1812
Sodium silicate, densities of aqueous solutions of, 183
Sodium sulfate, constructional materials for, 1515
densities of aqueous solutions of, 183
drying of, 866
Sodium sulfide, constructional materials for, 1516
densities of aqueous solutions of, 183
spray-drying of, 845
Sodium sulfite, constructional materials for, 1517
densities of aqueous solutions of, 183
Sodium sulfonate, drying of, 866
Sodium thiosulfate, constructional materials for, 1518
densities of aqueous solutions of, 183
Sodium thiosulfate, pentahydrate, densities of aqueous solutions of, 183
Soft solders, 453
Softening point of glasses, 1548
Softening temperature of coal, 1564
Solar day, apparent, 45
Solar evaporation, 505
Solar heat evaporators, 505
Solar time, 45
Solders and cements, 453
Solenoid(s), 1329, 1753
valves, 1329
Solid(s), clutch, 1671
cubical expansion of, 202
desiccants, 877
feeder, 954
feeding of, 1370
filling and weighing equipment, 1395
fuels, 1560*ff*.
geometry, 55–57
granular, mass and heat transfer in gases through, 547
heat capacity of, 298
fugacity of, 307
solubility of, in liquids, 319
heating and cooling of, 462
miscellaneous, specific heats of, 235
model charts, 100
organic, specific heats of, 230–232
plastic, 369
plastico-viscous, 369
properties of, 1110
solution of, in batch apparatus, 557
specific gravity and volume by immersion, 363
storage of, 1376
surfaces of, 58–59
volumes of, 58–59
Solid-cone nozzles, 1171
Solid-liquid separation, 994
Solubility(ies), data for gas absorption, 668
effect of, on plate efficiencies of bubble-tray absorption columns, 698
elements and inorganic compounds, 110–128
gases in liquids, 317
liquids in liquids, 318
gases in non-aqueous liquids, 676
solids in liquids, 319
in water, 673
of hydrocarbons in oil, 676
inorganic compounds in water, 196–199
of liquid hydrocarbons in furfural, 647
organic compounds, 129–148
Solution(s), alkaline SO_2, 679
aqueous, specific heats of, 233, 234
colloidal, definition of, 939
compositions and freezing points of, 1694
feeder, 954
heats of, 246–248, 299
ideal, 308
at infinite dilution of aqueous solution, 530
real, 308
rate of, 1058
Solution(s), of solids in liquids, batch apparatus for, 557
specific heats of non-freezing, 1696
true, definition of, 939
vapor pressures of, 166–172
water evaporated for change in concentration of (table), 501
Solvent, concentration, effect of, on vapor-liquid equilibrium, 644
effect of, on plate efficiencies of bubble-tray absorption columns, 698
content-concentration diagram, 724, 730, 731, 734
extraction, 713–753
recovery, drying with, 875
recovery and continuous drying, 815
selection of in gas absorption, 668
Sonic flocculation of dust, 1049
Sonic oscillator, 1169
Sorbitol production, 1796
Sorensen chlorine-cell, 1806
Sources of power, 1628
South America, weights and measures of, 44
Soxhlet, method of extraction, 717
Soybean oil-furfural system, 728
purification by adsorption, 890
refining, centrifuges in, 1008
Soybeans, milling of, 1148
Space furnaces, 1598
Space lattices, 1050
Space-time-yield, in flow systems, 329
Spargers, 1176
gas, 1203, 1217
Sparking, electrical, 1039
potentials, 1040
Sparking potential, minimum, 1747
Sparkler filter, 988
Spaulding precipitator, 947
Specific conductivity, 1782
Specific flame output, 1587
Specific gravity, of coals, 1561
elements and inorganic compounds, 110–128
equivalents °Bé., °Tw., lb./gal., lb./cu. ft., 38
of glasses, 1548
of liquid fuels, 1569
of liquid hydrocarbons, variation with temperature, 536
measurement of, 1296
organic compounds, 129–148
and relationship to degrees A.P.I., 1296, 1297
Baumé, 1296, 1297
Brix, 1296, 1297
Twaddell, 1297
of solids by immersion, 363
Specific heats, of air at high pressures, 233
effect of temperature and pressure in ratio of, 297
ratios of, at high pressures, 233
aqueous solutions, 233, 234
of coke, 1562
of gases, 229, 1588
ratios of at 1 atm., 233
of glasses, 1548
of liquefied gases, 233

Specific heats, of liquids, 228, 235
miscellaneous materials, 235
miscellaneous oils, 235
miscellaneous porcelains, 235
miscellaneous solids, 235
of non-freezing solutions, 1696
of organic liquids, 225–227
of organic solids, 230–232
of paraffin gases, 567
of petroleum oils, 567
of pure compounds, 219–232
Specific inductive capacity, 1732
Specific resistance, 1782
Specific surface of particles, 1113
Specific viscosity, 369
Specific volume of dry air, 811
of saturated air, 811
Specific volume of mixtures, 289
Specific volume of water below 0°C., 176
Specific volumes of pure substances, 177
Specific weight of water vapor, 758
Specification numbers of S.A.E. steels, 1667
Specifications for fuel oils, 1570
Specifications for piezometer taps, 397
Spectrometers, 1305
Spectrometry, infrared, 1305
ultraviolet, 1305
Spectrophotometers, 1302
Speed, floating, 1338
measurement and control, 1295
Speed control of motors, 1768
Speed of flame propagation, 1587
Speeds, peripheral, of rotary dryers, 832
Sphere(s), diameters, by hundredths, 34
drag coefficient for, 1018
target efficiencies of, 1022
hollow, volume of, 58
segments of, 35
solid, heating and cooling of, 462
surfaces, areas, 34
volumes, 34
Spherical geometry, theorems of, 57–58
Spherical particles, drag, 1018
coefficient for, 1018
charge and motion of, 1040
terminal velocities of, 1021
Spherical sector, 58, 74
Spherical segment, 74
Spherical surfaces, evaporation rates from, 546
Spherical triangles, 58, 74–75
area of, 74–75
Spherical trigonometry, 74–76
Spherical zone, 74
Spheronizing, 1188
Spices, grinding of, 1158
Spinner air classifier, 1138
Spinner external air classifier, 1135
Spiral bevel gear unit, 1670
Spiral-coil hygrometer, 1298
Spiral concentrators, 1079
Humphreys, 1079
Spiral element, 1279
Spiral tiles, costs of, 709
Splash feed for drum dryer, 864
Splash lubrication, 1681
Splint coal, 1563
Split clutch, 1671
Split-Flo homogenizer valve, 1168
Split-phase starting of motors, 1769
Split-stage compression, 1681
Spot, dead, 1338
Spray (*see* Dust)
Spray(s), absorber, desorption of oxygen from water in, 700
dryers, 838–848
Swenson, 842
Western Precipitation, 842
film feed for drum dryer, 864
nozzles, 1170
characteristics of for spray dryers, 840
for dehumidifiers, 783
pressure, 840
two-fluid, 840
and packed towers for extraction, 748
rate data for, 750
estimation of capacity coefficients, 751
ponds, 1647
tower as mixer, 1204, 1217
rates of absorption in, 699
type unit air conditioners, 784
Spraying, 1169
Spreader stokers, 1641
with furnace boiler, 1639
Sprinkler heads, friction loss through, 1850
Sprinkler systems, 1854
Sprout-Waldron attrition mill, 1123, 1143, 1144
Sprout-Waldron crusher, 1128, 1129
Sprout-Waldron rotary cutters, 1161
Square(s), circle construction, 60
measure, equivalents and table, 43
of numbers, 23
octagon construction in, 61
root(s), 23, 52
by logarithms, 50
by slide rule, 52
sides of, with diameters of circles of equal areas, 33
Square-edged orifices, expansion factor, frictional loss of, 389
Squarex clarifier, Dorr, 943
Squirrel-cage motor, 1768
Stability (floating bodies), 362
Stability of coals, 1562
Stackers, 1363
Stacks for power plants, 1642
Stage(s), determination of number of, 716
efficiency of, 716, 729
in extraction, definition of, 714
ideal, 716, 729
minimum, 736, 737
Stainless-steel, pipe, 417
tubing, 417
Stairs, 1861
Standard cells, 1772
cadmium, 1772
Weston, 1772
Standard cone crusher, 1126, 1127
Standard-cost methods, 1833
Standard free energy change, 309
Standard heats, of combustion and formation, units, 335
of reaction, 344
from heats of combustion, 345
from heats of formation, 344
Standard National print dryer, 850
Standard rating of refrigerating machine, 1676
Standard short tube, 407
Standard states, 307
Standard ton of refrigeration, 1676
Standard time belts of U.S., 45
Standard tray dryer, 818
Standard vertical-tube evaporator, 506
Standardization of thermometers, 1269
Stannic chloride, densities of aqueous solutions of, 183
Stannous chloride, densitics of aqueous solutions of, 183
Starch, by-product, drying of, 837
drying of, 826
filtration of, 988
grinding of, 1148
spray-drying of, 846
suspensions, inverted plasticity of, (chart), 1201
Starting devices for centrifuges, 1004
Stassano furnace, 1815
State, funicular, 800
pendular, 800
States, corresponding, theory of, 292
standard, 307
Static corrosion test, 1459
Static electricity, 1869
Static head, 337
Static head of liquids in pumps, 1415
Static pattern for concentric orifice plate, 1284
Static pressure, average, 397
absolute, 360
local, definitions of, 396
measurement of, 396
in other than straight passages, 397
specifications for piezometer taps, 397
Statics, fluid, 359–412
Statistical calculations of thermodynamic properties, 313
Stay-new dust filter, 1047
Steam-atomizing burners, 1573, 1574
Steam, boiler furnaces, heat transfer in, 496
over-all thermal resistances in surface condenser for (graph), 470
consumption in batch steam distillation, 584
diagrams, 1633
distillation, batch, 582
steam consumption in batch, 584
vaporization efficiency in, 583
engine diagram factors, 1645
types of, 1644
extra, in evaporation, 519
jets, 1682
condensers for, 1682
ejector, 1224, 1453
vacuum system, control of, 1333
power plants, 1633

Steam, power plants, equipment, life of, 1632
input-output curves, 1632
prime movers, 1644
requirements for steam jet vacuum cooling units, 1682
saturated, thermodynamic properties of, pressure table, 278
temperature table, 277, 278
superheated, thermodynamic properties of, 279, 280
surface condenser, 1648
temperature of, in evaporator operation, 509
turbines, 1645
condensing extraction, 1646
velocity for free expansion of, 1645
viscosity of (table), 370
Steam-tube rotary dryer, 860
Stearates, grinding of, 1160
Stearic acid, constructional materials for, 1519
Stedman column packing, 608
Stedman Disintegrator, 1140
Stedman Heavy-Duty Crushers, 1129
Stedman Type A 2-stage swing hammer grinders, 1140
Steel(s), action of gases on, 1235
for alcohol synthesis, 1235
American Iron and Steel Institute standard, chemical composition of, 1543*ff.*
for ammonia synthesis, 1235
for coal hydrogenation, 1235
decarbonization of, 1235
effect of temperature on tensile properties, 1236
elastic failure of, 1239
elongation of, 1235
enameled equipment, 1549
furnaces, electric, 1812
open-hearth, 1620
overstrained, 1236
pipe, 414
tubing, 417
for pressure vessels, 1234–1237
roller chains, 1672
sampling of, 1101
tensile strengths of, 1236
tests of, for high pressure use, 1235
vibrac, 1235
yield points of, 1235
Stefan-Boltzmann law, 484, 1274
Steffensen flotation machine, 1089
Stevedore trucks, 1363
Stilling box, 367
Stirrer-shaft packings, 1247
Stirrup feeder, 1370
Stobie furnace, 1815
Stoke (definition), 370
Stokers, coal, 1640
mechanical, 1567
Stokes-Cunningham law, 1019
Stokes-Einstein equation (dialysis), 754
Stokes's formula, 937
Stokes's law, 1019
correction factor for deviations from, 1020
Stokes McLeod gage, 1280
Stokes & Smith automatic packager, 1398
Stokes & Smith filling machines, 1401
Stokes & Smith wrapping machine, 1406
Stokes Universal fiber cutter, 1162
Stone, drying of, 833
Stoneware, 1547
(chemical) equipment, 1549
towers, costs of, 709
packing, pressure drop in, 396
Storage, of coal, 1564
of compressed gases in cylinders, 1869
of gases, 1380
at high pressures, 1242
of liquid fuels, 1573
liquids, bulk, 1378
of material, 1376–1381
in bins or bunkers, 1377, 1382
bulk material, 1376, 1382
containers, 1377, 1382
indoor pile, 1377
outdoor pile, 1376
silos, 1377
systems, floor loadings for, 1379
tanks for flammable liquids, 1868
Storage battery, 1788
Storage-bin construction, 1382
Stormer viscometer, 1197, 1200
classification of paints by, 1201
starch suspensions, measurements with, 1201
and viscous materials, 1201
Stoves, blast-furnace, 1623
heat transfer in, 463
Straight-blade fans, 1448
Straightening vanes for venturis, 1285
Strain-energy, maximum, of steels, 1239
Strainers for filters, 983, 984
Straub crushers, 1120
Straw as fuel, 1568
Streamline filter, 984
flow, 375
flow formulas, for open channels, 387
for pipes running full, 387
for straight passages, 385
economic pipe diameter for, 385
Strength, ultimate, of steels, 1235
Stress(es), distribution in thick-walled cylinders, improving, 1241
equations for, 1238
effect of temperature on, 1241
in mixing, 1197, 1198, 1199
in threads or shanks of bolts, 1244
in walls of pressure vessels, 1237
Strip-chart recorders, 1267
Strip, electric heaters, 1759
Stroboscope, 1296
Stuffing boxes, 1423
Sturtevant mill, 1137, 1144
Sturtevant mill classifiers, 1135
Sturtevant Rock-Emery mill, 1144
Sublimation, 660–665
auxiliary equipment for, 665
carrier, 660, 662
condensation, 665
definition, 660
Sublimation, design calculations, 662
desublimation, 660
drying by, 854
entrainer, 665
equipment, 663
examples, 663
of iodine, 663
operating diagram, 662
of salicylic acid, 664
simple, 660, 662
plant for, 664
snow point, 660
symbols for, 662
triple point, 660
uses of, 660
Submarine signal sonic oscillator, 1169
Submerged-coil condensers, 1688
Submerged-spiral classifier, 926
Substances, miscellaneous, linear expansion of, 201
Substitutes for integration, 89
Subtraction, 63
algebraic, 63
Suchar process for sugar refining, 903
characteristics of, 902
Sucrose solutions, viscosity of (table), 374
Suction box, 924
Suction pump, head, net positive, 1415
lift, 1415, 1652
limitations, 1415
Duplex, Dorrco, 946
Sugar, liquors, effect of vegetable carbons on, 902
mill, three-roll, 1074
refining, Suchar process for, 903
solutions, viscosity of (table), 374
Sulfate (sodium), constructional materials for, 1515
Sulfates, grinding of, 1150
Sulfide (sodium), constructional materials for, 1516
Sulfite (sodium), constructional materials for, 1517
Sulfite pulp liquor, evaporation of, 503
Sulfur, adsorption by magnesia, 916
constructional materials for, 1520
corrosion of metals and alloys by, 1526
grinding of, 1158
Sulfur dioxide, absorption in caustic, 706
absorption from flue gases in ammonium sulfite-bisulfite solution in spray tower, 702
absorption in sodium carbonate solution in cyclone spray tower, 701
absorption in water in cyclone spray tower, 701
from air, adsorption by silica gel, 912
alkaline solutions, 679
constructional materials for, 1521
household refrigerator, tests on, 1698
partial pressures of, over SO_2, aqueous, 167
radiation of heat from, 493

Sulfur dioxide, saturated, thermodynamic properties of, 275
superheated, thermodynamic properties of, 276
vapor-liquid equilibrium relations for, 572
in water, absorption of, 696
Sulfur dioxide–carbon dioxide, temperature-pressure diagrams for, 572
Sulfur trioxide, partial pressures of, over H_2SO_4, 169
Sulfuric acid, absorption of ammonia in, 706
aqueous, enthalpy-concentration diagram for, 277
aqueous, latent heats of vaporization for, 168
aqueous, normal boiling points of, 168
aqueous, vapor pressures of, 168
constructional materials for, 1522
densities of aqueous solutions of, 184, 185
freezing points of aqueous solutions of, 281
and nitric acid, aqueous, enthalpy-concentration diagram for, 277
partial pressures of, over H_2SO_4, 169
specific heats of aqueous solutions of, 234
Sulfuric acid–nitric acid, aqueous, vapor pressures of, 170
Sulfurous acid, constructional materials for, 1524
Summary of thermodynamic equations, 314
Sump pumps, 1419
Super-Jumbo Crusher, 1129
Superficial mass velocity, 374
Superfine classifier, 1135, 1136
Superheat, effect of, in pure vapor, on heat transfer, 477
Superheaters, 1641
Superimposed a.c. on d.c., 1788
Superior McCully crusher, 1126
Superpressures, design for, 1240
Supersaturation, 1054
Supervision of instruments, 1336
Supplementary angles, 53
Supports for packings, 710
pipe, 453
Surface, and area equivalents, 41
condenser, effect of, on heat transfer, 520
steam, 1648
over-all thermal resistances in (graph), 470
drop size for gas atomization of water, 840
evaporation rates from spherical, 546
heating, calculation of, 500
of particles, specific, 1113
screening, 958
of solids, 57
Surface-active agents in paint grinding, 1115
Surface temperature, 803
graph, 804
Surface temperature, tension, of common liquids (table), 363
and capillarity, 363
Eötvös equation for, 363
tension and mixing, 1202
Surface unit (S.U.), 931
Surface-water temperatures in U.S., 1687
Susceptance, 1732
Suspended centrifuge, 998
Suspensions, 1195, 1196, 1218, 1229
definition of, 939
flow of, 385
Sutherlands' constant, 370
formula (viscosity), 370
Sutton, Steele, and Steele dry table, 1079
Sutton, Steele, and Steele Electrofloat separator, 1094
Sweating, 1075
Sweden, weights and measures of, 44
Sweetland filter, 973
leaves, 974
Swellgrip bag holder, 1390
Swenson spray dryer, 842
Swenson-Walker crystallizer, 1063
Swindell furnace, 1813
Swing-hammer grinders, 1140
Swing-hammer mills, 1116
Swinging-plate feeder, 1372
Switches, mercury, 1324
Switzerland, weights and measures of, 44
Symbols, algebraic, 62
mathematical, 5
Symons cone crushers, 1126
Short Head, 1126
standard, 1121
Synchronized cam telemeter, Bailey, 1325
Synchronous condensers, 1767
converters, 1765, 1766
motor, polyphase, 1767
phase modifier, 1767
Synchroscope, 1739
Synthetic fiber filter mediums, 968
Synthetic fibers, hygroscopic moisture of, 779
Synthetic-resin ion-exchangers, 916
Syntron, hydraulic jolter, 1395
mechanical seal, 1424
vibrator for container packing, 1395
System, control, 1338
liquid water–air–water vapor, simultaneous heat and mass transfer in, 558
Systems, following Raoult's law, 526
multicomponent, gas absorption design for, 672
of power, 1628
of weights and measures of different countries, 44

T

Table(s), conductance separator, 1094
conversion, 36
Deister-Overstrom deck, 1079
dry, 1079
engineering, thermodynamic properties, explanation of, 249
Table(s), equivalents, 36
feeder, 1371
mathematical and weights and measures, 1–47
Sutton, Steele, and Steele dry, 1079
trigonometric, 71
wet, 1078
Tabling, agglomeration, 1079
Tabulating machines, 1828
Tabulations of thermodynamic data, 314
Tachometers, 1295
Tagliabue Celectray potentiometer circuit, 1278
Tagliaferri furnace, 1815
Talc, grinding of, 1150
Tanbark as fuel, 1568
Tangent(s), to circle, 55, 60
equation of, 78
to two circles, 60
Tank (chemical-), concrete for construction of, 1558
Tank crystallization, 1062
Tank ends, volume of, 1378
Tank furnaces, 1598
Tanks, capacities of horizontal, 1380
capacities of vertical cylindrical tanks, 1381
and pipe (wood), manufacturers of, 1553
volumes of, 1378, 1379
Tanning extract, drying of, 866
spray-drying of, 845
Tantalum tubing, 430
Taps, corner, 405
flange, 405
piezometer, specifications for, 397
pressure measurement, 404
radius, 405
vena contracta, 405
Tar in manufactured gas, 1578
oil as fuel, 1575
Target efficiency of spheres, cylinders, and ribbons, 1022
Task determination, 1835
Taylor, formula of, 83
bellows-type flowmeter, 1286
-Stiles precision cutters, 1162
volumetric pressure gage, 1280
Technical calculations, 333–358
Technical journals, abbreviations, 4
Technique, high-pressure, 1233–1262
(*See also* High pressure)
Telemeter, Bailey synchronized-cam, 1325
bridge-circuit, 1325
electric, 1325
Republic pneumatic, 1324
self-balancing bridge, 1325
Selsyn motor, 1325
Telemetering, 1324
electrical, 1324
hydraulic, 1324
pneumatic, 1324
of weight readings, 1292
Telescopic gas holders, 1380
Telsmith breaker crushers, 1126
cone crushers, 1126
Temperature, 464
absolute, 290
of automatic sprinklers, 1854

Temperature, cold-storage, 1698
composition diagrams at constant pressures, 578
constant, 282, 283
control for packing liquids, 1396
control system, 1382
open-kettle, 1330
critical, 291, 292, 565
difference, log mean, 464
distribution of, in multiple-effect evaporators, 509
drop in evaporators, 500
energy diagram, 331
equivalents, 42
as a factor in mixing, 1202
on equilibrium constant, 320
chart, 350
in high-pressure design, 1240
on liquid-activity coefficients, 529
on metal corrosion, 1459
on packed distillation towers, 619
on reaction velocity, 323
on solubility of gases in liquids, 317
on solubility of solids in liquids, 319
on thermal conductivity of metals and alloys (table), 456
on vapor pressure, 293
flame, 1588
for food preservation, 1699
in furnaces and kilns, 1599
high, 283
ignition, 1582
inversion, 333
in a limestone shaft kiln, 1614
lines of humidity chart, effective, 777
measurement of, 1269, 1762
differential, 1334
reduced, 568
regulator, self-operated, 1320
scales, 334, 1269
on ternary phase equilibrium, 721
true, of gas, calculation of, 473
true surface, 803
graph, 804
valve for elevated, 1251
water, in U.S., 1687
wet-bulb, 759
Temperature–vapor-pressure curves, 98
Tensile strength of glasses, 1548
of materials, 1539*ff.*
Tension, hoop, 362
interfacial, 363
effect in extraction, 747
surface, 363
Tenter dryer, 849*ff.*
Ternary azeotropic mixtures, 634
Ternary Margules equations, 528
Ternary systems, graphical representation, 719
Terminal settling velocity (of particles), 1019, 1021
Test-tube centrifuge, 992
Testing, sieve, 962
Testing of coal, 1560
Testing devices for flammable vapors, 1869
Tests, drying, 874
Tetragonal system of crystals, 1051
Textile bags, hand sewing, 1390
Textiles, hygroscopic moisture of, 778
Theisen disintegrator, 1038
Theorem, Bernoulli's, 375
Theorems, binomial, 64
plane geometry, 53
solid geometry, 55
spherical geometry, 57
Theoretical plates in countercurrent apparatus, 548, 554
Theory, general, of diffusional operations, 523–559
Theory of corresponding states, 292
Thermal conductivity, of alloys, 456
of building materials, 457, 458
cells, 1303
conversion factors for, 41, 456
of gases and vapors, 461, 1303
of glasses, 1548
of insulating materials, 458, 1693
of liquids, 459
of metals, 456
of petroleum oils (tables), 460
for refrigeration insulation (table), 458
of rubber, 463
of vapors (table), 461
of wood, 463
Thermal decomposition of coal, 1564
Thermal diffusivity of wood, 463
Thermal efficiency of kilns, 1611
Thermal efficiencies of tray and compartment dryers, 820
Thermal expansion, 200–202
of gases, coefficients of, 200
of glasses, 1548
Thermal flowmeter, 1288
gages, 1281
Thermal instruments, 1738
Thermal meters, 412
Thermal performance of power plants, 1630
Thermal precipitation of dust, 1050
Thermal radiation, 483
Thermal treatment of unstable gases, 1625
Thermionic amplifiers, 1743
Thermionic tubes, 1742, 1751
Thermix tube, 1028
Thermochemistry, 294
Thermocompressors, 519
Thermocouple(s), 1272
base-metal, 1270
calibrating, 1274
depth of immersion, 1273
installation of, 1274
life of, 1273
multiple, 1274
noble-metal, 1270
protecting tubes for, 1273
selection of, 1273
speed of response, 1273
Thermodynamic analyses of oxygen processes, 1705
Thermodynamic data tabulations of, 314
Thermodynamic efficiency of oxygen processes, 1706
Thermodynamic equations, summary of, 314
Thermodynamic functions, graphical representation, 330
Thermodynamic properties of air, 760
engineering, 249–281
of gases from statistical calculations, 313
of water, 770
Thermodynamic temperature scale, 1269
Thermodynamics, 294, 295, 303
first law of, 295
functions of, graphical treatment, 330–333
laws of, 1703
second law of, 303
third law of, 313
Thermoelectricity, 1742
Thermofor cracking units, 1618
Thermofor kiln, 895, 1612
Thermometer(s), bimetallic, 1270
constant-volume gas, 1269
gas-pressure, 1270, 1271
industrial, 1270
angle-type, 1270
liquid-in-glass, 1270
liquid-pressure, 1271
mercury-in-glass, 1270, 1267
mercury-pressure, 1270
pressure-filled, 1270
resistance, 1270, 1274
vapor-pressure, 1270, 1272
wet- and dry-bulb, 1299
Thermometric fixed points, 1269
Thermometry, wet- and dry-bulb, 557
Thermopile, 1275
Thermostat(s), 1762
bellows, 787
bimetal, 787
diaphragm, 787
duct, 787
Thermostatic control, bath fluids for, above zero degrees, 283
below zero degrees, 282
Thickener(s), Briggs filter, 949
Dorr Torq, 941
traction, 942
tray, 942
washing-tray, 953
filter, 947
Hardinge Auto-raise, 944
mechanical continuous, 941
mixing between, 954
non-mechanical continuous, 940
Oliver-Borden filter, 948
selection of type of, 947
single-compartment Dorr, 941
Thickening, continuous, 937
capacity, 939
mechanics of, 939
phases of intermittent, 938
Thickness, measurement and control, 1295
Thief sampling, 1100
Thiosulfate (sodium), constructional materials for, 1518
Third law of thermodynamics, 313
Thixotropic filter cakes, drying, 824
fluid, 1198
substance, 369

Thixotropy, 1197, 1198
definition of, 369
Thomas converter, 1617
Thompson e.m.f., 1272
Threads, buttress, 1244
screw, 1244
Three-component systems, 579
Three-drum Jeffrey-Steffensen magnetic separator, 1092
Three-phase Delta system, 1735
power, measurement of, 1738
Y system, 1735
Three-roll sugar mill, 1074
Three-way valve in process control, 1330
Throttle pressure, effect on by-product power cycle, 1635
Throttling, 333
controller, 1320
Through-circulation dryers, batch, 820
continuous, 823
drying data, 826
Thwing radiation pyrometer, 1275
Thyratrons, 1748, 1749
Tie lines (condensed systems), data for, 721–728
definition of, 719
effect of temperature on, 721
interpolation of, 724
in multicomponent systems, 728
Tilting-box meter, 1283
Tilting manometer, 366
Tilting-type manometer, 1286
Time, charts, 100
dead, 1338
floating, 1338
lag, finite, 1338
and its measurement, 45
and motion study, 1835
standard belts of the U.S., 45
Timekeeping, 1834
Timer, interval, 1326
Tin, electrolytic refining, 1804
Tin boxes, cartoning machine for, 1406
Tin pipe, 430
Titanium dioxide, drying of, 826
Titanium pigments, grinding of, 1157
Titone, drying of, 819
Titration coulometers, 1775
Tobaccos, hygroscopic moisture of, 780
Toboggan conductance separator, 1094
Toluene-air vapor mixtures, humidity chart for, 815
Toluene-benzene mixtures, partial pressure-composition diagram for, 577, 578
Toluene–water–benzoic acid system, extraction rate data for, 750
Toluidine red, drying of, 819
Tongue and groove joint, 1245
Toothed-roll crushers, 1127
Top slitters, 1162
Torq thickener, Dorr, 941
Torque control, high starting, 1673
of motors, 1766
Torque-type viscometer, 1301
Torus, volume of, 58
area, 58
Total head tube, 397
heat of liquid, 331
immersion test, 1458
reflux (distillation), 591, 623
Totalizing conveyor scales, 1294
Towers, costs of, 709
packed, flow through, 393
packed distillation, 616
rectifying, types of, 596
Townsend cell, 1808
Traction thickener, Dorr, 942
Tractors, industrial, 1362
-trailer train, 1364
and trailers, 1363
Trailer(s), caster, 1364
fifth-wheel, 1364
four-wheel-steer, 1364
industrial, 1362
tractor-trailer, 1364
and tractors, 1363
-train pallet transportation, 1367
Transcendental functions, calculus, 87
Transfer units, calculation of number required in countercurrent apparatus, 552
in countercurrent apparatus, 550
equations for, in extraction, 744
Transfer unit, length of in dryers, 810
use of, for extraction, 745
Transformer(s), 1762
connections, 1763
coupled amplifier, 1744
instrument, 1739
Transformer oil, acidity, reduction of, with Activated Alumina, 911
purification with centrifuges, 1010
Transition, heat of, 299
Transition-point location, 1272
Transition points of hydrates, 283
Translation velocity, 1058
Transmission, hydraulic, 1324
mechanical power, 1660–1674
pneumatic, 1333
radiant-heat, 483–498
heat, 455–498
remote, 1324
for remote control, pneumatic, 1333
Transmitter, pneumatic, 1324
Transparency of glasses, 1548
Transparent attritus in coals, 1562
Transparent plastic pipe, 437
Transportation as plant location factor, 1724
Transporting water vapor, cost of, 856
Trapezoid, 55
Trapezoidal rule for areas, by integration, 89
Traps, air-separating for centrifuges, 1003
cinder, 1643
and hatches, 1860
moisture, 991
Traveling-cushion bowls, 1001
Traveling mixer, 1205, 1219
Traveling paddle mixer, 1205
Traverse, for mean velocity, 399
Traxler and Baum elutriator, 1084
Tray dryers, 815
Tray thickener, Dorr, 942
Traylor crusher, 1126
gyratory crusher, 1126
Trays, capacities of, in bubble-cap towers, 600
Trenches and excavations, 1860
Triangle(s), congruent, 54
construction in circle, 60
equilateral, 54
oblique, solution of, 73
plane, 54
solution of (trigonometry), 72–74
polar, 58
right, 54
similar, 54
spherical, 58
Triangle Package Machinery Co. packers, 1402
Triangular diagrams, in extraction, 719, 729, 732, 737
Triangular graph paper, 99
Triangular phase diagram, application in extraction, 719–722, 729, 732, 737
with solid components, 725
Trichloroethane-water-acetone system, 721, 722, 723
Trichlorethylene, constructional materials for, 1525
saturated, thermodynamic properties of, 281
Trichloromonofluoromethane (F-11), saturated, thermodynamic properties of, 259
superheated, thermodynamic properties, of 260
Triclinic system of crystals, 1051
Triethanolamine-CO_2 equilibrium data, 677
Trigonal system of crystals, 1051
Trigonometric functions, 80
approximations of, 72
by curves, 80
hyperbolic, 76
by slide rule, 52
Trigonometric series, algebra, 64
Trigonometrical tables, 71
Trigonometry, 70–76
plane, 70
spherical, 74
Trilinear graph paper, 99
Triple point, 660
Trippers, for conveyors, 1356
Trisodium phosphate, costs of crystn. of, 1070
Tropical year, 45
Trough dryer, 862
Trouton's rule, 299
Troy and metric units, equivalents of, 1781
Troy weights and measures, 43
Truck(s), factory, 1363
floor, 1363
fork, 1366
hand electric platform, 1369
hand pallet, 1369
high-lift, 1365, 1366
industrial, 1362
jack-lift, 1364
low-lift platform, 1365, 1366
mechanical jack, 1364
pneumatic-tire, 1366
stevedore, 1363

True-boiling-point apparatus, 608
-boiling point distillations, 607
comparison with A.S.T.M., 607
True separator bowls, 1000
True solutions, definition of, 939
Trumpet-shaped entrance, 388
Tube(s), banks, heat transfer to fluids flowing areas, 472
pressure drop across, 390
flow in, 377
alignment chart, 379
friction factors for in-line (chart), 392
gaseous-discharge, 1748
impact, 397
mills, 1116, 1130
air classifiers with, 1135
closed-circuit, 932
grinding at various distances from feed end, 1117
pitot, 397
short, 407
static, 360
thermionic, 1742
total head, 397
Tubing, aluminum, 424
bimetallic, 430
block tin, 430
block tin lined, 430
columbium, 430
condenser, 425
copper and copper alloys, 424
copper water, 427
finned copper, 428
glass, 436
heat exchanger, 425
hypodermic needle, 417
laminated phenolic, 438
Lucite, 437
mechanical, 417
nickel and nickel alloy, 430
non-metallic, 431
pressure, 417
Saran, 438
seamless, dimensions (table), 417
stainless, 417
steel, 417
tantalum, 430
tin, 430
tolerances of cold-drawn seamless tubing, 417
(*See also* Pipe)
Tubular-bowl centrifuge, 1000
Tubular gage glasses, 436
Tumbling barrel mixer, 1212
Tumbling mills, 1130
performance of, 1132
Tungar, 1751
Tunnel dryers, 821
furnace, 1607
kilns, 1608
Turbidimeter, particle-size detn. by, 1113
Turbine(s), blower mixer, 1210
expansion line on Mollier diagram, 1635
gas, 1655
absorber, 1211
generator, expansion lines for, 1645
mixers, 1210, 1215, 1218, 1222
shell pressures, 1646
steam, 1645
Turboabsorber, 1216, 1217
Turboalternator, excitation curves of, 1765
Turboblowers, 1439, 1450
Turbocompressors, 1450
Turbodisperser, 1211, 1215
Turbodryers, vertical, 821
Turboexpanders, 1712
Turbo-Gas-Absorber, 1179, 1211, 1216, 1217
Turbomixer, 1217
Turbulaire spray dryer, 843, 845, 847
Turbulence, 375
mixers, 1203
Turbulent flow, 375, 384
through smooth pipes, 542
Turbulent motion in settling, 994
Turning for size reduction, 1163
Twaddell, degrees, °Bé, lb./gal., lb./cu. ft., sp. gr., 38
and sp. gr. relationship, 1297
Twin-drum dryers, 863
Two-fluid U-tube, 366
Two-position (on-and-off) controllers, 1320
Two-stage closed-circuit grinding, 932
compression, 1680
open-circuit grinding, 931
Tyler standard screen-scale sieves, 963
Type "B" Junior Hercules mill, 1137
Type of dryers, 815
Type "R" reduction gyratory crusher, 1122, 1126
Type S clarifier, Dorrco, 947
Type T Bulldog crusher, 1126
Type TY Reduction crusher, 1126
"Ty-Rock" screen, 957

U

U-leather packing, 1247
Ultimate analysis of coals, 1561
Ultrasonic precipitator, 1049
Ultraviolet spectrometry, 1305
Unaccomplished moisture change, 800
Unbound moisture, 800
Undercut-gate feeder, 1370
Underfeed stokers, 1641
Under-firing furnace, 1607
Underwood chart (extraction), 741
Unidan mill, 1133
Uniflow steam-engine cylinder, 1644
Unikom mill, 1133
Unimolecular reactions, at constant volume, 321
gases, at constant pressure, 322
Union special sewing heads, 1391
Unit(s), air conditioner, 784
conversion of, 334
of energy, 295
heat, types of, 335
latent heat, 335
selection of, in stoichiometry, 334
standard, of formation and combustion, 335
electrical and magnetic, 1733
electrochemical, 1772
Unit(s), filters, 1033
of heat transmission, 467
of refrigeration, 1676
British, 1677
U.S. Automatic Box Machinery Co. weigher and packer, 1403
U.S. standards, liquid volume equivalents, 40, 43
prime mover capacity, 1629
series of sieves, 963
time belts, 45
Weather Bureau weather data, 46
weights and measures of, 43
Univariant system, 315, 316
Universal Filling machine, 1401
Universal gas mask of U.S. Bureau of Mines, 908
Universal jaw crusher, 1125
performance data for, 1126
Universal Road Machinery classifiers, 1135
Unloaders for centrifuges, mechanical, 998
Unseeded crystallization, 1054
Unstable gases, thermal treatment of, 1625
"Unsupported area" joint, Bridgman, 1246
Updraft kilns, 1608
Urea resin, spray-drying of, 845
Utilization of activity coefficients, 532
Henry's law, 532
prediction of absorption equilibrium, 532
prediction of azeotropes, 532
prediction of distribution in liquid-liquid extraction, 533
prediction of heats of solution, 532
prediction of immiscibility, 533
U-tube, 364
closed, 365
differential, 364
inclined, 366
inverted differential, 364
manometer, 1279
open, 364
two-fluid, 366

V

Vacuums, control, 1333
measurement, 1279
system, control of steam-jet, 1333
crystallizer, 1065
drum dryers, 864
filler, General Mills Co., 1402
kneader, 1208
low-temperature, dryers, 854
manometric range, 655
Multi-vane air classifiers, 1136
optimum, in evaporation, 510
pump, displacement of, 521
use of, in evaporation, 510
packing containers, 1393
pan dryers, 857
pumps, 990, 1439
rotary dryer, 858
scraper condensers, 855
shelf dryers, 853
tests for filtration, 969

Valence crystals, 1050
Vallez filter, 974
Value, desired, 1338
Valve(s), 447
bags, sleeve, 1390
balanced, 1327
bevel-plug, 1328
bowls, 999
Brown relay, 1322
butterfly, 1328
characteristics, 1327
for close control of flow, 1251
check valve, 1251
high pressure, 1250 et seq.
for high temperatures, 1251
magnetic control, 1252
pressure-regulating, 1252
repackable under pressure, 1251
control, 1326
controlling a piston, slide, 1321
direct-acting diaphragm, 1328
reverse-acting diaphragm, 1328
double-seat V-port, 1327
double-seated parabolic plug, 1327
electrolytic, 1787
flapper-type pilot, 1320
gate, 1328
Hammel-Dahl small-flow, 1327
installation, 1327
jet-type pilot, 1321
materials of construction for, 1328
needle, 1327
for packing liquids, control, 1396
parabolic plug, 1327
pilot, 1321
and relay, 1321
pilot ball-type, 1321
pilot-operated reducing, 1320
positioner, 1329
in process control, three-way, 1330
ratio plug, 1327
rotary, 1328
Saunders, 1327
selection, 1327
single-seat V-port, 1327
single-seated bevel-disk, 1327
slide-valve pilot, 1320
solenoid, 1329
Split-Flo Homogenizer, 1168
Vol-U-Meter electrically controlled, 1396
V-port, 1328
Valve-bag-filling machine, St. Regis, 1386
Valve-bag packers, 1387
Van der Waal's equation, 290
Van Laar, binary, equations, 527
Van Tongeron cyclone, 1024, 1025
Vane anemometer, 398
Vane meter, 412
Vaned disk mixer, 1217
Vapor-compression refrigerating machine, 1677
Vapors, adsorption by activated carbon, 906
compression systems, refrigerants for, 1686
condensation of, in presence of non-condensable gas, 477
diffusion, 801
of liquid mixtures, 317, 318
pure condensation of, 476
Vapors, saturated heat capacity of, 297
thermal conductivity of (tables), 461
velocity in bubble-cap towers, 597
effect of, on plate efficiencies of distillation columns, 613
Vapors and gases, diffusion coefficients of, 539
effect of external pressure on, 293
estimation of, 293
explosion, 1868
Vapor-liquid equilibrium data in gas absorption, 668
Vapor-liquid equilibrium of mixtures, 537
Vapor-liquid equilibrium relations for CO_2-SO_2 at constant pressure, 572
Vapor pressures, of acetic acid aqueous, 170
of 1-butene, 647
Cox Chart for, 564
curves of liquids, 564
of diethylene glycol, aqueous, 170
effect of indifferent gas on, 293
effect of superheat in, 477
of furfural, 647
of inorganic compounds, above 1 atm., 149
of inorganic compounds, up to 1 atm., 150–152
of isobutane, 647
of liquid water and water ice, 149
of liquids, 293, 319
of moist air, water, 760
of NH_3, aqueous, 172
of organic compounds, above 1 atm., 165, 166
of organic compounds, up to 1 atm., 153–165
of phosphoric acid, aqueous, 167
of pure substances, 149–166
of refrigerants, 1687
of small drops, 293
of solutions, 166–172
of sulfuric acid, aqueous, 168
of sulfuric acid-nitric acid, aqueous, 170
temperature curves, 98
thermometers, 1270, 1272
of water and diethylene glycol over glycol-water solution, 680
Vaporization, continuous equilibrium, 585
of petroleum fractions, 587
efficiency in steam distillation, 583
estimation of heats of, 299
flash, 585
heat of, 299
latent heats of, 218
latent heats of, for sulfuric acid, aqueous, 168
methods of, 580
Vaporizer control, Dowtherm, 1335
Variable, controlled, 1338
Variable-speed drives, 1670
Variance accounting, 1841
Variation, of entropy with pressure, volume, and temperature, 304
of free energy with pressure and temperature, 305
Variation, of heat of reaction with temperature, 299
Varnish, filtration of, 990
and lacquer clarification, 1011
Varteressian-Fenske diagram (extraction), 733
V-belt drive, 1670
Vector representation, electrical, 1735
Vegetable(s), carbons, effect of on sugar liquors, 902
dried, grinding of, 1148
drying of, 821
glue, drying of, 866
oil refining, centrifuges in, 1008
products, milling of, 1147
seeds, drying of, 821
Velocity(ies), acoustic, 375
of approach, 401
factor for orifices, 1285
critical fluids, 375, 383, 384
distribution of, for circular cross section streamline flow, 385
effect of, on packed distillation towers, 619
equivalents and conversion table, 41
flooding for tower packings, 684
head, 377
of liquids, 1415
of light, 47
limiting, for absorption columns, 680
linear (fluids), 374
liquids in pumps, 1415
loading for tower packings, 683
local, 397
mass (definition of), 374
in piping systems, 1642
settling of particles, 1017, 1021
terminal, of spherical particles, 1021
translation, 1058
traverse for mean, 399
Vena contracta, 402
for orifice, 1284, 1285
taps for, 405
Ventilation, 1853
Venting, from evaporators, 511
Venturi, meter, 400, 406
working formulas, 403
nozzles, 1285
tubes, 401, 1283, 1285
coefficients of discharge, 407
Vernal equinox, 45
Versator, Cornell, 1168
Vertex dryer, 834
Vertical compressors, 1681
Vertical cylindrical tanks, capacities of, 1381
Vertical-flow plate precipitator, 1042
Vertical furnaces, 1598
Vertical pump, 1421
Vertical-tube evaporators, 501, 506
Vertical turbodryers, 821
Vessel closures, 1248
Vessels, for high pressures, reaction, closures for, 1243, 1248
Vibrating chutes, 1361
Vibrating conveyors, 1361
Vibrating dryer, 824
conveyor dryers, 863

Vibrating feeder, 1292, 1361, 1372
Vibrating mills, 1130
Vibrating screens, 956, 1361
Vibration of centrifuges, 1005
Vibration-type speed indicator, 1296
Vibrations, mechanical, 991
Vibrators for packing containers, 1394
Syntron, 1395
Vibrox packer, 1395
Vickers-Anderson closure, 1250
Victaulic coupling, 442
Vinegar, filtration of, 988
Vinyon filter mediums, 968
Virial coefficients, equation for, 537
Viscolizer, Cherry-Burrell, 1167
Viscometer(s), Brookfield, 1201
couette, 1201
Engler, formula for, 370
MacMichael, 1201
modified Stormer, 1200
Redwood, formula, 370
Saybolt, formula, 370
Viscose rayon, hygroscopic moisture of, 779
Viscosity(ies), 1197
absolute, 369
apparent, 1197, 1199
conversion chart, 1198
conversion of units for, 369
data, 369, 371, 372
definitions of, 369–370
dimensions of, 369
effect of, on plate efficiencies of bubble-tray absorption columns, 698
on plate efficiencies of distillation columns, 612
English unit of, 1197
falling sphere, 1199
Faxen, 1199
of gases (alignment chart), 371
effect of pressure, 372
index to data in "International Critical Tables," 369
kinematic, 370
of liquid fuels, 1570
of liquids (alignment chart), 373
application of Duhring rule, 372
effect of pressure, 370
generalized function of temperature, 370
of liquids in pumps, 1415
measurement of, 1300
of refrigerant solutions, 1697
of refrigerating solutions, 1696
relative, 369
specific, 369
of steam (table), 370
of sucrose solution (table), 374
true, 1197
variation of, in mixing, 1202
of various materials, 1193–1231
of viscous materials, 1201
of water (Bingham's formula) table, 374
Viscosity-gravity, constant, use of in extraction, 728
Viscous-drag-type gas-density meter, 1297
Viscous flow, 375
Viscous resistance to settling, 995
Vissac dryer, 824
Vitrain in coals, 1562
V-notch meters, 954
Void content, abnormal, of solids, 394
Volatile matter in coal, 1560
Volatility(ies), relative, 579
direct prediction of values of, 529
effect of, on plate efficiencies of distillation columns, 613
of ideal mixtures, 526
Volatilization, 887
Volt, 1732, 1772
box, 1740
Volt-ampere, 1732
reactive, 1732
Voltage (breakdown), effective of primary cell, 1784
of glasses, 1548
inverse, 1749
vectors, three-phase, 1765
Volume(s), additive, 291
atomic, 538
calculations involving, for gases, 301
and capacity equivalents, 40
coulometers, 1775
critical, 291, 292
of horizontal cylindrical tanks, 1380
humid, 759
of moist air, 760
saturated, 759
measure, table, 40
English system, 44
metric system, 43
molal, 289
of 1 g. of water between 0° and 40°C., 177
of 1 g. of water between 40° and 100°C., 177
partial, 291
percentage of water vapor in gases, 173
and pressure, 330
pressure diagram, 330
pressure-temperature diagram, 330
of solids, determination by immersion, 363
specific, of dry air, 811
of mixture, 289
of saturated air, 811
of spheres, 40
Vol-U-Meter electrically controlled valve, 1396
Volumetric efficiency, 1679
efficiency of pumps, 1431
meters, 1282
pressure gage, 1280
Volute-type pumps, 1419
"Vomit" tubes as mixers, 1203
Von Mises-Hencky formulas, 1239
Vorce cell, 1808
Vortex contactor, 1037
Votator, 1213
V-pen for recorders, 1267
V-port valves, 1328
double-seat, 1327
single-seat, 1327
V-type pump, Dorrco, 946
Vulcanized fiber, physical properties of, 1551

W

Wage control, 1834
job evaluation for, 1835
Wall-effect factor, 394
Wallace and Tiernan feeder, 1374, 1375
Ward pipe joint, 419
Wash-water feeders, 954
Washing, 715, 927
Washing-tray thickener, Dorr, 953
Waste heat from furnaces, 1641
Waste-heat recovery, 1623
Water, demineralizing with adsorbents, 915
density of, below 0°C., 176
between 0° and 40°C., 175
between 40° and 100°, 175
extraction rate data for, 750
filtration, activated carbon plant for, 903
flowing inside tubes, heat transfer coefficients for, 470
heat capacity of, 225
ice, vapor pressures of, 149
liquid, effect of in still on steam distillation, 583
vapor pressures of, 149
measure (table), 43
meter, rotary-disk, 411
mol percentages of, over NH_3, aqueous, 171
Mollier diagram for, 333
and nitric acid, absorption of nitrogen oxides in, 706
partial pressures of, over HBr, aqueous, 170
over HCl, aqueous, 166
over HNO_3, aqueous, 169, 170
over H_2SO_4, 169
over methyl alcohol, aqueous, 172
over NH_3, aqueous, 171
over SO_2, aqueous, 167
over sodium carbonate, aqueous, 172
over sodium hydroxide, aqueous, 173
phase diagram of, 316
as plant location factor, 1723
pumps, hot, 1416
rate curves for steam engines, 1645
removal from silica gel by air, 913
requirements of condensers, 1689
for evaporator, condensers, 521
for steam jet vacuum cooling units, 1682
sampling of, 1102
specific volume of, below 0°C., 176
supplies for fire protection, 1849
systems, drinking, 1853
temperature-entropy diagram for, 332
in U.S., 1687
thermodynamic properties of, 770
tube boilers, 1638
viscosity of, Bingham's formula, 374
table, 374

Water, volume of 1 g., between 0° and 40°C., 177
between 40° and 100°C., 177
Water–acetic acid–benzene system, 751
Water–acetone–trichloroethane system, 721, 722, 723
Water–butanol vapor–liquid equilibriums, 632
Water content of silica gel as function of temperature, 912
Water-cooled rheostats, 1752
Water–ether–acetic acid, 721
Water–ethyl alcohol, azeotropes, 631
mixtures, densities of aqueous solutions of, 189
specific gravity of solutions of, 190
solvent, 723
Water gas, as fuel, 1576
manufacture of, 1579
sets, 1617
Water-jet scrubber, 1035
Water–methyl isobutyl ketone system, 751
Water-power plants, 1655
Water-sealed gas meter, 1283
Water vapor, and air mixtures, humidity chart for, 811
content of gases, 173, 174, 338
of hydrogen, 173
of nitrogen, 173
N_2, compressed, effect of pressure on, 174
cost of transporting, 856
in gases, volume percentage of, 173
partial pressure of, 758
pressure of moist, 760
specific weight of, 758
Water wheel generator, 1656
Watson method, for estimation of critical constants, 291
of latent heats of vaporization, 300, 343
Watson and Nelson characterization factor, 565
Watt, 1732
Watt-hour, 1732
Wattmeter, 1738
polyphase, 1738
"Wave-ring" pressure joint, 1246
Wax, paraffin, decolorization of by adsorbents, 892
Waxes, grinding of, 1159
Weather data of U.S. and foreign countries (tables), 46
Weathering of natural gasoline, 581
Web slitting, Camachine, 1162
Weddle's rule for integration, 90
Wedge optical pyrometer, 1275
Weigh batching systems, 1293
Weighers, automatic, 1292
Weighing, 1382
and bagging, 1382
fillers for liquids, 1397
hand, 1385
limits of industrial scales, 1292
in packaging, 1382
and weight control, 1291
Weight(s), and masses, per unit length equivalents, 40
Weight(s), and measures of different countries, 44
English system, 44
metric system, 43
of water vapor, specific, 758
Weight coulometer, 1775
feeders, continuous, 1292
Weir(s), 408
broad crested, 408
Cipolletti trapezoidal, 409
curved sharp-edged, 410
downspouts, 600
evaporator, heat-transfer coefficients in, 503
formulas, 409
for gases, 409, 410
for liquids, 409
Francis formula, 409
head measurement, 409
inward flow, 410
meters, 1258
rectangular, 408
sharp-edged, 408
submerged, 409
suppressed, 409
trapezoidal, 409
triangular, 408, 409
V-notch, 409
Welded pipe joints, 441
autogenous or fusion, 1243
vs. brazing, 1244
and corrosion, 1459
Welds, types of, 1244
Well type manometer, 1279
Wellner-Jelinek evaporator, 501, 506
Western Precipitation spray dryer, 842
Weston cadmium cell, 1732
Weston standard cell, 1772
Westphal balance, 1296
Wet-bulb control system, 1335
Wet-bulb method of determining humidity, 777
Wet-bulb temperature, 759
lines for air-water vapor, 811
Wet compression, 1680
Wet condenser, 520
Wet- and dry-bulb thermometers, 1299
Wet- and dry-bulb thermometry, 557
Wet dust collection, 1034
Wet elutriation, 1084
Wet gas meter, 410
Wet vs. dry grinding, 1118
Wet grinding, 1115
of barytes and limestone, 1151
Wet tables, 1078
Wet-weight basis, 800
Wetherill magnetic separator, 1092
Wetted-glass-fiber filter, 1049
Wetted wall towers for extraction, 748
Wetting, ease of, and mixing, 1202
heats of, 301
of silica gel, 884
Weymouth formula, 383
Wheatstone's-bridge, 1739
principle, 1740
Wheel, impulse, meter, 412
Wheeler cell, 1808
Whey, drying of, 837
Whipper mixer, 1206
Whirlwind external air classifier, 1135
pulverizer, 1124, 1141
White lead, drying of, 826
production, 1796
Whiting, classification of, 936
closed-circuit grinding, 936
grinding of, 936
Whizzer external air classifier, 1135, 1136
Wickel band, 1243
Wien's law, 484, 1274
Wiley Laboratory mill, rotary knife cutter, 1161
Williams, hammer mills, performance of, 1148
hay cutter, 1148
Helix-Seal mill, 1139
lines for steam engines, 1645
patent crushers, 1129, 1139
classifiers, 1135
ring-roller mill, 1138
Williams and Hazen formula, 377
Window, cylindrical, for high pressures, 1252, 1253
Wine, filtration of, 988
Winterschall-Schmalfeldt pulverizing process, 1167
Wire, characteristics of nichrome, 1760
cloths, 960
covering, 1183
screens, 959
and sheet-metal gages, 39
table, 1754
tying bags, 1390
Wire-wound vessels for high pressure, 1241
Wiring, electrical, 1753, 1759
Witch of Agnesi, 81
Wolf rotary cutter, 1161
Wood(s), block hygrometer, 1298
for chemical equipment, physical properties of, 1552
dehydration by solutions, 1558
equipment, manufacturers of, 1537
for filters, 970
as fuel, 1568
heating of, 463
-lined pipe, 440
pipe, 440
stove, 440
and tanks, manufacturers of, 1553
thermal conductivity of, 463
Wool cloth filter mediums, 968
felt filters, 968
grease recovery, centrifuges in, 1008
hygroscopic moisture of, 778, 780
Work, of crushing, 931
and circulating load, 931
and feed rate, 931
of grinding, 931
and heat and energy, equivalents and conversion table, 40
of liquefaction and separation, 1703
maximum, 305
useful, 305
performed in pumping, 1416

Work, required for size reduction, 1114
reversible, 1703
Working capital, 1829
Worm-gear reduction units, 1669
Worsted, hygroscopic moisture of, 778
Wound rotors, 1768
Wrapping, package, 1406
Wrought-iron pipe, 413
Wrought-steel pipe, 413
Wulff-Bock crystallizer, 1062

X

X-ray diffraction for analytical controls, 1304
o-Xylene–air vapor mixtures, humidity chart for, 816

Y

Yankee dryers, 867
Yeast, spray-drying of, 845
Yeast filter cake, drying of, 837
Yield point, 1197
of materials, 1539*ff*.
of steels, 1235
Y system, three-phase, 1735

Z

Zeolite ion-exchangers, 915
Zinc, electroplating, 1799
electrowinning, 1805
production, 1820
slimes, drying of, 833
Zinc bromide, densities of aqueous solutions of, 185
Zinc chloride, densities of aqueous solutions of, 185
Zinc-dust feeder, 1376
Zinc nitrate, densities of aqueous solutions of, 185
Zinc oxide, hygroscopic moisture of, 779
Zinc stearate, drying of, 826
Zinc sulfate, densities of aqueous solutions of, 185
specific heats of aqueous solutions of, 234
spray-drying of, 845
Zone, dead, 1338
Zone of unsaturated surface drying, 806
Zone where internal liquid flow controls, 806
Zsigmondy classification of solid particles, 939

INTERNATIONAL ATOMIC WEIGHTS, 1949

Substance	Sym.	At. Wt.	At. No.	Substance	Sym.	At. Wt.	At. No.
Actinium	Ac	227	89	Molybdenum	Mo	95.95	42
Aluminum	Al	26.97	13	Neodymium	Nd	144.27	60
Americium	Am	241	95	Neon	Ne	20.183	10
Antimony	Sb	121.76	51	Neptunium	Np	237	93
Argon	A	39.944	18	Nickel	Ni	58.69	28
Arsenic	As	74.91	33	Niobium	Nb	92.91	41
Astatine	At	211	85	Nitrogen	N	14.008	7
Barium	Ba	137.36	56	Osmium	Os	190.2	76
Beryllium	Be	9.02	4	Oxygen	O	16.0000	8
Bismuth	Bi	209.00	83	Palladium	Pd	106.7	46
Boron	B	10.82	5	Phosphorus	P	30.98	15
Bromine	Br	79.916	35	Platinum	Pt	195.23	78
Cadmium	Cd	112.41	48	Plutonium	Pu	239	94
Calcium	Ca	40.08	20	Polonium	Po	210	84
Carbon	C	12.010	6	Potassium	K	39.096	19
Cerium	Ce	140.13	58	Praseodymium	Pr	140.92	59
Cesium	Cs	132.91	55	Promethium	Pm	147	61
Chlorine	Cl	35.457	17	Protactinium	Pa	231	91
Chromium	Cr	52.01	24	Radium	Ra	226.05	88
Cobalt	Co	58.94	27	Radon	Rn	222	86
Copper	Cu	63.54	29	Rhenium	Re	186.31	75
Curium	Cm	242	96	Rhodium	Rh	102.91	45
Dysprosium	Dy	162.46	66	Rubidium	Rb	85.48	37
Erbium	Er	167.2	68	Ruthenium	Ru	101.7	44
Europium	Eu	152.0	63	Samarium	Sm	150.43	62
Fluorine	F	19.00	9	Scandium	Sc	45.10	21
Francium	Fr	223	87	Selenium	Se	78.96	34
Gadolinium	Gd	156.9	64	Silicon	Si	28.06	14
Gallium	Ga	69.72	31	Silver	Ag	107.880	47
Germanium	Ge	72.60	32	Sodium	Na	22.997	11
Gold	Au	197.2	79	Strontium	Sr	87.63	38
Hafnium	Hf	178.6	72	Sulfur	S	32.066	16
Helium	He	4.003	2	Tantalum	Ta	180.88	73
Holmium	Ho	164.94	67	Technetium	Tc	99	43
Hydrogen	H	1.0080	1	Tellurium	Te	127.61	52
Indium	In	114.76	49	Terbium	Tb	159.2	65
Iodine	I	126.92	53	Thallium	Tl	204.39	81
Iridium	Ir	193.1	77	Thorium	Th	232.12	90
Iron	Fe	55.85	26	Thulium	Tm	169.4	69
Krypton	Kr	83.7	36	Tin	Sn	118.70	50
Lanthanum	La	138.92	57	Titanium	Ti	47.90	22
Lead	Pb	207.21	82	Uranium	U	238.07	92
Lithium	Li	6.940	3	Vanadium	V	50.95	23
Lutetium	Lu	174.99	71	Wolfram	W	183.92	74
Magnesium	Mg	24.32	12	Xenon	Xe	131.3	54
Manganese	Mn	54.93	25	Ytterbium	Yb	173.04	70
Mercury	Hg	200.61	80	Yttrium	Y	88.92	39
				Zinc	Zn	65.38	30
				Zirconium	Zr	91.22	40